Global Energy Assessment

Toward a Sustainable Future

Council Co-Presidents: Ged Davis and José Goldemberg

Executive Committee Co-Chairs: Thomas B. Johansson and Anand Patwardhan

Director: Nebojsa Nakicenovic

Associate Director: Luis Gomez-Echeverri

Convening Lead Authors: Rangan Banerjee, Sally M. Benson, Daniel H. Bouille, Abeeku Brew-Hammond, Aleh Cherp, Suani T. Coelho, Lisa Emberson, Maria Josefina Figueroa, Arnulf Grubler, Mark Jaccard, Suzana Kahn Ribeiro, Stephen Karekezi, Kebin He, Eric D. Larson, Zheng Li, Susan McDade, Lynn K. Mytelka, Shonali Pachauri, Anand Patwardhan, Keywan Riahi, Johan Rockström, Hans-Holger Rogner, Joyashree Roy, Robert N. Schock, Ralph Sims, Kirk R. Smith, Wim C. Turkenburg, Diana Ürge-Vorsatz, Frank von Hippel, and Kurt Yeager

Review Editors: John F. Ahearne, Ogunlade Davidson, Jill Jäger, Eberhard Jochem, Ian Johnson, Rik Leemans, Sylvie Lemmet, Nora Lustig, Mohan Munasinghe, Peter McCabe, Granger Morgan, Jürgen Schmid, Jayant Sathaye, Leena Srivastava, Youba Sokona, John Weyant, and Ji Zou

Secretariat: Martin Offutt, Mathis L. Rogner, Hal Turton, and Pat Wagner

Global Energy Assessment (GEA)

Editors

Thomas B. Johansson
Co-Chair

Nebojsa Nakicenovic
Director

Anand Patwardhan
Co-Chair

Luis Gomez-Echeverri
Associate Director

International Institute for Applied Systems Analysis
Schlossplatz 1, A-2361 Laxenburg, Austria
www.iiasa.ac.at

CAMBRIDGE UNIVERSITY PRESS
Cambridge, New York, Melbourne, Madrid, Cape Town,
Singapore, São Paulo, Delhi, Mexico City

www.cambridge.org
www.globalenergyassessment.org

First published 2012

Printed in the United States of America

A catalog record for this book is available from the British Library.

ISBN 9781 10700 5198 hardback
ISBN 9780 52118 2935 paperback

GEA, 2012: *Global Energy Assessment – Toward a Sustainable Future*, Cambridge University Press, Cambridge UK and
New York, NY, USA and the International Institute for Applied Systems Analysis, Laxenburg, Austria.

Cover photo:
Figure 1.2 | Global energy flows of primary to useful energy, including losses, in EJ for 2005. Source: adapted from Nakicenovic et al., 1998, based on IEA, 2007a; 2007b; 2010. Artwork by Anka James.

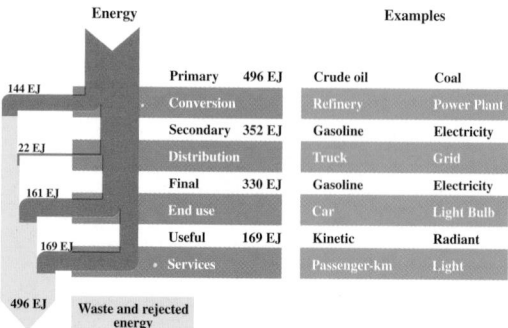

Global Energy Assessment (GEA) Council

Sponsoring Organizations

Austrian Development Agency

Climateworks Foundation

Deutsche Gesellschaft für Internationale Zusammenarbeit

First Solar Inc.

Global Environment Facility

Global Environment and Technology Foundation

Italian Ministry for the Environment and Territory

International Institute for Applied Systems Analysis

Petrobras

Research Council of Norway

Swedish Research Council for Environment, Agricultural Sciences and Spatial Planning/

Swedish Energy Agency

United Nations Development Programme

United Nations Environment Programme

United Nations Foundation

United Nations Industrial Development Organisation

United States Department of Energy

United States Environmental Protection Agency

World Bank/ESMAP

World Energy Council

Contents

Contents

Foreword and Preface

Foreword

Energy is central to addressing major challenges of the 21st Century, challenges like climate change, economic and social development, human well-being, sustainable development, and global security. In 2005, Prof. Bert Bolin, the founding Chair of the Intergovernmental Panel on Climate Change (IPCC), with other eminent scientists and policy-makers, identified that a comprehensive, science-based assessment of the global energy system was needed if these challenges were to be realistically addressed. The Global Energy Assessment (GEA) is the result of this shared vision.

Since the establishment of the GEA in 2006 by governing Council of the International Institute for Applied Systems Analysis (IIASA), 500 independent experts (about 300 authors and 200 anonymous reviewers) from academia, business, government, intergovernmental and non-governmental organizations from all the regions of the world have contributed to GEA in a process similar to that adopted by the IPCC.

The final GEA report examines: (a) the major global challenges and their linkages to energy; (b) the technologies and resources available for providing adequate, modern and affordable forms of energy; (c) the plausible structure of future energy systems most suited to addressing the century's challenges; and (d) the policies and measures, institutions and capacities needed to realize sustainable energy futures.

Undertaking such a massive assessment has required extraordinary leadership, intellectual input, support and coordination. Governance of the Assessment has been overseen by the GEA Council, led by two Co-Presidents, Ged Davis and José Goldemberg and comprising members of supporters and sponsors of the GEA, including international organizations, government agencies, corporations, and foundations and non-governmental organizations. Convening Lead Authors (CLAs) coordinated the 25 Chapters and the contributions of Lead and Contributing Authors. The GEA Executive Committee, led by two Co-Chairs, Thomas B. Johansson and Anand Patwardhan includes all CLAs. Review Editors were appointed by the GEA Council for each Chapter. They in turn appointed anonymous reviewers and guided the rigorous peer-review process.

Completion of GEA has involved dedication and sustained contributions from many colleagues around the world. Our thanks and gratitude go to: Leen Hordijk, the IIASA Director who initiated GEA at IIASA; Sten Nilsson, IIASA Acting Director and Deputy Director; and Detlof von Winterfeldt, the IIASA Director who provided personal and institutional support throughout. The resources and the encouragement they provided helped make GEA a reality. The GEA Organizing Committee and the GEA Council provided wise counsel and guidance throughout. Additionally the GEA Council solicited financial and in-kind resources without which GEA would not have been possible.

We are especially grateful for the contribution and support of the GEA Council, the Executive Committee, the Organizing Committee, the Secretariat, as well as the IIASA Council and management. As host organization for the GEA Secretariat, IIASA has provided substantial in-kind support to GEA over the past seven years.

The Co-Chairs Thomas B. Johansson and Anand Patwardhan of the GEA Executive Committee and the Associate Director, Luis Gomez-Echeverri, coordinated the work of multiple authors and provided intellectual leadership, the vision needed to conduct an assessment of this magnitude, and guidance consistent with the GEA Council resolutions.

It is a pleasure to acknowledge the contribution of the team of editors, Geoff Clarke, Esther Eidinow, Valerie Jones, Susan Guthridge-Gould, Karen Holmes, Gail Karlsson, Wendy Knerr, John Ormiston, Emily Schabacker, Misti Snow, Mark F. Speer, Jon Stacy, Linda Starke, Julia Stewart, Lloyd Timberlake, Michael Treadway, Thomas Woodhatch who patiently edited GEA manuscripts. Thanks to IIASA colleagues who worked with the GEA Secretariat – including Colin Adair, Brigitte Adamik, Marilyn Bernardo, Anita Brachtl, Claire Capate, Elisabeth Clemens, Katalin David, Susanne Deimbacher, Sanja Drinkovic, Linda Foith, Walter Foith, Amy Fox, Bill Godwin-Toby, Amnah Kasman, Martin Gugumuck, Margit Hegenbart, Anka James, Shari Jandl, Elizabeth Lewis, Monica Manchanda, Eri Nagai, Olivia Nilsson, Patrick Nussbaumer,

Sheila Poor, Leane Regan, Susan Riley, Michaela Rossini, Iain Stewart, Ingrid Teply-Baubinder, Mirjana Tomic, and Alicia Versteegh.

Finally we express our sincere gratitude to the GEA authors, whose knowledge and experience has made possible this unique and valuable volume. Behind these people are families who have generously foregone time such that GEA could be completed, we thank them also.

The publication of GEA in June 2012 and the importance of energy at Rio+20 is no coincidence. The UN General Assembly declared 2012 the year of "Sustainable Energy for All" and the UN Secretary General's office initiated a campaign for an Action Agenda to meet the world's energy challenges. The GEA shows that an energy transformation toward a sustainable future is possible with strong political commitment. It is our belief that this assessment will provide policy- and decision-makers around the world, with invaluable new knowledge to inform action and commitment towards achieving these goals and thereby resolving the 21st Century's greatest challenges.

Pavel Kabat
IIASA Director/CEO

Nebojsa Nakicenovic
GEA Director

Preface

Today the world of energy has many of the features established in the 20th century:

– Energy consumption grows on average at 2% per year, most of it (80%) originates in fossil fuels

– Energy growth is driven by population growth and economic growth, now predominantly in developing countries and high levels of consumption in the developed countries

– 3 billion people don't have access to basic energy services and have to cook with solid fuels

However, the present path of uninterrupted reliance on fossil fuels poses four challenges to sustainability:

– Soaring greenhouse gas emissions

– Decreasing energy security

– Air pollution at the local and regional levels with resulting health problems

– Lack of universal access to energy services

Most reviews of the energy system needed for the 21st century start with "business as usual" futures and then analyze the effectiveness of specific corrections of course. For many the preferred options are technological fixes such as such as carbon capture and storage (CCS), nuclear energy and even geo-engineering schemes. However, to achieve sustainable development all the needed attributes of energy services, that is availability, affordability, access, security, health, climate and environmental protection, must be met concurrently. The Global Energy Assessment (GEA) accepts this and is unashamedly normative, examining future energy pathways that point to new solutions. The aspirational goals in GEA are defined as:

– Stabilizing global climate change to 2°C above pre-industrial levels to be achieved in the 21st century

– Enhanced energy security by diversification and resilience of energy supply (particularly the dependence on imported oil),

– Eliminating household and ambient air pollution, and

– Universal access to modern energy services by 2030.

GEA's approach is the one adopted by policy planners and governments, that is to take a holistic view of the problems they faced, of which energy supply is only one of them. In such an approach externalities play a big role in determining choice among options. This is what governments do all the time, and is exemplified by the current debates on the future of nuclear energy, shale gas, the building of big dams or a large expansion of biofuels production. None of the preferred options can be established without an understanding of the wider policy agenda. For example, integrated urban planning leads to lower costs than a combination of non-integrated policies in building efficiency, compact layout and decentralized energy production.

The main purpose of GEA has been to establish a state-of-the-art assessment of the science of energy. This work examines not only the major challenges that all face in the 21st Century, and the importance of energy to each, but also the resources that we have available and the various technological options, the integrated nature of the energy system and the various enablers needed, such as policies and capacity development. Central to the integrated analysis of the energy system has been a novel scenario exercise exploring some 40 pathways that satisfy simultaneously the normative social and environmental goals outlined above.

Without question a radical transformation of the present energy system will be required over the coming decades. Common to all pathways will be very strong efforts in energy efficiency improvement for buildings, industry and transportation, offering much-needed flexibility to the energy supply system. But in implementing efficiency options there will be a need to avoid continued lock-in to inefficient energy demand patterns and obsolescent technologies. we will see an increased share of renewables (biomass, hydro, wind, solar and geothermal), which could represent by 2050 over a half of the global energy supply. The foundation is being put in place. For example, half the world's new electric generating capacity added during 2008–10 was renewable, the majority in developing countries. Global 2010 renewable capacity, with additions of ~66 GW, is larger than nuclear power's global installed capacity. In the European Union electric capacity additions have been over 40% renewables in each year between 2006 and 2010, and in Denmark 30% of the electricity produced in 2010 was renewable. Even though China is still building coal plants, its 2010 net capacity additions were 38% renewables.

This will come at a cost, increasing the 2% of global GDP investment currently spent in the energy sector, especially in the next 20 years. However, this should not constrain the drive for universal access, which could be achieved by 2030 for as little as $40 billion per year, less than 3% of overall yearly investment. This would build on successful programs for energy access in a number of developing countries, such as Brazil, Mexico and South Africa. And results have been dramatic. In Brazil, during the ten years prior to September 2011 14 million people were connected to the electricity grid, at a cost of some 10 billion dollars.

Although the required transformation of the energy system is substantial, it is not without precedent. Last century between the 1920's and the 1970's oil replaced coal as the dominant energy source despite the immense available coal reserves. This occurred due to oil, as a liquid, being superior to coal in many respects, particularly for transportation. Similarly energy efficiency and renewables can be an easier way to solve energy security than producing fossil energy at higher costs that usually exacerbate environmental problems.

There are many combinations of energy resources, end-use, and supply technologies that can simultaneously address the multiple sustainability challenges. There will be an increased role of electricity and gases as energy carriers, co-utilization of biomass with fossil fuels in integrated systems, co-production of energy carriers, electricity, and chemicals, and, CCS.

All GEA energy pathways to a more sustainable future represent transformative change from today's energy systems. Large, early, and sustained investments are needed to finance this change, and can be in part achieved through new and innovative polices and institutional mechanisms that should reduce risks and increase the attractiveness of early, upfront investments, that have associated low long-term costs.

The GEA pathways that meet the sustainability goals generate substantial benefits across multiple economic and social objectives. This synergy is advantageous and important, given that measures which lead to local and national benefits, e.g. improved local and immediate health and environment conditions, support the local economy, may be more easily adopted than those measures that are put forward primarily on the grounds of goals that are global and long-term in nature, such as climate mitigation. An approach that emphasizes the local benefits of improved end-use efficiency and increased use of renewable energy would also help address global concerns.

Policies and incentive structures that promote R&D should be key areas for intervention. Rationalizing and reallocating subsidies, including subsidies to fossil fuels and nuclear energy can create new opportunities for investment. A major acceleration of publicly financed R&D and its reorientation towards energy efficiency and renewable energy technologies is required. And to bring new technologies to market an integrated approach towards energy for sustainable development is needed; with policies in sectors such as industry, buildings, urbanization, transport, health, environment, climate, security, and others made mutually supportive.

The transition from coal to oil occurred without significant government regulations although subsidies played a role. However the transformation GEA envisages this century is more fundamental in character, and government policies are a key ingredient needed particularly in changing buildings codes, fuel efficiency standards for transportation and mandates for the introduction of renewables. A new found appreciation by policy-makers of the multiple benefits of sustainability options and their appropriate valuation will be critical for the transformation to occur.

The Global Energy Assessment's report establishes a benchmark for current understanding of the options for building a sustainable future for the energy system. But the Assessment consists more than just a report. Analytical tools have been developed to help translate the Assessment into actionable findings. Tools for decision making, that include global and regional scenarios, can be used to develop policy choices to address country-specific problems.

An important contribution to knowledge is the massive data base that is at the disposal of research and scientific community for their own use, and eventually analysis will be made available to the public at large.

Outreach has already started with the presentation of the early findings of GEA at the Vienna Energy Forum in June 2011. Importantly at that forum a Ministerial declaration, supported by the UNIDO leadership, endorsed the solutions offered by GEA, particularly:

 — Ensure universal access to moderns forms of energy for all by 2030

 — Reduce global energy intensity by 40% by 2030

 — Increase the share of renewables 30% by 2030

These three objectives are reflected in the Action Agenda of the UN Secretary General's High-Level Group on "Sustainable Energy for All".

The aim going forward is to ensure the widest dissemination of GEA's work that is possible, including both national and regional policy dialogues.

This opportunity to layout a new approach to the design and implementation of sustainable energy pathways would not be possible without the extraordinary effort of the 500 or so contributors, be they authors from various disciplines and walks of life, reviewers, editors or members of the secretariat, executive team and council. We thank you all.

Ged Davis and José Goldemberg
GEA Co-Presidents

Key Findings

The Global Energy Challenge

Since before the Industrial Revolution, societies have relied on increasing supplies of energy to meet their need for goods and services. Major changes in current trends are required if future energy systems are to be affordable, safe, secure, and environmentally sound. There is an urgent need for a sustained and comprehensive strategy to help resolve the following challenges:

- *providing affordable energy services for the well-being of the 7 billion people today and the 9 billion people projected by 2050;*

- *improving living conditions and enhancing economic opportunities, particularly for the 3 billion people who cook with solid fuels today and the 1.4 billion people without access to electricity;*

- *increasing energy security for all nations, regions, and communities;*

- *reducing global energy systems greenhouse gas emissions to limit global warming to less than 2°C above pre-industrial levels;*

- *reducing indoor and outdoor air pollution from fuel combustion and its impacts on human health; and*

- *reducing the adverse effects and ancillary risks associated with some energy systems and to increase prosperity.*

Major transformations in energy systems are required to meet these challenges and to increase prosperity.

The Global Energy Assessment (GEA) assessed a broad range of resources, technologies and policy options and identified a number of 'pathways' through which energy systems could be transformed to simultaneously address all of the above challenges

These are the Key Findings:

1. Energy Systems can be Transformed to Support a Sustainable Future: the GEA analysis demonstrates that a sustainable future requires a transformation from today's energy systems to those with: *(i)* radical improvements in energy efficiency, especially in end use, and *(ii)* greater shares of renewable energies and advanced energy systems with carbon capture and storage (CCS) for both fossil fuels and biomass. The analysis ascertained that there are many ways to transform energy systems and many energy portfolio options. Large, early, and sustained investments, combined with supporting policies, are needed to implement and finance change. Many of the investment resources can be found through forward-thinking domestic and local policies and institutional mechanisms that can also support their effective delivery. Some investments are already being made in these options, and should be strengthened and widely applied through new and innovative mechanisms to create a major energy system transformation by 2050.

2. An Effective Transformation Requires Immediate Action: Long infrastructure lifetimes mean that it takes decades to change energy systems; so immediate action is needed to avoid lock-in of invested capital into existing energy systems and associated infrastructure that is not compatible with sustainability goals. For example, by 2050 almost three-quarters of the world population is projected to live in cities. The provision of services and livelihood opportunities to growing urban populations in the years to come presents a major opportunity for transforming energy systems and avoiding lock-in to energy supply and demand patterns that are counterproductive to sustainability goals.

3. Energy Efficiency is an Immediate and Effective Option: Efficiency improvement is proving to be the most cost-effective, near-term option with multiple benefits, such as reducing adverse environmental and health impacts, alleviating poverty, enhancing energy security and flexibility in selecting energy supply options, and creating employment and economic opportunities. Research shows that required improvements in energy efficiency particularly in end-use can be achieved quickly. For example:

- retrofitting buildings can reduce heating and cooling energy requirements by 50–90%;

- new buildings can be designed and built to very high energy performance levels, often using close to zero energy for heating and cooling;

- electrically-powered transportation reduces final energy use by more than a factor of three, as compared to gasoline-powered vehicles;

- a greater integration between spatial planning and travel that emphasizes shorter destinations and enhances opportunities for flexible and diverse choices of travel consolidating a system of collective, motorized, and non-motorized travel options offer major opportunities;

- through a combination of increased energy efficiency and increased use of renewable energy in the industry supply mix, it is possible to produce the increased industrial output needed in 2030 (95% increase over 2005) while maintaining the 2005 level of GHG emissions.

A portfolio of strong, carefully targeted policies is needed to promote energy efficient technologies and address, *inter alia*, direct and indirect costs, benefits, and any rebound effects.

4. Renewable Energies are Abundant, Widely Available, and Increasingly Cost-effective: The share of renewable energy in global primary energy could increase from the current 17% to between 30% to 75%, and in some regions exceed 90%, by 2050. If carefully developed, renewable energies can provide many benefits, including job creation, increased energy security, improved human health, environmental protection, and mitigation of climate change. The major challenges, both technological and economic, are:

- reducing costs through learning and scale-up;

- creating a flexible investment environment that provides the basis for scale-up and diffusion;

- integrating renewable energies into the energy system;

- enhancing research and development to ensure technological advances; and

- assuring the sustainability of the proposed renewable technologies.

While there remain sound economic and technical reasons for more centralized energy supplies, renewable energy technologies are also well-suited for off-grid, distributed energy supplies.

5. Major Changes in Fossil Energy Systems are Essential and Feasible: Transformation toward decarbonized and clean energy systems requires fundamental changes in fossil fuel use, which dominates the current energy landscape. This is feasible with known technologies.

- CO_2 capture and storage (CCS), which is beginning to be used, is key. Expanding CCS will require reducing its costs, supporting scale-up, assuring carbon storage integrity and environmental compatibility, and securing approval of storage sites.

- Growing roles for natural gas, the least carbon-intensive and cleanest fossil fuel, are feasible, including for shale gas, if related environmental issues are properly addressed.

- Co-processing of biomass and coal or natural gas with CCS, using known technologies, is important for co-producing electricity and low-carbon liquid fuels for transportation and for clean cooking. Adding CCS to such coproduction plants is less costly than for plants that make only electricity.

Strong policies, including effective pricing of greenhouse gas emissions, will be needed to fundamentally change the fossil energy system.

6. Universal Access to Modern Energy Carriers and Cleaner Cooking by 2030 is Possible: Universal access to electricity and cleaner cooking fuels and stoves can be achieved by 2030; however, this will require innovative institutions, national and local enabling mechanisms, and targeted policies, including appropriate subsidies and financing. The necessary technologies are available, but resources need to be directed to meet these goals. Universal access is necessary to alleviate poverty, enhance economic prosperity, promote social development, and improve human health and well-being. Enhancing access among poor people, especially poor women, is thus important for increasing their standard of living. Universal access to clean cooking technologies will substantially improve health, prevent millions of premature deaths, and lower household and ambient air pollution levels, as well as the emissions of climate-altering substances.

7. An Integrated Energy System Strategy is Essential: An integrated approach to energy system design for sustainable development is needed – one in which energy policies are coordinated with policies in sectors such as industry, buildings, urbanization, transport, food, health, environment, climate, security, and others, to make them mutually supportive. The use of appropriate policy instruments and institutions can help foster a rapid diffusion and scale-up of advanced technologies in all sectors to simultaneously meet the multiple societal challenges related to energy. The single most important area of action is efficiency improvement in all sectors. This enhances supply side flexibility, allowing the GEA challenges to be met without the need for technologies such as CCS and nuclear.

8. Energy Options for a Sustainable Future bring Substantial Multiple Benefits for Society: Combinations of resources, technologies, and polices that can simultaneously meet global sustainability goals also generate substantial and tangible near-term local and national economic, environmental, and social development benefits. These include, but are not limited to, improved local health and environment conditions, increased employment options, strengthened local economies through new business opportunities, productivity gains, improved social welfare and decreased poverty, more resilient infrastructure, and improved energy security. Synergistic strategies that focus on local and national benefits are more likely to be implemented than measures that are global and long-term in nature. Such an approach emphasizes the local benefits of improved end-use efficiency and increased use of renewable energy, and also helps manage energy-related global challenges. These benefits make the required energy transformations attractive from multiple policy perspectives and at multiple levels of governance.

9. Socio-Cultural Changes as well as Stable Rules and Regulations will be Required: Crucial issues in achieving transformational change toward sustainable future include non-technology drivers such as individual and public awareness, community and societal capacities to adapt to changes, institutions, policies, incentives, strategic spatial planning, social norms, rules and regulations of the marketplace, behavior of market actors, and societies' ability to introduce through the political and institutional systems measures to reflect externalities. Changes in cultures, lifestyles, and values are also required. Effective strategies will need to be adopted and integrated into the fabric of national socio-cultural, political, developmental, and other contextual factors, including recognizing and providing support for the opportunities and needs of all nations and societies.

10. Policy, Regulations, and Stable Investment Regimes will be Essential: A portfolio of policies to enable rapid transformation of energy systems must provide the effective incentive structures and strong signals for the deployment at scale of energy-efficient technologies and energy supply options that contribute to the overall sustainable development. The GEA pathways indicate that global investments in combined energy efficiency and supply will need to increase to between US$1.7–2.2 trillion per year compared to present levels of about US$1.3 trillion per year (about 2% of current world gross domestic product) including end-use components. Policies should encourage integrated approaches across various sectors and promote the development of skills and institutional capacities to improve the investment climate. Examples include applying market-oriented regulations such as vehicle emissions standards and low carbon fuel standards and as well as renewable portfolio standards to accelerate the market penetration of clean energy technologies and fules. Reallocating energy subsidies, especially the large subsidies provided in industrialized countries to fossil fuels without CCS, and nuclear energy, and pricing or regulating GHG emissions and/or GHG-emitting technologies and fules can help support the initial deployment of new energy systems, both end-use and supply, and help make infrastructures energy efficient. Publicly financed research and development needs to accelerate and be reoriented toward energy efficiency, renewable energy and CCS. Current research and development efforts in these areas are grossly inadequate compared with the future potentials and needs.

* * * * *

The full GEA report is available for download in electronic form at www.globalenergyassessment.org. The website includes an interactive scenario database that documents the GEA pathways.

2 Summaries

SPM

Summary for Policymakers

Convening Lead Authors (CLA):
Thomas B. Johansson (Lund University, Sweden)
Nebojsa Nakicenovic (International Institute for Applied Systems Analysis and Vienna University of Technology, Austria)
Anand Patwardhan (Indian Institute of Technology-Bombay)
Luis Gomez-Echeverri (International Institute for Applied Systems Analysis, Austria)

Lead Authors (LA)
Rangan Banerjee (Indian Institute of Technology-Bombay)
Sally M. Benson (Stanford University, USA)
Daniel H. Bouille (Bariloche Foundation, Argentina)
Abeeku Brew-Hammond (Kwame Nkrumah University of Science and Technology, Ghana)
Aleh Cherp (Central European University, Hungary)
Suani T. Coelho (National Reference Center on Biomass, University of São Paulo, Brazil)
Lisa Emberson (Stockholm Environment Institute, University of York, UK)
Maria Josefina Figueroa (Technical University of Denmark)
Arnulf Grubler (International Institute for Applied Systems Analysis, Austria and Yale University, USA)
Kebin He (Tsinghua University, China)
Mark Jaccard (Simon Fraser University, Canada)
Suzana Kahn Ribeiro (Federal University of Rio de Janeiro, Brazil)
Stephen Karekezi (AFREPREN/FWD, Kenya)
Eric D. Larson (Princeton University and Climate Central, USA)
Zheng Li (Tsinghua University, China)
Susan McDade (United Nations Development Programme)
Lynn K. Mytelka (United Nations University-MERIT, the Netherlands)
Shonali Pachauri (International Institute for Applied Systems Analysis, Austria)
Keywan Riahi (International Institute for Applied Systems Analysis, Austria)
Johan Rockström (Stockholm Environment Institute, Stockholm University, Sweden)
Hans-Holger Rogner (International Atomic Energy Agency, Austria)
Joyashree Roy (Jadavpur University, India)
Robert N. Schock (World Energy Council, UK and Center for Global Security Research, USA)
Ralph Sims (Massey University, New Zealand)
Kirk R. Smith (University of California, Berkeley, USA)
Wim C. Turkenburg (Utrecht University, the Netherlands)
Diana Ürge-Vorsatz (Central European University, Hungary)
Frank von Hippel (Princeton University, USA)
Kurt Yeager (Electric Power Research Institute and Galvin Electricity Initiative, USA)

Introduction

Energy is essential for human development and energy systems are a crucial entry point for addressing the most pressing global challenges of the 21st century, including sustainable economic and social development, poverty eradication, adequate food production and food security, health for all, climate protection, conservation of ecosystems, peace and security. Yet, more than a decade into the 21st century, current energy systems do not meet these challenges.

A major transformation is therefore required to address these challenges and to avoid potentially catastrophic future consequences for human and planetary systems. The Global Energy Assessment (GEA) demonstrates that energy system change is the key for addressing and resolving these challenges. The GEA identifies strategies that could help resolve the multiple challenges simultaneously and bring multiple benefits. Their successful implementation requires determined, sustained and immediate action.

Transformative change in the energy system may not be internally generated; due to institutional inertia, incumbency and lack of capacity and agility of existing organizations to respond effectively to changing conditions. In such situations clear and consistent external policy signals may be required to initiate and sustain the transformative change needed to meet the sustainability challenges of the 21st century.

The industrial revolution catapulted humanity onto an explosive development path, whereby, reliance on muscle power and traditional biomass was replaced mostly by fossil fuels. In 2005, some 78% of global energy was based on fossil energy sources that provided abundant and ever cheaper energy services to more than half the people in the world. Figure SPM-1 shows this explosive growth of global primary energy with two clear development phases, the first

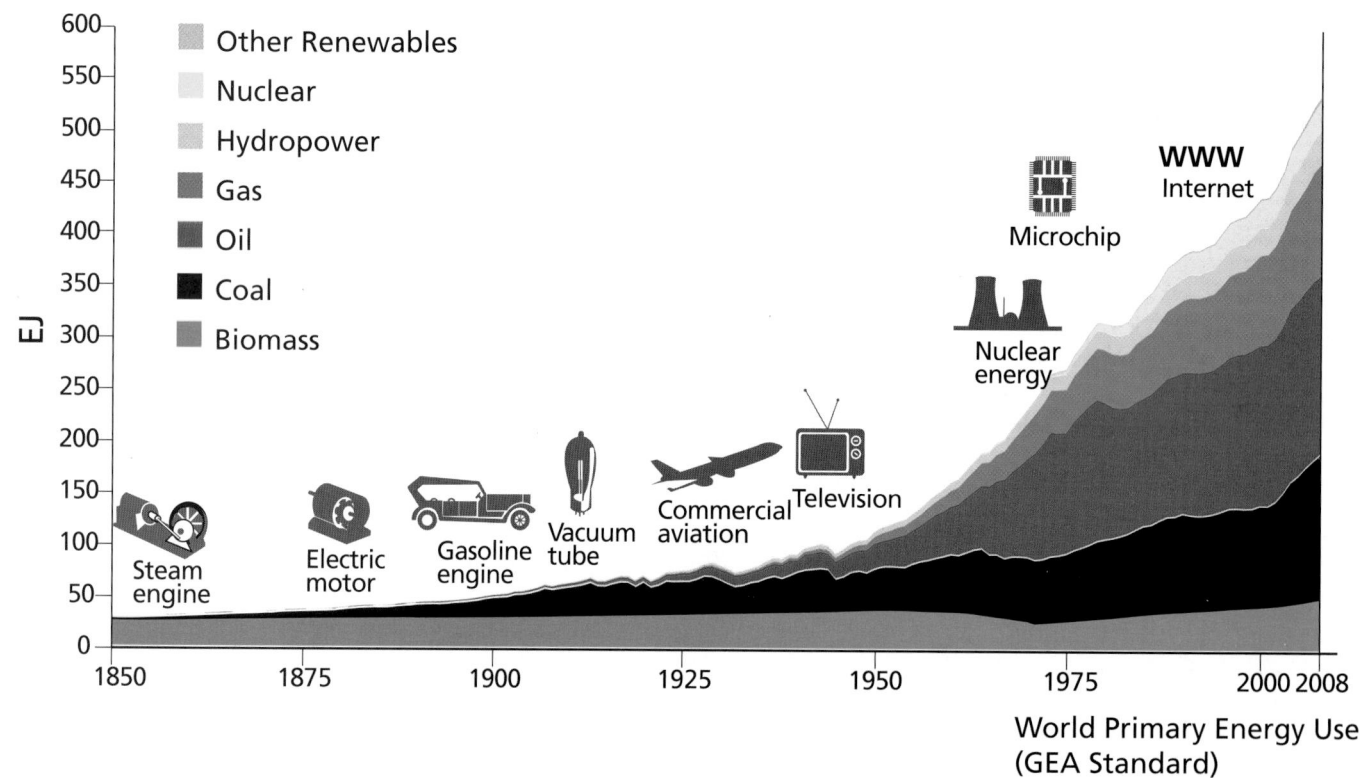

Figure SPM-1. | Evolution of primary energy shown as absolute contributions by different energy sources (EJ). Biomass refers to traditional biomass until the most recent decades, when modern biomass became more prevalent and now accounts for one-quarter of biomass energy. New renewables are discernible in the last few decades. Source: updated from Nakicenovic et al., 1998 and Grubler, 2008, see Chapter 1.[1]

---

1   **Nakicenovic, N., A. Grubler and A. McDonald (eds.), 1998:** *Global Energy Perspectives*. International Institute for Applied Systems Analysis (IIASA) and World Energy Council (WEC), Cambridge University Press, Cambridge, UK.

   Grubler, A., 2008: Energy transitions. In *Encyclopedia of Earth*. C. J. Cleveland (ed.), *Environmental Information Coalition, National Council* for Science and the Environment, Washington, DC.

characterized by a shift from reliance on traditional energy sources to coal and subsequently to oil and gas. Hydropower, biomass and nuclear energy during the past decades have a combined share of almost 22%. New renewables such as solar and wind are hardly discernible in the figure.

Despite this rapid increase in overall energy use, over three billion people still rely on solid fuels such as traditional biomass, waste, charcoal and coal for household cooking and heating. The resulting air pollution leads to over two million premature deaths per year, mostly of women and children. Furthermore, approximately 20% of the global population has no access to electricity. Addressing these challenges is essential for averting a future with high economic and social costs and adverse environmental impacts on all scales.

An energy system transformation is required to meet these challenges and bring prosperity and well-being to the 9 billion people expected by 2050. The encouraging news is that a beginning of such a transformation can be seen today in the rapidly growing investments in renewable energy sources, high-efficiency technologies, new infrastructures, near zero-energy buildings, electric mobility, 'smart' energy systems, advanced biomass stoves, and many other innovations. The policy challenge is to accelerate, amplify and help make the implementation of these changes possible, widespread and affordable. Initial experience suggests that many of these changes are affordable, although they may be capital intensive and require high upfront investments. However, in general they have lower long-term costs that offset many of the up-front added investment requirements. Many of these innovations also lead to benefits in other areas such as equity and poverty, economic development, energy security, improved health, climate change mitigation, and ecosystem protection.

This Summary for Policymakers expands on the GEA approach and the Key Findings. The Technical Summary provides further support for the key findings.

Goals Used in the Assessment and in the GEA Pathways Analysis

For many of the energy related challenges, different goals have been articulated by the global community, including, in many instances specific quantitative targets. Meeting these goals simultaneously has served as the generic framework for all assessments in the GEA. The GEA pathways illustrate how societies can reach global normative goals of welfare, security, health, and environmental protection outlined below simultaneously with feasible changes in energy systems.

The selection of indicators and the quantitative target levels summarized here is a normative exercise, and the level of ambition has, to the extent possible, been guided by agreements and aspirations expressed through, for example, the United Nations system's actions, resolutions, and from the scientific literature. This, of course, only refers to the necessary changes of the local and global energy systems; much more is required in other sectors of societies for overall sustainability to be realized.

In the GEA pathways analysis, global per capita gross domestic product (GDP) increases by 2% a year on average through 2050, mostly driven by growth in developing countries. This growth rate falls in the middle of existing projections. Global population size is projected to plateau at about 9 billion people by 2050. Energy systems must be able to **deliver the required energy services** to support these economic and demographic developments.

To avoid additional complexity, the GEA pathways assume one intermediate population growth pathway that is associated with uncertainty. Given that population growth has significant implications for future energy demand, however, it should be remembered that policies to provide more of the world's men and women the means to make responsible parental decisions (including safe contraception technologies) can significantly reduce the growth in population over the century as well as energy demand and CO_2 emissions. By increasing birth spacing, they would also bring benefits for maternal and child health.

Access to affordable modern energy carriers and end-use conversion devices to improve living conditions and enhancing opportunities for economic development for the 1.4 billion people without access to electricity and the 3 billion who still rely on solid and fossil fuels for cooking is a prerequisite for poverty alleviation and socioeconomic development.

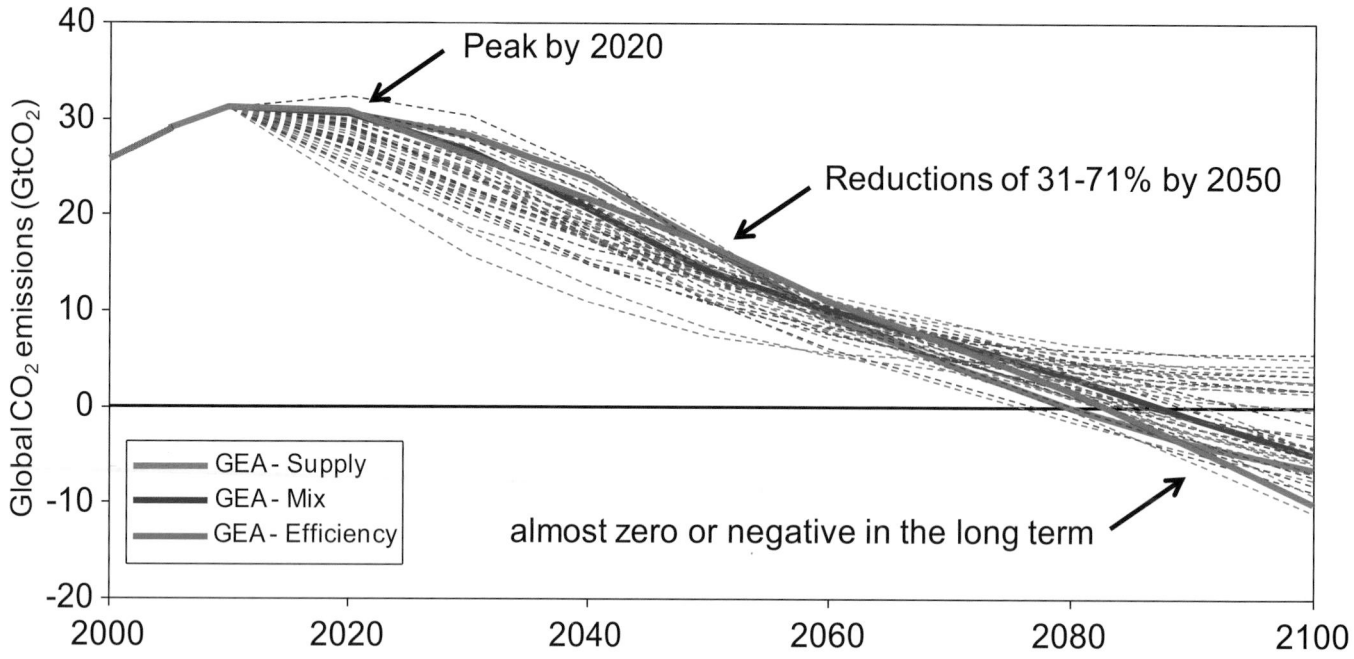

Figure SPM-2. | Development of global CO_2 emissions from energy and industrial sources to limit temperature change to below 2°C (with a success probability of >50%). Shown is that the emissions need to peak by around 2020 (or earlier) and decline toward zero during the following four to five decades. The later the peak occurs, the steeper the decline needs to be and higher the net "negative" emissions. The latter can be achieved through in the energy system through carbon dioxide capture and storage in conjunction with the use of sustainable biomass. Source: Chapter 17. For further details of the GEA pathways see the interactive web-based GEA scenario database hosted by IIASA: www.iiasa.ac.at/web-apps/ene/geadb.

Enhanced energy security for nations and regions is another key element of a sustainable future. Reduced global interdependence via reduced import/export balances, and increased diversity and resilience of energy supply have been adopted as key energy-related metrics. The targets for these goals were assessed ex-post through the GEA pathways analysis (Chapter 17), identifying the need for energy efficiency improvements and deployment of renewables to increase the share of domestic (national or regional) supply in primary energy by a factor of two and thus significantly decrease import dependency (by 2050). At the same time, the share of oil in global energy trade is reduced from the present 75% to below 40% and no other fuel assumes a similarly dominant position in the future.

The *climate change mitigation* goal is to contain the global mean temperature increase to less than 2°C above the preindustrial level, with a success probability of at least 50%. This implies global CO_2 emissions reductions from energy and industry to 30–70% of 2000 levels by 2050, and approaching almost zero or net negative emissions in the second half of the century (Figure SPM-2).

Health goals relating to energy systems include controlling household and ambient air pollution. Emissions reductions through the use of advanced fuels and end-use technologies (such as low-emissions biomass cookstoves) for household cooking and heating can significantly reduce human morbidity and mortality due to exposure to household air pollution, as well as help reduce ambient pollution. In the GEA pathways, this is assumed to occur for the vast majority of the world's households by 2030. Similarly, a majority of the world's population is also expected to meet the World Health Organization's (WHO) air quality guidelines (annual PM2.5 concentration < 10 µg/m³ by 2030), while remaining populations are expected to stay well within the WHO Tier I-III levels (15–35 µg/m³ by 2030). In addition, there needs to be a major expansion of occupational health legislation and enforcement in the energy sector.

Linkages between the energy system and the **environment** are at multiple levels and scales – from local to global. While the local environmental and ecological consequences of resource extraction, processing and energy conversion have been long recognized, attention is increasingly turning towards the growing evidence that humanity has reached a phase when anthropogenic pressures on Earth systems – the climate, oceans, fresh water, and the biosphere – risk irreversible disruption to biophysical processes on the planetary scale. The risk is that systems on Earth may then

Table SPM-1. | Global Burden of Disease, 2000 from Air Pollution and other Energy-related causes. These come from the Comparative Risk Assessment (CRA) published in 2004 by the World Health Organization (WHO). Estimates for 2005 in GEA for outdoor air pollution and household solid fuel use in Chapter 17 are substantially larger, but were not done for all risk pathways shown. Estimates for 2010 in the new CRA by WHO will be released in 2012 and will again include all pathways in a consistent framework.

	Total Premature Deaths – million	Percent of all Deaths	Percent of Global Burden in DALYs	Trend
Direct Effects [except where noted, 100% assigned to energy]				
Household Solid Fuel	1.6	2.9	2.6	Stable
Energy Systems Occupational*	0.2	0.4	0.5	Uncertain
Outdoor Air	0.8	1.4	0.8	Stable
Pollution				
Climate Change	0.15	0.3	0.4	Rising
Subtotal	2.8	5.0	4.3	
Indirect Effects (100% of each)				
Lead in Vehicle Fuel	0.19	0.3	0.7	Falling
Road Traffic Accidents	0.8	1.4	1.4	Rising
Physical Inactivity	1.9	3.4	1.3	Rising
Subtotal	2.9	5.1	3.4	
Total	5.7	10.1	7.7	

* One-third of global total assigned to energy systems.

Notes: These are not 100% of the totals for each, but represent the difference between what exists now and what might be achieved with feasible policy measures. Thus, for example, they do not assume the infeasible reduction to zero traffic accidents or air pollution levels.

Source: Chapter 4.

reach tipping points, resulting in non-linear, abrupt, and potentially irreversible change, such as destabilization of the Greenland ice sheet or tropical rainforest systems.

There are also a number of other concerns related to how energy systems are designed and operated. For example, activities need to be occupationally safe, a continuing concern as nano- and other new materials are used in energy systems. Other impacts such as oil spills, freshwater contamination and overuse, and releases of radioactive substances must be prevented (ideally) or contained. Waste products must be deposited in acceptable ways to avoid health and environmental impacts. These issues mostly influence local areas, and the regulations and their implementation are typically determined at the national level.

The world is undergoing severe and rapid change involving significant challenges. Although this situation poses a threat, it also offers a unique opportunity – a window of time in which to create a new, more sustainable, more equitable world, provided that the challenges can be addressed promptly and adequately. Energy is a pivotal area for actions to help address the challenges.

The interrelated world brought about by growth and globalization has increased the linkages among the major challenges of the 21st century. We do not have the luxury of being able to rank them in order of priority. As they are closely linked and interdependent, the task of addressing them simultaneously is imperative.

Energy offers a useful entry point into many of the challenges because of its immediate and direct connections with major social, economic, security and development goals of the day. Among many other challenges, energy systems are tightly linked to global economic activities, to freshwater and land resources for energy generation and food production, to biodiversity and air quality through emissions of particulate matter and precursors of tropospheric ozone, and to climate change. Most of all, access to affordable and cleaner energy carriers is a fundamental prerequisite for development, which is why the GEA places great emphasis on the need to integrate energy policy with social, economic, security, development, and environment policies.

Reaching the GEA goals simultaneously requires transformational changes to the energy system, in order to span a broad range of opportunities across urban to rural geographies, from developing to industrial countries, and in transboundary systems. The ingredients of this change are described in the following section.

Key Findings

The Global Energy Assessment (GEA) explored options to transform energy systems that simultaneously address all of the challenges above. A broad range of resources and technologies were assessed, as well as policy options that can be combined to create pathways[2] to energy for a sustainable future. These are the Key Findings:

1. Energy Systems can be Transformed to Support a Sustainable Future: the GEA analysis demonstrates that a sustainable future requires a transformation from today's energy systems to those with: (i) radical improvements in energy efficiency, especially in end use, and (ii) greater shares of renewable energies and advanced energy systems with carbon capture and storage (CCS) for both fossil fuels and biomass. The analysis ascertained that there are many ways to transform energy systems and many energy portfolio options. Large, early, and sustained investments, combined with supporting policies, are needed to implement and finance change. Many of the investment resources can be found through forward-thinking domestic and local policies and institutional mechanisms that can also support their effective delivery. Some investments are already being made in these options, and should be strengthened and widely applied through new and innovative mechanisms to create a major energy system transformation by 2050.

Humanity has the capacity, ingenuity, technologies and resources to create a better world. However, the lack of appropriate institutions, coordination mandates, political will and governance structures make the task difficult. Current decision making processes typically aim for short-term, quick results, which may lead to sub-optimal long-term outcomes. The GEA endeavors to make a compelling case for the adoption of a new set of approaches and policies that are essential, urgently required, and achievable.

The GEA highlights essential technology-related requirements for radical energy transformation:

- significantly larger investment in energy efficiency improvements especially end-use across all sectors, with a focus on new investments as well as major retrofits;

- rapid escalation of investments in renewable energies: hydropower, wind, solar energy, modern bioenergy, and geothermal, as well as the smart grids that enable more effective utilization of renewable energies;

- reaching universal access to modern forms of energy and cleaner cooking through micro-financing and subsidies;

- use of fossil fuels and bioenergy at the same facilities for the efficient co-production of multiple energy carriers and chemicals with full-scale deployment of carbon capture and storage; and

- on one extreme nuclear energy could make a significant contribution to global electricity generation, but on the other extreme, it could be phased out.

To meet humanity's need for energy services, comprehensive diffusion of energy and an increased contribution of energy efficiencies are required throughout the energy system – from energy collection and conversion to end use. Rapid diffusion of renewable energy technologies is the second but equally effective option for reaching multiple objectives. Conversion of primary energy to energy carriers such as electricity, hydrogen, liquid fuels and heat along with smart transmission and distribution systems are necessary elements of an energy system meeting sustainability objectives.

---

2   The GEA developed a range of alternative transformational pathways to explore how to achieve all global energy challenges simultaneously. The results of the GEA pathways are documented in detail at the interactive web-based GEA scenario database hosted by IIASA: www.iiasa.ac.at/ web-apps/ene/geadb.

The GEA makes the case that energy system transformation requires an iterative and dynamic transformation of the policy and regulatory landscape, thereby fostering a buildup of skills and institutions that encourage innovation to thrive, create conditions for business to invest, and generate new jobs and livelihood opportunities.

A major finding of the GEA is that some energy options provide multiple benefits. This is particularly true of energy efficiency, renewables, and the coproduction of synthetic transportation fuels, cooking fuels, and electricity with co-gasification of coal and biomass with CCS, which offer advantages in terms of supporting all of the goals related to economic growth, jobs, energy security, local and regional environmental benefits, health, and climate change mitigation. All these advantages imply the creation of value in terms of sustainability. This value should be incorporated into the evaluation of these and other measures and in creating incentives for their use.

One implication of this is that nations and corporations can invest in efficiency and renewable energy for the reasons that are important to them, not just because of a global concern about, for example, climate change mitigation or energy security. But incentives for individual actors to invest in options with large societal values must be strong and effective.

The GEA explored 60 possible transformation pathways and found that 41 of them satisfy all the GEA goals simultaneously for the same level of economic development and demographic changes, including three groups of illustrative pathways that represent alternative evolutions of the energy system toward sustainable futures.[3] The pathways imply radically changed ways in which humanity uses energy, ranging from much more energy-efficient houses, mobility, products, and industrial processes to a different mix of energy supply – with a much larger proportion of renewable energy and fossil advanced fossil fuel technologies (see Figure SPM-3).

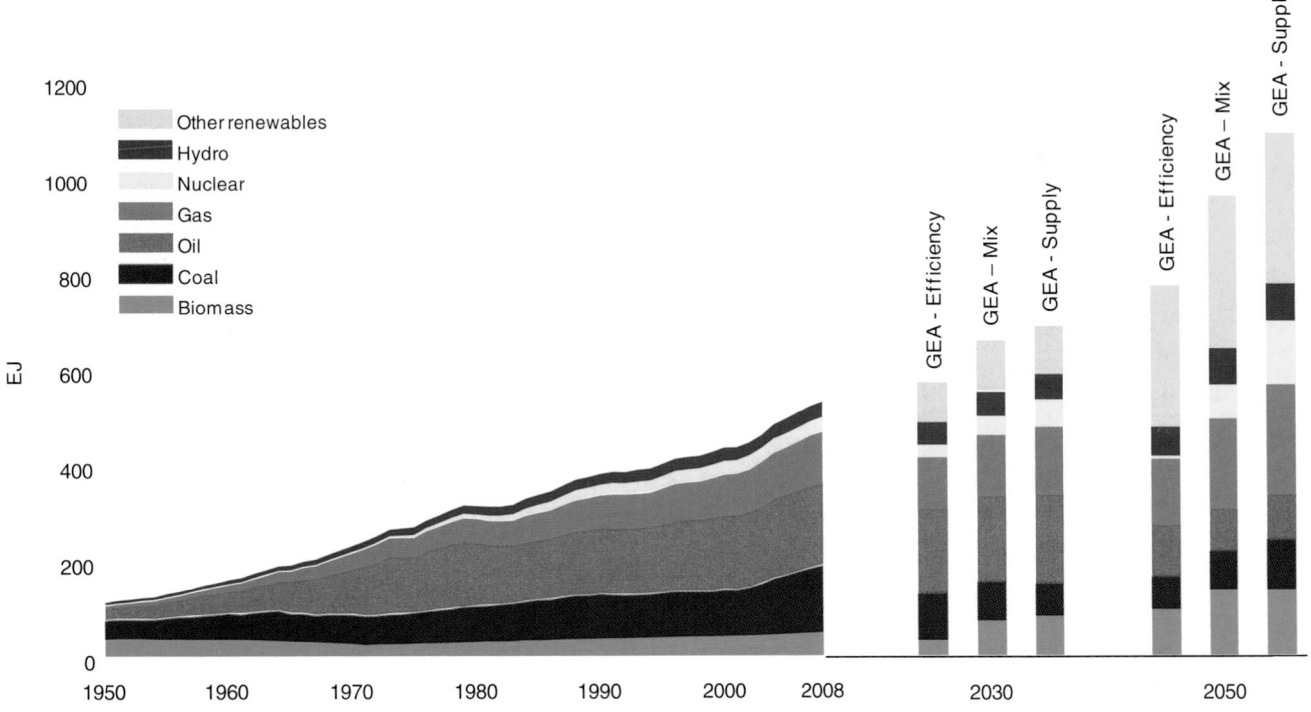

Figure SPM-3. | Development of primary energy to 2008 and in the three illustrative GEA pathways for the years 2030 and 2050. Source: based on Figures TS-24 and 17.13, Chapter 17. For further details of the GEA pathways see the interactive web-based GEA scenario database hosted by IIASA: www.iiasa.ac.at/web-apps/ene/geadb.

---

3   The pathways encompass eleven world regions, grouped into five GEA regions and energy sectors, including supply and demand, with a full range of associated social, economic, environmental and technological developments.

On the demand side, the three groups of GEA pathways pursue the energy efficiency options to a varying extent. On the supply side, the GEA pathways highlight the broad portfolio of technologies that will be needed to achieve the energy system transformation. Particularly important options are low-carbon energy from renewables, bioenergy, nuclear power, and CCS. In aggregate, at least a 60–80% share of global primary energy will need to come from zero-carbon options by 2050; the electricity sector in particular will need to be almost completely decarbonized by mid-century (low-carbon shares of 75–100%). Getting to that point requires major progress in several critical areas:

- Renewables: Strong renewable energy growth beginning immediately and reaching a global share of 30–75% of primary energy by 2050, with some regions experiencing in the high case almost a complete shift towards renewables by that time;

- Coal: A complete phase-out of coal power without CCS by 2050;

- Natural Gas: Natural gas acting as a bridging or transitional technology in the short to medium term and providing 'virtual' storage for intermittent renewables;

- Energy Storage: Rising requirement for storage technologies and 'virtual' systems (e.g., smart grids and demand-side management) to support system integration of intermittent wind and solar;

- Bioenergy: Strong bioenergy growth in the medium term from 45 EJ in 2005 to 80–140 EJ by 2050, including extensive use of agricultural residues and second-generation bioenergy to mitigate adverse impacts on land use and food production, and the co-processing of biomass with coal or natural gas with CCS to make low net GHG-emitting transportation fuels and or electricity;

- Nuclear: Nuclear energy as a choice, not a requirement. The GEA pathways illustrate that it is possible to meet all GEA goals even in the case of a nuclear phase-out. Nuclear energy can play an important role in the supply-side portfolio of some transition pathways; however, its prospects are particularly uncertain because of unresolved challenges surrounding its further deployment, as illustrated by the Fukushima accident and unresolved weapons proliferation risks;

- Carbon Capture and Storage: Fossil CCS as an optional bridging or transitional technology in the medium term unless there is high energy demand, in which case CCS may be essential. CCS technology offers one potentially relatively low-cost pathway to low carbon energy. CCS in conjunction with sustainable biomass is deployed in many pathways to achieve negative emissions and thus help achieve climate stabilization.

New policies would be needed to attract capital flows to predominantly upfront investments with low long-term costs but also low short-term rates of return.

The pathways indicate that the energy transformations need to be initiated without delay, gain momentum rapidly, and be sustained for decades. They will not occur on their own. They require the rapid introduction of policies and fundamental governance changes toward integrating global concerns, such as climate change, into local and national policy priorities, with an emphasis on energy options that contribute to addressing all these concerns simultaneously.

In sum, the GEA finds that there are possible combinations of energy resources and technologies that would enable societies to reach all the GEA goals simultaneously, provided that government interventions accommodate sufficiently strong incentives for rapid investments in energy end-use and supply technologies and systems.

2. An Effective Transformation Requires Immediate Action: Long infrastructure lifetimes mean that it takes decades to change energy systems; so immediate action is needed to avoid lock-in of invested capital into energy systems and associated infrastructure that is not compatible with sustainability goals. For example, by 2050 almost three-quarters of the world population is projected to live in cities. The provision of services and livelihood opportunities to growing urban populations in the years to come presents a major opportunity for transforming energy systems and avoiding lock-in to energy supply and demand patterns that are counterproductive to sustainability goals.

Given the longevity of the capital stock of energy systems and of the built environment, rates of change are slow and possible irreversibilities or 'lock-in' effects can have powerful long-lasting effects. Long-term transformations need to be initiated earlier rather than later. Therefore the time for action is *now*. Changes in current policies that are particularly critical in triggering longer-term transformations are technology, and urbanization.

Reflecting economic, social and environmental externalities in the market conditions is therefore a necessary first step to provide appropriate incentives for redirecting private sector investments. Such measures would include removal, or at least substantial reduction, of subsidies to fossil fuels without CCS and nuclear energy, stimulation of development and market entry of new renewable options, and emphasis on energy efficiency in all end-use sectors. According to the GEA pathway analysis, global energy systems investments need to increase to some US$1.7–2.2 trillion annually to 2050, with about US$300–550 billion of that being required for demand-side efficiency. This compares to about US$1 trillion supply-side investments and about $300 billion demand-side investments in energy components per year today. These investments correspond to about 2% of the world gross domestic product in 2005, and would be about 2–3% by 2050, posing a major financing challenge. New policies would be needed to attract such capital flows to predominantly upfront investments with low long-term costs but also low short-term rates of return.

Today about 3.5 billion people, about half the world population live in urban environments. Projections suggest that by 2050 an additional three billion people need to be integrated into the urban fabric. Housing, infrastructure, energy and transport services, and a better urban environment (especially urban air quality) are the key sustainability challenges for urban development.

Urban energy and sustainability policies can harness local decision-making and funding sources to achieve the largest leverage effects in the following areas:

* urban form and density (which are important macro-determinants of urban structures, activity patterns, and hence energy use, particularly for urban transport);

* the quality of the built environment (energy-efficient buildings in particular);

* urban transport policy (in particular the promotion of energy-efficient and 'eco-friendly' public transport and non-motorized mobility options); and

* improvements in urban energy systems through zero-energy building codes, cogeneration or waste-heat recycling schemes, where feasible.

There are important urban size and density thresholds that are useful guides for urban planning and policymaking. The literature review identified a robust density threshold of 50–150 inhabitants per hectare (5,000–15,000 people per square kilometer) below which urban energy use, particularly for transport, increases substantially and which should be avoided. There are also significant potential co-benefits between urban energy policies and environmental policies. However, they require more holistic policy approaches that integrate urban land use, transport, building, and energy policies with the more-traditional air pollution policy frameworks.

Policy coordination at an urban scale is as complex as potentially rewarding in sustainability terms. Institutional and policy learning needs to start early to trigger longer-term changes in urban form and infrastructures. A particular challenge is represented by small to medium sized cities (between 100,000 and 1 million inhabitants), as most urban growth is projected to occur in these centers, primarily in the developing world. In these smaller-scale cities, data and information to guide policy are largely absent, local resources to tackle development challenges are limited, and governance and institutional capacities are insufficient.

3. Energy Efficiency is an Immediate and Effective Option: *Efficiency improvement is proving to be the most cost-effective, near-term option with multiple benefits, such as reducing adverse environmental and health impacts, alleviating poverty, enhancing energy security and flexibility in selecting energy supply options, and creating employment* *and economic opportunities. Research shows that required improvements in energy efficiency particularly in end-use can be achieved quickly. For example:*

- *retrofitting buildings can reduce heating and cooling energy requirements by 50–90%;*

- *new buildings can be designed and built to very high energy performance levels, often using close to zero energy for heating and cooling;*

- *electrically-powered transportation reduces final energy use by more than a factor of three, as compared to gasoline-powered vehicles;*

- *a greater integration between spatial planning and travel that emphasizes shorter destinations and enhances opportunities for flexible and diverse choices of travel consolidating a system of collective, motorized, and non-motorized travel options offers major opportunities;*

- *through a combination of increased enegry efficiency and increased use of renewable energy in the industry supply mix, it is possible to produce the increased industrial output needed in 2030 (95% increase over 2005) while maintaining the 2005 level of GHG emissions.*

A portfolio of strong, carefully targeted policies is needed to promote energy efficient technologies and address, inter alia, direct and indirect costs, benefits, and any rebound effects.

Progress in accelerating the rate of energy efficiency improvement worldwide is critical to an energy system for sustainability. Quickly improving energy efficiency through new investments and retrofits requires focused and aggressive policies that support rapid innovation through more stringent regulations of energy efficiency, fiscal incentives for new technologies, and pricing GHG emissions. Combined with higher energy prices, a culture of conservation among consumers and firms, and an increase in urban density societies can realize a dramatic increase in energy efficiency.

A major challenge is to resolve the issue of split incentives, that is, the situation where those who would be paying for efficiency improvements and other energy investments are more oriented toward short-term rates of return than to the long-term profitability of the investments and, likewise, they are rarely the beneficiaries of reduced energy costs and other public benefits.

Regulations are essential elements of energy policy portfolios to drive an energy transition. Standards for building codes, heating and cooling, appliances, fuel economy, and industrial energy management are one of the most effective policy tools for improving energy efficiency and should be adopted globally. These regulatory policies are most effective when combined with fiscal incentives and attention-attracting measures such as information, awareness, and public leadership programs.

The GEA analysis provides considerable evidence of the ability of such policy packages to deliver major change. However, the results from three decades of experiences with energy efficiency policies in industrial countries also show other effects.

These cost factors and rebound effects mean that subsidies to encourage acquisition of energy-efficient devices are unlikely, on their own, to cause the dramatic energy efficiency gains called for in the GEA analysis. For these gains to be realized, carefully targeted policies are needed. For example, strong efficiency regulations have proven effective. These are updated regularly and have incentives to reward manufacturers who push technology designs toward advanced efficiency by using electricity tariffs that reward efficiency investments and conservation.

In the buildings sector, new and existing technologies as well as non-technological opportunities represent a major opportunity for transformative change of energy use. Passive houses that reduce energy use for heating and cooling by 90% or more, for example, are already found in many countries. Increased investments in a more energy-efficient building shell are in part offset by lower or fully eliminated investments in heating/cooling systems, with energy costs for operation almost avoided, making these new options very attractive. Passive house performance is possible also

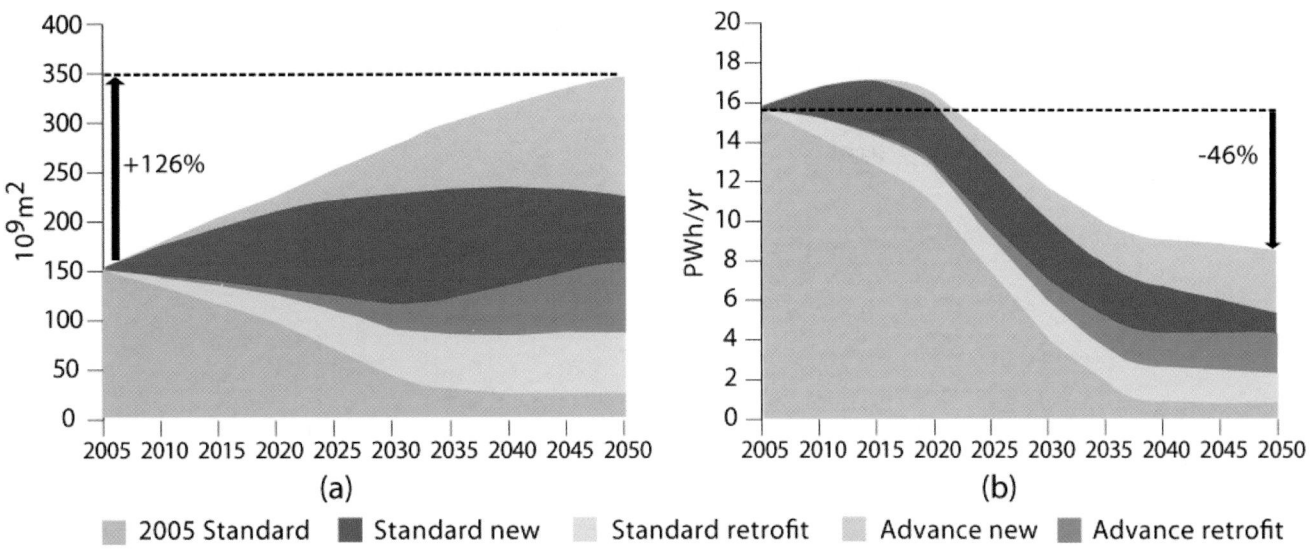

Figure SPM-4. | Global final thermal energy use in buildings (a) and global floor area (b) in the state-of-the-art scenario (corresponding approximately to the "GEA-Efficiency" group of pathways), 2005–2050. Source: Chapter 10.

Key: Explanations of efficiency categories: standard: today's stock; new: new buildings built to today's building code or anticipated new building codes (without additional policies); advance new: new buildings built to today's state-of-the-art performance levels; retrofit: assumes some efficiency gains, typically 35%; advanced retrofit: retrofit built to state-of-the-art levels.

for existing buildings, if it is included as a performance goal when major renovations are done. Energy Plus houses, delivering net energy to the grid over a year, have been constructed even in high latitudes. Building-integrated solar photovoltaics can contribute to meeting the electricity demand in buildings, especially in single-family homes, and solar water heaters can cover all or part of the heat required for hot water demand. However, requiring buildings to be zero-energy or net-energy suppliers may not be the lowest-cost or most sustainable approach in addressing the multiple GEA goals and typically may not be possible, depending on location.

Analysis carried out under the GEA pathway framework demonstrates that a reduction of global final energy use for heating and cooling of about 46% is possible by 2050 compared with 2005 through full use of today's best practices in design, construction, and building operation technology and know-how. This can be obtained even while increasing amenities and comfort and simultaneously accommodating an increase in global floor area of over 126% (Figure SPM-4).

There is, however, a significant risk of lock-in. If stringent building codes are not introduced universally and energy retrofits accelerate but are not subject to state-of-the-art efficiency levels, substantial energy use and corresponding GHG emissions can be 'locked-in' for many decades. This could lead to a 33% increase in global energy use for buildings by 2050 instead of a decrease of 46% (Figure SPM-5).

Wide adoption of the state-of-the-art in the buildings sector would not only contribute significantly to meeting the GEA's multiple goals, such developments would also deliver a wide spectrum of other benefits. A review of quantified multiple benefits showed that productivity gains through reduced incidence of infections from exposure to indoor air pollution score particularly high in industrial countries. Other benefits included increases in productivity, energy security, indoor air quality and health, social welfare, real estate values, and employment. The approximately US$57 trillion cumulative energy cost savings until 2050 in avoided heating and cooling energy costs alone substantially exceeds the estimated US$15 trillion investments that are needed to realize this pathway. The value of the additional benefits has also been shown to be substantial, often exceeding the energy cost savings. In several cases the multiple benefits are so significant, and coincide with other important policy agendas (such as improved energy security, employment, poverty alleviation, competitiveness), that they provide easier and more attractive entry points for local policymaking than climate change or other environmental agendas.

Influencing energy use in the transport sector involves affecting transport needs, infrastructure, and modes, as well as vehicle energy efficiency. Policies for urbanization will have a large impact on transport needs, infrastructure, and

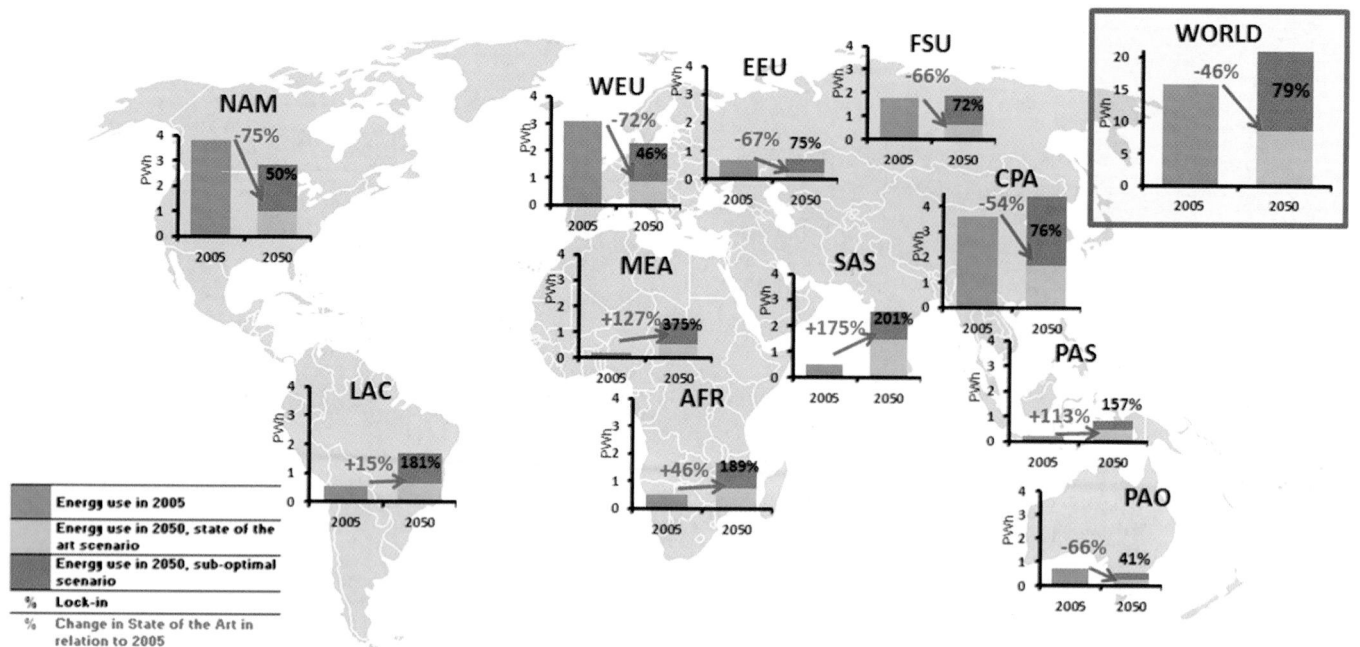

Figure SPM-5. | Final building heating and cooling energy demand scenarios until 2050: state-of-the-art (~corresponding roughly to the GEA-Efficiency set of pathways) and sub-optimal (~corresponding roughly to the GEA-Supply set of pathways scenarios, with the lock-in risk (difference). Note: Green bars, indicated by red arrows and numbers, represent the opportunities through the state-of-the-art scenario, while the red bars with black numbers show the size of the lock-in risk (difference between the two scenarios). Percent figures are relative to 2005 values. Source: Chapter 10.

the viability of different transport modes on the local scale. Both the decision to travel and the choice of how to travel affect fuel consumption. With a focus on urban road transport, a transition to sustainable transport can follow the framework known as 'avoid-shift-improve'. This considers three major principles under which diverse policy instruments are grouped, with interventions assuming different emphasis in industrial and developing countries. They need to focus on technological options, ('improve'), not only with respect to climate mitigation but also with respect to local environmental conditions and social concerns. The other two components – modal shift and avoiding travel influence the level of activity and structural components that link transport to carbon emissions.

A major transformation of transportation is possible over the next 30–40 years and will require improving vehicle designs, infrastructure, fuels and behavior. In the short term improving overall sector energy efficiency, introducing alternative low-carbon fuels and electricity, enhancing the diversification, quantity and quality of public modes of transport is necessary. Medium term goals require reducing travel distances within cities by implementing compact urban design that improves accessibility to jobs and services and facilitates use of non-motorized modes, and replacing and adopting vehicle and engine design (for trucks, airplanes, rail, and ships) following the best available technological opportunities for increasing efficiency and societal acceptability.

Transport policy goals for urbanization and equity include the adoption of measures for increasing accessibility and the affordable provision of urban mobility services and infrastructure that facilitates the widespread use of non-motorized options Cities can be planned to be more compact with less urban sprawl and a greater mix of land uses and strategic siting of local markets to improve logistics and reduces the distances that passengers and goods need to travel. Urban form and street design and layout can facilitate walking, cycling, and their integration within a network of public transport modes. Employers in many sectors can enhance the job-housing balance of employees through their decisions on where to be located and can provide incentives for replacing some non-essential journeys for work purposes with the use of information technologies and communication.

Modal share could move to modes that are less energy-intensive, both for passenger and freight transport. In cities, a combination of push and pull measures through traffic demand management can induce shifts from cars to public transit and cycling and can realize multiple social and health benefits. In particular, non-motorized transportation could be promoted everywhere as there is wide agreement about its benefits to transportation and people's health. Parking policies and extensive car pooling and car sharing, combined with information technology options can become key policies to reduce the use of cars. Efficient road capacity utilization, energy use and infrastructure costs for different modes could be considered when transport choices are made.

There are still many opportunities to improve conventional vehicle technologies. The combination of introducing incremental efficiency technologies, increasing the efficiency of converting the fuel energy to work by improving drive train efficiency, and recapturing energy losses and reducing loads (weight, rolling, air resistance, and accessory loads) on the vehicle has the potential to approximately double the fuel efficiency of 'new' light-duty vehicles from 7.5 liters per 100 km in 2010 to 3.0 liters per 100 km by 2050.

The emergence of electric drive technologies such as plug-in hybrid electric vehicles allows for zero tailpipe emissions for low driving ranges, up to around 50 kilometers in urban conditions. All-electric battery vehicles can achieve a very high efficiency (more than 90%, four times the efficiency of an internal combustion engine vehicle but excluding the generation and transmission of the electricity), but they have a low driving range and short battery life. If existing fuel saving and hybrid technologies are deployed on a broad scale, fleet-average specific fuel savings of a factor of two can be obtained in the next decade.

The aggregate energy intensity in the industrial sector in different countries has shown steady declines due to improvements in energy efficiency and a change in the structure of the industrial output. In the EU-27, for instance, the final energy use by industry has remained almost constant (13.4 EJ) at 1990 levels; 30% of the reduction in energy intensity is due to structural changes, with the remainder due to energy efficiency improvements.

In different industrial sectors, adopting the best achievable technology can result in savings of 10–30% below the current average. An analysis of cost-cutting measures in 2005 indicated energy savings potentials of 2.2 EJ for motors and 3.3 EJ for steam systems. The economic payback period for these measures ranges from less than nine months to four years. A systematic analysis of materials and energy flows indicates significant potential for process integration, heat pumps, and cogeneration.

Nevertheless, such a transformation has multiple benefits. Improved energy efficiency in industry results in significant energy productivity gains as a result, for example, in improved motor systems; compressed air systems; ventilation, heat recovery, and air conditioning systems; and improvements in comfort and the working environment through better lighting, thermal comfort, and reduced indoor air pollution from improved ventilation systems, and, in turn, improved productivity boosts corporate competitiveness.

4. Renewable Energies are Abundant, Widely Available, and Increasingly Cost-effective: The share of renewable energy in global primary energy could increase from the current 17% to between 30% to 75%, and in some regions exceed 90%, by 2050. If carefully developed, renewable energies can provide many benefits, including job creation, increased energy security, improved human health, environmental protection, and mitigation of climate change. The major challenges, both technological and economic, are:

- *reducing costs through learning and scale-up;*
- *creating a flexible investment environment that provides the basis for scale-up and diffusion;*
- *integrating renewable energies into the energy system;*
- *enhancing research and development to ensure technological advances; and*
- *assuring the sustainability of the proposed renewable technologies.*

While there remain sound economic and technical reasons for more centralized energy supplies, renewable energy technologies are also well-suited for off-grid, distributed energy supplies.

The GEA pathways show that renewable energies can exceed 90% of projected energy demand for specific regions. The GEA pathways analysis indicates that a significant increase in renewable energy supplies is technically feasible and necessary in order to meet the GEA goals.

Table SPM-2. | Renewable energy flows, potential, and utilization in EJ of energy inputs provided by nature.[a]

	Primary Energy 2005[b] [EJ]	Direct Input 2005 [EJ]	Technical potential [EJ/yr]	Annual flows [EJ/yr]
Biomass, MSW, etc.	46.3	46.3	160–270	2200
Geothermal	0.78	2.3	810–1545	1500
Hydro	30.1	11.7	50–60	200
Solar	0.39	0.5	62,000–280,000	3,900,000
Wind	1.1	1.3	1250–2250	110,000
Ocean	-	-	3240–10,500	1,000,000

[a] The data are direct energy-input data, not primary energy substitution equivalent shown in the first column. Considering technology-specific conversion factors greatly reduces the output potentials. For example, the technical 3150 EJ/yr of ocean energy in ocean thermal energy conversion (OTEC) would result in an electricity output of about 100 EJ/yr.

[b] Calculated using the GEA substitution method (see Chapter 1, Appendix 1.A.3).

Source: Chapter 7.

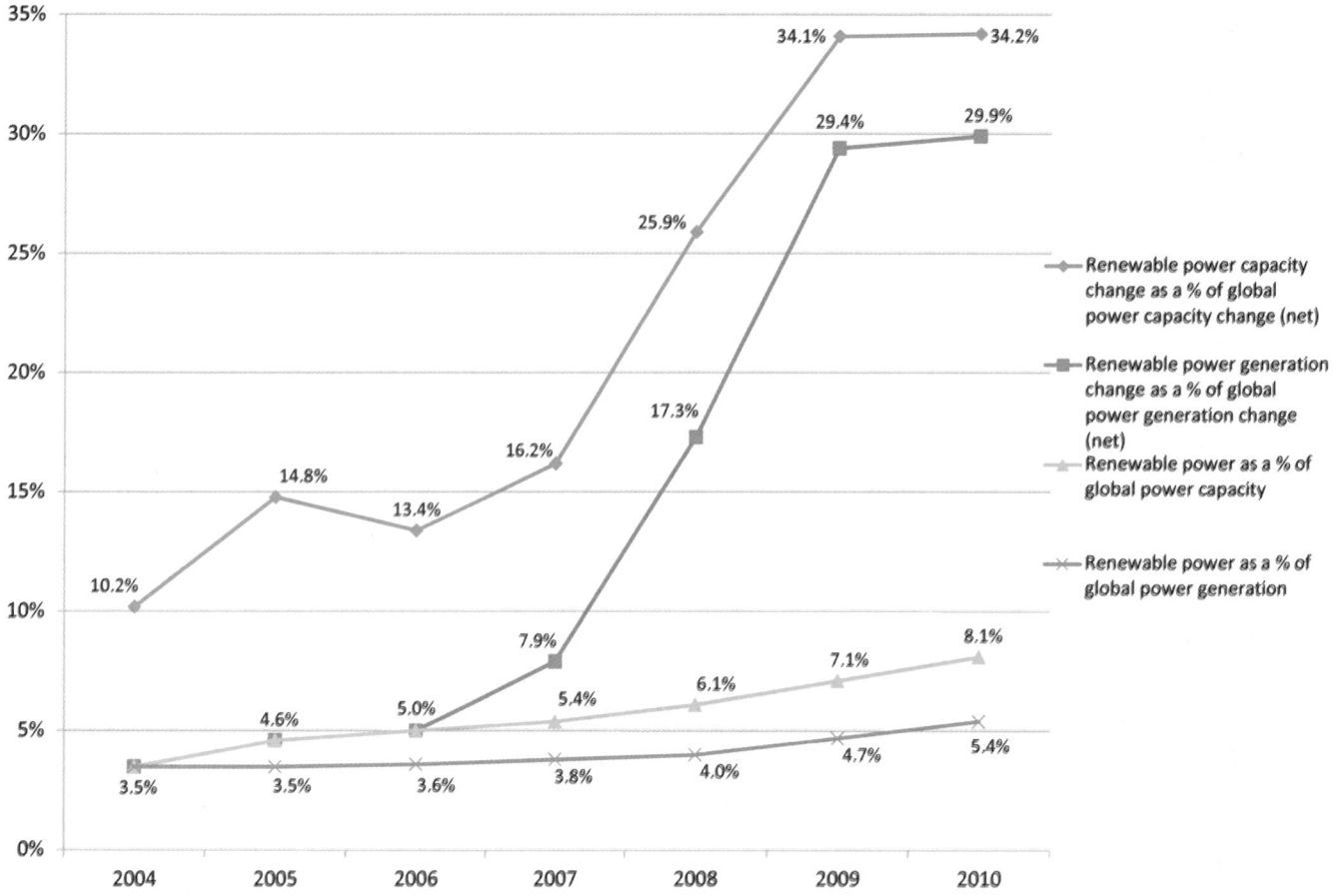

Figure SPM-6. | Renewable power capacity and generation (excluding large hydro) as a percentage of global capacity and generation, respectively, and their rates of change also in percent; 2004–2010. Source: UNEP and BNEF, 2011, see Chapter 11.[4]

The resource base is sufficient to provide full coverage of human energy demand at several times the present level and potentially more than 10 times this level (see Table SPM-2). Starting in 2007 renewable power generating capacity has grown fast in the world (see Figure SPM-6), and is now over 30% of total capacity expansion, excluding large scale hydropower. Figure SPM-7 shows a regional breakdown of the investments.

---

4   **UNEP and BNEF, 2011:** *Global Trends in Renewable Energy Investment 2011: Analysis of Trends and Issues in the Financing of Renewable Energy*. United Nations Environment Programme (UNEP), Nairobi, Kenya and Bloomberg New Energy Finance (BNEF), London, UK.

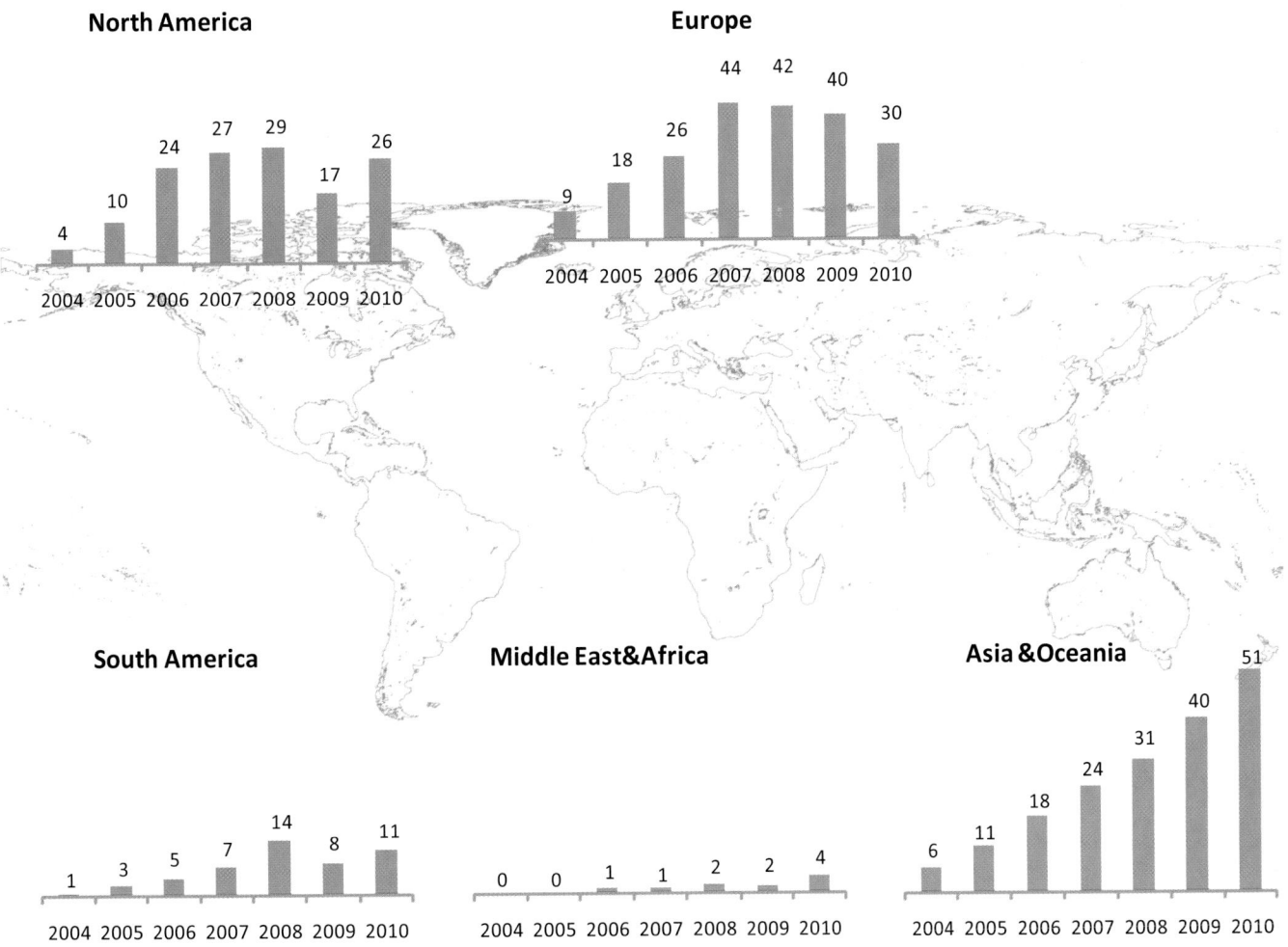

Figure SPM-7. | New financial investments in renewable energy, by region, 2004–2010 (US$$_{2005}$bn). New investment volume adjusts for reinvested equity; total values include estimates for undisclosed deals. This comparison does not include small-scale distributed energy projects or large-scale hydropower investments. Source: Chapter 11.

The rapid expansion in renewables, which has largely taken place in only a few countries, has usually been supported by incentives of different types or driven by quota requirements. Especially successful have been the feed-in tariffs used in the majority of EU countries, China, and elsewhere. Global investments in 2009 were slightly lower as a result of the financial crises (although with less reduction than for most other energy technologies), and in 2010 they rebounded. Both wind and solar PV electricity are nowadays cost-competitive in some markets and are projected to become so in many markets in the next 5–10 years without any public subsidy. However, renewables face resistance due to lock-in to conventional energies and substantial market barriers in the majority of markets.

The intermittent and variable generation of wind, solar and wave power must be handled within an electricity system that was not designed to accommodate it, and in which traditional base load-power from nuclear, geothermal and fossil power stations with restricted flexibility limit the system's ability to follow load variations. Energy systems have historically been designed to handle loads that vary over seconds, days, weeks, and years with high reliability. These systems are becoming increasingly able to accommodate increased quantities of variable generation through use of so-called smart systems with advanced sensing and control capabilities. With support from accurate and timely load forecasting, capacity management, and overall intelligent load and demand-side management, experience has shown that at least 20%, perhaps up to 50%, of variable renewable generation can be accommodated in most existing systems at low costs and that it is feasible to accommodate additional intermittent generation with additional investment in grid flexibility, low capital cost fuel-based generation, storage, and demand-side management (smart grids).

Safe and reliable improvements of interconnections between nations and across geographical regions will facilitate the compensation due to fluctuations in electricity generation from rapidly increasing shares of variable renewable energies in the system. Wind and solar PV and most hydrokinetic or ocean thermal technologies offer the unique additional attribute of virtually complete elimination of additional water requirements for power generation. Other renewable options, including bio-based options, geothermal, concentrating solar, and hydropower on a life-cycle basis, still require water for cooling of a steam turbine or are associated with large amounts of evaporation.

The development of high-voltage direct current (HVDC) transmission cables may allow the use of remote resources of wind and solar at costs projected to be affordable. Such cables have been installed for many years in sub-marine and on-shore locations, and demand is increasing (in the North Sea, for example). This is significant, as some of the best renewable energy resources are located far from load centers. In conjunction with energy storage at the generation location, such transmission cables can be used to provide base load electricity supply.

The GEA pathways analysis indicates that a significant increase in renewable energy supplies is technically feasible and necessary in order to meet the GEA objectives.

5. Major Changes in Fossil Energy Systems are Essential and Feasible: Transformation toward decarbonized and clean energy systems requires fundamental changes in fossil fuel use, which dominates the current energy landscape. This is feasible with known technologies.

- *CO_2 capture and storage (CCS), which is beginning to be used, is key. Expanding CCS will require reducing its costs, supporting scale-up, assuring carbon storage integrity and environmental compatibility, and securing approval of storage sites.*

- *Growing roles for natural gas, the least carbon-intensive and cleanest fossil fuel, are feasible, including for shale gas, if related environmental issues are properly addressed.*

- *Co-processing of biomass and coal or natural gas with CCS, using known technologies, is important for co-producing electricity and low-carbon liquid fuels for transportation and for clean cooking. Adding CCS to such coproduction plants is less costly than for plants that make only electricity.*

Strong policies, including effective pricing of greenhouse gas emissions, will be needed to fundamentally change the fossil energy system.

Table SPM-3. | Fossil and uranium reserves, resources, and occurrences.[a]

	Historical production through 2005 [EJ]	Production 2005 [EJ]	Reserves [EJ]	Resources [EJ]	Additional occurrences [EJ]
Conventional oil	6069	147.9	4900–7610	4170–6150	
Unconventional oil	513	20.2	3750–5600	11,280–14,800	> 40,000
Conventional gas	3087	89.8	5000–7100	7200–8900	
Unconventional gas	113	9.6	20,100–67,100	40,200–121,900	> 1,000,000
Coal	6712	123.8	17,300–21,000	291,000–435,000	
Conventional uranium[b]	1218	24.7	2400	7400	
Unconventional uranium	34	n.a.		7100	> 2,600,000

[a] The data reflect the ranges found in the literature; the distinction between reserves and resources is based on current (exploration and production) technology and market conditions. Resource data are not cumulative and do not include reserves.

[b] Reserves, resources, and occurrences of uranium are based on a once-through fuel cycle operation. Closed fuel cycles and breeding technology would increase the uranium resource dimension 50–60 fold. Thorium-based fuel cycles would enlarge the fissile-resource base further.

Source: Chapter 7.

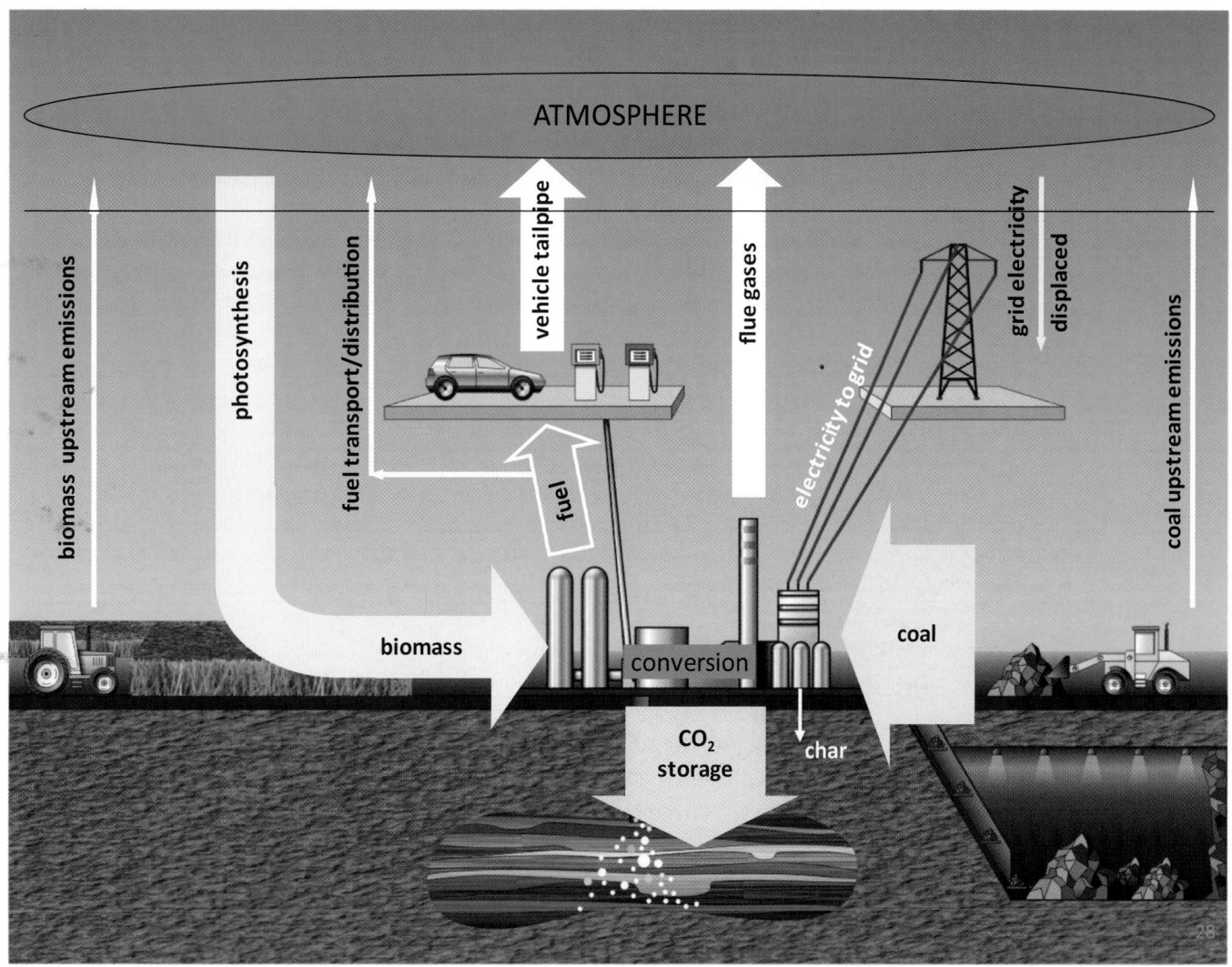

Figure SPM-8. | Carbon flows for conversion of coal and biomass to liquid fuels and electricity. For this system, when biomass is approximately 30% of the feedstock input (on a higher heating value basis), the net fuel cycle GHG emissions associated with the produced liquid fuels and electricity would be less than 10% of the emissions for the displaced fossil energy. Source: Larson et al., 2010, see Chapter 12.[5]

Continued use of coal and other fossil fuels in a carbon-constrained world requires strategies that deal with this reality. For industrial and developing countries, these strategies would differ in the short term but converge in the long term. For developing countries, the emphasis could be on increasing access to energy services based on clean energy carriers, building new manufacturing and energy infrastructures that anticipate the evolution to low-carbon energy systems, and exploiting the rapid growth in these infrastructures to facilitate introduction of the advanced energy technologies needed to meet sustainability goals. In industrial countries, where energy infrastructures are largely already in place, a high priority could be overhauling existing coal power plant sites to add additional capabilities (such as co-production of power and fuels) and CCS. Simply switching from coal to natural gas power generation without CCS will not achieve the needed carbon emission reductions.

Among the technologies that use fossil fuels, co-production strategies using coal plus biomass and CCS have the greatest ability to address all the major energy-related societal challenges. In the long term, hydrogen made from fossil fuels with CCS is an energy option, but infrastructure challenges are likely to limit this option in the near

5   **Larson, E. D., G. Fiorese, G. Liu, R. H. Williams, T. G. Kreutz and S. Consonni, 2010:** Co-production of Decarbonized Synfuels and Electricity from Coal + Biomass with CO₂ Capture and Storage: an Illinois Case Study. *Energy – Environmental Science*, **3**(1):28–42.

term. Co-production with CCS of electricity and carbon-based synthetic transportation fuels such as gasoline, diesel and jet fuel represent low-cost approaches for simultaneously greatly reducing carbon emissions for both electricity and transportation fuels and providing multiple benefits (Figure SPM 8): enhancing energy supply security, providing transportation fuels that are less polluting than petroleum-derived fuels in terms of conventional air pollutants, providing super-clean synthetic cooking fuels as alternatives to cooking with biomass and coal (critically important for developing countries), and greatly reducing the severe health damage costs due to air pollution from conventional coal power plants.

No technological breakthroughs are needed to get started with co-production strategies, but there are formidable institutional hurdles created by the need to manage two disparate feedstock supply chains (for coal and biomass) and provide simultaneously three products (liquid fuels, electricity, and CO_2) serving three different commodity markets.

6. Universal Access to Modern Energy Carriers and Cleaner Cooking by 2030 is Possible: Universal access to electricity and cleaner cooking fuels and stoves can be achieved by 2030; however, this will require innovative institutions, national and local enabling mechanisms, and targeted policies, including appropriate subsidies and financing. The necessary technologies are available, but resources need to be directed to meet these goals. Universal access is necessary to alleviate poverty, enhance economic prosperity, promote social development, and improve human health and well-being. Enhancing access among poor people, especially poor women, is thus important for increasing their standard of living. Universal access to clean cooking technologies will substantially improve health, prevent millions of premature deaths, and lower household and ambient air pollution levels, as well as the emissions of climate-altering substances.

Access to affordable modern energy carriers and cleaner cooking, improves well-being and enables people to alleviate poverty and expand their local economies. Enhanced access to modern energy carriers and cleaner cooking can become an effective tool for improving health for example, by reducing air pollution and can also help combat hunger by increasing food productivity and reducing post-harvest losses. Modern energy carriers and end-use conversion devices could improve education and school attendance by providing better lighting, heating, and cooling services. Electrifying rural health centers enables medical services to be provided at night, medicines to be preserved and more-advanced medical equipment to be used. Reduction of the proportional cost of energy services, particularly for rural poor people who spend a significant part of their time and disposable income on energy, is also important. This can liberate financial and human, especially women's, resources for other important activities or expenses, such as education, purchasing more and better-quality food, and expanding income-generating activities.

Several challenges exist to improving access to modern forms of energy and cleaner cooking. These include low income levels, unequal income distribution, inequitable distribution of modern forms of energy, a lack of financial resources to build the necessary infrastructure, weak institutional and legal frameworks, and a lack of political commitment to scaling up access. Even among households that have physical access to electricity and modern fuels, a lack of affordability and unreliable supplies limit their ability to use these resources, particularly for productive purposes. In addition to access to modern forms of energy, there must be access to end-use devices that provide the desired energy services. Those who can afford modern energy carriers may still not be able to afford the upfront costs of connections or the conversion technology or equipment that makes that energy useful.

While the scale of the challenge is tremendous, access to energy for all, electricity for all, and modern fuels or stoves for all by 2030 is achievable. This will require global investments of US$36–41 billion annually – a small fraction of total energy infrastructural investments required by 2030. It is expected that as households with public sector support gain access to modern energy and end-use devices and start earning incomes, their standard of living and ability to pay for the energy services utilized would successively expand.

Between 1990 and 2008 almost two billion people gained access to electricity, more than the corresponding population increase of 1.4 billion people over that time period (see Figure SPM-9). By 2030, the 1.4 billion people currently without access to electricity plus the projected population increase to 2030 of 1.5 billion people need to be connected to meet the GEA goal on electricity access (see Figure SPM-10). To achieve this, a multitrack approach is needed, combing grid extension with microgrids and household systems. Grid extension is currently the lowest cost per kWh delivered and also the preferred delivery form by most customers because of the capacity to deliver larger quantities of power for

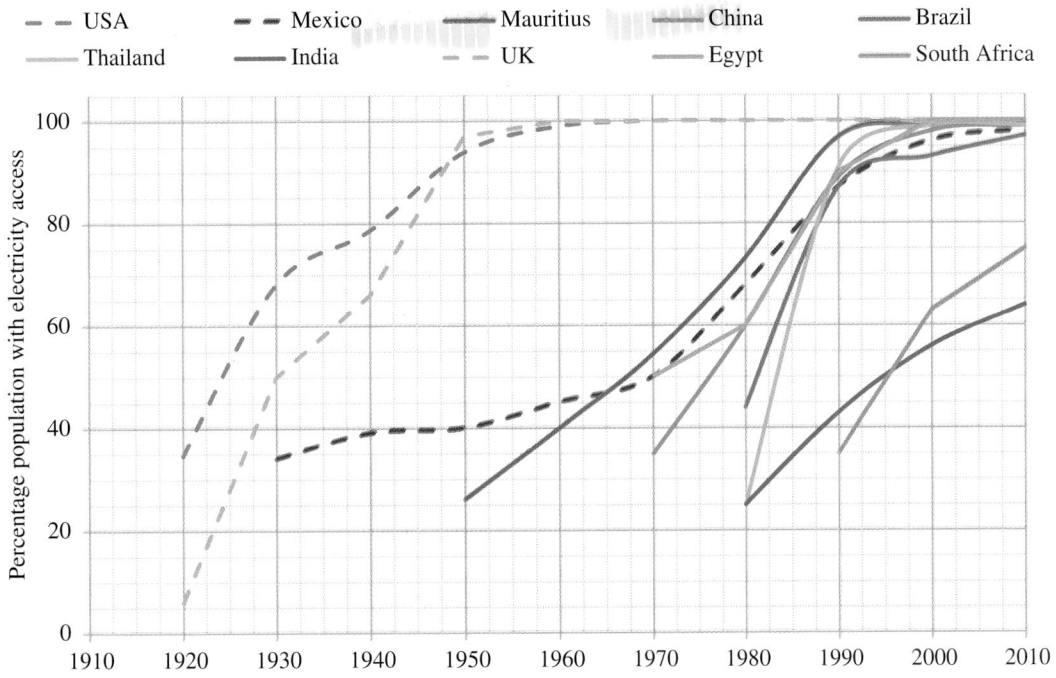

Figure SPM-9. | Historical experience with household electrification in select countries. Source: Chapter 19.

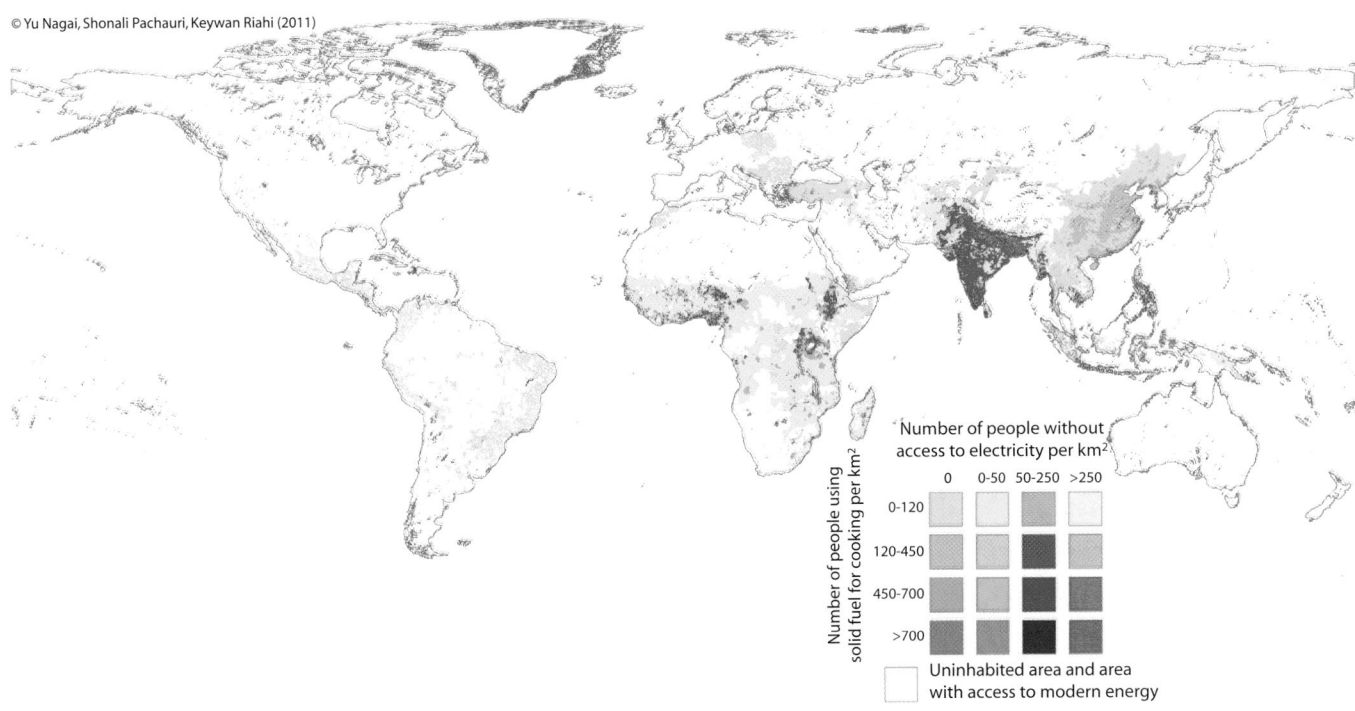

Figure SPM-10. | Density of population lacking access to modern energy carriers in 2005. Colored areas show people per km² without access to electricity and those that use solid fuels for cooking, e.g., dark blue and red areas show where people do not have access to electricity and cook predominately using solid fuels. Source: Chapters 17 and 19.

productive purposes. For many remote populations grid extension by 2030 will be highly unlikely and microgrids offer an alternative, based on local renewable energies or imported fossil fuels. An interesting approach to providing modern energy and development in remote villages is the multifunctional platform beginning to gain hold in West Africa. Household electrification is expanding rapidly in some countries, based on solar PV that are financed by micro-credits that has been done without increasing household expenses for energy (replacing candles and kerosene).

About 3 billion people rely entirely, or to a large degree, on traditional biomass or coal for cooking and heating. This number has not changed appreciably over the last decades, particularly among households in rural areas. Indeed, more people rely on these fuels today than any time in human history. Improving the cooking experience for these populations will require access to cleaner liquid or gaseous fuels, especially biogas, liquid petroleum gas (LPG), and ethanol, or alternatively access to advanced biomass stoves with efficiency and pollutants emissions similar to those of gas stoves. Transitioning to such fuels or stoves is not likely to have negative implications for climatic change. This is because transitioning to modern fuels (even in the case that these are fossil based) will displace large quantities of traditional biomass use. Current technologies that use traditional biomass are a factor 4–5 times less efficient than cooking with modern fuels such as LPG, and are associated with significant emissions of non-CO_2 Kyoto gases (e.g., CH_4, N_2O) and aerosols (e.g., BC, OC) due to incomplete combustion.

Providing universal and affordable access to electricity and cleaner cooking is possible if timely and adequate policies are put in place. Overall, and on the basis of successful experiences of increasing access to modern energy, no single approach can be recommended above the others. What is clear, however, is that the current institutional arrangements and policies have met with mixed success, at best. Reforms are needed, at global and country level, to strengthen the feasibility of energy projects for poor people, expand the range of players involved, open up the regulatory system, and allow for innovation. In the specific case of access to cleaner cooking, fuel subsidies alone will be neither sufficient nor cost-effective in terms of achieving ambitious energy access objectives (see Figure SPM-11). Financial mechanisms, such as micro-credit, will need to complement subsidies to make critical end-use devices such as cleaner cookstoves affordable for poor people.

A paradigm shift is needed in the approach to energy planning and policy implementation in order to facilitate access to modern forms of energy and cleaner cooking. Current supply-side approaches that simply take as their starting point the provision of electricity, petroleum, or gas, or of equipment of a particular type (solar technology, improved cookstoves, biogas, and other forms of bioenergy) are unable to reap the full potential of social and economic improvements that

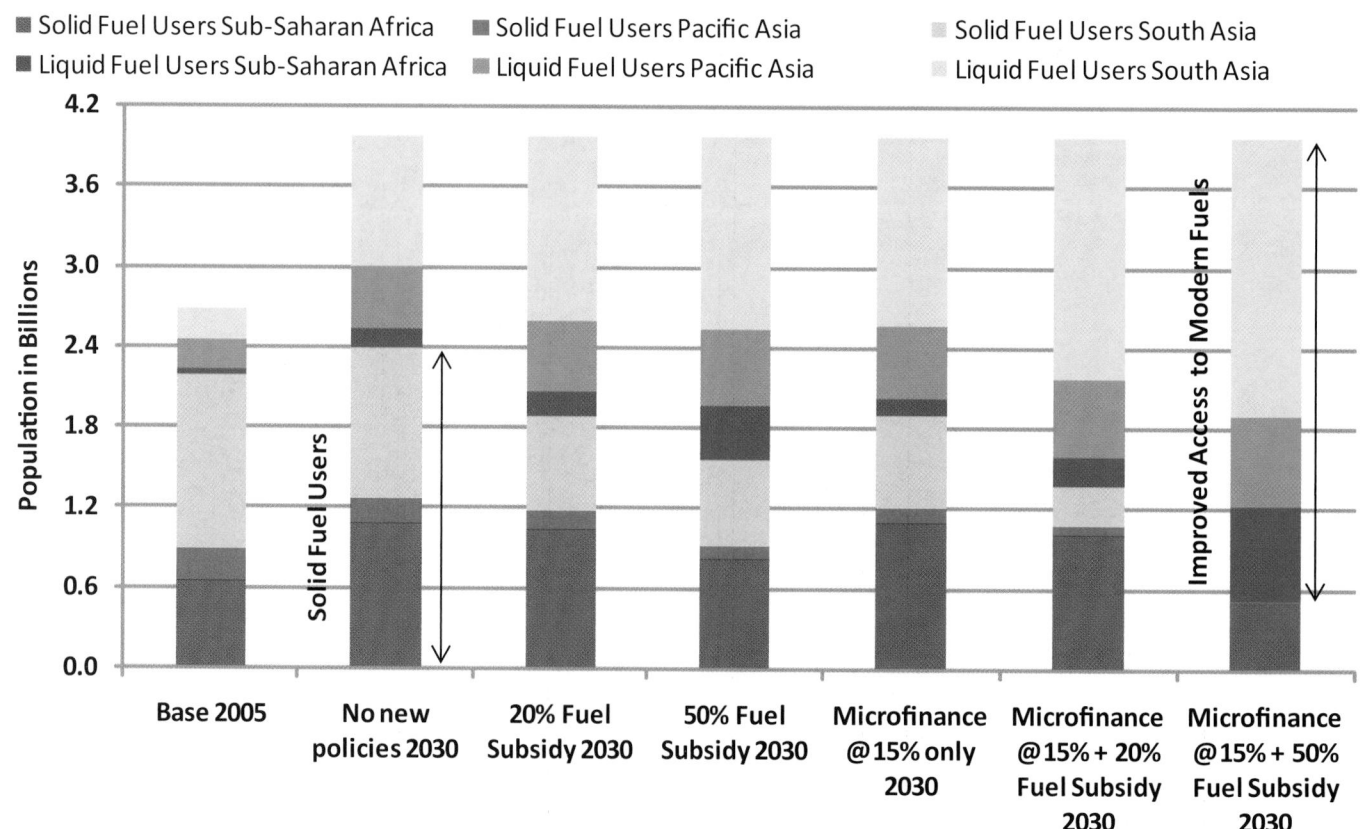

Figure SPM-11. | Impact of alternative policy scenarios on access to cleaner cooking fuels in three developing regions. Subsidies are relative to consumer price levels and are additional to existing subsidies. Source: Chapter 17.

follow from improved energy access and cleaner cooking. Leveraging funding and access to capital from public and private sources – for needed investments at the macro level and, at the micro level, for meeting costs for low-income households – is crucial in efforts to expand access to energy services for the poorest people. Creative financing mechanisms and transparent cost and price structures will be critical to achieving the required scale-up and quick roll-out of solutions to improve access.

Policy recommendations in the form of general ideas or guidelines are provided below. Regional and national contexts should be considered in defining strategies, instruments, and measures.

- A better understanding and a clearer diagnosis of the structure and functioning of energy systems, along with the needs (energy services) to be supplied, is needed. It has often been absent in the discussion of proposals and the role of public policies. Good policies need good diagnoses. Support and funds for diagnosis and information should be part of the strategies.

- Subsidies are generally justified as a response to inequality and social expectations in energy provision. However, their net effect can be positive or negative depending on the intended goals of the subsidy, and the way a subsidy is implemented. An effective tariff and subsidy regime has to be transparent and minimize administrative costs to avoid gaming of the system and to maximize the benefits that accrue to the intended recipients. Subsidies to energy should be complemented with funds toward solving the first-cost capital financing problem since up-front costs of equipment are, usually, the key barrier.

- Financing mechanisms are needed for every scale of energy intervention. Mobilizing affordable and genuine international, regional, national, and local funds is crucial.

- Energy access policy is part of a wider development policy and should be aligned with other sector policies and objectives. If these policies are misaligned, they can reduce the effectiveness of any given policy. Policy misalignments can occur when different energy policies work at cross-purposes or when government priorities that could benefit from an effective energy policy are not aligned. In particular, there is a need to link rural and peri-urban energy supply more closely with rural development. This would shift the focus from minimal household supply to a more comprehensive approach to energy that includes productive activities and other welfare-enhancing uses of energy. Ideally, the linkages between energy and other policy priorities, such as health, education, gender equality and poverty alleviation, should be recognized explicitly and local solutions that address these needs be encouraged and supported.

- Capacity development is needed, especially for the design and implementation of public policies oriented to poor people.

7. An Integrated Energy System Strategy is Essential: An integrated approach to energy system design for sustainable development is needed – one in which energy policies are coordinated with policies in sectors such as industry, buildings, urbanization, transport, food, health, environment, climate, security, and others, to make them mutually supportive. The use of appropriate policy instruments and institutions can help foster a rapid diffusion and scale-up of advanced technologies in all sectors to simultaneously meet the multiple societal challenges related to energy. The single most important area of action is efficiency improvement in all sectors. This enhances supply side flexibility, allowing the GEA challenges to be met without the need for technologies such as CCS and nuclear.

Energy-focused policies must be coordinated and integrated with policies addressing socioeconomic development and environmental protection in other sectors. Effective policy portfolios will require a combination of instruments, including regulatory frameworks and investment policies, as well as measures for strengthening capacity development, which stimulate innovation.

The main conclusion from the GEA pathways analysis is that energy efficiency improvements are the most important option to increase the flexibility of regional and sectoral energy end use and supply systems. In pathways with high rates of efficiency improvements, it was possible to achieve the GEA normative goals under any of the assumed supply portfolio restrictions and even without including nuclear energy and CCS technologies.

Energy systems differ between regions, between major economies, and between developing and industrial countries. Approaches to the necessary transitions to create energy systems for a sustainable future therefore vary, and policies that work successfully in one region may fail in another. Nevertheless, there are lessons to be learned from shared experiences. The evolution of energy systems will depend on how well technologies are implemented and how well policies are instituted to bring about the required changes.

Prevailing market and institutional structures in the energy sector have a significant influence on investments in different end-use and supply-side options. In countries with well-developed energy markets, spot-market energy trading is common and long-term contracts are becoming less frequent; and it is now more difficult to ensure long-term returns on large-scale investments. This is the main impediment to financing of large capital-intensive energy-supply projects.

Governments must recognize that policies promoting competition in the electricity sector must also prevent the short-term exercise of market power that results in unjustified excessive profits for some producers and speculators as well as price volatility for consumers, requiring continued regulation and public sector involvement in energy system planning and long-term contracting.

A regulatory framework is essential as it facilitates the creation and modernization of physical infrastructure and capital investments in energy end-use and supply systems. It is also necessary for economic development and poverty reduction.

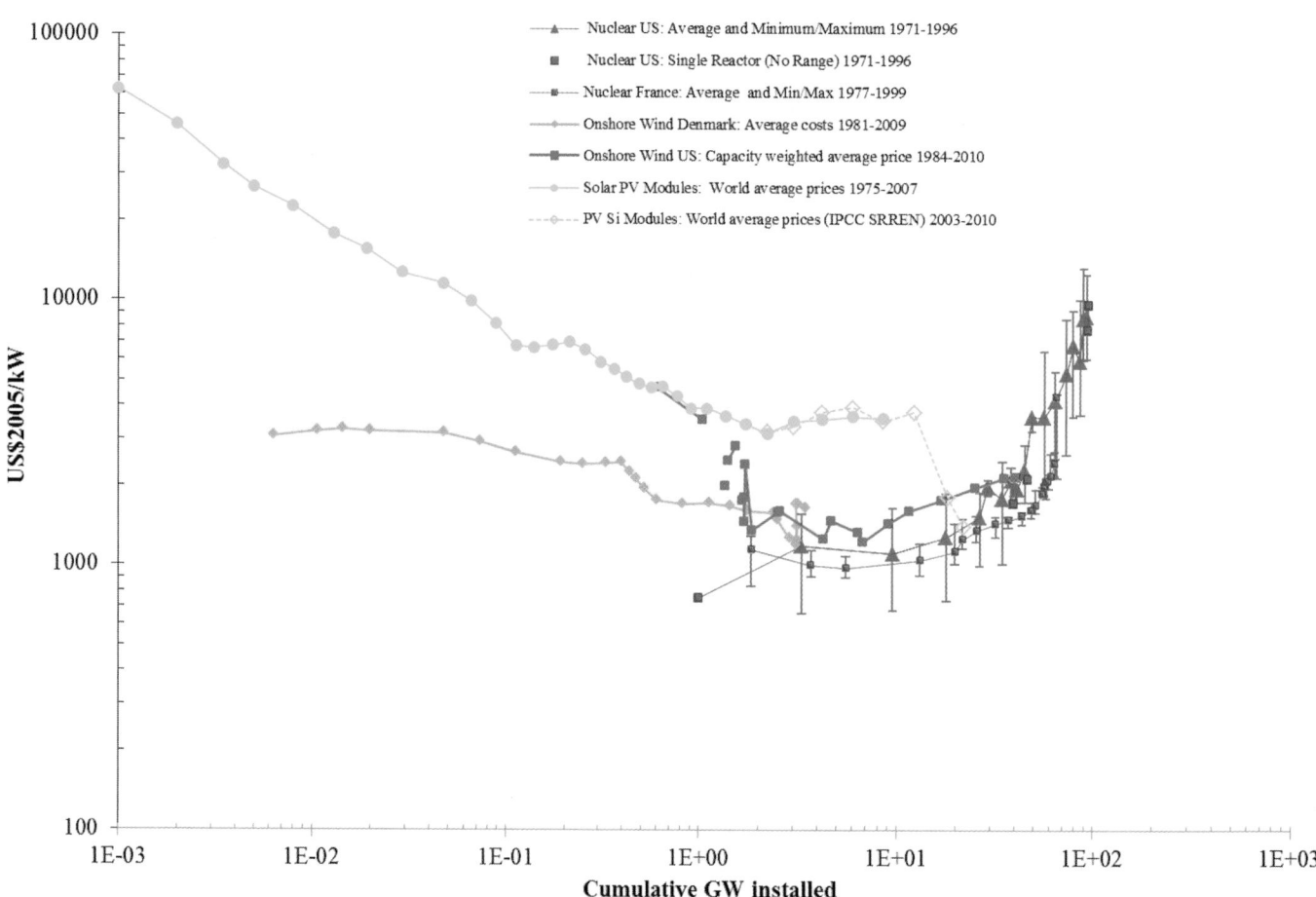

Figure SPM-12. | Cost trends of selected non-fossil energy technologies (US$2005/kW installed capacity) versus cumulative deployment (cumulative GW installed) Chapter 24 data have been updated with most recent cost trends (2010) available in the literature for PV Si Modules and US onshore wind turbines. Note that the summary illustrates comparative cost trends only and is not suitable for direct economic comparison of different energy technologies due to important differences between the economics of technology components (e.g. PV modules versus total systems installed), cost versus price data, and also differences in load factors across technologies (e.g., nuclear's electricity output per kW installed is three to four times larger than that of PV or wind turbine systems). Source: Chapter 24.

Success will depend on the implementation of robust public and private partnerships that can achieve unprecedented cooperation and integration between and among the public and private sectors, civil society, and academia.

A multiplicity of policies is required to address the potential impacts of the energy system on human health and the environment. A mix of regulations, information programs, and subsidies are needed, for example, to stimulate the rapid adoption of household energy-using devices that have virtually zero indoor emissions. Ambient air quality requires regulations on emissions from fuel combustion. Similarly, regional air quality must be protected by technology and emissions regulations or by direct emissions pricing.

Policies to foster energy from biomass should seek to minimize the trade-offs between biomass for food and biomass for fuel by encouraging the use of biomass residues and sustainable feedstocks as well as efficient conversion processes. Many developing countries import all or most of their current liquid fuels (in the form of oil and diesel) at increasingly higher costs and have at the same time large areas that are off-grid.

Greenhouse gas pricing policies will be essential in shifting energy systems toward low-carbon emission technologies, fuels, and activities. While there is disagreement on which pricing method is best – carbon taxes or cap and trade – the two approaches can be designed so that their effects are quite similar. The price certainty of a carbon tax can be approximated with cap and trade by setting a price floor and ceiling for permit prices. The revenues generated by a carbon tax can also be obtained by auctioning permits in the cap and trade approach

It is important to complement GHG pricing with a portfolio of other regulatory and market mechanisms. This is because different instruments are more effective in different sectors, geographic and cultural regions, as well as for different options. For instance, due to the magnitude and diversity of market barriers prevailing in the building sector, different regulatory and market-based instruments and their packages needed to be tailored to overcome specific barriers.

Strategic alliances and strong coordination among various policy fields will be able to lead to the realization of a larger share of technological potential by improving the economics of efficiency investments through the addition of further benefits to cost-efficiency considerations, such as security, employment, social welfare, and regional development. For example, policies for urban planning that encourage high density development with investments in public transport are likely to lead to lower long-term energy demand. Similarly, policies for renewable energy technologies could emphasize positive spillover effects on new venture and job creation. By actively seeking opportunities for such cross-sectoral integration; the required changes in the energy system may be accelerated. For example, a shift to clean cooking may be regarded as much a required change in the energy system, as an intervention to improve maternal and child health.

8. Energy Options for a Sustainable Future bring Substantial, Multiple Benefits for Society: *Combinations of resources, technologies, and polices that can simultaneously meet global sustainability goals also generate substantial economic, environmental, and social development benefits. These include, but are not limited to, improved local health and environment conditions, increased employment options, strengthened local economies through new business opportunities, productivity gains, improved social welfare and decreased poverty, more resilient infrastructure, and improved energy security. Synergistic strategies that focus on local and national benefits are more likely to be implemented than measures that are global and long-term in nature. Such an approach emphasizes the local benefits of improved end-use efficiency and increased use of renewable energy, and also helps manage energy-related global challenges. These benefits make the required energy transformations attractive from multiple policy perpectives and at multiple levels of governance.*

The energy systems illustrated by the GEA pathways meet the sustainability goals by design while generating substantial economic, environmental, and social benefits. For example, achieving near-term pollution and health objectives is furthered by investing in the same energy technologies that would be used to limit climate change. Policies to control emissions of greenhouse gases, or to increase access to cleaner cooking fuels could, in turn, bring significant improvements in pollution related health impacts. For example as the GEA pathways indicate, a saving of 20 million disability adjusted life years (DALYs) from outdoor air pollution and more than 24 million DALYs from household air pollution. In addition, universal access to electricity and cleaner cooking fuels opens up opportunities for education, for income generating activities, and significantly improved well-being.

This synergy is crucial and advantageous, given that measures which lead to local and national benefits (e.g., improved health and environment) may be more easily adopted than those measures that are put forward solely on the grounds of global goals. Many energy efficiency and renewable energy options enjoy such synergies and generate benefits across multiple objectives. Some of these advantages can be so substantial for certain investments and measures that they may offer more attractive entry points into policymaking than the climate or social targets alone. This is particularly the case where benefits are local rather than global. Seeking local benefits and receiving global benefits as a bonus is very attractive, and this is often the case for investments in energy efficiency and renewable sources of energy.

Therefore, even if some of these multiple benefits cannot be easily monetized, identifying and considering them explicitly may be important for decision-making. Cost-effectiveness (or cost-benefit) analyses evaluating energy options may fare differently when multiple benefits are considered.

The enhancement of end-use efficiency in buildings, transport and industry offers many examples of benefits across multiple environmental, social, and economic objectives:

* *inter alia* improved social welfare as a result of very high efficiency and thus very low fuel-cost buildings;

* reduced need for public funds spent on energy price subsidies for people living in poverty; health benefits through significantly reduced indoor and outdoor air pollution, often translating into commendable productivity gains;

* productivity gains and general improvements in operational efficiency in industry translate into strengthened competitiveness; and

* enhancing efficiency by increasing the rate of building retrofits can in addition be a source of employment and know-how.

Other benefits that are difficult to quantify include improved comfort and well-being, reduced congestion, new business opportunities, and better and more durable capital stock.

Rapid decarbonization of the energy system for climate protection also reduces the need for subsidies presently given to carbon-intensive petroleum products and coal. Subsidies for these fuels amount to approximately US$132–240 billion per year, and only 15% of this total is spent directly towards those with limited access to clean energy. However, GHG mitigation in the GEA pathways would, at the same time, reduce consumption of carbon-intensive fossil fuels, leading to a reduction in the need for subsidies for petroleum products and coal in the order of US$70–130 billion per year by 2050 compared to today.

Whether an impact is a benefit or a liability depends on the baseline and specific local situation. For example, while LPG causes major environmental and climate impacts in itself, it still has major advantages in many areas when it replaces traditional biomass as a fuel. Thus a unique novelty of the analysis is that it provides a new, additional framework for a well-founded assessment for individual decisions to choose among various energy alternatives, which complement financial appraisals. For example, in regions where access to modern forms of energy is a major energy policy goal, evaluations of "energy security" will play an essential role in ranking the different options available at comparable costs. In other areas, access or employment may be key secondary objectives of energy policy and these can play an important role in additional prioritization of options with comparable local costs.

There is a broad array of different benefits in the spectrum of policy target areas, which represent many potential entry points into policy-making. However, some options can have a wider range of multiple benefits than others, in particular renewable energies and improved energy end-use efficiency.

9. Socio-Cultural Changes as well as Stable Rules and Regulations will be Required: Crucial issues in achieving *transformational change toward sustainable future include non-technology drivers such as individual and public awareness, community and societal capacities to adapt to changes, institutions, policies, incentives, strategic spatial planning, social norms, rules and regulations of the marketplace, behavior of market actors, and societies' ability to introduce through the political and institutional system, measures to reflect externalities. Changes in cultures, lifestyles, and values are also required. Effective strategies will need to be adopted and integrated into the fabric of national socio-cultural, political, developmental, and other contextual factors, including recognizing and providing support for the opportunities and needs of all nations and societies.*

The complexity, magnitude, and speed of the changes envisaged in this transformation will necessitate a major shift in the way that societies analyze and define the concept of 'capacities' and the way in which they go about the important task of developing these capacities to meet the challenges of energy transitions. Different from some of the linear approaches to capacity development and to technology transfer and deployment used today, which often fail to appreciate the complexity of change processes, the concept of capacity development advanced by the GEA is intimately linked to the energy transitions perspective based on multilayered processes of system change.

In these processes, special attention is paid to the informal institutions that arise out of historically shaped habits, practices, and vested interests of players in the system already in place and to the tendency for path dependence, where past choices constrain present options. They are given special attention because they constitute potential impediments to needed change. In the transitions perspective, both learning and unlearning such habits, practices, and norms in the course of change are important.

Traditional habits, practices, and norms also shape the styles of communication in societies. Evidence shows that the more successful change processes take place in environments that tend to move away from top-down communication and consultation to more active and continuous dialogue practices. Capacity development has an important role to play in building mechanisms of support and capacities for interactive feedback, flexibility, and adaptive management and change. And because these traditional habits, practices, and norms are embedded in a broader social context, building capacities for dialogue at the local level is essential.

Market development and the role of feedback and flexibility at the local and project level are also essential in support of the diffusion of new energy technologies, but they are usually ignored in the design of capacity building initiatives. Also important is the need to build and strengthen capacities for local manufacture, repair, and distribution of new energy-related technologies, whether related to improved cookstoves, solar home systems, or other forms of early energy access initiatives, or to the introduction of more modern and decentralized forms of energy. Successful examples of energy technology development and diffusion also point to the need to develop and strengthen local research capacities, participating in collaborative research and development efforts and coordinating across sectors and disciplines.

But these new and emerging forms of knowledge networking, coupled with new and innovative forms of finance and technology research collaboration and development, require new and enhanced capacities for effective participation on the international level that many countries, particularly developing ones, do not have or are not well developed today. The increasingly complex and fast-paced world of energy and climate change finance is a good example of an area where present capacities fall far short of the need. The recent climate change negotiations alone have generated pledges of fast-start finance up to 2012 of some US$30 billion and promises to work collaboratively so that this funding can grow to some US$100 billion by 2020.

This is only a small part of the overall investment projections needed to meet the high growth in energy demand – some US$1.7–2.2 trillion per year are needed to 2050. The world of energy finance has always been a large and complex market. The difference today is that it is becoming even more complex, with new and innovative instruments of finance, including the carbon market, and with countries demanding more attention to the need to develop, introduce, and diffuse new technologies. Under these conditions, a multi-goal approach can both speed the diffusion of new energy technologies as well as stimulate the development and energy transition processes in developing countries.

10. Policies, Regulations, and Stable Investment Regimes will be Essential: *A portfolio of policies to enable rapid transformation of energy systems must provide the effective incentive structures and strong signals for the deployment at scale of energy-efficient technologies and energy supply options that contribute to the overall sustainable development. The GEA pathways indicate that global investments in combined energy efficiency and supply will need to increase to between US$1.7–2.2 trillion per year compared to present levels of about US$1.3 trillion per year (about 2% of current world gross domestic product) including end-use components. Policies should encourage integrated approaches across various sectors and promote the development of skills and institutional capacities to improve the investment climate. Examples include applying market-oriented regulations such as vehicle emissions standards and low carbon fuel standards and as well as renewable portfolio standards to accelerate the market penetration of clean energy technologies and fuels. Reallocating energy subsidies, especially the large subsidies provided in industrialized countries to fossil fuels without CCS, and nuclear energy, and pricing or regulating GHG emissions and/or GHG-emitting technologies and fuels can help support the initial deployment of new energy systems, both end-use and supply, and help make infrastructures energy efficient. Publicly financed research and development needs to accelerate and be reoriented toward energy efficiency, renewable energy and CCS. Current research and development efforts in these areas are grossly inadequate compared with the future potentials and needs.*

The GEA analysis has identified pronounced asymmetries in current incentive structures for the development, early deployment, and the widespread diffusion of energy end-use and supply technologies that need rebalancing. Current technology policy frameworks are also often fragmented and contradictory instead of integrated and aligned. Nowhere is this more apparent that in the continued subsidies for fossil fuels that amount to close to US$500 billion and are in direct contradiction with policy initiatives that promote increasing energy end-use efficiency and deployment of renewables. This assessment has also identified a marked mismatch between the critical needs for vastly improved energy efficiency and the under-representation of energy efficiency in publicly funded energy research and development and deployment (RD&D) and incentives for early market deployment of new technologies which are presently characterized by a distinct supply-side over-emphasis.

A first, even if incomplete, assessment of the entire global investments into energy technologies – both supply and demand-side technologies – across different innovation stages suggests RD&D investments of some US$50 billion, market formation investments (which rely on directed public policy support) of some US$150 billion, and an estimated range of US$1–5 trillion investments in mature energy supply and end-use technologies (technology diffusion). The GEA pathways estimate the current annual energy investments at about US$1.3 trillion per year. The difference to the estimated range up to US$5 trillion is related mostly to the magnitude of demand-side investments that is not included in the pathways. Demand-side investments are of critical importance, particularly because the lifetimes of end-use technologies can be considerably shorter than those on the supply side. Demand-side investments might thus play an important role in achieving pervasive and rapid improvements in the energy system.

Major developing economies have become significant players in global energy technology RD&D, with public- and private-sector investments approaching some US$20 billion – in other words, almost half of global innovation investments – which are significantly above OECD public-sector energy RD&D investments (US$13 billion).

Policies now need to move toward a more integrated approach, stimulating simultaneously the development as well as the adoption of efficient and cleaner energy technologies and measures. RD&D initiatives without simultaneous incentives for consumers to adopt the outcomes of innovation efforts risk not only being ineffective but also precluding the market feedbacks and learning that are critical for continued improvements in technologies.

Another area of near-term technology policy focus is the domain of enhancing the international cooperation in energy technology research and development as well as in the domains of technology standards. Through dynamic standard setting and international harmonization, predictable and long-term signals are provided to innovation players and markets. Ambitious efficiency standards are of particular urgency for long-lived capital assets such as buildings. Other end-use technologies such as vehicles or appliances turn over much more quickly, offering the possibility of more gradually phased in technology standards as long as clear long-term signals are provided.

Table SPM-4. | Energy investments needed between 2010 and 2050 to achieve GEA sustainability goals and illustrative policy mechanisms for mobilizing financial resources. GEA pathways indicate that global investments in combined energy efficiency and supply have to increase to about US$1.7–2.2 trillion per year compared with the present level of some US$1.3 trillion (2% of current gross world product). Given projected economic growth, this would be an approximately constant fraction of GDP in 2050.

Times	Investment (billions of US$/year)		Policy mechanisms			
		2010–2050	Regulation, standards	Externality pricing	Carefully designed subsidies	Capacity building
Efficiency	n.a.[a]	290–800[b]	*Essential* (elimination of less efficient technologies every few years)	*Essential* (cannot achieve dramatic efficiency gains without prices that reflect full costs)	*Complement* (ineffective without price regulation, multiple instruments possible)[c]	*Essential* (expertise needed for new technologies)
Nuclear	5–40[d]	15–210	*Essential* (waste disposal regulation and, of fuel cycle, to prevent proliferation)	*Uncertain* (GHG pricing helps nuclear but prices reflecting nuclear risks would hurt)	*Uncertain* (has been important in the past, but with GHG pricing perhaps not needed)	*Desired* (need to correct the loss of expertise of recent decades)[e]
Renewables	190	260–1010	*Complement* (feed-in tariff and renewable portfolio standards can complement GHG pricing)	*Essential* (GHG pricing is key to rapid development of renewables)	*Complement* (tax credits for R&D or production can complement GHG pricing)	*Essential* (expertise needed for new technologies)
CCS	<1	0–64	*Essential* (CCS requirement for all new coal plants and phase-in with existing)	*Essential* (GHG pricing is essential, but even this is unlikely to suffice in near term)	*Complement* (would help with first plants while GHG price is still low)	*Desired* (expertise needed for new technologies)[e]
Infrastructure[f]	260	310–500	*Essential* (security regulation critical for some aspects of reliability)	*Uncertain* (neutral effect)	*Essential* (customers must pay for reliability levels they value)	*Essential* (expertise needed for new technologies)
Access to electricity and cleaner Cooking[g]	n.a.	36–41	*Essential* (ensure standardization but must not hinder development)	*Uncertain* (could reduce access by increasing costs of fossil fuel products)	*Essential* (grants for grid, micro-financing for appliances, subsidies for clean cookstoves)	*Essential* (create enabling environment: technical, legal, institutional, financial)

[a] Global investments into efficiency improvements for the year 2010 are not available. Note, however, that the best-guess estimate from Chapter 24 for investments into energy components of demand-side devices is by comparison about US$300 billion per year. This includes, for example, investments into the engines in cars, boilers in building heating systems, and compressors, fans, and heating elements in large household appliances. Uncertainty range is between US$100 billion and US$700 billion annually for investments in components. Accounting for the full investment costs of end-use devices would increase demand-side investments by about an order of magnitude.

[b] Estimate includes efficiency investments at the margin only and is thus an underestimate compared with demand-side investments into energy components given for 2010 (see note a).

[c] Efficiency improvements typically require a basket of financing tools in addition to subsidies, including, for example, low- or no-interest loans or, in general, access to capital and financing, guarantee funds, third-party financing, pay-as-you-save schemes, or feebates as well as information and educational instruments such as labeling, disclosure and certification mandates and programs, training and education, and information campaigns.

[d] Lower-bound estimate includes only traditional deployment investments in about 2 GW capacity additions in 2010. Upper-bound estimate includes, in addition, investments for plants under construction, fuel reprocessing, and estimated costs for capacity lifetime extensions.

[e] Note the large range of required investments for CCS and nuclear in 2010–2050. Depending on the social and political acceptability of these options, capacity building may become essential for achieving the high estimate of future investments.

[f] Overall electricity grid investments, including investments for operations and capacity reserves, back-up capacity, and power storage.

[g] Annual costs for almost universal access by 2030 (including electricity grid connections and fuel subsidies for cleaner cooking fuels).

Some of the policies for energy for sustainability described above simply involve an improvement of existing policies, such as better management of the electricity sector or more responsible use of fossil fuel resource rents. But the dominant message of the GEA is that the global energy system must be rapidly modified and expanded to provide energy access to those who have none, and must quickly transform to an energy system more supportive of sustainable development. This transition will require considerable investments over the coming decades. Table SPM-4 indicates the necessary investments to achieve this as estimated by the GEA, and links these to the types of policies needed. It also assesses these policies in terms of their necessity and their ability to complement or substitute for each other. Although considerable, these investment levels can be compared to estimates of global fossil fuel subsidy levels on the order of US$500 billion a year, of which an estimated US$100 billion goes to producers.

Table SPM-4 compares the costs and policies for different technology options to those of promoting energy access. Different types of technologies and objectives will require different combinations of policy mechanisms to attract the necessary investments. Thus, the Table identifies 'essential' policy mechanisms that must be included for a specific option to achieve the rapid energy system transformation, 'desired' policy mechanisms that would help but are not a necessary condition, 'uncertain' policy mechanisms in which the outcome will depend on the policy emphasis and thus might favor or disfavor a specific option, and policies that are inadequate on their own but could 'complement' other essential policies.

GEA findings indicate that global investments in combined energy efficiency and supplies have to increase to about US$1.7–2.2 trillion per year compared with the present level of some US$1.3 trillion (2% of current gross world product). Given projected economic growth, this would be an approximately constant fraction of GDP in 2050.

For some objectives, such as energy access, future investment needs are comparatively modest. However, a variety of different policy mechanisms – including subsidies and regulation as well as capacity building programs – need to be in place. Regulations and standards are also essential for almost all other options listed in the Table, while externality pricing might be necessary for capital-intensive technologies to achieve rapid deployment (such as a carbon tax to promote diffusion of renewables, CCS, or efficiency). The GEA estimates that the investment requirements to transform energy systems are in the range of US$1.7–2.2 trillion per year through 2020. Capital requirements for energy infrastructure are only a small part of the overall investment projections, but among the highest priorities of the options listed. A multi-goal approach can both speed the diffusion of new energy technologies as well as stimulate the development and energy transition processes in developing countries.

Increasing investments in the energy system as depicted by the GEA pathways requires the careful consideration of a wide portfolio of policies in order to create the necessary financial incentives, adequate institutions to promote and support them, and innovative financial instruments to facilitate them The portfolio needs to include regulations and technology standards in sectors with, for example, relatively low price elasticity in combination with externality pricing to avoid rebound effects, as well as targeted subsidies to promote specific 'no-regret' options while addressing affordability. In addition, focus needs to be given to capacity development to create an enabling technical, institutional, legal, and financial environment to complement traditional deployment policies (particularly in the developing world).

In sum, the GEA finds that attainment of a sustainable future for all is predicated on resolving energy challenges. This requires the creation of market conditions, via government interventions, that invite and stimulate investments in energy options that provide incentives for rapid investments in energy end-use and supply technologies and systems.

Technical Summary

Convening Lead Authors (CLA):
Thomas B. Johansson (Lund University, Sweden)
Nebojsa Nakicenovic (International Institute for Applied Systems Analysis and Vienna University of Technology, Austria)
Anand Patwardhan (Indian Institute of Technology-Bombay)
Luis Gomez-Echeverri (International Institute for Applied Systems Analysis, Austria)

Lead Authors (LA):
Doug J. Arent (National Renewable Energy Laboratory, USA)
Rangan Banerjee (Indian Institute of Technology-Bombay)
Sally M. Benson (Stanford University, USA)
Daniel H. Bouille (Bariloche Foundation, Argentina)
Abeeku Brew-Hammond (Kwame Nkrumah University of Science and Technology, Ghana)
Aleh Cherp (Central European University, Hungary)
Suani T. Coelho (National Reference Center on Biomass, University of São Paulo, Brazil)
Lisa Emberson (Stockholm Environment Institute, University of York, UK)
Maria Josefina Figueroa (Technical University of Denmark)
Arnulf Grubler (International Institute for Applied Systems Analysis, Austria and Yale University, USA)
Kebin He (Tsinghua University, China)
Mark Jaccard (Simon Fraser University, Canada)
Suzana Kahn Ribeiro (Federal University of Rio de Janeiro, Brazil)
Stephen Karekezi (AFREPREN/FWD, Kenya)
Eric D. Larson (Princeton University and Climate Central, USA)
Zheng Li (Tsinghua University, China)
Susan McDade (United Nations Development Programme)
Lynn K. Mytelka (United Nations University-MERIT, the Netherlands)
Shonali Pachauri (International Institute for Applied Systems Analysis, Austria)
Keywan Riahi (International Institute for Applied Systems Analysis, Austria)
Johan Rockström (Stockholm Environment Institute, Stockholm University, Sweden)
Hans-Holger Rogner (International Atomic Energy Agency, Austria)
Joyashree Roy (Jadavpur University, India)
Robert N. Schock (World Energy Council, UK and Center for Global Security Research, USA)
Ralph Sims (Massey University, New Zealand)
Kirk R. Smith (University of California, Berkeley, USA)
Wim C. Turkenburg (Utrecht University, the Netherlands)
Diana Ürge-Vorsatz (Central European University, Hungary)
Frank von Hippel (Princeton University, USA)
Kurt Yeager (Electric Power Research Institute and Galvin Electricity Initiative, USA)

Contents

1   Introduction

Energy is essential for human development and energy systems are a crucial entry point for addressing the most pressing global challenges of the 21st century, including sustainable economic, and social development, poverty eradication, adequate food production and food security, health for all, climate protection, conservation of ecosystems, peace, and security. Yet, more than a decade into the 21st century, current energy systems do not meet these challenges.

In this context, two considerations are important. The first is the capacity and agility of the players within the energy system to seize opportunities in response to these challenges. The second is the response capacity of the energy system itself, as the investments are long-term and tend to follow standard financial patterns, mainly avoiding risks and price instabilities. This traditional approach does not embrace the transformation needed to respond properly to the economic, environmental, and social sustainability challenges of the 21st century.

A major transformation is required to address these challenges and to avoid potentially catastrophic consequences for human and planetary systems. The GEA identifies strategies that could help resolve the

multiple challenges simultaneously and bring multiple benefits. Their successful implementation requires determined, sustained, and immediate action.

The industrial revolution catapulted humanity onto an explosive development path, whereby reliance on muscle power and traditional biomass was replaced mostly by fossil fuels. In 2005, approximately 78% of global energy was based on fossil energy sources that provided abundant and ever cheaper energy services to more than half the world's population. Figure TS-1 shows two clear development phases in this explosive growth of global primary energy: the first characterized by a shift from reliance on traditional energy sources to coal and subsequently to oil and gas. During the past decades, hydropower, biomass, and nuclear energy have a combined share of almost 12%, while new renewables, such as solar and wind, are hardly discernible in Figure TS-1. These major transitions are also illustrated in Figure TS-2, which shows the shares of global primary energy and their changes over the period from 1850 to 2008. The dominance of biomass in the 1800s was overtaken by coal in the first half of the 20th century, giving way to oil around 1970. Oil still retains the largest share of global primary energy.[2]

Despite this rapid increase in overall energy use, over three billion people still rely on solid fuels such as traditional biomass, waste,

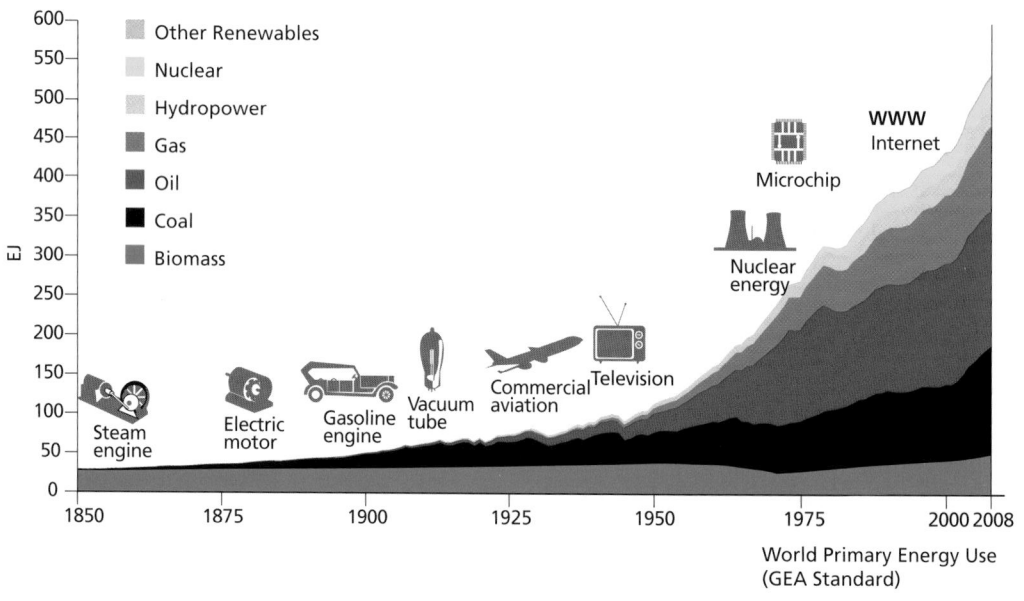

Figure TS-1 | Evolution of primary energy shown as absolute contributions by different energy sources (EJ). Biomass refers to traditional biomass until the most recent decades, when modern biomass became more prevalent and now accounts for one-quarter of biomass energy. New renewables have emerged in the last few decades. Source: updated from Nakicenovic et al., 1998 and Grubler, 2008, see Chapter 1.[1]

---

1   **Nakicenovic, N., A. Grubler and A. McDonald (eds.), 1998:** *Global Energy Perspectives.* International Institute for Applied Systems Analysis (IIASA) and World Energy Council (WEC), Cambridge University Press, Cambridge, UK.
    Grubler, A., 2008: Energy transitions. In *Encyclopedia of Earth.* C. J. Cleveland (ed.), Environmental Information Coalition, National Council for Science and the Environment, Washington, DC.

2   GEA convention on primary energy using primary energy substitution equivalent (see Chapter 1.A.3) is used throughout.

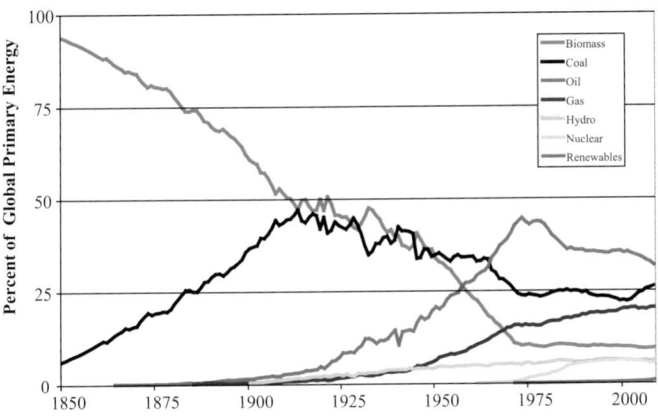

Figure TS-2 | Evolution of primary energy shown as shares of different energy sources. Source: updated from Nakicenovic et al., 1998 and Grubler, 2008; see Chapter 1.[3]

charcoal, and coal for household cooking and heating. The resulting air pollution leads to over two million premature deaths per year, mostly of women and children. Furthermore, approximately 20% of the global population has no access to electricity, making it difficult for children to study after sunset and impossible, for example, to keep vaccines cold, provide mechanical energy for agriculture and irrigation, and power the most simple machines for manufacturing and commerce. This situation undermines economic development and energy security, and causes indoor and outdoor air pollution and climate change. Addressing these challenges is essential to averting a future with high economic and social costs and adverse environmental impacts on all scales.

An energy system transformation is required to meet these challenges and bring prosperity and well-being to the nine billion people expected by 2050. The encouraging news is that the beginnings of such a transformation can be seen today, in the rapidly growing investments in renewable energy sources, high-efficiency technologies, new infrastructure, near zero-energy buildings, electric mobility, 'smart' energy systems, advanced biomass stoves, and many other innovations. The policy challenge is to accelerate, amplify, and help make the implementation of these changes possible, widespread, and affordable. Initial experience suggests that many of these changes are affordable, although they may be capital intensive and require high upfront investments. However, in general, they have lower long-term costs that offset many of the upfront added-investment requirements. Many of these innovations also lead to benefits in other areas such as equity and poverty, economic development, energy security, improved health, climate change mitigation, and ecosystem protection.

At the same time, the beginning of this grand transformation is, to a large extent, obscured by business-as-usual (BAU) thinking and behavior. Also, it tends to be blocked by current decision-making processes, institutions, consumption patterns, capital vintages, interests, and investment patterns that show a lock-in to old development pathways. However, there are excellent examples in many countries of transformational changes showing the opportunities available. The transition from the 'pervasive old' to the 'emerging new' will require continuous and major enhancements in awareness, knowledge, and skills, as well as new institutions, policies, and strategies.

GEA shows that while the local, regional, and global challenges, and their demands on energy systems, are enormous, they can all be met in a timely manner, effectively, and simultaneously – if societies want to do so. The assessment shows that a transformation toward energy systems supportive of sustainable development is possible. However, it will require decision makers to approach energy systems in an innovative and integrated way, to significantly strengthen their efforts domestically, and to coordinate their activities internationally.

GEA explored 60 alternative energy transformation pathways toward a sustainable future that simultaneously satisfy all the normative social and scientifically based environmental goals: continued economic development, universal access to modern energy carriers, climate and environment protection, improved human health, and better energy security.

The 60 pathways were grouped into three different approaches toward achieving the normative goals: GEA-Supply, GEA-Mix, and GEA-Efficiency. They were selected to represent three alternative evolutions of the energy system toward a sustainable future (details are in Section TS-4). A major conclusion is that many of these pathways satisfy all the GEA goals (see Sections TS-2.6 and TS-4).

This Technical Summary synthesizes and integrates the main findings from the individual chapters in the GEA report. It is structured as follows. Section TS-2 outlines the magnitude and orientation of the energy system change that is required, and forms the basis for specific goals expressed in terms of qualitative and quantitative indicators. Section TS-3 presents the building blocks of the transformation, such as energy resources, energy end-use and supply technologies, infrastructures, and systems. Section TS-4 describes the energy pathways for sustainable development and their implications. Section TS-5 presents policy tools and interventions to implement energy pathways that deliver on the goals for sustainable development.

2   The Need for Change

This section summarizes the major global challenges of the 21st century that require actions on energy systems in order to be resolved. GEA has developed energy-related indicators of sustainability that are discussed

---

3   **Nakicenovic, N., A. Grubler and A. McDonald (eds.), 1998:** *Global Energy Perspectives.* International Institute for Applied Systems Analysis (IIASA) and World Energy Council (WEC), Cambridge University Press, Cambridge, UK.
Grubler, A., 2008: Energy transitions. In *Encyclopedia of Earth.* C. J. Cleveland (ed.), Environmental Information Coalition, National Council for Science and the Environment, Washington, DC.

in this section, and used in Section TS-4 in the development of pathways toward a sustainable future.

2.1   Economic and Population Growth[4]

Energy access is fundamental to the growth and development of modern economies. Energy use is rising rapidly, driven by worldwide population growth, increased economic prosperity, an expanding middle class, and the lifestyles of the richest 1–2 billion people, as well as by the burgeoning use of ever more energy-intensive technologies in homes and workplaces around the world. This explosive growth in energy use is illustrated in TS-1. Even as the total energy use has increased rapidly, there is wide variation on a per capita basis (up to a factor of 10) in the final energy use among major world regions (see Figure TS-3).

On an individual basis, this variation is of the same order of magnitude as the disparity in global income distribution. Particularly striking on the regional level is how the sectoral shares in final energy vary from about equal percentage allocations at the higher levels of energy use to almost all energy being used in the residential (and commercial) sector at the lowest levels – meaning that almost no energy services are available to support production and development.

Several historic shifts are likely to fundamentally alter the global economy over the coming decades. First, as developing nations move from poverty to relative affluence, there will be a shift from agriculture to more energy-intensive commercial enterprises. Greater affluence has historically also been associated with an increase in meat consumption and other protein-rich diets, which multiply the stresses on the global environment due to the elevated need for water and land and increased greenhouse gas (GHG) emissions.

Currently, the process of industrialization is energy-intensive, as it requires high levels for material transformations. Rising incomes are expected to generate higher demands for private transport as well as for space and water heating, space cooling, and power-hungry household appliances. Demand for freight transport also implies increasing demand for energy transport services. It is unclear to what extent these demands are subject to saturation at higher income levels.

Second, for the first time in human history half the world population now lives in cities, and this urban fraction is growing faster than the overall population growth. The largest and fastest-growing urban centers are found in the world's poorer regions, where lack of energy access is most prevalent.

Access to affordable and sustainable energy services is fundamental to human development and economic growth. Economies lacking proper access to modern forms of energy (particularly electricity and other forms of mechanical energy for productive purposes, and also cleaner household combustion) cannot develop and contribute to improvements in well-being.

Figure TS-3 | Final energy (GJ) per capita versus cumulative population for 11 world regions sorted by declining per capita energy use (with final energy use disaggregated by sector and total, color bars) and final energy per capita for 137 countries in 2005 (black, solid line). Dashed horizontal line shows the average final energy per capita, which indicates that approximately 1.5 billion people are above and 5.5 billion below that level. Source: Chapters 1 and 19.

Energy for sustainable development must concurrently meet, without compromise, all dimensions of energy service requirements. These include availability, affordability, accessibility, security, health, climate, and environmental protection. The approach in GEA focuses on the energy options that deliver benefits for many, if not all, of these dimensions to avoid costly lock-in effects from a focus on a single dimension.

Being a multiplier of consumption, population growth remains a major driver of global impacts. Given the absolute limits of the planet, as illustrated by the need to limit concentrations of climate-altering pollutants, reductions in population growth trends can provide valuable additional decades to help resolve energy and other problems before reaching planetary limits. There is no coercion implied here, as studies show that hundreds of millions of women wish to control their family size but do not have access to modern contraceptive technologies. Models show, for example, that by providing such services, CO_2 emissions from energy use could be reduced by 30% in 2100 over what is otherwise projected. Providing reproductive health services to these women is also an equity issue – all women, not just those in rich countries, ought to have access to such services. It is also an important health issue, as spacing births, which, along with reducing the total number of births, is an effect of giving women access to contraception, has major benefits for child and maternal health.

2.2   Energy Access, Poverty, and Development[5]

Poverty is the most critical social challenge that faces developing and industrialized countries globally. Approximately three billion people live

---

4   Section TS-2.1 is based on Chapter 6.

5   Section TS-2.2 is based on Chapters 2 and 19.

© Yu Nagai, Shonali Pachauri, Keywan Riahi (2011)

Figure TS-4 | Density of population lacking access to modern energy carriers in 2005. Colored areas show people per km² without access to electricity and those that use solid fuels for cooking, e.g., dark blue and red areas show where people do not have access to electricity and cook predominately using solid fuels. Source: Chapters 17 and 19.

on less than US$2 a day, with about 1.4 billion living in extreme poverty, on less than US$1.25 a day. The number of people with no access to electricity is 1.4 billion. Around 2.7 billion people rely on traditional biomass, such as fuel wood, charcoal, and agricultural residues (including animal dung), for cooking and heating, and another 400 million cook and/or heat with coal, making a total of around three billion people who rely on solid fuels for cooking and heating (see Figure TS-4).

Providing access to modern energy carriers and end-use conversion devices, such as cleaner cookstoves, is a major step to enable people living in poverty to improve their lives and reach the United Nations Millennium Development Goals and beyond.

Enhanced access to electricity, fuels and cleaner cooking systems can be an effective tool for improving health, for example by reducing air pollution, and can combat extreme hunger by increasing food productivity and reducing post-harvest losses. The energy technologies needed are relatively affordable, can often be produced locally, and in many cases do not require large-scale centralized energy supply options or costly infrastructure. Modern energy carriers, such as electricity and cleaner burning fuels, and end-use conversion devices can also improve education and school attendance by providing better energy services, such as lighting, heating, and cooling services. Electrifying rural health centers enables medical services to be provided at night, medicines to be preserved, and more-advanced medical equipment to be used.

Modern carriers and end-use conversion devices also encourage investments in capital goods that use electricity, which, in turn, allows the establishment of advanced agro-processing industries in rural areas, such as sugar production, milk cooling, grain milling, and food preservation. Processing will help make more food edible for longer, keep more money in local communities if the processing is done locally, and, in some

cases, help farmers retain control of sales. These not only enhance rural incomes through increased sales and better prices, they also increase food production, and thereby contribute to reducing extreme hunger. Enhancing access among poor people, especially poor women, Is thus important for increasing their standard of living. Reducing the proportional cost of energy services is also important, particularly for the rural poor, who spend a significant part of their time and disposable income on energy. This can liberate financial and human resources for other important activities or expenses, such as education, purchasing more and better-quality food, and expanding income-generating activities.

2.3   Energy Security[6]

Energy security, that is, the uninterrupted provision of vital energy services, is critical for every nation. For many industrial countries, the key energy security challenges are dependence on imported fossil fuels and reliability of infrastructure. Many emerging economies have additional vulnerabilities, such as insufficient power generation capacity, high energy intensity, and rapid demand growth. In many low-income countries, multiple vulnerabilities of energy systems overlap, making them especially insecure.

Oil is at the center of contemporary energy-security concerns for most nations, regions, and communities. Oil products provide over 90% of transport energy in almost all countries. Thus, disruptions of oil supplies may have catastrophic effects, not only on personal mobility, but also on food production and distribution, medical care, national security, manufacturing, and other vital functions of modern societies. At the same time, conventional oil resources are increasingly concentrated in just a few

6   Section TS-2.3 is based on Chapter 5.

regions. The concerns over political stability affecting resource extraction and transport add to uncertainty. Moreover, the global production capacity of conventional oil is widely perceived as limited (see Section TS-3.1.1). Furthermore, the demand for transport fuels is steadily rising, especially rapidly in emerging Asian economies. Thus, for most countries, an ever higher share of their oil, or even all of it, must be imported. More than three billion people live in countries that import more than 75% of the oil and petroleum products they use (see Figure TS-5). An additional 1.7 billion people living in countries with limited domestic oil resources (including China) are likely to experience similarly high levels of import dependence in the coming decades.

The increasing concentration of conventional oil production, and the rapidly shifting global demand patterns, make some analysts and politicians fear a 'scramble for energy' or even 'resource wars'. These factors result in rising and volatile oil prices that affect all economies, especially low-income countries, almost all of which import over 80% of their oil supplies. The costs of energy imports (primarily oil products) exceed 20% of the export earnings in 35 countries that together are home to 2.5 billion people.

Import dependence is also common in countries that extensively use natural gas. Almost 650 million people live in countries that import over 75% of their gas. Most of these countries rely on a very limited number of gas suppliers (in many cases just one) and import routes. The risks of supply disruptions and price fluctuations are often the most serious energy security issues in such countries. The potential of recent shale gas technology developments to alleviate these concerns is at present uncertain.

Electricity systems in many low- and middle-income countries have inadequate generation capacity, low diversity of generation options, and low reliability of transmission and distribution systems. In over two-thirds of low-income countries, electricity supply is interrupted for at least one hour each day. Over 700 million people live in countries that derive a significant proportion of their electricity from only one or two major dams. Hydroelectric power production may also become insecure due to increased stress on global water supplies through increased population,

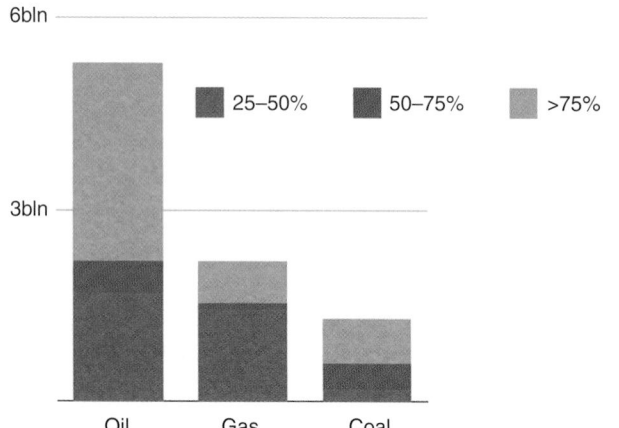

Figure TS-5 | Number of people in countries that are dependent on imported oil, gas and coal. Source: data from Chapter 5.

agriculture, energy production, and climate change, which may affect seasonal water availability.

Many countries using nuclear power are experiencing an aging of the reactor fleet and workforce, as well as problems obtaining the capital and technologies to renew or expand nuclear programs. Twenty-one out of 29 countries with nuclear power plants have not started to build a new reactor in the last 20 years, and in 19 of these countries the average age of nuclear power plants is over 25 years. Large-scale enrichment, reactor manufacturing, and reprocessing technologies are currently concentrated in just a few countries (see Figure 5.5 in Chapter 5). The spread of enrichment or reprocessing to a larger number of countries is opposed by concerns over nuclear weapons proliferation – one of the main controversies – and the risks and costs associated with nuclear energy.

Another energy security issue is 'demand security'. Vital energy export revenues play a major part in the economies of some 15–20 mainly low- or middle-income countries. In many cases these revenues are not expected to last for more than one generation, and in several cases they may cease in less than a decade. In addition, poor energy-exporting nations are at a high risk of the 'resource curse': economic and political instability eventually affecting human development and security. The present economic and social importance of energy export revenues should be recognized in international arrangements, while diversifying the economies of countries excessively dependent on energy exports is also a high priority in dealing with 'demand security'.

2.4 Environment[7]

Linkages between the energy system and the environment are seen on multiple levels and scales – from local to global. While the local environmental and ecological consequences of resource extraction, processing, and energy conversion have been long recognized, attention is increasingly turning toward the growing evidence that humanity has reached a phase when anthropogenic pressures on Earth systems – the climate, oceans, freshwater, and the biosphere – risk irreversible disruption to biophysical processes on the planetary scale. The risk is that systems on Earth may then reach tipping points, resulting in non-linear, abrupt, and potentially irreversible change, such as destabilization of the Greenland ice sheet or tropical rainforest systems.

The challenges are illustrated in Figure TS-6, showing planetary boundaries for nine Earth system processes, which together define a safe operating space for humanity (indicated by the green area), within which human development stands a good chance of proceeding without large-scale deleterious change. Estimates indicate that the safe levels are being approached or, in some cases, transgressed. Energy systems contribute to humanity's approach to many of the planetary boundaries,

---

7   Section TS-2.4 is based on Chapters 3 and 17.

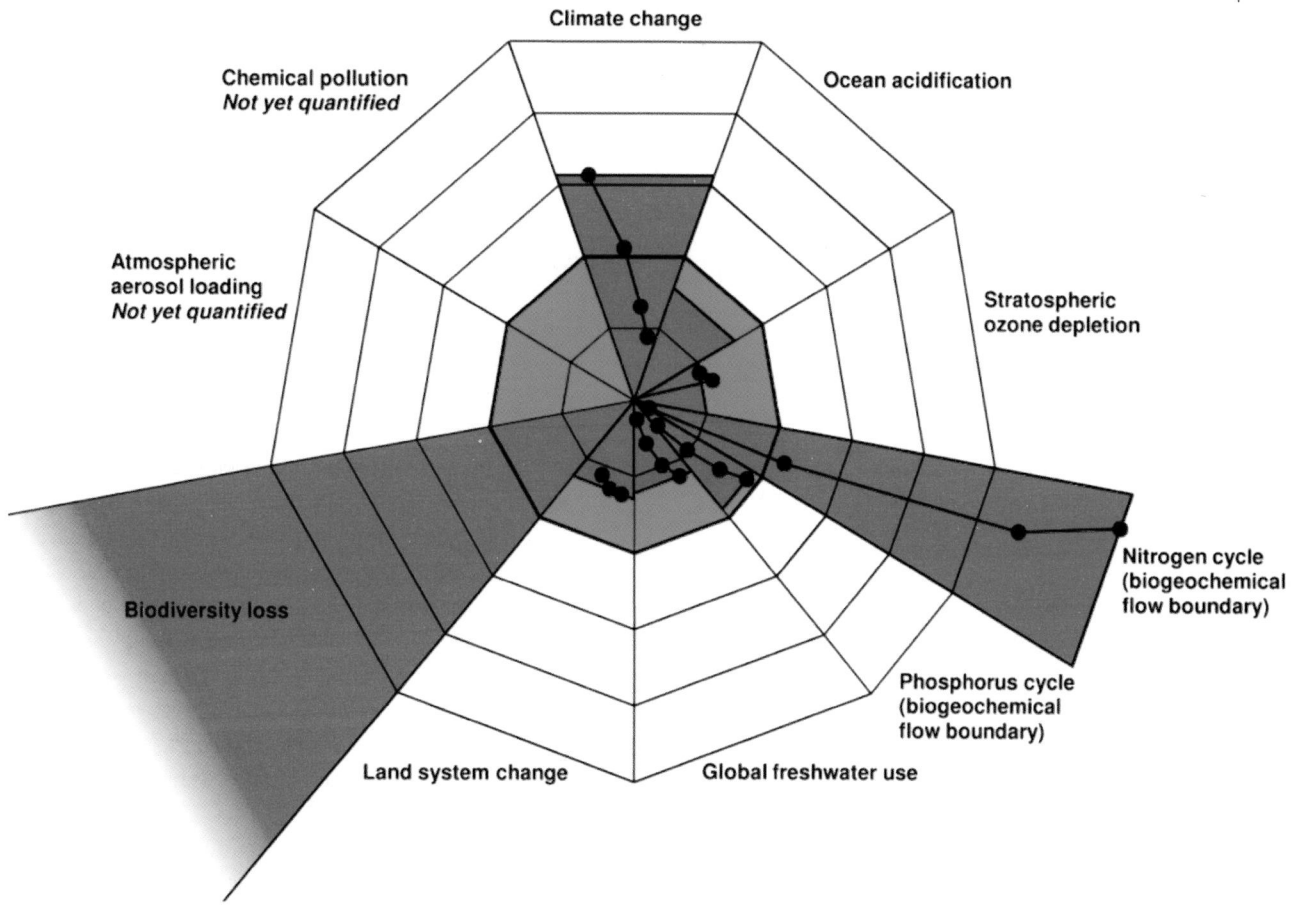

Figure TS-6 | Current global state of the world for the 10 proposed planetary boundaries. The green area denotes a "safe operating space" for human development, and red indicates the current position for each boundary process. The dots indicate evolution by decade from the 1950s. Source: Chapter 3.

and in particular climate change, aerosol loading, ocean acidification, biodiversity loss, chemical pollution, land system change, the nitrogen cycle, and fresh water use.

In 2005, energy supply and use contributed around 80% of CO_2 emissions and 30% of methane emissions (Chapter 1), as well as large fractions of other substances, such as black carbon, organic carbon, and aerosols that can either warm or cool the atmosphere, depending on their composition. Energy systems are furthermore tightly linked to land and freshwater use through dependence on water and land resources for energy generation. They are also linked to ecosystem services and air quality through emissions of particulate matter and atmospheric pollutants, such as nitrogen and sulfur oxides, and precursors of tropospheric ozone that can lead to acidification, eutrophication, and reduced net primary productivity. Consequently, the energy system has a critical part to play in achieving global sustainability.

One of the areas of concern most influenced by energy systems is climate change. Threats to agriculture, biodiversity, ecosystems, water supply in some areas and floods in others, sea levels, and many other environmental aspects will continue to worsen unless climate change

is curbed significantly. Moreover, degradation of land, biodiversity, and freshwater resources, which are closely related to energy generation and use, has accelerated climate change. In response to this challenge, the global community decided to take actions to limit anthropogenic warming to less than 2°C above pre-industrialized levels (UNFCC, Decision 1/CP.16). This stabilization target is the normative goal adopted by GEA and is reflected in all pathways toward a more sustainable future (see Section TS-4). GEA recognizes however, that even a 2°C target may lead to large adverse effects, including risks of reaching tipping points, thus highlighting the need for an even more ambitious target.

Limiting global temperature increase to less than 2°C above pre-industrial levels (with a probability of greater than 50%) requires rapid reductions of global CO_2 emissions from the energy sector with a peak around 2020 and a decline thereafter to 30–70% below 2000 emissions levels by 2050, finally reaching almost zero or even negative CO_2 emissions in the second half of the century (see Figure TS-7). Given that even a 2°C target will likely lead to significant impacts, assuring only a 50% chance of success is a rather low bar to set. A higher probability of meeting the 2°C target, or a lower temperature-increase target, would require higher emission reductions by 2050 and beyond. In particular, the later

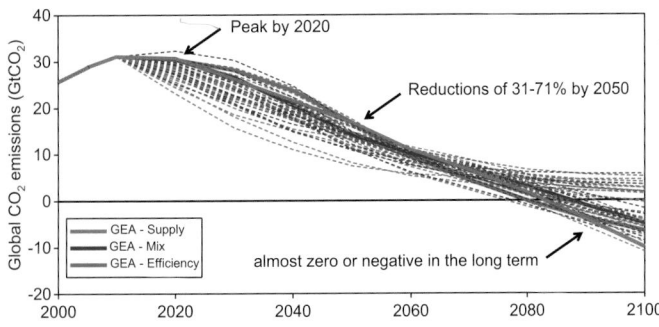

Figure TS-7 | Development of global CO$_2$ emissions from energy and industrial sources to limit temperature change to below 2°C (with a probability of >50%). Shown is that the emissions need to peak by around 2020 (or earlier) and decline toward zero during the following four to five decades. The later the peak occurs, the steeper the decline needs to be and higher the net 'negative' emissions. The latter can be achieved in the energy system through CCS in conjunction with the use of sustainable biomass. Source: Chapter 17. For further details of the GEA pathways see the interactive web-based GEA scenario database hosted by IIASA: www.iiasa.ac.at/web-apps/ene/geadb.

the emissions peak, the higher the need for net 'negative' emissions in the second half of the century, for example, by using biomass together with carbon capture and storage (CCS) (see Section TS-3.3.2.2).

Reducing emissions of both long-lived GHGs, such as CO$_2$, and short-lived climate forcers, such as ozone precursors and black carbon (both emitted, for example, from the combustion of diesel fuel and household biomass fuel), is essential. Reducing short-lived climate forcers is critical to slow the rate of near-term climate change and provides a far greater likelihood of achieving the 2°C target when coupled with aggressive measures to also bring down long-lived GHG emissions.

The focus on planetary-scale impacts does not reduce the importance of addressing local and regional environmental and ecological impacts. Atmospheric pollutants may limit the net primary productivity of ecosystems, and lead to the acidification and eutrophication of land and seascapes. Land is affected through loss or damage to ecosystems from land-use change and contamination from energy-related waste arising from activities such as mining, drilling, and the transport of fossil fuels. In addition, disaster mitigation systems need to be continually developed and implemented to avert energy-related environmental disasters, for example, from nuclear accidents, oil rig explosions, oil tanker spills, flooding from hydroelectric dam bursts, and so on.

Differences in energy provisioning systems around the world cause variability in the environmental challenges. Reducing such problems requires policy implementation targeted at the most threatening environmental effects specific to a nation and region. Problems may arise when the impacts considered a priority at the regional scale differs from those at the global scale.

In such situations, international cooperative approaches may be required to provide economic and social infrastructural support to address both

national and regionally perceived priorities, and more internationally driven concerns over environmental threats and the ability to remain within the safe operating limits of planetary boundaries.

2.5   Health[8]

Energy systems are currently responsible for a large proportion of the global burden of disease, which is in the order of five million premature deaths annually from air pollution and other energy-related causes and more than 8% of all ill health (lost healthy life years from both morbidity and premature mortality) (see Table TS-1).

Air pollution from incomplete combustion of fuels and biomass burning is a major contributor to ill health. As cooking fuel is the greatest source of household indoor air pollution, and a significant source of outdoor pollution, access to cleaner cooking (see Section TS-2.2) would provide significant improvement. Outdoor air pollution in both urban and rural areas accounted for 2.7 million premature deaths globally in 2005, while about 2.2 million premature deaths are estimated to occur annually from exposure to indoor air pollution in developing countries, mainly among women, the elderly, and young children. Other sources of outdoor air pollution include the transportation sector, industry, power plants, and space conditioning.

Occupational health impacts, particularly from harvesting/mining and processing biomass and coal, are currently the next most important impact on health from energy systems. Miners are exposed to collapsing mine shafts, fire and explosion risks, toxic gases (carbon monoxide), lung-damaging dusts (coal and silica), and hot work environments, as well as injury and ergonomics hazards. Oil and gas workers face injury risks, particularly during drilling, emergency situations, and work on offshore platforms, as well as exposure to toxic materials at refineries.

Unlike biomass and fossil fuels, nuclear power systems are not a significant source of routine health impacts, although they often garner considerable public and policy concern. Average radiation doses to workers in nuclear power industries have generally declined over the past two decades. For nuclear power facilities, as with large hydroelectric facilities, the major health risks lie mostly with high-consequence but low-probability accidents.

Climate change is beginning to have an important impact on health, causing an estimated 150,000 premature deaths in 2000, with more than 90% of these occurring among the poorest populations in the world. Both the direct health burden and the share of climate change impacts due to energy systems are expected to rise under current projections of GHG emissions and changing background health conditions in vulnerable populations. By 2010, this impact may have doubled.

8   Section TS-2.5 is based on Chapter 4.

Table TS-1 | Global burden of disease in 2000 from air pollution and other energy-related causes. These come from the Comparative Risk Assessment (CRA) published in 2004 by the World Health Organisation. GEA estimates for 2005 of outdoor air pollution and household solid fuel use in Chapter 17 are substantially larger, but were not done for all CRA risk pathways shown. Estimates for 2010 in the new CRA will be released in 2012 and will again include all pathways in a consistent framework.

	Total Premature Deaths – million	Percent of all Deaths	Percent of Global Burden in DALYs	Trend
Direct Effects [except where noted, 100% assigned to energy]				
Household Solid Fuel	1.6	2.9	2.6	Stable
Energy Systems Occupational*	0.2	0.4	0.5	Uncertain
Outdoor Air Pollution	0.8	1.4	0.8	Stable
Climate Change	0.15	0.3	0.4	Rising
Subtotal	2.8	5.0	4.3	
Indirect Effects (100% of each)				
Lead in vehicle Fuel	0.19	0.3	0.7	Falling
Road Traffic Accidents	0.8	1.4	1.4	Rising
Physical Inactivity	1.9	3.4	1.3	Rising
Subtotal	2.9	5.1	3.4	
Total	5.7	10.1	7.7	

* One-third of global total assigned to energy systems.

Notes: These are not 100% of the totals for each, but represent the difference between what exists now and what might be achieved with feasible policy measures. Thus, for example, they do not assume the infeasible reduction to zero traffic accidents or air pollution levels. DALYS = disability adjusted life years.

Source: Chapter 4

2.6   Goals Used in the Assessment and in the GEA Pathways Analysis

For many of the energy-related challenges different goals have been articulated by the global community, in many instances including specific quantitative targets. This sub-section summarizes the concrete goals in major areas that require changes to energy systems, based on Section TS-2. Meeting these goals simultaneously has served as the generic framework for all assessments in GEA. The GEA pathways, described and elaborated in Section TS-4, illustrate how societies can reach the global normative goals of welfare, security, health, and environmental protection outlined below, simultaneously with feasible changes in energy systems.

The selection of indicators and quantitative target levels summarized here is a normative exercise, and the level of ambition has, to the extent possible, been guided by agreements and aspirations expressed through, for example, the United Nations system's actions and resolutions, and from the scientific literature. This, of course, only refers to the necessary changes of the local and global energy systems; much more is required in other sectors of societies for overall sustainability to be realized.

In the GEA pathways analysis, the global per capita gross domestic product (GDP) increases on average by 2% per year through 2050, mostly driven by growth in developing countries. This growth rate declines in

the middle of existing projections. Global population size is projected to plateau at about nine billion people by 2050. Energy systems must be able to *deliver the required energy services* to support these economic and demographic developments.

Universal access to affordable modern energy carriers and end-use conversion (especially electricity and cleaner cooking)[9] by 2030 for the 1.4 billion people without access to electricity and the three billion people who still rely on solid and fossil fuels for cooking is a prerequisite for poverty alleviation and socioeconomic development.

Enhanced energy security for nations and regions is another key element of a sustainable future. Reduced global interdependence via reduced import/export balances and increased diversity and resilience of energy supply have been adopted as key energy-related metrics. The targets for these goals were assessed ex-post through the GEA pathways analysis (Chapter 17), identifying the need for energy efficiency improvements and deployment of renewables to increase the share of domestic (national or regional) supply in primary energy by a factor of two, and thus significantly decrease import dependency (by 2050). At the same time, the share of oil in global energy trade is reduced from the present 75% to below 40% and no other fuel assumes a similarly dominant position in the future.

---

9   See Chapter 2.2.

The *climate change mitigation* goal is to, at a minimum, contain the global mean temperature increase to less than 2°C above the pre-industrial level, with a probability of at least 50%. This implies global CO_2 emissions reductions from energy and industry to 30–70% of 2000 levels by 2050, and approaching almost zero or net negative emissions in the second half of the century.

Health and environment goals include controlling household and ambient air pollution, ocean acidification, and deforestation. Emissions reductions through the use of advanced fuels and end-use technologies for household cooking and heating can significantly reduce human morbidity and mortality due to exposure to household air pollution, as well as help reduce ambient pollution. In the GEA pathways, this is assumed to occur for the vast majority of the world's households by 2030. Similarly, a majority of the world's population is also expected to meet WHO air-quality guidelines (annual PM2.5 concentration[10] <10 µg/m^3), while the remaining population is expected to stay well within the WHO Tier I-III levels (15–35 µg/m^3) by 2030. In addition, there needs to be a major expansion of occupational health legislation and enforcement in the energy sector.

There are also a number of other concerns related to how energy systems are designed and operated. For example, activities need to be occupationally safe, a continuing concern as nano-technologies and other new materials are used in energy systems. Other impacts such as oil spills, freshwater contamination and overuse, and releases of radioactive substances must be prevented (ideally) or contained. Waste products must be deposited in acceptable ways to avoid health and environmental impacts. These issues mostly influence local areas, and the regulations and their implementation are typically determined at the national level. The analysis of indicators and pathways to sustainability in Section TS-4 assumes that such concerns and impacts are under control.

Reaching these goals simultaneously requires transformational changes to the energy system in order to span a broad range of opportunities across urban to rural geographies, from developing to industrial countries, and in transboundary systems. The ingredients of this change are described in the Section TS-3.

3   Options for Resources and Technologies

This section assesses the building blocks that can be used for transforming energy systems toward a sustainable future, including the available resources, whether fossil fuels, fissile material, or renewable energy flows. It assesses technology options on the demand and supply sides and concludes with some insights from a comparative evaluation of options. We begin with a brief description of current global and regional energy systems.

The ultimate purpose of energy systems is to deliver energy that either directly or indirectly provides goods and services to meet people's needs and aspirations. The *energy system* includes all steps in the chain – from primary energy resources to energy services (see Figure TS-8). The *energy sector* refers to the steps in the chain, from the extraction of primary energy resources through to the delivery of final energy carriers for use in end-use technologies that produce energy services or goods. In economic terms, the energy sector includes those businesses responsible for the different steps in this chain.

It is important to define that, for GEA purposes, energy services refer to illumination, information and communication, transport and mobility of people and goods, hot water, thermal comfort, cooking, refrigeration, and mechanical power. Electricity and kerosene are examples of energy carriers, not energy services. All goods and services are provided using energy and thus have energy embedded in them; however, they are not energy services *per se*.

The supply side of energy systems consists of energy resources and the technologies that convert them into energy carriers for final use. Increased demand for improved energy services, mostly in developing countries, and driven in part by population growth and socioeconomic development, is inevitable. Meeting the increased global demand will require a transformation of current energy supply systems (as well as of conversion and end-use systems, as previously described) globally. Such a transformation is not a new phenomenon. Figure TS-1 shows how the relative role of different sources has varied during the growth of global primary energy over the decades. The first transition was from biomass to coal, followed by the transition from coal to oil, which currently remains the largest source of primary energy, although natural gas is steadily increasing its share.

The energy supply situation in 2005 is illustrated by Figure TS-9. Fossil fuels dominate in all regions of the world, with oil having the largest share in the Organisation for Economic Co-operation and Development (OECD), Countries the Middle East and Africa (MAF), and Latin America and the Caribbean (LAC), while coal dominates in Asia and natural gas in Eastern Europe and the Former Soviet Union (REF).

The distribution of primary sources of energy in 2005 shows fossil fuels contributing over 78%, renewables (including large hydro) over 16%, and nuclear over 5% (Figures TS-9 and TS-10). In 2009, fossil fuels were used in 68% of electricity generation, hydropower contributed 16%, nuclear 13.5%, and other renewables contributed 2.6% (see Figure TS-17).

---

10 PM2.5 refers to particulate matter less than 2.5 micrometers in size.

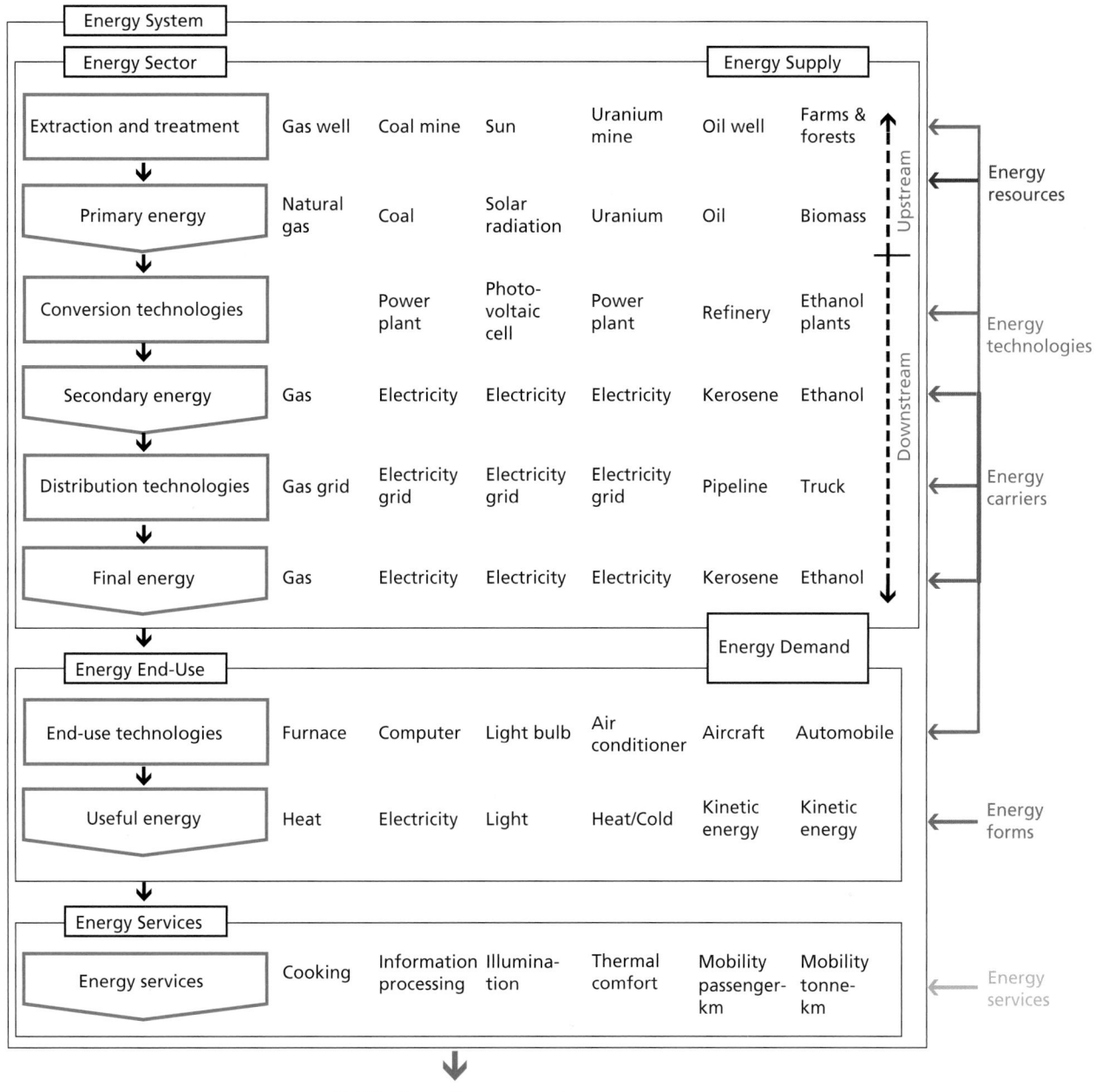

Figure TS-8 | Schematic diagram of the energy system with some illustrative examples of the energy sector, energy end-use, and energy services. The energy sector includes energy extraction, treatment, conversion, and distribution of final energy. The list is not exhaustive and the links shown between stages are not 'fixed'; for example, natural gas can also be used to generate electricity, and coal is not used exclusively for electricity generation. Source: adapted from Nakicenovic et al., 1996b; see Chapter 1.[11]

3.1   Energy Resources

An energy resource is the first step in the energy services supply chain. Provision of energy carriers is largely ignorant of the particular resource that supplies them, but the infrastructure, supply, and demand technologies, and fuels along the delivery chain, often depend highly on a

11 **Nakicenovic, N., A. Grubler, H. Ishitani, T. Johansson, G. Marland, J. R. Moreira and H.-H. Rogner, 1996b:** Energy primer. In *Climate Change 1995 – Impacts, Adaptations and Mitigation of Climate Change: Scientific-Technical Analyses, Contribution of Working Group II to the Second Assessment Report of the Intergovernmental Panel on Climate Change.* R. T. Watson, M. C. Zinyowera and R. H. Moss (eds.), Cambridge University Press, Cambridge, UK, pp.75–92.

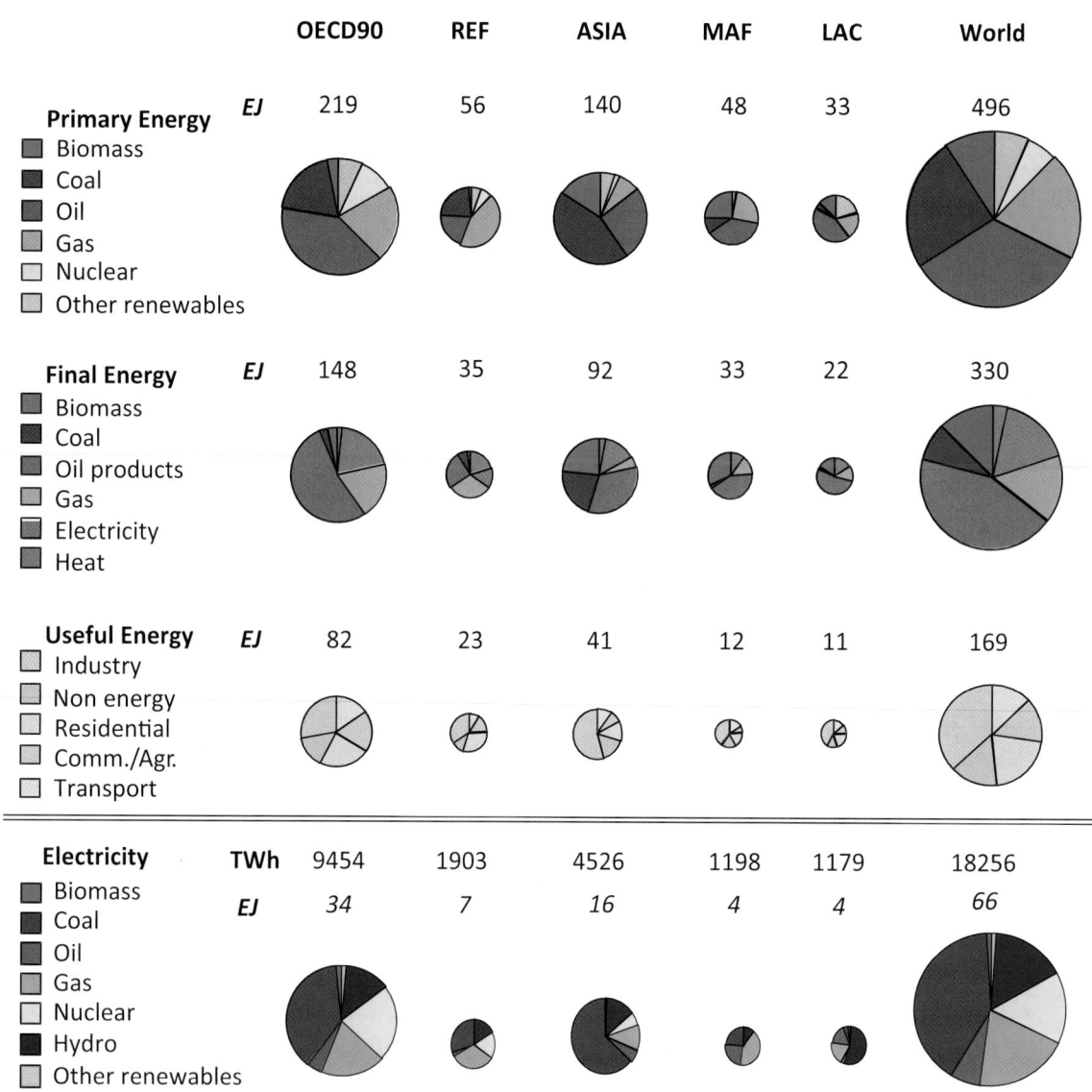

Figure TS-9 | World primary energy (by source), final energy (by energy carrier), useful energy (by sector/type of energy service), and electricity generated by different energy sources for 2005, World and five GEA regions.[12] Primary, final, and useful energy in EJ and electricity in TWh and EJ; feedstocks are included. Note: circle areas are proportional to electricity generated, but electricity graphs are not to the same scale as those for other energy forms; 1 TWh = 0.0036 EJ. Source: Chapter 1.

particular type of resource. The availability and costs of bringing energy resources to the point of end-use are key determinants of affordable and accessible energy carriers.

12   The five GEA regions consist of OECD90, which includes the UNFCCC Annex I countries; REF, which includes Eastern Europe and the Former Soviet Union; Asia, which excludes Asian OECD countries; MAF consisting of the Middle East and Africa; and LAC, which includes Latin America and the Caribbean.

The availability of energy resources *per se* poses no inherent limitation to meeting the rapidly growing global energy demand as long as adequate upstream investments are forthcoming for exhaustible resources in exploration, production technology, and capacity and, by analogy, for renewables in conversion technologies. However, exploitation of sufficient energy resources will require major investments and is not without significant environmental and other consequences.

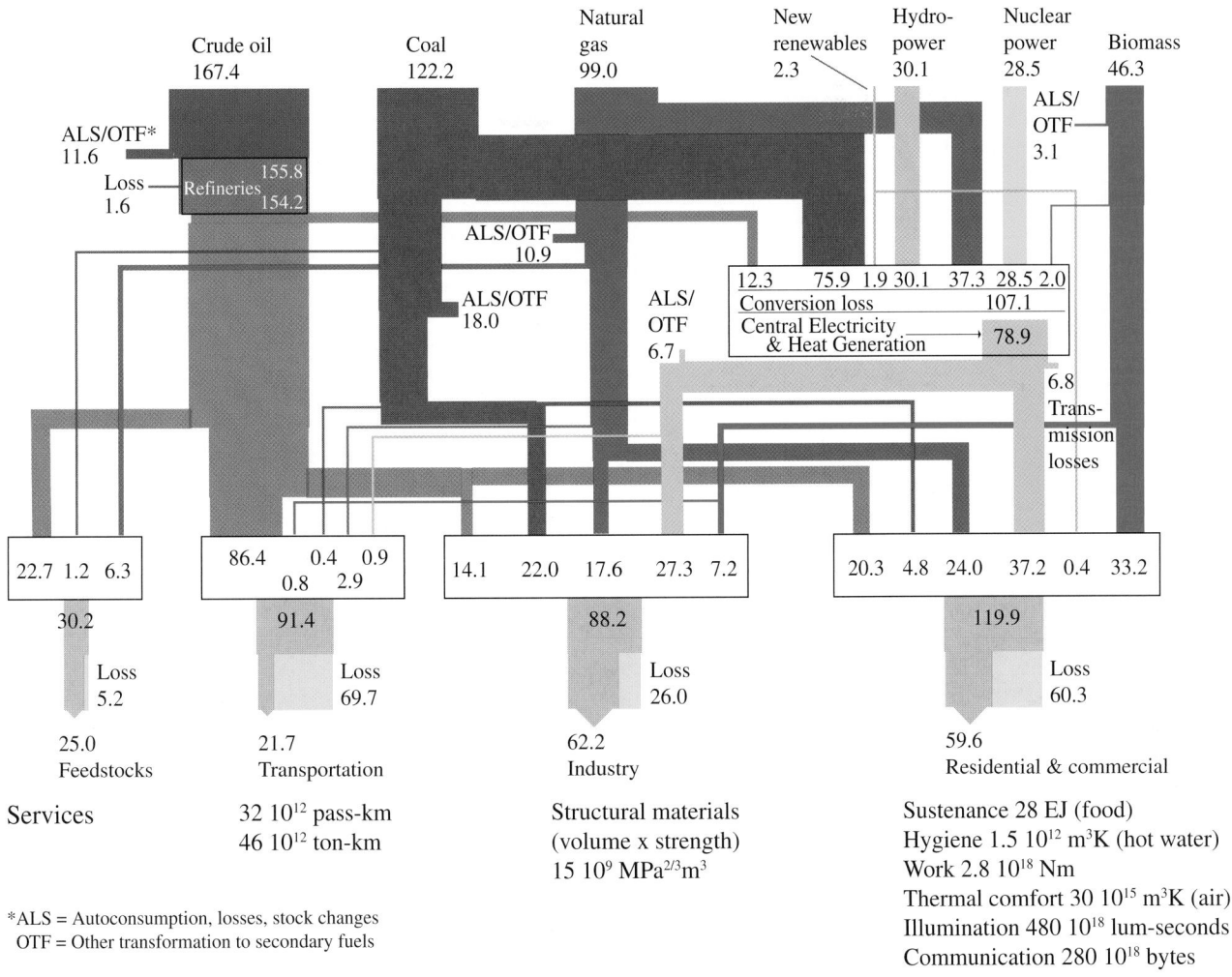

Figure TS-10 | Global energy flows (in EJ) from primary to useful energy by primary resource input, energy carrier (fuels), and end-use sector applications in 2005. Source: data from IEA, 2007a; 2007b (corrected for GEA primary energy accounting standard) and Cullen and Allwood, 2010, see Chapter 1.[13]

3.1.1   Hydrocarbons and Fissile Resources[14]

Hydrocarbons and fissile materials are plentiful in the Earth's crust, yet they are finite. The extent of ultimately recoverable oil, natural gas, coal, or uranium has been subject to numerous reviews, and there are wide ranges of estimates in the literature (see Table TS-2). For example, figures between 4900 and 13,700 exajoules (EJ) for conventional oil reserves and resources have caused continued debate and controversy. Such large ranges can be the result of varying boundaries

of what is included in the analysis of a finite stock of an exhaustible resource (e.g., conventional oil only, or conventional oil plus unconventional occurrences such as oil shale, tar sands, and extra-heavy oils). Uranium resources are a function of the level of uranium ore concentrations in the source rocks considered technically and economically extractable over the long run as well as the prospects of tapping into the vast amounts in sea water.

Oil production from difficult-to-access areas or from unconventional resources is not only more energy-intensive, it is also technologically and environmentally more challenging. Production from tar sands, shale oil, and gas, or the deep-sea production of conventional oil and gas, raise further environmental risks – ranging from oil spillages, ground- and freshwater contamination, and GHG emissions, to the release of toxic materials and radioactivity. A significant fraction of the energy gained needs to be reinvested into the extraction of the next unit, adding to already higher exploration and production costs.

---

13   **IEA, 2007a:** *Energy Balances of OECD Countries*. International Energy Agency (IEA), IEA/OECD, Paris, France.
   IEA, 2007b: *Energy Balances of Non-OECD Countries*. International Energy Agency (IEA), IEA/OECD, Paris, France.
   Cullen, J. M. and J. M. Allwood, 2010: The efficient use of energy: Tracing the global flow of energy from fuel to service. *Energy Policy*, **38**: 75–81.

14   Section TS-3.1.1 is based on Chapter 7.

Table TS-2 | Fossil and uranium reserves, resources, and occurrences.[a]

	Historical production through 2005	Production 2005	Reserves	Resources	Additional occurrences
	[EJ]	[EJ]	[EJ]	[EJ]	[EJ]
Conventional oil	6069	147.9	4900–7610	4170–6150	
Unconventional oil	513	20.2	3750–5600	11,280–14,800	>40,000
Conventional gas	3087	89.8	5000–7100	7200–8900	
Unconventional gas	113	9.6	20,100–67,100	40,200–121,900	>1,000,000
Coal	6712	123.8	17,300–21,000	291,000–435,000	
Conventional uranium[b]	1218	24.7	2400	7400	
Unconventional uranium	34	n.a.		7100	>2,600,000

[a] The data reflect the ranges found in the literature; the distinction between reserves and resources is based on current (exploration and production) technology and market conditions. Resource data are not cumulative and do not include reserves.

[b] Reserves, resources, and occurrences of uranium are based on a once-through fuel cycle operation. Closed fuel cycles and breeding technology would increase the uranium resource dimension 50–60 fold. Thorium-based fuel cycles would enlarge the fissile-resource base further.

Source: Chapter 7

Historically, technology change and knowledge accumulation have largely counterbalanced otherwise dwindling resource availabilities or steadily rising production costs (in real terms). They extended the exploration and production frontiers, which, to date, have allowed the exploitation of all finite energy resources to grow. The questions now are whether technology advances will be able to sustain growing levels of finite resource extraction, and what the necessary stimulating market conditions will be.

Resources first need to be identified and delineated before the technical and economic feasibility of their extraction can be determined. But having identified resources in the ground does not guarantee either that they can be technically recovered or their economic viability in the marketplace. The viability is determined by the demand for a resource (by the energy service-to-resource chain), the price it can obtain over time, and the technological capability to extract the resource efficiently.

Thus, timely aboveground investment in exploration and production capacities is essential in unlocking belowground resources. Private-sector investment is governed by expected future market and price developments, while public-sector investment competes with other development objectives. At least 10 years can elapse between investment in new production capacities and the actual start of deliveries, especially for the development of unconventional resources. Until new large-scale capacities come online, uncertainty and price volatility will prevail.

Nuclear fuel reserves are sufficient for approximately 100 years of consumption at today's production rates (see Table TS-2). The resources are much larger and, if uranium in seawater is included, practically sufficient for much longer, even with an expansion of global nuclear capacity.

There appears to be a consensus that there are sufficient fossil (for oil, see Figure TS-11) and fissile energy resources to fuel global energy

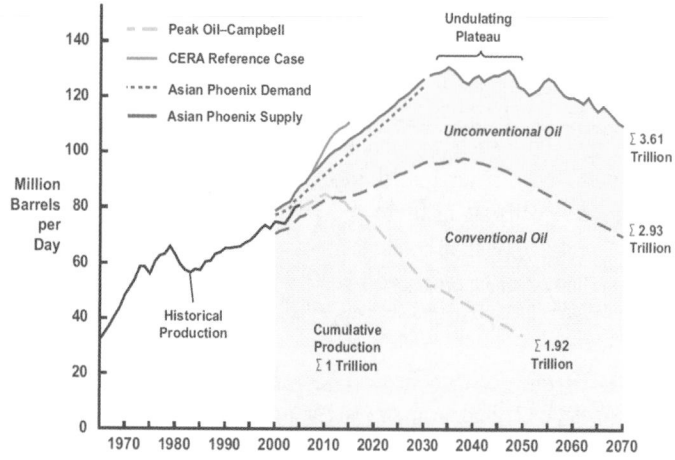

Figure TS-11 | The figure shows future oil production projections, comparing an undulating plateau with a peak oil projection. The Campbell peak oil projections is from Campbell (see Chapter 7) and is shown together with more optimistic projections of increased use of conventional oil resources as well as the use of unconventional oil resources. Source: Witze, 2007; see Chapter 7.[15]

needs for many decades. There is much less consensus as to their actual future availability in the marketplace. This availability is dependent on the balance between a variety of constraining and enabling factors. For example, the factors that can reduce the accessible stocks and flows from them include:

- smaller and smaller deposits in harsher and harsher environments, leading to rising exploration, production, and marketing costs;

- excessive environmental burdens;

15 **Witze, A., 2007:** Energy: That's oil folks…, *Nature*, **445**: 14–17.

Table TS-3 | Renewable energy flows, potential, and utilization in EJ of energy inputs provided by nature.[a]

	Primary Energy 2005[b] [EJ]	Utilization 2005 [EJ]	Technical potential [EJ/yr]	Annual flows [EJ/yr]
Biomass, MSW, etc.	46.3	46.3	160–270	2200
Geothermal	0.78	2.3	810–1545	1500
Hydro	30.1	11.7	50–60	200
Solar	0.39	0.5	62,000–280,000	3,900,000
Wind	1.1	1.3	1250–2250	110,000
Ocean	-	–	3240–10,500	1,000,000

[a] The data are energy-input data, not output. Considering technology-specific conversion factors greatly reduces the output potentials. For example, the technical 3150 EJ/yr of ocean energy in ocean thermal energy conversion (OTEC) would result in an electricity output of about 100 EJ/yr.

[b] Calculated using the GEA substitution method (see Chapter 1, Appendix 1.A.3).

Source: Chapter 7 (see also Chapter 11 for a discussion of renewable resource inventories and their differences). Note: MSW = municipal (and other) solid wastes.

- diminishing energy ratios;

- low rate of technological advances; and

- public intolerance of accident risks.

On the other hand, demand, high prices (plus associated investments), innovation, and technology change tend to increase stock sizes and flow rates. The question is: which combinations of these forces acting in opposite directions are going to govern fuels production in the mid to long term? It is likely that, due to the constraints, only a fraction of these resources may ever be produced.

In conclusion:

- Hydrocarbon resources are huge compared with conceivable future energy needs, but realizing any significant proportion of these available resources will require major investments and is not without significant consequences.

- Development of these resources and potentials is subject to many constraints, but not by a constraint on physical availability.

- The peak of oil and other fossil fuels is not caused by the lack of resources, but rather by other changes, such as the transformation toward sustainable futures and perhaps insufficient investments in the supply chains.

3.1.2   Renewable Energy Flows[16]

Renewable energy resources comprise the harvesting of naturally occurring energy flows. While these flows are abundant (see Table TS-3) and far exceed (by orders of magnitude) the highest future energy demand

imagined for global energy needs, the challenge lies in developing adequate technologies to manage the often low or varying energy densities and supply intermittencies and to convert them into usable energy carriers or utilize them for meeting energy demands.

Solar radiation reaching the Earth's surface amounts to 3.9 million EJ/yr and, as such, is almost 8000 times larger than the annual global energy needs of some 500 EJ. Accounting for cloud coverage and empirical irradiance data, the local availability of solar energy is 633,000 EJ. The energy carried by wind flows is estimated at about 110,000 EJ/yr, and the energy in the water cycle amounts to more than 500,000 EJ/yr, of which 200 EJ/yr could theoretically be harnessed for hydroelectricity. Net primary biomass production is approximately 2400 EJ/yr, which after deducting the needs for food and animal feed, leaves, in theory, some 1330 EJ/yr for energy purposes. The global geothermal energy stored in the Earth's crust up to a depth of 5000 meters is estimated at 140,000 EJ/yr, with the annual rate of heat flow to the surface of about 1500 EJ/yr. Oceans are the largest solar energy collectors on Earth, absorbing on average some one million EJ/yr.

The amounts of these gigantic annual energy flows that can be technically and economically utilized are significantly lower, however. Renewables, except for biomass, convert resource flows directly into electricity or heat. Their technical potentials are limited by factors such as geographical orientation, terrain, or proximity of water, while their economic potentials are a direct function of the performance characteristics of their conversion technologies within a specific local market setting. The data shown in Table TS-3 are the energy input potentials provided by nature.

3.2   Energy End-Use

The sectoral and regional distributions of per capita final energy use are shown in Figure TS-2. Globally, of a total final energy use in 2005 of

---

16 Section TS-3.1.2 is based on Chapters 7 and 11.

330 EJ, three sectors dominate – buildings (residential and public/commercial) is the largest sector with 34% (112 EJ), followed by transport at 28% (91 EJ), and industry with 27% (88 EJ). The relative shares of these sectors vary somewhat by region, with high-income countries having greater shares of residential, commercial, public, and transportation energy use, while residential and industrial sectors dominate the energy use in low-income countries. Therefore, this section reviews the options for enhancing efficiency in the industry, transportation, and buildings sectors.

3.2.1 Industry[17]

As noted, the industrial sector accounted for about 27% (88 EJ) of global final energy use in 2005. The production of materials – chemicals, iron and steel, non-metallic minerals (including cement), non-ferrous metals, paper and pulp, and mining – accounts for about 70% of global industrial final energy use. The final energy use of 88 EJ in 2005 excludes the energy use in coke ovens and blast furnaces, and feedstock energy use for petrochemicals. The addition of the energy inputs for these subsectors results in a final energy use of 115 EJ in 2005.

There has been a geographic shift in primary materials production, with developing countries accounting for the majority of production capacity. China and India have high growth rates in the production of energy-intensive materials like cement, fertilizers, and steel (12–20% per year after 2000). In other economies, the demand for materials is seen to grow initially with income and then stabilizes. For example, in industrial countries per capita use seems to reach saturation at about 400–500 kilograms for cement and about 500 kilograms for steel.

The aggregate energy intensity in the industrial sector in different countries has shown steady declines due to improvements in energy efficiency and a change in the structure of the industrial output. In the EU-27, for example, the final energy use by industry has remained almost constant (13.4 EJ) at 1990 levels despite output growth; 30% of the reduction in energy intensity is due to structural changes, with the remainder due to energy efficiency improvements.

In different industrial sectors, adopting the best achievable technology can result in savings of 10–30% below the current average costs. An analysis of cost-cutting measures in 2005 indicated energy savings potentials of 2.2 EJ for motors and 3.3 EJ for steam systems. The economic payback period for these measures ranges from less than nine months to four years. A systematic analysis of materials and energy flows indicates significant potential savings for process integration, heat pumps, and cogeneration.

An exergy analysis (the second law of thermodynamics) reveals that the overall global industry efficiency is only 30%. Clearly, there are major

energy efficiency improvements possible through research and development (R&D) in next-generation processes. The effective use of demand-side management can be facilitated by a combination of mandated measures and market strategies. To level the playing field for energy efficiency, a paradigm shift is required – with a focus on energy services, not energy supply *per se*. This requires a reorientation of energy supply, distribution companies, and energy equipment manufacturing companies.

Nevertheless, such a transformation has multiple benefits. Improved energy efficiency in industry results in significant energy productivity gains, for example, in improved motor systems; compressed air systems; ventilation, heat recovery, and air conditioning systems; and improvements in comfort and working environments through better lighting, thermal comfort, and reduced indoor air pollution from improved ventilation systems, and, in turn, improved productivity boosts corporate competitiveness.

Policies and capacity development to capture the opportunities are needed globally. New business models are also needed and are being deployed to deliver a transformation that shifts the focus to energy services. For example, energy service companies (ESCOs) are already a multibillion dollar market per year globally, and substantial new business opportunities await progressive enterprises and innovative technological and business initiatives.

A frozen efficiency scenario based on today's technologies (close to the GEA counterfactual pathway, see Section TS-4) has been constructed for industry between 2005 and 2030, which implies a demand for final energy of 225 EJ in 2030. This involves an increase of the industrial energy output in terms of manufacturing value added of 95% over the 2005 value. Owing to normal efficiency improvements from new technology designs over time, the BAU scenario results in a final energy demand being reduced from 225 EJ to 175 EJ in 2030.

An aggressive energy-efficient scenario (consistent with the GEA-Efficiency pathway, see Section TS-4) can result in a significant reduction in the energy intensity of the industrial sector. Such a scenario for 2030 has been constructed with the same increase in the manufacturing value added and only a 17% increase in final energy demand (to a total final energy demand for industry of 135 EJ) (see Figure TS-12.)

For existing industries, measures include developing capacity for systems assessment for motors, steam systems, and pinch analysis; sharing and documentation of best practices, benchmarks, and roadmaps for different industry segments; and enabling access to low-interest finance. A new energy management standard, ISO50001, for energy management in companies has been developed by the International Organization for Standardization. It will allow industries to systematically monitor and track energy efficiency improvements. To significantly improve energy efficiency, a paradigm shift is required – with a focus on energy services, not on energy supply *per se*. This requires a reorientation and new

---

17 Section TS-3.2.1 is based on Chapter 8.

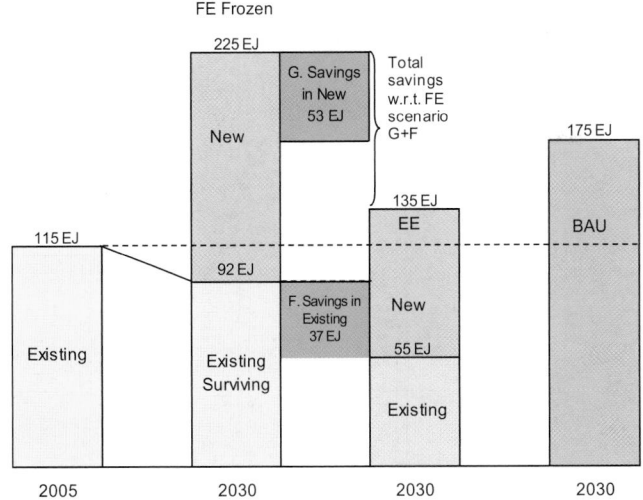

Figure TS-12 | Energy use in industry in 2005 and scenarios based on frozen efficiency and business-as-usual (BAU) scenarios for 2030 for an increase of manufacturing value added by 2030 of 95% over the 2005 level. The BAU and efficiency scenarios are consistent with GEA pathways. The surviving plants in 2030 from 2005 are subject to energy efficiency improvements (37 EJ) reducing their energy use from 92 EJ in 2005 to 55 EJ in 2030. Savings in the new capacity brings the frozen efficiency number from 225 − 92 = 133 EJ in 2005 to 80 EJ in 2030. Source: Chapter 8.

business models for energy supply, distribution companies, and energy equipment manufacturing companies.

Renewables currently account for 9% of the final energy use of industry (10 EJ in 2005). If an aggressive renewables strategy results in an increase in renewable energy supply to 23% in 2030 (23 EJ), it is possible to have a scenario of near-zero growth in GHG emissions in the industrial sector. Further reductions in overall energy use and emissions would be possible by dematerialization, redesign of products, and materials recycling.

Future industrial growth to 2030 is possible with net zero growth in GHG emissions provided there is a reorientation of the energy sector to focus on energy services, renewable energies (Chapter 11), and low-carbon fossil fuel use (Chapters 12 and 13).

3.2.2   Transport[18]

The transportation sector is responsible for approximately 28% (91 EJ) of global final energy demand. Road transport accounts for more than 70% of that total and 95% of transport energy comes from oil-based fuels. A major transformation of transportation is possible over the next 30–40 years and will require improving vehicle designs, infrastructure, fuels, and behavior. In the short term, improving overall sector energy efficiency, introducing alternative low-carbon fuels and electricity, and enhancing

18   Section TS-3.2.2 is based on Chapters 9 and 17.

the diversification, quantity, and quality of public modes of transport is necessary. Medium-term goals require reducing travel distances within cities by implementing compact urban design that improves accessibility to jobs and services and facilitates use of non-motorized modes, and replacing and adopting vehicle and engine design (for trucks, airplanes, rail, and ships) following the best-available technological opportunities for increasing efficiency and societal acceptability.

Transport policy goals for urbanization and equity include the adoption of measures for increasing accessibility and the affordable provision of urban mobility services and infrastructure that facilitates the widespread use of non-motorized options. Cities can be planned to be more compact with less urban sprawl and a greater mix of land uses and strategic siting of local markets to improve logistics and reduce the distances that passengers and goods need to travel. Urban form and street design and layout can facilitate walking, cycling, and their integration within a network of public transport modes. Employers in many sectors can enhance the job–housing balance of employees through their decisions on where to locate and can provide incentives for replacing some non-essential journeys for work purposes with the use of information technologies and communication.

Modal shares could move to modes that are less energy-intensive, both for passenger and freight transport. In cities, a combination of push-and-pull measures through traffic-demand management can induce shifts from cars to public transit and cycling and can realize multiple social and health benefits. In particular, non-motorized transportation could be promoted everywhere as there is wide agreement about its benefits to transportation and people's health. Parking policies and extensive car pooling and car sharing, combined with information technology options, can become key policies to reduce the use of cars. Efficient road-capacity utilization, energy use, and infrastructure costs for different modes could be considered when transport choices are made (see Figure TS-13).

Life cycle analyses (LCA), together with social and environmental impact assessments, are useful tools to compare different technologies. Significant uncertainties need to be addressed with respect to LCA system boundaries and modeling assumptions – especially in the case of biofuels and land use – and to future unknown technological advances. Hybrid electric vehicles (HEVs) can improve fuel economy by 7–50% over comparable conventional gasoline vehicles, depending on the precise technology used and on driving conditions (although comparable modern diesel engines can be equally fuel-efficient). Plug-in hybrid electric vehicles (PHEVs) allow for zero tailpipe emissions for low driving ranges, such as around 50 km in urban conditions. All-electric battery vehicles (BEVs) can achieve a very high efficiency (more than 90%, four times the efficiency of an internal combustion engine vehicle, but excluding the generation and transmission of the electricity), but they have a short driving range and battery life. Charging times are also, at present, significantly longer than fueling time for liquids. Consequently, BEVs have limited market penetration at present. If existing fuel saving

A)							
	2 000	9 000	14 000	17 000	19 000	22 000	80 000
B) MJ/p-km	1.65-2.45	0.32-0.91*	0.1	0.24*	0.2	0.53-0.65	0.15-0.35
C) €/p-km infrastructure	2 500-5 000	200-500	50-150	600-500	50-150	2 500-7 000	15 000-60 000
D) Fuel	Fossil	Fossil	Food	Fossil	Food	Electricity	Electricity

*Lower values correspond to Austrian busses, upper values correspond to diesel busses in Mexico city before introduction of BRT system.

Key:

A) Values are indicative for European and Asian cities and can vary significantly across cities, world regions, and particular situations. For example, BRT capacity can more than double with a second lane. Suburban rails in India can transport up to 100,000 passengers per hour.

B) Energy intensity in MJ per passenger km. SUVs can exceed depicted values for cars. Energy values for bus in the US are generally higher due to low ridership. While BRT systems have similar energy efficiencies as normal busses, they provide signficant systemic energy savings via modal shift, small bus substitution, and reduction in parallel traffic. BRT systems can also be converted from oil based fuels to renewable based electricity and hydrogen.

C) Estimated infrastructure costs in euros per passenger kilometer are highest for subway systems and heavy rail. Costs for bus system can be significantly lower than for individual motorized transport. Infrastructure costs for non-motorized transport are very cost competitive and can realize significant social benefits.

D) Dominant fuels are given for each mode.

Figure TS-13 | Comparative corridor capacity (people per hour), energy intensity per passenger kilometer (MJ/p-km), infrastructure cost (€/p-km), and main source of energy. Source: modified from Breithaupt, 2010; see Chapter 9.[19]

and hybrid technologies are deployed on a broad scale, fleet-average specific fuel savings of a factor of two can be obtained in the next decade (Figure 9.41, Chapter 9).

Increasing the performance of high-energy density batteries for PHEVs could lead to higher market penetration of BEVs. Hydrogen fuel cell vehicles (FCVs) could alleviate the dependence on oil and reduce emissions significantly. For HEVs and FCVs, the emissions are determined by the mode of production of hydrogen and electricity. Further technological

advances and/or cost reductions would be required in fuel cells, hydrogen storage, hydrogen or electricity production with low- or zero-carbon emissions, and batteries, including charging time. Substantial and sustained government support is required to reduce costs further and to build up the required infrastructure.

There are still many opportunities to improve conventional technologies. The combination of introducing incremental efficiency technologies, increasing the efficiency of converting the fuel energy to work by improving drivetrain efficiency, and recapturing energy losses and reducing loads (weight, rolling, air resistance, and accessory loads) on the vehicle has the potential to approximately double the fuel efficiency of 'new' light-duty vehicles from 7.5 liters per 100 km in 2010 to 3.0 liters per 100 km by 2050 (Figure 9.41, Chapter 9).

19 **Breithaupt, M., 2010:** *Low-carbon Land Transport Options towards Reducing Climate Impacts and Achieving Co-benefits.* Presented at the Fifth Regional Environmentally Sustainable Transport (EST) Forum in Asia, 23–25 August 2010, Bangkok, Thailand.

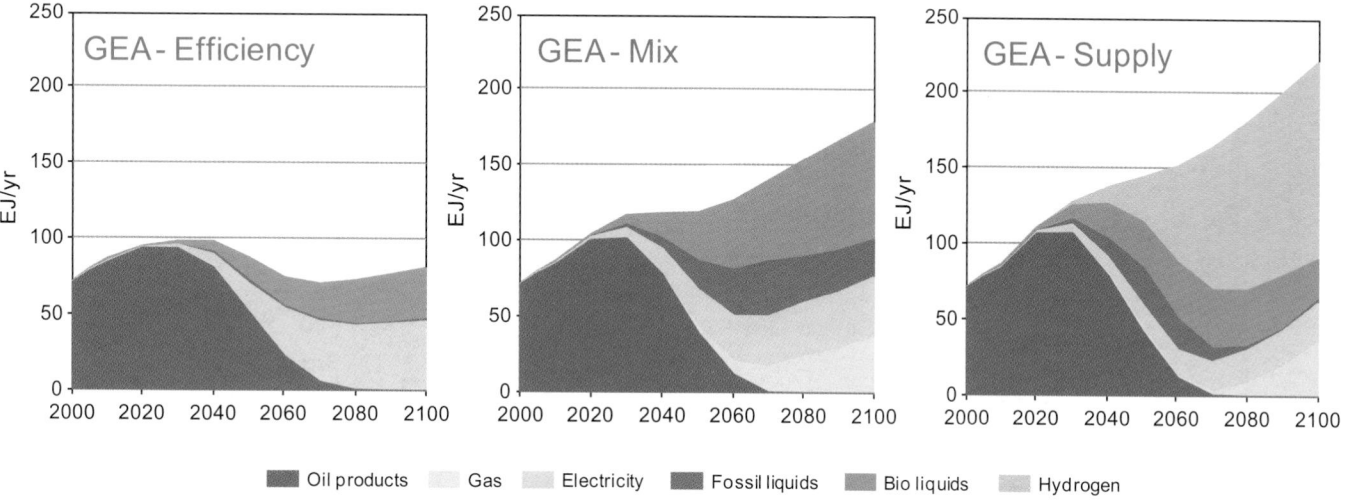

Figure TS-14 | Illustrative examples of fuel use in the transport sector for three GEA pathways. GEA-Supply features a strong technology push for new advanced technologies such as hydrogen, GEA-Efficiency features a strong reliance on regulation to reduce transport energy demand in combination with hybrid electric/biofuel technologies, and GEA-Mix features a co-evolution of both strategies, leading to regionally diverse transport systems. Source: Chapter 17 pathways and the GEA online database. For further details of the GEA pathways see the interactive web-based GEA scenario database hosted by IIASA: www.iiasa.ac.at/web-apps/ene/geadb.

Fuel economy standards have been effective in reducing fuel consumption and therefore could be adopted worldwide. The overall effectiveness of standards can be significantly enhanced if combined with fiscal incentives and consumer information. Taxes on vehicle purchase, registration, use, and motor fuels, as well as road and parking pricing policies, are important determinants of vehicle energy use and emissions.

Aviation transportation presents unique challenges owing to the requirement for very high density fuels. Studies indicate that fuel efficiency of aviation can be improved by 40–50% by 2050 through a variety of means, including technology, operation, and management of air traffic. As aviation's growth rate is projected to be the highest of the transport sub-sectors, such efficiency improvements will not be enough to keep overall energy use in the sector from increasing; thus, alternative low-carbon, high energy-density fuels will play a crucial role in decarbonizing emissions from aviation.

In the maritime sector, a combination of technical measures could reduce total energy use by 4–20% in older ships and 5–30% in new ships by applying state-of-the-art knowledge, such as hull and propeller design and maintenance. Reducing the speed at which a ship operates brings significant benefits in terms of lower energy use. For example, cutting a ship's speed from 26 to 23 knots can yield a 30% fuel saving.

GEA explored three distinctly different pathways for the transport sector (Figure TS-14), all of which satisfied the goals adopted for the GEA analysis (Section TS-4). In all pathways conventional oil is essentially phased out shortly after 2050. In the GEA-Efficiency pathway, electricity and biofuels dominate, while in GEA-Supply,

hydrogen plays a large role. In GEA-Mix, natural gas and fossil/biofuel liquids are also being used. The conclusion is that there are many combinations of energy carriers that would be able to fuel the transport sector.

3.2.3   Buildings[20]

Buildings are integrated systems that encompass and deliver multiple energy services and that require holistic approaches to achieve substantial reductions in energy demand and associated benefits. The sector, and activities in buildings themselves, are responsible for approximately 34% (112 EJ) of global final energy demand, with three-quarters of this amount for thermal purposes. Several energy-related problems in buildings (such as poor indoor air quality or inadequate indoor temperatures) affect the health and productivity of residents significantly.

New and existing technologies, as well as non-technological opportunities, represent a major opportunity for transformative change of energy use in buildings. Passive houses that reduce energy use for heating and cooling by 90% or more, for example, are already found in many countries. Increased investments in a more energy-efficient building shell are in part offset by lower or fully eliminated investments in heating/cooling systems, with energy costs for operation almost avoided, making these new options very attractive. Passive-house performance is also possible for existing buildings, if it is included as a performance goal when major renovations are done. Energy Plus houses, delivering net energy to the

---

20   Section TS-3.2.3 is based on Chapter 10.

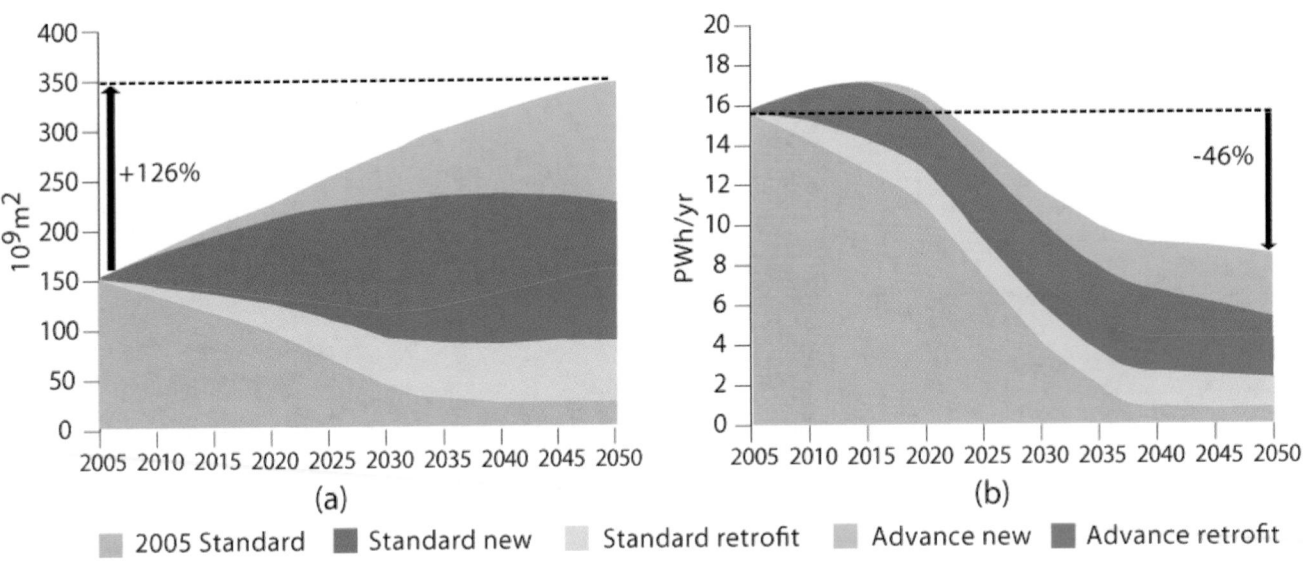

Figure TS-15 | Global final thermal energy use in buildings (a) and global floor area (b) in the state-of-the-art scenario (corresponding approximately to the 'GEA-Efficiency' group of pathways), 2005–2050. Source: Chapter 10.

Key: Explanations of efficiency categories: standard, today's stock; new, new buildings built to today's building code or anticipated new building codes (without additional policies); advance new, new buildings built to today's state-of-the-art performance levels; retrofit, assumes some efficiency gains, typically 35%; advanced retrofit, retrofit built to state-of-the-art levels.

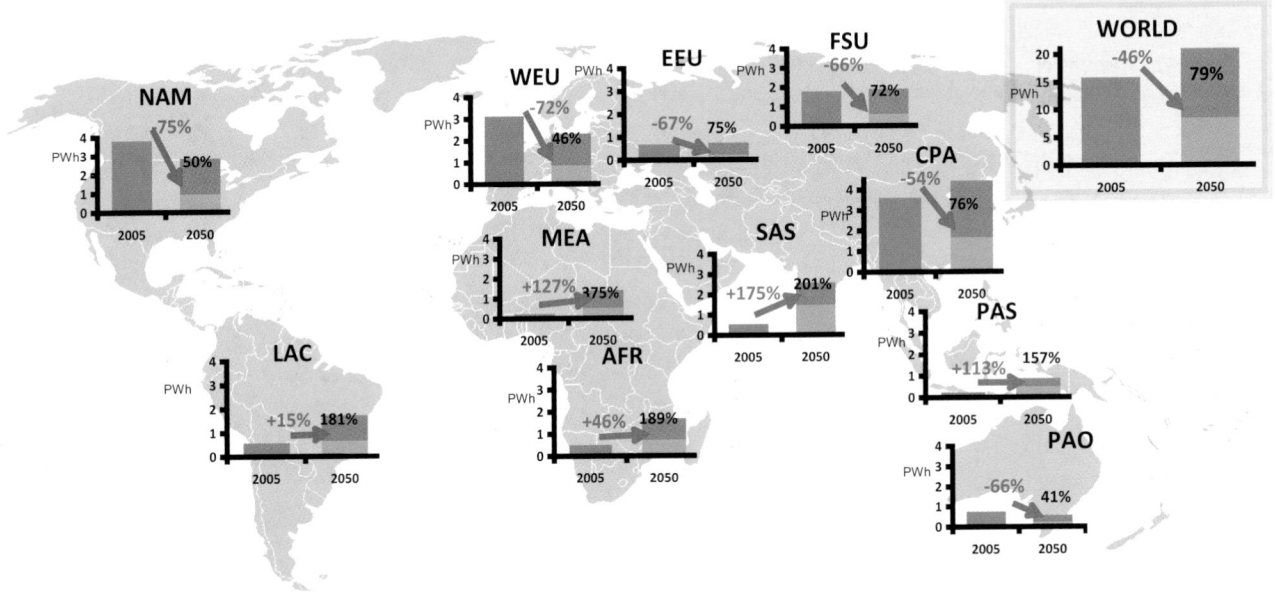

Figure TS-16 | Final building heating and cooling energy demand scenarios until 2050: state-of-the-art (~corresponding to the GEA-Efficiency set of pathways) and sub-optimal (~corresponding to the GEA-Supply set of pathways scenarios), with the lock-in risk (difference). Note: Green bars, indicated by red arrows and numbers, represent the opportunities through the state-of-the-art scenario, while the red bars with black numbers show the size of the lock-in risk (difference between the two scenarios). Percent figures are relative to 2005 values. Source: Chapter 10.

grid over a year, have been constructed even in high latitudes. Building-integrated solar photovoltaics (PVs) can contribute to meeting the electricity demand in buildings, especially in single-family homes, and solar water heaters can cover all or part of the heat required for hot water demand. However, requiring buildings to be zero-energy or net-energy suppliers may not be the lowest cost or most sustainable approach in

addressing the multiple GEA goals, and sometimes may not be possible, depending on location.

Analysis carried out under the GEA pathway framework demonstrates that a reduction of global final energy use for heating and cooling of about 46% by 2050 compared with 2005 is possible through the full use

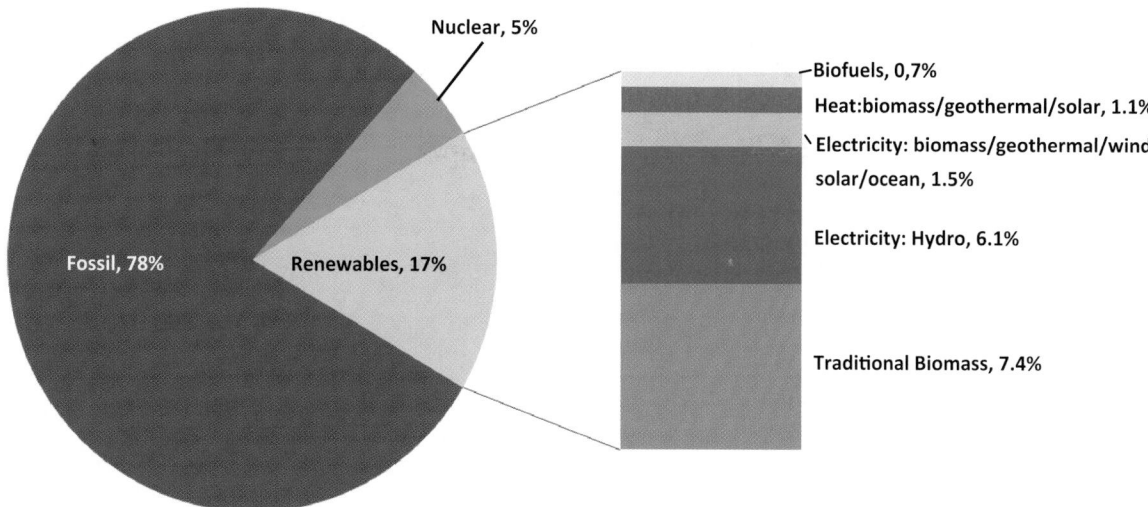

Figure TS-17 | Renewable share of primary energy use, 2009 (528 EJ). Source: Chapter 11.

of today's best practices in design, construction, and building operation technology and know-how. This can be obtained even while increasing amenities and comfort and simultaneously accommodating an increase in floor space of over 126% (see Figure TS-15.)

However, there is a significant risk of lock-in. If stringent building codes are not introduced universally and energy retrofits accelerate but are not subject to state-of-the-art efficiency levels, substantial energy use and corresponding GHG emissions can be 'locked in' for many decades. This could lead to a 33% increase in global energy use for buildings by 2050 instead of a decrease of 46% (see Figure TS-16).

Wide adoption of the state-of-the-art in the buildings sector would not only contribute significantly to meeting GEA's multiple goals, but such developments would also deliver a wide spectrum of other benefits. A review of quantified multiple benefits showed that productivity gains through reduced incidence of infections from exposure to indoor air pollution score particularly high. Other benefits included increases in productivity, energy security, indoor air quality and health, social welfare, real estate values, and employment. The approximately US$57 trillion cumulative energy-cost savings until 2050 in avoided heating and cooling energy costs alone substantially exceeds the estimated US$15 trillion investments that are needed to realize this pathway. The value of the additional benefits has also been shown to be substantial, often exceeding the energy cost savings. In several cases the multiple benefits are so significant and coincide with other important policy agendas (such as improved energy security, employment, poverty alleviation, competitiveness) that they provide easier and more attractive entry points for local policymaking than climate change or other environmental agendas.

A transition to a very low energy-use level for buildings requires a shift in the focus of energy-sector investment from the supply side to

an integrated system solution and services perspective, as well as the innovation and cultivation of new business models.

A broad portfolio of approaches is available and has been increasingly applied worldwide to capture the cost-effective efficiency potentials. Owing to the large number and diversity of market barriers, single instruments such as a carbon pricing will not unlock the large efficiency potentials. Policy portfolios tailored to different target groups and a specific set of barriers are needed. Nevertheless, deep reductions in building-energy use will not be possible without ambitious and strictly enforced performance standards, including building codes for new construction and renovation as well as appliance standards.

3.3   Energy Supply

3.3.1   Renewable Energy[21]

The potential to provide electricity, heat, and transport fuels to deliver all energy services from renewable energies is huge. The resource base is more than sufficient to provide full coverage of human energy demand at several times the present level and potentially more than 10 times this level.

In 2009, renewable energy sources contributed about 17% of world primary energy use, mainly through traditional biomass (7.4%) and large hydropower (6.1%), while the share from solar, wind, modern biomass, geothermal, and ocean energy was 3.3% (see Figure TS-17).

Many examples exist of hydropower plants, geothermal power plants, and biomass combustion for heat and for combined heat and power

---

21   Section TS-3.3.1 is based on Chapter 11.

Table TS-4 | Current status of renewable energy technologies as of 2009 (all financial figures are in US$_{2005}$).

Technology	Installed capacity increase in past five years (percent per year)	Operating capacity end 2009	Capacity factor (percent)	Secondary energy supply in 2009	Primary energy supply in 2009 (EJ/yr) based on the substitution calculation method	Turnkey investment costs ($/kW of output)	Current energy cost of new systems (¢/kWh, for biofuels $/GJ)	Potential future energy cost (¢/kWh, cost for biofuels $/GJ)
Biomass energy								
Electricity	6	54 GW$_e$	51[a]	~ 240 TWh$_e$	3.3	430–6200	2–22¢/kWh	2–22¢/kWh
Bioethanol	20	95 bln liter	80[a]	~ 76 bln liter	2.7	200–660	11–45 $/GJ	6–30 $/GJ
Biodiesel	50	24 bln liter	71[a]	~ 17 bln liter	0.9	170–325	10–27 $/GJ	12–25 $/GJ
Heat CHP	~ 3	~ 270 GW$_{th}$	25–80	~ 4.2 EJ	5.2	170–1000	6–12¢/kWh$_{th}$	6–12¢/kWh$_{th}$
Hydroelectricity								
Total capacity	3	~ 950 GW$_e$	30–80	~ 3100 TWh$_e$	32	1000–3000	1½-12¢/kWh$_e$	1½-10¢/kWh$_e$
Smaller scale plants (<10 MW)	~ 9	~ 60 GW$_e$	30–80	~ 210 TWh$_e$	2.2	1300–5000	1½-20¢/kWh$_e$	1½-20¢/kWh$_e$
Geothermal energy								
Electricity	4	~ 8 GW$_e$	70–90	~ 67 TWh$_e$	0.7	2000–4000	3–9¢/kWhe	3–9¢/kWh$_e$
Direct use of heat	12	~ 49 GW$_{th}$	20–50	~ 120 TWh$_{th}$	0.5	500–4200	2–19¢/kWh$_{th}$	2–19¢/kWh$_{th}$
Wind electricity								
Onshore	27	~ 160 GW$_e$	20–35	~ 350 TWh$_e$	3.6	1200–2100	4–15¢/kWh$_e$	3–15¢/kWh$_e$
Offshore	28	~ 2 GW$_e$	35–45	~ 7 TWh$_e$	0.07	3000–6000	7–25¢/kWh$_e$	5–15¢/kWh$_e$
Solar PV electricity	45	~ 24 GW$_e$	9–27	~ 32 TWh$_e$	0.33	3500–5000	15–70¢/kWh$_e$	3–13¢/kWh$_e$
Solar thermal electricity (CSP)								
Without heat storage	15	0.8 GW$_e$	30–40	~ 2 TWh$_e$	0.02	4500–7000	10–30¢/kWh$_e$	5–15¢/kWh$_e$
With 12h heat storage	–	–	50–65	–	–	8000–10,000	11–26¢/kWh$_e$	5–15¢/kWh$_e$
Low-temperature solar thermal energy	19	~ 180 GW$_{th}$	5–12	~ 120 TWh$_{th}$	0.55	150–2200	3–60¢/kWh$_{th}$	3–30¢/kWh$_{th}$
Ocean energy								
Tidal head energy	0	~ 0.3 GW$_e$	25–30	~ 0.5 TWh$_e$	0.005	4000–6000	10–31¢/kWh$_e$	9–30¢/kWh$_e$
Current energy	–	exp. phase	40–70	PM	–	5000–14,000	9–38¢/kWh$_e$	5–20¢/kWh$_e$
Wave energy	–	exp. phase	25	PM	–	6000–16,000	15–85¢/kWh$_e$	8–30¢/kWh$_e$
OTEC	–	exp. phase	70	PM	–	6000–12,000	8–23¢/kWh$_e$	6–20¢/kWh$_e$
Salinity gradient energy	–	R&D phase	80–90	–	–	–	–	–

[a] Industry-wide average figure; on plant level the CF may vary considerably.

Source: Chapter 11

(CHP) that have been fully competitive with fossil fuels for decades. In select locations, and for specific markets (such as remote locations), wind and solar (or hybrid systems) have also provided least-cost, highly reliable energy supplies. In broader markets, starting from an initially very low level, very rapid growth of wind and solar technologies has occurred in the last decade, strongly stimulated by public policies. Wind energy has grown globally by more than 25% per year for more than 15 years and solar PV by more than 50% per year for around 10 years. Costs have dropped by 50–90% on a dollar per megawatt-hour basis over the past decades, and they continue to decline rapidly. Technologies such as solar water heaters and wind farms on good sites are nowadays competitive with conventional energy technologies on standard economic terms (without including any external costs and benefits). An overview of the present status and future potential of renewable energy options is presented in Table TS-4.

The rapid expansion in renewables, which largely has taken place in only a few countries, has usually been supported by different types of incentives or driven by quota requirements. The feed-in tariffs (FITs) used in the majority of EU countries, China, and elsewhere have been especially successful. Global investments in 2009 were slightly lower as a result of the financial crises (although with less reduction than for most other energy technologies) (see Figure TS-18); however, they rebounded in 2010. Both wind and solar PV electricity are already

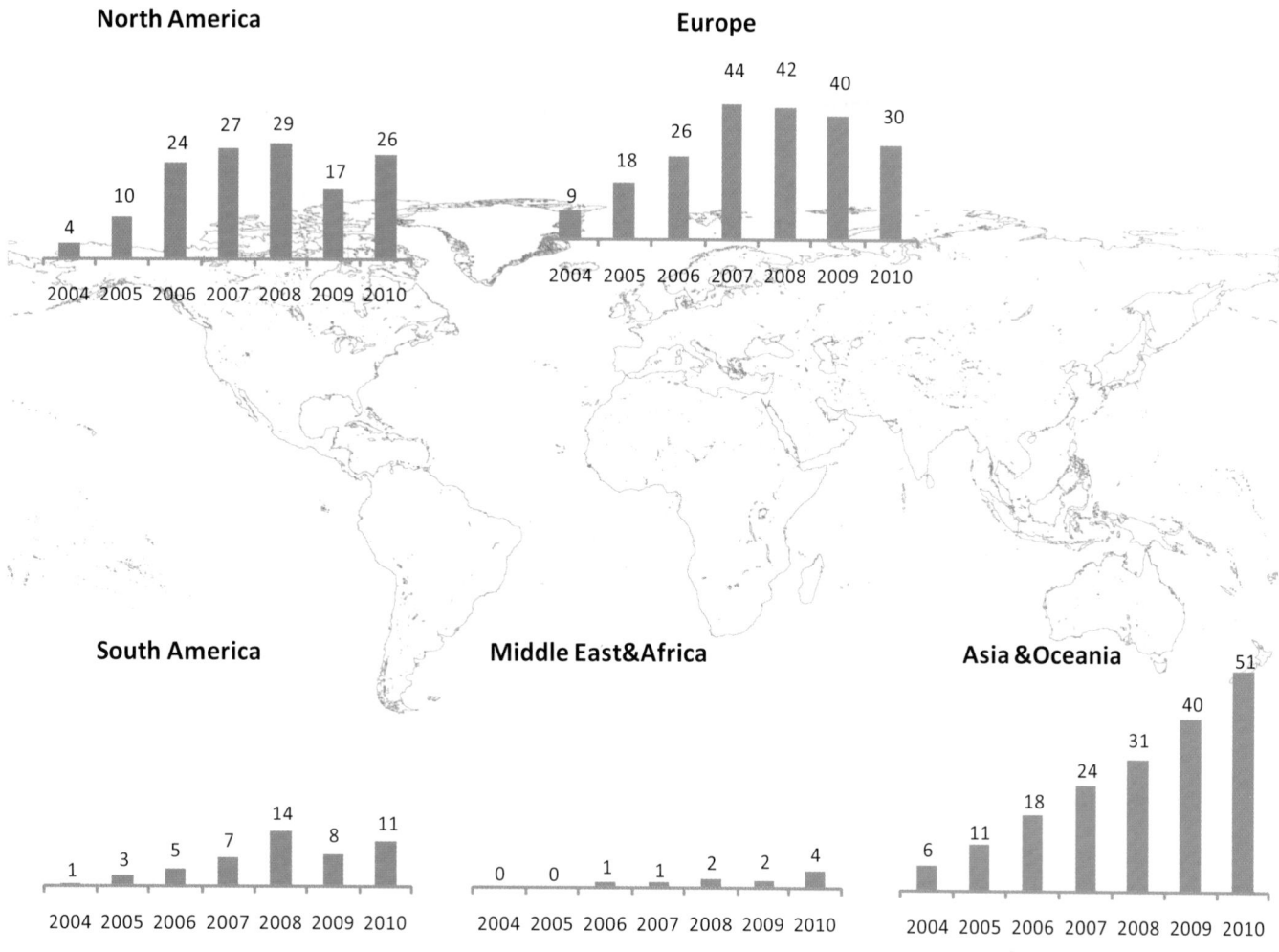

Figure TS-18 | New financial investments in renewable energy, by region, 2004–2010 (billion US$_{2005}$\$). New investment volume adjusted for reinvested equity; total values include estimates for undisclosed deals. This comparison does not include small-scale distributed energy projects or large-scale hydropower investments. Source: Chapter 11.

cost-competitive in some markets and are projected to become so in many more markets in the next 5–10 years without being favored by public policy. However, renewables face resistance due to lock-in to conventional energies and substantial market barriers in the majority of markets.

Renewable power capacity additions now represent more than one-third of all global power capacity additions (see Figure TS-19). While substantial on an annual basis, with a very large total installed capacity, renewables remain a relatively small contributor to global energy supply.

The intermittent and variable generation of wind, solar, and wave power must be handled within an electricity system that was not designed to accommodate it, and in which traditional base load-power from nuclear, geothermal, and fossil power stations with restricted flexibility limit the system's ability to follow load variations. Energy systems have historically been designed to handle loads that vary over seconds, days, weeks, and years with high reliability. These systems are becoming increasingly

able to accommodate increased quantities of variable generation through use of so-called smart systems with advanced sensing and control capabilities. With support from accurate and timely load forecasting, capacity management, and overall intelligent load and demand-side management, experience has shown that at least 20%, and perhaps up to 50%, of variable renewable generation can be accommodated in most existing systems at low costs, and that it is feasible to accommodate additional intermittent generation with additional investment in grid flexibility, low capital cost fuel-based generation, storage, and demand-side management (smart grids).

Intelligent improvement and increase of interconnection between states and across geographic regions will help maintain and increase reliability of energy systems in an environment with rapidly increasing shares of variable renewable energies in the system. Wind and solar PV, and most hydrokinetic or ocean thermal technologies, offer the unique additional attribute of virtually complete elimination of additional water requirements for power generation. Other renewable options, including bio-based options, geothermal, concentrating solar, and hydropower on

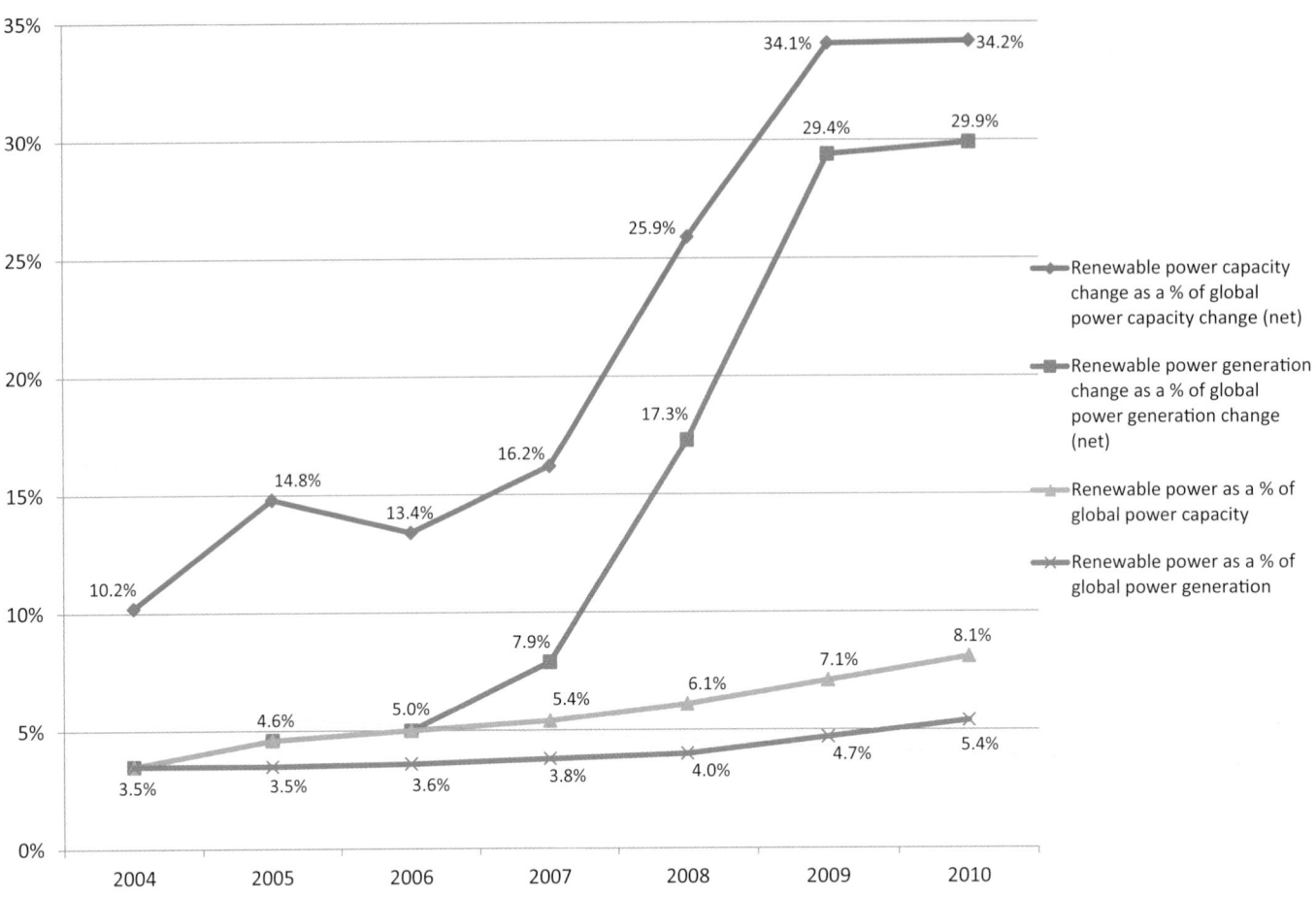

Figure TS-19 | Renewable power capacity (excluding large hydro) and generation as a percentage of global capacity and generation, respectively, and their rates of change also in percent; 2004–2010 Source: UNEP and BNEF, 2011; see Chapter 11.[22]

a life-cycle basis, still require water for cooling a steam turbine or are associated with large amounts of evaporation.

The development of high-voltage direct current transmission cables may allow the use of remote resources of wind and solar at costs projected to be affordable. Such cables have been installed for many years in submarine and on-shore locations, and demand is increasing (in the North Sea, for example). This is significant, as some of the best renewable energy resources are located far from load centers. In conjunction with energy storage at the generation location, such transmission cables can be used to provide base load electricity supply.

The GEA pathways show that renewable energies can meet up to 100% of projected energy demand for specific regions. The GEA pathways analysis indicates that a significant increase in renewable energy supplies from 17% of global primary energy use in 2009 up to 30–75% and would result in multiple benefits.

3.3.2   Fossil Energy Systems[23]

This section has two parts. The first part covers new opportunities in the conversion of fossil fuels to liquid energy carriers and electricity. Co-utilization of coal and biomass is highlighted in conjunction with CCS, which is the subject of the second part.

3.3.2.1   Fuels, Heat, and Electricity from Fossil Resources

A radical transformation of the fossil energy landscape is feasible for simultaneously meeting the multiple sustainability goals of wider access to modern energy carriers, reduced air pollution, enhanced energy security, and major GHG emissions reductions. The essential technology-related requirements for this transformation are continued enhancement of unit energy conversion efficiencies, the development of CO_2 capture and storage, use of both fossil and renewable energy in the same facilities, and efficient co-production of multiple energy carriers at the same facilities.

22   **UNEP and BNEF, 2011:** *Global Trends in Renewable Energy Investment 2011: Analysis of Trends and Issues in the Financing of Renewable Energy.* United Nations Environment Programme (UNEP), Nairobi, Kenya and Bloomberg New Energy Finance (BNEF), London, UK.

23   Section TS-3.3.2.1 is based on Chapter 12.

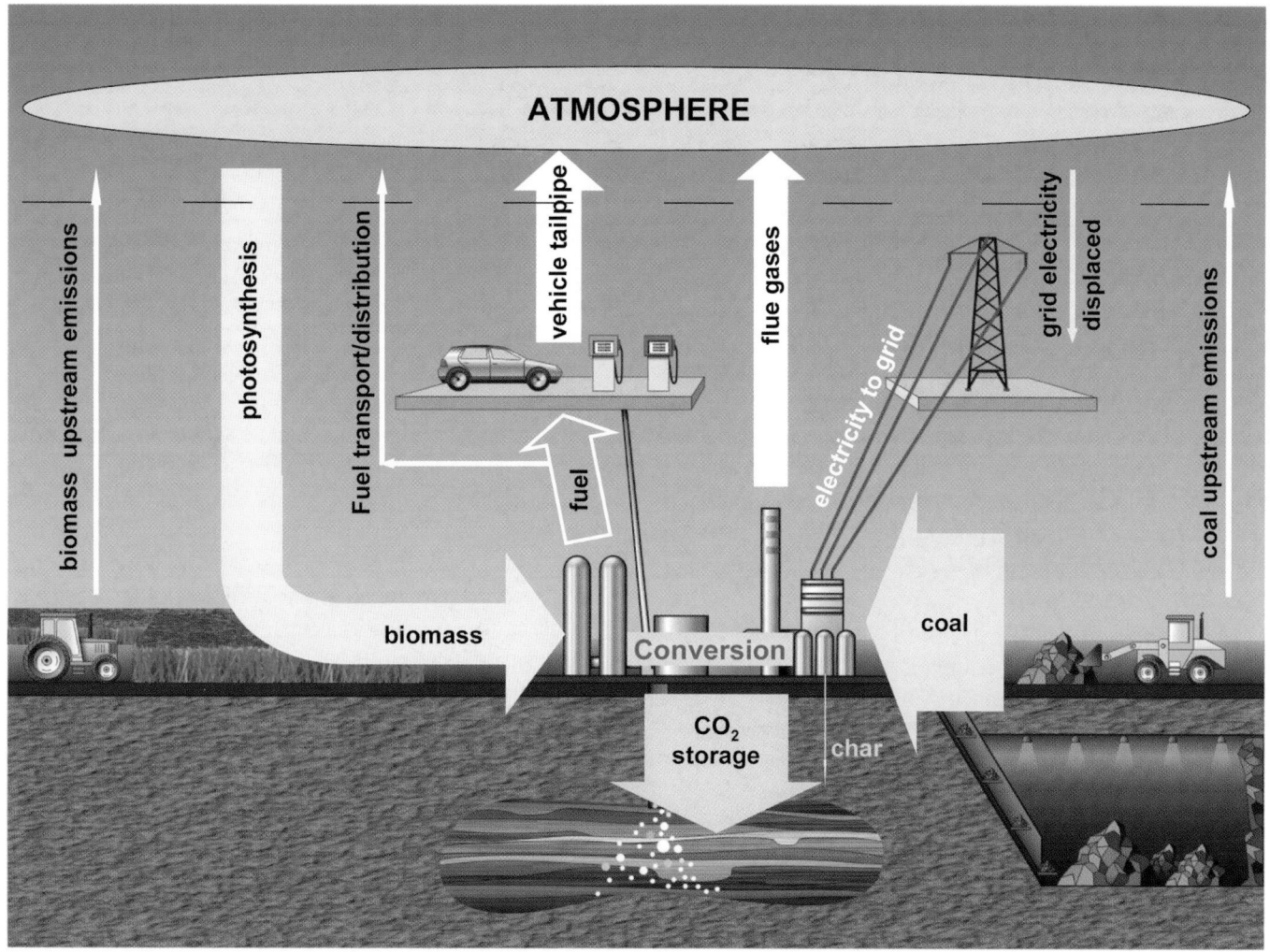

Figure TS-20 | Carbon flows for conversion of coal and biomass to liquid fuels and electricity. When biomass is approximately 30% of the feedstock input (on a higher heating value basis), the net fuel cycle GHG emissions associated with the produced liquid fuels and electricity would be less than 10% of the emissions for the displaced fossil energy. Source: Chapter 12.[24]

For developing and industrial countries alike, fossil fuels – which will dominate energy use for decades to come – must be used judiciously by designing energy systems for which the quality of energy supply is well matched to that required and also by exploiting other opportunities for realizing high efficiencies. Continued use of coal and other fossil fuels in a carbon-constrained world will increase the requirement for CO_2 capture and storage.

Since developing and industrial countries have different energy priorities, strategies for fossil energy development will vary in the short term, but they must converge in the long term. In developing countries, the emphasis could be on increasing access to energy services based on cleaner energy carriers, building new manufacturing and energy infrastructures that anticipate the evolution to low-carbon energy systems, and exploiting

the rapid growth in these infrastructures to facilitate introduction of the advanced energy technologies needed to meet sustainability goals.

In industrial countries, where energy infrastructures are largely already in place, a high priority could be overhauling existing coal power plant sites to add additional capabilities (such as co-production of electricity and liquid transport fuels) and CCS. (Simply switching from coal to natural gas power generation without CCS will not achieve the needed carbon emission reductions.)

Among the technologies that use fossil fuels, only co-production and co-processing strategies using biomass with fossil fuel and with CCS have the ability to achieve deep reductions in CO_2 through 'net negative' emissions. These technologies could begin to be deployed in the 2015–2020 time frame as nearly all of their components are already in commercial use. In the long term, hydrogen made from fossil fuels with CCS is a decarbonization energy option, but infrastructure challenges are likely to limit this option in the near term.

24 **Larson, E. D., G. Fiorese, G. Liu, R. H. Williams, T. G. Kreutz and S. Consonni, 2010:** Co-production of Decarbonized Synfuels and Electricity from Coal + Biomass with CO_2 Capture and Storage: an Illinois Case Study. *Energy & Environmental Science*, **3**(1):28–42.

Co-production with CCS represents a low-cost approach for simultaneously greatly reducing carbon emissions for both electricity and transportation fuels (such as gasoline, diesel, and jet fuel), enhancing energy supply security, providing transportation fuels that are less polluting than petroleum-derived fuels in terms of conventional air pollutants, providing clean synthetic cooking fuels as alternatives to cooking with biomass and coal (critically important for developing countries), and greatly reducing the severe health-damage costs due to air pollution from conventional coal power plants.

Co-processing biomass with coal or natural gas in co-production systems requires, at most, half as much biomass to provide low-carbon transport fuels compared with advanced fuels made only from biomass such as cellulosic ethanol. Co-production also represents a promising approach for gaining early market experience with CCS (because CO_2 capture is less costly than for stand-alone power plants) and can serve as a bridge to enabling CCS as a routine activity for biomass energy (with corresponding negative GHG emissions) after 2030 (see Figure TS-20).

No technological breakthroughs are needed to get started with co-production strategies, but there are formidable institutional hurdles created by the need to manage two disparate feedstock supply chains (for a fossil fuel and biomass) and simultaneously provide three products (liquid fuels, electricity, and CO_2) serving three different commodity markets.

Creative public policies that promote the needed changes in the fossil fuel landscape would include the setting of a price on GHG emissions, more stringent regulations on air pollution, performance-based support for the early deployment of promising technology, and an emphasis on cost reduction through accelerated learning. These actions would need to be supported by international collaboration and intellectual and financial assistance from industrial to developing countries for technology adoption and technological and institutional capacity building.

3.3.2.2   Carbon Capture and Storage[25]

Over the past decade there has been a remarkable increase in interest and investment in CCS. In 2011, 280 projects were in various stages of development, and governments have committed billions of dollars for R&D, scale-up, and deployment. Considering full life-cycle emissions, CCS technology can reduce CO_2 emissions from fossil fuel combustion from stationary sources by about 65–85%. CCS is applicable to many of these sources, including the power generation and industrial sectors. Applying CCS with bioenergy would open up a route to achieving negative emissions.

Although the technology for CCS is available today, significant improvements are needed to support its widespread deployment. CCS involves the integration of four elements: CO_2 capture (separation and compression of CO_2), transportation to a storage location, and isolation from the

25   Section TS-3.3.2.2 is based on Chapter 13.

atmosphere by pumping the CO_2 into appropriate saline aquifers, oil and gas reservoirs, and coal beds with effective seals that keep it safely and securely trapped underground.

Successful experiences with five ongoing projects (Weyburn-Midale, La Barge, In Salah, Sleipner, Snøhvit) demonstrate that, at least on a limited scale, CCS appears to be safe and effective for reducing emissions. Moreover, relevant experience from nearly 40 years of CO_2 utilization for enhanced oil recovery, currently at the aggregate rate of 40 Mt/yr, also shows that CO_2 can safely be pumped and retained underground.

Significant scale-up will be needed to achieve large reductions in CO_2 emissions through CCS. A five- to ten-fold scale-up in the size of individual projects is needed to capture and store emissions from a typical coal-fired power plant. A thousand-fold scale-up in CCS would be needed to reduce emissions by billions of tonnes per year.

Worldwide storage capacity estimations are improving, but more experience is needed. Estimates for oil and gas reservoirs are about 1000 billion tonnes (Gt) CO_2, saline aquifers are estimated to have a capacity ranging from about 4000 to 23,000 Gt, and coal beds have about 200 Gt. However, there is still considerable debate about how much sequestration capacity actually exists, particularly in saline aquifers. Research, geological assessments, and – most important – commercial-scale demonstration projects will be needed to improve confidence in capacity estimates.

Added costs and reduced energy efficiencies are associated with CCS. Costs for CCS are estimated to be from below US$30 to above US$200 per tonne of CO_2 avoided, depending on the type of fuel, the capture technology, and the assumptions about the baseline technology. And they would increase the cost of stand-alone power generation by 50–100%. Capital costs and parasitic energy requirements of 15–30% are the major cost drivers. Further R&D could help reduce costs and energy requirements. In addition, pursuing electricity generation via co-production with transportation fuels could also reduce costs for generating decarbonized electricity from coal (as described in Chapter 12).

Early CCS demonstration projects are likely to cost much more than projected long-term costs, but there are opportunities to keep costs down for demonstration by coupling to existing sources of low-cost CO_2 (e.g., coal-to-chemicals or fuels facilities in China) and/or to storage of anthropogenic CO_2 via enhanced oil recovery (as currently practiced in a large-scale CO_2 storage project in Canada).

Access to capital for large-scale deployment could be a major factor limiting the widespread use of CCS. Owing to the added costs, CCS will not take place without strong incentives to limit CO_2 emissions. Certainty about the policy and regulatory regimes will be crucial for obtaining access to capital to build these multibillion dollar projects.

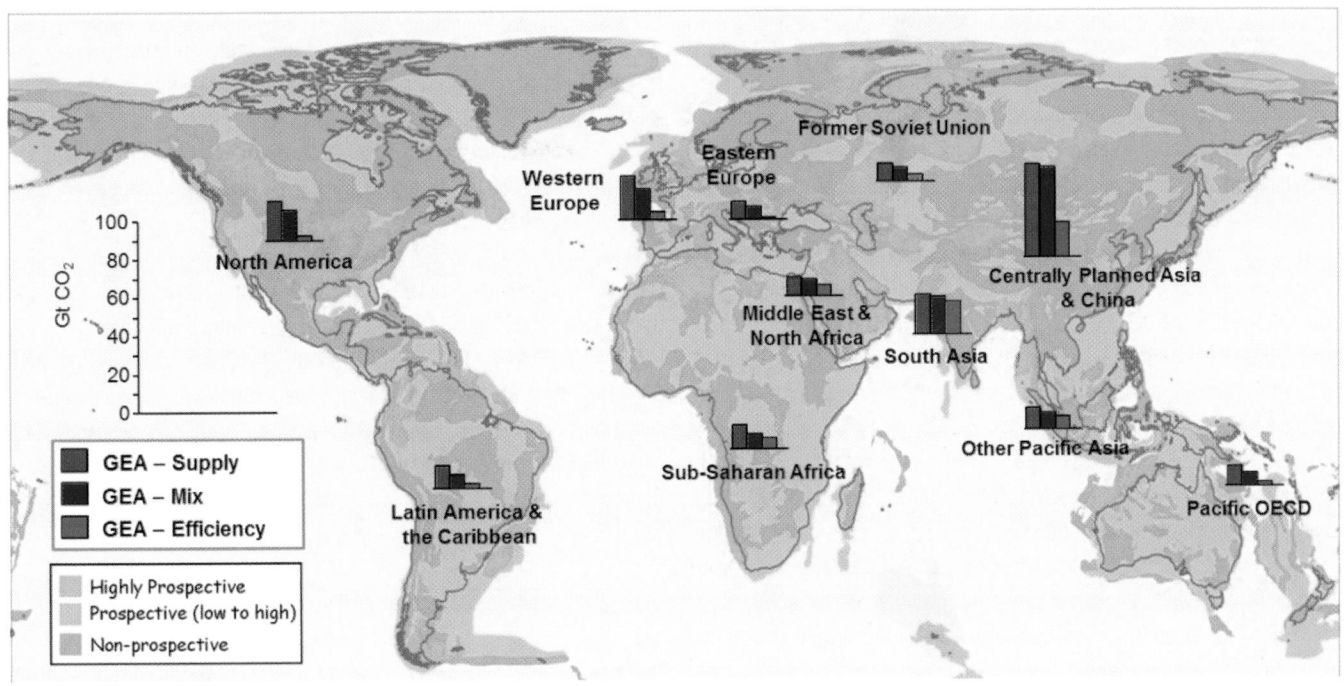

Figure TS-21 | Global map showing prospective geological carbon storage areas (Bradshaw and Dance, 2005)[26] superimposed on the estimated CO_2 storage requirements from CCS across the three illustrative GEA pathways (Chapter 17). Storage requirements in the illustrative GEA pathways are below 250 $GtCO_2$ by 2050, and below 1300 $GtCO_2$ by 2100. This is significantly less compared to the global geological CO_2 storage capacity, which includes saline aquifers ranging from about 4000 to 23,000 $GtCO_2$, and oil and gas reservoirs of about 1000 $GtCO_2$ (Chapter 13). For further details of the GEA pathways see the interactive web-based GEA scenario database hosted by IIASA: www.iiasa.ac.at/web-apps/ene/geadb.

To manage the environmental risks of CCS, clear and sufficient regulations are needed and enforced to ensure due diligence over the lifecycle of the project – particularly siting decisions, operating guidelines, monitoring, and closure of a storage facility.

Social, economic, policy, and political factors may limit deployment of CCS if not adequately addressed. Critical issues include ownership of underground pore space, long-term liability and stewardship, GHG accounting approaches, and verification and regulatory oversight regimes. Government support to lower barriers for early deployment is needed to encourage private-sector adoption. Developing countries will need support for getting access to technologies, lowering the cost of CCS, developing workforce capacity, and training regulators for permitting, monitoring, and oversight. CCS combined with biomass gasification has negative emissions, which are likely to be needed to achieve atmospheric stabilization of CO_2 and may provide an additional incentive for CCS adoption.

The assessment of future pathways suggests an overall requirement of CCS of up to 250 $GtCO_2$ of cumulative captured emissions by 2050. This is much less than the estimated storage capacity (see Figure TS-21).

3.3.3   Nuclear Energy[27]

The share of nuclear energy, currently 14% of world electricity, has declined in recent years. Nevertheless, there are 441 nuclear power reactors in the world with 374 GWe of generating capacity, and another 65 under construction, of which 45 have been launched in the past five years, 27 of them in China. New grid connections peaked at 30 GW/yr in the mid-1980s, and the last decade has witnessed a decline to an average of 3 GWe/yr in new nuclear capacity and 1 GWe/yr of retirements for a net increase of 2 GWe/yr. Increases in the capacity factors of existing units have, to a degree, compensated for a slower increase in installed capacity. Efforts are being made to extend the lives of existing plants and to encourage the construction of new ones with government loan guarantees, caps on liability for the consequences of accidents, and other subsidies, and thereby sustain, and even increase by a few percent, the share of nuclear energy in a growing global electric power sector. This is the most that the International Atomic Energy Agency believes is achievable by 2050, based on its review of national plans.

Although the momentum of global nuclear power expansion slowed considerably in recent decades, there are important differences between nations and regions. In OECD countries, home of 83% of global installed nuclear capacity, very little construction is under way. Costs per unit remain high, and may even be increasing in Western Europe and North

26 **Bradshaw, J. and T. Dance, 2005:** Mapping geological storage prospectivity of CO2 for the world's sedimentary basins and regional source to sink matching. *Proceedings of the 7th International Conference on Greenhouse Gas Control Technologies*.

27 Section TS-3.3.3 is based on Chapter 14.

America, which together account for 63% of global capacity. Among the reasons for cost increases in the 1980s and 1990s were increased stringency in safety requirements and construction delays in the United States after the Three Mile Island accident (see Figure TS-23). These factors can be expected to play a role again for some time after the accident at Japan's Fukushima Daiichi nuclear power plant in 2011. Costs are lower in East Asia and Russia, where most of the new construction is underway. There has been concern in China, however, about the availability of qualified workers and the adequacy of regulatory oversight. After the Fukushima accident, Germany, Italy, Switzerland, and Japan have decided, or announced plans, to scale back nuclear energy, and the United States, European Union, Japan and China announced comprehensive safety reviews.

Many developing countries have aspirations to build their first nuclear power plant, but a large fraction do not have the funds and currently have grid capacities that are too small to manage the large unit sizes of the currently available nuclear power plants. There is currently interest in the nuclear energy establishments of the developed countries, but in smaller reactors whose costs may be reduced through mass production.

Although over the past decades many proposals have been made for improving reactor safety and strengthening the barriers blocking the misuse of nuclear energy technologies for weapons purposes, it is still not clear how these problems will be dealt with. Most importantly, it has been understood since the end of World War II that the non-proliferation regime would be greatly strengthened if enrichment and plutonium were placed under international or multinational control. Plutonium separation and recycle persists in some countries, however, despite the fact that it is unlikely to be economic for the foreseeable future. Also countries continue to build national enrichment plants that could be misused to produce highly enriched uranium for weapons. Consequently, unless problems associated with proliferation and safety are effectively addressed, nuclear energy may not be a preferred climate change mitigation option even though it has low carbon emissions – other low-carbon electric power supply options may be more attractive.

Uncertainty characterizes the long-term role of nuclear energy in the GEA pathways. As with other energy technologies, it is an option, not a necessity, to meet future energy needs in a climate-friendly way. As discussed later, the scenarios in GEA demonstrate it is possible to meet all the GEA goals, including the climate goals, without nuclear power, even in the case of high demand scenarios. At the same time, in some scenarios with higher energy demand, it plays a large role in the energy mix. The resulting nuclear installed capacity in the scenarios ranges between 75 and 1850 GWe by 2050 with the lower bound resulting in a complete phase-out in the second half of the century. This uncertainty in the future of nuclear energy results from uncertainties in its future cost sund public concerns about reactor safety, the proliferation of weapons-grade fissile materials, and the absence of arrangements

in most countries for final disposal of spent fuel and/or the radioactive waste from spent-fuel reprocessing and plutonium recycle. Note that the high end of the above range is higher than the high projection put forward by the IAEA, based on the high ends of national projections before the Fukushima accident in 2011. In the past, nuclear growth has been far below the IAEA's high projection – and, until 2000, even below its low projections.

We have not considered nuclear fusion separately in the scenarios, given that fusion power is not likely to become a commercial energy option before the middle of the century at the earliest and would compete most directly with fission. However, pure fusion would have significant advantages relative to fission with regard to safety, proliferation resistance, and radioactive waste. Fusion–fission hybrids would not have these advantages.

3.4   Energy Systems[28]

The mechanisms by which energy is supplied to the final consumer are critical to the success of global and local economies. For the complex and diverse energy supply system to operate smoothly, many sources of energy – and their conversion to forms that can be delivered for use by consumers, from households to industry and commercial businesses – must operate in harmony within stable markets. Without the smooth operation of this complex interwoven system, the global economy cannot function (Chapter 15.1).

Energy supply systems differ between regions, between major economies, and between developing and industrial countries. Approaches to the necessary transitions[29] to create energy systems for a sustainable future therefore vary, and policies that work successfully in one region may fail in another. Nevertheless, there are lessons to be learned from shared experiences. The evolution of energy systems will depend on how well technologies are implemented and how well policies are instituted to bring about these changes.

Sustainable conversion from energy sources to energy carriers and efficient transmission and distribution for end-uses is crucial. This places particular emphasis on energy carriers such as electricity, hydrogen, heat, natural gas, biogas, and liquid fuels. All are key to transporting energy from more-remote production locations to growing urban population centers. Marketplaces will determine how much of each is used in any geographic region and when and how rapidly that use occurs.

As noted earlier, industry accounts for 27% of global final energy use, residential and public/commercial buildings use 34%, transport uses 28%, and agriculture/feedstocks/other uses account for 11%. Electricity

---

28   Section TS-3.4 is based on Chapters 15.

29   Transitions are covered more generally in Section TS-3.4.1 and in Chapters 16, 24, and 25.

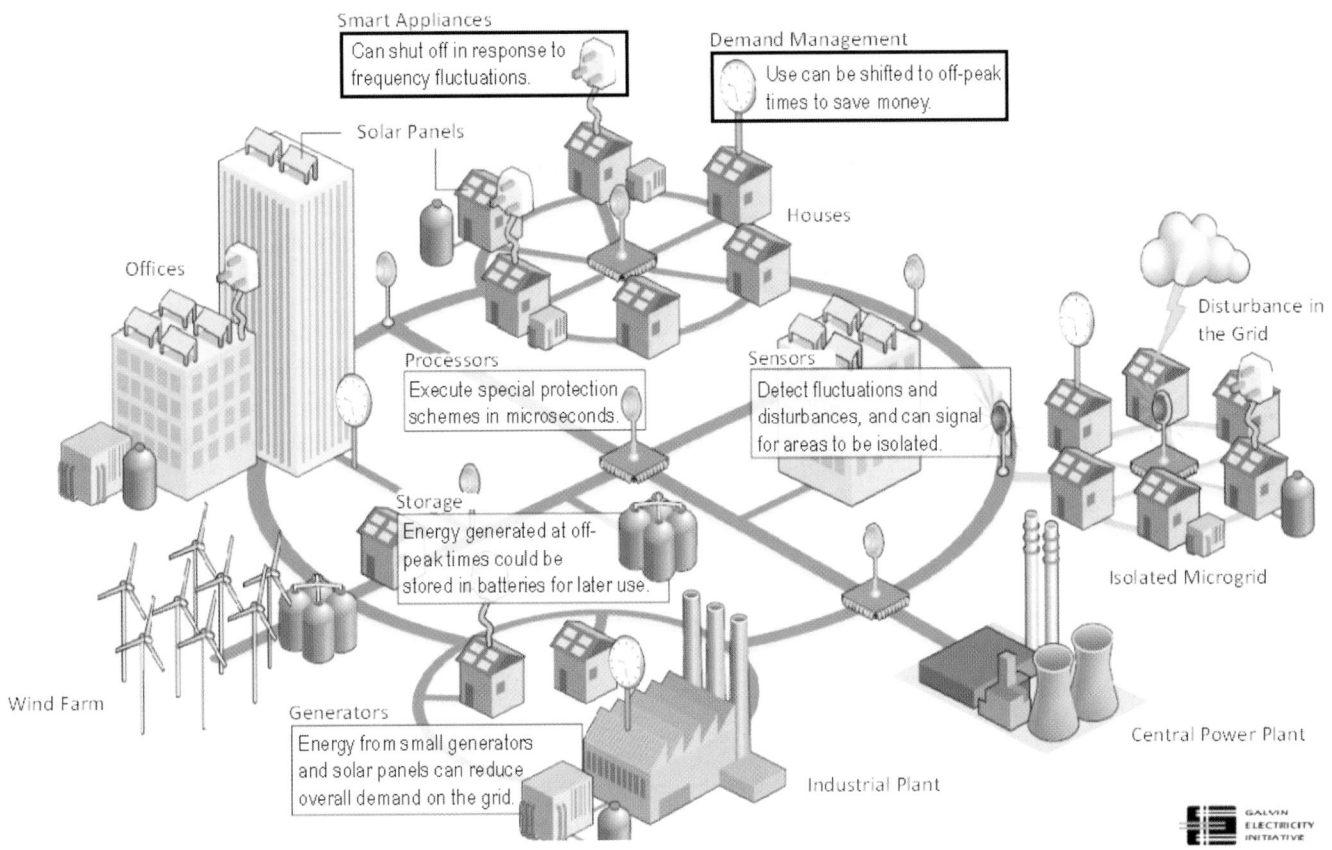

Figure TS-22 | Example of a smart grid, a network of integrated microgrids that can monitor and heal itself. Source: Amin, 2008; see Chapter 15.[30]

generation accounts for over one-third of the world's primary energy demand, with an average conversion efficiency of only around 36% (ranging from up to 90% for large hydro to less than 15% in some very old coal-fired power stations). Over half of all electricity generated is used in buildings, a category that includes households, services, and the public sector. More importantly, 40% of primary energy use results in heat, and much of this is wasted. Using more heat from thermal and geothermal power stations for industrial processes, district-heating schemes, and so on, would increase the overall efficiency of energy use (Chapter 15.2.1).

Providing integrated and affordable energy storage systems for modern energy carriers is essential. This is perhaps the largest and most perplexing part of the energy systems for a sustainable future that is needed for future economic security since costs are relatively high; pumped hydro storage being the lowest storage cost where feasible.

The entire energy grid for each energy source, from conversion to end-use, must be optimized into 'smart grids' using a digital system that continuously communicates between source and end-use and that integrates all energy carriers and their transmission and distribution systems (see Figure TS-22).

The potential of competitive electricity markets to most efficiently match energy supply and demand is a proven principle that is at the heart of

the industrial world's economic system. Based on that success, it is rapidly becoming the standard for economic systems around the world.

In countries with well-developed energy markets, spot-market energy trading has been introduced and long-term contracts are becoming less frequent. The result for countries that supply energy is generally negative, and it is now more difficult to ensure long-term returns on large-scale investments. This is a threat to the financing of large capital-intensive energy supply projects.

Effective approaches to improving energy systems will be led by the private sector – but it is essential that there be a stable governance framework, facilitation of physical infrastructure, capital investments, and the social cohesion necessary for economic development and poverty reduction. Success will depend on the implementation of robust global public/private partnerships that can achieve unprecedented cooperation and integration between governments, between businesses, and between governments and businesses. This needs to happen rapidly to achieve energy systems for a sustainable future envisioned in the GEA goals.

30 **Amin, M., 2008:** Interview with Massoud, Amin, "Upgrading the Grid". *Nature*, **454**: 570–573.

Crucial to improved energy systems to meet rapidly changing needs is the urgent requirement to boost development and investment in advanced systems (Chapter 15, Sections 15.7.5, 15.8.5, and 15.9). The lag time between research and large-scale commercial deployment is long, yet funding for energy system R&D and demonstrations has been declining for 20 years. This trend must be reversed and involves enhanced cooperation among and between private and public sectors.

3.4.1   Transitions in Energy Systems[31]

The beginning of transformative change may be seen in a number of innovations and experiments in the energy sector. These experiments include technology-driven innovations in generation and end-use; system-level innovations that could reconfigure existing systems; and business-model innovations centered on energy service delivery. Experiments in generation include hybrid systems, where combining multiple primary energy sources help address issues such as intermittency. Experiments in end-use include technology options for the simultaneous delivery of multiple energy services, or even energy and non-energy services. System-level experiments include innovations in storage, distributed generation, and the facilitation of energy efficiency by effectively monetizing savings in energy use and the creation of new intermediaries such as ESCOs. In some of these experiments, technology can lead to changing relationships between players or changing roles for players; for example, the process of consumers becoming producers as seen in small-scale biogas projects.

To generate a base of innovations and effectively support those that show promise, understanding of the dynamics of technology transitions is essential. The transitions literature suggests that large-scale, transformative change in technology systems involves a hierarchy of changes from experiments to niches to technology regimes, with linkages across different scales.

3.4.2   Opportunities in System Integration

It is important to focus energy-related solutions on providing the energy services needed rather than on energy supply *per se*. Examples include telecommuting and electronic/IT services (such as e-banking) that replace the need for many routine car trips, or a relaxation of summer dress codes in offices (saving the energy used for air conditioning).

Measures to improve efficiency on the end-use side of the energy chain typically save more primary energy and associated pollution than measures on the supply side. This is because 1 kilowatt-hour (kWh) saved on the end-use side can often reduce 2–3 kWh worth of primary energy use and associated emissions. The GEA pathways assessment (Section TS-4) concludes that strategies focusing more on improved

energy efficiency are by and large associated with lower costs for reaching the GEA goals than those focusing more on supply-side options.

GEA also finds that end-use-focused solutions reduce the risk that sustainability goals become unattainable. They increase the chances for multiple benefits, including improved local and indoor air pollution, and thus result in significant health gains, reduced congestion, reduced poverty, productivity gains, increased comfort and well-being, potential energy security improvements, new business opportunities, and sometimes enhanced employment opportunities.

The efficiency of end-use devices is also important when providing energy access. Lowest-cost devises, sometimes second-hand, can lock poor populations into using much more energy than needed; conversely, if efficient basic devices or appliances are also subsidized when providing access, consumers will be able to afford a higher level of energy services.

While individual system-, process-, and component-level efficiencies have improved significantly over the last few decades, the major opportunities for reducing energy intensity of economic activities lie in system optimization strategies rather than a focus on single components.

In industry, major advances have been achieved in process efficiency, with relatively fewer potentials remaining for such improvements. Still, most markets and policymakers tend to focus on individual system components (e.g., motors and drives, compressors, pumps, and boilers) with improvement potentials of 2–5%, while systems have much more impressive improvement potentials: 20% or more for motor systems and 10% or more for steam and process-heating systems.

Another way to reduce energy use in industry is to use fewer materials, as 70% of industrial energy use goes into the production of materials. A focus on increasing the rates of product reuse, renovation, remanufacture, and recycling has significant potential.

In buildings, novel approaches focusing on holistic methods that involve integrated design principles have been known to achieve as much as 90% energy reductions for heating and cooling purposes compared with standard practices (Section TS-3.2.3). Small-scale CHP has attracted interest for large buildings, and with the cost declines of PV, even if currently more expensive than grid electricity, the prospect of not being dependent on grid electricity has emerged. However, for feasibility, economic, and environmental reasons, requiring buildings to be zero-energy or net-energy suppliers is not likely to be the lowest cost or most sustainable approach in eliminating fossil fuel use, and is sometimes even impossible.

Energy use in densely built and populated areas, up to hundreds of watts per square meter (W/m^2) land area, typically significantly exceeds

---

31   Section TS-3.4.1 is based on Chapter 16.

annual average local renewable energy flows (typically below 1 W/m²). Therefore net zero-energy buildings are only feasible in low-density areas and in building types with low power or heat loads. High-rise or commercial buildings with high energy use, such as hospitals, cannot meet their entire energy demand through building-integrated renewable energy sources.

It is typically single-family or low-rise, lower-density multifamily residential neighborhoods that can become zero net-energy users for their residential energy needs (excluding transportation). Therefore, care needs to be exercised that zero-energy housing mandates do not incentivize further urban sprawl that leads to more automobile dependence and a growth in transport energy use, as efficient public transport systems are not economical in low-density urban areas.

Rather than aiming for buildings that use zero fossil fuel energy as quickly as possible, an economically sustainable energy strategy would implement a combination of the following: reduced demand for energy; use of available waste heat from industrial, commercial,

or decentralized electricity production; on-site generation of heat and electricity; and off-site supply of electricity. Many forms of off-site renewable energy are less expensive than on-site PV generation of electricity and are thus able to achieve more mitigation and sustainable energy supply per unit of expenditure. At the same time, PV costs are expected to become competitive for grid-connected small consumers in major grid markets during this decade. Furthermore, PV must compete against the retail rather than the wholesale cost of electricity produced off-site. On-site production of electricity also improves overall system reliability by relieving transmission bottlenecks within urban demand centers. All of these steps point to the importance of policy integration.

3.5   Evaluating Options

The previous sub-sections describe various supply-side and end-use technologies, as well as the primary energy sources and their availability, and the changes taking place in the entire energy system. A number of

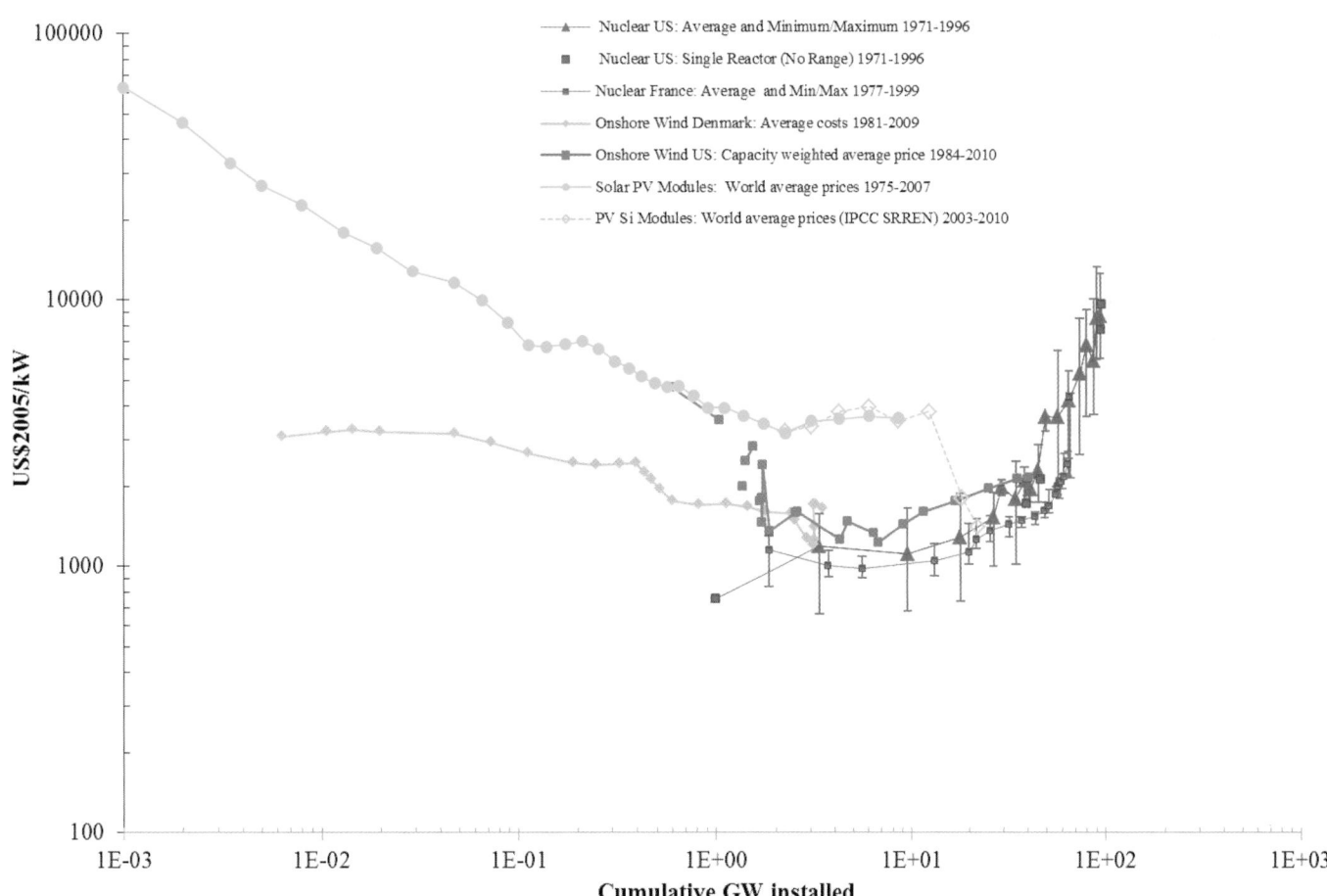

Figure TS-23 | Cost trends of selected non-fossil energy technologies (US2005$/kW installed capacity) versus cumulative deployment (cumulative GW installed) Chapter 24 data have been updated with the most recent cost trends (2010) available in the literature for PV Si Modules and US onshore wind turbines. Note that the summary illustrates comparative cost trends only and is not suitable for a direct economic comparison of different energy technologies due to important differences between the economics of technology components (e.g., PV modules versus total systems installed), cost versus price data, and also differences in load factors across technologies (e.g., nuclear's electricity output per kW installed is three to four times larger than that of PV or wind turbine systems). Source: Chapter 24.

attributes of these options need to be considered while evaluating their feasibility, appropriateness, and attractiveness. These include costs, benefits, environmental outcomes, and trade-offs across multiple objectives.

3.5.1　Costs and Environmental Performance of Options

Costs of Non-fossil Energy Supply Options

Figure TS-23 shows cost trends of selected non-fossil energy technologies. Despite a wide range in experiences of cost trend across technologies, two important observations stand out. First, there is a marked contrast between nuclear technologies, showing persistent cost escalations, versus the other non-fossil technologies, which generally show declining costs/prices with accumulated market deployment experience. Second, improvement trends are highly variable across technologies and also over time. For some technologies (e.g., wind in the United States and Europe) historical cost improvements were temporarily reversed after the year 2003–2004, suggesting possible effects of ambitious demand-pull policies in face of manufacturing capacity constraints and rising profit margins that (along with rising commodity and raw material prices) have led to cost escalations in renewable energy technologies as well.

Greenhouse Gas Emissions

One of the goals in the GEA analysis is to reduce emissions of GHGs. Tables TS-5 and TS-6 present the specific emissions per kWh for different technologies. Data are all life-cycle emissions of the major gases converted to CO_2 equivalent per GJ. The lowest emissions from renewables are from run-of-the-river hydropower, followed by hydro (reservoir), which displays a large variation depending on biological material in the reservoir, solar PV, wind, biomass, and geothermal. Nuclear energy has lower emissions than fossil-based generation, even if CCS is applied.

Combining fossil fuels with CCS reduces emissions significantly, although they are still significantly higher than renewables and nuclear. Combined use of fossil fuel and biomass with CCS can offer electricity and fuels with no net or negative emissions. As photosynthesis removes CO_2 from of the atmosphere, this results in negative emissions – that is, removal and storage of CO_2 when the biomass fuel is combined with CCS, which returns the CO_2 molecules to the ground (Table TS-7).

Quantifying emissions for electricity co-produced with liquid fuels (or for liquid fuels co-produced with electricity) is complicated by the issue of how to apportion emissions between the different products. The approach adopted here is to assign a percentage reduction (or increase) in emissions equal to the emission for the complete co-production system divided by emissions for a reference system consisting of separate conventional fossil fuel technologies (without CCS) that collectively produce the same amount of fuels and electricity.

Table TS-5 | Life-cycle emissions of GHGs for electricity generation. These numbers include direct emissions at the power plant and associated emissions upstream and downstream of the plant. The numbers here must be regarded as approximate as they depend on many assumptions, including the composition of the present energy systems.

Type	Conversion scheme	g CO_2-eq per kWh$_e$	
		w/o CCS	with CCS
Renewables	Solar photovoltaic[a]	16	
	Wind[b]	8	
	Hydro (reservoir)[c]	2–48	
	Hydro (run-of-river)	1–18	
	Geothermal[d]	~100	
	Biomass-IGCC[e]	25	−776
Fossil	Sub-critical coal[f]	896	187
	Super-critical coal[f]	831	171
	Coal-IGCC[g] Coal+biomass in coproduction with transportation fuels[h]	787–833 750	126–162 71
	NGCC[i] Natural gas + biomass in co-production with transportation fuels[h]	421 448	110 87
Nuclear	LWR, once-through fuel cycle[j]	38±27	

CCS = carbon capture and storage; CO_2-eq = carbon dioxide equivalents; HT = high-temperature; IGCC = integrated gasification combined cycle; LWR = light water reactor; MWh = megawatt-hour; and NGCC = natural gas combined cycle.

a) For an in-plane irradiation of 1700 kWh/m^2 and a value of 10 g CO_2-eq per kWh$_e$ or even less, see Chapter 11, Section 11.6.

b) Data from ExternE, vol 6, 1995.

c) Chapter 11, Section 11.3 gives estimates on emissions from hydropower.

d) High temperature brine, Chapter 11, Section 11.4. Note: emissions are very variable between existing geothermal power plants.

e) As-received biomass moisture content is 15% by weight (see Chapter 12, Table 12.1). This assumes zero emissions associated with indirect land-use change, as would be the case with utilization of most biomass residues or biomass grown on abandoned cropland.f) From Chapter 12, Table 12.6.

g) From Chapter 12, Table 12.7.

h) With co-production, how the life-cycle emissions for the system are allocated to each product is arbitrary. In Chapter 12 a Greenhouse Gas Emissions Index (GHGI) for co-production is defined as the system life-cycle emissions divided by life-cycle emissions for a reference system producing the same amount of electricity and transportation fuels. The reference system makes electricity in a supercritical coal-fired plant with CO_2 vented and transportation fuels from petroleum. The emissions rates given in this table are calculated as GHGI × (831 kgCO_2eq/kWh), where the GHGI values are from Table 12.27. The GHGI values will change with the fraction of fuel input to co-production that is biomass. For the systems in this table, biomass accounts for 29% of the higher heating value fuel input for the coal+biomass system and 34% for the natural gas+biomass system.

i) From Chapter 12, Table 12.8.

j) From Chapter 14.

Table TS-6 | Life-cycle emissions of GHGs for different liquid fuel supply technologies. Note: ILUC: Indirect Land Use Change; GHGI: Greenhouse Gas Index; CTL: Coal to Fischer-Tropsch liquid fuels; BTL: biomass to Fischer-Tropsch liquid fuels.

	Without ILUC	With ILUC	Source
	kgCO$_2$-eq/GJ LHV		
FOSSIL FUELS			
Gasoline from crude oil	91.6		Table 12.15, Note (c)
Diesel from crude oil	91.8		
Kerosene-type jet fuel from crude oil	87.8		
LPG	86		Table 12.15, Note (c)
FIRST GENERATION BIOFUELS			
US Midwest corn ethanol	69	90–210[a]	Chapter 9
US Midwest soy biodiesel	20	82[b]	
EU rape biodiesel	49		Chapter 11, average values in Figure. 11.20
EU sugarbeet ethanol	46		
Brazil sugarcane ethanol	19[c]	73	Chapter 20, ref. Macedo et al, 2008
SECOND GENERATION BIOFUELS (electricity is at most a minor byproduct)			
Cellulosic ethanol (farmed trees)	2	20	Chapter 20, ref. Macedo et al, 2008
Cellulosic ethanol (switchgrass)*	16		Chapter 12, Table 12.26
Cellulosic ethanol with CCS (switchgrass)*	−19		Chapter 12, Table 12.26
BTL (waste wood)	5.8		Chapter 11, Figure 11.20
BTL (switchgrass)*	6		Chapter 12, Table 12.15
BTL with CCS (switchgrass)	−87		
FOSSIL FUELS and FOSSIL/BIOMASS COMBINATIONS (electricity is a minor byproduct)			
CTL (coal to liquids)*	157		Chapter 12, Table 12.15
CTL with CCS*	81		
Coal+biomass with CCS to liquids, 43% biomass input*	2.7		
GTL (gas to liquids)	101		Chapter 12, Section 12.4.3.1
GTL with CCS	82		
CO-PRODUCTION OF LIQUID FUELS AND ELECTRICITY (electricity is a major co-product)			
CTL with CCS*	64		Chapter 12, Table 12.22
Coal+biomass to FTL with CCS*	8.5		
Natural gas+biomass to FTL with CCS*	9.6		Chapter 12, Table 12.27

* The emissions estimates for these systems are from Chapter 12, based on the GHG emissions index (GHGI), defined in Table 12.15, note (c). The emissions rates in this table are calculated as GHGI × Z, where the GHGI values are taken from Chapter 12 and Z is the lifecycle GHG emissions for petroleum-derived fuels that would be displaced by the fuels produced from coal, gas, and/or biomass.

a) Chapter 11 (Fig. 11.20) gives range from 73 to 210 for corn ethanol (without ILUC).

b) Chapter 11 (Fig. 11.20) gives 19 (without ILUC)

c) Chapter 11 (Fig. 11.20) gives 23 kgCO^2-eq/GJ w/o ILUC

3.5.2   Multiple Benefits

The GEA pathways that meet the sustainability goals also generate substantial economic and social benefits. For example, achieving society's near-term pollution and health objectives is greatly furthered by investing in the same energy technologies that would be used to limit climate change. Increased stringency of air pollution policies globally and increased access to cleaner cooking fuels would bring significant improvements in pollution-related health impacts as compared to currently planned air-quality legislations and access trends, with a saving of 20 million disability adjusted life years (DALYs) from outdoor air pollution and more than 24 million DALYs from household air pollution.

This synergy is advantageous and important, given that measures which lead to local and national benefits, for example, improved health and environment, may be more easily adopted than those measures that are put forward solely on the grounds of global goals. Many energy efficiency and renewable energy options enjoy such synergies and generate

Table TS-7 | Synthesis and taxonomy of multiple benefits related to sustainable energy options. The options are discussed in the different rows, while the policy goal where the multiple benefits occur is covered in the different columns. The first five columns summarize how the various options discussed contribute to (or occasionally compromise) reaching the multiple objectives of GEA. The remaining three columns identify further multiple benefits and impacts. The table focuses on the more sustainable sub-options within the rows, and acknowledges if there are major differences in the particular impact among the sub-options (such as with and without CCS). It is important to recognize, however, that this is a synthesis table and thus cannot be fully comprehensive.

Sustainable Energy Options	Access	Energy Security	Health	Environment	Climate Change (mitigation and adaptation)	Development and Economic benefits, poverty alleviation	Risks of large accidents	Employment (local)
Efficiency								
Industry	–	Energy import need reductions due to saved energy; a robust industry makes for stronger social and defense systems	Reduced health impacts from lower regional industrial and energy-related air pollution	Reduced energy-related emissions from saved energy; lower industrial pollution from material efficiency & recycling	Lower CO_2 emissions due to saved energy; lower non-CO_2 GHG emissions due to process/material efficiency	Productivity gains and increased competitiveness; For all end-use sectors: New business opportunities for efficiency implementation, e.g. ESCOs, Reduced investment needs in supply thus more funds for development	Lower risk of industrial accidents from more efficient, safer and fewer processes; Gains through displaced risky generation	New jobs in efficiency implementation, ESCOs, equipment production
Transport	Better and more equitable access to mobility services	Alternative fuels: Shift to non-oil dependent economy; lower oil consumption for efficiency, modal shift and non-motorized mobility	Health gains from lower urban air pollution; lower mortality and morbidity from reduced accidents	Significant improvement in urban air quality due to reduced specific emissions and transport volumes	Lower GHG emissions due to efficiency gains, alternative fuels, non-motorized and alternative mobility	Reduced economic damages from congestion; better access to economic activities; economic savings through lower transport costs due to efficiency and alternative mobility; more time for productive activity from improved mobility	Lower risk for oil spills due to lower oil consumption and trade	Local employment gains for public transport
Buildings (residential, public and services)	Access to higher energy service levels from same budget and production capacity through efficiency	Reduced needs for imports due to saved energy; more resilient energy systems from building-integrated distributed generation	Clean/efficient cooking; lower respiratory infectious morbidity in well-ventilated buildings; reduced noise exposure	Reduced energy-related emissions from saved energy; both local and regional	Reduced GHG emissions: CO_2 from saved energy, non-CO_2 from less cooling; more climate and heat resilience: adaptation gains	Increased social welfare: More disposable income through saved energy costs; potentially eliminated poverty; Reduced needs for tariff subsidies; productivity gains from reduced illnesses in well-ventilated buildings; increased value for real estate	Gains if energy savings are large enough to displace risky power generation	Large net local employment benefits, especially for retrofits; high employment intensity of energy savings' through efficiency
Systems and grids								
Advanced electricity and gas systems (possibly hydrogen in future)	Inexpensive and more linked systems provide easier access; Distributed generation provides access where needed	Smart systems provide redundancy through rapid deployment of energy, and enhanced use of alternatives. Microgrids offer autonomy, stability, flexibility	Smart systems are more efficient, reducing air pollution from sources	Smart systems reduce overall need for energy with less environmental impact; Can assimilate large amounts of variable renewable energy source (RES)	More efficient, use less energy and produce fewer GHGs	ICT, smart systems, microgrids, and distributed generation increase efficiency and productivity by providing stability and instant flexibility; Smart systems are less expensive and thus more ubiquitous.	Faster, more reliable and redundant systems reduce accident risks	Distributed generation and smart systems mean mostly local jobs
Supply								

Oil (with and without CCS)	Oil products such as kerosene and LPG provide clean solutions to the access to cleaner cooking goal, and would reduce GHG emissions from the cooking	Dependence from producers in politically unstable regions; control of oil sources as demonstrated cause of armed conflicts	Health benefits if cleaner fuels replace traditional biomass use	Lower indoor and local pollution if cleaner oil-based fuels replace traditional biomass	Lower black carbon emissions if cleaner oil-based fuels replace traditional biomass; with CCS: reduced CO_2 emissions for large point-sources	Major income source for many oil-rich countries; with CCS can provide low GHG energy source for many places and uses. Its price volatility can impact the economies; and affects mobility affordability	Risk of spills in the extraction and transport of oil and oil products	Mostly low labor intensity of extraction and use; mostly centralized
Coal (with and without CCS and biomass)	Can provide the fuels for access to cleaner cooking in many countries; provided technology routes such as analyzed in Chapter 12	Coal and biomass are geographically rather equally available; can be imported easily	Health impacts through mining and burning-related emissions need to be addressed along the lines in Chapter 12	Potential major landscape impacts from mining. Large particulate and other emissions, including radioactive; damages through mining, may be difficult to avoid	Highest specific emissions without CCS; with biomass co-gasification CCS enables CO_2-free power; and can even have negative emissions	With high oil prices low C electricity can be provided at costs that are much lower than for power-only plants with CCS via coal/biomass co-production with CCS systems that simultaneously offer low C fuels at lower costs than biofuels.	Risk of coal mine accidents, although not large mortalities; for CCS: risk of leakage	If the primary future role of coal is shifted from the present focus on generating electricity to co-production of liquid fuels and electricity with CCS from coal + biomass, there would be significant new industries associated with making these liquid fuels that would replace oil imports
Natural gas	Provides access to more energy services if available locally	Might reduce reliance on other imported energy if available locally; more equally distributed than oil	Little or no impact due to relatively perfect combustion	Lower emissions than other fossil fuels	High GHG emissions prices needed to induce CCS for power-only systems, a challenge that can be greatly mitigated if instead electricity and synfuels are co-produced with CCS; CH_4 emissions can be controlled	Can provide low-cost energy needs for economic development in regions where available; cleaner with CCS, the added cost of CCS can be mitigated in a world of high oil prices by making synthetic fuels + electricity instead of just electricity	Risks of gas leaks from storage and pipelines and exposures	Low labor intensity of extraction and use
Nuclear	Can provide the energy for access in countries with insufficient local energy sources	More ubiquitously available/importable than many other sources. Major security risks from facilitating nuclear-weapon proliferation	Large health benefits in comparison with fossil and traditional biomass energy	Occasional radioactive releases from whole fuel cycle; radioactive waste challenges	Much lower emissions of CO_2 than fossil-fueled power plants on a life-cycle basis	Can provide the energy for development in regions where insufficient local resources are available; though at what cost depends on many factors	Risk of large releases of radioactivity; risks from facilitating nuclear-weapon proliferation	Low employment intensity; centralized employment

Sustainable Energy Options	Access	Energy Security	Health	Environment	Climate Change (mitigation and adaptation)	Development and Economic benefits, poverty alleviation	Risks of large accidents	Employment (local)
Renewables								
Solar (CSP and PV)	PV: Provides interim power source until full access to grid is provided; Thermal: provides some hot water. Can be self-made	Security gains through reduced energy and fuel import needs	Improved health due to lower emissions and pollution	Reduced emissions from fossil fuels and resources depletion; lifecycle environmental impact for PV	Limited life-cycle GHG	Most distributed energy source: can provide the energy right where needed; High cost of solar electric technology, but no fuel expenditures once equipment is installed; low cost (potentially self-made) for solar thermal; new industrial opportunities	Risk reductions through avoided conventional risky generation	More employment intensive than large-scale power generation, but many related jobs can be 'exported'
Wind	Better access to electricity through local/regional power generation	Security gains through reduced energy import needs	Improved health due to lower emissions, pollution	Reduced emissions from fossil fuels, but noise and visual impacts; some ecological impacts	Limited life-cycle GHG emissions	Most competitive RES-E technology and thus can provide affordable power; new industrial opportunities; eliminated fuel costs: no fuel import costs eliminate price volatility	Risk reductions through avoided conventional risky generation	More employment intensive than large-scale power generation, but many related jobs can be 'exported'
Hydro	Better access to electricity through local power generation (e.g., small hydro)	Security gains through reduced energy import needs	Improved health due to lower emissions, pollution	Reduced emissions from fossil fuels, but large hydro has serious environment and social liabilities	Limited life-cycle GHG emissions	Large hydro is the most ubiquitous large RES, can provide significant energy for development in many countries. Can be very affordable; can help in flood regulation; new recreational space creation	Risk of dam break and catastrophic flooding for large hydro	Can be employment intensive, part of jobs created are local
Biomass (with and without CCS)	If biomass is used primarily to make synthetic transportation fuels via gasification, LPG which can be used as a clean cooking fuel will be an inevitable byproduct, and much lower cooking fuel use rate via LPG compared to burning biomass implies that the LPG could go a long way in meeting cooking fuel needs even if providing transportation fuel is the main objective. This is discussed in Chapter 12	Security gains through reduced energy import needs; biomass can be ubiquitously available and easily imported, land use competition issues can be avoided with adequate environmental legislation as done in Brazil	Traditional biomass burning can have very high indoor pollution and related health toll	Like any agricultural crop, needs to be sustainably produced to avoid deforestation and other major environmental impacts; risk of ecological damages through monocultures, which can be avoided with adequate environmental legislation such as fauna corridors and maintenance of local native forest.	Variable life-cycle GHG emissions; needs to be consciously minimized. *With CCS it can provide one of the few opportunities for negative GHG emissions*	Presently provides the most ubiquitous source of energy for the poorest, but with large impacts when not produced in a sustainable way; If sustainably produced, can fuel development in many regions, but potential competition with food production and thus impact on food prices	Risk reductions through avoided conventional risky generation	Very employment intensive if sustainably produced, most jobs created are local

benefits across multiple objectives. Some of these benefits can be so substantial for certain investments/measures that they may offer more attractive entry points into policymaking than climate or social targets. This is particularly the case where benefits are local rather than global. Seeking local benefits and receiving global benefits as a bonus is very attractive, and this is often the case for investments in energy efficiency and renewable sources of energy.

Therefore, even if some of these multiple benefits cannot be easily monetized, identifying and considering them explicitly may be important for decision making. Cost-effectiveness (or cost–benefit) analyses evaluating sustainable energy options may fare differently when several benefits are combined into one investment evaluation.

The enhancement of end-use efficiency in buildings, transport, and industry offers many examples of benefits across multiple social and economic objectives. Among the most important ones are: improved social welfare, including alleviated or eliminated poverty as a result of very high efficiency and thus very low fuel-cost buildings; reduced need for public funds spent on energy price subsidies or social relief for poor people; and health benefits through significantly reduced indoor and outdoor air pollution, often translating into commendable productivity gains. Productivity gains and general improvements in operational efficiency in industry translate into improved competitiveness. Improving efficiency by increasing the rate of building retrofits can be a source of employment generation. Other benefits that are non-quantifiable or difficult to account for include improved comfort and well-being, reduced congestion, new business opportunities, and improved and more durable capital stock.

Other economic benefits of rapidly decarbonizing the energy system are the reduced need for subsidies into carbon-intensive petroleum products and coal. At present, subsidies for these fuels amount to approximately US$132–240 billion per year[32] (Chapter 17). Only 15% of this total is spent directly toward poor people who have limited access to clean energy. As noted in Section TS-5.2.1, subsidies to poor people must be increased in order to achieve universal access. GHG mitigation in the GEA pathways would, however, at the same time reduce consumption of carbon-intensive fossil fuels by the rest of the population, leading to a reduction in the need for subsidies for oil products and coal in the order of US$70–130 billion per year by 2050 compared to today.

A review of the co-benefits related to the discussions in this assessment, that is, selected key routes through which the various sustainable energy options described earlier in this section and in Chapters 7–15 of the report contribute to different policy goals, is presented in Table TS-7. The purpose is to summarize the main routes of impacts through which the various options affect the multiple objectives set out in GEA, as well as to identify further benefits. However, whether an impact is a benefit or a liability depends on the baseline and local situation: for example,

while liquid petroleum gas (LPG) has major environmental and climate impacts in itself, it still has major advantages in many areas when it replaces traditional biomass burning. Thus, a key novelty of Table TS-7 is that it provides a new, additional framework for a well-founded assessment for individual choices among various energy alternatives that complements the financial appraisals. For example, in jurisdictions where access to modern forms of energy is a major energy policy goal, evaluations in the 'energy security' column may play a key role in ranking the different options available at comparable costs. In other areas, access or employment may be key secondary objectives of energy policy and these may play the chief role in additional prioritization of options with comparable local costs. The evaluations Table TS-7 are qualitative and, due to the challenges of comparing very different types of effects, are indicative only.

Table TS-7 demonstrates that there is a very broad array of different benefits in a large spectrum of policy target areas, representing many potential entry points into policymaking. However, some options can have a wider range of co-benefits than others, such as improved efficiency, system solutions, and some renewables, such as biomass with CCS and fossil fuel/biomass co-processing with CCS. The lowest levels of co-benefits arise from nuclear power and fossil-fuel related options, even with CCS. There are less-marked differences among the options with regard to policy goals, such as energy security: most of the sustainability options discussed in this report have positive impacts on security. In contrast, others, such as poverty and access, are harder to contribute to: it is mainly renewable forms of energy that can significantly contribute to these goals. Improved efficiency, at the same time, has positive impacts on almost all policy goal areas and often has the broadest range and largest co-benefits.

3.5.3   Trade-offs and Constraints[33]

Changes in food and energy use will not only have substantial environmental impacts, they will also influence each other in many ways. At the same time, the production of food and energy and their dependence on water resources will be affected by global environmental change, including climate change.

Population growth and economic growth are major factors contributing to increased demand for land and water. In addition, growth in incomes is strongly correlated with increased consumption of animal-derived food (meat, milk, eggs). This combination will increase pressure on land and water resources if not counteracted by environmentally sound land- or water-saving innovations.

Sustainability issues arising from competition and synergies between the future production of bioenergy and food are highly important in this context, but they can be avoided with adequate policies, as discussed in

---

32 'Tax-inclusive' subsidies are higher at a little below US$500 billion.

33 Section TS-3.5.3 is based on Chapter 20.

Chapter 20. The global bioenergy potential from dedicated energy crops, considering sustainability constraints (such as maintenance of forests, areas with high biodiversity value, or protected areas) as well as food demand and feed demand of livestock, was estimated to be 44–133 EJ/yr in 2050. Substantially higher or lower levels could also be possible, as there are large uncertainties with respect to many important factors: land and water availability; feedbacks between food, livestock, and energy systems (in particular, future crop yields and feeding efficiency of livestock); and climate change.

Food prices are a concern, particularly in poorer areas of the world. They are influenced by many factors, including temperature and precipitation, pest attacks, and oil and fertilizer prices. There is a concern that major bioenergy crop expansion would become a significant factor, but there are studies, as mentioned in Chapter 20, that show different points of view. The potential indicated above takes such concerns into consideration. However, land allocation for different purposes will have to be monitored and managed carefully, as free competition between food, bioenergy, and other markets (e.g., fiber) could lead to swings in agricultural prices and supply. Adequate policies for introducing bioenergy plantations could avoid adverse social, economic, and ecological effects, while best-practice examples suggest that bioenergy plantations could be highly beneficial in terms of sustainability if based on sound strategies. Monitoring, managing, and enforcing adequate policies are required to ensure sustainability of bioenergy production. Adequate land and water-use environmental zoning and planning should be implemented to consider specific environmental conditions of each region.

Policies that support biofuels expansion should carefully evaluate the price and related food-security implications of scenarios relevant to the situation in each country. They should ensure sustainable production for any agricultural product and give priority to the diversification of technologies and fuels, while identifying different options for the future, based on adequate environmental zoning, and sustainable policies, as well as considering the overall impacts of each fuel. This must occur through public policies that govern and regulate markets and stimulate efficient technologies. These policies include biofuels sustainability based, for example, on certification schemes adequate for each country

The impact of climate change on land use systems is, at present, rather imperfectly understood. In subtropical and tropical regions, changes in climate and the rainfall regime may change the agricultural suitability of a region significantly. Temperature change may require the migration of some crops and agricultural areas to regions with a more temperate climate or higher levels of soil moisture and rainfall. In general, crop productivity in the tropics may decline even with a 1–2°C increase in local temperature. This would also have significant impacts on renewable energy resources, for example cloud cover, rainfall, and wind speed.

Multiple uses of water – for human consumption, hydro and thermal power generation, manufacturing, agriculture, water security, bioenergy, and so on – are feasible and associated with environmental, social, and strategic aspects, as well as with potential trade-offs. Competition between food and energy crops may not always be over 'the same water'. Depending on the type of feedstock, it is possible to cultivate adequate bioenergy crops in areas where conventional food production is not feasible due to, for example, water constraints – that is, the 'water footprints' are of a different character.

3.5.4   Rethinking Consumption[34]

The well-being of the 'final consumer' drives the production of goods and services and consequent energy service demand. Is a lifestyle that has high throughput of energy and materials globally sustainable in the long run? The literature on ecological footprints shows the unsustainability of ever-growing consumption in a growth path led by economic well-being. Studies offer the potential to get life-cycle approaches into a decision-making context and open up the possibility of a diversification of the policy portfolio.

The growing constraints on people's time as they pursue economic well-being have led people to buy appliances that save labor but use more energy. Walking, cycling, jogging, and natural green spaces are being taken over by energy-guzzling health clubs and the like, and by highly irrigated green spaces, while small traditional retail stores are being replaced by high energy using air-conditioned shopping malls. A convergence in the high level of energy service demands across various cultures, geographies, and income classes is the dominant trend.

Technology, income levels, and lifestyles are causing important changes in both direct and indirect energy requirements of households. While energy efficiency through technological improvement is helping, energy use and GDP growth have not really been decoupled in many countries. Lifestyle changes are essential to realize the full benefits of the technical potential.

In the short term, for incremental changes it is advantageous to consider consumers as shoppers and purchasers in a marketplace. By controlling information, education, and so on, what people buy can be influenced to achieve the desired outcome. In the medium term, an approach that relies on human well-being in terms of sustainable development, on Millennium Development Goal indicators, and on the triple bottom line (with more emphasis toward environmentalism) can have a moderate dampening effect on energy use.

In the longer term, an ecological footprint index and the criterion of 'sufficiency' provide promising policy options in individualistic liberal

---

34   Section TS-5.4.4 is based on Chapter 21.

societies for increasing sustainability in the energy system and motivating the adoption of a new value system. A human well-being indicator needs to evolve beyond GDP and the Human Development Index to reflect responsible individual and community behavior, sufficiency, happiness, and social ecosystem balance. Transformational change in the social fabric that places individual and community actions in the proper context has a role to play in reaching a low-energy path.

Despite health alerts and religious taboos, meat consumption has increased due to the aggressive marketing strategies of producers and distributors, creating an association between wealthy people's diet and meat consumption. There is a lack of awareness that a reduction in per capita meat consumption, especially in industrial countries, could reduce numerous health risks as well as global energy use and GHG emissions.

Education systems in modern societies can promote the virtues of going beyond classical humanistic contents of individual freedom and dignity and instead emphasize more collective aspects. The role of the state is to ensure adoption of a rights-based policy line that can make the duty to 'do no harm' a global right that matches the right to not be harmed. Governance that evolves organically can shape the course of action that involves the state and various communities such as non-government organizations, corporations, communities, civil society, and religious institutions.

Formal, informal, ethical, public, and mass media systems of education could generate social values that redefine modernism through more cultural diversity and local specificities instead of homogenization. Responsible individual and community behavior that justifies sufficiency in liberal societies needs broader and faster dissemination through investments in various institutions.

4   GEA Transformational pathways[35]

GEA explored 60 alternative pathways of energy transformations toward a sustainable future that simultaneously satisfy all its normative social and environmental goals of continued economic development, universal access to modern energy carriers, climate and environment protection, improved human health, and higher energy security (see Section TS-2.6).

The pathways were divided into three different groups, called GEA-Supply, GEA-Mix, and GEA-Efficiency, representing three alternative evolutions of the energy system toward sustainable futures. The pathways within each group portray multiple sensitivity analyses about the 'robustness' of the three different approaches in mastering the transformational changes needed to reach more sustainable futures. Of the

60 pathways explored, 41 clearly simultaneously fulfilled all the normative goals – indicating that such futures are reachable from a resource, technology, and economic point of view.

For such a transformation to be achieved, the pathways presume that political commitment, the availability of necessary financing (coming forward in response to the right market signals), and technological learning and diffusion occur pervasively throughout the world. Achieving the goals assumed in GEA generates significant benefits. Realizing these benefits requires governments to provide market conditions that ensure appropriate investments are mobilized (see Section TS-5). Thus, planning for a future energy system requires going beyond pure economic costs and needs to factor in salient environmental and social externalities. Thus, the 41 pathways jointly indicate the plausibility of transformative changes if the externalities and benefits related to the GEA sustainability goals are appropriately accounted for.

Together, these 41 transformational pathways integrate the conclusions of individual GEA chapters on major challenges and options into a consistent framework of scenario analysis. The analysis included a narrative that constituted the initial platform for the three alternative sets of quantitative pathways, which were developed by two different integrated assessment modeling frameworks (MESSAGE and IMAGE).

The energy transformations captured by the pathways encompass 11 world regions, grouped into five GEA regions. They also include various energy sectors, including supply and demand, with a full range of associated social, economic, environmental, and technological developments. They result in radically changed ways in which humanity uses energy, ranging from much more energy-efficient houses, mobility, products, and industrial processes to a different mix of energy supply – with a much larger proportion of renewable energy and advanced fossil fuel technologies.

The pathways indicate that the energy transformations need to be initiated without delay, gain momentum rapidly, and be sustained for decades. They will not occur on their own. In fact, the pathways imply a significant departure from recent trends. Serious policy commitments are therefore required. Furthermore, it would require the rapid introduction of policies and fundamental governance changes toward integrating global concerns, such as climate change, into local and national policy priorities, with an emphasis on energy options that contribute to addressing all these concerns simultaneously.

Although energy transformations in the pathways are fundamental and rapid, they are not historically unprecedented. In the past, energy systems have experienced similar, or even more profound, transformations – for example, when coal replaced biomass in the 19th century, or when electricity was introduced in the first half of the 20th century. More recently, a number of countries have rapidly shifted from coal to natural gas. Transformational changes closer to the consumer,

---

35   Section TS-4 is based on Chapter 17.

such as the replacement of horse carriages by automobiles, occurred within three decades in many parts of the world. The need for sustained transformation is, however, new and unprecedented on the global scale.

The GEA pathways are based on the assumption that the advanced technologies known today, often commercially unviable under current market conditions, would be improved through vigorous R&D and deployment to achieve cost reductions and better technical performance through economies of scale and through learning by using and by doing. Various combinations of resources, technologies, and policies are incorporated across the different pathways. While there is some flexibility in the choice of specific policy mechanisms, achieving all the GEA normative goals simultaneously is an extremely ambitious task.

For each of the three groups of pathways, one 'illustrative' case has been chosen that captures the salient characteristics of the group. The illustrative pathways are not necessarily the average or median of the set, but rather pathways that capture the overall characteristics of the respective group. They depict salient branching points for change and policy implementation. The characteristics differ significantly and depend on choices about technologies, infrastructures, behaviors, and lifestyles, as well as on future priorities on supply and demand-side policies. These choices have, in turn, widespread implications for technology availability and scale-up, institutional and capacity requirements, and financing needs.

The main distinguishing dimensions of the three illustrative GEA pathways are as follows:

- *Demand versus supply focus.* While the assessment shows that a combination of supply-side and demand-side measures is needed to transform the energy system, emphasis on either side is an important point of divergence, as exemplified by GEA-Supply compared to GEA-Efficiency. A critical factor is thus how much the changes in demand for energy services together with demand-side efficiency measures can reduce the amount of energy required to provide mobility, housing, and industrial services. This dimension is one of the main distinguishing characteristics and motivates the naming of the three illustrative pathways.

- *Global dominance of certain energy options versus regional and technological diversity.* Once technological change is initiated in a particular direction, it becomes increasingly difficult to alter its course. Whether the transformation of the future energy system follows a globally more uniform or diverse path thus has important implications, given irreversibility, 'lock-in', and the path dependency of the system. GEA-Efficiency and GEA-Supply pathways depict worlds with global dominance of certain demand and supply options, while GEA-Mix pathways are characterized by higher levels of regional diversity.

- *Incremental versus radical new solutions.* Given that the GEA sustainability objectives are ambitious, the transformational changes

to realize them need to be introduced rapidly across all GEA pathways. For instance, all pathways feature decreasing shares of carbon-intensive supply options. The pathways differ, however, with respect to the emergence of new solutions. Some rely more heavily on today's advanced options (such as efficiency and renewables in GEA-Efficiency) and infrastructures (such as biofuels in GEA-Mix), while others depict futures with more radical developments (such as hydrogen or CCS in GEA-Supply).

The transformation can be achieved from different levels of energy demand as well as through alternative combinations of primary energy resources (see Box TS-1 and Figure TS-24).

4.1 Requirements for Achieving the Transformation

Despite the flexibility and choices available across the pathways regarding the direction and dynamics of the energy system transformations, also a large number of robust characteristics are common to all pathways. These commonalities are summarized below. They illustrate the magnitude of energy system changes that would need to be introduced to reach the GEA sustainability objectives.

Improvements to at least the historical rate of change in energy intensity are necessary to reduce the risk that the sustainability objectives become unreachable. Further improvements in energy intensity, entailing aggressive efforts to improve end-use efficiency, increase the flexibility of supply and improve the overall cost-effectiveness of the energy system transformation (see Box TS-2). 'Negawatts' provide more choices at lower costs than 'Megawatts'.[36]

A broad portfolio of supply-side options, focusing on low-carbon energy from renewables, bioenergy, nuclear power, and CCS, was explored, achieving at least a 60–80% share of zero-carbon options in primary energy by 2050. These include:

- Strong renewable energy growth beginning immediately and reaching between 165–650 EJ of primary energy by 2050. This corresponds to a global share of 30–75% of primary energy with some regions experiencing, in the high case, almost a complete shift toward renewables by 2050.

- Rising requirement for storage technologies and 'virtual' systems (e.g., smart grids and demand-side management) to support system integration of intermittent wind and solar.

---

36 Amory Lovins is well known for highlighting the need for achieving high energy efficiencies through "Negawatts rather than Megawatts".
   Lovins, A.B., 1990: The Negawatt Revolution, *Across the Board*, **27**(9) 23–29.
   http://www.thewindway.us/pdf/E90–20_NegawattRevolution.pdf.

Box TS-1 | The three groups of GEA pathways

All GEA pathways share common socioeconomic assumptions, including demographic and economic developments. They differ radically in the structure of the future energy systems.

GEA-Efficiency pathways emphasize efficiency. The global pace of energy intensity improvements thus double compared to the long-term historical average. For example, this implies that in the buildings sector, efficiency would be improved by a factor of four by 2050. This would require measures and policies to achieve the rapid adoption of best-available technology throughout the energy system, for example, to, retrofit existing plants, enhance recycling, improve life-cycle product design in the industry sector; reduce energy demand through aggressive efficiency standards, including electrification, a shift to public transport, and reduction of demand for private mobility. Emphasis in GEA-Efficiency is thus on demand-side R&D and solutions to limit energy demand for services. This results in a primary energy demand level in 2050 of 700 EJ, compared to the level of 490 EJ in 2005.

GEA-Efficiency also relies on increasing renewable energy, approaching 75% of primary energy by 2050, and further increasing its contribution to about 90% by the end of the century. Figure TS-24 shows the illustrative efficiency pathway with various sensitivity analyses, indicating the changes by 2050. In some of the efficiency pathways nuclear power is assumed to be phased out over the lifetime of existing capacities, whereas CCS provides an optional bridge for the medium-term transition toward renewables. In the illustrative pathway, coal use declines immediately while the oil peak is reached before 2030. Unconventional oil resources thus remain largely untapped, given the GEA environmental objectives to reduce GHG and air pollutant emissions. Natural gas contribution remains at about current levels as it is the least carbon-intensive of all fossil sources. In contrast, the role of renewables increases across all pathways.

GEA-Supply features a major focus on the rapid up-scaling of all supply-side options. A more modest emphasis on efficiency leads to energy intensity improvement rates roughly comparable to historical experience. Primary energy demand in the illustrative Supply pathway reaches about 1050 EJ in 2050. Massive up-scaling of energy supply R&D and deployment investments lead to new infrastructures and fuels (such as hydrogen and electric vehicles in the transportation sector). Renewables contribute about half of primary energy by the middle of the century. The GEA-Supply pathways show similar levels of expansion for renewables as those in the GEA-Efficiency pathways. As a result of relatively higher levels of energy demand, the share of renewables is, however, comparatively smaller in the GEA-Supply pathways. Further implications of the relatively higher energy demand is that fossil CCS is becoming an essential building block in the medium term to decarbonize the remaining fossil share of the supply system. In the long term, the contribution of fossil CCS declines as the transition toward zero-carbon options progresses. New nuclear power plants gain significant market share after 2030 in some of the supply pathways. This presupposes that issues related to weapons proliferation, nuclear waste, and other inherent risks of nuclear energy are satisfactorily resolved. However, supply pathways also include nuclear phase-out cases that imply vigorous increases of alternative energy sources given the relatively high energy demand. This is possible; however, it was not possible to formulate a GEA-Supply pathway without the CCS technology, indicating that CCS is a must with such high levels of demand for fossil fuels (Figure TS-24). In principle, fossil fuels can either be used with CCS at high demand levels or the energy demand level can be reduced to meet GHG emission reduction goals, as in GEA-Efficiency.

GEA-Mix pathways are intermediate with respect to many scenario characteristics, such as efficiency focus and the up-scaling of advanced and cleaner supply-side technologies discussed in Section TS-4. The primary energy demand level reaches 920 EJ in 2050. The main emphasis is on diversity of energy supply and technology portfolios, thus enhancing system resilience against innovation failures or technology shocks. Furthermore, large differences in regional implementation strategies reflect local choices and resource endowments. This results in the co-evolution of multiple fuels, particularly in the transport sector, where, for example, second-generation bio-liquids, fossil/bio-liquids with CCS, and electricity gain importance in different regions.

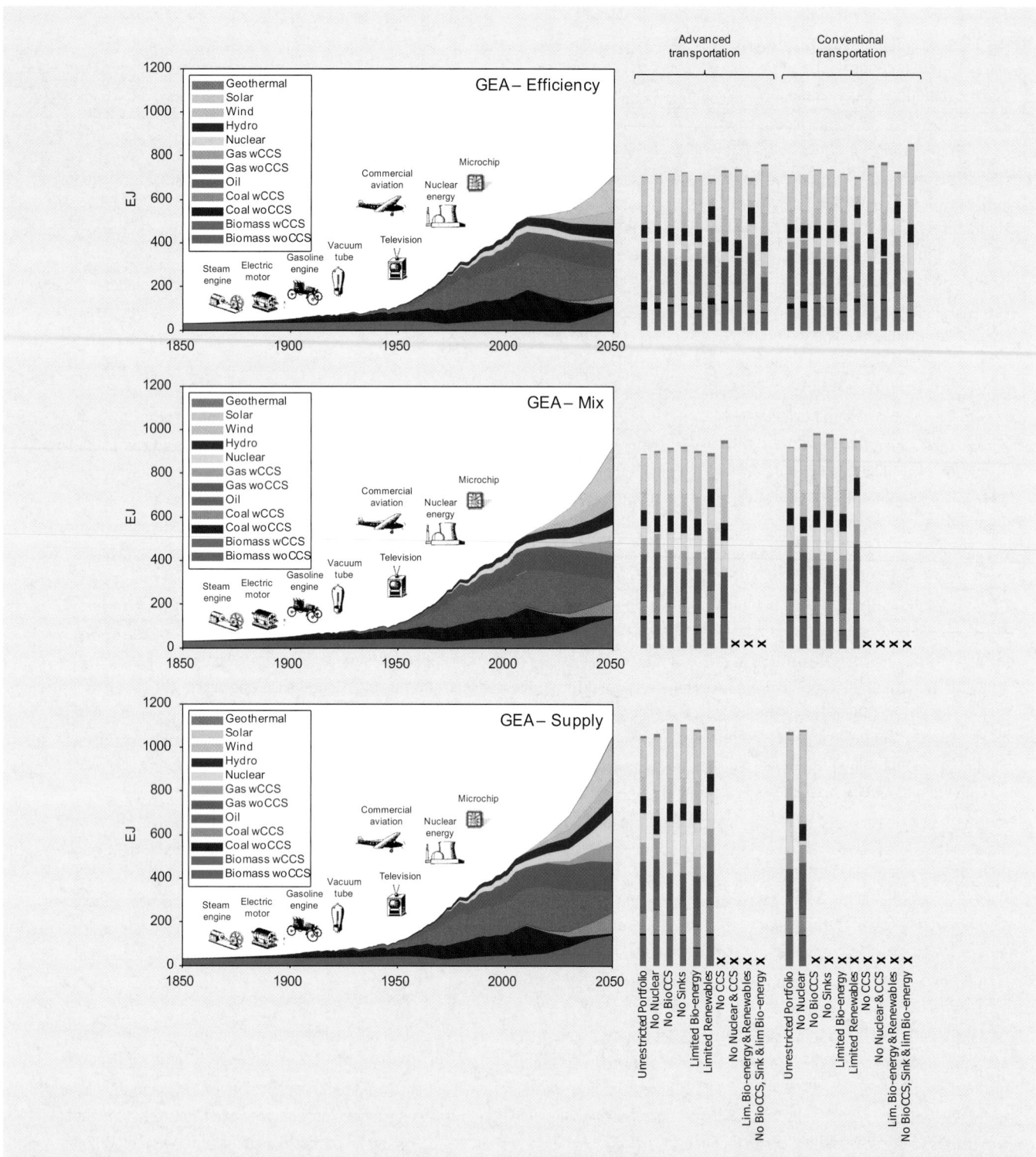

Figure TS-24 | Development of primary energy in the three sets of GEA pathways. Left-hand panels show the three illustrative GEA pathways, and right-hand bars give the 2050 primary energy mix of all 60 pathways explored. Conventional transportation refers to pathways that assume the continuation of a predominantly liquid-based transportation system, whereas advanced transportation refers to pathways that allow for fundamental changes in infrastructures – for example, high penetration of electric vehicles or other major breakthroughs in transportation technology such as hydrogen fuel cells. Pathways marked 'x' indicated the 19 cases where, under the specific combination of assumptions, the GEA normative goals could not be reached. For further details of the GEA pathways see the interactive web-based GEA scenario database hosted by IIASA: www.iiasa.ac.at/web-apps/ene/geadb.

- Strong bioenergy growth in the medium term from 45 EJ in 2005 to 80–140 EJ by 2050 (including extensive use of agricultural residues and second-generation bioenergy to mitigate adverse impacts on land use and food production).

- Nuclear power plays an important role in some of the pathways, while others assess the consequences of a nuclear phase-out. They illustrate that it is possible to meet the GEA normative goals even with a nuclear phase-out. The GEA-Efficiency pathways provide more flexibility in these cases. The range of nuclear in the GEA pathways is similar to the ranges found in other studies in the peer-reviewed scenario literature. In some GEA-Supply pathways, nuclear energy's contribution grows to 1850 GWe installed capacity, which is well above the IAEA's high projection of 1228 GWe. In the past, nuclear growth has been far below the IAEA's high projection – and, until 2000, even below the low projections. The nuclear contribution, however, is particularly uncertain across the pathways because of unresolved challenges surrounding its further deployment due to weapons proliferation risks and especially in the aftermath of Fukushima.

- Fossil CCS as an optional bridging or transitional technology in the medium term, unless there is high energy demand, in which case cumulative storage from CCS of up to 250 GtCO$_2$ by 2050 may be needed. In the pathways, it is a bridging technology that helps offset the gap between the need to vigorously decarbonize the energy system and the time needed to diffuse low-carbon options across the pathways. In the long term, CCS in conjunction with sustainable biomass is deployed in many pathways to achieve negative emissions and thus help achieve climate stabilization.

Vigorous decarbonization in the electricity sector leads to low-carbon shares in total generation of 75–100% by 2050, while coal power without CCS is phased out. Natural gas acts as a bridging or transitional technology in the short to medium term and provides 'virtual' storage for intermittent renewables.

The availability of energy resources by itself does not limit deployment on an aggregated global scale, but it may pose important constraints regionally, particularly in Asia, where energy demand is expected to grow rapidly.

Global energy systems investments need to increase to some US$1.7–2.2 trillion annually during the coming decades, with about US$300–550 billion of that being required for demand-side efficiency. This compares to about US$1 trillion supply-side investments and about $300 billion demand-side investments (energy components) per year currently. These investments correspond to about 2% of the world GDP in 2005, and would be about 2–3% by 2050, posing a major financing challenge. New policies would be needed to attract such capital flows to predominantly upfront investments with low long-term costs, but also low short-term rates of return.

4.2   Meeting Multiple Objectives

Universal access to electricity and cleaner cooking fuels requires the rapid shift from the use of traditional biomass to modern, cleaner, flexible energy carriers, and cleaner cooking appliances. This is achievable by 2030 provided that investments of US$36–41 billion per year are secured (half of which would be needed in Africa). About half of the investments would be required for a transition to modern fuels and stoves for cooking and the other half for electrifying rural populations.

Pollution control measures across all sectors need to be tightened beyond present and planned legislation so that the majority of the world population meets the WHO air-quality guideline (annual PM2.5 concentration <10 μg/m^3 by 2030), while the remaining population stays well within the WHO Tier I-III levels (15–35 μg/m^3 by 2030). This would lead to total annual air pollution control costs of about US$200–350 billion by 2030. The ancillary benefits of climate mitigation policies enacted in the pathways reduce the overall pollution control costs by about 50–65%.

Limiting global temperature increase to less than 2°C over pre-industrial levels (with a probability of >50%) is achieved in the pathways through rapid reductions of global CO$_2$ emissions from the energy sector, peaking around 2020 and declining thereafter to 30–70% below 2000 emissions levels by 2050, ultimately reaching almost zero or even 'net' negative CO$_2$ emissions in the second half of the century.

Enhanced energy security across world regions is achieved in the pathways by limiting dependence on imported energy and by increasing the diversity and resilience of energy systems. A focus on energy efficiency improvement and renewable deployment across pathways increases the share of domestic (national or regional) supply by a factor of two and thus significantly decreases import dependency. At the same time, the share of oil in global energy trade is reduced from the present 75% to below 40% and no other fuel assumes a similarly dominant position in the future.

5   Policy Tools and Areas of Action[37]

The previous sections describe the need for transformative change in the energy system (Section TS-2.6) and the different combinations of supply side and end-use technologies that could enable this need to be met in a timely and adequate fashion (Section TS-4). As noted in Section TS-4, a number of such different combinations meet the normative goals of access, security, climate protection, and health. These combinations are very different in terms of the magnitude and

---

37   Section TS-5 is based on Chapter 22–25 and specific policy discussions in Chapters 8–10 and 11–15, and 17.

Box TS-2 | Flexibility of Supply

The pathways explore a wide range of future energy transformations that are consistent with the GEA sustainability goals. Some of them explore future developments in which selected supply-side options were either limited or excluded completely. These pathways focus on overall questions of the 'feasibility' of such limitations and their economic and resource implications. In sum, 60 pathways were explored and 41 were found to be compatible with the GEA sustainability goals. They constitute a wide portfolio of sensitivity analyses regarding the nature and direction of future energy transformations.

The main conclusion from the GEA pathways analysis is that energy efficiency improvements are the single most important option to increase the flexibility of supply and the structure of the regional and sectoral energy systems. With high rates of efficiency improvements in GEA-Efficiency, it was possible to achieve the GEA normative goals under any of the assumed portfolio restrictions. Only in the GEA-Efficiency pathways was it possible to do so in the absence of both nuclear energy and CCS.

GEA-Supply, with high energy demands, requires the rapid and simultaneous growth of many advanced technologies, resulting in reduced flexibility on the supply side. Some more critical options are needed in all GEA-Supply pathways, such as bioenergy, non-combustible renewables (i.e., hydropower, intermittent renewables, and geothermal), and CCS. Excluding, or limiting, these options renders the high demand pathways infeasible. Nevertheless, the high demand pathways explore feasible energy transformations with limitations of some options, such as nuclear energy, as shown in Figure TS-25.

The transportation sector configuration has profound implications for supply-side flexibility. In the case of rapid penetration of advanced transportation technologies (electricity or hydrogen), the GEA-Supply group of pathways was found to still be feasible if any one of BioCCS, carbon sink enhancement, nuclear energy, the full bioenergy supply, or large-scale renewable energy deployment are excluded as options in the future. In the case of restricted penetration of advanced transportation technologies, essentially the full set of supply-side options is needed to keep the GEA sustainability targets within reach (in GEA-Supply).

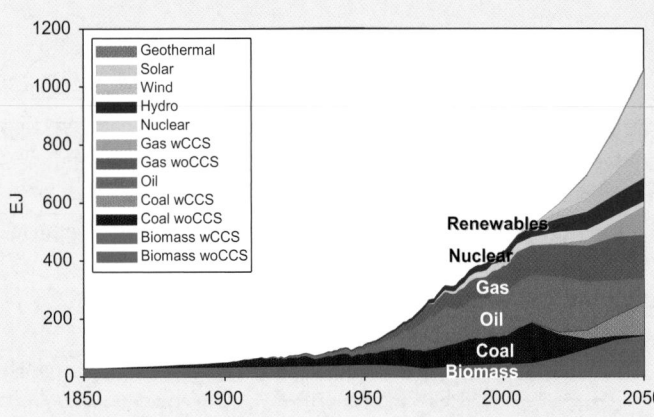

Figure TS-25 | Development of primary energy in the GEA-Supply pathway with a nuclear phase-out shortly after 2050. Source: Chapter 17 and the GEA online database.

orientation of investments required. They would therefore require, at least in part, different policy packages and institutional arrangements for implementation.

The design and formulation of such policy packages needs to reflect the diversity and heterogeneity of national circumstances, the need to consider issues of sociopolitical feasibility and acceptability, and the need to adequately address implementation challenges, including institutional design.

This section describes the possible approaches, basic elements, and policy tools and instruments that may be utilized. Effective policy portfolios will require a combination of instruments, including regulatory and investment policies, as well as measures for strengthening capacity, stimulating innovation, and guiding behavior and lifestyle changes. Moreover, the energy goals cannot be pursued in isolation. Energy-focused policies must be coordinated and integrated with non-energy policies for

socioeconomic development and environmental protection. These latter include, for example, policies that foster sustainable urban areas, preserve forested land and biodiversity, reduce poverty and inequality within and between countries, provide efficient and environmentally acceptable transportation, ensure vibrant rural areas, and improve human health.

The remainder of this section is organized as follows. Section TS-5.1 describes the main elements of, and areas for, policy intervention. We then consider two of the key challenges at the nexus of energy and development – those of universal access and of urbanization. Policies to address these challenges are discussed. Rapid improvement in energy efficiency, scaled up and accelerated development of renewables, and the modernization of fossil fuel systems (such as described in Section TS-3.3.2) form the key building blocks of all of the pathways meeting the sustainable goals. Policies to address these building blocks are discussed in Section TS-5.3. Section TS-5.4 examines the major areas of policy intervention – innovation, finance, and capacity building.

5.1 Framework for Policies and Policy Design[38]

If **universal energy access** is to be achieved by 2030, then energy policies must work in concert with economic development policies by harnessing the collective investment potential of markets, international organizations, central governments, regional governments, cooperatives, and local organizations. The global community must increase its support, including providing financial assistance for major clean-energy infrastructure, and leverage this where possible with private funds by creating a sound climate for sustainable energy-related investments. Bilateral cooperation and the role of Development Banks are key in supporting the necessary investments needed to achieve universal access.

Energy security can be enhanced by combined technically and strategically oriented policies, which various regions and countries will emphasize differently. The technical policies include steps that lead to timely energy network upgrades, greater interconnection and backup agreements between neighboring energy network operators, grid tariffs that induce short-term demand reductions in response to constrained market conditions, and long-term supply procurement strategies that foster local supplies and a diversity of external supplies. Strategically oriented policies would include international cooperation and agreements to reduce the risk of supply disruptions, such as coordination to protect international energy supply routes and to stockpile critical energy resources for release during acute market shortages.

Market power is a concern in much of the energy sector. Governments must recognize that policies to promote electricity competition must also prevent the short-term exercise of market power that results in unjustified excessive profits for some producers and speculators as well as price volatility for consumers. This will involve continued regulation and public involvement in energy system planning and long-term contracting to ensure operation of the system in a socially desirable manner.

The effective and transparent **management of valuable resources** is important in the energy sector. Oil and gas especially are valuable resource endowments that can provide great wealth if their exploitation is properly managed. Policies should maximize the collection of resource rents (via royalties, taxes, and, where applicable, national oil companies) for present and future generations, and should control the rate of exploitation to minimize inflationary harm to otherwise sustainable economic sectors. Depending on a country's current level of well-being, part of the resource rents should be streamed into sovereign wealth funds that are invested domestically to, among other goals, offset the negative impacts of resource exploitation at home, as well as abroad, in a balanced effort to maximize the benefits for future generations.

A multiplicity of policies is required to address the potential impacts of the energy system on **human health and the environment**. A mix of regulations, information programs, and subsidies are needed, for example, to stimulate the rapid adoption of household energy-using devices that have virtually zero indoor emissions. Subsidies should be applied to equipment such as zero-emission stoves and efficient light bulbs to replace the indoor combustion of kerosene and other fuels for lighting and cooking.

Ambient air quality requires regulations on emissions from fuel combustion in buildings, industry, vehicles, power plants, and other sources. Some regulations may be highly prescriptive – specifying combustion technologies such as the use of catalytic converters, for instance, or restricting certain fuels – while others may focus on the absorptive capacity of a given air shed for a particular pollutant. The latter case could involve the establishment of air shed emission limits, perhaps using a cap-and-trade system. Similarly, regional air quality must be protected by technology and/or emissions regulations or by direct emissions pricing.

Extractive activities and the various uses of land and water – coal mines, oil and gas fields, hydropower dams, reservoirs, nuclear plants, storage sites for radioactive wastes and captured CO_2, wind farms, solar electricity farms – should all face a **regulatory framework** that assesses their benefits against a precautionary consideration of their impacts and risks, but that also ensures a streamlined regulatory process for more favorable projects.

Policies to foster **energy from biomass** should seek to minimize the trade-offs between biomass for food and biomass for fuel by encouraging the use of biomass residues and only the most sustainable and productive feedstocks and efficient conversion processes. Subsidies to corn-based ethanol could be replaced, for example, by emissions charges or by regulations requiring sustainability standards and minimum biofuel content in gasoline or diesel, motivating competitive markets to find the most efficient processes for producing biofuels. Many developing countries import all or most of their current liquid fuels at increasingly higher costs, and have, at the same time, large areas that are off-grid. One very successful example is that of *Jatropha* intercropped with food crops in an off-grid small town in Mali – which uses the jatropha oil to product electricity for a local grid.

GHG pricing policies will be key in shifting energy systems toward low-carbon emission technologies, fuels, and activities. While there is disagreement on which pricing method is best – carbon taxes or cap-and-trade – the two approaches can be designed so that their effects are quite similar. The price certainty of a carbon tax can be approximated with cap-and-trade by setting a price floor and a ceiling for permit prices. The revenues generated by a carbon tax can also be achieved by auctioning permits in cap-and-trade. Cap-and-trade will be difficult to apply at a global level, but the process could start with a subset of countries (as Europe and some countries and subregions of countries have done) and eventually link systems by various mechanisms. Or there could be parallel systems of cap-and-trade in some jurisdictions

---

38   Section TS-5.1 is based on Chapter 22.

with carbon taxes in others. In the final analysis, environmental sustainability will need to be achieved at the lowest cost, and in an equitable manner.

In addition to GHG emissions pricing, other policies will be needed to develop, and then support, **new technologies** through the various stages from laboratory research to prototype demonstrations to wide-scale commercialization. These include R&D support, subsidies for large-scale technology demonstrations, and, for emerging favorites, market sales requirements or guarantees. Such policies will apply to initiatives such as CCS, new nuclear power technologies, renewable energy innovations, highly efficient energy-using devices, energy storage devices, grid management systems, and zero-emission vehicles and other transportation technologies.

It is important to complement GHG pricing with a portfolio of other regulatory and market mechanisms. This is because different instruments are most effective in different sectors, geographic/cultural regions, and for different options. For example, owing to the magnitude and diversity of market barriers prevailing in the building sector, different regulatory and market-based instruments and their packages are needed and tailored to overcome the specific barriers.

As a new technological pathway, CCS requires additional targeted policies to clarify pore space, property rights, storage site risk assessment, short- and long-term liabilities at storage sites, and measurement and crediting protocols to ensure that such projects are valued as emission reductions in GHG regulatory frameworks. Some CCS demonstrations should target low-cost CO_2 sources (existing or new coal-chemicals or coal-fuels plants) and enhanced oil recovery for CO_2 storage.

In cities, regulatory policies such as land use zoning, building codes, development permitting, and local emission standards must drive the shift toward low- and near-zero emission buildings, in some cases in concert with low- and near-zero emission decentralized energy supply. In addition to buildings, new urban developments should be required – with greater stringency in industrial countries initially – to be low- or near-zero emission (of local air pollutants and GHG emissions) through the local supply of renewable energy sources, where feasible, and through the import of energy from near-zero emission external sources. These requirements, which in part apply to rural areas as well, should also be gradually phased in to the retrofit of existing buildings and the redevelopment of existing urban areas.

Policy integration is especially important for unlocking energy-saving potentials due to the many barriers as well as multiple benefits. Strategic alliances and strong coordination among various policy fields will be able to capture a much larger share of technological potential by improving the economics of efficiency investments through the addition of further benefits to the cost-efficiency considerations, such as security, employment, social welfare, regional development, reduced congestion, and so on.

5.2   Policies for Meeting Energy-development Challenges

The importance of energy for achieving development goals is outlined above. Two challenges are of particular importance –providing universal access to modern forms of energy, and addressing urbanization and the provision of urban energy services.

5.2.1   Energy Access[39]

Access to affordable modern energy carriers and cleaner cooking improves well-being and enables people to alleviate poverty and expand their local economies. Even among people who have physical access to electricity and modern fuels, a lack of affordability and unreliable supplies limit their ability to use these resources. In addition to access to modern forms of energy, there must be access to end-use devices that provide the desired energy services. Those who can afford the improved energy carriers may still not be able to afford the upfront costs of connections or the conversion technology or equipment that makes that energy useful.

The lack of access to modern forms of energy is due to a number of factors. They include low income levels, unequal income distribution, inequitable distribution of modern forms of energy, a lack of financial resources to build the necessary infrastructure, weak institutional and legal frameworks, and a lack of political commitment to scaling-up access. Public policies should develop the strategies, and create the conditions, to overcome the mentioned barriers.

While the scale of the challenge is tremendous, access to energy for all, electricity for all, and modern fuels or stoves for all by 2030 is achievable. This will require global investments of US$36–41 billion annually – a small fraction of the total energy infrastructural investments required by 2030. It is expected that as households with public-sector support gain access to modern energy and end-use devices and start earning incomes, the standard of living and ability to pay for the energy services utilized would successively expand.

Access to Electricity

Between 1990 and 2008 almost two billion people gained access to electricity, more than the corresponding population increase of 1.4 billion people over that period (see also Figure TS-26). By 2030, the 1.4 billion people currently without access to electricity, plus the projected population increase to 2030 of 1.5 billion people, need to be connected to meet the GEA goal on universal electricity access. To achieve this, a multitrack approach is needed, combining grid extension with microgrids and household systems.

---

39   Section TS-5.2.1 is based on Chapters 2, 17, 19, and 23.

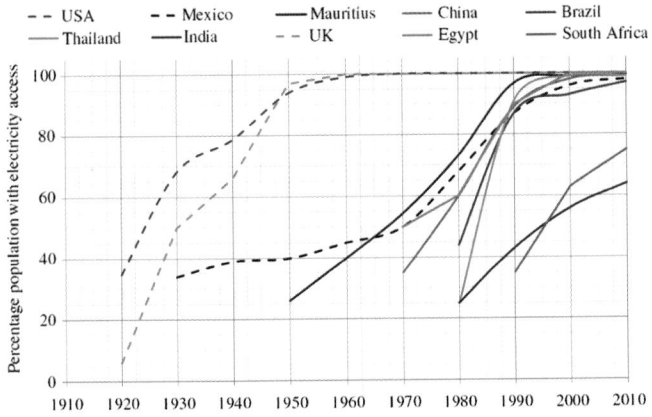

Figure TS-26 | Historical experience with household electrification in select countries. Source: Chapter 19.

Access to electricity can be arranged in several ways. In areas with high energy density, grid extension is currently the lowest cost per kWh delivered and also the preferred delivery form by most customers because of the capacity to deliver larger quantities of power for productive purposes. For many remote populations grid extension by 2030 will be highly unlikely and microgrids offer an alternative, based on local renewable energies or imported fossil fuels. An interesting approach to providing modern energy and development in remote villages is the multifunctional platform beginning to gain hold in West Africa (see Chapter 25). Household electrification is expanding rapidly in some countries, based on solar PVs that are financed by micro-credits and this has been done without increasing household expenses for energy (replacing candles and kerosene).

Experience shows that grid extension on commercial grounds is ineffective. Initial loads are small and provide insufficient income for the utilities. Privatizations of utilities have amplified this situation. The use of life-line rates or a free allocation of a small quantity of electricity per month per household has proven effective in starting a development process.

Grants should support equipment that uses energy more efficiently and more cleanly rather than subsidizing the energy used – except in the case of lifeline rates for the lowest-income customers, which can be achieved in part through cross-subsidies from other customers. Grants for high-voltage grid extensions and decentralized microgrids should involve competitive bidding to ensure the most cost-effective use of funds. Policies should support local participation in developing and managing energy systems, as this approach has been shown to have the best chance of providing a stable environment for new investment and reinvestment in increased energy access.

Cleaner cooking

About three billion people rely entirely, or to a large degree, on traditional biomass or coal for cooking and heating. This number has not changed appreciably over the last decades, particularly among households in rural areas. Indeed, more people rely on these fuels today than at any time in human history.

Access to cleaner cooking refers to liquid or gaseous fuels, especially biogas, LPG, ethanol, and others, or alternatively access to advanced biomass stoves with pollutants emissions similar to those of gas stoves. In many regions, current inefficient use of biomass fuels requires women and children to spend many hours per week collecting and carrying traditional biomass that is burnt in highly inefficient and polluting stoves. The resulting household air pollution leads to significant ill health (see Section TS-2.5).

Achieving universal access to modern energy is not likely to have negative implications for climate change. This is because transitioning to modern fuels (even in the case that these are fossil based) will displace large quantities of traditional biomass use. Current technologies that use traditional biomass are 4–5 times less efficient than cooking with modern fuels like LPG, and are associated with significant emissions of non-CO_2 Kyoto gases (e.g., CH_4, N_2O) and aerosols (e.g., black carbon, organic carbon) due to incomplete combustion.

Unlike the new biomass stoves, there is no feasible technology for burning coal cleanly at the household level. Attempts to provide 'clean coals' for household use attempted around the world over the last several decades had little success. Chimneys do not protect people sufficiently as they simply move the pollution from one place to another in the household environment. Country after country, therefore, has found that the only way to provide clean residential environments is to shift to other fuels. Only a few countries today, notably China, have significant household use of coal, which is a source of much ill-health, inefficient energy use, and high-climate impact per unit energy service delivered.

The observation that introducing cleaner cooking brings multiple benefits in terms of development (Section TS-2.2), situation of women, improved health from reduced exposure to household air pollution (Section TS-2.5), and reduced contributions to climate change (Section TS-2.4) should be very attractive for developing countries as well as for development cooperation organizations.

Policies for improving energy access in general

Providing universal and affordable access to electricity and cleaner cooking is possible if timely and adequate policies are put in place. Overall, and on the basis of successful experiences of increasing access to modern energy, no single approach can be recommended above others. What is clear, however, is that the current institutional arrangements and policies have met with mixed success, at best. Reforms are needed, on global and country levels, to strengthen the feasibility of energy projects for poor people, expand the range of players involved, open up the regulatory system, and allow for innovation.

Several examples demonstrate that the major features behind the success of programs include: political priority and government commitment; continued support and strengthening of programs through various administrations; effective mechanisms for targeting the available subsidies exclusively to poor families in need, thus ensuring highly efficient public expenditure; and integration of the energy-access program into governments' broader policies of social support for poverty alleviation.

Success depends on regional, national, and local circumstances. In some instances, a decentralized and participatory decision-making process and a holistic development approach is very important. This goes together with a strong community-mobilization process that focuses on organizational development, skills enhancement, capital formation, promotion of technology, environmental management, and empowerment of vulnerable communities.

Significant success has been achieved with small pilot projects to improve access in some rural areas and among poor communities in urban areas. But less thought has been focused on how to scale-up from these projects to market development and to meet the needs of the larger population.

A paradigm shift is needed in the approach to energy planning and policy implementation to facilitate access to modern forms of energy and cleaner cooking. Current supply-side approaches that simply take as their starting point the provision of electricity, petroleum, or gas, or of equipment of a particular type (solar technology, improved cookstoves, biogas, and other forms of bioenergy) are unable to reap the full potential of social and economic improvements that follow from improved energy access and cleaner cooking.

Several countries, including India, China, Argentina, Chile, Vietnam, Laos, and Brazil, have demonstrated that if effective political decisions are taken, the results are positive. This is not yet the situation in much of sub-Saharan Africa. Challenges and economic, sociocultural, and political barriers require more elaborate strategies and a higher global commitment to satisfy the GEA scenario objectives. Universal access in 20 years is not going to be available in Africa with micro actions and isolated measures unless they are integrated in long-term national programs with clear targets, dedicated and guaranteed funds, an adequate institutional framework, and robust strategies.

Policy recommendations in the form of general ideas or guidelines are provided below. Regional and national contexts should be considered in defining strategies, instruments, and measures.

- A better understanding and a clearer diagnosis of the structure and functioning of energy systems, along with the needs (energy services) to be supplied, are needed. These have often been absent in the discussion of proposals and the role of public policies. Good policies need good diagnoses. Support and funds for diagnosis and information should be part of the strategies.

- Subsidies are generally justified as a response to inequality and social expectations in energy provision. However, their net effect can be positive or negative depending on the intended goals of the subsidy, and the way a subsidy is implemented. An effective tariff and subsidy regime has to be transparent and minimize administrative costs to avoid gaming of the system and to maximize the benefits that accrue to the intended recipients. Subsidies to energy should be complemented with funds toward solving the first-cost capital financing problem, since upfront costs of equipment are, usually, the key barrier.

- Financing mechanisms are needed for every scale of energy intervention. Mobilizing affordable and genuine international, regional, national, and local funds is crucial.

- Energy-access policy is part of a wider development policy and should be aligned with other sector policies and objectives. If these policies are misaligned, they can reduce the effectiveness of any given policy. Policy misalignments can occur when different energy policies work at cross-purposes or when government priorities that could benefit from an effective energy policy are not aligned. In particular, there is a need to link rural and peri-urban energy supply more closely with rural development. This would shift the focus from minimal household supply to a more comprehensive approach to energy that includes productive activities and other welfare-enhancing uses of energy. Ideally, the linkages between energy and other policy priorities, such as health, education, and poverty alleviation, should be recognized explicitly and local solutions that address these needs should be encouraged and supported.

- Capacity development is needed, especially for the design and implementation of public policies oriented to poor people.

In the specific case of access to cleaner cooking, fuel subsidies alone will be neither sufficient nor cost-effective in terms of achieving ambitious energy-access objectives. Financial mechanisms, such as micro-credit, will need to complement subsidies to make critical end-use devices such as cleaner cookstoves affordable for poor people.

Leveraging funding and access to capital from public and private sources – for needed investments at the macro level and, at the micro level, for meeting costs for low-income households – is crucial in efforts to expand access to energy services for the poorest people. Figure TS-27 shows the estimated impacts of policy scenarios for cleaner cooking. It is only with the combined attention to fuel costs and equipment purchases that universal access is approached. Creative financing mechanisms and transparent cost and price structures will be critical to achieving the required scale-up and quick roll-out of solutions to improve access.

No single solution fits all in improving access to energy among rural and poor households. Programs need to be aware of local needs, resources, and existing institutional arrangements and capabilities. Diverse sources

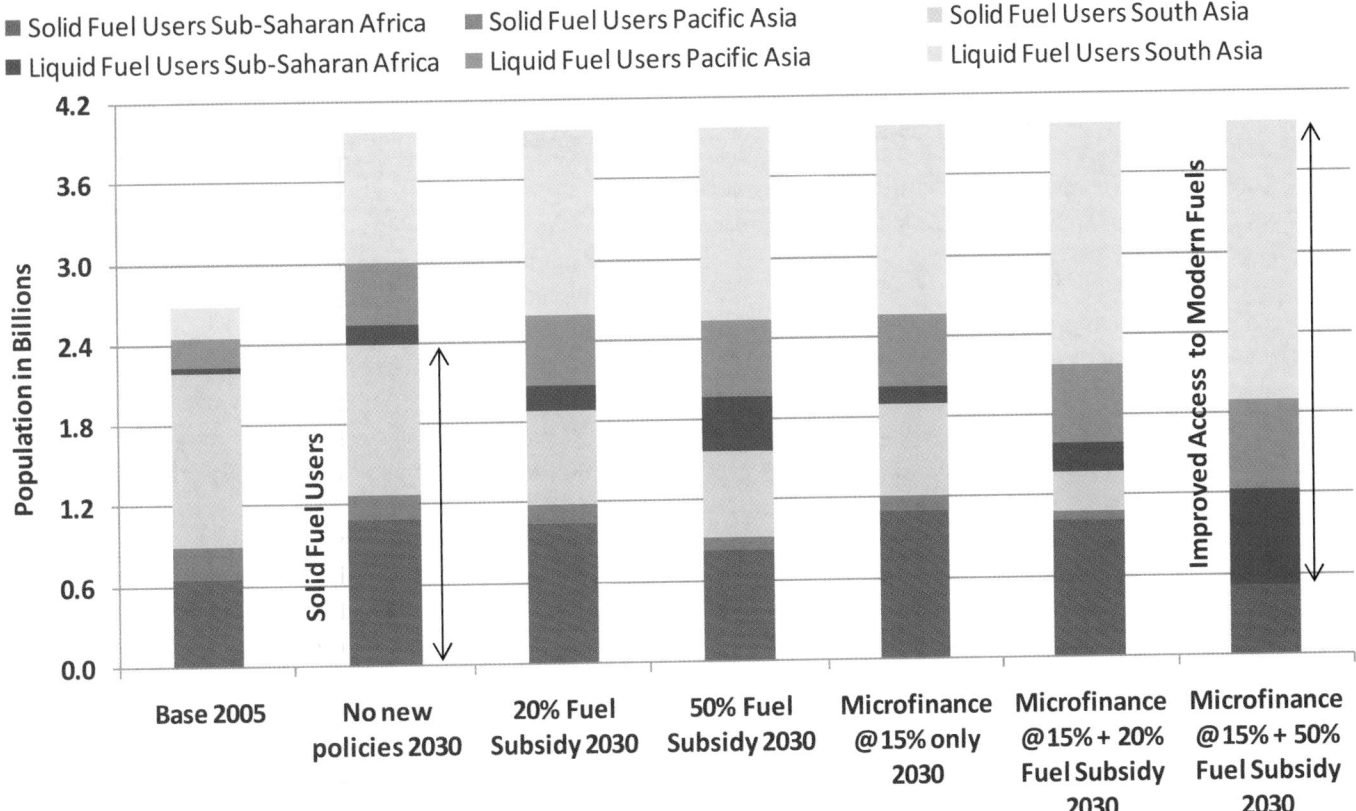

Figure TS-27 | Impact of alternative policy scenarios on access to cleaner cooking fuels in three developing regions. Subsidies are relative to consumer price levels and are additional to existing subsidies. Source: Chapter 17.

of energy supply (fossil and renewable), a wide portfolio of technologies, and a variety of institutional and innovative business models adapted to local circumstances are required to meet the challenge. An enabling environment shaped by sustained government commitment and enhanced capacity building at all levels is essential to ensure that access targets are met. Complementary development programs and enhancement of market infrastructure are needed to ensure sustained economic growth as well as steady employment and income generation for poor people to provide them with a means to pay for improved energy carriers.

As discussed in Chapter 12, establishing synthetic liquid transportation fuels in the market via gasification offers an innovative approach to expanding production of, and making more affordable access to, LPG, an inherent co-product of synthetic liquid transportation fuel manufacture. Such systems that co-produce electricity with CCS involving coal or the co-processing of coal and a modest amount of biomass (which could be deployed in this decade) would make available LPG with a significantly lower carbon footprint than LPG derived from petroleum sources, and with attractive system economics in a world with high oil prices.

Moreover, at GHG emissions prices that might be typical in the post-2030 period, small-scale systems making liquid transportation fuels from biomass with CCS would be able to provide substantial quantities of byproduct LPG for cooking at competitive costs in biomass-rich but regions lacking fossil fuels. Chapter 12 sketches out a plausible business

model for making this domestically produced LPG affordable without subsidy, even for the very poorest households in such regions. Near-term actions to facilitate exploitation of this future opportunity include assessments of CO_2 storage capacity in such regions, building the human and physical infrastructure capacities needed to support these new industrial activities, establishing carbon trading systems that would facilitate realization of the attractive economics for these bioenergy with CCS systems, and testing alternative business models for providing the LPG to poor households without subsidy.

5.2.2 Urbanization[40]

Currently, about half the world population lives in urban areas, which also account for an overly large share of global economic output and energy use (an estimated 60–80% of the global total). Projections invariably suggest that almost all future population growth of some three billion people by 2050 would be absorbed by urban areas, which would also account for a majority of economic and energy demand growth. By 2050 the global urban population is expected to approach 6.4 billion people – about the size of the entire global population in 2005. In contrast, the global rural population would plateau around 2020 at 3.5 billion people and decline thereafter.

---

40 Section TS-5.2.2 is based on Chapter 18.

Most urban growth would continue to occur in small- to medium-size urban centers (between 100,000 and one million inhabitants) in the developing world, which poses serious policy challenges. In these smaller-scale cities, data and information to guide policy are largely absent, local resources to tackle development challenges are limited, and governance and institutional capacities are weak. Housing, infrastructure, energy and transport services, and a better urban environment (especially urban air quality) are the key sustainability challenges for urban poverty alleviation.

Several hundred million urban dwellers in low- and middle-income nations lack access to electricity and are unable to afford cleaner, safer fuels, such as gas or LPG. In addition to poverty and poor urban energy infrastructures, poor people face political or institutional obstacles to obtaining cleaner energy carriers.

Given capital constraints, daring new architectural and engineering designs of 'eco' or 'zero-carbon' cities can serve as inspirational goals and as field experiments, but they are unlikely to play any significant role in integrating some three billion additional urban dwellers by 2050 into the physical, economic, and social fabric of cities. (Building 'zero-carbon' cities for three billion new urban citizens along the Masdar model in Abu Dhabi would require some US$1000 trillion, or some 20 years of current world GDP.)

Cities in OECD countries generally have lower per capita final energy use than their respective national averages. Conversely, cities in developing and emerging economies generally have substantially higher per capita energy use than the national average, primarily due to substantially higher income levels than those in rural areas.

Urban systems are, however, by definition inherently open systems: they are characterized by vast imports of resources and commodities and by vast exports of goods and services to their respective hinterlands and the rest of the world. 'Embodied' energy (and GHG emissions) is, as a rule, several fold larger than the direct energy uses in urban settings, at least for the handful of megacities for which data are available.

The overall design of cities and their components affect the energy use to a large degree. For buildings, energy use for thermal purposes can cost-effectively be reduced by 90% or more, as compared with current standard practice (see Figure TS-15, Section TS-3.2.3). Not incentivizing the adoption of available building-efficiency technologies and practices will lock cities into a much higher energy-use level than necessary. Figure TS-16 illustrates this for energy use in buildings. Next to buildings, urban density, form, and usage mix are also important determinants of urban energy use and efficiency, especially in transportation (see Figure TS-13). Avoiding spatial lock-in into urban sprawl and ensuing automobile dependence should, therefore, be another important urban policy objective.

Significant potential co-benefits between urban energy and environmental policies do exist. However, they require more holistic policy approaches that integrate urban land use, transport, building, and energy policies with the more-traditional air pollution policy frameworks.

Urban energy and sustainability policies could focus on where local decision making and funding also provides the largest leverage effects:

- urban form and density (which are important macro-determinants of urban structures, activity patterns, and hence energy use, particularly for urban transport);

- the quality of the built environment (energy-efficient buildings in particular);

- urban transport policy (in particular the promotion of energy-efficient and 'eco-friendly' public transport and non-motorized mobility options); and

- improvements in urban energy systems through cogeneration or waste-heat recycling schemes, where feasible.

Illustrative model simulations for a 'synthetic' city suggest improvement potentials of at least a factor of two each by buildings that are more energy-efficient and by a more compact urban form (at least medium density and mixed-use layouts), with energy system optimization through distributed generation and resulting cogeneration of electricity, heat, and air conditioning adding another 10–15% improvement in urban energy use (see Figure TS-28).

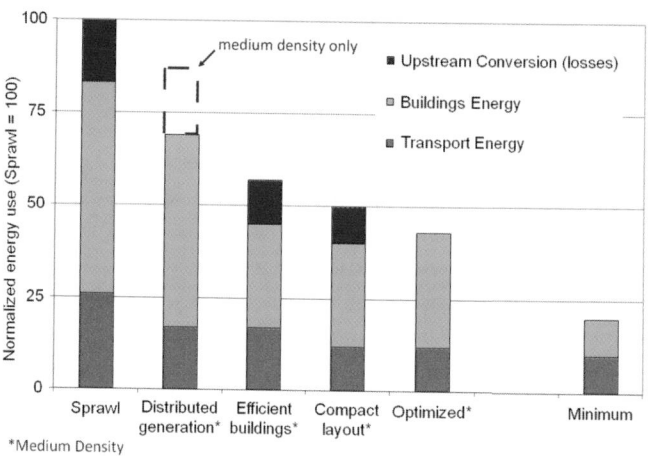

Figure TS-28 | Policy integration at the urban scale. Simulated energy use for an urban settlement of 20,000 inhabitants using the SimCity Model combining spatially explicit models of urban form, density, and energy infrastructures, with energy systems optimization. Individual policy options are first simulated individually and then combined in a total systems optimization. Baseline (index = 100) sprawl city corresponds to a secondary energy use of 144 GJ/capita; energy use is shown by major category: transport, buildings, and upstream energy conversion losses (which can be eliminated by local cogeneration of electricity and heat or by on-site energy systems). The potential for efficiency improvement of narrow energy sector-only policies (local renewables, cogeneration) at the urban scale is smaller than policies aiming at minimizing buildings energy use or at higher urban density and mixed uses, which minimize transport energy use. The largest improvement potentials can be realized by a combination of energy, building efficiency, and urban form and density policies. Source: Chapter 18.

There are important urban size and density thresholds that are useful guides for urban planning and policymaking. The literature review identified a robust density threshold of 50–150 inhabitants per gross hectare (5000–15,000 people per square kilometer) below which urban energy use, particularly for transport, increases substantially. Note that there is little empirical evidence to suggest substantial further energy efficiency gains at much higher densities. Energy-wise, there are pronounced diseconomies of scale of low urban densities (leading to lower efficiency and higher energy use), but no significant economy-of-scale effects beyond intermediary density levels.

5.3   Policies for Key Energy System Building Blocks

The GEA pathways describe the various combinations of transformations in energy systems required to meet the GEA's various goals and objectives simultaneously (see Section TS-4.) While they differ in terms of their relative proportions and the magnitudes of the various changes they involve, all of them include a dramatic increase in energy end-use efficiency, larger and more rapid deployment of renewables, decarbonization and modernization of fossil fuel systems, and the judicious use of nuclear energy. Policies to address the changes required in each of the building blocks are described below.

5.3.1   Energy Efficiency

Progress in accelerating the rate of energy efficiency improvement worldwide is critical to an energy system for sustainability. Quickly improving energy efficiency requires more focused and aggressive policies that: support rapid innovation; significantly tighten efficiency regulations in energy supply and demand; increase energy prices; create a culture of conservation among consumers and firms; change land use zoning to increase urban density; and integrate mixed land uses so that transportation needs decline and low-energy transportation modes flourish. In some cases, these policies will involve subsidies for new technologies, but these will not be effective unless they are combined with pricing of GHG emissions via taxes and/or cap-and-trade plus well-designed efficiency regulations.

Regulations, especially standards, are essential elements of energy policy portfolios of the transition. Building codes, appliance standards, fuel economy standards, and industrial energy management standards have proven to be very environmentally sound in improving efficiency and should be adopted globally. The combination of regulations, incentives (e.g., fiscal incentives), and measures to attract attention, (e.g., information, awareness, or public leadership programs) has the highest potential to increase energy efficiency. Policies encouraging the use of multi-generation and renewable energy in each end-use sector are important further components of energy policy portfolios.

The GEA analysis provides considerable evidence for the ability of such policy packages to deliver major change. However, the results from three decades of experience with energy efficiency policies in industrial countries also show other effects. For example, the adoption of energy efficiency devices has both a direct rebound effect (more-efficient fridges, with lower operating costs, encourage the adoption of larger fridges) and an indirect rebound effect (sometimes called a productivity rebound) that relates to the apparent causal link between energy efficiency breakthroughs and the development of new devices and new energy services (fridge efficiency improvements foster the development of new refrigeration devices, such as beer and wine coolers, water coolers, desk-top fridges and freezers, portable fridges, etc.). Evidence also shows that when estimating costs it is important to take into account all transaction costs and differences in technology risks and technology quality. Ideally, beyond transaction costs, all other indirect costs and benefits, including monetizable co-benefits, need to be integrated into cost-effectiveness assessments related to policy choices, as these can both be substantial and fundamentally alter final cost-effectiveness outcomes and thus instrument choices.

These cost factors and rebound effects mean that subsidies to encourage acquisition of energy-efficient devices are unlikely, on their own, to cause the dramatic energy efficiency gains called for in the GEA analysis. For these gains to be realized, a portfolio of stronger, carefully targeted policies is needed. Examples include: strong efficiency regulations that are updated regularly (say, every five years); incentives to reward manufacturers to push the technology design envelope toward advanced efficiency; increases in energy prices (because of direct or indirect emissions pricing); electricity tariffs that give high rewards to efficiency investments and behavior; land use planning and zoning that fosters efficient urban development and renewal; and public (and private) investments in efficient infrastructure such as mass transit, cycling paths, and CHP systems.

In the buildings sector, to be able to reduce final thermal energy use by over 40%, the goal in the GEA efficiency pathway, all jurisdictions need to introduce and strictly enforce building codes that mandate very low specific energy-use levels, equal or similar to passive-house levels. They also need to extend these requirements to renovations, and building retrofits will need to significantly accelerate the present rates. The remaining building energy needs can be met from locally generated renewable energy sources, where feasible, and economically and environmentally optimal – typically, low-density residential neighborhoods. Achieving the needed transformation in the buildings sector entails massive capacity-building efforts to retrain all the trades involved in the design and construction process, as well as the building owners, operators, and users.

Influencing energy use in the transport sector involves affecting transport needs, infrastructure, and modes, as well as vehicle energy efficiency.

Table TS-8 | Regulatory policies potential contribution to transport and GEA multiple goals.

GEA Overall Systemic Goals	Sustainable Transportation Systems Multiple Goals and Benefits	Aim: Establishing Clear Regulatory Framework								
		Vehicle Standards	Fuel Standards and Mandates	Reduce Travel Speed Limit in Urban Areas	Reduce Travel Speed/Volume of Freight Transport in Urban Areas	Reduce Speed of Airplanes	Reduce Speed of Commercial Maritime Transport	Improved Management Intelligent Transport System	Mandatory Vehicle Inspections	Traffic Safety Regulation
Economic Growth, Equity & Urbanization	Functionality, Efficiency									
	Accessibility									
	Affordability									
	Acceptability									
Health & Environmental Protection	Traffic Safety									
	Acces of less fit									
	Human Motion									
	Reduce Air Pollution									
	Reduce Noise									
	Reduce Congestion									
Climate	Reduce GHG									
Energy Security	Diversification Energy sources									
	Independence from Fossil fuels									

Legend: Role of Policies or potential contribution to attainment of goal according to literature
essential
uncertain
complementary

Source: Chapter 9.

Policies for urbanization will have a large impact on transport needs, infrastructure, and the viability of different transport modes on the local scale. Both the decision to travel and the choice of how to travel affect fuel consumption. With a focus on urban transport, a transition to sustainable transport can follow the framework known as 'avoid–shift–improve'. This considers three major principles under which diverse policy instruments are grouped, with interventions assuming different emphasis in industrial and developing countries. They need to focus on improving technological options, not only with respect to climate mitigation, but also with respect to local environmental conditions and social concerns. The other two components – modal shift and avoiding travel – influence the level of activity and structural components that link transport to carbon emissions.

This approach to urban transport would include policies and measures for developing alternatives to car use, reducing the need for travel, improving existing infrastructure use, and setting a clear regulatory framework (alternative fuels and efficient vehicles). In addition, policies targeting freight and long-distance travel (shipping, trucks, rail, and air) are needed. To illustrate the complexity of transportation policy, Table TS-8 shows some regulatory options and their potential impact.

For energy efficiency in industry it is useful to separate what can be achieved when a new plant is being built and what can be done in existing industry. Most of the new industrial growth will occur in developing countries. Under the business-as-usual scenario, a mix of technologies would be installed with varying levels of specific energy use. In addition to regulations and economic incentives, regional centers for industrial energy efficiency could be set up that help disseminate information related to specific energy use and best-available technologies for different processes. There could also be web-based facilities established where any industry that is being proposed can compare its design energy performance with the best available benchmark technologies. An incentive scheme should provide funding for energy performance analysis at the design stage. Governments could help provide financing of the incremental costs of energy-efficient technologies as low-interest loans through commercial banks.

In existing industries, realizing the potential for energy efficiency can be achieved through a combination of measures, including incentives for demand-side management. Regulatory commissions can provide regulations and standards for energy-using equipment and process improvements. Information gaps need to be reduced, especially the sharing and documentation of best practices. Capacity needs to be developed for systems assessment rather than individual components assessment.

In developing countries or jurisdictions with suppressed energy service levels, improved efficiency may lead to an increase in energy service levels rather than a decrease in energy demand. However, this should normally be the goal of efficiency policies in such jurisdictions. In industrial countries, such rebound effects need to be minimized through appropriate energy pricing and taxes that complement efficiency policies.

The transition into a very low energy future requires a shift in the focus of energy-sector investment from the supply-side to end-use capital stocks, as well as the cultivation of new innovative business models (such as performance contracting and ESCOs).

5.3.2   Renewable Energies

Increased use of renewable energy technologies can address a broad range of aims, including energy security, equity issues, and emission reductions, thereby linking beneficially with other policies related to poverty eradication, water provision, transport, agriculture, infrastructure development, industrial development, job creation, and development cooperation. For this to occur, policy measures must overcome the barriers within the current energy system that prevent wider uptake of renewables (see Table TS-9 for an overview). A key issue is how to accelerate the deployment of renewable energies so that their deep penetration into the energy system can be achieved quickly.

Given the enormous size and momentum of the existing global energy system, new technologies such as renewables face significant market barriers. To address these, policy measures should support a level playing field where renewables can compete fairly with other forms of energy; they should also support the development of renewables so that they can overcome additional hurdles to their deployment.

While competitive markets operate effectively for many goods and services, a number of failures need to be addressed in relation to energy. A central concern is the way that markets currently favor conventional forms of energy by not fully incorporating the externalities they are responsible for and by continuing to subsidize them – making it harder to incorporate new technologies, new entrants, and new services in the energy system. This both distorts the market and creates barriers for renewables.

Similarly, the potential benefits of renewables are also often not accounted for when evaluating the return on investment, such as increased energy security, access to energy, reduced economic impact volatility, climate change mitigation, and new manufacturing and employment opportunities. These issues are exacerbated by ongoing subsidies for fossil fuels that globally amount to hundreds of billions of US dollars per year, much more than the support renewables are receiving. It is through public policies that the values to society can be reflected in market conditions such that it will be advantageous for investors to seek out energy options that support and contribute to a sustainable future for all.

Using a portfolio of policies helps to increase successful innovation and commercialization, providing they complement each other. To expand renewable technologies, it is important to note that:

- market growth results from the use of combinations of policies;

- long-term, predictable policies are important;

- multi-level involvement and support from national to local players is important; and

- each policy mechanism evolves as experience of its use increases.

Policy approaches for renewable energy intend to address the innovation chain both technologically and socially, to pull technologies to the marketplace and commercialize them, and to improve the financial attractiveness and investment opportunities of renewables.

Of the market-pull policies, two are most common: a policy that sets a price to be paid for renewable energy and ensures connection to the grid and off-take (often known as a feed-in tariff or FIT), and a policy that sets an obligation to buy, but not necessarily an obligation on price (often known as a quota or obligation mechanism or a renewable portfolio standard). So far, FITs have been used for electricity only, although some countries, for example the United Kingdom, are now considering how to provide them for heat. Quotas have so far been used for electricity, heat, and transport. Biofuel quotas are now common globally.

A FIT that provides a strong, stable price for renewable electricity has proven successful in some countries for accelerating investment in renewables. Some jurisdictions prefer renewable portfolio standards that set a minimum, but growing, quota for renewable or low-emission electricity generation technologies. Although there is considerable debate between advocates of these two approaches, the detailed way in which they are implemented is the key to success. In addition, the GEA analysis for meeting climate stabilization goals shows that, currently, in industrial countries virtually no new investments in electricity generation should result in the new emission of GHGs. Unfortunately, such investments are still possible in countries with FITs, green certificate markets, or other renewable energy support schemes, and indeed this has been the case in most jurisdictions with such policies, although at a lower rate.

Table TS-9 | Summary of renewable energy policies.

Policy	Definition	End-use Sector		
		Electricity	Heat/Cooling	Transport
Regulatory Policies				
Targets	A voluntary or mandated amount of renewable energy (RE), usually a percentage of total energy supply	X	X	X
Access-related Policies				
Net metering	Allows a two-way flow of electricity between generator and distribution company and also payment for the electricity supplied to the grid	X		
Priority access to network	Allows RE supplies unhindered access to network for remuneration	X	X	
Priority dispatch	Ensures RE is integrated into the energy system before supplies from other sources	X	X	
Quota-driven Policies				
Obligation, mandates, Renewable Portfolio Standards	Set a minimum percentage of energy to be provided by RE sources	X	X	X
Tendering/bidding	Public authorities organize tenders for a given quota of RE supplies and ensure payment	X		
Tradable certificates	A tool for trading and meeting RE obligations	X	X	
Price-driven Policies				
Feed-in tariff (FIT)	Guarantees RE supplies with priority access, dispatch, and a fixed price per unit payment (sometimes declining) delivered for a fixed number of years	X	X	X
Premium payment	Guarantees RE supplies an additional payment on top of their energy market price or end-use value	X	X	
Quality-driven Policies				
Green energy purchasing		X	X	
Green labeling	Usually government-sponsored labeling that guarantees that energy products meet certain criteria to facilitate voluntary green energy purchasing	X	X	X
Fiscal Policies				
Accelerated depreciation	Allows for reduction in tax burden	X	X	X
Investment grants, subsidies, and rebates	One-time direct payments usually from government but also from other actors, such as utilities	X	X	X
Renewable energy conversion payments	Direct payment by government per unit of energy extracted from RE sources	X	X	
Investment tax credit	Provides investor/owner with an annual tax credit related to investment amount	X	X	X
Other Public Policies				
Research and development	Funds for early innovation	X	X	X
Public procurement	Public entities preferentially purchasing RE or RE equipment	X	X	X
Information dissemination and capacity building	Communications campaigns, training, and certification	X	X	X

Source: Chapter 11

The obvious next step is to require that all new investments for electricity generation are in near-zero emissions technologies, and some jurisdictions have done this. Since 2006, for example, British Columbia in Canada has a 100% clean electricity standard for all new investments.

5.3.3 Modernized Fossil Fuels

Low-, zero-, or negative-emission fossil fuel use will require a transition to systems that co-utilize fossil fuels with renewable energy and with CCS. Co-processing of biomass with coal or natural gas for the co-production of power, fuels, and chemicals with CCS, is especially promising. New policies are needed that encourage environmentally acceptable deployment of such systems. Some of the following leading policies have already been enacted on an experimental basis, but these efforts would need to be intensified significantly over the next decades to realize a dramatic shift. Governments or regulators could, among others:

- implement GHG emissions pricing via carbon taxes and/or cap-and-trade systems;

- reduce all subsidies to fossil fuels without CCS. This includes fuel price subsidies that promote increased energy use; subsidies to private vehicle use (e.g., untolled roads), and a host of subsidies to industrial, commercial, institutional, and other combustion uses of fossil fuels;

• provide demonstration and commercialization subsidies;

• offer to pay above-market rates for electricity, heat, or low-net GHG-emitting fuels provided via projects that co-process sustainable bio-mass and fossil fuel feedstocks in systems with CCS. This would be similar to the FIT for renewables;

• ban construction of new coal-fired electricity plants that lack CCS or are not CCS ready;

• require land use planning that facilitates socially and environmentally acceptable siting of underground carbon storage and CO_2 pipelines. There is also a need for land use planning to safeguard against potential impacts of carbon storage on other uses of the subterranean, such as geothermal energy, or at least consider a balance between the possible uses;

• legally clarify geological rights to underground pore spaces for CO_2 storage; and

• establish short- and long-term liabilities and risk management and monitoring responsibilities at CO_2 storage sites and on CO_2 pipeline right-of-ways.

5.3.4   Nuclear Energy

People's views on the value and risks of nuclear power differ greatly and are often polarized. Some people see nuclear power as a risky technology. These perceived threats from nuclear power include catastrophic accidents at nuclear plants (either through operational failures or terrorist attacks), the inability to safely transport and permanently store radioactive wastes, and the exploitation of civilian nuclear expertise for the proliferation of nuclear weapons.

Depending on the severity of these concerns about nuclear power, its regulatory burden (for design, permitting, operation, and decommissioning) can be such that nuclear power is a high-cost option for electricity generation. However, where public policy (local, national, international) is able to allay these concerns, then nuclear power can be a competitive energy option. However, everything hinges on risk preferences among the public and decision makers, particularly with respect to trading off the extreme event risks of nuclear power with the ongoing impacts and risks of its alternatives. The following policies therefore focus on how to ensure a safe use of nuclear power that is both real and perceived:

• At the international level, governments and the nuclear industry need to continue to improve their mechanisms for monitoring and controlling

the use of nuclear power and the reprocessing of nuclear fuel to prevent acquisition of expertise and materials for nuclear weapons production.

• Governments need to collaborate in the establishment of permanent storage sites for radioactive materials.

• By facilitating collaborative investments, governments can help the nuclear industry settle on two or three dominant designs that have the best chance of achieving regulatory approval and thus reducing regulatory costs, which have been very high in jurisdictions like the United States.

5.4   Elements of Policy Packages

The preceding sections describe a variety of policy instruments, tools, and approaches for different objectives, whether energy access or decarbonization through the use of renewables. Across the various domains of intervention, there are some common requirements for transformative change. For example, whether in the context of CCS or renewable energy technologies, accelerating the process of research, development, demonstration, and deployment is a common requirement. Similarly, it is necessary to enhance and reorient investment. Capacity building is essential to ensure that countries, regions, and policymakers are able to design and implement policies. It is possible that fundamental rethinking of lifestyles and consumption patterns may be required for sustainability. This may require new knowledge (such as green accounting practices) as well as a range of tools to influence public thinking, opinion, and behavior.

5.4.1   Innovation[41]

Innovation and technological change are integral to the energy system transformations described in the GEA pathways. Energy technology innovations range from incremental improvements to radical breakthroughs and from technologies and infrastructure to social institutions and individual behaviors. The innovation process involves many stages – from research through incubation, demonstration, (niche) market creation, and ultimate widespread diffusion. Feedback between these stages influences progress and likely success, yet innovation outcomes are unavoidably uncertain. Innovations do not happen in isolation; inter-dependence and complexity are the rule under an increasingly globalized innovation system.

A first, even if incomplete, assessment of the entire global investments into energy technologies – both supply- and demand-side technologies – across different innovation stages suggests RD&D investments of some US$50 billion, market formation investments (which rely on directed public policy support) of some US$150 billion, and an estimated range

---

41   **Grubler, A. and K. Riahi, 2010:** Do governments have the right mix in their energy R&D portfolios? *Carbon Management*, **1**(1):79–87.

of US$1–5 trillion investments in mature energy supply and end-use technologies (technology diffusion) are required. Demand-side investments are of critical importance, particularly as the lifetimes of end-use technologies can often be considerably shorter than those on the supply side. Demand-side investments might thus play an important role in achieving pervasive and rapid improvements in the energy system.

Major developing economies have become significant players in global energy technology RD&D, with public- and private-sector investments approaching some US$20 billion – in other words, almost half of global innovation investments – and are significantly above OECD public-sector energy RD&D investments (US$13 billion).

Policies now need to move toward a more integrated approach, simultaneously stimulating the development and adoption of efficient and cleaner energy technologies and measures. R&D initiatives without simultaneous incentives for consumers to adopt the outcomes of innovation efforts risk not only being ineffective, but also precluding the market feedbacks and learning that are critical for continued improvements in technologies.

Few systematic data are available for private-sector innovation inputs (including investments). Although some of the data constraints reflect legitimate concerns to protect intellectual property, most do not. Standardized mechanisms to collect, compile, and make data on energy technology innovation publicly available are urgently needed. The benefits of coupling these information needs to public policy support have been clearly demonstrated.

The energy technology innovation system is founded on knowledge generation and flows. Increasingly these are global, but need to be adapted, modified, and applied to local conditions. Long-term, consistent, and credible institutions underpin investments in knowledge generation, particularly from the private sector. Yet consistency does not preclude learning. Knowledge institutions have to be responsive to experience and adaptive to changing conditions; see, for example the discussion on open and distributed innovation, university-industry linkages, and knowledge networks (in the North, the South, and North–South) discussed in Chapter 25.7. Although knowledge flows through international cooperation and experience, sharing at present cannot be analyzed in detail; the scale of the innovation challenge emphasizes their importance alongside efforts to develop capacity to absorb and adapt knowledge to local needs and conditions.

Clear, stable, and consistent expectations about the direction and shape of the innovation system are necessary for innovators to commit time, money, and effort with only the uncertain promise of distant returns. To date, policy support for the innovation system has been characterized by volatility, changes in emphasis, and a lack of clarity. An example is the development of solar thermal electric (STE) technology in the United States (see Figure TS-29). After successful development during one decade, sudden policy changes in 1992 terminated interest in STE in the country. Now US interest has revived, with some projects underway in California

(although none completed yet) with all the knowledge and technology imported from Europe (Spain), as associated knowledge entirely depreciated in the United States after the 1992 sudden policy changes.

Policies have to support a wide range of technologies. However seductive they may seem, silver bullets do not exist without the benefit of hindsight. Innovation policies should use a portfolio approach under a risk hedging and 'insurance policy' decision-making paradigm. The portfolio approach is also emphasized in Chapter 25 as part of a capacity development approach, especially in developing countries. The whole energy system should be represented, not just particular groups or types of technology. The entire suite of innovation processes should be included, not just particular stages or individual mechanisms. Less capital-intensive, smaller-scale (that is, granular) technologies or projects are a lower drain on scarce resources, and failures have less-serious consequences.

Public technology policy should not be beholden to incumbent interests that favor support for particular technologies that either perpetuate the lock-in of currently dominant technologies or transfer all high innovation risks of novel concepts to the public sector.

Portfolios need to recognize that innovation is inherently risky. Failures vastly outnumber successes. Experimentation, often for prolonged periods (decades rather than years), is critical to generate the applied knowledge necessary to support the scaling-up of innovations into the mass market.

Public sector energy R&D as a function of total public sector financed R&D has declined since the early 1980s, with a small reversal in the trend over the last few years (see Figure TS-30). Spending on technology groups has been relatively constant over time. Nuclear energy has received the largest part of the funding.

Technology needs from the pathway analysis shows a very different picture (see Figure TS-31). Energy efficiency dominates this analysis which also shows a doubling or more for renewable energies, and a significant lower emphasis on nuclear energy. This historical energy R&D portfolio bias needs to be addressed urgently to stimulate the innovations needed for realization of the GEA transition pathways.

5.4.2   Finance[42]

Some of the policies for energy sustainability described above simply involve an improvement of existing policies, such as better management of the electricity sector or more responsible use of fossil fuel resource rents. But the dominant message of the GEA is that the global energy system must be rapidly modified and expanded to provide energy access to those who have none, and must quickly transform to an energy system more supportive of sustainable development. This transition will require considerable investments over the coming decades. Table TS-10 indicates the

---

42   Section TS-5.4.1 is based on Chapter 24.

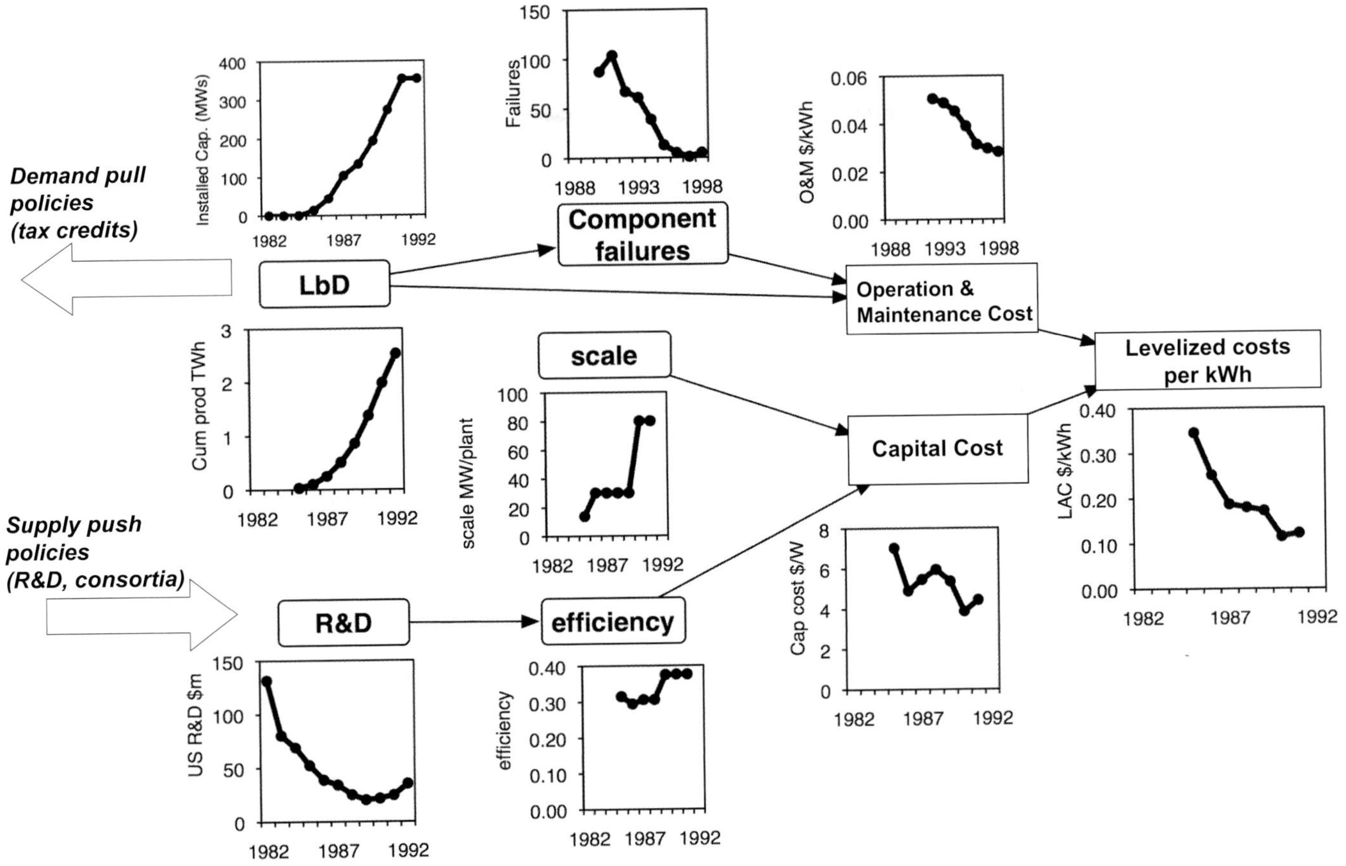

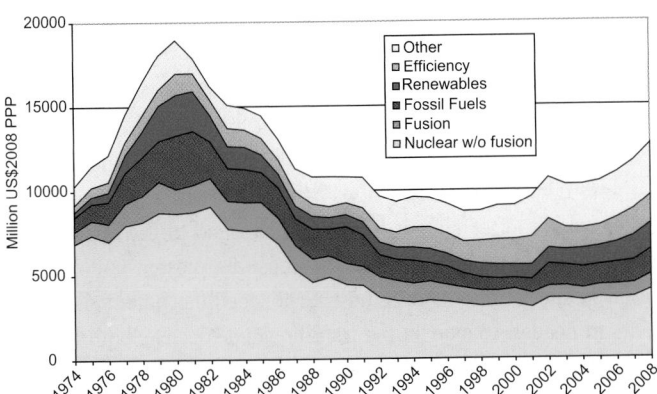

Figure TS-29 | History of the US solar thermal electricity program, 1982–1992. This shows a 'virtuous' technology development cycle as a result of well-coordinated policies. Demand pull policies enabled expanding market applications, which in turn enabled a scaling-up of the technology – reducing capital costs through economies of scale effects, learning by doing (LbD), and reductions in component failures (lowering operating costs). Supply push policies, such as R&D (even at declining budgets), led to technology improvements (efficiency) that further lowered capital costs. In the aggregate, levelized total costs per kWh declined by a factor of three over 10 years. This positive innovation development cycle came to an abrupt halt after 1992 with the sudden discontinuation of public policy support. Source: Chapter 24.

Figure TS-30 | Public sector energy RD&D in IEA Member countries by major technology group. Source: Chapter 24.[43]

necessary investments to achieve this, as estimated by the GEA, and links these to the types of policies needed. It also assesses these policies in terms of their necessity and their ability to complement or substitute for each other. Although considerable, these investment levels can be compared to estimates of global fossil fuel subsidy levels on the order of US$500 billion a year, of which an estimated US$100 billion goes to producers.

Table TS-10 compares the costs and policies for different technology options to those of promoting energy access. Different types of technologies and objectives will require different combinations of policy mechanisms to attract the necessary investments. Thus, Table TS-10 identifies 'essential' policy mechanisms that must be included for a specific option to achieve the rapid energy system transformation, 'desired' policy mechanisms that would help but are not a necessary condition, 'uncertain' policy mechanisms in which the outcome will depend on the policy emphasis and thus might favor or disfavor a specific option, and policies that are inadequate on their own but could 'complement' other essential policies.

The GEA findings indicate that global investments in combined energy efficiency and supplies have to increase to about US$1.7–2.2 trillion per

43   **IEA, 2009a:** *World Energy Outlook*. International Energy Agency, Organization for Economic Cooperation & Development, Paris.

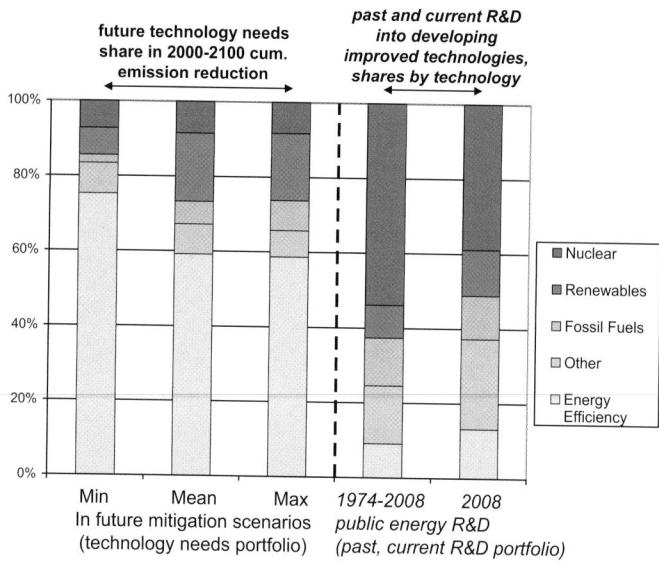

Figure TS-31 | Distribution of past (1974–2008) and current (2008) public sector energy technology R&D portfolios in member countries of the IEA (right) versus portfolios of future GHG mitigation needs (min/mean/max, left) derived from an extensive scenario uncertainty analysis. Source: Adapted from Grubler and Riahi, 2010, see Chapter 24.[44]

year compared with the present level of some US$1.3 trillion (2% of current world GDP). Given projected economic growth, this would be an approximately constant fraction of GDP in 2050.

For some objectives, such as energy access, future investment needs are comparatively modest. However, a variety of different policy mechanisms – including subsidies and regulation as well as capacity building programs – need to be in place. Regulations and standards are also essential for almost all other options listed in Table TS-10, while externality pricing might be necessary for capital-intensive technologies to achieve rapid deployment (such as a carbon tax to promote diffusion of renewables, CCS, or efficiency). Capital requirements for energy infrastructure are among the highest priorities of the options listed.

Increasing investments in the energy system as depicted by the GEA pathways requires the careful consideration of a wide portfolio of policies to create the necessary financial incentives and adequate institutions to promote and support them, and innovative financial instruments to facilitate them The portfolio needs to include regulations and technology standards in sectors with, for example, relatively low price elasticity in combination with externality pricing to avoid rebound effects, as well as targeted subsidies to promote specific 'no-regret' options while addressing affordability. In addition, focus needs to be given to capacity development to create an enabling technical, institutional, legal, and financial environment to complement traditional deployment policies (particularly in the developing world).

5.4.3   Capacity Development[45]

Wealthier countries need to improve mechanisms for supporting *capacity development* in developing countries, including financial support, technical training, and sharing of industry, trade, and institutional experiences. Any energy capacity strategy must, however, be tailored to the specific characteristics of a given country or region if it is to succeed in stimulating a rapid transition of the energy system to a more sustainable path. While this strategy must address basic needs for education and training, it must also be adapted to the local cultural norms and practices.

The transitions put forward in GEA require a transformation of energy systems that demand significant changes in the way energy is supplied and used today. These transitions are, by definition, long-term, socially embedded processes in the course of which capacities at the individual, organizational, and systems levels and the policies for capacity development themselves will inevitably change. From this perspective, capacity development can no longer be seen as a simple aggregation of individual skills and competences or the introduction of new 'technology'. Rather, it is a broad process of change in production and consumption patterns, knowledge, skills, organizational form, and – most important – established practices and norms of the players involved: in other words, a host of new and enhanced capacities. Energy transitions are thus innovative processes (Chapter 25.1).

The complexity, magnitude, and speed of the changes envisaged in these transitions will necessitate a major shift in the way that societies analyze and define the concept of 'capacities' and in the way in which they go about the important task of developing these capacities to meet the challenges of energy transitions. Different from some of the linear approaches to capacity development and to technology transfer and deployment used today, which often fail to appreciate the complexity of change processes, the concept of capacity development advanced by GEA is intimately linked to the energy transitions perspective based on multilayered processes of system change.

In these processes, special attention is paid to the informal institutions that arise out of historically shaped habits, practices, and vested interests of players in the system already in place and to the tendency for path dependence, where past choices constrain present options. They are given special attention because they constitute potential impediments to needed change. In the transitions perspective, both learning and unlearning such habits, practices, and norms in the course of change are important (Chapter 25.4).

Traditional habits, practices, and norms also shape the styles of communication in societies. Evidence shows that the more successful change processes take place in environments that tend to move away from top-down communication and consultation to more active and continuous

---

44 Section TS-5.4.2 is based on Chapter 6 and 17.

45 Section TS-5.4.3 is based on Chapter 25.

Table TS-10 | Energy investments needed between 2010 and 2050 to achieve GEA sustainability objectives and illustrative policy mechanisms for mobilizing financial resources.

Times	Investment (billions of US$/year)		Policy mechanisms			
		2010–2050	Regulation, standards	Externality pricing	Carefully designed subsidies	Capacity building
Efficiency	n.a.[a]	290–800[b]	*Essential* (elimination of less efficient technologies every few years)	*Essential* (cannot achieve dramatic efficiency gains without prices that reflect full costs)	*Complement* (ineffective without price regulation, multiple instruments possible)[c]	*Essential* (expertise needed for new technologies)
Nuclear	5–40[d]	15–210	*Essential* (waste disposal regulation and of fuel cycle, to prevent proliferation)	*Uncertain* (GHG pricing helps nuclear but prices reflecting nuclear risks would hurt)	*Uncertain* (has been important in the past, but with GHG pricing perhaps not needed)	*Desired* (need to correct the loss of expertise of recent decades)[e]
Renewables	190	260–1010	*Complement* (feed-in tariff and renewable portfolio standards can complement GHG pricing)	*Essential* (GHG pricing is key to rapid development of renewables)	*Complement* (tax credits for R&D or production can complement GHG pricing)	*Essential* (expertise needed for new technologies)
CCS	<1	0–64	*Essential* (CCS requirement for all new coal plants and phase-in with existing)	*Essential* (GHG pricing is essential, but even this is unlikely to suffice in near term)	*Complement* (would help with first plants while GHG price is still low)	*Desired* (expertise needed for new technologies)[e]
Infrastructure[f]	260	310–500	*Essential* (security regulation critical for some aspects of reliability)	*Uncertain* (neutral effect)	*Essential* (customers must pay for reliability levels they value)	*Essential* (expertise needed for new technologies)
Access to electricity and cleaner cooking[g]	n.a.	36–41	*Essential* (ensure standardization but must not hinder development)	*Uncertain* (could reduce access by increasing costs of fossil fuel products)	*Essential* (grants for grid, micro-financing for appliances, subsidies for clean cookstoves)	*Essential* (create enabling environment: technical, legal, institutional, financial)

[a] Global investments into efficiency improvements for the year 2010 are not available. Note, however, that the best-guess estimate from Chapter 24 for investments into energy components of demand-side devices is by comparison about US$300 billion per year. This includes, for example, investments into the engines in cars, boilers in building heating systems, and compressors, fans, and heating elements in large household appliances. Uncertainty range is between US$100 billion and US$700 billion annually for investments in components. Accounting for the full investment costs of end-use devices would increase demand-side investments by about an order of magnitude.

[b] Estimate includes efficiency investments at the margin only and is thus an underestimate compared with demand-side investments into energy components given for 2010 (see note a).

[c] Efficiency improvements typically require a basket of financing tools in addition to subsidies, including, for example, low- or no-interest loans or, in general, access to capital and financing, guarantee funds, third-party financing, pay-as-you-save schemes, or feebates, as well as information and educational instruments such as labeling, disclosure and certification mandates and programs, training and education, and information campaigns.

[d] Lower-bound estimate includes only traditional deployment investments in about 2 GW capacity additions in 2010. Upper-bound estimate includes, in addition, investments for plants under construction, fuel reprocessing, and estimated costs for capacity lifetime extensions.

[e] Note the large range of required investments for CCS and nuclear in 2010–2050. Depending on the social and political acceptability of these options, capacity building may become essential for achieving the high estimate of future investments.

[f] Overall electricity grid investments, including investments for operations and capacity reserves, back-up capacity, and power storage.

[g] Annual costs for almost universal access by 2030 (including electricity grid connections and fuel subsidies for cleaner cooking fuels).

dialogue practices. Capacity development has an important role to play in building mechanisms of support and capacities for interactive feedback, flexibility, and adaptive management and change. And because these traditional habits, practices, and norms are embedded in a broader social context, building capacities for dialogue at the local level is essential.

Market development and the role of feedback and flexibility at the local and project level are also essential in support of the diffusion of new energy technologies, but they are usually ignored in the design of capacity building initiatives. Also important is the need to build and strengthen capacities for local manufacture, repair, and distribution of new energy-related technologies, whether related to improved cookstoves, solar home systems or other forms of early energy-access initiatives, or to the introduction of more modern and decentralized forms of energy. Successful examples of energy technology development and diffusion also point to the need to develop and strengthen local research capacities, participating in collaborative R&D efforts and coordinating across sectors and disciplines.

Brazil's sustained research effort that led to its development of the biofuels industry and to multiple development goals ranging from energy-access improvements to lowering GHG emissions is a good example of this interaction and the success that it brings (see Chapter 25.6.1). Research and advisory services have also played an important role in the development of smallholder jatropha farms to produce oil for off-grid electricity production in Mali (Chapter 25.6.1). Other examples where bottom-up approaches have been critical to the successful introduction of new energy technologies include experiences in the introduction of small hydropower schemes in China and village power schemes in Bhutan (Chapter 25.6.2).

Because the need to transform energy systems applies to all economies – whether industrial, emerging, developing, or poorest – the new concept of capacity development for energy transitions in some of the examples just mentioned must also apply to all programs, whether they relate to small energy-access projects or major transitions and innovations across society and at the national level. The differences reside in the types of objectives and outcomes sought – ranging from countries where the main objective may be to attain the highest levels of cleaner, sustainable, and secure forms of energy to those where the goal is to provide access to cleaner and affordable modern forms of energy to the largest possible number of residents.

Making choices about transition pathways requires access to a wide range of knowledge and information as well as the capacities to use this knowledge in the policy process. Two new approaches have emerged recently from contemporary business practices that may have great relevance in future capacity development approaches. These have developed over the past several decades as production has become more knowledge-intensive, competition more globalized, and information technology more accessible to the population at large. These 'open innovation' and 'distributed innovation' systems require very different and complex approaches to capacity development, involving special skills for managing risks and for creating innovative partnerships that speed the development and diffusion of new energy technologies (Chapter 25.7).

Open innovation involves a network culture in which the world outside is used to generate knowledge inside, and knowledge flows in and out of the institution purposefully rather than at random. The main objective is to leverage existing knowledge rather than depend solely on intellectual property. Distributed innovation, in contrast, is more closely associated with the development of open source software such as Linux, but the innovation has spread and is being practiced in other fields, including the biosciences. In this case, existing practices are not just modified but disrupted. The innovation power comes from a collected set of individuals whose individual actions 'snap together' to create something new.

These approaches point to the importance of building very special capacities for networking and knowledge networks and for appreciating the increasing relevance of open and distributed systems. Brazil's systematic collaborative research since the early 1980s that led to the biofuels success and the Dutch use of 'transition platforms' to advance efforts toward a low-carbon economy, relying on bottom-up processes

and open networks involving business, the non-governmental sector, and government, illustrate the applicability of this approach for industrial as well as developing countries (Chapter 25.8.3). In these and other examples, the lesson is that access to information and the capacity to use such inputs are critical in making choices for energy transitions – for individual players, the community, or a national government.

But these new and emerging forms of knowledge networking, coupled with new and innovative forms of finance and technology research collaboration and development, require new and enhanced capacities for effective participation at the international level that many countries, particularly developing ones, do not have, or are not well-developed today. The increasingly complex and fast-paced world of energy and climate change finance is a good example of an area where present capacities fall far short of the needs. The recent climate change negotiations alone have generated pledges of fast-start finance up to 2012 of some US$30 billion and promises to work collaboratively so that this funding can grow to some US$100 billion by 2020.

This is only a small part of the overall investment projections needed to meet the growth in energy demand – some US$1.7–2.2 trillion per year are needed up to 2050. The world of energy finance has always been a large and complex market. The difference today is that it is becoming even more complex, with new and innovative instruments of finance, including the carbon market, and with countries demanding more attention to the need to develop, introduce, and diffuse new technologies. Under these conditions, a multi-goal approach can both speed the diffusion of new energy technologies as well as stimulate the development and energy transition processes in developing countries.

6   Conclusions

The world is undergoing severe and rapid change involving significant challenges. Although this situation poses a threat, it also offers a unique opportunity – a window of time in which to create a new, more sustainable, more equitable world, provided that the challenges can be addressed promptly and adequately. Energy is a pivotal area for actions to help address the challenges.

The interrelated world brought about by growth and globalization has increased the linkages among the major challenges of the 21st century. We do not have the luxury of being able to rank them in order of priority. As they are closely linked and interdependent, the task of addressing them simultaneously is imperative.

Energy offers a useful entry point into many of the challenges because of its immediate and direct connections with major social, economic, security, and development goals of the day. Among many other challenges, energy systems are tightly linked to global economic activities, to freshwater and land resources for energy generation and food production, to biodiversity and air quality through emissions of particulate matter and precursors of tropospheric ozone, and to climate change. Most of all, access to affordable and cleaner energy carriers is a fundamental prerequisite for development, which is why GEA places great emphasis on the need to integrate energy policy with social, economic, security, development, and environment policies.

The good news is that humanity has the resources, the ingenuity, and the technologies to create a better world. The bad news is that the lack of appropriate institutions, their interaction and integration, capacities, and governance structures makes the task difficult. Raising the level of political will to address some of these challenges could go a long way toward making significant progress in achieving multiple goals. This is a major task, however, given the tendency of current decision-making processes to aim for short-term, quick results. GEA endeavors to make a compelling case for the adoption of a new set of pathways – pathways that are essential, required urgently, and – most important – achievable.

GEA highlights essential technology-related requirements for radical energy transformation:

- significantly larger investment in energy efficiency improvements, especially end-use, across all sectors, with a focus on new investments as well as major retrofits;

- rapid escalation of investments in renewable energies: hydropower, wind, solar energy, modern bioenergy, and geothermal, as well as the smart and super grids that enable renewable energies to become the dominant sources of energy;

- reaching universal access to modern forms of energy and cleaner cooking through micro-financing and subsidies;

- use of fossil fuels and bioenergy at the same facilities for the efficient co-production of multiple energy carriers and chemicals;

- full-scale deployment of CCS; and

- on one extreme nuclear energy could make a significant contribution to the global electricity, but in the other, it could be phased out.

To meet humanity's need for energy services, comprehensive diffusion of advanced energy technologies and an increased contribution of energy efficiencies are required throughout the energy system – from energy collection and conversion to end-use. Rapid diffusion of renewable energies is the second, but equally most effective, option for reaching multiple objectives. Sustainable conversion to carriers such as electricity, hydrogen, and heat, along with smart transmission and distribution systems for the most important end-uses are crucial.

A major policy challenge is to resolve the current issue of split incentives, in the sense that those who would be paying for efficiency improvements and other energy investments are more oriented toward short-term rates of return than to the long-term profitability of the investments and, likewise, that they are rarely the beneficiaries of reduced energy bills and other public benefits.

GEA makes the case that energy system transformation is possible only if there is also an interactive and iterative transformation of the policy and regulatory landscape, thereby fostering a buildup of skills and institutions that encourage innovation to thrive, create conditions for business to invest, and generate new jobs and livelihood opportunities.

It is projected that, by mid-century, more than six billion people will live in urban environments. This underscores the importance for policymakers to consider the window of opportunity available in designing the urban landscape, specifically in terms of urban layout, transport structure, and individual buildings/structures and their energy use.

A major finding of GEA is that some energy options provide multiple benefits. This is particularly true of energy efficiency, renewables, and the co-production of synthetic transportation fuels, cooking fuels, and electricity with CCS, which offer advantages in terms of supporting all of the goals related to economic growth, jobs, energy security, local and regional environmental benefits, health, and climate change mitigation. All these advantages imply the creation of value. This value should be incorporated into the evaluation of these measures (and others) and in creating incentives for their use.

One implication of this is that nations and corporations can invest in efficiency and renewable energy for the reasons that are important to them, not just because of a global concern about, for example, climate change mitigation or energy security. But incentives for individual players to invest in options with large societal values must be strong and effective.

Finally, the GEA pathways describe the transformative changes needed to achieve development pathways toward a more sustainable future – a 'sustainable future' that simultaneously achieves normative goals related to the economic growth, energy security, health, and environmental impacts of energy conversion and use, including the mitigation of climate change.

In sum, GEA finds that attainment of a sustainable future for all is predicated on resolving energy challenges. This requires the creation of market conditions, via government interventions, that invite and stimulate investments in energy options that provide incentives for rapid investments in energy end-use and supply technologies and systems.

3 Cluster 1–4

CLUSTER 1

Chapter 1–6

Energy Primer

Lead Authors (LA)
Arnulf Grubler (International Institute for Applied Systems Analysis, Austria and Yale University, USA)
Thomas B. Johansson (Lund University, Sweden)
Luis Mundaca (Lund University, Sweden)
Nebojsa Nakicenovic (International Institute for Applied Systems Analysis and Vienna University of Technology, Austria)
Shonali Pachauri (International Institute for Applied Systems Analysis, Austria)
Keywan Riahi (International Institute for Applied Systems Analysis, Austria)
Hans-Holger Rogner (International Atomic Energy Agency, Austria)
Lars Strupeit (Lund University, Sweden)

Contributing Authors (CA)
Peter Kolp (International Institute for Applied Systems Analysis, Austria)
Volker Krey (International Institute for Applied Systems Analysis, Austria)
Jordan Macknick (National Renewable Energy Laboratory, USA)
Yu Nagai (Vienna University of Technology, Austria)
Mathis L. Rogner (International Institute for Applied Systems Analysis, Austria)
Kirk R. Smith (University of California, Berkeley, USA)
Kjartan Steen-Olsen (Norwegian University of Science and Technology)
Jan Weinzettel (Norwegian University of Science and Technology)

Review Editor
Ogunlade Davidson (Ministry of Energy and Water Resources, Sierra Leone)

Contents

1.1 Introduction and Roadmap

Life is but a continuous process of energy conversion and transformation. The accomplishments of civilization have largely been achieved through the increasingly efficient and extensive harnessing of various forms of energy to extend human capabilities and ingenuity. Energy is similarly indispensable for continued human development and economic growth. Providing adequate, affordable energy is a necessary (even if by itself insufficient) prerequisite for eradicating poverty, improving human welfare, and raising living standards worldwide. Without economic growth, it will also be difficult to address social and environmental challenges, especially those associated with poverty. Without continued institutional, social, and technological innovation, it will be impossible to address planetary challenges such as climate change. Energy extraction, conversion, and use always generate undesirable by-products and emissions – at a minimum in the form of dissipated heat. Energy cannot be created or destroyed – it can only be converted from one form to another, along a one-way street from higher to lower grades (qualities) of energy. Although it is common to discuss energy "consumption," energy is actually transformed rather than consumed.

This Energy Primer[1] aims at a basic-level introduction to fundamental concepts and data that help to understand *energy systems* holistically and to provide a common conceptual and terminological framework before examining in greater detail the various aspects of energy systems from challenges and options to *integrated* solutions, as done in the different chapters of the Global Energy Assessment (GEA). Different chapters will quite naturally emphasize different aspects and components of the global energy system, but they all share this basic common understanding of the importance of *integrating* all aspects related to energy into a common systems framework. Given the focus on assessing *current* energy systems as well as possible transformation pathways into *future* energy systems throughout this publication, the Energy Primer also aims at providing historical context that helps to understand how current energy systems have emerged and what characteristic rates of change are in these large-scale systems.

After an introduction and roadmap to Chapter 1 (Section 1.1), Section 1.2 introduces the fundamental concepts and terms used to describe global energy systems (Section 1.2.1) and then proceeds with an overview of the fundamental driver: the demand for *energy services* (Section 1.2.2), which is key in this assessment. Section 1.2.3 then summarizes the major links between energy services and primary energy resources at the global level for the year 2005. The section also contains a summary of major energy units and scales (with technical details given in Appendix 1.A).

Section 1.3 then turns to a historical perspective on energy transitions, covering both energy end-use demand and services (Section 1.3.1), as

well as energy supply (Section 1.3.2), and concludes with a brief introduction into the relationship between energy and economic growth (Section 1.3.3). A long historical perspective is important in understanding both the fundamental drivers of energy system transitions, as well as the constraints imposed by the typically slow rates of change in this large, capital-intensive system characterized by long-lived infrastructures (Grubler et al., 1999).

Section 1.4 then discusses the central aspect of energy efficiency, summarizing key concepts and measures of energy efficiency (Section 1.4.1), and estimates of global energy efficiencies based on the first (Section 1.4.2) and second law of thermodynamics (Section 1.4.3), as well as energy intensities (Section 1.4.4).

Section 1.5 provides a summary of key concepts (Section 1.5.1) and numbers of global *energy resources* that provide both key inputs and key limitations for energy systems. Fossil, fissile (Section 1.5.2), and renewable resources (Section 1.5.3) are covered comprehensively along with a basic introduction to energy densities, which are particularly critical for renewable energy (Section 1.5.4).

Section 1.6 provides a summary of major energy flows associated with production, use, and trade of energy (Section 1.6.2) and energy conversions (Section 1.6.3) that link energy resources to final energy demands. After an introduction and overview (Section 1.6.1), production, use, and trade of both direct (Section 1.6.2.1) and indirect "embodied" energy, (Section 1.6.2.2) are discussed, and all energy trade flows summarized in Section 1.6.2.3. The discussion of energy conversions is short, as it is dealt with in detail in the various chapters of this publication. After an introductory overview (Section 1.6.3.1), the electricity sector is briefly highlighted (Section 1.6.3.2).

Section 1.7 summarizes the main impacts of global energy systems on the environment in terms of emissions, including greenhouse gases (Section 1.7.2) and other pollutants where the energy sector plays an important role (Section 1.7.3). Emissions are central environmental externalities associated with all energy conversions.

Section 1.8 then complements the global synthesis of Chapter 1 by highlighting the vast heterogeneities in levels, patterns, and structure of energy use, by first introducing basic concepts and measures (Section 1.8.1), before addressing the heterogeneity across nations (Section 1.8.2), within nations (Section 1.8.3), as well as energy disparities (Section 1.8.4). This short section is of critical importance, especially in terms of a global assessment, as the inevitable top-down perspective involving Gigatonnes and Terawatts often glosses over differences in time, social strata, incomes, lifestyles, and human aspirations.

Section 1.9 provides a primer on basic economic concepts related to energy end-use and energy supply, using cooking in developing countries and electricity generation options as illustrative examples.

---

1   This text draws on, extends, and updates earlier publications by the authors including: Goldemberg et al., 1988; Nakicenovic et al., 1996b; 1998; Rogner and Popescu, 2000; Grubler, 2004; and WEA (World Energy Assessment), 2004.

Lastly, Section 1.10 leads into the full GEA, by providing an overview roadmap to the structure of GEA and its chapters.

Appendix 1.A returns to the rather technical, but nonetheless fundamental, aspect of units, scales, and energy accounting intricacies. This document uses uniformly the International System (SI) of (metric) units and has also adapted a uniform accounting standard for primary energy to achieve consistency and comparability across the different chapters. This is especially important in the energy field, that to date continues to use a plethora of vernacular units and accounting methods.

Appendix 1.B provides convenient summary tables of conversion and emission factors, and summarizes the various levels of regional aggregations used throughout GEA.

1.2 The Global Energy System

1.2.1 Description of the Global Energy System

The *energy system* comprises all components related to the production, conversion, and use of energy.

Key components of the energy system comprise: primary *energy resources* which are harnessed and converted to *energy carriers*[2] (such as electricity or fuels such as gasoline), which are used in end-use applications for the provision of *energy forms* (heat, kinetic energy, light, etc.) required to deliver final *energy services* (e.g., thermal comfort or mobility). The key mediator linking all energy conversion steps from energy services all the way back to primary resources are energy conversion *technologies*. Energy systems are often further differentiated into an *energy supply* and an *energy end-use* sector. The energy supply sector consists of a sequence of elaborate and complex processes for extracting energy resources, for converting these into more desirable and suitable forms of *secondary energy*, and for delivering energy to places where demand exists. The part of the energy supply sector dealing with primary energy is usually referred to as "*upstream*" activities (e.g., oil exploration and production), and those dealing with secondary energy as "*downstream*" activities (e.g., oil refining and gasoline transport and distribution). The energy end-use sector provides energy services such as motive power, cooking, illumination, comfortable indoor climate, refrigerated storage, and transportation, to name just a few examples. The purpose of the entire energy system is the fulfillment of demand for energy services in satisfying human needs.

Figure 1.1 illustrates schematically the architecture of the energy system as a series of linked stages connecting various energy conversion and transformation processes that ultimately result in the provision of goods and services. A number of examples are given for energy extraction, treatment, conversion, distribution, end-use (final energy), and energy services in the energy system. The technical means by which each stage is realized have evolved over time, providing a mosaic of past evolution and future options (Nakicenovic et al., 1996b).

Primary energy is the energy that is embodied in resources as they exist in nature: chemical energy embodied in fossil fuels (coal, oil, and natural gas) or biomass, the potential kinetic energy of water drawn from a reservoir, the electromagnetic energy of solar radiation, and the energy released in nuclear reactions. For the most part, primary energy is not used directly but is first converted and transformed into *secondary energy* such as electricity and fuels such as gasoline, jet fuel, or heating oil which serve as energy carriers for subsequent energy conversions or market transactions (Nakicenovic et al., 1996b).

Final energy ("delivered" energy) is the energy transported and distributed to the point of retail for delivery to final users (firms, individuals, or institutions). Examples include gasoline at the service station, electricity at the socket, or fuel wood in the barn. Final energy is generally exchanged in formal monetary market transactions, where also typically energy taxes are levied. An exception are so-called non-commercial fuels – i.e., fuels collected by energy end-users themselves such as fuel wood or animal wastes, which constitute important energy sources for the poor.

The next energy transformation is the conversion of final energy in end-use devices such as appliances, machines, and vehicles into *useful energy* such as the energy forms of kinetic energy or heat. Useful energy is measured[3] at the crankshaft of an automobile engine, by the mechanical energy delivered by an industrial electric motor, by the heat of a household radiator or an industrial boiler, or by the luminosity of a light bulb. The application of useful energy provides *energy services* such as a moving vehicle (mobility), a warm room (thermal comfort), process heat (for materials manufacturing), or light (illumination).

Energy services are the result of a combination of various technologies, infrastructures (capital), labor (know-how), materials, and energy forms and carriers. Clearly, all these input factors carry a price tag and, within each category, are in part substitutable for one another. From the consumer's perspective, the important issues are the quality and cost of energy services. It often matters little what the energy carrier or the "upstream" primary energy resource was that served as input. It is fair to say that most consumers are often unaware of the upstream activities of the energy system. The energy

---

2   In the literature (e.g. Rosen, 2010, Scott, 2007, Escher, 1983) also the term *energy currency* is used to highlight the fact that different energy carriers are to a degree interchangeable and can be converted to whatever form is most suitable for delivering a given energy service task. Like monetary currencies, energy currencies are also exchangeable (at both an economic and [conversion] efficiency price). In this assessment, the term energy carrier is used throughout. A concise compendium of energy-related concepts and terms is given in Cleveland and Morris, 2006.

---

3   Useful energy can be defined as the last measurable energy flow before the delivery of energy services.

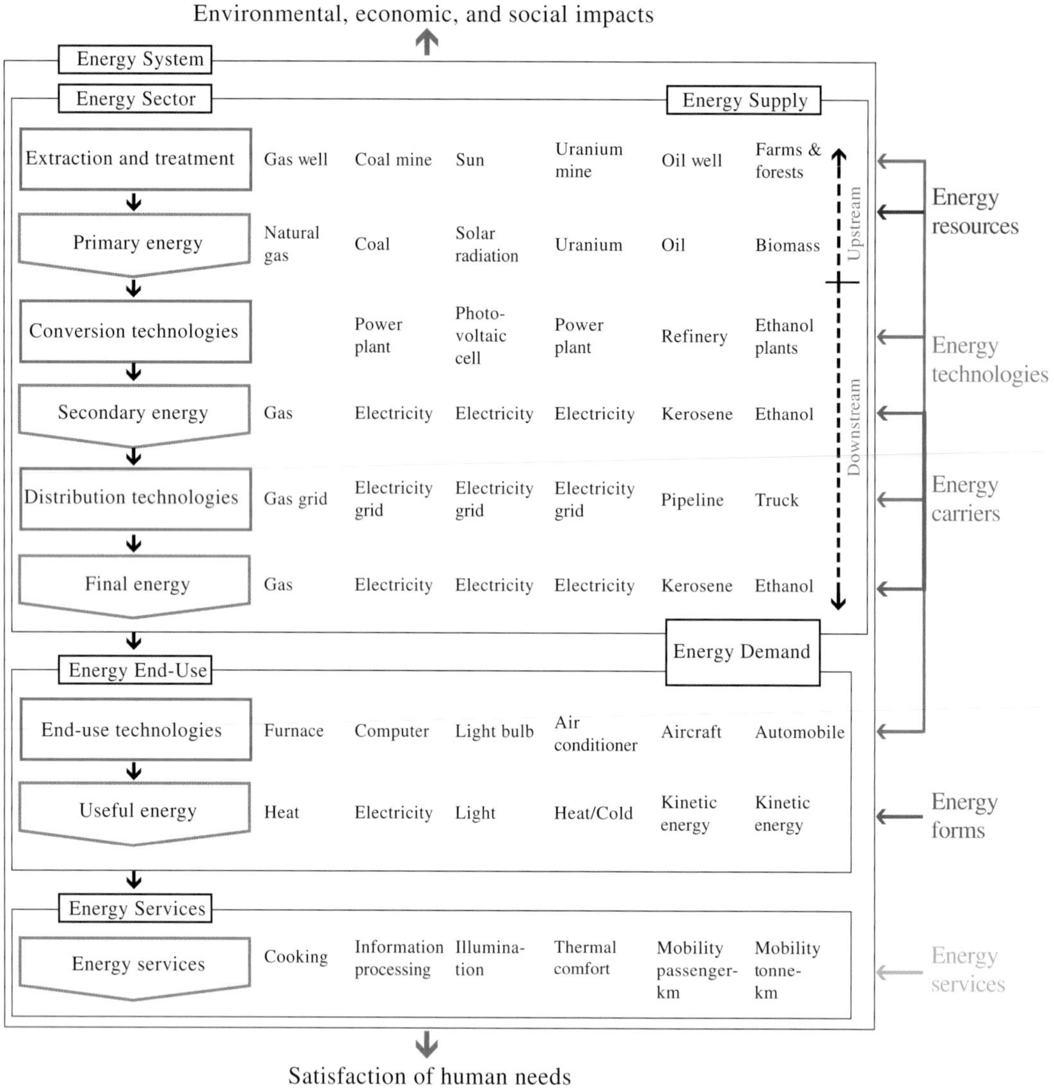

Environmental, economic, and social impacts
Energy System
Energy Sector
Energy Supply
Extraction and treatment
Gas well
Coal mine
Sun
Uranium mine
Oil well
Farms & forests
Primary energy
Natural gas
Coal
Solar radiation
Uranium
Oil
Biomass
Upstream
Energy resources
Conversion technologies
Power plant
Photo-voltaic cell
Power plant
Refinery
Ethanol plants
Energy technologies
Secondary energy
Gas
Electricity
Electricity
Electricity
Kerosene
Ethanol
Downstream
Distribution technologies
Gas grid
Electricity grid
Electricity grid
Electricity grid
Pipeline
Truck
Energy carriers
Final energy
Gas
Electricity
Electricity
Electricity
Kerosene
Ethanol
Energy Demand
Energy End-Use
End-use technologies
Furnace
Computer
Light bulb
Air conditioner
Aircraft
Automobile
Useful energy
Heat
Electricity
Light
Heat/Cold
Kinetic energy
Kinetic energy
Energy forms
Energy Services
Energy services
Cooking
Information processing
Illumina-tion
Thermal comfort
Mobility passenger-km
Mobility tonne-km
Energy services
Satisfaction of human needs

Figure 1.1 | The energy system: schematic diagram with some illustrative examples of the energy sector and energy end use and services. The energy sector includes energy extraction, treatment, conversion, and distribution of final energy. The list is not exhaustive and the links shown between stages are not "fixed"; for example, natural gas can also be used to generate electricity, and coal is not used exclusively for electricity generation. Source: adapted from Nakicenovic et al., 1996b.

system is *service driven* (i.e., from the bottom-up), whereas energy flows are driven by resource availability and conversion processes (i.e., from the top-down). Energy flows and driving forces interact intimately. Therefore, the energy sector should never be analyzed in isolation: it is not sufficient to consider only how energy is supplied; the analysis must also include how and for what purposes energy is used (Nakicenovic et al., 1996b).

Figure 1.2 illustrates schematically the major energy flows through the global energy system across the main stages of energy transformation, from primary energy to energy services, with typical examples. For an exposition of energy units see Box 1.1 below and Appendix 1.A.

Energy
Examples
Primary 496 EJ
Crude oil
Coal
144 EJ
Conversion
Refinery
Power Plant
Secondary 352 EJ
Gasoline
Electricity
22 EJ
Distribution
Truck
Grid
Final 330 EJ
Gasoline
Electricity
161 EJ
End use
Car
Light Bulb
Useful 169 EJ
Kinetic
Radiant
169 EJ
Services
Passenger-km
Light
496 EJ
Waste and rejected energy

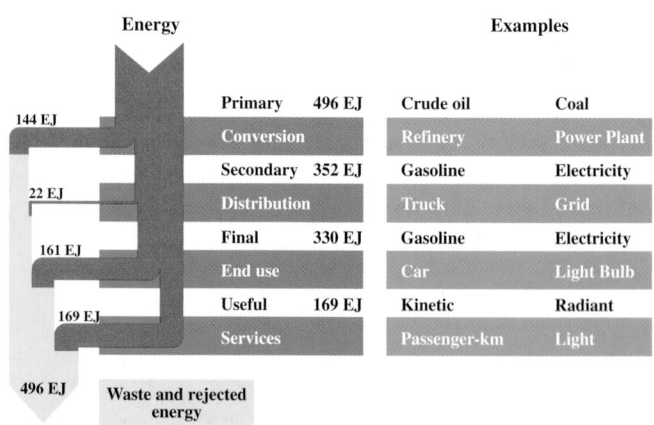

Figure 1.2 | Global energy flows of primary to useful energy, including conversion losses (waste and rejected energy), in EJ for 2005. Source: adapted from Nakicenovic et al., 1998, based on IEA, 2007a; 2007b; 2010.

Box 1.1 | Energy Units and Scales

Energy is defined as the capacity to do work and is measured in joules (J), where 1 joule is the work done when a force of 1 Newton ($1\ N = 1\ kg\ m/s^2$) is applied over a distance of 1 meter. Power is the rate at which energy is transferred and is commonly measured in watts (W), where 1 watt is 1 joule/second. Newton, joule, and watt are defined as basic units in the International System of Units (SI).[4]

Figure 1.3 gives an overview of the most commonly used energy units and also indicates typical (rounded) conversion factors. Next to the SI units, other common energy units include kilowatt-hour (kWh), used to measure electricity and derived from the joule (1 kWh – 1000 Watt-hours – being equivalent to 3600 kilo-Watt-seconds, or 3.6 MJ). In many international energy statistics (e.g., by the IEA and OECD) tonnes of oil equivalent (1 *toe* equals 41.87×10^9 J) are used. Some national energy statistics (e.g., in China and India) report tonnes of coal equivalent (1 *tce* equals 29.31×10^9 J).

The energy content of combustible energy resources (fossil fuels, biomass) is expressed based on either the so-called higher (HHV) or lower heating value (LHV). For non-combustible energy resources (nuclear, hydropower, wind energy, etc.) different conventions exist to convert those into primary energy equivalents. (For a detailed discussion see Appendix 1.A). In this publication non-combustible energies are accounted for using the so-called substitution equivalent method, with 1 kWh of nuclear/renewable electricity equivalent to some 3 kWh of primary energy equivalent, based on the current global average conversion efficiency of 35%. Combustible energies are reported based on the LHV of fuels.

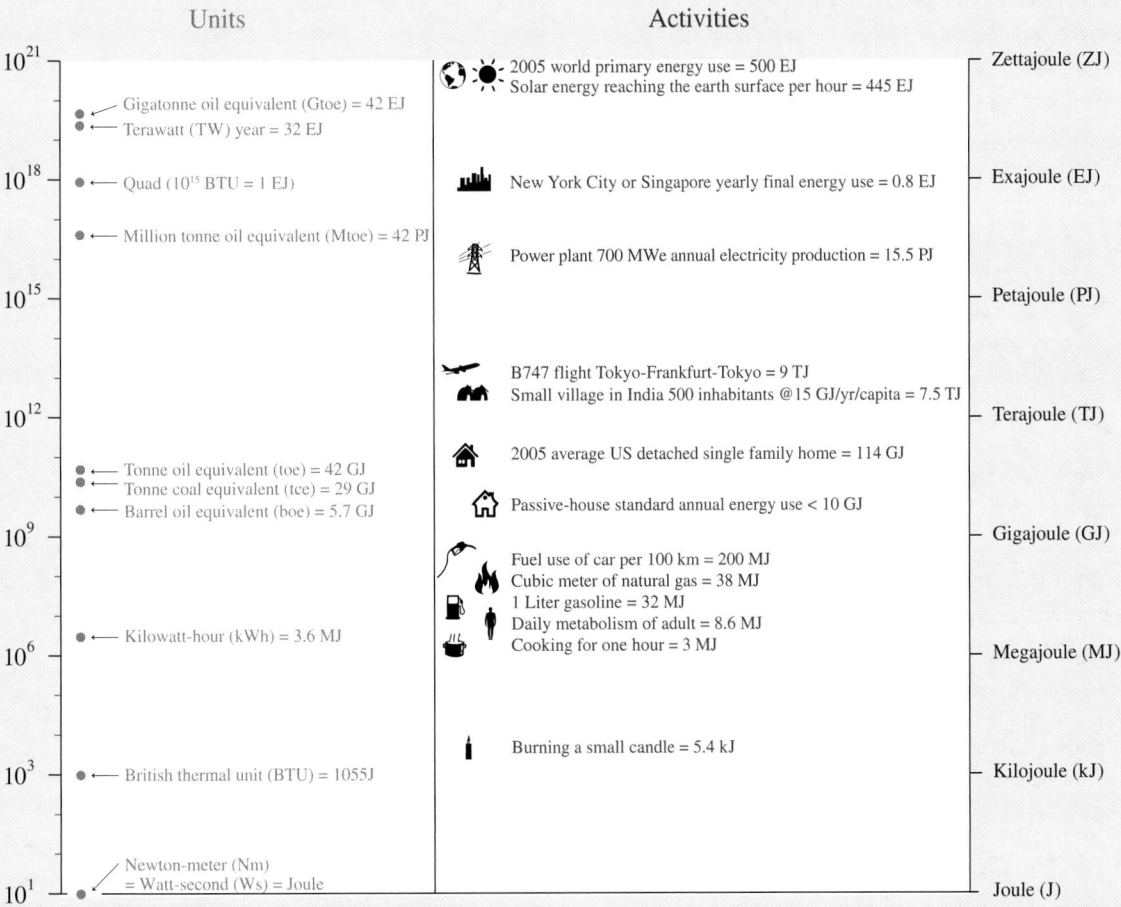

Figure 1.3 | Illustrative examples of energy units and scales used in the GEA.

---

4   International System of Units, *SI* from the French *le Système international d'unités*

1.2.2   Energy Services

Despite the centrality of energy services for the energy system, their measurement and statistical reporting is sparse. As the different types of energy services – from passenger and goods transport to illumination, to materials produced and recycled, to information communicated – are so diverse, activity levels are non-commensurable (i.e., cannot be expressed in common units). Hence energy service levels are often assessed via their required energy *inputs* (useful, final, or primary energy) rather than by their actual *outputs*. This can distort the picture quite substantially, as those energy services with the lowest conversion efficiency (and thus highest proportional energy inputs) are over-weighted in the energy accounts. Measuring services via inputs rather than outputs can also significantly mask the enormous efficiency gains which have historically characterized technological change in energy end-use applications (from candles to white diode lighting, or from horses to electric vehicles), and which generally go unnoticed in long-term estimates of economic productivity and welfare growth (see Nordhaus, 1998).

A notable global assessment of energy service provision is given by Cullen and Allwood (2010) and summarized in Table 1.1 below. The assessment used primary energy as a common energy metric, which is problematic for energy services due to the ambiguities of primary energy accounting conventions (see Appendix 1.A). Using primary energy inputs to characterize energy services also gives greater weight to lesser efficient energy service provision chains. A passenger-km traveled by car is accounted and weighted for by its much larger primary energy inputs (crude oil) compared to a passenger-km traveled by bicycle (food caloric intake). The multitude of energy services summarized here can be conveniently grouped into three broad categories and are assessed in separate chapters in this publication: Industry (Chapter 8), Transportation (Chapter 9), and Buildings (Chapter 10), which are the physical structures in which the remainder of energy services are provided.

It is useful to put these rather abstract engineering-type summary estimates of energy service levels into perspective – for example, on a per capita basis for a global population of 6.5 billion in 2005. These illustrative global average levels of energy service provision should not distract from the vast heterogeneity in levels of energy service provision between rich and poor, or between urban and rural populations (see Section 1.8 below).

Transport: The 46 trillion tonne-km and 32 trillion passenger-km translate into a daily average mobility of some 13 km/day/person, and transporting on average 1 tonne/day per capita over a distance of some 20 km.

Industry: The structural materials summarized in Table 1.1 translate in absolute terms into close to 2 billion tonnes (Gt) of cement, 1 Gt of crude steel, some 0.3 Gt of fertilizer, 0.1 Gt of non-ferrous metal ores processed, and over 50 million tonnes of plastics produced per year (UN, 2006a, 2006b). Estimates of the global total material flows reveal a staggering magnitude of the industrial metabolism (Krausmann et al., 2009). In terms

Table 1.1 | Estimated levels of energy services and corresponding shares in primary energy per service type for the year 2005.

Energy service	2005 levels	Units	As a percentage of pro-rated primary energy use (including upstream conversion losses)
Thermal comfort	30	10^{15} m^{3}K (degree-volume air)	19%
Sustenance (food)	28	10^{18} J (food)	18%
Structural materials	15	10^9 MPa$^{2/3}$m^3 (tensile strength × volume)	14%
Freight transport	46	10^{12} ton-km	14%
Passenger transport*	32	10^{12} passenger-km	14%
Hygiene	1.5	10^{12} m^{3}K (temperature degree-volume of hot water)	11%
	2.8	10^{18} Nm (work)	
Communication	280	10^{18} bytes	6%
Illumination	480	10^{18} lumen-seconds	4%

* The original passenger transport data have been corrected by adding non-reported categories provided in Chapter 9.

Source: adapted from Cullen and Allwood, 2010.

of tonnage, humankind uses each year (values for 2005) some 12 Gt of fossil energy resources, some 6 Gt of industrial raw materials and metals (ores and minerals), 23 Gt of construction materials (sand, gravel, etc.), and an additional 19 Gt of biomass (food, energy, and materials), for a total material mobilization of approximately 60 Gt/year, or more than 9 tonnes/year per capita on average. The use of around 10 Gt of energy thus enables the "leverage" of the mining, processing, refinement, and use of an additional 50 Gt of materials.

Buildings: The size of the residential and commercial building stock worldwide (2005 data) whose internal climate needs to be maintained through heating and cooling energy services is estimated to be about 150 billion m^2 (including some 116 billion m^2 residential and 37 billion m^2 commercial floorspace, see Chapter 10) which corresponds to approximately 20 m^2 per person on average.

Useful energy as a common energy input denominator minimizes distortions among different energy service categories, as it most closely measures the actual energy service provided. Chapter 1 has, therefore, produced corresponding useful energy estimates based on the 2005 energy balances published by the International Energy Agency (IEA, 2007a and 2007b) using typical final-to-useful conversion efficiencies available in the literature (Eurostat, 1988; Rosen, 1992; Gilli et al., 1996; BMME, 1998; Rosen and Dincer, 2007). This method has some drawbacks, as the available energy balances are based on an economic sectoral perspective, which does not always perfectly correspond with

Table 1.2 | Energy service levels, world in 2005, as estimated by their corresponding useful and final energy inputs (in EJ, and as share of total; see also Footnote 5).

Energy service	Final energy [EJ]	As percentage of total final energy [%]	Useful energy [EJ]	As percentage of total useful energy [%]
Transport				
Road	66.9	20.3	13.7	8.1
Rail	2.3	0.7	1.1	0.7
Shipping	9.0	2.7	3.0	1.8
Pipelines	2.9	0.9	0.9	0.5
Air	10.3	3.1	3.0	1.8
Total transport	91.4	27.7	21.7	12.9
Industry				
Iron and steel	14.4	4.4	11.5	6.8
Non-ferrous metals	4.0	1.2	1.9	1.1
Non-metallic minerals	11.1	3.4	4.5	2.7
Other	58.7	17.8	44.3	26.3
Total industry	88.2	26.8	62.2	36.9
Other sectors				
Feedstocks	30.2	9.2	25.0	14.8
Agriculture, forestry, fishery	7.5	2.3	3.0	1.8
Residential	81.0	24.6	35.6	21.1
Commercial and other	31.4	9.5	21.0	12.5
Total other sectors	150.1	45.5	84.6	50.2
Grand Total	**329.7**	100.0	**168.5**	100.0

Source: final energy: data from IEA, 2007a and 2007b; useful energy: Chapter 1 estimation.

particular energy service types.[5] It needs to be emphasized that different forms of useful energy (such as thermal versus kinetic energy) are not interchangeable, even when they are expressed in a common energy unit and aggregated. Global totals for useful and final energy inputs per energy service category are summarized in Table 1.2 (see also Figure 1.5 below), with regional details given in Figure 1.6 below.

The largest category of energy service demands arise in **industry** (62 EJ of useful energy in 2005), with the dominant energy service application being (high-temperature) industrial process heat associated with the processing, manufacturing, and recycling of materials. *Feedstocks* refer to non-energy uses of energy, where energy carriers serve as a raw material (e.g., natural gas used for the manufacture of fertilizers), rather than as an input to energy conversion processes proper. Feedstocks are also

associated with industrial activities (the chemical sector) and add another 25 EJ of useful energy to the 62 EJ of industrial energy service demands.

The residential and commercial sectors (some 57 EJ of useful energy in 2005) are dominated by the energy use associated with **buildings**, both in maintaining a comfortable indoor climate (heating and air conditioning), as well as various energy services performed *within* buildings such as cooking, hygiene (hot water), and the energy use of appliances used for entertainment (televisions) or communication (computers, telephones). Agriculture, forestry, and fisheries are comparatively minor in terms of useful energy (3 EJ) and are only summarily included in the "other sectors" category here.

Transport is comparatively the smallest energy service category when assessed in terms of useful energy, with an estimated level of 22 EJ (some 13% of total useful energy, but due to low conversion efficiencies, some 28% in total primary energy, see Table 1.1 above). Road transportation (cars, two- and three-wheelers, buses, and trucks) are the dominant technologies for providing mobility of people and goods. Due to the low final-to-useful conversion efficiency associated with internal combustion engines (some 20% only, with 80% lost as waste heat of engines and associated with friction losses of drive trains), road transport accounts for only 8% of useful energy but for approximately 20% of total final energy. This example once more highlights the value of an energy service perspective (Haas et al., 2008) on the energy system, by looking at service outputs rather than final or primary energy inputs that overemphasize the least efficient energy end-use applications. Nonetheless, it needs to be noted (see the discussion below) that transportation is one of the fastest growing energy demand categories. This adds further emphasis on efforts to improve transport energy efficiency, which has both technological (more efficient vehicles), as well as behavioral and lifestyle dimensions (changing mobility patterns, shifts between different transport modes – e.g., by using public transportation or bicycles instead of private motorized vehicles).

Global trends since 1971 for different energy service categories and in measuring final energy inputs are shown in Figure 1.4.

1.2.3   From Energy Services to Primary Energy

Figure 1.5 illustrates the interlinkages of global energy flows from useful energy up to the level of primary energy, and also shows major energy carriers and transformations. Different primary energies require different energy system structures to match the demand for type and quality of energy carriers and energy forms with available resources.

As a result, there is great variation in the degree and type of energy conversions among different fuels in the global energy system. At the one extreme, biomass is largely used in its originally harvested form and burned directly without intervening energy conversions. At the other extreme are nuclear, hydropower, and modern renewables that are not used in their original

---

5   For instance, transport energy use is reported by mode of transport (road, rail, sea, air) in the underlying IEA statistics, which does not allow differentiation between passenger and goods transport.

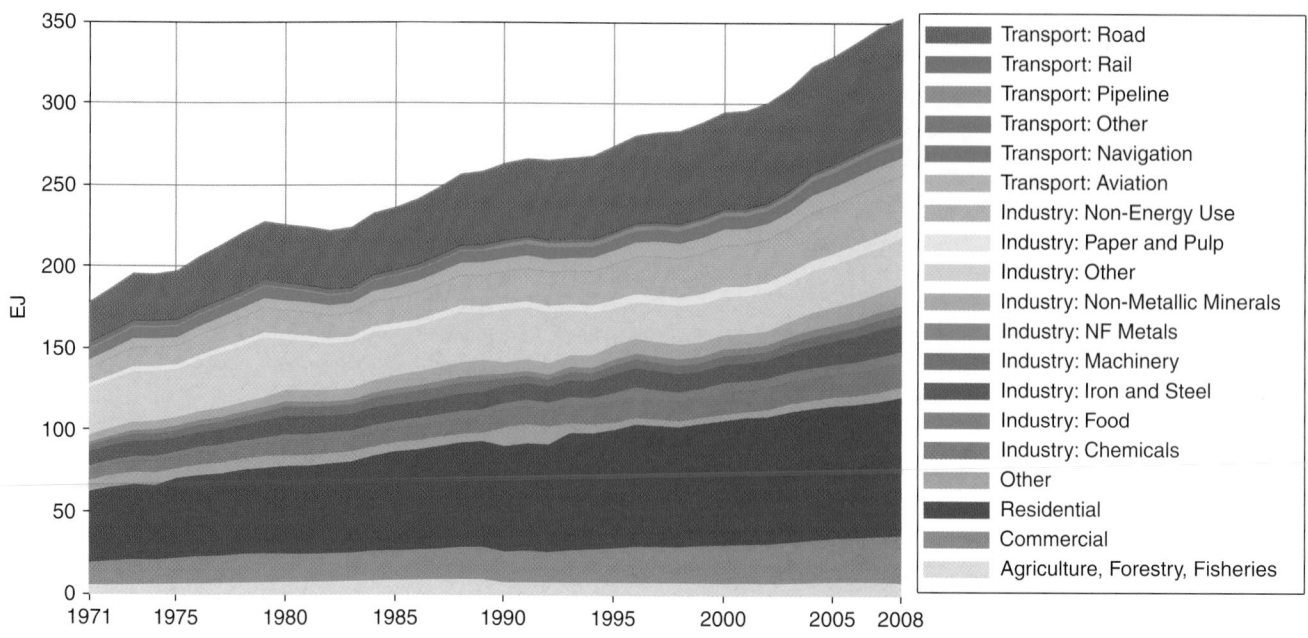

Figure 1.4 | Global final energy input into different energy services categories since 1971 (in EJ), by major energy service category. Source: adapted from IEA, 2010.

resource state but converted into electricity. Electricity is the energy carrier with the highest versatility of providing different energy forms required for various energy services (heat, light, mobility). Crude oil also needs to be converted (refined) to the liquid fuels required for energy end-uses (gasoline, diesel, kerosene, for cars, trucks, and aircraft), or for further secondary energy conversions (e.g., fuel oil-fired power plants generating electricity). Coal is a major input for electricity generation and for specific industrial uses (metallurgy) but is not often used in direct form outside these two applications (remaining uses for residential heating/cooking are declining rapidly due to air pollution concerns). Conversely, natural gas is a major energy carrier directly used as final energy and for end-uses, mainly due to its convenience (grid delivered, no combustion ashes to dispose of) and cleanliness. Natural gas is also increasingly being used in electricity generation, where the advent of highly efficient combined cycle power plants with flat economies of scale (i.e., costs per MW capacity are not significantly different across different plant sizes) allows fast construction of modular units. Due to the low emission characteristics of these highly efficient conversion processes, plants can also often be located in high demand density areas, thus opening up the possibility of using waste heat from electricity generation for industrial and residential customers, a scheme known as *cogeneration* or combined heat and power production (CHP).

From an energy systems perspective, the electricity sector assumes a special role (also the reason why it is discussed in greater depth in Section 1.6.3 on Energy Conversions below.) Electricity generation is the energy conversion process that can accommodate the greatest diversity of primary energy inputs. As shown in Figure 1.5, all primary energy carriers enter to different degrees into electricity generation, from biomass, to all fossils, nuclear, hydro, and new renewables. Electricity is also a very specific energy carrier: its absolute cleanliness at the point of end-use (not necessarily at the point of electricity generation, however)

and its high energy quality translate into the greatest versatility and flexibility in delivering whatever type of energy form and energy service required. However, electricity cannot be stored easily, which means that generation needs to follow the inevitable intertemporal variations of electricity demand over the seasons, during the day, even during minute-intervals.[6]

Overall, there is great variation in energy systems structures across different regions as a result of differences in the degree of economic development, structure of energy demand, and resource availability, among others. These differences are summarized at the level of useful, final, and primary energy respectively for the 5 GEA regions and the world in Figure 1.6.

1.3 Historic Energy Transitions

1.3.1 Transitions in Energy End-Use (United Kingdom)

Levels and structure of energy services have changed dramatically since the onset of the Industrial Revolution, reflecting population and income growth and, above all, technological change. Due to the "granular" nature of energy services, the measurement intricacies discussed above, and the traditional focus of energy statistics on (primary) energy supply, it is not possible to describe long-term transition in energy services and

---

6   The variation in electricity demand over time is enshrined in the concept of *load curves* that describe the instantaneous use of electric power (in Watts or typically rather GW) over time (on a daily, weekly, or monthly basis). A cumulative load curve over all of the 8760 hours of a year, sorted by declining GW load, yields a *load duration curve* (or cumulative load curve) that helps to design a whole electricity system and to dimension different types of power plants used for *peak*, *intermediate*, and *base load* electricity generation.

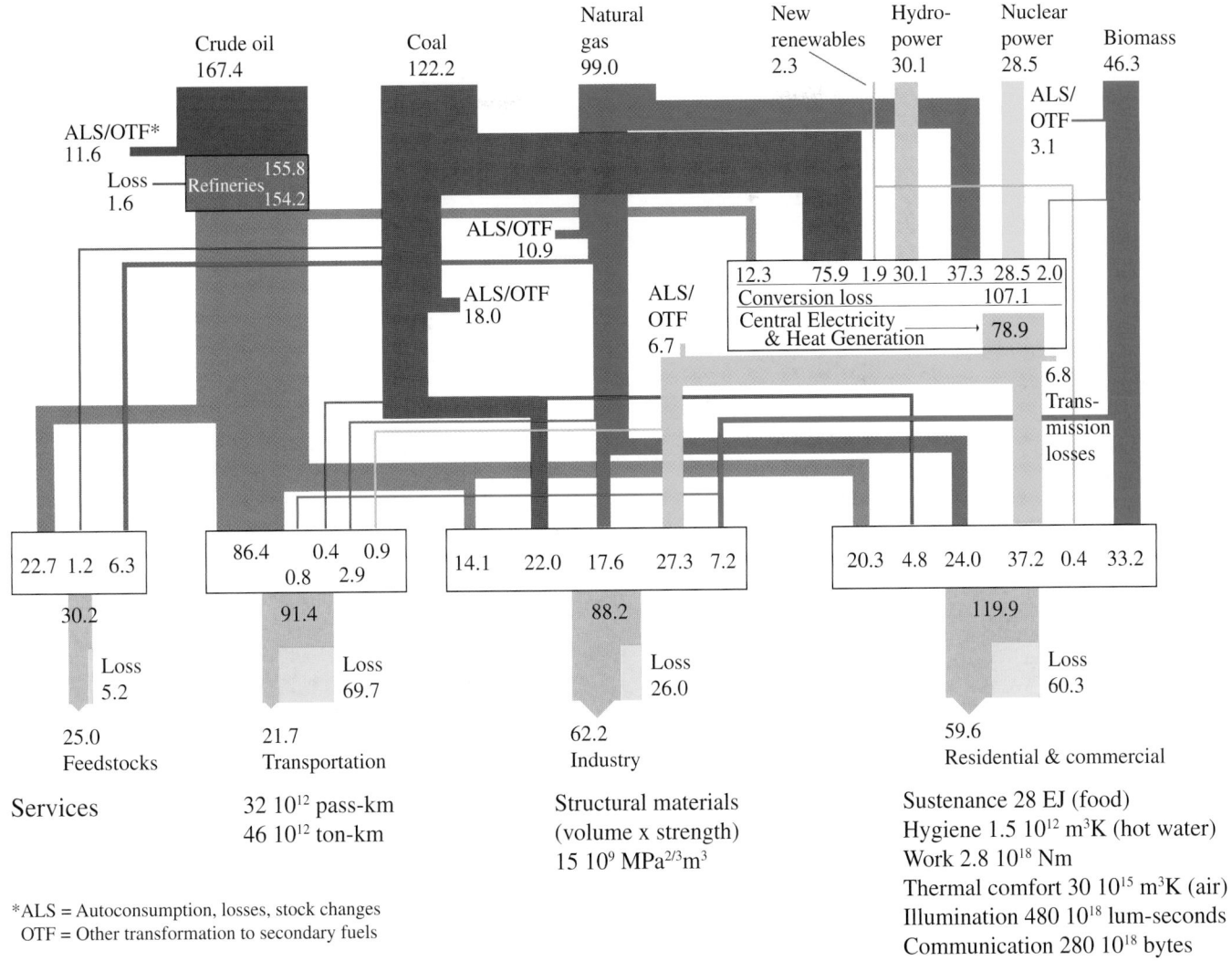

Figure 1.5 | Global energy flows (in EJ) from primary to useful energy by primary resource input, energy carriers (fuels) and end-use sector applications in 2005. Source: data from IEA, 2007a; b[7] (corrected for GEA primary energy accounting standard), and Cullen and Allwood, 2010.

7   Readers should note some small differences (1–5%) between the 2005 base year energy flows reported in Chapter 1 and the ones derived from Chapter's 17 scenario modeling and reported in the GEA Scenario Data Base. Whereas Chapter 1 is based exclusively on statistics as reported by the International Energy Agency (IEA), Chapter 17 and the GEA Scenario data base also include data revisions and draw at times different system boundaries for the accounting of energy flows, in line with standard energy modeling practices.

The largest global differences are for final energy (330 vs. 315 EJ in Chapters 1 and 17 respectively) related to: (a) new improved estimates of non-commercial energy use based on household surveys that have revised downwards the IEA statistics on residential, traditional biomass use; and (b) different accounting of energy use for pipeline transportation, and bunker fuels for international shipping which are accounted in Chapter 17 as energy-sector auto-consumption and at the international level only whereas in Chapter 1 they are accounted at the national and regional levels as final transport energy use. Differences in global primary energy are smaller (496 vs. 489 EJ between Chapters 1 and 17 respectively, for the same reasons as outlined above).

These small energy accounting differences are within the inevitable uncertainty range of international energy statistics (for a review see Appendix 1.A) and do not diminish the coherence of this Assessment.

energy end-use on the global scale. Long-term detailed national-level analyses are available for the United States (Ayres et al., 2003) and the United Kingdom (Fouquet, 2008), as well as (for shorter time horizons) in the form of useful energy balances for Brazil (BMME, 1998).

The long-term evolution and transitions in energy end-use and energy services is described below for the United Kingdom over a time period of 200 years. The United Kingdom is used as an illustrative example, not only due to the level of detail and time horizon of the original data available, but particularly because of its history of being the pioneer of the Industrial Revolution, which thus illustrates the interplay of industrialization, income growth, and technological change as drivers in energy end-use transitions.

Figure 1.7 illustrates the growth in energy service provision for the United Kingdom since 1800 by expressing the different energy services in terms of their required final energy inputs. Three main periods can be distinguished:

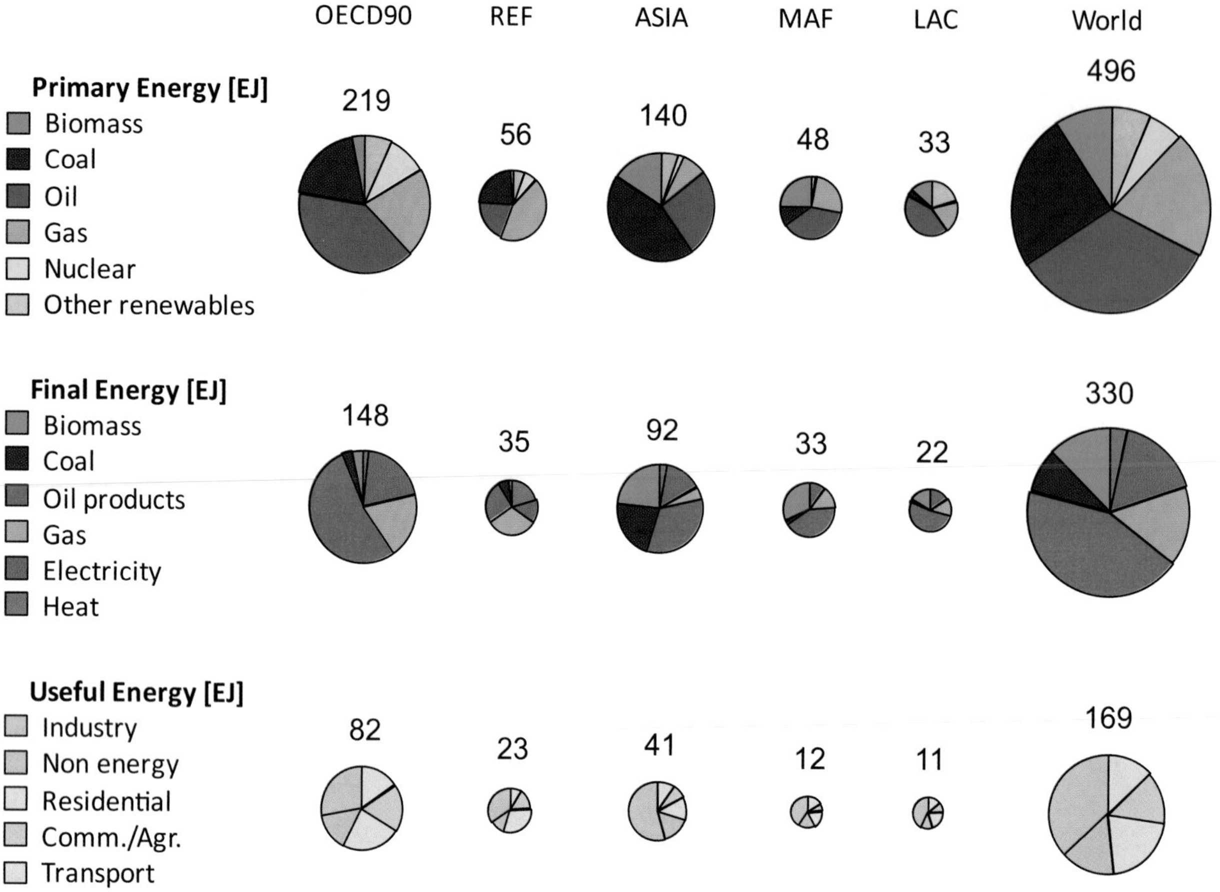

OECD90
REF
ASIA
MAF
LAC
World
Primary Energy [EJ]
Biomass
Coal
Oil
Gas
Nuclear
Other renewables
219
56
140
48
33
496
Final Energy [EJ]
Biomass
Coal
Oil products
Gas
Electricity
Heat
148
35
92
33
22
330
Useful Energy [EJ]
Industry
Non energy
Residential
Comm./Agr.
Transport
82
23
41
12
11
169

Figure 1.6 | World energy use: primary energy (by fuel), final energy (by energy carrier), and useful energy (by sector/type of energy service) for the world and five GEA regions for 2005 (in EJ). Source: based on IEA, 2007a and 2007b (corrected for GEA primary accounting standard, see also Footnote 5, above). For a definition of the GEA regions, see Appendix 1.B.

- a regular expansion of energy services in the 19th century that characterized the emergence of the United Kingdom as a leading industrial power, in which growth is dominated by industrial energy service demands and to a lesser degree by rapidly rising transportation services enabled by the introduction of steam-powered railways;

- a period of high volatility as a result of cataclysmic political and economic events (World War I, the Great Depression of 1929, and World War II) that particularly affected industrial production and related energy services; and

- a further (more moderated) growth phase after 1950, again punctuated by periods of volatility, such as the energy crisis of the 1970s characterized by the gradual decline of industrial energy services, compensated by strong growth in passenger transportation resulting from the diffusion of petroleum-based collective, and individual transport technologies (buses, aircraft, and cars).

Light
Freight-transport
Passenger-transport
(Mechanical) power
Heat-industry
Heat-domestic
EJ
8
6
4
2
1800
1850
1900
1950
2000

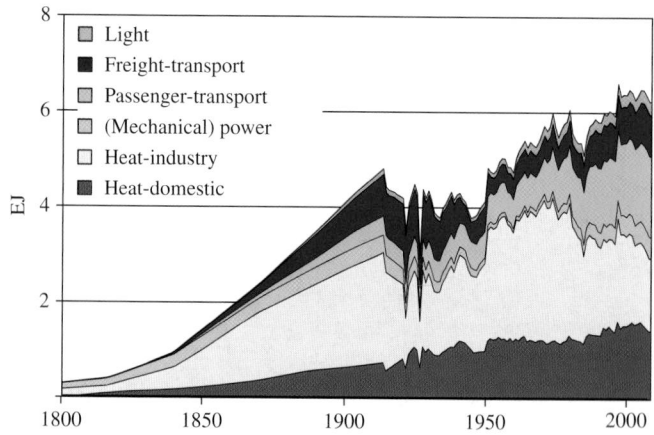

Figure 1.7 | Growth in energy service demand (measured by final energy inputs) United Kingdom since 1800, in EJ. Source: data from Fouquet, 2008. Updates after 2000 and data revisions courtesy of Roger Fouquet, Basque Centre for Climate Change, Bilbao, Spain.

At present, levels of energy services appear saturated at a level of above 6 GJ, or 100 GJ of final energy input equivalent per capita. Industry (with an ever declining share) accounts for about 30% of all energy services, residential applications (with a stable share) for another 30%, and transportation (with an ever growing share) for about 40% of total energy services.

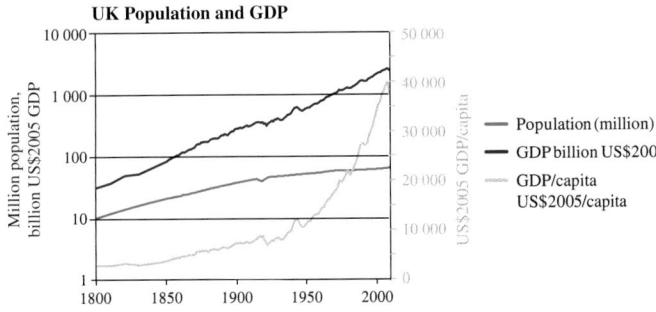

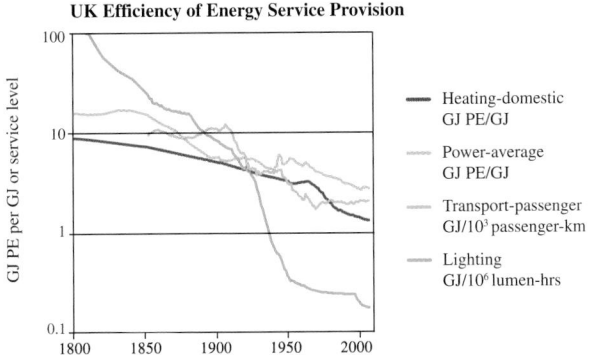

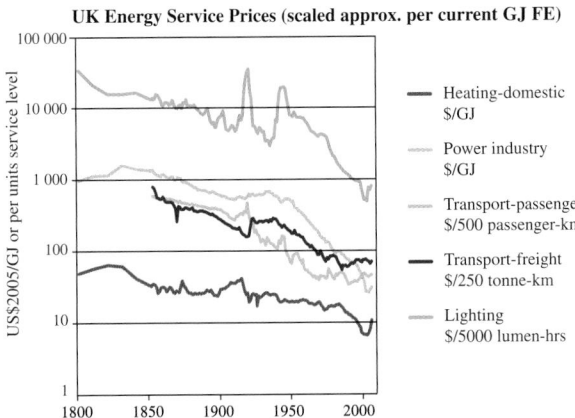

Figure 1.8 | Drivers of UK energy service demand growth: population, GDP and income per capita (panel 1); efficiency of energy service provision (per GJ service demand or service activity level – panel 2); and prices of energy services (per GJ service demand or activity level, activity level units have been normalized to approximately equal one GJ of current final energy use – panel 3). Source: data from Fouquet, 2008. Updates after 2000 and data revisions courtesy of Roger Fouquet, Basque Centre for Climate Change, Bilbao, Spain.

Figure 1.8 illustrates the evolution of the determinants of the growth in UK energy services and shows the mutually enhancing developments that led to the spectacular growth in energy services since 1800 (by a factor of 15 when measuring final energy inputs, and much more – perhaps as much as by a factor of 100 – when considering the significant improvements in the efficiency of energy service provision that have ranged between a factor of five for transportation, to up to a factor of 600 for lighting, see Fouquet, 2008). Population growth (from 10 million to 60 million people) and rising incomes (per capita Gross Domestic Product (GDP) has grown from some US$3000 at 2005 price levels and exchange rates in 1800, to close to US$40,000 at present) increase both

the demand for energy services[8] and the purchasing power of the population to afford traditional, as well as novel energy services.

Improvements in the energy efficiency of service provision and other technological improvements in turn are key factors contributing to the significant lowering of energy service prices, which have declined by a factor of under 10 for heating to over 70 for lighting since 1800. In short, more consumers that became more affluent enjoy increasingly energy-efficient and cheaper energy services, which fuels growth in energy service demand (a positive feedback loop in the terminology of systems science). A narrow interpretation of this dynamic process of increasing returns to adoption (e.g., costs of technologies and energy services decline, the higher their market application) as a simple "take-back"[9] effect, represent a static "equilibrium" perspective of energy systems evolution. The history of technological revolutions in energy services and in energy supply suggests rather a "dis-equilibrium" interpretation of major energy transitions: the transformation is so far-reaching that the ultimate future state of the system could have never been reached by incremental improvements in efficiency and costs of existing technologies and energy services. *"Add as many mail-coaches as you please, you will never get a railroad by so doing"* (Joseph A. Schumpeter, 1935).

1.3.2   Transitions in Energy Supply Systems (Global)

The history of energy transitions is a story of development interlaced with periods of crisis and shortages. The Neolithic revolution brought the first transformational change. Hunters and gatherers settled and turned to agriculture. Their energy system relied on harnessing natural energy flows, animal work,

---

8   There are both direct as well as indirect effects on energy service demands. A larger population translates into more food to cook, more people needing housing, etc., and a corresponding growth in related energy services. Higher incomes from economic growth imply growth in energy service demand in industrial and commercial activities and related services. This growth in energy service demand is "indirect" in the sense that production-related energy services are embedded in the private consumption of goods and services by private households and public services (schools, hospitals, etc.). Lastly, higher incomes make traditionally expensive energy services (such as air transportation) affordable for larger segments of society, an effect amplified by decreasing prices for energy services resulting from energy efficiency and other technology improvements.

9   The "take-back" (or "rebound") effect describes a situation where an improvement in energy efficiency leads to lower energy costs and hence consumer savings, which are often spent on (energy-intensive) consumption activities. Part of the energy savings is thus "taken back" by changed consumer expenditures. For example, a new, more energy-efficient car, with lowered fuel costs, can lead to driving more, or alternatively to spending the saved fuel bill on additional recreational air travel. This effect was first postulated by William Stanley Jevons in 1865 (and hence is referred to also as "Jevons Paradox," see also Binswanger, 2001). Empirical studies suggest that in high-income countries the take-back effect can be anywhere between 0% and 40% (see the 2000 special issue of *Energy Policy* 28(6–7) and the review in Sorell et al., 2009). If absolute reductions of energy use are on the policy agenda, compensating for take-back effects leads to increases in energy prices via taxes. Studies in developing countries (Roy, 2000) – e.g., on compact fluorescent lighting – suggest that take-back effects can approach 100%. In this case, the effect of energy efficiency improvements are less in reductions of total energy use but rather in vastly increased human welfare.

and human physical labor to provide the required energy services in the form of heat, light, and work. Power densities and availability were constrained by site-specific factors, with mechanical energy sources initially limited to draft animals and later to water and windmills. The only form of energy conversion was from chemical energy to heat and light – through burning fuel wood, for example, or tallow candles (Nakicenovic et al., 1998). It is estimated that early agricultural societies were based on annual energy flows of about 10–20 GJ per capita, two-thirds in the form of food for domesticated animals and humans, and the other third in the form of fuel wood and charcoal for cooking, heating, and early industrial activities such as ore smelting (Smil, 2010). China already experienced acute wood and charcoal shortages in the north of the country by the 13th century. In Europe, and particularly in the UK, domestic fuel wood became increasingly scarce and expensive as forests were overexploited without sufficient replanting or other conservation measures (Ponting, 1992).[10]

The fuel crisis was eventually overcome through a radical technological end-use innovation: the steam engine powered by coal.[11] The steam cycle represented the first conversion of fossil energy sources into work; it allowed the provision of energy services to be site-independent, as coal could be transported and stored as needed; and it permitted power densities previously only possible in exceptional locations of abundant hydropower (Smil, 2006). Stationary steam engines were first introduced for lifting water from coal mines, thereby facilitating increased coal production by making deep-mined coal accessible. Later, they provided stationary power for what was to become an entirely new form of organizing production: the factory system. Mobile steam engines, on locomotives and steam ships, enabled the first transport revolution, as railway networks were extended to even the most remote locations and ships were converted from sail to steam. While the Industrial Revolution began in England, it spread[12] throughout Europe, the United States and the world. Characteristic primary energy use levels during the "steam age," (the mid-19th century in England), were about 100 GJ/year per capita (Nakicenovic et al., 1998). These levels exceed even the current average global energy use per capita. By the turn of the 20th century, coal had become the dominant source of energy, replacing traditional non-fossil energy sources, and supplied virtually all of the primary energy needs of industrialized countries.

---

10   See also Perlin (1989) on the role of wood in the development of civilization. In fact, the first coal uses in the UK date back to Roman times, and coal was already being used for some industrial applications (e.g., brewing beer) before the Industrial Revolution. The absence of new and efficient end-use technologies for coal use (the later steam engine) implied only very limited substitution possibilities of traditional biofuel uses by coal before the advent of the Industrial Revolution.

11   Note, however, that the fuel wood crises did not cause or induce the numerous technological innovations including the steam engine that led to the Industrial Revolution. These were not caused by price escalation associated with an early "fuel wood peak," but rather resulted from profound transformations in the social and organizational fabric and incentive structures for science and entrepreneurship (see Rosenberg and Birdzell, 1986).

12   Quantitative historical accounts for major industrial countries are given in Gales et al., 2007, Kander et al., 2008, and Warr et al., 2010.

Figure 1.9 shows the exponential growth of global energy use at a rate close to 2%/yr since the advent of the Industrial Revolution. Figure 1.10 is based on the same data and shows relative shares of different primary energy sources. Substitution of traditional energy sources by coal characterized the first phase of the energy revolution – the "steam revolution" – a transformation that lasted until the early 1920s when coal reached its maximal share of close to 50% of global primary energy.

The second "grand" energy transformation also lasted for about 70 years. Primary energy demand increased even more rapidly, reaching 5% or even 6% growth annually, from the late 1940s to the early 1970s. This development phase was characterized by increasing diversification of both energy end-use technologies and energy supply sources. Perhaps the most important innovations were the introduction of electricity as an energy carrier which could be easily converted to light, heat, or work at the point of end-use, and of the internal combustion engine, which revolutionized individual and collective mobility through the use of cars, buses, and aircraft (Nakicenovic et al., 1998). Like the transition triggered by the steam engine, this "diversification transformation" was led by technological innovations in energy end-use, such as the electric light bulb, the electric motor, the internal combustion engine, and aircraft, as well as computers and the Internet, which revolutionized information and communication technologies.

However, changes in energy supply have been equally far-reaching. In particular, oil emerged from its place as an expensive curiosity at the end of the 19th century to occupy the dominant global position, where it has remained for the past 60 years. The expansion of natural gas use and electrification are other examples of important changes in energy supply in the 20th century. The first electricity generation systems were based on the utilization of small-scale hydropower, followed by a rapid expansion of thermal power-generating capacity utilizing coal, oil, and more recently, natural gas. Commercial nuclear power stations were increasingly put into operation in the period from 1970 to 1990. Renewable sources other than hydropower have become more intensively explored for electricity generation since the mid-1970s, with most of the new capacity being added during the past decade.

Despite these fundamental changes in the energy system from supply to energy end-use, the dynamics of energy system transformations have slowed down noticeably since the mid-1970s. Figure 1.10 shows that after oil reached its peak market share of some 40% during the early 1970s, the 1990s and the first decade of the 21st century saw a stabilization of the historical decline in coal's market share, and a significant slowdown in the market growth for natural gas and nuclear. Since 2000, coal has even experienced a resurgence, mostly related to the massive expansion of coal-fired power generation in rapidly developing economies in Asia.

The shift from fuels such as coal with a high carbon content to energy carriers with a lower carbon content such as natural gas, as well as the introduction of near-zero carbon energy sources such as hydropower

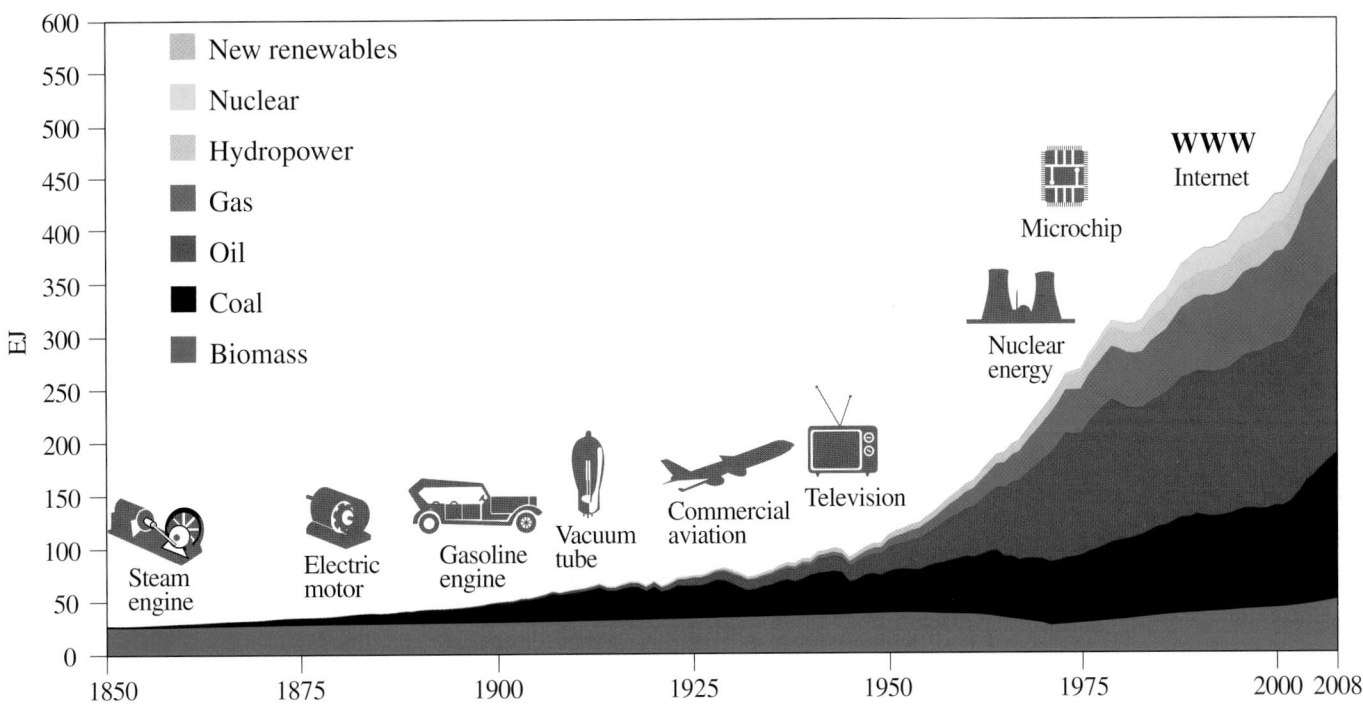

Figure 1.9 | History of world primary energy use, by Source (in EJ). Source: updated from Nakicenovic et al., 1998 and Grubler, 2008.

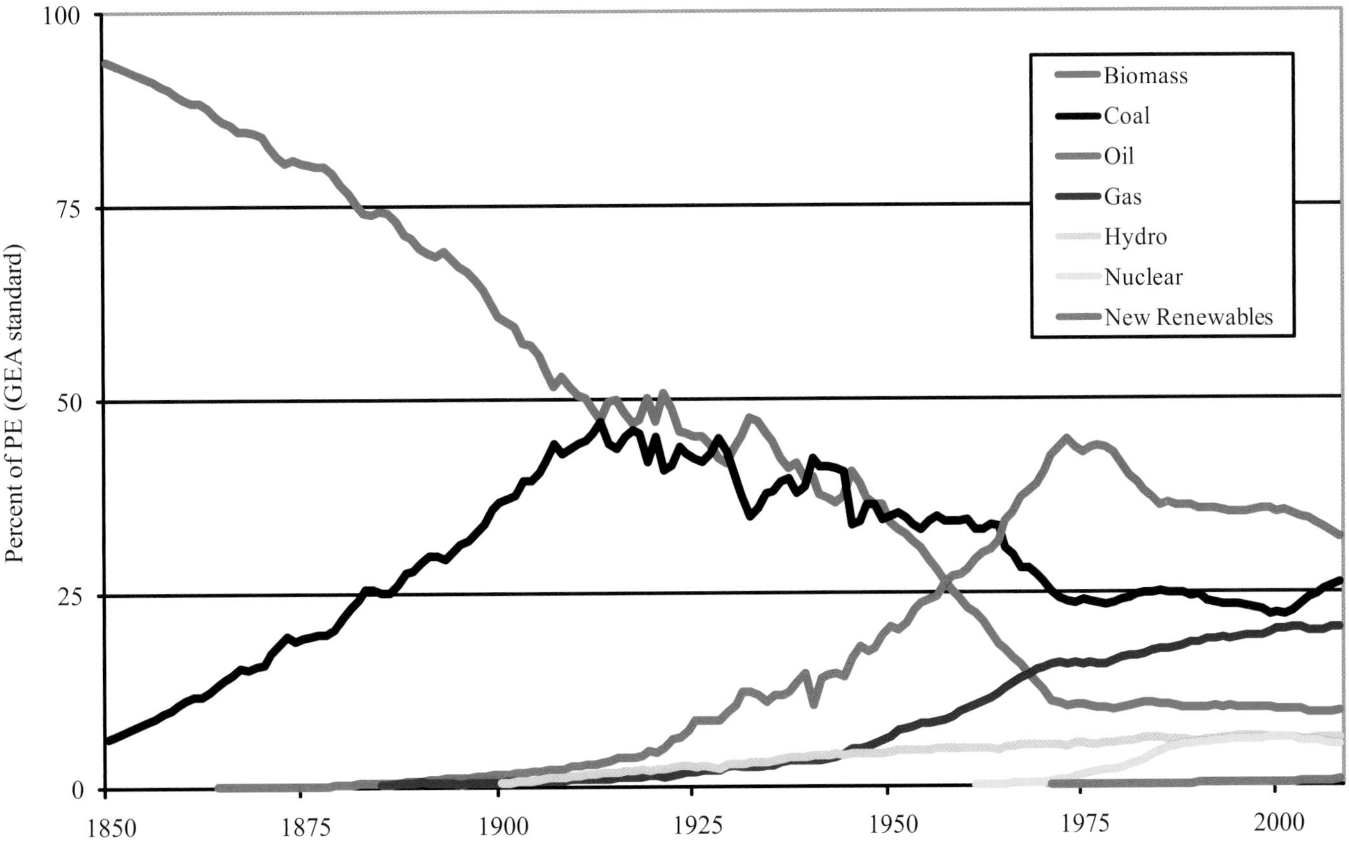

Figure 1.10 | Structural change in world primary energy (in percent). Source: updated from Nakicenovic et al., 1998 and Grubler, 2008.

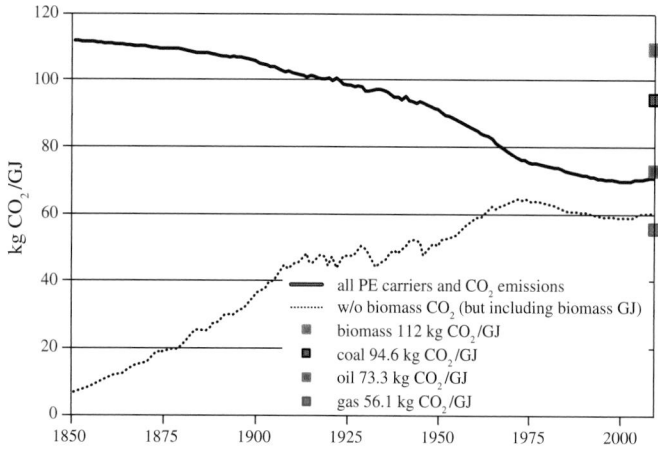

Figure 1.11 | Decarbonization of primary energy (PE) use worldwide since 1850 (kg of CO_2 emitted per GJ burned). Note: For comparison, the specific emission factors (OECD/IPCC default emission factors, LHV basis) for biomass (wood fuel), coal, crude oil, and natural gas are also shown (colored squares). See also discussion in text. Source: updated from Grubler and Nakicenovic, 1996.

and nuclear, has resulted in the *decarbonization* of energy systems (Grubler and Nakicenovic, 1996; Grubler, 2008). Decarbonization refers to the decrease in the specific emissions of carbon dioxide (CO_2) per unit of energy. Phrased slightly differently, it refers to the decrease in the carbon intensity of primary or any other energy form. Figure 1.11 illustrates the historical trend of global decarbonization since 1850 in terms of the average carbon emissions per unit of primary energy (considering all primary energy sources). The dashed line indicates the same trend but excluding biomass CO_2 emissions, assuming they have all been taken up by the biosphere under a sustainable harvesting regime (biomass regrowth absorbing the CO_2 released from biomass burning). Historically, emissions related to land-use changes (deforestation) have far exceeded[13] carbon releases from energy-related biomass burning, which suggests that in the past, biomass, like fossil fuels, has also contributed significantly to increases in atmospheric concentrations of CO_2.

The global rate of decarbonization has been on average about 0.3% annually, about six times too low to offset the increase in global energy

use of some 2% annually. Again, the significant slowing of historical decarbonization trends since the energy crises of the 1970s is noteworthy, particularly due to rising carbon intensities in some developing regions (IEA, 2009), and in general due to the slowed dynamics of the global energy system discussed above.

Decarbonization can be expected to continue over the next several decades as natural gas and non-fossil energy sources increase their share of total primary energy use. Some future scenarios (for a review see Fisher et al., 2007) anticipate a reversal of decarbonization in the long term as more easily accessible sources of conventional oil and gas become exhausted and are replaced by more carbon-intensive alternatives. Others foresee continuing decarbonization because of further shifts to low-carbon energy sources, such as renewables and nuclear energy. Nonetheless, virtually all scenarios foresee some increases in the demand for energy services as the world continues to develop. Depending on the rate of energy efficiency improvement,[14] this mostly leads to higher primary energy requirements in the future. As long as decarbonization rates do not significantly accelerate, this means higher carbon emissions compared to historical experience.

1.3.3 Energy and Economic Growth

The relationship between economic growth and energy use is multifaceted and variable over time. The relationship is also two-directional: provision of adequate, high-quality energy services is a necessary (even if insufficient)[15] condition for economic growth. In turn, economic growth increases the demand for energy services and the corresponding upstream energy conversions and resource use.

Figure 1.12 summarizes the long-term history of economic and energy development for a few countries for which such long-term data (since 1800) are available. To separate the impacts of population growth, both economic output (GDP) and (primary)[16] energy use are expressed on a per capita basis. Thereby, the usual temporal dimension of historical comparisons is replaced by an economic development metric in which countries are compared at similar levels of per capita incomes (GDP).

13 Cumulative emissions of fossil fuels between 1800 and 2000 are estimated to have released some 290 GtC (gigatonnes of elemental carbon – to obtain CO_2 multiply by 44/12, yielding 1060 GtCO_2), compared to land-use-related (deforestation, but excluding energy-related biomass burning) emissions of some 155 GtC. Total cumulative energy-related biomass carbon emissions are estimated at 80 GtC from 1800 to 2000 (all data from Grubler, 2002). Houghton (1999) estimates a net biospheric carbon flux (deforestation plus biomass burning minus vegetation regrowth) over the same time period (net emissions) of 125 GtC, which suggests that only a maximum (attributing – quite unrealistically – all residual net biospheric uptake to fuel wood) of 30 GtC (155 GtC deforestation release minus 125 GtC net biospheric emissions), or a maximum of 38% (30/80) from energy-related biomass burning has been absorbed by the biosphere historically. In the past, biofuel combustion for energy can, therefore, hardly be classified as "carbon neutral." Evidently, in many countries (at least in Northern latitudes) forests and energy biomass are harvested currently under sustainable management practices that in many cases (avoiding soil carbon releases from changing vegetation cover) will qualify as "carbon neutral." The extent of current net carbon releases of energy-related biomass burning in developing countries remains unknown.

14 The growth in emissions can be conveniently decomposed by the following identity (where annual percentage growth rates are additive) covering their main determinants of emissions and their growth: population, income, energy efficiency, and carbon intensity: CO_2 = Population x GDP/capita x Energy/GDP x CO_2/Energy (proposed by Holdren and Ehrlich, 1971, and applied for CO_2 by Kaya, 1990). Due to spatial heterogeneity in trends and variable interdependence, caution is advised in interpreting component growth rates of this identity.

15 Human (education) and social (functioning institutions and markets) capital as well as technology (innovation) are recognized as important determinants of economic growth (see Barro, 1997).

16 The most direct link between energy and economic activity is revealed at the level of final energy use. However, historical data are mostly available for primary energy use. For the United Kingdom, both primary and final energy (see Figure 1.7 above) are shown.

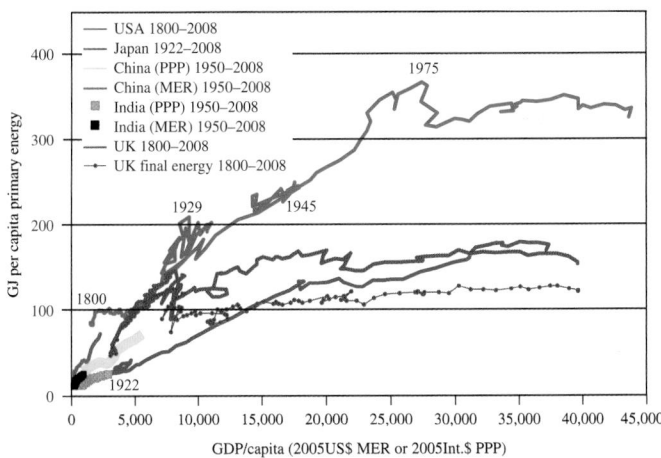

Figure 1.12 | Primary energy use (GJ) versus GDP (at market exchange rates (MER) in 2005US$) per capita. Source: USA, Japan: updated from Grubler, 1998, UK: Fouquet, 2008, India and China: IEA (2010) and World Bank (2010). Note: Data are for the United States (1800–2008), United Kingdom (primary and final energy, 1800–2008), Japan (1922–2008), China (1950–2008), and India (1950–2008). For China and India, also GDP at purchasing power parities (PPP, in 2005International$) are shown.

There are two ways of comparing GDP across different national economies depending on which exchange rate is used to convert a given national currency into a commensurable common currency (usually dollar denominated): at market exchange rates (MER) and in terms of purchasing power parities (PPP). The former are based on national accounts and official market (e.g., bank) exchange rates, while the latter are calculated based on relative prices for representative baskets of goods and services across countries denominated in an accounting currency of International$ (that equals the US$ in the United States). At present, differences between GDP rates denominated in MER and PPP exchange rates are comparatively minor among industrialized countries, and to simplify the exposition only MER-based GDP values are shown for the UK and Japan (MER and PPP GDPs are identical in the case of the US by definition). However, differences are significant in the case of developing economies (with PPP-based GDPs usually being larger than MER-based GDPs by a factor of two to three due to the much lower domestic price levels in developing countries –and hence the higher purchasing power of their population compared to industrialized ones), and, therefore, both GDP measures are shown in the case of China and India.

Three observations help to understand the relationship between economic and energy growth:
the importance of metrics;
the overall positive correlation, that is, however, variable over time; and
the distinctive differences in development paths among different countries and their economies.

First, both the starting points and the growth rates (the slopes of the trend lines shown in Figure 1.12) of economies are dependent on the economic metric chosen for comparing incomes across countries (MER or PPP). For instance, China's and India's GDP per capita in 1970 are

estimated to have been approximately US$170 and US$250, respectively (in US$2005), based on MER, and $700 and $1000 (in International $2005), respectively, when based on PPP, which compares to the GDP of the US of approximately US$1000 (at US$2005 rates) of 200 years ago, and to that of Japan in 1885.[17]

Thus, developing countries are by no means in a better position for economic "take-off"; they are not comparatively "richer" today than today's industrialized countries were some 100 or even 200 years ago, albeit enjoying unique development opportunities due to new technologies and improved communication and trade flows (Grubler, 2004). This illustrates the time dimension of economic development that entails many decades. Developing countries are today at the beginning of a long uphill development path that will require many decades to unfold and is also likely to include setbacks, as evidenced by the historical record of the industrialized countries. However, overall levels of energy use can be expected to increase as incomes rise in developing countries.

The overall positive correlation between economic and energy growth remains one of the most important "stylized facts" of the energy development literature, even if the extent of this correlation and its patterns over time are highly variable. Although the pattern of energy use growth with economic development is pervasive, there is no unique and universal "law" that specifies an exact relationship between economic growth and energy use over time and across countries. The development trajectory of the US illustrates this point. Over much of the period from 1800 to 1975, per capita energy use in the US grew nearly linearly with rising per capita incomes, punctuated by two major discontinuities: the effects of the Great Depression after 1929, and the effects of World War II (recognizable by the backward-moving "snarls" in the temporal trajectory of both income and energy use per capita shown in Figure 1.12). However, since 1975, per capita energy use has remained remarkably flat despite continuing growth in per capita income, illustrating an increasing decoupling of the two variables as a lasting impact of the so-called "energy crisis" of the early 1970s, an experience shared by many highly industrialized countries. It is also important to recognize significant differences in timing. During the 100 years from 1900 to 2000, Japan witnessed per capita income growth similar to that experienced by the US over 200 years (Grubler, 2004). This illustrates yet another limitation of simple inferences: notwithstanding the overall evident coupling between economic and energy growth, the growth experiences of one country cannot necessarily be used to infer those of another country, neither in terms of speed of economic development, nor in terms of how much growth in energy use such development entails.

Lastly, there is a persistent difference between development trajectories spanning all of the extremes from "high energy intensity" (the US) at one end of the scale to "high energy efficiency" (Japan) at the other (see also the discussion on energy intensities in Section 1.4.4 below).

---

17 Based on MER. Using PPP, Japan's GDP per capita in 1885 is estimated to have been well above $4000 (in 2005International$).

The relationship between energy and economic growth thus depends on numerous and variable factors. It depends on initial conditions (e.g., as reflected in natural resource endowments and relative price structures) and the historical development paths followed that lead to different settlement patterns, different transport requirements, differences in the structure of the economy, and so on. This twin dependency on initial conditions and the development paths followed to explain differences among systems is referred to as "path dependency" (Arthur, 1989). Path dependency implies considerable inertia in changing development paths, even as conditions prevailing at specific periods in history change – a phenomenon referred to as "lock-in" (Arthur, 1994). Path dependency and lock-in in energy systems arise from differences in initial conditions (e.g., resource availability and other geographical, climatic, economic, social, and institutional factors) that in turn are perpetuated by differences in policy and tax structures, leading to differences in spatial structures, infrastructures, and consumption patterns. These in turn exert an influence on the levels and types of technologies used, both by consumers and within the energy sector, that are costly to change quickly owing to high sunk investment costs, hence the frequent reference to "technological lock-in" (Grubler, 2004).

The concepts of path dependency and technological lock-in help to explain the persistent differences in energy use patterns among countries and regions even at comparable levels of income, especially when there are no apparent signs of convergence. For instance, throughout the whole period of industrialization and at all levels of income, per capita energy use has been lower in Japan than in the US (Grubler, 2004). The critical question for emerging economies such as China and India is, therefore, what development path they will follow in their development and what policy leverages exist to avoid lock-in in energy- and resource-intensive development paths that ultimately will be unsustainable, which puts *energy efficiency* at the center of the relationship between the economic and energy systems.

1.4   Energy Efficiency and Intensity

1.4.1   Introduction

Energy is conserved in every conversion process or device. It can neither be created nor destroyed, but it can be converted from one form into another. This is the First Law of Thermodynamics. For example, energy in the form of electricity entering an electric motor results in the desired output – say, kinetic energy of the rotating shaft to do work – and in losses in the form of heat as the undesired by-product caused by electric resistance, magnetic losses, friction, and other imperfections of actual devices. The energy entering a process equals the energy exiting. Energy efficiency is defined as the ratio of the desired (usable) energy output to the energy input. In the electric motor example, this is the ratio of the shaft power to the energy input electricity. Or in the case of natural gas for home heating, energy efficiency is the ratio of heat energy supplied to the home to the calorific value of the natural gas entering the furnace. This definition of energy efficiency is sometimes called *first-law efficiency* (Nakicenovic et al., 1996b).

A more efficient provision of energy services not only reduces the amount of primary energy required but, in general, also reduces costs and adverse environmental impacts. Although efficiency is an important determinant of the performance of the energy system, it is not the only one. In the example of a home furnace, other considerations include investment, operating costs, lifetime, peak power, ease of installation and operation, and other technical and economic factors (Nakicenovic et al., 1996b). For entire energy systems, other considerations include regional resource endowments, conversion technologies, geography, information, time, prices, investment finance, age of infrastructure, and know-how.

As an example of energy chain efficiency, Figure 1.13 illustrates the energy flows in the supply chain for illumination services (lighting). In this example, electricity is generated from coal in a thermal power station and transmitted and distributed to the point of end-use, where it is converted to light radiation by means of an incandescent light bulb. Only about 1% of the primary energy is transformed to illumination services provided to the end-user. In absolute terms, the majority of losses occur at the thermal power plant. The conversion of chemically stored energy from the coal into high-quality electricity comes along with the production of a significant amount of low-grade heat as a by-product of the process. Idle losses[18] at the point of end-use reflect the amount of time when the light bulb is switched on with the illumination service not being needed at that moment – for example, when the user is temporarily not present in the room.

In this example, abundant opportunities for improving efficiency exist at every link in the energy chain. They include shifting to more efficient fuels (e.g., natural gas) and more efficient conversion, distribution, and end-use technologies (e.g., combined cycle electricity generation, fluorescent or LED lighting technologies), as well as behavioral change at the point of end-use (e.g., reducing idle times). Integration of energy systems is another approach to reduce losses and improve overall system efficiency. An example of such system integration is combined heat and power production, where low temperature residual heat from thermal power production is utilized for space heating, a technique which can raise overall first-law fuel efficiency up to 90% (Cames et al., 2006). At the point of end-use, idle losses can be reduced through changed user behavior and control technology such as building automation systems that adapt energy services to the actual needs of the user.

---

18 Similar concepts are captured by the term "load factor" referring to the capacity utilization of plant and equipment. In typical commuting situations in industrialized countries there are no more than 1.2 passengers per automobile, which is a lower load factor than for 2-wheelers (bicycles and scooters) in most cities of developing countries.

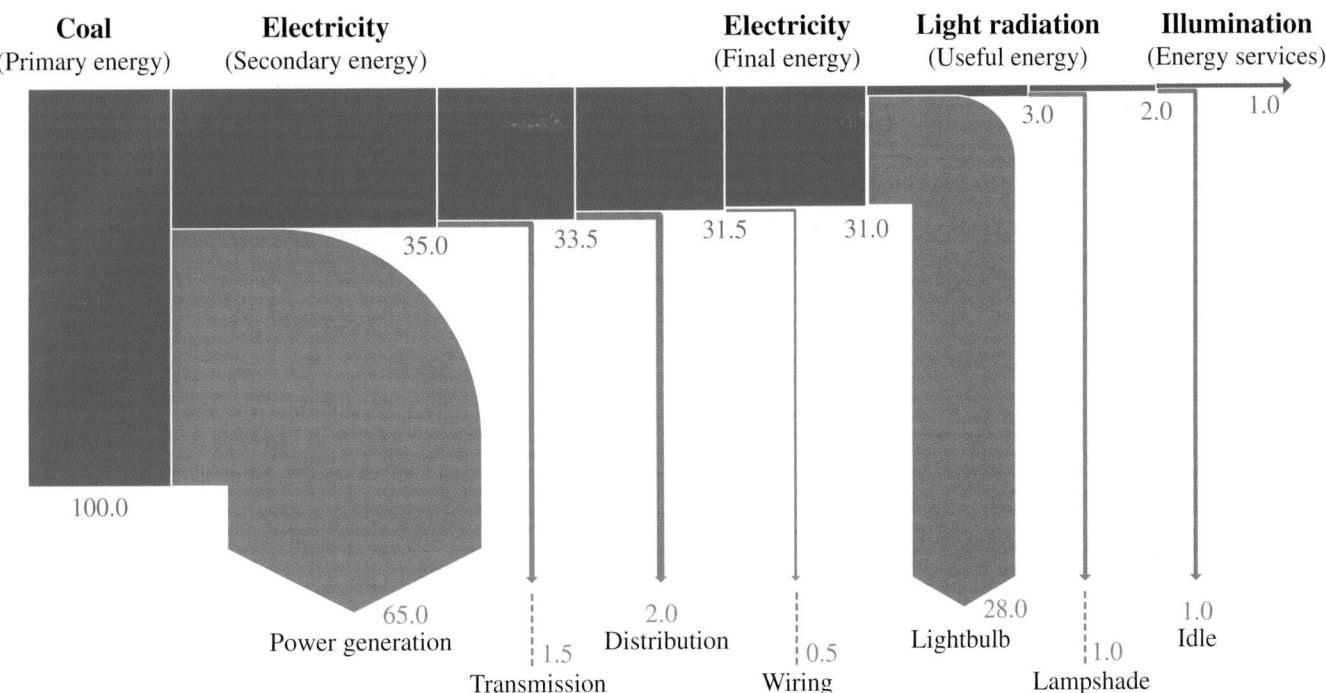

Figure 1.13 | Illustrative example of the compound First-Law efficiency of an entire energy chain to provide the energy service of illumination. Index: primary energy entering system = 100%.

1.4.2   First-Law Efficiencies

In 2005, the global efficiency of converting primary energy sources to final energy forms, including electricity, was about 67% (330 EJ over 496 EJ; see Figure 1.2 above). The efficiency of converting final energy forms into useful energy is lower, with an estimated global average of 51% (169 EJ over 330 EJ; see Figure 1.2). The resulting average global efficiency of converting primary energy to useful energy is then the product of the above two efficiencies, or 34%. In other words, about two-thirds of global primary energy use does not end up as useful energy input for providing energy services but is dissipated to the environment in the form of waste heat (or what is colloquially termed energy "losses"). The ultimate efficiency of the energy system in the provision of energy services cannot be determined by calculations based on the First Law of Thermodynamics but requires an extension of the discussion to the Second Law of Thermodynamics.

1.4.3   Second-Law Efficiencies and Exergy

How much energy is needed for a particular energy service? The answer to this question is not so straightforward. It depends on the type and quality of the desired energy service, the type of conversion technology, the fuel, including the way the fuel is supplied, and the surroundings, infrastructures, and organizations that provide the energy service. Initially, energy efficiency improvements can be achieved in many instances without elaborate analysis through common sense, good housekeeping, and leak-plugging practices. Obviously, energy service efficiencies improve as a result of sealing leaking window frames or the

installation of a more efficient furnace. Or if the service is transportation, getting to and from work, for example, using a transit bus jointly with other commuters is more energy-efficient than taking individual automobiles. After the easiest improvements have been made, however, the analysis must go far beyond energy accounting.[19]

Here the concept that something may get lost or destroyed in every energy device or transformation process is useful. This "something" is called "availability," which is the capacity of energy to do work. Often the availability concept is called "exergy."[20]

The following example should help clarify the difference between energy and exergy. A well-insulated room contains a small container of kerosene surrounded by air. The kerosene is ignited and burns until the container is empty. The net result is a small temperature increase of the air in the room ("enriched" with the combustion products). Assuming no heat leaks from the room, the total quantity of energy in the room has not changed. What has changed, however, is the quality of energy. The initial fuel has a greater potential to perform useful tasks than the resulting

---

19   This section updates and expands on material that was first published in Nakicenovic et al. (1996b).

20   Exergy is defined as the maximum amount of energy that under given (ambient) thermodynamic conditions can be converted into any other form of energy; it is also known as "availability" or "work potential." Therefore, exergy defines the minimum theoretical amount of energy required to perform a given task. The ratio of theoretical minimum energy use for a particular task to the actual energy use for the same task is called exergy or *second-law efficiency* (based on the Second Law of Thermodynamics). See also Wall, 2006.

slightly warmer air mixture. For example, one could use the fuel to generate electricity or operate a motor vehicle. The scope of a slightly warmed room to perform any useful task other than space conditioning (and so provide thermal comfort) is very limited. In fact, the initial potential of the fuel or its exergy has been largely destroyed.[21] Although energy is conserved, exergy is destroyed in all real-life energy conversion processes. This is what the Second Law of Thermodynamics says.

Another, more technical, example should help clarify the difference between the first-law (energy) and second-law (exergy) efficiencies. Furnaces used to heat buildings are typically 70% to 80% efficient, with the latest best-performing condensing furnaces operating at efficiencies greater than 90%. This may suggest that minimal energy savings should be possible, considering the high first-law efficiencies of furnaces. Such a conclusion is incorrect. The quoted efficiency is based on the specific process being used to operate the furnace – combustion of fossil fuel to produce heat. Since the combustion temperatures in a furnace are significantly higher than those desired for the energy service of space heating, the service is not well matched to the source and the result is an inefficient application of the device and fuel. Rather than focusing on the efficiency of a given technique for the provision of the energy service of space heating, one needs to investigate the theoretical limits of the efficiency of supplying heat to a building based on the actual temperature regime between the desired room temperature, and the heat supplied by a technology. The ratio of theoretical minimum energy use for a particular task to the actual energy use for the same task is called exergy or *second-law efficiency.*

Consider the following case. To provide a temperature of 30°C to a building while the outdoor temperature is 4°C requires a theoretical minimum of one unit of energy input for every 12 units of heat energy delivered to the indoors. To provide 12 units of heat with an 80% efficient furnace, however, requires 12/0.8, or 15 units of heat. The corresponding second-law efficiency is the ratio of theoretical minimum to actual energy use – i.e., 1/15 or 7%.

The first-law efficiency of 80% gives a misleading impression that only modest improvements are possible. The second-law efficiency of 7% says that a 15-fold reduction in final heating energy is theoretically possible by changing technologies and practices.[22] In practice, theoretical

maxima cannot be achieved. More realistic improvement potentials might be in the range of half of the theoretical limit. In addition, further improvements in the efficiency of supplying *services* are possible by task changes – for instance, in reducing the thermal heat losses of the building to be heated via better insulation of walls and windows.

What is the implication of the Second Law of Thermodynamics for energy efficiencies? First of all, it is not sufficient to account for energy-in versus energy-out ratios without due regard for the quality difference – i.e., the exergy destroyed in the process. Minimum exergy destruction means an optimal match between the energy service demanded and the energy source. Although a natural gas heating furnace may have a (First-Law) energy efficiency of close to 100%, the exergy destruction may be very high depending on the temperature difference between the desired room temperature and the temperature of the environment. The Second-Law efficiency, defined as exergy-out over exergy-in, in this natural gas home heating furnace example is some 7% – i.e., 93% of the original potential of doing useful work (exergy) of the natural gas entering the furnace is destroyed. Here we have a gross mismatch between the natural gas potential to do useful work, and the low temperature nature of the energy service space conditioning.

There are many examples for exergy analysis of individual conversion devices (e.g., losses around a thermal power plant) as well as larger energy systems (cities, countries, the entire globe). This literature is reviewed in detail in Nakicenovic (1996b). Estimates of global and regional primary-to-service exergy efficiencies vary typically from about 10 to as low as a few percent of the thermodynamically maximum feasible (see also Ayres, 1989, Gilli et al., 1996, and Nakicenovic et al., 1996a).

The theoretical potential for efficiency improvements is thus very large, and current energy systems are nowhere close to the maximum levels suggested by the Second Law of Thermodynamics. However, the full realization of this potential is impossible to achieve. First of all, friction, resistance, and similar losses can never be totally avoided. In addition, there are numerous barriers and inertias to be overcome, such as social behavior, vintage structures, financing of capital costs, lack of information and know-how, and insufficient policy incentives.

The principal advantage of second-law efficiency is that it relates actual efficiency to the theoretical (ideal) maximum. Although this theoretical maximum can never be reached, low exergy efficiencies identify those areas with the largest potentials for efficiency improvement. For fossil fuels, this implies the areas that also have the highest emission mitigation potentials. A second advantage of exergy efficiency is that the concept can be transferred to the assessment of energy service provision, which is not possible in first-law efficiency calculations. By comparing an actual configuration (a single driver in an inefficient car) with a theoretically ideal situation (a fuel-efficient car with five people in it), respective exergetic service efficiencies while maintaining the same type of energy service (i.e., not assuming commuting by bicycle) can be determined. This is important, especially as the available literature

---

21 Alternative example: In terms of energy, 1 kWh of electricity and the heat contained in 5 kg of 20°C (raised from 0°C) water are equal, i.e. 3.6 MJ. At ambient conditions, it is obvious that 1 kWh of electricity has a much larger potential to do work (e.g., to turn a shaft, provide light, or allow to run a computer) than the 5 kg of 20°C water that cannot perform any useful work.

22 For example, instead of combusting a fossil fuel, Goldemberg et al. (1988) give the example of a heat pump that extracts heat from a local environment (outdoor air, indoor exhaust air, ground water) and delivers it into the building. A heat pump operating on electricity can supply 12 units of heat for three to four units of electrical energy. The second-law efficiency then improves to 25–33% for this particular task – still considerably below the theoretical maximum efficiency. Not accounted for in this example, however, are the energy losses during electricity generation. Assuming a modern gas-fired combined cycle power plant with 50% efficiency, the overall efficiency gain is still higher by a factor of two compared to a gas furnace heating system.

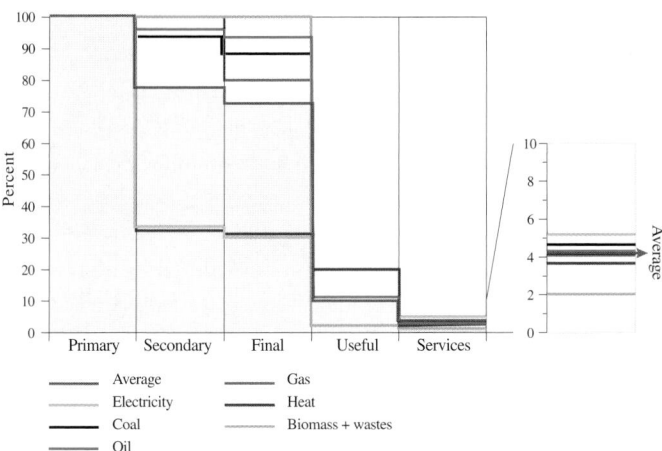

Figure 1.14 | Estimated exergy efficiencies (average for OECD countries) from primary exergy (= 100%) to useful exergy and to services by energy carrier (fuel). Source: adapted from Nakicenovic, 1993.

suggests that efficiencies in energy end-uses (in the conversion of final to useful energy and of useful energy to energy services) are particularly low (see Figure 1.14).

1.4.4 Energy Intensities

A related concept to that of energy efficiency is that of energy intensity. Instead of measuring input/output relations in energy terms, as is the case for energy efficiency, energy inputs are divided by a range of appropriate activity indicators that represent the energy service provided (such as tonnes of steel produced, vehicle-km driven, floorspace inhabited, monetary measures of output, number of employees, etc.) to yield energy intensity indicators. Such comparative benchmarking across countries, industries, or products, yields valuable insights into potentials for efficiency improvements related to various activities (comparing current intensities to best practice), and is applied widely in the corresponding energy efficiency improvement and greenhouse gas (GHG) mitigation literature (see Fisher et al., 2007; and the GEA end-use chapters 8, 9, and 10 in this publication). Extending this concept to entire energy systems and economies yields a widely used indicator of energy intensity, per unit of economic activity (GDP, which is the monetary quantification of all goods and services consumed in an economy in a given year subject to market transactions).[23] This parsimonious indicator is appealing because of its relative simplicity (usually a single number) and seeming ease of comparability across time and across different systems (global and/or national economies, regions, cities, etc.). However, its simplicity comes at a price.

First, the indicator is affected by a number of important measurement and definitional issues (see the discussion below). Second, the underlying factors for explaining differences in absolute levels of energy intensities across economies and their evolution over time requires detailed, further in-depth analysis using a range of additional explanatory variables. They cannot be distilled from an aggregate indicator such as energy intensity of the national or global GDP.

The literature on energy intensities, their trends, and drivers is vast (for useful introductory texts see, e.g., Schipper and Myers, 1992; Nakicenovic et al., 1996b; Greening et al., 1997; Schäfer, 2005; Baksi and Green, 2007; Gales et al., 2007). Apart from definitional, accounting, and measurement conventions, differences in energy intensities have been explained by a set of interrelated variables including demographics (size, composition, and densities – e.g., urban versus rural population), economics (size and structure of economic activities/sectors – e.g., the relative importance of energy-intensive industries versus energy-extensive services in an economy; per capita income levels), technology and capital vintages (age and efficiency of the production processes, transport vehicles, housing stock, etc.), geography and climate, energy prices and taxes, lifestyles, and policies, just to name the major categories.

In terms of energy and economic accounting, energy intensities are affected by considerable variation depending on which particular accounting convention is used (and which is often not disclosed prominently in the reporting reference). For energy, the largest determining factors are whether primary or final energy is used in the calculations, and if non-commercial (traditional biomass or agricultural residues, which are of particular importance in developing countries) are included or not. Another important determinant is which accounting method is used for measuring primary energy (see Appendix 1.A). For GDP, the largest difference in energy intensity indicators is the conversion rate used for expressing a unit of national currency in terms of an internationally comparable currency unit based on either MER or PPP exchange rates (see the discussion in Section 1.3.3 above).

Figure 1.15 illustrates some of the differences in the evolution of historical primary energy intensity for four major economies in the world: China, India, Japan, and the United States. It shows a number of different ways of measuring energy intensity of GDP. The first example can be best illustrated for the US (where there is no difference between the MER and PPP GDP measure by definition).

The (thin red) curve shows the *commercial* energy intensity. Commercial energy intensities increase during the early phases of industrialization, as traditional and less efficient energy forms are replaced by commercial energy. When this process is completed, commercial energy intensity peaks and proceeds to decline. This phenomenon is sometimes called the "hill of energy intensity." Reddy and Goldemberg (1990) and many others have observed that the successive peaks in the procession of countries achieving this transition are ever lower, indicating a possible catch-up effect and promising further energy intensity reductions in

---

23 Like energy, GDP is a *flow* variable and, therefore, does not measure wealth or welfare (which are *stock* variables).
  The measurement of GDP through market transactions (sales/purchases of goods and services) is at the same time a strength (measurability by statistical offices) and a weakness of the concept, as excluding non-market transactions (such as household and voluntary work that should increase GDP if valued monetarily) as well as environmental externalities (the negative impacts of pollution, congestion, etc. that would lower GDP).

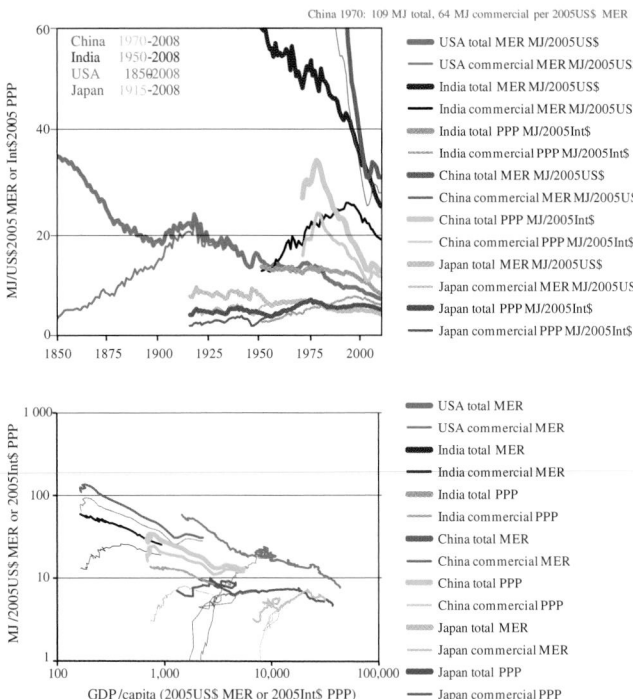

Figure 1.15 | Energy intensity improvements over time (top) and against per capita income (bottom) US (1800–2008), Japan (1885–2008), India (1950–2008), and China (1970–2008). Source: see Figure 1.12. Note: Energy intensities (in MJ per $) are always shown for total primary energy (bold lines) and commercial primary energy only (thin lines) and per unit of GDP expressed at market exchange rates (MER in 2005US$) and for China, India, and Japan also at purchasing power parities (PPP in 2005International$). For the United States, MER and PPP are identical.

developing countries that still have to reach the peak. In the US, for example, the peak of commercial energy intensity occurred during the 1910s and was higher than Japan's subsequent peak, which occurred in the 1970s (Nakicenovic et al., 1998). More important than this "hill" in commercial energy intensities is, however, a pervasive trend toward overall lower *total* energy (including also non-commercial energy) intensities over time and across all countries.

Figure 1.15 also shows energy intensities for China and India for two alternative measures of converting national GDP to an internationally comparable level: using MER or PPP exchange rates. In the cases of India and China, MER energy intensities are very high, resembling the energy intensities of the now industrialized countries more that 100 years ago (Nakicenovic et al., 1998). This gives the appearance of very low energy efficiency in producing a unit of economic output in China and India, and by implication in other emerging and developing countries. However, China and India's PPP-measured GDPs are much higher than official MER-based GDPs suggest (and resulting PPP-based energy intensities much lower) due to generally much lower prices in the two countries compared to industrialized countries. This translates into a more favorable PPP exchange rate of the local currency compared to MER (often by a factor of two to

three). Consequently, with the same dollar amount, a consumer can purchase more goods and services in developing countries than in more industrialized countries. PPP-measured energy intensities are thus generally much lower for developing countries, indicating substantially higher energy efficiencies in these countries than would be calculated using MER.

The substantially lower energy intensity of GDP when expressed in terms of PPP rather than MER should be contrasted with the much lower energy intensity improvement *rates* in terms of PPP compared to energy intensities based on MER. The differences can indeed be substantial. In 2005 the energy intensity in China was about 33 MJ/US$2005 for MER, with an average historical reduction rate of 3.3%/year since 1971, compared with about 14 MJ per 2005International$ for PPP for the same year and an improvement rate of 1.9%/year. Since 1971, China's per capita GDP in terms of MER has grown by some 7%/year, whereas the estimated per capita GDP in PPP terms has grown by some 5%/year, compared to a growth rate of per capita primary energy use of some 3%/year (from 20 GJ in 1971 to 57 GJ in 2005 and 71 GJ in 2008). Therefore, caution is needed when interpreting the apparent rapid energy intensity improvements, measured by MER-based GDPs, which are reported for some countries. In theory, as countries develop and their domestic prices converge toward international levels, the difference between the two GDP measures largely disappears (see the case of Japan in Figure 1.15).[24]

Adding traditional (non-commercial) energy[25] to commercial energy reflects total primary energy requirements and yields a better and more powerful measure of overall energy intensity. Total energy intensities generally decline for all four countries in Figure 1.15. There are exceptions, including periods of increasing energy intensity that can last for a decade or two. This was the case for the US around 1900 and China during the early 1970s. Recently, energy intensities are (temporarily) increasing in the economies in transition, due to economic slowdown and depression (declining per capita GDP). In the long run, however, the development is toward lower energy intensities. Data for countries with long-term statistical records show improvements in total energy intensities by more than a factor of five since 1800, corresponding to an average decline of total energy intensities of about 1%/year (Gilli et al., 1990; Nakicenovic et al., 1998; Fouquet, 2008). Improvement rates can be much faster, as illustrated in the case of China discussed above (2–3%/year for PPP- and MER-based energy intensities, respectively. Energy intensities in India have improved by 0.8%/year (PPP-based) to 1.5%/year (MER-based) over the period from 1970 to 2005. The much higher improvement rates of China compared to India reflects both a

---

24 As by definition an International$ used for PPP accounting is equal to one US$, no distinction is made between PPP- and MER-based intensities in the case of the US in Figure 1.15.

25 Traditional biomass fuels are often collected by end-users themselves and thus not exchanged via formal market transactions. Their collection costs in terms of effort and time can be substantial but are not reflected in official GDP estimates.

less favorable (less energy-efficient) starting point as well as much faster GDP per capita growth in China than in India. Faster economic growth leads to a faster turnover of the capital stock of an economy, thus offering more opportunities to switch to more energy-efficient technologies. The reverse side also applies, as discussed above for the economies in transition (Eastern Europe and the former Soviet Union): with declining GDP, energy intensities deteriorate – i.e., increase rather than decline.

It is also useful to look at long-term energy intensity trends using a more appropriate "development" metric than a simple calendar year. Even if in many aspects not perfect, income per capita can serve as a useful proxy for the degree of economic development. From this perspective, the vast differences in energy intensities between industrialized and developing countries are *development gaps* rather than inefficiencies in developing economies. For similar levels of income, energy intensities of developing countries are generally in line with the levels that prevailed in industrialized countries about a century ago, when these had similar low income levels (see lower graph, Figure 1.15).

However, such a perspective also reveals more clearly distinctive differences in development patterns spanning all the extremes between "high intensity" (e.g., the US) and "high efficiency" (e.g., Japan). The United States has had at all times significantly higher energy intensities than other countries, reflecting its unique condition of originally prevailing resource abundance,[26] coupled with a vast territory, and a comparative labor shortage that led to early mechanization and the corresponding substitution of human and animal labor by mechanical energy powered by (cheap) fossil fuels (David and Wright, 1996). The concepts of path dependency and lock-in (introduced above) describe these differences in development patterns and trajectories. Current systems are deeply rooted in their past development history. Initial conditions and incentives in place (such as relative prices) structure development in a particular direction, which is perpetuated (path dependent), ultimately leading to lock-in – i.e., the resistance to change of existing systems (due to, e.g., settlement patterns, industrial structure, lifestyles). From this perspective, a rapid convergence of levels of energy intensity and efficiency across all countries would indeed be a formidable challenge, notwithstanding that all systems can improve their energy intensities toward an "endless" innovation "frontier" in energy efficiency.

Energy intensity improvements can continue for a long time to come. As discussed above, the theoretical potential for energy efficiency and intensity improvements is very large; current energy systems are nowhere close to the maximum levels suggested by the Second Law of Thermodynamics. Although the full realization of this potential is impossible, many estimates reflecting the potential of new technologies and opportunities for energy systems integration indicate that the improvement potential might be large indeed – an improvement by a factor of

ten or more could be possible in the very long run (see Ayres, 1989; Gilli et al., 1990; Nakicenovic., 1993; 1998; Wall, 2006). Thus, reductions in energy intensity can be viewed as an endowment, much like other natural resources, that needs to be discovered and applied.

1.5 Energy Resources

1.5.1 Introduction

Energy *resources* – or rather *occurrences* – are the stocks (e.g., oil, coal, uranium) and flows (e.g., wind, sunshine, falling water) of energy offered by nature. Stocks, by definition, are exhaustible, and any resource consumption will reduce the size of the concerned stock. Flows, in turn, are indefinitely available as long as their utilization does not exceed the rate at which nature provides them. While the concept of stocks and flows is simple and thus intriguing, it quickly becomes complex and confusing once one is tasked with their quantification (the size of the "barrel") or recoverability ("the size and placement of the tap"). Crucial questions relate to the definition and characterization of, say, hydrocarbons in terms of chemical composition, concentration of geological occurrence, investment in exploration, or technology for extraction. Just by accounting for lowest concentration occurrences or lowest-density flow rates, stocks and flows assume enormous quantities. However, these have little relevance for an appreciation of which parts of the stocks and flows may be or become practically accessible for meeting societies' energy service needs. Private- and public-sector energy resource assessments, therefore, distinguish between *reserves* and *resources*, while occurrences are usually ignored for reasons of lack of technical producibility or economic attractiveness. Put differently, what is the benefit of knowing the size of the barrel when no suitable tap is available?[27]

Despite being used for decades, the terms energy *reserves* and *resources* are not universally defined and thus poorly understood. There are many methodological issues, and there is no consensus on how to compare reserves and resources across different categories fairly. A variety of terms are used to describe energy reserves and resources, and different authors and institutions have different meanings for the same terms depending on their different purpose.

The World Energy Council (WEC, 1998) defined resources as "the occurrences of material in recognizable form." For oil, it is essentially the amount of oil in the ground. Reserves represent a portion of resources and is the term used by the extraction industry. *Reserves* are the amount currently technologically and economically recoverable (WEC, 2007). *Resources* are detected quantities that cannot be profitably recovered with current technology but might be recoverable in the future, as well as those quantities that are geologically possible but yet to be found.

---

26 A similar case can be found in the development history of the former Soviet Union, whose long-term economic data are, however, too uncertain for cross-country comparisons of energy intensity.

27 This section updates and expands on material that was first published in Rogner et al. (2000).

Occurences include both reserves and resources as well as all additional quantities estimated to exist in the Earth's crust.

BP (2010a) notes that "proven reserves of oil are generally taken to be those quantities that geological and engineering information indicate with reasonable certainty, which can be recovered in the future from known reservoirs under existing economic and operating conditions." Other common terms include probable reserves, indicated reserves, and inferred reserves – that is, hydrocarbon occurrences that do not meet the criteria of proven reserves. Undiscovered resources are what remain and, by definition, one can only speculate on their existence. Ultimately recoverable resources are the sum of identified reserves and the possibly recoverable fraction of undiscovered resources, and generally include production to date.[28]

Then there is the difference between conventional and unconventional resources (e.g., oil shale, tar sands, coal-bed methane, methane clathrates (hydrates), uranium in black shale or dissolved in sea water). In essence, unconventional resources are occurrences in lower concentrations, different geological settings, or different chemical compositions than conventional resources. Again, unconventional resource categories lack a standard definition, which adds greatly to misunderstandings. As the name suggests, unconventional resources generally cannot be extracted with technology and processes used for conventional oil, gas, or uranium. They require different logistics and cost profiles and pose different environmental challenges. Their future accessibility is, therefore, a question of technological development – i.e., the rate at which unconventional resources can be converted into conventional reserves (notwithstanding demand and relative costs). In short, the boundary between conventional and unconventional resources is in permanent flux. Occurrences are in principle affected by the same dynamics, albeit over a much more speculative and long-term time scale. Technologies that may turn them into potential resources are currently not in sight, and resource classification systems, therefore, separate them from resources (often considering occurrences as speculative quantities that may not become technologically recoverable over the next 50 years).

In short, energy resources and their potential producibility cannot be characterized by a simple measure or single numbers. They comprise quantities along a continuum in at least three, interrelated, dimensions: geological knowledge, economics, and technology. McKelvey (1967) proposed a commonly used diagram with a matrix structure for the classification along two dimensions (Figure 1.16): decreasing geological certainty of occurrence and decreasing techno-economic recoverability (Nakicenovic et al., 1996b). The geological knowledge dimension is divided into identified and undiscovered resources. Identified resources are deposits that have known location, grade, quality, and quantity, or that can be estimated from geological evidence. Identified resources are further subdivided into demonstrated (measured plus indicated) and

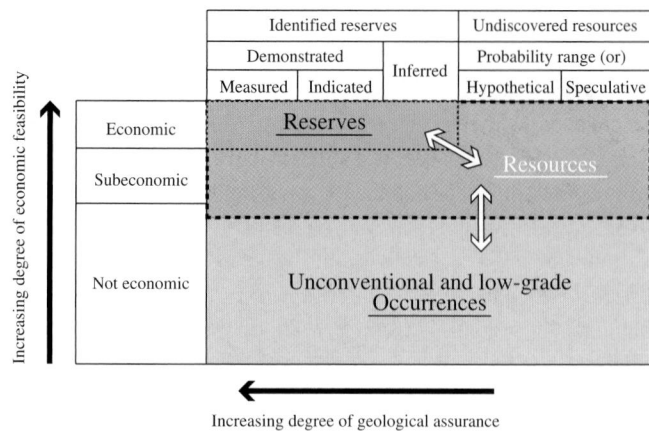

Figure 1.16 | Principles of resource classification, illustrating the definition of the three fundamental concepts: reserves, resources, and occurrences. Source: adapted from McKelvey, 1967.

inferred resources to reflect varying degrees of geological assurance. The techno-economic dimension accounts for the feasibility of technical recoverability and economic viability of bringing the resource to the market place. Reserves are identified resources that are economically recoverable at the time of assessment (see the BP definition above).

Undiscovered resources are quantities expected or postulated to exist under analogous geological conditions. Other occurrences are materials that are too low-grade, or for other reasons not considered technically or economically extractable. For the most part, unconventional resources are included in other occurrences.

Reserve and resource estimations, as well as their production costs, are subject to continuous revision for several reasons. Production inevitably depletes reserves and eventually exhausts deposits, while successful exploration and prospecting adds new reserves and resources. Price increases and production cost reductions expand reserves by moving resources into the reserve category and vice versa. Technology is the most important force in this process. Technological improvements are continuously pushing resources into the reserve category by advancing knowledge and lowering extraction costs. The outer boundary of resources and the interface to other occurrences is less clearly defined and often subject to a much wider margin of interpretation and judgment. Other occurrences are not considered to have economic potential at the time of classification. Yet over the very long term, technological progress may upgrade significant portions of occurrences to resources and later to reserves (Rogner et al., 2000).

In contrast, long-term supply, given sufficient demand, is a question of the replenishment of known reserves with new ones presently either unknown, not delineated, or from known deposits presently not producible or accessible for techno-economic reasons (Rogner, 1997; Rogner et al., 2000). Here the development and application of advanced exploration and production technologies are essential prerequisites for the

---

28 Physical and economic limitations of the *rates* of extraction do not enter the estimations of these stock variables.

Table 1.3 | Fossil and uranium reserves, resources and occurrences (in EJ).

	Historical production through 2005	Production 2005	Cumulative extraction GEA scenarios 2005–2100	Reserves	Resources	Additional Occurrences
	[EJ]	[EJ]	[EJ]	[EJ]	[EJ]	[EJ]
Conventional oil	6 069	147.9	6 600–10 000	4 900–7 610	4 170–6 150	
Unconventional oil	513	20.2	2–470	3 750–5 600	11 280–14 800	> 40 000
Conventional gas	3 087	89.8	7 900–11 900	5 000–7 100	7 200–8 900	
Unconventional gas	113	9.6	180–8 500	20 100–67 100	40 200–121 900	> 1 000 000
Coal	6 712	123.8	3 300–16 500	17 300–21 000	291 000–435 000	
Conventional uranium[b]	1 218	24.7	1 520–28 500	2 400	7 400	
Unconventional uranium[c]	n.a.	–			4 100	> 2 600 000

(a) The data reflect the ranges found in the literature; the distinction between reserves and resources is based on current (exploration and production) technology and market conditions. Resource data are not cumulative and do not include reserves.

(b) Reserves, resources, and occurrences of uranium are based on a once-through fuel cycle operation. Closed fuel cycles and breeding technology would increase the uranium reserve and resource dimensions 50–60 fold. Thorium-based fuel cycles would enlarge the fissile-resource base further.

(c) Unconvential uranium occurrences include uranium dissolved in seawater

Source: Chapter 7.

long-term resource availability. In essence, sufficient long-term supply is a function of investment in research and development (exploration and new production methods) and in extraction capacity, with demand prospects and competitive markets as the principal drivers.

For renewable energy sources, the concepts of reserves, resources, and occurrences need to be modified, as renewables represent (in principle) annual energy flows that, if harvested without disturbing nature's equilibria, are available sustainably and indefinitely. In this context, the total natural flows of solar, wind, hydro, geothermal energy, and grown biomass are referred to as *theoretical potentials* and are analogous to fossil occurrences. For resources, the concept of *technical potentials* is used as a proxy. The distinction between technical and theoretical potentials thus reflects the possible degree of use determined by thermodynamic, geographical, technological, or social limitations without consideration of economic feasibility.

Economic potentials then correspond to reserves – i.e., the portion of the technical potential that could be used cost-effectively with current technology and costs of production. Future innovation and technology change expand the techno-economic frontier further into the previously technical potential. For renewables, the technical and economic resource potentials are defined by the techno-economic performance characteristics, social acceptance, and environmental compatibility of the respective conversion technology – for instance, solar panels or wind converters. Like hydrocarbon reserves and resources, economic and technical renewable potentials are dynamically moving targets in response to market conditions, demand, availability of technology, and overall performance. Conversion technologies, however, are not considered in this discussion on resources. Consequently, no reserve equivalent (or economic potential) is given here for renewable resources. Rather, the deployment ranges resulting from the GEA pathways analyses (see Chapter 17) are compared with their annual flows.

1.5.2 Fossil and Fissile Resources

Occurrences of hydrocarbons and fissile materials in the earth's crust are plentiful – yet they are finite. The extent of the ultimately recoverable oil, natural gas, coal, or uranium has been subject to numerous reviews, and still there is a large range in the literature – a range that sustains continued debate and controversy. The large range is the result of varying boundaries of what is included in the analysis of a finite stock of an exhaustible resource – for example, conventional oil only, or conventional oil plus unconventional occurrences such as oil shale, tar sands, and extra heavy oils. Likewise, uranium resources are a function of the level of uranium ore concentrations in the source rocks considered technically and economically extractable over the long run.

Table 1.3 summarizes the global fossil and fissile reserves, resources, and occurrences identified in the GEA and contrasts these with the cumulative resource use (2005–2100) in the GEA pathways.

At the low end, cumulative global oil production in GEA pathways amounts to little more than total historical oil production up to 2005 – a sign of oil approaching peak production but also of a continued future for the oil industry. At the high end, future cumulative oil production is about 60% higher than past production without tapping unconventional oil in significant quantities.

1.5.3 Renewable Resources

Renewable energy resources represent the annual energy flows available through sustainable harvesting on an indefinite basis. While their annual flows far exceed global energy needs, the challenge lies in developing adequate technologies to manage the often low or varying

Table 1.4 | Renewable energy flows, potential, and utilization in EJ of energy inputs provided by nature.

	Primary Energy Equivalent in 2005	Utilization GEA pathways	Technical potential	Annual flows
	[EJ]	[EJ/yr]	[EJ/yr]	[EJ/yr]
Biomass, MSW, etc.	46	125–220	160–270	2200
Geothermal	1	1–22	810–1545	1500
Hydro	30	27–39	50–60	200
Solar	< 1	150–1500	62,000–280,000	3,900,000
Wind	1	41–715	1250–2250	110,000
Ocean	–	–	3240–10,500	1,000,000

Note: The data are energy-input data, not output. Considering technology-specific conversion factors greatly reduces the output potentials. For example, the technical potential of some 3000 EJ/yr of ocean energy in ocean thermal energy conversion (OTEC) would result in an electricity output of about 100 EJ/yr.

Source: Chapter 7 (see also Chapter 11 and IPCC, 2011 for a discussion of renewable resource inventories and their differences). Note: MSW = municipal (and other) solid wastes.

energy densities and supply intermittencies, and to convert them into usable fuels (see Section 1.5.4 below). Except for biomass, technologies harvesting renewable energy flows convert resource flows directly into electricity or heat. Their technical potentials are limited by factors such as geographical orientation, terrain, or proximity of water, while the economic potentials are a direct function of the performance characteristics of their conversion technologies within a specific local market setting.

Annual renewable energy flows are abundant and exceed even the highest future demand scenarios by orders of magnitude. The influx of solar radiation reaching the Earth's surface amounts to 3.9 million EJ/yr. Accounting for cloud coverage and empirical irradiance data, the local availability of solar energy reduces to 630,000 EJ. Deducting areas with harsh or unsuitable terrain leads to a technical potential ranging between 62,000 EJ/yr and 280,000 EJ/yr. By 2100 the GEA pathways, presented in Chapter 17, utilize up to 1500 EJ/yr of solar radiation (see Table 1.4). Note: The flows, potential, and utilization rates in Table 1.4 are given in terms of energy input – not as outputs (secondary energy or using any accounting scheme for equivalent primary energy – see Appendix 1.A). The production and utilization data, therefore, differ from the presentation in Chapter 17.

The energy carried by wind flows around the globe is estimated at about 110,000 EJ/yr,[29] of which some 1550 EJ/yr to 2250 EJ/yr are suitable for the generation of mechanical energy. The GEA pathways range of wind utilization varies between 41 EJ/yr and 715 EJ/yr. The energy in the water cycle amounts to more than 500,000 EJ/yr, of which 200 EJ/yr could theoretically be harnessed for hydroelectricity. The GEA pathways utilize between 27 EJ/yr and 39 EJ/yr compared to a technical potential estimated at 53 EJ/yr to 57 EJ/yr.

Net primary biomass production is approximately 2400 EJ/yr, which, after deducting the needs for food and feed, leaves in theory some 1330

EJ/yr for energy purposes. Accounting for constraints such as water availability, biodiversity, and other sustainability considerations, the technical bioenergy potential reduces to 160 EJ/yr to 270 EJ/yr, of which between 125 EJ/yr and 220 EJ/yr are utilized in the GEA pathways. The global geothermal energy stored in the Earth's crust up to a depth of 5000 meters is estimated at 140,000 EJ. The annual rate of heat flow to the Earth's surface is about 1500 EJ/yr, with an estimated potential rate of utilization of up to 1000 EJ/yr.

Oceans are the largest solar energy collectors on Earth, absorbing on average some 1 million EJ/yr. These gigantic annual energy flows are of theoretical value only, and the amounts that can be technically and economically utilized are significantly lower.

1.5.4   Energy Densities

The concept of energy density refers to the amount of energy generated or used per unit of land. The customary unit for energy densities is Watts per square meter (W/m^2), referring to a continuous (average) availability of the power of one Watt over a year. Typical energy densities for demand as well as supply are illustrated in Figure 1.17.

Energy demand and supply densities have co-evolved since the onset of the Industrial Revolution. In fact, one of the advantages of fossil fuels in the industrialization process has been their high energy density, which enables energy to be produced, transported, and stored with relative ease, even in locations with extremely high concentration of energy demand, such as industrial centers and rapidly growing urban areas. The mismatch between energy demand and supply densities is largest between urban energy use, which is highly concentrated, and renewable energies, which are characterized by vast, but highly diffuse energy flows.[30] The density of energy demand in urban areas is typically between 10 W/m^2 and 100

---

29 Wind, biomass, hydro, and ocean energy are all driven by the solar energy influx. Their numbers are, therefore, not additive to the solar numbers discussed above.

30 Exceptions are geothermal energy and urban (municipal) wastes, which are characterized by high energy density.

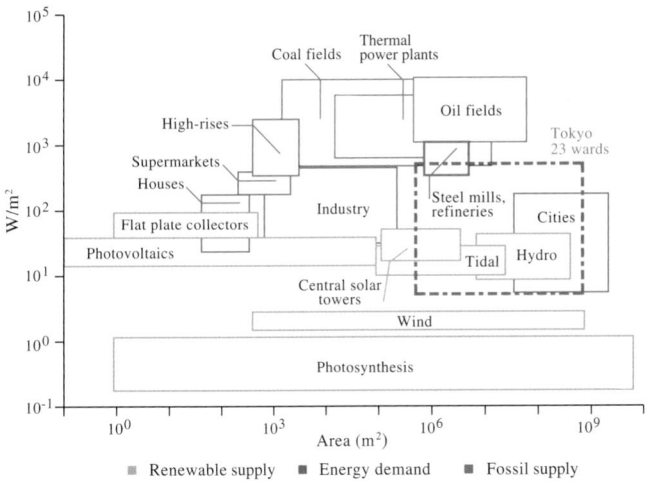

Figure 1.17 | Energy densities of fossil (grey) and renewable (green) energy supply and demand (red). Source: modified and adapted from Smil, 1991.

W/m^2 and can reach some 1000 W/m^2 in extreme locations, such as sky-scraper complexes and high-density business districts (see Figure 1.17 for an illustration of Tokyo's 23 wards [districts]). Conversely, renewable energies have typical energy supply densities of a few W/m^2 under ideal conditions (assuming all land can be devoted to harnessing renewable energy flows). Under practical conditions (considering competing land-uses for agriculture and human settlements) renewable energies can yield typically between 0.1 W/m^2 and 0.5 W/m^2 (see Chapter 18). As a result, locally harvested renewable energies in densely populated areas such as cities can only provide a very small fraction (some one percent) of urban energy demand. Therefore, renewable energies also have to be harvested in locations where land and favorable renewable resource potentials are available, and need then to be transported over longer distances to urban centers with their high energy demand densities.

1.6   Production, Trade, and Conversions

1.6.1   Introduction

The energy system consists of an intricate web of energy conversion processes linking primary energy resources to the provision of energy services. A first overview of energy conversion, therefore, can be gained by looking at the associated energy flows (see Figure 1.5 above). Due to the associated conversion losses, the energy flows get larger the further upstream the energy system one moves, which is the reason why global primary energy flows (496 EJ) are about three times larger than useful energy flows (169 EJ).

As the geographical location of energy resources and "downstream" components of the energy system are distributed very unevenly, this section begins by describing major global primary energy flows from production, use, and trade of energy.

1.6.2   Production, Use, and Trade

The sheer size of global energy flows, that dwarf energy storage capacities, implies a fundamental energy market identity: production of energy flows needs to equal demand, and vice versa. As energy demand and production capacities are distributed unevenly geographically, this basic market identity translates into vast flows of *energy trade*. Energy is traded in three forms:

direct energy flows of primary energy (coal, crude oil, and natural gas) and secondary energy (primarily refined oil products); and

indirect (embodied) energy flows, in which energy is traded in the form of (energy-intensive) commodities (aluminum, steel, etc.) and products (fertilizer, steel rails, cars, etc.).

The following sections summarize the status of primary energy production, trade, and *use* (defined as "Total Primary Energy Supply" – or TPES – in energy balances) for fossil fuels, as they are the dominant form of current global energy trade flows.[31]

1.6.2.1   Direct Energy

Table 1.5 summarizes primary energy production, trade, and use for nine regions and the world in 2005. From the TPES of some 390 EJ of fossil fuels in 2005, some 230 EJ (or close to 60%) are represented by energy imports. The share of traded energy (direct primary and secondary energy trade) in TPES is markedly different for different fuels: it is lowest for coal (18%), followed by natural gas (30%), and reaches 80% for crude oil. Including trade in refined oil products (secondary energy), oil-related energy trade flows (172 EJ) actually exceed the global TPES of oil products (167 EJ). This apparent paradox results from the fact that large importers of crude oil have corresponding large refining capacities, becoming in turn large exporters of refined petroleum products. The international division of labor in energy means that a barrel of crude oil can be traded various times and in various forms across national boundaries (not to mention the multiple "virtual" trades of the same barrel on speculative and futures markets). A good (even if extreme) illustration is provided in the case of Singapore: total fossil fuel imports equal a staggering 880 GJ/capita, of which 210 GJ/capita are used as primary energy input to the Singapore economy (with 120 GJ/capita final energy use), 450 GJ/capita are re-exported as oil products, and an additional 220 GJ/capita exported as bunker fuels for international

---

31 Renewables are dominated by traditional biomass use that is harvested and used locally without international trade. Modern renewables such as hydropower, solar, or wind, or for that matter also nuclear power, enter the energy system as secondary energy carriers (predominantly electricity, with some direct heat), which are generally not traded internationally. International trade in biofuels remains comparatively modest at some 0.2 EJ in 2005. International trade in electricity is also small: slightly above 2 EJ in 2005.

Table 1.5 | World trade flows between regions (in EJ) for 2005.

	Coal				Crude Oil				Oil Products				Natural Gas			
	Prod.	Exp.	Imp.	Stock Changes	Prod.	Exp.	Imp.	Stock Changes	Prod.	Exp.	Imp.	Stock Changes	Prod.	Exp.	Imp.	Stock Changes
Asia w/o China	14.4	−3.9	5.5	−0.2	7.6	−3.1	17.5	0.1	0.0	−6.9	6.8	0.1	9.4	−3.1	2.3	0.0
China	48.0	−2.3	0.9	−0.7	7.6	−0.3	5.3	0.0	0.0	−0.8	2.4	0.0	1.8	−0.1	0.1	0.0
EU27	8.5	−1.2	6.4	−0.1	5.4	−3.0	26.7	0.0	0.1	−11.9	13.2	−0.3	7.9	−2.5	13.3	0.0
Japan	0.0	0.0	4.7	0.0	0.0	0.0	9.0	0.0	0.0	−0.4	2.2	−0.1	0.1	0.0	2.8	0.0
LAC	2.2	−1.7	0.8	0.0	23.6	−11.0	3.0	0.0	0.2	−4.3	3.0	0.0	6.2	−1.1	1.0	0.0
MAF	6.0	−2.0	0.6	0.0	72.3	−52.1	2.6	0.0	0.0	−8.4	3.7	0.0	17.1	−5.5	0.4	0.0
rest-OECD	11.0	−7.5	1.0	0.1	12.2	−8.7	4.2	−0.1	0.8	−2.2	2.2	0.0	11.2	−7.1	1.4	0.3
REF w/o EU	10.3	−2.9	0.9	−0.1	24.1	−13.6	2.0	0.0	0.0	−4.9	0.5	0.0	27.7	−9.7	4.0	−0.1
USA	23.9	−1.2	0.8	0.1	13.3	−0.2	23.7	−0.2	0.2	−2.3	6.3	0.0	17.6	−0.7	4.2	0.1
World trade between regions	124.3	−22.8	21.7	−1.0	166.3	−92.1	94.0	−0.2	1.2	−42.0	40.3	−0.2	99.1	−29.8	29.5	0.2
World trade between countries (IEA data)	121.8	−21.6	21.8	0.1	167.8	−133.8	133.9	−0.5	−	−40.7	37.6	−	99.3	−30.0	29.4	0.3

Note: The five GEA regions have been expanded to nine to better represent international trade flows. These nine regions represent well the major international energy trade flows obtained from aggregating inter-country trade flows based on IEA statistics. Only for crude oil, the inter-regional trade flows cover only 70% of the true international trade in crude oil (the difference is intra-regional trade – e.g., within the EU27 countries, or within Latin America (LAC) countries.

Source: data from IEA, 2007a and 2007b.

shipping and aviation (Schulz, 2010). In addition, Singapore's energy trade is also characterized by vast energy flows embodied in exported products (petrochemicals) as well as in goods imported into this city state (see the discussion below).

The largest annual international trade flows (from aggregate country imports or exports) in 2005 were crude oil, with some 135 EJ, followed by oil products (40 EJ), natural gas (30 EJ), and coal (20 EJ).

In terms of regions, the largest[32] exporters for crude oil were the Middle East (MEA) region (some 50 EJ), the former Soviet Union (14 EJ), and Latin America and the Caribbean (LAC – 11 EJ), balanced from the oil import side with imports to Europe (27 EJ), the United States (24 EJ), and developing economies in Asia, excluding China, with 18 EJ. For gas trade, only exports from the former Soviet Union (10 EJ) and imports to Europe (13 EJ) are beyond the 10 EJ reporting threshold level adopted here. Inter-regional coal trade is comparatively small (with largest regional exports and imports of 8 EJ (Australia) and Europe (6 EJ), respectively).

Perhaps the least known aspect of international energy trade is the significant exports and imports of petroleum products. Europe, while a main crude oil importer (27 EJ), nonetheless exports 11 EJ of oil products, in order to import in turn a further 13 EJ of oil products. The trade in oil products to/from other regions is much smaller. The picture emerging from the international energy trade is thus less one of directed "source–sink" energy resource flows, but rather one of an increasingly complex "foodweb" in which energy is traded in primary and secondary forms across multiple boundaries.

1.6.2.2   Embodied Energy

The literature and statistical basis of embodied energy flows is thin, as existing studies almost invariably focus on embodied CO_2 emissions in international trade, without disclosing the underlying energy data. Notable exceptions are studies on embodied energy in the international trade of Brazil (Machado, 2000), China (Liu et al., 2010), and Singapore (Schulz, 2010). Current energy accounting and balances report direct energy flows, whereas *embodied energy* trade is quite under-researched and not reported systematically.

The only data source available for estimating embodied energy flows is the GTAP7 (Narayanan et al., 2008) database that contains data suitable for estimating the fossil fuel energy embodied in international trade flows by input-output analysis (Table 1.6). Important limitations and intricate methodological issues need to be considered when trying to estimate the energy embodied in internationally traded commodities

---

32   Flows below 10 EJ are not discussed separately. Details are given in Table 1.5. All data refer to the year 2005.

Table 1.6 | Trade in embodied energy between major regions (in EJ, only fossil primary energy) as derived from the GTAP7 Multi-Regional Input-Output Tables for 2005.

	EU	US	Japan	REF	Rest-OECD	Asia w/o China	China	LAC	Africa	Sum of Exports
EU		3.5	0.6	1.1	2.2	2.1	0.9	0.8	1.2	12.4
US	3.3		0.9	0.1	2.8	1.5	0.9	2.8	0.3	12.6
Japan	0.7	0.9		0	0.2	1.2	1.2	0.1	0.1	4.4
REF	6.1	1.3	0.3		1.1	1.5	1	0.4	0.3	12
Rest-OECD	2.4	4.1	0.4	0.2		1	0.5	0.2	0.2	9
Asia w/o China	5.1	3.6	3.5	0.3	1.3		4.9	0.7	1.1	20.5
China	5.1	5.4	3	0.4	1.3	4.8		1	0.5	21.5
LAC	1.4	4.1	0.2	0.1	0.4	0.5	0.4		0.2	7.3
Africa	2.4	0.9	0.2	0	0.3	0.6	0.3	0.2		4.9
Sum of imports	26.5	23.8	9.1	2.2	9.6	13.2	10.1	6.2	3.9	104.6

Source: Narayanan et al., 2008.

and products based on multi-regional input-output tables. The flows summarized in Table 1.6, therefore, need to be considered as order of magnitude estimates that await further analytical and empirical refinements. Nonetheless, even these "rough" data help to get a sense of proportion. GTAP estimates that (fossil) energy embodied in international trade amounts to some 100 EJ – i.e., some 20% of global primary energy use – compared to direct energy trade flows of some 190 EJ in 2005 when using the same regional aggregation[33] as reported in Table 1.6.

In other words, at least half of global primary energy use is traded among regions in either direct or indirect (embodied) form, which illustrates the multitude of interdependencies at play in the global energy system that go far beyond traditional concerns of oil import dependency. Assuming that the relative proportions of intra-regional to international trade flows hold for embodied energy flows in a similar way, as in the case of direct energy trade flows, then direct and embodied energy trade flows (of perhaps 400 EJ) approach the level of world primary energy use in the year 2005 (500 EJ). Evidently, these numbers must not be interpreted through the traditional lens of (additive) "net" energy trade flows. The nature of the international division of labor is precisely that a Joule of energy can be traded many times, hence the trade numbers discussed above include multiple double-counting. Consider two examples: Iran is a major oil producer and exporter but lacks sufficient domestic refining capacity. A barrel of oil exported to Singapore may be re-exported back to Iran in the form of gasoline, or it may be re-exported back in the form of plastic or chemical products. The same physical energy thus ends up being counted twice as an international energy trade flow. China is a major steel producer, Australia a major exporter of metallurgical coal

(used in the steel industry), and Germany a major car manufacturer. In our example, coal is exported from Australia to China, where it serves to produce steel, and this steel is exported from China to Germany, where manufacturers use it to produce German cars for export to China. Direct energy trade (coal) becomes embodied energy trade (steel), which in turn becomes embodied energy trade again (cars), with a physical Joule energy counted three times as international energy trade. This example also illustrates the great difficulties in comprehensive accounting of energy (or GHG emission flows) through multiple exchanges and trade flows. Who ultimately "owns" the corresponding energy or GHG "footprint": the Australian coal producer, the Chinese steel manufacturer, the German car company, or the Chinese consumer (car buyer)?

1.6.2.3   Energy Trade Flows

Figure 1.18 summarizes all direct (primary and secondary) and indirect fossil fuel-related international trade flows in the form of a map to demonstrate the high degree of energy interdependence worldwide. The term interdependence suggests that the energy system is much more integrated than conventional wisdom or energy security concerns would suggest. Not only do many countries critically depend on oil exports from the Middle East, the Middle East also depends on numerous other countries for its supply of food, consumer products, and investment goods that all embody (part of) the region's previous energy exports.

1.6.3   Conversions

1.6.3.1   Introduction and Overview

One way of looking at energy conversion processes is to consider the associated energy conversion capacity, which is a proxy of the aggregated size of energy conversion technologies and hence an indicator of

---

33 The difference between the 190 EJ intra-regional trade (nine regions) and the 230 EJ reported as international energy trade reflects the energy trade between countries within a given region (e.g., between Germany and France in the EU region, or between Indonesia and Bangladesh in the Asia-sans-China region) which is not counted in the regional trade flows but included in the global total trade numbers (summed from national statistics).

Crude oil and oil products

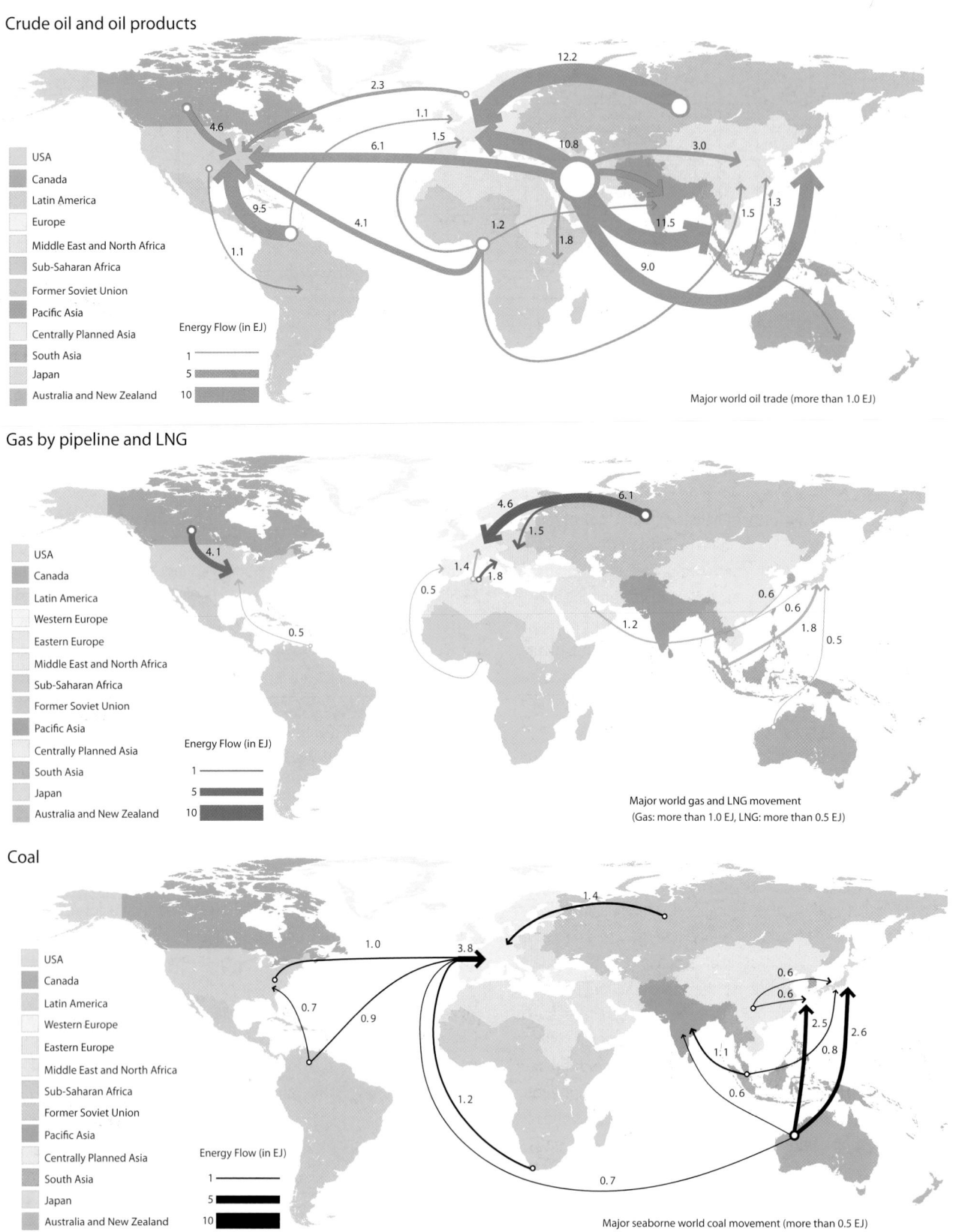

Figure 1.18a | World energy trade of fossil fuels: direct primary and secondary energy coal (black), oil and oil products (red) and gas (LNG light blue, pipeline gas: dark blue), in EJ. Source: Oil/gas energy trade for 2005 (BP, 2007), coal trade for 2008 (WCI, 2009).

Direct energy trade

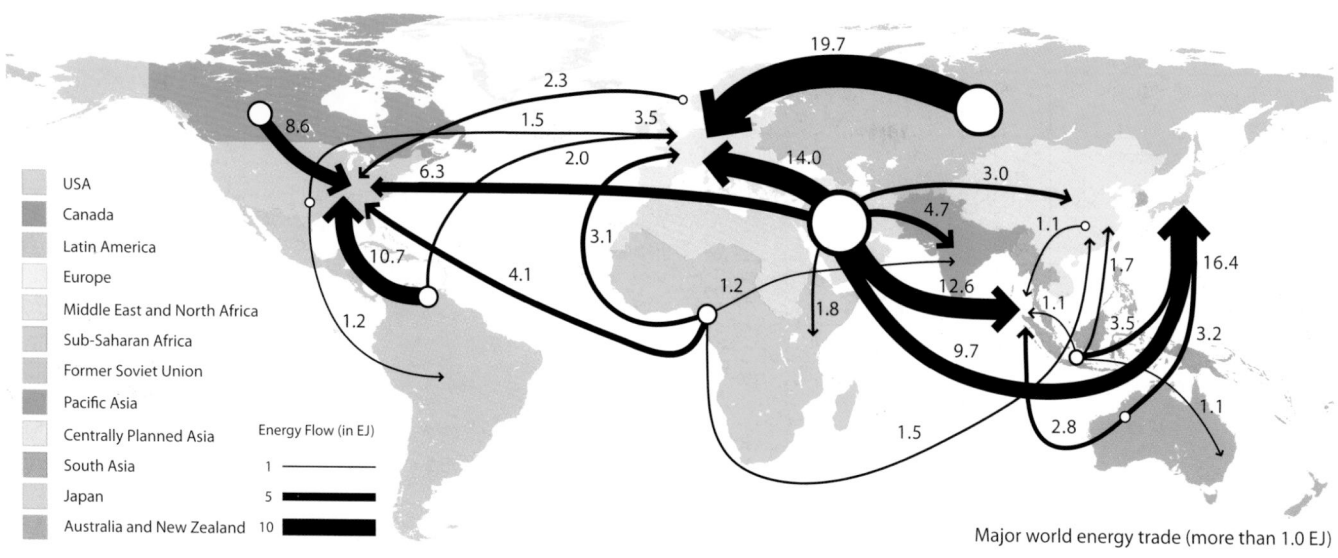

Major world energy trade (more than 1.0 EJ)

Embodied energy trade

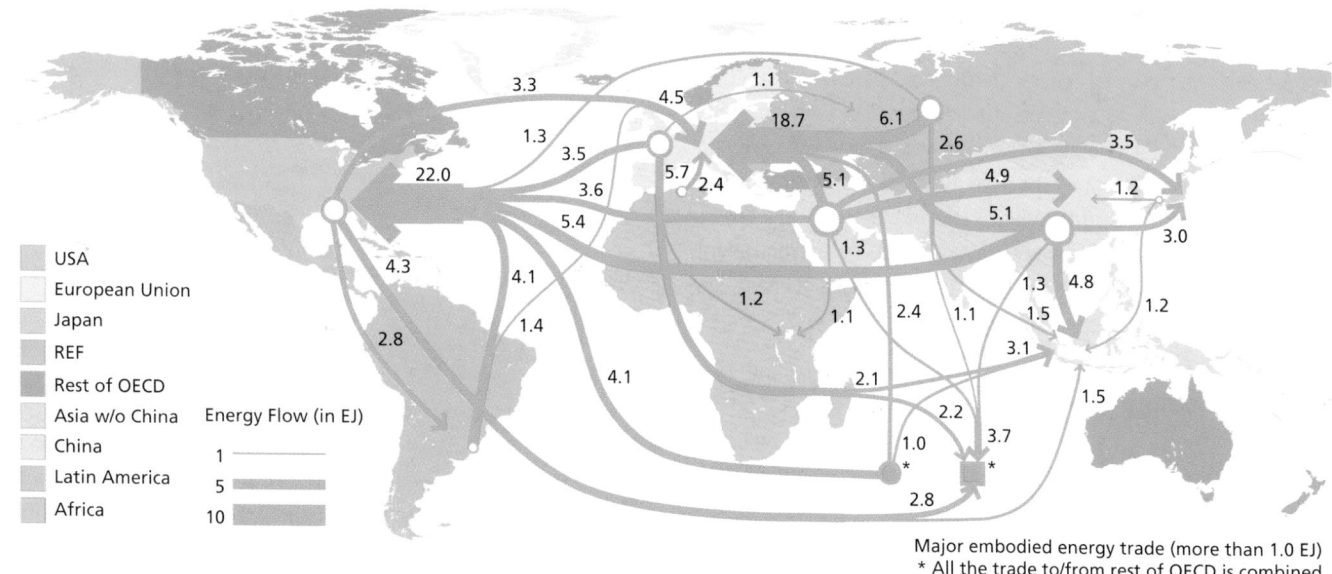

Major embodied energy trade (more than 1.0 EJ)
* All the trade to/from rest of OECD is combined

Figure 1.18b | World energy trade of total direct energy (black) as well as embodied energy in goods traded (grey), in EJ. Source: Embodied energy trade for 2004 (GTAP database, 2010), direct energy trade (BP, 2007; WCI, 2009).

the magnitude of technological change and capital replacement required for improving energy efficiency through the application of more efficient processes and technologies. Unlike the picture that emerges when looking at energy flows, the scale of energy conversion technologies portrays a different pattern in which energy end-use conversions dominate. Although global numbers are not available, this pattern of an increasing scale of energy conversion processes and devices revealed by the long-term history of the US energy system (Table 1.7) is quite characteristic of the global picture as well.

For instance, in 2000 the total installed capacity of all US energy conversion devices equaled a staggering 35 TW (that compares to a global

energy flow of some 16 TW-yr).[34] Energy supply-related conversion processes account for some 5 TW, with 30 TW in energy end-use, most notably in the form of automobiles (25 TW). Assuming all cars ran on zero-emission hydrogen fuel cells, the installed capacity of the existing car fleet would be about ten times larger than that of all electricity-generating power plants and could easily substitute the traditional utility-dominated

---

34 In other words, if all US energy conversion devices operated 24 hours a day, 7 days a week, they would transform energy flows twice as large as the entire world energy use. The fact that US primary energy of 100 EJ is equal to 20% of global primary energy use illustrates the comparatively low aggregate capacity utilization of energy conversion devices, particularly in energy end use. (Transportation surveys suggest, for instance, that on average a car is used only one hour per day).

Table 1.7 | Installed capacity of energy conversion technologies (in GW) for the United States, 1850 to 2000.

GW (rounded)		1850	1900	1950	2000
stationary end-use	thermal (furnaces/ boilers)	300	900	1900	2700
	mechanical (prime movers)	1	10	70	300
	electrical (drives, appliances)	0	20	200	2200
mobile end-use	animals/ships/trains/ aircraft	5	30	120	260
	automobiles	0	0	3300	25,000
stationary supply	thermal (power plant boilers)	0	10	260	2600
	mechanical (prime movers)	0	3	70	800
	chemical (refineries)	0	8	520	1280
TOTAL		306	981	6440	35,140

Source: Chapter 24 case studies, Appendix 24.B.

centralized electricity-generation model by an entirely decentralized generation system, powered by cars during their ample idle times. Such drastic transformations in electricity generation have been proposed (e.g., Lovins et al., 1996), especially as a means of accommodating vastly increased contributions from intermittent renewables such as wind, solar thermal, or photovoltaic systems without the need for centralized energy storage. Even if currently futuristic, such daring visions of technology are a useful reminder that the analysis of energy systems needs to look beyond energy flows only and to always consider both major components of energy systems: energy supply *and* energy end-use.

1.6.3.2 Electricity Generation

Electricity is growing faster as a share of energy end-uses than other direct-combustion uses of fuels. Between 1971 and 2008, world electricity production almost quadrupled from 19 EJ to 73 EJ of secondary energy (see Figure 1.19 below) – an absolute increase of 54 EJ. Some 60% of this growth (32 EJ) was in countries outside the Organisation for Economic Co-operation and Development (OECD).

Figure 1.19 depicts the fuel share in global electricity production. About 68% of global electricity is generated from the combustion of fossil fuels, with coal accounting for more than 40% of total production. The share of oil in power production has decreased considerably from 23% to 6% since the first oil crisis in 1973. On the other hand, the share of natural gas has increased from 12% to 21%. Renewable energy sources contribute about 18%, with hydropower accounting for more than 85% of this. Following a rapid expansion in the 1970s and 1980s, nuclear electricity generation has seen little growth since.

Figure 1.19 also shows electricity production for the GEA regions for the base year 2005. Fuel mixes vary widely, primarily reflecting the

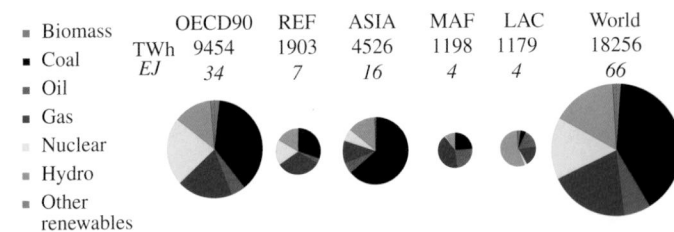

- Biomass
- Coal
- Oil
- Gas
- Nuclear
- Hydro
- Other renewables

	OECD90	REF	ASIA	MAF	LAC	World
TWh	9454	1903	4526	1198	1179	18256
EJ	34	7	16	4	4	66

Figure 1.19 | Electricity output by generating source in 2005: World and five GEA Regions in TWh and EJ (in italics). Source: IEA, 2007a and 2007b. Note: Circle areas are proportional to electricity generated.

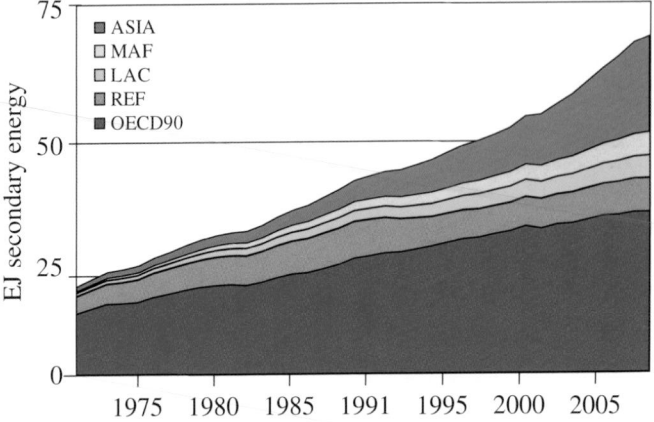

Figure 1.20 | World electricity generation by five GEA Regions, cumulative total (in EJ of secondary energy generated per year). Source: data from IEA, 2010.

availability of local energy resources and to some extent also reflecting past technical and financial capacity to invest in advanced technologies such as nuclear. Coal for electricity generation is most prominent in Asia, accounting for almost 70% of production. OECD and Africa also have significant shares of coal-based power generation. Nuclear energy is primarily used in OECD countries as well as in Eastern Europe and the countries of the former Soviet Union. It makes only a minor contribution in developing countries, except China, which currently has the most nuclear power under construction in the world. Hydropower is unevenly used, providing 66% of electricity in Latin America and the Caribbean. Non-hydro renewable energy in electricity production is low in all regions. However, as a result of various policy support mechanisms in a rapidly increasing number of countries (see Chapter 11), about half of current investments in power generation are in renewable generation.

Figure 1.20 shows regional trends in electricity output: growth trends are across heterogeneous regions. Most additional electricity production since 1971 was actually in the OECD countries (+22 EJ), slightly larger than in the Asia region[35] (+20 EJ/yr). More recent growth trends, however, change this picture dramatically. Since 1990, growth in electricity generation has focused heavily on Asia (most notably in China, an additional 16 EJ of electricity generated), followed by the OECD

---

35   For the definition of GEA regions, see Appendix 1.B.

Table 1.8 | Global GHG and pollutant emissions by source for the year 2005.

	Pollutant Emissions							Main Greenhouse Gases		
	Sulfur	NOx	BC	OC	CO	VOC	PM2.5	CO_2	CH_4	N_2O
	$TgSO_2$	$TgNO_2$	Tg	Tg	Tg	Tg	Tg	$PgCO_2$	Tg	Tg
Energy & Industry	110.0	106.5	5.1	12.2	561.0	131.1	34.6	26.5	105.2	-
international shipping	13.1	18.8	0.1	0.1	1.3	3.1	-	-	0.5	-
transport	3.4	34.6	1.2	1.3	162.0	28.5	2.9	-	1.0	-
industry	27.0	17.2	1.6	2.3	115.3	31.8	13.2	-	0.9	-
residential & commercial	8.8	9.6	2.1	8.2	261.3	38.6	15.7	-	14.3	-
energy Conversion	57.7	26.3	0.1	0.3	21.1	29.1	2.8	-	88.5	-
Non-Energy	4.1	20.8	1.6	23.6	475.3	81.8	32.2	6.8	233.4	-
agriculture (animals, rice, soil)	-	2.3	-	-	-	0.8	-	-	134.4	-
waste (landfills, wastewater, incineration)	0.1	0.3	-	-	4.1	1.5	-	-	72.6	-
waste (agricultural burning on field)	0.2	0.6	0.1	0.7	19.9	2.7	-	-	1.5	-
savannah burning	1.6	11.6	1.5	10.9	222.0	35.1	-	-	8.9	-
forest fires	2.2	6.0	-	12	229.3	41.7	-	-	16.0	-
TOTAL	114.1	127.3	6.7	35.8	1036.3	212.9	66.8	33.3	338.6	12.1

Sources: data from Lamarque et al., 2010; Smith et al., 2011; IPCC-RCP database[38] Houghton, 2008; GEA Chapter 17.

(+9 EJ) and all other developing countries combined (6 EJ). The REF region even experienced a slight drop in electricity output in the aftermath of its economic restructuring.

1.7 Environmental Impacts (Emissions)

1.7.1 Introduction

Energy extraction, conversion, and use are major contributors to GHG emissions and thus global warming. In addition, a host of energy-linked pollutant emissions, including suspended fine particles and precursors of acid deposition, contribute to local and regional air pollution and ecosystem degradation. Energy-related pollutants also result in adverse effects for human health. The largest single source of health impacts of energy is associated with indoor air pollution resulting from the use of traditional biomass in open fires or inefficient cooking stoves by poor people in developing countries. Its human health impacts are estimated to result in

about 2 million premature deaths, or about 42 million person-years of life (DALYs[36]) lost per year, due to respiratory and other diseases, affecting particularly women and children, making access to culturally acceptable, clean, and efficient cooking fuels a priority policy concern.

Table 1.8 summarizes the major sources of global GHGs and selected[37] pollutant emissions. The main pollutants emitted in the combustion of fossil fuels are sulfur and nitrogen oxides, carbon monoxide, and black and organic carbon, including suspended particulate matter. In addition, fossil fuel combustion in the energy sector produces more CO_2 than any other human activity, and contributes to about 30% of global methane (CH_4) emissions. Altogether, the energy sector is thus the biggest source of anthropogenic GHG emissions that are changing the composition of the atmosphere.

1.7.2 CO_2 and other GHGs

CO_2 emissions from fossil energy use in 2005 are estimated at 7.2 Pg C or 26.4 Pg CO_2 (Boden et al., 2010). This represents 80% of all anthropogenic sources of CO_2 in that year, with the remainder associated with land-use changes (deforestation) (Houghton, 2008).

---

36 Thus, on average, each premature death is associated with close to 20 life-years lost. Estimates for the health impact of outdoor air pollution suggest close to 3 million pre-mature deaths/year and some 23 million DALYS. The health impacts of indoor and outdoor air pollution are not additive. See Chapters 4 and 17 for a more detailed discussion. Note: DALYS: "Disability-adjusted Life Years are units for measuring the global burden of disease and the effectiveness of health interventions and changes in living conditions. DALYs are calculated as the present value of future years of disability-free life that are lost as a result of premature death or disability occurring in a particular year. DALY is a summary measure of population health and includes two components, years of life lost due to premature mortality and years lost due to disability" (WHO, 2011).

37 Only emissions of pollutants where energy plays an important role are highlighted here. As such, Chapter 1 does not suggest that other pollutants and emissions sources are not important, but rather that their assessment is beyond the scope of this energy focused précis.

38 See www.iiasa.ac.at/web-apps/tnt/RcpDb.

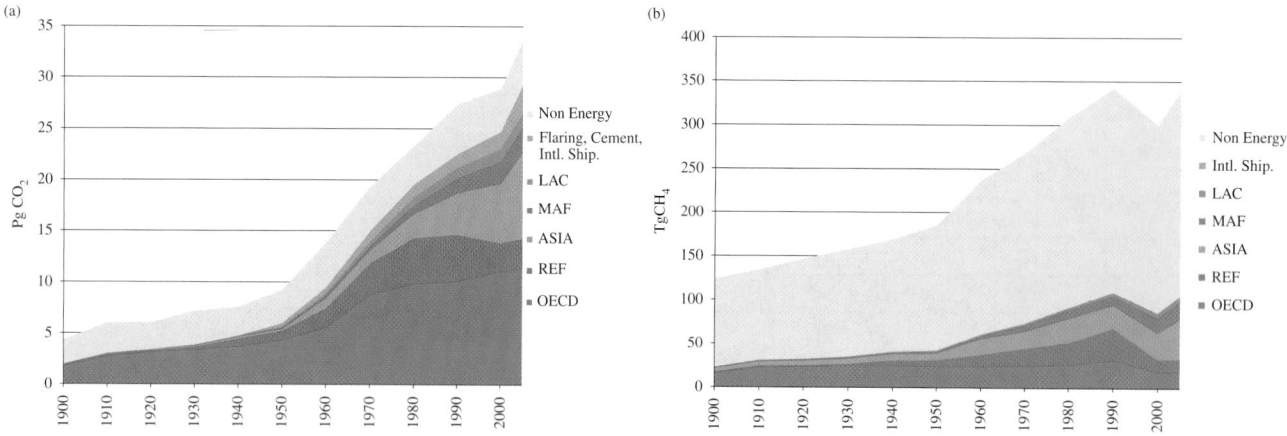

Figure 1.21 | Development of annual energy-related (a) CO_2 and (b) CH_4 emissions by region (compared to global non-energy sources) from 1900 to 2005 in Pg (Gt)(CO_2) and Tg (CH_4). Source: data from Boden et al., 2010; Houghton, 2008; IPCC-RCP database.[39]

Figure 1.21 shows the historical development of fossil energy CO_2 emissions by major world regions (compared to global non-energy-related sources of CO_2). Today's industrialized countries contribute most to the present global CO_2 emissions and have also emitted most of the historical emissions associated with the observed increase in atmospheric CO_2 concentrations. Although they are presently at lower absolute levels, emissions are growing more rapidly in developing countries. The largest source of energy-related carbon emissions are coal and oil (including oil products for feedstocks), with each about a 40% share, followed by natural gas, which represents about 20% of carbon emissions from the energy sector.

CH_4 is the second largest GHG contributing to anthropogenic global warming. Energy-related sources include coal production (where it is a major safety hazard), oil production (from associated natural gas), and natural gas production, transport, and distribution (leaks). Municipal solid waste, animal manure, rice cultivation, wastewater, and crop residue burning are the major non-energy-related sources of CH_4 emissions. While CH_4 emissions from energy accounted for only 30% of total CH_4 emissions in 2005, the relative share of the energy sector has been continuously increasing due to the rise of fossil fuel use throughout the 20th century (see Figure 1.21).

Other GHGs include nitrous oxide (N_2O), tetrafluoromethane (CF_4), sulfur hexafluoride (SF_6), and different types of ozone-depleting hydro-fluoro-carbons (HFCs). These gases are predominantly emitted from non-energy sectors. N_2O is the largest contributor to global warming among these other GHGs (IPCC, 2001). Important sources of N_2O include agricultural soil, animal manure, sewage, industry, automobiles, and biomass burning, with energy contributing about 5% to total N_2O emissions. CF_4, SF_6, and HFCs are predominantly emitted by various industrial sources, with only minor contributions from the energy sector (and are, therefore, not reported separately here).

1.7.3   Traditional Pollutants (SO_x, NO_x, Particulates, etc.)

Energy-related air pollution is responsible for a number of health effects including increased mortality and morbidity from cardio-respiratory diseases (Brunekreef and Holgate, 2002). Developing countries in particular face the greatest burden of impacts from air pollution, both outdoor and indoor. They tend to have high long-term levels of exposure from pollution sources such as forest fires, biomass burning, coal-fired power plants, vehicles, and industrial facilities, thus implying relatively high health impacts. In addition, indoor air pollution due to the lack of access to clean cooking fuels adds to exposure to air pollution, particularly in large parts of Asia and Africa. According to the *World Health Report 2002*, indoor air pollution is the second largest environmental contributor to ill health, behind unsafe water and sanitation (WHO, 2002).

Figure 1.22 shows the historical development of selected pollutant emissions by major world regions (compared to global non-energy-related sources). It builds upon the collaboration of major inventory experts (Lamarque et al., 2010; Smith et al., 2011).

Unfortunately for some important pollutants, such as lead or particulate matter, comparable global inventories with historical trends do not exist. Information for these pollutants is usually summarized at the regional, national, or city level only. Below, the trends for various pollutants are summarized, starting with those that are dominated by emissions from the energy sector.

Anthropogenic *sulfur emissions* have resulted in greatly increased sulfur deposition and atmospheric sulfate loadings and acidic deposition in and around most industrialized areas (Smith et al., 2011). High levels of ambient sulfur concentrations impact human health and cause corrosion. Sulfuric acid deposition can be detrimental to ecosystems, harming aquatic animals and plants, and is also damaging a wide range of terrestrial plant life. In addition, sulfur dioxide forms sulfate

---

39   See www.iiasa.ac.at/web-apps/tnt/RcpDb.

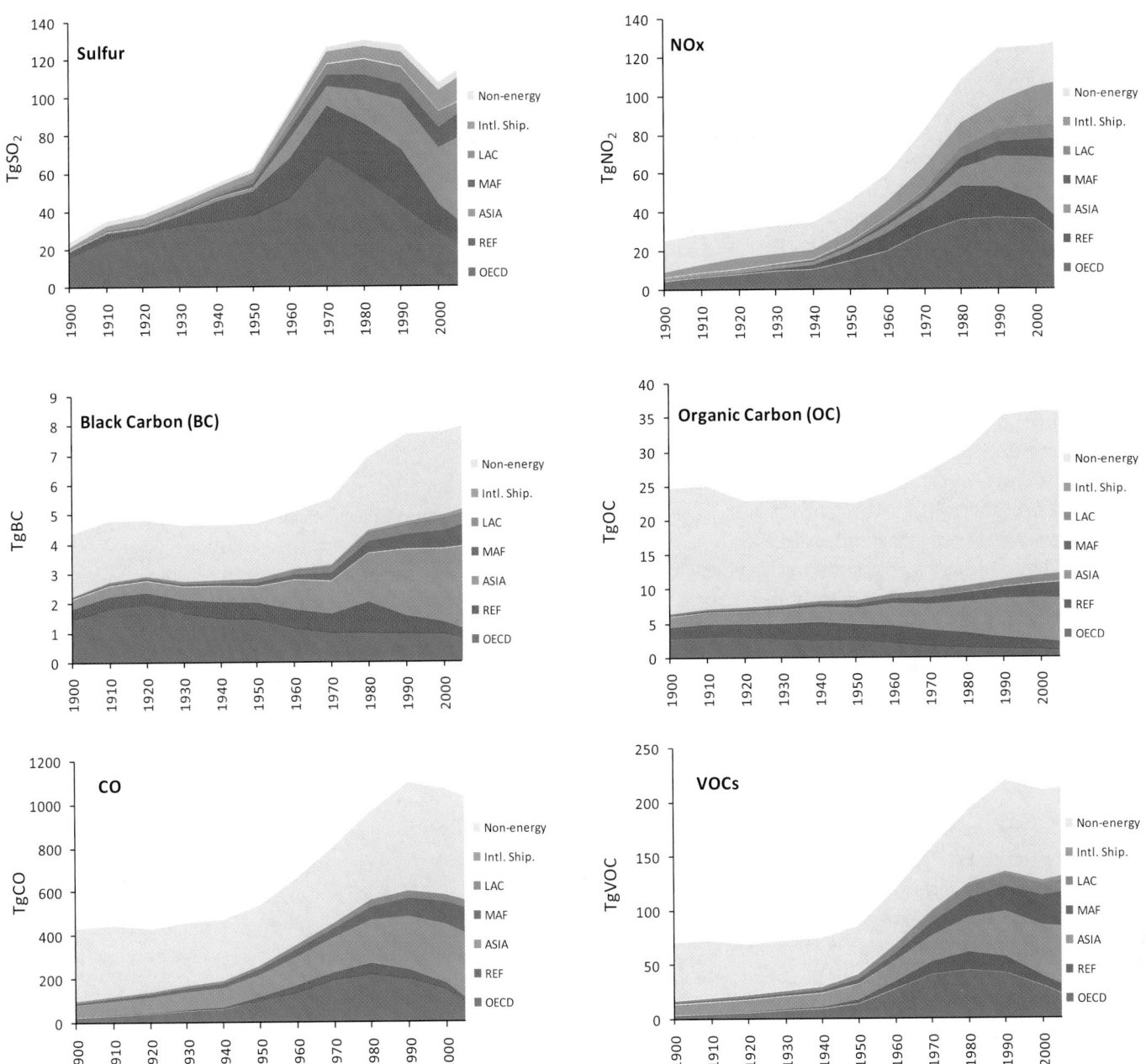

Figure 1.22 | Development of annual energy-related pollutant emissions in Tg: sulfur (SO₂), nitrogen oxides (NOₓ), black carbon (BC), organic carbon (OC), carbon monoxide (CO), and volatile organic compounds (VOCs) by annual region (compared to global non-energy sources) from 1900 to 2005.

aerosols that have a significant effect on global and regional climates. The effect on global climate change of sulfate aerosols may be second only to that caused by CO_2, albeit in the opposite direction (Forster et al., 2007). Stratospheric sulfate aerosols back-scatter incoming solar radiation, producing (regional) cooling effects that mask the global warming signal from increased atmospheric concentration of GHGs. Sulfur is ubiquitous in the biosphere and often occurs in relatively high concentrations in fossil fuels, with coal and crude oil deposits commonly containing 1–2% sulfur by weight (and much higher in some deposits). The widespread combustion of fossil fuels from the energy sector has, therefore, greatly increased sulfur emissions into the atmosphere, with the anthropogenic component now substantially greater than natural emissions on a global basis (Smith et al., 2001; 2011). More than 90% of present sulfur emissions are released from the energy sector. Historically, global emissions peaked in the early 1970s due to the tightening of air pollution legislation particularly in industrialized countries and were decreasing until 2000. Sulfur emissions have resurged since (see Figure 1.22), with increased coal-related emissions in China, international shipping (using heavy fuel or "bunker" oil that has a particularly high sulfur content), and developing countries in general (Smith et al., 2011).

Emissions from nitrogen oxides (NOₓ – predominantly nitrogen dioxide and nitric oxide) contribute to a wide variety of health and environmental problems (respiratory diseases such as asthma, emphysema, and bronchitis; heart disease; damage to lung tissue; acid rain). NO_x is also a main component of ground-level ozone and smog and thus contributes to global warming. Similar to sulfur, NO_x emissions are dominated by the energy sector, which accounts for more than 80% of total anthropogenic NO_x emissions. Emissions from NO_x have continuously been increasing with the use of fossil fuels at the global level. Emissions trends differ significantly, however, at the regional level. While control measures in industrialized countries have resulted in improved air quality and decreasing NO_x emissions since the early 1980s, the rapid increase in NO_x emissions in Asia and from international shipping have more than compensated for improvements elsewhere, leading to an overall global increase in emissions (see Figure 1.22).

The incomplete combustion of carbon-containing fuels (fossil as well as biomass) causes emissions of *carbon monoxide* and other pollutants, including *particulate matter, black carbon,* and *organic carbon.*[40] In addition, black carbon strongly absorbs solar radiation and is contributing to climate warming (although its net aggregated effect is subject to uncertainty), and its deposition is a significant contributor to Arctic ice-melt. In 2005, combustion from the energy sector contributed about 75% of the total anthropogenic emissions of black carbon, with forest fires and savannah burning accounting for the remainder. Due to relatively higher emissions coefficients of organic carbon and carbon monoxide from vegetation fires, the contribution of the energy sector is between 35% and 50% and thus smaller than for black carbon (see Table 1.8 above). Historically, industrialized countries were once the primary source of emissions from incomplete combustion. However, emissions of black carbon and organic carbon in the industrialized world have been declining since the 1920s, as have those of carbon monoxide since the 1980s. Major drivers of this trend are improved technology and the introduction of air quality legislation. Today, the majority of energy-related emissions from incomplete combustion occur in developing countries (see Figure 1.22), resulting in significant health risks, particularly from household combustion of solid fuels (mostly biomass) that affect between half and three-quarters of the population in most poor countries, particularly in rural areas.

Volatile Organic Compounds (VOCs) are emitted by a variety of sources, including industrial processes (solvents), on-road vehicles, refineries, vegetation fires, and residential wood burning, as well as emanations from a wide array of household products. Total global anthropogenic VOC emissions are estimated at about 220 Tg in 2005, with the energy and industry sectors accounting for about 60% of the total. VOCs contribute to the formation of ground-level ozone and include a variety of chemicals, some of which have short- and long-term adverse health effects. As for other pollutants, the energy and industrial emissions have been increasing substantially, and in the recent decades the major sources of VOCs have moved from the industrialized world to developing countries, which contribute about 75% of present energy and industrial VOC emissions.

1.8   Heterogeneity in Energy Use

1.8.1   Introduction

In addition to the temporal variations in global energy use described in earlier sections, there is a huge degree of cross-sectional heterogeneity in energy use evident across the globe today. While aggregate energy statistics are insightful for describing the energy system globally, regionally, or nationally, they often mask the large disparities in energy use both across and within national and regional boundaries. Heterogeneities are evident both in the quantities of energy used and in the structure of use across different nations and sub-populations. These disparities stem, for the most part, from differences in incomes or levels of economic affluence, production and consumption activities, and lifestyles. Yet a small part of the variations might also be on account of differences in climatic conditions and thus energy service needs across regions (e.g., heating/cooling). Differences also exist in the types of energy carriers that are predominantly used and the levels of access to these across countries and populations.

1.8.2   Heterogeneity in Energy Use across Nations

Akin to the uneven development of economies around the world, energy use and service varies significantly across countries. In 2005, the total final energy use was about 330 EJ globally, with the average per capita final energy use about 50 GJ. However, this global average conceals enormous differences in final energy use per capita across nations. The starkest disparity in average national final energy use per capita can be found by comparing Qatar, the country with the highest average in 2005 (445 GJ/capita), with Eritrea, that with the lowest (<5 GJ/capita), a difference of a factor of about 94. The OECD countries, with less than a sixth of the world's population, account for over 45% of total final energy use (see Figure 1.23). Developing countries, with about four-fifths of the world's population, account for just under 40% of this total. OECD countries on average consume over 16 times as much energy per capita than developing countries in South Asia and Africa.

Differences in the amounts of final energy use per capita are mirrored in variations in the structure of energy use across nations and regions. In general, countries with higher levels of energy use per capita also use a larger proportion of their total final energy for transport uses. For instance, the OECD countries use over a third of their final energy for transport. In contrast, in Africa and Asia, over 40% of final energy is for residential and commercial uses. Finally, in addition to variations

---

40 Black carbon: pure carbon (soot) emitted ("black smoke") from the combustion of fossil fuels, biofuels and other biomass (vegetation burning). It absorbs sunlight and reradiates heat into the atmosphere, thus producing a climate warming effect. Organic carbon: carbon combined with oxygen/hydrogen atoms (organic radicals) mainly arising from the incomplete combustion ("brown" or "white smoke") of biomass. Organic carbon aerosols (fine particles suspended in the atmosphere) tend to back-scatter sunlight, producing a cooling effect on climate.

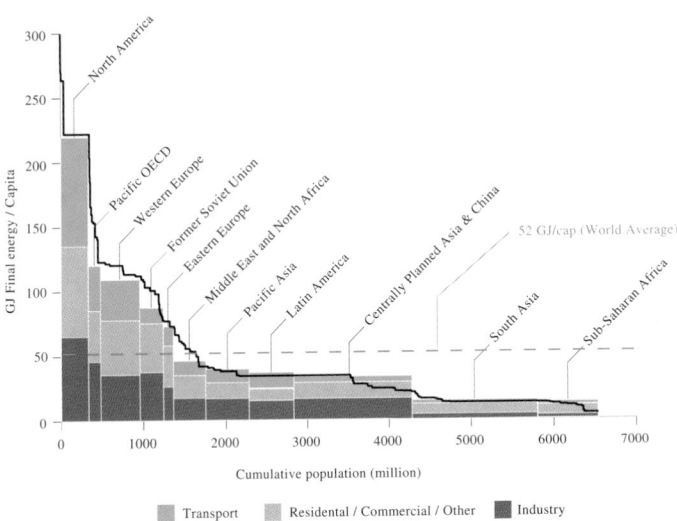

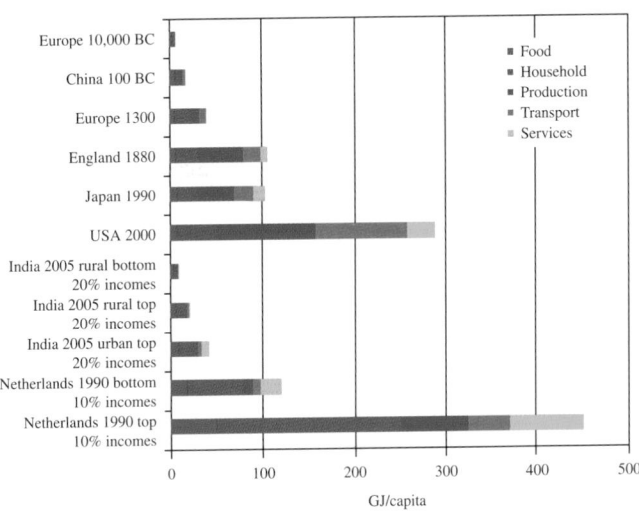

Figure 1.23 | World final energy (GJ) per capita versus cumulative population for 11 GEA regions sorted by declining per capita energy use (with final energy use disaggregated by sector and total, color bars); and total per capita final energy use for 137 countries in 2005 (black, solid line). Dashed horizontal line shows the average global final energy use per capita. Source: data from IEA, 2010.

Figure 1.24 | Per capita primary energy use by service category over time and across different populations. Source: historical estimates: Smil, 1991; Japan, United States: IEA, 2010; India: Pachauri, 2007; Netherlands: Vringer and Block, 1995.

in the levels and purposes of energy use across nations, countries also exhibit very divergent patterns of energy use. Nearly a third of all final energy use in developing countries is unprocessed biomass, with this fraction being close to 90% for some least-developed countries. In addition, about 30% of the population in developing countries lack access to, and so do not use, any electricity. In developed OECD countries, however, almost all final energy use is in the form of electricity or oil/gas products. These differences in patterns have implications for the level of energy services across nations, as some carriers like traditional biomass used in traditional end-use devices have very low efficiencies, and are associated with high emissions, and social externalities.

1.8.3   Heterogeneity in Energy Use within Nations

Often the variance in energy use within nations can be of the same or greater order of magnitude as that across nations. In such instances, aggregate national indicators disguise intra-national disparities, sometimes grievously. Within nations, substantial differences in energy use exist across geographical regions, rural versus urban residents, and among other socio-economic and demographic sub-groups of the population. Spatial patterns of economic development and industrial activity are reflected in variations in quantities and structures of energy use between regions. In many developing countries, one can find evidence of a dual economy with substantial disparities in quantities and types of energy use between rural hinterlands, with poor infrastructure and formal development, and urban metropolitan areas that are the centers of industrial production and economic activity. Thus, for instance, as shown in Figure 1.24, the poorest 20% of the rural population in India have per capita energy use levels comparable to those estimated for the pre-agrarian European population some 10,000 years ago. Even

the richest 20% of the rural population in India uses only about half as much energy per capita as the richest 20% of the urban population, with their energy use levels comparable to the estimates for China in 100 B.C. Some of this difference in the quantity of energy used can be explained by disparities in income levels across rural and urban regions. However, large disparities in the structure of energy use are also evident, both in terms of uses of energy and the types of energy used.

The starkest disparities in energy use within (and between) nations are those between rich and poor people. Thus, as Figure 1.24 illustrates, the richest decile of the Dutch population uses almost four times as much energy per capita as the poorest decile, which is about the same order of difference as between the richest and poorest urban Indian quintiles. The richest 20% of urban Indians use only a third as much of the energy used by the poorest 10% of the Dutch, albeit the richest 20% in India will include many examples of very wealthy individuals whose energy use vastly surpasses that of the average Dutch top 10% income class. As such, these illustrative numbers reflect the wide disparities in incomes and development levels across and within nations. The richest Dutch also use almost three times as much energy for food on average as their poorest compatriots. This, of course, does not imply that rich people eat three times as much as poor people in the Netherlands. However, the food habits and types of provisions consumed do differ. For instance, the rich Dutch eat more exotic fruits and vegetables (e.g., Kiwi fruit flown in from New Zealand) than the poor which explain their much larger food-related (embodied) energy use. The biggest differences in the structure of energy use between rich and poor people, both within and across nations, is the substantially larger share of energy used for transport and for the consumption of products and services. Poor people, by contrast, use the largest proportion of energy for basic necessities such as food and household fuels (cooking and hygiene). These differences illustrate the substantial variations in lifestyles and growing consumerism evident with rising incomes and retail market sophistication.

1.8.4 Disparities in Energy Use

While fairness and equity are normative, ethical concepts, several methodologies and metrics exist to measure dispersions and distributions which help to describe disparities in energy use. Lorenz curves and Gini indices or coefficients are widely used to measure inequalities in income and wealth. The Lorenz curve is a graphical representation of a cumulative distribution function, often with a ranked cumulative distribution of population on the x-axis versus a ranked distribution of cumulative value of a given variable such as income, wealth, or energy on the y-axis. A perfectly equal distribution is described by a straight line where y = x along the diagonal or along 45 degrees, where every given percentage of the population consumes or owns an equal percentage of the variable in question (e.g., energy, wealth, etc.). The greater the distance of the Lorenz curve from this diagonal, the greater the degree of inequality it represents. The Gini coefficient, also used as a measure of inequality, is mathematically represented as the ratio of the area between a Lorenz curve and the diagonal (or line of perfect equality) to the total area under the diagonal. The Gini coefficient can range from 0 to 1, with a value closer to 0 representing a more equal distribution. In addition to Lorenz curves and Gini coefficients, other measures of inequality commonly in use are ratios of percentiles, deciles, quintiles, or quartiles of the population.

Figure 1.25 illustrates inequality across nations by depicting the Lorenz curves for important energy and economic variables for the year 2000. The x-axis depicts the ranked cumulative distribution of population by nation, while the cumulative disposal of income (in PPP terms), final energy, and electricity are shown on the y-axis.

In terms of income and final energy use, the poorest 40% of the world's population only disposes of some 10% of global income and final energy use; the richest third disposes of two-thirds of global income and final energy. It is noteworthy that final energy use mirrors prevailing (vast) income inequalities closely. Energy and economic poverty and wealth thus go hand in hand. Access to electricity is even more inequitable. In 2005, some 23% of the world's population (1.4 billion people) had no access to electricity at all.

1.9 The Costs of Energy

1.9.1 Accounting Frameworks and Different Types of Costs

In one way or another, energy services carry a price tag. The price a consumer pays for a particular energy service based, for example, on electricity use is made up by a variety of components, the most important of which are generating costs, systems costs, rents, profits, taxes, subsidies, and externalities.

Generating costs are not only a key component determining the price of a service but also a central decision criterion for investment and operating decisions alike. Generating costs they consist of three major

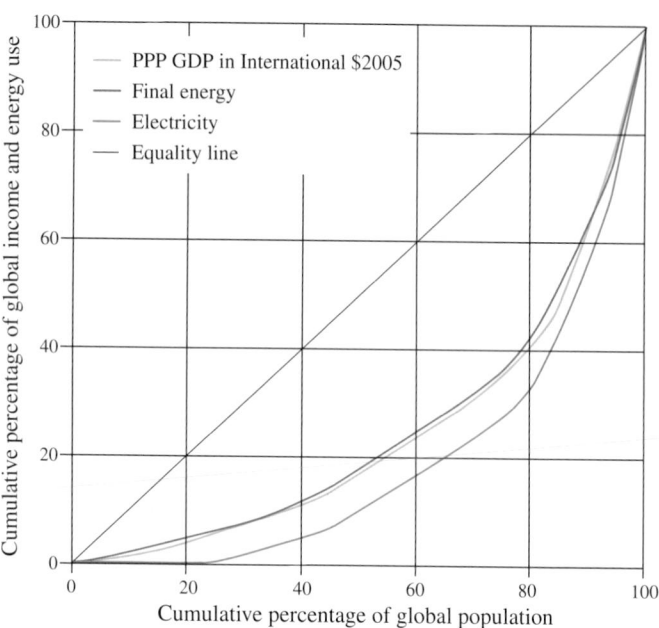

Figure 1.25 | Lorenz curves of energy inequality, measuring cumulative global population (in percent) disposing of corresponding fraction of cumulative income (in percent of PPP$ (green), final energy (blue), and electricity use (red) (in percent of Joules energy used) for the Year 2000. Source: IEA, 2010 and UNDP and WHO, 2009.

components: capital costs, fuel costs, and non-fuel operating and maintenance (O&M) costs. Capital costs are the costs associated with the construction/acquisition/purchase of a power plant, refinery, or home furnace. Fuel costs are the expenditures associated with the fuel supply for plant operation or service provision. O&M costs cover labor costs, insurance, consumables other than fuel, repairs, etc. More recently, capital costs also include decommissioning expenditures at the end of a plant's service life, while O&M costs may include waste disposal costs.

While fuel and O&M costs are largely incurred on a per-use basis, capital or investment costs occur upfront – for some technologies spread over several years of plant construction – before earning revenue or providing energy services for the investor. Capital costs must be recovered over the lifetime of the investment, reflecting the wear and tear of the plant (the investment) over its economic lifetime.

The levelized cost[41] of electricity (LCOE) is a widely used tool in policy analysis for comparing the generating costs of different technologies over their economic life. A critical parameter in the LCOE approach is the discount rate, which reflects the interest rate on capital (cost of capital or return) for an investor in the absence of specific market or technology risks. LCOE spreads the capital costs (including the finance costs) uniformly over the lifetime of an investment, accounts for the fuel and O&M costs, and calculates the specific costs per unit of energy delivered.[42]

---

41 For a review see Anderson, 2007. For a review quantitative cost estimates for electricity generation see Heptonstall, 2007 and NREL, 2010.

42 Note: LCOE assumes perfect knowledge about future fuel prices and interest rates several decades into the future. Scenarios of different price trajectories are commonly used to reflect uncertainty.

Table 1.9 | Total levelized costs of different electricity generation technologies (in percent using a 5% discount rate) and representative cost ranges in 2005US$/MWh as used in GEA. Note: These are direct energy (electricity generating) costs only, i.e. excluding externality costs; Data source: Chapters 12 and 17, and IPCC, 2011.

		Solar PV	Wind (onshore)	Nuclear[a]	Advanced coal	Adv. coal with CCS	Gas combined cycle
Capital	$/kWe	900–2800	900–1300	4000–6200	1100–1600	1700–2400	400–500
O&M	$/kWe	6–18	19–30	118–180	46–65	69–96	16–20
Fuel	$/GJ	0	0	0.7–0.9	1.3–2.8	1.3–2.8	2.6–6.5
Waste	$/MWh	0	0	1–2	0	6	0
Total generating costs	$/MWh	27–151	21–131	53–100	27–46	44–69	24–49

(a) Current (pre-2010) nuclear investment costs under construction in several developing countries range between 1800 and 2500 $/kWe.

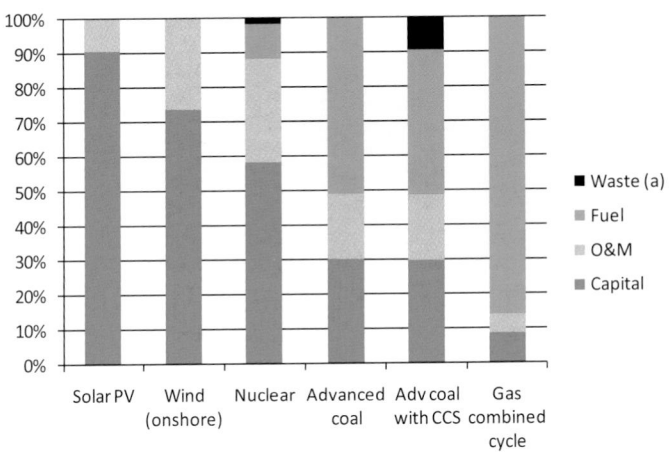

Figure 1.26 | Structure of total levelized costs of different electricity generation technologies (in percent using a 5% discount rate) and representative cost ranges for 2030 as used in the GEA pathways (cf. Chapter 17) for capital costs, operation and maintenance (O&M), fuel costs, as well as waste disposal costs for nuclear and advanced coal with carbon capture and sequestration (CCS). All values are given in 2005US$ and are shown in Table 1.9. Note: (a) Includes decommissioning costs for nuclear power and costs for transport and disposal of 90% of CO_2 emissions for advanced coal power plants with carbon capture and storage (CCS).

The relative structure of the various generating cost components varies significantly per unit of output for different generating options (see Figure 1.26 and Table 1.9, using a real annual discount rate of 5%), and the variation indicates the inherent risks associated with a particular option. For example, gas combined cycle technology (CCGT) has the lowest capital costs but the highest fuel costs of the options shown in Figure 1.26 and Table 1.9. Consequently, CCGT generating costs are almost all fuel costs. Any change in natural gas prices thus impacts its generating costs greatly. Conversely, nuclear power generation is dominated by high capital costs (>70%), with fuel cycle and O&M costs assuming approximately equal shares of the remaining costs.[43] The high share of capital costs exposes nuclear power projects to financial risks associated with rising interest rates and to cost escalation caused by delays in construction completion. Adding carbon

capture and storage (CCS) can also increase costs substantially: typically adding some 50 $/MWh levelized costs to pulverized coal fired power plants (and 20–30 $/MWh for IGCC or natural gas electricity generation), see Chapter 12.

In addition to the generating costs, the price of electricity for consumers then includes transmission and distribution (T&D) costs and taxes or subsidies. Taxes and subsidies are policy instruments to influence consumer behavior. Taxes can be used to discourage politically undesirable behavior patterns, while subsidies provide incentives to adopt a more desirable investment or consumption pattern. Subsidized electricity or gasoline prices are also an instrument for extending access to energy services to low-income families, supporting small rural business developments, or connecting rural areas to markets.

Figure 1.27 compares gasoline prices with and without taxes for a variety of countries. While prices without taxes vary by a factor of two, this doubles to a factor of four when taxes are included. The taxes imposed by countries reflect national policy objectives, e.g. revenue needs, trade balances, etc., and not necessarily the countries' endowment with oil resources. For example, oil-exporting Norway features the second highest gasoline taxes in this comparison (equivalent to a carbon tax of US$576/tonne of CO_2), while oil-importing US has the second lowest gasoline taxation (equivalent to US$56/tonne of CO_2). Other oil-exporting countries such as Kuwait (not shown in Figure 1.27) even subsidize[44] domestic gasoline use.

---

43   Unlike natural gas or coal-fired generation, the fuel cost of nuclear power generation is not dominated by the resource (uranium) input price but by enrichment and fuel fabrication costs. Uranium accounts for approximately 25% of fuel costs only.

44   In the most general definition, an energy subsidy is represented by the difference between (low) local and prevailing (high) world market prices (without taxes). When the local gasoline price is below the marginal costs of producing and refining crude oil, this represents a direct financial transfer/subsidy from energy producers (usually nationalized industries – i.e., from the government) to energy consumers (households, taxi companies). Beyond that, any difference between local and world market prices is best conceptualized as opportunity costs associated with foregone potential export revenues, and also classified as energy subsidy. The marginal costs of producing a barrel of crude oil in many oil-exporting countries can be as low as US$5/bbl. The difference to a world market price of say US$100/bbl (in economic theory determined by the global marginal [i.e., highest] production costs plus profits) is referred to in economics as "scarcity rent," leading to vast financial transfers and wealth to energy producers, which may, however, not always have only beneficial effects (the so-called "resource curse," Humphreys et al., 2007).

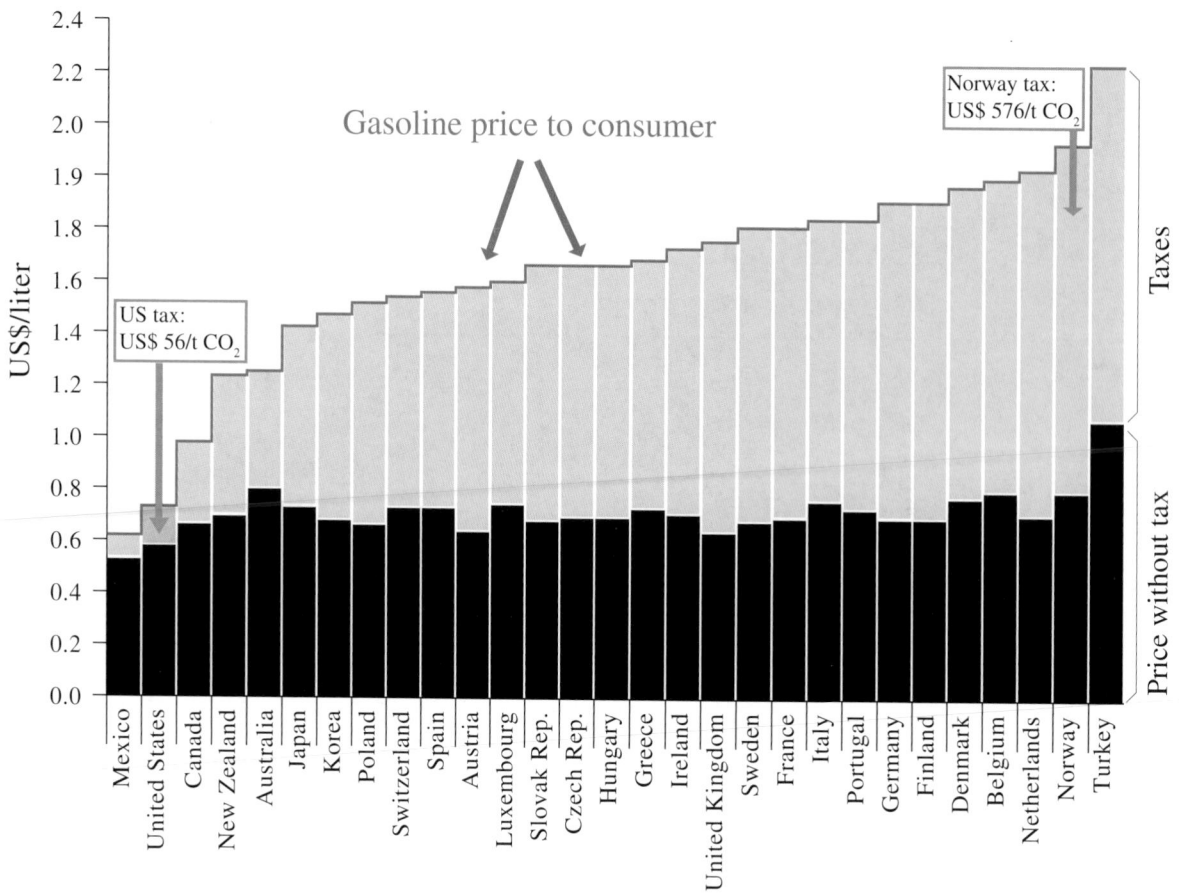

Figure 1.27 | Gasoline prices with and without taxes in US$/liter and implied price of carbon (US$/tCO$_2$) for 1st quarter of 2010. Source: data from IEA, 2011.

Finally there are cost elements caused by the conversion and use of energy and energy services which – although real – are not included in the price paid by the consumer but paid by society at large. Examples of such costs, called "externalities," are health and environmental damage costs resulting from air and water pollution from fossil fuel combustion or lower property values due to the proximity of a nuclear power plant or noise from wind converters. Ignoring externalities masks the true costs of energy and sends the wrong signal to the market place. Charges or taxes on carbon emissions or investment in carbon capture and storage (CCS) technology are ways to internalize externalities caused by GHG emissions. They also change the merit order of electricity generation favoring low-GHG emission technologies.

While investment decisions are guided by LCOE considerations, operating decisions and dispatch of an existing fleet of power stations are based on short-term marginal costs – in essence, fuel costs and possibly emission charges. Capital costs are no longer a decision criterion, as these are "sunk."

Figure 1.28 explains the inherent substitutability between capital and fuel costs using the example of providing heat for cooking. Higher-efficiency stoves are more capital-intensive but reduce fuel costs, which in a rural developing country context often mean time spent collecting wood for fuel. Shifting to more capital-intensive stoves (and higher-exergy fuels) reduces the time spent on fuel supply and at the same time improves indoor air quality through lower combustion-related emissions. The time released from gathering fuel is then available for more productive uses. This freed time, lower pollution exposure, and improved human health are important examples of positive externalities of moving to cleaner household fuels.

A transition to an improved cooking service can occur in one of two ways, as shown in Figure 1.28. A simple shift or substitution to higher-exergy energy carriers (e.g., from firewood to liquefied petroleum gas – LPG) will result in higher combustion efficiency, lower combustion-related emissions, lower time costs associated with fuel collection, but higher capital costs for stoves (and cash expenditure for commercial fuels). On the other hand, improvements in cooking services can also be achieved through the use of more

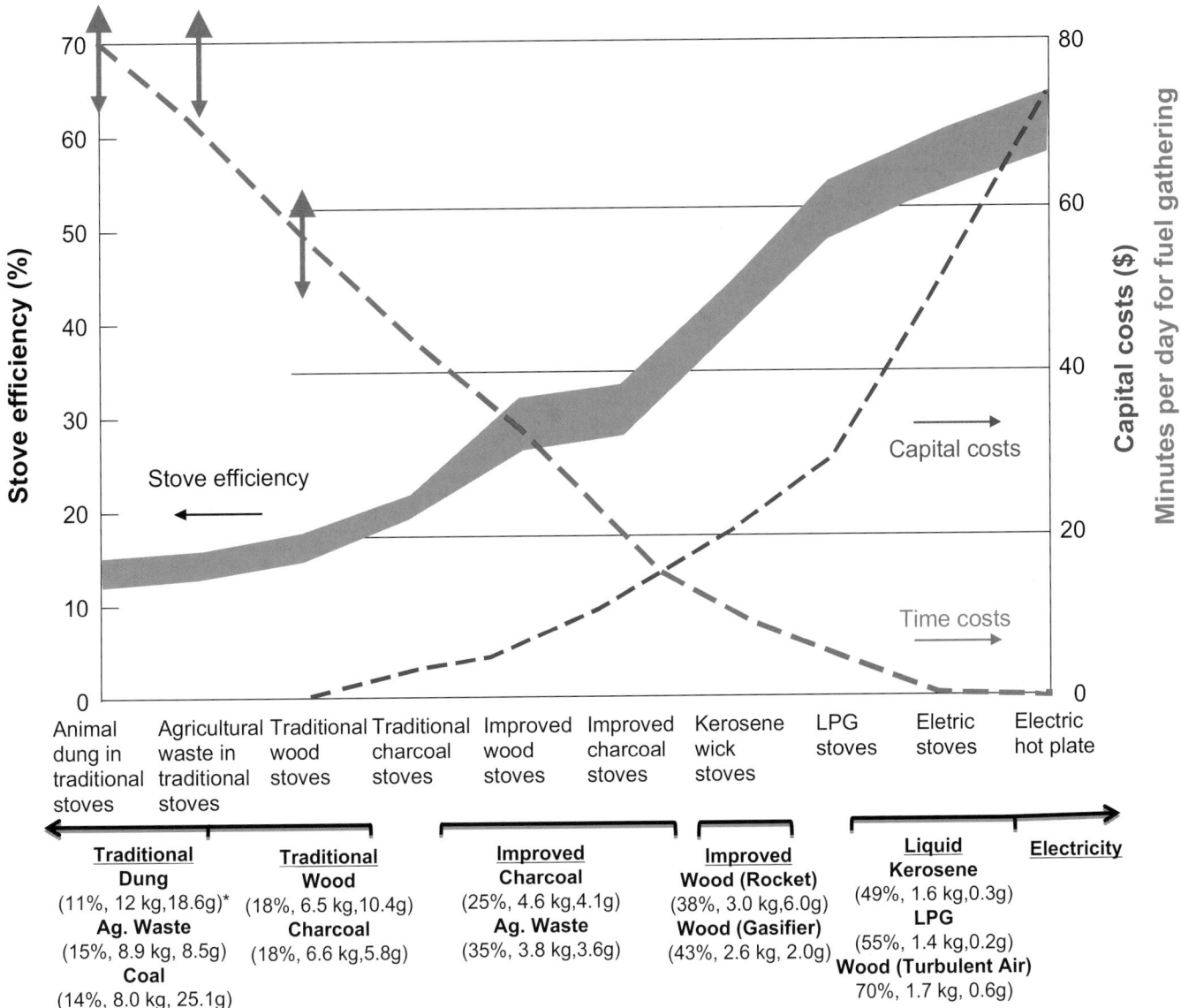

*Avg. thermal efficiency (%), GHG emissions (kgCO$_2$-e/day assuming 11 MJ delivered/day), Particulate matter emissions (g/day assuming 11 MJ delivered/day.

Figure 1.28 | Different costs of energy service provision. Example: cooking in developing countries. Notes: Red arrows show uncertainty ranges in time costs. GHG emissions are illustrative estimates and include all greenhouse gases and other emissions that affect radiative forcing. Emissions from fuel production and renewability of charcoal, wood, and agricultural residues are not included. Source: adapted from OTA, 1991 and Chapter 3.

capital-intensive improved technologies that continue to use traditional fuels (e.g., firewood and residues) but more efficiently (e.g., biogasifiers).

Further cost components related to Figure 1.28 are "inconvenience" or "opportunity" costs. Depending on the levelized costs of the heat for cooking, it might well be that using traditional fuel wood in an inefficient stove is the cheapest way to produce the required heat. However, factoring in alternative uses of the time spent for wood collection – for example, for other productive uses or just leisure activities – turns wood collection into an inconvenient task. A more efficient stove using commercial fuels reduces pollution and time spent gathering fuel wood, and hence reduces *inconvenience* costs. Likewise, the capital spent on a

more efficient stove may not be available for other investments – say, a pump for irrigation – and thus represents an *opportunity* cost.

1.10   Roadmap to the Chapters of the GEA

Earlier transitions of the world's energy system, from biomass to coal, to oil, and now to a mix of coal, oil, and natural gas as the dominating energy carriers were all driven by convenience and cost reductions. Coming energy transitions will occur in a world that has changed through the "great acceleration" (Chapter 3) that started in the 1950s and is still ongoing. The next transitions will have to consider these changes that in fact create demands on the performance of the coming energy systems in order for the world to develop in a sustainable manner.

The 25 chapters of the GEA are divided into four clusters.

Cluster I sets out to describe and assess the nature and magnitude of energy system changes required from a key set of conditions and concerns. Chapter 2 evaluates the role of energy in poverty alleviation and socio-economic growth and what will be required from the energy systems to make poverty a condition of the past, especially in terms of access to electricity and clean cooking fuels and practices. Chapter 3 reviews the environmental impacts of energy systems and what changes would be required, especially in terms of emissions reductions, to protect the environment as it is now known, including mitigating climate change. Chapter 4 addresses the health impacts of energy systems, especially indoor and outdoor air pollution, and health impacts of climate change. Chapter 5 analyzes energy security from several points of view, and Chapter 6 reviews the demands for energy services from a growing global population with increased standards of living, especially for the poorer parts of the world. Together, indicators defined and quantified for the purpose of the GEA in these five areas are used to define a "sustainable" state of the world from an energy systems perspective by 2050.

Cluster II reviews the resources and energy technologies available, or on the horizon, to address the energy sustainability challenges.

Chapter 7 evaluates reserves and resources of fossil fuels, fissile material, and renewable energy flows. Chapters 8 through 10 deal with energy end-use in industry, transport, and buildings, respectively, and Chapters 11 through 16 review energy supply-side options, including renewable energy technologies, fossil fuel technologies, carbon capture and storage, nuclear energy, energy systems operation, and transitions to new energy systems.

Cluster III then explores how the elements of Cluster II can be combined into *systems* that address all the concerns identified in Cluster I, all at the same time. Chapter 17 presents this back-casting (normative scenario) analysis and identifies a number of conceivable energy systems that would meet the goals from Cluster I. Special attention is then given to urbanization (Chapter 18), energy access for development (Chapter 19), trade-offs in land and water use (Chapter 20) and life-styles (Chapter 21).

Cluster IV deals with policies and institutions to bring about the sustainable energy systems that were identified in Cluster III. Chapter 22 reviews the overall implementation situation, Chapter 23 the implementation of options for access to modern energy carriers and clean cooking fuels, Chapter 24 technology innovation systems, and Chapter 25 the capacity development required in terms of policies, institutions, and people that will be the agents of change to make the next energy transition toward sustainability happen.

Appendix 1.A Accounting for Energy

1.A.1 Introduction

The discussion of energy systems above described how primary energy occurs in different forms embodied in resources as they exist in nature, such as chemical energy embodied in fossils or biomass, the potential kinetic energy of water drawn from a reservoir, the electromagnetic energy of solar radiation, or the energy released in nuclear reactions. A logical question is, therefore, how to compare and assess the potential substitutability of these energy "apples and oranges." This is the objective of this more technical section.

The primary energy of fossil energy sources and biomass is defined in terms of the heating value (enthalpy[45]) of combustion. Together, combustibles account for about 90% of current primary energy in the world, corresponding to some 440 EJ in 2005. There are two different definitions of the heat of combustion, the *higher* (HHV) and *lower heating values* (LHV – see the discussion below), but otherwise the determination of apple-to-apple primary energy comparisons among combustible energy sources is relatively straightforward.

The situation is more complicated for non-combustible primary energy sources such as nuclear energy and renewables other than biomass. In these cases, primary energy is not used directly but is converted and transformed into *secondary energy* (energy carriers) such as electricity as in the case of modern wind or photovoltaic power plants. The measurable energy flow is the secondary energy, whereas the primary energy input needed to generate electricity needs to be estimated. In the two examples of wind and solar photovoltaics, primary energy estimates of the kinetic energy of wind and the electromagnetic energy of solar radiation are needed to determine primary energy equivalences to other energy sources. There are various conventions that specify the appropriate conversion from different renewable energy forms based on the generated electricity. For these conventions, the type of energy flow and its technological characteristics – such as the efficiency of the wind converters or photovoltaic cells – are needed. These various important accounting issues are dealt with below, starting with units and heating values.

1.A.2 Energy Units, Scale, and Heating Values (HHV/LHV)

Energy is defined as the capacity to do work and is measured in joules (J), where 1 joule is the work done when a force of 1 Newton (1 N = 1 kg m/s²)

is applied over a distance of 1 meter. Power is the rate at which energy is transferred and is commonly measured in watts (W), where 1 watt is 1 joule/second. Newton, joule, and watt are defined as basic units in the International System of Units (SI).[46]

There is a wide variety of energy units which can be converted into each other. Figure 1.3 in Section 1.2.1 above, gives an overview of the most commonly used energy units and also indicates typical (rounded) conversion factors (see also Appendix 1.B). Typically, the choice of an energy unit depends on various factors such as the type of the energy carrier itself, the respective energy sector, as well as geographical and historical contexts. Next to the internationally standardized SI units, the most common energy unit used for electricity is the kilowatt-hour (kWh), which is derived from the joule (one kWh (1000 Watt-hours) being equivalent to 3600 kilo-Watt-seconds, or 3.6 MJ). In many international energy statistics (e.g., IEA and OECD) tonnes of oil equivalent (1 *toe* equals 41.87 x 10⁹ J) is used as a core energy unit, but it is not included in the SI system. Certain energy subsectors often use units that apply best to their respective energy carrier. For example, the oil industry uses barrels of oil equivalent (1 *boe* equals 5.71 x 10⁹ J or about 1/7 of a *toe*), the coal industry tonnes of coal equivalent (1 *tce* equals 29.31 x 10⁹ J), whereas the gas industry uses cubic meters of gas at a normalized pressure (1 m³ of methane equals 34 MJ – all numbers refer to LHV; see the discussion below). Some countries such as the US use the imperial system of units, which include British Thermal Units (1 BTU equals 1055 J) as a unit for energy, cubic feet (for natural gas, one ft³ equals about 1000 BTU, or 1 MJ), and barrels as volumetric energy units (bbl is another name for *boe*).

The *calorific value* or *heating value* of a fuel expresses the heat obtained from combustion of one unit of the fuel. It is important to distinguish between the higher heating value (HHV or gross calorific value) and the lower heating value (LHV or net calorific value). Most combustible fuels consist of hydrocarbon compounds that are primarily mixtures of carbon and hydrogen. When the hydrogen combines with oxygen, it forms water in a gaseous state, which is typically carried away with the other products of combustion in the exhaust gases. Similarly, any moisture present in the fuel will typically also evaporate. When the exhaust gases cool, this water will condense into a liquid state and release heat, known as *latent* heat, which can be captured and utilized for low-temperature heating purposes.

The *HHV* of a fuel includes the latent heat recovered from condensing water vapor from combustion. Modern condensing natural gas or oil boilers can capture this latent heat.[47] The *LHV* excludes the latent heat of the water formed during combustion.

---

45 Enthalpy – from the Greek "to warm/heat" – is the product of the mass of a fuel times its specific enthalpy, which is defined as the sum of its internal energy (from combustion) plus pressure times volume. Heating values per unit mass of a fuel are, therefore, defined for standardized pressure/volume conditions.

46 International System of Units – *SI* from the French *le Système international d'unités*.

47 Commercial advertisements often inappropriately refer to furnaces as "more than 100% efficient," which is thermodynamically impossible. The seeming paradox simply results from comparing apples and oranges in the form of LHV fuel energy inputs but HHV combustion energy releases.

The differences between LHV and HHV are typically about 5–6% of the HHV for solid and liquid fuels, and about 10% for natural gas (IEA, 2005). Typically, the LHV is used in energy balances, since most current energy conversion devices are still not able to recover latent heat. The distinction between HHV and LHV becomes important when comparing international energy statistics and balances (usually based on LHV, as in IEA or UN statistics) with national ones that can sometimes be based on HHV (as in case of the US Energy Information Administration, EIA). Care is also required when applying fuel-specific emission factors – for example, for CO_2 – that are specified separately per HHV or LHV to the corresponding heating value of the fuel as defined in the underlying energy statistics but not always spelled out prominently. As a precautionary measure to avoid accounting errors, literature sources on emission factors and energy use numbers that do not specify their underlying heating value concept definition should be avoided. In this publication both definitions are used, but the LHV is the default, as in most international energy statistics (e.g., UN or IEA).

1.A.3   Accounting for Primary Energy

As discussed above, the determination of the primary energy equivalent of combustible fuels (all fossils as well as biomass) is straightforward (only a consistent HHV or LHV reporting format needs to be adopted). For non-combustible energies (modern renewables such as wind or solar photovoltaics, geothermal, hydropower, and nuclear), there are different conventions that specify the appropriate conversion factors to account for primary energy equivalents: the *substitution*, the *direct equivalent*, and the *physical energy content* method (which is a hybrid combination of the substitution and direct equivalent methods). The share of non-combustible energy sources in total primary energy supply will appear to be very different depending on the method used (Lightfoot, 2007; Macknick, 2009):

The (partial) *substitution* method estimates the primary energy from non-combustible sources as being equivalent to the LHV or HHV of combustible fuels that would have been required in conventional thermal power plants to substitute the generated electricity or some other secondary energy form. Basically, this means that some average or representative efficiency of thermal power plants is applied to calculate the equivalent primary energy from the generated electricity from nuclear and renewables outside biomass.[48] This method is used, for example, by BP (2010a) and WEC (1993) and as the default method in the GEA (see Annex-II Technical Guidelines) to maintain a consistent accounting framework across different energy options.[49] Throughout

the GEA there is always a clear indication if another method is used. The difficulties with this method include choosing an appropriate thermal power generating efficiency factor and the fact that the method displays "hypothetical" transformation losses in energy balances which end up as reported primary energy use, but which do not have any physical basis.

The (direct) *equivalent* method counts one unit of secondary energy such as generated electricity from non-combustible sources as one unit of primary energy. This method is also often used in the literature – for example, by UN Statistics (2010) and in multiple IPCC reports that deal with long-term energy and emission scenarios (Watson et al., 1995; Nakicenovic and Swart, 2000; Morita et al., 2001; Fisher et al., 2007). The difficulties with this method are twofold: (i) an increase in the share of non-combustible energy sources results in the apparent efficiency improvement of the whole energy system because ever higher shares of primary energy have a definitional 100% "efficiency" of conversion into secondary forms, and (ii) actual conversion efficiencies even for these non-combustible sources of primary energy are substantially lower than 100% – for instance, the theoretical maximum efficiency (under optimal conditions) of converting wind kinetic energy into electricity is about 59%, but actual machines today achieve at best 47%.

The (physical) *energy content* method adopts a hybrid approach, using the direct equivalent approach for all energy sources other than those where primary energy is heat, such as nuclear, solar thermal, and geothermal energy sources. Thermal energy generated in a nuclear, geothermal, or solar power plant is considered primary energy equivalent. For example, in the case of nuclear energy, the heat released by fission is taken as primary energy, even though two-thirds are dissipated[50] to the environment through the turbine's condenser and the reactor cooling system and only one-third is actually delivered as electricity. This approach is identical to the case of fossil energy, for which the heat of combustion is taken as primary energy. In effect, the hybrid system leads to the following assumed primary energy accounting: (i) *substitution method* for heat from nuclear, geothermal, and solar thermal, and (ii) *direct equivalent method* for electricity from hydropower, wind, tide, wave, and solar photovoltaic energy. This hybrid method is used by the OECD, the International Energy Agency and Eurostat (IEA, 2005). The difficulty with this method is that it can result in confusion, as some energy forms such as hydropower are accounted for by the direct equivalent method, while for others such as nuclear conversion efficiencies are applied. Even though they both generate about the same electricity in the world, nuclear's primary energy equivalent is counted as three times larger than that of hydropower.[51]

---

48 Note, however, that different variants of the substitution method use somewhat different conversion factors. For example, BP applies 38% conversion efficiency to electricity generated from nuclear and hydro (BP, 2010), whereas the World Energy Council uses 38.6% for nuclear and non-combustible renewables.

49 In the GEA a uniform primary accounting equivalent of 35% conversion efficiency for electricity from non-combustible sources (equivalent to the global average of fossil-fuel power generation in 2005) and of 85% conversion efficiency for heat is applied.

50 In principle such waste heat could be "recycled" but would require a close co-location of nuclear power plants with main energy uses such as major cities, which raises issues of safety and public risk perception.

51 For example, in IEA/OECD (2005) the assumed conversion efficiency factor for hydropower, solar electricity, and wind is 100%, for nuclear power it is 33%, and for geothermal electricity it is 10%.

Table 1.A.1 | Comparison of global primary energy supply in 2005 using three different accounting methods for primary energy.

	GEA Substitution Method		Direct Equivalent Method		Physical Energy Content Method	
	EJ	%	EJ	%	EJ	%
Fossil fuels	389	78	389	85	389	81
Biomass	46	9	46	10	46	10
Nuclear	28	6	10	2	30	6
Hydro	30	6	11	2	11	2
Other Renewables	< 3	1	1	<1	3	1
Total	496	100	457	100	479	100

Source: data from IEA, 2010.

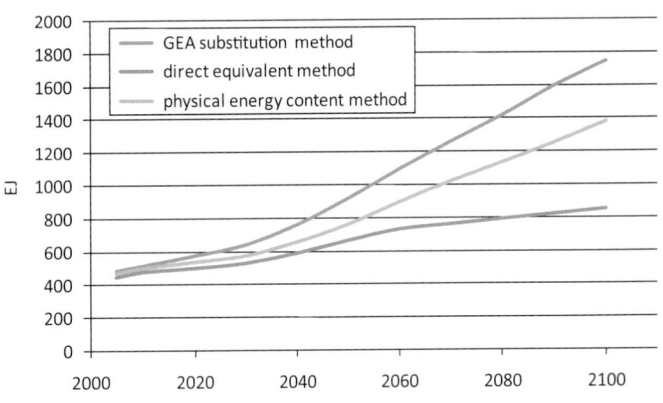

Figure 1.A.1. | Comparison of global total primary energy supply between 2005 and 2100 using three different primary energy accounting methods based on identical useful energy demands as quantified in the illustrative GEA-M set of pathways (see Chapter 17).

A detailed overview of differences in primary energy accounting from different energy statistics is described in Macknick (2009); see also Figure 1.A.2 below (the paper also contains a link to a data base where users can specify their own standardized accounting convention applied to the main international energy statistical data sources).

Table 1.A.1 compares the differences across the primary energy accounting methods for the world by energy source using the GEA primary substitution equivalent (see Technical Guidelines, Annex-II), the direct equivalent, and the physical energy content methods for the year 2005 based on IEA data (IEA, 2010). As is to be expected, the main differences in absolute terms across the methods are for nuclear, hydropower, and other renewables (except biomass). Great care is, therefore, advised when using and comparing reported primary energy across different statistical sources in general and in comparing the numbers reported in the GEA in particular.

1.A.4   Limitations of Primary Energy Accounting

The alternative primary energy accounting methods outlined above show significant differences in how non-combustible energy sources are presented in energy statistics. As the differences are significant for nuclear and renewables, the accounting method chosen has an impact on how the primary energy structure is interpreted. This in itself is an important limitation of the concept of primary energy. It is also a cause of considerable confusion in comparing different statistics, data sources, and analyses (and the ensuing emphasis on the importance of different energy options).

The differences of applying the three accounting methods to current energy use levels are relatively modest compared to those in scenarios of possible future major energy transformations where the structure of the global energy system changes significantly (see Chapter 17). The accounting gap between the different methods tends to become bigger over time as the share of combustible energy sources declines. The very

concept of a statistically defined primary energy that has no real physical equivalence is thus becoming more limited as more radical future energy systems depart from current ones.

Figure 1.A.1 illustrates this growing divergence across the three primary energy accounting methods for an otherwise identical scenario in terms of final and useful energy demand (based on the intermediary GEA-M set of pathways; see Chapter 17). As the structure of the global energy system changes, different accounting methods differ by more than a factor of two in terms of implied primary energy growth. No such significant accounting ambiguities affect secondary and final energy, which are thus preferable descriptors for radical, transformative changes in energy systems.

1.A.5   Main Energy Statistics and Data Sources

Four institutions regularly publish globally comprehensive statistics on energy use: British Petroleum (BP), the US Energy Information Administration (EIA), the International Energy Agency (IEA), and the United Nations (UN). As Table 1.A.2 shows, these energy statistics differ in terms of energy coverage ranging from primary energy (PE), primary and secondary energy (EIA, IEA, UN), to primary, secondary, and final energy (IEA).

Data are mainly collected through questionnaires and exchanges between the organizations as well as with others, including but not limited to publications from the Statistical Office of the European Communities (Eurostat), the International Atomic Energy Agency (IAEA), the Organization of the Petroleum Exporting Countries (OPEC), the Organización Latinoamericana de Energía (OLADE), etc.

Statistics differ in the extent to which they include non-commercial energy (use of traditional biomass), which is fully covered in the IEA statistics (all sources) and partially in UN (mainly fuel wood), as well as modern renewables (outside hydropower), with only IEA and UN

Table 1.A.2 | Overview of the four major data sources for Global Energy Statistics.

	BP	EIA	IEA	UN
Primary energy	X	X	X	X
Secondary energy		X	X	X
Final energy			X	–[1]
New renewables[2]		X	X	X
Traditional biomass[3]			X	X
Electronic availability	Online free	Online free	Online subscription ($)	Offline tape order ($)

[1] Not reported directly by UN but can be calculated from full data base statistics.
[2] New renewable refers to solar, wind, modern bioenergy, and geothermal.
[3] Traditional biomass refers to fuel wood, dung, and agricultural residues.

providing (near) full coverage (with BP reporting selected modern renewables). Different reporting organizations also use different methods for expressing the primary energy equivalent of non-combustible energies (see Section 1.11.3 on Primary Energy Accounting above) and in their use of heating values (see the discussion above). LHV are used by the UN and IEA (and unless otherwise specified in this report). HHV are used in US EIA statistics (which therefore tend to report systematically higher energy use compared to other data sources), with BP using a hybrid approach which is closer to UN/IEA statistical values).

Both the UN and IEA provide comprehensive energy statistics on the production, trade, conversion, and use of primary and secondary, conventional and non-conventional, and new and renewable sources of energy covering the period from 1970 onwards (UN, 2010; IEA, 2010).[52] IEA's energy balances represent convenient aggregates of all energy flows in a common (non-SI) energy metric in tonnes of oil equivalent, summarized from IEA's energy statistics and for global and regional aggregates as well as for individual countries. The IEA statistics cover approximately 130 countries (of 192 UN Member countries), which represent about 98% of worldwide energy use (IPCC, 2006). BP statistics focus on commercial and conventional energy carriers and exclude fuels such as wood, peat, and animal waste and energy flows of other renewables such as wind, geothermal, and solar power generation[53] (BP, 2010a). Its

statistics cover the period since 1965, are updated regularly,[54] and are available free of charge on the Internet. Cumulative installed renewable power capacity data are provided in BP's full workbook of historical statistical data from 1965–2009 (BP, 2010b). US EIA energy statistics, which are also freely available online, cover primary and secondary energy use by fuel category and per country since 1980, using the (non-SI) BTU as a common energy metric and based on HHV, which is different than other energy statistics.

As a result of differences in data collection sources, boundary conditions, methodologies, and heating values used in different statistics, global primary energy use numbers reported by these four organizations differ from 442 EJ (BP) to 487 EJ (EIA), or by some 10%, for the GEA base year 2005 and throughout their entire reporting horizon (see Figure 1.A.2). Adjusting[55] the different primary accounting conventions to the GEA standard and completing non-reported energies (non-commercial, traditional biomass using the IEA numbers) reduces this data uncertainty to a range from 495 EJ (IEA and BP) to 528 EJ (EIA[56]), or some 7%, with the UN statistics taking an intermediary position (506 EJ) for the GEA base year 2005 (see Figure 1.A.2). This assessment adopts a value of 495 EJ for the level of world primary energy use in the year 2005.

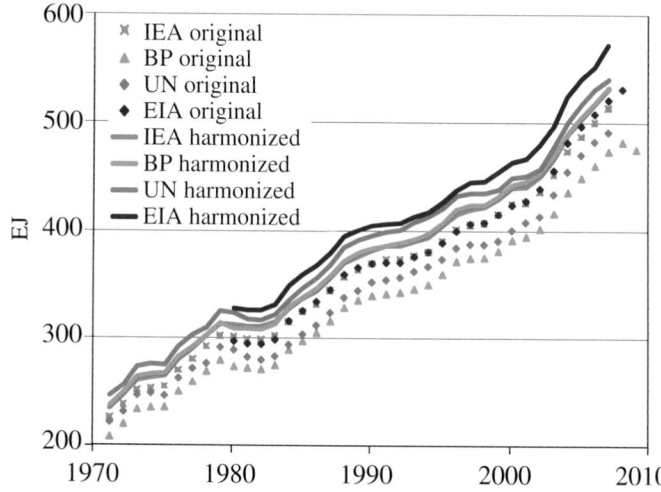

Figure 1.A.2. | World primary energy. Original data by four reporting agencies and harmonized primary energy equivalences. Source: data from Macknick, 2009.

52 Electronic data need to be purchased from the UN and processed with appropriate data base software tools as few aggregates are contained in the statistics. For instance final energy use is not reported directly by the UN, but can be calculated from a multitude of individual energy flows reported. The UN data portal allows free electronic access to statistics of individual energy flows as well as few aggregate energy indicators (primary energy use, electricity generation) from 1990 onwards. Full IEA energy balances, by energy flow, use, and sector since 1971 are available online to subscribers (including many universities) of the OECD iLibrary online publication and statistical query service: The statistics of the EIA and BP are available online free of charge but provide a somewhat more limited coverage as well as adopt differing accounting conventions to UN and IEA.

53 With exception of ethanol, only installed capacity data for geothermal, wind, and solar are reported by BP.

54 Updates are fastest among all energy statistics and available by September each year for the preceding year.

55 A software tool performing data comparison and adjustments to consistent and comparable accounting conventions for the 20 largest energy-using countries worldwide as well as the global total has been developed by Macknick (2009) and is available online: www.iiasa.ac.at/Research/TNT/WEB/Publications/Energy_Carbon_DataBase/.

56 Due to the use of HHV in the EIA statistics.

Appendix 1.B Conversion Tables and GEA Regional Definitions

Table 1.B.1 | Conversion factors.

FROM	TO->												
	MJ	TCE	btu	toe	boe	kWyr	kcal	TJ	Gcal	Mtoe	Mbtu	GWh	GWyr
MJ	1	3.4121E-05	947.8134	2.39E-05	0.000175	3.171E-05	238.8459	0.000001	0.00023885	2.3885E-11	0.00094781	2.7778E-07	3.171E-11
TCE	29307.6	1	27778140	0.7	5.131	0.9293379	7000000	0.0293076	7	0.0000007	27.77814	0.008141	9.2934E-07
btu	0.00105506	3.6E-08	1	2.52E-08	1.85E-07	3.346E-08	0.2519968	1.0551E-09	2.52E-07	2.52E-14	0.000001	2.9307E-10	3.3456E-14
toe	41868	1.428571	39683050	1	7.33	1.327626	10000000	0.041868	10	0.000001	39.68305	0.01163	1.3276E-06
boe	5711.869031	0.19489378	5413784.9	0.136426	1	0.1811222	1364256	0.00571187	1.364256	1.3643E-07	5.413786	0.00158663	1.8112E-07
kWyr	31536	1.07603488	29890240	0.753224	5.521135	1	7532244	0.031536	7.532244	7.5322E-07	29.89024	0.00876	0.000001
kcal	0.0041868	1.4286E-07	3.968305	1E-07	7.33E-07	1.328E-07	1	4.1868E-09	0.000001	1E-13	3.9683E-06	1.163E-09	1.3276E-13
TJ	1000000	34.12084	947813400	23.88459	175.074	31.70979	238845900	1	238.8459	2.3885E-05	947.8134	0.27777778	3.171E-05
Gcal	4186.8	0.14285714	3968305	0.1	0.733	0.1327626	1000000	0.0041868	1	0.0000001	3.968305	0.001163	1.3276E-07
Mtoe	41868000000	1428571	3.968E+13	1000000	7330000	1327626	1E+13	41868	10000000	1	39683050	11630.0004	1.327626
Mbtu	1055.06	0.03599953	1000000	0.0252	0.184714	0.0334557	251996.8	0.00105506	0.2519968	2.52E-08	1	0.00029307	3.3456E-08
GWh	3600000	122.835	3.412E+09	85.98452	630.2666	114.1553	859845200	3.6	859.8452	8.5985E-05	3412.128	1	0.00011416
GWyr	31536000000	1076034.88	2.989E+13	753224.4	5521135	1000000	7.5374E+12	31536	7532244	0.7532244	29910720	8759.99625	1

Source: Nakicenovic et al., 1998.

Table 1.B.2a | Typical calorific values of solid energy carriers.

	Gross calorific value HHV [MJ/kg]	Net calorific value LHV [MJ/kg]
Anthracite	29.65–30.35	28.95–30.35
Cooking coals	27.80–30.80	26.60–29.80
Other bituminous	23.85–26.75	22.60–25.50
Metallurgical coke	27.90	27.45
Gas coke	28.35	27.9
Low-temperature coke	26.30	25.4
Petroleum coke	30.5–35.8	30.0–35.3
Wood	15–19	-

Note: Detailed information on energy and chemical characteristics for a wide range of biomass fuels can be found at IEA Task 32 biomass database: http://www.ieabcc.nl;

Phyllis biomass database: http://www.ecn.nl/phyllis;

TU Vienna biomass database: http://www.vt.tuwien.ac.at/biobib.

Source: IEA/OECD/Eurostat, 2005.

Table 1.B.2b | Typical calorific values of liquid energy carriers.

	Gross calorific value HHV [GJ/tonne]	Net calorific value LHV [GJ/tonne]
Ethane	51.90	47.51
Propane	50.32	46.33
Butane	49.51	45.72
LPG	50.08	46.15
Naphtha	47.73	45.34
Aviation gasoline	47.40	45.03
Motor gasoline	41.10	44.75
Aviation turbine fuel	46.23	43.92
Other kerosene	46.23	43.92
Gas/diesel oil	45.66	43.38
Fuel oil, low sulphur	44.40	42.18
Fuel oil, high sulphur	43.76	41.57
Biodiesel	–	36.8–40.9
Biogasoline	–	24.0–37.4
Ethanol	29.4	26.87
Methanol	22.36	19.99
Dimethyl ether	30.75	28.62

Source: IEA/OECD/Eurostat, 2005; Agarwal, 2007; IEA, 2009b.

Table 1.B.2c | Typical calorific values of gaseous energy carriers per kg and m³.

	Gross calorific value HHV [MJ/kg]	Net calorific value LHV [MJ/kg]	Gross calorific value HHV [MJ/m³]	Net calorific value LHV [MJ/m³]
Methane	55.52	50.03	37.652	33.939
Natural gas (Norway)	–	–	39.668	–
Natural gas (Netherlands)	–	–	33.339	–
Natural gas (Russia)	–	–	37.578	–
Natural gas (Algeria)	–	–	42.000	–
Natural gas (United States)	–	–	38.341	–

Source: IEA/OECD/Eurostat, 2005 ; IEA, 2009a.

Table 1.B.3 | CO_2 emission factors on a net calorific basis.

Fuel type	IPCC default [kg/GJ]	Range from [kg/GJ]	to [kg/GJ]
Crude Oil	73.3	71.1	75.5
Motor Gasoline	69.3	67.5	73.0
Jet Gasoline	70.0	67.5	73.0
Jet Kerosene	71.5	69.7	74.4
Kerosene	71.9	70.8	73.7
Gas / Diesel Oil	74.1	72.6	74.8
Residual Fuel Oil	77.4	75.5	78.8
Liquefied Petroleum Gases	63.1	61.6	65.6
Ethane	61.6	56.5	68.6
Naphtha	73.3	69.3	76.3
Petroleum Coke	97.5	82.9	115.0
Anthracite	98.3	94.6	101.0
Coking Coal	94.6	87.3	101.0
Lignite	101.0	90.9	115.0
Oil Shale and Tar Sands	107.0	90.2	125.0
Brown Coal Briquettes	97.5	87.3	109.0
Natural Gas	56.1	54.3	58.3
Compressed Natural Gas	56.1	54.3	58.3
Liquefied Natural Gas	56.1	54.3	58.3
Municipal Wastes (non-biomass fraction)	91.7	73.3	121.0
Municipal Wastes (biomass fraction)	100.0	84.7	117.0
Industrial Wastes	143.0	110.0	183.0
Waste Oils	73.3	72.2	74.4
Peat	106.0	100.0	108.0
Wood / Wood Waste	112.0	95.0	132.0
Sulphite lyes (Black Liquor)	95.3	80.7	110.0
Other Primary Solid Biomass	100.0	84.7	117.0
Charcoal	112.0	95.0	132.0
Biogasoline	70.8	59.8	84.3
Biodiesels	70.8	59.8	84.3
Other liquid biofuels	79.6	67.1	95.3
Landfill Gas	54.6	46.2	66.0
Sludge Gas	54.6	46.2	66.0
Other Biogas	54.6	46.2	66.0

Note: Values represent CO_2 emissions that arise with 100 percent oxidation of fuel carbon content at the point of combustion. Life-cycle CO_2 emissions for various fuels can be higher or lower, due to emissions in the supply chain of the fuel and due to carbon absorbed during the growth phase of biomass feedstock.

Source: IPCC, 2006.

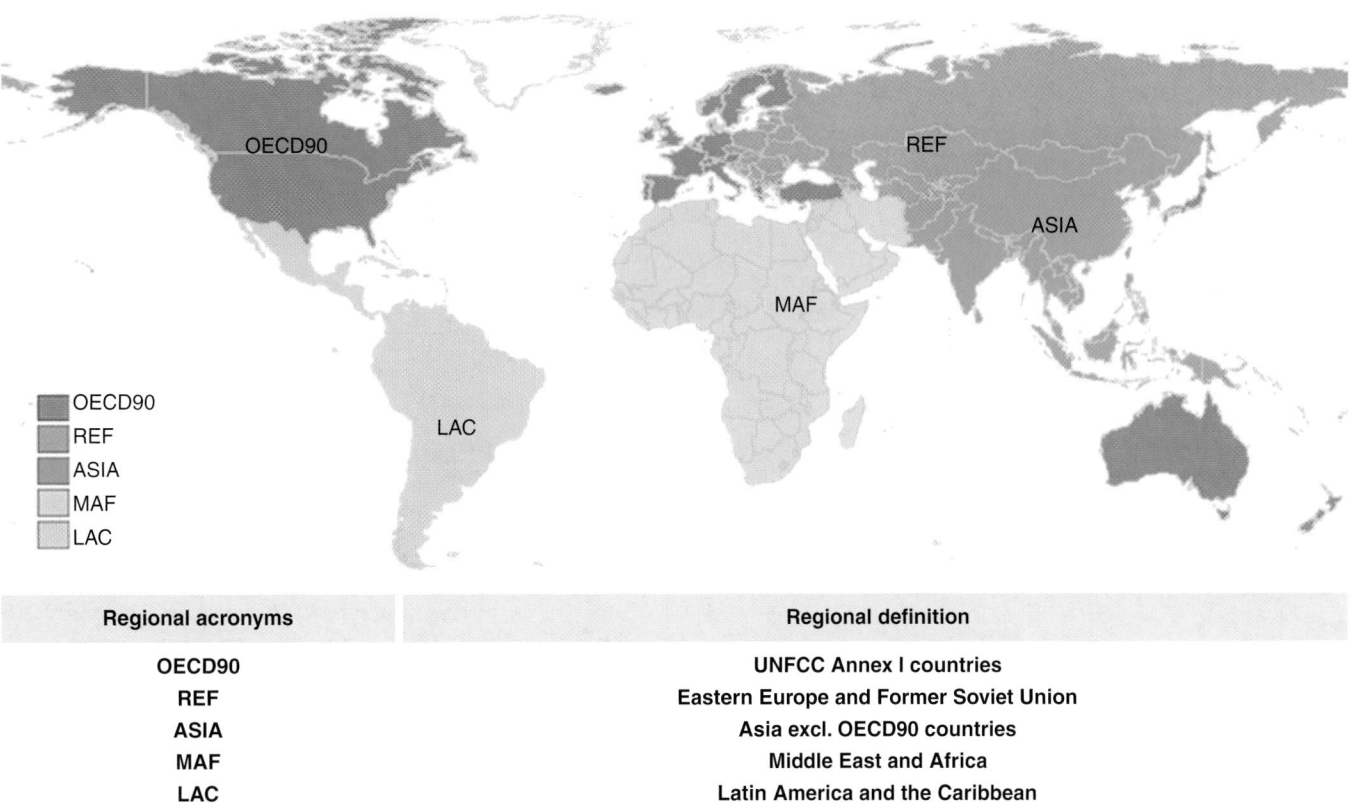

Regional acronyms	Regional definition
OECD90	UNFCC Annex I countries
REF	Eastern Europe and Former Soviet Union
ASIA	Asia excl. OECD90 countries
MAF	Middle East and Africa
LAC	Latin America and the Caribbean

OECD90 = OECD countries as of 1990 in Western Europe, North America, and Pacific Asia (and defined in UNFCCC as Annex-I countries)

REF = Countries undergoing economic reform, i.e. countries in Eastern Europe and the former Soviet Union

ASIA = non-Annex-I countries in Asia

MAF = Middle East and North, and Sub-Saharan Africa

LAC = Latin America and the Caribbean

For country listings and finer-resolution regional definitions see Annex-II.

Figure 1.B.1. | Definition of GEA regions, see also Annex-II Technical guidelines.

References

Anderson, D., 2007: Electricity generation costs and investment decisions: A review. UKERC Working Paper, February 2007. UK Energy Research Centre. ukerc.rl.ac.uk/UCAT/cgi-bin/ucat_query.pl?GoButton=Browse.

Arthur, W. B., 1989: Competing technologies, increasing returns, and lock-in by historical events. *The Economic Journal*, 99: 116–131.

Arthur, W.B., 1994: *Increasing Returns and Path Dependence in the Economy*. Michigan University Press, Ann Arbor, MI.

Ayres, R. U., 1989: Energy Efficiency in the U.S. Economy: A New Case for Conservation, RR-89–12, International Institute for Applied Systems Analysis (IIASA), Laxenburg, Austria.

Ayres, R. U, L. W. Ayres and B. Warr, 2003: Exergy, power and work in the US economy. *Energy*, 28: 219–271.

Baksi, S., and C. Green, 2007: Calculating economy-wide energy intensity decline rate: The role of sectoral output and energy shares. *Energy Policy*, 35: 6457–6466.

Barro, R. J., 1997: *Determinants of Economic Growth*. The MIT Press, Cambridge, MA.

Binswanger, M., 2001: Technological progress and sustainable development: What about the rebound effect? *Ecological Economics*, 36:119–132.

BMME, 1998: Useful energy balance – BEU, *Economy & Energy*, 2(10). Brazilian Ministry for Mines and Energy (BMME). www.ecen.com/eee10/beup.htm.

Boden, T. A., G. Marland and R. J. Andres, 2010: *Global, Regional, and National Fossil Fuel CO2 Emissions 1751–2007*. Carbon Dioxide Information Analysis Center. http://cdiac.ornl.gov/trends/emis/overview_2007.html.

Brunekreef, B. and S. T. Holgate, 2002: Air pollution and health. *Lacet*, 360(93441): 1233–1242.

BP, 2007: *BP Statistical Review of World Energy*. BP plc., London, UK.

BP, 2010a: *BP Statistical Review of World Energy*. BP plc., London, UK.

BP, 2010b: *Excel* Workbook of Historical Statistical Data from 1965–2009. BP plc., London, UK. www.bp.com/liveassets/bp_internet/globalbp/globalbp_uk_English/reports_and_publications/statistical_Energy_review_2008/STAGING/local_assets/2010_downloads/Statistical_Review_of_World_Energy_2010.xls (accessed 19 March, 2011).

Cames, M., C. Fischer, B. Praetorius, L. Schneider, K. Schumacher, J. P. Voß and M. Pehnt, 2006: *Micro Cogeneration – Towards Decentralized Energy Systems*. Springer, Berlin.

Cleveland, C. J. and C. Morris (eds.), 2006: *Dictionary of Energy*. Elsevier, Amsterdam, The Netherlands.

Cullen, J. M. and J. M. Allwood, 2010: The efficient use of energy: Tracing the global flow of energy from fuel to service. *Energy Policy*, 38: 75–81.

David, P. A. and G. Wright, 1996: *The Origins of American Resource Abundance*. WP-96–015, International Institute for Applied Systems Analysis (IIASA), Laxenburg, Austria.

Ehrlich, P. R. and J. P. Holdren, 1971: Impact of population growth. *Science*, 171: 1212–1217.

Escher, W. J. D., The case for solar/hydrogen energy. *International Journal of Hydrogen Energy*, 8(7): 479–498.

Eurostat, 1988: *Useful Energy Balances*. Statistical Office of the European Communities (Eurostat), Luxembourg.

Fisher, B. S., N. Nakicenovic, K. Alfsen, J. Corfee Morlot, F. de la Chesnaye, J.-C. Hourcade, K. Jiang, M. Kainuma, E. La Rovere, A. Matysek, A. Rana, K. Riahi, R. Richels, S. Rose, D. van Vuuren and R. Warren, 2007: Issues related to mitigation in the long term context. In *Climate Change 2007: Mitigation. Contribution of Working Group III to the Fourth Assessment Report of the Intergovernmental Panel on Climate Change*. Cambridge University Press, Cambridge, UK.

Forster, P., V. Ramaswamy, P. Artaxo, T. Berntsen, R. Betts, D. W. Fahey, J. Haywood, J. Lean, D. C. Lowe, G. Myhre, J. Nganga, R. Prinn, G. Raga, M. Schulz and R. Van Dorland, 2007: Changes in atmospheric constituents and in radiative forcing. In *Climate Change 2007: The Physical Science Basis. Contribution of Working Group I to the Fourth Assessment Report of the Intergovernmental Panel on Climate Change*. S. Solomon, D. Qin, M. Manning, Z. Chen, M. Marquis, K. B. Averyt, M.Tignor and H. L. Miller (eds.), Cambridge University Press, Cambridge, UK.

Fouquet, R., 2008: *Heat, Power and Light: Revolutions in Energy Services*. Cheltenham, UK.

GTAP7, 2010: *GTAP7 Data Base*. Global Trade, Assistance, and Production (GTAP7). www.gtap.agecon.purdue.edu/databases/v7/v7_doco.asp (accessed 22 April 2011).

Gales, B., A. Kander, P. Malanima and M. Rubio, 2007: North versus south: Energy transition and energy intensity in Europe over 200 years. *European Review of Economic History*, 11(2): 219–253.

Gilli, P.V., N. Nakicenovic, A. Grubler, F.L. Bodda, 1990: *Technischer Fortschritt, Strukturwandel und Effizienz der Energieanwendung – Trends weltweit und in Österreich*. Volume 6, Schriftenreihe der Forschungsinitiative des Verbundkonzerns, Vienna, Austria.

Gilli, P.V., N. Nakicenovic and R. Kurz, 1996: *First- and Second-law Efficiencies of the Global and Regional Energy Systems*. RR-96–2, *International Institute for Applied Systems Analysis (IIASA)*, Laxenburg, Austria.

Goldemberg, J., T.B. Johansson, A.K.N. Reddy, and R.H. Williams, 1988: *Energy for a Sustainable World*. Wiley Eastern Limited, New Dehli, India.

Greening, L. A., W. B. Davis, L. Schipper and M. Khrushch, 1997: Comparison of six decomposition methods: Application to aggregate energy intensity for manufacturing in 10 OECD countries. *Energy Economics*, 20(1): 43–65.

Grubler, A, 1998: *Technology and Global Change*. Cambridge University Press, Cambridge, UK.

Grubler, A., 2002: Trends in global emissions: Carbon, sulfur and nitrogen. In *Encyclopedia of Global Environmental Change*. T. Munn (ed.), Vol. 3, Wiley, Chichester, UK, pp.35–53.

Grubler, 2004: Transitions in Energy Use. In *Encyclopedia of Energy*. C. J. Cleveland (ed.), Vol.6, Elsevier, Amsterdam, The Netherlands, pp.163–177.

Grubler, 2008: Energy transitions. In *Encyclopedia of Earth*. C. J. Cleveland (ed.), Environmental Information Coalition, National Council for Science and the Environment, Washington, DC.

Grubler, A. and N. Nakicenovic, 1996: Decarbonizing the global energy system. *Technological Forecasting and Social Change*, 53: 97–100.

Grubler A., N. Nakicenovic and D. G. Victor, 1999: Dynamics of energy technologiesand global change. *Energy Policy*, 27(5): 247–280.

Haas, R., N. Nakicenovic, A. Ajanovic, T. Faber, L. Kranzl, A. Müller and G. Resch, 2008: Towards sustainability of energy systems: A primer on how to apply the concept of energy services to identify necessary trends and policies. *Energy Policy*, 36: 4012–4021.

Heptonstall, P., 2007: *A Review of Electricity Unit Cost Estimates*. UKERC Working Paper, December 2006, updated May 2007. United Kingdom Energy Research Centre (UKERC), UK. http://ukerc.rl.ac.uk/UCAT/cgi-bin/ucat_query.pl?GoButton=Browse.

Houghton, R. A., 1999: The annual net flux of carbon to the atmosphere from changes in land use 1850–1990. *Tellus*, 51(B): 298–313.

Houghton, 2008: Carbon flux to the atmosphere from land-use changes: 1850–2005. In TRENDS: A Compendium of Data on Global Change. Carbon Dioxide Information and Analysis Center, Oak Ridge, TN, USA.

Humphreys, M., J. D. Sachs and J. E. Stiglitz (eds.), 2007: *Escaping the Resource Curse*. Columbia University Press, New York, NY, USA.

IEA, 2005: *Energy Statistics Manual*. International Energy Agency (IEA) and Eurostat, IEA/OECD, Paris, France. www.iea.org/stats/docs/statistics_manual.pdf.

IEA, 2007a: *Energy Balances of OECD Countries*. International Energy Agency (IEA), IEA/OECD, Paris, France.

IEA, 2007b: *Energy Balances of Non-OECD Countries*. International Energy Agency (IEA), IEA/OECD, Paris, France.

IEA, 2009: *World Energy Outlook 2009*. International Energy Agency (IEA), IEA/OECD, Paris, France.

IEA, 2010: *IEA Data Services (Online Statistics)*. International Energy Agency (IEA), IEA/OECD, Paris, France.

IPCC, 2001: *Climate Change 2001: The Scientific Basis. Contribution of Working Group I to the Third Assessment Report of the Intergovernmental Panel on Climate Change (IPCC)*. Houghton, J.T., Y. Ding, D.J. Griggs, M. Noguer, P.J. van der Linden, X. Dai, K. Maskell, and C.A. Johnson (eds.), Cambridge University Press, Cambridge, UK.

IPCC, 2006: *2006 IPCC Guidelines for National Greenhouse Gas Inventories*. Intergovernmental Panel on Climate Change (IPCC) report prepared by the National Greenhouse Gas Inventories Programme, H. S. Eggleston, L. Buendia, K. Miwa, T. Ngara and K. Tanabe (eds.), IGES, Japan.

IPCC, 2011: IPCC Special Report on Renewable Energy Sources and Climate change Mitipetion, O,Edennofer, R/. Pichs-Madrnja, Y. Sohona et al. (eds.), Cambridge University Press, Cambridge, UK.

Jevons, W. S., 1865: *The Coal Question; An Inquiry Concerning the Progress of the Nation, and the Probable Exhaustion of Our Coal Mines*. MacMillan and Co., London.

Kander, A., P. Malanima and P. Warde, 2008: *Energy Transitions in Europe: 1600–2000*. CIRCLE Electronic Working Papers 2008/12, Center for Innovation, Research and Competences in the Learning Economy (CIRCLE), Lund University, Sweden.

Kaya, Y., 1990: *Impact of Carbon Dioxide Emission Control on GNP Growth: Interpretation of Proposed Scenarios*. Paper presented to the Intergovernmental Panel on Climate Change (IPCC) Energy and Industry Subgroup, Response Strategies Working Group, Paris, France.

Krausmann, F., S. Gingrich, N. Eisenmenger, K. H. Erb, H. Haberl and M. Fischer-Kowalski, 2009: Growth in global materials use, GDP and population during the 20th century. *Ecological Economics*, 68(10): 2696–2705.

Lamarque, J.-F., T. C. Bond, V. Eyring, C. Granier, A. Heil, Z. Klimont, D. Lee, C. Liousse, A. Mieville, B. Owen, M. G. Schultz, D. Shindell, S. J. Smith, E. Stehfest, J. van Aardenne, O. R. Cooper, M. Kainuma, N. Mahowald, J. R. Mc-Connell, V. Naik, K. Riahi and van Vuuren, D. P., 2010: Historical (1850–2000) gridded anthropogenic and biomass burning emissions of reactive gases and aerosols: methodology and application. *Atmos. Chem. Phys.*, 10: 7017–7039, doi:10.5194/acp-10–7017–2010.

Lightfoot, H. D., 2007: Understand the three different scales for measuring primary energy and avoid errors. *Energy*, 32: 1478–1483.

Liu, H., Y. Xi, J. Guo and X. Li, 2010: Energy embodied in the international trade of China: An input-output analysis. *Energy Policy*, 38: 3957–3964.

Lovins, A., M. Brylawski, D. Cramer and T. Moore, 1996: *Hypercars: Materials, Manufacturing, and Policy Implications*, Rocky Mountain Institute, Snowmass, CO, pp. 234–236.

McKelvey, V. E., 1967: Mineral resource estimates and public policy. *American Scientist*, 60: 32–40.

Machado, G. V., 2000: *Energy Use, CO2 Emissions and Foreign Trade: An IO Approach Applied to the Brazilian Case*. Proceedings XIII International Conference on Input-Output Techniques, 21–25 August 2000, Macerata, Italy.

Macknick, J., 2009: *Energy and Carbon Dioxide Emission Data Uncertainties*. IIASA Interim Report, IR-09–032, International Institute for Applied Systems Analysis (IIASA), Laxenburg, Austria.

Morita, T., J. Robinson, A. Adegbulugbe, J. Alcamo, D. Herbert, E. Lebre la Rovere, N. Nakicenivic, H. Pitcher, P. Raskin, K. Riahi, A. Sankovski, V. Solkolov, B.d. Vries, and D. Zhou, 2001: Greenhouse gas emission mitigation scenarios and implications. In: *Climate Change 2001: Mitigation; Contribution of Working Group III to the Third Assessment Report of the IPCC*. Cambridge University Press, Cambridge, pp. 115–166.Nakicenovic, N. (ed.), 1993: Long-term strategies for mitigation global warming. *Energy*, Special Issue, **18**(5): 401–609.

Nakicenovic, N, P. V. Gilli and R. Kurz, 1996a: Regional and global exergy and energy efficiencies. *Energy*, 21(3): 223–237.

Nakicenovic, N., A. Grubler, H. Ishitani, T. Johansson, G. Marland, J. R. Moreira and H.-H. Rogner, 1996b: Energy primer. In *Climate Change 1995 – Impacts, Adaptations and Mitigation of Climate Change: Scientific-Technical Analyses, Contribution of Working Group II to the Second Assessment Report of the Intergovernmental Panel on Climate Change*. R. T. Watson, M. C. Zinyowera and R. H. Moss (eds.), Cambridge University Press, Cambridge, UK, pp.75–92.

Nakicenovic, N., A. Grubler and A. McDonald (eds.), 1998: *Global Energy Perspectives*. International Institute for Applied Systems Analysis (IIASA) and World Energy Council (WEC), Cambridge University Press, Cambridge, UK.

Nakicenovic, N., and R. Swart (eds.), 2000: *IPCC Special Report on Emissions Scenarios*. Cambridge University Press, Cambridge, UK.

Narayanan, G. B. and T. L. Walmsley (eds.), 2008: *Global Trade, Assistance, and Production: The GTAP7 Data Base*. Center for Global Trade Analysis, Purdue University, IN, USA. www.gtap.agecon.purdue.edu/databases/v7/v7_doco.asp (accessed 2 November, 2010).

Nordhaus, W. D., 1998: *Do Real-output and Real-wage Measures Capture Reality? The History of Lighting Suggests Not*. Cowles Foundation Paper No. 957, New Haven, CT, USA.

NREL, 2010: *Energy Technology Cost and Performance Data*. National Renewable Energy Laboratory (NREL), Golden, Colorado. www.nrel.gov/analysis/costs.html (accessed 28 April 2010).

Pachauri, S., 2007: *An Energy Analysis of Household Consumption*. Springer, Dordrecht, the Netherlands.

Perlin, J., 1989: *A Forest Journey: The Role of Wood in the Development of Civilization*. W. W. Norton, New York.

Ponting, C., 1992: *A Green History of the World: The Environment and the Collapse of Great Civilizations*. St. Martin's Press: New York.

Reddy, A. K. N. and J. Goldemberg, 1990: Energy for the developing world. *Scientific American*, 263(3):110–118.

Rogner, H.-H., 1997: An assessment of world hydrocarbon resources, *Annual Review of Energy and the Environment*, 22: 217–262.

Rogner, H.-H., F. Barthel, M. Cabrera, A. Faaij, M. Giroux, D. O. Hall, V. Kagramanian, S. Kononov, T. Lefevre, R. Moreira, R. Nötstaller, P. Odell and M. Taylor, 2000: Energy resources. In *World Energy Assessment*. J. Goldemberg (ed.), United Nations Development Programme (UNDP) and World Energy Council (WEC), New York, pp.135–172.

Rogner, H.-H. and A. Popescu, 2000: An introduction to energy. *World Energy Assessment (WEA)*. United Nations Development Programme (UNDP), New York, pp.30–37.

Rosen, M. A., 1992: Evaluation of energy utilization efficiency in Canada. *Energy*, 17: 339–350.

Rosen, M. A., 2010: Advances in hydrogen production by thermochemical water decomposition: A review. *Energy*, 35: 1068–1076.

Rosen, M. A. and I. Dincer, 2007: *Exergy: Energy, Environment, and Sustainable Development*. Elsevier, Amsterdam, The Netherlands.

Rosenberg, N., and L. E. Birdzell, 1986: *How the West Grew Rich: The Economic Transformation of the Industrial World*. I. B. Tauris and Co., London, UK.

Roy, J., 2000: The rebound effect: Some empirical evidence from India. *Energy Policy*, 28(6–7): 433–438.

Schipper, L. and S. Myers, 1992: *Energy Efficiency and Human Activity: Past Trends, Future Prospects*. Cambridge University Press, Cambride, UK.

Schipper, L. (ed.), 2000: Special Issue of Energy Policy. *Energy Policy*, 28(6–7): 351–500.

Schäfer, A., 2005: Structural change in energy use. *Energy Policy*, 33(4): 429–437.

Schulz, N. B., 2010: Delving into the carbon footprints of Singapore – Comparing direct and indirect greenhouse gas emissions of a small and open economic system. *Energy Policy*, 38(9): 4848–4855.

Schumpeter, J. A., 1935: The Analysis of Economic Change. *The Review of Economics and Statistics*, 17(4): 2–10.

Scott, D. S., Energy currency crisis. *Nuclear Engineering International*, **52(64)**: 38–41.

Smil, V., 1991: *General Energetics*. John Wiley and Sons, New York.

Smil, V., 2006: Industrial revolution. In *Dictionary of Energy*. C. J. Cleveland and C. Morris (eds.), Elsevier, Amsterdam, The Netherlands.

Smil, V., 2010: *Energy Transitions: History, Requirements*, Prospects. Praeger, Santa Barbara, CA.

Smith, S. J., H. Pitcher, and T. M. L. Wigley, 2001: Global and Regional Anthropogenic Sulfur Dioxide Emissions, *Global Planet.Change*, 29(1–2): 99–119.

Smith, S. J., J. van Aardenne, Z. Klimont, R. Andres, A. C. Volke and S. Delgado Arias, 2011: Anthropogenic sulfur dioxide emissions: 1850–2005. *Atmos. Chem. Phys.*, 11: 1101–1116.

Sorell, S., 2007: Jevons' Paradox revisited: The evidence for backfire from improved energy efficiency. *Energy Policy*, 37(4): 1456–1469.

Sorell, S., J. Dimitropoulos and M. Sommerville, 2009: Empirical estimates of direct rebound effects: A review. *Energy Policy*, 37(4): 1356–1371.

UN, 2006a: *Industrial Commodity Statistics Yearbook 2006*, United Nations (UN),

UN, 2006b: *Industrial Commodity Production Statistics Dataset, 1950–2003*. United Nations (UN), Statistics Division, New York, CD-ROM.

UN, 2010: *UNData*, United Nations (UN). data.un.org/ (accessed 22 October 2011).

UNDP and WHO, 2009: The Energy Access Situation in Developing Countries: A Review Focusing on Least Developed Countries and Sub-Saharan Africa. Sustainable Energy Programme Environment and Energy Group Report, United Nations Development Programme (UNDP) and World Health Organization (WHO), New York.

Vringer, K. and K. Block, 1995: The direct and indirect energy requirements of households in the Netherlands. *Energy Policy*, 23: 893–910.

Wall, G., 2006: Exergy. In *Dictionary of Energy*. C. J. Cleveland and C. Morris (eds.), *Elsevier, Amsterdam*, The Netherlands.

Warr, B., R. U. Ayres, N. Eisenmenger, F. Krausmann and H. Schandl, 2010: Energy use and economic development: A comparative analysis of useful work supply in Austria, Japan, the United Kingdom and the US during 100 years ofeconomic growth. *Ecological Economics*, 69: 1904–1917.

Watson, R., M. C. Zinyowera and R. Moss, (eds.), 1995: *Climate Change 1995. Impacts, Adaptations and Mitigation of Climate Change: Scientific Analyses. Contribution of Working Group II to the Second Assessment Report of the Intergovernmental Panel on Climate Change*. Cambridge University Press, Cambridge.

WCI, 2009: *The Coal Resource: A Comprehensive Overview of Coal*. World Coal Institute (WCI), London, UK.

WEA, 2004: Basic Energy Facts. *World Energy Assessment Overview: 2004 Update*.World Energy Assessment (WEA), United Nations Development Programme (UNDP), New York, pp.25–31.

WEC, 1993: *Energy for Tomorrow's World*. WEC Commission Global Report, World Energy Council (WEC), London, UK.

WEC, 1998: *Survey of Energy Resources*. World Energy Council (WEC), London, UK.

WEC, 2007: *2007 Survey of Energy Resources*. World Energy Council (WEC), London, UK.

WHO, 2002: *The World Health Report 2002: Reducing Risks, Promoting Healthy Life*. World Health Organization (WHO), Geneva. www.who.int/whr/2002/en/ (accessed 17 October 2010).

WHO, 2011: *Global Burden of Disease*. World Health Organization (WHO), www.who.int/topics/global_burden_of_disease/en/ (accessed 3 March 2011).

World Bank, 2010: *World Development Indicators*. Online database, World Bank, Washington, DC.

2

Energy, Poverty, and Development

Convening Lead Authors (CLA)
Stephen Karekezi (AFREPREN/FWD, Kenya)
Susan McDade (United Nations Development Programme)

Lead Authors (LA)
Brenda Boardman (University of Oxford, UK)
John Kimani (AFREPREN/FWD, Kenya)

Review Editor
Nora Lustig (Tulane University, USA)

Contents

Executive Summary

There is often a two-way relationship between the lack of access to adequate and affordable energy services and poverty. The relationship is, in many respects, a vicious cycle in which people who lack access to cleaner and affordable energy are often trapped in a re-enforcing cycle of deprivation, lower incomes and the means to improve their living conditions while at the same time using significant amounts of their very limited income on expensive and unhealthy forms of energy that provide poor and/or unsafe services.

Access to cleaner and affordable energy options is essential for improving the livelihoods of the poor in developing countries. The link between energy and poverty is demonstrated by the fact that the poor in developing countries constitute the bulk of an estimated 2.7 billion people relying on traditional biomass for cooking and the overwhelming majority of the 1.4 billion without access to grid electricity. Most of the people still reliant on traditional biomass live in Africa and South Asia.

Limited access to modern and affordable energy services is an important contributor to the poverty levels in developing countries, particularly in sub-Saharan Africa and some parts of Asia. Access to modern forms of energy is essential to overcome poverty, promote economic growth and employment opportunities, support the provision of social services, and, in general, promote sustainable human development. It is also an essential input for achieving most Millennium Development Goals (MDGs) – a useful reference of progress against poverty by 2015 and a benchmark for possible progress much beyond that. Poverty alleviation and the achievement of the MDGs will not be possible as long as there are billions of people who do not have access to electricity and or to cleaner and better quality as well as adequate supplies of cooking fuels or with limited access to affordable and more efficient end-use energy devices such as improved cookstoves (those using traditional fuels but burning in a cleaner fashion), proper heating, more efficient lights, water pumps, low-cost agro-processing equipment as well as energy-efficient housing and transportation options.

The lack of modern and affordable forms of energy affects agricultural and economic productivity, time budgets, opportunities for income generation, and more generally the ability to improve living conditions. Low agricultural and economic productivity as well as diminished livelihood opportunities in turn result in malnourishment, low earnings, and no or little surplus cash. This contributes to the poor remaining poor, and consequently they cannot afford to pay for cleaner or improved forms of energy (often neither the fuels nor the equipment). In this sense the problem of poverty remains closely intertwined with a lack of cleaner and affordable energy services.

On the other hand, when the poor gain access to stable electricity supplies or cleaner fuels that can support job creation, trade, and value-adding activities within the family, they are able to accumulate the small levels of "surplus" or savings that facilitate access to education and health services, improved nutrition, or improved housing conditions that in turn enable them to gradually escape their poverty.

Cleaner affordable forms of energy and the energy services that they provide, reverses the aforementioned vicious cycle as they contribute to building a basis for supporting job creation, economic growth, agriculture, education, commerce, and health – the key areas that can contribute to overcoming poverty, as the following examples demonstrate:

*   Energy plays a vital role in enhancing the food security of the poor through greater use of energy technologies for irrigation and water pumping. A number of these technologies not only ensure food supply throughout the year but also generate additional income for poor households, particularly in rural areas.

*   Access to modern and affordable forms of energy is essential to creating employment, which can directly reduce poverty levels. Employment in formal and non-formal sector activities is positively correlated to increased access to cleaner energy options such as electricity, as is workers' productivity in value-adding processes.

*   Increased access to modern forms of energy contribute to the transformation of agriculture-based economies – where significant animate energy is used – into industry-based economies, where modern forms of energy play a key role in the more-advanced value-added activities that characterize more industrialized economies.

- Improved low-cost cookstoves reduce the amount of fuel used, which translates into direct cash savings. They also reduce respiratory health problems associated with smoke emission from traditional biomass stoves and offer a better home and working environment. Other benefits include the alleviation of the burden placed on women and children in fuel collection, freeing up more time for women to engage in other activities, especially income-generating endeavors.

- Modern forms of energy play an important role in improving access to safe water and sanitation in developing countries.

- Modern health services (the facilities to provide them, and the professional and health sector workers who deliver them) are greatly enhanced by access to modern forms of energy and the services they provide, not only electricity for lighting, refrigeration and modern medical equipment but also heat-related services often linked to the availability of cleaner and affordable fuels for institutional uses.

- Access to modern forms of energy greatly improves the quality and availability of educational services and increases the likelihood that children will attend and complete schooling. Rural electrification helps retain good teachers in rural areas – a key lever for enhancing the quality of rural education.

- Gender parity and school enrollment of girls improve with greater access to cleaner and affordable energy options, especially in rural areas. Access to mechanical power for threshing and grinding grains help older girls stay in school by minimizing school absenteeism or desertion, while cleaner modern cooking systems reduce the time spent by girls on fuelwood collection for their families.

Universal access to electricity and modern as well as cleaner forms of energy for cooking provide a wide range of high-quality energy services even when the amount of energy used is modest. Given the important contribution of access to cleaner and affordable energy options to the improvement of livelihoods of poor people in developing countries, special attention should be paid to extending universal access to electricity and modern forms of energy for cooking in developing countries.

From a policy and public financing point of view, there is a need to alleviate the negative income effect of time use and family expenditure related to the use of traditional fuels and to overcome the barriers related to household energy. In developing countries, where poor people are highly dependent on traditional fuels that have incomplete combustion in inefficient devices, much greater emphasis must be given to increasing the availability and affordability of liquid and gaseous fuels and associated stove technologies. Special attention should also be paid to mechanical power solutions and labor-saving appliances that can provide homestead-based pumping, agro-processing, milling, and grinding, among other services. These are also critically important for gender and health reasons, as discussed later in this chapter.

The absence of mechanical options in many rural areas of the developing world, to provide access to mechanical power for water pumping, milling, or other household activities is also related to the absence of affordable and cleaner fuels or electricity to drive engines, mills, and pumps to supplement human labor (see Chapters 19, 23 and 25). The absence of mechanical power affects not only the poorest families, particularly girls and women, but often entire geographic areas due to the distances from electricity supply and cleaner fuel delivery systems.

Prioritizing access to electricity and energy-using technology for commercial/productive activities (as opposed to only low-load household activities) could increase employment opportunities that can contribute to income generation and the fight against poverty, especially for the working poor. Access to more efficient, cleaner and affordable energy options is an effective tool for combating extreme hunger by increasing food productivity and reducing post-harvest losses.

Low-cost energy-efficient transportation for both passengers and goods is crucial for economic development. This is particularly important for long-distance transit of passengers and goods. Rail transport is considered the least expensive energy-efficient land transport option for long-distance passenger travel and has been developed extensively in industrial countries in the form of surface and underground rail networks. For freight transportation, with the exception of India and China, many developing countries make limited use of lower-cost energy-efficient rail transport mainly due to its lack of sustained investment.

There are several health-related benefits associated with access to modern, cleaner and affordable energy options. Electricity in rural health centers, allows the provision of medical services at night, greater use of more advanced medical equipment, helps retain qualified staff in rural health centers among other benefits. In this regard, one of Global Energy Assessment (GEA)'s goals is achieving universal access to electricity and cleaner cooking by 2030. It should, however, be emphasized that access to electricity and cleaner cooking fuels constitute only a part of the desired policy objective of reducing poverty – they are a prerequisite, but electricity supply and cleaner cooking fuels, on their own, are not sufficient to move out of poverty.

Access to modern and affordable forms of energy can play an important role in improving access to safe water and sanitation in developing countries. Some of the energy solutions available are relatively affordable for the poor and do not require electricity or fossil fuels. Low-cost mechanical options can play an important role in ensuring access to safe water. For example, hand pumps and wind pumps are robust technologies that can dramatically expand access to safe water for domestic use as well as for irrigation and watering livestock.

The challenge of providing safe water and sanitation facilities is still immense particularly in sub-Saharan Africa where more than half the urban population lives in slums – a sizable part of the more than 825 million people living in urban dwellings without improved sources of drinking water and sanitation. Solar, wind, handpumps and biogas options can play an important role in enhanced access to safe water and sanitation in low-income peri-urban and urban of sub-Saharan Africa.

Prevailing energy systems in developing countries and associated economic and welfare policies need to be redesigned to ensure an emphatic pro-poor orientation that will move toward universal access to cleaner and affordable forms of energy in key economic sectors that the poor rely on such as health, water, education, agriculture and transport.

Access to electricity supply (both grid and non-grid) in many developing countries is almost an exclusive service enjoyed by the non-poor in urban areas. Even after two decades of energy sector reforms, initial indications from a wide range of developing countries indicate that few of the initiatives have resulted in significant improvement in the provision of electricity to the world's poor.

Experiences in developing countries point to an overarching conclusion: when power sector reforms were introduced with the sole intention of improving the performance of utilities, the expected and hoped-for social benefits did not necessarily follow. Where governments maintained a role as instigator or at least regulator of improved access to electricity by the poor, tariffs for poor households tended to decrease and levels and rates of electrification tended to increase.

In some cases, wide-scale deployment of renewable energy systems may provide the best options for providing access to cleaner and modern energy options while ensuring long-term sustainability. In the near term, however, cleaner fossil fuels such as liquefied petroleum gas (LPG) combined with more efficient and low-cost end-uses devices such as LPG cookstoves reduce a host of social, economic, and environmental barriers to overcoming poverty due to higher combustion efficiency.

The ongoing search for win-win solutions that are completely reliant on renewable energy must also put a time premium on overcoming social injustice and the conditions of inequality that entrench poverty and reproduce underdevelopment. From a policy point of view, it is unrealistic to support arguments for "near-term renewables only" options that delay the satisfaction of the basic needs of people who remain poor today and who constitute a negligible contributor to global historical and current carbon emissions. It is, therefore, important to retain cleaner fossil fuel options such as LPG combined with more efficient low-cost end-use devices in near-term priority action plans that would set the stage for eventually moving the poor to a fully sustainable path that they are able to rely more heavily on renewable energy.

2.1   Introduction

This chapter focuses on the situation of poor people – especially in developing countries – paying particular attention to their access to cleaner and affordable energy options. Poverty, as discussed in this chapter, includes concepts of both lack of access to income and inequality as well as limited access to services (such as access to cleaner energy options), opportunities and social exclusion, which are often intimately linked to inequality.

The chapter calls for setting of targets for bringing cleaner and affordable energy options to a wider proportion of the poor at affordable prices – plus the policy frameworks and the public and private investment needed to make this a reality.

The issue of access to cleaner and affordable energy options is emphasized here in line with the GEA goal of providing universal access to affordable electricity and affordable cleaner forms of energy for cooking by 2030. However, this 2030 goal requires substantial investments as well as a transition from the use of traditional solid fuels such as biomass and coal to cleaner energy carriers such as gas, electricity and liquid fuels. These investments and transition require sufficient time for low-income countries to establish the right institutions, regulation and policies that would enable mobilization and appropriate deployment of the required capital investments.

While the term "access" can simply refer to physical proximity to modern energy carriers such as electricity, natural gas, liquefied petroleum gas (LPG), biogas and ethanol, access also implies the availability of affordable improved and more efficient end-use energy devices such as improved cookstoves (those using traditional fuels but burning in a cleaner fashion), more efficient lights, water pumps, low-cost agro-processing equipment as well as energy-efficient housing and transportation options. In a much broader sense, access refers to the affordable and stable services of cleaner energy options that are delivered in a reliable fashion and ensure consistent quality (see also Chapter 19).

GEA considers universal energy access to refer to access to affordable modern energy carriers (e.g., electricity, natural gas, LPG, biogas, ethanol) including a wide array of efficient low-cost end-use options used by the poor in agro-processing, small-scale value addition processes, water pumping, housing and transportation as well as energy-efficient devices such as improved cookstoves and affordable, advanced stoves that use traditional fuels for cooking, by 2030.

This chapter focuses on developing countries for several reasons. The bulk of poor people in the world live in developing countries and require energy to support poverty alleviation, economic growth, social inclusion and development. The chapter focuses first on the concept of poverty: who are poor people and where they live.

It goes on to examine the historic worldwide relationship between access to energy services and development, arguing that developing countries also need to have access to improved energy services in order to overcome poverty, especially the most extreme forms faced by the lowest fifth of income earners around the world. The specific role of energy services – those provided by modern energy carriers such as electricity and fuels such as LPG, biogas and ethanol – is discussed in this chapter in relation to both household and productive energy applications. The essential role that energy plays in supporting job creation, economic growth, agriculture, transport, and commerce is reviewed, as these are key factors for overcoming poverty.

The importance of energy as an input to support the delivery of services such as education, health, and other social services is examined, followed by an analysis of the particular hardships faced by women in relation to traditional energy systems and their impacts on women's time use, literacy, and health. The importance of energy for the achievement of the Millennium Development Goals (MDGs) is also briefly illustrated by showing that energy is an essential input to each MDG. The local, regional, and global environmental impacts of energy use in developing countries are discussed, with an emphasis on the extremely small historical and current contribution made to global green house emissions by the poorest people in the world.

The chapter concludes that special emphasis must be given to extending universal access to electricity and modern forms of energy for cooking in developing countries and that the switch from traditional solid fuels to cleaner liquid fuels and combustion technologies constitute an essential step for overcoming poverty and supporting economic growth and sustainable human development. It is recommended that near-universal access is needed in developing countries due to the range of high-quality energy services that this carrier provides, even when it is available only in small amounts.

Specifically, with regard to electricity, this chapter argues that access to electricity as a carrier is only part of the desired policy objective of reducing poverty – it is a prerequisite, but electricity supply, in itself, is not sufficient for poor households to move out of poverty. The ultimate, well-being of people depends on the actual consumption of goods and services that electricity as a carrier can facilitate. Equally important is access to cleaner or modern forms of energy for cooking and fueling a wide array of energy efficient and low-cost end-use options and devices used by the poor in agro-processing, small scale value-addition processes, water pumping, housing and transportation.

Looking toward 2030, the need to focus on improving access to energy services in developing countries becomes even more crucial due to their higher population growth and their potential for higher economic growth rates, implying that the decisions taken regarding the nature of emerging energy systems in terms of energy sources, conversion technologies, and distribution and financing systems are important not only to developing countries but to the world as a whole throughout this century.

2.1.1   Poverty and Development

Economic growth is an essential prerequisite for overcoming poverty. No country has achieved sustained economic growth without improving access to cleaner and modern forms of energy and the services that they provide. It is also globally recognized – based on the experiences of most industrial countries – that policies to "share" the benefits of growth are needed to address inequality and combat poverty. Energy services to support economic growth and energy policies to combat inequality in human welfare are thus both critically important.

Using an absolute poverty line set at US$_{2005}$\$1.25 per day per capita (purchasing power parities (PPP)), the number of poor people worldwide fell from 1.8 billion in 1990 to 1.4 billion in 2005 (see Figure 2.1) in spite of population growth – a tremendous achievement based on the economic growth experienced worldwide during this period (UN, 2010a). According to the World Bank, the financial and food crises of recent years unfortunately pushed another 64 million people back into extreme poverty (World Bank, 2010a).

Table 2.1, provides a comparison of poverty trends in 1999, 2002 and 2005. It shows a substantial drop in poverty levels in East Asia and Pacific (particularly China) and highlights the continued high-levels of poverty in Sub-Saharan Africa and South Asia.

Using the multidimensional poverty index (MPI) recently developed jointly by Oxford University and United Nations Development Programme (UNDP), it is estimated that, worldwide, a larger proportion of humanity, about 1.75 billion, remains poor. The MPI is an advanced method of measuring poverty by determining the extent of deprivation among poor people across three key dimensions, namely: health, education and living standards. MPI also takes into account access to essential services such as electricity, water and sanitation in order to measure the magnitude of poverty (UNDP, 2010; Alkire et al., 2011).

Table 2.1 | Proportion of people with incomes below US$_{2005}$\$1.25 per day PPP (%).

Region	Year		
	1999	2002	2005
Sub-Saharan Africa	58.4	55.0	50.9
South Asia	44.1	43.8	40.3
East Asia and Pacific	35.5	27.6	16.8
Latin America and Caribbean	11.0	10.7	8.1
Europe and Central Asia	5.1	4.6	3.7
Middle East and North Africa	4.2	3.6	3.6

Source: World Bank, 2011a.

Using the MPI methodology, available data (Figure 2.2) re-confirms the fact that the greatest concentrations of poor people live in the least developed countries of sub-Saharan Africa and South Asia.

Both income-defined and MPI poverty measurements leave out the fundamental issue of income distribution within countries. Today the World Bank, the United Nations system, and economists that work on poverty indicators at large argue that it is essential to look at not only the aspects of poverty expressed by income levels alone, but also at equity measures such as how the lowest 20% of the population (determined by income) fare in comparison to the top 20% in any given country. The more unequal the income distribution, the less likely it is that the poorest segment of the population will benefit from growth, although overall national statistics may show positive trends (UNDP, 2011).

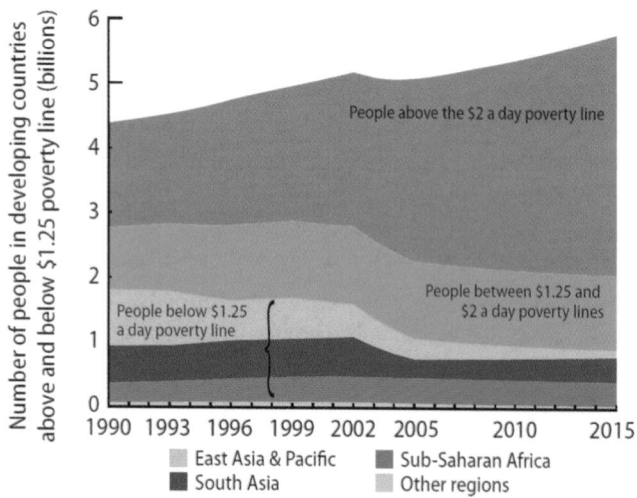

Figure 2.1 | Progress in poverty reduction. Source: World Bank, 2010a.

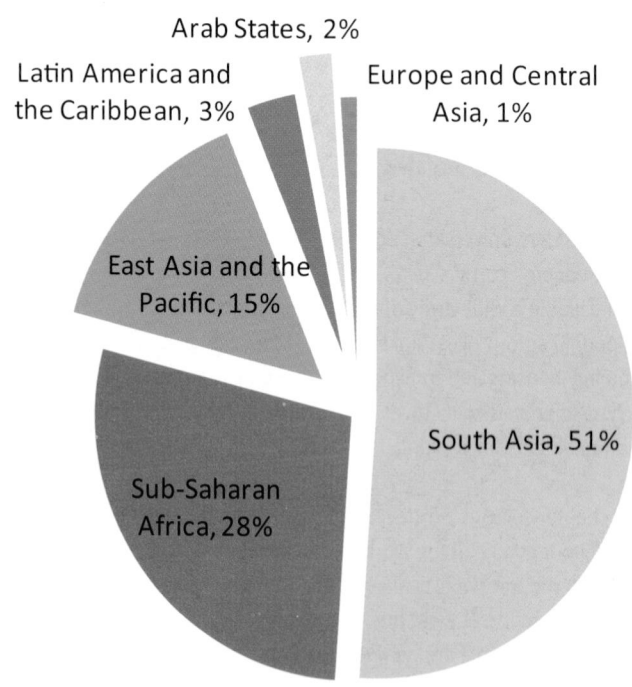

Figure 2.2 | Distribution of the poor (1.75 billion people) in developing countries (using multidimensional poverty measurement). Source: UNDP, 2010.

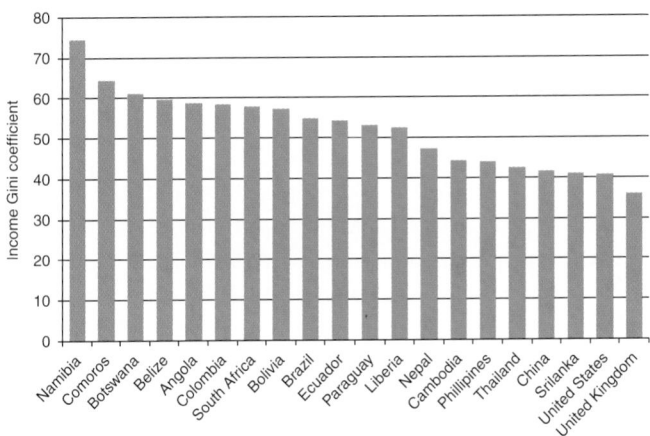

Figure 2.3 | Gini indices for selected developing countries (2000–2011*). Source: adapted from UNDP, 2010; 2011. *Data refer to the most recent year available during the period specified.

The Gini coefficient[1], which compares income levels among the richest and poorest segments of populations, is a useful indicator of inequality (see Figure 2.3). Comparing the lowest 20% of income earners to those who earn the most is also useful for understanding the role that inequality plays in defining country contexts. Inequality has been increasing alongside growth and economic development around the world triggering new development challenges – and therefore new energy challenges. Countries which have a higher coefficient, even in situations where energy access indicators show positive growth, the poorest segment of the population may not be benefiting from the growth. Currently, Latin America, the Caribbean region and Africa, are the most unequal regions in the world (Gasparin et al., 2009; Rosenthal, 2010; Gasparin and Lustig, 2011).

While many developing countries, particularly in Latin America, the Middle East, and Asia, have achieved significant growth and progress in development indicators over the last 20 years, progress has often not been evenly distributed across the population. As a result, inequality and increasingly unequal income distribution pushes more people into poverty thus creating new development as well as public financing and service delivery challenges for governments. Turkey, Gabon and South Africa, for instance, have made impressive advances in improving income per capita in the last decade, while levels of poverty and inequality have persisted (UNDP, 2010). As Andy Sumner notes in *Global Poverty and the New Bottom Billion*:

> The global poverty problem has changed because most of the world's poor no longer live in poor countries meaning low-income countries (LICs). In the past, poverty has been viewed as, predominantly, an LIC issue. Nowadays such simplistic assumptions/

classifications can be misleading because a number of the large countries that have graduated into the middle-income countries (MIC) category still have large number of poor people. In 1990, we estimate that 93 per cent of the world's poor people lived in LICs. In contrast, in 2007–8 we estimate that three-quarters of the world's approximately 1.3 billion poor people now live in MICs and only about a quarter of the world's poor – about 370 million people live in the remaining 39 low-income countries, which are largely in sub-Saharan Africa (Sumner, 2010).

The largest number of poor people who "moved" from poor countries to middle-income ones are found in India, Nigeria, and Pakistan (Sumner, 2011). The two-tiered concept is itself questionable as it divides the world into developing and developed countries, which is contradicted by the emergence of the BRICS group, the rapidly growing and industrializing countries of Brazil, Russia, India, China, and South Africa (Sumner, 2011). The recent creation of the G-20 group of countries with industrialized or industrializing economies includes the BRICS as well as Argentina, Indonesia, Mexico, South Korea and Turkey.

Yet these emerging economic powers continue to have millions of poor people, with conditions of wealth and poverty[2] often existing side by side – sometimes only neighborhoods away and sometimes defined by the rural-urban divide. Thus, in terms of poverty, development, and growth, the world is more heterogeneous now than it was in 1990. And looking forward to 2050, many emerging economic powers historically viewed as "developing" will take on new roles while entire segments of poor people remain a feature of their national makeup. This is evident in urban areas of emerging economies where high numbers of population live in deplorable dwellings, i.e., slums, without access to basic services thus constituting a major challenge to urban governance and design in developing countries (GNESD, 2008).

It should, however, be emphasized that the poor in emerging economic powers of Latin America and Asia have better prospects for moving out of the poverty largely due to the significant and abundant economic, institutional and skill resources that are available in these countries plus their growing clout in re-arranging international economic relationships to protect their national interests.

In contrast, the prospects for the poor in Sub-Saharan Africa are significantly less promising (see Table 2.1) because of the very limited economic, institutional and skill resources that sub-Saharan African countries have combined with modest and more importantly "diminishing" ability to influence global economic relationships that are disadvantageous to the sub-Saharan Africa region. As shown in Table 2.1,

---

1   Gini coefficient measures the extent to which the distribution of income, (at times it includes consumption expenditure) among people or households within an economy deviates from a perfectly equal distribution. The values vary between 0, which reflects complete equality and 1, indicating a complete inequality situation, where one individual has all the income or consumption, while all the others have none).

2   These countries poverty conditions are in terms of relative poverty: this refers to lack of the usual or socially acceptable level of resources or income as compared with others within a society or country. On the other hand absolute poverty refers to a common international poverty line of: population living on less than $1.25 at 2005 PPP.

well over half of sub-Saharan Africa's population is poor – higher than any other region of the world.

Virtually all future poverty scenarios show substantial reductions in poverty levels in the large emerging economic powers of Latin America and Asia with virtually the whole of sub-Saharan Africa and parts of South Asia continuing to be the epicenters of global poverty.

In conclusion, poverty includes concepts of low income and inequality as well as limited access to services (such as access to cleaner energy options), opportunities and social exclusion, which are often intimately linked to inequality.

2.2   Energy and Development

Access to modern forms of energy is an essential pre-requisite for overcoming poverty, promoting economic growth, expanding employment opportunities, supporting the provision of social services, and, in general, promoting human development. It is also an essential input for achieving most MDGs, which are a reference of progress against poverty by 2015 and a benchmark for possible progress beyond that date. As mentioned earlier, it is, however, important to underline that access to electricity and cleaner cooking constitutes only part of the desired policy objective of reducing poverty. They are a pre-requisite, but electricity supply and cleaner cooking fuels, on their own, are not sufficient for poor households to move out of poverty.

Table 2.2 illustrates how energy can contribute to the MDGs targets. It also shows clearly how the lack of access to energy can be a major constraint to achieving the MDGs (Modi et al., 2006; UN-Energy, 2005; UNDP, 2005, DFID, 2002). One of GEA's goals is the provision of universal access to affordable electricity and clean cooking by 2030 – a key target that would contribute to reducing poverty levels.

Human well-being, poverty reduction, social inclusion, and economic improvement cannot be advanced without access to electricity, fuels, mechanical power, and the range of services that they provide. Countries with the highest levels of poverty and underemployment tend to be those that also lack access to adequate levels of energy services and the modern conveniences that they provide. This is most pronounced in Africa and South Asia, where the number of people who depend on traditional biomass for heating and cooking and who lack access to electricity are the greatest. As shown in Figure 2.4, based on the Energy Development Index (EDI) developed by the International Energy Agency (IEA), all sub-Saharan African countries, with the exception of South Africa, feature in the bottom half of the EDI ranking.

The EDI was developed to track the progress in country's or region's transition to the use of modern energy and to better understand the role that energy plays in human development. It is computed by factoring in the UNDP's Human Development Index (HDI) and is composed of four

indicators; per capita commercial energy use, per capita electricity consumption, share of modern fuels in total residential sector energy use and share of population with access to electricity (Bazilian et al., 2010; IEA, 2010a; Nussbaumer et al., 2011).

Abundant cheap and highly polluting energy supplies, while providing essential inputs for economic growth in the short term, result in unsustainable local, regional, and global environmental and health effects over the medium and long term, as seen throughout the industrial world and now increasingly in China and India (UNDP, 2011). From a sustainable human development perspective, therefore, the key challenge is how to make available modern and cleaner forms of energy that can support economic growth and human well-being, helping to reduce poverty while not establishing long-term structural and negative environmental consequences in energy systems that will eventually undermine development itself.

For industrial countries, this challenge largely results in a focus on cleaner energy systems, energy efficiency, and the drive to diversify the energy supply mix to include more renewable sources to address global climate change concerns and rising or unstable global oil prices. For the least developed countries, which have significantly less installed energy infrastructure and where in many cases the vast majority of the population does not have access to either modern and cleaner forms of energy or electricity, the key near-term question is how to promote the development of energy systems that can provide accessible and affordable services using new models of production, financing, and distribution as well as introducing technologies that can avoid the limits to growth that traditional energy systems will eventually run into.

Energy as a key input to development is undeniable, but significant debate exists about how to go forward, and the models must differ from those put in place by industrial countries that have proved to be unsustainable. To move forward, it is essential to understand the multiple linkages between access to energy services and social, economic, and human development. These linkages shape emerging energy markets, define and often limit poor peoples' options to escape the cycle of poverty, and will determine the opportunities and limitations to key issues of pertaining to sustainability of development. Putting peoples' needs at the center of the development model that defines the energy services required to overcome poverty is a key starting point.

Currently there are almost one and a half billion people worldwide who do not have access to electricity (IEA, 2010a; see also Chapter 19). Most of these people live in Africa and South Asia. An estimated 3 billion people primarily rely on solid fuels (coal and traditional biomass) of which 2.7 billion people cook and heat their homes with traditional fuels and low-efficiency stoves (UNDP and WHO, 2009; IEA, 2010a; see also Chapter 19). For many millions of people, especially women and school-aged children, the lack of mechanical power for pumping water and grinding food grains results in hours of manual labor carrying fuel and water and undertaking repetitive pounding and grinding activities that pumps and mills could do so much more efficiently.

Table 2.2 | Energy and the MDGs.

Goal and target	Some direct and indirect contributions of cleaner/affordable energy options
MDG 1. Extreme poverty and hunger: To halve, between 1990 and 2015, the proportion of the world's people whose income is less than one dollar per day. To halve, between 1990 and 2015, the proportion of people who suffer from hunger.	• Cleaner burning fuels and electricity can reduce the large share of household income spent on cooking, lighting and heat. The bulk of staple foods (95%) need cooking before they can be eaten and need water for cooking. • Post-harvest losses can be reduced through improved electric-powered preservation (for example, drying and smoking) and chilling/freezing. • Energy technologies such as wind pumps and treadle pumps can be used for irrigation in order to increase food production and improve nutrition. Access to affordable energy options from gaseous and liquid fuels and electricity can assist enterprise development. • Electrically driven machinery can increase productivity and provide opportunities for income generation. • Local energy supplies can often be provided by small-scale, locally owned businesses creating employment.
MDG 2. Universal primary education: To ensure that, by 2015, children everywhere will be able to complete a full course of primary schooling.	• Lighting at homes (e.g., through solar lanterns) allows children to study after school hours, with a significant impact on learning outcomes. • Lighting in schools can assist in retaining teachers, especially if their houses are electrified. • Availability of electricity can enable access to educational media and communications in schools and, at home, can facilitate distance learning. • Access to energy can provide opportunities for using specialized equipment for teaching. • Cleaner energy systems and efficient building design can reduce heating/cooling costs and thus school fees. • Energy can help create more child-friendly environments, thus improving attendance at school and reducing dropout rates.
MDG 3. Gender equality and women's empowerment: To ensure that girls and boys have equal access to primary and secondary education, preferably by 2005, and to all levels of education no later than 2015.	• Availability of cleaner energy options can free girls' and young women's time from survival activities (gathering firewood, fetching water, etc.). • Good-quality lighting can facilitate home study and organization of evening classes for girls and women who are often housebound due to traditional family responsibilities. • Affordable and reliable energy options can broaden the scope for women's enterprises, thereby fostering employment and income generation among women. • National decision-making by women representatives, especially on energy use at household level, can be beneficial, hence improving energy access among the poor.
MDG 4. Child mortality: To reduce by two-thirds, between 1990 and 2015, the death rate for children under the age of five.	• GEA estimates for 2005 put the burden of disease caused by household air pollution at about 2.2 million premature deaths annually, mostly affecting children and women (see Chapters 4 and 17). Gathering and preparing traditional fuels exposes young children to health risks and can reduce time spent on childcare. • Cleaner energy options facilitate the provision of nutritious cooked food and space heating, while boiled water contributes to better health. • Improved energy options can provide access to better medical facilities for pediatric care, including vaccine refrigeration and equipment sterilization. • Energy can be used to purify water or pump clean groundwater locally, which can reduce the burden of water-borne diseases.
MDG 5. Maternal health: To reduce by three-quarters, between 1990 and 2015, the rate of maternal mortality.	• Clean cooking fuels and equipment can reduce pregnant women's exposure to indoor air pollution and improve health. • Improved energy options can provide access to better medical facilities for maternal care, including laboratory services, medicine refrigeration, equipment sterilization, and operating theatres, as well as safer caesarean sections. • Improved energy options can also help retain qualified medical personnel in remote rural areas. • Cleaner energy options can reduce excessive workloads and heavy manual labor (carrying heavy loads of fuelwood and water), which could adversely affect a pregnant woman's general health and well-being.
MDG 6. HIV/AIDS, malaria, and other major diseases: By 2015, to have halted and begun to reverse the spread of HIV/AIDS, malaria, and other major diseases that afflict humanity.	• Electricity in health centers can help provide medical services at night, retain qualified staff, and allow the use of more advanced medical equipment (e.g., sterilization). • Energy for refrigeration can facilitate vaccination and medicine storage for the prevention and treatment of diseases and infections. • Energy is needed to develop, manufacture, and distribute drugs, medicines, and vaccinations. • Electricity can enable access to health education media through information and communications technologies.
MDG 7. Environmental sustainability: To stop the unsustainable exploitation of natural resources. To halve, between 1990 and 2015, the proportion of people who are unable to reach or afford safe drinking water and sanitation.	• Increased agricultural productivity can be facilitated by the greater use of electric-powered machinery and irrigation, which in turn reduces the need to expand the amount of land under cultivation. • Increased renewable energy technology use can contribute greatly to alleviation of deforestation and reduction of green house emissions that lead to climate change. • Cleaner burning fuels can reduce greenhouse gas emissions, which contribute to climate change conversion technologies. • Simple cleaner energy solutions such as low-cost sterilization of drinking water can save many lives.
MDG 8. Global partnership for development	• Global and subregional partnerships are valuable for ensuring cross-border trade and exchange of skills in cleaner energy options as well as joint lower-cost development of transmission interconnections.

Source: adapted from Modi et al., 2006; DFID, 2002.

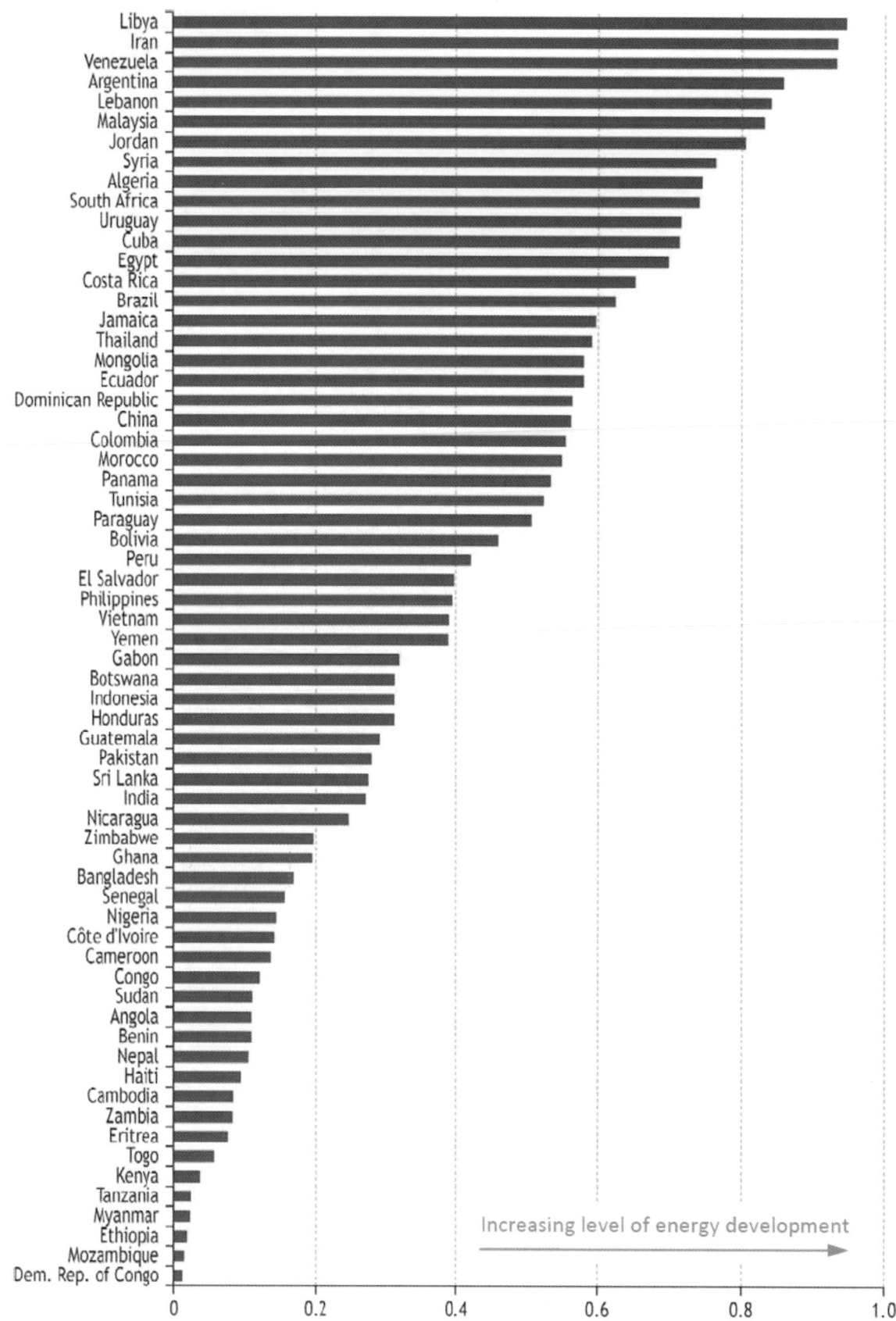

Figure 2.4 | Energy Development Index, 2009. Source: IEA, 2010a.

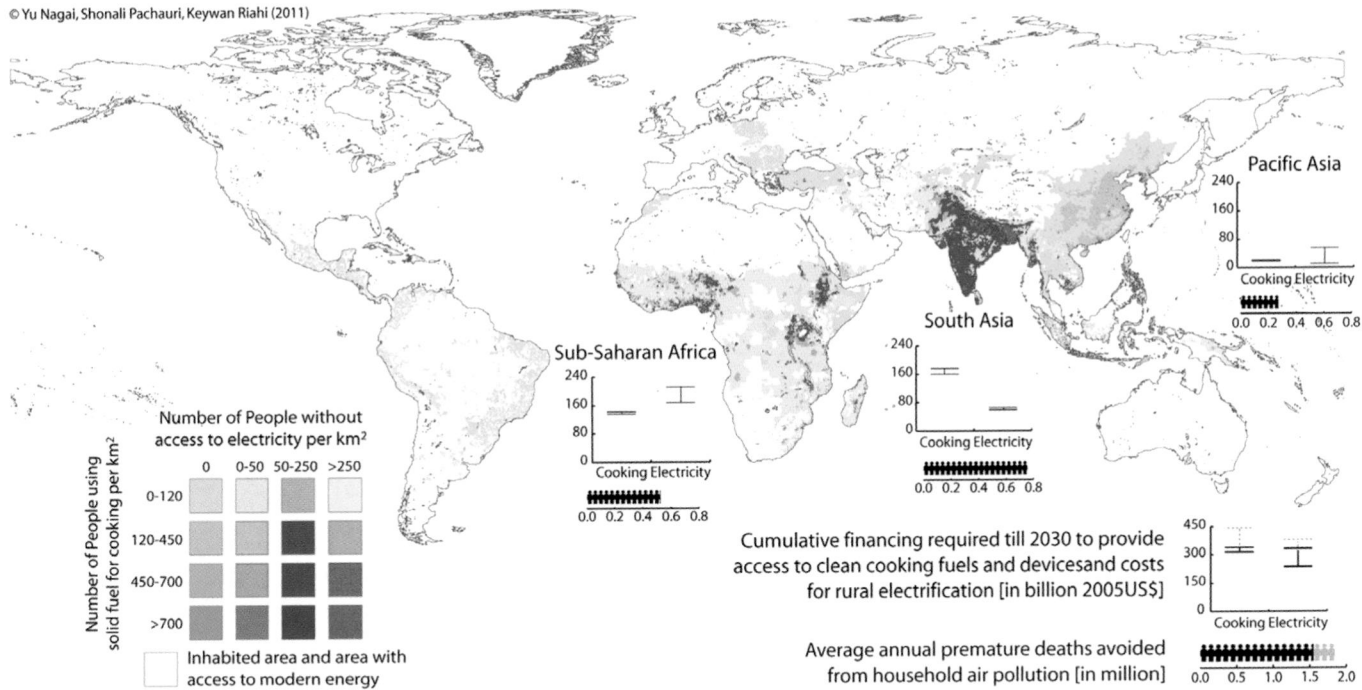

© Yu Nagai, Shonali Pachauri, Keywan Riahi (2011)

Figure 2.5 | Proportion of population without access to electricity and using solid fuels. Source: Compiled by IIASA using data from UNDP and WHO, 2009.

Figure 2.5 shows the proportion of people without electricity and using solid fuels for cooking. The figure indicates that, among developing countries, sub-Saharan Africa has the highest proportion of its population having a combination of low access to electricity and reliance on solid fuels (see also Table 2.4 and Table 2.6) while the absolute largest number of the people with limited access live in Asia. According to projections by IEA (2010a), the population without electricity will continue to rise in sub-Saharan Africa, unlike in other parts of the developing world, which still have pockets of underserved populations but which are projected to significantly increase access to electricity (e.g., Latin America and North Africa).

As shown in Table 2.3, within countries, there is a significant disparity in terms of access to electricity between rural and urban populations (IEA, 2010a).

As shown in Table 2.4, in many developing countries, especially those in sub-Saharan Africa and some parts of Asia, low electrification rates are compounded by very low levels of electricity consumption implying very limited use of this important and flexible energy service.

As mentioned earlier, while there are different approaches to defining poverty, there is a positive worldwide correlation between energy services and GDP per capita (approximately average personal income) as well as in relation to human development. As shown in Figure 2.6, there is a correlation between the level of well-being (shown by the HDI) and access to modern forms of energy (indicated by per capita electricity consumption). As the electricity consumption per capita increases, the HDI indicator increases, while an increase in income and

Table 2.3 | Urban and rural populations without access to electricity, 2009.

Region	Rural (million)	Urban (million)
Sub-Saharan Africa	465	120
China	8	–
India	381	23
Other developing countries in Asia	328	59
Latin America	27	4
Total	**1209**	**206**

Source: IEA, 2010a.

electricity access has a direct relationship in the increase of electricity consumption per capita.

As mentioned earlier, provision of and access to modern, cleaner and affordable energy options *per se* does not, in itself, alleviate poverty. However, there is growing consensus that enhancing access to modern, cleaner and affordable energy options can play a key contributing role to reducing poverty, especially for the poorest people (the lowest fifth in a population) and the poorest countries. Figure 2.7 illustrates the link between high poverty levels and low access to electricity. Energy services therefore, are a means to facilitate development given that energy is an essential input for productive, household and social sectors.

It is, however, important to note that at higher levels of incomes, this correlation begins to break down at a national level – an important indicator of wasteful energy use which highlights the importance of energy efficiency in richer industrialized nations.

Table 2.4 | Snapshot of electricity access and consumption in developing countries, 2008.

Country	Population with electricity access (%)	Electricity consumption per person (kWh)
Africa		
South Africa	75.0	4532
Nigeria	50.6	121
Cameroon	48.7	271
Cote d'Ivoire	47.3	203
Senegal	42.0	196
Sudan	35.9	114
Eritrea	32.0	51
Angola	26.2	202
Benin	24.8	91
Togo	20.0	111
Ethiopia	17.0	46
Kenya	16.1	147
Tanzania	13.9	86
Mozambique	11.7	453
Congo, Democratic Republic of	11.1	104
Asia		
Singapore	100.0	7949
China	99.4	2631
Malaysia	99.4	3614
Thailand	99.3	2045
Vietnam	97.6	918
India	75.0	597
Indonesia	64.5	590
Pakistan	62.4	449
Cambodia	24.0	131
Myanmar	13.0	104
Latin America		
Venezuela	99.0	3152
Chile	98.5	3283
Brazil	98.3	2206
Paraguay	96.7	1056
Dominican Republic	95.9	1358
Colombia	93.6	1047
Ecuador	92.2	1115
El Salvador	86.4	845
Peru	85.7	1136
Guatemala	80.5	548
Bolivia	77.5	558
Honduras	70.3	678

Source: IEA, 2009; 2011; World Bank, 2011a.

The correlation between access to cleaner energy and poverty reduction is corroborated by a number of recent World Bank studies. For example,

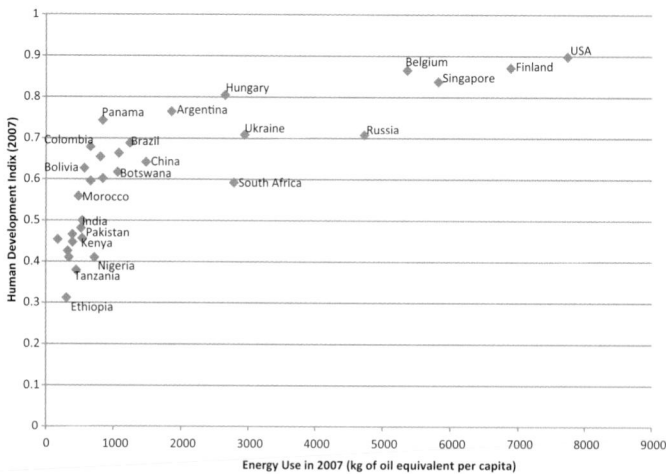

Figure 2.6 | Correlation between HDI and electricity consumption per capita. Source: adapted from UNDP, 2009; World Bank, 2007.

according to a study by Barnes et al. (2010), increased access to cleaner and affordable energy options contributes to monetary gains among the poor and leads to better quality of life, such as an improved diet and amount of food intake, the ability to afford better health and education facilities, etc.

There tends to be a two-way causal relationship between poverty and the lack of access to adequate and affordable energy forms. The relationship can be characterized as a vicious cycle because poor people who lack access to cleaner and affordable energy are often trapped in a repeating cycle of deprivation, limited incomes and the means to improve their living conditions while at the same time using significant amounts of their very scarce income on expensive and unhealthy forms of energy that provide poor or unsafe services such as dry cell batteries, rudimentary and inefficient kerosene lamps, charcoal and candles.

The lack of modern and affordable forms of energy affects agricultural and economic productivity, time budgets, opportunities for income generation, and more generally the ability to improve living conditions. Low agricultural and economic productivity as well as diminished livelihood opportunities in turn result in malnourishment, low earnings, and no or little surplus cash. This contributes to the poor remaining poor, and consequently they cannot afford to pay for cleaner or improved forms of energy (often neither the fuels nor the equipment). In this sense the problem of poverty remains closely intertwined with a lack of cleaner and affordable energy services.

On the other hand, when the poor gain access to stable electricity supplies or cleaner fuels that can support job creation, trade, and value-adding activities within the family, they are able to accumulate the small levels of "surplus" or savings that facilitate access to education and health services, improved nutrition, or improved housing conditions that in turn enable them to gradually escape their poverty (UNDP, 2006).

A number of developing countries have set targets for enhancing access to electricity, cleaner fuels, improved stoves, and mechanical power

(see Table 2.5). In overall terms, however, the total number of developing countries with such targets is relatively small.

As shown in Figure 2.8, nearly all 140 countries assessed in a UNDP-WHO study have some data on electricity access/supply, but none had data on mechanical power.

Table 2.5 | Developing countries with access targets for modern energy availability.

Energy Source/ Application	Number of Developing Countries with Target	Number of Sub-Saharan African Countries with Target
Electricity	68	35
Modern fuels	17	13
Improved cookstoves	11	7
Mechanical Power	5	5

Source: adapted from UNDP and WHO, 2009.

2.3   Household Energy

For the poorest people in the world, especially those who live in the poorest countries, the most inelastic segment of demand for energy is that for cooking and heating to ensure basic survival. This "household

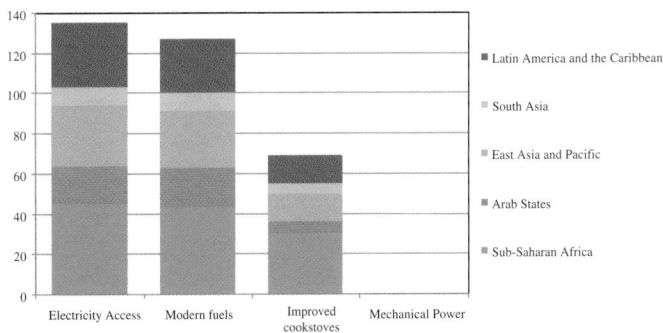

Figure 2.8 | Number of countries with data on energy sources and applications. Source: adapted from UNDP and WHO, 2009.

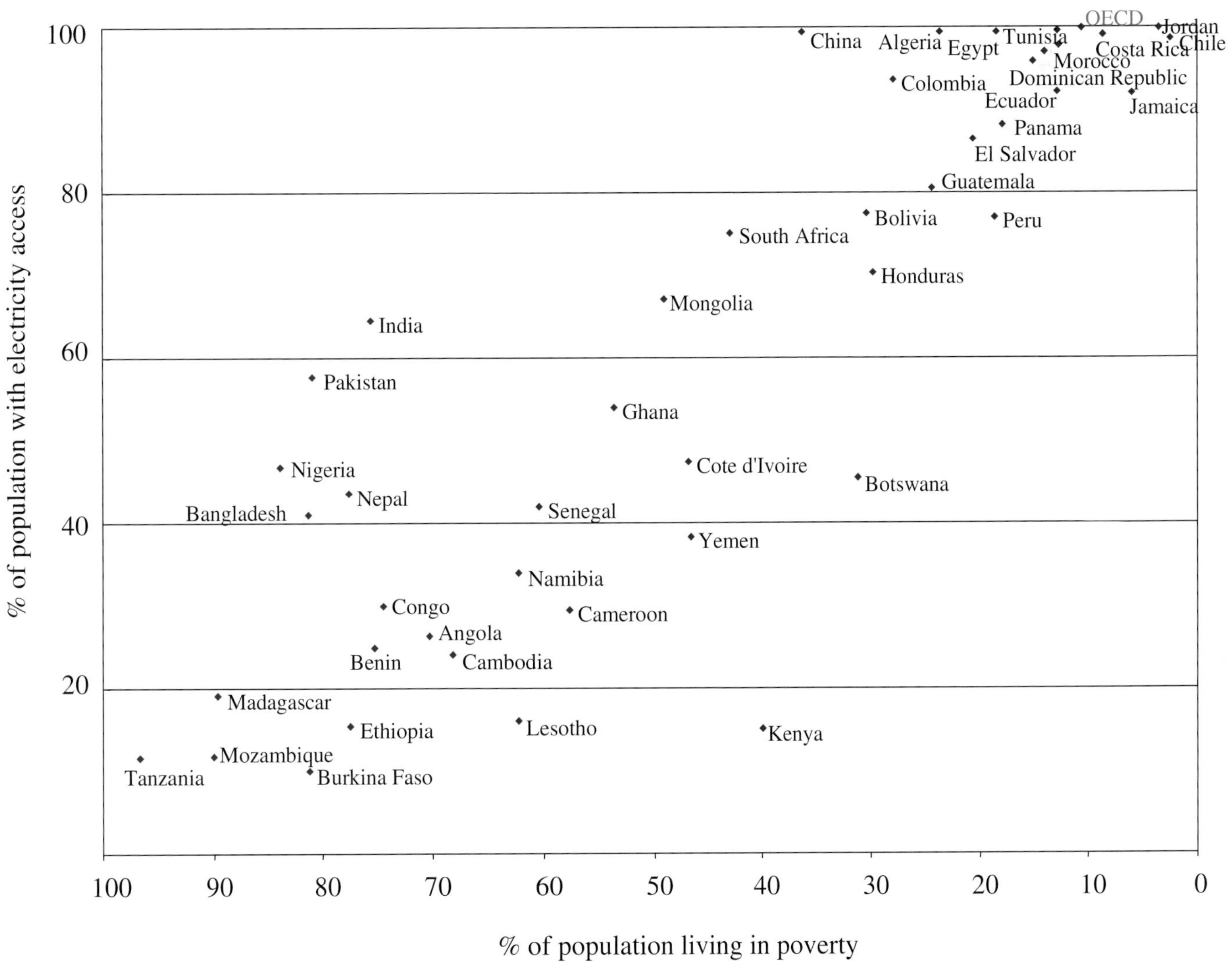

Figure 2.7 | Access to electricity versus poverty levels. Source: adapted from World Bank, 2011a; IEA, 2008.

energy" is often poorly understood by development planners at large, and within the energy sector it is often not considered in policies that historically have been focused on electricity supply rather than other household fuels. Even poor families do not have a choice of whether to use heat to cook their food because 95% of all basic staples must be converted to food through heat for cooking (ESMAP, 2004).

All families worldwide, rich or poor, depend on heat to cook their food, boil water, and, where the climate requires it, provide space heating. While traditional fuels such as wood, agricultural residues, or dung can be gathered locally, considerable time is spent collecting them (ESMAP, undated) (see Figure 2.19). As traditional fuels become scarcer due to overharvesting, agricultural decline, or increased competition among growing populations, more and more unpaid time is dedicated to fuel collection, leaving less time for income-earning activities.

In many poor countries, biomass, mainly for cooking, accounts for over 90% of household energy use. According to IEA (2010a) estimates, about 2.7 billion people rely on traditional biomass, such as fuelwood, charcoal, agricultural waste, and animal dung, to meet their energy needs for cooking. A significant proportion of these people are the majority of the poor, who live on less than US$_{2005}$2 a day (PPP).

As shown in the Table 2.6, sub-Saharan Africa and parts of Asia currently rely heavily on traditional biomass (IEA, 2010a). In all subregions of the developing world, people in rural areas account for the highest proportion of the population relying on traditional biomass, a key indication that rural areas in most developing countries have limited access to cleaner energy options, especially for cooking.

According to available data, if access to cleaner cooking fuels remains constrained, a large proportion of the population in sub-Saharan Africa and parts of South Asia will continue relying on traditional biomass for cooking for the next couple of decades. And as shown in Figure 2.9, in sub-Saharan Africa the number of people relying on traditional biomass for cooking is projected to increase at a much steeper pace than in other regions over the next two decades (Brew-Hammond, 2007) if current trends continue.

Unlike other regions and developing countries where traditional biomass energy use is expected to stagnate or decline, Africa's traditional biomass use is likely to increase sharply by 2030 unless there is a significant increase in access to cleaner and affordable energy carriers for cooking (IEA, 2006a). In the meantime, there is evidence that in areas where local prices of energy carriers have increased because of recent high international energy prices, the shift to cleaner and more efficient energy options for cooking has actually slowed and even reversed (IEA, 2006a; UNDP, 2007a; World Bank, 2010a).

Enhancing the poor's access to modern and cleaner forms of household energy is important due to its potential for increasing income levels for

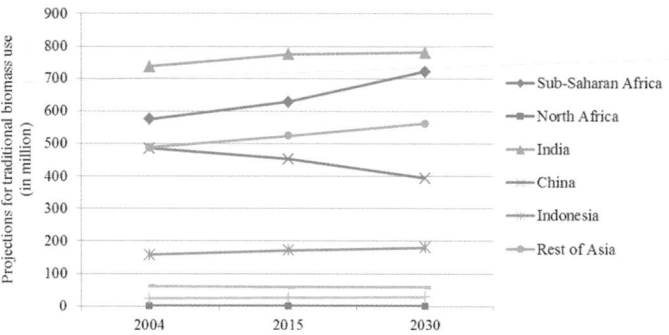

Figure 2.9 | Projections for traditional biomass use in developing regions (million people). Source: based on IEA, 2006a.

Table 2.6 | Rural and urban populations relying on traditional biomass as principal household energy option, by region and selected country.

Subregion	Total number of people relying on traditional biomass (million)			Proportions of population relying on biomass within subregion (%)		Population relying on biomass as share of subregion's population (%)
	Rural	Urban	Total	Rural	Urban	
Africa	481	176	657	73.2	26.8	67
Sub-Saharan Africa	477	176	653	73.1	26.9	80
Developing Asia	1694	243	1 937	87.5	12.5	55
China	377	47	423	89.1	11.1	32
India	765	90	855	89.5	10.5	75
Other Asia	553	106	659	83.9	16.1	63
Latin America	60	24	85	70.6	28.2	18
Developing countries*	**2235**	**444**	**2679**	**83.4**	**16.6**	**54**

*– Includes Middle East

Note: Traditional use of biomass refers to the basic technology used and not the resource itself. The people relying on traditional use of biomass refer to those households where biomass is the primary fuel for cooking. In addition to the number of people relying on biomass for cooking, some 0.4 billion people, mostly in China, rely on coal cooking.

Source: adapted from IEA, 2010a.

Table 2.7 | Share of household income expenditure on energy, selected countries (%).

Country	Date of data Publication	Bottom quintile	Top quintile
Armenia	2006	13.9 (electricity)	6.5 (electricity)
Djibouti	2005	18.2 (18.5)[a]	16.3 (13.4)[a]
Bolivia	2006	2.6 (LPG)	1.1 (LPG)
Georgia	2006	6.3 (electricity)	2.0 (electricity)
Ghana	2006	0.1 (petrol)	2.1 (petrol)
	2006	5.9 (kerosene)	1.6 (kerosene)
	2006	0.0 (LPG)	0.2 (LPG)
Hungary	2006	6.5 (electricity)	3.7 (electricity)
Jordan	2006	1.0 (kerosene)	0.3 (kerosene)
	2006	1.8 (LPG)	0.7 (LPG)
	2006	3.1 (electricity)	1.8 (electricity)
Kazakhstan	2006	0.9 (electricity)	0.6 (electricity)
Moldova	2006	6.3 (electricity)	3.0 (electricity)
	2006	0.4 (LPG)	0.4 (LPG)
Mongolia	2003	10.8 for the poor on heating (18% in winter months)	5.7% for the non-poor on heating (10.1% in winter months)
South Africa	2004	75.0 (housing, energy, clothing, and food)	33.3 (housing, energy, clothing, and food)
Poland	2006	5.8 (electricity)	2.9 (electricity)
Uganda	2002	15.0	9.5
Ethiopia	2002	10.0	7.0
India	2002	8.5	5.0
United Kingdom	2002	6.8	2.0

[a]   Numbers are calculated for Djibouti Ville and shown in parenthesis for other towns.

Source: World Bank, 2008a.

this group. Just as important, however, is the need to reduce the poor's expenditure on energy services. As shown in Table 2.7, the poor spend a higher share of their income on energy than the non-poor, which underscores the need to ensure affordable energy options for people with lower incomes.

Traditional fuels are typically used in low-efficiency stove technology – for the poorest people, nothing more than stones or very basic stoves. Relatively poor-quality heat is derived from the use of traditional fuels. This is not the nature of the fuels themselves but a byproduct of stove technology.

Global analysis conducted by World Health Organization (WHO) evaluated different intervention scenarios if the number of people cooking with solid fuels were halved by providing them with access to LPG by 2015. The analysis, conducted in 11 WHO subregions, showed a payback of US$_{2005}$$91 billion a year on an investment of US$_{2005}$$13 billion a year (See Table 2.8).

In poor countries, due to scarcity, rural people are unable to collect all the fuels they need (UNDP, 2005; UN-Energy, 2005). Increasingly, traditional

fuels including wood, charcoal, and other biomass are bought in commercial markets. The share of poor families' incomes spent on fuels is a significant portion of their total expenditures and can sometimes overtake other essential items such as schooling and health costs when local fuel prices rise (Modi et al., 2006; UNDP, 2005). This is also true for the urban poor, who rely on traditional fuels for cooking and heating and who must buy such fuels.

It is essential to put to rest two myths regarding the use of traditional fuels. The first myth is that all traditional fuels are free or available at no monetary cost (even those that are harvested locally for family consumption are not free as they involve a high opportunity cost based on the time spent on their collection). The second is that traditional fuels are mainly used in households in rural areas in terms of absolute numbers of consumers. In sub-Saharan Africa, much of the rural deforestation and overharvesting of fuelwood can, in part, be traced to trees cut down for feedstock for the charcoal industry. This is not to supply rural consumers with energy but to provide a steady supply of cheap fuels for cooking and household energy use for urban dwellers, as indicated by the thriving charcoal markets throughout the African continent (AFREPREN/FWD, 2006a).

Table 2.8 | Remarkable returns from investing in household energy. Benefits of household energy and health interventions (US$_{2005}$ million), by type of benefit, 2005.

	If 50% of the population cooking with solid fuels in 2005 switch to cooking with liquefied petroleum gas by 2015	If 50% of the population cooking with solid fuels in 2005 switch to cooking with modern forms of biofuels by 2015	If 50% of the population cooking with solid fuels in 2005 switch to cooking on an improved stove by 2015
Health care savings	384	384	65
Time savings due to childhood and adult illness prevented: school attendance days gained for children and productivity gains for children and adults	1460	1460	510
Time savings due to less time spent on fuel collection and cooking: productivity gains	43,980	43,980	88,100
Value of deaths averted among children and adults	38,730	38,730	13,560
Environmental benefits	6070	5610	2320
Total benefits	**90,624**	**90,164**	**104,555**

Note: Costs and benefits of different intervention scenarios were estimated using 2005 as the base year and a 10-year time horizon, taking into account demographic changes over this period. The analysis was conducted for 11 WHO subregions to reflect variations in (i) the availability, use and cost of different fuels and stoves; (ii) disease prevalence; (iii) health care seeking as well as quality and cost of health care; (iv) the amount of time spent on fuel collection and cooking; (v) the value of productive time based on Gross National Income per capita; and (vi) variations in environmental and climatic conditions. A 3% discount rate was applied to all costs and benefits.

Source: WHO, 2006.

A reduction in energy expenditure or time spent collecting fuel can liberate much-needed financial resources for other important expenses, such as increasing the quantity and quality of food or investing in new income-generating activities or expanding existing income-generating activities – thereby enhancing income as well as creating employment. For example, by switching from traditional to improved cookstoves, the poor can significantly reduce their expenditure on fuel and also save time spent in fuel collection (Galitsky et al., 2006). Improved charcoal cookstoves modelled on the Kenya Ceramic Jiko (KCJ) can cut charcoal consumption by up to half. The money thereby saved can be reinvested in more or better-quality food, partially contributing to minimizing extreme hunger, or in an income-generating activity. The KCJ reduces emissions and particulate matter, the latter of which contributes to acute respiratory infection among women and infants under the age of five. The relevance of the KCJ experience in Kenya is demonstrated by its introduction into Burkina Faso, Ethiopia, Ghana, Malawi, Mali, Rwanda, Senegal, Sudan, Tanzania, and Uganda, to mention just a few countries. Other successful improved biomass cookstoves as well as ethanol cookstoves are also being disseminated in Asia and Latin America.

Liquid and gaseous fuels can provide better cooking and heating options due to the combustion efficiency and quality of heat they provide. Unfortunately, fuels such as kerosene and LPG are often not available in local markets, especially in rural areas. In addition, due to their higher costs, they cannot compete with traditional biomass-based fuels. Poor people also face challenges associated with the upfront cost of stoves. For example, families who would be able to afford the recurring fuel costs of LPG versus charcoal may not be able to afford the requisite LPG stove, thereby eliminating the possibility of switching fuels (Quansah et al., 2003).

In most developing countries, there has been a significant increase in the use of LPG over the past decade (Karekezi et al., 2008b), but this has occurred among people with higher disposable incomes except in countries where active interventions to reach the poor are in place. In Senegal, for example, about 90% of Dakar's inhabitants rely on LPG for cooking (ANSD, 2006; Schlag and Zuzarte, 2008; Fall et al., 2008). To reach the poor, a number of LPG distributors have promoted smaller LPG cylinders, thus lowering upfront costs. India and Brazil have registered impressive progress in promoting LPG use through the use of subsidies to promote market wide changes in consumption patterns (Shankar, 2007; Coelho, 2009). However, world oil prices are highly variable, and in a time of increasing resource scarcity, a significant price increase is likely to erode some of the gains made in LPG use among the poor.

Socioeconomic determinants can be expected to have a significant influence on fuel or other intervention choices (such as improved stoves or improved household ventilation). Household fuel demands have been shown to account for more than half of the total energy demand in most countries with per capita incomes under US$_{2005}$1000 (PPP), while accounting for less than 2% in industrial countries (UNDP, 2007b). While it is known that households switch to cleaner, more-efficient energy systems for their domestic energy needs as per capita incomes increase (i.e., they move up the "energy ladder"), these changes have largely been due to increases in affordability, demand for greater convenience, and energy efficiency (IEA, 2010a). Figure 2.10 is a partial depiction of a typical "energy ladder" with regard to some of the fuels associated with cooking.

From a policy and public financing point of view, there is a need to alleviate the negative income effect of time use and family expenditure related to the use of traditional fuels and to overcome the barriers

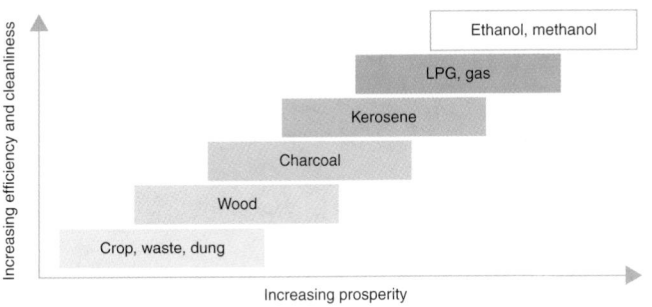

Figure 2.10 | Typical "energy ladder" for cooking fuels[3]. Source: UNDP, 2007b.

related to household energy. In developing countries, where poor people are highly dependent on traditional fuels that are poorly combusted in inefficient devices, much greater emphasis must be given to increasing the availability and affordability of liquid and gaseous fuels and associated stove technologies. Special attention should also be paid to mechanical power solutions and labor-saving appliances that can provide homestead-based pumping, agro-processing, milling, and grinding, among other services. These are also critically important for gender and health reasons, as discussed later in this chapter.

The absence of mechanical options in many rural areas of the developing world, to provide access to mechanical power for water pumping, milling, or other household activities is also related to the absence of affordable and cleaner fuels or electricity to drive engines, mills, and pumps to supplement human labor (see Chapters 19, 23, and 25). The absence of mechanical power affects not only the poorest families, particularly girls and women, but often entire geographic areas due to the distances from electricity supply and cleaner fuel delivery systems (UN Energy, 2005; UNDP, 2005). Overcoming the mechanical power deficit that affects the household energy balance is also related to the appliances (pumps, mills, and engines) that convert energy carriers into the service of mechanical power. In this sense, it has many similarities with the cooking and heating challenges in the household energy mix.

2.4   Energy and Production

Virtually all value-adding activities, both paid and unpaid, require energy as an input in the production process (Modi et al., 2006; UN-Energy, 2005). In its simplest form it can be human or animal energy transporting goods, doing manual work, or providing manual power through simple machines such as mills. As greater degrees of technology are applied in the production process, different forms of energy are used. This could be heat energy in manufacturing processes requiring ovens or kilns and therefore fuels, in which case the quality of the heat, access, and affordability of the fuels and efficiency of the ovens used all affect the final product (from bread to bricks). Other productive processes

require motors, machines, or other tools that often use liquid fuels (such as petrol for motors) or electricity (for sewing machines, welders, etc.) to power them. Table 2.9 illustrates various mechanical power technologies and applications that could help minimize the amount of human energy spent on different productive activities.

In addition to production activities in which energy and machines are applied to raw materials and other inputs to produce final products, service industries also require energy as inputs. This includes telephones, computers, other information technology, or air conditioners, specialized equipment, or security systems – all of which consume electricity.

In short, from the simplest manufacturing industries to the most complex technology-intensive ones, energy – whether electricity or fuels – is an essential input. It therefore would seem obvious that there is a link between energy and the development objectives of achieving full and productive employment and decent work for all. Yet while many countries have active programs to promote employment through workers' training or other skills-building activities, sources of energy for machines or financing schemes to get access to the energy-using equipment itself is often not part of the job creation agenda.

Many working conditions, especially for the poorest people, are hazardous or life-threatening. In countries where a work-related injury can greatly reduce the economic conditions of entire families due their vulnerability to a range of factors beyond their control (food and fuel price rises, access to affordable housing, increases in education or health service fees), access to decent work is a key determinant of the ability of people, communities, and entire countries to overcome conditions of poverty. Cleaner energy options can enhance working conditions and open opportunities to generate livelihoods, increase the number of jobs, and provide decent work for entire populations are directly related to access to energy services for value-adding processes, whether this be in manufacturing, the services sector, government, or social services (Modi et al., 2006; UN-Energy, 2005).

Employment in formal and non-formal sector activities is positively correlated to access to cleaner energy options such as electricity, as is workers' productivity in value-adding processes. In addition, increased access to cleaner energy contributes to the transformation of economies from being agriculture-based – where significant animate energy is consumed – to industry-based, where cleaner energy plays a key role in the production of commodities. Figure 2.11 illustrates that economies with relatively high cleaner energy use (due to high access levels) have lower contributions of agriculture to the GDP implying that reliance on agriculture is inversely proportional to cleaner energy use as more non-agricultural industries are established contributing to economic development. Conversely, economies with very low cleaner energy use show relatively high levels of contribution of agriculture to GDP, as their industrial sector is not developed, in part, due to an inadequate supply of cleaner energy options (Modi et al., 2006; UN-Energy, 2005; UNDP, 2005).

---

3   Electricity represents the highest rung on the 'energy ladder' and produces different energy services than fuels.

Table 2.9 | Mechanical power technologies and applications.

Activity	Service	Traditional technology	Mechanical power alternative
Water supply	Drinking	Container (bucket) for lifting/carrying water	Diesel pump
	Irrigation		Treadle pump
	Livestock watering		Rope pump
			Ram pump
			Persian wheel
			Hand pump
			River turbine
			Wind pump
Agriculture	Tillage/plowing	Animal-drawn tiller/hand hoe	Power tiller/ two-wheel tractor
	Harvesting	Scythe Animal-drawn mower Manual practices	Harvester
	Seeding	Hand planting	Bed planter Row planter Seed drill
Agro-processing	Milling/Pressing	Hand ground/Flail	Powered mill Oil expellers
	Cutting/Shredding	Knife	Sawmills Powered shredder
	Winnowing/Decorticating	Winnowing basket	Powered shaker Grinder
	Spinning	Manual spin	Powered spinner
	Drying	Handheld fan Sun drying	Powered fan Solar dryer
Natural resource extraction	Small-scale mining	Shovel Chisel Hammer Pickax	Manual percussion drill Petrol-powered drill Expandable tube with hydraulic pump
		Hand washing	Hand /fuel/water-powered water jet
		Hand screen	Hand/fuel/water-powered shaker
	Lumbering	Hand saw	Powered saw (saw mill, chain saw)
Small-scale manufacturing	Metal working	Hammer	Sheet metal/pipe bender Hole puncher
	Woodworking/Carpentry	Hand saw	Sawmill Treadle lathe
	Briquetting/brick pressing	None	Hand/foot-powered pressers
	Textile making	Hand-woven	Treadle loom
	Papermaking	Mould and deckle	Paper press Pulp mill
	Pottery	Hand powered potter's wheel	Treadle pottery's wheel
Lifting and crossing	Lifting	Manual labor (climbing, lifting)	Chain/rope hoist
	Crossing	Manual labor (swimming, walking)	Gravity ropeway Tuin (aerial tramway in Himalaya)

Source: adapted from Bates et al., 2009.

In the countries with the least access to electricity countrywide, or in specific geographic areas within national borders that have limited electricity access or an unreliable supply, limits to productive employment and income generation are also likely to be present. The cost, reliability, and ease of access to energy services are also factors in creating conditions for jobs creation, production, and decent work.

For example, lack of reliable electricity supply in sub-Saharan Africa has significant impacts on job security, especially among casual laborers in industry. On annual basis, cumulative time of electrical supply interruptions is equivalent to about three months of production time lost (IEA, 2010a), a significant duration that implies little or no production, thereby risking loss of jobs. In addition, it is estimated that businesses lose production worth 6.1% of their turnover due to electrical outages (IEA, 2010a). Table 2.10 summarizes the reliability of electricity supply in sub-Saharan Africa and developing countries in general.

At the macroeconomic level, lack of a reliable supply of electricity is estimated to have a significant impact on economic growth and productivity. Figure 2.12 shows the proportion of GDP lost to unreliable electricity supply in some countries in sub-Saharan Africa.

Table 2.10 | Summary of electricity supply reliability in sub-Saharan Africa and developing countries.

Indicator	Sub-Saharan Africa	Developing Countries
Cumulative electrical interruptions (days/year)	90.9	28.7
Lost Production (% of turnover)	6.1	4.4
Firms owning/sharing generators (% of total)	47.5	31.8
Number of days for new electrical connection	79.9	27.5

Source: World Bank & IBRD, 2008.

Access to modern and affordable forms of energy options is essential for creating employment, thereby contributing to a reduction in poverty levels. According to a survey of a rural Indian village by Hiremath and

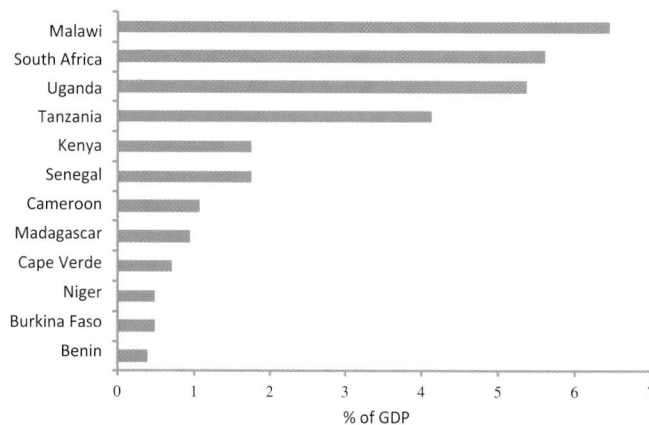

Figure 2.12 | Proportion of GDP lost due to unreliable power supply in 2007. Source: Vivien and Briceno-Garmendia, 2010.

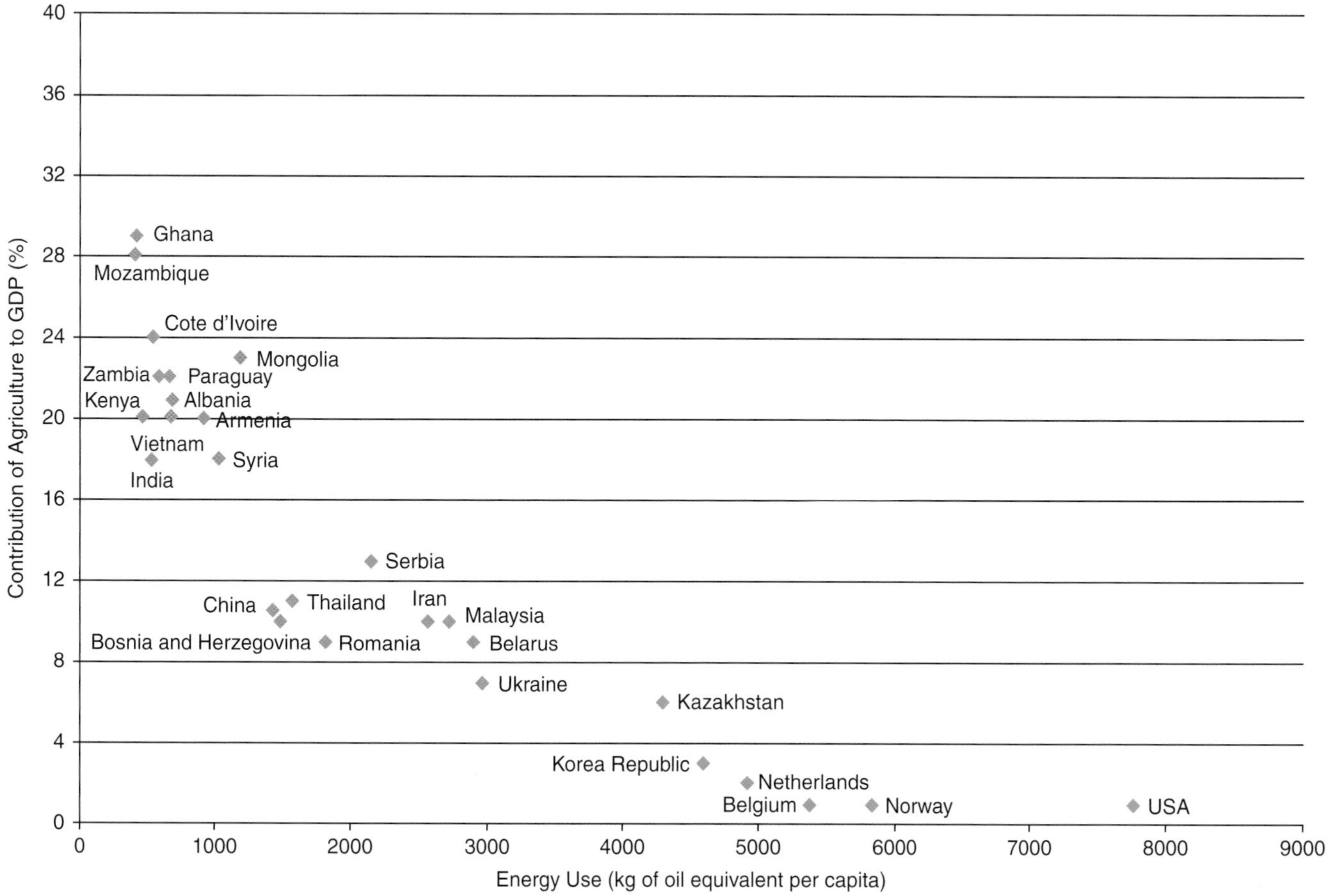

Figure 2.11 | Correlation between cleaner and modern forms of energy use and contribution of agriculture to GDP, 2007. Source: adapted from World Bank, 2011a.

Misra (2009), modern forms of bioenergy systems can create opportunities for 25–45 jobs in one village alone. The jobs would arise from developing forests for feedstock, preparing wood feedstock, collecting cattle dung or leaves, operating and maintaining biogas/gasifier systems, etc., as well as the operation, repair, and maintenance of associated equipment such as generators, pumps, flour mills, diesel engines, and domestic appliances.

World Bank studies in Asia have attempted to quantify the benefits of rural electrification. For example, a survey in Bangladesh revealed that rural electrification could lead to an increase of income in the range of 9–30% (Khandker et al., 2009). The value of benefits of rural electrification accrued by households in the Philippines is shown in Table 2.11 (World Bank, 2008b), which are substantial in comparison to Philippine's per capita GDP of US$146 in 2009.

Thus although the presence of electricity or fuels does not, on its own, guarantee full employment or enough value-adding jobs to ensure decent work and poverty elimination, access to electricity and fuels for formal and informal sector work expands the range of employment options in both urban and rural areas, with the most dramatic effects being seen in the countryside. This is not because of the electricity or the fuel itself but because of the expansion in the range of services for manufacturing, communication, mechanical power, and illumination that they provide.

Prioritizing access to electricity and energy-using technology for commercial activities (as opposed to only low-load household activities) could increase employment opportunities that can contribute to income generation and the fight against poverty, especially for the working poor.

2.5 Energy and Agriculture

Most of the caloric intake of the poorest people worldwide comes from basic grains such as rice, corn, millets, and wheat. In all cases not only

Table 2.11 | Quantifying electrification benefits for typical households in rural Philippines.

Benefits	Value of benefits (US$/month)	Beneficiary
Less expensive and expanded use of lighting	36.6	Household
Less expensive and expanded use of radio and television	19.6	Household
Improved returns on education and wage income	37.1	Wage earner
Time savings for household chores	24.5	Household
Improved productivity of existing home business	34.0	Business
Productivity of new home business	75.0	Business

Source: adapted from World Bank, 2008b.

do basic grains need to be cooked to be consumed, which requires heat energy, but their very production, harvest, and processing require energy inputs for cultivation, irrigation, transport, and, for some foodstuffs, preservation.

By facilitating irrigation, cleaner and affordable energy options can help ensure food security among the poor in spite of increasingly frequent drought conditions in many countries. As shown in Table 2.12, irrigation is still an embryonic practice in sub-Saharan African countries but it is widely practiced in Asia (FAO, 2011). Most of the Asian countries represented in Table 2.12 source more than half of their grain production from irrigated land. In comparison, in most of sub-Saharan African countries, grain production from irrigated land accounts for around 20% or less (FAO, 2011).

Energy can play a vital role in enhancing food security among the poor through technologies that can be used for irrigation and water pumping. Energy for mechanical power (water pumping or distribution) can come from grid-connected electricity, local motors using fuels, or renewable-energy-derived water-lifting devices.

Some irrigation and water pumping technologies have significant potential for not only ensuring food supply throughout the year but also generating additional income for households (Karekezi et al., 2005). Table 2.13 provides estimates of the initial investment required to buy and install various irrigation or water pumping technologies. However, where farmers are supplied with excessively subsidized or extremely low-cost electricity there are possibilities of misusing it, for example by use

Table 2.12 | Proportion of grain production from irrigated land (%).

Country	%
ASIA	
Japan	98.0
Vietnam	94.5
Tajikistan	84.5
Malaysia	70.0
China	67.0
Uzbekistan	61.5
India	56.0
Bangladesh	47.0
AFRICA	
Madagascar	67.0
Namibia	43.9
Mali	22.4
Gambia	19.6
Nigeria	14.2
Burkina Faso	3.2
Botswana	2.6
Malawi	2.0
Mozambique	2.0

Source: FAO, 2011.

Table 2.13 | Initial investment and specific cost of irrigation water, various technologies.

Component	Treadle pumps[4]	Hydrams (hydraulic ram pumps)	Wind pump	Diesel pump
Cost + installation (US$)	171	295	2981	2,000
Lifetime (years)	6	40	20	10
Rate of discharge (m³/hour)	4.3	4.5	7.5	10.0
Specific cost of water (US$/m³/year)	0.03	0.05	0.29	0.18

Note: The costs in the table assume that the requisite renewable energy resources are available at the site where the pump is located and also that the water source is within a reasonable distance for the pump to operate at its optimum.

Source: Karekezi et al., 2005; Karekezi et al., 2008a.

of inefficient pumps, which leads to increased electricity consumption as well as wasteful use of water which can lower water table levels.

As Table 2.13 indicates, not all technologies may be affordable to poor households due to high upfront cost. However, the treadle pump appears to have the least upfront cost and is likely to be the most attractive option for the poor. This technology – widely used in Asia and parts of sub-Saharan Africa – is reportedly very effective in enhancing food production.

Many sub-Saharan African farmers are still irrigating very small plots of land using bucket-lifting technologies, which are slow, cumbersome, and labor-intensive, especially for women. Small-scale irrigation is usually developed privately by farmers in response to family and local market requirements, often with limited government interventions.

Small-scale irrigation using treadle pumps is one of the success stories in many countries in Africa at a time when large-scale water developments have failed to meet expectations. Case studies from Kenya, Niger, Zambia, and Zimbabwe show that by using animate energy-driven treadle pumps instead of bucket irrigation, farmers can increase irrigated land, reduce work time, improve crop quality, grow new crops, and increase the number of cropping cycles (AFREPREN/FWD, 2006b). Treadle pumps make it easier for farmers, especially women, to retrieve water for their fields or vegetable gardens, and they are cheap and easy to handle. Treadle pump technology has enabled poor rural farmers, especially women, to increase their incomes by selling surplus produce in the local market.

In some cases, cropping intensity has been extended to three crops per year. In Zimbabwe, treadle pumps are mostly used for irrigation of small vegetable gardens. And because these pumps usually reach water only within five meters, they do not deplete valuable groundwater resources. It is noted that the local water table would drop only if a large number of farmers were operating in the same area.

---

4   A treadle pump is a human-muscle powered irrigation device.

In Kenya, treadle pumps were introduced in 1991 and have generated a wide range of benefits for users, manufacturers, promoters, and retailers. The low-income owners in Kenya purchase them mainly through their savings as well as from the sale of crops and livestock and use of retirement benefits. While men own 84% of the treadle pumps, women manage nearly three-quarters of them, which are mainly used for irrigating crops and to some extent for supplying water for household use and animals (Karekezi et al., 2005).

According to available data, the areas under irrigation increased among users by 700% from an average of 0.03–0.24 ha in 1999 to 0.59 ha in 2004 (Karekezi et al., 2005). For farmers not involved in irrigation previously, the number of crop cycles increased from 1.2 to an average of 2.3. One advantage associated with an increased number of crop cycles is the timing of cropping, in order to harvest when the crop fetches higher prices from the market.

If pumps are produced locally, they can also create jobs and income – the pumps cost less than US$_{2005}$200 and can be built with ordinary workshop equipment (AFREPREN/FWD, 2006b). Some 186,804 pumps are reported to have been sold across West, East, and Southern Africa since 1991 (Kickstart, 2011).

Wind pumping technologies can supply water for irrigation as well as for household use and livestock (Harries, 2002). The wind speeds in most regions (even in sub-Saharan Africa, which is located in a generally low wind-speed zone that straddles the equator) are sufficient for windpumps. In sub-Saharan Africa, South Africa and Namibia possess large numbers of wind pumps. An estimated 400,000 wind pumps are in operation in South Africa (Karekezi et al., 2009) and are believed to have played a major role in the past in supporting irrigated agriculture and transforming South Africa into a food-surplus country.

Another key area in which better energy options can help combat hunger is post-harvest losses. These are related to inadequate facilities for food harvest, storage, and transport; a lack of conversion technologies to preserve or otherwise extend the life of food products; and the lack of protection from pests, disease, and other threats to the quality and quantity to harvested food products.

FAO estimates that post harvest losses among food grains in developing countries account for 15–50 % (FAO, 2009). However, more alarming are the post-harvest losses of fruits and vegetables (such as tomatoes, bananas, and citrus fruit) as well as of sweet potatoes and plantains, which can be as high as 50% (Farm Radio International, 2006a). It is also estimated that between 20% and 50% of the fish caught in Africa are lost after capture (Farm Radio International 2006b).

The impact of high post-harvest losses on the poor is twofold: first, it implies that less nutritious food is available, which exposes the poor to undernourishment. Second, post-harvest losses equate lost potential income. This means that the poor end up with less income from the sale of food crops. In addition, lower income erodes any potential cushioning

Table 2.14 | Various options for low-cost options for minimizing post-harvest losses and the requisite energy inputs.

Fruit/vegetable	Post-harvest transformation	Type of energy inputs required
Most fruits & vegetables	Dried fruits and vegetables	Solar thermal using solar drier
African locust bean	Traditional mustard	Heat for cooking
Peanuts	Peanut butter	Heat for cooking and animate energy for grinding
Tomato	Tomato paste	Heat for cooking
Most fruits	Jams and sauces	Heat for cooking and sterilizing storage containers

Source: adapted from Farm Radio International, 2008.

against food price hikes as well as reduces poor people's ability to afford better-quality food.

Several modern, cleaner and affordable forms of energy options could reduce post-harvest losses of fruits and vegetables. One important technology directly linked to energy services is transportation, especially on-time transportation to get harvested foods to markets and processing centers. In rural areas with difficult terrain, the use of motorized transportation can be a significant challenge. Alternative and low-cost options for such areas include animate energy and the use of mechanical ropeways[5]. In parts of Nepal, ropeways have dramatically reduced transportation of farm produce from 3–4 hours to a mere five minutes (Bates et al., 2009).

Processing options involve transforming fruits and vegetables into other forms of foodstuff with longer life spans (see Table 2.14). Energy for cooking and drying (solar drying) is an important input for transforming fruits and vegetables into longer-lasting food products that can go on the shelf or to market, thereby extending consumers' access and increasing the income of small scale poor rural farmers. However, it is noteworthy that for more-sophisticated industrial-level food processing, during which food is transformed into high-value products such as canned food, a significant amount of modern forms of energy are required, such as electricity and steam. Due to the limited access to modern forms of energy in rural areas, where most food is grown, most advanced food processing plants in developing countries are located in urban areas, in some cases far away from where the crops are grown, thereby denying rural areas of employment opportunities.

Industrial-scale agricultural production, while in some cases providing less local employment, has also contributed to increasing overall food production. In the case of basic grains, the globalization of markets for rice, soya, wheat, maize, and other staples implies that the benefits of energy inputs in one country can benefit consumers in other countries due to global trading systems. Without exception, these systems have

relied on energy inputs for fertilizer production, irrigation, harvesting, and the transport of their products.

Energy technologies required to increase food production and reduce post-harvest losses, can be produced locally, are relatively affordable, and in many cases may not require electricity to operate. For example, pico and micro hydro for shaft power can be used to process agricultural produce and increase its value for a prolonged period, thereby improving food security. Low-cost efficient hand tools and animal-drawn implements can increase the agricultural productivity of rural areas in the developing world.

In conclusion, access to more efficient, cleaner and affordable energy options is an effective tool for combating extreme hunger by increasing food productivity and reducing post-harvest losses.

2.6   Energy and Transport

Mobility is an essential requirement that ensures the delivery of goods and services. For the poor, it is an essential factor in providing options for employment, production, and livelihoods. The more limited the transport options, often due to lack of access to energy, the more limited the livelihood options.

In recent years, the high cost of motorized transport due to rising oil prices is a major challenge, especially for the poor in developing countries who need to transport to take their agricultural produce to the market and to take up employment.

In rural areas, poor road conditions and networks limit the availability of motorized transport. Consequently, in some rural areas of sub-Saharan Africa, Asia, and Latin America, people walk 10 kilometers a day one way to attend to their chores (e.g., fuelwood or water collection, farm work) as well as attend school or visit health facilities (See Chapter 9).

In urban areas, the high cost of public transport has a negative impact on the poor. Table 2.15 shows the proportion of income spent by the poor on motorized transport in selected cities.

In poorer parts of the developing world, particularly sub-Saharan Africa and some parts of Asia, the urban poor cannot afford motorized transport and largely rely on walking. The relatively high cost of motorized transport for the urban poor is partially linked to the limited attention

Table 2.15 | Proportion of urban poor's income spent on motorized transport in four cities.

City	Share of income spent on motorized transport (%)
Colombo	21
Mumbai	15
Manila	14
São Paulo	8

Source: Chapter 9; Gomide, 2008; Kumarage, 2007; Baker et al., 2005.

---

5   A ropeway is a form of lifting device used to transport light goods across rivers or ravines.

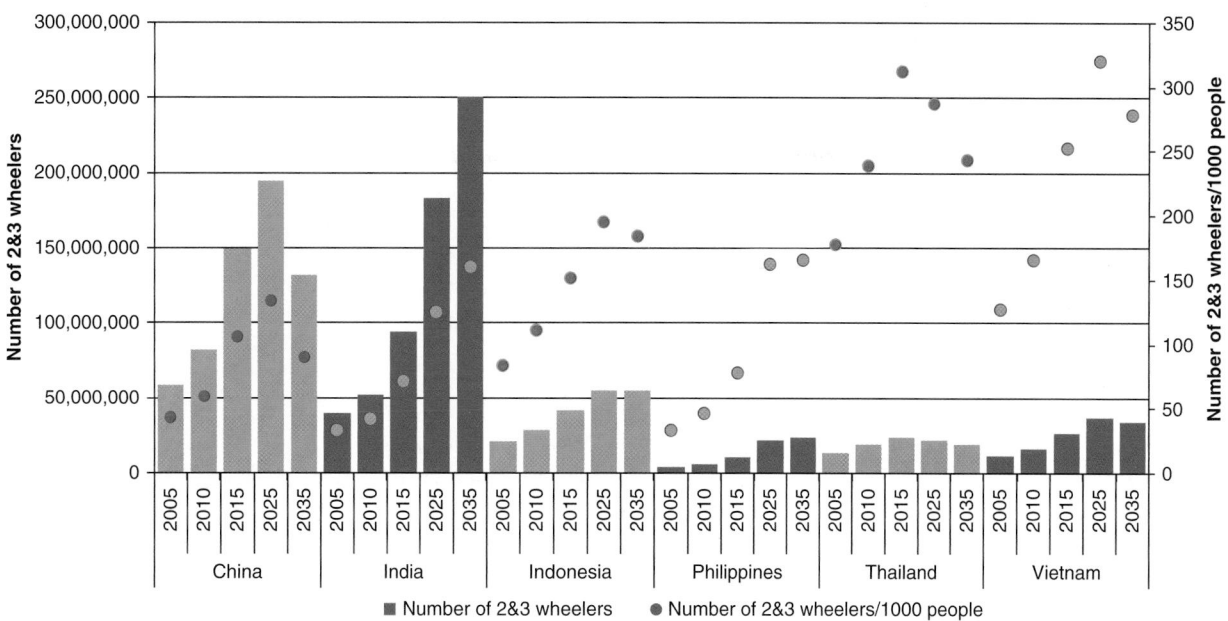

Figure 2.13 | Population of 2 and 3-wheelers in selected Asian countries[6]. Source: CAI-Asia, 2011.

being paid to the development of energy-efficient non-motorized transport options and of low-cost efficient mass transit systems in cities of developing countries. Most road network systems and infrastructure in these cities appear to favor the convenience of private motorized transport, although most people use either non-motorized transport or motorized public transport.

In the cities, the increased demand for low-cost motorized transport, especially for short trips, has led to a sharp increase of two- and three-wheelers in developing countries, especially in Asia (see Figure 2.13). In Asia more than two-thirds of the total vehicle population consists of two- and three-wheelers in many countries (Hook and Nadal, 2011).

The relatively high uptake of two- and three-wheelers in developing countries is mainly linked to their low upfront cost compared with cars. In certain parts of the developing world, however, such as in sub-Saharan Africa, two- and three-wheelers have received a mixed reception. Among the poor in both urban and rural areas these vehicles are perceived as a lucrative source of employment as well as an affordable and convenient form of motorized transport. But in urban areas of sub-Saharan Africa, the high rates of fatal accidents linked to undisciplined driving and absence of dedicated lanes has discouraged use of 2 and 3-wheelers in many sub-Saharan African countries.

Low-cost energy-efficient transportation for both passengers and goods is crucial for economic development. This is particularly important for long-distance transit of passengers and goods. Rail transport is considered the least expensive energy-efficient land transport option for

long-distance passenger travel and has been developed extensively in industrial countries in the form of surface and underground rail networks. For freight transportation, with the exception of India and China, many developing countries make limited use of lower-cost energy-efficient rail transport mainly due to its lack of sustained investment.

2.7   Energy and Education

While basic educational services and basic literacy can be achieved without the use of cleaner energy inputs, access to energy services can improve the quality and availability of educational services and increase the likelihood that children will attend and complete school (IEA, 2010a; UNDP, 2005; UN-Energy, 2005).

Global education goals place emphasis on the primary education due to its long-lasting impact on literacy and social inclusion. According to recent estimates of the World Bank (2010a), about 50 developing countries have already achieved the goal of universal primary education. Although seven others are likely to attain the goal by 2015, nearly 40 developing countries, mostly in sub-Saharan Africa, are very unlikely to achieve universal primary education by 2015 (World Bank, 2010a). This is clearly indicated in Figure 2.14. Sub-Saharan Africa has the highest proportion of countries seriously off track in making progress toward universal primary education.

Modern, cleaner and affordable energy options can help create a more child-friendly environment that encourages school attendance and reduces the significant dropout rates experienced in many low-income countries (Mapako, 2010; UNEP, undated). For example, cleaner and affordable energy can enhance access to clean water, sanitation,

---

6   In Asian cities 2 and 3-wheelers account for 50–90% of total vehicle fleet.

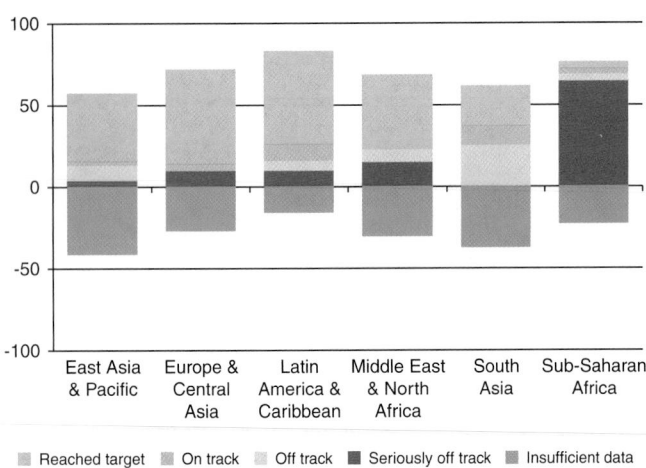

■ Reached target   ■ On track   ■ Off track   ■ Seriously off track   ■ Insufficient data

Figure 2.14 | Status of progress (%) on achieving universal primary education. Source: World Bank, 2010a.

Table 2.16 | Impact of rural electrification on children's study time in Bhutan (number of children).

Electrification status	Study hours of children in the evenings			Total (children)
	< 1 hour	1–2 hours	>2 hours	
Electrified	67	23	21	111
Non-electrified	71	32	7	110
Total (children)	138	55	28	221

Source: Bhandari, 2006.

lighting, space heating/cooling, and energy for cooking lunch (and other meals in the case of boarding schools).

There is growing evidence of the positive impact of rural electrification (particularly grid extension to rural schools) on the retention of the math and science teachers who are much sought after in rural areas. When the general quality of life increases due to the rural electrification, teachers are willing to relocate to rural schools thus mitigating the problem of shortage of teachers in rural areas (AllAfrica, 2004; Cabraal et al., 2005; World Bank, 2008a; Harsdorff and Peters, 2010).

Electricity can facilitate access to educational media and communications in schools and in homes. It can increase use of distance-learning modules. Access to electricity provides the opportunity to use more sophisticated equipment for teaching (such as overhead projectors, computers, printers, photocopiers, and science equipment), which allows wider access to more-specialized teaching materials and courses (Mapako, 2010).

In addition, cleaner energy systems and efficiently designed buildings reduce heating/cooling costs and thus school fees thus increasing the poor's access to education. As discussed in more detail in Chapter 6, education is the key for higher world economic growth. Education is a long-term investment associated with near-term costs, but in the long-term it is arguably one of the best investments societies can make for their future.

One of the indicators for monitoring progress in universal primary education is the proportion of pupils who start grade one and reach the last grade of primary school. Often children, especially girls, face family pressure to contribute to household energy supplies through the collection of wood and other fuels and of water for home heating and cooking purposes. Access to cleaner fuels, efficient stoves, and alternative fuelwood management practices can reduce fuel collection times significantly, which can translate into increased time for education of rural children.

When families make the hard decision as to which children to keep home from school to help with such activities as fuel and water fetching and basic cooking and cleaning, preference is often given to boys' education over that for girls, whose traditional roles mean these tasks fall disproportionately on them and they are withdrawn from school before completing the full primary cycle (UNDP, 2004). (See also Section 2.10 on Energy and Women.)

When girls who do go to school help their mothers in the kitchen, they lose valuable study time (Mapako, 2010). In addition, inhalation of kitchen smoke exposes them to illnesses associated with the respiratory system, which can lead to significant school absenteeism from poor health. Cleaner energy options, especially in rural areas, can free up time for girls to study and improve their academic performance.

For both boys and girls, modern energy options can provide quality lighting for comfortable night-time studying (Mapako, 2010). A study in Bhutan found that with modern forms of energy options such as electricity, children can study much later in the evening (see Table 2.16). In addition, good lighting reduces risks to children's eyesight (WHO, 2011a). In many rural areas of the developing world, energy for transportation is also essential if children are to reach schools that are beyond walking distance.

One of the suitable ways to enhance access to cleaner and affordable energy in schools and other institutions of learning is to piggy-back on existing school-related programs. For example, the World Food Programme has introduced "school feeding" initiatives in many developing countries. These focus on poverty-stricken areas as a means of encouraging primary school attendance through the incentive of onsite meals, thus combating malnutrition among the poor (WFP, 2010). Use of improved biomass stoves in the school feeding initiatives reduces cost as well as provides live demonstrations of improved cookstoves and encourages local people to adopt cleaner and more-efficient stoves.

To sum up, access to cleaner and affordable energy options can contribute to achieving universal primary education as well as enhance the quality of education (Meisen and Akin, 2008; Mapako, 2010; Harsdorff and Peters, 2010):

- Providing improved cookstoves that reduce the cost of school feeding programs, which is instrumental in maintaining high attendance levels in rural primary schools;

- Helping to create a more child-friendly environment (access to clean water, sanitation, lighting, and space heating or cooling), thus improving attendance at school and reducing dropout rates;

- Introducing cleaner energy systems and efficient building design that reduce heating/cooling costs and thus school fees which, in turn, expands the poor's access to education;

- Retaining qualified teachers in rural areas by providing electricity for their residences;

- Providing the opportunity to use advanced teaching aids and equipment (overhead projector, computer, and science equipment);

- Enabling access to educational media and communication in schools and at home, hence increasing education opportunities and allow distance learning;

- Allowing children to study after school hours, which can have a significant impact on learning outcomes; and

- Saving girls' time doing household chores, thereby giving them time to study at home and also attend school.

2.8   Energy and Health

Modern health services, the facilities to provide them, and the professional and health sector workers who deliver them require access to energy options in the form of electricity and cleaner and affordable fuels for both institutional and household use. Detailed energy-related health issues are covered in greater depth in Chapter 4.

Public expenditures on health care provision in developing countries have typically been low. When public-sector spending on health services increases, urban and peri-urban areas are often been given priority, leaving the rural poor with inadequate health care options that are distant and of poor quality as well as often unaffordable. Access to health care services varies based not only on the rural-urban divide but also on the income bracket of the people and families in question.

From a human development point of view, the most grievous negative health effects are loss of life, especially among the youngest and most vulnerable, as well as shortened life spans due to poor overall living conditions. Global efforts have prioritized reducing infant mortality in recent years. While deaths of children below the age of five dropped in developing countries by more than 25% between 1990 and 2008, the absolute number of children dying before age five increased (UN, 2010a; World Bank 2010a). Sub-Saharan Africa appears to have made the least progress in terms of reducing child mortality (see Figure 2.15) with one child in seven there dying before the age of five (World Bank, 2010a).

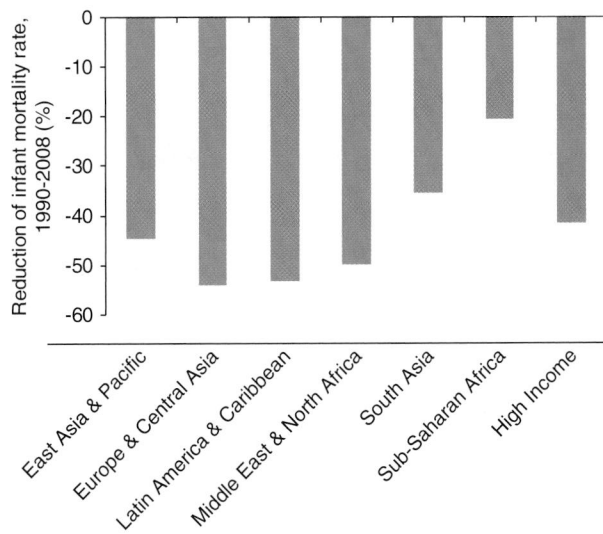

Figure 2.15 | Reduction of infant mortality rate. Source: World Bank, 2010a.

The leading causes of under-five child mortality worldwide (after neonatal causes) are diarrhea and pneumonia (14% each), both of which can be directly linked to inadequate access to cleaner and affordable energy options (UN 2010a).

In the case of waterborne diseases such as diarrhea that has a particularly detrimental impact on children's health, the contributing factors that are energy-related include lack of pumping systems from clean water sources, lack of energy or fuels to boil water, and lack of industrial processes for water treatment and purification.

Pneumonia cases among children under the age of five from poor households are mainly linked to household air pollution due to reliance on traditional biomass for cooking (WHO, 2006). The inefficient burning of solid fuels in an open fire or traditional stove indoors creates a dangerous cocktail of hundreds of pollutants, primarily carbon monoxide and small particles, but also nitrogen oxides, benzene, butadiene, formaldehyde, polyaromatic hydrocarbons, and many other health-damaging chemicals. Day in and day out, for hours at a time, women and small children breathe in amounts of smoke equivalent to consuming two packs of cigarettes per day. Where coal is used, additional contaminants such as sulfur, arsenic, and fluorine may also be present in the air – see Chapter 4 for a more detailed assessment of the health impacts of using solid fuels.

As shown in Figure 2.16, the prevalence of tuberculosis is highest in developing countries that predominantly use traditional biomass energy for cooking. As noted earlier, modern, cleaner and affordable energy options for cooking can play a significant role in reducing acute respiratory infections such as tuberculosis by improving the kitchen environment where women and infants spend a significant amount of time (WHO, 2006; see also Chapter 4). Examples of energy options

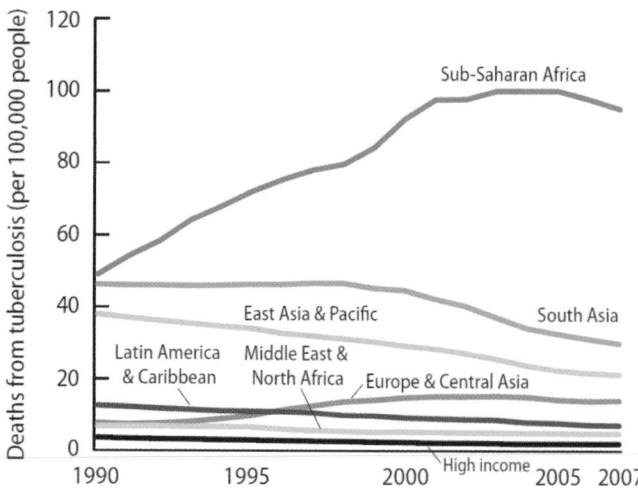

Figure 2.16 | Tuberculosis prevalence by region. Source: WHO, 2011b.

that could replace the traditional biomass energy for cooking include improved fuelwood and charcoal cookstoves, kerosene, LPG, ethanol, and biogas.

The quality of the medical care available in health care institutions can be determined by access to energy services. Examples include electricity for illumination and communication, to power diagnostic machines and provide refrigeration for medicines, fuels for hot water and sterilization processes to maintain sanitary conditions, and fuels to provide space heating or cooking.

One of the critical moments in a person's life for access to health care facilities is during pregnancy and birth. Every year, hundreds of thousands of women die from complications related to pregnancy or childbirth, most of which are preventable. About 99 percent of these deaths occur in developing countries. Indeed, childbirth and complications during childbirth are still the leading cause of death among women aged 20 to 45 in developing countries (UNFPA, 2005). In some areas, one in every seven pregnant women dies in childbirth (WHO, 2008).

Access to health care in the first trimester of pregnancy and having doctors attend births is directly correlated with improved maternal health and reduced infant mortality (UNFPA, 2009). In rural areas of sub-Sahara Africa, women often deliver without skilled care. About a third of maternal deaths are due to hemorrhage that could be stopped by skilled medical personnel.

Many complications associated with pregnancy can be detected and rectified early with the use of advanced technologies such as ultrasound equipment, which needs a reliable source of electrical power. The absence of electricity in rural areas denies rural women such essential health services. Thus both directly (through support the provision of health services) and indirectly (by removing barriers to attracting medical personnel), access to electricity in rural areas can contribute to averting needless maternal deaths.

In some sub-Saharan African countries, life expectancy has actually reverted to below 1970 levels over the last two decades. Largely due to deaths from Human Immunodeficiency Virus /Acquired Immune Deficiency Syndrome (HIV/AIDS), this is the case in the Democratic Republic of Congo, Lesotho, South Africa, Swaziland, Zambia, and Zimbabwe (UNDP, 2010). The HIV pandemic appears to be disproportionately affecting women, tragically highlighted by the phenomenon of AIDS orphans being raised by grandmothers – a common occurrence in many parts of sub-Saharan Africa (UN, 2011a). Although the global spread of HIV has finally begun to slow worldwide, combating this global pandemic requires energy services in a variety of applications. For example, safe disposal of used hypodermic syringes by incineration prevents re-use and the potential further spread of HIV/AIDS.

Education, prevention, condom use, and reproductive health services for women and men are key ingredients to combating new AIDS infections. These require communication campaigns, distribution routes for condoms, and health workers to reach at-risk populations, all of which benefit from energy services in terms of transportation and communication. When HIV infections occur, quality of life and longevity are directly related to access to medical services, anti-retroviral drugs, and overall health maintenance, including good nutrition. Clean energy options for the poor can contribute to improving access to health services and ensure better nutrition.

The challenging nexus of energy, poverty, and health can also be found in industrial countries. The symptoms of energy problems of the poor in richer industrialized economies include high levels of debt to the energy utilities, disconnections from supply because of debt, cold homes in winter, and an increase of cold-related ill health and death. During hot weather periods, energy is needed to power fans and home conditioning. Inadequate access can lead to increased ill health and death of elderly people. The 2003 heat wave in Western Europe led to 35,000 deaths (Bhattacharya, 2003).

A concerted international effort to meet the globally agreed development goals related to human health is starting to make inroads, but more resources need to be allocated to cleaner low-cost energy options for powering for pro-poor health services.

To sum up, there are several health-related benefits associated with access to modern, cleaner and affordable energy options. Electricity in health centers allows the provision of medical services at night, greater use of more advanced medical equipment as well as helps retain qualified staff in rural health centers. Energy for refrigeration can facilitate vaccination and medicine storage for the prevention and treatment of diseases and infections. Energy is needed to generate and maintain basic sanitary conditions to support health services. Finally, energy is needed to develop, manufacture, and distribute drugs, medicines, and vaccinations as well as to provide access to health education media through information and communications technologies.

2.9   Energy and Water

Sanitation and access to improved water in developing nations are key environmental concerns of the poor. Recent estimates indicate that about 1.2 billion people lack access to safe water (IAEA, 2007; UN Women, 2011) and about 2.6 billion people lack access to toilets, latrines, and other forms of improved sanitation in developing countries (UN, 2011). Progress has been slowest in South Asia and sub-Saharan Africa. To halve the proportion of people without basic sanitation by 2015, more than 1.3 billion people would have to gain access to an improved facility (World Bank, 2010a).

On the positive side, rapid progress is being made in improving access to safe water, especially in East Asia, where access to drinking water increased by 30% between 1990 and 2008 (UN, 2010a) to reach 89% of the population. Access to safe water in Latin America in 2008 reached 93%, while for developing countries as a whole the figure was 84% (UN, 2010a). These encouraging indicators of progress are marred by the appalling situation in sub-Saharan Africa where only 48% of rural households have access to piped clean water (World Bank, 2011a).

The higher safe water access that characterize richer developing countries hide a continuing rural-urban divide. Figure 2.17 compares access to improved water supply in rural and urban areas in selected developing countries. In most developing countries, a much larger proportion of the urban population enjoys access to improved sources of water (which includes household piped water connection, public standpipe, borehole, protected well or spring, and rainwater collection).

Inadequate supplies of safe water and poor sanitation, particularly in rural areas accounts for nearly 80% of all illnesses in developing countries (BPN, 2010). In addition, patients with waterborne ailments account for half of the people in hospital beds (TWP, 2010; BPN, 2010). Some of these ailments include cholera, typhus, and dysentery, which account for an annual death toll of 1.6 million deaths (SODIS, 2010b).

Access to water is key to not only improved health but also to increased agricultural output through irrigation

Lack of sanitation facilities in many developing countries contributes to a loss of a sense of dignity, especially among the poor. Inadequate access to sanitation has a disproportionate large impact on women and girls, especially during menstruation. Lack of adequate sanitation in schools in many developing countries has been reported as a key contributor to absenteeism and even school dropouts among one in ten girls at the puberty stage (IMW, 2008; WaterAid, 2009).

Access to modern and affordable forms of energy can play a vital role in improving access to safe water and sanitation in developing countries. Some of the energy solutions available are very low-cost and do not require electricity or fossil fuels. For example, solar water disinfection (SODIS) only requires a clear glass or plastic polyethylene terephthalate bottle, which is filled with water and placed on the roof in the sun for at least six hours to eliminate most disease-causing pathogens (Wegelin and Sommer, 1997; EAWAG/SANDEC, 2002; Foran, 2007; SODIS, 2010a).

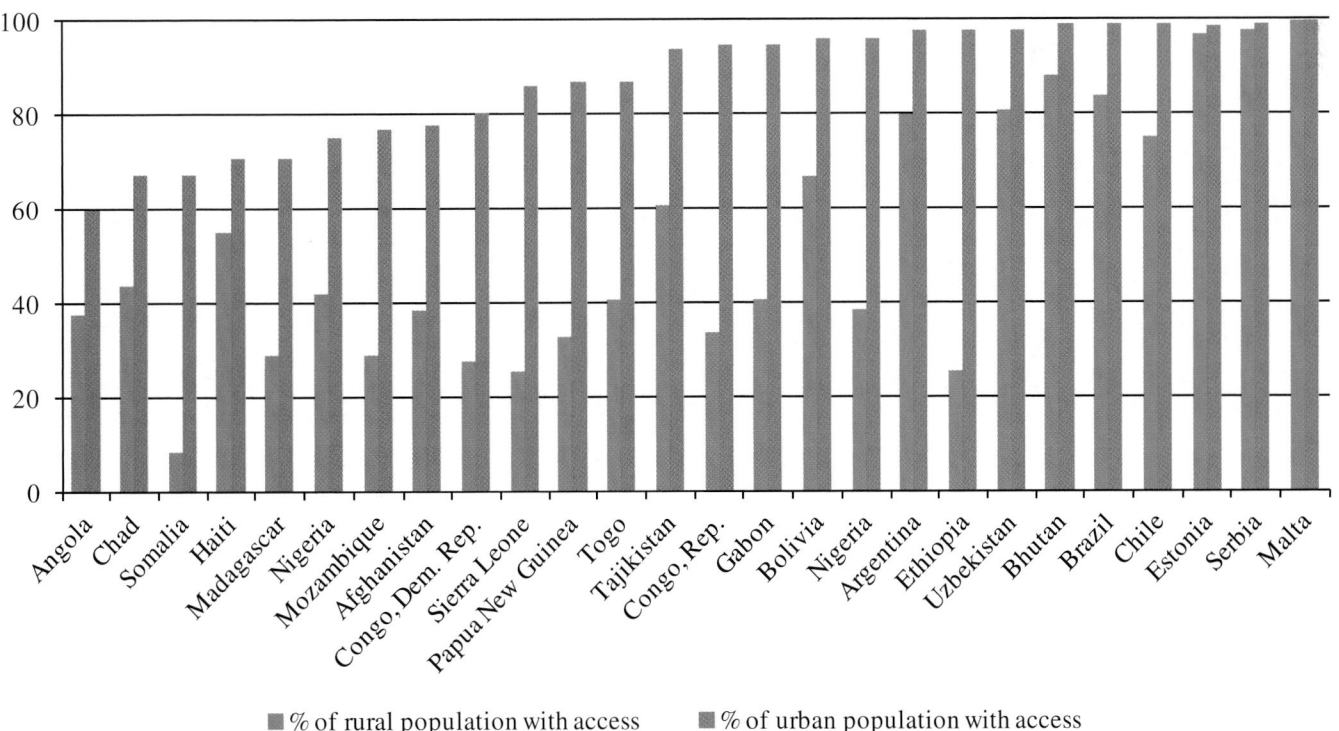

■ % of rural population with access   ■ % of urban population with access

Figure 2.17 | Access to improved sources of water (%), selected developing countries, 2008. Source: adapted from World Bank, 2011a.

Table 2.17 | Scientific results of solar water disinfection technology.

Bacteria	Disease caused by bacteria	Reduction with SODIS method (6h, 40°C)
Escherichia coli	Indicator for water quality & enteritis	99.999%
Vibrio cholera	Cholera	99.999%
Salmonella species	Typhus	99.999%
Shigella flexneri	Dysentery	99.999%
Campylobacter jejuni	Dysentery	99.999%
Yersinia enterocolitica	Diarrhea	99.999%
Virus	**Disease caused by virus**	**Reduction with SODIS method (6h, 40°C)**
Rotavirus	Diarrhea, dysentery	90%
Polio virus	Polio	99.99%
Hepatitis virus	Hepatitis	Reports from users
Parasites	**Disease caused by parasite**	**Reduction with SODIS method (6h, 40°C)**
Giardia species	Giardiasis	Cysts rendered inactive
Cryptosporidium species	Cryptosporidiasis	Cysts rendered inactive after >10h exposure
Amoeba species	Amibiasis	Not rendered inactive Water temperature must be above 50 °C for at least 1h to render inactive

Notes: 1) The SODIS method needs relatively clear water. The source of the water (well, surface water) does not matter. 2) Relative levels of pathogens used in tests.

Source: SODIS, 2009.

As shown in table 2.17, through heat and ultraviolet radiation from the sun, SODIS can reduce diarrhea-related cases in developing countries by nearly 40% (Wegelin and Sommer, 1997; EAWAG/SANDEC, 2002; Foran, 2007; Heaselgrave et al., 2006; SODIS, 2010a).

Low-cost mechanical options can play an important role in ensuring access to safe water. For example, hand pumps and wind pumps are robust technologies that can dramatically expand access to safe water for domestic use as well as for irrigation and watering livestock. Hand-pumps and wind-pumps are well suited for rural areas as they are robust and their maintenance is relatively simple. The ease-of-maintenance of hand-pumps has been demonstrated in Uttar Pradesh, India, where women have successfully repaired hand pumps, thereby ensuring a continuous flow of clean water (WaterAid, 2009).

There are other cleaner energy options but somewhat higher-cost cleaner energy options that can address problems of sanitation and at the same time provide energy. For example, a biogas digester linked to a toilet can be an effective means of treating sewage (Karekezi and Ranja, 1997), while the methane gas produced can be used for heating applications such as cooking or running a genset[7] to produce electricity. This

technology is widely used in Asian countries, especially in Bangladesh, Bhutan, China, India, Laos, Nepal, and Sri Lanka (SNV 2007).

Biogas technology can also be used to improve sanitary conditions in large institutions such as schools, hospitals, and prisons. In Africa, prisons in Rwanda and Kenya have used large-scale biogas plants resolve long-standing sanitation challenges (Ashden Awards for Sustainable Energy, 2005). The gas is used for cooking, thereby reducing woodfuel consumption by half. Similar applications, but at a smaller scale, have been reported in other African and Asian countries, where biogas technology has been adopted in schools and mainly used for cooking (Huba and Paul, 2007; AFREPREN/FWD, 2010).

The challenge of providing safe water and sanitation facilities is still immense particularly in sub-Saharan Africa where more than half the urban population lives in slums; a sizable part of the more than 825 million people living in urban dwellings without improved sources of drinking water and sanitation (World Bank, 2010a). Solar, wind, handpumps and biogas options can play an important role in enhanced access to safe water and sanitation in low-income peri-urban and urban of sub-Saharan Africa.

2.10   Energy and Women

Promoting gender equality and empowering women is an agreed national and global priority and is reflected in major global conferences and intergovernmental agreements on the need to enhance the status of women, including in economic and political life (CEDAW, 2011). In 1979, the UN General Assembly adopted the Convention on the Elimination of All Forms of Discrimination against Women to protect fundamental human rights and equality for women around the world (CEDAW, 2011).

Global studies indicate that a higher percentage of women than men live in poverty (UNECA, 2010). Gender-related studies of the poor in sub-Saharan Africa, parts of South Asia, and Latin America and the Caribbean demonstrate that poverty affects women and men differently, with women often experiencing the most severe levels of deprivation, in part, demonstrated by inadequate access to cleaner energy options (Karekezi et al., 2002).

Gender equality and empowering women are also important issues because they foster progress toward other development outcomes, such as reducing poverty, hunger, and disease and improving access to education and maternal health. Educational opportunities for girls have expanded since 1990. Enrollment patterns in upper-middle-income countries now resemble those in high-income countries, and those in lower-middle-income countries are nearing equity. However, gender gaps remain large in low-income countries, especially at the primary and secondary levels in sub-Saharan Africa (World Bank, 2010a).

Girls in poor households and rural areas are least likely to be enrolled in school. Cultural attitudes and practices also pose formidable obstacles

---

7   A "genset" is also known as an "engine-generator," which is a machine used to generate electricity.

to gender parity. Currently, 64 developing countries have achieved gender parity in primary schools and 21 others are on track to achieving the target. However, about 29 countries (more than two-thirds of them in sub-Saharan Africa) are seriously off track and unlikely to achieve parity by 2015 if current trends prevail (World Bank, 2010a).

Wider access to cleaner and affordable energy options can improve gender parity and school enrollment of girls. For example, access to cleaner energy options (electricity for lighting in schools and cleaner cooking fuels at home such as LPG) can extend studying hours for girls by reducing the time they spent collecting fuel. Access to pumped water can reduce the time that girls spend carrying water for household use.

Ensuring girls in developing countries are educated is important for their future life as mothers. Global evidence has shown that better-educated women will chose to have fewer children and have more access to family planning methods, while illiterate women overall are the least well able to gain access to health services and medical care at large (UNFPA, 2008; SIL International, 2011). The cycle of exclusion from education and literacy arising from preoccupation with household chores of collecting fuel, water, cooking, cleaning and looking after younger children can lead to early onset of reproduction and a higher risk for maternal health caused by frequent multiple pregnancies. According to Population Action International (2001), teenage mothers face twice the risk of dying from childbirth as women in their twenties. In addition, their children are more vulnerable to health risks.

Although most rural households use biomass, women in poor households will spend more time searching for firewood than those in households with higher incomes (Energia, 2008). In areas where there is decreased vegetation cover, women and children are required to walk longer distances to collect firewood (Ekouevi, 2001; Kammen et al., 2001; Ward, 2002; ESMAP, 2003; BIC, 2009). As shown in Figure 2.18, in Africa this can take up to four hours a day (Njenga, 2001; WHO, 2006) – and these distances increase as wood from stressed ecosystems.

The energy that women and girls expend fetching water, carrying fuel, and preparing food has received virtually no treatment in traditional energy sector policies because, as it is not traded, it has no "price" as determined by the market and therefore is considered "free." The time people spend carrying water that is not transported by pumping systems or fuels that have been harvested or bought locally, or pounding and grinding maize, wheat, rice, sorghum, or other basic staples, has an opportunity cost. It is time that could otherwise be spent on more economically productive or socially beneficial activities such as paid agricultural work, commercial activities, pursuit of education, or physical rest to preserve the health of women, especially the poorest women.

Access to mechanical power for threshing and grinding grains can help older girls stay in school by minimizing school absenteeism and drop-out rates among girls. It is difficult to change cultural practices where food preparation is assumed to be the responsibility of women and girls. With mechanical power, however, food preparation can be made easier and faster, which liberates valuable time for girls to study and attend school.

Lack of close proximity to safe water places a huge burden on women and girls in many developing countries. For example, on average, women travel 6 kilometers daily in search of water (IAEA, 2007; BPN, 2010). In rural areas, the distances walked by women can be as high as 16 kilometers; during drought, the distance could be at least twice as long (IMW, 2008).

The long distances travelled carrying a load of approximately 20kg pose serious long-term health risks for women, including spine and pelvic deformities that can create complications during childbirth (WaterAid, 2009). In addition, these distances imply that a significant amount of time is spent obtaining this precious commodity, especially in rural areas. According to WaterAid (2009), in rural Africa, about 26% of women's time is spent fetching water. In rural Ghana, a study by Costa et al. (2009) revealed that women spend far more time than men fetching water (see Figure 2.19).

As noted earlier, fuelwood is not always collected. Often it is a commercial commodity bought and sold through markets like any other fuel. This is especially the case in urban areas, where cleaner liquid fuels are either unavailable or unaffordable for poor families. Here, too, women must spend time on the purchase and transport of traditional fuels for household use. And this too has an opportunity cost for women.

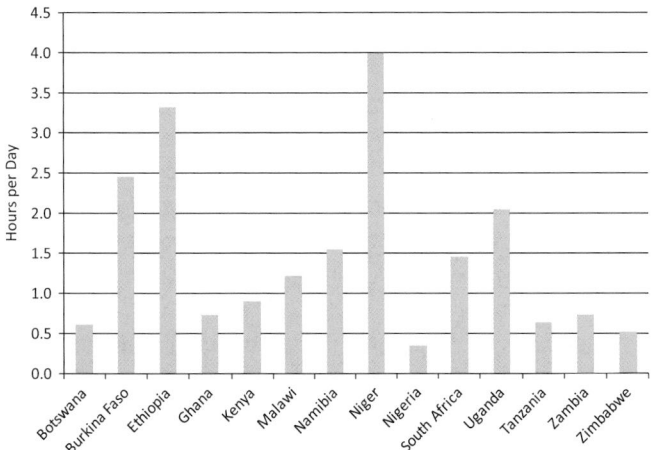

Figure 2.18 | Time ticking away: daily hours that women spend collecting fuel in different African geographical settings, by country, 1990–2003. Source: WHO, 2006.

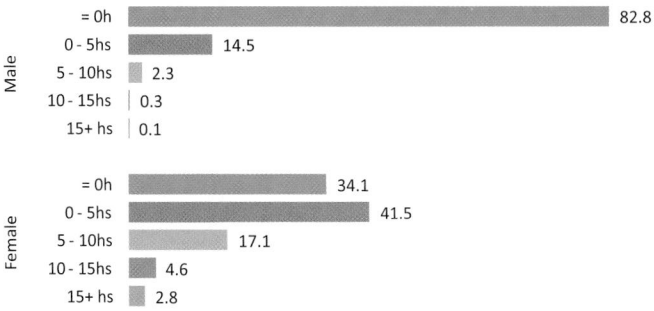

Figure 2.19 | Comparison of time spent by men and women fetching water in rural Ghana (hours/week). Source: Costa et al., 2009.

While improved biomass energy technologies such as cookstoves have several benefits for poor women, a more pronounced switch to cleaner fuels and more efficient cooking devices can lead to increased economic, health, and environmental benefits. As shown in Table 2.18, for example, in India, using a typical LPG stove for cooking can reduce cooking time by slightly over 1 hour compared with a conventional woodstove (Kanagawa and Nakata, 2005). Similar benefits can also be accrued from use of ethanol stoves.

The use of improved cookstoves can result in savings in the amount of fuel used, which translates into direct cash savings. They can also reduce respiratory health problems associated with smoke emission from biomass stoves and offer a better home and working environment for rural housewives and cooks/operators in institutions.

The production and dissemination of improved biomass energy technologies and associated sustainable fuel supply initiatives provide employment opportunities for a significant proportion of the population, particularly women. In Senegal, for example, the sustainable forestry subsector earns the country about US$_{2004}$\$12.5 million a year, with women in control of one-third of this revenue stream (World Bank, 2010a).

The development linkages between women and energy are not limited to cooking fuels and water collection challenges, although policy and analysis in development debates often stops there. Women are disproportionately affected by the lack of cleaner and affordable energy options for maintaining households and enterprises that can be a source of income (Energia, 2008). This is particularly perverse given that the best means to combat absolute poverty is through the stimulation and increase in women's income – either through productive activities or targeted state transfers aimed at women. Women have a relatively higher propensity to reinvest in family welfare expenditures such as food, education, and health, especially for their dependent children.

Women are key users of energy for productive activities, and as is the case for productive employment for men, access to labor-saving energy-using technologies allows the amount of time spent on non-paid work to be reduced in favor of time spent on paid work, which greatly improves the overall welfare of women, their families, and especially their children. The

limited access to cleaner energy and labor-saving technological options therefore has a double negative effect on women: first through limiting the opportunities to generate income and second through the forgone benefits that the additional income could bring to poor families when spent by women on inputs and services that benefit the family.

Affordable and reliable energy options can broaden the scope of women's enterprises. In West Africa, for example, mechanical power to grind grains and other agricultural products through a fee-for-service model not only was able to meet household grain consumption processing needs but also freed up women's time for other household and agricultural activities (Morris and Kirubi, 2009). In addition, the volumes of milled grain produced surpluses that could be marketed, thereby supplementing limited family cash income that was used to support children's schooling.

The upfront cost of cleaner and affordable energy options is in most cases beyond the reach of the rural poor. This affects women who are the major collectors, managers, and users of traditional energy sources for household activities (BIC, 2009). In developing countries, since many women lack the money to pay for energy services, small-scale food production and processing – a large component of which is often dominated by women – largely relies on human and animal energy (Clancy, 2004) – further discussion on this issue is found in Chapter 20.

An important approach for enhancing access to cleaner and affordable energy for women is to develop programs specifically aimed at financing the upfront costs of energy using equipment for household use as well as targeting income-generating activities that involve a high number of women or women's groups. A number of financing institutions in developing countries have emerged focusing solely on financing women. In Bangladesh, for example, the Grameen Bank has a credit program in which 98% of its borrowers are women (Giridharas and Bradsher, 2006). These funds could be used to mechanize farming and the water supply, hence reducing the burden on women, as well as improving lighting and cooking energy technologies at the household level.

Since in many areas women are the primary users of energy, especially for cooking, it makes sense for them to be involved in designing and implementing policies and projects to meet their own energy needs. Globally, as more women obtain positions of political leadership, it is anticipated that the increased political representation by women will have significant policy implications. For example, it is expected that women politicians are more likely to push for increased access to cleaner and affordable energy options that would benefit women and cleaner cooking fuels such as LPG, ethanol stoves, and improved biomass stoves. Other modern energy options that benefit women are summarized in Table 2.19.

To sum up, access to cleaner and affordable energy options can play an important role in gender equality and women empowerment:

- Available energy services can free girls' and young women's time from household activities (gathering firewood, fetching water, cooking inefficiently, crop processing by hand, manual farming work). This

Table 2.18 | Comparison of cooking times (in hours) of woodstove and LPG in India.

	Time spent on cooking-related activities		
	Fuel collection[a]	Food preparation and cooking	Total time
Woodstove	0.67	2.73	3.4
Gas Stove	-	2.30	2.30
Time saved by gas stoves	0.67	0.43	1.10

[a] The time spent taking gas cylinders for refilling is negligible when spread over the time the newly refilled gas cylinder is used.

Source: adapted from Kanagawa and Nakata, 2005.

Table 2.19 | Summary of some energy options favoring women.

Energy Option	Benefits for women
Electricity	Radio, TV, and telecommunications can improve access to the outside world and provide useful information for women.
	Street lighting improves safety for women, especially in slum areas, and enables them to trade after dark.
	In urban areas, electricity helps women to cope with modern lifestyles.
	Availability of electricity enables women to operate specialized enterprises such as hair dressing.
Mechanical power	Women can irrigate crops and water livestock, thereby enhancing food security and nutrition.
	Mechanized tillage and weeding can contribute to higher crop yields.
	Briquette presses can help women obtain cleaner and sustainable biomass fuel supplies.
	Oil presses can help women earn more from agriculture by selling a higher-value product.

Sources: adapted from Bates et al., 2009; UNDP, 2007a.

time can be productively used to generate more income or acquire education.

• Cleaner cooking fuels and equipment can reduce exposure to indoor air pollution and improve health.

• Street lighting can improve women's safety.

• Modern, cleaner and affordable energy options can broaden the scope for women's enterprises, thereby fostering employment and income generation among women.

• Representation of women at the national level can influence energy policies and investments to ensure wider access to cleaner and affordable energy options that benefit women.

• Specific policy initiatives that prioritize "women's" or household energy needs, especially for fuels and mechanical power, can provide a balance to the focus on rural electrification, bringing about a host of family welfare benefits.

2.11 Energy and Environment

This section briefly discusses energy and environmental sustainability in the context of development, while the broader issue is covered in depth in Chapters 3, 4, and 20.

Many of the world's poor depend on forests for their livelihoods as well as to meet their energy needs for cooking. In sub-Saharan Africa about 80% of the population relies on biomass (IEA, 2010a). Although the major cause of deforestation comes from land use change, excessive

harvesting of biomass fuel, especially from forests, also contributes to this deforestation. The loss of forests threatens the livelihoods of the poor, destroys habitats that harbor biodiversity, and eliminates important carbon sinks that offset global warming. Since 1990, forest losses have been substantial (more than 1.4 million square kilometers), especially in Latin America and the Caribbean, East Asia and the Pacific, and sub-Saharan Africa (World Bank, 2010a). Some of this deforestation is due to excessive harvesting of biomass fuel.

As mentioned earlier, much of the rural deforestation and overharvesting of fuelwood experienced in sub-Saharan Africa, can, in part, be traced to trees cut down for feedstock for the charcoal industry. This is not to supply rural consumers with energy but to provide a steady supply of cheap fuels for cooking and household energy use for urban dwellers, as indicated by the thriving charcoal markets throughout the African continent (AFREPREN/FWD, 2006a). Important options for addressing the charcoal and deforestation challenge include dissemination of proven improved cookstoves such as the charcoal-burning KCJ that can cut charcoal consumption by half and switching to LPG.

As shown in the Figure 2.20, biomass – mainly traditional biomass biofuels – accounts for about 44% of the world's final renewable energy use. The bulk of this supply is from developing countries.

Poor households using open fires and traditional biomass stoves for cooking, tend to be highly inefficient due to incomplete combustion of the biomass. Consequently, methane, a more potent greenhouse gas than carbon dioxide (CO_2), is released. The use of traditional and unsustainable biomass cooking energy fuels has significantly higher greenhouse gas (GHG) emissions compared to fuels such as LPG and biogas (see Chapter 19). Use of cleaner cooking fuels and technologies can reduce the GHGs of developing countries.

Although developing countries emit fewer emissions on a per capita basis, in gross terms and, if one ignores, historical emissions or per capita emissions, some emerging industrializing countries account for a significant share of current global GHGs emissions comparable to those of industrialized countries. For example, China's CO_2 emissions in 2008 exceeded those of the entire European continent and compare with those of the United States, Canada, and Mexico combined (see

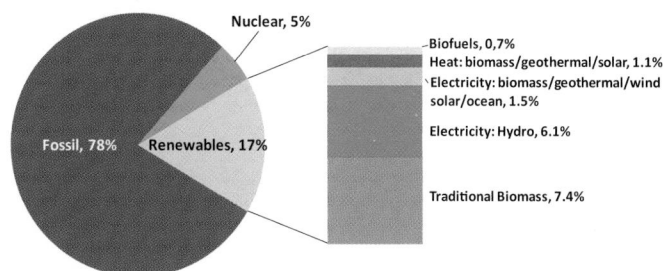

Figure 2.20 | Renewable share of final energy use in 2009 (528 EJ). Source: fossil and nuclear fuel use: IEA 2011b; renewables: Chapter 11.

Chapter 3). However, the picture changes dramatically if historical and per capita emissions are included. Wealthy western countries of Europe and North America account for a dominant share of historical emissions and are by far the highest emitters of GHGs on a per capita basis.

It can therefore be argued that the poor in developing countries should not be restricted when the current high GHGs emissions levels are still largely a result historical and of high per capita emissions of industrialized countries of North America and Europe.

On a per capita basis, poor people and poor countries in the world generate fewer GHGs than either the people who are well-off in poor countries or the inhabitants of industrialized countries. The poor contributions to global greenhouse gas emissions is negligible. For example, Africa has very low levels of access to GHGs-intensive energy options and, as a result, the continent generates only 3% of the global aggregate GHG emissions (UNIDO, 2009; IEA, 2010b). But lifting the poor out of poverty through improving access to jobs, social services, and improved living conditions requires greater energy inputs, as has been argued throughout this chapter, which would inevitably result in increasing emissions further from the developing countries. Replacement of traditional biomass (with incomplete combustion) with LPG could lead to lower overall GHG emissions especially when particulate matter is accounted for.

In some cases, wide-scale deployment of renewable energy systems may provide the best options for providing access to cleaner and modern energy options while reducing GHGs emissions. In the near term, however, cleaner fossil fuels such as LPG combined with more efficient and low-cost end-uses devices such as LPG cookstoves reduce a host of social, economic, and environmental barriers to overcoming poverty due to higher combustion efficiency.

The ongoing search for win-win solutions in low-income areas of poor developing countries that are completely reliant on renewable energy must also put a time premium on overcoming social injustice and the conditions of inequality that entrench poverty and reproduce underdevelopment. As mentioned earlier, from a policy point of view, it is unrealistic to support arguments for "near-term renewables only" options that delay the satisfaction of the basic needs of people who remain poor today and who constitute a negligible contributor to global historical and current carbon emissions. It is, therefore, important to retain cleaner fossil fuel options such as LPG combined with more efficient low-cost end-use devices in near-term priority action plans that would set the stage for eventually moving the poor to a fully sustainable path that relies more heavily on renewable energy.

There is, however, a growing interest on the part of energy decision makers in developing countries in examining efficient, environmentally sound, climate-friendly energy options that also deliver substantial development benefits and reduce the climate risk profile of their energy industries. Studies indicate that many of these options (such as demand-side management, renewable energy use and efficient energy end-user

devices such as tube lights, compact fluorescent lamps, biomass cogeneration, ethanol, etc.) are not as capital-intensive as conventional centralized energy options (DuPont, 2009).

2.12   Conclusions and the Way Forward

The intrinsic link and positive relationship between energy and overcoming poverty – whether income-defined poverty or poverty linked to inequality in income distribution within countries – is very strong in developing countries. It is clearly demonstrated by the fact that poor people constitute the bulk of the estimated 3 billion people primarily relying on solid fuels (coal and traditional biomass) of which 2.7 billion people cook and heat their homes with traditional fuels and low-efficiency stoves (UNDP and WHO, 2009; IEA, 2010a; see also Chapter 19) and almost one and a half a billion without access to electricity (IEA, 2010a; see also Chapter 19).

Access to electricity supply (both grid and non-grid) in many developing countries is almost an exclusive service enjoyed by the non-poor in urban areas. Even after two decades of energy sector reforms, initial indications from a wide range of developing countries indicate that few of the initiatives have resulted in significant improvement in the provision of electricity to the world's poor (GNESD, 2010; Karekezi and Sihag, 2004; IMF, 2008). Access to modern and cleaner forms of energy improves the livelihoods[8] of the poor in developing countries. As discussed in earlier sections, increased access to cleaner energy in rural areas opens up these regions to greater economic productivity.

In most cases, market-led reforms were primarily designed to improve the financial health of electricity companies and were introduced into countries where a large sector of the potential "market" consisted of very poor people not served by national grids or who resided in geographic areas with low levels of energy use. Expanding access to electricity to the poor meant attempting to service low-income consumers whose incomes may well be highly unstable and who often live in isolated areas that are difficult to access or in urban areas that are characterized by high degrees of informality. To provide electricity services, the "reformed" companies had to cover operating and investment costs (required by market-oriented reforms) while providing expensive transmission lines and connections as well as maintenance, billing, and collection services in a market where return on investment was far from assured. In the majority of the countries, these contradictory demands proved to be irreconcilable. It is therefore not surprising that newly privatized or reformed electricity companies tended to "cherry pick" the most lucrative markets (i.e., non-poor urban areas), raised their tariffs, and virtually ignored widening of their networks to poorer consumers.

---

8   Livelihood refers to "means of support, subsistence or securing the necessities of life." Livelihoods can be improved through increased agricultural production, better social welfare – including increased income among other factors.

Thus, experiences in developing countries point to an overarching conclusion: when power sector reforms were introduced with the sole intention of improving the performance of utilities, the expected and hoped-for social benefits did not necessarily follow. Where governments maintained a role as instigator or at least regulator of improved access to electricity by the poor, tariffs for poor households tended to decrease and levels and rates of electrification tended to increase.

The excessive reliance on private sector-driven approaches that have proven detrimental to widening access to electricity in many parts of the developing world, especially sub-Saharan Africa and some parts of Asia, are also becoming prevalent in efforts to disseminate improved cookstoves and scale up the use of renewables in developing countries. There is heavy emphasis on bottom-of-the-pyramid profit-driven approaches with limited attention being paid to the important role of public interventions that have the resources and long-term horizon to engineer the scale-up required to reach millions of the unelectrified homes in the developing world that currently rely on inefficient and traditional biomass-based cookstoves to meet their cooking and heating needs. A more balanced approach is needed that judiciously combines large-scale and long-term public initiatives with innovative pilot private sector based programs that rely on the bottom-of-the-pyramid profit-driven small and medium scale enterprises.

In order to ensure that modern, cleaner and affordable forms of energy are accessed by poor people, the right choice of energy supply has to be made. For example, large-scale renewable energy technologies have lower running costs, hence might be, in the long-term, the most attractive options. In addition, some fossil fuels such as LPG can also be attractive due to their cleaner combustion and higher efficiency characteristics.

Based on political and policy considerations, there is need to minimize any delays to the satisfying the basic need of the poor. Therefore, appropriate solutions to overcoming social injustice and the conditions of inequality that entrench poverty and reproduce underdevelopment must be achieved within the shortest possible time. It is for this reason that GEA shares a goal with the UN proposed targets of achieving universal access to electricity and cleaner cooking fuels by 2030 (UN, 2010b). Current estimates indicate that achieving the aforementioned goal will require significant investment (see Chapter 17) and political commitment (see Chapters 19 and 23). In addition, the relationship between poverty and energy requires a better understanding of the demand profile of this segment of the energy market and the recognition that poor people are energy users. This is demonstrated by the fact that, for example, at current electrification rates it is estimated that about 15% of the world population will not have access to electricity in 2030 (IEA, 2010a). As shown in Table 2.20, most of these people will be in sub-Saharan Africa and some parts of Asia. However, it is noted that the IEA baseline contrasts with GEA pathways of achieving universal access by 2030.

In order for national leaders in developing countries to deliver the much-needed political commitment required to address the access question, approaches to dealing with the issue need to be "homegrown." The rationale for this approach is that policymakers in developing countries have successfully used a similar quantified campaign strategy with the MDGs to accelerate the pursuit of globally agreed social indicators fundamental to overcoming underdevelopment and poverty. The community of nations should adopt global goals for minimum levels of energy services, especially universal access to electricity and cleaner cooking fuels, to address poverty and support sustainable human development.

It is worth pointing out that while promoting access to electricity is generally given more attention due to the services it typically provides, such as communications, lighting, refrigeration, and motor power, equally important is the role that nonelectric forms of energy, in particular fuels and mechanical power, play in both the household and the social sectors. This second set of energy issues has been given much less attention in energy sector planning, especially in developing countries. In order to ensure that modern, cleaner and affordable forms of energy are accessed by poor people, the right choice of energy supply has to be made. For example, large-scale renewable energy technologies have lower running costs, hence might be, in the long-term, the most attractive options. More challenging still is that while power sector reform has received much attention in developing countries, energy policies addressed at fuel switching and improving heating and cooking systems, especially in rural areas, have received very little policy attention in energy sector reform. Electricity is not synonymous with energy, a concept which is much broader. Some fossil fuels such as LPG can

Table 2.20 | Current and projected electrification levels in developing countries.

Region/Country	2009		2030	
	Unelectrified population (million)	Proportion of all unelectrified (%)	Unelectrified population (million)	Proportion of all unelectrified (%)
Latin America	31	2.2	10	0.8
Sub-Saharan Africa	585	40.7	652	53.8
North Africa	2	0.1	2	0.2
South & East Asia	812	56.5	549	45.3
China	8	0.6	0	0
Total	1438	100.0	1213	100.0

Source: adapted from IEA, 2010a; 2010c.

also be attractive due to their cleaner combustion and higher efficiency characteristics and therefore it is crucially important that all fuel options are given consideration when designing energy sector strategies. Finally many of the energy issues treated here fall outside the energy sector per se and are intimately linked with service and supply decisions taken in other sectors and policy frameworks including those in education, health, agriculture, water, housing etc.

As this chapter indicates, for the poorest segments of the population and especially for poor women and their children, the role of cleaner energy carriers for cooking, household use and homestead-based productive activities are disproportionately more important in reducing the day-to-day barriers to education, health, and family food security. Access to modern and cleaner forms of energy contribute to a general improvement in social welfare – including increased income – due to improved health, sanitation, education, etc. For example, in Bangladesh, an impact assessment of its rural electrification revealed that 63% of electrified households surveyed reported an increase of income as a direct result of electrification (Berthaud et al., 2004). In Lao PDR rural electrification using solar systems has demonstrated the viability of decentralized renewable energy systems in enhancing rural livelihoods through increased income, improved healthcare and access to information (Theuambounmy, 2007).

In short, to ensure the poor's access to cleaner and affordable energy options, energy policy must see the poor as a consuming market that, when better served, will generate income and opportunities for value-adding activities that can have benefits at the family, community and sector wide levels. However, the excessive reliance on private sector-driven approaches that have proven detrimental to widening access to electricity in many parts of the developing world, especially sub-Saharan Africa and some parts of Asia, should not become the only route to expand poor's access to cleaner energy options. The almost exclusive recourse to private sector-driven approaches to disseminate improved cookstoves and scale up the use of renewables in developing countries is of some concern. There is heavy emphasis on bottom-of-the-pyramid profit-driven approaches with limited attention being paid to the important role of public interventions that have the resources and long-term horizon to engineer the scale-up required to reach millions of the unelectrified homes in the developing world that currently rely on inefficient and traditional biomass-based cookstoves to meet their cooking and heating needs. A more balanced approach is needed that judiciously combine large-scale and long-term public initiatives with innovative pilot private sector based programs that rely on the bottom-of-the-pyramid profit-driven small and medium scale enterprises.

Prevailing energy systems in developing countries need transformational change in order to have a pro-poor orientation as well as to ensure universal access to modern forms of energy and energy services in key economic sectors providing livelihoods to poor people. In addition, universal access serves both to advance education and all other concerns, and to avoid the detrimental effects of dependence on traditional fuels in terms of women's time, health effects, pollution, global warming etc. Chapters 17, 19, and 23 discuss how universal access, especially of electricity and cleaner cooking fuels, could be achieved. All GEA pathways lead to universal access to electricity and cleaner cooking fuels by 2030.

References

AFREPREN/FWD, 2006a: *Report of the ADB FINESSE Training Course on Renewable Energy and Energy Efficiency for Poverty Reduction*, 19–23 June, Volume III: Write-Ups. Energy, Environment and Development Network for Africa (AFREPREN/FWD), Nairobi.

AFREPREN/FWD, 2006b: *Renewables for Poverty Reduction in Africa*. Energy. Environment and Development Network for Africa (AFREPREN/FWD), Nairobi.

AFREPREN/FWD, 2010: *Arusha Field Visit to SMES Dealing With Renewable Energy Report*. Energy. Environment and Development Network for Africa (AFREPREN/FWD), Nairobi.

Alkire, S., J.M. Roche, M.E. Santos, and S. Seth, 2011: *Multidimensional Poverty Index: 2011 Data*. Oxford Poverty and Human Development Initiative. www.ophi.org.uk/policy/multidimensional-poverty-index/ (accessed 4 March 2010).

AllAfrica, 2004: Zimbabwe: Infrastructure improvement key to teacher retention in rural areas – Chigwedere. *The Herald*, 21 May.

ANSD, 2006: *Sénégal: resultats du troisième recensement général de la population et de l'habitat (2002): rapport National de presentation*. Agence Nationale de la Statistique et de la Demographie (ANSD), Senegal.

Ashden Awards for Sustainable Energy, 2005: *Case-study: Large-scale biogas for sanitation. Kigali Institute of Science*, Technology and Management (KIST), London.

Baker, J., R. Basu, M. Cropper, S. Lall, and A. Takeuchi, 2005: *Urban Poverty and Transport: The Case of Mumbai*. Policy Research Working Paper No. 3693. World Bank, Washington, DC.

BIC, 2009: *Ugandan environmentalist speaks out on large dams, renewable energy, and poverty alleviation*. Bank Information Center (BIC) Update. Washington, DC.

Barnes, D. F., A. Hussain, A. H. Samad, and R. S. Khandker, 2010: *Energy Access, Efficiency and Poverty: How Many Households are Energy Poor in Bangladesh?* Policy Research Working Paper No. 5332. World Bank, Washington, DC.

Bates, L., S. Hunt, S. Khennas, and N. Sastrawinat, 2009: *Expanding Energy Access in Developing Countries: The Role of Mechanical Power*. Practical Action, Rugby.

Bazilian, M., P. Nussbaumer, A. Cabraal, R. Centurelli, R. Detchon, D. Gielen, H. Rogner, H. McMahon, V. Modi, N. Nakicenovic, B. O'Gallachoir, M. Radka, K. Rijal, M. Takada, and F. Ziegler, 2010: *Measuring Energy Access: Supporting a Global Target*. The Earth Institute, Columbia University, New York.

Berthaud, A., A. Delescluse, D. Deligiorgis, K. Kumar, S. Mane, S. Miyamoto, W. Ofosu-Amaah, L. Storm, and M. Yee. 2004: *Integrating Gender in Energy Provision: Case Study of Bangladesh*. International Bank for Reconstruction and Development/The World Bank, Washington, DC.

Bhandari, O., 2006: *Socio-Economic Impacts of Rural Electrification in Bhutan*. Master's thesis, Asian Institute of Technology, Bangkok.

Bhattacharya, S., 2003: *European heatwave caused 35,000 deaths*. New Scientist, 10 October.

BPN, 2010: *The Facts About the Global Drinking Water Crisis*. Blue Planet Network (BPN). blueplanetnetwork.org/water/facts (accessed 22 November 2010).

Brew-Hammond, A., 2007: *Challenges to Increasing Access to Modern Energy Services in Africa*. Background Paper Prepared for the Forum of Energy Ministers of Africa Conference on Energy Security and Sustainability, 28–30 March, Maputo, Mozambique.

Cabraal, R. A., D. F. Barnes, and S. G. Agarwal, 2005: Productive uses of energy for rural development. *Annual Review of Environment and Resources*, 30:117–44.

CAI-Asia, 2011: *Managing Two and Three-Wheelers in Asia*. Clean Air Initiative for Asian Cities (CAI-Asia), Manila and Partnership for Clean Fuels and Vehicle, Nairobi.

CEDAW, 2011: *About CEDAW*. Convention on the Elimination of All Forms of Discrimination against Women (CEDAW). www.cedaw2010.org/index.php/about-cedaw (accessed 3December 2011).

Clancy, J., 2004: *Urban Poor Livelihoods: Understanding the Role of Energy Services*. University of Twente, Enschede.

Coelho, S., 2009: *Energy Access in Brazil*. PowerPoint presentation at Taller Latinoamericano y del Caribe: Pobreza y el Acceso a la Energía. Brazilian Reference Center on Biomass (CENBIO), University of São Paulo. October 2009, Santiago.

Costa, J., D. Hailu, E. Silva, and R. Tsukada, 2009: *Water supply and women's time use in rural Ghana*. Poverty in Focus No. 18, International Policy Centre for Inclusive Growth (IPC-IG). Brasilia.

DFID, 2002: *Energy for the poor: Underpinning the Millennium Development Goals*. Department for International Development (DFID), Crown, London.

DuPont, P. 2009: *Case Study of Energy Efficiency in Asia and Internationally: Some Examples and Food for Thought*. USAID-Asia, PowerPoint presentation at First Mekong Energy and Ecology Training, 14 May, Bangkok.

EAWAG/SANDEC, 2002. *Solar Water Disinfection: A Guide for the Application of SODIS*. Swiss Federal Institute of Environmental Science and Technology (EAWAG)/Department of Water and Sanitation in Developing Countries (SANDEC), Duebendorf.

Ekouevi, K., 2001: An overview of biomass energy issues in sub-Saharan Africa. In *Proceedings of the African High-Level Regional Meeting on Energy and Sustainable Development for the Ninth Session on the Commission on Sustainable Development*. N. Wamukonya (ed.), United Nations Environment Programme, Risoe Centre, Roskilde, pp.102–08.

Energia, 2008: *Turning Information into Empowerment: Strengthening Gender and Energy Network in Africa*. TIE-ENERGIA Project.

ESMAP, undated: *Fighting Poverty through Decentralized Renewable Energy*. Energy Sector Management Assistance Program (ESMAP), World Bank, Washington, DC.

ESMAP, 2003: *Energy and Poverty: How Can Modern Energy Services Contribute to Poverty Reduction? Proceedings of a Multi-Sector Workshop*, 23–25 October, Addis Ababa.

ESMAP, 2004: *Renewable Energy for Development – The Role of the World Bank Group*. World Bank, Washington, DC.

Fall, A., S. Sarr, T. Dafrallah and A. Ndour, 2008: Modern Energy Access in Peri-Urban Areas of West Africa: The Case of Dakar, Senegal. *Energy for Sustainable Development*, 12(4): 22–37.

Farm Radio International, 2006a: Increasing post-harvest success for smallholder farmers. *Voices Newsletter*, No. 79. Ontario.

Farm Radio International, 2006b: *Three fishing ladies with a message about solar dryers*. Radio Scripts. Package 79, Script 6. www.farmradio.org/english/radio-scripts/79–6script_En.asp (accessed 17 February 2011).

Farm Radio International, 2008: *Food processing and storage*. Radio Scripts by Subject. www.farmradio.org/english/radio-scripts/food.asp (accessed 17 February 2011).

FAO, 2009: *Post-harvest Losses Aggravate Hunger – Improved Technology and Training Show Success in Reducing Losses*. Food and Agriculture Organization

(FAO) of the United Nations, Rome. www.fao.org/news/story/en/item/36844/icode/ (accessed 16 October 2011).

FAO, 2011: *AQUASTAT Online Database: Percentage of total grain production irrigated (%)*.www.fao.org/nr/water/aquastat/data/query/index.html (accessed 20 June 2011).

Foran, M. M., 2007: *An analysis of the time to disinfection and the source water and environmental challenges to implementing a solar disinfection technology (SolAgua)*. Harvard School of Public Health, Harvard University, Boston.

Galitsky, C., A. Gadgil, M. Jacobs, and Y. Lee, 2006: *Fuel efficient stoves for Darfur IDP camps: Report of field trip to North and South Darfur*. 16 November – 17 December, 2005. Lawrence Berkeley National Laboratory, Berkeley, CA.

Gasparini, L., and N. Lustig, 2011: *The Rise and Fall of Income Inequality in Latin America*. Tulane Economics Working Paper No. 1110. Tulane University, New Orleans.

Gasparini, L., G. Cruces, L. Tornarolli, and M. Marchionni, 2009: *A Turning Point? Recent Developments on Inequality in Latin America and the Caribbean*. Centro de Estudios Distributivos, Laborales y Sociales, Universidad Nacional de La Plata, Buenos Aires.

Giridharas, A., and K. Bradsher, 2006: Microloan pioneer and his bank win Nobel Peace Prize. *New York Times*, 13 October.

GNESD, 2008. *Clean Energy for the Urban Poor: An Urgent Issue*. Global Network on Energy for Sustainable Development (GNESD), Roskilde.

GNESD, 2010. *Energy Access – Making Power Sector Reform Work for the Poor*. Global Network on Energy for Sustainable Development (GNESD), Roskilde.

Gomide A., 2008: Mobility and the urban poor. *Urban Age*. Newspaper essay, South America, December 2008.

Harries, M., 2002: Disseminating wind pumps in rural Kenya: Meeting rural water needs using locally manufactured wind pumps. *Energy Policy*, 30(11–12).

Harsdorff, M., and J. Peters, 2010: *On-Grid Rural Electrification in Benin*. RWI Materialien, Germany.

Heaselgrave, W., N. Patel, S. Kilvington, S.C. Kehoe and K.G. McGuigan, 2006: Solar disinfection of poliovirus and Acanthamoeba polyphaga cysts in water—a laboratory study using simulated sunlight. *Letters in Applied Microbiology*, 43(2):125–130.

Hiremath, B. N., and H. Misra, 2009: *Management of livelihood centric rural development projects: A systemic view*. Second National Conference on Agro-Informatics and Precision Farming, University of Agricultural Sciences, Raichur, organized by INSAIT, Dharwad, 2–3 December.

Hook, W., and L. Nadal, 2011: *Motorized Two and Three Wheeler Design and Regulation*. PowerPoint presentation. Institute for Transportation & Development Policy. www.itdp.org/index.php/library/publications (accessed 27 November 2011).

Huba, E-M., and E. Paul, 2007: *National Domestic Biogas Programme Rwanda: Baseline Study Report*. Netherlands Development Organisation, German Technical Cooperation, and Ministry of Infrastructure.

IAEA, 2007: *Top Stories & Features: Women & Water*. International Atomic Energy Agency (IAEA). www.iaea.org/newscenter/news/2007/womenday2007.html (accessed 6 October 2011).

IEA, 2006: *World Energy Outlook 2006*. International Energy Agency (IEA), Paris.

IEA, 2008: *World Energy Outlook: Energy and Development*. International Energy Agency (IEA), Paris.

IEA, 2009: *The Electricity Access Database*. International Energy Agency (IEA), Paris. www.iea.org/weo/database_Electricity/electricity_access_database.htm (accessed 18 April 2011).

IEA, 2010a: *Energy Poverty – How to Make Modern Energy Access Universal? World Energy Outlook 2010*. International Energy Agency (IEA), Paris.

IEA, 2010b: *CO_2 Emissions from Fuel Combustion – Highlights*. International Energy Agency (IEA), Paris.

IEA, 2010c: *The Electricity Access Database*. International Energy Agency (IEA), Paris. www.iea.org/weo/database_Electricity10/electricity_database_web_2010.htm (accessed 27 November 2010).

IEA, 2011a: *Energy and Development Methodology*. International Energy Agency (IEA), Paris.

IEA, 2011b: *Energy Balances of Non-OECD Countries 2011*. International Energy Agency (IEA), Paris.

IMF, 2008: *Africa's power supply crisis: Unraveling the paradoxes*. International Monetary Fund (IMF) Survey Magazine, 22 May.

IMW, 2008: Water Woes: Women Fetch the World's Water. Online Exhibition. International Museum of Women (IMW). www.imow.org/wpp/stories/viewStory?storyId=1308 (accessed 16 May 2011).

Kammen, D. M., R. Bailis, and A. V. Herzog, 2001: *Clean Energy for Development and Economic Growth: Biomass and Other Renewable Energy Options to Meet Energy and Development Needs in Poor Nations*. University of California, Berkeley, CA.

Kanagawa, M. and T. Nakata, 2005: *Analysis of the Energy Access Improvement and Its Socio-Economic Impacts in Rural Areas of Developing Countries*. Tohoku University, Sendai.

Karekezi, S. and T. Ranja, 1997: *Renewable energy technologies in Africa*. Zed Books, London.

Karekezi, S., G.B., Khamarunga, and W. Kithyoma, and X. Ochieng, 2002: Improving Energy Services for the Poor in Africa – A Gender Perspectiv. *ENERGIA News* Vol. 5 Nr 4, Energia.

Karekezi, S., and A. Sihag, 2004: *Synthesis/Compilation Report. Energy Access Working Group*. Global Network on Energy for Sustainable Development, Risoe National Laboratory, Roskilde, Denmark.

Karekezi, S., J. Kimani, A. Wambille, P. Balla, F. Magessa, W. Kithyoma, and X. Ochieng, 2005: *The Potential Contribution of Non Electrical Renewable Energy Technologies (RETS) to Poverty Reduction in East Africa*. Energy, Environment and Development Network for Africa, Nairobi.

Karekezi, S., W. Kithyoma, K. Muzee, and J. Wangeci, 2008a: *The Potential for Small and Medium Scale Renewables in Poverty Reduction in Africa*. Energy, Environment and Development Network for Africa, Nairobi.

Karekezi, S., J. Kimani, and O. Onguru, 2008b: *Urban and Peri -Urban Energy Access Working Group – Thematic Study*. Energy, Environment and Development Network for Africa, Nairobi.

Karekezi, S., W. Kithyoma, K. Muzee, and A. Oruta. 2009: *Renewable Energy for Africa: Potential Markets and Strategies*. Energy, Environment and Development Network for Africa, Nairobi.

Khandker, S. R., D. F. Barnes, and H. A. Samad, 2009: *Welfare Impacts of Rural Electrification: A Case Study from Bangladesh*. Policy Research Working Paper No. 4859. World Bank, Washington, DC.

KickStart, 2011: *Our Impact*. www.kickstart.org/what-we-do/impact (accessed 9 June 2011).

Kumarage, A. S., 2007: Impacts of transportation infrastructure and services on urban poverty and land development in Colombo, Sri Lanka. *Global Urban Development Magazine*, 3(1).

Mapako, M., 2010: *Energy, the Millennium Development Goals and the Key Emerging Issues*. Department of Environmental Affairs, South Africa.

Meisen, P., and I. Akin, 2008: *The Case for Meeting the Millennium Development Goals through Access to Clean Electricity*. Global Energy Network Institute, San Diego, CA.

Modi, V., S. McDade, D. Lallement, and J. Saghir, 2006: *Energy and the Millennium Development Goals*. The International Bank for Reconstruction and Development/ The World Bank, Washington, DC, and the United Nations Development Programme, New York, NY.

Morris, E., and G. Kirubi, 2009: *Bringing Small-Scale Finance to the Poor for Modern Energy Services: What is the Role of Government?* United Nations Development Programme, New York, NY.

Nussbaumer, P., M. Bazilian, V. Modi and K. Yumkella, 2011: *Measuring Energy Poverty: Focusing on What Matter*. Oxford Poverty and Human Development Initiative, Oxford.

Njenga, B. K., 2001: *Upesi rural stoves project. In Generating Opportunities: Case Studies on Energy and Women*. United Nations Development Programme, New York, NY.

Population Action International, 2001: *A World of Difference: Sexual and Reproductive Health and Risks*. Washington, DC.

Quansah, S. K., P. Russell, and N-O. Mainoo, 2003: *Ghana LPG Rural Energy Challenge*. Workshop Report. Kumasi Institute of Technology and Environment, Kumasi.

Rosenthal, G., 2010: The financial and economic crisis of 2008 and its repercussions on economic thought. *CEPAL Review* 100:29–39.

Schlag, N. and F. Zuzarte. 2008: *Market Barriers to Clean Cooking Fuels in Sub-Saharan Africa: A Review of Literature. Working Paper*. Stockholm Environment Institute, Stockholm.

Shankar, K. K., 2007: Govt extends kerosene, LPG subsidy till 2010. *Indian Express Limited,* 12 October.

SIL International, 2011: *Women and Literacy*. www.sil.org/literacy/wom_lit.htm (accessed 4 April 2011).

SNV, 2007: *SNV and FMO Provide Bio-Digesters and Micro Loans to Farmers in Cambodia*. Netherlands Development Organisation (SNV). www.snvworld.org/en/aboutus/news/Pages/Bio%20digesters%20for%20Cambodian%20farmers.aspx (accessed 20 November 2011).

SODIS, 2009: *Microbiology*. Solar Water Disinfection (SODIS). www.sodis.ch/methode/forschung/mikrobio/index_en (accessed 17 March 2011).

SODIS, 2010a: *SODIS – Safe Drinking Water in 6 Hours*. www.sodis.ch/index_en (accessed 17 March 2011).

SODIS, 2010b: *Health*. www.sodis.ch/methode/forschung/gesundheit/index_en (accessed 17 March 2011).

Sumner, A., 2010: *Global Poverty and the New Bottom Billion: What if Three-quarters of the World's Poor Live in Middle-income Countries?* IDS Working Paper 349. Institute of Development Studies, University of Sussex, Sussex.

Sumner, A., 2011: *The New Bottom Billion: What If Most of the World's Poor Live in Middle-Income Countries?* Institute of Development studies and Centre for Global Development

Theuambounmy, H. 2007: *Status of Renewable Energy Development in the Lao People's Democratic Republic*. Paper presented at the Greening the Business and Making Environment and Business Opportunity. 5–7 June 2007, Bangkok.

TWP, 2010: *Water Scarcity and Its Effects*. The Water Project (TWP), thewaterproject.org/water_stats.asp (accessed 10 October 2011).

UN, 2010a: *The Millennium Development Goals Report*. United Nations (UN), New York, NY.

UN, 2010b: *Energy for a sustainable future: summary report and recommendations of the Secretary-General's Advisory Group on Energy and Climate Change (AGECC)*. United Nations (UN), New York, NY.

UN, 2011a: *Water for life Meeting Global Targets for Water and Sanitation*. United Nations (UN), New York, NY.

UNDP, 2004: *Reducing Rural Poverty Through Increased Access to Energy Services: A Review of Multifunctional Platform Project in Mali*. United Nations Development Programme (UNDP), New York, NY.

UNDP, 2005: *Energizing the MDGs: A Guide to Energy's Role in Reducing Poverty*. United Nations Development Programme (UNDP), New York, NY.

UNDP, 2006: *Energizing Poverty Reduction. A Review of Energy-Poverty Nexus in Poverty Reduction Strategy Paper*. United Nations Development Programme (UNDP), New York, NY.

UNDP, 2007a: *A Review of Energy in National MDG Reports by Minoru Takada and Silvia Fracchia*. United Nations Development Programme (UNDP), New York, NY.

UNDP, 2007b: *Delivering Energy Services for Poverty Reduction: Success Stories from Asia and the Pacific*. UNDP Regional Centre, Bangkok.

UNDP, 2009: *UNDP Human Development Indicators Report 2009*. United Nations Development Programme (UNDP), New York, NY.

UNDP, 2010: *UNDP Human Development Indicators Report 2010*. United Nations Development Programme (UNDP), New York, NY.

UNDP, 2011: *Human Development Report 2011. Sustainability and Equity: A better future for all*. United Nations Development Programme (UNDP), New York, NY. http://hdr.undp.org/en/media/HDR_2011_en_Complete.pdf

UNDP and WHO, 2009: *The Energy Access Situation in Developing Countries, A Review Focusing on the Least Developed Countries and Sub-Saharan Africa*. United Nations Development Programme (UNDP) and World Health Organization (WHO), New York, NY.

UNECA, 2010: *Economic Report on Africa Promoting High-Level Sustainable Growth to Reduce Unemployment in Africa*. United Nations Economic Commission for Africa (UNECA), Addis Ababa.

UNEP, undated: *Empowering Rural Communities by Planting Energy: Roundtable on Bioenergy Enterprise in Developing Regions*. United Nations Environment Programme (UNEP), Nairobi.

UNIDO, 2009: *Industrial Development Report 2009*. United Nations Industrial Development Organization (UNIDO), Vienna.

UNFPA, 2005: *State of World Population 2005: Journalists Press Kit*. United Nations Population Fund (UNFPA), New York, NY. www.unfpa.org/swp/2005/presskit/factsheets/facts_rh.htm (accessed 18 June 2011).

UNFPA, 2008: *Linking Population, Poverty and Development – Reducing Poverty and Achieving Sustainable Development*. United Nations Population Fund (UNFPA), New York, NY. www.unfpa.org/pds/poverty.html (accessed 18 June 2011).

UNFPA, 2009: *A Review of Maternal Health in Eastern Europe and Central Asia*. United Nations Population Fund (UNFPA), New York, NY.

UN Energy, 2005: *The Energy Challenge for Achieving the Millennium Development Goals*. United Nations, New York, NY.

UN Women, 2011: *Facts and Figures on Gender and Climate Change*. www.unifem.org/partnerships/climate_change/facts_figures.php (accessed 25 May 2011).

Vivien, F. and C. Briceno-Garmendia, (eds.), 2010: *Africa's Infrastructure – A Time for Transformation*. The International Bank for Reconstruction and Development and The World Bank, Washington, DC.

Ward, S., 2002: *The Energy Book for Urban Development in South Africa*. Sustainable Energy Africa, Noordhoek.

WaterAid, 2009: *Women's issues*. Issue Sheet, November 2009.

Wegelin, M. and B. Sommer, 1997: SODIS at the turning point – A technology ready for use. *SANDEC News* No. 3, EAWAG/SANDEC, October 1997, Ueberlandstrasse.

WFP, 2010: *WFP Activities: Tanzania*. World Food Programme (WFP), Rome. www.wfp.org/countries/Tanzania – United-Republic-Of/Operations (accessed 15 July 2011).

WHO, 2006: *Fuel for Life, Household Energy and Health*. World Health Organization (WHO), Geneva.

WHO, 2008: *Maternal Health – Fact Sheet*. World Health Organization (WHO), Geneva.

WHO, 2011a: Indoor Air Pollution: Multiple Links between Household Energy and the Millennium Development Goals. www.who.int/indoorair/mdg/energymdg/en (accessed on 18 May, 2011).

WHO, 2011b: *Global Tuberculosis Control: WHO Report 2011*. World Health Organization (WHO), Geneva.

World Bank, 2007: *World Development Indicators*. Washington, DC. data.worldbank.org/products/data-books/WDI-2007 (accessed 7 March 2011).

World Bank, 2008a: *Climate Change and the World Bank Group, Phase 1: An Evaluation of World Bank Win-Win Energy Policy Reforms*. Washington, DC, pp.114–119.

World Bank, 2008b: *Designing Sustainable Off-Grid Rural Electrification Projects: Principles*

World Bank, 2010a: *The Millennium Development Goals and the Road to 2015. Building on Progress and Responding to Crisis*. Washington, DC.

World Bank, 2010b: *World Development Report 2010*. Washington, DC.

World Bank, 2011a: *World Development Indicators*. Washington, DC.

World Bank & IBRD, 2008: *Africa development Indicators 2007*. The World Bank and International Bank for Reconstruction and Development (IBRD), Washington, DC.3   A ropeway is a form of lifting device used to transport light goods across rivers or ravines.

3

Energy and Environment

Convening Lead Authors (CLA)
Lisa Emberson (Stockholm Environment Institute, University of York, UK)
Kebin He (Tsinghua University, China)
Johan Rockström (Stockholm Resilience Centre, Stockholm University, Sweden)

Lead Authors (LA)
Markus Amann (International Institute for Applied Systems Analysis, Austria)
Jennie Barron (Stockholm Environment Institute, University of York, UK)
Robert Correll (Global Environment Technology Foundation, USA)
Sara Feresu (Institute of Environmental Studies, University of Zimbabwe)
Richard Haeuber (United States Environmental Protection Agency)
Kevin Hicks (Stockholm Environment Institute, University of York, UK)
Francis X. Johnson (Stockholm Environment Institute, Stockholm University, Sweden)
Anders Karlqvist (Swedish Polar Research Secretariat)
Zbigniew Klimont (International Institute for Applied Systems Analysis, Austria)
Iyngararasan Mylvakanam (United Nations Environment Programme)
Wei Wei Song (Tsinghua University, China)
Harry Vallack (Stockholm Environment Institute, University of York, UK)
Qiang Zhang (Tsinghua University, China)

Review Editor
Jill Jäger (Sustainable Europe Research Institute, Austria)

Contents

Executive Summary

Modern energy systems have been central to the development of human societies. They have perhaps been the single most important determinant of growth of our industrial societies and our modern economy. Unfortunately, they have also been a key driver of many of the negative environmental trends observed in the world today. For example, current energy systems are the predominant source of carbon dioxide (CO_2) emissions, accounting for 84% of total global CO_2 emissions and 64% of global greenhouse gas (GHG) emissions related to human activities. Past trends suggest that this percentage is likely to increase in the future if our energy needs continue to be met by fossil fuels.

The impact of GHG emissions on climate is arguably the most significant environmental impact associated with our energy systems, as the effects of such emissions are felt globally. However, these effects will not necessarily be equitable. Due to the realities of global and national economics, the areas that may suffer the greatest impacts from climate change may be those that have to date contributed the least in terms of GHG emissions. Our fossil fuel-based energy systems also emit substantial quantities of other atmospheric pollutants, for example sulphur dioxide (SO_2), nitrogen oxides (NO_x), primary particulate matter (PM), and non-methane volatile organic compounds (NMVOCs), which degrade air quality and cause damage to health and ecosystems through processes such as acidification, eutrophication, and the formation of ground-level ozone (O_3) and secondary PM. Biomass-based energy systems can also have substantial impacts on land and water resources.

Nevertheless, climate change and the reduction of CO_2 emissions are issues that need to be addressed immediately if we are to prevent irreversible environmental change on a planetary scale. This requires an international effort to develop alternative pathways that will enable our global energy systems to keep us within safe limits of environmental change whilst still providing the energy required for our human development. There has been much discussion of what such 'safe limits' might be for climate change, with scientists urging for constraining the temperature increase to 2°C, or even 1.5°C. The Global Energy Assessment (GEA) scenarios have been developed with a view towards achieving the former target by stabilizing concentrations of GHGs to less than 450 parts per million (ppm) CO_2 (equivalent). This target is considered to provide a reasonable chance of avoiding average global mean surface temperature (GMT) increases of above 2°C. However, it is important to realize that even if these alternative pathways are achieved in the future, there is still a significant risk of substantial environmental change, since it is not certain how the climate will respond to changes in GHG emissions. It is also important to note that although the focus of the GEA scenarios is on long-lived GHGs, short-lived climate forcers are always emitted along with CO_2 from combustion sources in energy systems, and have both warming (e.g., black carbon (BC)) as well as cooling properties (e.g., sulphur, particulate organic carbon (OC)). Therefore, energy policies that seek to mitigate climate change should also consider these other radiative forcers. Doing so could have co-benefits, since many of these short-lived climate forcers are air pollutants, impacting human health as well as the environment. Ideally, mitigation policy would target emission reductions that improve the situation for both climate and air quality, though such efforts may well be confounded by gaps in our scientific knowledge and by socioeconomic conditions and political priorities that vary by global region.

In view of such considerations, this chapter discusses the role of our energy systems in relation both to climate change and air quality. The latter is achieved by assessing current knowledge about air pollutants whose major sources are energy-related activities. The scale of the stresses placed on ecosystems from climate change and air quality vary by global region. This is in part due to the fact that regions are situated at different points along the energy-transition pathway, with transitional progress determining both their energy mix and associated emissions. For example, biomass energy may release carbon monoxide (CO), NMVOCs, and primary PM including BC and OC, but the net CO_2 emissions are typically small. Hence, economic regions that tend to rely on these more traditional types of energy (e.g., many African countries) will tend to contribute less to climate change and may suffer from more localized effects of poor indoor and outdoor air quality. In contrast, rapidly industrializing countries, such as those in Asia, rely heavily on fossil fuels (in particular coal), which leads to emissions of SO_2, NO_x, PM, and CO_2. As a result, such economies will be contributing to climate change as well as suffering the effects of eutrophication, acidification, elevated concentration of fine PM and ground-level O_3. Finally, those economies that have industrialized and are now putting efforts into renewable energy (including some economies in Europe) are seeing improvements in air quality, but still have a long way to go to achieve the reductions in CO_2 emissions that would be needed to achieve the 2°C limit to global warming.

Of course, changes in atmospheric composition are not the only way in which energy systems are causing environmental degradation. Energy systems, especially those reliant on biomass and hydropower, are also having substantial impacts on land and water resources. The potential threat posed by the trend towards increased biomass energy, in particular through the displacement of food crops and the added pressure this places on already scarce water resources, also needs to be considered when identifying future pathways to sustainable energy.

Finally, it is not enough to try to identify sustainability criteria that focus only on particular and individual environmental threats caused by energy systems. Rather, it is necessary to develop indicators that treat all threats in a holistic manner, recognizing the connections that exist between sources, processes, and impacts that result in environmental degradation. Only by understanding such connections will it be possible to identify pathways of change that will truly lead us towards energy systems that are able to meet demand while keeping within the safe limits of change of the Earth's biophysical processes. As such, an important conclusion of Chapter 3 is that a move towards redefining global sustainability criteria for energy following the 'planetary boundaries' approach may be advisable. Such a move could provide the holistic framework necessary to ensure that our global energy systems develop to achieve sustainability on a planetary scale.

3.1   Introduction

Growing evidence indicates that humanity has entered a global phase of development, in which anthropogenic pressures on Earth-system processes are at risk of reaching the limits of hardwired biophysical processes on a planetary scale (Crutzen, 2002; Steffen et al., 2007). The human pressures on the climate system, the oceans, the stratosphere, and the biosphere have now reached a point where the prospect of large-scale deleterious impacts on human development cannot be excluded (Lenton et al., 2008; Schaeffer et al., 2008; Gordon et al., 2008).

The Earth may be entering a new geological era, the Anthropocene, in which humanity constitutes the main driver of planetary change (Steffen et al., 2007; Zalasiewicz et al., 2010). This change may be threatening the environmental stability of the current geological era, the Holocene. The perception of these risks has been amplified by newly acquired knowledge that Earth systems may cross 'tipping points,' resulting in nonlinear, abrupt, and potentially irreversible change, such as the destabilization of the Greenland ice sheet or the tropical rainforest systems (Schellnhuber, 2009).

Energy systems play a critical role in determining our ability to achieve global sustainability in the short- and long-term, as they depend on natural resources and are among the most significant drivers of environmental impacts on the Earth's physical and living systems (Steffen et al., 2007). Easy access to energy was a prerequisite for the rapid improvements in human welfare and the rapid growth of the world's population, which began at the start of the industrial revolution. Empirical evidence shows that in the mid-1950s, humanity embarked on a 'great acceleration' (Steffen et al., 2007), characterized by an exponential exploitation of natural resources and ecosystems (Figure 3.1) and closely associated with a similar exponential growth in global population, economic growth, and energy use (see Chapter 1).

Increased energy use is also accompanied by heavy environmental costs. The true magnitude of these costs is only now being fully realized. Perhaps the most threatening of these environmental impacts is the marked increases in atmospheric concentrations of trace gases, in particular CO_2 (Keeling and Whorf, 2005), which have occurred in parallel with this rapid human development (Steffen et al., 2007). Observations of increases in global surface temperature (GMT) (IPCC, 2007) provide evidence that these increases in GHGs are causing climate change and that, when data are compared from the same source regions, this change has been rising as quickly in the past decade as in the previous two decades, and has been steadily increasing over the last century (Figure 3.2).

Other impacts associated with energy systems, historically viewed as acting at the local or regional level, are now also threatening at the global level. Increases in energy use per capita and exponentially increasing populations have led to an agglomeration of local impacts, which pose global threats (e.g., deforestation from the harvesting of wood fuel and limited availability of fresh water due to water extraction and pollution).

Additionally, environmental impacts resulting from extreme events such as nuclear accidents and oil spills can also have global implications, both in terms of environmental consequences and in terms of changes in public perception of 'safe' energy that would make up our future energy systems.

This chapter addresses the environmental impacts of energy systems, focusing on those that place pressure on the atmosphere, the terrestrial biosphere and the hydrosphere while health related impacts are discussed in Chapter 4. The chapter considers how these impacts, in particular those associated with atmospheric emissions related to energy systems, vary across the globe in Europe, the Americas, Africa, Asia and polar regions. The chapter then considers all these impacts in relation to current environmental sustainability criteria used in energy systems, giving thought to how these criteria could be modified to capture the full range of environmental impacts. Such a holistic approach is appropriate for efforts that seek to ensure planetary-scale sustainability.

3.2   The Atmosphere and Energy Systems

Energy-related activities are responsible for a major share of anthropogenic emissions of GHGs, other radiative forcing substances, and air pollutants into the atmosphere. For example, energy-related GHG emissions, mainly from fossil fuel combustion for heat supply, electricity generation, and transport, account for approximately 64% of total emissions, including carbon dioxide (CO_2), methane (CH_4) and some traces of nitrous oxide (N_2O) (IPCC, 2007a). These atmospheric pollutants are primarily associated with impacts caused either by their radiative forcing (RF) properties leading to climate change, by their deposition to sensitive ecosystems causing damage through processes such as acidification and eutrophication, or by high PM concentrations leading to health impacts. In some cases, the atmospheric pollutants may have a role in all these types of impacts. For example, ground-level ozone (O_3), which is a secondary pollutant formed from a series of chemical reactions involving nitrogen oxides, (NO_x), volatile organic compounds (VOC) and carbon monoxide (CO), may cause direct damage to human health, and vegetation and is a powerful GHG.

First, we begin by describing the most important energy-related atmospheric pollutants, their major emission sources, and how emissions have varied over recent decades (see Section 3.2.1). Emission trends are derived from the Emissions Database for Global Atmospheric Research (EDGAR) which provides data on global annual emissions for 1990, 1995, and 2000 for direct GHG, precursor gases, and acidifying gases (van Aardenne et al., 2001; Olivier et al., 2005).[1] We then describe how these emissions lead to impacts on climate, both in the long-term and short-term (Section 3.2.2) and on air quality leading to impacts on terrestrial and aquatic ecosystems resulting from acidic deposition and eutrophication

---

1   The GEA scenarios are based on more recent emission inventories from the Reference Concentration Pathways developed for the IPCC Fifth Assessment Report, to be published in 2013 and 2014. See www.iiasa.ac.at/web-apps/tnt/RcpDb/.

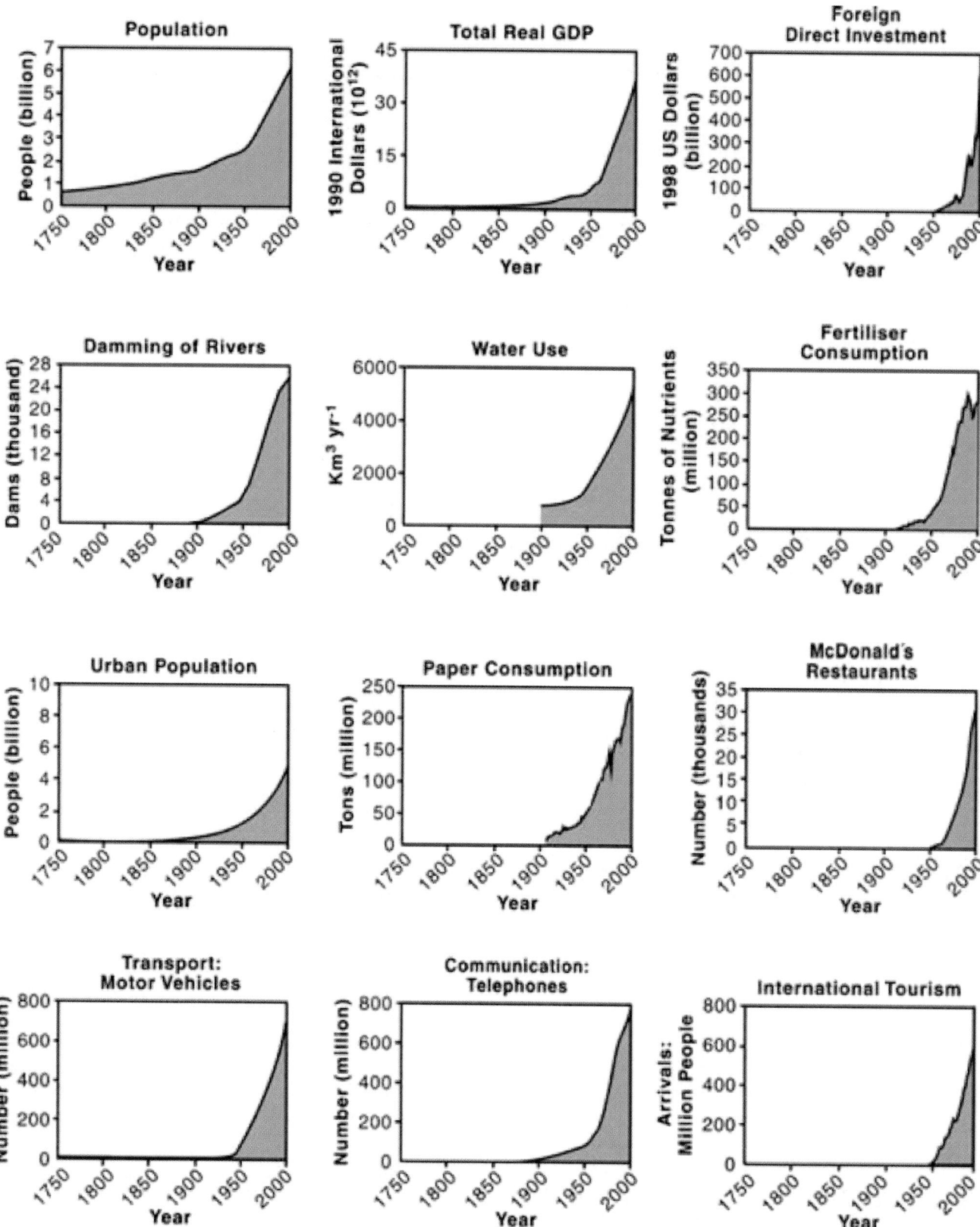

Figure 3.1 | Observed global trends of key environmental processes providing the evidence of the 'great acceleration' in human enterprise in the mid-1950s. Source: Steffen et al., 2007.

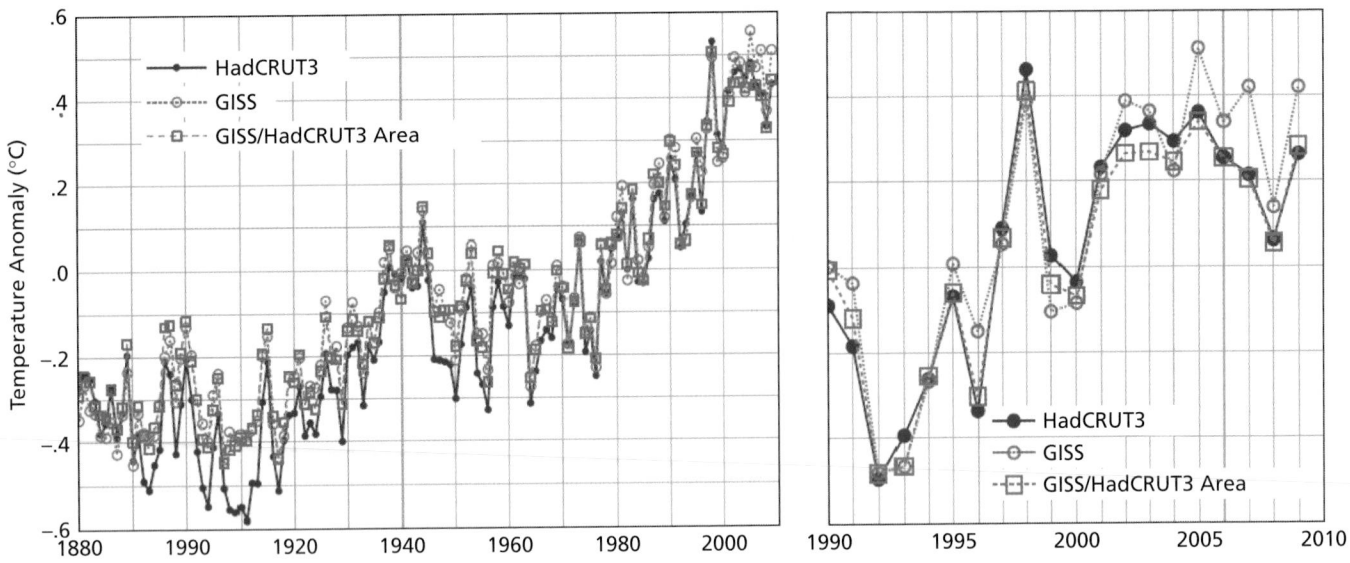

Figure 3.2 | Global surface temperature anomalies (°C) relative to 1961–1990 base period for three cases: HadCRUT, GISS, and GISS anomalies limited to the HadCRUT area. Source: Hansen et al., 2010.

(e.g., overfertilization with nitrogen), and through damage to ecosystems resulting from ground-level O_3 (Section 3.2.3); the health impacts from PM are discussed in Chapter 4. The regional variation in emission sources, climate change, and air quality impacts are discussed in Section 3.2.4.

3.2.1 Major Energy-related Sources of Atmospheric Pollution

3.2.1.1 Carbon Dioxide (CO₂)

Carbon dioxide released from anthropogenic sources has been identified as the key contributor to observed climate change (IPCC, 2007a). These sources include fossil fuel combustion, cement production, and emission flux from land-use changes. The largest sources of CO_2 emissions globally are related to energy, mainly from the combustion of fossil fuels such as coal, oil, and gas in power plants, vehicles, industrial facilities, and residential homes. In 2005, these energy-related activities accounted for 84% of total global anthropogenic CO_2 emissions (IEA, 2009).

Carbon dioxide emissions have increased continuously during the last century. The EDGAR database shows that global CO_2 emissions increased by 16% from 1990 to 2000 (Olivier et al., 2005), and a more recent analysis showed that global CO_2 emissions from fossil fuel combustion in 2008 have increased by 41% since 1990 (Le Quere et al., 2009). Although emissions increased in both developed and developing countries, the scale of the increase has varied between different global regions. For example, emissions of CO_2 in the United States increased by 16%, from 18.7 gigatonnes (Gt) of CO_2 in 1990 to 21.3 $GtCO_2$ in 2008 (US EPA, 2010a), while Chinese emissions increased by approximately 160%, from 9.2 $GtCO_2$ in 1990 to 23.8 $GtCO_2$ in 2007. However, when CO_2 emissions are expressed per capita, a different picture emerges. In 2007, the United

States was the biggest per capita emitter at 18.7 tonnes, compared to China, which emitted only 4.6 tonnes per capita (IEA, 2009).

3.2.1.2 Methane (CH₄)

Methane is a potent GHG with a global warming potential (GWP) 25 times greater than that of CO_2 over a 100-year time horizon (IPCC 2007a). Importantly, CH_4 has an atmospheric lifetime of just over 10 years, which is fundamentally different from CO_2 and N_2O. Decreases in CH_4 emissions will therefore lead to a relatively rapid reduction in atmospheric concentrations of this GHG. Given the high RF of this pollutant, decreasing CH_4 emissions may be an attractive mitigation option.

Although CH_4 is relatively unreactive compared with other VOCs, it has a highly significant role in determining background O_3 concentrations on large geographic scales. Thus, higher emissions of CH_4 lead to elevated ground-level O_3 concentrations. The anthropogenic sources of CH_4 include rice agriculture, livestock, landfill, biomass burning, and fossil fuel combustion. Natural CH_4 is emitted from sources such as wetlands, oceans, forests, wildfires, termites, and geological sources. Total global pre-industrial emissions of CH_4 were dominated by natural sources (approximately 90%), with anthropogenic sources accounting for the rest (IPCC, 2007a). In contrast, anthropogenic emissions dominate present-day CH_4 budgets, accounting for more than 60% of the total global budget. Estimates of the share of anthropogenic CH_4 emissions due to fossil fuel extraction and combustion range from 20–30% (IPCC, 2007a).

3.2.1.3 Suphfur Dioxide (SO₂)

Sulphur dioxide and its atmospheric products (e.g., sulphate aerosols, sulphuric acid) cause a number of environmental problems. They can act as atmospheric radiative forcers (with implications for climate change), as air pollutants causing acidic deposition and acidification of terrestrial and aquatic ecosystems, and as air pollutants adding to the atmospheric PM load (with impacts on human health). These pollutants also act at a range of spatial levels, from local to global, depending on atmospheric circulation patterns. Sulphur dioxide is emitted into the atmosphere from both anthropogenic sources (e.g., fossil fuel combustion, industrial process) and natural sources (e.g., volcanic eruptions). It is estimated that anthropogenic sources account for more than 70% of SO_2 global emissions, 85% of which are from fossil fuel combustion (Olivier et al., 2005). The largest anthropogenic sources of SO_2 emissions are related to fossil fuel combustion at power plants and other industrial facilities. Smaller anthropogenic sources of SO_2 emissions include industrial processes such as extracting metal from ore, and the burning of high sulphur-containing fuels by locomotives, large ships, and non-road equipment.

The EDGAR database shows that global anthropogenic SO_2 emissions in 2000 were 150 megatonnes (Mt) of SO_2, coming mainly from power generation (54 $MtSO_2$, 35.6% of total), industrial fossil fuel combustion (24 $MtSO_2$, 16.2%), nonferrous metals melting (21 $MtSO_2$, 14.2%), other fossil fuel (10 $MtSO_2$, 6.8%), residential fossil fuel (8 $MtSO_2$, 5.4%), and international shipping (7 $MtSO_2$, 4.8%), according to Olivier et al. (2005).

Global SO_2 emissions have risen dramatically over the last century, approximately in parallel with increased fossil fuel use. Emission sources have changed considerably with time, both geographically and by sector. For example, emissions in most industrialized countries have fallen over the past two decades, due to the implementation of sulphur controls and a shift to lower sulphur fuels (Smith et al., 2011). Emissions in the United States have decreased from a high of 28 $MtSO_2$ in 1970 to 10 $MtSO_2$ in 2008 (US EPA, 2009). European anthropogenic SO_2 emissions have also decreased over the last two decades, from 55 $MtSO_2$ in 1980 to 15 $MtSO_2$ in 2004 (Vestreng et al., 2007). Conversely, anthropogenic SO_2 emissions in China are of increasing concern; they contributed about one quarter of the global total and more than 90% of East Asian emissions during the 1990s. A recent study showed that from 2000 to 2006, total SO_2 emissions in China increased by 53%, from 21.7 $MtSO_2$ to 33.2 $MtSO_2$, at an annual growth rate of 7.3% (Lu et al., 2010).

3.2.1.4 Nitrogen Oxides (NO_x)

Nitrogen oxides ($NO_x = NO + NO_2$) play a key role in tropospheric chemistry. Nitrogen oxides can either be deposited directly to ecosystems through dry deposition or through wet deposition, caused when nitrates (NO_3^-) form in cloud and rain. Both processes can cause eutrophication of ecosystems. Nitrogen oxides also play a role in the production of O_3 in the troposphere, where the abundance of O_3 is controlled by atmospheric concentrations of NO_x and VOCs. Nitrogen oxides can also contribute to fine particle pollution through the formation of nitrate aerosols. Nitrogen oxides are therefore linked to climate change (through contributions to

RF atmospheric aerosols and tropospheric O_3 formation), to eutrophication (through wet and dry deposition), and ecosystem damage (again though tropospheric O_3 formation). Nitrogen oxides have also been linked to impacts on human health, as they can cause adverse effects on the respiratory system. They can also directly impact vegetation.

Like SO_2 emissions, global NO_x emissions have risen dramatically during the past century. It is estimated that global anthropogenic NO_x emissions increased five-fold between 1890 and 1990, from 6.9 MtN to 35.5 MtN (van Aardenne et al., 2001). After 1990, NO_x emissions from some industrialized regions began to decrease, mainly due to regulations in the transportation sector. Emissions in the United States decreased from 23 $MtNO_2$ in 1990 to 14.8 $MtNO_2$ in 2008, due to reductions in emissions from vehicles and power plants (US EPA, 2009). In Europe, road transport has been the dominant source of NOx emissions, accounting for 40% of the total emissions in 2005 (Vestreng et al., 2009). As a result of the combined control measures, the total NO_x emissions in Europe decreased by 32% between 1990 and 2005 (Vestreng et al., 2009). Nitrogen oxide emissions in China have continuously increased over the past two decades; the growth rate itself accelerated during this period. It is estimated that NOx emissions in China were 10.9 $MtNO_2$ in 1995 and 18.6 $MtNO_2$ in 2004, increasing by 70% during the period, at a 6.1% annual average growth rate (Zhang et al., 2007). More recent data estimates that nearly 21 $MtNO_2$ of NO_x have been emitted in China, compared to a value of 36.7 $MtNO_2$ of NO_x for the whole of Asia (Zhang et al., 2009).

3.2.1.5 Carbon Monoxide (CO)

Carbon monoxide is a significant air pollutant, capable of damaging human health as well as being an O_3 precursor. Carbon monoxide is emitted whenever fossil fuels and vegetation are incompletely combusted, whether in residential stoves, industrial boilers, vehicles, or through biomass burning. According to the EDGAR global inventory, in 2000 about 50% of CO emissions came from open biomass burning, 23% from biofuel combustion, and 22% from fossil fuel combustion, with the rest coming from industrial processes and waste treatment (Olivier et al., 2005).

Since the 1960s, great strides have been taken to reduce CO levels in the United States and Europe. However, in the developing world, few regulatory steps have been implemented. Emissions in the United States decreased from 185 MtCO in 1970 to 70 MtCO in 2008 (US EPA, 2009), while emissions in Asia increased from 207 MtCO in 1980 to 340 MtCO in 2003 (Ohara et al., 2007). A recent inventory study found that global CO emissions decreased by only 2–3% globally between 1988 and 1997, as increases in eastern Asia of 51% caused by rapid economic development were offset by declines in Europe and North America (Duncan et al., 2007). The largest decline, in Eastern Europe (45%), was largely caused by the economic contraction of the former Soviet Union (FSU). There were smaller declines in Western Europe (32%) and North America (17%), caused primarily by increasing levels of emissions control on vehicles (Duncan et al., 2007).

3.2.1.6   Non-methane Volatile Organic Compounds (NMVOCs)

Non-methane volatile organic compounds include a variety of chemicals that play an important role in atmospheric chemistry through tropospheric O_3 formation. There are many sources of NMVOCs in the atmosphere, including natural or 'biogenic' sources, such as trees and vegetation. Anthropogenic sources of NMVOCs include combustion of fossil and biofuels; biomass burning; the production, processing, and storage of liquid fuels; solvent use; and many other industrial processes. According to the EDGAR global inventory for 2000, about 42% of NMVOC emissions came from the combustion and processing of fossil fuels, 26% from open biomass burning, 16% from biofuel combustion, and 14% from solvent use and other industrial processes. Sectoral distribution might vary strongly between regions (e.g., Klimont et al., 2002; Wei Wei et al., 2008). Geographically, about 33% of the emissions came from Asia, 20% from Europe, 17% from Africa, 15% from Latin America, and 13% from North America (Olivier et al., 2005).

Global anthropogenic NMVOC emissions have risen dramatically during the past century. It is estimated that emissions increased from 24.8 MtNMVOC to 181 MtNMVOC between 1890 and 1990 (van Aardenne et al., 2001). After 1990, NMVOC emissions from some industrialized regions began to decrease. Emissions in the United States have decreased from 21.8 MtNMVOC in 1990 to 14.4 MtNMVOC in 2008 (US EPA, 2009), mainly due to emissions control on vehicles, as well as improved technologies related to solvent use. Meanwhile, emissions in Asia have increased from 21.9 MtNMVOC in 1980 to 45.5 MtNMVOC in 2003 (Ohara et al., 2007). As for NO_x, more recent estimates of 54 MtNMVOC for the whole of Asia for 2006 are provided by Zhang et al. (2009).

3.2.1.7   Black Carbon (BC) and Organic Carbon (OC)

Recent work has suggested that climate forcing by carbonaceous aerosols is probably a significant component of anthropogenic forcing. Forcing by BC from fossil fuel combustion ranges from about +0.1 to +0.3 W/m^2, hence having a warming influence on climate, and similar estimates for primary OC particles are −0.01 to −0.06 W/m^2 causing a cooling effect on climate. As global averages, these values, especially BC forcing, are significant relative to the average CO_2 forcing of about +1.5 W/m^2. Some studies suggest that regional aerosol forcings can be an order of magnitude greater than GHG forcings (Ramanathan and Carmichael, 2008). Carbonaceous aerosols are mainly produced during incomplete combustion of fossil fuels, as well as open biomass burning. It has been estimated that global annual emissions in 1996 were 8.0 Mt for BC and 33.9 Mt for OC (Bond et al., 2004). The contributions of fossil fuel, biofuel, and open biomass burning are estimated at 38%, 20%, and 42% respectively for BC, and 7%, 19%, and 74% respectively for OC (Bond et al., 2004). Emissions of BC and OC have increased

steadily over the past century. Global anthropogenic BC emissions increased from 2200 kilotonnes (kt) in 1900 to 4400 kt in 2000, and OC emissions increased from 5800 kt to 8700 kt during the same period (Bond et al., 2007).

Among 'contained' combustion sources (fossil fuel and biofuel), significant contributors to BC include the transportation, industry, and residential sectors, which account for 20%, 10%, and 25% respectively (Bond et al., 2004). Transportation is the most significant contributor to BC in developed regions, such as North America and Europe, with on-road and off-road diesels having approximately equal contributions. On the other hand, in developing regions like China, India, and Africa, the residential sector contributes the most to BC, though industry (e.g., coke making and brick kilns) and the transportation sector also makes significant contributions. For example, in China the residential and transport sector together comprise 63% and 82% of the total anthropogenic emissions respectively for BC and OC (Zhang et al., 2009). As a consequence of the poorer combustion in small devices, residential solid fuels (biofuel and coal) dominate 'contained' OC emissions in all regions except the Middle East and the Pacific. It is also estimated that residential solid fuels and transport contribute 20% and 4% to the global budget of OC respectively, if one considers all sources (e.g., open biomass burning (Bond et al., 2004)). The regional analyses for Asia presented in Zhang et al. (2009) and Klimont et al. (2009) highlight the dominant role of domestic combustion in OC emissions.

3.2.2   Climate Change and Energy Systems

The world's energy systems constitute an extremely important driver of climate change. This section reviews the current state of knowledge of climate change on consideration of energy system related emissions of GHGs and air pollutants.

3.2.2.1   Long-term Climate Change and Energy Systems

Key to understanding the physical mechanisms of climate change is the concept of energy balance and radiative forcing (RF) in the Earth's atmosphere (see Box 3.1). For the Earth's GMT to remain at an average of 15°C, the net incoming flux of solar radiation at the top of the atmosphere must equal the flux of long-wave radiation out to space. The chief physical mechanism by which the radiation imbalance arises as a consequence of human interference in the climate system is through increases in the atmospheric concentrations of long-lived GHGs. Long-lived GHGs, i.e., those gases that persist for periods of time ranging from decades to centuries, include CO_2, N_2O, and halocarbons. Other atmospheric trace gases are also crucial in determining the energy balance, including stratospheric O_3, which decreases RF (see Box 3.1); CH_4 and tropospheric O_3, which increase it; and aerosols, which on aggregate also decrease RF (IPCC, 2007a). The Intergovernmental Panel on Climate Change (IPCC, 2007a) (see Box 3.2) estimates that the combined net RF

Box 3.1 | Radiative Forcing, Climate Sensitivity, and Carbon Dioxide Equivalent

Radiative Forcing (RF)

Radiative forcing (RF) can be defined as the net change in the energy balance between the Earth and space (i.e., the difference in incoming solar radiation less outgoing terrestrial or long-wave radiation) at the tropopause. It is quantified as the rate of energy change per unit area of the globe as measured at the top of the atmosphere and is expressed in units of 'watts per square meter' (W/m^2). Radiative forcing is used to assess and compare the anthropogenic and natural drivers of climate change (IPCC, 2007) and can be linearly related to the global mean equilibrium surface temperature (GMT) change (ΔGMTs); $\Delta GMT = \lambda RF$, where λ is the climate sensitivity parameter (e.g., Ramaswamy et al., 2001).

Climate Sensitivity (λ)

Climate sensitivity (λ) is a measure of the responsiveness of equilibrated global mean surface temperature (GMT) to a change in the radiative forcing equivalent to a doubling of the atmospheric equivalent CO_2 concentration (CO_2-eq) (IPCC, 2007a). Climate sensitivity (λ) is hard to predict, since it needs to incorporate various couplings, feedbacks (particularly those related to clouds, sea ice, and water vapor), and interactions that occur within the climate system in response to any changes within the system.

Carbon Dioxide Equivalent (CO_2-eq)

Carbon dioxide equivalent (CO_2-eq) is a universal unit of measurement used to indicate the GWP of one unit of CO_2 over a 100-year time horizon. It is used to evaluate the releasing of different GHGs against a common basis. Thus, for methane (CH_4) the GWP is 25, and for nitrous oxide (N_2O) the GWP is 296 (IPCC, 2007a).

for all anthropogenic agents is +1.6 W/m^2 (with a 0.6–2.4 W/m^2 90% confidence range) indicating that, since 1750, it is extremely likely that humans have exerted a substantial warming influence on climate.

The contributions from CO_2 and CH_4 to this RF are 1.66 W/m^2 (with a range of $\pm$0.17 W/m^2) and 0.48 W/m^2 (with a range of $\pm$0.05 W/m^2) respectively. The energy sector is important in determining emissions of both these GHGs. Energy systems are the predominant source of CO_2 emissions, accounting for 84% of total global CO_2 emissions in 2005 and for 64% of global GHG emissions related to human activities (IEA, 2009).

Observations of climate change in response to these anthropogenic increases in radiative forcers are now clearly being recorded (IPCC, 2007a). Observations show GMT to have risen by 0.74°C $\pm$ 0.18°C when estimated by a linear trend over the past 100 years (1906–2005), with the rate of warming over the past 50 years almost double that of the past 100 years (IPCC, 2007a; see also Figure 3.2). This has lead to changes in Earth system climate. For example, long-term trends in precipitation amounts from 1900 to 2005 have been observed across many large regions, with precipitation significantly increasing in the eastern parts of North and South America, northern Europe, and northern and central Asia. In contrast, drying has been observed in the Sahel, the Mediterranean, southern Africa, and parts of southern Asia. Substantial increases in heavy precipitation events have been observed (IPCC, 2007a). In addition, during the 1961–2003 period the average rate of global mean sea-level rise was estimated to be 1.8 $\pm$ 0.5 mm/yr (IPCC, 2007a).

The contemporary climate has moved out of the envelope of Holocene variability, sharply increasing the risk of dangerous climate change. Observations of a climate transition include a rapid retreat of summer sea ice in the Arctic Ocean (Johannessen, 2008), the retreat of mountain glaciers around the world (IPCC, 2007a), the loss of mass from the Greenland and West Antarctic ice sheets (Cazenave, 2006), an increased rate of sea-level rise in the last 10–15 years (Church and White, 2006), a 4° latitude pole-ward shift of subtropical regions (Seidel and Randel, 2006), increased bleaching and mortality in coral reefs (Bellwood et al., 2004; Stone, 2007), a rise in the number of large floods (Milly et al., 2002; MEA, 2005b), and the activation of slow feedback processes like the weakening of the oceanic carbon sink (Le Quéré et al., 2007).

Box 3.2 | Scientific Assessments of Climate Change and the IPCC

The main source of scientific knowledge on climate change is contained in the assessment reports of the Intergovernmental Panel on Climate Change (IPCC). The IPCC was established in 1988 by two United Nations Organizations, the World Meteorological Organization (WMO) and the United Nations Environment Programme (UNEP), to assess "the scientific, technical and socioeconomic information relevant for the understanding of the risk of human-induced climate change." The First Assessment Report, or FAR (IPCC, 1990), informed the intergovernmental negotiations that led to the United Nations Framework Convention on Climate Change (UNFCCC). The Second Assessment Report, or SAR (IPCC, 1995a), informed the negotiations leading to the Kyoto Protocol in 1997. The Third Assessment Report, or TAR (IPCC, 2001), and the Fourth Assessment Report, or AR4 (IPCC, 2007), informed the process leading up to the 2009 United Nations Climate Change Conference in Copenhagen (COP-15), which was intended to create an extended or new regime in anticipation of the 2012 expiration of the first commitment period of the Kyoto Protocol. Each Assessment Report consists of three volumes, from Working Group I on the science of climate change, Working Group II on impacts and adaptation, and Working Group III on mitigation.

IPCC has also produced a series of Special Reports, including the Special Report on Emissions Scenarios (SRES) in 2000 (IPCC, 2000a), the Special Report on Methodological and Technological Issues in Technology Transfer, also in 2000 (IPCC, 2000b), and the Special Report on Carbon Dioxide Capture and Storage (SRCCS) in 2005 (IPCC, 2005).

Each of the IPCC Reports is peer-reviewed and assesses a vast number of scientific publications, and is the most authoritative assessment available. IPCC has recently been subject to criticism due to a few mistakes in the AR4. These mistakes have not affected the overall conclusions or their soundness.

How climate will change in the future under anthropogenic pressures will depend to a large extent on future GHG emissions, changes to the biosphere, and feedbacks in the Earth system. Even if emissions of all anthropogenic RF agents were to remain constant at today's levels, the Earth's climate system would continue to change. This is often referred to as 'committed warming,' and is largely due to the thermal inertia of the oceans and ice sheets and their long time-scales for adjustment. For example, the IPCC (2007a) estimates that committed climate change due to atmospheric composition in the year 2000 corresponds to a warming trend of about 0.1°C per decade over the next two decades, in the absence of large changes in volcanic or solar forcing. By 2050, about a quarter of the 1.3–1.7°C warming relative to 1980–1999 estimated using Special Report on Emission Scenerios, often referred to as the IPCC SRES 'marker Scenarios' (Nakicenovic et al., 2000; see also Box 3.2), would be due to committed climate change if all RF agents were to be stabilized at today's concentration levels.

It is extremely unlikely that RF agents will be held constant, as evidenced by the continued rise in GHG emissions. In order to assess the likely future trends in our climate, the IPCC (2007a) has assessed global climate change projected from six SRES scenarios of emissions of RF agents (see Figure 3.3). These scenarios represent a range of plausible future trajectories of population, economic growth, and technology change, in the absence of policies to specifically reduce emissions in order to address climate change. The assessment of climate change under these scenarios was made using a number of climate models of varying levels of complexity (and hence capable of incorporating different aspects of climate sensitivity; see Box 3.1), from simple climate

models to those that include ocean-atmosphere general circulation models and feedbacks between climate change and the carbon cycle (Betts et al., 2011). The IPCC (2007a) concluded that GMTs are likely to increase by between 1.1–6.4°C by the end of the 21st century relative to the 1980–1999 average. A key question is, What are these projected increases in GMT likely to mean for impacts associated with climate change? A related question is, How is this likely to affect our ability to stabilize GHG concentrations so as to "*prevent dangerous anthropogenic interference of the climate system*," as referred to in Article 2 of the United Nations Framework Convention on Climate Change (UNFCCC, 1992)? One of the core objectives of GEA is to answer these questions by assessing the implied constraints on future energy-related emissions of GHGs that would fulfill the stated objective of the UNFCCC (1992).

3.2.2.2 Impacts of Climate Change in the Future

How will climate change influence Earth systems, and what risks are involved? This has been continually assessed in the four IPCC Assessment Reports (IPCC, 1991a; 1995b; 2001a; 2007a) (see Box 3.2). Increases in GMT of the magnitude projected for 2100 as described in Figure 3.3 would be expected to have substantial global consequences both for near-term climate change and throughout the 21st century. Such consequences would include continued sea-level rise, changes in the cryosphere, decreases in snow cover, changes in global and regional patterns of temperature and precipitation, changes in extreme weather events such as heat-waves and drought, changes in the number and intensity of tropical cyclones, loss of genetic species and ecosystem

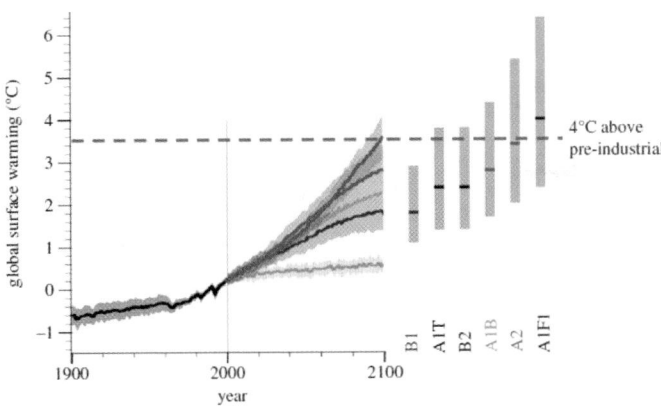

Figure 3.3 | Past changes in global mean surface temperature (GMT) (black curve), and projected future changes resulting from the IPCC SRES (Nakicenovic et al., 2000) 'marker scenarios' of GHG and aerosol emissions (colored curves and gray bars), relative to the 1980–1999 mean (Meehl et al., 2007). Climate changes under the A2, A1B, and B1 scenarios were projected with general circulation models (red, green, and blue lines, with plumes showing 5–95% range of model projections without uncertainties in climate-carbon cycle feedbacks). The full set of 'marker scenarios,' including a range of strengths of climate-carbon cycle feedbacks, were examined with simple climate models. Gray bars show the likely range of warming at 2090–2099 for each scenario, from expert assessments based on all available evidence from general circulation models, simple climate models, and observational constraints. The red dashed line marks warming of 3.5°C relative to 1980–1999, which represents 4°C relative to pre-industrial levels. Red line, A2; green line, A1B; blue line, B1; orange line, year 2000 constant concentrations; black line, 20th century. Source: IPCC, 2007a.

diversity, acidification of the oceans, and, perhaps most importantly, dangers of crossing tipping points that could lead to catastrophic ecological consequences (Schellnhuber et al., 2009). The uncertainties in assessments of these projections in climate change are many, and the IPCC process summarizes these uncertainties in its reports.

The IPCC (2001b) summarized an extensive analysis of the impacts of climate change. The IPCC identified 'reasons for concern' (RFCs), describing them in what has since become known as the 'burning embers' diagram. The diagram aimed to characterize the extent of the level of threat or risk associated with future projected anthropogenic climate change, defined as a change from 1990 levels of GMT.

The 'burning embers' diagram (IPCC, 2001b), as described in Smith et al. (2001), is shown in Figure 3.4, together with data updated by Smith et al. (2009). The comparison of the two diagrams suggests that the temperature range from which a consensus definition of dangerous anthropogenic interference might be drawn is getting lower, a result of advancements in our scientific insights regarding the functioning of the Earth system. There has been a growing consensus toward adopting a '2°C guardrail' approach, shown as a black dashed line in Figure 3.4 (Hare and Meinshausen, 2006). This approach has been adopted by the 'Copenhagen Accord' and the European Commission (United Nations Conference of the Parties, 2009). The 2°C barrier is based on recommendations by numerous scientific studies (Schneider and Mastrandrea, 2005; Fisher et al., 2007; Nakicenovic and Riahi, 2007; Hansen et al., 2008; Schellnhuber, 2008; Kriegler et al., 2009;

Meinshausen et al., 2009; Rockström et al., 2009), which suggest that global warming in excess of 2°C from pre-industrial times could trigger several climate tipping elements and lead to unmanageable changes (Smith et al., 2009). This target represents a clear guiding principle for acceptable limits of climate change. However, it has been recognized that establishing how the target will be met is rather complicated, since uncertainties associated with our knowledge of climate sensitivity, particularly the carbon cycle and climate response (see Box 3.1), complicate efforts to estimate the GHG emission reductions that would be necessary to remain below this warming target. Figure 3.4 indicates that even if this rather ambitious target is met, three out of the five 'reasons for concern' would still be at high risk of manifestation.

It is also worth emphasizing that significant risks of adverse climate impacts for society and the environment will have to be faced even if the 2°C line can be held (see also IPCC, 2007b; Richardson et al., 2009; WGBU, 2009). In view of this fact, it is possible that the 2°C barrier will be revised to lower values; efforts to make the target more stringent may be renewed as our understanding of regional consequences of climate change improves (Schneider and Mastrandrea, 2005; Hansen et al., 2008; Kriegler et al., 2009; Rockström et al., 2009).

3.2.2.3   Emission Scenario Requirements to Remain Below the 2°C 'Guardrail'

The UNFCCC Conference of the Parties (COP) in Copenhagen (COP 15) in December 2009 and the UNFCCC COP 16 in Cancun in December 2010 did not arrive at a legally binding agreement on how to proceed after the first commitment period of the Kyoto Protocol ends in 2012. However, the three-page "Copenhagen Accord," which was offered by a subgroup of Parties and taken note of by the COP, provides a consensus, however limited, on defining a 2°C GMT increase as a global 'guardrail' for human-induced climate change. This is the nearest expression of how to interpret dangerous climate change and of the level of mitigation desired. It is therefore taken as the normative goal for energy systems development and used to define the global sustainability criteria used in this GEA assessment (see Chapter 17).

The IPCC (Fischer et al., 2007) addressed the question of what the GHG emission reductions might need to be in order to provide a chance of stabilizing GMT below the 2°C 'guardrail.' The left-hand graph in Figure 3.5 shows the emission paths that are consistent with various stabilization levels, and the right-hand graph indicates that staying below a 2°C 'guardrail' with a 50% probability would require long-term GHG stabilization at around 440–450 ppm CO_2-eq. Figure 3.5 also indicates the uncertainties in climate-sensitivity estimates. For example, to increase the probability to around 90% would require stabilization below 400 ppm CO_2-eq, or essentially the maintenance of current concentrations throughout the century. Basically, global emissions need to decline almost immediately (within the next decade) to keep the goal of stabilizing at 2°C within reach. The higher the 'overshoot' of emissions,

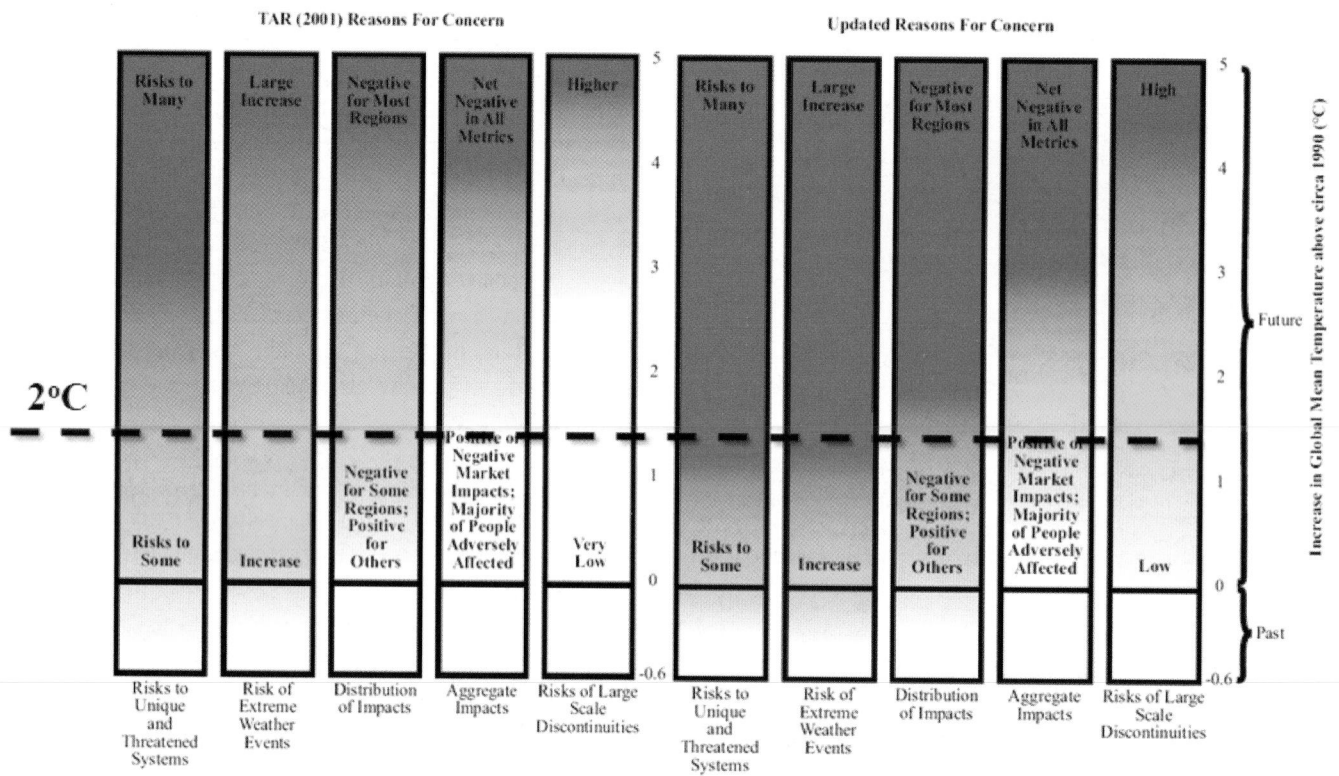

Figure 3.4 | Risks from climate change, by 'reason for concern' (Smith et al., 2001), compared with updated data (Smith et al., 2009). Climate change consequences are plotted against increases in GMT after 1990. The pre-industrial temperature level is also indicated. Each column corresponds to a specific reason for concern and represents additional outcomes associated with increasing GMT. The color scheme represents progressively increasing levels of risk. Both figures suggest that all stabilization levels, including the current atmospheric concentrations of GHGs, can be considered to be in principle dangerous, but it is important to note that the level of concern increases significantly with higher stabilization levels. Source: Smith et al., 2009.

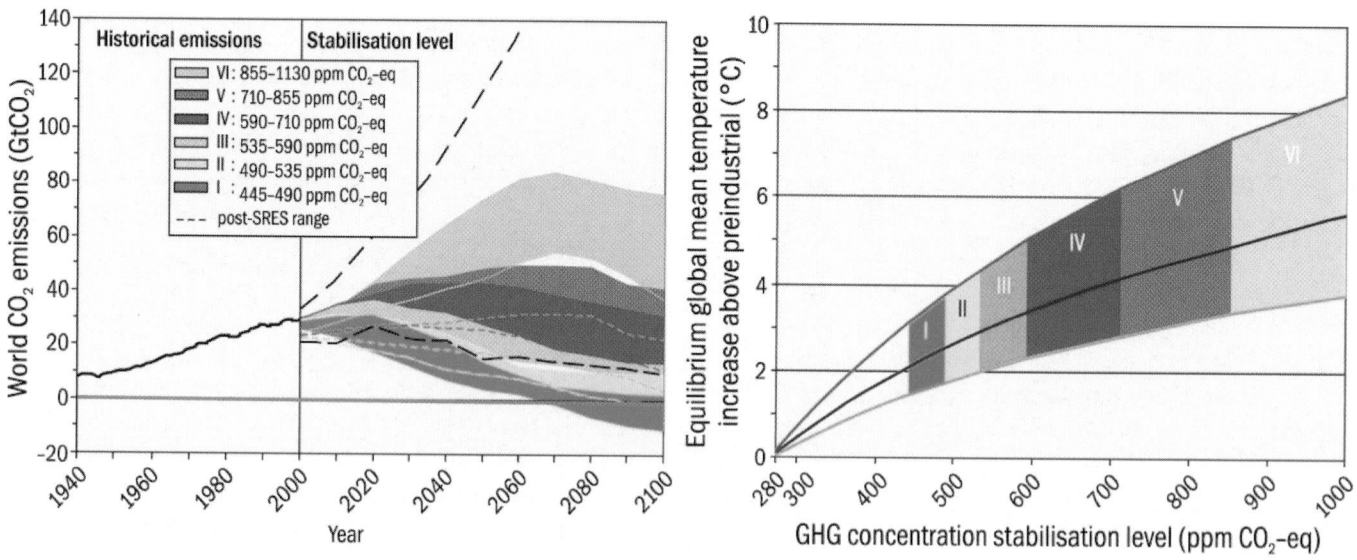

Figure 3.5 | The graph on the left shows the ranges of future emissions pathways for reaching different GHG emissions concentrations, expressed in terms of $GtCO_2$-eq. For example, the green range indicates the emissions trajectories that would lead to stabilization levels between 445–490 ppm $GtCO_2$-eq (as shown in the box within the figure). Note the need for net negative emissions post-2060 in case of the most stringent (green) trajectory. The graph on the right shows the equilibrium GMT increase above pre-industrial levels as a function of GHG stabilization level (ppm CO_2-eq). The middle black line indicates the most likely climate sensitivity, which is the most likely temperature increase at a certain GHG stabilization level. The red and blue lines indicate higher and lower climate sensitivity, that is higher or lower temperature increases for a given GHG stabilization level. Source: IPCC, 2007b.

the steeper the decline needs to be thereafter. At the moment, global GHG emissions are continuing to increase at close to historical rates without a sign of a reversal. A possible decline is being pushed further and further into the future with the recent failure to reach a 'global deal' in Copenhagen during the recent UNFCCC COP 15 (United Nations Conference of the Parties, 2009).

The temperature increase is in the first approximation a function of cumulative emissions. So far, humanity has emitted about 1000 $GtCO_2$-eq into the atmosphere, which has resulted in CO_2 concentrations increasing from about 280 to some 400 ppm today. In the case of lower stabilization levels, the remaining emissions 'endowment' is smaller than the cumulative historical emissions. Meinshausen et al. (2009) estimate that future cumulative emissions will be substantially lower than 1000 $GtCO_2$-eq. The exact amount will depend on the climate sensitivity to future emissions (which is not known with complete certainty) and the desired likelihood of not exceeding a particular stabilization level, say of 2°C. According to the German Advisory Council on Global Change (2009), the cumulative CO_2 emissions from 2010 to 2050 must not exceed 750 $GtCO_2$-eq in order to stay below a 2°C temperature increase with 67% probability. This assumes that there will not be any 'negative' emissions after 2050 to offset the excess emissions of the next several decades. Most of the 2°C stabilization scenarios do actually assume the possibility of negative emissions in the second half of the century (e.g., Fujino et al., 2006, Riahi et al., 2007; Van Vuuren et al., 2007; Wise et al., 2009).

The necessity of adopting such stringent emission reductions is evident, given the facts that a growing body of evidence suggests that the climate is changing more quickly than previously projected by the IPCC Assessment Reports (Jackson, 2009); that substantial climate impacts are occurring at lower GMTs (Smith et al., 2009); and that temperature changes may well be greater during this century than had been previously projected (Sokolov et al., 2009).

3.2.2.4   Near-term Changes of Radiative Forcing

Recent scientific studies indicate that short-term changes of RF play a significant role in climate change. For example, forcing of BC, a short-term radiative forcer, has been estimated to be 20–50% of CO_2 forcing, making it the second or third largest contributor to global warming (Wallack and Ramanathan, 2009). Black carbon and other short-term radiative forcers (e.g., O_3) will enhance warming, and therefore their mitigation would help prevent climate change. In the atmosphere, these short-term radiative forcers often co-occur with other short-lived pollutants such as sulphates, nitrates, OC, and other aerosols. These pollutants cool the climate through scattering and reflection of incoming solar radiation, and hence their mitigation would actually lead to a warming of the climate. These mixtures of anthropogenic particles and gases are sometimes referred to as atmospheric brown clouds (ABCs), especially when they occur in regions that particularly suffer from visible pollution. Such pollution may, for example, result from enhanced biomass burning, such as that occurring in south and southeast Asia (see also Box 3.3 and Ramanathan and Feng, 2008).

The realization of the substantial effect that these short-lived forcers can have on climate has led to a growing consensus regarding the need not only to mitigate those atmospheric agents responsible for long-term climate change, but also to manage the magnitude and rate of change of emissions of near-term radiative forcers and hence their RF (Ramanathan and Xu, 2010). Such mitigation of near-term climate change involves different pollutants, which often arise from different

---

Box 3.3 | Atmospheric Brown Clouds (ABCs)

What are ABCs

Atmospheric brown clouds (ABCs) are regional scale plumes of air pollution that consist of copious amounts of aerosols (tiny particles of BC, OC, sulphates, nitrates, fly ash) as well as many other pollutants including tropospheric O_3. The brownish color of ABCs is due to the absorption and scattering of solar radiation by BC, OC, fly ash, soil dust particles, and NO_2 gas. Typical background concentrations of aerosols are usually in the range of 100–300 particles/cm³, in polluted continental regions suffering ABCs, aerosol concentrations are in the range of 1000–10,000 particles/cm³.

ABCs start as indoor and outdoor air pollution consisting of particles (referred to as primary aerosols) and pollutant gases, such as NO_x, CO, SO_2, NH_3, and hundreds of organic gases and acids. These pollutants are emitted from anthropogenic sources, such as fossil fuel combustion, biofuel cooking, and biomass burning. Gases, such as NO_x, CO, and many VOCs, are important precursors of O_3 which is both an air pollutant and a strong GHG. Gases such as SO_2, NH_3, NOx, and referred to as aerosol precursor gases. These gases – over a period of a day or more – are converted to aerosols through the so-called gas-to-particle conversion process. Aerosols that are formed from gases through chemical changes (oxidation) in the air are referred to as secondary aerosols (Ramanathan et al., 2008).

Impacts of ABCs

Radiative forcing: Some components of ABCs, such as sulphate and nitrate aerosols, have a cooling effect on the climate system through reflection and scattering of incoming solar radiation. Others, such as BC, have a warming effect through absorption of solar radiation, which can lead to warming of the atmosphere or, where the BC is deposited on reflective snow- and ice-covered surfaces, can lead to surface warming and melting with implications for hydrological flows.

Glacial melting: ABCs solar heating (by BC) of the atmosphere is suggested to be as important as GHG warming in accounting for the anomalously larger warming trend observed in the elevated regions. In addition to the heating effect, deposition of BC on snow or ice can reduce the surface albedo and accelerate melting. Scientific studies suggest that ABC is one of the major contributing factors in glacier and sea-ice melting.

Water budget: ABCs change the cloud properties (cloud droplet numbers, size, albedo) and produce brighter clouds that are less efficient at releasing precipitation. Together, these effects can cause localised dimming (reduction of solar radiation reaching the Earth's surface) and lead to a alterations of the hydrological cycle.'

Human health: A large fraction of the aerosol particles that make up ABCs originate from emissions at the Earth's surface caused by the incomplete combustion of fossil fuels and biofuels. Humans are exposed to these particles both indoors and outdoors. Available information about the adverse health effects of airborne fine particles from studies conducted in many areas of the world suggests that ABC exposure is very likely associated with significant adverse health effects.

In summary, ABCs cause perturbations to regional climates, due to their comprising RF species. They also affect human health and agricultural productivity directly, through impacts resulting from the air pollution component (aerosols and ground-level O_3) of ABCs but also (particularly in the case of hydrology and agriculture) indirectly through their mediation of local climate. ABCs therefore represent a striking example of the interactions between climate change and air pollution, not only in relation to commonality in the atmospheric species causing both these environmental problems, but also in relation to the processes by which impacts are propagated. Hence, important lessons can be learned in relation to understanding, with a view to ultimately controlling, the adverse impacts associated with ABCs.

source activities compared with their long-term counterparts. This situation was eloquently described by Jackson (2009), who explored the contributions from near- and long-term climate forcers to climate change over a 20-year time frame, showing the relative contributions to RF from past emissions and from a variety of different pollutants (Figure 3.6). These pollutants are all sourced, albeit to varying extents, through energy-related processes.

Figure 3.6 highlights two important issues. The first is that positive RF resulting from the next 20 years of unrestrained human activity would exceed positive RF remaining from historical human activity after a couple of decades. The second issue is that short-lived pollutants (in particular BC, O_3, and CH_4) account for more than half (57–60%) of the positive RF generated in years 1 to 20.

Jackson also identified the 'top 10' pollutant-generating activities contributing to net RF, taking into account multiple pollutants from each source activity (Figure 3.7, see also Koch et al., 2007). From this information it is argued that the seven sources that appear on the left side (purple bars) would be overlooked by mitigation strategies focusing

exclusively on long-lived pollutants. There is therefore an urgent need for integrated mitigation strategies that include both the long- and short-term changes of RF; of these, gas and coal production and residential biofuel combustion are the categories which Jackson believed could be addressed by changes in future energy use and supply.

Raes and Seinfeld (2009) describe the current policy conundrum associated with these short-lived climate forcers. This relates to the fact that in addition to being radiative forcers, these species also play a role in air pollution, causing impacts on both human health (see Chapter 4) and ecosystems (see Section 3.2.3). To prevent and control these air pollutants, particularly in relation to human health, policies are already in place to reduce some of these pollutants (especially those classified as PM which include BC, OC, and other aerosols). These policies do not distinguish between positive or negative radiative forcers. Therefore, they may not improve the situation for climate change, or even actually enhance RF by reducing atmospheric concentrations of the negative forcers more effectively than the positive forcing species. The reality of this situation has been recently investigated by Penner et al. (2010), who argue that the short-term climate forcers need to be brought under

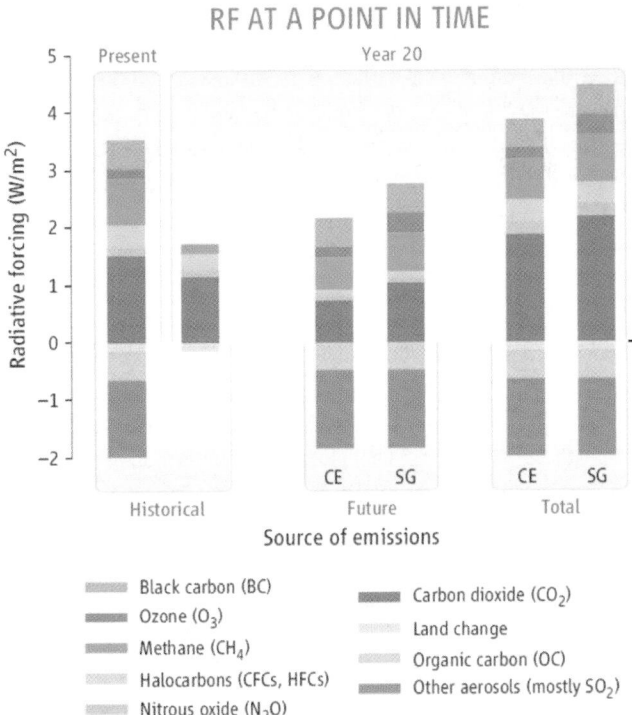

Figure 3.6 | Global radiative forcing (RF). The left-most bar shows RF attributable to historical human emissions (1750–2000), with the next bar representing historical RF that would remain after 20 years of atmospheric decay with zero additional human emissions. The next two bars represent 'future' RF in year 2020 resulting from human emissions. The two scenarios depicted are: emissions remain constant at year 2000 levels (CE), or emissions grow steadily at current rates (SG). The right-most columns show total RF experienced in year 2020 (historical + future emissions), again for both scenarios. For further details, see Jackson, 2009.

control within a few decades, and that the effect of this control on atmospheric composition and climate sensitivity needs to be monitored to provide an understanding of the warming and cooling contributions from CO_2 and short-lived air pollutants. Only with this information will it be possible to identify mitigation options that will afford the largest benefits in the alleviation of climate change while also addressing impacts on human health and ecosystems; a first attempt at identifying such options on a global scale has been achieved by a UNEP/WMO commissioned assessment on Black Carbon and Ozone (UNEP/WMO, 2011). This study found that considerable human health and crop productivity benefits could be realized through a number of technical and non-technical measures to limit emissions of BC and O_3 precursors and that these measures also provide a chance of constraining temperature increases below the 2°C, and even the 1.5°C, threshold if implemented in the very near future. This study helps to highlight that, especially in relation to energy systems, the most obvious policy option would seem to be identifying mitigation measures that would reduce both CO_2 and short-lived climate forcers that would otherwise lead to a warming of the climate systems.

3.2.3   Air Pollution and Energy Systems

As described previously, GHGs and RF agents that are produced from our energy systems can cause impacts other than climate change. Atmospheric emissions of SO_2, NO_x, and O_3 precursors are associated with eutrophication, acidification, and other types of direct ecosystem damage (Fowler et al., 2009). Much of the information describing these various air pollution impacts provided in the following sections is set within the context of an effects-based concept developed by the United Nations Economic Commission for Europe (UNECE) Convention

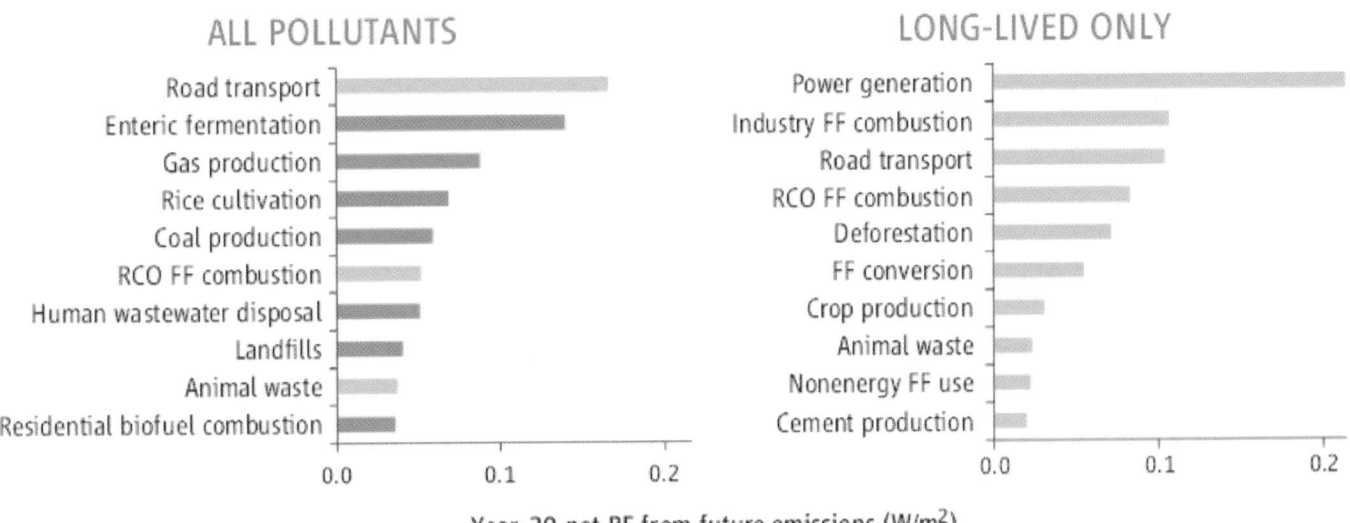

Figure 3.7 | Top-10 global sources of year 20 net radiative forcing (RF). Note: Long-lived pollutants (CO_2, N_2O) have only positive RF, whereas pollutants that are not long-lived have both positive RF (BC, O_3) and negative RF (OC, SO_2). Hence, a source may show a different RF on the left-hand versus the right-hand graph. See Jackson (2009) for further details.

on Long-Range Transboundary Air Pollutants (LRTAP). This approach has been successfully applied at the regional level in Europe to set critical limits below which ecosystems are unaffected by air pollution according to current knowledge (see Sliggers and Kakebeeke, 2004). These limits, which are referred to as critical loads and levels, have been established for different pollutants and impacts and have been used to define national and international air-quality guidelines (e.g., WHO, 2006a). They have also been used in European regional integrated assessment models to optimize air pollution mitigation policies, so that maximum benefits can be achieved at minimum cost (Schopp et al., 1999).

The following sections describe the processes by which these air pollutants impact ecosystems. The focus is on impacts by pollutants acting individually. However it is recognized, though poorly understood, that in reality these pollutants act together. For example, the potential for these pollutants to cause feedbacks on climate systems through perturbations to the terrestrial carbon sink strength has been identified as a major uncertainty which affects our ability to quantify the carbon cycle. Understanding interactions between pollutants is especially important here, given that impacts can both increase or decrease carbon sequestration.

3.2.3.1   Acidification of Soils and Freshwater

It is well documented in Europe and North America that sulphur(S) and nitrogen (N) deposition, predominantly from fossil fuel use, have caused widespread acidification of terrestrial and aquatic ecosystems. Acidic deposition has been linked to serious losses of fish stocks and other sensitive aquatic species, and it has been implicated as a potential cause of symptoms of forest decline (Rodhe et al., 1995; Menz and Seip, 2004) and effects on biodiversity (Bobbink et al., 1998). As emissions and deposition of S and N compounds increase in other parts of the world, particularly in Asia (see Section 3.2.4.4), there is a risk that acidification problems will become more widespread. However, the type and severity of acidification problems depend on many environmental factors, and evidence to date suggests that acidification problems in other parts of the world may not be as serious as they were in Europe (Hicks et al., 2008). It is also known that the reduction of atmospheric emissions of S and N to safe levels will not result in the immediate recovery of impacted ecosystems (e.g., Bishop and Hultberg, 1995), though there is evidence of recovery in European ecosystems (e.g., Vanguelova et al., 2010), where acidic deposition has been reduced significantly since the 1980s (Nakicenovic and Riahi, 2007; Vestreng et al., 2007).

3.2.3.2   Eutrophication

Increased release of reactive nitrogen (Nr) to the environment resulting from energy use is due mainly to increased NO_x emissions from fossil-fuel combustion processes that provide energy for transport, power generation, and industry. However, in terms of the total amount of Nr emitted to the Atmosphere, an equal or even greater amount is emitted from agricultural sources in the form of ammonia. Furthermore, Nr losses to ground and surface waters, mainly as nitrates, are dominated by agricultural activity; for example, in Europe, agriculture contributes 60% of these ground and surface water Nr losses, with the remainder largely made up of discharges from sewage and waste water treatment systems. As such, although the atmospheric emission of NO_x from fossil fuel combustion contributes to the eutrophication effects described below, these emissions can be dwarfed by agricultural flows depending on the location of the sensitive systems (Galloway et al., 2008, ENA, 2011).

Nitrogen is an essential nutrient that limits growth in many ecosystems; however, when applied in excess, it can cause eutrophication of terrestrial and aquatic ecosystems. Eutrophication can result in excessive plant growth such as algal blooms in aquatic systems. It can also cause changes in biodiversity. The eutrophying effect of N pollution at the global level is of particular concern (Gruber and Galloway, 2008). Molecular nitrogen (N_2) in the atmosphere is not usable by most organisms; it is only when it is fixed into reactive compounds (i.e., Nr) that it causes environmental impacts. Before the 20th century, the fixation of Nr occurred predominantly via a limited group of microorganisms and by lightning. This was sufficient until the demand for food, driven by rapid population growth, led to new ways of converting nonreactive gaseous N_2 into reactive forms for agricultural purposes, mainly through industrial production of fertilizers. Such Nr fixation requires substantial amounts of energy to break the strong triple bond of N_2; hence, increased anthropogenic Nr fixation became possible through increased access to fossil fuels. This has led to an increase in the use of inorganic fertilizers in agriculture and associated increases in Nr leaching and eutrophication when fertilizer is applied in excess of agricultural system requirements (Smith et al., 1997). A recent study on energy use in the fertilizer industry also showed that despite significant energy-efficiency improvements in fertilizer manufacture, these improvements have not been sufficient to offset growing energy demand, due to rising fertilizer consumption (Ramírez and Worrell, 2006).

At the global level, current Nr emission scenarios project that most regions will have increased rates of atmospheric N deposition by 2030 (Dentener et al., 2006), which may cause significant impacts to global plant biodiversity in sensitive ecosystems (Vitousek et al., 1997; Sala et al., 2000; Phoenix et al., 2006). A significant proportion of these emissions will be related to air pollution from energy sources (Galloway et al., 2008). In addition to the acidification effects of atmospheric Nr loading described above, deposits of Nr compounds act as a fertilizer to increase the productivity of terrestrial and aquatic ecosystems. However, if the supply of Nr continues to increase, a complex series of alterations to soil and biogeochemistry may affect productivity, competition, and microbial community structure (Bobbink et al., 2010). Many of the European arctic, boreal, and temperate terrestrial ecosystems have already been allocated effect thresholds or empirical critical N loads under the LRTAP Convention in the UNECE region, but there is a lack of information on impacts in other parts of the world. Bobbink et al. (2010) conclude that reductions in plant diversity as a result of increased atmospheric N deposition may be more widespread than first thought. They show that vulnerable regions outside Europe and North America include eastern and southern Asia (China, India), an important part of the Mediterranean ecosystems (California, southern Europe), and several subtropical and tropical parts of Latin America and Africa. However, to date the effects of N deposition on biodiversity are mostly only quantified for plant richness and diversity, and the impacts on animals and other groups are barely studied. Freshwater ecosystems are also affected, for example, across most of Europe nitrate levels in freshwaters greatly exceed a threshold above which water bodies may suffer biodiversity loss (ENA, 2011).

Nitrogen pollution is now considered to be the biggest pollution problem in coastal waters (Howarth et al., 2000; NRC, 2000; Rabalais, 2002), due to algal blooms and hypoxia (Levin et al., 2009) predominantly resulting from Nr fertilizer runoff. Human activities have severely altered many coastal ecosystems by increasing the input of anthropogenic Nr through sources such as rivers and groundwater, direct discharges from wastewater treatment, and atmospheric deposition, resulting in increasing eutrophication (Duce et al. 2008). In addition to runoff from land, atmospheric anthropogenic fixed Nr can enter the open ocean. Duce et al. (2008) found that this could account for approximately one third of the ocean's external (non-recycled) Nr supply and up to 3% of the annual new marine biological production.

In addition to impacts on biodiversity, N deposition and eutrophication in general will also have significant impacts on ecosystem services. Potential impacts include human health impacts of high nitrate levels in drinking water, carbon sequestration, GHG fluxes from soils, pollination, and cultural aspects, such as the loss of treasured species (MEA, 2005; GBO-3, 2010; NEA, 2011). Studies on the significance of these aspects are only just beginning in earnest around the world.

3.2.3.3   Vegetation Damage from Ground-level Ozone (O_3)

Tropospheric O_3 is a naturally occurring atmospheric trace gas. Historically, pre-industrial levels of the pollutant were in the region of 15–20 parts per billion (ppb) (Vingarzan et al., 2004). However, concentrations increased following the advent of industrialization and the burning of fossil fuels for transport, industry, and power generation. This increase was largely the result of increasing NO_x emissions, coupled with increases in NMVOC emissions from industry; these precursor pollutants combine under the action of sunlight through photochemically driven processes to form O_3. Heavily polluted regions can experience frequent occurrences of O_3 concentrations of approximately 40–50 ppb, especially during the summer periods, when the photochemical activity that drives the O_3 formation reactions is high (Royal Society, 2008).

Ozone concentrations vary locally, regionally, and seasonally across the globe (see Figure 3.8). Ozone concentrations can be high in both urban and rural locations (downwind from precursor pollutant sources).

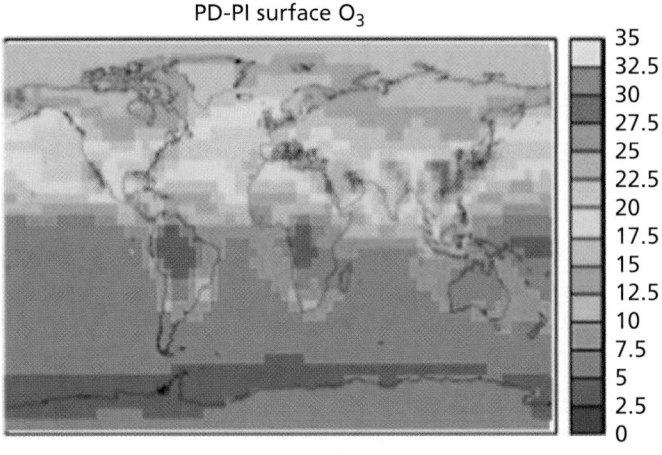

PD-PI surface O_3

Figure 3.8 | Modelled global increase in surface ozone (O_3) concentrations between pre-industrial times and the present day (PD-PI). Source: Royal Society, 2008.

However, in urban areas, episodes tend to be short-lived, as O_3 reacts with other chemical pollutants, namely nitric oxide (NO). In contrast, rural episodes of elevated O_3 tend to have an enhanced longevity, as the pollutants that destroy O_3 are not so prevalent. This makes O_3 an extremely important rural pollutant, which is heightened by the fact that it is strongly phytotoxic and capable of causing a wide variety of damage to various ecosystems; most notable are decreases in agricultural crop yields (Fuhrer 2003; Fuhrer and Booker, 2003), reductions in forest biomass (Matyssek and Sandermann, 2003), and changes in species composition of seminatural vegetation communities (Davison and Barnes, 1998; Ashmore, 2005). Ground-level O_3 also impacts human health through damage to the respiratory system, with reduced lung function and lung irritation among the most commonly experienced respiratory impairments; these issues are discussed in more detail in Chapter 4.

Although tropospheric O_3 is a relatively short-lived pollutant, with an average tropospheric lifetime of 22 (±2) days (Stevenson et al., 2006), attention has been drawn recently to the phenomenon of hemispheric transport of this air pollutant. This phenomenon is important, since it means that geographical regions (e.g., Europe, Asia, North America) are not in complete control of the air pollution load they experience (UNECE Task Force on Hemispheric Transport of Air Pollution, 2007). This concern is a result of the complex atmospheric chemistry associated with O_3, which can result in formation from precursors long after they have been emitted (Stevenson et al., 2006).

The changes in mean global O_3 concentration between pre-industrial times to the present day, shown in Figure 3.8, emphasise the historical importance of fossil fuel-based energy use and supply as the major source of anthropogenic O_3 precursor emissions, both in industrialized and developing countries. This is reflected in the distribution of the global 'hotspots' of elevated O_3 concentrations (in the United States, Europe, south asia, and east Asia) in regions that have experienced rapid industrialization over the past 200 or so years.

3.2.3.4   Combined Effects of Air Pollutants

Since fossil fuel combustion is an important source of N and S pollution and O_3 precursors, these chemicals will have a tendency to occur as a mix of pollutants in the atmosphere. This has consequences for impacts on receptor ecosystems, since damage by a single pollutant can be altered by the presence of other pollutants, frequently with synergistic (i.e., more than additive) effects. However, calculating response functions for each pollutant in the presence of others is extremely complex (Bell, 1985; Bender and Weigel, 1993). Additionally, the impacts on net primary productivity caused by these pollutants will also affect terrestrial carbon sequestration and hence create feedbacks to climate change.

Perhaps the best-studied of such feedbacks are climate change interactions with the nitrogen cycle. There is growing evidence that the increase in N deposition since industrialization may have been responsible for maintaining at least part of the current terrestrial carbon sink (Magnani et al., 2007; de Vries et al., 2008; Reay et al., 2008; Janssen and Luyssaert, 2009). The natural nitrogen cycle may also have been accelerated by climate change thereby increasing N availability to ecosystems. Nitrogen deposition can also lead to increased emissions of the potent GHG N_2O from soils and can increase soil emissions of NO, one of the important chemical precursors for O_3 formation (Prather et al., 1995). Nitrogen deposition can also enhance the growth rates of Nr-limited forests (Hungate et al., 2009), resulting in enhanced uptake/sequestration of carbon in terrestrial ecosystems where Nr is the limiting nutrient. As atmospheric CO_2 concentrations increase, such Nr limitation may become more common. However, where N deposition exceeds critical loads, adverse effects on growth and carbon sequestration can occur.

There is some evidence that a long-term trend of increased productivity in European forests is associated with such environmental factors, that including N deposition, as well as increased CO_2, and climate change, (Nabuurs et al., 2003), and is not simply due to improved management as is sometimes suggested. These factors can also ameliorate the resilience of trees to O_3. However, O_3 itself is considered a factor potentially capable of reducing the 'benefits' of CO_2 and nitrogen fertilization (King et al., 2005; Magnani et al., 2007). Seminatural grasslands are often limited by nutrients such as Nr or phosphorous. Alleviating such constraints, for instance by the addition of Nr, could decrease the sensitivity of the plant community to O_3 through increasing biochemical detoxification capacity, or increase the sensitivity though increased stomatal conductance. However, to date, evidence of these effects is limited and contradictory. Research into the nitrogen cycle is relatively less well developed than for the carbon cycle. Several authors have evaluated the effect of including C-N coupling in carbon and or climate models (Sokolov et al. 2008; Xu-Ri and Prentice, 2008). These studies suggest that the likelihood of greatly enhanced global CO_2 sequestration resulting from future changes in N deposition is low (Dolman et al., 2010).

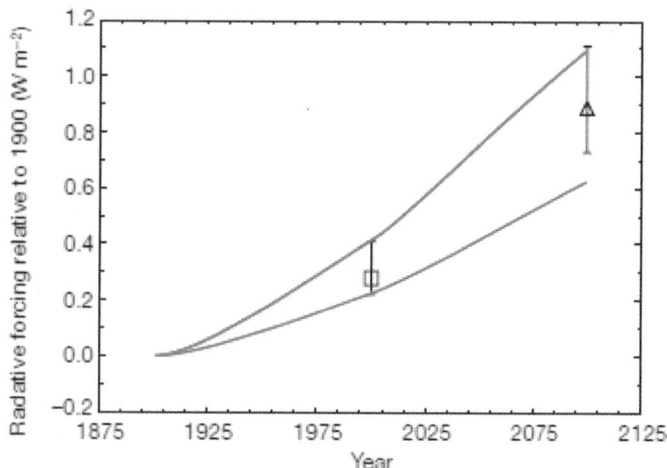

Figure 3.9 | Temporal changes in indirect radiative forcing (RF) due to O_3 for 'high' (red) and 'low' (blue) plant sensitivities to O_3. These results are diagnosed from model simulations using a fixed pre-industrial CO_2 concentration. For comparison, estimates of the direct RF forcing due to O_3 increases are shown by the bars. For further details, see Sitch et al. (2007).

The impact of tropospheric O_3 on ecosystem net primary productivity has also recently been used to estimate the indirect RF that would occur through O_3-induced alterations to carbon sequestration (see Figure 3.9). These estimates suggest that the indirect forcing by O_3 by 2100 would be 0.62 W/m^2 and 1.09 W/m^2 for 'low' and 'high' plant O_3 sensitivity respectively. This compares with a mean direct RF, estimated from 11 atmospheric chemistry models, of 0.89 W/m^2, highlighting the extreme importance of air pollution-dependant ecosystem feedbacks in relation to our understanding of how climate forcing is likely to change in the future.

3.2.4   Regional Impacts of Climate Change and Air Pollution

Energy systems are a significant contributor to climate change and atmospheric pollution in most parts of the world. In this section, the current state of environmental impacts that are related to atmospheric pollution across a number of regions (Europe, the Americas, Asia, Africa, and the polar regions) are discussed in detail. Climate change-related impacts have already been reviewed in great detail in the IPCC 2007 assessment, therefore only a short summary of these impacts is provided here. A more detailed account of relevant air quality impacts is given, since these tend to be less considered compared to climate change impacts, though they cause serious environmental degradation on a regional scale. Here we limit the discussion to environmental impacts; other critical issues, such as health-related impacts of outdoor and household air pollution, are discussed in other chapters, including Chapter 4 and Chapter 17.

3.2.4.1   Europe

Europe has a history of environmental policies that have targeted air quality, dating back to the 1950s. Following the region's increase in economic wealth, industrial activity, and use of fossil fuels, emissions of harmful air pollutants increased significantly in Europe, from the beginning of industrialization to the 1980s (see, for example, historical SO_2 emissions in Figure 3.10). However, the long-term growth trend of emissions reversed after the 1980s, triggered by public concern about the detrimental impacts of urban air pollution on public health, the dieback of forests in Central Europe, and the disappearance of fish and aquatic life in Scandinavia (UNECE, 2004). In numerous international agreements for harmonized emission reductions under the LRTAP Convention and the European Union, countries have agreed to substantially reduce their emissions. As of 2010, emissions of SO_2 and NO_x had declined by 70% and 50%, respectively, since their peaks, due to widespread application of dedicated end-of-pipe pollution-control equipment, as well as improved energy intensities of the European economies and changes in the composition of fuel consumption.

Polices to reduce emissions of GHGs to prevent climate change have only been recently established, even though the European Union (EU) prides itself on having been a driving force behind the international negotiations that led to the establishment of the UNFCCC in 1992 and the Kyoto Protocol in 1997. For the latter, the 15 EU member states signed up to reduce emissions in the 2008–2012 period to 8% below 1990 levels, a target that currently looks achievable. European Union leaders have also endorsed an integrated approach to climate and energy policy to make a transition to an energy-efficient, low-carbon economy. They have made a unilateral commitment in the 'Climate and Energy package', adopted in 2008. By adopting this strategy, the 27 nations of the EU-27 (Austria, Belgium, Bulgaria, Cyprus, Czech Republic, Denmark, Estonia, Finland, France, Germany, Greece, Hungary, Ireland, Italy, Latvia, Lithuania, Luxemburg, Malta, the Netherlands, Poland, Portugal, Romania, Slovakia, Slovenia, Spain, Sweden, and the United Kingdom) pledged to reduce overall emissions to at least 20% below their 1990 levels by 2020. The EU pledges that this emission reduction would be increased to 30% by 2020, if other major emitting countries in the developed and developing worlds take similar action.

However, there is ample and robust scientific evidence that even at present rates, Europe's emissions to the atmosphere pose a significant threat to human health, ecosystems, and the global climate (UNECE, 2007). Discussed below are some of the main environmental impacts associated with climate change and air pollution in Europe.

Climate change environmental impacts in Europe
Wide-ranging climate change impacts have been observed in Europe through warming trends (Jones and Moberg, 2003) and spatially variable changes in

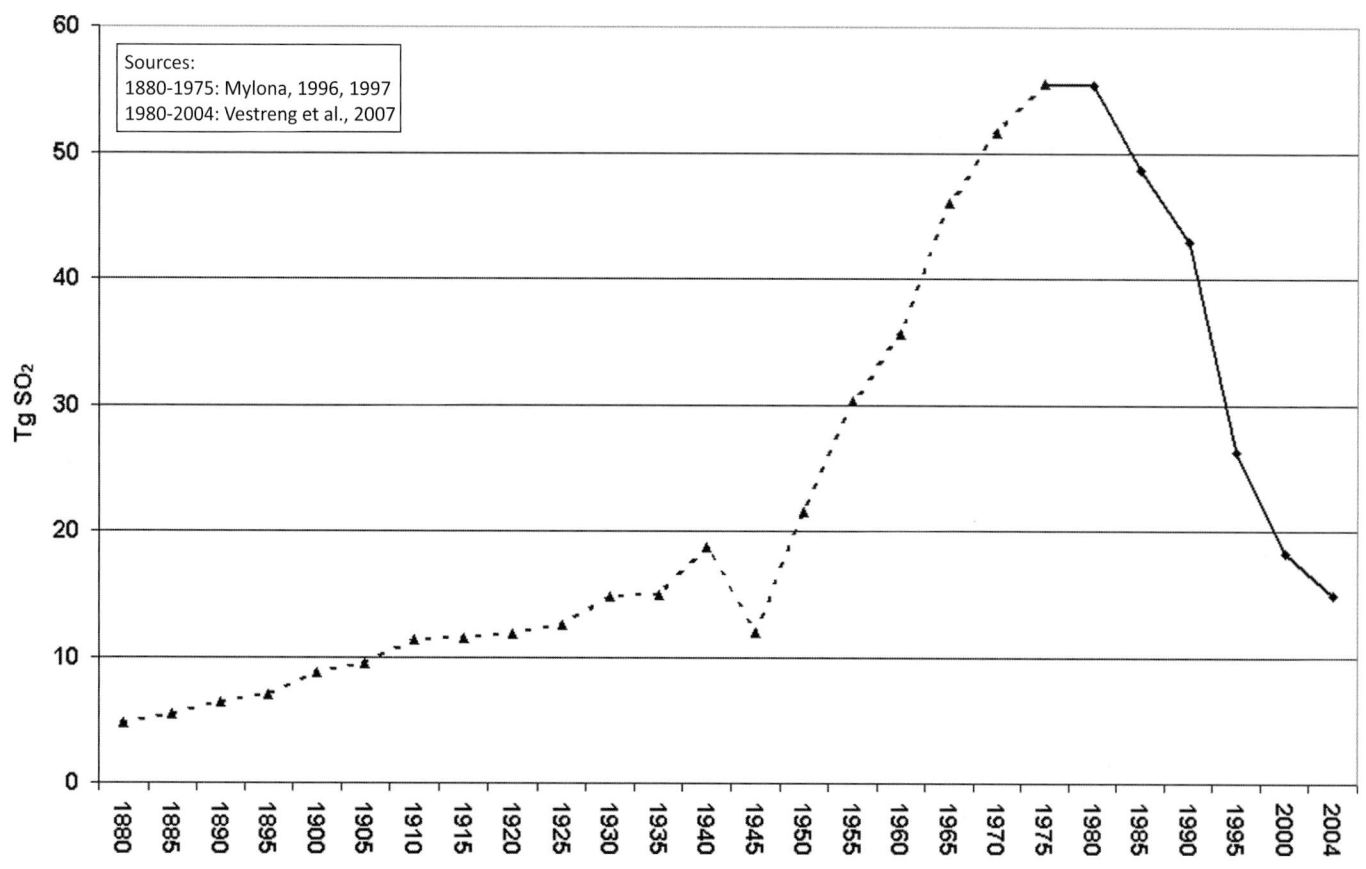

Figure 3.10 | Historical development of sulphur dioxide (SO_2) emissions in Europe (Unit: MtSO₂). Source: Mylona, 1996; 1998; Vestreng et al., 2007.

rainfall (Klein Tank et al., 2002). These impacts have affected the cryosphere, as seen in the retreat of glaciers (Hoelzl et al., 2003) and changes in the extent of permafrost (Frauenfeld et al., 2004). They have also impacted natural and managed ecosystems, as evidenced by changes in growing season length (Menzel et al., 2006) and species distribution (e.g., Walther et al., 2005), which have implications for biodiversity. Europe is considered most sensitive to climate change that causes extreme seasons, especially hot, dry summers and mild winters; the former would lead to more frequent and prolonged droughts (Schär et al., 2004) as well as to a longer fire season, especially in the Mediterranean (Moriondo et al.,2006). Climate-change projections indicate greater warming during winter in the north and during summer in the south. These projected changes have implications for crop suitability and production and for forest expansion and biomass growth, both of which are likely to increase in northern Europe and decrease in southern Europe (Olesen et al., 2007; Shiyatov et al., 2005; Metzger et al., 2004). The northward expansion of forests may reduce the extent of tundra regions (White et al., 2000). These changes may be accompanied by a shift in peak electricity demand from winter to summer, as demand for heating decreases and demand for cooling increases (Hanson et al., 2006). Water stress is projected to increase over central and southern Europe. This has important implications for energy supply, since the hydropower potential of Europe is expected to decline on average by 6% by the 2070s, and by 20–50% in the Mediterranean region (Lehner et al., 2005). The important role that water abstraction plays in energy supply is also evident from the fact that 31% of total water withdrawals from 30 European countries are used as cooling water in power stations (Flörke and Alcamo, 2005). Short-duration climate events such as windstorms and heavy rains may also increase in frequency, causing problems such as flooding (Christensen and Christensen, 2003). Longer-term changes in climate will also place considerable pressure on coastal areas through sea-level rise (Devoy et al., 2007). Further details of the observed and projected climate change impacts for Europe are provided by the IPCC in Alcamo et al. (2007).

Air Quality Environmental Impacts in Europe

Eutrophication in Europe

Many plant species are endangered as a result of eutrophication in terrestrial ecosystems (WHO, 2006a). Ecosystems that include meadows, forests, and bogs that are characterized by low nutrient content and species-rich, slowly growing vegetation adapted to lower nutrient levels are overgrown by faster growing and more competitive species-poor vegetation, like tall grasses, that can take advantage of unnaturally elevated Nr levels. As a result, certain habitats may be changed beyond recognition, and vulnerable species may be lost (Hettelingh et al., 2007); for example, the majority of orchid species in Europe are considered at risk from eutrophication (WHO, 2006a). For the year 2000, it was estimated that N deposition had significantly exceeded thresholds that would guarantee ecological sustainability (i.e., critical loads) in most of the European forests and grassland areas (Figure 3.11). It should be noted that energy-related NOx emissions in Europe only account for about half of the total N deposition, with the rest coming from agricultural sources, in particular intensive livestock rearing (Stigliani and Shaw, 1990).

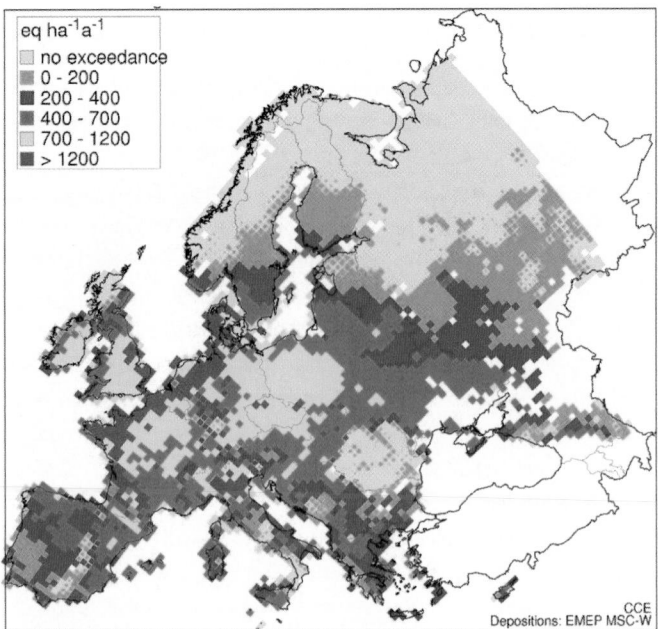

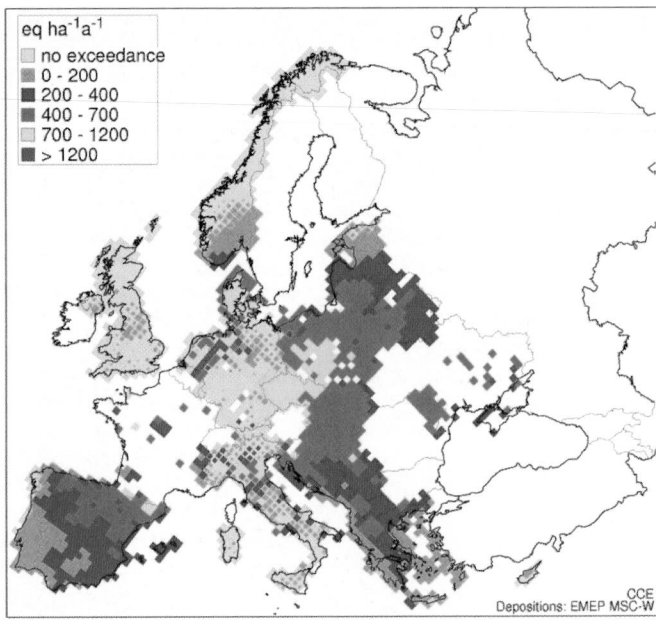

Figure 3.11 | Exceedance of critical loads for eutrophication for forest soils (top panel) and seminatural vegetation (grasslands, shrubs, etc., bottom panel) in the year 2000. The size of a colored grid cell is proportional to the fraction of the ecosystem area in the cell where critical loads are exceeded. Source: Hettelingh et al., 2008.

Methods have recently been developed that can be used to quantify the impacts, rather than merely identifying areas in exceedance, of excess N deposition on plant-species diversity (Hettelingh et al., 2008). These have estimated that current levels of Nr significantly degrade the species richness in many European ecosystems, leading to losses of up to 20% of the species in forests in northwest Europe (Figure 3.12).

Under the Natura 2000 program, the EU declared specific nature reserve areas to maintain and restore natural habitats. Among other stresses (e.g., from the fragmentation of habitats), ecosystems are under pressure

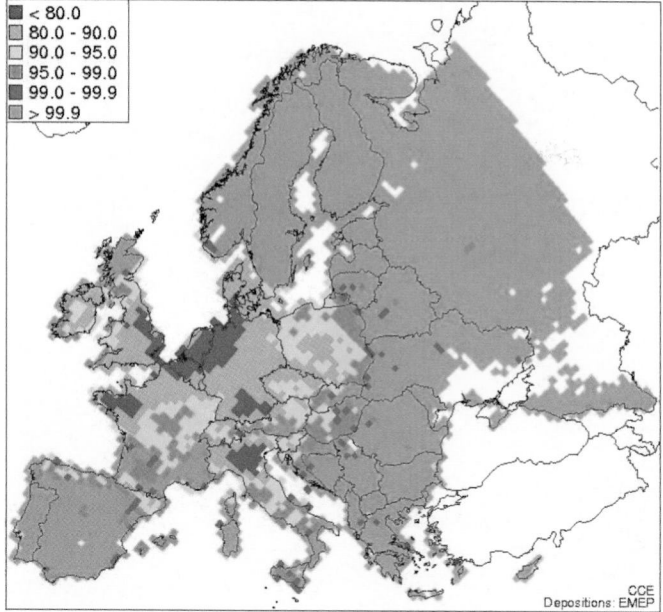

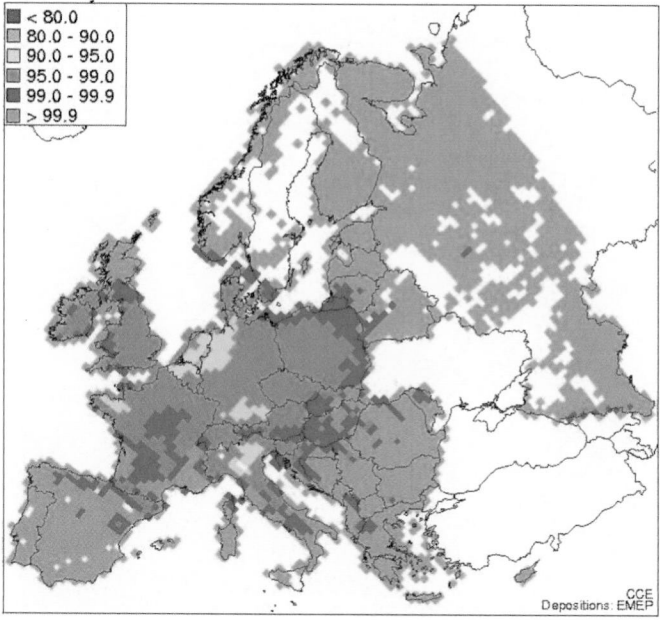

Figure 3.12 | Percentage of species richness in forests in Europe (top panel) and semi-natural vegetation (bottom panel) in the year 2000. Red-shaded areas indicate that the estimated biodiversity indicator percentages are lower than 80%, while green shadings indicate areas where this percentage is between 95–100%. Source: Hettelingh et al., 2008.

from excess deposition of atmospheric pollutants that affect plant species diversity and soil chemistry. For the year 2000, overly high deposition of N compounds constituted an important stress factor to Natura 2000 areas in Germany, the Netherlands, France, Poland, the Czech Republic, and Bulgaria, where deposition exceeded the tolerable levels of input (Figure 3.13).

Acidification in Europe

In the second half of the 20th century, the health of European forests, lakes, and rivers was heavily compromised by high acid deposition

resulting from emissions of SO_2, NO_x, and NH_3. As a consequence, soils in forests and freshwater catchment areas, as well as lakes in Scandinavia, experienced strong acidification that led to plant damage and the disappearance of fish and other aquatic fauna. Since then, steep reductions in SO_2 and NO_x emissions have reduced deposition levels considerably, and many areas are now gradually recovering from past acidification. Nevertheless, current deposition rates are still exceeding sustainable levels for large forest areas in central Europe and freshwater catchments in Scandinavia and the United Kingdom (Figure 3.14). It is also clear that the full recovery of acidified soils will require deposition to be below critical-load thresholds for a substantial period of time in order to replenish the buffering capacity of soils, which has been depleted over the last decades. Thus, full recovery of acidified ecosystems in Europe would require SO_2 and NO_2 emissions to decline by 80–90% below current levels.

Ground-level Ozone (O_3)

In Europe, ground-level O_3 has been found to cause impacts to agricultural crops. These impacts include visible injury, (particularly important for leafy salad crop species (Emberson et al., 2003); declines in yields of arable crops (Mills et al., 2007); and alterations to quality of crop yields, for example the nitrogen content of harvestable products (Pleijel et al., 1999).

Exposure to ground-level O_3 also causes negative effects on sensitive forest trees, including reduced photosynthesis, premature leaf shedding, and growth reductions (Skarby et al., 1998). Ozone-sensitive forest tree species, including birch, beech, Norway spruce, Sessile oak, Holm oak, and Aleppo pine, are present across large areas of Europe (Karlsson et al., 2007). Ozone effects on these species have important negative consequences for carbon sequestration, biodiversity, and other ecosystem services that are provided by forest trees. Such services include reducing soil erosion and decreasing flooding and avalanches.

By impacting growth, seed production, and environmental stress tolerance, O_3 also affects the vitality and balance of seminatural vegetation ecosystems and the ecosystem services they provide. These services include carbon storage, water storage, and biodiversity (Fuhrer et al., 2009). The floral diversity of this vegetation type makes it more difficult to generalize about effects and to establish critical levels applicable across Europe. Widespread effects on *Trifolium* (clover species), an important component of productive pasture, have been found in O_3-exposure experiments (Mills et al., 2011). Effects include reductions in biomass, forage quality, and reproductive ability at ambient and near-ambient concentrations in many parts of Europe.

In recent years, research has focused on identifying appropriate indicators to quantify the risk of vegetation damage from O_3 (Emberson et al., 2007). Originally, risks for damage have been associated in Europe with the AOT40 (accumulated O_3 exposure over a threshold of 40 parts per billion) indicator, which measures O_3 concentrations during daylight hours that exceed a 40 ppb threshold, accumulated over the entire vegetation period (Fuhrer et al., 1997). This indicator suggests that for the year 2000 the largest risk to forest trees was in the Mediterranean countries (Figure 3.15).

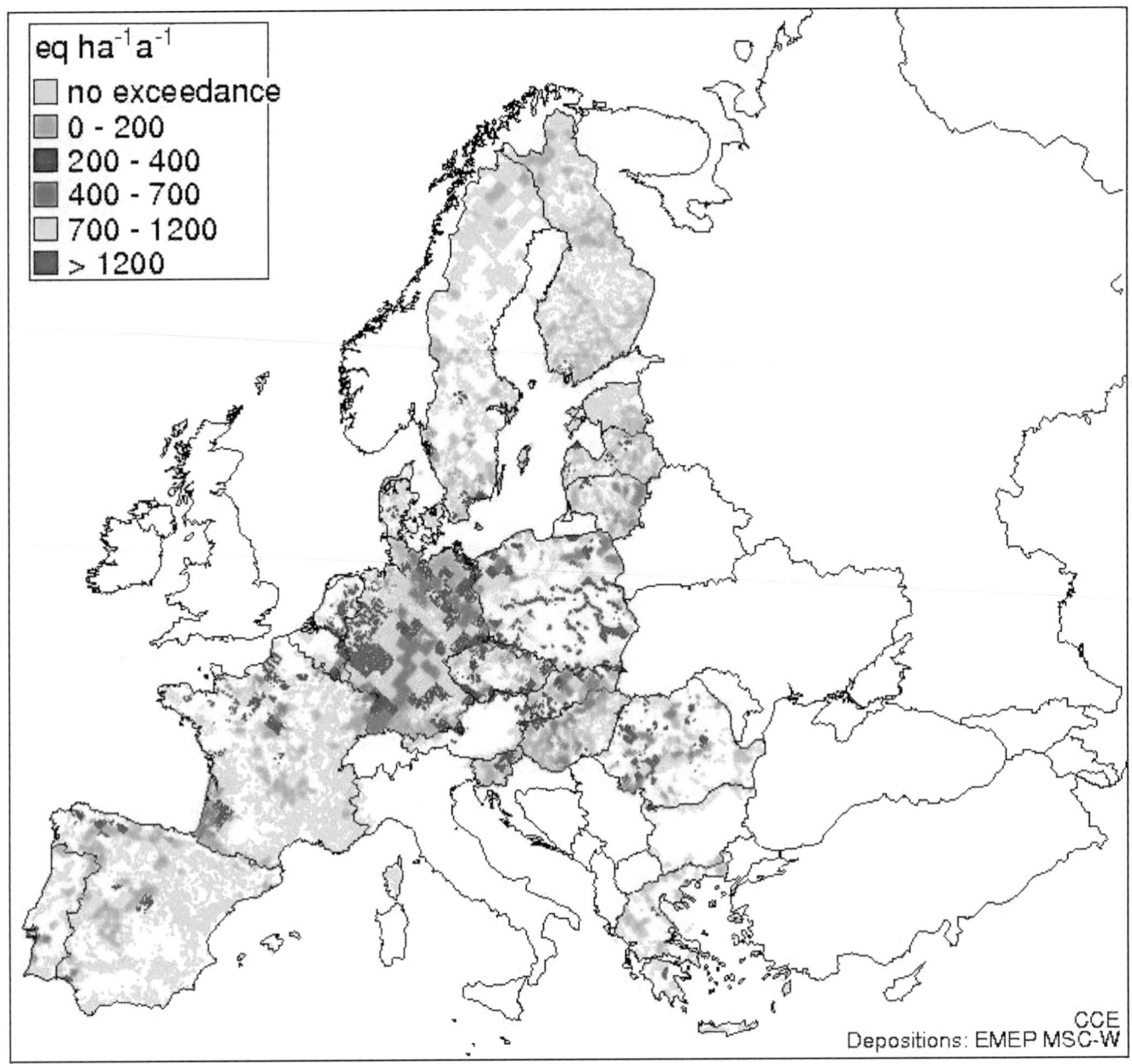

Figure 3.13 | Excess nitrogen deposition (Nr-eq/ha/yr) in Natura 2000 nature protection areas in 2000. Source: Hettelingh et al., 2008.

However, more recent work that associates actual vegetation damage with the O_3 dose that is absorbed by plants (often termed the 'ozone flux' approach) shows higher risks of O_3 in central Europe than indicated by AOT40. The areas at risk identified using this flux-based approach also bear a closer relation to those areas where vegetation damage has actually been found 'on the ground' (Mills et al., 2011).

Reductions in European precursor emissions of ground-level O_3 will certainly alleviate the pressure on vegetation. However, there is growing evidence of an increasing trend in hemispheric background concentrations of O_3 that could counteract the positive effects of measures within Europe (Royal Society, 2008). Effective response strategies that seek to

eliminate the risk of vegetation damage from O_3 will therefore need to address not only sources in Europe, but also sources from other continents that may well be contributing to these increasing background concentrations.

3.2.4.2   The Americas

North, Central, and South American countries suffer the full range of environmental threats due to atmospheric emissions to which energy systems remain a major contributor. Problems exist in both rural and urban areas, including increased human morbidity and mortality plus agricultural, forest, water, visibility, and other welfare damage. Air-

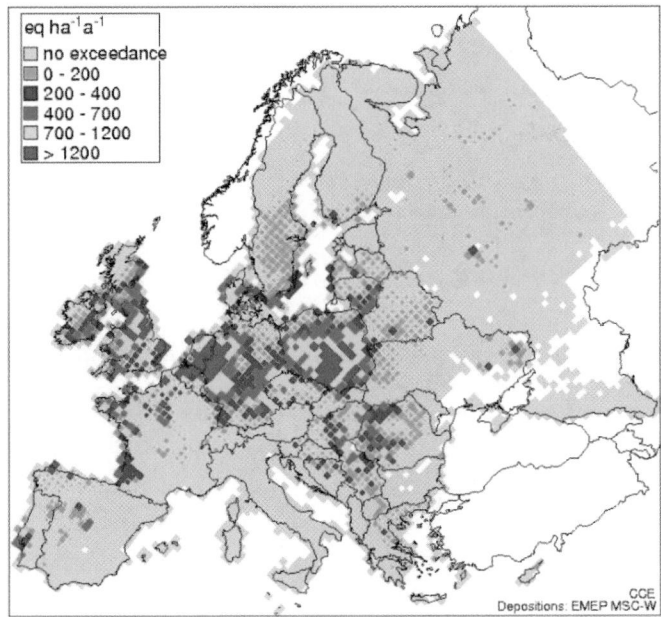

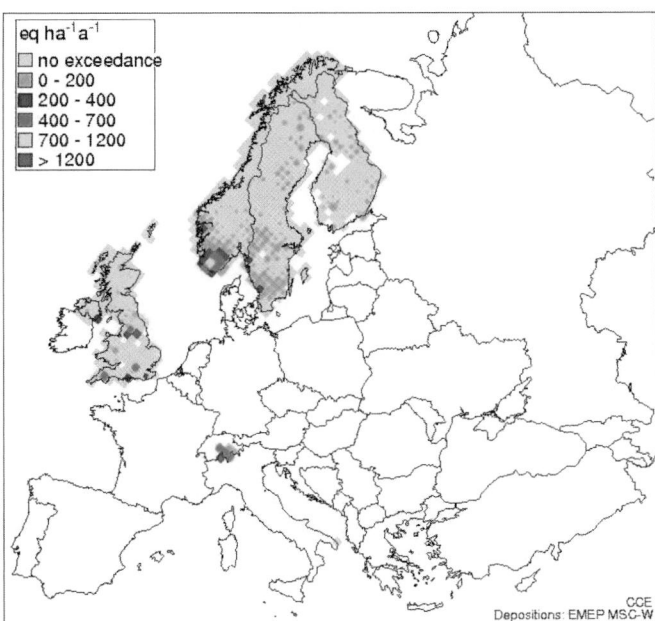

Figure 3.14 | Exceedance of critical loads of acidification (N-eq/ha/yr) in the year 2000 for forest soils (top panel) and freshwater catchment areas (bottom panel). The size of a colored grid cell is proportional to the fraction of the ecosystem area in the cell in which critical loads are exceeded. Source: Hettelingh et al., 2008.

the impacts from the release of toxic materials in all parts of North America. The Clean Air Act was established in 1970 to foster the growth of a strong American economy and industry while improving human health and the environment. Over the last 20 years, total emissions of the six principal air pollutants (also known as 'common' or 'criteria' pollutants) – PM, ground-level O_3, CO, sulphur oxides (SO_x), NO_x, and lead – have decreased by more than 41%. During the same period, GDP has increased by more than 64%.

Although significant progress has been made in improving the quality of the air in most US cities and communities, there is more to be done over the next 40 years. The development and use of energy in North America has been, and still remains, the prime source of environmental degradation. According to the Commission for Environmental Cooperation (CEC) in North America, there is a total of 3.17×10^{10} kilograms of criteria-related emissions emitted by the industrial sector in North America. Of these, 60% of the emissions are released by industrial sources in the United States, 26% by industrial sources in Mexico, and 14% by industrial sources in Canada (CEC, 2009).

Climate Change Environmental Impacts in the Americas

The Fourth Assessment Report of the IPCC states that there is high confidence that North America has experienced locally severe economic damage, plus substantial ecosystem, social, and cultural disruption from recent weather-related extremes, including hurricanes, other severe storms, floods, droughts, heat-waves, and wildfires (Field et al., 2007). There is also high confidence that climatic variability and extreme events have been severely affecting the Latin America region over recent years (Magrin et al., 2007).

Many coastal areas in North America are exposed to storm-surge flooding (Titus, 2005), especially those areas below sea-level. The breaching of New Orleans floodwalls following Hurricane Katrina in 2005 and storm-wave breaching of a dike in Delta, British Columbia, in 2006 demonstrate this vulnerability. Under El Niño conditions, high water levels combined with changes in winter storms along the Pacific coast have produced severe coastal flooding and storm impacts (e.g., Walker and Barrie, 2006). Significant impacts of projected climate change and sea-level rise are also expected for 2050–2080 on the Latin American coastal areas. With most of their population, economic activities, and infrastructure located at or near sea-level, coastal areas will be very likely to suffer floods and erosion, with high impacts on people, resources, and economic activities.

Changes in precipitation and increases in temperature are constraining over-allocated water resources, increasing competition among agricultural, municipal, industrial, and ecological uses across the Americas. In Latin America during the last decades, significant changes in precipitation and increases in temperature have been observed. Increases in rainfall in southeast Brazil, Paraguay, Uruguay, the Argentinean Pampas, and some parts of Bolivia have had impacts on land-use and crop yields, and have increased flood frequency and

quality deterioration in North American urban areas was initially noted in the first half of the 20th century, which resulted in the establishment of air-quality management programs in a number of the larger urban areas in the United States by 1950. Such programs were extended to Canada in the 1970s and to Mexico after 1980. Regional and global environmental threats were recognized in the late 1970s, when acid deposition was established as a significant problem in the northeastern part of the United States and in eastern Canada. By the 1980s, the potential threat of global climate change was recognized, along with

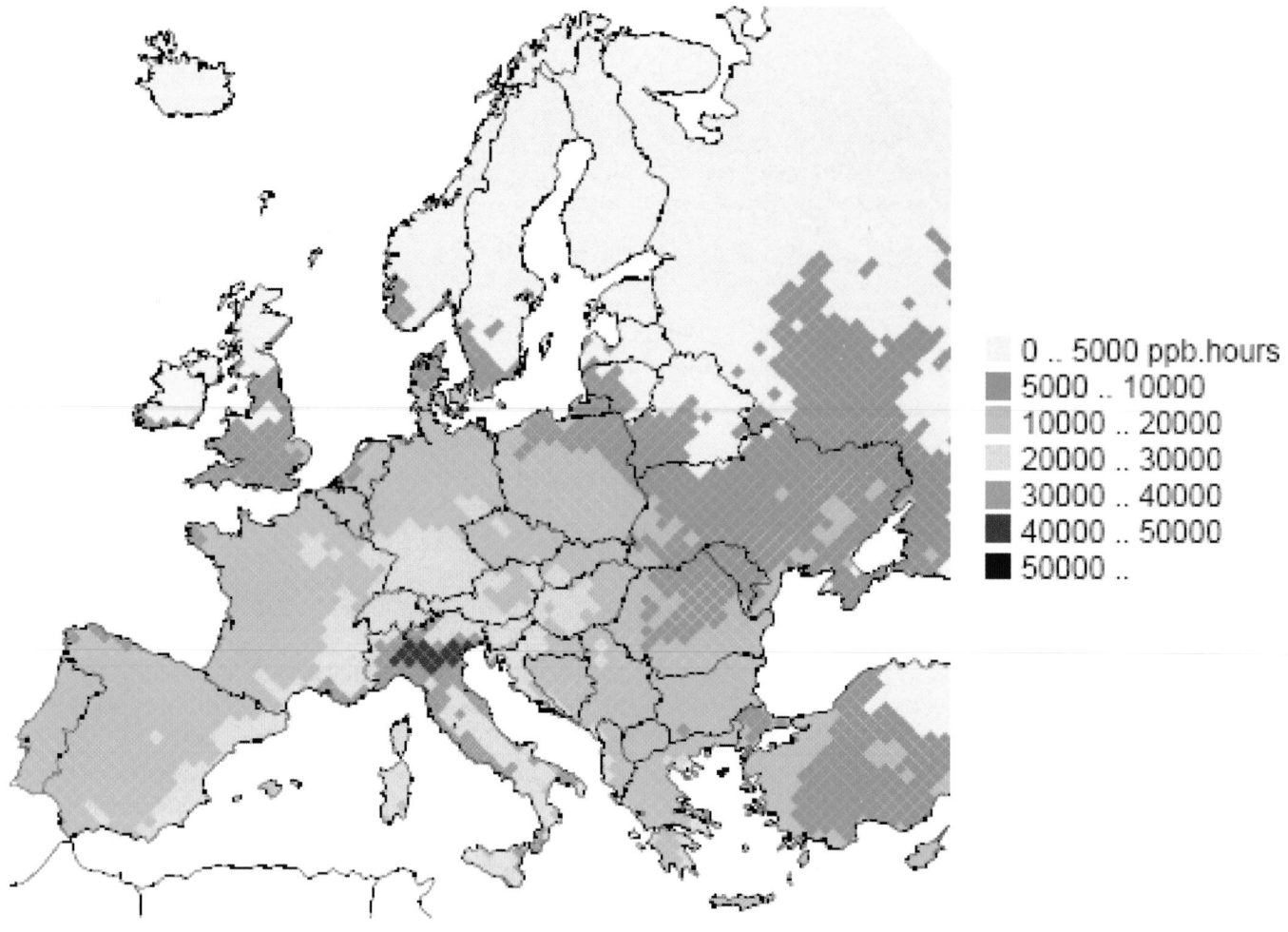

Figure 3.15 | An indicator of ozone damage to forest trees (AOT40) calculated for 2000. Source: Amann et al., 2011.

intensity. On the other hand, a declining trend in precipitation has been observed in southern Chile, southwest Argentina, southern Peru, and western Central America. As a consequence of temperature increases, the trend in glacier retreat is accelerating, with recent studies indicating that the volumes of most of the South American glaciers from Colombia to Chile and Argentina are decreasing at an accelerated rate (e.g., Leiva, 2006). During the next 15 years, intertropical glaciers are very likely to disappear, affecting water availability and hydropower generation. Hydropower production is known to be sensitive to total runoff, the timing of runoff, and to reservoir levels in North America. For example, during the 1990s, water levels in the Great Lakes fell as a result of a lengthy drought, and in 1999 hydropower production was down significantly both at Niagara and Sault St. Marie (CCME, 2003).

Climate change causes a risk of significant species extinctions in many areas of tropical Latin America. Up to 40% of the Amazonian forests could react drastically to even a slight reduction in precipitation. The tropical vegetation, hydrology, and climate system in South America could rapidly change to another steady state (Rowell and Moore,

2000). It is more probable that forests will be replaced by ecosystems that have more resistance to multiple stresses caused by temperature increases, droughts, and fires, such as tropical savannas. The replacement of tropical forest by savannas is expected in eastern Amazonia and the tropical forests of central and southern Mexico, along with the replacement of semiarid vegetation by arid vegetation in parts of northeast Brazil and most of central and northern Mexico, due to the synergistic effects of both land-use and climate changes (Magrin et al., 2007). By 2050, desertification and salinization will affect 50% of agricultural lands in Latin America and the Caribbean zone (FAO, 2004). Over the 21st century, pressure on species to shift north and to higher elevations will fundamentally rearrange North American ecosystems. Differential capacities for range shifts and constraints from development, habitat fragmentation, invasive species, and broken ecological connections will alter ecosystem structure, function, and services (Field et al., 2007).

Further details of the observed and projected climate change impacts for North America and Latin America and the Caribbean are provided by the IPCC in Field et al. (2007) and Magrin et al. (2007), respectively.

Air Quality Environmental Impacts in the Americas

Eutrophication

Although eutrophication was identified as an issue in North America in the 1970s, the problem is not ranked particularly highly within national environmental protection agencies. Nevertheless, the World Resource Institute, which carried out an assessment of eutrophication in North American coastal areas, found 131 areas that show symptoms of eutrophication (WRI et al., 2008).

Various studies in western North America (e.g., US EPA, 2008) demonstrate that some aquatic and terrestrial plant and microbial communities may be significantly altered by N deposition. For example, an accumulating weight of evidence has led some researchers to conclude that high-altitude watersheds in the Colorado Front Range show symptoms of ecological impacts, even at current N deposition levels. These effects include changes in alpine plant communities, elevated surface water nitrate concentrations, and changes in lake algal species communities. Further west, levels of stream water and groundwater nitrate in the San Gabriel and San Bernardino Mountains have been found to be strongly linked to the magnitude of N deposition in watersheds throughout the region. Stream water nitrate concentrations at Devil's Canyon in the San Bernardino Mountains and in chaparral watersheds with high smog exposure in the San Gabriel Mountains northeast of Los Angeles are the highest in North America for forested watersheds. Chronic N deposition and nitrate export from these watersheds contribute to the groundwater nitrate problems in the eastern San Gabriel Basin, where levels often exceed the federal drinking water standard.

A significant amount of research has been conducted since 1990 on the effects of N deposition on terrestrial ecosystems in the Los Angeles air basin in southern California (Fenn et al., 2003). Researchers have found that Nr enrichment, in combination with O_3 exposure, causes major changes in tree health by reducing fine root biomass and carbon allocation below ground and by greatly decreasing the life span of pine foliage. Nitrogen enrichment results in greater leaf growth, while O_3 causes premature leaf loss at the end of the growing season. The net result of these pollutant interactions is significant litter accumulation on the forest floor. Nitrogen cycling rates in soil are also stimulated by the high N inputs, resulting in large leachate losses of nitrate from these watersheds and elevated fluxes of NO gas from soil. Greenhouse and field studies indicate that N deposition may be one factor promoting the invasion of exotic annual grasses into coastal sage ecosystems occurring in low-elevation sites in the region.

One of the main adverse ecological effects resulting from N deposition, particularly in the Mid-Atlantic region of the United States, is the effect associated with nutrient enrichment in estuarine (Bricker et al., 2007) and coastal (Valigura et al., 2001) waters. Eutrophication in such ecosystems is associated with a range of adverse ecological effects, including low dissolved oxygen, harmful algal blooms, loss of submerged aquatic vegetation, and low water clarity. These changes disrupt aquatic habitats; cause stress to fish and shellfish (which in the short term can lead to episodic fish kills, and in the long-term can damage growth in fish and shellfish populations); and cause aesthetic impairments to estuaries. A recent assessment of 141 estuaries in the United States by the National Oceanic and Atmospheric Administration concluded that 64 estuaries (45%) suffered from moderately high or high levels of eutrophication due to excessive inputs of both Nr and phosphorus, with a majority of these estuaries being located in the coastal area from North Carolina to Massachusetts (Bricker et al., 2007). For estuaries in the Mid-Atlantic region, the contribution of atmospheric distribution to total Nr loads is estimated to range between 10% and 58% (Valigura et al., 2001).

Acidification and Visibility

Acid deposition and visibility reduction were noted as significant issues in the United States in the 1980s. In 1990, the US Clean Air Act was amended to specifically address these issues. The US program to address acid rain, which began in 1995, required US power plants to reduce SO_x emissions. A second phase of the SO_x reduction program began in 2000. Overall, the goal of the program has been to reduce SO_x emissions by 10 Mt/yr and cap emission levels from power plants at 8.95 Mt/yr in 2010. In addition to the requirement to reduce SO_x emissions, the US Environmental Protection Agency (US EPA) was required to adopt programs to reduce NO_x emissions by 2 Mt/yr beyond levels projected to occur without the program. Figure 3.16 indicates the emission reductions in NO_x and SO_2 in the United States between 1990 and 2009 (US EPA, 2010b) in relation to changes in electricity generation and retail price.

The success of the US program to address acid rain can be seen in monitoring data collected by the National Atmospheric Deposition Program/National Deposition Trends Network, or NADP/NTN (US EPA, 2010b). The data show significant improvements in the deposition of S and N across the United States (Figure 3.17). Reductions in N deposition recorded since the early 1990s have been less pronounced than those for sulphur.

Eastern Canada also suffers from acid deposition similar to that found in the northeastern United States. More than half of the acid rain in eastern Canada comes from emissions in the United States (Environmental Canada, 2010). It is estimated that 68% of the Canadian emissions are from industrial sources, while 67% of the US emissions are from electric power plants. In Mexico, there is concern that acid deposition is damaging the ancient Maya ruins that can be found in many parts of the country (Bravo et al., 2006).

Acidification Impacts on Aquatic Ecosystems

Acid deposition resulting from SO_2 and NO_x emissions is one of many large-scale anthropogenic impacts that negatively affect the health of lakes and streams in the United States (US EPA, 2010b). Surface-water chemistry provides direct indicators of the potential effects of acidic deposition on the overall health of aquatic ecosystems. Long-term surface water monitoring networks provide information on the chemistry of lakes and streams and on how water bodies are responding to changes in emissions. Since the 1980s, scientists measuring changes in a number

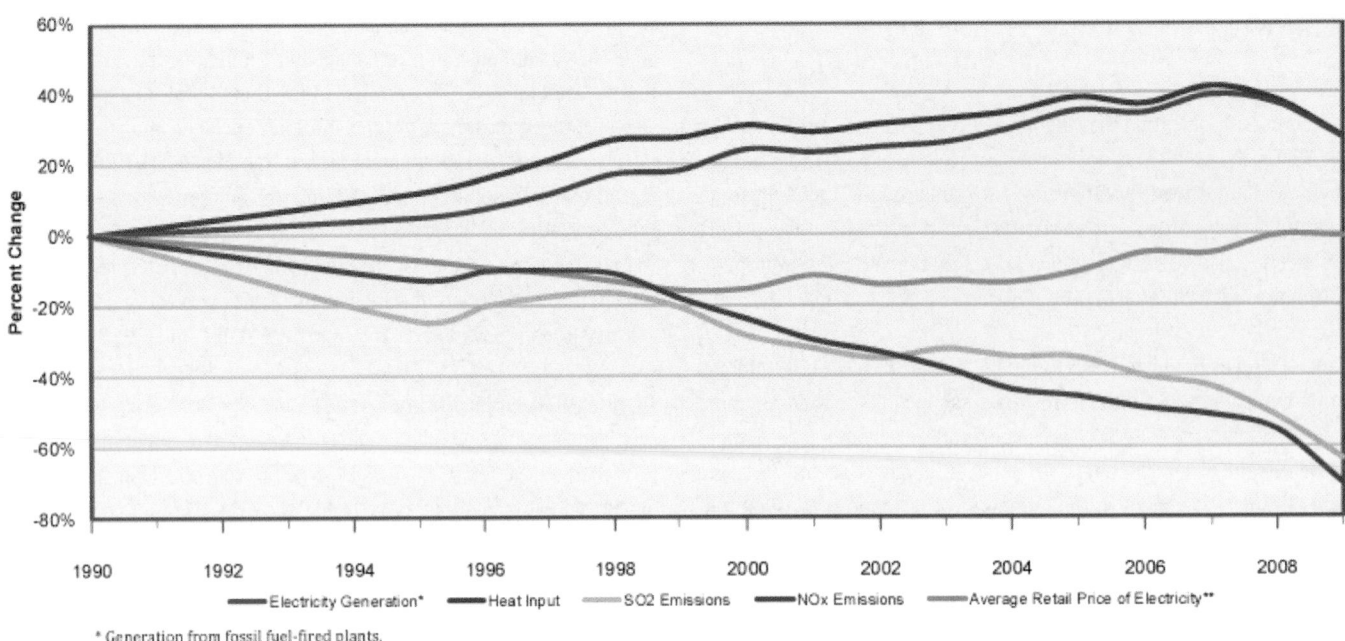

* Generation from fossil fuel-fired plants.
** Constant year 2000 dollars adjusted for inflation.

Figure 3.16 | Trends in SO_2 and NO_x emissions in relation to electricity generation and retail price from 1990 to 2009. Source: US EPA, 2010b.

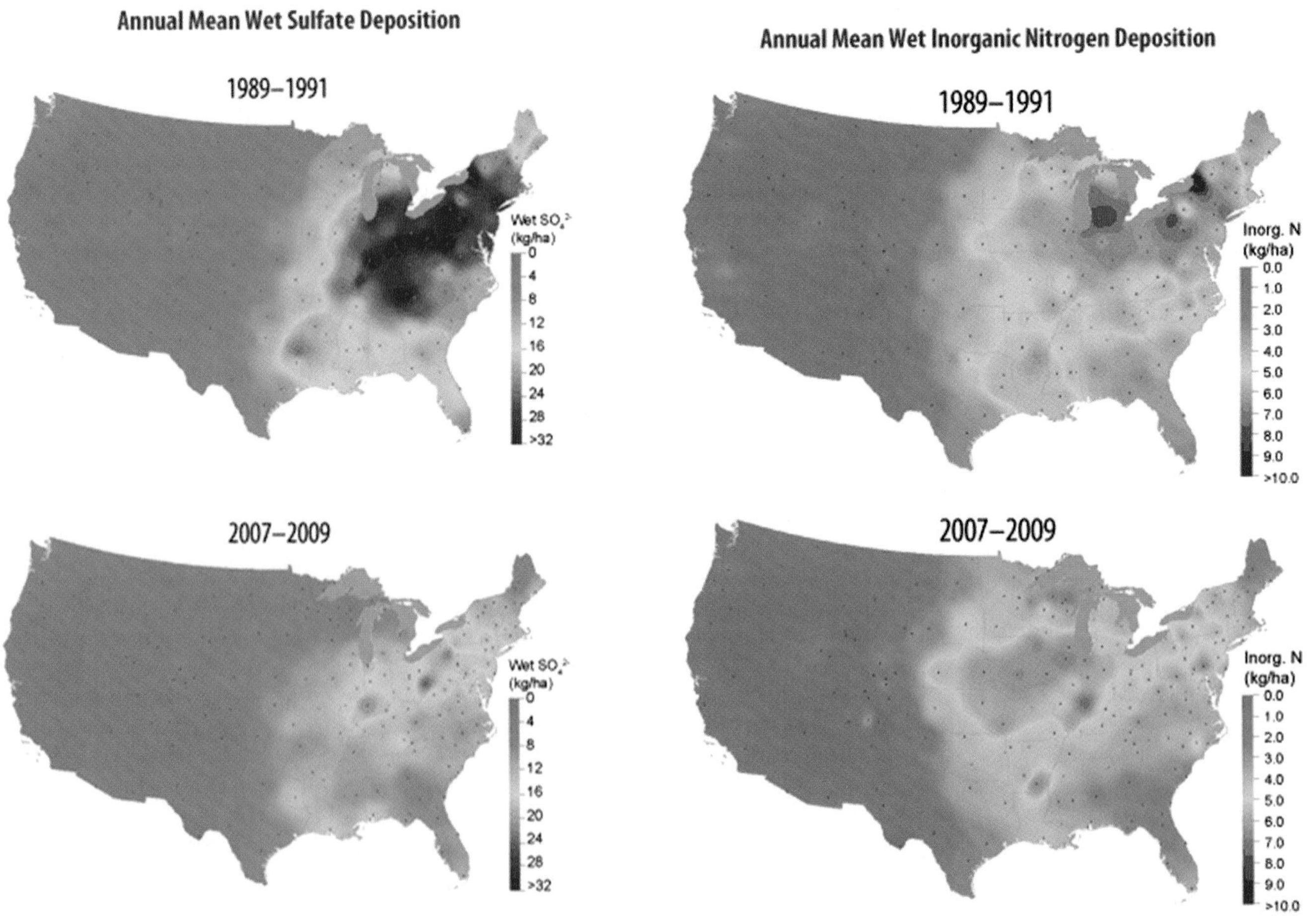

Annual Mean Wet Sulfate Deposition

1989–1991

Annual Mean Wet Inorganic Nitrogen Deposition

1989–1991

2007–2009

2007–2009

Figure 3.17 | Reductions in sulphate and nitrogen deposition in the United States between 1989 and 1991. Source: NADP/NTN monitoring data, US EPA, 2010b.

of lakes and streams in the eastern United States have found signs of recovery in many, but not all, of those bodies of water. These positive signs are a result of emission reductions that have occurred since the early 1990s (US EPA, 2010b).

The Long-Term Monitoring (LTM) program provides information on the effects of acid rain on aquatic systems. This program was designed to track the success of the 1990 Clean Air Act Amendments in reducing the acidity of surface waters in four regions: New England, the Adirondack Mountains, the Northern Appalachian Plateau, and the Central Appalachians (the Ridge and Valley and Blue Ridge Provinces). The surface-water chemistry trend data in the four regions monitored by the Long-Term Monitoring program (US EPA, 2010b) is essential for tracking the ecological response to emission reductions. One major finding of the program was that sulphate concentrations are declining at most sites in the Northeast (New England, Adirondacks, and the Catskills/Northern Appalachian Plateau). However, in the Central Appalachians, sulphate concentrations in some streams (21%) are increasing as a result of a decreasing proportion of the deposited sulphate being retained in the soil and an increasing proportion being exported to surface waters. Another important finding was that nitrate concentrations are decreasing in some of the sites in all four regions, but several lakes and streams indicate flat or slightly increasing nitrate trends. This trend does not appear to reflect changes in atmospheric pollutant deposition in these areas and is likely a result of ecosystem factors. A third finding was that acid-neutralizing capacity, as measured in surface waters, is increasing in many of the sites in the Adirondack and Catskills/Northern Appalachian Plateau regions, which can be attributed in part to declining sulphate deposition.

The critical load concept, which was first developed in Europe under the LRTAP Convention to assess the effects of acidification (see Section 3.2.3), has been used in Canada since the 1980s. The concept is currently being considered for use in the United States. The most recent analysis of the status of North American ecosystems with respect to acidification uses this critical load concept. This analysis is found in the Progress Report of the United States and Canada Air Quality Agreement, known as AQA (IJC, 2010). This report shows that in Canada, lakes that are highly acid-sensitive exist throughout northern Manitoba and Saskatchewan. Critical loads were set to protect 95% of the lake ecosystems. Critical-load exceedance occurs close to the base metal smelters in Manitoba and downwind of the oil sands operations in western Alberta. The exceedances were almost entirely due to sulphate deposition. Nitrogen inputs to the lakes, while significant, were virtually entirely retained within their catchments (lake water nitrate levels were below analytical detection in most cases), meaning that at present, N deposition is not an acidifying factor. It is estimated that lakes having critical loads as low as those observed in northern Manitoba and Saskatchewan will be threatened by long-term acid inputs. However, they do not presently exhibit obvious symptoms of chemical damage from anthropogenic acidic deposition (i.e., low pH and/or reduced alkalinity). Hence, there is still time to protect them from the acidification effects observed in many eastern Canadian lakes.

In the United States, the critical load approach is not yet officially accepted for ecosystem protection. For example, language specifically requiring a critical load approach does not exist in the Clean Air Act. Nevertheless, several federal agencies are now employing critical load approaches to protect and manage sensitive ecosystems. Modelling studies (US EPA, 2008) show that for the period 1989–1991, 56% of modelled lakes and streams received acid deposition greater than their estimated critical load, compared to approximately 36% in the 2006–2008 period. However, there are still regions where exceedances of critical loads occur. Areas with the greatest concentration of lakes in which acid deposition currently is greater than – or exceeds – estimated critical loads include the Adirondack Mountain region in New York; southern New Hampshire and Vermont; northern Massachusetts; northeast Pennsylvania; and the central Appalachian Mountains of Virginia and West Virginia (US EPA, 2008). Therefore, even though there has been improvement in acidic deposition rates over the past decade (US EPA, 2010b) (see Figure 3.17) application of the critical load concept and estimates from the scientific literature indicate that acid-sensitive ecosystems in the northeastern United States are still at risk of acidification at current deposition levels. As a result, additional reductions in acidic deposition from current levels might be necessary to protect these aquatic ecosystems (US EPA, 2008).

Acidification Impacts on Terrestrial Ecosystems

Certain ecosystems in the continental United States are potentially sensitive to terrestrial acidification, which is the greatest concern regarding N and S deposition (US EPA, 2008). Figure 3.18 depicts the areas across the United States that are potentially sensitive to terrestrial acidification. Current understanding of the effects of acid deposition on forest ecosystems has increasingly focused on the biogeochemical processes that affect plant uptake, retention, and cycling of nutrients within forested ecosystems. Research conducted during the 1990s indicated that decreases in base cations (e.g., calcium, magnesium, and potassium) from soils are at least partially attributable to acid deposition in the northeastern and southeastern United States (US EPA, 2008).

Ground-level Ozone (O_3)

Across North America, the approaches to establishing air quality guidelines for O_3 differ. In Canada, the federal government sets National Ambient Air Quality Objectives (NAAQOs) on the basis of recommendations from the National Advisory Committee and Working Group on Air Quality Objectives and Guidelines. Currently, 'desirable' and 'acceptable' NAAQOs for O_3 are defined as 50 ppb and 80 ppb respectively, expressed as an average O_3 concentration over a one-hour period. Provincial governments have the option of adopting these either as objectives or as enforceable standards, according to their legislation (Environment Canada, 2010).

In the United States, the US EPA is responsible for setting National Ambient Air Quality Standards (NAAQs), which are legally enforced. A revision of the NAAQs is currently ongoing as of 2011, and includes the possibility of introducing a more stringent secondary standard in addition to the primary standard that is designed to protect human health.

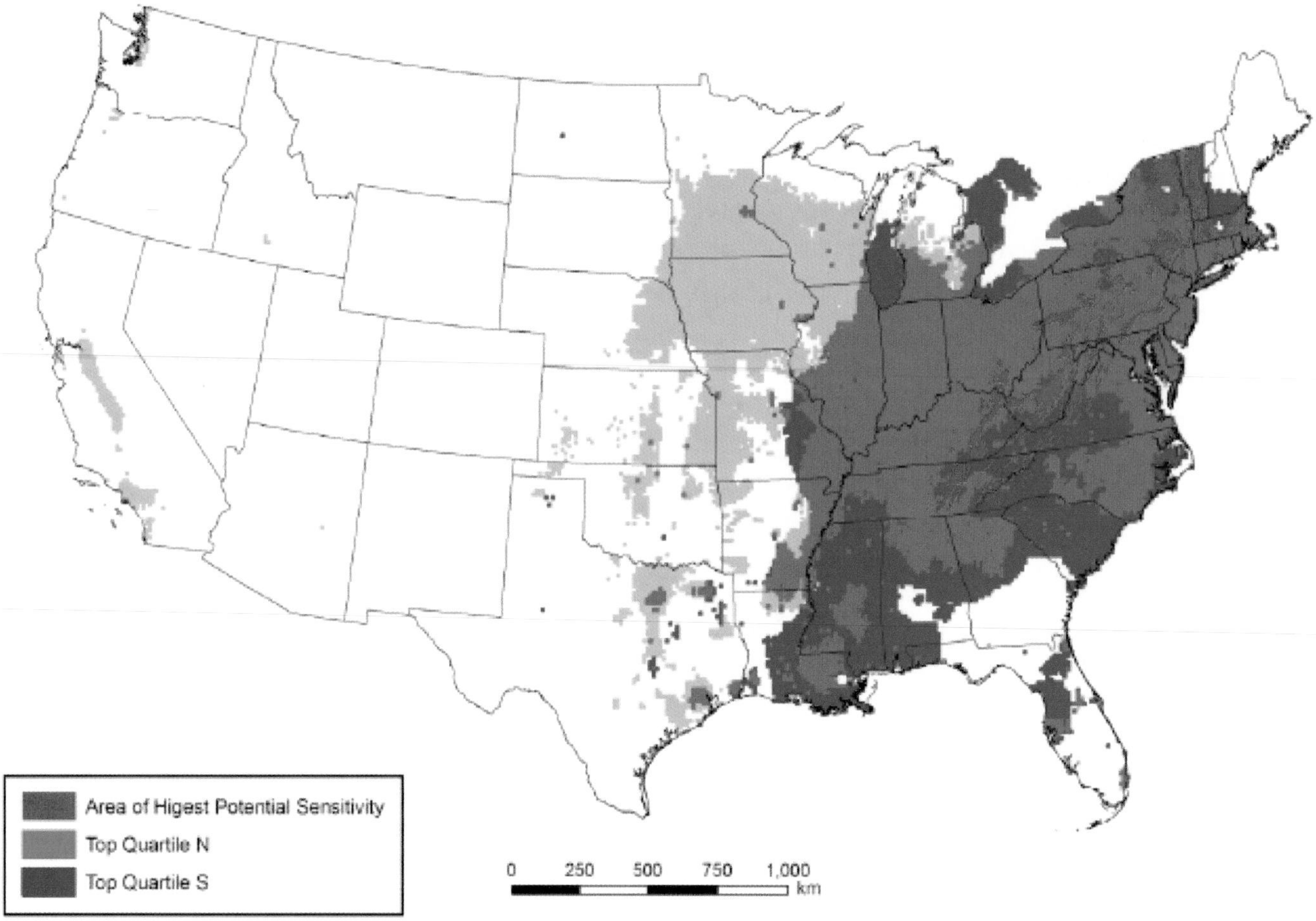

Figure 3.18 | Areas potentially sensitive to terrestrial acidification. Source: US EPA, 2011c.

The secondary standard protects against welfare effects (e.g., impacts on vegetation, crops, ecosystems, visibility, climate, man-made materials, etc.). Currently, the primary standard is 75 ppb averaged over an eight-hour period, or 120 ppb averaged over a one-hour period, with the secondary standard the same as the primary standard (US EPA, 2011a). Long-term trends in O_3 concentrations across the United States reflect notable decreases of approximately 29% in the second highest one-hour O_3 concentrations from 1980–2003, and of about 21% in the fourth highest eight-hour O_3 concentration during the same time period (see Figure 3.19). However, the effects of rising background O_3 concentrations (Vingarzen et al., 2004) and worsening chronic O_3 concentrations are uncertain in relation to ecosystem effects, since most experimental studies have used fumigation profiles that mimic episodic O_3 peaks.

The ecological effects of O_3 appear to be widespread across North America, based on biomonitoring studies and forest health surveys (US EPA, 2011b). The majority of existing experimental evidence comes from biomonitoring and Open Top Chamber studies that were conducted in North America under the National Crop Loss Assessment Network Programme (Heck et al., 1988). In recent decades, there has been a shift from chamber-based studies to field-based approaches to assess O_3

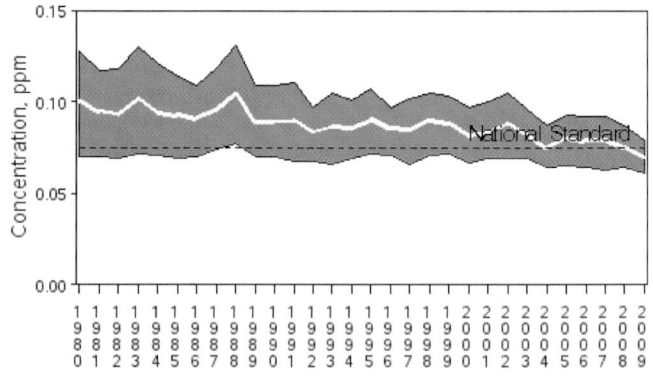

Figure 3.19 | Long-term trends in O_3 concentrations for the United States (1980–2003), based on 255 sites, represented as the fourth highest eight-hour O_3 concentration, with a 30% decrease in national average. Source: US EPA, 2011a.

impacts. In the United States, such studies have investigated forest trees (aspen, maple, and birch) in Wisconsin (Karnosky et al., 2005) and crops (soybeans) in Illinois (Morgan et al., 2006). These studies tend to confirm the detrimental effects of O_3 exposure on vegetation that where found in chamber studies (King et al., 2005; Long et al., 2005).

Extrapolation of data provided by these site-specific studies to the entire United States suggests serious risk to crops and forests from O_3 pollution. The 'cornbelt' in the United States produces 40% of the world's corn and soybean crops, and this region is already potentially losing 10% of its soybean production to O_3 (Tong et al., 2007). In the United States as a whole, agronomic crop loss to O_3 is estimated to range from 5–15%, with an approximate cost of $3–5 billion annually (Fiscus et al., 2005; US EPA, 2006). Despite the overwhelming evidence that current O_3 concentrations are causing yield losses, new O_3-tolerant crop cultivars are not being developed for a future world that could be characterized by higher levels of O_3 (Ainsworth et al., 2008). In addition to regional-scale O_3 risk assessment based on site-specific experimental data, epidemiologic methods have also been used for crop-loss assessments (Fishman et al., 2010). These studies have found that the influence of O_3 can be detected in regional-level production statistics and field trial data, though damage estimates have been found to differ from those obtained from risk assessments performed using empirically derived dose-response relationships. This may be due to the fact that these methods are most effective in those regions characterized by higher average O_3 concentrations (Fishman et al., 2010), where the ozone signal is strong enough to overcome the influence of confounding variables affecting yield.

For forest trees, O_3 has been shown to cause visible foliar injury, accelerate leaf senescence, reduce photosynthesis, alter carbon allocation, and reduce growth and productivity (Karnosky et al., 2007). These effects vary by forest tree species and genotype (Karnosky and Steiner, 1981). Since the mid 1950s, the San Bernardino mixed coniferous forest in southern California has been exposed to some of the highest O_3 concentrations in North America. This exposure has affected such species as ponderosa and Jeffrey Pines. When concentrations were at their height in the 1970s, annual tree mortalities of between 2–2.5% were being reported (Miller and McBride, 1999). Extrapolation of site-specific experimental data in order to perform regional assessments in the United States provides estimates of annual yield losses equivalent to US$80 million (Muller and Mendelsohn, 2007). Finally, although few experimental studies have been conducted investigating the effects of O_3 on seminatural ecosystems in the United States, those that do exist have highlighted O_3 impacts on the nutritive quality of forage crops (Krupa et al., 2004).

Unlike in North America, in South America there have been relatively few studies on the impact of O_3 on vegetation. There is also a dearth of observational evidence on the subject. The most notable exceptions are studies conducted in Mexico in the vicinity of Mexico City. These studies have found evidence of visible foliar injury occurring on forest species, especially in the Valley of Mexico and in Desierto de los Leones National Park (Hernandez and de Bauer, 1984; Hernandez and de Bauer, 1986).

3.2.4.3   Africa

Energy supply and demand in Africa is extremely heterogeneous. For example, northern Africa is a major supplier of oil and gas, while in sub-Saharan Africa more than 70% of the energy demand is met by biomass, mostly from wood and charcoal. (see Chapters 19 and 23 for discussions on energy access issues in Africa). Wood, including charcoal, is the most environmentally detrimental biomass energy source, as it leads to considerable deforestation (Okello, 2001), which is one of the most pressing environmental problems faced by many African nations. Forests cover about 22% of the region, but they are disappearing faster than anywhere else in the world (FAO, 2005). Many sub-Saharan African countries have suffered depletion of more than three quarters of their forest cover (Energy Information Administration, 2000). Africa lost 10.5 % of its forests during the 1980s. It is estimated that if current trends continue, many areas, especially those in the Sudano-Sahelian belt, will experience a severe shortage of fuelwood by 2025 (Energy Information Administration, 2000). It should also be noted that producing and using charcoal is less efficient than using fuelwood directly and leads to forest depletion in rural areas providing fuel to cities (WHO, 2002). Deforestation also causes other problems, such as increased erosion and loss of biodiversity, and has contributed to desertification. Other environmental impacts directly associated with energy systems in Africa include atmospheric emissions from gas 'flaring' and impacts of large-scale hydropower.

Climate Change Environmental Impacts

Although Africa's contribution to global GHG emissions is only about 3%, the continent is more vulnerable than many of the world's regions to the impacts of climate change (Davidson et al., 2007). This vulnerability stems from the fact that, on average, Africa is hotter and drier than most other regions of the world and has less dependable rainfall. Moreover, Huang et al. (2009) show a large-scale effect of aerosols on precipitation in the West African monsoon region. They report a statistically significant precipitation reduction associated with high aerosol concentrations near the coast of the Gulf of Guinea from late boreal autumn to winter. Aerosols originate from various African sources. For example, large quantities of desert dust and biomass-burning smoke are emitted during much of the year across the African continent.

Climate change and variability have the potential to impose additional pressures on water availability, water accessibility, and water demand in Africa (Boko et al., 2007). The population at risk of increased water stress due to climate change is projected to be 75–250 million and 350–600 million people by the 2020s and 2050s, respectively (Arnell, 2004). The impact on water resources is likely to be greatest in northern and southern Africa. In eastern and western Africa, however, more people will be likely to experience a reduction in water stress (Arnell, 2006). In the future, climate change may become a contributing factor to conflicts, particularly those concerning water scarcity (Fiki and Lee, 2004).

It has been estimated by Mendelsohn et al. (2000) that by 2100, parts of the Sahara are likely to emerge as the most vulnerable to climate change, showing likely agricultural losses of between 2–7% of GDP. Western and central Africa are also vulnerable, with impacts ranging from 2–4% of GDP. By contrast, it's estimated that northern and southern Africa will experience losses of 0.4–1.3% GDP.

It is predicted that a significant decrease in the extent of suitable rain-fed land and production potentials for cereals will occur by the 2080s as a result of climate change. Furthermore, for the same projection and time horizon, the area of semiarid land in Africa could increase by 5–8%. southern Africa would be likely to experience a notable reduction in maize production under possible increased El Niño/La Niña-Southern Oscillation (ENSO) conditions (Stige et al., 2006). In some countries, additional risks that could be exacerbated by climate change include greater erosion, deficiencies in yields from rain-fed agriculture of up to 50% during the 2000–2020 period, and reductions in the length of the crop-growing season (Agoumi, 2003). Other agricultural activities could also be affected by climate change and variability, including changes in the onset of rainy days and the variability of dry spells (Reason et al., 2005).

It has been suggested that climate change will have a range of impacts on terrestrial and aquatic ecosystems. For example, two climatic-change scenarios used to assess the sensitivity of African mammals in 141 national parks in sub-Saharan Africa suggest that climate change will have an impact on species diversity. Assuming no migration of species, Boko et al. (2007) projected that 10–15% of the species will fall within the International Union for Conservation of Nature (IUCN) Critically Endangered or Extinct categories by 2050, a figure that will increase to 25–40% by 2080. Assuming unlimited species migration, the results were less extreme, with these percentages dropping to approximately 10–20% by 2080.

In Africa, highly productive ecosystems (mangroves, estuaries, deltas, lagoons, and coral reefs), which form the basis for important economic activities such as tourism and fisheries, are located in the coastal zone. The projected rise in sea-level will have significant impacts on the coastal megacities of Africa because of the concentration of poor populations in areas that are especially vulnerable to such changes (Klein et al., 2002; Nicholls, 2004). By 2080, across a range of SRES scenarios and climate change projections, three of the five regions shown to be at risk of flooding in coastal and deltaic areas of the world are located in Africa (Nicholls and Tol, 2006). Sea-level rise, combined with increases or decreases in rainfall, will alter the penetration of salt water into estuaries and could induce overtopping and even destruction of the low barrier beaches that limit the coastal lagoons, while changes in precipitation could affect the discharges of rivers feeding them. These changes could also affect lagoonal fisheries and aquaculture (République de Côte d'Ivoire, 2000).

The Indian Ocean islands could also be threatened by potential changes in the location, frequency, and intensity of cyclones, while East African coasts could be affected by potential changes in the frequency and intensity of ENSO events and coral bleaching (Klein et al., 2002). Coastal agriculture (e.g., plantations of palm oil and coconuts in Benin and Côte d'Ivoire, shallots in Ghana) could be at risk of inundation and soil salinization. In Kenya, losses for three crops (mangoes, cashew nuts, and coconuts) could cost almost US$500 million for a 1 meter sea-level rise (Republic of Kenya, 2002). In Guinea, between 130–235 km^2 of rice fields (17–30% of the existing rice field area) could be lost as a result

of permanent flooding, depending on the inundation level considered (between 5–6 m) by 2050 (République de Guinée, 2002). In Eritrea, a 1 meter rise in sea-level could cost over US$250 million, as a result of the submergence of infrastructure and other economic installations in Massawa, one of the country's two port cities (State of Eritrea, 2001). These results confirm previous studies stressing the great socioeconomic and physical vulnerability of settlements located in marginal areas.

Initial assessments show that several regions in Africa may be affected by different impacts of climate change (Figure 3.20). Such impacts may further constrain attainment of the Millennium Development Goals in Africa.

Air Quality Environmental Impacts

Increased activities in key social and economic sectors (see Table 3.1) are contributing significantly to air pollution in Africa. Unsustainable patterns of use and supply of energy resources by industry, transport, and household sectors have been particularly important as sources of indoor and outdoor air pollutants. Generally, emission levels of NO_2 and SO_2 have increased significantly in many African countries due to the region's industrial activity. In addition, about 1.8 $MtSO_2$/yr are emitted from electricity generation alone (UNEP, 2006). However, Table 3.1 also shows the significant contribution to PM, NMVOCs, NO_x, and CH_4 pollution resulting from agricultural activities (i.e., not associated with energy systems).

Africa's rate of urbanization is the highest in the world. The rate is increasing rapidly, with urbanization rates in sub-Saharan Africa generally in excess of 4% (Clancy, 2008). This results in rapidly increasing energy use and pollutant emissions from industry, motor vehicles, and households, including SO_x, CO_2, and NO_x, as well as PM and other organic compounds (UNEP, 2006). The transportation sector is increasingly being recognized as the highest polluter in key African cities such as Cairo, Nairobi, Johannesburg, Cape Town, Lagos, and Dakar (UNEP, 2006). The transport systems in these cities are emitting tonnes of air pollutants, mainly NO_x, PM, and VOCs. Some of the challenges related to transportation are the use of old vehicles without emission controls; rapidly increasing fleets; an increase in two- and three-wheeler vehicles with 'dirtier' two-stroke engines; absent or improper vehicle maintenance; lack of cleaner burning fuels; and absent or poor regulatory frameworks for vehicle emissions and their enforcement (Schwela, et al., 2007). In addition, inadequate urban planning and poor road networks have led to traffic congestion in most African cities, with impacts on fuel wastage and air pollution. Traditional cooking is also a significant emitter of CO_2, CO, and NO_x (UNEP, 2006). Charcoal production emits significant levels of CH_4, CO, and other products of incomplete combustion and PM (UNEP, 2006).

However, most African countries do not have monitoring networks to measure air pollutants. Therefore, no comparisons of air quality can be made at the subregional level. Recognizing such limitations, the Air Pollution Information Network for Africa (APINA) has identified the lack of reliable emission inventories, the lack of experience in the use of atmospheric models, and the lack of data on measured impacts as important

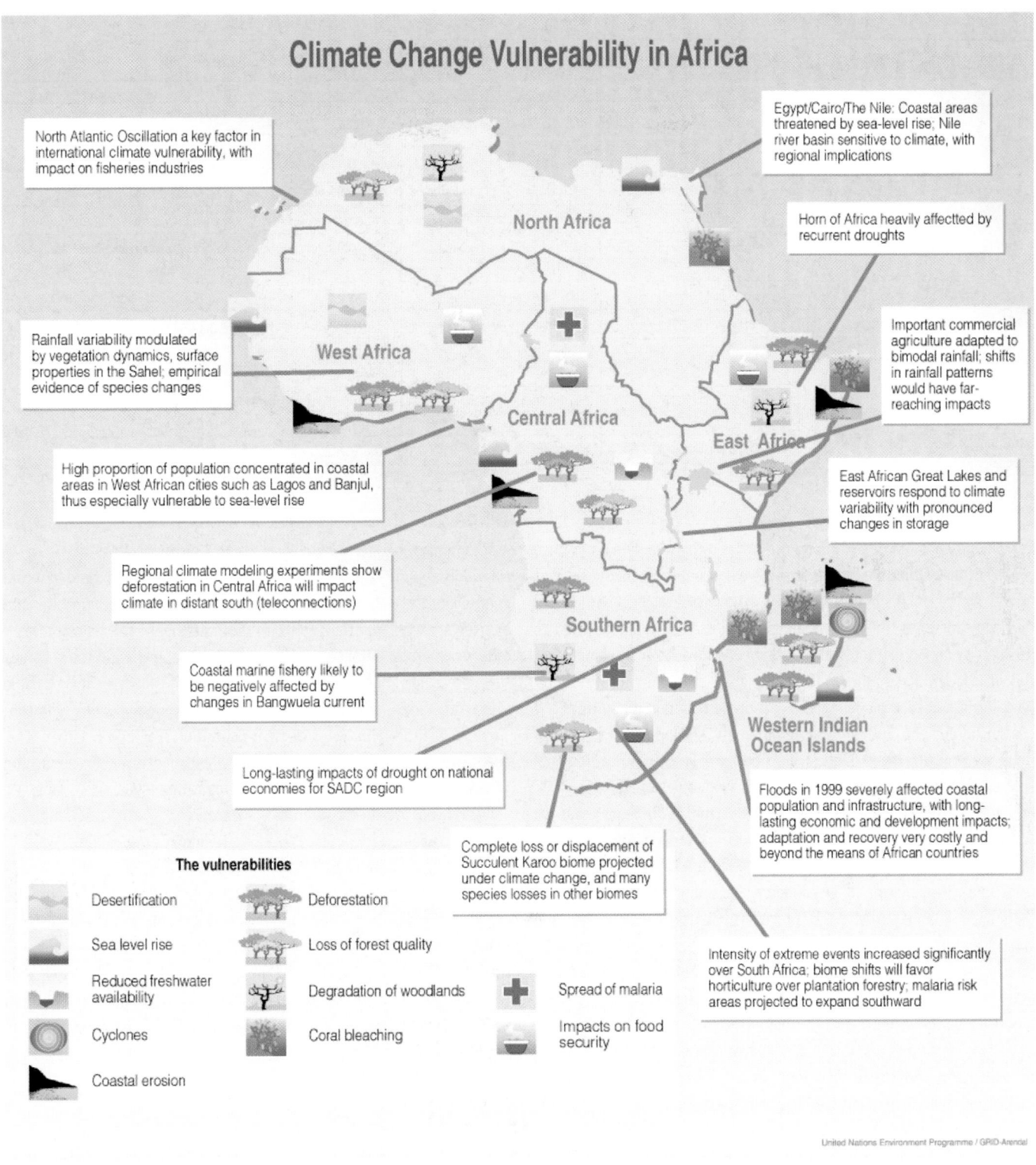

Figure 3.20 | The vulnerability of different African regions to the multiple stresses likely to result from climate change. These impacts include desertification, sea-level rise, reduced freshwater availability, cyclones, coastal erosion, deforestation, loss of forest quality, woodland degradation, coral bleaching, the spread of malaria, and impacts on food security. Source: UNEP/GRID-Arendal, 2005. Courtesy of Delphine Digout, Revised by Hugo Ahlenius, UNEP/GRID-Arendal.

Table 3.1 | Contribution of key social and economic sectors to air pollution in Africa.

Pollutant	Commercial energy supply	Traditional energy supply	Agriculture	Manufacturing, other
Particulate Matter (PM)	35% (fossil fuel fires)	10% (traditional fuel fires)	40% (agriculture fires)	15% (smelting, land clearing, municipal, etc.)
Lead (Pb)	41% (gasoline burning)	Negligible	Negligible	59% (metal processing, manufacturing, municipal)
Cadmium (Cd)	13% (fossil fuel burning)	5% (traditional fuel burning)	12% (agriculture fires)	70% (metal processing)
Mercury (Hg)	20%	1% (traditional fuel burning)	2% (agriculture fires)	77%
Non-methane hydrocarbons (NMVOCs)	35% (fossil fuel burning)	5% (traditional fuel burning)	40% (agriculture fires)	20% (non-agricultural land clearing)
Sulphur dioxide (SO_2)	85% (fossil fuel burning)	0.5% (traditional fuel burning)	1% (agriculture fires)	13% (smelting, municipal)
Nitrogen oxides (NO_x)	30% (fossil fuel burning)	2% (traditional fuel burning)	67% (fertilizer, agriculture fires)	1% (municipal)
Carbon dioxide (CO_2)	75% (fossil fuel burning)	3% (net deforestation for fuelwood)	15% (net deforestation for land clearing)	7% (net deforestation for lumber, cement manufacturing)
Methane (CH_4)	18% (fossil fuel burning)	5% (traditional fuel burning)	65% (rice fields, animals, land clearing)	12% (landfills)
Nitrous oxide (NO)	12% (fossil fuel burning)	8% (traditional fuel burning)	80% (fertilizer, land clearing)	Negligible

Source: UNDP, 2000.

knowledge gaps to understanding the status of air quality across Africa (UNEP, 2006). Other challenges include weak national energy policies as well as the lack of local air-pollution exposure data. Such data would be necessary to establish air-quality standards that could be used to drive policy making targeted at reducing emissions. At the regional level, APINA and UNEP, together with other global air pollution institutions, have facilitated subregional agreements on air pollution that include the Lusaka Agreement (2008) – Southern African Development Community Regional Policy Framework on Air Pollution; the Eastern Africa Regional Framework Agreement on Air Pollution (Nairobi Agreement–2008); and the West and Central Africa Regional Framework Agreement on Air Pollution (Abidjan Agreement–2009). These agreements detail the actions that are required to reduce air pollution, including emissions from energy use, but the financial means and political will to implement them are still limited. Such policy processes will be important, given the variety of the proved energy potential that exists in Africa. At the global level, most African countries are signatories to the UNFCCC and produce national communications that are GHG inventories. However, they do not produce inventories for the other pollutants.

Acidification and Eutrophication

Africa has many ecosystems with a very high biodiversity value and which provide important ecosystems services. There is potential for sensitive ecosystems to be detrimentally affected by acidifying and eutrophying deposition. However, further research is required in this area to assess the full extent of ecosystem damage.

Acidification is potentially a problem for sensitive terrestrial and aquatic ecosystems downwind of major emission sources of SO_x and NO_x. One such sensitive ecosystem is in Mpumalanga on the Highveld of South Africa, where there are eight coal-fired power stations (Josipovic et al., 2010). A concentration, distribution, and critical level exceedance

assessment of SO_2, NO_2, and O_3 in South Africa (Josipovic et al., 2010) shows that some critical levels for vegetation are only exceeded in the central area of the South African industrial Highveld where emission sources are high. Pollutant concentrations are below the critical thresholds for environmental damage in remote areas, including the sensitive forested regions of the Drakensberg escarpment. However, it is possible that critical loads for acidification may be exceeded in areas with sensitive soils and higher rates of total deposition (Fey and Dodds, 1998). There are also soils that are potentially sensitive to acidic deposition across western and central Africa, but deposition levels of S and N are still relatively low in these areas (Kuylenstierna et al. 2001), although they are projected to increase by 2030 (Dentener et al., 2006).

It is thought that between 2000 and 2030, deposition levels in Africa are likely to increase, in part due to increased intensity in the use of fossil fuel to meet increasing energy demand; NO_x emissions in sub-Saharan Africa resulting from fossil fuel combustion are already estimated to be 14 Mt/yr (Selman and Greenhalgh, 2009). Such enhanced deposition may seriously affect the integrity of protected areas, especially those in west and central Africa.

Ground-level Ozone (O_3)

Sunlight and heat stimulate O_3 formation. Therefore, the potential for elevated O_3 is high in many areas of Africa because of the combination of solar radiation and the high emission of precursors such as NO_x and VOCs, from both human activities and natural sources. There is a sharp increase of O_3 with altitude in the tropics, leading to a risk of vegetation damage in high mountainous regions (UNEP, 2006).

Studies in Africa have shown that current-day ambient air pollution concentrations can significantly damage human health, crops, local vegetation, and other materials. Ozone injury to turnips has been

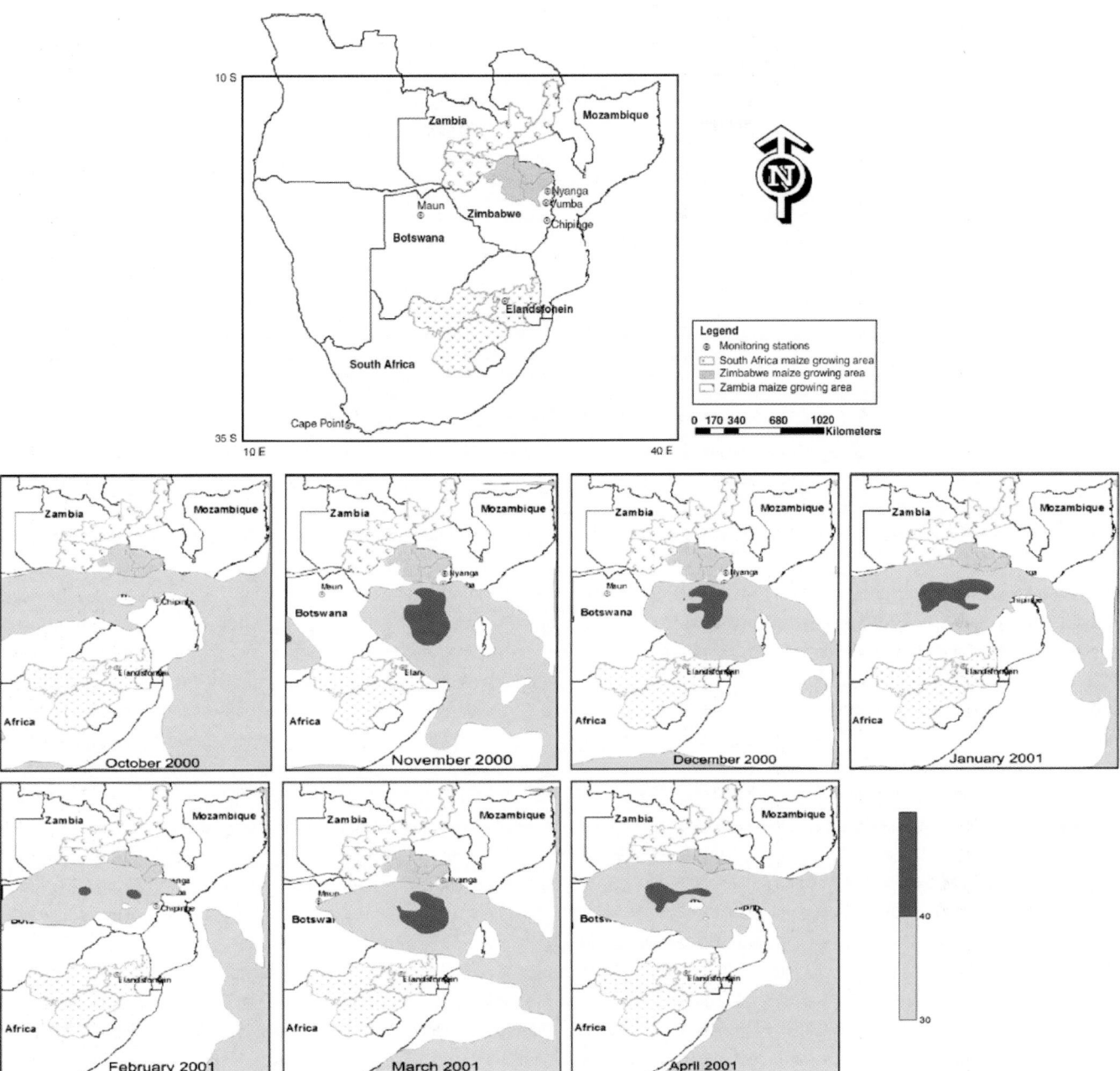

Figure 3.21 | Map of southern Africa showing the modelling domain, maize-growing regions, and locations of O_3-monitoring sites; the series of maps shows average daylight O_3 concentrations in ppb over the main maize-growing area in southern Africa for a five-day period (10th-14th) in each of the months from October 2000 to April 2001. Source: van Tienhoven et al., 2006.

reported in Egypt (Hassan et al., 1995). In parts of Africa, predicted increases in global background concentrations of O_3 combined with trends of increased emissions of O_3 precursors suggest that current and future O_3 impacts on crops and forests in these areas may be significant (Ashmore, 2005). Modelling studies based on 'current legislation' emission projections performed by the Royal Society (2008) suggest that O_3 concentrations are likely to increase in many regions by the year 2050, particularly over landmasses in the tropics, Asia, and Africa, largely because of climate change.

Even under current-day conditions, O_3 may be posing a threat to ecosystems across Africa. For example, photochemical modelling studies have suggested that levels of O_3 concentration that have resulted in impacts in Europe may be occurring in parts of Botswana, South Africa, and Zimbabwe in southern Africa (Figure 3.21). In addition, the maize-growing areas at risk from drought are similar to those modelled to be at risk from elevated O_3 concentrations (USAID, 2002). Therefore, maize may be suffering multiple stresses resulting from O_3 and drought, which may compromise crop productivity. Such stress may be magnified in the future, as increased air pollutant emissions lead to higher and more persistent O_3 concentrations. Additional studies are required to fully understand the potential threat posed by these combinations of stress.

3.2.4.4   Asia

Asia is experiencing rapid population growth. China and India alone account for approximately 2.5 billion of the global population; according to UN estimates (2008), this figure will have risen to over 3 billion by 2030. This population growth has been closely matched by strengthening economies in many of the most rapidly developing Asian nations. For example, over the past five years (2005–2010), China's annual GDP growth rate has been between 9–13%, while India's growth rate has fluctuated between 5–9 % (UN, 2011). This growth has in part been made possible by increased access and utilization of fossil fuels (van Ardenne et al., 1999), as evidenced by the present energy mix for the Centrally Planned Asia and South Asia regions. In these regions, approximately 85% of energy demand is met by fossil fuels, with coal being the principle energy source (summarized from the GEA database).

This heavy reliance on coal and other fossil fuels translates into substantial CO_2 emissions, with over 1600 MtC/year under current day (2005) conditions (as estimated from the GEA database). Such coal use also results in substantial emissions of SO_2. For example, in 2005 China emitted 25.5 $GtSO_2$, reversing a decreasing trend in SO_2 emissions that was evident prior to 2002 (Chan and Yao, 2008); this has implications for acidification across many parts of the country. In addition to increases in CO_2 and SO_2 emissions, NO_x emissions have also increased, due to combustion of fossil and biomass fuels, especially in the transport, industrial, and power-generation sectors (van Ardenne, 1999). The resulting increases in atmospheric NO_x concentrations will hasten environmental degradation caused by acid deposition, eutrophication, and ground-level O_3. The resulting high levels of air pollution across the Asian region are not a new phenomenon; atmospheric emissions have been steadily increasing over the past few decades, as countries in the region have experienced rapid industrialization and economic growth (Ohara et al., 2007). Even with this increase in economic growth in many Asian countries, large sectors of the population remain deprived of basic facilities and amenities. For example, around 500 million people still do not have access to electricity, and many more lack access to clean drinking water and modern energy fuels like liquid petroleum gas (LPG) and kerosene to meet their domestic cooking and lighting needs. This state of affairs has consequences for both indoor and outdoor pollution and human health (WHO, 2006b; for more information, see Chapter 4). Additionally, many poor people are reliant on agriculture for their livelihoods, and hence are also at risk from the threat of air pollutants that reduce the quality and quantity of crops as well as from droughts that are becoming more frequent with the onset of climate change.

Climate Change Environmental Impacts in Asia

The IPCC Fourth Assessment reported the increasing intensity and frequency of extreme weather events in Asia over the last century and into the 21st century (Cruz et al., 2007). It is likely that climate change will impinge on the sustainable development of most of the developing countries of Asia, as it compounds the pressures on natural resources and the environment associated with rapid urbanization, industrialization, and economic development.

Climate change is expected to affect forest expansion and migration, and exacerbate threats to biodiversity resulting from land-use, cover change, and population pressure in most of Asia (Cruz et al., 2007). Increased risk of extinction for many flora and fauna species in Asia is likely as a result of the synergistic effects of climate change and habitat fragmentation. In North Asia, forest growth and a northward shift in the extent of boreal forest is probable; the frequency and extent of forest fires in North Asia may well increase in the future. In southeast Asia, extreme weather events associated with El Niño were reported to be more frequent and intense in the past 20 years (Cruz et al., 2007). Significantly longer heat-wave duration has been observed in many countries of Asia, as indicated by pronounced warming trends and several cases of severe heat-waves. Generally, the frequency of occurrence of more intense rainfall events in many parts of Asia has increased, causing severe floods, landslides, and debris and mud flows, while the number of rainy days and the total annual amount of precipitation has decreased. However, there are reports that the frequency of extreme rainfall in some countries has exhibited a decreasing tendency. Increasing frequency and intensity of droughts in many parts of Asia are attributed largely to a rise in temperature, particularly during the summer and normally drier months, and during ENSO events. Recent studies indicate that the frequency and intensity of tropical cyclones originating in the Pacific have increased over the last few decades. In contrast, cyclones originating from the Bay of Bengal and Arabian Sea have decreased since 1970, but their intensity has increased. Damage caused by intense cyclones originating in both areas has risen significantly, particularly in India, China, Philippines, Japan, Vietnam, Cambodia, Iran, and the Tibetan Plateau.

Marine and coastal ecosystems in Asia may well be affected by sea-level rise and temperature increases (Cruz et al., 2007). Projected sea-level rise is very likely to result in significant losses of coastal ecosystems, and a million or so people along the coasts of south and southeast Asia may be at risk from flooding. Sea-water intrusion due to sea-level rise and declining river runoff is likely to increase the habitat of brackish water fisheries. Coastal inundation could seriously affect the aquaculture industry and infrastructure, particularly in heavily populated megadeltas. The stability of wetlands, mangroves, and coral reefs around Asia is likely to be increasingly threatened. Risk analysis of coral reefs suggests that 24–30% of the reefs in Asia may be lost during the next 10–30 years (Wilkinson, 2004), unless the stresses are removed and relatively large areas are protected.

Rapid thawing of permafrost and a decrease in the depths of frozen soils, due largely to rising temperatures, have threatened many cities and human settlements. They have led to an increase in the frequency of landslides, a degeneration of some forest ecosystems, and an increase in lake-water levels in the permafrost region of Asia (Cruz et al., 2007). In drier parts of Asia, melting glaciers account for over 10% of freshwater supplies. Glaciers in Asia are melting faster in recent years than before, as reported in Central Asia, Western Mongolia, and Northwest China, particularly the Zerafshan glacier, the Abramov glacier, and the glaciers on the Tibetan Plateau (Pu et al., 2004). Mudflows and avalanches have increased as a result of the rapid melting of glaciers, glacial runoff, and the frequency of glacial lake outbursts (WWF, 2005). A recent study in northern Pakistan, however, suggests that glaciers in the Indus Valley region may be expanding, due to increases in winter precipitation over western Himalayas over the past 40 years (Archer and Fowler, 2004).

The production of rice, maize, and wheat during the past few decades has declined in many parts of Asia, due to increasing water stress. This stress can be attributed in part to increasing temperatures, the increasing frequency of El Niño, and a reduction in the number of rainy days. In a study at the International Rice Research Institute, the yield of rice was observed to decrease by 10% for every 1°C increase in growing-season minimum temperature (Peng et al., 2004). A decline in potentially good agricultural land in East Asia and substantial increases in suitable areas and production potentials in currently cultivated land in Central Asia have also been reported (Fischer et al., 2002). Climate change could make it more difficult to step up agricultural production to meet the growing demands in Russia and other developing countries in Asia.

In Asia, water shortages have been attributed to rapid urbanization and industrialization, population growth, and inefficient water use, which are aggravated by changing climate and its adverse impacts on demand, supply, and water quality (Cruz et al., 2007). Overexploitation of groundwater in many countries of Asia has resulted in a drop in its level, leading to ingress of sea water in coastal areas, making the sub-surface water saline. India, China, and Bangladesh are especially susceptible to increasing salinity of their groundwater as well as surface water resources, especially along the coast, due to increases in sea-level as a direct impact of global warming. Increasing sea-level by 0.4–1.0 m can induce saltwater intrusion 1–3 km farther inland in the Zhujiang estuary. Increasing frequency and intensity of droughts in the catchment area will lead to more serious and frequent salt-water intrusion in the estuary and thus deteriorate surface and groundwater quality.

Air Quality Environmental Impacts in Asia

Eutrophication

Recent global modeling studies suggest that N deposition related to NO_x and NH_3 emissions are now as high in some parts of south, southeast, and southern East Asia as they are in Europe and North America (Figure 3.22), with much of the NH_3 emission coming from agricultural sources such as cattle rearing. There is growing scientific consensus that eutrophication impacts will be considerable in Asia, although local evidence is only just starting to emerge (Phoenix et al., 2006). For example, in China, manipulation experiments suggest that N deposition has the potential to influence the species richness of the understory of temperate and tropical forests. The UN Convention on

Biological Diversity has identified N deposition as an indicator of a threat to biodiversity in many Asian countries. In addition, effects on biodiversity have been linked to the type of forest decline that has occurred in some areas of China. This decline is the result of the direct effects of SO_2, via extremely acidic mist or rain events, and the effects of soil acidification, which is caused by S and N deposition as discussed in more detail in the following sectors.

Acidification

The extensive use of coal as a primary energy source across Asia has led to concerns over environmental degradation resulting from acidification. The main sectors contributing to SO_2 emissions are coal-fired electricity production and industry, which are responsible for approximately 45% and 36%, respectively, of SO_2 emissions (Hicks et al., 2008). However, other pollutants, especially NO_x and NH_3, are also involved in acidification via their transformations in the atmosphere and soil. Acid deposition is mainly a problem in the southern and southwestern regions of East Asia, where neutralization by alkaline desert dust in the atmosphere is relatively low. These acidifying emissions have resulted in parts of East Asia being identified as the third largest acid rain-prone region in the world. The variability in susceptibility to acid rain evident across this East Asian region would appear to exist across the whole of Asia, with modelling studies indicating that soil acidification effects are unlikely to be widespread, due to the insensitivity of soils and high concentrations of alkaline dust in the atmosphere. According to Hicks et al. (2008), areas at risk of acidification effects over the next 50 years are mainly restricted to areas of southern East and southeast Asia (Figure 3.23).

Various attempts have been made to estimate the societal costs of air pollution and acid rain in China. In 2004, the former Chinese State Environmental Protection Administration estimated the economic loss due to negative effects of acid rain on crops to be US$4.6 billion (World Bank, 2007). Although such estimates are highly uncertain, they make clear that the economic benefits of tackling acid rain may be considerable.

Policies to control SO_2 emissions in East Asia began in the early 1990s, when the concept of acid rain control zones was developed as the main framework for setting priorities on acid rain reduction policy (Shi et al., 2008). Mitigation actions focused on establishing emissions standards for coal- and oil-fired power plants, targeting the use of low-sulphur coals, the relocation of coal-fired power plants to rural areas, the installation of de-sulphurization technologies, increased chimney stack height, the closing of small coal-fired generating units and polluting industries, and the shifting to cleaner energy sources (Cofala and Syri, 1998). Intergovernmental cooperation has also been established through the Acid Deposition Monitoring Network in East Asia, or EANET (Tsunehiko et al., 2001). These measures, along with the economic recession of the late 1990s, resulted in significant decreases in SO_2 emissions. However, coal consumption continued to increase in East Asia, largely due to the continued expansion of electricity generation and heat provisioning coal-fired power plants (e.g., Fang et al., 2008).

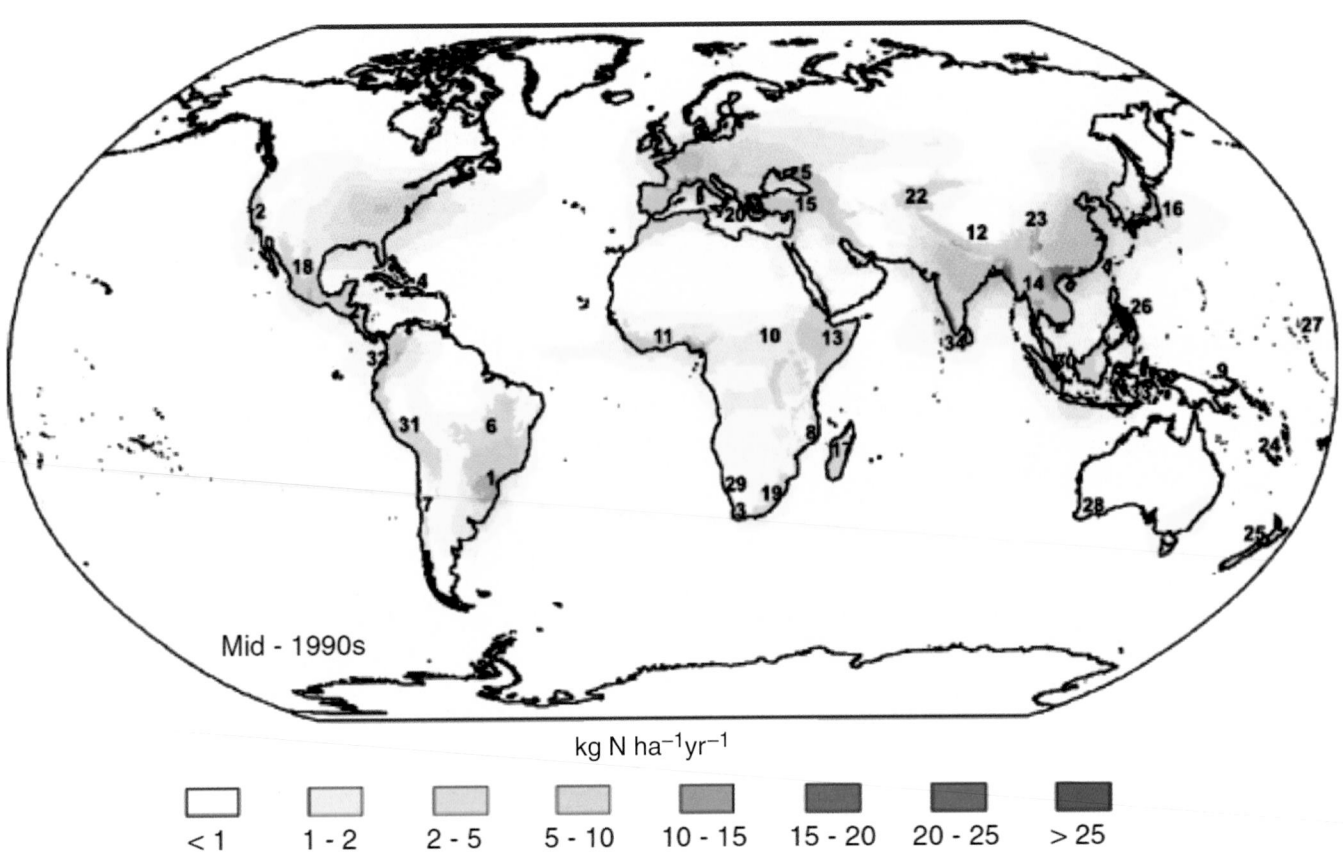

kg N ha^{-1}yr^{-1}

| <1 | 1 - 2 | 2 - 5 | 5 - 10 | 10 - 15 | 15 - 20 | 20 - 25 | > 25 |

Figure 3.22 | Distribution of nitrogen deposition in relation to biodiversity hotspot areas in the mid-1990s showing high nitrogen deposition values in southern China and parts of south and southeast Asia. Numbers on map identify World biodiversity hotspots. Source: Phoenix et al., 2006.

Ground-level Ozone (O₃) and Food Security

The air pollutant that is arguably of greatest concern due to its occurrence at elevated concentrations across large parts of rural Asia is ground-level O$_3$ (Emberson et al., 2009). Experimental evidence collected in Asia over the past few decades provides a strong indication of the threat posed by O$_3$ to agricultural production, with staples such as wheat, rice, and beans commonly showing substantial yield losses of 10–30% under ambient pollution concentrations (Emberson et al., 2009). Modelling studies provide an indication of the magnitude and geographical extent of O$_3$ risk to crop production. These studies indicate that the Indo-Gangetic Plain, one of the most important agricultural regions in the world, is particularly at risk from high O$_3$ concentrations. This potentially has important implications for agricultural production (Van Dingenen et al., 2009). For example, it is possible that O$_3$ may be a significant contributing factor to the yield gap that currently exists across much of Asia. It is also possible that O$_3$ may have played a role in recent reductions in the growth rate in yield of key staple crops, which have recently been a cause for concern across much of south Asia (Emberson et al., 2009).

Economic loss estimates that have been conducted across Asia suggest substantial effects due to yield losses caused by O$_3$. Estimates from East Asia suggest annual losses of US$5 billion, based on four key crops – wheat, rice, soybean, and maize (Wang and Mauzerall, 2004). A global

modelling study that translated production losses (see Figure 3.24, which shows global production loss estimates for wheat) into total global economic damage for the same four commodities, using world market prices for the year 2000, estimated an economic loss of US$16–30 billion per year. About 40% of this damage was found to occur in parts of China and India. For those Asian countries with economies largely based on agriculture, the O$_3$-induced damage was estimated to offset a significant portion (20–80%) of the increase in GDP in the year 2000 (Van Dingenen et al., 2009).

These modelling studies have relied on North American or European dose-response relationships to assess the yield losses caused by O$_3$, since equivalent Asian relationships do not exist. However, recent comparisons of Asian and North American O$_3$-response data strongly suggest that Asian crops (in this instance wheat, rice, and beans) and cultivars may well be more sensitive to O$_3$ concentrations when growing under Asian conditions (i.e., under Asian climates, on Asian soils, and under Asian management practices). This implies that current economic loss estimates may have substantially underestimated the yield, and subsequent production losses, in the region (Emberson et al., 2009).

The indirect effects that air pollutants may have on the climate system should also be considered within the context of agriculture. Work that

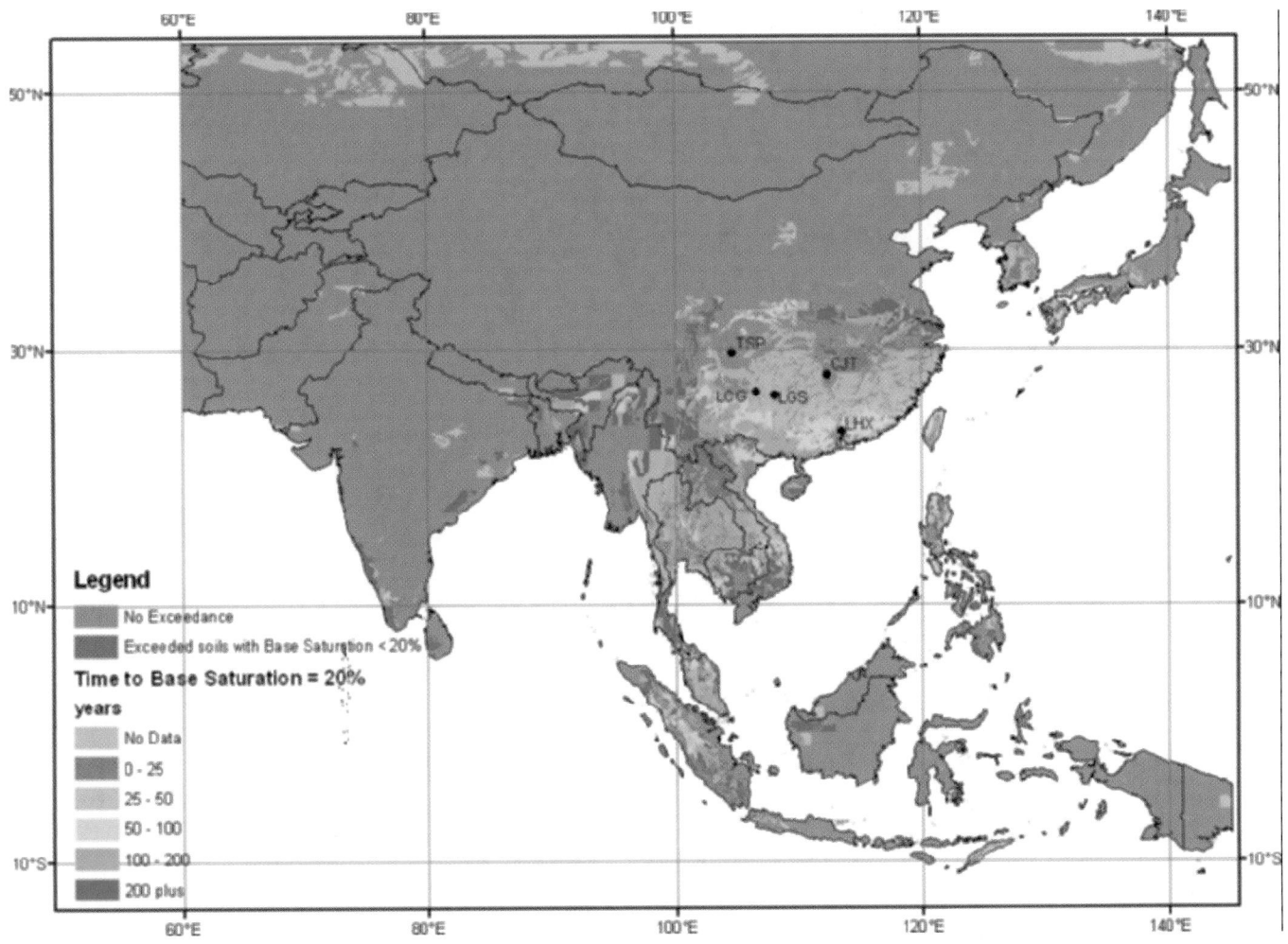

Figure 3.23 | Time development of soil acidification damage according to a modelling study for Asia using the best available data for soil and deposition parameters and deposition estimates obtained using a more pessimistic emission scenario (IPCC SRES A2) for 2030. The model calculates the time it takes for the neutralizing capacity of the soil (expressed as base saturation) to be reduced to a level where acidification effects are observed (i.e., approximately 20% base saturation). Source: Hicks et al., 2008.

has focused on understanding the role of ABCs on climate provides a particularly good example of such effects. The surface dimming (reduction in sunlight reaching the Earth's surface) associated with ABCs, described earlier in this chapter, is thought to impact agricultural yields in Asia, with estimates suggesting that 70% of the crops grown in China may have their yields suppressed by 5–30%. In addition, yields may be impacted through alterations to monsoon rainfall (UNEP, 2008). In particular, BC may play an important role in causing monsoonal shifts, as it dims the surface and warms the lower atmosphere, producing effects on the vertical atmospheric temperature profile, evaporation, atmospheric stability, and strength of convection. These changes in precipitation will also impact agricultural management and productivity across the region.

3.2.4.5   Polar Regions

The Earth's polar regions, the areas of the globe surrounding the North and South Pole, are dominated by polar ice caps. The fact that both of these polar regions are so remote from large-scale industrial activities might

suggest that they would remain as pristine environmental strongholds. However, over recent decades the particular sensitivity of these regions to environmental degradation and global environmental change has been striking. Examples of such degradation have involved the global-scale transport and accumulation of persistent organic pollutants (POPs) within ecosystems, the depletion of the stratospheric O_3 layer, the transport of radionuclide material and, most recently, observations of the occurrence of enhanced climate change above the global mean. In addition, the lack of obvious sovereignty of the Arctic region, coupled with this region's wealth of natural resources, threatens to lead to further exploitation of a region that is unprotected by international treaty. Here, the potential of the Arctic as a supply of fossil fuels and the increased demand for energy is of particular relevance. These issues are discussed in the following sections to highlight the possible consequences of unrestrained energy supply and demand for these sensitive regions.

The Sensitivity of the Polar Regions

In recent decades, the particular sensitivity of the polar regions has become worryingly apparent. One of the best examples of this is the

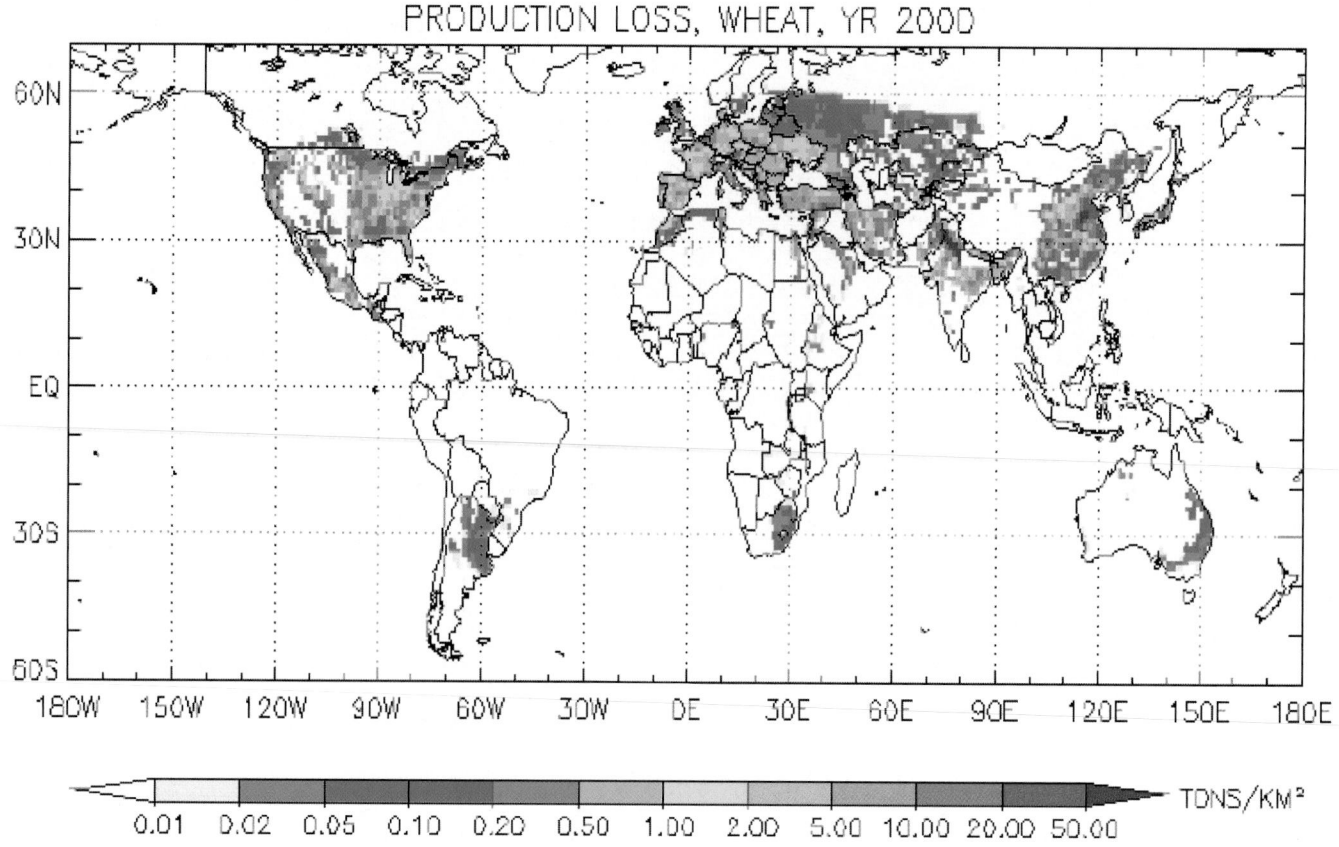

Figure 3.24 | Average wheat crop production losses due to O_3 estimated for the year 2000 using European and North American concentration-based exposure-response relationships. Source: Van Dingenen et al., 2009.

bioaccumulation and biomagnification of POPs in species and ecosystems of these regions. These processes refer to the accumulation of certain long-lived substances such as pesticides or heavy metals as they move up the food chain through a combination of ecosystem contamination and predation, concentrating the substances in tissues or internal organs. The first evidence of the occurrence of these processes was the measurement of the pesticide DDT and its derivatives in Adélie penguins in Antarctica in the 1960s (Sladen et al., 1964). This provided one of the first demonstrations of how human-made substances could spread and affect ecosystems on a global scale. The polar regions are particularly vulnerable to such processes, since their cold climate prevents the breakdown of contaminants, exacerbating their effects.

Another example of polar region sensitivity is that posed by the potential of ozone-depleting substances (ODS) to deplete the stratospheric O_3 layer, which was first recognized in 1974 (Molina and Rowland, 1974). Stratospheric O_3 depletion is important, because the O_3 layer in the stratosphere keeps 95–99% of the sun's ultraviolet-B radiation (UV-B) from striking the Earth. Consequences of increased levels of UV-B include genetic damage, eye damage, and damage to marine life. A decade after identification of the problem in the mid-1980s, the discovery of the O_3 hole over Antarctica (Farman et al., 1985) and the subsequent attribution to ODS (Soloman et al., 1986) further heightened concern

for the O_3 layer. This scientific evidence resulted in the establishment of the Montreal Protocol in 1987, which has led to successful restrictions on ODS. These restrictions have meant that net ODS production, consumption, and subsequent emissions have decreased. However, the long atmospheric lifetimes of many ODSs, particularly the chlorofluoro carbons, mean that the mixing ratios of ODS in the stratosphere are decreasing relatively slowly (Velders et al., 2007). Once again, the particular vulnerability of the polar regions is a consequence of their climatic characteristics. Atmospheric circulation patterns, coupled with the absence of solar radiation during the polar winter, allow the buildup of reservoirs of chemical species in the atmosphere. The reintroduction of solar radiation and the concurrent increases in temperature kick-start the photochemical processes that are able to act on the large reservoir of chemicals, allowing rapid destruction of the stratospheric O_3 layer.

Another example of polar sensitivity is that of radionuclide contamination. The Arctic Monitoring and Assessment Programme (AMAP, 2010) reported radionuclide transport to the Arctic, which could be traced back to the nuclear reprocessing facility at Sellafield in the United Kingdom. Discharges of radioactive substances such as Technetium-99 (^{99}Tc), a long-lived product of uranium fission, were found to have been transported by the Atlantic currents to the Norwegian Sea, continuing northwards along the Norwegian coastline to the Barents Sea by the

Norwegian Coastal Current. The Sellafield-to-Barents Sea transport is estimated to take four to five years. In an attempt to avoid similar contamination in the southern polar region, the Antarctic Treaty has declared the region a nuclear-free zone, where nuclear explosions and radioactive waste disposal are forbidden.

The fact that contaminants in water, air, and ecosystems that originate from outside the polar regions can be observed and measured within the region is a critical reminder of the impossibility of confining emissions and their impacts to source regions, regardless of the remoteness of the impacted environment. This points to the need to consider environmental change, and the limits to such change, at the planetary level. In recent years the particular threat posed to the polar regions by climate change has been recognized, as once again these regions seem particularly sensitive when compared to the global mean. The following section describes climate change in the polar regions in more detail, highlighting the physical, socioeconomic, and political considerations that are particularly relevant when considering the sustainability of future global energy systems.

Climate Change in the Polar Regions

Arctic mean temperatures were found to be rising at almost twice the rate of the rest of the Earth. Many parts of the Arctic have seen warming of 2–3°C in the last 50 years alone. In comparison, over the last century GMTs have increased by only 0.6°C (ACIA, 2004).

By 2100, climatologists project an additional mean temperature increase of 5–7°C, with regions closest to the North Pole projected to reach 10–12°C above their current temperatures in the winter months (ACIA, 2004).

Similarly, a number of stations in the Antarctic Peninsula have experienced a major warming over the last 50 years, with temperatures at Faraday/Vernadsky station having increased at a statistically significant rate of 0.56°C/decade over the year and 1.09°C/decade during the winter. Overlapping 30-year-trends of annual mean temperatures also indicate a strong tendency for the warming trend to be greater during the 1961–1990 period compared with 1971–2000 (Turner et al., 2005).

This 'polar amplification,' whereby the rate of regional temperature change far outpaces global changes, is related to a powerful set of thermodynamic and dynamic factors. A driver of this phenomenon is the ice-albedo feedback mechanism, whereby the melting of sea ice in the Arctic caused by warming associated with climate change exposes dark open water or melt ponds on top of ice. This can decrease the albedo (reflectivity of the surface) to between 0.2 for dark melt ponds to 0.6–0.7 for melting multiyear ice, with such effects occurring during the summer, when warming is most acute. In comparison, the albedo of unaffected multiyear ice is high at 0.8–0.9 (Perovich, 2002). The mechanism is a 'positive feedback,' since the absorbed energy resulting from the lower albedo raises the sea and air temperatures, thereby melting even more ice and slowing down the winter refreeze. A phenomenon

known as the Arctic haze (see next section) will also contribute to this enhanced warming of the North Polar Region.

Quantitative assessments of the extent of sea-ice melt suggest that sea-ice in September 2007 was half the typical ice extent in the same month between the 1950s-1970s (Stroeve et al., 2008). This area was 23% smaller than the previous ice extent minimum reached in 2005. Arctic ice also has an increased tendency to be younger and weaker. For example, while in 1987 ice older than five years comprised 57% of the central Arctic basin, now it has all but disappeared from this area (Maslanik et al., 2007). In addition, ice thickness has declined by 25% over the last two decades. This appears to be a critical parameter determining sea-ice extent, with evidence suggesting that a threshold may exist below which summer melt potential is greatly enhanced (Holland et al., 2006). The rapid melt of sea ice that has been observed in the Arctic has led some experts to predict that a 'tipping point' will be crossed, resulting in a seasonally ice-free Arctic. Based on observations of the 2007 ice melt, Stroeve et al. (2008) revised earlier projections, suggesting that rapid ice decays in Arctic sea-ice, caused by increases in ocean heat transport to the region, could, with a high degree of probability, realize a seasonally ice-free Arctic Ocean as early as 2030.

The Arctic Haze

The Arctic haze, a visible reddish-brown haze found in the atmosphere at high latitudes, results from the long-range transport of air pollution to the polar regions. This pollution remains in the polar atmosphere due to the slow rate of removal of pollutants in the cold polar air (Shaw, 1995). It was first observed in the 1950s by flight crews on weather reconnaissance missions, but only became public knowledge when Raatz (1984) reanalyzed and published data from these missions (known as the Ptarmigan flights). This reanalysis found that the most frequent reports of the Arctic haze occurred at about 75–80°N during late winter, under clear-sky conditions (Shaw, 1995).

Practically all pollution contributing to the Arctic haze originates from more southerly latitudes (Law and Stohl, 2007); with local pollution sources being small and limited to areas near the Arctic circle. These local pollution sources include volcanic emissions; anthropogenic emissions from urban centers, e.g., Murmansk; and industrial emissions, particularly in the northern parts of Russia, and emissions from the oil industry and shipping (AMAP, 2006). In terms of the long-range transported pollutants, Eurasian emissions are more important than those from North America, due to the closer proximity of Eurasian sources, atmospheric circulation patterns, and the extended distribution of the haze over Eurasia (Law and Stohl, 2007). Some studies have even suggested that Asian dust may be contributing to the Alaskan Arctic haze (Rahn et al., 1981), though this claim has been disputed (Stohl, 2006).

The Arctic haze is important, since it changes the short- and long-term radiation balance in the Arctic. Because large areas are affected, this can have significant effects, even though the concentrations of aerosols found in the Arctic are approximately an order of magnitude lower than those found in polluted and industrial locations (AMAP, 2006). These aerosols can also affect the physical properties of clouds, which will affect the radiation balance (Garrett et al., 2002). Black carbon will also cause heating in the haze layers (Quinn et al., 2007) and, on deposition to snow and ice, will result in a reduction in surface albedo. The resulting warming may lead to the melting of ice and may be contributing to earlier snowmelts on tundra in Siberia, Alaska, Canada, and Scandinavia (Foster et al., 1992).

In terms of future trends for this pollutant haze, there is evidence that ship traffic is already affecting the summertime Arctic atmosphere; increases have been seen in the deposition of BC from increased shipping following reductions in summer sea-ice (Corbett et al., 2010). The warming of northern latitudes could also cause boreal forest fires to become more frequent, thus increasing pollution transport to the Arctic (Stocks et al., 1998). Finally, the polar dome, which currently presents a barrier to pollution transport in the Arctic, may weaken in the future, as the Arctic continues to warm relatively faster than lower latitudes, allowing more efficient pollution import into the Arctic (Eckhardt et al., 2003).

Energy Issues in the Arctic

The physical changes in the Arctic environment will cause substantial impacts on the regions human populations and ecosystems. However, perhaps more worrying, at least from the point of view of enhancement of the level and rate with which impacts are occurring in the Arctic, are the economic and political implications of these physical changes. Importantly, the physical feedback of ice-melt has realized the potential for transport routes (with the apparent opening of the North West Passage for ice-free navigation which would substantially reduce shipping distance from Europe to Asia) and hence improved accessibility to energy sources (Powell, 2008). Johnson (2010) describes this situation as 'accumulation by degradation' where the previous combustion of fossil fuels generates geophysical changes that are exploited in order to identify, extract, and combust additional hydrocarbons that fuel development and lead to future CO_2 emissions and additional climate change; this process will lead to self-amplifying feedbacks of the geophysical world.

Fuelling the likelihood of such a feedback process are the claims of undiscovered hydrocarbon reserves in the Arctic region. There are estimates that as much as 25% of the World's undiscovered oil and gas lie in the Arctic (AMAP, 2007) though this figure has been contested (Powell, 2010). The US Geological Survey recently completed its first probabilistic assessment of all undiscovered oil and gas deposits of the Arctic Circle. It estimated that of the world's undiscovered resources the region holds 30% of natural gas, 20% of natural gas liquids and 13% of oil (US GS, 2008).

The potential hydrocarbon gains, coupled with climate change-driven ice melt, could substantially reduce the costs associated with hydrocarbon exploration and exploitation, including deep-water exploration and the transport of cargo, drilling equipment, and petroleum, and extend the seasonal period during which such operations are feasible.

These activities themselves will cause environmental degradation of the region by increasing the likelihood of oil spills and pollution (Casper et al., 2009).

3.3   Land-use and Energy Systems

Renewable energy sources often require more land than the fossil energy sources that they replace, which means that making energy systems more sustainable can place additional pressures on land-use. The transition towards sustainable energy thus tends to increase land pressures due in part to the fact that current energy systems to some extent substitute nonrenewable resources – fossil and nuclear – for land. It is clear that a critical component of a sustainable energy path – and indeed the sustainable use of natural resources in general – is to use land more effectively in the future, with the recognition that some of that land will have to substitute for the fossil resources that will no longer be available at acceptable levels of economic and environmental costs. Renewable resources that occur on water (e.g., tidal power) or land that is difficult to use for other human needs (e.g., mountain tops) will naturally also become more valuable in the sustainability transition.

In this section, we provide a brief overview on the *quantity* of land associated with energy systems, while of course recognizing that the quality of land-use is just as important, if not more important, than the quantity *per se*. The reason for focusing on the quantity of land required is to provide some rough idea of how priorities might be set where land-use pressures are a key concern. Some explanation and/or examples are also given here, in instances where there are significant qualitative land-use impacts that are obscured by simple quantitative comparisons.

3.3.1   Land-use Changes

Land is essentially always transformed in the initial establishment of any energy system. This transformation can be significant, in that the effects are not easily reversed. In the case of coal, hundreds of years may be needed for recovery or reclamation from the toxic effects of coal mining. In the case of nuclear power, the consequences of a serious accident result in land-use transformation that requires tens of thousands of years for recovery. Consequently, changes in land-use from toxic, nonrenewable resource uses should not be directly compared to land-use changes associated with biological or natural resources.

Indeed, the term 'land-use change' has a special meaning within the context of natural resources and climate, in that it refers to the transition from one type of land-use such as forestry to another, such as agriculture. Although also representing a transformation and involving energy use and GHG emissions, these land-use changes are qualitatively quite different from fossil or nuclear land-use impacts, in that there are generally no toxic effects and the land could potentially be converted back

Table 3.2 | Land-use intensity (LUIs) for various electric power production systems.

Type	Reference capacity (for energy conversion)	Land-use intensity for installed power (km²/GW)		Land-use intensity for electricity generation (km²/TWh/yr)		Assumptions and/or type of impact
		low	high	low	high	
Coal	85%	18.6	126.6	2.5	17	Area disturbed by mining
Hydropower	44%	62.2	354.8	16.1	92.1	Area submerged by lake
Biomass electricity	75%	2844	4294	432.9	653.6	Area for growing woody feedstock (willow)
Geothermal	85%	7.5	103.6	1.0	13.9	Area covered by plant and access infrastructure, fragmented habitat
Solar thermal	29%	25.9	51.8	10.2	20.4	Area covered by plant and access infrastructure, fragmented habitat
Solar photovoltaic (PV)	28%	51.8	129.5	21.1	52.8	Area covered by plant and access infrastructure, fragmented habitat
Onshore wind	35%	199.4	242.8	65.0	79.2	Area covered by turbine and access infrastructure, fragmented habitat
Nuclear	91%	3.02	4.78	1.9	2.8	Area covered by plant, as well as area for uranium mining and waste storage

Source: adapted from McDonald et al., 2009.

to its previous use or to some other use, especially if the previous uses were agricultural. In this respect, the impacts of the sheer quantity of land-use change associated with agriculture or bioenergy are tempered somewhat by the fact that they can be reversed in some cases without extensive damage. However, there can also be serious and irreversible losses when land of high carbon stock or land with high biodiversity value is converted for food, fiber, energy, or other commercial uses. The complexity and dynamic nature of land-use change places it beyond the scope of this brief review, which focuses mainly on the quantity of land required by energy systems.

3.3.2   Land-use Intensity

Land-use intensity (LUI) is a static measure of the land needed for continuous operation and that which is unavailable or otherwise impacted with respect to other uses. The LUIs of energy systems vary by several orders of magnitude. Nuclear and fossil fuels have fairly low land-use intensities, while renewable energy systems exhibit much more variation, ranging from modest requirements for geothermal and solar photovoltaics (PVs) to extremely high requirements for some types of biomass to electricity systems. The high value for biomass electricity is also partly due to the fact that biomass is most efficiently used and optimized for heat production in combination with electricity.

The land-use intensity of energy systems provides some first-order indication of the types of land-use pressures associated with the transition to sustainable energy systems. Land-use impacts are based on the amount of land that is rendered unavailable for other uses for at least the duration of the lifetime of the energy systems. It is important to note that such calculations include only the energy system in use and do not include land-use impacts due to accidents, which could be quite

serious in the case of nuclear power and could contaminate land for thousands of years. Accidents in most renewable energy facilities would generally not be serious, as there are very few potentially toxic elements and those that are used could not spread over large areas.

A representative comparison of LUIs with respect to installed power and electricity production in the United States is given in Table 3.2. The systems with the lowest LUI are nuclear, coal, and geothermal. Solar and wind have LUIs that are 1 order of magnitude higher, while biomass has LUIs that are 2 orders of magnitude higher. However, since biomass energy systems normally have coproducts and since the land can often accommodate other uses, the effect of the land impacts can be significantly less. In the case of geothermal energy, there are underground impacts that are not included. There are also some similarities in the land impacts of geothermal energy with those of oil and natural gas. In the case of geothermal energy supply that involve wells, only 5% of the associated impacts are from direct land-use effects; the remaining 95% are due to habitat fragmentation and species-avoidance behavior. Onshore wind has a comparable breakdown, in that 95–97% of the impacts are not due to the land occupied, but rather to fragmentation and the effects on birdlife (McDonald et al., 2009).

3.3.3   Power Density

Another way to compare energy use in relation to area is to use the power density, expressed in watts per square meter (W/m²) and representing the incident energy that can be delivered. Incident solar energy is typically in the range of 100–200 W/m² in Europe, but is much higher in the lower latitudes. The highest delivery of high-quality (electric) renewable energy per unit of land is likely to be from solar PV,

Table 3.3 | Land-use intensity and nitrogen intensity of various energy crops used for biofuel production.

	Land-use intensity (LUI)		Nitrogen intensity		Combined weighted ranking
	(ha/1000 GJ)	rank	(kg N/1000 GJ)	rank	
Sugarcane	2.3	2	110	2	0.02
Willow	5.3	6	90	1	0.03
Miscanthus	4.2	5	210	5	0.03
Sugar beet	1.9	2	460	8	0.03
Oil palm	3.0	4	440	8	0.04
Birch	6.8	8	160	3	0.04
Poplar	7.2	8	160	3	0.04
Switchgrass	6.5	8	300	6	0.05
Corn	4.9	6	490	8	0.05
Sweet sorghum	6.1	8	390	7	0.05
Algae	0.3	1	1100	11	0.05
Grain sorghum	16.2	12	1000	11	0.13
Rapeseed	16.5	12	1400	12	0.15
Soybean	20.2	13	3900	13	0.30

Source: Miller, 2009.

which could capture in the order of 10–20% or roughly 10–40 W/m^2 for European conditions when placed on south-facing surfaces. This might be compared with other energy systems in terms of how much electricity they can deliver, such as wind (2 W/m^2) or tidal (3 W/m^2), again based on conditions common in Europe (MacKay, 2008). The calculation of power density is somewhat more location-specific than the LUI measure used in Table 3.2, because it is based mainly on the energy that can be instantaneously extracted, whereas the LUI includes various land impacts that are then averaged across the power or energy that is available.

3.3.4   Biofuels, Land-use, and Nutrient Demand

The land-use associated with energy crops that are used to make liquid biofuels differ qualitatively from other types of energy supply in that they involve agricultural operations, and therefore other types of land issues arise. A particularly useful metric that can be compared alongside LUI is the nitrogen requirement; by combining this measure with LUI, one obtains a much better indicator for sustainability. Table 3.3 provides an example of such analysis, in which a combined ranking for two criteria was made, with equal weighting given to LUI and nitrogen-use intensity (Miller, 2009). An obvious conclusion from this comparison is that soybean, rapeseed, and grain sorghum are highly inefficient crops for biofuel production compared to almost any other option. Sugarcane scores highest in the initial ranking, and this held true even when sensitivity analysis was conducted for key parameters such as higher heating value (HHV), nitrogen and harvestable yield (Miller, 2009).

3.3.5   Multiple Uses, Hybrid Systems, and Land-use Efficiency

As the demand for food, feed, fiber, energy, and other products requiring significant land-use increases, greater land-use competition will be expected. Furthermore, since land is also valued for recreation, conservation, biodiversity, and many other uses, the 'productive' capacity of land must also be weighed against ecosystem services that are in fact highly productive, but whose value is poorly reflected in our socioeconomic systems. Consequently, the rise in importance of bioenergy has raised concerns that other valuable land-uses will suffer as a result of bioenergy expansion. One approach to reducing the LUI, while also improving the system design, can be hybrid systems that combine biomass with intermittent renewables such as wind. One example is a hybrid wind/biomass system that uses compressed air storage, which was estimated to require 70 m^2/MWh, a LUI that is just a fraction of the range given for biomass in Table 3.2 (Denholm, 2006).

The attention paid to bioenergy systems has also led to increased scrutiny of the land-use efficiency of agricultural and livestock practices. Greater land-use efficiency in agriculture, forestry, and other land-uses could free up some land for the renewable energy options that will be ultimately required to replace fossil and nuclear fuels. Current agricultural practices have tended to be land-intensive, as the low price of rural land in combination with fossil fuel infrastructure led to continual expansion. With precision farming methods and low tillage methods, land can be used more effectively while also reducing GHG impacts from fertilizer use, N$_2$O, and soil carbon release; this is true regardless of whether the land is used for food, feed, or fuel (Fischer et al., 2010).

Table 3.4 | Principal water uses for various energy systems.

Process requiring water	Type of energy produced	Environmental impacts	Energy eq. produced
Biomass growth	Traditional heat/cooking	Consumptive: less quantity of water for other uses, ecosystem services	36 EJ (traditional)
	Commercial biomass: heat, electricity biofuels	Non-consumptive: water quantity used released back into flow system, sometimes with degraded quality	9 EJ (commercial)
Direct energy generation	Hydropower	Non-consumptive, but can have large impacts on upstream and downstream habitats, depending on water flows affected by hydropower dams Consumptive: (unproductive) evaporation losses from dams, reservoirs	20% of total global electricity
Direct energy generation	Wave and tidal power (sea water)	Non-consumptive	
Energy processing	Thermal power	Non-consumptive	
Energy processing	Fossil fuel and nuclear power stations	Non-consumptive: for cooling; release of cooling water increases local habitat water temperatures	

Source: IEA, 2008.

3.4 Water Resources and Energy Systems

Water resources and the flow of freshwater through the biosphere are fundamental to supplying benefits to society, including various forms of energy. Fresh water is a finite global resource, with no substitute in biological processes. On a planetary scale, water stress and water pollution are already reaching alarming levels, though currently concentrated in local and regional 'hot spots.' With increasing demand for energy, the competition for appropriating more water is expected to increase, in particular for hydropower and for the production of bioenergy crops. Mining activities also have implications for water resources, as they can cause pollution and change local water flows with implications for groundwater reservoirs. Impacts on the quality and quantity of water resources often relate to local social, economic, and environmental settings. The decisions on how best to develop energy system strategies and investments therefore need to be made at all levels, from local to national, in a transparent manner and according to best-informed practices.

3.4.1 Water Resources in Energy Systems: Demand and Impacts

Unlike the atmospheric impacts of different energy systems, which can be considered globally or regionally, the impacts of energy systems on water resources are more meaningfully considered at smaller spatial scales, from local to national levels. This is due to the nature of water resources and the boundaries of water flows. The impacts of various energy systems on water resources must be considered in terms of quality as well as quantity, at local and regional scales, in landscapes, catchments, or river basins. There are large differences in the availability of water resources among regions in the world, with ample freshwater in temperate cold regions and humid tropical regions contrasted by water scarcity in arid, semiarid, and sub-humid tropical regions. As a result, assessments of aggregate water resource availability at the global level

may exaggerate actual access to water resources in societies and for specific ecosystem purposes.

Water has three key uses in energy systems (see Table 3.4). The first use is biomass production (rainfed and irrigated), including crops, plantations, and seminatural vegetation types, including forests and grasslands. The second use is processing (e.g., sugarcane, coal washing) and cooling to produce energy (e.g., fossil fuel and nuclear power stations). The third use is generating energy (e.g., hydropower, wave power, and tidal power).

3.4.2 Water Embedded in Biomass Production for Energy

A large part of the energy stemming from biomass is used for primary human energy needs such as heating and cooking. Only a small part of biomass is used for secondary energy supply, involving substantial postharvest processing, sometimes including additional water uses.

All biomass production, be it for crops or primary biomass production for energy, is associated with large consumptive use (water that does not return to the river or groundwater) of freshwater. Biomass production is the single largest freshwater-consuming process on Earth, with large volumes of water used to produce energy (e.g., a range of 60–110 m^3 per gigajoule (GJ), for major bioenergy crops), according to Geerben-Leenes et al. (2009).

The use of wood fuel for basic human needs of heating and cooking will continue to remain dependent on local resources, and thus the water input will depend on the local hydrological conditions prevailing at the source of production, which in this instance will be the same as the site of use and energy supply. Change in energy use, for example cooking facilities with better wood fuel energy conversion, can be translated into a 'water saving,' as less embedded water is used to

Table 3.5 | Water productivity (as actual evapotranspiration per GJ output of electricity for biofuel).

Crop	Biofuel	Energy water footprint (as tonne water per GJ output)
Rapeseed	Biodiesel	100–175
Sugarcane	Ethanol	37–155
Sugar beet	Ethanol	71–188
Corn	Ethanol	73–346
Wheat	Ethanol	40–351
Lignocellulosic crops, including salix, eucalyptus, switchgrass, miscanthus	Ethanol	11–171
Lignocellulosic crops, including salix, eucalyptus, switchgrass, miscanthus	Methanol	10–137
Lignocellulosic crops, including salix, eucalyptus, switchgrass, miscanthus	Hydrogen	10–124
Lignocellulosic crops, including salix, eucalyptus, switchgrass, miscanthus	Electricity	13–195

Source: adapted after Berndes, 2008.

produce the same amount of energy. Although 'water footprints' are a common way of accounting for water use embedded in biomass produce, care should be taken, as water-use efficiency values vary globally and locally, depending on production systems and agroclimatic conditions for the same species. Also, the calculations vary in regard to how the water consumption is related to the supply of energy. For example, is the full crop used for energy supply, or are only crop by-products used for energy supply? A synthesis by Berndes (2008) show large variations in embedded water in different types of energy systems (Table 3.5). Thus, good planning both in terms of location and type of energy can have large local impacts on landscape and river basin water use efficiency.

Practices that will cause changes in land-use, e.g., shifting biomass production systems, are likely to lead to changes in landscape hydrology, which in turn can affect the downstream availability of water. Bioenergy production upstream may thus impact on water-dependent provisioning and supporting or regulating ecosystem services downstream in a catchment or river basin. A meta-analysis of catchment land-use changes shows that transitions from annual crop and/or grassland to more water-demanding woody species can reduce stream flow (e.g., Jackson et al., 2005; Locatelli and Vignola, 2009). Other crops and habitats that depend on certain volumes and the timeliness of river flows may be impacted, with subsequent implications for livelihoods and economies depending on these ecosystem services. These changes in water flow, caused by changes in crop production for bioenergy or other land-use changes, may not be evident at the outset and may be difficult to anticipate (e.g., Gordon et al., 2008).

3.4.3   Impacts on Water Resources by Post-harvest Processing Biomass for Biofuel

Crops (annual or perennial) that are produced for commercial purposes and eventually processed into energy as heat, electricity, or biofuels, may have a production and processing location that is different from

the location of consumption. Thus, over extraction or pollution of water resources due to such commercial biomass production or to post-harvest processing is not necessarily co-located with the place of consumption of the end product. Two water-related issues, in addition to the previous discussion about water footprints, have also been documented: degradation of surface and groundwater by agrochemicals, and post-processing of biomass for bioenergy.

Degradation of surface and groundwater by agrochemicals: As with other intensive crop-production systems, biofuel crops have also been associated with pollution due to the intensive use of agrochemicals such as fertilizers, pesticides, and herbicides. These different chemical components have been shown to cause harm to both surface and groundwater resources in areas of intensive, large-scale use, such as maize (corn) in the midwestern United States for ethanol (e.g., Donner and Kucharik, 2008), and sugarcane in various parts of Brazil (e.g., Goldemberg et al., 2008).

A second pressure on water resources can be associated with post-processing of biomass for bioenergy. This water use is non-consumptive, and the water used will eventually be released back into the landscape. However, as has been seen, for example in association with the production of large amounts of sugarcane, the release can be associated with degraded quality, thus affecting downstream users and habitats.

3.4.4   Impacts on Water Resources by Utilization of Water for Cooling

In various energy systems, such as fossil fuelled power stations and nuclear power stations, water is used for cooling purposes. This use is non-consumptive, and water can be used downstream for other purposes. However, local impacts on water-related habitats have been identified, as the returned cooling water has a higher temperature than natural temperature cycles, especially during cold seasons. This can alter local habitats and species (e.g., Teixeira et al., 2009; Yi-Li et al., 2009; Svensson and Wigren-Svensson, 1992).

3.4.5   Impacts on Water Resources by Hydropower Generation

A major impact on global freshwater flows is the construction of dams and reservoirs for hydropower generation. Thus the environmental impacts, in terms of the hydropower energy chain, are almost entirely due to infrastructure construction activities. Although the energy supply is non-consumptive in terms of landscape water flows, several water-related impacts are associated with hydropower generation. The key ecosystem services that are affected by dams are: the change of flow patterns in time and space at site and downstream; the change of aquatic and riparian habitats; and river-system fragmentation.

The most important impact is the fragmentation of river systems (e. g., Nilsson et al., 2005). The construction of large dams and reservoirs change temporal water flow distribution in river systems, affecting downstream habitats and water quality. This has far-reaching impacts on various ecosystem services that benefit human activities, economies, and development. Although the World Commission on Dams (2000) estimated 40–80 million people to be affected by dam constructions, the numbers of people impacted downstream of dams are in the order of 500 million (Richter et al., 2010). The loss of functioning habitats such as wetlands and floodplains often means that benefits, both valued and non-valued, are irreversibly lost. A recent global assessment indicates that of 292 rivers, 172 are already seriously affected and fragmented by dams (Nilsson et al., 2005). The most fragmented rivers systems are also some of the most habitat-rich, biodiverse, and populated river systems of the world. On a global scale, dams and reservoirs have already had a significant impact on freshwater flow-related ecosystem services, and the societies, economies, and population relying on these ecosystem services (Meybeck, 2003). In addition, natural recreational areas may be destroyed by large-scale dam construction.

Sometimes the regulating function of dams is desired. For example, dams may protect valuable infrastructure and settlements from recurrent flooding events. Dams and reservoirs can also ensure the provision of a regular water supply to society and agriculture. However, the net outcome of a holistic assessment of dams and reservoirs is, more often than not, the loss of provisioning, supporting, and regulating ecosystem services that would directly benefit humans in terms of energy, water supply, and water regulation. In some cases, large dams are being deconstructed to restore natural river flows (e.g., Gosnell and Kelly, 2010). In view of the development demands of the world, it is expected that a number of large dams for hydropower purposes will be constructed in many countries in Latin America, Africa, and south Asia in the near future. However, even the World Bank recognizes the limited operational capacity that exists to ensure that ecosystem values are truly accounted for, despite the increased recognition of these values highlighted by the groundbreaking report of the World Commission of Dams in 2000 (World Bank, 2009).

GHG release from large dams and reservoirs: A recently recognized added impact of large manmade dams is the release of CH_4 to the atmosphere. Methane is released as a by-product by anaerobic breakdown of inundated biomass. Methane has GWP 25 times greater than that of CO_2 and thus will have a significant contribution to total effective GHG emissions from large hydropower energy systems (e.g., Chen et al., 2009; Guerin et al., 2006). According to International Rivers (2011), 23% of all manmade CH_4 emissions can be attributed to large dams.

The trapping of sediments in dams: a serious impact of reservoirs and dams is the reduced levels of sediments transported to various parts of the river systems. An earlier estimate suggested that globally, dams in river systems trap 4–5 Gt/yr (Vorosmarty et al., 2003), or approximately 25% of total sediment transport. The environmental issue related to the sediment trapping is the concentration of harmful compounds that are stored in these sediments. Sometimes this has unintended benefits, such as dams acting as storage for harmful substances. Some authors are warning that a restoration or unintended break in these dams would pose serious threats downstream due to the years of built-up polluted sediments.

3.5   Environmental Sustainability in Energy Systems

It is a major challenge, and an absolute necessity, to understand the demands that will be placed on future energy systems to develop means of ensuring that energy can be provided in a sustainable manner. One approach to achieving this is to develop energy systems within globally defined sustainability criteria for critical Earth-system processes.

Defining these criteria requires the inclusion of living and nonliving systems (ecosystems and natural resources) and the consideration of risks of abrupt nonlinear changes, or tipping points/thresholds, as a result of environmental overshoot. Attempts have been made in the past to develop global indicators for sustainability, most famously in the Limits to Growth World3 scenarios in the 1972 Club of Rome report (Meadows et al., 1972). These scenarios, later revised twice (Meadows et al., 1992; 2004), excluded ecosystems and focused on nonrenewable resources and persistent pollution. Later global assessments, such as the Millennium Ecosystem Assessment (MEA, 2005) and the IPCC (2007a), include analyses of environmental states and trends, with partial attempts to also define global sustainability criteria. The IPCC stabilization scenarios come closest to defining such criteria at the global scale for the climate system. These scenarios involve targets of maximum atmospheric CO_2 concentration, but the CO_2 concentrations are not linked to specific thresholds or other features of the climate system. Other atmospheric pollutants that are targeted for their impacts on poor air quality rather than climate change have been addressed at the regional level. Methods such as the effects-based approach developed by the UNECE Convention on Long-Range Transboundary Air Pollution aim to optimize improvements in air quality (see Sliggers and Kakebeeke, 2004) and have defined critical loads and levels for different pollutants to protect human health, materials, and a range of ecosystems from the impacts of air pollution impacts.

Other approaches to defining global sustainability criteria include the World Wildlife Fund Living Planet Index, which is an indicator of global biodiversity, and the global ecological footprint assessment, which attempts to compare humanity's demand for renewable provisioning ecosystem services (crops, fish, livestock, timber) in relation to the capacity of the planet to regenerate these services (Living Planet Report, 2010). Regulating services from ecosystems, such as climate stability and freshwater supplies, are only partially represented, through estimates of carbon footprints.

Perhaps coming closest to defining appropriate global sustainability criteria for the Anthropocene era are the scientific advancements made in assessing the risks of tipping elements in the climate system (Schellnhuber, 2009), and the 'tolerable windows approach' that has been developed to provide a climate policy guidance framework. In the tolerable windows approach, which was first proposed by the German Advisory Council on Global Change (WBGU, 2009), limits designed to avoid dangerous anthropogenic climate change are set against assessments of tolerability. A key feature of this approach, and most other attempts to develop sustainability criteria, is the compromise between limits set by nature on the one hand, and the demands and the 'tolerable' costs set by society on the other.

A number of modeling tools have also been developed to specifically represent the sustainability impacts of energy systems. These include life-cycle analysis tools that provide quantitative assessments that compare the potential for environmental impacts (GHG emissions, SO_2 emissions, NO_x emissions, direct land requirements) associated with different energy systems at different points along the energy chain (Evans et al., 2009). Other tools include natural resource management; ecologically sustainable development tools, and integrated assessment models. The latter have been specifically developed for research on climate change and air quality (e.g., IPCC, 2007) and incorporate, for example, the full cycle of anthropogenic GHG emissions, the options and costs of their mitigation, the resulting climate change, the impacts of climate change, and the related options and costs of adaptation (Toth, 2003). There is also increasingly a focus in models on the 'triple bottom line,' or sustainability in terms of ecological, sociological, and economic factors (Harris, 2002).

3.5.1   Environmental Sustainability in GEA

The GEA transition pathways (see Chapter 17) build on existing frameworks and use a range of modeling tools. These pathways propose the development of future energy systems under a number of sustainability criteria. Key environmental targets are formulated with respect to climate change and air pollution. These include limiting global mean surface temperature change to 2°C above pre-industrial levels with a

likelihood of more than 50% and achieving global compliance with WHO air-quality standards (PM2.5 < 35 μg/m^3) by 2030 (see Chapter 17 for more detail). The climate-change target is discussed in Section 3.2.2, while the air-quality target relates to human health and is considered in Chapter 4. Both targets consider current knowledge and policy in developing stringent environmental goals for energy systems.

The stringency of such goals implies significant environmental outcomes of these pathways measured with respect to atmospheric pollutants. As discussed in Chapter 17, the GEA transitions are most comparable to the most stringent IPCC scenarios. In these low pathways, GHG emissions may continue to increase only for a very short period of time (peaking between 2020 and 2030) before they must decline to reach levels at about zero or even negative over the long-term (beyond 2060). These scenarios are compatible with long-term atmospheric CO_2 concentrations stabilizing at below 400 ppm (see Figure 17.35 in Chapter 17).

The GEA pathways indicate that the nature and magnitude of future environmental impacts will greatly depend on the evolution of energy systems and other anthropogenic activities in the coming decades. A major global energy transition in the future as described by the GEA pathways will require on the one hand conditional convergence in incomes across regions, and on the other hand a combination of policies at the local and regional levels, with global availability of energy-efficient technologies and devices (see Chapter 17 for a detailed discussion). Future energy systems are expected to evolve differently across regions under such a major energy transition. In general, one can expect increases in fossil-energy systems equipped with carbon capture and storage (CCS) and increases in renewable energy (wind, solar, bioenergy) across all regions. The exact extent of such changes in energy systems will depend on availability of energy sources and other constraints (see Chapter 17, Section 17.3.4.4 for more regional discussion of energy systems). The changes in the energy system will have a major impact on GHG emissions and air pollutants in the future. In addition to global climate change policies, regional and local policies will play a major role in influencing future environmental impacts. Policies will need to be stringent in order to significantly control future environmental and other impacts. Ensuring universal access to clean cooking fuel and controlling levels of air pollution across all regions through stringent legislation are seen as being essential to ensuring improvements in environmental quality.

As summarized in Figure 3.25, the GEA 'Counterfactual' scenario indicates that in the absence of stringent policies in all regions, GHG emissions can be expected to increase across all regions, while the regional evolution of air pollutants (shown here are SO_2, NO_x, and $PM_{2.5}$) will depend on the effectiveness of currently legislated air-quality controls. While currently planned air quality legislation, as in the GEA Counterfactual scenario, will bring reductions in outdoor air pollution levels across Organisation for Economic Co-operation and Development (OECD) regions, for other regions, in particular Asia and sub-Saharan Africa, increasing populations and rapid expansions in energy systems

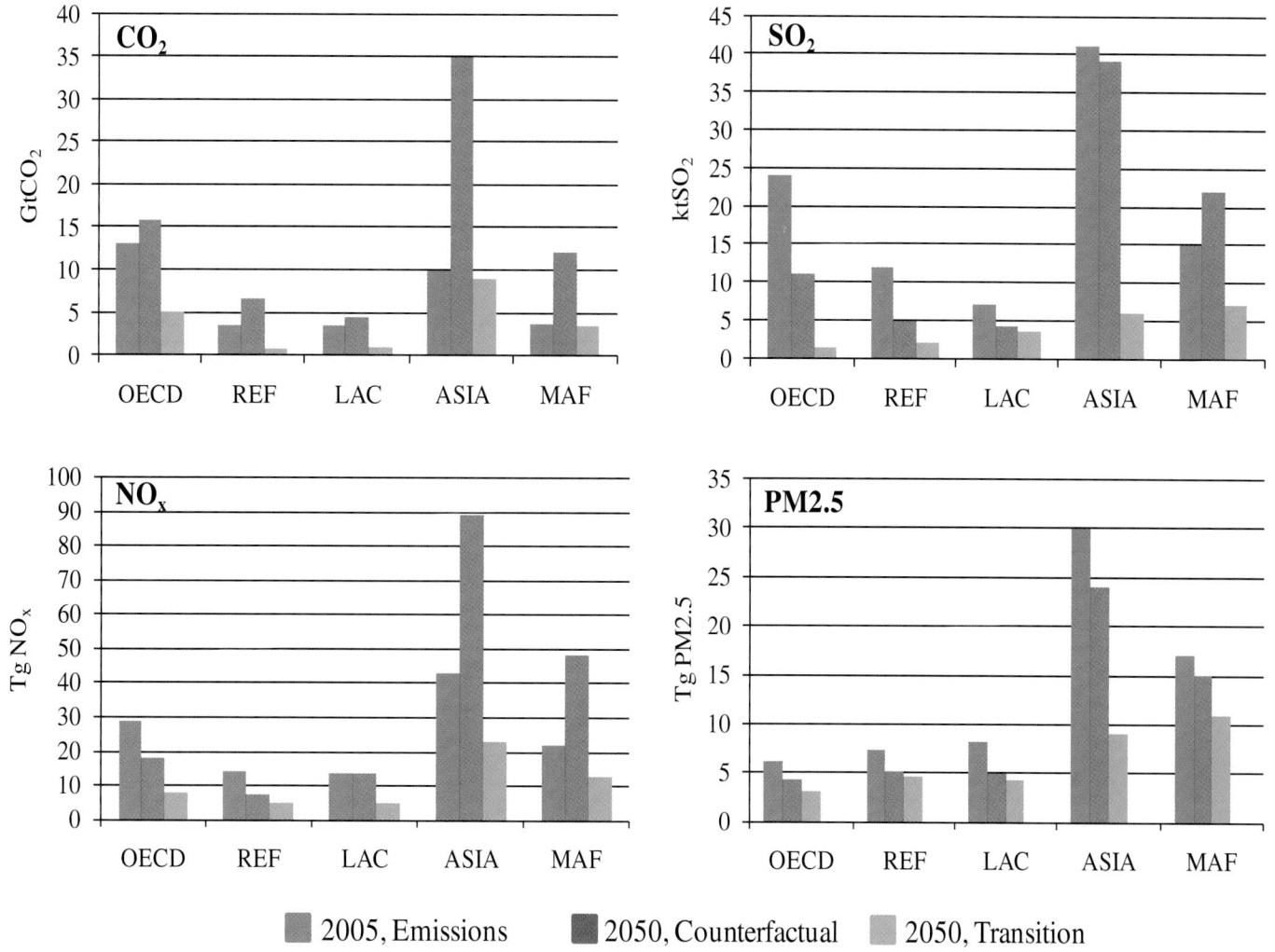

Figure 3.25 | CO_2, SO_2, NO_x, and $PM_{2.5}$ emissions in 2005 and 2050 in GEA Counterfactual and GEA Mix (Transition) Scenarios. Regional definition in Chapter 17.

will mean that emissions continue to increase. The inclusion of sustainability criteria, as in the GEA transition pathways (shown here is the GEA Mix; see Chapter 17 for a full description of pathways), and the subsequent increase in the stringency of policies, lead to significant declines in both GHG and pollution levels across regions.

These emission reductions from anthropogenic systems can be expected to bring significant environmental benefits across regions compared to current levels. While a detailed description of regional environmental impacts is beyond the scope of this discussion, such impacts will extend to a wide range of environmental concerns, as discussed earlier in the regional descriptions. These would include, for example, decreased sea-level rise; decreased frequency of droughts and flooding; decreased impacts of BC-related climate effects, decreased crop losses from O_3-related effects; and decreased acidification and eutrophication. In addition to environmental benefits, significant health-related benefits can be expected as a result of stringent pollution energy access policies (see Chapters 4 and 17 for discussions on the health-related impacts of air pollution).

3.5.2 Proposal for Alternative Framework for Environmental Sustainability of Energy Systems

Our assessment is that global sustainability criteria for the global energy systems in the Anthropocene era must be comprehensive, including all relevant environmental processes (e.g., carbon and nitrogen cycles) and systems on Earth that are affected by our energy systems (e.g., the climate system, the ocean system, and terrestrial and aquatic systems). Such criteria must consider the interactions and risks of abrupt, nonlinear change (tipping points and thresholds). They must also address the ability of systems on Earth to maintain their resilience, and thereby their capacity to remain in desired states conducive to human development in an era of rapid global change. The GEA sustainability indicators provide a systematic approach to defining specific quantitative goals for environmental sustainability with respect to the atmospheric impacts associated with particular scenarios involving air pollution and climate change (see Chapters 4 and 17). This approach will also need to extend to encompass the full range of Earth-system considerations.

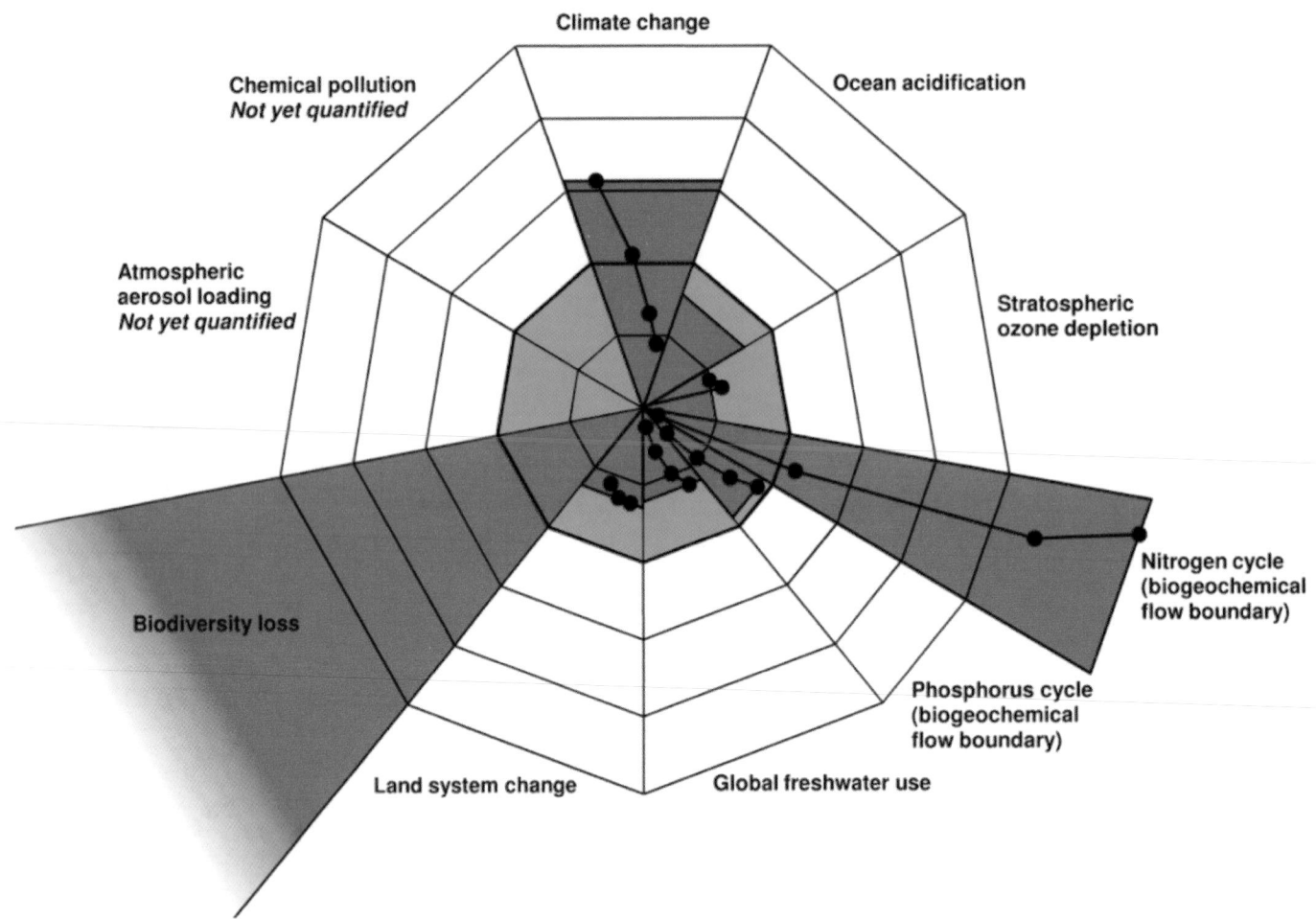

Figure 3.26 | Current global state of the world for the ten proposed planetary boundaries. The green areas denotes a "safe operating space" for human development, and red indicates the current position for each boundary process. The dots indicate evolution by decade from the 1950s.

Based on this conclusion and on the assessment of previous approaches to defining global sustainability criteria, the recently advanced planetary boundaries concept (Rockström et al., 2009) is considered here as a complementary way of framing the assessment of sustainability criteria for global energy systems. This framework, which builds on previous global sustainability approaches, advances in Earth system science, and resilience research, was developed as a means of addressing and establishing global sustainability criteria that recognize the global pressures from human enterprise and the risk of undermining the resilience of major biophysical systems on Earth. It complements existing environmental indicators, such as the critical loads and levels approach, global footprint analysis, and the 'tolerable windows approach,' by addressing the biophysical prerequisites for avoiding abrupt and undesired global change. It is an expansion of the 'guardrail' approach (Hare and Meinshausen, 2006) used to define safe climate-mitigation targets by also addressing other global environmental change processes. The approach is presented in Figure 3.26, which shows the planetary boundaries for nine Earth-system processes, which together define a safe operating space for humanity (indicated by the green area), within which human development

stands a good chance of proceeding without large-scale deleterious change. Estimates indicate that the safe levels are approached or in some cases surpassed. Energy systems contribute to humanity's approach to all the planetary boundaries, but in particular climate change, biodiversity loss, land-system change, atmospheric aerosol loading, and chemical pollution.

So how can the planetary boundary framework be redefined to be appropriate for an assessment of our energy systems? First this requires that the main Earth-system processes affected by our energy systems are defined (as already discussed in previous sections) and related to impacts that act at the planetary level. The planetary boundary framework already includes proposed boundary levels appropriate for the prevention of Earth system-threatening climate change (see Rockström et al., 2009). Conceptually, the framework may also facilitate efforts to understand the impacts of the more 'traditional' atmospheric pollutants on ecosystems, since a similar critical limits threshold-driven concept has long been used in scientific research to define loads and levels below which pollution would not threaten perturbations to environmental systems (Sliggers and

Table 3.6 | Summary of proposed global sustainability indicators and target levels for energy systems based on planetary boundaries.

Pressure	Receptor system	Planetary boundary	Indicator target or orientation value
Emission of radiative forcers (e.g., CO_2, PM, aerosols), leading to changes in global and regional climate	Atmosphere	Climate change; Rate of biodiversity loss; Ocean acidification; Atmospheric aerosol loading	350 ppm CO_2 or climate forcing less than 1 W/m² above the pre-industrial level, or less than 1000 GtCO₂-eq released in the period 2000–2050
			No more than 2°C above pre-industrial of human-induced global warming, with a probability greater than 50%
			Aim for 50% GHG emission reduction by 2030 and subsequent reductions to achieve the target
			Reduce by 2030 and eliminate by 2050 emissions of BC, OC, nitrogen and sulphur, and other PM that contribute to atmospheric aerosol loading, so that radiative forcing remains less than 1 W/m².
Emission of air pollutants (e.g., SO_2, No_x, O_3)	Atmosphere	Climate change, Rate of biodiversity loss, Nitrogen cycle, Chemical pollution; Atmospheric aerosol loading	Reduce by 2030 and eliminate by 2050 emissions of BC, OC, nitrogen and sulphur species, and other PM that contribute to atmospheric aerosol loading, so that radiative forcing remains less than 1 W/m².
			Reduce by 2030 and eliminate by 2050 emissions of air pollutants that contribute to human health and ecosystem damage – *further research required to define such limits*
Land requirement and degradation	Terrestrial biosphere	Change in land-use; Rate of biodiversity loss; Chemical pollution	No more than 15% of global land cover should be converted to cropland
Water resource requirement and impaired water quality	Freshwater and marine systems	Global freshwater use; Rate of biodiversity loss; Chemical pollution	Limit global freshwater use to no more than 4,000–6,000 km³/yr of consumptive use of accessible river flow

Kakabeeke, 2004). Therefore, for many of the environmental threats related to atmospheric emissions from energy, it might be considered relatively straightforward to use existing indicators and sustainability criteria to assess whether the planetary boundary has been crossed. The actual reality of using these indicators is discussed in more detail below. For the other environmental impacts, which are not directly related to atmospheric emissions (for example, land-use and water resources), different threshold criteria will have to be defined according to a full understanding of the potential impacts, their causes in relation to energy use and supply, and trends and variations within geographical regions.

Discussed below are proposed definitions for planetary boundaries for energy systems that include a range of environmental impacts. They build on earlier analysis by Rockström et al. (2009), and also include boundaries for air pollution, land-use, and water use, based on the summary of current impacts in Sections 3.2, 3.3, and 3.4. The proposal is summarized in Table 3.6.

3.5.2.1 Defining Planetary Boundaries for Climate Change

The planetary boundary GHG concentration-based target proposed by Rockström et al. (2009) calls for limiting concentrations to 350 ppm CO_2, or approximately 400 ppm CO₂-eq, with an uncertainty range of 350–500 ppm CO_2. The climate-change boundary proposed here aims at minimizing the risk of highly nonlinear, possibly abrupt and irreversible, Earth-system responses (National Research Council, or NRC, 2002; IPCC, 2007c). These responses may be related to one or more thresholds, the crossing of which could lead to the disruption of regional climates (Lenton et al., 2008), trigger the collapse

of major climate-dynamic patterns such as the thermohaline circulation (Clark et al., 2002), and drive other impacts that would be difficult for society to address, such as rapid sea-level rise. The risk of crossing such thresholds will rise sharply with further anthropogenically driven deviation from the natural variability of the Holocene climate.

This dual approach to defining the planetary boundary for climate change uses both atmospheric CO_2 concentration and RF as global-scale control variables. Boundary values of 350 ppm CO_2 and 1 W/m² above pre-industrial levels are suggested. The boundary is based on: an analysis of the equilibrium sensitivity of the climate system to GHG forcing; the behavior of the large polar ice sheets under climates warmer than those of the Holocene (Hansen et al., 2008); and the observed behavior of the climate system at a current CO_2 concentration of about 387 ppm and +1.6 W/m² (with a 0.6–2.4 W/m² 90% confidence range) net RF (IPCC 2007a).

3.5.2.2 Defining Planetary Boundaries for Air Pollution

How do the critical loads and levels currently defined by the UNECE LRTAP Convention relate to the planetary boundary concept? Firstly, it is useful to identify which boundaries presented in Rockström et al. (2009) are relevant for air pollution. Perhaps the most obvious is that for the nitrogen cycle, which proposes a boundary of 35 Mt/year of N_2 removed from the atmosphere for human use. How does this compare with LRTAP Convention critical loads? Current empirical critical loads for nutrient nitrogen (which protect against changes in plant growth, interspecific relationships, and soil-based processes) are provided for a wide variety of different natural and seminatural ecosystem habitats. These include forest habitats, heathlands, scrubland, tundra, grasslands, mires, bogs, fens, and inland surface water and coastal and marine habitats. The critical loads vary from 5–40 kg N/ha/year, with the range indicating the variability in ecosystem sensitivity to excess Nr input. Considering that there are similar variations in ecosystem sensitivity to acidifying nitrate and to ammonium deposition, which causes soil acidification, the difficulty of defining a single, planetary-scale boundary for N deposition becomes apparent.

The planetary boundary limit on N_2 fixation attempts to reduce Nr at the source, rather than at different points along the nitrogen cascade that will lead to varying types of ecosystem damage defined by Galloway et al. (2003). However, the difficulty comes in using scientific or expert judgment (the latter is necessary where there are gaps in our scientific understanding) to define a threshold for damage for the different processes through which excess Nr causes eutrophication and acidification. This is the case, since unless N_2 fixation rates equal denitrification rates, there will always be excess Nr in the system, which may cause damage. The geographical spatial heterogeneity of atmospheric Nr pollution and deposition as well as ecosystem sensitivity will mean that some ecosystems will be unprotected by a single planetary-scale boundary that is set to avoid stepping outside the resilience of the system as a whole. It is also not possible to exclude the possibility of cumulative impacts caused by steady increases in N deposition that may result in systems crossing tipping points, even under low pollution loads, if they persist for substantial periods of time. Therefore, we propose that the planetary boundary for N_2 fixation provides a useful guide to encourage more efficient use of Nr in our agricultural systems and more efficient combustion of fossil fuels, but that it should be used in concert with regionally based indicators that employ an effects-based approach to limit pollution impacts. Only then will control be sufficient to ensure protection of the most vulnerable systems. Similar arguments can be made for each of the regional-scale air pollutants discussed in this Chapter.

3.5.2.3   Defining Planetary Boundaries for Land-use

The planetary boundary for land-use proposes that no more than 15% of global land cover should be converted to cropland. Given the current level of malnourishment in the world and the expected 50% increase in world population by 2050, the demand for food and animal feed is expected to require a 70% increase in agricultural production by 2050 (FAO, 2006). Even if most of the increase comes through intensification and yield improvements, there remains a chance that land converted to agriculture could surpass 15% by 2050; this chance is increased by the likelihood that an increasing share of agricultural land will also be devoted to production of biofuels. However, with major investments in agricultural research and an emphasis on high-efficiency agricultural and integrated food-energy systems, expansion of land under agriculture could be significantly constrained (Johnson and Virgin, 2010). The conclusion is that expansion of land used for biofuels must accompanied by much greater investment in 'smart agriculture,' both for the sake of reducing land pressures and for improving food and energy security simultaneously.

3.5.2.4   Defining Planetary Boundaries for Water Resources

Actual freshwater availability is manifested at the local catchment or river-basin level. At the same time, there is increasing evidence that humanity faces global freshwater constraints due to the finite nature of freshwater resources, and the coupling of local water balances with the global hydrological cycle. Currently some 30% of the world's population faces water stress, and approximately 25% of the world's rivers dry out before reaching the ocean (Molden, 2007). The global freshwater cycle has entered the Anthropocene era (Meybeck, 2003), where humans now constitute the dominant driving force, altering river flows at the global level (Shiklimanov and Rodda, 2003) and the spatial patterns and seasonal timing of vapor flows (Gordon et al., 2005).

Global freshwater assessments show that the accessible global volume of runoff water (accessible base flow) is in the order of 12,500–15,000 km³/yr (Postel et al., 1998; deFraiture et al., 2001). Several analyses show that severe water scarcity is experienced on the regional scale when withdrawals of runoff exceed 40–60% of this stable freshwater resource. This provides an uncertainty range of sustainable global freshwater withdrawals of 5000–9000 km³/yr, beyond which negative implications for human societies are expected. However, not all of these withdrawals constitute consumptive use. Current withdrawals of approximately 4000 km³/yr (World Water Development Report, 2009), of which less than 3000 km³/yr is consumptive use. Based on these indicators of sustainability thresholds for freshwater use, a planetary boundary range for global freshwater use has been proposed at 4000–6000 km³/yr of consumptive use of accessible river flow. Evidence indicates that transgressing this boundary range leads to an overuse of freshwater in catchments and river basins where water-induced thresholds, e.g., the collapse of freshwater dependent ecosystems, can no longer be excluded (Rockström et al., 2009). This freshwater boundary is highly tentative, given the uncertainties associated with aggregating sustainable freshwater use at a global level, but it does provide an indicator of the magnitude of freshwater that can be used for bioenergy and other purposes before serious water-related problems occur.

3.6   Conclusion

The assessment in this chapter confirms earlier scientific findings that a global energy transformation is needed to address the growing risks associated with accelerated global environmental change. Anthropogenic pressures on the planet have reached a level where large-scale deleterious impacts, or even catastrophic ones, can no longer be excluded. Such impacts have the potential to undermine human development. This new global social-environmental predicament is closely associated with energy.

Atmospheric emissions from energy use contribute to multiple environmental impacts. In addition to climate change, atmospheric pollutants may limit net primary productivity of ecosystems, and lead to the acidification and eutrophication of land and seascapes. These impacts interact, reinforcing impacts on social and environmental systems, in complex ways that are not always well understood. This chapter confirms the necessity for the global energy system, which is the largest source of GHG emissions, to – as a minimum requirement – operate within the 2°C climate guardrail. In fact, based on the latest science, this chapter concludes that 1.5°C may be a more appropriate guardrail. This conclusion is based on the high likelihood that even small increases in global mean surface temperature will have extensive negative impacts on societies and ecosystems. This chapter also concludes that immediate action on reducing BC and tropospheric O_3 should be a high priority for short-term climate mitigation with associated benefits for human health. Action to control BC will particularly aid improvements in indoor air quality in the poor households of the world (see also Chapter 4).

This chapter also confirms the interconnectedness among all regions of the world, in terms of high dependency on fossil-energy sources and negative environmental impacts from the current energy mix. This energy mix generates impacts at the local, regional, and global levels. This chapter also confirms that there are winners and losers on the global energy scene – with the poorest tropical regions in the world being most vulnerable to the environmental impacts of unsustainable energy use, and the lowest energy-using regions, including the poorest developing countries and the polar regions, being highly affected by negative impacts originating from energy use in other regions.

This chapter confirms the intricate link between land and energy. Land is affected through loss or damage to ecosystems from land-use change and contamination from energy-related waste arising from activities such as mining, drilling, and the transport of fossil fuel raw materials. The alternatives to fossil fuel-based energy systems (e.g., nuclear power, hydropower, biomass-derived fuels and solar power) also lead to a variety of adverse environmental impacts on air, land, and water at various stages in the energy chain. In particular, the intensive use of land and fresh water by bioenergy systems has implications for meeting increased global food demands, as assessments increasingly indicate the existence of regional and global limitations to the expansion of agricultural land and water use for biomass production. It is concluded that the expansion of land used for biofuels must be accompanied by greater investment in 'smart agriculture,' both for the sake of reducing land pressures and for improving food and energy security simultaneously.

Water resources and aquatic ecosystems are also adversely impacted by various types of energy systems. Water may be diverted from other uses by biomass crop production or hydropower schemes. Aquatic ecosystems may be damaged by the interruption of hydrological flows (e.g., from dam construction or open-cast mining operations) as well as by contamination during coal and uranium ore mining, oil and gas drilling,

fossil fuel-processing and transportation, or thermal pollution from power stations.

Drawing upon the latest science, this chapter confirms earlier assessments (particularly within climate science) that atmospheric emissions, of both GHGs and air pollutants, constitute the core and most immediate environmental challenges within the energy sector. However, it is equally clear that energy impacts the biosphere; this calls for immediate attention to reduce the impacts of energy systems on land and water resources, including the use and flow of nitrogen; biodiversity loss; the toxic effects of tropospheric O_3 and other toxic chemical pollution.

Chapter 3 therefore concludes that there is a need for an integrated approach, in which all environmental impacts from energy use are considered, both in terms of climate and ecosystem change. An energy transformation would bring multiple benefits and would help humanity tunnel out of the current era of rapid global environmental change. Such a transformation would have benefits ranging from averting global climate change to reducing the burden of air pollution and ecosystem degradation. It would also require the integration of policy and development action on climate change, air pollution, and ecosystem management, from local to regional to global levels. Energy systems, climate change, and air pollution are strongly connected, in such a way that integrated decision making, coordinated at an international level, will be absolutely crucial to the development of viable options for mitigating the adverse environmental impacts of our energy needs.

A new framework is therefore needed to guide a global energy transformation. This chapter concludes that there is an urgent need for global sustainability criteria, within which the global energy system can operate and identifies the 'planetary boundary' approach as one means of defining global sustainability criteria that could help establish future sustainable energy pathways.

Acknowledgements

The Stockholm Environment Institute's contribution to the study presented here has been made possible through financial support provided by the Swedish International Development Cooperation Agency (Sida). However, Sida was not involved in the preparation of the chapter and does not necessarily support the views expressed. This paper has not been subjected to US EPA peer and administrative review; therefore, the conclusions and opinions contained here are solely those of the authors, and should not be construed to reflect the views of the US EPA. Thanks also to Mr. Richard Falk and Ms. Freya Forest, who helped with the layout and referencing of the Chapter.

References

ACIA, 2004: *Impacts of a Warming Arctic: Arctic Climate Impact Assessment.* Cambridge University Press, Cambridge, UK and New York, NY.

Agoumi, A., 2003: *Vulnerability of North African Countries to Climatic Changes: Adaptation and Implementation Strategies for Climate Change*. International Institute for Sustainable Development.

Ainsworth, E. A., et al., 2008: Targets for crop biotechnology in a future high-CO_2 and high-O_3 world. *Plant Physiology*, **147**: 13–19.

Alcamo, J., J. M. Moreno, B. Nováky, M. Bindi, R. Corobov, R. J. N. Devoy, C. Giannakopoulos, E. Martin, J. E. Olesen, and A. Shvidenko, 2007: *Climate Change 2007: Impacts, Adaptation and Vulnerability. Chapter 12: Europe. Contribution of Working Group II to the Fourth Assessment Report of the Intergovernmental Panel on Climate Change*, M. L. Parry, O. F. Canziani, J. P. Palutikof, P. J. van der Linden, and C. E. Hanson (eds.), Cambridge University Press, Cambridge, UK and New York, NY, pp.541–580.

AMAP, 2006: *AMAP Assessment 2006: Acidifying Pollutants, Arctic Haze and Acidification in the Arctic*. Arctic Monitoring and Assessment Programme (AMAP), Oslo, Norway.

AMAP, 2007: *Arctic Oil and Gas 2007 Oslo. Arctic Monitoring and Assessment Programme (AMAP), Oslo, Norway.AMAP, 2010: AMAP Assessment 2009: Radioactivity in the Arctic*. Arctic Monitoring and Assessment Programme (AMAP), Oslo, Norway.

Amann, M., I. Bertok, J. Borken-Kleefeld, J. Cofala, C. Heyes, L. Höglund-Isaksson, Z. Klimont, B. Nguyen, M. Posch, P. Rafaj, R. Sandler, W. Schöpp, F. Wagner, and W. Winiwarter, 2011: Cost-effective control of air quality and greenhouse gases in Europe: modeling and policy applications. *Environmental Modelling and Software, in press* doi 10.1016/j.envsoft.2011.07.012

Archer, D. R. and H. J. Fowler., 2004: Spatial and temporal variations in precipitation in the Upper Indus Basin, global teleconnections and hydrological implications. *Hydrology and Earth System Sciences*, 8: 47–61.

Arnell, N.W., 2004: Climate change and global water resources: SRES emissions and socio-economic scenarios. *Global Environmental Change*, 14: 31–52.

Arnell, N. W., 2006: *Global impacts of abrupt climate change: an initial assessment*. Working Paper 99, Tyndall Centre for Climate Change Research, University of East Anglia, Norwich, UK, pp.37.

Ashmore, M., 2005: Assessing the future global impacts of ozone on vegetation. *Plant, Cell & Environment*, 28(8): 949–964.

Bell, J. N. B., 1985: *SO2 effects on productivity of grass species, in Effects of SO2 on Plant Productivity*, Mooney and A. H. Goldstein (eds.), Stanford University Press, Palo Alto, CA, pp.209–266.

Bellwood, D., T. Hughes, C. Folke, and M. Nystrom, 2004: Confronting the coral reef crisis. *Nature*, 429: 827–833.

Bender, J. and H. J. Weigel, 1993: *Crop responses to mixtures of air pollutants, in Air Pollution and Crop Responses in Europe*, H. J. Jager et al. (eds.), CEC, Brussels, Belgium, pp.445–453.

Berndes, G., 2008: Future biomass energy supply: the consumptive water use perspective. *International Journal of Water Resources Development*. 24: 235–245.

Betts, R. A., M. Collins, D. L. Hemming, C. D. Jones, J. A. Lowe, and M. G. Sanderson, 2011: When could global warming reach 4⬜C? *Philosophical Transactions of the Royal Society A*. 369: 67–84.

Bishop, K. and H. Hultberg, 1995: Reversing acidifcation in a forest ecosystem – The Gardsjon covered catchment. *Ambio*, 24(2): 85–91.

Bleeker, A., W. K. Hicks, F. Dentener, J. Galloway, and J. W. Erisman, 2011: N deposition as a threat to the World's protected areas under the Convention on Biological Diversity. *Environmental Pollution* (in press).

Bobbink, R., K. Hicks, J. Galloway, T. Spranger, R. Alkemade, M. Ashmore, M. Bustamante, S. Cinderby, E. Davidson, F. Dentener, B. Emmett, J.W. Erisman, M. Fenn, F. Gilliam, A. Nordin, L. Pardo, and W. De Vries, 2010: Global assessment of nitrogen deposition effects on terrestrial plant diversity: a synthesis. *Ecological Applications*, 20(1): 30–59.

Bobbink, R., M. Hornung, and J. G. M. Roelofs, 1998: The effects of air-borne nitrogen pollutants on species diversity in natural and semi-natural European vegetation. *Journal of Ecology*, 86: 717–738.

Boko, M., I. Niang, A. Nyong, C. Vogel, A. Githeko, M. Medany, B. Osman-Elasha, R. Tabo, and P. Yanda, 2007: *Climate Change 2007: Impacts, Adaptation and Vulnerability. Chapter 9: Africa. Contribution of Working Group II to the Fourth Assessment Report of the Intergovernmental Panel on Climate Change*, M. L. Parry, O. F. Canziani, J. P. Palutikof, P. J. van der Linden, and C. E. Hanson (eds.), Cambridge University Press, Cambridge, UK and New York, NY, pp.433–467.

Bond, T., E. Bhardwaj, R. Dong, R. Jogani, S. Jung, C. Roden, D. Streets, and N. Trautmann, 2007: Historical emissions of black and organic carbon aerosol from energy-related combustion, 1850–2000. *Global Biogeochemical Cycles*, 21: GB2018.

Bond, T., D. Streets, K. Yarber, S. Nelson, J-H. Woo, and Z. Kilmont, 2004: A technology-based global inventory of black and organic carbon emissions from combustion. *Journal of Geophysical Research*, 109: D14203.

Bravo, H., A. R. Soto, R. Sosa, P. Sánchez, A. L. Alarcón, J. Kahl, and J. Ruíz, 2006: Effect of acid rain on building material of the El Tajín archaeological zone in Veracruz, Mexico. *Environmental Pollution*, 144: 655–660.

Bricker, S., B. Longstaff, W. Dennison, A. Jones, K. Boicourt, C. Wicks, and J. Woerner, 2007: *Effects of Nutrient Enrichment In the Nation's Estuaries: A Decade of Change. NOAA Coastal Ocean Program Decision Analysis Series No. 26*. National Centers for Coastal Ocean Science, Silver Spring, MD, USA.

Brohan, P., J. J. Kennedy, I. Harris, S. F. B. Tett, and P. D. Jones, 2006: Uncertainty estimates in regional and global observed temperature changes: A new data set from 1850. *Journal of Geophysical Research*, 111: D12106.

Casper, K. N., 2009: Oil and gas development in the Arctic: Softening of Ice demands hardening of international law. *Natural Resources Journal*, 49: 825–881.

Cazenave, A., 2006: How fast are the ice sheets melting? *Science*, 314, 1250–1252.

CCME, 2003: *Climate, Nature, People: Indicators of Canada's Changing Climate. Climate Change Indicators Task Group of the Canadian Council of Ministers of the Environment, Canadian Council of Ministers of the Environment Inc.*,Winnipeg, Canada.

CEC, 2009: *Taking Stock – 2005 North American Pollutant Releases and Transfers*, Commission for Environmental Cooperation, Montreal, Canada.

Chan, C. K. and X. Yao, 2008: Air pollution in mega cities in China. *Atmospheric Environment*, 42: 1–42.

Chen, H., Y. Wu, X. Yuan, Y. Gao, N. Wu, and D. Zhu, 2009: Methane emissions from newly created marshes in the drawdown area of the Three Gorges Reservoir, *Journal of Geophysical Research*, 114: D18301.

Christensen, J. H. and O. B. Christensen, 2003: Severe summertime flooding in Europe. *Nature*, 421, 805–806.

Church, J. A. and N. J. White, 2006. A 20th century acceleration in global sea-level rise. *Geophysical Research Letters,* **33**: L01602.

Clancy, J. S., 2008: Urban ecological footprints in Africa. *African Journal of Ecology,* 46(4): 463–470.

Clark, P., N. Pisias, T. Stocker, and A. Weaver, 2002: The role of thermohaline circulation in abrupt climate change. *Nature,* 415: 863–869.

Cofala, J. and S. Syri, 1998: *Sulfur Emissions, Abatement Technologies and Related Costs for Europe in the RAINS Model Database.* IR-98–035. IIASA, Laxenburg, Austria.

Corbett, J. J., D.A. Lack. J. J. Winebrake, S. Harder, J. A. Silberman, and M. Gold, 2010: Arctic shipping emissions inventories and future scenarios. *Atmospheric Chemistry and Physics,* 10: 10271–10311.

Crutzen, P. J., 2002: Geology of mankind the anthropocene. *Nature,* 415: 23.

Cruz, R. V., H. Harasawa, M. Lal, S. Wu, Y. Anokhin, B. Punsalmaa, Y. Honda, M. Jafari, C. Li, and N. Huu Ninh, 2007: *Climate Change 2007: Impacts, Adaptation and Vulnerability, Chapter 10: Asia. Contribution of Working Group II to the Fourth Assessment Report of the Intergovernmental Panel on Climate Change,* M. L. Parry, O. F. Canziani, J. P. Palutikof, P. J. van der Linden, and C. E. Hanson, (eds.), Cambridge University Press, Cambridge, UK and New York, NY, pp.469–506.

Davidson, O., M. Chenene, E. Kituyi, J. Nkomo, C. Turner, and B. Sebitosi B., 2007: *Sustainable Energy in sub-Saharan Africa.* Draft Science/Work plan, International Council for Science, Regional Office for Africa.

Davison A.W. and J. D. Barnes, 1998: Effects of ozone on wild plants. *New Phytologist,* 139(1): 135–151.

deFraiture, C., D. Molden, U. Amarasinghe, and I. Makin, 2001: PODIUM: Projecting water supply and demand for food production in 2025. *Physics and Chemistry of the Earth, Part B: Hydrology, Oceans and Atmosphere,* 26(11–12), 869–876.

de Vries, W., S. Solberg, M. Dobbertin, H. Sterba, D. Laubhahn, G. J. Reinds, G-J. Nabuurs, P. Gundersen, M. A. Sutton, 2007: Ecologically implausible carbon response? *Nature,* 451: E1-E3.

Denholm, P., 2006: Improving the technical, environmental and social performance of wind energy systems using biomass-based energy storage. *Renewable Energy,* 1: 1355–1370.

Dentener, F., S. Kinne, T. Bond, O. Boucher, J. Cofala, S. Generoso, P. Ginoux, S. Gong, J. J. Hoelzemann, A. Ito, L. Marelli, J. E. Penner, J. P. Putaud, C. Textor, M. Schulz, G. R. van der Werf, and J. Wilson, 2006: Emissions of primary aerosol and precursor gases in the years 2000 and 1750 prescribed data-sets for AeroCom. *Atmospheric Chemistry and Science,* 6: 4321–4344.

Devoy, R. J. N., 2007: Coastal vulnerability and the implications of sea-level rise for Ireland. *Journal of Coastal Research,* 24(2): 325–341.

Di Iorio, L. and C. W. Clark, 2010: Exposure to seismic survey alters blue whale acoustic communication. *Biology Letters,* 6(1): 51–54.

Dolman, A. J., G. R. van der Werf, M. K. van der Molen, G. Ganssen, G., J. W. Erisman, and B. Strengers, 2010: A carbon cycle science update since IPCC AR-4. *Ambio,* 39: 402–412.

Donner, S. D., and C. J. Kucharik, 2008: Corn-based ethanol production compromises goal of reducing nitrogen export by the Mississippi River. *Proceedings of the National Academy of Sciences of the United States of America,* 105: 4513–4518.

Duce, R. A., J. LaRoche, K. Altieri, et al., 2008: Impacts of atmospheric anthropogenic nitrogen on the open ocean. *Science,* 320(5878): 893–897.

Duncan, B., J. Logan, I. Bey, I. Megretskaia, R. Yantosca, P. Novelli, N. Jones, and C. Rinsland, 2007: Global budget of CO, 1988–1997: Source estimates and validation with a global model. *Journal of Geophysical Research,* 112: D22301.

Eckhardt, S., A. Stohl, S. Beirle, N. Spichtinger, P. James, et al., 2003: The north Atlantic Oscillation controls air pollution transport to the Arctic. *Atmospheric Chemistry and Physics,* 3: 1769–1778.

EEA, 2008: *Energy and environment indicator fact sheet, EN13 Nuclear waste production.* European Environment Agency (EEA), Copenhagen, Denmark.

Emberson, L. D., et al., 2003: *Air Pollution Impacts on Crops and Forests-A Global Assessment,* edited, Imperial College Press, London, UK.

Emberson, L. D., P. Bueker, and M. R. Ashmore, 2007: Assessing the risk caused by ground level ozone to European forest trees: A case study in pine, beech and oak across different climate regions. *Environmental Pollution,* 147: 454–466.

Emberson L. D., P. Büker, M. R. Ashmore, et al., 2009: A comparison of North American and Asian exposure-response data for ozone effects on crop yields. *Atmospheric Environment,* 43: 1945–1953.

ENA, 2011: *The European Nitrogen Assessment: Sources, Effects and Policy Perspectives.* Eds. Sutton. M.A., Howard, C.M., Erisman, J.W., Billen, G., Bleeker, A., Grennfelt, P., van Grinsven, H. and Grizzetti, B. Cambridge University Press. Avaialble at: http://www.nine-esf.org/ENA-Book

Energy Information Administration, 2000: Sub-Saharan Africa: Environmental Issues, November 2000.

Environment Canada, 2008: Canadian Environmental Sustainability Indicators. ec.gc.ca/indicateurs-indicators/default.asp?lang=en&n=7D7BDF1E-1 (accessed 23 March 2010).

Environment Canada 2010: www.ec.gc.ca/acidrain/acidfact.html (accessed 29 March 2010).

Environment Canada 2010: National Ambient Air Quality Objectives. www.ec.gc.ca (accessed 27 March 2010).

Evans, A., V. Strezov, and T. J. Evans, 2009: Assessment of sustainability indicators for renewable energy technologies. *Renewable and Sustainable Energy Reviews,* 13: 1082–1088.

Fang, Y. P., Y. Zeng, and S. M. Li, 2008: Technological influences and abatement strategies for industrial sulphur dioxide in China. *International Journal of Sustainable Development and World Ecology,* 15: 122–131.

FAO, 2004: *Seguridad alimentaria como estrategia de Desarrollo rural. Proc. 28ava Conferencia Regional de la FAO para América Latina y el Caribe,* Ciudad de Guatemala, Guatemala.

FAO, 2005: *State of the World's Forests 2005. Food and Agriculture Organization of the United Nations (FAO),* Rome.

FAO, 2006: *World agriculture: towards 2030/2050. Food and Agriculture Organization of the United Nations (FAO),* Rome.

Farman, J. C., R. J. Murgatroyd, A. M. Silnickas, and B. A.Thrush, 1985: Ozone photochemistry in the antarctic stratosphere in summer. *Quarterly Journal of the Royal Meteorological Society,* 111: 1013–1025.

Fenn, M. E., J. S. Baron, E. B. Allen, H. M. Reuth, K. R. Nydick, L. Geiser, W. D. Bowman, J. O. Sickman, T. Meixner, D. W. Johnson, and P. Neitlich, 2003: Ecological Effects of Nitrogen Deposition in the Western United States. *Bioscience,* 53(4): 404–420.

Fey, M. V. and H.A. Dodds, 1998: Classifying the sensitivity of soils of the South African highveld to acidification. *South African Journal of Plant and Soil,* 15: 99–103.

Field, C. B., L. D. Mortsch, M. Brklacich, D. L. Forbes, P. Kovacs, J. A. Patz, S. W. Running, and M. J. Scott., 2007: *Climate Change 2007: Impacts, Adaptation and Vulnerability. Chapter 14: North America. Contribution of Working Group II to the Fourth Assessment Report of the Intergovernmental Panel on Climate Change*, M.L. Parry, O.F. Canziani, J.P. Palutikof, P.J. van der Linden, and C.E. Hanson, (eds.), Cambridge University Press, Cambridge, UK and New York, NY, pp.617–652.

Fiki, C., and B. Lee, 2004: Conflict generation, conflict management and self-organizing capabilities in drought-prone rural communities in north-eastern Nigeria: A case study. *Journal of Social Development in Africa*, 19(2):25–48.

Fischer, G., S. Prielera, H. van Velthuizena, S. M. Lensink, M. Londo, and M. de Wit, 2010: Biofuel production potentials in Europe: Sustainable use of cultivated land and pastures, Part I: Land productivity potentials. *Biomass and Bioenergy*, 34: 159–172.

Fischer, G., M. Shah, and H. vanVelthuizen, 2002: *Climate change and agricultural vulnerability. Preprints, World Summit on Sustainable Development*, Johannesburg, pp.160.

Fiscus, E. L., et al., 2005: Crop responses to ozone: uptake, modes of action, carbon assimilation and partitioning. *Plant Cell Environment*, 28: 997–1011.

Fisher, B., N. Nakicenovic, K. Alfsen, J. Corfee Morlot, F. de la Chesnaye, J.-C. Hourcade, K. Jiang, M. Kainuma, E. La Rovere, A. Matysek, A. Rana, K. Riahi, R. Richels, S. Rose, D.P. van Vuuren, R. Warren, P. Ambrosi, F. Birol, D. Bouille, C. Clapp, B. Eickhout, T. Hanaoka, M.D. Mastrandrea, Y. Matsuoko, B. O'Neill, H. Pitcher, S. Rao, and F. Toth, 2007: Issues related to mitigation in the longterm context. In: B. Metz, O. Davidson, P. Bosch, R. Dave, L. Meyer (eds.), *Climate change 2007: Mitigation of climate change. Contribution of working group III to the fourth assessment report of the intergovernmental panel on climate change*. Cambridge University Press, Cambridge, UK and New York, NY, pp.169–250.

Fishman, J., et al., 2010: An investigation of widespread ozone damage to the soybean crop in the upper Midwest determined from ground-based and satellite measurements. *Atmospheric Environment*, 44(2010): 2248–2256.

Flörke, M., and J. Alcamo, 2005: European Outlook On Water Use. *Prepared for the European Environment Agency*, pp.83.

Foster, J. S., J. W. Winchester, and E. G. Dutton, 1992: The date of snow disappearance on the Arctic tundra as determined from satellite, meteorological station and radiometric in situ observations. *IEEE Transactions on Geoscience and Remote Sensing*, 30(4): 793–798.

Fowler, et al., 2009: Atmospheric composition change: Ecosystesm-Atmosphere interactions. *Atmospheric Environment*, 43: 5193–5267.

Frauenfeld, O. W., T. Zhang, R. G. Barry, and D. Gilichinsky, 2004: Interdecadal changes in seasonal freeze and thaw depths in Russia. *Journal of Geophysical Research*, 109: D05101.

Fuhrer, J., 2003: Agroecosystern responses to combinations of elevated CO2, ozone, and global climate change. *Agriculture, Ecosystems and Environment*, 97: 1–20.

Fuhrer, J., 2009: Ozone risk for crops and pastures in present and future climates. *Naturwissenschaften*, 96: 173–194.

Fuhrer, J. and F. Booker, 2003: Ecological Issues related to ozone: agricultural issues. *Environment International*, 29(2–3): 141–154.

Fuhrer, J., L. Skärby, and M. R. Ashmore, 1997. Critical levels for ozone effects on vegetation in Europe. *Environmental Pollution*, 97: 91–106.

Fujino, J., R. Nair, M. Kainuma, T. Masui, and Y. Matsuoka, 2006: Multigas mitigation analysis on stabilization scenarios using AIM global model. *Energy Journal*, 27: 343–354.

Galloway J., A. Townsend, J. W. Erisman, Z. Cai, F. Martinelli, S. Seitzinger, and M. Sutton, 2008: Transformation of the nitrogen cycle: Recent trends, questions and potential solutions. *Science*, 320: 889–892.

Galloway, J. N., J. D. Aber, J. W. Erisman, S. P. Seitzinger, R. W. Howarth, E. B. Cowling, and B. J. Cosby, 2003: The Nitrogen Cascade. *BioScience*, 53(4): 341–356.

Garrett, T. J., L. F. Radke, and P. V. Hobbs, 2002: Aerosol effects on the cloud emissivity and surface longwave heating in the Arctic. *Journal of the Atmospheric Sciences*, 59: 769–778.

Geerben-Leenes, W., A. Y. Hoekstra, and T. H. van der Meer, 2009: The water footprint of bioenergy. *Proceedings of the National Academy of Sciences of the United States of America*, 106(25): 10219–10223.

Global Biodiversity Outlook 3, 2010: Secretariat of the Convention on Biological Diversity, Montreal. www.cbd.int/gbo3/ebook/ (accessed 14 March 2010).

Gordon, L. J., G. D. Peterson, and E. B. Bennett, 2008: Agricultural modifications of hydrological flows create ecological surprises. *Trends in Ecology and Evolution*, 23(4): 211–219.

Gordon, L., W. Steffen, B. F. Jönsson, C. Folke, M. Falkenmark, and Å. Johannessen, 2005: Human modification of global water vapour flows from the land surface. *PNAS*, 105: 7612–7617.

Gosnell, H., Kelly, K., 2010: Peace on the river? Social-ecological restoration and large dam removal in the Klamath basin, USA. *Water Alternatives*, 3(2): 362–383.

Grennfelt, P. and Ø. Hov, 1995: Regional Air Pollution at a turning point. *Ambio*, 34(1): 2–10.

Gruber, N. and J. N. Galloway, 2008: An Earth-System perspective of the global nitrogen cycle. *Nature*, 451: 293–296.

Guerin, F., G. Abril, S. Richard, B. B. Burban, C. Reynouard, P. Seyler, and R. Delmas, 2006: Methane and carbon dioxide emissions from tropical reservoirs: significance of downstream rivers. *Geophysical Research Letters*, 33: L21407.

Hansen, J., R. Ruedy, M. Sato, M. Imhoff, W. Lawrence, D. Easterling, T. Peterson, and T. Karl, 2001: A closer look at United States and global surface temperature changes. *Journal of Geophysical Research*, 106: 947–963.

Hansen, J., M. Sato, P. Kharecha, D. Beerling, R. Berner, V. Masson-Delmotte, M. Pagani, M. Raymo, D. L. Royer, and J. S. Zachos, 2008: Target atmospheric CO2: Where should humanity aim? *Open Atmospheric Science Journal*, 2: 217–231.

Hansen, J., R. Reudy, M. Sato, and K. Lo, 2010: Global surface temperature change. *Reviews of Geophysics*, 48: RG4004.

Hanson, C. E., J. P. Palutikof, A. Dlugolecki, and C. Giannakopoulos, 2006: Bridging the gap between science and the stakeholder: the case of climate change research. *Climate Research*, 13: 121–133.

Hare, B. And M. Meinshausen, 2006: How much warming are we committed to and how much can be avoided? *Climatic Change*, 75(1–2): 111–149.

Harris, G., 2002: Integrated assessment and modelling: an essential way of doing science. *Environmental Modelling & Software*, 17(3): 201–207.

Hassan, I., M. Ashmore, and J. Bell, 1995: Effect of ozone on radish and turnip under Egyptian field conditions. *Environmental Pollution*, 89(1): 107–114.

Heck, W. W, O. C. Taylor, and D. T. Tingey (eds.), 1988: *Assessment of Crop Loss from Air Pollutants.* Barking, UK: Elsvier Applied Science.

Hernandez, T., and L. I. de Bauer, 1984: Evolucion del daño por gases oxidantes en Pinus hartwegii P. montezumae var. Lindleyi en el Ajusco. *D. F. Agrociencia*, 56:183–194.

Hernandez, T., and L. I. de Bauer 1986: Photochemical oxidant damage on Pinus hartwegii at the Desierto de los Leones, *D. F. Phytopathology*, 76(3): 377.

Hettelingh, F.-P., M. Posch, J. Slootweg, G. J. Reinds, T. Spranger, and L. Tarrason, 2007: Critical loads and dynamic modelling to assess Eruopean areas at risk of acidification and eutrophication. *Water, Air and Soil Pollution Focus,* 7: 379–384.

Hettelingh, J. P., M. Posch, and J. Slootweg, 2008: *Critical Load, Dynamic Modelling and Impact Assessment in Europe: CCE Status Report 2008. Coordination Centre for Effects, Netherlands Environmental Assessment Agency,* Bilthoven, Netherlands.

Hicks, W. K., J. C. I. Kuylenstierna, A. Owen, F. Dentener, H. M. Seip, and H. Rodhe, 2008: Soil sensitivity to acidification in Asia: Status and prospects. *Ambio,* 37: 295–303.

Hoelzle, M., W. Haeberli, M. Dischl, and W. Peschke, 2003: Secular glacier mass balances derived from cumulative glacier length changes. *Global and Planetary Change,* 36: 295–306.

Holland, M, C. Bitz, and B. Tremblay, 2006: Future abrupt reductions in the summer Arctic sea ice. *Geophysical Research Letters,* 33(L23503): 1–5.

Howarth, R., D. Anderson, J. Cloern, C. Elfring, C. Hopkinson, B. Lapointe, T. Malone, N. Marcus, K. McGlathery, A. Sharpley, and D. Walker, 2000: Nutrient pollution of coastal rivers, bays, and seas. *Ecological Society of America,* 7: 2–15.

Huang, J., C. Zhang, and J. M. Prospero, 2009: Large-scale effect of aerosols on precipitation in the West African Monsoon region. *Quarterly Journal of the Royal Meteorological Society,* 135: 581–594.

Hungate, B. A., et al., 2009: Nitrogen and climate change. *Science,* 203: 1512–1513.

IEA, 2009: *World energy outlook 2009. International Energy Agency, OECD Publication Service,* OECD, Paris.

IJC, 2010: *US and Canada Air Quality Agreement Progress Report*. International Joint Commission (IJC).

International Rivers, 2011: www.internationalrivers.org/ (accessed 24 March 2011).

IPCC, 1990: *First Assessment Report (FAR) – Climate Change 1990*. Cambridge University Press, Cambridge, UK and New York, NY.

IPCC, 1995a: *Second Assessment Report (SAR) – Climate Change 1995*. Cambridge University Press, Cambridge, UK and New York, NY.

IPCC, 1995b: *The Science of climate change. Contribution of Working Group I to the Second Assessment Report of the Intergovernmental Panel on Climate Change*. J. T. Houghton, L. G. Meira Filho, B. A. Callander, N. Harris, A. Kattenberg and K. Maskell (eds.), Cambridge University Press, Cambridge, UK, New York, NY, and Melbourne, Australia, pp.531.

IPCC, 2000a: Emission scenarios. N. Nakicenovic, and R. Swart, (eds.). *The Science of climate change. Contribution of Working Group I to the Second Assessment Report of the Intergovernmental Panel on Climate Change,* pp.570. Cambridge University Press, Cambridge, UK and New York, NY.

IPCC, 2000b: Methodological and technological issues in technology transfer. Metz, B., Davidson, O., Martens, J.-W., Van Rooijen, S. and Van Wie Mcgrory, L. (eds.), *The Science of climate change. Contribution of Working Group I to the Second Assessment Report of the Intergovernmental Panel on Climate Change,* Cambridge University Press, Cambridge, UK and New York, NY, pp.432.

IPCC, 2001a: *The Scientific Basis. Contribution of Working Group I to the Third Assessment Report of the Intergovernmental Panel on Climate Change*. Houghton, J. T., Y. Ding, D. J. Griggs, M. Noguer, P. J. van der Linden, X. Dai, K. Maskell, and C. A. Johnson (eds.), Cambridge University Press, Cambridge, UK and New York, NY, pp.881.

IPCC, 2001b: *Impacts, Adaptation, and Vulnerability. Contribution of Working Group II to the Third Assessment Report of the Intergovernmental Panel on Climate Change*. J. J. McCarthy, O. F. Canziana, N. A. Leary, D. J. Dokken, K. S. White, (eds.), Cambridge University Press, Cambridge, UK and New York, NY, pp.976.

IPCC, 2005: *Carbon dioxide capture and storage*, B. Metz, O. Davidson, H. de Coninck, M. Loos, L. Meyer, (eds.), Cambridge University Press, Cambridge, UK and New York, NY. pp.431.

IPCC, 2007a: *The Physical Science Basis. Contribution of Working Group I to the Fourth Assessment Report of the Intergovernmental Panel on Climate Change*, S. Solomon, D. Qin, M. Manning, Z. Chen, M. Marquis, K. B. Averyt, M. Tignor, H. L. Miller (eds.), Cambridge University Press, Cambridge, UK and New York, NY.

IPCC, 2007b: *Impacts, adaptation and vulnerability. Contribution of Working Group II to the Fourth Assessment Report of the Intergovernmental Panel on Climate Change*, M. L. Parry, O. F. Canziani, J. P. Palutikof, P. J. van der Linden, and C. E. Hanson (eds.), Cambridge University Press, Cambridge, UK and New York, NY.

IPCC, 2007c: *Mitigation of climate change. Contribution of Working Group III to the Fourth Assessment Report of the Intergovernmental Panel on Climate Change*, B. Metz, O. Davidson, P. Bosch, L. A. Meyer, (eds.), Cambridge University Press, Cambridge, UK and New York, NY, USA.

Jackson, R. B, G. J. Esteban, A. Roni, S. B. Roy, D. J. Barrett, C. W. Cook, K. A. Farley, D. C. le Maitre, B. A. McCarl, and B. C. Murray, 2005: Trading water for carbon with biological carbon sequestration. *Science,* 310(5756): 1944–1947.

Jackson, C., 2009: Parallel pursuit of near-term and long-term climate mitigation. *Science,* 326(5952): 526–527.

Janssens, I. A. and S. Luyssaert, 2009: Carbon cycle: Nitrogen's carbon bonus. *Nature Geoscience,* 2: 318–319.

Johannessen, O. M., 2008: Decreasing Arctic Sea Ice mirrors increasing CO_2 on Decadal time scale. *Atmospheric and Oceanic Science Letters,* 1: 51–56.

Johnson, F. X. and I. Virgin, 2010: Future Trends in biomass resources for food and fuel. In *Food versus Fuel: an informed introduction*. F. Rosillo-Calle and F. X. Johnson (eds.), ZED books, London.

Jones, P. D., and A. Moberg, 2003: Hemispheric and large scale surface air temperature variations: an extensive revision and an update to 2001. *Journal of Climate,* 16, 206–223.

Josipovic, M., H. J. Annegarn, M. A. Kneen, J. J. Pienaar, and S. J. Piketh, 2010: Concentrations, distributions and critical level exceedance assessment of SO_2, NO_2 and O_3 in South Africa. *Environmental Monitoring and Assessment,* 171: 181–196.

Karlsson, P. E., et al., 2007: Risk assessments for forest trees: The performance of the ozone flux versus the AOT concepts. *Environmental Pollution,* 146, 608–616.

Karnosky, D. F., and K. C. Steiner, 1981: Provenance variation in response of Fraxinus americana and Fraxinus pennsylvanica to sulfur dioxide and ozone. *Phytopathology,* 71: 804–807.

Karnosky, D. F., K. S. Pregitzer, D. R. Zak, M. E. Kubiske, G. R. Hendrey, D. Weinstein, M. Nosal, and K. E. Percy, 2005: Scaling ozone responses of forest trees to the ecosystem level in a changing climate. *Plant, Cell and Environment,* 28: 965–981.

Karnosky, D. F., et al., 2007: Perspectives regarding 50 years of research on effects of tropospheric ozone air pollution on US forests. *Environmental Pollution,* 147: 489–506.

Keeling, C. D. and T. P. Whorf, 2005: *Atmospheric CO^2 records from sites in the SIO air sampling network. In Trends: A compendium of data on global change. Carbon Dioxide Information Analysis Centre*, Oak Ridge National Laboratory, US Dept. of Energy, Oak Ridge, TN.

King, J. S., et al., 2005: Tropospheric O3 compromises net primary production in young stands of trembling aspen, paper birch and sugar maple in response to elevated atmospheric CO_2. *New Phytologist,* 168: 623–636.

Klein Tank, A. M. G., J. B. Wijngaard, G. P. Konnen, R. Bohm, G. Demaree, A. Gocheva, M. Mileta, S. Pashiardis, L. Hejkrlik, C. Kern-Hansen, R. Heino, P. Bessemoulin, G. Muller-Westermeier, M. Tzanakou, S. Szalai, T. Palsdottir, D. Fitzgerald, S. Rubin, M. Capaldo, M. Maugeri, A. Leitass, A. Bukantis, R. Aberfeld, A.F.V. VanEngelen, E. Forland, M. Mietus, F. Coelho, C. Mares, V. Razuvaev, E. Nieplova, T. Cegnar, J.A. López, B. Dahlstrom, A. Moberg, W. Kirchhofer, A. Ceylan, O. Pachaliuk, L.V. Alexander, and P. Petrovic, 2002: Daily dataset of 20th-century surface air temperature and precipitation series for the European Climate Assessment. *International Journal of Climatology,* 22: 1441–1453.

Klein, R. J.T., R. J. Nicholls, and F. Thomalla., 2002: *The resilience of coastal megacities to weather-related hazards. Building Safer Cities: The Future of Disaster Risk,* A. Kreimer, M. Arnold and A. Carlin, Eds., The World Bank Disaster Management Facility, Washington, DC, 101–120.

Klimont, Z. Z., Cofala, J., Xing, J., Wei, W., Zhang, C., Wang, S., Bhandari, P., Mathur, R., Purohit, P., Rafaj, P., Chambers, A., Amann, M., Hao, J., 2009: Projections of SO2, NOx and carbonaceaous aerosols emissions in Asia. *Tellus,* 61: 602–617.

Klimont, Z., Streets, D. G., Gupta, S., Cofala, J., Fu, L., Ichikawa, Y., 2002: Anthropogenic emissions of non-methane volatile organic compounds in China. *Atmospheric Environment,* 36(8): 1309–1322

Koch, D., T. C. Bond, D. Streets, N. Unger, 2007: Linking future aerosol radiative forcing to shifts in source activities. *Geophysical Research Letters,* 34, L05821

Kriegler, E., J. W. Hall, H. Held, R. Dawson, H. J. Schellnhuber, 2009: Imprecise probability assessment of tipping points in the climate system. *Proceedings of the National Academy of Sciences of the United States of America, USA,* 106: 5041–5046.

Krupa, S. V., et al., 2004: Effects of ozone on plant nutritive quality characteristics for ruminant animals. *The Botanica,* 54: 1–12.

Kuylenstierna J. C. I, H. Rodhe, S. Cinderby, and K. Hicks, 2001: Acidification in developing countries: Ecosystem sensitivity and the critical load approach on a global scale. *Ambio,* 30: 20–28.

Law, K. S., A. Stohl, 2007: Arctic air pollution: Origins and impacts. *Science,* 315: 1537–1540.

Le Quéré, C., M. Raupach, J. Canadell, G. Marland, et al., 2009: Trends in the sources and sinks of carbon dioxide. *Nature Geoscience,* 2: 831–836.

Le Quéré, C., C. Rodenbeck, E. Buitenhaus, T. Conway, R. Lagenfelds, A. Gomez, C. Labuschagne, M. Ramonet, T. Nakazawa, N. Metzl, N. Gillett, M. Heimann, 2007: Saturation of the southern ocean CO2 sink due to recent climate change. *Science,* 316: 1735–1738.

Lehner, B., G. Czisch, and S. Vassolo, 2005: The impact of global change on the hydropower potential of Europe: a model-based analysis. *Energy Policy,* 33: 839–855.

Leiva, J. C., 2006: *Assement climate change impacts on the water resources at the northern oases of Mendoza province, Argentina. Global Change in Mountain Regions,* M. F. Price (ed.), Sapiens Publishing, Kirkmahoe, Dumfriesshire, 81–83.

Lenton, T. M., H. Held, E. Kriegler, J. W. Hall, W. Lucht, S. Rahmstorf, and H. J. Schellnhuber, 2008: Tipping elements in the Earth's climate system. *Proc Natl Acad Sci USA,* 105: 1786–1793.

Levin, L. A., W. Ekau, A. J. Gooday, F. Jorissen, J. J. Middelburg, S. W. Naqvi, C. Neira, N. N. Rabalais, and J. Zhang, 2009: Effects of natural and human-induced hypoxia on coastal benthos. *Biogeosciences,* 6(10): 2063–2098.

Living Planet Report, 2010: Biodiversity, biocapacity and development, WWF. wwf. panda.org/about_our_earth/all_publications/living_planet_report/ (accessed 26 March 2011).

Locatelli, B. and R. Vignola, 2009: Managing watershed services of tropical forests and plantations: can meta-analyses help? *Forest Ecology and Management,* 258: 1664–1870.

Long, S. P., E. A. Ainsworth, A. D. B. Leakey, and P. B. Morgan, 2005: Global food insecurity. Treatment of major food crops with elevated carbon dioxide or ozone under large-scale fully open-air conditions suggests recent models may have overestimated future yields. *Philosophical Transactions of the Royal Society B,* 360: 2011–2020.

Lu, Z., D. G. Streets, Q. Zhang, S. Wang, G. R. Carmichael, Y. F. Cheng, C. Wei, M. Chin, T. Diehl, and Q. Tan, 2010: Sulfur dioxide emissions in China and sulphur trends in East Asia since 2000. *Atmospheric Chemistry and Physics,* 10: 6311–6331.

Lugina, K. M., et al., 2005: Monthly surface air temperature time series area-averaged over the 30-degree latitudinal belts of the globe, 1881–2004. In: *Trends: A Compendium of Data on Global Change. Carbon Dioxide Information Analysis Center,* Oak Ridge National Laboratory, US Department of Energy, Oak Ridge, TN, cdiac.esd.ornl.gov/trends/temp/lugina/lugina.html (accessed 3 April 2011).

MacKay, 2008: Sustainable Energy – without the hot air. UIT, Cambridge, UK. www. withouthotair.com/ (accessed 22 January 2011).

Magnani, F., M. Mencuccini, P. Borghetti, P. Berbigier, F. Berninger, S. Delzon, et al., 2007: The human footprint in the carbon cycle of temperate and boreal forests. *Nature,* 447: 849–851.

Magrin, G., C. Gay García, D. Cruz Choque, J. C. Giménez, A. R. Moreno, G. J. Nagy, C. Nobre, and A. Villamizar, 2007: *Climate Change 2007: Impacts, Adaptation and Vulnerability. Chapter 13: Latin America. Contribution of Working Group II to the Fourth Assessment Report of the Intergovernmental Panel on Climate Change,* M. L. Parry, O. F. Canziani, J. P. Palutikof, P. J. van der Linden, and C. E. Hanson (eds.), Cambridge University Press, Cambridge, UK and New York, NY, 581–615.

Maslanik, J., C. Fowler, J. Stroeve, S. Drobot, J. Zwally, D. Yi, and W. Emery, 2007: A younger, thinner Arctic ice cover: increased potential for rapid, extensive sea-ice loss. *Geophysical Research Letters,* 34: L24501.

Matyssek, R., and H. J. Sandermann, 2003: Impact of ozone on trees: An ecophysiological perspective. *Progress in Botany,* 64: 349–404.

McDonald, R. I., J. Fargione, J. Kiesecker, W. M. Miller, and J. Powell, 2009: Energy Sprawl or Energy Efficiency: Climate Policy Impacts on Natural Habitat for the United States of America. *PLoS ONE,* 4(8): e6802.

Meadows, D. H., and D. L. Meadows, 1972: *The Limits to Growth.* New American Library, New York, NY.

Meadows, D. H., D. L. Meadows, and J. Randers, 1992: *Beyond the limits: Confronting global collapse-envisioning a sustainable future.* Chelsea Green Publishing Company, Post Mills, VT.

Meadows, D. H., J. Randers, and D. L. Meadows, 2004: *Limits to Growth – the 30 year update.* Earthscan, London.

Meehl, G. A. et al., 2007: *Global climate projections. In Climate change 2007: the physical science basis. Contribution of Working Group I to the 4th Assessment Report of the Intergovernmental Panel on Climate Change.* S. Solomon, D. Qin,

M. Manning, Z. Chen, M. Marquis, K. B. Averyt, M. Tignor, and H. L. Miller (eds.), Cambridge University Press, Cambridge, UK and New York, NY.

Meinshausen, M., N. Meinshausen, W. Hare, S. C. B. Raper, K. Frieler, R. Knutti, D. J. Frmae, M. R. Allen, 2009: Greenhouse-gas emission targets for limiting global warming to 2°C. *Nature*, 458: 1158–1162.

Mendelsohn, R., A. Dinar, and A. Dalfelt, 2000: *Climate change impacts on African agriculture*. Preliminary analysis prepared for the World Bank, Washington, DC.

Menz, F. C. and H. M. Seip, 2004: Acid rain in Europe and the United states: an update. *Environmental Science and Policy*, 7(4): 253–265.

Menzel, A., T. H. Sparks, N. Estrella, E. Koch, A. Aasa, R. Ahas, K. Alm-Kübler, P. Bissoli, O. Braslavska, A. Briede, F. M. Chmielewski, Z. Crepinsek, Y. Curnel, Å. Dalh, C. Defila, A. Donnelly, Y. Filella, K. Jatczak, F. Måge, A. Mestre, Ø. Nordli, J. Peñuelas, P. Pirinen, V. Remišová, H. Scheifinger, M. Striz, A. Susnik, A. VanVliet, F.-E. Wielgolaski, S. Zach, and A. Zust, 2006: European phenological response to climate change matches the warming pattern. *Glob. Change Biol.*, 12: 1969–1976.

Metzger, M. J., R. Leemans, D. Schröter, W. Cramer, and the ATEAM consortium, 2004: *The ATEAM Vulnerability Mapping Tool*. Quantitative Approaches in System Analysis No. 27. Wageningen, C.T. de Witt Graduate School for Production Ecology and Resource Conservation, Wageningen, CD ROM.

Meybeck, 2003: M. Global analysis of river systems: from earth system controls to Anthropocene controls. *Philosophical Transactions of the Royal Academy, London B*, 358: 1935–1955.

MEA (Millennium Ecosystem Assessment) 2005: *Millennium Ecosystem Assessment* 2005, Synthesis, Island Press, Washington, DC, USA.

Miller, P. and J. McBride (eds.), 1999: *Oxidant air pollution impacts in the Montane forests of Southern California: The San Bernardino Mountain Case Study*. Springer-Verlag, New York, USA.

Miller, S., 2009: Minimizing Land Use and Nitrogen Intensity of Bio-energy. *Environmental Science & Technology*, 44: 3932–3939.

Mills, G., et al., 2007: A synthesis of AOT40-based response functions and critical levels of ozone for agricultural and horticultural crops, *Atmospheric Environment*, 41: 2630–2643.

Mills, G., et al., 2011: Evidence of widespread effects of ozone on crops and (semi-) natural vegetation in Europe (1990–2006) in relation to AOT40 and flux-based risk maps, *Global Change Biology*, 17: 592–613.

Milly, P. C. D., R. T. Wetherald, K. A. Dunne, and T. L. Delworth, 2002: Increasing risk of great floods in a changing climate. *Nature*, 415: 514–517.

Molden, D. (ed.), 2007. *Water for food, water for life: A Comprehensive Assessment of Water Management in Agriculture, Earthscan*, London, and IWMI, Colombo.

Molina, M. J. and F. S. Rowland, 1974: Stratospheric sink for chlorofluoromethanes: chlorine atomc-atalysed destruction of ozone. *Nature*, 249(5460): 810–812.

Morgan, P. B., et al., 2006, Season-long elevation of ozone concentration to projected 2050 levels under fully open-air conditions substantially decreases the growth and production of soybean. *New Phytologist*, 170: 11.

Moriondo, M., P. Good, R. Durao, M. Bindi, C. Gianakopoulos, and J. Corte-Real, 2006: Potential impact of climate change on fire risk in the Mediterranean area. *Climate Research*, 31: 85–95.

Muller, N. Z. and R. Mendelsohn, 2007: Measuring the damages of air pollution in the United States. *Journal of Environmnetal Economics and Management*, 54: 1–14.

Mylona, S., 1996: Sulphur dioxide emissions in Europe 1880–1991, *Tellus*, 48B: 662–689.

Mylona, S., 1997: Corrigendum to Sulphur dioxide emissions in Europe 1880–1991, *Tellus*, 49B: 447–448.

Nabuurs, G. J., et al., 2003: Temporal evolution of the European forest sector carbon sink from 1950 to 1999, *Global Change Biology*, 9: 152–160.

Nakicenovic, N. and R. Swart (eds.), 2000: *Emissions Scenarios: A Special Report of Working Group III of the Intergovernmental Panel on Climate Change*. Cambridge University Press, Cambridge, UK and New York, NY.

Nakicenovic, N. and K. Riahi (eds.), 2007: Integrated assessment of uncertainties in greenhouse gas emissions and their mitigation, *Technological Forecasting and Social Change*, 74(2007): 873–886.

NRC (National Research Council), 2002: Defining the mandate of proteomics in the post-genomics era: Workshop report. *Molecular and Cellular Proteomics*, 1(10): 763–780.

NRC (National Research Council), 2000: *Clean Coastal Waters: Understanding and Reducing the Effects of Nutrient Pollution*. National Academy Press, Washington, DC.

NEA (National Ecosystem Assessment), 2011: *The UK National Ecosystem Assessment: Synthesis of the Key Findings*. UNEP-WCMC, Cambridge, UK.

Nicholls, R., and R Tol, 2006: Impacts and responses to sea-level rise: a global analysis of the SRES scenarios over the twenty-first century. *Philosophical Transactions of the Royal Society A*, 364: 1073–1095.

Nicholls, R. J., 2004: Coastal flooding and wetland loss in the 21st century: changes under the SRES climate and socio-economic scenarios. *Global Environmental Change-Human and Policy Dimensions*, 14(1): 69–86.

Nilsson, C., C. A. Reidy, M. Dynesius, and C. Revenga, 2005: Fragmentation and flow regulation of the world's large river systems. *Science*, 308(5720): 405–408.

Ohara, T., H. Akimoto, J. Kurokawa, N. Horii, K. Yamaji, X. Yan, and T. Hayasaka, 2007: An Asian emission inventory of anthropogenic emission sources for the period 1980–2020. *Atmospheric Chemistry and Physics*, 7: 4419–4444.

Okello, B. D., T. G. O'Connor, and T. P. Young, 2001: Growth, biomass estimates, and charcoal production of Acacia drepanolobium in Laikipia, Kenya. *Forest Ecology and Management*, 142: 143–153.

Olesen, J. E., T. R. Carter, C. H. Díaz-Ambrona, S. Fronzek, T. Heidmann, T. Hickler, T. Holt, M. I. Mínguez, P. Morales, J. Palutikof, M. Quemada, M. Ruiz- Ramos, G. Rubæk, F. Sau, B. Smith, and M. Sykes, 2007: Uncertainties in projected impacts of climate change on European agriculture and terrestrial ecosystems based on scenarios from regional climate models. *Climatic Change*, 81: S123-S143.

Olivier, J., J. Van Aardenne, F. Dentener, L. Ganzeveld, and J. Peters, 2005: *Recent trends in global greenhouse gas emissions: regional trends and spatial distribution of key sources, in: Non-CO2 Greenhouse Gases (NCGG-4)*. A. van Amstel (ed.), Millpress, Rotterdam, the Netherlands. pp.325–330.

Pacnya, E., and J. Pacnya, 2002: Global emission of mercury from anthropogenic sources in 1995. *Water, Air & Soil Pollution*, 137(1–4): 149–165.

Pacnya, E., J. Pacyna, F. Steenhuisen, and S. Wilson, 2006: Global anthropogenic mercury emission inventory for 2000. *Atmospheric Environment*, 40(22): 4048–4063.

Peng, S., J. Huang, J. E. Sheehy, R. E. Laza, R. M. Visperas, X. Zhong, G. S. Centeno, G. S. Khush, and K. G. Cassman, 2004: Rice yields decline with higher night temperature from global warming. *Proceedings of the National Academy of Science of the United States of America*, 101: 9971–9975.

Penner, J. E., M. J. Prather, I. S. A. Isaksen, J. S. Fuglestvedt, Z. Klimont, and D. S. Stevenson, 2010: Short-lived uncertainty? *Nature Geoscience*, 3(2010): 587–588.

Perovich, D., T. Grenfell, and P. Hobbs, 2002: Seasonal evolution of the albedo of multi-year Arctic sea-ice. *Journal of Geophysical Research-Oceans,* 107: SHE20–21.

Phoenix, G. K., W. K. Hicks, S. Cinderby, J. C. I. Kuylenstierna, W. D. Stock, F. J. Dentener, K. E. Giller, A. T. Austin, R. D. B. Lefroy, B. S. Gimeno, M. R. Ashmore, and P. Ineson, 2006: Atmospheric nitrogen deposition in world biodiversity hotspots: the need for a greater global perspective in assessing N deposition impacts. *Global Change Biology,* 12: 470–476.

Pirrone, N., and R. Mason, 2008: *Mercury Fate and Transport in the Global Atmosphere, Measurements, Models and Policy Implications, Interim Report of the UNEP Global Mercury Partnership,* Mercury Air Transport and Fate Research Partnership Area.

Pleijel, H. et al., 1999: Grain protein accumulation in relation to grain yield of spring wheat (Triticum aestivum L.) grown in open-top chambers with different concentrations of ozone, carbon dioxide and water availability. *Agriculture, Ecosystems & Environment,* 72(3): 265–270.

Postel, S. L., 1998: Water for food production: will there be enough in 2025? *BioScience,* 48: 629–638.

Prather, M., R. Derwent, D. Ehhalt, P. Fraser, E. Sanheuza, and X. Zhou, 1995: *Other trace gases and atmospheric chemistry. In: Climate change 1994: Radiative forcing and climate change and an evaluation of the IPCC IS92 Emission Scenarios.* J. T. Houghton, et al. (eds.) Cambridge University Press, Cambridge, UK and New York, NY, pp.73–126.

Pu, J. C., T. D.Yao, N. L.Wang, Z. Su, and Y. P. Shen, 2004: Fluctuations of the glaciers on the Qinghai-Tibetan Plateau during the past century. *Journal of Glaciology and Geocryology,* 26: 517–522.

Quinn, P. K., Shaw, G., Andrews, E., Dutton, E. G., Ruoho-Airola, T., Gong, L., 2007: Arctic Haze: Current trends and knowledge gaps. *Tellus,* 59B: 99–114.

Raatz, K. A., 1984: Observations of Arctic haze during the Ptarmigan weather reconnaissance flights 1948–1961. *Tellus,* 36B: 126–136.

Rabalais, N. N., 2002: Nitrogen in aquatic ecosystems. *Ambio,* 31(2): 102–112.

Raes, F., and J. H. Seinfeld, 2009: New Directions: Climate change and air pollution abatement: a bumpy road. *Atmospheric Environment,* 43: 5132–5133.

Rahn, K. A., R. D. Borys, and G. E. Shaw, 1981: Asian desert dust over Alaska: anatomy of an Arctic haze Episode, Desert Dust. In: *Origin, Characteristic and Effect on Man,* T. Pewe (ed.), Special paper No. 186, The Geological Society of America, Boulder, Colorado, pp.37–70.

Ramanathan, V. and G. Carmichael, 2008: Global and regional climate changes due to black carbon. *Nature Geoscience,* 1: 221–227.

Ramanathan, V. and Y. Feng, 2008: On avoiding dangerous anthropogenic interference with the climate system: Formidable challenges ahead. *Proceedings of the National Academy of Sciences of the United States of America,* 105(38): 14245–14250.

Ramanathan, V. and Y. Xu, 2010: The Copenhagen Accord for limiting global warming: Criteria, constraints and available avenues. *Proceedings of the National Academy of Sciences of the United States of America,* 107(18): 8055–8062.

Ramaswamy, V., et al., 2001: *Radiative forcing of climate change. In: Climate Change 2001: The Scientific Basis. Contribution of Working Group I to the Third Assessment Report of the Intergovernmental Panel on Climate Change.* J. T. Houghton, et al., (eds.), Cambridge University Press, Cambridge, UK and New York, NY, pp.349–416.

Ramírez C. A. and E. Worrell, 2006: Feeding fossil fuels to the soil: An analysis of energy embedded and technological learning in the fertilizer industry. *Resources, Conservation and Recycling,* 46: 75–93.

Reason, C. J. C., and M. Rouault, 2005: Links between the Antarctic Oscillation and winter rainfall over western South Africa. *Geophysical Research Letters,* 32(L07705).

Reay, D., F. Dentener, P. Smith, J. Grace, and R. Feely, R., 2008: Global nitrogen deposition and carbon sinks. *Nature Geoscience,* 1: 430–437.

Republic of Kenya, 2002: *First National Communication of Kenya to the Conference of Parties to the United Nations Framework Convention on Climate Change.* Ministry of Environment and Natural Resources, Nairobi.

République de Côte d'Ivoire, 2000: *Communication Initiale de la Côte d'Ivoire.* Ministère de l'Environnement, de l'Eau et de la Forêt, Abidjan.

République de Guinée, 2002: *Communication Initiale de la Guinée à la Convention Cadre des Nations Unies sur les Changements Climatiques.* Ministère des Mines, de la Géologie et de l'Environnement, Conakry.

Riahi, K., A. Grübler, and N. Nakicenovic, 2007: Scenarios of long-term socioeconomic and environmental development under climate stabilization. *Technological Forecasting and Social Change,* 74: 887–935.

Richardson, K., W. Steffen, H. J. Schellnhuber, J. Alcamo, T. Barker, D. M. Kammen, R. Leemans, D. Liverman, M. Munasinghe, B. Osman-Elasha, N. Stern, and O. Wæve, 2009: *Synthesis Report: Climate Change- Global Risks, Challenges & Decisions.* International Alliance of Research Universities. www.climatecongress.ku.dk (accessed 14 April 2011).

Richter, B., S. Postel, C. Revenga, T. Scudder, B. Lehner, A. Churchill, and M. Chow, 2010: Lost in development's shadow: The downstream human consequences of dams. *Water Alternatives,* 3(2): 14–42.

Rockström, J., W. Steffen, K. Noone, A. Persson, F. S. Chapin, E. F. Lambin, T. M. Lenton, M. Scheffer, C. Folke, H. Schellnhuber, B. Nykvist, C. A. De Wit, T. Hughes, S. Van der Leeuw, H, Rodhe, S. Sörlin, P. K. Snyder, R. Costanza, U. Svedin, M. Falkenmark, L. Kalberg, R. W. Corell, V. J. Fabry, J. Hansen, B. Walker, D. Liverman, K. Richardson, P. Crutzen, and J.A. Foley, 2009: A safe operating space for humanity. *Nature,* 461: 472–475.

Rodhe, H., J. Langner, L. Gallardo, and E. Kjellstrom, 1995. Global scale transport of acidifying pollutants. *Water, Air, & Soil Pollution,* 85(1): 37–50.

Rowell, A., and P. F. Moore, 2000: *Global Review of Forest Fires.* WWF/IUCN, Gland, Switzerland.

Royal Society, 2008: *Ground-level ozone in the 21st century: future trends, impacts and policy implications.* Science Policy Report 15/08, The Royal Society, London, UK.

Schaeffer, S. M., J. B. Miller, B. H. Vaughn, J. W. C. White, and D. R. Bowling, 2008: Long-term field performance of a tunable diode laser absorption spectrometer for analysis of carbon isotopes of CO2 in forest air. *Atmospheric Chemistry and Physics,* 8(17): 5263–5277.

Schär, C., P. L. Vidale, D. Lüthi, C. Frei, C. Häberli, M.A. Liniger and C. Appenzeller, 2004: The role of increasing temperature variability in European summer heatwaves. *Nature,* 427: 332–336.

Schellnhuber, H. J., 2008: Global warming: Stop worrying, start panicking? *Proceedings of the National Academy of Sciences of the United States of America,* 105: 14239–14240.

Schellnhuber, H.J., 2009: Tipping elements in the Earth System. *Proceedings of the National Academy of Sciences of the United States of America Special Issue,* 106(49): 20561–20563.

Schneider, S. H. and M. D. Mastrandrea, 2005: Probabalistic assessment of 'dangerous' climate change and emissions pathways. *Proceedings of the National Academy of Sciences of the United States of America,* 102: 15728–15735.

Schopp, W., M. Amann, J. Cofala, C. Heyes, and Z. Klimont, 1999: Integrated assessment of European air pollution emission control strategies. *Environmental Modelling and Software*, 14(1): 1–9.

Schwela, D., G. Haq, C. Huizenga, W.-J. Han, H. Fabian, and M. Ajero, 2006: *Urban Air pollution in Asian Cities: Status, Challenges and Management.* Earthscan, London.

Seidel, D. J., and W. J. Randel, 2006: Variability and trends in the global tropopause estimated from radiosonde data. *Journal of Geophysical Research Atmospheres*, 111(D21101).

Selman, M., and S. Greenhalgh, 2009: *Eutrophication: Sources and drivers of nutrient pollution.* WRI Policy Note. Water Quality: Eutrophication Hypoxia No. 2. World Policy Institute, Washington, DC.

Shaw, G. E., 1995: The Arctic Haze Phenomenon. *American Meteorological Society*, 76: 2403–2413.

Shi, L., L. Xing, G. F. Lu, and J. Zou, 2008: Evaluation of rational sulphur dioxide abatement in China. *International Journal of Environmental Pollution*, 35: 42–57.

Shiklomanov, I. A., and J. C. Rodda, 2003: *World Water Resources at the Beginning of the 21st Century.* United Nations Educational, Scientific and Cultural Organization (UNESCO) and Cambridge University Press, Cambridge, UK and New York, NY.

Shiyatov, S. G., M. M. Terent'ev, and V. V. Fomin, 2005: Spatiotemporal dynamics of forest-tundra communities in the polar Urals. *Russian Journal of Ecology*, 36, 69–75.

Sitch, S., P. M. Cox, W. J. Collins, and C. Huntingford, 2007: Indirect radiative forcing of climate change through ozone effects on the land-carbon sink. *Nature*, 448: 791–794.

Skarby, L., et al., 1998: Impacts of ozone on forests: A European perspective. *New Phytologist*, 139: 109–122.

Sladen, W. J. L., C. M. Menzie, and W. L. Reichel, 1966: DDT residues in Adélie penguins and a crabeater seal from Antarctica. *Nature*, 210: 670–673.

Sliggers, J. and W. Kakebeeke, 2004: *Clearing the Air: 25 years of the Convention on Long-Range Transboundary Air Pollution.* United Nations ECE/EB.AIR/84.

Smith, J. B. et al., 2001: Impacts, Adaptation, and vulnerability. In *Climate Change 2001.* J. McCarthy, O. Canziana, N. Leary, D. Dokken, K. White (eds.), Cambridge University Press, Cambridge, UK and New York, NY, pp.913–967.

Smith, J. B., H. Schneider, M. Oppenheimer, G. W. Yohe, W. Hare, M. D. Mastrandrea, A. Patwardhan, I. Burton, J. Corfee-Morlot, C. H. D. Magadza, H.-M. Füssel, A. B. Pittock, A. Rahman, A. Suarez, and J.-P. van Ypersele, 2009: Assessing dangerous climate change through an update of the Intergovernmental Panel on Climate change (IPCC) "reasons for concern." *PNAS*, 106: 4133–4137.

Smith, K., 1997: The potential for feedback effects induced by global warming on emissions of nitrous oxide form soils. *Global Change Biology*, 3: 327–338.

Smith, T. M. and R. W. Reynolds, 2005: A global merged land and sea surface temperature reconstruction based on historical observations (1880–1997). *Journal of Climate*, 18: 2021–2036.

Smith, S. J., van Aardenne, J., Klimont, Z., Andres, R., Volke, A., Delgado Arias, S., 2011: Anthropogenic sulfur dioxide emissions: 1850–2005. *Atm. Chem. & Phys.*, 11(3): 1101–1116

Sokolov, A., D. Kicklighter, J. Melillo, B. Felzer, C. Schlosser, and T. Cronin, 2008: Consequences of considering carbon-nitrogen interactions on the feedbacks between climate and the terrestrial carbon cycle. *Journal of Climate*, 21(15): 3776–3796.

Sokolov, A., P. Stone, C. Forest, C. Prinn, M. Sarofim, M. Webster, S. Paltsev, and C. Schlosser, 2009: Probabilistic Forecast for Twenty-First-Century Climate Based on Uncertainties in Emissions (Without Policy) and Climate Parameters. *Journal of Climate*, 22: 5175–5204.

Solomon S., R.R. Garcia, F.S. Rowland, and D. J. Wuebbles, 1986: On the Depletion of Antarctic Ozone. *Nature*, 321(6072): 755–758.

State of Eriteria, 2001: *Eritrea's Initial National Communication under the United Nations Framework Convention on Climate Change (UNFCCC).* Ministry of Land, Water, and Environment. Asmara.

Steffen, W., J. Crutzen, and J. R. McNeill, 2007: The Anthropocene: are humans now overwhelming the great forces of Nature? *Ambio*, 36(8): 614–621.

Stevenson, D. S., F. J. Dentener, M. G. Schultz, K. Ellingsen, T. P. C. van Noije, O. Wild, G. Zeng, M. Amann, C. S. Atherton, N. Bell, D. J. Bergmann, I. Bey, T. Butler, J. Cofala, W. J. Collins, R. G. Derwent, R. M. Doherty, J. Drevet, H. J. Eskes, A. M. Fiore, M. Gauss, D. A. Hauglustaine, L. W. Horowitz, I. S. A. Isaksen, M. C. Krol, J. F. Lamarque, M. G. Lawrence, V. Montanaro, J. F. Muller, G. Pitari, M. J. Prather, J. A. Pyle, S. Rast, J. M. Rodriguez, M. G. Sanderson, N. H. Savage, D. T. Shindell, S. E. Strahan, K. Sudo, and S. Szopa, 2006: Multimodel ensemble simulations of present-day and near-future troposheric ozone. *Journal of geophysical research-atmospheres*, 111(8): D08301.

Stieb et al., 2002: Meta-analysis of time-series studies of air pollution and mortality: Effects of gases and particles and the influence of cause of death, age and season. *Journal of the Air and Waste management association*, 52(4): 470–484.

Stige, L., J. Stave, K. Chan, L. Ciannelli, N. Pettorelli, M. Glantz, H. Herren, and N. Stenseth, 2006: The effect of climate variation on agro-pastoral production in Africa. *PNAS*, 103(9): 3049–3053.

Stigiiani, W. M., and R. W. Shaw, 1990: Energy use and acid deposition: The view from Europe. *Annual Review of Energy*, 15: 201–216.

Stocks, B. J. et al., 1998: Climate change and forest fire potential in Russian and Canadian boreal forests. *Climatic Change*, 38: 1–13.

Stohl, A., 2006, Characteristics of atmospheric transport into the Arctic troposphere. *Journal of Geophysical Research-Atmospheres*, 111: D11306.

Stone, R., 2007: A world without corals? *Science*, 316(5825): 678–681.

Stroeve, J., M. Serreze, S. Drobot, S. Gearheard, M. Holland, J. Maslanik, W. Meier, and T. Scambos, 2008: Arctic sea ice extent plummets in 2007. *EOS Transactions of the American Geophysical Union*, 89: 13–14.

Svensson, R., and M. Wigren-Svensson, 1992: Effects of cooling water discharge on the vegetation in the Forsmark Biotest Basin, Sweden. *Aquatic Botany*, 42(2): 121–141.

Teixeira, T., L. Neves, and F. Araújo, 2009: Effects of a nuclear power plant thermal discharge on habitat complexity and fish community structure in Ilha Grande Bay, Brazil. *Marine Environmental Research*, 68(4): 188–195.

Titus, J. G., 2005: Sea-level rise effect. *Encyclopaedia of Coastal Science*. M. L. Schwartz (ed.), Springer, Dordrecht, pp.838–846.

Tong, D., et al., 2007: The use of air quality forecasts to assess impacts of air pollution on crops: methodology and case study. *Atmospheric Environment*, 41: 8772–8784.

Toth, F., 2003: Climatic policy in light of climate science: The iclips project. *Climatic Change*, 56: 7–36.

Tsunehiko, O., F. Norio, L. Hu, H. Hiroshi, S. Hiroyuki, S. Massahi, and S. Katsunori, 2001: Quality control and its constraints during the preparatory-phase activities of the acid deposition monitoring network in East Asia. *Water, Soil and Air Pollution,* 131: 1613–1618.

Turner, J., S. R. Colwell, G. J. Marshall, T. A. Lachlan-Cope, A. M. Carleton, P. D. Jones, V. Lagun, P. A. Reid, and S. Lagovkina, 2005: Antarctic climate change during the last 50 years. *International Journal of Climatology,* 25: 279–294.

UN, 2008: World Population Prospects, the 2008 revision. United Nations Department of Economic and Social Affairs. esa.un.org/unpd/wpp2008/index.htm (accessed 10 June 2011).

UN, 2011: National Accounts database. United Nations Department of Economic and Social Affairs unstats.un.org/unsd/snaama/introduction.asp (accessed 9 September 2010).

UNDP, 2000: *World Energy Assessment: Energy and the Challenge of Sustainability.* UNDP, New York.

UNECE, 2004: *25 Years of the Convention on Long-Range Transboundary Air Pollution.* United Nations Economic Commission for Europe (UNECE), New York.

UNECE 2007: *Strategies and Policies for Air Pollution Abatement. 2006 Review prepared under The Convention on Long-range Transboundary Air Pollution.* United Nations Economic Commission for Europe (UNECE), Geneva.

UNECE Task Force on Hemispheric Transport of Air Pollution, 2007: Hemispheric Transport of Air Pollution 2007, Air pollution studies No. 16. United Nations Economic Commission for Europe (UNECE), Geneva.

UNEP, 2006: Report on Atmosphere and Air Pollution. Africa Regional Implementation Review for the 14th Session of the Commission on Sustainable Development (CDS-14). United Nations Environmental Programme (UNEP), Nairobi.

UNEP/GRID-Arendal, 2005: Climate change vulnerability in Africa. In UNEP/GRID-Arendal Maps and Graphics Library. maps.grida.no/go/graphic/climate_change_vulnerability_in_africa (accessed 22 July 2011).

UNFCCC, 1992: *United Nations Framework Convention on Climate Change.* FCCC/INFORMAL/84. United Nations (UN), New York and Geneva.

USAID, 2002: USAID response to drought in southern Africa. www.usaid.gov/press/releases/2002/fs020510.html (accessed 12 March 2011).

US EPA, 2006: *Air quality criteria for ozone and related photochemical oxidants (final).* US Environmental Protection Agency (US EPA). Washington, DC.

USEPA 2008. Integrated Science Assessment for Oxides of Nitrogen and Sulfur – Ecological Criteria National (Final Report). National Center for Environmental Assessment, Research Triangle Park, NC. US Environmental Protection Agency (US EPA). EPA/600/R-08/139. cfpub.epa.gov/ncea/cfm/recordisplay.cfm?deid=201485 (accessed 22 November 2010).

USEPA 2010a: AirNow web site. US Environmental Protection Agency (US EPA). www.epa.gov/aircompare/compare-trip.htm (accessed 22 November 2010).

USEPA 2010b: Acid Rain and Related Programs: 2009 Environmental Results. US Environmental Protection Agency (US EPA). Washington, DC. www.epa.gov/airmarkets/progress/ARP09_3.html (accessed 20 November 2010).

USEPA 2011a: National Ambient Air Quality Standards. US Environmental Protection Agency (US EPA). www.epa.gov/air/criteria.html (accessed 20 November 2011).

USEPA 2011b: Ozone National Ambient Air Quality Standards: Scope and Methods Plan for Welfare Risk and Exposure Assessment. Office of Air Quality Planning and Standards. US Environmental Protection Agency (US EPA).

USEPA 2011c: Regulatory Impact Analysis for the Federal Implementation Plans to Reduce Interstate Transport of Fine Particulate Matter and Ozone in 27 States; Correction of SIP Approvals for 22 States. Office of Air and Radiation (US EPA). http://www.epa.gov/crossstaterule/pdfs/FinalRIA.pdf (accessed 20 November 2011).

Ramanathan, V., M. Agrawal, H. Akimoto, M. Auffhammer, H. Autrup, L. Barregard, P. Bonasoni, M. Brauer, B. Brunekreef, G. Carmichael, W.-C. Chang, U. K. Chopra, C. E. Chung, S. Devotta, J. Duffus, L. Emberson, Y. Feng, S. Fuzzi, T. Gordon, A.K. Gosain, S. I. Hasnain, N. Htun, M. Iyngararasan, A. Jayaraman, D. Jiang, Y. Jin, N. Kalra, J. Kim, M. G. Lawrence, S. Mourato, L. Naeher, T. Nakajima, P. Navasumrit, T. Oki, B. Ostro, Trilok S. Panwar, M.R. Rahman, M. V. Ramana, H. Rodhe, M. Ruchirawat, M. Rupakheti, D. Settachan, A. K. Singh, G. St. Helen, P. V. Tan, S.K. Tan, P. H. Viet, J. Vincent, J. Y. Wang, X. Wang, S. Weidemann, D. Yang, S. C. Yoon, J. Zelikoff, Y. H. Zhang, and A. Zhu, 2008: *Atmospheric Brown Clouds: Regional Assessment Report with Focus on Asia.* Published by the United Nations Environment Programme, Nairobi, Kenya.

Valigura, R. A., R. B. Alexander, M. S. Castro, T. P. Meyers, H. W. Paerl, P. E. Stacy, and R. E. Turner, 2001: *Nitrogen Loading in Coastal Water Bodies: An Atmospheric Perspective.* American Geophysical Union, Washington, DC.

van Aardenne, J. A., F. J. Dentener, J. G. J. Olivier, C. G. M. K. Goldewijk, and J. Lelieveld, 2001: A 1 degrees x 1 degrees resolution data set of historical anthropogenic trace gas emissions for the period 1890–1990. *Global Biogeochemical Cycles,* 15 (4): 909–928.

van Aardenne, J.A., G. R. Carmichael, H. Hiram Levy II, D. Streets, and L. Hordijk, 1999. Anthropogenic NOx emissions in Asia in the period 1990–2020. *Atmospheric Environment,* 33: 633–646.

van Dingenen, R., F. J. Dentener, F. Raes, M. C. Krol, L. D. Emberson, and J. Cofala, 2009: The global impact of ozone on agricultural crop yields under current and future air quality legislation. *Atmospheric Environment,* 43: 604–618.

van Tienhoven, A. M., M. Zunckel, L. Emberson, A. Koosailee, and L. Otter, 2006: Preliminary assessment of risk of ozone impacts to maize (Zea mays) in southern Africa. *Environmental Pollution,* 140: 220–230.

van Vuuren, D. P., M. den Elzen, P. Lucas, B. Eickhout, B. Strengers, B. van Ruijven, S. Wonink, and R. van Houdt, 2007: Stabilizing greenhouse gas concentrations at low levels: an assessment of reduction strategies and costs. *Climatic Change,* 81: 119–159.

Vanguelova, E. I., S. Benham, R. Pitman, A. J. Moffat, M. Broadmeadow, T. Nisbet, D. Durrant, N. Barsoum, M. Wilkinson, F. Bochereau, T. Hutchings, S. Broadmeadow, P. Crow, P. Taylor, and T. D. Houston, 2010: Chemical fluxes in time through forest ecosystems in the UK – Soil response to pollution recovery. *Environmental Pollution,* 158(5): 1857–1869.

Velders, G. J. M., S. O. Andersen, J. S. Daniel, D. W. Fahey, and M. McFarland, 2007: in *The Montreal Protocol, Celebrating 20 years of environmental progress,* D. Kaniaru (ed.), Cameron May, London, UK, pp.285–290.

Vestreng, V., G. Myhre, H. Fagerli, S. Reis, and L. Tarrason, 2007: Twenty-five years of continuous sulphur dioxide emission reduction in Europe. *Atmospheric chemistry and physics,* 7(13): 3663–3681.

Vestreng, V., L. Ntziachristos, A. Semb, S. Reis, I.S.A. Isaksen, and L. Tarrason, 2009: Evolution of NOx emissions in Europe with focus on road transport control measures. *Atmospheric Chemistry and Physics,* 9: 1503–1520.

Vingarzan, R., 2004: A review of surface ozone background levels and trends. *Atmospheric Environment,* 38: 3431–3442.

Vitousek, P. M., H. A. Mooney, J. Lubchenco, and J.M. Melillo, 1997: Human domination of Earth's ecosystems. *Science,* 277: 494–499.

Vörösmarty, C., M. Meybeck, B. Fekete, K. Sharma, P. Green, and J. Syvitski, 2003: Anthropogenic sediment retention: major global impact from registered river impoundments. *Global and Planetary Change*, 39(1–2): 169–190.

Wafula, E., M. Kinyanjui, L. Nyabola, and E. Tenambergen, 2000: Effect of improved stoves on prevalence of acute respiration infection and conjunctivitis among children and women in a rural community in Kenya. *East African Medical Journal*, 77(1): 37–41.

Walker, I. J., and J. V. Barrie, 2006: Geomorphology and sea-level rise on one of Canada's most 'sensitive'coasts: northeast Graham Island, British Columbia, Canada. *Journal of Coastal Research SI*, 39: 220–226.

Wallack, J. S., and V. Ramanathan, 2009: The other Climate Changers: Why black carbon and ozone also matter. *Foreign Affairs*, 88(5): 105–113.

Walther, G.-R., S. Beissner and C.A. Burga, 2005: Trends in upward shift of alpine plants. *Journal of Vegetation Science*, 16, 541–548.

Wang X., and D. L. Mauzerall, 2004: Characterizing distributions of surface ozone and its impact on grain production in China, Japan and South Korea: 1990 and 2020. *Atmospheric Environment*, 38: 4383–4402.

Wei Wei., Wang, S., Chatani, S., Klimont, Z., Cofala, J., Hao, J., 2008: Emission and speciation of non-methane volatile organic compounds from anthropogenic sources in China. *Atmospheric Environment* 42(20): 4976–4988

WGBU (German Advisory Council on Global Change), 2009: *Solving the Climate Dilemma: The Budget Approach*, WGBU, Berlin, Germany.

White, A., M. G. R. Cannel, and A. D. Friend, 2000: The high-latitude terrestrial carbon sink: a model analysis. *Global Change Biology*, 6: 227–246.

WHO, 2002: *Addressing the links between indoor air pollution, household energy and human health*. World Health Organization (WHO), Switzerland.

WHO 2006a: *Air quality guidelines: global update 2005, particulate matter, ozone, nitrogen dioxide and sulphur dioxide*. World Health Organization (WHO) Regional Office for Europe: Copenhagen, Denmark.

WHO 2006b: *Health risks of particulate matter from long-range transboundary air pollution. European Centre for Environment and Health, Report E88189*, World Health Organization (WHO), Switzerland.

Wilkinson C. R. (ed.), 2000: *Global Coral Reef Monitoring Network: Status of Coral Reefs of the World in 2000*. Townsville, Queensland: Australian Institute of Marine Science.

Wise, M., A. Calvin, A.Thomson, L. Clarke, B. Bond-Lamberty, R. Sands, S. J. Smith, A. Janetos, and J. Edmonds, 2009: Implications of limiting CO2 concetrations for land use and energy. *Science*, 324: 1183–1186.

World Commisssion on Dams, 2000: *Dams and development: a new framework for decision-making*. Report of the World Commission on Dams, London: Earthscan.

World Resources Institute (WRI), United Nations Development Programme, United Nations Environment Programme, and World Bank, 2008. **World Resources 2008**: Roots of Resilience – Growing the Wealth of the Poor. Washington, DC.

World Bank, 2007: Cost of pollution in China: Economic estimates of physical damages. The World Bank, State Environmental Protection Administration. P.R. China. www.worldbank.org/eapenvironment (accessed 20 February 2011).

World Bank Group, 2009: *Directions in hydropower*. The World Bank, Washington, DC.

World Water Assessment Programme, 2009: *The United Nations World Water Development Report 3: Water in a Changing World*. United Nations Educational, Scientific and Cultural Organization (UNESCO), Paris, and Earthscan, London.

WWF, 2005: *An overview of glaciers, glacier retreat, and subsequent impacts in Nepal, India and China*. World Wildlife Fund (WWF), Nepal Programme.

Xu-Ri, and I. Prentice, 2008: Terrestrial nitrogen cycle simulation with a dynamic global vegetation model. *Global Change Biology*, 14(8): 1745–1764.

Yi-Li, C., Y. Hsiao-Hui, and L. Hsing-Juh, 2009: Effects of a thermal discharge from a nuclear power plant on phytoplankton and periphyton in subtropical coastal waters. *Journal of Sea Research*, 61(4):197–205.

Zalasiewicz, J., M. Williams, W. Steffen, and P. Crutzen, 2010: The new world of the Anthropocene. *Environmental Science & Technology*, 44: 2228–2231.

Zhang, Q., D. G. Streets, G.R. Carmichael, K. He, H. Huo, H., A. Kannari, Z. Klimont, I. S. Park, S. Reddy, J. S. Fu, D. Chen, L. Duan, Y. Lei, L. T. Wang, and Z. L. Yao, 2009: Asian emissions in 2006 for the NASA INTEX-B mission. *Atmospheric Chemistry and Physics*, 9: 5131–5153.

Zhang, Q., D. Streets, K. He, Y. Wang, A Richter, J. Burrows, I. Uno, C. Jang, D. Chen, Z. Yao, and Y. Lei, 2007: NOx emission trends for China, 1995–2004: The view from the ground and the view from space. *Journal of Geophysical Research*, 112(D22306).

4

Energy and Health

Convening Lead Author (CLA)
Kirk R. Smith (University of California, Berkeley, USA)

Lead Authors (LA)
Kalpana Balakrishnan (Sri Ramachandra University, India)
Colin Butler (Australian National University)
Zoë Chafe (University of California, Berkeley, USA)
Ian Fairlie (Consultant on Radiation in the Environment, UK)
Patrick Kinney (Columbia University, USA)
Tord Kjellstrom (Umea University, Sweden)
Denise L. Mauzerall (Princeton University, USA)
Thomas McKone (Lawrence Berkeley National Laboratory, USA)
Anthony McMichael (Australian National University)
Mycle Schneider (Consultant on Energy and Nuclear Policy, France)
Paul Wilkinson (London School of Hygiene and Tropical Medicine, UK)

Review Editor
Jill Jäger (Sustainable Europe Research Institute, Austria)

Contents

Executive Summary

Despite providing significant benefits for human health, energy systems also negatively affect global health in major ways today, causing directly perhaps as many as five million premature deaths annually and more than 5% of all ill-health (measured as lost healthy life years). Air pollution from incomplete combustion of fossil fuels and biomass fuels is by far the single major reason that energy systems negatively affect global health, although ash, sulfur, mercury, and other contaminants in fossil fuels also play a role. Effects on workers in energy industries are the second biggest health impact globally.

The largest exposures to energy-related air pollution occur in and around households, particularly in developing countries where unprocessed biomass (wood and agricultural wastes) and coal are used for cooking and heating in simple appliances.

This chapter does not focus on differences in impacts among alternative energy systems that have *minor* impacts on global health; rather, the focus is on the most significant impacts of energy systems on health. The important positive impacts of energy systems on health are mostly addressed in Chapter 2.

Given the importance of avoiding climate change, there is secondary focus on the ways that mitigating climate change through changes in energy systems might achieve important health improvements: co-benefits.

Unless major policy interventions are introduced, energy systems are expected to continue contributing significantly to the global burden of disease for years to come.

Household air pollution: GEA estimates for 2005 put the burden of disease caused by household air pollution at about 2.2 million premature deaths annually. These deaths occur mainly among women and young children in developing countries because they receive the highest exposures to household air pollution from cooking and heating with solid fuels. Although the fraction of households relying on solid fuels is slowly declining, the absolute numbers are still rising among the world's poorest populations.

The only way to ameliorate this health risk is through encouraging as many households as possible to use clean-burning gases and liquids made from biomass or petroleum fuels while initiating widespread promotion of new generations of advanced combustion biomass stoves. These stoves reduce biomass emissions to nearly the level of emissions from clean fuels, by using small blowers and other technical innovations.

Outdoor air pollution: Outdoor air pollution from incomplete combustion and other emissions from fuel use is also an important health risk globally. GEA estimates that it was responsible for some 2.7 million premature deaths in 2005. Outdoor air pollution affects not only urban areas, but many regions between cities, due to long-range transport of pollutants. Dominant sources of outdoor air pollution include combustion of fossil fuels in industry and transportation. In addition, poor household fuel combustion is a significant contributor to outdoor pollution in many parts of the world, for example South Asia and China, which means that the goals for clean household energy and general ambient air pollution are linked. GEA scenarios project the magnitude of the improvement that could be achieved with (1) implementation and enforcement of strict emissions controls, (2) universal access to clean cooking fuels, and (3) shifting of energy to non-combustion energy sources and efficiency.

Occupational health impacts: Occupational injuries and diseases, particularly in biomass and coal harvesting and processing (such as coal mining and transport), are the next most important impact on health from energy systems. Only strict adherence to international norms for worker health and safety can blunt this impact. Despite substantial gains over the past 50 years at some energy-related workplaces, health and safety systems do not exhibit best practices in many countries. The advent of novel nano- and other engineered materials for advanced solar and other energy systems potentially pose risks to workers and the public that need to be carefully investigated before widespread deployment to avert health impacts before they occur.

Nuclear power impacts: Unlike biomass and fossil fuels, nuclear power systems are not a significant source of routine public health impacts globally, although they often garner considerable public and policy concern. This said, reprocessing facilities release significantly larger amounts of radioactivity than power reactors and uranium mining leads to considerable environmental effects and health risks to workers. Average radiation doses to workers in nuclear power industries have generally declined over the past two decades, but tend to be concentrated in workers under less regulatory control. For nuclear power facilities, as with large hydroelectric facilities, the major health risks lie mostly with high-consequence but low-probability accidents. These risks are difficult to compare to the impacts that occur day-to-day. (See Chapter 14 for more discussion of nuclear accidents).

Energy efficiency and health: Although energy efficiency is usually found to be the best overall first strategy to improve the sustainability of energy systems, there can also be downsides if done without care. This has been seen, for example, in programs to improve the energy efficiency of buildings without considering the impacts on indoor air quality of reducing ventilation or use of improper materials.

Climate change, energy, and health: Climate change to date, to which energy systems are a significant contributor, is starting to have an important impact on health, perhaps exceeding 200,000 premature deaths annually by 2005, more than 90% among the poorest populations in the world. It is expected, however, that the health burden due to climate change will rise under current projections of greenhouse gases and other climate-altering pollutants (CAP) and background health conditions in vulnerable populations, which are largely in developing countries. Well over 20 million people along the vulnerable coastal regions of Bangladesh, Egypt, and Nigeria are estimated to be at risk of inundation from a one-meter sea level rise, not accounting for inevitable population growth. Other health impacts from climate change include the effects of extreme weather events, heat waves or sustained periods of extreme heat, malnutrition, spreading infectious diseases, and resource-related conflicts.

Protecting the climate system and improving health: There are a number of opportunities to reduce the current burden of disease while also reducing the pressure on global climate. Some relate simply to reducing energy use and its associated health and climate impacts through efficiency improvements and increased use of non-carbon based fuels, particularly renewable energy sources. Others, however, take advantage of the relatively high health and climate risk per unit emission posed by targeting specific short-lived greenhouse pollutants produced by energy systems, in particular black carbon and the precursors to ozone such as methane. Some of the energy-associated and health-damaging pollutants, such as sulfate and organic carbon particles, however, have cooling characteristics that create potential trade-offs between health and climate goals when controlling certain sources.

Overarching concepts: Per unit of useful energy, the health and climate benefits of emission reduction interventions rise with the fraction of incomplete combustion; and, also per unit of useful energy, the health benefits of emission reduction interventions rise as the combustion is closer to the population, increasing the proportion of emissions inhaled by the population.

From a health standpoint, in addition to reducing risk factors such as air pollution and climate change, there is equal importance to reducing vulnerability by improving background health conditions, particularly among the world's poor. Meeting the Millennium Development Goals (MDGs) as soon as possible will be critical, and bringing modern energy services to the world's people living in poverty is a necessary, if not sufficient, condition in doing so.

4.1   Introduction

Humanity requires energy to bring the benefits of health care, adequate food, education, protection against the elements, and the many other activities of society that enhance population health. On the other hand, dirty, dangerous, and environmentally disruptive energy supplies can lead to disease, injury, and premature death that is significant on local and global scales. Health impacts have thus been a major consideration in the promotion of some types of energy supplies and the avoidance of others.

Energy services provide important direct health benefits; for example, refrigeration preserves food and allows storage of vaccines. Energy for lighting enables health clinics to operate after dark. Energy for heating and cooling helps avoid heat stroke, hypothermia, and other health impacts of extreme conditions. It also improves general quality of life. These benefits are described in detail in Chapter 2. The lack of sufficient energy services, therefore, can be considered a health risk. However, whereas the impacts of lack of energy access are discussed elsewhere in the report, this chapter focuses on the health consequences of the energy society *does* use.

Two principal categories of risk lead to human health impacts associated with energy systems: risks resulting from routine operation (e.g., air pollution, occupational accidents) and risks from low-probability, high-consequence events (e.g., accidents at nuclear power plants and dams). Although the former category has the greatest impact on health year in and year out, the risk of major, if infrequent, events creates uncertainty and unease. Even as society seeks to limit such risks, there is sometimes a disconnect between calculated risk and perceived risk, the latter being influenced by the dread of disastrous, though rare, events associated with energy systems.

In addition to the public at large, many workers in the energy sector, especially in low-income countries, also bear health burdens from the world's energy supply systems. Although the total burden (mortality and morbidity) is larger in the public as a whole, the risk per person is usually greatest among workers, raising equity issues.

Within the general public, the health impacts of energy systems are also distributed inequitably. Not surprisingly, people living in poverty bear a disproportionate burden due to, for example, reliance on poor-quality combustion in households and living in the polluted areas of cities. They suffer more both because of their higher exposures and because of their greater vulnerability and co-risk factors, such as malnutrition and poor access to health care. The health effects of climate change also burden the poorest and most vulnerable populations disproportionately.

As with the environmental impacts presented in Chapter 3, the largest health impacts globally from energy systems are associated with the combustion of carbonaceous fuels, whether fossil or biomass. Chapter 3 presented the patterns of pollutant emissions from energy systems across the world and the resultant ambient air pollution levels. These are inputs into an assessment of health impacts but are insufficient by themselves. Firstly, there are substantially different types and magnitudes of health

impacts from different types of pollutants. Secondly, health impacts are not a direct function of emissions or environmental concentrations but of human *exposure* to them. For instance, pollution over the ocean may cause impacts on marine ecosystems and the climate but no appreciable impact on human health because no one is there to breathe it. On the other hand, a relatively small amount of pollution emitted inside households may have major impacts on human health, but a small effect on ambient (outdoor) pollution levels. The exposure implications of different emissions sources has come to be termed the "intake fraction" – that is, the fraction of the material emitted to which the population is actually exposed (Bennett et al., 2002). This can vary by three orders of magnitude among energy systems and thus is crucial when considering health effects. Thus, no one-to-one relationship between the rankings of impacts in Chapter 3 and this chapter exists, even when the same sources and pollutants are considered (see Box 4.1).

In simple combustion situations, such as household stoves and small boilers in developing countries, combustion is far from complete, thus releasing a significant amount of the fuel carbon in products of incomplete combustion (PICs). These include a range of toxic compounds, such as carbon monoxide (CO), polyaromatic hydrocarbons (PAHs), benzene, and formaldehyde. Even when burned under excellent conditions, fossil fuels vary in quality. Some supplies contain significant toxic contaminants, such as sulfur, ash, and mercury, which are released in large amounts unless emissions are controlled after combustion. Finally, even when burned completely without any release of contaminants, carbon dioxide (CO_2) from fossil fuel use is the primary source of greenhouse gas (GHG) emissions, which threaten health through climate change. In general, consideration of climate change aside, health damage of all sorts is greatest with solid fuels, intermediate with liquid fuels, and least with gaseous fuels.

Health impacts from energy supply systems do not stem solely from the final conversion step – the power plant or automobile, for example. A common framework for evaluating the environmental and health implications of energy systems is the fuel cycle, which starts with harvesting of the raw fuel and proceeds through processing, transport, perhaps more than one conversion stage, and through to final waste disposal. In addition, there may be important impacts from the construction as well as the operation of facilities, together called "life-cycle assessment" (LCA) (see Box 4.4). Although the majority of health risk is usually exerted at the final conversion stage, a full understanding and comparison of energy systems requires such LCAs.

There is essentially no human endeavor that does not affect and is not affected by both health and energy in some way. The importance of and knowledge about the connections, however, vary dramatically. Here in the limited space available, we must focus only on the largest and most well understood of these relationships and thus set a fairly conventional system boundary for most of our analysis, i.e., energy supply systems.

As in the World Energy Assessment (Holdren and Smith, 2000) this chapter is organized by spatial scale, starting with what happens inside

Box 4.1 | Intake fraction

Health impacts of airborne and other pollutants depend on the level of exposure to the population of concern, which in turn depends not only on the amount and toxicity of emissions, but also on the proximity of the emissions to the population. Thus, if no one is downwind to breathe it, a pollutant released far from populations may not affect health significantly, while a pollutant released in the direct proximity of people can have a major effect, even if it is released in relatively small amounts. The metric used to compare such situations is called "intake fraction" (iF) which, for airborne pollutants, is simply the amount inhaled by the population divided by the amount released (Bennett et al., 2002). The difference between major categories of pollution can be several orders of magnitude, as illustrated in Figure 4.1.

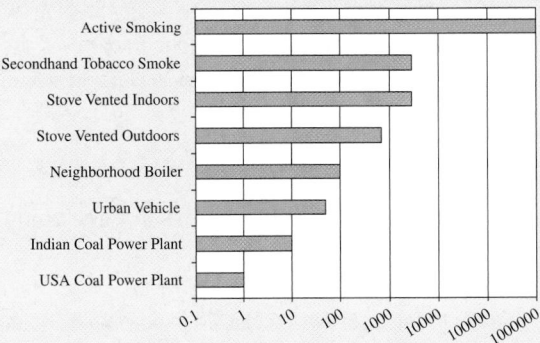

Figure 4.1 | Approximate intake fractions for typical sources of air pollution. Expressed as grams of pollutant inhaled for every tonne emitted. Thus, about one gram per tonne (one in a million emitted) is breathed by someone downwind of a typical power plant in the USA, while about 3000 grams are inhaled per tonne released indoors from a stove. Source: data from Smith et al., 1993.

Note that the iF for active smoking is 1.0 (one million grams inhaled for every million emitted): by definition, all the released pollutant is breathed in. At the other end of the iF spectrum in this figure are remotely sited US power plants, where only one millionth of what is released is breathed by someone. For pollutants released indoors, iFs are a few thousands per million, by comparison, while outdoor neighborhood sources range in between. The iFs vary by local situations, of course; for example, for indoor sources they depend on ventilation (Nazaroff, 2006). In cities, the iF of traffic pollution tends to be larger for bigger cities because the pollution is carried across the breathing zones of more people on its way downwind (Marshall et al., 2005). iF is generally used for primary pollutants, i.e., those actually emitted, and it must be supplemented by other analyses when addressing secondary pollutants, e.g., particles created downwind from sources emitting gases such as sulfur dioxide (SO_2).

The large difference in iFs across pollutant categories goes a long way toward explaining why relatively small amounts of pollution in the form of cigarette smoke or smoke from indoor cookfires can have such significant health impacts compared to the much larger amounts of pollution released from outdoor sources.

households and workplaces, moving to community and regional impacts, and ending with health risks from global changes due to energy use. In addition, we take up some special topics, including the health impacts of the nuclear fuel cycle, emerging technologies, and energy efficiency. Throughout, we start with a summary of what is known about health impacts today and then describe how these might evolve under different pathways of energy supply. As impacts on climate are a central theme throughout GEA, we end with a discussion of the potentials for "co-benefits," i.e., ways in which the evolution of energy systems can be directed to simultaneously achieve climate and health protection.

4.2   Household Energy Systems

4.2.1   Key Messages

- Human exposure to health-damaging air pollution from household combustion of fuels is widespread, affecting about 40% of the world population, and larger fractions in most developing countries.

- Women and children are likely to bear the largest share of the health risk burden from these exposures. According to GEA estimates in Chapter 17, household use of solid fuel for cooking resulted in 2.2 million premature deaths, or 41.6 million DALYs (disability-adjusted life years), in 2005, from exposures around the households.[1]

- Although only gaseous fuels and electricity offer truly clean performance, an emerging set of advanced biomass stove technologies

---

1   The most authoritative source of such estimates is the international Comparative Risk Assessment (CRA) which is part of the Global Burden of Disease Project. An extensively revised edition of the CRA is planned for publication in 2012, which will have updated authoritative estimates of the use of household solid fuels by region and the associated burden of disease as well as exposures and burdens from outdoor air pollution. These estimates are the most useful available for global policy because they are done across a range of risk factors, both environmental and other, in a consistent fashion. Thus they provide reliable information about the relative importance of interventions to improve health across sectors with minimal concern about differential assessment methods. Unfortunately, the updated CRA results were not available before GEA was completed but are scheduled to be available on the World Health Organization website in 2012: www.who.int/healthinfo/global_burden_disease/en/index.html.

provides intermediate performance that is more affordable and nearly as clean. More research and development is needed, however, to assess the field performance and acceptance of these stoves in different parts of the world, as well as the development of quantitative international standards for performance.

- Poverty and education deficits are likely to further exaggerate exposure potentials for vulnerable groups. To help meet the Millennium Development Goals (MDGs), opportunities to include household fuel and consequent air pollution issues in the mainstream public health agenda should be identified.

4.2.2 Background

A great majority of people living in poverty in developing countries have limited or no access to modern energy services, including access to electricity and modern fuels for cooking. As discussed in Chapter 1, GEA estimates that in 2005 about 2.8 billion people, mostly in the least developed and developing countries, relied on solid fuels such as biomass (wood, agricultural residues, and animal dung), charcoal, and coal for cooking and other household energy needs. Solid fuels in these households are often used in inefficient, poorly vented combustion devices, which results in the bulk of the fuel energy being emitted and wasted as toxic products of incomplete combustion. Further, the use of traditional stoves in small and poorly ventilated kitchens – in close proximity to household members on a daily basis – leads to exposures that are significantly detrimental to the health of family members, particularly to women and children, who spend the most time in or near the kitchen. Very young children are especially at risk, as they receive some of the highest exposures during vulnerable periods of growth. The scale of the exposures (spread across many countries), the complexity of the exposure situation (with multiple household-level determinants influencing frequency, duration, and magnitude of exposure), and the limited availability of data on exposures and health outcomes have resulted in a somewhat belated recognition of this risk factor as a major contributor to the disease burden at the global, regional, and national scales. Health impacts surrounding this risk factor are thus often neglected in global energy discussions.

With persuasive evidence for health impacts from the Comparative Risk Assessment (CRA) exercises conducted by the World Health Organization (WHO) (WHO, 2002), new, revised estimates of the burden of disease are being prepared, and an increasing body of scientific literature indicates substantial co-benefits for health and climate change from household energy interventions. This section provides an overview of patterns of household fuel use across world regions; concentrations and exposures experienced within household micro-environments; linkages between exposures and select health outcomes; and selected intervention options that are available and/or are being evaluated for broader application within countries (Household fuel systems of developed countries are not elaborated in this section but are discussed in Section 4.9 in the context of energy efficiency).

4.2.3 Patterns of Household Fuel Use in Developing Countries

Hundreds of demographic surveys conducted in developing countries, especially over the last decade, have collected information on household fuel use. According to a recent United Nations Development Programme review (UNDP, 2009), some 129 countries have data available on fuel use, including access to modern fuels and fuels used for cooking (see Table 4.1 and Figure 4.2). India and China together account for nearly half the global population that uses solid fuels for cooking (27% and 25%, respectively); sub-Saharan Africa also accounts for a significant share. Wood is the predominant type of solid fuel used, although in many surveys the distinction between wood, woodchips, and agricultural residues is not clearly made. Among developing countries generally, nearly 40% rely on modern fuels; however, in the poorest, least developed countries, gas use is uncommon. Use of other fuels is concentrated in certain countries, e.g., charcoal in sub-Saharan Africa, coal in China, dung in India, kerosene in Djibouti, and electricity in South Africa.

Accurate information on fuel use at the national and regional scales is a critical input for calculating the attributable burden of disease. Traditionally, the risk estimates from epidemiological studies have

Table 4.1 | Number of people relying on solid and modern fuels for cooking for all developing countries, least-developed countries, and sub-Saharan Africa.

	Number of people relying on solid fuels (millions)			Number of people with access to modern fuels (millions)
	Traditional biomass	Coal	Total	
Developing Countries	2564	436	2999	2294
LDCs	703	12	715	74
Sub-Saharan Africa	615	6	621	132

Notes: Based on UNDP's classification of developing countries, and the UN's classification of least developed countries (LDCs). There are 50 LDCs and 45 sub-Saharan African countries, with 31 countries belonging to both categories. Traditional biomass includes wood, charcoal, and dung. Wood includes wood, wood chips, straw, and crop residues. Modern fuels refer to electricity, liquid fuels, and gaseous fuels such as LPG, natural gas, and kerosene.

Source: data estimated by UNDP, 2009 for 2007.

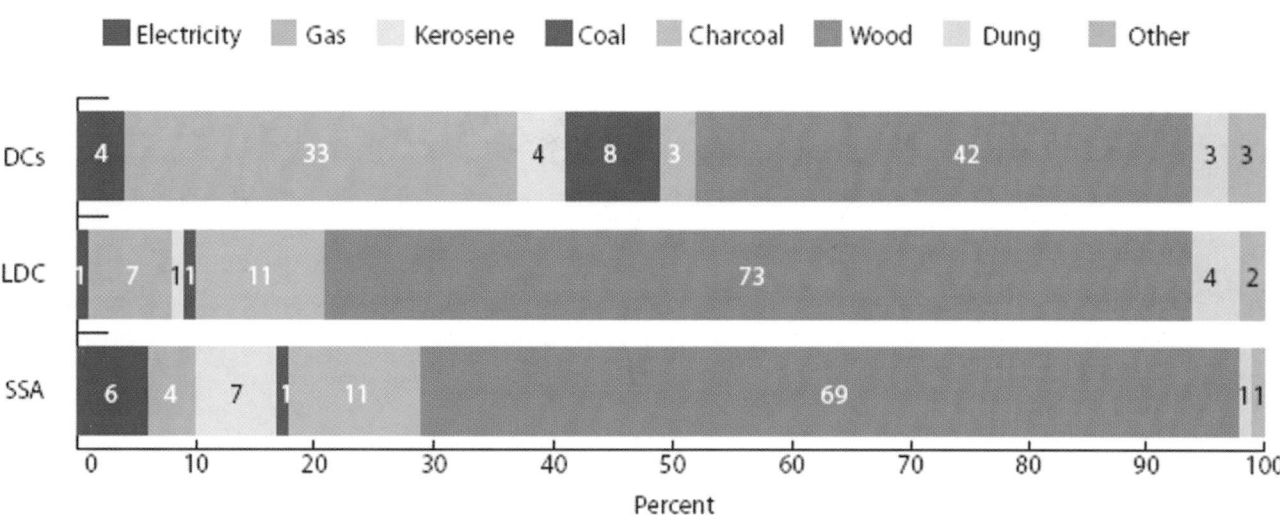

Figure 4.2 | Share of population relying on different types of cooking fuels in all developing countries, least-developed countries, and sub-Saharan Africa. Source: data estimated by UNDP, 2009 for 2007.

Notes: Based on UNDP's classification of developing countries (DCs), and the UN's classification of least developed countries (LDCs). There are 50 LDCs and 45 sub-Saharan countries (SSA), of which 31 countries belong to both categories. Gas includes natural gas, LPG, biogas, and ethanol. Kerosene includes kerosene and paraffin. Coal includes coal dust and lignite. Wood includes wood, wood chips, straw, and crop residues. "Other" includes missing data, "no cooking in the house," and other fuels.

been based on whether solid fuels are used for cooking or heating (WHO, 2002).[2]

Socio-economic factors significantly influence household fuel choice. In most countries with per capita incomes under US$1000, household fuel demands account for more than half of the total primary energy demand for cooking and heating; in contrast, such demands account for less than 2% of total primary energy use in industrialized countries (UNDP, 2000). As per capita incomes increase, households switch to cleaner, more efficient energy systems for their household energy needs, i.e., they move up the "energy ladder" (Hosier and Dowd, 1987; UNDP, 2000). With technological progress, the income levels at which people make the transition to cleaner modern fuels has fallen. However, availability of cleaner fuels at the national level does not guarantee availability of supply in rural areas (Masera et al., 2000), due to issues related to transport, reliability of supply, and socio-cultural preferences. Moreover, in every poor country the income disparities between rural and urban people are large, with most people in rural areas pursuing subsistence livelihoods. Household fuel generation, distribution, and consumption are thus closely related to the overall status of energy, environment, and development in the respective countries.

Several household factors directly influence patterns of human exposure to cooking fuel smoke, which occurs both inside and around households using poor combustion. Fuel type, kitchen location, use and maintenance of stoves, household layout and ventilation, time-activity profiles of individual household members, and behavioral practices (such as where children are located when cooking is being done) have been shown to influence pollution levels and individual exposures. Countries with low gross domestic product (GDP) also typically experience greater gender inequities in terms of income, education, access to health care, social position, and socio-cultural preferences, all of which could potentially influence exposures for vulnerable groups, such as women and children.

Geographic variables can also significantly affect pollution intensity and duration. Extreme temperature differentials between seasons, rainfall, altitude, and even meteorological factors (such as wind speed, wind direction, and relative humidity) could determine whether fuels are used for both cooking and heating, as well as whether they are affecting aerosol dispersion and/or deposition. Patterns of vegetation (e.g., tropical rain forests vs. scrub) could contribute to household decisions on seeking alternative energy sources. Easy availability of wood or other biomass at little or no cost is likely to encourage continued use, especially among people living in poverty.

Although the available literature does not allow a detailed attribution of exposures to each of these variables, these can be expected to make varying contributions and should be considered while creating local or regional profiles of the exposure situation. A schematic showing the potential determinants is shown in Figure 4.3.

2   Only recently has it also been possible to estimate the exposure-response relationships for a few important diseases at exposure levels typical for populations using solid fuels for cooking. In addition, estimates of household exposures across major populations have been done. These will be available in the new CRA. See Footnote 1. The estimates in GEA, however, are based on the traditional approach: yes-no epidemiological relationships in which the average difference in disease between people living in polluted and less polluted households has been determined based on cookfuel type.

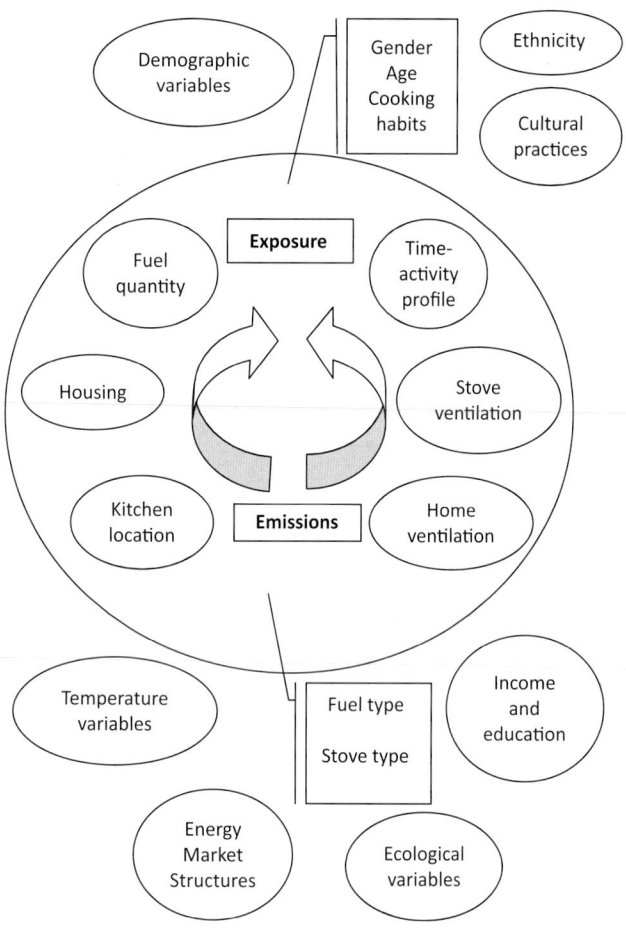

Figure 4.3 | Macro and micro-environmental determinants of exposure to solid fuel smoke.

In a recently concluded assessment by the International Agency for Research on Cancer, emissions from household combustion of coal have been classified as being "carcinogenic to humans" (Group 1 carcinogen, based on sufficient human and experimental evidence), while indoor emissions from household combustion of biomass fuel (mainly wood) has been classified as "probably carcinogenic to humans" (Group 2A carcinogen, on the basis of limited evidence of carcinogenicity of biomass combustion emissions, mainly from wood, in humans and experimental animals; sufficient evidence of carcinogenicity of wood-smoke extracts in experimental animals; and strong evidence of mutagenicity) (IARC, 2010).

4.2.5 Exposures to Health-Damaging Pollutants from Combustion of Household Fuels

Well over 100 studies over the last two decades have assessed levels of indoor air pollutants in households using solid fuels. The methods employed range from the collection of questionnaire-based information to quantitative measurements of household-level exposures. A global database documenting results from these measurements across some 70 studies in the developing countries of Asia, Latin America, Africa, and China (if reported in English) is available from WHO (2005). Although a great majority of studies have performed single-pollutant measurements on a cross-sectional sample of households, select studies have examined temporal, spatial, or multi-pollutant patterns, in addition to day-to-day or seasonal variability in concentrations and exposures (Saksena et al., 1992; Ezzati et al., 2000; Balakrishnan et al., 2002; He et al., 2005). A few have also developed models to examine the differential contributions of household determinants and validate the use of simpler household indicators (that are relatively easy to collect) as a proxy for actual household exposures (Balakrishnan et al., 2004; Bruce et al., 2004).

Data from more than 70 studies are available in the global indoor air pollution databases (Saksena et al., 2003), with some 30 new studies added in a recent update. Nearly one-third of the studies are from South Asia; Latin America, Africa, East Asia, and Eastern Mediterranean are also well represented. On a country-by-country basis, India reported the most studies (27), followed by China (13), Guatemala (10), and Mexico (10). Other countries generally reported fewer than five studies. While most studies reported results from rural areas, 34 studies reported on urban households, indicating a substantial use of solid fuels in urban settings of developing countries. These measurements cover approximately 13,000 households worldwide.

Wood was the most commonly reported solid fuel, with many studies reporting wood, wood chips, and agricultural residues in aggregate or as "mixed fuels." Similarly, gaseous and liquid fuels were often reported as "mixed clean fuels" rather than individually. Prior to 2000, most studies reported measurements from households using traditional solid fuel stoves, while more recently nearly 30 studies have reported results from households using self-styled "improved (solid fuel) stoves."

4.2.4 Emissions from Combustion of Household Fuels

A majority of households in developing countries using solid fuels for cooking burn them in poorly functioning mud or metal stoves or use open pits, usually without a chimney or other arrangement to vent the smoke from the area. Under ideal conditions, complete combustion of carbon would produce only CO_2 and water. However, virtually all traditional ways of burning household biomass fuel emit substantial quantities of PICs, since conditions for the efficient combustion of these fuels are difficult to achieve in typical household stoves. PICs include: small respirable particles, gases such as carbon monoxide and nitrogen, phenols, quinones/semi-quinones, chlorinated acids such as methylene chloride, and dioxins. Combustion of coal may release, in addition to the above pollutants, sulfur oxides, heavy metal contaminants, arsenic, and fluorine. On average, a typical solid fuel stove converts 6–20% of the fuel into toxic substances. At least 28 pollutants present in smoke from solid fuel use have been shown to be toxic in animal studies; some 14 carcinogenic compounds and four cancer-promoting agents have been identified (Table 4.2).

Table 4.2 | Toxic pollutants from incomplete biomass combustion.

Pollutant	Known Toxicological Characteristics
Particulates (PM $_{10}$, PM $_{2.5}$)	Bronchial irritation, inflammation increased reactivity, reduced mucociliary clearance, reduced macrophage response
Carbon monoxide	Reduced oxygen delivery to tissues due to formation of carboxyhemoglobin
Nitrogen dioxide (relatively small amounts from low temperature combustion)	Bronchial reactivity, increased susceptibility to bacterial and viral lung infections
Sulfur dioxide (relatively small amount from most biomass)	Bronchial reactivity (other toxic end points common to particulate fractions)
Organic air pollutants	
Formaldehyde / 1,3 butadiene / Benzene / Acetaldehyde / Phenols / Pyrene, Benzopyrene / Benzo(a)pyrene / Dibenzopyrenes / Dibenzocarbazoles / Cresols	Carcinogenicity / Co-carcinogenicity / Mucus coagulation, cilia toxicity / Increased allergic sensitization / Increased airway reactivity

Source: Naeher et al., 2007.

Table 4.3 | Distribution of particulate matter concentrations in households using solid fuels in traditional stoves.

Type of fuel	Number of households	Mean (µg/m³)	Median (µg/m³)	Standard Deviation
Wood	3269	2097	858	3893
Dung	159	7756	2619	11200
Charcoal	231	3926	1190	8396
Coal	134	820	533	579
Mixed solid fuel	3506	591	575	456
Fuel unspecified	341	537	298	566
Kerosene	283	270	100	422
LPG/ Methane /Electricity	5821	316	139	371
Mixed Liquid/ Gaseous fuel	76	376	398	149

Notes: Data include all size fractions and sampling durations.

Source: author's calculations based on data from 108 studies in the revised WHO Global Indoor Air Pollution Database (WHO, 2005).

Particulate matter of various sizes was the most commonly measured pollutant, followed by carbon monoxide. Other air toxics were seldom reported. While earlier studies measured total suspended particulate matter (TSP) over short sampling times (e.g., during cooking), many recent studies report results of PM$_{10}$ or PM$_{2.5}$ over 24-hour periods. Table 4.3 shows the distribution of measured values across studies that used alternative fuels.[3]

Median values in households using liquid or gaseous fuel are observed to be 5- to 10-fold lower than those reported for households using solid fuel. When compared with more carefully done studies, however, the levels in households using gaseous fuel is often 10- to 20-fold lower. Peak values in dung- and coal-using households have occasionally been observed to be 100-fold higher than levels in corresponding gas-using households.

Households using solid fuels experience indoor concentrations many times higher than WHO Air Quality Guidelines (AQG) for PM$_{10}$ and PM$_{2.5}$ (WHO, 2006). Even gas-using households in rural areas barely meet the strictest AQG values, presumably due to a high background

3   It should be noted that PM$_{10}$ and PM$_{2.5}$ are generic particle classes defined simply by the upper limit of the particle diameters contained within that class (i.e., 10 or 2.5 micrometers). These metrics encompass particles of any chemical composition whose aerodynamic diameters fall below the respective size, potentially including sulfates, nitrates, metals, elemental and organic carbons.

Table 4.4 | Mortality and morbidity attributable to indoor air pollution from solid fuel use, by global region, 2004.

	Attributable deaths per year		Attributable DALYs per year	
	Number ('000)	Per 1 million population	Number (in millions)	Per 1 million population
Developing countries	1,994	378	40.5	7,878
LDCs	577	771	18.4	24,606
Sub-Saharan Africa	551	781	18	25,590
South Asia	662	423	14.2	9,075
Arab States	35	114	1.1	3,489
East Asia and Pacific	665	341	6.5	3,308
Latin America and Caribbean	29	54	0.7	1,334
World	1,961	305	41	6,374

Notes: Numbers and rates of deaths and disability-adjusted life years (DALYs) for all causes – i.e., child pneumonia, adult COPD, and adult lung cancer.

Source: UNDP, 2009.

of outdoor air pollution resulting from the venting and dispersion of solid fuel smoke from neighboring households.[4]

Collectively, the evidence from these studies shows that rural women, children, and men in households using solid fuels experience extremely high levels of exposure to particulate matter and toxic gases. Often these exposures are an order of magnitude or even higher than levels generally considered safe. Further, some emissions from coal combustion (e.g., arsenic and fluorine) have additional, non-inhalational exposure routes (such as deposition on food and contamination of drinking water sources), compounding health effects (He et al., 2005).

4.2.6   Health Effects Associated with Household Fuel Combustion

The amount of disease resulting from the use of solid fuels has been estimated (Smith et al., 2004) as part of WHO's CRA (WHO, 2002; Ezzati et al., 2004). The burden of disease is calculated by combining information on the increased risk of a disease resulting from exposure with information on how widespread the exposure is in the population (in this case, the percentage of people using solid fuels). This allows for the calculation of the "population attributable fraction" (PAF), which is the fraction of the disease seen in a given population that can be attributed to a particular exposure – in this case, solid fuel use. Applying this fraction to the total burden of disease (e.g., child pneumonia expressed as deaths or DALYs), gives the total number of deaths and DALYs that result from use of solid fuels.

For computing disease burdens associated with solid fuel use, estimates of relative risk obtained from all published epidemiological studies have been combined together in a process called meta-analysis into a single best estimate of the risk. In the 2004 CRA, three diseases were felt to have

sufficient evidence bases to be included in the final estimate of the health burden of solid fuel use: pneumonia in children aged under five years, chronic obstructive pulmonary disease (COPD), and lung cancer (only for use of coal). The 2004 WHO-led CRA exercise estimated that annually about 1.6 million deaths and over 38.5 million lost healthy life years (also measured in DALYs) were attributable from household solid fuel use in 2000 (Table 4.4) (Smith et al., 2004).[5] Cooking with solid fuels is thus responsible for a significant proportion, about 3%, of the global burden of disease. In many poor countries, such as India, solid fuel use was found to be the third most important risk factor for ill health, exceeded only by malnutrition and unsafe water and sanitation (Smith and Ezzati, 2005).

The regional or national burden of disease attributable to solid fuel use varies considerably and is based not only on differences in fuel use but also on underlying disease rates. For example, among developing countries, attributable deaths from COPD are greater in India and China as compared to least developed countries (LDCs) and sub-Saharan Africa, while attributable deaths from child pneumonia are highest in the latter because of different patterns of background disease rates. It is to be emphasized that child pneumonia accounts for some 75% of DALYs attributable to solid fuel use in developing countries overall (compared with less than half of all premature deaths) and an even greater share of DALYs in the LDCs and sub-Saharan Africa. In India, for example, pneumonia accounts for more than half of all DALYs attributable to solid fuel use, but only about a third of premature deaths. Because the DALY measure captures the many years of life lost due to deaths from child pneumonia, it shows an even stronger impact of solid fuel use on health in the poorest countries, where pneumonia is a major cause of death in children under five years of age. As expected, the poorest are at the greatest risk of exposures and disease burdens wherever they are found.

At the time of the preparation of the CRA (~2002), the evidence for attributable disease burden was sufficient only for the three health outcomes

---

4   The completed global indoor air pollution measurement database is expected to be available on the WHO website in 2012: www.who.int/indoorair/health_impacts/databases_iap/en/

5   As noted in Footnote 1, a completely revised update will be available on the WHO website in 2012.

mentioned above (i.e., child pneumonia, COPD, and lung cancer). An emerging body of evidence, however, indicates additional burdens from other diseases and conditions, including new studies on cataracts, low birth-weight, and tuberculosis, as well as lung cancer associated with biomass fuel use (Lin et al., 2007; Pope et al., 2010; Hosgood III et al., 2011). Although not yet conclusive, there is also growing evidence of other cancers and burns related to biomass fuel use. Evidence on the original health outcomes is growing as well; see Dherani et al. (2008) for information on pneumonia, Kurmi et al. (2010) on COPD, and the IARC (2010) on lung cancer from coal. In addition, results will soon be available from the first randomized controlled trial for an intervention related to solid fuel use, through the use of chimney stoves in Guatemala (Smith et al., forthcoming). Although there have not been any studies of heart disease in these settings, substantial evidence from other exposures to combustion particles (outdoor air pollution, environmental tobacco smoke, and active smoking) together provide strong, although indirect, evidence of a probable major impact from household fuels as well (Pope et al., 2009; Smith and Peel, 2010).[6]

For the future scenarios used in GEA, a simplified health assessment was conducted using the same relative risks published in the 2004 CRA, as well as new relative risks for ischemic heart disease published in Wilkinson et al. (2009), but with changing exposures as advanced stoves and fuels come into use in the next decades (in line with GEA assumptions). GEA estimated premature mortality from household fuels in 2005 using this approach was 2.2 million, which is somewhat higher than what was found for 2000 in the CRA. The methods used are discussed in detail in Chapter 17.[7]

4.2.7 Intervention Effectiveness

Household energy interventions to date have largely centered on improving fuel efficiency, either by using better fuels and stoves or using improved stoves with the same fuels (Barnes et al., 1994). Considerable evidence is available to indicate that households using gaseous (liquified petroleum gas or biogas) or liquid fuels experience considerably lower pollution levels as compared to homes using solid fuels (Albalak et al., 2001; Balakrishnan et al., 2002; Balakrishnan et al., 2004). A limited number of studies have also shown significant reductions with the use of electricity (Rollin et al., 2004). Improved biomass stoves, which burn more efficiently and vent emissions outside the home, have been an intervention option for more than two decades. Improved stove programs have been implemented in many countries, most notably in India and China. Although designed to conserve fuel, these programs have

had some impacts on reducing household emissions of health-damaging pollutants and CAPs (US EPA, 2000; Edwards et al., 2004).

Somewhat lower indoor concentrations using stove models equipped with chimneys have also been documented in many regions, including China (Sinton et al., 2004), India (Smith et al., 1983; Ramakrishna et al., 1989), Nepal (Reid et al., 1986; Pandey et al., 1990), Latin America (Brauer and Bartlett, 1996; Albalak et al., 2001; Bruce et al., 2004), Mexico (Riojas-Rodríguz et al., 2001), and sub-Saharan Africa (UNDP, 2009). Interventions that reduce exposures either through behavioral interventions or through improved ventilation have also been described (Barnes et al., 2004; UNDP, 2009).

More recently, advanced combustion stoves – using traditional woodfuel but burning much more cleanly – have become available. One promising approach involves so-called "gasifier" stoves that achieve very high combustion efficiency through designs that facilitate two-stage combustion. The most reliable of these use small electric blowers to stabilize the combustion. Where there is no reliable electricity available, inexpensive thermal-electric generators (TEGs) are now being incorporated into stoves to generate the needed power from the stove heat itself.

Although programs promoting the use of improved stoves have not always proved successful, some have achieved remarkable penetration. For example, the Chinese National Improved Stove Programme was able to provide stoves to some 180 million rural households during the 1980s and 1990s (Barnes et al., 1993; Smith et al., 1993; Sinton et al., 2004). Several programs are underway in India that attempt to promote penetration of improved stoves using market-based approaches, in contrast with earlier, government-subsidized efforts. While substantial reductions in emissions have been achieved with many of these improved stove models, the residual levels of pollutants are still high compared to WHO's health-based AQGs. More recent programs increasingly emphasize not only the provision of stoves, but also support for installation and routine maintenance, training and education, and use of market mechanisms to continuously assess user preferences. These innovations are expected to expand coverage and improve performance, leading to further sustained exposure reductions across large populations.

Some evidence suggests that health benefits can accrue even with modest reductions in exposures. In the trial with chimney stoves in Guatemala, a 50% reduction in smoke exposures resulted in an 18% reduction in physician-diagnosed childhood pneumonia (Bruce et al., 2007; Smith et al., forthcoming). To achieve this, however, required nearly a 90% reduction in indoor levels, which is difficult to achieve reliably in practice without advanced combustion stoves. In those households achieving a 90% reduction in exposure, on the other hand, children had only half as much pneumonia, an improvement greater than achieved by available vaccines and nutrition supplements, which are the other major interventions for this major killer of children. To obtain this much reduction, however, will require extremely clean burning stoves used regularly over long periods.

---

6 See Footnote 2.

7 As noted in Footnote 1, readers interested in the burden of disease attributable to household air pollution or outdoor air pollution in 2005 or 2010 are advised to refer to the updated CRA on the WHO website, as it is part of a larger consensus assessment in which fair comparisons were made across risk factors. Please note that the CRA does not attempt to project changes in burden of disease over future years as does GEA in Chapter 17.

Table 4.5 | Health benefits of the Indian Improved Stove Program, 2010–2020 (estimated).

	Deaths from ALRI	Deaths from COPD	Deaths from IHD	Total DALYs for these diseases
Avoided in 2020 (%)	30.2	28.2	5.8	17.4
Annual number in 2020 without stoves (x10⁶)	0.14	1.00	1.77	63.0
Total avoided 2010–20 (x10⁶)	0.24	1.27	0.56	55.5

Note: ALRI = acute lower respiratory infections. COPD = chronic obstructive pulmonary disease. IHD = ischaemic heart disease. DALY = disability-adjusted life year.

Source: Wilkinson et al., 2009.

In a striking set of analyses, simulations to estimate percent use of improved stoves under baseline and assumed enhanced rates of dissemination for the case of India have shown that nearly 12,500 DALYs could be avoided annually per million people. The national burden of disease (DALYs) in 2020 from these three major diseases is estimated to be about a sixth lower than it would have been without the stove program, which is equivalent to the elimination of nearly half the entire cancer burden in India in 2020 (Table 4.5 and Wilkinson et al., 2009). What is not well understood yet, however, is how well such advanced stoves actually perform in field conditions and how well they are accepted by large populations.[8]

4.2.8 Conclusion

The UNDP report of 2009 assesses that while many developing countries have clear targets for providing access to electricity, only a handful (17 for access to modern cooking fuels and 11 for provision of improved stoves) have set targets to reduce household solid fuel use. Evidence of health benefits from incremental improvements could be expected to provide relevant cost-effectiveness information to policy makers, greatly facilitating the acceleration of intervention efforts. To effectively implement stoves that truly lower exposures, however, there is a need to define more quantitatively what is meant by various levels of "improvement," probably separately in terms of expected emissions/ exposure and fuel use per meal. In the past, unfortunately, almost any new stove could be claimed to be "improved" when many actually did little. Fortunately, the WHO has embarked on the process to develop formal health-based guidelines, slated for publication in 2012.[9] Another factor stove programs need to address directly is adoption/usage – it does no good to disseminate a high-performance stove that no one uses, something that has happened in many past programs. New monitoring technologies, however, are making it possible to keep track of actual household usage inexpensively (Ruiz-Mercado et al., 2011). This will allow programs to better understand the stove designs, training, and incentives that enhance usage.

---

8   In 2010, a number of non-governmental organizations; foundations; international, bilateral, and national agencies and companies announced the formation of Global Alliance for Clean Cookstoves. Its goal is to engender the introduction of 100 million clean-burning cookstoves globally by 2020. See http://cleancookstoves.org/the-alliance/.

9   For more information, see www.who.int/topics/air_pollution/en/.

4.3 Occupational Health Effects of Energy Systems

4.3.1 Key Messages

- Energy systems involve a large number of workers, particularly in low- and middle-income countries, and entail significant occupational health problems.

- Solid fuel systems (i.e., biomass and coal) tend to have higher occupational risks, in both absolute and per-worker terms.

- Despite substantial gains over the past 50 years at some energy-related workplaces, health and safety systems do not exhibit best practices in many countries. Health risks could be reduced with the adoption of modern occupational health management and prevention programs, in accordance with international conventions and guidance from the International Labour Organization (ILO).

- Renewable energy systems, such as hydropower, wind power, and photovoltaic power, likely involve lower occupational health risks. These risks occur mainly during construction work.

- Occupational health hazards for each type of energy source need to be assessed at the local level, followed by the creation of a plan for effective health protection. Such assessments should also include analysis of occupational health co-benefits of changing to more renewable energy sources.

4.3.2 Background

Energy production, processing, transport, waste disposal, and end use involves millions of workers around the world. In addition, the construction work required to build the energy supply units requires large numbers of additional workers. The health hazards they are exposed to during work activities are important concerns for the energy industry. Many of these jobs are hazardous, with particular risks of injuries, dust diseases, poisoning, noise/induced hearing loss, heat stroke, and radiation effects.

Energy-related work for rural and urban people living in poverty primarily involves the collection of biomass fuel for daily household cooking and heating needs. Their work involves injury risks, heavy load bearing and heat exposure in bright sun during the hot season, the latter being exacerbated by global climate change. In some areas, risks of physical violence, snake bite, and leeches accompany biomass collection for women. Biomass fuel collection may also occur in dumps and landfill sites, which increases the risks of injuries and chemical exposures for the collectors. Much of the work needed for household energy supply in the developing countries is carried out as a household task that does not figure in the national statistics as an "occupational" issue.

Coal as an energy source involves coal mining, one of the most hazardous occupations on Earth. Miners are exposed to collapsing mine shafts, fire and explosion risks, toxic gases (carbon monoxide), lung-damaging dusts (coal and silica), and hot work environments, as well as injury and ergonomics hazards. Coal also requires major transport arrangements for coal and waste that create health hazards. Oil and gas workers face injury risks, particularly during drilling and emergency situations and work on offshore platforms, as well as exposure to toxic materials at refineries. The Gulf of Mexico oil platform disaster in 2010 is an example of the extreme hazards involved.

Hydropower electricity production on a large scale requires major dam, tunnel, and building construction, which also entails high injury risks and exposure to noise and heat. Mini- and micro-hydro schemes do not require such major infrastructure.

Wind power, photovoltaic solar power, and solar water heating involve more limited occupational hazards during construction and are associated with risks of manufacturing steel, cement, and other materials, while nuclear power has its own special hazards associated with potential radiation exposure and the disposal of radioactive waste (see Sections 4.7 and 4.8).

This section will analyze the health issues based on the type of energy source and give examples of how the effects have been documented in different countries. The ILO Encyclopedia of Occupational Safety and Health (ILO, 1998) was published more than 10 years ago but is still a very useful source of information on occupational health aspects of energy systems. This section therefore refers to "older" sources from that period, but that does not mean that the health information is out of date. Ideally, there would be data on the occupational health impacts of different energy sources expressed as "burden of disease and injury per megawatt-hour of energy supply," but these are not yet available.

The analysis of occupational health impacts of different energy systems should include a full life-cycle analysis of the extraction, transport, and processing of raw materials; the production of machinery and buildings required; and the transportation and disposal of waste from the process, as well as the actual energy medium (see Box 4.4 (p 205)).

4.3.3   Biomass

Wood, agricultural plant waste, cow dung, etc., are common sources of cooking and heating energy in poor households in developing countries. Wood is still also widely used in developed countries for space heating, in some cases promoted in the interest of reducing CAP emissions (UNDP, 2000). Biomass constitutes approximately 9% of the global primary energy supply. A detailed review of the occupational health and safety issues of forestry can be found in Poschen (1998a). Additional and updated information can be searched, for instance, on the website "Atlantic Network for Research in Forestry Occupational Health, Safety and Ergonomics."[10]

The majority of wood and agricultural waste, etc. is collected by women and children in local fields and forests (Sims, 1994). This is a part of the daily survival activities, which also include water hauling, food processing, and cooking. An analysis of time use for these activities in four developing countries (Reddy et al., 1997) showed that women spent 9–12 hours per week doing this, whereas men spent 5–8 hours. Women's role in firewood collection was most prominent in Nepal (2.4 hours for women and 0.8 hours for men). Updated similar information about sub-Saharan Africa is now available (Blackden and Wodon, 2006). Firewood collection may be combined with harvesting of wood for local use in construction and small-scale cottage industry manufacturing. This is subsistence work, often seasonal, unpaid, and unrecorded in national economic accounts. Globally, about 16 million people were involved in forestry in the 1990s (Poschen, 1998b), more than 14 million of them in developing countries and 12.8 million of them in subsistence forestry (UNDP, 2000).

A number of health hazards are associated with the basic conditions of the forest. The workers have a high risk of suffering from insect and snake bites, stings from poisonous plants, cuts, falls, and drowning. In tropical countries, the heat and humidity create great strain on the body (Wasterlund, 1998; Kjellstrom et al., 2009), whereas in temperate countries the effect of cold climate is a potential hazard. The work is outside, and, in low-latitude countries, high levels of ultraviolet radiation can be another health hazard, as it increases the risk of skin cancer and cataracts (WHO, 1994). All forestry work is hard physical labor, often causing repetitive stress injuries, such as painful backs and joints, as well as fatigue, which in turn increases the risk of injuries from falls, falling trees, or equipment (Poschen, 1998b). Women and children carrying very heavy loads of firewood are a common sight in areas with subsistence forestry (Sims, 1994). Children may also drop out of school due to the need to help with family forestry work. In addition, the living conditions of forestry workers are often poor quality, and workers may be spending long periods in simple huts in the forest with limited protection against the weather and poor sanitary facilities (UNDP, 2000).

Urbanization leads to the development of a commercial market for firewood and larger-scale production of firewood from logs or from smaller waste material left over after the logs have been harvested. Energy forestry then becomes more mechanized and the workers are exposed to additional

10   www.safetynet.mun.ca/forestry/index.htm

hazards associated with commercial forestry (Poschen, 1998b). Motorized hand tools (e.g., the chain saw) become more commonly used, which leads to high injury risk, as well as noise-induced hearing loss and "white finger disease" caused by vibration of the hands. In addition, synthetic fertilizer and pesticides become a part of the production system, with the potential for pesticide poisoning of workers (UNDP, 2000).

As the development of forestry progresses, more and more of the logging becomes mechanized with very large machinery involved, which reduces the direct contact between worker and materials. Workers in highly mechanized forestry have only 15% of the injury risk of highly skilled forestry workers using chainsaws (Poschen, 1998b). However, firewood production remains a hazardous operation because it requires manual handling of the product close to cutting tools (UNDP, 2000).

Another health aspect of wood-based energy is the risk of burning wood that has been treated against insect damage with copper-arsenic compounds or painted with lead paint. Such wood may be more difficult to sell, and may therefore be used to a greater extent by the firewood production workers themselves in stoves and open fires. When burnt, poisonous arsenic and lead compounds, as well as dioxins and furans, are emitted with the smoke and can lead to ill-health when inhaled (UNDP, 2000; Tame et al., 2007).

4.3.4   Coal

Coal is a major energy source, constituting approximately 25% of total primary energy (see Chapter 1). Coal can be produced through surface mining ("open cast") or underground mining. Both operations are inherently dangerous to workers. About 1% of the global workforce is engaged in mining, but they account for 8% of the global fatal occupational accidents (about 15,000 per year) (Jennings, 1998). Since 1900, over 100,000 people have been killed in coal mining accidents in the United States, and in China, underground mining accidents cause 3,800–6,000 deaths annually (Epstein et al., 2011). A detailed review of occupational health and safety issues in coal mining and other mining is available in Armstrong and Menon (1998).

Underground coal miners are exposed to the hazards of excavating and transporting materials underground. This includes injuries from falling rocks and falls into mine shafts, as well as injuries from machinery used in the mine. There are no reliable global data on injuries of this type from developing countries (Jennings, 1998), but in developed countries miners have some of the highest rates of compensation for injuries. The situation is likely to be worse in developing countries. In addition, much of the excavation involves drilling into silica-based rock, which creates high levels of silica dust inside the mine. Occupational lung disease (silicosis) is therefore a common health effect in coal miners (Jennings, 1998). In addition, coal miners have an increased risk of lung cancer (Donoghue, 2004).

Other health hazards specific for underground coal mining include the coal dust, which can cause "coal workers' pneumoconiosis" or anthracosis (black lung disease), often combined with silicosis (Ross and Murray, 2004). Since 1900, coal workers' pneumoconiosis has killed over 200,000 people in the United States; in the 1990s alone, over 10,000 former US miners died from the disease. The prevalence has more than doubled since 1995 (Epstein et al., 2011).

The coal dust is explosive, and explosions in underground coal mines are an ever-present danger for coal miners. Coal is also inherently a material that burns, and fires in coal mines are not uncommon. Once such a fire has started it may be almost impossible to extinguish it. Apart from the danger of burns, the production of smoke and toxic fumes will create great health risks for the miners. Even without fires, the coal material will produce toxic gases, such as carbon monoxide, carbon dioxide, and methane, when it is disturbed (Weeks, 1998). Carbon monoxide is extremely toxic, as it binds to hemoglobin in the blood, blocks oxygen transport, and causes "chemical suffocation" (WHO, 2000). It is a colorless and odorless gas, giving no warning before symptoms such as drowsiness, dizziness, headache, and unconsciousness occur. Carbon dioxide can sometimes displace oxygen in the underground air and can cause suffocation. Another health hazard in mining is exhaust fumes from diesel engines used in machinery or transport vehicles underground. These exhausts contain very fine particles, nitrogen oxides, and carbon monoxide, all of which can create serious health hazards (WHO, 2000).

Work in coal mines is also uncomfortably hot during the warmer seasons of the year (Kampmann and Piekarski, 2005). In both Germany and South Africa, heat impacts on mine workers increase during and just after the summer season. Ongoing climate change is expected to increase the occupational health risks related to coal mining.

Surface coal mining avoids the hazards of working underground but does still involve injury risk from machinery, falls, and falling rocks. In addition, coal mining is very physically intensive work, and heat, humidity, and other weather factors can affect workers' health (Kjellstrom, 2009). The machinery used is noisy and hearing loss is a common effect in miners (Armstrong and Menon, 1998). Another health hazard is the often squalid conditions under which many coal workers in developing countries live, creating particular risk for the diseases of poverty (Jennings, 1998). In addition, many modern coal mines involve mountaintop removal and strip mining, which adds to ecological damage, causes mental stress among nearby communities, leads to ammonia releases, and contaminates water sources with waste emissions (Epstein et al., 2011).

After extraction the coal needs to be processed and transported to the sites where it will be used, including residential areas, power stations, and factories. This creates other types of occupational hazards (Armstrong and Menon, 1998). For instance, coal for residential use is often ground and formed into briquettes. This work involves high levels of coal dust, as well as noise hazards. Loading, transportation, and offloading of large amounts of coal involves ergonomic hazards, noise, and injury hazards (UNDP, 2000).

The use of coal on a large scale in power stations or industry creates yet more hazards. One of the more specific hazards is the conversion of coal to coke in steel production. This process, though not entirely associated with energy supply, distills off a large number of volatile polycyclic aromatic hydrocarbons in coal, the so-called "coal tar pitch volatiles" (Moffit, 1998). Coke oven workers have twice the lung cancer risk of the general population (IARC, 1984). Additional health hazards for workers are created when the large amounts of ash produced in power stations or by industry need to be transported and deposited. The health hazards of power generation workers have been reviewed by Crane (1998).

4.3.5   Oil and Gas

Oil and gas exploration, drilling, extraction, processing, and transportation involve a number of hazards, including heavy physical labor, ergonomic hazards, injury risk, noise, vibration, and chemical exposures (Kraus, 1998). This type of work is often carried out in isolated geographic areas with inclement weather conditions. Long-distance commuting may cause fatigue, stress, and traffic accident risks (UNDP, 2000).

The ergonomic hazards lead to risk of back pain and joint pain. Injuries include burns and those caused by explosions. Skin damage from exposure to the oil itself and from chemicals used in the drilling processes creates a need for well-designed protective clothing (Kraus, 1998). In addition, many oil and gas installations have used asbestos in insulating cladding on pipes and equipment. This creates the hazard of inhaling asbestos dust during the installation and repair of such equipment. This in turn creates a risk of lung cancer, asbestosis and mesothelioma (WHO, 1998).

Much exploration and drilling for oil and gas now occurs offshore. The Gulf of Mexico catastrophe in 2010 highlighted the hazards involved, including underwater diving work, which is inherently dangerous. In addition, the weather-related exposures can be extreme, particularly as the work often requires continuous, around-the-clock operations (Kraus, 1998). Transport to and from the oil rig is another hazard (Gardner, 2003).The work is also stressful due to its shift-work character, characterized by long work shifts and extended periods in cramped, crowded living conditions (Knutsson, 2003).

4.3.6   Hydropower

This type of energy generation has its own set of occupational health hazards (McManus, 1998). Constructing a hydroelectric power station usually means building a large dam, excavating underground water channels, and erecting large structures to house the electricity generator. McManus (1998) lists 28 different hazards involved in the construction and operation of these power stations, including chemical exposures from paints, oils, and

polycholrinated biphenyls (PCBs); asbestos exposure; diesel fumes; welding fumes; work in confined spaces or awkward positions; drowning; electrocution; noise; heat; electromagnetic fields; and vibration (UNDP, 2000).

Much of the construction of new hydropower stations occurs in low- and middle-income countries (see International Energy Agency databases), where occupational health management and prevention may be laxer than in high-income countries. Thus, expanded use of hydropower, as an element of policies to reduce GHG emissions, must entail the application of up-to-date occupational health standards.

4.3.7   Nuclear

Nuclear power generation has its own particular hazards due to the radiation hazards involved in mining, processing, and transporting uranium, as well as the radiation present in the power station itself. These hazards are addressed in Section 4.7 below.

4.3.8   Other Electricity (Wind, Solar, Waste)

The manufacture of equipment used in the wind power and solar power industries involves the hazards typical of manufacturing industries, including injuries, noise, and chemical exposures. In addition, the technologies for solar electricity generation involve new chemical compounds, some based on rare metals, with poorly understood toxic properties. These risks are discussed in Section 4.8 below.

4.3.9   Number of Workers and Quantitative Health Effects Comparisons

Much of the comparative analysis of the human impacts of various energy systems has been carried out as economic analysis (e.g., Pearce, 2001). Energy policy analysis often assesses "externalities" of different types, but occupational health issues are not always included and the "life-cycle" approach mentioned earlier (Sorensen, 2003) is not always applied. A comparison of wind and coal as energy sources in the United States (Jacobson and Masters, 2001) concluded that wind was at a great advantage, with one key rationale being the high occupational mortality in coal mining. Another more comprehensive analysis (Rabl and Dreicer, 2002) highlighted the lower health and economic impacts of renewable energy systems compared to coal.

Comparisons of the occupational health aspects of energy systems require data on the number of workers involved, as well as the level of occupational health risks in each system and the different elements of the lifecycle of that system. The number of people necessary for each type of energy system to meet the energy requirements of a community is difficult to estimate accurately. As mentioned earlier, much of the work done to supply the energy needs within the poorest communities is carried out by family members, particularly women, who are not employed in the formal sense. In addition, much of this work is carried out by small enterprises, and is not always recorded in national employment statistics.

In the ILO Encyclopedia on Occupational Health (Hamilton, 1998) a summary table outlines the occupational health and general public health impacts of selected energy systems used in electricity generation (Table 4.6). Each of these systems has important occupational health effects; it is unfortunate that a comparison with renewable energy sources was not included.

Mining is a particularly dangerous occupation and miners are a large occupational group in the international statistics (e.g., the UN Demographic Yearbooks and the Laborsta database at the International Labor Organization). They represent up to 2% of the economically active

population in certain developing countries (Kjellstrom, 1994), but the breakdown by different types of mining work is not always reported. Coal mining is, however, a common type of mining in certain developing countries.

Table 4.7 highlights the high occupational mortality rates in mining in the official statistics of a number of countries. It is likely that in some developing countries a number of occupational deaths are excluded from these statistics due to the limited scope of reporting and assembling statistical information. The table also shows that workers in the electricity (and other infrastructure occupations) have a higher mortality rate than the country averages.

It should be emphasized that these risk estimates are based on the current situation and the risks in many countries and for several energy systems can be significantly reduced if appropriate occupational health

Table 4.6 | Significant health effects of technologies for generating electricity.

Technology	Occupational Health	Public Health
Biomass	Trauma from accidents during gathering and processing Exposure to hazardous chemical and biological agents	Air pollution health effects Diseases from exposure to pathogens Trauma from house fires
Coal	Black lung disease Trauma from mining accidents Trauma from transport accidents	Air pollution health effects Trauma form transport accidents
Oil	Trauma from drilling accidents Cancer from to refinery organics	Air pollution health effects Trauma from explosions and fires
Oil shale	Brown lung disease Cancer from exposure to retorting emissions Trauma from mining accidents	Cancer from exposure to retorting emissions Air pollution health effects
Natural gas	Trauma from drilling accidents Cancer from exposure to refinery emissions	Air pollution health effects Trauma from explosions and fires
Tar sands	Trauma from mining accidents	Air pollution health effects Trauma from explosions and fires

Source: Hamilton, 1998.

Table 4.7 | Fatal occupational injury rates per 100,000 workers; crude rates for all occupations and rate ratio between mining and all occupations.

Country	Crude Occupational Mortality Rate, all occupations (per 100,000 workers)	Rate Ratio, mining workers vs. all occupations
Argentina	15	3
Nicaragua	10	6
El Salvador	10	4.5
South Korea	12	12
Hong Kong	7.5	11
Zimbabwe	7	3.5
Japan	2	8
Sweden	1.5	6
United Kingdom	0.7	13

Notes: Based on rates per 100,000 insured workers or rates per million work hours and 2000 hours per worker and year. Recent years, 2006 or average of five recent years until 2006, approximate figures, men and women combined. These data most likely underestimate the true rates and should be seen as indicative, of the additional risks in mining. The rates for electricity workers are in a number of countries similar to the rates for mining. Most countries provide incomplete information to ILOs website.

Source: ILO website, Laborsta, 2009.

standards are applied. The differences in mining mortality risks in Table 4.7 between countries such as Sweden or Japan and the other countries, even developed countries, are likely to indicate the impacts of good preventive occupational health programs.

4.3.10   Conclusion

Biomass and fossil fuel systems have a number of occupational health problems and involve very large numbers of workers, particularly in low- and middle-income countries. Major successful efforts to reduce occupational risks in the energy supply chain can be demonstrated during the last 50 years, but in many mines, construction sites, and power stations, the health and safety systems are not commensurate with best practice. The health risks can be further reduced if modern occupational health management and prevention programs are applied in all countries and all energy supply sites. International conventions and guidelines from the ILO can be used as a basis for such applications.

Renewable energy systems such as hydropower, wind power, and photovoltaic solar power most likely involve less occupational health risks, while nuclear power has special hazards with low probability but enormous health impact if a major accident occurs. Further comparative analysis of the occupational health risks in different energy systems would be of great value.

Each energy investment project should include a detailed occupational health and safety impact assessment as well as a strategy and program to implement prevention. This program should identify the initial and ongoing costs for maintenance of the program, and these should be included in future budgeting.

4.4   Community Effects of Energy Systems

4.4.1   Key Messages

- Air pollution emissions from fuel combustion is a major risk factor for morbidity and mortality in communities, particularly for people living in poverty, who often rely on low-quality energy sources and technologies. Within and across communities, vulnerability to pollution-related health risks varies as a function of age and income.

- Community air quality is relatively good and improving in most developed countries, and relatively poor and deteriorating in many developing countries.

- Exposure to outdoor particle air pollution, largely due to fuel combustion, is estimated by GEA to have been responsible for about 2.7 million premature deaths in 2005.[11]

- In developed-world communities, transportation emissions are a particularly important cause of health burdens, due both to the nature of the emitted pollutants and the close proximity of emissions to vulnerable populations. Energy use in buildings also makes a significant contribution.

- In developing-world communities, the situation is more complex, with contributions from vehicles, industry, trash burning, household biomass fuels, agricultural burning, and other sources. Health burdens associated with community energy use are largest in developing-world cities, especially in China and India.

- Experiences from developed countries indicate that outdoor pollution levels can be greatly reduced through the application of control technologies on combustion and shifts to non-combustion sources of energy.

4.4.2   Background

Communities are residential groupings of people ranging in size from small rural villages to large, densely populated megacities. Patterns of energy use and resulting health impacts vary with size, and also with economic development. Within communities, vulnerability to pollution exposures and health impacts varies considerably across the population, with the young, old and those living in poverty often at greatest risk. Community-scale use of energy and associated environmental health consequences will increase in importance as populations continue to concentrate in urban areas, particularly in developing countries.

For the purposes of this section, the wide spectrum of communities measured on scales of density and economic development has been simplified to three representative categories: large developing-world cities, large developed-world cities, and low-density, peri-urban communities in the developed world. This simplification helps to organize the discussions that follow. However, it should be recognized that tremendous heterogeneity exists within these categories. Where particularly relevant, variations within categories are highlighted. (We exclude the category of rural villages in developing countries because energy use in this setting is usually dominated by solid fuel combustion for cooking, an issue dealt with in Section 4.2.)

Energy-related health impacts can arise in community settings either when fuel storage, processing, or combustion occurs within the community itself (e.g., due to exposures to fuel or combustion products), or when it occurs someplace else (e.g., due to transported pollutants). In this section, we focus primarily on the former situation. Section 4.5 addresses regional scale energy use and health impacts.

Large, developing-world cities are often characterized by rapid growth, increasing traffic congestion, the intermingling of industrial, commercial, agricultural and residential zones, and often inadequate sanitary

---

11   See Footnote 1 (p 263).

and solid waste disposal systems. In the absence of technology interventions, urbanization in this setting typically leads to higher outdoor air pollution concentrations (Smith and Ezzati, 2005), a substantial portion of which could be attributed to increasing energy use in buildings, transport, and industry as population and consumption rise. At the same time, household energy use for cooking and space heating may shift towards cleaner fuels and technologies, reducing indoor exposures from those sources. Depending on climate, development level, and fuel availability, building-related energy use for heating, cooling, lighting, and cooking ranges widely across and within cities in the developing world (Kandlikar and Ramachandran, 2000).

Income inequalities within developing-world cities may lead to large variations in exposures to, and resulting health impacts of, energy use for transport, buildings, or industrial production. Asthma and other chronic disease rates are on the rise worldwide, adding to vulnerability. In addition, children may be at special risk due to their activity patterns and developing respiratory systems (Selevan et al., 2000). Over coming decades, rapid population growth is likely to continue as people seek economic opportunities in cities. Technological interventions that reduce emissions per unit of energy used will be needed to avoid worsening health impacts. Recent successes in reducing transport emissions in Delhi and Dhaka serve as useful examples.

Most large cities in developed countries have relatively stable populations, adequate infrastructure for transport and waste disposal, and relatively stable or declining air pollution levels. Improvements in air quality are due in part to emissions controls and in part to slow economic and population growth rates. Per capita energy use in developed-world cities is high compared with developing cities, but low compared with the peri-urban communities that surround them. For example, a 2005 energy use survey reported that urban households use on average less than 80% of the energy used by suburban households in the United States (US EIA, 2005). Transportation and the heating and cooling of buildings are the two major categories of local energy use. In contrast to the situation in developing cities, industrial production is usually not a major energy sector in developed cities. Per capita transport energy demand varies considerably depending on urban density and the availability of public transportation. While pollution emissions per unit of energy used may be relatively low, population exposures can be elevated owing to high population density and the proximity of people to emissions sources, especially motor vehicles. High and increasing rates of asthma result in a large pool of vulnerable people. In the next few decades, continuing efforts to reduce energy use and pollution emissions have the potential to accelerate improvements in environmental health in large cities in the developed world.

Low-density, peri-urban communities (i.e., "suburbs") have been growing rapidly in developed countries for many decades (Frumkin et al., 2004). Residents of many such communities depend almost completely on automobiles for moving people between home, work, school, and commercial services. Thus peri-urban communities tend to have high per capita energy use for transportation. Because dwellings and other buildings are typically large, detached, and surrounded by high-maintenance landscapes, these communities also have relatively high per capita building-related energy use. Accordingly, emissions of pollutants from fuel combustion may be high on a per capita basis, but relatively low per unit of land area, resulting in lower ambient air pollution levels than in nearby cities. Local air pollution hot-spots may still occur where high-traffic roadways pass close to residences, schools, or other places where people congregate.

4.4.3 Community Health Risks From Energy-Related Air Pollutants

Fuels (including wood, coal, oil, gasoline and diesel, liquified petroleum gas, and natural gas) may be combusted in communities to supply energy for transportation, to heat and cool buildings, and for waste processing and disposal. Fuels may also be burned to produce electricity for lighting, though this combustion often occurs outside of city limits. In addition, solid fuels (e.g., dung, wood, charcoal, coal) are often used for cooking in developing-world cities. While the household impacts of solid fuel use for cooking are discussed above (see Section 4.2), here we consider the implications of solid fuel combustion for ambient air quality in urban areas.

Depending on the fuel used and the way it is burned, combustion may produce a wide spectrum of solid and gaseous pollutants that can impact community air quality. It is useful to distinguish between primary and secondary pollutants. Primary pollutants are emitted directly at the source and thus have the potential for very local air pollution health impacts. Secondary pollutants form via reactions of primary pollutants in the atmosphere and thus have the potential for wider, more regional impacts. Pollutants emitted within communities will be experienced by community members mainly as primary pollutants; communities downwind of the source are more likely to experience them as transformed secondary pollutants. Thus, isolated communities (i.e., those not downwind from many other communities) will be exposed mainly to primary pollutants generated locally. On the other hand, communities that are downwind of many other communities will experience both locally generated primary pollutants and transported secondary pollutants. As population and economic activity rise, the relative importance of secondary pollutants is likely to rise. Indeed, a relatively easy way to reduce local primary pollution impacts is to build taller smokestacks on industrial and power facilities, though secondary pollutant impacts in downwind communities may increase.

Primary pollutants emitted by fuel combustion include many of the "criteria" pollutants that are often measured and regulated (i.e., SO_2, NO_2, CO, and some of PM_{10} and $PM_{2.5}$, including lead if used as a gasoline additive), as well as many other less-familiar or less-measured "non-criteria" pollutants such as trace metals, elemental carbon, and a wide range of solid and vapor-phase organic compounds. Secondary pollutants include ozone, some components of PM_{10} and $PM_{2.5}$, and organic vapors. Among secondary particles, sulfates, nitrates, and organic and elemental carbon represent important components. The levels and character of the pollutants emitted from fuel combustion depend greatly on the type of fuel used, the conditions of combustion, and on any post-combustion pollutant capture technologies. Across these dimensions, communities tend to utilize cleaner options as economic development advances, which is likely an important reason for the trend towards lower community-scale emissions as one moves up the economic development scale (Smith and Ezzati, 2005).

Available knowledge regarding the effects on human health of the pollutants that derive from fuel combustion is robust for only a subset of these pollutants (Kunzli et al., 2010). The knowledge base is most extensive for the criteria pollutants, owing to their widespread measurement in developed countries. In developed-world cities, scientific evidence for health impacts at observed concentrations levels suggests that particulate matter (either PM_{10} or $PM_{2.5}$) and ozone remain of greatest concern among the criteria pollutants.

Laboratory studies have shown that ozone can cause acute, reversible drops in lung function, increases in non-specific bronchial responsiveness, and pulmonary inflammation (Kim et al., 2011). In addition, epidemiology studies demonstrate associations with worsened asthma, emergency room visits, hospital admissions, and deaths. In particular, time-series analyses have shown that ozone is associated with an increased risk of premature mortality. Daily changes in ambient ozone concentrations have been found to be significantly associated with daily changes in the number of deaths, on average, across 98 US communities. Any anthropogenic contribution to ambient ozone, however slight, presents an increased risk for premature mortality (Smith et al., 2009). NO_x, CO, or volatile organic compounds (VOCs) emissions can contribute to ozone concentration both near and far from emission sources. Populations most at risk include children and adults who are active outdoors, especially those with asthma.

Epidemiology studies addressing health effects of PM_{10} and $PM_{2.5}$ have reported associations with both acute and chronic mortality in urban areas, as well as increases in hospitalizations and respiratory symptoms and decreases in lung function. Long-term studies in the United States have found strong associations between human exposure to fine particulate matter and adverse impacts on human health, including lung cancer and deaths from cardiopulmonary disease (Pope et al., 2002). Time-series studies have found similar concentration-response relationships in the developing countries of Asia (HEI, 2010). Because recent research indicates there is no well-defined threshold below which adverse health impacts from PM do not occur (Schwartz et al., 2008), incremental increases in PM concentration can increase rates of premature death in relatively clean communities as well as polluted ones. Populations at greatest risk of PM effects include the elderly and those with pre-existing cardiopulmonary disease. Not yet clear from the available evidence is whether specific chemical components of PM are more important than others in the observed health effects.

Assessing health risks associated with energy-related air pollution emissions in communities is challenging. First, there are many relevant pollutants that cause adverse health impacts, and choices must be made as to which to include in the analysis. A common approach, particularly in developing-country settings, is to use particulate matter as a proxy for the entire mix of pollutants. This is a reasonable strategy, given the strong evidence base for premature mortality effects of PM. Another key challenge in developing countries is the limited availability of data – including air pollution exposures, baseline health statistics, and the exposure-response relationship linking pollution to health risk– for use in risk assessments. Also, not all air pollution in communities arises from energy use, so it can be challenging to parse out the energy-related component of health risk.

WHO tackled many of these challenges in its analysis of the adverse health burden related to urban air pollution for world cities with populations of 100,000 or more in the year 2000 (Cohen et al., 2004). This assessment used particulate matter to represent all urban air pollution, and estimated the numbers of deaths associated with $PM_{2.5}$ levels in each city as compared to a reference concentration of 7.5 ug/m³. It focused on three categories of cause of death: cardiopulmonary and lung cancer deaths among adults, and acute respiratory infections in children from birth to age four. The largest mortality burden was estimated to occur in the more polluted, rapidly growing cities of the developing world. On a global basis, it was estimated that urban PM caused approximately 3% of mortality due to cardiopulmonary disease, about 5% of mortality from lung cancer, and about 1% of mortality from acute respiratory infections.[12]

Adult mortality attributable to $PM_{2.5}$ was estimated using a concentration-response function derived from the US-based American Cancer Society epidemiologic study (Pope et al., 2002). Since the effect of PM pollution was not well known above 50 ug/m³ $PM_{2.5}$, mortality impacts were truncated at that level. In spite of these uncertainties, this work provided a valuable snapshot of the total health burden in 2000 attributed to urban air pollution, which totaled 0.8 million premature deaths annually. Note that not all outdoor fine particle pollution is due to the operation of energy systems; they are, however, probably responsible for about 85% globally.

As part of the current GEA effort, estimates of the global burden of years of life lost and DALYs due to ambient particulate matter in 2005 are calculated based on the WHO (2004) (Cohen et al., 2004) methodology using estimated exposures and most recent baseline mortality and DALYs from WHO (2008). Globally, there were 23 million DALYs and 2.75 million premature deaths due to outdoor air pollution from energy

---

12   As noted previously, this WHO assessment is being extensively revised for 2012 publication.

systems. These new estimates are higher than those reported earlier in the WHO work discussed above, due in part to higher pollution concentration estimates in China and India, but also because all exposures, urban and rural, were included. The reader is referred to Chapter 17 for a full description of this assessment. The distribution of estimated $PM_{2.5}$ levels is shown in Figure 17.45.

It is also worth noting that, when monetary values are assigned to the health damage caused by air pollution, the economic benefits of air quality improvements can be very substantial. Nemet et al. (2010) and colleagues reported that air pollution-related health co-benefits can substantially offset control costs for GHG emissions reductions.

4.4.4   Exposure vs. Emissions

An important consideration in assessing the potential health significance of air pollutants emitted by fuel combustion is the temporal-spatial relationship between pollutant concentrations and the people living in those communities. No matter how much pollution is emitted, people must breathe air pollution of sufficient concentrations for sufficient periods of time to elicit adverse health effects (see Box 4.1). Digging deeper, we may examine this question for population subgroups of potentially greater vulnerabilities, including the young, the old, and those with pre-existing diseases. Although clear understanding of which levels produce how much ill health in each population is not yet available, exposures that are higher and/or longer will be of greater health significance.

Spatial relationships between source and receptor can vary widely for different types of fuel combustion sources. Also, human activity patterns play an important role in determining the extent to which individuals encounter air pollution. In developing-world cities, for example, poor individuals often live in more polluted locations, and are more likely to commute by walking or cycling along congested roadways. People living in poverty are thereby exposed to elevated levels of air pollution, while also experiencing high lung doses due to enhanced activity and breathing rates. This is particularly true for children, who have higher metabolic rates than adults (Selevan et al., 2000). Wealthier individuals often live in cleaner neighborhoods and are more likely to commute by car, and their pollution doses may be lower due to less time spent on the roads and slower resting breathing rates. In developed-world cities and suburbs, similar patterns are seen, although a far greater proportion of the population commutes in vehicles. Housing design and climate factors influence the extent of penetration of ambient air pollution indoors. The intake fraction (Box 4.1) is a metric to integrate over these domains and provides a relative ranking of human impacts of different sources per unit of emissions. Kandlikar and Ramachandran (2000) estimated values of a related metric, exposure efficiency, for a range of sources across the development gradient. Exposure efficiency ranges from 100 Exposure Units [EU = (ug/m^3)-person-year] per tonne for a US power plant to 100,000–200,000 EU/tonne for indoor combustion in urban India.

4.4.5   Building-Related Air Pollution Emissions

Residential and commercial buildings are an important source of energy-related air pollution emissions in communities, due to the combustion of fuels for heating, cooling, and cooking. The nature and magnitude of the impacts on air quality vary with the type and amount of fuel used, how it is burned, the density of buildings, and local topographic and meteorological features. Emissions and local concentrations generally diminish as one moves from coal and wood to oil and to natural gas. Major air pollution episodes such as London's killer Great Smog of December 1952 have been attributed to dense, urban household coal combustion combined with stagnant "inverted" meteorological conditions, favoring the buildup of particulate matter and other pollutants in the lower atmosphere. Transition to oil and other clean fuels for household heating led to substantial improvements in air quality in London and other developed-world cities in the 20th century, essentially eliminating the periodic mortality peaks that had previously been observed. In Dublin, Clancy et al. (2002) documented a 10% drop in cardiovascular deaths and a 15% drop in respiratory deaths in the years following the 1990 elimination of household coal combustion. Black smoke concentrations in Dublin fell by about two-thirds following the coal ban. Oil is far cleaner than coal, and more refined grades of oil are cleaner than residual oil. Recent studies in New York have highlighted the pollution impacts of residual oil combustion for heating of buildings in winter and for hot water and steam generation throughout the year (Peltier and Lippmann, 2009). Peltier et al. (2008) show that commercial burning of residual fuel oil, mainly for space heating, leads to high concentrations of trace metals (e.g., nickel) that may have particular health significance.

In addition to fuel switching, technology improvements can lower emissions while also reducing energy use and GHG emissions. Many cities are developing GHG emissions inventories, and the building sector can be a dominant source of such emissions. For example, the 2005 New York City emissions inventory found that over 60% of CO_2-equivalent emissions were from residential, commercial, and institutional buildings (City of New York, 2007; Bloomberg and Aggarwala, 2008). Substantial health co-benefits can be achieved via technology strategies whose primary objective is to reduce GHG emissions, because they reduce air pollution emissions (Wilkinson et al., 2009). The monetary value of these health benefits could partially or completely compensate for the costs of the new technologies.

In developing-world cities, emissions from cooking with solid fuels like wood, dung, and charcoal contribute significantly to urban air pollution. The proportional use of wood and dung is usually lower in urban than in rural areas (Kandlikar and Ramachandran, 2000), but the density of emissions is far higher. In Africa, charcoal often becomes the dominant cooking fuel in urban areas. Though emitting fewer particles at the stove than wood, the production of charcoal in low-technology kilns is highly polluting, although fortunately usually occurring outside cities.

4.4.6 Transportation Emissions

Energy use for transport is a major source of air pollutants in all communities, but with wide variations depending on vehicle densities, congestion, fuels, and engine technologies. Vehicle emissions have special significance from a human health perspective because they occur in close proximity to people, enhancing the fraction of emissions that is inhaled. In developed cities, where industrial and uncontrolled point source combustion is relatively rare, vehicle emissions can be the dominant local air pollution source (Qin et al., 2006). The situation in developing cities is far more complicated, with much higher levels of air pollution emissions overall, from a wide range of sources. However, vehicles play an important and probably increasing role in urban air pollution in developing cities Kinney et al. (2011). In a study of four Indian megacities, Chowdhury et al. (2007) found that gasoline and diesel vehicle emissions together represented 20–50% of $PM_{2.5}$ concentrations, depending on the city and season.

Though there are few reliable data on vehicle emissions in developing cities, emissions per kilometer are estimated to be at least an order of magnitude higher than in developed cities, and perhaps far higher (Kandlikar and Ramachandran, 2000). Reasons for higher per vehicle emissions in developing countries include higher proportions of diesel vehicles, adulteration of fuels with inexpensive alternatives like kerosene, and the lack of catalytic control of engine emissions. Several large developing-world cities have achieved major air quality improvements by enacting laws requiring the use of cleaner fuels in commercial vehicle fleets (Chelani and Devotta, 2007; Begum et al., 2008; Reynolds and Kandlikar, 2008).

Reducing per vehicle emissions, either through fuel or technology interventions, is a relatively fast and economical approach for achieving significant improvements in urban air quality. Another challenging but ultimately more sustainable solution is to address growing road congestion by providing public transportation options. This has the potential to reduce the rapid rise in private vehicle use being seen in

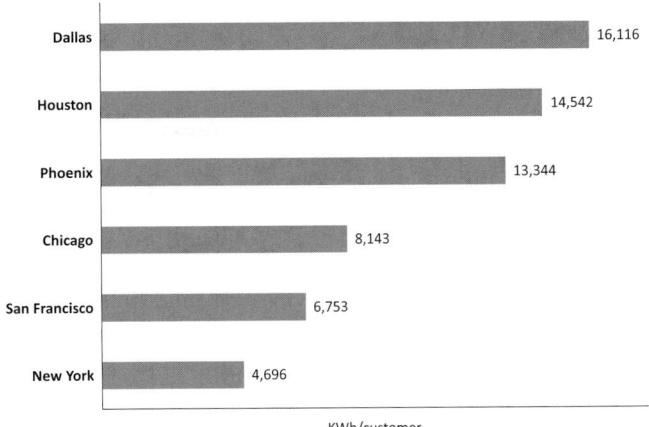

Figure 4.4 | Average annual residential electricity usage by city, 2000–2005. Source: adapted from City of New York, 2007.

many developing cities, which is far outpacing road infrastructure (see Table 4.8). Many developing-world cities are examining urban transport reform, with some success stories (e.g., Brazil or Bangkok). Dedicated lanes for bus rapid transit can reduce both congestion and air pollution levels, while also providing affordable transportation for urban residents. Developed cities (e.g., London, Milan) have been successful at reducing congestion by imposing automated congestion fees within the downtown zone. While many developed cities are dependent on private vehicles for workforce commuting, there are wide variations even within countries. For example, the proportion of commuters using public transportation in six US cities ranges from 5% to over 50% (Figure 4.4).

Vehicles can affect air quality in other ways besides their engine emissions. Road dust can be a serious nuisance in developing-world cities, and often carries health risks due to toxic materials re-suspended with the dust, including asbestos from brake linings, lead from tire weights, and oil from leaking tanks.

Table 4.8 | Composition of vehicle population in India.

Year end March	Mopeds, Motorcycles etc.	Cars, Jeeps etc.	Buses	Goods Vehicle	Others	Total (Million)
	(as % age of total vehicle population)					
1951	8.8	52.0	11.1	26.8	1.3	0.31
1961	13.2	46.6	8.6	25.3	6.3	0.66
1971	30.9	36.6	5.0	18.4	9.1	1.86
1981	48.6	21.5	3.0	10.3	16.6	5.39
1991	66.4	13.8	1.5	6.3	`11.9	21.37
2001	70.1	12.8	1.2	5.4	10.5	54.99
2002	70.6	12.9	1.1	5.0	10.4	58.92
2003	70.9	12.8	1.1	5.2	10.0	67.01
2004	71.4	13.0	1.1	5.2	9.4	72.72
2005	72.1	12.7	1.1	4.9	9.1	81.5
2006	72.2	12.9	1.1	4.9	8.8	89.61

Source: Government of India, 2009.

4.4.7 Community Impacts of Transported Regional Pollutants Related to Energy Supply

As described below in Section 4.5, community air quality can be impacted by pollutants transported from other regions. In fact, upwind sources may often dominate local sources as contributors to community air quality, particularly for communities that are downwind of major source regions, such as cities in the northeastern United States, and those downwind of the industrial heart of China, such as Hong Kong. The scale over which pollution can be transported may be hundreds or even thousands of kilometers.

4.4.8 Energy-Related Pollutants in Water and Soils

Health can also be negatively impacted by energy-related pollutants in water and soils. Spills are common near oil drilling and handling operations. Spills from ageing or sabotaged pipelines have increased in Arctic Russia, the Niger Delta, and the northwestern Amazon, leading to contaminated drinking water supplies, fishing areas, and agricultural fields (Jernelov, 2010). In a preliminary analysis of health impacts in the months after the BP oil spill in the Gulf of Mexico, Solomon and Janssen (2010) report that 300 people sought treatment in Louisiana for symptoms typical of hydrocarbon and/or hydrogen sulfide exposures, including nausea, vomiting, cough, and respiratory distress. Mining has been associated with offsite contamination of domestic groundwater supplies, even after mine-site reclamation (Palmer et al., 2010). Exposure can also come from contact with surface water (Besser et al., 1996) or re-suspended dust (Ghose and Majee, 2007). Hazardous compounds in dust were elevated near surface coalmining operations in India (Ghose and Majee, 2007). Elevated rates of cardiac, pulmonary and kidney diseases, and hypertension were observed in the general population living near coal mining operations in the US state of West Virginia (Hendryx and Ahern, 2008). Health damage from fossil fuel extraction processes can be reduced through improved mining practices, but also by efforts targeting energy demand, such as improved energy efficiency and use of renewable sources.

4.4.9 Other Sources

Other energy-related sources that can play an important role in community air quality include electric utility facilities (as well as smaller distributed energy sources), industrial operations, port facilities, and uncontrolled combustion of solid waste. The latter is a widespread source of toxic pollutants in developing cities. Though not strictly an energy issue, the energy contained in solid waste could be utilized for energy supply while at the same time improving air quality and health. Waste-to-energy facilities may represent a long-term solution to this growing problem.

Electric utilities and small power generators often are sited in or near urban areas. Depending on the fuel burned, stack height, and emission control technology employed, emissions and local impacts vary widely. Peak energy loads during summer hot spells are often supplied by diesel

generators, with higher impacts than plants fired by natural gas. Large electric utilities require large amounts of water for their routine operations, and can adversely impact downstream water quality.

Industrial operations have been largely eliminated from developed cities but still remain significant sources of air pollution in many developing cities (Kandlikar and Ramachandran, 2000; He et al., 2002). While moving dirty industries away from populated areas would be the most sustainable solution, improved fuels and emissions control technologies can provide important short-term gains. There is also growing recognition of the air quality impacts on local communities of emissions from port operations (e.g., Newark, New Jersey; Los Angeles).

4.4.10 Conclusions

Air pollution emissions from fuel combustion are a major risk factor for morbidity and mortality in communities. Key energy-related emissions sectors include transportation, building heating, cooling and cooking, electricity generation, industrial processes, and waste combustion. Transportation emissions are particularly influential in health burdens, due both to the nature of the emitted pollutants and to the close proximity of emissions and vulnerable populations. Within and across communities, vulnerability to pollution-related health risks varies as a function of age and income. While health burdens due to community air pollution exist everywhere, these burdens are largest and in developing-world cities, particularly in China and India.

4.5 Regional and Transboundary Impacts

4.5.1 Key Messages

- Air pollutants can be transported long distances before being removed from the atmosphere and thus health impacts from emissions can occur locally, regionally, and even globally, following intercontinental

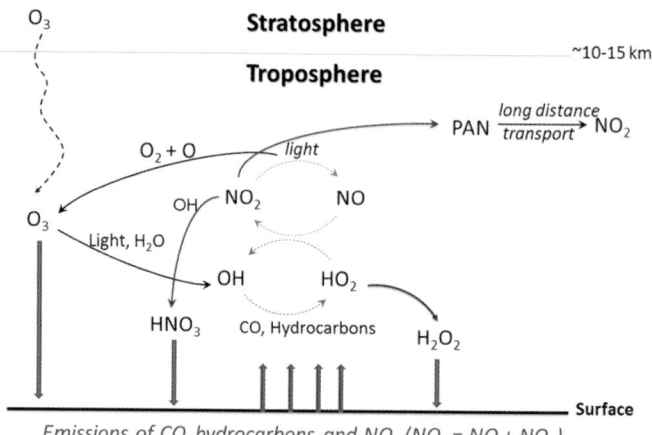

Figure 4.5 | Ozone production in the troposphere.

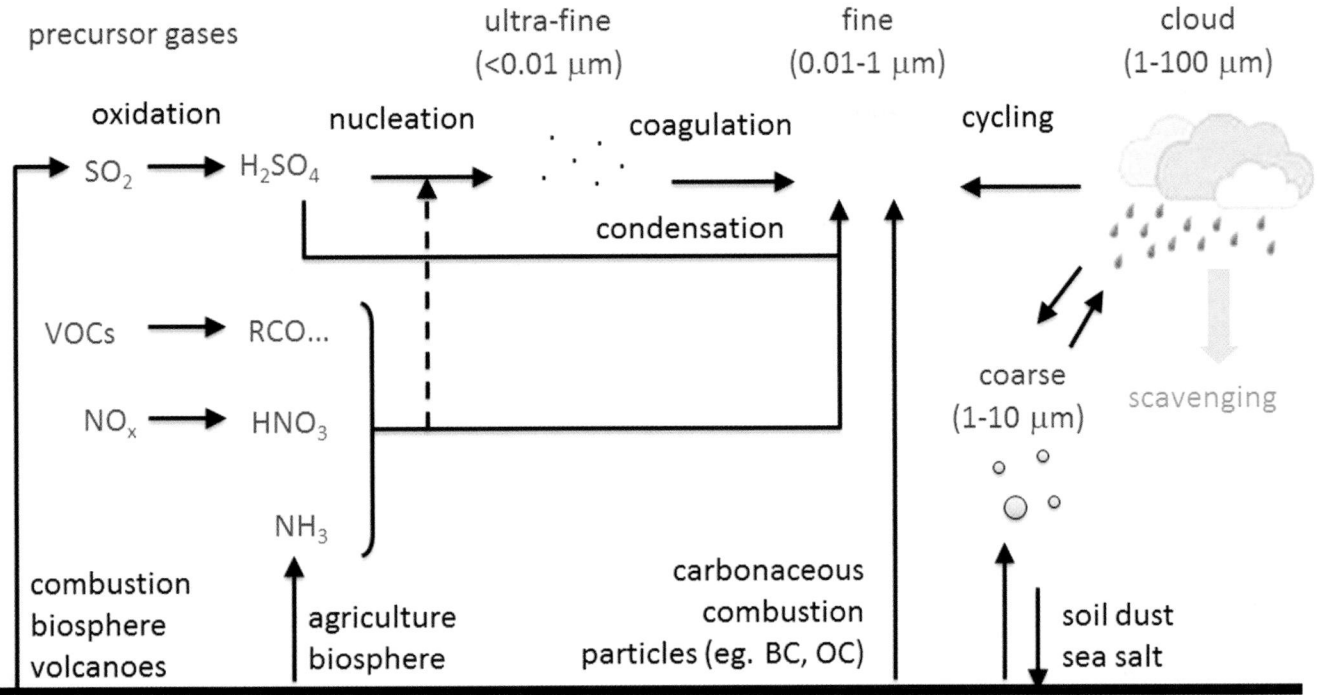

Figure 4.6 | Particulate matter (PM) sources and processes.

transport. In general, health damage due to ambient air pollution from the combustion of fossil and biomass fuels is greatest close to the source of emissions and within population centers.

- Emissions of particulate air pollutants have a significant effect on climate change. Externally mixed sulfate aerosols decrease radiative forcing, while black carbon (BC) aerosols increase it. Hence, although reductions in particle concentrations benefit public health, reductions in emissions of sulfur dioxide (the precursor to sulfate) will likely warm climate and reductions in the emissions of BC will likely cool climate.

- Long-range transport can bring black carbon to glaciers and snow-covered regions (Kopacz et al., 2011) where it may increase melting rates, affecting water supplies and health.

4.5.2   Contribution of Energy Systems to Air Pollution

Combustion of both fossil fuels and biomass releases a myriad of gaseous and particulate air pollutants in large quantities that have direct adverse impacts on public health. These pollutants include nitrogen oxides (NO_x); VOCs including methane, carbon monoxide (CO), sulfur dioxide (SO_2); and black carbon and organic carbon (OC) particulates. In addition, a variety of persistent organic pollutants (POPs) and heavy metals, including mercury (Hg), are also emitted. The reaction of some of these compounds in the atmosphere in the presence of sunlight forms secondary pollutants. In particular, NO_x, CO, and VOCs can react to form

ozone (O_3) (see Figure 4.5). Sulfur dioxide and nitrogen oxides can form aerosols. (See Figure 4.6 for a summary of the sources and processes of particulate matter.)

The lifetime of air pollutants varies from hours (NO_x) to days/weeks (aerosols, VOCs) to months (O_3, CO, and Hg) and even years (POPs). The lifetime of the pollutant has a large influence on the distance it can travel before being removed from the atmosphere. Methane –produced by both natural and anthropogenic sources is released by the incomplete combustion of fossil fuels and, in larger amounts, as leaks from oil refineries, pipelines, coal mines, and other parts of energy systems – has a lifetime of approximately 12 years. As a result, it is relatively well mixed in the atmosphere. Oxidation of methane contributes to the production of tropospheric ozone. However, due to methane's long lifetime in the atmosphere, ozone may be produced far from the source of methane emissions, contributing to the observed rise in global background ozone levels (Fiore et al., 2008).

4.5.3   Mechanisms of Regional and Intercontinental Transport

Historically, air pollution was viewed as an urban problem. The Great London Smog of December 1952, caused by the combustion of coal, mainly for residential heating, during an atmospheric inversion that trapped the pollutants close to the surface, caused the death of thousands of people. This resulted in increased recognition of the toxic effects of air pollution on the urban scale.

In the 1970s, several studies confirmed that acid deposition could occur hundreds of miles from where SO_2 was emitted as a result of the oxidation of SO_2 to sulfuric acid (H_2SO_4) and sulfate. Starting in the 1980s, there was increasing recognition that other pollutants can be transported hundreds of miles and affect air quality far downwind of emission sources. The Long Range Transboundary Air Pollution Convention and its protocols (www.unece.org/env/irtap) were signed and ratified by nations of Europe and North America in the 1980s and 1990s to control emissions of sulfur, nitrogen oxides, VOCs, heavy metals, and POPs, and to abate acidification, eutrophication, and ground-level ozone concentrations. The Convention was the first international effort at coordinated abatement of air pollutant emissions.

In the late 1990s, efforts began in the United States to develop interstate emission controls on NO_x in order to reduce the long-range transport of ozone. NO_x emitted in cities and then transported to rural areas can react with VOCs emitted from vegetation to form ozone on a regional scale (see Figure 4.5). Elevated ozone concentrations can impact human health, ecosystems, and agriculture on a regional to global scale (Bell et al., 2004; Mauzerall and Wang, 2001).

Transport of air pollutants across the North American or European continent takes approximately one week, while circumpolar transport of pollution at northern mid-latitudes takes approximately one month. Mixing between the tropics and the high latitudes of the northern hemisphere requires approximately three months. Starting in the late 1990s, intercontinental transport of air pollutants, particularly ozone, mercury, and POPs, was recognized by the research community as potentially affecting the ability of some regions to meet their own air quality goals (Fiore et al., 2002).

Although aerosol transport from Asia to the United States has been observed by satellite as taking less than one week, average transport time is calculated to be two to three weeks (Liu and Mauzerall, 2005). Pollutants with lifetimes of a few days or less tend to remain within a limited area, while those with lifetimes of several weeks or more can cross oceans and influence air quality on downwind continents. Pollutants with lifetimes less than several years, however, have their largest concentrations close to the source of emission.

Intercontinental transport also influences surface ozone concentrations. Ozone has a lifetime of days in the continental boundary layer (the surface to approximately 2 km) but several weeks in the free troposphere (approximately 2–10 km above the surface). Ozone can be transported in the free troposphere and then subside to the surface, where it has adverse effects on human health, ecosystems, and agriculture. Observations at northern mid-latitudes have shown background ozone concentrations to be rising from preindustrial concentrations of approximately 10 ppbv (parts per billion by volume) (Marenco and Gouget, 1994). Model simulations define background ozone

concentrations to be the ozone that results when emissions of ozone precursors from one region are turned off while emissions from the rest of the world are maintained at current levels. Present background concentrations in the United States are in the range 20–40 ppbv, at least half of which is of anthropogenic origin (Fiore et al., 2002). Rapid increases in combustion of fossil fuels for industry and vehicles in Asia has led to large increases in the emission of NO_x (Ohara et al., 2007) and resulting increases in hemispheric background concentrations of ozone, despite simultaneous decreases in NO_x emissions in the United States and Europe.

4.5.4 Implications of the Lack of a Threshold below which Air Pollutants No Longer Damage Health

As noted in Section 4.4, there is good evidence of health effects of small particles, even at low levels. In addition, both observational and modeling studies show that concentrations are influenced by long-range transport, as well as by local anthropogenic and natural emissions (Jaffe et al., 1999; Park et al., 2004). As a result, even relatively modest increases in PM concentration due to regional and intercontinental transport can increase rates of premature mortality in both polluted and relatively clean regions. A recent evaluation of the global health impact of intercontinental transport of fine aerosols found that nearly 380,000 premature deaths globally of adults age 30 and older are associated with exposure to particulates originating from foreign continents, with approximately 90,000 of these deaths attributable to fine, non-dust aerosols (Liu et al., 2009).

4.5.5 Effect of Ozone on Agricultural Yields

Episodes of elevated ozone are frequently observed in suburban and rural regions due to NO_x outflow from urban centers reacting with hydrocarbons from local vegetation. Elevated ozone concentrations can reduce agricultural yields (Mauzerall and Wang, 2001). Surface ozone in East Asia in 1990 is estimated to have reduced agricultural production of wheat, rice, and corn in China, Korea, and Japan by 1–9% and of soybeans by 23–27% (Wang and Mauzerall, 2004). In 2020, assuming no change in agricultural production practices, grain loss due to increased levels of ozone pollution is projected to increase to 2–16% for wheat, rice and corn and to 28–35% for soybeans (Wang and Mauzerall, 2004). More recent studies have examined the global impact of ozone on agricultural crop yields in the years 2000 and 2030 (van Dingenen et al., 2009; Avnery et al., 2011a; Avnery et al., 2011b), and found 2000 global relative yield reductions to be approximately 9–14% for soybean, 4–15% for wheat, and 2–6% for maize; in 2030, the projected yield losses were 5–26% for wheat, 15–19% for soybean, and 4–9% for maize. Similar results were obtained by van Dingenen (2009). Developing countries are likely to experience simultaneous increases in population and in emissions of ozone precursors, with resulting reductions in agricultural yields. This will have an indirect effect on health via escalating commodity prices, hunger, and malnutrition in some parts of the world.

exposure to MeHg in the United States. Fetal exposure via fish consumption by pregnant women can pose neurodevelopmental risks (NAS, 2000).

4.5.6 Mercury Emissions: Transport and Intake that Affects Health

Mercury, largely emitted from coal combustion, has long been recognized as a global pollutant by the scientific community (Selin, 2005). Mercury is largely emitted in the elemental form Hg(0), which is oxidized in the atmosphere to Hg(II) and subsequently deposited. The atmospheric residence time of Hg(0) is about a year, which permits its global transport (Selin et al., 2007). Microbiological action converts Hg(II) into methylmercury (MeHg), an organic form of mercury. MeHg is highly toxic and can accumulate up the food chain in aquatic systems, leading to high concentrations in predatory fish. Consumption of contaminated fish is the major source of human

4.5.7 Effects of Air Quality on Climate Change and Vice Versa: Resulting Effects on Health

Some air pollutants have a significant effect on climate. As shown in Figure 4.7, tropospheric ozone and black carbon together have a positive global radiative forcing larger than methane, the second most important CAP (Forster and Ramaswamy, 2007). In contrast, sulfate and organic carbon have a negative radiative forcing due to both direct effects (reflection of incoming solar radiation) and indirect effects on clouds (smaller droplets and hence whiter clouds with higher albedo) (Forster and Ramaswamy, 2007). Table 4.9 summarizes the climate, health, and environmental issues associated with emissions leading to sulfate and BC (see also Box 4.2).

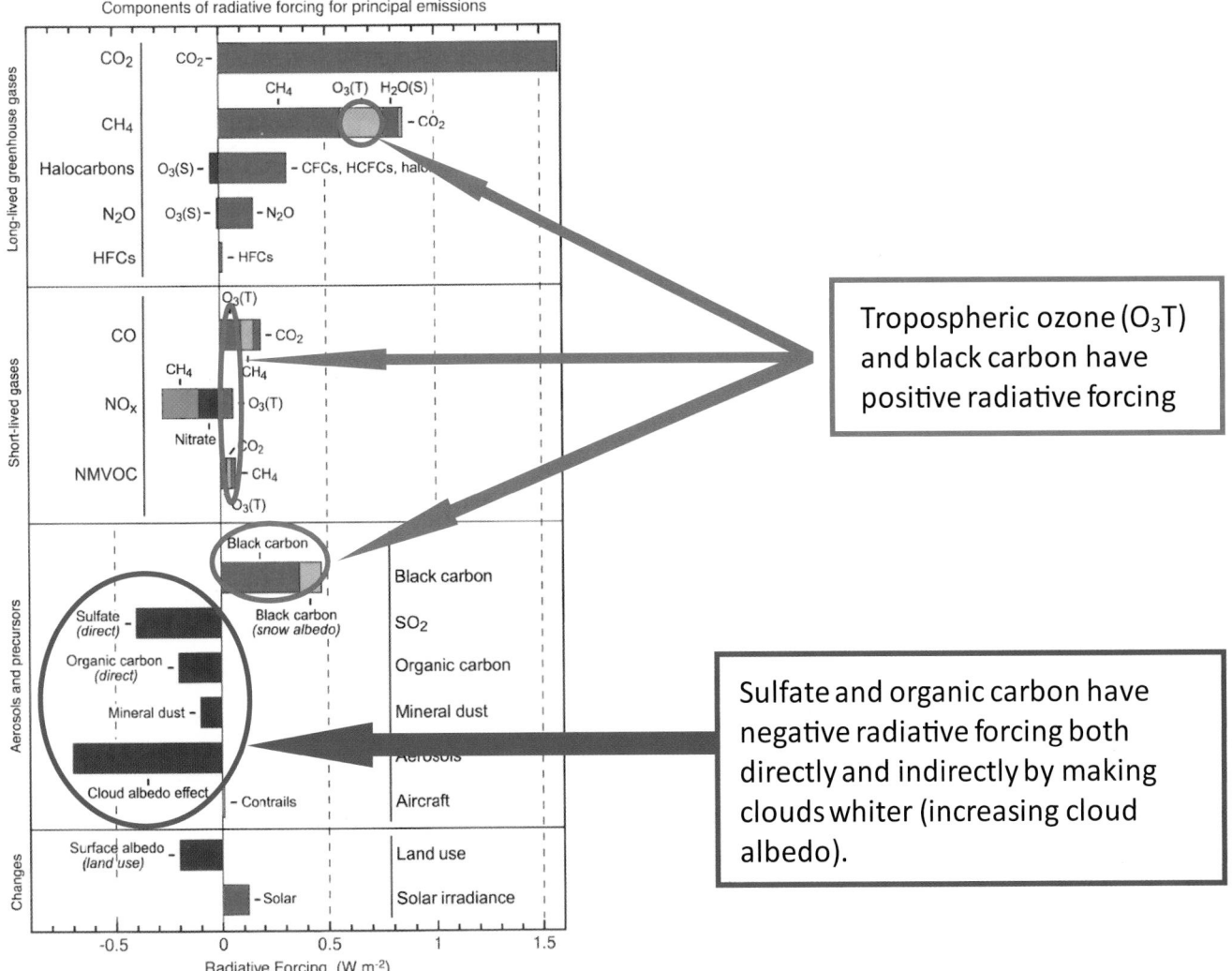

Figure 4.7 | Global radiative forcings due to emissions changes, 1750–2005: important contributions of air pollutants. Source: adapted from Forster and Ramaswamy, 2007.

Table 4.9 | Summary of sulfate and black carbon, climate, health, and environment issues.

	Source	Measurement	Health	Climate	Environment	Confounders
Sulfate	Mainly power and industry, but also transport	Little uncertainty although emissions are mainly SO_2 gas, with calculation of transformation to sulfate necessary	May be less damaging than the average PM, but uncertain	Relatively little uncertainty on direct and indirect climate cooling effects, but cloud physics still poorly understood	Acid precipitation – relatively little uncertainty but wide difference in impact by location	Sulfur control can reduce mercury emissions from power plants but has minor interaction with other types of control
Black Carbon	Mainly diesel engines and household fuel	Because BC is co-emitted with other carbonaceous aerosols basic measurement methods and metrics not well characterized	May be more damaging than the average PM, but uncertain	Major uncertainties but high radiative forcing – complicated by location and short lifetime	Major melting impact if falls on ice or snow, particularly in Arctic and Himalayan glaciers. Not well understood	Except for emissions from diesel engines, it's difficult to control BC alone without also reducing organic aerosol emissions, which are generally cooling

Source: Smith et al., 2009.

Box 4.2 | Health and climate primer for energy-related climate-active pollutants

Long-lived (hundreds of years)

Carbon dioxide (CO_2) poses a low direct health hazard and indeed is also the weakest greenhouse pollutant by mass. However, because of its magnitude of emissions, it is the most important overall. It also has a much longer lifetime in the atmosphere than any of the other energy-related pollutants – most is gone in 100 years or so, but a portion of emissions is thought to remain in the atmosphere for thousands of years. Even without special measures to burn fuels more cleanly, measures that reduce fossil fuel use or increase its efficiency will have co-benefits, through associated reduction of CO_2 emissions and the health-damaging pollutants noted below that accompany fuel burning. About 78% comes from the combustion of fuels.

Medium-lived (tens of years and thus globally mixed)

Methane (CH_4) is the second most important greenhouse pollutant. About one-third of global emissions come directly from energy-related sources, including leakage from oil/gas facilities and coal mines as well as incomplete combustion of biomass and fossil fuels. Its main sources, however, are from agriculture and poor management of wastes. Although not directly health-damaging, methane is a primary precursor to the global rise in tropospheric ozone levels. Although methane has a shorter atmospheric lifetime than carbon dioxide, a tonne of methane will have a much bigger warming impact than a tonne of CO_2 for the first few decades after emission, because of its large direct and indirect impacts on warming.

Short-lived (days to weeks, and thus effects depend on local conditions)

- Carbon monoxide (CO) is mainly a product of incomplete combustion. Although it does not have a direct climate effect, it acts to sweep up hydroxyl (OH) radicals in the atmosphere, thus effectively increasing the lifetime of methane and adding to tropospheric ozone. The impacts of CO on methane and tropospheric ozone are both potentially climate warming. About two-thirds comes from the energy sector, the rest from forest/savannah/agricultural fires.

- Non-methane volatile organic compounds (NMVOCs) come from several human-generated sources, including incomplete combustion and evaporation from fuels. They also have primarily indirect rather than direct impacts on climate. They play an important role in urban ozone formation as well. NMVOC emissions reduce the oxidizing capacity of the troposphere, increasing the lifetime of methane and adding to tropospheric ozone. Many of these compounds also have direct health effects on humans. About one-third is due to human energy use.

- Nitrogen oxides (NO_x) derive from fuel combustion and have a complex relationship with and indirect impact on both climate warming and cooling by affecting ozone, methane, and particle levels. NO_x emissions act to decrease the oxidizing capacity of the troposphere and increase the lifetime of methane, but also are a major precursor to tropospheric ozone. Nitrate particles, like those of sulfate, are lighter colored and thus generally cooling. There also seems to be a small increase in carbon capture in natural ecosystems due to the eutrophication from deposited nitrate. About half comes from the energy system.

- Sulfur oxides (SO_x), which derive mainly from combustion of fuels, partly convert to sulfate (SO_4^{2-}) aerosols in the atmosphere, which although potentially health-damaging, are generally thought to exert a net cooling effect on the climate. Along with organic carbon (OC, see below), can sometimes be coated with black carbon (BC, see below) to create "brown carbon" with warming potential. Essentially all human emissions derive from fuel use (see Figure 4.10).

- Black carbon (BC), which is fine particulate matter of dark color containing a large fraction of elemental carbon, is derived exclusively from incomplete combustion. They are strongly warming in the atmosphere and increase heat absorption if deposited on ice and snow, such as on Himalayan glaciers or in the Arctic. About two-thirds of human emissions come from energy systems (Figure 4.10).

- Organic carbon (OC) aerosol, which is less dark carbonaceous particulate matter produced, like BC, largely from incomplete combustion, but also from secondary processes involving biogenic VOCs. Although not well characterized and sometimes coated around BC, it is thought generally to produce a net cooling effect globally, although with much local variation. It is a major form of health-damaging small particles globally. About half comes from energy systems.

Very short lived (hours)

- Tropospheric ozone (O_3): is a secondary pollutant formed through complex photochemical reactions involving nitrogen oxides and volatile organic compounds including methane in the presence of sunlight. Worldwide, background ozone has more than doubled since pre-industrial times and continues to rise (Unger et al., 2008). Although plants and other natural sources such as forest fires contribute to ozone levels, the major sources are the rise in methane emissions and burning of fuels with subsequent increased emissions of nitrogen oxides and VOCs. In the Intergovernmental Panel on Climate Change (IPCC) assessments, as shown in Figure 4.7, taken all together, ozone is the third most important greenhouse gas (after carbon dioxide and methane), although itself created from other gases. Stratospheric ozone generally has different sources and, although also warming, protects Earth's surface from health- and ecosystem-damaging ultraviolet radiation.

Source: Smith et al., 2009.

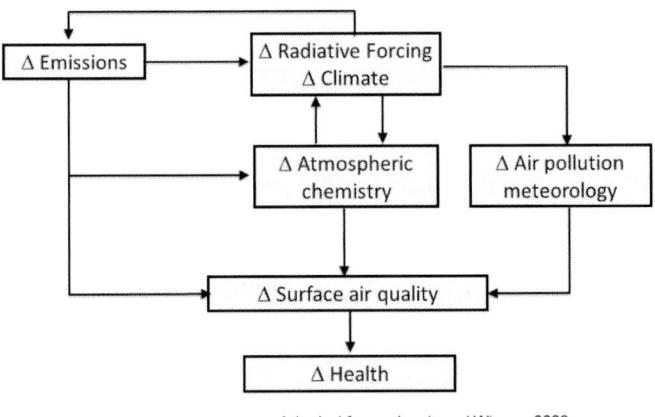

Air Pollution – Climate Interactions

Adapted from: Jacob and Winner, 2009.

Figure 4.8 | Overview of air pollution – climate interactions from emissions to surface air quality to health. Source: adapted from Jacob and Winner, 2009.

Figure 4.8 summarizes interactions and feedbacks of air pollution, climate change, and health, which are in general poorly understood. Changes in emissions of CAPs affect radiative forcing and climate change (ozone, PM, methane), which in turn influences atmospheric chemistry, surface air quality (ozone and PM), and public health. Climate change affects meteorology and hence transport and mixing of air pollutants and natural emissions (biosphere, dust, fires, lightning) with implications for surface air quality (Jacob and Winner, 2009), health, and agricultural and ecosystem impacts. A recent study concluded that, "[c]oupled GCM (General Circulation Model)–CTM (Chemical Tracer Model) studies find that climate change alone will increase summertime surface ozone in polluted regions by 1–10 ppb over the coming decades, with the largest effects in urban areas and during pollution episodes" (Jacob and Winner, 2009). This climate penalty means additional impacts on public health unless stronger emission controls are implemented.

Sulfate, BC, and OC all contribute to suspended particulate matter less than 2.5 micrometers in diameter ($PM_{2.5}$). Figure 4.6 describes the sources and processes for forming, coagulating, and scavenging PM from the atmosphere (see also Box 4.2).

BC is emitted during the incomplete combustion of fossil fuel and biomass. BC is generally co-emitted with OC, another type of carbonaceous (carbon-based) aerosol. Because OC has a cooling effect on climate, the net warming effect of carbonaceous aerosol emissions decreases as their OC:BC ratio increases (Kopp and Mauzerall, 2010). This ratio is relatively high for biomass combustion and quite low for diesel engines. The warming effect of BC varies among source types and regions due to differences in co-emitted aerosols, transport, and deposition location. In addition to causing warming when lofted in the atmosphere, BC also enhances melting rates when it is deposited onto snow and ice. It thus contributes significantly to the warming of the Arctic, where it accelerates melting of sea and land ice. When it is emitted from South and East Asia, a major source area, it accelerates Himalayan glacier and snowpack melting, with resulting impacts on water supplies.

BC predominantly falls within the PM_{10} category (particulate matter with a diameter less than 1 micrometer). Both diffuse and concentrated BC particles, like all fine particles, have adverse impacts on human health, including premature mortality. There is a perception that particles from diesel engines, elemental, and organic carbon are the "most toxic" constituents of $PM_{2.5}$ (Cooke et al., 2007). The few long-term studies, however, do not indicate that BC particles are significantly more toxic than the "average" ambient particle (Smith et al., 2009).

As long as the BC/OC ratio is sufficiently high, reduction in emissions of combustion particles will result in rapid, short-term reductions in radiative forcing, hence slowing warming significantly in the near term while simultaneously improving public health (UNEP, 2011).

4.5.8   Conclusion

Poor combustion of both fossil and biomass fuels releases gaseous and particulate air pollutants in large quantities, nearly all of which have adverse impacts on public health. Pollutants with lifetimes of a few days or less tend to remain within a limited area, while those with lifetimes of several weeks or more can cross oceans and influence air quality on downwind continents. Because research indicates that there is no threshold of health effects at ambient concentrations of fine particulates, incremental increases in particle concentration due to regional and intercontinental transport can increase rates of premature mortality in both polluted and relatively clean regions. The secondary formation of ozone has direct adverse health impacts and also decreases agricultural yields, hence reducing food availability in some regions. Long-range transport of BC to glaciers may increase melting rates, thus influencing water supplies and health.

4.6   Global Health Impacts from Climate Change

4.6.1   Key Messages

* Of the many impacts of climate change on human health, most will be adverse and will particularly affect the world's people living in poverty and otherwise vulnerable populations. These health impacts will occur largely by exacerbating existing health problems. This will impede efforts to reduce longstanding public health impacts, particularly in low-income countries. Climate change will also confer some benefits to health in some populations, at least in the early stage of the process.

* There is broad agreement among researchers that, as climate change progresses, indirect effects on health are likely to account for the largest population health burden, compared to events with direct impacts such as heat waves. These indirect effects include the risks of malnutrition, altered patterns of infectious diseases, and the consequences of conflict, disruptions, and displacement due to climate-exacerbated resource shortages.

* The WHO's estimate of the burden of disease attributable to incipient climate change in the year 2000 identified malnutrition as the preeminent component of health loss. Most of that loss (i.e., premature deaths, stunting, and susceptibility to infection) was in young children in developing countries.

* The existing systems for managing public health problems should provide a foundation for CAP dealing with most of the health impacts of climate change. Therefore, it is likely that most of the required public health adaptations will entail incremental and complementary changes in those systems.

* Actions taken by governments and communities to reduce emissions should, in general, confer localized health benefits as a bonus (i.e., "health co-benefits"). This adds further incentive to undertake those mitigation actions.

4.6.2   Background

The science of climate change, including the influence of human use of energy, is now robust (see Chapter 3) and provides a base for anticipating and estimating human health consequences. Climate change affects social institutions, economic activities, and human health both directly and indirectly. Many major health problems in populations around the world, particularly in the poorer and more geographically vulnerable regions, are climate-sensitive. Hence, a change in climatic conditions will inevitably affect the rates and patterns of those diseases and disorders, including diarrheal disease, malaria, and under-nutrition – all

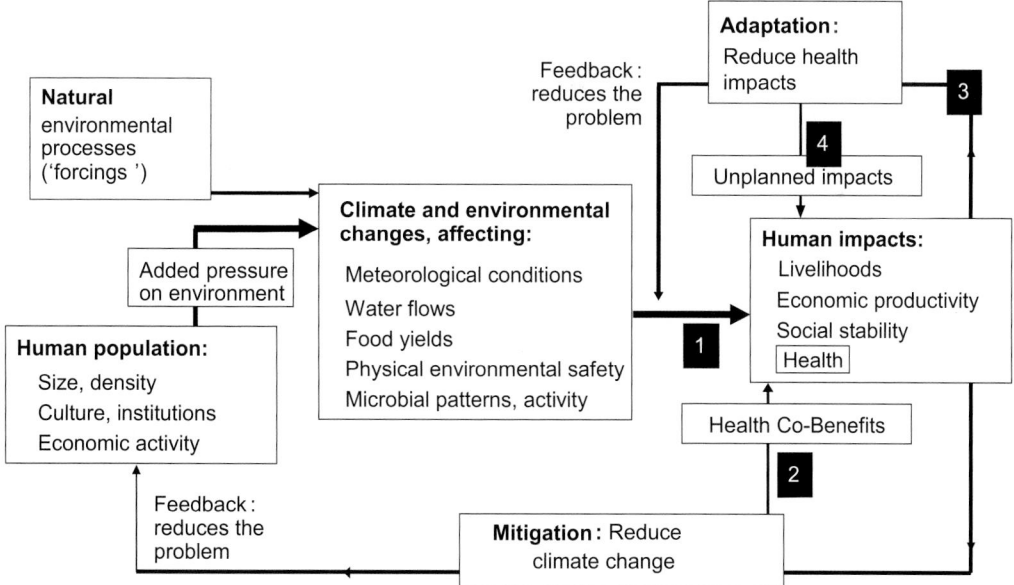

Figure 4.9 | Major pathways leading to impacts on population health associated with climate change and social responses. The four numbered paths refer to major, distinct, areas of research enquiry: 1 = identifying and estimating risks to health from climate change; 2 = estimating the collateral health benefits to local populations as they take action to mitigate climate change (mostly to reduce greenhouse emissions); 3 = research on the effectiveness and equity of adaptive strategies to protect population health against adverse climate change impacts; 4 = monitoring and estimating the collateral impacts of adaptive strategies on (mostly local) population health.

major killers of infants and young children in most low-income countries (Akachi et al., 2009).

Substantial inequalities in material and social conditions, and hence in health status and life expectancy, persist between subgroups, national populations, and geographic regions (Commission for the Social Determinants of Health, 2008). Many of these existing health inequities will be exacerbated by the environmental and social consequences of climate change. Climate-related exposures often impinge with differing intensity between locations (Kesavan and Swaminathan, 2006), and the resultant variations in adverse impacts on well-being, health, and survival will often be compounded by differences in human and financial resources, coping capacity, and social resilience.

The Fourth Assessment Report of the Intergovernmental Panel on Climate Change (IPCC) summarized the published scientific evidence, both direct and inferential, of current and future risks to human health from climate change (Confalonieri et al., 2007). It is not yet possible to confirm and quantify all of the many potential impacts of climate change on health. Those with limited available evidence include the risks to health from heightened (especially local/regional) food shortage and food insecurity, from sea level rise, and from population displacement and conflict. Further, climate change may cause widespread economic disruption, leading to severe health consequences – again, especially in poor and vulnerable populations (see also Costello et al., 2009).

Emerging evidence strongly suggests that some health impacts that are reasonably attributable to climate change have already occurred (Costello et al., 2009; McMichael and Bertollini, 2011). This evidence includes reports of an uptrend over recent decades in deaths, injuries, and other adverse health impacts from cyclones, storms, wildfires, and flooding; an increase in annual deaths from heat waves in several countries; shifts in the range and seasonality of some climate-sensitive infectious diseases; adverse mental health consequences in farming communities affected by drying; and impairment of food yields (and hence increased risk of malnutrition) in some already food-insecure populations. These are discussed in more detail below.

Recognition of the risks to human health strengthens the rationale for the rapid abatement of human-driven climate change (see Section 4.6). That is, those risks, properly understood, signal that climate change is beginning to weaken and disrupt nature's life support systems – the systems that are prerequisite for the attaining and sustaining of high levels of human population health (Raven, 2002; McMichael, 2009a). Meanwhile, appreciation of the range and likely ubiquity of these health risks provides an evidence base for adaptive interventions to protect population health against those climate change-related risks that already exist or are apparently unavoidable, given the as-yet-unrealized climate change from extant emissions.

4.6.3   Diverse Health Risks; Multiple Causal Pathways

Weather variations and changes in climatic conditions affect human well-being, safety, health, and survival in many ways (see Figure 4.9). Some health impacts are direct-acting, familiar, and easily understood in relation to future climate change. This category includes the diverse health consequences of weather disasters (which will typically increase in frequency and intensity under climate change) and the health consequences of heat extremes within both the general community and segments of the workforce. Other health impacts are less immediate, occur via more complex pathways, and may be less obvious and more difficult to attribute (Butler and Harley, 2010). This category includes the changes in patterns of various infectious diseases, especially those transmitted via vector organisms, and the nutritional consequences of impairments in regional food availability, in part due to alterations in pests, aflatoxin contamination, and crop diseases (Butler, 2010). Climate change also harms ocean acidity (Doney et al., 2009), phytoplankton production (Boyce et al., 2010), and corals (Hoegh-Guldberg and Bruno, 2010).

Further, since climate change is part of a larger set of contemporary, human-induced global environmental changes, most of them of unprecedented scale, many of the health impacts of climate change will be modulated, and often amplified, by these other coexistent environmental changes. A ready example is that of food yields and human nutrition. In many parts of the world, the productivity of food systems is being jeopardized by the combination of soil exhaustion, chemicalization, diminished supplies of fresh water, local losses of biodiversity (e.g., pollinating organisms), coastal salination (rising sea level), and climate change (McIntyre et al., 2007).

Climate change entails a complex of environmental, ecological, and social changes. It is therefore not some new and distinctive "risk factor," nor is it likely to generate *new* diseases or health disorders. Instead, via that complex of changes, many existing population health problems will be exacerbated. It is important, in a policy context, to note that this threat of exacerbation heightens the already strong practical and moral rationale for reducing pre-existing disease rates as quickly as possible – something that the world is striving to achieve via the MDGs. An obvious focus of efforts to protect population health against further incursions by climate change is that of childhood malnutrition and infection, especially acute respiratory infection and diarrheal diseases, which currently kill around nine million children annually.

The following subsections consider first the health impacts of climate change that occur via apparently direct, relatively simple, pathways, followed by a discussion of those health impacts that arise via less direct, and often diffuse and deferred, causal pathways.

4.6.4   Direct Health Impacts: Temperature Extremes, Weather Disasters, and Health Impacts

Heat waves (extreme temperature events relative to local average, and of, typically, around three to seven days duration) will become more frequent and more severe as background temperatures rise and weather variability increases. It is well established that heat waves can kill people and increase the incidence of heart attacks and strokes, especially in the elderly, the frail, and those with underlying chronic diseases (Kovats and Hajat, 2008).

Temperature extremes that exceed physiological coping capacity also affect bodily functioning, mood, and behavior (Kjellstrom, 2009). This usually happens at temperatures above 30°C, depending on humidity, wind movement, and heat radiation. This has particular relevance to segments of the workforce exposed to extreme thermal stress. Poor people without access to household or workplace cooling devices are likely to be most affected, in hot, high-income countries as well as in hot, low- and middle-income countries. This climate change impact will primarily affect adults 20–60 years of age. Heat also affects other daily physical activities that are unrelated to work for all age groups.

Physically active workers in low- and middle-income tropical countries are particularly vulnerable, since many are engaged in heavy physical work, outdoors or indoors, without effective cooling. If high work intensity is maintained in excessively hot workplaces, serious health effects can occur, including heat stroke, organ damage, and death. Meanwhile, depending on the type of occupation, the required work intensity, and the level of heat stress, as temperature extremes rise over time, working people will need to work more slowly to reduce internal body heat production and the risk of heat stroke. Hence, without preventive interventions to reduce heat stress on workers, both their individual health and economic productivity will be impaired (Kjellstrom et al., 2009).

Hotter weather also increases the formation and concentration of various noxious ambient air pollutants, especially ozone, in large, motorized, industrial cities. Ozone is a well-established risk factor for respiratory tract damage and cardiovascular disease. Much of the huge mortality excess caused by the August 2003 heat wave in Western Europe may have been due to the coexistent high levels of ozone in and around some of the big cities – high levels that, indeed, may have been partly attributable to the unusually high temperatures (Dear et al., 2005; Kinney, 2008).

Weather disasters – such as the cyclones that have struck vulnerable, mostly poor, coastal populations of Myanmar, Haiti, and Viet Nam in recent years – kill, injure, dispossess, impoverish, and cause mental health burdens. They also predispose people to outbreaks of infectious diseases and cause damage to crops. Among vulnerable populations, droughts variously cause hunger, loss of farming jobs, impoverishment, population displacement, and misery; suicide rates may rise in response to these hardships.

4.6.5   Direct Health Impacts: Sea Level Rise

The average global rate of sea level rise is now approximately 3 mm/year (Rahmstorf et al., 2007; Allison et al., 2009) – that is, approximately 50% higher than in the 1980s. A growing proportion of the rise is due to the melting of polar ice and (less certainly) alpine glaciers, on top of the contribution from the ongoing thermal expansion of ocean water. Increasingly, scientists expect that the sea level may rise by a meter or more this century (Hansen, 2007; Allison et al., 2009) – a figure that is noticeably higher than the cautious forecast contained in the IPCC's Fourth Assessment Report in 2007 (IPCC, 2007b). Well over 20 million people along the vulnerable coastal regions of Bangladesh, Egypt, and Nigeria are estimated to be at risk of inundation from a one-meter sea level rise, without also factoring in the inevitable population growth (Perch-Nielsen et al., 2008).

Sea level rise is already endangering food yields, freshwater supplies, and physical safety in several low-lying small island states (Barnett and Adger, 2003; Kelman, 2006; White et al., 2007). In some cases, these problems are exacerbated by high population density and continuing high fertility rates (Ware, 2005). India, with its low-lying and densely populated coastline extending for over 7000 km, is highly vulnerable to sea level rise. This includes the threat of inundation and salinization of many coastal paddy fields (Kesavan and Swaminathan, 2006). Various less direct risks to health (physical and psychological) will result from likely displacement of populations and the breakup of families and communities due to sea level rise.

4.6.6 Indirect Pathways

The health of human populations depends fundamentally on various environmental and ecological systems and processes that determine the cycling and flow of water, food yields, and the natural geographic and seasonal constraints on infectious agents. As human-induced climate change progresses, these essentially indirect pathways are likely to account for an increasing and substantial proportion of the population health burden attributable to climate change (Confalonieri et al., 2007; Costello et al., 2009; Butler and Harley, 2010). These impacts will occur particularly via increased malnutrition, altered range and seasonality of infectious disease, and the consequences of conflict, social disruptions, and displacement due to resource shortages.

4.6.6.1 Water Insecurity

Climate change will exacerbate water insecurity in many regions, via alterations in the seasonality and intensity of rainfall, via geographic shifts in rainfall systems, and by increased rates of evaporation. As is evident in the paleoclimatic record, global warming displaces rainfall systems in subtropical regions towards the poles. Such displacement appears to have emerged recently in southern Australia, southern Africa, southern Canada, southern Spain, and Italy, for example. Further, stratospheric ozone depletion, as a concomitant global environmental change, also contributes to such displacement of rainfall systems (Cai, 2006).

By 2050, the projected changes, from medium-to-high scenarios of climate change, in rainfall, surface runoff, and depleted flows from reduced glacier masses will expose an estimated additional 2–3 billion people to severe water stress (IPCC, 2007a). Water scarcity poses multiple risks to health, including water-borne infectious diseases (e.g., cholera, other diarrheal organisms, cryptosporidium, etc.), vector-borne diseases associated with water storage, exposure to higher concentrations of salt and chemical contaminants in water, and impaired food yields. Sufficient water also reduces the risk of water-washed diseases, including diarrhea (Curtis and Cairncross, 2003), trachoma (Mecaskey et al., 2003), and scabies. Water stress seriously constrains sustainable development, particularly in savannah regions, which represent two-fifths of the world's land area. It may also lead to conflict over diminishing supplies, as well as the many adverse health consequences that flow from conflict, property loss, bereavement, and population displacement (McMichael and Bertollini, 2011).

4.6.6.2 Food Yields, Food Insecurity, and Health

Many aspects of climate change and its environmental consequences affect food yields, on land and at sea. Impairment of food yields by climate change, including the impacts of an increase in extreme weather events and heightened risks of infestations and infectious diseases, pose a great threat to population health. Meanwhile, food insecurity persists widely, and appears to have increased in absolute numbers over the past decade. In 2009, over one billion persons were classified as undernourished (FAO, 2009), representing around a one-fifth increase on the corresponding estimate at the start of the decade (Butler, 2009).

During 2008, as food prices escalated and shortages emerged, concerns were expressed about contributory climatic influences on food yields (Sheeran, 2008). Indeed, climate change is increasingly viewed as a likely contributor to altered food yields in many regions (Lobell and Field, 2007; Lobell et al., 2008; Battisti and Naylor, 2009). This superimposed threat to food supplies from climate change looms greatest in food-insecure regions where high levels of malnutrition and child stunting already exist (Lobell and Field, 2007).

Modeling studies consistently project that climate change will, overall, have a negative impact on global food yields (Nelson et al., 2009). However, those impacts will occur unevenly; some temperate regions may benefit, particularly where the soil is fertile. In general, countries in the tropics and subtropics, where both warming and reduced rainfall are likely to occur, are at greatest risk. Many studies indicate that South Asia is particularly vulnerable, and likely to experience declines in total cereal grain yields of the order of 10–20% by later this century (Fischer et al., 2005). In temperate regions and at high latitudes, agricultural productivity could initially increase, but elsewhere smallholder and subsistence farmers, especially in the tropics, are at particular risk (Morton, 2007).

Many such model-based estimates are likely to be conservative, particularly since they are unable to take into account the episodic, perhaps "surprise," events that will greatly increase the damage to yields and harvests. Extreme weather events (storms, floods, fires, etc.) can wreak disaster. This happened in 2010 with the extreme wildfires in Russia, which, in combination with preceding prolonged drought, resulted in the loss of almost one-third of the nation's annual wheat crop. Climatic conditions also affect the probability of damage by plant and livestock pests and pathogens. The northern movement of the blue-tongue virus in Southern Europe, in association with warming over the past decade, has extended the region of an economically catastrophic threat to cattle and sheep populations (Purse et al., 2005). Warming affects a range of pests; flooding favors fungal growth, whereas drought encourages aphids, whiteflies, and locusts.

It is clear from the above that estimating the current and future contributions of climate change to local or regional food availability per capita is necessarily an inexact and incomplete science. There are further methodological difficulties in translating food availability into an estimation of the actual burden of disease and functional impairment (including intellectual development in children) due to malnutrition. One major reason is the considerable variation in the role of climatic conditions as direct determinants of food availability, as well as other factors in the causation of stunting and wasting, such as diarrhea, tropical enteropathy (Humphrey, 2009), and other infections.

In remote and food-insecure regions, climate-related downturns in yields of crops and pastures can quickly result in hunger, under-nutrition, starvation and, on occasion, conflict. For example, a field assessment in western Sudan by the United Nations Environment Programme (UNEP) concluded that tensions between traditional farmers and nomadic herders over declining pasture and evaporating water holes, during a protracted period of declining rainfall, may have contributed to recent conflicts (UNEP, 2007). Inevitably, the cause of such conflict is complex (McMichael, 2009a). Other authors have noted the lag of several decades between a stepped decline in rainfall in the region and the outbreak of conflict, and have assigned a primary causal role in the region discrimination against the population of Darfur by the central Sudanese government (Kevane and Gray, 2008).

Concerns over the risk of conflict due to the environmental consequences of climate change should be set against other evidence indicating that genuine cooperation can eliminate conflict (Salehyan, 2008). Indeed, care should be taken to ensure that the implication of environmental factors is not used to evade political responsibility for the occurrence of conflict (Butler, 2007). Nevertheless, many examples from human history suggest that the long-term prospect of conflict arising in part from resource insecurity, both real and perceived, is genuine – especially when a tipping point is exceeded. Unabated climate change will reduce many kinds of resources, including food, fertile coastal land, and habitable areas within low-lying cities. The nurturing of protective human institutions to reduce this threat is vital – but so, too, are the

technological developments and the forms of social reorganization that are required to accelerate the energy and sustainability transitions (Walker et al., 2009).

4.6.6.3 Infectious Diseases

Many infectious diseases – whether food-borne, water-borne, or vector-borne – are sensitive to climatic conditions (Dobson, 2009; Wilson, 2009). Various combinations of warmer temperatures, increased rainfall (affecting both humidity and surface water bodies and flows), changes in wind patterns, altered profiles of vegetation, and the aftermath of weather disasters can affect a wide variety of infectious disease agents, vectors, and the pathogen's intermediate (non-human) host species.

The rate at which bacteria multiply in food and in nutrient-loaded water increases as temperature rises. Research in Australia and elsewhere has shown a clear positive relationship between weekly or monthly temperatures and the incidence of (reported) salmonella food poisoning (D'Souza et al., 2004; Kovats et al., 2004). Changes in rainfall can affect local flooding, well-water quality, river flows and general sanitary conditions, thereby affecting the spread of diarrheal diseases, including cholera. Research based on extensive systematic historical records in southern India shows that cholera outbreaks occur most often either when conditions are very dry (and vibrio bacteria are therefore concentrated) or during times of flooding and crowding when person-to-person contact increases and hygienic conditions deteriorate (Ruiz-Moreno et al., 2007). Furthermore, especially in temperate and tropical areas, flooding can lead to increased moulds in and around dwellings, and hence additional respiratory illnesses such as asthma (Rao et al., 2007).

The range and seasonality of many vector-borne infections (diseases transmitted by mosquitoes, other insects, or rodents) are sensitive to temperature, rainfall, humidity, and wind. Pathogens incubating within mosquitoes (e.g., the malaria plasmodium and dengue virus) mature and replicate more rapidly as temperatures rise (within limits – excessive heat is harmful to insects, as it is to humans); and mosquitoes feed more frequently (McMichael, 2009a). Minimum temperature is often particularly important, often as a threshold. As predicted by climate change modeling, minimum daily temperature has been rising more rapidly around the world in recent decades than has maximum daily temperature.

Surface water patterns influence mosquito breeding; humidity and temperature affect mosquito survival. Modeling vector-borne disease transmission shows that a small temperature rise can substantially increase transmission probability. Hence, where the geographic range of malaria has increased alongside gains in climate-determined suitability the geographic range of for transmission – as in parts of eastern Africa recently – it is reasonable to assume that the climate trend partly explains the observed increases in disease (Pascual and Bouma, 2009).

The occurrence of infectious diseases (many of them vector-borne) that spill over sporadically into human populations from animal sources – that is, 'zoonotic' infections – is often influenced by climate-related changes in the population density and geographic range of the "reservoir" animal species (Harley et al., 2011). Examples include: West Nile Fever (now in the United States and Canada: host animal species are birds), bubonic plague (sub-Saharan Africa, western USA, Central America: wild rodent species), Lyme disease (south-eastern Canada: deer), Ross River Virus (Australia: kangaroos and wallabies), and Rift Valley Fever (Kenya: cattle).

In recent years, the geographic range of some vector-borne infectious diseases appears to have increased in association with documented regional warming (Alonso et al., 2011; McMichael and Lindgren, 2011). The explanation for any one, disease-specific observation is usually contestable. However, the overall pattern, across multiple settings and continents, indicates the likely emergence of a climate "signal." This set of changes includes shifts in patterns of occurrence of malaria in some eastern African highlands, tick-borne encephalitis in Sweden and the Czech Republic, and Lyme disease in Canada.

Recent warming in eastern China has been accompanied by the northwards extension of the temperature-limited transmission zone for schistosomiasis (for which the temperature-sensitive water snail is intermediate host), putting an estimated 21 million more persons at risk (Zhou et al., 2008).The first reported outbreaks of dengue in the Himalayan foothills have occurred recently – in Bhutan in 2004 and in Nepal in 2006 (TDR, 2009). Meanwhile, in Europe, there has been recent evidence of warming-induced northwards extensions of the bluetongue virus disease in livestock, and of its midge vector (Purse et al., 2005). In 2006, a further northerly extension occurred, with an outbreak 900 km north of its previous limit (Carpenter et al., 2009).

4.6.7 Social, Economic, and Cultural Disruption and Health

The disruptive effects of climate change on social conditions, relationships, and economic circumstances will affect human health and well-being in many ways. For example, adverse impacts on freshwater supplies and food yields can cause social and economic disruptions. These, in turn, may result in the displacement of people, which then often leads to various health risks, including malnutrition, mental health problems, and exposures to infectious diseases.

These direct and indirect risks to health may be further amplified by accompanying changes to other health-related behaviors, such as the (economic) need to resort to transactional sex, or by increases in the consumption of tobacco, alcohol, and perhaps illicit drugs. An illustration of the complex and sometimes extended pathways by which climate change may increase the burden of disease is in relation to HIV/AIDS (see Figure 4.10).

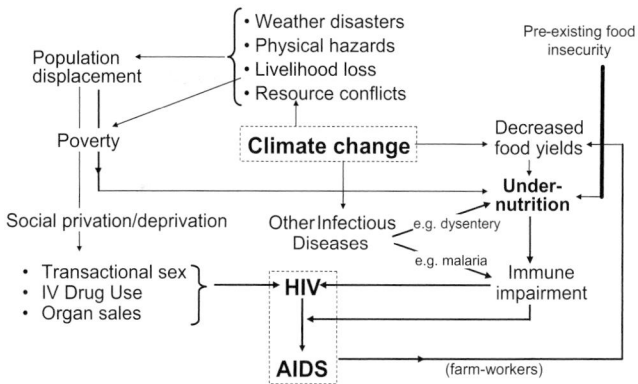

Figure 4.10 | Multiple pathways for impacts of climate change on risks of HIV/AIDS.

This category is also well illustrated by the impacts of the relatively rapid ongoing warming in the Arctic region. The resultant loss of sea ice and permafrost is already adding further stress to already disturbed traditions of living, hunting, and eating patterns in the Inuit communities of northern Canada. One consequence has been a reduction in physical activity, an increased reliance on imported energy-dense processed foods, and hence potentially amplified risks of obesity, cardiovascular disease, and diabetes (Nickels et al., 2005). The mental health consequences of these social and cultural disruptions, and of associated perceptions of future threats, pose an increasingly important risk to health.

4.6.7.1 Population Displacement

In addition to the above-mentioned general health risks associated with population displacement, changes in climate will often intensify the stresses and health risks of displacement and may erode environmental resources at the site of resettlement.

Estimates vary as to the likely number of persons displaced by climate change, with mid-century totals often of the order of several hundred million (McMichael and Bertollini, 2011). Much climate change-related displacement is likely to occur in developing regions where public health resources are lacking or inadequate (Carballo et al., 2008). Increased flooding, water shortages, and drought are likely to amplify rural-to-urban migration in many such countries, yet many poor urban communities are situated in parts of cities that are themselves at high risk for climate change impacts. Hence, people migrating into these settings may face continued environmental and, hence, health threats.

4.6.8 Burden of Human Disease Attributable to Climate Change

The task of risk assessment and risk estimation, in relation to either current or future climate change, is a very heterogeneous one; it depends on type of health outcome, type of population, and the time of interest (present or future). Some estimations are relatively straightforward; for example, for a given plausible temperature increase, and with no future change in population biomedical profile nor in adaptive circumstances, how will annual age-specific death rates from heat waves change?

Identifying and quantifying the climate-attributable impact on various other specified population health outcomes (e.g., child stunting due to malnutrition) are inherently difficult, especially at the relatively earlier stages of climate change. The often diverse, coexistent non-climate influences, upon the same health outcomes, of cultural attributes, behaviors, and exposures can easily obscure any early real signal of a climate-related health impact. Child stunting is, for example, influenced by many factors other than agricultural production or even food intake. These include co-infection with parasites, caloric demand due to ambient temperature and workload, and the frequency of other infections, including malaria, diarrhea, and tropical enteropathy (Humphrey, 2009).

Another area of difficulty relates to estimating how and how much climate change affects mental health – a health risk in many of Australia's rural communities, for example, who face increased warming, drying, and fires; declining incomes; and possible future relocation. Similarly, inhabitants of low-lying coastal areas and many small island states live with the growing insecurity, somewhat analogous to people who live in river valleys, that they are destined for future inundation (Jackson and Sleigh, 2000).

It is possible, however, to estimate the health impacts of climate change via statistical modeling, given sufficient prior knowledge of exposure-response relationships. Indeed, such an exercise was carried out in the early 2000s as part of a systematic international CRA project. That project estimated and compared portions of disease burdens (globally and regionally) attributable to a nominated list of major risk factors, including climate change. The estimation for climate change (McMichael et al., 2004) was necessarily limited to just the few health outcomes for which there was sufficient direct epidemiological evidence (from at least several geographic regions for each outcome) to derive an estimated average exposure-response relationship. The five such conditions were: diarrheal disease, malaria, malnutrition, deaths due to flooding, and (in developed countries) cardiovascular events and deaths due to thermal stress.

The modeling of attributable disease burdens from those specific health outcomes was carried out both for the year 2000 and for various future years (including 2020 and 2030). The best estimate of the average exposure-response relationship was derived from published literature. That relationship was then applied, using the estimated region-specific level of climate change "exposure" for 2000, and 2030 (from global climate models), to the project-generated estimates of current and future disease burdens from the health outcomes of interest. The modeling thus yielded estimates, at both the global level and by geographic region, of burdens of (chronic/disabling) disease and premature mortality attributable to climate

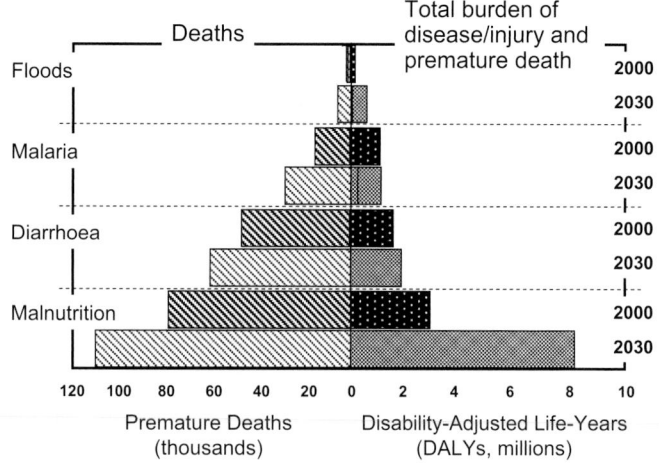

Figure 4.11 | Deaths and total burden of disease/disability/death attributable to climate change for the four selected health outcomes shown, in the years 2000 and (projected) 2030. Climate change is assessed relative to the average (reference) climate during 1961–1990. Estimations were done by major geographic region, in relation to region-specific estimates of climate change (2000 and 2030) and then summed. Estimates for 2030 were based on the projected region-specific disease and death rates for the year 2000. Source: data based on McMichael et al., 2004.

change that had already occurred by 2000 (or was projected to occur). These were expressed and summed as total loss of DALYs.[13]

The main results for the global level are summarized for 2000 and 2030 in Figure 4.11. The exercise yielded a modest estimate of the climate change-attributable burden of disease for the year 2000 (~0.4% of DALYs; 0.3% of deaths). Inevitably, that small figure reflected the marginal change in world climate over the course of the preceding several decades. In addition, only a few health outcomes that also had great relevance to the developing world (where the brunt of the impacts from malnutrition, infection, and flooding would be expected) could be included. An estimated 88% of the climate change-attributable loss of DALYs worldwide (and more than 85% of deaths) was in children under age five in poor countries. In 2000, the climate change-attributable burden of disease was estimated to increase by around 35%, compared to 2000, with a global population increase of only 25%. Malnutrition was considered to be responsible for the largest fraction of this burden, and also to have the greatest proportional increase. Beyond that time, population growth will further slow but the impact of climate change is likely to be disproportionately more, perhaps 40–50% greater than in 2000. Many synergies are likely; for example, heavier rainfall events that cause flooding, drowning, injury, and psychological trauma also contribute to diarrhea, reduced harvests (and resultant food prices rises), social disruption, and outbreaks of vector-borne diseases, including malaria, in some locations. Poor communities are likely to have the highest burden of disease, but high-income countries will not be fully spared.

---

13 The DALY was developed for this international assessment as a universally applicable metric enabling the project to provide the first-ever systematic assessment of disease burdens, within which comparisons could be made across risk factors, ages, sexes, and regions.

Box 4.3 | Mitigation and adaptation actions: Implications for health risks

If mitigation is to *avoid the unmanageable* (the types of changes and risks to health that would be seriously different, worse and perhaps uncontrollable, compared to current conditions), then adaptation is to *manage the unavoidable* risks to health (i.e., those that have already begun to occur, that are impending and probably unstoppable as further climate change occurs, or are fully realized with the physical, biological, and social environments).

Reducing emissions of climate-active pollutants (i.e., "mitigation") is the first-order task. Many mitigation actions will, in their own right, affect the health of localized populations; encouragingly, most of those consequences should be beneficial, and they are thus sometimes called "co-benefits" (see Section 4.10).

Adaptive strategies are also needed, however, given that no feasible mitigation measures can fully prevent human-caused climate change. Some strategies will entail improving or extending existing policies and practices, such as mosquito control, better sanitation, nutritional supplementation programs, and stronger flood controls. Others will require innovative actions, such as community-wide early warning systems for impending weather extremes (e.g., heat waves, storms, and cyclones), crop substitution, water harvesting, and (if feasible) genetic modification of mosquito populations (McMichael, 2009a).

Unless international cooperation and resultant mitigation actions increase markedly, the need for adaptive responses to lessen the impacts on health will stretch into the distant future. However, any such extended and prolonged reliance on adaptation – particularly if seen as a substitute for mitigation – would increase the moral dilemma, since adaptation will, in general, be more achievable and affordable in the higher-income countries (McMichael, 2009a). To constrain the risks to health (and to other environmental and social assets), to do so in the framework of *sustainability*, and to facilitate the achievement of a more equitable world, mitigation must remain the centerpiece of national and international climate policy.

The above comparative risk assessment exercise is almost a decade old. It may well be that its estimates for future health impacts are too conservative. For example, world food prices reached a fresh peak in 2010. A rising trend of extreme weather events, contributed to by climate change, appears to be a major cause of this peak. Higher food prices inevitably harm the nutrition and health of the poorest on a very wide scale.

4.6.9   Differences in Vulnerability

The adverse health impacts of climate change today primarily affect low-income, poorly resourced, and geographically vulnerable populations – just those who have not contributed historically to GHG emissions (Patz et al., 2007). This pattern will persist as climate change intensifies (Confalonieri et al., 2007; Costello et al., 2009; McMichael and Bertollini, 2011). Many of the low-income countries in tropical and subtropical regions, especially their poor slum-dwelling populations, will be at particular risk. Bangladesh, for example, is vulnerable on multiple counts: widespread poverty and food insecurity, high rates of infectious diseases associated with tropical climates and with poverty and crowding, a low-lying coastal population vulnerable to cyclones and storm surges (Mitchell et al., 2006), and threats to river water flows (Himalayan glacier retreat and the probability of upstream damming and diversion of rivers by China and India).

Higher-income countries also can have marked differences in vulnerability. In the United States, for example, the impacts of the 1995 heat wave in Chicago (Semenza et al., 1996) and the 2005 Hurricane Katrina in New Orleans differed markedly between ethnic and socio-economic groups. In Australia, the population groups considered to be particularly vulnerable to climate change include (McMichael et al., 2009):

- rural communities exposed to long-term drying conditions;
- older and frailer persons, especially in relation to heat waves;
- coastal communities facing storm surges, coastal erosion, and greater risks from cyclones;
- remote indigenous communities facing heat, drying, water shortages and loss of traditional food species; and
- persons living in regions where climate-sensitive infectious diseases may tend to spread, including communities in northern Australia with greater exposure to mosquito-borne infections.

In principle, for reasons of both moral obligation and achievable social benefit, adaptive strategies should be weighted towards high-vulnerability groups and subpopulations. In practice, such interventions could take many forms, depending on the scale (global, regional, local), the type of health risk, the timeframe, and the resources available. This has implications for the ways in which intervention options are selected, and the level of specificity at which they are evaluated.

Meanwhile, human-induced climate change is already occurring and will increase over the coming decades, no matter what action humankind takes today. Therefore, adaptation is required, at least transitionally, to reduce otherwise unavoidable adverse impacts on health. This will be particularly important in populations where underlying disease rates (e.g., child diarrhea, malnutrition) are already high and will therefore rise further due to multiplier effects caused by a change in climate (Smith and Desai, 2002).

Many adaptive strategies to reduce health risks require planning and coordinated action across diverse research disciplines, sectors of government, and community interests. This collaboration extends beyond the skills and experience of the formal health sector – and reflects the fact that a society's ways of building, moving, living, producing, sharing, and consuming are the prime determinants of population health. The typical 'health' sector plays a more limited, focused, and mostly reactive role to states of health and disease. However, in addition to joining the above collaboration, there remains a major distinct role for the health sector in adaptation strategy. Institutions (hospitals, ambulance services, etc.) will need added capacity to cope

with impacts of more severe weather disasters. Health professional education and training should incorporate new learning about climate-related health risks likely to apply within any particular population. Public health programs such as vaccination, child nutrition, sanitation, and community education will require further fortification. Primary health care, as the front line of the health, will be important in educating, counseling, and detecting changing health outcome patterns (Smith, 2008).

4.6.10 Climate Change and Health: Attuning Public Health Capacities

Climate change will affect human health by modulating (mostly intensifying and extending) existing climate-related risks to health. Novel health outcomes are unlikely. Therefore, the existing social and public health systems should provide a foundation for the strengthening of adaptive strategies, via incremental and complementary changes in existing health risk management programs (Ebi et al., 2006; McMichael, 2009b). Such changes also include:

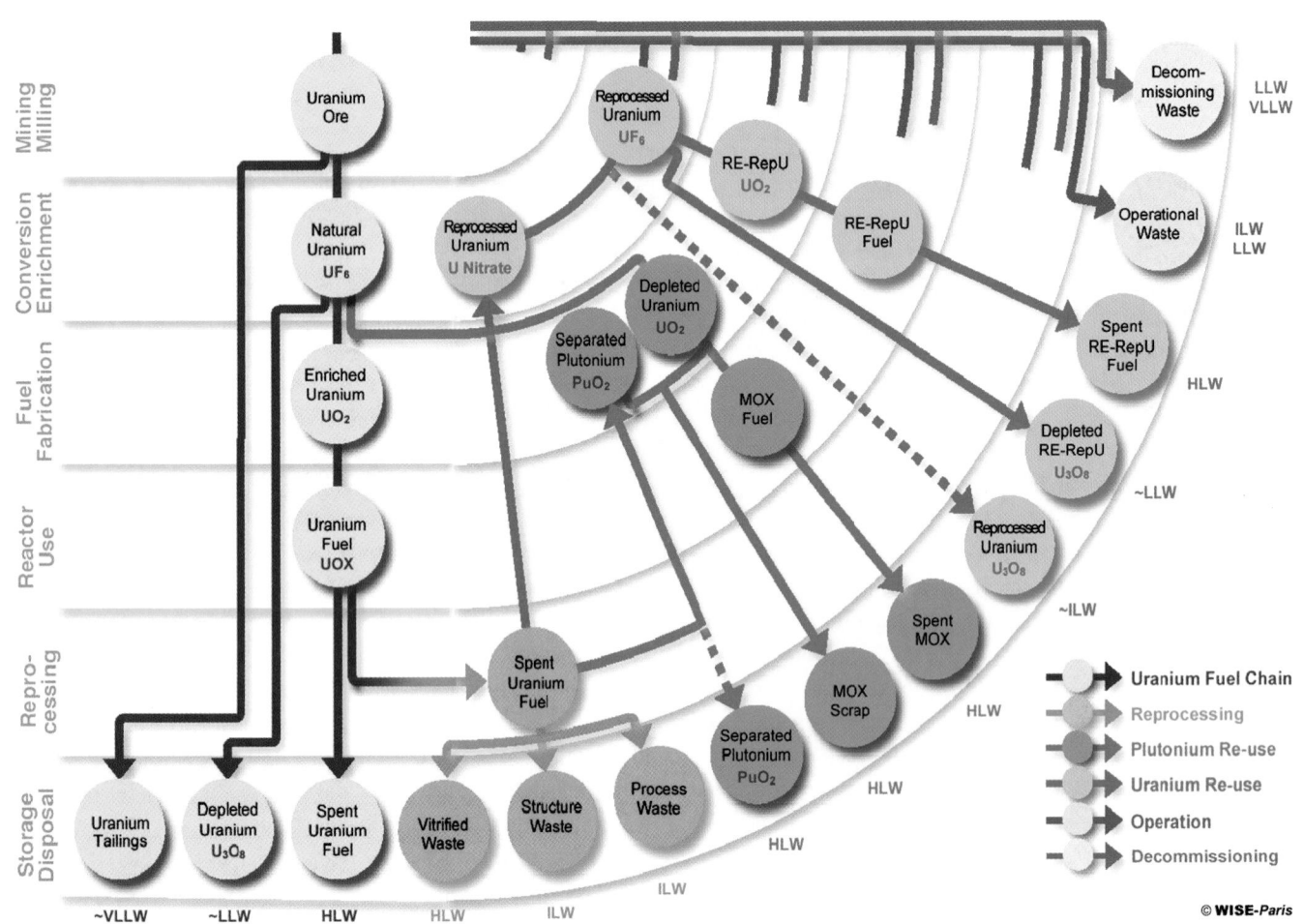

Figure 4.12 | The nuclear fuel chain. Source: WISE-Paris.

- Orienting the existing structure, staffing, knowledge base and awareness of the health system and its infrastructure to the altered and impending risks to health needs (Jackson and Shields, 2008)

- Translation of policies and knowledge from other countries or regions, to address changes in the geographic range of some health risks and diseases

- Restoration and strengthening of surveillance, maintenance and prevention programs that may have been neglected or abandoned

- Development of new policies and strategies to address any apparently new threats to health

4.7 Routine Health Impacts of the Nuclear Power Fuel Chain

4.7.1 Key Messages

- Based on current evidence, the main health effect from routine radiation releases from nuclear power facilities is the increased incidence of childhood leukemia – a rare disease. The resulting burden of disease is small compared to routine operation of fossil fuel facilities, but child cancer garners considerable public and policy concern.

- Radioactive releases from reprocessing plants are greater than from reactors. In addition, concerns exist about the significant environmental effects of uranium mining and milling which are often excluded from assessments of nuclear power.

- Average radiation doses to full-time permanent workers in nuclear power industries have generally declined over the past two decades. An increasing share of the collective dose is received by the workers of subcontractors and operators of fuel chain facilities.

- As with large hydroelectric facilities, there are difficult-to-quantify risks due to high-consequence low-probability accidents that affect workers and the public, which are also difficult to compare with day-to-day impacts (see Chapter 14).

4.7.2 Overview

The nuclear fuel supply system, from uranium mining to radioactive waste disposal, is frequently called the nuclear fuel "cycle." A strategy that leads to the direct disposal of spent fuel is referred to as "open" or "once-through," while a reprocessing strategy that separates plutonium and uranium from fission products and other wastes is called a "closed fuel cycle."

The use of such terminology suggests a circular system that confines most of the materials. In fact, each step in the system leads to specific radioactive and chemical emissions and specific new waste streams. A more accurate term would be the nuclear fuel "chain." In a typical light water reactor, with a direct disposal scheme (i.e., no reprocessing), the uranium is mined and milled before it is converted into uranium hexafluoride (UF6). In this gaseous form, the percentage of the fissile isotope U-235 is increased by diffusion or centrifugation. It is then converted back to uranium oxide, shaped into fuel pellets, and inserted into tubes in nuclear fuel assemblies. After the fuel has undergone fission in a reactor for three to four years, it becomes "spent" and is put into cooling ponds, in most cases, for at least 10 to 20 years. The term "spent" is slightly misleading, as this fuel remains extremely radioactive and dangerous for hundreds of years. The ultimate fate of spent fuel is a vexed issue and remains unresolved at this date. In many countries, spent fuel is being left in cooling ponds for increasingly lengthy periods due to the absence of technical solutions and to political opposition to plans for nuclear waste facilities. This unsafe and unsatisfactory situation poses considerable challenges for future nuclear waste management.

The decision to separate plutonium and uranium from spent fuel in reprocessing plants greatly increases the complexity of the system and its impact on the environment and health (see Figure 4.12). For these reasons, most future plans for nuclear power do not include reprocessing. (See Chapter 14 for more discussion.)

4.7.3 The Nuclear Fuel Chain

This assessment of the health impact of nuclear power systems distinguishes between effects on workers and the public. Average doses to workers in nuclear countries have generally declined over the past two decades. However, a larger share of the collective dose is received by workers of nuclear subcontractors and operators of fuel chain facilities. In France, for example, in 2008 about 31% of the nuclear workers under radiological surveillance belonged to subcontractor companies, but they received around 49% of the collective dose (calculation based on IRSN (2009)).

In 2005, a retrospective study (Cardis et al., 2005) on over 400,000 nuclear industry workers in 15 countries, the largest study of nuclear workers ever conducted, concluded that there were increased risks of solid cancers and leukemia, "even at the low doses and dose rates typically received by nuclear workers in this study" (Cardis et al., 2005). This finding is consistent with other studies of radiation health effects at low doses (NRC, 2006).

4.7.4 Uranium Mining and Milling

Canada, as the world's largest producer of uranium, illustrates issues in these stages of the nuclear fuel chain. After mining, uranium is separated from its ore by acidic leaching, as occurs in many uranium mines around the world. Because the ore usually contains only about 0.1%

uranium, leaching results in extremely large volumes of mine wastes. Canadian mine companies, for example, had accumulated about 213 million tonnes of uranium mine tailings (LLRWMO, 2004) and 109 million tonnes of waste rock, as of the end of 2003. In addition, this process has generated about 400 million cubic meters of contaminated process water. The resulting tailings, waste water, and runoff from waste rock piles contain over 80% of the initial radioactivity in the ore, and thus constitute serious low-level, but long-lived, radioactive hazards.

In addition, as in other types of ore processing, large volumes of sulfuric acid are used to mobilize heavy metals such as copper, zinc, nickel, and lead, which are highly toxic to aquatic and terrestrial wildlife. Severe contamination of groundwater constitutes a permanent risk. The government health ministry, Health Canada, states: "There is a serious possibility that the food chain can be contaminated unless appropriate mitigation is instituted. Fish, wildlife, vegetation, country foods, and drinking water are all at risk should spills or leakages occur. The need to manage the water from waste management areas is important, particularly if there are drinking water sources in the vicinity" (Health Canada, 2004). However, leakages, mostly small, do occur frequently. In the Elliot Lake area in Ontario, for example, over 30 failures of tailings dams have been reported.[14]

Recent research suggests that the health detriments of ingested and inhaled uranium may be greater than previously estimated, mainly because of the likely synergism between the carcinogenic chemical and radiation effects of uranium exposures (UNIDIR, 2008). There is also evidence that, like many other chemicals, uranium may have an endocrine-disrupting function in the body, implying that those exposed should be followed for increased risk of fertility problems and reproductive cancers (Raymond-Whish et al., 2007). As a result of such evidence, the German government recently decided to lower the regulatory limit to 10 micrograms of uranium per liter of drinking water (Associated Press, 2008).

The most significant health impacts from uranium mining probably result from the release of radioactive radon gas, which is particularly significant in underground uranium mines. Various studies have documented significant excesses in lung cancer among uranium miners (e.g., NRC, 1999; Grosche et al., 2006).

4.7.5   Uranium Conversion and Enrichment

The health risks associated with uranium conversion and enrichment are primarily due to the inhalation or ingestion of uranium in various chemical and physical forms. The uranium concentrate from the milling process (U_3O_8), which is termed yellowcake, is converted into uranium hexafluoride (UF_6), a highly volatile and toxic chemical. Airborne UF_6

immediately reacts with water vapor to form hydrofluoric acid, which is extremely toxic and can cause pulmonary irritation, edema, corrosion of the lining of the lungs, seizures, and even death.

The large stockpiles of depleted uranium accumulated as a byproduct of uranium enrichment, over 700,000 tonnes in the United States alone, pose challenges for future safe management. Depleted uranium is presently stored in ageing UF_6 containers at the sites of the enrichment plants, mostly in the United States. The conversion back to more stable uranium oxide involves the same type of risks as the original conversion from oxide to UF_6.

4.7.6   Nuclear Power Reactors and Nuclear Reprocessing Plants

During normal operation, nuclear power reactors routinely release radioactive gases to the atmosphere and radioactive liquids to the sea or rivers. In addition, when reactors are depressurized prior to being opened for refueling, larger gaseous emissions occur over short time periods. The main radioactive releases are tritium (hydrogen-3, half-life of about 12 years), carbon-14 (5,700 years), krypton-85 (11 years), argon-41 (1.8 hours), and a number of iodine isotopes (including iodine-129, 16 million years).

Commercial nuclear fuel reprocessing is carried out primarily in the United Kingdom and France. During reprocessing, spent but still highly radioactive nuclear fuel is cut up and dissolved in boiling nitric acid. Plutonium and uranium isotopes are then separated out from the fission products in the fuel. The recovered plutonium is fissile and can be used in nuclear weapons. Beyond the environmental and safety problems, it therefore constitutes a serious security and proliferation risk and is stored in guarded vaults. Part of the plutonium is mixed with uranium oxide to form mixed oxide (MOX) fuel, which is used in some reactors in Europe. However, the use of this fuel entails several economic, operational, and waste management disadvantages. (See Chapter 14 for more detail.)

The recovered uranium has no commercial value as it is contaminated with isotopes (U-234, U-236) that make further use unattractive. Most of it is currently stored in steel drums in the United Kingdom and France or has been exported to Russia. The remaining fission products, mainly cesium-137 and strontium-90, are stored in solution at high temperatures in dozens of huge tanks. These tanks, containing large unstable quantities of radioactivity, require continuous cooling by various independent cooling systems. Originally, these wastes were to be vitrified online and stored in the safer and more manageable solid form. However, such processes have proved complex and some of these (and future) wastes have remained in liquid form for many years. Again, this situation poses challenges for future nuclear waste management.

Annual nuclide releases from reprocessing plants are several orders of magnitude larger than those from nuclear power stations (European

---

14 This is not new. Thirty years ago, the Ontario Royal Commission on Electric Power Planning stated that, "The mining and milling of uranium ore produces very large volumes of long-lived, low-level radioactive tailings which have leached into waterways in the vicinity of Elliot Lake, Ontario, thereby posing serious health and environmental problems" (Ontario Royal Commission, 1978).

Parliament, 2001). The main releases are krypton-85, hydrogen-3, carbon-14, and a number of iodine isotopes (European Parliament, 2001). The global collective dose resulting from the discharges of the La Hague reprocessing facility for its remaining years of operational life – truncated[15] at 100,000 years – would cause over 3,000 additional cancer deaths globally, if the linear no-threshold theory of radiation is applied.

Although nuclides are routinely released from nuclear facilities, the health consequences among populations living downwind remain a subject of controversy. In the 1990s, several studies found increased incidences of childhood leukemia near UK nuclear facilities. However, official estimated doses from released nuclides were too low, by two to three orders of magnitude, to explain the increased leukemias. Recent epidemiological studies have reopened the child leukemia debate. Baker and Hoel (2007) carried out a meta-analysis of 136 nuclear sites in the United Kingdom, Canada, France, United States, Germany, Japan, and Spain and found cancer death rates for children were elevated by 5–24% depending on their proximity to nuclear facilities. Hoffmann et al. (2007) found 14 leukemia cases between 1990 and 2005 in children living within 5 km of the Krümmel nuclear plant in Germany, significantly exceeding the 0.45 predicted cases.

Most important, however, is the KiKK study, (*Kinderkrebs in der Umgebung von Kernkraftwerken* = Childhood Cancer in the Vicinity of Nuclear Power Plants) (Kaatsch et al., 2008; Spix et al., 2008) which found a 60% increase in solid cancer risk and a 120% increase in leukemia risk among young children living within 5 km of all German nuclear reactors. The findings are significant because KIKK was a large, scientifically rigorous study: the German government, which commissioned it, has confirmed its findings. The KiKK results are presently the subject of much research and it is too early to provide a firm explanation for the increased cancers although radiation exposures are thought to be involved. One hypothesis (Fairlie, 2009) proposes that infant leukemias are mainly a teratogenic effect from in utero radiation exposures due to radionuclide intakes during pregnancy. Whatever the explanation(s), the epidemiological evidence points to raised cancer risks among children living near nuclear reactors.

In recent years, a number of official reports have been published on the hazards of tritium, the radioactive form of hydrogen that is released in large quantities from all nuclear facilities, in particular from spent fuel reprocessing plants and heavy water reactors. Extremely large amounts of tritium would also be released from any fusion facility, should one ever be operated successfully. In the past, this isotope had been regarded as being only weakly radiotoxic, but this view is now changing among governments and international agencies concerned with radiation exposures. Reports have been published to date by radiation safety agencies in the United Kingdom (AGIR, 2007), Canada (CNSC, 2010a; CNSC, 2010b), and most recently in France. The French Nuclear Safety Authority (ASN, 2010) has published a comprehensive white paper on tritium and the French Institute of Radioprotection and Nuclear Safety (IRSN, 2010a; IRSN, 2010b; IRSN, 2010c; IRSN, 2010d; IRSN, 2010e; IRSN, 2010f) has published six major reports on tritium. These reports draw attention to various hazardous properties of tritium, including its extremely rapid distribution in the environment and its heterogeneous distribution within tissues.

4.7.7   Decommissioning and Waste Management

The collective dose of atmospheric discharges during the decommissioning processes of nuclear facilities in the European Union in 2004 have been estimated to be nearly 100 times less than from the operation of nuclear facilities (European Commission, 2005).[16] In total today, however, both are small by comparison to other human-caused radiation exposures, which are currently dominated by medical procedures and radon in buildings (Culling and Smith, 2010; see also Section 4.9).

Nuclear waste storage and disposal has led to significant environmental pollution in the past. In France, the Centre de Stockage de la Manche (CSM) repository stores 527,000 cubic meters of low- and intermediate-level waste and was filled and closed in 1994. The CSM contains considerable amounts of long-lived radioisotopes, including at the minimum about 100 kg of plutonium and 200,000 kg of uranium. It also contains large amounts of heavy metals, including about 20,000 tonnes of lead and 1,000 kg of mercury. Tritium leaks from the site have polluted the local water tables and streams. Contamination levels measured near the site exceeded six million becquerels per liter in 1983. Similar problems have been raised in Germany concerning the Asse nuclear waste disposal site that, in addition, is struggling with massive water intrusion and structural instability. The Federal Radiation Protection Office has recently recommended taking out all of the 126,000 drums with radioactive waste.

High-level waste (HLW) in the form of spent nuclear fuel or vitrified waste from reprocessing contains more than 90% of the radioactivity in wastes from nuclear electricity generation. However, there is no operating HLW repository in the world, and none is expected in the foreseeable future. This means that estimates of the radiological impacts of HLW disposal remain highly speculative, but HLW still poses the key question of intergenerational liability and justice. Although unlikely to pose health risk to populations approaching that of fossil alternatives, the difficulty of assuring the long-term integrity of HLW disposal sites and the difficulty of siting them due to local opposition is a major bottleneck to expansion of nuclear power in some countries (see Chapter 14).

---

15   The authors of the present report consider that, while the use of truncated collective doses might be justified in some comparisons, untruncated collective doses are necessary to treat present and future generations equally, a key requirement of the principle of sustainable development. Untruncated doses are of course larger than truncated ones.

---

16   These collective dose estimates were truncated by the European Commission at 500 years.

4.7.8   Transport

As can be seen from Figure 4.12, the nuclear fuel chain involves many stages, which inevitably results in a great deal of transportation between the various steps. These include uranium ore mining, milling, leaching, yellowcake production, U-235 enrichment, fuel fabrication, reactor operations, defuelling, temporary storage of spent fuel, encapsulation of spent fuel, and final disposal of spent fuel. This is for the direct once-through path; the reprocessing path involves significantly more stages.

Few recent studies have examined the routine health impacts of these transportation stages, as opposed to the health impacts of transport accidents that, by their nature, are not routine. Perhaps the most likely step where radiation doses may be a matter for consideration is the transport of spent nuclear fuel, in particular spent plutonium fuels (MOX fuels) by ship, rail, and road from reactors to reprocessing facilities. Older studies indicate that members of the public can receive small radiation doses from prolonged standing near spent fuel railway flasks. The result has been, at least in the United Kingdom, that railway transports of spent fuel are not allowed to stop in areas to which members of the public may have access or where they may be present.

4.7.9   Summary

Nuclear power facilities have been associated with the risk of some cancers in selected sub-populations, such as workers, but nevertheless do not represent a significant fraction of radiation exposure or the burden of disease in any population. Leaving aside the important potential of weapons-grade material being diverted from commercial nuclear fuel systems and major nuclear accidents, all major studies indicate that well-run modern nuclear power systems pose much less risk than fossil power systems, even modern ones. Nevertheless, because of the character and history of radiation, the risks of nuclear power systems are of particular concern to the public and policy makers and thus do significantly influence energy policy decisions in many countries. In addition, as the size and scale of nuclear facilities expand, such risks may also rise unless careful measures are taken. Finally, as discussed in Chapter 14, the difficult-to-quantify risks of proliferation from some nuclear fuel technologies may overwhelm those from all other parts of the system, including routine operation, accidents, and waste disposal.

4.8   Emerging Energy Systems

4.8.1   Key Messages

- Energy derived at a large scale from sunlight and wind is not free of health and environmental impacts. Issues of land use, maintenance, materials inputs, and energy storage raise concerns about environmental, occupational, and community health impacts.

- The health and environmental impacts of advanced materials such as nanotechnology materials that are used in some advanced energy technologies are not well characterized. There are a number of concerns about both occupational and environmental exposures from the widespread use of these materials and the resulting impact on disease burden

4.8.2   Background

Accessible and affordable energy has facilitated the industrialization, strengthening, and growth of the world economy over the past century. But rising energy use has also brought recognition to the human health and environmental impacts of different systems for producing, transforming, transporting, and consuming energy. As fossil fuel use developed over the last century, a variety of associated health and environmental effects began to appear. To address these problems, ongoing investments have been made in alternative energy sources and technologies, often with an implicit assumption that alternate sources will reduce impacts relative to existing technologies. The flaw in this assumption, however, is that lack of knowledge about impacts is not the same as absence of impact.

Changes from one energy source to another are likely to have different impacts through the whole usage chain (or fuel life cycle), from resource capture through to conversion, distribution, and end use. Moreover, because the full impact burden of extant energy systems has not been adequately characterized, there is no appropriate baseline against which to compare the relative benefits of new systems. Therefore, as world leaders consider options for changing the portfolio of future energy sources, it is important to examine each of the options for associated impacts and strategies to address those impacts.

The history of technology development (e.g., pesticides, vaccines, drugs, nuclear power, transportation, fuel additives) reveals that all technologies have both benefits and drawbacks (health, environmental, financial, and social). Not all benefits and drawbacks can be accurately characterized in advance of technology deployment, but technology assessment has proven useful in avoiding strategic errors that can derail a promising technology. Prior to widespread technology deployment is the time to confront issues such as demands on resources and materials, occupational and environmental health burdens, impacts on water supply and climate, and land-use offsets. As emerging energy technologies move from concept to deployment, there is a need to go beyond a simple current carbon benefit or net energy balance assessment and include other impacts such as human health benefits and impacts.

One tool to evaluate the cumulative impacts of existing technologies and anticipate the impacts of future technologies is lifecycle assessment (LCA). LCA provides a framework for comparing services and products (or product-related emissions) according to their total estimated environmental impact summed over all chemical emissions and activities associated with the product's life cycle. Box 4.4 below illustrates life-cycle impact as it applies to both existing and emerging energy systems.

Box 4.4 | Life-cycle assessment

Life-cycle assessment (LCA) has evolved over the last three decades as an effective strategy for organizing a comprehensive assessment of energy system impacts. The purpose of LCA is to quantify and compare environmental flows of resources and pollutants (to and from the environment) associated with an industrial system, such as an energy supply and use system, over the entire life cycle of the system – from resource extraction to energy use. LCA evaluates a broad range of requirements and impacts for technologies, industrial processes, and products in order to determine their propensity to consume natural resources or generate pollution. The term "life cycle" refers to the need to include all stages of a process – raw material extraction, manufacturing, distribution, use, and disposal, including all intervening transportation steps – so as to provide a balanced and objective assessment of alternatives. In the case of electrical energy, this means tracking the system from resource extraction (i.e., collection of sunlight, extraction of tar sands, collection of wind energy, etc.) through conversions, storage, transmission, and end use. In the case of transportation energy, LCA tracks the system of resource extraction (biofuel production, oil shale extraction, oil well production, etc.) through conversions, storage, and transport to the point of energy release and use in a vehicle (auto, bus, train, airplane, etc.).

An LCA includes three types of activities: collecting life-cycle inventory data on materials and energy flows and processes; conducting a life-cycle impact assessment (LCIA) that provides characterization factors to compare the impacts of different product components; and life-cycle management, which is the integration of all this information into a form that supports decision making. As an illustration of the LCA approach, Figure 4.14 shows a comprehensive LCA for transportation biofuels that addresses cumulative impacts to human health and the environment from all stages, impacts from alternative materials, and impacts from obtaining feedstocks and raw materials.

"To assess human toxicity impact, the LCIA practitioner considers for each chemical involved the cumulative exposure associated with the mass released to a defined (indoor, urban, regional, etc.) environment by multiplying the release amount by a measure of toxic impact to characterize the likelihood of health effects and their potential consequences. The SETAC Life Cycle Impact Assessment (LCIA) Working Group on Human Toxicity (Krewitt et al., 2002) has classified measures of toxic impact into two broad categories: potency-based characterization factors that are used to assess the likelihood of a disease or effect (cancer, death, reproductive failure, etc.) and severity-based characterization factors or damage factors that, in addition to the qualitative or quantitative likelihood of disease, reflect population consequences of the disease in terms of years of life lost or some other measure of societal impact" (McKone et al., 2006).

Below we discuss a set of emerging energy systems, along with the new and/or potentially worrisome health issues surrounding these systems. The health impacts from energy systems include those that can be measured through occupational or community health tracking and those that are estimated using tools such as risk assessment. Most of the health impacts from emerging systems, and many of the health impacts from existing systems, are in the latter category. So, a key challenge of addressing health impacts from emerging systems is the lack of large empirical databases or well-established assessment methods. To address this gap, we first consider the baseline impact from existing systems and then consider how emerging systems will perform to reduce and/or redistribute this burden.

4.8.3   Baseline Impact Assessment

For both transportation fuels and electricity production, an assessment of the current disease impact associated with a base metric, such as 100,000 vehicle km traveled and with a GWe-yr of electricity produced, provides baselines for the comparative assessment of emerging energy systems. It is difficult to obtain accurate and consistent assessments of the disease burden of energy in these terms, but some key studies offer insight regarding the magnitude of the disease burden. In its extensive study of energy externalities, the US National Research Council (NRC) (2010) noted that among alternate approaches employed to address baseline assessments, some are "top-down," others "bottom-up." The top-down approach makes use of morbidity and mortality statistics for a specific population, such as the inhabitants of a country or a population group, and postulates a link to a specific source, such as transportation or power plant emissions or to an activity (smoking). The bottom-up approach begins with hazard identification, defining sources for pollutants of concern and then follows these sources from a point of release into a receptor population to assess exposure and damage. According to the NRC (2010), "Top-down assessments for air pollution have been carried out for many regions such that it is possible to provide a disease burden estimate for air pollution. But allocation to specific energy

Table 4.10 | Examples of the categories of impacts that have been associated with pollutant emissions from energy supply and the specific types of health effects and other damages associated with these pollutants.

Impact Category	Pollutant/Burden	Effects
Human health – mortality	PM_{10}, SO_2, NO_x, O_3	Reduction in life expectancy
	Benzene	Cancers
	Benzo-[a]-pyrene	
	1,3-butadiene	
	Diesel particles	
	Noise	Loss of amenity, impact on health
	Accident Risk	Fatality risk from traffic and workplace accidents
Human health – morbidity	PM_{10}, O_3, SO_2	Respiratory hospital admissions
	PM_{10}, O_3	Restricted activity days
	PM_{10}, CO	Congestive heart failure
	Benzene	
	Benzo-[a]-pyrene	Cancer risk (non-fatal)
	1,3-butadiene	
	Diesel particles	
	O_3	Asthma attacks
		Symptom days
	Noise	Myocardial infarction
		Angina pectoris
		Hypertension
		Sleep disturbance
	Accident Risk	risk of injuries from traffic and workplace accidents
Building material	SO_2 Acid deposition	Ageing of galvanized steel, limestone, mortar, sand-stone, paint, rendering, and zinc for utilitarian buildings
	Combustion particles	Soiling of buildings
Crops	NO_x, SO_2	Yield change for wheat, barley, rye, oats, potato, sugar beet
	O_3	Yield change for wheat, barley, rye, oats, potato, rice, tobacco, sunflower seed
	Acid deposition	Increased need for liming
Global warming	CO_2, CH_4, N_2O, N, S	World-wide effects on mortality, morbidity, coastal impacts, agriculture, energy demand, and economic impacts due to temperature change and sea level rise
Amenity losses	Noise	Amenity losses due to noise exposure
Ecosystems	Acid deposition, nitrogen deposition	Acidity and eutrophication (avoidance costs for reducing areas where critical loads are exceeded)

Source: European Commission, 2005.

systems cannot be resolved, because the top-down approach usually lacks the spatial and temporal resolution needed to track impacts to specific technologies." To address the damages associated with specific technologies requires the bottom-up approach such as was used in the comprehensive ORNL/RFF (ORNL/RFF, 1992–1998) and ExternE (European Commission, 2005) studies.

Both the ORNL/RFF and ExternE studies employ a bottom-up approach, in which environmental benefits and costs are determined using a pathway analysis that follows the energy-technology pollutant emissions from the source, then tracking dispersion in air, soil, and water and ultimately to health and environmental impacts. Both ORNL/RFF

and ExternE used a life-cycle approach to characterize externalities associated with electricity production from a variety of technology options. In order to permit comparisons across multiple technologies, the ORNL/RFF and ExternE characterized externalities in monetized units, that is dollars of damage per kWh. This involved the use of detailed models to predict the dispersion from energy supplying facilities of primary pollutants as well as the atmospheric formation of secondary pollutants, specifically ozone and fine particulate matter (see Table 4.10). The ExternE study also considered transportation systems. More recently, Hill et al. (2009) assessed life-cycle impacts from air emissions attributable to transportation fuels. They used a life-cycle approach to assess the potential impacts of emerging

Box 4.5 | US National Research Council study on the "hidden costs" of energy

In 2008, the US Congress asked the National Research Council (NRC) of the US National Academy of Science to prepare a report that defined and evaluated "key external costs and benefits – related to health, environment, security, and infrastructure – that are associated with the production, distribution, and use of energy but not reflected in market prices or fully addressed by current government policy." The NRC released this report in late 2009 (NRC, 2009b). To assess damages, the committee used a life-cycle approach with relatively high spatial resolution for characterizing emissions and exposures attributable to energy used in the United States by electricity generation, transportation systems, process heat, and building heating/cooling. They applied these methods to a year close to the present (2005) for which data were available, and also to a future year (2030) so as to gauge the impacts of possible changes in technology. In addition to impacts in the near term (health damages), the committee considered damages in the future attributable to emissions of atmospheric greenhouse gases.

The report found that, in 2005, the unpriced impacts of energy supply amounted to US$120 billion. This amount is due mainly to the negative impact of air pollution on health. Of this amount, US$62 billion was attributable to electricity (mainly coal) and about US$56 billion was attributable to transportation, with the remainder due to process heat and comfort control. The long-term damages from climate change were estimated to be comparable (~US$120 billion) and security costs were not quantified. Coal-fired plants, which produce about half of US electricity, had an impact of about US¢3.2 of "nonclimate" damages for every kilowatt-hour (kWh) generated. In contrast, natural gas electricity had an impact of US¢0.16. In the transportation sector, which accounts for 30% of US energy use, the estimated unpriced health damages of US$56 billion works out to US¢0.7–1 per vehicle km traveled. The estimated air pollution damages associated with electricity generation in 2030 will depend on many factors, such as the future mix of generation technologies, end-use efficiency, air pollution regulations, etc. With regard to transportation impacts, the committee noted that substantially reducing nonclimate damages related to transportation by 2030 would require major breakthroughs, such as cost-effective technologies to create cellulosic biofuels and to capture and store carbon from coal-fired power plants. In addition, great increases in renewable or other forms of electricity generation (for electric vehicles) with lower emissions would be needed. Further enhancements in fuel economy will help, especially for emissions from vehicle operations, although they are only about one-third of the total life-cycle emissions (NRC, 2010).

biofuels relative to gasoline. The impact pathways included in ExternE are listed below.

The NRC (2010) report on the unpriced consequences in the United States considered impacts from all forms of energy supply, distribution, and use. The approach and findings of this report are summarized in Box 4.5. Although these results are most relevant to North America, they provide key insights about the relative magnitude of the health and environmental impacts from existing and emerging energy technologies. This report makes clear that, even for relatively clean energy technologies, the upstream inputs can have a significant impact because they often rely on existing technologies for energy inputs. For example, the production of fuels, whether from biomass or from petroleum, requires a significant amount of energy that could be coming from a coal-fired power plant.

4.8.4   Biofuels

Biofuels have become the focus of both governments and industry efforts to find technically feasible and carbon-neutral fuels for transportation and (to a lesser degree) electricity production. The challenge for biofuels is to compete with fossil-based fuels in terms of cost and impact. The enormous world demand for liquid fuels makes it inevitable that a biofuel economy will involve significant changes in agricultural, fuel-production, and transportation infrastructure. Building and operating this infrastructure will involve many complex technical and policy choices. Addressing these complex choices requires research to anticipate the nature and scale of the supply/demand systems that will arise in large-scale deployment of biofuels (Energy Biosciences Institute, 2010). However, the current methods used to measure, evaluate, and regulate human health impacts and benefits of biofuels suffer from a number of information gaps and have not been fully integrated into an LCA framework. Moreover, the call for biofuels gives rise to environmental and ethical questions regarding the impacts in rural communities from deforestation, food price changes, and loss of ecosystem services that can accrue when monoculture crops replace extant systems.

Ethanol derived from corn or sugarcane as an additive to or substitute for gasoline, along with biodiesel derived from vegetable oils or waste animal fat, are currently the largest and most economically viable biofuels options. However, biofuels can be produced from a wide variety of plant materials and wastes. The biofuel feedstocks receiving the most attention include corn stover (stalks and cobs), perennial grasses (switchgrass and miscanthus), sorgum, algae, trees and dedicated woody crops, and even garbage. Because plants absorb CO_2 during growth and may

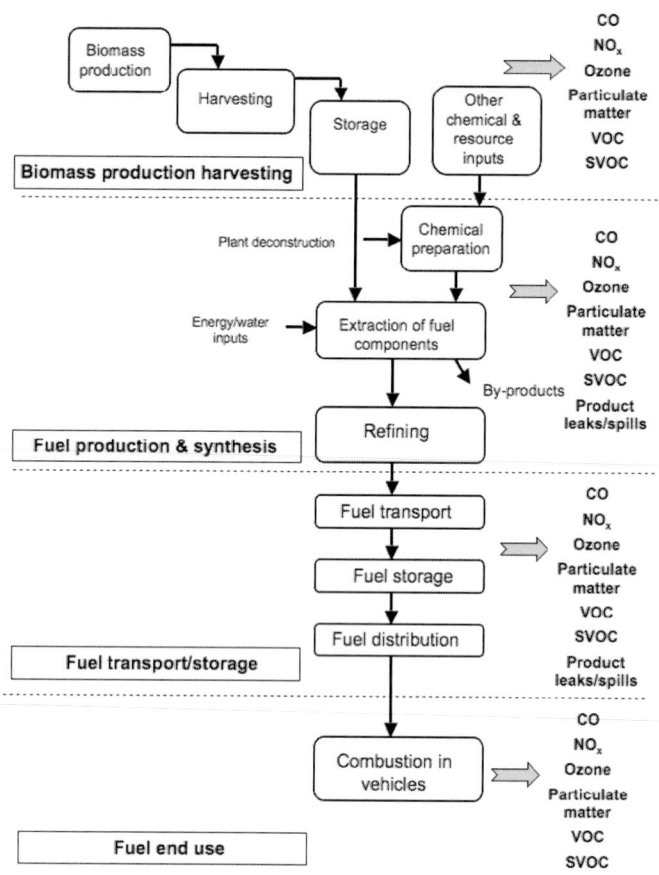

Figure 4.13 | Pollutants associated with biofuel production, transport and use.

In characterizing the life-cycle health impacts associated with liquid transportation fuels from biomass, one needs to consider the emissions generated at each of the following stages (see Figure 4.14):

- production of the feedstock (farm or forest externalities);
- transportation of the feedstock to the processing facility;
- processing of the feedstock into liquid fuels;
- transportation of the fuel to distribution endpoints; and
- downstream effects of using the fuel.

The last two steps are identical to any liquid transportation fuel and, at least for air emissions, have been addressed by air quality studies and models. But these studies have significant uncertainties, so it remains difficult to make an accurate characterization of the overall disease burden from gasoline, much less to project future scenarios. The upstream impacts of biomass feedstock effects will be quite location specific, as different feedstocks will be economically viable in different locations (e.g., corn stover and switchgrass in the corn belt, miscanthus in warmer climates, trees and forestry in the southeast, etc.). In addition, the health impacts associated with any given feedstock are also likely to vary by specific field/watershed within a region (depending on climate, land use history, soils, slope of the land, proximity to water bodies, etc.) and can be attenuated by farming practices (use of conservation tillage, nutrient management of both fertilizer and manure applications, placement of buffers or wetlands, etc.). Finally, transportation of feedstocks to processing facilities is expected to remain expensive, even after technological improvements, so that numerous, small processing facilities located throughout the landscape is a likely configuration of the industry. This means that health impacts associated with production and transportation of the feedstock and liquid fuels will be both site-specific and widespread.

Hill et al. (2009) estimated that for "each billion ethanol-equivalent gallons of fuel produced and combusted in the USA, the combined climate-change and health costs are US$469 million for gasoline, US$472–952 million for corn ethanol depending on biorefinery heat source (natural gas, corn stover, or coal) and technology, but only US$123–208 million for cellulosic ethanol depending on feedstock (prairie biomass, miscanthus, corn stover, or switchgrass)." However, their analysis does not distinguish mortality from morbidity or define the equivalent disease burden associated with these numbers. Moreover, their approach focuses on air emissions and excludes the impacts of toxic air pollutants and releases to soil and water. Future life-cycle assessments are needed to address a full range of air emissions, as well as emissions to surface water, soil, and groundwater, which have complex and significant pathways from release to human intake (e.g., food pathways).

The debate concerning the net GHG effects of biofuels centers on indirect land use changes. Indirect land use refers to crop allocation changes occurring indirectly as a result of biofuels policies and the effect of such

increase stores of soil organic carbon (Tilman et al., 2006), biofuels may reduce GHG emissions relative to petroleum-derived fuels. Many studies (see, for example, the review in NRC, 2010) conclude that there are at least small, positive life-cycle GHG benefits from most biofuels. But there continues to be debate concerning the net GHG effects from first-generation biofuels, particularly corn-based ethanol.

In addition to potential reductions in GHG emissions, vehicle tailpipe emissions of many air pollutants harmful to human health will be different and may be lower with biofuels (McCormick, 2007). Questions remain regarding how shifts from petroleum-based fuels to biofuels will affect the combustion-phase emissions of air pollutants. Evidence for lower primary particle emissions must be weighed against evidence for potentially higher VOC emissions, which could increase ozone and secondary particle formation. Whether and how much biofuels reduce human health impacts (see Figure 4.13) relative to fossil fuels depends strongly on how they are produced, distributed, and combusted (Robertson et al., 2008). Energy inputs, fertilizer use, biomass yields, conversion efficiencies, pollution control technologies, and direct and indirect damages to water supplies and land all determine the cumulative impacts on human health from both biofuels and fossil fuels (Hill et al., 2006; Fargione et al., 2008; Searchinger et al., 2008).

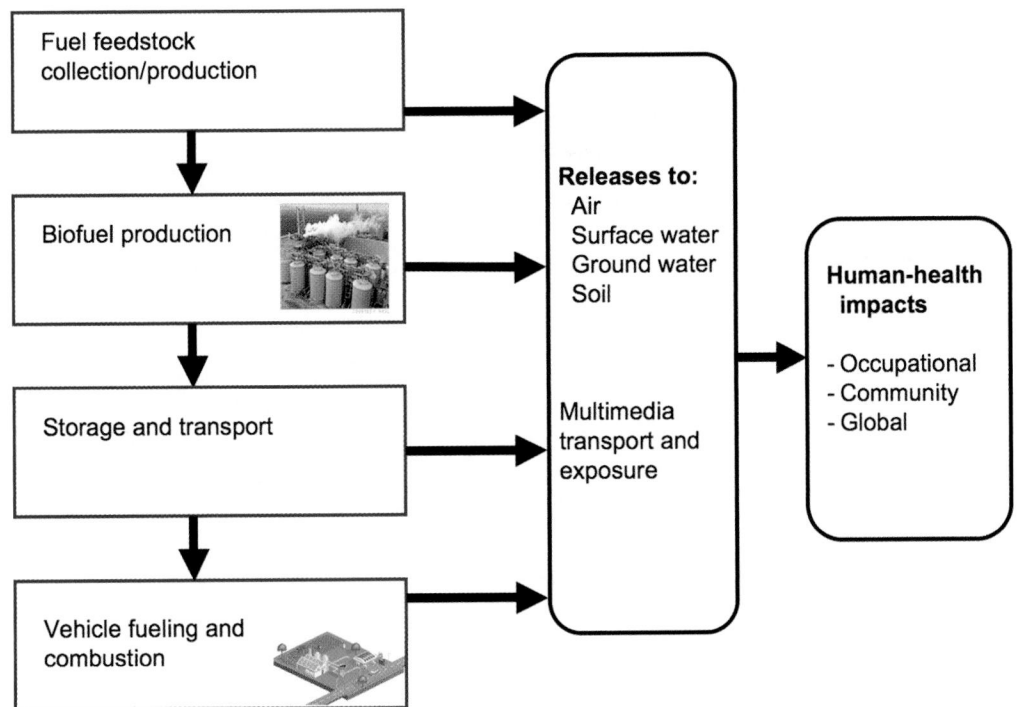

Figure 4.14 | An illustration of the life-cycle approach showing the link of human health impacts to the major life stages of biofuel production and use.

changes on GHG emissions. It has been a major source of discussion since Searchinger et al. (2008) published a paper that explains how increasing demand for corn or farmland causes crop prices to increase, making it profitable for farmers to increase their acreage. If this increased acreage comes from plowing up land that has not been in agricultural production and/or is particularly environmental sensitive (e.g., rainforests in Brazil, pristine ecosystems in the United States), this could actually increase CAP emissions (since burning rainforests would release lots of carbon) and have other detrimental environmental concerns. There is continuing study of this issue. A recent report by the US Environmental Protection Agency considered a range of biomass production scenarios and found significant variability with regard to indirect land use (US EPA, 2009). In some cases, their analysis suggests that significant reduction in CAP emissions relative to gasoline could be largely offset by indirect land use changes. In other cases, the effect is small. These types of results focus attention on a formal treatment of uncertainties in the life-cycle assessment.

4.8.5 Synfuels

Synfuels are considered emerging technologies because most synthetic fuels technologies are either not yet cost-competitive with existing fossil fuels or because they are still in a research and development phase. Synthetic fuels can be made from coal, oil shale, or tar

sands. Various fuel gases, such as substitute natural gas and synthesis gas (made from CO and hydrogen) are also included in this category. Synthetic crude oil (often called "syncrude") is a complex mixture of hydrocarbons somewhat similar to petroleum. It is obtained from coal (liquefaction), from synthesis gas (a mixture of carbon monoxide and hydrogen), or from oil shale and tar sands. Gaseous fuels can be produced from sources other than petroleum and natural gas. Syncrude mixtures have a number of key differences in composition relative to petroleum. For example, syncrude from coal usually contains more aromatic hydrocarbons than petroleum. The most important current source of synthetic crude oil is the tar sand deposit that occurs in northeastern Alberta, Canada. Tar sand is a common term for oil-impregnated sediments that can be found in almost every continent. The routes by which synthetic fuels can be prepared from coal involve either gasification or liquefaction.

Atmospheric emissions from synfuel production are generated from mining, processing, retorting, and waste disposal. Large emissions of PM can arise from mining, blasting, conveying, crushing, and on-site retorting operations. The on-site liquefaction/gasification needed to mobilize synfuels also poses the potential for large emissions of oxides of sulfur and nitrogen, mercury and other metals, and carbon monoxide, as well as large numbers of organic compounds, including alkanes, benzene, and polycyclic aromatic hydrocarbons (PAHs) to the local airsheds.

Production of synthetic fuels from geological formations such as oil sands or oil shale also have the potential for impacting human health through significant reductions in water quality. The impacts on water quality are both direct and indirect. Direct impacts arise from the release of contaminants to surface and groundwater from both routine and off-normal operations. Synthetic fuel operations produce large amounts of wastewater that must be stored, treated, and moved through pipelines. Even under ideal conditions, the treatment process can introduce trace amounts of metals and toxic organic compounds to surface and groundwater. Moreover, off-normal events such as leaks in pipes and containment ponds, which can be frequent, will add even more of these substances to both surface and groundwater. Indirect effects are those associated with large demands on limited water supply and impacts from disrupting water-bearing formations. For example, extracting one barrel of oil from an oil sand requires many barrels of water. The demand on surface and/or groundwater for synfuel mining and processing will limit the amount of water available for diluting the pollutants from these processes. Lack of available fresh surface or groundwater can be a disease burden in and of itself. The alteration of geologic formations due to mining operations can limit the storage and transmission of both surface and groundwater, leading to a reduction of freshwater and the health consequences of this privation.

Significant human health impacts are associated with the production and use of synfuels (NRC, 2010). These technologies have documented impacts on occupational health as well as the health of local, regional, and global populations. The life-cycle impacts of synfuels include both upstream processes for extraction, transportation, and reefing the fuel-use stage. The upstream stages have potentially significant local and even regional impacts on air, surface water, and groundwater with resulting impacts on human health. In particular, the need to liquefy or gasify the fuels from the formations that supply them will result in CAP emissions that impose both local and global disease burdens. In the refining and combustion process, the impacts of synfuels are likely to exceed those of an energy-equivalent quantity of gasoline; however, there are currently insufficient data available to accurately make this assessment (NRC, 2010).

4.8.6 Large-Scale Solar and Wind Power Systems

Solar energy has been deployed in both small-scale (mainly rooftop) applications and in large-scale electrical production. A number of large-scale solar power plants are already in operation worldwide, with plans slated for others. Three large-scale technologies transform solar radiation into a form that provides electricity: photovoltaic cells, which generate electricity directly; concentrating solar power thermal systems, which use an absorption medium such as water or oil to transfer absorbed heat to a steam generator that drives a turbine; and solar towers, which are effectively chimneys where rising hot air powers turbine generators.

Wind power or wind energy is the process by which the wind is used to generate mechanical power or electricity and is one of the fastest-growing forms of electricity generation in the world. Archer and Jacobson (2007) produced a global wind-energy resource map that estimated the global potential for wind-generated energy at 72 terawatts, 40 times the worldwide demand in 2000.

Large-scale solar and wind systems are not free of health and environmental impacts. Issues of water demand, land use, maintenance, materials inputs, and energy storage raise concerns for environmental, occupational, and community health impacts. Interest is growing in the large-scale use of nanomaterials in solar applications. Energy storage is an issue currently being addressed with new technologies that may require large quantities of materials whose toxic properties are not well studied. Constructing new transmission lines to areas with the best wind and solar resources also poses significant environmental problems. There is no indication, however, that the overall health burden of these systems will be even close to those from traditional fossil fuel systems. Moreover, the health burdens for large wind power and solar power systems are likely to be lower than even advanced fossil and nuclear power systems. Nevertheless, it is important to anticipate risks in advance in order to head off any unpleasant surprises as new systems come into widespread use.

4.8.7 Geo-engineering Schemes

Concerns about the environmental impacts of GHG emissions and the persistent pollutants from energy technologies have led to a call for large-scale "geo-engineering" projects for mitigating the impacts of global pollutants. Geo-engineering is the deliberate modification of Earth's environment on a large scale to provide for human needs and promote habitability. Keith (2001) emphasizes that it is the deliberateness that distinguishes geo-engineering from other large-scale, human impacts on the global environment, such as those that result from large-scale agriculture, global forestry activities, or fossil fuel combustion. Originally, the term geo-engineering referred to a proposal to collect CO_2 at power plants and inject it into deep ocean waters. The concept of geo-engineering now also includes:

- underground storage of CO_2;

- fertilization of oceans with urea, a nitrogen-rich substance, or with iron, a limiting nutrient, to encourage growth of plankton;

- deflection of sunlight from the Earth through the use of a giant space mirrors;

- injection of sulfur particles into the stratosphere in order to modify the Earth's albedo;

- cloud-seeding by spraying seawater in the atmosphere to increase the reflectiveness of clouds; and

- windscrubbers to filter carbon dioxide from the air.

Each of these concepts carries with it some risk of health and environmental impacts that have not been fully identified or evaluated. In addition, most of these concepts are energy intensive and will thus have an impact that depends on the quantity of energy required and the source of that energy.

4.8.8   Advanced Materials in Energy Systems

Advanced materials such as nanotechnology materials and advanced semiconductor technology play a growing role in both energy supply and energy use. Nanotechnology is increasingly impacting the US and world energy balance, on both the supply and demand sides. On the supply side, nanotechnology is being used to optimize supply from existing energy sources (e.g., crude petroleum) and to exploit new sources, such as heavy oil, liquefied coal, and solar energy (including using solar energy to produce hydrogen). Nanotechnology is also improving and opening new possibilities for the transmission and storage of energy, especially electricity and possibly hydrogen in the future. On the demand side, nanotechnologies have the potential to reduce energy use by making it possible to manufacture lighter and/or more energy-efficient automobiles and appliances. Nanotechnologies can also be used to improve energy efficiency in buildings.

The health and environmental impacts of advanced materials such as nanotechnology materials are not well characterized. There are a number of concerns about both occupation and environmental exposures to the widespread use of these materials and what impact that will have on disease burden (NRC, 2009a).

4.9   Energy Efficiency

4.9.1   Key Messages

- Increases in combustion efficiency have the potential to greatly improve human health, especially in households that use solid fuels for cooking.

- Measures to increase energy efficiency in buildings can improve health by reducing cold-related morbidity and mortality and by increasing psychosocial well-being.

- In homes with relatively low levels of air exchange, energy efficiency measures can have some negative impacts on health by increasing exposure to dust, mold, radon, volatile organic compounds, and indoor tobacco smoke.

- Energy efficiency projects require capital. Fuel taxes or tariffs may be levied to account for these expenditures, while subsidies for poorer consumers may be needed to counteract the effect of increased fuel prices.

4.9.2   Energy Efficiency, Energy Use, and Health

Energy efficiency is often looked upon as the most desirable approach to limiting energy use and as potentially the most cost-effective means of achieving net health benefits. Obtaining the same level of energy service with less primary energy input should entail lower fuel cycle risks, reduce atmospheric emissions of pollutants, limit demands on infrastructure, lower costs, and contribute to economic competitiveness. Although all these factors are generally true, the relationship between energy efficiency and health is more complex.

To a large degree, in fossil fuel combustion, efficiency equates with cleanliness of burning and hence with lower emissions of health-damaging pollutants for a given quantity of fuel and a specified level of energy service. This has most immediate and direct relevance to the indoor combustion of biomass for cooking and heating in low-income households (see Section 4.2), where efficiency of stove technology may appreciably influence exposure to indoor air pollutants, especially for women and children.

Some argue that, at a societal level, increasing efficiency (and hence improving the affordability of energy services) can promote greater overall use of the technology in question than otherwise would occur, resulting in a rise in total per capita energy use, and potentially greater community-level pollutant emissions (Smil, 2005). Among poor populations, an increase in energy services due to greater energy efficiency can have major welfare benefits, including better health. This set of issues is sometimes called the "takeback" or "rebound" effect and is discussed in more detail in Chapter 10.

Increased energy efficiency is a factor in the web of socio-technological influences that contribute to (industrialized) patterns of supply and demand, which have many detrimental (as well as beneficial) consequences for health in rich countries. Among the risks are those associated with increasing dependence on motorized transport (in relation to risk of road injuries and deaths; physical activity with its consequences for obesity, diabetes, cancer risks; and many other health effects; as well as adverse effects on the quality of urban living); overconsumption of energy-dense, processed foods high in salt and saturated fats, derived from intensive agriculture; and diets with a high proportion of meat and dairy products.

Energy efficiency is seen as one means of helping to limit the vulnerability to price volatility in global energy markets and of establishing greater energy security – an argument that applies at national and household levels (see Chapter 5). Though opinion about the imminence and consequences of peak oil remains divided (Deffeyes, 2001; Frumkin et al., 2007), there is no doubt that the world's low-cost energy reserves are being steadily depleted. Over time, the costs of extraction will rise in financial, environmental, and energy terms, as deposits become less and less accessible. There are justifiable concerns that this trend will be a driver for economic change and restructuring in many countries, with adverse effects on health and life expectancy. Reducing energy dependence through greater efficiency may help to limit these adverse impacts.

Among the various methods by which energy efficiency may be improved is the use of more distributed energy supply systems based on micro-generation and combined heat and power systems. Though broadly desirable, significant uncertainties remain about the adverse impacts that multiple microgeneration facilities may have on air quality in some settings, relative to centralized energy supply and distribution systems. This is due in part to the higher intake fraction (see Box 4.1) of the emissions from such facilities.

Increasingly, energy efficiency interventions – including programs of home insulation and heating system upgrades, as well as building regulation – are being used to tackle GHG emissions. While such interventions are likely to be broadly beneficial for health (e.g., resulting in warmer indoor temperatures in winter, lower fuel use/cost, improved use of space with various psychosocial benefits, etc.), there can be important unintended consequences. One method to increase energy efficiency is to limit the degree of uncontrolled ventilation by improving the air tightness of windows, doors, and other parts of the building fabric. While this may have little effect if the relative reduction in ventilation is small, it may have more critical impact in dwellings that already have fairly modest ventilation rates. The potential adverse health consequences include hazards relating to dampness and mold, indoor air quality (especially increases in radon levels), and in some circumstances, risk of adverse heat-related effects.

There is limited epidemiological evidence about many of these connections, which often are modified by local circumstances. The impacts of energy efficiency therefore cannot be quantified with any precision, despite their importance to a range of energy-related policy decisions.

4.9.3   Household Energy Efficiency

Household energy efficiency merits particular discussion because it is increasingly becoming a target for energy policy in the context of climate change (Stern, 2007; Friedlingstein, 2008) and recent fuel price rises. In developed countries, it is commonly being addressed through a combination of building regulation (e.g., minimum energy efficiency standards for new dwellings) and initiatives aimed at the refurbishment/upgrading of the much larger pool of existing housing stock. The scope and cost-effectiveness of energy efficiency gains is

generally greatest for new dwellings, where efficiency measures can be incorporated at the design stage. However, in nearly all settings, new dwellings account for a tiny fraction of the overall housing stock, so that in the short term, the larger potential for changing collective energy efficiency arises in relation to the upgrading and retrofitting of existing dwellings. For such dwellings, the types of upgrades may be limited (older housing stock may not have cavity walls, for example), the basic building fabric may be inherently inefficient from a thermal perspective, and the health benefits and financial returns are correspondingly lower.

The cost and benefits of specific building standards and efficiency upgrades depend on numerous factors, including, crucially, fuel type and price. It is also a function of occupancy levels and occupant behaviors. To date, few systematic attempts have been made to assess the associated costs and benefits to health, though such assessments are now beginning based on risk assessments of a range of housing-related hazards. Energy efficiency upgrades have a number of (mainly unquantified) adverse health costs as well as benefits.

4.9.4   Potential Adverse Health Effects

The main concerns about adverse health effects arise from reduced air exchange, which may increase indoor pollutants. In the most air-tight dwellings, these problems can be alleviated by mechanical ventilation and heat recovery (MVHR) systems, which allow greater ventilation while maintaining energy efficiency by recovering heat from the vented air. However, because MVHR systems require high levels of air tightness for effective operation, it is critical that they are correctly installed and maintained. In dwellings reliant on passive ventilation, pollutants derived from indoor or 'under-house' sources tend to rise as air exchange is reduced.

4.9.4.1   Radon

Radon has received surprisingly little attention despite strong theoretical arguments about its detrimental health effects if ventilation rates are lowered. Radon is a naturally occurring, colorless, odorless radioactive inert gas, with a half-life of 3.8 days, formed from the decay of radium. It is a significant contaminant of indoor air worldwide. Radon results from radioactive decay in rock formations beneath buildings or in certain building materials themselves. As a very heavy gas, it tends to accumulate at floor level.

Exposure to radon increases the risk of lung cancer (NRC, 1999; Darby et al., 2005). A common action level for remediation is 200 becquerels per m^3, a level that corresponds to a 3% lifetime excess risk of lung cancer. In many countries, it is the second most important cause of lung cancer deaths after cigarette smoking, accounting for around 10% of all lung cancer cases in the United Kingdom, for example.

Reduction in air exchange because of energy efficiency may give rise to an appreciable shift in the distribution of indoor concentrations, and hence alter risks of lung cancer. A variety of cost-effective engineering solutions are available to deal with high radon levels. The concern is that dwellings with elevated levels of radon may go unrecognized, or the energy efficiency interventions give rise to a more subtle shift in the population average exposure.

4.9.4.2   Dampness/Mold and House Dust Mite

Dampness and mold are primarily determined by indoor (relative) humidity. With energy efficiency, there is a tradeoff between higher temperatures (which tend to lower humidity, especially in winter months, and thus to decrease mold risk) and lower ventilation, which tends to increase humidity (Oreszczyn et al., 2006).

Although there is insufficient evidence, as yet, of a causal link between the presence of visible mold and the many putative health effects, several epidemiological studies show associations between visible mold (≥ 1 m²) and respiratory and other symptoms, including asthma, rhinitis, cough, wheeze, nausea, vomiting, and general ill health (Bornehag et al., 2004; Davies et al., 2004). The most vulnerable groups are young children, the elderly, allergy sufferers, and immuno-compromised individuals.

It is difficult to generalize about the patterns of ventilation and mold, which depend on climatic factors, quality of housing stock, and multiple other factors, but mold is a common problem in most settings. At reasonably high levels of air exchange (> one per hour), temperature improvements associated with energy efficiency may tend to lower mold risks. However, concerns arise towards the low end of air exchange in developed countries, i.e., less than 0.5 per hour.

4.9.4.3   Other Indoor Pollutants

Lower ventilation rates may also be important for a range of other pollutants arising from indoor sources (COMEAP, 2004).

- *Volatile Organic Compounds (VOCs)* include over 200 component chemicals typically existing in indoor air, including formaldehyde, benzene, and benzo(a)pyrene (BaP). Potential health effects are complicated: biological effects vary according to the specific chemical components present in the mixture, as well as with environmental factors such as temperature and humidity. Indoor sources are varied and include foam insulation (urea formaldehyde foam insulation), particle board used for construction, furniture, furnishings, and household cleaning agents. Indoor levels in developed countries are generally approximately 10-fold higher than ambient levels. Increased concentrations tend to be associated with newer and/or recently decorated houses. Health effects are poorly characterized but principally relate to allergic responses (Mendell, 2007), comfort and well-being, and some (theoretical) genotoxic/carcinogenic risk.

- *Products associated with combustion, especially nitrogen dioxide (NO_2) and particles* are a significant component of indoor air pollution. The major source of indoor NO_2 is from cooking with gas. On average, concentrations encountered indoors are not high enough to produce serious acute effects but are thought to be sufficient to cause a reduction in lung function and an increase in response to allergens that can cause narrowing of the airways in some sensitive individuals. Other possible health effects of indoor NO_2 include: increased likelihood of respiratory infection, symptoms, or illness in children living in homes using gas for cooking and heating; reduction in lung function in women living in homes using gas for cooking and heating (de Bilderling et al., 2005); impaired respiratory health associated with long-term exposure to NO_2; and a possible effect of NO_2 on the response to allergens in specifically sensitized individuals. Recent work by Chauhan et al. (2003) has shown that exposure to nitrogen dioxide increases the likelihood of viral infections in children and increases the bronchoconstrictor response seen in asthmatic subjects so infected (COMEAP, 2004). This is an important finding, as it is known that such infections play an important role in triggering attacks of wheezing in young children.

- *Carbon monoxide* is an under-recognized risk associated with poorly ventilated or maintained boilers and burners. In relation to energy efficiency, the main impact is more likely to be *reduced* likelihood of exposure (from more modern and efficient combustion devices installed as part of energy efficiency upgrades) than higher levels of exposure from reduced air exchange.

- *Tobacco smoke* poses a significant risk in households where one or more family members are smokers. In dwellings with lower air exchange rates, the known adverse effects of environmental tobacco smoke may be greater for children or other family members.[17]

4.9.4.4   Heat Risks

Greater energy efficiency has the potential to improve protection against heat-related risks by buffering against peak temperatures and smoothing out the diurnal fluctuations in indoor temperatures during heat waves. However, energy efficiency may be detrimental where solar gain is high (e.g., dwellings with large areas of glass facing the sun), as the high level of insulation in the building fabric may then act to maintain and accentuate high indoor temperatures. This may be a particular problem with high energy efficiency, low thermal mass dwellings. Traditional

---

17   Risks from secondhand tobacco smoke exposures are quantified in the latest CRA. For further information, see Footnote 1.

Table 4.11 | Effects of cold temperature on cardiovascular mortality, and modification by standardized indoor temperature.

Quintile of standardized indoor temperature. Range (mean) in degrees celsius	No. of deaths	% increment (95% CI) in mortality per degree celsius fall in outdoor temperature	
		Unadjusted	Adjusted for region
CARDIO-VASCULAR			
1 <14.8 (13.3)	2648	1.7 (0.5, 2.9)	2.2 (0.6, 3.9)
2 14.8- (15.7)	2555	1.1 (-0.1, 2.3)	1.1 (-0.5, 2.8)
3 16.6- (17.2)	2314	1.2 (-0.0, 2.4)	1.2 (-0.5, 2.9)
4 18.4- (18.7)	2523	1.4 (0.2, 2.6)	1.3 (-0.4, 3.0)
5 19.4- 27.0 (21.0)	2963	0.2 (-1.0, 1.3)	-0.1 (-1.7, 1.5)
Trend (change per degree celcius increase in hall temperature)		-0.15% (-0.28, -0.03)	-0.13% (-0.26, -0.00)

Source: Wilkinson et al., 2009.

dwellings in hotter climates typically have high thermal mass, small solar gain (small outward facing windows and doors, inner courtyards), and orientations to promote natural ventilation at night. Although there is reasonable understanding of how dwellings can be adapted to heat, countries that currently do not have hot climates may be optimized for cold rather than the hotter summer conditions that may prevail under climate change.

4.9.5 Health Benefits

Despite the list of potential adverse consequences, improvements in energy efficiency are likely to lead to net benefits through various routes, at least in communities where there are appreciable cold-related impacts and high energy costs.

4.9.5.1 Reduced Cold-Related Morbidity and Mortality

Burdens of cold-related mortality and morbidity are substantial in most temperate climates and are measurable in almost all populations studied to date (see, for example, McMichael et al., 2008). Much uncertainty remains about the degree to which housing quality affects this vulnerability, but evidence from the United Kingdom, which has a comparatively large burden of winter excess death, provides the most relevant quantification (Wilkinson et al., 2001), broadly consistent with other published work (The Eurowinter Group, 1997; WHO, 2007). It shows that the magnitude of cold- and winter-related excess death is greater in people living in dwellings that have low winter indoor temperatures. Specifically, there is evidence to quantify the risk of cold-related mortality in relation to adequacy of home heating, as reflected by the standardized indoor temperature – the indoor temperature measured when the outdoor temperature is 5°C (Table 4.11). There is less evidence regarding the relationship between housing characteristics and health impacts other than mortality.

4.9.5.2 Psychosocial and Mental Health Benefits of Warmer Indoor Environments

A growing body of evidence indicates that improved household energy efficiency is accompanied by important changes in the use of space, social interaction, and related psychosocial well-being (Gilbertson et al., 2006). What is less clear is the contribution of different effects: sense of control, reduced costs of home heating (for low-income families), or effects that are a direct function of warmer indoor environments.

4.9.5.3 Indirect Effects on Environmental Pollution and Climate Active Pollutant Emissions

Although household energy efficiency is likely to have its greatest impact on health through direct effects on householders, the impact on wider environmental emissions of toxic pollutants and CAPs is also important. However, the evidence is mixed whether energy efficiency gains translate into lower emissions of CAPs. To date, the evidence suggests that much of the benefit of energy efficiency is taken as higher indoor temperatures in winter rather than as energy savings.

4.9.6 Conclusion

The goal of reducing CAP emissions should assume should central importance in the energy policies of many national and local administrations. A range of policy instruments and interventions are could be used to try to limit energy use and drive energy efficiency, as well as to support a modal shift towards renewable energy sources. However, most energy efficiency improvements require appreciable capital investment, and renewable energy sources currently remain more expensive than conventional (fossil) fuels. Carbon taxes or tariffs to support market shifts therefore generally act to increase fuel prices. This may be damaging to poor groups in higher-income countries already unable to pay for their fuel or to those in developing countries if households are driven down the "energy ladder" to polluting biomass fuels by higher fuel prices. To avoid the adverse health consequences of such policies, it is necessary to provide direct support to low income groups through price subsidies ("winter fuel payments" and the like), though this in part negates the rationale for the carbon tax, or to combine carbon taxation with targeted infrastructure investment to improve energy efficiency for disadvantaged groups.

From a health perspective, however, it is not always clear that such investments represent the most cost-effective gains in public health, as when they are set, for example, against other potential forms of health care expenditure. A consequence of the imperative of tackling the challenges of climate change is that substantial resources will need to be transferred from richer to poorer populations to assist them in attaining and using efficient, low-carbon technology.

4.10 Co-benefits for Climate Change Mitigation and Health Promotion

4.10.1 Key Messages

- For long-term climate protection, CO_2 emissions need to be reduced through energy efficiency and shifts away from fossil fuels, both of which may reduce health-damaging pollutants even if no special emission controls are used.

- Reducing most CAPs, which have substantial direct and indirect health impacts, will lead to substantial health and climate protection over the next decades.

- Other important co-benefits opportunities for health and climate include providing access to reproductive services for women, working to lower meat consumption, and actions to redesign cities and transport systems to reduce physical inactivity.

- Reduction of the CAPs, nitrogen and sulfur emissions, although important for health and ecosystem protection, unfortunately slightly aggravates global warming by reducing the cooling effects of the light-colored particles that are created from current emissions.

There are a number of potential co-benefits opportunities, which ameliorate both health and climate, in the wake of changes in the world's energy system to reduce CAPs.[18] For example, reducing fossil fuel combustion in power production will improve air quality (Markandya et al., 2009), preventing respiratory and cardiovascular disease; and an increase in use of mass transit, cycling, and walking will increase physical

activity, reduce obesity, and stimulate social contacts (Woodcock et al., 2009). In high-income countries, where average daily intake of red meat typically exceeds nutritional needs, a reduction in meat production and consumption – especially from ruminant animals, including cattle, sheep, and goats – would confer several health and environmental gains, partly through reduction in energy use from processing, storage, and transport (Friel et al., 2009). Another major set of co-benefits can accrue by promoting reproductive rights for women around the world. Studies show that giving women control over their fertility will lower both maternal and child mortality, which are both important parts of the unfinished health agenda worldwide, as targeted in the MDGs (Smith and Balakrishnan, 2009). Helping countries reach replacement fertility earlier than would otherwise occur will, in turn, reduce energy use and climate impacts (O'Neill et al., 2010). Finally, nearly any improvement in energy efficiency or shift to different sources that reduces carbonaceous fuel combustion will generally have commensurate health advantages, although care must be exercised that other risks are not substituted, for example, from increased indoor air pollution when building insulation is enhanced (Wilkinson et al., 2009).

Although there are a number of examples of important co-benefits to be gained within these classes of interventions, designing acceptable policies to take advantage of them will require consideration of other factors, among which cost is probably most important. It is not straightforward, however, to convert either health or climate protection benefits into economic terms, let alone in a consistent way across both classes of benefits, in order to determine which projects should be undertaken first. There is no space in GEA to provide such analyses across the large range of alternative projects, countries, and impacts, but we have provided some entry into the literature in Boxes 4.2 and 4.6.

Not all CAPs are created equal from either a climate or a health standpoint. Some are substantially more damaging than others and, in the case of climate impacts, some actually have beneficial (cooling) characteristics, although all are damaging to health. This creates a complex landscape for planning interventions that have benefits for both health and climate.

To inform this strategy, in Box 4.2 we provide a brief primer of the climate and health characteristics of the major shorter-lived CAPs emitted from the energy sector, all of which have implications for health.[19] Figure 4.15 summarizes what is understood about them at present (Smith et al., 2009). Figure 4.7, which is taken directly from the IPCC Fourth Assessment (Forster and Ramaswamy, 2007), shows the estimated contributions from these agents of global warming in the year 2005 due to all human emissions since 1750. The CAPs are compared by radiative forcing (RF), which is the metric used for the comparisons of the strength of different human and natural agents in causing climate change (Smith et al., 2009).

---

18   See the special series on co-benefits in the *Lancet* published November 2009, in particular the summary policy paper (Haines et al., 2009).

19   Not discussed are CAPs such as N_2O, H_2, and NH_3, for which fuel combustion is a minor contributor.

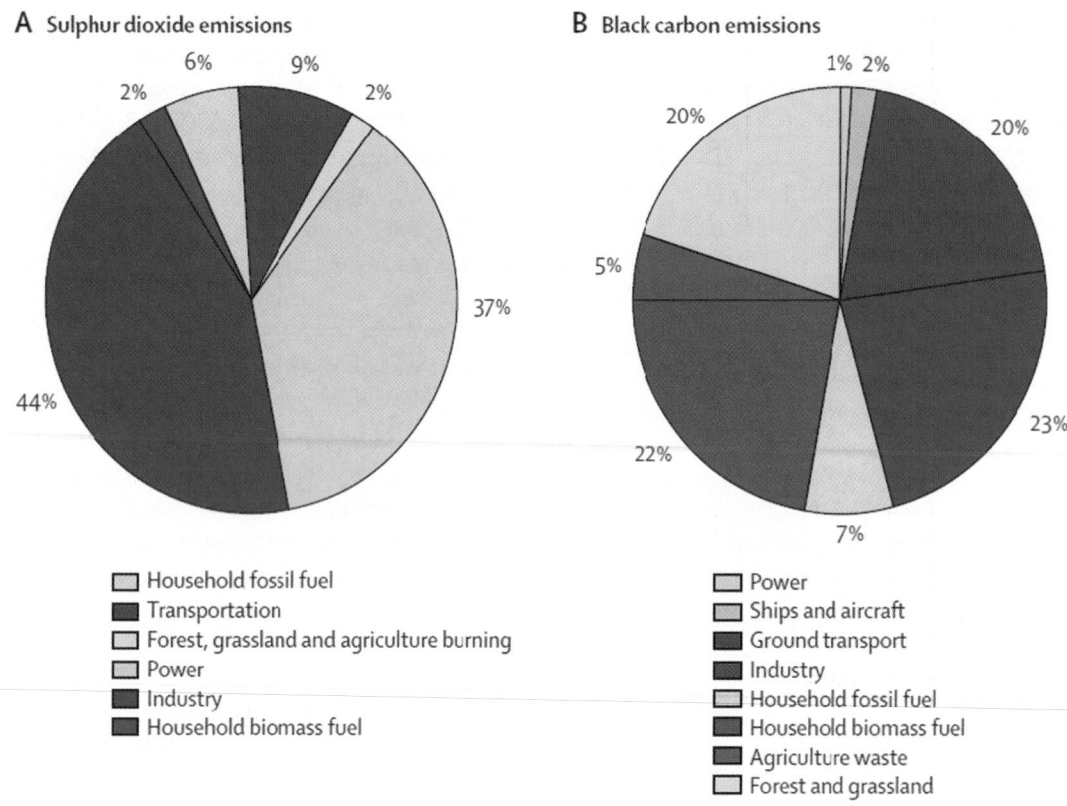

A Sulphur dioxide emissions

B Black carbon emissions

Household fossil fuel
Transportation
Forest, grassland and agriculture burning
Power
Industry
Household biomass fuel

Power
Ships and aircraft
Ground transport
Industry
Household fossil fuel
Household biomass fuel
Agriculture waste
Forest and grassland

Figure 4.15 | Sulfur dioxide and black carbon emission sources. Source: Smith et al., 2009.

Box 4.6 | Economic evaluations of co-benefits

There is a growing literature attempting to quantify the economic implications of CAP reductions for health and climate protection. Progress has been made in establishing scoping methods (Smith and Haigler, 2008). There are also important and often not well understood atmospheric interactions that require detailed local information and modeling for estimating co-benefits. Nevertheless, studies of growing sophistication, using local data, have been done in Asia and elsewhere to pin down the potentials.[20]

As an illustration of one approach, Figure 4.16 below from Smith and Haigler (2008) compares the cost-effectiveness of a range of interventions in terms of both climate and health protection. For metrics, it uses international US dollars per tonne CO_2-equivalent, which applies a global warming potential to tonnes of methane and nitrous oxides reduced to be able to add them to the tonnes of CO_2 reduced. For health, it uses the approach recommended by the World Health Organization (WHO) for cost-effectiveness, i.e., international US dollars per saved DALY (Murray et al., 2000). WHO recommends that any intervention costing less than three times the local GDP per capita be considered seriously for implementation (WHO, 2003). The price of CO_2-equivalents in early 2012 on the international market was about US$ 9 per tonne CO_2-eq (Carbon Positive, 2012). The projects shown in the figure are only illustrative examples from the literature and are not meant to be best or worst options. Rather, the Figure illustrates one approach for such analysis (see article for details).

---

20 A good example is the set of studies done in cities around the world as part of the Integrated Environmental Strategies Project of the USA EPA (en.openei.org/wiki/EPA-Integrated_ Environmental_Strategies). Many other references are found in Smith and Haigler (2008).

In general, household fuel interventions in poor countries have the highest returns because of the inefficiency by which traditional fuels are currently used, leading to high health-damaging CAP emissions, and the high intake fractions, leading to large health effects.

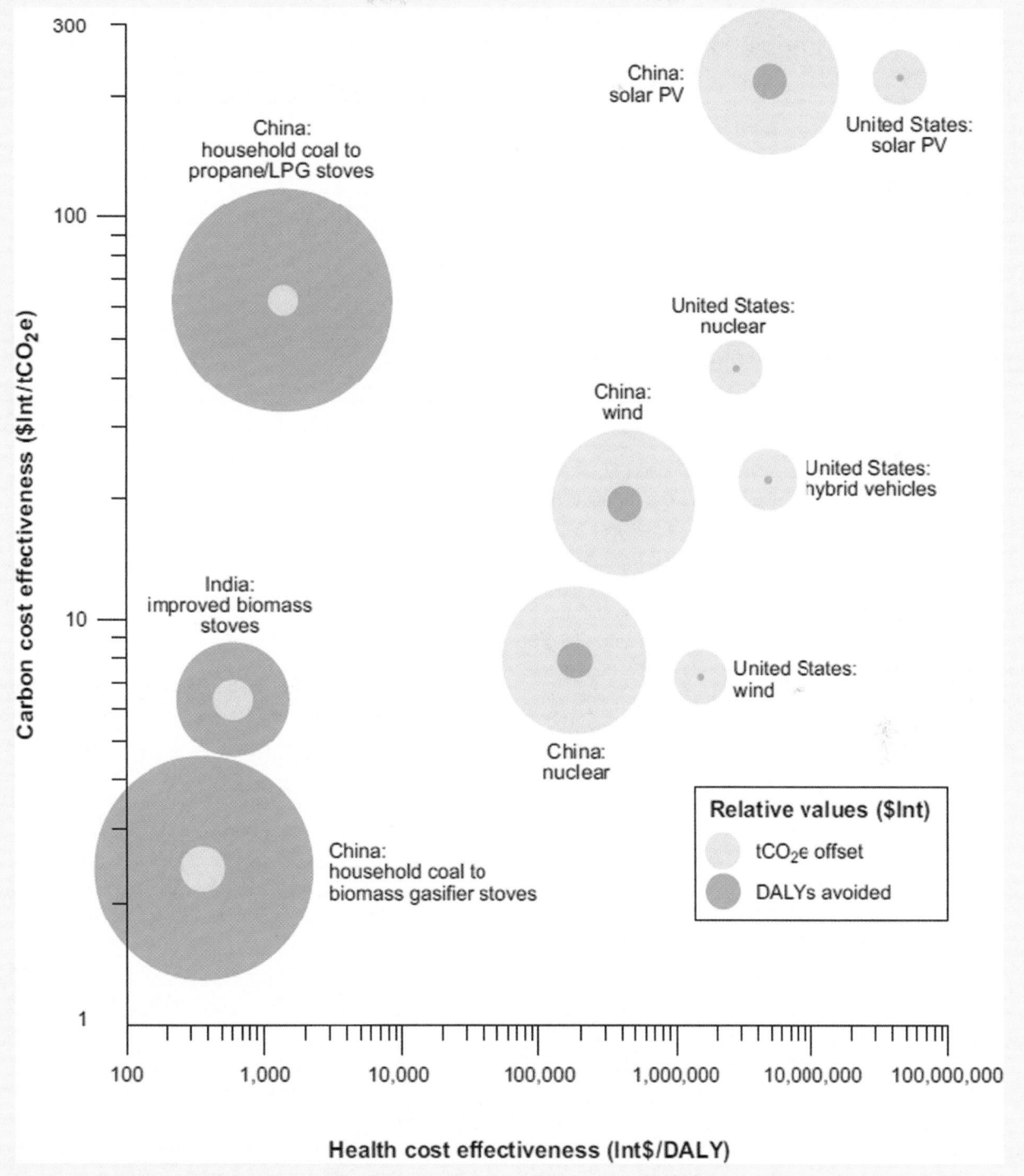

Figure 4.16 | Illustrative examples of co-benefits analysis: comparison of the health and climate cost-effectiveness of household, transport, and power sector interventions. Source: Smith and Haigler, 2008.

Aerosols influence RF directly through the reflection and absorption of solar and infrared radiation (Smith et al., 2009). Some aerosols, such as BC, cause a positive RF (i.e., climate warming) whereas others, such as sulfates, cause a negative RF (i.e., climate cooling) (see Figure 4.7).

The direct RF summed over all aerosol types is thought to be negative. Aerosols also indirectly cause a negative RF through the changes they cause in cloud properties, which are not shown in Figure 4.7. It has not been possible to explain the observed global temperature records for the 20th century with global climate models without a substantial post-World War II cooling term from the burning of sulfur-containing fossil fuels to partially offset the global warming from the CAPs. As nitrogen and sulfur oxides are reduced globally due to health and acid-precipitation concerns, their cooling effect will also be reduced, thus unmasking the climate warming due to other CAPs that they are currently counteracting.

In addition, as it is not possible to control BC emissions without simultaneously controlling the associated OC particle emissions from the same combustion sources, the global climate benefits of carbon-particle control measures, including improved combustion efficiency, depend on the ratio of BC to OC of each source. More recent observations and modeling (Ramanathan and Carmichael, 2008), however, indicate that the estimated warming of BC may be higher than those in the most recent IPCC estimates indicated in Figure 4.7. This would have policy implications because a much broader range of combustion sources would consequently have net climate benefits from combined control of BC and OC (Smith et al., 2009).

It is important to distinguish between direct and indirect RF. Direct RF results when the emitted substance is a greenhouse pollutant itself, such as CO_2. Indirect RF results when the emitted pollutant is not a greenhouse pollutant but takes part in chemical reactions to form a greenhouse pollutant or to change the global distribution of another greenhouse pollutant. SO_2, for example, is transformed in the atmosphere to form aerosol sulfates that act to produce a negative RF. NO_x emissions act to increase the oxidizing capacity of the troposphere, reducing methane (negative RF), but adding to tropospheric ozone (positive RF), while methane, carbon monoxide (CO), and non-methane volatile organic carbons (NMVOCs) contribute to O_3. Although not emitted directly, nevertheless tropospheric ozone is the third most important human-influenced CAP in the atmosphere, as shown in Figure 4.7 (IPCC, 2007a).

As methane and carbon dioxide are well mixed globally, it is possible to treat emissions in all places and seasons as essentially equal, which led to the deployment of so-called "global warming potentials" by the IPCC, used to gauge the relative importance of emissions of different GHGs in treaties, inventories, and international carbon-offset programs. The complexity, short life, and local dependence of the short-lived CAPs, however, makes it difficult if not impossible to establish official "global" warming potentials for policy making. At present, therefore, they are not included in most international climate policy deliberations, although they are subject to much research and media attention and featured prominently in scientific assessments (Smith et al., 2009).

This gap is increasingly important as the world seeks to reduce global warming risks in ways that are both cost-effective and compatible with other goals, such as protection from outdoor air pollutants. In addition, control of the CAPs will not only have immediate health benefits, but much quicker climate benefits than control of CO_2. Not

having a way to systematically look at co-benefits across the CAPs hampers these efforts. In addition, although the aerosols seem to have a net cooling effect, they can still impose regional climate impacts in South and East Asia, for example, and thus should not be considered benign (or beneficial) for the climate in addition to their important health consequences. Some have suggested, therefore, that regional climate disruption potentials might someday be applied to weight the relative importance of emission reductions for these CAPs.

4.11 Conclusions

4.11.1 Key Messages

- Most health impacts of today's energy systems come from the extraction and combustion of solid fuels. Some impacts are among workers collecting and processing coal and biomass. About 40% of the people in the world are exposed to household air pollution from the poor combustion of solid fuels used for heating, cooking, and lighting. This air pollution also contributes substantially to outdoor air pollution.

- Health impacts from most new and renewable energy sources are likely to be much smaller, but vigilance is needed to be sure these energy sources are managed carefully.

- Nuclear power poses low routine health risks to workers and the public if properly regulated. See Chapter 14 for a discussion of accident and proliferation risks.

- For long-term climate protection, CO_2 emissions need to be reduced through energy efficiency measures and a general shift away from fossil fuels. Shifting away from fossil fuel combustion may also proportionally reduce associated health-damaging pollutants, even if no special pollutant emission controls are used.

- Although energy efficiency measures are generally desirable, there are potential health impacts if done poorly, for example as a result of reduced home air exchange.

- Human-engendered climate change, which is largely but not entirely caused by energy use, seems already to be imposing health impacts, particularly among poor populations. Health impacts from climate change can be expected to steadily grow in the next decades unless major mitigation and adaptation efforts are undertaken.

- There are large health benefits from reducing carbon-containing aerosols, but the climate benefits are blunted somewhat because both BC aerosols (which are primarily warming) and OC aerosols (which are primarily cooling) usually decline together.

- There are climate, as well as health, benefits from reducing BC emissions, however, that may land on snow or ice, even if OC emissions are reduced as well.

- Per unit useful energy, the health benefits of emission reduction interventions rise as the combustion is closer to the population, i.e., as the intake fraction rises.

- Per unit useful energy, the health and climate benefits of emission reduction interventions rise with the fraction of incomplete combustion.

- Given the previous two conclusions and the widespread use of solid fuels in households, the largest co-benefit opportunities for health and climate lie in substituting away from solid fuels or increasing the solid fuel combustion efficiency in households. This is illustrated in Box 4.6.

- To capture the potential co-benefits for health and climate of changes in energy systems, society needs to make sure actions do not just shift the hazards from one population to another, e.g., shift the health burden from coal burning to a potentially even higher burden from nuclear proliferation.

- Using a life-cycle approach to evaluate energy sources, especially new and renewable fuels, is important to fully understand the health

and climate costs and benefits across production, storage, transport, and end-use processes.

- The rush to find and implement ways to mitigate CAP emissions should not result in significant relaxation of the environmental and safety controls on new or existing energy systems, including those affecting outdoor air pollution, which were developed over the years to protect health.

- ILO and WHO guidelines for the protection of worker and public health should be adopted into law and enforcement practices worldwide.

- Control of primary and secondary aerosols may, in some cases, yield negative effects for either health or climate goals. For this reason, there is a need for multidisciplinary approaches to assessing and establishing aerosol policy goals, such that health and climate co-benefits can be maximized.

Energy systems play important roles in the burden of disease globally. GEA's estimates are that household air pollution from solid fuels was responsible for 2.2 million premature deaths in 2005 and outdoor air pollution was responsible for about 2.7 million premature deaths in the same year. Part of the latter can also be attributed to poor household

Table 4.12 | Rough contribution of energy systems to the global burden of disease, 2000.

Direct Effects [except where noted, 100% assigned to energy]	Total Premature Deaths – million	Percent of all Deaths	Percent of Global Burden in DALYs	Trend
Household solid fuel	1.6	2.9	2.6	Stable
Energy Systems Occupational*	0.2	0.4	0.5	Uncertain
Outdoor Air Pollution	0.8	1.4	0.8	Stable
Climate Change	0.15	0.3	0.4	Rising
Subtotal	2.8	5.0	4.3	
Indirect Effects (100% of each)				
Lead in vehicle fuel	0.19	0.3	0.7	Falling
Road traffic Accidents	0.8	1.4	1.4	Rising
Physical inactivity	1.9	3.4	1.3	Rising
Subtotal	2.9	5.1	3.4	
Total	5.7	10.1	7.7	

* One-third of global total assigned to energy systems

Notes: These are not 100% of the totals for each, but represent the difference between what exists now and what might be achieved with feasible policy measures. Thus, for example, they do not assume the infeasible reduction to zero traffic accidents or air pollution levels.

Source: Ezzati 2004.

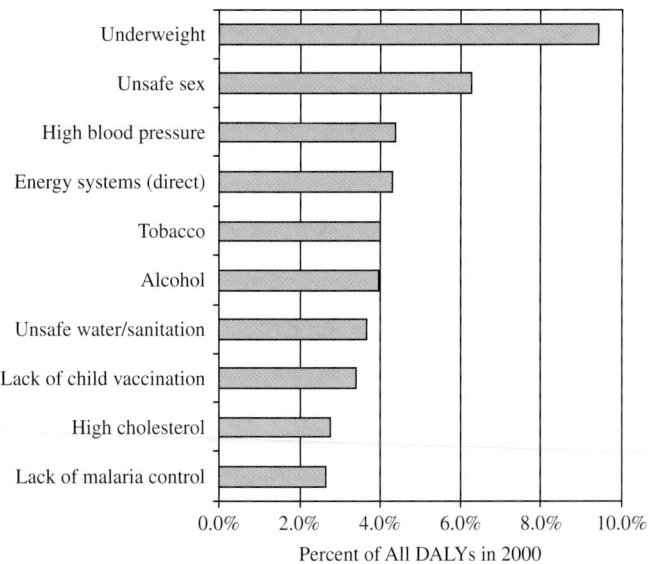

Figure 4.17 | Comparison of global burden of disease from major risk factors, including direct impacts of energy systems around the year 2000. Source: Table 4.12 and WHO Global Burden of Disease database (WHO, 2008).

Note: All expressed as a percentage of the total global burden in lost healthy life years (DALYs).

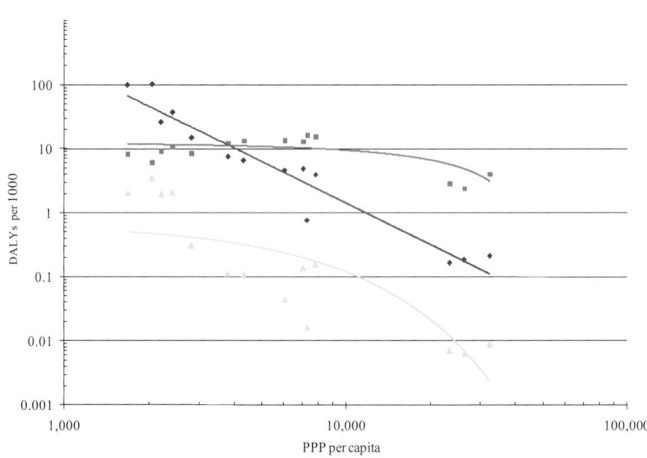

Figure 4.18 | Environmental risk transition: experienced burden of disease from environmental risk factors around 2000. Source: data based on Smith & Ezzati, 2005.

Notes: Household environmental risks (in blue), including pollution from solid fuel use, tend to decline with economic development, while community air pollution and traffic risks from energy use (red) first rise and then fall. Experienced risks from climate change (yellow) also tend to decline with development status. Vertical axis is in lost healthy life years (DALYs) per capita in each of 14 WHO epidemiological world sub-regions distributed by average income per capita. Note log scale on axes.

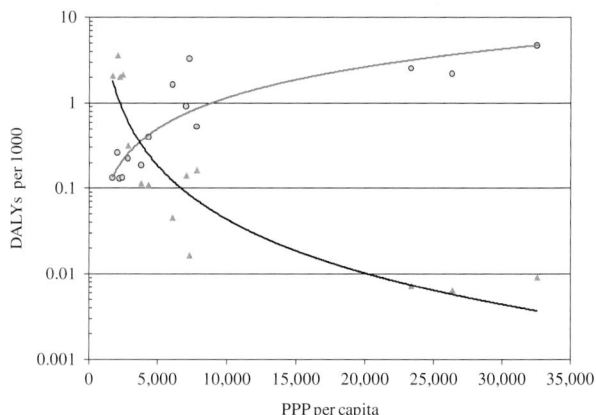

Figure 4.19 | Distribution of health impacts from climate change, experienced versus imposed, 2000. Source: data based on Smith & Ezzati, 2005.

Notes: Based on cumulative CO_2 emissions. Imposed health impacts from climate change (red line) rise with economic development. This is inverse of the pattern for experienced health impacts (black line), resulting in large inequities between rich and poor regions globally. Vertical axis is in lost healthy life years (DALYs) per capita in each of 14 WHO epidemiological world sub-regions distributed by average income per capita.

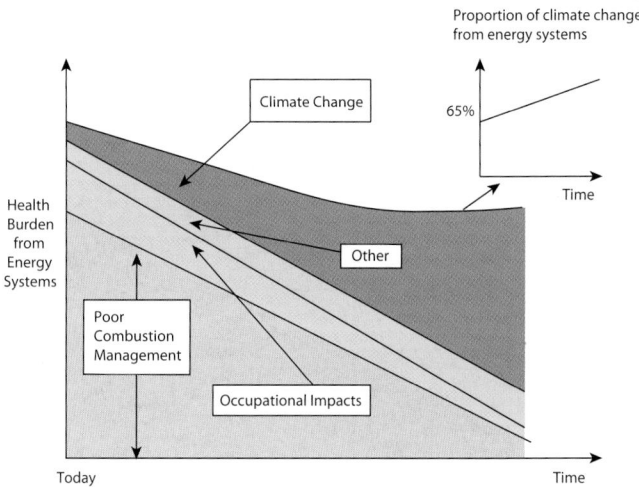

Figure 4.20 | Current trends in health impacts from energy systems.

Note: This represents a pessimistic future projection, in which combustion efficiency improves but effective climate policies do not exist.

fuel combustion. (Details on the calculations used to determine these figures can be found in Chapter 17.) Inclusion of energy's indirect role in lead pollution and occupational risks would probably add another 10–20% to that figure. Although outside the system boundaries of energy supply systems used in GEA, inclusion of road traffic accidents and energy's role in physical inactivity would roughly double these values. Recognizing the uncertainties of such calculations, the direct effects alone put energy systems above the global health impact of all other risk factors except malnutrition and unsafe sex and roughly equal to the global impacts of tobacco, alcohol, and high blood pressure based on WHO estimates for 2000 (see Table 4.12 and Figure 4.17). As noted in this chapter, the vast part of the direct impact comes from the poor management of fuel combustion.

Note these are not 100% of the totals for each, but represent the difference between what exists now and what might be achieved with

314

feasible policy measures. Thus, for example, they do not assume the infeasible reduction to no traffic accidents or zero pollution levels.

Figure 4.18 shows how these energy system risks vary across the world by income level, illustrating what has been called the "Environmental Risk Transition" (Smith and Ezzati, 2005). Household-level risks tend to fall with economic development because they are so heavily influenced by household solid fuel use. Community (urban) risks, here including occupational impacts, form a more complex pattern by rising first during economic growth and then falling, roughly following what has been called the Environmental Kuznets Curve. Note, however, that there is much variation across the world in this pattern due to earlier or later introduction of policy measures. Experienced global risks due to climate change fall heavily on those living in poverty and thus also decline with development.

Figure 4.19 shows, however, that if global risks are distributed not by who experiences them but rather by who imposes them, the pattern is inverted. Poor countries receive much more impact than they have imposed in terms of their cumulative CO_2 emissions and rich countries impose much more than they experience. This has been called the "largest regressive tax in history."

Most of the health impacts of energy systems today are due to poor management of the combustion of fuels through emissions of incomplete combustion products and contaminants such as sulfur and nitrogen oxides. Second in importance are occupational impacts, particularly in the harvesting and processing of solid fuels. As world energy planning reorients toward reducing greenhouse emissions, there is need to take close account of the potential for doing so in ways that also continue to move society toward other important goals, such as those codified in the MDGs. Energy systems in general, not only have close links to CAP emissions, but in some cases to MDGs as well. In particular, household fuels are the system in which the two are most closely linked, with consequent opportunity to move both climate mitigation and health protection forward together. This is why providing access to clean household

fuels and technologies is one of the major goals of the GEA scenarios in Chapter 17.

The health impact of climate change from energy use is relatively minor at present, partly because only about two-thirds of human-caused warming comes from energy systems (see Figure 4.20). The Figure assumes that the goals for the health indicators in the scenarios (Chapter 17) are met, which will mean very much lower health impacts from these two sources within a few decades. The impacts of climate change, however, are more refractory, partly because considerable climate change has already been committed and also because there is no sign at present that world society has found a way to grapple with greenhouse emissions effectively. In addition, as shown in the figure, the fraction of climate change due to energy systems will slowly rise under current trends. Shown in Figure 4.20, therefore, is a pessimistic outcome in which total health impacts from energy systems fall for some period as air pollution and occupational impacts come under control, but then start to rise as the health impacts from climate change begin in earnest. This outcome also depends, however, on the progress of society in dealing with the vulnerability of poor populations, since climate change operates mainly to enhance existing patterns of ill health. Thus, other scenarios than the one shown here are not only desirable but possible, albeit at considerable effort across several sectors.

Acknowledgements

11 April 2012

The authors are grateful for research assistance provided by Dr. Alfred Körblein and the, logistical assistance provided by Pat Wagner at IIASA. Zoë Chafe's participation in the IIASA Young Scientists Summer Program was enabled, in part, by a grant from the National Academy of Sciences Board on International Scientific Organizations, funded by the National Science Foundation (Grant No. OISE-0738129).

References

AGIR, 2007: *Review of Risks from Tritium*. Report of the independent Advisory Group on Ionising Radiation (RCE-4), The Advisory Group on Ionising Radiation (AGIR) on behalf of the United Kingdom Health Protection Agency, Chilton, UK.

Akachi, Y., D. Goodman and D. Parker, 2009: *Global Climate Change and Child Health: A Review of Pathways, Impacts and Measures to Improve the Evidence Base.* UNICEF Innocenti Research Centre.

Albalak, R., N. Bruce, J. P. McCracken, K. R. Smith and T. de Gallardo, 2001: Indoor Respirable Particulate Matter Concentrations from an Open Fire, Improved Cookstove, and Lpg/Open Fire Combination in a Rural Guatemalan Community. *Environmental science and technology*, **35**(13): 2650–2655.

Allison, I., N. L. Bindoff, R. A. Bindschadler, P. M. Cox, N. d. Noblet, M. H. England, J. E. Francis, N. Gruber, A. M. Haywood, D. J. Karoly, D. Kaser, C. L. Quéré, T. M. Lenton, M. E. Mann, B. I. McNeil, A. J. Pitman, S. Rahmstorf, E. Rignot, H. J. Schellnhuber, S. H. Schneider, S. C. Sherwood, R. C. J. Somerville, K. Steffen, E. J. Steig, M. Visbeck and A. J. Weaver, 2009: *The Copenhagen Diagnosis. Updating the World on the Latest Climate Science.* The University of New South Wales Climate Change Research Centre (CCRC), Sydney, Australia.

Alonso, D., M. J. Bouma and M. Pascual, 2011: Epidemic Malaria and Warmer Temperatures in Recent Decades in an East African Highland. *Proceedings of the Royal Society B*, **278**(1712): 1661–1669.

Archer, C. L. and M. Z. Jacobson, 2007: Supplying Baseload Power and Reducing Transmission Requirements by Interconnecting Wind Farms. *Journal of Applied Meteorology & Climatology*, **46**(11): 1701–1718.

Armstrong, J. and R. Menon, 1998: Mining and Quarrying. In *Encyclopaedia of Occupational Health and Safety*. J. M. Stellman (ed.), International Labour Office (ILO), Geneva, Vol. 3, Chapter 74, pp. 74.71–74.55.

ASN, 2010: *White Paper on Tritium.* Autorite de Securite Nucleaire (ASN), Paris.

Associated Press, 2008: *Bund Und Länder Einig Über Uran-Grenzwert Für Trinkwasser.* Associated Press (AP), New York.

Avnery, S., D. L. Mauzerall, J. Liu and L. W. Horowitz, 2011a: Global Crop Yield Reductions Due to Surface Ozone Exposure: 1. Year 2000 Crop Production Losses, Economic Damage, and Implications for World Hunger. *Atmospheric Environment*, **45**: 2297–2309.

Avnery, S., D. L. Mauzerall, J. Liu and L. W. Horowitz, 2011b: Global Crop Yield Reductions Due to Surface Ozone Exposure: 2. Year 2030 Potential Crop Production Losses, Economic Damage, and Implications for World Hunger under Two Scenarios of O3 Pollution. *Atmospheric Environment*, **45**: 2284–2296.

Baker, P. J. and D. G. Hoel, 2007: Meta-Analysis of Standardized Incidence and Mortality Rates of Childhood Leukaemia in Proximity to Nuclear Facilities. *European Journal of Cancer Care*, **16**(4): 355–363.

Balakrishnan, K., J. Parikh, S. Sankar, R. Padmavathi, K. Srividya, V. Venugopal, S. Prasad and V. L. Pandey, 2002: Daily Average Exposures to Respirable Particulate Matter from Combustion of Biomass Fuels in Rural Households of Southern India. *Environmental Health Perspectives*, **110**(11): 1069.

Balakrishnan, K., S. Sambandam, P. Ramaswamy, S. Mehta and K. R. Smith, 2004: Exposure Assessment for Respirable Particulates Associated with Household Fuel Use in Rural Districts of Andhra Pradesh, India. *Journal of Exposure Analysis & Environmental Epidemiology*, **14**: S14–S25.

Barnes, B. R., A. Mathee, L. Krieger, L. Shafritz, M. Favin and L. Sherburne, 2004: Testing Selected Behaviors to Reduce Indoor Air Pollution Exposure in Young Children. *Health Education Research*, **19**(5): 543–550.

Barnes, D., F., K. Openshaw, K. Smith, R. and R. van der Plas, 1993: The Design and Diffusion of Improved Cooking Stoves. *The World Bank Research Observer*, **8**(2): 119–142.

Barnes, D. F., K. Openshaw, R. S. Kirk and R. van der Plas, 1994: *What Makes People Cook with Improved Biomass Stoves? A Comparative International Review of Stove Programs.* World Bank Technical Paper (242). Energy Series, World Bank, Washington, DC, USA.

Barnett, J. and W. N. Adger, 2003: Climate Dangers and Atoll Countries. *Climatic Change*, **61**: 321–337.

Battisti, D. S. and R. L. Naylor, 2009: Historical Warnings of Future Food Insecurity with Unprecedented Seasonal Heat. *Science*, **323**: 240–244.

Begum, B., S. Biswas and P. Hopke, 2008: Assessment of Trends and Present Ambient Concentrations of Pm2.2 and Pm10 in Dhaka, Bangladesh. *Air Quality, Atmosphere & Health*, **1**: 125–133.

Bell, M. L., A. McDermott, S. L. Zeger, J.M. Samet and F. Dominici, 2004: Ozone and Short-term Mortality in 95 USA Urban Communities, 1987–2000. *Journal of the American Medical Association*, **292**(19):2372–2378.

Bennett, D. H., T. E. McKone, J. S. Evans, W. W. Nazaroff, M. D. Margni, O. Jolliet and K. R. Smith, 2002: Peer Reviewed: *Defining Intake Fraction. Environmental Science & Technology*, **36**(9): 206A–211A.

Besser, J. M., J. P. Giesy, R. W. Brown, J. M. Buell and G. A. Dawson, 1996: Selenium Bioaccumulation and Hazards in a Fish Community Affected by Coal Fly Ash Effluent. *Ecotoxicol Environ Saf*, **35**(1): 7–15.

Blackden, C. M. and Q. Wodon, 2006: *Gender, Time Use, and Poverty in Sub-Saharan Africa. World Bank Working Paper no. 73*, Washington, DC, USA.

Bloomberg, M. R. and R. T. Aggarwala, 2008: Think Locally, Act Globally: Howe Curbing Global Warming Emissions Can Improve Local Public Health. *American Journal of Preventive Medicine*, **35**(5): 414–423.

Bornehag, C. G., J. Sundell and T. Sigsgaard, 2004: Dampness in Buildings and Health (Dbh): Report from an Ongoing Epidemiological Investigation on the Association between Indoor Environmental Factors and Health Effects among Children in Sweden. *Indoor Air*, **14**(s7): 59–66.

Boyce, D. G., M. R. Lewis and B. Worm, 2010: Global Phytoplankton Decline over the Past Century. *Nature*, **466**: 591–596.

Brauer, M. and K. Bartlett, 1996: Assessment of Particulate Concentrations from Domestic Biomass Combustion in Rural Mexico. *Environmental Science & Technology*, **30**(1): 104–110.

Bruce, N., J. McCracken, R. Albalak, M. Schei, K. R. Smith, V. Lopez and C. West, 2004: Impact of Improved Stoves, House Construction and Child Location on Levels of Indoor Air Pollution Exposure in Young Guatemalan Children. *Journal of Exposure Analysis & Environmental Epidemiology*, **14**: S26–S33.

Bruce, N., M. Weber, B. Arana, A. Diaz, A. Jenny, L. Thompson, J. McCracken, M. Dherani, D. Juarez, S. Ordonez, R. Klein and K. R. Smith, 2007: Pneumonia Case-Finding in the Respire Guatemala Indoor Air Pollution Trial: Standardizing Methods for Resource-Poor Settings. *Bulletin of the World Health Organization*, **85**(7): 535–544.

Butler, C. D., 2009: Food Security in the Asia-Pacific: Malthus, Limits and Environmental Challenges. *Asia Pacific Journal of Clinical Nutrition*, **18**(4): 577–584.

Butler, C. D., 2010: Climate Change, Crop Yields, and the Future. *SCN News*, **38**: 18–25.

Butler, C. D. and D. Harley, 2010: Primary, Secondary and Tertiary Effects of Eco-Climatic Change: The Medical Response. *Postgraduate Medical Journal*, **86**(1014): 230–235.

Butler, D., 2007: Darfur's Climate Roots Challenged. *Nature (London)*, **447**(7148): 1038–1038.

Cai, W., 2006: Antarctic Ozone Depletion Causes an Intensification of the Southern Ocean Super-Gyre Circulation. *Geophysical Research Letters*, **33**: L03712.

Carballo, M., C. B. Smith and K. Pettersson, 2008: Health Challenges. *Forced Migration Review*, (**31**): 32–33.

Carbon Positive, 2011: *Carbon Positive*. www.carbonpositive.net (accessed 31 December, 2010).

Cardis, E., M. Vrijheid and E. Gilbert, 2005: Risk of Cancer after Low Doses of Ionising Radiation – Retrospective Cohort Study in 15 Countries. *BMJ: British Medical Journal*, **331**(7508): 77–81.

Carpenter, S., A. Wilson and P. S. Mellor, 2009: Culicoides and the Emergence of Bluetongue Virus in Northern Europe. *Trends in Microbiology*, **17**(4): 172–178.

Chauhan, A., H. M. Inskip, C. H. Linaker, S. Smith, J. Schreiber, S. L. Johnston and S. T. Holgate, 2003: Personal Exposure to Nitrogen Dioxide (No2) and the Severity of Virus-Induced Asthma in Children. *Lancet (British edition)*, **361**(9373): 1939–1944.

Chelani, A. B. and S. Devotta, 2007: Air Quality Assessment in Delhi: Before and after Cng as Fuel. *Environmental Monitoring and Assessment*, **125**: 257–263.

Chowdhury, Z., M. Zheng, J. J. Schauer, R. J. Sheesley, L. G. Salmon, G. R. Cass and A. G. Russell, 2007: Speciation of Ambient Fine Organic Carbon Particles and Source Apportionment of Pm2.5 in Indian Cities. *Journal of Geophysical Research*, **112**(D15): D15303.

City of New York, 2007: *PlaNYC: A Greener, Greater New York*. Office of the Mayor, New York.

Clancy, L., P. Goodman, H. Sinclair and D. W. Dockery, 2002: Effect of Air-Pollution Control on Death Rates in Dublin, Ireland: An Intervention Study. *Lancet (British edition)*, **360**(9341): 1210–1214.

CNSC, 2010a: *Health Effects, Dosimetry and Radiological Protection of Tritium*. Canadian Nuclear Safety Commission (CNSC), Ottawa, Canada.

CNSC, 2010b: *Tritium Studies Project Synthesis Report*. Canadian Nuclear Safety Commission (CNSC), Ottawa, Canada.

Cohen, A. J., H. R. Anderson, B. Ostro, K. D. Pandey, M. Krzyzanowski, N. Künzli, K. Gutschmidt, C. A. P. III, I. Romieu, J. M. Samet and K. R. Smith, 2004: Urban Air Pollution. In *Comparative Quantification of Health Risks: Global and Regional Burden of Disease Attributable to Selected Major Risk Factors Geneva*. M. Ezzati, A. D. Lopez and C. J. L. Murray (eds.), World Health Organization, pp. 1353–1433.

COMEAP, 2004: *Guidance on the Effects on Health of Indoor Air Pollutants*. Committee on the Medical Effects of Air Pollutants, UK Department of Health, London, UK.

Commission for the Social Determinants of Health, 2008: *Closing the Gap in a Generation*. World Health Organization (WHO), Geneva, Switzerland.

Confalonieri, U., B. Menne, R. Akhtar, K. L. Ebi, M. Hauengue, R. S. Kovats, B. Revich and A. Woodward, 2007: *Human Health. In Climate Change 2007: Impacts, Adaptation and Vulnerability*. Contribution of Working Group II to the Fourth Assessment Report of the Intergovernmental Panel on Climate Change, M. L. Parry, O. F. Canziani, J. P. Palutikof, P. J. van der Linden and C. E. Hanson, Eds., Cambridge University Press, Cambridge, UK. pp. 391–431.

Cooke, R. M., A. M. Wilson, J. T. Tuomisto, O. Morales, M. Tainio and J. S. Evans, 2007: A Probabilistic Characterization of the Relationship between Fine Particulate Matter and Mortality:Elicitation of European Experts. *Environmental Science & Technology*, **41**(18): 6598–6605.

Costello, A., M. Abbas, A. Allen, S. Ball, S. Bell, R. Bellamy, S. Friel, N. Groce, A. Johnson, M. Kett, M. Lee, C. Levy, M. Maslin, D. McCoy, B. McGuire, H. Montgomery, D. Napier, C. Pagel, J. Patel, J. A. Puppim de Oliveira, N. Redclift, H. Rees, D. Rogger, J. Scott, J. Stephenson, J. Twigg, J. Wolff and C. Patterson, 2009: Managing the Health Effects of Climate Change. *The Lancet*, **373**(9676): 1693–1733.

Crane, M., 1998: Power Generation and Distribution. In *Encyclopaedia of Occupational Health and Safety*. J. M. Stellman (ed.), International Labour Office (ILO), Geneva, Vol. 3, Chapter 76, pp. 76.71–76.17.

Culling, H. and K. R. Smith, 2010: Better Radiation Exposure Estimation for the Japanese Atomic Bomb Survivors Enables Us to Better Protect People from Radiation Today. *Journal of Exposure Science and Environmental Epidemiology (JESEE)*, **20**(7): 575–576.

Curtis, V. and S. Cairncross, 2003: Effect of Washing Hands with Soap on Diarrhoea Risk in the Community: A Systematic Review. *The Lancet Infectious Diseases*, **3**(5): 275–281.

D'Souza, R. M., N. G. Becker, G. Hall and K. B. A. Moodie, 2004: Does Ambient Temperature Affect Foodborne Disease? *Epidemiology (Cambridge, Mass.)*, **15**(1): 86–92.

Darby, S., D. Hill, A. Auvinen, J. M. Barros-Dios, H. Baysson, F. Bochicchio, H. Deo, R. Falk, F. Forastiere, M. Hakama, I. Heid, L. Kreienbrock, M. Kreuzer, F. Lagarde, I. Makelainen, C. Muirhead, W. Oberaigner, G. Pershagen, A. Ruano-Ravina, E. Ruosteenoja, A. S. Rosario, M. Tirmarche, L. Tomasek, E. Whitley, H. E. Wichmann and R. Doll, 2005: Radon in Homes and Risk of Lung Cancer: Collaborative Analysis of Individual Data from 13 European Case-Control Studies. *British Medical Journal*, **330**(7485): 223.

Davies, M., M. Ucci, M. McCarthy, T. Oreszczyn, I. Ridley, D. Mumovic, J. Singh and S. Pretlove, 2004: A Review of Evidence Linking Ventilation Rates in Dwellings and Respiratory Health – a Focus on House Dust Mites and Mould. *The International Journal of Ventilation*, **3**(2): 155–168.

de Bilderling, G., A. J. Chauhan, J. A. R. Jeffs, N. Withers, S. L. Johnston, S. T. Holgate and J. B. Clough, 2005: Gas Cooking and Smoking Habits and the Risk of Childhood and Adolescent Wheeze. *American Journal of Epidemiology*, **162**(6): 513–522.

Dear, K., G. Ranmuthugala, T. Kjellström, C. Skinner and I. Hanigan, 2005: Effects of Temperature and Ozone on Daily Mortality During the August 2003 Heat Wave in France. *Archives of Environmental & Occupational Health*, **60**(4): 205–212.

Deffeyes, K. S., 2001: *Hubbert's Peak: The Impending World Oil Shortage*. Princeton University Press, Princeton, NJ.

Dherani, M., D. Pope and M. Mascarenhas, 2008: Indoor Air Pollution from Unprocessed Solid Fuel Use and Pneumonia Risk in Children Aged under Five Years: A Systematic Review and Meta-Analysis. *Bulletin of the World Health Organization*, **86**(5): 390–402.

Dobson, A., 2009: Climate Variability, Global Change, Immunity, and the Dynamics of Infectious Diseases. *Ecology*, **90**(4): 920–927.

Doney, S., V. Fabry, R. Feely and J. Kleypas, 2009: Ocean Acidification: The Other Co2 Problem. *Annual Review of Marine Science*, **1**: 169–192.

Donoghue, A. M., 2004: Occupational Health Hazards in Mining: An Overview. *Occupational Medicine*, **54**(5): 283–289.

Ebi, K. L., J. Smith, I. Burton and J. Scheraga, 2006: Some Lessons Learned from Public Health on the Process of Adaptation. *Mitigation and Adaptation Strategies for Global Change*, 11(3): 607–620.

Edwards, R., D., K. Smith, R., J. Zhang and Y. Ma, 2004: Implications of Changes in Household Stoves and Fuel Use in China. *Energy Policy*, 32(3): 395–411.

Energy Biosciences Institute, 2010: Life-Cycle Environmental and Economic Decision-Making for Alternative Biofuels.

Epstein, P. R., J. J. Buonocore, K. Eckerle, M. Hendryx, B. M. Stout Iii, R. Heinberg, R. W. Clapp, B. May, N. L. Reinhart, M. M. Ahern, S. K. Doshi and L. Glustrom, 2011: Full Cost Accounting for the Life Cycle of Coal. *Annals of the New York Academy of Sciences*, 1219(1): 73–98.

European Commission, 2005: *Radioactive Effluents from Nuclear Power Stations and Nuclear Fuel Reprocessing Sites in the European Union, 1999–2003.* Radiation Protection 143, S. Van der Stricht and A. Janssens, Directorate-General for Energy and Transport, Luxembourg.

European Parliament, 2001: Possible Toxic Effects from the Nuclear Reprocessing Plants at Sellafield (Uk) and Cap De La Hague (France) – a First Contribution to the Scientific Debate. STOA unit, European Parliament, Directorate General for Research, Luxembourg.

Eurowinter Group, 1997: Cold Exposure and Winter Mortality from Ischaemic Heart Disease, Cerebrovascular Disease, Respiratory Disease, and All Causes in Warm and Cold Regions of Europe. *Lancet (British edition)*, 349(9062): 1341–1346.

Ezzati, M., H. Saleh and D. M. Kammen, 2000: The Contributions of Emissions and Spatial Microenvironments to Exposure to Indoor Air Pollution from Biomass Combustion in Kenya. *Environmental Health Perspectives*, 108(9): 833–839.

Ezzati, M., S. V. Hoorn, A. D. Lopez, G. Danaei, A. Rodgers, C. D. Mathers and C. D. L. Murray, 2004: Comparative Quantification of Mortality and Burden of Disease Attributable to Selected Risk Factors. In *Global Burden of Disease and Risk Factors*. A. D. Lopez , C. D. Mathers , M. Ezzati , D. T. Jamison and C. D. L. Murray (eds.), World Bank, Washington, DC.

Fairlie, I., 2009: Childhood Cancers near German Nuclear Power Stations: Hypothesis to Explain the Cancer Increases. *Medicine, Conflict and Survival*, 25(3): 206–220.

FAO, 2009: *The State of Food Insecurity in the World: Economic Crises – Impacts and Lessons Learned.* Food and Agriculture Organization of the United Nations, Rome, Italy.

Fargione, J., J. Hill, D. Tilman, S. Polasky and P. Hawthorne, 2008: Land Clearing and the Biofuel Carbon Debt. *Science*, 319(5867): 1235–1238.

Fiore, A. M., D. J. Jacob, I. Bey, R. M. Yantosca, B. D. Field and A. C. Fusco, 2002: Background Ozone over the United States in Summer: Origin, Trend, and Contribution to Pollution Episodes. *J. Geophys. Res.*, 107(D15): 10.1029.

Fiore, A. M., J. J. West, L. Horowitz, V. Naik and M. D. Schwarzkopf, 2008: Characterizing the Tropospheric Ozone Response to Methane Emission Controls and the Benefits to Climate and Air Quality. *Journal of Geophysical Research*, 113.

Fischer, G., M. Shah, F. N. Tubiello and H. van Velhuizen, 2005: Socio-Economic and Climate Change Impacts on Agriculture: An Integrated Assessment, 1990–2080. *Philosophical Transactions of the Royal Society B: Biological Sciences*, 360(1463): 2067–2083.

Forster, P. and V. Ramaswamy, 2007: Changes in Atmospheric Constituents and in Radiative Forcing. In *Climate Change 2007: The Physical Science Basis* , S. Solomon , D. Qin , M. Manning , M. Marquis , K. Averyt , M. M. B. Tignor , H. L. Miller Jr. and Z. Chen, Eds., Intergovernmental Panel on Climate Change, New York. pp. 129–234.

Friedlingstein, P., 2008: A Steep Road to Climate Stabilization. *Nature (London)*, 451(7176): 297–298.

Friel, S., A. D. Dangour, T. Garnett, K. Lock, Z. Chalabi, I. Roberts, A. Butler, C. D. Butler, J. Waage, A. J. McMichael and A. Haines, 2009: Public Health Benefits of Strategies to Reduce Greenhouse-Gas Emissions: Food and Agriculture. *The Lancet*, 374: 2016–2025.

Frumkin, H., L. Frank and R. Jackson, 2004: *Urban Sprawl and Public Health: Designing, Planning and Building for Healthy Communities. Island Press*, Washington, DC.

Frumkin, H., J. Hess and S. Vindigni, 2007: Peak Petroleum and Public Health. *JAMA*, 298(14): 1688–1690.

Gardner, R. O. N., 2003: Overview and Characteristics of Some Occupational Exposures and Health Risks on Offshore Oil and Gas Installations. *The Annals of Occupational Hygiene*, 47(3): 201–210.

Ghose, M. K. and S. R. Majee, 2007: Characteristics of Hazardous Airborne Dust around an Indian Surface Coal Mining Area. *Environmental Monitoring Assessment*, 130(1–3):17–25.

Gilbertson, J., M. Stevens, B. Stiell and N. Thorogood, 2006: Home Is Where the Hearth Is: Grant Recipients' Views of England's Home Energy Efficiency Scheme (Warm Front). *Social Science and Medicine*, 63(4): 946–956.

Government of India, 2009: *Road Transport Year Book 2006–2007.* R. T. Ministry of Shipping, and Highways, Transport Research Wing, New Delhi.

Grosche, B., M. Kreuzer, M. Kreisheimer, M. Schnelzer and A. Tschense, 2006: Lung Cancer Risk among German Male Uranium Miners: A Cohort Study, 1946–1998. *British Journal of Cancer*, 95(9): 1280–1287.

Haines, A., A. J. McMichael, K. R. Smith, I. Roberts, J. Woodcock, A. Markandya, B. G. Armstrong, D. Campbell- Lendrum, A. D. Dangour, M. Davies, N. Bruce, C. Tonne, M. Barrett and P. Wilkinson, 2009: Public Health Benefits of Strategies to Reduce Greenhouse-Gas Emissions: Overview and Implications for Policy Makers. *The Lancet*, 374(9707): 2104–2114.

Hamilton, L. D., 1998: Energy and Health. In *Encyclopaedia of Occupational Health and Safety*. J. M. Stellman (ed.), International Labour Office (ILO), Geneva, Vol. 3, Chapter 53, pp. 53.19–53.22.

Hansen, J. E., 2007: Scientific Reticence and Sea Level Rise. *Environmental Research Letters*, 2(2): 024002–024002.

Harley, D., A. Swaminathan and A. J. McMichael, 2011: Climate Change and the Geographical Distribution of Infectious Diseases. In *A Geographic Guide to Infectious Diseases*. E. Petersen , L. H. Chen and P. Schlagenhauf (eds.), Wiley-Blackwell, Oxford, pp. 414–423.

He, G., B. Ying, J. Liu, S. Gao, S. Shen, K. Balakrishnan, Y. Jin, F. Liu, N. Tang, K. Shi, E. Baris and M. Ezzati, 2005: Patterns of Household Concentrations of Multiple Indoor Air Pollutants in China. *Environmental science and technology*, 39(4): 991–998.

He, K., H. Huo and Q. Zhang, 2002: Urban Air Pollution in China: Current Status, Characteristics, and Progress. *Annual Review of Energy and the Environment*, 27: 397–431.

Health Canada, 2004: *Canadian Handbook on Health Impact Assessment.* Volume 4: Health Impacts by Industry Sector, A Report of the Federal/Provincial/Territorial Committee on Environmental and Occupational Health.

HEI, 2010: *Outdoor Air Pollution and Health in the Developing Countries of Asia: A Comprehensive Review.* Health Effects Institute (HEI), Boston.

Hendryx, M. and M. M. Ahern, 2008: Relations between Health Indicators and Residential Proximity to Coal Mining in West Virginia. *Am J Public Health*, **98**(4): 669–671.

Hill, J., E. Nelson, D. Tilman, S. Polasky and D. Tiffany, 2006: Environmental, Economic, and Energetic Costs and Benefits of Biodiesel and Ethanol Biofuels. *Proceedings of the National Academy of Sciences of the United States of America*, **103**(30): 11206–11210.

Hill, J., S. Polasky, E. Nelson, D. Tilman, H. Huo, L. Ludwig, J. Neumann, H. Zheng and D. Bonta, 2009: Climate Change and Health Costs of Air Emissions from Biofuels and Gasoline. *Proceedings of the National Academy of Sciences of the United States of America*, **106**(6): 2077–2082.

Hoegh-Guldberg, O. and J. F. Bruno, 2010: The Impact of Climate Change on the World's Marine Ecosystems. *Science*, **328**: 1523–1528.

Hoffmann, W., C. Terschueren and D. B. Richardson, 2007: Childhood Leukemia in the Vicinity of the Geesthacht Nuclear Establishments near Hamburg, Germany. *Environmental Health Perspectives*, **115**(6): 947–952.

Holdren, J. P. and K. R. Smith, 2000: Chapter 3: Energy, the Environment, and Health. In *World Energy Assessment.* United Nations Development Programme, New York, NY, USA.

Hosgood III, H. D., H. Wei, A. Sapkota, I. Choudhury, N. Bruce, K. R. Smith, N. Rothman and Q. Lan, 2011: Household Coal Use and Lung Cancer: Systematic Review and Meta-Analysis of Case-Control Studies, with an Emphasis on Geographic Variation. *International Journal of Epidemiology*.

Hosier, R.H., and J. Dowd, 1987: Household fuel choice in Zimbabwe – An empirical test of the energy ladder hypothesis, *Resources and Energy*, **9**: 347–361.

Humphrey, J. H., 2009: Child Undernutrition, Tropical Enteropathy, Toilets, and Handwashing. *The Lancet*, **374**: 1032–1035.

IARC, 1984: *Polynuclear Aromatic Compounds, Part 3, Industrial Exposures in Aluminium Production, Coal Gasification, Coke Production, and Iron and Steel Founding.* IARC Monographs on the evaluation of Carcinogenic Risks to Humans, Volume 34, International Agency for Research on Cancer (IARC) of the World Health Organization (WHO), Lyon, France.

IARC, 2010: *Household Use of Solid Fuels and High Temperature Frying.* Monographs on the Evaluation of Carcinogenic Risks to Humans, International Agency for Research on Cancer, World Health Organization, Lyon.

ILO, 1998: Encyclopaedia of occupational health and safety. Fourth edition. J. M. Stellman, Ed. International Labour Office, Geneva.

ILO, 2009: *International Labour Office Database on Labour Statistics.* laborsta.ilo.org (accessed 31 December, 2010).

IPCC, 2007a: *Contribution of Working Group I to the Fourth Assessment Report of the Intergovernmental Panel on Climate Change.* In Climate Change 2007: The Physical Science Basis. S. Solomon , D. Qin , M. Manning , Z. Chen , M. Marquis , K. B. Averyt , M. Tignor and H. L. Miller (eds.), Cambridge, UK and New York, NY, USA.

IPCC, 2007b: Contribution of Working Group II to the Fourth Assessment Report of the Intergovernmental Panel on Climate Change. In *Climate Change 2007: Impacts, Adaptation and Vulnerability.* M. Parry , O. Canziani , J. Palutikof , P. van der Linden and C. Hanson (eds.), Cambridge, Cambridge, UK and New York, NY, USA.

IRSN, 2009: *La Radioprotection Des Travailleurs – Bilan 2008 De La Surveillance Des Travailleurs Exposés Aux Rayonnements Ionisants En France.* Institute de Radioprotection et Surete Nucleaire (IRSN), Fonteney-aux-Roses, Paris.

IRSN, 2010a: *Sources of Production and Management of Tritium Produced by Nuclear Plants.* Institute de Radioprotection et Surete Nucleaire (IRSN), Fonteney-aux-Roses, Paris.

IRSN, 2010b: *Tritium in the Environment – Review of the Irsn.* Institute de Radioprotection et Surete Nucleaire (IRSN), Fonteney-aux-Roses, Paris.

IRSN, 2010c: *Tritium in the Environment – a View from the Irsn on the Key Issues and Avenues of Research and Development.* Institute de Radioprotection et Surete Nucleaire (IRSN), Fonteney-aux-Roses, Paris.

IRSN, 2010d: *Elements of Reflection on the Health Risk Posed by Tiritium.* Institute de Radioprotection et Surete Nucleaire (IRSN), Fonteney-aux-Roses.

IRSN, 2010e: *Tritium: Limits of Relseases and Impact.* Institute de Radioprotection et Surete Nucleaire (IRSN), Fonteney-aux-Roses, Paris.

IRSN, 2010f: *Tritium and Ospar.* Institute de Radioprotection et Surete Nucleaire (IRSN), Fonteney-aux-Roses, Paris.

Jackson, R. and K. N. Shields, 2008: Preparing the Us Health Community Climate Change. *Annual Review of Public Health*, **29**: 57–73.

Jackson, S. and A. C. Sleigh, 2000: Resettlement for China's Three Gorges Dam: Socio-Economic Impact and Institutional Tensions. *Communist and Post Communist Studies*, **33**: 223–241.

Jacob, D. J. and D. A. Winner, 2009: Effect of Climate Change on Air Quality. *Atmospheric Environment*, **43**(1): 51–63.

Jacobson, M. Z. and G. M. Masters, 2001: Exploiting Wind Versus Coal. *Science*, **293**(5534): 1438.

Jaffe, D., T. Anderson, D. Covert, R. Kotchenruther, B. Trost, J. Danielson, W. Simpson, T. Bernsten, S. Karlsdottir, D. Blake, J. Harris, G. Carmichael and I. Uno, 1999: Transport of Asian Air Pollution to North America. *Geophys. Res. Lett. 26*, **26**: 711–714.

Jennings, N. S., 1998: Mining: An Overview. In *Encyclopaedia of Occupational Health and Safety.* J. M. Stellman (ed.), International Labour Office (ILO), Geneva, Vol. 3, Chapter 74, pp. 74.72–74.74.

Jernelov, A., 2010: The Threats from Oil Spills: Now, Then, and in the Future. *Ambio*, **39**(5–6): 353–366.

Kaatsch, P., C. Spix, R. Schulze-Rath, S. Schmiedel and M. Blettner, 2008: Leukaemia in Young Children Living in the Vicinity of German Nuclear Power Plants. *International Journal of Cancer*, **122**(4): 721–726.

Kampmann, B. and C. Piekarski, 2005: Assessment of Risks of Heat Disorders Encountered During Work in Hot Conditions in German Hard Coal Mines. In *Environmental Ergonomics: The Ergonomics of Human Comfort, Health, and Performance in the Thermal Environment*. Ergonomics Book Series. 3, Y. Tochihara and T. Ohnaka, Eds., Elsevier, Amsterdam, the Netherlands, pp. 79–84.

Kandlikar, M. and G. Ramachandran, 2000: The Causes and Consequences of Particulate Air Pollution in Urban India: A Synthesis of the Science. *Annual Review of energy and the environment*, **25**: 629–684.

Kelman, I., 2006: Island Security and Disaster Diplomacy in the Context of Climate Change. *Les Cahiers de la Sécurité*, **63**: 61–94.

Kesavan, P. C. and M. S. Swaminathan, 2006: Managing Extreme Natural Disasters in Coastal Areas. *Philosophical Transactions of the Royal Society A: Mathematical, Physical and Engineering Sciences*, **364**(1845): 2191–2216.

Kevane, M. and L. Gray, 2008: Darfur: Rainfall and Conflict. *Environmental Research Letters*, **3**(3): 034006–034006.

Kim, C. S., N. E. Alexis, A. G. Rappold, H. Kehrl, M. J. Hazucha, J. C. Lay, M. T. Schmitt, M. Case, R. B. Devlin, D. B. Peden and D. Diaz-Sanchez, 2011: Lung

Function and Inflammatory Responses in Healthy Young Adults Exposed to 0.06 Ppm Ozone for 6.6 Hours. *American Journal of Respiratory and Critical Care Medicine*, **183**(9): 1215–1221.

Kinney, P. L., 2008: Climate Change, Air Quality, and Human Health. *American Journal of Preventive Medicine*, **35**(5): 459–467.

Kinney, P. L., M. Gatari Gichuru, N. Volavka-close, N. Ngo, P. K. Ndiba, A. Law, A. Gachanja, S. M. Gaita, S. N. Chillrud, and E. Sclar, 2011: Traffic Impacts on PM$_{2.5}$ Air Quality in Nairobi, Kenya. Environmental Science and Policy, **14**: 369–378.

Kjellstrom, T., 1994: *Issues in the Developing World*. In *Textbook of Clinical Occupational and Environmental Medicine*. L. Rosenstock and M. Cullen (eds.), W.B. Saunders & Co., Philadelphia, pp. 25–31.

Kjellstrom, T., 2009: Climate Change, Direct Heat Exposure, Health and Well-Being in Low and Middle-Income Countries. *Global Health Action*, **2**: 1–4.

Kjellstrom, T., I. Holmer and B. Lemke, 2009: Workplace Heat Stress, Health and Productivity – an Increasing Challenge for Low and Middle-Income Countries During Climate Change. *Global Health Action*, **2**.

Knutsson, A., 2003: Health Disorders of Shift Workers. *Occupational Medicine*, **53**(2): 103–108.

Kopacz, M., D. L. Mauzerall, J. Wang, E. M. Leibensperger, D. K. Henze, and K. Singh, 2011: Origin and Radiative Forcing of Black Carbon Transported to the Himalayas and Tibetan Plateau. Atmos. Chem. Phys. 11:2837–2852.

Kopp, R. E. and D. L. Mauzerall, 2010: Assessing the Climatic Benefits of Black Carbon Mitigation. *Proceedings of the National Academy of Sciences*.

Kovats, R. S., S. J. Edwards, S. Hajat, B. G. Armstrong, K. L. Ebi and B. Menne, 2004: The Effect of Temperature on Food Poisoning: A Time-Series Analysis of Salmonellosis in Ten European Countries. *Epidemiology and Infection*, **132**: 443–453.

Kovats, R. S. and S. Hajat, 2008: Heat Stress and Public Health: A Critical Review. *Annual Review of Public Health*, **29**(1): 41–56.

Kraus, R. S. (1998). Oil Exploration and Drilling. Encyclopaedia of occupational health and safety. J. M. Stellman. *Geneva, International Labour Office*. **3**: 75.71–75.14.

Krewitt, W., D. W. Pennington, S. I. Olsen, P. Crettaz and O. Jolliet, 2002: Indicators for Human Toxicity in Life Cycle Impact Assessment. In *Life Cycle Impact Assessment: Striving Towards Best Practice*, SETAC Press, Stuttgart, Germany, pp. 123–148.

Kunzli, N., L. Perez and R. Rapp, 2010: *Air Quality and Health. European Respiratory Society*, Lausanne, Switzerland.

Kurmi, O. P., S. Semple, P. Simkhada, W. C. S. Smith and J. G. Ayres, 2010: Copd and Chronic Bronchitis Risk of Indoor Air Pollution from Solid Fuel: A Systematic Review and Meta-Analysis. *Thorax*, **65**(3): 221–228.

Lin, H.-H., M. Ezzati and M. Murray, 2007: Tobacco Smoke, Indoor Air Pollution and Tuberculosis: A Systematic Review and Meta-Analysis. *PLoS Medicine*, **4**(1): 1–17.

Liu, J., D. L. Mauzerall and L. W. Horowitz, 2009: Evaluating Inter-Continental Transport of Fine Aerosols:(2) Global Health Impact. *Atmospheric Environment*, **43**(28): 4339–4347.

Liu, J. F. and D. L. Mauzerall, 2005: Estimating the Average Time for Inter-Continental Transport of Air Pollutants. *Geophysical Research Letters*, **32**(11).

LLRWMO, 2004: *Inventory of Radioactive Waste in Canada*. A report prepared for Natural Resources Canada by the Low-Level Radioactive Waste Management Office, Ottawa, Canada.

Lobell, D. and C. Field, 2007: Global Scale Climate–Crop Yield Relationships and the Impacts of Recent Warming. *Environmental Research Letters*, **2**(1): 014002.

Lobell, D. B., M. B. Burke, C. Tebaldi, M. D. Mastrandrea, W. P. Falcon and R. L. Naylor, 2008: Prioritizing Climate Change Adaptation Needs for Food Security in 2030. *Science*, **319**: 607–610.

Marenco, A. and H. Gouget, 1994: Evidence of a Long-Term Increase in Tropospheric Ozone from Pic Du Midi Data Series: Consequences: Positive Radiative Forcing. *Journal of Geophysical Research-Atmospheres*, **99**(D8): 16617–16632.

Markandya, A., B. G. Armstrong, S. Hales, A. Chiabai, P. Criqui, S. Mima, C. Tonne and P. Wilkinson, 2009: Public Health Benefits of Strategies to Reduce Greenhouse-Gas Emissions: Low-Carbon Electricity Generation. *The Lancet*, **374**(9706): 2006–2015.

Marshall, J. D., S. K. Teoh and W. W. Nazaroff, 2005: Intake Fraction of Nonreactive Vehicle Emissions in Us Urban Areas. *Atmospheric Environment*, **39**(7): 1363–1371.

Masera, O. R., B. D. Saatkamp and D. M. Kammen, 2000: From Linear Fuel Switching to Multiple Cooking Strategies: A Critique and Alternative to the Energy Ladder Model. *World Development*, **28**(12): 2083–2103.

Mauzerall, D. L. and X. Wang, 2001: Protecting Agricultural Crops from the Effects of Tropospheric Ozone Exposure: Reconciling Science and Standard Setting in the United States, Europe, and Asia. *Annual Review of energy and the environment*, **26**: 237–268.

McCormick, R., 2007: The Impact of Biodiesel on Pollutant Emissions and Public Health. *Inhalation Toxicology*, **19**(12): 1033–1039.

McIntyre, B. D., H. R. Herren , J. Wakhungu and R. T. Watson, Eds. 2007: *Synthesis Report – a Synthesis of the Global and Sub-Global Iaastd Reports*. International Assessment of Agricultural Knowledge, Science and Technology for Development (IAASTD), Washington, DC.

McKone, T. E., A. D. Kyle, O. Jolliet, S. I. Olsen and M. Hauschild, 2006: *Dose-Response Modeling for Life Cycle Impact Assessment: Findings of the Portland Review Workshop* Lawrence Berkeley National Laboratory, Berkeley, California.

McManus, N., 1998: Hydroelectric Power Generation In *Encyclopaedia of Occupational Health and Safety*. J. M. Stellman (ed.), Inernational Labour Office (ILO), Geneva, Vol. 3, Chapter 76, pp. 76.72–76.76.

McMichael, A., H. J. Weaver, H. Berry, P. J. Beggs, B. Currie, J. Higgins, B. Kelly, J. McDonald and S. Tong, 2009: *National Climate Change Adaptation Research Plan: Human Health*. National Climate Change Adaptation Research Facility, Gold Coast, Australia.

McMichael, A., J., P. Wilkinson, R. S. Kovats, S. Pattenden, S. Hajat, B. Armstrong, N. Vajanapoom, M. Niciu Emilia, H. Mahomed, C. Kingkeow, M. Kosnik, O. N. M. S, I. Romieu, M. Ramirez-Aguilar, L. Barreto Mauricio, N. Gouveia and B. Nikiforov, 2008: International Study of Temperature, Heat and Urban Mortality: The "Isothurm" Project. *International Journal of Epidemiology: Official Journal of the International Epidemiological Association,* **37**(5): 1121–1131.

McMichael, A. J., D. Campbell-Lendrum, S. Kovats, S. Edwards, P. Wilkinson, T. Wilson, R. Nicholls, S. Hales, F. Tanser, D. L. Sueur, M. Schlesinger and N. Andronova, 2004: Global Climate Change. In *Comparative Quantification of Health Risks: Global and Regional Burden of Disease Due to Selected Major Risk Factors*. M. Ezzati , A. D. Lopez , A. Rodgers and C. J. L. Murray (eds.), World Health Organization, Geneva, pp. 1543–1649.

McMichael, A. J., 2009a: Climate Change and Human Health. In *Commonwealth Health Ministers' Update 2009*, Pro-Book Publishing, London, UK, pp. 11–20.

McMichael, A. J., 2009b: *Climate Change in Australia: Risks to Human Wellbeing and Health.* Mapping Causal Complexity in Climate Change: Austral Special Report 09–03S, R. Tanter et al., RMIT Press, Melbourne.

McMichael, A. J. and R. Bertollini, 2011: Risks to Human Health, Present and Future. In *Climate Change: Global Risks, Challenges and Decisions*. K. Richardson , W. Steffen , D. Liverman , T. Barker , F. Jotzo , D. Kammen , R. Leemans , T. Lenton , M. Munasinghe , B. Osman- Elasha, J. Schellnhuber , N. Stern , C. Vogel and O. Waever (eds.), Cambridge University Press, Cambridge, pp. 114–116.

McMichael, A. J. and E. Lindgren, 2011: Climate Change: Present and Future Risks to Health – and Necessary Responses. *Journal of Internal Medicine*.

Mecaskey, J. W., C. A. Knirsch, J. A. Kumaresan and J. A. Cook, 2003: The Possibility of Eliminating Blinding Trachoma. *The Lancet Infectious Diseases*, **3**(11): 728–734.

Mendell, M. J., 2007: Indoor Residential Chemical Emissions as Risk Factors for Respiratory and Allergic Effects in Children: A Review. *Indoor Air*, **17**(4): 259–277.

Mitchell, J. F. B., J. Lowe, R. A. Wood and M. Vellinga, 2006: Extreme Events Due to Human-Induced Climate Change. *Philosophical Transactions of the Royal Society A: Mathematical, Physical and Engineering Sciences,* **364**(1845): 2117–2133.

Moffit, A., 1998: Iron and Steel. In *Encyclopaedia of Occupational Health and Safety*. J. M. Stellman (ed.), International Labour Office (ILO), Geneva, Vol. 3, chapter 73, pp. 73.71–73.18.

Morton, J. F., 2007: The Impact of Climate Change on Smallholder and Subsistence Agriculture. *Proceedings of the National Academy of Science (USA)*, **104**(50): 19680–19685.

Murray, C. J. L., D. B. Evans, A. Acharya and R. M. P. M. Baltussen, 2000: Development of Who Guidelines on Generalized Cost-Effectiveness Analysis. *Health Economics*, **9**(3): 235–251.

Naeher, L. P., M. Brauer, M. Lipsett, J. T. Zelikoff, C. D. Simpson, J. Q. Koenig and K. R. Smith, 2007: Woodsmoke Health Effects: A Review. *Inhalation Toxicology*, **19**(1): 67–106.

NAS, 2000: *Toxicological Effects of Methylmercury.* C. o. L. Sciences, National Academy of Sciences, The National Academies Press, Washington, DC.

Nazaroff, W. W., 2006: Inhalation Intake Fraction of Pollutants from Episodic Indoor Emissions. *Building and Environment*, **43**(3): 269–277.

Nelson, G. C., M. W. Rosegrant, J. Koo, R. Robertson, T. Sulser, T. Zhu, C. Ringler, S. Msangi, A. Palazzo, M. Batka, M. Magalhaes, R. Valmonte-Santos, M. Ewing and D. Lee, 2009: *Climate Change: Impact on Agriculture and Costs of Adaptation.* Food Policy Report, International Food Policy Research Institute, Washington, DC, USA.

Nemet, G. F., T. Holloway and P. Meier, 2010: Implications of Incorporating Air-Quality Co-Benefits into Climate Change Policymaking. *Environmental Research Letters*, **5**(January-March 2009): 014007.

Nickels, S., C. Furgal, M. Buell and H. Moquin, 2005: *Unikkaaqatigiit – Putting the Human Face on Climate Change: Perspectives from Inuit in Canada*. Joint publication of Inuit Tapiriit Kanatami, Nasivvik Centre for Inuit Health and Changing Environments at Université Laval and the Ajunnginiq Centre at the National Aboriginal Health Organization, Ottawa, Canada.

NRC, 1999: *Health Effects of Exposure to Radon.* BEIR VI, National Research Council (NRC), Natoinal Academies Press, Washington, DC.

NRC, 2006: *Health Risks from Exposure to Low Levels of Ionizing Radiation.* BEIR VII Phase 2, National Research Council (NRC), National Academies Press, Washington, DC.

NRC, 2009a: *Review of the Federal Strategy for Nanotechnology-Related Environmental, Health, and Safety Research.* National Research Council Committee for Review of the Federal Strategy to Address Environmental, Health, and Safety Research Needs for Engineered Nanoscale Materials, National Research Council, Washington, DC.

NRC, 2009b: *Review of the Federal Strategy Fornanotechnology-Related Environmental, Health, and Safety Research.* Committee for Review of the Federal Strategy to Address Environmental, Health, and Safety Research Needs for Engineered Nanoscale Materials, National Research Council (NRC), National Academies Press, Washington, DC.

NRC, 2010: *Hidden Costs of Energy: Unpriced Consequences of Energy Production and Use.* Committee on Health, Environmental, and Other External Costs and Benefits of Energy Production and Consumption, N. A. o. Sciences, National Research Council (NRC), Washington, DC.

O'Neill, B. C., M. Dalton, R. Fuchs, L. Jiang, S. Pachauri and K. Zigova, 2010: Global Demographic Trends and Future Carbon Emissions. *Proceedings of the National Academy of Science.*

Ohara, T., H. Akimoto, J. Kurokawa, N. Horii, K. Yamaji, X. Yan and T. Hayasaka, 2007: An Asian Emission Inventory of Anthropogenic Emission Sources for the Period 1980–2020. *Atmospheric Chemistry and Physics*, **7**(16): 4419–4444.

Ontario Royal Commission, 1978: *Ontario Royal Commission on Electric Power Planning, Race against Time.* Interim Report on Nuclear Power, Ontario Royal Comission, Toronto.

Oreszczyn, T., I. Ridley, H. Hong Sung and P. Wilkinson, 2006: Mould and Winter Indoor Relative Humidity in Low Income Households in England. *Indoor and Built Environment*, **15**(2): 125–135.

ORNL/RFF, 1992–1998: *External Costs and Benefits of Fuel Cycles: Reports 1–8.* Oak Ridge National Laboratory (ORNL) and Resources for the Future (RFF), Washington, DC.

Palmer, M. A., E. S. Bernhardt, W. H. Schlesinger, K. N. Eshleman, E. Foufoula-Georgiou, M. S. Hendryx, A. D. Lemly, G. E. Likens, O. L. Loucks, M. E. Power, P. S. White and P. R. Wilcock, 2010: Science and Regulation. Mountaintop Mining Consequences. *Science*, **327**(5962): 148–149.

Pandey, M. R., R. P. Neupane, A. Gautam and I. B. Shrestha, 1990: The Effectiveness of Smokeless Stoves in Reducing Indoor Air Pollution in a Rural Hill Region of Nepal. *Mountain Research and Development*, **10**(4): 313–320.

Park, R. J., D. J. Jacob, B. D. Field, R. M. Yantosca and M. Chin, 2004: Natural and Transboundary Pollution Influences on Sulfate-Nitrate-Ammonium Aerosols in the United States: Implications for Policy. *J. Geophys. Res.*, **109**: D15204, doi:15210.11029/12003JD004473.

Pascual, M. and M. J. Bouma, 2009: Do Rising Temperatures Matter? *Ecology*, **90**(4): 906–912.

Patz, J. A., H. K. Gibbs, J. A. Foley, J. V. Rogers and K. R. Smith, 2007: Climate Change and Global Health: Quantifying a Growing Ethical Crisis. *EcoHealth*, **4**(4).

Pearce, D., 2001: *Energy Policy and Externalities: An Overview*. Paper prepared for OECD Nuclear Energy Agency, Keynote address to Workshop on Energy Policy and Externalities: the Life Cycle Analysis Approach, Paris, France.

Peltier, R. and M. Lippmann, 2009: On the Sources of Ambient Respirable Particles Laden with Trace Metals: An Investigation of Extensive Sources of Aerosol with High Fractions of Transition Metals in New York City. *Epidemiology (Cambridge, Mass.)*, **20**(6): S201-S201.

Peltier, R. E., S.-I. Hsu, R. Lall and M. Lippmann, 2008: Residual Oil Combustion: A Major Source of Airborne Nickel in New York City. *J Expos Sci Environ Epidemiol*, **19**(6): 603–612.

Perch-Nielsen, S. L., M. B. Bättig and D. Imboden, 2008: Exploring the Link between Climate Change and Migration. *Climatic Change*, **91**(3): 375–393.

Pope, C. A., III, R. T. Burnett, M. J. Thun, E. E. Calle, D. Krewski, K. Ito and G. D. Thruston, 2002: Lung Cancer, Cardiopulmonary Mortality and Long-Term Exposure to Fine Particulate Air Pollution. *JAMA: Journal of the American Medical Association*, **287**(9): 1132–1141.

Pope, C. A., III, R. T. Burnett, D. Krewski, M. Jerrett, Y. Shi, E. E. Calle and M. J. Thun, 2009: Cardiovascular Mortality and Exposure to Airborne Fine Particulate Matter and Cigarette Smoke: Shape of the Exposure-Response Relationship. *Circulation : official journal of the American heart association*, **120**(11): 941–948.

Pope, D. P., V. Mishra, L. Thompson, A. R. Siddiqui, E. A. Rehfuess, M. Weber and N. G. Bruce, 2010: Risk of Low Birth Weight and Stillbirth Associated with Indoor Air Pollution from Solid Fuel Use in Developing Countries. *Epidemiologic Reviews*, **32**: 70–81.

Poschen, P., 1998a: Forestry. In *Encyclopaedia of Occupational Health and Safety*. J. M. Stellman (ed.), International Labour Office (ILO), Geneva, Vol. 3, Chapter 68, pp. 68.61–68.41.

Poschen, P., 1998b: General Profile (Forestry). In *Encyclopaedia of Occupational Health and Safety*. J. M. Stellman (ed.), International Labour Office (ILO), Geneva, Vol. 3, Chapter 68, pp. 68.62–68.66.

Purse, B. V., P. S. Mellor, D. J. Rogers, A. R. Samuel, P. P. C. Mertens and M. Baylis, 2005: Climate Change and the Recent Emergence of Bluetongue in Europe. *Nature Reviews Microbiology*, **3**(2): 171–181.

Qin, Y., E. Kim and P. K. Hopke, 2006: The Concentrations and Sources of Pm2.5 in Metropolitan New York City. *Atmospheric Environment*, **40**: 312–333.

Rabl, A. and M. Dreicer, 2002: Health and Environmental Impacts of Energy Systems. *International Journal of Global Energy Issues*, **18**(2/3/4): 113–150.

Rahmstorf, S., A. Cazenave, J. A. Church, J. E. Hansen, R. F. Keeling, D. E. Parker and R. C. J. Somerville, 2007: Recent Climate Observations Compared to Projections. *Science*, **316**(5825): 709–709.

Ramakrishna, J., M. B. Durgaprasad and K. R. Smith, 1989: Cooking in India: The Impact of Improved Stoves on Indoor Air Quality. *Environment International*, **15**(1–6): 341–352.

Ramanathan, V. and G. Carmichael, 2008: Global and Regional Climate Changes Due to Black Carbon. *Nature geoscience*, **1**(4): 221–227.

Rao, C. Y., M. A. Riggs, G. L. Chew, M. L. Muilenberg, P. S. Thorne and D. V. Sickle, 2007: Characterization of Airborne Molds, Endotoxins, and Glucans in Homes in New Orleans after Hurricanes Katrina and Rita. *Applied and Environmental Microbiology*, **73**(5): 1630–1634.

Raven, P. H., 2002: Science, Sustainability, and the Human Prospect. *Science*, **297**(5583): 954–958.

Raymond-Whish, S., L. P. Mayer, T. O'Neal, A. Martinez, M. A. Sellers, P. J. Christian, S. L. Marion, C. Begay, C. R. Propper, P. B. Hoyer and C. A. Dyer, 2007: Drinking Water with Uranium Below the U.S. Epa Water Standard Causes Estrogen Receptor-Dependent Responses in Female Mice. *Environmental Health Perspectives*, **115**(12): 1711–1716.

Reddy, A. K. N., R. H. Williams and T. B. Johnasson, 1997: *Energy after Rio: Prospects and Challenges.* United Nations Development Programme (UNDP), New York.

Reid, H. F., K. R. Smith and B. Sherchand, 1986: Indoor Smoke Exposures from Traditional and Improved Cookstoves Comparisons among Rural Nepali Women. *Mountain research and development*, **6**(4): 293–304.

Reynolds, C. C. O. and M. Kandlikar, 2008: Climate Impacts of Air Quality Policy: Switching to a Natural Gas-Fueled Public Transportation System in New Delhi. *Environmental Science & Technology*, **42**(16): 5860–5865.

Riojas-Rodríguz, H., P. Romano-Riquer, C. Santos-Burgoa and K. R. Smith, 2001: Household Firewood Use and the Health of Children and Women of Indian Communities in Chiapas, Mexico. *International Journal of Occupational and Environmental Health*, **7**(1): 44–53.

Robertson, G. P., V. H. Dale, O. C. Doering, S. P. Hamburg, J. M. Melillo, M. M. Wander, W. J. Parton, P. R. Adler, J. N. Barney, R. M. Cruse, C. S. Duke, P. M. Fearnside, R. F. Follett, H. K. Gibbs, J. Goldemberg, D. J. Mladenoff, D. Ojima, M. W. Palmer, A. Sharpley, L. Wallace, K. C. Weathers, J. A. Wiens and W. W. Wilhelm, 2008: Agriculture: Sustainable Biofuels Redux. *Science*, **322**(5898): 49–50.

Rollin, H. B., A. Mathee, N. Bruce, J. Levin and Y. E. R. von Schirnding, 2004: Comparison of Indoor Air Quality in Electrified and Un-Electrified Dwellings in Rural South African Villages. *Indoor Air*, **14**(3): 208–216.

Ross, M. H. and J. Murray, 2004: Occupational Respiratory Disease in Mining. *Occupational Medicine*, **54**(5): 304–310.

Ruiz-Mercado, I., O. Masera, H. Zamora and K. R. Smith, 2011: Adoption and Sustained Use of Improved Cookstoves. *Energy Policy*, **39**(2011): 7557–7566.

Ruiz-Moreno, D., M. Pascual, M. Bouma, A. Dobson and B. Cash, 2007: Cholera Seasonality in Madras (1901–1940): Dual Role for Rainfall in Endemic and Epidemic Regions. *EcoHealth*, **4**(1): 52–62.

Saksena, S., R. Prasad, R. C. Pal and V. Joshi, 1992: Patterns of Daily Exposure to Tsp and Co in the Garhwal Himalaya. *Atmospheric Environment Part A, General Topics*, **26**(11): 2125–2134.

Saksena, S., L. Thompson and K. R. Smith, 2003: *Database of Household Air Pollution Studies in Developing Countries, Protection of the Human Environment.* World Health Organization, Geneva.

Salehyan, I., 2008: From Climate Change to Conflict? No Consensus Yet. *Journal of Peace Research*, **45**(3): 315–327.

Schwartz, J., B. Coull, F. Laden and L. Ryan, 2008: The Effect of Dose and Timing of Dose on the Association between Airborne Particles and Survival. *Environmental Health Perspectives*, **116**(1): 64–69.

Searchinger, T., R. Heimlich, R. A. Houghton, F. Dong, A. Elobeid, J. Fabiosa, S. Tokgoz, D. Hayes and T.-H. Yu, 2008: Use of U.S. Croplands for Biofuels Increases Greenhouse Gases through Emissions from Land-Use Change. *Science*, **319**(5867): 1238–1240.

Selevan, S. G., C. A. Kimmel and P. Mendola, 2000: Identifying Critical Windows of Exposure for Children's Health. *Environ Health Perspect*, **108 Suppl 3**: 451–455.

Selin, N. E., 2005: Mercury Rising: Is Global Action Needed to Protect Human Health and the Environment? *Environment*, **47**(1): 22–35.

Selin, N. E., D. J. Jacob, R. J. Park, R. M. Yantosca, S. Strode, L. Jaegle and D. Jaffe, 2007: Chemical Cycling and Deposition of Atmospheric Mercury: Global Constraints from Observations. *Journal of Geophysical Research*, **112**(D02308).

Semenza, J. C., C. H. Rubin, K. H. Falter, J. D. Selanikio, W. D. Flanders, H. L. Howe and e. al., 1996: Heat-Related Deaths During the July 1995 Heat Wave in Chicago. *New England Journal of Medicine*, **335**(2): 84–90.

Sheeran, J., 2008: The Challenge of Hunger. *The Lancet*, 371(9608): 180–181.

Sims, J., 1994: *Anthology on Women, Health and Environment.* WHO/EHG/94.11, World Health Organization, Geneva, Switzerland.

Sinton, J. E., K. R. Smith, J. W. Peabody, L. Yaping, Z. Xiliang, R. Edwards and G. Quan, 2004: An Assessment of Programs to Promote Improved Household Stoves in China. *Energy for Sustainable Development*, 8(3): 33–52.

Smil, V., 2005: Energy at the Crossroads: Global Perspectives and Uncertainties. Massachusetts Institute of Technology, Massachusetts.

Smith, K. R., A. L. Aggarwal and R. M. Dave, 1983: Air Pollution and Rural Biomass Fuels in Developing Countries: A Pilot Village Study in India and Implications for Research and Policy. *Atmospheric Environment*, 17(11): 2343–2362.

Smith, K. R., G. Shuhua, H. Kun and Q. Daxiong, 1993: One Hundred Million Improved Cookstoves in China: How Was It Done? *World Development*, 21(6): 941–961.

Smith, K. R. and M. Desai, 2002: The Contribution of Global Environmental Factors to Ill-Health. In *Environmental Change, Climate, and Health: Issues and Research Methods.* P. Martens and A. T. McMichael (eds.), Cambridge Univ. Press, pp. 52–95.

Smith, K. R., S. Mehta and M. Maeusezahl-Feuz, 2004: Indoor Air Pollution from Household Use of Solid Fuels. In *Comparative Quantification of Health Risks: Global and Regional Burden of Disease Attributable to Selected Major Risk Factors*. M. Ezzati , A. D. Lopez , A. Rodgers and C. J. L. Murray (eds.), World Health Organization, Geneva, Switzerland, pp. 1435–1494.

Smith, K. R. and M. Ezzati, 2005: How Environmental Health Risks Change with Development: The Epidemiologic and Environmental Risk Transitions Revisited. *Annual Review of Environment & Resources*, 30(1): 291-C-298.

Smith, K. R., 2008: Comparative Environmental Health Assessments. *Annals of the New York Academy of Sciences*, 1140: 31–39.

Smith, K. R. and E. Haigler, 2008: Co-Benefits of Climate Mitigation and Health Protection in Energy Systems: Scoping Methods. *Annual Review of Public Health*, 29(1): 11–25.

Smith, K. R. and K. Balakrishnan, 2009: Mitigating Climate, Meeting the Mdgs, and Blunting Chronic Disease: The Health Co-Benefits Landscape. In *Commonwealth Health Ministers' Update 2009*, Woodgridge, UK. pp. 59–65.

Smith, K. R., M. Jerrett, H. R. Anderson, R. T. Burnett, V. Stone, R. Derwent, R. W. Atkinson, A. Cohen, S. B. Shonkoff, D. Krewski, C. A. Pope, M. J. Thun and G. Thurston, 2009: Public Health Benefits of Strategies to Reduce Greenhouse-Gas Emissions: Health Implications of Short-Lived Greenhouse Pollutants. *The Lancet*, 374(9707): 2091–2103.

Smith, K. R. and J. Peel, 2010: Mind the Gap. *Environmental Health Perspectives*, 118(12): 1643–1645.

Smith, K. R., J. P. McCracken, M. W. Weber, A. Hubbard, A. Jenny, L. Thompson, J. Balmes, A. Diaz, B. Arana and N. Bruce, 2011: Effect of Reduction in Household Air Pollution on Childhood Pneumonia in Guatemala (RESPIRE): a randomised controlled trial. *The Lancet*, 378 : 1717–1726.

Solomon, G. M. and S. Janssen, 2010: Health Effects of the Gulf Oil Spill. *JAMA: The Journal of the American Medical Association*, 304(10): 1118–1119.

Sorensen, B., 2003: *Total Life-Cycle Assessment of Pem Fuel Cell Car.* Roskilde University, Energy & Environment Group, Roskilde, Denmark.

Spix, C., S. Schmiedel, P. Kaatsch, R. Schulze-Rath and M. Blettner, 2008: Case-Control Study on Childhood Cancer in the Vicinity of Nuclear Power Plants in Germany 1980–2003. *European Journal of Cancer*, 44(2): 275–284.

Stern, N. H., 2007: *The Economics of Climate Change: The Stern Review.* Cambridge University Press, Cambridge.

Tame, N. W., B. Z. Dlugogorski and E. M. Kennedy, 2007: Formation of Dioxins and Furans During Combustion of Treated Wood. *Progress in Energy and Combustion Science*, 33(4): 384–408.

TDR, 2009: *Dengue Guidelines for Diagnosis, Treatment, Prevention and Control.* Special Programme for Research and Training in Tropical Deseases and World Health Organization, Geneva, Switzerland.

Tilman, D., J. Hill and C. Lehman, 2006: Carbon-Negative Biofuels from Low-Input High-Diversity Grassland Biomass. *Science*, 314(5805): 1598–1600.

UNDP, 2000: *World Energy Assessment: Energy and the Challenge of Sustainability.* United Nations Development Programme, New York, NY, USA.

UNDP, 2009: *The Energy Access Situation in Developing Countries: A Review Focused on Least Developed Countries and Sub-Saharan Africa.* United Nations, Nairobi, Kenya.

UNEP, 2007: *Sudan: Post-Conflict Environmental Assessment.* United Nations Environment Programme (UNEP), Nairobi, Kenya.

UNEP, 2011: *Integrated Assessment of Black Carbon and Tropospheric Ozone: Summary for Decision Makers.* B. Ullstein, United Nations Environmental Programme (UNEP) and World Meterological Organization (WMO), Nairobi, Kenya.

Unger, N., D. T. Shindell, D. M. Koch and D. G. Streets, 2008: Air Pollution Radiative Forcing from Specific Emissions Sectors at 2030. *Journal of Geophysical Research-Atmospheres*, 113(D02306).

UNIDIR, 2008: *The Health Hazards of Depleted Uranium.* Plublished in Disarmament Forum, I. Fairlie, United Nations Institute for Disarmament Research (UNIDIR), Geneva, Switzerland.

US EIA, 2005: *Average Consumption by Fuels Used, 2005.* 2005 Residential Energy Consumption Survey: Energy Consumption and Expenditures Tables, Energy Information Administration (US EIA), Washington, DC, USA.

US EPA, 2000: *Report on Greenhouse Gases from Small-Scale Combustion Devices in Developing Countries: Household Stoves in India.* United States Environmental Protection Agency (US EPA), Ann Arbor, MI.

US EPA, 2009: *Epa Lifecycle Analysis of Greenhouse Gas Emissions from Renewable Fuels.* United States Environmental Protection Agency (US EPA), Ann Arbor, MI.

van Dingenen, R., F. Raes, M. C. Krol, L. Emberson and J. Cofala, 2009: The Global Impact of O3 on Agricultural Crop Yields under Current and Future Air Quality Legislation. *Atmospheric Environment*, 43: 604–618.

Walker, B., S. Barrett, S. Polasky, V. Galaz, C. Folke, G. Engström, F. Ackerman, K. Arrow, S. Carpenter, K. Chopra, G. Daily, P. Ehrlich, T. Hughes, N. Kautsky, S. Levin, K.-G. Mäler, J. Shogren, J. Vincent, T. Xepapadeas and A. d. Zeeuw, 2009: Looming Global-Scale Failures and Missing Institutions. *Science*, 325(5946): 1345–1346.

Wang, X. P. and D. L. Mauzerall, 2004: Characterizing Distributions of Surface Ozone and Its Impact on Grain Production in China, Japan and South Korea: 1990 and 2020. *Atmospheric Environment*, 38(26): 4383–4402.

Ware, H., 2005: Demography, Migration and Conflict in the Pacific. *Journal of Peace Research*, 42(4): 435–454.

Wasterlund, D. S., 1998: A Review of Heat Stress Research with Application to Forestry. *Applied Ergonomics*, 29(3): 179–183.

Weeks, J. L., 1998: Health Hazards of Mining and Quarrying. In *Encyclopaedia of Occupational Health and Safety.* J. M. Stellman (ed.), International Labour Office (ILO), Geneva, Vol. 3, Chapter 74, pp. 74.51–74.55.

White, I., T. Falkland, T. Metutera, E. Metai, M. Overmars, P. Perez and A. Dray, 2007: Climatic and Human Influences on Groundwater in Low Atolls. *Vadose Zone Journal*, **6**(3): 581–590.

WHO, 1994: *Ultraviolet Radiation: An Authoritative Scientific Review of Environmental and Health Effects of Uv, with Reference to Global Ozone Layer Depletion.* Environmental health criteria no. 160, World Health Organization, Geneva, Switzerland.

WHO, 1998: *Chrysotile Asbestos.* World Health Organization, Geneva, Switzerland.

WHO, 2000: *Air Quality Guidelines for Europe, Second Edition.* WHO Regional Publications, European Series, No. 91, World Health Organization, Regional Office for Europe, Copenhagen, Denmark.

WHO, 2002: *World Health Report 2002: Reducing Risks, Promoting Healthy Life.* World Health Organization, Geneva, Switzerland.

WHO, 2003: *Making Choices in Health: Who Guide to Cost-Effectiveness Analysis.* World Health Organization, Geneva, Switzerland.

WHO, 2004: *Health Aspects of Air Pollution: Results from the Who Project "Systematic Review of Health Aspects of Air Pollution in Europe".* World Health Organization (WHO), Geneva, Switzerland.

WHO, 2005: *Global Indoor Air Pollution Database.* www.who.int/indoorair/health_impacts/databases_iap/ (accessed 31 December, 2010).

WHO, 2006: *Air Quality Guidelines: Global Update for 2005.* World Health Organization Regional Office for Europe, Copenhagen.

WHO, 2007: *Housing, Energy and Thermal Comfort: A Review of 10 Countries within the Who European Region.* World Health Organization, Regional Office for Europe, Copenhagen, Denmark.

WHO, 2008: *The Global Burden of Disease: 2004 Update.* World Health Organization, Geneva, Switzerland.

Wilkinson, P., M. Landon, B. Armstrong, S. Stevenson, S. Pattenden, M. McKee and T. Fletcher, 2001: *Cold Comfort: The Social and Environmental Determinants of Excess Winter Deaths in England, 1986–96.* Bristol, UK.

Wilkinson, P., K. R. Smith, M. Davies, H. Adair, B. G. Armstrong, M. Barrett, N. Bruce, A. Haines, I. Hamilton, T. Oreszczyn, I. Ridley, C. Tonne and Z. Chalabi, 2009: Public Health Benefits of Strategies to Reduce Greenhouse-Gas Emissions: Household Energy. *The Lancet*, **374**(9705): 1917–1929.

Wilson, K., 2009: Forum: Climate Change and the Spread of Infectious Ideas. *Ecology*, **90**(4): 901–903.

Woodcock, J., P. Edwards, C. Tonne, B. G. Armstrong, O. Ashiru, D. Banister, S. Beevers, Z. Chalabi, Z. Chowdhury, A. Cohen, O. H. Franco, A. Haines, R. Hickman, G. Lindsay, I. Mittal, D. Mohan, G. Tiwari, A. Woodward and I. Roberts, 2009: Public Health Benefits of Strategies to Reduce Greenhouse-Gas Emissions: Urban Land Transport. *The Lancet*, **374**(9705): 1930–1943.

Zhou, X.-N., G.-J. Yang, K. Yang, X.-H. Wang, Q.-B. Hong, L.-P. Sun, J. B. Malone, T. K. Kristensen, N. R. Bergquist and J. Utzinger, 2008: Potential Impact of Climate Change on Schistosomiasis Transmission in China. *American Journal of Tropical Medicine and Hygiene*, **78**(2): 188–194.

5

Energy and Security

Convening Lead Author (CLA)
Aleh Cherp (Central European University, Hungary)

Lead Authors (LA)
Adeola Adenikinju (University of Ibadan, Nigeria)
Andreas Goldthau (Central European University, Hungary)
Francisco Hernandez (Lund University, Sweden)
Larry Hughes (Dalhousie University, Canada)
Jaap Jansen (Energy Research Centre of the Netherlands)
Jessica Jewell (Central European University, Hungary)
Marina Olshanskaya (United Nations Development Programme)
Ricardo Soares de Oliveira (Oxford University, UK)
Benjamin Sovacool (National University of Singapore)
Sergey Vakulenko (Cambridge Energy Research Associates, USA)

Contributing Authors (CA)
Morgan Bazilian (United Nations Industrial Development Organization)
David J. Fisk (Imperial College London, UK)
Saptarshi Pal (Central European University, Hungary)

Review Editor
Ogunlade Davidson (Ministry of Energy and Water Resources, Sierra Leone)

Contents

Executive Summary

Uninterrupted provision of vital energy services (see Chapter 1, Section 1.2.2) – energy security – is a high priority of every nation. Energy security concerns are a key driving force of energy policy. These concerns relate to the robustness (sufficiency of resources, reliability of infrastructure, and stable and affordable prices); sovereignty (protection from potential threats from external agents); and resilience (the ability to withstand diverse disruptions) of energy systems. Our analysis of energy security issues in over 130 countries shows that the absolute majority of them are vulnerable from at least one of these three perspectives. For most industrial countries, energy insecurity means import dependency and aging infrastructure, while many emerging economies have additional vulnerabilities such as insufficient capacity, high energy intensity, and rapid demand growth. In many low-income countries, multiple vulnerabilities overlap, making them especially insecure.

Oil and its products lack easily available substitutes in the transport sector, where they provide at least 90% of energy in almost all countries. Furthermore, the global demand for transport fuels is steadily rising, especially rapidly in Asian emerging economies. Disruptions of oil supplies may thus result in catastrophic effects on such vital functions of modern states as food production, medical care, and internal security. At the same time, the global production capacity of conventional oil is widely perceived as limited. These factors result in rising and volatile prices of oil affecting all economies, especially low-income countries, almost all of which import over 80% of their oil supplies. The costs of energy (primarily oil) imports exceed 20% of the export earnings in 35 countries with 2.5 billion people and exceed 10% of gross domestic product (GDP) in an additional 15 countries with 200 million people.

The remaining conventional oil resources are increasingly geographically concentrated in just a few countries and regions. This means that most countries must import an ever-higher share or even all of their oil. More than three billion people live in 83 countries that import more than 75% of the oil products they consume. This does not include China, where oil import dependency is projected to increase from the current 53% to 84% in 2035. The increasing concentration of conventional oil production and the rapidly shifting global demand patterns make some analysts and politicians fear a "scramble for energy" or even "resource wars."

Import dependency is also common in countries that rely on natural gas to provide heat and generate electricity. Almost 650 million people live in 32 Eurasian countries that import over 75% of their gas. Many of these countries are landlocked and import gas through a limited number of pipelines. The interregional trade in natural gas is projected to significantly increase, with yet more uncertain consequences for energy security. Developments in unconventional gas extraction and liquefied natural gas (LNG) technologies may have a defining influence in this regard.

Vulnerabilities of electricity systems are not limited to power plants relying on imported fossil fuels (which currently provide over 50% of electricity in some 39 countries with 600 million people). Hydroelectric power production, especially from major dams located on internationally shared rivers, is often perceived as insecure, particularly in light of climate change affecting seasonal water availability. Over 700 million people live in 31 countries that derive a significant proportion of their electricity from just one or two major dams and thus are vulnerable to failures of these dams.

Many countries using nuclear power are experiencing an aging of the reactor fleet and workforce, as well as difficulties in accessing capital and technologies to renew, expand, or launch new nuclear programs. Twenty of the 31 countries with nuclear power programs have not started building a new reactor in the last 20 years, and in 19 countries the average age of nuclear power plants is over 25 years. Large-scale enrichment, reactor manufacturing, and reprocessing technologies and capacities are currently concentrated in just a few countries. Transfer of these technologies and capacities to a larger number of countries is constrained by serious concerns over nuclear weapons proliferation, which is one of the main controversies and risks associated with nuclear energy. If nuclear energy can address energy security challenges, it will only happen in a few larger and more prosperous economies.

Various vulnerabilities in electricity supply are often made worse by demand-side pressures. Some 4.2 billion people live in 53 countries that will need to expand the capacity of their electricity systems massively in the near future because they have either less than 60% access to electricity or an average demand growth of over 6% over the last decade. Both fuels and infrastructure for such an expansion will need to be provided without further compromising the sovereignty or resilience of national electricity systems. The reliability of electricity supply is a serious concern, especially in developing countries. In almost three-quarters of low-income countries blackouts are on average for more than 24 hours per month, and in about one-sixth of low-income countries blackouts average over 144 hours (six days) a month. In over one-half of low-income countries blackouts occur at least 10 times a month.

The energy sector also provides vital export revenues for some 15–20 countries. In the majority of these oil- and gas-exporting countries the revenues are not expected to last for more than one generation, and in several cases they may cease in less than a decade. In addition, poor energy-exporting nations are at a high risk of the "resource curse": economic and political instability eventually affecting human development and security.

Almost all countries associate enhanced energy security with higher diversity of energy sources (especially in the transport sector), lower energy intensity of national economies, and reduced import dependency by relying on domestic energy sources. International regimes fostering cooperation between exporters and importers of energy and interacting with global governance arrangements for climate change and energy access are important for achieving these energy security goals. Energy security under sustainable energy transitions will be determined by the dynamics of phasing out fossil fuels and their substitution by new energy sources, as well as by new technologies in the end-use sector. A quantitative analysis of such developments is conducted in Chapter 17.

Table 5.1 | Summary of energy security issues in the world.

Energy sector and its significance	Energy security concerns and the population affected	
	Shorter term	**Longer term**
Oil (125 countries, 5.9 billion)*	>75% import dependency (**3 billion**) consumption growth >5%/year (**1.8 billion**)	Reserves/Consumption <15 years (**1.7 billion**)
Gas (78 countries, 2 billion)*	>75% import dependency (**650 million**)	Reserves/Consumption <16 years (**780 million**)
Coal (45 countries, 4.5 billion)*	>80% import dependency (**300 million**)	
Nuclear (21 countries, 1.3 billion)**		Average age of nuclear power plants >25 years (**1.9 billion**) Start of last plant construction >20 years (**1.4 billion**)
Hydro (58 countries, 1.5 billion)***	Low diversity (one or two major dams) (**730 million**)	
Electricity (all countries)	>50% dependency on imported fossil fuels (**600 million**) low diversity (one or two fuel sources) (**450 million**)	annual demand growth >6%/year and/or access rate <60% (**4.2 billion**)
Transport	>50% dependency on imported fuels (**4.9 billion**)	annual consumption growth >8% (**1.7 billion**)
Industry (>25% of GDP in 60 countries; 4.5 billion)	>50% dependency on imported fuels (**800 million**)	
Residential and commercial (all countries)	>50% dependency on imported fuels (**500 million**)	Reliance on traditional biofuels for >80% of the residential sector energy (**700 million**)
Cross-sectoral energy supply (all countries)	>50% overall import dependency (**700 million**) low diversity of PES (one or two dominant sources) (**1 billion**) cost of energy imports >20% of export earning (**2.5 billion**); cost of energy imports >10% of GDP (**200 million**)	energy intensity >50% of world average (**400million**) consumption growth >6% (**1.8 billion**) consumption per capita <30 GJ/year (**3 billion**)

Notes: PES – primary energy sources;

Numbers in brackets indicate the number of people who live in countries with the indicated energy security conditions;

* – more than 10% in total energy supply; ** – more than 10% in electricity generation; *** – more than 20% in electricity generation

5.1   Introduction

Energy systems are closely entangled with national and human security. Concerns over the reliability of vital energy services have shaped public opinions and political agendas, eventually affecting broader security issues ranging from risks of armed conflicts, to viability of national economies, and to integrity and stability of political systems. Policies developed in the quest for energy security have been – and are likely to remain – a key driving force in the transformations of energy systems.

Our analysis of energy security in the world is based on a rich tradition of addressing this topic in political, professional, and academic circles. Historically, the notion of energy security emerged in the first half of the 20th century as a concern over the secure supply of fuels (coal and oil) for naval fleets and armies. Political and military leaders sought to ensure security of fuel supplies through diversifying suppliers, substituting foreign imports with domestic production (e.g., synthetic aviation fuel in Germany), restricting non-essential uses of fuels (e.g., rationing of gasoline in the United States) and, finally, seeking military control over energy resources and infrastructure (military campaigns in Indonesia, Caucasus, and other theaters during World War II) (Yergin, 2006).

In the second half of the 20th century, oil became increasingly important not only for the military but also for sustaining such vital functions of industrialized societies as transport, mechanized agriculture, electricity generation, and the heating of buildings. At the same time, the global oil trade dramatically increased as major economies such as the United States became dependent on imported rather than domestic resources. The oil embargoes in the 1970s brought energy security to the forefront of political attention in industrialized countries. Energy security strategies prompted by these embargoes included establishing emergency stocks and joint response mechanisms in Organisation for Economic Co-operation and Development (OECD) countries, substituting oil by other energy sources (natural gas, nuclear, coal, etc.) in heating and electricity generation, investing in oil reserves outside the Organization of the Petroleum Exporting Countries (OPEC) – for example, in Alaska and the North Sea – and promoting energy efficiency to decrease oil intensities of economies.

By the end of the 20th century, many of these strategies bore fruit so that the fear of global oil supply disruptions subsided. At that time, other concerns at the interface of energy systems and national security emerged. One was the security of electricity transmission and generation systems. Vulnerability of nuclear power plants was exposed by the Three Mile Island, Chernobyl, and Fukushima accidents, with Chernobyl virtually halting the construction of new nuclear reactors in the world for two decades. Large-scale blackouts due to failures of generation and transmission exposed the vulnerability of modern societies to even short-term disruptions of electricity supply. At the same time, with the collapse of the Soviet Union, it became clear that the economies of many oil-exporting nations are not viable without steady energy export revenues. This prompted a shift in attention towards "demand security," another aspect of energy security.

Nowadays, the energy security debate is a mixture of many concerns, as "[i]n the background – but not too far back – is the anxiety over whether there will be sufficient resources to meet the world's energy requirements in the decades ahead" (Yergin, 2006). More immediate concerns include high and volatile oil prices, especially painful for lower-income countries; the predicted "plateau" of conventional oil production apparently falling short of the rising demand (see Chapter 7, Figure 7.4); the increasing geographic concentration of conventional oil and gas resources in just a few countries and regions; the shift of oil demand to India and China; and the fear of tensions and conflicts as new and old consumers "scramble" for the remaining, increasingly concentrated resources. The present energy security concerns also include the recent conflicts over deliveries of Russian natural gas to Eastern Europe, and fears over the excessive dependency of some European countries on a very limited number of energy supply options. In light of the September 2001 terrorist attacks and the disruption brought about by Hurricane Katrina, there are also serious worries over the vulnerability of critical energy infrastructure to terrorist attacks or extreme natural events. Moreover, energy security concerns are now closely entangled with other critical energy issues, most notably energy access and climate impacts of energy systems.

Throughout history, energy security has been viewed as protection from disruptions of essential energy systems. The notion of "essential energy systems" evolved from supplies of oil for military purposes to encompass various energy sources, infrastructure, and end-use sectors. The idea of "protection from disruptions" has also evolved from securing military or political control over energy resources to setting up complex policies and measures of strategically managing risks that affect all elements of energy systems.

Though it is possible to discuss energy security at household, community, and other levels, most political concerns and scholarly research are about energy security of individual countries. This is because nation-states have a historic responsibility for security, national energy systems provide appropriate units of analysis of key risks and vulnerabilities, and the majority of policy interventions to maintain energy security occur at the national level. Consequently, the analysis presented in this chapter primarily concentrates on the national level of energy security.

In line with the existing tradition, our analysis defines a nation's energy security as protection from disruptions of energy systems that can jeopardize nationally vital energy services. Numerous definitions of energy security (Box 5.1) are largely elaborations of this broader concept, particularly on the notions of "protection from disruptions" and "vital energy systems."

Box 5.1 | Defining Energy Security

The analysis of energy security in academic literature recognizes that its meaning varies from one country or one context to another. Thus, universal definitions of energy security are less frequently attempted than contextualized discussions of its various aspects or dimensions. One of the most frequently quoted definitions is the "availability of sufficient supplies at affordable prices" suggested by Yergin (2006). It is preceded by the European Commission's (2000) definition of energy security as the "uninterrupted physical availability on the market of energy products at a price which is affordable for all consumers." Energy insecurity is defined by (Bohi and Toman, 1996) as "the loss of economic welfare that may occur as a result of a change in the price or availability of energy."

These definitions contain notions of "availability," "sufficiency," "affordability," "welfare," "energy products" (or "supplies"), and "interruptions," which are open to wide interpretations. For example, Yergin (2006) discusses the different meaning of energy security – within his given definition – for several different countries. This concept of variability of the notion of energy security is also stressed by Müller-Kraenner (2008), Kruyt et al. (2009), and Chester (2010).

In its analysis of energy security, scholarly literature draws different boundaries for energy systems and subsystems. These boundaries differ between how many and which fuels are considered, as well as how far up- and downstream boundaries are drawn within that system. In terms of fuel-related boundaries, studies range from focusing on a specific fuel (generally oil) (Kendell, 1998; Gupta, 2008; Greene, 2010); looking at all fossil fuels (Le Coq and Paltseva, 2009); analyzing the security of an electricity system (Stirling, 1994) or critical energy infrastructure (Farrell et al., 2004) to evaluating the security of the whole primary energy system (Neff, 1997; Jansen et al., 2004; Jansen and Seebregts, 2010). Within each of these divisions, some studies focus only on the supply side, while others integrate supply and demand aspects and indicators.

With respect to the characteristics of energy systems that are associated with their security, various studies propose and discuss different dimensions of energy security. The simplest discussion uses two dimensions of energy security: the "physical" and "economic" dimensions (Kendell, 1998; Gupta, 2008). Another commonly used taxonomy is the 4 A's, or: "availability" (i.e., physical availability of resources), "accessibility" (geopolitical aspects associated with accessing resources), "affordability" (economic costs of energy), and "acceptability" (social and often environmental stewardship aspects of energy) (Kruyt et al., 2009). Other dimensional classifications include "economic, environmental, social, foreign policy, technical and security" (Alhajii, 2007) dimensions, as well as "energy supply, economic, technological, environmental, social-cultural, and military security" dimensions from von Hippel et al. (2009) and others.

Contemporary literature on energy security considers risks linked to natural (e.g., resource scarcity, extreme natural events), technical (e.g., aging of infrastructure, technological accidents), political (e.g., intentional restriction of supplies or technologies, sabotage and terrorism), and economic (e.g., high or volatile prices) factors. Correspondingly, "protection from risks" is defined by various authors as independence, reliability, resilience, availability, accessibility, affordability, or sustainability of energy systems.

This analysis considers three perspectives on energy security linked to distinct policy strategies and rooted in specific scholarly disciplines: robustness, sovereignty, and resilience (Cherp and Jewell, 2011) (see Box 5.2):

- **Robustness** is focused on protection from disruptions originating from predictable and "objective" natural, technical, and economic factors such as resource scarcity, rapid rise of demand, aging of infrastructure, or rising energy prices.

- **Sovereignty** is focused on protection from disruptions originating from intentional actions of various actors (such as unfriendly political powers and overly powerful market agents). Sovereignty implies the ability to control the behavior of energy systems and is often linked to much-discussed "energy independence."

- **Resilience** is focused on protection from disruptions originating from less predictable factors of any nature, such as political instability, game-changing innovations, or extreme weather events.

Box 5.2 | Three Perspectives on Energy Security

For assessing energy security, when it is defined as protection from disruptions of energy systems, it is necessary to understand the nature of such disruptions and adequate protection mechanisms. Energy security policies as well as scholarly literature present three distinct but complementary perspectives on this issue.

Historically, the oldest – "sovereignty" – perspective on energy security focuses on disruptions potentially arising from actions of "external" actors, be it hostile powers or terrorists, "unreliable" exporters, "foreign" energy companies, or overly powerful market agents. Protection from such disruptions is seen in increasing control over energy systems, be it by military, political, economic or technical means. In a broader sense, the sovereignty perspective focuses on interests, intentions, power, and the space for maneuver of various energy actors and institutions. In its most familiar and most simplistic form, a sovereignty strategy is the quest for energy independence.

The "robustness" perspective on energy security focuses on risks that arise from predictable and largely controlled characteristics of energy systems rather than from malevolent actions. Scarcity of energy resources, failures of infrastructure, or inadequate capacity to cope with the rising demand are examples of the issues addressed within this perspective. The main strategies for minimizing the risks of disruptions within this perspective are switching to more abundant and accessible energy sources, investment in infrastructure to minimize the risks of technical failures, and decreasing energy intensity to reduce vulnerability to high prices.

In contrast, the "resilience" perspective emphasizes unpredictable factors affecting energy security. Precisely due to the complexity and unpredictability of energy systems, this perspective focuses on diversity of energy options as the main strategy to cope with the potential threats.

Figure 5.1 shows the three perspectives on energy security in relation to fundamental assumptions about the nature of potential disruptions. The robustness strategies are most prominent when disruptions of energy systems are both controllable and predictable. Sovereignty strategies are focused on disruptions arising from forces outside of our control (these can be more or less predictable). Resilience strategies work in situations of unpredictability, independently of whether or not we have control over energy systems.

Figure 5.1 also shows that many widespread energy security strategies are defined by more than one perspective. For example, increasing diversity of energy suppliers or maintaining competitive energy markets addresses both resilience and sovereignty concerns, since it reduces the power of individual agents to disrupt energy systems and provides a diversity of options which may be useful in the face of unpredictable disruptions. Investing in redundant capacities responds to both robustness and resilience concerns, whereas maintaining emergency stocks and selecting "trusted" partners is an overlap of sovereignty and robustness strategies.

Source: Cherp and Jewell, 2011

"Vital energy systems" that should be protected to ensure energy security were historically interpreted as a supply of oil and subsequently included other fuels, energy infrastructure, and eventually energy services. The literature identified "vital" as linked to "national values and objectives" (Yergin, 1988), "affecting welfare" (Bohi and Toman, 1996), and, in some cases, not associated with unacceptable environmental and social impacts (Sovacool and Brown, 2010) of energy systems.

In this analysis, "vital energy services" mean those that are necessary for the stable functioning of modern societies. Inherent in this definition is the notion that governments and the public perceive the interruption of such services as a national security concern. Such services vary from one country to another but commonly include transportation and energy for buildings. Most countries also depend upon a supply of energy for industrial purposes, and for some countries uninterrupted energy exports provide vital revenues. Security of these vital services is linked to vulnerability of energy sources (such as oil, gas, coal, hydro, and nuclear energy) and infrastructure for energy conversion and transmission (such as power plants, fuel reservoirs, and pipelines). Thus, our analysis is not limited to energy supply but also extends to energy conversion and distribution infrastructure and to the demand aspects within the vital energy sectors. We also analyze the connections between supply, demand, and infrastructure to understand how vulnerabilities propagate within energy systems so that if one element is at risk, other connected elements may also be affected.

Table 5.2 | Summary of the three perspectives on energy security.

Perspective	Sovereignty	Robustness	Resilience
Historic roots	War-time oil supplies and the 1970s oil crises	Large technological accidents, electricity blackouts, concerns about resource scarcity	Liberalization of energy systems.
Key risks for energy systems	Intentional actions by malevolent agents	Predictable natural and technical factors	Diverse and partially unpredictable factors
Primary protection mechanisms	Control over energy systems. Institutional arrangements preventing disruptive actions	Upgrading infrastructure and switching to more abundant resources	Increasing the ability to withstand and recover from various disruptions

Source: Cherp and Jewell, 2011.

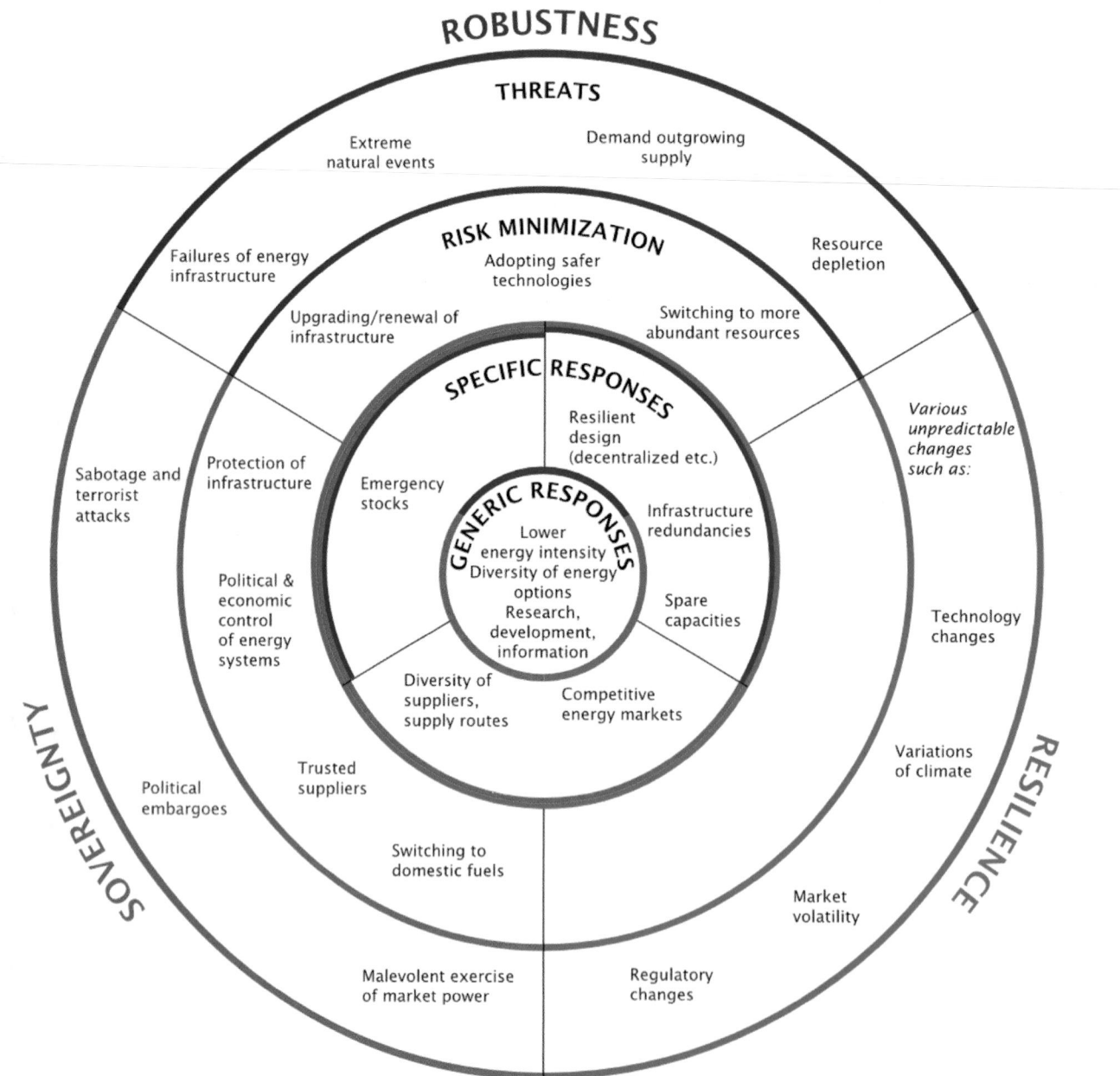

Figure 5.1 | Three perspectives on energy security. With their roots in political studies, natural science, engineering and economics, the three perspectives on energy security focus on different types of threats and risk minimization strategies. Source: Cherp and Jewell, 2011.

This chapter does not consider environmental and social impacts of energy systems as part of energy security, although such consideration is sometimes encountered in academic literature. For example, among 91 scholarly articles on energy security reviewed by Sovacool and Brown (2010), about one-quarter include environmental performance of energy systems in their definition of energy security. Our decision to separate the analysis of environmental and social impacts from the more conventional analysis of energy security is guided by two considerations. First, such impacts are analyzed in other chapters in this publication (Chapters 2, 3, and 4). Second, policies in the majority of countries consider energy security and climate change as distinct, although related, concerns. Thus, an analysis focused on energy security in a more narrow sense may be more policy relevant, as it appeals to a distinct policy community. Nevertheless, we acknowledge strong linkages between energy security and climate change, as explained in Box 5.3 and discussed throughout the chapter.

Box 5.3 | Energy Security and Other Energy-Related Concerns

Energy security concerns overlap with other concerns related to energy systems. This box briefly discusses this overlap with respect to climate and other environmental impacts, energy access and poverty, and energy affordability.

1. **Climate change and energy security.** Limiting the impacts of energy systems on climate change and ensuring energy security are clearly distinct policy concerns, as evidenced by many countries (especially in the developing world) that rigorously pursue energy security agendas without strong climate change agendas. Energy security and climate change concerns are associated with different policy communities, paradigms, and discourses. Yet there are strong links between the two policy areas. First, in some cases changes in climate present energy security risks affecting, for example, availability of water for hydro and thermal power generation or stability of permafrost in areas of oil and gas exploration. In some other cases, climate change results in new energy options such as hydropower from the melting glaciers of Greenland, or additional hydrocarbon resources available for exploration or marine transportation routes opened as a result of a retreat of Arctic sea ice. Second, climate and energy security policies may interact in a synergistic or conflicting manner. For example, reducing reliance on imported fossil fuels and promoting distributed generation from renewable energy sources has both climate change and energy security benefits. On the other hand, switching from imported natural gas to domestic coal advocated on energy security grounds is clearly disadvantageous from the climate change perspective. New technological dependencies, market arrangements, and international regimes associated with climate change governance may also have energy security implications. A similar argument also applies to other environmental impacts of energy systems and energy security.

2. **Energy access and energy security.** Whereas climate change is mostly a global concern, energy security is primarily a national concern, and energy access is often a local concern. Concerns over access to modern energy primarily focus on rural poor people, who are often not a high priority for statesmen dealing with national-level energy security, which concentrates on vital sectors often located in more prosperous urban centers. Likewise, poverty is often tolerated by national policymakers insofar as it does not affect the stability of national political or economic systems. Yet there are several strong links between energy security and energy access. First, enhancing access to modern energy services usually imposes a need to find additional resources and investments, which may exacerbate already existing energy security challenges. Second, lack of access to affordable energy *can* actually catalyze political or economic instability and thus become an energy-related security threat (for example, the uprising that resulted in the change of regime in Kyrgyzstan in April 2010 was allegedly initially triggered by high electricity prices and unreliable supply). Thus, our analysis considers lack of access an indicator of energy insecurity.

3. **Energy security and energy affordability.** Affordability of energy is often quoted as an important aspect of energy security. There is extensive literature analyzing and even defining energy security in purely economic terms. This is understandable, since some disruptions of energy systems are often nothing other than rapid increases in energy prices and can even, in other cases, be translated into economic losses. Yet, there is a clear distinction between "affordability" of energy and energy security. The former is a measure expressing the cost of energy relative to other economic parameters (GDP, income per capita, etc.). Thus energy affordability can be increased or decreased by changes outside energy systems, such as increases in income levels. Affordability also primarily addresses the relative cost of energy in the situation of economic equilibrium. In contrast, energy security deals with price disruptions that are outside economic equilibrium and induced by changes in energy systems (such as supply disruptions) rather than by general economic developments.

This chapter includes three main sections. The next section contains an assessment of energy security in over 130 countries carried out within a framework designed specifically for this purpose. The goal of this assessment is not to compare or rank the nations, but rather to identify common energy security concerns affecting significant parts of the world's population. The following section discusses energy security strategies pursued by individual countries and embodied in various international institutions. The last section provides an outlook for energy security in the future, connecting to Chapter 17 (Energy Pathways for Sustainable Development).

5.2 Energy Security Conditions in the World

This section presents a global overview of energy security in more than 130 countries using a National Energy Security Assessment Framework ("the Framework") specifically developed for this purpose. The Framework is presented in the first subsection and used in subsequent subsections to identify and map vulnerabilities of national energy systems globally.

5.2.1 National Energy Security Assessment Framework

There is extensive scholarly literature on measuring energy security. The methods and indicators used for this purpose vary depending upon the chosen definition of energy security, the intended use of the results of the analysis, the selection of boundaries of energy systems and time-horizons, the assumptions about the nature of potential risks, and the availability of data (Cherp and Jewell, 2010). Table 5.3 summarizes the key choices made in this study as compared to other quantitative evaluations of energy security.

The Framework used in this study proceeds from the definition of energy security as the protection from disruptions of nationally vital energy services. Our analysis aims to identify globally predominant national

energy security concerns rather than to compare or rank countries as more or less secure.

This definition and the purpose of assessment leads to certain choices regarding the boundaries of energy systems analyzed within the Framework (see Figure 5.2). At the center of the analysis are national energy systems subdivided into subsystems of primary energy supply (oil, natural gas, hydro, nuclear energy, biomass, and "new renewables"); systems for the generation and transmission of electricity; and energy end-use sectors. This national-level analysis is supplemented by a global analysis of the vulnerabilities of internationally traded fossil fuels (oil, natural gas, and coal) electricity systems, and the nuclear fuel cycle.

The significance of energy end-use sectors analyzed in this chapter varies across countries. The first three of such sectors, discussed in Chapters 8, 9, and 10, are energy for (1) industry, (2) transport, and (3) commercial and residential buildings. The fourth important energy end-use sector is energy exports, which provide vital revenues to certain countries. By analyzing vulnerabilities of export revenues, we address "demand security," a serious concern of energy-exporting countries.

The Framework takes into account the propagation of vulnerabilities from energy sources to electricity systems and to energy end-uses. Electricity generated from less secure primary energy sources is considered more vulnerable. In turn, vulnerability of end-uses depends upon the vulnerabilities of primary energy sources and electricity used in a given sector. With respect to time-horizons we look into short- to medium-term concerns (up to 15–20 years), which are typically predominant in policy agendas.

The second set of choices concerns the nature of potential risks of disruptions of energy systems. The Framework assesses three dimensions of energy security: robustness (protection from disruptions due to predictable natural and technical factors), sovereignty (protection from disruptions originating from external actors), and resilience (the ability to withstand shocks and disruptions of various natures). We have chosen these three dimensions rather than other attributes of energy systems

Table 5.3 | Goals, choices and assumptions for measuring energy security made in this chapter as compared to other studies of energy security.

Choices and assumptions	Existing studies quantifying energy security	Chapter 5 of the Global Energy Assessment
Assessment goals	Rank groups of countries	Identify energy security concerns affecting a large number of countries and people globally
	Identify vulnerabilities of particular countries	
Definition of energy security	See Box 5.1	Low risk of disruptions of nationally vital energy systems
Energy security concerns	See Box 5.1	As reflected in sovereignty, robustness and resilience of energy systems, see Table 5.2. Exclude environmental and social issues
Energy systems and subsystems	Depending on the assessment purpose; often focused on energy supply, especially oil	Global systems for internationally traded fuels and nuclear fuel cycle; national energy systems including subsystems for energy sources, electricity generation and end-use sectors: transport; industry, residential and commercial energy, and energy export revenues
Aggregation of data	Compound or disaggregated energy security indicators	Aggregated for fuel systems, electricity and end-use sectors at the national and, where appropriate, the global level

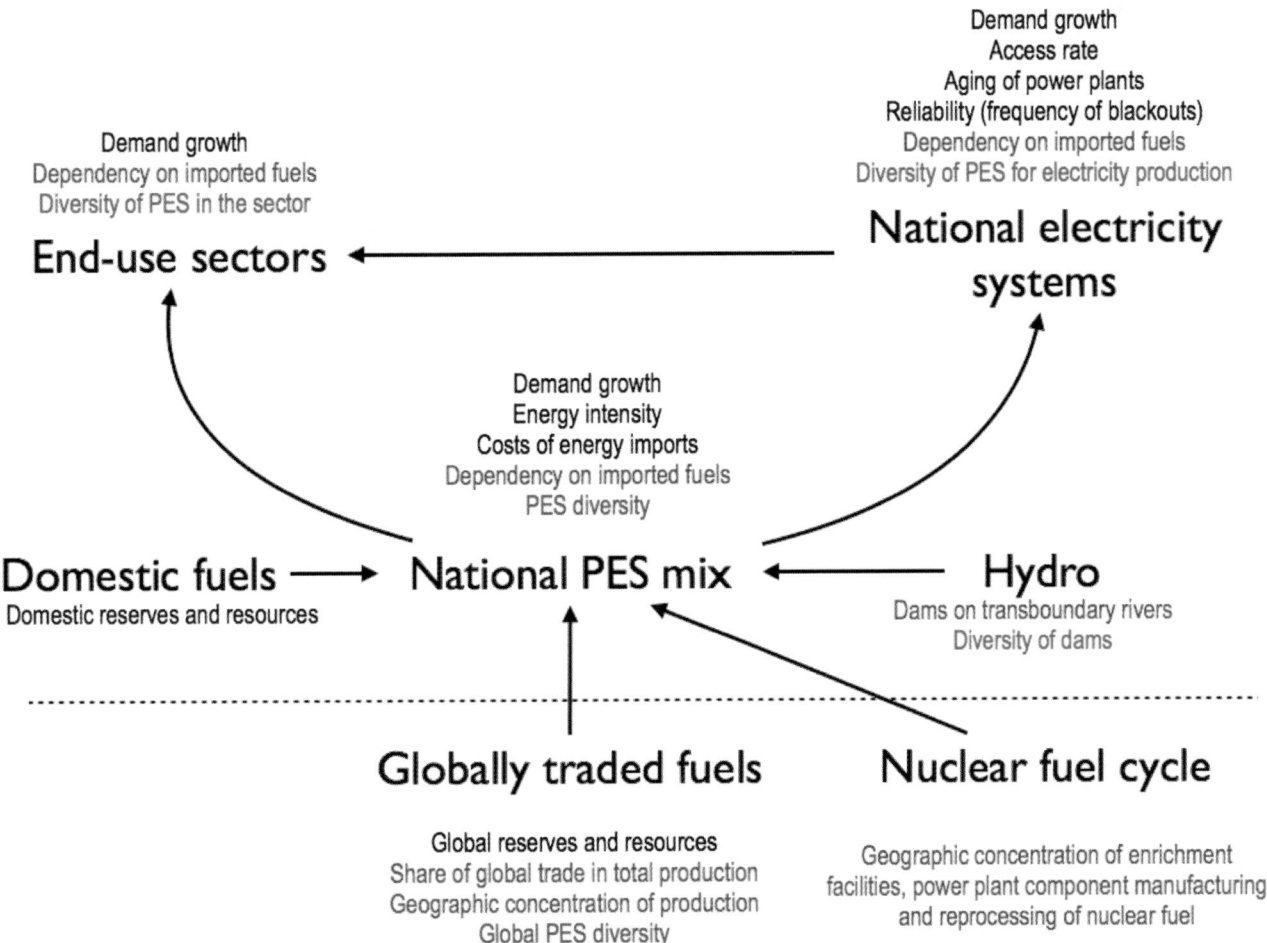

Figure 5.2 | Energy security assessment framework. The framework addresses the security of national energy systems with focus on energy sources, carriers and nationally vital energy services. The figure shows the elements of energy systems and their interconnections accounted for in the framework. The blue, red and green text indicates concerns related to the three perspectives on energy security (see Figure 5.1) – robustness, sovereignty and resilience respectively. The dotted line separates external and domestic factors.

proposed in the literature because they are linked to distinct policy mindsets and strategies, and thus we believe that the analysis along these dimensions will have the most policy relevance, as explained in Figure 5.1 and Box 5.2.

The four types of energy systems and the three dimensions of energy security are combined in a matrix defining key energy security concerns (Table 5.4).

The application of the Framework consists of several elements, as shown in Figure 5.1. This figure and the text below it provide only a brief description of the sample indicators. Further explanation of the choice of indicators and a qualitative analysis are given in the main text of the chapter.

The first element is the analysis of the global vulnerabilities of internationally traded fuels. The main robustness concerns regarding such fuels are available resources and reserves, as well as predicted supply-demand balances. Although these estimates are surrounded by large uncertainties, they nevertheless allow us to compare the major fuels and explain policy concerns about their vulnerabilities. The sovereignty concerns related

to globally traded fuels are the share of international trade in total fuel production (which roughly indicates the degree of global dependency on imported fuels) and the geographic concentration of the fuel production (which indicates the market power of individual producing countries). Finally, the resilience concerns of the global primary energy sources system can be measured by the diversity of the global fuel mix.

The nuclear fuel cycle is also first evaluated at the global level, where we look at the concentration of capacities for uranium enrichment, manufacturing of nuclear power plant components, and nuclear fuel reprocessing.

The rest of the analysis is conducted at the national level for each of the 134 countries where the data are available in the IEA database (IEA, 2010b). With respect to the national primary energy supply, the Framework looks into such robustness indicators as domestic reserves and resources of non-renewable fuels, demand growth, and energy intensity. It uses import dependency (by fuel and for primary energy sources as a whole) as indicators of sovereignty concerns. Finally, the overall diversity of primary energy sources at the national level and the diversity of hydroelectric dams are used as an indicator of resilience of energy supply.

Table 5.4 | Energy security concerns analyzed in this chapter.

Energy subsystems	Energy security dimensions		
	Robustness	Sovereignty	Resilience
Global level			
Globally traded fuels: oil, coal, and gas	Availability of resources and reserves	Share of international trade in the overall production	Dominance (share) of a fuel in the total global PES mix
		Geographic concentration of fuel production	
Nuclear fuel cycle		Geographic concentration of uranium enrichment, manufacturing of nuclear power plant components, and reprocessing of nuclear fuel	
National level			
Energy sources			
Fossil fuels: oil, natural gas, and coal	Available domestic reserves (R/C ratio)		
	Demand growth for a particular fuel	Import dependency	Diversity of import routes
Hydro energy	Climate change effects on water availability and variation	Usage of transboundary water resources	Diversity of hydroelectric dams (see also electricity generation)
Electricity generation and transmission	Age of power plant fleet Growth in consumption of electricity Reliability (frequency of blackouts) Access rate	Reliance on imported fuels	Diversity of fuels used for electricity production Diversity of power plants
End-use sectors: industry, transport, residential and commercial, energy exports	Growth (decline*) in energy demand for the sector	Reliance on imported fuels within the sector	Diversity of energy sources used in the sector
National energy systems (cross-sectoral)	Energy intensity Growth in overall energy consumption Energy consumption per capita	Overall import dependency	Overall diversity of PES used in the national energy system

Notes: Concerns quantified by indicators are highlighted in bold. * – for energy exports

With respect to national electricity systems, the robustness indicators are the age of power plants, demand growth, and the reliability (blackout) statistics. The sovereignty indicator is the dependency of electricity systems on imported electricity and fuels. The resilience indicator is the diversity of primary energy sources used in electricity production.

With respect to end-use sectors, the Framework uses the rate of demand growth as the robustness indicator, the share of imported fuels in the sectoral primary energy sources mix as the sovereignty indicator, and the diversity of primary energy sources as the resilience indicator for a particular sector.

The final set of choices in evaluating national energy security concerns the aggregation of indicators and interpretation of results. This chapter recognizes that the nature of energy security risks varies among countries and in time, and thus a unified indicator is not plausible or relevant. The Framework seeks to identify, map, and explain the varying conditions of energy insecurity encountered around the world. It uses simplification and aggregation, necessary for mapping these concerns, rather than producing a single universal energy security index.

The Framework used in this analysis is naturally not capable of evaluating all important energy security concerns. Some of the issues that were intentionally left outside the analysis are discussed in Box 5.4. The method and some of the key indicators used in the Framework are

subsequently deployed in analyzing energy security within the GEA Scenario in Chapter 17. The method of such analysis is explained both in Chapter 17 and in the last section of this chapter.

5.2.2   Security of Primary Energy Sources

This section considers the security of key primary energy sources. It starts by analyzing major internationally traded fuels: oil, natural gas, and coal. For each of these fuels, the analysis of their global vulnerabilities is followed by the analysis of national vulnerabilities in individual countries. The second subsection discusses vulnerabilities of nuclear energy: both at the global and the national level. The remaining three subsections discuss vulnerabilities of hydro energy, biomass, and "new renewables." The final subsection contains an analysis of the total primary energy supply of individual countries by aggregating the data on individual sources at the national level. The indicators used for the analysis in this section are summarized in Table 5.5; they cover the robustness, sovereignty, and resilience perspectives on each of the energy sources.

5.2.2.1   Globally Traded Fuels

The three major fuels traded on an international and global scale are oil and its products, natural gas, and coal. Global trade in these fuels means

Box 5.4 | Limitations of Current Analysis

The analysis presented in this chapter does not consider several issues that are occasionally addressed in energy security literature. These issues have been scoped out to maintain the focus of the assessment as well as due to time, space, and data limitations. The limitations relate to the types of energy systems considered in this assessment, time-horizons, and the nature of risks and disruptions analyzed.

First, the assessment is focused on key energy security issues that presently dominate policy concerns in a large number of countries. This means that only major traditional mainstream energy security systems are analyzed. For example, wind, solar, geothermal, and tidal energy at the moment do not make for a large share of energy supply in many countries and thus are not analyzed in detail. Yet, availability of alternative energy sources – to say nothing about technologies and finances – may be key factors in shaping national energy security in certain countries.

For largely the same reason, the analysis in this chapter is focused on short- to medium-term concerns (up to 15–20 years) that typically dominate policy agendas. We believe that energy security in a longer-term future will largely depend on policy choices and may unfold in the context of radical energy transitions. If these policy choices are guided by economic, environmental, and social sustainability goals, energy systems may evolve along pathways described in the GEA Scenario (Chapter 17). Energy security under such pathways is analyzed in Chapter 17 using the assumptions, methods, and indicators developed in this chapter.

Third, energy security is affected by a large number of complex and often intangible factors that are difficult to identify, quantify, or compare on the global scale. Some of these factors are subjective, such as trust between various actors in energy systems, consistency and predictability of policies, or reflexivity of market price-setting. Other factors are related to vulnerabilities of complex technical systems ranging from technological interdependencies (with respect to materials, equipment, expertise, and capacities related to energy systems) to cyber-security of critical energy infrastructure, especially electricity networks. Another critical group of factors relates to financing options, availability and effectiveness of investment, and capacities to develop various energy options. None of these factors could be extensively analyzed in this chapter, which does not make them less important in specific national contexts.

Despite these limitations, we believe that this chapter fulfills its goal of mapping the major energy security concerns of today's world and provides a methodological and factual basis for analyzing the potential evolution of these concerns under sustainable energy transitions.

that their vulnerabilities can be analyzed within the global and national systems using the three perspectives on energy security: robustness, sovereignty, and resilience.

Robustness of globally traded fuels means their physical availability, technological accessibility, and economic affordability (Kruyt et al., 2009). Although each of these characteristics is affected by a number of complex and often uncertain factors, for the purposes of this analysis they can be described by relatively simple proxy indicators such as the global reserves-to-production (R/P) ratio, supply and demand projections, and price dynamics.

At the national level, robustness of a particular fuel source can be characterized by national R/P ratios. Although such ratios are notoriously fluid (because the estimates of "resources" and "reserves" often change in time with economic and technological changes), they still signal important vulnerabilities to policymakers, especially if they are relatively short-term (under 10–20 years). A rapidly growing demand for a particular fuel signals a pressure on resources and infrastructure. Other national-level indicators of robustness of a particular fuel supply are the fuel intensity of a national economy (higher fuel intensity means that it is more difficult to adjust to potential disruptions of supply) and the proportion of national

gross domestic product (GDP) or export revenues that is spent on imports of this fuel.

The main sovereignty concern related to internationally traded fuels at the global level is market power of dominant actors, which relates to their potential ability to disrupt prices or even physical supply of that fuel. Proxy indicators of this concern include the share of internationally traded fuel in the total fuel production and the geographic concentration of fuel production in particular countries or regions. These indicators are more informative where there is a single global market for the fuel (as in the case of oil and, partially, coal). In the case of gas, such indicators should be used in each of the separate regional markets.

The main sovereignty concern at the national level is import dependency on a particular fuel. This is probably the most widely used metric of energy security. Import dependency is important for policymakers because it makes their energy supply vulnerable to (a) global price volatility determined by factors beyond their control; (b) market power of major exporters, which may in extreme cases be manifested in direct physical supply disruptions; and (c) exposure to other disruption factors (including in transit countries) along import routes. In addition, many nations are concerned about the security of their energy imports if these originate in countries considered

Table 5.5 | Overview of concerns and indicators of national energy security addressed in this chapter.

Energy sector	Energy security concerns (indicators)*	
	Shorter term >	**Longer term**
Oil	**Exposure to the global oil market** (import dependency, cost of imports, **Demand-side vulnerabilities** (annual growth in oil consumption, *oil intensity*) **Domestic availability of oil** (R/C) Environmental acceptability of oil production and use	Global conventional oil scarcity ("peak oil")
Gas	**Exposure to the global and international gas markets** (import dependency, cost of imports **Demand-side vulnerabilities** (*gas intensity*) **Domestic availability of gas** (R/C) Environmental acceptability of gas production and use	
Coal	**Exposure to the global coal market** (import dependency) **Domestic availability of coal** (R/C) Environmental and health acceptability of coal production and use	
Nuclear	Seasonal water availability **Aging infrastructure** (average age of nuclear power plants) **Capacity to replace existing fleet** (start of last plant construction) Access to capital, enrichment, reactor manufacturing, reprocessing Environmental, safety and security acceptability of nuclear power	
Hydro	Reliance on dams which are shared (*transboundary dams* or *dams on transboundary rivers*) Seasonal water availability Aging and silting of dams and other infrastructure **Exposure to risk of dam failure or sabotage** (diversity of dams)	Effects of climate change on water patterns and availability
Electricity	**Exposure to imported fuels** (dependency on imported fuels) **Exposure to a single fuel market** (low diversity of energy sources used for electricity production) **Adequate capacity** (annual demand growth rate, access rate) Underinvestment and aging infrastructure	
Transport	**Exposure to imported fuels** (dependence on imported fuels) **Demand-side vulnerabilities** (annual consumption growth rate)	
Industry	**Exposure to imported fuels** (dependence on imported fuels) **Demand-side vulnerabilities** (industrial energy intensity)	
Residential and commercial	**Exposure to imported fuels** (dependency on imported fuels) **Demand-side vulnerabilities** (annual consumption growth rate) **Adequacy of provision** (reliance on traditional biofuels in the residential sector)	
Energy for export	**Exposure to price fluctuations** (revenue from energy exports as share of GDP) "Security of demand" (diversity of export routes and destinations) "Dutch disease" and "resource curse"	
Cross-sectoral	**Exposure to imported fuels** (overall import dependency, cost of energy imports compared to GDP, cost of energy imports compared to export earnings) **Overall resilience of primary fuels** (diversity of PES) **Exposure to energy price volatility** (overall energy intensity) **Demand-side pressure** (annual growth rate in consumption, consumption per capita)	

Notes: R/C – reserves to consumption ratio; PES – primary energy sources

unfriendly, unstable, unreliable, politically unacceptable, or potentially exercising asymmetrical market power. In other situations, imports come from culturally and politically close and trusted partners and thus lead to few security concerns. In the global context, we do not systematically quantify these idiosyncrasies but comment on them in particularly relevant cases (e.g., in regional energy security discussions).

Finally, the main resilience metric with respect to energy supply is the diversity of primary energy sources, either nationally or globally.

Strategic reserves of oil and, to a lesser extent, gas and coal support both resilience and robustness of energy supply, but their presence was not quantitatively analyzed in this chapter due to data limitations.

Global oil supply vulnerabilities

Oil is a non-renewable resource massively traded on the global scale. It is the largest single primary energy source worldwide and dominates the transport sector, where it lacks easily available substitutes. The global demand for oil is steadily rising, particularly as a result of increasing motor fuel consumption in emerging economies. Given limited new discoveries, the global production of conventional oil is predicted to "peak" or "plateau" in the first half of this century (see Chapter 7 for details) in spite of globally rising demand. This supply-demand imbalance will lead to higher oil prices in the medium to long term, as exploration and production will turn to progressively more demanding, and thus more expensive, sources (such as oil in deep-sea water or the Arctic). The costs are rising even in established production regions. The Kashagan oilfield in the Caspian Sea, one of the biggest discoveries in decades, was due to enter production in 2005. Now the target is 2012, and the costs could exceed US$100 billion.[1] The 2010 disaster at BP's well in the Gulf of Mexico may increase costs further as environmental regulations are tightened.

In addition to this expected medium- to long-term cost increase, global oil prices have been increasingly volatile in the last decade for several reasons. First, the global demand for oil has been growing faster than production capacity, and as a result the spare production capacity (the difference between what is possible to produce in the short term and what is actually produced to meet the demand) has significantly decreased. With the smaller spare capacity, even relatively minor disruptions of supply – whether they are due to natural, economic or political causes – may knock the demand and supply off balance, thus signaling rocketing prices or even a physical scarcity of fuel at least in the short term,[2] or the reverse.

Furthermore, the global oil price has been increasingly affected by highly volatile market expectations through speculation. Future contracts and other derivates have opened the oil market to speculative money, which is blamed as a contributing factor in oil price volatility. Due to the rapid increase in the number of actors and improvements in communication tools, the oil markets have become highly reflexive – i.e., driven by market sentiments and expectations. This reflexivity does not blend well with the decreasing and often uncertain spare capacity. A speculative market combined with low spare capacity and high uncertainty easily results in price bubbles and longer-term volatility.

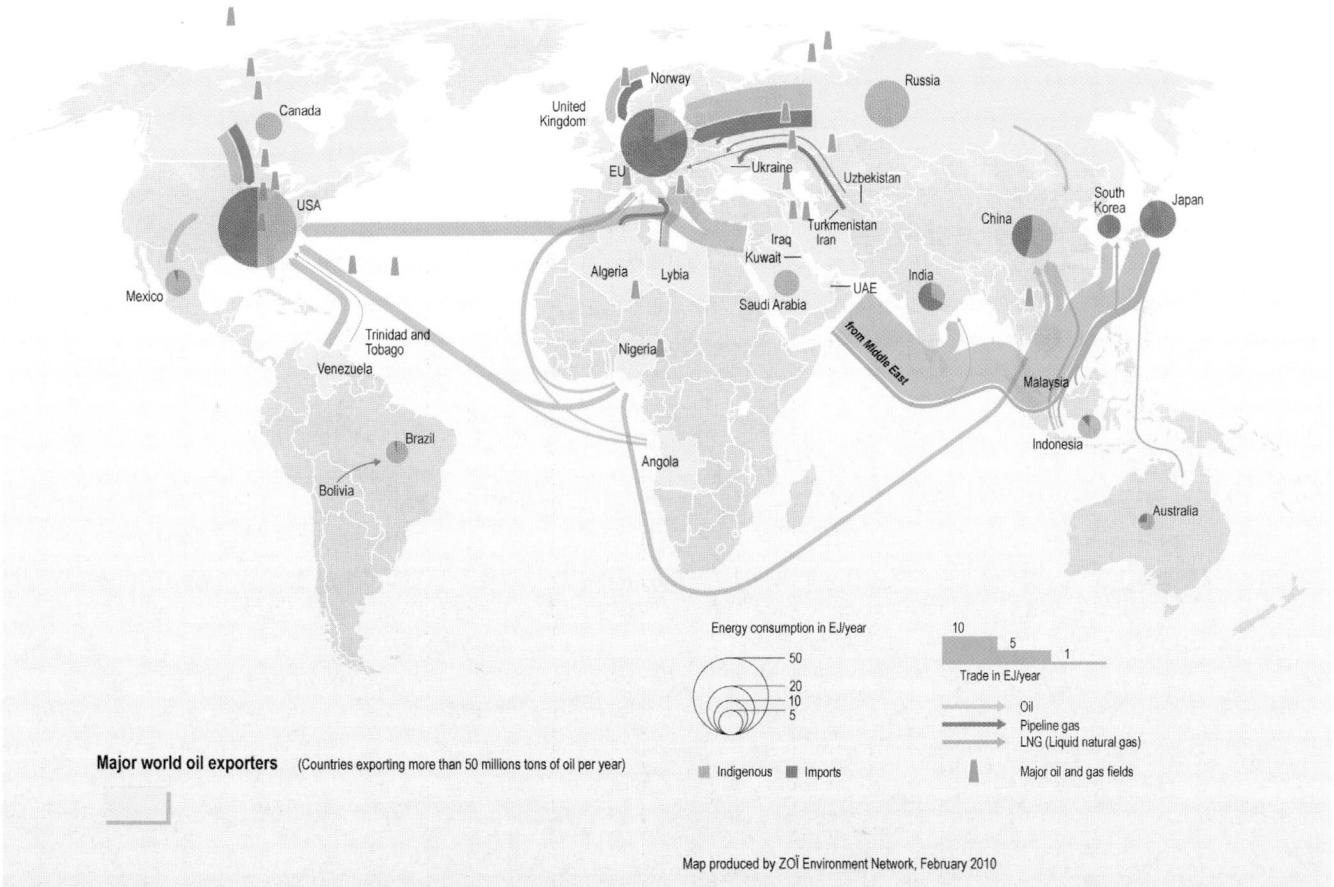

Figure 5.3 | Global oil and gas production and trade. The figure shows the concentration of global oil and gas production in a limited number of regions and large volumes of trade in these fossil fuels. Source: BP, 2009; IEA, 2009a; b.

---

1   According to the Cambridge Energy Research Associates' (CERA) Upstream Capital Costs Index (UCCI), prices of drilling technology, skilled labor, and equipment have soared. Even in spite of recent economic recession, exploration and development costs in oil and gas have risen by 200% between 2000 and 2011 (IHS, 2011) adding a significant cost to crude supply.

2   When BP had to shut down a single refinery in 2006, world crude prices jumped 2% (Sovacool, 2009).

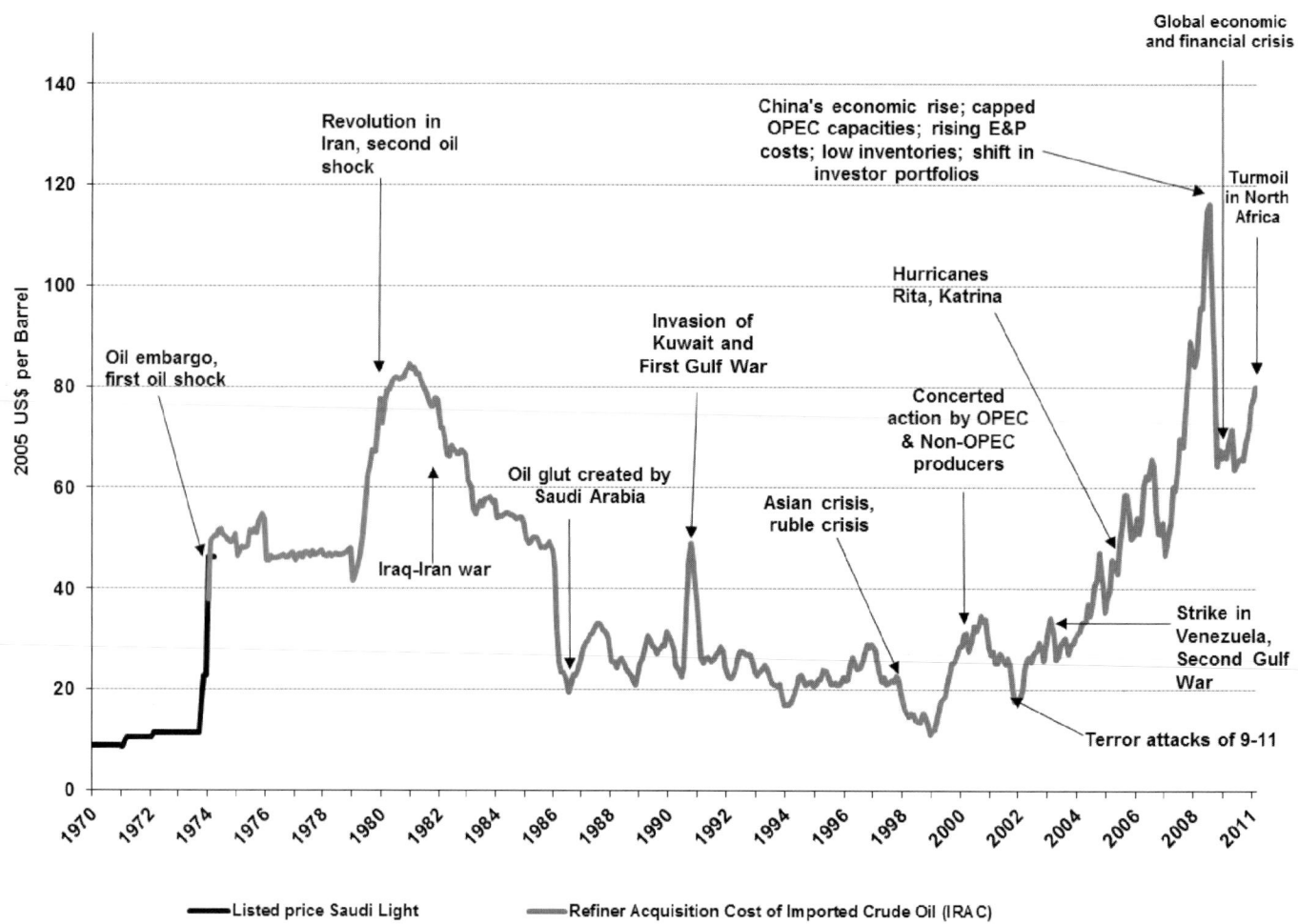

Figure 5.4 | Price of oil (in US$_{2005}$$) and major political events 1970–2011. Source: modified from US EIA, 2011b. Note: Imported Refiner Acquisition Cost (IRAC) is the volume-weighted average price of all crude oil imported to the United States.

Moreover, the volatility of the global oil market is exacerbated by the fact that, whereas oil is traded globally, its production and reserves are heavily concentrated in just a few countries and regions (see Figures 5.3 and 5.4) away from the major consumption centers. In 2006 the Middle East accounted for 62% of the world's identifiable proved liquid reserves and 31% of the output. The world's top 10 producers represent approximately 62% of global output. Eight of the top 10 producers are net oil exporters that collectively supply 34% of global oil demand. On the other hand, seven of the world's 10 largest consuming countries lack sufficient oil production capacity to meet their internal consumption, importing more than 35% of the world's demand (Lehman Brothers, 2008). According to the IEA (2010a), the concentration of oil production and the market power of major producers are expected to increase by 2030, not only in the Reference Scenario but also in the presence of strong climate policies (the "450 scenario"). The interregional trade in oil will increase by over 30% and comprise almost one-half of the total volume of global production by 2035 (IEA, 2010a) in the New Policies Scenario.

This means that global oil supply is becoming increasingly more vulnerable to conditions in oil-producing countries and regions, as well as the demand dynamics and expectations of major consumers. There is

an extensive record of political, economic, and natural events, military conflicts, and deliberate acts of sabotage that have either physically disrupted global oil flows or resulted in price hikes. According to (Jones et al., 2004), quoted in (Farrell et al., 2004), 24 oil supply shocks between 1950 and 2003 averaged eight months or 3.7% of the global supply. The 1990 Gulf War, for instance, resulted in production loses of 2.8 million barrels per day (mbd) of Iraqi and 1.4 mbd of Kuwaiti oil – some 7% of global supply (BP, 2007). The strike at Petróleos de Venezuela, SA during 2002/2003 brought to a halt the country's entire oil sector and temporarily reduced global oil supplies by 2.3 mbd, or 3% of total world supply (US Government Accountability Office, 2006). As it coincided with unrest in oil producer Nigeria and the looming US invasion of Iraq, the strike contributed to pushing the nominal price of oil to new highs (Shore and Hackworth, 2003). For an overview linking oil price developments and major political events since the early 1970s, see Figure 5.4.

Thus, countries relying on imported oil are at risk of facing medium- to long-term price increases combined with volatility of oil prices in the short term. Physical supply disruptions similar to the Arab OPEC oil embargo of the early 1970s are less likely nowadays because alternative suppliers can normally be found, given the "liquid" character of the global oil market, and strategic oil reserves present in most major economies can smooth out such impacts. However, infrastructural limitations (e.g., pipeline capacities or refineries "tuned up" to deal only with a particular kind of crude) can still make physical disruptions very painful for particular countries. For example, several landlocked countries in Central and Eastern Europe receive the majority of their oil supplies from Russia through the Druzhba pipeline and are, therefore, vulnerable to disruptions affecting this particular supplier and the transportation route. Globally, 64% of the global oil supply flows through just 10 supply chain "choke" points, with the top three accounting for 46% of supply (Lehman Brothers, 2008). In addition to the Druzhba pipeline, these include the Straits of Hormuz, Strait of Malacca, Abqaiq processing facility, Suez Canal, Bab el-Mandab, Bosporus/Turkish Straits, Mina al-Ahmadi terminal (Kuwait), Al Basrah oil terminal (Iraq), and LOOP (United States).

In summary, the global oil supply is vulnerable from robustness, sovereignty, and resilience perspectives. The main robustness concerns are the increasing demand in the face of limited conventional production capacities, rising costs of production, and high and volatile prices. The main sovereignty concerns are the increasing concentration of production away from major demand centers and the increasing market power of major producers. The resilience concerns include not only the dominance of oil in the world's energy use and the lack of easily available substitutes for oil in the transport sector (as we will be discussing further later in this chapter), but also the limited diversity of global oil supply routes and the vulnerability of the "choke" points.

National oil supply vulnerabilities

Oil is the only primary energy source that plays an important role in all of the national energy systems except in a few less-developed countries. Virtually every country (125 countries with 5.9 billion people)[3] has at least 10% of its primary energy derived from oil, while over 5.5 billion people live in 112 countries which rely on oil for more than 18% of their total primary energy supply. Moreover, some 370 million people live in 32 countries that use oil for more than half of their primary energy supply.

The majority of countries import most or even all of the oil and petroleum products they need. Over three billion people live in 83 countries that import more than 75% of the oil and petroleum products they consume, and 101 countries with 5.3 billion people import over 25% of the oil they consume.

The number of people living in countries significantly dependent on oil imports is likely to rise in the near future. At the moment there are 3.6 billion people in countries that import more than half of their oil, but a further 1.7 billion people live in countries (including Argentina, China, Indonesia, and the United Kingdom) where the ratio of domestic oil

reserves to domestic annual oil consumption is under 15 years.[4] Even though some countries (e.g., Ghana) may become less dependent on imported oil due to new discoveries and developments, this is unlikely to reverse the overall trend of the rapid increase of oil import dependency in the world.

Several of the highly import-dependent countries have additional demand-side vulnerabilities. In 16 such countries (with 1.8 billion people), the demand for petroleum products grew at 5–10% per year on average from 1996–2006. In addition, five of these (including several African countries and Vietnam) have outstanding fuel intensities that make them highly vulnerable to market price disruptions.

China and India – which are together home to about 2.5 billion people – tell a story of their own. Oil consumption in China more than quadrupled between 1980 and 2009. It has grown at the annual rate of over 7% a year in the last 10–15 years. In the WEO (World Energy Outlook)'s New Policies Scenario, it is expected to almost double between 2009 and 2035, increasing on average by 2.4% per year. In this scenario, China accounts for 48% of the global rise in demand for oil. The majority of this increase will likely come from imports, as China already imports over 40% of its oil supply and its domestic reserves/consumption ratio is between seven and eight years. India already imports over 75% of its oil, and its domestic resource/supply ratio is under six years, with consumption growing on average 4.6% per year and projected to grow faster than in China at 3.6% between 2009 and 2035, accounting for a further 30% of the increase in the global demand for oil (IEA, 2010a).

It is worth commenting that all low-income countries (over 600 million people) except Uzbekistan,[5] Myanmar,[6] and Yemen import over 80% of their oil and petroleum products. In sub-Saharan Africa, apart from South Africa, Gabon, Nigeria, Angola, Sudan, Congo, Cameroon, and Côte d'Ivoire, all of the countries are completely dependent on the import of petroleum products!

Global natural gas supply vulnerabilities

Natural gas is the fastest growing fuel of choice for electricity generation in the Western hemisphere and has also gained an increasing importance in emerging economies. In addition, it is extensively used in the residential sector (for heating and cooking) and in many industrial applications. Similarly to oil, gas is a non-renewable resource with reserves and production concentrated in a few countries and regions, and most nations relying on imported supplies. There are, however,

---

3   According to IEA (2007), oil and its products represented less than 10% of total final energy use in only nine countries: Mozambique, Tanzania, Democratic People's Republic of Korea, Zimbabwe, Nepal, Ethiopia, Democratic Republic of Congo, and Nigeria.

4   The concepts of reserve/production or reserve/consumption ratios are notoriously unreliable when used for projecting the date of "running out of oil." This is because the estimates of reserves are constantly updated and the rates of production/consumption also change. However, the very low (<10–15 years) ratio usually signals serious energy security concerns and the need to develop new reserves, imports, or other energy sources.

5   Uzbekistan has the resource/supply ratio of about 16 years.

6   Currently, Myanmar is a net crude oil exporter, but it imports more than 50% of its petroleum products and has the resource/supply ratio of just 3.75 years.

Table 5.6 | Vulnerabilities of primary energy sources.

Energy security perspectives	Robustness		Sovereignty		Resilience
	Global R/P ratio	Projected demand growth 2008–2035*	Share of international trade in global production in 2009	Number of people (billions) in countries with import dependencies over 25/50/75%	Diversity of global producers by region, SWDI
Globally traded fuels					
Oil	30 yr.	15%	66%	5.3/3.6/3.1	1.63
Gas	80 yr.	44%	29%	2.2/0.75/0.65	1.84
Coal	150 yr.	19%	14%	1.3/1.1/0.70	1.92
Other energy sources					
Nuclear	Aging of nuclear power plants; sensitivity to political interventions		Concentration of enriched uranium and reactor manufacturing technologies; nuclear fuel cycle controlled for non-proliferation reasons		Generally large facilities; difficult to substitute in case of failure
Hydro	Sensitivity to water availability; vulnerability to climate change in some regions.		Hydroelectric facilities located on internationally shared rivers		In certain cases extremely large facilities providing majority of electricity of certain countries
NRES	High initial costs; intermittency of supply		Technological dependencies; potential import dependencies for biofuels		Generally assumed to be higher than in the case of traditional sources due to distributed generation and more diverse energy mix

Source: see main text; *– New Policies Scenario (IEA, 2010a).

important differences between energy security concerns associated with oil and natural gas as globally traded energy sources.

With respect to globally available resources, the production of conventional gas is likely to peak (or plateau) several decades later than conventional oil in most scenarios. In addition, technological advances have recently added significant quantities of unconventional (primarily shale) gas to global reserve estimates (see Box 5.5). There are, generally speaking, fewer concerns about the risk of rapid increases in gas prices than in the case of oil (Söderbergh et al., 2009; Söderbergh et al., 2010). In other words, "peak gas" may not be as significant a concern as "peak oil."

Yet the global demand for natural gas is projected to increase faster than that for oil (by 44% in the WEO's New Policies Scenario, compared to 18% for oil). A quarter of this rise in global demand is set to come from China, where the consumption of natural gas is projected to increase on average by 6% per year up to 2035 (IEA, 2010a).

On the other hand, similarly to oil, a globally sufficient supply of natural gas depends upon adequate investments in exploration, production, and transportation. Such investment is all but granted, especially in connection with uncertainty over shale gas and other factors affecting gas prices (see Box 5.5). According to the IEA (2007), the total upstream and infrastructure investment needs in natural gas amount to US$4.2 (US$_{2005}$$ 4.0) trillion until 2030.

The difference between oil and natural gas is also in the volumes of international trade. Whereas about two-thirds of the globally produced oil is traded, this share is only about one-third for natural gas (see Table 5.6). However, interregional trade in natural gas is projected to increase very significantly (by 80% between 2008 and 2035 in the WEO's New

Policies Scenario (IEA, 2010a) to account for over one-quarter of the total production). China's imports of natural gas are set to grow some 40-fold in this scenario, accounting for some 40% of the total growth in the interregional gas trade over this period (IEA, 2010b).

In contrast to oil, there is no unified global market for natural gas, primarily because the regional price differentials do not exceed the still high costs of interregional transportation (Stevens, 2010). Natural gas is transported via pipelines or between liquefied natural gas (LNG) installations tied with long-term contracts. Natural gas is traded on several largely independent regional markets, most notably the Eurasian, North American, and Asia-Pacific markets. While the North American natural gas market is liquid, regionally integrated, and deregulated, the Eurasian gas market is dominated by long-term bilateral contracts, entailing off-take agreements and destination clauses. Consequently, the Eurasian gas market, though considerably larger in volume than the North American one, is characterized by a comparably lower liquidity and a still marginal, although growing, role of spot markets. The emergence of a global gas market was widely predicted by experts and signaled by considerable investments in LNG infrastructure but delayed by the economic crisis and shale gas developments.

Natural gas is traded in regional markets by pipelines and globally as LNG. This has two major implications for energy security concerns associated with this fuel. On one hand, it prevents the emergence of globally powerful market actors, be it individual producers or cartels (like OPEC in the case of oil). On the other hand, it significantly increases the market power of regionally dominant market actors, such as Russia in Eurasia. Regional markets dominated by long-term contracts may also be more protected from price volatility, especially if the prices of gas are decoupled from those of oil, which has been a long-term trend and expectation (Stern, 2009; Stevens, 2010).

Box 5.5 | Shale Gas Revolution and Energy Security

Over the last several years the production of unconventional shale gas in the United States has dramatically increased (from 1% of total energy supply in 2000 to some 20% in 2009) and its share of total gas reserves increased by 50%. This has been heralded as the "shale gas revolution," with significant implications for energy security not only in the United States but also in the rest of the world, particularly in Western and Central Europe, which may become significantly less dependent on Russian gas imports.

At the same time, many uncertainties have been quoted with respect to shale gas developments. Even in the United States, cost estimates vary widely, and there are concerns over shorter life and low recovery rates of shale gas fields as well as environmental impacts of shale gas exploration. More uncertainties are associated with replication of North American success on other continents. Obstacles to shale gas development in Europe include different geology, lack of tax breaks, lack of technologies and expertise, more sensitive natural and cultural environment, potential opposition of local communities, and different land/natural resource structure than in the United States. Despite the potential shale gas revolution, the IEA New Policies Scenario predicts the overall decline in natural gas production in Western Europe, with the average rate of −1.5% per year up to 2035. Globally, the share of unconventional gas is predicted to increase from the current 12% to 15% in 2030, much of this increase being in North America.

In spite of these potentially positive (although less than certain) effects of the shale gas revolution, there may be some negative effects as well. So far, increasing shale gas production has resulted in decreasing gas prices and cancellation of investment projects in LNG and natural gas infrastructure. If the developments in shale gas do not follow the most optimistic forecasts, there may be a significant lack of capacity to meet the global demand. The shale gas revolution has also slowed down the development of spot trading and the emergence of a global gas market, leading to a temporary "glut" in LNG production and capacity. Investments in renewable energy alternatives to natural gas have also been slowed by this decrease in price.

Another concern associated with shale gas may be technological dependency. Chemicals, drilling technologies, and the expertise required for shale gas production are primarily available in the United States, where they have been developed over decades of trial and error. These will need to be exported, at least in the short term, to enable a similar increase in production in other countries.

Source: Stevens, 2010; IEA, 2010a. See also Chapter 7 for the discussion of technical issues related to shale and other unconventional natural gas.

Thus, sovereignty concerns related to natural gas are highly region-specific. They are most prominently highlighted by the recent gas disputes between Russia and the Ukraine (in 2006 and 2009) which led to interruptions in the supply of gas to members of the European Union (EU). This interruption most severely affected such countries as Slovakia, which is significantly dependent on Russian gas and lacks emergency gas stocks to cope with such short-term disruptions of supply. In another similar case, the Georgian gas supply in 2005/2006 was struck by a sabotaged Russian gas pipeline, causing severe energy shortages during that winter. Subsequently, the EU's energy security debate has become largely framed by the need to diversify the routes and origins of gas imports (as well as to substitute gas by other energy sources). The tense political relations between some Central and Eastern European countries and Russia played an important role in sustaining this debate. Outside Europe such deliberate disruptions of gas pipeline deliveries have been rarer, although not unknown.

Finally, in contrast to oil, which currently lacks readily available substitutes in the transport sector, natural gas can be more easily replaced by other sources such as oil, coal, nuclear power, or imported electricity (not to mention biogas and other renewable sources). Naturally, such a replacement might require substantial infrastructure investment.

In summary, the global vulnerabilities of the natural gas supply are largely similar to those of oil but with several important distinctions. With respect to robustness, it has larger conventional reserves and probably more optimistic expectations concerning non-conventional reserves, although it still remains a limited resource and upstream investment requirements are very high. In the case of natural gas, sovereignty concerns are strongly articulated on the regional level, particularly in Eurasia. With respect to resilience, the substitutability of natural gas is its major advantage, but for many markets the low diversity of origin of imports and supply routes presents a potential vulnerability to natural or political disruptions.

National natural gas supply vulnerabilities

Significantly fewer countries rely on natural gas than on oil. Fifty-seven countries with a population of almost two billion people use gas for 20% or more of their primary energy supply. Among those, 16 countries (with a combined population of over 350 million people) use natural gas for more than half of their energy needs. A further 850 million people in 21 countries rely on natural gas for between 10% and 20% of their energy. Further analysis in this section relates to these 78 countries (which notably exclude India and China, where the consumption of natural gas is still not very high, although it is projected to grow rapidly).

Most of the countries using natural gas have to rely on imports. Almost 650 million people live in 32 countries that import over 75% of all their gas needs. All of these countries are in Eurasia: most are in Europe, Turkey, and the former Soviet Union, but they also include Jordan, Singapore, Korea, and Japan. Remarkably, only seven of these highly import-dependent countries have LNG facilities. Of those, Korea and Japan only import natural gas through LNG facilities, and the remaining countries use pipeline imports as well. The remaining 25 highly import-dependent countries do not have LNG re-gasification terminals and rely exclusively on pipelines for gas imports. Eleven of those countries are landlocked and thus have no prospects of benefiting from LNG trade (Coutsoukis, 2008). At the same time, global LNG markets are rapidly expanding (IGU, 2010).

There are 35 countries, with a combined population of over 750 million people, which import more than 50% of their gas needs, while the number of people living in countries which import over 25% of their gas needs is almost 2.2 billion.

Among the countries that import less than 50% of their natural gas needs, including those relying on domestic reserves (such as the United Kingdom, the United States, Bangladesh, Mexico, Thailand, Brazil, and Argentina), 12 (780 million people) have a domestic reserves/supply ratio under 16 years, which signals the likely increase in import dependency in the near future.

Thus, over 1.5 billion people live in countries that are either seriously dependent on imported natural gas or are likely to experience such dependency soon. In addition, the majority of the 37 countries (with over 2.5 billion people) relying on natural gas for more than 10% of their supply experienced a growth rate of over 6% per year over the last decade, which is likely to put further pressures on their natural gas supply.

Global coal supply vulnerabilities

Coal is the world's fastest growing fossil fuel energy source, currently providing about one-third of the global primary energy supply. Similarly to oil and gas, coal is a non-renewable fuel that is traded on the global market. However, coal is different than oil and gas in several important aspects, which affect the vulnerability of coal supplies.

First of all, the worldwide reserves of coal are larger than those of conventional oil and gas, although they are subject to large uncertainties, as explained in detail in Chapter 7. The global R/P ratio of coal varies between different organizations and years in which estimates are provided but is generally believed to exceed 130 years.[7] Global coal production is expected to increase by some 14%, growing on average at 0.6% per year until 2035 in the WEO's New Policies Scenario. All of this growth is projected to come from non-OECD countries, with over 90% concentrated in China, India, and Indonesia (IEA, 2010a).

Global coal reserves are not as geographically concentrated as those of oil and natural gas. The United States, China, India, and the former Soviet Union together account for some 80% of global hard core reserves. Southern Africa and Australia account for 13% of the remaining 17% of reserves, with the remainder being split among the rest of the world. However, the existing coal reserves and production capacities are largely located in the same countries that consume or are expected to consume the majority of coal. For example, the WEO projects that China will account for half of global coal production in 2035 but will also consume all of this coal, being the largest world consumer and producer at the same time (IEA, 2010a). International trade in coal was only 16% of its total consumption in 2009 (compared to 66% of oil and 29% of natural gas). It may nevertheless increase in both the short and medium term. The main driver of this increase will be consumption and production of coal in China. In 2009, China's imports of coal tripled, but their growth in the future will be determined by the competitiveness of imported coal against the coal domestically produced and transported from China's western provinces. In the WEO's projections the share of globally traded coal will remain at approximately today's 16% of global demand, whereas the absolute volume of trade may grow some 15%. In this scenario the largest exporters will be Australia and Indonesia, and the largest importer, India.

The geographic distribution of coal reserves means that supplies of coal are less likely to be disrupted for such "geopolitical" reasons that are much feared in the cases of oil and gas. However, coal production may be slowed down because of its severe environmental and health costs. (Coal has the highest greenhouse gas emission factor of all fuels and also contributes significantly to local air pollution). Global climate change policies may also affect coal affordability (if carbon capture and storage becomes a requirement). If coal becomes a globally traded commodity, fluctuations of domestic currencies and a host of other factors can also affect its price.

As with other fuels, domestic availability of coal does not automatically mean that it is easily accessible to consumers. A case in point is China, whose reserves are mainly found in its western provinces, while consumption is concentrated on its east coast. Given China's rapidly rising consumption levels, its transport infrastructure faces heavy capacity challenges (see Box 5.6).

---

7   The BP *Statistical Review of World Energy 2007* gives a coal R/P ratio as 133 years at the end of 2007 (BP, 2007). The *World Energy Outlook 2010* (IEA, 2010a) notes a 1:150 R/P ratio.

Box 5.6 | Coal Use in China

China relies heavily on coal as a primary fuel for industrial use and electricity generation. Coal combustion provided 65% of national electricity in 1985 but ballooned to more than 80% in 2006 (though it shrank to 71% in 2008). From 2002 to 2007, demand for electricity in China grew by about 12%, and more than 70,000 MW of capacity were brought online to meet it. A majority of this capacity was coal-fired, and China is currently constructing the equivalent of two 500 MW coal-fired plants per week, or a capacity comparable to the entire power grid in the United Kingdom every year. During this time, every week to 10 days over the course of five years, a coal-fired power plant opens somewhere in China big enough to serve all of the households in Dallas or San Diego. More than half of China's total coal use is in the non-electricity sector. Coal provides 60% of Chinese chemical feed-stocks, 55% of industrial fuel, and about 45% of China's national railway capacity is devoted to the transport of coal. Coal is, therefore, China's most abundant and widely used fuel, and China is the world's largest coal producer (mining about 2.3 billion tons per year compared to just 1.1 billion tons in the United States). Put another way, coal production and consumption account for more than 65% of China's total energy supply and use. China already uses more coal than the European Union, Japan, and the United States combined.

In 2009, coal imports of China tripled, and for the first time it became a net coal importer. This import dependency may be reversed if coal production capacities of remote western provinces are utilized. For example, the province of Xinjiang holds about 40% of China's total coal resources. According to the expectations of its regional government reported by the IEA (2010a), the planned upgrade of the railway line linking Xinjiang to the east coast may help to increase coal production there by more than 10-fold so that its share of global coal production will be double the share of current oil production of the world's biggest oil field, Ghawar in Saudi Arabia.

Source: IEA, 2010a; Sovacool and Khuong, 2011.

National coal supply vulnerabilities

At present, 4.2 billion people live in 45 countries where coal represents more than 10% of the primary energy supply. About 3.4 billion people live in 28 countries (including China, India, Japan, and the United States) where coal represents more than 20% of the total primary energy supply. About 1.4 billion people live in seven countries (including China) where coal accounts for more than one-half of the total primary energy supply. In Mongolia, South Africa, and the Democratic People's Republic of Korea, more than 70% of the primary energy supply is derived from coal.

Only seven of these 28 countries (approximately 300 million people) import more than 80% of the coal they consume. These include Japan, which imports 100% of its coal. Some of these countries (e.g., Morocco, Slovakia, and the Republic of Korea) may be considered especially vulnerable, as either their coal consumption has been growing at over 5% per year or their economies have very high coal intensities. Despite its rapidly growing coal extraction, India is a net coal importer and is projected to increase its imports of coal more than five-fold between 2008 and 2035 (IEA, 2010a).

Most of the countries that significantly rely on domestic coal (net exporters and those that import less than 50% of their consumption) have a domestic resource/consumption ratio of over 30 years. The only exception is Vietnam, for which this ratio is 8.5 years. This makes the situation for coal very different than that for oil and gas, where import dependency is likely to significantly increase in a large

number of countries even if the current levels of consumption do not notably grow.

5.2.2.2   Nuclear Power

Global nuclear energy supply vulnerabilities

Whereas the energy security concerns related to fossil fuels are primarily related to the supply and demand of resources, in case of nuclear power the primary concerns relate to nuclear energy infrastructure and technologies. Unlike fossil fuels, the fuel of nuclear energy (uranium) has a fairly high security of supply, offers protection from fuel price fluctuations, and is possible to stockpile. In comparison to oil and gas, uranium is abundant and more geographically distributed, with a third of proven reserves in OECD countries (NEA, 2008). Recent estimates indicate that even in the face of a large expansion of nuclear energy, proven uranium reserves would last at least a century (Macfarlane and Miller, 2007; NEA, 2008).[8] Furthermore, electricity produced from nuclear energy offers a greater protection from fluctuations in raw commodity prices; while doubling uranium prices leads to a 5–10% increase in generating cost for nuclear power, doubling the cost of coal and gas leads to a 35–45% and 70–80% increase, respectively (IAEA, 2008). Uranium is also a relatively easy fuel to stockpile. The refueling of a nuclear power plant generally provides fuel for two to three years of operation (Nelson and Sprecher, 2008), and it is possible to store up to a 10-year supply of

---

8   Chapter 7 contains a more extensive discussion of uranium resource availability.

nuclear fuel (IAEA, 2007b). In contrast, oil and gas emergency reserves, where they exist, are measured in days, weeks, or – in exceptional cases – months, not years.

At the same time, there are significant energy security risks associated with technological, economic, and institutional characteristics of nuclear power production. As the most capital-intensive electricity-generation technology, it is economically difficult for nuclear energy to compete in liberalized markets where the investor has to assume the financial risk of investment. As a result, strong government backing is necessary for the development of nuclear power (Finon and Roques, 2008). Such political backing depends on the public support of nuclear power, which has been very uneven. In particular, public opinion is swayed by nuclear accidents such as the ones at Three Mile Island in the United States in 1979, Chernobyl in the USSR in 1986, and Fukushima in Japan in 2011. Each such change of public opinion and the resulting change in the government policy may affect energy security both in the short term (e.g., as a result of shutting down nuclear power plants immediately affected by the accident[9] and those deemed unsafe) and in the longer term (through complicating the investment climate). Unlike other energy sources and electricity-generating technologies, for nuclear energy the risks associated with accidents extend beyond the plant level or national level to the entire nuclear power plant fleet. Thus, nuclear power globally faces the systemic risk of nuclear accidents.

Additionally, in most countries nuclear power plants are aging and often reaching the end of their licenses. The mean age of nuclear power plants worldwide is 26 years (calculated from IAEA, 2010). Since the standard lifespan of nuclear power plants is 30–40 years, many plants are nearing the end of their planned operational period. The IAEA has recently begun efforts to create a dialogue on effective management and safety enhancements to extend the lifespan of many of the world's nuclear power reactors (IAEA, 2007a). The power plants are not the only part of the industry that is old; in many countries the industry faces an aging workforce and a dearth of young workers to replace retiring nuclear engineers and plant operators (Sacchetti, 2008).

Nuclear power and other thermal plants are also subject to heat waves and water shortages. In 2006, France, Spain, and Germany had to shut down or scale back electricity production in several of their nuclear power plants due to low water levels. With growing concerns over water availability due to increasing pressure from uses and climate change, thermal power plants could face problems involving water supply more frequently. In addition to these robustness concerns, there are also sovereignty issues associated with nuclear power since capacities for fuel enrichment and nuclear reactor construction are concentrated in relatively few countries. Only six countries currently possess large-scale enrichment plants, and seven countries possess small-scale enrichment

facilities (see Figure 5.5).[10] The fact that several countries (including Australia, Brazil, and South Africa) are considering constructing enrichment facilities indicates that even though countries can relatively easily stockpile nuclear fuel, national governments may feel too vulnerable if they rely solely on foreign suppliers. In addition to the concentration of nuclear fuel enrichment, construction capacity for new nuclear power plants is concentrated in just 12 companies in eight countries (see Figure 5.5). The number of countries holding the ability to forge the bottleneck component of large LWR pressure vessels is currently even more restricted.

National nuclear energy supply vulnerabilities

Currently, 29 countries with a total population of 4.4 billion people operate nuclear reactors. Nuclear power is located in middle- and high-income countries that are almost all relatively stable (see Figure 5.5). Nuclear energy comprises more than 10% of the electricity supply in 21 countries with a population of 1.3 billion people. Of these, only 200 million people live in 13 countries that rely on nuclear energy for at least 30% of their electricity generation, and about 80 million people live in three countries that rely on nuclear energy for more than 50% of their electricity production.

The most pressing energy security concerns for nuclear power in most countries are robustness concerns related to the age and obsolescence of their nuclear power programs combined with a lack of recent investment. Twenty-one out of the 29 countries with nuclear power (with a combined population of 1.3 billion people) have not started constructing a new nuclear power plant in the last 20 years. Without new nuclear power plants, the nuclear industry in these countries lacks the vitality of recent activity. This can, in turn, lead to a lack of dynamic capacity needs of the industry, from both a human resources and a manufacturing perspective.

There is also clear evidence that many of the countries with stagnating nuclear power programs face imminent human resources shortages. A nuclear industry institute in the United States reports that in the next five years as much as 35% of the nuclear workforce may reach retirement (NEI, 2010). The United Kingdom and Germany also face a dearth of young qualified workers: over 75% of nuclear employers in the United Kingdom report that they have trouble filling scientific and engineering positions, and in Germany in many recent years not a single person has graduated in a nuclear discipline (Sacchetti, 2008).

Nineteen of the 29 countries that use nuclear energy (with a combined population of 1.4 billion) have nuclear power plants with an average age greater than 25 years (see Table 5.7). While countries can extend the operating licenses of their existing nuclear power plants, this raises

---

9   For example, in the immediate aftermath of the March 2011 earthquake in Japan, 10.5 GW of nuclear power capacity was shut down (Nakano, 2011).

10   There are conventional security concerns linked with uranium enrichment due to the link between civilian enrichment capabilities and nuclear weapons. This topic is beyond the scope of this chapter but is discussed in detail in Chapter 14 (Nuclear Energy).

Figure 5.5 | Nuclear power worldwide. The figure shows the concentration of nuclear power and related capacities in a few industrialized countries in the Northern Hemisphere. It also indicates the aspirations of a large number of new countries, predominately in the developing world to deploy nuclear energy. Source: NEA, 2008; IAEA, 2011; Jewell, 2011; World Nuclear Association, 2011a; b; c.

Table 5.7 | The average age of nuclear power plants and the date of the start of the most recent construction.

Start of the most recent construction (years prior to 2010) Average age of NPPs (from the time of completion), years	Less than 22	22 or more
Less than 25	China, Brazil, India, Rep. of Korea	Czech Republic, Bulgaria, Ukraine, Slovakia, Romania, Mexico
More than 25	Russia, Japan, France, Pakistan, Finland	UK, USA, Canada, Germany, Argentina, Sweden, Belgium, Spain, Hungary, South Africa, Slovenia, the Netherlands, Switzerland, Armenia

safety concerns, especially in the wake of the Fukushima nuclear accident, which occurred at an older power plant. In Vermont, in the United States, the state legislature recently voted down extending the license of its nuclear power plant, which currently meets 73% of electricity demand, due to safety concerns over the 28-year-old plant.

Countries with aging plants will face the decision of whether to invest the required resources to jump-start a stale industry or redirect resources to fill the gap that aging nuclear power plants will open up. Such countries as the United States, the United Kingdom, and South Africa have recently expressed interest in restarting their nuclear power programs. However, the ability of these plans to get off the ground remains to be

seen. This fact, combined with the lack of recent construction experience in all but eight countries (Table 5.7), indicates that nuclear power faces significant robustness challenges in almost three-quarters of the countries with nuclear power.

Due to safety concerns regarding the obsolescence of the operating technology, Bulgaria, Lithuania, and Slovakia were forced to close their early Soviet-designed plants as a precondition for accession to the EU. Since the Chernobyl-style nuclear power plant in Lithuania, the most recent EU Member State to close its nuclear reactor (at the end of 2009), met about 70% of its electricity demand, the government has had to seriously reconsider its energy strategy. The only remaining countries that operate Chernobyl-type reactors are Armenia and Russia. Although not an EU candidate country, Armenia has come under significant pressure from both the EU and neighboring Turkey to shut down its nuclear power plant. However, as the nuclear power plant currently meets almost 40% of its electricity generation, there is resistance.

In addition to the 29 countries currently operating nuclear power, an additional 52 countries, often citing growing energy security concerns and electricity demand, have expressed interest in starting a national nuclear power program (Rogner, 2009). While nuclear energy can affect energy security in both types of countries, in so-called "newcomer countries" the energy security risks pertain to the *potential of* and *capacity for* nuclear energy rather than risks to current generating capacity (Jewell, 2011). These newcomer countries will be discussed in Section 5.3.2, which explores the future of energy security.

5.2.2.3   Hydroelectric Power

Global hydropower supply vulnerabilities
Although hydropower is a renewable resource and hydroelectricity is not associated with massive global trade, as in the case of fossil and nuclear fuels, there are still related energy security concerns from the robustness, sovereignty, and resilience perspectives.

With respect to robustness, hydropower is vulnerable to seasonal and annual variability in hydrological regimes, as well as local weather conditions, temperature, and precipitation in the catchment area. These factors affect not only the availability of water for hydropower production, but also the amount of pressure on water resources in a region from competing uses.

The pressures on hydrological resources are likely to be exacerbated by climate change and shifting hydrological patterns. According to the Intergovernmental Panel on Climate Change (IPCC) (Bates et al., 2008), climate change is likely to alter river discharge, and thus water availability, for hydropower generation, although the exact extent of these effects is difficult to forecast. Furthermore, intensified glacial melting, while it initially has a positive impact on hydropower production, will eventually lead to reduced water flows and could seriously affect regions with glacier-fed rivers used for power production.[11] Additionally, an increase in the frequency of extreme weather events (both floods and droughts), which is predicted by the IPCC to accompany climate change (Bates et al., 2008), could alter the temporal availability of hydropower resources and place a greater stress on hydroelectric dam infrastructure.

11   Rain rather than snow in the winter will lead to less snow pack acting as a reservoir for the spring and summer months, leading to summer shortages. Higher temperatures can lead to more evaporation and less generation. Higher summer temperatures lead not only to water scarcity but also to higher electricity demand for air conditioning. This puts an additional pressure on electricity generation.

In addition, changes in the available discharge of a river might have a direct influence on the economic and financial viability of a hydropower project, since hydropower plants have a life of more than 50 years. Even if a significant part of the changes takes place in the distant future, the impacts are inevitable. For instance, in North America, potential reductions in the outflow of the Great Lakes could result in significant economic losses as a result of reduced hydropower generation both at Niagara and on the St. Lawrence River. For a Coupled Global Climate Model (CGCM1) projection with 2°C global warming, Ontario's Niagara and St. Lawrence hydropower generation would decline by 25–35%, resulting in annual losses of C$240–350 million at 2002 prices (Buttle et al., 2004).

Changing hydrology, but also possible extreme events such as floods and droughts (which are expected to increase as a result of climate change), pose sediment risks. More sediment, along with other factors such as a changed composition of water, could raise the probability that a hydropower project suffers greater exposure to turbine erosion. When a major destruction actually occurs, the cost of recovery will be enormous. An unexpected amount of sediment will also lower turbine and generator efficiency, resulting in a decline in energy generated (Iimi, 2007).

In addition to these relatively gradual pressures that hydropower faces, there is the risk of sudden failure either from terrorism, faulty construction, or aging infrastructure. A report by the American Society of Civil Engineers (2009) recently highlighted the risk of dam failure as hydropower dams age.

From the sovereignty perspective, energy security risks associated with hydroelectricity primarily arise in situations where multiple countries share rivers feeding hydropower dams. In such cases, competing uses on different sides of a national border can be exacerbated by the fact that upstream countries can significantly affect the quantity and quality of hydrological resources that a downstream country receives. For example, China's plans to build a series of dams on the Mekong has raised significant concerns downstream in Vietnam, where scientists worry that the planned dams would significantly accelerate the disappearance of the Mekong Delta, thus decimating fisheries and livelihoods of locals in the region. Another example of water scarcity and conflicting uses of water between sectors is manifested in the ongoing tension in Central Asia, where downstream countries (Kazakhstan, Uzbekistan, and Turkmenistan) prefer to use water for irrigation in summer, while upstream countries (Tajikistan and Kyrgyzstan) need it for generating hydroelectricity in winter.

Finally, from the resilience perspective, energy security of hydropower may be considered to be under threat where a nation's electricity is supplied from just one or a few major hydroelectric facilities and thus is vulnerable to failures of these facilities.

National and regional hydropower supply vulnerabilities
Hydropower accounts for at least 20% of electricity generation in 58 countries with a combined population of 1.5 billion people and for at

least 10% of electricity generation in 74 countries with a combined population of 4.6 billion people. About 800 million people live 35 countries that derive over half of their electricity from hydropower. The risks to hydroelectric production are highly contextualized and subject to local factors. This brief assessment considers the regional vulnerability of hydropower to climate change and the diversity of hydropower facilities in a particular country.

The IPCC's projections (Bates et al., 2008) for the impact of climate change on hydropower are available for the following regions:

In Europe, the IPCC's estimates state that the total European hydropower potential is projected to decrease by 7–12% by 2070s. The highest decreases of 20–50% are expected in Portugal, Spain, Ukraine, and Bulgaria. Some countries in Southern Europe are already vulnerable to water shortages. Albania, where 92% of power comes from hydro resources, has experienced severe shortages of power for a few consecutive summers. At the same time, in some parts of Europe, such as Scandinavia and northern Russia, the electricity production potential of hydropower plants is expected to increase by 15–30% by the end of the century.

In Africa, where little of the continent's hydropower potential has been developed, climate change simulations for the Batoka Gorge hydroelectric scheme on the Zambezi River projected a significant reduction in river flows and declining power production (a decrease in mean monthly production from 780 GWh to 613 GWh).

In Asia, changes in runoff could have a significant effect on the power output of hydropower-generating countries such as Tajikistan, which predominately relies on hydropower for its electricity needs and is among the larger hydroelectricity producers in the world. Climate change will also result in an increase in the frequency and duration of extreme climate events, i.e., floods and droughts. For instance, a hydrological model indicates a great risk of substantial increases in (mean) peak discharges in the three major rivers in Bangladesh: the Ganges, Brahmaputra, and Meghna (Iimi, 2007).

In Latin America, where hydropower is one of the main electrical energy sources, expected further glacier retreat is projected to negatively impact the generation of hydroelectricity over the long term in countries such as Colombia, Bolivia, and Peru (Bates et al., 2008). The consequences of droughts and increased energy demand caused a virtual breakdown of hydroelectricity in most of Brazil in 2001, which contributed to a GDP reduction of 1.5% (Kane, 2002). More recently, because of a shift in precipitation patterns due to El Niño, Venezuela experienced a water (and hydroelectricity production) shortage. President Chavez ordered a 20% reduction in electricity consumption in December 2009 (Cancel, 2009).

In 10 countries (with 100 million people), hydropower is derived from one dam; another 11 countries (with 290 million people) rely on one major dam and one to three significantly smaller dams (with a combined capacity less than half the size of the main dam). In another 10 countries, 330 million people rely on two major dams and one to four significantly smaller dams. Ten countries with 560 million people rely on at least three main dams for electricity supply, and 16 countries with 260 million people have four or more dams.

5.2.2.4   Traditional Biomass

There are several difficulties in assessing the vulnerabilities of biomass supply. First, biomass comes in a variety of very different forms ranging from dung, firewood, and charcoal to biodiesel, ethanol, palm oil, and specially grown plant species. Secondly, the flows of "traditional" biomass (which is not commercially produced, processed, and distributed) are often difficult to estimate. The available statistics often do not distinguish between "modern" and "traditional" *solid* biomass, especially when "modern" is used in a decentralized fashion. Yet different types of biomass are associated with different risks and vulnerabilities. We discuss traditional biomass in this subsection and further deal with modern *liquid* biomass (i.e., biofuels) in the next subsection.

Traditional biomass is primarily used for heating and cooking in residential buildings. It is estimated that 2.5 billion people rely on such energy in their daily lives. Over half of these people live in India, China, and Indonesia. In some countries, mainly in sub-Saharan Africa, these sources account for 90% of total household energy use. The main threats to traditional biomass availability are resource scarcity resulting from the growing demand (population growth, migration and other demographics, growth in consumption) and depletion of resources (change of land use, climate change). Though demand for wood-fuel is unlikely to deplete or remove forest cover on a large scale (Arnold et al., 2006), localized wood-fuel scarcities may indeed be projected to occur in parts of Southeast Asia and Africa.

Access to traditional energy resources may be further restricted for political or economic reasons. In many countries large portions of forest areas are controlled by the government, and access and use of firewood might therefore be limited. For example, in India in the mid-1990s about 55% of household needs for firewood were collected for free from common pool resources (CPRs). The share of CPRs has been shrinking at a five-yearly rate of 1.9%; coupled with decreased productivity of what remains, this has seriously restricted the CPRs' ability to meet the growing demand (Chopra and Dasgupta, 2000). Formal and informal privatization of land holdings in Africa have similarly reduced the areas available as CPRs, leading to local scarcity and shortages of firewood (Arnold et al., 2006). Seasonal variability is also an important factor that might undermine the availability of resources: bio-waste is abundant in the time of harvest; without appropriate storage capacity it is difficult to sustain the supply through the year.

Traditional biomass energy systems are typically inefficient, unreliable, and unsafe, with high environmental, social, and health externalities. They often fail to provide the level of services associated with modern energy systems (e.g., adequate heating in winter) and may not be equally available to vulnerable groups (elderly, disabled, landless, etc.). Therefore, residential energy sectors relying on traditional biomass are in most cases considered inadequate and need to be replaced by modern systems. We return to this issue in Section 5.3.3.

5.2.2.5   New Renewable Energy Sources

New renewable energy sources (NRES) include "modern" biomass (in solid, liquid, and gaseous forms), wind, solar photovoltaic, solar thermal, geothermal, tide, waves, and ocean current energy. NRES can provide most of the nationally vital energy services by generating electricity or heat, or providing liquid fuels for transportation. Despite being labeled together as "renewables," these are very different energy sources potentially associated with distinct energy security concerns. At present, these sources significantly contribute to energy supply in only a small number of countries, so there is much less empirical evidence on the nature and scale of their potential disruptions than for traditional energy sources summarized in the previous sections. Therefore, the discussion in this section is limited to general considerations and a brief overview of available quantitative data.

The main robustness concern related to fossil fuels – finite resources – is not applicable to NRES. Instead, the main concerns about the robustness of NRES are costs, intermittency, and sufficient availability in locations where they are needed. With respect to costs, vulnerabilities of NRES are different from vulnerabilities of fossil fuels because the running (fuel) costs account for a much smaller share in the total cost of energy (except solid and liquid biomass). This means that, although NRES are less vulnerable to price fluctuations, they may be more expensive in terms of total lifetime costs per unit of energy. Intermittency of NRES (with the exception of biomass and geothermal) is another concern: it is connected with daily, seasonal, or other variations in natural factors (sun, wind, tides). The impacts of intermittency can be reduced if NRES sources are integrated in larger networks where energy generated in one part of the system can compensate for the shortage of energy in another part. Developing technological solutions concerning electricity and heat storage may be another answer to the challenge of intermittency. More details on these solutions are provided in Chapter 11 (Renewable Energy).

Sovereignty concerns are in general less pronounced in the case of NRES. This is because most NRES are found locally and do not need to be imported as other fuels. The exception is liquid biofuels, which are already traded on a global scale, with Brazil and the United States accounting for over 90% of world production in 2007 (Balat and Balat, 2009). At the same time, most NRES require cutting-edge technologies that are not readily available for the majority of countries. There are

signals that dependencies in renewable energy technologies may be perceived as energy security concerns by some countries (Burnett and Dwyer, 2011).

NRES also seem to be associated with better resilience than mainstream energy sources. This is because energy systems based on renewables are typically more diverse with respect to energy sources and more distributed in space. It has been argued (Lovins and Lovins, 1982) that such diversity means better protection from human or natural disruptions. Farrell et al. (2004) partially confirm this conclusion, but indicate that much more research needs to be done concerning the resilience of distributed energy generation.

Thus, NRES are widely perceived as providing better energy security, particularly from the sovereignty and resilience perspectives. That is why promotion of NRES figures prominently in almost all national energy security strategies. Although the number of countries where NRES comprise a significant share in energy supply remains small, some of these (notably the United States, Germany, Spain and China) have witnessed significant growth in NRES in recent years. Detailed information on the growth in the use of renewable energy resources in the world is provided in GEA Chapter 11.

In the residential sector, modern biomass (including waste residues) comprises over 10% of energy use in such high-income countries as Austria, Denmark, Finland, Estonia, France, Croatia, Slovenia, and Greece. Biofuels are used for transport in Brazil (15.4% of the overall transport energy use), Germany (7.2%), Cuba (6.3%), France (3.2%), and the United States (2.3%), as well as in insignificant amounts in smaller countries (IEA, 2010b).

As this discussion shows, energy security concerns associated with NRES are difficult to quantify empirically because the share of NRES in the total energy supply remains limited in the majority of countries. If large shares of renewables were to be used around the world, it might change the energy security landscape considerably and new energy security threats might emerge (e.g., demand for water and arable land). For example, it may turn out that trade in biofuels will acquire many characteristics of the current global oil market, which, as we argued, has many vulnerabilities. Answering these concerns about the security of future energy systems where NRES play a large role is beyond the scope of this chapter. Instead, they are systematically addressed in Chapter 17, which analyses energy security implications of major energy transformations.

5.2.2.6   Overall Energy Supply Security

The sectoral energy security analysis presented in the previous sections makes the most sense for identifying vulnerabilities to relatively minor disruptions unfolding in the short or medium term, when the elements and connections within energy systems remain largely fixed. Under such conditions we can make certain assumptions concerning the propagation of risks and vulnerabilities. For example, we assume that if the supply of petroleum products is disrupted, the transport system will be affected. However, over a longer term or under much stronger shocks, the energy system may become reconfigured. Liquid fuels may be produced from coal or biomass, or electrically-driven trains may take more cargo and passengers. System-wide, cross-sectoral indicators may reflect the more general vulnerabilities of national energy systems to systemic shocks and long-term threats.

We use both supply and demand indicators of cross-sectoral vulnerabilities of energy systems. On the supply side, we consider the overall import dependency, the diversity of primary energy sources, and the cost of imported fuels as a proportion of GDP or export earnings. On the demand side, we consider the total energy intensity, the overall rate of demand growth, and energy use per capita.

With regard to the supply side, 46 countries (700 million people) have a total energy import dependency higher than 50%. Fifty-eight countries (one billion people) have a low diversity (one or two dominant sources) of primary energy supply. For 15 countries (200 million people), the cost of imported fuels is higher than 10% of GDP, and for 35 countries (2.5 billion people), the cost of energy imports exceeds export earnings by 20% or more. In total, there are 102 countries (3.8 billion people) that are vulnerable in at least one of the abovementioned aspects.

In relation to the demand side, in 24 countries with 400 million people energy use per capita intensity exceeds the world average by more than 50%. In 18 countries (1.8 billion people), including China, the average annual growth in energy use has been higher than 6% over the last decade. There are 55 predominately low- and lower middle-income countries (three billion people) with energy use per capita of 30 GJ/year or lower. Such levels of energy use – at least three times less than even in the least energy-intensive developed countries such as Israel, Italy, or Japan – are likely to signal pressures on energy systems to rapidly and radically increase production to meet the future demand. There are 76 countries (4.7 billion people), primarily low- or lower middle-income, subject to at least one demand-side vulnerability mentioned above.

Cross-sectoral demand- and supply-side vulnerabilities overlap in 63 countries with 2.8 billion people, of which 47 are either low- or lower middle-income countries. Additionally, three billion people live in 52 countries that have either cross-sectoral demand or cross-sectoral supply vulnerabilities. Only 19 countries with 600 million people can be considered free from overall energy supply vulnerabilities.

5.2.3   Security of Electricity Systems

5.2.3.1   General Considerations

Due to the large and growing importance of electricity systems, the risks of their potential disruption are increasingly prominent on energy

security agendas. The importance of secure delivery of electricity is growing in every country due to:

- the increasing dependency on electricity (for example, in the information and communications technology [ICT] sectors, and eventually in the transportation sector) of modern economies;

- the expansion in the coverage of electricity grids in developing nations;

- the increasing "electrification" of energy services in emerging economies where the entry into the middle class is often signaled with the arrival of domestic electrical appliances; and

- the advance of "new" energy systems relying on distributed generation, renewable energy sources, and possibly electric propulsion for vehicles.

In contrast to other energy carriers, electricity cannot at present be stored cheaply and easily and should in many critical cases be supplied without any interruption, closely matching the changing demand. This makes electricity systems vulnerable even to short-term disruptions and imposes particularly stringent requirements on electricity generation, conversion, and transmission. The vulnerability of electricity systems can be considered from robustness, sovereignty, and resilience perspectives.

In this analysis we use four indicators to estimate the robustness of electricity systems. Two of them relate to the supply side: the age of electric power plants and the frequency of blackouts. The two others relate to the demand side: the rate of access to electricity and the growth in demand for electricity. Rapid demand growth increases the pressure on existing electricity systems and stimulates countries to seek new sources of energy.[12] Another demand-side vulnerability is the inadequate provision of electricity, reflected in a low rate of access to electricity. The low access rate is untenable for most governments, and therefore results in pressures on the energy system to expand. Thus, all else being equal, a country with low access or higher growth will be less secure, as it will need to expand rather than merely maintain its electricity supply at the present level. Our analysis also considered the third demand-side vulnerability factor: the electricity intensity of the economy. In principle, more electricity-intensive economies should be more sensitive to energy price fluctuations and other disruptions of electricity systems. However, electricity intensity was not found to be a sufficiently discriminating factor in the final analysis.

The sovereignty perspective on the security of electricity systems is reflected in the reliance on imported fossil fuels for electricity generation and on imports of electricity. Finally, a proxy indicator for the resilience

---

12 Jewell (2011) demonstrates that the rapid growth in electricity demand has correlated with countries launching nuclear energy programs.

of national electricity systems used in this analysis is the diversity of fuels used for electricity generation.

In many industrialized countries, electricity subsystems are undergoing transformations. These include distributed decentralized generation, "smart grids," and possibly an increasing role of international interconnectedness and electricity imports. Energy security factors associated with these new developments will play increasingly important roles in the security of electricity generation, distribution, and transmission in the future. For example, it is often argued that distributed generation of electricity will result in higher reliability and resilience of energy provision (Lovins and Lovins, 1982). At the same time, the empirical evidence that distributed systems are more robust than centralized ones seems to be inconclusive (Farrell et al., 2004). Key technological and institutional developments that may affect the reliability of electricity systems are discussed in Chapter 15.

5.2.3.2 Vulnerability of National Electricity Systems

The primary energy security concerns from the perspective of robustness of electricity systems include the aging of power plants, rapid growth in demand and/or lack of adequate provision of electricity, and the reliability of electricity systems. The aging of power plants is particularly a problem for industrialized countries with established and stagnant fleets. Especially problematic is the aging of nuclear reactors because they typically provide large shares of national electricity and their replacement with other nuclear reactors may be problematic for economic and political reasons (see a more detailed discussion of this point, as well as statistics on the aging of nuclear reactors, in the previous section). Non-nuclear generating facilities are also aging in many countries. For example, in the United Kingdom, the aging of power stations is one of the three major national energy challenges (the other two being decarbonization and an increasing reliance on imported natural gas – see Figure 5.6).

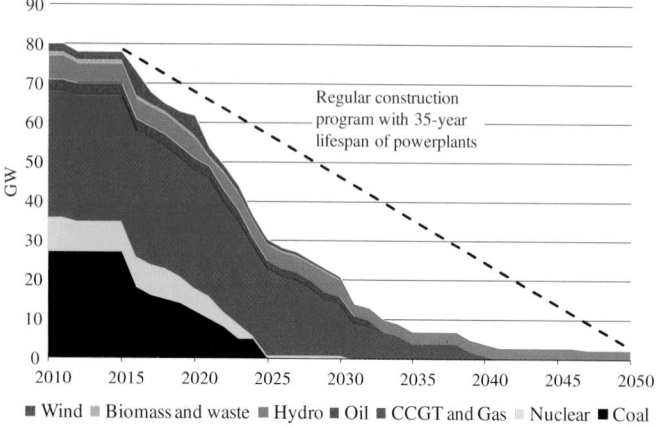

Figure 5.6 | Projected retirement of electricity-generating capacity in the United Kingdom. The figure shows the decline in capacity much faster than if the regular pace of construction was maintained in recent years (as shown by the dotted line). Source: Parsons Brinckerhoff, 2009.

The reliability of electricity infrastructure is addressed in Box 5.7. The scale of the problem is dramatically higher in low- and middle-income countries, although it is high on the political agenda of industrialized nations as well because their economies are increasingly sensitive to even minor disruptions of electricity supply.

There are energy security concerns from the robustness perspective related to the demand side of electricity systems. There are 32 countries with a total population of some 2.4 billion people that have less than 60% access to electricity. In 35 countries with some 2.6 billion people, the electricity demand has been growing at over 6%. The need to increase electricity-generation capacities is especially dramatic in larger low-income countries. For example, Ibitoye and Adenikinju (2007) estimate that under the assumption that Nigeria becomes a middle-income economy and meets its Millennium Development Goals by 2030, it will need to achieve a 25-fold increase in its electricity-generating capacities (from the current 6.5 GW to 164 GW). All in all, 53 countries with some 4.2 billion people experience either one or both of these demand-side vulnerabilities.

The importing of fuels for electricity production or the importing of electricity itself are considered an energy security concern in some countries. Almost 1.3 billion people live in 62 countries that rely on imported energy for more than 25% in their electricity, almost 600 million live in 39 countries importing over 50%, and almost 200 million live in 21 countries that import over 75% of energy for electricity production. Most of this energy is imported in the form of fuels for electricity production rather than directly as electricity imports.[13]

From the resilience perspective, the security of electricity systems may be measured as the diversity of fuels used for electricity generation. The Shannon-Weiner Diversity Index[14] of such fuels is less than 0.2 in 30 countries, which are home to some 360 million people. This means that virtually all of their electricity comes from a single source (often hydropower or natural gas-fired power plants). Even the diversity index of 0.4 (applicable to some 450 million people in 35 countries) can still be considered low.

All in all, some 990 million people live in 62 countries where electricity systems are vulnerable with respect to supply, because they are either significantly (over 40%) based on imported fossil fuels and/or have a low (<0.4) diversity of electricity generation.

---

13 There are only 12 relatively small countries (six in Europe and six in sub-Saharan Africa) that import more than 25% of the electricity they consume.

14 The Shannon-Wiener Diversity Index (SWDI) is a most common measure of diversity of options or categories. It is based on the idea that the greater the variety and the more even the spread the greater is the diversity. The index is calculated as: $SWDI = -\Sigma_i p_i \ln p_i$, where pi represents the share of each category/option in the energy mix (Stirling, 1998; Jansen et al., 2004).

Box 5.7 | Reliability of Electricity Supply

Challenges for ensuring adequacy and reliability of electricity grids exist for both developed and developing countries. For developing countries, extending electricity networks and increasing generating capacity is the main challenge. Even for households connected to electricity grids, planned outages due to the lack of generating capacity are not uncommon. At the same time, an increasing number of emerging economies and all industrialized countries have sufficient network coverage and generating capacity. However, precisely because these societies rely heavily on permanent access to electricity, any disruptions of the systems are potentially very harmful. The increasing scale, complexity, age, and interconnectedness of modern electricity grids present a serious investment and energy security challenge worldwide.

Advanced electricity distribution systems are spatially distributed and technically complex. The spatial extent of electricity grids means that they face "distributed threats," which are difficult to predict and manage. Moreover, the scale and the degree of interconnectedness of modern grids mean that even a minor failure can have dramatic consequences throughout the system – the so-called "cascading failure." This is especially plausible when the system operates near its critical load. Paradoxically, some measures to control small-scale blackouts may actually increase the probability and severity of large-scale blackouts, since they increase the interconnectedness and thus the size of the system that may fail (Dobson et al., 2007).

There are no comparative international statistics on the number and intensity of disruptions of electricity supply also known as blackouts. However, detailed data that exist on some industrialized countries can be not only interpreted but also compared with analysis of the situation in developing countries.

Several comprehensive studies analyze blackouts in the United States. The analysis of the US data from 1984–2006 by Hines et al. (2008) indicates that the frequency of large blackouts was not decreasing with time despite the introduction of advanced data management, reliability standards, and many other measures. A particularly catastrophic blackout on August 14, 2003 affected approximately 50 million people in the northeastern United States and parts of Canada and resulted in US$3 billion in insurance claims alone. An assessment of major electricity blackouts conducted by the Institute of Electrical and Electronics Engineers estimates that no less than 17 major blackouts have affected more than 195 million residential, commercial, and industrial customers in the United States, with seven of these major blackouts occurring in the past 10 years. Sixty-six smaller blackouts (affecting between 50,000 and 600,000 customers) occurred from 1991 to 1995, and 76 occurred from 1996 to 2000. The costs of these blackouts are monumental: the US Department of Energy (DOE) estimates that power outages and power quality disturbances cost customers as much as US$206 billion annually, or more than the entire nation's electricity bill for 1990 (Hines et al., 2008).

According to Hines et al. (2008), the single most important reason for blackouts in the United States was technical failure (nearly 30% of all events), which together with operator failure, volunteer reduction, and weather conditions accounted for over 85% of all blackouts. Intentional attacks were responsible for some 1.6% of blackouts. Larger blackouts also resulted from earthquakes and hurricanes or tropical storms.

In the report on the reliability of the US electricity system presented to the Assistant Secretary of State for Energy Efficiency and Renewable Energy, Osborn and Kawann (2001) indicate that reserve capacity margins (for both transmission and generation) have been decreasing in most US regions for the past two decades and were projected to decrease until 2020, which may be the single most important factor explaining the lack of a downward trend in the number of blackouts in the United States. The decrease in this margin can be explained by both increasing demand and falling investment in transmission and electricity grid infrastructure. The report connects this fall in investment with institutional and market failures of the electric power system.

The frequency and severity of blackouts are reportedly less in Western Europe than in the United States. Nevertheless, severe blackouts in September 2003 affected some four million people in southern Sweden and eastern Denmark, and the nationwide blackout in Italy was the worst in the history of the nation (Andersson et al., 2005).

Despite concerns that electricity blackouts cause in developed countries, their intensity is dramatically less than in the developing world. The three largest blackouts in world history occurred in developing nations, including the 2005 Java-Bali blackout in Indonesia, which affected 100 millllion people (Donnan, 2005), the 1999 South Brazil blackout, which affected 75 million people (Yu and Pollitt, 2009), and the November 2009 Brazil-Paraguay blackout, which originated in the world's second largest hydroelectric plant and left tens of millions of people in Brazil and the whole of Paraguay without electricity (Reuters, 2009).

World Bank's Enterprise Survey data (World Bank, 2010) clearly show that both the number and severity of electricity blackouts are highest for low-income countries, lower for lower middle-income countries and the lowest for upper middle-income countries in the survey. For most lower-income countries, blackouts are not measured in hours per year (as in the United States and Europe), but rather in hours per *day*. In almost three-quarters of low-income countries, blackouts are on average for more than 24 hours per month, and in about one-sixth of low-income countries blackouts average over 144 hours (six days) a month. There are more than 10 blackouts every month in more than half of the surveyed low-income countries. Companies in Albania, Guinea, Kosovo, Nepal, Pakistan, and Yemen experienced 30 or more blackouts in a month, whereas firms in Bangladesh experienced, on average, 101 blackouts per month. The worst statistics of blackouts were shown in South Asia, followed by sub-Saharan Africa. The results of this analysis should be interpreted with an understanding that companies may, in fact, experience less frequent outages than private customers, as they tend to be located in urban areas with better infrastructure.

In 28 countries with over 500 million people, high supply- and demand-side vulnerabilities overlap. These electricity systems, pressed from both sides, can be considered especially insecure. The majority of the world's population (some 4.3 billion in 65 countries) lives in countries that are vulnerable in at least one of the supply or demand aspects. Finally, some 1.6 billion people live in about 38 countries with relatively secure energy systems, which rely primarily on diverse domestic resources and have high access to electricity and moderate rates of growth in demand.

5.2.4   Vulnerability of End-Use Sectors

A modern state cannot function without several vital services provided by national energy systems. For the purpose of this analysis, we distinguish between four such services. The first three are the "end-uses" discussed in Chapters 8, 9, and 10, i.e., energy in (1) transport, (2) industry, and (3) buildings. The last energy "service" is (4) revenue from energy exports. It is only relevant to a relatively small number of countries whose economies rely significantly on such revenues.

This subsection evaluates the vulnerabilities of each of the vital energy services: transportation, residential and commercial (R&C), industrial, and export-revenue generation. Such vulnerabilities are generally divided into those associated with the energy service itself (demand-side vulnerabilities) and those linked to the properties of the energy system in which a particular service is embedded, particularly the security of relevant energy sources.

5.2.4.1   Energy for Transport

Transportation is vital for almost every aspect of modern societies, ranging from personal mobility to food production and distribution, trade, availability of goods and services, and, not least of all, military security. Transport accounts for over 20% of energy use in virtually all developed countries. It is still under 15% in many developing countries, including China and India (where it is rapidly growing). Only in Least Developed Countries (LDCs) does transport account for under 10% of their total energy use. Modern transport systems rely on motorized vehicles that share fuels with agricultural, construction, and other machinery. Thus, this discussion of energy security for transportation is also applicable for the security of fuels for agricultural production and related sectors.

Among all nationally vital energy services, transport is beyond doubt the most vulnerable. This vulnerability is primarily a result of the reliance on imported oil as transport fuel. The absolute bulk of energy for transport is derived from oil and its products.

Only nine countries (Algeria, Argentina, Belarus, Brazil, Moldova, Russia, Slovakia, Ukraine, and Uzbekistan) rely on oil for less than 85% of their transport energy needs. In all these countries except Brazil, natural gas is the main alternative to oil products in transport. Whereas Algeria, Argentina, Russia, and Uzbekistan have their own natural gas reserves (with the R/S ratio of only 11 years in the case of Argentina), the other countries import most of their natural gas.

This means that the security of energy for transport largely reflects the security of oil supply discussed in Section 5.1.2.1. Around 3.2 billion people live in 85 countries that import[15] 70% or more, and 4.9 billion live in 93 countries (including China)[16] that import one-half or more of their transport fuels. Finally, 101 countries (5.3 billion people) import more than 25% of the fuel used for transport. Two other countries (Argentina and the United Kingdom), which currently import less than 25% of their transport fuels, may face import dependency in the near future, since their oil R/S ratio is under 15 years.

Transport energy systems are also subject to demand-side vulnerabilities. In particular, rapid growth in transport energy use signals a pressure on transport. About 1.7 billion people live in some 17 countries (including China) where transport energy use has grown faster than 8% annually over 1998–2007, and 29 countries with 1.9 billion people

---

15 Dependency on imported fuels for transport considers all fuels used in transport. In the case of oil, it includes petroleum products and crude oil imports.

16 In 2007, China's dependency on imported fuels for transport was 47.4%. The resource/consumption ratio of domestic oil in China is 7.23 years, and the annual rate of growth of transport energy was 8.8% in 1998–2007.

have experienced a growth rate of over 6%. This is in contrast with some developed countries, notably the United States, where the car fleet shrank by four million vehicles in 2009 and the consumption of gasoline peaked in 2007 (NADA, 2009, quoted in Brown, 2010).

Supply- and demand-side vulnerabilities of transport energy systems overlap in some 15 countries (including China) with 1.7 billion people, which depend on imported fuels for one-half or more of their energy use and at the same time experience over 6% growth in demand. An additional 84 countries with 3.2 billion people experience either supply- or demand-side vulnerabilities. Only about 1.2 billion people live in 31 countries with relatively import-independent and slowly growing transport systems.

5.2.4.2   Energy for Industry

The industrial use of energy – mainly in the form of heat and electricity – varies between countries. Virtually all developed countries use at least 15% of their energy in the industrial sector. In 60 countries with 4.5 billion people, industry accounts for over 25% of their energy use, and in 12 countries with 1.7 billion people (which include Brazil, China, and Ukraine), it accounts for over 40% of their energy use. Many smaller developing and emerging economies are dominated by one or few industries relying on distinct energy systems, which are therefore critical for the energy security of those societies.

The industrial sector is less based on imported fuels than the transport sector. In 23 countries (400 million people) industry relies on fuels that are 75% or more imported, in 46 countries (800 million people) on fuels that are 50% or more imported, and in 77 countries (1.5 billion people) on fuels that are 25% or more imported.

Demand-side vulnerabilities should also be taken into account. Although many economies experience growth in industrial energy use, this cannot be considered as permanent and pressing, as in the case of the transport and residential sectors, because industrial growth may be reversed. Industrial energy intensity, on the other hand, is an important characteristic that can make the industrial sector relatively more or less vulnerable to price volatility and other energy supply disruptions. In 34 countries (including China, the Russian Federation, Brazil, Ukraine, and several other emerging economies) with 2.8 billion people, industrial energy intensity exceeds the world average by 25% or more (159% in Ukraine, 49% in China, 45% in Russia, 35% in Brazil). In the vast majority of these countries industrial products also account for a large portion of GDP (49% in China, 38% in Russia, 37% in Ukraine), which makes them even more vulnerable to industrial energy supply disruptions.

Supply- and demand-side vulnerabilities of industrial energy systems overlap in only two countries: Jordan and Moldova. About 80 countries in the world, with a combined population of some 3.6 billion people,

experience either high import dependency of industrial fuels or high industrial energy intensity.

5.2.4.3   Energy in the Residential and Commercial Sector

The R&C sector in all countries vitally depends upon a supply of electricity that is used for lighting, cooking, heating, and other appliances. Other forms of energy are also used in this sector, particularly for heating, which is an important national priority in cold climates. In many countries, prevention of failures to provide adequate heating during cold seasons are rightly considered a matter of national priority.

Another key feature of the R&C sector is that it significantly relies on traditional biomass in many developing countries. Energy statistics typically designate this source as "renewables and combustibles" without making the distinction between "modern" (e.g., pellets, straw boilers, modern heaters) and "traditional" (e.g., firewood or dung used in open cooking stoves) uses of biomass. However, reliance on traditional biomass in the R&C sector represents a national energy security issue because such use is generally considered unacceptable on health, environmental, and developmental grounds. This means that the massive use of traditional biomass in the R&C sector, much like the low rate of access to electricity, is untenable for modern nation-states. When such use is widespread, the energy policy and the energy system are under pressure to find new sources of energy to replace traditional biomass, which can be combined with other pressures to exacerbate a nation's energy vulnerabilities.

The patterns of energy use in the R&C sector differ greatly between developing and industrialized countries. Lower-income countries typically have a higher proportion of R&C energy in the total energy use and a higher share of (mostly "traditional") biomass in R&C energy use.[17] In the following analysis we make a distinction between three groups of countries that differ in the degree to which they rely on traditional biomass in their R&C sector:

- 25 countries (700 million people) rely on combustibles and renewables for over 80% of their R&C energy, most of them low- or lower middle-income countries. These countries are not included in our analysis of import dependency of fuels for the R&C sector.[18]

- 22 countries (3.3 billion people) rely on combustibles and renewables for between 33% and 80% of their R&C energy, of these two-thirds are low- or lower middle-income countries and these rest are

---

17 All of the 27 countries with R&C energy representing more than 40% of total energy use are either low- or lower middle-income countries. The share of biomass in R&C energy use of those 27 countries is higher than 70%.

18 The rationale for this approach is that the overarching vulnerability for those countries is an excessive reliance on traditional biomass in their R&C sectors rather than import dependency of the relatively insignificant remaining share of their R&C energy.

upper middle-income countries. This group also includes India and China. For these countries we analyze import dependency only for the share of their R&C energy, which does not include combustibles and renewables.[19]

- The remaining 86 countries (2.3 billion people) rely on combustibles and renewables for less than 30% in their R&C energy. About 75% of these countries are high- or higher middle-income. For these countries we analyze vulnerabilities of the R&C sector.

The R&C sector is generally less dependent on imported fuels than transport, but more dependent than the industrial sector. Twenty countries (400 million people) import more than 75% of their energy for R&C. In 39 countries (500 million people), the energy imported for this sector is higher than 50%. Finally, 52 countries (2.3 billion people) rely on more than 25% of imported energy for the R&C sector.

There are also demand-side pressures on R&C energy use. In 28 countries (400 million people), the growth rate of energy use in this sector exceeded 5% per year. In many low-income countries per capita energy use is far below the world average, which probably indicates an inadequate provision of energy services in this area and signals a potential vulnerability. This question will be explored further.

Only three small countries (Portugal, Cyprus, and Albania) have an import dependency on energy for R&C higher than 50% and annual demand growth higher than 6%. Forty-seven countries (1.1 billion people) have either dependency on energy higher than 50% or demand per capita for R&C higher than 6%. These vulnerabilities should be considered in addition to 25 countries (with 700 million people) that primarily rely on traditional biomass in their R&C sector.

5.2.4.4   Energy for Export and Resource Curse

Energy exports are central for several countries where export revenues provide core funding for the public sector and where energy export industries sustain employment and a large part of the economies. To articulate the role of energy exports, these countries sometimes formulate them as "demand security" to contrast with the "supply security" of energy-importing nations. We consider export revenues as another energy sector vital for the stability of economic, political, and social order in exporting countries.

Yet energy exports are as vulnerable to various disruptions as other energy services and can also be considered from the robustness, sovereignty, and resilience perspectives. First of all, energy exports are only

possible when sufficient domestic energy resources are available. This presumes both physical availability and accessibility of resources, as well as a reasonably satisfied (or artificially constrained) national demand for such resources. Second, resources can only be exported if there are adequate extraction, conversion, and transportation technologies and infrastructure. Third, any export depends upon safe export routes and destinations. Thus, exports security depends upon the physical safety of export routes (from adverse natural events and deliberate disruptions) as well as upon functioning institutional arrangements (e.g., markets) to match energy supply and demand and ensure sufficient and predictable revenues.

Energy exports can be considered a vital national service for only a few of all energy-exporting countries. There are about 15 countries where energy exports constitute 20% or more of GDP. About half of these countries are LDCs; most of the rest are middle-income countries.

The first insecurity stems from the very fact that a particular national economy extensively relies on oil exports. Worldwide, such dependency is highly correlated with slower long-term rates of economic growth and political problems. The economic and political risks of significantly relying on oil exports are particularly pronounced for poor countries and are extensively analyzed in the literature on the so-called "resource curse" discussed in Box 5.8. Furthermore, domestic oil reserves in some exporting countries are quite limited and will not last longer at current rates of production unless new resources are found or put in production. For example, the R/P ratios for Angola, Cameroon, and Malaysia in 2007–2009 were estimated at 12–20 years and for Argentina at about 10 years (BP, 2007; 2009; US EIA, 2011b).

5.2.5   Energy Security in Selected Regions

The above analysis primarily concerns energy security in individual countries, as well as relevant global considerations. At the same time, energy security issues have a strong regional dimension because countries located in geographic proximity often experience similar conditions with respect to access to energy resources and the structure of energy use, and may be engaged in more intensive mutual trade and transit of energy carriers. This subsection outlines energy security issues from the robustness, sovereignty, and resilience perspectives in the major world sub-regions: Africa, Asia, Europe, and North and South America.

5.2.5.1   Energy Security in Africa[20]

Energy security is an important policy issue in Africa. A number of factors contribute to rising energy security challenges. The rising economic

---

19 The rationale for this approach is that such countries typically have two distinct residential sectors: one "modern," which can be subject to standard import vulnerability analysis, and the other "traditional," where the vulnerabilities associated with traditional fuels are most prominent.

20 Source: Adenikinju, 2005; 2008; Anyanwu et al., 2010; US EIA, 2011a; Wohlgemuth, 2008.

Box 5.8 | Energy Exports and Resource Curse

Exports of energy resources – primarily oil and gas – have strongly influenced national economies and political systems in different historical periods in many countries. It has arguably supported economic development in such industrialized countries as Norway, the United Kingdom, Canada, and the United States. It has fueled economic development in Malaysia, Indonesia, the United Arab Emirates, and other emerging economies. Yet it has been blamed for slower economic growth, poor governance, political instability, and conflict in several low- and middle-income countries. This latter phenomenon has been termed the "resource curse" or "natural resource trap." There is an extensive body of scholarly literature that attempts to explain the resource curse and find appropriate remedies.

Poor economic performance of energy exporters, observed by many researchers, has been blamed on macroeconomic instability induced by the volatility of energy prices. More often than not, producers that face a season of windfalls spend extravagantly and have difficulty curbing this spending during lower price periods. In addition, some energy exporters tended to unsustainably borrow in times of high oil prices and were saddled with unbearably heavy debts in times of lower prices. Added to this, many low-income energy exporters are faced with money flows that far surpass their absorptive capacity, resulting in overheated economies and rampant inflation.

Another reason for slower economic growth of energy exporters is the so-called "Dutch disease," an economic phenomenon named after the negative impact on the Dutch economy of natural gas-related revenues in the 1970s resulting from over-valuation of the national currency and a set of other factors which inhibit the development of non-energy branches of an economy producing "tradable goods." The result of the Dutch disease is thus a steady decrease in non-oil productive activity and, in the face of the eclipse or lack of a non-oil fiscal base, the growing dependency of state coffers on petroleum rents.

The political consequences of oil wealth have been shown to be equally negative for many countries. Indeed, it is primarily at institutions, the mentalities of decision-makers, and the quality of governance rather than at macroeconomic trends that one must look to account for the predicament of the petro-state. An immediate consequence of oil revenues is the increase in state power and the absolute social, economic, and political centrality it acquires. The boom phase characteristically includes a growth in construction and proliferation of employment in the state sector and the services sector. As the key actor of the domestic economy and main supporter of the lingering private sector through handouts and public contracts, the state is everywhere and in constant expansion. The oil economy is meanwhile managed under a shroud of secrecy. Production levels and financial amounts transacted are matters of speculation; budgets are fictitious; and money flows are the object of unheard-of detours and misappropriations in a context of no public accountability (Ross, 2001). It is, therefore, unsurprising that of the top 15 oil exporters only one (Norway) is a well-functioning democracy and four others (Mexico, Venezuela, Russia, and Iraq) can be considered "imperfect democracies." This lack of accountability extends to the international level. More often than not, oil producers with questionable domestic records are nonetheless courted and protected by oil-consuming states in the industrial world, in a demonstration that oil, on account of its centrality for the functioning of modern economies, is a commodity like no other.

Easy oil money means that the state does not need to expand the domestic fiscal base. In fact, OPEC members' oil revenues represent an average of 42% of total government revenues. In cases where the building of state institutions runs parallel with the availability of oil wealth, no fiscal administrations are created at all. Oftentimes, the state does not seek to pursue this option anyway, for it faces articulate national constituencies that see themselves as legitimate beneficiaries of the national wealth rather than as net contributors to it. This fiscal aloofness strengthens the executive but also cuts its link with society as well as the minimal social contract the link might imply.

These arrangements, however, become unstable at times of low or decreasing oil prices. At such times, a petro-state becomes highly indebted, with a vast unemployed urban working class and a restive youth, a large and intermittently paid civil service, a neglected countryside, and an inequitable pattern of wealth distribution. Recent research shows that, faced with a powder keg of distorted economies and dissatisfied populations, oil states are far more likely than non-oil states to suffer from civil war or separatist bids from resource-rich regions. This political instability and poor governance can feed on each other, creating a vicious circle. In already unstable contexts such as West and Central Africa, for instance, oil is clearly a factor that exacerbates conflict and perpetuates the control of the state by self-seeking elites.

Researchers disagree on the root causes of the resource curse. The term itself tends to underplay two important factors. Regarding the first, it is becoming clear that poor institutions are largely responsible for economic under-performance, and vice versa: poor non-diversified economies are obstacles to improving governance. The second factor is the character of the economy, institutions, and political culture in oil-rich states prior to the arrival of oil rents. OECD oil-producing states had developed the institutions (strong and accountable governments, a free press, a robust political party system, etc.) that proved instrumental in steering oil revenues in a constructive direction decades before oil materialized. In much the same way, many of the problems found in oil-rich-but-poor states were everyday occurrences before oil: fragile economies, despotic rulers, and weak and inefficient institutions. In this context, oil accentuated pathologies that existed already.

More detailed recent analysis shows that endowment with national resources may be a curse or a blessing, depending on many other conditions. The larger the endowment of natural resources, the stronger are the institutions needed to overcome the risk of resource curse. However, under certain conditions – prevalent in many developing countries – the opposite occurs. The presence of natural resources actually weakens the institutions, which in turn are unable to manage the natural resource wealth; this leads to more resource-dependency, further weakening of the institutions, etc. Thus, suffering from resource curse or falling into a natural resource trap is not inevitable. However, it is a matter of very high risk and affects many developing oil-exporting nations. While analysts dispute what is at the root of the disappointing performance of most petroleum exporters – ranging from the single-minded determinism of a resource curse to more nuanced institutional and sociological analyses – there is a near consensus that few states have gone up the developmental ladder by virtue of their oil endowments.

As evidence that the resource curse does not always occur, one recent study looked at oil and gas production in Indonesia, Malaysia, Myanmar, and Thailand from 1987–2007. In each of these countries, including Myanmar, economies have become more diversified, per capita incomes have risen, and life expectancies and living standards have improved along with increased energy supply. Political transparency and accountability have remained constant in Indonesia, Malaysia, and Thailand, and oil and gas have contributed less to government revenues over time in Indonesia and Thailand. The situation in Southeast Asia suggests that extractive resources themselves are neither a curse nor a blessing. Instead, it is how they are used and the particular socio-political environment in which they evolve that determines whether they contribute to peace and prosperity or risk conflict and environmental degradation. The successful practice of managing energy export revenues in Chile included avoiding excessive spending in boom times, allowing deviations from a target surplus only in response to output gaps and long-lasting commodity price increases, as judged by independent panels of experts rather than politicians. Other measures to avoid the resource curse proposed by experts include hedging export proceeds in commodity futures markets, denominating debt in terms of commodity prices, allowing some nominal currency appreciation in response to an increase in world prices of the commodity, but also adding to foreign exchange reserves, especially at the early stages of the boom when it may prove to be transitory.

Source: Fardmanesh, 1991; Auty, 1993; 1994; Chaudhry, 1997; Karl, 1997; 1998; Leite and Weidmann, 1999; Fearon and Laitin, 2003; Katz et al., 2004; Karl, 2005; Soares de Oliveira, 2006; Collier, 2007; Soares de Oliveira, 2007; Brunnschweiler and Bulte, 2008; EIU, 2009; Sovacool, 2010.

prosperity of the continent after decades of economic underperformance has led to an increasing demand for modern energy. The rising costs of energy imports constitute a threat to economic resurgence of the continent. Weak investments in energy infrastructure, poor domestic policies that favor poorly targeted subsidies, and limited transportation connectivity across Africa also hamper the extent and pace at which trade in energy can occur between energy-surplus and energy-deficient parts of the continent. Most of the countries in the continent continue to depend on imports from outside the continent to meet the rising domestic demand.

Over the years, energy prices have risen rapidly and become highly volatile. While the upwards trend in global energy prices has led to a massive accumulation of reserves in many energy-rich African countries, net energy importers in the continent have borne the costs in terms of slower economic growth, rising trade deficit, faster depletion of foreign reserves, and rising inflation.

Affordability of and access to modern energy are very poor in Africa though in total, the continent produces more energy than it consumes. In 2009, extraction of crude oil, natural gas, and coal were 3.2, 2.2, and 1.34 times higher than their respective use, whereas the overall total primary energy supply was more than twice the amount of primary energy used. However, these aggregate figures mask a high degree of intra-African diversity in energy demand and supply, which largely reflects the diverging regional resource endowments.

The TPES in Africa grew by a mean of 4% between 2005 and 2009, mainly as a result of robust growth in crude oil production in North and West Africa. The share of biomass, while still significant in parts of Africa, especially in sub-Saharan Africa, decreased from 62% in 1971 to less than 45% in 2009. However, Africa's use of combustible renewable energy (mainly wood-fuel) still remains significantly higher than the world average. Total primary energy use in the same time period rose by an annual average of 3.7%.

Though the continent exports energy, it is unable to meet the energy needs of its population, especially in sub-Saharan Africa. The systemic energy problem in Africa is especially evident within its power sector. Africa is estimated to require 7000 MW of new power generation capacity each year, but has been installing only 1000 MW annually. Outside South Africa, power consumption averages 124 kWh (4.46 GJ) per person per year. Only one-fifth of the population of sub-Saharan Africa has access to electricity, compared with one-half in South Asia and four-fifths in Latin America. At the same time, the cost of electricity is higher in Africa. Frequent power outages force firms to rely on expensive back-up generators that cost up to US$0.40 per kWh. Many countries rely on inefficient, expensive, small-scale, oil-based power generation. Africa is well endowed with large-scale, cost-effective energy resources, but they tend to be located a long distance from the major demand centers and their development is often too expensive for the countries where they are found.

Energy resource ownership and use vary widely across Africa. For instance, three countries – South Africa, Egypt, and Algeria – account for two-thirds of total primary energy use. Algeria and Egypt account for 73.6% of total natural gas consumption. South Africa and Egypt jointly account for 61% of the continent's hydroelectricity consumption. South Africa alone accounts for 93% of total African coal consumption and 100% of its nuclear energy generation.

Nigeria, Algeria, Angola, and Libya are the major producers of oil, although oil is also produced in Cameroon, Chad, Congo, Côte d'Ivoire, Egypt, Equatorial Guinea, Gabon, Mauritania, Sudan, and Tunisia. Natural gas reserves are located in four major countries: Nigeria, Algeria, Egypt, and Libya.

There is a very limited amount of intra-Africa trade in energy. Only 1.8% of the continent's oil exports go to other African countries. LNG trade from Algeria, Egypt, Equatorial Guinea, Libya, and Nigeria go to countries in North America, Europe, and Asia/Pacific. Three countries – Egypt, Mozambique, and South Africa – are responsible for nearly all the coal

exports in Africa. Most of the coal exports go to other African countries – Algeria, Congo Kinshasa, Kenya, Morocco, Mauritius, Namibia, and Senegal. There is also some electricity trading within the continent, especially within the various sub-regions. A good example is the West African Power Pool, designed to interconnect national grids across 5000 km of most West African countries. Another important initiative in West Africa is the West African Gas Pipeline. The US$635 million project was initially developed to utilize some of the gas currently being flared in Nigeria for power generation in Benin, Togo, and Ghana.

The presently low level of energy trade among countries has made it difficult to leverage the huge energy surplus produced in the continent to meet the needs of energy-poorer countries. Hence, significant scope exists for energy cooperation across sub-regions on the continent. Although there are a number of pan-African cooperative energy initiatives, the pace of implementation of these initiatives has been slow.

The diversity of energy resources in Africa is relatively low. Oil is the dominant fuel in most African countries. Four-fifths of the countries derive over 50% of their energy use from oil. Over one-third of the refined petroleum consumed in the region is imported. While the amount may vary, all the countries import some proportion of their product consumption. Over 63% of the countries import more than 90% of the refined products they consume. Paradoxically, Nigeria, the leading producer and exporter of crude oil in Africa, imports 69.8% of its product requirements and accounts for 17.7% of the continent's total imports. Dependency on imported energy is a risk to energy security and makes the country vulnerable to high and volatile world oil prices.

In some African countries, energy sources other than oil dominate consumption. Countries significantly dependent on coal are South Africa (76.4%), Zimbabwe (53.4%), Botswana (41%), and Swaziland (34.7%). Countries with high natural gas dependency are Equatorial Guinea (92.6%), Algeria (64.4%), Tunisia (50.5%), Egypt (49%), Côte d'Ivoire (44.5%), and Nigeria (41.7%). Countries heavily dependent on hydroelectric energy are Mozambique (82.1%), Zambia (75.2%), Congo Kinshasa (68.2%), Malawi (45.3%), and Cameroon (41.6%). Countries that depend almost exclusively on a single primary energy source are Benin, Chad, Djibouti, The Gambia, Seychelles, Sierra Leone, Somalia, and Togo.

The generally low energy diversity in Africa relates to low capacity and a weak technological base in most countries. Poverty and a lack of appropriate technologies have hampered diversification into renewables such as solar, wind, and small hydro.

Several options are open to African countries to address the huge energy gap in the region. These include the use of renewable energy to supplement the non-renewable energy sources that currently dominate the energy mix, the reduction of huge energy losses and wastages and adoption of energy efficiency technologies, increased investment in new technologies and energy supply infrastructure, and an enhanced level of intra-regional energy cooperation. The latter is required because many African countries are very small and may not be able to finance the huge investment costs needed to develop alternative energy sources. Africa must engage the private sector in addressing its energy supply inadequacies. While energy subsidies may not disappear for a while, such subsidies must be smart and better targeted to meet specific societal objectives.

5.2.5.2 Energy Security in Europe[21]

Europe[22] is among the world's largest energy-consuming regions. Oil, natural gas, and coal constitute the dominant fuels in Europe's energy mix. In 2009, coal accounted for 17% of Europe's total primary energy demand, oil accounted for 35%, and gas for 25%. That year, Europe represented around 18% of the global demand for oil, some 17% of the global demand for natural gas, and 9% for coal. Nuclear made up 14% and renewables 9% of primary energy use in Europe. Overall, Europe, representing 7% of the world's population, accounted for 15% of the world's primary energy demand.

Europe is also a key supplier of energy. In 2009, it produced 2.6% of global oil output, 5.7% of global gas output, 4.6% of global coal output and 11% of global renewable energy. Still, Europe is crucially relying on energy imports to satisfy its needs. In 2009, the EU imported more than half of its energy from non-EU countries. This number has increased in recent years, rising from less than 40% of gross energy use in 1980 to 55% at present. If the present trends continue, European import dependency is likely to steeply increase in the near future, a function of policies aimed at reducing carbon-intensive fuels such as coal with comparably less carbon-heavy fuels such as gas, but also due to declining domestic production, notably in the North Sea.

European energy imports are dominated by a few producers of energy. Some two-thirds of EU-27 imports of natural gas stem from only three countries: Norway, Algeria, and Russia. Russia is also Europe's dominant supplier of crude oil, accounting for some 30% of the bloc's imports in 2008. Even in hard coal, Russia plays an important role, supplying around 24% of European imports. High European import dependency is not a concern for all fuels. The coal market is relatively small in volume, and global reserves are relatively evenly distributed. The oil market is global, liquid, integrated, and therefore unlikely to give a dominant producer much political leverage over the consumer region supplied. Gas markets, however, still remain by and large regional and bilateralized in nature. Given the predominantly pipeline-bound infrastructure in natural gas, alternative suppliers are hard to find in the short to medium term. In light of this, concerns have been expressed over some European countries' heavy reliance on gas imports from Russia,

reaching up to 100% of total demand in Central and Eastern Europe, especially the Baltic States. Recent disputes over gas transit between Russia and neighboring Ukraine, and also Belarus, have added to this concern and sparked political initiatives to reduce overall European import dependency on Russia. The degree to which Russian gas actually poses a security problem is, however, disputed. Russia's share of European natural gas imports has in fact declined steadily over the last two decades, from some 75% in 1990 to 31.5% today. Current developments in unconventional gas production (see Box 5.5), coupled with the increasing global capacity in LNG, may possibly also change existing market structures and contribute to a higher integration of regional markets and more gas-to-gas competition in the European market (IGU, 2010). The planned construction of the Nord Stream gas pipeline under the Baltic Sea, connecting Russia directly to Germany and Denmark, is intended to diversify gas trade routes and reduce dependency on transit countries.

Still, policy initiatives center on reducing gas import dependency on Russia by promoting diversification strategies, including pipelines in the "southern corridor" aimed at bringing Caspian gas to European households. A key European initiative in this context is the planned Nabucco pipeline, a 31 bcm per year interconnector for Azeri and possibly Turkmen or even Iraqi gas, via Turkey. Since available gas volumes remain uncertain, the pipeline has so far not left the planning stage. Recent Azeri pledges to commit parts of the gas output generated in the Shah-Deniz II field (planned to come on stream in the next years) may constitute a breakthrough for the project.

In addition to import dependency, European energy security challenges may, however, also be of domestic origin. In particular, they may stem from Europe's commitment to pursue low-carbon energy transition. Policy packages such as the EU's 20–20–20 initiative, aimed at reducing emissions of greenhouse gases by 20% by 2020 and flanked by reaching 20% of renewable energy in total energy use, will put demand-side pressure on European energy generation and systems. Replacing coal in power generation by bridge fuels (notably natural gas) before phasing them out to the benefit of renewables will pose unprecedented challenges with regards to finance, infrastructure, and technology. Adding to this challenge is a rapidly aging infrastructure in European nuclear energy, coupled with stiff public opposition to new nuclear power plants in most European countries.

5.2.5.3 Energy Security in North America

The North American continent has been endowed with immense energy wealth. The United States is among the world's top ten producers of coal, oil, natural gas, and electricity from nuclear and hydroelectricity, while Canada is in the top ten for oil, natural gas, and electricity from nuclear and hydroelectricity production, and Mexico ranks in the top ten for oil production (IEA, 2010a). Despite this, each country has its own set of energy security problems.

---

21 Sources: BP, 2010; IEA, 2010a; Eurostat, 2011.

22 The term "Europe" is used here to designate European OECD member countries or the European Union Member States.

Probably the most dominant and well known of these problems is that being faced by the United States and its dependence on foreign supplies of crude oil. Every US president, from Nixon to Obama, has set targets, put forward proposals, commissioned reports, and signed legislation in an effort to stem crude oil imports and improve energy security (US DOE, 2010). Today, over 60% of US demand for crude oil is met from imports (US EIA, 2010).

Support for the US transportation system is the driving force behind all energy security legislation put forward in the United States. For example, the 2007 Energy Independence and Security Act (EISA) calls for, amongst other things, reducing vehicular fuel consumption through increased CAFE (Corporate Average Fuel Economy) standards, replacing gasoline with ethanol, and requiring auto manufacturers to develop a new generation of vehicles to operate on electricity (EISA, 2007).

EISA has had unintended consequences. The push for ethanol from cornstarch means that a significant percentage of US farmland is being diverted from food into fuel production; this has had an impact on world corn supplies, indirectly affecting food security in countries such as Mexico (Roig-Franzia, 2007).

The increasing demand for electricity in general, and the inevitable reliance on mains electricity to meet the energy needs of plug-in electric vehicles in particular, will have an impact on (electrical) energy security. At present, about 50% of the electricity in the United States is produced from domestic coal, followed by natural gas and nuclear (about 20% each), hydroelectricity (5%), and a mix of renewables (2.5%) (US EIA, 2011a). Demand pressures are forcing electricity suppliers to plan for new generation capacity and, if climate change is ever addressed seriously by the US Congress, it will be necessary to develop generation facilities that emit little or no carbon. However, the supply mix is only part of the problem – the US electrical grid is showing its age and must be refurbished if it is to meet the expected future reliance on electricity. The costs of new generation facilities (whether or not the United States addresses the issue of climate change) and grid upgrades are estimated in the trillions of dollars – the price of ensuring the availability of the electrical supply.

Until the middle of the last decade, it was assumed that domestic supplies of natural gas in the United States had peaked and, like crude oil before it, would make the United States increasingly reliant on imports of natural gas. To ensure (natural gas) energy security, plans were drawn up for dozens of liquefied natural gas (LNG) facilities around the continental United States (McAleb, 2005). Today, things look considerably different – the use of horizontal drilling fracking is making shale gas available as a replacement for declining stocks of conventional natural gas (Grape, 2006). Shale gas is rich in natural gas liquids (NGLs), meaning it can also improve US energy security by offsetting imports of crude oil (Sandrea, 2010). Optimistic reserve projections have industry analysts suggesting that the United States could soon start exporting LNG (PennEnergy, 2010); this would not appear to be in the long-term

energy security interests of the United States. There are also concerns over the environmental impacts associated with the extraction of shale gas (Doggett, 2010); time will tell whether it is considered an acceptable source of natural gas that will improve the energy security of the United States.

Two of the countries on which the United States depends for its energy are its nearest neighbors, Mexico and Canada; both countries are exporters of crude oil and other refined petroleum products to the United States, while Canada also exports natural gas and electricity. The United States' reliance on Mexico and Canada for its energy has politicians and analysts in all three countries talking about North American, or continental, energy security (Angevine, 2010).

North America's energy security is governed by chapter six of NAFTA, the North America Free Trade Agreement, which outlines the rules and regulations regarding the trade of energy and petrochemicals. NAFTA requires a signatory to maintain its energy exports; short of war, any reduction in exports must be met by a proportional reduction in supply within the exporting nation. Mexico is exempt from this provision; Canada is not (NAFTA, 2002).

Mexico is facing energy security challenges of its own. Its most important oil field, Cantarell (in the Gulf of Mexico), is in decline and further exploration is hampered by the Mexican constitution that restricts oil and natural gas development to the state oil company, Pemex.

Canada, unlike Mexico, has few restrictions on international players exploiting its crude oil and natural gas. Despite the availability of these resources, not all Canadians have access to them; for example, although Canada is self-sufficient in crude oil, over 60% of it is exported to the United States, meaning that Canada meets almost 50% of its crude oil needs from imports (Hughes, 2010). Canada is also self-sufficient in natural gas, yet almost 60% is exported to the United States (US EIA, 2011a). Not only is Canada exporting energy that could improve its future energy security, but also it has compounded the problem by failing to develop the pipeline infrastructure to connect parts of eastern Canada with the oil and natural gas fields in western Canada (Hughes, 2010).

Although Canada's production of conventional crude oil and natural gas has peaked, the tar sands (euphemistically referred to as the "oil" sands in the United States) are seen as essential to continental energy security. Canada's current prime minister has gone so far as to call Canada an "energy superpower" with respect to the development of unconventional energy resources such as the tar sands, shale gas, and Arctic oil and natural gas for export to the United States (Hester and Welsh, 2009).

Canada is one of the few countries with the capacity to improve its energy security with its own energy resources. Despite this, Canada's trade and energy policies have evolved to the point where much of the energy that could be used for its own energy security is, instead, contributing to the improvement of energy security in the United States.

5.2.5.4   Energy Security in Asia[23]

Asia – meant here to encompass the big four energy consumers of China, India, Japan, and South Korea, as well as the developing economies including Southeast Asia and South Asia – faces a series of daunting energy security challenges that crisscross the three themes of robustness, sovereignty, and resilience.

Growth in energy use, both in terms of per capita use and total use in aggregate, is expected to rise dramatically in the next few decades. As a whole, Asia Pacific's per capita electricity demand was only about 1300 kWh in 2005, compared to the world average of more than 2500 kWh. Under a business-as-usual scenario, between 2005 and 2030 energy demand is expected to grow at 2.4%, whereas the world average during the same period will be 1.5%. Net imports of fossil fuels in Asia Pacific are expected to more than double. The region's oil dependency will increase from 57.5% to 66.4%. The region will also need between US$7 trillion and US$9.7 trillion of cumulative investment in the energy sector during this period, of which about two-thirds will be in electricity generation, transmission, and distribution. The 10 countries that comprise the Association of Southeast Asian Nations, for example, will likely experience an annual growth rate in energy demand of 2.5% between 2010 and 2030. If that projection holds true, regional demand for energy will equal the current combined total demand of Japan, Australia, South Korea, and New Zealand. Yet, although Southeast Asia is home to 8.5% of the world population (530 million people), the region possesses about 1% each of the world's oil and coal stocks and less than 4% of total natural gas reserves.

Security of supply has thus become a key economic and political concern. In China, Beijing had to ration its gas supply to shopping malls and supermarkets in January 2010 as a result of extreme winter weather. In 2008, India walked out of the deal to build an Iran-Pakistan-India (IPI) gas pipeline – on which discussions were conducted over 13 years – over security issues and the inability of Pakistan to agree to provide penalties for supply disruptions. Japan buys nearly 90% of its oil from the Middle East, making it vulnerable to disruptions of even a few days on the Strait of Hormuz or through shipping routes from the Middle East.

Threats need not be international or external. Laborers of India's public-sector petroleum company, Oil and Natural Gas Corporation Limited (ONGC), went on a three-day strike in early 2009, shutting down the Hazira plant that processes oil and gas from offshore operations and threatening to create shortages of compressed natural gas used for public transportation in Gujarat. Large parts of China also had to confront

energy shortages in 2010 due to a combination of weather and infrastructure factors: the difficulty of transporting coal in the snow, less hydropower output due to freezing temperatures, and reduced coal supplies from Shanxi province due to mine closures. In 2008, shortages of gasoline and diesel occurred in Bali, Indonesia, when oil tankers had trouble accessing the island during a series of storms, and in Kalimantan long lines formed at petrol stations due to a shortfall of 10,000 liters of gasoline. In Jakarta and Java, as well, shortages of premium gasoline and LPG occurred after a refinery had maintenance problems, and disruptions of electricity hit every Indonesian province in both 2007 and 2008.

Trade in energy is another essential challenge. Apart from Indonesia, Malaysia, Vietnam, and Brunei, all other Asian countries are currently net energy importers. This means that the promotion of trade is instrumental to building energy markets so that countries can improve access to multiple sources of energy. Without such access, buyers must negotiate directly with producing nations such as those in the Middle East. Several "energy-poor" nations are relying on trade to overcome their energy shortages. The Bangladesh Power Development Board is currently holding a road show around the world to encourage foreign investors to help them erect about 3500 MW of new power plants and a terminal for LNG. The country suffers from an acute shortage of power, especially during the hot summer months. Pakistan is also facing severe energy shortages, and searching for private foreign investment by offering incentives related to upstream and downstream hydrocarbon development.

Some countries, such as China, India, and Japan, have begun aggressively investing overseas to then export energy fuels back to their mainland. The China National Offshore Oil Corporation (CNOC), after a well-publicized yet unsuccessful takeover bid for Unocal in the United States in 2005, took over Canadian-based PetroKazakhstan in 2006, and since then has won contracts in politically volatile places such as Angola and Nigeria and strengthened ties with Sudan, Cuba, Venezuela, Iran, Sudan, Kazakhstan, and Myanmar. India's government aimed to produce 20 million barrels of equity oil and gas abroad by 2010, and the overseas arm of ONGC has already acquired properties in Vietnam, Russia, and Sudan. Japan, in its quest to produce more "Hinomaru" oil (oil developed and imported through domestic producers), has integrated its key oil companies – Inpex and Teikoku Oil – under a joint holding company, to make them more competitive against foreign oil companies.

5.2.5.5   Energy Security in Latin America and the Caribbean

The Latin America and the Caribbean region (LAC) has considerable energy resources, including oil, natural gas, biomass, and hydro energy, and is a net energy exporter, producing about 8.4% of global energy and consuming about 6.3%. Venezuela and Mexico are among the top global oil exporters, whereas Brazil is the largest ethanol exporter, accounting for half of the world's bio-ethanol exports (US EIA, 2011a). However, these resources are unequally distributed. For example, more

---

23 Sources: Asian Development Bank, 2009; IEA, 2010a; Bambawale and Sovacool, 2011.

than 90% of proven oil reserves in LAC are concentrated in three countries: Venezuela (which also holds about two-thirds of the region's natural gas reserves), Brazil, and Mexico (BP, 2009).

Moreover, utilization of these resources requires considerable investment and capacities, which can often only be mobilized at the international level. For example, significant investment will be required to develop the giant Lula (Tupi) and other "pre-salt" oil and gas fields recently discovered in Brazil or to implement the ambitious plans for expanding production of hydroelectricity where only 22% of the regional potential is currently used (SESEM-CFT, 2005). The current underinvestment in energy infrastructure is sometimes explained by legal and political uncertainty and insecurity, including changing the rules and nationalization of energy assets in several countries (Iranzo and Carrasco, 2008).

The region is also diverse with respect to capacities of individual countries, some of which are both too small and poor to address their energy challenges. For example, Nicaragua, one of the Heavily Indebted Poor Countries, is dependent on imported oil for almost 40% of its primary energy supply, including electricity generation (IEA, 2010b). Ecuador, another low-income country, relies on oil exports for almost half of its total export earnings and one-third of all tax revenues. Despite its large oil production, Ecuador must still import refined petroleum products due to a lack of sufficient domestic refining capacity. As a result, the country cannot always benefit from high oil prices, which increase its export revenues but also increase its refined product import bill (US EIA, 2009). Almost all of the electricity in Paraguay relies on one hydroelectric plant (Itaipu).

Energy infrastructure in politically unstable LAC countries is often the target of sabotage. Only in 2001, 170 attacks were registered on one of the most important oil pipelines in Colombia, Cano Limón (CEPAL, 2007). The TransAndino pipeline connecting Colombia and Ecuador has also occasionally been the target of rebel forces in Colombia, and an attack in March 2008 shut the system down for several days. Similarly, another oil pipeline in Ecuador, Sistema Oleducto Trans-Ecuatoriano (SOTE), has suffered from natural disasters that severely disrupted Ecuador's oil production. In March 2008, landslides damaged SOTE, shutting operations for several days. In 1987, an earthquake destroyed a large section of SOTE, reducing Ecuador's oil production for that year by over 50% (US EIA, 2009).

In LAC, regional integration is often viewed as a means to both redistribute uneven resources and to pool forces for infrastructure development and needed investment. Several energy integration organizations were created in LAC as long ago as the 1960s and 1970s:

- OLADE (Latin America Energy Organization), formed in 1973 by 26 LAC countries as an umbrella organization promoting the political, institutional, and technical integration of energy systems as well as energy efficiency;

- ARPEL (Latin America and Caribbean Regional Association of Oil and Natural Gas Companies), created in 1976 by 27 public and private companies and organizations that account for 90% of total upstream and downstream operations in the region; and

- CIER (Commission for Regional Electricity Integration), created in 1964 and including all South American countries except Surinam and Guyana.

Energy security has received high-level political attention in the last decade. In the Caracas Declaration (made in 2005), energy ministries of South America agreed to seek energy integration and cooperation. In April 2007, the first Presidential Energy Summit in South America resulted in a common energy strategy known as the "Margarita Declaration" (CEPAL, 2007), which advocates for a stronger role of the state in energy issues and promotion of renewables, especially biofuels. In November 2008, the Energy Ministers of OLADE member countries issued the Buenos Aires Declaration, which stated that energy security (defined as "safe and reliable energy resources availability") is a priority of the region (OLADE, 2007a; 2007b).

Another regional energy integration effort is Venezuela-backed Petroamérica, which provides a framework of cooperation initiatives in the areas of oil and gas supply and infrastructure (PDVSA, 2009). Petroamérica is divided into sub-regional frameworks: Petrocaribe, Petroandina, and Petrosur.

Petrocaribe includes 14 Caribbean countries, as well as Venezuela and Surinam. Within Petrocaribe, Venezuela directly sells oil and products to these countries under favorable financing conditions (CEPAL, 2007; US EIA, 2008; PDVSA, 2009). In 2007, 10 Petrocaribe members signed an energy security agreement that promotes the expansion of refinery capacity, ethanol production, and LNG infrastructure, as well as energy efficiency measures (PDVSA, 2009). Many of the Caribbean countries import oil from Mexico and Venezuela under favorable terms. Under the San Jose Pact, Barbados, the Dominican Republic, Haiti, and Jamaica receive oil and refined products from those two countries.

Petroandia includes Bolivia, Colombia, Ecuador, Peru, and Venezuela. It is an alliance of state oil and energy organizations to promote electric and gas interconnection, mutual energy supply, and joint investments (PDVSA, 2009). As part of Petroandia, Venezuela operates joint ventures in oil exploration, production, and capacity with Bolivia and Ecuador (PDVSA, 2009). Petrosur is made up by Brazil, Argentina, Uruguay, and Venezuela. Activities within Petrosur include the construction of joint Venezuelan-Uruguayan and Venezuelan-Brazilian refineries and the creation of Petrosuramérica, a joint Venezuelan-Argentinean company (Comesaña, 2008).

One area of regional energy integration is jointly constructed infrastructure for transporting natural gas. An agreement on joint construction of "the southern gas pipeline" for transporting natural gas from Venezuela to Brazil and Argentina was signed in 2005. However, a later discovery of large oil and gas reserves in Brazil decreased that country's interest in the project. The needed investment (around US$25 billion) could not be secured, and the project was shelved in 2009 (Hidrocarburosbolivia.com, 2009). A more recent project, called the "Energy Ring," would connect three gas exporters (Ecuador, Bolivia, and Peru) to four (potential) importers (Brazil, Chile, Argentina, and Uruguay). However, the project was stopped due to the withdrawal of Bolivia and political differences between Chile and Peru. Instead, such countries as Chile have chosen to expand their LNG infrastructure to increase the diversity of gas import energy and routes.

There are also efforts to integrate electricity infrastructure such as a large integration initiative, the "Mesoamerican Integration and Development Project" (Proyecto Mesoamérica, 2009). The Electrical Interconnection System for Central America (SIEPAC), a 1,800 km power line that links six countries (Panama, Costa Rica, Honduras, Nicaragua, El Salvador, and Guatemala), was almost completed at the time of writing and a Regional Electrical Market was being put in place. This infrastructure is complemented by the existing electrical interconnection between Guatemala and Mexico, and a project to interconnect Panama and Colombia (IADB, 2010).

5.2.6   Summary: Energy Security in the World

This section explores energy security conditions in the world with respect to vulnerabilities of primary energy sources at the national and, where appropriate, global level, security of national electricity systems, and vulnerabilities of vital end-use sectors: transport, residential and commercial, industrial, and energy export revenues.

The results of the analysis of energy sources are summarized in Table 5.6. It shows that all primary energy sources have vulnerabilities from the robustness, sovereignty, and resilience perspectives. Globally, oil stands out as the most vulnerable in all three aspects among the internationally traded fuels, although natural gas may develop equally strong vulnerabilities in the near future. Rising demand plays as strong a role as supply limitations. Although the vulnerabilities of nuclear and hydro energy are not directly comparable to those of fossil fuels, they still affect hundreds of millions of people in dozens of countries.

Many countries using nuclear power experience aging of the reactor fleet and workforce as well as difficulties in accessing capital and technologies to renew, expand, or launch new nuclear programs. Of the 31 countries with nuclear power programs, 20 have not started constructing a new reactor in the last 20 years, and 19 countries have nuclear power plants with an average age of over 25 years. Large-scale enrichment, reactor manufacturing, and reprocessing technologies and capacities are currently concentrated in just a few countries. The transfer of these technologies and capacities to a larger number of countries is

constrained by serious concerns over nuclear weapons, which is one of the main controversies and risks associated with nuclear energy.

Hydroelectric power production, especially from major dams located on internationally shared rivers, is often perceived as insecure, particularly in light of climate change affecting seasonal water availability. Over 700 million people live in 31 countries that derive a significant proportion of their electricity from just one or two major dams, and are thus vulnerable to failures of these dams.

The analysis also identifies vulnerabilities of national electricity systems. First of all, electricity systems inherit the vulnerabilities of energy sources used for electricity generation described above. For example, power plants relying on imported fossil fuels currently provide over 50% of electricity in some 39 countries with 600 million people. Some 450 million people in 35 countries primarily rely on just one source of energy for generating electricity, which is a concern from the resilience perspective.

In addition, electricity systems in developed countries often bear risks associated with aging power plants (especially pronounced in the case of nuclear reactors, which have not been renewed in most industrialized countries in the last 25 years) and other infrastructure. In developing and emerging economies, electricity systems are under strong demand-side pressures. The majority of the world population – some 4.2 billion people – lives in 53 countries that will need to massively expand the capacity of their electricity systems in the near future because they either have less than 60% access to electricity or demand growth averaging 6% or more over the last decade. Both fuels and infrastructure for such an expansion will need to be provided without further compromising the sovereignty or resilience of national electricity systems.

Finally, reliability of electricity supply is a serious concern, especially in developing countries. In almost three-quarters of low-income countries blackouts are on average for more than 24 hours per month, and in about one-sixth of low-income countries blackouts average over 144 hours (six days) a month. In over one-half of low-income countries blackouts occur at least 10 times a month.

With respect to nationally vital end-use energy services, transport is globally the most vulnerable. The absolute majority of countries rely on oil products for most of their transport energy and, as we have seen, in most of the world this oil has to be imported. Around 4.9 billion people live in 93 countries that import more than one-half of their transport energy requirements. This supply-side vulnerability is made worse by demand-side pressures: in some 17 developing countries with 1.7 billion people, transport energy use was growing faster than 8% annually from 1998 to 2007.

The energy sector also provides vital export revenues to some 15–20 countries. In the majority of these oil- and gas-exporting countries the revenues are not expected to last for more than one generation, and in several cases they may cease in less than a decade. In addition, poor energy-exporting nations are at a high risk of the resource curse: economic and political instability eventually affects human development and security.

If rapidly growing demand for energy and high risk of resource curse are considered vulnerabilities of energy systems, there will be very few, if any, countries in the world that do not experience significant energy security challenges. The next section considers how national governments and international institutions attempt to deal with such challenges.

5.3   Energy Security Actors and Institutions

The key to understanding energy security is not only in the quantitative analysis of energy systems, but also in examining perspectives and strategies of energy security actors, primarily nation-states. These perspectives and strategies shape energy security risks by posing or dissuading real and perceived threats and by determining responses to likely disruptions. The energy security perspectives and strategies reflect the objective conditions but are also influenced and limited by cognitive and political factors that shape the views of policymakers and by capacities to enact these views. In response to perceived energy security threats, nations initiate energy security strategies, as we describe in this section. We also discuss the interaction of national strategies at the international level through both conflict and cooperation, including within various international institutions.

5.3.1   National Perspectives and Strategies

National energy security strategies exist in an increasing number of countries and focus on minimizing the risks for the energy end-uses (transport, residential and commercial energy, industry or energy export revenues) most vital for a given country. Some strategies are rooted in specific energy security perspectives – robustness, sovereignty, or resilience – whereas other strategies present more generic responses to multiple threats (see Figure 5.1). This section first discusses the historically earlier and more prominent strategies and then reviews robustness and resilience strategies.

5.3.1.1   Sovereignty Strategies

The essence of sovereignty strategies is increasing control over energy systems vis-à-vis "foreign" or "external" agents. Such threats are often perceived as more imminent and more easily catching public attention. They have also historically been at the center of energy security concerns.

In many cases, sovereignty-driven strategies focus on attaining or increasing control over existing energy resources. One example is the so-called "resource nationalism" of energy exporters that has recently attracted significant attention following the re-nationalization of the energy sector in Venezuela in 2003; another example is the transfer of several major oil and gas projects from international oil companies (IOCs) to state companies in Russia in the last decade. Resource nationalism and the resulting dominance of national oil companies (NOCs) (see Table 5.8) is not exactly a new phenomenon (see Box 5.11). It is not uncommon for the states to seek sovereign control over technology and infrastructure, as well as natural resources. For example, there was significant resistance in the UK against Russian Gazprom acquiring Centrica, a gas distribution company, in 2006–2008, while the US blocked the attempt of the China National Offshore Oil Corporation (CNOOC) to buy a US oil company, Unocal, in 2005.

Resource nationalism is not only typical in energy-exporting countries. Most notably the emerging major importers, such as India and China, are supporting their NOCs in acquiring energy assets around the world. Even OECD countries such as Japan and Korea have shown a renewed interest in the idea of securing supply through a state-owned corporation. A plethora of existing bilateral energy deals, long-term contracts, and joint projects reflect the increasing interests of nation-states to enhance their influence over energy resources, if not to full control, then at least to management through a "trusted partner."

As a result of this trend towards state actors in the global energy market, NOCs have come to dominate the scene, as shown in Table 5.8. In fact, only two of the world's top 15 energy companies in terms of reserves are private, while only one is headquartered in an OECD country.

In fact, and contrary to widespread perception, the trend towards a greater role of NOCs in global energy does not fundamentally affect supply, although it increases the market power of certain states. A strong role of NOCs in global energy does not necessarily imply "less market." Producer NOCs supply the same global market as their private counterparts, while major consumer NOCs tend to sell significant volumes of their equity oil on global markets rather than shipping it back home (IEA, 2007).

Similarly to oil, and with the exception of North America, state companies also dominate the supply of natural gas. In Eurasia, these are, among others, Russia's Gazprom, Algeria's Sonatrach, Norway's Statoil and Azerbaijan's State Oil Company of Azerbaijan Republic (SOCAR).[24] State-owned companies also tend to dominate the downstream gas market in Asia (with Malaysia's Petronas, China's China National Petroleum Company [CNPC], and India's GAIL being prominent examples). At the same time, most European downstream markets have been privatized. Due to restricted access of gas producers to attractive European downstream markets, private retail companies – such as Germany's EoN

---

24   Where IOCs are admitted to exploration and production endeavors (as in the case of the Azerbaijan International Operating Company [AIOC] developing Azeri gas), state companies are at least party to all of the international consortia developing new gas projects.

Table 5.8 | Oil reserves of top 15 companies.

Rank	Company	Type	Proven reserves (bbl)	Percentage of world total (%)
1	Saudi Aramco	NOC	264.3	21.9
2	National Iranian Oil Co.	NOC	137.5	11.4
3	Iraq National Oil Co.	NOC	115.0	9.5
4	Kuwait Petroleum Corp.	NOC	101.5	8.4
5	Abu Dhabi National Oil Co.	NOC	92.2	7.6
6	Petróleos de Venezuela S.A.	NOC	80.0	6.6
7	National Oil Corp of Libya	NOC	41.5	3.4
8	Nigerian National Petroleum Corp.	NOC	36.2	3.0
9	Lukoil (Russian G8 Presidency)	IOC	16.1	1.3
10	Qatar Petroleum	NOC	15.2	1.3
11	Gazprom	NOC	13.8	1.1
12	Pemex (Mexico)	NOC	12.9	1.1
13	Petrobras (Brazil)	NOC	12.2	1.0
14	China National Petroleum Corp.	NOC	11.5	1.0
15	Chevron (USA)	IOC	8.0	0.7
	Total, top 15		957.9	79.3

Source: BP, 2007; Klare, 2008.

Ruhrgas, Spain's Gas Natural, or France's GDF Suez[25] – enjoy a generally powerful position vis-à-vis producers.

Other examples of moves to secure control over resources include an increasing number of deals that involve China, India, Russia, the United States, and other countries to secure access to uranium deposits in Mongolia, Kazakhstan, and other Central Asian countries (Pistilli, 2009).

It should also be noted that asserting full or partial control over fossil fuel resources is not an option for the majority of countries that both lack such resources domestically and do not have economic, political, or military power to project their influence internationally. Their primary sovereignty strategies are to increase their reliance on domestic resources. Among resources available to a majority of countries – and thus playing an increasingly important role in national energy security strategies – are renewables (hydro, wind, solar, and modern biomass) as well as, to a lesser extent, peat and coal. One of the most celebrated examples of increasing energy security by switching to indigenous fuels is the Brazilian ethanol program, which resulted in the replacement of a large share of imported oil as a transport fuel by domestically produced ethanol. Developing renewable energy resources is seen not only as addressing more pressing and immediate concerns with the volatilities and uncertainties of global fossil fuel markets, but also a more systemic and long-term pressure of the perceived scarcity of fossil fuels and in order to reduce climate impacts of energy systems.

25 The French government holds a 37.5% stake in GDF Suez.

Launching or expanding national nuclear energy programs may also be viewed as a sovereignty strategy. Although few states can build and manage a nuclear power plant and the related nuclear fuel cycle on their own, they typically feel that there are fewer uncertainties beyond their control once the facility is up and running. Nuclear power can also be considered a diversification strategy for states relying on fossil fuels. For example, several Gulf States are import-independent but excessively relying on oil and gas for their electricity generation (Jewell, 2010). Another example is Belarus, whose electricity sector almost entirely depends on imported Russian natural gas. Belarus' planned nuclear power plant will be manufactured from Russian parts and most likely use Russian fuel and expertise, thus not reducing the country's dependency on its neighbor. However, it will provide the much-needed diversity in terms of related technologies, markets, and institutions so that disruptions of natural gas supply will not necessarily be devastating for the country's electricity sector.

5.3.1.2 Robustness and Resilience Strategies

Robustness strategies focus on minimizing predictable and manageable risks within energy systems. For example, many industrialized countries have extensive standards concerning the reliability of electricity transmission and generation (European Parliament, 2006; North American Electric Reliability Corporation, 2010). Robustness strategies may also be focused on constraining demand. For example, following the oil crisis of the early 1970s, Japan, the United States, and eight Western European countries (including energy-exporting Norway) reduced their energy intensity by 30–34% (Geller et al., 2006). In its current energy strategy, Russia aims to reduce the energy intensity of its economy by 40% by 2020, as compared to 2007 (Ministry of Energy of the Russian Federation, 2010). However, no single country in modern times has been able to deliberately reduce or even stabilize its overall energy use over a long term, though demand could be reduced temporarily in response to short-term disruptions (Meier, 2005).

Resilience of an energy system is its ability to provide critical energy services in the face of disruptions. The concept of resilience has recently been brought into the public energy security debate, in particular in the United States, where an influential public figure commented in 2008:

> Our aim should not be total independence from foreign sources of petroleum. That is neither practical nor necessary in a world of interdependent economies. Instead, the objective should be developing a sufficient degree of resilience against disruptions in imports. Think of resilience as the ability to absorb a significant disruption, bigger than what could be managed by drawing down the strategic oil reserve (Grove, 2008).

Enhancing the ability to cope with short-term disruptions that do not alter the fundamental character of energy systems is often more prominent in national energy security strategies. For example, emergency fuel stocks, which are currently maintained by all developed and many developing countries, address the risks of unexpected short-term interruptions of oil supply or price volatility. Some European countries now have emergency storage of natural gas, which serves the same purpose and proved instrumental in dealing with the shutdown of Russian gas deliveries to Europe in 2006 and 2009. Russia is exploring constructing its own gas storage facilities to deal with potential interruptions of delivery, be it for economic, political, or technical reasons.

Another resilience strategy is increasing diversity. Diversity can ensure that an energy system is able to adjust to more systemic disruptions. It is important to distinguish between various types of "diversity" that can relate to individual elements or aspects of energy systems.

Many energy security strategies contain elements of diversification. These range from increasing the number of import/export routes, origins, or destinations, to increasing the number of actors in the energy sector or the number and types of energy facilities and primary energy sources. Some of the diversity strategies only address relatively limited threats. For example, diversification of import and export routes does not address global price volatility or resource scarcity. In order to deal with more systemic and long-term risks to energy security, more profound diversity strategies are needed. For example, many energy-exporting states are aware of their long-term vulnerability to price fluctuations, eventual resource depletion, and the resource curse. Some of them, most notably the Gulf States and Russia, proclaim strategic focus on diversification of their economies away from reliance on energy exports.

Diversity can also be fostered by certain types of institutions. Although the key rationale for introducing energy markets is economic efficiency, markets may also foster diversity. The existence of the global oil market, where the ability of any single actor to significantly disrupt the supply is limited, is an example of enhancing security through diversity. Naturally, markets (and international trade in general) may also have negative effects on resilience and sovereignty, as further discussed below.

5.3.1.3   Limitations of National Strategies

The majority of countries in the world pursues or at least declares energy security strategies that have both resilience and sovereignty dimensions. These strategies, however, have generally not been effective, which is best evidenced by the presentation of the dire situation of energy security in the world. The reasons for this ineffectiveness are manifold. First of all, the strategies may be internally inconsistent or otherwise poorly designed.

It has proven to be very difficult to strategically reconcile various aspects of energy security. Some measures to boost short-term energy security have had negative impacts in the long term. Certain resilience strategies have adversely affected sovereignty, and vice versa. For example, liberalized markets may have increased the diversity of supply options

but lead to increased price volatility and reduce the (real or perceived) control of critical resources by the government.

Similarly, some sovereignty strategies negatively affect resilience. Asserting control may increase confidence but also increase the objectively measured risks and vulnerabilities. For example, nationalization of oil and gas sectors has in some cases resulted in decreased government income from energy revenues due to increasing inefficiencies.[26] Squeezing foreign companies out of energy projects has also resulted in underinvestment and thus the deterioration of infrastructure, as national governments have not possesed the necessary capital, technology, and know-how.

The next reason for the mixed record of national sovereignty strategies lies in the fact that energy policies become too entangled with economic interests, foreign policy, and even conventional security imperatives. Foreign bids by Sinopec, PetroChina, and the CNOOC, for instance, tend to be accompanied by Chinese state aid projects; Gazprom's efforts to make the Ukraine pay "market prices" serves both Gazprom's economic and Russian state foreign policy interests. In addition, NOCs do not necessarily rely on financial markets when financing their exploration and production endeavors, and also enjoy a compelling lender of last resort: the state. Such "political" components of the NOCs' operation almost by definition compromise their energy security performance. What is worse, there is a dangerous trend of viewing energy security policies as an extension of conventional security policies, which leads to the discourse of geopolitics considered in more detail in the next section.

The other line of NOCs' evolution is perhaps more promising, even though it goes against the hope of asserting their parent states' control over energy resources. Observers note that maturing NOCs behave more and more like private companies, especially when operating in global or international markets. This is because they are subject to the same market rules and pressures. In particular, and ironically, some NOCs are affected by the same "resource nationalism" of host countries that resulted in their emergence in the first place. The case in point is Petrobras, a Brazilian semi-public NOC, having its assets nationalized in Bolivia. As any globally operating company, NOCs seek a stable and transparent regulatory environment, a level playing field, and well-protected property rights.[27]

The final reason for the lack of success may lie in the fundamental limitations to conceive and implement an energy security strategy by a single nation-state acting alone. It is quite obvious that small economies are rarely, if ever, able to implement energy system transformations on their own, since they simply do not possess the necessary financial,

---

26   A case in point here is Venezuela's PDVSA, which has experienced a strong downward trend both in output and overall efficiency after Chavez's re-nationalization of the energy sector in 2003.

27   That is why Russian Gazprom is, for example, domestically championing corporate social responsibility and other "good business" causes and the CNOC is reportedly considering joining the Extractive Industries Transparency Initiative (EITI).

technological, and human resources. For example, a study shows that launching a national nuclear power program relying on their own resources may be out of reach for at least 28 of the 52 countries that expressed an interest in nuclear energy based on their energy security imperatives (Jewell, 2011).

Even larger countries face serious limitations in ensuring their own energy security. This is not only due to lack of capacity, but also to the natural reflexivity of energy security. If states start acting alone or in closed groups, other states may perceive their actions as threats to their own energy security. In no time, energy security becomes a "zero-sum game" dominated by "geopolitical" and other discourses drawn from the vocabulary of diplomacy and military security. We turn to such developments in the next section.

5.3.2   Energy, Geopolitics, and Confrontation: the Specter of Energy Wars

A comprehensive understanding of energy security includes perceptions and perspectives of nation-states alongside "objective" indicators. A peculiar aspect of such perceptions is reflexivity: their ability to dynamically influence each other. If the position or actions of a country are perceived as a "threat" to energy security, other countries may start responding in such a way that is also perceived as threatening, causing another round of threatening responses, etc. The situation may escalate into a zero-sum game, with the energy security of certain nations only being achieved at the expense of other nations. Eventually, such developments may result in a lack of much-needed confidence, disrupted cooperation, increased tensions, or even conflicts over energy resources or infrastructure.

The risk of conflicts over energy resources is a significant concern on the global security agenda. The extreme form of such confrontation is a much-feared "resource war" – an inter-state armed conflict aimed to secure access to energy resources. In 1980, the US President Jimmy Carter proclaimed that the United States would use military force in the Persian Gulf region to defend its national interests, specifically "the free movement of Middle Eastern oil" (Carter, 1980). The fear of resource wars has considerably grown in recent years, particularly prompted by the rise of the "new consumers" of globally tradable energy resources (India and China), concerns over inadequate capacity to meet the increasing demand for oil, rising oil prices, and a series of high-profile disputes involving Russian gas supply to Europe. An influential geopolitical school of thought (e.g., Klare, 2008) points to numerous factors increasing the risk of such a confrontation in the near future.

Other researchers (Jaffe et al., 2008) note that there have been very few – if any – resource wars in the recent past[28] (save the Iraq-Kuwait

war of 1990) and downplay the risks of such conflicts in the future. The arguments about the future probability of oil conflicts are difficult to resolve. On the one hand, the risk of resource wars significantly depends upon non-energy factors such as the capacity of international and bilateral regimes and institutions; on the other hand, the configuration of global oil production, trade, and use are undoubtedly major factors determining such risks. Moreover, there are several forms of confrontations involving energy resources that are only marginally less worrisome than resource wars.

First, energy resources, particularly oil, have played an important role (sometimes as a weapon) in past inter-state confrontations, including armed conflicts. The "tanker war" linked to the Iran-Iraq conflict, the Arab oil embargo related to the Arab-Israeli war, and other modern examples are listed in Table 5.9. It may be argued that the presence of oil (or many other natural resources, for that matter) have made some armed conflicts more prolonged or destructive.

Second, energy resources and infrastructure have shaped inter-state relations, prompting either collaboration or confrontation or, more commonly, a mixture of both. The most prominent examples include:

- the dispute over borders in the potentially oil- and gas-rich Arctic (but also possible collaboration over the exploration of oil and gas in the Arctic); similar disputes have also been documented in other regions;[29]

- debates over gas and oil pipeline routes in Eurasia (such as the Nord Stream gas pipeline bringing Russian gas to Western Europe and the Baku-Tbilisi-Ceyhan (TBC) pipeline transporting Caspian oil to the Mediterranean); these were closely linked to several heated disputes over gas and oil supplies between Russia and former Soviet states in Eastern Europe (Belarus and Ukraine), as well as in the Caucasus;

- military and development support from major oil importers (notably the United States and China) to some oil-producing and transit countries in Africa (e.g., Sudan and Nigeria), the Gulf Region (e.g., Saudi Arabia), the Caucasus (Georgia and Azerbaijan), and Central Asia (e.g., Kazakhstan); which is often linked to:

- the struggle for influence in the remaining few oil-producing regions among major oil importers (the United States, Western Europe, China, and Japan).

Third, increasingly strained supplies of energy resources justify the growing deployment of military overseas to "protect" oil infrastructure (for example, terminals, tanker routes, etc.) against real and perceived threats and disruptions. The recent creation of the US Navy command

---

28 On the other hand, historically more wars can be linked to some resource issues. The very devastating Paraguay-Argentinean war (1932–1935) for Gran Chaco was for presumed oil reserves in that area (that never materialized). Many aspects of World War II (e.g., Japanese occupation of the Dutch East Indies or the German-Soviet battle over the Caucasus) also had an oil sub-text.

---

29 The Spratly Islands are claimed by China, Taiwan, and Vietnam; parts of them are claimed by Malaysia, the Philippines, and Brunei (Klare, 2008). The Falkland Islands (Malvinas) are claimed by the United Kingdom and Argentina (Luft, 2010).

Table 5.9 | Major inter-state conflicts and tensions related to oil and gas systems (since the end of World War II).

Year	Resource/system in question	Security event or measure
1950	US and other oil exports to China	The Western bloc's Coordinating Committee for Multilateral Export Controls placed China under an oil embargo during the Korean War of 1950.
1956	Saudi oil reserves/production	Saudi oil embargo against France and the United Kingdom following the Suez crisis
1967	Middle Eastern oil embargo	Imposed by Arab nations on the USA, the UK or in relation to all oil exports after the beginning of the Six-Day War.
1973–1974	Oil production/reserves of Arab oil exporting countries	OPEC and Arab oil embargo, generating the first "oil price shock"
1979	Oil exports of Iran	Iranian revolution
1980	Crude oil exports of Iran/Iraq	"Tanker War" between Iraq and Iran
1981	Algerian gas supply	"Gas Battle" between Algeria, Italy, the United States and others
1990–1991	Kuwait oil reserves	Iraq invasion of Kuwait eventually repelled by the United States and allies
2003	Russian crude oil delivery/pipeline infrastructure	Cut-off in Russian oil supplies to Latvia
2003	Iraq and Middle East oil reserves	US invasion of Iraq
2005	Pricing mechanism of Russian gas	Gas dispute/cut-off in Russian gas supplies to Georgia
2006	Pricing mechanism of Russian gas	Cut-off in Russian gas supplies to Ukraine and Moldova
2006	Russian crude oil delivery/pipeline infrastructure	Cut-off in Russian oil supplies to Lithuania
2007	Pricing mechanism of Russian oil deliveries to Ukraine	Russian interruption of the Druzhba oil pipeline
2009	Pricing mechanism of Russian gas	Cut-off in gas supplies to Western Europe, causing side unclear (Ukraine or Russia)

for Africa (AFRICOM) and Russia's plans to increase its military presence in the Baltic to protect the Nord Stream gas pipeline (as well as the Swedish response of increasing its own forces in the Baltic) are quoted as some examples. US national security and oil imports, as well as the implications of various energy issues for the US Navy, have been the subject of two recent RAND reports (see Box 5.9).

Not only do energy resources (particularly oil and gas) shape inter-state relations, but they may also affect internal security issues, especially in poorer countries that face nation-building challenges. Current inter-ethnic conflicts in Iraq – largely related to oil – are most prominent, but energy resources shape and color internal tensions in many other countries. Civil wars in Angola, Colombia, the Republic of Congo, Indonesia (Aceh), Morocco, and Sudan, as well as a simmering conflict in the Niger Delta in Nigeria (see Box 5.10), have also been linked to oil. Whether the political challenges of allocating revenues from oil and gas production will lead to instability and conflict depends on the quality of governance and institutions as well as key international actors.

Finally, there have been concerns about the connection between energy resources and international terrorism, particularly the argument that oil revenues help to fund terrorist activities. However, very few, if any, facts have been found to support this assertion except circumstantial evidence that certain terrorist organizations and individuals come from the Arab-speaking countries which also happen to have significant oil reserves. On the contrary, many researchers point out that terrorism is a "low-cost activity" that does not depend on oil revenues. Moreover, the emergence of contemporary terrorist networks (such as al-Qaeda) occurred in the mid-1990s when oil revenues were at their lowest. On the other hand, the resource curse – i.e., economic and political problems in many poor oil-exporting countries, described earlier – has contributed to dissatisfaction and disenfranchisement of individuals and social groups that eventually support or engage in terrorist activities.[30]

Box 5.9 | Energy and US Defense Costs

Due to heavy dependency of the United States on imported oil, access to foreign oil remains a top priority driving the country's strategy and defense policy. US military forces have been present in the Persian Gulf since the 1970s to protect access to Middle Eastern oil. US efforts to ensure secure access to foreign oil also include, since the 1990s, deepened ties (economic, political, and military) with oil-producing states in Central Asia, South America, and West Africa.

The presence of US military forces to maintain the security of international oil flows for the global market undoubtedly incurs substantial costs. However, to date, there has been no comprehensive, accurate, publicly available US government study of the costs. Nevertheless, several attempts have been made to quantify them. In general, the analyses addressed the costs incurred by US Central Command (USCENTCOM)[31] in its mission to protect the maritime transit of oil supplies in the Persian Gulf and Indian Ocean, as well as to assist in

---

30   This hypothesis explains why terrorism often emerges during low rather than high oil prices when the level of popular dissatisfaction may be highest in those economies that primarily rely on oil revenues for their welfare programs.

31   USCENTCOM is a military force created in 1979 to be available for worldwide contingencies. However, its focus quickly tilted heavily toward the Persian Gulf region.

the defense of United States-friendly oil-producing governments. Estimates in these studies have varied from as low as US_{2005}\$12 billion up to as high as US_{2005}\$130 billion of military spending per year, representing, respectively, 2.5% and 27% of the defense budget in 2006. This range in estimates reflects the complexity of how US forces are planned and operated and, thus, the difficulty of being very specific in allocating precise costs to this mission.

Recently, two RAND research teams estimated the incremental costs that the US government would likely avoid if it were to entirely drop the mission of ensuring the secure production and transit of oil from the Persian Gulf for the global market. Estimates of potential annual savings amounted to between US\$67.5 billion and US\$83 billion – respectively, 12% and 15% of the US defense budget. However, these estimates do not claim to be precise. Moreover, RAND's analysis does not argue that a partial reduction of the US dependency on imported oil would automatically lead to a proportional reduction in US spending that is focused on this mission (Crane et al., 2009). Another RAND study predicts that the defense budget is likely to decrease due to future increased social spending for the United States' growing numbers of elderly citizens, which, in turn, will likely affect the incremental amount available for protecting oil supply and transit (Gordon et al., 2008).

Two other academic assessments produce similarly startling figures. Researchers at the Oak Ridge National Laboratory in the United States estimated that from 1970 to 2004 American dependency on foreign supplies of oil has cost the country US\$5.6–14.6 trillion in macroeconomic shocks and unnecessary transfers of wealth.[32] When the numbers are adjusted to 2007 dollars, the amount is greater than the costs of all wars fought by the country going back to the Revolutionary War, including both invasions of Iraq. Researchers from the University of California-Davis and University of Alaska-Anchorage calculated that US defense expenditures exclusively to protect oil in the Persian Gulf amount to about US_{2005}\$28 billion to US_{2005}\$75 billion per year (Delucchi and Murphy, 2008).

Box 5.10 | Resource Wealth, Civil Conflict and Disruption of Oil Supplied in Nigeria

The complex interplay between resource wealth, political and inter-ethnic conflict, and its impact on production facilities, infrastructure, and output is evident in the case of political conflict in the energy-rich Niger Delta. Nigeria is one of the world's top 10 oil exporters and produced around 3% of the world's total output in 2006 (BP, 2007). Yet traditional tension between different ethnic groups in the Niger Delta, one of the most densely populated areas in the world, turned into violent conflict as oil revenue started to pour into the country. The behavior of large oil corporations and poor capacities of the local government is largely regarded as having contributed to raising the degree of violence and to prolonging conflict. Violence has become a major cause of slowing production and causing interruptions in Nigerian crude deliveries. Attacking installations, kidnapping oil corporations' personnel, and siphoning off crude from pipelines have become common features. In 2006, militant attacks on oil facilities resulted in a shutdown of almost 500,000 barrels per day, or 20% of Nigeria's oil output. In addition, oil tankers have been stolen, with the crude being sold on foreign markets (Vesely, 2004).

5.3.3   Energy Alliances, Institutions, and Markets

Despite the rhetoric of geopolitics, the zero-sum game, and the resource wars, the actual international interaction in the field of energy security has so far been largely dominated not by conflict but by cooperation, albeit imperfect. This section examines the existing cooperation mechanisms as well as their successes and shortcomings.

At the moment there is no overarching global energy institution, but rather a plethora of alliances, multi- and bilateral deals, and arrangements. The most significant international energy alliances unite major exporters and importers of energy (primarily oil). First and foremost, these include OPEC, established in 1960 in Baghdad, with the aim of regulating global oil production. OPEC seeks to influence oil prices by adjusting production levels through the use of a quota system. At present, OPEC member states control around 80% of global oil reserves and almost half of global production. Gas-producing countries established the Gas Exporting Countries Forum (GECF) in 2001, which remains a rather loose gas club that might – depending on the development of the take-or-pay-dominated market – potentially become a future gas cartel similar to OPEC.

32 Source: Greene and Ahmad, 2005. Numbers have been adjusted to US_{2007}\$.

The International Energy Agency (IEA) is a watchdog for energy-importing members of the OECD. It has developed rules concerning strategic petroleum reserves and a supply shock emergency response mechanism. The IEA can draw on the International Energy Program (IEP, established in 1974) for larger supply interruptions, and on the coordinated emergency response mechanism (CERM), which applies for smaller emergencies.

Another type of energy alliance is formed on the basis of regional proximity. One example is the EU, whose prototype, the European Community for Coal and Steel, was created to govern access to coal resources, the then-dominant energy source. The EU aims to operate a single energy market. For that purpose, it fosters an integrated energy infrastructure (such as in electricity and natural gas) and liberalized cross-border trade of energy services governed by common rules and policy agenda. The EU has also sought to develop a common energy policy towards third parties, encompassing energy security and other energy-related goals as well as environmental and climate-related topics. Vis-à-vis main producers, the EU has initiated steps towards linking up with neighboring regions via energy partnerships within the realm of various agreements, including the Partnership and Cooperation Agreement with Russia, or via efforts targeting the Maghreb/Mashrek region. Project proposals also exist for the large scale import of renewably generated electricity from northern Africa (Komendantova et al., 2009).

Another regional club is the Shanghai Cooperation Organization (SCO) uniting Russia, Central Asian countries (all net energy exporters), and China (an energy importer). Whether the energy role of the SCO will extend beyond declarations is so far not clear. Regional "clubs" with energy agendas also exist in other regions. For example, in West Africa, common arrangements feature the West African Power Pool, West Africa Gas Pipeline projects, and the West Africa Regional Energy Access initiative, all designed to reduce energy insecurity in the region. Similar arrangements are being tried with the East African Power Pool and the Southern Africa Power Pool.

Some energy alliances have also been established based on nations' economic, rather than regional, characteristics. For example, the historically prominent G7, created in the aftermath of the first oil shock in 1973, gathered the largest economies of that time (Germany, Italy, Japan, the United Kingdom, the United States, France, and Canada) to coordinate its members' energy-related and macroeconomic policies. The G7 did not noticeably address energy issues during the almost two-decade-long low-price period on the oil market from the early 1980s to the early 2000s. It was joined by Russia in 2004, becoming the G8, and has recently sought to include other large economies including major consumers and producers of energy (China, India, Brazil, South Africa, and Mexico) to become the G20. However, the G8 and G20's energy activities so far have been limited to declarations rather than to establishing permanent effective institutions.

The relationships between various different alliances are sometimes portrayed as competition or confrontation.[33] In our view, a more accurate description would be that of "non-engagement." As already noted, there are very few global energy rules or institutions extending beyond specific alliances. The International Energy Forum (IEF), an organization that seeks to unite energy exporters and importers, has not yet resulted in any tangible institutions or arrangements. Perhaps the only widely applicable organizing principle is that of a free market for certain energy products, most importantly oil.

Establishing a global oil market was facilitated by the advances in transportation and communication technologies, as well as the end of the Cold War. As a result, many security concerns of the 1970s and the earlier era have subsided. The producer countries are no longer at the mercy of international oil corporations dictating their conditions (see also Box 5.11 on IOCs and NOCs). The consumer countries can seamlessly switch and mix suppliers and not worry about excessive dependency on a single supplier. The global oil trade system can assimilate minor shocks of disruptions in particular countries or regions.

However, the free market has its limitations. For example, it does not function well under conditions of imperfect information or monopoly, which often arise naturally in the case of grid based carriers (compare to Chapter 22). Also the global oil market has increasingly developed some of these features. Information about reserves and production capacities has been severely impeded by an explicit non-reporting policy of OPEC and other exporters. Moreover, due to the increasing geographic concentration of oil reserves and the lack of suitable substitutes, the global supply develops the characteristics of a monopoly. In addition, as discussed above in relation to national energy security, markets tend to under-provide such public goods as sufficient investment in production capacity and infrastructure.

This last feature has been empirically observed in the case of global oil markets. A prime example includes the current global shift from a "strategically planned," vertically integrated approach in the energy industry, practiced even in private companies, towards decision-making based on immediate economic objectives such as return on capital. This shift, partially encouraged by the financial investment community and seen especially in the low oil price environment of the 1990s, led to a decrease in spare capacity in production, storage, and transportation, thus increasing the effects of even small disruptions. Markets in their current shape tend not to reward players for maintaining spare capacity and do not have adequate mechanisms for charging for capacity on a pay-as-you-go basis, which tends to enhance booms and busts. Price hikes effectively allow some of those market players who did invest for extraordinary circumstances to obtain a return on their investments, albeit in a one-off fashion, as opposed to a constant stream of income.

---

33 For example, the SCO is sometimes considered a "geopolitical bloc" aimed at undermining the US presence in Central Asia. In the eyes of some pundits, such new geopolitical constellations will be increasingly framed by energy issues and will compete with each other (e.g., the SCO competing with the United States and Japan in the Western Pacific). See, among others, Klare, 2008.

To buffer these shortcomings, institutional arrangements need to be strengthened to provide for a greater degree of information, notably data collection and exchange. Here, the IEA's data-generating activities (though notoriously criticized on analytical fronts and short on information on producing countries) and institutions of producer-consumer cooperation such as the IEF in Riyadh become essential. The IEF's Joint Oil Data Initiative (JODI) is a particularly promising step.

Finally, there are vocal concerns that market arrangements might crumble in the face of rapidly unfolding scarcity or another severe energy security crisis. According to Klare (2008):

> Oil will cease to be primarily a trade commodity, to be bought and sold on the international market, becoming instead the preeminent strategic resource on the planet, whose acquisition, production, and distribution will increasingly absorb the time, effort, and focus of senior government and military officials.

While it is difficult to judge the validity of such concerns, history definitely provides many examples of when market arrangements were either heavily modified or entirely replaced by other rules in times of severe crisis. This leads many observers to believe that markets should not be viewed as the only or the most effective mechanism for providing energy security.

To summarize this brief overview of the most notable international institutions, the world governance of energy security has been ineffective for the following two reasons:

First, there has been little success in creating institutions that would serve and include key actors and stakeholders. There is no effective organization that would unite exporters and importers of energy. Some of the largest consumers, China and India, are not part of the "importers club," the IEA. The majority of countries are simply "too small" to qualify for various memberships, although their energy security concerns are no less significant than those of larger countries.

Second, the existing energy security institutions largely focus on oil and partially on naturally gas. This is understandable from the point of view of the present concerns, where these resources are at the front and center of energy security. However, this also shapes the expertise and the frame of reference of the existing institutions and largely predetermines their inability to govern seriously systemic energy transitions involving various supply, infrastructure, and demand elements.

Third, the international arena governing energy security should be more strongly connected to the international arenas governing climate change and supporting the provision of access to modern energy. These three arenas have historically developed in isolation from each other, but it is no longer possible for them to operate independently in the world where energy challenges are increasingly entangled (Cherp et al., 2011).

Potential pathways for the future evolution of global energy security institutions are touched upon in Section 5.3.

Box 5.11 | IOCs, NOCs and the Global Oil Market

The origins of the current global oil market date back to the 1930s. At that point, the industry became dominated by a small number of transnational corporations, controlling most of the sales of oil products. Most oil-rich nations lacked the domestic capacity to develop their reserves and were thus dependent on these large, integrated oil companies, also since the latter controlled the downstream assets in the main Western markets. The notions of the "Seven Sisters" or "Big Oil" still reflect the unprecedented power of the Western-dominated international oil companies (IOCs) in that era.[34] US policies supporting the "one base" or "Gulf plus freight" formula (later revised to the "two base" oil pricing formula) helped maintain a single price for all oil consumers outside the United States and secured enormous rents for this cartel. Until the 1960s, the United States was a major producer and a net exporter of oil. Thereafter, the decline of US mainland production, coinciding with the process of decolonialization and a growth in nationalism in producing countries shifted the balance of (market) power to the countries actually owning the resources. OPEC, initially established to enable producing countries to enhance their share in oil revenues vis-à-vis the dominating Western IOCs, is a result of this process, and profoundly changed the global oil market by introducing production quotas, a tool to influence oil prices. In response to OPEC's cartelization efforts in the 1970s, industrialized countries established the IEA, a consumers' club intended to strengthen the market power of importing nations; in addition, they pushed for energy efficiency improvements and the development of advanced offshore production techniques in the 1970s and the 1980s, leading to the opening up of new hydrocarbon provinces. As a result, the global oil market became more liquid and highly integrated, reducing the power of OPEC to set prices. OPEC defended the high price level established in 1980 by cutting production

---

34 The "Seven Sisters" consisted of Exxon (or Esso, Humble, Standard of NJ), Royal Dutch Shell, BP (originally British Petroleum, Burmah Oil + Anglo-Iranian), Gulf, Texaco, Mobil (Standard of NY or Socony-Vacuum), and Chevron (Standard of California).

every year until 1986, the long-term trends of world oil demand and non-OPEC supply eventually resulting in longer prices. However, in the new millennium, due to economic globalization and a steep rise in demand from "emerging consumers" (e.g., India and China), the pendulum has again swung back. In addition, financial market actors and mechanisms have entered the global oil market, which is now characterized by a variety of hedging instruments, swap and future contracts.

As this brief historical survey suggests, IOCs – i.e., private players – dominated the oil market for a long time. Historically, many of them (BP and the predecessors of Total, ENI) were fully or majority-owned by Western governments, and even private companies enjoyed a high level of support from their home states. In that respect, their role was somewhat comparable to the national oil companies (NOCs) from emerging importing nations such as India and China: a secure energy supply for their home nations abroad. With a more integrated global oil market, these companies have become fully fledged profit-driven enterprises. IOCs see their core competence in the ability to mobilize the large amounts of capital, equipment, and manpower required to develop and manage large projects; to use their technical expertise and know-how to reach even increasingly difficult reserves; and to live with price and resource market uncertainties and to manage political risks even in "frontier" environments through large investment portfolios. IOCs usually received a share in the upstream (equity oil) of the projects in which they were involved. More recently, and particularly against the backdrop of rising resource nationalism and market pressure to reduce costs, a second set of private players including integrated service providers (ISPs) such as Halliburton and Schlumberger have entered the scene. These companies bring in their specialized expertise and act as contractors to both IOCs and NOCs. In contrast to IOCs, they usually do not engage in exploration and production projects with their own capital.

5.4   The Future of Energy Security

This section considers possible future developments in energy security. It discusses the implications of projected or likely demographic, economic, natural resource, technological, and institutional developments. It also lays the foundation for a quantitative analysis of energy security under sustainable energy transition pathways modeled in GEA. This analysis is presented in Chapter 17.

5.4.1   Technology and Resource Developments

The role of oil on the global energy security scene will likely become even more important in the short- to medium-term future. Oil production will probably become more geographically concentrated, and demand for oil will continue to increase, primarily in Asia. Several present-day exporters (for example, the United Kingdom and Argentina) will likely become importers, and many countries will need to import more in both absolute and relative terms. The *World Energy Outlook 2010* (IEA, 2010a) predicts that both supply and demand of oil will become less responsive to prices, which will likely lead to long-term price increases for this fuel. At one point in the future, perhaps as soon as in one or two decades, the global production of conventional oil will likely "peak" or "plateau."

Many concerns have been expressed in connection with this imminent "peak oil." Some predictions paint a collapse of organized oil markets with catastrophic economic, and possibly political and military, consequences (Korowicz, 2010). It seems more likely that peak oil will have many more localized effects on those countries (mostly in the developing world) that lack the capacity to cope with steep increases in energy prices. Whether the international community will be able to mitigate the shock for these and other countries depends not only on other technological and resource developments but also upon the presence, focus, and effectiveness of global energy governance institutions in the future.

The production of natural gas is also likely to become more concentrated, but it is unlikely to peak in the nearest decades. The main developments affecting the situation with natural gas will most likely be advances in transportation technologies and infrastructure. A much more extended network of gas pipelines is likely to emerge in Eurasia, linking Russian and Central Asian producers not only to Europe but also to China and the rest of Asia. The development of LNG infrastructure will lead to the sharp increase in intercontinental trade. Growing LNG trade, while contributing to the emergence of a truly global natural gas market, may imply similar supply security patterns as the global oil market, including an increasing dependency on less reliable producer countries, transportation choke-points, and global supply-demand balances. Technological developments affecting the accessibility of shale and other unconventional gas may also alter the global gas market landscape, although the exact extent of this much-heralded "shale gas revolution" are difficult to predict.

A likely consequence of the dynamics of global oil and natural gas production will be a shift away from these energy sources, which is already vigorously pursued by many countries. It is unlikely that a new globally dominant source of energy, such as oil, will be found. Instead, there will be a shift to diverse sources, appropriate to national and regional contexts. One such source may be coal, especially if technologically and economically acceptable ways of low-carbon utilization of coal are found.

Nuclear energy will clearly be another alternative to fossil fuels considered by many countries. While the global scale of future nuclear expansion is not clear, it is likely that nuclear energy will be able to address energy security challenges only in a relatively limited number of countries. At the moment, the lion's share of the world's nuclear capacity is concentrated in the United States, whereas almost all short-term growth occurs in China, other Asian economies, France, and Russia. For most of the other countries with existing nuclear energy programs, such programs may become a liability rather than a solution to their energy security challenges.

There are over 50 "newcomer countries" that are interested in launching nuclear energy programs for the first time, by and large, to meet their energy security needs. According to Jewell (2011), safe deployment of nuclear power is likely in only 10 such countries under present conditions. Others will need very significant international help, including possibly forming energy partnerships among themselves, or dramatically altered nuclear energy markets and policies (see Figure 5.7).

Finally, oil, natural gas, coal, and nuclear power may be partially displaced by renewable energy sources. One such source is biomass. If the future biofuels or biomass systems assume production and transportation patterns similar to that of present-day fossil fuels (i.e., concentrated production

regions away from consuming regions and large centralized refineries), they may also become subject to similar risks and vulnerabilities. Furthermore, the availability of biomass feedstocks may be disrupted by climate change as well as policies guided by conflicting uses (e.g., food production or ecosystems preservation). Given the emerging state of the present-day biofuels systems, the scale of these risks is difficult to estimate.

Replacing with other renewables (such as wind and solar) will require technological innovation and creation of new infrastructure, including (as penetrations rise significantly) new systems for the storage and distribution of energy and for possible new propulsion systems for vehicles (e.g., electricity or hydrogen). There are possibilities that the storage requirements associated with large-scale electricity system transitions towards distributed renewable resources may be offset by synergies between parallel emerging smart management procedures for distributed energy storage in electric vehicles. Relatively simple co-ordination systems enabling the remote control of certain consumer products also offer an important resource in cost-effective management of intermittency and other electricity security challenges. Specific technologies for carbon capture at centralized fossil fuel plants may also present opportunities for improving capacity to manage intermittency. These developments introduce a potentially significant level of effective aggregate electricity system storage capacity as a side effect. However, the task of

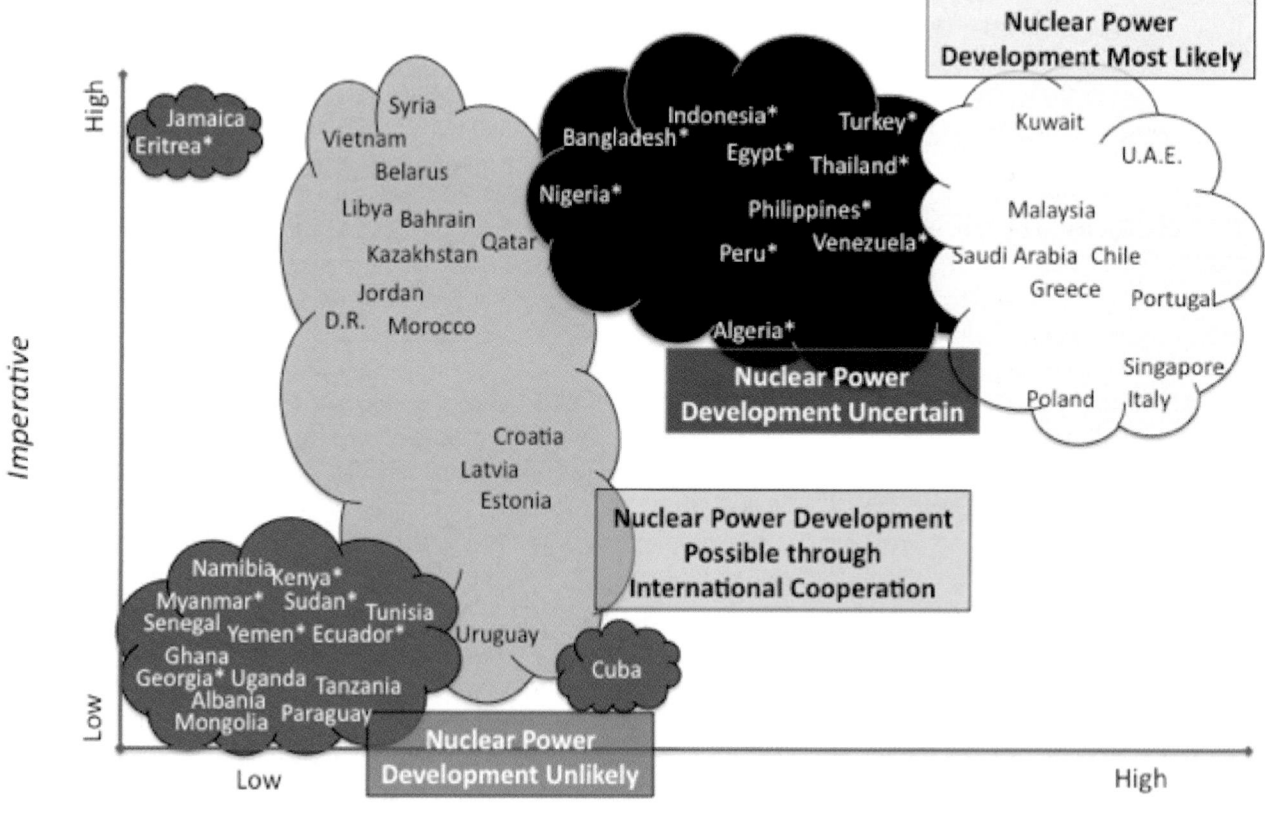

Figure 5.7 | Feasibility and uncertainty of nuclear power adoption by new countries. Source: Jewell, 2011.

Table 5.10 | Factors in future energy scenarios related to national energy security strategies and international institutions.

Focus of national strategies International institutions	Sovereignty	Robustness and Resilience
Less effective	Fragile markets and geopolitics; national state-driven transitions to centralized solutions	Self-organizing markets; decentralized and diverse transitions
More effective	Strong rules and powerful alliances; regulated markets	More uniform transitions at several levels

realizing the positive synergies of these technological innovations raises many challenges for associated infrastructural, organizational, contractual, and regulatory innovation. Wind, solar, and hydro energy are more likely to be produced domestically and thus not invoke traditional sovereignty concerns. However, there are also plans (such as Desertec) to generate solar energy for intercontinental trade.

Another imminent development affecting energy security in the future is the increasing role of electricity in energy systems. Electricity will play a more important role due to the likely advent of plug-in electric propulsion vehicles and the continuing spread of ICTs and other consumer technologies requiring electricity, as well as to the increasing use of electricity by the rising middle class in emerging economies.

An increasing reliance on electricity will mean that reliability of its production and distribution is likely to come to the forefront of energy security concerns in the future. The complexity of future electricity systems is likely to increase to incorporate:

- technologies for storage of electricity;

- "smart grids" including "active load" devices;

- "super grids" for transferring large quantities of electricity over long distances with minimal losses (e.g., through high-voltage direct current lines) when localized distributed systems are not feasible or sufficient; and

- hybrid systems,[35] to increase reliability of power generation and distributed generation such as modular small-scale systems with improved or increased energy storage capacity.

Some of these approaches may reduce the risk of "cascading failures" inherent in modern complex centralized grids. It is possible that other approaches to increasing reliability will evolve as a result of a combination of electric and information technologies. As the role of electricity in energy systems increases, the factors affecting energy security will have increasingly more to do with institutional structures and capacities than with more traditional issues of access to natural resources.

5.4.2   Institutional Uncertainties in Future Energy Security

Energy security perspectives and strategies are both a key driving force of and a central strategic uncertainty in future energy transformations. Security has been a main driver of past technological and political transformations, and it is also a prominent consideration in most, if not all, of the global long-term scenarios. It is not certain which of the perspectives on energy security will prevail in national energy agendas and discourses. Another, separate uncertainty is whether new international energy institutions will emerge and be able to function effectively (see Table 5.10).

If the resilience and robustness perspectives gain more prominence at the national level, countries are likely to promote diverse energy solutions. The focus of national policies may become decentralized and distributed energy generation, improvement of electricity grids, and demand-side measures such as energy efficiency. Regulated but generally liberalized markets may become more common national energy arrangements. Many such measures will also improve sovereignty – i.e., reduce import dependency – although nations driven by these two perspectives may remain open to international trade and investments.

An additional uncertainty in this case may arise at the international level. If effective international energy institutions are created, they will be able to help countries with technology transfer, ensuring investments and well-functioning markets to enhance energy resilience. These global energy institutions may be able to oversee new patterns of intensive international energy trade (e.g., in electricity, hydrogen, or biofuels). This will be a more favorable scenario.

It will be more difficult for most countries to pursue resilience and robustness strategies in the absence of effective international institutions. The pace of transformation to diverse and more robust energy systems may be significantly slowed down, especially in smaller and poorer economies. More disparity in energy security may emerge as a result. In fact, the absence of functioning international institutions may push many countries towards the sovereignty perspective.

If the sovereignty perspective in national energy security policies prevails, countries may view international trade, foreign capital, and even domestic private actors with a degree of suspicion. Energy solutions are likely to be less diverse, more centralized, and relying on state support (e.g., nuclear power, or coal with carbon capture and storage). If international energy institutions continue to be weak, the global energy markets may start failing, especially in the case of increasingly scarce "strategic"

---

35   For example, hybrid systems using wind or solar photovoltaics to provide emergency back-up to conventionally distributed electricity.

commodities such as oil and natural gas. Geopolitics may become the rule of the day, with stronger and larger countries gaining better "access" to scarce resources. This may, in turn, push national governments towards surprisingly rapid transformations of energy systems. There is naturally no guarantee that such transformations will always be safe or successful.

Finally, one can imagine strong international institutions in the world where most countries pursue sovereignty strategies. The role of such institutions will be more limited than in the resilience scenario, but they will still be able to set the rules of the game (such as market institutions) to prevent geopolitical manipulations from exploding into open confrontation. Regionally or ideologically based "energy clubs" based on "mutual trust" may also become increasingly prominent.

5.4.3 Energy Security under Sustainable Energy Transitions

The GEA argues that energy systems can support such global goals as poverty reduction and the stability of Earth's climate only if they are radically and rapidly transformed. Chapter 17 (Energy Pathways for Sustainable Development) defines multiple pathways for such sustainable energy transitions.

The question is: what are the energy security implications of these pathways? Due to the radical transformations of energy systems, the energy security concerns identified in this chapter cannot be simply projected into the future. While some of these concerns may disappear or become irrelevant, new vulnerabilities may emerge. For example, oil, an energy source that is at the heart and center of most energy security concerns today, may be phased out and replaced by other fuels. But will using these new fuels on the global scale result in similar energy security concerns? Will they be produced in just a few countries and regions that will set terms for the global market? Will they dominate end-use sectors and lack easily available substitutes so that any disruption may be catastrophic? Will poorer countries and regions spend a significant portion of their revenues to procure these new fuels?

This section summarizes generic energy security considerations that form the basis of assessing energy security under sustainable energy transitions. We presume that, although energy security concerns might change in the future, the perspectives on energy security will remain the same. This means that nations will still be concerned about the robustness (i.e., the protection from certain known risks), the sovereignty (i.e., the protection from actions by external actors), and the resilience (i.e., the ability to withstand various disruptions, both known and unknown) of their energy systems.

Applying these three perspectives to energy sources, carriers, end-use services, and regions[36] of the future will answer the following questions:

---

36 The GEA model cannot forecast energy developments in individual nations, and the analysis is, therefore, concentrated on the regional level. The caveats of such an analysis of energy security are explained in the relevant section of Chapter 17.

- With respect to *energy sources*: what will be the energy sources of the future? Will any of them dominate the global energy supply to the same degree that oil, gas, and coal dominate the present-day energy landscape? Will they be based on non-renewable and hence limited resources or utilize renewable energy? Which of these sources will be traded on the global scale? Will their global trade be equally as intensive as it is currently for oil? Will their production be concentrated in only a few regions or be more evenly spread around the world?

- With respect to *energy carriers* and *end-use sectors*: will they be based on diverse energy sources or dominated by one fuel, like the transport sector today? Will they primarily rely on imported or domestic energy sources? Will they experience very rapid and destabilizing growth or decline during any periods of the transition?

- With respect to *countries and regions*: will any of them be significantly dependent on imported energy? Will any of them rely on just a small number of energy sources? Will energy intensity and hence vulnerability to changing energy prices increase or decline?

The above questions are answered in Chapter 17 with the help of a quantitative projection of selected energy security indicators. The overall conclusion of this analysis is that in most pathways of the GEA Scenario, energy systems become more secure from all perspectives and in most of the regions. At the same time, certain vulnerabilities may emerge in particular regions or globally, and these need to be taken into account and mitigated while managing sustainable energy transitions.

5.5 Conclusions

Adequate protection from disruptions of vital energy systems – "energy security" – is one of the most politically prominent energy-related concerns. Disruptions of energy systems may result from both short-term shocks, such as natural events, technical failures, malfunctioning markets, or deliberate sabotage, and slowly unfolding but more permanent threats, such as resource scarcity, aging of infrastructure, and unsustainable demand growth. Such disruptions may affect broader security issues ranging from the viability of national economies and stability of political systems to the danger of armed conflicts. This means that policies developed in the quest for higher energy security have been, and are likely to remain, a key driving force in the transformations of energy systems.

Although energy security concerns differ from one country to another, they typically relate to the robustness, sovereignty, and resilience of energy systems. Robustness means minimizing risks arising from well-defined natural, technical, and economic factors. It is associated with sufficient energy resources, reliable infrastructure, energy efficiency, and managed demand. Sovereignty means the protection from disruptions to energy systems by external agents. Reliance on domestic resources and technologies, stable prices, control over infrastructure, and trusted institutions are typically associated with sovereignty. Resilience means the ability of energy systems to withstand diverse and uncertain threats. Resilience is linked to diversity of supply and infrastructure options, redundancies and spare capacities, and institutions capable of adequately adjusting to disruptions, as well as flexibility in demand.

This chapter examines robustness, sovereignty, and resilience concerns with respect to energy sources, electricity systems, and the primary end-uses of energy in over 130 countries. Its main conclusion is that significant energy security concerns from at least one of these perspectives affect the absolute majority of countries and are likely to become more serious in the short- to medium-term if the present trends are allowed to continue. Whereas the primary energy security concern of most industrialized countries is import dependency and aging infrastructure, many emerging economies have additional vulnerabilities such as insufficient capacity, high energy intensity, and rapid demand growth. In many low-income countries supply and demand vulnerabilities overlap, making them especially insecure.

Most globally prominent energy security concerns relate to the oil sector. Oil is a significant source of energy in almost every nation, whereas the majority of countries must import most or even all of the oil and petroleum products they need. Over three billion people live in 83 countries that import more than 75% of the oil and petroleum products they consume. This number does not yet include China, with its 1.3 billion people, where oil consumption has been rising on average by 7% per year over the last decade, and the current oil import dependency of 53% is projected to reach 84% by 2035 (IEA, 2010a). Virtually all of the world's low-income countries import over 80% of their oil supplies. Those that do not have another type of dependency: on oil export revenues to sustain their economies.

Furthermore, the recent rise in the global demand for oil has not been matched by an increase in the supply capacity, which – together with concerns over eventual scarcity – has made markets more volatile. A longer-term issue, the much-discussed "peak oil" – a rapid forced decline in production and consumption – also cannot be dismissed, although it is likely to be initially experienced as painful disruptions of vital energy services in low-income countries rather than as a global energy crisis. Peak oil is covered in more detail in Chapter 7 (Energy Resources and Potentials).

Oil is the main, but not the only, energy source causing widespread energy security concerns. Natural gas accounts for over 10% of primary energy supply in 78 countries with a combined population of some 2.9 billion people. Among those, almost 650 million people live in 32 Eurasian countries that import over 75% of their gas needs. In addition, 12 countries with some 780 million people have a domestic reserves/consumption ratio of natural gas under 16 years and are thus likely to experience significant import dependency in the future.

For many countries, coal is a potential solution to the energy sovereignty problem, since its resources are more abundant and more evenly geographically distributed. Only a small number of countries (12) currently significantly depend on coal imports, and this number is not likely to significantly increase in the near future. However, the use of coal may be subject to environmental and health constraints, as discussed in other parts of this publication.

These vulnerabilities of energy sources affect the security of national electricity systems. For example, almost 600 million people live in 39 countries where over 50% of electricity production is based on imported fossil fuels. In many developing and transition countries these sovereignty concerns are aggravated by low robustness and resilience: inadequate generation capacity, as well as low diversity of electricity generation options. For example, in 35 countries with a population of 450 million people, the absolute majority of electricity production comes from just one or two major power plants. In addition, both developing and industrialized countries suffer from disruptions of electricity systems due to natural and technical failures exacerbated by increasing complexity, poor maintenance, aging infrastructure, and insufficient investment.

In addition to fossil fuels, many countries rely on nuclear power for significant parts of their electricity supply. The sovereignty aspects of nuclear power are access to enrichment, reactor manufacturing, and reprocessing technologies and capacities, which are currently concentrated in just a few countries. For example, large-scale uranium enrichment plants exist in only six countries, and commercial reprocessing facilities in only five. One of the main problems with currently operating nuclear power programs is the aging of reactors and workforce. There are 21 countries with 1.3 billion people with existing nuclear power programs that have not started constructing a new reactor in the last 20 years, and 19 countries (1.4 billion people) have nuclear power plants with an average age of over 25 years. The nuclear power programs in these countries would need to be either "re-launched" or phased out. Access to capital and technology will be critically important for nuclear power expansion and renewal.

An additional 52 countries have expressed their intention to start nuclear power programs for the first time. However, only about one-quarter of them are likely to be able to securely deploy nuclear power with their own resources. In the remaining "newcomers," access to capital and creating necessary institutional arrangements for secure deployment of nuclear power will present serious challenges. For the same reasons, nuclear power will remain beyond the reach of the majority of less-developed countries.

Another significant source of electricity supply in many countries is hydropower. In many regions of the world, notably in Southern Europe, Africa, and South Asia, its long-term security may be affected by shifting patterns of water availability due to climate change. Many existing hydroelectric plants are located on internationally shared rivers, with divergent water use interests of the riparian states in some cases threatening the secure supply of hydroelectricity. Finally, over 700 million people live in 31 countries that derive a significant proportion of their electricity from just one or two major dams and thus are vulnerable to a variety of natural and technical factors affecting these dams.

Reliability of electricity systems is a source of concern for both industrialized and developing countries. The economies of the former are increasingly sensitive to even minor disruptions of electricity supply, so even relatively short blackouts (typically not exceeding a few hours per year) translate into major economic costs. The scale and frequency of blackouts in most developing countries is at least one to two orders of magnitude higher. Companies in the majority of low-income countries experience 10 or more blackouts averaging 24 hours or more every month, and in some cases the number of blackouts exceeds 100 a month and their total duration approaches 100 hours.

In addition to these supply-side vulnerabilities, electricity systems in developing countries are also under strong demand-side pressures. The majority of the world's population – some 4.2 billion people – live in 53 countries that will need to massively expand the capacity of their electricity systems in the near future because they either have less than 60% access to electricity or average annual demand growth of over 6% over the last decade. Both fuels and infrastructure for such an expansion will need to be provided without further compromising the sovereignty or resilience of national electricity systems.

Insecurities of energy sources and electricity systems affect vulnerabilities of vital energy services: transport, industry, the residential and commercial sector, and providing energy for export.

Insecurities of oil supply affect first and foremost the transport sector, where petroleum products provide at least 90% of energy in the majority of countries. Almost five billion people live in countries that import over 50% of fuels for their transport sector. The situation could get especially serious in 17 countries (with 1.7 billion people) where transport energy use has grown on average more than 8% per year over the last decade.

In the industrial sector, 46 countries with some 800 million people, and in the residential and commercial sector, 39 countries with some 500 million people rely on imported fuels for over 50% of energy use. This does not include 25 countries with some 700 million people that use traditional biomass for over 80% of their residential energy use.

Some 15–20 national economies significantly rely on revenues from energy (primarily oil) exports. In the majority of these oil-exporting countries the revenues are not expected to last for more than one generation, and in several cases they are likely to cease in less than a decade. In addition, poor energy-exporting nations are at a high risk of the resource curse: economic and political instability eventually affecting human development and security.

National energy security strategies throughout the world seek to address those of the above-listed issues that are most prominent in a particular national context. Such strategies generally seek to increase the robustness, sovereignty, or resilience of national energy systems. With respect to robustness, the main strategies are to switch to more abundant and affordable energy sources, stimulate investments in infrastructure, and manage energy demand. With respect to resilience, many states maintain emergency stocks of critical fuels and seek to increase reliability of energy infrastructure by securing necessary investment. Energy exporters seek to achieve resilience by establishing Sovereign Welfare Funds and promoting diversification of their economies. One generic resilience strategy is to increase the diversity of various elements in energy systems: import or export transportation routes, production options and facilities, primary energy sources, or actors on the energy market.

In addition, most nation-states pursue sovereignty strategies that range from switching to domestic energy resources and entering long-term contractual arrangements with trusted partners to nationalizing energy-related assets ("resource nationalism"), establishing nationally controlled energy companies to secure energy resources abroad, and in some cases projecting economic, political, or military power to secure access to energy resources. Domestically, sovereignty strategies may mean increasing state control over the supply and use of energy.

Many nation-states lack the capacity to implement coherent and effective energy security strategies. One reason is that they often focus on shorter-term issues and solutions such as potential disruptions of supply, especially by "hostile" actions of "foreign" actors. Thus, politically it is often challenging to strike the right balance between short-term sovereignty and longer-term resilience strategies. Moreover, energy security strategies need to be reconciled with broader foreign policy and economic strategies that do not necessarily favor the most secure energy solutions. Finally, many countries simply lack financial and other resources to implement the required energy security measures.

Resource nationalism and some other sovereignty strategies may be beneficial when they mobilize additional resources and capacities to energy systems but may turn self-defeating in the international context, where the pursuit of energy security may become a zero-sum game, where states seek to achieve their own energy security at the expense of each other. Thus, in some situations concerns over energy security – especially in relation to oil and to a lesser extent natural gas – translate into broader geopolitical concerns. Although all-out "resource wars" are highly unlikely in the foreseeable future, the "energy security factor" already plays a significant role in US defense outlooks and may increase the tensions between states and make existing confrontations more protracted.

However, the present international interaction in the field of energy is largely dominated by cooperation rather than conflict. Unfortunately, the plethora of existing energy institutions and alliances have not always been successful in reducing national energy security concerns. This may be largely due to their narrow focus (on a particular energy sector, region, or a group of countries) not reflecting the systemic nature of the energy security risks and their connection with other energy issues such as access and climate change.

The future of energy security will be affected by a variety of technological, economic, and natural factors. It is likely that the production of conventional oil will reach its maximum in the next few decades, whereas the resources of both oil and natural gas may become more geographically concentrated as the center of consumption will shift towards Asia, especially China and India. New nuclear energy programs – despite their large costs – may be able to alleviate energy security concerns in some of the larger and more prosperous economies. At the same time, the inevitable phasing-out of some of the existing nuclear programs will result in new energy security issues. Many countries are likely to expand energy supply from domestic resources such as coal and renewables (including biofuels). The prominence of energy security concerns related to electricity will undoubtedly increase. Whereas the overall demand for energy, and especially for electricity, will grow dramatically, especially in developing countries, there may

also be significant gains in energy efficiency, reducing this demand to more manageable levels.

The energy security landscape of the future will depend critically on both the direction of national strategies and the nature of international energy institutions. In the scenario where national strategies focus on sovereignty concerns and international institutions are weak, one can expect centralized but not internationally integrated energy infrastructure and fragile markets subordinated by resource nationalism and geopolitics. In the opposite scenario, when the national strategies focus on resilience under strong international institutions, it may be possible to support transitions to more secure energy systems even in those countries that lack the capacity to do it on their own.

Under sustainable energy transitions modeled in the GEA Scenario in Chapter 17, the energy security landscape will change so significantly that many of the current energy security threats may disappear and new ones may emerge. To assess energy security in the future, it is important to know how diverse and geographically concentrated the future global energy supply will be, what the diversity of fuels used in key end-use sectors will be, and whether some regions will continue to be seriously dependent on imported energy sources.

References

Adenikinju, A., 2005: *Analysis of the Cost of Infrastructure Failures in a Developing Economy: The Case of the Electricity Sector in Nigeria*. AERC Research Paper 148, African Economic Research Consortium (AERC), Nairobi, Kenya.

Adenikinju, A., 2008: *West African Energy Security Report. University of Ibadan, the Centre for Energy Economics at the University of Texas at Austin, and the Kumasi Institute of Energy*, Technology and Environment, Ibadan, Nigeria; Austin, TX; Kumasi, Ghana.

Alhajii, A. F., 2007: What is energy security? (4/5). *Middle East Economic Survey*, L(52).

Andersson, G., P. Donalek, R. Farmer, N. Hatziargyriou, I. Kamwa, P. Kundur, N. Martins, J. Paserba, P. Pourbeik, J. Sanchez-Gasca, R. Schulz, A. Stankovic, C. Taylor and V. Vittal, 2005: Causes of the 2003 Major Grid Blackouts in North America and Europe, and Recommended Means to Improve System Dynamic Performance. *IEEE Transactions on Power Systems*, **20**(4):1922–1928.

Angevine, G., 2010: *Towards North American Energy Security: Removing Barriers to Oil Industry Development*. Studies in Energy Policy, Fraser Institute, Vancouver, BC.

Anyanwu, J. C., k. Abderrahim and A. Feidi, 2010: *Crude-oil and Natural Gas Production in Africa and the Global Market Situation*. Commodities Brief, Vol. 1, Issue 4., African Development Bank, Abidjan, Côte d'Ivoire.

Arnold, M., G. Kohlin and R. Persson, 2006: Wood fuels, Livelihoods, and Policy Interventions: Changing Perspectives. *World Development*, **34**(3):596–611.

ASCE, 2009: *Dams*. Available at www.infrastructurereportcard.org/fact-sheet/dams (accessed 24 March, 2010).

Asian Development Bank, 2009: *Energy Outlook for Asia and the Pacific.* Asian Development Bank, Manila, Philippines.

Auty, R. M., 1993: *Sustaining Development in Mineral Economies: The Resource Curse Thesis.* Routledge, London, UK and New York, NY.

Auty, R. M., 1994: Industrial policy reform in six large newly industrializing countries: The resource curse thesis. *World Development*, **22**(1):11–26.

Balat, M. and H. Balat, 2009: Recent trends in global production and utilization of bio-ethanol fuel. *Applied Energy*, **86**(11):2273–2282.

Bambawale, M. J. and B. K. Sovacool, 2011: China's Energy Security: The Perspective of Energy Users. *Applied Energy*, **88**(5):1949–1956.

Bates, B. C., Z. W. Kundzewicz, S. Wu and J. P. Palutikof, (eds.), 2008: *Climate Change and Water*. Technical Paper of the Intergovernmental Panel of Climate Change (IPCC). IPCC Secretariat, Geneva.

Bohi, D. R. and M. A. Toman, 1996: *The Economics of Energy Security.* Kluwer Academic Publishers, Norwell, Massachusetts.

BP, 2007: *Statistical Review of World Energy 2007.* BP plc., London, UK.

BP, 2009: *Statistical Review of World Energy 2009*. BP plc., London, UK.

BP, 2010: *Statistical Review of World Energy 2010*. BP plc., London, UK.

Brown, L., 2010: *U.S. Car Fleet Shrank by Four Million in 2009 – After a Century of Growth, U.S. Fleet Entering Era of Decline*. Available at www.earth-policy.org/index.php?/plan_b_updates/2010/update87 (accessed 7 April, 2010).

Brunnschweiler, C. N. and E. H. Bulte, 2008: The Resource Curse Revisited and Revised: A Tale of Paradoxes and Red Herrings. *Journal of Environmental Economics and Management*, **55**(3):248–264.

Burnett, S. H. and W. Dwyer, 2011: *Will Green Energy Make the United States Less Secure?* Brief Analyses No. 739, National Center for Policy Analysis (NCPA), Dallas, TX.

Buttle, J., J. T. Muir and J. Frain, 2004: Economic Impacts of Climate Change on the Canadian Great Lakes Hydro-electric Power Producers: A Supply Analysis. *Canadian Water Resources Journal*, **29**:89–109.

Cancel, D., 2009: *Venezuela Orders 20% Reduction in Electricity Use (Update2)*. Available at www.bloomberg.com/apps/news?pid=newsarchive&sid=ah2q612 aVM58 (accessed 22 December, 2009).

Carter, J., 1980: *State of the Union Address Delivered Before a Joint Session of the Congress*. President of the United States of America.

CEPAL, 2007: *La Seguridad Energética en America Latina y el Caribe en el Contexto Mundial (Energy Security in Latin America and the Caribbean in the Global Context)*. Santiago, Chile, Comisión Económica para Lationamerica y el Caribe de Naciones Unidas (United Nations Economic Commission for Latin America and the Caribbean [ECLAC])

Chaudhry, K. A., 1997: *The Price of Wealth: Economies and Institutions in the Middle East.* Cornell University Press, Ithaca, NY.

Cherp, A. and J. Jewell, 2010: Measuring energy security: From universal indicators to contextualized frameworks. In *The Routledge Handbook to Energy Security*. B. Sovacool, (ed.), Routledge.

Cherp, A, and J. Jewell, 2011: The Three Perspectives on Energy Security: Intellectual History, Disciplinary Roots and the Potential for Integration. *Current Opinion in Environmental Sustainability*, **3**(4):202–212

Cherp, A., J. Jewell and A. Goldthau, 2011: Governing Global Energy: Systems, Transitions, Complexity. *Global Policy*, **2**(1):75–88.

Chester, L., 2010: Conceptualising Energy Security and Making Explicit its Polysemic Nature. *Energy Policy*, **38**(2):887–895.

Chopra, K. and P. Dasgupta, 2000: Common Property Resources and Common Property Regimes in India: a Country Report. *Institute of Economic Growth, New Delhi (Draft)*.

Collier, P., 2007: *The Bottom Billion: Why the Poorest Countries are Failing and What Can be Done About It.* Oxford University Press, Oxford, UK.

Comesaña, F., 2008: *Integración energética en América Latina (Energy Integration in Latin America)*. Available at www.economias.com/2008–03–27/569/integracion-energetica-en-america-latina/ (accessed 3 April, 2011).

Coutsoukis, P., 2008: *World Pipelines Maps*. Available at www.theodora.com/pipelines/index.html (accessed 24 February, 2011).

Crane, K., A. Goldthau, M. Toman, T. Light, S. E. Johnson, A. Nader, A. Rabasa and H. Dogo, 2009: *Imported Oil and US National Security.* RAND Corporation, Santa Monica, CA.

Delucchi, M. A. and J. J. Murphy, 2008: US Military Expenditures to Protect the Use of Persian Gulf Oil for Motor Vehicles. *Energy Policy*, **36**(6):2253–2264.

Dobson, I., B. A. Carreras, V. E. Lynch and D. E. Newman, 2007: Complex systems analysis of series of blackouts: cascading failure, critical points, and self-organization. *Chaos*, **17**(2):026103.

Doggett, T., 2010: *EPA Begins Study on Shale Gas Drilling*. Reuters. Washington, DC

Donnan, S., 2005: *Financial Times*. Indonesian Outage Leaves 100m Without Electricity

EISA, 2007: *Energy Independence and Security Act (EISA) of 2007*. Public Law 110–140, 110th United States Congress, Washington, DC.

EIU, 2009: *Index of Democracy 2008*. Economist Intelligence Unit, London, UK.

European Commission, 2000: *Towards a European Strategy for the Security of Energy Supply*. Green Paper, Communication from the Commission to the European parliament, The Council, The European Economic and Social Committee and the Committee of the Regions, European Commission, Brussels.

European Parliament, 2006: *Directive 2005/89/EC of the European Parliament and of the Council of 18 January 2006 Concerning Measures to Safeguard Security of Electricity Supply and Infrastructure Investment.* Official Journal of the European Union, European Parliament and the Council of the European Union, Strasbourg.

Eurostat, 2011: *Statistical Office of the European Communities.* Available at epp.eurostat.ec.europa.eu/portal/page/portal/eurostat/home/ (accessed 24 April, 2011).

Fardmanesh, M., 1991: Dutch Disease Economics and Oil Syndrome: An Empirical Study. *World Development,* **19**(6):711–717.

Farrell, A. E., H. Zerriffi and H. Dowlatabadi, 2004: Energy Infrastructure and Security. *Annual Review of Environment and Resources,* **29**(1):421–469.

Fearon, J. D. and D. D. Laitin, 2003: Ethnicity, Insurgency, and Civil War. *American Political Science Review,* **97**(01):75–90.

Finon, D. and F. Roques, 2008: *Financing Arrangements and Industrial Organisation for New Nuclear Build in Electricity Markets.* Cambridge Working Papers in Economics, Faculty of Economics, University of Cambridge, Cambridge, UK.

Geller, H., P. Harrington, A. H. Rosenfeld, S. Tanishima and F. Unander, 2006: Polices for Increasing Energy Efficiency: Thirty Years of Experience in OECD Countries. *Energy Policy,* **34**(5):556–573.

Gordon, J., R. W. Button, K. J. Cunningham, T. I. Reid, I. Blickstein, P. A. Wilson and A. Goldthau, 2008: *Domestic Trends in the United States, China, and Iran. Implications for US Navy Strategic Planning.* RAND Corporation, Santa Monica, CA.

Grape, S. G., 2006: *Technology-Based Oil and Natural Gas Plays: Shale Shock! Could There Be Billions in the Bakken?* Washingtonk, DC, United States Energy Information Administration (US EIA)

Greene, D. L. and S. Ahmad, 2005: *Costs of US Oil Dependence: 2005 Update.* ORNL/TM-2005/45, United States Department of Energy (USDOE), Washington, DC.

Greene, D. L., 2010: Measuring energy security: Can the United States achieve oil independence? *Energy Policy,* **38**(4):1614–1621.

Grove, A. S., 2008: *An Energy Policy We Can Stick To.* The Washington Post

Gupta, E., 2008: Oil Vulnerability Index of Oil-importing Countries. *Energy Policy,* **36**(3):1195–1211.

Hester, A. and J. Welsh, 2009: *Superpower? Oil could make Stephen Harper a Superhero.* The Globe and Mail. Toronto, ON

Hidrocarburosbolivia.com, 2009: *Oficial: Gasoducto del Sur al archivo" (Official: South Gas Pipeline Shelved).* Available at www.hidrocarburosbolivia.com/index.php?option=com_content&view=article&id=16570:oficial-gasoducto-del-sur-al-archivo&catid=75:brasil&Itemid=98 (accessed 24 April, 2011).

Hines, P., J. Apt and S. Talukdar, 2008: Trends in the history of large blackouts in the United States. *Power and Energy Society General Meeting – Conversion and Delivery of Electrical Energy in the 21st Century, IEEE,* 20–24 July, Pittsburgh, PA.

Hughes, L., 2010: Eastern Canadian crude oil supply and its implications for regional energy security. *Energy Policy,* **38**(6):2692–2699.

IADB, 2010: *Mesoamerica Renews Push Towards Integration.* Available at www.iadb.org/en/news/webstories/2010–10–25/integration-mesoamerica-idb,8234.html (accessed 3 April, 2011).

IAEA, 2007a: *Extending the Operational Life Span of Nuclear Plants.* Available at www.iaea.org/NewsCenter/news/2007/npp_Extension.html (accessed 7 April, 2010).

IAEA, 2007b: *Considerations to Launch a Nuclear Power Programme.* International Atomic Energy Agency (IAEA), Vienna, Austria.

IAEA, 2008: *Financing of new nuclear power plants.* No. NG-T-4.2, International Atomic Energy Agency (IAEA), Vienna, Austria.

IAEA, 2010: *Power Reactor Information System.* International Atomic Energy Agency (IAEA), Vienna, Austria.

IAEA, 2011: *Integrated fuel cycle information system.* Available at www-nfcis.iaea.org (accessed 25 July, 2011).

Ibitoye, F. I. and A. Adenikinju, 2007: Future Demand for Electricity in Nigeria. *Applied Energy,* **84**(5):492–504.

IEA, 2007: *World Energy Outlook 2007.* International Energy Agency (IEA) of the Organisation for Economic Co-operation and Development (OECD), Paris, France.

IEA, 2009a: *Natural gas information 2009.* IEA Statistics, International Energy Agency (IEA) or the Organisation for Economic Co-operation and Development (OECD), Paris, France.

IEA, 2009b: *Oil information 2009.* IEA Statistics, International Energy Agency (IEA) of the Organisation for Economic Co-operation and Development (OECD), Paris, France.

IEA, 2010a: *World Energy Outlook 2010.* International Energy Agency (IEA) of the Organisation for Economic Co-operation and Development (OECD), Paris, France.

IEA, 2010b: *International Energy Agency: Statistics and Balances.* International Energy Agency (IEA) of the Organisation for Economic Co-operation and Development (OECD), Paris, France.

IGU, 2010: *World LNG Report.* Petronas and International Gas Union (IGU), Kuala Lumpur, Malaysia and Oslo, Norway.

IHS, 2011: *IHS Indexes.* Available at www.ihsindexes.com/ (accessed 24 April, 2011).

Iimi, A., 2007: *Estimating Global Climate Change Impacts on Hydropower Projects: Applications in India, Sri Lanka and Vietnam.* Policy Research Working Paper 4344, The World Bank, Washington, DC.

IPCC, 1998: *The Regional Impacts of Climate Change, An Assessment of Vulnerability.* A Special Report of Working Group II of the Intergovernmental Panel on Climate Change, Cambridge University Press, Cambridge, UK.

Iranzo, S. and M. C. Carrasco, 2008: *La situación energética en Latinoamérica. (Energy Status in Latin America).* Boletín Económico, Banco de España, Madrid.

Jaffe, A. M., M. T. Klare and N. Elhefnawy, 2008: The Impending Oil Shock: An Exchange. *Survival: Global Politics and Strategy,* **50**(4):61–82.

Jansen, J. C., W. G. van Arkel and M. G. Boots, 2004: *Designing Indicators of Long-term Energy Supply Security.* ECN-C-007, Energy Research Centre of the Netherlands (ECN), Petten, Netherlands.

Jansen, J. C. and A. J. Seebregts, 2010: Long-term Energy Services Security: What is it and How Can it be Measured and Valued? *Energy Policy,* **38**(4):1654–1664.

Jewell, J., 2011: A Nuclear-powered North Africa: Just a Desert Mirage or is There Something on the Horizon? *Energy Policy,* **39**(8): 4445–4457.

Jewell, J. 2011: Ready for Nuclear Energy?: An Assessment of Capacities and Motivations for Launching New National Nuclear Power Programs. *Energy Policy,* **39**(3):1041–1055.

Jones, D. W., P. N. Leiby and I. K. Paik, 2004: Oil Price Shocks the Macroeconomy: What Has Been Learned Since 1996. *Energy Journal,* **25**(2):1–32.

Kane, R. P., 2002: Precipitation Anomalies in Southern America Associated with a Finer Classification of El Nino and and La Nina Events. *International Journal of Climatology,* **22**:357–373.

Karl, T. L., 1997: *The Paradox of Plenty, Oil Booms and Petro-State.* University of California Press, Berkley.

Karl, T.L., 1998: State Building and Petro Revenues. *The Geopolitics of Oil, Gas, and Ecology in the Caucasus and Caspian Sea Basin*, M. Garcelon, E. W. Walker, A. Patten-Wood and A. Radovich, (eds.), UC Berkeley: Berkeley Program in Soviet and Post-Soviet Studies., Berkely, CA.

Karl, T.L., 2005: Understanding the Resource Curse. In *Covering Oil: A Guide to Energy and Development*, Revenue Watch, Open Society Institute, New York, NY, pp.21–27.

Katz, M., U. Bartsch, H. Malothra and M. Cuc, 2004: *Lifting the Oil Curse: Improving Petroleum Revenue Management in Sub-Saharan Africa.* International Monetary Fund (IMF), Washington, DC.

Kaufmann, D., A. Kraay and M. Mastruzzi, 2008: *Governance Matters VII: Aggregate and Individual Governance Indicators 1996–2007.* Policy Research Working Paper 4654, Development Research Group and Global Governance Program, The World Bank, Washington, DC.

Kendell, J. M., 1998: *Measures of Oil Import Dependence.* United States Department of Energy (US DOE), Washington, DC.

Klare, M. T., 2008: *Rising Powers, Shrinking Planet: The New Geopolitics of Energy.* Metropolitan Books, New York, NY.

Komendantova, N., A. Patt, L. Barras and A. Battaglini, 2009: Perception of risks in renewable energy projects: The case of concentrated solar power in North Africa. *Energy Policy*, **In Press, Corrected Proof**.

Korowicz, D., 2010: *Tipping point: Near-Term Systemic Implications of a Peak in Global Oil Production: An Outline Review.* The Foundation for the Economics of Sustainability, Dublin.

Kruyt, B., D. P. van Vuuren, H. J. M. de Vries and H. Groenenberg, 2009: Indicators for Energy Security. *Energy Policy*, 37(6):2166–2181.

Le Coq, C. and E. Paltseva, 2009: Measuring the Security of External Energy Supply in the European Union. *Energy Policy*, 37(11):4474–4481.

Lehman Brothers, 2008: *Global Oil choke Points: How Vulnerable is the Global Oil Market?*, Lehman Brothers Inc., New York, NY.

Leite, C. A. and J. Weidmann, 1999: *Does Mother Nature Corrupt? Natural Resources, Corruption, and Economic Growth.* Working Paper No. 99/85, International Monetary Fund (IMF), Washington, DC.

Lovins, A. B. and L. H. Lovins, 1982: *Brittle Power: Energy Strategy for National Security.* Brick house Publishing Company, Andover, MA.

Luft, G., 2010: The Falkland Islands – A New Frontier in the 21st Century Resource War? *Journal of Energy Security*, (March 2010).

Macfarlane, A. M. and M. Miller, 2007: Nuclear Energy and Uranium Resources. *ELEMENTS*, 3(3):185–192.

McAleb, W., 2005: *The Future of the US LNG Marketplace.* Business Briefing: LNG Review, R.W. Beck and Touch Briefings, McLean, VA and London, UK.

Meier, A., 2005: *Saving Electricity in a Hurry: Dealing with Temporary Shortfalls in Electricity Supplies.* International Energy Agency (IEA) of the Organisation for Economic Co-operation and Development (OECD), Paris, France.

Ministry of Energy of the Russian Federation, 2010: *Energy Strategy of Russia for the Period up to 2030.* Moscow Institute of Energy Strategy, Moscow.

Müller-Kraenner, S., 2008: *Energy Security: Re-measuring the World.* Earthscan, London, UK.

NADA, 2009: *Economic Impact of America's New-car and New-truck Dealers.* Available at www.nada.com/nadadata (accessed 4 April, 2010).

NAFTA, 2002: Chapter Six: Energy and Basic Petrochemicals. In *North America Free Trade Association Treaty*.

Nakano, J., 2011: *Japan's Energy Supply and Security since the March 11 Earthquake.* Available at csis.org/publication/japans-energy-supply-and-security-march-11-earthquake (accessed 3 March, 2011).

NEA, 2008: *Nuclear Energy Outlook.* Nuclear Energy Agency (NEA) of the Organisation for Economic Co-operation and Development (OECD), OECD Publishing, Paris, France.

Neff, T., 1997: *Improving Energy Security in Pacific Asia: Diversification and Risk Reduction for Fossil and Nuclear Fuels.* Project Commissioned by the Pacific Asia Regional Energy Security (PARES) Project, Massachusetts Institute of Technology, Cambridge, MA.

NEI, 2010: *Nuclear Industry's Comprehensive Approach Develops Skilled Work Force for the Future.* Fact Sheet, Nuclear Energy Institute (NEI), Washington, DC.

Nelson, P. and C. M. Sprecher, 2008: *What determines the extent of national reliance on civil nuclear power?* NSSPI-08–014, Nuclear Security Science and Policy Institute, College Station, TX.

North American Electric Reliability Corporation, 2010: *Reliability Standards for the Bulk Electric Systems of North America.*, Princeton, NJ.

OLADE, 2007a: *Energy Statistics. Energy Economic Information System.* Latin American Energy Organisation (OLADE), Quito, Equador.

OLADE, 2007b: *Eficiencia Energética: Recurso no Aprovechado (Energy Efficiency: a Forgotten Resource).* Latin American Energy Organisation (OLADE), Quito, Ecuador.

Osborn, J. and C. Kawann, 2001: *Reliability of the US Electricity System: Recent Trends and Current Issues.* LBNL-47043, Ernest Orlando Lawrence Berkeley National Laboratory, University of California, Berkeley, CA.

Parsons Brinckerhoff, 2009: *Powering the Future: Mapping Out Low-carbon Path to 2050.* Parsons Brinckerhoff, Newcastle upon Tyne, UK.

PDVSA, 2009: *Petroamerica.* Available at www.pdvsa.com/index.php?tpl=interface.en/design/readmenuprinc.tpl.html&newsid_temas=46 (accessed 24 April, 2011).

PennEnergy, 2010: *Chesapeake Energy wants to export LNG.* Available at qa.pennenergy.com/index/petroleum/display/6421909998/articles/oil-gas-financial-journal/unconventional/chesapeake-energy.html (accessed 23 January, 2011).

Pistilli, M., 2009: *Uranium Resource Competition Heats Up.* Available at www.u3o8.biz/s/MarketCommentary.asp?ReportID=363571&_Title=Uranium-resource-competition-heats-up (accessed 7 April, 2010).

Proyecto Mesoamérica, 2009: *Proyecto Integracion y Desarrollo Mesoamerica (Mesoamerican Integration and Development Project).* San Salvador, El Salvador, Proyecto Mesoamérica

Reuters, 2009: *Brazil's Largest Cities Hit by Blackout*, Reuters

Rogner, H.-H., 2009: Nuclear Power in the World today: Today and in the 21st Century. *IAEA Regional Asia and the Pacific Seminar on Providing Decision Support for Nuclear Power Planning and Development*, International Atomic Energy Agency (IAEA),, Chengdu, China.

Roig-Franzia, M., 2007: *A Culinary and Cultural Staple in Crisis: Mexico Grapples With Soaring Prices for Corn – and Tortillas.* Washington Post

Ross, M. L., 2001: Does Oil Hinder Democracy? *World Politics*, 53(3):325–361.

Sacchetti, D., 2008: Generation Next. *IAEA Bulletin*, 49(2):64–65.

Sandrea, R., 2010: *U.S. Shale Gas – A Key to Energy Security.* Available at www.pennenergy.com/index/articles/display/2551165565/articles/pennenergy/micro-blogs/rafael-sandrea/us-shale-gas-a-key-to-energy-security.html (accessed 23 January, 2011).

SESEM-CFT, 2005: *A Review of the Power Sector in Latin America and the Caribbean, Evolution in the Market and Investment Opportunities for CFTs*. Contract No. NNE5–2002–96, Securing Energy Supply and Enlarging Markets through Cleaner Fossil Technology (SESEM-CFT), Deutsche Montan Technologie GmbH (DMT), Latin American Energy Organisation (OLADE) and the Spanish Centre for Energy, Environment and Technology (CIEMAT), Essen, Germany; Quito, Equador and Madrid, Spain.

Shore, J. and J. Hackworth, 2003: *EIA Petroleum Feature Articles,*. Impacts of the Venezuelan Crude Oil Production Loss, Energy Information Agency, (ed.) US Department of Energy, Washington, DC.

Soares de Oliveira, R., 2006: Context, Path Dependency and Oil-Based Development in the Gulf of Guinea. In *Dead Ends of Transition: Rentier Economies and Protectorates*. M. Dauderstädt and A. Schildberg, (eds.), Campus Verlag, Frankfurt.

Soares de Oliveira, R., 2007: *Oil and Politics in the Gulf of Guinea.* Columbia University Press, New York, NY.

Söderbergh, B., K. Jakobsson and K. Aleklett, 2009: European Energy Security: The Future of Norwegian Natural Gas Production. *Energy Policy*, **37**(12):5037–5055.

Söderbergh, B., K. Jakobsson and K. Aleklett, 2010: European Energy Security: An Analysis of Future Russian Natural Gas Production and Exports. *Energy Policy*, **38**(12):7827–7843.

Sovacool, B. and M. Brown, 2010: Competing Dimensions of Energy Security: An International Perspective. *Annual Review of Environment and Resources*, **35**(1):77–108.

Sovacool, B. and M. V. Khuong, 2011: Energy Security and Competition in Asia: Challenges and Prospects for China and Southeast Asia. In *The Dragon and the Tiger Cubs: China, ASEAN, and Regional Integration*. D. S. L. Jarvis and A. Welch, (eds.), Palgrave MacMillan, New York, NY.

Sovacool, B. K., 2009: Sound climate, energy, and transport policy for a carbon constrained world. *Policy and Society*, **27**(4):273–283.

Sovacool, B.K., 2010: The political economy of oil and gas in Southeast Asia: heading towards the natural resource curse. *Pacific Review*, **23**(2):225–259.

Stern, J. P., 2009: *Continental European Long-Term Gas Contracts: Is a Transition Away from Oil Product-linked Pricing Inevitable and Imminent?*, Oxford Institute for Energy Studies, Oxford.

Stevens, P., 2010: *Chatham House Report.* the Royal Institute of International Affairs, London, UK.

Stirling, A., 1994: Diversity and Ignorance in Electricity Supply Investment: Addressing the Solution Rather than the Problem. *Energy Policy*, **22**(3):195–216.

Stirling, A., 1998: *On the Economics and Analysis of Diversity*. Paper No. 28, Science Policy Research Unit, University of Sussex, Sussex, UK.

US DOE, 2010: *History of the Department of Energy*. Available at www.energy.gov/about/history.htm (accessed 23 January, 2011).

US EIA, 2008: *Caribbean*. Analysis Briefs, United States Energy Information Administration (US EIA), Washington, DC.

US EIA, 2009: *Ecuador*. Country Analysis Brief, United States Energy Information Administration (US EIA), Washington, DC.

US EIA, 2010: *U.S. Crude Oil Imports by Country of Origin*. Available at www.eia.doe.gov/dnav/pet/pet_move_impcus_a2_nus_Epc0_im0_mbbl_m.htm (accessed 23 January, 2011).

US EIA, 2011a: *International Energy Statistics*. Available at www.eia.doe.gov/countries/ (accessed 24 April, 2011).

US EIA, 2011b: *International Petroleum (Oil) Prices and Crude Oil Import Costs*. Independent Statistics and Analysis, United States Energy Information Administration (US EIA), Washington, DC.

US Government Accountability Office, 2006: *Energy Security: Issues Related to Potential Reductions in Venezuelan Oil Production*. Report to the Chairman, Committee on Foreign Relations, U.S. Senate., Washington, DC.

Vesely, M., 2004: The Vanishing Oil Tankers. *African Business*, **November**(303):44.

von Hippel, D., T. Suzuki, J. H. Williams, T. Savage and P. Hayes, 2009: Energy Security and Sustainability in Northeast Asia. *Energy Policy*, **In Press, Corrected Proof**.

Wohlgemuth, N., 2008: *Powering Industrial Growth – The Challenges of Energy Security for Africa*. Working Paper, United Nations Industrial Development Organization (UNIDO), Vienna, Austria.

World Bank, 2010: *Infrastructure*. Available at www.enterprisesurveys.org/ (accessed 1 August, 2010).

World Nuclear Association, 2011a: *Processing of used nuclear fuel*. Available at www.world-nuclear.org/info/inf69.html (accessed 25 July, 2011).

World Nuclear Association, 2011b: *Uranium enrichment*. Available at www.world-nuclear.org/info/inf28.html (accessed 25 July, 2011).

World Nuclear Association, 2011c: *World uranium mining*. Available at www.world-nuclear.org/info/inf23.html (accessed 25 July, 2011).

Yergin, D., 1988: Energy Security in the 1990s. *Foreign Affairs*, **67**(1):110–132.

Yergin, D., 2006: Ensuring Energy Security. *Foreign Affairs*, **85**(2):69–82.

Yu, W. and M. Pollitt, 2009: *Does Liberalisation cause more electricity blackouts? Evidence from a global study of newspaper reports*. EPRG Working Paper 0827, Cambridge Working Paper in Economics 0911, Electricity Policy Research Group (EPRG), Faculty of Economics, University of Cambridge, Cambridge, UK.

6

Energy and Economy

Convening Lead Author (CLA)
Kurt Yeager (Electric Power Research Institute and Galvin Electricity Initiative, USA)

Lead Authors (LA)
Felix Dayo (Triple "E" Systems Inc., USA)
Brian Fisher (BAEconomics, Australia)
Roger Fouquet (Basque Centre for Climate Change, Spain)
Asmerom Gilau (Triple "E" Systems Inc., USA)
Hans-Holger Rogner (International Atomic Energy Agency, Austria)

Contributing Authors (CA)
Marianne Haug (University of Hohenheim, Germany)
Richard Hosier (World Bank, USA)
Alan Miller (International Finance Corporation, USA)
Sabine Schnitteger (BAEconomics, Australia)

Review Editor
Nora Lustig (Tulane University, USA)

Contents

Executive Summary

The three most basic drivers of energy demand are economic activity, population, and technology. Longer-term trends in economic growth for a particular economy depend on underlying demographic and productivity trends, which in turn reflect population growth, labor force participation rate, productivity growth, national savings rate, and capital accumulation (USEIA, 2011).

Several historic shifts are likely to fundamentally alter global demographics over the coming decades. First, as developing nations move from poverty to relative affluence, there is a fundamental shift from agriculture to more energy-intensive but much more productive commercial enterprises. Second, labor forces in the developed countries are aging considerably, which has implications on many fronts, including energy use and employment structures. Third, for the first time the majority of the world's population has become urbanized, with the largest urban centers emerging in developing regions where energy access is a serious constraint. All of these will have immense impacts on the level and quality of energy demand and on concerns about energy security.

Global energy security and sustainability in the twenty-first century will depend less on the total global population than on incomes and their distribution. This in turn will depend to a large extent on how effectively the lack of energy services, which now limit economic opportunities in the less developed regions, is addressed. In addition, energy security will depend on the ability of countries to maintain reliable sources of energy to meet their needs.

As economies develop, countries' energy needs and priorities change. The evolution of demand at different stages of economic development changes. As economies develop, as happened with industrialized countries, the tendency is to adopt more efficient technologies for the provision of energy services, and the composition of economic activities change with energy intensity tending to decline over time.

Prices play several essential roles in economic production and demand. Most importantly prices send signals to buyers and sellers. Yet it is important to distinguish between prices and costs. There are four types of costs: monetary costs, opportunity costs, environmental (and health) costs, and sociopolitical costs. Most consumers are predominantly exposed to monetary costs and less to the other ones, although these are also important.

Renewable energy technologies, energy efficiency, advanced energy technologies and their associated products and services have been among the most rapidly growing sectors for investment in recent years, with major developing countries becoming investment leaders rather than simply technology transfer followers. In spite of this progress, the total public and private funding of energy-related research, development, and deployment remains much less than the amount needed for the transition to a sustainable, climate-constrained world.

Due to their importance as major contributors to job creation and economic growth, small and medium enterprises are potential leaders in business model transformation in many parts of the world.

6.1   Introduction

The primary role of Chapter 6 is to define the nature and magnitude of the demands on local and global energy systems arising from economic activity. Energy is not an end in itself but rather the means for providing energy services. The energy system is driven by the demand for energy services – a demand that in turn is driven by population and demographic trends, by the level of economic activity and income, and by technological and structural changes. In essence, providing energy services involves investment, operating costs and, if applicable, fuel costs.

Energy fuels the economy, which provides for the establishment of necessary energy infrastructures – from resource and material extraction to the technologies producing electricity as well as other energy carriers and end-use equipment to deliver the desired energy services. The economy is also the financier of energy systems and of its components and energy flows. A central question is: how much energy do economies need to function smoothly and thus being able to augment social development and well-being.

At the same time, economic and demographic developments play a fundamental role along with other drivers (e.g. technology) and GEA sustainability goals (e.g., access, security, and environmental and climate protection) in determining the energy needs and structure of energy systems. Chapter 6 focuses on these two drivers central to GEA. Both were assumed exogenously so as to be in the middle of the range in the literature (see Chapter 17). However, the convergence of per capita gross domestic product (GDP) was assumed to be stronger than in other median economic projections. In order not to subtract from the energy focus of GEA, single, median GDP and population projections were chosen.

An important reason is that economic and demographic developments like other GEA goals are in themselves necessary dimensions of sustainability. The relative GDP convergence enhances the achievement of other GEA goals and enables the transformation of energy systems and achievement of sustainable development. For example, economic development furthers technological change through higher investments and more rapid capital turnover.

Given the long time frames under consideration, serious attention to incorporating highly energy efficient end-use technologies has the decisive potential to address the major energy related challenges addressed in GEA (Goldemberg et al., 1985).

Infrastructures such as power plants, roads, railways, buildings, and so on are inherently long-lived, with service times counted in decades to half-centuries and more. Longevity means stability and inertia at the cost of short-term flexibility. Still, energy systems are constantly in flux – at rates often difficult to detect in the short run at the level of supply. Rates can be much faster on the energy demand side, as energy-using appliances, plants, and equipment have much shorter lifetimes than supply-side infrastructures. The shorter lifetimes are directly related to the growth and changing mix of goods and services provided by the economy.

The evolving energy system epitomizes technology change and innovation. Technology is the crucial tether between the energy system and the economy, especially the modern economy[1]. Energy and economy evolve in tandem, and technology defines which energy carriers and services the system can provide and which the economy can demand. The industrial revolution was powered by coal, which provided industries and households with a much more concentrated fuel. This enabled a higher productivity with respect to wood fuels and which boosted economic progress and urbanization. In the nineteenth century, abundant access to coal increased productivity and stimulated economic development.

Today access to modern forms of energy or rather secure, clean, affordable energy carriers fundamentally defines the modern economy. Electricity is key in this regard, as it is most compatible with the needs of the modern economy. It is more than just an energy carrier in the strict physical sense. Electricity enables all kinds of transactions – from information exchange to transportation. Its productivity in economic production and consumption as well as cleanliness at the point of use is second to none. The factors contributing to income disparities within and between countries can be traced to many reasons that vary across countries and regions. A lack of access to modern energy carriers and services is one of these contributing factors (Modi et al., 2005).

The world is now in the midst of an unparalleled period of dramatic growth in multiple parameters, also known as "the great acceleration" (see also Chapter 3). Major populous developing countries are actively and successfully pursuing industrialization and socioeconomic development with concomitant growth in demand for energy-intensive goods and services. As a consequence, energy demand is rising rapidly compounded by population trends. If current trends continue, human beings will use more energy over the next half-century than in all of recorded history. Energy demand on this scale will put increasing pressure on global energy resources and distribution networks. This is unsustainable without a fundamental transformation of the energy system. The unsustainability, however, has many other elements than resource availability, including energy services for economies and poverty alleviation, as well as addressing various social and environmental and health dimensions, security and peace (see also Chapters 2–5).

The dominant fossil energy resources today, especially oil, are concentrated in only a few regions. Energy security – that is, the potential disruption of supply – is viewed by many countries as a potential threat to their economic well-being (see also Chapter 5).

---

1   Technology is more than just hardware and includes a range of factors from cultural aspects to education and training (Arthur, 2009).

Providing access to energy services involves the conversion of primary energy resources as well as the manufacturing (and construction) of required technologies and distribution networks. These activities take material from the environment and inevitably split them into desirable products and wastes of various forms. The latter are returned to the environment, increasingly at levels beyond the carrying capacities of ecosystems, and threatens to have environmental damages undermining economic gains.

Achieving the partly conflicting objectives of environmental protection and economic gains will require substantial input from the economy in the form of finance as well as research, development, and deployment (RD&D). The *2009 World Energy Outlook* of the International Energy Agency (IEA, 2009a) estimated that US$20 trillion will be required over the next 25 years just to meet the projected increase in global energy demand by 2030. Similarly, the GEA pathways from Chapter 17 show that the transformation of the energy system would require dedicated efforts to increase global energy-related investments to between US$1.7 trillion and US$2.2 trillion annually, compared with about US$1.3 trillion in annual investments today (see also Chapter 24). Out of this total, about US$300 to US$550 billion of efficiency-related investments are required on the demand-side of the pathways. This includes only the efficiency-increasing part of the investment to improve energy intensity beyond historical improvement rates. The full demand-side investments into all energy components of appliances might thus be significantly higher. Total investments into energy supply and efficiency-related investments at the demand-side correspond in sum to a small fraction (about 2%) of global GDP. Future transitions with a focus on energy efficiency achieve the targets at more modest cost and thus represent the lower bound of the investment range (see Chapter 17 for further details).

Current modest levels of investment in clean energy facilities and energy efficiency measures contrast starkly with these immense investment requirements. Furthermore, significantly strengthened innovation will also be essential to support the continued development of new solutions to these critical energy challenges, as governments and industry struggle to expand new energy resources and new ways to use existing resources in a sustainable manner.

The demands of a changing energy paradigm have wider institutional implications. The institutions created over the past several decades are struggling to remain relevant in the face of profound geopolitical and economic changes (outside the domain of their membership). New institutional frameworks more suited to the needs of today are urgently needed. These institutions would need to support the delivery on the goals of access to affordable modern energy carriers and end-use conversion, enhanced energy security, climate change mitigation, and health and environment.

In the twenty-first century, a global energy system for sustainable societies must reflect multiple objectives that include energy availability,

affordability, security, and consistency with climate change goals (see also Chapters 2–5). These have been further complicated by the usual market implementation issues that have been aggravated by the current global economic crisis.

The required energy system transformation will be difficult to accomplish without some transformation of the world economy – a process that will be complex and characterized by marked clashes of interest. This transformation will therefore require a long-term vision and sustained cooperation among a large array of diverse stakeholders at both the national and international level, coupled with strong public policy support and private-sector engagement.

6.2 Basic Drivers of Energy Demand

The three most basic drivers of energy demand are population, economic activity per capita, and technology performance. Based on these fundamentals, this century is likely to see major shifts in energy demand and development. According to a recent UN report (UNFPA, 2011), global population by 2050 is projected at 9.3 billion, a revision upwards from previous reports (UNDESA, 2004). Virtually all this projected population growth will occur in the developing world. By comparison, the present world population is 7 billion, and was only 2 billion as recently as 1930.

This unprecedented massive global population growth over little more than a century is arguably one of the most defining events of our era. The past century also represents a period of intense technological expansion, which has fundamentally increased humanity's ability to harness energy. This has contributed to changing the historic equilibrium between human fertility and mortality. As a result, average life expectancy worldwide more than doubled over the last century from 32 to 67 years, and it continues to increase steadily.

6.2.1 Demographic and Income Changes

Extending population projections beyond 2050 depends on uncertain fertility, mortality, and migration assumptions. Researchers at the International Institute for Applied Systems Analysis (IIASA) have addressed this uncertainty constraint by developing probabilistic population projections that reflect a realistic range of uncertainty (Lutz and Samir, 2010). Table 6.1 outlines these projections through 2100 for 13 major world regions. The results suggest that the world population will most probably peak by 2050 at slightly less than 9 billion[2] and remain above 8 billion through at least the remainder of this century. Yet the probabilistic shifts in population distribution among the continents

---

2   A more recent study by the UNFPA (2011) has adjusted this figure slightly upward to 9.3 billion as mentioned above.

Table 6.1 | Population projections.

	Total Population (millions)		
	2010	2050	2100
North Africa	208	307 (+48%)	324–346 (+56%–+66%)
Sub-Saharan Africa	799	1,617 (+102%)	2,074–2,247 (+160%–+180%)
North America	339	427 (+26%)	421–468 (+24%–+38%)
Latin America	595	834 (+40%)	909–977 (+53%–+64%)
Central Asia	65	96 (+48%)	101–108 (+55%–+66%)
Middle East	215	359 (+67%)	392–417 (+82%–+94%)
South Asia	1,625	2,289 (+41%)	2,016–2,140 (+24%–+32%)
China & CPA	1,468	1,342 (–9%)	829–881 (–56%—40%)
Pacific Asia	542	699 (+29%)	649–689 (+20%–+27%)
Pacific OECD	152	137 (–10%)	85–103 (–44%—32%)
Western Europe	462	449 (–3%)	320–364 (–31%—21%)
Eastern Europe	120	94 (–22%)	54–60 (–55%—50%)
Former Soviet Union	228	169 (–26%)	103–111 (–55%—51%)
WORLD	6,816	8,816 (+29%)	8,280–8,920 (+21%–+31%)

Source: Lutz et al., 2008.

during this time are dramatic. Europe and China are projected to have population reductions on the order of 40–50% by the end of the century, while Africa and the Middle East are likely to double their populations and Latin America and Central Asia are expected to have population increases on the order of 50%.

Four historic shifts, revealed in the most recent United Nations population data, are likely to fundamentally alter global demographics over the next several decades (UNFPA, 2011). First, the relative demographic weight of the world's industrialized nations is forecast to decline by at least 25%, with a corresponding shift of economic power to developing nations. As these nations move from poverty to relative affluence, there is a fundamental shift from agriculture to more energy-intensive commercial enterprises. Second, the labor forces in the industrialized countries will age considerably. Third, most of the world's expected population growth will be concentrated in today's poorest countries, which typically lack employment, capital, and educational opportunities.

Finally, the majority of the world's population is becoming urbanized, with the largest urban centers found in the world's poorer regions, where energy access typically remains a serious economic constraint. All of these shifts point to substantial growth of demand in developing countries where energy systems are notoriously underdeveloped and therefore open early and sizable prospects for an effective energy system transformation. Expanding systems simply offer more opportunities for market penetration of new technologies than stagnating or shrinking systems. Being a multiplier of demand for goods and services including energy, however, population remains a major driver of, especially adverse, impacts (Campbell et al., 2007). Given the absolute limits of the planet, as illustrated by the need to limit concentrations of climate-altering pollutants, reductions in population growth trends can give valuable additional decades to resolve energy and other problems before these planetary limits are reached. There is no coercion implied here, as studies show that there are hundreds of millions of women wishing to control their family size who do not have access to modern contraceptive technologies (Cleland et al., 2006). Analyses show, for example, that doing so could lower CO_2 emissions from energy use by 30% in 2100 over what is projected otherwise (O'Neill et al., 2010). Providing reproductive health services to these women is also an equity issue as all women, not just those in rich countries, ought to have access to such services (Prata, 2009). It is also an important health issue as spacing births, which, along with reducing growth rates, gives men and-women access to contraception, which has major benefits for both child and maternal health (Smith and Balakrishnan, 2009).

In light of these enormous demographic challenges, the global economy is significantly underperforming mainly in industrialized countries, while developing economies performances are not sufficient to provide for everybody. Given the annual addition to the world of about 40 million workers and the average levels of GDP per worker, the world economy has to generate at least $500 billion in additional output each year just to employ new workers (Martin, 2005). At current growth rates, tens of millions of workers will remain unemployed. Unfortunately, the issue of how to achieve sustainable world economic growth is still not being effectively addressed. Education is the key. As Lutz et al. (2008) point out, "education is a long-term investment associated with near-term costs, but in the long term it is one of the best investments societies can make in their future."

Nearly 80% of the people in the world continue to have inadequate purchasing power parity (PPP) to afford basic needs and thus live in poverty (Chen and Ravallion, 2008). Lack of access to modern forms of energy is a constraint on the economic and social progress for a large fraction of this population. Indeed, due to extreme poverty[3], huge rural and urban populations are being largely excluded from the social and economic development processes in which markets are a part. Unless this growing demand for energy is met by cleaner, safer, and more efficient energy technologies, the negative impacts on global health and environmental will continue to grow.

The economic prosperity of the world's population is, however, rapidly increasing. Barring a cataclysmic crisis, global economic output is projected to increase at the rate of 3–4% over the next several decades (IEA, 2011; USEIA, 2011). As a result, and as was the case in the last century, global income will most likely increase far faster than population. Indeed, stabilization of the world's population has been shown to be a direct result of increased prosperity. In 2000, the poorest half of the world's population owned only about 1% of total global wealth (Davies et al., 2008). This extreme poverty is decreasing as nations develop and their standards of living steadily improve.

The rise in world affluence holds promise for improved well-being but also comes with significant ecosystem risks if prevailing patterns of energy

demand, supply, and waste persist. Since the Industrial Revolution, economic development and expansion have been tied to increased energy use, and this link remains strong today. Without the grand transformation in the global energy system as explored in GEA, fossil fuels will remain the dominant energy source through at least the middle of the century under any achievable circumstances – a definite challenge for achieving greenhouse gas (GHG) stabilization at 450 ppm. Fossil fuels with carbon capture and storage (CCS) or moving beyond fossil fuels to cleaner, renewable and other non-carbon energy sources will require a much larger and sustained commitment as well as accelerated upfront investments. In 2009, non-fossil fuel energy resources provided approximately 22% of the world's primary energy supply (see Chapter 11). Providing adequate energy supplies at minimum GHG emissions begins with the large-scale adoption of currently best available technologies (BAT) and practices (Pacala and Socolow, 2004; IPCC, 2007) but eventually will require a major upscaling of this BAT. In either case this only occurs when people have the incentives to adopt innovative technologies and are willing to forego excessive current consumption for future benefit. Investing in BAT and technological progress is a major concern, as the demands for short-term returns on investment constrain the strategic development opportunities on which those returns ultimately depend.

The challenges facing rural regions over the next 50 years are significant. Fertility rates are much higher than in urban areas, where health care and education are relatively advanced and available. Not only will these rural regions be the source of a large portion of the world's population growth, but they must be able to help feed the world. Only about 13% of global land area is arable, and the average population density in the developing world is expected to increase from about 60 people/km^2 to around 96 people/km^2 by 2050. This compares to an industrial-world stable population density of about 25 people/km^2 and poses unprecedented problems of land use and preservation in the developing world (UNDESA, 2004).

Countries' energy priorities and services demand change dramatically as their economies develop. Thus global energy security and sustainability in the 21st century will depend less on the total global population than on population incomes and how that income is distributed. Nations are very different with respect to their incomes. For example, member countries of the Organisation for Economic Co-operation and Development (OECD) currently account for approximately 18% of the world's population but for almost three quarters of GDP. By comparison, China and India represent some 37% (20% and 17%, respectively) of the world's population and about 11% of global GDP in terms of market exchange rates (MER) (World Bank, 2011ba). Even if expressed in units of Purchasing Power Parity (PPP), their share of global GDP was below 25% in 2010 (GEA scenario database, see www.globalenergyassessment.org).

Figure 6.1 depicts the disparity of income measured in terms of GDP per capita across 11 world regions in 2005. The level of primary energy use appears to mirror the level of affluence in a region, and there is a

---

3   Currently defined as per capita income of less than $US_{2005}$$ 1.25 per day (Chen and Ravallion, 2008).

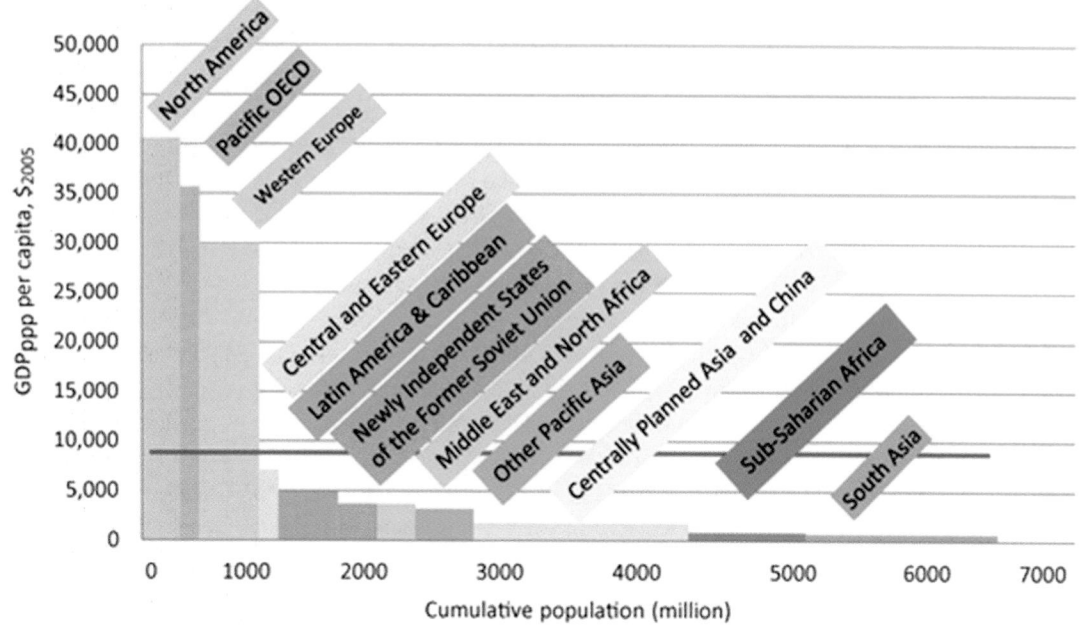

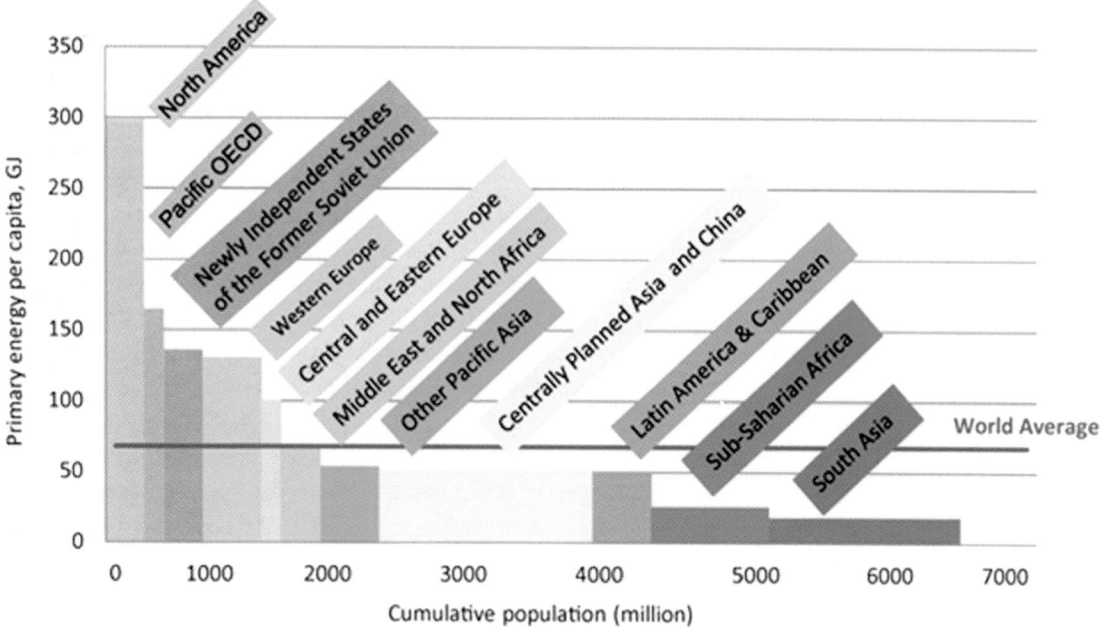

Figure 6.1 | Regional disparities in income and energy use per capita in 2005.[4]

common understanding that the access to clean and affordable energy is critical for achieving the Millennium Development Goals (MDGs)s and enabling sustainable economic development. The UN Advisory Group on Energy and Climate Change (AGECC) defines energy access[5] as

"access to clean, reliable and affordable energy services for cooking and heating, lighting, communications and productive uses" (AGECC, 2010).

6.2.2 The Move to Modern Energy Carriers—Advances in Electricity

Electricity provides the essential key to energy access and enables technical innovation and productivity growth – the lifeblood of a modern

---

4   Figure 6.1 depicts the large income and energy use disparities between major regions and countries. However, it is important to note that access disparities within nations are typically greater than between nations.

5   For definitions of energy access, see Chapters 2 and 19.

society. Electricity's central role in achieving economic and environmental progress is broadly accepted throughout the world, depending on how the electricity is produced (see Chapters 3 and 4). The determining factor is whether nations will find the leadership and make the commitment necessary to harness and expand electricity's unique capabilities. Ironically, the electronic technological revolution that electricity has enabled over the past decades has largely been ignored by the electricity supply industry itself (Galvin and Yeager, 2009).

The pacing factor is the necessary investment for the innovative technologies that are needed to replace today's generally centralized and aging electricity infrastructure and establish a truly intelligent global electricity infrastructure spanning from generation to end use. The accelerated growth of intermittent renewable technologies combined with the introduction of smart grids may make it possible to transform them into useful and valuable fuels. Of equal importance is that electricity provides a continuous efficiency improvement opportunity, i.e., the inherent intelligence that comes from the incorporation of modern digital electronic monitoring and control technology throughout the electricity delivery and utilization networks.

While renewable technologies are desirable for providing clean energy carriers to the unconnected, especially in rural areas, one must not ignore the ongoing urbanization trend and industrialization aspirations of developing countries. Cities have a much higher energy demand density per hectare than rural settlements. Energy-intensive industries represent large-scale off-take nodes. While supply densities of renewables or distributed electricity generation are consistent with the demand densities and offer opportunities in cities as well, metropolitan areas will continue to require a certain share of baseload electricity from large central conversion facilities such as large scale hydro, nuclear, or fossil fuels with CCS.

Today, global electrification is distributed very unevenly (see Chapters 2 and 19). The highest proportion of people without electricity is concentrated in rural sub-Saharan Africa and South Asia – regions that are also projected to have very high population growth this century. A realistic and universally achievable goal over the next decades is therefore to eliminate the electrification gaps for the 1.4 billion people without access today and to even more by 2030 under business-as-usual conditions (see also Chapters 17, 19, and 23). This goal is not just a matter of equity but an essential prerequisite for eliminating extreme poverty and stabilizing population in fast-growing developing regions. In order to keep pace with the world's growing population and to provide the foundation for the corresponding economic development, electricity must reach at least an additional 100 million people per year for at least the next 30 years. This is more than twice the current rate.

6.2.3 Economic Production Processes

In the context of the economic production process, the output of which is measured in terms of GDP, energy is but one production factor. It is

a necessary but not sufficient input[6] to the production process.[7] Other essential production factors are capital, labor, land, and materials. A simple production function then would take on the following form: GDP = f(capital, labor, energy, land, materials and know-how).[8] The contribution of each factor (or input quantity) to output depends on the state of development of an economy, physical conditions and location (e.g., climate, geography, land), relative factor prices, and factor productivity as well as social and cultural conditions. Within certain limits production factors can substitute each other.

The ease of substitution is a matter of relative production factor prices, availability of factors, vintage of technologies, and lock-in in existing infrastructures, timelines, etc. For example, as an agriculture-based economy advances, higher incomes make labor more expensive (or more productive elsewhere in the economy), and rural workers are progressively replaced by machines (capital), energy (fuel for the machines), and materials. In the case of rising energy prices, substitution of capital and materials for energy can lower the physical energy input to the economy (although the economic value of the input may not change due to the higher prices of fuel and technology). Efficiency improvements are largely a substitution of capital and know-how for energy. Using the production function approach, the optimal energy input and thus energy demand can be determined by a first-order derivation of the production function where the marginal productivity equals the factor price.

Figure 6.2 presents three capital-energy combinations – all of which produce identical outputs (isoquant) both in terms of quantity and quality (labor and materials remain constant). Option A is an energy-intensive process but uses little in terms of capital (investment). Option B is much more energy-efficient but requires a more capital-intensive production structure. The most economically efficient option then is determined by the combined effect of the respective factor costs – that is, the price of fuel p_E, capital p_k (interest and depreciation), labor p_L (wages), and materials p_M. If factor costs, say for energy, were to include externalities (which generally is not yet the case), this would make energy more expensive, hence change the slope of the optimal factor combination (dashed line in Figure 6.2), i.e., from point C to B (lower energy – higher capital input). Therefore policies enforcing the internalization of externalities would encourage capital for energy substitution (for identical energy services) or change the nature of energy services demanded. Similarly energy subsidies reduce the incentive to deploy capital and energy efficiency measures resulting in higher energy use than otherwise.

---

6 According to the laws of thermodynamics nothing can be changed without energy input (or rather exergy consumption).

7 While long recognized for its many limitations, economic output is predominantly measured as GDP, either at market exchange rates or purchasing power parities even though alternative measures have been proposed.

8 Production functions found in the economic literature are way more complex than the illustrative example presented here. A more detailed discussion of production functions is beyond the scope of this chapter.

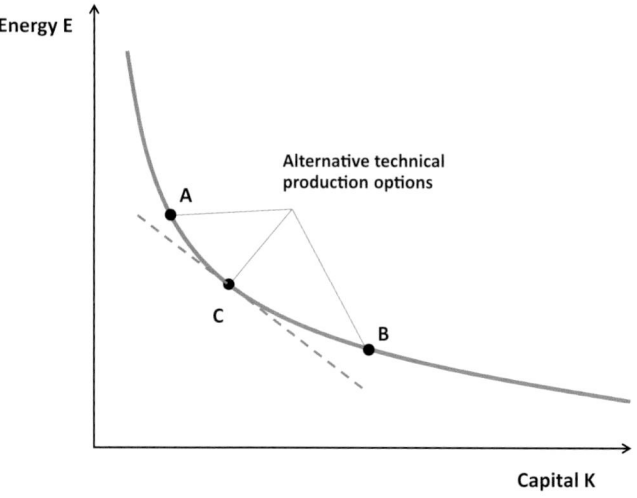

Figure 6.2 | Capital-energy substitution frontier.

Production costs $C = p_k K + p_E E + p_L L + p_M M$

The optimal factor combination results when

$dC = 0 = p_k dK + p_E dE$ (as dL and dM are zero by assumption) or

$$\frac{dE}{dK} = -\frac{p_K}{p_E}$$

which at constant factor costs leads to the dashed straight line with the slope $-p_K/p_E$ shown in Figure 6.2. The cost optimal factor combination results where the line touches the output isoquant – that is, Option C.

Generally it takes longer for substitution processes to materialize for a variety of reasons, ranging from system inertia and lock-in to the availability of capital and finance costs. The vintage structure of the capital stock and thus the natural rate of capital replacement due to wear and tear are ideal dates for the introduction of more-efficient plant and equipment or building stock. In the short run, a recently opened new factory or apartment building is unlikely to modify its heating and cooling equipment in response to, say, higher natural gas prices. If prices remain at elevated levels for extended periods of time, displacement of natural gas by inter-fuel substitution or efficiency improvements become economically attractive. Refurbishment of an aging building stock, the early retirement of boilers in a factory, or relocation of energy-intensive manufacturing process to a lower energy cost area are routine economic decision processes based on cost functions and relative factor prices.

6.3   Challenges of Projecting Energy Demand

In generating projections of future energy requirements, a key difficulty arises because the demand for energy derives from the demand for various services required by the industrial production of goods and services as well as buildings and energy-using appliances, such as refrigerators or cars. Energy demand therefore depends on evolving personal preferences and behavior, including how the demand for energy services evolves at different stages of economic development (Haas et al., 2008). A variety of approaches have been used to estimate energy demand on the basis of the use of different types of energy-dependent services. They include:

- "bottom-up" approaches that evaluate the mix of economic activities and the energy intensity of these activities for each sector of the economy (Schipper and Meyers, 1992);

- approaches that consider the relationship between the demand for energy services and efficiency improvements in service provision (Ayers and Warr, 2009); and

- detailed studies of the price of energy services (such as lighting), economic development, and the consumption of energy services (Nordhaus, 1997; Fouquet and Pearson, 2006; Fouquet, 2008).

Since the early 1970s, a large number of empirical studies of energy demand have also sought to identify the effect of real incomes on energy demand (Griffin, 1993; Espey, 1998; Hunt and Ninomiya, 2003; Espey and Espey, 2004). In general, the estimates from these statistical analyses show that the income elasticity of the demand for energy is positive (i.e., rising incomes will lead to increased energy use). However, the magnitude of these elasticities[9] differs by stage of development. In industrialized countries, the estimated income elasticities tended to be less than one (e.g., a 10% increase in income would result in an increase in energy use of less than 10%). Judson et al. (1999) analyzed data from OECD and non-OECD countries and concluded that income elasticities are lower at low levels of economic development, rise substantially at medium levels, and fall at higher levels. Hence the state of economic development and the standard of living of populations in a given region strongly influence the link between economic growth and energy demand (USEIA, 2011).

Based on empirical evidence, the evolution of demand at different stages of economic development can be described as follows (Fouquet, 2008). The demand for energy rises rapidly at early stages of economic development as an economy evolves from an agrarian economy and industrializes, with an associated increase in mining and manufacturing activities. Industrialization is also associated with a rapid rise in demand for heating and for freight transport. As the industrialized economy matures, the demand for energy by the industrial sector continues to grow in absolute terms but declines relative to the household and service sectors, in part because energy-intensive imports are substituted for domestic production. Household demand for energy continues to grow in absolute terms and rises with disposable incomes because of increased demand for space, heating and cooling, and individual transportation.

---

9   Income elasticity of energy demand is the ratio of percentage change of energy demand to the percentage change of income.

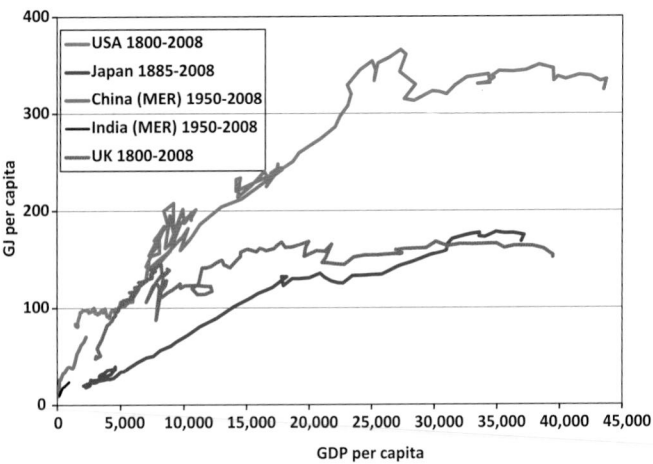

Figure 6.3 | Primary Energy Use per capita (GJ) versus GDP (at market exchange rates in US$_{2005}$$) per capita. Data sources: US, Japan: updated from Grubler, 1998, UK: Fouquet, 2008, India and China: IEA, 2010b and World Bank, 2010. Note: Data are for the United States (1800–2008), United Kingdom (primary and final energy, 1800–2008), Japan (1922–2008), China (1950–2008), and India (1950–2008). For China and India, also GDP at purchasing power parities (PPP, in International$_{2005}$$) are shown.

While it might be expected that the demand for energy will reach a saturation point at high levels of development, the evidence available to date suggests that if new, more-efficient technologies reduce the amount of energy required to operate appliances and equipment, the implied cost reduction often serves, everything else being equal, to stimulate additional energy-dependent activities. This is referred to as the rebound effect and arises because improvements in energy efficiency effectively decrease the cost of energy services, leading to additional disposable income and so potentially increases demand (see Chapter 22). Consumers and businesses change their behavior – they may raise thermostat levels in the winter, cool their buildings more in the summer, buy more appliances and operate them more frequently, or drive their vehicles more (IEA, 2005). However, as the price of energy services is what matters, higher fuel prices may well outweigh the benefits of more-efficient appliances, plants, equipment, and infrastructures. Data for the United States and Japan suggest that once annual per capita income levels exceed US$30,000, per capita energy use no longer increases with GDP (see Figure 6.3) and that total energy demand is largely a function of population and prices. However, it can be argued that per capita energy use in the United States is higher than shown in Figure 6.3 as it excludes embedded energy, i.e., the energy content of imported goods and services as energy-intensive materials, mining, processing and manufacturing occur abroad. Likewise, accounting for embedded energy would reduce per-capita energy use of export-driven Japan.

A simple approach to present the energy – economy link is by way of demand functions such as the aggregate top-down relation:

$$E = POP \times \frac{GDP}{POP} \times \frac{E}{GDP}$$

where energy demand (E) is the product of population (POP), per capita income (GDP/POP), and the energy intensity of the economy (E/GDP), usually measured in megajoules per unit of economic output or GDP.

For all countries, population growth is a modest to strong driver of energy demand. The income effect in terms of GDP and GDP per capita is positive and generally larger in developing countries, where small improvements in per capita income translate into an over-proportional use of energy services. Energy intensities vary from region to region depending on the stage of economic development, geographic and climate conditions, and energy prices, but intensities have been declining in most countries and regions and thus are chiefly responsible for keeping energy demand growth in check globally. Figure 6.4 shows the average growth rates of primary energy, population, primary energy per capita, primary energy and electricity intensities of GDP for four consecutive 10-year periods in the World, OECD, Reforming Economies, Middle East and Africa, Asia, Latin America and Caribbean.

As described above, the relationship between energy availability and economic growth has been the subject of many studies, but no clear causality has emerged. Instead, the research suggests that the direction of causality differs as energy efficiency becomes a higher priority among countries, especially between industrialized and developing countries. In industrialized countries, the link now appears to be relatively weak, with energy use decoupled from economic growth. In developing countries, energy demand and economic growth have been more closely correlated in the past, with demand tracking or exceeding the rate of economic expansion (USEIA, 2011). These countries are still in the process of building their human and physical capital, which is inherently more energy-intensive than the operation of well-developed infrastructures.

6.4 Relationship Between Economic Growth and Energy Demand

As noted earlier, economic growth and energy demand are linked, but the strength of that link varies among regions and their stages of economic development. The state of economic development and the standard of living of individuals in a given region strongly influence the link between economic growth and energy demand. Advanced economies with high living standards have a relatively high level of energy use per capita, but they also tend to be economies where per capita energy use is stable or changes very slowly. In industrialized countries, there is a high penetration rate of modern appliances and motorized personal transportation equipment. To the extent that spending is directed to energy-using goods, it often involves purchases of new equipment to replace old capital stock. The new stock is often more efficient than the equipment it replaces, resulting in a weaker link between income and energy demand (USEIA, 2011).

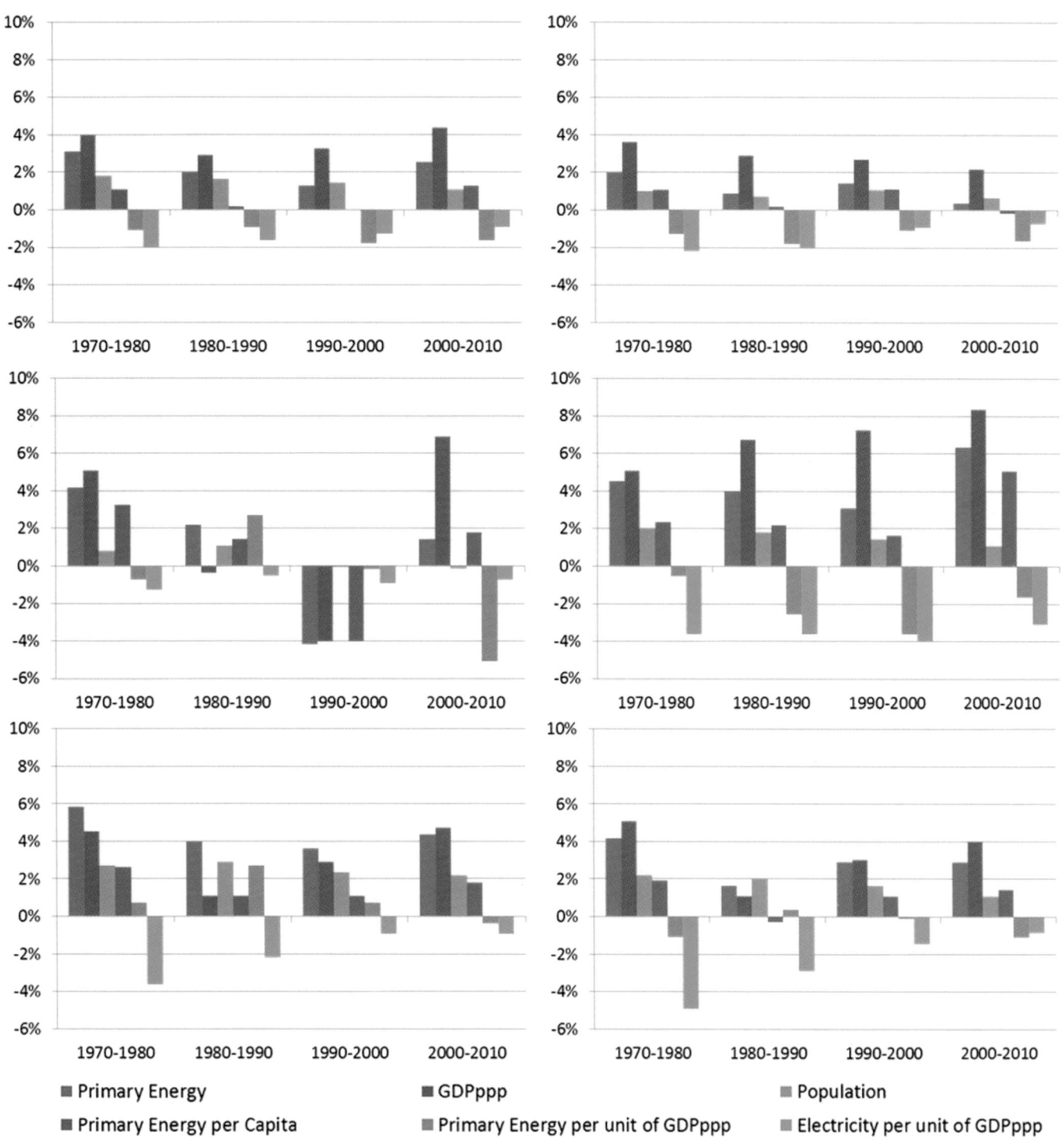

Figure 6.4 | Annual growth rates of indicators for primary energy demand, GDP, population changes and selected indicators for four ten-year time-periods and different regions (World, OECD, REF: reforming economies, ASEA, MAF: Middle East and Africa, LAC: Latin America and Caribbean). Source: data from GEA database.

6.4.1   Industrialized Countries

Historically, and beginning with the industrial revolution, increased energy use has fueled economic development in advanced industrialized societies (Fouquet, 2008). The first industrial revolution, which began in the eighteenth century, merged into the second industrial revolution around 1850, when technological and economic progress gained momentum with the development of steam-powered ships, railways, and, later in the nineteenth century, the internal combustion engine and electrical power generation. Although a number of social and non-energy factors were responsible for the productivity increases of that era, there is agreement that the transition to the use of available energy sources such as coal coupled with technology innovations accounted for the impressive growth results achieved during this era (Landes, 1969; Ayers and Warr, 2009).

A commonly observed characteristic of this development is that advanced industrialized countries use less energy per unit of economic output and more energy per capita than poorer societies do (Toman and Jemelkova, 2003). Figure 6.5 show the energy intensity of various regional economies in 2005. In contrast to energy use per capita, OECD countries have generally lower energy intensities than developing and transition countries. The latter are less efficient in their economic production, which indicates a large potential for efficiency improvements as these regions industrialize and adopt the best available technologies.

The stylized trend in energy intensity is for a rapid increase of industrialization, reaching a high plateau, and then a decline, especially when services begin to dominate the economic production. Changes in the demand for energy services associated with structural shifts in the economy are especially important in explaining the upward trend in energy intensity with industrialization. This rise is often exaggerated, however, by the lack of estimates of non-commercial energy sources. When wood fuel is included in statistical studies, and even human and animal power, there appears to be a more gradual rise in energy intensity (Schurr et al., 1960; Nilsson, 1993; Fouquet, 2008).

A more reliable observation is that as industrialized countries adopt more-efficient technologies for energy supply and use and as the composition of economic activities changes, energy intensity tends to decline over time (Nakićenović, 1996). In the last 30 years, the dominant driver for the declining trends in energy intensity have been technical improvements (Liu and Ang, 2007). Also, despite variability across countries, energy intensities tend to converge (Markandya et al., 2006; Liu and Ang, 2007). Even then, total energy use and energy use per capita continue to rise in most industrialized countries.

One of the first major studies of the relationship between energy use and economic growth in industrialized economies was undertaken in the United States for the period from 1947 to 1974. Although that study found that economic growth had a causal effect on energy demand (Kraft and Kraft, 1978), subsequent studies (Yu and Choi, 1985; Erol and Yu, 1987; Abosedra and Baghestani, 1989; Hwang and Gum, 1991; Cheng, 1995) alternately confirmed or rejected these conclusions (Soytas and Sari, 2003).

More recent studies suggest that the direction of any causality is country-specific. Soytas and Sari (2003) re-examined the relationship between economic growth and energy supply in the top 10 emerging economies and G7 countries and found that growth of GDP raised energy demand in the United States and in Italy, but that energy use raised GDP growth in Turkey, France, Germany, and Japan. The authors' conclusion was that in the long run, restriction on energy use, which reduces productivity, may harm economic growth, and that this may in fact have occurred in Turkey, France, Germany, and Japan. Another recent study of about 100 countries tested for causality between energy and GDP and found that causality from energy to GDP is more prevalent in OECD countries than in developing countries (Chontanawat et al., 2008).

At the aggregate level, the OECD region has seen steadily declining population growth rates over 40 years, caused by and large by an aging population and low fertility rates (see Figure 6.4). GDP per capita growth rate also declined steadily from 2.5% during the 1970s to less than 1.5% between 2000 and 2010 (GEA Database). Everything else being equal, population and per capita income together would have caused total primary energy demand to expand on average by 2.7% annually between 1970 and 2010. In reality, primary energy demand

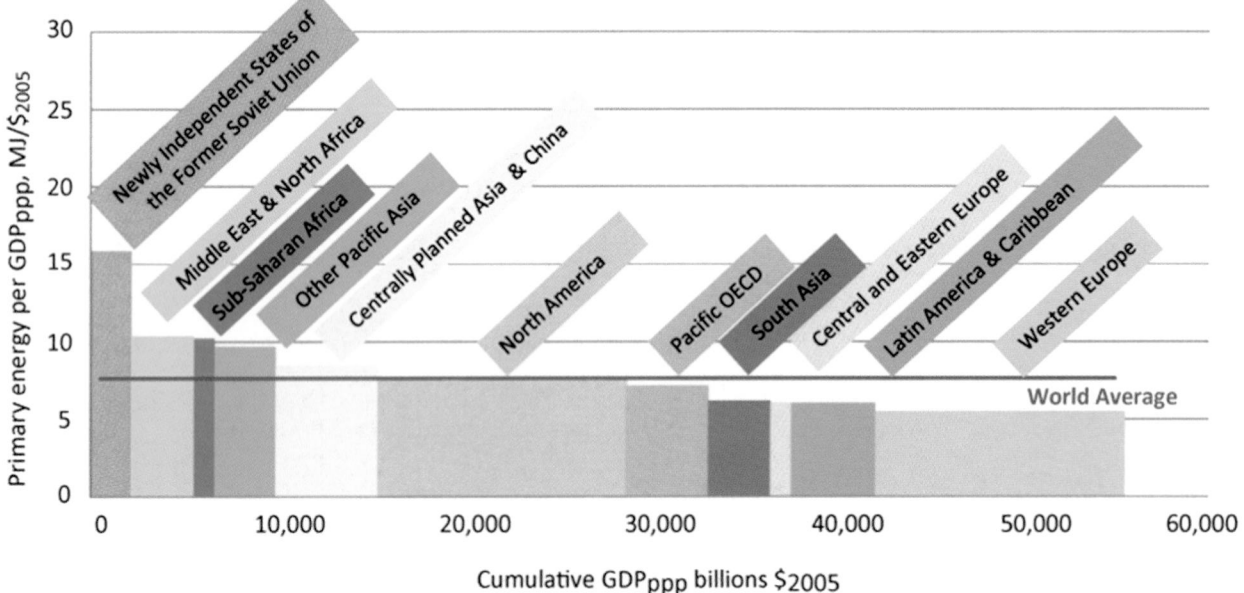

Figure 6.5 | Primary energy intensities across regions in 2005. Source: data from GEA Scenario Database.[10]

---

10   See www.globalenergyassessment.org

grew only by little more than 1.1% per year as a result of energy innovation, efficiency improvements, and structural economic change but also due to policy and energy price responses, all of which reduce the average energy input required for the production of one unit of GDP. The OECD graph clearly depicts how during the 1990s low energy prices slowed down energy intensity improvements with energy demand growth rebounding (see Figure 6.4). The first decade of the this century is marked by a very modest growth in primary energy due to higher prices, policy incentives stimulating efficiency improvements, and consequent lower energy intensities leading to negative growth rates of per capita energy and electricity demand.

6.4.2   Developing Countries

Despite fluctuations in global economic growth, some countries are consistent regional economic growth leaders[11]. The dynamics and influence of these countries, particularly China, India, Brazil, and South Africa, will be critical in terms of shaping not only the developing world but global energy, economic, and climate change mitigation trends. The enormous investment needs in energy infrastructure by these countries in the coming decades will provide a unique window of opportunity for sustainable, low-carbon energy development. At the same time, they will continue to face the challenges of sustaining economic growth and eliminating poverty.

A proper understanding of the type and direction of causality between energy use and economic growth in developing countries would help illuminate the role of energy in the future evolution of these counties' economies. Soytas and Sari (2003) concluded that there is bi-directional causality in Argentina and causality running from GDP to energy use in Korea. An earlier paper tested for co-integration between total energy use and real income of six Asian nations: India, Pakistan, Malaysia, Singapore, Indonesia, and the Philippines. Masih and Masih (1996) concluded the existence of unidirectional causality from energy to income for India, unidirectional causality from income to energy for Indonesia, and mutual causality for Pakistan. The pertinent inferences that were drawn by the authors from these conclusions were that improving living conditions and providing goods and services will require more energy services, and therefore an energy carrier is necessary. How much of this energy carrier is needed will depend on the energy end-use efficiency achieved in converting the energy carrier to energy services.

Another paper, based on data sets from 18 developing countries, refuted the neutrality hypothesis and concluded that energy use generally causes GDP growth and not vice versa (Lee, 2005). A recent paper focusing on China used a co-integration analysis and an error-correction model to examine the long-term equilibrium relationship between GDP

and energy supply and use in 1980–2005 (Wang et al., 2008). It concluded that the two variables are co-integrated but specifically that the growth of GDP forcefully drives the growth of energy supply to increase while energy has a little effect on GDP. Based on this analysis, they concluded that if scientific actions were taken regarding development in the energy sector by: keeping a reasonable economic growth rate; optimizing industrial structures; exploring the use of high-efficiency energy utilization, and developing energy technology, then reducing the country's energy intensity by 20% in five years is achievable.

Despite not many publications being available on the causality between energy and economic development in Africa, Wolde-Rufael (2009) provided a good introduction to this issue and examined the relationship between energy use and economic growth for 17 African countries in a multivariate framework that included labor and capital as additional variables. He found the existence of causality between energy use and economic growth in 15 of the 17 countries. In Kenya and Zambia, energy relative to labor and capital was the most important determinant of economic growth, but causality running from energy use to economic growth was marginally rejected. In 11 countries, energy was not even the second most important factor when compared with capital and labor. Even though energy's contribution relative to output was not so high as that of capital or labor, its contribution to output growth was still relatively high in Algeria (29%), Cameroon (41%), South Africa (23%), and Tunisia (44%). Energy made the least contribution to economic growth in Côte d'Ivoire, Gabon, Senegal, Sudan, and Zimbabwe (Wolde-Rufael, 2009).

An important observation is that for most African countries energy appears to be a smaller factor in economic growth and not as important as labor and capital. In contrast, for many industrialized and developing Asian economies energy is a relatively important contributing factor (Soytas and Sari, 2003; Soytas and Sari, 2006). This conclusion is consistent with the economic growth, energy supply, and use realities of most African countries. Many of these nations are characterized by low economic development, which is reflected in their limited energy development and consumption. In many cases energy supplies are unreliable and the infrastructure needed to meet the needs and demands are lacking.

There is also the issue of consumers lacking access to commercial energy markets. All these characteristics of the energy sector tend to decouple energy supplies and use as drivers for economic growth and development. What is clear is that African countries must endeavor to find ways to direct investments into energy infrastructure development as well as to reduce inefficiencies in energy supplies and use in order to stimulate and promote sustainable economic development. In recent years, a number of African nations have become oil producers, boosting GDP and offering a source of funding for the development of energy infrastructure. Nevertheless, the proximity of oil production does not imply that more energy will get to African households and smaller businesses. Instead, coherent policies will be needed to ensure that the infrastructure and supply reach them.

---

11   Defined as large emerging economies with global impacts on demand and supply balances.

Despite the many studies which estimate the effect of energy on economic development in developing economies, the research is fraught with difficulty. First, many low-income economies are predominantly agrarian. Much of their energy services are provided through muscular effort of human and animal power, which is not included in energy statistics. Similarly, a great deal of the heating, either for cooking or warming space and water, is fueled using biomass, which is also omitted from most estimates of energy use. Few econometric studies of the causality of energy on GDP (or for that matter, of GDP on energy) incorporate these traditional fuels. Also, time series studies in developing countries are based on few data points, and cross-sectional analysis suffers from a great deal of economic, political, institutional, social, and cultural factors that are hard to include as variables. Thus, conclusions about the smaller influence of energy on GDP in developing economies compared with post-industrial ones must be taken with some caution (Chontanawat et al., 2008).

6.4.3   Small and Medium Enterprises

In both developing and developed countries, small and medium enterprises (SMEs) are major contributors to job creation and economic growth (UNCTAD, 2001). The figure and facts below show the important role of SMEs and the need to take them into account when examining the energy and economy nexus in any economy.

SMEs account for over 90% of the number of businesses in the world and for 50–60% of worldwide employment (USAID, 2009). Their contribution to GDP and employment share increases when national economies mature and branch out. In an evolutionary view to the economic structure, they are crucial in providing diversity ("mutations") in goods and services offered, from which the evolving preference structure of consumers choose. They are the backbone of innovative and dynamic technology development, often crucial in bridging the "valley of death" between lab and the marketplace. In emerging economies and liberalized markets they can also be important vehicles of social mobility. Evidence shows that SMEs have played a major role

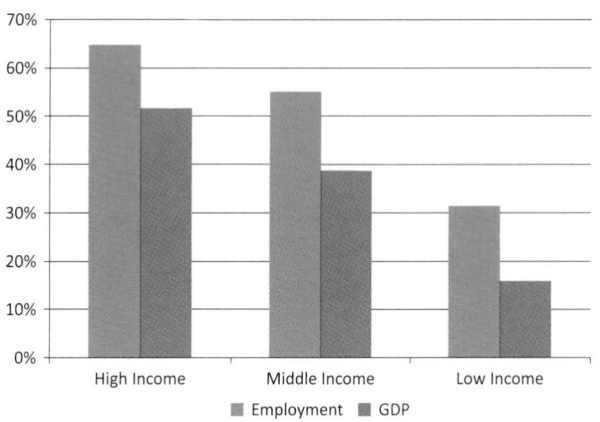

Figure 6.6 | SME contribution to employment and GDP. Source: Yago, 2007.

in the growth and development of all the leading economies in Asia (UNCTAD, 2001).

Access to reliable, affordable, high-quality modern forms of energy is vital for SMEs to operate efficiently and profitably. Large enterprises can afford back-up generators and often benefit from preferential tariff structure and supply security from the utilities. SMEs in contrast pay the highest electricity and modern fuel prices (USAID, 2009), and energy expenses of SMEs in developing countries commonly range from ten to more than 65 percent of the total costs of production. The estimated costs of intermittent electricity provision in several African economies ranges in the order of magnitude of 5–10% of GDP (World Bank, 2011a); the dominant share of this is due to effects on SMEs

Even within the energy sector, SMEs can provide important technology initiatives. The commercialization of formerly unconventional tight and shale gas deposits in the US was largely led by small and medium size energy companies, and significantly changed the supply security of the US gas market. Also in Europe (particularly in Denmark and Germany) small and medium sized enterprises contributed to technology development and reshuffling of the energy markets. A condition was the effective liberalization (and unbundling of supplier and network operator for grid based energy carriers) and democratization of access to energy markets, e.g., through feed-in-tariff structures.

6.5   The Prices and Costs of Energy

6.5.1   Prices as Market Signals

Prices play several essential roles in the economic production and consumption process. Most important, prices send signals to buyers and sellers. When market prices change, they provide incentives to consumers and producers, and demand and supply of goods and services adjust accordingly. Higher prices may signal scarcity and provide incentives for buyers to purchase less of that good or service or look for alternatives. Higher prices can also adversely affect the affordability of basic services and thus extend poverty. For producers, higher prices stimulate additional supplies and increase sales. In short, prices regulate the quantities of goods and services supplied and consumed. If prices increase because of demand exceeding available supplies, they also are a way to let producers know what consumers demand.

Prices also determine income and profit. For producers, price times the quantity sold results in the total revenues, which – when corrected for production costs and taxes – become income. For consumers prices determine affordability, and price time the quantity purchased governs the disposable income left for other activities.

It is important to distinguish between prices and costs. While the price of a commodity or service represents indeed costs to the buyer (the cash outlay at the time of purchase), from the perspective of the consumer there are other economically relevant "cost" factors – ranging from inconvenience and individual preference to regulation and externalities. For example, in terms of direct costs, heating a home with coal would still be the cheapest way in many jurisdictions. When affordable, however, most home owners prefer fuels such as natural gas. It is simply more convenient (no shuffling of coal, easy temperature control, no extra space for coal storage), cleaner, and more benign for the environment. Moreover, coal combustion in residential areas is banned in many communities. In essence, the price includes a component of willingness-to-pay for the avoidance of inconvenience. In the absence of monopolistic supply situations, price in a competitive market is to a large part a matter of opportunity costs.

Producers or sellers determine the price of providing the fuel, with production cost usually lower than the sales price and with the difference being profit. Sellers test the market and explore what prices consumers are willing to bear and adjust supply in response to demand.

6.5.2   Costs Associated with Energy Supply and Use

The economic literature often distinguishes four types of costs: monetary costs, opportunity costs, environmental and health costs, and sociopolitical costs (Schipper and Meyers, 1992). Most consumers are predominantly exposed to monetary costs – that is, the money they pay for the goods and services they consume, such as the monthly electricity bill or the price at the pump when filling up the car. Enterprises purchasing raw materials, compensating workers (paying wages), servicing debt and dividends, or paying property taxes are also paying monetary costs.

Opportunity costs represent the value of alternative uses of investments, labor, materials, and so on if used for other economic purposes instead of for the supply or use of energy. For example, what is the cost of sequestering carbon by growing trees on a tract of fallow public land? In particular, what is the value of the land, which could well be zero "cost" because the land currently does not earn any rent? In terms of opportunity costs, the cost of the land is to be measured as the value of the output that could be received from that land if used for other activities, such as a shopping mall or soccer pitch.

Health and environmental costs include the premature deaths, injuries, and illnesses suffered by workers in energy supply industries as well as the public at large as a result of effluents and accidents associated with energy supply; damages to buildings, infrastructures, agriculture, and forestry productivity, tourism, and so on; climate change, overloading the carrying capacities of ecosystems, and loss of biodiversity; and nuisance from noise and odor, congestion, or visual blight (Schipper and Meyers, 1992).

Sociopolitical costs include adverse impacts on energy security, geopolitical relations, income distribution, or land use patterns resulting from

energy supply and use. Also included are undesired impacts on cultural, community, and family values.

While the monetary costs associated with energy supply and use are fully paid by the producer or user, this is only partially the case for the opportunity, health and environmental, and sociopolitical costs. Insofar as these are not incorporated into the monetary costs, they are called "externalities."

6.5.3   Externality Costs

Externalities arise when an economic agent enjoys benefits or imposes costs without having to make a payment for doing so. As such, externalities can be positive or negative. For example, the adverse health and environmental damages (hidden costs) caused by fossil-sourced electricity generation that are not compensated by the producer are negative externalities. At the same time, the cheaper electricity (without externalities) enjoyed by consumers and that contribute to overall welfare generation represent positive externalities. Factoring external costs into the market price of energy ("internalization") would raise prices. It would send correct pricing signals to the marketplace and thus change the merit order of investment and operating decisions as well as reduce demand and emissions, with subsequent lower externalities.

Negative externalities are often associated with using public goods provided by the environment for free (e.g., air, soil, water, landscape, ecosystem services). To reduce these, a utility needs to take countermeasures, such as installing pollution abatement equipment or making compensation payments for the damages caused. Identifying, measuring, and monetizing externalities are particularly important steps for the quantification of hidden costs and assessment of the effectiveness of policy instruments aimed at internalizing external costs. Since private enterprise normally does not incorporate external costs in investment decision-making, government intervention is necessary to "internalize externalities" resulting from energy supply and use. According to the International Energy Agency, "governments are best positioned to assess, on a broad scale, the social and environmental costs and benefits associated with power generation, as well as the energy security aspects of, for example, a high dependence on natural gas imports destined to the power sector" (IEA/NEA, 2010).

With the exception perhaps of climate change, the internalization of externalities has been inextricably linked with socioeconomic development. As incomes and welfare grow, parts of the health and environmental impact costs have been increasingly internalized through regulation, such as mandatory emission abatement, caps on effluents or waste charges, higher wages compensating for occupational risks, employer-paid insurance schemes, and user fees. As regards climate change, cap-and-trade arrangements such as the European Emission Trading System, carbon taxes, mandatory performance standards, etc. are tools for internalizing the costs of using the atmosphere as a carbon dioxide waste repository. Their individual effectiveness depends on several factors ranging from levels of caps, tax rates and assessed penalties to rigor to implementation (enforcement) and possibilities of leakage.

The evaluation or monetization of externalities is highly controversial. While the evidence that externalities are real is generally unquestioned, their quantification has been fraught with uncertainties arising from issues of boundaries (what to include) to the valuation of loss of life. Externalities attributed to emissions from energy supply and use have been assessed using the impact pathway approach – that is, the pathway from emissions through dispersion, exposure, physical impact, and damage to the monetization of the damage costs to individuals or society at large (Rabl and Spadaro, 1999; Ricci, 2010). Controversy usually emerges in the last steps of this chain.

Other methodologies include the willingness to pay (WTP) and willingness to accept (WTA) approaches (Markandya and Boyd, 2002). Both WTP and WTA are closely related to the valuation of opportunity costs. WTP/WTA can be interpreted as the willingness to pay for the avoidance of a damage cost or the financial compensation for any damage inflicted that is deemed acceptable. Paying tolls for the use of a highway to avoid congestion and reduce commuting time is an example for WTP. (Note: the congestion charge presents a positive externality for motorists who stay on the regular roads that now are less congested.) Accepting a higher wage for a risky construction job is an example for WTA.

Several studies have attempted to quantify externalities, most of which focus on electricity generation (EU, 2003; Ricci, 2010). The latest systematic analysis of external costs of various electricity supply technologies and their associated chains is available from the European Commission's CASES project (Markandya et al., 2010). Figure 6.7 provides the summary results for the 27 countries of the European Union (EU), estimated for the period 2005–2010.

Human health impacts due to classic air pollutant emissions and the adverse consequences of GHG emissions dominate the external costs across all technologies. On a life-cycle basis, renewables and nuclear power emit only a few grams of GHGs per kWh. The full technology chain for nuclear energy includes the front end of the fuel cycle, power plant construction and operation, the back end of the fuel cycle (including reprocessing), the construction of interim and final repositories, and plant decommissioning. The main contributors to GHG emissions are plant construction (emissions from cement and material production and component manufacturing) and, in the case of nuclear power, enrichment of uranium (depending on enrichment technology and fuel mix used for the electricity input) (Rogner, 2010).

Due to the higher amount of GHG emissions and air pollutants along their respective energy chains from resource extraction, conversion, transmission and distribution to end-use, fossil-based electricity generation followed by biomass-sourced technologies have considerably

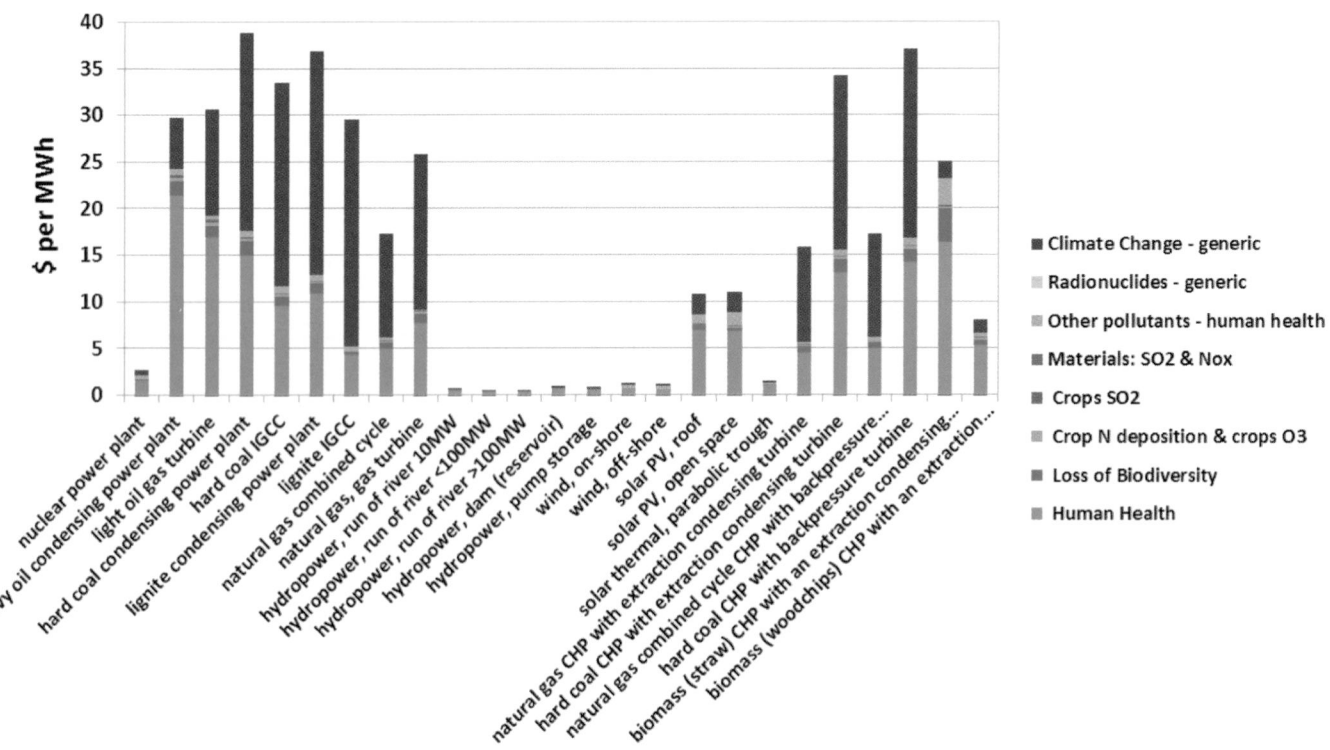

Figure 6.7 | Average external costs for the European Union. Source: Markandya et al., 2010.

higher external costs than renewables and nuclear power, generally ranging between 8.1 and 39.0 US$/MWh which is more than nuclear power and most non-biomass renewables technologies. Externalities are also location dependent as well as how wide the boundaries of the analysis are drawn, i.e., what is included in the analysis and what is not, especially with respect to indirect factors.

Finally, a few words of caution: In addition to the uncertainties associated with quantifying externalities, the external cost data discussed here are dynamic over time and highly nonlinear. Technology change affects the overall economic and environmental performance of energy conversion technologies (higher efficiency, better abatement equipment, etc.). Effluent reductions often result from capital turnover of retired plant and equipment by new technology. Next, the costs are highly nonlinear because of either saturation or threshold effects, which have different implications:

- An example of a saturation effect is found in certain disease risks from air pollution in which an increase in pollution causes a much larger increase in health burden for populations living in relatively clean environments than the same pollution increase causes in populations already living with high pollution levels.

- An example of a threshold effect is found in the impact of acid precipitation on some ecosystems in which the environment is able to absorb extra acid up to a point without much damage but above that level (sometimes called a "tipping point"), the damage increases dramatically.

Moreover, external costs vary considerably from location to location, depending on population density, geography, land use patterns, wind speeds and direction, regulations, and so on.

6.5.4   Resource Depletion as an Externality

Resource economics finds its roots in the perception that natural nonrenewable resources are being extracted too quickly and sold too cheaply for the good of future generations – that is, their excessive cheapness has given rise to wasteful use and inefficiency. As Harold Hotelling noted in 1931, "contemplation of the world's disappearing supplies of minerals, forests and other exhaustible assets has led to demands for regulation of their exploitation" (Hotelling, 1931).

A dynamically efficient allocation occurs when the present value of marginal profit or marginal scarcity rent for the last unit produced is equal across various time periods. Therefore, for resource markets to be in equilibrium, the marginal profit from resource sales must rise at the rate of interest (see Figure 6.8). Then owners of resource stocks are indifferent between extracting the marginal unit of the resource and leaving it in the ground, since the return on holding the resource stock as an asset is equal to the return on alternative interest-bearing assets (SAUNER,

2000). It also implies that if there is no substitute for a resource that is essential to society, there is no maximum resource price. The efficient extraction path may then be identified by noting that the resource stock must reach zero in infinite time.

The rationale behind what is known as Hotelling's Rule is that as a finite resource is produced, less will be available in the future – causing scarcity rent, and thus the resource price, to rise. Increasing prices reduce the quantities demanded by the market and preserve the resource for future use.

In the presence of an alternative technology or a substitute for the resource (usually referred to as backstop technology), the resource price rise is capped at the level when the backstop becomes economically viable (backstop price). In Figure 6.8, the backstop price is the flat part of the price path (P_b), which, once reached, causes discontinuation of production. At this point (T), the resource is said to be "economically exhausted" even though there is plenty of the resource remaining in the ground (Q_T).

However, Hotelling's approach to efficient resource extraction trajectories ignored production costs, technical progress in exploration and

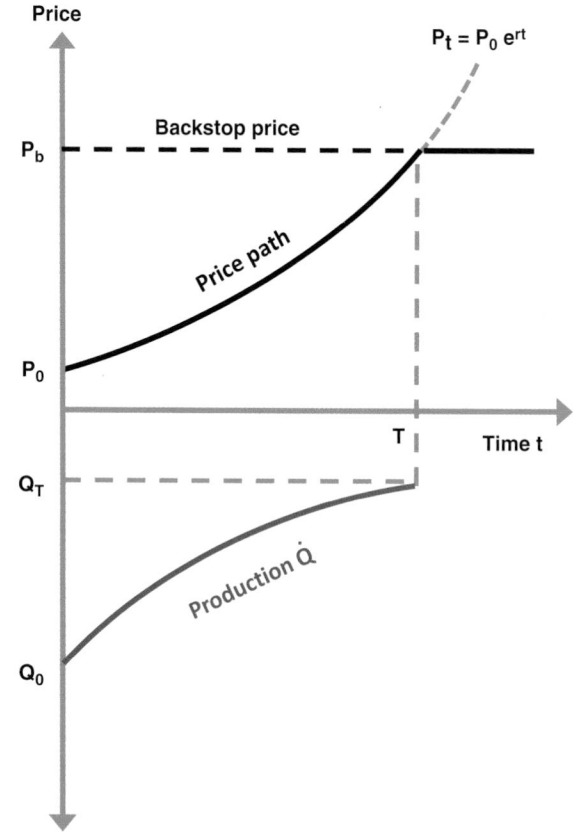

Figure 6.8 | The role of backstop technologies in capping resource prices.

production technologies, and new discoveries. Over the years, several authors have extended the analysis and included these essential real-life aspects in the analysis (e.g., Hartwick and Hageman, 1993; Krautkraemer, 1998, for extraction costs; Dasgupta and Heal, 1979, for exploration and new discoveries).

The questions of interest are whether the historically observed resource prices bear any relation to the Hotelling rule and, more important in relation to the "peak oil" debate, to what extent the observed price paths are an indication of resource growing scarcity over time.

Barnett and Morse (1963) analyzed mineral resource prices between 1870 and 1957, finding that prices fell. They attributed this to new discoveries, to technological progress in resource extraction and processing, and to substitution of alternative materials. Periods of increasing prices have been noted, such as oil prices during the 1970s, but they were attributed to geopolitical tension and Organization of Oil Exploiting Countries supply restrictions (Barnett and Morse, 1963). Lack of investment in exploration and production capacity by producer countries and an unforeseen strong demand growth have been blamed for the oil price hikes during the first decade of this century (IEA, 2008b).

Berck and Roberts (1996) offer three possible explanations for stagnant or decreasing resource prices: technological progress in resource exploration, extraction, and processing; high natural abundance (so scarcity is not yet an economic problem); and environmental or policy constraints limiting production or use where the physical occurrence of the resource is not the limiting factor and the resource is not therefore economically scarce (SAUNER, 2000). Norgaard (1988) attributes the failure of resource prices to rise to the shortsightedness of markets. Increased scarcity is not a sufficient indicator as long as markets are not capable of reflecting it. Markets are simply not farsighted enough for the intertemporal arbitrage function required for price paths to be determined by resource scarcity.

The expectation that innovation and technical progress may reduce the cost of the backstop technology below the scarcity rent trajectory well before existing stocks approach depletion is another explanation for prices staying below scarcity rents. The threat of a cheap backstop encourages depletion of reserves more quickly rather than restricting production and causing prices to rise.

Regarding projections of future energy resource price paths that usually reflect Hotelling's rising scarcity rents (Manne and Schrattenholzer, 1986). Berck and Roberts (1996) found that the probability of rising rents depends on the econometric model used and the assumption about the size of resource stocks.

A final issue on trends in resource prices is the divergence between long run prices of energy and of energy services. Nordhaus (1997) and Fouquet (2011) showed that over decades or more these price trends tend to diverge. This implies that studying long run trends in energy

prices, and the apparent resource scarcity, indicates only partially the prices and incentives facing consumers, who benefit from technological development. Indeed, resource scarcity and higher energy prices tend to encourage the adoption of energy efficiency measures that after accounting for all implementation and transaction costs involved, lower energy service prices. So, despite some threats of resource scarcity, the long run trends in energy service prices have been downwards since the Industrial Revolution (Fouquet, 2011). Thus, although there may be short- and even medium-term scarcity related to individual fuels, it is questionable whether consumers are likely to face rising long run trends in energy service prices (Nordhaus, 1997; Fouquet, 2011).

6.5.5 Discounting and Discount Rates

Generally, the discount rate accounts for the time value of money. It reflects the general attitude that money available today is worth more than the same amount in the future. It is expected that deferred consumption, say by investing in bonds issued by a utility, should earn interest. The discount rate then determines the present value of future interest payments received by the investor. The discount rate also reflects the risk and uncertainties associated with an investment – that is, higher risk projects command higher returns, hence higher discount rates. In short, discounting is a critical step in determining whether a project or investment is desirable and is used as a tool to make costs and benefits with different time paths or different risks to the investor comparable. Arguments for why costs and benefits with different time profiles are not comparable typically make several points (Dasgupta and Pearce, 1972; Arrow et al., 1996). The first is that individuals expect their level of consumption to increase in the future, so that the marginal utility of consumption can be expected to diminish. Alternatively, individuals can have a positive pure time preference either because they are generally impatient or myopic or because they perceive a risk associated with not being alive in the future. A second argument for discounting future costs and benefits takes the perspective that capital is productive and that resources acquired for a particular project can be invested elsewhere, generate returns, and so have an opportunity cost.

Discount rates are key elements in most energy investment decisions, especially when the choice is between high upfront investments but low operating costs versus low upfront and higher operating costs (essentially all technologies potentially involved in large scale energy system transformation share the common characteristic of higher upfront investment costs and lower operating costs than traditional forms of energy service supplies). Public policy considerations affect the choice of the social discount rate and opportunity costs more than private sector entities where competitive returns on investment commensurate with vthe (perceived) risks determine discount rates.

Choosing the appropriate social discount rate has long been contentious. Setting the social discount rate too high could preclude many socially desirable projects or investments from being undertaken, while setting it too low risks encouraging a lot of economically inefficient investments. Further, a relatively high social discount rate, by attaching less weight to benefit and cost streams that occur in the distant future, favors projects with benefits occurring at earlier dates, while a relatively low social discount rate favors projects with benefits occurring at later dates. The choice of the social discount rate affects not only the ex-ante decision of whether a specific project deserves funding but also the ex post evaluation of its performance (Zhuang et al., 2007).

At an individual level, the discount rate affects inter-temporal consumption decisions. Insofar as the choice of energy requires the purchase of energy-using capital equipment, the amount of investments consumers choose to make will depend on the perceived financial profitability. There is evidence that consumers use high implicit discount rates, hindering the adoption of energy-efficient technologies. Specific causes of high implicit discount rates include a lack of information about the cost and benefits of efficiency improvements, a lack of knowledge on how to use available information, uncertainties about the technical performance of investments, a lack of sufficient capital to purchase efficient products (or capital market imperfections), a low income level, high transaction costs for obtaining reliable information, and risks associated with investments (Train, 1985; Lutzenhiser, 1993; Jaffe and Stavins, 1994b; Jaffe and Stavins, 1994a; Howarth and Sanstad, 1995).

In the case of low-income populations, capital scarcity, liquidity constraints, and high transactions costs associated with borrowing can result in households having particularly high implicit discount rates. Hausman (1979) and Train (1985) also argue that implicit discount rates vary inversely with income. Evidence from India suggests that these high implicit discount rates among poor consumers can hinder the adoption of cleaner-combusting, more efficient and convenient fuels and technologies (Reddy and Reddy, 1994; Ekholm et al., 2010). Due to the fact that the use of high implicit discount rates may be a function of asymmetric information, bounded rationality, low incomes, and/or transaction costs, it is argued that policy instruments may affect the implicit discount rate used by consumers by targeting those market imperfections (Howarth and Sanstad, 1995). In the study by Ekholm et al. (2010), the availability of easy and cheap microfinance, for instance, influenced consumer choices regarding cooking fuels and technologies by lowering the costs of borrowing and making upfront capital stove purchases more affordable. In many richer industrialized countries as well, studies have shown that financial incentives that lower the upfront cost of investments can influence individuals' technology adoption and investment decisions (WEC, 2008).

6.5.6   The Impact of Long-term and Short-term Energy Demand

The price of energy is but one element in determining the price of an energy service. It is the price of the service – the combination of the prices of service technology (building, heating system, vehicle, or light

bulb) and the fuel (and convenience) – rather than the price of the fuel alone that matters to consumers.

Demand for a particular fuel then derives from the demand for energy services provided by that fuel, the price of the fuel, and the techno-economic performance of the conversion technology and related infrastructure. For heating services this would include the building envelope, the boiler, the heat distribution system (in short, labeled "service technology"), and possibly any environmental compliance plus, of course, climatic conditions and the preferred indoor temperature.

Here the time horizon plays an essential role. In the short term the service technology is "fixed" and represents a "sunk" cost, which explains the focus on day-to-day oil price movements or the adjusted structure of electricity rates. In the longer run, services technologies as well as infrastructure are replaced ("capital turn-over") as they reach the end of their service times (see Chapter 1). This opens the opportunity of replacing the aged and underperforming equipment with more-efficient and cleaner models and designs. How quickly the energy-related capital stock can be replaced and to what degree best-available technologies will be adopted depends, however, on several constraints, ranging from access to finance, individual preferences to information, policy incentives, and market transparency (IEA, 2010b). Moreover, consumers attempt to optimize their utility of using energy for meeting the desired energy services over time. Especially, the market penetration of service technologies with high upfront investment costs is largely determined by the underlying individual discount rate described earlier.

Figure 6.9 depicts the interplay between short-term and long-term demand responses to a generic one-time energy price rise. The point of

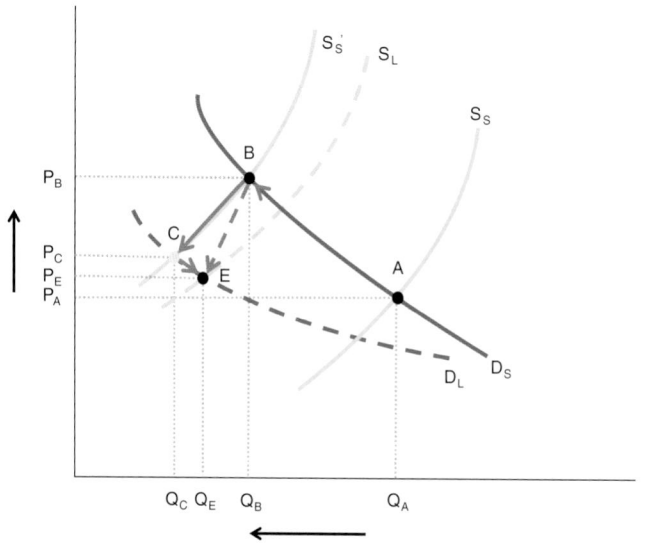

Figure 6.9 | Short and long-term demand/supply responses to a price increase due to a loss of supply capacity. Source: adapted from Samuelson et al., 1988.

departure is a market equilibrium situation at point A (that is, demand and supply functions intersect for demand Q at the market clearing price P_A). A price increase to P_B caused by a loss of supply capacity due to a geopolitical conflict induces short-term demand responses. A steep rise in gasoline prices leaves most consumers with little choice in the short run: pay the high price at the pump, drive less, or shift to other modes of transportation (behavioral adaptation). All this results in a movement from point a along the short term demand curve to point b. The option of purchasing a more efficient car usually arises when the current vehicle approaches the end of its economic service life. In the graph, demand moves along the short-run demand curve D_S, characterized by the fixity of existing capital infrastructure. The loss of supply capacity shifts the supply curve S_S to $S_S{}'$.

Although demand is reduced to point Q_B, consumers pay an overall higher energy bill than before despite the cutback in use. In the longer run, when more-efficient appliances, vehicles, and capital in general penetrate the marketplace or when buildings are progressively better insulated, demand declines further to Q_C, now determined by the new long-term demand curve D_L, which represents the efficiency, technology, and infrastructure adjustments undertaken in response to the price increases. The shift of the demand curve and lower demand reduce supply requirements, and prices drop to P_C. In the interim, higher prices also simulate energy supply- side measures, (e.g. additional exploration, additional mining capacity, etc.) which effectively easing the pressure on prices. The combination of demand and supply-side responses leads to the new equilibrium E. Market clearing now occurs at demand Q_E (lower than Q_B but higher than without the supply side response Q_C) and at a market price P_E.

The production function represents the energy supply to economy link needed for providing key services, including information exchange, transportation, heating and cooling. In contrast, the demand function represents the economic income effect on the demand for energy services influenced by economic growth and development. Regardless of the methodology applied, the influence of economic activity on energy demand is unquestioned. Equally unquestioned is the importance of energy input for economic production, process, and economic development, but the relationship between energy and economic growth is not unidirectional. A major source of uncertainty, however, concerns the dominance over time of relationships between the economy and energy requirements to fuel the economy, on the one hand, and energy as the driver of economic (GDP) growth, on the other. As such, energy plays a vital role in economic development and growth, and it underpins countries' ability to provide employment for their populations. Increased production requires more energy inputs, while greater disposable income increases the demand for heating, cooling and mobility. More generally, the demand for energy is likely to vary at different stages of economic development, with the availability of resources and technologies, according to economic, social, and institutional structures and policies, and across cultures.

6.5.7   Energy Markets and Energy Price Subsidies

While markets are playing a growing role in funding energy investment, significant impediments continue to fundamentally constrain progress in moving toward an energy system and supply patterns for a more sustainable society. Key issues include energy prices that are not cost-reflective, and energy subsidies which distort markets. Energy subsidies and cross-subsidies have profound effects on countries' ability to lift incomes and economic growth by making energy affordable. Feed-in tariffs for low-carbon electricity generation are policy tools for guiding action and channeling investment into socio-politically desirable directions, assisting technology learning (cost buy-down) and reducing investor risks. In essence, subsidies absorb risks and costs private sector entities are not willing to bear for reasons of competitiveness and profitability. However, if not time-bound and performance based, subsidies in the longer run do not create incentives for energy suppliers to recover their costs and invest in a timely manner, discourage new entrants and private-sector investment, and undermine the pursuit of energy efficiency objectives.

Many hundreds of statistical studies have been undertaken to assess the responsiveness of the demand for different forms of energy – including electricity, natural gas, oil and oil products, coal, or total demand for all forms of energy – to changes in prices. Irrespective of the form of energy considered, the demand for energy is typically price-elastic, so that energy use declines as the price of energy increases (and vice versa). A recent study of energy use expenditure in the United States between 1970 and 2006, for instance, found that use of all forms of energy declines in response to energy price increases (Kilian, 2008), although there are important differences across different forms of energy. The study identified a short-run (over one year) price elasticity of gasoline of –0.48 (so that a 10% price increase in gasoline will reduce consumption by 4.8%), and a price elasticity of electricity of –0.15 (so that a 10% electricity price increase will reduce consumption by 1.5%). Elasticities of energy demand with respect to price are typically higher over the longer term, as users have time to adjust, either by switching to alternative fuels or by conserving energy.

In the normal course of events, rising energy prices associated with a scarcity of energy resources would therefore tend to discourage use, but at a minimum it would ensure that energy is not "wasted" and that GHG emissions are not created needlessly. The prevalence of energy price subsidies, in particular subsidies of fossil fuels, however, fundamentally undermines this objective. In some countries (mostly industrialized market economies) energy markets have been deregulated so that prices for energy carriers and services provided to consumers are broadly cost-reflective. Many countries, however, including many developing nations, have administered energy pricing regimes that are not cost-reflective and that entail significant cross-subsidies from one sector (for instance, industry) to others, such as the residential or agricultural sectors. Thus a recent extensive survey found that 37 countries account for over 95% of global subsidized fossil fuel consumption, leading to higher levels of consumption than would occur without price distortions IEA, 2010a; World Bank, 2010). Overall subsidies amounted in 2008 to US$557 billion (an increase of US$215 billion from 2007).

Phasing out these subsidies would provide a clear incentive to use energy in a more efficient manner and would facilitate the switch from fossil fuels to less GHG-intensive energy sources. Modeling undertaken by the IEA (2010b) indicates that phasing out energy subsidies between 2011 and 2020 would:

- cut global energy demand by 5.8% by 2020;

- cut global oil demand by 6.5 million barrels per day in 2020, predominately oil used in the transport sector; and

- reduce carbon dioxide emissions by 6.9% by 2020, the equivalent of 2.4 Gt of carbon dioxide.

Subsidized energy prices that do not reflect the costs of producing and supplying that energy (including the cost of the very significant infrastructure required to do so) encourage wasteful consumption, but they will also not provide sufficient compensation to producers. Timely investments to meet growing demand are therefore not undertaken by existing utilities; the incentives for new investors to enter the market are similarly removed. While low energy prices may promote access to modern energy carriers in the short term, in the longer term such policies discourage investment in the energy sector, thereby constraining supplies and economic development.

Policies to phase out subsidies for kerosene, liquefied petroleum gas (LPG), and electricity must be carefully designed not to restrict access to essential energy services, as these fuels often are used to meet the basic needs of the poor people. IEA (2010b) analysis indicates that today 1.4 billion people around the world are still without access to electricity and around 2.7 billion people rely on traditional biomass as their primary source of energy (see Chapter 2). However, subsidies to kerosene, LPG, and electricity in countries with low levels of modern energy access (that is, with electrification rates below 95% or access to modern fuels below 85%), represented just 11% of the US$557 billion in consumption subsidies in 2008.

Furthermore, while fuel subsidies are often justified on the grounds that they help address income inequality and provide assistance to low-income groups, in practice such subsidies represent inadequately targeted transfers, with most of the benefit accruing to the largest consumers of oil products, who are typically not the poorest members in the society (World Bank, 2010). A reduction in subsidies combined with a more effective system of taxation would enable government revenues to be better targeted at income transfers, with the broader aim of supporting investments in education, health, and physical infrastructure to assist in economic development (Barnes et al., 2008).

6.6   Investments in Energy

6.6.1   Investment Characteristics and Trends

The transition to modern energy systems is characterized by increasing investment levels in technology and infrastructure. Investments are necessary for energy resource extraction, energy conversion to usable fuels, transmission and distribution systems, and end-use infrastructures. In principle, there are two major categories of investments that are intimately interrelated: investment in the expansion of (and replacement of retired) technologies and infrastructures under competitive or regulated market conditions and investment in innovation, development, and commercialization, including market formation. While the first category is important for supporting energy system growth – from mobilizing upstream exploration and resource extraction to energy conversion and supporting access to energy services, especially in the short-run for balancing demand and supply – it is the second category that enables energy system evolution and transformation, and hence progress in terms of socioeconomic development and environmental protection. The second category is also dependent on policy support, especially market formation.

Market-formation investments include public and private investments in the early stages of technological diffusion and are sometimes also referred to as "niche market" investments. In the energy domain, these investments include policies with respect to certain technologies (such as feed-in tariffs or production tax credits) and public procurement. They also include private investments that may take advantage of markets created by government policies, such as renewable performance standards or price instruments like carbon taxes (UNDESA, 2011).

Market-formation investments in the energy sector as a whole are difficult to track, because many transactions are unreported, the ways of measuring market-formation investments are not yet harmonized internationally, and efforts to track such investments are only relatively recent.

Investment in infrastructure growth and innovation are interlinked, as investment in innovation is a prerequisite for the development of improved and better-performing technologies and processes. Investment in growth and replacement of plant and equipment creates market opportunities for the diffusion of innovative technologies, whereas growth generally offers larger penetration opportunities than mere replacement in an otherwise stagnant market. Using the natural rate of capital turnover for the introduction of innovative technologies has lower transaction costs.

Investment is deferred consumption, and all the features of risk, discount rates commensurate with the risk, present value considerations, opportunity costs, and of course demand discussed earlier apply equally here as well. Energy sector investment time horizons are long-term: investments are made over periods of up to 15 years before the first revenues are received, and capacities are built to last for 15–60 years and more. The long lifetime of energy sector capital means a slow turn-over of its capital stock – a limiting factor to speedy energy system transformation. The recent financial economic crises and the uncertainty in financial markets about the indebtedness of key industrialized countries have made energy infrastructure a highly risky proposition, resulting in underinvestment in key energy sector areas (exploration, production capacity, transmission, and distribution grids). Nationalistic policy solutions focusing on the short run without a global vision (such as of energy system transformation or a global international environmental agreement) further aggravate uncertainty for private-sector investors and result in lock-in effects for the little investment that takes place. The risk premium on such investment and the widening demand-supply gap lead to higher energy prices until additional investment is forthcoming. However, absent solid and predictable energy policy objectives and long-term policy targets which are hugely important ingredients for investor confidence, such investment is unlikely forthcoming.

Since 1980, global total annual investments have fluctuated between 21 and 24% of gross world product with at times a significantly higher share in developing countries and a slightly lower share in the industrialized world (UNCTAD, 2007; IMF, 2011). The share of capital formation allocated to the energy sector is estimated at about 4–8% of total investment, or 1.0–1.8% of GDP. These figures exclude energy-related investments at the end use of the energy system (buildings, heating systems, cars, refrigerators, etc.) that are delivering the energy services that consumers demand.

At the country level, especially for small but energy-exporting countries, the share of energy supply–related investments of GDP can be much higher than the global share and at times can amount to 5–10% of GDP or more.

6.6.2   Energy Supply Investments

Data on energy supply investments are extremely limited, so the literature typically relies on limited surveys or on model estimates (multiplying statistical data and/or estimates on capacity expansion with average technology-specific investment costs to derive total energy supply investments). Energy supply modeling studies have been available since the mid-1990s in academia (e.g., Nakicenovic and Rogner, 1996; Nakicenovic et al., 1998; Riahi et al., 2007) as well as from the work of the IEA, particularly the *World Energy Investment Outlook* (IEA, 2003); the *Energy Technology Perspectives* (IEA, 2006b; IEA, 2008a); and the recurrent projections of the *World Energy Outlook* (WEO) (e.g., IEA, 2006a; IEA, 2007; IEA, 2008b; IEA, 2009a; IEA, 2010a; IEA, 2011), which also contain unique survey data on energy supply investments, particularly in the oil and gas industry.

A common feature (and drawback) of all modeling studies is that energy sector investments are not reported for their corresponding base year values but instead as cumulative totals of the projection period of typically

30 years. The absence of published base year input data for energy sector investment projections not only reduces the credibility of the studies, it also makes an assessment of current investment levels and structure and a comparison among the different studies almost impossible.

The assessment in this section summarizes available information by drawing on the only modeling study that has disclosed its underlying base year energy investment numbers (Riahi et al., 2007) and the surveys reported in IEA's WEO (IEA, 2006a; IEA, 2007; IEA, 2008b; IEA, 2009a; IEA, 2010a; IEA, 2011). Because of the significant price escalation observed for energy sector investments (particularly for oil and gas since 2004), the Riahi et al. (2007) estimate (which refers to year 2000 investments and price levels) can be considered a lower bound, assuming recent price escalations will not remain permanent. Conversely, the IEA numbers can be considered as an upper-bound estimate of investments in energy supply. Comparing and making sense of investment estimates or quotes from different sources is fraught with uncertainty as boundaries of what is included in a particular estimate and what is not, which price basis and exchange rate has been applied, etc. are rarely clearly specified and documented.

Despite differences and uncertainties in estimated supply-side investments per supply category, the available data suggest a range of energy supply-side investment during the mid 2010s of US$700 billion a year to some US$840 billion a year (in 2005 dollars). Investments are dominated by electricity generation and by transport and distribution (T&D), at some US$500 billion. Fossil fuel supply, particularly the "upstream" component (exploration and production), accounts for US$250–400 billion annually, mostly for oil and gas.

Renewables are still relatively minor players under current energy market conditions despite substantial subsidy support for market formation. Liquid and gaseous biofuels account for US$20 billion, including US$8 billion for Brazilian ethanol (UNEP/SEFI/NEF, 2009). Large-scale hydropower (approximately US2005$40 billion for annual capacity additions of between 25 and 30 gigawatts) accounts for some 17% of current supply-side investments. Nevertheless, it should be noted that investments in renewable investments are increasing significantly in recent years. According to REN21 Global Status Report (2011), investments in renewables amounted to 211 billion in 2010 up from 160 billion from the previous year and five times the size of similar investments in 2004.

Exploration and resource extraction are the prime components of upstream investment in fossil fuel supply, although detailed data are difficult to come by due to proprietary issues and clear separation from extraction investment. Major differences also exist for electricity transport and distribution infrastructure investments, for which only modeling study data are available, and estimates differ by about a factor of three. The IEA WEO (2008b) projection of average annual electricity T&D infrastructure investments of US$230 billion over the period 2007–2015 appears high, possibly due to large replacement investments in OECD countries, and is comparable to corresponding electricity generation capacity expansion investments.

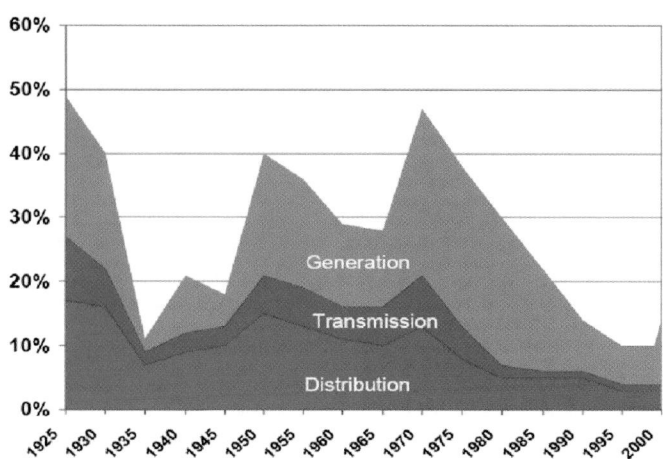

Figure 6.10 | Declining investments (as share of revenues) in the US electricity sector, 1925–2000. Source: Modified from EPRI, 2003.[12]

Given the scarce data available on time trends of supply-side energy investments, an intriguing empirical finding from the United States, however, shows a significant decline in energy supply-side investments for electricity generation in the second half of the twentieth century (see Figure 6.10.) The declining investments (as a share of revenues) in the US electricity sector suggest a substantial thinning of resources available for capital turnover and diffusion of new technologies as a twin result of slowing demand growth and energy sector deregulation and liberalization. At present, it remains unclear if this trend is a specific phenomenon of OECD countries or of US electricity supply (an increasingly deregulated sector), but the example supports the conclusion that better current and longitudinal data on energy sector investments are needed for both in-depth analysis and improved decision-making.

6.6.3   Energy End-use Investments

Investment in energy efficiency, appliances, heating systems, and related infrastructure (buildings, factories) throughout the economy is a necessary prerequisite for the energy system transformations needed.

The decentralized nature of these investments by private households (and their corresponding classification as consumer expenditures rather than investments) and by firms (whose energy-specific investments go unrecorded) explains the absence of energy end-use investment numbers in the literature. The small-scale nature and formidable definitional challenges of these numbers also contribute to their absence.[13] This lack

of data, or even of model estimates, introduces a serious challenge in both energy modeling and policy, because the potentially largest opportunity for energy demand (and emissions) reduction is either entirely ignored or assumed to cost nothing.[14] Customary energy and climate policy models deal with energy end-use costs by either "assuming away" missing data by exogenous (and policy-independent) autonomous energy efficiency trends or by considering investment costs for the incremental component of energy end-use investments related to improved energy efficiency, which in itself provides a formidable definitional and data challenge.

GEA addressed the gap and presents the first global, bottom-up estimate of total investment costs in energy end-use technologies (see Chapter 24). Volume data (production, delivery, sales, and installations) and cost estimates to approximate total investment costs in 2005 are estimated for both end-use technologies and their specific energy-using components.[15] Low and high sensitivities around central estimates are included, taking into account uncertainties in both volume and cost assumptions. The intention is to provide a first-order, educated guess benchmark for comparison with supply-side investments. Supporting data and a discussion text are posted on the GEA Chapter 24 website to document the assumptions underlying the estimates here, to solicit feedback and comments, and to invite further research in this critical area.[16]

To ensure comparability between supply-side and demand-side investments, a clear definition unit of analysis and boundaries is needed. Supply-side investments are quantified at the level of the power plant, refinery, or liquefied natural gas terminal. These are complex, integrated technological systems with energy conversion technologies at their core. These energy-conversion components are configured within their corresponding technological system to provide a useful service to intermediate users (e.g., utilities, fuel distributors, pipeline, or shipping companies).

The demand-side analogues are the aircraft, vehicle, refrigerator, or home heating systems. Although generally less complex, each of these technological systems similarly has an energy conversion technology at its core (e.g., the jet engine, internal combustion engine, compressor, boiler, or refrigerator). In addition, each is configured to provide a useful service to final users.

---

12 More recent data indicates that the trend persists to today. According to the Edison Electric Institute (EEI), for example, which publishes annual utility revenue and investment reports, there has not been any significant change in the percentage of revenues being invested in infrastructure over the past 10 years.

13 For instance, it is far from trivial to discern the energy component in the total investments of a new building. Depending on where the systems boundary is drawn, one could look at the heating and air conditioning system, including that part of the building structure that determines its energy use (insulation, windows). Indeed, the entire building structure may be considered.

14 Some studies include incremental energy end-use technology investments associated with additional energy efficiency gains above a typical "business as usual" scenario (e.g., IEA, 2009b). Apart from introducing additional definitional ambiguities (i.e., what constitutes incremental investments?), the modeling is usually only done for a few technologies (e.g., transport), which limits its usefulness for informed policymaking.

15 Available data do not allow a further disaggregation into those subcomponents of investments on energy efficiency improvements, which remain an important future research task.

16 See chapter 24 at www.globalenergyassessment.org

With demand-side technologies, however, this definition of the unit of analysis is problematic. Investments in (and performance of) end-use technologies are dependent on investments in associated infrastructure, such as airports, roads, and buildings. Is it meaningful to quantify the investment cost of a home heating system without quantifying the investment cost of the home and the insulation level that determine the dimensioning of the home heating system in the first place? Is the end-use technology to consider the furnace or the building?

Although the same issue exists on the supply side, it is largely addressed by additionally quantifying investment costs in associated T&D infrastructures in policy models, as comprehensive statistics are also lacking on the supply side. The problem on the demand side is that the same approach would result in a sum of the total investment costs in all building structures, roads, railways, ports, airports, industrial machinery, equipment, and appliances. Such an exercise would amount to a *reductio ad absurdum*.

A pragmatic pathway out of this system boundary ambiguity is to provide a range of estimates for a range of system boundaries of energy end-use technologies. An initial broader definition and data set describe end-use technologies as the smallest (or cheapest) discrete units that can be purchased by final consumers. This implies boilers and air conditioning units, not houses, and dishwashers and ovens, not kitchens. A second, narrower definition and data set describe the specific energy-using components of these end-use technologies. This implies engines in cars and light bulbs in lighting systems. Table 24.12 (see Chapter 24) summarizes these distinctions for the technologies analyzed. In some cases (e.g., industrial motors, mobile heating appliances), a distinct energy-using component was not identified.

The investments in 2005 in end-use technologies are estimated at US$1–3.5 trillion; the estimate in 2005 in the energy-using components of these end-use technologies is on the order of US$100–700 billion.

Given these definitional problems, the appropriate point of comparison for estimates of supply-side investment costs is a range spanning the narrow category of "energy-using components" at the lower end to the broader category of "end-use technologies" at the upper end. Taking also into account the extent of end-use technologies missing from this analysis, the range of demand-side investment costs is conservatively on the order of US$300 billion–4 trillion. This compares with the range of supply-side investment costs on the order of US$950 billion a year.

The upper bound of demand-side investment outlays is five times higher than the supply-side equivalent whereas the latter is likely a (potentially substantial) underestimate. It is noteworthy that the GEA findings align well with the IEA estimation that demand-side investment needs exceed supply-side investment needs by a factor of 4.5 in climate policy scenarios (IEA, 2008b).

6.7 Financing

6.7.1 The Constraints

The need for a shift in emphasis to higher efficiency and low emissions technologies implies the need for accelerated and up-front energy-related investments over the next several decades. However, even existing low-cost clean technologies in energy supply as well as at end use, particularly when they entail long-lived capital assets and infrastructures, will take decades to fully penetrate the energy sector. Given the long lead times for new technology development, deploying advanced technologies on a large scale after 2025 requires a strong policy response and creation of market incentives today, especially in order to facilitate innovation in such technologies. For example, market incentives to move from obsolete electromechanical to real time electronic, internet-based control of energy delivery and end-use devices would have a major positive impact on energy efficiency.

6.7.2 Energy efficiency investment constraints

Achieving even cost effective improvements in energy efficiency is commonly hindered by a range of market and non-market barriers and failures and many of them can be described as principal-agent problems. The extraordinary large number of actors involved, fragmented institutional framework and often comparatively small size of individual investment needs for energy efficiency improvements and long payback time make this market little attractive for individual investors. Regional energy agencies and energy service companies (ESCOs), together with developing banks, gathered experience with energy efficiency contracting. Almost always, both financiers and end users require some degree of independent assessment. For example, where a trusted ESCO might be able to fully meet the needs of both parties, usually the financier or the end user still wish to have some level of independent technical assessment. Choices then need to be made concerning the degree of outsourcing. Among end users, major industrial enterprises often may conduct technical assessments largely in-house, with perhaps only some very specialized expertise acquired from outside. Building owners, on the other hand, usually outsource nearly all of the project development and assessment effort. The situation among financiers also varies: some development finance institutions may have quite sophisticated in-house technical assessment capacity, whereas many commercial banks will contract out such work to trusted partners. In all cases, energy efficiency investment financing mechanisms must include efficient and cost-effective organizational and institutional arrangements for delivering marketing and technical assessment requirements in which incentives of all the parties are properly aligned. In each respective economic environment, this is likely to include differing combinations of in-house expertise and outsourcing arrangements.

Box 6.1 | Barriers to Energy Efficiency Improvements

Barriers to achieving cost saving investments in energy efficiency include:

Low or underpriced energy. Low energy prices undermine incentives to save energy.

Regulatory failures. Consumers who receive unmetered heat lack the incentive to adjust temperatures, and utility rate-setting can reward inefficiency.

A lack of institutional champion and weak institutional capacity. Energy-efficiency measures are fragmented. Without an institutional champion to coordinate and promote energy efficiency, it becomes nobody's priority. Moreover, there are few energy-efficiency service providers, and their capacity will not be established overnight.

Absent or misplaced incentives. Utilities make a profit by generating and selling more electricity, not by saving energy. For most consumers, the cost of energy is small relative to other expenditures. Because tenants typically pay energy bills, landlords have little or no incentive to spend on efficient appliances or insulation.

Consumer preferences. Consumer decisions to purchase vehicles are usually based on size, speed, and appearance rather than on efficiency.

Higher up-front costs. Many efficient products have higher up-front costs. Individual consumers usually demand very short payback times and are unwilling to pay higher up-front costs. Preferences aside, low-income customers may not be able to afford efficient products.

Financing barriers and high transaction costs. Many energy-efficiency projects have difficulty obtaining financing. Financial institutions usually are not familiar with or interested in energy efficiency, because of the small size of the deal, high transaction costs, and high perceived risks. Many energy service companies lack collateral.

Products unavailable. Some efficient equipment is readily available in high-and middle-income countries but not in low-income countries, where high import tariffs reduce affordability.

Limited awareness and information. Consumers have limited information on energy-efficiency costs, benefits, and technologies. Firms are unwilling to pay for energy audits that would inform them of potential savings.

Lack of capacity in terms of trained people and regulatory frameworks in the energy using sectors (Chapters 8–10 and Chapter 25).

6.7.3   Financing Energy Efficiency

Success in capturing a bigger share of the large numbers of financially attractive energy efficiency retrofit projects has proven stubbornly difficult, primarily because the intrinsic nature of the projects and their broader setting make it hard for effective markets to develop naturally. In some countries, price distortions may undermine incentives, but in most sectors in Brazil, China, and India, and many other countries, this is not the case, as project financial returns are high in most instances. Flow of information about energy efficiency opportunities is far from perfect, but it has improved. In some countries, the required technical or managerial expertise is lacking, but in the case of Brazil, China, and India the issue is more how to bring existing strong expertise to bear. For energy efficiency investments to be made, energy efficiency concepts must be marketed to enterprises, and specific projects must be identified, designed, and appraised. This requires marketing, project development, and technical assessment skill, typically provided by local energy efficiency experts. Human and organizational capacity is needed to define target markets and market outreach strategies, identify project opportunities, design appropriate project packages at end-user facilities, assess financial returns and the risks influencing delivery of the project cost savings cash flow, and understand the incentives to participate by each of the designated parties. For countries such as Brazil, China, and India, the main issue is how to most efficiently access existing project development capacity (Taylor et al., 2008).

Keeping transaction costs reasonable is often a major challenge, especially given the relatively small size of energy efficiency loans. Design of programs to achieve this requires creativity and innovation. For example, for their general and energy efficiency lending to SMEs, Indian banks have relied on new geographical and industry-specific clustering approaches. For investment delivery mechanisms integrating project development and financing to be successful in increasing energy efficiency project investment, they should build upon the following principles: Delivery mechanisms need to be customized, based on a careful diagnostic review of the local institutional environment, including the financial sector, local capacities for technical assessment, the energy efficiency market, and the role of government. Such diagnostic review critically requires local expertise.

End users need to face commercial terms for the financing and technical services being provided as the best foundation for the creation of an energy efficiency market. End-user subsidies tend to ultimately undermine sustainable market development, because they are usually short-lived and can create market distortions and unrealistic expectations. However, concessional financing has often proven valuable to help buy down the high costs and risks of starting up new commercially oriented programs, build necessary new capacity, and assume risks with new approaches. Appropriate incentives must be included for the various actors in each mechanism to participate. Particularly important are incentives to generate deal flow. Combined with the last point, this implies a focus on organizational and institutional arrangements ("deal structuring") that deliver positive incentives for all actors without relying upon long-term market-distorting subsidies (Taylor et al., 2008).

6.7.4   Financial Sources and Instruments

Following are some of the main financial instruments used to support energy efficiency policies:

A. State or Municipal Bonds: Given the magnitude of needs, the high degree of public ownership in the energy sector of the project countries and the limited funds available from external sources, this is one of the most important financial sources.

B. Grants: Many bilateral and multilateral partners provide grants under different programs for financing energy efficiency. Some of the most important international organizations include the United Nations Development Program, the United Nations Environment Program, the World Bank and the Global Environment Facility (GEF) – the financial mechanism of the United Nations Framework Convention on Climate Change (UNFCCC). In the Global Environment Facility, climate change activities are currently divided into four areas: (a) removing barriers to energy efficiency and energy conservation; (b) promoting the adoption of renewable energy by removing barriers and reducing implementation costs; (c) reducing the cost of low greenhouse gas emitting technologies and (d) supporting the development of sustainable transport.

C. Loans: These are mostly provided by many actors including banks – both private and public development banks, international financial institutions, and private investors. International Financial Institutions (IFIs) have increased their role in this important category of financial instrument. Their involvement can provide political motivation to pursue energy sector reform and promote investment. IFIs can play a particularly important role in low income developing countries by building up a standardized risk data base and enhancing the quality of governance. This would both facilitate the use of appropriate investment instruments and help the international investment community achieve the confidence needed to invest in low income countries. This underscores the catalytic role that IFIs can play in promoting policies that stimulate and stabilize internal cash generation, and attract substantially higher levels of commercial and private investment.

D. Equity financing: Under these arrangements, investors take a whole or fractional share of ownership, and thus sharing both risks and benefits from that investment. The types of investors under this category are very diverse including institutional, government, and private investors.

E. Leasing: Under a leasing contract, a lessor conveys to a lessee the right to use a piece of property for an agreed period of time against payment or a series of payments. It is sometimes chosen to be the mechanism, which provides the project owner with necessary equipment under performance contracting.

F. Tax and customs tariffs incentives: Another option for countries to promote measures in energy efficiency is through stimulating utilities or the market for technology distribution and providers of energy efficiency services. This may also be called a form of indirect financing. Yet another is to diminish or abolish customs tariffs on imported energy efficiency equipment.

G. Revolving fund: This is a financial scheme aimed at establishing sustainable financing for a row of investment projects. The fund may include loans or grants and aims at becoming self-sustainable after its first capitalization. The objective is to invest in profitable projects with short payback time, be repaid, and use the same fund to finance new projects. It can be established as a bank account of the owner or as a separate legal entity. There are several parties in a revolving fund: The owners can be either public or private companies, organizations, institutions or authorities. The operator of the fund can be either its owner or an authority appointed by him. External donors and financiers provide contributions to the fund in the form of grants, subsidies, loans or other types of repayable contributions. The borrowers can be either the project owners or contractors. According to the conditions of the revolving fund, savings or earnings gained from projects should be paid back to the fund within a fixed period of time, at certain time intervals. The revolving fund, as financial instruments has its advantages and drawbacks.

H. Venture Capital: Venture capital is particularly important for investments in new technology. By providing early up-front investment finance to companies that have high potential but also high risk, venture capital plays an important role in innovation. It could also play an important role in the setting of new businesses with little or not operating history or that are too small to raise their own funding. The trust placed by venture capital to launch many of these could be essential.

I. Energy Service Companies: Although not a financial instrument *per se*, ESCOs constitute an indirect way to attract capital to investments or third party financing. The ESCOs may fulfill the functions of project identification, planning, implementation and financing. Financing is obtained through contractual relations between the ESCO and the project owner; the investment is carried out by the ESCO and is financed from the costs savings achieved. There are two types of contracts. The first is a guaranteed savings contract, in which payments from the project owner are made according to savings achieved. The second might not include the ESCO in the contract but it makes the project owner liable to cover the capital outlays; payments to the ESCO from the project are not related to the actual savings achieved.

Box 6.2 | Financing Mechanisms

Very large investments will be required to develop new energy technologies and also to increase their uptake throughout the world. Almost all growth in energy demand over the coming four decades will occur in non-OECD countries, where financing capacity is weakest with the exception of the large emerging economies like China and Brazil. The availability of finance and transfer of clean energy technologies into these markets is often considered essential. IFIs and Multilateral Development Banks are expected to play a central role in the process of technology transfer. According to a recent study, multilateral development banks are increasing their role in energy finance in developing countries significantly (BNEF, 2010). From 2008 to 2009, loans to developing countries increased three-fold to some $21.1 billion. Some national development banks, both in OECD countries which invest in developing countries, as well as some in developing countries such as the BNDES, Brazilian Development Bank, have also become major actors and suppliers of energy finance. Over the past five years, KfW of Germany has had financial cooperation commitments for energy projects in developing countries amounting to €3.8 billion KfW, 2011.

The overall contribution of the World Bank Group to energy investments over the past 20 years had been varying considerably, with a lowest point at about US$1 billion in 2004, but had been increasing since then to about US$ 7 billion in 2009 and over US$10 billion in 2010, of which $4.9 billion went to renewables. Compared to the overall portfolio of World Bank investments, the energy sector contributed about 15% of World Bank lending in that time period. Compared with the global investment in the energy sector of many hundred billion and exceeding several trillion if end use-investments are included, these contributions are just a small share of the total investment capital needed (World Bank, 2011a).

The GEF – an operating entity of the financial mechanism of the UNFCCC – has been a key contributor to helping countries eliminate market barriers to the introduction of new technologies and catalyzing energy investments in developing countries. Up to 2009, the GEF has allocated some US$ 2.7 billion to support climate change mitigation projects in developing countries and economies in transition. It has also leveraged an additional $17.2 billion in project co-financing, most of which are energy-related projects. This has resulted in more than 1 billion tons of greenhouse gas emission being avoided between 1991 and 2009, thanks to the support of the GEF. The portfolio of GEF projects is diverse. The energy efficiency portfolio ranges widely from district-heating to efficient lighting, industrial energy efficiency and pioneering guarantees for energy efficiency investments. Through its life-time, the GEF has provided support to over 40 different technologies, many of which were new to the countries into which they were introduced.

The Climate Investment Fund is another more recent and specialized financing mechanism of various development banks set up to provide climate change funding pending the setting up of the Green Climate Fund being established under the UN Framework Convention on Climate Change. It contains two funds of direct relevance to the funding of energy. The Clean Technology Fund which provides concessional finance at as significant scale to help developing countries formulate projects that can be included in the National Appropriate Mitigation Actions emerging from the climate change negotiations and a second one for scaling up renewable energy in low-income countries.

Lastly, it is also important to mention that as a result of the recent financial and economic crisis, many countries such as the Republic of Korea, China, and many others have set up stimulus packages with a significant focus on green investments and the promotion of investments in low carbon technologies. Many of these middle income countries are also playing an increasingly important role in trilateral arrangements as well as south-south cooperation for technology transfer and finance.

6.8   Technology Innovation and Diffusion

In its sixteenth meeting of the Conference of the Parties of the UN Framework Convention on Climate Change, and more specifically in its Cancun Agreements (UNFCCC, 2010), the Parties recognized that "deep cuts in global greenhouse emissions are required…with a view to reducing greenhouse gas emissions so as to hold the increase in global temperature below 2 degrees above pre-industrial levels…". The Intergovernmental Panel on Climate Change have also called for reductions of 50% or greater in global GHG emissions to keep the concentration of these gases below 450 ppm (IPCC, 2007). According to the IEA, achieving this objective would require "major improvements in efficiency and rapid switching to renewables and other low carbon technologies, such as carbon capture and storage (CCS)" and "deployment and development of technologies still under development, whose progress and ultimate success are hard to predict" (IEA, 2008a). Such a shift, "if achievable, would certainly be unprecedented in scale and speed of deployment" IEA, 2008a). The World Bank similarly states that addressing climate change "requires … widespread diffusion of renewable energy technologies… and breakthroughs in technologies from batteries to carbon capture and storage" (World Bank, 2010).

While there is a general recognition that new technologies will need to be developed for a transition to a less energy-intensive and emissions-intensive world, neither public nor private funding of energy-related research, development, and deployment is remotely close to what is required (World Bank, 2010). (See Box 6.3.)

In absolute terms, public energy-related RD&D expenditures in the OECD have declined since 1980 and fell by almost half from then to 2000, to around US$10 billion, before rebounding to US$15 billion in 2008. The expenditures reported for 2009 reach an unprecedented high of US$23 billion (in 2009 prices and exchange rates), with fossil fuels and technologies absorbing the lion's share (US$3.8 billion) of the increase of US$8 billion from 2008 (IEA, 2011).

Private energy RD&D, estimated at US$40–60 billion a year, far exceeds public spending (Chapter 24). At 0.5% of revenue, private expenditures on energy RD&D remain small, however, for instance, compared with 8% of revenues invested in RD&D in the electronics industry and 15% of revenues in the pharmaceuticals sector (World Bank, 2010). (Note: the latter are largely non-energy-related.) Taken overall, the energy-related figures pale in comparison to estimates of required RD&D expenditures in the order of US$100–700 billion a year (World Bank, 2010).

Within the OECD, Japan spends 0.08% of GDP on research and development (R&D), compared with the 0.03% of GDP average for the high- and upper-middle-income members of the OECD (World Bank, 2010). By contrast, in the United States the share of public and private R&D fell from 10% to 2% between 1980 and 2006 (Weiss and Bonvillian, 2009). Total energy R&D now accounts for less than 1% of the annual revenues of the US energy sector. Weiss and Bonvillian offer several explanations for this decline:

- deregulation of energy markets, which has increased competition and led to cutbacks in discretionary expenditures, including R&D;

- an extended period of relatively low energy prices; and

- a mature and cost-competitive sector, which has generally tended to deter new entrants and potential competitors.

Box 6.3 | The Role of Entrepreneurial Innovators

The path to global sustainability will require nearly US$2 trillion annually of primarily private sector investment, plus making hundreds of thousands of sites available for often locally controversial energy facilities and training hundreds of thousands of energy workers at all levels each year. Governmental institutions that support innovation are generally unprepared for this challenge. The energy sector is also dominated by large, risk-averse corporations with strong interests in preserving their comfortable status quo and with a resulting history of underinvestment in innovation.

The role of entrepreneurial innovators will thus be crucial. However, the energy sector has not been very encouraging to entrepreneurial innovators for several decades.[17] Also the priority of avoiding risk at all costs of most governmental organizations has deterred investors in entrepreneurial firms from entering energy markets. This is particularly true for governmental regulatory agencies that have typically put protecting the incumbent industries they regulate ahead of innovation goals. This has inhibited the reallocation of economic resources from the established to the new entrepreneurial organizations that necessarily lead transformative innovation (see also Chapters 24 and 25).

---

17   Entrepreneurs in wind and solar PV have been successful in some countries, including: Germany, Spain, Denmark and Bangladesh.

The process of technology deployment and adoption is widely seen as being about much more than the technical aspects or performance of some piece of equipment. This is particularly true in developing countries, where many factors influence technology adoption and where capacity building will be crucial to enable developing countries to adopt new technologies (Tomlinson et al., 2008). Technology "diffusion" then covers the process of understanding, using, and replicating technologies, as well as adapting them to local conditions and integrating them with other technologies (IPCC, 2000).

The increasing importance of investment flows to promote technology is the focus of current research to promote climate-friendly development. The majority of global investment and technology diffusion occurs via the private sector in the form of corporate R&D, venture capital or asset financing arrangements, or funds raised in public markets (UNEP/SEFI/NEF, 2009). Studies assessing technology transfer and diffusion in developing countries have accordingly noted that openness to trade is a necessary prerequisite for successful transfer.

A second key factor for facilitating technology transfer (see also Chapter 25), which relates directly to the incentives for private firms to undertake R&D, is intellectual property rights (IPRs). Predictable and clearly defined IPRs can stimulate technology transfer from abroad, while weak IPR enforcement discourages foreign firms from investing in R&D activities, licensing new technologies, and investing in domestic enterprises abroad (World Bank, 2010). For example, foreign subsidiaries of global wind equipment producers have registered very few patents in Brazil, China, India, or Turkey, all of which have weak IPR regimes.

At the same time, there are trade-offs. IPRs may hamper innovation if a patent blocks other useful inventions by being too broad in scope, or they can hamper technology transfer if firms refuse to license their technology. To date, overly restrictive IPRs have not been identified as a material barrier to transferring renewable energy production capacity to middle-income countries (ICHRP, 2011), but this situation may change if patenting activity accelerates in photovoltaics and biofuels and if equipment supplier consolidation continues in the wind sector (World Bank, 2010). The World Bank (2010) therefore highlights a role for high-income countries in ensuring that:

- excessive industry consolidation in the renewable energy sectors does not reduce incentives to license technology to developing countries;

- national policies do not prevent foreign firms from licensing publicly funded research for clean technologies of global importance; and

- concerns over IPRs and the transfer and innovation of clean technologies are considered in international treaties such as those of the World Trade Organization.

The importance of the RD&D effort for developing new low-emissions technologies and the observed shortfall in activity suggests that there is an ongoing role for government in this area. The role of governments is threefold and extends beyond energy RD&D: first, in general support for knowledge creation (education, support for international science and technology cooperation, and information exchange); second, in supporting basic and applied energy technology R&D via direct public R&D expenditures as well as by creating and maintaining appropriate incentives for private sector R&D; and third, creating favorable market deployment incentives as well as removing existing barriers for the adoption of more-efficient and cleaner energy technologies.

Broadly speaking, government has a role in RD&D where there are specific "market failures" that will inherently limit private RD&D, but it also needs to put in place a framework in which research activities and innovation are facilitated (IPCC, 2007). There are many market failures that can prevent technology development and deployment. A fundamental failure occurs when prices do not reflect the full costs of the service rendered. Subsidized consumer prices are one example.

Another prominent example, as described earlier, is the failure to internalize the health and environmental damage costs from fossil fuel combustion, especially the costs caused by climate change. Failing to charge the full costs sends the wrong signal to the marketplace and results in overuse of services and underinvestment in more-efficient and cleaner technologies and processes. The net result is reduced demand for climate-friendly technologies and private-sector incentives stimulating innovation. Failures of this type can be addressed by establishing a corresponding market mechanism, for instance in the form of a carbon price, or by removing subsidies.

Additional market failures include monopolistic market structures, different incentives between short-run first-cost minimization (e.g., apartment buildings) by contractors and operating costs over the longer term, access to finance for technologies with higher up-front investment and transaction costs, lack of institutional support, and limited awareness and information.

More generally, new technologies usually face a range of technical and market hurdles and associated uncertainties about entering into widespread commercial use, including innovation uncertainty, technology performance, cost, financial risks, lengthy timescale for deployment, and very large sunk costs. Here governments have a role to play by putting in place the enabling "infrastructure" required to support RD&D and innovation, including the rule of law, open markets, the protection of intellectual property, and the movement of goods, capital, and people.

Several promising efficient and environmentally benign technologies are in their infancy and thus require public RD&D support, while others are more mature and need primarily market incentives for deployment and diffusion. Providing effective deployment incentives is therefore another area for public-sector involvement. These include corrections of the market failures just mentioned, financial incentives (tax credits, carbon prices, fuel taxes, technology rebates, time-limited subsidies), a reliable and predictable regulatory framework, institutional arrangements (e.g., energy service companies), public procurement, promotion, and education.

6.9   Institutional Change

In the face of the profound demographic changes occurring throughout the world, the institutions that were created over the past three decades to help ensure energy security are struggling to remain relevant. The rapid changes of the last few decades point to the need for a more nimble mechanism or system that could help address the complexity of security, global environmental and social and economic concerns, and so on. In order to be effective, this mechanism or system would need to balance the interests of governments, importers, and exporters while aligning with the needs of the private and state-owned firms that provide most of the energy infrastructure investment.

A broad-based cooperative leadership coalition for positive change is the indispensable but missing ingredient needed to transform the energy systems that sustainable societies would require. There is no shortage of governance institutions in today's energy markets. What is missing is a practical strategy for setting effective norms for governing the global energy economy. The basic problem lies in the massive economic and political risks inherent in new projects, particularly those that supply energy across national borders and thus face a multitude of uncertainties. Longtime antagonists must work together to create a shared vision of, and an implementation commitment to achieving, a new, sustainable global energy strategy. This means setting clear goals to address the pivotal challenges of energy access, security, poverty, health, climate change and environment and crafting the necessary policies and practices to achieve these goals. The key challenges can be overcome through a blend of carefully targeted policy initiatives that build on the power of the market, plus public-private partnerships for financing and technology development

Support for new "green" or low-carbon technologies is a second area where a governance vacuum has made progress difficult. Firms are not likely to invest the trillions of dollars needed to develop energy infrastructure in the coming decades without credible signals that governments are serious about establishing and maintaining policies that enable the private sector to confidently cash in on their investments. Based on experience, a sharper focus on energy investments is likely to be much more successful than the more typical broader multilateral governmental agreements on foreign investments.

6.10   Conclusion

Following are the main conclusions and key messages of the chapter:

- Energy is not an end in itself but a prerequisite for economic development (including for the achievement of the Millennium Development Goals), and for the achievement of growth.

- Energy is crucial for the necessary transition to a more equitable and sustainable world and one where all have access to the energy services required for comfort and for a secure and healthy livelihood.

- Energy service demand is a function of population and income, as well as technology. While more affluence may lead to a demand for more energy intensive services it could also lead to a demand for cleaner energy carriers.

- The immense global population increase of the past century has been matched by a period of intense technological expansion that has increased the capacity to harness energy more efficiently and effectively. The reality is that while technology has expanded dramatically, for the most part it is not yet being commercially implemented to meet the 21st Century needs.

- Electricity provides the essential key to energy access, and is the energy prime mover enabling technical innovation and productivity growth. Filling the global electrification gaps to an ever-growing 1.4 billion people today is an essential requirement for eliminating extreme poverty and global security threats.

- A healthy economy is needed to ensure that the energy demands are met, that investments and infrastructure work is carried out and that resources for Research and Development flow to meet the needs and requirements for a sustainable future.

- Most clean technologies are capital-intensive however; they also lower energy demand and fuel consumption. Proper incentives and financial schemes to promote their development are essential.

- Prices play an essential role in the production and consumption process by sending important signals to sellers and buyers. But it is important to distinguish between prices and costs.

- Ideally, the life-cycle cost is what matters when assessing the costs of energy carriers (rather than the cost of fuels or capital costs being accounted for separately as is often the case).

- While monetary costs associated with energy supply and use are fully paid by the producer and/or user, this is only partially the case for other costs associated with "externalities."

- Energy, capital, labor and materials are, within limits, substitutes of each other and the optimal mix is a matter of their relative factor prices.

- Demand responds to price changes – slowly in the short-run because lock-in effect leaves little room for immediate fuel switching and efficiency improvements but in the longer run, it does respond more profoundly through capital replacement (e.g., buildings, refurbishment, new technologies, process adaptation, etc.) and market penetration effects.

- As incomes rise energy demand grows but eventually the tendency is for this growth to take place slower than GDP.

- Investment in R&D drives innovation and is the key for technology improvements, for new technologies to emerge and for a lower energy intensive production of GDP.

- Required investments in clean energy systems are staggering and require investors with a long-term vision – usually with the support of governments which would ideally come in to provide the strategic policy certainty and level playing field for private sector involvement.

- The shift to higher efficiency and more sustainable forms of energy require accelerated and up-front energy-related investments over the next several decades that will need to be sustained and supported by coherent and coordinated policy and regulatory frameworks to mitigate the many existing constraints.

- Although investments in clean energy systems in recent years have been impressive and continue growing, much more is required for the energy transitions that GEA necessitates, including to avoid lock-in effects associated with obsolete energy infrastructures with little short-term flexibility.

- Subsidies change the relative merit order based on private cost only, and can, if well designed, reflect the externalities in market conditions. Subsidies are required for transitions to energy systems supporting a sustainable future.

- Overall energy system performance (intensities, cleanliness, affordability) is more dependent on investments in end-use infrastructures and technologies than on traditional energy sector investment although the latter are also important.

- Energy transitions do not happen in isolation. They require a robust public policy framework and an adequate institutional infrastructure to help make things happen. The evidence shows that major policy and institutional reforms are necessary to lead us into the energy transitions that the GEA necessitates.

References

Abosedra, S. and H. Baghestani, 1989: New Evidence on the Causal Relationship between U.S. Energy Consumption and Gross National Product. *Journal of Energy Development*, **14**(2): 285–292.

AGECC, 2010: *Energy for a Sustainable Future*. The United Nations Secretary-General's Advisory Group on Energy and Climate Change (AGECC), New York, NY, USA.

Arrow, K. J., W. R. Cline, K.-G. Maler, M. Munasinghe, R. Squitieri and J. E. Stiglitz, 1996: Intertemporal Equity, Discounting, and Economic Efficiency. In *Climate Change 1995: Economic and Social Dimensions of Climate Change*. J. P. Bruce, L. H. and E. F. Haites (eds.), Second Assessment Report of the Intergovernmental Panel on Climate Change (IPCC), Cambridge University Press, Cambridge, UK and New York, NY, USA., pp. 125–144.

Arthur, W. B., 2009: *The Nature of Technology: What It Is and How It Evolves*. The Free Press, New York, NY, USA.

Ayers, R. U. and B. Warr, 2009: *The Economic Growth Engine: How Energy and Work Drive Material Prosperity*. Edward Elgar Publishing Ltd., Cheltenham, UK.

Barnes, J., D. Brumberg, M. E. Chen, D. B. Cook, J. Elass, M. El-Gamal, M. Gillis, J. González-Gómez, P. R. Hartley, D. Hertzmark, A. M. Jaffe, Y. J. Kim, N. Lane, D. Li, D. R. Mares, K. Matthews, K. B. Medlock, R. Soligo, L. Smulcer, R. Stoll and X. Xu, 2008: *The Global Energy Market: Comprehensive Strategies to Meet Geopolitical and Financial Risks—the G8, Energy Security, and Global Climate Issues*. Baker Institute Policy Report #37, James A. Baker III Institute for Public Policy of Rice University, Houston, TX, USA.

Barnett, H. and C. Morse, 1963: *Scarcity and Growth: The Economics of Natural Resource Availability*. Johns Hopkins University Press for Resources for the Future, Baltimore, MD, USA.

Berck, P. and M. Roberts, 1996: Natural Resource Prices: Will They Ever Turn Up? *Journal of Environmental Economics and Management*, **31**(1): 65–78.

BNEF, 2010: *Weathering the Storm – Public Funding for Low-Carbon Energy in the Post Financial Crisis Era*. United Nations Environment Programme Sustainable Energy Finance Alliance (UNEP SEF Alliance) and Bloomberg New Energy Finance (BNEF), London, UK.

Campbell, M., J. Cleland, A. Ezeh and N. Prata, 2007: Return of the Population Growth Factor. *Science*, **315**(5818): 1501–1502.

Chen, S. and M. Ravallion, 2008: *The Developing World Is Poorer Than We Thought, but No Less Successful in the Fight Against Poverty*, Policy Research Working Paper No. 4703, The Work Bank, Washington, D.C., USA.

Cheng, B. S., 1995: An Investigation of Cointegration and Causality between Energy Consumption and Economic Growth. *Journal of Energy Development*, **21**(1): 73–84.

Chontanawat, J., L. C. Hunt and R. Pierse, 2008: Does Energy Consumption Cause Economic Growth?: Evidence from a Systematic Study of over 100 Countries. *Journal of Policy Modeling*, **30**(2): 209–220.

Cleland, J., S. Bernstein, A. Ezeh, A. Faundes, A. Glasier and J. Innis, 2006: Family Planning: The Unfinished Agenda. *The Lancet*, **368**(9549): 1810–1827.

Dasgupta, A. K. and D. W. Pearce, 1972: *Cost-Benefit Analysis: Theory and Practice*. Palgrave Macmillan, Basingstoke, UK.

Dasgupta, P. and G. M. Heal, 1979: *Economic Theory and Exhaustible Resources*. Cambridge University Press, New York, NY, USA and Cambridge, UK.

Davies, J. B., S. Sandström, A. Shorrocks and E. N. Worlff, 2008: *The World Distribution of Household Wealth*. Discussion Paper No. 2008/3, United Natoins University World Institute for Development Economics Research (UNU-WIDER), Helsinki, Finland.

Ekholm, T., V. Krey, S. Pachauri and K. Riahi, 2010: Determinants of Household Energy Consumption in India. *Energy Policy*, **38**(10): 5696–5707.

EPRI, 2003: *Electricity Technology Roadmap: Meeting the Critical Challenges of the 21st Century*. Project Number: 1010929, Electric Power Research Institute (EPRI), Palo Alto, CA, USA.

Erol, U. and E. S. H. Yu, 1987: On the Casual Relationship between Energy and Income for Industrialized Countries. *Journal of Energy Development*, **13**(2): 113–122.

Espey, J. A. and M. Espey, 2004: Turning on the Lights: A Meta-Analysis of Residential Electricity Demand Elasticities. *Journal of Agricultural and Applied Economics*, **36**(1): 65–81.

Espey, M., 1998: Gasoline Demand Revisited: An International Meta-Analysis of Elasticities. *Energy Economics*, **20**(3): 273–295.

EU, 2003: *World Energy Technology and Climate Policy Outlook 2030 (WETO)*. Directorate-General for Research, Energy, European Commission, Brussels, Belgium.

Fouquet, R. and P. J. G. Pearson, 2006: Seven Centuries of Energy Services: The Price and Use of Light in the United Kingdom (1300–2000). *Energy Journal*, **27**(1): 139–177.

Fouquet, R., 2008: *Heat, Power and Light: Revolutions in Energy Services*. Edward Elgar Publishing Ltd., Cheltenham, UK and Northampton, MA, USA.

Fouquet, R., 2011: Long Run Trends in Energy-Related External Costs. *Ecological Economics*, **70**(12): 2380–2389.

Galvin, R. and K. Yeager, 2009: *Perfect Power – How the Microgrid Revolution Will Unleash Cleaner, Greener, and More Abundant Energy*. McGraw-Hill, New York, NY, USA.

Goldemberg, J., T. B. Johansson, A. K. N. Reddy and R. H. Williams, 1985: Basic Needs and Much More with One Kilowatt Per Capita. *Ambio*, **14**(4/5): 190–200.

Griffin, J. M., 1993: Methodological Advances in Energy Modelling: 1970–1990. *The Energy Journal*, **14**(1): 111–124.

Grubler, A., 1998: *Technology and Global Change*. Cambrdige University Press, Cambridge, UK.

Haas, R., N. Nakicenovic, A. Ajanovic, T. Faber, L. Kranzl, A. Müller and G. Resch, 2008: Towards Sustainability of Energy Systems: A Primer on How to Apply the Concept of Energy Services to Identify Necessary Trends and Policies. *Energy Policy*, **36**(11): 4012–4021.

Hartwick, J. M. and A. P. Hageman, 1993: Economic Depreciation of Mineral Stocks and the Contribution of El Serafy. In *Toward Improved Accounting for the Environment*. E. Lutz (ed.), World Bank, Washington, D.C., USA.

Hausman, J. A., 1979: Individual Discount Rates and the Purchase and Utilization of Energy-Using Durables. *Bell Journal of Economics*, **10**(1): 33–54.

Hotelling, H., 1931: The Economics of Exhaustible Resources. *Journal of Political Economy*, **39**(2): 137–175.

Howarth, R. B. and A. H. Sanstad, 1995: Discount Rates and Energy Efficiency. *Contemporary Economic Policy*, **13**(3): 101–109.

Hunt, L. C. and Y. Ninomiya, 2003: *Modelling Underlying Energy Demand Trends and Stochastic Seasonality: An Econometric Analysis of Transport Oil Demand in the Uk and Japan*. School of Economics Discussion Papers 107, Surrey Energy Economics Centre (SEEC), School of Economics, University of Surrey, Surrey, UK.

Hwang, D. and B. Gum, 1991: The Causal Relationship between Energy and Gnp: The Case of Taiwan. *Journal of Energy Development*, **16**(2): 219–226.

ICHRP, 2011: *Beyond Technology Transfer – Protecting Human Rights in a Climate-Constrained World*. International Council on Human Rights Policy (ICHRP), Geneva, Switzerland.

IEA, 2003: *World Energy Investment Outlook 2003*. International Energy Agency (IEA) of the Organisation for Economic Cooperation and Development (OECD), Paris, France.

IEA, 2005: *The Experience with Energy Efficiency Policies and Programmes in Iea Countries – Learning from the Critics*. IEA Information Paper, H. Geller and S. Attali, International Energy Agency (IEA) of the Organisation for Economic Cooperation and Development (OECD), Paris, France.

IEA, 2006a: *World Energy Outlook 2006*. International Energy Agency (IEA) of the Organisation for Economic Cooperation and Development (OECD), Paris, France.

IEA, 2006b: *Energy Technology Perspectives 2006 – Scenarios & Strategies to 2050*. International Energy Agency (IEA) of the Organisation for Economic Cooperation and Development (OECD), Paris, France.

IEA, 2007: *World Energy Outlook 2007*. International Energy Agency (IEA) of the Organisation for Economic Cooperation and Development (OECD), Paris, France.

IEA, 2008a: *Energy Technology Perspectives 2008 – Scenarios & Strategies to 2050*. International Energy Agency (IEA) of the Organistion for Economic Cooperation and Development (OECD), Paris, France.

IEA, 2008b: *World Energy Outlook 2008*. International Energy Agency (IEA) of the Organisation for Economic Cooperation and Development (OECD), Paris, France.

IEA, 2008b: *Energy Technology Perspectives: Energy Technology Perspectives to 2050*. International Energy Agency – OECD, Paris.

IEA, 2009a: *World Energy Outlook 2009. International Energy Agency (IEA) of the Organisation for Economic Cooperation and Development (OECD)*, Paris, France.

IEA, 2009b: *World Energy Outlook*. International Energy Agency, Organization for Economic Cooperation & Development, Paris.

IEA, 2010a: *World Energy Outlook 2010*. International Energy Agency (IEA) of the Organisation for Economic Cooperation and Development (OECD), Paris, France.

IEA, 2010b: *Energy Statistics and Balances: Online Database. International Energy Agency (IEA) of the Organization for Economic Co-operation and Development (OECD)*, Paris, France http://www.iea.org/stats/index.asp (accessed 25 August, 2011).

IEA, 2011: *World Energy Outlook 2011*. International Energy Agency (IEA) of the Organisation for Economic Cooperation and Development (OECD), Paris, France.

IEA/NEA, 2010: *Projected Costs of Generating Electricity, 2010 Edition*. International Energy Agency (IEA) of the Organisation for Economic Cooperation and Development and the Nuclear Energy Agency (NEA), Paris, France.

IPCC, 2000: *Ipcc Special Report – Methodological and Technological Issues in Technology Transfer*. Special Report of IPCC Working Group III, World Meteorological Organisation (WMO), United Nations Environment Programme (UNEP), and the Intergovernmental Panel on Climate Change (IPCC), Geneva, Switzerland.

IPCC, 2007: *Mitigation of Climate Change*. Contribution of Working Group III to the Fourth Assessment Report of the Intergovernmental Panel on Climate Change, 2007 B. Metz, O. R. Davidson, P. R. Bosch, R. Dave and L. A. Meyer, Cambridge University Press, Cambridge, UK and the Intergovernmental Panel on Climate Change (IPCC), Switzerland, Geneva.

Jaffe, A. B. and R. N. Stavins, 1994a: The Energy Paradox and the Diffusion of Conservation Technology. *Resource and Energy Economics*, **16**(2): 91–122.

Jaffe, A. B. and R. N. Stavins, 1994b: The Energy-Efficiency Gap What Does It Mean? *Energy Policy*, **22**(10): 804–810.

Judson, R. A., R. Schmalensee and T. M. Stoker, 1999: Economic Development and the Structure of the Demand for Commercial Energy. *The Energy Journal*, **20**(2): 29–58.

KfW, 2011: *Action by Kfw Entwicklungsbank*. KfW Banking Group, Frankfurt, Germany http://www.kfw-entwicklungsbank.de/ebank/EN_Home/Sectors/Energy/Action_by_KfW_Entwicklungsbank/index.jsp (accessed 1 August, 2011).

Kilian, L., 2008: The Economic Effects of Energy Price Shocks. *Journal of Economic Literature*, **46**(4): 871–909.

Kraft, J. and A. Kraft, 1978: On the Relationship between Energy and Gnp. *Journal of Energy Development*, **3**(2): 401–403.

Krautkraemer, J. A., 1998: Nonrenewable Resource Scarcity. *Journal of Economic Literature*, **36**(4): 2065–2107.

Landes, D., 1969: *The Unbound Prometheus*. Cambridge University Press, Cambridge, UK.

Lee, C.-C., 2005: Energy Consumption and Gdp in Developing Countries: A Cointegrated Panel Analysis. *Energy Economics*, **27**(3): 415–427.

Liu, N. and B. W. Ang, 2007: Factors Shaping Aggregate Energy Intensity Trend for Industry: Energy Intensity Versus Product Mix. *Energy Economics*, **29**(4): 609–635.

Lutz, W., J. C. Cuaresma and W. Sanderson, 2008: The Demography of Educational Attainment and Economic Growth. *Science*, **319**(5866): 1047–1048.

Lutz, W. and K. C. Samir, 2010: Dimensions of Global Population Projections: What Do We Know About Future Population Trends and Structures? *Philosophical Transactions of the Royal Society B: Biological Sciences*, **365**(1554): 2779–2791.

Lutzenhiser, L., 1993: Social and Behavioral Aspects of Energy Use. *Annual Review of Energy and the Environment*, **18**: 247–289.

Manne, A. S. and L. Schrattenholzer, 1986: International Energy Workshop: A Progress Report. *OPEC Review*, **10**(3): 287–320.

Markandya, A. and R. Boyd, 2002: *Airpacts Economic Valuation*. University of Bath, UK, for the International Atomic Energy Agency (IAEA), Vienna, Austria.

Markandya, A., S. Pedroso-Galinato and D. Streimikiene, 2006: Energy Intensity in Transition Economies: Is There Convergence Towards the Eu Average? *Energy Economics*, **28**(1): 121–145.

Markandya, A., A. Bigano and R. Porchia, Eds. 2010: *The Social Costs of Electricity: Scenarios and Policy Implications*. Edward Elgar Publishing Limited, Cheltenham, UK.

Martin, P., 2005: *Migrants in the Global Labor Market*. Prepared for the Policy Analysis and Research Programme, Global Commission on International Migration (GCIM), New York, NY, USA.

Masih, A. M. M. and R. Masih, 1996: Energy Consumption, Real Income and Temporal Causality: Results from a Multi-Country Study Based on Cointegration and Error-Correction Modelling Techniques. *Energy Economics*, **18**(3): 165–183.

Modi, V., S. McDade, D. Lallement and J. Saghir, 2005: *Energy Services for the Millenium Development Goals*. Energy Sector Management Assistance Programme (ESMAP), United Nations Development Programme (UNDP), UN Millenium Project, and the World Bank, New York, NY, USA.

Nakicenovic, N. and H.-H. Rogner, 1996: Financing Global Energy Perspectives to 2050. *OPEC Review*, **20**(1): 1–23.

Nakicenovic, N., A. Grubler and A. McDonald, Eds. 1998: *Global Energy Perspectives*. Cambridge Universitiy Press Cambridge, UK.

Nakićenović, N., 1996: Freeing Energy from Carbon. *Daedalus*, **125**(3): 95–112.

Nilsson, L. J., 1993: Energy Intensity Trends in 31 Industrial and Developing Countries 1950–1988. *Energy*, **18**(4): 309–322.

Nordhaus, W. D., 1997: *Beyond the Cpi: An Augmented Cost of Living Index (Acoli)*. Cowles Foundation Discussion Papers 1152, Cowles Foundation for Research in Economics, Yale University, New Haven, CT, USA.

Norgaard, R. B., 1988: Sustainable Development: A Co-Evolutionary View. *Futures*, **20**(6): 606–620.

O'Neill, B. C., M. Dalton, R. Fuchs, L. Jiang, S. Pachauri and K. Zigova, 2010: Global Demographic Trends and Future Carbon Emissions. *Proceedings of the National Academy of Sciences*.

Pacala, S. and R. Socolow, 2004: Stabilization Wedges: Solving the Climate Problem for the Next 50 Years with Current Technologies. *Science*, **305**(5686): 968–972.

Prata, N., 2009: Making Family Planning Accessible in Resource-Poor Settings. *Philosophical Transactions of the Royal Society B: Biological Sciences*, **364**(1532): 3093–3099.

Rabl, A. and J. V. Spadaro, 1999: Damages and Costs of Air Pollution: An Analysis of Uncertainties. *Environment International*, **25**(1): 29–46.

Reddy, A. K. N. and B. S. Reddy, 1994: Substitution of Energy Carriers for Cooking in Bangalore. *Energy*, **19**(5): 561–571.

REN21, 2011: *Renewables 2011 – Global Status Report*. Renewable Energy Policy Network for the 21st Century (REN21), Paris, France.

Riahi, K., A. Grübler and N. Nakicenovic, 2007: Scenarios of Long-Term Socio-Economic and Environmental Development under Climate Stabilization. *Technological Forecasting and Social Change*, **74**(7): 887–935.

Ricci, A., 2010: *Policy Use of the Needs Results*. New Energy Externatlities Development for Sustainability.

Rogner, H.-H., 2010: Innovating for Development: Nuclear Power and Sustainable Development. *Journal of International Affairs*, **64**(1): 137–163.

Samuelson, P. A., W. D. Nordhaus and J. McCallum, 1988: *Macroeconomics – Sixth Canadian Edition*. McGraw-Hill Ryerson Ltd, Toronto, Canada.

SAUNER, 2000: *Sustainability and the Use of Non-Renewable Resources (SAUNER) – Summary Final Report*. Contract ENV4-CT97–0692, University of Bath, UK; University of Suttgart, Germany and Montanuversität Leoben, Austria.

Schipper, L. and S. Meyers, 1992: *Energy Efficiency and Human Activity – Past Trends, Future Prospects*. Stockholm Environment Institute, Stockholm, Sweden and Cambridge University Press, Cambridge, UK.

Schurr, S. H., B. C. Netschert, V. F. Eliasberg, J. Lerner and H. H. Landsberg, 1960: *Energy in the American Economy, 1850–1975*. Johns Hopkins Press, Baltimore, MD, USA.

Smith, K. R. and K. Balakrishnan, 2009: Mitigating Climate, Meeting Mdgs, and Moderating Chronic Disease: The Health Co-Benifts Landscape. In *Commonwealth Health Ministers' Update 2009*, Commonwealth Secretariat, London, UK.

Soytas, U. and R. Sari, 2003: Energy Consumption and Gdp: Causality Relationship in G-7 Countries and Emerging Markets. *Energy Economics*, **25**(1): 33–37.

Soytas, U. and R. Sari, 2006: Energy Consumption and Income in G-7 Countries. *Journal of Policy Modeling*, **28**(7): 739–750.

Taylor, R. P., C. Govindarajalu, J. Levin, A. S. Meyers and W. A. Ward, 2008: *Financing Energy Efficiency – Lessons from Brazil, China, India, and Beyond*. Report No. 42529, Energy Sector Management Assistance Program (ESMAP) and the World Bank, Washington, D.C., USA.

Toman, M. T. and B. Jemelkova, 2003: Energy and Economic Development: An Assessment of the State of Knowledge. *The Energy Journal*, **24**(4): 93–112.

Tomlinson, S., P. Zorlu and C. Langley, 2008: *Innovation and Technology Transfer*. Chatham House and E3G, London, UK.

Train, K., 1985: Discount Rates in Consumers' Energy-Related Decisions: A Review of the Literature. *Energy*, **10**(12): 1243–1253.

UNCTAD, 2001: *Growing Micro and Small Enterprises in Ldc – the "Missing Middle" in Ldcs: Why Micro and Small Enterprieses Are Not Growing*. UNCTAD/ITE/TEB/5, United Nations Conference on Trade and Development (UNCTAD), Geneva, Switzerland.

UNDESA, 2004: *World Population to 2300*. ST/ESA/SER.A/236, Population Division, United Nations Department of Economic and Social Affairs (UNDESA), New York, NY, USA.

UNDESA, 2011: *World Economic and Social Survey 2011 – the Great Green Technological Transformation*. E/2011/50/Rev.1 ST/ESA/333, United Nations Department of Ecnomic and Social Affairs (UNDESA), New York, NY, USA.

UNEP/SEFI/NEF, 2009: *Global Trends in Sustainable Energy Investment 2009 – Analysis of Trends and Issues in the Financing of Renewable Energy and Energy Efficiency*. United Nations Environment Programme (UNEP), Sustainable Energy Finance Initiative (SEFI) and New Energy Finance (NEF), New York, NY, USA.

UNFCCC, 2010: *The Cancun Agreements – an Assessment by the Executive Secretary of the United Nations Framework Convention on Climate Change*. United Nations Framwork Convention on Climate Change (UNFCCC), Geneva, Switzerland.

UNFPA, 2011: *State of the World Population 2011 – People and Possibilities in a World of 7 Billion*. United Nations Population Fund (UNFPA), New York, NY, USA.

USAID, 2009: *Energy and Small and Medium Enterprise*. United States Agency for International Development (USAID), Washington, D.C., USA.

USEIA, 2011: *Interational Energy Outlook 2011*. United States Energy Information Administration (USEIA), Washington, D.C., USA.

Wang, Y., J. Guo and Y. Xi, 2008: Study on the Dynamic Relationship between Economic Growth and China Energy Based on Cointegration Analysis and Impulse Response Function. *China Population, Resources and Environment*, **18**(4): 56–61.

WEC, 2008: *Energy Efficiency Polices around the World: Review and Evaluation*. World Energy Council, London, UK.

Weiss, C. and W. B. Bonvillian, 2009: *Structuring and Energy Technology Revolution*. The MIT Press, Cambridge, MA, USA.

Wolde-Rufael, Y., 2009: Energy Consumption and Economic Growth: The Experience of African Countries Revisited. *Energy Economics*, **31**(2): 217–224.

World Bank, 2010: *State and Trends of the Carbon Market*. A. Kossoy and P. Ambrosi, Carbon Finance Unit of the World Bank, Washington, D.C., USA.

World Bank, 2011a: *Energy Portfolio Data*. World Bank, Washington, D.C., USA http://go.worldbank.org/ERF9QNT660 (accessed 25 August, 2011).

World Bank, 2011b: *World Development Indicators 2011*. World Bank, Washington, D.C., USA.

Yago, G., 2007: *Financing the Missing Middle – Transatlantic Innovations in Affordable Captial*. Milken Institute, Wasington, D.C., USA.

Yu, E. S. H. and J. Y. Choi, 1985: Causal Relationship between Energy and Gnp: An International Comparison. *Journal of Energy Development*, **10**(2): 249–272.

Zhuang, J., Z. Liang, T. Lin and F. de Guzman, 2007: *Theory and Practice in the Choice of Social Discount Rate for Cost-Benefit Analysis: A Survey*. ERD Working Paper Series No. 94, Economic and Research Department, Asian Development Bank, Manila, Phillipines.

7

Energy Resources and Potentials

Convening Lead Author (CLA)
Hans-Holger Rogner (International Atomic Energy Agency, Austria)

Lead Authors (LA)
Roberto F. Aguilera (Curtin University, Australia)
Cristina L. Archer (California State University and Stanford University, USA)
Ruggero Bertani (Enel Green Power S.p.A., Italy)
S.C. Bhattacharya (International Energy Initiative, India)
Maurice B. Dusseault (University of Waterloo, Canada)
Luc Gagnon (HydroQuébec, Canada)
Helmut Haberl (Klagenfurt University, Austria)
Monique Hoogwijk (Ecofys, the Netherlands)
Arthur Johnson (Hydrate Energy International, USA)
Mathis L. Rogner (International Institute for Applied Systems Analysis, Austria)
Horst Wagner (Montan University Leoben, Austria)
Vladimir Yakushev (Gazprom, Russia)

Contributing Authors (CA)
Doug J. Arent (National Renewable Energy Laboratory, USA)
Ian Bryden (University of Edinburgh, UK)
Fridolin Krausmann (Klagenfurt University, Austria)
Peter Odell (Erasmus University Rotterdam, the Netherlands)
Christoph Schillings (German Aerospace Center)
Ali Shafiei (University of Waterloo, Canada)

Review Editor
Ji Zou (Renmin University of China)

Contents

Executive Summary

An energy resource is the first step in the chain that supplies energy services (for a definition of energy services, see Chapter 1). Energy services are largely ignorant of the particular resource that supplies them; however, often the infrastructures, technologies, and fuels along the delivery chain are highly dependent on a particular type of resource. The availability and costs of bringing energy resources to the market place are key determinants to affordable and accessible energy services.

Energy resources pose no inherent limitation to meeting the rapidly growing global energy demand as long as adequate upstream investment is forthcoming – for exhaustible resources in exploration, production technology, and capacity (mining and field development) and, by analogy, for renewables in conversion technologies.

Hydrocarbons and Nuclear

Occurrences of hydrocarbons and fissile materials in the Earth's crust are plentiful – yet they are finite. The extent of the ultimately recoverable oil, natural gas, coal, or uranium is the subject of numerous reviews, yet still the range of values in the literature is large (Table 7.1). For example, the range for conventional oil is between 4900 exajoules (EJ) for reserves to 13,700 EJ (reserves plus resources) – a range that sustains continued debate and controversy. The large range is the result of varying boundaries of what is included in the analysis of a finite stock of an exhaustible resource, e.g., conventional oil only or conventional oil plus unconventional occurrences, such as oil shale, tar sands, and extra-heavy oils. Likewise, uranium resources are a function of the level of uranium ore concentrations in the source rocks that are considered technically and economically extractable over the long run.

- Oil production from areas that are difficult to access or from unconventional resources is not only more energy intensive, but also technologically and environmentally more challenging. The production of oil from tar sands, shale oil, natural gas from shale gas or the deep-sea production of conventional oil and gas raises further environmental risks, ranging from oil spillages, groundwater contamination, greenhouse gas (GHG) emissions, and water contamination to the release of toxic materials and radioactivity. A significant fraction of the energy gained needs to be reinvested in the extraction of the next unit, thus further exacerbating already higher exploration and production costs.

- Historically, technology change and knowledge accumulation have largely counterbalanced otherwise dwindling resource availabilities or steadily rising production costs (in real terms). They extended the exploration and production frontiers, which to date have allowed the production of all finite energy resources to grow. The question now is whether technology advances will be able to sustain growing levels of finite resource production and what will be the necessary stimulating market conditions?

- Resources need first to be identified and delineated before the technical and economic feasibility of their extraction can be determined. However, having identified resources in the ground does not guarantee its prerequisite technical producibility or its economic viability in the market place. The viability is determined by:

  - the demand for a resource (by the energy service-to-resource chain);
  - the price it can obtain; and
  - the technology capability, economic performance and environmental limitations. The last is becoming increasingly difficult to accomplish (see Sections 7.2.6 and 7.3.9).

- Thus, timely above-ground investment in exploration and production capacities is essential to unlocking below-ground resources. Private sector investment is governed by the expected future market and price developments, while public sector investment competes with other development objectives. The time lag between investment in new production capacities and the actual start of deliveries can be up to 10 years and more, especially for the development of unconventional resources. Until new large-scale capacities come online, uncertainty and price volatility will prevail.

Table 7.1 | Fossil and uranium reserves, resources, and occurrences.[a]

	Historical production through 2005	Production 2005	Reserves	Resources	Additional occurrences
	[EJ]	[EJ]	[EJ]	[EJ]	[EJ]
Conventional oil	6069	147.9	4900–7610	4170–6150	
Unconventional oil	513	20.2	3750–5600	11,280–14,800	> 40,000
Conventional gas	3087	89.8	5000–7100	7200–8900	
Unconventional gas	113	9.6	20,100–67,100	40,200–121,900	> 1,000,000
Coal	6712	123.8	17,300–21,000	291,000–435,000	
Conventional uranium[b]	1218	24.7	2400	7400	
Unconventional uranium	34	n.a.		7100	> 2,600,000

[a] The data reflect the ranges found in the literature; the distinction between reserves and resources is based on current (exploration and production) technology and market conditions. Resource data are not cumulative and do not include reserves.

[b] Reserves, resources, and occurrences of uranium are based on a once-through fuel cycle operation. Closed fuel cycles and breeding technology would increase the uranium resource dimension 50–60 fold. Thorium-based fuel cycles would enlarge the fissile-resource base further.

- There appears to be general consensus that the occurrence of fossil and fissile energy resources is large enough to fuel global energy needs for many centuries. There is much less consensus as to their actual future availability in the market place. While the 'barrels' of exhaustible resources may well be humongous, the sizes of their taps that enable the flow from the barrels to the market are subject to a variety of components, including:

  - smaller and smaller deposits in harsher and harsher environments;
  - rising exploration, production, and marketing costs;
  - excessive environmental impacts;
  - diminishing energy ratios;
  - rate of technology change; and
  - environmental policy.

These factors inherently reduce accessible stocks (size of the barrel) and flow rates, while demand, high prices (plus associated investments), innovation, and technology change tend to increase stock sizes and flow rates. The question arises, which of these opposing forces is going to prevail in the mid-to-long term? It suffices to say that because of these constraints only a fraction of these resources is likely to be produced.

Renewables

Renewable energy resources represent the annual energy flows available through sustainable harvesting on an indefinite basis. While their annual flows far exceed global energy needs, the challenge lies in developing adequate technologies to manage the often low or varying energy densities and supply intermittencies, and to convert them into usable fuels.

Annual renewable energy flows[1] are abundant and exceed even the highest future demand speculations by orders of magnitude (Table 7.2). The influx of solar radiation[2] that reaches the Earth's surface amounts to 3.9 million EJ/yr. Accounting for cloud coverage and empirical irradiance data, the availability of solar energy reduces to 630,000 EJ/yr. The energy carried by wind flows is estimated at about 110,000 EJ/yr and the energy in the water cycle amounts to

---

1   The numbers presented here are different than in other reports; please see Table 11.3 for an explanation.

2   A good graphical summary of renewable energy flows can be found in Sorensen (1979).

Table 7.2 | Renewable energy flows, potential, and utilization in EJ of energy inputs provided by nature.[a]

	Utilization 2005	Technical potential	Annual flows
	[EJ]	[EJ/a]	[EJ/a]
Biomass, MSW, etc.	46.3	160–270	2200
Geothermal	2.3	810–1545	1500
Hydro	11.7	50–60	200
Solar	0.5	62,000–280,000	3,900,000
Wind	1.3	1250–2250	110,000
Ocean	–	3240–10,500	1,000,000

[a] The data are energy-input data, not output. Considering technology-specific conversion factors greatly reduces the output potentials. For example, the technical 3150 EJ/yr of ocean energy in ocean thermal energy conversion (OTEC) would result in an electricity output of about 100 EJ/yr.

more than 500,000 EJ/yr, of which 200 EJ/yr theoretically could be harnessed for hydro electricity. Net primary biomass production is approximately 2200 EJ/yr, which, after deducting the needs for food and feed, leaves in theory some 1100 EJ/yr for energy purposes. The global geothermal energy stored in the Earth's crust up to a depth of 5000 m is estimated at 140,000 EJ/yr. The annual rate of heat flow to the Earth's surface is about 1500 EJ/yr. Oceans are the largest solar energy collectors on Earth absorbing on average some 1 million EJ/yr. These gigantic annual energy flows are of theoretical value and the amounts that can be utilized technically and economically are significantly lower. Renewables, except for biomass, convert resource flows directly into electricity or heat. Their technical potentials are thus a direct function of the performance characteristics of their respective conversion technologies as well as of factors such as geographic location and orientation, terrain, supply density, distance to markets or availability of land and water, while the economic potentials of renewables depend on their competitiveness within a specific local market setting.

7.1   Introduction

This chapter reviews the world's endowment of exhaustible and renewable energy occurrences. It foremost attempts to clarify what nature has to offer, what it may cost to make its resource stocks and flows accessible to the market place, and what the social and environmental implications of their extraction are. It does not *per se* speculate whether, how, or how much of these resources will be utilized – this is the subject of Chapter 17 (Scenarios) and Chapters 8–16, which cover energy-conversion technologies throughout the energy system to the supply of energy services (for a definition of energy services, see Chapter 1).

This is not to say that demand is irrelevant. On the contrary, without a demand dimension any resource assessment is a futile undertaking. Indeed, nature's offerings become relevant resources only in the presence of demand and the existence of an appropriate technology for mining and harvesting at affordable costs. Therefore, in the presence of demand, technology and technology change (innovation) are fundamental in bringing energy resources to the market place. This chapter restricts its assessment to technologies necessary for mining and mobilizing hydrocarbon or fissile materials, for improving land productivity, or for damming up water. Here, advances in geosciences, exploration techniques, mining methods, or biotechnology are examples that will shape our knowledge about resource dimensions, producibility, costs, and potential adverse consequences.

This approach works for finite resources, but not for most renewables – the degree of their future use is rather a question of the anticipated technological and economic performance of technologies that feed on these natural flows and not the magnitude of the flows themselves, as these are undeniably enormous.

The natural question arises of how to reconcile the apparent contradiction between the approach adopted here and the above statement that resources are determined by demand, technology, and costs (relative to alternatives). To begin, the assessment takes stock of the material volumes contained in the Earth's crust, the magnitudes of renewable flows, and the land available for energy-crop production. Next, the quantified stocks and flows are divided into separate resource categories or classes to reflect the different degrees of quality or technological challenge of extraction and/or harvesting, whenever possible by way of cost tags. For example, aggregate supply cost curves are developed for oil and coal, while wind energy resources are presented in categories of specific wind speed, general geographical condition, etc.

Finally, supply cost curves and resource categories serve as an input to the scenario and technology chapters. These chapters determine the overall call on resource utilization (the demand) depending on the relative merit order of the various conversion technologies (technology and costs) and associated energy inputs (resource categories). Finally, feedback from the technology and scenario chapters helps to refine the resource categories and supply cost curves of this chapter.

7.1.1   Definitions and Classifications

There is no consensus on the exact meanings of the terms reserves, resources, and occurrences. Many countries and institutions have developed their own expressions and definitions, and different authors and institutions have different meanings for the same terms. This lack of consistent definitions is one cause of confusion. Another cause is rooted in the fact that most reserves quantities, estimated as deposits, are often located several kilometers below the surface. The estimates are based on inherently limited information and the geological data derived from exploration activities are subject to interpretation and judgment. "Reserves estimation is a bit like a blindfolded person trying to judge what the whole elephant looks like from touching it in just a few places. It is not like counting cars in a parking lot, where all the cars are in full view" (Hirsch, 2005).

Principally, this chapter adopts the concept of the McKelvey box (Figure 7.1), which presents resource categories in a matrix that shows increasing degrees of geological assurance and economic feasibility. This scheme, developed by the US Bureau of Mines and the US Geological Survey (USGS), is reflected in the international classification system used by the United Nations (USGS, 1980; UNESC, 1997; ECE, 2010).

In this classification system, 'resources' are defined as "concentrations of naturally occurring solid, liquid, or gaseous material in or on the Earth's crust in such form that economic extraction is potentially feasible." The geological dimension is divided into 'identified' and 'undiscovered' resources.

'Identified' resources are deposits that have a known location, grade, quality, and quantity – or that can be estimated from geological evidence. Identified resources are further subdivided into 'demonstrated' and 'inferred' resources, to reflect varying degrees of geological assurance, or lack thereof. 'Undiscovered' resources are quantities expected or postulated to exist based on materials found in analogous geological

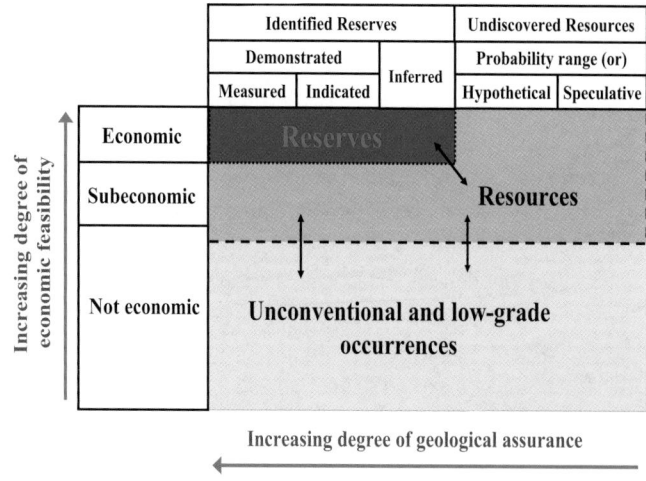

Figure 7.1 | Principles of resource classification. Source: McKelvey, 1967.

433

conditions. 'Other occurrences' are materials that are too low grade or for other reasons not considered technically or economically extractable. For the most part, unconventional resources are included in this category.

The boundary between 'reserves' and 'resources' is the current or expected profitability of exploitation, governed by the ratio of future market prices to the long-run cost of production. Price increases and production cost reductions expand reserves by moving resources into the reserve category and vice versa. Production costs of reserves are usually supported by actual production experience and feasibility analyses, while cost estimates for resources are often inferred from current production experience, adjusted for specific geological and geographical conditions.

Reserve outlooks reported by the media are usually framed within a short-term market perspective, which focuses on prices, who is producing from which fields, where might spare capacity exist to meet short-term demand peaks, the politics of oil, and how this all balances with demand.

In contrast, long-term supply, given sufficient demand, is a question of the replenishment of known reserves with new ones presently either unknown, not delineated, or from known deposits currently not producible or accessible for technoeconomic reasons (Rogner, 1997; Rogner et al., 2000). Here the development and application of advanced exploration and production technologies are essential prerequisites for the long-term resource availability. Technologies are continuously shifting resources into the reserve category by advancing knowledge and lowering extraction costs. The outer boundary of resources and the interface to other occurrences is less clearly defined and often subject to a much wider margin of interpretation and judgment. Other occurrences are not considered to have economic potential at the time of classification. Production inevitably depletes reserves and eventually exhausts deposits, yet over the very long term, technological progress may upgrade significant portions of occurrences to resources and later to reserves. In essence, sufficient long-term supply is a function of investment in research and development (R&D) of exploration and new production methods and in extraction capacity, with demand prospects and competitive markets as the principal drivers.

More precisely, this assessment uses the following definitions in decreasing order of certainty and economic producibility:

- 'Occurrences' are hydrocarbon or fissile materials contained in the Earth's crust in some sort of recognizable form (WEC, 1998).

- 'Resources' are detected quantities that cannot be profitably recovered with current technology, but might be recoverable in the future, as well as those quantities that are geologically possible, but yet to be found. Undiscovered resources are what remain and, by definition, one can only speculate on their existence.

- 'Reserves' are generally those quantities that geological and engineering information indicate with reasonable certainty that can be

recovered in the future from known reservoirs under existing economic and operating conditions (BP, 2010).[3]

Another major factor in estimating future availabilities of oil, gas, coal, and uranium is the difference between 'conventional' and 'unconventional' occurrences (e.g., oil shale, tar sands, coal-bed methane [CBM], methane clathrates, and uranium in black shale or dissolved in sea water). Again, terms that lack a standard definition are often used, which adds greatly to misunderstandings, especially in the debates on peak oil, gas, or coal. As the name suggests, unconventional resources generally cannot be extracted with technology and processes used for conventional oil, gas, or uranium. They require different logistics and cost profiles, and pose different environmental challenges. Their future accessibility is, therefore, a question of technology development, i.e., the rate at which unconventional resources can be converted into conventional reserves (notwithstanding demand and relative costs). In short, the boundary between conventional and unconventional resources is in permanent flux.

This chapter is based on a comprehensive literature review which revealed a wide range of resource quantifications with particularly high variability for unconventional oil and gas over short time intervals at the national or regional levels. Here the responsible author(s) used their expert judgment on which data to report in this assessment.

For renewable energy sources, the concepts of reserves, resources, and occurrences need to be modified as renewables represent, at least in principle, annual energy flows that, if their flows are harvested without disturbing nature's equilibria, are available indefinitely. In this context, the total natural flows of solar, wind, hydro, and geothermal energy, and of grown biomass are referred to as theoretical potentials and are analogous to 'occurrences.' 'Resources' of renewable energy are captured by using the concept of 'technical potential' – the degree of use that is possible within thermodynamic, geographical, or technological limitations without a full consideration of economic feasibility.

'Reserves' of renewable energy would then correspond to the portion of the technical potential that could be utilized cost-effectively with current technology. Future innovations and technology will change and expand the technoeconomic frontier further, so 'reserves' will move dynamically in response to market conditions, demand, and advances in conversion technologies. For example, the economic potential of solar is largely a matter of the cost of photovoltaic or concentrated solar

---

3 The industry associates proved reserves (so-called 1P reserves) with quantities recoverable with at least 90% probability (P90) under existing economic and political conditions and using existing technology. Reserves based on median estimates and at least a 50% probability (P50) of being produced are referred to by industry as probable or 2P (proved plus probable) reserves. 3P (proved plus probable plus possible) reserves characterize occurrences with at least a 10% probability of being produced. 2P and 3P reserves reflect the inherent uncertainties of the estimation process caused by varying interpretations of geology, future market conditions, and future recovery methods (SPE, 2005).

conversion systems, electricity system integration, and local market conditions, and not of the overall amount of solar radiation.

Conversion technologies are outside the scope of this chapter, but are extensively dealt with in Chapters 8–16. Chapter 17 (Energy Pathways for Sustainable Development) integrates demand, resources, and all technologies throughout the energy system and balances supply and demand. However, some basic assumptions regarding the current and future performance of conversion technologies to harvest renewable energy flows as well as system aspects, distances to demand centers, etc., are necessary to quantify 'resources.' Therefore, the term 'practical potential' is used in this assessment as proxy for renewable resources.

The reserve, resource, and potential estimates of this chapter serve as inputs to the later chapters that present different energy future pathways and technology reviews. At the same time, feedback from the technology and scenario chapters has helped refine the resource categories and supply cost curves in this chapter. In this context, it is important that there is one fundamental difference between this resource chapter and the technology and scenario chapters. The latter report potentials or utilization in terms of output (e.g., kilowatt-hours [kWh] or megajoules [MJ]), while this chapter presents resources and potential in terms of inputs (e.g., the kinetic energy of the wind hitting the turbine blades of a wind power plant).

7.1.2   The 'Peak Debate'

How much oil, gas, coal, or uranium does the Earth's crust hold? This question has preoccupied resource analysts since the dawn of the 20th century and the answers provided reflect a deep divide between representatives of different disciplines, i.e., between geologists and economists. Over time, the divide appears to have widened rather than narrowed, resulting in what is now termed the 'peak debate.' Traditionally, its focus has been on the availability of conventional oil (e.g., Aleklett et al., 2010), but more recently the notions of peak coal (Heinberg and Fridley, 2010), peak gas (Laherrère, 2004), and peak uranium (EWG, 2006) also entered the debate. The arguments brought forward in support or rejection of an imminent peak are largely the same for each resource. Therefore, the following paragraphs summarizing the peak-oil debate are representative for all resources.

An increasing number of analysts expect the production of oil to peak in the near future, i.e., over the next 10 to 20 years. Some even argue that 'peak oil is now' (EWG, 2007). They base their projections on the fact that large oil discoveries ('super giants') ended in the mid-1960s, followed by a substantial decline in the discoveries of new reserves (Figure 7.2). Between 1980 and 2009, slightly more than 65% of global oil production was reported to be offset by new additions of oil reserves (Figure 7.3) (BP, 2010). However, it has been argued that the increased levels of reported oil since 1980 are merely the result of belated corrections to previous oil-field estimates. Backdating the revisions to the years in which the fields were discovered reveals that

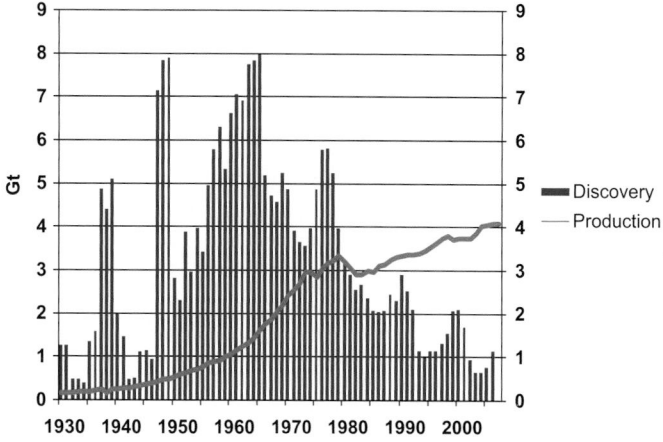

Figure 7.2 | Oil discoveries and oil production. Source: adapted from Earth Policy Institute, 2007.

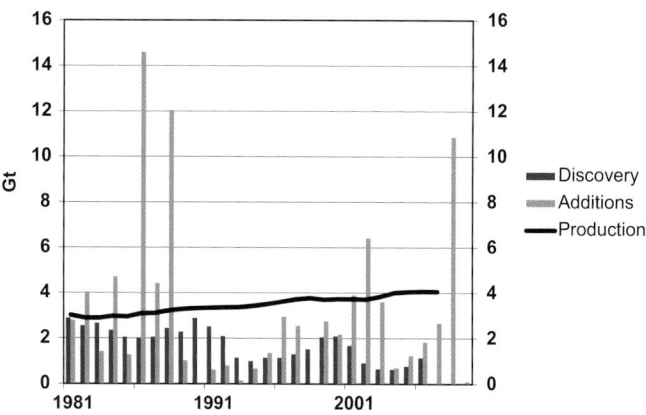

Figure 7.3 | Oil production, oil reserve discoveries, and oil reserve additions. Source: adapted from Earth Policy Institute, 2007; BP, 2010.

reserves have been falling because of a steady decline in truly new-found oil deposits. On this account, for each barrel of oil newly listed as discovered some 4–6 barrels are removed from the ground (Campbell and Laherrère, 1998; Hirsch, 2005). For example, the reserve additions in 2008, reported by BP, are "primarily due to an upward revision in Venezuela of 73 billion barrels (10 Gt)" (Rühl, 2010).

Continuously removing more oil than is offset by new reserve additions will eventually result in reaching a level of "peak oil" at approximately the time when half of the oil reserves have been used. After peak oil, the global availability of oil will decline year after year at a rate depending on the rate of production. The assumed ultimate global oil reserve endowment is therefore a critical parameter in determining both the level of peak production and the point in time when it will occur.

'Estimated ultimate recoverable oil'[4] (EUR) refers to estimates of the total amount of the world's conventional oil – that which has already

---

4   EUR includes past production.

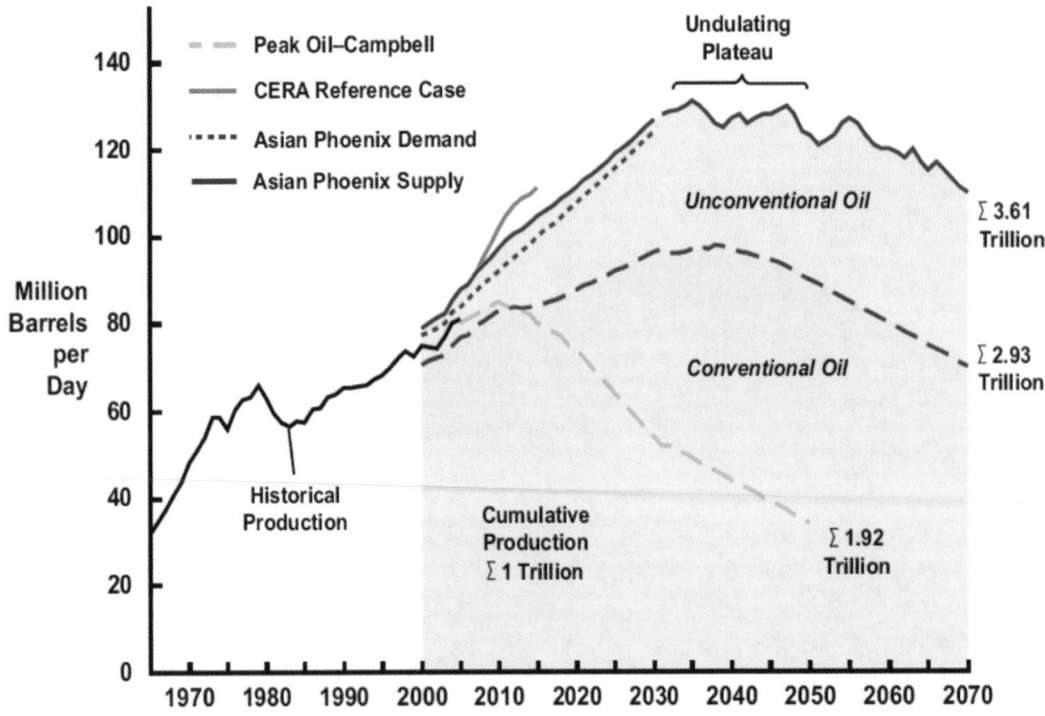

Figure 7.4 | Undulating plateau versus peak oil. Source: Witze, 2007.

been taken out of the ground and that which remains. Since the 1950s, approximately 100 different estimates have been made. The majority lie in the range 12.6–16.7 ZJ (300–400 Gt[5]). By the end of 2008, cumulative oil production amounted to some 6.5 ZJ (156 Gt). For the lower estimates, this means the half-way production mark (the peak) has almost been reached. Using the higher estimates would shift the peak only by a decade or so.

The term 'recoverable' in 'estimated ultimate recoverable oil' means that it is not a measurement of the total amount of oil in place, but refers only to the portion that is recoverable, taking into account geological complexities and economic limitations. Estimates, therefore, include certain (usually tacit) assumptions about costs, markets, and technology. It is this interplay between technology (innovation and knowledge) and prices, but also with demand, that prompts some analysts, especially with an economics background, to be more sanguine about EUR estimates. Not that the economists ignore the ultimate finiteness of exhaustible resources or that peak oil production will eventually take place – this is not the issue. In their view, human ingenuity has kept ahead of reserve depletion and there is no obvious reason why this should change (Bentley and Smith, 2004) as technological innovation will continue to unlock additional reserves currently not identified or understood, or not economically extractable with existing technology and market conditions.

In terms of economic factors, higher prices not only push the frontier of marketable resources (exploitation of smaller fields, higher recovery rates, ability to operate in more challenging environments, etc.), but also stimulate upstream technology R&D for expanding exploration and production activities. At the same time higher prices generally suppress demand, lowering pressure on supply. Claims that recent skyrocketing oil prices reflect that the peak has arrived ignore these market fundamentals. As was observed in the second half of 2008, a slight reduction in global oil demand started a precipitous drop in prices along the entire oil chain.

'Unconventional' oil resources are not included in EUR estimates until they become economically and technically producible. The inclusion of unconventional oil resources in the standard future production profiles of the 'peak oil' analysts would radically change the shape of the global oil production profile (Odell, 2004; McKenzie-Brown, 2008). In fact, the notion of 'peak oil' is misleading. When total (conventional and unconventional) oil production approaches a maximum level, production is likely to be characterized by an 'undulating plateau' (see Figure 7.4) rather than by a peak followed by a sharp drop-off in output (IHS-CERA, 2006; Witze, 2007). However, the 'peak oil' proponents would counter that even if the resource base of non-conventional oil is tapped, production would be constrained by high specific investment and production costs as well as environmental regulations. As rising costs continue to push prices, consumers

---

5   1 t = 42 MJ = 7.33 barrels of oil (bbl).

would eventually turn to cheaper non-oil alternatives leaving plenty of untapped oil in the ground.

Both sides do agree that conventional oil production is going to peak in the foreseeable future, e.g., sometime between now and 2040, with a peak production volume between 4.1–4.5 Gt/yr (82–95 Mbbl/day). Differences in EUR estimates and the role of technology and price explain the variations in time and volume.

Both sides see a role for unconventional oil in future oil supply. There is agreement that the overall 'barrel' of conventional and unconventional oil resources is large indeed, but there is disagreement about the size of the tap, i.e., the rate at which the barrel's (finite) contents be developed, and on the related economic and environmental costs. Potential financial, environmental, and sociopolitical constraints that could limit exploitation of unconventional oil include:

- the high capital intensity of bringing unconventional oil to the market;

- the extraction of unconventional oil is more energy intensive than conventional oil (up to 30% of net output, with corresponding increases in GHG emissions); and

- enormous local environmental burdens (severe soil and water contamination by chlorinated hydrocarbons and heavy metals) from processing unconventional oil into marketable oil.

Large-scale exploitation of unconventional oil necessitates conditions of low financial, environmental, and geopolitical risk averseness, which the 'peak oil' school simply does not see forthcoming. Indeed, most of the unconventional oil in place may never reach the market place. However, this is less a resource-existence issue, but more a result of sociopolitical choice. Then again, leaving aside such constraints, unconventional oil may well postpone the overall decline of oil into the second half of the 21st century.

7.1.3   Units of Measurement

This chapter uses the International System of Units (SI) and reports on energy resources in exajoules (1 EJ = 10^{18} joules). Other units are used intermittently for reasons of comparison with the units commonly used by the different trades.

Tonnes are metric tonnes (1 t = 10^6 g). The energy resource-related industry and energy-related statistics or resource surveys report oil occurrences in gigatonnes (1 Gt = 10^9 tonnes) or gigabarrels using the energy equivalent of 42 GJ/tonne of oil equivalent (GJ/toe) and 5.7 GJ/bbl, respectively (see Table 7.3).

Gas resources are usually reported in teracubic meters (1 Tm^3 = 10^{12} cubic meters) and are converted to EJ using 37 GJ/1000 m^3.

Table 7.3 | Energy conversion factors and prefixes of the metric system used in Chapter 7.

	Tonnes of oil (toe)	Barrel (bbl or boe)	Cubic feet of gas (CFG)	Cubic meter of gas (m³)	Tonnes of coal equivalent (tce)	Megajoule [MJ]	Brtish Thermal Unit (BTU)
Tonnes of oil (toe)	1	7.33	40,000	1124	1.429	41,868	40×10^6
Barrel (bbl or boe)	0.136	1	5414	153.3	0.195	5712	5.414×10^6
Cubic feet of gas (CFGc)	0.25×10^{-6}	0.185×10^{-3}	1	0.028	35.997×10^{-6}	1.055	1000
Cubic meter of gas (m³)	0.890×10^{-3}	6.523×10^{-3}	35.31	1	1.271×10^{-6}	37	31.35×10^3
Tonnes of coal equivalent (tce)	0.7	5.131	$27.78 * 10^3$	786.634	1	29,000	27.78×10^6
Megajoule (MJ)	24×10^{-6}	0.175×10^{-3}	0.948	0.0268	34.121×10^{-6}	1	947.867
British Thermal Unit (BTU)	25×10^{-9}	0.185×10^{-6}	0.001	28.32×10^{-6}	35.997×10^{-9}	1.055×10^{-3}	1

Milli (m)	10^{-3}
Centi (c)	10^{-2}
Kilo (k)	10^3
Mega (M)	10^6
Giga (G)	10^9
Tera (T)	10^{12}
Peta (P)	10^{15}
Exa (E)	10^{18}
Zeta (Z)	10^{21}

Coal resources are usually accounted for in natural units, although the energy content of coal may vary considerably within and between different coal categories. The Bundesanstalt für Geowissenschaften und Rohstoffe (BGR; German Federal Institute for Geosciences and Natural Resources) in Hannover (Germany) is the only institution that converts regional coal occurrences into tonnes of coal equivalent (1 tce = 29.3 GJ). Thus coal resource data comes from the BGR.

Uranium and other nuclear materials are usually reported in tonnes of metal. The thermal energy equivalent of 1 tonne of uranium in average once-through fuel cycles is about 589 terajoules (IPCC, 1996).

7.2 Oil

7.2.1 Overview

Oil is one of the most important sources of global energy because of its high energy density, high abundance and easy transportability at standard temperatures and pressures. In 2009, oil was responsible for over one-third (34.8%) of the world's total primary energy supply. Almost 70% of global oil production is used for transportation and petrochemistry (IEA, 2010a).

Analysts fear resource depletion of oil will produce significant supply scarcities in the near future. Currently, annual oil production exceeds added new oil reserves. Resource estimation is by no means an exact science and is dynamic by nature (see Section 7.1.1). As a result of advances in exploration and production technologies, the borders that distinguish reserves from resources and resources from occurrences are increasingly blurred. To a large extent, this explains the variability of reserve and resource estimates over time, as well as those of different authors. For example, the USGS estimates of oil resources have nearly doubled since the early 1980s. Figure 7.5 shows the evolution of oil

resource estimates since 1940. However, it is important that new additions of resources or reserves, especially in recent years, may not be the result of new discoveries, but of improved extraction technologies or processes, deposit reevaluations, changing economic conditions, or underreporting in the first place.

7.2.1.1 Classifying Conventional and Unconventional Oil

Oil occurrences can be divided into many different subclasses and categories, depending on the physical properties of the oils. Some oils are classified as heavy or extra heavy, which implies that they have high viscosities and do not flow easily. Other oils are sour, which signifies that they contain a certain amount of sulfur. Other oils are considered unconventional, which implies that they are not recoverable using the standard technologies used to extract conventional low-viscosity oils from subsurface reservoirs.

The term crude oil typically refers to conventional oils, but many countries or organizations have their own criteria to define the boundary between conventional and unconventional oils.

The BGR defines conventional oil as oil having a specific gravity of less than 1.0 g/cm^3. This is equivalent to the American Petroleum Institute (API) definition, which specifies API gravity greater than 10°.[6] However, the USGS defines conventional oil as oil having API gravity greater than 15°. 'Conventional oil' generally also includes gas condensates and natural gas liquids.

Unconventional oil includes oils of high viscosity, as well as the oils extracted from kerogen-rich shale (oil shale). The term 'viscous oil' is used generically for all oil with viscosities greater than 100 mPa·s. Kerogen is semisolid organic matter that can be directly converted into oil through pyrolysis. While other definitions exist (e.g., Head et al., 2003; Meyer and Detzman, 1979), the definitions used in this report for the various subclasses of oils are given in Table 7.4.

Many also consider unconventional oil to include products such as biomass oil, synthetic oil from conversion of natural gas, and coal liquefaction (Farrell, 2008; IEA, 2008a). While these are, indeed, potential sources of liquid hydrocarbon fuels, they are not considered unconventional oil in this assessment. Natural gas, coal and biomass are covered in Sections 7.3, 7.4, and 7.7, respectively.

The distinction between conventional and unconventional oil is further distorted when regarding deep offshore or Arctic oil. Despite having API gravity greater than 10°, some assessments consider these as 'unconventional' because of the novelty of deep-sea drilling or Arctic production. It is also important that some unconventional oils are cheaper to recover than some high-end conventional oils, e.g., oil sands versus arctic oil.

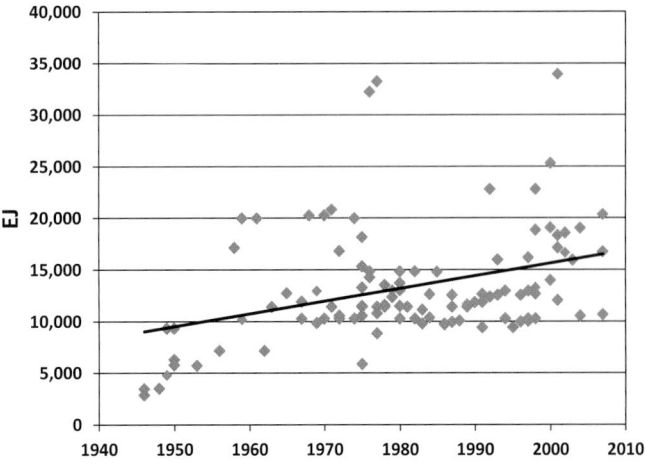

Figure 7.5 | World oil resource (reserves and resources) estimates, 1940–2007. Sources: adapted from Ahlbrandt and Klett, 2005; NPC, 2007; IEA, 2008; BGR, 2009.

6   10° = 1.0 g/cm^3

Table 7.4 | Definitions of oil terms.

	Term	Definition	Physical properties
Oil type	Conventional oil	Oil that is mobile in situ and can be produced economically using conventional methods	μ < 100 mPa·s[a] Low S content
	Viscous oil	Inclusive term for the next three categories below	μ > 100 mPa·s
	Heavy oil	Oil that is only slightly mobile in situ, usually requiring stimulation techniques to improve mobility	μ 100–10,000 mPa·s S content <1%
	Extra-heavy oil	Oil of exceptionally high gravity (ρ >1.0) that has some mobility because of high in situ temperatures (particularly used to refer to Orinoco, Venezuela oils)	μ <10,000 mPa·s, ρ >1.0 S content >1%
	Bitumen	Oil that is immobile in situ so that large viscosity reductions or mining methods are needed	μ >10,000 mPa·s ρ >1.0[b] S content >1%
Rock strata	Oil sands	Sand strata > 25% porosity, containing extra-heavy oil or bitumen, more viscous than heavy oil	k >0.5 Darcy, usually μ > 1000 mPa·s
	Heavy-oil NFCRs	Naturally fractured carbonate reservoirs (NFCRs) containing viscous oil	μ >100 mPa·s Porosity 10–20%
	Oil shale	Kerogenous shales and marls that produce more than 50 l/t of product during Fischer Assay tests	Porosity >15–20%

[a] μ = measure of viscosity in mPa·s = cP = 0.01 g/cm·s.

[b] ρ = specific gravity in g/cm^3.

The boundary is further blurred with respect to currently mined deposits. Using standard conventional extraction technologies, only around 20–40% of oil in place is recoverable by natural depletion of the reservoir. To extract further quantities, improved oil-recovery technologies, which can extract 30–60% of the oil in place, need to be applied (DOE, 2010). The goal of enhanced oil recovery is to alter the original properties of the oil, to restore formation pressure, or to improve oil-displacement efficiency or fluid-flow patterns in the reservoir. There are three major types: chemical flooding (e.g., alkaline flooding or polymer injection), miscible displacement (e.g., carbon dioxide [CO_2] or solvent injection), and thermal processes (e.g., steam flooding or in situ combustion). The application of each technology depends on reservoir temperature, pressure, depth, permeability, net pay, residual oil and water saturation, porosity, and fluid viscosities (Schlumberger, 2010).

7.2.2   Estimates of Conventional Oil

Estimates of the amount of oil are derived from only a limited number of sources. While both production and reserves data are collected by

government agencies at several levels, reserves data are less commonly published than production statistics. Several private companies (e.g., IHS Energy Inc., Wood Mackenzie, BP p.l.c., and others), industry journals (e.g., *Oil and Gas Journal*), and government agencies (e.g., International Energy Agency [IEA], US Energy Information Administration [US EIA]) compile and publish these data.

Reserve estimates are burdened with uncertainty as some countries hold such data as confidential. Estimates published in databases and in the literature are often made by the authors themselves. Growth potentials for undiscovered resources and reserves understandably suffer from greater uncertainties than those of proven reserves. Most reports provide estimates for yet-to-be-discovered resources, but only a few give estimates for reserves-growth potential. Some reports give aggregate estimates for the world as a whole, others for individual countries or regions, but few sources cover the entire world in detail, which adds a further difficulty to a general synthesis of reserves estimations. The USGS and BGR provide the most comprehensive set of estimates.

Table 7.5 compares recent estimates of conventional oil reserves. The estimates vary between 116.3 Gt (EWG, 2008) and 181.7 Gt (BP, 2010). Importantly, some estimates include reserves of oil sands, which are, by definition, unconventional oils.

The low estimate of the Energy Watch Group stems from their suspicion that the reserve data reported by the governments of the Middle East are politically motivated and hence unrealistically high. The USGS estimate, which dates back to the year 2000, appears low as well. However, the USGS reports much higher oil resources than the recent BGR (2009) assessment (see Table 7.6), which suggests that some resources of 2000 became reserves in 2009.

As previously mentioned, terms like 'resources' and 'reserves' are used with varying definitions by different authors. Further confusion arises when resource statistics are interpreted to be more comprehensive than in actuality. While reserves reporting in the United States require a 90% (1P) probability of recovery under existing economic, technological, and political conditions, other reporting bodies typically declare reserves at a median, 50%, (2P), probability. Despite this, reserve estimates given at 1P have tended to increase over time. While some growth may be legitimate because of better extraction or surveying technology, this cannot explain all the reserves growth experienced. Some reports claim that reserves have been deliberately underreported so that estimates could be revised upwards over time to give a comforting yet misleading image of steady growth (WEC, 2007).

Considering oil reserves in a narrow sense, largely conventional oil, cumulative production to date is roughly equal to the remaining proven reserves – for the proponents of 'peak oil' a clear indication of the imminent peak.

Table 7.6 summarizes the conventional oil reserves and resource quantities for the 18 GEA regions. An additional region is added to account

Table 7.5 | Comparison of estimates of conventional oil reserves.[a]

Region	OGJ [Mt]	EWG [Mt]	EIA [Mt]	EXXON [Mt]	BP [Mt]	BGR [Mt]	OPEC [Mt]	USGS [Mt]
Europe	1942	3469	1977	1913	1849	2264	2164	4632
CIS	13,452	20,952	16,784	13,453	16,808	17,543	17,450	22,773
Africa	15,622	17,007	15,192	15,366	17,404	17,276	16,268	9,973
Middle East	101,808	49,252	98,301	101,610	102,803	102,366	100,893	70,866
Asia	4673	7007	4893	4628	5749	5600	5208	7167
North America	28,737	11,429	7921	28,442	8392	6121	5111	5221
Latin America	14,946	7143	9600	15,225	28,706	9854	16,369	10,174
World	181,180	116,259	154,668	180,637	181,712	161,024	163,463	130,806
Oil sands	23,665			23,665				
World w/o oil sands	157,515	116,259	154,668	156,972	181,712	161,024	163,463	130,806

OGJ = Oil and Gas Journal (2007); EWG = Energy Watch Group (2008); EIA = Energy Information Agency (2008); Exxon (2008), BP = British Petrol (2010), OPEC = Organization of Petroleum Exporting Countries (2008), USGS = U.S. Geological Survey (2000).

[a] Conventional oil reserves include NGLs.

Source: adapted from BGR, 2009; 2010.

Table 7.6 | Conventional oil reserves and resources.[a]

Region	Oil production 2009 [EJ]	Historical production till 2009 [EJ]	Reserves BP [EJ]	Reserves BGR [EJ]	Reserves USGS [EJ]	Resources BGR [EJ]	Resources USGS [EJ]	Reserves + Resources BGR [EJ]	Reserves + Resources USGS [EJ]
USA	15.00	1246	162	162	183	420	476	582	659
CAN	6.70	200	189	28	36	101	21	129	57
WEU	8.98	329	74	88	179	186	492	275	671
EEU	0.28	47	4	6	15	13	11	19	26
FSU	27.64	1017	704	735	953	1008	952	1743	1906
NAF	10.38	336	389	388	252	184	158	573	410
EAF	0.00	0	0	4	0	13	7	17	7
WCA	6.07	214	263	254	142	302	375	556	517
SAF	3.78	48	77	77	24	150	97	227	121
MEE	50.78	1823	4308	4286	2967	889	1654	5175	4621
CHN	7.90	220	85	84	142	97	95	181	237
OEA	1.02	11	26	26	0	32	1	58	1
IND	1.57	46	33	33	40	17	18	50	58
OSA	0.14	4	4	2	3	13	11	15	13
JPN	0.01	2	0	0	0	0	0	1	0
OCN	1.20	41	25	24	94	44	108	69	202
PAS	4.90	203	68	65	22	88	63	153	86
LAC	20.30	862	1203	479	426	614	853	1093	1279
Circum-Arctic							768		768
Total	166.68	6647	7615	6742	5477	4172	6161	10914	11,638

[a] Includes natural gas liquids (NGLs). USA = United States of America; CAN - Canada; WEU = Western Europe, incl. Turkey; EEU = Central and Eastern Europe; FSU = Former Soviet Union; NAF = Northern Africa; EAF = Eastern Africa; WCA = Western and Central Africa; MEE = Middle East; CHN =China; OEA = Other East Asia; IND = India; OSA = Other South Asia; JPN =Japan; PAS = Other Pacific Asia; OCN = Australia, New Zealand, and other Oceania; LAC = Latin America and the Caribbean

Sources: author's estimate; BP, 2010; USGS, 2000; 2008; BGR, 2009; 2010.

for resources located within the Arctic Circle (USGS, 2008). The table compiles reserves estimates from BP, BGR, and the USGS – the three organizations that regularly assess global oil reserves or resources. While reserve estimates exhibit only slight variance, less than 15% between the highest (7615 EJ by BP) and lowest estimates (6635 EJ by BGR), resource estimates show almost a 50% difference (4170 EJ by BGR and 6150 EJ by USGS).

The discrepancies arise because of different definitions, boundaries, and classifications of different oil types. While reserves estimation is somewhat better defined, resource estimation has very few guidelines and is thus subject to greater institutional subjectivity. For example, the USGS resource estimate includes oil occurrences in the Arctic (768 EJ). Furthermore, estimates of resources in undiscovered fields, despite their inherent ambiguity, have also grown over time. This is mainly because technology changes that have either shifted resources from the unconventional to conventional category, or have opened new territories to exploration (e.g., deep-water areas).

Figure 7.6 displays the regional distribution of conventional oil reserves and resources. Almost two-thirds of global conventional oil is shared between the Middle Eastern countries (48.6%) and the Former Soviet Union (FSU) (16.7%), while the remaining regions, except Latin America (9.0%) and the United States (5.7%) hold less than 5% (BGR, 2009).

The regional distribution of oil resources has significant implications for rapidly developing economies such as China and India. China's ability to provide for its own needs is limited because its proven oil reserves are small compared to its consumption. Despite recent attempts at diversifying its oil supply, China will inevitably become more dependent on Middle Eastern oil to fuel its economic development.

7.2.3   Types of Unconventional Oil

7.2.3.1   Viscous Oil

Anaerobic biodegradation of light oil is recognized as the main process responsible for the large viscous oil deposits around the world (Atlas and Bartha, 1992; Head et al., 2003). Aitken et al. (2004) investigated the hydrocarbon degradation of 77 oil samples from around the world, including Canadian tar sands deposits, and concluded that the hydrocarbon biodegradation must have been an anaerobic process, at least at some point, and this is certainly the case for viscous oil deposits at depths where aerobic biodegradation was unlikely. Thus, light oil generated deep in megasynclinal structures migrated up-dip because of density differences and hydrodynamic forces until geochemical conditions suitable for biodegradation were encountered. Archaeobacteria then fed on the light oil, generating viscous oil and CH_4. Furthermore, at shallow depths, some aerobic effects may occur, and light-oil fractions can also be removed by hydrodynamic washing and diffusion, further increasing the viscosity. As a result of increasing temperature and the biodegradation origins of viscous oil, it is rarely found below depths of ~2000 m; probably greater than 85% of the resource base is shallower than 1000 m (Figure 7.7).

Many of the molecules of high molecular weight, such as asphaltenes, that give the oil its high viscosity are the remnants of the cell walls (lipids) of the bioorganisms. The oil contains sulfur, as well as heavy

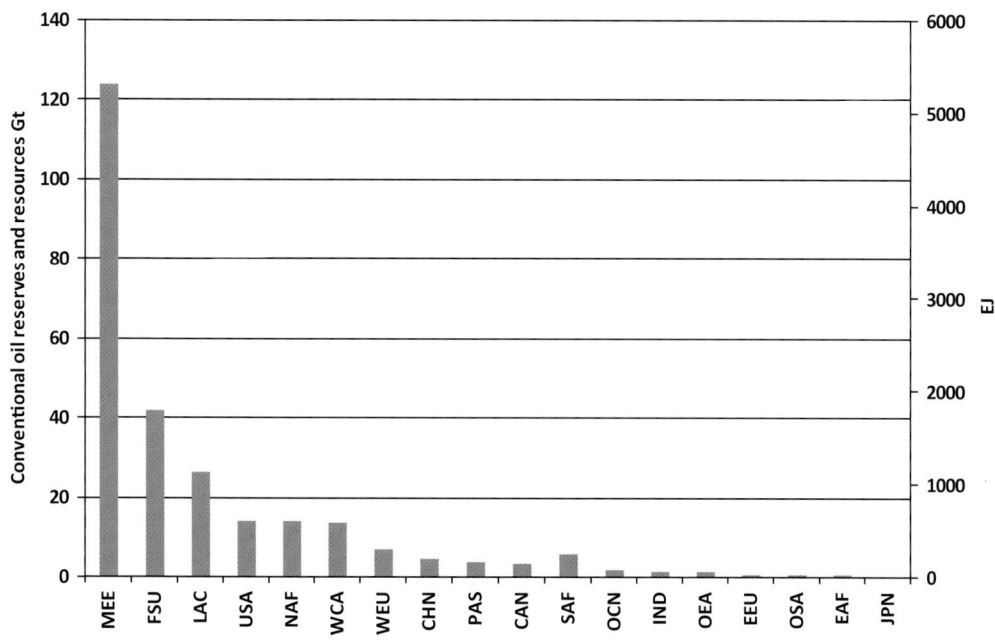

Figure 7.6 | Regional distribution of conventional oil reserves and resources. Source: BGR, 2010.

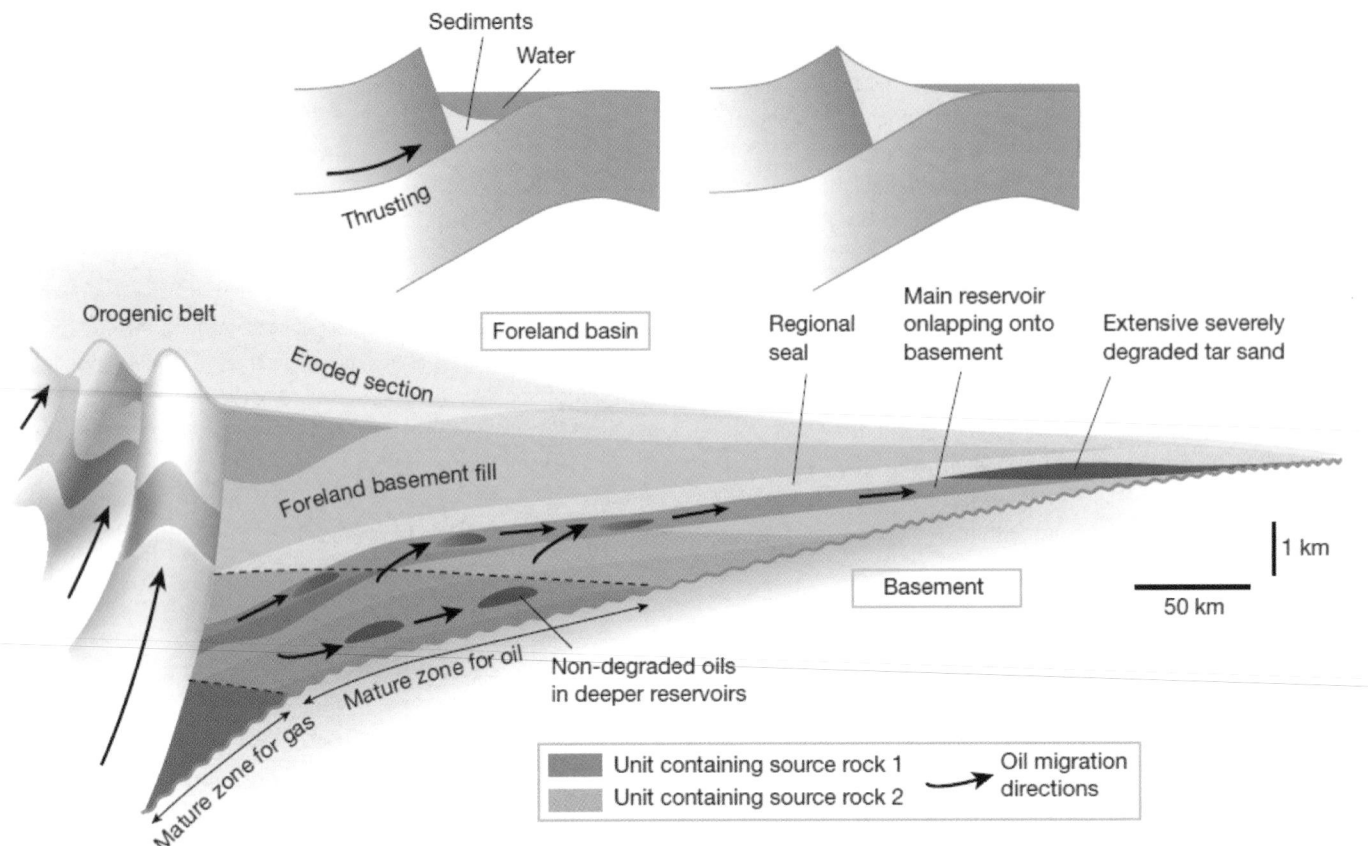

Figure 7.7 | The 'standard' model for viscous oil formation. Source: Head et al., 2003.

metals such as vanadium and nickel. The concentration of these elements in viscous oil is a function of density: oils of $\rho > 1.0$ can have sulfur concentrations greater than 4% and heavy metal contents as high as 500 parts per million (ppm). Viscous oils with $\rho < 0.92$–0.93, by contrast, typically have less than 25% of such concentrations. In the production of these resources, the elements S, V, Ni, Fe, and the high oil viscosity lead to problems in processing and environmental management (Head et al., 2003).

Bitumen is the heaviest, thickest form of petroleum. The bitumen contained in shallow oil (tar) sands can be recovered using surface mining techniques. However, it takes about two tonnes of oil sands mined (0.9 m³) and processed to obtain one barrel of synthetic crude oil. About 90% of mined bitumen can be recovered by way of hot water and surfactant mixing and separation from the sand in special agitation-flotation tanks. After oil extraction, solids and water are returned to the mine, which is eventually reclaimed. Separated bitumen is then shipped or upgraded to synthetic crude oil.

Deposits with overburdens of more than perhaps 60 m must be exploited by in situ methods, e.g., injecting steam into the deposits to reduce the bitumen's viscosity and increase reservoir pressure (which also induces fracturing), and thereby enable bitumen to flow (Dusseault, 2002). The advantage of in situ extraction is avoidance of

the massive movement of the oil's host material and attendant surface disruption. The disadvantage is its water and energy intensity (about 3–4 barrels of steam per barrel of oil produced). Two processes are currently applied: cyclic steam simulation (CSS) (Nasr and Ayodele, 2005) and steam-assisted gravity drainage (SAGD) (Butler and Stephens, 1980; 1981). CSS involves the injection of a charge of high-pressure steam into a well over some time, and during the production cycle the heat and pressure mobilize bitumen flow to the same well, along with hot water. The cycle is repeated when the rate of production flow declines and a well may have a lifespan of 8–12 years. SAGD involves twin parallel horizontal wells separated vertically by about 4–5 m. Steam is injected into the top well, the mobilized bitumen drains into the bottom well and is lifted to the surface. Other in situ extraction processes being tested include the use of chemical solvents, electric heating, and partial combustion, none of which are currently considered commercially viable.

In situ production of heavy and extra-heavy oils utilizes methods similar to those for the production of oil sands, except that if the viscosity is modest and the permeability high, conventional pumping and water flooding may result in modest recovery factors (5–20%) without steam stimulation. If the sand is unconsolidated and has good solution-gas content, in many cases encouraging sand influx has proven to be a viable and economic production process.

Not only are viscous oils difficult to extract, they are difficult to process and transport. These oils may be viewed as deficient in hydrogen (carbon rich) and must be upgraded to produce useful products. Hydrogenation and carbon rejection (coking) are used to improve the hydrogen-to-carbon ratio, with CH_4 used as the source of hydrogen during upgrading. The heavy metals lead to difficulties in processing because they rapidly poison catalyst beds. Large volumes of sulfur and coke are by-products of typical upgrading operations. Once upgraded, the synthetic crude oil is sent to conventional refineries for further beneficiation to create marketable products. If local upgrading is not used, pipelining viscous oil requires blending with diluents, light oil, or synthetic crude oil.

At the present time (2010), only nine countries in the world produce more than 250,000 bbl/day or 12.5 Mt/yr of non-conventional (viscous) oil: Canada (~90 Mt/yr), Venezuela (~50 Mt/yr), the United States (~20 Mt/yr), Mexico (~18 Mt/yr), and each of Russia, China, Oman, Iran, and Brazil at about 15 Mt/yr.

7.2.3.2   Oil Shales (Kerogen)

Kerogen is a semisolid organic compound formed from algal residues. It is found in shales – sedimentary rocks that were buried rapidly, which led to low oxygen conditions that preserved the lipids and other similar cellular debris from the algae. Over geological time, kerogen buried to depths greater than 3 km and at temperatures greater than 80–90°C is broken down and converted into light oil. Economically interesting kerogen deposits are found at relatively shallow depths, less than 1500–2000 m. Kerogen in oil shales is viewed as 'immature oil' and, like viscous oil, is deficient in hydrogen (Dyni, 2006; Altun et al., 2006; WEC, 2007).

Near-surface deposits can be produced with standard surface and underground mining techniques using surface retorting (pyrolysis),

and the shale-oil product (a mixture of gases and liquids) used directly as boiler fuel or as hydrocarbon products. The generation of synthetic crude oil from kerogen requires pyrolytic retorting with collection and treatment of the product. In contrast to most viscous oil, kerogen in deeper reservoirs cannot be mobilized, but must be converted physically through pyrolytic approaches. For example, kerogenous shale can be exposed to a combustion process (oxypyrolysis) that literally cooks the shale by partial combustion, generating valuable lighter products.

In situ or surface anoxic pyrolysis requires temperatures greater than 325°C to decompose the complex molecules into liquid and gaseous products that can be recovered, leaving behind perhaps 30% of the organic matter as elemental carbon. Anoxic in situ extraction technologies are based on conductive heating of shale beds using thermal energy (electrical heating or hot fluid circulation), with separate wells to collect generated fluids (e.g., Biglarbigi et al., 2009; Crawford et al., 2008, 2009).

Unlike heavy oil, kerogen does not contain sulfur and heavy metals. Thus, the products from pyrolysis are light 'sweet' gases and liquids that are processed relatively easily.

7.2.4   Estimates of Unconventional Oil

There are about 1780 Gt (74.8 ZJ) of liquid petroleum (shale oil, heavy oil, extra-heavy oil, and bitumen) trapped in sedimentary rocks in several thousand basins around the world (Table 7.7). Oil-shale resources are estimated at about 382–450 Gt (16–18.9 ZJ) (Dyni, 2006; WEC, 2007). However, these figures, particularly for oil shale, are somewhat conservative because of the lack of detailed exploration for these resources, particularly in countries with large conventional oil resources. For example, Libya, Iraq, Saudi Arabia, Iran, Oman, Kuwait, the United Arab Emirates, and other countries

Table 7.7 | World oil resource estimates.

Oil type	Definition	Gt	ZJ
Conventional	Original oil <100 cP viscosity in place, including sandstones and carbonates	614	25.7
Produced to date	Estimated cumulative production to date, 98% of which has been conventional oil <100 cP in situ	150–164	6.3–6.9
Remaining conventional	Total remaining in place Technically recoverable, current technology Ultimately recoverable, including above	464 150 205	19.4 6.3 8.6
Viscous oil	Original oil in place, all rock types	1350	56.4
Viscous oil in sandstones	All viscous oil in sandstone reservoirs Technically recoverable, current technology Ultimately recoverable, including above	~1023 109 273	42.8 4.6 11.4
Viscous oil in NFCRs	All viscous oil in NFCRs Technically recoverable, current technology Ultimately recoverable, including above	205–300 <14 41	8.6–12.5 <0.59 1.7
Shale oil converted into liquid oil (excluding gas)	Estimated barrels in place, >50 l/t shale oil Technically recoverable, current technology Ultimately recoverable, including above	382–450 <14 41	16–18.8 <0.59 1.7

Sources: quantitative data from authors' estimates; WEC, 2007; USGS, 2000; 2008; IEA, 2008a; BGR, 2009.

Table 7.8 | Unconventional oil reserves and resources.

Region	Oil Sands						Heavy & Extra Heavy Oil					
	Resources in place		Reserves		Cumulative production		Resources in place		Reserves		Cumulative production	
	[Mt]	[EJ]	[Mt]	[EJ]	[Mt]	[EJ]	[Mt]	[EJ]	[Mt]	[EJ]	[Mt]	[EJ]
USA	5905	247			95	4.0	108,542	4537	1590	66	1250	52
CAN	245,380	10,257	27,450	1147	2980	125	61,300	2562	1	0.04		
MEX							18,400	769			100	4.2
WEU	106	4.4	33	1.4			580	24	26	1.1	450	19
EEU	35	1.5					12	0.5	6	0.3	1	0.04
FSU	180,744	7555	17,869	747	20	0.8	258	11	21	0.9	490	20
NAF							380	16	8	0.3	420	18
EAF	111	4.6	35	1.5								
WCA	306	13	97	4.1								
SAF	233	10	74	3.1								
MEE							87,620	3663	1	0.04	1390	58
CHN	89	3.7					5587	234	119	5.0	460	19
OEA							142	6			35	1.5
IND							430	18				
OSA											52	2.2
JPN												
OCN												
PAS	222	9.3	70	2.9	4	0.2						
LAC	219,000	9154			3610	151	100	4.2	6423	268	213	8.9
TOTAL	652,131	27,259	45,628	1907	6709	280	283,351	11,844	8195	343	4861	203

Note: Viscous oil production from oil sands mining, Canada only, for the period of 1967–2009 is about 4 Gbbl (CAPP, 2009). The shale oil production for the period of 1880–2009 is estimated at 1.5 Gbbl assuming an average shale oil content of 100 l/t (WEC, 2007).

Sources: Authors' estimates; IEA, 2008a; USGS (i.e., see Meyer and Dietzman, 1981; Meyer and Duford, 1989; Meyer and Attanasi, 2004; Meyer et al., 2007; Schenk et al., 2009); Laherrère, 2005; CAPP, 2009; BGR, 2009; WEC, 2010; Dusseault and Shafiei, 2011.

have extensive deposits of viscous oil, but the volumes are not well delineated largely because of lack of interest. As conventional oil becomes more difficult to exploit, greater interest in unconventional oil will result in better delineation of the world resource base.

The definition of what constitutes a resource remains contentious. A 3 m thick conventional oil zone with good permeability is exploitable under many conditions, but not a 3 m thick bitumen bed, or a deeply buried 10 m thick oil shale, or oil shale with less than 7–8% organic content by mass. Viscous oil in thinly bedded, low-permeability strata at depth is not likely to be included in a resource survey.

Tables 7.8 and 7.9 contain estimates of unconventional oil resources for the 18 GEA regions. The figures are approximate because the category of technically recoverable reserves is a moving target as prices change, costs rise, and technologies are developed. Technically 'recoverable oil' implies the use of currently commercialized methods, whereas 'ultimately recoverable oil' assumes that currently known but non-commercial methods achieve commercial status. Additional oil will also become accessible with innovative, but presently unknown, extraction technology.

7.2.5   Oil Supply Cost Curves

As seen in Table 7.6, conventional oil reserves and resources, more than half of which are in the Middle East and North Africa, amount to about 11,000 EJ (~2 Tbbl). To put this amount of oil into perspective, cumulative past oil production amounts to around 6500 EJ.

Figure 7.8 shows an aggregate (of the 18 GEA regions) global oil supply cost curve. The curve plots the potential long-term contributions from conventional resources (Table 7.6) and non-conventional resources (Table 7.8) against their 2007 and projected 2050 production costs. The costs do not include taxes or royalties. The projected productivity gains in upstream oil operations vary between 0.25% and 0.75% per year. While the 2050 supply cost curve accounts for a constant oil production rate of 4000 Mtoe/yr until 2050, it does not reflect the added knowledge brought about by geological work between now and 2050, and the volumes underlying the 2050 curve are very much on the conservative side.

The supply curve suggests some 4000 EJ of conventional oil can be produced for 2 $/GJ (~12 $/bbl) or less at 2007 prices and exchange

Table 7.8 | (continued)

Region	Shale Oil						All Unconventional Oil					
	Resources in place		Reserves		Cumulative production		Resources in place		Reserves		Cumulative production	
	[Mt]	[EJ]	[Mt]	[EJ]	[Mt]	[EJ]	[Gt]	[EJ]	[Mt]	[EJ]	[Mt]	[EJ]
USA	301,566	12,605	32,700	1367	20	0.8	416.0	17,389	34,290	1433	1365	57
CAN	2200	92					308.9	12,911	27,451	1147	2980	125
MEX							18.4	769	0	0.0	100	4.2
WEU	13,532	566	137	5.7	275	11	14.2	594	196	8.2	725	30.3
EEU	66	2.8	2	0.1			0.1	4.7	8	0.3	1.0	0.04
FSU	46,652	1950	244	10	1250	52	227.7	9516	18,134	758.0	1760	74
NAF	8983	375	150	6.3			9.4	391	158	6.6	420	17.6
EAF	5	0.2					0.1	4.8	35	1.5	0.0	0.0
WCA	14,310	598					14.6	611	97	4.1	0.0	0.0
SAF	19	0.8			5	0.2	0.3	11	74	3.1	0.0	0.0
MEE	5796	242	1320	55			93.4	3905	1321	55.2	1390	58.1
CHN	2290	96	191	8.0	1000	42	8.0	333	310	13.0	1460	61.0
OEA	42	1.7					0.2	8	0	0.0	35	1.5
IND							0.4	18	0	0.0	0.0	0.0
OSA							0.0	0	0	0.0	52	2.2
JPN							0.0	0	0	0.0	0.0	0.0
OCN	4534	190	518	22	6	0.3	4.5	190	518	21.7	6.0	0.25
PAS	1202	50	243	10			1.4	60	313	13.1	4.0	0.17
LAC	11,794	86.1	134	1.0	500	21	230.9	9244	6557	269.5	4323	180.7
TOTAL	412,991	16,856	35,639	1485	3056	127.7	1,348	55,959	89,462	3734.9	14,621	611

rates (these supply costs are not the actual prices of the fuel in the market place).

The total long-term potential oil resource base assessed here amounts to some 34,000 EJ (810 Gt). This includes 11,600 EJ of conventional oil reserves and resources (USGS, 2000, 2008), 3700 EJ of unconventional oil reserves, and 33% or 18,700 EJ of unconventional oil resources (Table 7.8).

The cost of producing conventional resources typically ranges from less than 5 $/bbl in the Middle East up to 40 $/bbl. Deep-water and Arctic production could drive costs to 70 $/bbl.

The production costs of heavy oils and oil sands range from 15–80 $/bbl. A lack of major commercial production experience means the costs of shale-oil recovery are uncertain and estimated to range between 60–140 $/bbl. Continued demand for oil-based energy services throughout the 21st century is expected to induce technology change, and production cost levels could be 25% lower by 2050.

Environmental considerations could adversely affect oil-production costs, especially for unconventional resources that leave a large environmental footprint, including through the GHGs emitted during the extraction and upgrading processes. GHG-emission penalties would change the shape of the cost curve, as unconventional oil would become relatively more expensive, and enhanced oil recovery based on CO_2 injection potentially cheaper (IEA, 2008a).

Figure 7.9 presents the uncertainties regarding future liquid-fuel availability and associated production costs (Farrell, 2008).

7.2.6 Environmental and Social Implications

The upstream oil industry has a blemished reputation in environmental protection. There are four major factors behind this. First, spending on waste management has been viewed as a direct revenue loss. For example, rehabilitation of drill sites prevents erosion and long-term soil and water degradation, but brings no financial returns and is thus often avoided or poorly managed.

Second, in exploration and production activity, there has been little commitment to local communities. Company employees are usually from

Table 7.9 | Unconventional oil reserves and resources.

Region	Oil sands			Heavy & extra heavy oil			Shale oil			All unconventional oil		
	Amount in place	Resources	Reserves	Amount in place	Resources	Reserves	Amount in place	Resources	Reserves	Amount in place	Resources	Reserves
	[Mt]	[Mt]	[Mt]	[Mt]	[Mt]	[Mt]	[Mt]	[Mt]	[Mt]	[EJ]	[EJ]	[EJ]
USA	5905	2065	0	415	76	3	301,566	73,030	35,970	12,870	3142	1504
CAN	272,000	81,853	27,450	0	2	1	2200	0	0	11,462	3422	1147
WEU	336	106	33	2312	419	26	13,532	307	151	676	35	9
EEU	100	35	0	61	12	6	66	4	2	9	2	0
FSU	180,744	57,005	17,869	1434	258	21	46,652	545	268	9565	2416	759
NAF	0	0	0	79	14	8	8983	335	165	379	15	7
EAF	352	111	35	0	0	0	5	0	0	15	5	1
WCA	971	306	97	0	0	0	14,310	0	0	639	13	4
SAF	739	233	74	0	0	0	19	0	0	32	10	3
MEE	0	0	0	1	0	1	5796	2948	1452	242	123	61
CHN	253	89	0	1411	254	119	2290	427	210	165	32	14
OEA	0	0	0	0	0	0	42	0	0	2	0	0
IND	0	0	0	0	0	0	0	0	0	0	0	0
OSA	0	0	0	0	0	0	0	0	0	0	0	0
JPN	0	0	0	0	0	0	0	0	0	0	0	0
OCN	0	0	0	0	0	0	4534	1156	569	190	48	24
PAS	708	222	70	0	0	0	1202	543	267	80	32	14
LAC	260	92	0	240,740	46,820	6423	11,794	298	147	10,567	1973	275
TOTAL	462,368	142,117	45,628	246,453	47,855	6608	412,991	79,592	39,202	46,892	11,268	3822

Source: BGR, 2009.

distant communities, spending a limited time in the area. Again, drill-site rehabilitation serves as a useful example. In Wytch Farm (UK), only a few kilometers from expensive beach properties, drill sites are fully rehabilitated, reforested, and invisible from all recreational sites. However, in southeastern Ecuador, 30-year-old drill sites in the jungle, including drilling waste pits and oil spills, continue to erode and cause silting and local contamination (San Sebastian et al., 2001).

Third, there is the joint problem of poor regulatory enforcement and occasionally associated corruption. Environmental protection requires strong, transparent, and consistent enforcement of regulatory guidelines, often in remote areas. Human fallibility and greed, combined with industry 'capture' of regulatory bodies, has led to unfortunate cases of collusion and corruption, invariably with negative consequences. Neutral third-party annual audits with full publication of results constitute an approach to reducing this problem.

Fourth, ignorance of environmental impacts has led to a general reactive attitude to environmental management rather than to proactive avoidance. Once a severe problem happens, mitigation costs and reputational loss often far exceed the initial investment in avoidance that should have been made. Increasing corporate attention to employee education is reducing this problem.

Recent and continued progress (Marika et al. 2009) suggests that goals of 'minimal impact-quasi zero emissions' are technically feasible and economically attainable, as well as socially desirable.

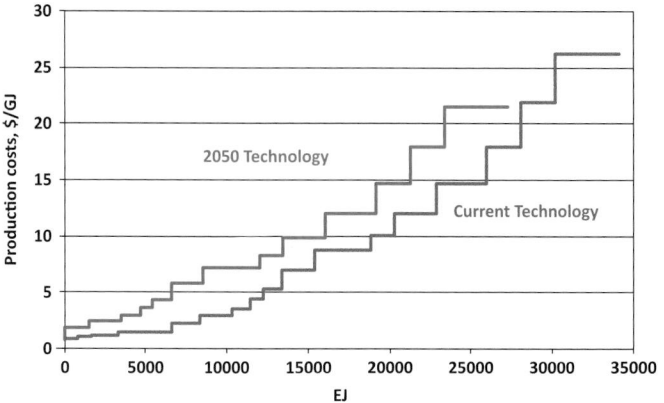

Figure 7.8 | Long-term oil supply curve[7] – combined global conventional and unconventional oil reserves and resources. Source: USGS, 2000; 2008; IEA, 2008a; Deutsche Bank, 2009; Aguilera et al., 2009.

7.2.6.1   Environmental Issues in Oil Development

Oil development involves, among others, the following activities: seismic exploration, exploratory and development drilling, infrastructure generation,

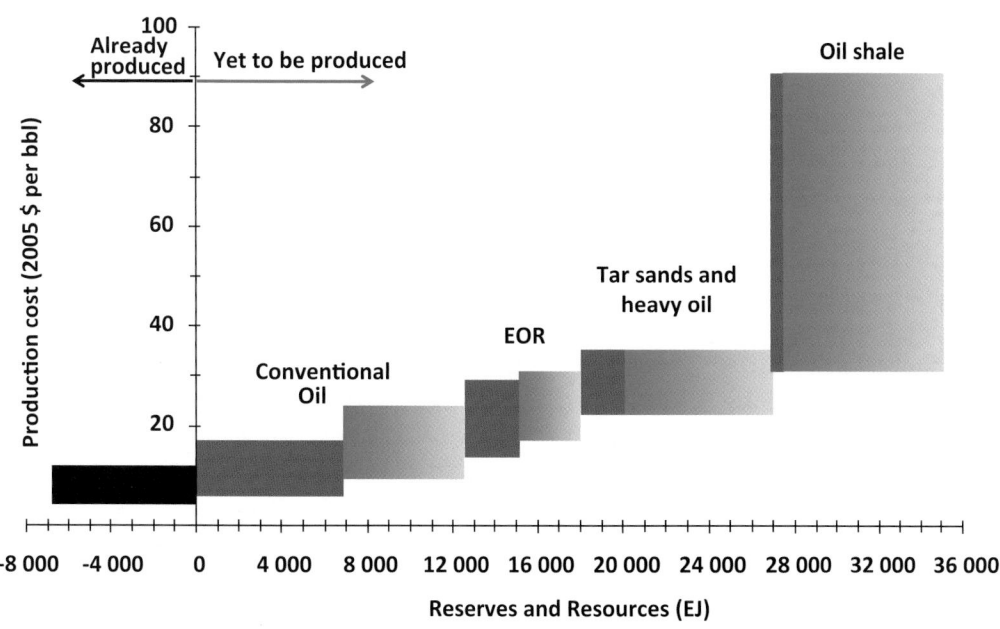

Figure 7.9 | Liquid fuel supply potentials and production costs. The green bars represent the economic cost of extracting fuels. The black bar to the left of the vertical axis represents cumulative past oil production (about 157 Gt). The height of the bars represents the degree of uncertainty of the production costs. The widths of the bars represent how plentiful the resource is, and the lightness of the bars reflects the degree of uncertainty of resource availability. Source: adapted from Farrell, 2008.

---

7   The 2050 supply curve is net of a hypothetical cumulative production between 2005 and 2050 based on the historically observed production growth rate between 1995 and 2007. The resources exploited during the period up to 2050 are no longer available after 2050. Despite upstream innovation and technology change, the 2050 supply cost curve is higher as resource extraction until 2050 is expected to exploit lower cost occurrences first.

long-term production, transportation (generally pipelines), siting for large upgrading, refining and petrochemical facilities, and postactivity cleanup.

Current liabilities that reside in previously contaminated sites are large, as the industry efforts to rectify the impact of poor historical practices. The full economic implications of poor environmental management is little understood and under heated debate. While issues common to oil and gas development are discussed in this section in a somewhat positive light, with the notion that the industry is doing better, it is clear that it can and must continue to improve.

With regard to climate-change issues, the major CO_2 emissions related to oil products come from their end-use combustion, and these numbers overwhelm the emissions from oil exploration and production activity (Charpentier et al., 2009; IHS-CERA, 2010).

However, there are other major impacts on land and water resources. Seismic work requires access to land areas, generally by cutting paths to allow the use of all-terrain vehicles. Sensitive areas include permafrost regions, forests, and steep terrain where vegetation disturbance leads to permafrost degradation or soil erosion. Low-impact methods reduce damage in these areas by hand-cutting narrow lines, and using helicopters and vehicles with small contact pressure for transport.

Exploratory drilling with light helicopter rigs has reduced road-triggered erosion, and better cuttings and waste-disposal practices, combined with improved site rehabilitation, have significantly reduced waste abandonment and erosion. Directional pad drilling in the North Slope of Alaska and winter-only transport have reduced the surface area needs to about 13%, compared to 30 years ago. Similar improvements have been made in jungle areas (Peru) and in farming regions (Saskatchewan and Alberta). Fossil-fuel exploitation can be compatible with agriculture and sensitive ecosystems.

Vast volumes of produced water (greater than three tonnes of water per tonne of oil worldwide) are now systematically reinjected for water flooding or disposed into deep saline aquifers (Veil and Quinn, 2008). In development plans for the Marañon Basin in Peru, companies are contractually committed to full reinjection without diversion to river courses. Landfills should be avoided for any solid waste that contains species potentially harmful to subsurface potable water because of long-term security issues: all landfills leak or will eventually leak. Other produced liquid wastes, emulsions, and solid wastes can be slurried and reinjected safely, virtually to eliminate environmental liability. After exploitation and beneficiation, sites and spills must be properly cleaned and disposed, and corporations are requested more and more to post bonds (insurance) against future problems. Cleanup must be an ongoing and regulated activity, not reserved for when production is fully terminated.

Pipeline and flowline installations in jungle areas now increasingly use forest canopy preservation, and the Aleyska pipeline from the Alaska

North Slope lies largely on elevated beams placed on thermo-siphon piles that do not degrade the permafrost. In built-up areas, pipeline burial, and reuse of rights-of-way as farmland is the rule, and pipeline companies have trained response teams for accident management. Aggressive reseeding of cleared areas using rapidly growing grasses and plants reduces erosion, and in many areas actually provides enhanced grazing for ungulates. Forty years ago all these practices only took place in developed areas; now they are basically the rule everywhere, but regulatory vigilance is vital. In particular, as the pipeline infrastructure ages, better inspections and corrosion-detection methods are needed to identify problems before they become breaks and spills (OGJ, 2010).

Environmental difficulties at large processing centers attract a great deal of media attention, which forces operators to become more vigilant and proactive. Nevertheless, a long history of site contamination blights the reputation of fossil-fuel companies. Many sites around the United States and the Mexico Gulf Coast are in a state of severe contamination (hydrocarbons, chlorinated hydrocarbons, and heavy metals); in some cases, sites (and responsibilities) have been abandoned for several decades. Deep slurried solids injection provides a method for cleaning these and other sites.

During offshore developments, the key to protecting marine environments near production and exploration activities is to plan and enforce zero liquid and solid discharge policies, as adopted by the Government of Norway, for example (Ekins et al., 2007). This is achieved by shipping waste to shore for treatment and by the injection of solids and liquid waste into oil-free saline aquifers. Spill risks must be managed properly and emergency response measures put in place. Old platforms are salvaged or sunk to provide rich environments for marine life, as iron is a limited nutrient in the ocean.

The recent (April 2010) Deep Water Horizon blowout in the Gulf of Mexico is having profound effects on safety practices, both to prevent and to respond to such incidents (OGJ, 2010). At the present time operators in the region have invested large sums to create a joint industry-response approach, and regulatory aspects are being examined carefully. This accident will affect all future offshore practices, and the cost to BP (US$40 billion projected is the most recent estimate) has already triggered improvements in procedures manuals, operational quality control, regulatory management, and response plans. These changes will reduce environmental risk for offshore operations around the world.

7.2.6.2 Oil Sands Mining
In Canadian surface mining of viscous oil, the issues are atmospheric emissions (sulfoxides [SO_x], nitrogen oxides [NO_x], CO_2, and particulates), water use (water sources, tailings ponds, seepage into groundwater), and solid-waste management (sand-tailings management, mature fine-grained oily sludges, long-term coke and sulfur storage, disposal of solid wastes such as pipe scale, tank bottoms, and emulsions). Concerns over massive surface changes can be mitigated through reclamation

efforts with the resultant terrain often more ecologically productive than before. Furthermore, the area amenable to surface mining in Alberta, about 3400 km^2, is no larger than 60% of the greater Los Angeles area; 95% of Alberta's boreal forests will remain unaffected ($\sim$470 km^2 of boreal forest has been cleared or mined to date).

One cubic meter of ore weighs 2.2 tonnes and averages 0.68 m^3 of silica sand, 0.23 m^3 of oil, 0.06 m^3 of water, and 0.03 m^3 of fine-grained silt and clay. The products are approximately 0.2 m^3 (0.171 t) of raw bitumen, 1.1 m^3 of bulked sand (35% porosity) that contains water, fine-grained minerals, and residual oil in the pores, 1.6 m^3 of water, much of which is recycled for further extraction, and 0.20 m^3 of oily aqueous sludge containing 60% of the clay minerals in the original raw ore. In addition, for each cubic meter of ore, 0.3 m^3 of overburden is removed, organic soil is stored for future reclamation, and 0.01 m^3 each of coke and sulfur are produced and stockpiled. CO_2 emissions for mining and upgrading are about 700–750 kgCO_2 per tonne of oil, exclusive of transportation and further refining (IHS CERA, 2011).

Of these materials, the large volumes of pond sludge constitute the major waste stream that presents a significant long-term problem, although tailings dams, coke, and sulfur may have local groundwater impacts. Since mining of oil sand in Alberta began 40 years ago, the volumes of sludge per barrel of bitumen produced have halved, and research continues to develop less water-intensive extraction methods. General groundwater drainage toward the large local river (Athabasca) means that management of contamination is feasible through interception and dilution. The recent focus of industry on sludge rehabilitation has resulted in large-scale experiments by corporations (Shell, Suncor, Syncrude) into methods to reduce sludge generation volumes and solidify the remaining sludge.

The incentive arises from a tightening of licensing practices by the Alberta government, but progress remains slow. Some low-volume, difficult wastes are also being disposed into deep salt caverns (e.g., Veil et al., 1999) in Alberta and Saskatchewan to achieve permanent isolation. Nevertheless, final pond closure, elimination of mature fine-grained tailings sludge, and reestablishing a stable productive landscape represent challenges to be met.

Recovery processes for in situ viscous oil

Steam for heavy oil recovery requires heat and water sources, as well as water recycling. Natural gas is by far the main source of heat, but asphaltene and coke (or coal) combustion for heat and H_2 generation will become more common. On average, recovery of one tonne of viscous oil by steam methods requires three tonnes of water as steam at 200–300°C. Steam generation requires water treatment and attendant sludge-disposal needs. Production of viscous oil also generates emulsions that are challenging to treat and are best disposed of through deep injection. Saline boiler subfeed water and other contaminated water can be injected, along with all other contaminated water, into saline aquifers. CO_2 emissions (extraction, upgrading, transportation, and refining only) are 990–1100 kg/t of oil for SAGD methods, and

1170–1390 kg/tCO_2 per tonne oil for cyclic steam stimulation (IHS CERA 2010). About 85% of emissions are allocated to steam generation, H_2 generation, power, and pumping inputs, ranked by size. For comparison, synthetic crude oil from mining requires about 950–1000 kg/tCO_2 to bring the product from the ground to the service station.

In situ combustion processes, by contrast, are anhydrous; CO_2 emissions are estimated at 880–1000 kg/t, generally less than steam projects based on natural gas. Furthermore, lack of significant liquid and solid wastes make in situ combustion processes more benign in terms of environmental impact. However, combustion methods remain far from commercialization at this time.

Oil-shale development

Anoxic in situ pyrolysis of oil shales at high temperatures generates no surface wastes or significant atmospheric emissions. Anoxic surface pyrolysis is also a process with low atmospheric emissions, but solid wastes are generated. If combustion-based in situ pyrolysis or surface retorting is used, CO_2 emissions are likely to be about 1170–1470 kg/t, not counting additional energy sources for heating or transportation. If ore is mined and retorted, spent shale to be disposed of has a volume bulking factor of about 20%, but it is an inert substance without residual oil or environmentally damaging substances. In the vast oil-shale deposits in the United States, as well as for some other oil-shale deposits in Jordan, China, and Australia, an arid climate means that reclamation of mined land is challenging, although the land-use value of these arid regions is low. Water requirements for oil-shale development are low because steam will not be used and retorting is anhydrous. It is unlikely that large-scale exploitation of the vast American oil-shale deposits will ever occur by surface mining methods because of the environmental impacts, solids management costs, and reclamation problems.

Zero emissions targets

Suppressing stack emissions of SO_x, NO_x, and particulates to near-zero levels is currently achievable, but CO_2 emissions will remain an environmental concern for the foreseeable future (see also Chapters 12 and 13). Produced solids, scale, cuttings, pond sludges, tank bottoms, emulsions, and treatment fluids, as well as many other noxious and hazardous wastes, can be slurried with untreated waste water and disposed of safely through deep-well injection. The slurry is injected at depths greater than 400 m into saline aquifers with adequate seals against upward flow. Solids are permanently retained in the stratum once the pressures dissipate, which happens rapidly in permeable strata. Costs are generally far less than those of chemical treatment or washing processes, which often generate other waste streams. An example of a large heavy-oil field approaching zero surface discharge is Duri, Indonesia, where deep-injection practices are used for the great majority of waste streams, including wastes from previous surface pits (Marika et al., 2009). Finally, as noted above, zero saline water discharge is now being practiced widely, and should be mandated in many other areas.

It appears that reasonable cost technologies exist to achieve near-zero discharge in oil and gas exploitation. Regulatory practices and enforcement are key to achieving the environmental goals that are now technically feasible. Good industry and regulatory practices must be promoted in the less-developed world, and the responsibility for this resides both with corporations and with non-industry agencies (governmental and non-governmental organizations) that promote good governance.

7.2.7   Summary

Cumulative global oil production to date is about equal to the current remaining conventional oil reserves. Based on conventional reserves alone, oil production will peak soon and steadily decline shortly thereafter. However, reserves and resources of all types of oil are sufficiently large to meet ever-rising demands for one to two decades until production reaches its peak.

Only the massive development of unconventional oil occurrences would shift peak liquid-fuel supply from oil-based resources to 2050 or beyond. This would require ramped up investments in upstream R&D, followed by the commercialization of new exploration, production, and upgrading techniques. The extraction of unconventional oil is more energy intensive than that of today's conventional oil production, claiming up to 20–30% of the oil's energy content rather than 10–15%, and leaving an emissions footprint that is about 10% higher than conventional oil on a well-to-wheels total CO_2 emissions basis (IHS-CERA 2010).

The investment share in extraction and upgrading of viscous oil is expected to increase slowly, compared to that of conventional oil, raising the issue of finance and timely investment to avoid future shortfalls and excessive upward pressure on prices. The lead time for large-scale viscous oil projects can be 3–5 years for in situ projects and 7–10 years for large mining projects.

Decisions on oil investments depend on several factors, ranging from aggregate demand, the nature of the markets served, to the availability of competitive alternative transportation fuels, as well as issues such as climate change and geopolitics.

7.3   Natural Gas

7.3.1   Overview

Natural gas is a mixture of combustible and non-combustible gases. Its chief combustible component is CH_4; other energy-relevant components include butane, ethane, and propane. Typical non-combustible components of natural gas are nitrogen, CO_2, and hydrogen sulfide (H_2S). Like oil, typically natural gas resources are trapped in porous underground rock formations, predominantly composed of sandstone. It is generally accepted that CH_4 is the result of the anaerobic decomposition of organic material.[8] Organic matter is ubiquitous in all sediments and so is CH_4. Although abundantly available, most of these occurrences are too diffuse to warrant commercial recovery.

Gas reservoirs differ greatly, with varying physical conditions affecting reservoir performance and recovery rates. Over geological time, almost all natural gas migrates upwards through the Earth's crust and eventually leaks to the atmosphere. If its migration is blocked by a geological trap – porous reservoir rock sealed above by impermeable cap rock – commercial quantities of gas can accumulate. This gas usually contains little admixtures of other hydrocarbons and is termed non-associated gas or dry gas.

Commercial amounts of gas can accumulate also as a gas cap above an oil pool or, with high reservoir pressures, dissolved in the oil. Such natural gas is referred to as 'associated gas' because its recovery is generally a by-product of oil production. The associated gas recovered along with oil is separated at the surface. Depending on location, field size, geology, and gas in place, associated gas is either recovered for revenue generation, reinjected for field pressurization and prolonged oil recovery, or flared. Approximately 17% (~135 billion m^3 or 5 EJ) of total recovered associated gas is currently flared because of the lack of harvesting infrastructure, especially for remote or small fields that do not warrant a commercial gas collection and transportation system.

Non-associated natural gas reservoirs are much more abundant than reservoirs with both oil and gas. When there are no significant liquid hydrocarbon components, a larger part of the in-place gas can be recovered by dropping reservoir pressures. Reservoir pressure, however, is often maintained by encroachment of water in the sedimentary rock formation and some of the gas will be trapped by capillarity behind the incoming water. Therefore, in practice, only approximately 60–80% of the in-place gas can be recovered (IEA, 2009).

Compared to oil, the remaining resources of natural gas are abundant. Reserve additions consistently outpace production volumes and resource estimations have increased steadily since the 1970s (IEA, 2010a). Figure 7.10 shows the evolution of natural gas resource estimates since 1950. The biggest uncertainty for future gas supply is thus a question of whether sufficient and timely investment will be made in developing these resources and associated transmission infrastructures to bring them to the market place at competitive costs.

7.3.2   Classifying Conventional Gas and Unconventional Gas

Like oil, natural gas resources are termed 'conventional' when recovery is possible with standard extraction technologies.

---

8   There exists a different theory (see Section 7.3.7).

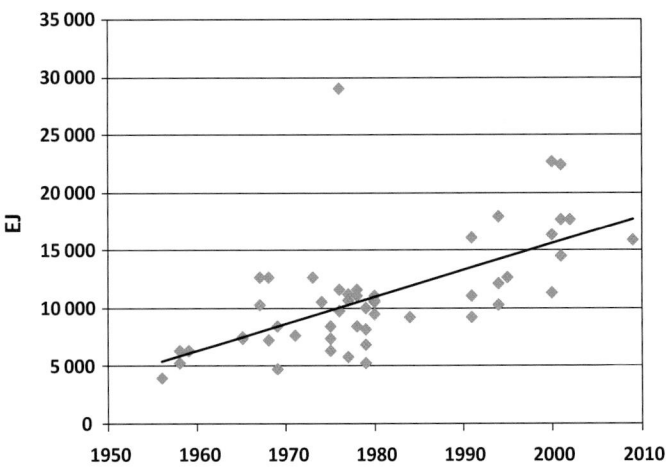

Figure 7.10 | World gas resource (reserves plus resources) estimates, 1950–2007. Source: adapted from Ahlbrandt and Klett, 2005; NPC, 2007; BGR, 2009.

In 'unconventional' gas deposits, additional technical measures are required for a production well to result in commercially viable flow rates and volumes – because of either the low permeability of the reservoir or the gas does not exist in a free gaseous phase. Hydraulic fracturing, horizontal drilling, or multilateral narrow-spaced wellbores are typical methods used to produce commercial flow rates and volumes. Still, per well, an unconventional gas reservoir will produce less gas over longer periods than a conventional higher permeability well (only 10–30% of total gas in place) (IEA, 2009). Thus, reservoir permeability is a key parameter that separates conventional and unconventional gas deposits. International practice suggests an average permeability of 0.1 milli-Darcy (mD) as the upper limit for unconventional gas (IGU, 2003; BGR, 2009).

Typically, unconventional gas can be found in low-permeability reservoirs that consist of sandstone (tight sands), shales (fine-grained sedimentary rock rich in organic material), carbonates, and coal deposits. It can also be found dissolved in groundwater, geopressured aquifers, saline brines, and gas hydrates. These fundamentally different geological settings require distinctly different production methods. Thus, unconventional gas occurrences can be classified further by applying simple criteria based on the host geology and the physics of natural gas in the reservoir.

The International Gas Union suggests grouping unconventional gas sources into two categories:

- 'Really unconventional gas sources,' i.e., where the non-free (solid, liquid) gas content of the reservoir is larger than 5%[9] of the total gas content of the reservoir (excluding conventional reservoirs). This category includes shale gas, CBM, water-dissolved (aquifer) gases,

and natural gas hydrates. Independent of actual gas flow rates, the exploitation of these sources requires new technologies and processes.

- 'Pseudo-unconventional gas sources,' i.e., where less than 5% of the total gas content is in a form different from the free state, but the gas is not economically feasible for development for geological or technical reasons. Tight reservoir gases, deep gas (basin-centered gas systems), and permafrost gas fall into the pseudo-unconventional category and can essentially be produced with adapted conventional technology.

The boundary between conventional and unconventional gas is dynamic and more blurred than that for oil. What was unconventional yesterday may, through some technological advance, ingenious new process, regulation, or dramatically different market conditions, become conventional tomorrow. For example, the latest reserve assessments, e.g., by Cedigaz (2009), also include gas from unconventional sources. In fact, the term 'unconventional' is becoming a misnomer as gas from these sources increasingly supplements conventional production, especially in the United States where 'unconventional gas' already accounts for over half its domestic gas supply (IEA, 2009). The recent surge of shale gas production in the United States, mainly through new technology applications (hydraulic fracturing) dramatically lowering recovery costs, was instrumental in bringing wellhead gas prices down from 11 $/GJ in 2008 to around 4 $/GJ in 2010 (US EIA, 2010).

7.3.3 Conventional Natural Gas Reserves and Resources

There have been only a few new assessments of global natural gas resources since 2000. These estimates generally point to a steadily growing resource base and are largely in agreement regarding reserves (see Table 7.10), but diverge for total resource potentials.

While the recent USGS appraisal of the oil and gas resource potential north of the Arctic Circle has boosted assessed resource potentials, it understandably has not yet affected reserves estimates.

Conventional natural gas reserves assessed by different organizations between 2007 and 2009 converge around 180–192 trillion m³ (6700 EJ to 7100 EJ).[10] More importantly, reserves estimations show a continuous upward trend with reserves expanding faster than production (see Figure 7.11). Reserve growth, new discoveries, and reclassification from resources to reserves have been instrumental in this regard. It is important, though, that approximately 22% of global conventional gas reserves exist in the form of 'associated gas' (IEA, 2009), with its availability thus dependent on future oil production.

---

9   5% is the approximate maximum share of natural gas adsorbed by a non-organic mineral surface in a reservoir.

10   Total gas production in 2009 was 3.0 trillion m³.

Table 7.10 | Estimates of conventional natural gas reserves.

Region	OGJ (2007) [bcm]	EIA (2008) [bcm]	Cedigaz (2009) [bcm]	BP (2010) [bcm]	BGR (2010) [bcm]	OPEC (2008) [bcm]	USGS (2000) [bcm]
Europe	4872	4976	5472	4368	5239	6232	7762
CIS	57,059	60,510	54,902	58,725	63,551	58,112	46,930
Africa	13,866	14,181	14,774	14,758	14,753	14,542	9559
Middle East	72,191	72,361	75,149	76,119	75,359	73,559	45,438
Asia	11,764	14,101	15,139	16,242	16,107	15,166	11,582
North America	8018	8124	9485	8682	9310	8018	6280
Latin America	7414	6858	7478	8534	7592	7542	7222
World	175,184	181,111	182,400	187,429	191,911	183,171	134,773

Source: OGJ, 2007; EIA, 2008; Cedigaz, 2009; Exxon, 2008; BP, 2010; BGR, 2010; OPEC, 2008; USGS, 2000.

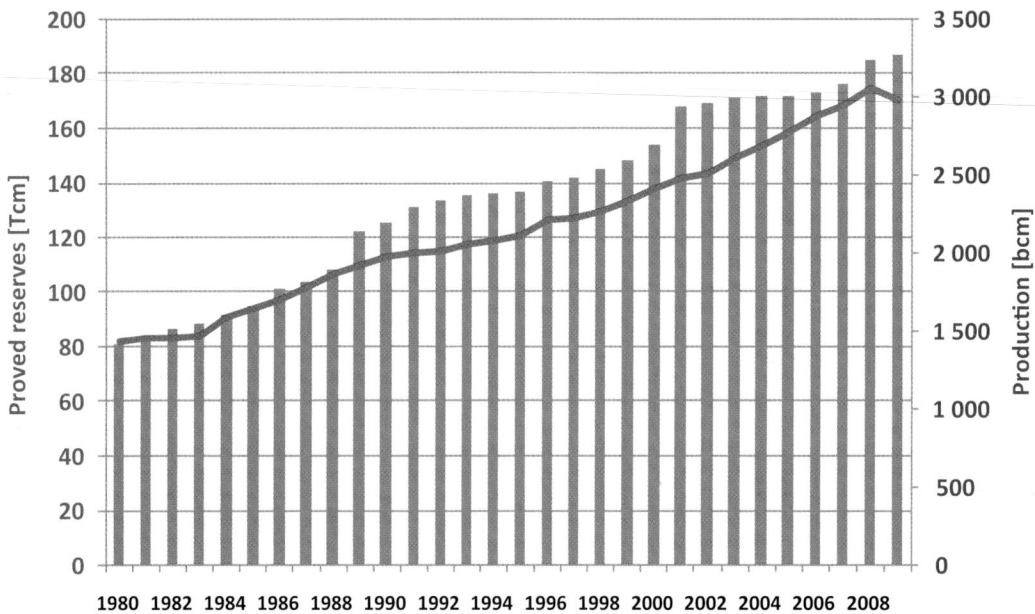

Figure 7.11 | Development of global natural gas reserves and production, 1980–2009. Source: BP, 2010.

While reserves have been steadily increasing, the increases typically result from revised reservoir estimations rather than from new discoveries. With total reserves still growing faster than production, there is little incentive for additional gas-exploration efforts, as the focus for exploration is, in many geographical areas, e.g., Africa, still on oil. Furthermore, knowledge about potential deepwater oil and natural gas resources is evolving fast. Deepwater deposits are basically conventional resources in an unconventional setting, and their potential exploitation will depend mainly on future technology developments.

Table 7.11 summarizes the cumulative historical production of natural gas up to 2009, the production in 2009, and the reserve and resource estimates of the BGR and USGS for the 18 GEA regions (the USGS data are from the year 2000, except for the Circum-Arctic area which was taken from USGS, 2008).

Resource estimations are not as accurate as reserve estimations. Unlike reserve assessments, which are mandatory legal requirements in many jurisdictions, resources are only assessed periodically. The most recent comprehensive assessment by the USGS dates back to 2000 and reflects the resource situation as it existed in 1995 (BGR, 2009). Since then, the USGS has added a comprehensive assessment of the Arctic (USGS, 2008) and updated its assessment for Afghanistan and parts of the Middle East. The gas-resource data in Table 7.11 may be conservative, as it may no longer adequately reflect the advances in geological knowledge and gas-production technologies.

Overall production in 2008 reached an all-time record of 3061 billion m³ (115.5 EJ), but declined to 2987 billion m³ (112.7 EJ) in 2009 because of the global economic slowdown and consequent lower demand. Cumulative gas production by 2008 approached 48% of proven

Table 7.11 | Conventional gas production, reserves, and resources.

	Natural gas production 2009	Historical production till 2009	Reserves BP	Reserves BGR	Reserves USGS	Resources BGR	Resources USGS	Reserves + resources BGR	Reserves + resources USGS
	[EJ]	[EJ]	[EJ]	[EJ]	[EJ]	[EJ]	[EJ]	[EJ]	[EJ]
USA	22.68	1153	258	262	180	740	552	1002	732
CAN	6.08	194	65	65	54	259	26	324	79
WEU	9.57	343	135	162	263	313	394	475	657
EEU	0.56	69	27	31	26	23	14	54	41
FSU	26.53	953	2188	2352	1748	4332	1633	6684	3382
NAF	6.01	102	307	310	244	324	119	634	363
EAF	0.74	1	48	1	0	22	11	23	11
WCA	0.94	13	196	213	99	183	198	396	297
SAF	0.00	2	0	21	13	70	49	91	61
MEE	15.17	205	2836	2788	1693	1309	1348	4097	3041
CHN	3.21	36	91	91	36	370	90	461	126
OEA	0.30	2	25	26	0	60	3	86	3
IND	1.48	21	42	41	24	33	32	75	56
OSA	2.17	35	47	49	37	81	81	130	117
JPN	0.00	4	0	1	0	0	0	1	0
OCN	1.75	36	115	115	238	78	214	193	452
PAS	7.61	133	285	274	96	304	119	577	216
LAC	7.90	166	318	299	269	402	562	700	831
Circum-Arctic							1748		1748
TOTAL	112.71	3467	6983	7101	5021	8902	7193	16,002	12,214

Sources: Author's estimate; BP, 2010; USGS, 2000; 2008; BGR, 2009; 2010.

reserves, which has led some analysts to conclude that peak gas production is imminent. When resources are added, however, that percentage drops to less than 22%, and during the first decade of the 21st century reserve additions outpaced production. In contrast to oil, gas resources exceed reserves by almost 30%, which suggests a much more favorable resource position for natural gas.

The geographical distribution of gas resources shows Russia (FSU) in the lead, followed by the Middle East and the United States. These three regions account for more than 70% of global conventional gas reserves and resources (BGR, 2010).

7.3.4 Unconventional Gas

7.3.4.1 Shale Gas

Shale is a common sedimentary rock consisting of clay, quartz and other minerals. Most shales are not suitable sources of natural gas as they have insufficient permeability to allow significant fluid flow to a well bore. In gas-bearing shales, the rock acts as source and reservoir for the gas. The gas occurs in three states: in the pore spaces of the shale, in vertical fractures (joints) which break through the shale, and adsorbed on mineral grains and organic materials. The bulk of recoverable gas is contained in the pore spaces.

In the absence of fractures, it is difficult for gas to escape from the pore spaces, because they are tiny and poorly connected. Gas from naturally fractured shale has been produced for over 100 years in the Appalachian Illinois Basins. However, well production was often marginal. The rate of production increases with well stimulation, such as hydraulic fracturing and horizontal drilling, but because of these extra expenses the gas tends to cost more to produce than gas from conventional wells.

The recent shale-gas boom in North America is the result of technology advances that create extensive artificial fractures around horizontal, rather than vertical well bores. High natural gas prices provided incentives for engineering advances in hydraulic fracturing and horizontal well drilling. The share of shale gas in US gas supplies rose from 1.6% (0.32 EJ) in 1996 to 10% (2.1 EJ) in 2008 (with a growth rate between 2005 and 2008 of 40% per year), as the surprise economic success of the Barnett Shale operations in Texas generated a rush for other sources of shale gas across the United States and Canada.

To date, almost all successful shale-gas wells have been in rocks of Paleozoic age, but shales of other ages are being evaluated, particularly Cretaceous shales in Rocky Mountain basins. In other parts of the world shale gas has not yet been produced commercially because of a limited geological knowledge about shale gas and host reservoirs, as well as higher technical and economic costs. However, since 2006 accelerated shale-gas exploration has occurred in both Europe and Asia.

7.3.4.2   Coal-bed Methane

CBM is gas that is adsorbed into the solid matrix of the coal. Its lack of H_2S content means the gas is considered a 'sweet gas.' It is located in various forms in natural fractures (free form), coal pores (free and adsorbed form), and coal structures (absorbed form). The presence of this gas is well known from its occurrence in underground coal mining, where it presents a serious safety risk. CBM is located at depths that range between 300–2000 m and can be extracted by multi-leg, horizontal wells or wells with massive hydraulic fracturing. As a result of low pressures and low well-head flow rates, the production of CBM is economically feasible only in the vicinity of gas-demand centers. Thus, its production is feasible in countries with considerable coal basins and where substantial populations exist on the territory of these basins: the United States, Canada, Australia, Russia, Ukraine, China, India, and approximately another 35 countries. CBM is produced commercially in the United States from some 40,000 coal-bed gas wells, and accounts for approximately 9% (1.8 EJ) of total domestic gas production.

The development of CBM in countries outside the United States continues to be slow for several reasons, including unfavorable reservoir characteristics, inadequate infrastructure, lack of operating knowledge, and competition from conventional gas. In some cases, leases have changed ownership several times before an operator with the right combination of corporate size, technical know-how, and contractual terms has been able to achieve a successful project.

7.3.4.3   Tight Reservoir (Tight Sands) Gas

Tight sands reservoirs are conventional gas resources located in sedimentary basins with less than 0.1 mD permeability, and at a depth of up to 4500 m. These geological structures can be found practically everywhere in the world. Still, countries with large conventional resources in very permeable reservoirs have had a tendency to pay little or no attention to tight reservoirs. Gas production from these reservoirs is developing in countries with mature gas industries (e.g., the United States, Canada, Great Britain, and Russia) and in countries poorly endowed with conventional natural gas resources (e.g., Japan and China). In fact, the distinction between conventional and unconventional tight gas is increasingly blurred, especially in North America. For example, the US EIA no longer lists the production of tight gas under unconventional gas, and it is now accounted for under conventional gas.

7.3.4.4   Gas in Deep Reservoirs (Basin-centered Gas Systems)

Deep reservoirs are deep sedimentary basins located at depths greater than 4500 m and are usually characterized by high pressures and high temperatures, and the presence of significant acid components (IGU, 1994; 2003; BGR, 2009). Basin-centered gas systems are typically further characterized by regionally pervasive accumulations that are gas saturated, abnormally pressured, commonly lack a down-dip water contact, and have low-permeability reservoirs.

Production from depths larger than 4500 m is technologically challenging. Low porosity and permeability limit gas flow rates to the well bore and require reservoir fracturing and complex horizontal and multilateral production wells. Drilling in tight formations with conventional technology at depths approaching 4500 m becomes increasingly prohibitive, and the extractable gas per well is lower than that for conventional gas at shallower depths. Deep gas has been explored in North America, North Europe, Russia, and some other regions. Exploration and production is technically challenging and there are only a few examples of deep-gas production: in the United States (on old exhausted fields and fields close to the consumer), in the North Sea (an old gas-producing region), and in Russia in the super-giant Astrakhan gas field. In all these cases production is constrained by technical, safety, and environmental restrictions.

7.3.4.5   Water-dissolved (Aquifer) Gas

'Aquifer gas' refers to CH_4 dissolved and dispersed in groundwater. This can be found practically everywhere as almost all porous rock formations below groundwater tables contain small amounts of CH_4. As a result of the low solubility of CH_4 in water, the gas content of water at depths less than 1000 m is low (0.3–3 m^3/m^3 of water) and production is economically unattractive. At greater depths, the CH_4 content can reach 10–15 m^3/m^3. In areas under high tectonic stress, gas concentrations of up to 90 m^3/m^3 can be found (BGR, 2009). The gas contained in aquifers exceeds reserves of conventional gas by two orders of magnitude, but even with new extraction technology only a small share of this gas is expected to become commercially viable in the long run.

Unlike conventional and other unconventional gases, the production of aquifer gas hinges upon the coextraction of the CH_4 substrate water. The process of water lifting and aboveground separation of typically low-concentration CH_4, as well as subsequent water treatment and disposal/recycling, generally has a low energy payback and thus lacks economic attractiveness. The economics may improve, for example in regions with high freshwater demand where gas production would be a convenient by-product, or in cases where geopressure and geothermal aquifers could be used for combined geothermal energy and gas supply.

7.3.5 Resource Estimates for Unconventional Gas Resources (Excluding Gas Hydrate)

Table 7.12 summarizes the estimated unconventional gas resources in the 18 GEA regions, except for gas hydrate, which is not yet produced commercially. Gas hydrate resource estimates are addressed in Section 7.3.6. The resource potential (omitting gas hydrate) amounts to 40,000 EJ, of which 20,000 EJ are considered potentially recoverable reserves.

7.3.6 Gas Hydrate

Gas hydrate is a solid crystalline substance composed of water and natural gas (primarily CH_4) in which water molecules form a cage-like structure around the gas molecules. The cage structure of the hydrate molecule concentrates the component gas so that a single cubic meter of gas hydrate will yield approximately 160 m^3 of gas and 0.8 m^3 of water at standard pressure and temperature (0.1 MPa, 20°C). Gas hydrate forms under conditions of moderately high pressure and

moderately low temperature (Figure 7.12) and is widespread in marine sediments of outer continental margins and in sediments in polar regions. In the marine environment, the pressure and temperature conditions for gas hydrate stability occur at water depths greater than 500 m at mid- to low-latitudes and greater than 150–200 m at high latitudes (Max et al., 2006). At these water depths, gas hydrate may occur within a zone of hydrate stability that extends into the sediment to depths of up to hundreds of meters beneath the seafloor. The thickness of the hydrate stability zone varies with temperature and pressure, typically increasing in deeper water as a result of increasing pressure. The base of the hydrate stability zone is determined largely by the local geothermal gradient – the rate at which temperature increases with depth. At some depth beneath the seafloor, the temperature increases to a point at which gas hydrate is no longer in a stable phase (Figure 7.13). As the geothermal gradient varies considerably within and between depositional basins, the thickness of the hydrate stability zone is highly variable.

In Arctic sediments, gas hydrate may occur within and beneath permafrost zones, with the upper boundary of the hydrate stability zone dependent upon local temperatures and pressures. As with the hydrate stability zone in marine environments, the base of this zone in polar environments is largely determined by the geothermal gradient.

Table 7.12 | Unconventional gas occurrences (without hydrate).

Region	Coalbed methane		Deep gas		Shale gas		Tight gas		Total	
	Resource potential	Reserves	Resource potential	Reserves	Resource potential	Reserves	Resource potential	Reserves	Resource potential	Reserves
	[EJ]	[EJ]	[EJ]	[EJ]	[EJ]	[EJ]	[EJ]	[EJ]	[EJ]	[EJ]
USA	1677	931	1677	745	4098	1863	1416	1118	8867	4657
CAN	559	261	373	186	373	186	820	559	2124	1192
WEU	559	261	186	112	559	224	186	149	1490	745
EEU	186	75	186	112	559	224	186	149	1118	559
FSU	1863	745	1863	1118	5402	2235	1304	1043	10,432	5141
NAF	373	149	559	373	373	149	373	298	1677	969
EAF	186	75	186	112	186	75	186	149	745	410
WCA	186	75	559	261	745	298	559	447	2049	1080
SAF	186	75	186	75	186	75	186	112	745	335
MEE	186	75	559	261	373	149	745	559	1863	1043
CHN	1490	559	186	75	186	75	186	112	2049	820
OEA	37	0	37	0	186	75	186	112	447	186
IND	559	261	186	75	186	75	186	112	1118	522
OSA	112	37	373	186	559	224	373	224	1416	671
JPN	112	37	0	0	0	0	0	0	112	37
OCN	373	186	186	75	373	149	186	149	1118	559
PAS	112	37	186	75	186	75	373	224	857	410
LAC	559	224	559	224	373	149	559	224	2049	820
TOTAL	9314	4061	8048	4061	14,903	6296	8010	5738	40,275	20,156

Source: IGU, 2003; Ananenkov, 2007; USGS, 2008; BGR, 2009.

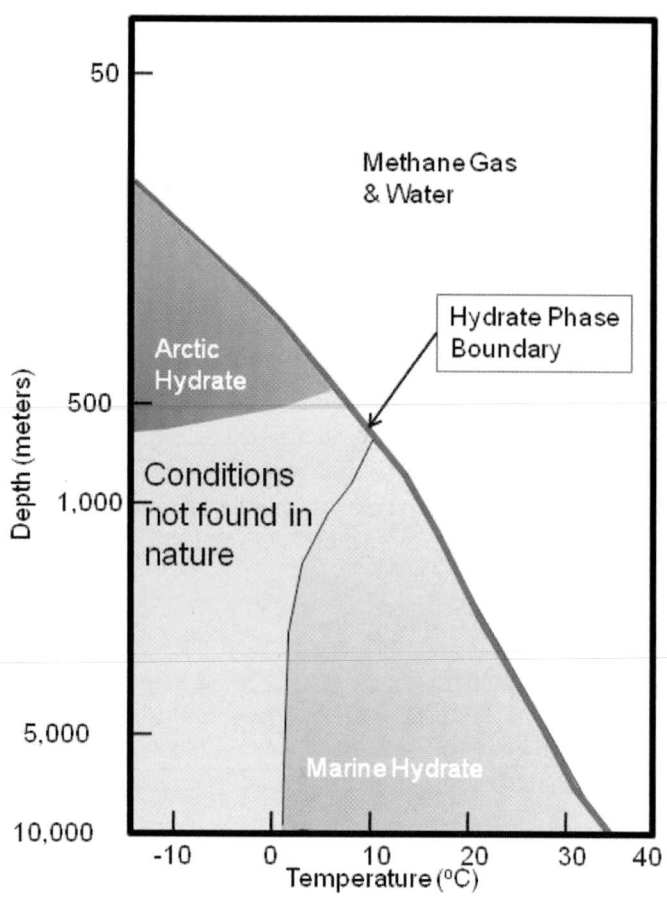

Figure 7.12 | Gas hydrate stability zone.

An important factor in the formation of high-grade gas hydrate deposits is the lithology (rock type) of the host sediment. The highest concentrations of gas hydrate occur in sediment with high porosity and permeability, such as sands and gravels. Cores recovered from such sediments have yielded hydrate concentrations up to 85%. In contrast, fine-grained sediments, such as shales, have concentrations of less than 10% of the bulk volume. The total volume of gas hydrate in fine-grained sediments represents the greatest proportion of the world's gas hydrate occurrences, but the prospects for commercial development of natural gas from such a highly disseminated resource are very poor without a major paradigm shift in technology.

Even in locations with excellent reservoir conditions, gas hydrate will not be present without an adequate supply of a hydrate-forming gas such as CH_4 generated from either biogenic or thermogenic sources. Evaluations indicate that insufficient microbial CH_4 is generated internally within the gas hydrate stability zone to account alone for the gas content of most hydrate accumulations. Most sediments that bear gas hydrate-bearing sediments have never been sufficiently heated or deeply buried to form thermogenic gas. Thus, it is likely that most of the gas that has formed hydrate has migrated into the hydrate stability zone from deeper sediments (Collett et al., 2009).

Current scenarios for the production of gas from hydrate-bearing sediments involve 'dissociating' the hydrate (converting it into its components – water and gas) by depressurization, thermal stimulation, or injection of an inhibitor such as methanol or glycol into the reservoir.

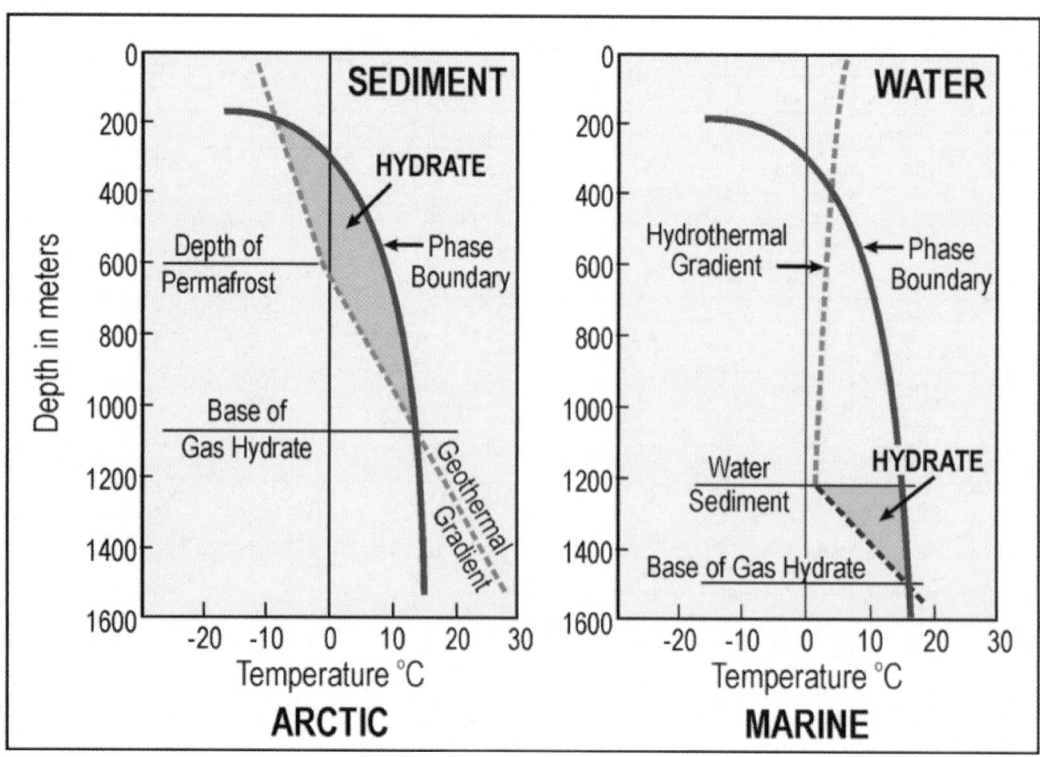

Figure 7.13 | Factors affecting the thickness of the hydrate stability zone.

The method utilized will depend on specific reservoir conditions, including hydrate saturation, porosity, and permeability. Natural gas may also be produced from gas hydrate through a chemical exchange, such as a substitution of CO_2 for CH_4. This approach has the advantage of sequestering CO_2 as well as yielding commercial gas production (Graue et al., 2006).

In summary, the volume of natural gas contained in the world's gas hydrate accumulations greatly exceeds that of known gas reserves (Collett, 2002), although a substantial proportion of that gas hydrate is in low-grade accumulations that are unlikely to be developed commercially. Ongoing research programs in the United States, Japan, India, and elsewhere have made great strides in understanding the formation of gas hydrate, identifying potential reservoirs where hydrate is concentrated, and developing production technologies for commercial exploitation. These programs have identified hydrate-bearing sediments in dozens of locations (Figure 7.14). Thus far, the vast energy potential of gas hydrate resources has not been proven commercially viable. There is, however, growing evidence that natural gas can be produced from high-grade gas hydrate accumulations with existing conventional oil and gas production technology (Moridis et al., 2008).

Estimates of the gas hydrate resource potential for each of the 18 GEA regions was undertaken by first segregating each region into separate subregions based on the local depositional setting. For marine gas hydrate, a range of values for the volume of sediment within the gas hydrate stability zone (corrected for sulfate reduction of CH_4 near the seafloor) was calculated using the model developed by Wood and Jung (2008). This volume was multiplied by a range of parameter estimates of the percentage of sand within the hydrate stability zone, the percentage of those sands that would be hydrate-bearing, sandstone porosity, hydrate saturation of the pore space, and the percentage of the gas that could be recovered from hydrate-bearing sands. This calculation provides an estimate of the technically recoverable gas hydrate resource.

As each of these parameters is poorly constrained in most of the world's depositional basins, the resulting resource estimates extend over several orders of magnitude. A narrower range of values will be obtainable in the future as data are collected. Where detailed analyses have been conducted, the results are integrated into this report.

For Arctic sediments, the estimate of technically recoverable gas hydrate was determined using recent analyses, such as USGS (2008), and extrapolating the range of results to areas where the parameters for a petroleum-systems approach are not available.

The economic resource potential is estimated to be between one and two orders of magnitude smaller than the estimate of technically recoverable gas hydrate because depositional basins typically contain a substantial fraction of thin and/or discontinuous sands and because of the energy required to dissociate hydrates that are at temperature and/or pressure conditions at some distance from the phase boundary. In addition, economic viability is strongly influenced by the presence of existing conventional infrastructure and proximity to markets. The theoretical resource potential is calculated with the inclusion of hydrate deposits that fill veins and fractures.

The combination of lithology and CH_4 flux required for extractable deposits of concentrated hydrate removes most of the world's hydrate from consideration as an energy resource (Figure 7.15). However, the energy potential of the gas hydrate deemed theoretically, technically, or economically recoverable is still extremely large (Table 7.13). In addition to the 18 GEA regions defined for use in this assessment, separate resource assessments are included for the Arctic Ocean without regard for national boundaries, and for the Southern Ocean (from the coast of Antarctica north to 60° south latitude).

Research programs undertaken by international consortia, government programs, and academic institutions have identified proven and

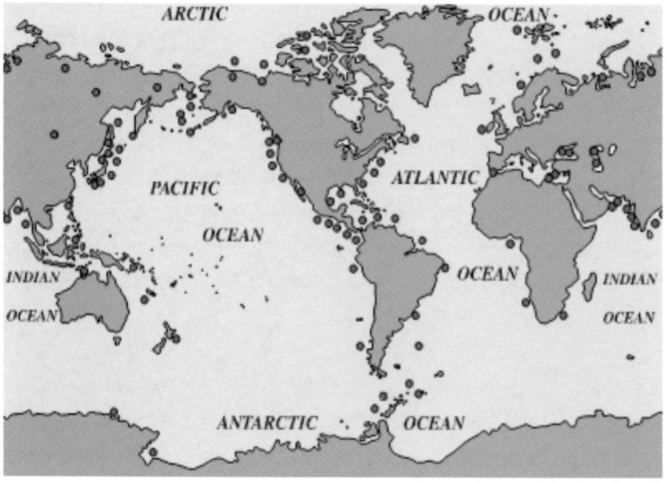

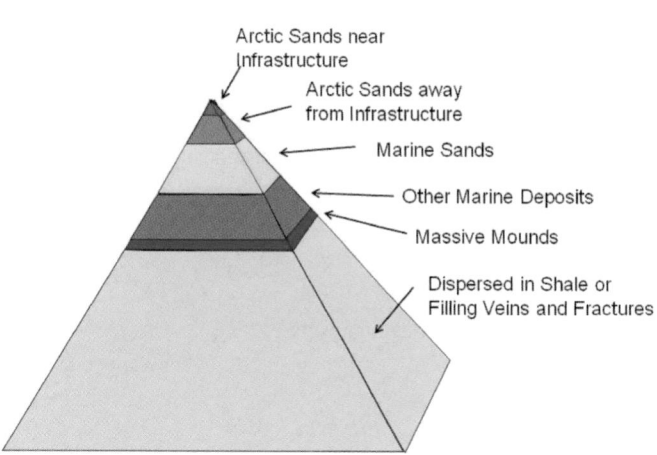

Figure 7.14 | Global gas hydrate locations.

Figure 7.15 | Magnitude of gas hydrate potential by type of deposit.

Table **7.13** | Recoverable gas hydrate resource potentials.

Region	Theoretical potential	Technical potential	Economic potential
	[EJ]	[EJ]	[EJ]
USA	782–122,128	395–12,213	0–1229
CAN	298–71,049	142–71,049	0–712
WEU	37–490,153	11–11,758	0–1177
EEU	0–820	0–83	0
FSU	820–80,997	402–8100	0–808
NAF	0–30,439	0–1449	0–71
EAF	0–155,883	4–15,588	0–779
WCA	37–211,061	22–21,106	0–1054
SAF	75–208,676	34–20,868	0–1043
MEE	0–30,439	7–3044	0–30
CHN	0–14,158	4–1416	0–142
OEA	0–21,386	4–2139	0–212
IND	37–49,589	11–4959	0–496
OSA	0–27,682	7–2768	0–276
JPN	37–3726	19–373	0–37
OCN	37–53,427	11–5343	0–533
PAS	37–205,323	15–20,532	0–2053
LAC	149–198,766	67–19,877	0–1986
SOO*	75–357,779	37–35,782	0
ARC**	75–439,409	45–43,926	0
Total	2,496–2,772,889	1237–238,428	0–12,638

* Southern Ocean

** Arctic Ocean

Table **7.14** | Global unconventional gas resources.

Really unconventional		
Water-dissolved (aquifer) gas (depth up to 4.5 km)	8000–10,000 Tm³	300,000–370,000 EJ
Gas of hydrates (including permafrost metastable hydrates)	2500–21,000 Tm³	90,000–780,000 EJ
Shale gas	380–420 Tm³	14,000–15,500 EJ
CBM (up to depth 4.5 km)	200–250 Tm³	7500–9300 EJ
Pseudounconventional		
Dense reservoirs (tight sands) gas (depth up to 4.5 km)	180–220 Tm³	6700–8200 EJ
Deep reservoirs (depth 4.5–7.0 km) gas	200–300 Tm³	7500–11,200 EJ

Source: Ananenkov, 2007.

7.3.7 Abiogenic Gas Theory

While petroleum (oil and gas) occurrences are commonly thought of as being derived from biological substances, i.e., by the thermal decomposition of organic matter or by microbial processes (Lollar et al., 2002), since the very beginning of the 'petroleum age' there has been a substantial amount of skepticism voiced about its exclusively biological origin (e.g., Mendeleev, 1877; Kudryavtsev, 1951; Gold and Soter, 1980; 1982; Gold, 1999; Kenney et al., 2009). According to the proponents of an abiogenic origin of hydrocarbons, CH_4 and oil are also, perhaps even predominantly, formed well below the usual depths of oil and gas deposits.

By comparing the carbon and hydrogen stable isotope signatures of CH_4 seeping from hard rock mines of Canadian and Fennoscandian shields, with microbial and thermogenic CH_4, Lollar et al. (2002) confirmed the abiogenic formation of hydrocarbons within the Earth's crust. Abiogenic hydrocarbon formation is attributed to the reduction of CO_2 in hydrothermal water-rock interactions similar to Fischer-Tropsch reactions and serpentinization of ultramafic rock. However, the lack of abiogenic isotope signatures in current conventional gases means these authors doubt that abiogenic formation is a significant hydrocarbon source. In contrast, the 'modern Russian-Ukrainian' theory of the origins of hydrocarbons postulates that petroleum is a primordial material of deep origin which is transported at high pressures via 'cold' eruptive processes into the crust of the Earth (Kenney et al., 2002). Much of this theory has developed through chemistry and chemical thermodynamics. One conclusion from the theory is that the origin of hydrocarbon molecules (except CH_4) from biogenic ones in the temperature and pressure regimes of the Earth's near-surface crust is in violation of the second law of thermodynamics (Kenney, 1996).

The abiogenic theory has been applied extensively across the former Soviet Union and has revealed 80 oil and gas fields in the Caspian district with production from the crystalline basement rock. Other examples

probable gas hydrate locations along continental margins and in polar regions throughout the world, although characterization of deposits is limited to relatively few locations.

7.3.6.1 Overall Unconventional Gas Resources (Including Gas Hydrate)

Table 7.14 summarizes the current understanding of global unconventional gas resources. Gas hydrate and water-dissolved gas are the most abundant, and are also relatively evenly distributed geographically. CBM, shale gas, and deep formation gas appear to be less evenly distributed, but this may well be the consequence of limited geological knowledge through historically low interest. Gas is likely to be found in the tighter formations of all sedimentary basins of the world, but actual quantities can only be confirmed and delineated through dedicated and costly exploration activities.

have been found in the western Siberian cratonic-rift sedimentary basin with numerous fields producing from the crystalline basement; on the northern flanks of the Dneiper-Donotz basin; and elsewhere in Azerbaijan, Tatarstan, and Asian Siberia.

The theory also postulates that many of the world's oil and gas fields are being continuously recharged from deeper horizons with abiogenic hydrocarbons, so making the fields 'effectively inexhaustible,' and therefore oil and gas fields now considered exhausted should be thoroughly investigated to ascertain the quantities of oil and gas that may have accumulated since the fields were shut-in. In view of the above considerations, Kenney et al. (2009) believe that petroleum abundances are limited by little more than the quantities of its constituent elements that were incorporated into the Earth at the time of its formation and the petroleum industry is only now "entering its adolescence."

Gold (1988) argues that hydrocarbons are of primordial origin without biological derivation. The presence of hydrocarbons in the planetary system, e.g., the atmospheres of Jupiter, Saturn, or Neptune contain enormous amounts of CH_4 and other hydrocarbons, while Titan, a satellite of Saturn, has a CH_4 and ethane atmosphere that is unlikely the result of biological processes. Rather, carbonaceous chondrite meteorites are thought to have brought carbon to the Earth during its formation. Subjected to the appropriate heat and pressure domains present at great depths, e.g., in the vicinity of magma cooling, the carbonaceous material would produce hydrocarbons, chiefly CH_4, as well as hydrogen and CO_2, which is subsequently outgassed to the upper layers of the Earth's crust. Such an environment is also quite amenable for the formation of heavier hydrocarbon molecules.

The discovery of microbial life in ocean vents too deep for photosynthesis is used by Gold (1999) in support of the abiogenic gas theory. Instead of photosynthesis, the large amounts of CH_4 in these hydrothermal vents migrating from deeper levels supply the energy required for life in chemical form. In addition, the existence of anaerobic bacteria in rock structures and CH_4 environments at depths of more than 4 km is another indication of abiogenic CH_4. In turn, bacteria brought along the upward-migrating hydrocarbons explain the existence of certain biomarkers in extracted petroleum commonly associated with the decomposition of organic matter.

The abiogenic origin of hydrocarbon debate dates back to the mid-19th century and since then the theory has undergone extensive development, refinement, and application. More than 4000 articles on the theory have been published (Odell, 2010) and numerous scientific conferences have been held to debate and evaluate the theory. The existence of abiogenic formation of CH_4 and hydrocarbon gases within the Earth has been confirmed (Lollar et al., 2002), but their substantial or exclusive contribution to, and replenishment of, current gas and oil deposits making hydrocarbons a quasi renewable source as proposed by the Russian-Ukrainian theory remains a matter of controversy.

7.3.8 Natural Gas Supply Cost Curves

The aggregate global gas supply cost curve shown in Figure 7.16 plots the potential long-term contributions from conventional and unconventional gas reserves and resources (see Tables 7.11 and 7.12) against their estimated production costs. The costs do not include taxes or royalties, nor do they include external environmental or social costs associated with gas production.

Conventional gas reserves and resources amount to about 12,000 EJ (~300 trillion m³). In this analysis, the costs of producing conventional resources range from 0.50–3.50 $/GJ. The supply curve suggests some 5000 EJ of conventional gas can be produced for around 1 $/GJ or less at 2007 prices and exchange rates (these supply costs are not the actual prices in the market place). To put this amount of gas into perspective, the cumulative past gas production amounts to 3467 EJ.

Unconventional gas reservoirs are very diverse in their geological structure, location, and accessibility and each deposit requires a specifically tailored extraction technology. Production cost estimates vary considerably, from less than 2 $/GJ for shale gas from very permeable reservoirs, CBM, or onshore gas hydrate to more than 13 $/GJ for offshore deep gas reservoirs. The supply cost curve of Figure 7.16 includes 28,000 EJ of unconventional gas (reserves plus 20% of resources assumed to become resources before 2050) and 5000 EJ of hydrate gas. The production costs of 40% of unconventional reserves are estimated at 2–5 $/GJ, while the remaining 60% of reserves plus 20% of the resources can eventually be recovered for 5–13 $/GJ.

Almost all unconventional resources are characterized by low permeability, and their extraction requires intensive reservoir stimulation through various techniques, e.g., horizontal drilling, multi-leg

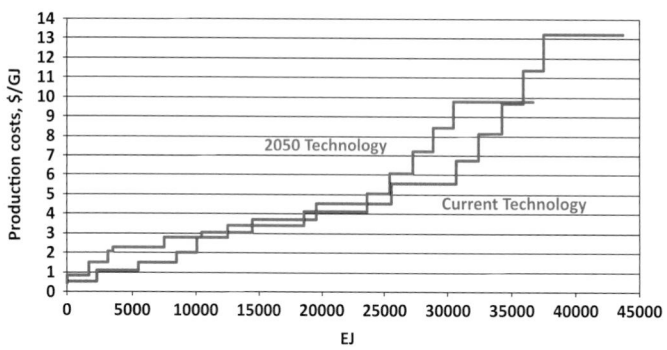

Figure 7.16 | Aggregate world-supply cost curve[11] for conventional and unconventional natural gas. Source: Rogner, 1997; Rogner et al., 2000; USGS, 2000; 2008; Aguilera et al., 2009; BGR, 2009.

---

11   See footnote 7.

completion, hydraulic fracturing, reservoir acid treatment (for carbonates), and thermal stimulation (for hydrate).

Regarding actual production experience with unconventional gas, the US gas industry is very much in the lead globally. Faced with perpetually declining production from conventional gas reserves, the industry sought alternatives. National fiscal policy created incentives for technological advances in exploration and production of unconventional gas. This led to a 'quiet revolution' in North America, where techniques, such as horizontal drilling and hydraulic fracturing, are allowing access to deposits of unconventional tight and shale gas, and of CBM at much lower costs than thought possible five years ago. In some cases, costs are now lower than those in conventional gas projects.

Costs for the production of gas from hydrate have yet to be delineated and to date no commercial production experience exists. Costs will depend on a variety of factors ranging from geological conditions (e.g., permeability of sediments, reservoir pressure, location, and size of hydrate deposits) to the necessary selective adaptation of exploration and extraction technology. For example, free gas trapped by a hydrate deposit may well be producible with conventional production technology for natural gas. Generally, site development will be more complex, and hence costlier, than that for conventional natural gas. The estimates of economically recoverable gas quantities from hydrate are based inherently on the expected competition with conventional and other unconventional gas occurrences and their respective future costs, and a range of 8–12 \$/GJ for gas from hydrate appears plausible.

7.3.9   Environmental and Social Implications

In addition to concerns similar to those relating to oil development, natural gas development raises several other issues.

CH_4, the chief constituent of natural gas, has a very high global warming potential and special care is needed to minimize leaks to the atmosphere. Leakage of gas behind casings of abandoned and 'sealed' oil and gas wells leads to gas entering shallow aquifers, thus degrading potable water resources. This is a significant environmental issue, identified by the US Environmental Protection Agency (US EPA) as a major concern. It can be addressed through better well-abandonment regulations, as well as enforcement of collection or flaring of small and low-grade CH_4 emissions.

Shale-gas development involving deep hydraulic fracturing in the United States has caused serious concerns about drinking water contamination and land degradation (Arthur et al. 2008). The US EPA has undertaken a scientific study to investigate the potential risks to drinking water from hydraulic fracturing and will consider the need for additional health and safety regulations.

Exploitation of sour gas resources requires special attention to escape of H_2S during production and SO_x during processing. Long-term stockage of elemental sulfur from gas plants is of environmental concern in areas such as Kazakhstan and western Canada because of distance-to-sea transportation (costs) and the oncoming world sulfur surplus as more sulfurous feedstocks are processed. For example, Tengizchevroil in Kazakhstan has been fined repeatedly for failure to meet contractual obligations in sulfur management. However, the environmental impact of elemental sulfur is small because it is inert.

Production of gas from hydrate involves potential adverse environmental consequences comparable to those of conventional natural gas. Appropriate environmental management and safety measures exist to minimize potential harmful effects. However, two threats are hydrate specific. First, destabilization of the seabed may cause subsidence and land slippage, including blow outs. Second, gas hydrate utilization may adversely affect the habitat of microorganisms and other forms of biocoenosis systems whose metabolisms depend on CH_4 from hydrate and which form an essential part of the food chain for mussels and beard worms (BGR, 2009). Both threats are generally considered of a local nature and manageable.

7.3.10   Summary

The global occurrence of natural gas is enormous and a significant portion of the total volume occurs in technically exploitable reservoirs. With regard to unconventional gas resources, the recent boom of shale-gas production in the United States is one example of how quickly technical change stimulated by declining conventional reserves and high market prices can reverse the overall supply outlook and price expectations.

The term 'unconventional' is becoming a misnomer, especially with respect to tight gas, shale gas, and CBM. Still, enormous technical challenges exist in the development of innovative exploration and production technology.

Broad-ranging technology impacts can take decades from initial concept to large-scale implementation. In addition, the energy needed to produce unconventional gas is usually higher than that for conventional gas. In the case of gas hydrate, this can reduce the net energy gain by up to 20% or more. Gas hydrate represents the largest of the unconventional natural gas resource options. Yet it is currently unclear how much gas hydrate can be exploited, both technically and economically. Field testing of marine and Arctic reservoirs is planned for the near term, and the results of these tests will determine their commercial viability. The development of gas hydrate has important implications, as it is widespread along the margins of every continent and, as such, includes areas that lack conventional oil and gas resources.

It is important, however, that progress in exploiting unconventional gases will depend on environmental regulation and suitable management practices, since many of these resources are in remote and ecologically sensitive areas.

7.4 Coal

7.4.1 Overview

Coal is a combustible, sedimentary, organic rock composed mainly of carbon, hydrogen, and oxygen. It is a fossil fuel formed from vegetation lying between rock strata and altered by the combined effects of pressure and heat over millions of years. Coal deposits can be found in sedimentary basins of various geological ages. Dependent on age, depositional conditions, and geological history, the nature of coal deposits and the quality and characteristics of coal vary in terms of heat value, reaction characteristics, and impurities.

The degree of change undergone by coal as it matures from peat to anthracite – known as coalification – defines its physical and chemical properties and is referred to as the 'rank' of the coal. Low-rank coals, such as lignite and sub-bituminous coals, are characterized by high moisture levels, low carbon content, and a consequently low energy content. These coals are typically used locally, predominantly power generation. Higher rank hard coals, i.e., anthracite and bituminous coals, have a higher carbon content and lower moisture content, which result in higher energy quantities per unit mass of coal. Hard coals are produced both for domestic and export markets, including power generation, iron and steel making processes, and chemical industries (WCI, 2008; IEA, 2008a). The large variety of different coal characteristics and specific energy values mean that assessments of reserve and resource are rather challenging.

Efforts to estimate global coal reserves and resources can be traced back to the beginning of the 20th century (Table 7.15). While there has

Table 7.15 | Historical development of hard coal resource estimates.

Source	Reserves (Gt)	Resources (Gt)
International Geological Congress (IGC) 1913	312	4399
World Power Conference (WPC) 1936	623	5269
World Energy Conference (WEC) 1974	998	7065
Federal Institute of Geo-Sciences and Mineral Resources (BGR) 2007	736	8818
Federal Institute of Geo-Sciences and Mineral Resources (BGR) 2010	723	17,167[a]
World Energy Council (WEC) 2010	861	_[b]

[a] BGR has included in the 2010 global hard coal resources figures, hypothetical coal resources in Canada and the United States, which have almost doubled the hard-coal resources figures published in 2007.

[b] The World Energy Council does not publish global coal resources figures.

Source: adapted from Fettweis, 1976.

been a steady increase in resource estimates, there has been considerable fluctuation in the estimates of proven hard-coal reserves.

7.4.2 Coal Production

Global hard-coal production amounted to 2.24 Gt in 1950 and has tripled since then, with 6.94 Gt produced in 2009. Three distinct periods occurred. A steady annual increase of about 64 Mt between 1950–1986, a period of stagnation from 1988–2000, and a dramatic increase of 260 Mt/yr or 4.7% from 2000–2009 (BP, 2010). The first period was characterized by a slight growth of coal consumption in the Organisation of Economic Co-Operation and Development (OECD) countries, a static situation in the FSU and marked increases in China, the rest of Asia, and Africa. The period of stagnation coincides with the collapse of centrally controlled economies. The unprecedented 50% increase in coal production since 2000 is primarily because of the economic upsurge of China and East Asia (see Table 7.16). Since 1981, coal production in the European Union (EU) has decreased by more than half and by one-third in the countries of the FSU; it has remained virtually constant in the OECD countries.

The reason for the marked increase in coal production lies in the relative abundance and broad geographical distribution of coal reserves and the comparatively low cost of coal production from shallow coal deposits. Coal is a domestically available energy source in many countries, and offers relative energy independence and foreign exchange savings.

Table 7.16 details the 2008 coal production for the 18 GEA regions. A heat value of 16.5 MJ/kg demarcates hard coal (bituminous and high-energy sub-bituminous coals) and soft brown coal (lignite and low-energy sub-bituminous coals). In 2008 world coal production amounted to 6799 Mt: 5774 Mt of hard coal and 1025 Mt of lignite/brown coal. Table 7.16 shows that:

- China and the United States together account for 56% of world coal production.

- Latin and South America and Africa together produce less than 6% of world coal.

- Asia has developed into the major coal-producing region in the world.

7.4.3 Reserves

Universally applicable, geological, quality, and technological criteria have yet to be uniformly applied, so assessments of global coal reserves and resources are subject to uncertainty. Most importantly, without a clear distinction between the quality or rank of coals and their specific energy contents, which can vary between 5–30 MJ/kg, aggregate

Table 7.16 | Coal production in 2008 by region.

Country/region	Hard coal [Mt]	Lignite/brown coal [Mt]	Total [Mt]	2008/1981	Hard coal [EJ]	Lignite/brown coal [EJ]
USA	994.3	69	1062.9	1.42	25.11	0.82
CAN	58.2	10	68.1	1.69	1.47	0.12
WEU	53.7	323	376.2	0.52	1.42	2.97
EEU	103.0	234	337.1	0.42	2.73	2.16
FSU	426.7	89	516.1	1.20	11.40	0.99
NAF	0.1	0	0.1	0.69	0.00	0.00
EAF	0.0	0	0.0	0.00	0.00	0.00
WCA	0.3	0	0.3	0.69	0.01	0.00
SAF	241.1	0	241.1	1.89	5.66	0.00
MEE	1.6	1	2.2	1.10	0.04	0.01
CHN	2646.4	115	2761.4	4.51	64.28	1.11
OEA	72.2	18	89.7	2.08	1.75	0.17
IND	489.5	32	521.7	3.94	11.89	0.31
OSA	3.8	0	3.8	1.05	0.09	0.00
JPN	1.2	0	1.2	0.07	0.03	0.00
OCN	330.1	73	402.8	3.15	8.02	0.70
PAS	255.0	57	311.6	11.39	6.20	0.55
LAC	96.5	6	102.6	6.85	2.39	0.06
World	5774	1025	6798.8	1.77	142.50	9.96

Source: BGR, 2009.

resource assessments will remain fraught with ambiguity. Without a precise definition of the energy contents used, it is impossible to compare any two sets of coal reserves or resources estimates.

Coal deposits that clear the threshold of economic producibility are referred to as mineable reserves. These quantities need to be corrected for the amounts of coal that are likely to be lost during the mining process – on average 10% for surface mining and 50% for underground mining (Fettweis, 1990; Daul, 1995).

It is important to define the separation point between higher and lower rank coals in any estimation. The most comprehensive information on coal reserves and resources is provided by the World Energy Council (WEC) and BGR. Unfortunately, the two agencies use different classifications of higher and lower rank coals that are not directly comparable. WEC publishes data for the categories (a) bituminous coal and anthracite, (b) sub-bituminous coal, and (c) lignite, whereas the breakdown used by the BGR is hard coals (>16.5 MJ/kg) and soft brown coals (<16.5 MJ/kg). Figure 7.17 clarifies the two definitions. A further quality criterion includes the content in the coal of environmentally harmful substances, e.g., sulfur, but is evaluated on a case-by-case basis. Nevertheless, harder coals typically have lower sulfur concentrations.

Coal reserves data also say little about the nature of the coal seams and the associated mining difficulties that determine the actual cost of production. Therefore, a further classification system based on geomining conditions is required. Geomining conditions include the depth of coal seams, the angle of dip of seams, number of seams and degree of disturbance of coal seams, and faulting (Wagner, 1998). Table 7.17 details the geomining descriptions used to classify coal fields (IEA Coal, 1983).

With increasing depth, rock stress conditions become more severe and require more complex and costly mine infrastructures; CH_4, heat, and water hazards also increase. Category D1 (Table 7.17) defines the current depth limit of the widely used drag-line operations in surface mining under conditions of competent overburden strata. The angle of dip adversely affects the operation of trackless coal mining equipment, because of gravity, and hazards associated with mine depth increase with steeper seams. Ground disturbances and faulting adversely affect the extent of mechanization in the coal mining operations, as well as mine safety, design, and layout. Highly mechanized and productive coal mining operations are presently confined to geologically relatively undisturbed coal deposits. Factors such as exceptional gas conditions or permafrost have to be considered as well.

The geological environments and geomining conditions of hard- and lignite/brown-coal deposits tend to be very different. Sandstone and shale formations are the typical rock strata in which hard-coal deposits are located, while soft-coal deposits are typically located in soft rocks and soils.

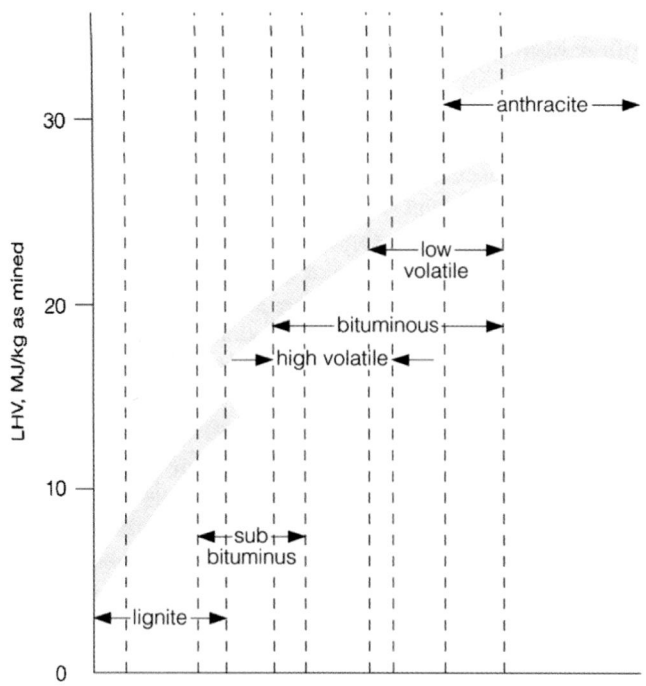

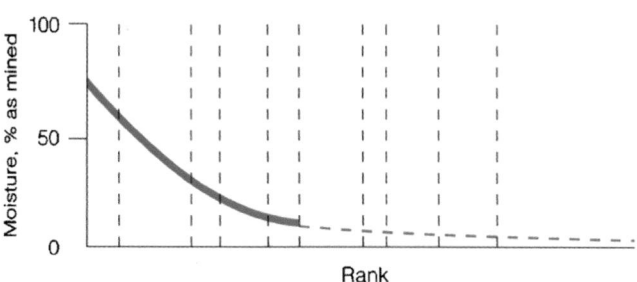

Figure 7.17 | Calorific value and moisture content of various coal ranks. Note: Volatility determines the burn rate of the coal. High volatile coals are easy to ignite but have lower energy content per unit volume than low volatile coals. Source: Couch, 1988.

Table 7.17 | Geomining characteristics of coal production categories.

Depth (D)	Angle of dip (A)	Degree of disturbance/ faulting (G)
D1, <70 m	A1, <15°	G1, undisturbed
D2, 70–300 m	A2, 15–30°	G2, disturbed
D3, 300–500 m	A3, >30°	G3, highly disturbed, faulted
D4, 500–800 m		
D5, >800 m		
Production category	**Hard-coal deposits**	
CI	Shallow coal seams (D1, D2); favourable conditions (A1, G1)	
CII	Shallow coal seams (D1, D2); less favourable conditions (A2, G1, G2) Intermediate depth (D3); favourable conditions (A1, A2; G1)	
CIII	Intermediate depth (D3); less favourable conditions (A2, G2) Great depth (D4), favourable conditions (A1,G1)	
CIV	Great depth (D4), unfavourable conditions (A2, A3; G2)	
	Lignite/brown-coal deposits	
BI	Few shallow, thick and level coal seams (D1, D2), (A1, G1)	
BII	Several shallow coal seams (D1, D2), (A1, A2), (G1, G2)	
BIII	Several coal seams at greater depths (D2,D3), (A1, G1, G2)	

Source: IEA Coal, 1983.

of categories CI and CII will be exploited first before extracting deposits of categories CIII and CIV.

It is also realistic to assume that coal reserves in categories CI and CII, and in the case of lignite/brown coal in category BI, will be exploited more rapidly – resulting in a change in future coal-reserves distribution toward the more difficult geomining categories. The impact of this development is most profound in geopolitical regions with limited CI and CII reserves, such as EEU, WEU, FSU, and Latin America and the Caribbean (LAC). As information on the coal reserve situation of China and India is uncertain, there could be potential longer term coal reserves difficulties in these regions.

Proven hard-coal reserves amount to 18.2 ZJ, while cumulative past production since the industrial revolution is estimated at close to 7 ZJ. Although there is a relatively broad geographical spread of proven hard-coal reserves, there is a concentration in four geopolitical regions (United States, China, India, and FSU), which together account for 84% of global reserves. OCN (essentially Australia), EEU, and southern Africa (SAF, essentially South Africa) account for 75% of the remaining 13% of hard coal reserves. Proven lignite/brown-coal reserves amount to 2.8 ZJ and are concentrated in the FSU, EEU, WEU, OCN, and the United States. These five regions account for 88% of proven lignite/brown-coal reserves.

7.4.4 Resources

Resources are coal occurrences that are either known to exist but are currently not economically recoverable or amounts that can be expected to exist based on geological indicators (BGR, 2009).

Table 7.18 shows that approximately 60% of CI and CII hard-coal reserves are situated in the United States and China, whereas the bulk of reserves in Europe and Japan fall into geomining categories CIII and CIV. The geomining conditions in different regions are clearly mirrored by the coal production trend from 1981–2008 shown in Table 7.16. According to Table 7.18, the bulk (categories BI and BII) of soft-coal reserves can be extracted by surface-mining methods and are concentrated in six geopolitical areas; in descending order: FSU, Western European Union (WEU), United States (USA), Oceania (OCN), Eastern European Union (EEU), and China (CHN). Slightly more than half of the soft-coal reserves fall in the most favorable geomining category (BI).

The future development of coal reserves depends on overall coal demand and the rate at which technology change and market conditions shift deposits presently listed in the coal resources category to coal reserves. Furthermore, the easily accessible and favorable coal deposits

Table 7.18 | Hard-coal and lignite/brown-coal reserves in terms of surface and mining potential and geomining conditions.

Region	Hard coal							Lignite/brown coal				
	Total reserves [EJ]	Surface [EJ]	Underground [EJ]	Category [EJ]				Total reserves [EJ]	Surface [EJ]	Categrory [EJ]		
				C I	C II	C III	C IV			B I	B II	B III
USA	5856	2342	3514	3924	1347	586	0	371	371	186	186	0
CAN	110	82	27	64	29	18	0	27	27	21	5	0
WEU	67	0	67	0	11	34	22	421	379	253	126	42
EEU	457	32	425	32	32	197	197	198	159	99	60	40
FSU	3306	694	2612	298	926	1455	628	1037	933	622	311	104
NAF	2	1	1	0	1	1	0	0	0	0	0	0
EAF	0	0	0	0	0	0	0	0	0	0	0	0
WCA	9	4	4	2	2	2	2	0	0	0	0	0
SAF	757	144	613	287	212	204	53	0	0	0	0	0
MEE	10	2	8	0	3	4	3	0	0	0	0	0
CHN	4387	965	3422	1579	1623	965	219	106	64	32	32	43
OEA	119	45	74	38	19	34	27	20	20	10	10	0
IND	1856	612	1243	612	1021	223	0	42	42	29	13	0
OSA	11	7	4	9	2	0	0	28	28	14	14	0
JPN	9	1	8	0	2	4	3	0	0	0	0	0
OCN	982	236	747	609	373	0	0	426	426	213	213	0
PAS	55	48	8	48	2	3	3	47	47	24	24	0
LAC	252	43	209	48	126	66	13	51	51	26	26	0
TOTAL	18,246	5259	12,987	7550	5730	3796	1170	2775	2547	1528	1018	228

Source: adapted from Wagner, 2010.

Resources are shown as in situ amounts; the eventually extractable quantities will be significantly lower. As a result of the large reserve base and the considerable costs associated with the exploration and delineation of resources that may potentially be at the cross-over threshold to reserves, there is limited economic incentive to expand the reserve base beyond a certain range, commonly expressed in terms of years of reserves at current production levels. In the case of coal this is more than 100 years (Wellmer, 2008).

The global in situ coal resource data in EJ of Table 7.19 were derived from regional estimates of physical in situ quantities by applying the heating values of coal reserves of these regions areas published by the BGR (Rempel et al., 2007).

While coal resources are 20 times higher than known extractable coal reserves, there is uncertainty about the minable portion of these in situ quantities. Information on the geomining conditions of coal resources is insufficient for a reliable production assessment. For example, most of the better delineated coal resources are situated at greater depths and thus belong to geomining categories CIII and CIV. Extraction ratios in geologically difficult coal deposits can be below 40% (Kundel, 1985; Daul, 1995; USGS, 2009). Since many of the 'in situ' hard-coal resource deposits are in narrow seams at depths of more than 1000 m, an overall recovery rate of 20% may well be achievable practically. For example, 60% of coal resources in China are found at depths deeper than of 1000 m (Pan, 2005; Minchener, 2007). Without new extraction methods, a 20% recovery rate puts the portion of coal resources that eventually could become available as reserves to 87,154 EJ.

7.4.5 Coal Mining Technology

The past 50 years have witnessed remarkable improvements in coal mining technology and productivity. Mechanization and the adoption of engineering-systems approaches resulted in a fourfold increase in labor productivity between 1980 and 2000. In surface mining, bucket-wheel excavators are employed under soft overburden conditions, while drag lines are used under hard overburden conditions, where the overburden strata has to be blasted to expose the coal. In underground mining, manual labor is becoming extinct, except in marginal micromines in developing countries. Coal is now mined by powerful continuous-mining machines in room and pillar mining, and pillar recovery operations. As a result of mechanization, safety in underground coal mining has improved and productivities have increased substantially. Output from individual longwall faces has increased more than 10 times over the past 30 years.

Table 7.19 | Hard-coal and lignite/brown-coal resources.

Country/ Region	Hard coal [EJ]	Lignite/ brown coal [EJ]
USA	163,816	16,382
CAN	3560	610
WEU	7416	449
EEU	5192	2405
FSU	76,947	14,204
NAF	10	0.4
EAF	22	0.4
WCA	72	2.6
SAF	1178	0.0
MEE	1125	0.0
CHN	121,693	2966
OEA	1298	3085
IND	3954	329
OSA	83	1753
JPN	297	11
OCN	2696	1720
PAS	941	551
LAC	753	202
World	391,052	44,671

However, these developments have not been achieved without costs. Higher degrees of mechanization make mining systems less flexible and more vulnerable to changes in geological conditions. While mechanization has contributed to substantial improvements in productivity, it has also resulted in significantly lower reserve assessments. Mining operations now increasingly concentrate on the most favorable coal seams in easy geological environments, which results in numerous previously mined coal seams being ignored. In the German Ruhr coal-mining region, inclined coal seams and faulted areas are no longer mined because of low productivity, prohibitive costs, and drastically reduced subsidies (Kundel, 1985; Gesamtverband Steinkohle, 2007). Once closed, it is extremely difficult, dangerous, and costly to reopen and extract any coal remnants. At present there are no mining methods available or known to overcome these difficulties. A recent revaluation of German hard-coal reserves showed a decline of more than 90% from previous assessments (Rempel et al., 2007).

Overall, labor productivity in shallow underground mines that operate under favorable geomining conditions (category CI, see Table 7.17) now approaches 10,000 tonnes per person and year. Yet in deep coal mines operating under difficult geomining conditions (CIII, CIV), overall labor productivities are below 1000 tonnes per person and year, and have improved only marginally over the past 20 years (Ritschel and Schiffer, 2007; Turek et al., 2008; Statistik der Kohlenwirtschaft, 2009). As the depth of mining progressively increases, operations will have to move into more difficult areas, so it is realistic to assume that the cost of coal extraction and reserve losses will increase in the future.

To harvest deep coal deposits, new and innovative technologies are required, including technologies for coping with geological disturbances. Depth-related challenges include the need for an underground roadway support that can withstand high rock pressures and effective cooling, ventilation, and gas-drainage systems. Whereas depth-related challenges are expected to be met by innovation, it is less clear how to solve the problem of extracting coal from highly disturbed and faulted areas. Some promising possibilities are highly flexible mining systems based on conventional mechanized and automated mining technology, or novel concepts utilizing underground coal gasification (UCG).

UCG, the controlled gasification of coal seams in situ, permits the exploitation of deposits where mining is no longer technically feasible. CO_2 from the process can be returned safely to the gasified seam, resulting in zero emissions and very little ground disturbance. Feasibility studies and demonstration projects are ongoing in the United Kingdom, Russia, China, South Africa, New Zealand, and elsewhere. These studies suggest the use of UCG could potentially increase world coal reserves by 500–600 Gt (12.3–14.8 ZJ) (WEC, 2007).

CBM is a relatively large and undeveloped resource in most regions and refers to CH_4 that is adsorbed to the solid matrix of coal (see Section 7.3.4).

7.4.6 Coal Mining Costs

The competitive nature of the international coal market means that detailed coal production costs are difficult to find. Data collected from coal mines in Australia, Canada, Columbia, Indonesia, South Africa, and Venezuela show a wide range of mine operating and capital costs. In 2007, mine costs for steam coal ranged from 15–65 US$/t, with the costs of underground coal some 10% to 20% higher than those of surface-mined coal (Ritschel and Schiffer, 2007). It is important that these cost figures refer to relatively shallow and favorable coal mining conditions, and deep-coal mining costs can be significantly higher (Turek et al., 2008).

As noted earlier, the cost of coal production is strongly influenced by geomining conditions, especially mining depths. Data for major European deep-coal mines reveal rapid cost increases at greater depths, mainly because of the higher costs of supporting and maintaining mine development, mine ventilation, and, in the case of the very deep mines, mine cooling.

Long-term projections of coal production costs need to account for the change in the ratio of surface to underground mining. After coal seams situated close to surface have been exploited, there will be a shift toward underground mining and toward deeper mines within underground mining. This means increased specific investment costs for new mine infrastructure as well as higher mine-operating expenditures. Technology change and innovation will be needed not only to keep short-term cost increases in check, but also to reduce the costs of long-term marginal coal production.

Highly productive multiple-entry longwall mining, which has been practiced successfully to depths of about 600 m, will reach its technical limits at greater depths and will need to be replaced by single-entry longwall development, which is limited in terms of ventilation and operational efficiency. Geological disturbances further impede longwall mining. Mining losses in faulted areas are expected to lower recovery rates. Without technology changes and innovation, these factors are likely to drive up long-term coal mining costs at a faster rate than those of oil or gas production costs.

7.4.7   Coal Supply Cost Curves

Future supply cost curves are a function of the energy content of coal deposits, geomining conditions, technology changes, overall coal demand, and inflation. Table 7.20 summarizes the current and projected cost of producing 1 TJ of in situ coal under different geomining conditions at constant US$_{2007}$ prices and exchange rates. The 2050 costs reflect the technological improvements between 2007 and 2050.

Further dividing the cost range per geomining category into four classes and applying these classes to the coal reserves of each of the 18 regions

(Table 7.18) in a fixed proportion (60%, 25%, 10%, and 5%) leads to a global hard-coal supply cost curve. Figure 7.18 shows two supply curves – one for the current reserves based on performance, productivity, and costs of current mining technology and one for the reserves, performance, and mining technology expected by 2050. The differences between the two supply cost curves reflect the tendency to extract coal frist from deposits having more favorable geomining conditions which over time will result in a reserve shift towards more unfavorable geomining conditions. The reserves in 2050 exclude any replenishment from coal resources potentially made possible by technology change and different market conditions but reflect the coal produced between 2007 and 2050 (assuming constant 2007 production levels).

7.4.8   Environmental and Social Implications

Major environmental and social problems can develop during mining operations, and also after the the coal has been exploited. However, concerns about adverse environmental and social impacts of coal as a primary energy source are often subordinated to the pressures that result from growing energy needs, especially in developing countries.

Surface mining requires the removal of overburden strata for access to the coal seams. Top soil can be destroyed and groundwater tables affected adversely by this removal process. Damage to land and water resources from the overburden removal method of coal mining has particularly destructive impacts on people living in the affected area. Many coal deposits are located in remote, isolated areas, but when

Table 7.20 | Hard- and lignite/brown-coal production costs for different geomining conditions at 2007 US$/TJ.

Geomining category	Production costs 2007		Technology change in %/yr	Production costs 2050	
	Lower bound	Upper bound		Lower bound	Upper bound
Hard coal					
CI	600 $/TJ (15 $/t)	1000 $/TJ (25 $/t)	1.5	310 $/TJ (7.75 $/t)	525 $/TJ (13 $/t)
CII	1000 $/TJ (25 $/t)	1600 $/TJ (40 $/t)	1	650 $/TJ (16.25 $/t)	1040 $/TJ (26 $/t)
CIII	1600 $/TJ (40 $/t)	2800 $/TJ (70$/t)	0.5	1290 $/TJ (32.25 $/t)	2260 $/TJ (56.50 $/t)
CIV	2800 $/TJ (70 $/t)	6000 $/TJ (150 $/t)	0.5	2260 $/TJ (56.50 $/t)	4900 $/TJ (122.50 $/t)
Lignite/brown coal					
BI	680 $/TJ (7 $/t)	1160 $/TJ (12 $/t)	1.5	350 $/TJ (3.60 $/t)	610 $/TJ (6.20 $/t)
BII	1160 $/TJ (12 $/t)	1750 $/TJ (18 $/t)	1.5	610 $/TJ ($ 6.20/t)	910 $/TJ (9.40 $/t)
BIII	1900 $/TJ (20 $/t)	3880 $/TJ (40 $/t)	0.5	1560 $/TJ (16.40 $/t)	3130 $/TJ (32.30 $/t)

The costs quoted are typical 2007 mining costs (Ritschel and Schiffer, 2007; Turek et al., 2008; Landsmann, 2009) and include operating and capital costs, but exclude costs of coal benefication and cost of mine-to-market transportation. The cost ranges reflect differences in geomining conditions within each class. Lignite/brown-coal mining costs are strongly influenced by environmental costs.

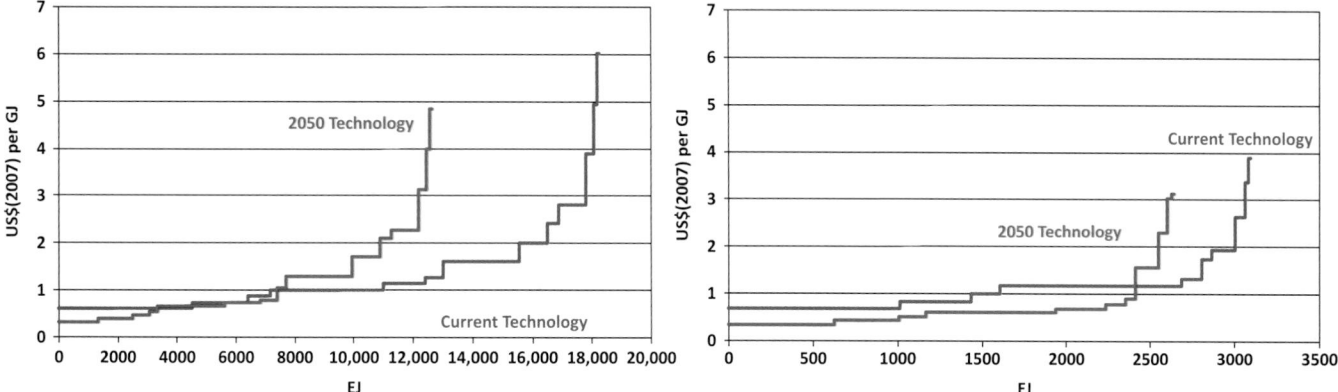

Figure 7.18 | Global coal supply cost curves[12] for 2007 and 2050.

mining operations are undertaken in densely populated areas, resettlement of the population can be a costly and politically sensitive issue. Differences in the cost of lignite mining in Europe and overseas are, to a considerable degree, linked to environmental legislation and the cost of resettlement.

Other environmental impacts of surface mining are ground vibrations and dust caused by overburden blasting. Once production has ceased, the overburden strata is returned to fill the void created by the coal extraction and the land surface is reshaped and rehabilitated. Many jurisdictions mandate rehabilitation to conditions at least equal to those prior to mining operations. As a result, the cost of land rehabilitation can be 20–30% of total mining cost and depends on climatic conditions, prior land use, and applicable mining regulations.

Underground coal mining is generally associated with less-severe environmental impacts, but presents significant health and safety hazards for workers, including mine fires, CH_4 and coal dust explosions, ground instability, and flooding. Coal mining requires a well-qualified and highly specialized workforce. In many countries such skills are not sufficiently available. Training and education of coal-mining personnel and continued mining R&D are essential as the extraction of deep coal resources in difficult geological environments requires new mining systems and technologies. With the demise of the nationalized coal-mining industries in most countries, mining research is left to equipment manufacturers and academic institutions (Wagner and Fettweis, 2001). As a result, research of coal-mine safety has suffered.

Mine subsidence is a critical issue. In the case of room and pillar mining, subsidence is a distinct possibility even many years after completion of mining operations. Caving of the roof strata tends to be less critical with longwall mining, particularly at great mine depths and moderate seam thicknesses. Mine subsidence associated with longwall mining is rather predictable and can be prevented by appropriate engineering

approaches. In sensitive areas, mine subsidence can be minimized by applying backfill (stowing).

Mine fires in relatively shallow underground mines can cause major environmental and safety hazards, especially where the sulfur content of coal is high. These fires can develop many years after completion of coal mining operations as a result of air entering through fractures in the overburden strata. Control of mine fires has been a serious challenge in many mining countries, notably China. High sulfur contents in coal seams can also give rise to problems of acidic mine drainage.

Many coal resources in the northern hemisphere are located in areas of permafrost. This creates unique environmental problems that not only impact the environment, but also can give rise to unique technical and social problems. Other environmental concerns include CH_4 emissions from shallow coal-mine workings, and transportation of coal from the remote mining areas to the industrialized areas.

7.4.9   Summary

Proven hard- and lignite/brown-coal reserves are abundant, i.e., 21,000 EJ, of which 18,000 EJ are hard coal. Estimated coal resources exceed proven reserves more than 20 times. Even on the basis of a very conservative assumption of a 20% in situ recovery rate, coal is sufficiently available to support annual coal production increases of 2% for more than 100 years. However, many coal deposits are located in remote areas with little or no infrastructure and often harsh climatic conditions. Bringing these coals to the market will pose immense challenges.

Coal mining technology and productivity have increased significantly in recent years and secured the position of coal as the fossil energy source of lowest cost. The expected increase in the average depth of coal mining and increased environmental regulation of coal mining operations will exert upward pressure on production costs, while advances in mining technology are expected to at least partly offset these increased costs.

---

12   See footnote 7.

7.5   Nuclear Resource Materials

7.5.1   Overview

Nuclear resource materials include the fuels required for the operation of both nuclear fission and fusion reactors. (For a discussion on nuclear energy technologies, see Chapter 14.)

The primary fuel, uranium, is a naturally occurring element that can be found in minute concentrations in all rocks, soils, and waters, always combined with other elements. It is found in hundreds of minerals, including uraninite (the most common uranium ore), lignite, monazite, and phosphates (Table 7.21).

The average uranium concentration in the continental Earth's crust is about 2.8 ppm, while the average concentration in ocean water is 3 parts per billion (ppb) (Bunn et al., 2003). The theoretically available uranium in the Earth's crust has been estimated at 100 teratonnes (Tt) uranium, of which 25 Tt occur within 1.6 km of the surface (Lewis, 1972). The amount of uranium dissolved in seawater is estimated at 4.5 Gt. However, these occurrences do not represent practically extractable uranium resource (at least not in the short run and without substantial R&D). For that, what matters are abnormal uranium concentrations in the Earth's crust, at levels that can be extracted, both technically and economically (currently about 100 ppm U).

A potential alternative fission fuel is thorium. Thorium is about three to four times more abundant than uranium, and is also found in seawater, soils, and rocks. It occurs in minerals on all continents.

Lithium is a valuable resource for nuclear fusion. Lithium does not occur in an elemental form because it is highly reactive. It is found in all geological strata and is the 25th most abundant element, but is still considered a comparatively rare element. Only a few deposits are concentrated enough to be of any actual or potential commercial value.

While the resource situation of uranium is currently well delineated, thorium and lithium resource profiles are, for different reasons, unclear. Thorium has limited uses and its natural ubiquity makes further exploration unprofitable. Lithium occurrences are concentrated mainly in Central and South America and China.

Table 7.21 | Typical uranium concentrations.

Occurrence	Average concentration [ppm U]
High-grade ore	20,000
Low-grade ore	1000
Granite	4
Sedimentary rock	2
Earth's continental crust	2.8
Seawater	0.003

Source: Bunn et al., 2003; WNA, 2009.

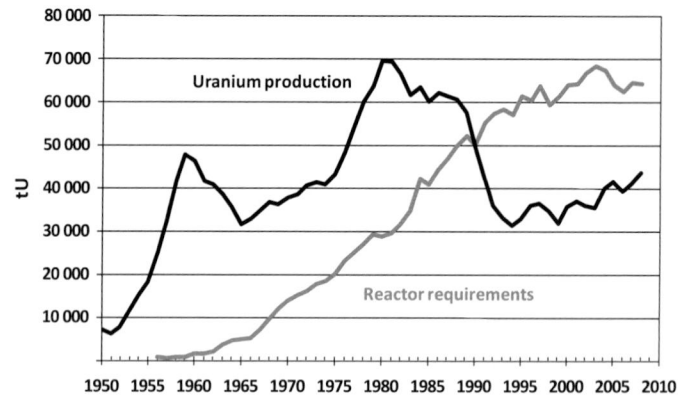

Figure 7.19 | Global annual uranium production and reactor requirements, 1945–2009. Source: adapted from NEA/IAEA, 2010.

7.5.2   Uranium

7.5.2.1   Historical Uranium Production

The uranium market has been characterized by a large disparity between global reactor requirements and mine production since the early 1990s when, after decades of production exceeding requirements by an unusually wide margin, mine output slipped below annual reactor requirements (see Figure 7.19).[13] The appearance of so-called secondary supplies, i.e., reactor fuel derived from warheads, military and commercial inventories, re-enrichment of depleted uranium tails, as well as enriching at lower tail assays, reprocessed uranium, and mixed oxide fuel, reduced the demand for fresh uranium. In addition, new entrants to the world uranium market, e.g., the Russian Federation, Kazakhstan, and Uzbekistan, further exerted competitive pressures. As a result of uncertain and low demand plus excess capacity, uranium prices fell, except for short-term aberrations.

As low prices seemed to suggest plentiful supply, utilities began to hold lower uranium inventories. This suppressed both production and prices even further. Requirements were increasingly met by accumulated past production and not from operating capacities, to the point that global operational mining output capacity dropped below the annual reactor requirements. In late 2000, uranium prices reached a historical low of 7.10 $/lb U_3O_8 (triuranium octoxide) or 18.45 $/kg U (uranium metal), threatening the economic survival of many mines. At the same time, global production had declined to less than 60% of annual reactor requirements. Low prices and a doubtful future for nuclear energy combined to make uranium exploration and investment in mine capacity a highly unpopular and risky business proposition. Uranium exploration almost totally ceased, mines closed, and institutions routinely reporting

---

13 The geopolitical landscape caused by the collapse of the Soviet Union accelerated the declining trend in global uranium production. Prior to 1990, information about uranium production from the Soviet Union's area of influence was not publicly available.

uranium reserves and resources became reluctant to do so in a comprehensive manner.

The situation changed dramatically starting in 2000 for four main reasons. First, the license extension of nuclear power plants in the United States generated another 20 years of unexpected uranium demand. Second, the emergence in the uranium market of China and India, potentially major customers for uranium, generated additional demand with consequent pressure on supply capacity.

Third, in the light of volatile fossil-energy prices, energy security concerns, and concerns about climate change, a growing number of countries began to reevaluate the nuclear option thus further adding to potential supply shortfalls. Fourth, when two of the world's top producing mines greatly reduced output because of fire and flooding incidents, the supply/demand imbalance began to impact uranium markets (PCA, 2006). Spot prices instantly shot through the roof and continued climbing, approaching 350 $/kgU in 2007. Exploration expenditures also increased by an order of magnitude, from US$100 million in 2002 to over US$1 billion in 2008. By April 2009 spot prices had dropped to one-third of the peak level or about 110 $/kgU and have not changed markedly since.

7.5.2.2   Conventional Uranium Reserves and Resources

Conventional uranium resources are defined as those occurrences from which uranium is recoverable as a primary product, a co-product, or an important by-product. Uranium reserves are periodically estimated by the OECD's Nuclear Energy Agency (NEA) together with the International Atomic Energy Agency (IAEA), and also the World Nuclear Association (WNA), WEC, and numerous national geological institutions. NEA-IAEA estimates have the widest coverage, so the reserves reported in their latest survey (the so-called Red Book) are given here (NEA/IAEA, 2010). The two organizations define 'reserves' as those deposits that could be produced competitively in an expanding market. This category is called identified resources[14] and, until 2008, included uranium recoverable at less than 130 $/kgU. Stimulated by high spot prices of up to 350 $/kg in 2007, the 2010 edition of the Red Book extended the cost ranges to 260 $/kgU.

What is important for a good comprehension of the long-term availability of uranium is the impact of demand prospects and prices on the volume of reported reserves and resources. The recent uranium-price hike stimulated exploration, prompted institutions to report occurrences long labeled irrelevant, and led companies to consider reopening high-cost mines. The RAR category in the Red Books increased by 27% between 2001 and 2009 (or 43% if reserves up to 260 $/kg production cost are included). In addition, the category of 'uranium that could be expected

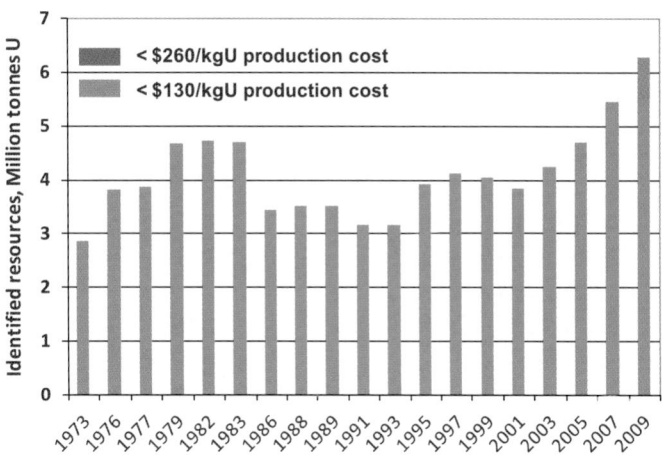

Figure 7.20 | Development of identified uranium resources at less than 130 $/kgU production costs, 1973–2009, and less than 260 $/kgU for 2009. Source: NEA/IAEA, 2010.

to be found based on the geological characteristics of known resources grew by 5%. Figure 7.20 shows the historical development of identified resources since the early 1970s.

Table 7.22 shows the identified uranium resources by region reported in the Red Book 2009 (NEA/IAEA, 2010). Altogether, there are 6.3 MtU (or 3.7 ZJ)[15] of conventional uranium resources available at extraction costs of less than 260 $/kgU. Vast additional uranium occurrences can be mobilized at costs larger than 260 $/kg.

Table 7.23 shows uranium occurrences yet to be discovered based on knowledge derived from already discovered and delineated deposits, as well as from regional geological analogies and mapping activities. Reports on 'prognosticated resources' are usually supported by some hard evidence. These deposits are generally located in known uranium provinces. Data on 'speculative resources' are not supported by direct evidence, but rather based on analogy and other considerations. Hence both categories require substantial additional exploration efforts before their existence can be confirmed and their overall qualities and quantities delineated.

Reporting on undiscovered resources (Table 7.23) is sketchy and incomplete, and recently the number of reporting countries has declined. Several reasons account for this decline, including funding constraints from governmental offices, lack of interest from industries, and plentiful known and accessible uranium occurrences.

7.5.2.3   Unconventional Uranium Resources

The distinction between conventional and unconventional resources is not clearly defined, but generally unconventional uranium resources

---

14   Until 2003 'Identified Resources' were reported as known conventional resources (KCR), i.e., the sum of reasonably assured resources (RAR) and estimated additional resources category-I (EAR-I). After 2003 Identified resources are the sum of RAR and inferred resources.

15   The thermal energy equivalent of 1 tonne of uranium in average once-through fuel cycles is about 589 TJ (IPCC, 1996).

Table 7.22 | Conventional uranium reserves and resources.

Regions	Theoretical up to 1.6 km depth [ZJ]	RAR (reserves)				Inferred (resources)				Total (reserves and resources)			
		<$40/ kg [EJ]	<80/kg [EJ]	<$130/ kg [EJ]	<$260/ kg [EJ]	<$40/ kg [EJ]	<$80/ kg [EJ]	<$130/ kg [EJ]	<$260/ kg [EJ]	<$40/ kg [EJ]	<$80/ kg [EJ]	<$130/ kg [EJ]	<$260/ kg [EJ]
USA	1,054,800	0	23	121	276	0	0	0	0	0	23	121	276
CAN	1,091,065	156	197	211	226	58	65	73	92	214	261	283	318
WEU	41,298	0	4	28	36	0	0	15	23	0	4	43	59
EEU	870,093	0	0	3	6	0	0	5	13	0	0	7	19
FSU	490,952	10	250	391	475	19	202	398	539	29	452	789	1014
NAF	146,720	0	0	11	11	0	0	0	1	0	0	11	12
EAF	2,418,526	0	0	3	3	0	0	0	2	0	0	3	4
WCA	1,048,701	10	25	151	153	0	18	18	19	10	43	169	173
SAF	252,887	85	167	281	282	46	103	153	153	130	271	434	435
MEE	474,121	0	26	26	26	0	40	40	40	0	65	65	67
CHN	359,212	14	32	36	36	7	16	18	18	21	48	53	53
OEA	201,465	0	22	22	22	0	3	7	10	0	24	29	33
IND	359,971	0	0	32	32	0	0	15	15	0	0	47	47
OSA	1,289,684	0	0	0	0	0	0	0	0	0	0	0	0
JPN	760,157	0	0	4	4	0	0	0	0	0	0	4	4
PAS	2,240,347	0	679	687	689	0	262	290	292	0	942	977	981
OCN	902,522	0	0	3	3	0	0	0	1	0	0	3	3
LAC	598,936	82	97	99	100	0	46	77	77	82	143	176	178
TOTAL	14,601,457	357	1522	2109	2381	130	754	1107	1294	487	2276	3215	3675

Source: Adapted from NEA/IAEA, 2010.

are those not rich enough to justify mining under current and expected market conditions; for example, where uranium is only recoverable as a minor by-product, or where the occurrence of uranium is at a level of concentration well below the capacity of current technology. Most unconventional uranium resources reported to date are associated with uranium in phosphate rocks, but seawater and black shale are other potential sources.

In some cases when uranium is produced as a by-product, however, it is considered a conventional resource, for example in the multimineral (copper and gold) Olympic Dam mine in Australia, where the average ore grade is about 280 ppm uranium.

Historically, unconventional uranium has been produced predominantly from phosphate deposits. The world average uranium content in phosphate rock is estimated at 50–200 ppm. Marine phosphorite deposits contain averages of 6–120 ppm, and organic phosphorite deposits up to 600 ppm. Although the concentration is of similar magnitude as that found in gold deposits, typically uranium recovered from phosphate rocks has been classified as coming from 'unconventional' resources.

Between 1975 and 1999, Moroccan phosphate rock processed in Belgium produced 690 tU. About 17,150 tU were recovered in the United States from Florida phosphate rocks between 1954 and 1962, and as much as 40,000 tU were recovered from processing marine organic deposits in Kazakhstan (NEA/IAEA, 2010). Estimated production costs for a 50 tU/yr project, including capital and investment, ranged between 40 $/kgU and 115 $/kgU in the United States during the 1980s (NEA, 2006). The estimates of uranium in phosphates vary widely globally, from 7–22 million tU, in large part reflecting the lack of interest in this resource, which is reported only by a few countries on a regular basis.

Other unconventional occurrences can be found in non-ferrous ores, carbonatites, black schists, and lignite. Table 7.24 summarizes the most recent ranges of unconventional uranium occurrences reported in NEA (2008). Even with the limited reporting, these unconventional occurrences represent between 6.6–7.4 ZJ.

Uranium is also an integral component of coal. For the majority of global coal deposits, uranium concentrations range from slightly below 1 ppm to 4 ppm, which is similar to the concentrations found in a variety of

Table 7.23 | Undiscovered conventional uranium occurrences.

Regions	Prognosticated resources						Total prognostic and speculative [EJ]	Unconventional [EJ]
	<$80/kgU [EJ]			<$130/kgU [EJ]				
USA	490	744	744	501	501	282	1526	2642
CAN	29	88	88	409	409	0	496	0
WEU	4	4	4	29	29	55	88	183
EEU	0	28	28	2	2	105	134	0
FSU	221	456	457	158	245	527	1229	34
NAF	0	0	0	0	0	0	0	3890
EAF	0	0	0	0	0	0	0	0
WCA	8	14	14	0	0	0	14	0
SAF	20	77	77	15	15	650	742	0
MEE	40	52	52	50	58	0	110	106
CHN	2	2	2	2	2	0	4	0
OEA	0	5	5	870	870	76	951	0
IND	0	0	37	0	0	10	47	10
OSA	0	0	0	0	0	0	0	0
JPN	0	0	0	0	0	0	0	0
OCN	0	0	0	0	0	0	0	0
PAS	0	0	0	9	9	0	9	1
LAC	179	197	197	138	138	398	733	156
TOTAL	994	1666	1705	2183	2279	2102	6086	7022

Source: adapted from NEA/IAEA, 2010.

common rocks and soils (see Table 7.21). Since uranium is not combustible, it remains in the ash at concentrations ten times and more than that in the original coal. Current global electricity-generation accumulates on average 3400 tU in the ashes per year.

Of specific interest are the ashes from lignite-burning coal plants located in areas where the lignite resources contain higher than normal levels of uranium. For example, the fly ash at the Xiaolongtang coal power plant in Yunnan Province, China, averages 160–180 ppm, which is only about a quarter of the concentration considered commercially viable for uranium extraction by in situ leaching (ISL), a conventional mining and extraction process. However, coal-ash piles are easier to drill in. In addition, it may be easier to protect local groundwater from contamination. There might also be some monetary value in reducing coal-ash radiotoxicity if regulatory requirements restrict the ash use or require its cleanup. The cumulative ash volume at the Xiaolongtang site is estimated at 5.3 Mt, and contains between 850–950 tU.

Seawater is also a possible unconventional source of uranium. The total mass of uranium in seawater is enormous and amounts to about 4.5 GtU, dwarfing any other exhaustible energy resource. However, because of the low concentration level (3–4 ppb), it is estimated that processing of about 350,000 tonnes of water would be required to produce one kilogram

of uranium. In the 1970s and 1980s, recovery from seawater was on the agenda of several countries. Today research is effectively ongoing only in Japan and France. A braid-type recovery system directly moored to the ocean floor recovered about 1.5 g U over a 30-day test period (Tamada et al., 2006; 2009), but the recovery costs of such a system are over 700 $/kgU, several times the current market price (NEA/IAEA, 2008). Other studies quote recovery costs between 260 $/kgU and 1700 $/kgU (Nobukawa et al., 1994; Kato et al., 1999; Bunn et al., 2003).

"The performance of current [uranium] adsorbents is highly dependent on temperature, and they are thus effectively limited to warm surface waters. However, horizontal and vertical mixing of the ocean would make seawater uranium accessible in warm surface waters at essentially constant concentrations for many centuries, so long as the rate of extraction did not exceed ~2 MtU/yr (30 times current global reactor requirements)" (Bunn et al., 2003).

These cost estimates for uranium from seawater do not include the value of the other metals (vanadium, cobalt, titanium, molybdenum) that would be co-recovered with the uranium (Sugo et al., 2001). Depending on the future long-term market prices of these co-recovered materials, their values might be sufficient to reduce substantially the net uranium recovery cost per kilogram of uranium.

Table 7.24 | Unconventional uranium resources based on limited reporting by countries.

Unit: EJ	Phosphates	Non-ferrous ores	Carbonatite	Black schist, lignite
USA	8.2–19.3	1.1		2336–2920
CAN				
WEU	0.3	1.5	1.5	179.3
EEU				
FSU	33.9			
NAF	3832	58.4		
EAF				
WCA				
SAF				
MEE	93.4–118.8			
CHN				
OEA				0.3
IND	1.0–1.5	3.9–13.4		
OSA				
JPN				
OCN				
PAS	0.3–0.9			
LAC	123.0–165.8	3.9–4.5	7.6	
TOTAL	4092–4456	10.8–28.1	9.1	2515–2925

Source: adapted from NEA, 2006.

7.5.2.4   Extraction Technologies and Production Costs

The extraction costs of uranium reserves are usually reported in categories up to 130 \$/kgU. Clearly, this production cost limit prevents the vast majority of low concentration and unconventional occurrences from being reported and accounted. In fact, from a current demand and supply perspective there is little interest in spending funds on a better delineation of these occurrences or on the development of technologies for their extraction on a commercial scale.

Figure 7.21 summarizes the technologies associated with the extraction of conventional uranium reserves and resources. Underground mining is expected to continue to be the dominant extraction process. At lower ore concentrations (and generally higher production costs), ISL becomes the technology of choice. Not surprisingly, mining of uranium as a by-product has the largest potential at the lowest production costs. Small amounts of uranium are also recovered from mine-water treatment and environmental restoration.

Uranium resources are divided into three production cost categories: less than 40 \$/kgU, less than 80 \$/kgU, and less than 130 \$/kgU. The third level is viewed as a price level "not likely to be seen for many decades, if not longer" (Bunn et al., 2003). This continues to be the general perception, even though spot prices in 2007 were more than twice that high for several months in 2007.

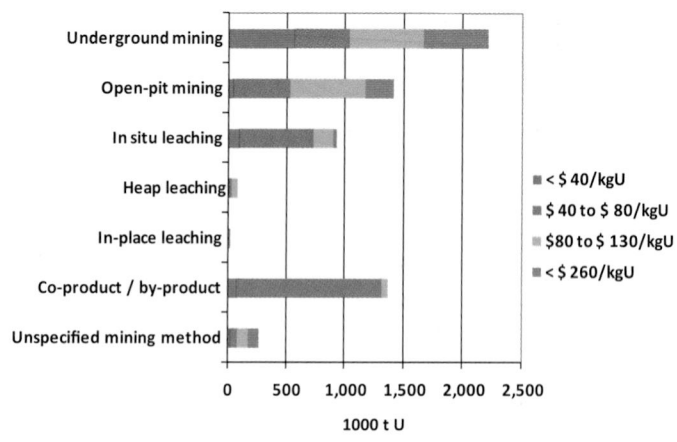

Figure 7.21 | Distribution of uranium reserves and resources by production method. Source: adapted from NEA/IAEA, 2010.

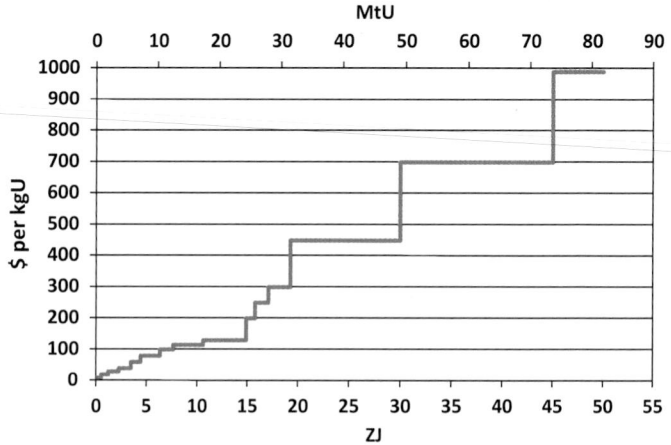

Figure 7.22 | Long-term uranium supply cost curve, in ZJ and MtU.

Figure 7.22 shows a resource cost curve developed from a limited number of indirect observations on the production cost distribution within each category. Costs higher than 130 \$/kgU are representative for uranium production from unconventional resources, and for the lowest ore concentration occurrences. Estimated extraction costs of uranium from seawater are now between the optimistically low 200 \$/kgU and 1000 \$/kgU (Schneider and Sailor, 2008). Although there is abundant uranium in seawater, it is unlikely ever to be produced because at costs consistently higher than 130 \$/kgU, reprocessing of spent fuel may become an attractive alternative (particularly in view of waste-disposal considerations).

7.5.3   Thorium

Thorium is a naturally occurring, slightly radioactive metal. It has been considered as an alternative nuclear fuel to uranium, especially by countries with limited domestic uranium resources. Until the mid-1970s, there was interest in the development of thorium deposits and thorium-based nuclear fuel cycles, in large part because of high growth expectations for nuclear energy and a limited understanding of the global uranium resource endowment. The initial enthusiasm cooled quickly once new discoveries of uranium deposits expanded the availability of uranium well beyond projected reactor requirements.

The average content of thorium in the Earth's outer crust amounts to three to four times the average concentration of uranium. Thorium is widely distributed in rocks and minerals and is found, to some extent, in virtually every continent of the world. There are three principal types of thorium deposits with concentrations of commercial interest: the rare earth phosphate mineral monazite in heavy-mineral, sand placer deposits, thorite ores in vein deposits, and thorium recovered as a by-product of uranium mining.

Current thorium production is based on phosphate mineral monazite deposits with a thorium concentration of 8–10%. Thorium-containing monazite occurs in Africa, Antarctica, Australia, Europe, India, North America, and Latin America (NEA, 2006).

Reliable data on current thorium production is not available, but thorium is produced in small quantities as a by-product of refining monazite for its rare-earth content. Monazite itself is recovered as a by-product of processing heavy-mineral sands for titanium and zirconium minerals. In 2007 Brazil, India, and Malaysia produced an estimated 6780 tonnes of thorium, 80% of which occurred in India. In addition, China, Indonesia, Nigeria, North Korea, the Republic of Korea, and countries of the Commonwealth of Independent States may produce some monazite (Jaskula, 2008).

Thorium is used as an alloying medium in magnesium, in the manufacture of high refractive index glass, in filaments for gas lamps, and as a catalyst. In the absence of demand for thorium as a reactor fuel, its use is expected to decline as the manufacturing industries fear the extra costs and potential liabilities associated with processing radioactive materials (Jaskula, 2008).

The present knowledge of the world's thorium resource base is poor and incomplete. When the global uranium resource situation improved markedly in the 1980s, the nuclear industry lost interest in thorium and thorium exploration effectively ceased, except for some low-key exploration efforts. Most thorium resource assessments date back to the 1970s and 1980s, except perhaps for India and more recently the United States.

There are two sources of information for world thorium reserve and resource estimates, the US Geological Service and the joint reports of the NEA-OECD and the IAEA. The two sources are not in agreement about thorium estimates in several countries. Differences in these estimates are the result of the divergent methodological approaches used, e.g., different costs and degrees of geological assurance, etc. Resource estimates may

Table 7.25 | Thorium resources.

Region	Identified resources (tTh) at <$80/kgTh	Prognosticated resources (tTh)
USA	400,000	32,000
CAN	44,000	129,000
WEU	540,000	574,000
EEU		
FSU	75,000	1,575,000
NAF	100,000	195,000
EAF	1500	
WCA	2500	NA
SAF	18,000	97,000
MEE	1500	8500
CHN	2000	13,000
OEA	1000	
IND	319,000	NA
OSA	500	
JPN		
PAS	4500	4500
OCN	452,000	NA
LAC	612,000	876,500
WORLD	2,573,500	3,504,500

Source: adapted from NEA, 2006; 2010.

also vary erratically from one reporting period to another. For example, recent reports have upgraded India's identified thorium deposits from approximately 300 kt to 650 kt without any explanation (IAEA, 2005). Also the resource estimates exclude data from much of the world. The sparse data reported are often based on assumptions and surrogate data for mineral sands, not direct geological evidence. The data in Table 7.25 indicates an identified thorium resource availability of more than 2.5 Mt at production costs of less than 80 $/kg Th. In addition, there are at least further 4 Mt of yet to be discovered thorium occurrences (NEA, 2008).

7.5.4 Fusion Materials

Lithium is a convenient source material for the deuterium-tritium fusion process. (For a discussion of nuclear fusion see Chapter 14).

The non-renewable part of nuclear fusion rests on the global endowment of lithium. The process requires deuterium (^{2}H) and tritium (^{3}H) which are naturally occurring isotopes of hydrogen. Deuterium is a stable isotope and as such is universally available in all water bodies at an average concentration of 154 ppm.[16] However, tritium is in short supply; it is radioactive, with a half-life of 12.3 years and is produced by cosmic

---

16 Deuterium recovery from water, however, is an energy-intensive process based on water electrolysis, laser application, or the Girdler sulfide process.

rays interacting with gases in the atmosphere. Tritium can be bred from lithium by neutron capture within the fusion reactor in an exothermic reaction (Harms et al., 2000; IEA, 2006b).

The availability of lithium, although more limited than that of deuterium, is still large enough to potentially supply the world's energy demand for thousands of years (Ongena and Van Oost, 2004).

Lithium is used in a number of products and processes, such as batteries, ceramics, glass, aircraft alloys, pharmaceuticals, lubricants, and air conditioning. Its use in rechargeable batteries has expanded significantly in recent years, driven by portable electronic devices and electrical tools. Today, batteries for electric cars constitute the largest and fastest-growing market for lithium compounds.

Lithium does not occur 'free,' i.e., purified or uncombined, in nature. It is found joined (ionized) in various salts in nearly all igneous rocks (especially pegmatites), in sedimentary rocks, and hydrated in brines and in seawater. Brines constitute almost two-thirds of the world's lithium resources, followed by pegmatites (26%).

In natural seawater, lithium is found hydrated at low levels, usually ranging from 140–250 ppb. Only near hydrothermal vents under the oceans does natural seawater contain elevated levels of lithium, where it approaches 7000 ppb. Altogether, an estimated 250 Gt of lithium dissolved in seawater represents an enormous resource (Chitraka et al., 2001). Japan and South Korea lead the development of lithium recovery technology from seawater, but these technologies are in their infancy and lithium from seawater is thus not included in most resource estimations.

As in many resource estimations, lithium reserve and resource estimates vary wildly and lack consistency. Recent resource estimates vary between 22 Mt (Tahil, 2007) and 62 Mt (Yaksic and Tilton, 2009) of lithium globally, with the bulk located in Latin America (Bolivia and Chile account for more than 40% of global resources). The discrepancies typically result from varying assumptions, methodologies, use of terminology, and inclusion of deposits.

Table 7.26 shows the geographical distribution of lithium, resources plus reserves, across the 18 GEA regions and at the same time shows the large range in estimations. It was necessary to combine reserve and resources in this table, as different authors use different definitions of reserves. Table 7.27 summarizes some of the reserve and resource estimations.

7.5.5   Environmental and Social Implications

Like all mining, uranium mining affects the environment and carries certain risks. In many respects these are much the same as those of the mining of any other mineral and, like any other mineral, uranium has unique features that must be managed with regard to its specific properties.

Table 7.26 | Estimates of lithium reserves plus resources in 1000 tonnes (kt).

Regions	Evans (2008) [kt]	Tahil (2007) [kt]	Yaksic and Tilton (2009) [kt]	Clarke and Harben (2009) [kt]
USA	5940	450	5735	6620
CAN	260	540	375	1073
WEU	115		125	221
EEU	850			957
FSU	1000		1160	2480
NAF				
EAF				
WCA	2300		2325	1145
SAF	60	50	70	57
MEE			2000	
CHN	3350	3800	5180	6173
OEA				
IND				
OSA				
JPN				
OCN	280	420	340	1603
PAS				
LAC	15,600	16,700	44,515	19,043
Total	29,755	21,960	61,825	39,372

Table 7.27 | Estimations of lithium reserves and resources in megatonnes (Mt).

Reserves [Mt]	Reference	Resources [Mt]	Reference
6.8	Tahil (2007)	15	Tahil (2007)
9.9	USGS (2010)	25.5	USGS (2010)
24.9	Yaksic and Tilton (2009)	29.9	Evans (2008)
		36.9	Yaksic and Tilton (2009)
		38.3	Gruber and Medina (2010)
		39.3	Clarke and Harben (2009)

One special property of uranium is its radioactive nature. Uranium is mildly radioactive comparable to the levels found in granite. Concern about radioactivity is not unique to uranium mining, but is also a feature of oil and gas production (radium and radon) or the mining of mineral sands (thorium). Although uranium itself is not very radioactive, the ore which is mined contains radium and radon and needs special attention to occupational health and safety protection. This is increasingly regulated by the International Organization for Standardization environmental management system.

Most of the radioactivity is in the tailings – the solid waste product left after uranium oxide has been extracted by milling of the uranium-bearing ore. Tailings include most of the original ore, including all the radium, and contain most of the radioactivity (WNA, 2009).The tailings are deposited in dam-like structures, and during mine operation the tailing dam is usually covered by water to confine surface radioactivity and radon emissions. After mine closure, the tailings are covered with clay, rock, topsoil, and vegetation. The radioactivity of the tailing is about 80–85% of the ore that initially contained the ore body (AUA, 2009) and the cover further reduces potential human exposure.

Run-off water from the mine stockpiles and process water from milling operations contain traces of radium and some other metals, hence this water needs to be prevented from entering biological systems. These liquids are collected in secure retention ponds for isolation and recovery of any heavy metals or other contaminants. The liquid portion is disposed of either by natural evaporation or recirculation to the milling operation. Several mines have begun to adopt a 'zero discharge' policy for any pollutants (WNA, 2009).

Historically, uranium production has principally involved open-pit and underground mining. However, over the past two decades, ISL mining, which uses either acid or alkaline solutions to extract the uranium directly from the deposit (the ore body stays in the ground), has become increasingly important. The uranium-dissolving solutions are injected into, and recovered from, the ore-bearing zone using a system of wells. In 2008, production by ISL exceeded production by open-pit mining and a larger reliance on ISL is expected to continue.

The main environmental consideration with ISL is avoiding the contamination of any groundwater away from the ore body, and leaving the immediate groundwater no less useful than it was initially (WNA, 2009). Apart from groundwater considerations, rehabilitation of ISL mines is very straightforward and much easier to accomplish than that of open-pit mining, making this a technique with lower environmental impact. Upon decommissioning, wells are sealed or capped, process facilities are removed, and evaporation ponds revegetated (WNA, 2009).

7.5.6   Summary

Fissile material resources are plentiful and lack of resource occurrences does not limit the future expansion of nuclear energy. However, a lack of timely investment in exploration and new mining capacities can result in supply shortages and market price volatility, as witnessed in 2007. A sudden shift in demand, such as the unexpected license extensions in the United States and elsewhere, can cause short-term aberrations if uranium producers fail to plan accordingly. Unconventional uranium occurrences (i.e., low concentration ores) exist, but are only economically extractable as by-products of other mining processes, or from coal ash.

Thorium and lithium are both available in large quantities, but require more comprehensive resource assessments and explorations activities. Still, the development of extraction and processing technologies for their large-scale commercial production will only be undertaken if demand prospects warrant such investments. Lithium as a fuel for nuclear fusion will most likely be preceded by its use in batteries for electric vehicles, while thorium, although an alternative to uranium with great potential, will be considered by countries lacking natural endowments of uranium primarily.

7.6   Hydropower

7.6.1   Overview

Hydropower exploits the energy of falling or flowing water by converting it into electricity. It is the most developed and mature renewable technology globally. Current electricity generation by hydropower utilizes some 12.8 EJ/yr of its potential and kinetic energy, from which 3208 TWh were produced in 2008. Hydropower accounts for about 16% of global electricity generation and over 90% of electricity from renewable sources (IEA, 2010a).

The global hydrological cycle is the result of solar energy evaporating water, with oceans having a larger evaporation per unit area than land. Winds transfer the water vapor from oceans to land through precipitation. A global water balance requires that the water precipitated on land eventually returns to the oceans as run off through rivers.

Hydropower projects can be classified by either storage capacity or by purpose. These classifications, however, are not mutually exclusive (IEA, 2000). Run-of-river projects, which have limited storage capacity, generate electricity according to the available hydrological fluctuations of the site. Reservoir-type projects involve damming water and creating reservoirs with significant storage capacity, which allows for the regulation of water flow and electricity production.

Hydropower resources may be designed for electricity generation alone, or for multiple purposes. Multipurpose projects typically have significant reservoir capacity and can provide services such as irrigation, freshwater supply, flood control, and recreation. These uses may affect the volume of water available for electricity generation. In fact, 30–40% of world irrigation is supplied by reservoirs (Lempérière, 2006). This can distort quantitative assessments of the economic potential of hydropower for energy purposes.

Pumped storage projects are used to provide efficient storage of energy, especially for intermittent renewable resources. As they do not generate net electricity, they are excluded from the assessment of hydropower potentials.

7.6.2   Estimating Hydropower Potentials

Assessment methods for hydropower are different to those used for other renewables. For renewable sources such as wind or solar, potentials are assessed by beginning with a theoretical estimate, which is then reduced, by the application of constraints, to define a technical potential, which is further reduced to an economical potential.

Hydropower potential, however, is generally assessed by adding up the potential of individual well-known sites, and many other possible sites are omitted for various reasons. As a result the theoretical potential is underestimated. The inherent subjectivity in determining site suitability plays a significant part in assessments. For example, the US Rocky Mountains are considered 'off-limits,' even though they are similar to the Alps, which have seen large developments. Smaller sites, as well as existing dams not used for hydropower, also were not considered in the United States until recently (UNWWAP, 2006). Finally, multi-purpose issues are not considered. Reservoirs often provide gravity fed irrigation, which is in itself an energy service, as it circumvents the electricity intensive pumping of water.

As a result of social and environmental constraints, it is prudent to assume that not all realistic or economic potential can be developed. It is important, however, that these potentials still be included in the theoretical and technical potentials. Social and political situations are subject to change over time. Climate change, new environmental policies, changed social preferences, and new technologies, especially for transmission, may make remote sites accessible, affect energy supply preferences and demand levels, and lead to a reassessment of previously excluded hydropower locations (and vice versa).

The assessment and reassessment of hydropower sites is a costly affair. Thus in many countries, where many well-known sites are still to be developed, there is less impetus to find additional sites.

7.6.3   Hydropower Potentials

About 1,260,000 EJ (40,000 TW) of the total solar flux of 2,800,000 EJ/yr (89,000 TW) reaching the surface of the Earth is actually used to evaporate water and drive the water cycle (Tester et al., 2005). In contrast to many renewable sources, such as wind or solar, which are characterized by flows of diffuse energy, hydropower benefits from the fact that the water cycle has already concentrated the energy in the form of a high-density flow of water (Shepherd et al., 2003). The theoretical potential energy in the water cycle is thus calculated as 40% of the solar energy used to evaporate water (Tester et al., 2005). Consequently, the theoretical potential of energy in the water cycle is 504,000 EJ (see Table 7.28).

A portion of the energy in the water cycle is held in flowing rivers – the part of the hydrological cycle useable for electricity generation – and several countries have tried to quantify this gross hydropower potential. It is based on average river flows, multiplied by the relative change in altitude of each river (Lehner et al., 2001). This results in an annual theoretical hydropower potential of 200 EJ globally.

The maximum technical potential is based on two sets of data. A few countries assess the total potential of their main rivers, while other countries add up the potentials of all sites, assuming that virtually all the energy at the site can be harnessed. Despite the fact that some sites may not be included in this data, this global estimate of 140–145 EJ is still huge. However, it represents a mix of technical and theoretical potentials.

Another approach to estimating technical potential is based on adding up site potentials, adopting a realistic use factor for each site using detailed information on the location and geographical factors of run-off water (such as available head and flow volume per unit of time). The resulting potential then could be called the 'practical technical potential.' Even if countries do not attribute a development cost to each site, it

Table 7.28 | Estimates of world hydropower potential.

Estimation method	Comments	Hydrootential [EJ/yr]
Energy in the water cycle (Tester et al., 2005)	40,000 TW of instant solar power serving to evaporate water 40% of the time	504,000
Theoretical potential (Lehner et al., 2001)	For most rivers: mass of runoff × gravitational acceleration × height	200
Maximum technical potential, based on rivers and or sites[a]	Technical potential of known sites, assuming a very high use factor	140–145
Technical potential, based on sites at 2–20¢ per kWh[b]	Portion of technical potential, with a realistic use factor, that is sufficiently promising to justify a site assessment	50–60
Economical potential, based on sites at 2–8¢ per kWh[c]	Portion of technical potential, with a realistic use factor, that is competitive with large thermal power plants	30

[a] Few countries provide information on the total potential of main rivers. When these data are not available, the technical potential of known sites, at 100%-use factor, is included.

[b] The 20¢ per kWh threshold is a global estimate.

[c] Economic potential is based on official country assessments. The threshold could be lower in some countries, as the main competitor is often coal fired generation at approximately 5¢/kWh.

is reasonable to assume that this potential is available at a cost between 20–200 $/MWh. This threshold of 200 $/MWh is justified in that a site that would cost more than 200 $/MWh would be considered a very poor site by an engineer, and would likely not be included in the assessment. For many countries, the threshold could be lower, such as 150 $/MWh. It is also important that, in many countries, environmentally sensitive sites were previously excluded from any assessment.

Finally, to assess the economic potential, the expected costs of sites are assessed and compared to the costs of other major sources of electricity, such as coal, gas, or nuclear. If the cost of the hydropower sites is lower or similar, the potential is considered economically viable; if it is not, it is excluded from the estimate. A cost estimate between 20–80 $/MWh is reasonable for this potential, as coal is often the main competitor at 40–60 $/MWh. The economic hydropower potential is estimated at around 20 EJ/yr (see Chapter 11).

7.6.3.1 Regional Data on Technical and Economic Potentials

Table 7.29 presents the maximum technical and practical technical potential, as assessed by each country. These country assessments can be made using different methods. Some countries exclude small hydropower or any development in national or state parks. Moreover, the assessment of remote sites can be expensive and thereby not included or only poorly estimated. These methodological issues do allow for a general conclusion: the data on hydropower potential is probably underestimated by a wide margin.

7.6.4 Environmental and Social Implications

The environmental issues surrounding hydropower development are numerous. This is unavoidable when creating a reservoir. However, it can be misleading to make general conclusions, as hydropower plants vary greatly and raise different environmental issues. This section does not review these issues, but discusses a few major constraints that can significantly affect future potential.

Ecosystem impacts usually occur downstream from hydropower sites and range from changes in fish biodiversity and in the sediment load of the river to coastal erosion and pollution. For comparable electricity outputs, GHG emissions associated with hydropower are one or two orders of magnitude lower than those from fossil-generated electricity, but can be non-negligible in cases where sites inundate large areas of biomass and consequent CH_4 releases to the atmosphere.

The land use of hydropower projects (per unit of electricity generated) can be very low for run-of-river plants and very high for plants with large reservoirs and multipurpose water uses. Large hydropower projects requiring large reservoirs and extensive relocation of communities increasingly encounter public resistance and, as a result, face higher costs.

Table 7.29 | Theoretical and technical potentials for the 18 GEA regions.

	Hydropower potential, based on known sites (EJ)			
	Production (2008)	Maximum technical potential[b]	Technical potential at 20–200 $/MWh	Economical potential at 20–80 $/MWh
USA	0.97	7.34	4.82	1.35
CAN	1.26	7.44	2.98	1.93
WEU	1.81	11.65	4.12	2.88
EEU	0.22	1.25	0.59	0.35
FSU	0.86	12.74	8.11	4.65
NAF	0.06	0.71	0.28	0.25
EAF	0.03	3.73	1.67	0.81
WCA	0.11	7.79	3.91	1.22
SAF	0.15	1.86	0.74	0.23
MEE	0.1	2.48	1	0.44
CHN	1.71	21.9	8.91	6.31
OEA	0.15	2.44	0.82	0.38
IND	0.44	9.5	2.38	1.59
OSA	0.16	6.8	2	0.28
JPN	0.33	2.58	0.49	0
OCN	0.14	1.69	0.64	0.11
PAS	0.14	11.26	2.89	0.44
LAC	2.35	30.22	11.07	6.61
World	10.96	143.41	57.41	29.84

[a] When statistics on technical or economic potentials are available only in MW, PJ was estimated by multiplying the MW by the typical use factor of hydro plants within that region.

[b] When the maximum technical potential is not provided, the realistic potential is used instead, assuming a use factor of 100%.

Source: adapted from Hydropower and Dams World Atlas, 2008.

Population density is a major constraint for future development. If a project requires resettlement, the high costs and uncertainty make planning quite difficult. Despite this, creating reservoirs does not mean that future hydropower sites are impossible to develop. With proper planning, consultation, and analysis, hydropower projects can be implemented with much smaller environmental and social impacts. The decision to include a reservoir may well depend on the demand for other services, such as irrigation or flood mitigation.

The effects of climate change may change precipitation patterns in various ways, as the water cycle will be intensified in some regions, causing a higher probability of extreme events and flooding. More reservoirs may be needed for flood impact mitigation, or reservoirs may lack sufficient water to provide enough electricity to meet demand. Evaporation may increase, so some regions may experience more frequent and more intense droughts. Construction of new reservoirs may be needed to adapt to climate change and prepare for additional storage needs. More storage may be needed for irrigation also, as rain-fed agriculture may become less reliable in some regions.

7.6.5   Summary

There is a large global potential of hydropower for electricity generation, with the annual economic potential estimated at around 30 EJ. Hydropower has very good development potential, in both short and long terms, as it is well adapted to meet future global challenges. The coupling of hydropower with other uses makes it a very versatile option, especially in mitigating and responding to the effects of climate change.

Although most of the suitable sites for large hydropower implementation in OECD countries have already been developed, the potential still remains for further small-scale plants as well as large scale developments in China, Africa, and especially Latin America.

7.7   Biomass energy

7.7.1   Overview

Biomass is biological material derived from living, or recently living, organisms. Biomass as an energy resource can be derived from agricultural crops, forest products, aquatic plants, crop residues, animal manures, and wastes, such as municipal solid waste (MSW). Biomass was the main energy source for humans until about the third quarter of the 19th century when the increasing availability of fossil fuels progressively reduced its dominance. Still, in many of the least-developed countries biomass remains the dominant source of energy and can contribute as much as 80% or more to these countries' energy supplies.

The primary process through which biomass becomes available on Earth is photosynthesis. Plants use solar energy to produce energy-rich organic materials from inorganic inputs (CO_2, H_2O, and plant nutrients such as nitrogen and phosphorous). Globally, current net biomass production of green plants (net primary production, NPP) in terrestrial ecosystems amounts to approximately 118 billion tonnes of dry matter per year (Gt/yr) with a gross calorific value (GCV) of 2190 EJ/yr. Additional biomass is produced by plants in freshwater and ocean ecosystems. A considerable proportion of the NPP is allocated to below-ground parts of plants, most of which cannot be harvested. Aboveground terrestrial NPP currently amounts to 67 Gt/yr or 1241 EJ/yr (Table 7.30).

The amount of biomass produced by terrestrial ecosystems depends on climate, soil, and human management (land use). Rising atmospheric CO_2 levels tend to raise plant productivity, partly as a result of changes in temperature and precipitation, and partly through the CO_2-fertilization effect which is, however, still poorly understood (Müller

et al., 2006; Long et al., 2006; Tubiello et al., 2007). This trend might be reversed in the future, because of possible changes in precipitation patterns that might limit productivity. Given unlimited water supply, higher temperatures boost plant growth, but if water supply is limited, rising temperatures may also reduce productivity through increased water stress (Sitch et al., 2008).

Land use may increase or reduce the biomass production of ecosystems. In areas where water availability limits biological productivity, irrigation can considerably increase biomass production per unit area and year. In temperate zones, intensive use of fertilizers and agricultural technologies may boost productivity compared to potential vegetation, i.e., the resulting vegetation in the absence of land use, but generally land use reduces NPP. Globally, agroecosystems, settlement areas, and soil degradation have reduced the biomass produced by green plants in terrestrial ecosystems by almost 10% (Haberl et al., 2007a). As plants in agroecosystems allocate more biomass to aboveground NPP than those in natural ecosystems, the human impact on aboveground productivity is smaller than that on total productivity (Table 7.30).

Globally, humans currently harvest or destroy about 20.1 Gt/yr of biomass (373 EJ/yr), i.e., 17% of the yearly biomass production of terrestrial ecosystems. Above the ground, however, humans currently harvest or destroy 30% of the annual biomass production of terrestrial ecosystems (Table 7.30).[17] About two-thirds of the biomass harvested and destroyed is actually used by humans: one-third is either burned in human-induced fires or left in the ecosystem (below-ground biomass, unused crop residues, and felling losses in forests). Table 7.30 shows that crops and crop residues account for 52% of the biomass harvested by humans. Almost one-third is directly taken up by grazing animals on the world's grazing lands. Wood removal from forests accounts for the rest. Livestock consumes almost 60% of the total amount of used biomass.

Bioenergy accounts for almost half of the final amount of biomass. A considerable fraction of the bioenergy stems from biogenic wastes, by-products, and residues. Fuelwood harvest in the year 2000, according to Food and Agricultural Organization (FAO) figures, had a GCV of 22 EJ/yr. Primary crops used for biofuels were negligible at that time, i.e., contributed about 0.8 EJ/yr in 2000, but since have increased to 3.1 EJ in 2008 (Chum et al., 2011).

7.7.2   The Theoretical Potential for Land-based Bioenergy Production

The point of departure for the estimation of the theoretical production potential for land-based bioenergy is the NPP data of Table 7.30. The

---

17   This value represents a cautious estimate because it only includes wood harvests according to FAO. Alternative studies suggest that wood harvests could be up to 33% higher than assumed here.

Table 7.30 | Global yearly biomass flows around the year 2000.

	USA	CAN	WEU	EEU	FSU	NAF	EAF	WCA	SAF	MEE	CHN	OEA	IND	OSA	JPN	OCN	PAS	LAC	World
										[EJ/yr]									
Total potential NPP [b]	157.4	138.7	87.3	26.1	331.6	41.5	73.0	243.1	150.2	16.9	129.7	40.1	64.0	19.4	8.8	121.3	165.7	608.9	2423.7
Abovegr. potential NPP [b]	85.7	76.4	48.9	15.3	179.7	20.8	39.8	133.0	79.5	6.4	70.1	21.2	34.1	9.5	5.1	60.0	92.0	331.4	1309.0
Total current NPP	140.9	132.0	79.5	20.2	298.9	36.0	59.1	217.7	136.3	15.2	123.0	35.4	50.6	19.8	8.1	114.1	140.6	563.6	2191.0
Aboveground current NPP	85.1	73.4	50.8	13.6	165.1	19.7	32.8	120.8	73.2	7.5	76.0	19.2	34.8	12.2	5.0	58.3	80.9	312.5	1241.0
Harvested biomass	25.7	4.6	20.4	5.2	10.4	5.3	6.9	10.9	6.4	2.6	29.2	3.3	24.3	8.2	1.0	6.2	13.3	40.4	224.6
Harvested primary crops	10.2	1.3	7.7	2.2	4.5	1.0	0.6	1.9	0.9	0.8	10.0	0.9	6.2	1.7	0.4	1.9	4.2	7.0	63.5
Harvested crop residues	5.8	0.8	3.9	1.7	2.8	1.0	0.9	2.9	1.4	0.8	9.2	0.8	7.5	2.1	0.2	0.6	4.2	7.5	54.3
Biomass grazed	4.8	0.6	6.0	0.4	1.3	2.7	3.5	3.0	2.7	1.0	7.1	1.0	7.1	3.4	0.2	3.3	1.7	21.1	71.0
Wood removals (FAO)	4.9	1.9	2.8	0.9	1.8	0.5	2.0	3.1	1.3	0.0	2.9	0.6	3.4	1.0	0.2	0.5	3.1	4.9	35.9
Biomass destroyed	9.8	1.8	7.1	2.5	8.1	7.5	5.7	21.4	14.2	0.7	14.1	3.4	10.9	2.4	0.4	3.9	10.6	23.5	148.0
Human-induced fires	0.3	0.0	0.3	0.2	3.6	6.3	4.3	17.7	12.5	0.0	0.9	2.1	2.3	0.5	0.0	2.6	4.5	14.7	72.8
Belowground biomass	4.7	1.1	3.2	1.1	2.2	0.4	0.8	1.9	0.9	0.3	5.3	0.6	3.9	0.9	0.2	0.6	2.8	4.4	35.5
Unused cropland residues	3.9	0.5	3.2	1.0	1.9	0.7	0.3	0.8	0.4	0.3	5.9	0.3	3.0	0.7	0.2	0.7	1.4	2.4	27.7
Felling losses in forests	0.8	0.2	0.4	0.2	0.5	0.1	0.2	0.9	0.4	0.0	2.0	0.3	1.7	0.3	0.0	0.1	1.9	1.9	12.1
Use of harvested biomass	23.8	3.3	21.0	5.0	9.4	5.8	6.9	10.5	6.3	3.3	29.5	3.2	23.6	8.2	2.4	5.5	12.5	38.6	219.1
Plant-based food	1.4	0.1	2.1	0.5	1.2	0.9	0.5	1.3	0.6	0.8	6.3	0.6	4.5	1.4	0.5	0.1	2.6	2.3	27.8
Feed (grazing, residues)	12.2	1.6	12.8	2.2	4.9	4.0	4.3	4.7	3.9	2.0	14.7	1.5	15.4	5.2	1.0	4.8	4.1	29.8	129.2
Wood	5.2	1.0	3.2	0.8	1.2	0.6	2.0	3.1	1.3	0.1	3.5	0.6	3.5	1.0	0.8	0.3	3.0	4.8	36.0
Other uses	4.9	0.6	2.9	1.5	2.1	0.4	0.1	1.5	0.5	0.3	5.1	0.4	0.1	0.6	0.1	0.3	2.9	1.7	26.1
Final biomass use	8.7	1.1	9.0	1.6	2.4	1.8	1.8	6.1	3.0	1.0	18.5	1.9	11.4	3.5	1.8	0.7	7.3	7.9	89.5
Food for humans	2.3	0.2	3.2	0.8	1.6	1.0	0.5	1.4	0.7	0.8	7.7	0.6	4.8	1.5	0.6	0.2	2.8	3.0	33.8
Timber and paper	3.3	0.4	2.9	0.4	0.4	0.1	0.1	0.2	0.2	0.1	1.4	0.1	0.2	0.1	0.9	0.2	0.7	0.9	12.4
Bioenergy (IEA)	2.6	0.5	2.3	0.4	0.3	0.6	1.2	4.4	2.1	0.0	8.9	1.1	6.2	1.8	0.2	0.2	3.6	3.4	40.2
Other uses	0.5	0.0	0.6	0.1	0.1	0.1	0.0	0.1	0.0	0.1	0.4	0.0	0.2	0.1	0.1	0.0	0.2	0.6	3.2

[a] Gross calorific value. 1 t dry matter biomass = 0.5 t carbon = 18.5 GJ.

[b] NPP of the vegetation assumed to exist in the absence of human land use under current climate conditions.

Sources: Haberl et al., 2007a; IEA, 2007a, 2007b, 2008b; Krausmann et al., 2008; Lauk and Erb, 2009.a

Table 7.31 | Estimates of the theoretical potential for global biomass production for bioenergy.[a]

	Total terrestrial surface area	Aboveground NPP of potential vegetation (NPP_0)	Aboveground NPP of current vegetation (NPP_{act})	Global human biomass harvest for food, feed, fiber	Theoretical total bioenergy potential	Theoretical (practical) bioenergy potential[b]
	[1000 km²]	[EJ/yr]	[EJ/yr]	[EJ/yr]	[EJ/yr]	[EJ/yr]
USA	11,367	86	85	22	64	48
CAN	9331	76	73	4	73	52
WEU	9178	49	51	17	32	25
EEU	1159	15	14	5	11	7
FSU	21,614	179	165	9	170	117
NAF	7984	21	20	4	17	12
EAF	3254	40	33	5	35	21
WCA	4440	133	121	9	124	84
SAF	6859	79	73	5	74	51
MEE	5169	7	8	2	5	5
CHN	9351	70	76	26	44	37
OEA	2411	21	19	3	18	12
IND	3147	34	35	18	16	12
OSA	1908	10	12	5	5	5
JPN	394	5	5	1	4	3
OCN	7913	60	58	5	55	40
PAS	4317	92	81	12	80	52
LAC	20,295	331	312	31	299	210
TOTAL	130,091	1308	1241	183	1126	793

[a] NPP and harvest values refer to the aboveground compartment only and exclude belowground biomass.

[b] This version of the theoretical potential was calculated by assuming that all (100%) aboveground NPP of current vegetation except NPP harvested for food, feed and fiber would be used as feedstock for bioenergy.

estimate of the theoretical bioenergy potential is based on the following assumptions:

- Significant increases in NPP (i.e., by >20%) over its natural potential are hardly possible without massive direct and indirect energy inputs (that would result in unfavorable energy returns on investment and high GHG emissions). Moreover, the global productivity of croplands is currently 35% lower than their potential productivity, despite all fertilizer and other inputs (Haberl et al., 2007a). Field et al. (2008) and Campbell et al. (2008) maintain that potential NPP is an upper limit for the large-scale average productivity of energy crop plantations.

- Only those parts of plants aboveground are harvested (for practical reasons).

- Land demand for all other human uses except bioenergy remains constant, i.e., expected future increases in demand for food, animal feed, and fiber have no impact on land allocation. This seems justified because a large part of the projected future agricultural biomass

production is expected to come from yield increases rather than from expansions of cropped area (FAO, 2006b; IAASTD, 2009).

- Primary bioenergy (i.e., bioenergy not derived from wastes, by-products, or residues) accounts for approximately 30% of the global final use of biomass (Krausmann et al., 2008).

- 100% of the potential aboveground NPP (NPP_0) not required for food, animal feed, or fiber production can be used to produce bioenergy.

The theoretical bioenergy potential amounts to some 1100 EJ/yr[18] (Table 7.31) and would leave no NPP for heterotrophic food chains

---

18 The estimate is based on a GCV of dry-matter biomass of 18.5 MJ/kg and assumes a carbon content of dry matter biomass of 50%. The GCV of woody biomass is 19.5 MJ/kg and that of herbaceous biomass is 17.5 MJ/kg. The net calorific value is approximately 10% lower than the GCV. The values refer to the total amount of biomass that might be harvested without consideration of how the biomass will be used as an energy source.

(animals, fungi, microorganisms), with consequent catastrophic impacts on diversity, resilience, and sustainability of ecosystems. As this would entail the clearing of all forests (no biomass left in the ecosystems to build up and maintain long-lasting carbon stocks), the resulting GHG balance would also be very unfavorable. The human appropriation of NPP – approximately 30% of aboveground NPP (Haberl et al., 2007a) – has already contributed to a global reduction in the terrestrial ecosystem's ability to deliver essential ecosystem services (MEA, 2005).

The theoretical potentials of Table 7.31 do not include the potential production of algae in coastal seawaters, industrial installations, and open-sea devices. Their productivity is not limited by land or freshwater availability and could exceed natural NPP per unit area and year by orders of magnitude.

7.7.3 Technical Biomass Potentials

7.7.3.1 Principal Biomass Flow Pathways Relevant for Bioenergy

The necessity to produce food and fiber, the area demand of infrastructure, and sustainability criteria such as biodiversity conservation and the GHG balance, limit the amount of bioenergy that can be produced sustainably. The technical potential is assessed for energy crops (see Chapter 20), as well as for agricultural and forestry residues.

Figure 7.23 shows the main synergies and competition between the demand for bioenergy, food, and fiber. If the demand for food and biomass-based materials is large, then less land is available for energy crops. However, during the food and material production process, residues become available that also can be used for energy applications.

Figure 7.23 distinguishes two types of energy crops plus residues from agriculture and from forestry production. The residues are further split into three categories: primary, secondary, and tertiary residues. Primary residues become available during the harvesting process, e.g., straw or wood from forest thinning. Secondary residues become available during the processing of food, wood, or other biomass, e.g., bagasse or sawdust. Tertiary residues become available after the use of the food or the material, e.g., MSW or waste wood.

The following sections assess the technical biomass potentials for three categories:

- agricultural residues, wastes, and by-products;

- biomass from forestry, including forestry residues; and

- aquatic biomass and/or algae.

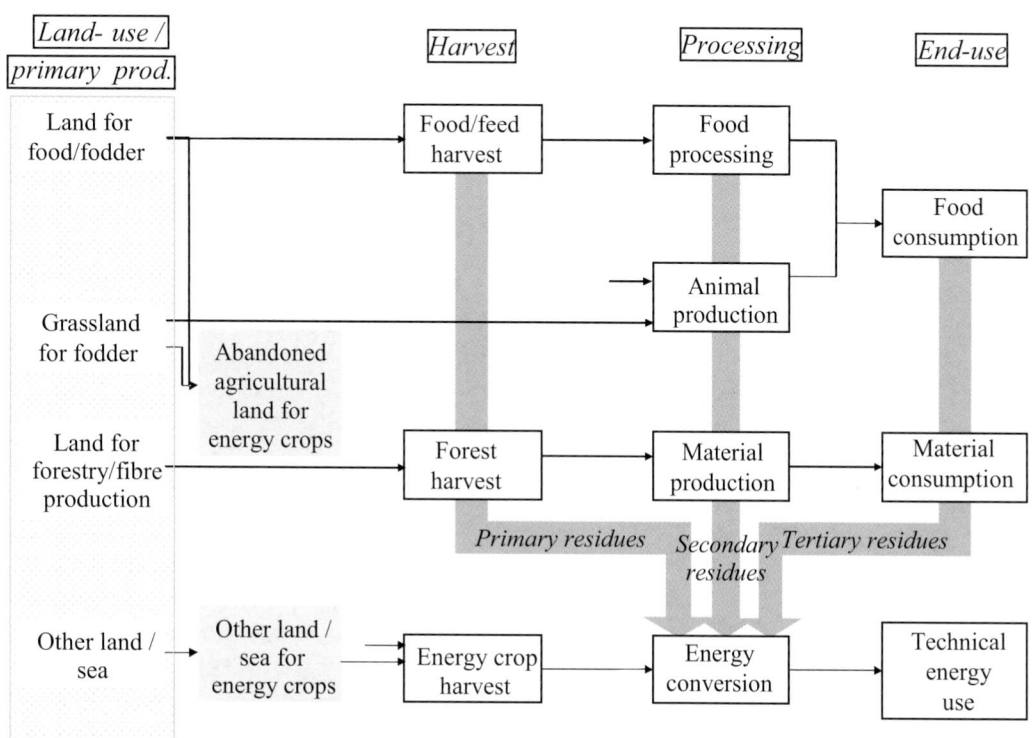

Figure 7.23 | Competition and synergy between different biomass feedstocks.

The energy potential from energy crops on cropland, abandoned agricultural land, or grazing land is discussed in Chapter 20.

7.7.3.2   Crop Residues, Municipal Solid Wastes, and Animal Manures

Overview

Large quantities of organic residues and wastes are produced annually, including crop residues, animal manures, and MSW. Rational utilization of these can produce energy, normally cost-effectively, and minimize environmental impacts that are often caused by other management and/or disposal methods.

Crop residues

Two types of residues are associated with crop production. These are primary, i.e., harvest or field residues, and secondary, i.e., processing residues. The recoverable fraction (RF) of a residue depends on the amount that could be collected practically and, in the case of field residues, on the amount that should be left in the field for maintaining soil quality.

Annual crop production in 2005 was estimated from the FAO online database (FAO, 2009). Values of residue to product ratio (RPR) for different crops are taken from the open literature (Fischer et al., 2007; Koopmans and Koppejan, 1998; Bhattacharya, 1991). The RPR values for primary residues of cereal crops depend on the level of technology used for cereal production. Higher levels of technology, including irrigation and fertilization, result in higher yields per hectare and lower values of RPR; RPR values selected in this study for 2050 correspond to higher levels of technology (Fischer et al., 2007). For the year 2005, region-specific RPR values were selected and assumed based on literature reviews; values corresponding to the average of higher and lower levels of technology are assumed for the regions for which no RPR values could be found in the literature.

Residue generation from cereal crop production for the year 2050 was estimated using projected production growth rates of cereals in developing, industrial, and transition countries from FAO (2006a). For estimating residue generation from the other crops, their production growth rates in different country categories are assumed to be the same as the projected rates for all food and non-food commodities obtained from FAO (2006b).

Table 7.32 shows the estimated energy potentials of crop residues in 2005 and 2050 for the 18 GEA regions. The total world technical potential of crop residues in 2050 is estimated to be 49.4 EJ.

Assuming RF values for different crops in the range 0.5–0.75, Hakala et al. (2009) estimated the world technical potential of field residues in 2050 to be 38–41 EJ; adding the process residue potential of 16 EJ (Smeets et al., 2007), the total technical potential of crop residues would be 54–57 EJ. These estimates for 2050 may be compared with the range 10–32 EJ/yr derived by Hoogwijk (2004) and 46–66 EJ by Smeets et al. (2007).

Differences in crop-residue energy potential estimated by different authors mainly result from differences in their assumed RPR and RF values and crop-production projection methodologies.

Municipal solid wastes

MSW consists of biogenic (i.e., renewable) and non-biogenic (i.e., non-renewable) constituents. The energy potential of the biogenic components was estimated from their energy values (US EIA, 2007), the total generation of MSW, assumed values for the recoverable fraction from MSW for energy, and the MSW composition. MSW composition and generation per capita for different countries/regions for the year 2000 were obtained from IPCC (2006); country/region-specific values were obtained, in some cases, from the literature.

MSW generation per capita is known to increase with growth in gross domestic product (GDP). Adhikari et al. (2006) established a relationship between MSW generation per capita and GDP per capita to project future MSW generation from expected growth in GDP per capita; this approach was used in this study. National GDP values were taken from World Bank (2008a) and the projected future national and/or regional growth rates obtained from IEA (2006a); population data for the years 2005 and 2050 were obtained from United Nations Population Division (2008).

Since an increase in MSW generation is not expected to continue indefinitely, Monni et al. (2006) projected a maximum future annual MSW generation of 900 kg/capita; this value was used in this assessment.

In this study, estimations for recycling of MSW organic matter throughout the world by 2050 are guided by a scenario for the EU-15 countries developed by Smith et al. (2001) for the year 2020: 60% paper recycling and 60% separation of food and yard-trimming wastes (two-thirds of the latter is assumed to be composted) and the remaining one-third used for anaerobic digestion. Thus, 60% of food and yard waste is available for energy purposes.

In 2005, leather and rubber recycling in the United States was 14.3%, while recycling of textiles was 15.3%; 9.5% of wood in MSW in the United States was recycled. It is assumed that 15% of both leather and textiles would be recycled in 2050 in all countries; the amount of wood recycled is assumed at 10%.

The theoretical potential of MSW is assumed to be the energy potential of all MSW produced in a country without any recycling, while technical potential is assumed to be the same as the recoverable potential.

As shown in Table 7.32, the total world recoverable potential of MSW in 2050 is estimated to be 11.02 EJ; this value can be compared with 1–3 EJ in 2050 as estimated by Hoogwijk (2004) and around 17 EJ as estimated by Smeets et al. (2007).

Table 7.32 | Energy potentials of crop residues, MSW, and animal wastes in EJ/yr.

Region	Crop residues				MSW				Animal wastes			
	2005		2050		2005		2050		2005		2050	
	Theoretical	Technical	Theoretical	Technical	Theoretical	Technical	Theoretical	Technical	Theoretical	Technical	Theoretical	Technical
USA	7.54	2.68	9.32	3.28	2.16	0.92	3.19	1.33	3.19	1.7	5.67	3.03
CAN	1.32	0.55	2.18	0.6	0.27	0.14	0.3	0.12	0.5	0.27	0.88	0.47
WEU	5.55	2.2	6.39	2.56	2.05	0.86	3.38	1.28	3.34	1.74	5.94	3.09
EEU	2.5	0.91	1.75	0.66	0.43	0.18	0.56	0.21	0.56	0.31	1	0.55
FSU	4.07	1.62	4.51	1.92	0.77	0.34	1.16	0.41	1.85	0.93	3.28	1.65
NAF	1.98	0.79	2.04	0.93	0.28	0.1	0.58	0.2	1.59	0.75	2.84	1.33
EAF	1.41	0.57	1.98	0.91	0.31	0.18	0.79	0.19	2	0.94	3.56	1.67
WCA	4.15	1.58	5.29	2.22	0.57	0.16	1.49	0.55	1.83	0.85	3.25	1.51
SAF	1.88	1.05	3.41	1.63	0.32	0.08	0.9	0.27	1.32	0.64	2.34	1.14
MEE	1.28	0.54	2.15	0.95	0.39	0.18	1.11	0.46	0.78	0.36	1.38	0.64
CHN	12.08	6.25	14.87	7.81	2.09	0.59	4.41	1.54	5.6	2.8	9.96	4.99
OEA	1.36	0.55	1.55	0.66	0.24	0.05	0.39	0.13	0.56	0.26	1	0.46
IND	9.31	4.08	14.15	6.75	0.96	0.15	2.58	0.77	7.08	3.47	12.59	6.17
OSA	3.04	1.27	4.25	1.95	0.5	0.09	1.09	0.33	1.66	0.82	2.95	1.46
JPN	0.34	0.14	0.33	0.14	0.56	0.19	0.87	0.31	0.17	0.1	0.29	0.17
PAS	7.29	3.18	9.49	4.52	0.93	0.32	2.63	1.09	1.46	0.76	2.6	1.35
OCN	1.52	0.66	1.47	0.67	0.16	0.09	0.2	0.09	1.32	0.59	2.34	1.65
LAC	11.41	5.69	21.64	11.28	1.44	0.78	3.99	1.75	8.87	4.44	15.79	7.9
WORLD	78.03	34.32	106.78	49.45	13.15	5.4	29.63	11.02	43.66	21.71	77.68	39.21

The energy potentials of harvest or field residues and processing residues were estimated as follows:

Residue energy potential (EJ/yr) = bone dry residue production (Mt/yr) × recoverable fraction of residue production × gross heating value of bone-dry residue (EJ/Mt).

Bone dry residue production (Mt/yr) = dry matter of crops produced (Mt/yr) × residue to product (or crop) ratio on dry basis.

Includes residue energy potentials of the following crops:

Cereals: wheat, rice, barley, maize, rye, oats, millets, sorghum.

Sugar crops: sugar beet, sugar cane.

Pulses: peas, chick peas, cow peas, pigeon peas, beans, broad beans, lentils, pulses of minor relevance.

Oil crops: groundnuts, rapeseed, soybeans, sunflower seed, safflower seeds, linseed, sesame seed, castor bean, other oilseeds of minor relevance.

Tree nuts: almonds, brazil nuts, cashew nuts, chestnuts, hazel nuts, kola nuts, pistachios, walnuts.

Fruits: apples, apricots, avocados, bananas, cashew apples, cherries, citrus fruit, dates, figs, fruit grapefruit (including pomelos), grapes, kiwi fruit, lemons and limes, mangoes, mangosteens, guavas, oranges, papayas, peaches and nectarines, pears, persimons, pineapples, plantains, plums and sloes, quinces, raspberries, sour cherries, stone fruit, tangerines, mandarins and clementines.

Tuber crops: cassava, potatoes, sweet potatoes, yams.

Other crops: coffee, cotton, jute.

Notes:

The energy potential of manure generated annually by an animal type i is estimated as follows:

Total volatile solid produced annually by an animal of a given type i, VS_i (kg/yr) = 365 × animal population (head) × average volatile solids production per head for animal of type i per day (kg volatile solid/head/day).

Total dry manure produced annually by an animal of a given type i, TS_i (kg/yr) = VS_i/volatile fraction

Total annual energy potential = TS_i × heating value of manure on dry basis.

Animal manures

The number of animals in 2005 was obtained from the FAO database on live animals (FAO, 2009). The number of all animals was assumed to increase at rates corresponding to the projected annual growth in world livestock production (FAO, 2006c) – these are 1.6% during the period up to 2030 and 0.9% in 2030–2050. Values of kilogram of volatile solid/head/day were obtained from the IPCC (2006); the amounts of total dry solids produced were estimated from these, assuming suitable values of volatile fraction. Values of RFs of manures were obtained from TERI (1985) and NRCS (1995). Based on literature reviews, the heating values of dry animal manures were assumed to be 17 MJ/kg for swine: 12.9 MJ/kg chicken, and 16 MJ/kg for other animals.

As shown in Table 7.32, the energy potential of recoverable manures was estimated to be 39.2 EJ in 2050. Johansson et al. (1993) estimated the heating value of annual recoverable manures in 2050 to be 25 EJ. Based on a literature review, Hoogwijk (2004) estimated the recoverable potential of animal manures to be 9–25 EJ.

Differences in the energy potential of animal manures reported by different authors appear to result mainly from differences in projected manure-generation values and assumed RFs, as well as heating values.

Technical potential of forestry residues

Figure 7.23 distinguishes three categories of forestry residues: primary (available from additional fellings or as residues from thinning or final fellings), secondary (available when processing the forest products, e.g., sawdust), and residues (available after end use, i.e., waste wood).

Various studies have assessed the future potential for forestry residues (e.g., Berndes et al., 2003; Smeets and Faaij, 2007; IPCC, 2007; Anttila et al., 2009). The Anttila et al. (2009) study estimates a primary forestry residue potential ranging from 5 EJ/yr to 9 EJ/yr globally, including logging residues from present cuttings, the stemwood from supplementary cutting, and the logging residues from supplementary cutting. This range is low compared to other estimates for the year 2050, e.g., 28 EJ/yr (Smeets et al., 2007) or 12–74 EJ/yr (IPCC, 2007). The difference largely results from the inclusion or not of secondary and tertiary residues. Secondary and tertiary residues can be 3–5 times higher than the primary residues Smeets et al. (2007). To estimate the regional potential for all forestry residues, the data from Anttila et al. (2009) were increased by a factor of four to include the other residues (see Table 7.33).

7.7.3.3 Aquatic Biomass

Land use constraints, competition with food, and demand for biomass with high caloric value have made marine-based biomass an attractive alternative supply option over past years. However, land-based biomass systems are far more developed than sea-based systems whose economic viability has yet to be determined. In addition, estimates of algae energy potentials are scarce and quite uncertain, as most are extrapolations from small-scale (pilot) projects.

Aquatic biomass can be grouped into three categories: microalgae, seaweed, and sea grass. The production of aquatic biomass depends on various factors, such as irradiation, limpidity, temperature, sea conditions, and nutrients. Naturally, regional variations can be significant. Promising concepts for aquatic biomass include (1) land-based open ponds for microalgae, (2) horizontal lines between offshore infrastructure, e.g., wind farms for macroalgae (seaweed), (3) vertical lines near shore in densely used areas and nutrient-rich areas for macroalgae, and (4) macroalgae colonies at open sea up to 2000 km offshore (Florentinus et al., 2008). The total potential at a global scale is assessed at over

Table 7.33 | Estimate of technical potential of forestry residues in comparison with the forestry residue estimates from IPCC (2007), Anttila et al. (2009), and Smeets et al. (2007). Anttila et al. (2009) only includes primary residues.

Regions	Technical potential		IPCC AR4		Anttila et al. (2009)		Smeets et al. (2007)	
	Low	High	Low	High	Low	High	Wood harvest	All forestry residues
	[EJ]							
North America	5.8	11.9	3	11	1.5	3	2	10
OECD Europe	3.8	6.7	1	4	1	1.7	1	5
Japan + Australia + NZ	1.1	1.8	1	3	0.4	0.7	0	2
FSU + Eastern Europe	3	5.4	2	10	0.8	1.3	1	3
Latin America	1.6	3.8	1	21	0.4	0.9	1	3
Africa	0.6	1.4	1	10	0.1	0.3	0	0
Centrally planned Asia	2	3	1	5	0.5	0.7	2	6
Other Asia	0.7	1.3	1	8	0.1	0.1	1	1
Middle East	0	0.1	1	2	0	0	0	0
World	18.8	35.1	12	74	4.7	8.7	8	30

6000 EJ/yr. For the first three production categories the potential is 235 EJ/yr. Algae production at open sea has by far the largest potential, but is also the most complicated.

7.7.4 Summary of Global Technical Bioenergy Potential Estimates

Table 7.34 summarizes the various bioenergy supply potentials. The total global technical bioenergy potential for the year 2050 varies between 162–267 EJ/yr – a range considerably lower than previous analyses, which suggested global bioenergy potentials around 400 EJ/yr by 2050. This lower estimate largely results from lower expectations about the overall potential to grow dedicated bioenergy crops. In general, these potentials are highly uncertain. However, it is increasingly recognized that land demand for food production and feed supply, urban and infrastructure areas, biodiversity conservation, and the need to maintain a favorable GHG emission balance (i.e., no deforestation) pose definite limits on land availability for bioenergy production. Moreover, new studies suggest that the productive potential of land areas available for bioenergy is much lower than previously thought (Johnston et al., 2009).

7.7.5 Economics of Bioenergy Production and Supply Cost Curves

7.7.5.1 Supply Cost curves of Bioenergy Crops

Costs for energy crops include capital costs for equipment and infrastructures, plant and seed material, land rent, labor costs, energy (e.g., for drying and transport), material expenditures (e.g., fertilizer and water), and storage. Bottom-up studies indicate that costs for drying

Table 7.34 | Summary of global technical bioenergy supply potentials in 2050.

Resource	MIN [EJ]	MAX [EJ]	Comments
Dedicated bioenergy crops	44	133	High uncertainty, depends on yields, diets, technology, and climate change
Crop residues	49		Soil conservation issues need to be addressed; GHG balance might depend on soil carbon balance (currently poorly understood)
Manures	39		Relatively small uncertainty and few, if any, environmental issues
Municipal solid waste	11		Relatively small uncertainty and few environmental issues
Forestry	19	35	Competition for other uses may reduce availability of residues
Total, excluding aquatic	162	267	

and storage are the least certain and therefore are often excluded (de Wit and Faaij, 2010). Future production costs will depend on productivity developments, labor costs, land rental, and experience. Lignocellulosic energy crops have the lowest cost of all energy crops because of the low input requirements and relative high productivities. Costs in Europe range from 1.5–4.5 €/GJ,[19] while other crops, such as starch, sugar, and oil crops, cost between 5–15 €/GJ (de Wit and Faaij, 2010). Similar results are found for the United Kingdom (E4Tech, 2009), where costs for energy crops in 2030 are generally estimated at 2.0–3.5 £/GJ. The cost curves for Europe and the United Kingdom as reported by de Wit and Faaij (2010) and E4Tech (2009) are all relatively flat in the longer term. In the EU about 90% of lignocellulosic crops are available at costs between 1.5–3 €/GJ by 2030, with the remaining amount between 3–6 €/GJ (de Wit and Faaij, 2010). For the United States projected costs for delivered biomass in 2025 are very low for forest residues (US$_{2005}$1.0–2.0/GJ) for limited amount (up to 1.5 EJ/yr). For larger supplies (up to 7 EJ/yr) energy crops have costs around US$_{2005}$3.5–4.0/GJ (Chum et al, 2011).

Projected costs for energy crops include capital costs for equipment and infrastructures, plant and seed material, land rent, labor costs, energy (e.g., for drying and transport to conversion facilities), material expenditures (e.g., fertilizer and water), and storage. Bottom-up studies indicate that costs for drying and storage are the least certain and therefore are often excluded (de Wit and Faaij, 2010). Future production costs will depend on productivity developments, labor costs, land rental, and experience. Lignocellulosic energy crops have the lowest cost of all energy crops because of the low input requirements and relative high productivities. Estimated costs in Europe range from 1.5–4.5 €/GJ and a production volume of about 7 EJ/yr, while other crops, such as starch (3 EJ/yr), sugar (6 EJ/yr), and oil crops (3 EJ/yr), cost between 5–15 €/GJ (de Wit and Faaij, 2010). Similar results are found for the United Kingdom (E4Tech, 2009), where costs for energy crops in 2030 are generally estimated at 2.0–3.5 £/GJ. The lignocellulosic cost curves for Europe and the United Kingdom as reported by de Wit and Faaij (2010) and E4Tech (2009) are all relatively flat in the longer term. In the EU about 90% of lignocellulosic crops are available at costs between 1.5–3 €/GJ by 2030, with the remaining amount between 3–6 €/GJ (de Wit and Faaij, 2010).

Long-term global economic supply potentials for dedicated energy crops were assessed by Hoogwijk et al. (2009). This study analyzed four distinct land-use scenarios using a production function approach with costs for labor, capital, land, and transport to the main distribution centers. Regional differences reflect differences in productivity, labor, and land costs. The bioenergy production cost ranges shown in Figure 7.24 are based on the production cost distribution of that land-use scenario with its cost distribution representing the middle of all the scenario distributions and applied to the technical potentials

---

19   Note that the market for lignocellulosic crops in particular is still small and costs in many literature sources are based on desktop data.

Table 7.35 | Technical and economic potential of energy crops.

	Technical potential		Potential cost categories			
	Low [EJ]	High [EJ]	<1 $/GJ [%]	<2 $/GJ [%]	<4 $/ GJ [%]	>4 $/ GJ [%]
USA	1.6	3.3	0	57	80	20
CAN	2.73	9.87	0	77	85	15
WEU	1.7	4.85	0	44	94	6
EEU	0.69	2.74	0	89	89	11
FSU	3.09	8.61	0	77	79	21
NAF	0.72	2.6	0	50	50	50
EAF	0.34	0.4	20	40	40	60
WCA	11	34.34	17	67	83	17
SAF	0.06	1.12	0	0	0	100
MEE	0.61	1.52	0	0	33	67
CHN	2.76	7.38	0	0	46	54
OEA	1.74	4.58	0	17	50	50
IND	0.12	0.65	0	17	50	50
OSA	1.45	7.11	0	17	50	50
JPN	3.09	6.89	0	44	94	6
OCN	4.12	18.06	20	80	83	17
PAS	5.64	10.91	20	80	83	17
LAC	2.23	7.74	0	27	68	32
TOTAL	43.69	132.67				

Source: Adapted from Hoogwijk et al., 2009.

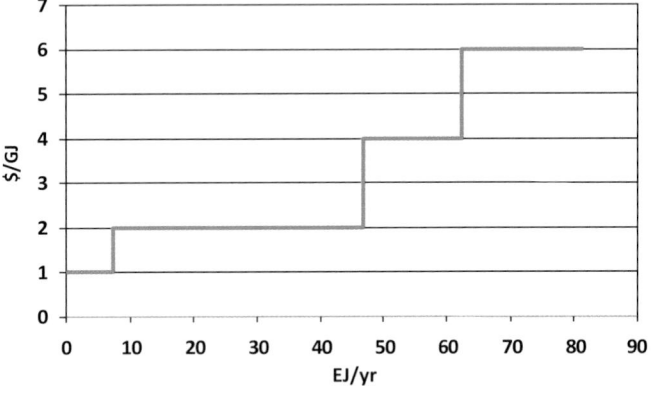

Figure 7.24 | Bioenergy crops supply cost curve (average biomass supply of Table 7.35).

summarized in Table 7.35. (The ranges are lower than those reported in the aforementioned studies.)

Figure 7.24 shows the aggregate global biomass supply cost curve based on the 'mean' regional supply volumes and the production-cost distribution data of Table 7.35.

7.7.5.2 Supply Cost Curves of Agricultural Residues and Municipal Solid Waste

In general, the cost of biomass residues is site-specific. Approximate values of residue costs are assumed for the purpose of this study. Based on Haq and Easterly (2006), the following cost values, including transportation and payment for a farmer premium, are assumed to be valid for the United States and all other developed countries and regions for the year 2005: 46 $/t for corn stover, wheat straw, and field residues of sorghum, barley, oats, rye, and cotton field trash, and 41 $/t for rice straw and processing residues. The cost of all other crop residues is assumed to be 46 $/t.

Based on Purohit et al. (2002), the cost of residues in India in 2005, including transportation, is estimated to be 25 $/t. The average cost of cellulosic biomass, including transportation, is assumed to be 22 $/t in China (Yang et al., 2005). The estimated cost of field residues in Nigeria has been reported as 23.1 $/t (Jekayinfa and Scholz, 2007). Based on the cost values of India, China, and Nigeria, the average cost of all agricultural crops in all developing countries, except India and China, in the year 2005 has been assumed to be 23.5 $/t.

The farm-gate cost of residues in United States mostly reflects the cost of their collection, i.e., delivering them in bales at the edge of the field. It has been projected that improvements in collection technology will significantly reduce the collection cost of residues below the 2005 level (Walsh, 2008). Farmer premiums, however, are likely to increase in the future. Based on the above considerations, it is estimated that the cost of residues in 2050 will remain at the 2005 level for all developed countries.

In developing countries, residues collection in 2050 is assumed to be based on machines similar to those used in developed countries; considering that the cost of labor involved in collection and transportation would remain lower in developing countries, the costs of residues in these countries in 2050 is assumed to be 80% of the estimated costs for developed countries.

MSW may have negative costs because of tipping fees. The average fee per tonne in the United States was $33.74 in 2002 and $34.29 in 2004 (Repa, 2005); the tipping fee was assumed to be 35 $/tonne in 2005. Based on the fees of six sites in Canada (Statistics Canada, 2005), the average fee for Canada is about 49 $/t. The following average 2005 tipping fees per tonne were assumed: Europe, $100; Japan, $150; OCN region, $32; China, $20 (based on the range 10–30 $/t, reported by Themelis and Themelis (2007)); developing countries of Asia, $5; and Latin America, $15; while India and the remaining regions have no tipping fees.

The tipping fee to breakeven in case of plasma gasification of MSW in Canada has been estimated to be $35 (Young, 2006); the value is likely to be less in the future because of improvements in technology. For this

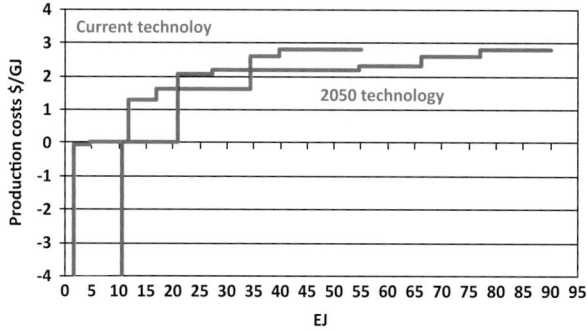

Figure 7.25 | Aggregate supply cost curve[20] for MSW, animal wastes, and crop residues. Negative costs are avoided tipping fees.

assessment, uniform tipping fees are $US_{2005}\$40/t$ in 2050 and have been assumed for all countries and regions by 2050, while animal manures are assumed to be available at zero cost. The supply cost curve for agricultural residues and MSW is shown in Figure 7.25.

7.7.5.3 Economic Potential of Bioenergy from Forestry

Forest residues are currently traded internationally, mostly in the form of wood pellets. The dominant raw material for wood-pellet production has traditionally been processing residues as sawdust. The prices of wood pellets for Europe range from 170–270 €/t (10–16 €/GJ) in a two-year period (UNECE/FAO, 2009) and for the Russian market from 95–165 €/t (6–10 €/GJ) (Junginger et al., 2009). The pellet market is integrated in the entire wood system and therefore influenced by the economic crisis, i.e., some wood-pellet producers have had problems with raw material supply (UNECE/FAO, 2009).

No global potential estimates include the production costs of forestry residues or pellets. Some assessments of the costs in Europe indicate levels ranging from 2.2–7.4 $US_{2000}\$/GJ$ (Lindner et al., 2005). Extensive cost curves at a country level from (Asikainen et al., 2008) taking transport distance as the main varying parameter, result in comparable levels of around 2.5–7.5 €/GJ. For Denmark, costs have been assessed below 1 $US_{2000}\$/GJ$ (Nord-Larson and Talbot, 2004). Estimates for Japan indicate costs in the order of 10–35 $/GJ (Yoshioka et al., 2006).

No studies include cost reductions caused by technological progress. Some cost curves for residues are known at the country level, e.g., for the United States (Walsh, 2008) and for some European countries (Asikainen et al., 2008). In general, costs are largely a function of transport distance, but a detailed breakdown is not provided. For the global potential it is assumed that 100% can be delivered at costs below 50 €/GJ and 10% at costs below 1 €/GJ.

7.7.5.4 Cost Estimates of Aquatic Biomass

The costs for algae cultivation are still high. Florentinus et al. (2008) estimate costs for biomass in the range of 300–700 €/t or even 1000 €/t of dry matter, or 18–60 €/GJ assuming a lower heating value of 16 GJ/t and significant technological improvements.

7.7.6 Social and Environmental Aspects of Bioenergy Use

Bioenergy is either produced by planting dedicated bioenergy crops or by using by-products, residues, and wastes from agriculture, forestry, food processing, and other economic activities (see Figure 7.23). These pathways can have environmental effects that are fundamentally different (Cherubini et al., 2009). The two pathways are therefore discussed in turn.

7.7.6.1 Environmental and Social Effects of Bioenergy Crop Plantations

Growing dedicated bioenergy crops can have positive and negative environmental and social impacts. For example, growing bioenergy crops increases the demand for agricultural products and creates income in the agricultural sector. However, demand for bioenergy crops may lead to rising agricultural prices (World Bank, 2008b) and reduce affordable food supply, in particular for the poor.

Likewise, bioenergy can help to reduce GHG emissions, but when emissions from direct and indirect land-use changes (e.g., deforestation) are included, GHG emissions of bioenergy can be large, indeed even higher than those of fossil-fuel based alternatives (Searchinger et al., 2008; WBGU, 2008).

To establish how much bioenergy can be produced without harming the environment, and without adverse social and economic impacts undermining gains, requires a systems approach that takes into account the relevant interactions between various land uses and socioeconomic functions of biomass (food, fiber/materials, energy).

The environmental effects of bioenergy plantations are species-specific as well as site-specific. Perennial grasses such as switchgrass, miscanthus, and short-rotation coppice are ecologically less demanding than food crops in terms of impacts on soils, soil erosion, biodiversity, nutrient leaching, pesticide application, etc. (Cherubini et al., 2009). Conversion of grasslands or forests for bioenergy production may cause dire ecological consequences. The environmental impacts of bioenergy depend only to some extent on the specific bioenergy plant and more on the previous use of the land on which it is planted (Gibbs et al., 2008).

Land demand is a central environmental issue associated with the expanding use of bioenergy crops (Sagar and Kartha, 2007; Firbank,

---

20 See footnote 7.

2008). Once considered a local environmental issue, it is increasingly recognized that land use has become a pervasive driver of global environmental change (Foley et al., 2005). The Millennium Ecosystem Assessment (MEA, 2005) concluded that land use has already reduced the ability of ecosystems to deliver vital ecosystem services. Large-scale bioenergy plantations would increase humanity's pressures on global terrestrial ecosystems and contribute to biodiversity loss (Haberl et al., 2009).

Bioenergy plantations usually reduce the amount of carbon stored in ecosystems compared to undisturbed ecosystems (Schimel, 1995; Watson et al., 2000), while land conversion and biomass harvest can lead to loss of soil organic matter and significant net emissions (Pulleman et al., 2000; Lal, 2004; see also Chapter 20). In addition to the emissions from land use, GHG emissions also result from the energy required to produce bioenergy, and from the production and use of fertilizers, pesticides, and all other activities of the full process chain. Gibbs et al. (2008) concluded that the expansion of bioenergy into carbon-rich ecosystems (e.g., forests) leads to 'carbon payback times' of decades to centuries, whereas GHG avoidance is almost instantaneous if degraded or already cultivated land is used. For areas with lower embedded carbon such as grasslands or savannahs and for high yield biomass feedstocks payback periods can be less than a decade.

Water demand of bioenergy crops may cause environmental and social problems. Humans already use or regulate more than 40% of all freshwater resources globally (MEA, 2005). Global water demand grows by 10% per decade adding to water supply stress in many regions. Agriculture currently demands some 70% of all freshwater use (UNEP, 2009), of which 2% was used for bioethanol production (WBGU, 2008).

The production of bioenergy uses 70–400 times more water per unit of energy than other primary energy carriers, excluding hydropower, and ranges from 24–143 m^3/GJ, with Jatropha having the largest water footprint among 12 bioenergy crops analyzed in a recent study (Gerbens-Leenes et al., 2009). The amount of water per unit of energy is highly dependent on the crops used, the efficiency of the cropping system, and the local hydrological and soil conditions. On a positive note, bioenergy plantations in marginal areas may alleviate water-related problems, i.e., local water harvesting and run-off collection may reduce water-related erosion (Berndes, 2008).

Bioenergy produced on currently grazed lands can have large-scale impacts on livestock-rearing subsistence farmers. These may be positive or negative, depending on the implementation strategy. Large-scale bioenergy plantations owned and operated by international, vertically integrated corporations tend not to benefit the local farming communities as most of the revenue is generated in the production stage that involves sophisticated biochemical conversion technologies (Sagar and Kartha, 2007). However, small-scale locally owned and operated plants,

as well as sustainability certification systems, might help ensure that benefits accrue to the local farming communities (Lewandowski and Faaij, 2006).

7.7.6.2 Environmental and Social Implications of Using By-products, Residues, and Wastes

In contrast to bioenergy crops, bioenergy production from agricultural by-products, residues, and wastes does not (1) require additional land or land use changes, (2) compete with food or fiber production, (3) affect agricultural and food prices, and (4) require large amounts of additional scarce inputs such as freshwater (Berndes, 2008). On the contrary, using biomass residues may help alleviate energy shortages, reduce landfill requirements, and create employment opportunities. However, there may also be negative environmental effects, depending on the respective biomass flow as well as on technology and management.

The agricultural residue straw can deliver substantial amounts of energy. However, straw plays a vital role for soil fertility, soil carbon pools, and the mitigation of water and wind erosion (Lal, 2005, 2006; Wilhelm et al., 2007). WBGU (2008) assumes about half of all crop residues could be used to produce bioenergy without compromising soil fertility. Still, the science underlying such assumptions is weak and more research is required. Removal of crop residues for energy production could also affect the GHG balance of cropping systems.

The removal of biomass from forests, including forest residues, may affect forest ecosystems, e.g., the coarse woody debris is essential for biodiversity and ecosystem functioning (Krajick, 2002; Shifley et al., 2006), and forest conservation objectives – the use of fuelwood and forestry residues – have to be jointly optimized.

Well-managed use of animal manures for biogas production can have significant positive environmental and social impacts. It reduces CH_4 emissions,[21] while returning most plant nutrients and parts of the carbon back to the soil, thereby mitigating land degradation and helping to maintain soil fertility (Rajabapaiah et al., 1993; Stinner et al., 2008). Moreover, energy from biogas can help to substitute for traditional biomass energy that has tremendously negative health and environmental effects and currently contributes to millions of premature deaths from respiratory diseases that result from indoor pollution (Jaccard, 2005).

Using MSW for energy production lowers CH_4 emissions from waste deposits (landfills). In the absence of effective air-pollution control technology, however, incineration of MSW can result in large

---

21   Conversion of animal manures into biogas plants and subsequent application of the residues as fertilizer reduce CH_4 compared to the storage and direct application of manures (Bhattacharya et al., 1997; Clemens et al., 2005).

emissions of toxic pollutants such as dioxin. Tight air-pollution regulation that vigorously enforces the use of the most advanced abatement technologies to reduce toxic emissions is required to minimize possible negative environmental effects from the combustion of MSW (McKay, 2002).

7.7.7   Summary

The global theoretical biomass is undoubtedly large at 1100 EJ/yr, not including the potential from aquatic biomass. However, while this value is more than double the current global energy consumption, harvesting even half of this biomass potential would result in severe adverse impacts on biodiversity, resilience, and sustainability of the Earth's ecosystems, humans included. How much biomass can or should be used for energy purposes is therefore less a question of the available theoretical potential than of ecological sustainability and socioeconomic desirability.

Thus, to quantify the harvestable portion of biomass potential, it is important to understand (a) the annual rates at which biomass for energy becomes available and (b) the competing land uses, such as food production, settlement, and infrastructures. Regarding biomass flows, these can be in the form of crops grown specifically for energy use (on various land types), residues (agricultural and forestry), wastes (agricultural and municipal), and by-products. The global technical potential ranges between 162–267 EJ/yr. Although the greatest portion of this technical potential is provided by dedicated bioenergy crops, it is also these crops that potentially have the greatest socioeconomic effects. However, use of residues, manures, and wastes could still have adverse effects on ecosystems, although with good management should be able to provide more benefits than costs.

Regarding competing land uses, it is important that both the technical and economic estimations of potential are significantly lower than previous assessments, as it is increasingly recognized that land demand for food and feed production, urban and infrastructure areas, and biodiversity conservation, as well as the need to maintain a favorable GHG emission balance, pose definite limits on land availability for bioenergy production.

As aquatic biomass is not limited by freshwater availability, this biomass feedstock could exceed terrestrial-based systems by orders of magnitude. However, as aquatic biomass systems are still in their infancy, this study omits their potentials in the final estimates.

Biomass is the most diverse energy feedstock, and it is also the most integrated into our everyday lives. For this reason, the potential exploitation of biomass for energy relies much more on the interplay between the demand for affordable energy, food and water, our ability to adapt to climate change, and social development.

7.8   Wind

7.8.1   Overview

While the total kinetic energy of the Earth's winds is enormous, the exploitable technical and economic potentials of wind energy depend on technology development on the ground, at sea and potentially at high altitudes. It is impossible to estimate a renewable resource potential without including explicit assumptions on the technology's technoeconomic performance profile. These technologies are discussed further in Chapter 11.

The theoretical potential delineates the total kinetic energy within the troposphere. The technical potential defines the upper limit of wind energy that can be harnessed effectively by technologies. Constraints include the theoretical maximum efficiency of power extraction by a wind turbine (the Betz conversion limit), as well as identifying locations where the average wind speed is strong enough for wind turbines to operate.

Finally, the economic potential takes into account limitations on the potential locations for wind converters, including costs, conflicting land uses, and the distance limitations of electricity transmission. As wind-technology performance has been quite dynamic, with substantial technology learning over recent years, wind's economic potential is certainly a moving target (see Chapter 11).

7.8.2   Theoretical Potential

The total energy in the winds at a given instant is defined as the sum of the kinetic energy of each air molecule in movement in the troposphere, from the surface all the way up to the top of the tropopause. This has been estimated as 11.8–13.9 J/m^2, corresponding to 604–711 EJ on average at each instant in time (Peixoto and Oort, 1992; Li et al., 2007).

What actually constrains wind-power extraction is the natural rate at which kinetic energy is dissipated in the atmosphere via friction. This is a difficult number to calculate directly, but it can be derived from the global energy budget of the Earth and other theoretical considerations. Estimates vary by orders of magnitude, from 113.5 ZJ/yr (3600 TW) (Lorenz, 1967) to 11,700 EJ/yr (370 TW) (Hubbert, 1971). A more recent estimate (Sørensen, 2004) suggests a natural dissipation of 63–160 ZJ/yr (2000–5100 TW).

7.8.3   Technical Potential

The estimation of technical potential adds further constraints based on the fundamental limitations of the application of turbine technologies. The first technical limit is a safety factor that prevents the significant modification of the Earth's climate, i.e., the fraction of the theoretical potential which can be exploited without adversely affecting natural global equilibria. Not all the energy available in wind can actually be extracted, as otherwise the air flow would stop and the air mass would pile up where the wind turbines are. Gustavson (1979) suggests a factor of no more than 10%, which, using Sørensen's (2004) theoretical potential estimate, would bring an initial technical potential maximum to 200–510 TW, or 6300–16,000 EJ/yr.

The Betz limit, the theoretical maximum efficiency of power extraction by a wind turbine, is 59% of the total power available in the wind going through the area swept by the turbine blades (Burton et al., 2001). The most efficient wind turbines nowadays can achieve efficiencies of about 50%. Even taking the Betz limit into account, the technical wind power output potential is 3700–9500 EJ/yr.

A further limitation is determining locations, both on the horizontal and vertical plane, where sufficient wind energy exists to turn wind turbines. The power in a wind stream is proportional to two factors: the cube of its speed (V^3) and the air density of the stream (ρ). Thus, doubling the wind speed increases the power by a factor of eight and decreasing the air density by 50% decreases the power by 50%. To take both factors – wind speed and air density – into account, the variable generally used is wind power density ($0.5 \, \rho V^3$), measured in W/m^2, which indicates how much power is available at a given site to drive a wind turbine, per unit area swept by blades perpendicularly to the wind.

Wind power density is not evenly distributed in the atmosphere. It varies by altitude and geographical location. In general, more power is available at high altitudes than near the ground. Also, near the surface more power is available over water than over land. Lastly, wind varies not only with aboveground height and terrain type, but also with the time of day and season. Figure 7.26 shows a map of median wind power density over a year at 80 m and 10,000 m above the ground. These two altitudes were selected because the former is representative of traditional technologies (i.e., wind turbines 80 m high), and the latter is of interest for emerging wind power technologies, e.g., kite- or rotor-based high-altitude devices (Canale et al., 2007; Roberts et al., 2007; Archer and Caldeira, 2009).

As wind energy is linearly proportional to the area swept by the turbine blades, and because wind speed generally increases with height, modern wind turbines have become taller. Hubs are now typically located at 80–100 m, and have longer blades (over 70 m in diameter) than those of 10 years ago, when the standard wind turbine was 50 m high with 30 m blades. Such large, tall, and heavy blades cannot be turned by weak winds, but typically need at least a wind speed of 3–4 m/s. As such, wind power is technically feasible (for large-scale applications) only in locations with average wind speeds at 80 m of 6.9 m/s or greater, corresponding to wind power class 3 or greater (Table 7.36). Thus, a threshold of 6.9 m/s is applied in this report to identify 'windy' sites. Classes of wind power density for three standard wind-turbine hub heights (10, 50, and 80 m) are listed in Table 7.36.

The global technical wind power potential from windy sites near the surface (i.e., 80 m wind speed ≥ 6.9 m/s) over land and offshore has been estimated as 72 TW (Archer and Jacobson, 2005), corresponding to a global annual energy from winds of 2256 EJ.

To extract this potential, a total area of about 16.5 million km^2 would be required (i.e., 1.5 times the area of China), covered with six 1.5 MW wind turbines per square kilometer. Figure 7.27 indicates locations where 80 m wind speeds exist in Europe and North America. Similar maps were also developed by the US National Renewable Energy Laboratory for land-based sites worldwide at 50 m above ground level (Denholm and Short, 2006).

Using the same wind speed data as Archer and Jacobson (2005), 8144 reporting sites are considered in this report, with the assumption that they are representative of the wind conditions in their areas. The fractions of windy sites can be used as proxies for the fractions of windy land in each region. Because the technical potential depends on the size of the turbines used, two modern turbines are used as benchmarks in this report, namely a 1500 kW (77 m diameter) turbine and a 5 MW (126 m diameter) turbine. Their rated power and blade size are used to calculate the yearly average wind power output as a function of the yearly average wind speed via the method in Jacobson and Masters (2001). Table 7.37 summarizes the values of relevant parameters in each of the 18 regions, including the range of technical potentials obtained from the two wind turbines. The data coverage, i.e., the number of reporting sites per unit area, can be used as a proxy for reliability of the calculated potentials. The greater the data coverage, the more representative are the data and wind power estimates. Japan and Europe have the best data coverage (more than three reporting sites per Mha), whereas Africa and some regions in South Asia have the lowest (less than 0.3 per Mha).

Small-scale wind turbines (~10 kW rated power) that can extract winds at modest wind speeds for single household applications are not considered in this study. Also, the intermittency of the wind resource is not considered in these estimates.

7.8.4 Practical Potential

The practical wind potential is calculated from the technical potential taking into account the following limitations:

- conflicting land uses – urban areas, protected natural areas, military exclusion areas, etc., – cannot be covered with wind turbines;

- remoteness – too remote locations (i.e., mountainous areas) cannot be reached via transmission or distribution electric lines.

All these constraints depend on the geographical, political, and socio-economic conditions of each location and are therefore difficult to

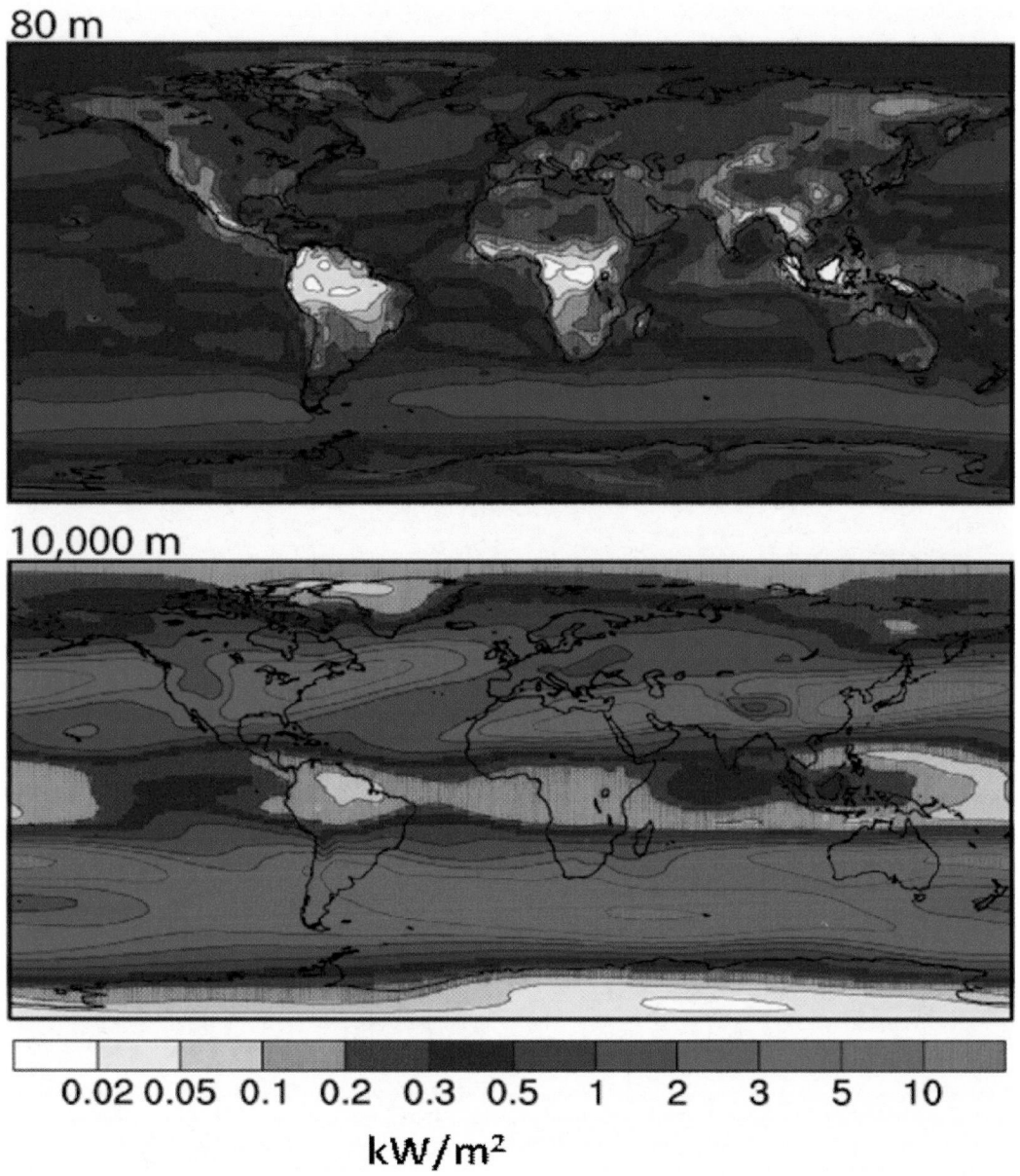

Figure 7.26 | Map of the median wind power density available in the winds near the surface (at 80 m) and near the jet streams (10,000 m) during 1979–2006. Source: Archer and Caldeira, 2009.

evaluate on a global scale. Hoogwijk et al. (2004) used suitability factors, i.e., the fraction of areas that can be devoted to wind harnessing, between 0% (bioreserves and tropical forests) and 90% (savannah). They calculated a practical onshore wind potential of 96 PWh/yr, or 346 EJ/yr, corresponding to about 20% of the technical output potential and over five times the total electricity production worldwide of ~73 EJ in 2008 (IEA, 2010a). The low-end values of the ranges for the practical potential shown in the last column of Table 7.37 were obtained by including fewer windy sites (determined with a cost analysis).

Offshore energy is excluded in this study because insufficient wind speed data are available to justify a proper analysis. Furthermore, suitable locations for offshore wind are dependent on factors such as sea conditions and shipping lanes, etc. Nevertheless, previous studies have assessed the global offshore potential at 37 PWh/yr at 50 m height (Hoogwijk, 2004).

Both the technical and the practical potentials of high-altitude wind power are zero at the moment, because it is still a prototype technology.

7.8.5   Environmental and Social Implications

The exploitation of wind energy has impacts on the environment that depend on the conversion technologies and their locations. These include disturbances to delicate ecosystems and impacts on birds and

Table 7.36 | Definition of wind power classes.[a]

Wind power class	10 m above ground		50 m above ground		80 m above ground	
	Wind power density	Wind speed	Wind power density	Wind speed	Wind power density	Wind speed
	[W/m²]	[m/s]	[W/m²]	[m/s]	[W/m²]	[m/s]
1	<100	<4.4	<200	<5.6	<250	<5.9
2	100–150	4.4–5.1	200–300	5.6–6.4	250–375	5.9–6.9
3	150–200	5.1–5.6	300–400	6.4–7.0	375–500	6.9–7.5
4	200–250	5.6–6.0	400–500	7.0–7.5	500–625	7.5–8.1
5	250–300	6.0–6.4	500–600	7.5–8.0	625–750	8.1–8.6
6	300–400	6.4–7.0	600–800	8.0–8.8	750–1000	8.6–9.4
7	>400	>7.0	>800	>8.8	>1000	>9.4

[a] Wind speed is assumed to have a Rayleigh distribution and to increase with height according to the power law with a friction coefficient of 1/7 and air density of 1.225 g/m³.

Source: adapted from AWEA, 2010.

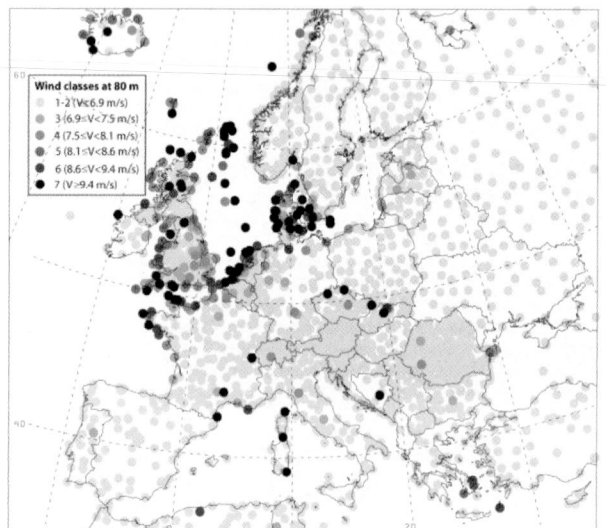 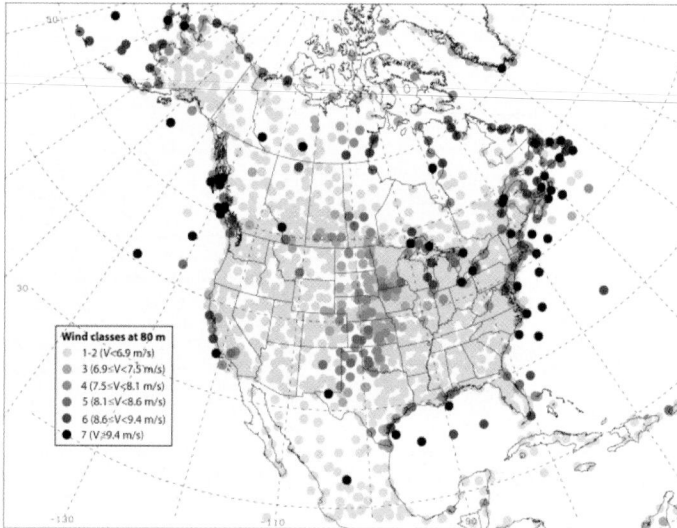

Figure 7.27 | Maps of annual average 80 m wind speed derived from observations in Europe and North America. Only locations in class 3 or greater (green and black dots) are windy enough for wind power. Source: Archer and Jacobson, 2005.

other wildlife. In addition, there are social concerns about noise pollution and landscape aesthetics. However, as previously mentioned, the exploitation of wind energy resources depends entirely on conversion technologies. The environmental and social impacts of wind turbines are discussed further in Chapters 3, 11, and 20.

7.8.6 Summary

The theoretical potential of wind energy was determined to lie in the range 110,000 ± 50,000 EJ/yr. Adding technical constraints, including fundamental conversion maximums, as well as wind power density minimums, a technical potential was estimated as 1500–5700 EJ/yr. Including the winds blowing across Antarctica adds approximately another 2100 EJ/yr. Finally, practical limitations on the technical potential

were imposed, which took converter performance characteristics, land use conflicts, and electricity transmission into account. The practical potential for onshore wind utilization was thus found to be 250–1200 EJ/yr (or about 20,000–100,000 TWh of electricity per year).

7.9 Solar

7.9.1 Overview

Solar energy is by far the most abundant energy resource on Earth and is ubiquitous over the Earth's surface. To put in perspective the enormity of the sun's energy, the average irradiation that hits the Earth's surface in one hour is about equal to that of the energy consumed by all human activities in a year (IEA, 2010c). Solar energy conversion technologies

Table 7.37 | Technical and practical wind (input) potentials for the 18 GEA regions.

Region	Area[a]	Fraction of area in class ≥3[b]	Average wind speed at sites in class ≥3[b]	Number of reporting sites[b]	Technical potential, land and offshore	Practical potential,[a] land only
	[Mha]	[%]	[m/s]		[EJ]	[EJ]
USA	925	17.2	8.04	1583	202.1–216.6	10.8–75.6
CAN	950	27.4	8.46	580	358.2–388.6	28.8–68.4
WEU	372	19.1	8.63	1459	100.8–109.9	3.6–14.4
JPN	37	8.3	7.87	266	3.8–4.0	0–0.4
OCN	838	24.1	8.68	531	289.1–315.5	3.6–50.4
EEU	116	3.1	9.12	449	5.6–6.1	0–1.4
FSU	2183	3.1	7.83	799	83.4–88.8	7.2–57.6
MEE	592	0.5	10.36	182	6.0–6.8	0–7.2
NAF	574	4	7.7	174	27.5–29.1	0–10.8
EAF	583	7.1	8.68	70	59.6–65.0	0–10.8
WCA	1127	4.8	8.71	126	77.2–84.4	0–0.7
SAF	676	4.6	8.68	197	44.2–48.2	0–0.7
PAS	442	2.2	8.23	462	12.6–13.6	0–0.2
CHN	960	2.8	8.04	434	33.8–36.2	n/a
OEA	243	1	7.63	100	2.8–3.0	n/a
IND	329	<0.01	n/a	97	<0.1	n/a
OSA	179	<0.01	n/a	52	<0.1	n/a
LAC	2030	9.4	8.33	583	257.5–278.5	18.0–36.0
World	11,990	12.3	8.39	8144	1564–1694.3 [c]	72.0–345.3

[a] From Hoogwijk et al. (2004).
[b] Derived from data in Archer and Jacobson (2005).
[c] This value does not include Antarctica. Including Antarctica (55 reporting sites), the world technical potential is 2256.2 EJ (or 7650 EJ input equivalent).

have the capability to provide electricity generation as well as a variety of energy services, including heating, cooling, and natural lighting.[22]

As the sun's radiation travels through the Earth's atmosphere toward the surface, it is reduced because of the reflection, scattering, and absorption of particles in the atmosphere. The fraction of radiation reflected back into space is considered the atmospheric albedo, or reflection coefficient, and is estimated to be between 30–35%. Thus, sunlight hits the Earth's surface both directly and indirectly. On a clear day, the direct irradiation accounts for between 80–90% of total irradiation, while on a foggy day, the direct irradiation approaches 0% and the ambient light is made up of indirect, i.e., diffused or scattered, light. As a result of the ubiquity of direct and diffuse radiation, photovoltaics (PV) and solar thermal collection systems can literally be placed almost anywhere on the surface of the Earth to generate electricity and heat, respectively.

As with other renewable resources, the availability of solar energy does not determine its role in the global energy spectrum; rather, it is a matter of the conversion technologies and their market competitiveness. However, as it is impossible to assess technical or economic potentials without a basic presumption of technology performance, this chapter aggregates technical potentials with respect to two technologies that directly contribute to the capture and application of solar energy: concentrating solar power (CSP) and PV providing electricity (see Chapter 11). Conversion of solar energy into heat can be quite straightforward, as any object placed in the sun will absorb some thermal energy, however, it is important to note that certain techniques and technologies exist to maximize absorbed energy and minimize losses. The wide variety of different techniques and the diffuse nature of their employment make it very difficult to quantify a technical potential for such technologies.

7.9.2 Theoretical Potential

The theoretical potential is defined as the total solar irradiation reaching the Earth's surface in a year. The incoming solar electromagnetic radiation upon reaching the Earth's atmosphere, the solar constant, is 1366 W/m² (Iqbal, 1983). The Earth's albedo and atmospheric absorption mean only approximately 51% of the incoming radiation reaches the Earth's

22 For a more in depth discussion on the various solar energy conversion technologies, see Chapter 11.

surface. The average irradiance thus reaching the surface is 697 W/m², which multiplied by the Earth's total land surface area results in a theoretical potential estimate of 3,300,000 EJ/yr. Including the oceans, the total solar irradiation reaching the Earth is over 11,000,000 EJ/yr. These estimates, however, do not take into account localized weather conditions. Hoogwijk (2004) estimated the theoretical potential using a bottom-up approach with average irradiance data from the Climate Research Unit, which records empirical irradiation data for the past 30 years. This study estimates the theoretical input potential of solar energy to be lower, at around 630,000 EJ/yr.

7.9.3 Technical Potential

Several factors reduce the practically harvestable potential of this vast energy source. The amount of solar energy available at a given location is subject to daily and seasonal variations. Geographical variation is another important factor. Areas near the Equator generally receive more radiation than those at higher latitudes. Weather (atmospheric) conditions are typically the strongest factors that influence solar energy availability. While solar tracking systems exist to reduce the impact of geographical variations, these can only harvest direct sunlight, which is most affected by weather conditions. As irradiation is often diffuse, large-scale generation of solar energy can carry significant land requirements.

Siting issues combine with all three of the above factors. Not all surfaces are suitable for solar energy conversion, even if they have suitable geography and weather conditions. Land use conflicts (man-made infrastructure, agriculture, or forests), geomorphology, topology, and protected or restricted areas pose siting constraints for larger solar installations. However, building structures provide interesting local siting possibilities for small-scale solar energy use.

These siting issues cause discrepancies in the various estimations of the global technical potential for solar energy, as different models have different criteria and methods for assessing site suitability.

7.9.3.1 Direct Irradiation Potential

The German Aerospace Center (DLR) models the optical transparency of the atmosphere to calculate the direct normal irradiance (DNI) on the ground at any time and any site, by detecting and quantifying those atmospheric components that absorb or reflect the sunlight (clouds, aerosols, water vapor, ozone, and gases). The DNI serves as a reference for CSP systems and is defined as the solar radiation received per unit area by a surface that is always held normal, or perpendicular, to the rays that come in a straight line from the direction of the sun at its current position in the sky (DLR, 2009).

CSP systems track the sun using mirrors and lenses to concentrate the solar energy for the operation of steam cycles. Diffuse solar energy

cannot be used, as it has no uniform direction. Heat loss means that for CSP systems to operate efficiently, they require a minimum input of direct sunlight (IEA, 2010b).

Thus, to plan a CSP plant, both site-specific radiation data and historical data are required to classify the actual ground-measurement radiation, taking year-to-year variations into account. Suitable areas, i.e., with high DNI, are usually found in arid and semiarid areas, between 15° and 40° latitude, and at higher altitudes, where both the air density and scattering absorption are lower (IEA, 2010b). In locations close to the equator it is too wet and cloudy during the rainy seasons, and nearer the poles it is often too cloudy. Figure 7.28 shows the DNI for 12 annual irradiation levels for the entire world.

Figure 7.28 shows that northern and southern Africa, the Middle East, parts of Chile, the southern United States, Mexico, and Australia are all suitable locations for CSP projects regarding DNI. The next constraint, then, is to find suitable land availability for the CSP systems. This is assessed by excluding land areas that are unsuitable because of ground structure, land cover, water bodies, slope, shifting sand, protected or restricted areas, forests, agriculture, urban areas, etc. The exclusion criteria can be strict and non-negotiable, or optional and thus subject to competition. Nevertheless, geographical characteristics and specific exclusion criteria, including competition for land use, jointly determine the viability of potential CSP project sites (Broesamle et al., 2001; Trieb et al., 2009).

Land areas with potentially suitable CSP sites, combined with the local DNI data, generate different classes of areas with a specific level of DNI. Table 7.38 shows the land area available for 13 DNI levels ranging from 1500–2800 kWh/m²/yr for all 18 GEA regions.

Most of the world's regions, except for CAN, EEU, JPN, and PAS, have significant areas with good CSP potential. Although developers typically set the bottom DNI threshold for CSP at 1900 kWh/m²/yr, locations with higher DNIs, everything else being equal, will be more economically attractive. OCN, NAF, MEE, SAF, and OSA are regions where more than 50% of the CSP-suitable areas have DNI levels larger than 2300 kWh/m²/yr. Nevertheless, the global technical potential is found to be 277,000 EJ/yr. This is considered the available input energy for CSP systems worldwide.

In total, 33.6 million km² of land is technically suitable to host CSP systems, corresponding to about 26% of the world's total land area. Without storage, CSP plants require around 2 ha/MWe, depending on the DNI and the technology. Even though the Earth's 'sunbelts' are relatively narrow, the technical potential for CSP is huge. If fully developed for CSP applications, the potential in the southwestern United States alone would meet the electricity requirements of the entire United States several times over. Potential in the Middle East and North Africa would cover about 100 times the current demand of the Middle East, North Africa, and the EU combined (IEA, 2010b).

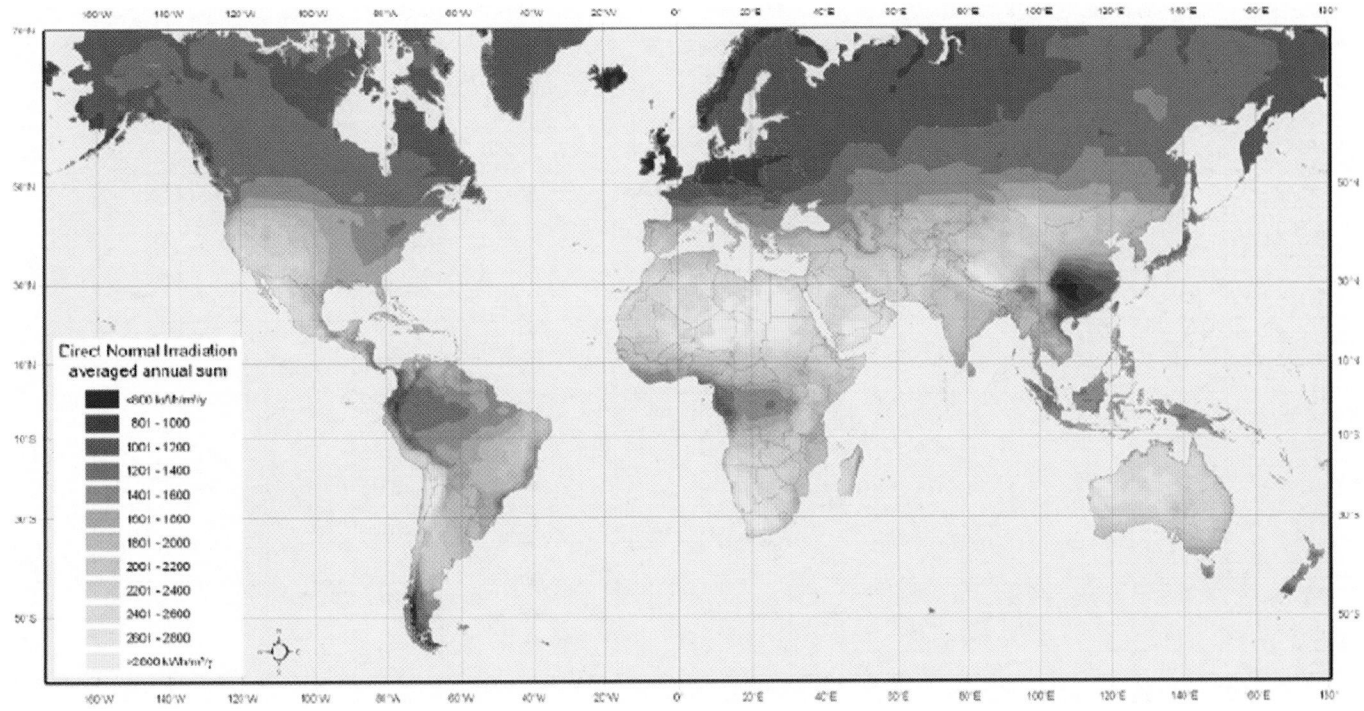

Figure 7.28 | Global map of DNI. Source: data from NASA radiation dataset SSE 6.0 (Trieb et al., 2009).

7.9.3.2   Direct and Diffuse Irradiation Potential

PV systems use semiconductors to convert solar energy directly into electricity. As these systems are able to exploit both direct and diffuse solar radiation, they are much more versatile in their application and a much lower irradiance boundary determines their effectiveness.

There are two types of PV systems, centralized (grid connected) and decentralized (stand alone). Centralized systems can also be defined as large scale (>10 Kilowatt peak (kWp) capacity), are installed on the ground, and are subject to competing land uses. Decentralized (100 Wp to 10 kWp) systems are typically for domestic energy supply and are installed close to settlements, utilities, or industry, and installation surfaces can include rooftops and facades (Hoogwijk, 2004). Although PV systems use diffuse and direct light, their orientation still plays an important role in both centralized and decentralized applications, as they cannot track the sun. Centralized systems can require large land footprints. For small-scale installations, shading from neighboring buildings as well as orientation of the roof are taken into account.

There are numerous studies on the future potentials of PV systems in local, national, and subregional contexts. Only three major studies assess the global technical potential of PV (Hofman, 2002;

Hoogwijk, 2004; de Vries et al., 2007). A comprehensive assessment of the technical and economic potentials at a level close to the 18 GEA regions could not be found in the literature except for that by Hoogwijk. Even in this case the data for several regions had to be supplemented from national assessments (Table 7.39). The potentials are influenced heavily by the definitions and boundaries of the suitability factors, which for centralized systems were based, in part, on Sørensen (1999), and for decentralized systems on Alesma et al. (1993) and IEA (2001).

The input energy technical potential for diffuse and direct irradiation was found to be 12,300 EJ/yr.

7.9.3.3   Economic Potential of Large-scale Solar Plants

Constructing a solar energy plant is dependent on a variety of other economic and market factors other than the irradiance intensity and suitability areas. Factors include the distance of the plant from demand centers, competition from alternative energy sources, and land use competition. Nonetheless, the economics of solar energy depend not only on the energy capturing and conversion technologies, but also on the ability to transport this energy to demand centers. A key challenge for solar energy, as well as other renewable

Table 7.38 | Land areas suitable for CSP by direct normal irradiance.

DNI classes [kWh/m²/y]	1501–1600	1601–1700	1701–1800	1801–1900	1901–2000	2001–2100	2101–2200	2201–2300	2301–2400	2401–2500	2501–2600	2601–2700	2701–2800+	Total
	[EJ/yr]	[EJ/yr]	[EJ/yr]	[EJ/yr]	[EJ/yr]	[EJ/yr]	[EJ/yr]	[EJ/yr]	[EJ/yr]	[EJ/yr]	[EJ/yr]	[EJ/yr]	[EJ/yr]	[EJ/yr]
USA	539	151	383	1003	1173	1100	1371	1669	1280	1872	637	183		11,360
CAN	2	0.3												2
WEU	6	16	27	89	174	169	82	66	14	9	6	2	3	664
EEU	4	0.3	0.4	0.4	0.1	0.4	1	0.2						7
FSU	2508	1684	542	1018	1655	1109	24	29	14	5				8587
NAF			224	552	1052	1472	2157	2613	2556	7357	8787	7176	14,764	48,709
EAF	25	56	142	864	1560	1571	2040	2146	1506	1883	471	74	130	12,468
WCA	830	1245	1434	1805	2724	3914	5776	3906	5067	3985	2757	3736	8435	45,614
SAF	74	107	218	460	891	1023	1301	1781	1918	3201	3984	3024	3846	21,827
MEE	2	5	34	9	56	167	1098	2875	4711	5588	2742	2534	2894	22,716
CHN	346	524	84	177	297	651	1637	3150	2226	878	889	171	242	11,271
OEA	120	619	880	134	29	0	42	377	1700	441	163	20		4526
IND	2	18	99	250	534	616	89	43	60	33	1	9	1	1756
OSA	0.0	7	13	4	12	338	364	466	703	1,070	183	0.0	11	3171
JPN	9	0.4	0.3	0.0										10
PAS	173	118	123	140	64	13	9	3	1	1				645
OCN	19	60	38	155	538	517	1570	2753	6871	11,601	16,299	11,184	3897	55,501
LAC	1300	2492	3082	5246	3820	2598	1925	2114	2077	1732	848	523	1075	28,831
Total	5958	7102	7324	11,908	14,579	15,258	19,486	23,990	30,705	39,654	37,768	28,637	35,298	277,666

Source: Trieb et al., 2009.

resources, is that electricity demand centers are not always situated close to the best sources.

7.9.4 Environmental and Social Implications

The environmental and social implications of solar energy depend on the technologies used and the amount of land covered. There is the potential concern of a significantly altered terrestrial albedo, particularly in regions where a lot of solar energy capture would exist. Land use conflicts, particularly affecting soils and biodiversity (Dubreuil et al., 2007), and landscape aesthetics are social issues that could spark debate.

For CSP there are concerns over impacts on fragile desert ecosystems, although some systems could have a positive impact on the surrounding microclimate through enhanced shading (Tsoutsos et al., 2005). There is also the possibility of thermal pollution to water resources, especially in arid regions.

7.9.5 Summary

Solar energy is not only in itself directly the most abundant energy resource on Earth, but it is also indirectly responsible as the ultimate source for hydro, wind, and, of course, biomass. The theoretical potential of solar energy, taking weather conditions into account, is enormous at 630,000 EJ/yr, while the practical potential is estimated at 12,300 EJ/yr for PV applications and at 277,000 EJ/yr for CSP. However, offshore solar potential was not taken into account in this assessment, and neither was the potential of small PV (<10 kWp), as these systems have almost no land suitability issues and their potential is thus very difficult to characterize. The economic potential of solar energy relies heavily on conversion technologies, as well as localized factors such as siting and transmission issues, and these are not addressed in this assessment.

7.10 Geothermal Energy

7.10.1 Overview

Geothermal energy, in the broadest sense, is the natural heat of the Earth due to primarily the decay of the radioactive isotopes of uranium, thorium, and potassium. Immense amounts of thermal energy are generated and stored in the Earth's core, mantle, and crust. At the base of the continental crust, temperatures range from 200–1000°C, and at the center of the Earth the temperatures may be in the range of 3500–5500°C (Alfé et al., 2001). The heat is transferred from the

Table 7.39 | Average irradiance, suitability factors, and solar energy input for PV for the 18 GEA regions.

Region	Average irradiance	Area	Theoretical potential	Suitability factor centralized	suitability factor decentralized	Suitable area centralized	Suitable area decentralized	Technical potential centralized PV	Technical potential decentralized PV
	[W/m²]	[million km²]	[EJ/yr]	[%]	[%]	[km²]	[km²]	[EJ]	[EJ]
USA	127.4	9.2	37,003	0.92%	0.08%	84,640	7360	340.4	29.6
CAN	93.6	9.5	28,042	0.50%	0.01%	47,500	950	140.2	2.8
WEU	108.8	3.7	12,695	0.69%	0.26%	25,530	9620	87.6	33.0
EEU	124.4	1.2	4708	0.63%	0.08%	7560	960	29.7	3.8
FSU	95.8	21.8	65,861	0.92%	0.01%	199,470	2180	605.9	6.6
NAF	203.1	5.7	36,700	4.50%	0.01%	256,500	570	1651.5	3.7
EAF	184.1	11.3	65,605	2.10%	0.00%	237,300	283	1377.7	0.0
WCA	195.3	5.8	35,845	2.71%	0.00%	157,180	145	971.4	0.0
SAF	180.2	6.8	38,643	2.10%	0.01%	142,800	680	811.5	3.9
MEE	198.1	5.9	36,859	3.32%	0.03%	195,880	1770	1223.7	11.1
CHN [a]	167.5	8.3	43,843	2.14%	0.06%	178,155	4995	938.2	26.3
OEA	165.9	1.1	5886	0.51%	0.05%	5610	550	30.0	2.9
IND	200.7	2.8	17,722	2.14%	0.06%	59,385	1665	379.3	10.6
OSA	193.0	5.1	31,041	1.92%	0.05%	97,920	2550	596.0	15.5
JPN	126.4	0.4	1594	0.23%	1.21%	920	4840	3.7	19.3
OCN	188.5	8.4	49,934	3.32%	0.01%	278,880	840	1657.8	5.0
PAS	148.8	3.3	15,579	0.51%	0.05%	16,830	1650	79.5	7.8
LAC	164.2	20.3	105,351	1.11%	0.03%	185,100	4600	1169.4	31.6
World	153.7	130.6	632,912	1.69%	0.11%	2,177,160	46,208	12,093.5	213.4

[a] Irradiation data from Green Cross International (2009).

Source: adapted from Hoogwijk, 2004.

interior toward the surface, mostly by conduction, and it this conductive heat flow that causes a rise in temperature with increasing depth at a rate, on average, of 25–30°C/km. This is called the geothermal gradient. In the most promising locations, typically along fault lines, the average gradient can be up to 5–10 times the average, with very significant increases of temperature even at shallow depths.

Although the total heat content of the first few km under the Earth's surface is immense, only a fraction can be utilized. Essentially, the Earth's crust acts as an insulating blanket, which can best be accessed through fluid conduits, magma, water, etc., to exploit the underground energy and transfer it to the surface.

Despite the inherent uncertainties in developing estimates of geothermal potential, it is possible to identify a range of estimations, also taking into consideration the possibility of new technologies, such as permeability enhancements, drilling improvements, enhanced geothermal system (EGS) technology, low temperature electricity production, and the use of supercritical fluids (Ledru et al., 2007).

7.10.2 Present Utilization of Geothermal Resources

Water from the deep wells and/or cold water from the surface is transported through this deep reservoir using injection and production wells, and recovered as steam or hot water or both. Injection and production wells, and surface installations, complete the circulation system. The utilization of geothermal energy is divided into two categories, electricity production and direct application, depending on fluid temperature.

Electricity is currently produced by geothermal energy in 24 countries. Five of these countries obtain 15–22% of their national electricity production from geothermal (Costa Rica, El Salvador, Iceland, Kenya, and the Philippines). There is ample opportunity for an increased use of geothermal energy both for direct applications and electricity production (Gawell et al., 1999).

Table 7.40 shows the recently observed rapid expansion in the utilization of geothermal energy for 18 world regions. In 2010, geothermal energy is expected to supply some 67 TWh/yr of electricity and 428 PJ/yr of heat for thermal applications (see Table 7.40).

Table 7.40 | Utilization of geothermal energy in the 18 GEA regions.

GEA Region	Electricity (GWh/yr)			Direct thermal (TJ/yr)		
	2000	2005	2010	2000	2005	2010
USA	14,000	16,840	16,603	20,302	31,239	56,552
CAN				1023	2546	8873
WEU	3852	7124	10,931	55,407	110,867	204,553
EEU				13,519	21,921	21,805
FSU	28	85	441	13,038	13,056	7825
NAF				1760	2651	2181
EAF	366	1088	1440	10	94	169
WCA						
SAF						115
MEE				3253	4500	4812
CHN	100	96	150	31,403	45,373	75,348
OEA				1077	469	2260
IND				2517	1606	2545
OSA				22	51	74
JPN	1722	3467	3064	5836	5161	15,698
OCN	2353	2775	4056	7375	10,054	9787
PAS	8307	15,357	20,363	83	112	162
LAC	7307	8877	10,200	5384	9840	15,301
World	38,035	55,709	67,246	162,009	259,540	428,060

Source: Bertani, 2003, 2005, 2007; 2010; Lund et al., 2005; Lund, 2010.

7.10.3 Theoretical Potential

The notion of geothermal resources refers to all of the thermal energy stored beneath the Earth's surface and is in the order of 10^{31} J (Fridleifsson et al., 2008). Even just the energy stored in the Earth's crust up to a depth of 5000 meters, estimated at 140×10^6 EJ (WEC, 1994) is still an enormous theoretical base. A detailed estimation of the heat stored 3 km deep under the continents by the EPRI (1978) applied an average geothermal temperature gradient of 25°C/km for normal geological conditions and accounted separately for diffuse geothermal anomalies and high-enthalpy regions located near plate boundaries or recent volcanism. High-enthalpy regions cover about 10% of the Earth's surface. The total amount of available heat was estimated to be about 42×10^6 EJ.

This number can be the cause of some confusion about the nature of geothermal resource, as it represents the static component of energy stored under the Earth's surface. The dynamic component, i.e., the terrestrial energy component or recharge of the crust's energy through thermal conduction from deeper, higher temperatures regimes, is much lower. The weighted average of conductive heat flow was estimated to be 87 mW/m² globally (Pollack et al., 1993), which corresponds to about 1500 EJ/yr (of which about 315 EJ on terrestrial surfaces), taking into account the additional flow rate caused by volcanic activity (Sigurdsson, 2000).

Using the definition of renewable energies, this can be assumed as the theoretical potential. Importantly, heat flow through the continents is much lower (65 mW/m²) than that at the ocean floor (101 mW/m²). However, convection processes at tectonic plate boundaries restore energies much faster, and thermal recharge is thus a localized factor.

7.10.4 Technical Potential

Until recently, geothermal energy was considered exploitable only in areas where naturally occurring water or steam is found concentrated at depths less than 3 km, and at temperatures above 180°C (Cataldi, 1999; Fridleifsson, 1999). This view has changed in the past two decades with the development of technologies that can economically utilize lower-temperature resources (around 100°C) and the emergence of ground-source heat pumps using the Earth as either a heat source or as a heat sink, depending on the season (Curtis et al., 2005). Furthermore, EGS technologies have become more viable. EGS, also known as hot dry rock geothermal energy, precludes the condition of naturally occurring water in place. It also allows for the potential use of more efficient thermodynamic fluids.

The assessment of technical potential must take the following limitations into account. Not all the available stored heat can be extracted from the rock through fluid circulation (whether natural or artificially induced). The greater part of the heat remains in rocks with low porosity and not connected to the fracture network. Not all surfaces are suitable for industrial development, because of inaccessibility of the area (mountains, desert, remoteness), presence of infrastructure (urban developments, grids, etc.), and restrictions through environmental protection. Finally, the localized nature of the geothermal gradient characteristics means that not all heat will be available at an economic temperature at reasonable depth.

A technical potential was evaluated (Stefansson, 2005), starting from a general correlation between the existing geothermal high-temperature resources and the number of volcanoes, inferring a total heat input potential suitable for electricity generation of 190 EJ/yr. However, this value considers only traditional geothermal resources, and does not account for emerging technologies, which allow for a more general global exploitation of the geothermal gradient.

The distribution between wet and dry systems and the recent statistical analysis of the lognormal heat distribution were used to calculate the technical potential of a 70% probability of adding up to 1000 MWe of EGS (Goldstein et al., 2009). According to the effective efficiency in the transformation from heat to electricity for different temperature classes (10% for 120–180°C, 20% for 180–300°C, 5% for EGS system), it is possible to evaluate the temperature-weighted average of the amount of equivalent heat extracted per year of about 660 EJ/yr. This assumes that locations with suitable conditions for electricity generation will be used for that purpose.

From the distribution of the geothermal resources over different temperature regimes, it is possible to estimate the low temperature potential (for direct thermal utilization only) using an empirical function reported in Stefansson (2005). The value of 153.5 EJ/yr has been split among the 18 GEA regions accordingly to the amount of low-temperature areas as described by the EPRI (1978). This sums to a total technical potential (for electricity generation and direct thermal uses) of 810 EJ/yr.

Finally, given the enormity of the heat stored within the first 10 km of the Earth's crust, the technical potential is not practically limited by the annual heat replenishment of 1500 EJ/yr and much higher technical potentials (than the annual rate of replenishment) have been reported in the literature (WEA, 2000).

7.10.5   Economic Potential

Over the short-to-medium term, the economic potential of geothermal energy can be estimated based on sites that are known and characterized by current drilling or by geochemical, geophysical, and geological evidence. Despite the promising prospect of EGS, the effective geothermal economic potential for the year 2050 is about 4.5 EJ/yr. EGS technologies are only now becoming commercially viable, and the first commercial plant only went online in Germany in 2010.

However, it is assumed that, in the coming years, EGS will become a leading technology, enabling the dissemination of geothermal electricity all around the world. Using the aforementioned statistical analysis of the lognormal distribution of the underground heat (Goldstein et al., 2009), and assuming for 2050 the exploitation of at least additional 70 GWe from EGS with 95% probability, the final value of 66 EJ/yr in 2050 has been evaluated. The economic potential of direct geothermal heat utilization depends to a large part on the technology associated with its utilization (heat pumps, binary cycles, etc.). By 2050 the global potential is estimated at 10 EJ/yr.

The results of the geothermal potentials (theoretical, technical, and economic) are presented in Table 7.41.

7.10.6   Cost Structure of a Typical Geothermal Electricity Project

Three major activities determine the cost structure of a geothermal project (GEA, 2005): exploration, resource confirmation and characterization (drilling and well testing), and site development (facility construction).

Table 7.41 | Geothermal heat supply potentials for the 18 GEA regions.

GEA region	Theoretical potential	Technical potential		Economic potential	
		Heat for direct utilization	Heat for electricity	Heat for direct utilization	Heat for electricity
	[10⁶ EJ]	[EJ/yr]	[EJ/yr]	[EJ/yr]	[EJ/yr]
USA	4.738	17.5	75	1.215	34.9
CAN	3.287	12.0	52	0.099	0.307
WEU	2.019	7.5	32	4.311	6.216
EEU	0.323	1.3	5.1	0.852	1.243
FSU	6.607	24.8	104	0.508	3.097
NAF	1.845	7.0	29	0.103	0.0
EAF	0.902	3.3	14	0.004	0.918
WCA	2.103	8.0	33	0.0	0.0
SAF	1.233	4.5	19	0.0	0.0
MEE	1.355	5.0	21	0.175	0.612
CHN	3.288	11.8	52	1.764	1.856
OEA	0.216	0.8	3.4	0.018	0.0
IND	0.938	3.5	15	0.062	0.613
OSA	2.424	9.3	38	0.002	0.0
JPN	0.182	0.5	2.9	0.201	0.612
OCN	1.092	3.5	17	0.004	7.424
PAS	2.304	8.8	36	0.391	1.568
LAC	6.886	24.8	109	0.383	6.216
World	41.743	153.5	657	10.1	65.6

Source: Adapted from Bertani, 2009.

Exploration is the process of locating geothermal fields, identifying fields at distinct levels of depths, and evaluating the overall prospects. It comprises several surface exploration surveys and eventually drilling of slim holes and/or geothermal gradient wells.

Site development covers all the remaining activities for the commercialization of a geothermal resource: drilling, plant permitting, gathering system, etc. Site development costs vary considerably from project to project depending on the well productivity and/or temperature and the depth to the geothermal reservoir. Recently, drilling costs have increased significantly (in some cases doubled) because of rising steel prices and competition from oil and gas activities for drilling-rig availability. In addition to factors common to exploration and confirmation, drilling costs are dominated by others, such as the permeability of the rock, depth of the reservoir, chemistry of geothermal fluid, resource temperature and pressure, and the liquid and steam gathering system.

The European Geothermal Energy Council (EGEC) collected the average of (all included) geothermal generation costs, in €/MWh, for different geothermal resource and sites, both for the present and projections for the year 2050. The EGEC data were reviewed for this publication and are presented here in Table 7.42. However, they have been modified to omit surface-equipment costs.

Table 7.42 | Geothermal production (heat supply) cost.

Category	Relative weight	Present cost [€/MWh]	Expected 2050 cost [€/MWh]
EGS	50%	120	25
Low temperature (100–180 °C)	33%	60	25
Medium temperature (180–250 °C)	12%	50	15
High temperature (>250 °C)	5%	30	10

Note: the relative weight has been taken from the temperature distribution of the geothermal resources of Stefansson (2005). This cost is only for the mining component. Plant and other surface equipment (approximately 50% of the total cost) are not taken into account.

Source: adapted from EGEC, 2009.

Table 7.43 | Geothermal direct utilization generation heat supply cost; the cost is for the mining only (approximately 10% of the total cost).

Category	Relative weight	Present cost [€/MWh]	Expected 2050 cost [€/MWh]
District heating	30%	5.00	4.50
Other shallow resources	70%	2.00	1.50

Source: modified from EGEC, 2009.

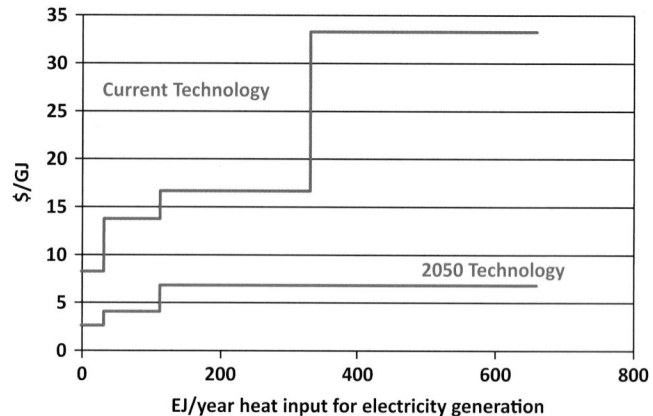

Figure 7.29 | Calculated geothermal heat supply curve suitable for electricity generation with present and expected (2050) technology.

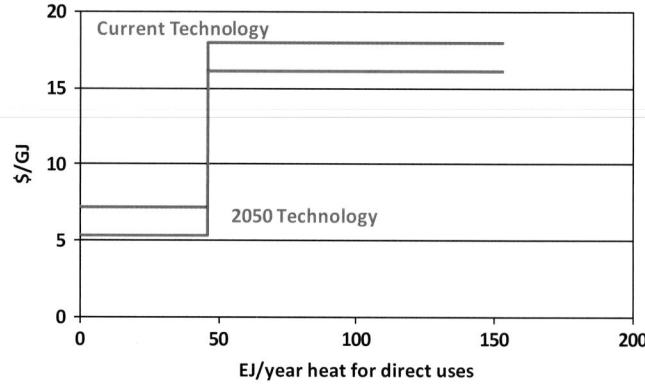

Figure 7.30 | Calculated geothermal direct heat utilization supply curve with present and expected (2050) generation cost.

Using the data of Table 7.42 it is possible to arrange the energy cost supply curve (Figure 7.29). The amount of heat available for each category according to its relative weight was considered, with lowest cost categories taking priority.

For direct utilization, the EGEC publication presents a very wide average value of present expected generation cost; the data are given in Table 7.43, and a cost curve is shown in Figure 7.30.

7.10.7   Environmental and Social Implications

Compared to other energy resources, the exploitation of geothermal energy exhibits a relatively low environmental footprint. Potential impacts range from the drilling of boreholes and of exploratory and production wells to some gaseous pollution releases during drilling and field testing. The installation pipelines may incur similar environmental impacts to those of drilling.

During plant operation, geothermal fluids (steam or hot water) usually contain gases as well as dissolved chemicals, the concentrations of which usually increase with temperature – with H_2S being one of the main pollutants. Various processes exist, however, that can reduce emissions significantly. Waste waters from geothermal plants also have higher temperatures than the surrounding environment and are therefore a potential thermal pollutant. The total reinjection of the used geothermal fluid is able to reduce strongly any impact of any potential liquid pollution. Adopting closed-loop systems essentially avoids any gaseous emissions.

The extraction of large quantities of fluids from geothermal reservoirs may give rise to subsidence phenomena, i.e., a gradual sinking of the land surface. This is a slow process distributed over vast areas and represents a low risk, if any. It takes years before such effects can be detected, but they need to be monitored systematically, as subsidence could damage the stability of buildings in the vicinity. In many cases subsidence can be prevented or reduced by reinjecting the geothermal waste waters.

The withdrawal and/or reinjection of geothermal fluids may trigger or increase the frequency of seismic events. However, these are microseismic events that can only be detected by means of instrumentation and to date no serious occurrence has been reported.

The high-pitched noise of steam traveling through pipelines and the occasional vent discharge are further potential adverse impacts. Finally, the scenic view can be affected adversely, although in some areas such as Larderello, Italy, the network of pipelines crisscrossing the countryside and the power-plant cooling towers have become an integral part of the panorama and are, indeed, a famous tourist attraction.

7.10.8   Summary

Geothermal energy is a renewable energy source that has been utilized economically in many parts of the world for decades. A great potential for an extensive increase in worldwide geothermal utilization exists, because new technologies allow for the exploitation of geothermal energy in many more locations. Still, the limiting factor for geothermal development rests in its geographical distribution, as high temperature gradients are available only in limited locations, even if the increasing role of binary fluids can strongly enlarge the possibility of using medium-to-low temperature areas.

The theoretical geothermal potential is not considered as the static heat content within the Earth's crust, but the conductive heat flow from deeper, higher temperature areas, and is estimated at 1500 EJ/yr. Extracting more than this theoretical potential would result in a

progressive cooling of the Earth's crust, and would no longer meet the principles of a renewable energy source. The technical potential was estimated to be 720 EJ/yr. This value approaches the upper limits of previous assessments of geothermal potential, as aggregated in Bertani (2003). This is because the current estimation assumes the inclusion of dry-rock system extraction technologies, which can exploit the general geothermal gradient and are not dependent on existing, naturally occurring fluids. The economic potential, 75 EJ/yr, takes into account the limited number of areas suitable for geothermal extraction due to geological constraints.

7.11   Ocean energy

7.11.1   Overview

Ocean energy refers to the kinetic energy carried by the waves, currents, and tides, as well as the potential energy stored in the ocean's salinity and temperature differences. The oceans are a tremendous source of energy, especially as many of the world's most concentrated populations are located close by. Ocean energy resources have an advantage in that they are more reliably available, especially compared to solar and wind.

7.11.2   Ocean Thermal Energy Conversion

Ocean thermal energy conversion (OTEC) takes advantage of the temperature difference that results from solar radiation warming the surface water and cooler deeper ocean currents. By using a heat engine, this natural temperature difference can be exploited to produce electricity.

The oceans are the world's largest solar energy collector and energy storage system. Although the average solar irradiation absorbed by the ocean is around 240 W/m², Vega (1995) used an estimate of the average solar flux as 95 W/m², which corresponds to the latent heat flux of the ocean. Using this average over the 360 million square-kilometer area of the Earth's oceans, the total theoretical potential of ocean thermal energy is estimated to be 1,000,000 EJ/yr.

However, as with any heat engine, larger temperature differences generate greater conversion efficiencies. A temperature difference of about 20 °C between the surface and 1000 m depth is required for OTEC systems to produce significant quantities of power (Binger, 2010). Figure 7.31 shows a world map indicating the ocean temperatures gradients.

The temperature irradiance generally increases with decreasing latitude, i.e., it is greatest near the equator. Various studies have estimated

---

23 These potentials are the input energy and do not include electricity-conversion efficiencies.

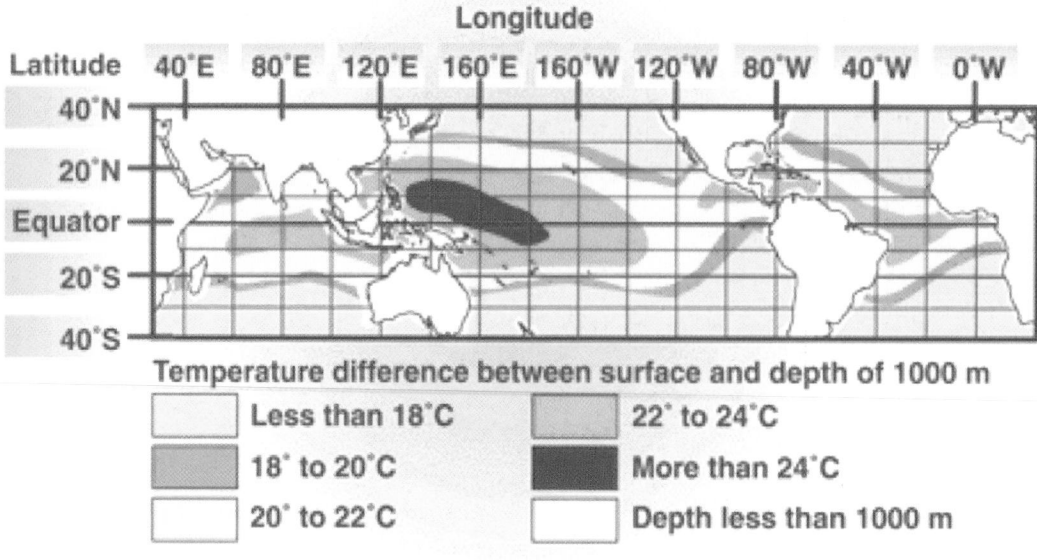

Longitude

Figure 7.31 | Ocean temperature differences between surface and 1000 m depth. Source: NREL, 2011.

the technical potential of ocean thermal energy. Taking just the tropics into account, Penney and Bharathan (1987) and Masutani et al. (2000) estimated a potential between 100,000–575,000 EJ/yr.[19] Other estimations have built upon these previous studies and assessed the technical potential as the maximum thermal energy extractable from the oceans without incurring significant environmental effects, with values between 15,500 EJ/yr (Pelc and Fujita, 2002) and 30,000 EJ/yr (Avery and Wu, 1994). One of the main discrepancies in these assessments is the viable ocean area for OTEC systems, as some estimations use 60 million km^2, while others use 100 million km^2. Finally, Nihous (2005) assesses the technical potential to be much lower than other studies, at 3150 EJ/yr. Nihous takes into account the disruption of the vertical thermal structure of the oceanic water columns that would likely occur if large-scale OTEC systems were put in place. Thus, the maximum thermal energy available for OTEC systems is on the same order of magnitude of the average upwelling of cold water in the oceans (Nihous, 2005).

7.11.3   Wave Resources

Wave energy is generated by the wind passing over the ocean surface, so that energy is transferred through frictional mechanisms into propagating instabilities in the free surface. It is these traveling instabilities that we refer to as waves.

Wind resources are greatest where the predominant wind directions pass over substantial ocean areas prior to heading toward coastal regions. The strongest winds are found in temperate latitudes, north and south, between 40° and 60°, and on the eastern ocean boundaries (Pelc and Fujita, 2002). The resource, which can be measured in terms of kW/m, is typically greater offshore than it is close to the coastline. However, the technology for harnessing the resource at distances beyond a few

tens of meters from the coast is more complex than that for coastal applications.

Pelc and Fujita (2002) estimate that up to 2 TW electricity could be provided by ocean waves. This corresponds to 68 EJ/yr, which, assuming an 80% conversion efficiency, leads to a technical potential of 78 EJ/yr.

7.11.4   Tidal Energy

The tides are cyclic variations in the level of the seas and oceans. Water currents accompany these variations in sea level, which, in some locations, such as the Pentland Firth to the North of the Scottish mainland, and the Bay of Fundy in Nova Scotia, Canada, can be extreme. Tidal energy is even more reliable than wave resources.

The availability of tidal energy, although predictable, is very site specific, and requires locations where the tides are amplified by topographical characteristics of the location. For large-scale plants, i.e., tidal barrages, this is an essential factor. However, tidal fences and turbines (basically underwater wind turbines) can be placed anywhere where the tide causes reliable currents greater than 2 m/s. The total worldwide technical potential of tidal energy amounts to approximately 3.4 EJ/yr (Pelc and Fujita, 2002).

7.11.5   Osmotic Energy

Osmotic energy, also known as salinity gradient energy, is the energy available by harnessing the osmotic pressure difference between seawater and freshwater, i.e., their different salt concentrations. Osmotic energy is available wherever significant amounts of freshwater and seawater come into contact, e.g., estuaries. The global discharge of freshwater to the seas is about 44,500 km^3/yr. Using a potential energy of 2.35 MJ/s/m^3 of freshwater (van den Ende and Groeman, 2007), a theoretical potential of 105 EJ/yr is determined. Van den Ende and Groeman (2007) assume that about 10% would be considered technically feasible, leading to a technical potential of 10 EJ/yr.

7.11.6   Environmental and Social Implications

The exploitation of ocean energy, as with any energy resource, is not without its downsides. For very large OTEC systems, thermal pollution, toxic releases, and impingement and entrainment of organisms on the plant itself pose potentially harmful sources. While the magnitude of thermal pollution would be small, even 3°C shifts in temperatures are known to cause high mortality to corals and fishes (Hoegh-Guldberg et al., 2007). Sensitive ecosystems could be affected further through nutrient loading caused by the drawing of deep coldwater. OTEC systems also release a small amount of CO_2, because gas solubility is reduced in warmer water. OTEC systems have the greatest potential for small island developing states, which need both a reliable energy source and freshwater (Binger, 2010).

Large-scale wave energy systems have the potential to disturb ocean ecosystems and harm marine life. They may also have a significant impact on ocean-atmospheric interactions. Wave energy devices alter the ocean's currents, typically slowing them down and calming the sea. While this could be used in conjunction with wave-break systems, especially around harbors, it will cause reduced mixing of ocean layers,

and could reduce food supplies, particularly for bottom feeders. Finally, dampened waves could alter erosion patterns of the coastline, in either a positive or negative way, depending on location (Boehlert et al., 2008).

Tidal barrages could pose threats to sensitive ecosystems, especially in estuaries, by altering the flow of saltwater. Ecosystems could be further affected by slower currents, reduction of the intertidal area, and limited ocean layer mixing. Tidal turbines, however, are more benign. Although they could be potentially dangerous to fish and crustaceans, slow-turning turbines could allow fish to swim through, and larger sea-life could be protected by undersea fences or nets (Boehlert and Gill, 2010).

As the main discharge waste of osmotic energy systems is brackish water, discharges of brackish water could cause salinity fluctuations outside those typical of the ecosystems, and so lead to loss of biodiversity. There is also the danger of impingement and entrainment of marine and estuarine organisms.

7.11.7   Summary

As the ocean covers almost 75% of the Earth's surface, it is Earth's largest collector of solar energy. This thermal potential accounts by far for the largest share of theoretical and technical ocean potentials. Ocean resources have the advantage over other renewables in that they are often more reliable and predictable. While the resource potential may be more than sufficient to meet global requirements (total technical potential, 3240 EJ/yr), special care must be given with regard to ocean ecosystems.

References

Adhikari, B. K., S. Barrington and J. Martinez, 2006: Predicted growth of world urban food waste and methane production. *Waste Management Resources,* **24**(5): 421–433.

Aguilera, R. F., R. G. Eggert, G. C. C. Lagos and J. E. Tilton, 2009: Depletion and the Future Availability of Petroleum Resources. *The Energy Journal,* **30**(1): 141–174.

Ahlbrandt, T. and T. Klett, 2005: Comparison of Methods Used to Estimate Conventional Undiscovered Petroleum Resources: World Examples. *Natural Resources Research,* **14**(3): 187–210.

Aitken, C. M., D. M. Jones, and S. R. Larter, 2004: Anaerobic hydrocarbon biodegradation in deep subsurface oil reservoirs. *Nature,* **431**(7006): 291–294.

Aldhous, P., P. McKenna, and C. Stier, 2010: Gulf Leak: Biggest Spill may not be Biggest Disaster. *New Scientist,* 2764.

Aleklett, K., M. Höök, K. Jakobsson, M. Lardelli, S. Snowden, and B. Söderbergh, 2010: The Peak of the Oil Age – Analyzing the world oil production Reference Scenario in World Energy Outlook 2008. *Energy Policy,* **38**(3): 1398–1414.

Alfè, D., M. J. Gillan, and G. D. Price, 2002: Composition and temperature of the Earth's core constrained by combining ab initio calculations and seismic data. *Earth and Planetary Science Letters,* **195**(1–2): 91–98.

Allsema, E. and M. van Brummelen, 1993: Het Potentieel van PV-Systemen in OECD Landen. Utrecht University, Department of Science, *Technology and Society*: pp. 53.

Altun, N. E., C. Hicyilmaz, J.-Y. Hwang, A. S. Bagci, and M. V. Kok, 2006: Oil shales in the world and Turkey; reserves, current situation and future prospects: a review. Oil Shale. *A Scientific-Technical Journal (Estonian Academy Publishers),* **23**(3): 211–227.

Ananenkov, D. G., 2007: *Role of Gas in Development of Global and Russian Fuel and Energy Industry in XXI Century, Paper P-1. In Proceedings of the International Conference on World Gas Resources and Reserves and Advanced Development Technologies*, VNIIGAZ, Moscow.

Antilla, P., T. Karjalainen, and A. Asikainen, 2009: Global Potential of Modern Firewood. Finnish Forest Institute, Vantaa. www.metla.fi/julkaisut/workingpapers/2009/mwp118.htm (accessed January 16, 2011).

Archer, C. L. and M. Z. Jacobson, 2005: Evaluation of global wind power. *J. Geophys. Res.,* **110**(D12): D12110.

Archer, C. L. and K. Caldeira, 2009: Global Assessment of High-Altitude Wind Power. *Energies,* **2**(2): 307–319.

Arthur, J. D., B. Bohm, and M. Layne, 2008: *Hydraulic Fracturing Considerations for Natural gas Wells of the Marcellus Shale*. In The Ground Water Protection Council 2008 Annual Forum, Cincinnati, OH.

Asikainen, A., H. Liiri, S. Peltola, T. Karjalainen, and J. Laitila, 2008: Forest energy potential in Europe (EU 27). Finnish Forest Institute, Vantaa. www.metla.fi/julkaisut/workingpapers/2008/mwp069.htm (accessed November 25, 2010).

Atlas, R. M. and R. Bartha, 1992: *Hydrocarbon Biodegradation and Oil Spill Bioremediation In Advances in Microbial Ecology*. K. C. Marshall, (ed.), Plenum Press, London, 13, pp. 282–338.

AUA (Australian Uranium Association), 2009: Uranium Stewardship – Best Practice Guidelines for Uranium Exploration Date Accessed: 2010, April, 10, www.aua.org.au/Content/ExplorationGuidelines.aspx.

Avery, W. H. and C. Wu, 1994: *Renewable Energy From the Ocean–A Guide to OTEC, Johns Hopkins University Applied Physics Laboratory Series in Science and Engineering*, J. R. Apel ed., Oxford University Press, NY.

AWEA (American Wind Energy Association), 2010: Basic Principles of Wind Resource Evaluation. www.awea.org/faq/basicwr.html (accessed on: March 30, 2010).

Bentley, R. W. and M. R. Smith, 2004: World Oil Production Peak – A Supply-Side Perspective. *International Association for Energy Economics Newsletters,* **13**(Third Quarter): 25–28.

Berndes, G., M. Hoogwijk, and R. van den Broek, 2003: The contribution of biomass in the future global energy supply: a review of 17 studies. *Biomass and Bioenergy,* **25**(1): 1–28.

Berndes, G., 2008: *Water demand for global bioenergy production: trends, risks and opportunities*. In Welt im Wandel: Zukunftsfähige Bioenergie und nachhaltige Landnutzung, WBGU (Wissenschaftlicher Beirat der Bundesregierung Globale Umwelt Veränderungen, Berlin.

Bertani, R., 2003: What is Geothermal Potential? *IGA News,* (**53**): 1–3.

Bertani, R., 2005: World geothermal power generation in the period 2001–2005. *Geothermics,* **34**(6): 651–690.

Bertani, R., 2007: *World Geothermal Power Generation in 2007. In Proceedings of the European Geothermal Congress*. 30 May – 1 June, Unterhaching, Germany.

Bertani, R., 2009: *Long-term Projections of Geothermal-electric Development in the World. GeoTHERM Expo & Congress*. March 5–6, 2009, Offenburg, Germany.

Bertani, R., 2010: *Geothermal Power Generation in the World: 2005–2010: Update Report. International Geothermal Association*. Bochum, Germany.

BGR, 2007: *Reserves, Resources and Availability of Energy Resources 2006*. Annual Report, Federal Institute for Geoscience and Natural Resources (BGR), Hannover, Germany.

BGR, 2009: *Energierohstoffe 2009 – Reserven, Ressourcen, Verfügbarkeit*. Annual Report, Federal Institute for Geoscience and Natural Resources (BGR), Hannover, Germany.

BGR, 2010: *Reserves, Resources and Availability of Energy Resources*. Annual Reports, Federal Institute for Geoscience and Natural Resources (BGR), Hannover, Germany.

Bhattacharya, S. C., 1991: *Characterization of Selected Agro-forestry Residues*. In Biocoal Technology and Economics, Bhattacharya S.C. and R. M. Shrestha, (eds.), Regional Energy Resources Information Center (RERIC), Bangkok.

Bhattacharya, S. C., J. M. Thomas, and P. Abdul Salam, 1997: Greenhouse gas emissions and the mitigation potential of using animal wastes in Asia. *Energy,* **22**(11): 1079–1085.

Biglarbigi, K., H. Mohan and J. Killen, 2009: USA, World Possess Rich Resource Base. *Oil & Gas Journal,* **107**(3): 56–61.

Binger, A., 2004: *Potential and Future Prospects for Ocean Thermal Energy Conversion (OTEC) In Small Islands Developing States (SIDS)*. UNESCO (UN Education, Scientific and Cultural Organisation).

Boehlert, G. W., G. R. McMurray, and C. E. Tortorici, (eds.), 2008: *Ecological Effects of Wave Energy in the Pacific Northwest. NOAA Technical Memo NMFS-F/SPO-92*. US Department of Commerce, Washington, DC.

Boehlert, G. W. and A. B. Gill, 2010: Environmental and Ecological Effects of Ocean Renewable Energy Development. *Oceanography,* **23**(2): 68–81.

BP, 2010: *Statistical Review of World Energy*. BP, London.

Broesamle, H., H. Mannstein, C. Schillings, and F. Trieb, 2001: Assessment of solar electricity potentials in North Africa based on satellite data and a geographic information system. *Solar Energy,* **70**(1): 1–12.

Bunn, M., S. Fetter, J. P. Holdren, and B. van der Zwaan, 2003: *The Economics of Reprocessing vs. Direct Disposal of Spent Nuclear Fuel. Project on Managing the Atom. DE-FG26–99FT4028, Belfer Center for Science and International Affairs,* John F. Kennedy School of Government, Harvard University, Cambridge, MA.

Burton, T., D. Sharpe, N. Jenkins, and E. Bossanyi, 2001: *Wind Energy Handbook.* John Wiley & Sons, Hoboken, NJ.

Butler, R. M. and D. J. Stephens, 1980: *The Gravity Drainage of Steam-Heated Heavy Oil to Parallel Horizontal Wells; Paper CIM # 80–31–31.* In Presented at the 31st Petroleum Society of CIM Annual Technical Meeting, 25–28 May, Calgary.

Butler, R. M. and D. J. Stephens, 1981: The Gravity Drainage of Steam-heated Heavy Oil to Parallel Horizontal Wells. *Journal of Canadian Petroleum Technology,* **2**(2): 90–96.

Campbell, C. J. and J. H. Laherrère, 1998: The End of Cheap Oil. *Scientific American,* **278**(3): 78–83.

Campbell, J. E., D. B. Lobell, R. C. Genova, and C. B. Field, 2008: The Global Potential of Bioenergy on Abandoned Agriculture Lands. *Environmental Science & Technology,* **42**(15): 5791–5794.

Canale, M., L. Fagiano, and M. Milanese, 2007: Power Kites for Wind Energy Generation EEE Control Systems Magazine, pp. 25–38.

CAPP (Canadian Association of Petroleum Producers), 2009: Crude Oil: Forecast Markets & Pipline Report. www.capp.ca/getdoc.aspx?DocId=173003 (accessed May 5, 2011).

Cataldi, R., 1999: Geothermal development in Europe to the year 2020: prospects or hopes? *Technika Poszukiwań Geologicznych,* **38**(4–5): 48–60.

Cedigaz, 2009: *Natural Gas in the World.* A. Lecarpentier, (ed.) Institut Français du Pétrole, Rueil-Malmaison, France.

Charpentier, A. D., J. A. Bergerson, and H. L. MacLean, 2009: Understanding the Canadian oil sands industry's greenhouse gas emissions. *Environmental Research Letters,* **4**(1): 014005.

Cherubini, F., N. D. Bird, A. Cowie, G. Jungmeier, B. Schlamadinger, and S. Woess-Gallasch, 2009: Energy- and greenhouse gas-based LCA of biofuel and bioenergy systems: Key issues, ranges and recommendations. *Resources, Conservation and Recycling,* **53**(8): 434–447.

Chitrakar, R., H. Kanoh, Y. Miyai, and K. Ooi, 2001: Recovery of Lithium from Seawater Using Manganese Oxide Adsorbent (H1.6Mn1.6O4) Derived from Li1.6Mn1.6O4. *Industrial & Engineering Chemistry Research,* **40**(9): 2054–2058.

Chum, H., A. Faaij, J. Moreira, G. Berndes, P. Dhamija, H. Dong, B. Gabrielle, A. Goss Eng, W. Lucht, M. Mapako, O. Masera Cerutti, T. McIntyre, T. Minowa, K. Pingoud, 2011: *Bioenergy. In IPCC Special Report on Renewable Energy Sources and Climate Change Mitigation,* O. Edenhofer, R. Pichs-Madruga, Y. Sokona, K. Seyboth, P. Matschoss, S. Kadner, T. Zwickel, P. Eickemeier, G. Hansen, S. Schlömer, C. von Stechow (eds.), Cambridge University Press, Cambridge, United Kingdom and New York, NY, USA.

Clarke, G. M. and P. W. Harben, 2009: Lithium Availability Wall Map. www.lithiumalliance.org/about-lithium/lithium-sources/85-broad-based-lithium-reserves (accessed January 3, 2011).

Clemens, J., M. Triborn, P. Weiland, and B. Amon, 2005: Mitigation of greenhouse gas emissions by anaerobic digestion of cattle slurry. *Agriculture, Ecosystems & Environment,* **112**(2–3): 171–177.

Collett, T. S., 2002: Energy Resource Potential of Natural Gas Hydrates. *AAPG Bulletin,* **86**(11): 1971–1992.

Collett, T. S., A. Johnson, C. C. Knapp, and R. Boswell, 2009: *Natural Gas Hydrates – A Review. In Natural Gas Hydrates – Energy Resource Potential and Associated Geologic Hazards.* T. S. Collett, A. Johnson, C. C. Knapp and R. Boswell, (eds.), American Association of Petroleum Geologists; Memoir 89, pp. 137.

Couch, G. R., 1988: *Lignite Resources and Characteristics, IEACR/13.* IEA Coal Research, London, UK.

Crawford, P. M., K. Biglarbigi, A. R. Dammer, and E. Knaus, 2008: *Advances in World Oil Shale Production Technologies, SPE # 116570 In Presented at the 2008 Annual Technical Conference and Exhibition,* 21–24 September, Denver, CO.

Crawford, P. M., K. Biglarbigi, E. Knaus, and J. Killen, 2009: New Approaches Overcome Past Technical Issues. *Oil & Gas Journal,* **107**(4): 44–49.

Curtis, R., J. Lund, B. Sanner, L. Rybach, and G. Hellström, 2005: *Ground Source Heat Pumps – Geothermal Energy for Anyone, Anywhere: Current Worldwide Activity. In Proceedings of the World Geothermal Congress.* April 24–29, Antalya.

Daul, J., 1995: *Untersuchung über die Verteilung und Veränderung der Steinkohlevorräte im Ruhrgebiet und deren Ausnutzung.* PhD-Thesis, Montanuniversität Leoben, Leoben.

de Vries, B. J. M., D. P. van Vuuren and M. M. Hoogwijk, 2007: Renewable energy sources: Their global potential for the first-half of the 21st century at a global level: An integrated approach. *Energy Policy,* **35**(4): 2590–2610.

de Wit, M. and A. Faaij, 2010: European biomass resource potential and costs. *Biomass and Bioenergy,* **34**(2): 188–202.

Denholm, P. and W. Short, 2006: *Documentation of WinDS Base Case. Version AEO 2006 (1).* National Renewable Energy Laboratory, Golden, CO.

Deutsche Bank, 2009: *The Cost of Producing Oil. FITT Research.* Deutsche Bank A.G., London.

DLR (Deutsches Zentrum für Luft- und Raumfahrt), 2009: *Characterisation of Solar Electricity Import Corridors from MENA to Europe – Potential, Infrastructure and Cost.* DLR, Stuttgart.

DOE (US Department of Energy), 2010: Enhanced oil Recovery/CO2 Injection. http://fossil.energy.gov/programs/oilgas/eor/ (accessed March 20, 2011).

Dubreuil, A., G. Gaillard, and R. Müller-Wenk, 2007: Key Elements in a Framework for Land Use Impact Assessment Within LCA (11 pp). *The International Journal of Life Cycle Assessment,* **12**(1): 5–15.

Dusseault, M. B., 2002: *New Oil Production Technologies. SPE Distinguished Lectures Series 2002–2003.* Paper SPE # 101463 DLP.

Dusseault, M. B. and A. Shafiei, 2011: *Tar Sands. In Ullmann's Encyclopedia of Chemical Engineering,* Wiley, Hoboken, NJ.

Dyni, J. R., 2006: *Geology and Resources of Some World Oil-Shale Deposits. Scientific Investigations Report 2005–5294,* US Geological Survey, Reston, VA.

E4Tech, 2009: *Biomass Supply Curves for the UK.* E4Tech, London.

Earth Policy Institute, 2007: Is World Oil Production Peaking? www.earth-policy.org/Updates/2007/Update67_printable.htm (accessed November 30, 2010).

ECE (Economic Commission for Europe), 2010: *United Nations International Framework Classification for Fossil Energy and Mineral Reserves and Resources 2009.* The ECE Energy Series, No. 39, Economic Commission for Europe, Geneva.

EGEC (European Geothermal Energy Council), 2009: *The Future of Geothermal Development. EGEC -Renewable Energy House,* Brussels, Belgium.

Ekins, P., R. Vanner, and J. Firebrace, 2007: Zero emissions of oil in water from offshore oil and gas installations: economic and environmental implications. *Journal of Cleaner Production*, **15**(13–14): 1302–1315.

EPRI (Electric Power Research Institute), 1978: Geothermal energy prospects for the next 50 years. Special Report ER-611-SR. www.osti.gov/energycitations/servlets/purl/5027376-y14BJ4/ (accessed February 12, 2011).

Evans R. K., 2008: An Abundance of Lithium. www.che.ncsu.edu/ILEET/phevs/lithium-availability/An_Abundance_of_Lithium.pdf (accessed April 20, 2010).

EWG (Energy Watch Group), 2006: *Uranium Resources and Nuclear Energy, EWG-Series No 1/2006*. Zittel W. And J. Schindler (eds.). Ludwig Bölkow Systemtechnik GmbH, Ottobrunn/Achen, Germany.

EWG (Energy Watch Group), 2007: *Crude Oil: The Supply Outlook Report to the Energy Watch Group. EWG-Series No 3/2007*, Ludwig-Bölkow-Systemtechnik GmbH, Ottobrunn.

EWG (Energy Watch Group), 2008: *Zukunft der weltweiten Erdölversorgung*. Ludwig-Bölkow-Systemtechnik GmbH, Ottobrunn.

Exxon, 2008: *Öldorado 2008*. Jubiläum 50 Jahre, ExxonMobil, Hamburg.

FAO, 2006a: *World agriculture: towards 2015/2030 – Interim report. Prospects for food, nutrition, agriculture and major commodity groups*. Food and Agricultural Organization (FAO), Rome.

FAO, 2006b: *World agriculture: towards 2030/2050 – Interim report. Prospects for food, nutrition, agriculture and major commodity groups*. Food and Agricultural Organization (FAO), Rome.

FAO, 2006c: *Gridded Livestock of the World*. Food and Agricultural Organization (FAO), Rome. http://www.fao.org/ag/AGAinfo/resources/en/glw/default.html (accessed September 21, 2010).

FAO, 2009: *FAOSTAT 2009*. Food and Agricultural Organization (FAO), Rome. http://faostat.fao.org/site/573/default.aspx#ancor (accessed September 21, 2010).

Farrell, A. E., 2008: Energy Notes. *News from the University of California Energy Institute*. Vol. **6**: p. 3.

Fettweis, G. B., 1976: *Weltkohlenvorräte. Eine vergleichende Analyse ihrer Erfassung und Bewertung*. Verlag Glückauf, Essen.

Fettweis, G. B., 1990: *Der Produktionsfaktor Lagerstätte*. In Bergwirtschaft. S. v. Wahl, (ed.), Verlag Glückauf, Essen.

Field, C. B., J. E. Campbell, and D. B. Lobell, 2008: Biomass energy: the scale of the potential resource. *Trends in Ecology & Evolution*, **23**(2): 65–72.

Firbank, L., 2008: Assessing the Ecological Impacts of Bioenergy Projects. *BioEnergy Research*, **1**(1): 12–19.

Fischer, G., E. Hizsnyik, S. Prieler, and H. van Velthuizen, 2007: *Assessment of biomass potentials for bio-fuel feedstock production in Europe: Methodology and results. REFUEL Project*. International Institute for Applied Systems Analysis (IIASA), Laxenburg.

Florentinus, A., C. Hamelinck, S. de Lint, and S. van Iersel, 2008: *Worldwide Potential of Aquatic Biomass Ecofys*, Utrecht.

Foley, J. A., R. DeFries, G. P. Asner, C. Barford, G. Bonan, S. R. Carpenter, F. S. Chapin, M. T. Coe, G. C. Daily, H. K. Gibbs, J. H. Helkowski, T. Holloway, E. A. Howard, C. J. Kucharik, C. Monfreda, J. A. Patz, I. C. Prentice, N. Ramankutty, and P. K. Snyder, 2005: *Global Consequences of Land Use. Science*, **309**(5734): 570–574.

Fridleifsson, I. B., 1999: Worldwide Prospects for Geothermal Energy in the 21st Century. *Technika Poszukiwań Geologicznych*, **38**(4–5): 28–34.

Fridleifsson, I. B., R. Bertani, E. Huenges, J. Lund, A. Ragnarsson, and L. Rybach, 2008: *The Possible Role and Contribution of Geothermal Energy to the Mitigation of Climate Change. In IPCC Scoping Meeting on Renewable Energy Sources*. 21–25 January, Lübeck.

Gawell, K., M. Reed, and P. M. Wright, 1999: *Geothermal Energy: The Potential for Clean Power from the Earth*. Geothermal Energy Association, Washington, DC.

GEA (Geothermal Energy Association), 2005: *Factors Affecting Costs of Geothermal Power Development. Report for the US Department of Energy*, US Department of Energy, Washington, DC.

Gerbens-Leenes, W., A. Y. Hoekstra, and T. H. van der Meer, 2009: The water footprint of bioenergy. *Proceedings of the National Academy of Sciences*, **106**(25): 10219–10223.

Gesamtverband Steinkohle, 2007: *Zahlen der Betribsstatistik*. Annual Report, Gesamtverband Steinkohle, Essen.

Gibbs, H. K., M. Johnston, J. A. Foley, T. Holloway, C. Monfreda, N. Ramankutty, and D. Zaks, 2008: Carbon payback times for crop-based biofuel expansion in the tropics: the effects of changing yield and technology. *Environmental Research Letters*, **3**(3): 034001.

Gold, T. and S. Soter, 1980: The Deep-earth Gas hypothesis. *Scientific American*, **242**(6): 155–161.

Gold, T. and S. Soter, 1982: Abiogenic Methane and the Origin of Petroleum. *Energy Exploration & Exploitation*, **1**(2): 89–104.

Gold, T., 1988: *Origin of Petroleum: Two Opposing Theories and a Test in Sweden. In The Methane Age*. T. H. Linden, D. A. Dreyfus and T. Vasko, (eds.), Kluwer Academic Publishers, Dordrecht.

Gold, T., 1999: *The Deep Hot Biosphere*. Springer, New York.

Goldstein, B. A., A. J. Hill, A. Long, A. R. Budd, F. Holgate, and M. Malavos, 2009: *Hot Rock Geothermal Energy Plays in Australia. In Proceedings of the 34th Workshop on Geothermal Reservoir Engineering*. 9–11 February, Stanford University Stanford, CA.

Graue, A., B. Kvamme, B. A. Baldwin, J. Stevens, J. Howard, G. Ersland, J. Husebo, and D. R. Zornes, 2006: *Magnetic resonance imaging of methane – carbon dioxide hydrate reactions in sandstone pores*. Paper Number SPE 102915. In Proceedings of the SPE Annual Technical Conference and Exhibition, 24–27 September, San Antonio, TX.

Green Cross International, 2009: *Global Solar Report Cards*. Green Cross International, Geneva.

Gruber P. and P. Medina, 2010: *Global Lithium Availability: A Constraint for Electric Vehicles? MSc Thesis*, Department of Natural Resources and Environment, University of Michigan.

Gustavson, M. R., 1979: Limits to Wind Power Utilization. *Science*, **204**(4388): 13–17.

Haberl, H., K.-H. Erb, F. Krausmann, W. Loibl, N. Schulz, and H. Weisz, 2001: Changes in ecosystem processes induced by land use: Human appropriation of aboveground NPP and its influence on standing crop in Austria. Global Biogeochem. *Cycles*, **15**(4): 929–942.

Haberl, H., K. H. Erb, F. Krausmann, V. Gaube, A. Bondeau, C. Plutzar, S. Gingrich, W. Lucht, and M. Fischer-Kowalski, 2007a: Quantifying and mapping the human appropriation of net primary production in earth's terrestrial ecosystems. *Proceedings of the National Academy of Sciences*, **104**(31): 12942–12947.

Haberl, H., K. H. Erb, C. Plutzar, M. Fischer-Kowalski and F. Krausmann, 2007b: *Human Appropriation of Net Primary Production (HANPP) as Indicator for pressures on Biodiversity. In Sustainability Indicators: A Scientific Assessment (SCOPE)*. T. Hak, B. Moldan and A. L. Dahl, (eds.), Island Press, Washington, DC.

Haberl, H., V. Gaube, R. Díaz-Delgado, K. Krauze, A. Neuner, J. Peterseil, C. Plutzar, S. J. Singh and A. Vadineanu, 2009: Towards an integrated model of socioeconomic biodiversity drivers, pressures and impacts. A feasibility study based on three European long-term socio-ecological research platforms. *Ecological Economics,* **68**(6): 1797–1812.

Hakala, K., M. Kontturi and K. Pahkala, 2009: Field Biomass as Global Energy Source. *Agricultural and Food Science,* **18**: 347–368.

Haq, Z. and J. Easterly, 2006: Agricultural residue availability in the United States. Applied *Biochemistry and Biotechnology,* **129**(1): 3–21.

Harms, A. A., D. R. Kingdon, K. F. Schoepf, and G. H. Miley, 2000: *Principles of Fusion Energy*. World Scientific Publishing Company, London.

HDWA (Hydropower and Dams World Atlas), 2008: *Hydropower and Dams World Atlas*. The International Journal on Hydropower and Dams, Sutton: Aquamedia Publications, Surrey, UK.

Head, I. M., D. M. Jones, and S. R. Larter, 2003: Biological activity in the deep subsurface and the origin of heavy oil. *Nature,* **426**(6964): 344–352.

Heinberg, R. and D. Fridley, 2010: The end of cheap coal. *Nature,* **468**(7322): 367–369.

Hirsch, R. L., 2005: *Peaking of World Oil Production: Impacts, Mitigation, & Risk Management. Association for the Study of Peak Oil and Gas*. National Energy Technology Laboratory.

Hoegh-Guldberg, O., P. J. Mumby, A. J. Hooten, R. S. Steneck, P. Greenfield, E. Gomez, C. D. Harvell, P. F. Sale, A. J. Edwards, K. Caldeira, N. Knowlton, C. M. Eakin, R. Iglesias-Prieto, N. Muthiga, R. H. Bradbury, A. Dubi, and M. E. Hatziolos, 2007: Coral Reefs Under Rapid Climate Change and Ocean Acidification. *Science,* **318**(5857): 1737–1742.

Hofman, Y., D. de Jager, E. Molenbroek, F. Schilig, and M. Voogt, 2002: *The Potential Of Solar Electricity to Reduce CO2 Emissions*. Ecofys, Utrecht.

Hoogwijk, M., 2004: *On the global and regional potential of renewable energy sources*. Universiteit Utrecht, Utrecht.

Hoogwijk, M., B. de Vries, and W. Turkenburg, 2004: Assessment of the global and regional geographical, technical and economic potential of onshore wind energy. *Energy Economics,* **26**(5): 889–919.

Hoogwijk, M., A. Faaij, B. de Vries, and W. Turkenburg, 2009: Exploration of regional and global cost-supply curves of biomass energy from short-rotation crops at abandoned cropland and rest land under four IPCC SRES land-use scenarios. *Biomass and Bioenergy,* **33**(1): 26–43.

Hubbert, M. K., 1971: Energy Resources of the Earth. *Scientific American,* **224**(3): 60–70.

IAASTD, 2009. *Agriculture at a Crossroads. International Assessment of Agricultural Knowledge, Science and Technology for Development (IAASTD)*, Global Report. Island Press, Washington, DC.

IAEA (International Atomic Energy Agency), 2005: *Thorium Fuel Cycle – Potential Benefits and Challenges*. IAEA-TECDOC-1350, Vienna.

IEA (International Energy Agency), 2000: *Hydropower and the Environment: Present context and guidelines for future action. Subtask 5 Report, Volume II: Main Report*, International Energy Agency of the Organisation of Economic Co-Operation and Development, Paris.

IEA, 2001: *Potential for building integrated photovoltaics*. International Energy Agency of the Organisation of Economic Co-Operation and Development, Paris.

IEA, 2006a: *World Energy Outlook 2006*. International Energy Agency of the Organisation of Economic Co-Operation and Development, Paris.

IEA, 2006b: *Energy Technology Perspectives 2006: Scenarios & Strategies to 2050*. International Energy Agency of the Organisation of Economic Co-Operation and Development, Paris.

IEA, 2007a: *Energy Balances of Non-OECD Countries, 2004–2005 – 2007 Edition*. CD-ROM. International Energy Agency of the Organisation of Economic Co-Operation and Development, Paris.

IEA, 2007b: *Energy Balances of OECD Countries, 2004–2005 – 2007 Edition*. CD-ROM. International Energy Agency of the Organisation of Economic Co-Operation and Development, Paris.

IEA, 2008a: *World Energy Outlook 2008*. International Energy Agency of the Organisation of Economic Co-Operation and Development, Paris.

IEA, 2008b: *Renewables Information 2008*. International Energy Agency of the Organisation of Economic Co-Operation and Development, Paris.

IEA, 2009: *World Energy Outlook 2009*. International Energy Agency of the Organisation of Economic Co-Operation and Development, Paris.

IEA, 2010a: *World Energy Outlook 2010*. International Energy Agency of the Organisation of Economic Co-Operation and Development, Paris.

IEA, 2010b: *Technology Roadmap: Concentrating Solar Power*. International Energy Agency of the Organisation of Economic Co-Operation and Development, Paris.

IEA, 2010c: *Technology Roadmap: Solar Photovoltaic Energy*. International Energy Agency of the Organisation of Economic Co-Operation and Development, Paris.

IEA Coal, 1983: *Concise Guide to World Coal fields*. International Energy Agency Coal Research, London.

IGC (International Geological Congress), 1913: *The Coal Resources of the World. Enquiry made upon the initiative of the Executive Committee of the Twelfth International Geological Congress*, Morang & Co., Toronto.

IGU (International Gas Union), 1994: *Report of IGU Committee A (Production, Treatment and Underground Storage of Natural Gas) 9.2 Production from Deep Fields*. IGU/A-94. In Proceedings of the19th World Gas Conference, 20–23 June, Milan.

IGU, 2003: *Report of IGU Working Committee 1: Exploration, production and treatment of natural gas*. In 22nd World Gas Conference, 1–5 June, Tokyo.

IHS-CERA (Cambridge Energy Research Associates), 2006: *Why the "Peak Oil" Theory Falls Down Myths, Legends, and the Future of Oil Resources*. P. Jackson (ed.), CERA, Cambridge, MA.

IHS-CERA, 2010: *Oil Sands, Greenhouse Gases, and US Oil Supply. Special Report*, IHS-CERA Inc., Cambridge, MA.

IHS-CERA, 2011: *Oil Sands Technology, Past, Present and Future. Special Report*, IHS CERA Inc., Cambridge, MA.

IPCC, 1996: *Climate Change 1995. In Adaptations and Mitigation of Climate Change: Scientific-technical Analysis. Contribution of Working Group II to the Second Assessment Report of the Intergovernmental Panel on Climate Change (IPCC)*. R. T. Watson, M. C. Zinyowera, and R. H. Moss, (eds.), Cambridge University Press, Cambridge, UK.

IPCC, 2006: *IPCC Guidelines for National Greenhouse Gas Inventories*. Institute for Global Environmental Strategies, Japan.

IPCC, 2007: *Mitigation of Climate Change. Contribution of working group III to the Fourth Assessment report of the IPCC*, B. Metz, O. R. Davidson, P. R. Bosch, R. Dave and L. A. Meyer, (eds.), Cambridge University Press, Cambridge, UK.

Iqbal, M., 1983: *An Introduction to Solar Radiation*. Academic Press, New York.

Jaccard, M., 2006: *Sustainable Fossil Fuels: the Unusual Suspect in the Quest for Clean and Enduring Energy*. Cambridge University Press, Cambridge, UK.

Jacobson, M. Z. and G. M. Masters, 2001: Exploiting wind versus coal. *Science,* **293**(5534): 1438.

Jaksula, B. W., 2008: *US Geological Survey Minerals yearbook – 2007*. US Geological Survey, Washington, DC.

Jeglic, F., 2004: *Analysis of Ruptures and Trends on Major Canadian Pipeline Systems*. National Energy Board of Canada, Calgary, AB.

Jekayinfa, S. O. and V. Scholz, 2007: *Assessment of availability and cost of energetically usable crop residues in Nigeria. In Conference on International Agricultural Research for Development*. 9–11 October, University of Göttingen.

Johansson, T.B., Kelly H., Reddy A.K.N. and H. Williams, (eds), 1993: *Renewable Energy: Sources for Fuels and Electricity*. Washington, DC: Island Press.

Johnston, M., J. A. Foley, T. Holloway, C. J. Kucharik and C. Monfreda, 2009: Resetting global expectations from agricultural biofuels. *Environmental Research Letters,* **4**(1): 01 4004.

Junginger, M., R. Sikkema and A. P. C. Faaij, 2009: *Analysis of the global pellet market. Including major driving forces and possible technical and non-technical barriers*. Report for the Pellet@las project, Copernicus Institute, Utrecht.

Kato, T., K. Okugawa, Y. Sugihara, and T. Matsumura, 1999: Conceptual Design of Uranium Recovery Plant from Seawater. *Journal of the Thermal and Nuclear Power Engineering Society,* **50**(1): 71–77.

Kenney, J. F., 1996: *Considerations about recent predictions of impending shortages of petroleum evaluated from the perspective of modern petroleum science*. Special Edition on "The Future of Petroleum" in Energy World, British Institute of Petroleum, London.

Kenney, J. F., V. A. Kutcherov, N. A. Bendeliani, and V. A. Alekseev, 2002: The evolution of multicomponent systems at high pressures: VI. The thermodynamic stability of the hydrogen-carbon system: The genesis of hydrocarbons and the origin of petroleum. *Proceedings of the National Academy of Sciences of the United States of America,* **99**(17): 10976–10981.

Kenney, J. F., V. I. Sozanksy, and P. M. Chepil, 2009: On the Spontaneous Renewal of Oil and Gas Fields. *Energy Politics,* **XVII**(Spring **2009**).

Koopmans, A. and J. Koppejan, 1998: *Agricultural and forest residues – generation, utilization and availability*. In Regional Consultation on Modern Applications of Biomass Energy, 6–10 January, Kuala Lumpur.

Krajick, K., 2001: Defending Deadwood. *Science,* **293**(5535): 1579–1581.

Krausmann, F., K.-H. Erb, S. Gingrich, C. Lauk, and H. Haberl, 2008: Global patterns of socioeconomic biomass flows in the year 2000: A comprehensive assessment of supply, consumption and constraints. *Ecological Economics,* **65**(3): 471–487.

Kudryavtsev, N. A., 1951: Against the Organic Hypothesis of Oil Origin. *Oil Economy Journal,* (**9**): 17–29.

Kundel, H., 1985: *Face technology in German coal mines in 1984. In German Longwall Mining – Facts and Figures. In Glückauf Bergbau Handbuch Nr. 32.*, Verlag Glückauf, Essen.

Laherrère, J. H., 2004: *Future of Natural Gas Supply. In Contribution to the 3rd International Workshop on Oil&Gas Depletion*. 24–25 May, Berlin.

Laherrère, J. H., 2005: Review on Oil Shale Data. http://www.oilcrisis.com/laherrere/OilShaleReview200509.pdf.

Lal, R., 2004: Agricultural activities and the global carbon cycle. *Nutrient Cycling in Agroecosystems,* **70**(2): 103–116.

Lal, R., 2005: World crop residues production and implications of its use as a biofuel. *Environment International,* **31**(4): 575–584.

Lal, R., 2006: Soil and Environmental Implications of Using Crop Residues as Biofuel Feedstock. *International Sugar Journal,* **108**(1287): 161–167.

Landsmann, H., 2009: Personal communication on soft brown coal mining cost ranges.

Lauk, C. and K.-H. Erb, 2009: Biomass consumed in anthropogenic vegetation fires: Global patterns and processes. *Ecological Economics,* **69**(2): 301–309.

Ledru, P., D. Bruhn, P. Calcagno, A. Genter, E. Huenges, M. Kaltschmitt, C. Karytsas, T. Kohl, L. Le Bel, A. Lokhorst, A. Manzella, and S. Thorhalsson, 2007: *Enhanced Geothermal Innovative Network for Europe: The State-of-the-Art*. Geothermal Resources Council Bulletin, 36.

Lehner, B., G. Czisch, and S. Vassolo, 2001: *Europe's Hydropower Potential Today and in the Future*. EuroWasser, Center for Environmental Systems Research, University of Kassel, Kassel.

Lemperiere, F., 2006: The Role of Dams in the XXI Century: Achieving a Sustainable Development Target. *International Journal on Hydropower & Dams,* **13**(3): 98–108.

Lewandowski, I. and A. P. C. Faaij, 2006: Steps towards the development of a certification system for sustainable bio-energy trade. *Biomass and Bioenergy,* **30**(2): 83–104.

Lewis, W. B., 1972: *Energy in the Future: the Role of Nuclear Fission and Fusion*. Proceedings of the Royal Society of Edinburgh, Sect. A.

Li, L., A. P. Ingersoll, X. Jiang, D. Feldman and Y. L. Yung, 2007: Lorenz energy cycle of the global atmosphere based on reanalysis datasets. *Geophysical Research Letters,* **34**(16): L16813.

Lindner, M., J. Meyer, T. Eggers and A. Moiseyev, 2005: *Environmentally enhanced bioenergy potential from European forests*. Report commission by the European Environmental Agency, European Forest Institute, Paris.

Lollar, B. S., T. D. Westgate, J. A. Ward, G. F. Slater, and G. Lacrampe-Couloume, 2002: Abiogenic Formation of Alkanes in the Earth's Crust as a Minor Source for Global Hydrocarbon Reservoirs. *Nature,* **416**(6880): 522–524.

Long, S. P., E. A. Ainsworth, A. D. B. Leakey, J. Nösberger, and D. R. Ort, 2006: Food for Thought: Lower-Than-Expected Crop Yield Stimulation with Rising CO2 Concentrations. *Science,* **312**(5782): 1918–1921.

Lorenz, E. N., 1967: *The nature and theory of the general circulation of the atmosphere*. World Meteorological Organization, Geneva. eapsweb.mit.edu/research/Lorenz/publications.htm (accessed November 29, 2011).

Lund, J. W., D. H. Freeston and T. L. Boyd, 2005: Direct Application of Geothermal Energy: 2005 Worldwide Review. *Geothermics,* **34**(6): 691–727.

Lund, J. W., 2010: Direct Utilization of Geothermal Energy. *Energies,* **3**(8): 1443–1471.

Marika, E., F. Uriansrud, R. Bllak, and M. B. Dusseault, 2009: *Acheiving Zero Discharge using Deep Well Disposal Paper 2009-WHOC09–350. In Proceeds of the World Heavy Oil Congress*, Marguarita, Venezuela.

Mastuani, S. M. and P. K. Takahashi, 2000: *Ocean Thermal Energy Conversion. In Encyclopedia of Electrical and Electronics Engineering*. J. G. Webster (ed.), Wiley-Interscience, New York.

Max, M. D., A. Johnson, and W. P. Dillon, 2006: *Economic Geology of Natural Gas Hydrate*. Springer-Verlag, Berlin.

McKay, G., 2002: Dioxin characterisation, formation and minimisation during municipal solid waste (MSW) incineration: review. *Chemical Engineering Journal*, **86**(3): 343–368.

McKelvey, V. E., 1967: Mineral Resource Estimates and Public Policy. *American Scientist,* (**60**): 32–40.

McKenzie-Brown, P., 2008: Colin Campbell and the Cracks of Doom. languageinstinct.blogspot.com/2008/03/colin-campbell-and-crack-of-doom.html (accessed January 17, 2009).

MEA (Millennium Ecosystem Assessment), 2005: *Ecosystems and Human Well-Being – Our Human Planet*. Summary for Decision Makers, Island Press, Washington, DC.

Mendeleev, D., 1877: L'Origine du Petrole. *Revue Scientifique*, **2**(8): 409–416.

Meyer, R. F. and W. D. Dietzman, 1979: *World Geography of Heavy Crude Oils*. In UNITAR Future of Heavy Crude and Tar Sands Conference, 4–12 July, Edmonton, AB.

Meyer, R. F. and W. D. Dietzman, 1981: *Future of Heavy Crude and Tar Sands*. In The Future of Heavy Crude and Tar Sands. R. F. Meyer, C. T. Steele, and J. C. Olson, (eds.), McGraw Hill New York.

Meyer, R. F. and J. W. Duford, 1989: *Resources of heavy oil and natural bitumen worldwide*. In Fourth United Nations Institute for Training and Research/United Nations Development Program International Conference on Heavy Crude and Tar Sands, Edmonton.

Meyer, R. F. and E. D. Attanasi, 2004: *Natural Bitumen and Extra Heavy Oil, Chapter 4*. In World Energy Council – Survey of Energy Resources J. Trinnaman and A. Clarke, (eds.), Elsevier, Amsterdam, pp. 93–117.

Meyer, R. F., E. D. Attanasi and P. A. Freeman, 2007: Heavy oil and natural bitumen resources in geological basins of the world. Open-File Report 2007–1084, US Geological Survey, Washington, DC. pubs.usgs.gov/of/2007/1084/ (accessed November 29, 2010).

Minchener, A., 2007: *Coal Supply Challenges for China*. CCC/127, IEA-Clean Coal Centre, London.

Monni, S., R. Pipatte, A. Lehtilä, I. Savolainen and S. Syri, 2006: *Global climate change mitigation scenarios for solid waste management*. Technical Research Centre of Finland (VTT), Epoo.

Moridis, G. J., T. S. Collett, R. Boswell, M. Kurihara, M. T. Reagan, E. D. Sloan and C. Koh, 2008: *Toward production from gas hydrates: assessment of resources and technology and the role of numerical simulation*. SPE 114163. In Proceedings of the 2008 SPE Unconventional Reservoirs Conference. 10–12 February, Keystone, CO.

Müller, C., A. Bondeau, H. Lotze-Campen, W. Cramer and W. Lucht, 2006: Comparative Impact of Climatic and Nonclimatic Factors On Global Terrestrial Carbon and Water Cycles. *Global Biogeochemical Cycles,* **20**(4): GB4015.

Nasr, T. N. and O. R. Ayodele, 2005: *Thermal Techniques for the Recovery of Heavy Oil and Bitumen*. SPE# 97488. In the SPE International Improved Oil Recovery Conference in Asia Pacific. 5–6 December, Kuala Lumpur.

NEA (OECD Nuclear Energy Agency), 2006: *Forty Years of Uranium Resources, Production and Demand in Perspective – The Red Book Perspective*. NEA No.6096, OECD, Paris.

NEA/IAEA, 2008: *Uranium 2007: Resources, Production and Demand*. NEA No.6098, OECD/IAEA (OECD Nuclear Energy Agency and International Atomic Energy Agency), Paris.

NEA/IAEA, 2010: *Uranium 2009: Resources, Production and Demand*. OECD/IAEA (OECD Nuclear Energy Agency and International Atomic Energy Agency), Paris.

Nihous, G. C., 2005: An Order-of-Magnitude Estimate of Ocean Thermal Energy Conversion Resources. *Journal of Energy Resources Technology*, **127**(4): 328–333.

Nobukawa, H., M. Kitamura, S. A. M. Swylem, and K. Ishibashi, 1994: *Development of a floating type system for uranium extraction from seawater using sea current and wave power*. In Proceedings of the 4th International Offshore and Polar Engineering Conference. 10–15 April, Osaka.

Nord-Larsen, T. and B. Talbot, 2004: Assessment of forest-fuel resources in Denmark: technical and economic availability. *Biomass and Bioenergy,* **27**(2): 97–109.

NPC, 2007: *Hard Truths – Facing the Hard Truths about Energy*. NPC (National Petroleum Council), Washington, DC.

NREL, 2011: What is Ocean Thermal Energy Conversion? National Renewable Energy Laboratories (NREL). www.nrel.gov/otec/what.html (accessed 29th December, 2011).

Odell, P., 2004: *Why Carbon Fuels will Dominate the 21st Century's Global Energy Economy*. Multi-Science Publishing Co. Ltd, Brentwood, UK.

Odell, P., 2010: Refereed Papers: The Long-Term Future for Energy Resources' Exploitation. *Energy & Environment,* **21**(7): 785–802.

OGJ (Oil & Gas Journal), 2007: Worldwide Look at Reserves and Production. *Oil & Gas Journal*, **105**(48).

OGJ, 2010: *Various articles between April and December 2010 related to the BP Macondo Well blow-out in the Gulf of Mexico*. Various articles between July and December 2010 related to the Enbridge pipeline rupture in Michigan, USA.

Ongena, J. and G. van Oost, 2004: Energy for Future Centuries – Will Fusion be an Inexhaustible, Safe, and Clean Energy Source? *Fusion Science and Technology*, **45**(2T): 3–14.

OPEC (Organization of Petroleum Exporting Countries), 2008: *Annual Statistical Bulletin 2007*. OPEC, Vienna.

Pan, K., 2005: *The depth distribution of Chinese coal resources. In Presentation at the School of Social Development and Public Policy*. Fudan University, Shanghai.

PCA (The Parliament of the Commonwealth of Australia), 2006: *Australia's uranium – Greenhouse friendly fuel for an energy hungry world. A case study into the strategic importance of Australia's uranium resources for the Inquiry into developing Australia's non-fossil fuel energy industry*. House of Representatives, Standing Committee on Industry and Resources, Canberra.

Peixóto, J. and A. H. Oort, 1984: Physics of Climate. *Reviews of Modern Physics*, **56**(3): 365–429.

Pelc, R. and R. M. Fujita, 2002: Renewable energy from the ocean. *Marine Policy*, **26**(6): 471–479.

Penney, T. R. and D. Bharathan, 1987: Power from the Sea, *Sci. Am.* **256**(1) 86–92.

Pollack, H. N., S. J. Hurter, and J. R. Johnson, 1993: Heat Flow from the Earth's Interior: Analysis of the Global Data Set. *Rev. Geophys.,* **31**(3): 267–280.

Pulleman, M. M., J. Bouma, E. A. van Essen, and E. W. Meijles, 2000: Soil Organic Matter Content as a Function of Different Land Use History. *Soil Science Society of America Journal*, **64**(2): 689–693.

Rajabapaiah, P., S. Jayakumar, and A. K. N. Reddy, 1993: *Biogas Electricity – the Pura Village Case Study. In Renewable Energy, Sources for Fuels and electricity*. T. B. Johansson, H. Kelly, A. K. N. Reddy and R. H. Williams, (eds.), Island Press, Washington, DC.

Rempel, H., S. Schmidt, and U. Schwarz-Schampera, 2007: *Reserven, Ressourcen und Verfügbarkeit von Energierohstoffen – Jahresbericht 2006*. Bundesanstalt für Geowissenschaften und Rohstoffe, Hannover.

Repa, E. W., 2005: 2005 *Tip Fee Survey*. National Solid Wastes Management Association, Washington, DC.

Ritschel, W. and H.-W. Schiffer, 2007: *Weltmarkt für Steinkohle*. RWE-Rhein Braun A.G., Essen.

Roberts, B. W., D. H. Shepard, K. Caldeira, M. E. Cannon, D. G. Eccles, A. J. Grenier, and J. F. Freidin, 2007: Harnessing High-Altitude Wind Power. *IEEE Transactions on Energy Conversion*, 22(1): 136–144.

Rogner, H. H., 1997: An Assessment of World Hydrocarbon Resources. *Annual Review of Energy and the Environment*, 22(1): 217–262.

Rogner, H. H., F. Barthel, M. Cabrera, A. P. C. Faaij, M. Giroux, D. O. Hall, V. Kagramanian, S. Kononov, T. Lefevre, R. Moreira, R. Nötstaller, P. Odell and M. Taylor, 2000: Energy Resources. In World Energy Assessment J. Goldemberg, (ed.), *United Nations Development Program, World Energy Council*, New York, pp. 135–172.

Rühl, C., 2010: *Statistical Review of World Energy 2010 – What's Inside*. BP, London.

Sagar, A. D. and S. Kartha, 2007: Bioenergy and Sustainable Development? *Annual Review of Environment and Resources*, 32(1): 131–167.

San-Sebastiàn, M., B. Armstrong, J. A. Córdoba, and C. Stephens, 2001: Exposures and Cancer Incidence near Oil Fields in the Amazon Basin of Ecuador. *Occupational and Environmental Medicine*, 58(8): 517–522.

Schenk, C. J., T. A. Cook, R. R. Charpentier, R. M. Pollastro, T. R. Klett, M. E. Tennyson, M. A. Kirschbaum, M. E. Brownfield, and J. K. Pitman, 2009: *An estimate of recoverable heavy oil resources of the Orinoco Oil Belt, Venezuela*. US Geological Survey Fact Sheet 2009–3028.

Schimel, D. S., 1995: Terrestrial ecosystems and the carbon cycle. *Global Change Biology*, 1(1): 77–91.

Schlumberger Inc., 2010: Oilfield Glossary. www.glossary.oilfield.slb.com/Display.cfm?Term=enhanced%20oil%20recovery (accessed March 20, 2011).

Schneider E.A. and W.C. Sailor, 2008: Long-Term Uranium Supply Estimates. *Nuclear Technology*, 162: 379–387.

Searchinger, T., R. Heimlich, R. A. Houghton, F. Dong, A. Elobeid, J. Fabiosa, S. Tokgoz, D. Hayes, and T.-H. Yu, 2008: Use of U.S. Croplands for Biofuels Increases Greenhouse Gases through Emissions from Land-Use Change. *Science*, 319(5867): 1238–1240.

Shepherd, W., 2003: *Energy Studies*. Imperial College Press, World Scientific Publication Co., London.

Shifley, S. R., F. R. Thompson, W. D. Dijak, M. A. Larson, and J. J. Millspaugh, 2006: Simulated effects of forest management alternatives on landscape structure and habitat suitability in the Midwestern United States. *Forest Ecology and Management*, 229(1–3): 361–377.

Sigurdsson, H., 2000: *Volcanic Episodes and the Rate of Volcanism*. In Encyclopaedia of Volcanoes. J. Stix, (ed.) Academic Press, Burlington, MA.

Sitch, S., C. Huntingford, N. Gedney, P. E. Levy, M. Lomas, S. L. Piao, R. Betts, P. Ciais, P. Cox, P. Friedlingstein, C. D. Jones, I. C. Prentice and F. I. Woodward, 2008: Evaluation of the terrestrial carbon cycle, future plant geography and climate-carbon cycle feedbacks using five Dynamic Global Vegetation Models (DGVMs). *Global Change Biology*, 14(9): 2015–2039.

Smeets, E. and A. Faaij, 2007: Bioenergy potentials from forestry in 2050. *Climatic Change*, 81(3): 353–390.

Smeets, E. M. W., A. P. C. Faaij, I. M. Lewandowski, and W. C. Turkenburg, 2007: A bottom-up assessment and review of global bio-energy potentials to 2050. *Progress in Energy and Combustion Science*, 33(1): 56–106.

Smith, A., K. Brown, S. Ogilvie, K. Rushton and J. Bates, 2001: *Waste Management Options and Climate Change*. Final Report to the European Commission, Brussels.

Sørensen, B., 1979: *Renewable Energy*, Academic Press, London, UK.

Sørensen, B., 1999: *Long Term Scenarios for Global Energy Demand and Supply. Four Global Greenhouse mitigation Scenarios*. Roskilde University.

Sørensen, B., 2004: *Renewable Energy: Its Physics, Engineering, Use, Environmental Impacts, Economy and Planning Aspects*. Elsevier Academic Press, London.

SPE (Society of Petroleum Engineers), 2005: *Glossary of Terms Used in Petroleum Reserves/Resources Definitions*, Richardson, Texas, USA.

Statistics Canada, 2005: *Human Activity and the Environment. Statistics Canada-Environment Accounts and Statistics Division*, Ottawa.

Statistik der Kohlenwirtschaft, 2009: *Der Kohlenbergbau in der Energiewirtschaft der Bundesrepublik Deutschland im Jahre 2008* Statistik der Kohlenwirtschaft E.V., Essen-Köln.

Stefansson, V., 2005: *World Geothermal Assessment*. In Proceedings of the World Geothermal Congress 2005. 24–29 April, Antalya

Stinner, W., K. Möller and G. Leithold, 2008: Effects of biogas digestion of clover/grass-leys, cover crops and crop residues on nitrogen cycle and crop yield in organic stockless farming systems. *European Journal of Agronomy*, 29(2–3): 125–134.

Sugo, T., M. Tamada, S. Tadao, S. Takao, U. Masaki and K. Ryoichi, 2001: Recovery System for Uranium from Seawater with Fibrous Adsorbent and its Preliminary Cost Estimation. *Journal of the Atomic Energy Society of japan*, 43(10): 1010–1016.

Tahil, W., 2007: *The Trouble with Lithium*. Meridian International Research, Martainville, France.

Tamada, M., N. Seko, N. Kasai, and T. Shimizu, 2006: *Cost Estimation of Uranium Recovery from Seawater with System of Braid Type Adsorbent*. JAEA Takasaki Annual Report 2005, JAEA-Review 2006–042, Takasaki Advanced Radiation Research Institute, Japan.

Tamada, M., 2009: *Current status of technology for collection of uranium from seawater*. In Erice Seminar. Erice, Italy.

TERI (Tata Energy Research Institute), 1985: "Biogas Technology – an information package," Tata Energy Documentation and Information Centre, Mumbai.

Tester, J. W., E. M. Drake, M. J. Driscoll, M. W. Golay and P. W. Peters, 2005: *Sustainable Energy Choosing Among Options*. MIT Press, Cambridge, MA.

Themelis, N. J. and L. N. Themelis, 2007: *Trip of Nickolas Themelis to China. Waste-to-Energy Research and Technology Institute (WTERT), Chongqing*. Columbia University, New York.

Trieb, F., C. Schillings, M. O'Sullivan, T. Pregger, and C. Hoyer-Klick, 2009: *Global Potential of Concentrating Solar Power*. In Solar Places Conference. Berlin.

Tsoutsos, T., N. Frantzeskaki and V. Gekas, 2005: Environmental impacts from the solar energy technologies. *Energy Policy*, 33(3): 289–296.

Tubiello, F. N., J.-F. Soussana, and S. M. Howden, 2007: Crop and pasture response to climate change. *Proceedings of the National Academy of Sciences*, 104(50): 19686–19690.

Turek, M., K. Skryzynksi and A. Smolinksi, 2008: *Structure and Changes of Production Costs in 1998–2005 in the Polish Hard Coal Industry*. Glückauf Essen.

UNECE/FAO, 2009: *Forest Products Annual Market Review, 2008–2009*. United Nations Economic Commission for Europe and Food and Agriculture Organization of the United Nations (UNECE/FAO), New York/Geneva.

UNEP, 2009: *Assessing Biofuels – Towards Sustainable Production and Use of Resources*. United Nations Environment Programme (UNEP), Division of Technology Industry and Economics, International Panel for Sustainable Resource Management, Paris.

UNESC (United Nations Economic and Social Council), 1997: "United Nations International Framework Classification for Reserves/Resources." Economic Commission for Europe, Geneva, Switzerland.

United Nations Population Division, 2008: *World Urbanization Prospects: The 2007 Revision Population Database*. United Nations Economic & Social Affairs, New York. esa.un.org/unup/ (accessed March 23, 2011).

UNWWAP, 2006: *Water: A Shared Responsibility*. New York: United Nations World Water Assessment Program (UNWWAP).

US EIA, 2007: *Methodology for allocating municipal solid wastes to biogenic energy*. US Energy Information Administration (US EIA), US Department of Energy. Washington, DC.

US EIA, 2008: *International Energy Statistics. US Energy Information Administration (US EIA)*, US Department of Energy. tonto.eia.doe.gov/cfapps/ipdbproject/IEDIndex3.cfm?tid=3&pid=3&aid=6 (accessed May 1, 2011).

US EIA, 2010: *US Natural Gas Wellhead Price. US Energy Information Administration (US EIA)*, US Department of Energy. tonto.eia.gov/dnav/ng/hist/n9190us3m.htm (accessed May 1, 2011).

USGS, 1980: Principles of a Resource/Reserve Classification for Minerals. US Geological Survey (USGS) Circular 831.

USGS, 2000: *World Petroleum Assessment*. CD-ROM. Washington, DC, US Geological Survey (USGS).

USGS, 2008: *Circum-Arctic Resource Appraisal*. Fact Sheet 2008–3049, US Geological Survey (USGS), Washington, DC.

USGS, 2009: *Chapter D: Availability, Coal Resource Recoverability and Economics Evaluation in the United States- A Summary*. Professional Paper 1625 F. In The National Coal Resource Overview. B. S. Pierce and K. O. Donnen, (eds.), US Geological Survey (USGS), Washington, DC.

USGS, 2010: *Mineral commodity Summaries 2010 (Lithium)*. US Geological Survey (USGS), Washington, DC.

Van den Ende, K. and F. Groeman, 2007: Blue Energy. Leonardo Energy. www.leonardo-energy.org/drupal/book/export/html/2243 (accessed April 20, 2011).

Vega, L. A., 1995: *Ocean Thermal Energy Conversion. In Encyclopedia of Energy Technology and the Environment*. A. Bisio and S. Boots, (eds.), Wiley-Interscience, Hoboken, NJ.

Veil, J. A., K. P. Smith, D. Tomasko, D. Elcock, E. L. Blunt, and G. P. Williams, 1999: Disposal of NORM-Contaminated Oil Field Wastes in Salt Caverns. Contract W-31–109-Eng-38, U.S. Department of Energy Office of Fossil Energy, National Petroleum Technology Office, Washington, DC.

Veil, J. A. and J. J. Quinn, 2008: *Water Issues Associated with Heavy Oil Production*. Report ANL/EVS/R-08/4 Argonne National Laboratory for the U S. Department of Energy, National Petroleum Technology Office, Washington, DC.

Wagner, H., 1998: *Zur Frage der Wirtschaftlichen Nutzung von Vorkommen Mineralischer Rohstoffe*. In Energievorräte und Mineralische Rohstoffe: Wie Lange Noch? J. Zemann, (ed.), Verlag der Österreichische Akademie der Wissenschaften, Vienna, 12, pp. 149–175.

Wagner, H. and G. B. L. Fettweis, 2001: About science and technology in the field of mining in the Western world at the beginning of the new century. *Resources Policy*, **27**(3): 157–168.

Walsh, M. E., 2008: *U.S. cellulosic biomass feedstock supplies and distribution*. M & E Biomass, Oak Ridge, TN.

Watson, R. T., I. R. Noble, B. Bolin, N. H. Ravindranath, D. J. Verardo, and D. J. Dokken, 2000: *Land Use, Land-Use Change, and Forestry. Special Report of the IPCC*, Cambridge University Press, Cambridge, UK.

WBGU, 2008: *Welt im Wandel. Zukunftsfähige Bioenergie und nachhaltige Landnutzung*. Wissenschaftlicher Beirat der Bundesregierung Globale Umweltveränderungen (WBGU), Berlin.

WCI, 2008: *The Coal Resource: A Comprehensive Overview of Coal*. World Coal Institute (WCI), London.

WEC, 1974: *World Energy Conference Survey of Energy Resources 1974. 9th World Energy Conference, Detroit, September 1974*. World Energy Council (WEC), US National Committee of the World Energy Conference, New York.

WEC, 1994: *New Renewable Energy Resources*. World Energy Council (WEC), London.

WEC, 1998: *Survey of Energy Resources (18th Edition)*, World Energy Council (WEC), London.

WEC, 2007: *Survey of Energy Resources*. World Energy Council (WEC), London.

WEC, 2010: *Survey of Energy Resources*. World Energy Council (WEC), London.

Wellmer, F. W., 2008: Reserves and resources of the geosphere, terms so often misunderstood. Is the life index of reserves of natural resources a guide to the future? *Zeitschrift der Deutschen Gesellschaft für Geowissenschaften*, **159**(4): 575–590.

Wilhelm, W. W., J. Johnson, D. Karlen, and D. Lightle, 2007: Corn Stover to Sustain Soil Organic Carbon Further Constrains Biomass Supply. *Agronomy Journal*, **99**(6): 1665–1667.

Witze, A., 2007: Energy: That's oil folks…, *Nature*, **445**: 14–17.

WNA, 2009: *Environmental Aspects of Uranium Mining*. World Nuclear Association (WNA). www.world-nuclear.org/info/inf25.html (accessed April 20, 2010).

Wood, W. T. and W. Y. Jung, 2008: *Modelling the Extend of the Earth's Marine Methane Hydrate Cryosphere*. In Proceedings of the 6th International Conference on Gas Hydrates. July 6–10, Vancouver, BC.

World Bank, 2008a: *Global Purchasing Power Parities and Real Expenditures*. The World Bank, Washington, DC.

World Bank, 2008b: *World Development Report 2008: Agriculture for Development*. The World Bank, Washington, DC.

WPC (World Power Conference), 1936: *Statistical yearbook of the World Power Conference*. No.1 1933–1934, Central Office of the World Power Conference, London.

Yaksic, A. and J. E. Tilton, 2009: Using the cumulative availability curve to assess the threat of mineral depletion: The case of lithium. *Resources Policy*, **34**(4): 185–194.

Yang, B., Y. Lu, J. Sun, and S. Donghai, 2005: *Potential for Cellulosic Ethanol Production in China*. In International Symposium on Alcohol Fuels. San Diego, CA.

Yoshioka, T., K. Aruga, T. Nitami, H. Sakai, and H. Kobayashi, 2006: A case study on the costs and the fuel consumption of harvesting, transporting, and chipping chains for logging residues in Japan. *Biomass and Bioenergy,* **30**(4): 342–348.

Young, G. C., 2006: Zapping MSW with Plasma Arc: An economic evaluation of a new technology for municipal solid waste treatment facilities. *Pollution Engineering,* **38**(11): 26–29.

8

Energy End-Use: Industry

Convening Lead Author (CLA)
Rangan Banerjee (Indian Institute of Technology-Bombay)

Lead Authors (LA)
Yu Cong (Energy Research Institute, China)
Dolf Gielen (United Nations Industrial Development Organization)
Gilberto Jannuzzi (University of Campinas, Brazil)
François Maréchal (Swiss Federal Institute of Technology Lausanne, Switzerland)
Aimee T. McKane (Lawrence Berkeley National Laboratory, USA)
Marc A. Rosen (University of Ontario Institute of Technology, Canada)
Denis van Es (Energy Research Centre, South Africa)
Ernst Worrell (Utrecht University, the Netherlands)

Contributing Authors (CA)
Robert Ayres (European Institute of Business Administration, France)
Marina Olshanskaya (United Nations Development Programme)
Lynn Price (Lawrence Berkeley National Laboratory, USA)
Değer Saygin (Utrecht University, the Netherlands)
Ashutosh Srivastava (Indian Institute of Technology-Bombay)

Review Editor
Eberhard Jochem (Fraunhofer Institute for Systems and Innovation Research, Germany)

Contents

Executive Summary

The industrial sector accounts for about 30% of the global final energy use and accounts for about 115 EJ of final energy use in 2005.[1] Cement, iron and steel, chemicals, pulp and paper and aluminum are key energy intensive materials that account for more than half the global industrial use.

There is a shift in the primary materials production with developing countries accounting for the majority of the production capacity. China and India have high growth rates in the production of energy intensive materials like cement, fertilizers and steel (12–20%/yr). In different economies materials demand is seen to grow initially with income and then stabilize. For instance in industrialized countries consumption of steel seems to saturate at about 500 kg/capita and 400–500 kg/capita for cement.

The aggregate energy intensities in the industrial sectors in different countries have shown steady declines – due to an improvement in energy efficiency and a change in the structure of the industrial output. As an example for the EU-27 the final energy use by industry has remained almost constant (13.4 EJ) at 1990 levels. Structural changes in the economies explain 30% of the reduction in energy intensity with the remaining due to energy efficiency improvements.

In different industrial sectors adopting the best achievable technology can result in a saving of 10–30% below the current average. An analysis of cost cutting measures for motors and steam systems in 2005 indicates energy savings potentials of 2.2 EJ for motors and 3.3 EJ for steam. The payback period for these measures range from less than 9 months to 4 years. A systematic analysis of materials and energy flows indicates significant potential for process integration, heat pumps and cogeneration for example savings of 30% are seen in kraft, sulfite, dairy, chocolate, ammonia, and vinyl chloride.

An exergy analysis (second law of thermodynamics) reveals that the overall global industry efficiency is only 30%. It is clear that there are major energy efficiency improvements possible through research and development (R&D) in next generation processes.

A comparison of energy management policies in different countries and a summary of country experiences, program impacts for Brazil, China, India, South Africa shows the features of successful policies. Energy management International Organization for Standardization (ISO) standards are likely to be effective in facilitating industrial end use efficiency. The effective use of demand side management can be facilitated by combination of mandated measures and market strategies.

A frozen efficiency scenario is constructed for industry in 2030. This implies a demand of final energy of 225 EJ in 2030. This involves an increase of the industrial energy output (in terms of Manufacturing Value Added (MVA)) by 95% over its 2005 value. Due to normal efficiency improvements the Business as Usual scenario results in a final energy demand of 175 EJ. The savings possibilities in motors and steam systems, process improvements, pinch, heat pumping and cogeneration have been computed for the existing industrial stock and for the new industries. An energy efficient scenario for 2030 has been constructed with a 95% increase in the industrial output with only a 17% increase in the final energy demand (total final energy demand for industry (135 EJ)). The total direct and indirect carbon dioxide emissions from the industry sector in 2005 is about 9.9 $GtCO_2$. Assuming a constant carbon intensity of energy use, the business as usual scenario results in carbon dioxide (CO_2) emissions increasing to 17.8 $GtCO_2$ annually in 2030. In the energy efficient scenario this reduces to 11.6 $GtCO_2$. Renewables account for 9% of the final energy of industry (10 EJ in 2005). If an aggressive renewables strategy resulting in an increase in renewable energy supply to 23% in 2030 is targeted (23 EJ), it is possible to have a scenario of constant greenhouse gas (GHG) emissions by the industrial sector (at 2005 levels) with a 95% increase in the industrial output.

---

1   This includes energy for coke ovens, blast furnaces and feedstock for petrochemicals.

Several interventions will be required to achieve the energy efficient or constant GHG emission scenario. For the existing industry measures include developing capacity for systems assessment for motors, steam systems and pinch analysis, sharing and documentation of best practices, benchmarks and roadmaps for different industry segments, access to low interest finance etc. A new energy management standard has been developed by ISO for energy management in companies. Its adoption will enable industries to systematically monitor and track energy efficiency improvements. In order to level the playing field for energy efficiency a paradigm shift is required with the focus on energy services not on energy supply per se. This requires a re-orientation of energy supply, distribution companies and energy equipment manufacturing companies.

Planning for next generation processes and systems needs the development of long term research agenda and strategic collaborations between industry, academic and research institutions and governments.

8.1   Introduction

The industrial sector is an important end-use sector, since all industrial processes require energy for the conversion of raw materials into desired products. The objective of this chapter is to assess the end-use efficiency of different industrial processes and systems. Earlier assessments include the End-Use Efficiency chapter of the World Energy Assessment (UNDP, 2000), the Industry chapter of Working Group III to the Fourth Assessment Report of the Intergovernmental Panel on Climate Change (IPCC) (2007) and the Energy Technology Perspectives 2008 scenarios and strategies for 2050 (IEA, 2008a).

The present analysis uses 2005 as the base year. We document time-series trends as well as regional variations in industrial energy use. The aim is to provide insights to understand parameters that affect global industrial energy use, review technological options, and identify potential for energy efficiency improvements. A review of industrial energy efficiency policies is also included.

Based on the status review, an energy efficiency scenario for 2030 is developed and the savings in energy and carbon dioxide (CO_2) emissions compared with respect to a frozen efficiency scenario and a business-as-usual (BAU) scenario.

8.2   Analysis of Industrial Energy Use Trends

The industrial sector accounted for 27% of the total global energy use in 2005 (IEA, 2008a). The total energy use by industry in 2005 was about 115 EJ (excluding traditional biomass and wood, which may add another 17 EJ). The share of final energy use by different industrial sectors in the world is shown in Figure 8.1.

8.2.1   Trends in Material Usage

Industry produces several products that are used by society on a daily basis. These products contain materials extracted from the environment. The conversion of the extracted feedstocks consumes large amounts of energy. A small number of key materials – cement, iron and steel, chemicals (plastics, fertilizer), pulp and paper, and aluminum – account for half of the global industrial energy use. Figure 8.2 shows trends in the global production of these materials.

Today, developing countries produce the majority of primary materials such as cement, steel, and fertilizers for infrastructure development. China alone produces about 46% of all the cement and 31% of the iron and steel in the world. As the industrial sectors of developing countries continue to grow, the same trend is likely to occur for other materials. Table 8.1 shows the comparison of production quantities of key energy-intensive materials in different countries. Among the developing countries, China and India show much higher growth rates and would

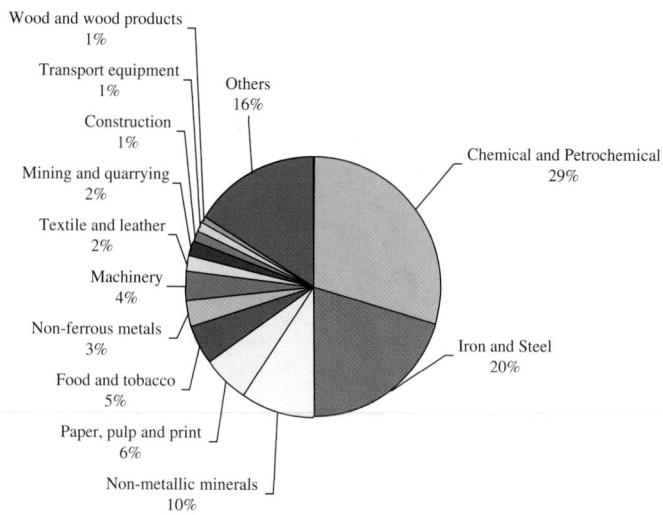

Figure 8.1 | Share of industrial final energy use in 2005. Source: data based on IEA, 2008a.

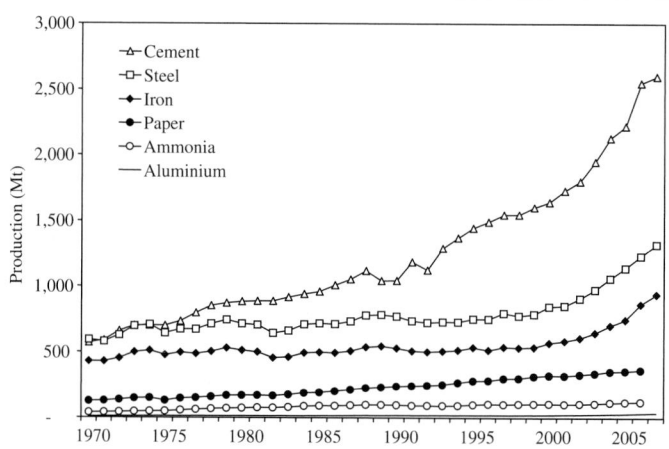

Figure 8.2 | Global production of key materials from 1970–2007. Source: data based on USGS, 2005; 2007; 2008 (cement, iron, steel, aluminum); IFA, 2009 (ammonia); FAO, 2009.

be responsible for the increased demand for materials in the future. Understanding future industrial energy use is based on future trends in material consumption and production. Generally, per capita materials demand increases with economic development and income, and is assumed to stabilize at a given level (following a so-called Kuznets curve; Mills and Waite, 2009). However, differences in the material intensity of different economies and regions suggest the potential to improve the efficiency with which we use materials.[2] Figures 8.3 and 8.4 depict the material intensity for cement and steel of various world regions.

---

2   Note that data availability also affects the material intensities. Consumption figures are given as apparent consumption, which equals domestic production plus imports minus exports of the material. Trade in products containing these materials (e.g., steel in a car) is not included in the apparent consumption. Hence, national-level data should be interpreted carefully. At a regional level, the data may provide a more consistent result.

Table 8.1 | Comparison of material production and growth rates in selected countries (2000–2007).

Regions	Material production in 2007 (Mt)									
	Steel	CAGR	Cement	CAGR	Paper and board	CAGR	Ammonia	CAGR	Primary aluminum	CAGR*
US	98	-0.5%	97	1.1%	84	-0.4%	9.5	-5.7%	2.6	-5.0%
Europe	202	1.1%	263	2.4%	100	4.2%	15.9	-0.3%	2.8	0.8%
South Korea	52	2.6%	57	1.5%	11	2.3%	0.1	-17.0%	N/A	N/A
Japan	120	1.8%	68	-2.5%	29	-1.4%	1.4	-3.3%	0.0	0.0%
China	495	21.4%	1361	12.5%	78	12.3%	51.6	6.3%	12.6	24.0%
India	53	10.2%	170	8.7%	4	1.4%	13.4	1.2%	1.2	9.6%
Brazil	34	2.8%	46	2.4%	9	4.7%	1.2	0.4%	1.7	3.8%
South Africa	9	1.0%	14	8.0%	3	5.7%	0.6	-2.2%	0.9	4.2%
World	1351	6.9%	2811	7.8%	386	2.6%	160	2.9%	38.0	6.5%

* CAGR = Compound Annual Growth Rate

Source: IISI, 2008; FAO, 2009; USGS, 2011.

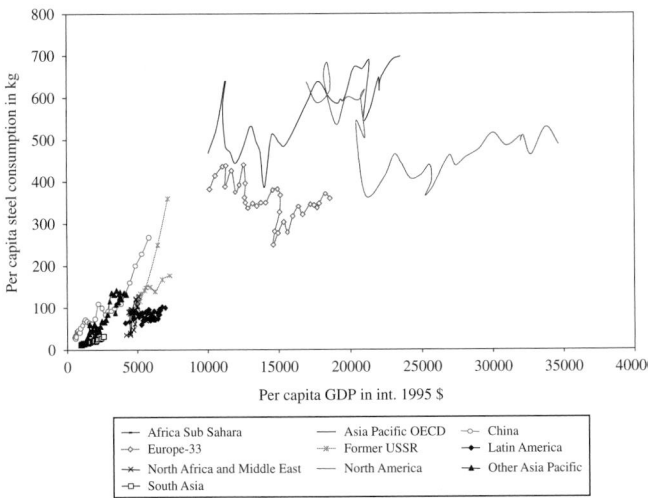

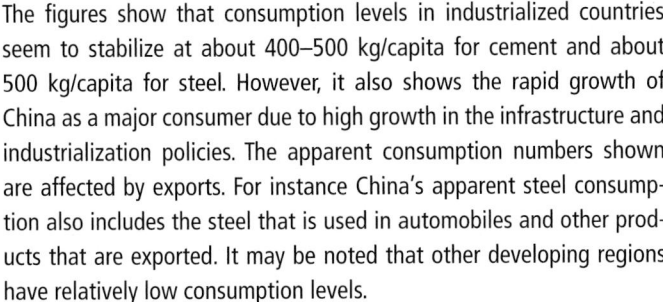

Figure 8.3 | Apparent steel consumption (expressed as kg/capita/yr) as a function of income (expressed as US$_{1995}$/capita) for different regions in the world.

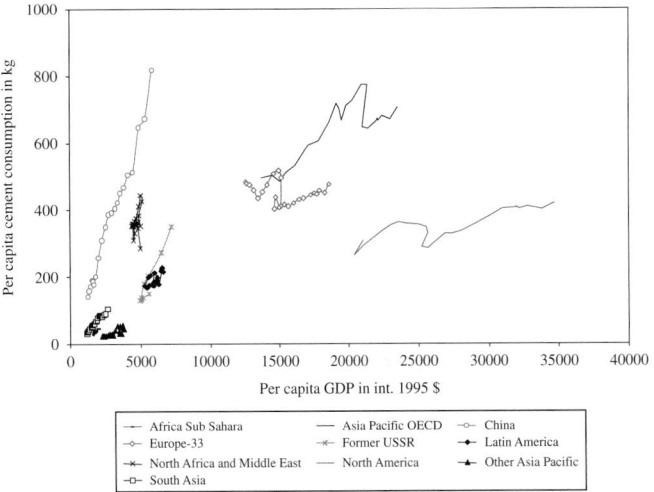

Figure 8.4 | Apparent cement consumption (expressed as kg/capita/yr) as a function of income (expressed as US$_{1995}$/capita) for different regions in the world.

The figures show that consumption levels in industrialized countries seem to stabilize at about 400–500 kg/capita for cement and about 500 kg/capita for steel. However, it also shows the rapid growth of China as a major consumer due to high growth in the infrastructure and industrialization policies. The apparent consumption numbers shown are affected by exports. For instance China's apparent steel consumption also includes the steel that is used in automobiles and other products that are exported. It may be noted that other developing regions have relatively low consumption levels.

Table 8.2 shows the conclusions obtained by Jänicke et al. (1992) for bulk material consumption and production per capita for different countries as a function of per capita Gross Domestic Product (GDP). De Vries et al. (2006) analyze trends in the per capita use of bulk materials including paper and board (see Figure 8.5), ammonia, bricks, polymers, and aluminum, in addition to cement and steel. For materials such as

paper and aluminum, there does not appear to be a saturation level. This could be due to an increase in the growth of the information and communications technology and aircraft sector.

The structure of the GDP and growth of the service sector for industrialized countries affect the overall trends. De Vries et al. (2006) conclude that there is no general trend for decoupling between physical and economic growth for industry.

The case of China is atypical. For example, China's 2005 production of cement of 1064 million tonnes (Mt) corresponds to a per capita production of 806 kg/capita. The cement industry in China is growing at more than 10%/yr. This is probably due to the high share of industry in China's GDP, high growth rates of infrastructure, and its export-oriented industrial development strategy. It is unlikely that other developing countries will reach this level of consumption/growth, as illustrated by the trends

Table 8.2 | General trends of per capita bulk materials production and consumption for 32 industrialized countries 1970–1990.

Product	Parameter	General trend
Paper and Paperboard	per capita production	increasing production at all income levels
Cement	per capita production	increasing producing until per capita GDP levels of US$5000–8000 generally decreasing production at higher GDP levels
Chlorine	per capita production	increasing production at all income levels
Pesticide	per capita production	increasing production at all income levels
Fertilizer	per capita production	increasing production until per capita GDP levels of US$9000 generally stabilizing production at higher GDP levels
	per capita consumption	increasing consumption until per capita GDP levels of US$8000 stabilizing consumption at higher GDP levels
Crude Steel	per capita production	increasing production until per capita GDP levels of US$6000–10000 decreasing or stabilizing production at higher GDP levels
	per capita consumption	increasing consumption until per capita GDP levels of US$5000–9000 Stabilizing or slightly decreasing consumption at higher GDP levels
Aluminum	per capita production	increasing production at all income levels
	per capita consumption	strong increase of consumption at all income levels

Source: Jänicke et al., 1992.

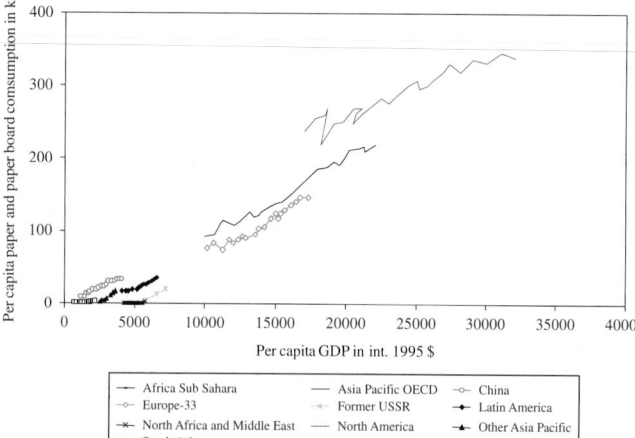

Legend:
— Africa Sub Sahara — Asia Pacific OECD ○ China
○ Europe-33 × Former USSR ◆ Latin America
✳ North Africa and Middle East — North America ▲ Other Asia Pacific
□ South Asia

Figure 8.5 | Paper and paper-based board – per capita consumption. Source: De Vries et al., 2006; Chateau et al., 2005.[3]

for India and Brazil, unless they focus on increasing their manufacturing capacity of cement for exports.

Despite this, the implication of these trends is the likely predominance of developing countries as major consumers (and producers) of energy-intensive materials during the next few decades. Investments in new, energy-efficient processes and plants, and efficient material usage (dematerialization) in developing countries will be important for managing the global industrial energy use.

8.2.2   Regional Variations

Table 8.3 shows a comparison of industrial energy use for select countries of the world. The Manufacturing Value Added (MVA) per capita,

total primary energy supply, and final energy and electricity use by industry are compared for different countries.

Different countries have different mixes of energy supply and sectoral energy use patterns. One of the factors affecting future global industrial energy use patterns is the growth of industry in developing countries. This is exemplified by the high growth rate of China in the production of energy-intensive materials, as shown in Table 8.4.

The trends of growth in developing countries and saturation in the energy-intensive industries of developed countries have implications on the future energy mix.

8.2.3   Structural Change

Overall in the economy there are structural shifts from agriculture to manufacturing to services. As countries develop, these structural shifts also result in changes in the overall energy intensity. Many countries have achieved a significant reduction in energy intensity in the industrial sector. Some of this effect has been due to a change in the structure of the industry, with a shift to less energy-intensive industry. These effects can be separated by decomposition analysis.

In the European Union (EU), the final energy use of industry in the 27 EU countries has remained almost constant since 1990 at 320 Mtoe (13.4 EJ). This has been possible through a 30% improvement in energy efficiency from 1990 to 2007 (2.1%/yr). For the EU-27, about 30% of the reduction has been due to structural changes. There are differences in the EU Member States. Figure 8.6 shows the changes in energy intensity for EU Member States from 2000–2007.

The example of the former Soviet Union illustrates the impact of structural change on industrial energy use. Olshanskaya (2004) revealed

---

3   This figure is from De Vries et al. 2006 which is based on (cites) the VLEEM Project report of Chateau et al., 2005.

Table 8.3 | Comparison of industrial energy use in selected countries for 2005.

	TPES (EJ)	Final Energy Use by Industry[2] (EJ)	Electricity use by Industry (EJ)	Industrial Share of GDP	MVA/capita[2]
World	478.9	115	22.2	32%	1014
Brazil	9.1	3.3	0.6	15%	594
China	72.7	24.6	4.9	34.1 %	492
India	22.5	5.5	0.8	14.1%	80
S. Korea	8.9	3.2	0.66	40.3%	187
Germany	14.4	2.38	0.83	21.4%	5090
UK	9.8	1.33	0.43	13.6%	3683
France	11.5	1.37	0.5	13.94%	3291
Japan	22.1	6.3	1.2	22.1%	8608
Russia	27.4	7.2	1.2	19.0%	461
South Africa	5.3	1.2	0.4	16.4%	550
USA	97.9	16.6	3.3	15.3%	5604

1 includes feedstocks (non-energy use); see Chapter 1, Section 1.2.2.

2 in constant US2000$ prices.

Source: IEA Database, 2011; UNIDO Database, 2011.

Table 8.4 | Production of energy-intensive materials in China, 2000–2005, in 10,000 tonnes.

Material	2000	2001	2002	2003	2004	2005	CAGR% (2000–2005)
Steel	128.5	151.6	182.2	222.2	272.8	352.4	22.4
Finished Steel	131.5	157	192.5	241.1	297.2	396.9	24.7
Nonferrous	7.8	8.8	10.1	12.3	14.3	16.4	15.8
Included Copper	1.4	1.5	1.6	1.8	2.2	2.6	13.7
Aluminium	3.0	3.6	4.5	6.0	6.7	7.8	21.1
Cement	597	661	725	862.1	966.8	1064	12.3
Flat glass	183.5	209.6	234.5	277	300.6	350	13.8
Ethylene	4.7	4.8	5.4	6.1	6.3	7.6	10.0
Synthetic ammonia	33.5	34.3	36.8	37.9	42.2	45	6.1
Caustic soda	6.7	7.9	8.8	9.5	10.6	12.6	13.6
Soda	8.3	9.1	10.3	11.3	13	14.7	12.0
Paper and paper board	30.5	37.8	46.7	48.5	54.1	54.0	12.1

Note: CAGR = Compound Annual Growth Rate.

Source: China Energy, 2009.

that in the Russian industrial sector there were changes toward a more energy-intensive industry between 1994–1997 which contributed positively to the increase in industrial energy intensity in that period. The trend reversed in the late 1990s, and until 2002 the aggregated contribution of structural changes within industry on industrial energy intensity was insignificant. Resulting positive changes in industrial energy intensity may be attributed to improvements in industrial energy efficiency per se (see Figure 8.7).

A decomposition analysis by Howarth et al. (1991) showed that the energy intensity of manufacturing declined by 45% in Japan during 1973–1987 (11.5% decline due to structure; 36.4% due to energy efficiency improvements), while for the United States the decline was 44.3% (14.8% due to structure; 32.4% due to energy efficiency improvements). An analysis of energy intensity trends in the US economy (Huntington, 2010) between 1997 and 2006 shows that structural change (within industries) accounted for more than half of the total energy intensity reduction in the United States.

In most economies there is a structural change where the share of energy-intensive industries is reducing in the total industrial mix. In order to account for this in an aggregate analysis, the decomposition analysis can

Table 8.5 | Energy use in the chemical and petrochemical industry, 2004 (excluding electricity).

	Amount	LHV	Feedstock Energy Needed	Fuel		Total Fuel + Feedstock
	Mt/yr	GJ/t	EJ/yr	GJ/t	EJ/yr	EJ/yr
Ethylene	103.3	47.2	4.9	13	1.3	6.2
Propylene	65.3	46.7	3.0	13	0.8	3.9
Butadiene	9.4	47.0	0.4	13	0.1	0.6
Butylene	20.3	47.0	1.0	10	0.2	1.2
Benzene	36.7	42.6	1.6	7	0.3	1.8
Toluene	18.4	42.6	0.8	7	0.1	0.9
Xylenes	33.7	41.3	1.4	7	0.2	1.6
Methanol	34.7	21.1	0.7	10	0.3	1.1
Ammonia	140.0	21	2.9	19	2.7	5.6
Carbon black	9.0	32.8	0.3	30	0.3	0.6
Soda ash	38.0	0.0	0.0	11	0.4	0.4
Olefins processing excl. polymerization	100.0	0.0	0.0	10	1.0	1.0
Polymerization	50.0	0.0	0.0	5	0.3	0.3
Chlorine and Sodium Hydroxide	45.0	0.0	0.0	2	0.1	0.1
Total			17.0		8.2	25.2

Source: IEA, 2007a.

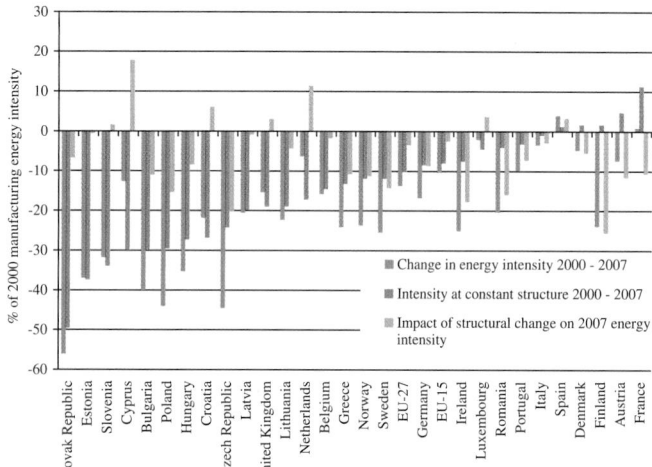

Figure 8.6 | Structural change impact for the EU. Source: Odyssee, 2009.

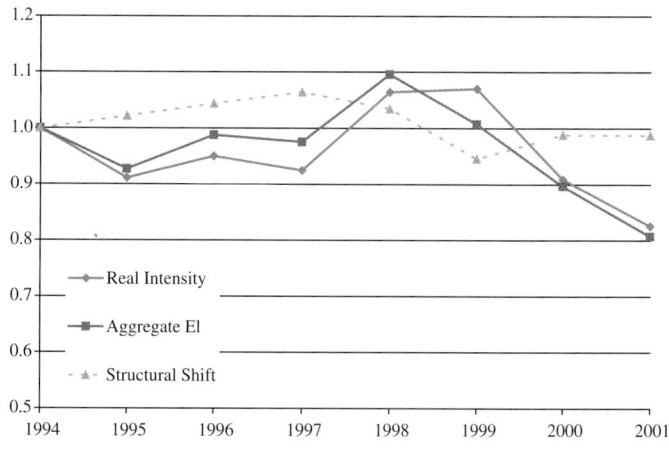

Figure 8.7 | Fisher ideal indices[4] for structural shift and real intensity in Russia, 1994 = 1. Source: Olshanskaya, 2004.

be used to reveal the actual impact of energy efficiency improvements in industry, as illustrated by the example shown in this section.

8.3 Consumption and Opportunities: Key Sectors

Industrial processes have significant variations in the energy use per unit of output depending on the vintage (age), process technology employed, quality of input new materials, and scale. Revamping old process plants often requires significant capital investment. In most industrial

processes there is a learning curve effect with newer plants being more energy efficient than earlier process plants. A few decades ago, developing countries often only had access to second hand plants and outdated technologies, and hence were often more inefficient than process plants in developed countries. This generalization is no longer possible with the world's most efficient aluminum smelters located in Africa and the most efficient cement plants located in India. Globalization has also resulted in new capital stock being of larger capacity with cutting edge

---

4   The Fisher ideal index is the geometric average of the Laspayre's and Paasche's price indices (Boyd and Roop, 2004).

Table 8.6 | Process-specific energy efficiency opportunities in ammonia production.

Measures
Highly integrated primary and secondary reformers
Improvements in reformers
Pre-reformer installation
Low-pressure ammonia synthesis
Highly efficient catalysts
Physical absorption CO_2 removal
CO_2 recovery with improved solvents and other improvements
Hydrogen recovery
Improved process control
Process integration

Source: FEMA, 2000; Nieuwlaar, 2001; Rafiqul et al., 2005; EC, 2007; EC, 2007; Worrell et al., 2008.

Table 8.7 | Revamp investments in natural gas-fueled steam reforming plants.

Retrofit measure	Average improvement	Range	Uncertainty Parameter	Cost	Applicability		
	(GJ/t)	(GJ/t)	(%)	(€ per t/yr)	EU (%)	US (%)	India (%)
Reforming large improvements	4.0	±1.0	17	24	10	15	10
Reforming moderate improvements	1.4	±0.4	20	5	20	25	20
Improvement CO_2 removal	0.9	±0.5	33	15	30	30	30
Low pressure synth	0.5	±0.5	67	6	90	90	90
Hydrogen recovery	0.8	±0.5	50	2	0	10	10
Improved process control	0.72	±0.5	50	6	30	50	30
Process integration	3.0	±1.0	23	3	10	25	20

Source: Rafiqul et al., 2005.

technologies. In some countries, (e.g., Russian Federation and Ukraine) the existing process plants that are inefficient have not been modernized due to the lack of capital investments (IEA, 2007a). This section provides an overview of chemicals and fertilizers, iron and steel, cement, pulp and paper, and aluminum, and discusses the factors affecting the energy use in these sectors.

8.3.1   Chemicals and Fertilizers

The chemical industry is highly diverse, with thousands of companies producing tens of thousands of products in quantities varying from a few kilograms to thousands of tonnes (t). Due to this complexity, reliable data on energy use are not available (Worrell et al., 2000a). However, a small number of (intermediate) products make up a large share of energy use in this sector – e.g., ammonia, chlorine and alkalines, ethylene, and other petrochemical intermediates. The chemicals and petrochemicals sector has a large number of products. Table 8.5 (IEA, 2007a) lists the major products that account for about 80% of the total energy use of the chemicals and petrochemicals sector.

Ethylene is a basic chemical that is used in the production of plastics and other chemical products. This is produced by steam cracking of hydrocarbon feedstocks. During this process several by-products are obtained like hydrogen, methane, propylene and other heavier hydrocarbons. Steam cracking consumes about 65% of the total energy used in ethylene production (Worrell et al., 2000a; Ren et al., 2006). Technology options like improved furnace and cracking tube materials, and cogeneration using furnace exhaust can result in 20% of total energy savings (IPCC, 2007). Improved separation and compression techniques (e.g., absorption technologies for separation) can result in 15% of total energy saving. Instead of steam cracking, alternative processes have been developed for converting methane in natural gas to olefins. However state of the art stream cracking of naphtha is more efficient than these processes (Ren et al., 2006).

Global ammonia (NH_3) production (mainly for fertilizer production) was estimated at 125 Mt in 2007. The main producers are China, Russia, India, the United States, Trinidad and Tobago, Indonesia, and Ukraine. The fertilizer industry accounts for about 1.2% of world energy use, and more than 90% of this energy is used in the production of ammonia. Modern

Table 8.8 | Summary of process-specific energy efficiency opportunities.

Process Specific Measures	
Process	**Measures**
Ethylene	More selective furnace coils
	Improved transfer line exchangers
	Secondary transfer line exchangers
	Increased efficiency cracking furnaces
	Pre-coupled gas turbine to cracker furnace
	Higher gasoline fractionator bottom temperature
	Improved heat recovery quench water
	Reduced pressure drop in compressor inter-stages
	Additional expander on de-methanizer
	Additional re-boilers (cold recuperation)
	Extended heat exchanger surface
	Optimization steam and power balance
	Improved compressors
Aromatics	Improved product recovery systems
Polymers	Low pressure steam recovery
	Gear pump to replace extruder
	Online compounding extrusion
	Re-use solvents, oils and catalysts
Ethylene Oxide / Ethylene Glycol	Increased selectivity catalyst
	Optimal design EO/EG-sections
	Multi-effect evaporators (Glycol)
	Recovery and sales of by-product CO_2
	Process integration
Ethylene Dichloride / Vinyl Chloride Monomer	Optimize recycle loops
	Gas-phase direct chlorination of ethylene
	Catalytic cracking EDC
Styrene	Condensate recovery and process integration

Source: Neelis et al., 2008.

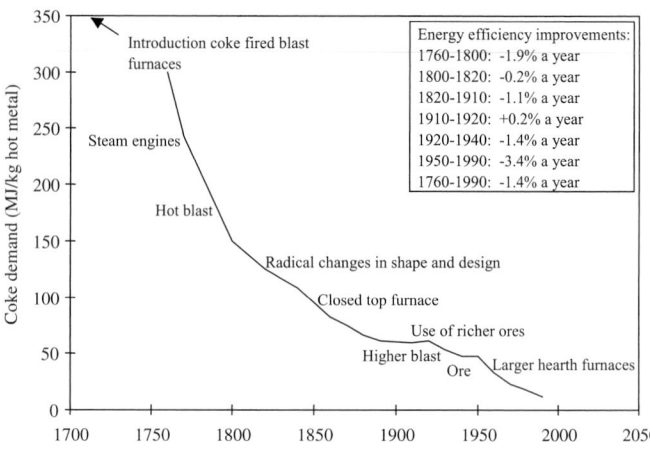

Figure 8.8 | Change in coke demand due to efficiency improvement in the blast furnace process. Source: de Beer et al., 1998.

Chinese coal-based production is about 53 GJ/t, compared with a global average of 41.4 GJ/t (Saygin et al., 2009). A summary of process-specific options for ammonia is shown in Table 8.6, while Table 8.7 from Rafiqul et al. (2005) shows the investments and energy-saving possibilities available from revamping ammonia plants.

A summary of process-specific energy efficiency opportunities in the petrochemical industry is shown in Table 8.8 below. The selection is limited to commercially available technologies and excludes emerging and cross-cutting technologies. Process integration offers significant scope for energy savings and is discussed in a subsequent section.

The use of nanocomposites as a filler material can help in reducing the energy use in polymer manufacture by 20% (Roes et al., 2010).

8.3.2   Iron and Steel

Steel is an important metal. The total global production of steel in 2007 was about 1350 Mt. The major steel producers were China (36% of global steel production), EU25 (15%), Japan (9%), and US (7%) (IISI, 2008). The main route used for steel making is the blast furnace route using coke or coal to reduce iron-ore oxides in a blast furnace to molten iron that is then processed to steel. About 60% of the global steel production is from this route (IPCC, 2007). Another important route accounting for 32% of steel production is the production of steel from melting scrap steel in an electric arc furnace (EAF). Since the raw material used in this route is scrap steel, the specific energy use in this process is only 30–40% of the blast furnace steel process route.

An alternative route is the use of natural gas or coal to produce direct reduced iron (DRI) that can be used in an electric arc furnace. DRI use and production is expected to grow as the share of electric arc furnaces grows in industrialized countries and globally. At present DRI accounts for only about 3% of total steel production.

ammonia plants are designed to use about half the energy per tonne of product than those designed in the 1960s, with energy use dropping from over 60 GJ/t of ammonia in the 1960s to 28 GJ/t of ammonia in the most recently designed plants (Worrell et al., 2009). Benchmarking data indicate that the best-in-class performance of operating plants ranges from 28.0 GJ/t to 29.3 GJ/t of ammonia (Chaudhary, 2001; PSI, 2004). Individual differences in energy performance are mostly determined by feedstock (natural gas compared with heavier hydrocarbons) and the age and size of the ammonia plant (Phylipsen et al., 2002; PSI, 2004).

Ammonia plants that use natural gas as a feedstock have an energy efficiency advantage over plants that use heavier feedstocks, and a high percentage of global ammonia production capacity is already based on natural gas. China is an exception, in that 67% of its ammonia production is based on coal (CESP, 2004) and small-scale plants account for 90% of the coal-based production. The average energy intensity of

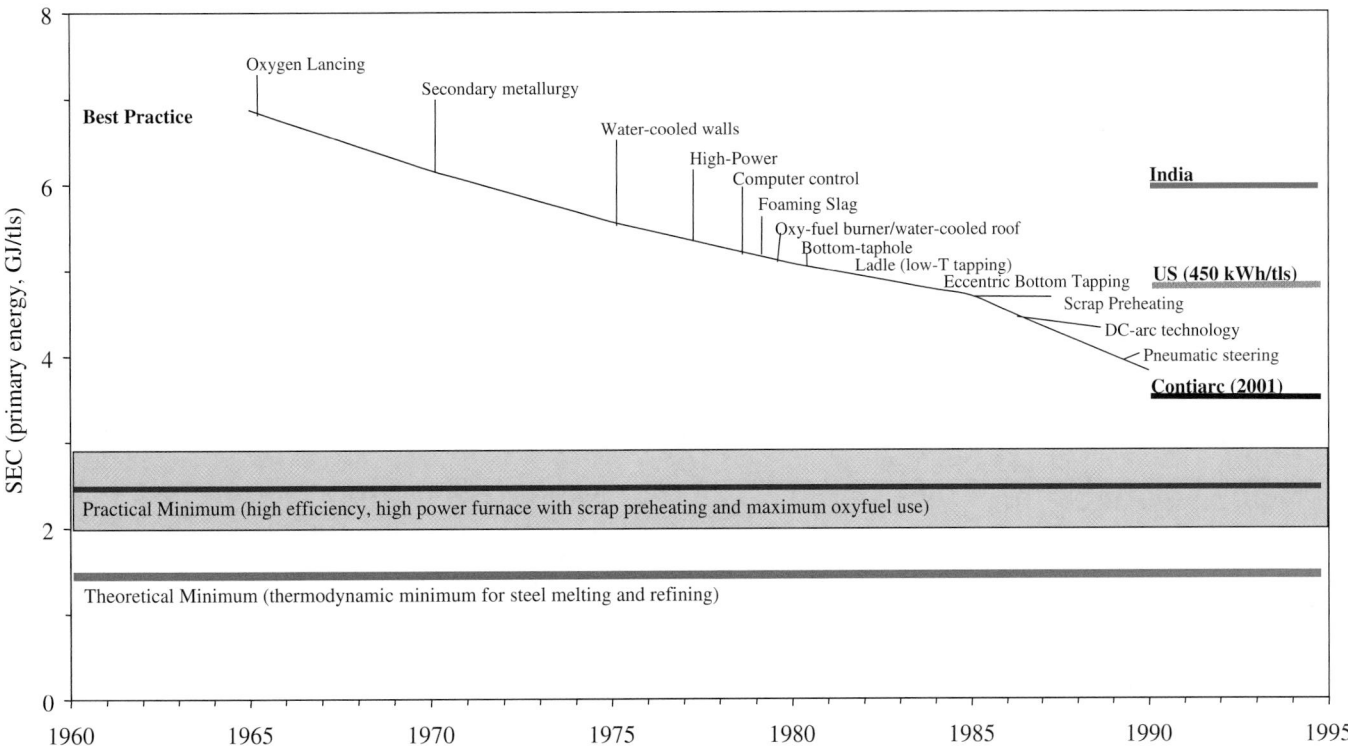

Figure 8.9 | Time-series trend of specific electricity consumption (SEC) values for the EAF process. Note: The Contiarc electric arc furnace was introduced in 2001 and provides an energy efficient option for the production of cast iron. tls = tonne of liquid steel.

Iron and steel making traditionally includes several batch processes. The introduction of continuous casting in steel making in the 1970s and 1980s resulted in significant energy and material savings. Continuous casting now accounts for about 93% of the world's steel production (IISI, 2008). Some major energy efficiency measurement adopted by the steel industry are enhancing continuous production processes to reduce heat loss, increasing recovery of waste energy and process gases, and efficient design of electric arc furnaces – for example, scrap preheating, high-capacity furnaces, foamy slagging, and fuel and oxygen injection. The effect of efficiency improvements on coke demand in the blast furnace process is shown in Figure 8.8. A time-series trend of specific energy use improvement is also shown for the EAF process in Figure 8.9.

Energy savings can be achieved by a combination of stock turnover and equipment retrofit. An analysis of electric arc furnaces in the US steel industry from 1990–2002 showed an efficiency improvement of 1.3%/yr (0.7% due to stock turnover and 0.5% due to equipment retrofit) (Worrell and Biermans, 2005).

Process modifications like near-net shape casting and smelt reduction, which integrates ore agglomeration, coke making and iron production in a single process, offering an energy-efficient alternative at small to medium scales (de Beer et al., 1998) offer scope for further improvements in energy efficiencies.

POSCO, a Korean steel producer, has developed a technology to replace the blast furnace (FINEX process technology) and constructed

a demonstration plant with a capacity of 600,000 t/yr in 2003. The coal consumption is about 770 kg/t of hot metal. (Siemens VAI, 2009).

A summary of process-specific energy opportunities in the iron and steel industry is shown in Table 8.9.

8.3.3 Cement

Cement is needed in the construction sector and is important for the growth of any economy. Cement is produced in almost all countries of the world. Developing countries account for about 73% of the global cement production (2811 Mt in 2007). China (1361 Mt in 2007) accounts for almost half of the global cement production (USGS, 2011).

Cement production is also highly energy- and CO_2-intensive. Clinker is the output of the cement kiln. Depending on the type of cement to be manufacture the clinker is further processed in a set of finishing operations. The production of clinker, the principal component of cement, consumes virtually all the fuel and emits CO_2 from the calcination of limestone. The major energy uses are fuel for the production of clinker and electricity for grinding raw materials and the finished cement. Coal dominates in clinker making.

The technical potential for energy efficiency improvements is about 40% (Worrell et al., 1995; Kim and Worrell, 2002b). An analysis of the US cement industry identified 30 opportunities for energy saving in the

Table 8.9 | Summary of process-specific energy opportunities in the iron and steel industry.

Iron Ore and Ferrous Reverts Preparation (Sintering)	
Heat recovery from sintering and sinter cooler	Use of waste fuel in sinter plant
Reduction of air leakage	Improve charging method
Increasing bed depth	Improve ignition oven efficiency
Emission Optimized Sintering (EOS®)	Other measures

Coke Making	
Coal moisture control	Coke dry quenching (CDQ)
Programmed heating	Coke oven gas (COG)
Variable speed drive coke oven gas compressors	Next generation coke making technology
Single Chamber System (SCS)	

Iron Making – Blast Furnace	
Injection of pulverized coal	Recovery of blast furnace gas
Injection of natural gas	Top gas recycling
Injection of oil	Improved blast furnace control
Injection of plastic waste	Slag heat recovery
Injection of coke oven gas and basic oxygen furnace gas	Preheating of fuel for hot stove
Charging carbon composite agglomerates (CCB)	Improvement of combustion in hot stove
Top-pressure recovery turbines (TRT)	Improved hot stove control

Steelmaking – Basic Oxide Furnace	
Recovery of BOF gas and sensible heat	Improvement of process monitoring and control
Variable speed drive on ventilation fans	Programmed and efficient ladle heating
Ladle preheating	

Steelmaking – EAF	
Increasing power	Refractories using engineering particles
Adjustable speed drives (ASDs)	Direct current (DC) arc furnace
Oxy-fuel burners/lancing	Scrap preheating
Post-combustion of flue gases	Waste injection
Improving process control	Airtight operation
Foamy slag practices	Bottom stirring/gas injection

Casting and Refining	
Integration of casting and rolling	Tundish heating
Ladle preheating	

Shaping	
Use efficient drive units	Installation of lubrication system
Gate Communicated Turn-Off (GCT) inverters	

Hot Rolling	
Recuperative or regenerative burners	Integration of casting and rolling
Flameless burners	Proper reheating temperature
Controlling oxygen levels and variable speed drives on combustion air fans	Process control in hot strip mill
Avoiding overload of reheat furnaces	Heat recovery to the product
Insulation of reheat furnaces	Waste heat recovery from cooling water
Hot charging	

Cold Rolling	
Continuous annealing	Inter-electrode insulation in electrolytic pickling line
Reducing losses on annealing line	Automated monitoring and targeting systems
Reduced steam use in the acid pickling line	

Source: Worrell et al., 2010.

cement industry with an economic potential of 11% savings in energy and 5% savings in emissions (Worrell et al., 2000b; Worrell and Galitsky, 2005). Blending of clinker with alternative cementitious materials like blast furnace slags, fly ash from coal fired power plants and natural pozzolanes can result in reduced energy and CO_2 emissions (IPCC, 2007). Worrell et al. (1995) and Humphreys and Mahasenan (2002) estimate that the use of blended cement has the potential to reduce CO_2 emissions by more than 7%.

Geo polymers and other alternatives to limestone-based cement are being studied (Humphreys and Mahasenan, 2002; Gartner, 2004) but are currently not economical for widespread deployment.

The energy use of the cement industry in China in 2005 was about 50% of energy consumption of the building materials industry, and became the largest energy consumer in the industry. From 2000 to 2005, the cement industry's energy consumption dropped from 5.0 GJ/t in 2000 to 4.36 MJ/t in 2005 (as shown in Table 8.10).

The small-scale cement industries in India had an average fuel consumption of 3.7 GJ/t of clinker and an average electricity consumption of 104 kWh/t of cement, while the average fuel consumption for large cement industries was 3.29 GJ/t and electricity consumption was 92 kWh/t (Bhushan and Hazra, 2005). The Indian cement industry is among the most efficient in the world. But there is still considerable scope for improvement in the energy use per tonne of output compared to the world's best, due to the potential of more blending in the cement, as shown in Figure 8.10. The blending of fly ash in the cement results in a reduction in the specific energy use. The figure also shows a high clinker content compared to the world's most efficient cement industries. The energy-efficient practices and technologies in cement production are shown in Table 8.11.

An analysis of ten large cement plants in India that account for 16% of total production has been carried out based on data from projects implemented between 2001 and 2006 (Bureau of Energy Efficiency Awards, 2006). The measures have been grouped into different categories, and the conservation supply curve is shown in Figure 8.11. About 8% of annual electricity consumption has been saved by these measures. The cost of saved energy (CSE) is computed by annualizing the cost of the measure and dividing by the annual electricity saving. The CSE varies from INR0.1–1.7/kWh, which is lower than the average price of electricity (INR4.5/kWh or US$0.10/kWh).

Table 8.10 | Specific energy use for cement in China.

	unit	2000	2001	2002	2003	2004	2005	Descending rate/yr
Cement energy use	GJ/t	5.0	4.9	4.7	4.6	4.5	4.4	2.8%

Source: Xiong, 2007.

Table 8.11 | Energy-efficient practices and technologies in cement production.

Raw Materials Preparation	
Efficient transport systems (dry process) Slurry blending and homogenization (wet process)	
Raw meal blending systems (dry process)	
Conversion to closed circuit wash mill (wet process)	
High-efficiency roller mills (dry process)	
High-efficiency classifiers (dry process) Fuel Preparation: Roller mills	

Clinker Production (Wet)	Clinker Production (Dry)
Energy management and process control Seal replacement Kiln combustion system improvements	Energy management and process control Seal replacement Kiln combustion system improvements
Kiln shell heat loss reduction	Kiln shell heat loss reduction
Use of waste fuels	Use of waste fuels
Conversion to modern grate cooler	Conversion to modern grate cooler
Refractories	Refractories
Optimize grate coolers	Heat recovery for power generation
Conversion to pre-heater, pre-calciner kilns Conversion to semi-dry kiln (slurry drier) Conversion to semi-wet kiln	Low pressure drop cyclones for suspension pre-heaters Optimize grate coolers Addition of pre-calciner to pre-heater kiln
Efficient kiln drives	Long dry kiln conversion to multi-stage pre-heater kiln
Oxygen enrichment	Long dry kiln conversion to multi-stage pre-heater, pre-calciner kiln
	Efficient kiln drives
	Oxygen enrichment

Finish Grinding	
Energy management and process control Improved grinding media (ball mills)	
High-pressure roller press	
High efficiency classifiers	

General Measures	
Preventative maintenance (insulation, compressed air system, maintenance)	
High efficiency motors	
Efficient fans with variable speed drives	
Optimization of compressed air systems Efficient lighting	
Product & Feedstock Changes	
Blended Cements Limestone cement	
Low Alkali cement	
Use of steel slag in kiln (CemStar®) Reducing fineness of cement for selected uses	

Source: Worrell and Galitsky, 2008.

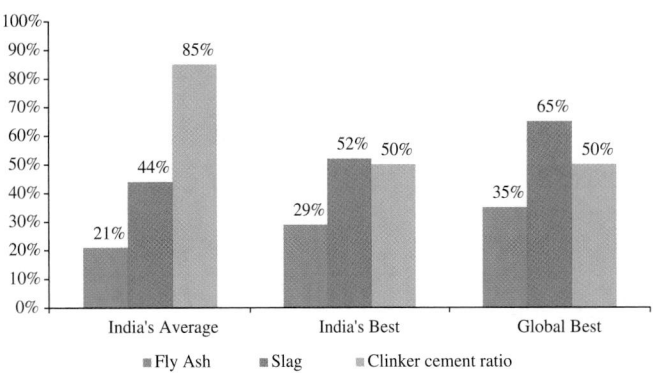

Figure 8.10 | India cement blend ratio – cement clinker ratio comparison with global best, 2005. Source: Bhushan and Hazra, 2005.

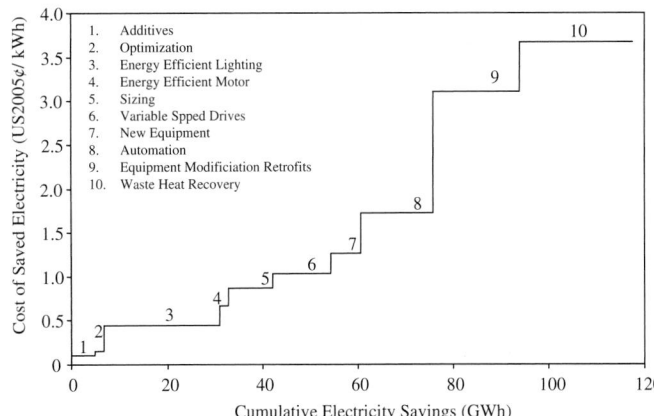

Figure 8.11 | Conservation supply curve for electricity savings in the Indian cement industry. Source: Rane, 2009.

The graph shown in Figure 8.12 covers 80% of global cement production. Note that the lower energy intensity of cement production is the effect of energy efficiency and the use of additives to blend cement. The low energy intensity does not necessarily mean that a country is more energy efficient. The Colombia Kiln (Zeman and Lackner, 2008) proposes a reduced-emission oxygen kiln for cement production. The concept is to use oxyfuel combustion and integrate with carbon capture and storage to reduce CO_2 emissions from the plant by 90%.

8.3.4 Aluminum

Global primary aluminum production was estimated at 38 Mt in 2007 and has grown by an average of 5%/yr over the last 10 years. The key producing countries are China, Russia, Canada, Australia, Brazil, India, and Norway.

Table 8.12 | Energy use in the Brazilian aluminum industry, 2002–2006.

		2002	2003	2004	2005	2006
Production	(10³ t)	1318.4	1380.6	1457.4	1497.6	1603.8
Domestic consumption	(10³ t)	715.5	666.0	738.5	802.3	837.6
Electricity consumption	(GWh)	19474.5	20758.9	22076.7	22939.6	23973.8
Fuel oil	(t)	58300	61000	62400	59100	54200

Source: data based on ABAL, 2008.

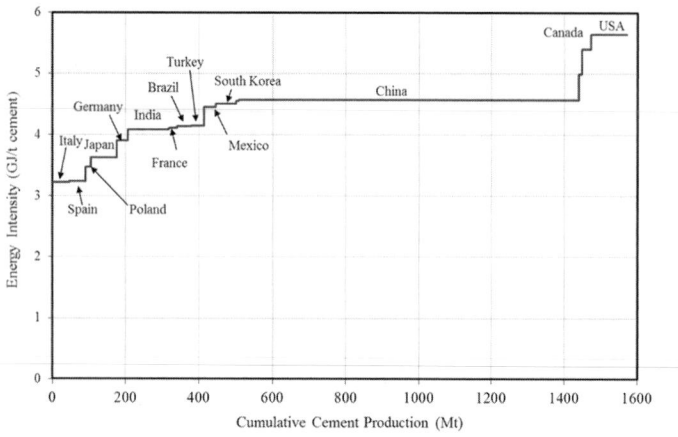

Figure 8.12 | Energy intensity of cement production in selected key cement-producing countries, expressed as primary energy (GJ/t).

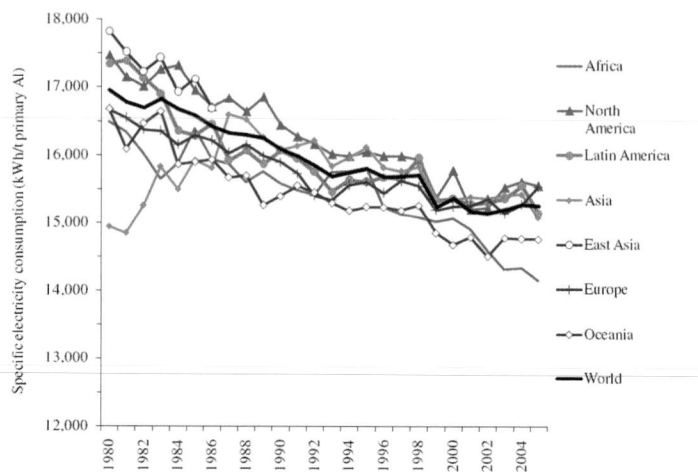

Figure 8.14 | Specific electricity consumption of the aluminum industry (kWh/t) by Region and for the world (black), 1980–2005. Source: data from IAI, 2007. Al = aluminum.

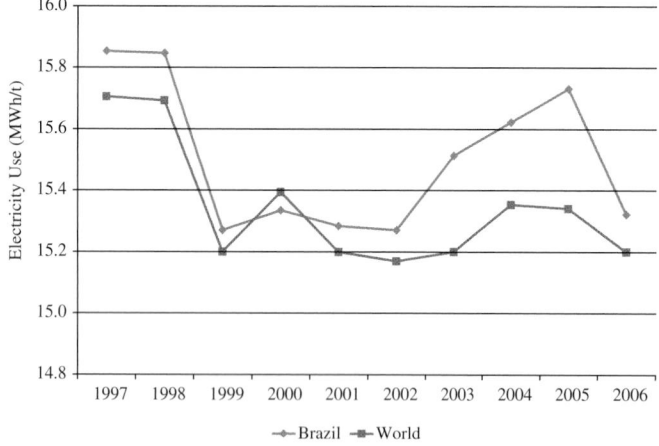

Figure 8.13 | Specific electricity consumption of the aluminum industry (MWh/t) – Brazil and world average, 1997–2007. Source: data based on ABAL, 2008.

Aluminum is produced by the electrolytic reduction of alumina (Al_2O_3). The process is energy intensive, using electricity. Apart from the CO_2 emissions associated with the electricity used, the process also results in emissions of perfluorocarbons (PFCs), carbon tetrafluoride (CF_4), and hexafluoroethane (C_2F_6) (IAI, 2007), which are all greenhouse gases (IAI, 2007). The International Aluminum Institute, a group of aluminum producers (accounting for 70% of the global production) committed to reducing their smelting energy use by 10% between

1990 and 2010 (IAI, 2007), achieved an actual reduction of 6% by 2004 (IPCC, 2007).

Additional energy efficiency improvements are possible through increased penetration of state-of-the-art, point feed, prebake smelter technology (replacing Söderberg cells), process control, and an increase in recycling rates for old scrap (IEA GHG, 2001). Figure 8.13 shows the trend in the specific electricity consumption in the aluminum industry in Brazil along with the world trend. Table 8.12 shows the trend in aluminum production and domestic consumption in Brazil. Almost 50% of production is for the export market. The time-series trend of electricity intensity of the aluminum industry across the regions of the world is shown in Figure 8.14.

Ongoing research to develop an inert anode is expected to reduce the energy used for anode baking and electrolysis. Though inert anodes are currently not viable, it is projected that commercially viable designs may be developed by 2020 (IAI, 2011).

Figure 8.15 shows the average specific electricity consumption for aluminum production in different regions of the world. Note that the International Aluminium Institute (IAI) data do not cover China completely. Hence, Figure 8.15 underestimates the relatively high specific electricity consumption for aluminum production in China. Europe includes the EU, Russia, and other countries. The high specific electricity consumption in Europe is due to Russian production capacity.

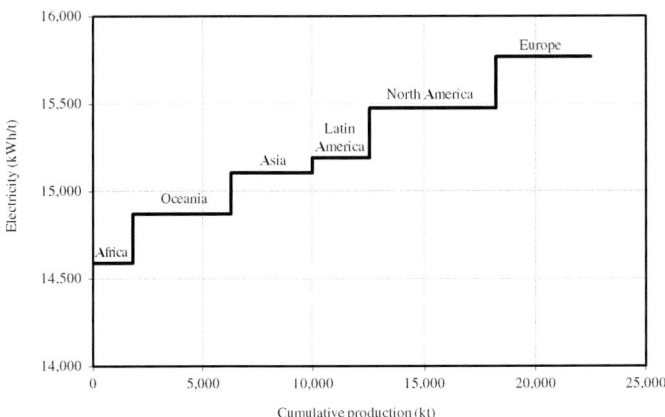

Figure 8.15 | Specific electricity consumption vs. production for world regions. Source: Production data based on USGS (2005; 2007; 2008) and electricity consumption data based on IAI, 2007.

8.3.5 Pulp and Paper Industry

The global pulp and paper industry is an important industry in many countries, from both an economic and an energy use perspective, consuming globally around 6.5 EJ (including printing). This makes the sector one of the largest energy-using sectors in industry after chemicals, iron and steel, and cement. The four largest paper-producing regions (the EU, the US, China, and Japan) account for 80% of energy use and CO_2 emissions. Despite recent changes in the drivers for paper demand, global paper demand is still growing at rates of over 3%/yr over the past 40 years.

The industry is also unique in its reliance on biomass as the key feedstock (besides recycled paper) and primary energy source. This means that while the energy intensity of the sector is high, the CO_2 intensity is far less. The most important processes are pulping (both mechanically and chemically) and papermaking. Energy is used in the pulping of the wood to prepare the fiber, which is processed in the paper machine, the other key energy-using process. About half of the energy is used in pulping, while the other half is used in papermaking. Energy use in the paper machine varies with the paper grade produced. Paper can be made in integrated mills (pulping and papermaking), in standalone pulp (for market pulp) or in paper mills (using imported pulp and recycled paper). Most energy is used in the form of heat (steam) and power. This makes the sector a large user of cogeneration (both using biomass as fossil fuels), but it still also provides a large opportunity for energy efficiency improvements.

Benchmarking and other studies (see, e.g., IEA, 2007a) have demonstrated a substantial potential for efficiency improvement, if best practice technology would be used (see, e.g., Worrell et al., 2008), both in heat use (varying between 0–40%) and electricity use (globally around 20–30%). Combined heat and power (CHP) use varies from 20% of the share of power use to highs of 60% or more (e.g., the United Kingdom, the Netherlands). The countries with the largest potential for energy efficiency improvement typically operate small-scale mills (e.g., China, India) or outdated process equipment (e.g., the United States), while energy-efficient

countries operate modern, large-scale mills (e.g., Japan, Scandinavia). China is an interesting example. Just a few decades ago, the majority of the paper industry consisted of very small, inefficient and polluting mills mainly using straw as the main fiber source. Today, China's share of global production is rapidly increasing, and this expansion is based on large, modern paper machines using (imported) recycled paper.

Table 8.13 provides a summary of process-specific energy efficiency opportunities (based on Martin et al., 2000; Kramer et al., 2009). Beyond these opportunities, cross-cutting options exist in motor systems and steam generation and distribution. New technology is being developed, of which black liquor gasification is the most important in pulping, and various new drying technologies are under development for papermaking. Moreover, paper recycling is an important option to reduce energy use (reducing the need for wood pulping) and save resources. Some paper-producing countries rely almost completely on the use of recycled fiber as feedstock (e.g., in Europe).

8.3.6 Small- and Medium-Sized Enterprises

The definition of Small- and Medium-Sized Enterprises (SMEs) varies by country. In some countries it is based on the value added, in others, it is based on the number of employees. Typically these are companies with up to a few hundred employees and a turnover of less than US$100 million. Some of the SME activities are energy intensive (see also Chapter 6). Substantial amounts of energy are used for the production of ferrous and non-ferrous foundries, ceramics, bricks, glass, lime, concrete, wood processing, food and beverages, small-scale pulp and paper mills, cement kilns, steel production and steel rolling mills, and DRI production. Reliable statistics in terms of economic activity and energy use are lacking. However, it is possible to make a rough estimate based on the physical production volume and the typical energy use per unit of product (Table 8.14 and Figure 8.16).

SMEs make economic sense in several sectors where there are no economies of scale. SMEs are adaptable and a source of technology innovation. Rapidly growing economies usually have a large share of SMEs. Access to large-scale production technology is an issue in certain countries. In countries that are members of the Organisation for Economic Co-operation and Development (OECD), where capital is cheap, labor expensive, and technology development has been targeting upscaling for decades, SMEs play a secondary role. However, in many developing countries they are the cornerstone of industrial development. In the context of the changing mix of global industrial output, SMEs in developing countries deserve special attention.

Estimated current final energy use of selected SMEs and small-scale clusters of the manufacturing industry is between 18–32 EJ. This is equivalent to 14–25% of the total final energy use of the manufacturing sector including feedstock use in 2007 (127 EJ), and 17–30% of the total process energy use when feedstock use is excluded (106 EJ).

Table 8.13 | Summary of process-specific energy efficiency opportunities in the pulp and paper industry.

Raw Material Preparation	
Cradle debarkers	Automatic chip handling and screening
Replace pneumatic chip conveyors with belt conveyors	Bar-type chip screening
Use secondary heat instead of steam in debarking	Chip conditioning

Chemical Pulping	
Pulping	
Use of pulping aids to increase yield	Digester blow/flash heat recovery
Optimize the dilution factor control	Heat recovery from bleach plant effluents
Continuous digester control system	Improved browstock washing
Digester improvement	Chlorine dioxide (ClO2) heat exchange
Bleaching	
Heat recovery from bleach plant effluents	Chlorine dioxide (ClO2) heat exchange
Improved brownstock washing	
Chemical Recovery	
Lime kiln oxygen enrichment	Improved composite tubes for recovery boiler
Lime kiln modification	Recovery boiler deposition monitoring
Lime kiln electrostatic precipitation	Quaternary air injection
Black liquor solids concentration	

Mechanical Pulping	
Refiner improvements	Increased use of recycle pulp
Refiner optimization for overall energy use	Heat recovery from de-inking plant
Pressurized groundwood	Fractionation of recycled fibers
Continuous repulping	Thermopulping
Efficient repulping rotors	RTS pulping
Drum pulpers	Heat recovery in thermomechanical pulp

Papermaking	
Advanced dryer controls	Waste heat recovery
Control of dew point	Vacuum nip press
Energy efficient dewatering – rewetting	Shoe (extended nip) press
Dryers bars and stationary siphons	Gap forming
Reduction of blow through losses	CondeBelt drying
Reduction air requirements	Air impingement drying
Optimizing pocket ventilation temperature	

Source: Kramer et al., 2009.

Policymakers can affect the sector's energy use. Interventions can be in the form of energy pricing, energy cost information systems, energy audits, workshops and conferences organized in cooperation with industry associations, technology cooperation schemes with technical universities and research institutes, and energy technology knowledge systems (centers, books, curricula, etc.).

One program considered successful at providing technical assistance comes from the US Department of Energy (US DOE). The program targets

Table 8.14 | Estimated total final energy use of the selected SMEs and small-scale clusters worldwide, 2007.

SMEs and small-scale clusters	Final Energy (PJ/Year)
Ferrous and non-ferrous metals	950 – 1750
Non-metallic minerals	7400 – 12,500
Bio-based chemical products	200 – 400
Food and beverage	2150 – 4400
Textiles and leather	950 – 1800
Building and construction	1450 – 2500
Wood processing	1200 – 2000
Energy transformation processes	925 – 1800
Small-scale energy-intensive sectors in developing countries	2450 – 5000
Total final energy use of SMEs and small-scale clusters	**17,675 – 32,150**

SMEs and has created a number of Industrial Assessment Centers housed within US universities. Engineering students from the centers are "seconded" to SMEs to provide relevant technical assistance, such as conducting energy audits and assessing potential energy efficiency projects (Mallett et al., 2010). SMEs like the program, as there are no costs involved on their part, and it also provides practical experience for the students. Many participating firms undertake the energy efficiency opportunities presented to them by the students, and some firms hire the students to continue working at their firm after graduation. An assessment of the program found that it helped to overcome informational barriers – there were significant changes in decision-making on energy efficiency within a relatively short period of time (Mallett et al., 2010).

Table 8.15 shows the summary of a study by the Confederation of Indian Industry and Forbes Marshall of fuel, electricity, and water in several SMEs in India. It is clear that significant savings are possible with respect to the best performance in each sector. The options considered in this study do not include process changes. An analysis of brick kilns shows significant potential for savings by introducing energy-efficient vertical shaft brick kilns. There is a need for increased efforts for benchmarking and analytical studies for energy efficiency in SMEs.

8.3.7 Industrial Benchmarking: A Tool for Realistic Assessment of Energy Efficiency Potentials

Benchmarking is a management tool that is used to compare similar plants. This is done for many operational aspects such as energy use and energy efficiency.

Benchmarking is primarily a tool that helps plant managers to gauge their improvement potential. However, if it is done for a representative set of plants or for a significant share of the total production volume, it can be used to estimate the improvement potential for the whole sector compared to best process technology. This is valuable information for policymakers.

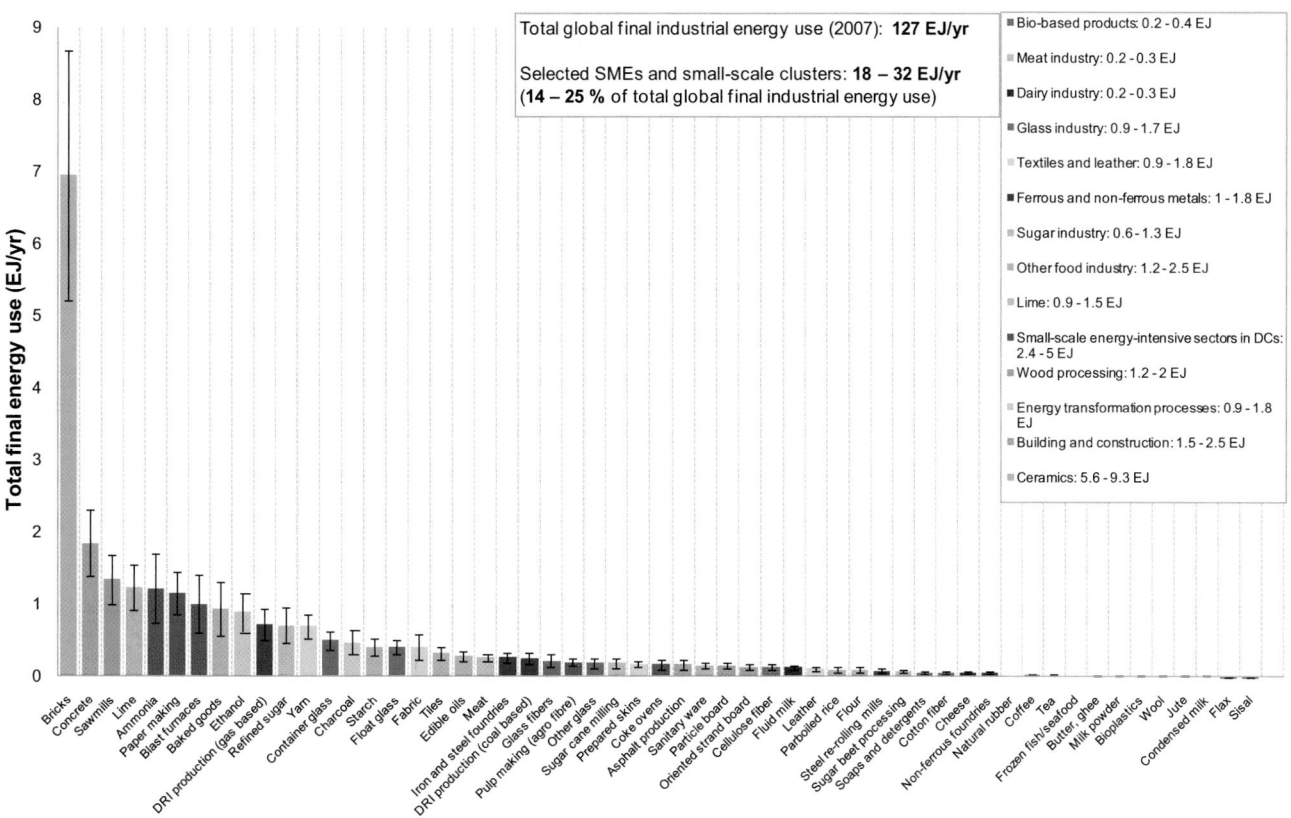

Figure 8.16 | Estimated total final energy use of selected SMEs and small-scale clusters worldwide, 2007.

Table 8.15 | Specific energy use savings for SMEs in India.

SMEs in India	Unit for Fuel	Fuel			Electricity (kWh)			Water (m³)		
		Average	Best	Savings	Average	Best	Savings	Average	Best	Savings
Breweries	Fuel L/kL Beer	58	44	24%	156	100	36%	9.1	7.9	13%
Beverage	Fuel L/kL Beverage	9.35	5.29	43%	–			–		
Tire	Fuel kg/t Finished Tire	210	162	23%	872	780	11%	8.4	4.8	43%
Textile	Coal kg/1000 Mt	390	168	57%	195	44	77%	10.15	7.43	27%
Soya	Coal t/t Seed Crushed	63	47	25%	40	21	48%	–		
Rice bran	Husk t/t Seed Crushed	111	100	10%	27	25	7%	–		
Paper	Coal kg/t Paper	360	259	28%	–			–		

Source: CII and Forbes Marshall Study, 2005.

The fact that benchmarking only considers measured data avoids a lengthy discussion about best available technology. As technology evolves over time and new technologies are gradually introduced, there is always a grey area between proven technology that can be applied in the short term in practice and technologies that are not yet fully mature or commercially available.

A challenge for benchmarking is the comparability of individual units. For example: feedstock quality may differ, the product quality may not be exactly the same and local climate conditions or opportunities for process integration may differ. Therefore, care must be taken

to compare "like with like." For benchmarking curves, a widely used approach is one where the 10th percentile is used to define the best available technology. Typically the benchmarking curves show a virtually linear rise from the 10th to the 90th percentile. This allows a comparatively straightforward estimate of the improvement potential: the average efficiency/energy use of the 10th and 90th percentile is the average for the whole group of plants, and the improvement potential is the percentage gap between this average and the 10th percentile.

This approach has been applied for primary aluminum, ammonia, cement clinker, and ethylene, as shown in Figure 8.17. A similar effort

Table 8.16 | Comparison between sectoral average energy intensities, best values, and potential savings.

Regions	Specific Energy Consumption in 2005 (GJ/t)			
	Steel	Cement	Paper	Aluminum
China	22.3	3.9	30.7	51.5
India	22.8	3.3	26.7	94.7
Brazil	26.6	3.9	22.0	61.6
World average	19.4	4.0	18.4	103
Thermodynamic Min (GJ/t)	6.9[1]	1.76		21.6[2]
Best Available Technology	16.3	2.9	17.6	70.6
Saving Potential %	16%	28%	4%	31%

Source: Worrell and Galitsky, 2008; IEA, 2008a;

1 IISI, 2008;

2 IEA, 2008a.

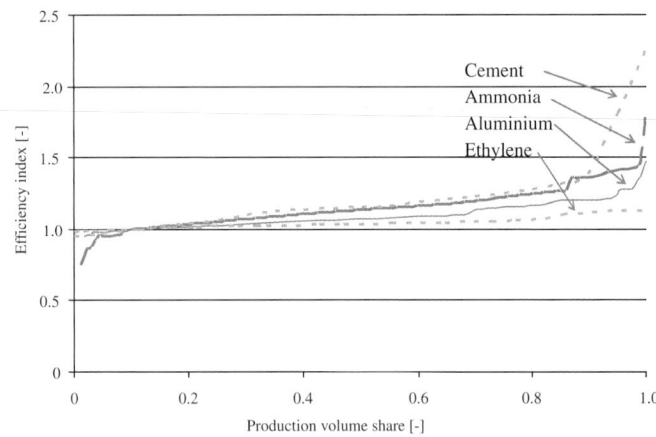

Figure 8.17 | Benchmarking curves for aluminum (2007), ammonia (2006/7), cement clinker (2006), and ethylene (2005). Source: IAI, 2007; IFA, 2009; CSI, 2006.

is ongoing for iron and steelmaking as part of the Task Force under the Asia-Pacific Partnership.

Typically the benchmarking curves suggest an average 10–20% efficiency potential. However, it should be considered that the coverage varies. China is missing from these datasets, and the average efficiency in China is relatively low. This raises the efficiency potential by a quarter to a half.

But what these numbers suggest is that the efficiency potentials based on best available technology in key energy-consuming sectors are significantly lower than other parts of the economy, such as buildings (typically with an average improvement potential of over 50%; see Chapter 10) and the power sector. Table 8.16 shows average energy intensities (GJ/t) in select industries and comparisons with international best averages.

Benchmarking curves do not account for the efficiency potential based on new technology. In all these sectors there are efforts to improve efficiency further. Inert anodes in combination with drained cells for aluminum smelters, new gas separation membranes for ammonia

plants, low-temperature heat recovery for cement kilns, and gas turbines and new separation systems for steam crackers are examples of such developments. The theoretical minimum energy use is typically half of the global average today. This does not mean that this minimum can be reached, but it indicates that further improvements can be expected as technology improves. Most benchmarking studies are based on statistical techniques by comparing existing plants. An alternative approach is model-based benchmarking (Sardeshpande et al., 2007) applied to industrial furnaces for glass manufacture. The model is used for predicting an achievable minimum energy use for a given furnace configuration based on design and operating parameters. This approach provides a rational basis for target setting and energy performance improvements for existing processes and can be extended to other industrial processes in metallurgical, cement, paper, petrochemical, and textile industries.

However, the large gains will not come from narrow process efficiency improvement but from the application of broader systems optimization strategies. Use of electricity outside the plant boundaries is often excessive, and here significant savings can be achieved. Also, options such as heat integration, cogeneration, recycling, and a change of process inputs can contribute to savings. Improved materials use efficiency does not contribute to savings per tonne of materials produced, but it reduces the materials production volume. Benchmarking curves do not capture all these improvement options or may do so only partially. Data are sketchier, but typically they can raise the average efficiency potentials by 5–10 percentage points. However, the economics of these improvements are not well established and they may vary widely.

The energy-saving potentials based on benchmarking and indicator data for OECD and non-OECD regions are shown in Table 8.17.

In 2007, the global manufacturing industry used 127 EJ of final energy (40% in industrialized countries and 60% in developing countries). More than half of the industrial energy use is due to the activities of

Table 8.17 | Energy efficiency improvement potentials in the manufacturing industry based on benchmarking and indicator data, 2007.

	Improvement potential (%)		Total Savings Potential (EJ/yr)		Global Subtotal (EJ/yr)
	Industrialized countries	Developing countries	Industrialized countries	Developing countries	
Chemical and petrochemical					
High value chemicals	15–25	25–30	0.4	0.3	2.3
Ammonia, methanol	10–15	15–30	0.1	1.4	
Non-ferrous metals					
Alumina production	30–40	40–55	0.1	0.5	1
Aluminium smelters	5–10	5	0.2	0.2	
Cast non-ferrous and other non-ferrous	35–60				
Ferrous metals					
Iron and steel	10–15	25–35	0.7	5.4	6.1
Cast ferrous	25–40				
Non-metallic minerals					
Cement	20–25	20–30	0.4	1.8	2.8
Lime	10–40	20–50	0.4	0.2	
Glass					
Ceramics					
Pulp and paper	20–30	15–30	1.3	0.3	1.6
Textile					
Food, beverages and tobacco	25–40		0.9	1	1.9
Other sectors	10–15	25–30	2.5	8.7	11.2
Total	10–20	30–35	7.2	20.1	27.3
Total (excl. feedstock)	15–20	30–35			

Source: Saygin et al., 2010.

the energy-intensive sectors: chemicals and petrochemicals (selected processes in Table 8.17), non-ferrous and ferrous metals, non metallic minerals, and pulp and paper (66 EJ/yr including feedstocks). According to benchmarking and indicator data, best practice technologies can reduce energy-intensive industry's final energy use by 11–17 EJ. This is equivalent to an improvement potential of 17–26% including feedstock use.

Additional energy efficiency potentials in light industries (e.g., textiles, food, beverages, and tobacco, etc.) are estimated at 12–16 EJ. This adds up to a total industrial energy-saving potential of 22–31% EJ if all industrial processes were to adopt best practice technologies (or 22–31 EJ improvement potentials excluding feedstock use).

Approximately three-quarters of this energy-saving potential are located in developing countries (17–23 EJ), with the estimated improvement potentials higher than worldwide, between 30% and 35%. The remaining 6–9 EJ of the potential is in industrialized countries. In the coming decades, industrial energy use is projected to increase much more in developing countries than in industrialized

countries. Given the high improvement potentials in developing countries and the future growth projections, improving energy efficiency at process level is a key measure to reduce energy demand and related carbon emissions.

Benchmarking has grown as an industrial management tool. Its use for sectoral agreements or for target setting raises new needs. For example, today in all benchmarks individual plant data are confidential for anti-trust and competitiveness reasons. Also, participation is voluntary, coverage is incomplete, and the process is driven by consultancies that have a natural interest to keep information confidential. These aspects need to be addressed to make the benchmarking tool more useful for the climate policymaking process.

Industries have recognized the importance of the benchmarking tool for a rational decision-making process. Certain sectors such as the European Chemicals Industry have devised innovative schemes for integrating benchmarking with emissions trading. In recent years efforts have been increased in iron and steel, cement, pulp and paper, and other sectors. More attention is needed for the use of benchmarking in the

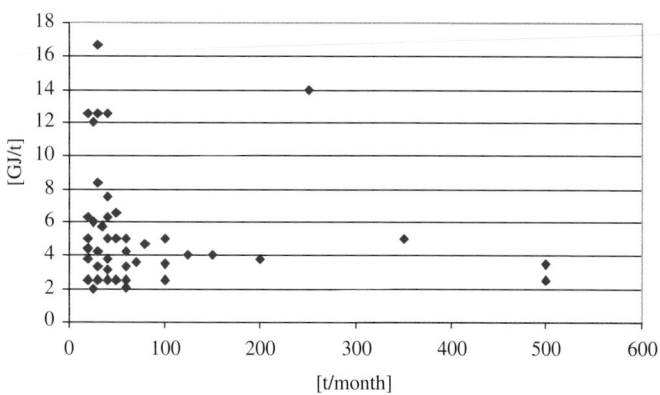

Figure 8.18 | Typical efficiency distribution for an iron SME cluster in India. Source: Gielen, 2010.

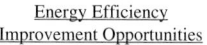
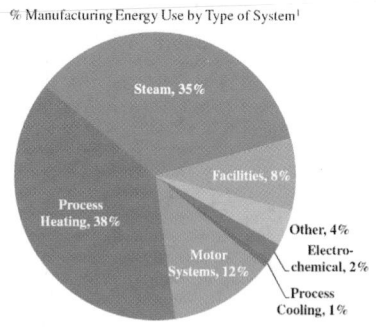

Figure 8.19 | Share of manufacturing energy use by system type. Source: data based on US DOE, 2004b.

1 Does not include offsite losses.

2 2002 MECS – plants indicated energy management activities for 6.3% steam, 16.6% compressed air, 7.5% process heating systems.

SME sector. The inclusion of broader system boundaries is also a theme that deserves policy attention.

The use of benchmarking curves is less established for sectors such as pulp and papermaking, SMEs, and less energy-intensive sectors. Typically these sectors account for 25–50% of total global industrial energy use. The issues are often very different than for large companies. Small-scale operations are in many cases operated intermittently, often based on outdated equipment and without much attention to energy use. As a consequence, the efficiency potentials in percentage terms tend to be much higher than for large industries.

As an example, a benchmarking effort for an Indian iron casting cluster is shown in Figure 8.18. While the larger plants in this cluster tend to be more efficient than the smaller ones, at the same time the small plants show a wide range of efficiencies, reflecting differences in operational practices. A shift to large plants would be an option to increase efficiency, but such an approach would be politically and socially unacceptable. An alternate approach would be to focus on energy efficiency improvements in smaller plants. Such considerations must be taken into account when estimating the improvement potential.

8.4 Consumption and Opportunities: Cross-Cutting End-Uses

8.4.1 Industrial Systems: Overview

System energy efficiency affords industrial facilities the opportunity to readily identify energy efficiency projects that can contribute to continuous improvement for energy management (UNIDO, 2007). However at present, most markets and policymakers tend to focus on individual system components (e.g., motors and drives, compressors, pumps, boilers) with an improvement potential of 2–5% – which can be seen, touched, and rated – rather than systems. While systems have impressive improvement potentials – 20% or more for motor systems and 10% or more for steam and process heating systems (see Figure

8.19) – achieving this potential requires engineering and measurement (US DOE, 2004a; IEA, 2007a).

Though equipment manufacturers are steadily increasing equipment efficiencies, the effect on the system efficiency needs greater attention.

A study of 41 completed industrial system energy efficiency improvement projects in the US between 1995 and 2001 documented an average 22% reduction in energy use. In aggregate, these projects cost US$16.8 million and saved US$7.4 million and 106 million kWh (0.38PJ), recovering the cost of implementation in slightly more than two years (Lung et al., 2003).

8.4.2 Motor Systems

Motor systems account for about 60% of industrial electricity use and about 15% of global final manufacturing energy use (17 EJ in 2005). The International Energy Agency (IEA) estimates global potential from motor system energy efficiency to be of the order of 2.6 EJ annually of final energy use for motor electricity consumption of 13 EJ (IEA, 2007a). Motor systems lose on average approximately 55% of their input energy before reaching the process or end-use work (US DOE, 2004c). Some of these losses are inherent in the energy conversion process – for example, a compressor typically loses 80% of its input energy to low-grade waste heat as the incoming air is converted from atmospheric pressure to the desired system pressure (US DOE, 2004d). Other losses can be avoided through the application of commercially available technology combined with good engineering practices (see Table 8.18) (UNIDO, 2007b). These improvements are cost-effective, with costs typically recovered in two years or less. Figure 8.20 provides an illustration of how a system approach can result in a substantial improvement in system operating efficiency.

Table 8.18 | Energy savings potential by compressed air improvement.

Compressed Air System Improvement Option	% Potential Energy Savings
Replace current compressor with more efficient model	2
Reconfigure piping to reduce pressure loss	20
Add compressed air storage	20
Add small compressor for off-peak loads	2
Add, restore, upgrade compressor controls	30
Install or upgrade distribution control system	20
Rework or correct header piping	20
Add, upgrade or reconfigure air dryers	1
Replace or repair air filters	10
Replace or upgrade condensate drains	5
Modify or replace regulators (controls at the process)	20
Improve compressor room ventilation	1
Install or upgrade (ball) valves in distribution system	10

Note: Does not account for interactions or inappropriate use.

Source: US DOE, 2004d.

In some industrial processes and utility compressed air systems where the delivery pressures are high (greater than 12 bar) it is better to opt for multi-stage compressors. Multi-stage compressors are a move toward the ideal isothermal compression process and results in significant electricity savings.

Even modern, well-maintained industrial systems can benefit from optimization. For example, the Canadian utility, Manitoba Hydro, offers industrial facilities system assessments through its PowerSmart program. System optimization projects completed and documented in 2004 reduced the energy requirements of compressed air systems at a milk plant and a garment manufacturer by more than 60%. One compressed air system was only nine years old with well-maintained, energy-efficient equipment (see Manitoba Hydro, 2011).

The total savings potential in motor systems is expected to range between 15–25%. It is estimated that this will result in a 3.5 EJ of final energy saving from the existing industry. The study by de Keulenaer et al. (2004) estimated annual savings of 202 TWh (0.72 EJ) in the EU with an investment of US$500 million and annual savings of US$10 billion. This would imply an investment requirement of US$2.4 billion for energy efficiency in motors globally for the 3.5 EJ savings.

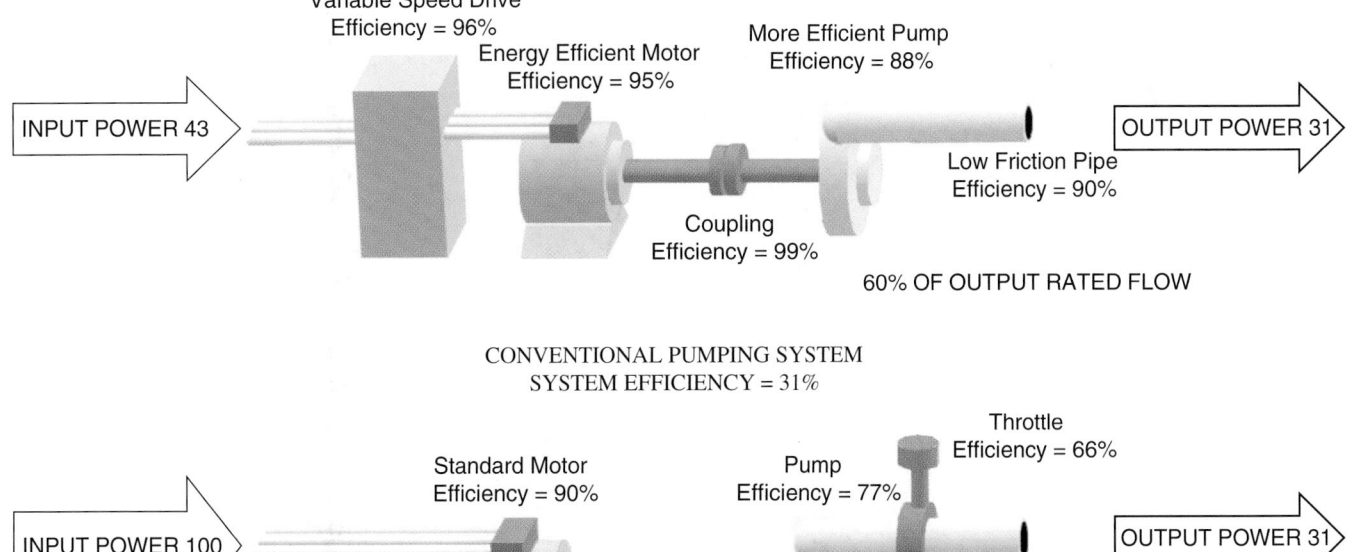

Figure 8.20 | Reconfiguration of pumping system to improve efficiency. Source: Almeida et al., 2005.

Table 8.19 | Total annual electricity saving and CO_2 emission-reduction potential in industrial pump, compressed air, and fan systems.

	Total Annual Electricity Saving Potential in Industrial Pump, Compressed Air, and Fan System (GWh/yr)		Share of Saving from Electricity use in Pump, Compressed Air, and Fan Systems in Studied Industries in 2008 (%)		Total Annual CO_2 Emission Reduction Potential in Industrial Pump, Compressed Air, and Fan System (ktCO$_2$/yr)	
	Cost-Effective	Technical	Cost-Effective	Technical	Cost-Effective	Technical
United States	71,914	100,877	25%	35%	43,342	60,798
Canada	16,461	27,002	25%	40%	8185	13,426
European Union	58,030	76,644	29%	39%	25,301	33,417
Thailand	8343	9659	43%	49%	4330	5013
Vietnam	4026	4787	46%	54%	1973	2346
Brazil	13,836	14,675	42%	44%	2017	2140
Total (sum of six countries)	172,609	233,644	28%	38%	85,147	117,139

* In calculation of energy savings, equipment 1000 hp or greater are excluded

Source: UNIDO, 2010a.

Motor system efficiency supply curves have been developed for the United States, Canada, the EU, Thailand, Vietnam, and Brazil (UNIDO, 2010a). Table 8.19 shows a summary of the results. Cost-effective savings of 172,600 GWh/yr are estimated in these six regions, accounting for 28% of the electricity use. Figures 8.21, 8.22, and 8.23 show the EU's pumping system efficiency curve. The discount rate used is 10%. Similar conservation supply curves have been drawn up for all the six regions studied.

8.4.3   Steam and Process Heating Systems

Steam systems are estimated to account for 38% of global final manufacturing energy use or 44 EJ in 2005. The IEA estimates global potential from steam system energy efficiency to be of the order of 3.3 EJ annually of final energy use for steam energy use of 33 EJ (IEA, 2007a). For steam systems, the losses are only marginally better than motor systems, with 45% of the input energy lost before the steam reaches point of use (US DOE, 2004c).

The best option for improving the energy efficiency of a steam system (Figure 8.24 shows a typical steam system) is through a CHP system. Particularly in more mature industries, excess steam production is fairly common and may be a cost-effective source for on-site generation. Higher-efficiency boilers currently under development offer the promise of higher efficiencies.[5] Sometimes other processes can be used in lieu of steam to perform the same work; for example, in recent decades the chemical industry has successfully developed new catalysts and process routes that reduce the need for steam.

In 2006, the US DOE created the Save Energy Now program based on more than a decade of experience in industrial system energy efficiency. The program offers system assessments to companies with an

energy use of 1 TBtu (1.1 PJ) or more annually (over 50% of US industrial energy use). DOE energy experts work with plant energy teams to identify opportunities for improving steam, process heating, pump, or compressed air systems through Energy Savings Assessments. There is a focus on transferring skills to plant personnel and identifying specific energy efficiency opportunities from one system type. The first year of program implementation focused on steam and process heating systems, with motor systems added in the second year. A total of 717 assessments have been completed, with implemented energy savings of US$135 million and planned energy savings of US$347 million. Recommended energy-saving projects totaled 87.2 TBtu (92 PJ), with more than US$937 million in energy cost savings and total potential reductions in CO_2 emissions of 7.9 $MtCO_2$. The payback periods from the measures are very cost-effective, as illustrated in Figure 8.25.

A study conducted in India in several industry segments revealed a potential for steam savings of 25% in breweries, 36% in paper, 28% in starch, 23% in tires, 28% in textiles, and 21% in starch. It is estimated that there is a savings potential of 20% or 8.8 EJ from cross-cutting measures in steam systems from the existing industry in 2005.

8.4.4   Barriers to Improving System Efficiency

The use of Energy efficient components does not guarantee the efficiency of the overall industrial system. Oversizing and incorrect application of energy efficient equipment such as Variable speed drives is common. System assessment to identify the end uses and optimization to determine the best configurations to meet these requirements can help improve energy efficiency (UNIDO, 2007).

System assessment services can provide high value to an industrial facility both in terms of lower operating costs and greater reliability.

---

5   US DOE is supporting the development of a 94% efficient boiler.

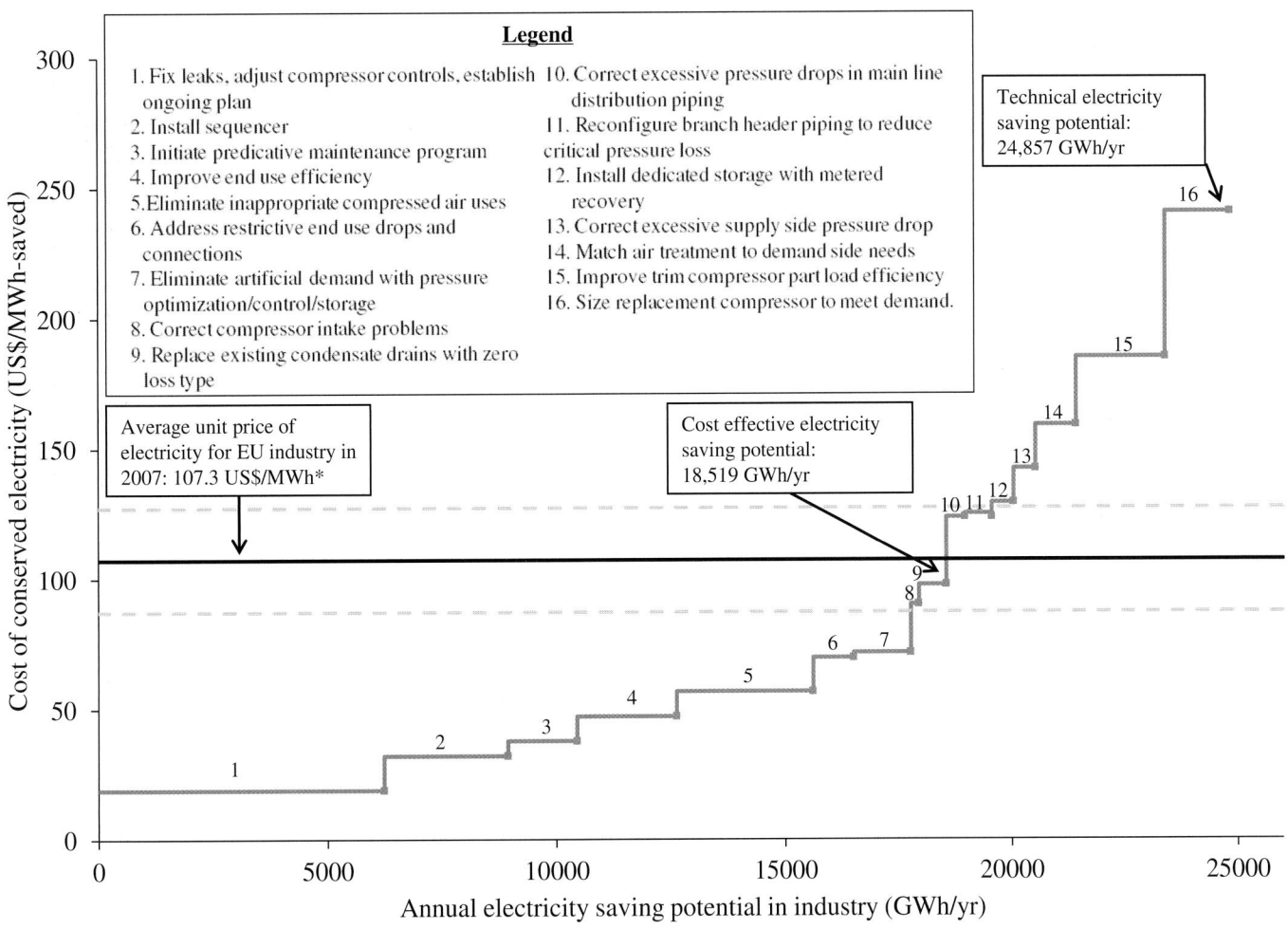

Legend

1. Fix leaks, adjust compressor controls, establish ongoing plan
2. Install sequencer
3. Initiate predicative maintenance program
4. Improve end use efficiency
5. Eliminate inappropriate compressed air uses
6. Address restrictive end use drops and connections
7. Eliminate artificial demand with pressure optimization/control/storage
8. Correct compressor intake problems
9. Replace existing condensate drains with zero loss type
10. Correct excessive pressure drops in main line distribution piping
11. Reconfigure branch header piping to reduce critical pressure loss
12. Install dedicated storage with metered recovery
13. Correct excessive supply side pressure drop
14. Match air treatment to demand side needs
15. Improve trim compressor part load efficiency
16. Size replacement compressor to meet demand.

Technical electricity saving potential: 24,857 GWh/yr

Average unit price of electricity for EU industry in 2007: 107.3 US$/MWh*

Cost effective electricity saving potential: 18,519 GWh/yr

* The dotted lines represent the range of price from the sensitivity analysis.

NOTE: this supply curve is intended to provide an indicator of the relative cost-effectiveness of system energy efficiency measures at the regional level. The cost-effectiveness of individual measures will vary based on site-specific conditions.

Figure 8.21 | EU compressed air system efficiency supply curve. Source: UNIDO, 2010a.

However, it is difficult for plant personnel to easily identify quality services at market-appropriate prices. A typical facility engineer does not have time to research the detailed system-specific information necessary for an informed, but infrequent, purchase of these services. The lack of market definition also creates challenges for providers of quality system assessment services to distinguish their offerings from others that are either inadequate to identify energy efficiency opportunities, or thinly-veiled equipment marketing strategies.

Furthermore, little data are available to support trending of performance for motor, steam, and process heating systems. Measuring the energy efficiency of the components (motors, furnaces, and boilers) is reasonably straightforward and well documented, allowing that some differences in the testing and rating of components still exist. The same is not true in the measurement of system energy efficiency, where most of the energy efficiency potential resides. Few industrial facilities can quantify the energy efficiency of motor, steam, or process heating systems without the assistance of a systems expert. Even experts can fail to identify large savings potentials if variations in loading patterns are not

adequately considered in the assessment measurement plan. If permanently installed instrumentation such as flow meters and pressure gauges are present, they are often non-functioning or inaccurate. Often orifice plates or other devices designed to measure flow actually result in restricting flow, as they accumulate scale with age (UN Energy, 2010).

8.4.5   Realizing System Energy Efficiency Potential

Expert knowledge based on assessments carried out in industrial systems in the United States, United Kingdom, and Canada have resulted in the identification of best practices. These assessment techniques have been further refined in recent years in the United States.[6]

---

6   US DOE's Energy Savings Assessments and Industrial Assessment Center Programs and the Compressed Air Challenge™. As an example, the Compressed Air Challenge's Best Practices for Compressed Air Systemscontains an appendix entitled "Detailed Overview of Levels of Analysis of Compressed Air Systems."

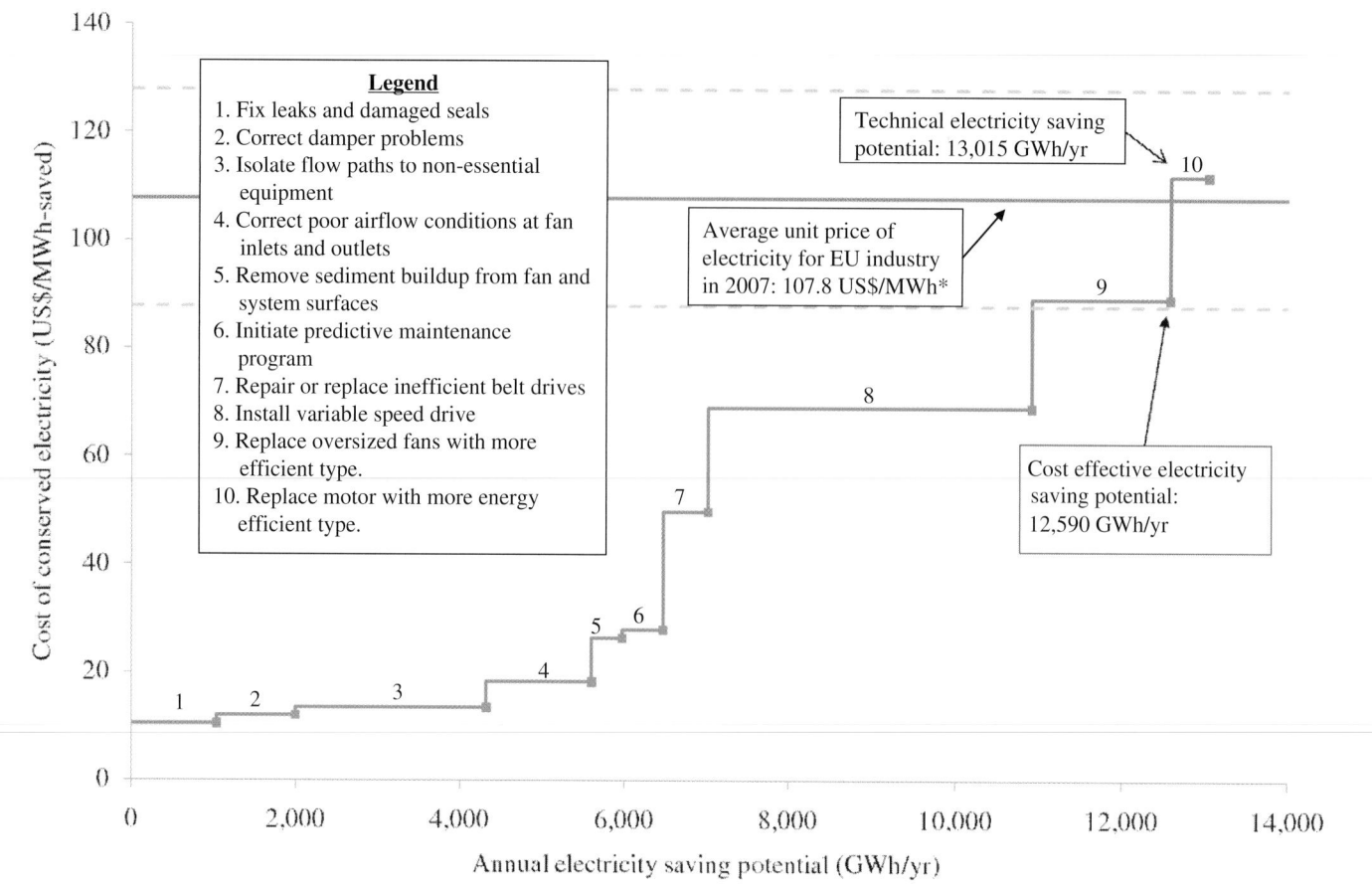

Legend
1. Fix leaks and damaged seals
2. Correct damper problems
3. Isolate flow paths to non-essential equipment
4. Correct poor airflow conditions at fan inlets and outlets
5. Remove sediment buildup from fan and system surfaces
6. Initiate predictive maintenance program
7. Repair or replace inefficient belt drives
8. Install variable speed drive
9. Replace oversized fans with more efficient type.
10. Replace motor with more energy efficient type.

Technical electricity saving potential: 13,015 GWh/yr

Average unit price of electricity for EU industry in 2007: 107.8 US$/MWh*

Cost effective electricity saving potential: 12,590 GWh/yr

* The dotted lines represent the range of price from the sensitivity analysis.

NOTE: this supply curve is intended to provide an indicator of the relative cost-effectiveness of system energy efficiency measures at the regional level. The cost-effectiveness of individual measures will vary based on site-specific conditions.

Figure 8.22 | EU fan system efficiency supply curve. Source: UNIDO, 2010a.

Best practices that contribute to system optimization are system-specific but normally include:

- evaluating work requirements and matching system supply;

- eliminating or reconfiguring inefficient uses and practices (throttling, open blowing);

- identifying and correcting maintenance problems;

- upgrading ongoing maintenance practices; and

- documenting these practices.

These systems remain inefficient primarily due to a series of institutional and behavioral barriers. Some of the important barriers are the limited awareness of the energy efficiency opportunities by industry, consultants, and suppliers, lack of understanding on how to implement energy efficiency improvements; and, most importantly, absence of a consistent organizational structure within most industrial facilities to effectively manage energy use.

Since energy use is rarely measured at the system level, there are few data available. Without performance indicators that relate energy use to production output, it is difficult to document improvements in system efficiency. If the facility also uses energy as a feedstock, even large system energy efficiency improvements can be lost in the "white noise" of overall plant energy usage, especially if production levels vary. System energy efficiency offers large improvement potential but is complex; a "one size fits all" approach will not work (McKane et al., 2007).

The Superior Energy Performance partnership, a collaboration involving the US DOE's Industrial Technologies Program, industrial companies, the American National Standards Institute (ANSI), nonprofit organizations, the US Environmental Protection Agency (US EPA), and the US Department of Commerce's National Institute of Standards and Technology, is facilitating the development of a market-based program for certifying industrial plants for energy efficiency (UN Energy, 2010). As part of this effort, a portfolio of System Assessment Standards has been published by the American Society of Mechanical Engineers (ASME) for compressed air, process

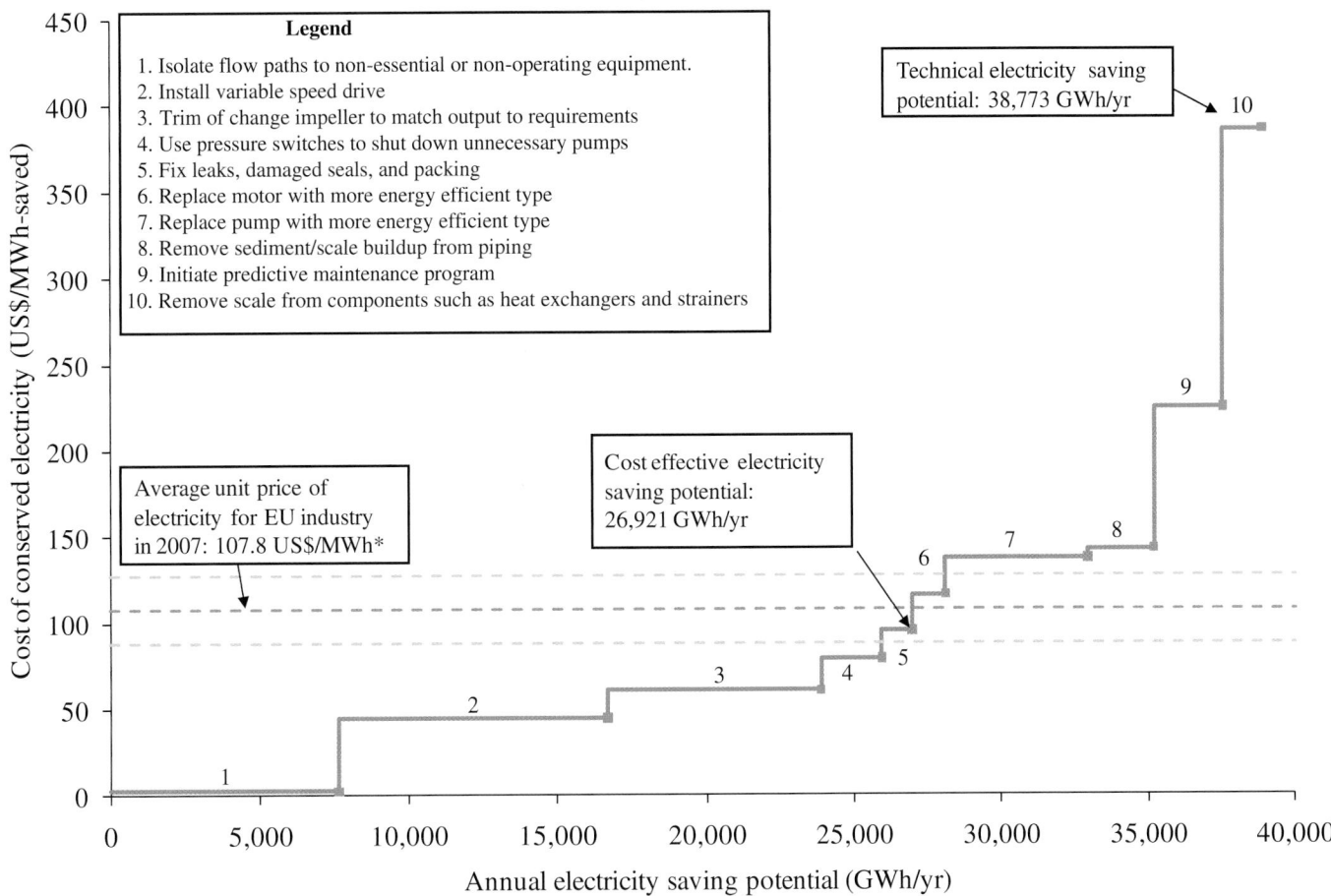

Legend

1. Isolate flow paths to non-essential or non-operating equipment.
2. Install variable speed drive
3. Trim of change impeller to match output to requirements
4. Use pressure switches to shut down unnecessary pumps
5. Fix leaks, damaged seals, and packing
6. Replace motor with more energy efficient type
7. Replace pump with more energy efficient type
8. Remove sediment/scale buildup from piping
9. Initiate predictive maintenance program
10. Remove scale from components such as heat exchangers and strainers

Technical electricity saving potential: 38,773 GWh/yr

Average unit price of electricity for EU industry in 2007: 107.8 US$/MWh*

Cost effective electricity saving potential: 26,921 GWh/yr

Figure 8.23 | EU pumping system efficiency supply curve. Source: UNIDO, 2010a.

heating, and pumping. Collaboration has already begun for one system type (compressed air) to develop an International Organization for Standardization (ISO) standard modeled on the approach used in the ASME standard.

The System Assessment Standards are designed to create a market threshold for industrial system energy efficiency assessments from the current body of expert knowledge and techniques. The standards will provide a common framework for conducting assessments of industrial systems that will help define the market for both users and providers of these services. By establishing minimum requirements and guidance for scope, measurement, and reporting, these standards will offer greater transparency and higher value to industrial facilities by providing assurance to plant managers, financiers, and other non-technical decision-makers that a particular assessment meets or exceeds a best practice threshold for accuracy and completeness. The existence of System Assessment Standards will also assist in training graduate engineers and others desiring a higher level of skill in the area of system optimization for energy efficiency. To assist industrial firms in identifying individuals with the necessary skills to apply the standards correctly, the US initiative will also include the creation of a professional credential with the working title of Certified Practitioner for each system type.

8.4.6   Process Integration, Heat Pumps, and Cogeneration

Industrial processes are systems (see Figure 8.26) in which raw materials are converted into products and by-products. Such transformations are realized by a succession of process units that use water or solvents and in which energy is the transformation driver. To realize the transformation, energy resources are first converted before being distributed and used in the process units. Applying mass and energy balances to the system shows that what are not useful products, goods, or services leave the system as waste in solid, liquid, or gaseous form, or as waste heat radiated to the environment.

By systematically analyzing the materials and energy conversion processes in the different unit operations in the system and representing their possible interactions, process integration aims at identifying synergies between the process units by allowing the following processes:

- *Recycling of materials and energy*: reuse materials – for example, by converting a waste into a product, to recycle waste streams, or to convert waste streams into useful energy.

- *Heat recovery*: utilization of the heat of the hot streams (streams to be cooled down) to heat up cold streams (streams to be heated up).

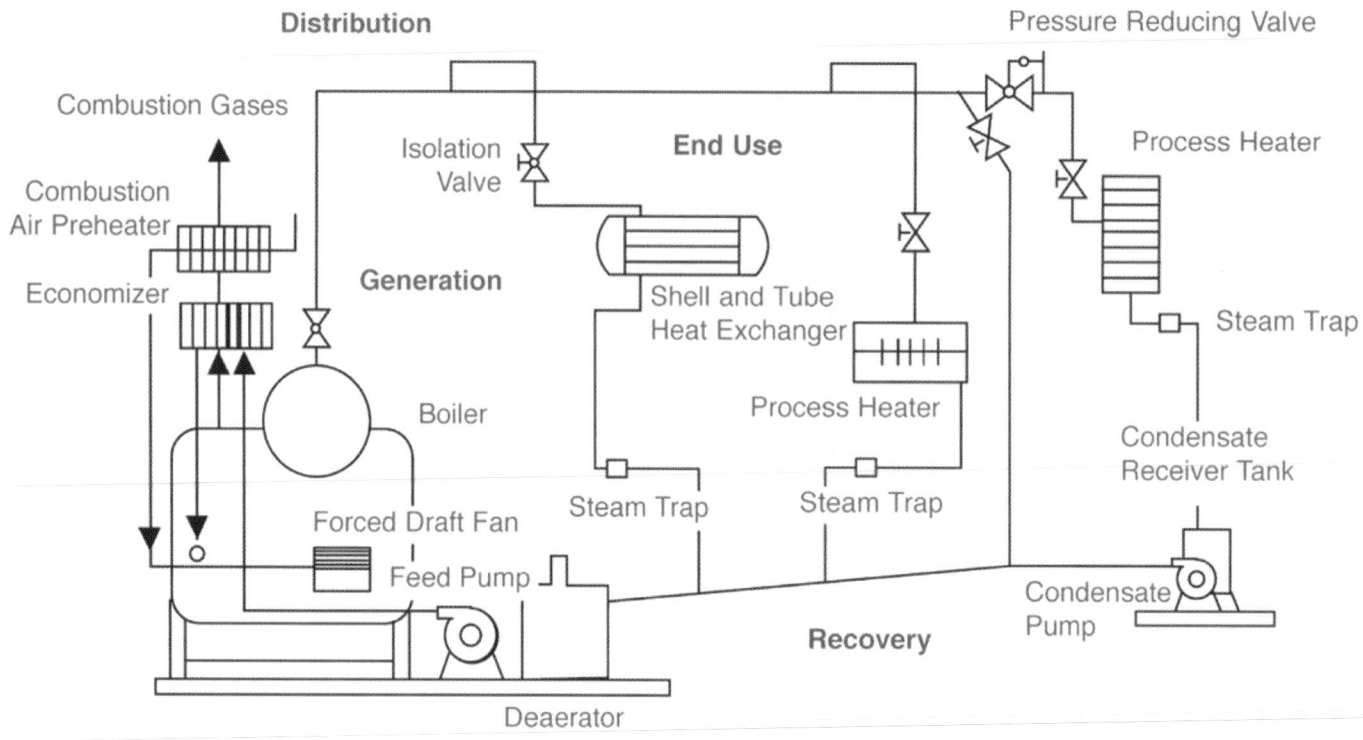

Figure 8.24 | Steam system schematic. Source: based on US DOE, 2002.

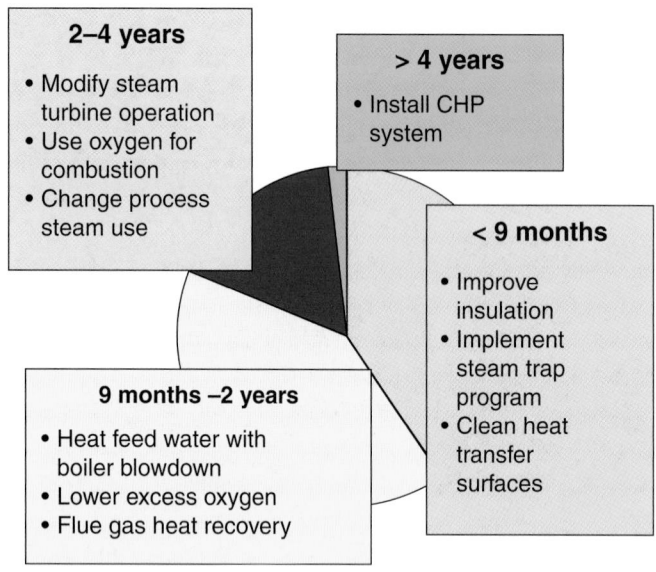

Figure 8.25 | Estimated payback periods for steam-based systems in 2006.

- *CHP production*: CHP – or, more generally, polygeneration – production is typically used to convert fuels into useful heat/cooling and electricity as a by-product or to convert the available exergy in the heat exchange system into useful mechanical power.

- *Waste heat utilization/upgrade in the system*: when the temperature level of waste heat is sufficient and the heat support media is

available (e.g., hot water), waste heat can be converted into mechanical power by thermodynamic cycles or can be upgraded by increasing its temperature level using heat pumping systems.

- *Waste heat upgrade*: by extending the system boundaries to other energy users, such as other processes or district heating systems.

Although there is a hierarchy between these different energy efficiency options, they have to be considered simultaneously within a systemic framework and applied considering local conditions and boundaries of the technically and economically accessible system. Extending the system boundaries may change the impact of the energy-saving actions and lead to different solutions. For example, if the waste heat of a process can be reused to heat up other processes in the surroundings, the use of a heat pump to recover waste heat for the reference process will become counterproductive at the system level.

8.4.6.1 Pinch Analysis

Recovering heat from hot streams to preheat cold streams is constrained by the temperature levels of the heat required, by topological constraints, and by the investment in heat exchangers. Pinch analysis is used to estimate the maximum heat recovery that can be realized in a system (without limitations on the number of streams

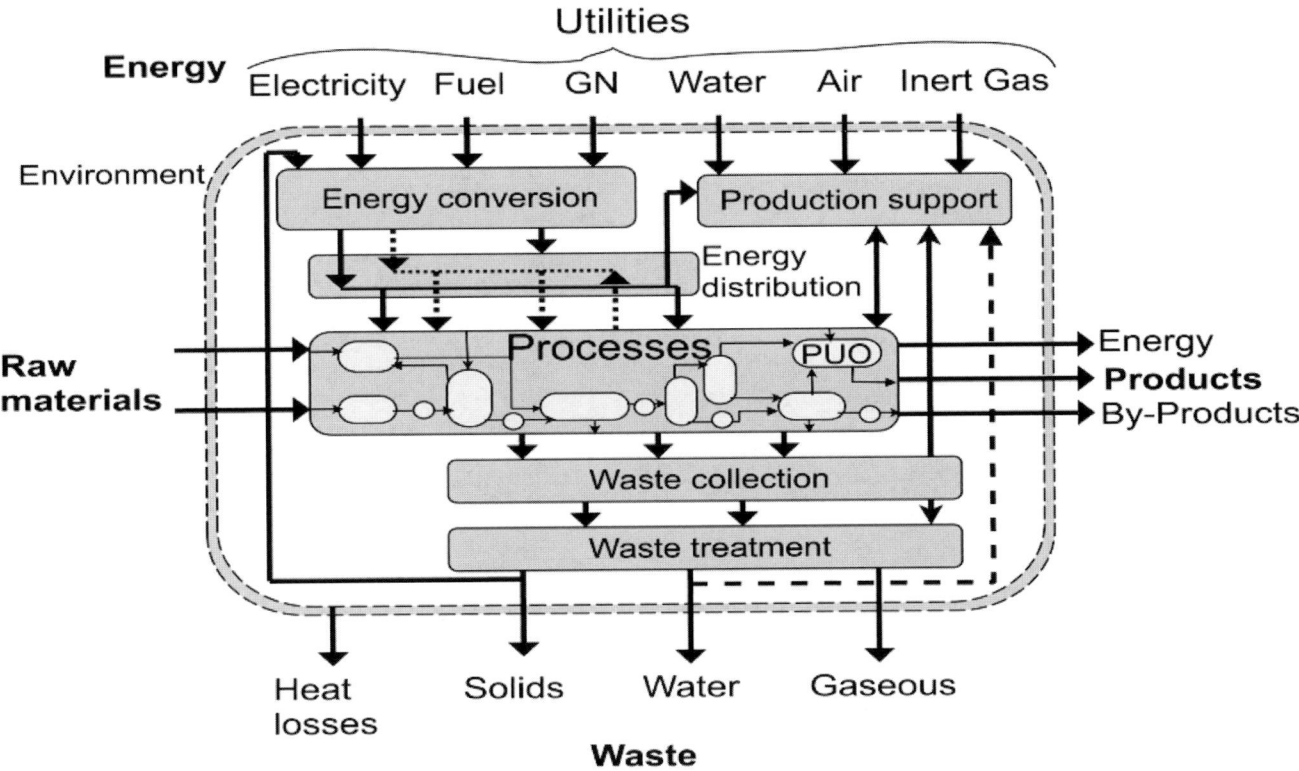

Figure 8.26 | The systemic vision of an industrial process.

and the size of the system) and without having to practically define the heat exchange interconnections. Pinch analysis takes into account the temperature levels of the heat requirement and uses a parameter to represent the energy recovery/capital trade-off that limits the heat exchanger cost by setting a minimum temperature difference (ΔT_{min}) between the hot and the cold streams considered for the heat recovery. The maximum heat recovery is obtained by calculating the integral of the heat available in the hot streams of the system and of the heat required by the cold streams as a function of the temperature. This draws respectively the hot and cold composite curves (see Figure 8.27) that represent the system as one overall hot and overall cold streams between which counter current heat exchange will be used to recover heat.

The maximum heat recovery is obtained by considering that the temperature difference between the two curves must always be higher than the ΔT_{min}. The point at which the curves are the closest is the pinch point. It divides the system into two subsystems: above the pinch point, the process features a deficit of heat (heat sink), and below it, the process has a surplus of heat (heat source). Having defined the maximum heat recovery, the minimum energy to be supplied and removed from the system is obtained by energy balance, while the heat cascade defines their corresponding temperature levels in the grand composite curve. Comparing the minimum energy requirement with the present energy use defines the amount of energy that is directly transferred from the

resources to the environment without having any real use in the process, as if this energy were brought to heat up the environment.

8.4.6.2   Optimizing Process Operating Conditions

The process composite curves can be used to adapt the process operating conditions in order to maximize the heat recovery potential. Based on the location of the pinch point location, the goal is to transfer heat excess from below the pinch point to above the pinch point, or heat demands from above to below the pinch by changing the process operating conditions. Important unit operations concerned are chemical reactors and evaporation systems (distillation or evaporators), in which the pressure can be modified to change the temperature of the evaporation and/or condensation. Multi effect evaporation is a good example of this mechanism: changing the operating pressure of the evaporation allows the recovery of the condensation heat to evaporate water in an effect with a lower pressure. Staging the evaporation in multi stages allows one to reduce the energy use for evaporation. For example, a three stage evaporation process will allow the reduction of 60% of the heat needed for evaporation. When considered along with process integration, the use of multi effect evaporators has to be considered with a holistic vision, considering the possible heat recovery and use elsewhere in the process, leading to even higher energy savings. Examples show that a well-integrated multi-effect

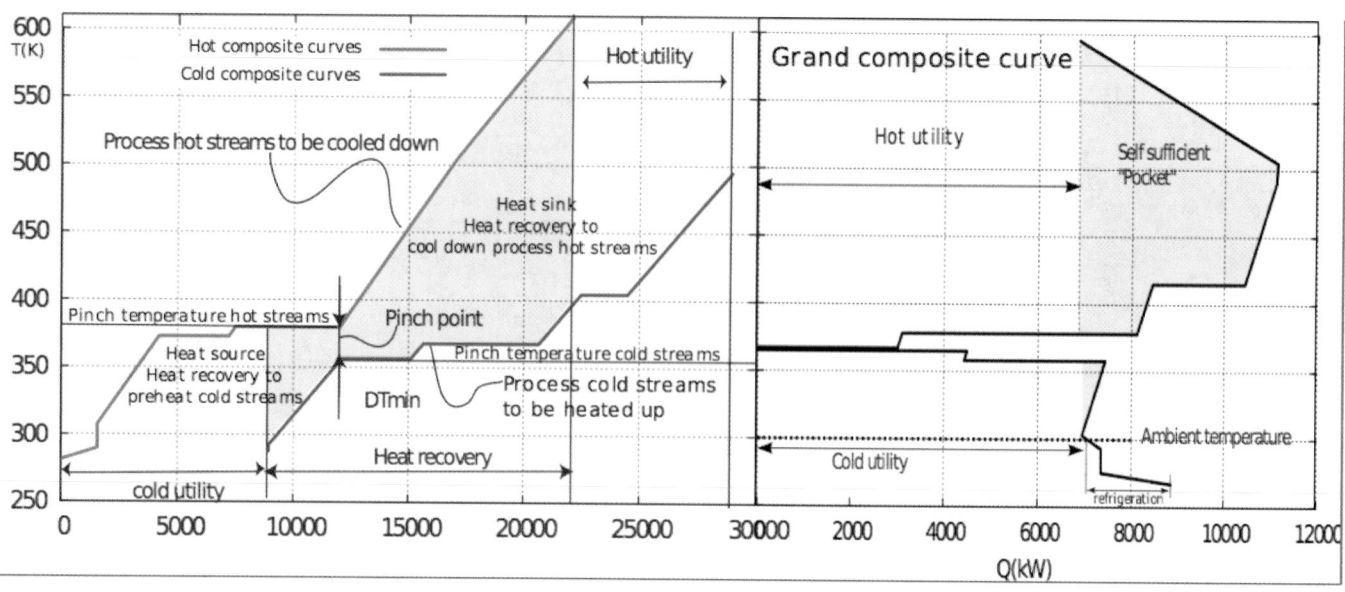

Figure 8.27 | Hot and cold composite curves and possible heat recovery in the system.

evaporation system can save an additional 30% of the remaining heat required.

8.4.6.3   Combined Heat and Power Production

In a CHP production system, energy resources are converted into useful process heat, and electricity. As the heat is supplied to the process by heat transfer, it must have a temperature level compatible with the heat requirement of the process as defined by the grand composite curve (see Figure 8.27) resulting from the pinch analysis. Only the heat of cogeneration that is compatible with the heat sink above the pinch point will create a cogeneration effect.

The different type of cogeneration units are given in Table 8.20. The specific investment, and the electrical (η_e) and thermal (η_{th}) efficiencies depend on the size of the unit. The primary energy savings related to the integration of a cogeneration unit depends on how the co-produced electricity is accounted. One may distinguish the avoided electricity import that substitutes a centralized electricity production facility and that is accounted with the electricity mix efficiency (η_{grid}) from the electricity export that could be considered as an additional electricity production unit. In addition, one may also consider the fuel substitution effect. When the overall heat available from the cogeneration unit is used in the process, the primary energy saving (see Equation 1) is calculated by comparing the resource consumed in the cogeneration unit with the one that would have been consumed for the separate production of heat in a boiler and of electricity from the grid.

$$\text{Energy savings [\%]} = 1 - \frac{1}{\dfrac{\eta_e}{\eta_{grid}} + \dfrac{\eta_{th}}{\eta_b}} \qquad (1)$$

with η_e – the electrical efficiency of the cogeneration unit

η_{th} – the thermal efficiency of the cogeneration unit

η_{grid} – the efficiency of the grid electricity production [38%]

η_b – the efficiency of the boiler replaced by the cogeneration unit [90%]

Rankine cycle-based CHP systems have several industrial applications. A typical example is the steam network of industrial processes, where high-pressure steam is produced in boilers or in the process, and is expanded in steam turbines to produce mechanical power before being condensed to supply heat to the process. Rankine cycles are also used to upgrade waste heat below the pinch point. In this case, the heat excess of the process is converted into mechanical power, and the environment is used as a heat sink. When operated under low temperature conditions (typically below 100°C), the fluid used in the cycle is an organic fluid such as the one used in refrigeration or heat pumping.

It is important to mention that the CHP unit will produce its benefit only when it is located entirely above (in the heat sink sub-system) or below (in the heat source sub-system) the pinch point. The heat that crosses the pinch point has no useful CHP effect. In addition, the CHP system allows one to convert the exergy available in the process hot streams.

8.4.6.4   Heat Pumping

In industrial processes, heat pumps are used to upgrade the temperature level of a heat source. Referring to the pinch analysis, heat pumping will be profitable only if it allows transfer of heat from below to above the pinch point, i.e., if it transforms excess heat from the system heat source into useful heat for the system heat sink. All the other situations do not lead to an overall saving. When operated only above the system

Table 8.20 | Examples of cogeneration units to supply heat to processes.

Type	Typical size	η_e	η_{th}	Investment	Energy Savings
	MWe	%	%	US$/kWe	%
Gas turbine	[1:50]	[30:40]	[40–45]	[1000:2500]	[29:36]
Gas turbine combined cycle	[10:100]	50	35		[41]
Engines	[0.1:10]	[30:50]	[55:40]	[800:4000]	[29:43]
Steam turbine	[0.1:15.0]	[20:25]	[74:69]	[69–71]	[26:30]
SOFC fuel cell	0.1	40	40	[NA]	[33]
Hybrid Gas turbine fuel cell	0.2	70	15	[NA]	[50]

pinch point, the heat pump is an expensive electrical heater for the system, while when operated only below the pinch point, the heat pump is an electrical heater that heats up the environment. The integration of a heat pump must therefore be studied with care.

Heat pumping systems may also use the environment as a heat source and the process as a heat sink. The different types of heat pumps are:

- *Mechanical vapor recompression (MVR):* a vapor stream of the process in the heat source is compressed before being condensed to supply heat to the heat sink. MVR heat pumps are typically used in evaporation, distillation, or drying processes.

- *Mechanical heat pumps:* a fluid (typically a refrigerant) is evaporated using heat from the process heat source and compressed before being condensed to supply heat to the process heat sink.

- *Absorption heat pumps:* instead of using mechanical power, the absorption heat pumps are tri-therm systems. High-temperature heat is used as a driver to raise the temperature of a fluid that is at a lower temperature in the heat source. The heat is sent back to the process at a medium temperature in the process heat sink.

- *Heat transformers:* heat transformers use the same principle as the absorption heat pump, but the driver in this case is at a medium temperature (evaporation from the process heat source) and sends the heat back at a higher temperature (in the heat sink), while the remaining heat is sent back at the lowest temperature (typically to the environment).

Especially in mechanical heat pumps, the temperature level of the heat delivered and the temperature lift define the heat pump's performance. The coefficient of performance (COP) of a heat pump represents its amplifying effect and is used to compute its primary energy saving. Considering constant temperature for the heat source ($\bar{T}_{source}$) and for the heat sink ($\bar{T}_{sink}$), the COP of the heat pump is given by Equation 2, where η_{COP} is the efficiency of the heat pump with respect to the theoretical COP. Typical values for η_{COP} are around 50%.

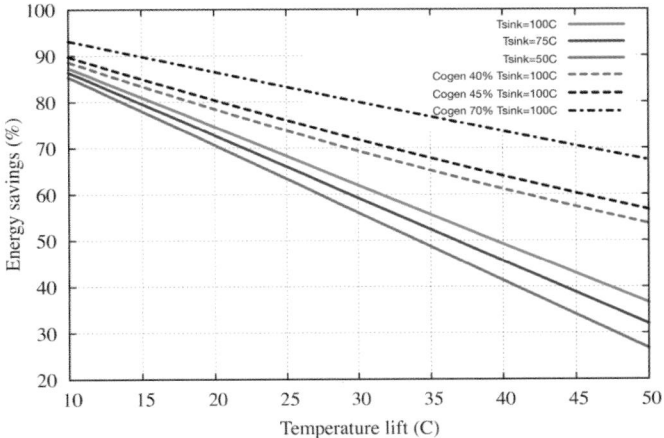

Figure 8.28 | Primary energy savings by the integration of heat pumps (η_{grid} = 38%, η_{boiler} = 90%. η_{COP} =50%) and of a cogeneration unit and heat pumps ($\eta_{electrical}$ = 40%, 45% and 70% with $\eta_{electrical} + \eta_{thermal}$ = 85%). Where η_{COP}= efficiency of the heat pump with respect to the theoretical COP.

$$COP = \frac{\dot{Q}_{th}}{\dot{E}_{hp}} = \eta_{COP} \frac{\bar{T}_{sink}(K)}{\bar{T}_{sink}(K) - \bar{T}_{source}(K)} \qquad (2)$$

In integrated systems, the flow of the heat pump and heat recovery are limited by the heat which is available and required by the process. Heat pumps are mainly used in processes with low to medium temperatures (up to 120°C), especially in processes with evaporation systems such as in the food industry or in drying processes. On the grand composite curves, the flow limitation corresponds to the activation of a new pinch point in the system and, therefore, introduces the possibility of cascading heat pumps.

The primary energy saving of a heat pump depends on the heat pump COP and on the efficiency of the grid electricity production, as shown in Figure 8.28. When the temperature level of the CHP permits, then the integration of heat pumps together with CHP units is an attractive option. In this case, the electricity required by the heat pump is produced by the CHP unit that supplies an additional amount of heat to the process. The primary energy savings in this case are even higher (see Figure 8.28).

Table 8.21 | Case studies of process integration, polygeneration, and heat pumping. In this Table the numbers are normalised so that the fossil fuel use in the present situation is taken as the base (100). All other quantities are computed with reference to the fossil fuel use. The savings for the improved scenarios are compared as a percentage of the existing scenario.

Kraft process	Present situation	Process modification	Heat exchangers
Fossil fuel	100	28.5(71.47%)	2.3 (97.74%)
Biomass	269.2	269.2	269.2
Electricity	353.1	289.8 (17.92%)	322.9 (8.56%)
Primary energy	1029.2	791.2 (23.12%))	851.9 (17.22%)
Cogeneration	5.6	68.9 (x12.3)	35.9 (x6.4)
Primary energy cogen.	354.3	116.4 (67.16%)	177.1 (50.03%)
Sulfite process	**Present situation**	**Process integration**	
Fossil Fuel	100	13 (85%)	
Biomass	11	11	
Electricity	48	36 (25%)	
Primary energy	237	110 (53.4%)	
Cogeneration	7	19 (x2.7)	
Primary energy cogen.	104	−23 (120%)	
Dairy	**Present situation**		**Heat pump + cogen.**
Fossil Fuel	100		21 (79%)
Electricity	15		10 (36%)
Primary energy	140		47 (67%)
Chocolate	**Present situation**	**Process integration**	**Heat pump + cogen.**
Fossil Fuel	100	53 (46%)	30 (70%)
Electricity	9	3 (66%)	0 (100%)
Primary energy	123	61 (50%)	30 (75%]
Ethylene	**Present situation**	**Process integration**	
Fossil Fuel	100	55 (45%)	
Electricity	18	7 (59%)	
Primary energy	147	74 (49%)	
Ammonia	**Present situation**	**Process integration**	**High temp. cogen.**
Fossil Fuel	100	38 (62%)	77 (33%)
Electricity	0	13	−3
Primary energy (38%)	100	71 (29%)	70 (30%)
Primary energy (58%)	100	60 (40%)	73 (27%)

8.4.6.5   Case Studies

The Kraft process is the dominant paper making process based on chemical pulping using sulfate. Case studies of a typical kraft pulp and paper process, the food industry (chocolate and dairy), and industrial chemicals (ethylene and ammonia) show benefits of applying process integration concepts, heat pumps, and cogeneration.

Table 8.21 shows the savings obtained against existing practice. It is clear that savings of more than 30% are possible through process integration, polygeneration, and heat pumping. It is, however, difficult to extrapolate the potentials, since most of the results refer to specific systems and conditions.

8.5   Renewables in Industry

In the present mix of energy use in industry, renewables account for about 9% of the final energy. This is mainly due to bagasse and rice husk in sugar and other traditional industries, biogas from sewage and farms, and black liquor in pulp and paper. It is technically feasible to plan a significant share of renewables for industry for process heating, cooling, and power. An analysis by UNIDO (2010b) estimates that renewable energy use in industry is likely to grow to about 50 EJ/yr of final energy by 2050. In 2006, cement manufacturers reported that 10% of total fuel use was from alternative fuels, of which 30% was biomass (CSI, 2006). IEA's analysis (IEA, 2006) reveals that more than 80 solar thermal plants with total collector areas of about 34,000 m² were being used for industrial heating, mainly in textiles, the food industry, chemical plants, car washing, and metal treatment facilities.

At present, solar-based thermal energy for low-grade steam and biomass-based thermal energy are feasible for replacing oil. In India, successful examples of thermal applications of biomass gasifiers are in steel rolling mills, the ceramic tile industry, and CO_2 manufacture. Solar thermal systems have been installed in dairies (a 160 m² dish replaced the 1 t/hour oil-fired boiler in a dairy in western India). The payback periods for thermal applications for gasifiers is one to two years, while for solar-based heating or steam systems for oil replacements the payback period ranges between four and six years. Biomethanation plants have also been applied in dairies, brewery waste, distillery waste, etc. These have been used for power generation; for example, the biomethanation plant at Haebowal Dairy in Ludhiana, Punjab, India (MNRE, 2010) generates a steady power output of about 1 MW of electricity based on the collection and use of 235 tonnes of animal waste per day. The cost of the plant installed in 2004 was INR134 million (US$3 million). This provides for cogeneration of heat and power and also provides stabilized organic manure. In sugar factories, bagasse-based cogeneration systems are cost-effective and can provide significant surplus power to the grid.

It is possible to increase the share of renewables in industry. In order to do this, it is proposed that a number of demonstration systems be financed in different industrial segments. This would facilitate the dissemination of renewables in the industrial sector. Bio-based renewables can also be used as a feedstock for chemicals (Hermann et al., 2007; Shen et al., 2010), resulting in energy and emissions reductions in the industrial sector.

8.6 Thermodynamic Limits

8.6.1 Analysis of the Global Industrial Sector

In this report, the focus is on the global industrial sector, which is composed of many industries. The energy used by each of these industries in 2005 is shown in Table 8.22. The most significant industries, based on quantity of energy used in 2005, are seen in Table 8.22 to be iron and steel; chemicals and petrochemicals; non-metallic minerals; paper, pulp, and printing; and food and tobacco. With this and other data, energy and exergy utilization in the global industrial sector is evaluated and analyzed. Note that all of the energy forms in Table 8.22 are primary

Table 8.22 | Energy input for the global industrial sector by industry and energy type, 2005 (in PJ).

Industry[1]	Coal and coal products	Crude, NGL and feedstocks	Petroleum products	Natural gas	Geothermal	Solar, wind, other	Combustible renewables and waste	Electricity	Heat	Total
Iron and steel	7910.3	0.2	635.0	2452.8	0.0	0.0	268.6	3259.0	494.3	15020.2
Chemical and petrochemical	1832.1	2.0	2550.3	4742.8	0.0	0.0	95.3	3570.5	1447.1	14240.1
Non-ferrous metals	486.1	0.0	326.2	605.4	0.0	0.0	4.9	2127.3	86.5	3636.3
Non-metallic minerals	5716.9	1.3	1520.9	2103.9	0.0	0.0	210.4	1346.8	105.7	11005.8
Transport equipment	150.6	0.0	126.3	422.8	0.0	0.0	0.6	589.7	136.8	1426.8
Machinery	416.2	0.4	472.9	879.0	0.0	0.0	2.5	2091.2	191.7	4054.0
Mining and quarrying	303.5	0.0	570.7	402.8	1.3	0.0	0.4	824.9	104.4	2208.0
Food and tobacco	794.2	1.5	1085.4	1367.7	0.2	0.0	1107.7	1284.1	356.7	5997.6
Paper, pulp and printing	794.9	0.0	601.1	1068.3	5.7	0.0	2069.0	1701.0	210.2	6450.3
Wood and wood products	87.1	0.3	140.4	121.0	0.0	0.0	414.6	348.6	213.3	1325.3
Construction	215.1	1.1	837.4	155.9	0.0	0.0	6.1	215.1	49.1	1479.9
Textile and leather	450.6	0.5	366.7	364.7	0.0	0.0	10.0	793.1	239.2	2224.8
Non-specified industry	2364.4	155.7	4214.8	3406.7	5.0	5.1	3322.6	4112.5	965.4	18,552.2
Total	**21,522.1**	**162.9**	**13,448.2**	**18,093.8**	**12.3**	**5.2**	**7512.7**	**22,263.9**	**4600.4**	**87,621.3**

Note: For the overall global industrial sector and all industry categories within it, there are no inputs of nuclear energy, or hydro, or heat production from non-specified combustion fuels. Units are petajoules (PJ), which is equal to 1015 Joule (see Chapter 1, Figure 1.3).

1 excludes feedstocks (non-energy use), see Chapter 1, Section 1.2.2.

Source: IEA, 2007a and 2007b.

energy resources or refined versions of them, except for electricity and heat, which are secondary energy carriers.

8.6.1.1 Methodology and Energy Data for the Global Industrial Sector

Each industry category in the industrial sector is analyzed separately, and then combined in a comprehensive assessment of the sector. To simplify the analysis of energy and exergy flows and efficiencies for a complex sector such as the one considered here, the most significant industries, in terms of proportion of total sector energy use, may be taken as representative of the overall sector. Many industries have been assessed individually (Brodyanski et al., 1994).

In the global industrial sector, most of the input energy is used to generate heat for production processes, mechanical drives, lighting, and air conditioning. Heating processes for each industry can be further divided into low-, medium-, and high-temperature categories. This differentiation is important in exergy analysis, as the temperature at which heat is supplied and used greatly affects the exergy associated with the heat.

Several steps are used to derive the overall efficiency of the sector:

- Energy and exergy efficiencies are obtained for process heating for each of the product-heat temperature categories.

- Mean heating energy and exergy efficiencies for the main industries are calculated using a two-part procedure:

  (i) weighted mean efficiencies for electrical heating and fuel heating are evaluated for each industry; and

  (ii) weighted mean efficiencies for all heating processes in each industry are determined with these values, using as weighting factors the ratio of the industry's energy use (electrical or fuel) to the total consumption of both electrical and fuel energy.

- Efficiencies are determined for other processes (i.e., heating, mechanical drives, and other processes).

- Weighted mean overall efficiencies for each industry are evaluated using as the weighting factor the fractions of the total sector energy input for heating, mechanical drives, and other processes (Dincer et al., 2004).

- To determine the industrial sector's efficiencies, weighted means for the weighted mean overall energy and exergy efficiencies for the major industries in the industrial sector are obtained, using as the weighting factor the fraction of the total industrial energy demand supplied to each industry (Utlu and Arif, 2008).

In the present analysis, a simplified approach is taken to evaluate exergy parameters. Here, we utilize global energy data for the industrial sector in 2005 as provided by IEA (2007a; b), which provides energy inputs, in terms of energy type, to each industry category in the global industrial sector. We then incorporate the energy and exergy efficiencies for the utilization of the different energy commodities in the industry sector, as determined in a previous global energy and exergy analysis (Nakicenovic et al., 1996). The assumption incorporated here is that efficiencies have not changed significantly on a global scale over the last ten to 15 years for the different industries in the global industrial sector. This assumption may not introduce significant inaccuracies, because although the efficiencies of technologies utilized in highly developed countries may have risen over the last decade, the same phenomenon may not be true in many developing countries, where the focus has been on increased energy use to drive economic development. In addition, the observation of Nakicenovic et al. (1996) that heat input to the industrial sector is predominantly for low- and medium-temperature heating, as well as process heating, is used, so the exergy of the heat input to the sector is thus taken to be 28% of the energy.

8.6.1.2 Energy and Exergy Flows and Efficiencies for the Global Industrial Sector

The energy and exergy inputs and outputs for the overall global industrial sector are presented in Table 8.23. For simplicity, exergy and energy values are assumed equal for commodities that normally exhibit an exergy-energy ratio of approximately one (e.g., most fossil fuels). Also, it is assumed for biofuels that the energy-exergy ratio is unity, and that the energy-exergy ratio for biofuels is representative of that for all renewables. In the present analysis, a reference environment which emulates the actual physical environment is utilized.

Energy and exergy flow diagrams for the overall global industrial sector for 2005 are presented in Figures 8.29 and 8.30, respectively. The input energy to the 2005 global industrial sector of 87.6 EJ shown in the present energy and exergy analyses (Figures 8.29 and 8.30) is less than 115 EJ reported in Section 8.1 (IEA, 2008a). The difference is due to the exclusion of coke ovens, blast furnaces and feedstock energy for petrochemicals from the industrial sector in the present analysis. The addition of the energy inputs for these sub-sectors provides a final energy input of 115 EJ in 2005.

This difference does not affect the overall results and conclusions as the efficiencies and fractional conversions of energy and exergy in the present analysis do not change significantly if coke ovens, blast furnaces and feedstock energy for petrochemicals are included. These figures illustrate the variations in flows shown on the basis of energy or exergy. In these figures, losses and wastes of energy involve only emissions, while losses and wastes of exergy involve emissions and internal destructions (with internal destructions normally being the most significant). It can be seen that the energy and exergy values of most of the inputs (except

Table 8.23 | Energy and exergy flows (in EJ) in the global industrial sector by energy commodity type, 2005.

Energy commodity	Input energy[1]	Product energy	Input exergy	Product exergy
Coal and coal products	21.5	9.3	21.5	3.3
Crude, NGL and feedstocks (including petroleum products)	13.6	3.4	13.6	3.0
Natural gas	18.1	10.2	18.1	3.7
Electricity	22.3	15.7	22.3	12.7
Heat	4.6	4.6	1.3	1.3
Renewables (including combustible renewables and waste, geothermal, and solar/wind/other)	7.5	1.5	7.5	1.0
Total	87.6	44.6	84.3	25.1

Note: Columns may not sum exactly due to round-off errors.

1 excludes feedstocks (non-energy use), see Chapter 1, Section 1.2.2.

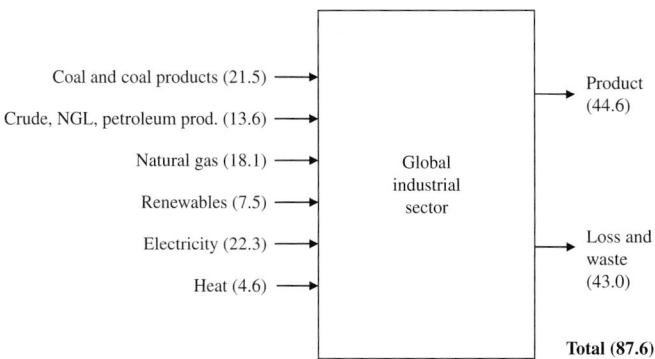

Figure 8.29 | Energy flows (in EJ) for the global industrial sector, 2005. Note: Final energy data excludes feedstocks (non-energy use), see Chapter 1, Section 1.2.2.

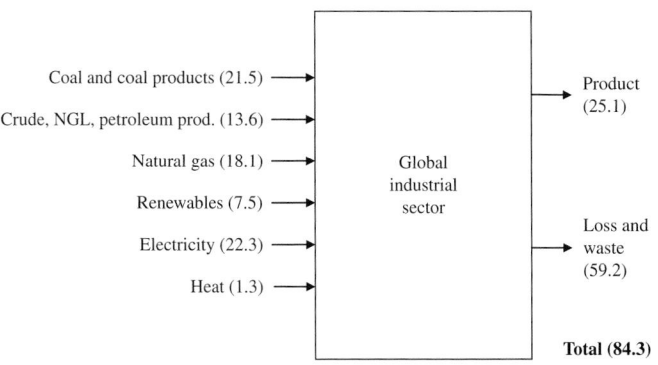

Figure 8.30 | Exergy flows (in EJ) for the global industrial sector, 2005.

heat) are similar, but the product and waste flows have significantly different energy and exergy values.

The overall efficiencies for the global energy sector, evaluated as the ratio of product to input using values from Table 8.23, are found to be 51% based on energy and 30% based on exergy. Consequently, exergy analysis indicates a less efficient picture of energy use in the global industrial sector than energy analysis does. A larger margin for improvement exists from an exergy perspective, compared to the overly optimistic margin indicated by energy.

An energy analysis of energy utilization in the global industrial sector does not provide a true picture of how well energy resources entering it are utilized. An assessment based on energy can be misleading because it often indicates the main inefficiencies to be in the wrong sectors, and a state of technological efficiency higher than actually exists. To accurately assess the true efficiency of energy utilization, exergy analysis must be used. Exergy parameters provide a powerful tool for indicating to industry and government where emphasis should be placed in programs to improve the use of the exergy associated with the main energy resources (Dincer et al., 2004). This analysis provides important insights about potential priorities for future research and development (R&D) initiatives and directions.

It is instructive to compare energy and exergy utilization for the industrial sector with that for other sectors, based on previous analyses of all sectors. Two key points are noted:

- Significant variations are usually exhibited between energy and exergy efficiencies in the residential/commercial and industrial sectors, but not in the utility and transportation sectors. This observation is mainly attributable to the high degree of heating and cooling processes that occur in the residential/commercial and industrial sectors.

- The industrial sector is relatively efficient on an exergy basis compared to other sectors. The reason for the low exergy efficiencies in some sectors is inefficient utilization of the quality of the input energy. The residential/commercial sector, in particular, has a notably low exergy efficiency, mainly because much of the primary use of energy is to produce cold or heat at near environmental temperatures, whereas in the industrial sector high-quality energy is used to produce higher-temperature heating and lower-temperature cooling (which are higher-quality energy forms). Thus the high quality of energy inputs, as reflected by their exergy values, is better utilized in the industrial sector. The production of low-quality products from a fossil fuel or electricity leads to a loss in energy quality that is only reflected properly with exergy analysis. Exergy methods clearly demonstrate that the closer the temperature of the environment to the temperature of the heat produced, the lower the exergy efficiency of the process.

Clearly, exergy provides important insights into the behavior and efficiency of the industrial sector in terms of efficiency limits and margins for improvement. Energy and exergy losses can be viewed as representing, respectively, perceived and actual inefficiencies. It is seen that actual inefficiencies in the residential/commercial and industrial sectors are much higher than the perceived inefficiencies. For the transportation and utility sectors, the actual inefficiencies are lower than the perceived inefficiencies.

8.7 Industrial Energy Efficiency and the Economy

8.7.1 Industrial Energy Efficiency and Economic Growth

At first sight, increased industrial energy efficiency should encourage economic growth. In fact, there is quite a lot of evidence that it has done so in the past by cutting costs, leading to lower prices of goods and, ultimately, increased demand for those goods or for new products and services that did not exist previously. Nielson's use of preheated air made blast furnaces more efficient and cut the cost of pig iron dramatically in the early 19th century. This, in turn, made wrought iron rails much cheaper and helped in the spread of railroads. Later, Bessemer's clever way of decarbonizing molten iron by blowing air through it to make steel was a tremendous gain in efficiency and made steel much cheaper than the older processes. The result was to substitute steel rails for iron rails, and to create immense new markets for steel products, especially in construction, ship-building, and, later, the manufacture of automobiles.

The invention of steam turbines in the 1880s made steam power cheaper and more efficient. The main application was for electric power generation. Cheaper electricity resulted in the rapid spread of electric lighting, replacing gas lights, candles, and acetylene lamps, and the substitution of electric motors for steam engines in factories and on railways. Cheap electric power also created new markets, including the aluminum and chlorine industries, the use of electric furnaces to refine metals with high melting points such as chromium and nickel (for stainless steel), tungsten (for light bulbs), and synthetic carbides to permit high-speed grinding machines that are essential for manufacturing complex automobile engine parts such as cam-shafts and crank-shafts. Electrification was arguably the single most important driver of economic growth in the first half of the 20thcentury.

Aluminum followed a similar trajectory. In the mid 19th century, aluminum was an expensive luxury metal used, for example, to put the cap on the Washington Monument in Washington DC. After the introduction of the more efficient electrolytic process, however, the price of aluminum

dropped rapidly, creating new markets such as pots and pans, window frames and beer cans, electrical transmission lines (replacing copper), as well as aircraft and truck bodies. Cheap plastics from petrochemicals have also created vast new markets for packaging materials, furniture, toys, and so on.

Plastics derived from petrochemicals have contributed significantly to economic growth in the second half of the 20thcentury. However, exergy efficiencies at the process level are now approaching 50% for some primary products, such as ammonia. While some further gains can be expected, the best that can be hoped for in the basic materials subsectors is typically an improvement in the range of 15–30% beyond the current best available technology. Gains of that magnitude will not lead to cost or price reductions sufficient to create significant new products, and only minor increases can be expected in price-driven demand for existing industrial products.

The most promising approach to reducing industrial energy use in the future is to sharply increase recycling (in the case of metals, glass, paper, and some plastics), and to use secondary materials, such as fly-ash, in the manufacture of cement. Another promising approach, albeit several decades in the future, will be to find ways of skipping intermediate process steps and/or to develop plastic substitutes from waste biomass, such as ligno-cellulose. Recycled metals from scrap are far less energy intensive than virgin metals. In the case of aluminum and copper the difference is more than a factor of ten. Unfortunately, there has been little progress in remanufacturing or recycling technology in recent decades (notwithstanding their potential importance) because of a lack of sustained support for the necessary collection and sorting activities by governments. The major exception seems to be some progress in recovering precious metals (mainly gold and silver) from electronic waste. But even in this case, most electronic waste is processed inefficiently by unskilled labor, with little or no concern about pollution or public health hazards.

However, all things considered, the energy intensity of new manufactured products cannot be expected to decrease significantly in the coming decades. The only way to cut energy use by industry more than marginally is to use much less of the products of industry and to sharply increase the rate of product reuse, renovation, remanufacturing, and recycling. This is technically feasible but will be politically difficult. The case of ammonia, the primary source of all nitrogen fertilizers, offers a possible example. The excessive use of nitrogen fertilizers is responsible for several environmental problems, such as pollution of ground water and eutrophication of rivers and streams. Precision agriculture offers the potential for actually cutting nitrogen fertilizer use by reducing waste and simultaneously reducing environmental harm.

8.7.2 Energy, Exergy and Economic Growth

As increased efficiency cuts costs and prices, it also follows that shortages and price increases will have a negative impact on economic growth, at least in the short and medium term. There have been many observations of growth slowdowns following energy price spikes (Hamilton 1983; 2003; 2005). While it is true that shortages induce invention and innovation, this is a slow process. The slow introduction of renewables (especially wind and solar photovoltaics), which may be the long-term substitutes for fossil fuels, illustrates the problem. Renewables only account for a tiny fraction of global electricity generation, and still cost considerably more per unit of delivered power than conventional coal-burning power plants. Introduction would be faster if the price of fossil fuels reflected the unpaid environmental damages they cause, but governments seem generally unwilling to impose those costs on the producers.

Parenthetically, it seems very likely that the high oil prices of the first half of 2008 hastened the end of the US real estate price boom by squeezing household expenditures at a time when many people already had credit card debts and no savings. This, in turn, caused an increase in foreclosures. That may have triggered the financial meltdown, which followed from the realization that mortgage-based securities could not be priced realistically, which meant that many financial institutions and banks were over-leveraged.

More importantly, in the long run, the forthcoming advent of "peak oil," whether it has already happened or whether it occurs ten or 20 years in the future, must have a significant negative impact on future global economic growth, at least until energy-related innovation creates a new boom (or "bubble"). The reason is that energy in general, and oil in particular, are essential to virtually all economic activity, with marginal productivity (output elasticity) far greater than its still small – though increasing – cost share. As the price of oil (and oil substitutes, such as they are) rises, the demand for energy-intensive products will fall, as happened in late 2008. That brings the price of oil temporarily back down, which encourages renewed consumption but discourages investment in energy conservation measures that depend on higher prices. This, in turn, delays needed economic adjustment while accelerating the onset of the next crisis.

Skipping over the details, in our multi-sector, multi-product economy there is econometric evidence that the output elasticity of an essential (non-substitutable) input, such as petroleum, or more generally, energy, tends to be much larger than its cost-share, whereas the output elasticity for labor in the industrialized countries tends to be much smaller than its cost-share. Simply put, it can be argued that raw (unskilled) labor is over-priced in modern economies, whereas flows of energy, especially petroleum, have been relatively underpriced up to now. The econometric evidence for this would take us far afield. However, the simple observation that most firms are able to increase profits by reducing employment seems to suggest that employment is being kept artificially high, partly for social and political reasons, and partly as a way of supporting consumption.

The non-equality of output elasticities and cost-shares has important consequences for the standard theory of economic growth. The first implication is that the Cobb-Douglas production function must be discarded, because it assumes that output elasticities are equal to cost-shares and that the latter are constant. Dropping this assumption implies that the output elasticities of factor inputs must be functions of all the input variables, namely capital, labor, and energy or energy services. Kuemmel et al. (2008) have shown that the simplest functional form for a production function that allows for non-constant output elasticities, takes into account the energy flows in a physically plausible way, and permits an explicit parametric formulation of the constraints is the so-called LINEX production function (Kuemmel et al., 2002; 2008; Ayres and Warr, 2005). Its mathematical characteristics have been discussed elsewhere and need not be recapitulated here.

When growth theory is suitably modified to reflect the true importance of energy as an input, it turns out that the primary driver of growth, apart from capital deepening, is the increasing supply of "useful work" (mechanical work, chemical work, electrical work, etc.) in the economy (Ayres et al., 2003; Ayres and Bergh, 2005; Ayres and Warr, 2005; 2009). This has been a consequence of two past trends: (1) the discovery of huge oil and gas reserves, and (2) the increasing efficiency of conversion of primary energy (fossil fuels) into various forms of useful work, such as electric power and motive power.

The advent of peak oil means that, as the supply of oil and gas cannot be expected to continue to increase in the future, driving energy prices down – as it did for most of the last two centuries – future economic growth will depend more than in the past on technological progress, especially in the area of increasing energy (exergy) efficiency in the economy. Yet, the rate of exergy efficiency increase (in the United States, at least) has been slowing down since the 1970s. The bottom line here is that either US economic growth will slow down permanently (with global consequences) or effective measures to increase the rate of increase of exergy efficiency must be undertaken to compensate for the coming decline in the availability of natural resources. Such an acceleration of technological progress vis-à-vis exergy efficiency is technically possible but politically difficult due to resistance from entrenched special interests, especially the electric utility monopolies.

8.8   Realizing the Opportunities – Policies and Programs

The principal business of an industrial facility is production, not energy efficiency. High energy prices or constrained energy supply will motivate industrial facilities to try to secure the amount of energy required for operations at the lowest possible price. However, price alone will not build awareness within the corporate management culture of the potential for energy use and cost savings, maintenance savings, and production benefits that can be realized from the systematic pursuit of industrial energy efficiency. It is this lack of awareness and the corresponding failure to manage energy use with the same attention that is normally provided to production quality, waste reduction, and labor costs, i.e., at the root of the opportunity.

To be effective, energy efficiency programs need to engage industry at the management level as well as facilities engineering. Since industrial decision-making is largely driven from the top, failure to engage management results in missed opportunities for energy efficiency improvement, even when the technical staff is educated and aware of the opportunities.

A number of countries have demonstrated the value of effective industrial energy efficiency programs. The IEA's (UN Energy, 2010) World Energy Outlook Policy Database has compiled information on industrial energy efficiency programs from the IEA Climate Change Mitigation Database, IEA Energy Efficiency Database, IEA Global Renewable Energy Policies and Measures Database, the European Conference of Ministers of Transport, and contacts in industry and government (IEA, 2008b). The IEA's Energy Efficiency Database contains 170 industrial energy efficiency policies and measures in 32 countries and the EU (IEA, 2008c).

Barriers to improved energy efficiency include lack of information regarding energy efficiency, limited awareness of the financial or qualitative benefits arising from energy efficiency measures, inadequate skills to implement such measures, capital constraints and corporate culture leading to more investment in new production capacities rather than energy efficiency, and greater weight given to addressing upfront (first) costs rather than recurring energy costs, especially if these costs are a small proportion of production costs (Monari, 2008).

In addition, for developing countries the marginal cost of adopting an industrial policy can be substantially greater than in a developed country that already has supportive institutions in place (Monari, 2008). For new technologies, the slow rate of capital stock turnover in many industrial facilities (Worrell and Biermans, 2005), coupled with the perceived risks of adopting new technologies, can hamper adoption. Table 8.24 (below) provides an overview of industrial needs and goals addressed by industrial policies and programs.

8.8.1   Energy Management

The implementation of successful national energy management programs is dependent on legislation, incentives and policies, and the institutional mechanisms for energy efficiency.

8.8.1.1   International Energy Management Standards

The purpose of an energy management standard is to provide guidance for industrial facilities to integrate energy efficiency into their management practices, including fine-tuning production processes and improving the energy efficiency of industrial systems (UN Energy, 2010). Energy management seeks to apply to energy use the same culture of continuous improvement that has been successfully used by industrial firms to improve quality and safety practices. An energy management standard

is needed to influence how energy is managed in an industrial facility, thus immediately reducing energy use through changes in operational practices, as well as creating a favorable environment for adopting more capital-intensive energy efficiency measures and technologies. Although the focus of this chapter is industrial energy efficiency, it is important to note that the energy management standards mentioned here are equally applicable to commercial, medical, and government facilities.

An energy management standard requires a facility to develop an energy management plan. In companies without a plan in place, opportunities for improvement may be known but may not be promoted or implemented because of organizational barriers. These barriers may include a lack of communication among plants, a poor understanding of how to create support for an energy efficiency project, limited finances, poor accountability for measures, or a perceived change from the status quo. Without performance indicators that relate energy use to production output, it is difficult to document improvements in energy intensity (UNIDO, 2007).

Companies that have voluntarily adopted an energy management plan (a central feature of an energy management standard) have achieved major energy intensity improvements. Some examples include:

- Dow Chemical achieved a 22% improvement (saving US$4 billion) between 1994 and 2005, and is now seeking another 25% from 2005 to 2015.

- United Technologies Corp. reduced global GHG emissions by 46% per dollar of revenue from 2001 to 2006, and is now seeking an additional 12% reduction from 2006 to 2010.

- Toyota's North American Energy Management Organization has reduced energy use per unit by 23% since 2002; company-wide energy efficiency improvements have saved US$9.2 million in North America since 1999.

Denmark, Sweden, Ireland, South Korea, Spain, Thailand, and the United States have national energy management standards. Japan has a legal requirement for its more energy-intensive industrial facilities to have an energy manager and an energy management plan. These requirements also extend to large commercial facilities and parts of the transportation sector. The Netherlands has an energy management specification closely linked to long-term agreements. The European Committee for Standardization and the European Committee for Electrotechnical Standardization developed a common standard for the EU in mid-2009. China and Brazil both have national energy management standards under development.

Table 8.25 compares the elements of the energy management standards in seven countries or regions with existing energy management standards (or specifications), two under development, and one country for

Table 8.24 | Industrial energy efficiency needs and goals addressed by policies and programs.

	Target-Setting Voluntary Agreements	Industrial Energy Management Standards	Capacity-Building for Energy Management and Energy-Efficiency Services	Delivery of Energy-Efficiency Products and Services	Equipment & System Assessment Standards	Certification and Labeling of Energy-Efficiency Performance	Financial Mechanisms and Incentives
ENERGY EFFICIENCY (EE) INFORMATION AND TOOLS							
Increased information on EE technologies and measures			X	X	X	X	
Increased information on EE standards			X	X	X	X	
Improved access to high-quality energy auditing services and assessment tools			X	X		X	
Access to training and tools for energy management (EM)			X			X	
Increased tracking of EE/GHG emissions: GHG inventories, product life-cycle and supply chain energy/GHG assessments	X		X			X	
Robust measurement, monitoring, and verification	X	X	X	X	X	X	
Development of high-quality EE data for analysts, policymakers	X					X	
International best practice information	X	X	X	X	X	X	X
SKILLED PERSONNEL							
Increased EE training at the college level			X	X			
Technical assistance providers for energy management			X				
Improved capability of EE service providers – assessment and EE services		X	X	X	X		
Increased EE focus of equipment suppliers and vendors		X	X	X	X		
Increased and enhanced skills of independent measurement and verification experts (GHG, EM, EE)		X	X	X	X	X	
Increased capacity for energy management at industrial facilities	X	X	X	X		X	
INCREASED MANAGEMENT ATTENTION TO EE							
Increased upper management support for energy efficiency/GHG mitigation investments	X	X				X	X
Management commitment to an energy management system	X	X				X	
Sustained, continuous improvement in EE/GHG mitigation	X	X				X	
EE/GHG MITIGATION COSTS AND FINANCING							
Improved access to capital for EE/GHG mitigation investments	X			X			X
Reduce transaction costs associated with smaller EE projects				X			
Improved understanding of among investors and financiers of potential financial returns		X				X	
Training in preparing project and loan request documents				X			
Pricing of energy to reflect actual costs, encourage EE efficiency							X
Reduce risks associated with assessing and securitizing revenues generated through using less energy				X		X	X

which energy management is a legislated practice for many industries. In all instances, the standard has been developed to be entirely compatible with the ISO quality management program (ISO 9001:2000) and environmental management program (ISO 14001).

Typical features of an energy management standard include:

- a strategic plan that requires measurement, management, and documentation for continuous improvement for energy efficiency;

- a cross-divisional management team led by a representative who reports directly to management and is responsible for overseeing the implementation of the strategic plan;

  - policies and procedures to address all aspects of energy purchase, use, and disposal;

  - projects to demonstrate continuous improvement in energy efficiency;

- creation of an energy manual, a living document that evolves over time as additional energy-saving projects and policies are undertaken and documented;

- identification of key performance indicators, unique to the company, that are tracked to measure progress; and

  - periodic reporting of progress to management based on these measurements.

As shown in Table 8.25, the existing energy management standards have many features in common. ISO now identifies energy management as one of the top five fields meriting the development and promotion of international standards.[7] Energy management received this priority focus based on its enormous potential to save energy and reduce GHG emissions worldwide.

The United Nations Industrial Development Organization (UNIDO) recognized the industry's need to mount an effective response to climate change and to the proliferation of national energy management standards. In March 2007, UNIDO hosted a meeting of experts, including representatives from the ISO Central Secretariat and nations that have adopted energy management standards. That meeting led to the submission of a formal request to the ISO Central Secretariat to consider undertaking work on an international energy management standard.

In February 2008, ISO's Technical Management Board approved the establishment of a new project committee to develop the new ISO Management System Standard for Energy. ANSI and the Associação Brasileira de Normas Técnicas are jointly serving as the secretariat to lead development of ISO PC 242, Energy Management.

The new ISO 50001 establishes an international framework for industrial plants or companies to manage energy, including all aspects of procurement and use. The standard provides organizations and companies with technical and management strategies to increase energy efficiency, reduce costs, and improve environmental performance (ISO 50001, 2011). Based on broad applicability across national economic sectors, the standard could influence up to 60% of the world's energy demand (US EIA, 2007). Corporations, supply chain partnerships, utilities, energy service companies (ESCOs), and others are expected to use ISO 50001 as a tool to reduce energy intensity and carbon emissions in their own facilities (as well as those belonging to their customers or suppliers) and to benchmark their achievements. ISO 50001, published in June 2011, is expected to promote industrial energy efficiency globally.

A successful program in energy management provides an organizational framework for a company to respond effectively through a program of continuous improvement to a national program that establishes energy intensity improvement and/or GHG reduction targets. UNIDO has identified a package of policies described as the "Industrial Standards Framework" that outlines the policy relationships among target-setting agreements, energy management standards, system optimization, and their intended impact on industrial markets (see Table 8.26).

UNIDO is currently engaged with ten countries in the development of industrial energy efficiency programs based on the framework described in Table 8.26.

8.8.1.2 Country Experiences in Energy Efficiency Implementation

A review of the national experiences of a few selected countries on industrial energy efficiency policies illustrates the state of energy policies in the world. The countries selected are Brazil, China, India, South Africa, and the United States. These countries account for almost half of total industrial consumption. Table 8.27 provides a summary of the country experiences for a few select countries in different regions of the world in energy efficiency implementation and impacts.

Brazil

Although a national electricity conservation program (Programa Nacional de Conservação de Energia Elétrica (PROCEL)) was implemented in 1985, energy policies have never addressed energy efficiency

---

7 Priorities also include calculation methods, biofuels, retrofitting and refurbishing, and buildings.

as a relevant issue. In particular, there has been very little attention to industrial energy efficiency policies, despite industry being the most important energy-consuming sector.

More comprehensive energy efficiency policies were only implemented in the country after a severe electricity supply crisis occurred during 2001–2002. Two main regulatory and legislative achievements which took place in the late 1990s and after the energy crisis were important to establish basic instruments for further advancement of energy efficiency in the country: the creation of a public benefit wire charge, and the energy efficiency law created after the energy crisis.

The public benefit wire charge sets aside 1% of electricity revenues to investments in energy R&D and energy efficiency. Since 1998 about R$2 billion (US$980 million) has been invested in energy efficiency utility programs. However, only 10% of these resources have been dedicated to the industrial sector so far. Energy efficiency programs are implemented by utilities under the regulator's oversight and also by a public fund called CTenerg.

During the crisis a series of measures were taken which saved 26 TWh (of a total national consumption of 284 TWh) and 13,000 MW peak, compared to PROCEL's estimated savings during 1994–2000 of 10.7 TWh and 640 MW peak (Maurer et al., 2005). In 2001, a national law was approved that introduced legal instruments to establish energy efficiency standards for appliances, buildings, and motor cars commercialized in the country. Electrical motors (three-phase) were the first equipment to receive mandatory minimum energy efficiency standards. More stringent standards and concentrated efforts in R&D (with the use of the public benefit funds) have the potential to accelerate the improvement of energy efficiency in the country, particularly in the industrial sector.

In general, energy efficiency has not been a priority for the industrial sector. There have been very few energy efficiency programs initiated by the local industries themselves using their own investment capital. In most cases, they are part of corporate policies related to environmental protection and improving the quality of their products, and have been restricted to larger corporations.

There is a preference for investments with a short payback (less than two years). Energy efficiency achieved has been the result of changes in production lines and products. However, there is a lack of expertise in industry in general, particularly in SMEs. Projects that have combined public and private funds have produced significantly better results. In some cases, there has been an increase in the specific energy use in energy-intensive industries due to stricter environmental rules and quality demand in the international market.

Since 1998, utilities have been required by the National Regulator to invest part of their revenues in energy efficiency programs. Up until 2007, a little less than US$1 billion was invested in energy efficiency programs,

but only 10% of this was targeted at the industrial sector, achieving an estimated savings of 376 GWh/yr. Figure 8.31 shows a supply curve for end-use efficiency in the electricity sector for Brazil in 2020.

Recent studies in Brazil have estimated opportunities to introduce energy efficiency measures and fuel substitution away from fossil fuels. A supply curve of conserved electricity for 2020 (base year 2004) is presented in WWF-Brazil, 2007 and shows a potential of 55 TWh at a cost of R$130/MWh (approximately US$80/MWh) with the replacement of motor systems (Figure 8.31). This was the largest potential depicted by this study.

Energy subsidies persist in the industrial sector and distort the cost-effectiveness of existing opportunities. For the most part, industrial energy costs are still irrelevant when compared to other inputs, taxes, and personnel costs. National energy conservation programs have focused more on the residential and public sectors, rather than the industrial sector. There are significant opportunities for more efficient technologies and processes in industry that have not been implemented on a significant scale. Opportunities to combine water and energy efficiency seem to be a better way to accelerate the improved use of resources in the industrial sector, since water usage is increasingly coming under the scrutiny of both policymakers and the population. Regulatory barriers still exist to impede energy companies from investing in efficiency programs for their end-use customers.

China

In 2005, China's government raised a strategic target to reduce its energy use per GDP by 20% by 2010. Industrial energy efficiency is expected to provide 80% of the reduction. To achieve this target, the Chinese government developed a series of initiatives:

- It emphasized regulation of the industrial structure. The government issued policies to import equipment and keep free import and export tariffs and VAT; release loans and lands dependent on energy efficiency standards; and phase out inefficient production capacities in iron and steel, cement, power generation, coal mining, etc. It is expected that 50 million kW in small thermal power units, 100 Mt of iron, 55 Mt of steel, and 250 Mt of cement will be phased out during China's 11th Five-year Plan.

- It established a responsibility mechanism to ensure that disaggregated energy conservation targets are achieved in each province. The central government signed energy-saving target responsibility agreements with 30 provincial governments and the top 1000 enterprises, which consumed energy of over 180,000 tonnes of coal equivalent (tce) annually (approximately 5.28 PJ), and set up a performance assessment system to track and evaluate energy-saving activities under the responsibility agreement mechanism. In May 2008, the National Development and Reform Commission and six other ministries implemented on-site assessment and evaluation of the performance of energy conservation target implementation for 30 provincial governments and released the results. The central

Table 8.25 | Comparison of national energy management standards.

Participating Countries	Management Commitment Required	Develop energy management plan	Establish energy use baseline	Management Appointed Energy Representative	Establish Cross-Divisional Implementation Team	Emphasis on Continuous Improvement
Existing						
Denmark	yes	yes	yes	yes	yes	yes
Ireland	yes	yes	yes	yes	yes	yes
Japan[3]	yes	yes	yes	licensed	implied	yes
Korea	yes	yes	yes	yes	yes	yes
Netherlands[5]	yes	yes	yes	yes	yes	yes
Sweden	yes	yes	yes	yes	unclear	yes
Thailand	yes	yes	yes	yes	implied	yes
United States	yes	yes	yes	yes	yes	yes
(Under Development)						
CEN (EU)	yes	yes	yes	yes	implied	yes
China	yes	yes	yes	yes	yes	yes

1 Certification is required for companies participating in voluntary agreements (also specified interval in Sweden). In Denmark, Netherlands & Sweden linked to tax relief eligibility.

2 As of 2002, latest date for which data is available

3 Japan has the Act Concerning the Rational Use of Energy, which includes a requirement for energy management

4 Korea invites large companies that agree to share information to join a peer-to peer networking scheme and receive technical assistance and incentives

5 Netherlands has an Energy Management System, not a standard, per se, developed in 1998 and linked to Long Term Agreements in 2000.

6 800 companies representing 20% of energy use have LTAs and must use the Energy Management System. The 150 most energy intensive companies, representing 70% of the energy use, have a separate, more stringent, bench marking covenant and are typically ISO 14000 certified, but are not required to use the EM System.

7 Thailand has made the energy management standard is mandatory for large companies, linked it to exisitng ISO-related program activities, coupled with tax relief; program evaluation not yet available

8 To date, the US government has encouraged energy management practices, but not use of the standard. A program was initiated in 2008 to address this which also includes validation; program evaluation results anticipated in 2011.

NOTE: National standards and specifications were used as source documents to develop this table Source: McKane, 2007 as updated by the author in 2008

Source: McKane et al, 2007.

government will reward the best-performing provinces and penalize the provinces which underperformed.

- It set up an energy conservation supervision and monitoring mechanism. Energy Conservation Supervision Centers were established in more than 19 provinces and cities. The centers are responsible for ensuring that enterprises and buildings follow the updated Energy Conservation Law, as well as mandatory energy efficiency standards and codes.

- It launched the Top 1000 Enterprises Energy Conservation Action. The 1000 enterprises are those 998 enterprises with annual energy use at or over 180,000 tce (5.28 PJ). According to statistics, comprehensive energy use of the 1000 enterprises in 2004 was 670 million tce (19.6 EJ) – 33% of the nation's total energy use and

47% of total industrial energy use. The purpose of the program is to encourage enterprises to carry out energy audits, work out energy saving plans, strengthen the implementation of energy efficiency standards, implement energy efficiency benchmarks, and to achieve the target of 100 million tce (2.9 EJ) of energy savings in 2010.

- Due to these policies, the following results were obtained:

- Energy use per GDP was reduced by 1.37% by 2006 and continued to decline (as shown in Figure 8.32).

- The specific energy use for key products from energy-intensive sectors has declined since 2006. Specific energy use per unit of thermal

Document Energy Savings	Establish Performance Indicators & Energy Saving Targets	Document & Train Employees on Procedural/ Operational Changes	Specified Interval for Reevaluating Performance Targets	Reporting to Public Entity Required	Energy Savings Externally Validated or Certified	Year Initially Published	Approx Market Penetration by Industrial Energy Use
yes	yes	yes	suggests annual	yes	optional[1]	2001	60%[2]
yes	yes	yes	industry sets own	yes	optional[1]	2005	25%
yes	yes	yes	yes, annually	yes	yes	1979	90%
yes	yes	yes	yes, annually	optional	optional[4]	2007	data not yet available
yes	yes	yes	yes	yes	optional[1]	2000	20–90%[6]
yes	yes	yes	yes[1]	yes	optional[1]	2003	50%elect
yes	yes	yes industry sets own	yes	evaluation plan	2004	not known[7]	
yes	yes	yes	annual recomm	no	no[8]	2000	<5%[8]
yes	yes	yes	industry sets own	national schemes	national schemes		
yes	yes	yes	industry sets own	not available	not available		

Table 8.26 | Industrial standards framework.

Policy Objective	Policy Response	Market Response
Establishing National Goals for GHG Reduction	Voluntary or Target-setting Agreements; Tax incentives	Companies commit to energy intensity reduction targets
Capacity Building	System Optimization Training of plant engineers/consultants/suppliers/ESCOs	Trained experts conduct plant assessments, sell system services
Integrating Energy Efficient Practices	Energy Management Standard, Guidance, Training	Plants actively manage energy like other resources
Identifying Energy Saving Projects	-Trained System Experts -System Optimization Library -Standardized Assessments	Plant managers used trained experts to identify projects
Implementing Energy Efficiency Projects	Financial incentives, loan guarantees & subsidies, energy efficiency credits, ESCOs	Plants implement more projects, buy system services, accrue credits
Documenting for Sustainability	-Energy Management Plan -System Optimization Library -Measurements & Verification	Energy savings continue through project lifetime & are tradable as credits
Market Recognition	Recognition Programs, Energy Efficiency Credits, Certification	Companies & financial institutions value energy efficiency

Source: UNIDO, 2010a.

power, crude steel, cement, petro-processing, crude copper, alumina, sodium carbonate, and ethylene declined by 3–10.5% by 2006. The energy intensity of steel and cement declined by about 3% by 2007, and thermal power-generating units over 6000 kW consumed 335 gce per kWh – 10 g lower than in 2006.

India

The Energy Conservation Act 2001 enacted by the government mandated the creation of the Bureau of Energy Efficiency (BEE) which promotes energy efficiency. The BEE has taken the following initiatives in the industrial sector:

- *National awards scheme* – an annual awards scheme which provides a mechanism to reward industries in different sectors based on their performance and initiatives in energy efficiency. This resulted in the creation of a database on the BEE website of energy conservation measures and their costs, encouraging other industries to adopt similar measures.

- *Designated consumers* – eight sectors have been designated as energy-intensive sectors: power plants, steel, cement, fertilizers, pulp and paper, chlor-alkali, textiles, and railways. For these sectors, all industrial units with total energy use above specified limits have to provide annual data regarding their energy performance. Designated consumers need to have certified energy managers employed in their plant and have

Table 8.27 | Type of industrial energy efficiency programs in selected countries.

Country	IN-Informational Programs; TP-Tax policies (incentives and/or penalties); REG-Regulations for energy efficiency; TSA-Target-setting Agreements w/ industry; FEII-Focus on Energy-Intensive Industries; EMS-Energy Management Standard; SA-Subsidized Energy Assessments or Audits; FEEP-Financial assistance for Energy Efficiency Project Implementation; TREM-Training for Energy Managers; TRSA-Training on System Assessments; IEES-Industrial Equipment Energy Efficiency Standards; RP-Recognition Program											
	IN	TP	REG	TSA	FEII	EMS	SA	FEEP	TREM	TRSA	IEEES	RP
Argentina	✓					(a)	✓					
Brazil	✓		✓			(a)	✓				✓	
Canada	✓		✓	✓	✓	(a)	✓			✓	✓	✓
Chile	✓				✓	(a)						
China	✓	-	✓	✓	✓	✓	✓	✓	✓	-	✓	✓
Colombia	✓					(a)	✓			✓		✓
Denmark	✓	✓	✓	✓	✓	✓	✓	✓	✓		✓	✓
Egypt	✓							✓	pend	pend		
Finland	✓				✓		✓	✓		✓		
France	✓	✓	✓		✓	(a)					✓	✓
Germany	✓	✓	✓		✓	(a)					✓	✓
India	✓		✓	pend	✓	(a)			✓			
Ireland	✓	✓	✓	✓	✓	✓	✓		✓	✓	✓	✓
Japan	✓	✓	✓		✓	(a)	✓	✓	✓	✓	✓	✓
Korea	✓	✓	✓	✓	✓	✓	✓	✓	✓	✓	✓	✓
Mexico												
Netherlands	✓	✓	✓	✓	✓	(a)	✓	✓	✓	✓	✓	✓
Norway	✓	✓	✓		✓		✓	✓				✓
Philippines	✓								pend	pend		✓
South Africa	✓			some	✓	pend		DSM	private		✓	
Spain	✓	✓				✓						
Sweden	✓	✓	✓	✓	✓	✓	✓		✓		✓	✓
Thailand	✓	✓	✓		✓	✓	✓		pend	pend		✓
United Kingdom	✓	✓	✓	✓	✓	(a)	✓				✓	✓
United States	✓	-	-	new	✓	✓	✓	-	-	✓	1	✓

audits carried out at specified intervals. BEE plans to establish benchmarks in the future for energy use for designated consumers.

- *National energy managers and auditors certification* – BEE initiated a process for the creation of certified energy managers and energy auditors. A syllabus has been drawn up and an annual examination carried out. About 4500 certified energy managers qualified through this process (3400 of them are also certified energy auditors). This process is expected to provide the skilled manpower needed to implement energy efficiency schemes in the country. Apart from these schemes, BEE plans to establish state energy conservation funds to facilitate and strengthen the implementation of energy efficiency by the state nodal agencies. BEE also plans to launch an initiative for small- and medium-sized industries.

India launched a National Mission on Enhanced Energy Efficiency (NMEEE) in April 2010 as a part of the National Climate Change Action Plan. The NMEEE includes a market-based mechanism to enhance energy efficiency in large energy-intensive industries and facilities (the Perform, Achieve and Trade scheme). This involves setting goals for specific energy use for each plant for reduction below the baseline. Industries are expected to meet their reduction targets within a three-year period. Industries which exceed their targets will be credited with tradable energy permits. Industries that fail to meet targets can either buy energy permits or pay penalties. The NMEEE will set up two fiscal instruments – the Partial Risk Guarantee Fund (PRGF) and the Venture Capital Fund for Energy Efficiency (VCFEE). The PRGF will provide commercial banks with partial coverage of risk exposure against loans made for energy efficiency projects. The VCFEE will facilitate the

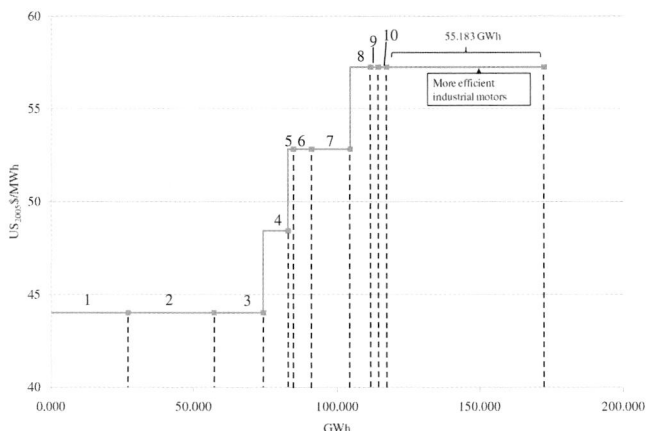

Figure 8.31 | Supply curve of saved electricity for 2020 (base year 2004). Source: adapted from WWF-Brazil, 2007.

Notes: 1- solar water heating, 2- efficient lighting (public and commercial), 3- other appliances (residential and commercial), 4- residential lighting, 5- air conditioning (residential), 6- air conditioning (commercial), 7- direct heat (industrial), 8- efficient refrigerator (residential), 9- freezer (commercial), 10- refrigerator (commercial).

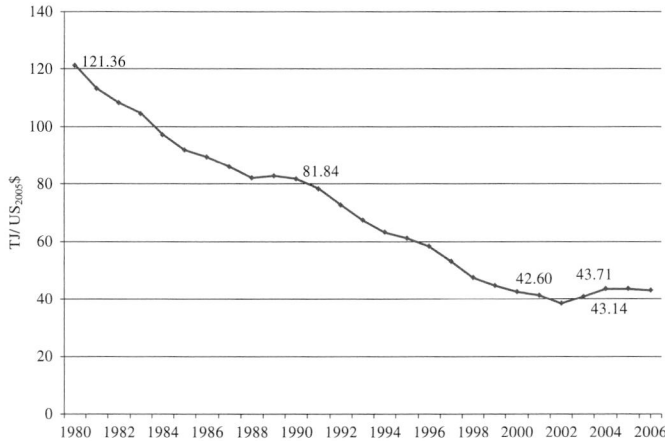

Figure 8.32 | Energy use of GDP per unit in China (TJ/US$_{2005}$$), 1980–2006. Source: adapted from China Energy, 2009.

availability of venture capital to energy supply companies and other companies investing in energy efficiency and demand-side management (DSM).

The Indian government has announced the establishment of Energy Efficiency Services Limited (EESL) with equity of INR1.9 billion (about US$43 million) from four public-sector companies: National Thermal Power Corporation, Power Finance Corporation, Rural Electrification Corporation, and Power Grid. EESL will work as an ESCO and lead the market-related actions of the NMEEE.

South Africa

In 2005, the South African government in the form of the Department of Minerals and Energy (DME) launched the "Energy Efficiency Strategy of the Republic of South Africa." This strategy addresses all sectors within South Africa and targets an overall improvement in energy efficiency of 12% by 2015. Specifically, the industry target is 15%. These targets do not refer to a baseline at a particular date. Instead, they refer to reductions against the projected use in 2015 if energy efficiency measures are not employed.

Although the strategy was divided into phases for implementation over the period, little to no funding, encouragement, or incentives were put in place to ensure a significant take-up. The DME did not have the resources to "make things happen."

There is generally low awareness in South Africa of the benefits of energy efficiency, and the cost of energy as a fraction of business expense is usually very small and, consequently, does not attract management's attention to reduce it. During times of economic well-being (as in recent years), business owners do not see the value in releasing additional profit in this way. Yet energy efficiency is cost-effective and very quickly pays back the initial investment.

In 2005 an (Industrial) Energy Efficiency Accord was signed between the DME and 30 of the country's largest energy users and business associations. No such accord was mentioned in the strategy, but it effectively became the mechanism for encouraging energy efficiency in industry. Many of the participants regularly attended meetings, particularly the Technical Committee meetings. The very large users of energy had an inherent incentive to reduce energy use, but the only financial incentive available to less energy-intensive users was the one generally available to all under the Energy Efficiency and Demand-Side Management (EEDSM) program. The rules were published by the National Electricity Regulator (subsequently the National Energy Regulator of South Africa) in 2004.

The EEDSM program is administered by Eskom, a vertically integrated national electricity monopoly. The energy efficiency component attracted a subsidy of 50% of the cost of implementation up to a maximum value set at a fraction of the cost of constructing a new power station. The approval process at Eskom became unacceptably long for many potential customers, and few Energy Efficiency Accord members ultimately used this route. The programme has been recently restructured and a number of options are now available in an attempt to simplify and facilitate adoption The National Energy Efficiency Agency was established in 2006. It was initially envisaged that it would oversee the implementation of DSM and energy efficiency projects undertaken by Eskom and other entities in the country. However, the agency was never appropriately resourced and did not grow beyond having a single staff member.

South Africa is now in a position where electricity demand often exceeds the supply capability, and a power curtailment program is in place. Legislation has recently been enacted that will require more attention to energy-saving and efficiency measures.

The national electricity utility, Eskom, has been piloting DSM for a number of years. In 2001 it began to formalize the process and procedures. The program is specified and monitored by the National Energy Regulator of South Africa, and Eskom acts as the facilitator. The early mechanism was to uses ESCOs as agents to "pull through" projects which the ESCOs had found. Measurement and verification teams were established at several universities with the express purpose of providing independent auditing of the results of the ESCO projects. Progress was very slow initially because there are not enough skilled ESCOs in South Africa. Many ESCOs have come from the realms of lighting suppliers, as there has been considerable attention paid to replacing the incandescent lamps with compact fluorescent lamps. Many of the linear fluorescent fittings with electro-magnetic ballasts have been replaced with electronic ballasts and small-diameter tubes.

Eskom lacks the capability to manage the program, and the extremely slow rate of project approval has frustrated many ESCOs, some of which no longer serve the program. It can be argued that a program that forces a supplier to curtail its supply without compensation is not properly located within a utility.

The DSM program is financed from public funds in the form of a levy on the tariff. Initially the need was to reduce or shift peak demand, but the emphasis has now moved to energy efficiency. A Rand/MW hurdle rate was established for various interventions, and ESCO proposals had to fall below this rate to be eligible for further consideration. Funding is at 100% of the project value up to the hurdle rate for load-reduction projects and at 50% for energy-reduction projects.

United States

Since 1993, the US DOE has been developing and offering an extensive array of technical training and publications to assist industrial facilities in becoming more energy efficient through its Best Practices program. As a result of these program activities, the US has developed a great deal of technical capability in industrial energy efficiency, especially motor, steam, and process heating systems (UNIDO, 2007). Under the program name of Save Energy Now, the US DOE initiated a series of program activities beginning in 2006 with the Energy Saving Assessments previously described in Section 8.3.3 (US DOE, 2010) also includes the Industrial Assessment Centers, a university-based program with a successful 30-year track record of training engineering students while conducting approximately 500 walk-through energy assessments annually, primarily of small- to medium-sized industries.

In 2002, the US EPA began a voluntary program called "Climate Leaders," which works with companies to develop long-term comprehensive climate change strategies. Using the GHG emissions protocol developed by the World Resources Institute and the World Business Council for Sustainable Development, 59 companies have set and report progress on a corporate-wide GHG reduction goal to be achieved over five to ten years. These goals are evaluated against the projected performance of the relevant sector. In 2003, the US EPA began offering information on

energy management guidelines and benchmarking as part of its ENERGY STAR for Industry program.[8] The US ENERGY STAR has developed a benchmarking tool called the Energy Performance Indicator for the cement, corn refining, and motor vehicle assembly industries that ranks a facility among its peers based on energy use, normalizing for specific activities or factors that influence energy use (UN Energy, 2010).

In 2007, the US DOE and EPA joined together with industry, ANSI, and the National Institute of Standardization and Technology to develop Superior Energy Performance, a collaborative program to certify plants for energy efficiency based on implementation of an energy management system and improvements in energy intensity measured against a baseline.[9] This program is centered on ANSI MSE 2000:2008, the national energy management standard developed by the Georgia Institute of Technology, which will be supplanted by ISO 50001. The Superior Energy Performance program creates a framework for fostering energy efficiency at the plant level and a methodology for measuring and validating energy efficiency/intensity improvements in a process that is voluntary, performance-based, and technically sound. The proposed approach can be integrated into existing corporate management systems, such as ISO 9001:2000 and 14001:2004. Certification will also position plants to be recognized by the financial community for superior energy management practices and their contribution to climate change mitigation. The strategic goals of Superior Energy Performance are:

* to foster an organizational culture of continuous improvement in energy efficiency;

* to develop a transparent system to validate energy intensity improvements and management practices; and thus

* create a verified record of energy source fuel savings and carbon emission reductions with potential market value that could be widely recognized both nationally and internationally.

Use of the ASME System Assessment Standards (see Section 8.3.5) is not required for participation in Superior Energy Performance, but the standards provide a clearly defined pathway for quickly achieving energy savings. Superior Energy Performance underwent an 18-month pilot period in Texas and launched in 2010.

8.8.2   Demand-side Management

Industries can participate in utility DSM programs to reduce their energy costs and contribute to the efficient operation of the energy supply system (see Chapter 15 for a description of DSM and its implementation).

---

8   See www.energystar.gov/index.cfm?c=in_focus.bus_industries_focus.

9   For more information, see www.superiorenergyperformance.net.

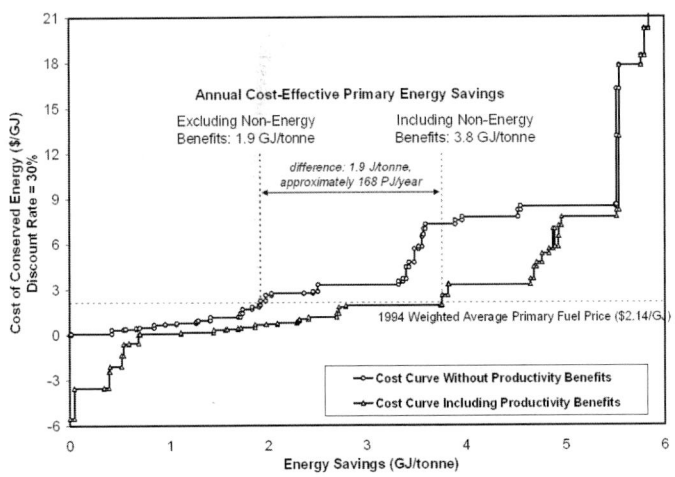

Figure 8.33 | Conservation supply curves for the iron and steel industry without productivity benefits and including productivity benefits. Source: Worrell et al., 2003

Table 8.28 | Additional demand-side investment in industry, 2005–2030, US2005$ billions.

	OECD	Non-OECD	World
APS	210	152	362
of which electrical equipment	121	74	195

Source: IEA, 2006.

Industrial programs can result in energy efficiency, load shifting and peak clipping, cool storage, cogeneration, deferrable interruptible load, and viable options for the industrial sector. Utility incentives can help facilitate widespread adoption of industrial DSM programs and can be part of the least-cost power plan for society.

8.8.3   Co-benefits for Industrial Energy Efficiency

Most industrial energy efficiency improvements also have additional benefits (co-benefits). These include reduced emissions and waste, improved product quality, increased product output and reduction of operation and maintenance costs (Worrell et al., 2003; IPCC, 2007).

Pye and McKane (1999) found that industrial efficiency projects adopted through the "Motor Challenge" program resulted in improved operations, extended lifetime of system components, and reduced expenditures and capital costs.

Worrell et al. (2003) quantified the monetary value of productivity benefits from 52 case studies. This revealed that the payback period for the measures based on energy savings only was 4.2 years. This reduced to 1.9 years when non-energy benefits were included.

Figure 8.33 shows the conservation supply curve for 14 measures in steelmaking with only energy benefits and including productivity benefits.

8.8.4   Financing

Energy efficiency is a largely invisible and nascent market. The financial benefits of energy efficiency accrue to end-users, representing a cost savings, rather than a financial return. Cost savings are difficult to collateralize, which makes it difficult to secure external financing for energy efficiency projects. Therefore, industrial energy efficiency is normally financed internally and is not generally identified as an investment or structured as a separate project.

In 2006, US$1.1 billion was invested in energy efficiency technologies, compared with US$710 million in 2005 (UNEP and New Energy Finance, 2007). These figures include both supply- and demand-side energy efficiency measures across all sectors. This is only a small fraction of the total investment in the industrial sector in 2005 of US$1379 billion (UNFCCC, 2007).

8.8.4.1   IEE Financing Requirements

Based on the IEA World Energy Outlook 2006 Alternative Policy Scenario (APS),[10] industrial energy demand in 2030 will be 337 million tonnes of oil equivalent (Mtoe) (9% lower than in the Reference Scenario). Over half of global energy savings in the industry sector can be achieved as the result of more energy-efficient production of iron and steel, chemicals, and non-metallic products.

The additional demand-side investment in APS amounts to US$360 billion to be financed by various industrial end-users, including about three-quarters to purchase more energy-efficient electrical equipment (see Table 8.28).

The UNFCCC (2007) estimates that an additional US$19.1 billion[11] of annual investments will be needed in 2030 to stabilize energy-related CO_2 emissions at the 2005 level (as set out in the IEA Beyond Alternative Policy Scenario (BAPS)).[12] This additional investment on the demand side can, however, is offset on the supply side by a decreased need for investment in new power-generation capacity and fossil fuel (US$60 billion less). The BAPS assumes that to achieve the GHG stabilization targets, industrial energy efficiency would need to improve by a further 7% compared to APS.

According to the UNFCCC (2007), most projected industrial energy efficiency measures can be achieved, as they assume very short payback periods (less than four years), and further additional investment needs

---

10 The APS considers how the global energy market can evolve by 2030 if countries are to adopt all policies they are considering related to energy security and energy-related CO_2 emissions (note: in APS global GHG emissions are 8Gt higher in 2030 than in 2006).

11 US$11.5 billion in OECD and US$8 billion in non-OECD countries.

12 The BAPS considers policies and changes in global energy market which need to be effected to stabilize GHG emissions at the 2004 level of 26.1 $GtCO_2$.

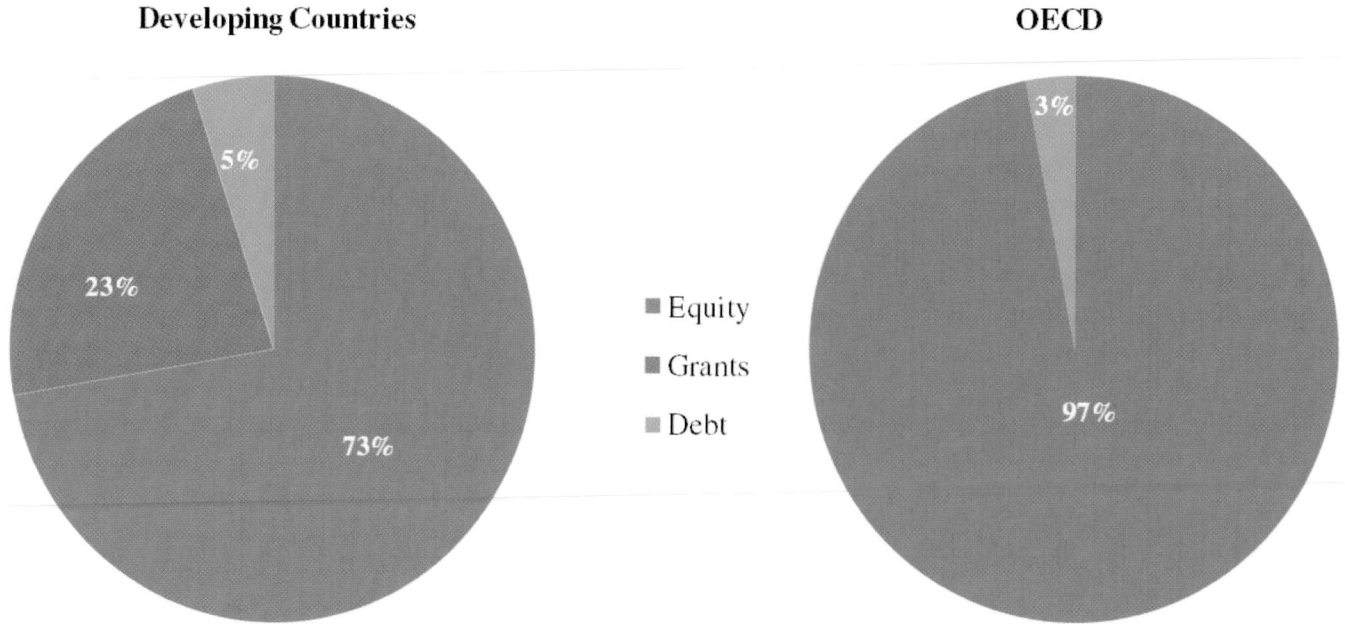

Figure 8.34 | Sources of energy efficiency investment in OECD and developing countries. Source: data based on UNFCCC, 2007.

constitute a small share (1%) of the total projected investment in the industrial sector.

The key bottleneck is that 75% of savings are projected to occur in non-OECD countries, where availability of capital is scarce, political and economic risks are high, financial markets are not sufficiently developed, and an enabling policy environment and technical skills are lacking.

8.8.4.2 Traditional Sources of Financing for Industrial Energy Efficiency

Patterns of investment in industrial energy efficiency (see Figure 8.34) generally mirror those for investment in the industrial sector in general, where the largest part (72%) comes from domestic sources, particularly in developing and transition economies. Foreign direct investment provides 22% of the global total, but more in OECD countries (up to 37% in North America). Debt plays a small role, while Official Development Assistance (ODA) barely registers as a source of industrial investment (UNFCCC, 2007).

Private equity/venture capital – as for energy efficiency financing, venture capital (primarily in OECD countries), and private equity (in developing and transition countries) provide the largest share of the total investment flows. Venture capital funds backed by the public sector can also be found in the United States, the United Kingdom, and Australia. For example, roughly one-third of the technology start-ups incubated by the United Kingdom's Carbon Trust operate in energy efficiency (UNEP and New Energy Finance, 2007).

Self-financing – in developing and transition economies, investment in industrial energy efficiency is mostly undertaken with companies' own funds. Key limitations in this respect are: energy efficiency is not part of the companies' core business and hence it has a low priority and needs to compete for scarce capital with other strategic projects; a lack of internal capacity for energy audits, project design, and implementation; and a lack of an enabling policy environment and motivation stemming from low/subsidized energy prices or an absence of mandatory energy performance targets.

Debt finance plays a very small role in financing industrial energy efficiency projects. This is due to the high cost of debt finance; the lack of long-term funds in the financial sector to invest in energy efficiency projects; the lack of understanding of how to evaluate energy efficiency investments on the part of the financial institutions and hence a higher perception of risk; and a lack of experience in structuring energy efficiency investment projects by companies, combined with a scarcity of competent local consultants and/or other intermediaries (e.g., ESCOs – see below) who could assist potential clients.

International Financial Institutions are an important source of funding for energy efficiency in developing and transition economies. In 2006, the World Bank (WB) committed more to energy efficiency projects (US$447 million) than to renewable energy (US$412 million) (WB, 2008). More than half of the WB's energy efficiency investment went into Central and Eastern Europe. Other important players are the European Bank for Reconstruction and Development, the Asian Development Bank, and the Global Environment Facility (GEF).

Public funds and other types of state financial support to industrial energy efficiency play a critical role in promoting investment in industrial energy efficiency in both developed and developing countries.

Following are examples of public support schemes and their impacts in developing countries.

It is very unlikely that initiatives in energy efficiency and energy (R&D) in Brazil would have taken place without the regulators' enforcement of compulsory programs in 1998 and later with the implementation of Law 9.991/00 by the National Congress. This initiative allocated 1% of annual utilities' revenues to energy efficiency and R&D programs. In 2000, a national law was approved by the congress that changed the allocation of the resources from the 1% obligation and created a national fund called CTEnerg, responsible for investing in energy efficiency and energy R&D in the public interest. Reforms in the power sector in Brazil provided the opportunity to enhance support and, in fact, increase significantly the level of funding in these areas. While PROCEL, the national electricity conservation program initiated in 1985, invested an annual average of US$14 million during 1994–2003, utilities' compulsory investments averaged US$57 million/yr during 1998–2004.

8.8.4.3   New Financing Sources for Industrial Energy Efficiency

ESCOs can be effective models for private-sector delivery of energy-efficient technologies and services. They had a global market volume of approximately US$2.5 billion in 2000 (Goldman et al., 2005; Vine, 2005). The role of an ESCO in structuring financing is to take on (fully or partially) energy efficiency project risks by guaranteeing with its own assets that a certain level of energy and cost saving will be achieved, thereby reducing project risks vis-à-vis potential financiers (a bank or the company itself). The ESCO model proved particularly successful in the United States, which still accounts for 75% of global ESCO operations (Goldman et al., 2005), and is gaining momentum in the EU (Bertoldi et al., 2005), but its prospects in developing and transition countries are limited by several factors: weak financial markets, companies without credit histories, the absence of risk-hedging instruments, and the absence of supportive governmental policies, which were critical for the success of ESCO business in the United States and the EU.

ODA is needed to overcome numerous barriers preventing cost-effective industrial energy efficiency measures from materializing in developing and transition economies (where countries lack resources and capacities to do so on their own). These include supporting governments in designing and implementing energy efficiency policies; building capacity of companies and other market participants to identify, structure finance for and implement projects; and promote technology transfer from developed to developing countries. All in all, the main role of ODA is to reduce risks of investment in energy efficiency, thereby making projects more attractive to financiers (see Figure 8.35). Figure 8.35 shows the effect of ODA in reducing the risk associated with a project (movement in the horizontal direction along the x-axis). The additional benefit of carbon finance results in an improved rate of return (movement in the vertical direction along the y-axis). The

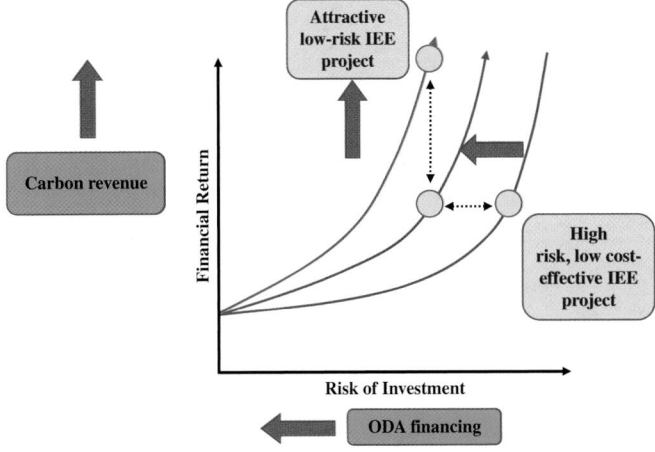

Figure 8.35 | Use of ODA and carbon finance for IEE projects. Source: modified from Glemarec, 2011.

combination of these two effects can result in changing a low return high risk industrial efficiency project to a high return low risk viable project. The GEF is the largest source of ODA for industrial energy efficiency projects.

Carbon finance is a new and rapidly growing market for clean energy financing and was valued at US$64 billion in 2007 (WB, 2008), including US$8 billion for projects under the Clean Development Mechanism (CDM). In contrast to ODA, the role of carbon markets in securing financing for industrial or other types of clean energy projects is to increase project profitability (i.e., Internal Rate of Return) or decrease the investment payback period by adding an additional revenue stream to the project (and in the case of industrial energy efficiency, the only cash revenue stream) through commercialization of CO_2 reductions associated with the project (see Figure 8.33). Despite early criticism, the volume of energy efficiency projects[13] constituted 44% of all projects in the CDM pipeline in 2007, compared to only 1% in 2005. Still, industrial energy efficiency projects represent a very tiny share of the market both in terms of size and volume (see Table 8.29).

Why is the carbon market slow to deliver the expected boost for investment in industrial energy efficiency? There are a number of barriers: first and foremost, industrial energy efficiency projects are less cost-effective than other alternatives for GHG mitigation. Average CO_2 abatement costs for industrial energy efficiency amount to €22/Certified Emissions Reduction (CER), while current prices for CDM in various market segments are between €10/CER and €20/CER (PointCarbon, 2008). In addition, projects are normally small, while transaction costs of structuring CDM projects are high (especially if new monitoring methodology needs to be designed), which prohibits their wide-scale application.

---

13 Including both energy-efficient supply- and demand-side measures in all economic sectors.

Table 8.29 | Industrial energy efficiency projects in the clean development mechanism pipeline.

CDM projects in the pipe-line	Projects, number (%)	kCERs till 2012 (%)
Cement	36 (1)	35,484 (1)
EE in industry	159 (5)	30,868 (1)
TOTAL	3498 (100)	2,639,741 (100)

Source: UNEP Risø, 2008.

In the post-2012 (post-Kyoto) scenario there are several carbon finance schemes to help carbon trading at the national and international level. For example, the European Investment Bank (EIB) has a Multilateral Carbon Credit Fund in Europe and Central Asia; the EIB/ KfW Carbon Purchase Program to help SMEs to comply with the EU's emissions trading schemes, the WB-EIB Carbon Fund for Europe, and the Fonds Carbone Capital Maroc (in French-speaking Africa) (Garcia and Roberts, 2008), with a total fund in these programs of more than €600 million.

Global carbon markets were worth about €40 billion in 2007 (PointCarbon, 2008), with the EU emissions trading scheme contributing to a trading volume of 1.6 GtCO$_2$-eq of carbon and a value of €28 billion, while the CDM market saw a volume of 0.95 GtCO$_2$-eq and a value of €12 billion.

Energy efficiency financing is a largely "invisible" market that represents a tiny share of global investment flows to the industrial sector but mirrors its structure. To move to a sustainable energy pathway, investment flows need to shift from west to east (from OECD countries to the developing world) and from the supply to the demand side. Projected increases in investment flows will be offset by decreased needs on the supply side (due to decreased energy demand compared to BAU). Given the variety and magnitude of risks facing industrial energy efficiency investment projects in developing countries, more attention needs to be paid to creating an enabling policy framework and support mechanisms to reduce these risks.

8.8.5 Technological development, R&D, and Technology Transfer

The thermodynamic analysis clearly reveals that industrial energy systems have relatively low exergetic efficiencies and significant potential for improvement. R&D can help in designing more efficient processes and utility systems. Most national energy efficiency programs focus on identifying and replicating "best practices." This is important from an implementation perspective, as it will result in significant energy savings in the short term. In addition to this, there should also be a focus on the evolution of "next practices" or future generation equipment and processes resulting in drastic reductions in energy use. A combined strategy that combines moving from existing processes to best practices

and also invests in R&D to evolve next practices is desirable. This will combine medium- and long-term energy efficiency needs.

The major portion of industrial energy use is accounted for by the production of a few energy-intensive materials. Demand for these materials in developed countries has become saturated, and the growth in consumption is mainly in developing countries. Developing countries such as China and India account for the largest share of these materials. If future capacity is to be more energy efficient, it is important that developing countries have access to these technologies.

In process plants, the energy use targets depend on local conditions such as ambient temperatures, raw materials, and scale. It is essential to develop the capability for R&D in developing countries for benchmarking and setting ambitious energy efficiency targets. Innovation in energy-efficient equipment or processes needs a critical amount of R&D funding. Most of the intellectual property and new technology know-how is available in the private sector, predominantly in developed countries. In the United States, the EU, and Japan there have been government-funded R&D programs to support industrial innovation to develop energy-efficient and clean technology.

Many countries have bilateral grant and assistance programs in the area of clean technologies. However, these are usually tied to the promotion or licensing of technologies developed by the donor country. For example, the Japanese Green Assistance plan resulted in the transfer of dry coke quenching technology from Japan to China. However, the technology was not to be adapted or indigenized by Chinese industry for a period of ten years.

Privately funded research initiatives usually have a short-term focus. Long-term initiatives for new materials, equipment, and process design need to involve researchers at universities. The evolution of road maps for the development of energy-efficient technology needs strategic partnerships between competing industries and academic and research organizations.

The challenge is to develop mechanisms to provide access to new technologies and build R&D capacity in the developing countries where the majority of the future industrial energy growth is likely to occur.

8.8.5.1 Dematerialization, Substitution and Eco Design

As material usage saturates, it is expected that there will be trends towards dematerialization. It is also expected that energy-intensive materials will be substituted by less energy-intensive materials. Table 8.30 shows an example of materials substitution in automobiles in the United States. The weight of the car decreased from 1663 kg in 1997 to 1524 kg in 2003 (8.4% dematerialization). The shares of aluminum, plastics, and composites have increased, while the share of conventional steel has decreased.

Table 8.30 | Examples of weights of materials used in cars.

Material	1977		1987		2003	
	kg	%	kg	%	kg	%
Conventional steel	904.9	54.4%	661.8	45.9%	614.6	40.3%
High-strength steel	56.7	3.4%	103.4	7.2%	171.9	11.3%
Stainless steel	11.8	0.7%	14.5	1.0%	25.6	1.7%
Other steel	25.4	1.5%	25.2	1.7%	12.0	0.8%
Iron	244.9	14.7%	208.7	14.5%	148.8	9.8%
Aluminum	44.0	2.6%	66.2	4.6%	125.9	8.3%
Rubber	68.0	4.1%	61.5	4.3%	67.6	4.4%
Plastics/composites	76.2	4.6%	100.5	7.0%	115.9	7.6%
Glass	39.7	2.4%	39.0	2.7%	44.7	2.9%
Copper	17.5	1.1%	20.9	1.4%	22.7	1.5%
Zinc die casting	17.2	1.0%	8.2	0.6%	3.9	0.3%
Powder metal parts	7.0	0.4%	8.8	0.6%	18.1	1.2%
Fluids n Lubricants	90.7	5.5%	83.0	5.8%	89.8	5.9%
Magnesium parts	58.1	3.5%	1.1	0.1%	4.3	0.3%
Other materials	0.5	0.0%	38.8	2.7%	57.8	3.8%
Total	1662.9	100%	1441.5	100%	1523.6	100%

Source: American Metal Market, 2003.

There are no aggregate estimates of the impacts of dematerialization and substitution on the energy required in the industrial sector. An analysis of buildings and selected industrial products would be useful to understand the potential for dematerialization and substitution.

A study by Fraunhofer and Eichammer (2010) shows that industrial energy use in Germany could be reduced by 13% through material efficiency. Germany has set up an agency for material efficiency[14] that document several case studies of targeted and achieved improvements in material and energy efficiency in German industry. Case studies include the use of structured composites, nano-coatings in automobile components, and biogenic raw materials.

The eco design process in the EU was mandated by a directive of the European Parliament in October 2009. This establishes a framework for setting eco design requirements for energy-related products. The commission supports life cycle thinking and has currently included more than 30 energy-using product categories with an eco-design regulation[15] that sets minimum energy efficiency requirements and environmental performance norms based on a life cycle approach. The systematic information dissemination mechanism followed for the introduction of eco design can be extended to other countries and regions and for other industrial products.

8.8.6   Capacity-building for Energy Efficiency

Energy systems and institutions are supply focused. The dominant principle is increased affluence, which requires increased consumption and, therefore, production of materials and products. The industrial system projects new manufacturing capacity and new utility systems to cater to increasing future demand. This results in an increasing amount of fuel and electricity used.

To level the playing field for energy efficiency, a paradigm shift is required with the focus on energy services – not on energy supply *per se*. This requires a re-orientation of energy supply, distribution companies, and energy equipment manufacturing companies.

What are the skills required to identify and implement the energy efficiency potential in industry? It is important to promote a systems approach and thinking. Motors, steam systems, cogeneration, and process integration need systems analysis capabilities. The US approach to developing systems assessment training modules is a replicable model. Several countries have evolved mechanisms for training and certification of energy auditors and energy managers (e.g., India's BEE has set up a syllabus, examination, and certification for energy auditors and managers). The development of ESCOs is also important for the success of industrial energy efficiency. ESCOs should be able to provide a complete energy efficiency solution to industry and should be able to take on the performance and financing risks of energy efficiency projects. ESCOs have not taken off in many markets. Most of the future growth in industry is expected in developing countries, which lack capital. Monitoring

14 See www.demea.de.

15 See www.inforse.org/europe/eu_Ecodesign.htm#Products.

and verification of savings for performance contracting also needs independent assessment, measurement, uncertainty analysis, quantification of effect of production changes and product mix, and clearly enforceable contracts for benefit-sharing.

Many countries have developed energy efficiency cells in regions or provinces (e.g., China has more than 400 energy management cells in provinces). These cells need access to information, training, research results, and support to plan their audits and energy management plans.

Regulators and government officials who deal with energy supply sectors and energy planning for the future need to be trained to integrate DSM and energy efficiency into the future supply mix.

There needs to be training for industry personnel at different levels. Short (half-day or one-day) workshops for top management (Chief Executive Officers and other key decision-makers) should ensure that energy efficiency and sustainability receive the necessary attention in the boardroom. Hands-on training modules need to be developed for technicians to inculcate the necessary skills for efficient operating and maintenance practices (steam trap maintenance, leakage reduction in compressed air systems, etc.) and retrofitting for energy efficiency. Plant engineers and managers should be familiarized with life cycle costing and sustainability analysis.

Planning for next-generation processes and systems needs the development of a long-term research agenda and strategic collaborations between industry, academic and research institutions, and governments.

International best case studies of new energy-efficient technologies and systems should be publicized and made available for industry. Searchable databases with information on plant-specific measures should be provided with translations into major languages.

8.8.7   Implementation Strategies

Investments in energy efficiency of even 1.6% of the present global fixed capital investment annually up to 2020 would provide an annual average return of 17%/yr. Investments of US$170 billion would result in US$900 billion a year in energy cost savings in 2020 (Farrell and Remes, 2008; UN Energy, 2010).

Country examples illustrate the importance of national energy efficiency action plans. Proprietary energy-efficient technologies and policies should be identified and methods to facilitate their access and deployment in developing countries should be supported. Capacity-building and information dissemination needs to be strengthened. Adoption of global ISO energy efficiency standards should be encouraged.

Policies need to address and overcome multiple barriers that exist at different levels. (Brunner et al. 2009) present an analysis of barriers to business at the sector level, to manufacturers at the original equipment

manufacturer level, and regarding wholesale planning, engineering, investment and energy management for electric motors. Figure 8.36 shows an approach to having multiple policy instruments in different parts of the product life cycle for electric motors. The adoption of life cycle costing would be an important step in promoting industrial energy efficiency, as the annual energy cost predominates in the life cycle cost for most energy utilization equipment in industry (motors, boilers, furnaces) but gets hidden in the conventional simple payback period analysis.

A locally organized energy efficiency network was created in Switzerland in the 1990s (Jochem and Gruber, 2007). This was facilitated by a stimulus from the Swiss Energy Agency for Industry to exempt participating industries from a fossil fuel surcharge of CHF 25/tCO$_2$. Participating companies agree to reduce energy-related CO$_2$ emissions to a negotiated target and undergo yearly evaluations.

The Swiss experience was replicated by an energy efficiency learning network in Baden-Württemberg in Germany in 2002 with 17 companies. The total energy use of the companies was 731 TJ in 2001. Participants agreed to 7% energy savings and 8% CO$_2$ emission reductions within four years (by 2005). Figure 8.37 shows the target and achievements of the companies (Jochem and Gruber, 2007). Encouraging the establishment of learning networks locally in different countries can help to facilitate sharing of experiences, increasing industrial energy efficiency, and setting group wide voluntary targets. National emissions trading schemes may provide a mechanism for funding and incentivizing industrial energy efficiency.

8.9   World Industrial Energy Projections up to 2030

8.9.1   Business-as-usual Scenario

Industrial energy use depends on the output of the industry sector. The Manufacturing Value Added (MVA) is used as a proxy to measure this output. The MVA for industry is disaggregated into developing and developed countries. The average annual growth rate in MVA of developing countries during 2000 to 2005 was 6.1%, and an average growth rate of 6%/yr was used to project the MVA in 2030. Zero growth rate in MVA has been assumed for industrialized countries (i.e., saturation).[16]

Energy intensity is the industrial energy use per unit of MVA. It is seen that in the past ten years the energy intensity value declined at a rate of 1.6%/yr. An annual decline of 1% in energy intensity is assumed as the BAU trend till 2030. Based on these assumptions, the industrial energy demand of the world in the BAU case is 175 EJ for 2030, as shown in Table 8.31.

---

16   The MVA values have been taken from the UNIDOdatabase (UNIDO Database, 2011) and the industrial energy consumption values from the IEA database (IEA Database, 2011).

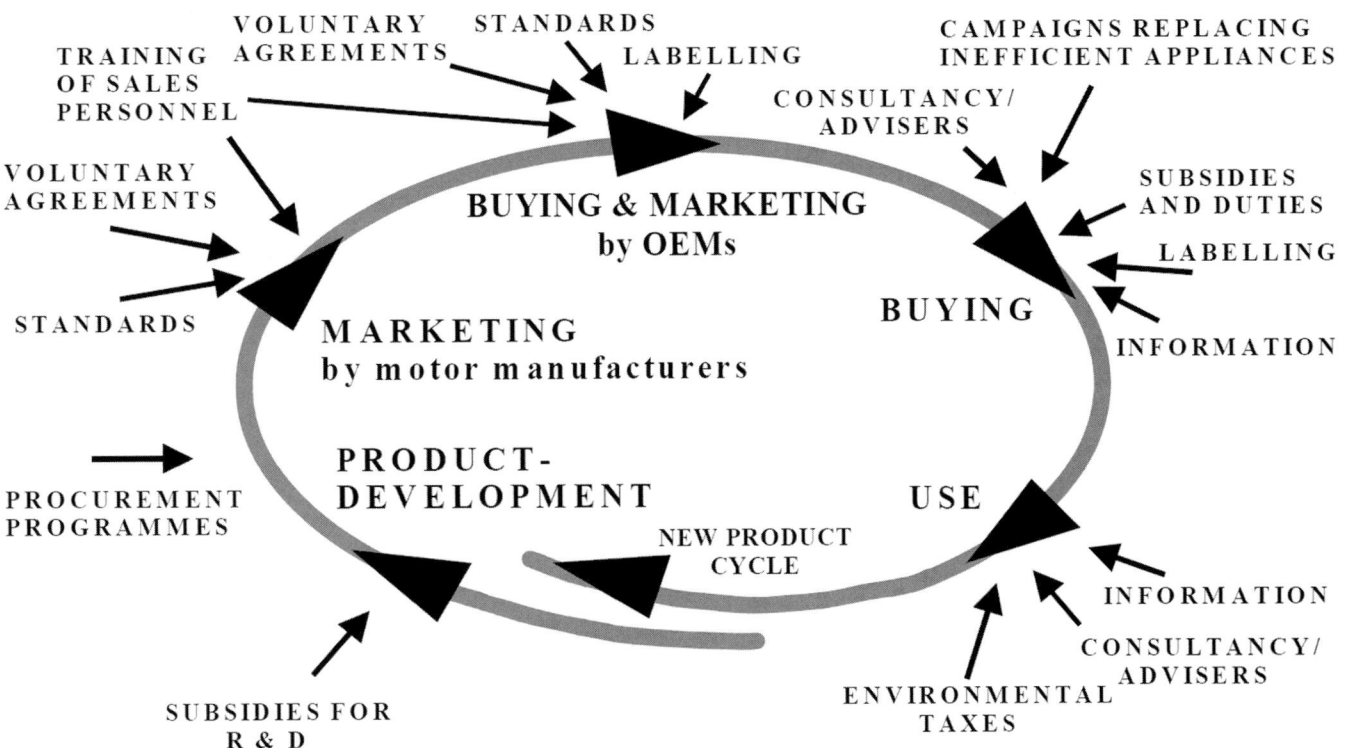

Figure 8.36 | Obstacles influencing the diffusion of highly efficient electrical motors. Source: IEA, 2011. ©OECD/International Energy Agency 2011.

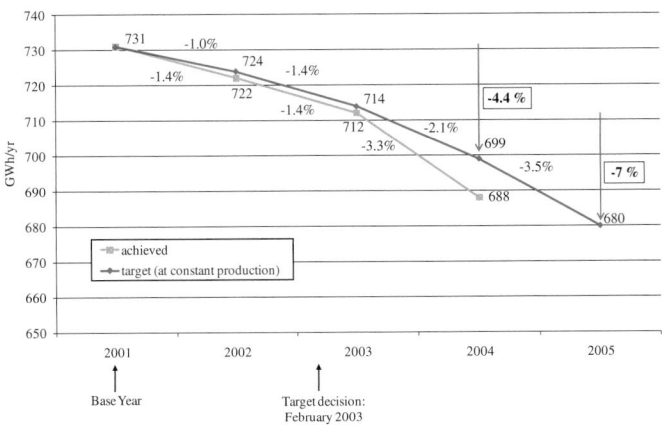

Figure 8.37 | Joint energy efficiency target path of 17 companies of the efficiency table and achieved energy savings (temperature adjusted). Source: Jochem and Gruber, 2007.

The country estimates are obtained from national plans and included for comparison. The forecasts for 2030 for the world are given in Table 8.32.

Predicted industrial final energy use and its share of total energy demand for selected countries in 2030, according to government estimates, are summarized in Table 8.33.

Material production growth estimates for 2030 by country, according to the countries' official projections, are listed in Table 8.34. The

selected industries presently account for about 40% of the world's energy use.

The specific energy consumptions for the countries are projected up to 2030. The data are given in Table 8.35, based on government estimates.

The CO_2 emission projections for 2030 for the world are given in Table 8.36. The figures for 2030 are based on the assumption of no improvement in specific energy consumption and the same fuel mix for material production.

8.9.2 Energy-efficient Scenario

The exergy analysis shown earlier reveals that industrial processes are not near the thermodynamic limits and significant potential for improvements exist. The exergy efficiency of 30% reveals the potential for further process improvements. Based on this potential in 2030 it is assumed that there is a 40% reduction in the energy intensity for the new stock with respect to the existing frozen efficiency (exergy efficiency of 50% for the new processes).

To compute the energy-efficient scenario, we differentiate between new stock (NS) and existing stock in 2005 (ES_{2005}). Of the existing stock in 2005, some fraction (f) will be retired or replaced. We consider a fraction

Table 8.31 | Basis for industrial energy demand in the BAU scenario.

Country	MVA-2005 (US$ billion)	CAGR (2000–2030)	MVA-2030 (US$ billion)	EI-2005 (MJ/$)	EI-2030 (MJ/$)	Industrial Energy (EJ)	Individual Country Estimates (EJ)
World	6536.6	–	12766	17.6	13.7	175	–
United States	1657.4	0%	1657.4	9.9	7.7	12.8	12.6
Japan	1096.8	0%	1096.8	5.7	4.4	4.9	–
China	641	3.5%	1514.8	38.2	29.7	45.0	46.4
India	68.2	8%	467.1	80.5	62.6	29.2	30.6
Brazil	81.4	5%	275.6	41.0	31.9	8.8	8.1
South Africa	24.4	6%	104.7	43.5	33.8	3.5	3.4

Note: CAGR- Compound Annual Growth Rate, EI- Energy Intensity

Table 8.32 | MVA, EI, and industrial final energy estimates for 2030.

MVA (Billion $ 2000)	2000	2005	CAGR (00–05)	CAGR (05–30)	2030
World	5774.3	6536.6	2.5%	2.7%	12766
Industrialized	4369.9	4644.1	1.2%	0%	4644
Developing	1404.4	1892.5	6.1%	6%	8122
Energy Intensity (MJ/$)	**1995**	**2005**	**CAGR (95–05)**	**CAGR (05–30)**	**2030**
World	20.6	17.6	−1.6%	−1%	13.7
Industrial Final Energy in 2030 (EJ)	175				

Note: CAGR- Compound Annual Growth Rate

Source: Based on UNIDO Database, 2011; IEA Database, 2011.

Table 8.33 | Projected final energy use by country up to 2030.

		US	China	India	Brazil	SA	World
Final Energy use by Industry1 (EJ)	2005	16.6	24.6	5.7	2.9	1.23	115
	2030 (B)	12.6	66.8	30.6	8.1	3.38	175
Industry % of Total Final Energy demand	2005	34%	51%	34%	45%	42%	30%
	2030 (B)	32%	54%	40%	38%	–	–

1 includes feedstocks (non-energy use), see Chapter 1, Section 1.2.2.

Source: Digest of South African Energy Statistics, 2006; Brazil Ministry of Mines and Energy, 2007; IEA, 2008a; TERI, 2008; China Government Estimates, 2011.

of 20% as the retirement rate up to 2030 (approximately 1%/yr). The output in 2030 (MVA_{2030}) would be from both existing equipment and new equipment, thus:

$$MVA_{2030} = MVA(ES_{2005}) * f + MVA(NS) \qquad (3)$$

Hence the existing industries (from 2005) account for US$5229 billion of MVA in 2030. The remaining MVA of US$7537 billion is produced from new stock.

Table 8.37 shows the details of the frozen efficiency scenario, BAU scenario and the energy-efficient scenario. The frozen efficiency scenario is computed assuming that the same energy intensity (17.6 MJ/US$) will continue in 2030. This results in a total final energy use for industry in 2030 of 225 EJ. For the existing surviving industry we compute the potential for energy efficiency in motor drive systems, steam systems, process improvements in energy-intensive industries and SMEs, and pinch and process integration. The basis and numbers are shown in Table 8.37. A total saving of 37 EJ is possible from the existing surviving industry. For the new industry an energy intensity improvement of 40% over the existing frozen efficiency scenario is assumed (corresponding to an energy intensity of 10.6 MJ/$). This includes process improvements, pinch and process integration, and efficiency in motors and steam systems. A saving of 53 EJ in the new stock is possible compared to the frozen efficiency scenario. This results in a total final energy use of 135 EJ in the energy-efficient scenario.

Table 8.34 | Material consumption growth rates country-wise until 2030.

Material Production		US	EU	China	India	Brazil	SA	World
Crude Steel	2005	95	196	355	45	31.6	9.5	1146
	05–30 B CAGR	−0.50%		3.70%	8.60%		5.20%	
Cement	2005	99	298	1060	153	36.7	13	2310
	05–30 B CAGR	−0.50%		2.50%	8.40%			
Paper & Paperboard	2005	88	98	62	7	10	4.6	361
	05–30 B CAGR	0.00%		3.30%	8.10%		1.30%	
Ammonia	2005	8	14.5	46.3	12	0.95	0.5	151
	05–30 B CAGR	0.00%		1.80%	1.90%			
Aluminum	2005	2.5	5.2	8.5	0.9	1.4	0.85	32
	05–30 B CAGR	0.60%		3.20%	7.90%		6.90%	

Note: Total Aluminum (Primary + Secondary) growth figures.

Source: American Forest and Paper Association, 2006; IISI, 2008; TERI, 2008; IEA, 2008a; USGS, 2011.

Table 8.35 | Projected specific electricity consumption by country up to 2030.

SEC (GJ/t)		US	EU	China	India	Brazil	SA	World
Steel	2005	15.4	17.1	22.3	22.8	26.6	31.3	19.4
	2030 (B)	13.7		19.3			25.2	
Cement	2005	4.1	3.1	3.9	3.3	3.9	4.5	4.0
	2030 (B)	3.9		3.4				
Paper	2005	30.9	16.6	30.7	26.7	22	28	18.4
	2030 (B)	27.4		26.1				
Ammonia	2005	37	52	48.2	38			
	2030 (B)	36.1		40.7				
Aluminum	2005	47*	35.7	51.5*	94.7	61.6*	50*	103
	2030 (B)	46.7*		47.5*			39.5*	

Source: TERI, 2008.

Table 8.36 | World industrial CO_2 emission quantities.

Industry (Million Tonnes of CO_2)	2005	2030
Iron and steel	1992	3598
Non-metallic minerals	1770	3287
Paper, pulp and print	189	262
Non-ferrous metals	110	189
Industry total – direct and process CO2 emissions	6660	12,031
Industry total – direct and indirect CO2 emissions	9860	17,812

Source: IEA, 2008a.

If industrial output were to increase by 95% of its 2005 value in 2030, in the frozen efficiency scenario this would require an input of 225 EJ of final energy input. In the BAU scenario, due to normal efficiency improvements, the final energy input required is 175 EJ. Hence industry would require 60 EJ in 2030 more than current consumption. Under the energy-efficient scenario an additional 40 EJ can be obtained through energy efficiency. Hence it is possible to target 95% growth in industrial output by 2030 (in MVA terms) with only a 17% growth in final energy input. Figure 8.38 shows the existing, frozen efficiency, and BAU scenarios and the savings in 2030.

Total direct and indirect CO_2 emissions from the industrial sector in 2005 were about 9.9 Gt (IEA, 2008a). Assuming the same carbon intensities for the industrial sector, under the BAU scenario total CO_2 emissions would increase to 17.8 Gt, and under the energy-efficient scenario to 11.6 Gt, in 2030. It is possible to stabilize the CO_2 emissions from the industry sector at 2005 numbers by a combination of the energy-efficient scenario and an increase in the share of renewables in the industrial mix. Renewables currently account for 9% of the total or about 10 EJ of final energy supply. This needs to be increased to about 32 EJ in 2030 to account for 23% of the final energy supply to the industrial sector. This would imply a compound annual growth rate of 4.8%/yr.

Table 8.37 | Industrial energy demand estimates for 2030 in the EE scenario.

World	2005	Fraction retirement 2030 (f)	2030 (FE)	2030 (BAU)
MVA (billion US$2000)	6536.6	20%	12,766	12,766
EI (MJ/$)	17.6	-	17.6	13.7
Energy (EJ)[1]	115	-	225	175
MVA (2005 surviving in 2030)	5229	US$ billions		
MVA (New Stock)	7537	US$ billions		
Energy-Efficient Scenario (EE)				
Energy required by existing surviving stock (in FE in 2030)			92 EJ	
End-uses	**Share**	**Potential Savings**	**EJ**	
A.	Motor systems	15%	20%	2.8	
B.	Steam systems	38%	20%	7.0	
C.	Process improvements in energy-intensive industries	66%	9%	5.5	
D.	Process improvements in SMEs	34%	10%	3.1	
Total Savings (A+B+C+D)			18.3	
Energy use in existing stock after A+B+C+D			73.7	
E.	Saving through pinch	100%	25%	18.4	
F. Total savings in existing stock (A to E)			37	
New Stock energy consumption			133	
G. Reduction in New Stock vs. FE scenario (EI – 10.6 MJ/US$)		40%	53	
Total savings obtained			90	
Energy required in EE scenario			135	

1 includes feedstocks (non-energy use), see Chapter 1, Section 1.2.2.

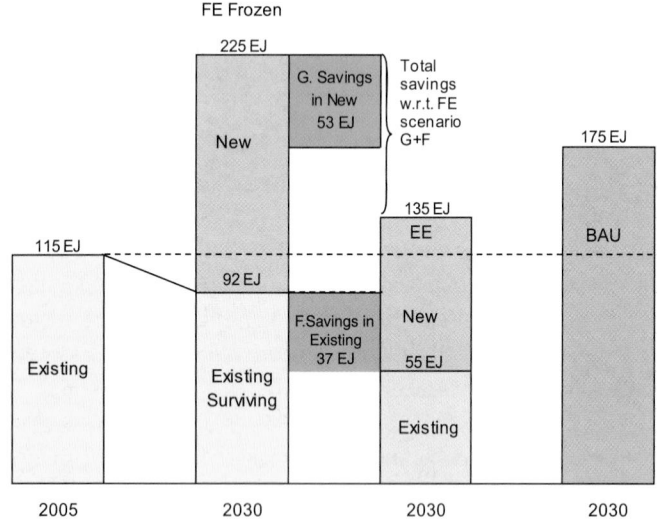

Figure 8.38 | Existing, frozen efficiency and BAU scenarios and the savings in 2030. Note: Final energy data includes feedstocks (non-energy use), see Chapter 1, Section 1.2.2.

8.10   Conclusions and Recommendations

The key messages emerging from the analysis of end-use efficiency for the industrial sector are:

- An energy-efficient scenario is possible which results in the required industrial growth till 2030 (95% increase over 2005 values) with only a 17% increase in the final energy required.

- If the energy-efficient scenario is coupled with a growth in the use of renewables for industry, it is possible to meet the growth in industrial output without any increase in the total CO_2 emissions of the industrial sector in 2030.

- Most of the growth will occur in developing countries. At present, developing countries account for 29% of industrial output; in 2030 they will account for 64%. Most investments in future industrial production capacities are likely to be in developing countries.

- Several interventions will be required if the energy-efficient scenario or the zero CO_2 growth scenario for industry are to be achieved, and the following interventions are suggested:

I.   Realizing the potential for energy efficiency in existing industry:

(a)   Incentivize DSM: regulatory commissions can provide incentives for motor efficiency programs and process improvements.

(b) Information gaps to be reduced: promote sharing and documentation of best practices.

(c) Develop capacity for system assessment: motors, steam, pinch, process. Recent efforts by ASME to develop System Assessment Standards for motors and steam systems in the US need to be extended to other regions of the world.

(d) Provide access to low-interest finance: investments in energy efficiency compete with investments in process improvements and enhancements in plant capacity. Separate credit lines should be provided especially for funding industrial energy efficiency in developing countries.

(e) Special efforts to focus on industry clusters of SMEs – for example, steel rolling, brick making: SMEs normally do not have the engineering capability to design and implement energy efficiency programs. Interventions that facilitate building this capability would be useful. Training workshops and energy efficiency manuals should be developed and tailor-made for different industry clusters. Sharing of best practices across industry groups is likely to result in new ideas for efficiency improvements. Multilateral agencies such as UNIDO can help facilitate this by creating regional information centers and funding dissemination workshops.

(f) National Energy Conservation Funds: governments should be encouraged to create National Energy Conservation Funds. These could be provided by a tax on all new supply (power plants, refineries). Funds should be used to provide a level playing field for energy efficiency vis-à-vis new supply.

(g) Energy management standard: industry should be encouraged to adopt the new ISO energy management standard.

(h) Benchmarking efforts: initiatives from industry organizations, such as Cement Sustainability Initiative (CSI) for cement, World Steel Association for steel, and IAI for aluminum, to draw up road maps and benchmarks for industry need to be encouraged.

(i) Risk guarantee and venture capital: there should be a mechanism to encourage commercial banks to invest in energy efficiency projects in industry. This could be in the form of guarantees for risk or special funding for venture capital.

II. New industry – Most of the new industrial growth will occur in developing countries. Under the BAU scenario, a mix of technologies would be installed with varying specific energy consumption.

It is suggested that regional centers for industrial energy efficiency be set up to disseminate information related to specific energy consumption and best available technologies for different processes. There should be a (web-based) facility where any industry that is being proposed can compare its designed energy performance with the best available technologies. Consultants can be provided to undertake energy integration and efficiency improvement studies at the design stage itself. An incentive scheme should provide funding for energy performance analysis at the design stage. Financing of the incremental costs of energy-efficient technologies should be provided as low-interest loans through commercial banks.

III. Dematerialization and substitution – To encourage studies and analysis for new products, it is proposed that design challenges be announced for some important products that may result in reduced usage of material and an overall reduction of energy-intensive materials. The potential for substitution and dematerialization needs to be studied in different sectors and end-uses.

IV. Next-generation processes and technologies – Most of the energy-intensive materials produced by industry have not reached their limits for efficiency improvements. To move from the best available technologies to next-generation technologies, it is essential to facilitate R&D in new processes. Industry groups such as CSI, World Steel Association, and IAI can play a role in bringing together industry and researchers to facilitate this. Pre-competitive consortia of industries and research institutions should be facilitated by funding from government and multilateral agencies. The funding for R&D for energy-efficient processes is sub-critical and needs to be enhanced. Financing of next-generation technologies may occur from national emissions trading schemes.

V. Increased use of renewables in industry – Funding should be available for demonstration and pilot projects for innovative applications of renewables in industry. Biomass- and solar-based process heating, cogeneration, cooling and power-generation applications need to be designed, implemented, and assessed for different industrial applications. Dissemination of information related to renewable case studies in industry should be facilitated internationally. Focused attempts to bring down the costs of renewable systems would imply setting up initiatives for technology development and consortium approaches.

Acknowledgements

We are grateful to Mr. Balkrishna Surve for assistance in preparation of the manuscript.

References

ABAL, 2008: Anuário Estatístico ABAL 2008 Statistical yearbook of the Brazilian Association of the Aluminum Industry. Associação Brasileirado Alumínio (ABAL), Sao Paulo, Brazil.

Ahiduzzaman, M. and A. K. M. Sadrul Islam, 2009: Energy utilization and environmental aspects of rice processing industries in Bangladesh. *Energies* 2002(2):134–149.

Alexeev, M. and W. Pyle, 2003: A note on measuring the unofficial economy in the former Soviet Republics. *Economics of Transition* 11(1):153–175.

Almeida, A., F. Ferreira and D. Both, 2005: Technical and Economical Considerations to Improve the Penetration of Variable Speed Drives for Electric Motor Systems. Institute of Electrical and Electronics Engineers (IEEE) Transactions on Industry Applications, January/February 2005, Washington DC.

American Forest and Paper Association, 2006: *Global Pulp and Paper Fact and Price Book.* Resource Information Systems Inc. (RISI), Washington, DC.

American Metal Market, 2003: Examples of Weights of Materials used in Cars. American Metal Market. www.amm.com/ref/carmat98.htm (accessed February 11, 2011).

Ayres, R. U. and B. S. Warr, 2005: Accounting for growth: The role of physical work. *Structural Change and Economic Dynamics*, 16(2):181–209.

Ayres, R. U. and B. S. Warr, 2009: *The Economic Growth Engine: How Energy and Work Drive Material Prosperity.* Edward Elgar, Cheltenham, UK, and Northampton, MA.

Ayres, R. U. and J. C. J. M. van den Bergh, 2005: A theory of economic growth with material/energy resources and dematerialization: Interaction of three growth mechanisms. *Ecological Economics*, 55(1):96–118.

Ayres, R. U., L. W. Ayres, and B. S. Warr, 2003: Exergy, power and work in the US economy, 1900–1998. *Energy*, 28(3):219–273.

BEE, 2010: National Mission on Enhanced Energy Efficiency 2010. Bureau of Energy Efficiency (BEE), New Delhi, India.

Bertoldi, P., S. Rezessy, and D. Urge-Vorsatz, 2005: Tradable certificates for energy savings: Opportunities, challenges, and prospects for integration with other market instruments in the energy sector. *Energy and Environment*, 16(6):959–992.

Bhushan, C. and M. Z. Hazra, 2005: *Concrete Facts: The Life Cycle of the Indian Cement Industry.* Centre for Science and Environment, New Delhi, India.

Brazil Ministry of Mines and Energy, 2007: Brazilian Energy Balance 2007. Ministério das Minas e Energia, Empresa de Pesquisa Energética.Balanço Energético Nacional, Rio de Janeiro, Brazil.

Brodyanski, V. M., M. V. Sorin, and P. Le Goff, 1994: The Efficiency of Industrial Processes: Exergy Analysis and Optimization. Elsevier, Amsterdam, the Netherlands.

Brunner, C. U., J. Martin, M. Martin, and E. Jochem, 2009: *Electric Motor Systems. Information paper, International Energy Agency (IEA)*, Paris, France.

Boyd, G. A., and J. M. Roop, 2004: "A note on the Fisher Ideal Index Decomposition for Structural Change in Energy Intensity." *The Energy Journal* 25(1):87–101.

CESP, 2004: *China's sustainable energy scenarios in 2020.* China Environmental Science Press (CESP), Beijing, China.

Chateau, B., M. Biberacher, U. Birnbaum, T. Hamacher, P. Lako, D. Martinsen, M. Patel, W. Pospischil, N. Quercia, and K. Smekens, 2005: VLEEM 2 (Very Long Term Energy-Environment Model). Report prepared for the European Commission, EC/DG Research, Grenoble, France.

Chaudhary, T. R., 2001: Technological Measures of Improving Productivity: Opportunities and Constraints. Proceedings of the Fertilizer Association of India Seminar "Fertiliser and Agriculture Future Directions," 6–8 December 2001, New Delhi, India.

China Energy, 2009: Statistical Yearbook. Department of Industry and Transport Statistics, National Bureau of Statistics of P. R. China, and Energy Bureau of National Development and Reform Commission of P. R. China, China Statistics Press, Beijing, China.

China Government Estimates, 2011: National Bureau of Statistics of China. www.stats.gov.cn/english/ (accessed May 9, 2011).

CII and Forbes Marshall Study, 2005: Energy Conservation-Time to Get Specific. Confederation of Indian Industries (CII), Forbes Marshall, Pune, India.

CSI, 2006: Formation and Release of POPs in the Cement Industry. Second edition. Cement Sustainability Initiative (CSI), World Business Council for Sustainable Development (WBCSD), Geneva, Switzerland.

de Beer, J. G., E. Worrell, and K. Blok, 1998: Future technologies for energy efficient iron and steelmaking. *Annual Review of Energy and Environment*, 23:123–205.

de Beer, J. G., J. Harnisch, and M. Kerssemeeckers, 2000: *Greenhouse Gas Emissions from Iron and Steel Production. U.S.* Environmental Protection Agency International Energy Agency (IEA) Greenhouse Gas R&D Programme, Cheltenham, UK.

De Keulenaer, H., R. Belmans, E. Blaustein, D. Chapman, A. de Almeida, B. De Wachter, and P. Radgen, 2004: Energy Efficient Motor Driven Systems: Can save Europe 200 billion kWh of electricity consumption and 100 million tonne of greenhouse gas emissions a year. Report prepared by the European Copper Institute (ECI), Brussels, Belgium.

de Vries, H. J. M., K. Blok, M. K. Patel, M. Weiss, S. Joosen, E. de Visser, and J. Sijm, 2006: Assessment of the Interaction between Economic and Physical Growth. EPIST project report UCE-34, prepared by Netherlands Research Programme on Scientific Assessment and Policy Analysis for Climate Change, June 2006, University of Utrecht, the Netherlands.

Dincer, I., M. M. Hussain and I. Al-Zaharnah, 2004: Analysis of sectoral energy and exergy use of Saudi Arabia. *International Journal of Energy Research*, 28(3):205–243.

Digest of South African Energy Statistics, 2006: *Pretoria: Department of Minerals and Energy*, Republic of South Africa, Pretoria.

EC, 2007: Reference Document on Best Available Techniques for the Manufacture of Large Volume Inorganic Chemicals – Ammonia, Acids and Fertilisers. European Commission (EC), Brussels, Belgium.

EFMA, 2000: Best Available Techniques for Pollution Prevention and Control in the European Fertilizer Industry. Booklet No. 1 of 8: Production of Ammonia. European Fertilizer Manufacturers Association (EFMA), Brussels, Belgium.

Eichhammer, F. and W. Eichhammer, 2010: Is energy efficiency enough to achieve the EU sustainability targets? Material efficiency in the industry sector. Proceedings of the Efonet Workshop 4.3: Increasing energy efficiency in industrial processes. February 19, 2010, Fraunhofer Institute for Systems and Innovation Research, Berlin, Germany.

FAO, 2009: ForestSTAT. Food and Agriculture Organization of the United Nations (FAO). www.faostat.fao.org/site/630/default.aspx (accessed May 30, 2011).

Farrell, D. and J. K. Remes, 2008: How the World should Invest in Energy Efficiency. *The McKinsey Quarterly*, July 2008.

Garcia, B. and E. Roberts (eds.), 2008: *Carbon Finance: Environmental Market Solutions to Climate Change. Report No. 18, Center for Business and the Environment*, Yale School of Forestry and Environmental Studies, New Haven, CT.

Gartner, E., 2004: Industrially interesting approaches to "low-CO2" cements. *Cement and Concrete Research*, **34**:1489–98.

Gielen, D., 2010: Industrial Energy Efficiency for the 21st Century. Presentation at International Workshop, Department of Energy Science and Engineering, Indian Institute of Technology (IIT)-Bombay, 22 February 2010, Mumbai, India.

Glemarec, Y., 2011: Catalysing Climate Finance: A Guidebook on Policy and Financing Options to Support Green, Low-Emission and Climate-Resilient Development. United Nations Development Programme (UNDP), New York, NY.

Goldman, C., N. Hopper, and J. Osborn, 2005: Review of US ESCO industry market trends: An empirical analysis of project data. *Energy Policy*, **33**:387–405.

Hamilton, J. D., 1983: Oil and the macroeconomy since World War II. *Journal of Political Economy*, **91**:228–248.

Hamilton, J. D., 2003: What is an oil shock? *Journal of Econometrics*, **113**:363–398.

Hamilton, J. D., 2005: Oil and the macroeconomy. In The New Palgrave: A Dictionary of Economics. J. Eatwell, M. Millgate and P. Newman (eds.), Macmillan Publishers, London, UK.

Hermann, B. G., K. Blok, and M. K. Patel, 2007: Producing bio-based bulk chemicals using industrial biotechnology saves energy and combats climate change. *Environmental Science and Technology*, **41**:7915–7921.

Howarth, R. B., L. Schipper, P. A. Duerr, and S. Strøm, 1991: Manufacturing energy use in eight OECD countries: Decomposing the impacts of changes in output, industry structure and energy intensity. *Energy Economics*, **13**:135–142.

Humphreys, K. and M. Mahasenan, 2002: Towards a Sustainable Cement Industry – Substudy 8: Climate Change. World Business Council for Sustainable Development (WBCSD), Geneva, Switzerland.

Huntington, H. G., 2010: Structural change and US energy use: recent patterns. *The Energy Journal*, **31**(3).

IAI, 2007: The Aluminium Industry's Sustainable Development Report. International Aluminium Institute (IAI), London, UK.

IAI, 2011: International Aluminium Institute. International Aluminium Institute (IAI), London, UK. www.world-aluminium.org (accessed May 30, 2011).

IEA Database, 2011: Statistics and Balances. International Energy Agency (IEA). www.iea.org/stats/ (accessed May 30, 2011).

IEA GHG, 2001: Abatement of emissions of other greenhouse gases: Engineered Chemicals. International Energy Agency (IEA) Greenhouse Gas (GHG) R&D Programme, J. Harnisch, O. Stobbe and D. de Jager (eds.), Cheltenham, UK.

IEA, 2006: IEA World Energy Outlook 2006. International Energy Agency (IEA), Paris, France.

IEA, 2007a: Tracking Industrial Energy Efficiency and CO2 Emissions. International Energy Agency (IEA), Paris, France.

IEA, 2007b: Energy Balances of Non-OECD Countries, 2004–2005. 2007 Edition. International Energy Agency (IEA), Paris, France.

IEA, 2008a: Energy Technology Perspectives 2008. International Energy Agency (IEA), Paris, France.

IEA, 2008b: World Energy Outlook Policy Database. International Energy Agency (IEA), Paris, France. www.iea.org/Textbase/pm/?mode=weo (accessed March 28, 2011).

IEA, 2008c: Energy Efficiency Policies and Measures. International Energy Agency (IEA), Paris, France.

IEA, 2011. Energy-Efficiency Policy Opportunities for Electric Motor-Driven Systems. Working Paper. International Energy Agency (IEA), Paris, France.

IFA, 2009: http://www.fertilizer.org/ifa/HomePage/STATISTICS/Production-and-trade (accessed July 25, 2011)

IISI, 2008: Steel Statistical Yearbook 2008.Worldsteel Committee on Economic Studies, International Iron and Steel Institute (IISI), Brussels, Belgium.

IPCC, 2007: Climate Change 2007 – Mitigation of Climate Change. Contribution of Working Group III to the Fourth Assessment Report of the Intergovernmental Panel on Climate Change (IPCC).Cambridge University Press, Cambridge, UK.

Jänicke, M., H. Mönch, and M. Binder, 1992: Umweltentlastung Durch Industriellen Strukturwandel? – Eine Explorative Studie über 32 Industrieländer (1970–1990). Rainer Bohn Verlag, Berlin, Germany.

Jochem, E. and E. Gruber, 2007: Local learning-networks on energy efficiency in industry: Successful initiative in Germany. *Applied Energy*, **84**:806–819.

Kim, Y. and E. Worrell, 2002a: International comparison of CO2 emissions trends in the iron and steel industry. *Energy Policy*, **30**:827–838.

Kim, Y. and E. Worrell, 2002b: CO2 emission trends in the cement industry: An international comparison. *Mitigation and Adaptation Strategies for Global Change* **7**:115–33.

Kramer, K.-J., E. Masanet, T. Xu, and E. Worrell, 2009: Energy Efficiency Improvement and Cost Saving Opportunities for the Pulp and Paper Industry: An ENERGY STAR Guide for Energy and Plant Managers. Lawrence Berkeley National Laboratory, Berkeley, CA.

Kuemmel, R., J. Henn, and D. Lindenberger, 2002: Capital, labor, energy and creativity: Modeling innovation diffusion. *Structural Change and Economic Dynamics*, **13**:415–433.

Kuemmel, R., J. Schmid, R. U. Ayres, and D. Lindenberger, 2008: Cost Shares, Output Elasticities and Substitutibility Constraints. Report EWI WP 08.02. University of Cologne, Germany.

Lung, R. B., A. McKane, and M. Olzewski, 2003: Industrial Motor System Optimization Projects in The US: An Impact Study. Proceedings of the 2003 American Council for an Energy-Efficient Economy (ACEEE) Summer Study on Energy Efficiency in Industry, Washington, DC.

Mallett, A., S. Nye, and S. Sorrell, 2010: Policy Options for Overcoming Barriers to Industrial Energy Efficiency. Report prepared by the Sussex Energy Group, Science and Technology Policy Research (SPRU) for the United Nations Industrial Development Organization (UNIDO), University of Sussex, Brighton, UK.

Manitoba Hydro, 2011: Power Smart Savings, Rebates and Loans. Manitoba Hydro. www.hydro.mb.ca/power_smart_for_industry/index.shtml (accessed May 30, 2011).

Martin, N., N. Anglani, D. Einstein, M. Khrushch, E. Worrell, and L. K. Price, 2000: Opportunities to Improve Energy Efficiency and Reduce Greenhouse Gas Emissions in the U.S. Pulp and Paper Industry. Report LBNL-46141. Lawrence Berkeley National Laboratory, Berkeley, CA.

Maurer, L., M. Pereira, and J. Rosenblatt, 2005: Implementing Power Rationing in a Sensible Way: Lessons Learned and International Best Practices. ESMAP Report 35/05, Energy Sector Management Assistance Program (ESMAP), Washington, DC.

McKane, A., R. Williams, W. Perry, and T. Li, 2007: Setting the Standard for Industrial Energy Efficiency. Proceedings of Energy Efficiency in Motor Driven Systems (EEMODS), 2007, Beijing, China. www.industrial-energy.lbl.gov/ (accessed March 28, 2011).

McKane, A., P Scheihing, and R. Williams, 2007: Certifying Industrial Energy Efficiency Performance: Aligning Management, Measurement, and Practice to Create Market Value, Lawrence Berkeley National Laboratory (LBNL), Berkeley, CA, USA.

Mills, J.H. and T.A. Waite, 2009: Economic prosperity, biodiversity conservation, and the environmental Kuznets curve. *Ecological Economics*, **68** (7): 2087–2095.

MNRE, 2010: The Ministry of New and Renewable Energy (MNRE), Government of India. www.mnre.gov.in (accessed June 12, 2008).

Monari, L., 2008: Energy Efficiency in Industry: Experience, Opportunities and Actions. Presentation at the UN-Energy Expert Group Meeting on Advancing Industrial Energy Efficiency in the Post-2012 Framework, September 22–23, 2008, Washington, DC.

Nakicenovic N., P. V. Gilli, and R. Kurz, 1996: Regional and global exergy and energy efficiencies. *Energy*, **21**:223–237.

Neelis, M., E. Worrell, and E. Masanet, 2008: Energy Efficiency Improvement and Cost Saving Opportunities for the Petrochemical Industry. An ENERGY STAR Guide for Energy and Plant Managers. LBNL-964E.Lawrence Berkeley National Laboratory, Berkeley, CA.

Nieuwlaar, E., 2001: ICARUS-4 Sector Study for the Chemical Industry. Report No. NWS-E-2001–100, Utrecht University, Utrecht, Netherlands.

Odyssee, 2009: Energy Efficiency Trends and Policies in the EU 27, Results of the ODYSSEE-MURE project, October, 2009.

Olshanskaya, M., 2004: Comparative analysis of energy intensity in Russia and Slovakia during economic transition of the 1990s. Diss. Department of Environmental Sciences and Policy, Central European University, Budapest, Hungary.

Phylipsen, D., K. Blok, E. Worrell, and J. de Beer, 2002: Benchmarking the energy efficiency of Dutch industry: an assessment of the expected effect on energy consumption and CO2 emissions. *Energy Policy*, **30**: 663–679.

PointCarbon, 2008: PointCarbon News. *CDM & JI Monitor*, **6**(14):9 July.

Price, L., J. Sinton, E. Worrell, D. Phylipsen, H. Xiulian, and L. Ji, 2002: Energy use and carbon dioxide emissions from steel production in China. *Energy*, **5**:429–446.

PSI, 2004: Energy Efficiency and CO_2 Emissions Benchmarking of IFA Ammonia Plants (2002–2003 Operating Period). Plant Surveys International, Inc (PSI). General Edition. Commissioned by the International Fertilizer Industry Association (IFA), Paris, France.

Pye, M. and A. McKane, 1999: Enhancing shareholder value: making a more compelling energy efficiency case to industry by quantifying on-energy benefits. In Proceedings of the 1999 Summer Study on Energy Efficiency in Industry. American Council for an Energy-Efficient Economy (ACEEE), Washington DC, pp.325–36.

Rafiqul, I., C. Weber, B. Lehmann, and A. Voss, 2005: Energy efficiency improvements in ammonia production – perspectives and uncertainties. *Energy*, **30**:2487–2504.

Rane, M., 2009: Impact of Demand Side Management on Power Planning. M.Tech. Diss., Department of Energy Science and Engineering, Indian Institute of Technology, Bombay, India.

Ren T., M. Patel, and K. Blok, 2005: Steam Cracking and Natural Gas-to-olefins: A Comparison of Energy Use and Economics. Proceedings of the American Institute of Chemical Engineers (AIChE), spring meeting 2005.Atlanta, GA and New York, NY.

Ren, T., M. Patel, and K. Blok, 2006: Olefins from conventional and heavy feedstocks: Energy use in steam cracking and alternative processes. *Energy*, **31**:425–451.

Roes, A. L., L. B. Tabak, L. Shen, E. Nieuwlaar, and M. K. Patel, 2010: Influence of using nano objects as filler on functionality-based energy use of nanocomposites. *Journal of Nanoparticle Research* **12**:2011–2028.

Sardeshpande, V., U. N. Gaitonde and R. Banerjee, 2007: Model based energy benchmarking for glass furnace. *Energy Conversion and Management*, **48**(10): 2718–2738.

Saygin, D., M. K. Patel, and D. Gielen, 2010: Global Industrial Energy Efficiency Benchmarking: An Energy Policy Tool, Working paper. November 2010, United Nations Industrial Development Organization (UNIDO), Vienna, Austria.

Saygin, D., M. K. Patel, C. Tam, and D. J. Gielen, 2009: Chemical and Petrochemical Sector: Potentials of Best Practice Technology and Other Measures. International Energy Agency (IEA) Information Paper. September 2009, Paris, France.

Shen, L., J. Haufe, and M. Patel, 2010: Present and future development in plastics from biomass. *Biofuels, Bioproducts and Biorefining*, **4**(1):25–40.

Siemens VAI, 2009: *The SIMETAL CIS Corex/Finex Process*, Siemens Iron Making, www.siemens-vai.com (accessed February 8, 2012).

TERI, 2008: *National Energy Map for India, Technology Vision 2030*. TERI Press, New Delhi, India.

UN Energy, 2010: Policies and Measures to Realize Industrial Energy Efficiency and Mitigate Climate Change. Report prepared by the UN Energy Energy Efficiency Cluster, 2010.

UNDP, 2000: World Energy Assessment. United Nations Development Programme (UNDP), United Nations Department of Economic and Social Affairs, and World Energy Council, New York, NY.

UNEP and New Energy Finance Ltd., 2007: Global Trends in Sustainable Energy Investment 2007. Analysis of Trends and Issues in the Financing of Renewable Energy and Energy Efficiency in OECD and Developing Countries. United Nations Environment Programme (UNEP), Nairobi, Kenya.

UNEP, 2008: UNEP Risø CDM/JI Pipeline Analysis and Database. United Nations Environment Programme (UNEP), Risø Centre on Energy, Climate, and Sustainable Development. www.cdmpipeline.org (accessed March 28, 2011).

UNFCCC, 2007: Investment and Financial Flows to Address Climate Change. United Nations Framework Convention on Climate Change (UNFCCC), Bonn, Germany.

UNIDO, 2010a: Motor Systems Efficiency Supply Curves. United Nations Industrial Development Organisation (UNIDO), Vienna, Austria.

UNIDO, 2010b: Renewable Energy in Industrial Applications, An assessment of the 2050 potential. United Nations Industrial Development Organization (UNIDO), Vienna, Austria.

UNIDO Database, 2011: United Nations Industrial Development Organization (UNIDO),Vienna, Austria. www.unido.org (accessed May 31, 2011).

US DOE, 2002: Steam System Opportunity Assessment for the Pulp and Paper, Chemical Manufacturing, and Petroleum Refining Industries. Main Report. Report prepared by Resource Dynamics Corporation for the U.S. Department of Energy (US DOE), Washington, DC.

US DOE, 2004a: Improving Steam System Performance: A Sourcebook for Industry. Lawrence Berkeley National Laboratory and Resource Dynamics Corporation, DOE/GO-102004–1868.U.S.Department of Energy (US DOE), Washington, DC. www1.eere.energy.gov/industry/bestpractices/techpubs_steam.html (accessed March 28, 2011).

US DOE, 2004b: Energy Loss Reduction and Recovery in Industrial Energy Systems. Prepared by Energetics, Inc., for the U.S. Department of Energy, Washington, DC.

US DOE, 2004c: Energy Use, Loss, and Opportunities Analysis: U.S. Manufacturing & Mining. Prepared by Energetics, Inc. and E3M, Inc. for the U.S. Department of Energy, Washington, DC.

US DOE, 2004d: Evaluation of the Compressed Air Challenge® Training Program. Lawrence Berkeley National Laboratory and XENERGY, Inc., DOE/GO-102004–1836. Washington, DC. www1.eere.energy.gov/industry/bestpractices/tech-pubs_compressed_air.html (accessed March 28, 2011).

US DOE, 2010: Save Energy Now. Industrial Technology Program (ITP), United States Department of Energy (US DOE). www1.eere.energy.gov/industry/saveenergy-now/ (accessed May 31, 2011).

US EIA, 2007: International Energy Outlook 2007, industrial and commercial world energy use. United States Energy Information Administration (US EIA), Washington, DC.

USGS, 2005: Mineral Resources Program 2005. United States Geological Survey (USGS), Reston, VA.

USGS, 2007: Mineral Resources Program 2007. United States Geological Survey (USGS), Reston, VA.

USGS, 2008: Mineral Resources Program 2008. United States Geological Survey (USGS), Reston, VA.

USGS, 2011: Mineral Resources Program 2011. United States Geological Survey (USGS), Reston, VA. www.minerals.usgs.gov/minerals/pubs/myb.html (accessed May 27, 2011).

Utlu, Z., and H. Arif, 2008: Energetic and exergetic assessment of the industrial sector at varying dead (reference) state temperatures: a review with an illustrative example. *Renewable and Sustainable Energy Reviews*, **12**(5):1277–1301.

Vine, E., 2005: An international survey of the energy service company (ESCO) industry. *Energy Policy*, **33**:691–704.

WB, 2008: State and Trends of The Carbon Market, 2008. Karan Capoor: Sustainable Development Operations, World Bank (WB) and Philippe Ambrosi: Climate Change Team, Washington, DC.

Worrell, E., D. Smit, K. Phylipsen, K. Blok, F. van der Vleuten, and J. Jansen, 1995: International comparison of energy effici1ency improvement in the cement industry. In Proceedings of the American Council for an Energy-Efficient Economy (ACEEE) Summer Study on Energy Efficiency in Industry. Washington, DC, Vol. 2, pp.123–134.

Worrell, E., D. Phylipsen, D. Einstein, and N. Martin, 2000a: Energy use and energy intensity of the US chemical industry. Berkeley, CA, Lawrence Berkeley National Laboratory (LBNL-44314).

Worrell, E., N. Martin, and L.K. Price, 2000b: Potentials for energy efficiency improvement in the US cement industry. *Energy*, **25**:1189–1214.

Worrell, E., J. A. Laitner, M. Rugh, and H. Finman, 2003: Productivity benefits of industrial energy efficiency measures. *Energy*, **28**:1081–1098.

Worrell, E. and G. Biermans, 2005: Move over! Stock turnover, retrofit and industrial energy efficiency. *Energy Policy*, **33**:949–962.

Worrell, E. and C. Galitsky, 2008: Energy Efficiency Improvement and Cost Saving Opportunities for Cement Making: An ENERGY STAR Guide for Energy and Plant Managers. Report LBNL-54036-Revision. Lawrence Berkeley National Laboratory, Berkeley, CA.

Worrell, E., L. Price, M. Neelis, C. Galitsky, and Z. Nan, 2008: World Best Practice Energy Intensity Values for Selected Industrial Sectors. Report LBNL-62806. Lawrence Berkeley National Laboratory, Berkeley, CA.

Worrell, E., L. Bernstein, J. Roy, L. Price and J. Harnisch, 2009: Industrial energy efficiency and climate change mitigation. *Energy Efficiency*, **2**(2):109–123.

Worrell, E., P. Blinde, M. Neelis, E. Blomen, and E. Masanet, 2010: Energy Efficiency Improvement and Cost Saving Opportunities for the U.S. Iron and Steel Industry: An ENERGY STAR Guide for Energy and Plant Managers. Lawrence Berkeley National Laboratory, Berkeley, CA.

WWF-Brazil, 2007: Sustainable Electricity Agenda 2020: Study of Scenarios for an Efficient, Safe and Competitive Brazilian Electrical Sector, Technical Series, Vol. 12, March 2007.

Xiong, H., 2007: Energy saving status, major problem and suggestions to china building materials industry during the 10th five-year plan. China Energy No.6.

Zeman, F. and K. Lackner, 2008: *The Reduced Emission Oxygen Kiln.* Lenfest Center for Sustainable Energy, Earth Institute at Colombia University, New York, NY.

Zhou, D., Y. Dai, C. Yi, Y. Guo, and Y. Zhu, 2003: *China's Sustainable Energy Scenarios in 2020.* China Environmental Science Press, Beijing, China.

9

Energy End-Use: Transport

Convening Lead Authors (CLA)
Suzana Kahn Ribeiro (Federal University of Rio de Janeiro, Brazil)
Maria Josefina Figueroa (Technical University of Denmark)

Lead Authors (LA)
Felix Creutzig (Technical University of Berlin, Germany)
Carolina Dubeux (Federal University of Rio de Janeiro, Brazil)
Jane Hupe (International Civil Aviation Organization, Canada)
Shigeki Kobayashi (Toyota Central R&D Laboratories, Japan)

Contributing Authors (CA)
Luiz Alberto de Melo Brettas (Brazilian National Civil Aviation Agency)
Theodore Thrasher (International Civil Aviation Organization, Canada)
Sandy Webb (Independent Consultant, USA)

Review Editor
Ji Zou (Renmin University of China)

Contents

Executive Summary

The world's demand of fuels for transportation has multiplied over the last decades due to the concurrent fast expansion of population, urbanization, and global mobility. The global transport sector is responsible for 28% of total final energy demand. The majority of the energy used in transportation – 70% – is utilized on the movement of passengers and goods on roads locally, nationally, and across regions. Transportation weighs heavily on climate, energy security, and environmental considerations, as 95% of transport energy comes from oil-based fuels. Transportation is the cause of other critical challenges due to its supporting role in local and global economies, as well as the implications of increasing transportation on human health and social interactions. The immense and multi-faceted challenges of a global transportation system deeply rooted in fossil fuels are compounded by the quickly evolving aspirations of a worldwide population that is increasingly on the move and has learned to regard mobility, in particular by motorized modes, as an important component of the modern lifestyle they have or are seeking to attain.

This chapter evaluates the roots of these challenges and outlines the options for a feasible major transformation of the global transportation system over the next 30–40 years. The goal of this transformation is the development of a robust path for the consolidation of transportation systems around the world that can deliver the mobility services needed to support growing economic and social activity while also creating the conditions for enhanced energy security, rigorous climate change mitigation, improved human health, better environment, and urban and social sustainability. This approach is in line with the GEA normative understanding of the consensus embodied in international negotiated agreements and plans of action, and in the more generally accepted aspirations of the international community. The goals of transforming transportation presented in this chapter cut across technical, planning, and policy issues of interest to public and private sector stakeholders and decision makers. A summary of key findings from the assessment are:

- The transformation of transportation systems locally and globally must start now. Time is of the essence. Immediate steps must be taken to define a multi-goal, multi-level, and multi-actor framework of action to identify context-specific "levers of change," or areas that are subject to policy intervention and can help sustain a positive rate of change for this transformation. Achieving multiple goals is possible, as transportation impacts all other areas in society. This will require strengthening institutional capacity and the consolidation of knowledge-based decision support systems to inform policy and decision making about impacts, cost, benefits, and potential trade-offs. The strengthening of institutional capacity will be necessary to deal with solutions that require merging concepts to improve energy and transport systems interaction, technology, urban and regional spatial planning, infrastructure development, and economic and social innovation.

- To remain consistent with the GEA multiple objectives, the transformation of the transport sector, requires that oil products peak before 2030 and are phased out over the long term. This is made possible in the medium term by adopting a policy mix to minimize fossil fuel use in transport, as well as rapidly introducing alternative sources such as renewable-based electricity transportation technologies. Technological choice and lifestyle changes in the long term are more uncertain and depend critically on the nature and direction of technological breakthroughs, but they must be made with the intention of complementing policies and in consideration of future climate impacts.

- Transportation goals for reducing fossil fuel consumption need to be pursued while simultaneously increasing and maintaining the provision of satisfactory economical and social levels of transportation services. Technological improvement is vital, but it is equally important to secure a timely and uninterrupted policy and decision making framework for action. This policy framework is necessary to establish clear conditions so that investments can be planned. Similarly, this framework should also provide context-specific feasibility information to assess local opportunities for creating a sustainable interaction between the transport and energy systems. The framework must also adequately assess conditions for the provision of a diversified set of choices and intelligent mobility services that can improve the efficiency of the transportation network for passengers and goods within all cities and regions.

- The many chances available today for improving conventional technologies require sustained attention. Enhancing the energy efficiency of all modes of transportation can help reduce transport fossil fuel use. Increasing efficiency can be effectively and immediately pursued through widespread adoption of current best available technologies and practices. In the longer term, this can be achieved through the subsequent systematic and comprehensive adoption of a range of new vehicle and other modal technologies. For example, introducing incremental efficiency technologies, improving drive train efficiency, and recapturing energy losses and reducing loads (e.g., weight, rolling, and air resistance and accessory loads) on the vehicle can approximately double the fuel efficiency of "new" light-duty vehicles by 2050. Fuel economy standards have been effective in reducing fuel consumption, and therefore should be tightened and adopted worldwide. The overall effectiveness and political feasibility of standards can be significantly enhanced if combined with fiscal incentives and consumer information. Taxes on vehicle purchase, registration, use, and motor fuels, in combination with the revamping of road and parking pricing policies, can influence vehicle energy use and emissions.

- The fuel efficiency of aviation can be improved by a variety of means, including technology, operation, and management of air traffic. Technology developments might offer a 40–50% improvement by 2050. As aviation's growth rate is projected to be the highest of the transport subsectors, such improvements will not be enough to keep energy use from increasing.

- In the maritime sector, a combination of technical measures could reduce total energy use by 4–20% in older ships and 5–30% in new ships by applying state-of-the-art knowledge, such as hull and propeller design and maintenance. Reducing the speed at which a ship operates brings significant benefits in terms of reduced energy use. For example, cutting a ship's speed from 26 to 23 knots can yield a 30% fuel saving. A similar rationale can be applied to aircraft. Improving the efficiency of aircraft on a global scale is essential.

- Reducing the use of fossil fuel energy in transport can potentially be achieved through the adoption of alternative energy sources such as advanced biofuels, fuel cells, and electric vehicles. Most of the barriers relate to cost, although there are also substantial performance barriers, especially for advanced batteries and infrastructure. Strong policies can ensure rapid uptake and full use of these technologies and will require encouraging sensible changes in lifestyle and behaviors. Life cycle assessment (LCA), together with social and environmental impact assessments, are useful tools to establish a level playing field comparison between different technologies. Significant uncertainties corresponding to modeling choice of system boundaries and modeling assumptions prohibit straightforward policy implications – especially in the case of biofuels. The future biomass potential strongly depends on production efficiency, the development of advanced techniques, costs, and competition with other uses of land. Current biofuels development needs to resolve its many sustainability challenges. Advanced biofuels are considered to have much greater potential for the future, not only for road transport but also for aviation and shipping.

- Electricity produced from renewable sources can provide a significant stronghold for the global transformation of the transportation system. Plug-in hybrid electric vehicles (PHEVs) allow for zero-tailpipe emissions for small vehicle driving ranges, e.g., in urban conditions. Hybrid electric vehicles (HEVs) can improve fuel economy by 7–50% compared to comparable conventional vehicles, depending on the precise technology used and driving conditions. All-electric or battery-powered electric vehicles (BEVs) can achieve a very high efficiency (up to four times the efficiency of an internal combustion engine vehicle) but have a low driving range and short battery life. If existing fuel-saving and hybrid technologies are deployed on a broad scale, significant fleet-average fuel savings can be obtained within the next decade. Increasing the performance of high-energy batteries for PHEVs could subsequently lead to higher market penetration of BEVs. Hydrogen fuel cell vehicles (FCVs) could alleviate the dependence on oil and reduce emissions significantly. For BEVs and FCVs, the emissions are determined by the production of hydrogen and electricity. Further technological advances and/or cost reductions would be required in fuel cells, hydrogen storage, hydrogen or electricity production with low- or zero-carbon emissions, and batteries. Substantial and sustained government supports are required to reduce costs further and to build up the required infrastructure.

- Reducing transport energy use can also be achieved by favoring those modes that are less energy-intensive, both for passenger and freight transport. Strong local and regional urban planning policies, practices, and implementation should aim to enhance the diversification and quality of public modes of transport. Spatial local plans aimed at reducing both the need for travel and the distances traveled can reduce energy demand and also improve the quality of urban life through improved accessibility, affordable mobility services, and improved traffic safety. In cities worldwide, a combination of push and pull measures and traffic demand management can induce a modal shift from cars to the more prevalent use of public transit and cycling, which has multiple benefits. In particular, use of non-motorized transportation (NMT) can improve the efficiency of the transport system and overall people's health. Parking policies and extensive car-sharing options can become key policies to reduce the use of private cars. Cities can be planned more compactly, with less urban sprawl and a greater mix of land uses and local markets. Neighborhood design and street layout can encourage walking and public transport access to schools, hospitals, shopping centers, and other places that attract people on a regular basis. Employers in many sectors can locate themselves strategically, with the goals of enhancing the job-housing balance for their employees and providing incentives for substituting non-essential work-related travel with information and communication technologies.

9.1 Introduction: Transportation and Energy – A Global Perspective

Broad accessibility to services, people, and goods is an essential and basic human need and a precondition to economic well-being in modern societies. The petroleum-fueled motor vehicle and the airplane are technologies that have presented opportunities for greatly increased mobility, flexibility and reduced travel times. Automobility, as a self-reproducing system, has created unprecedented flexibility and fostered profound changes in lifestyles and in the physical landscape of cities worldwide (Whitelegg, 1997; Urry, 2007). The enhanced potential for travel created by these transport innovations resulted in an unprecedented rate of growth of the volume of personal travel and the volume of goods moved (WBCSD, 2002; Schäfer et al., 2009; Sperling and Gordon, 2009).

At the close of the 20th century, the world witnessed both unprecedented urbanization in the developing world and the suburbanization of many cities in the developed world (for more, see Chapter 18). Motorized forms of mobility, particularly automobility, were favored as the principal means for personal and social accessibility in urban areas. Along with technological shifts and economic development, transport activities scaled up in distance and participants. This growth, and the reliance on individual motorized forms of urban transport, has resulted in many local social and environmental consequences, including congestion, air pollution, noise, olfactory and visual intrusion, disruption of ecosystems and landscapes, water and soil degradation, ozone depletion, social and urban fragmentation, road deaths and injuries, asthma, and obesity (Rothengatter, 2003; Whitelegg and Haq, 2003; Gilbert and Pearl, 2007, Schäfer et al., 2009).

The scale of growth in the movement of people and goods around the globe has been immense. But not all populations and geographic regions have participated evenly in the technological advances that facilitated this development. The average citizen in wealthy nations and urban areas can travel faster and further than ever before and overcome long distances in comfortable and reliable ways (van Wee et al., 2006). Conversely, average citizens in poor nations, living in areas with differentiated urban growth characterized by large differences in income levels and social disparities, face hurdles to meet their most basic transportation needs. These citizens witness an urban transport situation that, for the most part, is unbearable and only deteriorates over time (Newman and Kenworthy, 1999; Gwilliam, 2005; Tiwari, 2006, Figueroa, 2010).

The global expansion of mobility encompasses great innovations that have linked transportation and intelligent communications systems, transforming the way in which people organize their travel and communication considerably. The interplay of these systems has redefined the core of social interaction and urban life (Castells, 2001; Sheller and Urry, 2003; Castells 2004). The evolving transport mobility of the last century has converged into a dynamic system that includes new forms of virtual communications and is firmly rooted in a number of key components,

including motorized modes and the automobile, aviation, rail, ships and oil industry, consumerist lifestyles, energy and environmental resource use, global procurement of oil, spatial and infrastructure planning, urban and street design, and societal values that embrace mobility as part of what constitutes high quality of life standards (Urry 2004; 2007).

As economies grow, transport activity continues to increase around the world. In many areas of the developing world where globalization is expanding, trade flows and personal incomes are rising, leading to an increased demand for enhanced personal mobility. Mobility of people and goods has significant global economic benefits. A prime example is global travel for tourism, which constitutes the largest industry in the world. It is worth US$6.5 trillion and directly and indirectly accounts for 8.7% of world employment and 10.3% of world gross domestic product (GDP) (Urry, 2007).

9.1.1 The GEA Approach to Transportation and Energy: Acting on Multiple Goals to Realize Multiple Benefits

An efficient transportation system is crucial for economic development and an asset for the growing integration of international markets. This assessment follows GEA's normative approach in highlighting the necessity of realizing certain concrete goals that reflect part of the shared aspirations of the international community. These goals are related to economic growth and equity, improved environmental health, climate change mitigation, and enhanced energy security (see Chapters 2–6, as well as Chapter 17, for more detailed definitions). This assessment asserts that a global transformation of the transportation and other energy-using systems can help meet the major GEA goals. A complementary relationship exists between these goals, as summarized in Table 9.1. Two premises follow this approach. First, to meet the larger objectives, multiple goals must be addressed simultaneously and effectively. Strong and immediate actions need to be initiated and sustained at different scales to effectively achieve a radical, global transformation of the transportation and energy systems. Secondly, as transportation is deeply linked to other sectors, actions undertaken to achieve one goal can have directly calculable, and also unintended, effects on the other goals (Goodwin, 2003). Multiple benefits can be calculated in reference to how policies that promote clean air, walking and biking can contribute to control obesity, limit chronic disease and reduce air pollution emissions including greenhouse gases (Alliance for Biking & Walking, 2010; Nazelle et al., 2011). Facilitating walking and biking creates benefits for improved accessibility and energy security. Consequently, extensive follow up assessments to policy implementation are necessary to adequately measure these ripple effects. A framework that clearly outlines the multiple goals and benefits of a sustainable transformation can offer a variety of options and policy leverages for major stakeholders, governmental or otherwise, to take direct and immediate action. The policies, actions and actors are context dependent, but the long-term goals and expectations for this transformation are then made clear.

Table 9.1 | GEA framework for sustainable transportation multiple goals and benefits.

GEA systemic goal for Transport	Multiple goals and benefits of sustainable transportation systems (a)	Indicative approach to assess progress toward goal
Economic Growth and Equity	Functionality, Efficiency	Reduced travel time, cost and uncertainty, secures operationality and resilience
	accessibility	ease of reaching opportunities throughout the urban area (jobs, shops, medical and social activities); reduced access time, travel cost, and trip uncertainty
	affordability	% of income used to pay for transport services (<20%)
	acceptability	public acceptability; participation in decision-making process
Health and Local Ecosystems Protection	road safety	reduction of traffic deaths and injuries
	universal access	access provided for the elderly, very young, and people with disabilities
	personal physical activity	increased public health benefits from more physical activity
	reduce air pollution	air pollution reduction and public health benefits of reduced air pollution
	reduce noise	noise reduction
Climate Change	reduce GHG	GHG mitigation
Energy Security	diversification of fuel sources	use of alternative fuels
	independence from fossil fuels	% reduction in use of fossil fuels

(a) The general GEA goals for the global energy system are here adapted in attention to the specific services and goals that transportation systems deliver.

This multi-goal, multi-benefit approach aligns with the early approach labeled as co-benefits or ancillary benefits in the literature. It requires that a transport system is not only devised with the aim of maximizing mobility, but is also concerned with reaching other goals simultaneously. The concept of co-benefits has been pushed mostly in combination with climate change mitigation goals (Krupnick et al., 2000; Aunan et al., 2004; Bollen et al., 2009, Nazelle et. al., 2011), and has been applied particularly to urban transport (Creutzig and He, 2009; Bongardt et al., 2010; Creutzig et al., 2011b; UNEP, 2011), serving as the foundation for the systematic assessment of future scenarios and policy instruments.

Climate change mitigation is an important driver of envisaged improvements to transport systems. Fortunately, appropriate policies can be designed such that actions to reduce greenhouse gas (GHG) emissions also deliver other environmental and social benefits (EEA, 2008a). In addition to climate change mitigation, sustainable transportation solutions must also consider the multiple goals listed in Table 9.1. This perspective is also presented in other studies (EEA, 2008a, EEA, 2008b; Creutzig and He, 2009; Bongardt et al., 2010). The multi-goal, multi benefit approach maintains that meeting the social needs of growing urban populations requires affordable transportation services that efficiently facilitate accessibility to work, study, and leisure activities, and suggests that this can be designed with attention to operational efficiency to support economic development, low-carbon, energy security and environmental goals and for improving social and healthy living.

9.1.2 Overview of Energy use in Transportation

The concurrent fast expansion of population, urbanization, and global mobility has multiplied the world's demand for fuels for transportation and city-wide energy services. As shown in Figure 9.1, energy use in the transport sector in 2007 was high, 28% of total final energy use.

During the last several decades, the energy use of transport sectors in both Organisation for Economic Co-operation and Development (OECD) and non-OECD countries have increased substantially, as shown in Figure 9.2. In developing countries, the increase in recent years has become more prominent due to rapid urbanization and motorization. And, although a major increase in energy use was caused by road

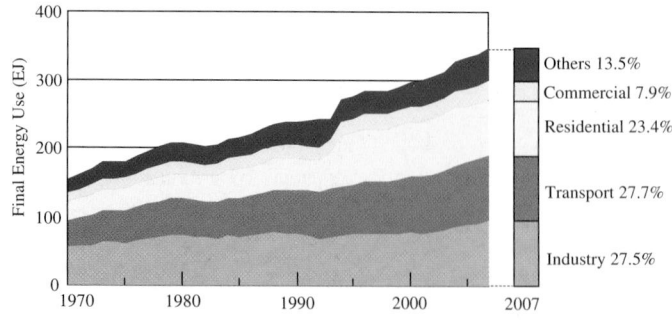

Figure 9.1 | Global energy use in various end-use sectors (others include agriculture/forestry, fishing, and non-energy use). Source: based on IEA, 2009b.

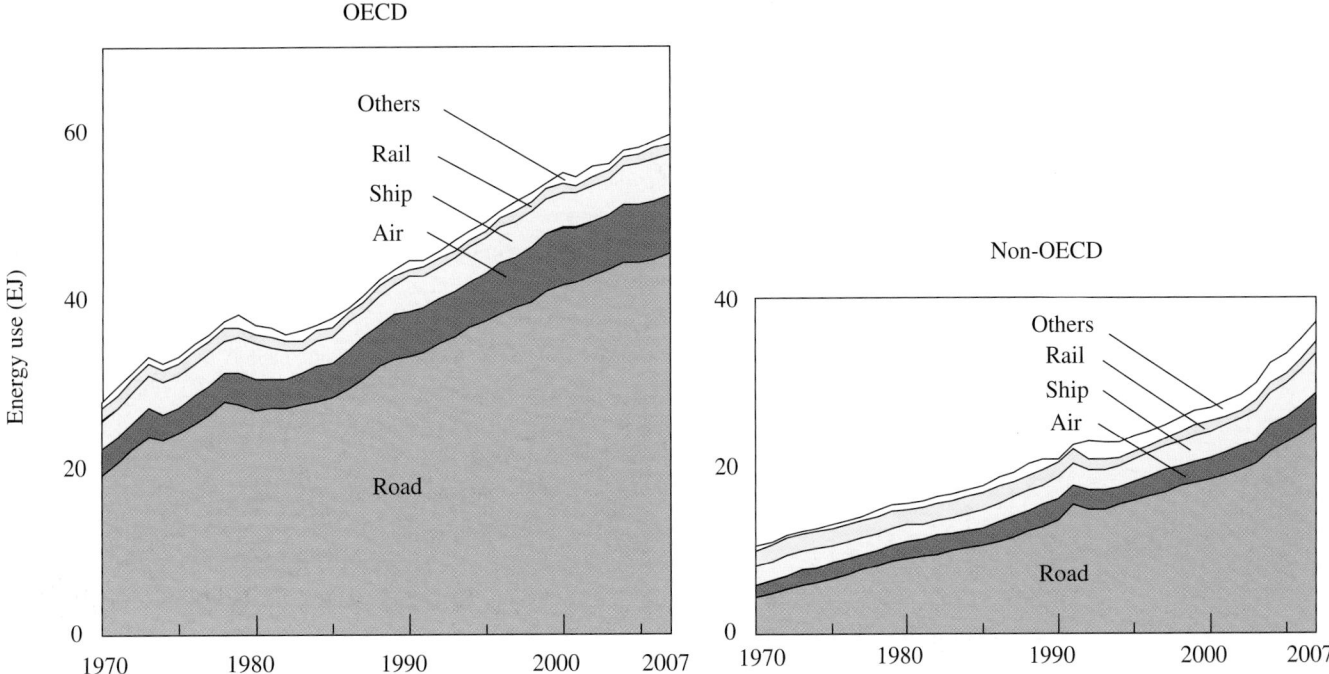

Figure 9.2 | Transport energy use in OECD and non-OECD countries by mode (others include pipeline and non-specified). Source: based on IEA, 2009b.

transport, the actual phenomenon of motorization is quite different between OECD and non-OECD countries. Non-OECD countries started later and still show motorization rates significantly below OECD countries. However, the speed of their concurrent urbanization and motorization is unrivalled, especially in China, and puts significant demands on adapting transport infrastructures.

A single fossil resource- petroleum- supplies 95% of the total energy used by world transport. This dependence results in two major areas of global concern: the long-term security of energy supplies and the fast-rising contribution of the transport sector to greenhouse gas (GHG) emissions (IEA 2009a; Stern, 2007). The carbon dioxide (CO_2) emissions and energy use of different transport sub-sectors are proportional (see Figure 9.3.).

The transport sector has the highest rate of growth in energy use and related CO_2 emissions of all final end-user sectors. This rate is expected to increase up to 1.7% a year between 2004 and 2030 (IEA 2009a).[1] In 2007, the global transport sector produced 6.6 $GtCO_2$ emissions, corresponding to 23% of world energy-related CO_2 emissions and, road transport, mostly passenger transport, accounts for 73% of this total.

A much higher rate of growth of 3.7%/year (between 1990–2003) corresponds to freight transport, this trend is expected to continue (see Figure 9.4) (McKinsey Global Institute, 2009).

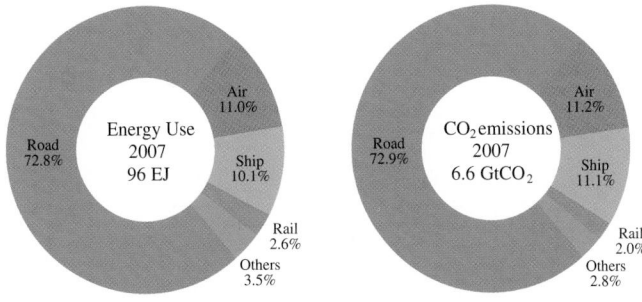

Figure 9.3 | Modal share of global energy use and CO_2 emission in transport sector (2007). Others include pipeline transport and non-specified use. Source: based on IEA, 2009a; 2009b.

Urbanization has been extremely rapid in the past 60 years, with a 2.6% annual average growth rate (UN, 2009). More than half of the world now lives in urban areas (UN, 2007; UN, 2008). In 2010, twenty one cities reported having a population over 10 million compared with two cities in 1950 (UNDP, 2010). Rapid growth in suburban areas and the rise of "edge cities" in the outer suburbs has been a common form of development facilitated by the rise of personalized motor transportation (see Chapter 18). The greater distances replicated through the low-density development discourage walking and bicycling as a share of total travel and are not easily served by public transport (WBCSD, 2002; 2009). A growing demand for travel and a declining share in the use and quality of public transportation services have been the observed result across developed and developing cities alike (Gwilliam, 2005; Tiwari, 2006, Hidalgo and Carrigan, 2010; Buehler and Pucher, 2011).

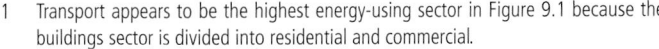

1   Transport appears to be the highest energy-using sector in Figure 9.1 because the buildings sector is divided into residential and commercial.

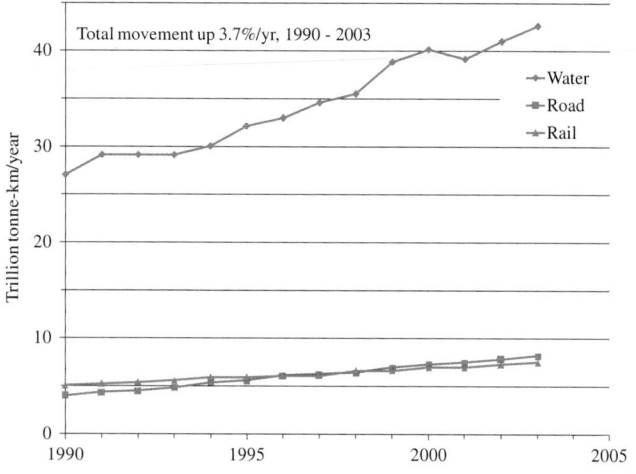

Figure 9.4 | Global freight movement in trillion tonne-km/year (t-km/yr). Source: Gilbert and Pearl, 2007.

The rapid motorization of many of the world cities has not been a global phenomenon. Many low-income urban areas in developing countries do not even have access to motorized public transport services of any sort (Tiwari, 2006; Vasconcellos, 2010). In developing country rural areas in sub-Saharan Africa, but also in parts of Asia and Latin America, walking more than 10 km/day each way to collect water or fuels, work on farms, and attend schools and health services is not unusual. Walking can be also prevalent for the urban poor, when commuting by public transport becomes very costly. For example, commuting accounts for 14% of the income of a poor family in Manila compared with 7% of the income of a higher income family in the same city (World Bank, 1996; Gwilliam, 2002).

The extent of global motorization explains only part of the increasing trends in energy use and carbon emissions from transport worldwide. Other contributing factors relate to consumer preferences for larger, heavier and more powerful passenger vehicles, particularly in the industrialized world, which have resulted in larger energy use for transport, despite the sustained technological improvement on car fuel efficiency attained on new passenger vehicles during the last decades. By 2008, more than half of all vehicles purchased in the United States were SUVs (sport utility vehicles) or light trucks. A weight reduction of the average car sold in the United States to European levels would bring about a 30% reduction in fuel intensity (Schipper, 2007). The average fuel economy of the new light-duty vehicle (LDV) fleet[2] in the United States in 2005 would have been 24% higher had the fleet remained at the weight and performance distribution it had in 1987 (Khan Ribeiro, et al., 2007). In addition a drop of fuel consumption rates of more than 1%/year would have resulted during

that same period, contrary to the fast rate of increase that has continued until today. (Heavenrich, 2005; Sperling and Gordon, 2009).

Another contributing factor is the switch to relatively faster speed modes of travel, particularly motorized vehicles and air transportation, which has led to an increase in the total yearly distance travelled (Schäfer, 2000). The switch to faster speed modes has been facilitated by a considerable decline in the share of income spent in transportation during the last decades (Berri et al., 2010). Faster modes and larger transport infrastructural investments enabled the expansion of urban areas and the process of suburbanization, allowing people to maintain the average time budget for their daily travel roughly constant (van Wee et al., 2006). Simultaneously, increasing income allowed commuters to choose the more expensive modes, e.g., motorized vehicles. In developing countries, where income increases have been less significant, a relative decrease in car purchase cost in the early 1990s, and a relaxation of restrictions or even the imposition of minimal restrictions on imports of second-hand cars has sustained a progressive diffusion of motorized modes of transport (Berri et al., 2010). Countries with different land use and transportation policies, including fuel prices, have different rates of auto mobility per unit of GDP per capita (Millard-Ball and Schipper, 2011; see Figure 9.5). Vehicle use (km per capita) differ for example between the United States and Japan by a factor of about 3 (Shipper, 2009).

If these historic trends continue, there will be a vast worldwide expansion of motorization and a resulting increase in fossil fuel use and GHG emissions from transport. The local impacts of congestion, noise, air pollution, health, safety, and energy security are of immediate concern. Globally, climate change – the atmosphere's sink capacity – and limited resources pose huge inter-temporal challenges. However, as the least developed areas become economically mature and their populations' incomes rise, the evidence suggests that individuals will tend to favor private motorized vehicles. This is in part because these vehicles promise to provide a faster and more flexible, convenient, and reliable form of travel than the available local and intercity public transport, but it is also because the car's

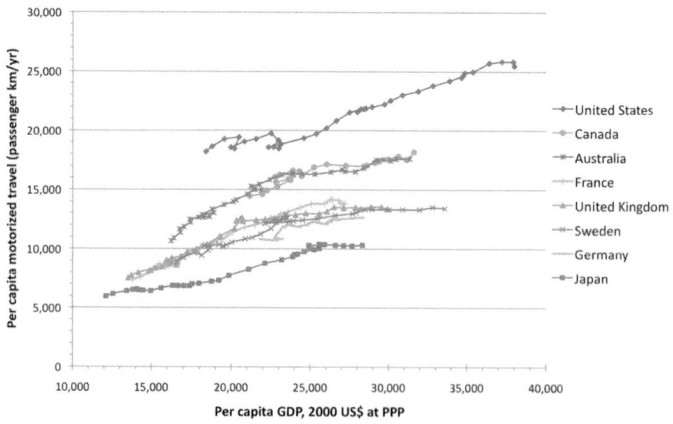

Figure 9.5 | Passenger motorized travel in some OECD countries. Passenger travel levels off with income. Developing countries, such as China and India (see Figure 9.8) start from low levels but are rapidly motorizing. Source: Millard-Ball and Schipper, 2011.

---

2   In the United States, fuel economy refers to the number of miles (distance) travelled/gallon of fuel (miles/gallon). The expected trend is for fuel economy to increase. In Europe, and elsewhere, the equivalent metric is the fuel intensity measured in liters of fuel used/100 km (liters/100 km); the expected trend is for the fuel intensity of the vehicle fleet to decrease over time.

attractiveness and seductiveness is deeply embodied in the culture and psychology of modernity (Whitelegg, 1997; Thomsen, et al., 2005; Urry, 2007). The car industry is strategically positioned in quickly emerging markets, notably in China and India, to tempt these new potential customers with simpler and affordable models (Sperling and Gordon, 2009).

In addition, the largely unregulated massive scale importation of used vehicles to developing countries, and the fact that these vehicles are kept running with great ingenuity for longer than the manufacturer's estimated lifetime, aggravates the energy requirements, local environmental problems, and carbon emissions contributions from the sector. Finally, a number of influential converging factors such as economic policies that maintain fuel subsidies and planning practices that incentivize suburban residential developments, large malls, and retail centers with extensive free parking all can play an important role in increasing motorization and energy use in transport.

The world auto fleet increased from about 50 million vehicles in 1950 to 580 million vehicles in 1997 (WBSCD, 2002). China's vehicle sales have already overtaken car sales in the United States by a huge margin increasing from 2.4 million in 2001 to around 17 million by 2010 (ADB, 2009). Two-wheeled scooters and motorcycles are important in the developing world and in parts of Europe. (WBSCD, 2002; Tiwari, 2006; Zegras and Gakenheimer, 2006; ADB, 2009).

Non-motorized transportations (NMTs) are dominant in developing countries. Walking accounts for 20–40% of all trips in many cities (WBCSD, 2002), while bicycles are important means of transportation in Asia and also in industrialized cities like Amsterdam and Copenhagen. Research and data for walking and cycling pale in comparison to research on motorized transport (Methorst et al., 2010).

Public transportation, especially buses, though declining in share in favor of private cars in the industrialized world and some emerging economies, have a high modal share elsewhere. For example, it is estimated that buses make up 61% of the motorized trips in Santiago, Chile (Estache and Gómez-Lobo, 2005). The public transportation share in 2007 in 15 cities in Latin America is 43% of the total trips, and 60% of the motorized trips. Of total public transport, 15% is by rail, 52% in formal buses, and 33% in informal microbuses and jitneys[3] (CAF, 2010). Public transport in many developing cities is characterized by hundreds of separate bus companies. For instance, it is estimated there are 200 operators on average for a single minibus route in Lagos, Nigeria (Gwilliam, 2005).

Intercity and international travel is growing rapidly and is dominated by auto and air modes. In Europe, Japan and in China intercity passenger travel is combined with fast rail travel. Worldwide passenger air travel is growing by 5% annually, a faster rate of growth than any other travel mode (WBCSD, 2002).

---

3   Jitneys are vans or small buses that transport passengers on a route for a small fare, follow somewhat regular routes (but generally not on a published schedule), and often allow for ad hoc route deviations to accommodate passenger needs (Small and Verhoef, 2007).

Industrialization and globalization have also stimulated freight transport, which now uses 35% of all transport energy (WBCSD, 2004a). The truck-transport sector has a continuously high demand for petroleum, particularly diesel. Truck-transport accounted for 12.5% of global petroleum demand and 45% of global diesel demand in 2006 (McKinsey Global Institute, 2009). Freight transport is more conscious of energy efficiency considerations than passenger travel because of pressure on shippers to cut costs (Larsen and Peterson, 2009). However, the historic rapid growth of truck-transport energy use reflects in large part that the truck sector has limited efficiency improvements opportunities for diesel engines (McKinsey Global Institute, 2009). The increasing demand for fast, reliable, smaller, "just-in-time" and door-to-door shipments contributes to the rapid growth of energy use for freight transport. The result has been similar to the case of passenger vehicles in that, although the energy efficiency of specific modes has been increasing, the movement to the faster and more energy-intensive modes translates into faster rates of energy use. The opportunities for switching from truck to rail apply to only a small number of truck shipments (those involving sufficiently large volumes and long distances), which represent a small percentage of truck-transport volume; for example, this will be an option for only 4% of shipments in Europe (McKinsey Global Institute, 2009). Worldwide the trends show rail and domestic waterways' shares of total freight movement declining, and highways' shares increasing, while air freight, though remaining a small share, growing rapidly. Furthermore, environmental harm continues to increase as fuel efficiency improvements are offset by increased freight kilometers.

Freight transport can be summarized as follows:

- the majority of international freight transported worldwide comes from a few countries. In 2008, more than three-quarters of exported freight were from only 25 countries (US DOT, 2010);

- international freight is dominated by ocean shipping; the bulk of international freight is carried aboard extremely large ships carrying bulk dry cargo (e.g., iron ore), container freight, or fuel and chemicals (i.e., tankers);

- regional freight is dominated by large trucks, with bulk commodities carried by rail and pipelines, and some water transport;

- national or continental freight is carried by a combination of large trucks on higher speed roads, rail, and ship;

- urban freight is dominated by trucks of all sizes (Khan Ribeiro, et al, 2007); and

- from mid-2008 to mid-2009, as global economic activities slowed, goods transported worldwide by ocean carriers and airlines fell (US DOT, 2010).

Transporting freight around the world depends on geography, available infrastructure, and economic development. All modes participate in the United States' freight transport system, which has the highest total traffic in the world. Russia's freight system is dominated by rail and pipelines. Europe's freight systems are dominated by trucking and shipping with a market share of 47% and 42% (in tonne-kilometers [t-km]), respectively, in European Union-27 countries, while rail's market share is only 11%, despite its extensive network (EC, 2008; EC, 2010). This rather small share of freight on rail in Europe is the result of priority given to passenger transport, divergent requirements for operability of urban rail systems, and market fragmentation between rival national rail systems, affecting the overall intermodality and a consolidation for rail freight in the European market (EC, 2008). China's freight system uses shipping as its largest carrier, with substantial contributions from rail (UNESCAP, 2004). Figure 9.6 shows the share of modes for passenger and freight.

9.2 Energy Use in Different Modes of Transportation

9.2.1 Road Transportation

From 1971 to 2007, global transport energy use rose steadily, with an average growth rate of 2.5%/year, which closely paralleled growth in economic activity around the world. The road transport sector (including

both LDVs and trucks) used the most energy and grew most in absolute terms (IEA, 2009b).

For passengers, road transport represents the most important mode (see Figure 9.7). The sums of passenger kilometers (p-km) for two- and three-wheelers, cars, minibuses, and buses are much higher than other modes in every single region of the world. For freight, trucking used about 23% of total energy used by the transport sector in 2005. Data on surface freight movement in many countries is poor, but most freight transport moved by road and rail appears to be domestic rather than international. In the European Union, for example, available data for 2005 indicates that only around 30% of all road freight (in terms of tonne kilometers) crossed an international border. The corresponding figure for rail freight, which accounted for 19% of all surface freight in the same year, was 51%.

Travel surveys and fuel use statistics indicate that passenger car travel per capita is approaching saturation[4] levels in most OECD countries (Cresswell, 2006; Dennis and Urry, 2009; Millard-Ball and Schipper, 2011), and that distances traveled by each vehicle each year may be declining as the total number of vehicles on the road and levels of congestion increase. Car ownership rates have risen above one LDV for every two people on average in OECD countries, nearly 1.5 vehicles for every two people in the United States, and an increasing number

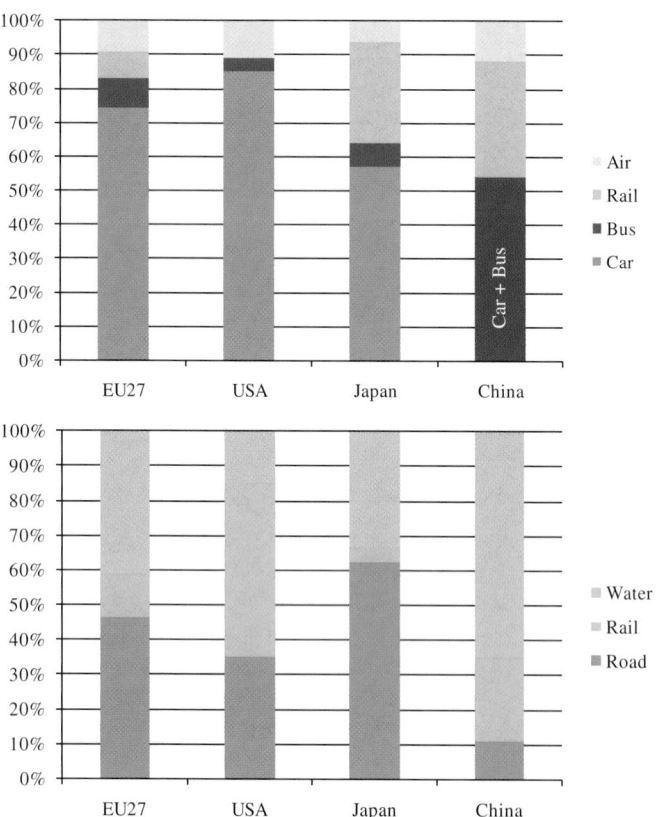

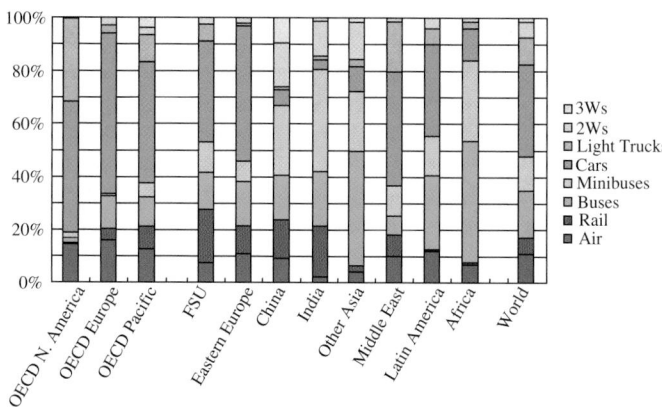

Figure 9.7 | Estimated motorized passenger travel split by mode (2005). Source: IEA, 2009c. ©OECD/International Energy Agency 2009.

Figure 9.6 | Modal split of passenger (top) and freight transport (bottom; air freight transport not included) in terms of passenger-km (p-km) and t-km, respectively (2005). Source: based on EU, 2010.

4   Saturation here refers to several observed trends: a) a slow turnover and penetration of vehicles due to low population growth in OECD countries. An equivalent slow turnover in developing countries can be attributed to long vehicle life plus extensive market penetration of second-hand vehicles; b) little or no variation in the number of trips and the time people use for travel daily. The value across travel surveys in OECD countries over time has been remarkably stable – between 2, 7 and 3 trips per day. Similarly, the per capita "time budget" shows a stable trend at about one hour/day (Lyons and Urry, 2005; van Wee et al., 2006). Comparatively, in developing countries, time series travel survey statistics are lacking, but the reported values vary within a similar range from 2.2 trips/day (Padam and Singh, 2001) to 4.0 trips/day and an average time between 70–90 minutes/day (Hu and Reuscher, 2004); c) Finally, the average distance travelled/day/ person (work related) has also remained unchanged, around 10–60 km/day globally (see Chapter 18 for a similar discussion).

Energy End-Use: Transport

of households own two or more vehicles. This leads to a flattening of, or even a decrease in, the average distance travelled by each vehicle each year (IEA, 2009c). For long-distance travel, faster modes such as air travel are growing more rapidly than passenger car travel, accounting for the majority of the increase in total distance travelled per year (Axhausen et al., 2003).

There is a close, bidirectional interaction between urban transport, land use, and urban form (Gordon et al., 1989; Anas et al., 1998; Wegener and Fürst, 1999; Small and Verhoef, 2007). Cities of similar wealth often have very different levels of motorization, reflecting the effect of other factors such as urban form, design, and density on transportation modes (Næss, 2006; Merriman, 2009; for more see Chapter 18 on urbanization). Modal shares vary dramatically across cities. The share of trips by walking, cycling, and public transport is 50% or higher in most Asian, African, and Latin American cities, and even in Japan and Western Europe. The use of bicycles in cities like Amsterdam and Copenhagen demonstrate that coordination of land use and transport planning can be key to maintaining high levels of safe use of NMT (Beatley, 2000; Spinney 2009).

The prevalence of NMT in most developing country regions and in Eastern Europe is partially the result of low levels of economic growth and infrastructure investment. Data for Africa, Asia, and Latin America shows walking to be the major transport mode among the poorest city residents (World Bank, 1996; Gwilliam, 2002; Tiwari, 2006; Vasconcellos, 2010). Starting in the 1970s, Western Europe has made a conscious effort, to restrict urban sprawl and motorization, and to invest in improved NMT infrastructure (Banister, 2005).

Motorization does not necessarily imply the increase of passenger four-wheeled automobiles. Motorized two- and three-wheelers are among the most popular means of transport in Asian countries and are important elements of motorization in parts of Africa (Nagai et al., 2003). Most cities in India have a rate of private motor vehicle ownership similar to the wealthier cities of Latin America, where motorized four-wheeled vehicles are predominant. By this same account, cities like Kuala Lumpur and Bangkok are already exhibiting motorization levels close to those of Western European countries (ADB, 2006; see Figure 9.8).

The increase in the number of two- and three-wheeled vehicles in developing countries is much more accelerated as compared with the increase of four-wheeled vehicles (Gwilliam, 2003; Zegras and Gakenhaimer, 2006). Once individuals have gained mobility with a two- or three-wheeler motorized vehicle, it is likely that many will shift to four-wheeled vehicles when they attain higher levels of income. If this condition is fulfilled, it can be expected that the ownership and number of four-wheeled vehicles in developing countries, particularly in China and great parts of Asia, will increase dramatically (Nagai et al., 2003; Wang et al., 2006; ADB, 2009). Reasons cited by these authors for the motorcycle motorization phenomenon in many countries include lower capital and operating costs than automobiles coupled with lower levels of real purchasing power; superiority in time and door-to-door

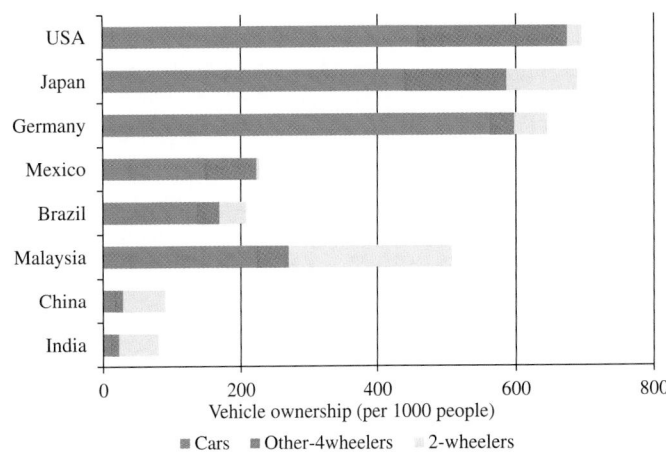

Figure 9.8 | Vehicular penetration in 2006 of several developed and developing countries (other four-wheelers include busses and trucks). Source: based on MoRTH, 2009.

convenience relative to automobiles in congestion; and superiority to an often deteriorating public transport system. Despite their prevalence, rapid growth and safety considerations, until recently motorcycles have been ignored both in planning and infrastructure implementation, including traffic counts and pollution control efforts, but this is changing (see, i.e., the Clean Air Initiative for Asian Cities, 2011). As motorcycles and three-wheelers are light vehicles, their overall fossil fuel consumption is not as large as conventional cars. Furthermore, new technologies such as electric and hybrid engines are being introduced, further reducing their fossil fuel consumption. Nevertheless, traffic safety concerns need to be thoroughly addressed.

For example, electric bikes (e-bikes) in China emerged from virtual non-existence in the 1990s to achieve annual domestic sales of 13.1 million in 2006, which is equal to the sales of gasoline two-wheelers (ADB, 2009). As it is likely that electric two-wheelers will continue to substitute for bicycles and public transport as incomes rise in China, appropriate policy initiatives may lead to wider electrification of the Chinese transport system (ADB, 2009). The rapid parallel development of new traffic safety rules and regulations to accommodate e-bikes is necessary, and it can serve as inspiration for other Asian cities experiencing the same trends.

9.2.2 Railways

Railway traffic systems are unique in terms of location and technology. Railway traffic systems are highly concentrated in a few world regions. Approximately 90% of all freight and passenger railway traffic can be found in North America, Russia (freight oriented), Japan (passengers), China, India, and Europe (see Figure 9.10). In total, in 2007 there were slightly over 960,000 km of rail lines globally, carrying over 28 billion passengers (2495 billion p-km) and 11.4 billion tonnes (8845 billion t-km) of freight (Thompson, 2010). The top five passenger rail systems

Box 9.1 | Case Study of Urban Mobility in Developing Countries

The World Business Council for Sustainable Development (WBCSD) Mobility for Development project has researched the state of mobility in rapidly growing cities at various stages of economic development (WBCSD, 2009). With three billion people surviving on less than US$2/day and not adequately served by existing mobility systems, the urgent challenge is to expand transport benefits to those currently excluded from urban transport systems and reduce transport's environmental impact. The project set out on a process of research, dialogue, and learning in four cities: Bangalore in India, Dar es Salaam in Tanzania, Sao Paulo in Brazil, and Shanghai in China.

Each of the cities has experienced rapid urban and economic growth, accompanied by growth in transportation, both passenger and freight, public and private – but in quite different ways. Public transport remains a major and sometimes overwhelmingly dominant provider of personal mobility. In these cities, the public transport share of all motorized trips ranged from 45% in Sao Paulo to 71% in Dar es Salaam. In Bangalore and Dar es Salaam, informal paratransit is an important provider of access (see Figure 9.9). Indeed, in the latter, public transport is currently being provided almost exclusively by approximately 9,000 privately-owned and operated minibuses known as *dala dalas*. It is also apparent that non-motorized modes still provide a major share of personal mobility in each city. In Shanghai in 2004, trips by foot and by bicycle accounted for 31% and 25% of the daily total, compared to 31% and 33%, respectively, in 1995. In Sao Paulo in 2002, 37% of daily trips were by foot and 1% was classified as "other" (it is assumed that this category includes trips by bicycle).

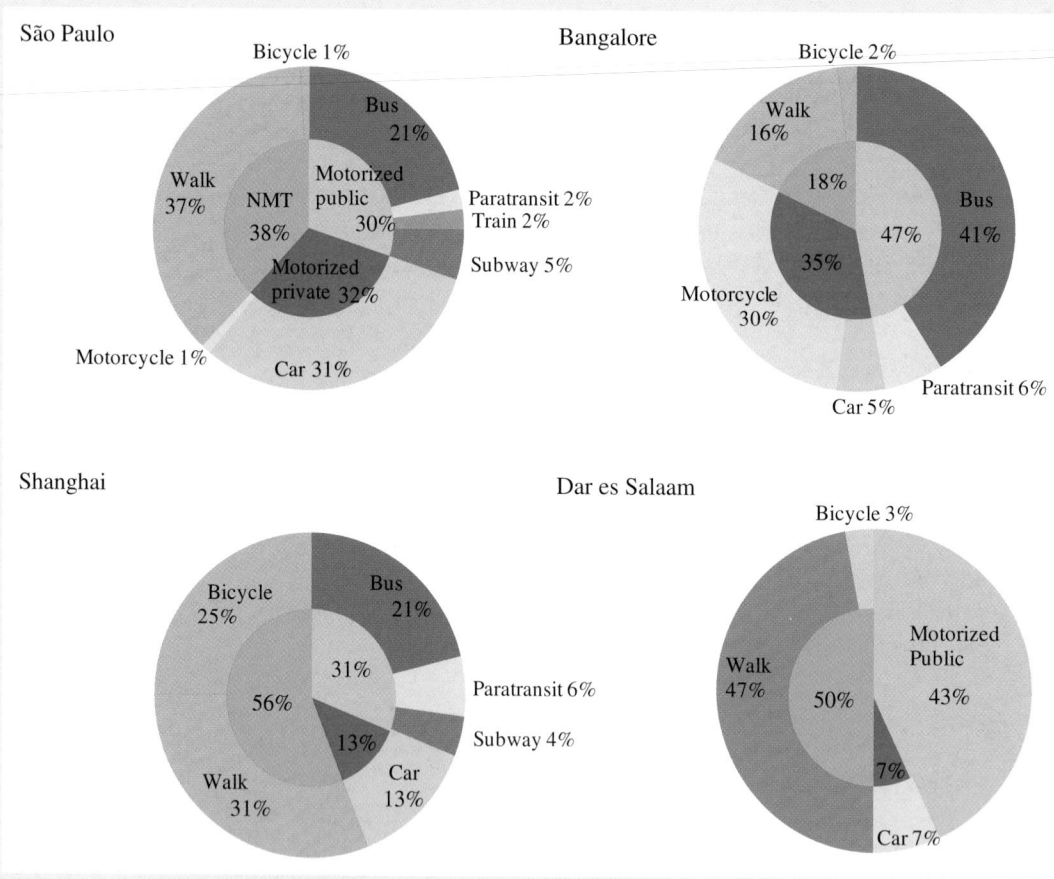

Figure 9.9 | Passenger transport modal shares of several cities in developing countries. Source: based on WBCSD, 2009.

carry 85% of the world's rail passenger traffic. Only three railway systems account for 79% of the world's railway t-km (Statistics Bureau of Japan, 2010).

Rail's main activities are high speed passenger transport between large cities, high-density commuter transport in the city, and freight transport over long distances. Heavy rail transit systems are generally

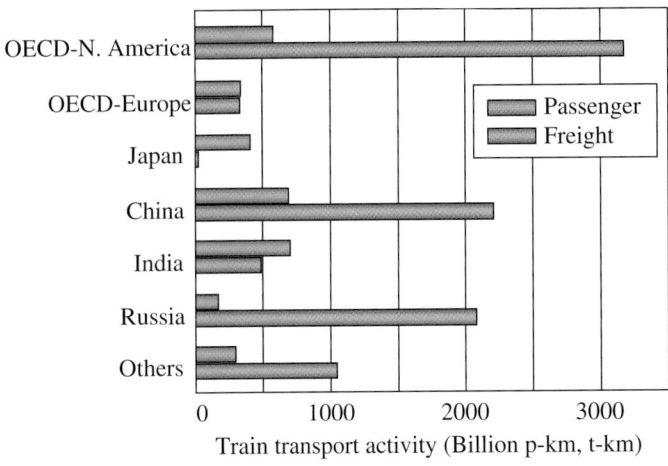

Figure 9.10 | Railway traffic in the major regions. Source: based on Statistics Bureau of Japan, 2010.

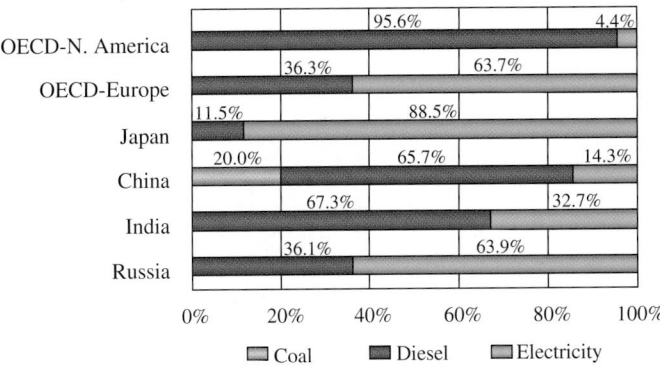

Figure 9.11 | Energy use in railway transport systems in the major regions. Source: based on IEA, 2009b.

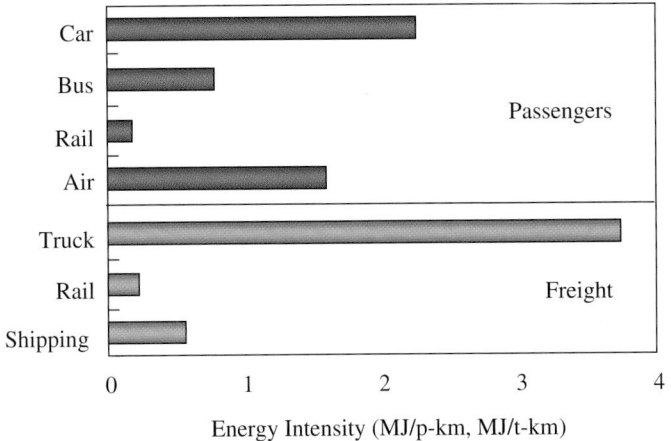

Figure 9.12 | Energy intensity of domestic transport system in Japan (2008). Source: based on ANRE, 2010.

found only in the largest, densest cities of the industrialized world and a few developing-world cities, like Beijing, Delhi, Mexico City, and Sao Paulo, among others (Kahn, Ribeiro et al, 2007).

In Europe, Japan, and Russia, electricity is a major energy source for rail, while diesel dominates in North America, China, and India (see Figure 9.11). Electric power for rail systems comes from a variety of sources, predominantly from coal. Coal is also directly used as locomotive fuel in some developing countries.

Trains are more efficient when compared with trucks, and the same applies for passenger trips when compared to all other motorized options (Figure 9.12). Efficiency is strongly dependent on the loading factor. Hence, the economic and environmental viability of public transport crucially depends on urban density.

9.2.3 Aviation

Since 1960, passenger air traffic grew 2.4 times faster than the global GDP growth rate, enabling unprecedented global mobility and causing total aircraft GHG emissions to rise (IPCC, 1999). From 1985–2005, total scheduled traffic, measured in terms of tonne-km, grew at an average annual rate of 5.5% (see Table 9.2).

In 2005, over 3940 billion domestic and international passenger-km were logged by the world's scheduled airlines (Figure 9.13). In the same year, nearly 150 billion t-km of domestic and international freight were transported by scheduled airline services (Figure 9.14). It is estimated that in 2005, the world's airlines carried over 2.1 billion passengers and some 40 million tonnes (Mt) of freight on scheduled services. During the same year, airlines performed on scheduled services 3940 billion p-km (equivalent to 365 billion t-km), some 150 billion freight t-km and 4.6 billion mail t-km (ICAO, 2007).

Today, international air traffic represents just over 60% of the total scheduled passenger air traffic and about 85% of global freight air traffic (ICAO, 2010). The demand for international flights has increased almost four times from 1985 to 2005, while domestic demand has only doubled (see Figure 9.15). The lower growth of domestic demand when compared to international demand can be attributed to competition with other modes by land.

The regional share of international air passenger traffic changed in the period 1985–2005. Europe, Asia/Pacific, and the Middle East increased their shares in total air passenger-km while North America, Latin America, and Africa have decreased (see Figure 9.15).

The share by region of international freight traffic changed in the period between 1985 and 2005 (see Figure 9.16). Asia/Pacific and the Middle East increased their shares in total freight-t-km considerably. North America did not vary much. Latin America, Africa, and Europe decreased (see Figure 9.17).

Aviation contributes to climate change in a number of different ways, and in more complex ways than most other sectors (IPCC, 1999). Aviation

Table 9.2 | Trends in total scheduled air traffic. World (1975–2005).

Scheduled Services	Average annual growth (%)		
	1975–1985	1985–1995	1995–2005
Passenger-km	7.0	5.1	5.2
Freight tonne-km	7.5	7.6	5.5
Mail tonne-km	4.3	2.5	-1.9
Total tonne-km	7.1	5.8	5.2

Source: ICAO, 2007.

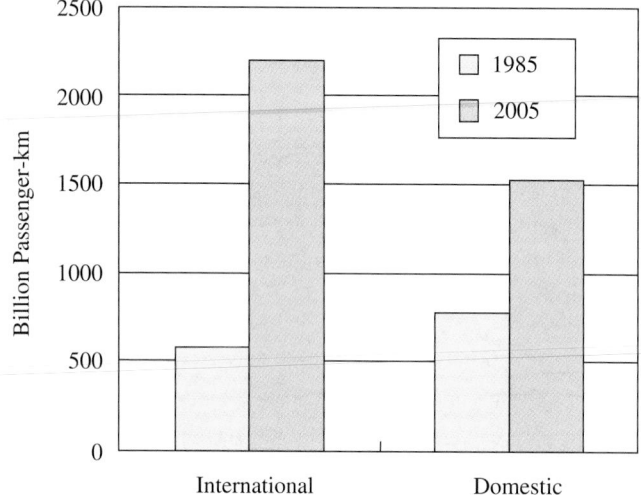

Figure 9.13 | Trends in schedule international and domestic traffic – world (1985–2005). Source: ICAO, 2007. ©ICAO, reproduced with permission.

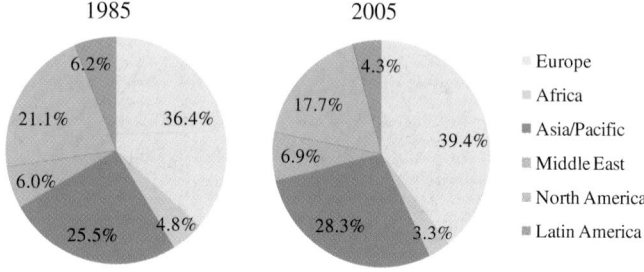

Figure 9.14 | Regional distribution of air passenger transport (p-km). Source: ICAO, 2007. ©ICAO, reproduced with permission.

used 252 Mtoe of energy in 2007, mostly jet fuel, which is around 11% of all transport energy used. International travel accounts for 62% of aviation, and is continuously increasing (IEA, 2009b). The total energy demand is expected to triple to about 750 Mtoe in 2050 in the IEA 2009 baseline scenario (IEA, 2009c). In the same study's high baseline scenario, this reaches nearly 1000 Mtoe (IEA, 2009c). Estimating the impact of aviation's GHG emissions is complicated by a number of uncertainties.

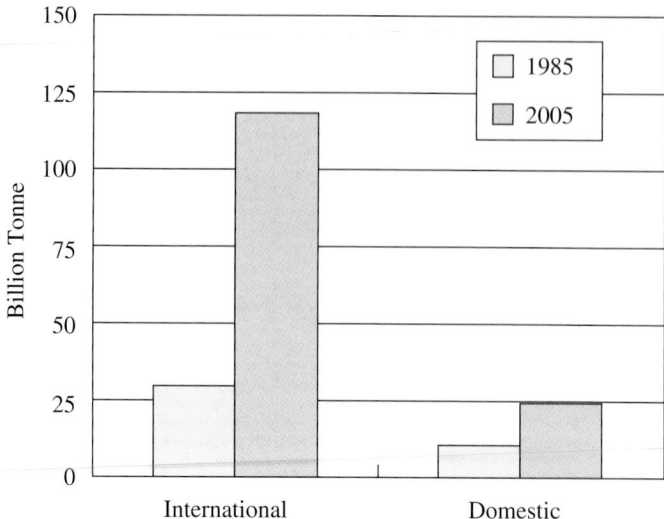

Figure 9.15 | Trends in scheduled international and domestic air freight traffic – world (1985–2005). Source: ICAO, 2007. ©ICAO, reproduced with permission.

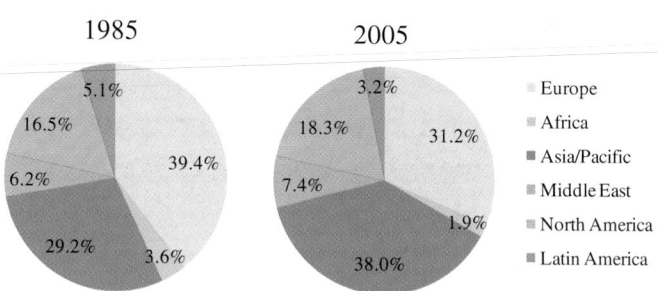

Figure 9.16 | Regional distribution of air transport – freight t-km. Source: ICAO, 2007. ©ICAO, reproduced with permission.

Total aircraft fuel consumption is expected to grow at a rate of 3 and 3.5 %/year, as shown in Figure 9.17. This is far less than the predicted 4.8%/year growth rate in air traffic. During the 37th Session of the Assembly in October 2010, the International Civil Aviation Organization (ICAO) adopted the first globally harmonized Resolution on international aviation and climate change. The resolution was adopted with some States expressing reservations and calling upon the ICAO Council to continue its work on specific aspects of the agreement. The resolution aims to collectively achieve global aspirational goals of improving fuel efficiency by 2% per year and stabilizing CO_2 emissions at 2020 levels. The Assembly also agreed on the guiding principles for market-based measures and decided to explore a global scheme for international aviation. A global CO_2 certification Standard for aircraft is being developed aiming for 2013.

Some measures, such as reducing overall airplane travel volumes, may reduce emissions across the board. But other measures, such as improving aircraft engine efficiencies, may result in trade-offs between different pollutants; for example, a reduction in CO_2 emissions might be

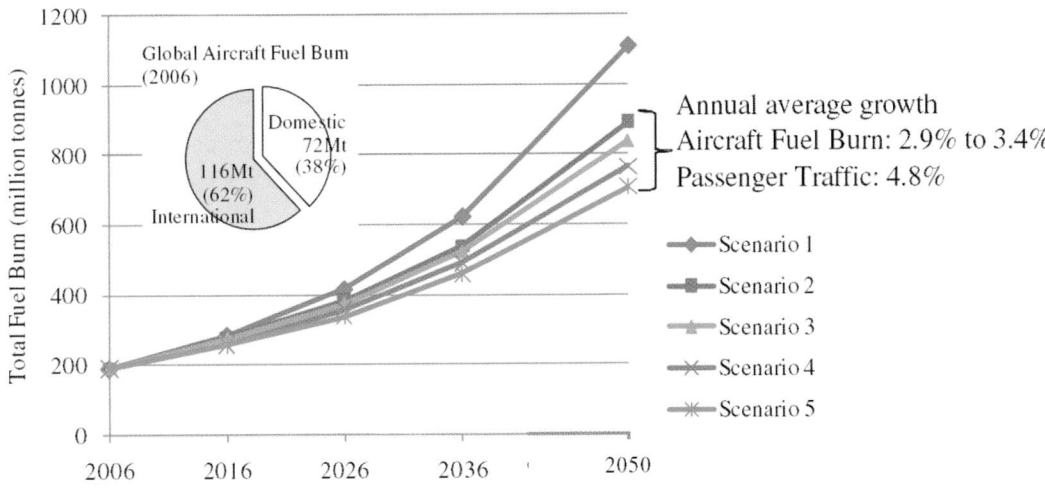

Figure 9.17 | Global aircraft fuel burn, international and domestic traffic combined (2006–2050) (Details of scenario; see Box 9.2). Source: modified from ICAO, 2010.

achieved at the cost of higher NO_x emissions. The levels of scientific understanding of the impacts of each of the contributors to radioactive force from aviation varies, and research is urgently needed to better understand and quantify the potential impacts of aviation emissions and the net effects of various mitigation strategies (IEA, 2009c). Box 9.2 details technology-specific scenarios developed by the ICAO (ICAO, 2010). Policies that also affect total air traffic demand and address air transport externalities are explained in Section 9.6.

9.2.4   Shipping

According to the United Nations Conference on Trade and Development (UNCTAD), more than 80% of global trade by volume is carried by sea (IMO, 2009; UNCTAD, 2010). Throughout the last century, the shipping industry has seen a general trend of increases in total trade volume. Increasing industrialization and the liberalization of national economies have fuelled free trade and a growing

Box 9.2 | The Future of Aviation

The ICAO (2010) estimated average annual passenger traffic growth of between 4.0–5.2 % through 2036. In order to accommodate this growth, the size of the global commercial aircraft fleet would need to grow by nearly 250% by 2036. The complete forecast is presented in the table below.

A range of potential technology and operational scenarios were considered in order to estimate global aviation fuel consumption through 2036, and they were then extrapolated to 2050. Described below as in ICAO (2010), these scenarios include a business-as-usual (BAU) sensitivity case, which was not considered to be likely, given the investments currently being made in operational improvements and the increasingly aggressive levels of operational and technological improvement.

Description of ICAO/CAEP (2010) Future Scenarios

* *Scenario 1 (Sensitivity)*: This scenario includes the operational improvements necessary to maintain current operational efficiency levels, but does not include any technology improvements beyond those available in current (2006) production aircraft.

* *Scenario 2 (Low Aircraft Technology and Moderate Operational Improvement)*: In addition to including improvements associated with migration of the latest operational initiatives, e.g., those planned in NextGen and Single European Sky ATM Research (SESAR) (Scenario 1), this scenario includes fuel burn improvements of 0.96 %/year for all aircraft entering the fleet after 2006 and prior to 2015, and 0.57 %/year for all aircraft entering the fleet beginning in 2015 out to 2036. It also includes additional fleet-wide moderate operational improvements by region.

- *Scenario 3 (Moderate Aircraft Technology and Operational Improvement)*: In addition to including improvements associated with migration of the latest operational initiatives, e.g., those planned in NextGen and SESAR (Scenario 1), this scenario includes fuel burn improvements of 0.96 %/year for all aircraft entering the fleet after 2006 out to 2036, and additional fleet-wide moderate operational improvements by region.

- *Scenario 4 (Advanced Aircraft Technology and Operational Improvement)*: In addition to including improvements associated with migration of the latest operational initiatives, e.g., those planned in NextGen and SESAR (Scenario 1), this scenario includes fuel burn improvements of 1.16 %/year for all aircraft entering the fleet after 2006 out to 2036, and additional fleet-wide advanced operational improvements by region.

- *Scenario 5 (Optimistic Aircraft Technology and Advanced Operational Improvement)*: In addition to including improvements associated with migration of the latest operational initiatives, e.g., those planned in NextGen and SESAR (Scenario 1), this scenario includes an optimistic fuel burn improvement of 1.5 %/year for all aircraft entering the fleet after 2006 out to 2036, and additional fleet-wide advanced operational improvements by region, as shown in Table 9.3 under "upper bound." This scenario goes beyond the improvements based on industry recommendations.

Table 9.3 | CAEP/8 passenger traffic growth rate forecast – Most Likely, High, and Low Scenarios.

Scenario/Sector	Average annual growth rate of passenger traffic (%)				
	2006–2016	2016–2026	2026–2036	2006–2026	2006–2036
High Scenario (Optimistic)					
International	5.9	5.5	5.0	5.7	5.5
Domestic	5.0	4.7	4.4	4.9	4.7
Global [Int. + Domestic]	5.5	5.2	4.8	5.4	5.2
Most Likely Scenario (Central Forecast)					
International	5.4	5.0	4.6	5.2	5.0
Domestic	4.5	4.3	4.1	4.4	4.3
Global [Int. + Domestic]	5.1	4.8	4.4	4.9	4.8
Low Scenario (Pessimistic)					
International	4.8	4.4	4.0	4.6	4.4
Domestic	3.6	3.2	2.8	3.4	3.2
Global [Int. + Domestic]	4.3	4.0	3.6	4.2	4.0

Source: ICAO, 2010.

The results of this analysis are shown in Figure 9.17, which estimates that global fuel consumption will increase from approximately 200 Mt in 2006 to between 711 and 897 Mt in 2050. This translates to between a 2.9 % and 3.4 % annual average growth rate in fuel consumption over the period. On a per flight basis, efficiency is expected to improve over the period; however, in absolute terms, GHG emissions are expected to rise significantly relative to 2006 emissions. Further, market-based policy instruments may possibly be required to achieve a notable contribution from the aviation sector to achieve 2050 climate mitigation goals (see Section 9.6).

demand for consumer products. In 2006, seaborne trade reached over 48 trillion t-km; this represents an increase of 49% compared to trade in 1996, as shown in Figure 9.18 (Singapore Maritime Careers, 2011)

Also according to the Singapore Maritime Careers (2011), the three main types of goods transported by sea are dry bulk, oil, and containerized cargo. Dry bulk (iron ore, grain, coal, bauxite/alumina, phosphate, etc.) accounted for 38% of the world's seaborne trade in 2006. The

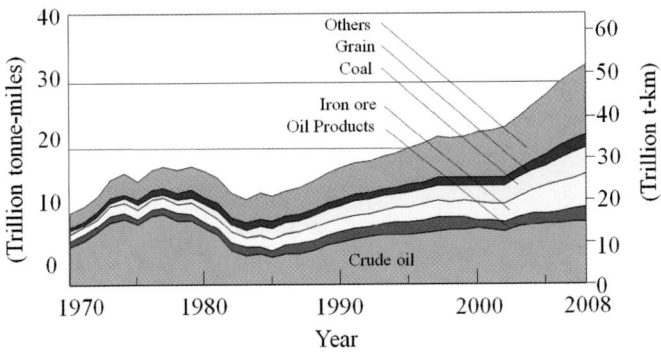

Figure 9.18 | World seaborne trade. Source: based on UNCTAD, 2010.

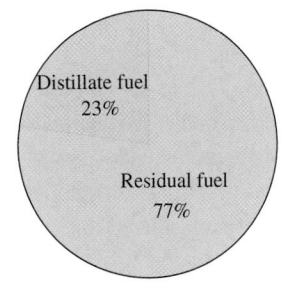 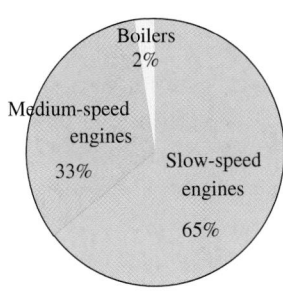

Figure 9.19 | Share of seaborne shipping fuel consumption by fuel type and engines in 2007. Source: data based on IMO, 2009.

second largest share of cargo was oil trade, which accounted for 36%, and containerized cargo contributed 15%. These trades are carried by more than 20,000 merchant ships to various ports around the world, leading to a global cargo throughput of 14.8 Gt and container throughput of 440 million twenty-foot equivalent units (TEUs).[5]

Ships transport food (grains, rice, maize, meat, fish, sugar, vegetables, vegetable oils, etc., and also fertilizers to increase crop productivity), energy (crude oil, refined oil products, ethanol, coal, and gas), raw materials (iron ore, scrap iron, minerals, lumber, wool, rubber, cotton, etc.) and finished products of all sorts. Besides this regular transportation, special ships perform other tasks like offshore service activities, cable laying, pipe laying and dredging, fishing, and exploration and research, among others.

Shipping fuel consumption is mainly residual fuel. Residual fuels are the cheapest fuels with the highest boiling points. They usually contain relatively high amounts of pollutants, in particular sulphur, and they contaminate water and air along coastlines. Estimates for 2007 show total fuel consumption was 333 Mt, as shown in Figure 9.19. The greatest demand for fuel (83%) comes from international routes.

Emissions from fuel in seaborne transport were estimated to amount to 1065 MtCO$_2$-eq in 2007, more than that of rail and aviation combined. However, black carbon emissions are not measured. Also, this figure does not include hydrofluorocarbon emissions from refrigerating systems and fugitive emissions of CH$_4$ from oil transport (IMO, 2009). Advances in technology have increased shipping's efficiency. For example, cargo vessels are estimated to emit 2–30 gCO$_2$/t-km, compared to a range of ca. 200 to 800 gCO$_2$/t-km for trucks with trailers (IMO, 2009; IEA, 2009d). However, efficiency gains are outweighed by significantly higher increases in overall volume.

Shipping activity is driven by the world economy. Understanding this mechanism for seaborne transport and other shipping activities is vital to estimate future energy demand and emissions from this mode. The huge growth rates of world seaborne trade figures were mostly driven by international trade, and the globalization of supply chains. Shipping is not immune to economic downturns – a notable fall in trade occurred, for example, during the worldwide economic recession of the early 1980s, during the Asian financial crisis of the late 1990s and, more recently, the financial crisis of 2008/2009 led to a small reduction in seaborne transport in 2009 that was already more than compensated for in 2010 (estimate by UNCTAD, 2010).

9.3 Transportation Trends and their Relation to Major Global Issues

Transport has a strong two-way interaction with the major global issues addressed by the Global Energy Assessment: urbanization, equity, climate, energy security, health, and environmental protection. This section explores this interaction, stressing that the results of implementation of transportation policies and investments can be favorable or detrimental to the attainment of goals in any of these areas. Therefore, gaining further understanding of these opportunities, challenges and potential co-benefits is necessary. Goals for sustainable transportation consider reducing the need to travel (to have fewer and shorter trips), encouraging a shift to more energy-efficient and safer modes of travel, and improving the technologies and operations of different transport modes (Banister, 2008; Dalkmann and Sakamoto, 2011). The last five years have seen the emergence of climate concerns as a flag point with the reporting by IPCC on the role of transport in CO$_2$ emissions (IPCC, 1999; Kahn Ribeiro et al., 2007; Stern, 2007; Bakker and Huizenga, 2010).

The GEA proposes combining the goals regarding global and local issues and those for sustainable transportation into a multi-goal framework approach that highlights normative goals and creates the opportunity of multiple entry points for decision makers (Chapters 2–6). In this way it envisions the synergy of effective action toward selected goals producing a catalyzing effect of change toward a larger systemic transport and energy transformation. The ability to impact

---

5   TEU is a unit of cargo capacity used to describe the capacity of container ships and container terminals. It is based on the volume of a 20-foot-long (6.1 m) intermodal container.

transportation activity and trends is paramount to achieving goals in any of the major global and energy issues that GEA addresses; this requires the assessment of the purpose and goals of specific transport interventions and their impact on larger goals. The methodological approach followed in this assessment begins with a review of the relationship between current transport trends and their impact on major global issues. Understanding the specific terms of this relationship can help to pinpoint the variables that serve to measure progress toward these goals. Transport sustainability is matched in this assessment with goals linking transport, urbanization, and equity, as well as those that link health and environmental protection (e.g., quality of life, reduced air pollution, noise). Climate mitigation focuses on emissions reduction goals, while energy security relates to fossil fuel independence (fuel diversification and net reduction of fossil fuels use for transportation).

Enabling conditions for change include increasing the level of investment, capacity building, technology transfer, institutional building, and general public awareness and education. The assessment of potential approaches at regional, national, and local levels are comprehensive but different approaches are necessary by region and locality, appropriate pricing policies for fuels, vehicles, and congestion will be needed across the globe and are further discussed in Section 9.6.

9.3.1 Economic Growth and Equity

Transportation systems are the backbone of urban social and economic development and are a determinant of the quality of urban life. As global urban populations are rapidly increasing, land-use planning and technologies and practices for smart urban transport become essential for urban development. Technological advances and motorization have brought major improvements in our ability to cover longer distances, become flexible, and make use of available space.

However, in most cities, urban travel and mobility has deteriorated over time. For instance, the yearly delay in urban areas in the United States with more than three million inhabitants has increased from 31 hours per auto commuter in 1982 to 50 hours in 2010 (Schrank et al., 2010). In most cities in developing countries, rapid urbanization and motorization are coupled with the inability of local governments to respond adequately to the demand for growth in infrastructure and public transport services (Vasconcellos, 2001; World Bank, 2008a). Simultaneously, the rate of car ownership growth supersedes the extension of road capacity. The combined rapid population and motorization growth and slower pace of economic development results in urban spatial impoverishment (see Chapter 18), making the provision of an adequate level of safe, reliable, and efficient transport services almost impossible. Further, as vehicle ownership and road use increase, traffic congestion, noise, and air pollution worsen (Vasconcellos, 2010).

Urbanization and urban form, or the spatial shapes of cities, have a complicated and interwoven bidirectional relationship with transportation (Small and Verhoef, 2007). Large-scale urbanization is only feasible with reduced transport costs (Krugman, 1991). Urbanization produces economics of agglomeration, a precondition to global integration. This generally leads to rising incomes and greater service provision, which, in turn, usually increases rates of ownership of motorized forms of transportation. Urban form, on the other hand, determines transportation needs (Næss, 2006). For example, dense cities and mixed land-use patterns create short distance destinations enabling walking and non-motorized transport options, whereas sprawling cities perpetuate dependence on motorized modes. In turn, highway infrastructure and inexpensive cars and fuels contribute to sprawled, auto-oriented urban development (Newman and Kenworthy, 1999; Banister, 2005). [6]

Accessibility is a useful measure to evaluate the quality of urban transport for users. It can be defined as ease of reaching opportunities throughout the urban area, such as jobs, shopping, medical services, and social activities (Cervero, 2005). Poor accessibility, quality of transportation services, and the question of transport service affordability are strong indicators of the links between transportation, urbanization, and social equity. Improving accessibility levels depends upon the urban form, distribution of land uses, transportation infrastructure, and the transport service level measured in travel time. Accessibility can be improved by proximity or mobility. Proximity is a function of urban form and land-use mix (i.e., dense, mixed-use development with a good jobs-housing balance) and mobility is determined by the level of transportation services, system operations, and vehicle performance characteristics within a particular urban area.

Cities with low auto-mobility could have high levels of accessibility due to the proximity of the urban opportunities in the city core if sustainable transport modes are used (Cherry, 2007). Cities with high auto-mobility might have low levels of accessibility due to the distance of decentralized activities in sprawling suburban communities. Lack of job accessibility, especially for vulnerable populations – those with low income, women, the elderly, and persons with disabilities – has been identified as a major contributor to poverty in developing countries and is an essential consideration in the development of any transportation policy (Gwilliam, 2002; Vasconcellos, 2001; 2010).

The provision of sustainable transport services, including low-carbon vehicles, can be hampered by lack of affordability. Affordability of transport means the ability to purchase access to basic goods and activities (e.g., medical care, basic shopping, school, work, other social services and socializing), usually expressed as the percentage of monthly income devoted to transport by poor families compared to a

---

6   For more information on the urban energy system see Chapter 18.

benchmark considered "affordable" for households (Serebrisky, et al, 2009). According to Litman (2007), this benchmark is achieved when households spend less than 20% of their income on transport and less than 45% on transport and housing combined. However, other authors (Serebrisky, et al, 2009) have observed that transport expenditures may be low in poor households due to a suppression of trips or a prevalence of walking and non-motorized modes; therefore subsidies for public transport, a common policy, may not reach the targeted population. These authors consider that the percentage of income used in transport can be seen as a good approximation of the hardship faced by certain population groups and an indication that deeper analyses are necessary.

A more effective use of resources within cities to aid affordability may be investing to increase accessibility improving sidewalks, crossing bridges and other non-motorized infrastructure (Serebrisky, et al, 2009). Urban planners and decision makers can target improving accessibility through integrated transport and urban design favoring mixed use, short distances, and modal integration, and by planning and implementing safe infrastructure for a number of low-cost, low-tech, non-motorized modes of transport (see further discussion in Section 9.6).

Transportation affordability is also affected by the number of vehicles that a household possess, the costs of using each vehicle, indirect costs such as residential parking, and the quality and costs of using alternative modes such as transit, rail, ridesharing, car-sharing, taxi services, cycling, and walking (Litman, 2007). The goal of introducing more efficient vehicles and low-carbon fuel technologies in developing countries will raise other dimensions to the concept of transport affordability at national and regional levels.

Finally, public acceptability is essential to transport sustainability, as it creates the conditions for implementation and improvement. The concept of acceptability will be further elaborated as part of the enabling conditions for sustainable transport implementation in Section 9.6.8. Improving the balance between transportation-urbanization and equity issues requires policies for enhancing the three A's: accessibility, acceptability, and affordability.

9.3.2   Public Health and Local Ecosystem Impacts

Transportation activity, particularly in cities, has specific impacts on public health and local ecosystems. According to Dora (2011), health risks from transport come from outdoor urban air pollution (1.2 million deaths in 2005), physical inactivity (3.2 million deaths and 19 million healthy life years lost in 2005), traffic injuries (1.3 million deaths in 2005), traffic noise (e.g., stress, memory loss, and analytical impairment), climate change (150,000 deaths), and lack of access to vital goods and services, social networks, equity, and cohesion, which is profound and under-reported. [7]

---

7   For more information on energy and health, see Chapter 4.

Road traffic injuries have become a global health and developmental problem (WHO, 2009). Higher road fatalities and air pollution are common in developing countries, while health problems associated with lack of physical activity are more common in industrialized nations. Over 90% of the world's fatalities on the roads occur in low- and middle-income countries, and almost half of those who die in road traffic crashes are pedestrians, cyclists, or users of motorized two-wheelers. These individuals are collectively known as vulnerable road users, and this proportion is higher in the poorer economies of the world (WHO, 2009). The death rates caused by road traffic accidents have been declining over the last decades in many high-income countries due to the adoption and proper enforcement of traffic laws, as well as persistent allocation of human and financial resources to build up effective and sustainable enforcement activities. However, even in these countries, road traffic injuries remain an important cause of death, injury and disability (WHO, 2009).

Dependence on car mobility and lack of physical activity foster obesity (Woodcock et al., 2009, 2011). The effects of physical inactivity and obesity on population health are more detrimental than those produced by traffic injuries and deaths when comparing the leading causes of death (heart attacks, heart diseases) and Burden of Disease studies by the WHO (WHO, 2009; Dora, 2011). The cost of urban road transport and air pollution on public health can be in the order of 3% of GDP (Creutzig and He, 2009).

Other significant indirect effects on health from transport are related to stress (e.g., noise-induced), which compound the dramatic effects that lack of exercise and car dependence have had in developed countries and result in other potentially debilitating health conditions. Both short- and long-term exposure to air pollution affects human health adversely. Air pollution is caused by emissions from mobile (vehicles) and stationary sources and can have significant local effects during stagnant weather conditions (WHO, 2006; see Box 9.3 and Figure 9.20). Even low concentrations increase the risk of adverse health impacts. The air quality guidelines of the WHO recommend 20 μg/m3 annual mean (indicated in Figure 9.24 by a dashed line), which is exceeded even for most cities in OECD countries. Limit values that are in force in several countries are typically 40–60 μg/m3 (the EU limit value is also indicated in the Figure 9.20 by a dashed line) (WHO, 2006). Air pollutant emissions can be directly linked to the fuel combustion, energy production, and industrial activities taking place in a particular urban environment. However, air pollution is also trans-boundary, as air pollutants can travel considerable distances from their sources.

A sustainable transport agenda needs to understand transportation as a key component of health in urban development. It thus requires considerable further research, particularly in non-OECD countries where most data or reliable statistics on traffic safety, air pollution, and noise are still lacking (WHO, 2006).

Box 9.3 | PM$_{10}$ and PM$_{2.5}$ as Indicators for Suspended Particulate Matter and Typical Values Worldwide

Globally, the most frequently used indicator for suspended particles in the air has been PM$_{10}$ – particles with an aerodynamic diameter <10 µm, but recent reviews by WHO, the USEPA, and European health agencies have pinpointed PM$_{2.5}$ as a better indicator of health risk as it penetrates more deeply into the respiratory tract. **Studies show correlation of health outcomes with either PM$_{10}$, PM$_{2.5}$ or both, although PM$_{2.5}$ better represents combustion sources, which are of the greatest concern in this report.** Here we present data for PM$_{10}$ only; however, since PM$_{2.5}$ is not yet measured routinely in many parts of the world. **PM$_{2.5}$ levels are typically about 10–60% lower than PM$_{10}$ depending on the location and season, however its temporal and geographical variations are often well correlated with PM$_{10}$** as shown for example in the recent global modeling study for the Global Burden of Disease project (Brauer et al., 2011).

An overview of typical annual average PM$_{10}$ concentrations in selected cities around the world is presented in Figure 9.20. The data selected for this presentation demonstrate that the general levels of suspended particles in Asia and Latin America are higher than those in Europe and North America. About 70% of the cities selected from these regions have annual average PM$_{10}$ concentrations above 50 µg/m^3.

In general, the highest concentrations of PM$_{10}$ were reported in Asia. This region also experiences relatively high background concentrations owing to forest fires and local emissions of particles from use of poor-quality fuels. A well-known springtime meteorological phenomenon throughout East Asia is the Asian Dust cloud, which originates from windblown dust from the deserts of Mongolia and China and adds to the general level of PM in the region.

Chinese cities experience very high airborne particle concentrations due to primary particles emitted from coal and biomass combustion and motor vehicle exhaust, as well as secondary sulphates formed by atmospheric chemical reactions from the sulphur dioxide emitted when coal is burned. Typical annual average PM$_{10}$ concentrations were reported to be as high as 140 µg/m3 in Beijing. In many areas of the world, massive and prolonged forest fires have caused significant increases in PM concentrations.

For a discussion of air pollution, including particulates, in the GEA pathways, see chapter 17, Section 17.5.2.

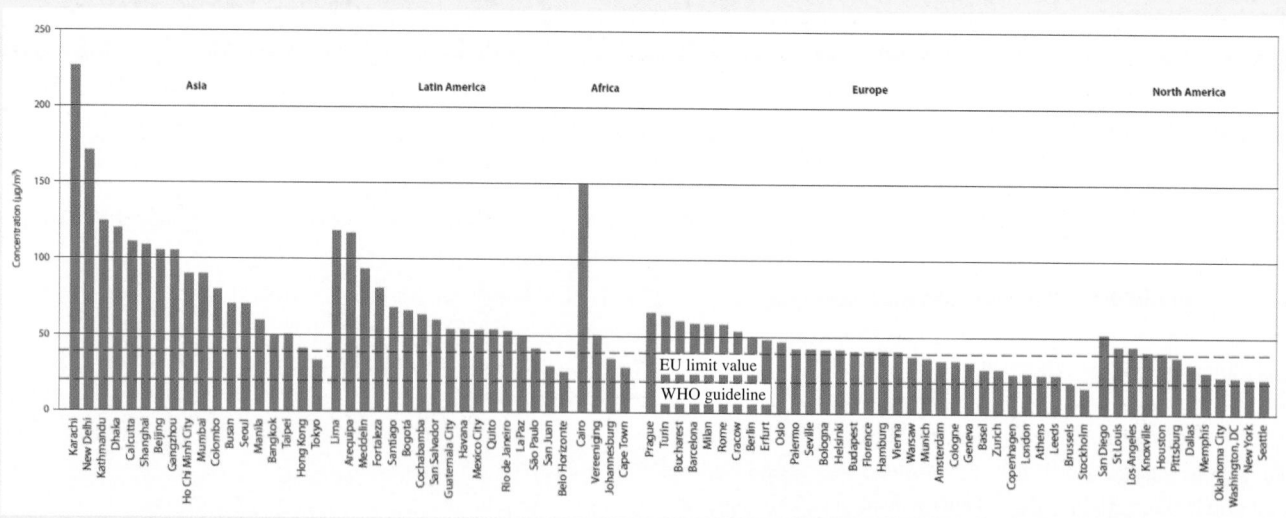

Figure 9.20 | Annual average PM$_{10}$ concentrations observed in selected cities worldwide. Even low concentrations increase chances of adverse health impact. The air quality guidelines of the WHO recommend 20 µg/m^3 annual mean, below levels of even most cities in OECD countries. Source: WHO, 2006.

Source: WHO, 2006; Brauer et al., 2011.

In addition, local and regional pollutants (e.g., acidifying substances, ozone precursors, and particulate matter) damage ecosystems. Releases of oil or chemical substances into the environment by trucks and tankers during routine work or accidents adds to the pollution of soils, rivers, and the sea. The expanse of land area covered by transportation infrastructure excludes it from other uses, cutting through ecosystems and impacting fauna and flora.

9.3.3   Climate Change

The reduction of GHG emissions in all sectors, including transportation, is of paramount importance to reduce the effects of climate change (IPCC, 2007; IEA 2008). While international climate negotiations and treaties (i.e., Kyoto in 1997) were mostly quiet on the transport sector, policy makers at local, national, and regional levels, as well as manufacturers of transport vehicles and systems, have recognized the climate challenge related to transportation.[8]

GHG emissions from transport can be calculated through decomposition of transport demand, energy intensity, and carbon intensity, using a similar approach to the Kaya identity (Creutzig and Edenhofer, 2010; Creutzig et al., 2011a). Transport demand can be divided into activity (trips) and modal share. Energy and carbon intensity are associated with particular modes of travel, as indicated in the ASIF model (Schipper and Marie-Lilliu, 1999).

Of all transport modes, LDVs contribute the highest percentage of GHG emissions (see Figure 9.21), followed by significant contributions by freight trucks, air traffic, and water-borne. Transport activity, and particularly the use of LDVs, generally increases with economic growth; hence the trends indicate increasing emissions if no action is taken. The recent 2008/2009 recession in OECD countries had the effect of reducing the total amount traveled in all transport modes, especially for international shipping; however, overall global transport activity, and hence transport-related emissions, continued increasing.

In terms of GHG intensity, or grams of CO_2 equivalent emitted per unit of freight or passenger transport (tank-to-wheel vehicle emissions during operation), there is a large difference between modes. Rail and shipping are the most efficient modes for both passenger and freight transport (see Figure 9.22).

With respect to CO_2 emissions, IEA estimates that, driven by increases in all modes of travel, but especially in passenger LDVs and aviation, the baseline projection[9] of GHG emissions in transport increases by nearly 50% by 2030 and 80% by 2050 (IEA 2009; see Figure 9.22). In a high

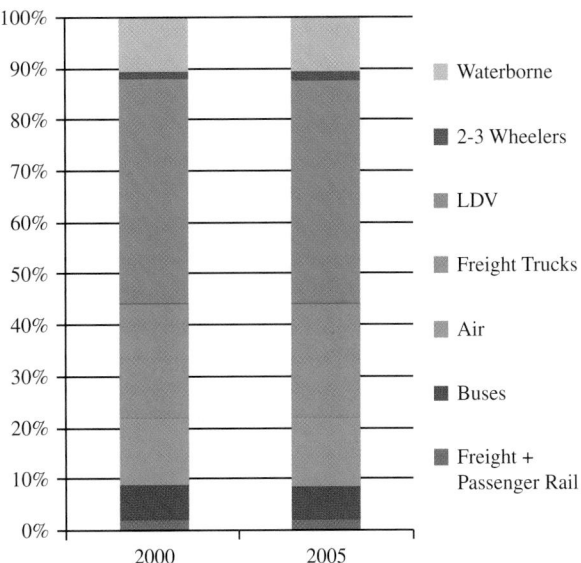

Figure 9.21 | Modal shares in world vehicles in CO_2 emissions (%). Source: based on ITF, 2008.

baseline scenario,[10] it increases by 130%. CO_2-eq emissions increase at even faster rates, due to increased use of high carbon fuels such as coal-to-liquids after 2030. GHG emissions from transport nearly double from about 7.5 Gt in 2006 to about 14 Gt in 2050 in the baseline scenario and to 18 Gt in the high baseline scenario.

Figure 9.23 shows a comparison of CO_2 emissions from various combinations of powertrain and fuel on a well-to-wheels (WTW) basis for LDVs. It is interesting to note that from a climate change mitigation perspective, it may be better to have an internal combustion engine-hybrid electric vehicle (ICE-HEV) than an electric vehicle in areas where the electricity is produced mainly in coal-powered plants. This is an example of why a life-cycle analysis is a crucial approach to inform policymaking.

9.3.4   Energy Security

Transportation demand is expected to grow at a rapid pace for the foreseeable future (IEA, 2009d). However, the way transport needs will be served depends on several factors that are directly related to energy security[11].

First, it is in every nations interest to promote and support transportation activity but as transport is dominated by one feedstock, petroleum, supply disruptions can jeopardize the essential continuity of transport activity (for a further discussion, see Chapters 7 and 17). Second, the primary driver of transport activity is economic

---

8   See also Chapter 3.

9   Baseline: follows the IEA World Energy Outlook 2008 Reference Case to 2030 and then extends it to 2050. It reflects current and expected future trends in the absence of new policies. These values represent expert opinions and are subject to high uncertainties.

10   High baseline: this scenario considers the possibility of higher growth rates in car ownership, aviation, and freight travel over the period to 2050 than occur in the baseline.

11   See also Chapter 5.

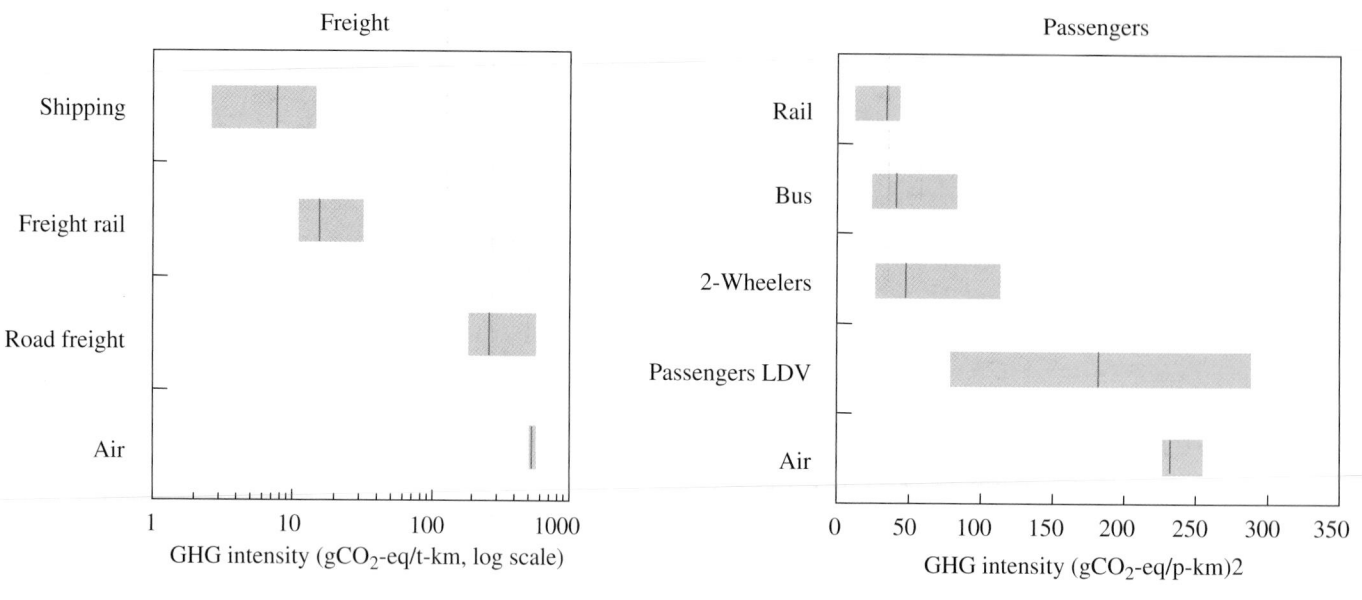

Figure 9.22 | GHG intensity per mode. Source: IEA, 2009c. ©OECD/International Energy Agency 2009.

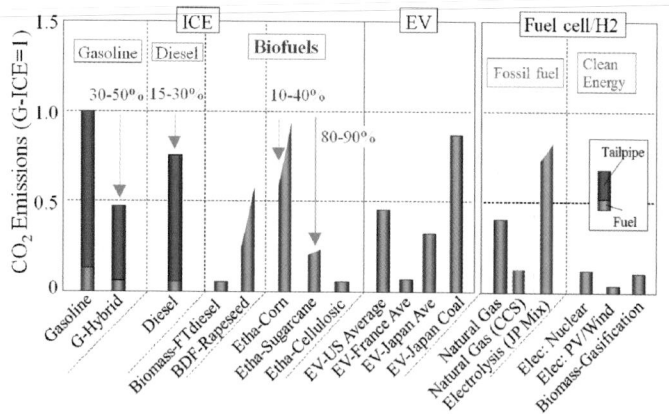

Figure 9.23 | Well-to-Wheel (WTW) CO$_2$ emissions by vehicle engine type. Arrows indicate the reduction compared with the conventional gasoline vehicles, and the range of data is expressed by slant bars. Source: based on data from Toyota/Mizuho, 2004.

In the aggregate, these trends indicate that energy use in transport will continue to increase and that fossil fuels are expected to continue to be the dominant propulsion source. Nevertheless, countries have choices in policy and infrastructure that may affect the key drivers of energy use and propulsion fuels: transport activity, modal choice, and technologies.

Most projections of transportation energy use and GHG emissions have developed reference cases that project future conditions in the case of governments continuing current policies, which mainly favor automobile use powered by fossil fuels. These reference cases establish a baseline that can be used to compare the implications of different policies and programs.

Widely-cited projections of world transportation energy use include reference cases by the International Energy Agency (IEA, 2008; 2009a; 2009c) and the WBCSD (2004b), whose forecast was undertaken by the IEA.

In contrast with the GEA pathways—which does not include a business as usual (BAU) path, the transportation projections from the mentioned scenario studies include business-as-usual forecasts under the assumptions that world oil supplies will be sufficient to accommodate the large projected increases in oil demand, and that world economies will continue to grow without significant disruptions. Attending to these fundamental assumptions, the BAU scenarios of each of the cited studies forecast robust growth in world transportation energy use over the next few decades, at a rate of around 2%/year. If these projections are correct, it will mean that transportation energy use in 2030 will be about 80% higher than in 2002 (see Figure 9.24 for the most recent estimates available). In addition for these studies, most of the new energy use

development, as global economic conditions fluctuate and financial capitals move from regions to regions so will the flows of passengers and goods globally. High economic development in industrialized countries, and also in China, India, Brazil guarantees that transport demand levels will remain high. If the rest of Asia, Latin America and Africa fulfill their economic potential, a fast growth of transport demand may take place over the next several decades. Finally, current trends in developing countries point toward an increasing dependence on private cars. This will require substantial increases in petroleum use even when transportation technology, the energy efficiency of different vehicles and fuels, as well as their cost and desirability, is expected to continue improving.

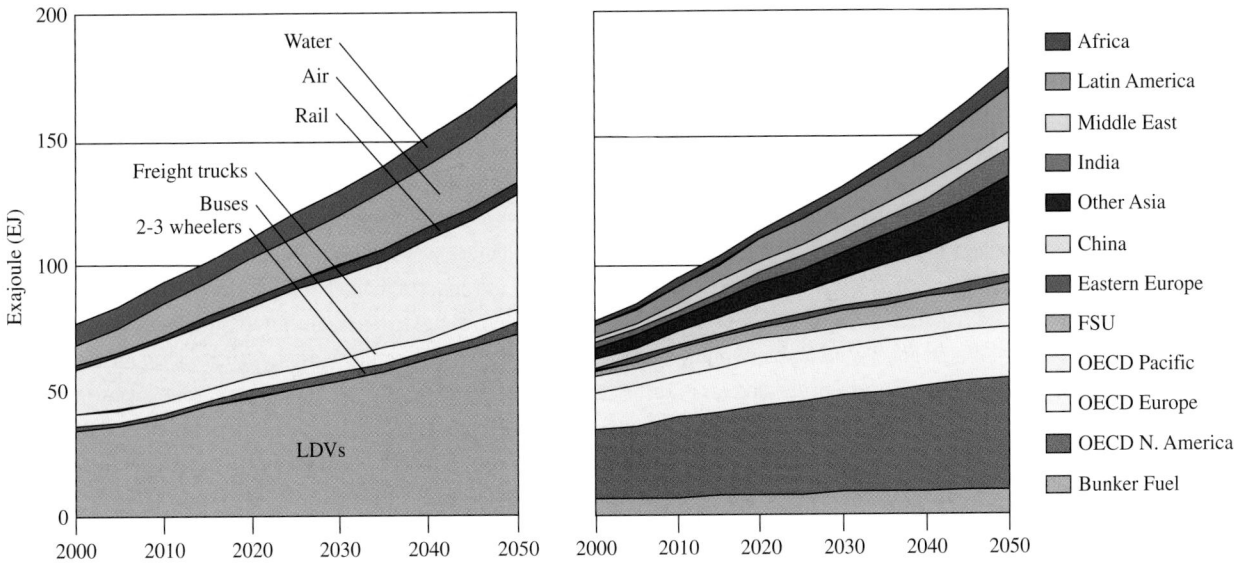

Figure 9.24 | Projection of transport world energy use by mode and region. Source: based on IEA, 2008.

is projected to be in petroleum fuels at 93–95% of transport fuel use over the period (see Figure 9.25). Nevertheless, the BAU scenarios imply increasing prices of oil; and the IEA predicts an average cost above US$130/barrel in 2030 (IEA, 2010a). All these estimates and assumptions are in danger of encountering further difficulties due to disruptive changes in resource supply (Bundeswehr Transformation Center, 2010), which would amplify energy security concerns. The GEA pathways (discussed in Chapter 17) and the GEA-Transport scenarios in this chapter include assumptions and considerations from a global energy security analysis.

With respect to energy security, it is important to point that the rise of consumption in emerging economies will lead to a shift in transport energy use. For example, in China, where the number of cars is growing at a rate of 20%/year, even under a slower projected growth of 6%/year, China's transportation energy use would increase by a factor of four between 2002 and 2025, from 4.3 EJ in 2002 to 16.4 EJ. India's transportation energy use is projected to grow at 4.7%/year during this period, and in countries such as Thailand, Indonesia, Malaysia, and Singapore, growth would be above 3%/year. The same path would be observed in the Middle East, Africa, and Central and South America, with growth rates at or near 3%/year. The current share of non-OECD countries in global transport energy use is 36% and would increase to 46% by 2030 (Kahn Ribeiro et al., 2007).

Developed countries transportation energy use is expected to grow at lower rates of 1.3%/year. This modest figure is attributable to more efficient engines and to the use of alternative fuels as a result of current policies. In North America, growth would be even lower, with a rate of 1%/year due to moderate growth in population and travel but only modest gains in energy efficiency. In Western Europe, transportation energy use would have an increase of 0.6%/year as a consequence of

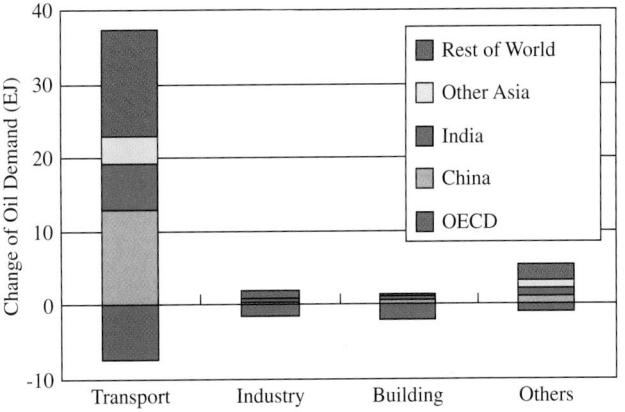

Figure 9.25 | Reference scenario: projected changes in oil demand, 2006–2030. Source: adapted from IEA, 2010a.

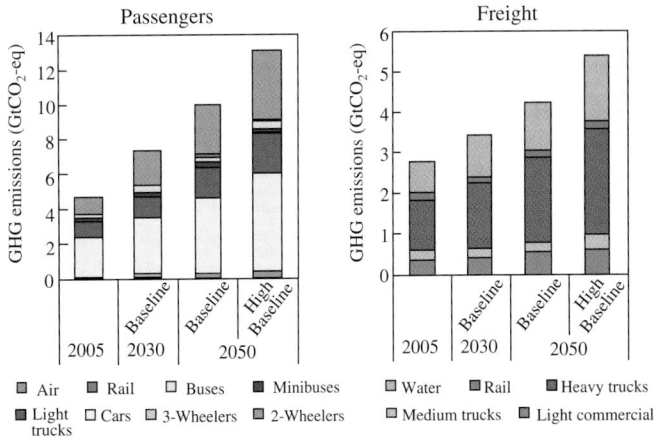

Figure 9.26 | Passenger and freight transport GHG emissions baselines by mode. Source: adapted from IEA, 2009d.

slower population growth, high fuel taxes and improvement in vehicle energy efficiency. Pacific OECD is also expected to grow at a rate of 0.6%/year. In total, developed economies' share of world transportation energy would be reduced from 40% to 31% from 2005 to 2030 (Kahn Ribeiro et al., 2007).

Worldwide expected growth of transport energy use means only that diversification of fuels for transportation is a key component of transport sustainability in terms of energy security. To reduce oil prices, strong policies to reduce demand for fossil fuels are required. IEA (2010a) projects prices in the range of US$110/oil barrel if new policies are in place and US$90/oil barrel in 2035 for policies that achieve the goal of keeping carbon concentration in the atmosphere below 450 parts per million (ppm). In the long term, global transportation systems will need to rely on greatly diversified and non-fossil sources of energy. The provision of a high level of transportation services with less dependence on fossil fuel and progressively more reliance on alternative sources of energy (i.e., electricity produced from renewable or non-fossil sources) is feasible worldwide, in combination with the implementation of avoid-shift-improve policies, as indicated by the United Nations Environment Programme (UNEP, 2011).

9.4 Assessing Energy Resources and Technology Options

The need for energy security, avoiding dangerous climate change, and other sustainability concerns, such as better air quality in cities, requires a combined approach (see UNEP, 2011 for more on the avoid-shift-improve framework). This section concentrates on assessing alternative and cleaner fuels and technologies for sustainable transport. Fuels that currently dominate are gasoline and diesel for road transport, electricity for rail, bunker fuels for shipping, and kerosene for air traffic. All modes could benefit from lower carbon technologies, including electric-powered rail, which needs decarbonization of electricity (as discussed in Chapter 11). As road and air are the dominant modes of transport, and are a major source of transport sector GHG emissions, the focus below is on alternative fuels for these modes.

Alternative fuels, including electricity, hydrogen, biofuels, and natural gas, are expected to gain increasing market shares in the coming decades (IEA, 2009d). An important indicator of fuel environmental quality is provided by global warming potential, i.e., GHG content per unit of energy, as estimated by full lifecycle analysis. Crucially, global warming potential can vary enormously for all fuel categories, depending on upstream production process. Hence, electricity, hydrogen, and biofuels constitute low-carbon fuels only if their respective supply chains are low carbon.

Fuel supply chain lifecycle is also important for fossil fuels. For example, upstream emissions of extraction increase significantly for oil tar sands and deep water resources, now entering the market due to increased oil prices (IEA, 2010a). From this perspective, Arabian oil resources perform better than unconventional fossil fuel resources.

Beyond their environmental footprint, novel fuels may have infrastructure constraints, at least initially. While some fuels, such as biofuels and biomethane, can rely on existing distribution infrastructures, electric, hydrogen and hybrid concept cars need completely new or partially new infrastructures. Conversely, from an energy generation standpoint, electric cars can use the capacity existing in current and planned power plants, whereas biofuels and hydrogen would have to scale up production facilities. Building new facilities and approving infrastructure investments necessary to support the market expansion of specific alternative fuels entail making complex decisions. Alternative fuels may function with a certain degree of complementarity, such as when biomass is used to produce electricity. But the most likely case involves trade-offs between alternative sources. Once a decision to create and develop a market for one source is made, it is likely that the chain of decisions and investments will lead to path dependency. Research and development of conventional fuels and powertrains represent sunken investments, as they build on a century of experience, learning curves, and economics of scale. Barriers to the entry of alternative fuels and technologies that hinder their widespread adoption include huge up-front costs for infrastructure and affordable vehicle technologies. This is particularly true for hydrogen, for which investments in infrastructure are considered to be in the range of US$200–500 billion for its use to be feasible in the United States (Hammerschlag and Mazza, 2005). Another uncertainty regarding future benefits and costs, these investments could constitute another technological lock-in. The issue has a lesser impact for electric cars, as the existing electric grid already provides a crucial part of the infrastructure requirement.

Multiple equilibria, corresponding to different fuels, are possible. Some of them could be far from the global optimum. In fact, several authors indicate that a hydrogen economy could result in a suboptimal equilibrium (Keith and Farrell, 2003; Odgen et al., 2004; Hammerschlag and Mazza, 2005). Taking this into consideration, it is useful to evaluate different fuels based on current knowledge of their respective lifecycle analyses. For more discussion on infrastructures, see Chapter 24.

9.4.1 Alternative Fuels for Road Transportation

9.4.1.1 Natural Gas and Biomethane[12]

Natural gas could rely on the existing internal combustion engine platform of vehicles and existing distribution infrastructures, subject to minor changes. While whole-scale infrastructure investments are not needed, a new pump system for gas fuel is required. Natural gas appears as compressed natural gas (CNG, 200–1500 psi) and liquid natural gas (LNG, liquefied at -160°C). Natural gas and biomethane are methane-based (CH_4) fuels and have about 20% lower carbon content than other fossil-based liquid fuels. As such, they are candidates for lowering the

---

12 See also Chapters 12 and 13

carbon intensity of transportation. Natural gas vehicles emit very low levels of particulate matter (PM) – NO_X, and carbon monoxide (CO) – and can reduce carbon emissions by up to 25% for heavy-duty vehicles (US DOE AFAVDC, 2009).

LNG is more than twice as energy dense as CNG, but has only 60% of the energy density of diesel. Vehicle technology, which uses large compressed gas storage tanks and spark ignition ICEs (internal combustion engines), is particularly mature for heavy-duty applications, which are less constrained by the low energy density of natural gas.

Biomethane is a biofuel refined from biogas, a natural byproduct of organic decay. Major sources of biogas include landfills, manure lagoons, and other agricultural systems. For end-use purpose, biomethane can be treated as natural gas. Fuel-graded biomethane can be directly injected into CNG pipelines or liquefied with LNG and be used in all the same applications. Using biomethane provides the opportunity to reduce GHG emissions, especially if it is sequestered from existing facilities such as landfills, manure lagoons, and rice paddies (unsequestered biomethane has 20 times the global warming potential of CO_2). Methane comprises over 10% of all GHG emissions for Annex I countries in terms of CO_2 equivalence, and around 26% for non-Annex I countries (IPCC, 2007). Harnessing and utilizing biogas is a key strategy for reducing these emissions.

9.4.1.2   Biofuels

Biofuels have been seen as a promising way forward to energy security and climate change mitigation. In some world regions, they are also regarded as a new form of revenue for farmers. However, market-mediated deforestation – and other so-called indirect land-use effects – can negate any GHG abatement and add pressure on global food security and biodiversity. The two major biofuel-producing countries are the United States and Brazil; 46% of global production occurs in the United States, 42% in Brazil, 4% in Europe, and 8% in the rest of the world (World Bank, 2008b). The European Union and the United States have ambitious relative or absolute targets for biofuel market shares by 2020 (see Section 9.6). Key questions for biofuels are their climate mitigation potential, and their (indirect) effects on competitive land uses, such as food production and ecosystem functioning (see also Chapter 20).

Biofuels were originally regarded as zero- or low-carbon fuels. This is based on the fact that biofuels release only CO_2 upon combustion that plants previously absorbed from the atmosphere via photosynthesis. However, GHG gases are also emitted during cultivation (by fertilizer and nitrate application) and harvest of the feedstocks, as well as subsequent processing, refining, distribution, or the various transportation requirements throughout the production stages. Moreover, direct and indirect land-use changes associated with growing the feedstock often change the carbon stored above or below ground.

The lifecycle GHG emissions of biofuels have been studied comprehensively, with emphasis usually placed on bioethanol (Quirin et al., 2004; Niven, 2005; Farrell et al., 2006; von Blottnitz and Curran, 2007; see also Chapter 11). The earlier literature suggests that biofuel systems show moderate to strong fossil fuel substitution potential and GHG savings when compared with conventional petroleum-based fuels. GHG savings vary for different biofuel systems, based on differences in growing methods, climate, and feedstock characteristics. Tropical sugar crops, in particular ethanol from Brazilian sugar cane, were shown to be the most productive feedstocks (e.g., Goldemberg, 2009) – a result that is disputed in the more recent literature (Lapola et al., 2010). Second generation biofuels from lignocelluloses are usually suggested to be the most abundant and potentially most sustainable feedstocks (von Blottnitz and Curran, 2007). A number of studies find GHG performance is worse for current biofuels than for fossil fuel systems (Patzek, 2004; Patzek and Pimentel, 2005; Hertel et al., 2010). Some variation is due to model assumptions regarding the system boundaries, the way co-products are accounted for, and the differences in input parameters and secondary data sources used (Farrell et al., 2006). More recently, three crucial qualifications of GHG benefits of biofuels have pointed out that:

- CO_2 emissions from direct and indirect land-use change are often neglected, tend to worsen GHG benefits, and add considerable uncertainties over the net GHG benefits of biofuels (e.g., Searchinger et al., 2008; Creutzig and Kammen, 2009a; Plevin et al., 2010).

- If land-use change is involved, emissions are better accounted for by temporally explicit lifecycle analysis that addresses upfront land-use emissions, long-term GHG savings from biofuels, and potential soil sequestration (O'Hare et al., 2009).

- Nitrous oxide emissions are estimated simplistically in most studies and have high uncertainties, which can have strong negative effects on the GHG emission balance of biofuels. For example, the IPCC default emission factor for nitrous oxides related to biofuels may underestimate emissions by a factor of 3–5 (Crutzen et al., 2008).

Uncertainties are a major issue, in particular with respect to indirect land-use emissions and nitrous oxide emissions. For example, US corn ethanol reduces GHG emissions per kilometer driven by 20% compared to conventional gasoline, as long as land-use change-related emissions are excluded from the calculations. However, once land-use emissions are accounted for, corn ethanol has twice as high lifecycle emissions as gasoline accounted over a 30-year production period (Searchinger et al., 2008). Within a 95% confidence margin, indirect land-use charge emissions from US corn ethanol expansion range from 21–142 gCO_2-eq/ MJ, i.e., from small, but not negligible, to several times greater than the lifecycle emissions of gasoline (Plevin et al., 2010).

In Europe, there is mainly permanent grassland that could be diverted to biofuel feedstock production. If this conversion takes place, it may take 20 to 110 years (+/- 50%) to recover the soil carbon through GHG emission savings from using the biofuel grown on the same land (de Santi et al., 2008). The same detailed bottom-up study reveals that emissions from nitrous oxides may vary by a factor of 100 depending on the site-specific conditions, notably soil organic content.

A summary of estimates for a variety of different biofuels is given in Figure 9.27. Note that these estimates are subject to major uncertainties (e.g., Plevin et al., 2010). In summary, indirect land-use change-related emissions can worsen the effect of biofuels significantly. For further discussion of the uncertainty of direct and indirect land-use emissions, see Chapter 11.

The performance of biofuels from different feedstock is very mixed across other environmental impact categories (von Blottnitz and Curran, 2007). Biofuel impact is often more harmful compared to conventional fossil fuels (Quirin et al., 2004). A comprehensive LCA of a broad variety of biofuel feedstocks suggests a trade-off between fewer GHG emissions and higher environmental disadvantages of different kinds compared to conventional petroleum-based fuels (Zah et al., 2007). Biofuels might be a greater overall public health risk than conventional petroleum-based fuels due to ozone effects. For a more comprehensive assessment on issues such as water use and biofuel production, see Chapter 20.

Crucially, agricultural crop production for biofuels competes with food production. The current production of biofuels based on conventional crops – first generation biofuels – has already put pressure on global food prices. While the World Bank estimates that biofuel production has been responsible for roughly 75% of the increase in food prices between 2002 and 2008, the overall literature is ambiguous on the magnitude of the biofuel-induced price hike (see Chapter 20 for a complete discussion). Food price hikes and induced food insecurity have severe consequences for the livelihoods of people in poor countries. The Food and Agricultural Organization of the United Nations (FAO, 2008) suggests that an additional 75 million people were pushed into undernourishment during the global food price crisis in 2007, bringing the total to nearly one billion hungry people in the world. Achieving the aggressive biofuel targets in the United States and Europe could potentially have much larger impacts, which justifies the need for questioning current biofuels strategies.

Second and third generation biofuels have been put forward as biofuels with low lifecycle emissions. Second generation biofuels are derived from non-food crops, and important variants such as cellulosic bioethanol are suggested to have very low lifecycle GHG emissions (Farrell et al., 2006). Third generation biofuels are produced from algae and have much lower land demand. However, more recent studies question the economic viability of second generation biofuels and point out that these biofuels compete with food production, too (Sanderson, 2009). For further discussion, see Chapter 11, Section 11.2.

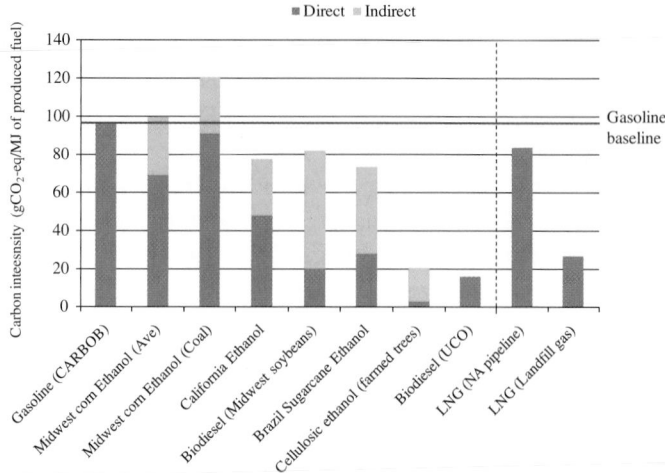

Figure 9.27 | Carbon intensity of different biofuels as compared to gasoline. UCO refers to used cooking oil which is generally assumed to have no indirect effects, and hence, has relatively low carbon intensity. Source: based on CARB, 2009; Creutzig et al., 2011a.

In summary, with a few exceptions[13] biofuels currently available in the market place have questionable impact on GHG emissions, food security, and other land-related issues, such as deforestation and biodiversity. Large uncertainty on indirect land-use changes and third level impact requires further research. Meanwhile, further large-scale biofuel production must be considered with caution. The accumulation of niche-based solutions, however, may be promising (Tilman et al., 2009). Chapter 11 and 20 suggest that the deployment of advanced biofuels with low GHG footprint is possible and can be recommended if managed in the strict context of sustainable development and if food security remains uncompromised.

9.4.1.3   Electricity and Hydrogen

Electricity and hydrogen as fuels for road transport both represent intermediate energy storage mediums that need to be produced by other sources of energy (e.g., coal, nuclear, or wind in the case of electricity, and natural gas or electricity itself in the case of hydrogen). Hence, they are discussed together. First, common features are presented and both fuels are compared. Then specific issues for each fuel are raised.

The limiting factor for battery electric cars is the total energy that can be delivered and the limited driving range, whereas the constraint for hydrogen fuel cell vehicles is the amount of power they can produce and whether this will be sufficient to meet vehicle acceleration requirements (Eaves and Eaves, 2004). Battery storage has higher WTW efficiency (75–85%) than hydrogen (30–50%) and lesser known variants such as compressed-air storage (<30%) (Schaber et al., 2004; Eaves and Eaves, 2004; Creutzig et al., 2009), translating into the more efficient

---

13   Such as Brazil Sugarcane Ethanol, cellulosic ethanol, Biodiesel from used cooked oil (UCO), and advanced biofuels.

economic and environmental performance of electric cars than hydrogen cars (Hammerschlag and Mazza, 2005). For electric cars, costs scale according to battery capacity. As most trips would be short distances, plug-in hybrid electric vehicles (PHEVs) provide the best cost-benefit ratio in the medium term (APS, 2008). Research and development (R&D) also needs to comprise demand-side technologies, as the current electric grid capacity is sufficient to fuel 20–50% of the European Union or the United States' fleets when demand-side technologies and policies are deployed (Denholm and Short, 2006; Kintner-Meyer et al., 2007; EC DGJRC, 2009).

Electricity

For electric mobility, electricity production and the required vehicle technology and recharging infrastructure need to be assessed. Since electricity generation and its distribution infrastructure are already widely used in most countries, there is no significant barrier to its use for road transport. Efficiency is discussed in Chapter 9, Section 9.3.2. Hence, the focus here is on electricity production. The key factor determining the lifecycle emissions of electric vehicles (EV) is the carbon intensity of the electricity generation.

Renewably produced electricity comprised 18% of global electrical consumption in 2007 and is projected to rise further (EIA, 2010, see also Chapter 11). The difference is mainly due to uncertainty in ambition of global climate change mitigation policies. In 2008, renewables comprised over 50% of all newly added electrical capacity in the United States and Europe combined, or 276.7 GW (Martinot and Sawin, 2009). With the current electricity mix in countries such as the United States and Germany, electric cars can have about the same global warming impact as conventional cars with otherwise the same characteristics (e.g., Creutzig et al., 2009). With an increasing share of renewable energies, the global warming impact of electrics is expected to decrease.

EVs do not necessarily reduce net emissions, but they remove them from inner cities where air pollution has the highest public health impact (i.e., intake fraction is highest). Hence, an electrification of current car fleets would produce public health advantages in most cities.

Electric mobility requires a new infrastructure. Electric vehicles face barriers to market entry, mostly related to range anxiety. The challenges that remain for all-electric vehicles (e.g., BEVs, PHEVs) are technological and logistical.

First, the requirements of fueling infrastructure and capacity of the electric grid to recharge EV batteries are important considerations. The fueling infrastructure for EVs is less costly to deploy than for hydrogen, requiring only recharging stations with conventional outlets connected to the existing electrical grid, which are estimated to cost around US$2000 per commercial charging station (Morrow et al., 2008). Another option would be to have fueling stations remove depleted batteries from vehicles and provide fully charged batteries to drivers, thereby helping them avoid lengthy recharging times. A study in the United States showed there is sufficient off-peak electrical capacity to charge 185 million EVs overnight, or 75% of the current US passenger vehicle fleet (Schneider et al., 2008). Policies and smart infrastructures are required to make use of off-peak capacity. While using off-peak electricity will result in overall increased grid efficiencies, it will still result in greater wear and tear of the grid, especially for transformers, resulting in higher maintenance costs (Blumsack et al., 2008).

Second, the batteries themselves are problematic components of EVs. The weight, size, cost, and lifetime of batteries must be optimized to improve the marketability and all-electric range (AER) of EVs. Larger batteries provide longer AERs and weigh from 23 kg (50 lbs) for limited-AER PHEVs to 450 kg (1000 lbs) for highway capable 320 km (200mile) AER BEVs (Shiau et al., 2009). Given the trade-off between battery weight and vehicle fuel efficiency, researchers are trying to reduce battery weight while maintaining desired AER. Most of today's EV models use nickel-metal hydride (NiMH) batteries, but the next generation will use Li-ion batteries, which are more compact and lightweight but also more expensive.

Third, the incremental cost of EVs compared to gasoline ICEs and Gs is significant. In the long term, PHEVs with an AER of 64 km (40 miles) are estimated to be at least US$11,000 more expensive than gasoline ICE vehicles (Simpson, 2006). The battery is the most expensive component of EVs, with a cost of US$30,000 or more for BEVs (US BTS 2001). Lower battery costs and higher petroleum costs will make them more competitive, along with other government incentives such as tax rebates and subsidies (Simpson, 2006). When taking into account the fuel cost savings, PHEVs could be up to US$11,000 cheaper than conventional ICE vehicles over their lifetime. However, the high sticker price for highway capable BEVs, as much as US$100,000 for some current models, requires substantial reduction before they can be mass-marketed (Tesla Motors, 2009).

Hydrogen

Hydrogen has been used in ICEs of cars, trains, ships, and other applications prior to the 1930s. Hydrogen gas burns cleanly in an ICE with minimal PM, NO_x, and CO_2 emissions. As a second possible technology, hydrogen-fuel cell vehicles emit only water as a byproduct.

Using hydrogen as a transportation fuel enables decentralized fuel production and refueling. Similar to electric vehicles, hydrogen vehicles allow for the displacing of fossil fuels if the hydrogen is produced from renewable energy. Hydrogen vehicles can also act as renewable energy storage devices. As an advantage over electric vehicles, hydrogen has higher energy density, and hence long-range driving is possible. However, WTW efficiency is lower than for electric cars. If not produced by electrolysis where the electricity is provided by renewable energies, hydrogen cars tend to have higher GHG emissions than electric cars.

The largest barrier to hydrogen vehicle use is the present economic infeasibility of mass-marketing the technology. Fuel cells, hydrogen

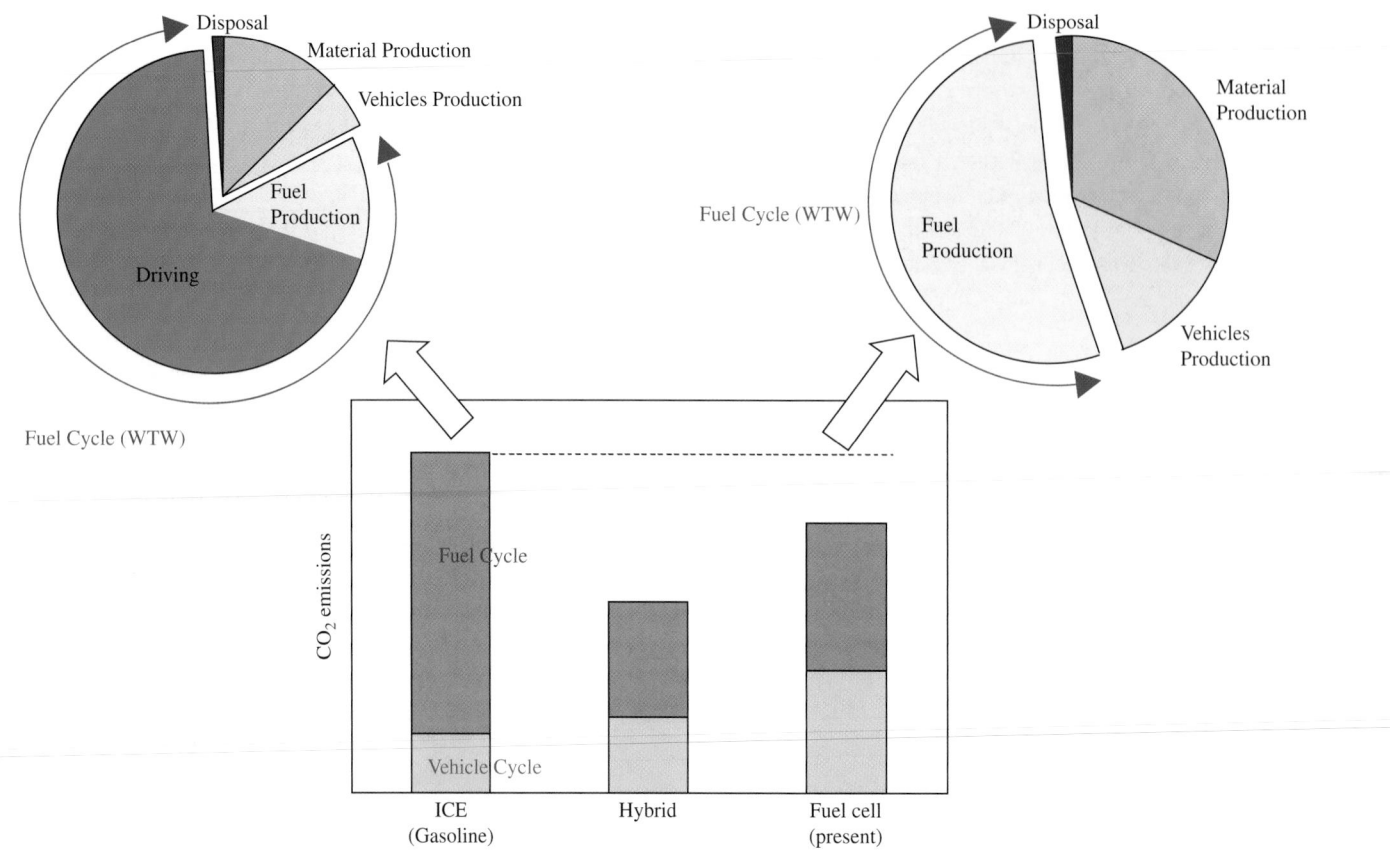

Figure 9.28 | LCA for passenger vehicles. (Pie charts show the details of CO$_2$ emissions from both fuel and vehicle cycles. Pie chart for HEV is not shown, but general features are between ICE and FCV.) Source: based on Toyota, 2004.

storage devices, fueling infrastructure, and hydrogen production are all currently very costly. Overcoming these kinds of economic and logistical barriers will require substantial political will and public funding into the foreseeable future, another impediment in its own right. However, should hydrogen transportation become economically viable, it provides an opportunity for significantly reducing GHG emissions.

9.4.2   Improving the Efficiency of Road Vehicles

Fuel use is a function of vehicle use (km/yr) and fuel economy (fuel/km). This section addresses potential for improvement for fuel economy.

9.4.2.1   Lifecycle Analysis of Motorized Vehicles

For automobiles, lifecycle analysis can be divided into the two parts; fuel cycle (extraction of crude oil, fuel processing, fuel distribution, and fuel use during vehicle operation) and vehicle cycle (material production, vehicle manufacturing, vehicle distribution, and disposal at the end of life). For a typical ICE vehicle, 70–90% of energy use and GHG emissions take place during the fuel cycle, depending on vehicle efficiency, driving

mode, and lifetime driving distance, as shown in Figure 9.28 (Toyota, 2004).[14] This indicates that the most effective CO$_2$ reduction measures would target the fuel cycle, making vehicle fuel economy an important aspect of vehicle technologies.

Although the vehicle cycle contributes 10–30% to the overall emissions in conventional cars, fuel cell and hybrid cars have higher levels of vehicle cycle emissions because more energy is needed to make battery, fuel cell stack, and electronic parts, such as motors and power control units (Toyota, 2004).

Recent research has summarized results of fuel cycle analyses (e.g., IPCC, 2007; CARB 2009; Creutzig et al., 2011a), especially in relation to biofuel and hydrogen production. Regional studies of the WTW CO$_2$ emissions of conventional and alternative fuels and vehicle propulsion concepts include a General Motors/Argonne National Laboratory (GM/ANL, 2005) analysis for North America, EU-CAR/CONCAWE/JRC (2006) for Europe, and Toyota/

---

14   For other transportation modes such as train and ship, very similar pictures can be observed where the CO$_2$ emission is dominant during operation, as shown in Figure 9.33 (Aihara and Tsujimura, 2002; Kameyama et al., 2005). CO$_2$ emissions during operation are very sensitive to total length, frequency of travel and other operational conditions.

Mizuho (2004) for Japan. These LCAs display small but significant differences for ICE-gasoline and ICE-D (diesel), reflecting the difference in oil producing regions and regional differences in gasoline and diesel fuel requirements and processing equipment in refineries (Khan Ribeiro, et al., 2007).

The WTW CO_2 emissions shown in Figure 9.30 are for the following four groups of vehicle/fuel combinations; ICE/fossil fuel, ICE/CNG, EV, FCVs and a number of hybrids. The full WTW CO_2 emissions depend on not only the drive train efficiency (tank-to-wheel or TTW) but also the emissions during fuel processing (well-to-tank, or WTT). ICE-D has 16–24% lower emissions due to the high efficiency of the diesel engine. The results

for hybrids vary among the analyses due to different assumptions of vehicle efficiency and different driving cycles. Toyota's analysis of the Prius, using a Japanese 10–15 driving cycle, demonstrates a potential for CO_2 reductions of 30–50% (Khan Ribeiro, et al, 2007).

The lifecycle emissions of ICE vehicles using biofuels and fuel cell vehicles are extremely dependent on the fuel pathways. For ICE-Biofuel, the CO_2 reduction potential is often assumed to be very large (30–90%) (e.g., IPCC, 2007). As the discussion in Section 9.3.1 demonstrated, recent research estimates that current biofuels possibly have a higher global warming potential than gasoline, mostly due to taking indirect land use change into account (see Figure 9.23).

The potential of EV is strongly dependent on the CO_2 emission factor of power generation. In regions where coal-fired plants dominate, switching to EV increases CO_2 emissions. The GHG reduction potential for natural gas-sourced hydrogen FCV is moderate, but lifecycle emissions could be dramatically reduced by using CCS (carbon capture and storage) technology during H_2 production. Using renewable energy such as C-neutral biomass as a feedstock or electricity from renewables as an energy source also will yield very low emissions (Khan Ribeiro, et al, 2007).

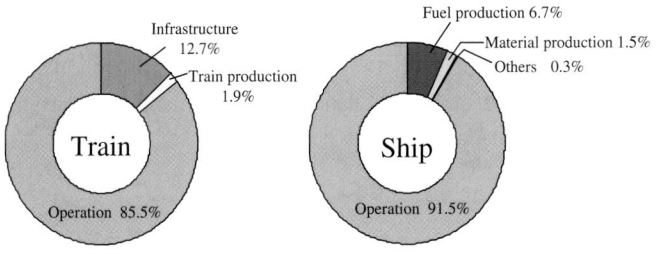

Figure 9.29 | LCA for trains and ships. Source: based on Aihara and Tsujimura, 2002; Kameyama et al., 2005.

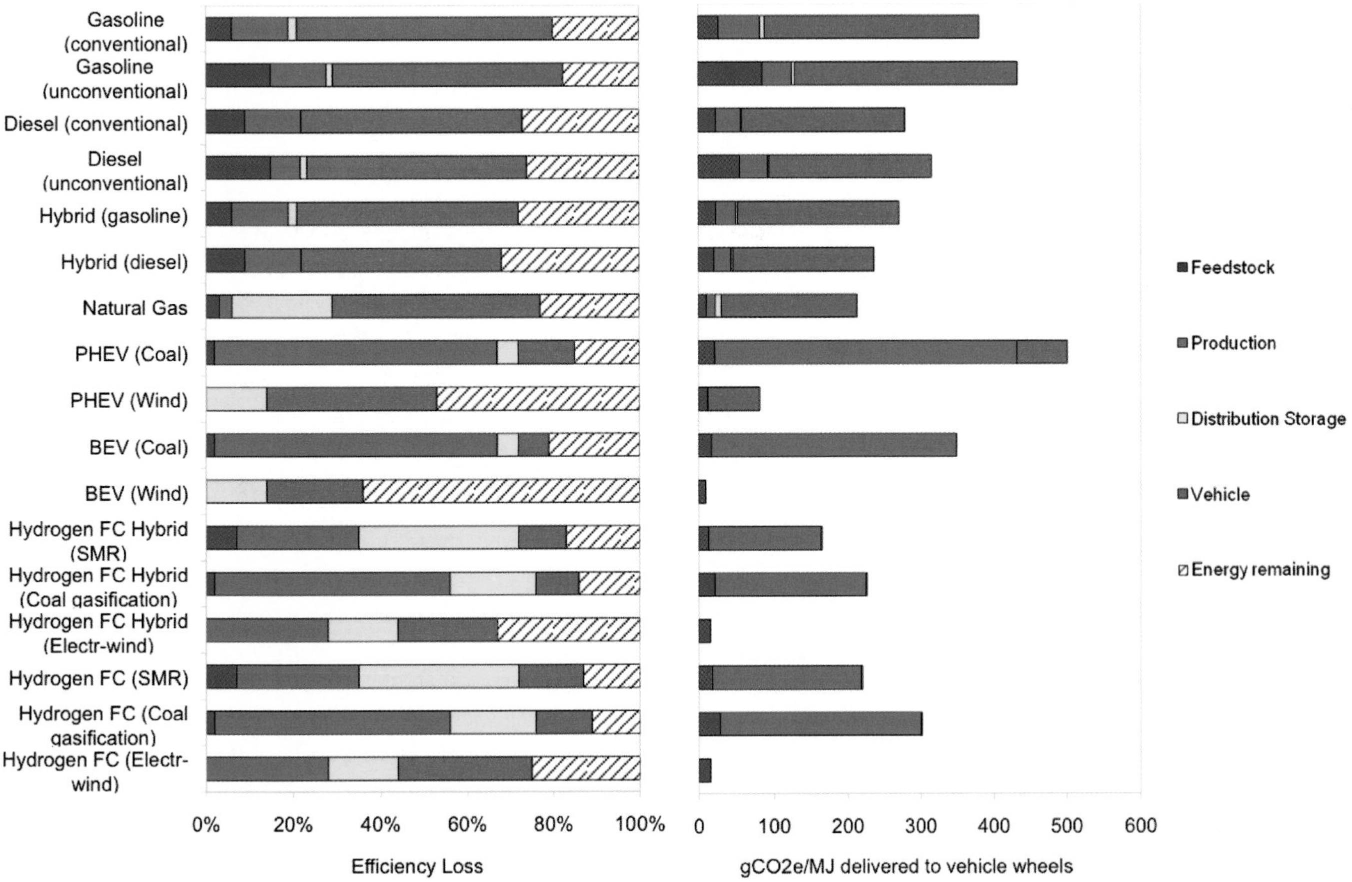

Figure 9.30 | Relative amount of energy required for different supply chains (the shaded area corresponds to the energy remaining at the wheel) and well-to-wheel GHG emission for various powertrain/fuel combinations. Source: based on Creutzig et al., 2011a.

9.4.2.2   Fuel Economy: Minimizing Energy Losses

There are significant opportunities to improve energy efficiency of road transport. Only about 20% of the energy from the fuel put into passenger vehicles gets used to operate the vehicle. The rest of the energy is lost in the engine and drive train and during idling (OTA, 1995; US DOE, 2010) as seen in Figure 9.31.

There is great potential to improve vehicle energy efficiency through improvements in vehicle technology. Fuel use of road transportation can be reduced by increasing the conversion efficiency of the fuel energy to work by improving drivetrain efficiency and recapturing energy losses; and reducing vehicle loads such as weight, rolling, air resistance, and accessory loads, thus reducing the work needed to operate it.

Engine losses

In gasoline-powered vehicles, more than 60% of the fuel energy is lost in the engine due to engine friction, pumping air into and out of the engine, and wasted heat. These losses can be reduced by advanced engine technologies such as variable valve timing and lift, turbocharging, direct fuel injection, and cylinder deactivation. Also, diesel engines have about 15–25% higher efficiency than gasoline engines. As will be discussed later, fuel cell vehicles and electric vehicles are future options for substantial improvement in powertrain efficiency.

Idling losses

During urban driving, idling at stop lights or in traffic causes significant energy loss. These losses can be reduced by automatically turning the engine off when the vehicle comes to a stop and restarting it instantaneously when the accelerator is pressed. This is a standard technology in current hybrid vehicles.

Drivetrain losses

Energy is also lost in the drivetrain and it is improved by technologies such as automated manual transmission and continuously variable transmission.

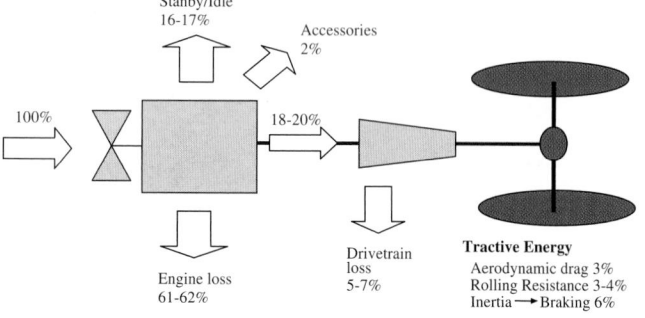

Figure 9.31 | Energy use in a typical gasoline passenger vehicle (mid-size, urban driving mode). Source: based on Kobayashi et al., 2009.

9.4.2.3   Fuel Economy: Reducing Vehicle Loads and Weight

Rolling resistance

Rolling resistance is the resistive forces between the tires and the road, and is directly proportional to weight. It amounts to up to about 30% of total tractive force. It can be reduced by tire design (tread and shoulder designs) and tire materials, and redesigning wheel bearings and seals. For passenger cars, a 5–7% reduction in rolling resistance leads to a 1% improvement of fuel efficiency. However, there are trade-off such as traction, durability, and noise.

Aerodynamic drag

About 30% of total tractive forces is lost by aerodynamic drag, and it become higher during fast highway driving, since drag is directly proportional to the square of speed. Drag forces are reduced by reducing the frontal area of the vehicle, smoothing out body surfaces and adjusting the body's basic shape and so on.

Accessories

Accessories such as air conditioning, power steering, windshield wipers, and audio systems use energy generated from the engine. This becomes substantial in battery-powered vehicles during hot or cold weather.

Braking losses

Upon braking, kinetic energy of vehicles is lost. The regenerative braking system can capture the energy as electricity or a kinetic force.

Vehicle weight

Energy to overcome the vehicle's inertia is directly related to its weight. The less a vehicle weighs, the less energy required to start and drive it. Weight can be reduced by lightweight materials and vehicle design. Lightweight materials in BEVs can increase the driving range – one of the crucial constraints of electric vehicles. For example, carbon fiber materials will be used for BMW's 2013 Mega City Vehicle, but with a penalty of cost.

9.4.2.4   Vehicle Efficiency through Hybrid Technology and Alternative Fuel Engines

Hybrid drivetrains

Typical hybrid-electric drive trains is a combination of a fuel-driven power source, such as a conventional internal combustion engine with an electric drivetrain – electric motor/generator and battery (or ultracapacitor) – in various combinations. In most current hybrids, the battery is recharged only by regenerative braking and engine charging, without external charging system from the grid. Since the 1997 introduction of the Toyota Prius hybrid in the Japanese market, hybrid electric drivetrain technology has advanced substantially. The markets for hybrid vehicles have expanded, presently providing a variety of improved systems in alternative forms and different combinations of costs and benefits.

Hybrid systems now cover a range of technologies from simple belt-drive alternator-starter systems, offering perhaps 7–8% fuel economy benefit under US driving conditions, to "full hybrids" such as the Prius, offering up to perhaps 40–50% fuel economy benefits (IPCC, 2007). Hybrids recover the kinetic energy by regenerative braking system and also save energy by idle-stop, therefore they can improve fuel efficiency in congested urban driving conditions. This might be particularly useful for taxis and vehicles dedicated to urban use as well as for some heavier-duty applications, including urban buses and urban delivery vehicles (IPCC, 2007). FedEx has claimed a 57% fuel economy improvement for its E700 diesel hybrid delivery vehicles (FedEx, 2004). Hybrid sales have expanded rapidly. In the United States, sales were about 7,800 in 2000 but rose to 350,000 in 2007. Worldwide cumulative Prius sales alone hit 500,000 by the end of April 2006, 1,000,000 by the end of April 2008, and 2,000,000 by the end of September 2010.

Plug-in hybrids, or PHEVs, are conventional hybrids with a capability of charging from the grid, and can be seen as a merging of hybrid electric and battery electric vehicles. PHEVs get some of their energy from the electricity grid. Plug-in hybrid technology could be useful for both LDVs and for a variety of medium-duty vehicles, including urban buses and delivery vehicles. Substantial market success of PHEV technology is, however, likely to depend strongly on further battery development, in particular on reducing battery cost and specific energy, as well as increasing battery lifetimes.

By using electricity to "fuel" a substantial portion of distance driven the potential of PHEVs to reduce oil use is clear (IPCC, 2007). The US Electric Power Research Institute (EPRI, 2001) estimates that 30 km hybrids (those that have the capability to operate up to 30 km solely on electricity from the battery) can substitute electricity for gasoline for approximately 30–40% of miles distance driven in the United States. With larger batteries and motors, the vehicles could replace even more mileage. However, their potential to reduce GHG emissions compared with more than that achieved by current hybrids depends on their sources of electricity. For regions that rely on relatively low-carbon electricity for off-peak power, e.g., natural gas combined cycle power, GHG reductions over the PHEV's lifecycle will be substantial. In contrast, PHEVs in areas that rely on coal-fired power could have increased lifecycle carbon emissions. In the long-term perspectives, transition to a low-carbon electricity sector could allow PHEVs to play a major role in reducing transport sector GHG emissions (IPCC, 2007). Technically, the development of high-performance batteries for PHEVs remains a big challenge.

Hydrogen/fuel cells

During the last decade, FCVs have attracted growing attention and made significant technological progress. Hydrogen fuel cells decouple carbon from power generation and emit only water vapor. Growing urban air pollution, climate impacts and energy security and the potential to provide new desirable attributes for vehicles (low noise, new designs) are main drivers for development of FCVs. Hydrogen can be produced from a wide range of sources including fossil, nuclear and renewable. However, the chicken-and-egg challenge to provide vehicles

and required infrastructure at the same time is more pronounced for hydrogen vehicles than it is for electric vehicles, reducing the short-term likelihood of large-scale deployment. However, a study about the US Hydrogen Initiative states that hydrogen can be produced in a "distributed" fashion at the refueling station by reforming natural gas and renewable fuels like ethanol utilizing existing delivery infrastructure to meet initial lower volume needs with the least capital investments, therefore fuel reformers would not require a substantial hydrogen transport and delivery infrastructure (Chalk, et al., 2008). While production is workable, hydrogen must be compressed, stored and dispensed at refueling stations or stationary power facilities. The same study considers that "hydrogen costs and availability and fuel-cell durability and cost are still formidable challenges to be solved" (Chalk, et al., 2008). Although there are several types of FCVs, such as direct-drive and hybrid powertrain architectures and fuelled by pure hydrogen, methanol, and hydrocarbons (gasoline, naphtha), nearly all auto manufacturers are now focused on the pure hydrogen FCV. Significant technological progress has been made for these several years, including improved fuel cell efficiency and durability, cold start (sub-freezing) operation, increased range of operation, and dramatically reduced costs. (Although FCV drive train costs remain at least an order of magnitude greater than ICE drivetrain costs).

A recent US National Research Council assessment (NRC, 2008) concludes:

- Concentrated efforts by private companies, together with governments around the world, have resulted in significant progress toward a commercially viable hydrogen fuel cell vehicle.

- Fuel cell costs have been reduced significantly over the past 4–5 years. Costs projected for high-volume (500,000 units/year) automotive fuel cell production are approximately US$100/kW for relatively proven technologies and US$67/kW for newer laboratory-based technologies.

- A significant market transition to FCVs could start around 2015 if supported by strong government policies to drive early growth.

The GHG reduction potential of FCVs strongly depends on the hydrogen production pathways and the efficiency of vehicles. At the present technology level, with FCV TTW efficiency of about 60%, and where hydrogen can be produced from natural gas at 60–80% efficiency, WTW CO_2 emissions can be reduced by 40–65% compared to current conventional gasoline vehicles. In the future, if hydrogen is produced by water electrolysis using renewable electricity such as solar and wind, or nuclear energy, the entire system from fuel production to end-use in the vehicle has the potential to be a true zero emissions system. The same is almost true for hydrogen derived from fossil sources with CCS system, in which as much as 90% of the CO_2 produced during hydrogen manufacture is captured and stored.

Table 9.4 | Increase of fuel efficiency due to various vehicle technologies.

Engine Technologies	Increase (%)	Transmission Technologies	Increase (%)
Variable Valve Timing & Lift	1–9	Continuously Variable Transmissions (CVT)	3–8
Cylinder Deactivation	7–7.5	Automated Manual Transmissions (AMT)	7–9
Turbochargers	2–7.5		
Idle Stop	0.5–8		
Direct Fuel Injection	3–15	Others	Increase (%)
Reduce Engine Friction	2–6	Weight+Drag+Time	7–18
Variable Compression Ration	10	Accessories	2–3

Source: Kobayashi et al., 2009.

Table 9.5 | Projections of vehicle efficiency improvement (%) relative to the base gasoline vehicle.

	2010	2020	2030–35	2050
Gasoline (G)	6–15	29	37–53	28–45
Diesel (D)	15–29	48	46–60	32–47
Hybrid -G	17–57	57	64–69	40–52
Hybrid -D	36–59	63	72	40–55
Fuel Cell	53–58	71–74	63	
Fuel Cell – Hybrid	52–73	76–78		
Electric Vehicle	82		80	
Ref #	1–4	5	6–8	9

Source: (1) Wurster, 2002; (2) GM/ANL, 2005; (3) EU CAR/CONCAWE/JRC, 2006: (4) JHFC, 2006; (5) Heywood and Weiss, 2003; (6) Bodek and Heywood, 2008; (7) Cheah et al., 2007; (8) Kromer and Heywood, 2008; (9) IEA, 2008.

Electric vehicles

Battery-powered electric vehicles operating today are powered by electricity acquired from the grid and stored on board in batteries. Electric vehicles have high efficiencies of more than 90%, but the big challenges are left to be overcome; short driving range and short battery life. Although the potential for CO_2 reduction strongly depends on the power mix, WTW CO_2 emission can be reduced by more than 50% compared to a conventional gasoline ICE (JHFC, 2006).

Vehicle electrification requires a more powerful, sophisticated, and reliable energy-storage component than lead-acid batteries. NiMH batteries currently dominate the hybrid market and Li ion batteries are expected to replace the market. The energy density has been increased to 170 Wh/kg and 500 Wh/L for small-size commercial Li ion batteries and 130 Wh/kg and 310 Wh/L for large-size EV batteries. The major hurdle left for Li ion batteries is high cost (IPCC, 2007).

Ultracapacitors offer long life and high power but have low energy density and a high current cost. The energy density of ultracapacitors has increased to 15–20 Wh/kg (Power System, 2005), compared with 40–60 Wh/kg for Ni-MH batteries. The cost of these advanced capacitors is in the range of several 10s of dollars/Wh, about one order of magnitude higher than Li batteries (IPCC, 2007).

9.4.2.5   Quantifying the Efficiency Improvement of Technologies

Table 9.4 summarizes the impact of vehicle technologies on vehicle fuel efficiency (Kobayashi et al., 2009). Most of these technologies are already commercially utilized, and the range of improvement by the use of these technologies is still relatively large. There are some trade-offs, such as emission control and safety issues. Employment of these technologies in new vehicles leads to an increase of vehicle production costs, although part (and, in some cases, all) of this cost increment is offset by fuel savings. Car manufacturers will select unique packages of technologies, based on the level of improvement required, economy viewpoint, and also their expertise.

Based on these technology developments, vehicle efficiency presented here is estimated along the timeline from 2010–2050 (Kobayashi et al., 2009). Since base vehicles, and assumptions about the employment of technologies and their effectiveness are different, the range of estimates among the studies are rather large, and the estimates of MIT (Ref. No. 5–8 in Table 9.7) are in the higher range compared with others. This is partly due to a more optimistic view, but also due to the fact that the potentials for improvement of vehicle efficiency in the US market is larger. As can be seen in Table 9.5, baseline gasoline ICE vehicles with conventional drivetrains could be continuously improved up to 2050, and achieve close to a 50% increase in fuel efficiency with advanced technologies. At the same time, diesel ICEs will gain higher efficiency, but their margin over gasoline vehicles will shrink (Kobayashi et al., 2009)

Several technical hurdles remain to be overcome for wide marketplace acceptance of hybrids, fuel cells, and electric vehicles. As mentioned above, the cost increase due to new technology is very important when gauging potential market penetration (Kobayashi et al., 2009). As in the case of efficiency improvement, there are large differences among studies. In particular, cost estimates for HEVs in EUCAR/CONCAWE/JRC (2004) are significantly higher than other studies. This may be explained by their assumption of very high HEV battery requirement with 6kWh, which allows the 20 km of full EV operation.

In Figure 9.32, most of the data lies below US$6000. Calculating how many years customers need to pay off the increment of vehicle price by saving the fuel cost provides some implication for the market (Kobayashi et al., 2009). Starting with a typical situation in the United States, that a vehicle has a fuel efficiency of 10 km/L (24 MPG) and travels 19,000 km (12,000 miles) yearly, results in an annual fuel consumption of 1900 L (500 gallons). If fuel costs US$0.80/l (US$3/gallon), the annual fuel cost is US$1500. If the fuel efficiency improvement is 50%, customers need 12 years to pay off the price increment. In Japan, where fuel prices are almost twice as high, but distance traveled per year is around 14,000 km, only six years are needed to pay off the price increment. Although this does not necessarily mean that customers will be willing to buy

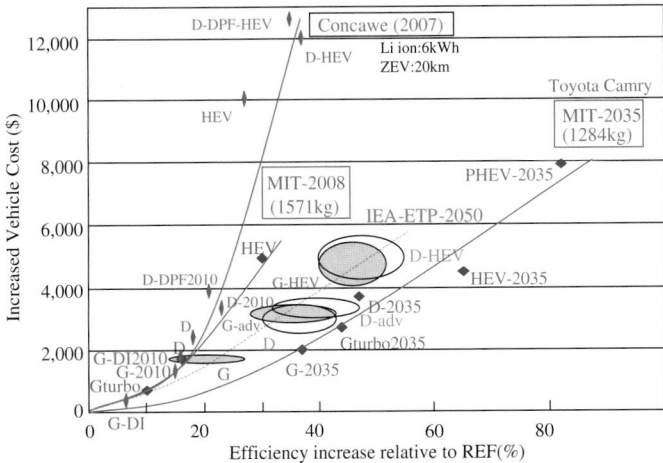

Figure 9.32 | Price increase of vehicle due to efficiency increase. Data for IEA are shown as ellipses indicating the range of data. (G: gasoline, D: diesel, DI: direct-injection, DPF: particulate filter) Source: based on data from EUCAR/CONCAWE/JRC, 2006; Bandivadekar, 2008; IEA, 2008.

the vehicle with such an increased price, this fuel savings/vehicle cost trade-off may be commercially allowable if future consumers are highly concerned about fuel economy.

9.4.3   Aviation

9.4.3.1   Alternative Fuels for Aviation

Sustainable alternative fuels produced from biomass or renewable oils offer the potential to reduce lifecycle GHG emissions from aviation. As these fuels contain significantly lower fuel sulphur, they may also help to reduce emissions of particulate matter and lessen aviation's impact on air quality. Aviation is heavily dependent over the short- and medium-term on drop-in liquid fuels. Sustainable alternative aviation fuels are not available today in significant quantities, and considerable doubts remain on their lifecycle GHG emissions. In fact, the lifecycle GHG emissions of biofuels may be higher than previously estimated when indirect effects are included (see Section 9.3), putting doubts on the straight-forward realization of win-win opportunities.

In the short term,[15] sustainable alternative fuels for aircraft may be available in limited quantities and have a lifecycle CO_2 footprint equal to, or less than, conventional jet fuel. It will be necessary to blend these alternative fuels with conventional jet fuel at up to a maximum of 50% to produce a drop-in fuel. Drop-in jet fuels are completely interchangeable with conventional jet fuel, and so will not require the modification of fuel handling and distribution systems.

In the medium term, it is possible that sustainable alternative fuels for aircraft will be available in much larger quantities. Given sufficient demand or incentive, significant supplies of jet fuel that offer a significant reduction

in lifecycle CO_2 emissions could be available in the mid-term. Significant research and development activities currently underway are expected to lead to a number of commercial scale production facilities. Also, the development of new sustainable alternative fuels for aircraft could be able to reduce costs of fuels to compete with conventional jet fuel in the mid-term.

In the long term, the aviation industry may explore more radical fuels that require redesigned engines and airframes. Fuels such as liquid hydrogen and liquid methane might be used to significantly reduce GHG emissions.

The following goals for alternative fuels are relevant for aviation:

- The US Consortium for Continuous Low Energy, Emissions, and Noise (CLEEN) has set a goal that 20% of jet fuel available for purchase by US commercial airlines and cargo carriers will be alternative fuels by 2016.

- The European Union has established a requirement that lifecycle GHG emission savings from the use of biofuels shall be at least 50% by 2017 and at least 60% by 2018 (however, indirect land-use emissions are not included in this accounting so far, as of April 2011).

- The European Union set a target of 10% use of renewable energy sources in transport by 2020.

Alternative aviation fuels are discussed individually below.

Natural gas and electricity

For ground operations in ground power units (GPUs) and energy backup systems, natural gas and electricity are already being used in some airports, mainly as substitutes for diesel. These measures are for short term implementation. Even though the cost of retrofit or equipment replacement can be expensive, these fuels still may be practicable to use in the aviation industry.

Fuel cells

There are ongoing studies on the use of fuel cells to power small generators and aircraft auxiliary power units when on the ground. The cost of these cells remains very high, and they are seen as a long-term measure for reducing emissions at airports.

Hydrogen

For aircraft operations, the most difficult option from a technological point of view is hydrogen. Great changes in aircraft design will be necessary for using this fuel that releases water vapor and NO_x (which also have radiative forcing) when burned. As it is a radical change of paradigm, this is considered a long-term solution. Since it requires changes in structures for fuel storage and transport, as well as new design technology and aircraft building, the economic impacts must be significant for the aviation industry, and society as a whole, to accept such a radical change.

Synthetic liquid fuel

Synthetic liquid fuels are obtained from different kinds of sources such as coal, natural gas, biomass, oil shale, or tar sand. One problem related to

---

15   Short term is approximately 5–10 years; medium term is 11–20 years; long term is over 20 years.

Box 9.4 | Alternative Fuels for Aviation Demonstration Cases

The international aviation industry has successfully demonstrated the technological feasibility of using sustainable alternative jet fuel blends in flight tests without affecting safety. The following test flights using alternative aviation fuels have been conducted during the past two years, which supported the approval of the ASTM D-7566 Standard Specification for Aviation Turbine Fuel Containing Synthesized Hydrocarbons, the first new jet fuel specification in more than 20 years.

Examples of aviation alternative fuel demonstrations:

- Airbus flew its A380 test aircraft with one of its four engines running on a 40% blend of Gas-To-Liquid (GTL) fuel with conventional jet fuel on 1 February 2008.

- Virgin Atlantic flew a Boeing 747–400 on 23 February 2008 with one engine operating on a 20% biofuel mix produced from babassu oil and coconut oil.

- Air New Zealand flew a Boeing 747–400 with one engine on 50% jatropha-derived hydro-treated renewable jet (HRJ) biofuel and 50% kerosene on 30 December 2008.

- Continental Airlines flew a Boeing 737–800 with one engine using 50% jet fuel and 50% algae and jatropha mix on 7 January 2009.

- JAL flew a Boeing 747–300 with 50% HRJ biofuel (derived from camelina, jatropha, and algae) and 50% kerosene mix on 30 January 2009.

- Qatar Airways performed the first revenue flight with alternative fuel on 12 October 2009. An A340–600 flew from London to Doha with its four engines running on a 48.5% blend of GTL with conventional jet fuel.

- KLM flew a Boeing 747–400 using a mix of 50% HRJ biofuel (derived from camelina) and 50% conventional Jet A1 on 23 November 2009.

Source: UNFCCC, 2010.

this type of fuel is its low lubricity due to the absence of sulphur in its composition. A second problem is the high cost of production, because it is still predominantly based on the Fischer and Tropsch process (coal-, gas-, or biomass-to-liquids), which requires an intense use of energy for precursor gasification and CO polymerization in the presence of water. Therefore, synthetic fuels are energy and CO_2 intensive. In spite of that, they are broadly compatible with current fuel systems, can produce a clean burn, and tend to produce less particulate matter. There are little technological barriers towards the introduction of synthetic liquid fuel technology

Biofuels

In the aviation industry, biofuels are seen as the most promising candidate. Biofuels are being tested by several companies. However, biofuels perform unfavorably in terms of global warming potential, as seen from a lifecycle perspective. For lifecycle accounting of biofuels, see Section 9.3.1.1.

There is an important difference between biofuels for road transport and biofuels for aviation. Because aviation fuels must have a high energy density and chemical stability as well as a low freezing point, conventional biofuels are not suitable. Therefore, biofuels for aviation must be drop-in (hydrocarbon) fuels that can be produced by hydro-treating the bio oil, gasification/F-T

process and other new biochemical pathways, such as sugar-to-diesel process. These production processes are still in the R&D phase.

9.4.3.2   Improving Energy Efficiency in Aviation

The aviation sector's efficiency track record is solid. Aircraft being built today are greater than four times more efficient than those built 40 years ago. According to Greener by Design (2011), "fuel efficiency in aviation improved by 70% between 1960 and 2000 – a better record than any other transport mode. Between 1990 and 2000 alone, aviation fuel efficiency improved by 20%. Improvements of an additional 20% are projected by 2015 and 40–50% by 2050, relative to today's aircraft. Current research programs have goals of reducing landing and take-off NO_x emissions by up to 70% over today's regulatory standards, while also improving fuel consumption by 8–10% over the most recent production engines by about 2010."

ICAO's Assembly Resolution A37–19 is aimed at improving the outlook for aviation's future climate footprint. The resolution recognizes the potential of technological and operational improvements, market-based measures, and the use of sustainable alternative fuels to reduce aviation's climate impact. The resolution aims to collectively achieve global aspirational goals of improving fuel efficiency by 2% per year and stabilizing CO_2 emissions at 2020 levels. However, with respect to projected growth rate in aviation (Schäfer et al., 2009), GHG emissions from aviation will challenge ambitious GHG mitigation targets.

ICAO (2009) has identified a basket of measures for addressing GHG emissions from international aviation:

- *Aircraft related technology development (including alternative fuels)*: Measures in this category include purchase of new aircraft, retrofitting and upgrade improvements on existing aircraft, new designs in aircraft/engines, fuel efficiency standards, and alternative fuels. Some of these measures have the potential for very high gains in fuel efficiency/emissions reduction, but the costs are likely to be high and there will be a long timeframe for implementation.

- *Improved air traffic management and infrastructure use*: More efficient air traffic management planning, ground operations, terminal operations (departure and arrivals), en route operations, airspace design and usage, and air navigation capabilities. These are measures with the potential for relatively short- to medium-term gains, although the scale of potential relative gains is low to medium. More efficient planning and use of airport capacities, construction of additional runways and enhanced terminal facilities, and clean fuel-operated ground support equipment. These can be implemented in the short to medium term, but potential emission reduction gains are likely to be low. Increased airport capacity may also encourage increased emissions from aircraft unless appropriate actions are taken to address the emissions.

- *More efficient operations*: Measures include minimizing weight, improving load factors, reducing speed, optimizing maintenance schedules, and tailoring aircraft selection to use on particular routes or services. This area is essentially a matter for aircraft operators and can be further incentivized by market-based instruments.

9.4.4 Shipping and Railways

9.4.4.1 Improving Energy Efficiency in Shipping

Giving the global volume of seaborne trade, the shipping industry is increasingly expected to share the burden of reducing global emissions of CO_2. Shipping is already a very efficient form of transport in terms of CO_2 emissions per tonne-km of goods moved. It compares favorably with rail transport, is more efficient than road transport, and is considerably more efficient than air freight (see Figure 9.12). Bunker fuels used in shipping are highly polluting for international seas and coasts and they challenge marine ecological systems.

In the past increases in fuel oil prices have encouraged improvements in ships fuel efficiency. During 2008, many ship operators, particularly of container ships, adopted a policy of "slow steaming," i.e., reducing the speed at which a ship operates. Reduction in a ship's speed from 26 knots to 23 knots can result in a 30% fuel savings.

Marine diesel engines are already highly efficient, operating near the theoretical maximum thermal efficiency. Almost 50% of the energy

Table 9.6 | Potential efficiency improvements from technology/design measures for existing ships (HFO: Heavy Fuel Oil; MDO: Marine Diesel Oil).

Improvement	Potential Fuel Saving (%)
Optimized hull shape	3–5
Propeller maintenance	1–3
Fuel injection improvements	1–2
Fuel switching (HFO to MDO)	4–5
Efficiency rating	3–5
Turbocharger upgrade	5–7
Potential total saving	**4–20**

Source: IMO, 2008.

Table 9.7 | Potential efficiency improvement from technology/design measures for new builds.

Improvement	Potential Fuel Saving (%)
Optimized hull shape	5–20
Optimized selection of propeller	5–10
Optimized propulsion efficiency	2–12
Fuel switching (HFO to MDO)	4–5
Planting improvements (e.g., waste heat recovery)	4–6
Machine monitoring	0.5–1
Potential total saving	**5–30**

Source: IMO, 2008.

Table 9.8 | Potential efficiency improvements from operational measures.

Improvement	Potential Fuel Saving (%)
Fleet planning (ship size, type, etc.)	5–40
"Just in time" voyage planning	1–5
Weather routing	2–4
Operating at constant engine speed	0–2
Operating with optimized vessel trim	0–1
Operating with minimum ballast	0–1
Optimizing propeller pitch	0–2
Reduced port time for cargo handling operations	1–5
Reduced time for mooring/anchoring operations	1–2
Potential total saving	**1–40**

Source: IMO, 2008.

Table 9.9 | Potential efficiency improvements from technological options.

Technology	Time Horizon	General Criteria			Energy Efficiency Potential	
		Barriers	Type of Traction	Market Potential	Single Vehicle (%)	Fleet (%)
Medium-frequency transformer	long-term	medium	Electric – AC	highly uncertain	2–5	1–2
HTSC transformer	long-term	high	electric – AC	highly uncertain	5–10	2–5
Single-axle bogies	mid-term	medium	electric – DC, AC – diesel	medium	5–10	2–5
Bogie fairings	mid-term	low	electric – DC, AC – diesel	medium	5–10	1–2
Common rail	mid-term	low	diesel	medium	> 10	1–2
Double-layer capacitors (storage technology)	mid-term	medium	electric – DC	low	> 10	not applicable
CO_2-based demand control for coach ventilation	mid-term	medium	electric – DC, AC – diesel	high	2–5	1–2
Driving advice systems in suburban operation	mid-term	medium	electric – DC, AC – diesel	medium	>10	> 5
Driving advice systems in main line operation	mid-term	medium	electric – DC, AC – diesel	high	5–10	2–5
Optimization of train operation by control center	mid-term	high	electric – DC, AC – diesel	medium	5–10	> 5
Wide-body stock	mid-term	high	electric – DC, AC – diesel	medium	> 10	2–5
Switch-off of traction group	mid-term	medium	electric – DC, AC – diesel	not applicable	2–5	1–2
On-board use of braking energy in diesel-electric stock	mid-term	low	diesel	medium	2–5	1–2
Control of comfort functions in parked trains	mid-term	medium	electric – DC, AC – diesel	high	2–5	2–5
Regenerative braking in DC systems	mid-term	medium	electric – DC	high	5–10	> 5
Modification of target temperature in passenger coaches	short-term	medium	electric – DC, AC – diesel	not applicable	< 2	1–2
Double-decked stock	short-term	high	electric – DC, AC – diesel	high	> 10	> 5
Multiple units vs. loco- hauled trains	short-term	low	electric – DC, AC – diesel	high	5–10	1–2
IGBT	short-term	low	electric – DC, AC – diesel	high	5–10	> 5
Ventilation control (in new stock)	short-term	medium	electric – DC, AC – diesel	high	2–5	–
Regenerative braking in 16.7 Hz, 15 kV systems	short-term	medium	electric – AC	high	5–10	> 5
Energy efficient driving strategies	short-term	medium	electric – DC, AC – diesel	not applicable	5–10	> 5
Energy efficient driving by low-tech measures	short-term	low	electric – DC, AC – diesel	not applicable	5–10	2–5
Re-engining of diesel stock (replacement of engine)	short-term	low	diesel	medium	> 10	–
Regenerative braking in 50 Hz, 25 kV systems	short-term	medium	electric – AC	medium	5–10	> 5
Aluminum car body	short-term	low	electric – DC, AC – diesel	medium	2–5	1–2

Source: UIC, 2011.

in the fuel is converted into mechanical work in a large marine diesel engine. Effective recovery of waste heat from the engine can further raise the efficiency to nearly 55%.

Because there is currently little or no scope for improving the efficiency of the marine diesel engine itself, improvements will have to come from other elements of ship design and operation. The International Maritime Organization (IMO) is currently debating which measures could be introduced to limit or reduce shipping's CO_2 emissions. Measures can be divided into operational measures, i.e., those that deal with how the world fleet is managed and operated, and technology/design measures. The latter can be further subdivided: some measures can be introduced only at the design/building stage for a ship, whereas others can be retrofitted to existing ships. Tables 9.6, 9.7, and 9.8 list some of the measures considered by IMO, giving an indication of what might be achievable.

9.4.4.2   Improving Energy Efficiency in Railways

Rail is more energy efficient than most other transport modes, For example, in Germany, rail-based passenger transport has about one-third of the GHG emissions per passenger of air traffic, and half of car transport (IFEU, 2010). Despite this, enhancement of energy efficiency is an important issue in order for railways to reap cost savings and to enlarge competition advantages, as well as to reduce railway's contributions to climate change.

One key means of improving energy efficiency in railways is to deploy advanced technologies. The International Union of Railways funded a project in which all relevant railway energy-saving technologies have been analyzed and evaluated. These evaluated technologies are summarized in Table 9.9. For electric railways, decarbonization can be achieved by switching from fossil fuels (e.g., coal fired power plants) to renewable energies.

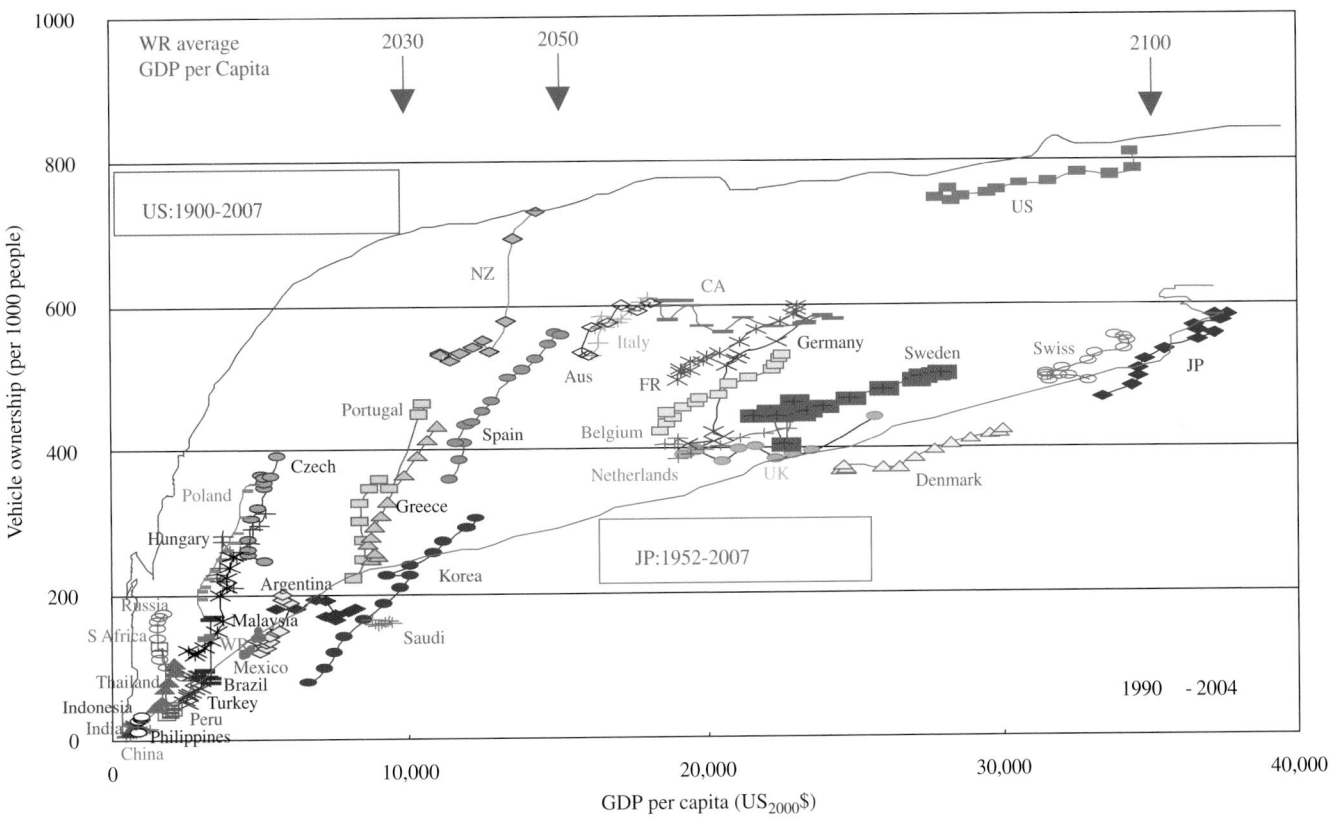

Figure 9.33 | Historical relationship between vehicle ownership and GDP per capita. World average values of GDP per capita are indicated by arrows. Note: plotted years vary by country depending on data availability. Source: based on World Bank, 2010.

9.5   Alternative Images of Future Transport Scenarios

9.5.1   GEA-Transport Model Overview

To conduct projections and policy analysis for GEA-Transport[16], we utilized the IEA's global transport spreadsheet model, which was developed for the WBCSD's Sustainable Mobility Project (SMP) (WBCSD, 2004a). The IEA model has the capability of making a scenario through 2050, which was expanded up to 2100 for this GEA transport study, which is a bottom-up model and differs from the one in Chapter 17. The WBCSD/SMP transport spreadsheet model is designed to handle all transport modes and most vehicle types. It produces projections of vehicle stocks, travel, energy use, and other indicators through 2050 for a reference case and for various policy cases and scenarios. It is designed to have some technology-oriented detail and to allow for fairly detailed bottom-up modeling. The model does not include any representation of economic relationships (e.g., elasticities), nor does it track costs. Rather, it is an accounting model, anchored by the ASIF identity:

- **A**ctivity (passenger and freight travel)
- **S**tructure (travel shares by mode and vehicle type)
- **I**ntensity (fuel efficiency)
- **F**uel type – fuel use by fuel type (and CO_2 emissions per unit fuel use).

For these factors in road transport, vehicle sales and stocks are most important variables to affect the final results. As shown in Figure 9.33, historical vehicle ownership is closely related with growth of income level per capita. Around the threshold of US$5000/person, vehicle ownership grows very rapidly, but the growth rate for specific regions is quite different and is affected by various factors, such as geographic conditions, infrastructures, policies, and so on.

As outlined from the long-term historical data for the United States and Japan, this growth pathway is important to determine vehicle stock at a certain economy level. If this growth rate remains at the lower level by some measures in the future, the vehicle stock in developing countries can be kept low compared with those in developed countries.

In the GEA transport model, the future level of vehicle ownership in developing countries is assumed to be lower compared to those for most of

---

16   The GEA-Transport model differs from the GEA model in Chapter 17 in that it is a bottom-up model. The departing approaches resulted in differences that were considered acceptable within the range of allowance.

Box 9.5 | Storyline of the Bottom-up Scenarios for Energy and Transport Development

Under the reference scenario, which is produced with the WBCSD/SMP model, energy use of the transport sector increases significantly when it is associated with economic development, especially in developing countries. Car ownership rate, use of more energy-intensive and faster modes of transport, and freight activity increase with an increase in economic level (see Section 9.5 where the transport model is discussed).

At the same time, more efficient powertrain technologies and more efficient operation reduce relative fuel consumption, and more use of low carbon fuel reduces CO_2 emissions from the transport sector. However, as demand increases with economic development, reductions due to technological advancement are offset, leading to an absolute increase in CO_2 emissions from the transport sector.

There are many potential technologies and measures to reduce energy use and CO_2 emissions from the transport sector. CO_2 emissions can be reduced significantly under three CO_2 constraint pathways: GEA-mix, GEA-supply, and GEA-efficiency. In Chapter 17, each pathway has a transport-specific branching point between liquid and advanced transportation fuels. In contrast, the pathways presented in Chapter 9 all assume the introduction of advanced technologies, but with different compositions and to varying degrees.

Under the GEA-mix pathways, the share of biofuels[17] increases in every transport sector using gasoline, diesel and jet fuel, and energy intensity of aviation and shipping are also improved over the reference scenario. While the share of hybrids in passenger and freight vehicles increases, new vehicle types such as electric vehicles and fuel cell vehicles are progressively put into the market. Efficient operation of trucks reduces the fuel consumption of freight transport. The characteristic of this pathway is a balanced use of all possible reduction options.

In the GEA-supply pathways, the negative emission of power generation through CCS, as discussed in Chapter 12, further reduces CO_2 emissions, especially from road transport. The modal shift from personal vehicle to public transport in passenger transport is enhanced by policy measures in both GEA-efficiency and GEA-mix pathways. The efficiency of vehicles is improved progressively toward 2100, but the rate of improvement is highest in the GEA-efficiency set of pathways. At the same time, electric vehicles, including Plug-in HEVs, are aggressively introduced into the market under the GEA-efficiency pathways.

Table 9.10 | Level of importance on each reduction measure for GEA-Transport scenarios.

Scenario		Efficiency	Mix	Supply
LDV	efficiency	***	*	*
	demand	****	**	*
	EV	****	**	*
	FCV	**	***	****
Trucks	efficiency	**	**	**
	demand	***	**	*
	new technology	**	**	**
Aviation	efficiency	**	**	*
	demand	***	**	**
Shipping	efficiency	***	**	**
BF		***	****	***
Electricity	emission factor	**	**	****
Modal shift		***	**	*
notes		FE+Modal shift+Plug-in	BioFuels	FCV/H2+Negative Emission

developed countries. This could be caused by a number of factors, including inhomogeneous structure of countries due to greater urbanization and lower suburbanization, greater income disparities between the wealthy and the poor, and limits on the infrastructure needed to support large numbers of vehicles, as well as future policies to reduce the need for motorized transport. Good examples of the role of policy measures to reduce personal vehicle ownership are discussed in the policy section. Based on GDP per capita, most countries presently classified as developing countries are not classified as developing countries in 2100.

9.5.2 Assumptions

9.5.2.1 Assumptions for Population and GDP

In Figure 9.34, global population reaches 10 billion in 2100, which is close to the level of SRES B2 scenario. Most of the growth occurs in

---

17 The values of emission factor for each biofuel are taken from the various studies, such as shown in Figure 9.27.

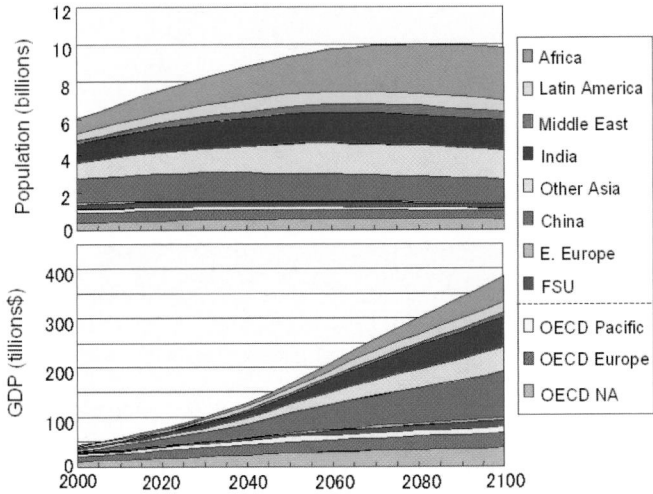

Figure 9.34 | Growth of population and GDP by region.

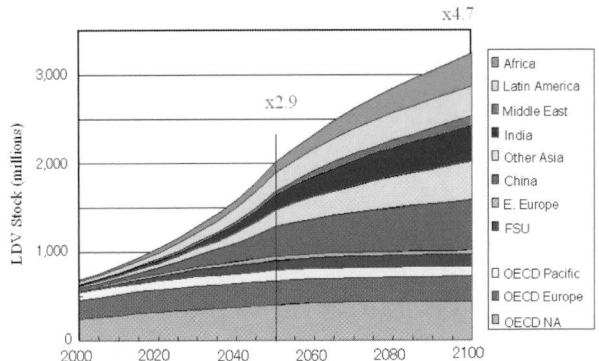

Figure 9.35 | Total stock of LDVs by region. They will grow 2.9, and 4.7 times higher than the present level by 2030, and 2050 respectively. Source: based on WBCSD, 2004b.

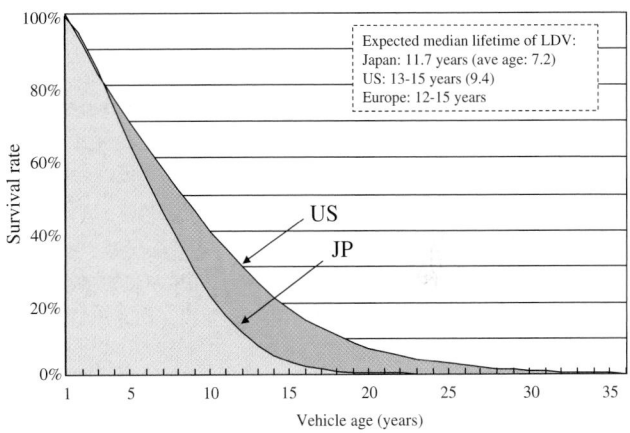

Figure 9.36 | Comparison of survival rate for passenger LDVs. Source: based on ORNL, 2010.

developing countries. Global GDP grows significantly up to around US$400 trillion toward the end of the century,[18] and GDP per capita also increases in most developing countries. As GDP is the strongest driver of car ownership, transport activities, and car ownership are expected to increase accordingly (see also Schäfer et al., 2009).

9.5.2.2 Assumptions for Vehicle Stock by Region and Fuel Economy

Three sectors are primarily propelling worldwide transport energy growth: LDVs, freight trucks, and air travel (IPCC, 2007). The Mobility 2030 study projects that these three sectors will be responsible for 38%, 27%, and 23%, respectively, of the total 100 EJ growth in transport energy that it foresees in the 2000–2050 period. The WBCSD/SMP reference case projection indicates the number of LDVs will grow to about 1.3 billion by 2030 and to just over 2 billion by 2050, almost three times higher than the present level (see Figure 9.35). Nearly all of this increase will be in the developing world (WBCSD, 2004b).

The present trend reveals a dominance of road transport and private cars, as well as an increase in aviation. As a consequence, it will be hard to curb fuel demand in any scenario timeline. LDVs and OECD countries will continue to dominate transport energy use for a long time, even with the growing participation of developing countries, mainly China.

One important factor affecting levels of car ownership is vehicle life, which determines the rate of replacement of older cars with new ones. As shown in Figure 9.36, the vehicle life is quite different from country to country. In terms of CO_2 reduction, vehicle efficiency is progressively improved, so the total level of CO_2 reduction is not strongly affected by the difference of longer tail at lower survival rate. However, this

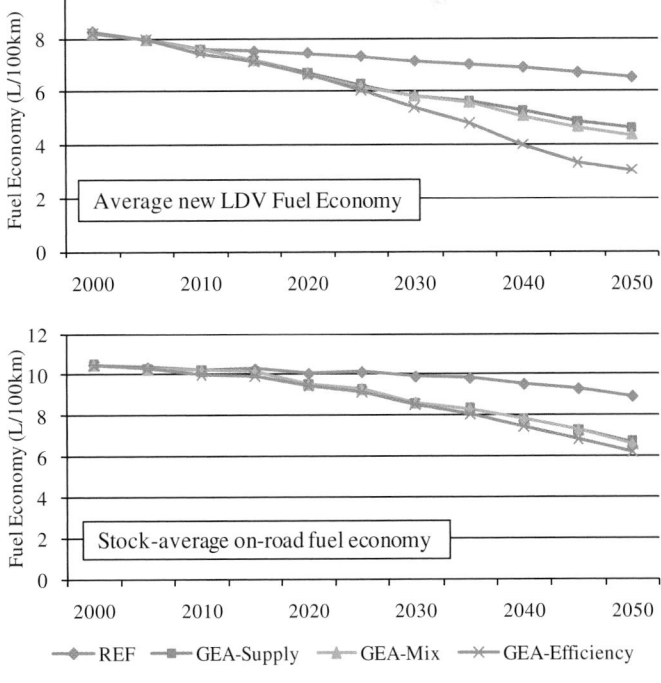

Figure 9.37 | Trend of new vehicle and stock-average on-road fuel economy.

---

18   The assumptions for population and GDP growth are slightly higher than those of Chapter 17, but GDP per capita data, which are only used for the scenario analysis, are consistent with each other.

becomes very important when tail pipe emissions are considered. Very few, but very old cars could be responsible for a large part of local air pollution, because these few cars might emit orders of magnitude more pollutants.[19]

Figure 9.37 shows the trend of new vehicle and stock-average on-road fuel economy values. Fuel economy values for three GEA scenarios show 40–50% improvement of new car fuel economy over 50 years, while stock-average fuel economy is improved by only 30% or so, because of the delay in replacement by older existing vehicles.

9.5.3 Key Results

9.5.3.1 Fuel Use Projection in the Reference and Pathways to Realize the GEA Scenario[20]

Along with an increase in population and GDP, as shown in Figure 9.34, vehicle stocks will increase significantly by a factor of 2.9 and 1.6 between 2000/2050 and 2050/2100, respectively, in the reference and GEA-supply scenarios (see Figure 9.35). LDV stocks in 2100 are 4.7 times larger than that in 2000. In 2050, the stocks in Asia are already bigger than that in the United States now, and are almost twice as large as that in all OECD countries in 2100. At the same time, truck stocks increase as fast as those for LDVs. The modal shift from passenger cars to trains will reduce ownership in passenger cars (LDV) by 40% and 20% in 2100, in the GEA-mix and GEA-efficiency pathways, respectively (see Figure 9.38). Truck stocks are reduced by 20% in 2100 in GEA-efficiency pathways due to a modal shift from trucks to freight train.

As the economy and population grow, other subsector activities also increase, as shown in Figures 9.39 and 9.40. Among passenger transport, the growth rate of aviation is largest. Aviation in 2100 is more than ten times larger than in 2000 in the reference scenario[21] and the GEA-supply and GEA-mix pathways. In the GEA-efficiency pathway, growth rate in aviation is slightly smaller.

In the GEA-efficiency and GEA-mix pathways, passenger train activity grows largely in 2100 because of modal shift from passenger cars. On the other hand, passenger car activity stays almost the same after 2050. In terms of magnitude, the share of LDV and aviation is significant and this is also true in terms of fuel consumption as shown in Figure 9.41.

In order to reduce fuel consumption and CO_2 emissions simultaneously, a variety of options for LDV can lead to the same goal (e.g., Figure 9.23),

19 These high-emitter vehicles were a hot issue in California in the 1990s.

20 See also Chapter 17.

21 Reference scenario is produced based on the WBCSD/SMP model and extended from 2050 to 2100. See Box 9.5.

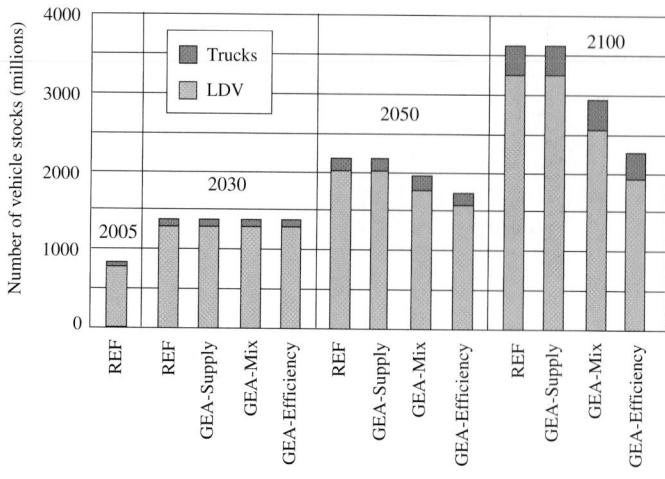

Figure 9.38 | Growth of stocks of LDVs and trucks.

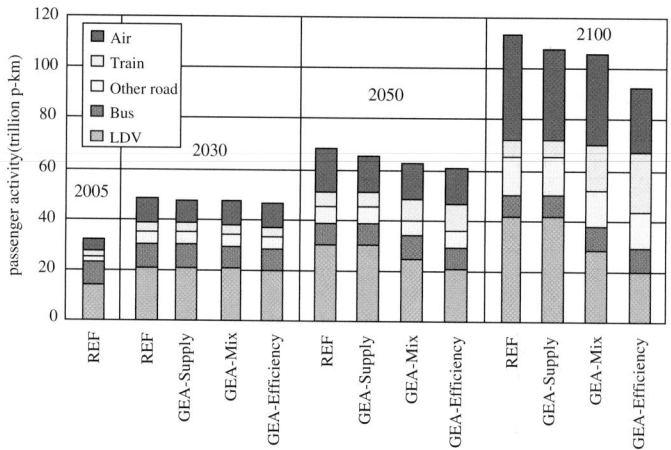

Figure 9.39 | Growth of passenger transport activity.

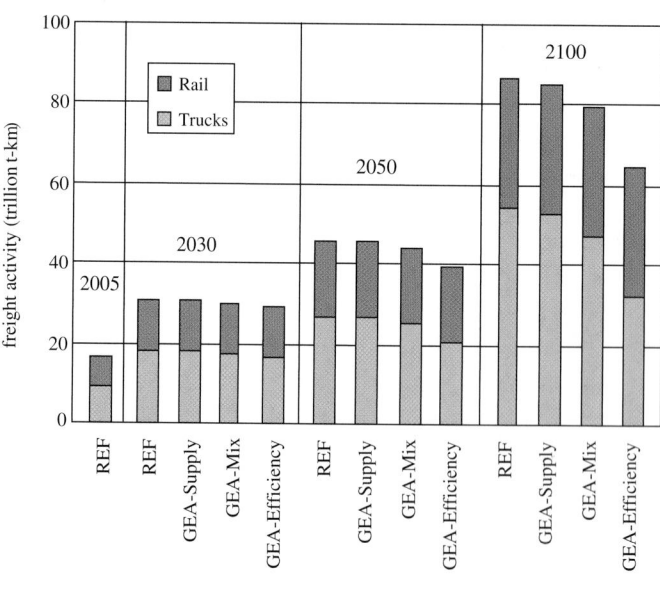

Figure 9.40 | Growth of freight transport activity.

but measures for aviation are limited. The most promising measures are efficiency improvements and biofuel; but there are many issues to overcome to utilize the benefit of biofuels, including the sustainability concern (see Section 9.3.1.1).

In modeling freight transport to project fuel consumption, the starting point was activity data for both trucks and freight trains. But for shipping, the growth rate of fuel consumption is directly projected from the first place of modeling because of its insignificant contribution to overall freight transport.

As shown in Figure 9.40, the growth of truck and freight train activity is significant and their activities in 2100 are four to seven times, and five times larger, than 2000, respectively. In terms of fuel consumption, the share used by trucks is most important, but the actual amount of fuel consumption differs very much depending on the pathways (see Figure 9.41).

In the reference scenario projection, global transportation fuel use increases by a factor of nearly 2.3 between 2000 and 2050 and slows down slightly afterwards, leading to an increase of a factor of around 1.5 between 2050 and 2100 (see Figure 9.42). Use of gasoline, diesel fuel, and jet fuel grows substantially, while other fuels retain a tiny share. Alternative fuels and advanced vehicles do not penetrate significantly in the reference scenario. Total fuel use by mode will grow significantly for all modes but buses, with the biggest growth occurring for LDVs, freight trucks, and air travel. It should be noted that most of the increase in fuel use occurs in developing countries.

Fuel efficiency improvement and introduction of new technologies significantly reduce fuel use by 35%, 40%, and 60% in 2100 for pathways GEA-supply, GEA-mix, and GEA-efficiency, respectively, compared with the reference scenario (see Figure 9.46). Reduction is most prominent in the LDV subsector, while air and trucks remain major players, along with LDVs (see Figures 9.41 and 9.42).

As shown in Figure 9.42, energy use decreases in the order of GEA-supply, GEA-mix, and GEA-efficiency pathways. This is mainly caused by different improvement rates of fuel efficiency and reduction rates of vehicle driving distance per year.

In the reference scenario, conventional fuels make up the major part of fuel consumed. In the GEA pathways, usage of biofuels, especially advanced biofuels, such as cellulosic ethanol and BTL diesel and jet fuel, increase with time. In 2100, the share of biofuels becomes 20–45%, depending on the pathway. Having assumed that advanced biodiesel is widely used in all vehicles with diesel engines, including trucks and buses, the share of advanced biodiesel becomes relatively large compared with that of advanced ethanol in 2100. Preconditions for wide-scale application of advanced biofuels are stringent policy instruments that discriminate against the mostly environmentally

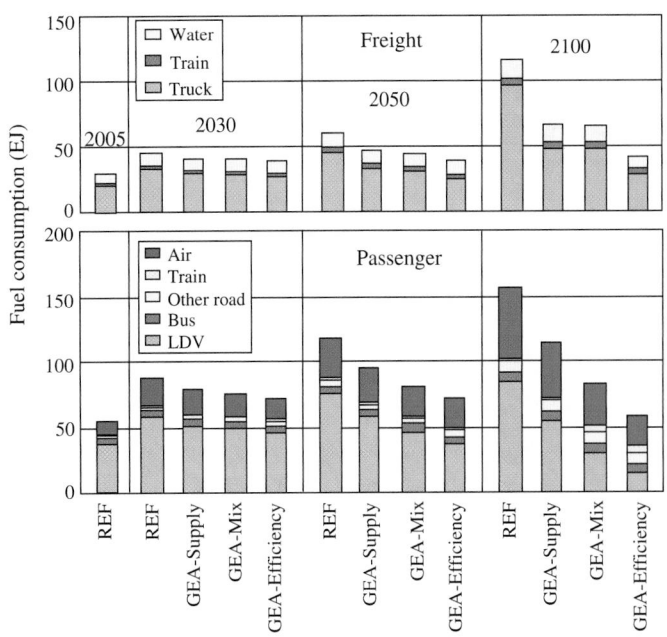

Figure 9.41 | Fuel consumption projection for both passenger and freight transport by mode.

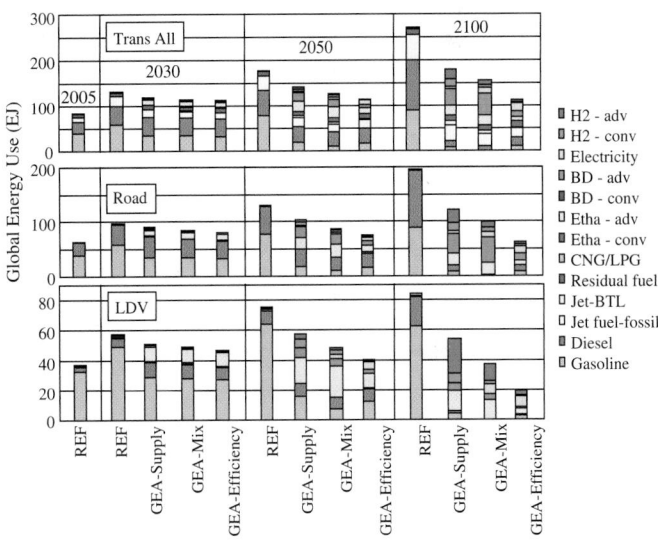

Figure 9.42 | Worldwide transport fuel use for the whole sector, road transport and LDV only by fuel type.

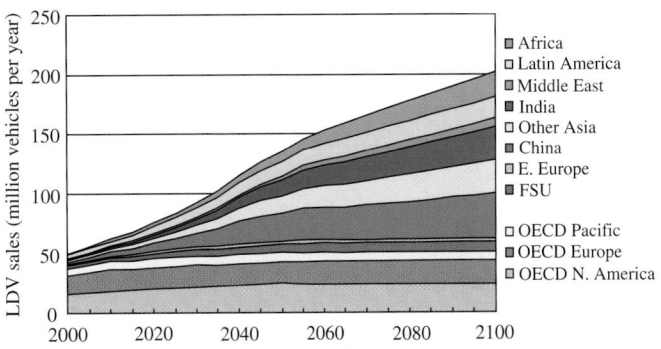

Figure 9.43 | Worldwide LDV sales by region (Reference and GEA-supply pathways).

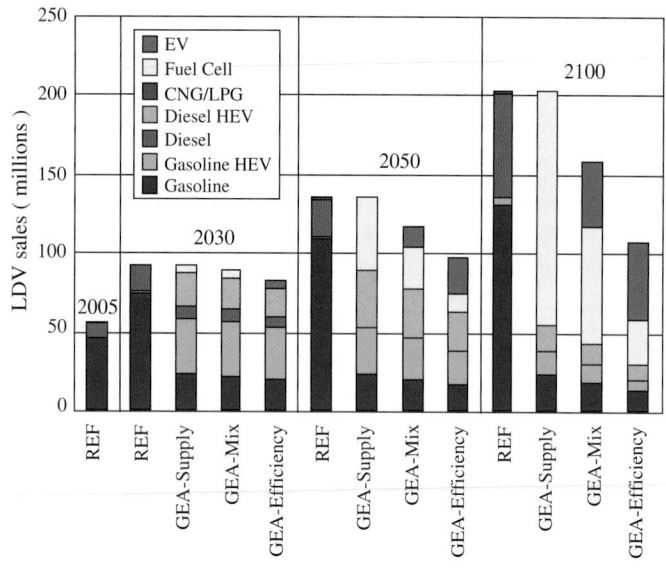

Figure 9.44 | Worldwide LDV sales by vehicle type.

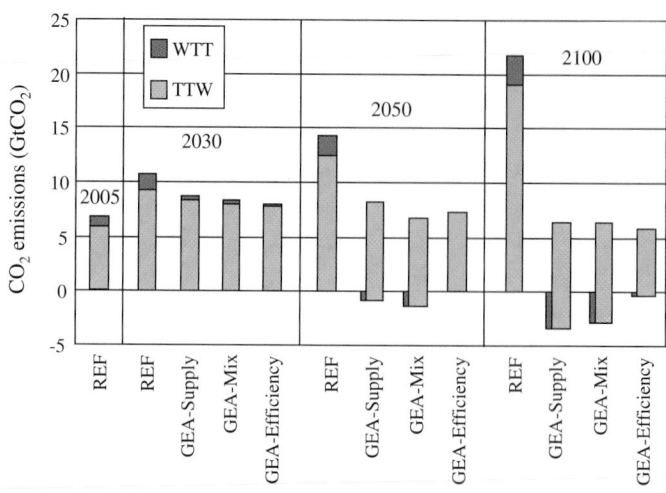

Figure 9.45 | Well-to-wheel CO_2 emissions from transport sector for reference and GEA pathways: tank-to-wheel (TTW), well-to-tank (WTT).

harmful, high-carbon but more price-efficient agrofuels that are currently produced, notably US corn ethanol. Lifecycle analysis of these biofuels is discussed earlier in this chapter, and appropriate policy instruments will be discussed later.

9.5.3.2   Light-Duty Vehicles

In the reference and GEA-supply pathways, new vehicle sales increase by a factor of 2.7 and 1.5 between 2000/2050 and 2050/2100, respectively. Although the new car market in OECD countries is almost saturated, the market in developing countries, especially in Asia (including China and India), will grow significantly (see Figure 9.43).

In the GEA-efficiency and GEA-mix pathways, new car sales decrease significantly because of a modal shift to trains. In 2100, new car sales in the GEA-efficiency and GEA-mix pathways are 44% and 22% lower, respectively, than that for the reference and GEA-supply pathways.

As shown in Figure 9.44, the share of HEVs increases with time, but these are replaced with fuel cell vehicles and electric vehicles in 2050 and 2100. In 2100, the share of FCVs and EVs is around 70%. Since there is a large uncertainty about the evolution of new vehicle technologies, FCV shares over EVs varied significantly among the GEA pathways.

9.5.3.3 CO_2 Projections

Global transport CO_2 emissions from vehicles tank-to-wheel (TTW) are projected to increase by a factor of 2.3, from about 5.4 Gt in 2000 to 12.4 Gt in

2050 and by 3.5 to 18.9 Gt in 2100 in the reference scenario. Also, like fuel use, the vast majority of CO_2 increase will be in non-OECD (i.e., developing) regions. In the GEA pathways, those emissions are significantly reduced to around 6–7 Gt in 2100. As shown in Figure 9.45, the well-to-wheel (WTW) emissions increase up to 21.6 Gt in 2100 in the reference scenario, but are reduced to 6–7 Gt due to the large use of alternative fuels, such as biofuels, electricity, and hydrogen in the GEA pathways.

As shown in Figure 9.45, the contribution of the TTW and WTT (well-to-tank) component to the reduction of CO_2 emissions is different among the three GEA pathways. In GEA-supply and GEA-mix pathways, the reduction of the WTT component is more important, because the negative emission from power generation, biofuels, and H_2 production is effectively utilized. In the GEA-efficiency pathway, the TTW component contributes more to the reduction of CO_2 emissions through the greater improvement of vehicle fuel efficiency.

Large fuel efficiency improvements and introduction of alternative fuels in LDVs and trucks in the GEA pathways contribute to the greater reduction of CO_2 emissions compared to the reference scenario. In aviation, the large growth of activity offsets CO_2 emission reductions due to efficiency improvement and biofuel use.

Analysis of the contribution of major measures on emission reduction in the LDV subsector is seen in Figure 9.46. It clearly shows that net reduction of CO_2 emissions from the current level can be achieved by the introduction of new technologies, such as EVs and fuel cell (FC) vehicles, although improvement of fuel efficiency in conventional vehicles (FE) and hybrid vehicles (HV) also contribute significantly to the reduction. It should be noted that for market penetration of these new emerging technologies, strong, continuous support from governments is needed from the R&D phase through to the commercialization phase. In the GEA-efficiency and GEA-mix pathways, the contribution of modal shift to trains is also important.

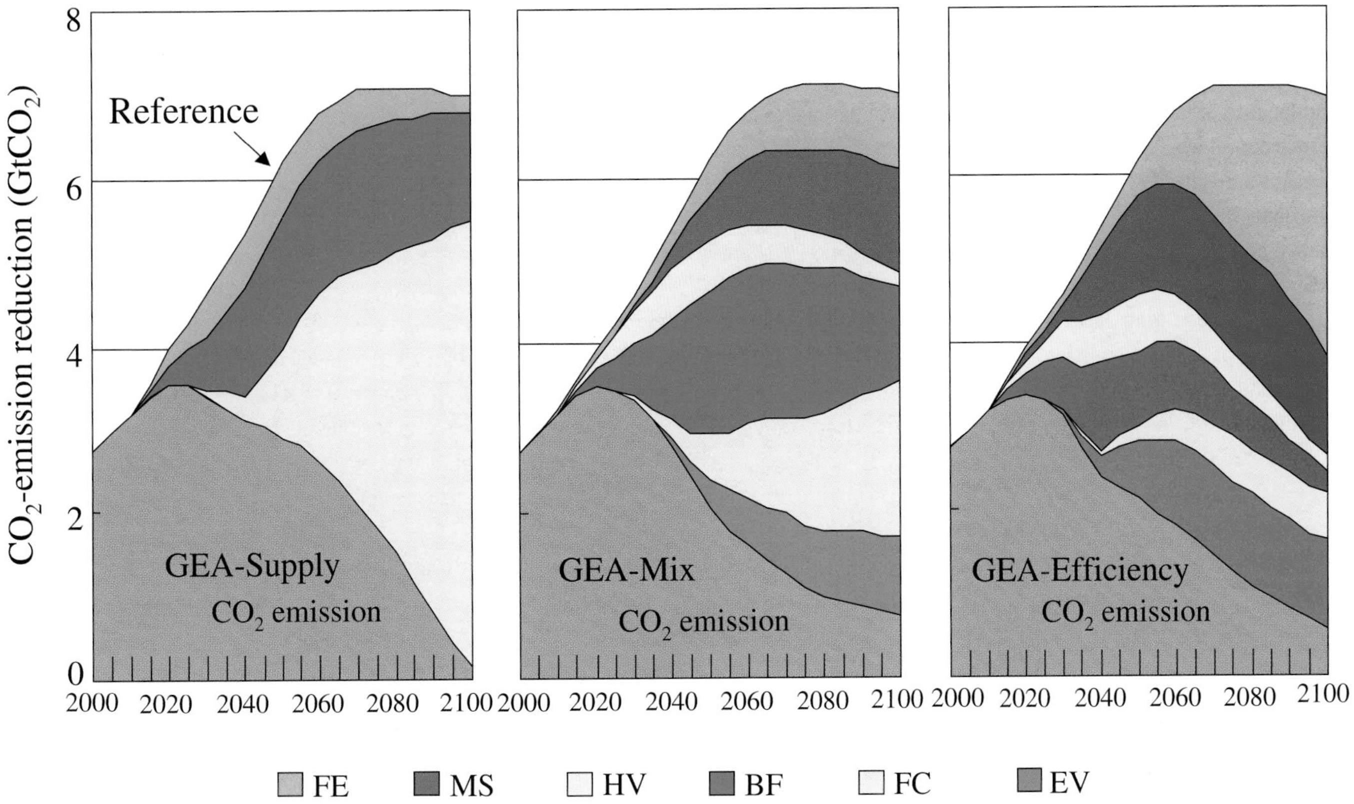

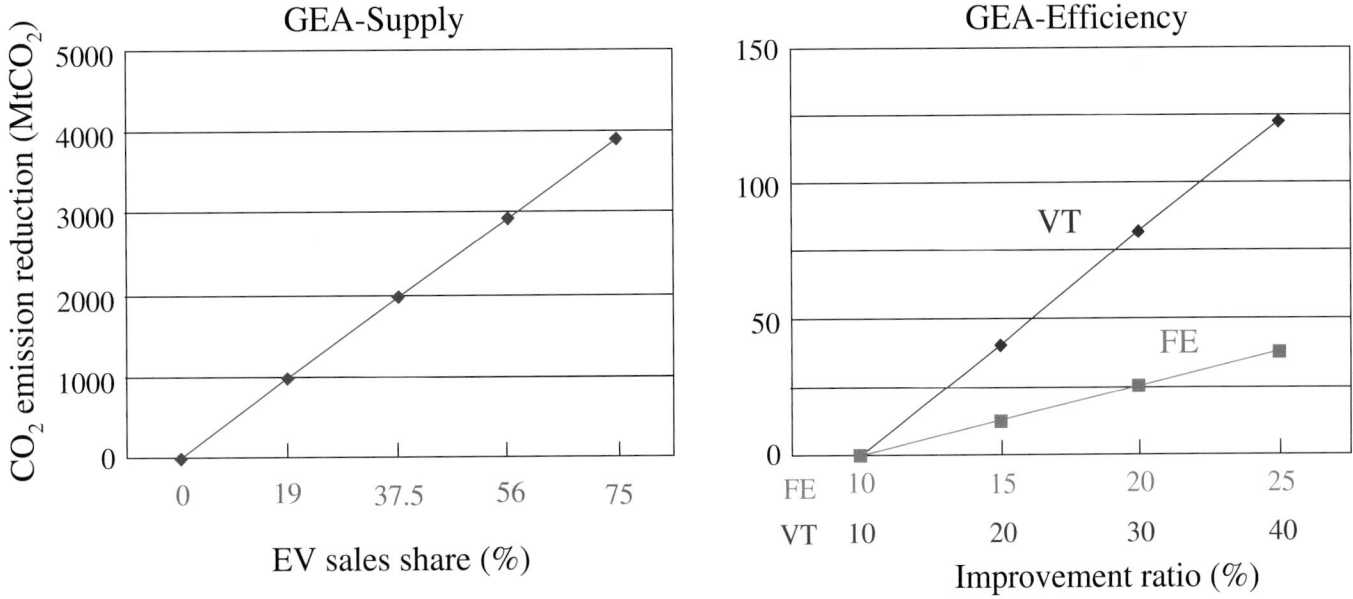

Figure 9.46 | Reduction of well-to-wheel CO_2 emissions for LDVs from reference scenario in GEA (FE: Fuel efficiency, MS: Modal shift, HV: Hybridization, BF: Biofuel, FC: Fuel cell, EV: Electric vehicle).

Figure 9.47 | Sensitivity analyses for the EV share in the GEA-supply pathway, and the reduction rate of vehicle driving distance per vehicle (VT) and fuel efficiency (FE) in the GEA-efficiency pathways.

Since perspectives of technology evolution are highly uncertain, especially toward 2100, some sensitivity analyses were performed for the LDV subsector, including the share of EVs, improvement of the fuel consumption rate, or fuel efficiency (FE), and reduction of the vehicle driving distance per vehicle (VT). In the GEA-Efficiency scenario, the share of EV and FCV sales is fixed to be around 75%, and CO_2 emissions from electricity generation factors for FCVs and EVs are nearly zero-emission and negative-emission, respectively. As shown in Figure 9.47, EV share in the GEA-Supply scenario was changed from 0 to 75%; accordingly, FCV share was changed from 75 to 0%. With the increase of EV share, the contribution of reduced emission of electricity generation using fossil fuels becomes larger and CO_2 reduction increases up to 4 Gt, which can be compared with the default emission level of the GEA-supply pathway – about 6 Gt (Figure 9.49).

In the reference scenario, LDV annual VT is assumed to be constant up to 2100, while in the GEA-efficiency scenario, it decreases after 2050 and is 13–17% lower than the reference scenario in 2100. The FE is assumed to be improved 2–3%/year in the reference scenario. In the GEA-efficiency pathway, this is further decreased after 2050 up to 13–17% in 2100, compared with the reference scenario.

The impact of these parameters on the reduction to CO_2 emissions is much smaller than that of the EV share mentioned above. However, this does not mean that the improvement of these parameters is not important, because they strongly affect the total amount of the fuel use in the transport sector. Figure 9.47 shows how a reasonable range of variation of these factors affect the estimated variations quite significantly.

9.5.4 Interpretations of How the Pathways Meet the GEA Goals

Measures to achieve the GEA goals should be distinguished by the timeframe on which one is focusing. Priorities among the many options may change over time. In the short and mid-term, most developing countries will still be in the transition stage of motorization, and available technological measures to meet the goals may be limited. In the long term, some developing countries will be in the middle of motorization, and various advanced technologies, such as advanced biofuels, fuel cell vehicles, and electric vehicles might be widely available in the market.

9.5.4.1 Fuel Efficiency

Improving efficiency is the primary option to reduce energy use and CO_2 emissions simultaneously from now until 2100. Many countries have imposed fuel economy standards. They have shown that these are very effective policy instruments, requiring car manufacturers to make more efficient cars. Additional measures, such as taxes and incentives, are required to encourage customers to buy these more efficient cars.

Fuel efficiency improvement itself is not good enough to reduce tail pipe emissions, such as NO_x, CO, and hydrocarbons. After-treatment devices for exhaust gases are required to significantly reduce pollutants being expelled from the tail pipe, which can be enhanced by emission standards.

9.5.4.2 Modal Shift

Switching passenger transport from LDVs to trains, buses, and non-motorized means can be very effective for reducing energy use. For dispersed suburban living, such as in many parts of the United States, this may not be an easy task. Infrastructure investment, integrated land use and transport planning, and sustainable transport policies are needed to shift personal travel to more energy-efficient modes such as walking, cycling, and public transportation.

A co-benefit of such a modal shift is reduction of traffic volumes, especially in urban areas, which contributes to a reduction in congestion and local air pollution. Since increased traffic volumes correlate to an increase in road traffic fatalities, reducing congestion through a modal shift has another co-benefit of improving road safety.

9.5.4.3 Advanced Biofuels

The merit of biofuels is their additive natures to other improvement measures in conventional vehicles, such as fuel efficiency, and therefore quick effects can be expected. With a careful choice of production pathways, biofuels can reduce CO_2 emissions significantly, as well as some tail pipe emissions.

9.5.4.4 Fuel-Cell Vehicles

FCVs can reduce energy use and drastically reduce emissions with the proper choice of pathways. The biggest barriers to FCV introduction are high costs and infrastructure issues. To build up infrastructure takes a long time and strong government support is needed.

9.5.4.5 Electric Vehicles

Hybrid vehicles and plug-in hybrid electric vehicles offer a cost-effective and smooth transition to electric mobility. PHEV, could expand the sale share to compromise the high cost of batteries. However, since only a small fraction of consumers may accept the limited driving range and high cost of a pure battery electric vehicle, battery electric vehicles may be limited to a small market in urban areas.

One of the major merits of FCVs and EVs is zero emissions from tail pipes. This is especially beneficial in highly-populated urban areas.

However, the GHG reduction potential of electric vehicles depends significantly on the carbon intensity of the electricity generation process. Since hydrogen and electricity can be produced by various energy sources, use of a secondary energy can reduce oil use and increase energy security.

Although advanced technologies have some barriers to overcome, new technologies are needed to reduce CO_2 emissions below the current level (Figure 9.46).

9.6   Policies and Measures to Meet Multiple Goals for Sustainability and toward a Global Energy Transition[22]

The transport pathways described in the previous section illustrate alternative paths and aspirations for a progressive transformation of the global transport sector in line with the GEA goals toward attaining a sustainable global energy transition within the next 50 to 100 years. The pathways presented are technologically optimistic but still display insufficient potential to decarbonize the transport sector. GEA emphasizes that the sustainable global energy transition needs to be anchored in socially equitable, environmentally sustainable, and economically effective development goals. The scale and scope of the transition indicates that nothing less than a sustainable systemic change is required. Hence, we also include shifts to non-motorized transport and public transport modes, and integrated land use and transport planning in our analysis of suitable policies towards a desired transition.

The world is now predominantly urban, and most future population increases will be absorbed by urban areas (see Chapter 18). Transport plays a fundamental role in the development and economic prosperity of urban areas because commercial organization, the location of industry, housing, and all other general services are transport dependent. At the present rate of world urbanization, cities will require increased transport services to make accessible the supplies needed for their physical expansion and to support economic development. The challenge of developing sustainable low-carbon transport systems will define the possibility of guaranteeing life in urban places as economically viable, socially constructive, environmentally safe, and, in general, qualitatively enjoyable spatial configurations. Hence, a large part of the policies and measures discussed in this section focus on urban areas.

From a climate change mitigation perspective, one can categorize emissions from transport into carbon intensity of energy, energy intensity of transport, and total transport demand (Schipper and Marie-Lilliu, 1999; Creutzig et al., 2011a). Both the decision to travel or not and the modal choice for this travel affects fuel consumption, and therefore carbon emissions. With a focus on urban road transport, a transition to sustainable transport can follow the "Avoid, Shift, Improve" framework (GTZ, 2007;

Bongardt et al., 2010). This framework considers three major principles under which diverse policy instruments (Planning, Regulatory, Economic, Information, Technological) are grouped interventions to mitigate GHG emissions from transport, assuming different emphases for developed and developing countries (Dalkmann and Brannigan, 2009). "Avoid" and "Shift" influence the level of activity and structural components that link transport to carbon emissions. "Improve" focuses on technological options, not only with respect to climate mitigation but also taking into account local environmental conditions and social concerns.

9.6.1   Reduce the Need to Travel

A transition toward sustainable transport goes beyond technology. In fact, the quality of transport can be improved in a broader sense by a multitude of individual measures and combined policy packages, particularly in urban areas. Reducing or avoiding the need to travel can be achieved through encouragement of greater use of transit systems and non-motorized transport, as well as through increasing urban density, mixed-use urban spaces, and greater emphasis on communications technologies.

While technological pathways and scenarios allow us to envisage large-scale changes that can help decarbonize the transportation sector, measures such as those mentioned above that reduce congestion, air pollution, and motor noise and that increase the accessibility of shopping and local services for pedestrians and cyclists ultimately increase quality of life in a very tangible manner.

9.6.1.1   Enhancing Accessibility

Enhancing accessibility for transport users often reduces overall transport demand in terms of distance while improving the quality of movements. Analysis focuses on aspects of mixed land use and density, modal interconnectedness, and transit-oriented development. The relation between mobility and urban structure is interdependent. While cities generate the transportation systems they require in the development process, mobility demands and transport system responses also shape urban form.

GEA has developed a full module on urbanization and energy use in Chapter 18 that explores the relationships between patterns of spatial, population, and economic growth of cities. It looks at interactions with the patterns generated by urban transport systems, and the resulting effects on total energy use. This section will only focus on general public policies for managing and shaping transport and land use urban activities, and recommend interested readers to take full advantage of the thorough review provided in Chapter 18.

The best opportunities for directing transportation, land-use planning, and urban development toward sustainability may reside in urban

---

22 See also Chapter 22.

centers of small and medium sizes, where most of the future population and economic growth are expected. Urban populations continue to expand at more than 6%/year in many developing countries. In some of these countries, per capita motor vehicle ownership and use continue to grow by up to 15–20%/year. Sprawling cities make the journey to work excessively long and costly for the very poor. This is overwhelming the operational sustainability of urban transport systems and is one of the causes of social, economic, and environmental imbalances and social inequities. The challenge for towns and cities everywhere is to enhance accessibility and reduce congestion, accidents, and pollution.

Traditional transportation planning developed largely in parallel with a post-WWII explosion of individual motorization in the United States, in an already wealthy society going through a period of sustained economic growth. The initial policy response in the United States, based on a predict-and-provide planning framework, was to accommodate the private car to the maximum extent possible, while retaining public transport modes for the dwindling market of those without cars.

An entirely different strategic approach was taken in the former communist countries, opting for public transport as the primary urban mode and developing cities accordingly. In both these cases, the common element has been the strength of institutions and the importance of public policies for the provision of infrastructure, the supply of services, and, consequently, the modal choices.

9.6.1.2   Enhancing Modal Interconnectedness

Smart Growth (also called New Community Design) is a general term for policies that integrate transportation and land use decisions, as an alternative to urban sprawl. Smart growth and mixed land use is closely related by multimodal interconnectivity. For example, public transit ridership crucially depends on design of pedestrian facilities, enabling smooth and rapid access to stations. Park-and-ride is ambivalent. On the one hand, park-and-ride facilities allow car-dependent commuters to participate in public transit in inner city areas. On the other hand, park-and-ride may encourage further urban sprawl and medium-length commuting to park-and-ride facilities. To increase ridership and improve the economic viability of transit systems, transit stations need to be accompanied with mixed-use development, i.e., including both employment opportunity and housing, and market-rate parking charges (Cervero, 1994). In contrast, large-scale free parking makes the area surrounding transit stations unattractive and increases insecurity. Easy access to car sharing can also have two effects. On the one hand, it can deter car purchase, and by this reduce overall car ownership. On the other hand, it may encourage car driving by those that usually rely exclusively on environmentally friendly modes. However, it is estimated that the overall effect of car sharing is beneficial for mobility and the environment.

9.6.1.3   Enhancing Regional Transit-Oriented Developments

The most consistent approach, which is by now state-of-the art in progressive cities, is to integrate land use and transportation planning to achieve a low-carbon infrastructure, reduce travel distances and travel time and increase accessibility to jobs, shops, and leisure facilities (Bongardt et al., 2010). The bottom line is that new mixed-use, high density developments must be close to public transit, and vice versa, and that the cost of using a car must be relatively increased in terms of money and time, to avoid high levels of motorization. Transit-oriented developments enable people to access housing, jobs, services by walking, biking, or riding transit, thus avoiding the necessity of driving to meet these needs and increasing the accessibility of goods and services. The case of Curitiba in Brazil offers a good example of how a coherent set of policies linking urban density, and a highly efficient bus rapid transit public transport system can transform a city (Bongardt et al., 2010)

9.6.1.4   Enhancing Mobility Management

Reducing the need to travel by car can be approached by implementing mobility management practices. The goal is to enhance and create transport services that promote a safe interconnection of modes and infrastructure for the use of sustainable modes such as walking, cycling, and public transport. Mobility management is a targeted approach that focuses directly on changing behavior at the individual or group level (e.g., company employees). It requires information, organization, coordination, and effective marketing and promotion to complement traditional traffic system management and infrastructure-orientated transport planning.

Mobility management can be seen as a new way of managing urban transport as a whole. It pulls together traditional approaches with public transport improvements and traffic management measures (including congestion charging and traffic calming). The main aims of mobility management are to:

- improve accessibility for all users and the conditions (social attractiveness, safety, economic efficiency) for an effective use of sustainable travel modes;

- improve the integration of activities and new land uses with sustainable travel modes and services covering the entire urban transport system;

- reduce traffic growth by limiting the number, length, and need for motorized vehicle trips; and

- improve interchange between transport modes and facilitate the interconnection of existing transport networks.

Mobility management has a demand-orientated approach that considers alternative transport modes – public transport, collective transport, carpooling, cycling, walking, etc. – as "products" that have to be marketed. It works with specific "clients" or "client groups," defined according to trip nature and purpose (home-school, home-work, shopping, leisure), and 'traffic generators' or the sites or activities that attract the traffic (city centers, companies, schools, tourist attractions, events, shopping centers, residential areas, etc.). Because each site is different, mobility management generally works with site-specific mobility plans.

Good marketing is necessary to convey and promote the positive results and benefits on the use of sustainable modes. Changing travel behavior is often associated with restrictions and sacrifices, or with critiques for placing unfair restrictions on car use. A well-planned mobility management plan ultimately enhances mobility opportunities for everybody, those who accept to shift to alternative modes and those who continue to drive.

9.6.2 Develop Alternatives to Car Use

9.6.2.1 Enhance Public Transportation

Public transportation carries a large proportion of passenger trips in most of the heavily populated urban areas of industrialized and developing regions in the world. Many developing cities, where the modal share for bus transit alone can be in upper range of 65–80%, are

A)		2 000	9 000	14 000	17 000	19 000	22 000	80 000
B)	MJ/p-km	1.65-2.45	0.32-0.91*	0.1	0.24*	0.2	0.53-0.65	0.15-0.35
C)	€ p-km infrastructure	2 500-5 000	200-500	50-150	600-500	50-150	2 500-7 000	15 000-60 000
D)	Fuel	Fossil	Fossil	Food	Fossil	Food	Electricity	Electricity

*Lower values correspond to Austrian busses, upper values correspond to diesel busses in Mexico city before introduction of BRT system.

Key:

A) Values are indicative for European and Asian cities and can vary significantly across cities, world regions, and particular situations. For example, BRT capacity can more than double with a second lane. Suburban rails in India can transport up to 100,000 passengers per hour.

B) Energy intensity in MJ per passenger km. SUVs can exceed depicted values for cars. Energy values for bus in the US are generally higher due to low ridership (Chester and Horvath, 2010). While BRT systems have similar energy efficiencies as normal busses, they provide signficant systemic energy savings via modal shift, small bus substitution, and reduction in parallel traffic (Schipper et al., 2009). BRT systems can also be converted from oil based fuels to electricity and hydrogen.

C) Estimated infrastructure costs in euros per passenger kilometer are highest for subway systems and heavy rail. Costs for bus system can be significantly lower than for individual motorized transport. Infrastructure costs for non-motorized transport are very cost competitive and can realize significant social benefits.

D) Dominant fuels are given for each mode.

Figure 9.48 | Corridor capacity of different modes of transportation (people/hr on a 3.5 mile-wide lane). Source: Modified from Breithaupt, 2010.

characterized by hundreds of separate bus companies. For example, a single minibus route in Lagos, Nigeria is estimated to be served by 200 mini enterprises (Gwilliam, 2005; Small and Verhoef, 2007). The provision of high-quality public transport connecting urban and suburban centers, with integrated fare systems and efficient intermodal/interface linkages, is paramount to enhancing the public transport services, equal accessibility, and other social aspects related to the provision of equitable services of these transport systems.

Bus Rapid Transit (BRT)

Bus Rapid Transit (BRT) is a high-quality bus-based transit system that delivers fast, comfortable, and cost-effective urban mobility (Wright, 2006). It is one of the most important transportation initiatives today and is increasingly being used by cities looking for cost-effective mass transport solutions.

BRT systems can enhance bus efficiency through segregated bus lanes, designs that make boarding and exiting buses quicker, bus priority at intersections, and effective coordination at stations and terminals. Political backing is a key ingredient for success in all BRT systems (Wright, 2006). However, dedicating road space to exclusive use by public transport can be politically difficult, especially given the relative political strength of private motorists. BRT systems can achieve higher speeds than conventional buses[23] and capacities that equal or exceed rail transit at lower costs as they usually take over an existing road lane (about 35,000 passengers can be carried in each direction; see Figure 9.48) (Hidalgo and Carrigan, 2010; ITDP, 2007). BRT has the flexibility to provide limited stop and express services.

There are currently 47 BRT systems operating worldwide and 16 more under construction. Most systems are in Europe (17, of which 10 are in France), while widely visible systems are in Latin America – in Curitiba, Brazil and Bogota, Colombia (Wright, 2006). By providing high-capacity and high-speed service, BRT systems attract more riders and provide service more efficiently than conventional bus services operating in mixed traffic (ITDP, 2010). BRT systems have achieved a certain amount of success in providing reasonable fare levels without the intervention of operating subsidies. Fares in the range of US$0.25–0.70 are typical with the subsidy-free systems in Latin America (Hidalgo and Carrigan, 2010).

Not all BRT strategies, technical and operational elements are transferable from one city to another. Availability of institutional, technical and management skills are vital to whether a BRT system will work effectively in its operations (handling of passenger boarding and alighting efficiently, with little delay); system signal prioritization (avoiding disruptions to traffic flow on major cross streets); vehicle designs (affecting system performance, cost, and appearance, both external and internal). These aspects are contributors to the overall success of any BRT system.

Light Rail Transit (LRT)

LRT uses electric rail cars operating along a dedicated lane or track, separated from other traffic most typically for density reasons in city centers, or in the case of trams, they often operate in mixed traffic without an exclusive lane. Light rail systems can carry 10–20000 passengers per hour down one corridor (Wright and Fjellstrom, 2005).

There are currently over 400 light rail systems operating worldwide, of which around 300 are tram systems. Most are in Europe, with a few in developing countries. Advances in electronics, software, and materials are playing a major role in making LRT systems more attractive to operators. This has been a major factor in the expansion of existing systems and the construction of entirely new systems during the past several decades (Wright and Fjellstrom, 2005).

LRT is seen as an extremely diversified mode that can be used for short urban lines, as well as long regional lines with various levels of speed and capacity, using alignments that vary from streets to fully separated tunnels, viaducts, and intercity railway tracks. LRT is also seen as an element of urban economic development; in fact, economic theory suggests that the increased value of land close to LRT is best used to finance public transit. Improvements in track and car design have reduced noise levels to a point where LRT protagonists consider LRT systems quieter than diesel buses (Wright and Fjellstrom, 2005).

Increasingly, BRT is challenging the development of new LRT systems on the grounds of cost. BRT systems are becoming more and more sophisticated and capable of emulating the service levels of LRT at much lower investment and operating costs.

There is considerable debate over the capacity of LRT systems compared with BRT systems. With the use of articulated and bi-articulated buses in Curitiba and Bogota, the capacity of BRT has equaled – and exceeded – that of comparable LRT systems. The development of LRT systems is therefore likely to be limited to cities where tram operations exist, and where they can be cost-effectively upgraded and enhanced.

A negative development limiting applications of LRT has sometimes been over-design. Instead of economical designs that allow for the construction of large networks, a number of projects have been upgraded step by step, resulting in very high costs.

BRT and LRT should be considered as complementary modes. LRT is seen as being suitable for situations demanding heavy passenger volumes, use of tunnels in high-density urban centers, and direct service in pedestrian zones.

Heavy Rail Transit

Heavy rail transit systems comprise Metro and Commuter Rail systems operating in exclusive rights-of-way without grade crossings, with high platform stations.

---

23 Conventional buses usually operate under significant levels of congestion. For example while conventional buses were operating at 10 km/h, the BRT Transmilenio System in Bogota operates at an average commercial speed of 27 km/h (ITDP, 2010)

Metro is the term for mostly underground (subway) heavy rail transit, which are the most expensive form of MRT/km, but have the highest theoretical capacity 50,000 passenger per hour in one direction (even 80,000) (Wright and Fjellstrom, 2005, Salter, et.al, 2011). The option of constructing an underground capital-intensive Metro transport system has been supported by many elected officials and voters with promises that it will help reduce congestion, pollution, transform social culture and even become the gem of a cosmopolitan city (Figueroa, 2010; Global Mass Transit, 2010). This despite evidence of extremely high costs compared to buses and a record of under-predicting the cost on this type of project (Flyvbjerg et al., 2004; Flyvbjerg et al., 2009).

Heavy rail transit systems are primarily found in large, dense cities of the industrialized world. However, there are also quite a few in upper-tier developing world cities, such as Tehran, Mexico City, Guangzhou, and Santiago. Bangkok has a new train system that reaches 60 km/hr, averaging 25–45 km/hr compared with the Bangkok traffic speed of 14 km/hr (Salter, et. al, 2011).

Intraurban – long distance – transport is most efficiently served by railways than it is today when car transport and aviation are the prevalent choices. High-speed railways trips are competitive in time with car and plane trips. Notably in some European countries and Japan, for up to 1000 km (Madrid-Barcelona, Paris-Marseille, Hannover-Munich, Kyoto-Tokyo). China puts efforts into replicating and even superseding this success story. Forty-two high-speed, mass transit train lines have been operationalized in China in the last years. The Chinese bullet train connects Ghangzhou to Wuhan, traveling 1036 km (664 miles) in a little more than three hours. The new superfast trains introduced in China are economically competitive and create economies of large-scale production for another big export industry.

9.6.2.2   Enhance Other Means of Transportation

There are many other alternatives to private car use that can be researched and considered.[24]

Enhance and Facilitate the Use of Non-Motorized Modes

The vast majority of all trips made daily worldwide are on foot. In the United States, for example, according to the 2001 US National Household Survey, 48% of trips were 3–4 km or less, and 24% were 1.4 km or less. These trips are the most suitable to the use of non-motorized modes. Many people rely on NMT, i.e., trips made on foot or by wheelchair, bicycle, tricycle, skateboard, handcart, cycle rickshaw, or other non-motorized vehicles, to meet their daily transportation and mobility needs. Facilitating the use of NMT is important because they are the least carbon-intensive mode of transport available and have

been shown to have health benefits. Most people worldwide have the potential to use NMT.

The presence of pedestrians and safe cyclists are a commonly-used gauge of a successful, vibrant urban space. Pedestrian environments provide safe and salubrious public spaces where people can meet and interact. Active NMT can be part of the solution to increasing public health and confront obesity concerns, particularly in developed countries.

Planning for NMT extends beyond the physical infrastructure to include psychological factors and institutional stasis hindering the adoption of non-motorized modes, variability in the facility needs of each of mode, safety, equity concerns, and land-use practices conducive to NMT. Below a brief discussion on the ramifications of these concerns, planning strategies applicable to non-motorized vehicles (NMV) and to pedestrians are presented.

Equity

In many places, NMT is viewed as the option of last resort if it is not for recreation. Low-income and transportation-disadvantaged persons (elderly, youth, people with disabilities) often rely and gain independent access to services from improvements on NMT facilities, and so does the community at large. The presence of sidewalks and curb ramps, free from surface irregularities and of adequate width, not only give people with disabilities a way of travelling independently but also benefits all pedestrians. Such infrastructure also benefits parents with prams, tourists with luggage, the old, the young, the pregnant, the visually impaired, wheelchair users, users with crutches, etc. Such universal and equitable access principals promote designs that accommodate the widest range of users possible.

NMT can also provide economic benefits for the poor. For many poor, private vehicle ownership of a bicycle, animal carts, or cycle rickshaws actually indicates a considerable economic achievement or a sign for upward economic mobility.

In urban areas, NMTs are not only relevant for the movement of people, but also for the transport of goods. NMVs such as cycle rickshaws and handcarts are an essential means of transporting people and goods in cities and towns in Asia and Africa. In many African towns, sellers or small-scale service providers use handcarts to transport goods to and from markets. In Asian cities, rickshaws designed for passengers often transport goods. NMT also facilitates indirect economic benefits sparked by the demand for NMTs, spare parts, and services such as repair, vehicle rent, and paid parking.

Safety

Road safety is a critical issue. Increased NMT can beget economic mobility and other equity benefits, but it is irresponsible to promote NMT without addressing various safety concerns associated with these relatively vulnerable road users. The speed and volume of cars, trucks, and buses are particularly threatening to cyclists and pedestrians. Data from several cities shows a "safety in numbers" phenomenon – that increasing the

---

24 For further references regarding other private providers of public services such as digital hitchhiking, hitchhiking, jitney, share taxis, shuttle services, slugging, taxis, vanpooling, see GTZ, 2002; ITDP, 2010.

number of bicyclists reduces the rate of accidents or injuries of bicyclists (accidents per 1000 bicyclists, for instance). This suggests that without changing the volume of vehicular traffic, introducing more cyclists to the road makes bicycling safer. There are several reasons and theories, including the argument that drivers become more familiar with seeing bikes on the road, and thus come to expect and look for them (Brandt, 2011).

Proper infrastructure for pedestrians and bicyclists is critically important for supporting increased NMT mode share. Sidewalks, pedestrian crossings, and bicycle lanes are only the most basic provisions needed. At a minimum, basic, unobstructed, and safe infrastructure is needed to accommodate and promote NMT. Desirable infrastructure would include secure bicycle parking at transit stations, physically segregated bike paths, bike boxes, dedicated bicycle turning signals, and audible and tactile pedestrian crossing signals. The quality of NMT infrastructure will largely determine the NMT mode share.

The presence of so-called "eyes on the street" NMT tends to increase perceptions of safety and can help pedestrians feel safer on less-frequented streets. Adequate light levels at night are important, particularly for women, to enhance safety. Separate facilities and proper lighting are also essential because NMVs and pedestrians may be difficult for drivers to see, particularly when traffic speeds are moderate or high. A concerted effort to increase safety for NMT users is crucial in its own right, but as an ancillary benefit, an attractive and safe pedestrian environment creates more livable communities.

Overcoming Barriers

Even if safe, extensive, convenient facilities for NMT were available, there remain various psychological, educational, geographic, and institutional barriers.

There is a widespread lack of respect for non-motorized users even in developing countries where NMT are dominant. In general, NMT tends to be stigmatized. Some people consider walking and cycling outdated, unsophisticated, and unexciting compared with motorized modes. Some even consider them as symbols of poverty and failure. These are one of the biggest psychological impediments for a modal shift away from motorized modes (UNEP, 2011; Salter, et al., 2011).

Constraints placed by physical barriers and excessive detours to overcome these barriers also impede use of NMT. In many cities, the environment is seen as unattractive for NMT and pedestrian use due to pollution, traffic volumes, and poor quality walkable areas.

Bicyclists report many reasons why they are encouraged to cycle, chiefly among them are cost-effectiveness, health, status. (Alliance for Biking and Walking, 2010). Potential cycle commuters can be however discouraged by a lack of shower and parking facilities at work. NMVs in general, and bicycles in particular, can be easy to steal. Provision of safe and secure bicycle and other NMV parking, in integration with public transportation stations and stop places are therefore essential ingredients

to promote NMV use. Creation of safe routes and parking facilities and simple education and training for people to gain confidence or road skills to cycle contribute to the widespread adoption of NMT (Alliance for Biking and Walking, 2010). Geography can be a factor that constrains NMV in many places.

There are institutional barriers and lack of technical staff to plan NMT facilities and manage a large increase of their use. It is often difficult to agree on prioritizing and secure funding for infrastructure and non-infrastructure projects for NMT due to finite funds, other transportation priorities, and underestimation of the magnitude of NMT. Travel surveys and traffic counts usually ignore or under-count short trips, non-work travel, travel by children, recreational travel, and non-motorized links. Many conventional user surveys attribute the entire trip to a single mode, regardless of the extent to which additional modes are involved and ignoring the fact that virtually every trip begins and ends with walking.

Increasing Non-motorized Transport Mode Share

There are many specific ways to encourage mode shift from other modes to NMT. A short list of suggestions is presented here modified from (Litman, 2010):[25]

- Planning for NMT is an important part of transport planning. An integrated transport policy is required that defines the role of NMT and NMVs within the total transport system and in relationship to other modes, such as private cars, public transport, and walking. Action plans need to include measures that will help overcome the barriers to NMT outlined above and better integrate NMT. Policies to encourage NMT usually form part of the mobility management and transport demand management measures.

- Improve pavements (sidewalks, cycle tracks), crossings (crosswalks), paths, and bicycle lanes and correct specific road hazards (sometimes called "spot improvement" programs).

- Improve the management and maintenance of NMT facilities, including reducing conflicts between users and maintaining cleanliness and enhancing safety.

- Accommodate people with disabilities and other special needs (universal design).

- Develop pedestrian-oriented land use and building design (new urbanism or transit-oriented development) and increase road and path connectivity, with special NMT shortcuts, such as paths between cul-de-sacs and mid-block pedestrian links.

- Introduce and maintain NMT related street furniture (e.g., benches) and design features (e.g., human-scale streetlights).

---

25 For a full overview on walking and cycling encouragement, see Victoria Transport Policy Institute, 2011.

- Use traffic calming, speed reductions, vehicle restrictions, and road space reallocation.

- Introduce safety education, law enforcement, and encouragement programs.

- Integrate NMT and public transport facilities (bike/transit integration).

- Provide adequate and secure bicycle parking and address security concerns of pedestrians and cyclists. Create multi-modal access guides, including maps and other information, on how to walk and cycle to particular destinations.

Municipalities and sustainable transport agents can market and promote activities to provide people with information, skills, and positive examples or role models that will promote the popularity of non-motorized vehicles.

Improve Pedestrian Infrastructure

While all the recommendations above also apply to pedestrians, pedestrians are inherently different from NMVs. Pedestrians generally travel more slowly than any other mode – about 4.5 km/hr. Many people will only walk about one kilometer, leaving pedestrians particularly sensitive to detours, roadway conditions, street aesthetics, and perceptions of street crime.

There are many ways to improve facilities, depending on existing infrastructure for pedestrians. A needs assessment can measure indicators of walkability, including factors such as land use mix, street connectivity, residential density, street and building design, scale and nature of place near homes, retail floor area ration, access to public transport, availability and quality of sidewalks, degree of separation from moving traffic and the degree to which moving traffic constitutes a barrier, pedestrian crossings, accessible and direct routes, and air quality.

Additionally, walkability can be evaluated at various scales (Litman, 2010):

- *Site level*: walkability is affected by the quality of pathways, access to buildings, and related facilities.

- *Street or neighborhood level*: the existence of footpaths (pavements or sidewalks) and pedestrian crossings (crosswalks), as well as roadway conditions (road widths, traffic volumes and speeds).

- *Community level*: accessibility, such as the relative location of common destinations and the quality of connections between them.

9.6.2.3   Enhance use of Telecommuting and Communication Technology

Technological change has fundamentally influenced the function and forms of cities and the way people and goods can travel in the city. The convergence of computing and information communications technologies allows a user-friendly interface for many actors and transactions, for information, education, business and a wide range of daily life social activities (Banister, et al., 2004). Communication networks and applications can improve mobility systems in several ways for example: by substituting some types of physical movement with Information Technology based solutions (for example: e-health & e-care; e-learning, e-working, e-culture & e-media, e-economy, e-mobility, e-government & e-democracy); by unleashing significant traffic efficiency via urban traffic managing systems that increase logistic opportunities and safety, and by facilitating opportunities to create a better connected and integrated array of public transport services.

There is a potential for telecommuting to substitute some work trips, particularly for jobs in which greater work flexibility is possible or where rules are made more flexible. Access to information and communication technologies may also have a dual effect where readily available information on new and exciting destinations and shopping-related entertainment stimulate travel and limit the scope of the substitution of virtual and physical travel. For the most part a synergistic relationship can be expected. That is, the more one or another form of telecommunication takes place, the more all forms of communication and travel are stimulated, while some types of travel are substituted (Mokhtarian, 1990; Banister, et.al., 2004). The full impacts of this dual relationship are still being determined. Some studies report that there is no clear certainty that activities such as online-shopping can result in environmental gains from a reduction in passenger transport (Cullinane, 2009). Similar evidence can be found regarding the effects on freight transport indicating that "internet shopping requires carriers to deliver goods to end users rather than to intermediaries, resulting in overall growth in direct shipments to customers. Together, these trends have increased the number of shipments, particularly small shipments, that carriers handle, and expanded the number of links in the freight supply chain that are needed to deliver goods to their final destination" (US DOT 2010).

9.6.3   Improve Use of Existing Infrastructure

Road traffic growth outpaces population and economic growth in most urban areas of the world. In developing countries, motorization can be as low as ten cars per thousand people (parts of China, India) and rarely exceeds 200 cars/1000 people, while two-wheeler ownership can be high. Motorization is higher than 600 cars/1000 people in the United States. Even with the current low levels of motorization in many developing countries, major challenges are associated with the unprecedentedly fast pace of urbanization, which imposes a high demand for adequate transport and the provision of other urban services and infrastructure. As more than one-half of the developing world's population and between one-third and one-half of its poor will be living in cities within the course of one generation, a substantial increase in the demand for energy services can be expected from most urban areas (Gwillian, 2002).

Further construction of major road infrastructure has done little to tame congestion or to reduce overall travel time. High-profile examples in large metropolitan areas of developing countries are Beijing, Bangkok, and Mumbai, where a decade of unprecedented investment in transport transformed an entire urban landscape with congestion becoming ever more problematic.

Institutions and policies that either accommodate the car or favor public transport have not been adequate, which resulting in greater congestion and poor accessibility and mobility. With well-known exceptions like Curitiba, Brazil and Hong Kong, China neither urban nor state-based institutions in developing countries have been strong and/or funded enough to accommodate rapid rates of population and motorization growth, exacerbated by the presence of sharp income inequalities.

The World Bank is actively working to carry out a retrospective/prospective study on urban transport policies and activities and is financially supporting construction of several urban transport systems. In addition, several initiatives of a global scale like the global Transport Knowledge Partnership, the Sustainable Urban Transport Project and the Partnership for Sustainable Low Carbon Transport are providing the best available international experience and expertise to help develop sustainable and efficient transport in developing and transition countries.

9.6.3.1   Parking Management Policies

Every car on the road needs a place to be parked; it is a key issue in almost all urban areas (GTZ, 2009). For an average of 23 hours of the day cars are parked, and if used for all journeys they would need a parking space at both ends of every trip – meaning that many spaces are required for every car. A parked car takes up around eight square meters when parked and often the same again in maneuvering space; this is a huge amount of space in dense urban areas where land is limited and expensive.

In many cities around the world, parking is unregulated, in great supply, and provided free of charge. Abundant parking makes it easier and cheaper to drive, and pandemic parking lots increase the distance between development, making cars more necessary (Shoup, 2005).

While drivers may pay very little to park a car, the actual cost of the parking spaces can be quite high. The cost of surface parking (on street or off-street in parking lots) depends on the price of the land and therefore varies significantly by location. Each space in a parking structure costs at least US$125/month (Shoup, 2005). Despite the high cost of providing parking spaces, in 1995, 95% of US automobile commuters parked free at work (Shoup, 2005).

Free or heavily subsidized parking reduces the costs borne by drivers, biasing individuals' travel choices toward more driving (Shoup, 2005). Shoup estimates that free parking reduces the cost of automobile commuting in the United States by 71%. Since many commuters base travel choices on the relative direct costs of each mode, charging for parking would make a substantial difference in travel behavior (Small and Vehoef, 2007). Removing subsidies for workplace parking in the United States would be equivalent to increasing the price of gasoline to US$4.44/ gallon (Shoup, 2005).

An environmental impact report for a new parking structure built at University of California, Los Angeles in 2003 estimated that each parking space would generate 1170 vehicle-km (727 vehicle-miles travelled) per month, and have associated external costs from increased congestion and emissions of nearly US$117/month (Shoup, 2005). Parking should be priced so that the people utilizing the parking internalize its costs.

On street, curb parking is often similarly underpriced. Where there is a parking meter, the price does not cover the actual costs of the parking. Economic theory says that under-pricing curbside parking creates a shortage. This shortage of curbside parking leads to "cruising" – drivers circling the block in search of available spaces. Studies have demonstrated that in some New York City neighborhoods, 30–45% of traffic is from cruising (Transportation Alternatives, 2006; 2007). A study by Transportation Alternatives (2008) estimated that each year, drivers circling for parking on a 15-block section of Columbus Avenue in New York City's Upper West Side drove 589,000 km (366,000 miles) and emitted 325 tonnes of CO_2.

A GTZ report reviews crucial parking effects and policies worldwide (GTZ, 2009), recommending, among others, the following:

* Recognize the role of parking in creating or limiting car demand in transport policy documents and actions.

* Implement maximum parking standards for new developments, paying careful attention to limit parking provision near public transit.

* Legislation is needed to set a framework for parking charges and fines, and to put liability for any fine with the owner of the car.

* Legislation should give local authorities the power to enforce parking regulations and use the parking revenue to improve a sustainable transport system.

* Manage parking demand by introducing paid parking; consider dynamic parking pricing that varies by time of day and parking availability.

* Parking tariffs should be higher for on-street than off-street, to encourage people to use the latter.

- All changes to parking must be communicated well in advance; a positive approach toward working with the public may increase compliance with parking regulations.

9.6.3.2 Intelligent Traffic/Infrastructure Systems

Traffic management via computerized control of traffic signals and traffic segregation is an important urban transport strategy for increasing the movement of people and goods with higher quality and a safer way as well as to enhance urban environment conditions. Segregation in the form of dedicated bus lanes, bicycle lanes and pedestrians also favors safety and attractiveness of public transport and it contributes even more to a healthy environment and to poverty alleviation.

Traffic management requires expertise and a strong local management agency (with strong regulatory and enforcement capacities) to be implemented, maintained and improved. The lack of these requirements might explain why many cities have not yet introduced such measures or benefited from them. According to the global Transport Knowledge Partnership fragmentation of responsibilities between agencies and lack of inter-agency coordination, coupled with a lack of staff and resources, reduces the effectiveness of many traffic management schemes."

9.6.3.3 Traffic Calming

Traffic calming strategies aim to reduce the speed and volume of traffic to improve safety for pedestrians and cyclists, as well as to improve environmental conditions. Similar to other sustainable transport policies, this involves more than just physical changes; it represents a process of social change requiring extensive community participation. Traffic calming measures are comprised of volume control measures that reduce through traffic by blocking certain movements and diverting traffic to other streets, as well as speed control measures that slow down traffic by changing vertical or horizontal alignment or narrowing the roadway.

Traffic calming measures range from relatively affordable (simple speed humps) to very expensive (designs that affect drainage patterns, utility pole locations or underground services).

Although largely beneficial, traffic calming measures can have drawbacks. Traffic calming measures such as speed humps or raised crosswalks are unsuitable on bus routes or streets that are used frequently by emergency vehicles. Such measures can cause delays up to 10 seconds per measure. Slow traffic on one street can displace vehicles onto adjacent streets, creating unwanted spillover that impacts inhabitants of neighboring streets. Measures that prohibit through traffic can make it more difficult for residents to reach their homes and for visitors to reach local businesses and institutions.

9.6.4 Policies for Alternative Fuels and Efficient Vehicles

9.6.4.1 New Vehicle and Fuel Economy Standards

Fuel economy standards are an effective and efficient policy instrument to reduce GHG emissions in the road transport sector. Fuel economy standards are mandated in many important world regions in order to curb fuel consumption and GHG emissions in new vehicles (An et al., 2007; IEA, 2009d; Creutzig et al., 2011a). Fuel economy standards, however, do not apply for the used vehicles that are often exported to developing countries. Fuel efficiency standards can also complement price instruments, such as carbon taxes and emission trading, that are not fully effective due to dynamic market failures (see also Plotkin, 2008; Flachsland et al., 2011). The European Union and Japan have the most ambitious fuel efficiency standards, with fleet-averaged targets of around 130 gCO$_2$-eq/km. Although still to be enforced, the European Union adopted an even more ambitious long-term target of 95 gCO$_2$-eq/km for 2020. China requires ca. 129 gCO$_2$-eq/km for new vehicles in 2015. The United States implemented a 153 gCO$_2$-eq/km target for 2016.

The development of fuel efficiency standards in different world regions is summarized in Figure 9.49. This figure is an update from An et al., (2007) with new significant European Union, US, and Chinese regulations (2007). The data is displayed in MJ/km – a suitable measure of energy efficiency, as it applies to all fuels (including electricity/biofuels) and is irrespective of supply-chain GHG emissions (Creutzig et al., 2011a).

An intensity reduction in terms of lower CO$_2$-eq/MJ is not necessarily equivalent to an absolute reduction in GHG emissions. Two different rebound effects could compromise the desired outcome. First, car

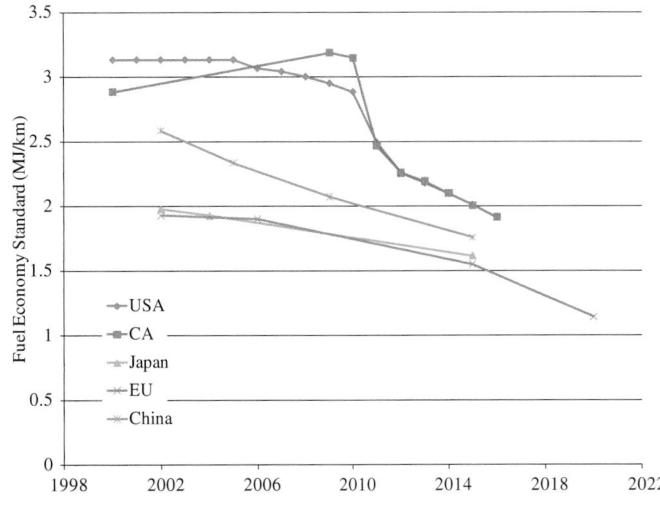

Figure 9.49 | Energy intensity standards of passenger cars extrapolated from current fuel efficiency standards (1 liter gasoline = 32 MJ). Source: An et al., 2007; Creutzig et al., 2011a.

Table 9.11 | Overview of fuel efficiency standards in some world regions.

European Union	CO₂ emissions	gCO₂/km	weight-based fleet standard	NEDC
California	GHG emissions	gCO₂-eq/mile	Absolute fleet standard for LDT1/LDT2	FTP 75
United States	Fuel economy and GHG	mpg and gCO₂-eq/mile	Footprint-based fleet standards for cars/ light trucks	FTP 75
Japan	Fuel economy	km/l	Weight-based fleet standard	Japan 10–15
China	Fuel economy	l/100km	Weight-based fleet standard	NEDC

Source: Creutzig et al., 2011a.

drivers could use the reduction in marginal cost from lower fuel use to increase total travel distance. Based on a review of 22 studies, Greening et al., (2000) suggest a potential rebound effect in the transport sector of between 10–30%. But they highlight the existence of unmeasured components, such as changes in automotive attributes related to shifts toward increases in weight, horse power, and acceleration of cars purchased. The rebound effect generally decreases with income and level of congestion and increases with fuel costs, suggesting a short- and long-run rebound effect of 4.5% and 22.2%. (Small and Van Dender, 2007; Hymel et al., 2010). The sharp rise in oil prices in 2008 might therefore have led to stronger rebound effects than previously observed. A review suggests that total travel volume in OECD countries started to level off between 2000 and 2010 (Millard-Ball and Schipper, 2011). As another kind of rebound effect, market forces could induce a higher production of fuel efficient cars without inducing a simultaneous reduction in gas guzzlers. Despite moderate rebound effects, total expected GHG abatement by fuel efficiency standards is significant and may be – as part of a broad policy mix – the single most effective climate policy in the transport sector (Creutzig et al., 2011a).

Furthermore, fuel efficiency standards are also considered to be relatively cost effective. Their relative contribution to climate change mitigation corresponds to that of other sectors, and costs are approximately in the same price range (e.g., Blom et al., 2007). One should ask, then, which specific design of fuel efficiency standards is most efficient? For example, some world regions have fleet average requirements (e.g., the European Union), whereas other world regions have targets for each car of a specific weight class (e.g., China; see Table 9.8). Given the same level of overall ambition, the first rule is more cost-efficient, as it gives flexibility to car manufacturers in determining where to invest in fuel efficiency. Effectiveness is not impacted if the fleet average target remains the same. Attribute-based standards redistribute the abatement burden among car manufacturers and consumers, and usually follow political rationale, while decreasing overall efficiency.

The unit of fuel efficiency deserves particular attention. Currently, the most common unit of fuel efficiency is volume per distance or its inverse. With alternative fuels and vehicles taking an increasing share of the market, additional units of fuel efficiency (TTW measures) are suggested to convert to energy or GHG emission per distance, e.g., MJ/km (Creutzig et al., 2011a). This choice of unit would provide a level playing field in a more diverse fuel and vehicle market.

The IEA recommended 25 different efficiency measures, including mandatory fuel economy standards, for LDVs and trucks (IEA, 2008). Four measures are recommended for worldwide implementation:

- Introduce, strengthen, and harmonize, where appropriate, mandatory fuel efficiency standards for LDVs.

- Introduce fuel efficiency standards and related policies, such as labeling and financial incentives, for heavy-duty vehicles.

- Introduce measures that promote proper inflation of tires and adequate tire maintenance and adoption of labeling and other efficient vehicle accessories such as headlights, internal lighting, and air conditioning systems.

- Provide incentives to encourage drivers to adopt safe driving techniques, the promotion of eco-driving, including driver training and deployment of in-car feedback instruments.

9.6.4.2   Used Vehicle Emissions Standards

Imports of used vehicles from developed countries contribute significantly to the vehicle fleet in developing countries. In 2005 alone, Mexico imported 1.3 million vehicles from the United States that were more than 10 years old (Johnson et al., 2009). Strict new vehicle fuel economy standards that increase vehicle prices further encourages the import and sale of older used cars in Mexico and other developing countries. Any fuel economy improvements achieved with new vehicle restrictions could be diminished or negated if older polluting vehicles introduced into the fleet reduce the overall on-road fuel economy.

It is therefore important for any new vehicle fuel economy standards to be accompanied by measures that discourage the purchase and ownership of fuel-inefficient used vehicles, such as inspection and maintenance requirements and border inspection interventions (Johnson et al., 2009).

The World Bank's *Low Carbon Development for Mexico* (Johnson et al., 2009) suggested that border vehicle inspections require imported vehicles to meet minimum environmental standards in order to indirectly regulate the efficiency of used imported vehicles. For instance, vehicles exceeding a certain emissions threshold would be banned from the country. Similarly, municipal or statewide vehicle inspection programs could restrict vehicle registration or licensing to those vehicles meeting certain tailpipe emissions standards.

9.6.4.3 Reducing the Carbon Intensity of Fuels

Biofuels have been discussed, and partially misunderstood, as low- or zero-carbon sources of energy for transportation, and as a suitable strategy for reducing oil dependency. Hence, the development of biofuels has been supported by a range of policy instruments, including volumetric targets or blending mandates, tax incentives or penalties, preferential government purchasing, government-funded research, development, and deployment (RD&D), and local business incentives for biofuel companies. For example, biodiesel production in Germany jumped with the introduction of a tax break, and slumped again with introduction of a new tax rate.

Another powerful tool that has been introduced into the policy arena over the past decade is renewable fuel mandates. Renewable fuel mandates require fuel producers to produce a pre-defined amount, or share, of biofuels and blend them with gasoline. The mandates aim to reduce the carbon intensity of transportation fuels by entering larger amounts of low-carbon fuels into the market without setting particular intensity targets. Some renewable fuel mandates are non-discriminatory in that they do not differentiate between different types of biofuels. From an environmental perspective, this assumes that any renewable fuel source is less carbon-intensive than conventional gasoline. However, recent evidence shows that this might not be the case (see Crutzen et al., 2008; Searchinger et al., 2008). Discriminatory renewable fuel mandates, on the other hand, apply to only a selection of biofuels or introduce quotas for the least carbon-intensive biofuels.

In its directive on the promotion of the use of energy from renewable sources (EC, 2009a), the European Union mandates 10% renewable fuels used in transportation by 2020. In the United States, the Renewable Fuel Standard 2 (RFS2) program will increase the volume of renewable fuel required to be blended into transportation fuel from 9 billion gallons in 2008 to 36 billion gallons by 2022, corresponding to around 9–13% of total fuels (EPA, 2010a).

However, renewable fuel standards are insufficient for a number of reasons (Creutzig et al., 2011a):

* Only biofuels, and no other alternative fuels, can contribute to achieve this goal. Hence, this is a technology-specific regulation.

* Only some, but not all, biofuels are subject to meeting threshold values. Regulation should be proportional to carbon content to be effective.

* In the current regulation, lifecycle accounting is not accurate as it is performed by relying on hypothetical technologies. It

underestimates indirect land-use emissions and ignores epistemic and highly relevant uncertainties related to land-use change (e.g., Plevin et al., 2010).

A Low Carbon Fuel Standard (LCFS) is different from renewable fuel standards in that it: a) mandates a specific overall decrease in the average carbon intensity of all fuels; and b) accounts for the carbon emissions of each individual fuel, including non-conventional fossil fuels. The primary purpose of a LCFS is to reduce the carbon intensity of fuels for LDVs. As such, a LCFS provides a level playing field across all fuels, rather than mandating specific fuels of the RFS. A LCFS has been implemented in California that requires a 10% reduction in fuel carbon intensity by 2020 (Schwarzenegger, 2007; CARB, 2009). In the European Union the Fuel Quality Directive COM-2007–18 requires a 6% reduction in CO_2-eq of transportation fuels from 2010 to 2020 (EC, 2009a). The US RSF2 indicated threshold values (EPA, 2010a), and a LCFS-like instrument is under discussion in China.

LCFS can remedy some of the insufficiencies in renewable fuel standards. However, a LCFS is also criticized for leakage and shuffling (Stoft, 2009). As an intensity-based instrument, a LCFS suffers from perverse incentives that may, in effect, increase emissions (Holland et al., 2009). Furthermore, the epistemic uncertainty in lifecycle analysis is pervasive, and is not only a matter of research accuracy but is of systemic nature (Plevin et al., 2010). Addressing transport biofuel emissions in a context of a total cap of GHG emissions by introducing fuel and feedstock accounting standards has been suggested to avoid perverse incentives and to decrease the reliance on subjective and unreliable lifecycle accounting to some degree (deCicco, 2009; Creutzig et al., 2011a).

9.6.5 Creating Economic Incentives or Disincentives

Price instruments in transport can cover several purposes. First, price instruments can be understood as user charges to recover investments in infrastructures. Second, price instruments can potentially aim to internalize the huge social costs of transportation (e.g., EEA, 2008b). Third, price instruments have a steering effect, and can reduce or shift demand that produces high social costs. Economic theory predicts that price instruments that address social costs comprehensively are immune to rebound effects. However, if only one category of social costs is priced (e.g., GHG emissions), other social costs of transport will not necessarily improve (e.g., congestion and space consumption of motorized transport in inner cities).

9.6.5.1 Fuel Taxation and Carbon Pricing

In many countries, fuel taxes are an important source of revenue for financing the transport sector. Fuel taxes reclaim the costs of transport infrastructure from road users. As a crude benchmark, US$0.10/liter may yield the financial resources necessary to maintain the road system, and an additional US$0.03–0.05/liter can be a source of finance for urban transport (GTZ, 2009).

Fuel taxes can be evaluated according to four categories of cost recovery and taxation purposes (GTZ, 2009), and countries can be sorted according to these four categories:

- In the first category, fuel prices do not reflect the costs of production and distribution. Some countries, notably Iran and Saudi Arabia, subsidize road users so strongly that not even the costs of oil production and distribution are covered. IEA estimated global fossil fuel consumption subsidies in 2008 alone at about US$557 billion (IEA, 2010b).[26]

- The second category recovers road infrastructure costs. Here, the relatively low fuel taxes of the United States are suggested as a conservative benchmark, i.e., the minimum level of fuel taxes to recover road construction and maintenance costs.

- The third category comprises the countries that attempt to internalize the external costs of road transport, i.e., tentatively including issues like consumption of urban space and air pollution.[27] Usually, countries that explicitly try to internalize the external costs of road transport have the highest fuel taxes (e.g. European Union countries, Hong Kong).

- In the fourth category, fuel taxes are used as a general source of revenue. Fuel taxes are easier to collect than income taxes or value-added taxes, and can constitute a major source of income for core state functions such as health services and education. This is particularly relevant in countries where tax code enforcement is limited by weak institutional capacity.

Fuel taxation is a straightforward instrument and a good proxy to achieve a number of objectives. However, it is crude with respect to localized externalities, such as congestion and air pollution. Localized road user charges (e.g., congestion charges) can remedy this gap (see below). For the current fleet of vehicles, because GHG emissions are proportional to gasoline and diesel consumption, fuel taxes can address GHG emissions of transport sufficiently accurately. However, with increasing shares of alternative fuels and technologies, fuel supply chains diversify, by this shifting GHG emission upstream in varying supply chains (Creutzig et al, 2011a). GHG emissions of transport are better addressed by upstream emission trading or carbon taxes (Flachsland et al., 2011).

---

26 www.iea.org/files/energy_subsidies.pdf.

27 Data suggests (e.g., EEA, 2008b) that even the relatively high fuel tax levels in the European Union insufficiently address external costs of road transport. The European Union intends to raise specific road user charges (e.g., EC, 2006).

9.6.5.2   Vehicle Taxation and Subsidies

Vehicle taxation can be used to regulate total car ownership. As most infrastructure and social costs are related to car usage, but not car ownership, vehicle taxation is commonly seen as a sub-optimal price instrument. However, a well designed vehicle tax can successfully mimic non-available first-best pricing instruments (Fullerton and West, 2000). High vehicle taxation limits motor vehicle penetration. For example, Denmark has higher vehicle taxes and lower car ownership than the United States. As another example, cities with dense urban agglomeration, such as Singapore or Shanghai, regulate vehicle ownership by auctioning licenses.

9.6.5.3   Road User Charging

Road user charging, or road pricing, means charging for the use of roads in a way that reflects the actual costs of using them, i.e., paying more when roads are congested and less when traffic is light. Congestion charging is a form of demand management that aims to reduce motor vehicle travel, or shift it to the most sustainable modes, for passenger or freight in peak hour traffic in order to reduce congestion, improve travel times, and reduce emissions (Pigou, 1920; Prud'homme and Bojacero, 2005). A city toll would use road pricing in cities not only for reducing congestion, but also to be an efficient measure to account for other externalities, including air pollution, noise pollution, accidents, and GHG emissions (Creutzig and He, 2009). Road pricing works best when applied in parallel with other measures, such as public transport improvements and provisions for cyclists and pedestrians, to support mode shifts. Technically, the joint application of push (pricing) and pull (investments into NMT and public transit) produces synergies via reduced demand elasticities, reduces opportunity costs for car drivers, and increases public welfare gains (Creutzig and He, 2009). Communication and the involvement of key stakeholders are vital to the success of a pricing scheme, and must be consulted effectively to raise the level of awareness and support.

Key examples of road pricing schemes are found in London, Singapore, Milan, and Stockholm. The London Congestion Charging Scheme, in place since 2003, led to a reduction of vehicle-km traveled of about 20%, significant travel time savings, 16% reduction in CO_2 emissions, substantial reduction of air pollution, and altogether higher life quality in London (Transport for London, 2007). The Singapore Area Licensing Scheme, introduced in 1975, is part of a comprehensive policy package that makes Singapore a leading city in urban transportation worldwide. The updated scheme, Singapore Electronic Road Pricing, is adaptive in pricing so that traffic is always fluid, maintaining speeds of 45–65 km/hr on expressways and 20–30 km/hr on arterial roads. The ECO-PASS scheme in Milan, in place since 2008, is relatively young and has better air quality as one of its primary objectives. Money raised will go toward buses, cycle paths, and green vehicles. The Stockholm congestion pricing scheme is particularly interesting for its political implementation. A trial period was introduced to persuade road users of the congestion relief benefits of this instrument. Following the trial period, in September 2006, a referendum was held in which a majority of voters, comprising city inhabitants but not outer districts, voted in favor of the congestion charging scheme. Carbon dioxide emissions dropped by 10–14%, which is very significant for a single policy instrument. Smaller cities, such as Durham, England and Valetta, Malta, have also had positive experiences with congestion charging.

Both design and distribution effects play a major role in the political success of road pricing and city tolls. A congestion charging scheme in Hong Kong was abandoned partly because people thought their movements might be tracked. Congestion charging is likely to be more accepted if the revenues subsidize alternative modes of transport, including public transit and NMT, and by this reduce opportunity costs for transport users. In contrast to Stockholm, a congestion charge referendum was rejected in Edinburgh. Car users were vehemently opposed and emphasized the costs to them, and neither car users nor public transit users were strong proponents of future benefits (loss aversion). A solid communication strategy with clear objectives was missing. Other crucial success criteria include a single implementation agency and a political champion and figurehead of the process. Design should correspond to the spatial scope of authority and voters. Both consultation and promotion of the project are required and, accordingly, the project must be kept simple (e.g., Gaunt et al., 2006).

9.6.6 Establish Policies Targeting Freight and Long-Distance Transport: Shipping, Trucks, Rail, and Air

Freight transportation continues to show growth rates that mirror development in an economy or exhibit even a slightly faster growth. It is likely that the recent economic downturn will have a short-term impact on the pace of growth, but this will not change the general trend toward long-term growth of freight transport.

9.6.6.1 Heavy-Duty Vehicles

The demand for freight transportation is driven by economic considerations and the development of logistics in the private sector (EEA, 2008b). Road transport dominates the inland freight transport market. Heavy-duty vehicles dominate inland freight transport at the expenses of rail and inland waterway transport, even in countries where a well-developed and extensive rail and water transport network exists, such as in Europe. Policies and measures to reduce the energy use of road freight are therefore focused on the short to medium term.

Three measures with energy saving potential that can also improve the fuel energy efficiency of the road freight sector include: "training of drivers in energy-efficient or eco-driving techniques, allowing larger vehicles, and reducing average driving speed" (McKinsey Global Institute,

2009). Lowering the speed of trucks has to be assessed vis-à-vis overall traffic safety considerations.

Measures such as the internalization of freight transport external costs (i.e., traffic-based air pollution, noise, and congestion), even if not directly aimed at reducing climate emissions, will contribute to that end goal.

A combination of regulatory policies, such as mandated fuel efficiency improvements for trucks, combined with a steadily increasing carbon tax, could achieve significant reductions in energy use per tonne-km of transport. Shifting freight and passenger transport to rail can be encouraged through government support for R&D of new technologies combined with a steadily increasing carbon tax. At present, neither aviation nor international shipping fuel is subject to any tax or regulation and is not subject to any restrictions under the Kyoto Protocol.

9.6.6.2 Rail

Encouraging a modal shift from road freight transport to rail and inland water transport and short-sea shipping are important measures, as the latter are less polluting modes for freight transportation. However, there are limits to how much a rail network can be expanded and this limits the potential for a substantial shift from trucks to rail.

9.6.6.3 Shipping

Inclusion of international transport emissions, or more generally, all sorts of environmental pollution attributable to international shipping, within a global climate policy framework has proven to be difficult, primarily because the responsibility for reducing emissions does not fall directly within the jurisdiction of any single country. Due to the global nature of shipping and aviation, sectoral approaches may be more appropriate for tackling emissions reductions in international transport. In fact, the shipping industry favors a global treatment of GHG emissions to avoid a multitude of regional regulations (UNCTAD, 2010). There are currently two main types of policy for GHG reduction considered by the IMO: market-based instruments and efficiency requirements. Market-based instruments include emissions trading schemes, fuel levies, energy efficiency credit trading schemes, and other hybrid schemes. Regulating emissions from maritime and air transport could potentially generate resources to finance climate change adaptation and mitigation measures in developing countries (Monkelbaan, 2010).

9.6.6.4 Aviation

A wide variety of economic, market-based measures have been identified for the aviation industry to make progress in reducing net emissions.

These include voluntary carbon offsetting, emissions trading schemes, emissions charges, and positive economic incentives.

The most advanced policy is the inclusion of aviation into the European Union emission trading scheme (ETS) (Directive 2008/101/EC; see European Commission, 2006), understood to be the most cost-efficient and environmentally effective option for controlling aviation emissions. Like industries, airlines will receive tradable allowances covering a certain level of CO_2 emissions per year. After each year, airline operators must surrender a number of allowances equal to their actual emissions in that year. From the start of 2012, emissions from all domestic and international flights – from or to anywhere in the world – that arrive at or depart from a European Union airport will be covered by the European Union ETS. The European Union ETS can thus form the basis for global action. Airline operators are expected to buy certificates from non-aviation abatement options. Due to the relatively weak cap of the European Union ETS, expected low carbon prices until 2020, and limited demand price elasticity of passengers, the overall reduction in demand growth and carbon emissions is expected to be insignificant (Anger and Köhler, 2010). Nonetheless, the European Union ETS directs attention to the growing emissions of the aviation sector and is expected to serve as a role model for other countries and world regions considering similar national or regional schemes, and to link these to the European Union scheme over time.

Possible regulatory and other measures include aircraft movement caps/slot management, enhanced weather forecasting, transparent carbon reporting, and education and training programs. Each of these can contribute to an overall action plan by individual Contracting States.

9.6.7 Assessment of Policy Contributions to Meeting the Multiple GEA and Transport Sustainability Goals

No single policy, actor, or technology has the silver bullet for an energy and sustainability transition in transportation. Achieving multiple goals within a relatively limited time frame requires action on all spatial scales in order to address technologies, urban design, and demand management alike. Governments of nations and cities around the world have different reasons for acting on transport energy sustainability. Appropriate policies will vary according to city, country, and circumstances but should be integrated across scales (Corfee-Morlot et al., 2011). The role of government is pivotal for an energy sustainability transition. The GEA approach considers a multiple-goal approach that requires multiple policies to be implemented simultaneously as policy packages.

Forming multi-policy packages as a way to frame transport policy is a strategy that is receiving increasing attention by researchers and policy makers (OPTIC, 2010). The motivation is the realization that traditional approaches based on single-goal policies have limitations when trying to resolve complex transport problems related to environment and climate. A multi-policy package has been defined as: "a combination of individual policy measures, aimed at addressing one or more policy goals with the objective of improving the impacts of the individual policy measures, minimizing possible negative side effects, and/or facilitating the measure's implementation and acceptability" (OPTIC, 2010).

Transportation is a sector bound by multiple direction causalities with land use and behavioral parameters that lead to major policy paradoxes. These multi-direction causalities affect transport policy implementation. One crucial example is "induced or latent traffic demand," in which building new road infrastructures induces additional traffic (e.g., Goodwin, 1996; Cervero, 2008).

The rebound effect states that efficient vehicle technologies make driving cheaper, and hence will increase the amount of driving and the external costs associated with it. The effectiveness of a fuel economy standard in reducing aggregate fuel consumption then is undermined if the rebound effect is large (Small and Van Dender, 2007; Van Dender, 2009). Two arguments contend that while the rebound effect is real, it should not deter advances in energy efficiency. First, trends clearly indicate that people favor a move toward faster modes of travel, and total distance travelled continues to grow globally. Given these preferences, efficiency improvements will mitigate but not limit the rise in fuel use that would be occurring anyway as trends unfold. Improving efficiency is necessary but not sufficient to meet climate, environmental protection, and energy security goals in transport. Second, the demand for driving and the distances driven are increasing because they generate net benefits to the driver, meaning there are benefits and not just external costs that need to be considered. Finally, while better fuel economy does lead to more driving, it appears to increase less when there is congestion, because congestion itself is a deterrent to driving (Van Dender, 2009).

The rebound effect and overall rising demand is difficult to tackle with fuel efficiency regulations and technological improvements alone and requires appropriate and encompassing demand policies, such as pricing, to be in place. In the absence of pricing policies, consumer preferences for faster, heavier, and stronger performance LDVs, induced and latent demand, and the rebound effect will result in total travel demand such that social benefits are heavily outweighed by social costs, such as congestion at the local level and climate change on a global scale. To effectively and efficiently deal with sub-optimally high demand, focused demand management (e.g., based on pricing policies) is a necessary complement to technological advancement.

A qualitative assessment pairs the sustainable goals outlined in this chapter with policies grouped according to the following categories: Planning, Regulatory, Economic, Information, and Technology. The contribution is rated as essential when policy implementation is a step toward the attainment of a particular goal and when this is well documented in the literature. The contribution is rated as uncertain when no conclusive study has proven the effectiveness between policy implementation and the goal. Finally, the contribution is categorized as complementary when the expected effect can indirectly or in addition to other policies contribute to creating some synergy to the attainment of that goal.

9.6.7.1　Planning Policies (Table 9.12)

- Two essential planning instruments can be adopted in all countries and cities to meet sustainability goals. First, design urban spaces to improve walkability and facilitate the use of non-motorized modes of transportation. This, in addition to parking policies (pricing, location, etc.) and all other policies to reduce car use and accommodate human-powered modes, are policies that serve most goals. Second, increase the speed, frequency, and coverage of public transport to improve the use of public transportation.

- The role of urban densification is essential to improving accessibility to services and jobs, while impact on congestion can be ambivalent. Increasing density is a planning policy that needs to be seen in a city-wide, or even regional, context when transportation corridors are designed (Cervero, 2008). Density needs to be a complementary policy to other land-use and transport planning policies, such as mixed use zoning, walkability, non-motorized modes, housing policies, etc. (Banister, 2008; Wheeler, 2004).

9.6.7.2　Regulatory Policies (Table 9.13)

- Regulatory policies have a clear role regarding traffic safety, air pollution, noise, climate mitigation, and diversification of energy sources.

- Gaps in regulatory implementation can diminish the effect of otherwise effective instruments. Therefore, capacity for implementation is also essential.

Table 9.12 | Planning policies – potential contribution to attainment of GEA and transport major goals.

GEA Overall Systemic Goals	Multiple Goals and Benefits for a Sustainable Transport System	Policies Aim: Developing Alternatives to Car Use and Reducing Need for Travel										
		Compact, Mixed Use Development	Regional Transit Oriented Development	Urban Design for Walkability	Create safe conditions for use of Non-motorized modes	Create Car-Free Zones. Calming-Parking	Improve Public Transport Access-Reliability	Increase Services of Low-Cost Mass Transit System (BRT)	Management of Urban Traffic System ITS-	Modal Shift Air to Rail long distance	Improve Logistics road freight transport	Shift Intercity Freight to rail and water transport
Economic Growth, Equity & Urbanization	Functionality, Efficiency											
	Accessibility											
	Affordability											
	Acceptability											
Health & Environmental Protection	Traffic Safety											
	Acces of less fit											
	Human Motion Promotion											
	Reduce Air Pollution											
	Reduce Noise											
	Reduce Congestion											
Climate	Reduce GHG											
Energy Security	Diversification Energy sources											
	Independence from Fossil fuels											

(dark gray)	essential
(light gray)	uncertain
(white)	complementary

Table 9.13 | Regulatory policies – potential contribution to transport and GEA multiple goals.

GEA Overall Systemic Goals	Sustainable Transportation Systems Multiple Goals and Benefits	Aim: Establishing Clear Regulatory Framework								
		Vehicle Standards	Fuel Standards and Mandates	Reduce Travel Speed Limit in Urban Areas	Reduce Travel Speed/Volume of Freight Transport in Urban Areas	Reduce Speed of Airplanes	Reduce Speed of Commercial Maritime Transport	Improved Management Intelligent Transport System	Mandatory Vehicle Inspections	Traffic Safety Regulation
Economic Growth, Equity & Urbanization	Functionality, Efficiency	essential	uncertain	uncertain	uncertain	uncertain	uncertain	essential	uncertain	uncertain
	Accessibility	uncertain	uncertain	complementary	complementary	uncertain	uncertain	uncertain	uncertain	uncertain
	Affordability	uncertain	uncertain	uncertain	complementary	uncertain	uncertain	uncertain	complementary	uncertain
	Acceptability	complementary	complementary	uncertain	uncertain	complementary	complementary	uncertain	uncertain	uncertain
Health & Environmental Protection	Traffic Safety	essential	uncertain	essential	uncertain	uncertain	uncertain	essential	essential	essential
	Acces of less fit	uncertain	uncertain	uncertain	uncertain	uncertain	uncertain	uncertain	complementary	uncertain
	Human Motion	uncertain	uncertain	uncertain	uncertain	uncertain	uncertain	uncertain	uncertain	uncertain
	Reduce Air Pollution	essential	essential	uncertain	uncertain	uncertain	uncertain	essential	essential	uncertain
	Reduce Noise	essential	uncertain	uncertain	uncertain	uncertain	complementary	essential	uncertain	uncertain
	Reduce Congestion	uncertain	uncertain	uncertain	complementary	uncertain	uncertain	essential	uncertain	uncertain
Climate	Reduce GHG	essential	essential	uncertain	uncertain	essential	essential	essential	complementary	uncertain
Energy Security	Diversification Energy sources	complementary	essential	complementary	complementary	uncertain	uncertain	complementary	uncertain	complementary
	Independence from Fossil fuels	complementary	essential	complementary	complementary	uncertain	uncertain	complementary	complementary	complementary

Legend:
- essential (dark gray)
- uncertain (light gray)
- complementary (white)

- Regulatory policies affecting one transport subsector may have little impact on other subsectors, demonstrating the need for instruments or sets of instruments creating comprehensive coverage.

- Although the impact of regulations on urbanization and equity goals related to transport as they are presented here can be seen as uncertain, the positive benefits of the stringent enforcement of safety, noise, and air pollution rules are without a doubt beneficial to the quality of life in a city for all its inhabitants.

9.6.7.3 Economic Policies (Table 9.14)

- Pricing policies are essential to address environmental, climate, and energy security goals. They play a complementary role in terms of sustainability and equity goals. Pricing policies can create incentives for the introduction of new transportation technologies and the use of new infrastructure. Depending on the design and revenue recirculation scheme, they can have regressive or progressive distributional effects.

- Pricing policies are considered in many respects to be more flexible and effective than regulatory and planning instruments, the latter of which are often viewed as "command and control measures." However, pricing policies may receive little or no acceptability by society and have high political, transaction, implementation, and enforcement costs.

- Economic policies are essential to achieve climate mitigation and diversification of energy sources and have mostly a complementary effect on achieving the other goals.

Table 9.14 | Economic policies – potential contribution to meet transport sustainability and multiple GEA goals.

GEA Overall Systemic Goals	Sustainable Transportation Systems Multiple Goals and Benefits	Land Taxation Pricing and Incentives	Vehicles Purchasing Taxes	Alternative Fuel Vehicle Tax Incentive	Vehicle Registration Taxes	Fuel Taxes	Reform of Fuel Subsidies	Distance Based Vehicle Taxation (ldv/trucks)	Congestion Charging (Inner City)	Heavy Goods Vehicles Tolls	Road Pricing (Passenger Vehicles- Inter City)	Road Pricing (Trucks Freight)	Parking Pricing Charges	Feebates for fuel economy standards
Economic Growth, Equity & Urbanization	Functionality, Efficiency													
	Accessibility													
	Affordability													
	Acceptability													
Health & Environmental Protection	Traffic Safety													
	Acces of less fit													
	Human Motion (avoid obesity/heart diseases)													
	Reduce Air Pollution													
	Reduce Noise													
	Reduce Congestion													
Climate	Reduce GHG													
	Reduce Fossil Fuel Use													
Energy Security	Diversification Energy sources													
	Independence from Fossil fuels													

essential
uncertain
complementary

- Road pricing effectiveness depends on the alternatives. If drivers are induced to divert their journeys to non-toll roads, there is only a very localized reduction in air pollution. If trips are taken with other modes of transportation that are more energy efficient, the effects are larger.

- Feebates for fuel economy standards may induce people to drive more (rebound effect), as the marginal cost is cheaper.

- Alternative fuels may not improve air pollution or climate mitigation if they are produced using fossil fuels.

9.6.7.4   Policies to Improve Information (Table 9.15)

Improving information is essential for a low-carbon transition and to contribute to larger goals for transport and urbanization, health, environmental protection, and climate change mitigation (Banister, 2008).

9.6.7.5   Contribution of New/Improved Technologies (Table 9.16)

Measures improving technology and inducing technological shifts contribute to energy security, climate, and environmental goals.

9.6.8   Enabling Conditions to Facilitating a Sustainable, Low-carbon Transition

9.6.8.1   Improving Institutional Capacity

A tension exists between the need for policy and decision makers to commit to a full systemic change and the need to get things done using existing policies, institutions, capacities, and decision-making frameworks. Institutional and human resource capacity is a prerequisite for the implementation of sustainable transport policies. As a rule of thumb, an unregulated transport sector tends to produce a number of inefficient environmental and social externalities. One important factor is

Table 9.15 | Assessment of policies for improving information.

GEA Overall Systemic Goals	Sustainable Transportation Systems Multiple Goals and Benefits	Aim: Improving Information, Inducing Behavioral Change								
		Enhancing Driving Efficiency (LDV Passengers)	Enhancing Driving Efficiency (Trucks)	Traffic Calming Information	Information facilitating Efficient Intermodal Travel	Improve Public Transport Service Information	Improve Information on Safety Rules for Passenger Tavel	Information on Safety of Use of Non-Motorized modes	Information on Vehicle Maintenance/Inspections	Encourage use of small efficient LDVs
Economic Growth, Equity & Urbanization	Functionality, Efficiency									
	Accessibility									
	Affordability									
	Acceptability									
Health & Environmental Protection	Traffic Safety									
	Acces of less fit									
	Human Motion									
	Reduce Air Pollution									
	Reduce Noise									
	Reduce Congestion									
Climate	Reduce GHG									
Energy Security	Diversification Energy sources									
	Independence from Fossil fuels									

Legend:
- essential
- uncertain
- complementary

institutional integration and coordination (Banister 1998, 2005; May et al., 2006).

9.6.8.2 Improving Acceptability and Public Information

Sustainable transport is not only characterized by a number of goals or co-benefits, but also by procedures such as participation and open communication. In fact, improving the acceptability of changes of the transport system often is a precondition for successful implementation. Such is its importance, that information itself has been identified as a key component of sustainable transport (Banister, 2008). In Table 9.17 key insights of information for transport transitions are identified.

9.7 Conclusion

Achieving sustainability goals in transport that help induce economic growth, energy security, equitable access and better health, and contribute to poverty alleviation and to meeting low emissions goals, within the limited time frame of a few decades requires transformative actions and political commitment from the local communities to the supra-national and global level. Actions must be taken to change the landscape of cities to favor multiple short-distance destinations, to enhance accessibility by foot and non-motorized modes, and to enhance and extend a variety of smart-integrated public transport services. These kinds of structural and demand-side policies are necessary complements to technological innovation for energy efficiency and alternative fuel deployment.

Table 9.16 | Potential contribution of new/improved technologies to transport and GEA goals.

GEA Overall Systemic Goals	Sustainable Transportation Systems Multiple Goals and Benefits	Improving Efficiency					Widespread Intake and Use of New Vehicles and Fuels											
		Improving Vehicle Efficiency	Improving efficiency of Trucks	Efficiency Technologies Rail	Efficient Technologies for Aviation	Efficient Technologies Shipping	Hybrids gasoline-electric	Hybrids fossil - biofuels	Natural Gas/ Biogas LDV	Electric Plug-in / Battery LDV	Hydrogen LDV	Natural Gas-Biogas Buses	Electric Buses	Hydrogen Buses	Hydrogen Service Vehicles (Industry)	Biofuels Freight (Trucks)	Biofuels in Aviation	Biofuels for Maritime Freight Transport
Economic Growth, Equity & Urbanization	Functionality, Efficiency																	
	Accessibility																	
	Affordability																	
	Acceptability																	
Health & Environmental Protection	Traffic Safety																	
	Acces of less fit																	
	Human Motion Promotion																	
	Reduce Air Pollution																	
	Reduce Noise																	
	Reduce Congestion																	
Climate	Reduce GHG																	
Energy Security	Diversification Energy sources																	
	Independence from Fossil fuels																	

	essential
	uncertain
	complementary

Globally, transportation policy makers, technicians, and manufacturers must purposefully participate in climate change mitigation and actively contribute to reduce energy security concerns. Locally, the livability of human settlements, environmental pollution, health, and spatial equity can be dramatically improved by sustainable transport practices.

Achieving this vital transition requires extensive investment in high-quality public transportation and non-motorized transport that encourages a shift to these modes from private automobiles. There needs to be an overall reduction in VMT, a radical shift in urban development from car-centric to people-centric models that prioritize walkable, dense, mixed-use urban spaces. Also required is the decarbonization of transportation fuels, mainstreaming of near-zero GHG emissions vehicle technologies, such as plug-in hybrid vehicles, full battery electric vehicles, and hydrogen fuel cell vehicles. Altogether, this comprehensive set of policies will affect not only transport use but also societal interaction, e.g., via land-use patterns and personal choices and preferences.

The task of transforming transportation will not be possible without sustained effort and deep involvement over the upcoming years and decades by industry, governments, and consumers (Lutsey and Sperling, 2007). Achieving a multitude of local transitions for sustainable urban

transport is a key challenge as cities for some additional 3 billion urban dwellers will need to be constructed by 2050. As discussed in Chapter 18 most of this urban growth will take place in medium size cities. This represents a window of opportunity to use many of the approaches covered in this Chapter, in particular, for giving a renewed focus on addressing issues of accessibility, density, land use and urban design to favor walking and bicycling, public transport service affordability, local equity and environmental concerns. The future sustainability of cities requires solid policy actions essential to decrease barriers for slow-mode traffic, reduce air pollution and noise nuisances, and to make cities safe and comfortable places for transport and living.

The required planning practices, policies, and technological innovation for this transition cannot be limited to improving the energy efficiency of personal vehicles (LDVs) and fuels. It must aim to upgrade the energy efficiency of all subsectors within all transport systems – including heavy-duty vehicles, buses and trucks, trains, airplanes, ships, agricultural vehicles, and off-road transport – on a scale from the local to the national to the international.

This transformation must embrace a long-term vision. It presents challenges that need to be overcome. Chief among them are societal barriers, limited institutional capacities, and status (e.g., Urry, 2007; Creutzig et al., 2011b). Another challenge is that fuel technologies that may be

Table 9.17 | Key elements for promoting public acceptability for a transport transition.

Information	• Education, awareness campaigns, and promotion through media and social pressure are an essential starting point • Explanation of the need for a transition toward sustainable, low-carbon mobility, emphasizing positive economic, social, and health benefits to individuals and businesses
Involvement and communication	• The process must be inclusive, with clear aims and an understanding of the consequences by those who the strategy will impact • Design to gain support and understanding, so that stakeholders can buy into proposals • Raise levels of consistency between expectations and outcomes
Packaging	• Push and pull policies measures need to be combined in mutually supporting packages • Policies restricting car use or raising its costs should be accompanied by well-publicized programs to improve availability and attractiveness of alternatives to driving alone, including car pooling, public transport, cycling and walking, all financed by dedicated revenues from car pricing measures
Selling the benefits	• It is necessary to widely publicize the benefits, even if there are costs, inconvenience, and sacrifice • Car drivers support funding of alternative modes to reduce congestion on the roads they drive on • Overweight or obese individuals would directly benefit from better walking and cycling conditions • Everyone benefits from cleaner air and safer traffic conditions • More walking, cycling, and public transport use would help relieve parking shortages. These are important and direct impacts that all individuals can support
Adopt controversial policies in stages	• Support needs to be built up in terms of positive outcomes and measurable improvements in the quality of life • Politics is about reflecting prevailing preferences and also forming opinions • Acceptance of responsibilities and commitment to change through actions is the key to success
Consistency between different measures and policy sectors	• Some measures (e.g., pricing) are common to all futures, and such measures need to be implemented now, even though the impacts may not be immediate • Regulations, standards, subsidies, and tax incentives should all be used to encourage manufacturers and other transport suppliers to develop and adopt the most energy-efficient and environmentally friendly technology possible • The precautionary principle should be followed, particularly on the global warming effects of transport emissions, and actions should be consistent over the longer term • Many problems created for the transport system do not emanate from the transport sector, but from other sectors. Thus, a more holistic perspective is needed that integrates decision making across sectors and widens public discourse
Adaptability	• Decisions today should not unnecessarily restrict the scope for future decisions, so that the adaptive behavior of individuals and agencies can be assessed • There is no prescription or blueprint for the correct procedures to follow. Each situation requires separate analysis and implementation, including flexibility to change policy measures if intentions and outcomes do not match. Assessment of risk and reversibility are both strong components of sustainable mobility • Adaptability is not an excuse for inaction or weak action. It is an argument for clear decision making and leadership, supported by analysis and monitoring to check the effectiveness of policy action

Source: adapted from Banister, 2008.

best suited to help with the transition are not presently at the commercialization stage. Furthermore, current marginal improvement in technologies can constitute further long-term technological lock-in, and may constitute a barrier for the deployment of transformational technologies.

Transforming transportation toward sustainability requires the implementation of a portfolio or aggregated bundle of policies and hybrids of regulations and market mechanisms directly linked to targets that enhance mobility while protecting the environment. This includes considerations of air pollution, GHG emission reduction, carbon tax policies, quality of accessibility, affordability of urban mobility, safety, and acceptance from urban residents. A transition must build in long-term signals to fuel producers, and needs to incentivize the development and dissemination of new, cleaner, and affordable technologies. To be successful, a transition must be accompanied by the effective communication of sustainable transport goals (Jaccard, 2006). Pioneer regions and cities in sustainable transport can provide

benchmark policies and lower the behavioral barriers in other regions and cities.

It is necessary to immediately start intelligently designing low-carbon transportation systems in developing countries that can also bring co-benefits, such as clean air or a reduction of congestion, that are necessary for sustainable development (Bongardt et al., 2010). Such measures could be supported by international climate mitigation efforts as administered by the UNFCCC, an intervention fostered by the Partnership on Sustainable Low Carbon Transport.

An important near-term approach is to identify strategies for decarbonizing key subsectors like aviation, marine transport, and heavy-duty vehicles. Transport intensity needs to be addressed in all road transport subsectors through urban planning and transportation demand management. The strategy for dealing with automakers is to bring highly efficient, alternative-fuel vehicles to the market with policies that focus on the link between new product commercialization and the mass dissemination necessary to realize substantial cost reductions from economies of scale and economies of learning (Jaccard, 2006).

Finally, with the multiplicities of sustainable transportation goals and different spatial scales involved, the cost of transforming transportation can vary dramatically across industries, countries, and regions. With this in mind, it is imperative that local governments gain experience and expertise even before national and global initiatives are in place. Lower-level government policy structures need not preclude – and can even advance – multiple goals toward transforming transportation and sustainability, integrating policy measures across goals while differentiating between geographic settings (Lutsey and Sperling, 2008; Creutzig and Kammen, 2009b).

References

ADB, 2006: *Summary of Country/City Synthesis Reports across Asia*. Asian Development Bank (ADB), Manila.

ADB, 2009: *Electric Bikes in the People's Republic of China: Impact on the Environment and Prospects for Growth*. Asian Development Bank (ADB), Mandaluyong City, Philippines.

Alliance for Biking &Walking 2010: Biking and Walking in the United States, Benchmarking Report, Release 1.1, The Library of Congress, Washington D.C.

Aihara, N. and T. Tsujimura, 2002: *Basic Study on Environmental Aspects of Tokaido Shinkansen Line by LCA Methods*. RTRI Report 16(10), prepared for Railway Technical Research Institute (RTRI), pp.13–18.

An, F., D. Gordon, H. He, D. Kodjak and D. Rutherford, 2007: *Passenger Vehicle Greenhouse Gas and Fuel Economy Standards: A Global Update*. The International Council on Clean Transportation (ICCT), Washington, DC.

Anas, A., R. Arnott and K. Small, 1998: Urban spatial structure. *Journal of Economic Literature*, **36**(3):1426–1464.

Anger, A. and J. Köhler, 2010: Including aviation emissions in the EU ETS: Much ado about nothing? A review. *Transport Policy*, **17**(1): 38–46.

ANRE, 2010: *FY 2009 Energy Supply and Demand Report*. Agency for Natural Resources and Energy (ANRE), Japanese Ministry of Economy, Trade and Industry, Tokyo.

APS, 2008: *Think Efficiency: How America Can Look Within to Achieve Energy Security and Reduce Global Warming*. American Physical Society (APS), USA.

Aunan, K., J. Fang, H. Vennemo, K. Oye and H. M. Seip, 2004: Co-benefits of climate policy – lessons learned from a study in Shanxi, China. *Energy Policy*, **32**(4): 567–581.

Axhausen, K. W., J.-L. Madre, J. W. Polak and P. L Toint, 2003: *Capturing Long-Distance Travel*. Research Studies Press, Ltd., Baldock, Herefordshire, England.

Bakker, S. and C. Huizenga, 2010: Making climate instruments work for sustainable transport in developing countries. *Natural Resources Forum*, **34**(4): 314–326.

Bandivadekar, A., K. Bodek, L. Cheah, C. Evans, T. Groode, J. Heywood, E. Kasseris, M. Kromer and M. Weiss, 2008: *On the Road in 2035 – Reducing Transportation's Petroleum Consumption and GHG Emissions*. Report No. LFEE 2008–05 RP, prepared for Laboratory for Energy and the Environment, Massachusetts Institute of Technology (MIT), Cambridge, MA.

Banister, D., 1998: Barriers to the implementation of urban sustainability. *International Journal of Environment and Pollution*, **10**(1): 65–83.

Banister, D., and D. Stead, 2004: Impact of information and communications technology on transport. *Transport Reviews* 24 (5), 611–632.

Banister, D., 2005: *Unsustainable Transport – City Transport in the New Century*. Routledge, London.

Banister, D.,2008: The sustainable mobility paradigm. *Transport Policy*, **15**(2): 73–80.

Banister, D., R. Hickman and D. Stead, 2008: Looking over the horizon: visioning and backcasting. In *Building Blocks for Sustainable Transport*. Vol. 1., A. Perrels, V. Hillanen and M. L. Gosselin (eds.), VATT and Emerald, Helsinki.

Beatley, T., 2000: *Green Urbanism: Learning from European Cities*. Island Press, Washington, DC.

Berri, A., S. Lyk-Jensen, I. Mulalic and T. Zachariadis, 2010: *Transport Consumption Inequalities and Redistributive Effects of Taxes: A Comparison of France, Denmark and Cyprus*. Working Paper 159, Society for the Study of Economic Inequality (ECINEQ), Palma de Mallorca, Spain.

Blom, M. J., B. E. Kampmann and D. Nelissen, 2007: *Price Effects of Incorporation of Transportation into EU ETS*. Committed to the Environment (CE) Delft Report, Netherlands.

Blumsack, S., C. Samaras and P. Hines, 2008: *Long-Term Electric System Investments to Support Plug-in Hybrid Electric Vehicles*. Power and Energy Society General Meeting – Conversion and Delivery of Electrical Energy in the 21st Century, Institute of Electrical and Electronics Engineers (IEEE), 20–24 July 2008, Pittsburgh, PA, pp.1–6.

Bodek, K. and J. Heywood, 2008: *Europe's Evolving Passenger Vehicle Fleet – Fuel Use and GHG Emissions Scenarios through 2035*. Report No. LFEE 2008–03 RP, prepared for Laboratory for Energy and the Environment, Massachusetts Institute of Technology (MIT), Cambridge, MA.

Bollen J., B. van der Zwaan, C. Brink and H. Eerens, 2009: Local air pollution and global climate change: A combined cost-benefit analysis. *Resource and Energy Economics*, **31**(3): 161–181.

Bongardt, D., M. Breithaupt and F. Creutzig, 2010: *Beyond the Fossil City: Towards Low Carbon Transport and Green Growth*. Working Paper, German Technical Cooperation (GTZ), Eschborn, Germany.

Brandt, S., 2011: As bicycle use climbs, rate of crashes with vehicles falls. *Star Tribune*, Minneapolis, MN, 6 February 2011. www.startribune.com/local/minneapolis/115449079.html?source=error (accessed 30 May 2011).

Brauer M, Amann M, Burnett RT, Cohen A, Dentener F, Ezzati M, et al. 2011. Exposure assessment for estimation of the global burden of disease attributable to outdoor air pollution. Environmental Science and Technology, in press.

Breithaupt, M., 2010: *Low-carbon Land Transport Options towards Reducing Climate Impacts and Achieving Co-benefits*. Presented at the Fifth Regional Environmentally Sustainable Transport (EST) Forum in Asia, 23–25 August 2010, Bangkok, Thailand.

Buehler, R., Pucher, J., 2011: *Making Public Transport Financially Sustainable*, Transport Policy, Elsevier, Vol 18, (2011), 126–138

Bundeswehr Transformation Center, 2010: *Peak Oil: Implications of Resource Scarcity on Security* (in German). Bundeswehr Transformation Center, Strasberg, Germany.

CAF, 2010: *Observatorio de Movilidad Urbana para América Latina*. Corporación Andina de Formento (CAF) (ed.), Caracas, Venezuela.

CARB, 2009: *Proposed Regulation to Implement the Low Carbon Fuel Standard. Vol. 1: Staff Report: Initial Statement of Reasons*. California Air Resources Board (CARB), Sacramento, CA.

CARB, 2010: California completes its commitment to a national greenhouse gas standard for cars. California Air Resource Board (CARB), Press release, 25 February 2010. Sacramento, CA.

Castells, M., 2001: *The Internet Galaxy*. Oxford University Press, Oxford, UK.

Castells, 2004: Informationalism, networks, and the network society: A theoretical blueprint. In *The Network Society*. M. Castells (ed.), Edward Elgar, Cheltenham, UK.

Cervero, R., 1994: Transit-based housing in California: Evidence on ridership impacts. *Transport Policy*, **1**(3): 174–183.

Cervero, R., 2005: *Accessible Cities and Regions: A Framework for Sustainable Transport and Urbanism in the 21st Century*. Working Paper UCB-ITS-VWP-2005–3, Institute Transportation Studies, University of California, Berkley, CA.

Cervero, R., 2008: Road expansion, urban growth and induced travel. *Journal of the American Planning Association*, **69**(2): 145–163.

Chalk, S., and Miller, James, 2008: The US Hydrogen Fuel Initiative. In *Making Choices About Hydrogen: Transport Issues for Developing Countries*. International Development Research Center, United Nations University Press, Canada.

Chandler, K., E. Eberts and L. Eudy, 2006: *New York City Transit Hybrid and CNG Transit Buses: Interim Evaluation Results*. Report NREL/TP-540–38843 prepared for National Renewable Energy Laboratory (NREL), Golden, CO.

Cheah, L., C. Evans, A. Bandivadekar and J. Heywood, 2007: *Factor of Two: Halving the Fuel Consumption of New US Automobiles by 2035*. Report No. LFEE 2007–04 RP, prepared for Laboratory For Energy and the Environment, Massachusetts Institute of Technology (MIT), Massachusetts, MA.

Cherry, C., 2007: *Electric Two-Wheelers in China: Analysis of Environmental, Safety and Mobility Impacts*. PhD Diss., Graduate Division of Civil and Environmental Engineering, University of California, Berkeley, CA.

Clean Air Initiative for Asian Cities, 2011: *Clean Air Portal*. Clean Air Initiative – Asia Center (CAI-Asia Center) and Asian Development Bank (ADB). http://clean-airinitiative.org/portal/index.php (accessed 30 May 2011).

Corfee-Morlot, J., I. Cochran, S. Hallegatte and P.-J. Teasdale, 2011: Multilevel risk governance and urban adaptation policy. *Climatic Change*, **104**(1): 169–197.

Creutzig, F. and D. He, 2009: Climate change mitigation and co-benefits of feasible transport demand policies in Beijing. *Transportation Research Part D*, **14**(2): 120–131.

Creutzig, F. and D. Kammen, 2009a: Getting the carbon out of transportation fuels. In *Global Sustainability – A Nobel Cause*. H. J. Schellnhuber, M. Molina, N. Stern, V. Huber and S. Kadner (eds.), Cambridge University Press, Cambridge, UK.

Creutzig, F. and D. Kammen, 2009b: The post-Copenhagen roadmap towards sustainability: Differentiated geographic approaches, integrated over goals. *Innovation*, **4**(4): 301–321.

Creutzig, F., A. Papson, L. Schipper and D. Kammen, 2009: Economic and environmental evaluation of compressed-air cars. *Environmental Research Letters*, **4**(4): 044011.

Creutzig, F. and O. Edenhofer, 2010: Mobilität im wandel – wie der klimaschutz den transportsektor vor neue herausforderungen stellt. *Internationales Verkehrswesen*, **62**(3):1–6.

Creutzig, F., E. McGlynn, J. Minx and O. Edenhofer, 2011a: Climate policies for road transport revisited (I): Evaluation of the current framework. *Energy Policy*, **39**(5): 2396–2406.

Creutzig, F., A. Thomas, D. Kammen and E. Deakin, 2011b: Transport demand management in Beijing, China: Progress and challenges. In *Low Carbon Transport in Asia: Capturing Climate and Development Co-benefits*. E. Zusman, A. Srinivasan and S. Dhakal (eds.), Earthscan, London.

Cresswell, T., 2006: *On the Move: Mobility in the Modern Western World*. Routledge, New York.

Crutzen, P., A. Mosier, K. Smith and W. Winiwarter, 2008: N_2O release from agrobiofuel production negates global warming reduction by replacing fossil fuels. *Atmospheric Chemistry and Physics*, **389**(8): 389–395.

Cullinane, S. 2009: From Bricks to Clicks: The Impacts of Online Retailing on Transport and the Environment. *Transport Reviews*, **26**: 6, 759–776

Dalkmann, H. and C. Brannigan, 2009: Module 5e: Transport and climate change. In *A Sourcebook for Policy-Makers in Developing Cities*. Gesellschaft für Technische Zusammenarbeit (GTZ), Eschborn, Germany.

Dalkmann, H. and K. Sakamoto, 2011: Transport: Investing in energy and resource efficiency. In *Towards a Green Economy: Pathways to Sustainable Development and Poverty Eradication – A Synthesis for Policy Makers*. United Nations Environment Programme (UNEP), St-Martin-Bellevue, France, pp.374–407.

DeCicco, J. M., 2009: *Addressing Biofuel GHG Emissions in the Context of a Fossil-based Carbon Cap*. Discussion Paper, October 2009, School of Natural Resources and Environment, University of Michigan, Ann Arbor, MI.

Denholm, P. and W. Short, 2006: *An Evaluation of Utility System Impacts and Benefits of Optimally Dispatched Plug-in Hybrid Electric Vehicles*. Technical Report NREL/TP-620–40293, prepared for National Renewable Energy Laboratory, US Department of Energy, Golden, CO.

Dennis, K. and J. Urry, 2009: *After the Car*. Polity Press, Cambridge, UK.

De Santi, G. (ed.), R. Edwards, S. Szekeres, F. Neuwahl and V. Mahieu, 2008: *Biofuels in the European Context: Facts and Uncertainties*. Joint Research Centre (JRC), Institute for Energy, Netherlands.

Dora, C., 2011: Measuring the impacts of transport systems on health. *Transforming Transportation*, slide presentation, 27 January 2011, World Health Organization (WHO).

Eaves, S. and J. Eaves, 2004: A cost comparison of fuel-cell and battery electric vehicles. *Journal of Power Sources*, **130**(1–2): 208–212.

EC, 2006: *European Energy and Transport: Trends to 2030 – Update 2005*. European Commission (EC), Office for Official Publications of the European Communities, Luxembourg.

EC, 2008: *Towards a European Rail Network for Competitive Freight*, (EC) Directorate General for Energy and Transport, Memo. Office for Official Publications of the European Communities, Luxembourg.

EC, 2009a: DIRECTIVE 2009/30/EC amending Directive 98/70/EC as regards the specification of petrol, diesel and gas-oil and introducing a mechanism to monitor and reduce greenhouse gas emissions and amending Council Directive 1999/32/EC as regards the specification of fuel used by inland waterway vessels and repealing Directive 93/12/EEC. Official Journal of the European Union, European Commission (EC), 23 April 2009.

EC, 2010: *European Union Energy and Transport in Figures*, Statistical Pocketbook 2010. Publications Office of the European Union, Luxemborug.

EEA, 2008a: *Success Stories within the Road Transport Sector Reducing Greenhouse Gas Emission and Producing Ancillary Benefits*. Report No. 2/2008, prepared for European Environment Agency (EEA), Office for Official Publications of the European Communities, Luxembourg.

EEA, 2008b: *Greenhouse Gas Emission Trends and Projections in Europe 2008: Tracking Progress Towards Kyoto Targets*. Report No. 5/2008, prepared for European Environment Agency (EEA), Office for Official Publications of the European Communities, Luxembourg.

EEA, 2008c: *Transport at a Crossroads. TERM 2008: Indicators Tracking Transport and Environment in the European Union*. Report No. 3/2008, prepared for European Environment Agency (EEA), Office for Official Publications of the European Communities, Luxembourg.

EIA, 2010: *International Energy Outlook 2010*. Energy Information Administration (EIA), US Department of Energy, Washington, DC.

EPA, 2010a: *Renewable Fuel Standards*. United States Environmental Protection Agency (EPA). www.epa.gov/otaq/fuels/renewablefuels/index.htm (accessed 31 December 2010).

EPRI, 2001: *Comparing the Benefits and Impacts of Hybrid Electric Vehicle Options*. Report 1000349, prepared for Electric Power Research Institute (EPRI), **Palo Alto, USA**.

Estache, A. and A. Gomez-Lobo, 2005: The limits to competition in urban bus service in developing countries. *Transport Reviews*, **25**(2): 139–158.

European Commission, 2006: Emissions trading: Commission decides on first set of national allocation plans for the 2008–2012 trading period. *Europa: Press Releases Rapid*, European Commission, **29** November 2006, Brussels. http://europa.eu/rapid/pressReleasesAction.do?reference=IP/06/1650&format=HTML&aged=0&language=EN&guiLanguage=en (accessed 25 May 2011).

EU, 2010: *EU Energy and Transport in Figures: Statistical Pocketbook 2010*. European Union (EU), Publications Office of the European Union, Luxembourg.

EUCAR/CONCAWE/JRC, 2006: *Well-to-Wheels Analysis of Future Automotive Fuels and Powertrains in the European Context*. European Council for Automotive R&D (EUCAR), Conservation of Clean Air and Water in Europe (CONCAWE), and the Joint Research Center (JRC)/Institute for Environment and Sustainability, Scientific Technology Options Assessment (STOA) Workshop: The Future of European Long Distance Transport.

FAO, 2008: *FAOSTAT Agriculture*. Food and Agriculture Organization (FAO), Rome, Italy.

Farrell, A., R. Plevin, B. Turner, A. Jones, M. O'Hare and D. Kammen, 2006: Ethanol can contribute to energy and environmental goals. *Science*, **311**(5760): 506–508.

FedEx, 2004: *FedEx Hybrid Update: 10 More to NYC*. Green Car Congress. www.greencarcongress.com/2004/10/fedex_hybrid_up.html (accessed 31 December 2010).

Figueroa, M. J., 2010: Global south mobility, democracy and sustainability. In *A New Sustainability Agenda*. Ashgate, London.

Flachsland, C., S. Brunner, O. Edenhofer and F. Creutzig, 2011: Climate policies for road transport revisited (II): Closing the policy gap with cap-and-trade. *Energy Policy*, **39**(4): 2100–2110.

Flyvbjerg, B., M. Garbujo and D. Lovallo, 2009: Delusion and deception in large infrastructure projects: Two models for explaining and preventing executive disaster. *California Management Review*, **51**(2): 170–193.

Flyvbjerg, B., M. Skamris and S. Buhl, 2004: What causes cost overrrun in transport infrastructure projects? *Transport Reviews*, **24**: 3–18.

Fullerton, D. and S. West, 2000: *Tax and Subsidy Combinations for the Control of Car Pollution*. NBER Working Paper No. 7774, prepared for National Bureau of Economic Research (NBER), Cambridge, MA.

Gaunt, M., T. Rye and S. Ison, 2006: Gaining public support for congestion charging: Lessons from referendum in Edinburgh, Scotland. *Transportation Research Record*, **1960**(1): 87–93.

Gilbert, R. and A. Pearl, 2007: *Transport Revolutions: Moving People and Freight without Oil*. Earthscan, London.

Global Mass Transit, 2010: *Global Mass Transit Report*. http://globalmasstransit.net/index.php (accessed 11 November 2010).

GM/ANL, 2005: *Well-to-Wheels Analysis of Advanced Fuel/Vehicle Systems – A North American Study of Energy Use, Greenhouse Gas Emissions, and Criteria Pollutant Emissions*. General Motors/Argonne National Laboratory (GM/ANL), Argonne National Laboratory, Argonne, IL.

Goldemberg, J., 2009: The Brazilian experience with biofuels. *Innovations*, **4**(4): 91–107.

Goodwin, P., 1996: Empirical evidence on induced traffic: A review and synthesis. *Transportation*, **23**: 35–54.

Goodwing, P., 2003: *Unintended effects of Policies*. Chapter 33, In Hensher, D., and Button, K., (eds) *Handbook of Transport and the Environment*, Handbooks in Transport Volume 4, Elsevier, UK.

Gordon, P., A. Kuman and H. Richardson, 1989: The influence of metropolitan spatial structure on commuting time. *Journal of Urban Economics*, **26**(2): 138–151.

Greener by Design, 2011 – http://www.greenerbydesign.co.uk/faq/index.php (accessed 29 July 2011)

Greening, L. A., D. L. Greene and C. Difiglio, 2000: Energy efficiency and consumption – the rebound effect – a survey. *Energy Policy*, **28**(6–7): 389–401.

GTZ, 2002: Introductory module: Sourcebook Overview, and Cross-cutting Issues of Urban Transport. *Sustainable Transport: A Sourcebook for Policy-makers in Developing Cities*. Deutsche Gesellschaft für Technische Zusammenarbeit (GTZ), Eschborn, Germany.

GTZ, 2007: Module 5e, Transport and Climate Change. *Sustainable Transport: A Sourcebook for Policy-makers in Developing Cities*. Deutsche Gesellschaft für Technische Zusammenarbeit (GTZ), Eschborn, Germany.

GTZ, 2009: *International Fuel Prices 2009*. 6th Edition. Deutsche Gesellschaft für Technische Zusammenarbeit (GTZ). www.gtz.de/fuelprices (accessed 31 December 2010).

Gwilliam, K., 2002: *Cities on the Move: A World Bank Urban Transport Strategy Review*. World Bank, Washington, DC.

Gwilliam, K., 2003: *Urban Transport in Developing Countries*, *Transport Reviews*, **23**: 2, 197–216

Gwilliam, K., 2005: *Bus Franchising in Developing Countries: Some Recent World Bank Experience*. Revised keynote paper, 8th International Conference on Ownership and Regulation of Land Passenger Transport, June 2003, Rio de Janeiro, Brazil.

Hammerschlag, R. and P. Mazza, 2005: Questioning hydrogen. *Energy Policy*, **33**: 2039–2043.

Heavenrich, R. M., 2005: *Light-Duty Automotive Technology and Fuel Economy Trends: 1975 through 2005*. Office of Transportation and Air Quality, US Environmental Protection Agency, (US EPA), Washington, DC.

Hertel, T. W., A. A. Golub, A. D. Jones, M. O'Hare, R. J. Plevin and D. M. Kammen, 2010: Global land use and greenhouse gas emissions impacts of US maize ethanol: Estimating market-mediated responses. *BioScience*, **60**(3): 223–231.

Heywood, J. B. and M. A. Weiss, 2003: *The Performance of Future ICE and Fuel Cell Powered Vehicles and Their Potential Fleet Impact*. Report LFEE 2003-004 RP, prepared for the Laboratory for Energy and the Environment (LFEE), Massachusetts Institute of Technology (MIT), Cambridge, MA.

Hidalgo, D., Carrigan, A. 2010: *Modernizing Public Transportation: Lessons Learned From Major Bus Improvement in Latin America and Asia*. EMBARQ Report, 2010. World Resources Institute http://www.embarq.org/sites/default/files/EMB2010_BRTREPORT.pdf (accessed 8 June 2011).

Holland, S. P., C. R. Knittel and J. E. Hughes, 2009: Greenhouse gas reductions under low carbon fuel standards? *American Economic Journal: Economic Policy*, **1**(1): 106–146.

Hu, P. S. and T. R. Reuscher, 2004: *Summary of Travel Trends, 2001 National Household Travel Survey*. Prepared for the US Department of Transportation, Federal Highway Administration, Washington, DC.

Hymel, K. M., K. A. Small and K. Van Dender, 2010: Induced demand and rebound effects in road transport. *Transportation Research Part B*, **44**(10): 1120–41.

IFEU, 2010: *Fortschreibung und Erweiterung Daten- und Rechenmodell: Energieverbrauch und Schadstoffemissionen des motorisierten Verkehrs in Deutschland 1960–2030*. Institut für Energie- und Umweltforshung (IFEU), TREMOD, Version 5, Heidelberg, Germany.

ICAO, 2007: *ICAO Environment Report 2007*. International Civil Aviation Organization (ICAO), Montreal.

ICAO, 2009: *Report from the Group on International Aviation and Climate Change (GIACC)*, Montreal.

ICAO, 2010: *ICAO Environment Report 2010*. International Civil Aviation Organization (ICAO), Montreal.

IEA, 2008: *Energy Technology Perspectives 2008*. International Energy Agency (IEA), Paris.

IEA, 2009a: *CO_2 Emissions from Fuel Combustion*. International Energy Agency (IEA), Paris.

IEA, 2009b: *Energy Balances of Non-OECD Countries*. International Energy Agency (IEA), Paris.

IEA, 2009c: *Transport, Energy and CO2: Moving Toward Sustainability*. International Energy Agency (IEA), Paris.

IEA, 2009d: *World Energy Outlook*. International Energy Agency (IEA), Paris.

IEA, 2010a: *World Energy Outlook*. International Energy Agency (IEA), Paris.

IEA, 2010b: *Energy Subsidies-Getting Prices Right*. Office of the Chief Economist, International Energy Agency (IEA), Paris.

IMO, 2008: *Opportunities for Reducing Greenhouse Gas Emissions from Ships*. Report IMO MEPC 58/INF.21, prepared for International Maritime Organization (IMO), London, UK.

IMO, 2009: *Second IMO GHG Study*. International Maritime Organization (IMO), London, UK.

IPCC, 1999: *Special Report on Aviation and the Global Atmosphere*. Intergovernmental Panel on Climate Change (IPCC), Cambridge University Press, UK.

IPCC, 2007: *Fourth Assessment Report AR4, Contribution of Working Group III to the Fourth Assessment Report of the Intergovernmental Panel on Climate Change*. B. Metz, O. Davidson, P. Bosh, R. Dave and L. Meyer (eds.), Intergovernmental Panel on Climate Change (IPCC), Cambridge University Press, UK.

ITDP, 2007: *Bus Rapid Transit Planning Guide*. http://www.itdp.org/documents/Bus%20Rapid%20Transit%20Guide%20-%20complete%20guide.pdf (accessed 14 June 2011).

ITDP, 2010: *Promoting Sustainable and Equitable Transportation Worldwide*. Institute for Transportation and Development Policy (ITDP). www.itdp.org (accessed 30 May 2010).

ITF, 2008: International Transport Forum (ITF), Transport Outlook 2008: Focusing on CO_2 Emissions from Road Vehicles, Discussion Paper No. 2008–13.

Jaccard, M., 2006: Mobilizing producers toward environmental sustainability: The prospects for market-oriented regulations. In *Sustainable Production: Building Canadian Capacity*. G. Toner (ed.), University of British Columbia Press, Vancouver, pp.154–177.

Jacobson, M., 2007: Effects of ethanol (E85) versus gasoline vehicles on cancer and mortality in the United States. *Environmental Science and Technology*, **41**(11): 4150–4157.

JHFC, 2006: *Well-to-Wheel Efficiency Analysis Results*. Japan Hydrogen and Fuel Cell Development Project (JHFC), Japan.

Johnson, T. M., C. Alatorre, Z. Romo and L. Feng, 2009: *Low-Carbon Development for Mexico*. World Bank, Washington, DC.

Kahn Ribeiro, S., S. Kobayashi, M. Beuthe, J. Gasca, D. Greene, D. S. Lee, Y. Muromachi, P. J. Newton, S. Plotkin, D. Sperling, R. Wit and P. J. Zhou, 2007: Transport and its infrastructure. In *Climate Change 2007: Mitigation. Contribution of Working Group III to the Fourth Assessment Report of the Intergovernmental Panel on Climate Change*. B. Metz, O. R. Davidson, P. R. Bosch, R. Dave, L. A. Meyer (eds.), Intergovernmental Panel on Climate Change (IPCC), Cambridge University Press, UK.

Kameyama, M., K. Hiraoka, A. Sakurai, T. Naruse and H. Tauchi, 2005: Development of LCA software for ships and LCI analysis based on actual ship-building and operation. In *Proceedings of the 6th International Conference on Ecobalance*, 25–27 October 2005, Tsukuba, Japan, pp.159–162.

Keith, D. W. and A. Farrell, 2003: Rethinking hydrogen cars. *Science*, **301**(5631): 315–316.

Kintner-Meyer, M., K. Schneider and R. Pratt, 2007: *Impacts Assessment of Plug-In Hybrid Vehicles on Electric Utilities and Regional US Power Grids, Part 1: Technical Analysis*. Report PNNL-SA-53523, prepared for Pacific Northwest National Laboratory, Richland, Washington.

Kobayashi, S., S. Plotkin and S. Kahn Ribeiro, 2009: Energy efficiency technologies for road vehicles. *Energy Efficiency*, **2**(2): 125–137.

Kromer, M. A. and J. B. Heywood, 2008: *A Comparative Assessment of Electric Propulsion Systems in the 2030 US Light-Duty Vehicle Fleet*. SAE 2008–01–0459, prepared for Society of Automobile Engineers (SAE) World Congress, 14–17 April 2008, Detroit, MI.

Krugman, P., 1991: Increasing returns and economic geography. *Journal of Political Economy*, **99**(3): 483–99.

Krupnick, A., D. Burtraw and A. Markandya, 2000: The ancillary benefits and costs of climate change mitigation: A conceptual framework. In *Ancillary Benefits and Cost of Green House Gas Mitigation*. Intergovernmental Panel on Climate Change (IPCC) and Organisation for Economic Co-operation and Development (OECD), Paris.

Lapola, D. M., R. Schaldach, J. Alcamo, A. Bondeau, J. Koch, C. Koelking and J. A. Priess, 2010: Indirect land-use changes can overcome carbon savings from biofuels in Brazil. *Proceedings for the National Academy of Science*, **23**: 3388–3339.

Larsen, H. and L. S. Peterson (eds.), 2009: How do we convert the transport sector to renewable energy and improve the sector's interplay with the energy system? Workshop on Transport, 17–18 March 2009, Technical University of Denmark, Lyngby, Denmark.

Litman, T., 2007: *Transportation Affordability: Evaluation and Improvement Strategies*. Victoria Transport Policy Institute, Victoria, Canada.

Litman, T., 2010: *Evaluating Non-Motorized Transportation Benefits and Costs*. Victoria Transport Policy Institute, Victoria, Canada.

Lutsey, N. and D. Sperling, 2008: Greenhouse gas mitigation supply curve for the United States for transport versus other sectors. *Transportation Research Part D*, **14**(3): 222–229.

Lyons, G. and J. Urry, 2005: Travel time use in the information age. *Transportation Research Part A*, **39**(2–3): 257–76.

Martinot, E. and J. Sawin, 2009: Renewables global status report 2009 update. *Renewable Energy World Magazine*, **9** September 2009.

May, A. D, C. Kelly and S. Shepherd, 2006: the principles of integration in urban transport strategies. *Transport Policy*, **13**(4): 319–327.

McKinsey Global Institute, 2009: *Averting the Next Energy Crisis: The Demand Challenge*. McKinsey Global Institute, San Francisco, CA.

Merriman, P., 2009: Automobility and the geographies of the car. *Geography Compass*, **3**(2): 586–599.

Methorst, R., H. Monterde i Bort, R. Risser, D. Sauter, M. Tight and J. Walker, 2010: *COST 358: Pedestrians' Quality Needs Final Report*. European Cooperation in Science and Technology (COST), Cheltenham, UK.

Millard-Ball, A. and L. Schipper, 2011: Are we reaching peak travel? Trends in passenger transport in eight industrialized countries. *Transport Reviews*, **31**(3): 357–378.

Mokhtarian, P., 1990: A typology of Relationships between Telecommunications and Transportation" Transportation Research-A. Vol. 24A. No. 3. PP. 231–242. Great Britain.

Monkelbaan, J., 2010: *COP16: Negotiating Bunker Fuels in Cancun*. Sustainable Shipping. www.sustainableshipping.com/forum/blogs/Joachim-Monkelbaan/98543/COP16-Negotiating-bunker-fuels-in-Cancun (accessed 21 May 2011).

Morrow, K, D. Karner and J. Francfort, 2008: *Plug-In Hybrid Electric Vehicle Charging Infrastructure Review – Final Report*. Report INL/EXT-08–15058, prepared for the US Department of Energy (US DOE), Batelle Energy Alliance, Idaho National Laboratory, ID.

MoRTH, 2009: *Road Transport Year Book, 2006–2007*. Ministry of Road Transport and Highways (MoRTH), India.

Naess, P., 2006: *Urban Structure Matters: Residential Location, Car Dependence and Travel Behaviour*. Routledge, Oxfordshire, UK.

Nagai, Y., A. Fukuda, Y. Okada and Y. Hashino, 2003: Two-wheeled vehicle ownership trends and issues in the Asian region. *Journal of the Eastern Asia Society for Transportation Studies*, **5**: 135–146.

Nazelle, A., Nieuwenhuijsen, M., Antó, J., Brauer, M., Briggs, D., Braun-Fahrlander, C., Cavill, N., Cooper, A., Desqueyroux, H., Fruin, S., Hoek, G., Panis, L., Janssen, N., Jerrett, M., Joffe, M., Andersen, Z., van Kempen, E., Kingham, S., Kubesch, N., Leyden, K., Marshall, J., Matamala, J., Mellios, G., Mendez, M., Nassif, H., Ogilvie, D., Peiró, R., Pérez, K., Rabl, A., Ragettli, M., Rodríguez, D., Rojas, D., Ruiz, P., Sallis, J., Terwoert, J., Toussaint, J., Tuomisto, J., Zuurbier, M., Lebret, E. 2011: *Improving health through policies that promote active travel: A review of evidence to support integrated health impact assessment*. Environment International, Elsevier, **37**, (2011), 766–777

Newman, P. and J. Kenworthy, 1999: *Sustainability and Cities: Overcoming Automobile Dependence*. Island Press, Washington, DC.

Niven, R., 2005: Ethanol in gasoline: Environmental impacts and sustainability review article. *Renewable and Sustainable Energy Reviews*, **9**(6): 535–555.

NRC, 2008: *Transitions to Alternative Transportation Technologies – A Focus on Hydrogen*. National Research Council (NRC), The National Academies Press, Washington, DC.

O'Hare, M., R. J. Plevin, J. I. Martin, A. D. Jones, A. Kendall and E. Hopson, 2009: Proper accounting for time increases crop-based biofuels' greenhouse deficit versus petroleum. *Environmental Research Letter*, **4**: 024001.

Ogden, J. M., R. H. Williams and E. Larson, 2004: Societal lifecycle costs of cars with alternative fuels/engines. *Energy Policy*, **32**(1): 7–27.

OPTIC, 2010: *Inventory of Measures, Typology of Non-intentional Effects and a Framework for Policy Packaging*. Institute of Transport Economics (TØI), Optimal Policies for Transport in Combination (OPTIC), Deliverable 1, EU Seventh Framework Program, 17 December 2010, Norway.

ORNL, 2010: *Transportation Energy Data Book*. 29th Edition. Oak Ridge National Laboratory (ORNL), US Department of Energy (US DOE), Oak Ridge, TN, USA.

OTA, 1995: *Advanced Automotive Technology – Visions of a Super Efficient Family Car*. Report OTA-ETI-638, prepared for US Congress, Office of Technology Assessment (OTA), US Government Printing Office, Washington, DC.

Padam, S. and S. K. Singh, 2001: *Urbanization and Urban Transport in India: The Sketch for a Policy*. Central Institute of Road Transport, Pune, India.

Patzek, T., 2004: Thermodynamics of the corn-ethanol biofuel cycle. *Critical Reviews in Plant Sciences*, **23**(6): 519–567.

Patzek, T. and D. Pimentel, 2005: Thermodynamics of energy production from biomass. *Critical Reviews in Plant Sciences*, **24**(5): 327–364.

Pigou, A. C. 1920: *Wealth and Welfare*. MacMillan, London.

Plevin, R. J., M. O'Hare, A. D. Jones, M. S. Torn and H. K. Gibbs, 2010: The greenhouse gas emissions from biofuels' indirect land use change are uncertain, but potentially much greater than previously estimated. *Environmental Science and Technology*, **44**(21): 8015–8021.

Plotkin, S., 2008: Examining fuel economy and carbon standards for light vehicles. In *The Cost and Effectiveness to Reduce Vehicle Emissions*. International Energy Agency (IEA) and International Transport Forum (ITF), Paris.

Power System, 2005: Development of high power and high energy density capacitor (in Japanese). Press release, 27 June 2005, Japan.

Prud'homme, R. and J. P. Bocajero, 2005: The London congestion charge: A tentative economic appraisal. *Transport Policy*, **12**(3): 279–287.

Quirin, M., S. Gärtner, M. Pehnt and G. Reinhardt, 2004: *CO_2 Mitigation through Biofuels in the Transport Sector: Status and Perspectives*. IFEU, Heidelberg.

Rothengatter, W., 2003: Environmental Concepts – Physical and Economic. In *Handbook of Transport and the Environment*. D. Hensher and K. Button (eds.), Elsevier, Oxford.

Salter, R., Dhar, S., Newman, P., (eds) 2011: *Technologies for Climate Change Mitigation*. United Nations Environment Programme, Risoe Centre on Energy, Climate and Sustainable Development (UNEP-Risoe), Roskilde, Denmark.

Sanderson, K., 2009: Wonder weed plans fail to flourish. *Nature*, **461**: 328–329.

Sausen, R., I. Isaksen, V. Grewe, D. Hauglustaine, D. S. Lee, G. Myhre, M. O. Kohler, G. Pitari, U. Schumann, F. Stordal and C. Zerefos, 2005: Aviation radiative forcing in 2000: An update on IPCC (1999). *Meteorologische Zeitschrift*, **114**: 555–561.

Schaber, C., P. Mazza and R. Hammerschlag, 2004: Utility-scale storage of renewable energy. *Electricity Journal*, **17**(6): 21–29.

Schafer, A., 2000: Regularities in travel demand: An international perspective. *Journal of Transportation and Statistics*, **3**(3): 1–31.

Schafer, A., J. B. Heywood, H. Jacoby and I. Waitz, 2009: *Transportation in a Climate Constrained World*. MIT Press, Cambridge, MA.

Schipper, L. and M. Marie-Lilliu, 1999: *Transport and CO2 Emissions: Flexing the Link – A Path for the World Bank*. Environment Department Paper No. 69, World Bank, Washington, DC.

Schipper, L., 2007: *Automobile Fuel, Economy and CO2 Emissions in Industrialized Countries: Troubling Trends through 2005/6*. EMBARQ, the World Resources Institute Center for Sustainable Transport, Washington, DC.

Schipper, L., 2009: Fuel economy, vehicle use and other factors affecting CO2 emissions from transport. *Energy Policy*, **37**(10): 3711–3713.

Schneider, K., C. Gerkensmeyer, M. Kintner-Meyer and R. Fletcher, 2008: Impact assessment of plug-in hybrid vehicles on Pacific Northwest distribution systems. In *Power and Energy Society General Meeting – Conversion and Delivery of Electrical Energy in the 21st Century*, Institute of Electrical and Electronics

Engineers (IEEE), 20–24 July 2008, Pacific Northwest National Laboratory, Richland, WA.

Schrank, D., T. Lomax and S. Turner, 2010: *TTI's 2010 Urban Mobility Report*. Texas Transportation Institute (TTI) and The Texas A&M University System, TX.

Schwarzenegger, A., 2007: *Executive Order S-01–07: Low Carbon Fuel Standard*. Sacramento, CA.

Searchinger, T., R. Heimlich, R. A. Houghton, F. Dong, A. Elobeid, J. Fabiosa, S. Tokgoz, D. Hayes and T.Yu, 2008: Use of US croplands for biofuels increases greenhouse gases through emissions from land-use change. *Science*, **319**(5867): 1238–1240.

Serebrisky, T., Gómez-Lobo, A., Estupiñán, N., Muñoz-Raskin, R., 2009: Affordability and Subsidies in Public Urban Transport: What Do We Mean, What Can Be Done?. *Transport Reviews*, **29**(6): 715–739

Sheller, M. and J. Urry, 2003: Mobile transformations of public and private life. *Theory, Culture and Society*, **20**(3): 107–25.

Shiau, C., C. Samaras, R. Hauffe and J. Michalek, 2009: Impact of battery weight and charging patterns on the economic and environmental benefits of plug-in hybrid vehicles. *Energy Policy*, **37**(7): 2653–2663.

Shoup, D., 2005: *The High Cost of Free Parking*. Planners Press, Chicago.

Simpson, A., 2006: *Cost-Benefit Analysis of Plug-In Hybrid Electric Vehicle Technology*. 22nd International Battery, Hybrid and Fuel Cell Electric Vehicle Symposium and Exhibition, 23–28 October 2006, National Renewable Energy Laboratory (NREL), Yokohama, Japan.

Singapore Maritime Careers, 2011 – www.maritimecareers.com.sg/maritime_industry_shipping_article1.html (accessed in 29 July 2011)

Small, K. A. and K. Van Dender, 2007: Fuel efficiency and motor vehicle travel: The declining rebound effect. *Energy Journal*, **28**(1): 25–51.

Small, K. A., and E. Verhoef, 2007: *The Economics of Urban Transportation*. Routledge, New York.

Sperling, D. and D. Gordon, 2009: *Two Billion Cars Driving toward Sustainability*. Oxford University Press, New York.

Spinney, J., 2009: Cycling the city. *Geography Compass*, **3**(2): 817–835.

Statistics Bureau of Japan, 2010: *The World Statistical Handbook 2010*. Statistical Research and Training Institute, MIC (ed.), Statistics Bureau of Japan, Tokyo.

Stern, N., 2007: *The Economics of Climate Change: The Stern Review*. Cabinet Office, HM Treasury, Cambridge University Press, New York.

Stoft, S., 2009: *The Global Rebound Effect Versus California's Low-Carbon Fuel Standard*. Research Paper No. 09–04, prepared for Global Energy Policy Center, Berkeley, CA.

Tesla Motors, 2009: *Tesla Showroom*. www.teslamotors.com/buy/buyshowroom.php (accessed 31 December, 2010).

Tilman, D., R. Socolow, J. A. Foley, J. Hill, E. Larson, L. Lynd, S. Pacala, J. Reilly, T. Searchinger, C. Somerville and R. Williams, 2009: Beneficial biofuels – The food, energy, and environment trilemma. *Science*, **325**(5938): 270–271.

Tiwari, G., 2006: *Urban Passenger Transport: Framework for an Optimal Mix*. India Resident Mission Policy Brief Series No. 1, Asian Development Bank (ADB), New Delhi, India.

Thompson, L., 2010: *A Vision for Railways in 2050*. Forum Paper 2010–4, prepared for *Transport Innovation: Unleashing the Potential*, International Transport Forum (ITF), 24–26 May 2010, Leipzig, Germany.

Thomsen, T., L. Drewers and H. Gudmundsson, 2005: *Social Perspectives on Mobility*. Aldershot, Ashgate, London.

Toyota, 2004: *Environmental & Social Report 2004*. Toyota Motor Corporation, Japan.

Toyota/Mizuho, 2004: *Well-to-Wheel Analysis of Greenhouse Gas Emissions of Automotive Fuels in the Japanese Context*. Mizuho Information and Research Institute, Tokyo.

Transport for London, 2007. *Impacts Monitoring: Fifth Annual Report*. Transport for London, Mayor of London, London.

Transportation Alternatives, 2006: *Curbing Cars: Shopping, Parking and Pedestrian Space in SoHo*. Transportation Alternatives, New York, NY.

Transportation Alternatives, 2007: *No Vacancy: Park Slope's Problem and How to Fix It*. Transportation Alternatives, New York, NY.

Transportation Alternatives, 2008: *Driven to Excess: What Under-priced Curbside Parking Costs the Upper West Side*. Transportation Alternatives, New York, NY.

UIC, 2011: *International Union of Railways Homepage*. International Union of Railways (UIC). http://www.uic.org/ (accessed 31 December 2010).

UITP, 2010. *Public Transport in Sub-Saharan Africa: Major Trends and Case Studies*. International Association of Public Transport (UITP), Trans-Africa Consortium, Brussels.

UN, 2007. *World Population Prospects: The 2006 Revision*. Population Division of the Department of Economic and Social Affairs of the United Nations Secretariat, New York.

UN, 2008. *World Urbanization Prospects: The 2007 Revision*. Population Division of the Department of Economic and Social Affairs of the United Nations Secretariat, New York.

UN, 2009. *World Urbanization Prospects: The 2009 Revision*. Population Division of the Department of Economic and Social Affairs of the United Nations Secretariat, New York.

UNCTAD, 2010: *Review of Maritime Transport*. United Nations Conference on Trade and Development (UNCTAD), Switzerland.

UNDP, 2010: *Human Development Report 2010 – The Real Wealth of Nations: Pathways to Human Development*. United Nations Development Programme (UNDP), Palgrave, Macmillan, New York.

UNEP, 2011: *Towards a Green Economy: Pathways to Sustainable Development and Poverty Eradication*. United Nations Environment Programme (UNEP), Nairobi, Kenya.

UNESCAP, 2004: *Manual on Modernization on Inland Water Transport for Integration within a Multimodal Transport System*. United Nations Economic and Social Commission for Asia and the Pacific (UNESCAP). www.ruralwaterways.org/en/case/case.php (accessed 20 May 2011).

UNFCCC, 2010: *Information relevant to emissions from fuel used for international aviation and maritime transport*. In 32nd session of the subsidiary body for scientific and technological advice, United Nations Framework Convention on Climate Change (UNFCCC), Bonn, Germany.

US BTS, 2001: *Highlights of the 2001 National Highway Household Travel Survey, Daily Passenger Travel*. United States Bureau of Transportation Statistics (US BTS), Research and Innovative Technology Administration (RITA). www.bts.gov/publications/highlights_of_the_2001_national_household_travel_survey/html/section_02.html (accessed 31 December 2010).

US DOE, 2010: *Advanced Technologies and Energy Efficiency*. United States Department of Energy (US DOE). http://www.fueleconomy.gov/feg/atv.shtml (accessed 30 May 2010).

US DOE AFAVDC, 2009: *Natural Gas Vehicle Emissions*. United States Department of Energy, Alternative Fuels and Advanced Vehicles Data Center (US DOE AFAVDC), www.afdc.energy.gov/afdc/vehicles/natural_gas_Emissions.html (accessed 31 December 2010).

US DOT, 2010: *Freight Transportation Global Highlights*, U.S. Department of Transportation, Research and Innovative Technology Administration, Bureau of Transportation Statistics, Washington, DC.

Urry, J., 2004: The system of automobility. *Theory, Culture and Society*, **21**(4–5): 25–39.

Urry, J., 2007: *Mobilities*. Polity Press, Cambridge, UK.

Van Dender, K., 2009: Energy policy in transport and transport policy. *Energy Policy*, **37**(10): 3854–3862.

Van Wee, B., P. Rietveld and H. Meurs, 2006: Is average daily travel time expenditure constant? In search of explanations for an increase in average travel time. *Journal of Transport Geography*, **14**: 109–22.

Vasconcellos, E. A., 2001: *Urban Transport Environment and Equity: The Case for Developing Countries*, Earthscan Publications Ltd.

Vasconcellos, E. A., 2010: *Análisis de la Movilidad Urbana, Espacio, Medio Ambiente y Equidad*. Corporación Andina de Fomento (CAF), Bogota, Columbia.

Victoria Transport Policy Institute, 2011: *Walking and Cycling Encouragement: Strategies That Encourage People To Use Non-motorized Transportation*. Victoria Transport Policy Institute, Victoria, Canada. http://www.vtpi.org/tdm/tdm3.htm (accessed 30 May 2011).

von Blottnitz, H. and M. A. Curran, 2007: A review of assessments conducted on bioethanol as a transportation fuel from a net energy, greenhouse gas, and environmental life cycle perspective. *Journal of Cleaner Production*, **15**(7): 607–619.

Wang, M., H. Huo, L. Johnson and D. He, 2006: *Projection of Chinese Motor Vehicle Growth, Oil Demand, and CO2 Emissions through 2050*. Report ANL/ESD/06–6, prepared for the Energy Systems Division, Argonne National Laboratory, Argonne, IL.

WBCSD, 2002: *Mobility 2001: World Mobility at the End of the Twentieth Century and Its Sustainability*. World Business Council for Sustainable Development (WBCSD), Geneva.

WBCSD, 2004a: *IEA/SMP Model Documentation and Reference Projection*. World Business Council for Sustainable Development (WBSCD) and International Energy Agency (IEA), Geneva.

WBCSD, 2004b: *Mobility 2030 – Meeting the Challenges to Sustainability*. World Business Council for Sustainable Development (WBSCD), Geneva.

WBCSD, 2009: *Mobility for Development*. World Business Council for Sustainable Development (WBSCD), Geneva.

Wegener, M. and F. Fürst, 1999: *Land Use Transport Interaction: State of the Art*. Deliverable 2a of the project Integration of Transport and Land Use Planning (TRANSLAND) of the 4th RTD Framework Programme of the European Commission, Institut für Raumplanung, Universität Dortmund, Germany.

Wheeler, S., 2004: *Planning for Sustainability Creating Livable, Equitable and Ecological Communities*. Routledge, New York.

Whitelegg, J., 1997: *Critical Mass*. Pluto Press, London.

Whitelegg, J. and G. Haq, 2003: *The Earthscan Reader in World Transport Policy and Practice*. Earthscan, London.

WHO, 2006: *Air Quality Guidelines, Global Update 2005: Particulate Matter, Ozone, Nitrogen Dioxide and Sulfur Dioxide*. World Health Organization (WHO) Regional Office for Europe, Copenhagen, Denmark. Figure 9.20: p38.

WHO, 2009: *Global Status Report on Road Safety: Time for Action*. World Health Organization (WHO), Geneva.

Woodcock, J., P. Edwards, C. Tonne, B. Armstrong, O. Ashiru, D. Banister, S. Beevers, Z. Chalabi, Z. Chowdhury, A. Cohen, O. Franco, A. Haines, R. Hickman, G. Lindsay, I. Mittal, D. Mohan, G. Tiwari, A. Woodward and I. Roberts, 2009: Public health benefits of strategies to reduce green-house gas emissions: Urban land transport. *The Lancet*, **374**(9705): 1930–1943.

Woodcock, J., O. Franco, N. Orsini and I. Roberts, 2011: Non-vigorous physical activity and all-cause mortality: Systematic review and meta-analysis of cohort studies. *International Journal of Epidemiology*, **40**(1): 121–138.

World Bank, 1996: *Sustainable Transport: Priorities for Policy Reform*. World Bank, Washington, DC.

World Bank, 2008a: *A Framework for Urban Transport Projects, Operational Guidance for World Bank Staff*. Transport Sector Board, World Bank, Washington, DC.

World Bank, 2008b: *World Development Report 2008*. World Bank, Washington, DC.

World Bank, 2010: *World Development Indicators*. World Bank, Washington, DC.

Wright, L., Fjellstrom, K. 2005: Mass Transit Options. In *Sustainable Transport: A Sourcebook for Policy-makers in Developing Cities*. Gesellschaft für Technische Zusammenarbeit (GTZ), Eschborn, Germany.

Wright, L. 2006: Module 3b: Bus rapid transit. In *Sustainable Transport: A Sourcebook for Policy-makers in Developing Cities*. Gesellschaft für Technische Zusammenarbeit (GTZ), Eschborn, Germany.

Wurster, R., 2002: *Well-to-Wheel Analysis of Energy Use and Greenhouse Gas Emissions of Advanced Fuel/Vehicle Systems – A European Study*. Prepared for General Motors/The Ludwig-Bölkow-Systemtechnik GmbH (GM/LBST), presented at the 15th World Hydrogen Energy Conference (WHEC), 28 June 2004, Yokohama, Japan.

Yang, J., C. Feng and G. Cao, 2007: Land and transportation development in china: economic analysis of government behavior. *Transportation Research Record*, **(2038)**: 78–83.

Zah, R., H. Boeni, M. Gauch, R. Hischier, M. Lehmann and P. Wäger, 2007: *Life-Cycle Assessment of Energy Products: Environmental Assessment of Biofuels*. Empa Science and Technology Lab, Bern.

Zegras, C. and R. Gakenhaimer, 2006: *Driving Forces in Developing Cities' Transportation Systems: Insights from Selected Cases*. Working Paper, Department of Urban Studies and Planning, Massachusetts Institute of Technology (MIT), Cambridge, MA.

Zhang, T., 2000: Land market forces and government's role in sprawl. *Cities*, **20**: 265–78.

Zhang, X. and X. Hu, 2003: *Enabling Sustainable Urban Road Transport in China: A Policy and Institutional Perspective*. Working Paper 2003:02, prepared for the Center for International Climate and Environmental Research (CICERO), Oslo.

10

Energy End-Use: Buildings

Convening Lead Author (CLA)
Diana Ürge-Vorsatz (Central European University, Hungary)

Lead Authors (LA)
Nick Eyre (Oxford University, UK)
Peter Graham (University of New South Wales, Australia)
Danny Harvey (University of Toronto, Canada)
Edgar Hertwich (Norwegian University of Science and Technology)
Yi Jiang (Tsinghua University, China)
Christian Kornevall (World Business Council for Sustainable Development, Switzerland)
Mili Majumdar (The Energy and Resources Institute, India)
James E. McMahon (Lawrence Berkeley National Laboratory, USA)
Sevastianos Mirasgedis (National Observatory of Athens, Greece)
Shuzo Murakami (Keio University, Japan)
Aleksandra Novikova (Climate Policy Initiative and German Institute for Economic Research, Germany)

Contributing Authors (CA)
Kathryn Janda (Environmental Change Institute, Oxford University, UK)
Omar Masera (National Autonomous University of Mexico)
Michael McNeil (Lawrence Berkeley National Laboratory, USA)
Ksenia Petrichenko (Central European University, Hungary)
Sergio Tirado Herrero (Central European University, Hungary)

Review Editor
Eberhard Jochem (Fraunhofer Institute for Systems and Innovation Research, Germany)

Contents

Executive Summary

Buildings are key to a sustainable future because their design, construction, operation, and the activities in buildings are significant contributors to energy-related sustainability challenges – reducing energy demand in buildings can play one of the most important roles in solving these challenges. More specifically:

- The buildings sector[1] and people's activities in buildings are responsible for approximately 31% of global final energy demand, approximately one-third of energy-related CO_2 emissions, approximately two-thirds of halocarbon, and approximately 25–33% of black carbon emissions.

- Several energy-related problems affecting human health and productivity take place in buildings, including mortality and morbidity due to poor indoor air quality or inadequate indoor temperatures. Therefore, improving buildings and their equipment offers one of the entry points to addressing these challenges.

- More efficient energy and material use, as well as sustainable energy supply in buildings, are critical to tackling the sustainability-related challenges outlined in the GEA. Recent major advances in building design, know-how, technology, and policy have made it possible for global building energy use to decline significantly. A number of low-energy and passive buildings, both retrofitted and newly constructed, already exist, demonstrating that low level of building energy performance is achievable. With the application of on-site and community-scale renewable energy sources, several buildings and communities could become zero-net-energy users and zero-greenhouse gas (GHG) emitters, or net energy suppliers.

Recent advances in materials and know-how make new buildings that use 10–40% of the final heating and cooling energy of conventional new buildings cost-effective in all world regions and climate zones. Holistic retrofits[2] can achieve 50–90% final energy savings in thermal energy use in existing buildings, with the cost savings typically exceeding investments. The remaining energy needs can be met at the building- and community-level from distributed energy sources or by imported sustainable energy supply. The mix of energy-demand reductions, on-site renewable energy generation, and off-site renewable energy supply that corresponds to the most sustainable solution and minimizes the total cost needs to be evaluated case by case, applying a full system life cycle assessment. Net zero-energy buildings and communities[3] are possible only for select building types and settlement patterns, mainly low-rise buildings and less densely populated residential areas. However, their economics are presently typically unfavorable, as opposed to high-efficiency buildings. Meanwhile, compact medium-rise and high-rise developments offer many advantages, such as reduced surface-to-volume ratios and typically lower energy service demands due to the higher density and concentration of building uses.

The scenarios constructed by the GEA buildings expert team, in concert with the GEA main pathways, demonstrate that a reduction of approximately 46% of the global final heating and cooling energy use in 2005 is possible by 2050 (see Figure 10.1). This is attainable through the proliferation of today's best practices in building design, construction, and operation, as well as accelerated state-of-the-art retrofits. This is achievable while increasing amenity and comfort and without interceding in economic and population growth trends and the applicable thermal comfort and living space increase. It goes hand in hand with the eradication of fuel poverty – i.e., supplying everyone with sufficient thermal

---

1   The GEA refers to energy use in the buildings sector as all direct energy use in buildings, including appliances and other plug loads, and accounting for all electricity consumption for which activities in buildings are responsible. Embodied energy use, emissions of the production of building materials, and their transport to the construction site, and other equipment are not included.

2   Holistic retrofit refers to a major renovation of a building involving a complex of various energy efficiency measures. It is the opposite to a stepwise renovation, when, first, some parts of the building are renovated (e.g., windows), later other parts (e.g., insulation), etc.

3   Net zero energy buildings (communities) are buildings (communities) that consume as much energy as they produce from renewable energy sources within a certain period of time (usually one year)

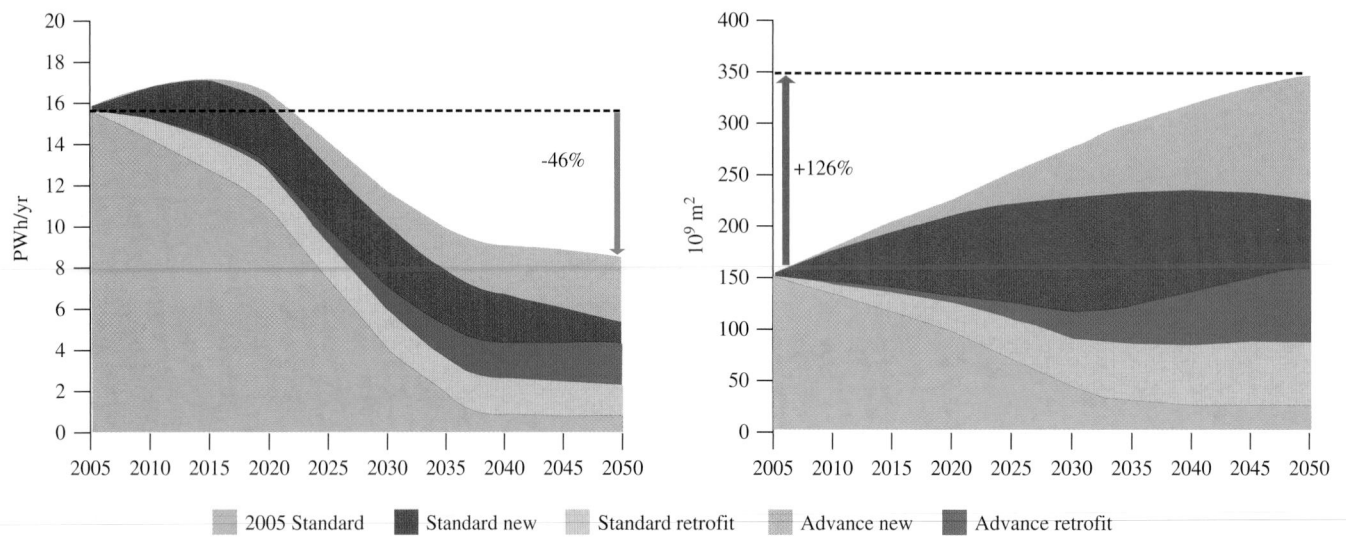

Figure 10.1 | Global final building heating and cooling energy use[4] until 2050 in the state-of-the-art scenario (corresponding roughly to the "GEA Efficiency" set of pathways) (left), contrasted to global floor area (on the right) projections.

comfort. Reaching these state-of-the-art energy efficiency levels in buildings requires approximately US$14.2 trillion in undiscounted additional cumulative investments (US$18.6 trillion with no technology learning) until 2050. However, these investments return substantially higher benefits, e.g., approximately US$58 trillion in undiscounted energy cost savings alone during the same period.

Present and foreseen cutting-edge technologies can reduce energy use of new appliances, information and communication technology (ICT), and other electricity-using equipment in buildings by 65% by 2020, as compared to the baseline. Longer-term projections of technology improvements are speculative, but likely to provide significant additional improvement. Through lifestyle, cultural, and behavioral changes, further significant reductions could be possible.

However, the scenario work also demonstrates that there is a significant lock-in risk. If building codes are introduced universally and energy retrofits accelerate, but policies do not mandate state-of-the-art efficiency levels, substantial energy use and corresponding GHG emissions will be "locked in" for many decades. Such a scenario results in an approximately 33% increase in global building energy use by 2050 compared to 2005, as opposed to a 46% decrease – i.e., an approximately 79% lock-in effect relative to 2005. This points to the importance of building shell-related policies being ambitious about the efficiency levels they mandate (or encourage). Figure 10.2 illustrates opportunities offered by a state-of-the-art scenario as well as the lock-in risk for the 11 GEA regions.

A future involving highly energy-efficient buildings can result in significant associated benefits, typically with monetizable benefits at least twice the operating cost savings, in addition to non-quantifiable or non-monetizable benefits now and avoided impacts of climatic change in the future. One of the most important future benefits is mitigation of the building sector's contribution to climate change. Other benefits include: improvements in energy security and sovereignty; net job creation; elimination of or reduction in indoor air pollution-related mortality and

4   In Chapter 10 energy use is measured in kWh, as it is the most commonly used metrics for the buildings sector. In order to convert kWh to kJ, please, follow the rule: 1 kWh = 3600 kJ.

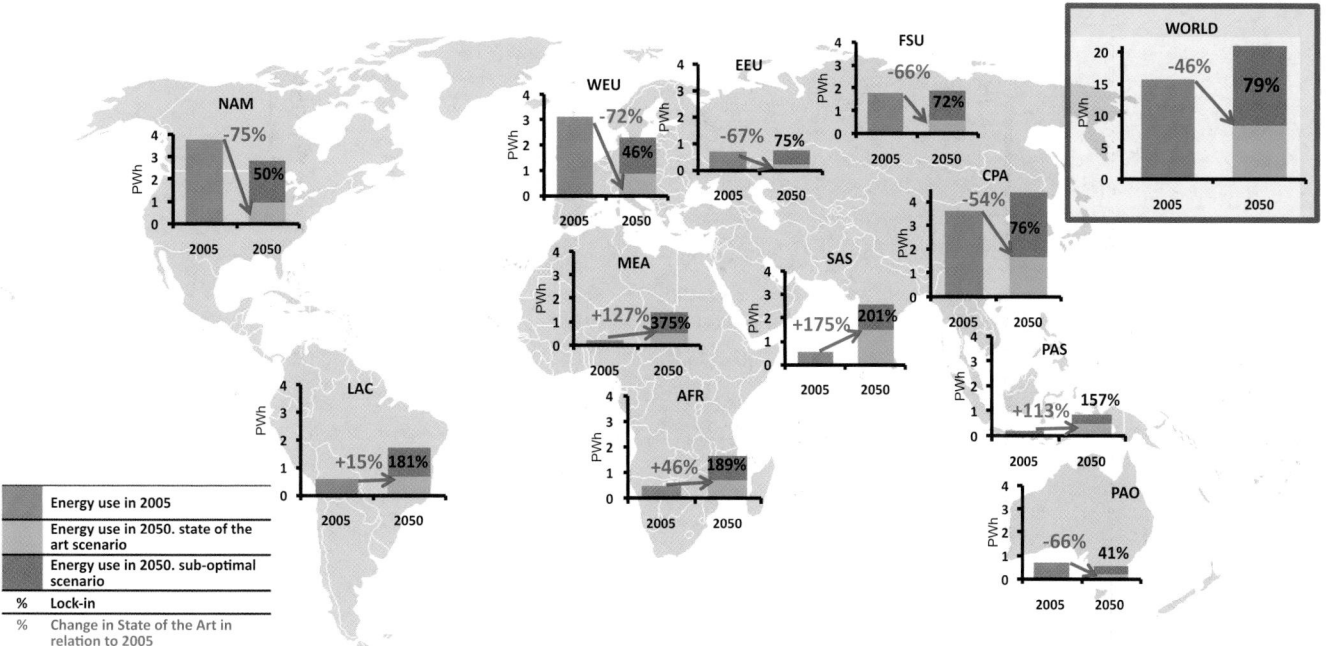

Figure 10.2 | Final building heating and cooling energy demand scenarios until 2050: state-of-the-art (~corresponding roughly to the GEA Efficiency set of pathways) and sub-optimal (~corresponding roughly to the GEA Supply set of pathways) scenarios, with the lock-in risk (difference).

Note: *Green bars, indicated by red arrows and numbers*, represent the opportunities through the state-of-the-art scenario, while the red bars with black numbers show the size of the lock-in risk (difference between the two scenarios). Percent figures are relative to 2005 values.

morbidity; other health improvements and benefits; alleviation of energy poverty and improvement of social welfare; new business opportunities, mostly at the local level; stimulation of higher skill levels in building professions and trades; improved values for real estate and enhanced ability to rent; and increased comfort, well-being, and productivity.

A survey of quantitative evaluations of such multiple benefits shows that even a single energy efficiency initiative in buildings in individual countries or regions has resulted in benefits with values ranging in the billions of dollars annually, such as health improvement-related productivity gains and cost aversions. At the same time, the market-based realization of significant, mostly cost-effective efficiency opportunities in buildings is hampered by a wide range of strong barriers. These barriers are highly variable by location, building type, and culture, as well as by stakeholder groups, such as planners, architects, craftsmen, investors, house and building users, and supervisors. Technological and human capacities of change need to be considered together, as it is through individual and organizational decisions that technologies are provided, adopted, and used. Analysis and examples in this chapter show that most of these barriers can be overcome or mitigated through policies, measures, and innovative financing schemes. A broad portfolio of instruments is available and has been increasingly applied worldwide to capture cost-effective efficiency and conservation potential and to tap other sustainable energy opportunities. Due to the large number and diversity of barriers, single instruments such as a carbon price will not unlock the large efficiency potential, but policy portfolios, tailored to different target groups and tailored to a specific set of barriers, are necessary to optimize results.

Among policy instruments, stringent, continuously updated, and well-enforced building and appliance standards, codes, and labeling – applied also to retrofits – are particularly effective in achieving large energy savings, mostly highly cost-effectively. In order to achieve the major building energy use reductions that have been shown to be possible in this chapter, an urgent introduction of strong building codes mandating near-zero-energy performance levels and progressively improving appliance standards, as well as the strong promotion of state-of-the-art efficiency levels in accelerated retrofits in existing building stocks, are crucial. In contrast, net-zero-energy building mandates are not the most sustainable, cost-effective, or even feasible solutions in many cases, such as dense urban zones or large commercial buildings, and may only encourage urban sprawl. Thus the introduction of such mandates and commitments should be carefully analyzed and, in some cases, re-examined.

For ICT and entertainment appliances, regulation also needs to tackle the durability of the equipment in addition to its operational energy use due to the high-embodied GHG emissions. Appropriate energy pricing is fundamental, and taxation provides the impetus for a more rational use of energy sources. In poor regions or population segments, subsidies enabling a highly energy efficient capital stock can be more effective in tackling energy poverty than energy price subsidies. Carefully designed subsidies enabling investments may be needed to bridge the discount rate gap between society and private decision-makers, and the availability of financing for building owners and users is often a crucial precondition. Innovative financing schemes, such as performance contracting, are paramount for groups with limited access to financing. Carbon prices need to be very high (above US$60/tCO$_2$) and sustained over a long period to achieve noticeable demand effects in the buildings sector. However, in order for energy price signals to be effective and sensible, energy price subsidies need to be removed so that the technology and fuel pricing environment provides a level playing field for sustainable energy options to be feasible. Awareness campaigns, education, and the provision of more detailed and direct information, including smart metering, enhance the effectiveness of other policies and enable behavioral changes.

A combination of sticks (regulations), carrots (incentives), and tambourines (measures to attract attention such as information or public leadership programs) has the greatest potential to increase energy efficiency in buildings by addressing a broader set of barriers. Achieving a transformation in the buildings sector that is in concert with ambitious climate stabilization targets by the mid-century entails massive capacity building efforts to retrain all trades involved in the design and construction process, as well as consumers, building owners, operators, and dwellers.

A transition into a very low building energy future requires a shift in focus of energy sector investment from the supply-side to end-use capital stocks, as well as the cultivation of new innovative business models, such as performance contracting and Energy Service Companies.

Novelties in this chapter, as compared to previous assessments, include (1) a focus on energy services, as well as life cycle approaches accounting for trade-offs in embodied vs. operational energy and emissions; (2) applying a holistic framework toward building energy use that recognizes buildings as complex, integrated systems; (3) presenting new global and regional building energy use scenarios until 2050, using a novel performance-based global building thermal energy model; (4) recognizing the importance of the lock-in effect and quantifying it; (5) in-depth attention to non-technological opportunities and challenges; (6) a large database on quantified and monetized co-benefits; and (7) a critical assessment of zero-energy buildings and related policies.

10.1   Setting the Scene: Energy Use in Buildings

10.1.1   Key Messages

Almost 60% of the world's electricity is consumed in residential and commercial buildings. At the national level, energy use in buildings typically accounts for 20–40% of individual country total final energy use, with the world average being around 30%. Per capita final energy use in buildings in a cold or temperate climate in an affluent country, such as the United States and Canada, can be 5–10 times higher than in warm, low-income regions, such as Africa or Latin America.

10.1.2   The Role of Buildings in Global and National Energy Use

Energy services in buildings – the provision of thermal comfort, refrigeration, illumination, communication and entertainment, sanitation and hygiene, and nutrition, as well as other amenities – are responsible for a significant share of energy use worldwide. The exact figure depends on where system boundaries are drawn. The global direct total final energy use in buildings was 108 EJ in 2007 and resulted in emitting 8.6 $GtCO_2e$ (IPCC, 2007), 33% of global energy-related CO_2 emissions (IEA, 2008a). Globally, biomass is the most important energy carrier for energy use in buildings, followed by electricity, natural gas, and petroleum products. Almost 60% of the world's electricity is consumed in residential and commercial buildings (IEA 2008a). In addition to the energy consumed directly in buildings, primary energy is lost in the conversion to electricity and heat and petroleum products, and the transport and transmission of energy carriers cost energy. In addition, the construction, maintenance and demolition of buildings requires energy, as do the manufacturing of furniture, appliances, and the provision of infrastructure services such as water and sanitation. The use of this indirect or embodied energy is influenced by the level and design of energy service provision in buildings. While comprehensive global statistics on indirect energy cost of buildings do not exist, regional data are presented below.

At the national level, direct energy use in buildings typically accounts for 20–40% of individual country's total final energy use (see Table 10.1), with the world average being 31%. In terms of absolute amounts, there is a significant variance among different world regions. Per capita final energy use in buildings in a cold or temperate climate in affluent countries, such as the United States and Canada, can be 5–10 times higher than in warm, low-income regions, such as Africa or Latin America (Table 10.1). Figure 10.A.1 in the online appendix and Figure 10.3 provide further information on the characteristics of building energy use by region or representative countries. Figure 10.4 shows total final energy use in buildings per capita in different world regions, according to the International Energy Agency (IEA) statistics. Figure 10.5 shows final energy use per square meter for thermal comfort by world region and building type, according to input data

collected from different sources for the model presented in Section 10.6. Because sources of building energy vary greatly, e.g., significant amounts of coal and biomass burned on site in China and India and a much higher share of electricity in other countries, this results in large differences in primary energy use because of the additional energy demands of power generation and distribution.

However, policies to address sustainability challenges of energy services rendered in buildings can often only be designed optimally if a life cycle approach is used for energy accounting and not only the direct energy use is optimized. For instance, there are trade-offs between minimizing operational energy use and embodied energy in building materials; these trade-offs in greenhouse gas emissions can be even larger. For example, reducing CO_2 emissions through increased Styrofoam insulation increases hydrochlorofluorocarbon (HCFC) emissions, potentially resulting in increased rather than decreased overall greenhouse gas emissions when measured in CO_2 equivalents. Further trade-offs exist in cooking energy use and embodied energy in foodstuffs. Reduction in certain energy service demands in buildings results in the reduction of energy use of other sectors, such as electricity transformation losses, transportation (such as for building materials, water, food, etc.), or industrial energy use (needed for products and appliances in buildings). Therefore, building-related energy services can only be optimized if a systemic, life cycle approach is used to reduce associated total primary energy use and associated environmental impacts. Unfortunately, global building energy use and emission data using a life cycle approach do not exist, but smaller-scale data on life cycle building energy use is presented below. As buildings are the end-point of a large share of our energy using activities – for example, a large share of products manufactured in industry are ultimately for the purpose of providing various services in buildings and many goods being transported are being used in buildings – reducing service needs requiring energy input in buildings is key to achieve a reduction in overall primary energy use. When a life cycle approach is applied to understand the energy services demanded, the importance of buildings grows substantially.

10.1.3   The Demand For Different Energy Services In Buildings And Their Drivers

10.1.3.1   Key Messages

Energy is used in buildings to provide a variety of services, including comfort and hygiene, food preparation and preservation, entertainment, and communications. The type and level of service and the quantity and type of energy required depend on the level of development, culture, technologies, and individual behavior. Global trends are toward electrification and urbanization, including toward multi-family from single-family dwellings. At all levels, large variations in cultural attitudes, individual behaviors, and the selection of construction materials and practices, fuels, and technologies contribute to a wide range of energy services and energy use.

Table 10.1 | Contribution of the buildings sector to the total final energy demand globally and in selected regions in 2007.

World regions	Share of the residential sector in %	Share of the commercial sector in %	Share of the total buildings sector in %	Residential and commercial energy demand per capita, MWh/capita-yr.
USA and Canada	17%	13%	31%	18.6
Middle East	21%	6%	27%	5.75
Latin America	17%	5%	22%	2.32
Former Soviet Union	26%	7%	33%	8.92
European Union-27	23%	11%	34%	9.64
China	25%	4%	29%	3.20
Asia excluding China	36%	4%	40%	2.07
Africa	54%	3%	57%	3.19
World	**23%**	**8%**	**31%**	**4.57**

Source: IEA online statistics, 2007.

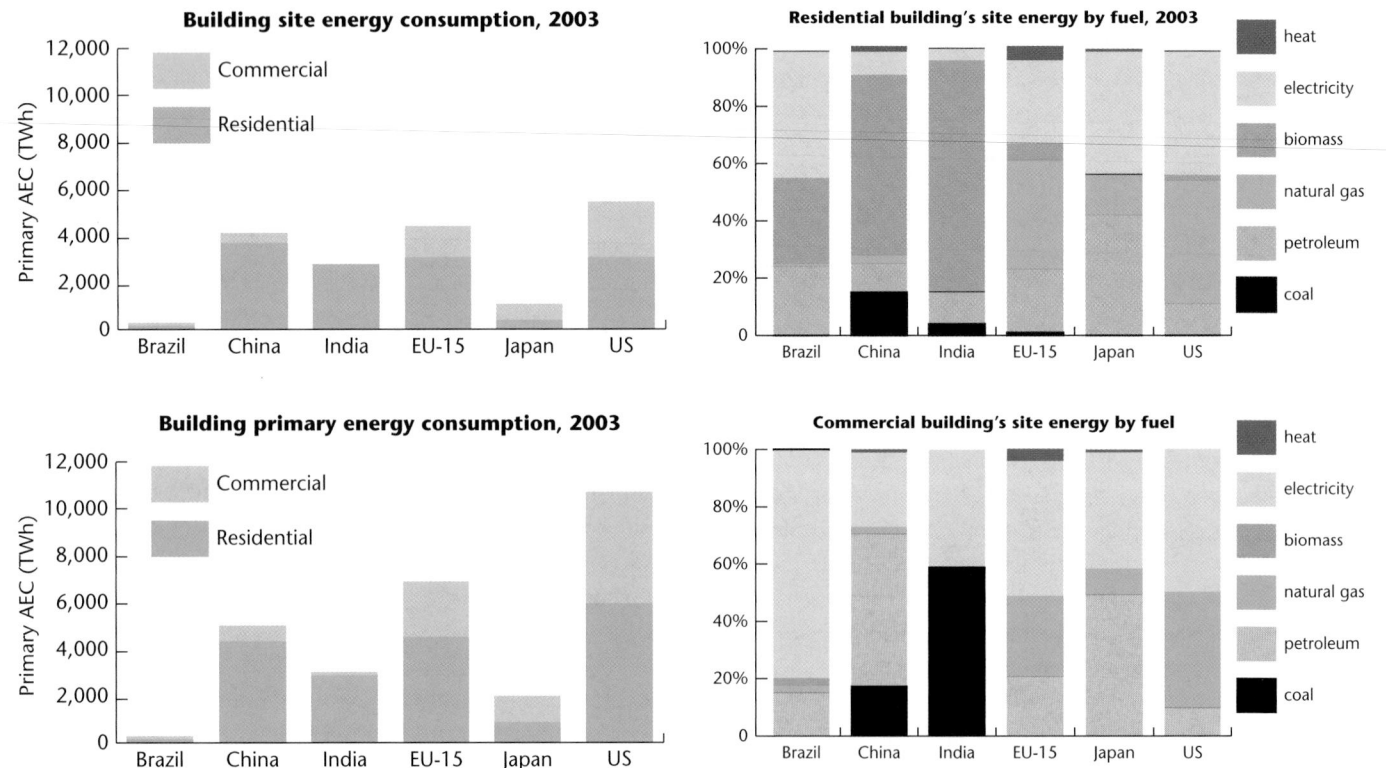

Figure 10.3 | Building final and primary energy use in selected countries in 2003; AEC = annual energy consumption. Source: WBCSD, 2008.

10.1.3.2   Building Energy Demand by Service Type

The type and level of service and quantity and type of energy required depend on a large number of factors, including culture, technologies, and individual behavior. This section includes a review of national and regional assessments conducted to understand the importance of different energy services in buildings. No global systematic studies have been performed to understand the importance of different energy services in buildings or other sectors, and therefore this section covers a selection of

national and regional assessments. Figure 10.6 shows the breakdown of primary energy use in commercial and residential buildings by end-use services in the United States. The figure demonstrates that five energy services accounted for 86% of primary energy use in buildings in 2006. These were: (1) thermal comfort – space conditioning that includes space heating, cooling and ventilation – 36%; (2) illumination – 18%; (3) sanitation and hygiene, including water heating, washing and drying clothes, and dishwashing – 13%; (4) communication and entertainment – electronics including televisions, computers, and office equipment – 10%;

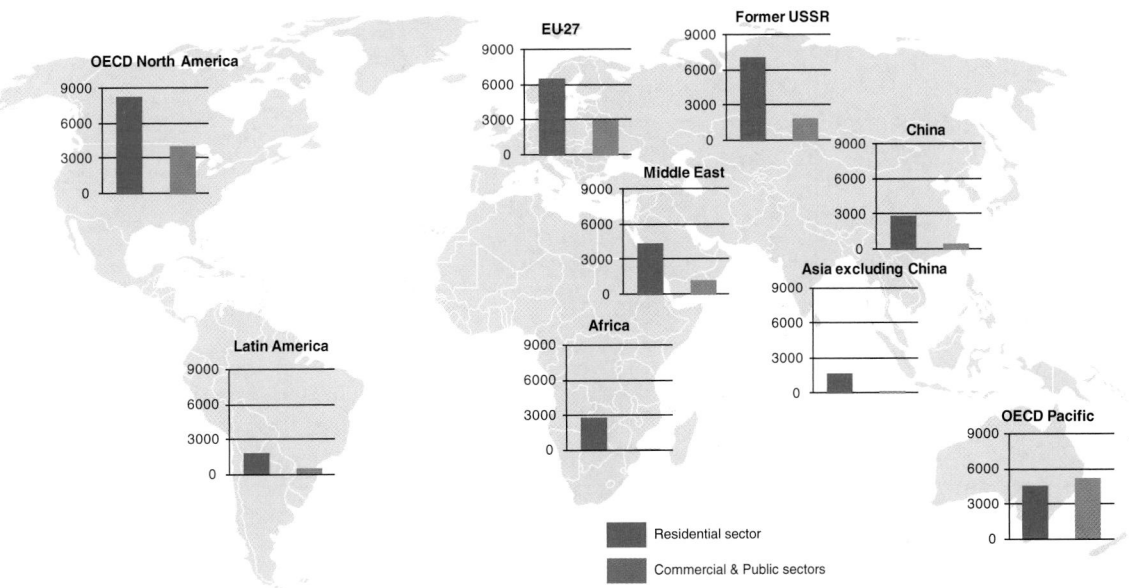

Figure 10.4 | Total annual final energy use in the residential and commercial/public sectors, building energy use per capita by region and building type in 2007 (kWh/capita/yr). Source: data from IEA Online Statistics, 2007.

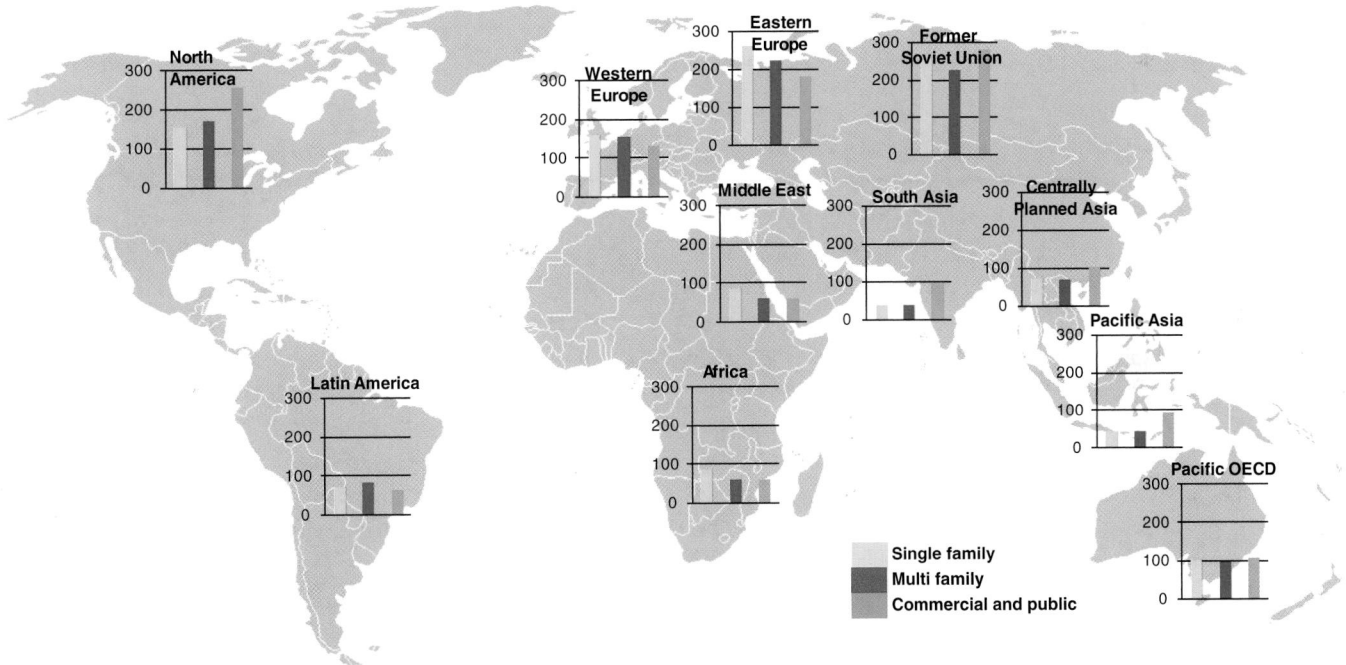

Figure 10.5 | Final heating and cooling specific energy consumption by region and building type in 2005 (kWh/m²/yr). Source: Model estimations (see Section 10.7).

and (5) provision of food, refrigeration and cooking – 9% (US DOE, 2008). The remaining 14% includes residential small electric devices, heating elements, motors, natural gas outdoor lighting, and commercial service station equipment, telecommunications equipment, medical equipment, pumps, and combined heat and power in commercial buildings.

Recently, McNeil et al. (2008) made an estimate of the current and projected end-use energy demand in buildings for ten separate regions covering the world. In the OECD member states, it was found that the five energy services listed above use 76% of the electricity and 69% of the

fuel final energy[5] in buildings. In developing countries (non-OECD member states), they account for 93% of site electricity and 78% of fuel use, respectively. According to the best available figures (IEA, 2006), household energy use in developing countries contribute almost 10% of the world primary energy demand. Household use of biomass in developing countries alone accounts for almost 7% of world primary energy demand.

---

5 Includes natural gas, bottled gas (LPG), and fuel oil. Does not include coal or biomass, and excludes district heating, which is significant in China, Europe, and the Former Soviet Union.

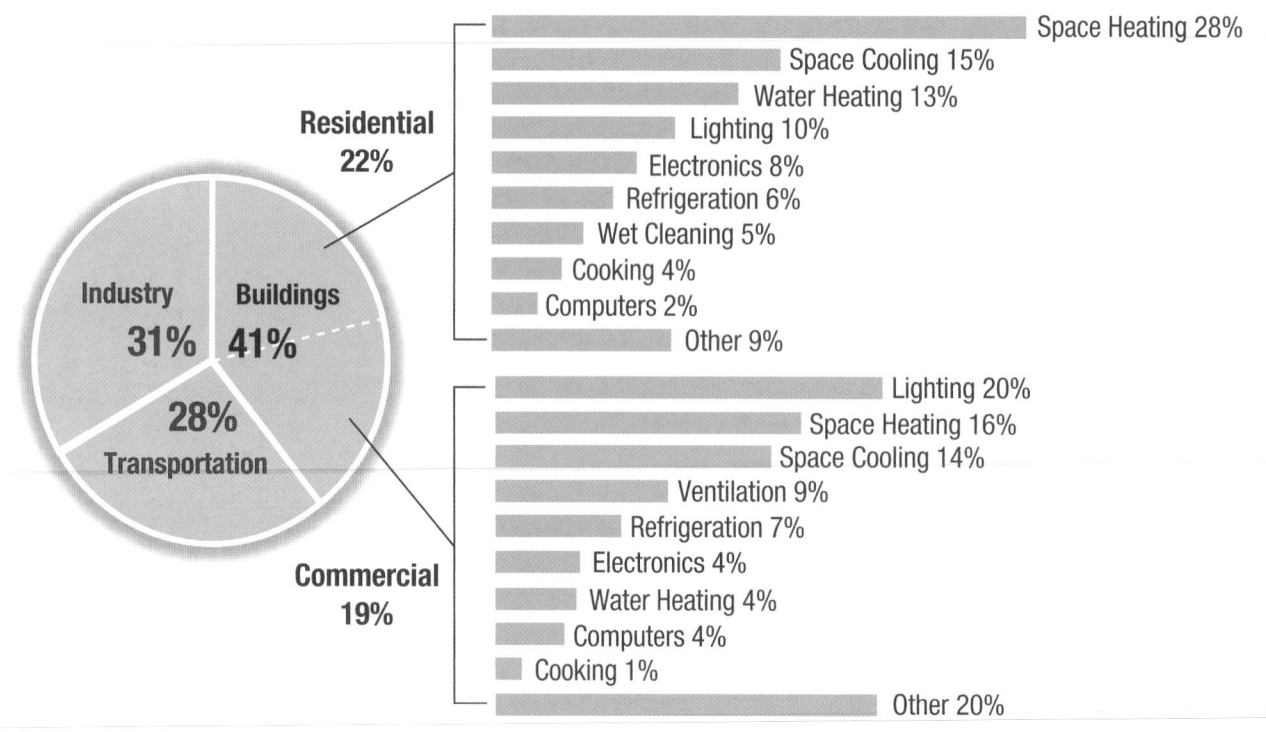

Figure 10.6 | Primary energy use in US commercial and residential buildings in 2010. Source: US EIA, 2011; 2012.

Valuable time and effort are devoted to fuel collection instead of education or income generation. While a precise breakdown is difficult, the main use of energy in households in developing countries is for cooking, followed by heating and lighting. Because of geography and climate, household space- and water-heating needs are small in these countries.

A review of national level studies of household energy services, analyzed on a life cycle basis, is presented in Figure 10.7. Buildings-related energy use contributes 60–70% of the total household energy use in OECD countries (Hertwich, 2005b) and up to 90% in India (Pachauri and Spreng, 2002; see also Box 10.1 and Figure 10.8). The remainder of household energy use is mostly related to mobility. On average across studies for a selected number of countries where data was available, buildings-related energy use, including the primary energy required to produce the energy carriers used in the household, accounts for 32% of the total household energy requirements. "Other shelter," which includes water and waste treatment utilities and construction and maintenance of buildings and furniture, accounts for 11%. Mobility accounts for 24%, food for 14%, recreation for 7%, clothing for 4%, and other for 9%. Variation in the importance of different categories, however, is substantial. Additional life cycle effects of building energy use are considered in Section 10.1.4.

10.1.3.3   Variations in Energy Service Needs and Key Drivers

The following factors are major contributors to changing energy service demands: (1) population growth; (2) urbanization; (3) shift from biomass to commercially available energy carriers, especially electrification

(percent of population having access to electricity); and (4) income, which is a strong determinant of the set of services and end-uses for which commercial energy is used and the quantity and size of energy-using equipment; (5) level of development; (6) cultural features; (7) level of technological development; and (8) individual behavior. Availability and financial aspects of technologies and energy carriers are also important.

The demand for energy services in buildings varies among regions according to geography, culture, lifestyle, climate, and the level of economic development. It also varies by the type of use, type of ownership, age, and location of buildings (e.g., residential or commercial, new or existing buildings, private or public, rural or urban, leased or owner-occupied) (Chakravarty et al., 2009). There are also significant differences in energy services among commercial subsectors – such as offices, retail, restaurants, hotels, and schools – and between single- and multifamily residential buildings. Different approaches, standards and technologies to how the buildings are sited, designed, constructed, operated, and utilized strongly affect the amount of energy used within buildings.

The level of economic development is a main driver of the global differences in energy use in buildings as set out in the previous sections. Table 10.1 shows that energy use per capita in buildings is up to an order of magnitude higher in North America than in most of Asia, Africa, and Latin America. This section sets out some more detailed differences, and their drivers, among countries as well as within individual countries.

Figure 10.9 shows there are differences in per capita energy use among six developed countries, at similar affluence levels (IEA, 2007b). The data

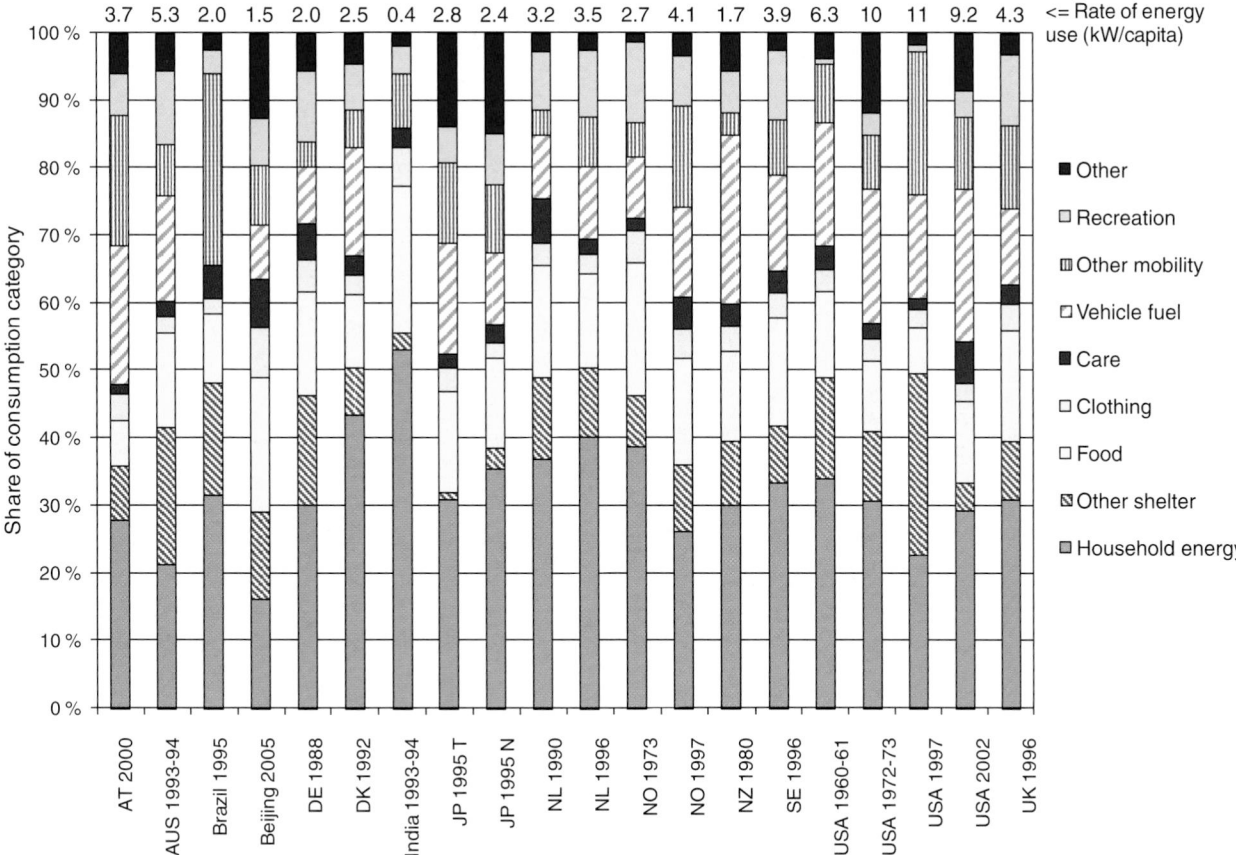

Figure 10.7 | Share of consumption categories in total energy use based on life-cycle analysis or input-output calculation and rate of annual energy use in kilowatt per capita (numbers on top of the columns). Source: Hertwich, 2011.

are "degree day normalized," i.e., corrected for the key climatic driver of heating demand. The overall energy use is higher in the United States, even by developed country standards, but lower in Japan, and intermediate in Europe and Canada. The variation between end-uses is large. For instance, space-heating demand follows the same broad pattern as the total use and is driven largely by per capita floor area and building internal temperatures (IEA, 2007b). Other end-uses are relatively less important, but there are substantial differences, notably the higher energy use in lighting and appliances in North America. The figure also illustrates the role of culture in determining energy efficiency. Japan uses less than a third of the energy used in the United States for space heating, even when normalized for climate. This is due to more compact living, as well as focusing on providing thermal comfort through alternative means rather than universal heating of all living space. The Japanese *kotatsu*, a table frame heat source which is combined with blankets and adaptive clothing, is still a common alternative despite high affluence levels.

10.1.3.4 Cultural, Social and Behavioral Drivers

Culture, values, and individual habits significantly influence consumption, as does the choice of technologies. The section above provided an example for heating energy use determined by culture in Japan.

This section provides more examples and highlights further issues. For instance, cooking energy demand varies largely with dietary choices. The use of refrigerated, packed, and tinned food is very limited in developing countries which leads to larger energy use for cooking. In rural China, similar to many other developing countries, cooking is the largest energy demand item for 60% of families.

A major source of variation in energy services and energy use is the impact of habits and behavior (see Section 10.4.8), irrespective of level of development. Within all countries, high-income groups contribute disproportionately to energy demand (Chakravarty et al., 2009). However, income is not the only factor. Lenzen et al. (2006) investigated the variation of energy use with household income and size, type of house, urban versus rural location, education, employment status, and age of the householder in several countries. Higher income social groups use more energy per capita, with the elasticity of energy expenditure ranging from 0.64 in Japan to 1.0 in Brazil. Lenzen et al. (2006) also found that energy use differs across countries even after controlling for the main socioeconomic and demographic variables, including income. This result confirms previous findings that the characteristics of personal energy use are partly determined by distinctive features such as historical events – for example, energy supply shortages or the introduction of taxes, socio-cultural norms, behavior, and present market conditions, as well as energy and environmental policy measures.

Box 10.1 | Energy Use in Buildings in India

In India, energy end use in buildings varies largely across income groups, building construction typology, climate, and several other factors. The energy sources utilized by the residential sector in India mainly include electricity, kerosene, liquefied petroleum gas (LPG or propane), firewood, crop residue, dung, and other renewable sources such as solar energy. The per capita energy use in the residential sector is as low as 2560 kWh/yr for India.

Traditional fuels are predominantly used for cooking. In the rural areas of the country, the three primary sources of energy for cooking are firewood and chips, dung cake, and LPG. In urban areas, cooking energy sources are primarily LPG and piped natural gas in select cities. Figure 10.9 shows seasonal differences in energy demand for Delhi, India (Manisha et al., 2007). In summer months, air-conditioners and refrigerators each account for about 28% of total monthly electricity consumption, while lighting accounts for about 8–14% of annual electricity consumption. In winter, major uses of electricity are refrigerators (44%), water heating ("Geyser" 18%), and lighting (14%).

Based on CPWD (2004), in India's commercial sector 60% of the total electricity is consumed for lighting, 32% for space conditioning, and 8% for refrigeration. End use consumption varies largely with space conditioning needs. In a fully air conditioned office building, about 60% of the total electricity consumption is accounted for by air conditioning followed by 20% for lighting.

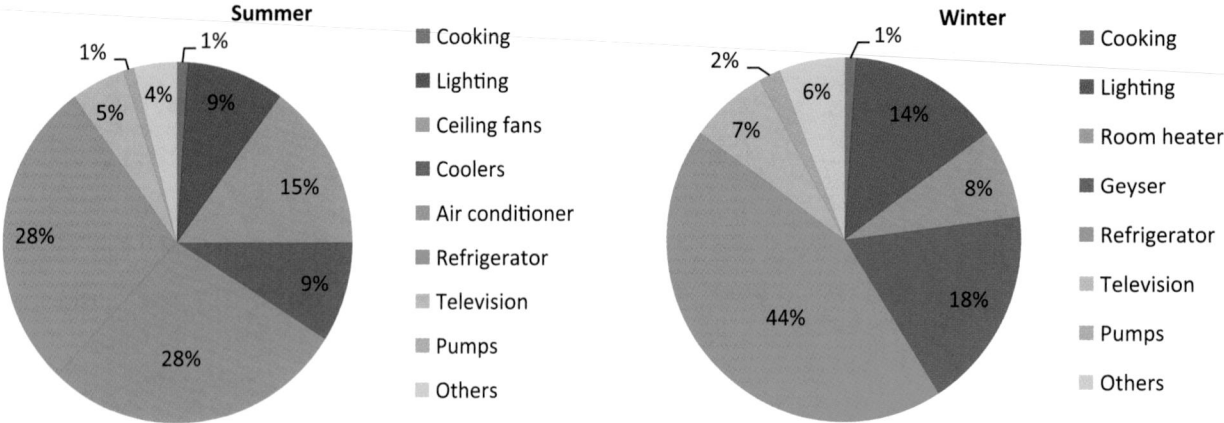

Figure 10.8 | Share of end-uses (by appliance) in electricity consumption in Delhi, India. Source: Manisha et al., 2007.

Note: "Others" include washing machine, computer and irons.

A study carried out by Socolow (1978) demonstrated energy use variations of more than a factor of two between houses that were identical but had different occupants. Gram-Hanssen (2004) found a similar variation of household electricity use in much larger sample of identical flats in Denmark. This is consistent with the findings of the World Business Council for Sustainable Development/Energy Efficiency in Buildings (WBCSD/EEB) that show that people can improve energy efficiency in buildings by around 30% by behavioral changes without any extra costs, or they can compromise a building's performance by up to 60% by behavioral effects (WBCSD, 2009).

A similar result is illustrated by a study of energy use for home air conditioning in a residential building in Beijing (Zhang et al., 2010). The building consists of 25 home units, all with similar income characteristics. Although each flat is fully occupied, the difference in energy use can be as large as a factor of 40 (see Figure 10.10). While the average is 2.3kWh/m², the range is from near zero to over 14 kWh/m². The real income of the high-energy users is lower than those of the low-energy users.

Social choices about cooling technology will prove increasingly important. For instance, a very large stock of residential, institutional, and commercial buildings in India is still designed to be non-air-conditioned and to use only ceiling fans to provide thermal comfort during hotter periods, while in other countries electric chilling is the standard in commercial buildings. Occupants of Indian buildings are acclimatized to higher temperature and humidity conditions without feeling uncomfortable. This translates to very low energy usage in such buildings, with a very limited scope for enhancing energy efficiency, but with a high potential for reducing future air conditioning needs.

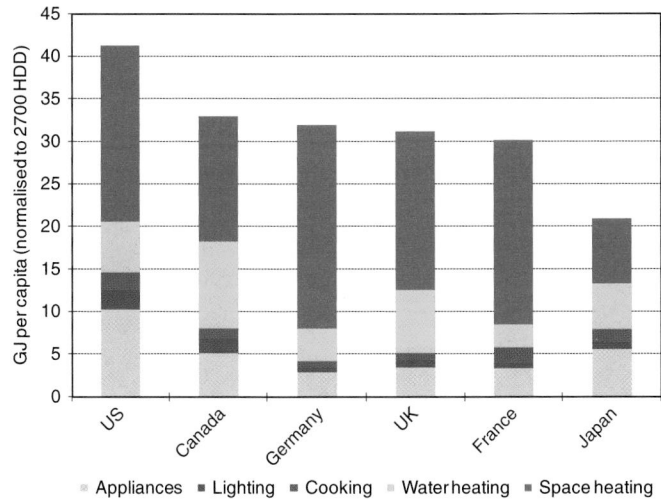

Figure 10.9 | Residential energy use in different developed countries. Source: adapted from IEA, 2007b. HDD = heating degree day.

10.1.3.5 The Drivers of Changing Demand for Building Energy Services

The share of energy use in buildings in the total energy use increases with the level of economic development. In India, with a near-consistent 8% annual rise in annual energy demand in the residential and commercial sectors, building energy use has seen an increase from 14% in the 1970s to nearly 33% of total primary energy use in 2004–2005 (authors' calculation based on the data from IndiaStat, 2010).

In addition to the determinants of building energy services discussed above, additional factors are major contributors to changing energy service demands: (1) population growth; (2) urbanization; (3) shift from biomass to commercially available energy carriers, especially electrification (percent of population having access to electricity); and (4) income, which is a strong determinant of the set of services and end-uses for which commercial energy is used and the quantity and size of energy-using equipment; (5) level of development; (6) cultural features; (7) level of technological development; and (8) individual behavior. Availability and financial aspects of technologies and energy carriers are also important.

While energy use in buildings is influenced by income, specific energy use does not necessarily continue growing at an equal rate at higher income levels. For instance, Figure 10.11 shows the trend of specific building energy use in the United States during the second half of the twentieth century, for Japan since 1970, and the trend for China since the mid-1990s. The most significant increase can be observed during the first two decades in this period. While gross domestic product (GDP) continued to increase in the second part of the period, improvements in technological efficiency have kept energy growth trends at bay. Chinese-specific building energy demand figures currently are in the same range as the United States in the 1950s. Whether China will follow trends of

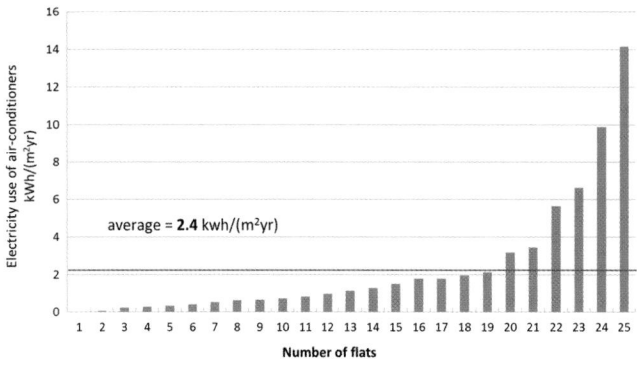

Figure 10.10 | Electricity consumption of air-conditioning in 25 flats of a residential building in Beijing. Source: based on Zhang et al., 2010.

the United States or will be able to decouple the increase in wealth from specific building energy use at an earlier stage is an important determinant of global future energy use.

Building location, form, and orientation are integrally related to urban/rural design, which, in turn, also influences energy use necessary for transporting people and products to buildings, as well as the feasibility of certain sustainable energy supply options such as district heating and cooling, and community-scale renewable energy generation. Therefore, urban design, building energy use, and urban transport energy use are integrally related (see Chapter 9 and Chapter 18).

In India and China, urban households tend to have higher energy requirements than rural households (Lenzen et al., 2006; Peters et al., 2007). In China, moving from a rural to an urban life currently increases household demand for energy by about a factor of three (Table 10.2), while in developed countries, urban households tend to have lower energy requirements. By 2020, both rural and urban demand for energy will increase due to a combination of urbanization, a shift from biomass to commercial energy carriers, and increased income. Thus, Chinese urban energy use per household in 2020 is expected to be five times the amount of rural energy per household today.

Building size and building floor space per person are also important factors that depend upon income and demographics. Households with more occupants tend to have lower per capita energy use. Older and wealthier individuals are more likely to occupy larger dwellings with fewer occupants. Often, improved energy efficiency is offset over time by bigger floor space per person or per household.

In sum, population growth, urbanization, the shift from biomass to commercial fuel carriers including electricity, and income growth are contributing to increasing demand for energy services. Technologies, practices, and policies toward increasing energy efficiency are offsetting growth in some locations and offer large future potential for reducing the quantity of energy required for energy services. Individual choices of lifestyle and specific behavior may greatly increase or decrease the demand for energy services.

Table 10.2 | Increasing energy intensities when moving from rural to urban life in China.

	China annual energy per household in the North China (kWh)						
	Year 2000		Year 2020		Ratio as percent		
	Rural	Urban	Rural	Urban	Urban/Rural 2000	Urban/Rural 2020	2020 Urban/ 2000 Rural
Space heating	631	4990	4638	9027	791%	195%	1431%
Water heating	1108	1001	1499	1579	90%	105%	143%
Lighting	155	189	220	488	122%	222%	315%
Cooking	277	250	375	395	90%	105%	143%
Other	50	100	150	420	200%	280%	840%
TOTAL	2221	6530	6882	11909	294%	173%	536%

Source: Zhou et al., 2007.

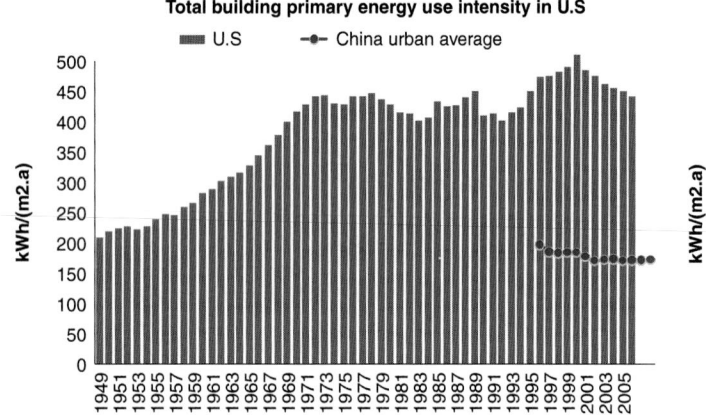
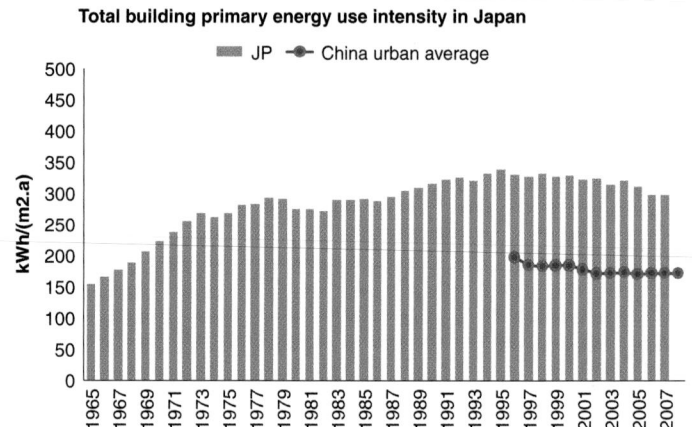

Figure 10.11 | Trend of total building final energy use per m² in the United States (1949–2006) and in Japan (1965–2007) (kWh/m²/yr) as compared to China. Sources: US Census Bureau, 2000; US Census Bureau, 2008; Energy Data and Modeling Center, 2008; US DOE, 2008; Building Energy Research Center, 2010.

10.1.3.6 Energy Carriers Used to Satisfy Energy Service Needs in Buildings

In developing countries, biomass, coal, oil products, and natural gas are mostly used to satisfy energy service needs in buildings because in rural areas people have easier access to biomass compared to people living in cities, and because even many urban building occupants do not have access to electricity, (Shepherd and Zacharakis, 2001; Melichar et al., 2004; see also Chapter 2). The progression from traditional biomass fuels to more convenient and cleaner fuels has traditionally been explained by the "energy ladder" model, suggesting that increasing affluence is the key to the transition.

More recently, the multiple fuel model was developed to explain household decision making in developing countries under conditions of resource scarcity or uncertainty taking into account the following factors: (1) economics of fuel and stove type and access to fuels; (2) technical characteristics of cookstoves and cooking practices; (3) cultural preferences; and (4) health impacts (Masera et al., 2000). In addition to urbanization, Pachauri and Jiang have identified income, energy prices, energy access, and local fuel availability as key drivers of the transition from traditional to modern fuels in China and India (Pachauri and Jiang, 2008).

In developed countries, electricity and natural gas are the most frequently used energy carriers, with electricity taking an increasing share. For instance, while buildings in the United States use only 39% of the country's primary energy, they use 71% of electricity and 54% of the natural gas supply.

Figure 10.12 presents the distribution of fuels in total final energy use in the residential and commercial sectors worldwide. Note that residential energy use (81.3 EJ) exceeds commercial and public sector energy use (27.5 EJ) by a factor of three.

10.1.4 Indirect Energy Use from Activities in Buildings in Detail Using the Life Cycle Assessment Approach

10.1.4.1 Key Messages

The life cycle approach is necessary to optimize the total energy required to provide energy services in buildings because the importance of indirect energy use can increase as more energy efficient technologies are applied. Depending on climate and energy efficiency, the construction of a building contributes as much as 25% to total indirect energy use, with

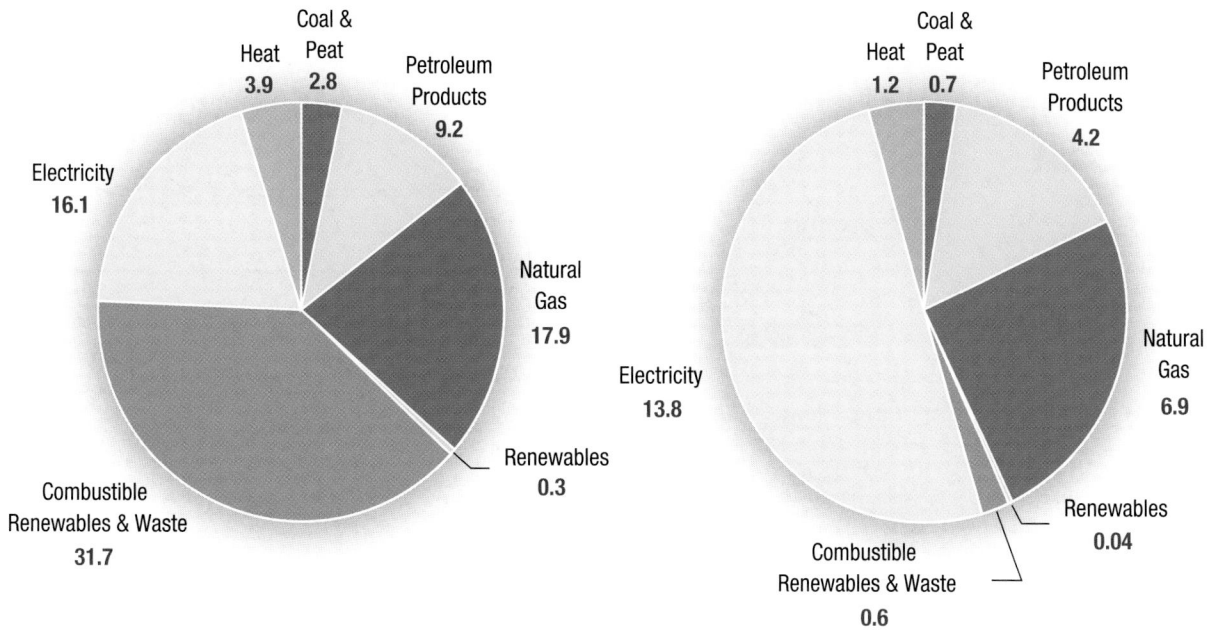

Figure 10.12 | World total final energy consumption by fuel in the residential (left) and commercial and public (right) sectors in 2007 (EJ). Source: IEA Online Statistics, 2007.

higher values for very high efficiency or energy self-sufficient buildings. Even though commonly pursued efficiency strategies increase the energy embodied in the building, in cold and temperate climates this invested energy is typically recovered with an energy payback time below one year. Building-integrated PV systems, however, require at least five years to recover the energy invested in their construction and may, from a life cycle perspective, not be the cleanest option of supplying electricity. The environmental impacts of different building materials and designs depend on a number of factors with wood offering an advantage in terms of carbon storage and potential energy recovery after demolition. A refurbishment of existing buildings to increase efficiency can offer savings in total life cycle energy use compared to demolition and new construction. While optimal solutions will be site- and case-dependent, in general significant reductions in environmental impacts can be obtained with energy efficient building designs, a wise choice of building materials, and renewable energy sources integrated in buildings.

10.1.4.2   Introduction to the Life Cycle Approach

A life cycle approach is necessary to optimize the total energy use required to provide energy services in buildings, because, for instance, the importance of indirect energy use can increase as more energy-efficient technologies are applied (see also earlier sections). In addition to direct energy use, a life cycle approach takes into account the energy used to produce the materials for constructing the buildings, energy losses associated with the provision of electricity and fuels to the buildings, energy used in the construction and maintenance of a building, and energy used in manufacturing and supplying building equipment – ranging from lighting and TV sets to heating and cooling equipment

(Treloar et al., 2000). This indirect energy use has been variously referred to as embodied, grey, or upstream energy.

This section provides an overview of embodied energy, including the trade-offs between embodied and operational energy, and examines it in detail for important cases: construction, heating, and energy embodied in water consumption in buildings. Indirect energy use is strongly affected by choices made during building construction, operation and/or use, as well as dietary choices.[6]

10.1.4.3   Embodied Energy

This section provides some general insights from life cycle assessment (LCA), and presents results of life cycle studies for building materials and buildings. Section 10.1.3 presented the direct use of energy for different energy services in the US residential sector. Figure 10.13 provides an overview of the direct and indirect energy use of the average household in the United States in 2002, from the life cycle perspective. Indirect energy use is split into energy losses and indirect energy connected to the purchase of all other goods and services. The largest category is private transport. The second largest category is "utilities," which includes direct energy use and the provision of water and wastewater treatment. The third largest category is the indirect energy embodied in food purchased by households.

---

6   Embodied GHG emissions cover a broader set of issues than just embodied energy. They also include process-based CO_2 emissions from clinker production, carbon storage in wood, and the non-CO_2 GHGs (mainly fluorinated gases) used in the production of certain construction materials and in the operation of some equipment, such as chillers. IPCC (2007) discussed non-CO_2 emissions related to buildings in detail, and thus this section focuses on indirect energy use.

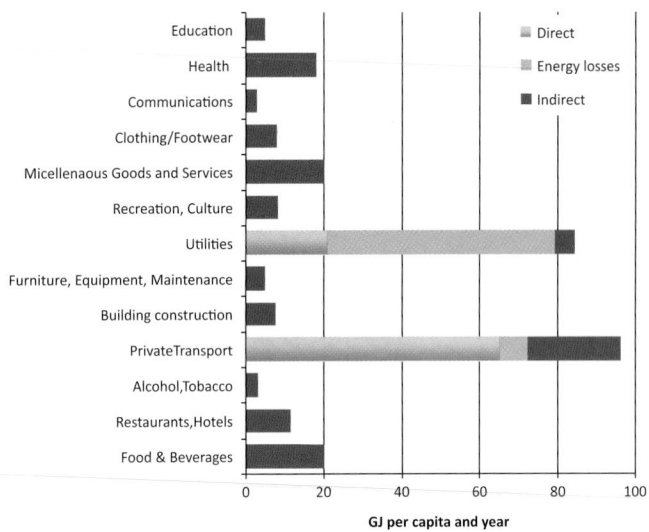

Figure 10.13 | Direct and indirect primary energy use of the average US citizen in 2002 for different consumption purposes. Source: based on Weber, 2009.

There is a trade-off between direct and indirect energy use in a number of areas, for example in thermal comfort and handling of foods. In the United States about 50% of direct energy use in buildings goes to ensuring thermal comfort (see section 10.1.3). While the building structure itself has other functions as well, its main energetic function is to provide thermal comfort. Clothing also functions partially to provide thermal comfort. Overall, however, the energy use in buildings and the cost of providing that energy dominate the total energy cost of providing thermal comfort in the United States. This issue is revisited in the remainder of Section 10.1.4.

Refrigeration and cooking requires about 10% of the direct household energy use, which is clearly less than the indirect energy required to grow and process the food. The importance of indirect energy used for manufactured goods used in buildings increases with increasing wealth, and hence overtime (Lenzen et al., 2006).

10.1.4.4 The Life Cycle Impact of Building Materials and Design

There is a distinction between the construction, operations and maintenance, and demolition of buildings. For most buildings, the bulk of energy use is in the operations phase, and energy conservation efforts have appropriately focused on reducing this energy use through smarter design, better insulation material, and improved building technology. However, for short-lived or highly efficient buildings, construction is responsible for a substantial share of the total energy use. Demolition offers an opportunity to recover some of the energy, either by combusting elements with high heating value or by reusing building materials and components, thus avoiding energy-intensive primary production of new materials. In construction, and especially demolition, energy for transport is an important consideration, constraining remanufacturing and recycling of building components and materials.

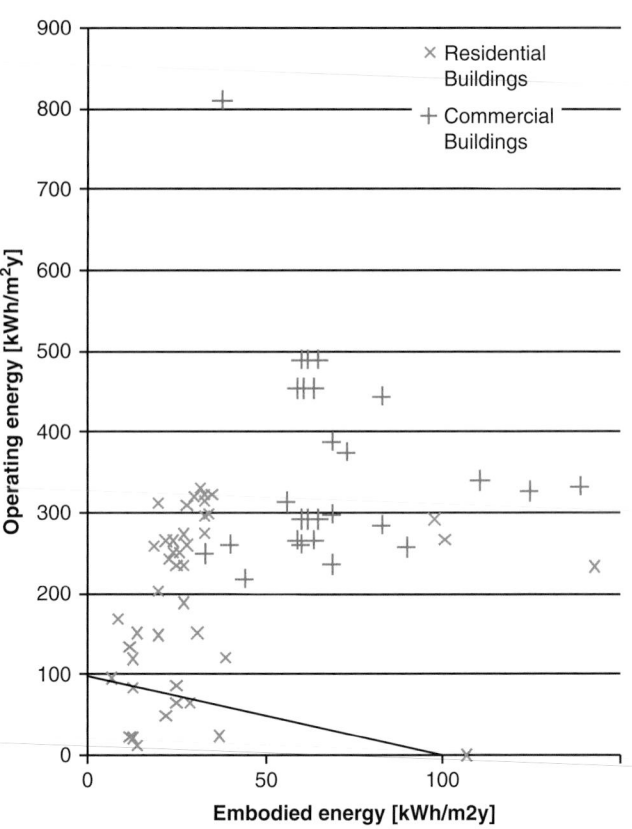

Figure 10.14 | Primary embodied and operating energy of buildings from a review of case studies mostly from northern/central Europe and Asia, including most efficient designs. Source: Sartori and Hestnes, 2007; Ramesh et al., 2010.

Note: the lines indicate points of equal life-cycle primary energy use.

Ramesh et al. (2010) reviewed life cycle energy studies of 73 buildings, including 60 buildings in OECD countries from Sartori and Hestnes (2007). The operating and embodied energy are presented in Figure 10.14. The embodied energy dominates only in three cases, two in climates not requiring heating or cooling and the other a self-sufficient solar home in a Nordic country. In most cases, the embodied energy contributes to 5–25% of the total energy over the lifecycle. The embodied energy, however, varies between 9 and 140 kWh/m^{2}y, given building lifetimes of 30–100 years. Similar results have been obtained for the United Kingdom (Monahan and Powell, 2011).

An analysis of different alternatives for the main building material in Sweden demonstrated that wood is preferable over concrete, especially if the wood is used as fuel after demolition or reused in buildings (Borjesson and Gustavsson, 2000; Lenzen and Treloar, 2002; Gustavsson and Sathre, 2006). Using a detailed input-output analysis of the entire Swedish construction sector, Nässen et al. (2007) show that building materials account for a little more than half of the energy use in the production of new detached and multifamily buildings. The energy use for excavation of the site and transport is important, as is the sum of construction and service inputs. This sector-wide study estimates energy use in the production phase of buildings for Sweden at about 25% of the total energy used for buildings (Nässen et al., 2007).Studies usually only account for energy, not

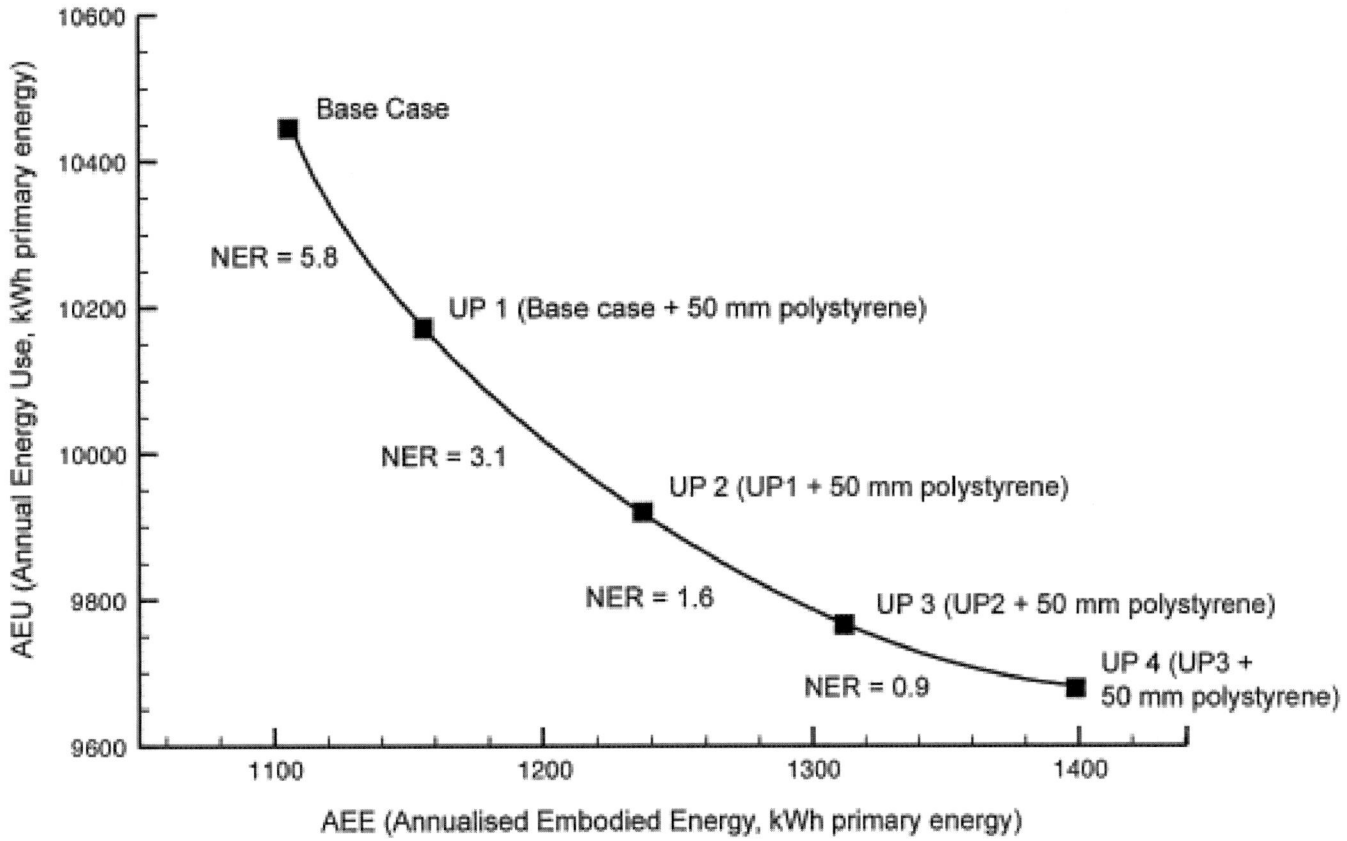

Figure 10.15 | Annualized operational energy use vs. annualized embodied energy use for polystyrene insulation of differing thickness for a case study of low-energy residential building in a maritime climate. Source: Hernandez and Kenny, 2010.

Note: The Net Energy Ratio (NER; also called the return on energy investment) indicates the benefit of each step of additional insulation.

environmental impacts. Depending on the energy mix chosen by various actors along the life cycle, a lower life cycle energy use does not necessarily result in lower environmental impacts (Brunklaus et al., 2010).

10.1.4.5   Highly Efficient Buildings and Active Components

For highly energy efficient buildings and the use of active[7] building technology, the trade-offs between embodied energy and environmental impacts and operational energy and environmental impacts becomes critical and requires a life cycle assessment to ensure that measures are not counterproductive.

Figure 10.15 illustrates an investigation of the trade-off between embodied energy and operating energy. Increasing the thickness of a fairly energy-intensive insulation material (polystyrene) in a house that already has a passive design (15 kWh/m²y) reduces energy use up to a point. The last step, from UP3 to UP4, leads to an increase of life cycle energy use, because the net energy ratio (energy return on investment) is smaller than one (Hernandez and Kenny, 2010).

Reviewed case studies of highly efficient buildings indicate that efficient design depends on higher initial investments of embodied energy. The embodied energy cost of efficiency, however, is small and the energy payback period is on the order of months (Feist, 1996; Winther and Hestnes, 1999). Similar gains can be made by retrofitting existing buildings in a cold climate (Dodoo et al., 2010). An environmental assessment of a passive house in France indicates that the passive design leads to a reduction in environmental impacts in 10 out of 12 impact categories investigated, by 28% on average over a conventional design (Thiers and Peuportier, 2008). Blengini and Di Carlo (2010) report similar results for a passive house in Italy. Environmental gains, however, significantly depend on the energy source and conversion technology for heat and electricity. A study in Sweden indicates that a passive house supplied entirely by electricity can lead to higher environmental impacts than a standard building supplied by district heat (Brunklaus et al., 2010). No comparable studies for hot and dry or hot and humid climates were found.

The introduction of active energy-generating components such as solar collectors and photovoltaic (PV) modules leads to a substantial increase in embodied energy and environmental impacts (Winther and Hestnes, 1999). Published net energy analyses indicate an energy payback time of solar hot water heaters from half a year to two years (Crawford and

---

7   By active building technology this section refers to components that generate energy, mostly power.

Treloar, 2004), while the energy payback time of building-integrated PV systems is five years and up (Lu and Yang, 2010; Leckner and Zmeureanu, 2011). Local circumstances, such as insolation at the site and orientation of the cells, as well as electric grid properties, determine whether building-integrated PV cells are beneficial from an environmental perspective. A solar building in Spain featuring solar collectors, an absorption cooling tower, and PV arrays was found to lead to substantial reductions in emissions of greenhouse gases, acidifying gases, and ozone precursors, while causing substantial increases in water use and the emissions of human toxicants and no change in other life cycle impacts (Batlles et al., 2010).

10.1.4.6   Demolition vs. Retrofitting

The question often arises about how far it is worth pursuing the retrofitting of existing poor quality buildings from a sustainable energy perspective rather than demolishing them and replacing them with state-of-the-art new construction. There is no single answer. Various aspects of building renovation, replacement, and demolition have been investigated. Building lifetime and the choice of demolition technique are important for the life cycle energy use of different building materials. If the building structure can be made to fit new purposes, retrofitting is often the more environmentally friendly option, because it preserves material and reduces transport needs. Itard and Klunder (2007) have investigated the case of two larger residential projects using life cycle assessment. According to their results, maintenance or transformation of the existing stock has lower impacts in both cases than demolition and new construction. These results are confirmed by studies of individual building components. Retrofitting to high energy efficiency standards is fully possible and often environmentally desirable (Dodoo et al., 2010). If a demolition is necessary, the embodied energy in the building material can be preserved through the reuse of building components or recycling of the material, or it may be recovered through incineration. The environmentally preferable option depends on local circumstances, and transportation required for alternative solutions is an important factor (Bohne et al., 2008).

10.1.4.7   Life Cycle Energy and Emissions of Residential Appliances

The electricity used by electric and electronic products used in buildings is ultimately converted to heat and either contributes to heating the building or needs to be removed through a cooling system, depending on the climate, building, and heat load. Such energy use in office buildings can be up to several 100kWh/m^2/yr, while appliance-related electricity consumption in residential buildings in OECD countries is typically around 50kWh/m^2/yr.

Life cycle assessments of large appliances indicate that operations-phase electricity use is the dominant source of environmental impacts (Cullen and Allwood, 2009). For personal computers, however, the production causes significant impacts (Williams, 2004). In what is to our knowledge the first study of life cycle impacts of household appliances and

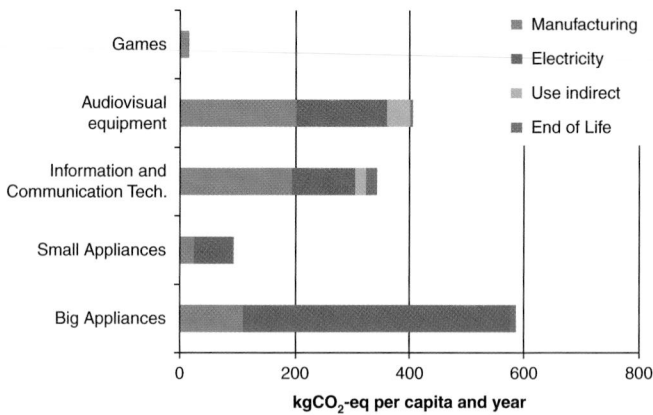

Figure 10.16 | GHG emissions associated with the purchase, use and disposal of electric and electronic equipment in Norwegian households, assuming 0.56 kgCO$_2$-eq/kWh for use-phase electricity (EU average). Source: Roux, 2010.

Note: Big appliances comprise cold appliances, wet appliances and big cooking appliances; small appliances include vacuum cleaners and microwaves; ICT includes computers, phones and peripherals; and audiovisual equipment include TVs, video equipment and audio equipment.

electronic equipment, Roux (2010) shows that the greenhouse gas emissions caused by the production of information and communication technology and audiovisual equipment purchased by Norwegian households is larger than the emissions caused by the electricity this equipment uses, even assuming a relatively polluting electricity mix (Figure 10.16). Taking the manufacturing of the equipment and the use of networks and content of ICT and audiovisual equipment into account, the GHG emissions caused by this equipment are equal to or larger than those caused by washing machines, driers, refrigerators, and freezers taken together. While these research results cannot be generalized, they indicate that the growing, rapid turnover of household electronics is an emerging problem that comes on top of the energy use of traditional household appliances. It may also stress the importance of regulating the durability of such equipment as a key energy- and GHG-saving strategy for residential emissions, potentially with comparable reductions as direct energy-saving policies.

10.1.4.8   Energy Embodied in Buildings-Related Water Use

Another commodity through which energy is embodied in buildings is water. The trade-offs between energy and the environment are discussed in detail in Chapter 20; however, this section narrows its focus only to the significant interactions of water and energy associated with their usage in buildings. These interactions are (1) water use for energy supply that is later used in buildings; (2) energy use to produce water consumed later in buildings; and (3) water used in energy consuming equipment and appliances installed in buildings as a special subcategory of interaction number two.

First, energy supply requires water. For instance, in the United States around 40% of freshwater withdrawals are used as cooling water for thermal power plants that are generating electricity (Huston et al., 2004).

Thus, reducing electricity demand in buildings has the potential to reduce freshwater withdrawals and associated environmental impacts. Second, water used in buildings[8] requires energy to extract, pump, transport, treat, and dispose the potable water and wastewater and to heat water for the final domestic and commercial use. For example, in California, these water-related energy uses for all sectors (buildings, industry, and agriculture) account for about 19% of the state's electricity, 32% of its natural gas, and 88 million gallons (3%) of diesel fuel every year (John et al., 2005). Barry (2007) estimated that 2–3% of the world's energy use is used only on pumping and treating water for civil and industrial supply. Water losses increase the amount of energy required to deliver water to the consumer. Barry (2007) provides case studies from Mexico, Brazil, and South Africa that exemplify that water loss ranges from one-third to one-half of the volume of water produced due to leaks and system inefficiencies.

Energy can be saved through water consumed in buildings by techno-logically reducing water demand at the point of its final use – such as through efficient washing machines, low flow faucets, toilets, and shower heads, increasing efficiency of water heating (see Section 10.4 for details), and eliminating water losses during the process of water extraction, treatment, transportation, distribution, and wastewater cleaning. Another cost-effective opportunity is the promotion of water-saving habits and lifestyles. Use of treated wastewater for other applications, such as for landscaping and flushing, is also predominant and even mandatory in India for several building typologies under environmental clearance norms. This, in turn, saves energy for pumping and transportation of water. The Alliance to Save Energy (2008) estimated that municipalities can cost-effectively save at least 25% of energy and money through water systems alone.

Finally, a special subcategory of energy saving measures is associated with water use in energy-using appliances, such as water heaters, cooking devices, and dish- and clothes-washing machines and in cooling and chilling applications. Energy savings can occur by means of water saving considerations in this equipment. Evaporative cooling is still largely used in India, for example, and provides a low-energy means of providing cool air in many parts of the world. For air conditioning applications, water-cooled chillers are more energy efficient than air-cooled ones. There is a dichotomy of choice between the use of water-efficient vis-à-vis energy-efficient chilling machines. Use of potable fresh water for space conditioning is prohibited in several states of India, especially where fresh water supplies are constrained. Also, existing environmental clearance norms for large-scale building projects in India encourage the use of recycled and treated wastewater or harvested rain water for space conditioning purposes. These measures not only address an environmental challenge – clean water conservation – but also reduce water-related energy use in connection with cooling.

---

8   According to UNEP, the global environmental footprint of the building sector represents 25% of total water use (UNEP SBCI, 2009).

10.1.5   The Impact of a Changing Climate on Building Energy Service Demand

A warming and changing climate has a strong influence on energy use in the building sector worldwide. While cooling demand increases as the climate warms, passive cooling approaches (e.g., overnight ventilation, shadowing) become less effective or do not achieve acceptable indoor temperatures. On the other hand, heating demand decreases in cold zones and allows acceptable winter comfort to be achieved more easily. In temperate climate areas such as much of Europe, Japan, South Africa, or the United States, both impacts on winter heating and summer cooling demand can be observed.

The net impact of warming depends on a complex set of factors. These include the choice of fuel and conversion efficiencies for heating fuels and power generation, building design, efficiency, and operation. Cooling loads will depend strongly on the market penetration of air conditioning, which itself will be dependent on income, building design, culture, and increasing internal loads of buildings by office automation, as well as external temperature. In addition, the cooling demand is exacerbated by the urban heat island effect (see Section 10.2.6) and the escalation of service demand for cooling. In some moderate climate regions, by contrast, heating loads may decrease substantially, or may even become unnecessary due to the combined effect of advanced know-how in building construction, building insulation performance, and an increase in internal heat loads. In contrast, the load on refrigeration equipment increases and its efficiency decreases with rising internal temperatures. The overall global effect is very likely to be an increase in electricity use, due to additional cooling demand in warmer continents and regions, despite a reduction in direct heating fuel use, with a net impact on primary energy that depends on a range of factors (Hunt and Watkiss 2011).

Changes in summer temperatures will also tend to increase the maximum load on electricity systems that already have summer peak demand, and therefore increase the need for power generation capacity. There are also implications for cooling strategies in buildings in some regions. In those regions with cold moderate climates where residential building over-heating is currently not a significant issue, it may become so, and passive cooling techniques currently associated with warmer climates will need to be incorporated into building design. On the other hand, in some arid climates, as diurnal average temperatures rise, existing passive cooling techniques may become inadequate, leading to greater reliance on active cooling. In general, buildings will need to be designed to allow comfortable conditions to be maintained in the range of climates they are expected to face over a building's lifetime. To the extent this is not done, there are increased mortality and morbidity risks from heat stress, although health impacts of cold will fall.

For much of Europe, increases in electricity demand for air conditioning and cooling will be outweighed by reductions in the need for heating energy up to 2050 (Jochem et al., 2009) while warming is moderate and no thermally advanced buildings are considered. The individual changes may be quite large: for example, 10% per degree change in winter temperature for residential space heating in the United Kingdom (Summerfield, 2010). The aggregate impact of the different effects is smaller. The net result in a 4°C scenario[9], for instance, is a reduction in final energy demand by 2050 of 3.3% (Jochem et al., 2009) for the European Union-27 plus Norway and Switzerland. However, electricity used in space cooling is presently more carbon intensive in many countries than energy used for heating (e.g., gas, heating oil, wood fuels) in Europe and other countries in moderate climates. Depending on the final energy mix for heating and cooling, and the mix of primary energy for electricity supply, net CO_2 emissions in quite a few countries could still slightly increase even though overall energy demand decreases (Cartalsi et al., 2001; Frank, 2005; Aebischer et al., 2007; Rivière et al.; 2008).

The total electricity demand in the buildings sector is projected to slightly decrease in Nordic and Baltic countries (0.5%) by 2050. However, a 7% increase in the electricity demand by 2050 is expected in southern Greece, Malta, Cyprus, Southern Italy, Spain, Bulgaria, and Romania in a 4°C Scenario (Mirasgedis et al., 2007; Jochem et al, 2009). This points to distributional issues regarding adaptation or mitigation policies between northern and southern European countries. It may also lead to a greater need to balance summer electricity flows via the trans-European electricity grid, particularly during extreme heat waves. Similar effects can also be assumed for northern and southern states in the United States, Russia, or provinces in China.

According to Mansur et al. (2005), the combination of climate warming and fuel switching – from fuels to electricity – in buildings in the United States results in increases in the overall primary energy demand. There is likely to be a significant growth in the installation of air conditioning, but, with low utilization, additional energy demand may remain modest (Henderson, 2005), as confirmed by Jochem et al. (2009) for European countries north of the Alps.

There will be smaller impacts on other uses of energy in buildings. While the energy required to supply the same amount of hot water slightly decreases in a warmer climate, any increased demand for showers and bathing in warmer weather and additional heat waves will offset this reduction.

Climate change will increase electricity and primary energy demand in most emerging and developing countries, in contrast to a small or even beneficial effect in more temperate industrialized countries of the developed world such as Scandinavian countries, Russia, or Canada.

If left to the market alone, i.e., in the absence of specific government interventions, the responses to these climate changes are likely to be based on incremental short-term considerations – for example, the

purchase of inefficient types of room air conditioning during hot summers. This may lock in the existing set of inefficient technologies (Unruh, 2000) in situations where innovative solutions would be more desirable (e.g., passive ventilation, passive buildings, cooling by absorption technology, shadowing by building elements and trees, white roofs and surfaces, vegetation in urban areas). Given the long life of building stock, it is clearly a priority that policies consider climate mitigation and adaptation of the building stock together.

10.2   Specific Sustainability Challenges Related to Energy Services in Buildings

10.2.1   Key Messages

This section focuses on major cross-cutting, building-specific issues that often present challenges and require trade-offs when pursuing sustainable energy goals; additional to those covered by other chapters in GEA. The key message that is valid across various subsections is that most of such challenges can be overcome by a high-efficiency, state-of-the-art building (from new and retrofits) and energy using equipment stock. The section on fuel poverty shows that high-efficiency buildings may eradicate fuel poverty and through this can also improve general social welfare.

10.2.2   Indoor Air Quality and Health Impacts of Air Tightness

Improving air tightness is an important method to reduce heating and cooling energy demand. However, it is also important to secure adequate ventilation, so as to maintain a healthy indoor environment due to the variety of chemicals used in interior materials, furniture, and daily goods. While this is not an issue with advanced buildings described in this chapter, since by design they operate with ventilation rates that result in very high indoor air quality, this is still an issue with existing or sub-optimally designed, inefficient future buildings.

One way to improve indoor air quality in increasingly airtight buildings is to reduce the use of materials emitting high levels of volatile organic compounds and ensure adequate ventilation. Health problems caused by airtight buildings without adequate ventilation, the so-called sick building syndrome,[10] were first identified as a result of reducing air change rate as an energy conservation measure. In order to ensure proper air quality, the American Society of Heating, Refrigerating and Air-Conditioning Engineers (ASHRAE) proposed a range of allowable CO_2 concentrations from 1000 ppm to 2500 ppm, as well as ventilation requirements of 10 liters/sec per person for office spaces. In addition,

---

9   The 4°C scenario assumes that global average temperature will rise by 4°C compared to pre-industrial levels (van Vuuren et al., 2008).

10   Sick building syndrome describes the situations in which building occupants experience health and comfort effects that appear to be linked to time spent in a building, but no specific illness or cause can be identified (US EPA, 1991).

case control studies of allergy symptoms in 390 Swedish households (Bornehag et al., 2004) showed that the air change rate of the case group that had allergy symptoms was lower than 0.5/h[11], the same value under which the sick building syndrome and other air infectious diseases have also been shown to increase (Seppanen et al., 1999). In Japan, measurement of the concentration of chemicals in 2800 households (Osawa and Hayashi, 2009) showed that concentrations of formaldehyde in 27.3% of the households were higher than the guideline value. As a result, the revised Building Standard Law stipulates that air change rate in the living rooms of new constructed buildings must be secured at at least 0.5/h.

10.2.3 Household Fuels vs. Environmental Health

As discussed in Chapter 4 and other Chapters of the GEA, indoor air pollution arising from poor quality fuels burnt in inefficient devices has a health toll of mortality and morbidity measured by hundreds of millions each year. Since the issue is treated in detail throughout this document (including, but not limited to, Chapters 2, 3, 4, 11, and 19), this section only brings the importance of the issue to the fore and highlights some key relevant aspects.

Traditional biomass fuels have been the single most important energy source in buildings for centuries. They still account for approximately 10% of global total primary energy use concentrated primarily in developing countries. Approximately 60% of all biomass is used in solid unprocessed forms such as firewood, agricultural waste, and dried animal dung burnt in crude and inefficient stoves and open fires for cooking and heating (IEA, 2008b). Chronic Obstructive Pulmonary Diseases, to which pollution from poor combustion of biofuels indoors contribute, are predicted to become the world's third largest cause of death by 2030 (WHO Statistics, 2008).

Women and children in rural and urban areas of developing countries are most at risk due to their daily, close proximity exposure. Providing chimneys and efficient wood-burning stoves have been shown to reduce health risks by up to 50% in some areas (Romieu et al., 2009). Facilitating access to clean fuels such as biogas, solar thermal energy, liquefied petroleum gas, or electricity could reduce health risks, particularly in urban areas where commercial energy is more widely available. Facilitating a "multiple fuel and clean technology" by simultaneously making a more efficient and cleaner use of biomass for cooking and increasing access to other modern fuels has wider potential ecological and economic benefits due to reduced forest degradation and the time spent, mostly by women, in collecting fuel (García-Frapolli et al., 2010). This challenge is explored in more detail in Section 10.7.3 and new developments on advanced stoves are reported in Section 10.4.3.

10.2.4 Fuel and Energy Poverty

Even if access to modern energy carriers has been enabled, many population segments still may not be able to afford sufficient amounts of energy to meet their basic needs. The problem exists even in the most affluent countries, in many of which significant shares of the population cannot afford adequate heating, or are forced to spend disproportionate shares of their income on meeting basic thermal comfort needs. Since, as the section below argues in detail, this is often not due to generic poverty; or is in cases cause of other poverty, this problem is intimately related to the sustainable development goals of GEA.

While there are several definitions of fuel poverty, this document's use of the term is broader than that in many other sources. According to Tirado Herrero (in preparation), "A household is in fuel/energy poverty when it is unable to afford an adequate amount of energy services to satisfy its basic domestic needs – particularly sufficient thermal comfort – or is forced to spend a disproportionate share of its income on them". This phenomenon, called "energy poverty" and "fuel poverty," was introduced in Chapter 2, and its health impacts were discussed in Chapter 4. This chapter elaborates further on these health and social consequences, and discusses how the solution to the problem goes hand in hand with sustainable energy goals in buildings.

Fuel poverty originates from a combination of three main causes: household income, energy prices, and domestic energy efficiency. In many cases the problem can be substantially alleviated, sometimes even eliminated, by significantly improved efficiency – thus providing a strong synergy with sustainable energy goals. Box 10.A.1(see the GEA Chapter 10 online appendix) provides a case study in India of a project to provide solar lighting for approximately 886 million rural residents.

Fuel poverty is often the long-term consequence of measures that were introduced to provide sufficient access: i.e., subsidized energy prices for the poor. Artificially low energy tariffs provide the wrong economic signals and thus result in the acquisition of inefficient equipment and occupation of energetically poor buildings. When consumers are weaned from the subsidized prices, they find themselves locked into disproportionally high energy expenditures. An example of this is the formerly communist countries of Central and Eastern Europe and the former Soviet Union, where highly subsidized energy pricing policies have resulted in a very poor efficiency building stock and highly inefficient infrastructure. After the fall of communism, the introduction of market-based energy pricing resulted in significant shares of the population now living in fuel poverty and not being able to afford adequate heating. Since they can especially not afford investments in improving the efficiency of their energy using stock or buildings, poor population segments may turn out to pay significantly more for lower levels of energy services than the more affluent who can afford higher levels of efficiency.

Fuel poverty is an insufficiently researched and reported problem, especially in certain regions like the former Soviet Union and Central and

---

11 Exchange rate of the total room air volume per hour.

Table 10.3 | Incidence of fuel poverty in selected countries and regions.

Country/Region	Main estimates	Reference
UK	- Number of fuel poverty households in the UK ranging between 2 and 6.5 million (1996/2007)	DECC (2009a)
Ireland	- 17.4% of households unable to adequately heat the home (2001)	Healy and Clinch (2002)
EU	- Average percentage of households unable to heat home adequately (1994/97) in EU14: 16.9% [max: 74.4% in Portugal; min. 1.6% in Germany]	Healy (2004)
CzechRepublic	- Less than 10% of households suffering from domestic energy deprivation	Buzar (2007)
Macedonia	- More than 50% of households suffering from domestic energy deprivation	Buzar (2007)
New Zealand	- Between 10% and 14% of households in fuel poverty using the UK definition of adequate indoor temperatures (2001).	Lloyd (2006)
Hungary	- The average Hungarian household allocated 9.7% of its net income to energy expenses (2000/07) - 15% of the population (1.5 million) declared to be unable to afford to keep their homes adequately warm (2005/07)	Tirado Herrero and Ürge-Vorsatz (2010)

Source: own elaboration after references consulted.

Eastern Europe, where it is suspected to be widespread (Buzar, 2007; Boardman, 2010). Though slowly gaining priority in some policy and research agendas (Friel, 2007), it is still far from being a common issue of concern (see Table 10.3). Even in economically and socially advanced regions like the European Union, few Member States have come up with specific strategies or policy frameworks to address the issue. In fact, few countries – only the United Kingdom (BERR, 2001; DEFRA/BERR, 2008), Ireland (MacAvoy, 1997), the United States (Power, 2006) and New Zealand (Chapman et al., 2009) – have started any significant action to tackle fuel poverty. In the United Kingdom, the government has set as a goal that by 2018 no British household should be spending more than 10% of its income on energy (DEFRA/BERR, 2008). The likely failure in meeting this target in the United Kingdom, as foreseen by Boardman (2010), evidences the scale of the challenge and provides arguments for jointly tackling fuel poverty, climate change mitigation, and energy security challenges. In fact, since domestic energy efficiency solutions allow bringing households out of fuel poverty while capping or reducing their energy use levels (Milne and Boardman, 2000), eradicating fuel poverty will certainly have positive effects on those related challenges.

There is evidence that inadequate indoor temperatures cause excess winter mortality (EWM) (Eurowinter Group, 1997), with most western countries reporting relative EWM rates ranging from 5% to 30% (Healy, 2004) (see Table 10.4). Based on a cross-country comparison, taking Norway as a control case, it is estimated that 44% of the cardiovascular- and respiratory-disease excess winter deaths registered in Ireland in 1986–1995 can be associated with poor housing standards (Clinch and Healy, 1999). Fuel poverty is also linked to certain illnesses (Morrison and Shortt, 2008; Roberts, 2008), with particularly negative physical and mental health effects being recorded for vulnerable populations, such as the elderly and children (de Garbino, 2004; Howieson, 2005; Liddell and Morris, 2010).

A common policy tool for alleviating fuel poverty has been subsidies aimed at reducing the energy bills or increasing the disposable income of low income households (DEFRA/BERR, 2008; Scott et al., 2008; Tirado Herrero and Ürge-Vorsatz, 2010). Such support schemes have, however, been criticized because, even though they succeed in reducing fuel poverty temporarily, in the long run they lock these households into fuel poverty by creating disincentives to improving the efficiency of energy using equipment and buildings. Healy (2004) has argued that the saved income most likely will be spent on more energy rather than invested in improving the quality of the dwellings. In addition, direct support schemes are often poorly targeted, distort the market, and constrain government budgets (Scott, 1996; IEA, 2007b; Fülöp, 2009).

A more sustainable and long-term solution is the retrofitting or replacement of inefficient equipment and building stock of these populations by high-efficiency ones. As this requires substantial investments that those experiencing fuel poverty themselves will not be able to afford, it may be necessary to (re)allocate public funds and financing to such purposes. For instance, since large sums of public funds, comparable to the investment costs of high energy efficiency retrofit programs that may fully eliminate the fuel poverty problem are devoted yearly to social energy price subsidies and temporary fuel poverty alleviation measures, a progressive substitution of the latter by the former can substantially contribute to the solution of the problem.

In addition, since a high-efficiency building and appliance stock contributes to the solution of many other problems – such as GHG and other environmental emissions, energy security, and employment – policy integration can result in the availability of funds that can more effectively reach those goals through improved efficiency, especially if sources from these several fields are combined. For instance, Ürge-Vorsatz et al.

Table 10.4 | Incidence of excess winter mortality (EWM) in selected countries.

Country	Period	EWM		Reference
		Relative[1]	Absolute	
Austria	1988–1997	14%	n.a	Healy, 2004
Belgium	1988–1997	13%	n.a	Healy, 2004
Denmark	1988–1997	12%	n.a	Healy, 2004
Finland	1988–1997	10%	n.a	Healy, 2004
France	1988–1997	13%	n.a	Healy, 2004
Germany	1988–1997	11%	n.a	Healy, 2004
Greece	1988–1997	18%	n.a	Healy, 2004
Hungary	1995–2007	12.71%	5566 deaths	Tirado Herrero and Ürge-Vorsatz, 2010
Ireland	1988–1997	21%	n.a	Healy, 2004
Italy	1988–1997	16%	n.a	Healy, 2004
Luxembourg	1988–1997	12%	n.a	Healy, 2004
Macedonia	1995–2004	n.a	885deaths	WHO, 2007
Netherlands	1988–1997	11%	n.a	Healy, 2004
New Zealand	1980–2000	18%	1,600	Davie et al., 2007
Poland	1991–2002	n.a	14,680deaths	Morgan, 2008
Portugal	1988–1997	28%	n.a	Healy, 2004
Romania	1991–2004	n.a	17,358deaths	Morgan, 2008
Spain	1988–1997	21%	n.a	Healy, 2004
UK	1988–1997	18%	n.a	Healy, 2004

Source: own elaboration after references consulted.

Notes: 1) Percentage of additional deaths in the cold season in comparison with the warm season

(2010) have suggested that Hungary could cover the bill of deep renovation of its entire inefficient building stock, and thus the complete elimination of its fuel poverty, from the redirection of existing budgets, while still reaching the objectives of those budget items.

10.2.5   Health Problems Caused by Intermittent Local Heating

Household heating in some cold regions, including Japan and parts of Europe, is often limited to the occupied space, causing large temperature differences between heated and unheated spaces.

In Japan, measurements of indoor air temperatures in residential buildings located in cold regions indicate that indoor air temperatures are maintained around 20°C in the heated rooms, while the temperatures of bedrooms, bathrooms, and toilets without heating can be as low as outdoor air temperatures (Yoshino et al., 1985). The average temperature difference between heated rooms and not heated rooms is often about 20°C. It is thought that blood pressure overshoots caused by such large temperature differences are one of the causes of high death rates from apoplexy in these districts. Moreover, a large percentage of deaths from accidental drowning in bathtubs in Japan is due to the low temperatures

in unheated bathrooms (Tochihara, 1999): the sudden change in blood pressure before and after bath might also be the cause of death from apoplexy or anemia. High-efficiency, state-of-the-art buildings advocated in this chapter could help overcome this problem. With minimal or no energy investments, they can provide full thermal comfort and thus reduce such mortality and morbidity.

10.2.6   Urban Heat Islands vs. Resilient Buildings

The outdoor air temperature in hot weather in thermally massive built environments with surfaces of low albedo is increasing due to the urban heat island phenomenon (Oke, 1982; Akbari et al., 1990; inter alia). It is becoming a major reason for the increase in energy use, and is exacerbated by the measure that is supposed to reduce the impact: the air conditioning of buildings. Air conditioners transport indoor heat to the outdoors, adding to the heat generated by air conditioners themselves, thereby contributing to the urban heat island effect in areas with mechanical cooling. The heat island effect occurs when surfaces of the built environment absorb sunlight, which is released as heat during cooler periods, such as nighttime, and keeps the air temperature warm. It can raise a city's temperature by up to 3–4°C and catalyzes smog formation and other health hazards (US EPA, 2007).

Table 10.5 | Energy saving for HVAC loads due to green roofs in different regions[12].

Percentage Energy Savings for HVAC loads (rounded to the nearest integer)	Number of Floors	City, Country, Latitude
73%	1	Toronto, Canada, Latitude: 43°41′ N
29%	2	
18%	3	
50%	5	Singapore, Latitude: 1°22′ N
26%	2	Athens, Greece, Latitude: 37°58′ N
25%	1	Madrid, Spain, Latitude: 40°23′ N
6%	8	
32% (non-insulated)	2 1 (top floor only)	Athens, Greece, Latitude: 37°58′ N
14% (insulated)		
48% (non-insulated)		
32% (insulated)		

Source: adapted from Ahrestani, 2010; based on Martens et al., 2008; Wong et al., 2003; Spala et al., 2008; Saiz et al., 2006; Santamouris et al., 2007.

Chapter 4 reviews some health impacts of heat stress. This section presents a few further examples. Narumi et al. (2007) carried out investigations on the relationship between infection and air temperature in Osaka, Japan. The number of daily reports of disease increased when the average outdoor temperature was higher than 22°C. Six out of fifteen types of infections: (1) hand-foot-and-mouth disease; (2) herpangina; (3) pharyngoconjunctival fever; (4) enterohemorrhagic escherichia coli; (5) infectious gastroenteritis; and (6) epidemic keratoconjunctivitis – showed a positive correlation with temperature.

Genchi et al. (2007) studied the increase of tropical nighttime temperatures caused by the urban heat island phenomenon, and quantified the impact on sleep disorders. The results show that when the temperature is higher than 26.7°C, about 10% of residents suffer from sleep disorders, with 1°C increase of air temperature at midnight. It was estimated that the economic losses due to insomnia was about 305 billion yen (US$ 3.53 billion).

Among the strategies to address the urban heat island effect are "cool roofs" and roof and vertical "greening." Cool roofs are solar reflective roofs that absorb less sunlight than conventional roofs. The greater reflectivity is achieved by utilizing a light color of roof surface and special highly reflective[13] and emissive[14] materials, which can reflect at least 60% of sunlight instead of 10–20%, reflected by traditional dark-colored roofs (US EPA, 2007). Standard black asphalt roofs can reach 74–85°C at midday during the summer. The surface temperature of bare metal or metallic roofs can increase up to 66–77°C. Cool roofs reach peak temperatures of only 43–46°C, even in full summer. Conventional roofs can be 31–47°C hotter than the air on any given day, while cool roofs tend to stay within 6–11°C of the background temperature.

Cool roofing materials cost 5–20% more than conventional ones, but in the long run they can provide considerable cost and energy savings, reduce GHG emissions, and improve human health. Human health improvements include reducing heat-related illnesses and reducing deaths in buildings without air conditioning (US EPA, 2007). Energy savings vary greatly from one building to another between 10 and 70% of total cooling energy use (Wang, 2008). Preliminary estimates of the global emitted CO_2 offset potentials for cool roofs and cool pavements by Akbari et al. (2008) are in the range of 24 Gt of CO_2 and 20 Gt of CO_2, respectively, giving a total global emitted CO_2 offset potential range of 44 Gt of CO_2.

Green roofs and walls also mitigate the heat island effect, improve urban air quality (Yang et al., 2008), reduce CO_2 concentrations, and reduce the need for winter heating (Takebayashi and Moriyama, 2007; Li et al., 2010). Green roofs have also been shown to provide thermal insulation to buildings through a combination of the reduced thermal conductivity of the roof structure, and the evapotranspiration of the plants (Martens et al., 2008). A number of studies have shown that insulation provided by green roofs can reduce energy use of heating, ventilation, and air conditioning (HVAC) systems. However, the energy savings reported in the literature are usually the results of simulations rather than real measured data, therefore, the range of presented values is very wide. For example Sailor (2008) finds that for a 2-story office building in Chicago and Houston a green roof with 0.2m thick soil reduces total electricity use by 2% in both cities and reduces natural gas use by about 9% in Chicago and by about 11% in Houston compared to a conventional membrane roof. Table 10.5 demonstrates the results from several studies and shows that modeling results of HVAC energy saving from green roofs vary between 6–72% depending on the climate zone and number of floors affected.

---

12 All energy savings presented in the Table are the results of simulations and not measured data

13 Solar reflectance, or albedo, is the percentage of solar energy reflected by a surface.

14 Thermal emittance is the amount of heat a surface material radiates per unit area at given temperature, i.e., how readily a surface gives up heat.

As the urban and global climate changes – and warms – the ability of buildings to continue to provide healthy and thermally-comfortable environments for inhabitants will be further challenged. A combination of efficiency (via passive solar design), bio-climatic design (where buildings are greened and integrated into their natural settings rather than set apart from them) (Yeang, 1994), and design for adaptability (when buildings are designed for simple retrofitting to enhance resilience to environmental climatic extremes) (Graham, 2005) is necessary to cope with climatic challenges (Bornstein and Lin, 1999). Designers, developers, regulators, and financiers – both government and private – urgently need to be made aware of this deteriorating situation. Occupants also need to be provided with a greater choice of strategies, including energy feedback and occupancy monitoring systems, in order to tune buildings to changing climatic conditions.

Among the primary purposes of buildings is the provision of thermal comfort. Thermal comfort is a dynamic quality based on the interaction of people's metabolism, sensory perceptions, expectations, and acclimatization experiences, as well as the human body's interaction with the surrounding interior building materials. The material nature of the urban environment is in a constant dynamic relationship with the urban climate (Santamouris, 2001), which is in turn embedded in the regional and ultimately the global climate – a relationship still little understood and appreciated. A change in any of these parameters can change perceptions and experiences of comfort, which exacerbates the demand for energy for heating or cooling. Integrating into a building the potential to naturally resist climatic extremes and especially temperature excesses in urban settings is a fundamental advantage of bio-climatically appropriate design. Both sustainability and livability are enhanced as a consequence.

10.3   Strategies Toward Energy-sustainable Buildings

This section briefly reviews the key strategies that can be applied to move towards buildings that use energy in more sustainable way. The sections to follow "unpack" these strategies, and translate them into concrete technological and non-technological options.

The key to achieving sustainability in the building stock is to reduce the energy requirements in operating buildings to the point where building energy needs can be met entirely through renewable and non-greenhouse gas emitting energy sources, while maintaining indoor air quality and avoiding hazardous chemicals. The extent to which present building energy requirements need to be lowered in order to be satiable by renewable energy depends on the overall energy demand in society and hence on the success of measures to reduce energy demands in other sectors of the economy. It also depends on the achievable and sustainable energy supply from renewable energy sources which, in the case of biomass energy, depends on a number of still uncertain biophysical and climatic factors, as well as on diet through its impact on the

availability of land for bio-energy crops (see Chapter 20). The building energy intensity (annual energy use per unit of floor area) that can be regarded as sustainable depends on the human population and the floor area per person, which together determine the total building floor area. Given limits – either physical, economic, or practical – to the renewable energy supply, a larger global building floor area will require lower energy use per unit of floor area. In the GEA pathways (Chapter 17) where future energy systems meet the key environmental, social, and economic objectives, energy intensity is reduced by the factor of three to four (and even larger in some regions).

A hierarchy of options for achieving reductions in the energy intensity of buildings of this magnitude is presented in Section 10.4, beginning with urban-scale energy systems, building-scale energy systems, and finally, individual energy using devices in buildings. A least-cost approach to achieving sustainable energy use in buildings will be to implement energy saving measures – either in the construction of new buildings or the retrofitting of existing buildings – up to the point where the cost of the next measure exceeds the cost of the least expensive renewable energy supply option. The least expensive renewable energy supply option might involve the on-site generation of electricity and thermal energy, or it might entail the provision of locally produced biomass or the provision of C-free electricity from distant but high quality wind, solar, or other renewable electricity sites. The relative costs of achieving a given low building energy intensity, of on-site generation of renewable energy, and of off-site supply of renewable energy will vary regionally, over time, and with the type of building under consideration. Some forms of renewable energy supply, such as passive heating, ventilation and daylighting, can be regarded as energy demand reduction measures rather than energy supply measures, but in any case, they tend to be low cost and so will be early choices in a hierarchy of increasingly stringent demand or supply measures.

10.4   Options to Reduce Energy Use in Buildings

10.4.1   Key Messages

Deep – 75% or more – reductions in the gross energy requirements of new buildings compared to the performance of most types of recent new buildings can be achieved in most or all jurisdictions in the world through application of the Integrated Design Process and of the principles discussed here. It is also possible to achieve significant – 50% or more – reductions in the energy use of existing buildings. Once gross energy requirements have been reduced by a factor of two to four, it is sometimes possible to supply most or all of the remaining energy requirements with on-site renewable energy generation such as active solar technologies (photovoltaic, solar thermal) mounted on or integrated into the building envelope, thereby reducing the net energy requirements to zero or achieving net energy generation.

Through drastic reductions in the net energy requirements of buildings, it will be significantly easier than otherwise to eventually supply all of the remaining energy requirements from off-site renewable energy sources (such as wind, solar and biomass), whether local or distant but of high quality.

10.4.2 Urban-Scale Energy Systems, Urban Design, and Building Form, Orientation, and Size[15]

10.4.2.1 Key Messages

Urban design influences energy use by buildings in several ways. Shape, height and orientation significantly affect heating and cooling loads and opportunities for passive ventilation. The density of urban developments influences the economic viability of district heating and cooling systems.

10.4.2.2 Role of Street Orientation and Width

Traditional houses and streets in many parts of the world used to be laid out so as to provide a significant amount of self-shading. The spacing of buildings close enough to provide significant self-shading will diminish wind strength near the ground, reducing the potential for ventilation, although daytime ventilation is not always useful. Close spacing also reduces solar access in winter, but such access will not be needed in those hot-summer regions with mild winters. Bourbia and Awbi (2004) measured temperatures at the 1.5 m height in traditional (narrow) and contemporary (wide) streets in a city in Algeria. Traditional streets are about five degrees cooler than contemporary streets, whether oriented north-south or east-west. This is due both to greater shading of traditional streets, which reduces direct solar irradiance, and the smaller sky viewing angles, which reduces diffuse solar irradiance. In hot-dry desert climates of India, the urban scape is defined by narrow roads banked by tall and compact houses with thick walls and small openings – all of these strategies help keep heat out of buildings.

10.4.2.3 Role of Building Shape, Form, and Orientation

Building shape (the relative length, width, and depth), form (small-scale variations in the shape of a building), and orientation are architectural decisions that have significant impacts on heating and cooling loads, as well as on daylighting and opportunities for passive ventilation, passive solar heating and cooling, and for active solar energy systems. For instance, in temperate climates, the optimal orientation for rectangular buildings is the long axis running east-west, as this simultaneously maximizes passive solar heating in the cold season and minimizes solar heating in the warm season. Traditional houses in warm climates in India were mostly designed

around courtyards and front courts (Aangan) that served as congregation spaces for families and for sleeping during nighttime, as these are naturally cool outdoor spaces. Developing countries, such as India, have a rich legacy of architecture that uses traditional low- or no-energy techniques to ensure thermal and visual comfort. Old forts and havelis (mansions) had deployed several innovative techniques of natural lighting, ventilation, and natural cooling to achieve desired comfort levels.

Roof design is another feature that can influence energy use. An unconditioned space between inhabited rooms and the roof is a traditional insulating technique, and one that allows significant improvements in thermal performance without loss of useful space, by installing additional insulation. An overhanging roof also provides passive cooling via shading, as well as protecting walls from rain and snow. A reflective (white or cool) roof reduces heat gains, as discussed earlier.

The options discussed in this section, as well as many other sections, may influence the aesthetics of the building and neighborhood. Nevertheless, by today most, if not all, sustainable energy solutions can be implemented in buildings that do not need to compromise on aesthetics.

10.4.2.4 Role of Building Size

Building size is an important factor in energy use. Increased size tends to reduce surface to volume ratio, thereby reducing thermal losses and gains relative to floor area. However, total surface area and hence total energy use will increase unless the envelope is sufficiently improved. In the United States, the living area in dwellings per family member increased by a factor of three between 1950 and 2000 (Wilson and Boehland, 2005). This is due in part to declining average family size (from an average of 3.67 to 2.62 members) and in part due to larger houses (from an average of 100 m^2 to 217 m^2). A moderately insulated 3000ft^2 (~300 m^2) house in Boston requires more heating and cooling energy than a poorly insulated 1500 ft^2 house in the same location. The larger house also requires substantially more materials. According to a designer-builder quoted by Wilson and Boehland (2005), the growth in house size is due to: (1) the loss of a sense of community and public life, so that the house becomes more of a fortress that needs to provide multiple forms of entertainment instead of basic shelter; (2) the promotion of the idea that "bigger is better" by the building industry; and (3) the diminishing craftsmanship in house construction and design, leading to a substitution of greater size to counteract the sterility of modern homes. Wilson and Boehland (2005) list various strategies to make more efficient use of space, so that smaller houses provide the same services.

10.4.2.5 Multi-unit Housing, Office and Retail Space

Multi-floor, multifamily housing is significantly more energy efficient than single-family housing, especially one-floor, single-family housing.

---

15 This section is a highly condensed discussion drawn from Harvey (2006, 2009, 2010).

This is due to the sharing of walls and reduction in roof area, with concomitant reduction in heat loss. Stacking housing units vertically, or designing single-family houses as two- or three-story houses rather than as one-story houses will increase the opportunities for passive ventilation in the summer by exploiting the buoyancy of warm internal air, and protects the lower floors from solar heat gain.

Analyses carried out by Smeds and Wall (2007) indicate that over twice the thickness of insulation is needed in a single-family house as in a multi-unit apartment, along with substantially better windows, in order to achieve the same reduction in heat loss. Conversely, adoption of about the same insulation levels and window performance in an apartment building as in high performance houses in Sweden reduces the annual energy use to one-third of that of the high performance house and to about one-tenth of that of conventional single-family Swedish houses.

Multifamily housing has a smaller surface to volume ratio than single-family housing, which reduces the building cost per unit of floor area by reducing the relative importance of the external envelope to the total cost. Construction material requirements are also reduced, while public transit, walking, and cycling are enhanced and land is spared because a more compact urban form can be created. Thus, multifamily and multi-unit office and retail buildings simultaneously reduce energy use and investment costs and enhance possibilities for alternatives to automobile use.

Another benefit of multi-unit housing and mixed (housing and commercial) development is that the connection to district heating and cooling grids is more likely to be economically justifiable, as explained later. However, large-scale office and retail buildings exceeding 30 meters in depth need more energy for lighting, ventilation, and cooling than small-scale office buildings. This is because natural lighting cannot reach the center of the building, so artificial lighting has to be relied upon; natural ventilation cannot provide enough outdoor air, causing a reliance on mechanical ventilation; and heat generated inside cannot be released through the envelope, so more cooling is needed. Different shapes – e.g., U-shaped or E-shaped rather than rectangular – provide better natural lighting and ventilation, but with an increase in exterior walls.

10.4.2.6   District Heating and Cooling

A district heating system consists of a network of underground pipes carrying steam or hot water from a centralized heating facility or heat source to individual buildings, while a district cooling system is a network of pipes to carry chilled water. District heating systems provide an energy savings if they make use of heat that would otherwise be wasted. The most common source of waste heat is heat produced from the generation of electricity in fossil fuel or biomass power plants. Conversely, district heating supplied entirely from centralized boilers does not save any energy, and may in fact increase energy use, compared to the use of on-site condensing boilers. System efficiency is maximized if heat from the cogeneration of electricity is supplied at the lowest possible temperature, as this minimizes the reduction in electricity generation caused by withdrawing useful heat from a steam turbine, maximizes the fraction of waste heat used, and minimizes heat losses during distribution. This, in turn, requires buildings with a high performance thermal envelope and ideally with radiant floor or ceiling heating systems, which permit low heat delivery temperatures, as discussed later. However, the heat load in this case might be so low that a district heating network cannot be economically justified unless the building density is very high.

District cooling can be supplied from large, dedicated centralized electric chillers or from absorption chillers that are driven with steam from steam turbines for electricity generation. The latter is referred to as "trigeneration," as it involves the concurrent production of electricity, hot water, and chilled water. In principle, district cooling from large, centralized chillers can provide significant (up to 45%) savings compared to the use of separate chillers in individual buildings (Dharmadhikari et al., 2000). This rate of savings is due to the larger full-load efficiency of large chillers compared to small chillers, and the ability to operate each chiller in a centralized system at, or close to, its maximum efficiency. Further savings are possible if the centralized system can make use of heat sinks, such as sewage or lake, river, or sea water that would not be available to chillers in individual buildings. However, in practice there may be no savings or even an increase in energy use if unfavorable behavioral changes – such as switching from cooling only individual rooms as needed, to cooling the entire building all of the time – accompany the switch to district cooling, as already highlighted in earlier sections of the chapter.

The total cost of district cooling systems can be less than the total cost of equipping individual buildings with their own chillers. This is due to low unit costs for large chillers, the need for less total capacity in centralized systems – because the peak cooling loads in individual buildings do not all occur at the same time – and the need for less backup capacity (IEA, 1999; Harvey, 2006). District cooling systems also eliminate the need for rooftop cooling towers, thereby freeing up roof space for other purposes, such as rooftop gardens or solar panels.

District heating networks can be coupled with the large-scale underground storage of heat that is collected from solar thermal collectors during the summer and used for space heating and hot water requirements during the winter (Schmidt et al., 2004; Harvey, 2006). Heat can also be supplied with biomass, as part of a biomass cogeneration system or from geothermal heat sources. If both heat and coldness are stored, then heat pumps can be used to recharge the thermal storage reservoirs or to directly supply heat or coldness to the district heating and cooling networks during times of excess wind energy. This, in turn, permits the sizing of wind systems to meet a larger fraction of total electricity demand without having to discard as much, or any, electricity generation potential during times of high wind and/or low demand.

10.4.3   Options Related to Building-Scale Energy Systems and to Energy Using Devices[16]

10.4.3.1   Key Messages

The energy use of buildings depends to a significant extent on how the various energy using devices (pumps, motors, fans, heaters, chillers, and so on) are put together as systems, as well as the efficiencies of individual devices. Savings opportunities at the system level are generally many times what can be achieved at the device level, and these system-level savings can often be achieved at net investment-cost savings. Significant savings are also possible for business and household plug loads.

The following subsections briefly explain how extraordinary savings can be achieved. Examples are presented of exemplary buildings from around the world, spanning a wide range of climates, followed by information on the initial investment cost of low energy buildings compared to conventional buildings. Much more detailed information can be found in Harvey (2006).

10.4.3.2   Integrated Design Processes

The key to achieving deep reductions in building energy use is to analyze the building as an entire integrated system, rather than focusing on incremental improvements to individual energy using devices. This requires a new approach to building design, referred to as the Integrated Design Process (IDP) (Lewis, 2004). IDP requires setting ambitious energy efficiency goals at the very beginning of a project, and requires an early brainstorming session involving all the members of the design team to develop a number of alternative concepts for achieving the energy target. The integrated design process also entails two-way interaction between the design team and the contractors. Simulation is an important tool in an IDP process. As a building will be operated over a large range of outdoor climates and indoor states, simulation can tell what happens in a part-load situation and help the design to achieve high efficiency during part-load conditions. It often happens that the building and its system perform very well during the hot and cold season (the design states), but very poorly during transitional seasons.

As Harvey (2006) discusses, the steps in the most basic IDP are to: (i) consider building orientation, form, thermal mass; (ii) specify a high performance building envelope and other measures to reduce heating and cooling loads; (iii) maximize passive heating, cooling, ventilation, and daylighting; (iv) install efficient systems to meet remaining loads; (v) ensure that individual energy using devices are as efficient as possible and properly sized; and (iv) ensure the systems and devices are properly commissioned.

By focusing on building form and a high performance envelope, heating and cooling loads are minimized, daylighting opportunities are maximized, and mechanical systems can be greatly downsized. This generates cost savings that can offset the additional cost of a high performance envelope and the additional cost of installing premium (high efficiency) equipment throughout the building. These steps alone can usually achieve energy savings on the order of 35–50% in new commercial buildings, while utilization of more advanced or less conventional approaches has often achieved savings on the order of 50–80%.

10.4.3.3   Reducing Heating and Cooling Loads

The term "heat load" refers to the rate of heat loss from a building during the heating season. This heat has to be replaced by the heating system in order to maintain a steady indoor temperature, and so represents a load on the heating system. The term "cooling load" refers to the rate of unwanted heat gain during the cooling season, heat that must be removed in order to maintain a steady indoor temperature.

Heating loads can be dramatically reduced through the use of a high performance thermal envelope, consisting of: (i) high levels of insulation in the walls, ceiling, and basement; (ii) avoidance of thermal bridges; (iii) windows and doors with a very high resistance to heat loss; and (iv) a high degree of airtightness, combined with mechanical ventilation with heat exchangers to recover heat or coldness from exhaust air and possibly waste water, depending on the season.

The heating energy requirement is the difference between heat losses and useable internal and passive solar heat gains. High levels of insulation, combined with high performance windows and airtightness and coupled with mechanical ventilation and heat recovery, can readily reduce heating energy requirements by a factor of 4–10 compared to current practices in cold climate regions. In areas with mild winters where previous practice was for no insulation, rather moderate levels of insulation can substantially reduce heating energy requirements, as well as reduce summer cooling energy use by a factor of two or more (Florides et al., 2002).

Heat loss through high performance windows is so small that perimeter heating units, usually placed below windows to prevent drafts, can be eliminated, even when winter temperatures dip to -40°C (Harvey and Siddall, 2008). When perimeter heating is eliminated, ductwork or hot water piping can be made shorter, as all the radiators can be located closer to the central core of the building, with associated cost savings but also savings in fan and pump size and energy use. If the default design involves floor-mounted fan-coil units, their elimination will increase the amount of useable floor space.

Options to reduce the cooling load include:

* orienting a building to minimize the wall area facing east or west;

---

16 Subsections 10.4.3 to 10.4.9 are a condensed discussion drawn from Harvey (2006, 2009, 2010).

- clustering buildings to provide some degree of self-shading, as in many traditional communities in hot climates;

- using high reflectivity building materials; for example, Parker et al. (2002) found that houses in Florida with white reflective roofs have a cooling-energy demand about 20–25% lower and peak power demand about 30–35% lower than houses with dark shingles;

- increasing insulation; for example, Florides et al. (2002) found that for a one-story house in Cyprus, adding 5cm of polystyrene insulation to the roof reduces the cooling load by 45% and the heating load by 67%, while addition of 5cm of polystyrene insulation to the walls reduces the remaining cooling load by about 10% and the remaining heating load by 30%;

- providing fixed or adjustable shading, as external shading devices block 90% of incident solar heat, compared to 50% for internal devices (Baker and Steemers, 1999);

- using windows with a low solar heat gain – as low as 25%, compared to 70% for a clear double glazed window – and avoiding excessive window area, particularly on east- and west-facing walls;

- using highly efficient lighting and household appliances, electronics, and office equipment to reduce internal cooling loads;

- utilizing thermal mass to minimize daytime interior temperature peaks, combined with nighttime cooling; and

- using vegetation to directly shade buildings and to indirectly reduce cooling loads by reducing ambient air temperature through evapotranspiration. Vegetation integrated into building surfaces, such as walls and roofs, also contributes to cooling by reducing heat gains and through evapotranspiration.

Thermal mass does not reduce the heat gain by a building and so does not represent a reduction in cooling load (as defined here). However, a high thermal mass reduces the temperature increase for a given heat gain and, for short temperature spikes, can eliminate the need for air conditioning. Porta-Gándara et al. (2002) simulated the cooling load for housing built with traditional adobe bricks and modern hollow concrete blocks (having minimal thermal mass) in Baja California, and found the air conditioner load of the former to be one-fourth that of the latter during the hottest summer months. However, unless temperatures drop sufficiently at night to remove the heat that enters the thermal mass by day, the temperature of the thermal mass will build up over a period of days, so it will become less and less effective in limiting daytime temperatures. The nighttime removal of daytime heat can be enhanced through deliberate nighttime ventilation of the building with outside air when the outside air is sufficiently cool, as discussed in the next subsection. External insulation will inhibit daytime penetration of outside

heat into the thermal mass while leaving it exposed to cool air during nighttime ventilation.

Thermal mass can also be provided through phase change materials (PCM), the most common being a paraffin wax that melts at around 25°C. The PCM can be embedded in drywall or plaster inside 50-μm capsules. The waxes will not rise in temperature above their melting point as they melt, just as ice will not rise above 0°C as it melts. Air in contact with the plaster or spheres will rise only a few degrees above the melting point of the wax. At night the waxes refreeze if they can be cooled to below their melting point with cool night air, releasing the heat that they absorbed during the day as they melted.

The combination of switching to a high albedo surface and planting shade trees can yield dramatic energy savings. Rosenfeld et al. (1998) calculated the impact on cooling loads in Los Angeles of increasing the roof albedo of all five million houses in the Los Angeles basin by 0.35 (a roof area of 1000 km²), increasing the albedo of 250 km² of commercial roofs by 0.35, increasing the albedo of 1250 km² of paved surfaces to 0.25 (by using whiter, limestone-based aggregates in pavement whenever roads are resurfaced), and planting 11 million additional trees. In the residential sector, they computed a total savings of 50–60%, with a 24–33% reduction in peak air conditioning loads. Akbari et al. (2008) estimate that a net albedo increase for urban areas of about 0.1, which they consider achievable by increasing both roof and pavement albedos by about 0.25 and 0.15 respectively, can achieve the equivalent of offsetting about 44 Gt of CO_2 emissions on a global scale. At the same time, these measures would induce significant savings in cooling costs, with an estimated savings potential in excess of US$1 billion/yr in the United States. Growing vegetation on building walls can also provide important reductions in cooling energy use; simulations by Kikegawa et al. (2006) indicate a savings of 10–30% for residential buildings in Tokyo.

In hot-humid climates, the energy required to dehumidify air can represent a significant fraction of the total cooling load. This portion of the cooling load will not be reduced through measures such as shading, external insulation, or use of thermal mass and windows with low-solar heat gain, so these measures will provide a smaller percentage savings in overall cooling loads. However, materials that absorb moisture can be placed at the internal surface of rooms so as to maintain nearly constant relative humidity inside. On dry days, the moisture can be released back to the air through ventilation. This can greatly reduce the energy required for dehumidification.

Thermal mass will be less effective in reducing daytime temperature-related cooling loads in humid climates because of the smaller day-night temperature difference in hot-humid climates than in hot-dry climates. In hot-humid climates, it is more appropriate to employ urban and building forms that promote air movement between and through buildings, in order to employ low thermal mass to minimize the storage of heat so that buildings can cool quickly whenever temperatures decreases (Koch-Nielsen, 2002).

Table 10.6 | Major strategies for reducing different energy uses in buildings in different climate zones.

	Climate Zone			
	Cold Moderate	**Warm Moderate**	**Arid**	**Tropical**
Heating	• High levels of insulation and air tightness with heat-recovery ventilation • Windows with high thermal resistance and high solar heat gain	• High levels of insulation and air tightness with heat-recovery ventilation • Windows with high thermal resistance and low solar heat gain	• Modest insulation • Windows with low solar heat gain	• Minimal or no insulation • Windows with low solar heat gain
Cooling	• Earth pipe • Shading • Thermal mass	• Earth pipe • Shading • Reflective surfaces • Thermal mass with external insulation and night ventilation	• Shading • Reflective surfaces • Thermal mass with external insulation and night ventilation • Compact form and self shading • Evaporative cooling	• Shading • Open structure with plenty of ventilation • Minimal thermal mass • Solar-driven desiccant dehumidification and evaporative cooling
Ventilation	• Hybrid passive-mechanical ventilation	• Hybrid passive-mechanical ventilation		

High performance thermal envelopes can readily reduce heating energy requirements by 75–90% in cold climates, while modest levels of insulation may eliminate the need for winter heating altogether in regions with mild winters. Modest levels of insulation are also effective in reducing cooling loads by about half in hot climates. Thermal mass, combined with external insulation and nighttime ventilation, can largely eliminate cooling requirements in hot-dry climates. In hot-humid climates, the potential to reduce cooling loads is smaller, due to the greater importance of dehumidification loads and the smaller diurnal temperature range. In both hot-dry and hot-humid climates, however, the remaining cooling loads can be handled through a variety of low-energy systems.

Table 10.6 summarizes the features of low-energy buildings that depend on the climate where the building is situated. These features largely pertain to building form and envelope, and the applicability of earth pipe, evaporative, or desiccant cooling systems. These and other building features and internal energy loads are discussed in the following subsections.

10.4.3.4   Passive and Passive-low-energy Heating and Cooling

Having reduced the cooling load through the techniques described above – often by a factor of two or more – the next strategy in priority is to use passive and/or passive-low-energy cooling strategies. A purely passive cooling technique requires no mechanical energy input at all. Other techniques involve small inputs of mechanical energy to enhance what are largely passive cooling processes. Some examples of passive and passive-low-energy cooling techniques are described below.

Natural Ventilation
Natural ventilation has a cooling effect whenever the outdoor air temperature is below the indoor air temperature. It reduces the perceived temperature due to the greater ability of moving air to remove heat from a warmer body, and increases the acceptable air temperature due to enhanced psychological adaptation to warmer conditions in naturally-ventilated buildings compared to buildings with mechanical ventilation.

Natural ventilation can be achieved through:

• cross ventilation and wind, a technique that has been widely employed in traditional architecture around the world and in passive ventilation stacks that are commonly used in north European residential buildings;

• solar chimneys, which consist of a tower in which air is heated and rises, drawing cooler outside air through the building. A striking example is the Building Research Establishment offices in Garston, United Kingdom, which is illustrated in Figure 10.A.3. in the GEA Chapter 10 online appendix;

• atria, which can induce natural ventilation through proper placement of air inlets and outlets, along with shading controls or passive measures, such as the geometry of laser-cut glazing, to avoid overheating in summer;

• cool towers, in which water is pumped into a honeycomb medium at the top of a tower and allowed to evaporate, thereby cooling the air at the top of the tower, which then falls through the tower and into an adjoining building under its own weight. These have been used in the Visitor Center at Zion National Park, United States (Torcellini et al., 2002), and at the Torrent Pharmaceutical Research Centre in Ahmedabad, India (Ford et al., 1998); and

• airflow windows, which are designed to facilitate the passage of outgoing exhaust air or incoming fresh air between the glazing in a double glazed window. In the Tokyo Electric Power Company

(TEPCO) research and development (R&D) center, constructed in Tokyo in 1994, this technique reduced the inward heat transfer by one-third to two-thirds compared to double glazed windows with interior or built-in blinds, and eliminated the need for perimeter air conditioning (Yonehara, 1998).

Nighttime Passive and Mechanical Ventilation

Where the day-night temperature variation is at least 5–7 degrees, cool night air can be mechanically forced through hollow core ceilings or through the occupied space to cool the building prior to its use the next day. Where artificial air conditioning is still needed by day, external air can be pre-cooled by passing it through the ceiling that has been ventilated at night. Effective night ventilation requires a high exposed thermal mass, an airtight envelope, minimal internal heat gains, and a building configuration that induces natural airflows so that minimal fan energy is required. In such buildings in the southern United Kingdom, as well as in Kenya, cooling energy savings of 30–40% can be achieved this way, as simulations for both places have shown (Kolokotroni, 2001). External insulation should be used in order to inhibit the inward penetration of daytime outside heat while leaving the thermal mass exposed to the cooling effect of nighttime ventilation and free to absorb internal heat during the day.

For Beijing, da Graça et al. (2002) found that thermally- and wind-driven nighttime ventilation eliminates the need for air conditioning of a six-unit apartment building during most of the summer, when an extreme outdoor peak of 38°C produces a 31°C indoor peak. Simulations by Springer et al. (2000) indicate that nighttime ventilation is sufficient to prevent peak indoor temperatures from exceeding 26°C over 43% of California's geography in houses that include improved wall and ceiling insulation, high performance windows, extended window overhangs, tight construction, and modestly greater thermal mass compared to standard practice in California.

Where mechanical air conditioning is used in combination with night ventilation, the energy savings from night ventilation depend strongly on the daytime temperature setpoint. For a three-story office building in La Rochelle, France, Blondeau et al. (1997) found through computer simulation that night ventilation with a 26°C setpoint requires only 9% of the cooling energy as the case with a 22°C setpoint and no night ventilation for this particular building and climate.

The combination of external insulation, thermal mass, and night ventilation is particularly effective in hot-dry climates, as there is a large diurnal temperature variation in such climates, and placing the insulation on the outside exposes the thermal mass to cool night air while minimizing the inward penetration of heat into the thermal mass. As previously noted, low thermal mass and an open design with plenty of cross ventilation are normally recommended in hot humid climates, although Tenorio (2007) finds that in humid tropical areas of Brazil, thermal mass combined with night ventilation and selective use of air conditioning can reduce cooling energy use in a two-story house by up to 80% compared to a fully air conditioned house.

Evaporative Cooling

Evaporation can cool water down to the wet bulb temperature (T_{wb}).[17] The difference between T_{wb} and the ambient temperature is greater the lower the absolute humidity, so the potential cooling effect of evaporative cooling is greater in arid regions, although the availability of water could be limiting. Evaporative cooling can provide comfortable conditions most of the time in most parts of the world (see Harvey, 2006). A number of residential evaporative coolers are on the market in the United States. Energy is required to operate the fans, which draw outside air through the evaporative cooler and directly into the space to be cooled, or into ductwork that distributes the cooled air. Simulations for a house in a variety of California climate zones indicate savings in annual cooling energy use of 92–95%, while savings are somewhat less (89–91%) for a modular school classroom (DEG, 2004). Evaporative cooling has been widely used in western China (such as XinJiang and Gansu Provinces) and some regions in India. It can provide very good cooling for office buildings, hotels, and shopping malls with outdoor temperatures as high as 38°C. As the energy savings would be much less in humid climates, a better approach is to enhance the evaporative cooling capacity using desiccants, as described later.

Underground Earth Pipe Cooling

Outside air can be drawn through a buried coil, cooled by the ground, and used for ventilation purposes. Simulations by Lee and Strand (2008) indicate that earth pipes can meet 70% of the June-August cooling load in Illinois and 65% in Spokane, Washington. The performance of such a system can be characterized by the ratio of the rate of heat removal by the air exchange to the power used by the fans – analogous to the coefficient of performance (COP) of a heat pump or air conditioner. The measured COP of a ground loop for a building in Germany is 35–50 (Eicker et al., 2006). Argiriou et al. (2004) built and tested an earth pipe that was coupled to a photovoltaic array on a building in Greece that directly powers a 370W DC motor, thereby avoiding the need for DC to AC conversion normally associated with PV power. The fan speed increases as the incident solar radiation increases, matching the need for increased cooling. The measured average COP (based on DC power output) was about 12. In climates with warmer mean annual temperatures, the ground temperature will be warmer, so the benefits of earth pipe cooling will be smaller.

Water can also be circulated through underground pipes and pre-cooled or pre-heated. This is ideal in conjunction with radiant floor heating or radiant ceiling cooling, and has been used in Europe, usually with a heat pump to enhance the heat extraction from or transfer to the ground.

---

17 Wet bulb temperature is a type of temperature that reflects the physical properties of a system with a mixture of a gas and a vapor, usually air and water vapor. It is the lowest temperature that can be reached by the evaporation of water only (Hart & Cooley Inc. 2009)

Desiccant Cooling and Dehumidification

Solid or liquid desiccants[18] can be used to remove moisture from air, with the desiccant subsequently regenerated using solar thermal heat such as can be provided by flat-plate solar thermal collectors. Desiccants combined with conventional heating for regeneration are sometimes used in supermarkets today, where their substantial drying capacity is an advantage over traditional dehumidification techniques. If the air is over-dried, evaporative cooling can be used to bring the air temperature close to the desired final temperature, with only supplemental cooling with mechanical air conditioners. Heat is required to regenerate the desiccant. A great advantage of desiccant cooling systems is that they avoid overcooling the air and then reheating it for dehumidification purposes. However, the COP of a desiccant system is typically only about 1.0, compared to typical values of four to six for electric chillers, so primary energy use may increase or decrease, depending on the efficiency of the electric power plant that supplies the displaced electricity and the extent of overcooling. However, in the hot and humid climate of Hong Kong, solid desiccant systems can reduce overall energy use for cooling and dehumidification by 50% if solar thermal energy is used for regeneration of the desiccant (Niu et al., 2002).

Passive heating techniques

Passive heating refers to the simple absorption of solar radiation inside a building, preferably by elements with a high thermal mass such as stone walls or concrete walls and floors, thereby minimizing overheating during the day and providing the opportunity to slowly release heat at night. Passive heating can occur directly, through absorption of solar radiation within the space to be heated, or indirectly, through thermal mass located between the sun and living space.

10.4.3.5   Heating equipment

Commercial buildings, multi-unit residences, and many single-family residences, especially in Europe, use boilers that produce hot water or, in some exceptional cases, steam, that is circulated, generally through radiators. Efficiencies (ratio of heat supplied to fuel use) range from 75% to 95%, not including distribution losses. Modern residential furnaces, which are used primarily in North America and produce warm air that is circulated through ducts, have efficiencies ranging from 78% to 96%, again not including distribution system losses. Old equipment tends to have an efficiency in the range of 60–70%, so new equipment can provide substantial savings. Space heating and hot water for consumptive use, e.g., showers, can be supplied with heat from small wall-hung boilers with an efficiency in excess of 90%.

Heat pumps are another option for heating. They transfer heat from something that is relatively cold, such as the outdoor air, ground, or outgoing exhaust air, to the warmer ventilation air or hot water heating

18 A desiccant is a substance that induces or sustains a state of dryness (desiccation) in its local vicinity in a moderately well-sealed environment. In this context desiccants are used to dehumidify air in a physical or chemical way rather than through air conditioning.

system. They make very effective use of electricity, as the ratio of heat supplied to electricity used (the COP) is at least three for ground source heat pumps and at least six to seven for exhaust air heat pumps (Halozan and Rieberer, 1997). Heat pumps provide one means of decarbonizing building heating, once the electric grid itself is decarbonized. If a building has a high performance thermal envelope, so that it loses heat very slowly when the heating system is turned off, heat pumps – like air conditioners – can serve as a dispatchable electricity loads that can, to some extent, be turned on and off to match variations in wind or solar generated electricity supply.

State-of-the-art biomass pellet boilers, with efficiencies of 86–94%, are another option for heating with renewable, carbon-neutral energy.

10.4.3.6   Heating, Ventilation, and Air Conditioning (HVAC) Systems

In the crudest HVAC systems, heating or cooling is provided by circulating enough air at a sufficiently warm or cold temperature to maintain the desired room temperature. The volume of air circulated in this case is normally much greater than what is needed for ventilation purposes, in order to remove contaminants and provide fresh air. The energy required to transport a given quantity of heat or coldness by circulating water is 25–100 times less than the energy required by circulating air. Thus, by using chilled or hot water for temperature control and circulating only the volume of air needed for ventilation purposes – that is, separating the ventilation and heating or cooling functions – considerable energy savings are possible. This is a system level change.

With regard to residential buildings in cold climates, distributing heat by circulating warm water rather than warm air can reduce the energy used to distribute heat by a factor of 10 to 15 and eliminates the infiltration of outside air through pressure differences induced by unbalanced airflow in ductwork (Harvey, 2006; see also Chapter 4). In radiant floor systems, the entire floor serves as a radiator by circulating warm water through pipes embedded in the floor. In this way, heat can be delivered at the coolest possible temperature – at 30°C rather than at 70–90°C – which improves the efficiency of furnaces or boilers and especially improves the efficiency of heat pumps. It would also improve the efficiency of district heating and cogeneration systems by allowing a lower temperature of the hot water provided to such systems. Ventilation requirements can then be met with a much smaller airflow that, during heating or cooling seasons when windows are closed, is circulated with fans. Heat exchangers allow 80–95% of the heat in the outgoing exhaust air to be transferred to the incoming fresh air.

In commercial buildings, three features of advanced HVAC systems with significant energy savings potential are (1) chilled ceiling cooling, (2) displacement ventilation, and (3) demand-controlled ventilation. Chilled ceiling (CC) cooling, which was pioneered in Europe in the 1980s, involves circulating water at a temperature of 16–20°C through radiant panels that cover a large fraction of the ceiling area, or through hollow concrete ceiling slabs. Stetiu and Feustel (1999) carried out simulations of buildings with all-air and combined air/CC cooling systems for a variety of climates in the United States, assuming the same rate of intake of outside air for the two cases. They found that radiant cooling reduces energy use by an amount ranging from 6% in Seattle to 42% in Phoenix. The savings are smaller in hot-humid or cool-humid climates than in hot-dry climates because relatively more of the total air conditioning energy is used for dehumidification, which is not affected by the choice of all-air versus air/CC chilling.

In displacement ventilation (DV), fresh air is introduced from many holes in the floor at a temperature of 16–18°C, is heated by heat sources within the room, and continuously rises and displaces the pre-existing air. Compared with conventional ventilation systems, which rely on the turbulent mixing of air from ceiling outlets, the required airflow is reduced because of the greater ventilation effectiveness of a given air flow. DV, like CC cooling, was first applied in northern Europe. By 1989, it had captured 50% of the Scandinavian market for new industrial buildings and 25% for new office buildings (Zhivov and Rymkevich, 1998). In a system where most of the cooling is done with chilled water, the airflow is reduced to that needed only for fresh air purposes, meaning that it will be completely vented to the outside and replaced with 100% fresh air after one circuit. This is referred to as dedicated outdoor air supply, or DOAS. Because the air in a DV rises from the occupants to the ceiling, and from there directly to the outside, heat picked up at ceiling level does not need to be removed by the chillers. Calculations by Loudermilk (1999) indicate that for an office in Chicago about one-third of the total heat gain – including 50% of the heat gain from electric lighting – can be directly rejected to the outside in this way. An overall savings in combined cooling and ventilation energy use of 30–60% can be achieved through a combination of DV and CC cooling (Bourassa et al., 2002; Howe et al., 2003).

Having decoupled the ventilation and heating or cooling functions of an HVAC system using some hydronic cooling method, preferably CC, one is free to vary the ventilation rate based on actual and changing ventilation requirements, rather than using a fixed ventilation rate or varying it according to some inflexible schedule. This is referred to as demand-controlled ventilation (DCV). Depending on the kind of building and occupancy schedule, DCV can save 20–30% of the combined ventilation, heating, and cooling energy use in commercial buildings (Brandemuehl and Braun, 1999).

Introducing high performance air conditioners is another way to save energy. Cooling and heating individual rooms by using air conditioners is as common in small- and medium-size non-residential buildings as in homes. The coefficient of performance (COP) – the ratio of heat removed to energy used – of package air conditioners for residential use is 4.9 for cooling, and 5.4 for heating. This COP value has significantly improved in Japan by recent technology development (Figure 10.17). In contrast,

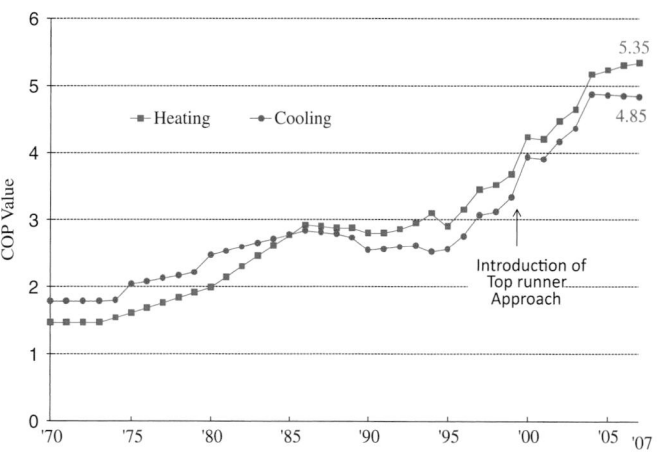

Figure 10.17 | COP values of air conditioners in Japan over time. Source: Jyukankyo Research Institute, 2007.

typical COPs of air conditioners available today in North America and Europe are 2–3.

In very humid climates, the task of air conditioning is to reduce both humidity and temperature. To remove humidity by condensing water vapor, 5–7°C chilled water is needed. However, for temperature control alone, water at 16–18°C is adequate, and this can often be obtained naturally or produced with a chiller at very high COP (up to 10). Liquid desiccants eliminate the need for producing water at 5–7°C because they can cool and dehumidify the air directly to the desired final conditions without overcooling and reheating. As noted earlier, savings of up to 50% are possible if the desiccant is regenerated with solar thermal energy.

10.4.3.7   Domestic Hot Water

The use of non-renewable energy to make hot water can be reduced by: (1) reducing the amount of hot water needed and used; (2) heating water as much as possible with solar energy using hot water panels; and (3) heating water more efficiently. Options for reducing the amount of hot water demand include using low flow showerheads (up to 50% savings), using more efficient washing machines and dishwashers, which require less water as well as less electricity, and washing using cold(er) water.

Where there are simultaneous inflows and outflows of water, as during showers, more than 50% of the heat in hot wastewater can be captured and used to preheat cold incoming water or air (Vasile, 1997).

If hot water is stored in a tank between periods when it is being used – the normal situation in North America – standby heat losses, which can account for one-third of total hot water energy use, will occur. Large losses can also occur from pipes that deliver hot water to where it is needed. These losses can be largely eliminated through tankless point-of-use water heaters, which are common in Europe and Asia. However, electrical tankless water heaters can amplify peak electrical loads. Solar water heaters can supply 50% or more of hot water requirements in most parts of the world and require a storage tank.

Heat pumps using CO_2 as a working fluid are ideal for hot water applications, as they can more readily reach the required 55–60°C water temperature. Compared to conventional combustion type water heaters, they enable about 30% energy conservation and about 50% reduction in CO_2 emissions given the electricity mix of Japan.[19]

Finally, substantial energy savings are often possible by upgrading existing hot water boilers. Older hot water boilers have efficiencies as low as 60%, compared to 80–85% for a modern non-condensing boiler and 92–95% for a condensing boiler.

10.4.3.8 Lighting

Lighting energy use constitutes 25–50% of the total electricity use in commercial buildings in OECD countries, and about 10% in residential buildings (see Figure 10.9 and Figure 10.A.2). Lighting energy use can be reduced by 75% or more through: (1) better design of lighting systems, including provision for task lighting; (2) maximizing the use of daylight, with light and occupancy sensors to add electric lighting only as needed; and (3) use of the most efficient lighting equipment (IEA, 2006).

Advances in window technology make it possible to increase window area to up to 40–60% of wall area without increasing heat loss in winter, after accounting for solar heat gain, and with minimal impact on cooling loads in summer. Detailed measurements and/or simulations demonstrate annual savings of 30–80% from daylighting of perimeter offices in commercial buildings (Rubinstein et al., 1998; Jennings et al., 2000; Reinhart, 2002; Bodart and De Herde, 2002; Li and Lam, 2003; Atif and Galasiu, 2003). The economic benefit of daylighting is enhanced by the fact that it reduces electricity demand most strongly when the sun is strongest, which is when the daily peak in electricity demand tends to occur during summer. Daylighting can also reduce cooling loads. This is because the luminous efficacy (the ratio of light to heat) of natural light is 25–100% greater than that of electric light systems.

In retrofits of fluorescent lighting systems, 30–50% lighting electricity savings can be routinely achieved. With considerable effort, 70–75% savings in retrofits are possible. In new construction, 75% and larger savings compared to current standards can be readily achieved. These remarkable energy savings could be increased yet further through advances in the efficiency and performance of the individual compo-

nents of the lighting system (described in Rubinstein and Johnson, 1998).

Light output is measured in lumens (lm), which takes into account differences in the sensitivity of our eyes to different wavelengths of light.[20] The ratio of lumens of light output to watts of energy used by a lamp is called the efficacy. At present,

- compact fluorescent lamps (CFLs) are about four times as efficacious as incandescent lamps, two to three times as efficacious as halogen infrared reflecting (HIR) lamps, and three times as efficacious as halogen lamps – all of which can be replaced in almost all applications;

- the 80-series T8 and T5 fluorescent tubes are about 60% more efficacious than T12 tubes used in old lighting and 25% more efficacious than standard (70-series) T8 tubes; and

- the ceramic metal halide lamp is about twice as efficacious as the HIR lamp, which it can replace.

Light-emitting diodes (LEDs) have the potential to be substantially more efficacious than any of the above. Commercially available LEDs currently have an efficacy of about 30 lm/W, compared to 10–17lm/W for incandescent lamps, 50 lm/W for CFLs, 105lm/W for the T5 fluorescent tubes, and up to 140lm/W for high-pressure sodium lamps. However, laboratory research LEDs have achieved an efficacy of up to 152lm/W at low current (Den Baars, 2008), and it is thought that efficacies of 150–200lm/W will be eventually achieved in commercial products. This would reduce electricity requirements by up to a factor of two compared to the best fluorescent tubes, a factor of four compared to CFLs, and a factor of 20 compared to the least efficacious incandescent lamps, which are still widely used. The effectiveness of LEDs is further improved by the fact that LEDs produce light that can be directed into only the directions where it is needed. By selective lighting, LEDs will be able to achieve energy savings compared to other lamps even before they achieve parity on a lamp-efficacy basis.

10.4.3.9 Commissioning, Control Systems, and Monitoring

Commissioning is the process of systematically checking that all of a building's systems – security, fire, life and safety, HVAC, lighting, electric, etc. – are present and function properly. It also involves adjusting the system and its controls to achieve the best possible performance. Commissioning costs about 1–3% of the HVAC construction cost, but in

---

19 Calculation based on 43°C quantity of hot water conversion (by Institute for Building Environment and Energy Conservation in Japan); primary energy and CO_2 emission intensity for electricity are based on Japanese Law.

20 Light is a form of energy. However, the efficiency of a lamp in producing light cannot be specified as the ratio of watts of light output to watts of electrical energy input. This is because not all watts of light are equal; our eyes are more sensitive to some wavelengths of radiation than to others.

the United States, only 5% of new buildings are commissioned (Roth et al., 2002). Consequently, the control systems never operate as intended in many buildings. Even if a building is commissioned after construction, it is important to continue adequate monitoring of the HVAC system. In a program involving over 80 buildings mentioned by Piette et al. (2001), improved controls reduced total-building energy cost by over 20% and combined heating and cooling energy costs by over 30%.

10.4.3.10   Appliances, Consumer Electronics, and Office Equipment

Residential appliances and office equipment are important uses of energy in their own right and, for office equipment and electronics in particular, can be an important source of internal heat gain that needs to be removed by the cooling system during summer from perimeter offices and year round from internal offices.

Appliances, consumer electronics, and office equipment in buildings are responsible for a large share of a building's electricity consumption and a share of natural gas use. In the United States they account for 40% of the electricity consumption and 12% of the natural gas use of buildings (US DOE, 2008; see also Chapter 1).

The energy use of most appliances used in buildings can be significantly reduced through increased equipment efficiency. Energy efficiency has increased gradually over time, as manufacturers develop new efficiency design options and incorporate them into their products as an enhancement of quality and performance.

New systems designs will reduce the energy requirements of energy services. For example, daylighting and controls or passive building shell design measures reduce the energy required for thermal comfort. Further significant reductions in energy requirements for energy services can be identified by examining specific technologies (IEA, 2008a; National Research Council, 2009). Some cross-cutting technologies offer significant promise for reducing energy use for devices providing a range of energy services. Examples include: electronics, computing and office equipment, display technologies (e.g., organic LEDs), motors (e.g., brushless DC permanent magnet rotors or variable reluctance motors), domestic refrigeration, and ceiling fans (Garbesi and Descroches, 2010). Sensors and controls for energy using devices deserve special attention, since they have many applications that can eliminate waste by limiting energy use to times and places when occupants are present and in need of the specific energy service.

Electronics and Computing and Office Equipment

Since the product life cycles of electronic products are typically months to years, much shorter than appliances or HVAC equipment, rapid improvements in energy efficiency of these products is possible. Currently, large amounts of energy are consumed while only a small capacity of devices is being utilized. Power management can reduce energy use by turning off unneeded capacity. For example, implementing sleep modes in

computers can reduce power consumption by as much as 5W to 65W. Ultimately, energy use can be minimized by proxying, in which a variety of internet connected devices maintain an internet presence while in sleep mode, unlike the current situation where those devices are kept in a higher power mode only in order to remain connected to the internet (Nordman and Christensen, 2010). Servers are designed for maximum demand, and thus are usually underutilized. Virtualization involves software that allows one server to take on the functions of several, while the others reduce power. Virtualization seems applicable to 80% of servers, with potential energy savings of 70%. Computer components are also expected to become more efficient. Laptops are efficient in order to maximize battery life. Replacing desktop computers with laptop technology provides large energy savings. Replacing disk drives with solid state drives could reduce that component's energy use by 70% or more.

Display Technologies (e.g., televisions and monitors)

Energy savings of 50–70% are possible from active power management, such as controlling the brightness of display technologies based on ambient lighting, only operating in the presence of a viewer, and lower brightness for dark colors (color content). In plasma TVs, energy use can be reduced with new phosphors, improved cell design, improved gas mixtures, and optimized electronic circuits. In current TV designs, significant light loss occurs from the backlight to the viewer. Future designs may eliminate backlighting completely, using organic light emitting diodes (OLEDs) that produce light themselves, currently in use in some mobile phones. Prototypes larger than 50 inches (diagonal screen size) have been demonstrated. Alternative technologies under development include efficient electronic circuits, quantum dots, and laser display phosphors.

Motors

Motors come in many sizes and have many residential and commercial applications, making them one of the most ubiquitous components in energy-using equipment. Most motors are single-speed induction. The most efficient on the market are brushless DC permanent magnet motors (BDCPM), which avoid rotor magnetization losses and have lower heat losses, providing a secondary benefit of reduced cooling energy. BDCPM motors require less material and operate at higher efficiency when under low loads than single-speed motors. Major reductions in energy use by motors could be achieved with a BDCPM having interior magnets in the motor, advanced core design, low resistance conductors, and low friction bearings. Further savings would result by including variable speed in the design.

Domestic Refrigeration

From 1974 to 2006, electricity consumption per new top-mount refrigerator-freezers in the United States declined by 70% due to technological improvements and mandatory energy efficiency standards. Changes included using polyurethane foam instead of fiberglass for insulation, improved compressors, improved heat exchangers, and better controls. Notably, after each major efficiency improvement, new approaches identified even further opportunities to reduce energy use. Today, the most efficient models available use 30% less energy than the maximum allowed. Research is underway on alternatives to vapor compression, including magnetic and thermo-acoustic refrigeration. The potential energy savings are not yet known.

Ceiling Fans

Fans suspended from the ceiling can provide thermal comfort in summer in some climates without compressor-driven air conditioning. Most fans have a light attached. A typical fan with a shaded pole motor has a power draw of 35W, with incandescent lighting requiring an additional 120W. An Energy Star[21] fan requires 30W, and its fluorescent lights require 30W. The most efficient model has a motor that requires 10W. A conceivable design with a DC motor, airfoil shaped blades and LED lights could draw as little as 5W for the fan, and 10W for the lights.

In large spaces, mostly commercial applications, substituting one large fan for several smaller ones results in higher efficiency. The same volume of air can be moved with larger blades at lower speed, and a more-than-linear decrease in energy use.

10.4.3.11   Energy and Greenhouse Gas Mitigation From Advanced Biomass-Based Cooking Technologies

Improved cookstoves have been disseminated since the 1980s to reduce demand for solid biomass fuels and to reduce health hazards related to cooking with open fire. Much experience was gained from early programs, which were generally not very successful. An exception is China, which put in place 250 million improved cookstoves as part of one of the largest social programs in the world. As a result of this learning process, a whole new generation of advanced biomass-based cookstoves, and dissemination approaches have been developed in the past ten years and the field is now bursting with innovations. An estimated 820 million people in the world are currently using some sort of improved cookstove (WHO, 2009). This new generation of cookstoves shows clear reductions in fuel use, indoor air pollution, and GHG emissions compared to open fires.

Innovations in the biomass cooking field relate to the: (1) technology (cooking devices); (2) dissemination approaches; and (3) monitoring and evaluation of the impacts.

Technology

Advanced biomass stoves (ABSs) for household cooking include: (1) direct combustion stoves; (2) gasifier stoves, (3) biogas, ethanol or other type of processed biomass fuels; and (4) hybrid models, which provide heat for cooking and water heating or other needs.

Direct combustion ABSs are cookstoves that directly burn biomass fuels. They include cookstoves designed to burn fuelwood, agricultural residues, and charcoal. The stoves fall typically into either portable, low-mass stoves (like the Rocket, which Envirofit disseminated in Africa and Asia, and the Darfur stove aimed at refugee camps in Africa) and high-mass chimney stoves (like the Patsari and Onil stoves disseminated in Latin America, and most of the Chinese models). These include improvements in the combustion chamber (such as the Rocket "elbow"), insulation materials, heat transfer ratios, stove geometry, and air flow (Still, 2009). Some models include a fan. As a result, combustion efficiency is greatly improved compared to open fires. The cost ranges between US$10 or less for the simpler models, to more than US$100 for the more sophisticated models. Fuel savings reach from 30% to 60%, measured in field conditions (Berrueta et al., 2008). Indoor air pollution levels are reduced up to 80–90% compared to the open fires in models with chimneys (Masera et al., 2007; Smith et al., 2007). Carbon mitigation has been estimated to range from 1–2 tCO_2-eq/yr for the simpler models, and have been measured in the field to range between 3–9 tCO_2eeq/yr for more sophisticated models such as the Patsari model in Mexico (Johnson et al., 2008).

Gasifier stoves have gone through a major development in the last five years. Stoves using wood-chips as fuel, with or without an electric blower, are commercially available in China and India (Bhattacharya and Leon, 2004). These stoves deliver 1–3kW of power with an efficiency of 35–40%. Major corporate enterprises have been involved in research and testing gasifier stoves – such as Shell and British Petroleum – and Philips is disseminating a model in India (Hegarty, 2006). Available data from lab testing indicates the potential for significant decreases in emissions of GHGs and health damaging pollutants. Programs are initially targeting urban areas, as the capital cost is still high and the stoves need pre-prepared fuel such as chips or pellets. A major challenge for their widespread dissemination so far is the need for standardizing currently used biomass into chips or pellets.

Biogas cookstoves, and stoves using other forms of processed biomass, provide clean combustion, and using animal dung as a feedstock have been disseminated worldwide, particularly in China, India, and Nepal on a large scale (Dutta et al., 1997). These cookstoves have the distinct co-benefits of enhancing the fertilizer value of the dung and serving to reduce the pathogen risks of human waste. Economic barriers include a high initial cost, which can run up to US$300 for some systems, including the digestion chamber unit. However, new designs of biogas digesters reduce the digestion time, increase the specific methane yield, and make use of alternate or multiple feedstocks, such as leafy material and food wastes. This substantially reduces the size and cost of the digestion unit (Lehtomäki et al., 2007).

Ethanol stoves have also been developed and tested in India, Nigeria, and other African countries (Rajvanshi et al., 2007). The Indian stove uses low-concentration ethanol (50%) and the Nigerian stove uses an ethanol-gel to minimize risks of explosion.

---

21   Energy Star means that products meet strict energy efficiency guidelines set by US Environmental Agency. Energy Star qualified ceiling fan/light combination units are over 50% more efficient than conventional units and use improved motor and blade designs (US EPA 2011).

Figure 10.18 | A stove with a thermoelectric unit includes a 2–4W power unit, which helps generate 100 lumens with a LED-based lantern. Source: BioLite, 2011.

Hybrid, or multiple-service, stoves serve multiple energy end-uses. There is currently a rapid development of these cookstoves. Commercially, stoves that provide both cooking and water heating are well established in Brazil (e.g., Ecofogao) and China (many models). A new exciting innovation is the development of thermo-electric modules, which will allow the stove's blower to generate electricity for lighting, either through CFLs or LED lanterns. Pilot units have been tested in Nepal. A modified Rocket model (Figure 10.18) shows significant improvements in stove performance as a result of the blower and is expected to sell for US$30 (BioLite, 2009).

Table 10.7 shows cooking costs for different types of stoves used in India. It can be seen that all the stoves are cheaper than LPG on an annualized cost basis. Some have high capital costs and thus need financial mechanisms or subsidies to foster dissemination.

Impact of Current Dissemination Programs
A second major innovation in the field of ABSs has to do with the production and marketing of the stoves. State-of-the-art manufacturing facilities are now in place that aim for disseminating stoves on a mass scale while at the same time assuring a quality product. Improved stoves, designed to appeal to consumers in various market segments and to be suitable for microfinance mechanisms, have been developed. As a result, several companies now produce over 100,000 stoves/yr. More than 70 stove programs are currently in place worldwide, with recently launched large-scale national programs in India and Mexico, as well as large donor-based programs in Africa.

The market for ABSs has also benefited from the recognition that using multiple fuels, rather than complete fuel switching to LPG or electricity, is the norm in many developing countries (Masera et al., 2000). Therefore, even households that already have access to LPG and kerosene continue using woodfuels, and many times an early-adopter market for ABS

devices provide substantial benefits for the adopting families (Masera et al., 2000).

Overall, ABSs can provide substantial benefits in terms of energy, health, and climate change mitigation. Approximately 2.4 billion people, representing 600 million households, cook with solid biofuels worldwide. Assuming fuel savings from 30–60%, and energy use of 40 GJ/household/yr for cooking with open fires, the technical energy mitigation potential ranges from 7 to 14 EJ/yr. Also, using a unit GHG mitigation of one to four tCO_2-eq/stove/yr compared to traditional open fires, the global mitigation potential of ABSs reaches between 0.6–2.4 $GtCO_2$-eq/yr, without including the effect of the potential reduction in black carbon emissions. Actual figures will depend on renewability of the biomass used for fuel, the characteristics of the fuel and stove, and the actual adoption and sustained used of the ABS.

Future Needs
Critical needs for ABSs include increased R&D, particularly for new insulating materials, as well as robust designs that endure several years of rough use. More field testing and stove customizing for user needs is required. There is also the need for strict product specifications and testing and certification programs. Finally, it is important to better understand the patterns of stove adoption given multiple devices and fuels, as well as mechanisms to foster long-term use.

10.4.4   Incorporation of Active Solar Energy into Buildings

10.4.4.1   Key messages

Potential active solar energy systems include PV panels for generation of electricity, solar thermal collectors for production of domestic hot water, the production of hot water or heating of air for space heating, the production of hot water to operate desiccant or other thermally-driven cooling systems, and the active collection and concentration of sunlight for daylighting. PV panels on all suitable rooftops and façades of existing buildings in a range of developed countries could supply 15–60% of current total electricity demand in these countries.

10.4.4.2   Active solar technologies

The design of low energy buildings incorporates passive solar energy in many forms – passive solar heating, passive ventilation and cooling, and daylighting – and is part of the package that can achieve 75% or greater savings in overall space conditioning energy use compared to conventional local practice. This level of energy use is so low that much or all of the remaining energy demand, or even more, can be met through active solar energy features such as photovoltaic panels on roofs and façades and solar thermal collectors. Thermal energy from solar collectors can be used for space heating, domestic hot water, or solar air conditioning, the

Table 10.7 | A comparison of annualized cost of cooking energy per household.

	Capital cost (US$)	Fuel Price (US$/kg)	Annualized cost for cooking (US$/ household)
Traditional stoves	1.25	0.025	30
Improved combustion stoves	25	0.025	23
Gasifier stoves	40–75	0.075	40–64
Biogas (Family unit)	300	0.0	47
LPG (subsidised)	64	0.49	65
LPG (non-subsidised)	64	0.8	100

Source: Venkataraman and Maithel, 2007.

latter either driving absorption chillers or desiccant-based dehumidification and cooling systems.

Active solar energy systems, while sometimes driving the use of off-site energy by the building to zero and even making the building a net exporter of energy, tend to be expensive at present. This is in contrast to the system level measures discussed above, which can deliver the first 75% or even more of the transition to zero-energy buildings at very low cost or even with a net savings in investment cost (as discussed below).

The potential energy generation from building integrated photovoltaics (BiPV) is large. Gutschner et al. (2001) estimated the potential power production from BiPV in member countries of the International Energy Agency, taking into account architectural suitability (based on limitations due to construction, historical considerations, shading effects, and the use of the available surfaces for other purposes) and solar suitability (restricting the useable roof and façade area to those elements where the solar irradiance is at least 80% of that on the best elements in a given location, defined separately for roof and façade elements). They estimate that 15% to almost 60% of current total national electricity demand could be provided using all available roof and façade surfaces, depending on the country. Thus, systematic incorporation of BiPV into new buildings, and into old buildings when they are renovated and whenever this is feasible, can make an important contribution to electricity supply.

PV modules typically cost between below US$2 and 5 per peak watt of output (US$2–5/$W_p$, see Chapter 11.7), with total system costs including installation typically running from US$4/$W_p$ to US$9/$W_p$ (IEA, 2008c). However, a number of studies have identified ways in which the cost of modules might eventually reach US$1/$W_p$ (Hegedus, 2006; Swanson, 2006) or even US$0.2–0.3/$W_p$ (Zweibel, 2005; Green, 2006). Keshner and Arya (2004) have presented perhaps the most aggressive scenario for future cost reductions, with an installed cost of US$1/$W_p$ or less for various thin film modules. At 5% financing over 20 years, a US$1/$W_p$ installed cost translates into an electricity cost of about US¢4/kWh for 12% efficient modules in sunny locations (2000 kWh/m²/yr irradiance) and US¢6.7/kWh for 7% efficient modules.

BiPV provides a number of benefits beyond the mere provision of electricity. This should be taken into account in deciding whether or not to incorporate PV into a building (Eiffert, 2003). These benefits include the role of BiPV as a façade element, replacing conventional materials and providing protection from UV radiation; providing thermal benefits such as shading or heating; augmenting power quality by serving a dedicated load; serving as backup to an isolated load that would automatically separate from the utility grid in the event of a line outage or disturbance; and reducing power transmission bottlenecks and the need for peaking power plants. The grid and load saving benefits to the power utility are worth about US$100–200/yr/kW of peak power produced. Inasmuch as BiPV is used as part of an aesthetic design element in a building, it can replace rather expensive building materials. The cost of PV modules ranges from US$400–1300/m². By comparison, the costs of envelope materials in the United States that PV modules can replace range from US$250–350/m² for stainless steel, US$500–750/m² for glass wall systems, to at least US$750/m² for rough stone and US$2000–2500/m² for polished stone (AEC, 2002).

10.4.4.2   Net-zero-energy buildings

The 'net-zero-energy building' is taken here to mean a building that generates onsite and exports to the grid an amount of electricity sufficient to offset the amount of electricity drawn from the grid at other times plus other energy (such as fuels for heating) that is supplied to the building. Requiring buildings to be zero-net-energy is not likely to be the lowest cost or most sustainable approach in eliminating fossil fuel use, and is sometimes impossible. Net-zero-energy buildings are feasible only in certain locations and for certain building types and uses.

Highly energy-efficient residential and commercial buildings have a total energy intensity of 50–100 kWh per m² of floor area/yr (compared to a typical value of 200–400 kWh/m²/yr today), which corresponds to an annual average energy flow of 5.7–11.4 W/m². Annual average solar irradiance ranges from 160 W/m² (middle latitudes) to 250 W/m² (in the sunniest regions). If 80% of the roof area is covered with PV modules and converted to electricity with a net sunlight-to-AC electricity conversion efficiency of 10%, and 20% of the roof area collects solar energy that is converted to useful heat (which can be used for space heating, production of domestic hot water, and desiccant cooling and dehumidification) with an efficiency of 50%, the overall capture of solar energy per unit of roof area is in the order of 30–45 W/m². Although additional electricity could be generated from some PV panels on the equator-, east-, or west-facing façades if shading is not an issue, total building energy demand will greatly exceed the solar energy supply in buildings that are more than a few stories high, even if the buildings are highly energy efficient. It is typically single-family or low-rise, lower-density multifamily residential neighborhoods that can become zero net energy users for their energy needs, excluding transportation. Therefore, care needs to be exercised that ill-conceived, uniform, zero-energy housing mandates do not incentivize further urban sprawl that leads to further automobile dependence and growth in transport energy use.

As discussed in Chapter 11, many forms of off-site renewable energy are less expensive than on-site PV generation of electricity, and are thus able to achieve more mitigation and sustainable energy supply from the same limited resources. Although PV electricity is currently expensive, as noted above, there are good prospects for a factor of two reduction in the installed cost of PV systems in buildings. Furthermore, PV must compete against the retail rather than the wholesale cost of electricity produced off-site. On-site production of electricity also improves overall system reliability by relieving transmission bottlenecks within urban demand centers.

In aiming for zero fossil fuel energy use as quickly as possible, an economical energy strategy would implement some combination of: (1) reduced demand for energy; (2) use of available waste heat from industrial, commercial, or decentralized electricity production; (3) on-site production of energy; and 4) off-site supply of C-free energy, taking into account all the costs and benefits and the reliability of various options.

10.4.5 Worldwide Examples of Exemplary High-Efficiency and Zero-energy Buildings[22]

10.4.5.1 Key Messages

Numerous examples exist worldwide of residential and commercial buildings that have annual energy use that is up to two to four times less than that of recently-built conventional buildings in the same jurisdictions. The most dramatic energy savings are seen in heating energy use, where many buildings have been built with heating requirements that are four times less than that for recent conventional new buildings and up to ten times less than that of the existing stock average.

10.4.5.2 Advanced Residential Buildings

Hamada et al. (2003) summarize the characteristics and energy savings for 66 advanced houses in 17 countries. For the 28 houses where the

savings in heating energy use are reported, the savings compared to the same house built according to conventional standards range from 23% to 98%, with eight houses achieving a savings of 75% or better.

Several thousand houses that meet the Passive House Standard – a house with an annual heating requirement of no more than 15kWh/m²/yr, irrespective of the climate – have been built in Europe (Passivehaus Institut, 2009). By comparison, the average heating load of new residential buildings is about 60–100kWh/m²/yr in Switzerland and Germany, but about 220kWh/m²/yr for the average of existing buildings in Germany and 250–400kWh/m²/yr in central and eastern Europe (Enerdata, 2009). Thus, Passive Houses represent a reduction in heating energy use by a factor of four to five compared to new buildings, and by a factor of 10–25 compared to the average of existing buildings. Technical details, measured performance, design issues, and occupant response to Passive Houses in various countries can be found in Krapmeier and Drössler (2001), Feist et al. (2005), Schnieders and Hermelink (2006), and Hastings and Wall (2007a, 2007b), with full technical reports at www.cepheus.de.

Holton and Rittelmann (2002), Gamble et al. (2004), and Rudd et al. (2004) have shown how a series of modest insulation and window improvements can lead to energy savings of 30–75% in a wide variety of climates in the United States. In all three studies, alterations in building form to facilitate passive solar heating, use of thermal mass combined with night ventilation to meet cooling requirements, where applicable, or use of features such as earth pipe cooling, evaporative coolers, or exhaust air heat pumps are not considered. Thus, the full potential is considerably greater. Demirbilek et al. (2000) found, through computer simulation, that a variety of simple and modest measures can reduce heating energy requirements by 60% compared to conventional designs for two-story single-family houses in Ankara, Turkey.

For single-family houses in the hot and relatively humid climate of Florida, Parker et al. (1998) show how a handful of very simple measures – attic radiant barriers, wider and shorter return air ducts, use of the most efficient air conditioners with variable speed drives, use of solar hot water heaters, efficient refrigerators, lighting, and pool pumps – can reduce total energy use by 40–45% compared to conventional practices. These savings are achieved while still retaining black asphalt shingle roofs that produce roof surface temperatures of up to 82°C. Further significant reductions in cooling energy requirements can be achieved through increasing the albedo surface of roofs (Akbari et al., 2008).

10.4.5.3 Commercial Buildings

Many commercial buildings in North America, Europe, and Asia have achieved a reduction of 50% or greater in overall energy use compared to current local conventional practice. A recent survey of such buildings can be found in Harvey (2009). The National Renewable Energy Laboratory (NREL) in the United States extracted key energy-related parameters from a sample of 5375 buildings in a 1999 commercial

---
22 Sections 10.4.5–10.4.7 is a highly condensed discussion drawn from Harvey (2006, 2009, 2010).

buildings energy use survey. They then used energy models to simulate the buildings' energy performance (Torcellini and Crawley, 2006). The results of this exercise were as follows:

- Average total energy use as built (including thermal and electric loads) is 266 kWh/m²/yr;

- Average energy use if complying with the ASHRAE 90.1–2004 standard is 157kWh/m²/yr, a savings of 41%; and

- The potential average energy use in new buildings is 92kWh/m²/yr with improved electrical lighting, daylight, overhangs for shading, and elongation of the buildings along an east-west axis, a savings of 65%.

With the implementation of technological improvements expected to be available in the future, the gross energy use is so small that PV panels can generate more energy than the buildings use, so that many buildings could serve as a net source of energy.

In the United Kingdom, the energy use guidelines indicate that energy use for office buildings is about 300–330kWh/m²/yr for standard mechanically-ventilated buildings, 173–186kWh/m²/yr with good practice (a savings of about 40–45%), and 127–145kWh/m²/yr for naturally-ventilated buildings with good practice (a savings of 55–60%) (Walker et al., 2007).

Voss et al. (2007) presented data on the measured energy use in 21 passively cooled commercial and educational buildings in Germany. The passive cooling techniques involve earth-to-air heat exchangers (nine cases), slab cooling directly connected to the ground via pipes in boreholes or connected to the groundwater (nine cases), and some form of night ventilation (16 cases), along with a limited window-to-wall ratio

(0.27–0.43) and external sun shading. The buildings also have a high degree of insulation and many have triple glazed windows. Nine of the buildings have total onsite energy use of 25–55kWh/m²/yr and 10 had 55–110kWh/m²/yr energy use, compared to 175kWh/m²/yr for conventional designs, so the rate of savings is up to a factor of seven. Three buildings have a heating energy use less than 20kWh/m²/yr and eight have a heating energy use of 20–40kWh/m²/yr, compared to a typical heating energy use of 125 kWh/m²/yr.

In north China, represented by Beijing, typical energy demand of high standard office buildings is 60–80kWh/m²/yr for heating and 30–100 kWh/m²/yr electricity for air conditioning, lighting, and plug loads. In south China, represented by Shenzhen, energy demand is 60–120kWh/m²/yr (all electricity) for all energy uses including office equipment. Design studies have also shown the feasibility of obtaining 50% savings through the use of relatively simple features in office buildings in Beijing (Zhen et al., 2005) and Malaysia (Roy et al., 2005). The measures in both cases involve insulation, shading, advanced windows, energy efficient lighting, and, in the Beijing case, natural ventilation. But measures such as displacement ventilation, chilled ceiling cooling, and desiccants were not considered, so the potential savings are even larger.

The Energy Base building in Vienna, illustrated in Figure 10.19, is another example of energy use in good practice. The glazing on the south façade is slightly overhanging to increase the proportion of diffuse to direct sunlight entering the room, while the incorporation of PV panels and reflective blinds enhances daylighting. In winter, the building combines solar preheating of ventilation air with heat recovery in a novel way, as illustrated in Figure 10.20. Ventilation air flows laterally from the north side to the south side of the building, then is overheated – to above the desired indoor temperature – by the space next to the glazing, which

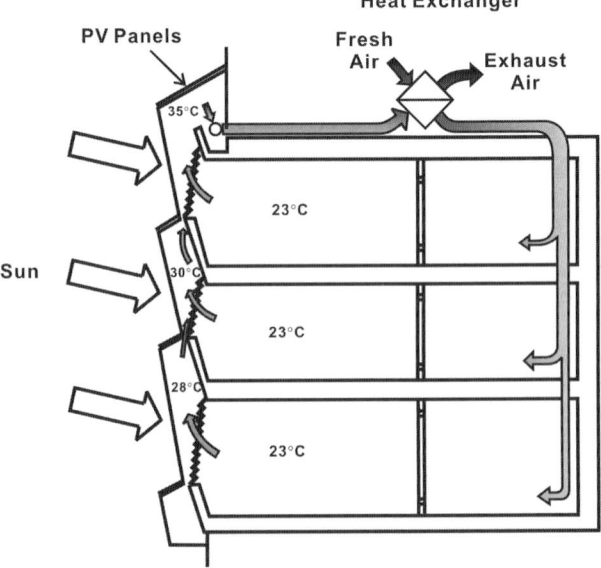

Figure 10.19 | Photograph of the energy base building in Vienna (left) and schematic diagram of the ventilation, solar-preheating, and heat recovery system (right). Source: Danny Harvey (photo); Ursula Schneider, Pos Architekton, Vienna.

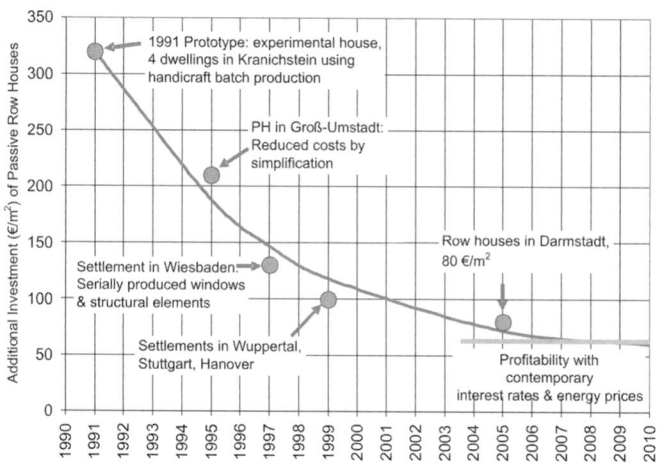

Figure 10.20 | Learning curve showing the progressive decrease in the incremental cost of meeting the passive house standard for the central unit of row houses. Source: Harvey, 2009.

functions as a solarium. This overheated air passes through a heat exchanger, where it is particularly effective in warming the incoming fresh air. At night, the system still has the benefit of heat recovery, unlike other systems for solar preheating of ventilation air. Additional heat is supplied by a ground source heat pump, which cools the ground sufficiently that the ground can be used for passive cooling (i.e., without operating a heat pump) of ventilation air during the summer. Solar regenerated desiccants are used for dehumidification during the summer.

10.4.6   Cost of New High Performance and Zero-energy Buildings

10.4.6.1   Key Messages

The additional cost of residential buildings in Central Europe that meet the Passive House standard has been steadily falling over the past two decades, and is now to the point where the additional costs are insignificant as compared to standard practice in Europe: in the range of 5–8% of the standard construction costs. In the case of high-performance commercial buildings in Europe, North America, and perhaps elsewhere, there is sometimes no additional cost or even cost savings compared to conventional buildings, because the extra cost of the high-performance envelope is offset by reduced costs for mechanical and electrical systems.

10.4.6.2   Residential Buildings

Figure 10.20 shows that in central Europe there is a progressive decline in the cost of the additional investment required to meet the Passive House standard, which uses four to eight times less heating energy than conventional new housing. Through learning, costs have fallen to the point where the incremental cost can be justified based on 2005 energy prices and interest rates. Schnieders and Hermelink (2006) report that the additional

cost averaged over 13 Passive House projects in Germany, Sweden, Austria, and Switzerland is 8% of the cost of a standard house. When amortized over 25 years at 4% interest and divided by the saved energy, the cost of saved energy averages 6.2 €cent/kWh (app. 7.9 cent(US$_{2005}$\$)/kWh; the range is 1.1–11 €cent/kWh – 1.4–14.1 cent(US$_{2005}$\$)//kWh). This is somewhat more than the present cost of natural gas to residential consumers in most European countries, which ranges from 2–8 €cent/kWh (app. 2.7–11.1 cent(US$_{2005}$\$)/kWh) (IEA, 2004). Audenaert et al. (2008) estimate extra costs of 4% for low-energy houses and 16% for Passive Houses in Belgium, having energy savings of 35% and 72%, respectively, relative to current standard houses in Belgium.

10.4.6.3   Commercial buildings

In the case of commercial buildings, the first (initial) cost of highly efficient buildings is sometimes less than the first cost of conventional buildings. This is due to the downsizing of mechanical systems that is possible in energy efficient buildings. As an example of the cost savings with advanced, energy efficient designs (Table 10.8) gives a breakdown of capital costs for commercial buildings in Vancouver, Canada, having conventional windows (double glazed, air filled, low-e with U=2.7 W/m^2/K and SHGC=0.48) and a conventional heating/cooling system, and for buildings with moderately high-performance windows (triple glazed, low-e, argon filled with U=1.4 W/m^2/K and SHGC=0.24) and radiant slab heating and cooling. The high performance building is 9% less expensive to build than a comparable conventional building, while using about half the energy.

Larger energy savings can cost less than smaller energy savings, as indicated by a survey of the incremental cost and energy savings for 32 buildings in the United States by Kats et al. (2003). These buildings meet various levels of the Leadership in Energy and Environmental Design (LEED) standard. Summary results are given in Table 10.9. The energy savings are broken into reductions in gross energy demand and reductions in net energy demand including on-site generation – by, for example, PV modules – which tends to be expensive. The cost premium is the total cost premium required to meet the various LEED standards and so includes the cost of non-energy features as well. Nevertheless, average incremental costs are less than 2% of the cost of the reference building and are smaller on average for buildings with 50% savings in net energy use than for buildings with 30% savings.

Measured performance information on ten buildings in the German SolarBau program where at least one year of data were available by 2003 is given in Wagner et al. (2004). Five of the ten buildings achieved the 100kWh/m^2/yr primary energy target, compared to 300–600 kWh/m^2/yr for conventional designs, but no building used more than 140 kWh/m^2/yr of primary energy. Additional costs are reported to be comparable to the difference in cost between alternative standards for interior finishings.

The final example presented of the beneficial economics of energy efficient buildings is one of the first buildings to be built on the new Oregon

Table 10.8 | Comparison of component costs for a building with a conventional VAV mechanical system and conventional (double-glazed, low-e) windows with those for a building with radiant slab heating and cooling and high-performance (triple-glazed, low-e, argon-filled) windows, assuming a 50% glazing area/wall area ratio. Prices given in 2001 Canadian dollars.

Building Component	Conventional Building	High-performance Building
Glazing	$140/m^2	$190/m^2
Mechanical System	$220/m^2	$140/m^2
Electrical System	$160/m^2	$150/m^2
Tenant Finishings	$100/m^2	$70/m^2
Floor-to-floor Height	4.0 m	3.5 m
Total	**$620/m^2**	**$550/m^2**
Energy Use	180 kWh/m^2/yr	100 kWh/m^2/yr

Source: McDonell, 2003; 2004.

Table 10.9 | Energy savings relative to ASHRAE 90.1–1999 and cost premium for buildings meeting various levels of the LEED standard in the USA.

LEED Level	Sample Size	% Energy Savings, Based on		Cost Premium
		Gross Energy Use	Net Energy Use	
Certified	8	18	28	0.66 %
Silver	18	30	30	2.11 %
Gold	6	37	48	1.82 %

Source: Kats et al., 2003.

Health and Science University, River Campus, and completed in 2006. This 16-story building is expected to achieve an energy savings of 60%, with a reduction in total construction costs of US$3.5 million out of an original budget of US$145.4 million and an operating cost savings of US$600,000/yr (Interface Engineering, 2005).

10.4.7 Renovations and Retrofits of Existing Buildings

10.4.7.1 Key messages

Comprehensive retrofits of existing residential buildings can usually reduce energy requirements, excluding plug loads, by a factor of at least two, with savings in heating energy requirements, which is the dominant energy use in cold climates, by up to a factor of ten. There are many examples of residential buildings, especially multi-unit residential buildings, which have been retrofitted to meet the Passive House standard. Fifty to 70% or more savings in non-plug energy use have been achieved through retrofits of commercial buildings throughout the world.

10.4.7.2 Residential Buildings

Energy use of residential buildings can be reduced through, among other things, upgrading windows, adding internal insulation to walls during renovations, adding external insulation to walls, adding insulation to roofs at the time that roofs need to be replaced, and through taking measures to reduce uncontrolled exchange of inside and outside air and introducing controlled ventilation with heat recovery. Some examples of modest retrofit measures and the corresponding energy savings are:

- the sealing of ductwork alone in houses in the United States saves an average of 15–20% of annual heating and air conditioning energy use (Francisco et al., 1998);

- retrofits of 4003 homes in Louisiana, including the switch from natural gas to a ground source heat pump for space and water heating, eliminated natural gas use and still decreased electricity use by one-third (Hughes and Shonder, 1998);

- an upgrade of multi-unit housing in Germany using, among other measures, External Insulation and Finishing Systems (EIFSs) achieved a factor of eight reduction in heating energy use;

- an envelope upgrade of an apartment block in Switzerland reduced the heating requirement by almost a factor of three (Humm, 2000);

- retrofits of houses in the York region of the United Kingdom reduced heating energy use by 35% through air sealing and modest insulation upgrades, while a 70% savings was projected with more extensive measures (Bell and Lowe, 2000); and

- a comprehensive retrofit of an old apartment block in Zurich, including the replacement of the roof, achieved an 88% savings in heating energy use measured over a two-year period (Viridén et al., 2003).

There are many examples where old buildings that have been retrofitted to very high energy performance standards. Table 10.10 includes cases where 90% or more savings in heating energy use have been achieved. A striking example is the retrofit of a ten-story panel

Box 10.2 | The Solanova Project

The SOLANOVA (Solar-supported, integrated eco-efficient renovation of large residential buildings and heat-supply-systems) project of the European Commission began in January 2003 (see www.solanova.eu). The project goal was to provide best-practice examples of the renovation of large residential buildings in Eastern Europe which, at present, are being renovated with only minimal improvements in energy intensity. In 2005, one seven-story panel building in the Hungarian town of Dunaújváros was renovated as part of this project. Heating energy demand decreased from 220 kWh/m^2/yr before the retrofit to a measured demand of 30 kWh/m^2/yr over a two-year period (a reduction of 86%). Overheating in the summer was one of the worst characteristics of the original building, so triple-glazed windows with an internal Venetian blind were installed on the south- and west-facing walls, resulting in a dramatic reduction in solar heat gain through the windows. Mechanical ventilation was provided to each individual flat with a real heat recovery of 82%. The investment cost was 240 €/m^2 + VAT. The time to pay back the initial investment, based on energy-cost savings only and at current energy prices, is 17 years. However, an unattractive and uncomfortable building was turned into an attractive and comfortable building at the same time. In Eastern Europe, about 100 million people live in large panel buildings, but results demonstrated in Dunaújváros can be transferred to the large stock of Western European panel buildings as well.

Table 10.10 | Documented examples of deep savings in heating energy use through renovations of buildings.

Building and Location	Year Built	Year Renovated	Energy intensity (kWh/m^2/yr)		
			Before	After	Metric
Apartment buildings in Ludwigshafen, Germany	1950s	2001	250	30 (m)	System
Villa in Purkersdorf, Vienna, Austria	Late 19th century	2008	---	20 (m)	System
2 apartment buildings on Tevesstrasse, Frankfurt	1950s	2005	290	17 (c) 13.6 (m)[a]	Load
18-unit apartment block, Brogärden, Sweden	1970	2009	115	30 (c)	System
24-unit apartment block, Zirndorf, Germany	1974	2009	116	35 (c)	Load
Apartment block, Ludwigshafen, Germany	1965	2006	141	18 (m)	Load
50-unit apartment, Linz, Austria	1958	2006	179	13.3 (c)	Load
Apartment block on Magnusstrasse in Zurich	~1900	~2000	165	19 (m)	System
Single-family house, Pettenbach, Austria		2005	280	14.6 (c)	Load
10-story apartment block, Dunaújváros, Hungary		2005	220	20–40	System

[a] Adjusted to an indoor temperature of 20°

building in the Hungarian town of Dunaújváros, of which hundreds of thousands of this similar type building exist in Eastern Europe and the former Soviet Union (see Box 10.2). Various studies indicate that the energy demand in old buildings in western Europe (EU-15) can be reduced by more than 50% with no additional cost over a thirty-year lifetime, and by up to 85% in new countries of the EU-27 (Petersdorff et al., 2005a, 2005b). Today's advanced solutions do not even need to compromise aesthetics: for instance, several historic and heritage buildings have also been already retrofitted to passive house standards.

In apartment buildings with balconies, the balcony slabs are a conduit for heat loss. Glazing the balconies so that they serve to preheat ventilation air, and integrating the balcony with the ventilation system of the apartments, can turn a thermal liability into an asset. Other solar options are transpired solar air collectors over vertically extensive equator facing walls, transparent solar insulation, construction of a second (glass) façade over the original façade, and installation of conventional solar air thermal collectors. Savings of 60–70% in old (per-1950) buildings and 30–40% in new (1970 or later) buildings in Europe have been obtained these ways (Boonstra et al., 1997; Haller et al., 1997; Voss, 2000).

A number of single-family and multi-unit residential buildings have been upgraded to the Passive Standard in Europe. In the case of an old detached house (Haus der Zukunft, 2006) the renovation reduced the heating energy use from 280kWh/m^2/yr to 14.6 kWh/m^2/yr at 16% greater cost than a conventional renovation, but the impact of the extra cost on mortgage payments is less than the energy cost savings. In the case of a 50-unit residential building (Haus der Zukunft, 2007), heating energy use was reduced from 179kWh/m^2/yr to 13.3kWh/m^2/yr at 27% greater renovation cost.

10.4.7.3   Commercial Buildings

Measures that can be taken to reduce energy use in existing commercial buildings include upgrades to the thermal envelope such as the reduction in air leakage, or the complete replacement of curtain walls, the replacement of heating and cooling equipment, the reconfiguration of HVAC systems, the implementation of better control systems, lighting improvements, and the implementation of measures to reduce the use of hot water. The quantitative savings from specific measures depend on the pre-existing characteristics, climate, internal heat loads, and occupancy pattern for the particular building in question. However, large (50–70% or more) savings in energy use have been achieved through retrofits of commercial buildings throughout the world.

Examples of savings achieved through relatively simple measures are:

- projected savings of 30% of total energy use in 80 office buildings in Toronto through lighting upgrades alone (Larsson, 2001);

- realized savings of 40% in heating, cooling, and ventilation energy use in a Texas office building through the conversion of the ventilation system from constant airflow to variable air flow (Liu and Claridge, 1999);

- realized savings of 40% of heating energy use through the retrofit of an 1865 two-story office building in Athens, where low energy was achieved through some passive technologies that required the cooperation of the occupants (Balaras, 2001);

- projected savings of more than 50% of heating and cooling energy for restaurants in cities throughout the United States by optimizing the ventilation system (Fisher et al., 1999);

- projected 51% savings in cooling and ventilation energy use in an institutional building complex in Singapore through upgrades to the existing system (Sekhar and Phua, 2003);

- realized savings of 74% in cooling energy use in a one-story commercial building in Florida through duct sealing, chiller upgrade, and fan controls (Withers and Cummings, 1998);

- realized fan, cooling, and heating energy savings of 59%, 63%, and 90%, respectively, at a university in Texas, roughly half due to a standard retrofit and half due to adjustment of the control system settings to optimal settings (Claridge et al., 2001);

- average realized savings of 68% in natural gas use after conversion of ten schools in the United States from non-condensing boilers producing low pressure steam to condensing boilers producing low temperature hot water, and an average savings of 49% after conversion of ten other schools from high to low temperature hot water and from non-condensing to condensing boilers (Durkin, 2006);

- projected savings of 30–60% in cooling loads in an existing Los Angeles office building by operating the existing HVAC system to make maximum use of night cooling opportunities (Armstrong et al., 2006);

- projected savings of 48% from a typical 1980s office building in Turkey through simple upgrades to mechanical systems and replacing existing windows with low-e windows having shading devices, with an overall economic payback of about six years (Çakmanus, 2007); and

- projected savings of 36–77% through retrofits of a variety of office types in a variety of European climates, with payback periods generally in the one to 30 year range (Hestnes and Kofoed, 1997, 2002; Dascalaki and Santamouris, 2002).

It should be emphasized that comprehensive retrofits of buildings are generally done for many reasons in addition to reducing energy costs. Thus, measures that are extensive enough to significantly reduce energy use may not pay for themselves in terms of energy cost savings alone, but may be feasible when complementing the regular renovation cycle of the building. In addition, accelerated retrofits may make economic sense if done to capture other non-energy benefits such as improved comfort or productivity, energy security, reduced greenhouse gas emissions, and increased employment (Ürge-Vorsatz et al., 2010).

A significant potential area for reduced energy use in existing buildings is through the replacement of existing curtain walls or upgrades of existing insulation and windows. Given the current trend of constructing nearly all-glass buildings, yet not using high-performance glazing, replacing existing glazing systems and curtain walls will be an essential future activity if deep reductions in heating and cooling energy use are to be achieved. Recently, the curtain walls were replaced on the 24-story 1952 Unilever building in Manhattan (SOM, 2010), which indicates that it is technically possible to completely replace curtain walls on high-rise office buildings.

In the case of brick or cement façades, one option is to construct a second, glazed façade over the first to create a double-skin façade, which opens up opportunities for passive ventilation and reduced cooling loads through the provision of adjustable external shading devices. This has often been done in Europe. A North American example of the construction of a second façade over the original façade is provided by the TELUS headquarters building in Vancouver. In this case, the second façade was constructed as part of seismic retrofitting. Construction of a second façade can also be undertaken as a measure to preserve original façades that are deteriorating due to moisture problems related to defects in original construction.

10.4.8   Professional and Behavioral Opportunities and Challenges

As shown in previous sections, energy use is profoundly affected by building design, social norms, and occupant behavior. This section draws on socio-technical studies to focus on two dimensions of the relationship between people, energy, and buildings. It describes how organizations and individuals are central to:

- providing energy efficient buildings and technologies;
- using buildings and energy using technologies in appropriate ways.

For technical solutions to affect energy use, they must be preceded by human decisions about design, purchase, installation, and use. Reducing the amount of energy used in buildings therefore depends on these factors to varying degrees.

The broader relationship between energy service demands and socio-economic factors, particularly in developing countries, is discussed in Chapter 21. This section focuses more closely on the opportunities and challenges of reducing energy used to create comfort, visibility, cleanliness, and convenience in buildings.

Delivering a global transformation to a low energy building stock raises a number of social challenges, including in the professions responsible for the built environment, wider society, and individual building owners and users. A major effort to train the construction sector workforce will be needed in all countries. Lifestyle and management practices consistent with low energy buildings will need to be encouraged. Programs will be required to deliver appropriate education, information, and advice to building designers, constructors, and users. Better feedback to users via smart meters has an important role to play, but alone it is insufficient.

10.4.8.1   Providing Energy Efficient Buildings: a Professional Challenge

Professionals and practitioners in the building industry are essential agents of the transformation towards energy efficient new builds and refurbishments. Section 10.6 sets out scenarios for the global transformation of new and existing buildings to very high efficiency standards that are currently demonstrated but not widely used. The number of buildings requiring transformation implies a huge challenge for the construction sector. This challenge is particularly difficult in the area of housing refurbishment, which is generally fragmented in small- and

medium-sized enterprises. The WBCSD (2009) suggests that a new "system integrator" profession is needed to develop the workforce capacity to save energy. The United Kingdom is training domestic energy assessors to draw up Energy Performance Certificates (Banks, 2008), while the Australian government is vigorously supporting the development of a new profession of in-home energy advisors (Berry, 2009). These efforts are essential in achieving the technical potential described above.

10.4.8.2   Using Buildings and Technologies Differently: A Personal Challenge

This section addresses how people can reduce energy by using buildings and technologies differently. It considers changing lifestyles, habits, norms, and practices, and increasing awareness.

Lifestyles
Substantial reduction, and in some cases even elimination, of energy use is possible by changing lifestyle through changing energy service demands – for example, through higher building occupancies, the use of internal temperatures closer to ambient temperatures, lower lighting levels, and the natural drying of clothes. In some cases, the change can be considered a loss of energy service (or utility), rather than providing the same service more efficiently. But levels of energy service that are considered normal, or even desirable, are a function of culture and lifestyle. Varying lifestyles require varying levels of energy service, both in different societies (Wilhite et al., 1996) and within the same society (Lutzenhiser, 1993). The variation is not only dependent on cost and income, but also on cultural practices.

The threat of climate change adds a new motivation for lifestyle change and has already generated a large number of new information sources, especially carbon footprinting web tools (Bottrill, 2007). Available evidence finds this has yet to have a major effect on most consumers (Lorenzioni et al., 2007), although there is some evidence of it providing a catalyst for community-based activity (Burgess, 2003; Darby, 2006a). However, as the thermal performance of buildings improves due to technical measures, the scope increases for occupant behavior to lead to 'in use' performance deviating (in percentage terms) from 'as designed'.

There are few quantitative studies on the impact of lifestyle change on energy demand in buildings. However, those that exist show potential for modest rates of lifestyle change to produce substantial energy use reductions in the long term, through changes in the use of energy coupled with higher propensities to adopt low energy technologies. Scenarios involving this type of lifestyle change can reduce energy use in buildings by 50% from existing levels in both Japan (Fujino et al., 2008) and the United Kingdom (UKERC, 2009). Dietz et al. (2009) examined the reasonably achievable potential for near-term reductions by altered adoption and use of available technologies in homes and non-business travel in the United States. They found that the implementation of these interventions could save an estimated

20% of household direct emissions or 7.4% of US national emissions, with little or no reduction in household well-being. Similar absolute reductions are not possible in developing countries where energy service demand will grow. However, the rate of growth of energy demand in buildings can be reduced by lower demand lifestyles (Wei et al., 2007). Although lifestyles vary across any society, lifestyle change is strongly affected by social interactions (Jackson, 2009) and therefore may be affected by policy. Chapter 21 contains a further discussion of this topic.

Changing Habits, Practices, and Norms

Even without major lifestyle changes, it is possible in high consumption societies for energy demand to be reduced through minor changes in behavior and conscious control of energy use. The reduction or elimination of wasteful behaviors generally requires increased awareness.

Behavioral changes that might take place include allowing higher/lower internal temperatures within acceptable comfort levels; the better use of shutters, blinds, or other artificial or natural shading (e.g., trees) to prevent unnecessary heat gains; reducing unnecessary air change (e.g., open windows) when heating or cooling; using showers rather than baths; using low water temperatures for clothes washing; naturally drying clothes; not overfilling pans and kettles; switching off lights in unoccupied space; using off, or other low power down states, for unused electronics. Box 10.3 gives two examples in residential and commercial buildings that further highlight social and cultural dimensions of energy use.

As shown in Section 10.1.3, conditioning living space to an acceptable temperature and humidity is generally the largest use of energy. Adaptation to higher/lower internal temperatures is therefore the most important single behavioral issue. Thermal comfort is subjective, variable, and to some extent influenced by previous experience. Recent studies have shown that people in different countries consider themselves comfortable at very different temperatures; that they will accept higher temperatures in naturally ventilated buildings than in mechanically-cooled buildings; and that indoor seasonal and diurnal temperature fluctuations may not be a bad thing. Box 10.4 discusses the concept of adaptive comfort in further detail.

Box 10.3 | Encouraging Adaptive Thermal Comfort in Japan

The importance of culture in determining buildings energy use is well demonstrated by the case of Japan. Japan is one of the most affluent countries in the world, but per capita space heating thermal energy demand per degree-day (heating degree day, or HDD) is significantly lower than in most other developed countries – about 8 GJ/capita compared to an OECD average of ~20 GJ/capita at 2700 HDD (IEA, 2007b). Japanese demand for space-heating is approximately one-third of that of other developed countries.

Beyond the typically much lower per capita floor area (29 m^2 per person compared to the OECD average of 46 m^2 per person (IEA, 2007b)), the reason is that most Japanese homes do not heat the entire living space, but use a modification of a traditional method to provide thermal comfort. In traditional Japanese homes, fire pits or charcoal braziers were generally used to warm the body. The "kotatsu," a direct body-warming apparatus unique to Japan, was common in Japanese houses. A kotatsu is a low table covered with a futon (a heavy quilted cover) placed on a tatami (floor mat). The inside of the futon is warmed by an electric heater attached to the bottom surface of the table. People sitting on the tatami put their feet under the table for direct body warmth. People can live comfortably in a room using a kotatsu even if the room air temperature is low; in many cases, rooms equipped with a kotatsu are heated only to a low temperature, or no heating is used at all. According to a detailed survey of residential energy use, the annual energy use of a house that uses a kotatsu as the main heating apparatus is approximately 40% less than the average use of all the houses studied (Sugihara et al., 2003). Over recent years, Japanese residences have been changing from a low-energy direct body-warming system such as kotatsu to a space heating system. This is one reason why energy use in the Japanese residential sector has been increasing.

Another example of a non-technological measure to reduce energy use is the Japanese "Cool Biz" program. Recognizing the fact that thermal comfort is highly dependent on clothing, which in turn is often determined by culture and dress codes, Japan attempted to change its existing dress code culture to a more sustainable one. In 2005, the Japanese Ministry of the Environment (MOE) promoted office building air conditioning settings of 28°C during summer. As a part of this campaign, MOE has been promoting a new standard for summer dress codes, "Cool Biz," to encourage business people to wear cool and comfortable clothes to work in summer rather than the traditional multi-layered, heavy and dark standard attire that often results in air conditioners having to be set to low temperatures. MOE estimated that CO$_2$ emissions had been reduced by approximately 460 thousand tonnes as a result of the campaign. MOE will continue to promote Cool Biz and higher summer temperatures in offices (MOE, 2006).

Box 10.4 | Focus on Adaptive Comfort

Warmer interior temperatures are acceptable on hot days and colder interior temperatures are acceptable on cold days, if an individual knows what to expect and is accustomed to it. In fact, surveys have shown that individuals typically find acceptable indoor temperatures to be several degrees lower in the heating season than in the cooling season (de Dear and Brager, 1998). The psychological adaptation to warmer/cooler temperatures is enhanced if an individual can *control* his or her environment by being able, for example, to open or close windows, or to activate or deactivate a fan. Research in Denmark indicates that a temperature of 28°C with personal control over air speed is overwhelmingly preferred to a temperature of 26°C with a fixed air speed of 0.2 m/s (de Dear and Brager, 2002). In Thailand, Busch (1992) found that the maximum temperature accepted by 80% of survey respondents is about 28°C in air conditioned offices and 31°C in naturally ventilated offices. Despite this evidence, most air conditioned buildings are operated to maintain a temperature in the lower part of the 23–26°C range, irrespective of outdoor conditions.

Although the percentage savings from increasing the thermostat setting for air conditioning diminish the warmer the outdoor temperature, the implications are substantial. Increasing the thermostat from 24°C to 28°C in summer will reduce annual cooling energy use by more than a factor of three for a typical office building in Zurich and by more than a factor of two in Rome (Jaboyedoff, Roulet et al., 2004), and by a factor of 2–3 if the thermostat setting is increased from 23°C to 27°C for night-time air conditioning of bedrooms in apartments in Hong Kong (Lin and Deng, 2004).

Increasing Awareness

Many countries and regions now have energy efficiency agencies that promote "easy" behavioral actions (e.g., ADEME, 2009; Efficiency Vermont, 2009; EST, 2009). There are a number of approaches that are designed to assist energy users voluntarily to change their own energy using behavior by increasing their awareness of energy issues. The types of measures that fall within this category are: feedback, education, information, and advice.

a.   Feedback – Billing and Metering

Feedback is the provision of information about personal energy use. This may be retrospectively with fuel bills, or in real time through metering and display technology. Reviews by Darby (2006a, 2006b) found that bills and other forms of indirect feedback can produce savings of 0–10%; savings from metering and direct feedback are typically in the range of 5–15%. The persistence of the effect is not well established. The use of smart metering for building energy management is growing. There is no agreed definition of smart metering. In some cases, programs of automated meter reading (AMR), i.e., remote meter reading, have been described in this way (Morch et al., 2007). These are already used extensively in some places, e.g., Italy, Sweden, and the Canadian province of Ontario (IBM, 2007; Haney et al, 2008) and provide clear benefits for energy suppliers through more timely and accurate information (Morch et al., 2007). Energy efficiency benefits within buildings are likely to be limited to those resulting from users making better decisions based on more accurate and frequent bills and incentives like dynamic pricing structure.

Greater involvement of building occupants demands two-way information flows. This is described as automatic meter management (AMM) or advanced metering infrastructure (AMI) (NETL, 2008). Information may be transferred through a range of communication channels, e.g., power lines, mobile phone text message, or radio, with results provided in real time within the building, e.g., via TV screens or internet (Wood and Newborough, 2003; Darby, 2008).

The costs of smart metering are expected to fall and the functionality to improve. Future technology will potentially use electrical harmonics, load profile data, and learn to identify, and feedback individual appliance consumption from aggregate meter data. A driver for smart metering is the need for smart grids to deal with the greater use of intermittent generation in low carbon electricity systems (see Chapter 11). These require electricity retailers and their customers to consider rescheduling of loads when possible, i.e., load switching or demand response measures(Hartway et al., 1999; Vojdani, 2008).This could be via building users responding to time-dependent price signals or by electricity supplier control over loads that are not time critical, e.g., cold appliances. In either case, smart meters potentially facilitate greater involvement of buildings in balancing electricity systems.

To effectively influence user behavior – whether to reduce demand or encourage load switching – information will need to be relevant and easy to assess. While the principle is clear, the exact format of the information that will be most effective is less obvious. Metrics could include energy, cost, or carbon and representations could be numeric or graphical, and could be based on comparisons with past use or other users. Different forms and levels of information will be required for professional users and householders, and may need to reflect other user differences, including culture and building type. More detailed research on the interaction between technology and behavior is required to maximize potential benefits.

b.   Education

Education is mainly targeted at young people. The objective is to provide the knowledge and skills about energy use that will allow them to make informed choices as energy users. Energy and environment form part of the curriculum in many countries, but not in all countries and not yet at adequate levels. A recent review of environmental programs around the world found that, although environmental education is growing, energy and energy efficiency are under-represented in national and international environmental educational programs (Harrigan and Curley, 2010). There is some evidence that education can be effective in the short term in influencing child and parent behavior (Heijne, 2003), but the lack of longitudinal studies makes firm conclusions about long-term impacts difficult.

c.   Information

Information is the provision of material on energy saving opportunities via government campaigns, the media, and other means, including energy company billing. Information may be combined with motivational content design to change behavior, generally focusing on either cost savings or environmental benefits. There is some evidence that information from public and not-for-profit sources is more trusted and effective than from energy companies (DEFRA, 2005).

d.   Advice

Advice differs from generally available information, as it targets the needs of the individual based on personal circumstances. There are a number of variants including face-to-face, telephone, and, more recently, internet-based advice systems. There is experience of this for energy use in businesses and households (Darby, 2003). Cost effectiveness of advice-oriented programs is generally very good. In one case, cost savings for consumers were 40 times the cost of the program (DEFRA, 2006b). Face-to-face advice is the most effective but least cost effective (Sadler, 2002). Much of the benefit resulting from advice leading to technical change persists longer than simple behavioral change.

10.5   Barriers Toward Improved Energy Efficiency and Distributed Generation in Buildings

10.5.1   Key Messages

Technologies and practices that are cost effective from an engineering-economic perspective are often not widely adopted in practice. The barriers include: lack of or imperfect information, transaction costs, limited access to capital, externalities, energy subsidies, risk aversion, principal agent situations, fragmented market and institutional structures, lack of feedback, administrative and regulatory barriers, and lack of enforcement. This section categorizes the barriers as financial costs and benefits, hidden costs and benefits, market failures, and behavioral, cognitive, and organizational barriers.

Solutions for the observed barriers must address many principal actors and their intermediaries and include increased education and training of professionals and consumers, improved information, pricing policies, and regulations (e.g., building codes and energy efficiency standards).

10.5.2   Introduction

The previous sections demonstrated that there is a broad spectrum of opportunities in buildings to significantly reduce energy demand without compromising the energy service delivered. This chapter attests that many opportunities offer net private or societal benefits. Subsidized energy prices (Kosmo, 1989; Lin and Jiang, 2010) and specific characteristics of the buildings sector – their occupants; agents who relate to construction, operation, maintenance, and use of buildings; and market characteristics – limit the "perfect" and "rational" energy efficiency function of buildings (de T'Serclaes, 2007). The IPCC (2007) concluded that the barriers that prevent many cost-effective energy efficiency and building-integrated distributed generation investments from being captured by market forces in present economic and political environments are especially strong in the built environment. Recent research identifies possible approaches to increase uptake of many of these opportunities (Brown et al., 2008; Brown et al., 2009; US DOE, 2010).

This section reviews barriers and solutions based on literature published since previous assessments, such as the IPCC (2007) and the World Energy Assessment (WEA) (2000; 2004). While the literature has been extensive in accounting for and explaining these imperfections, recent demands have become stronger for the quantification of these barriers to better inform private and public decision making. For instance, expenditure-based, climate change target setting, quantification of GHG reduction potentials at various cost levels, as well as several other policy goals require an understanding of the quantified importance of these barriers, the monetized impact of some of them, and an assessment of how much of these indirect costs can be prevented by policies. This section also aims to summarize quantitative estimates of these barriers so that their impacts can be incorporated into estimates of energy saving potential, in addition to incorporating co-benefits of energy efficiency (see Section 10.7). Different barriers are grouped according to typology as presented by the IPCC (2007) and summarized in Table 10.11.

10.5.3   Financial Costs and Benefits

Energy issues remain a low priority to most building owners and occupiers because energy is a relatively small part of the total costs in commercial and residential sectors (WBCSD, 2009). While other investments are subject to risk assessment, energy budgets for buildings are rarely analyzed (Jackson, 2008). Financial barriers to the penetration of energy efficiency and building-integrated distributed generation technologies include factors that increase the investment costs and/or decrease savings resulting from the improvement. These factors result in prolonging the payback time and downsizing the internal rate of return on investments. These factors include high initial costs of advanced technologies, costs associated with risks of implementation and financial operations (transaction costs), high discount rates for households and commerce,

Table 10.11 | Taxonomy of barriers that hinder the penetration of energy efficient technologies/practices in the buildings sector.

Barrier categories	Definition	Examples
Financial costs/benefits	Ratio of investment cost to value of energy savings	Higher up-front costs for more efficient equipment Lack of access to financing Energy subsidies Lack of internalization of environmental, health, and other external costs
Hidden costs/benefits	Cost or risks (real or perceived) that are not captured directly in financial flows	Costs and risks due to potential incompatibilities, performance risks, transaction costs etc.
Market failures	Market structures and constraints that prevent the consistent trade-off between specific energy-efficient investment and the energy saving benefits	Limitations of the typical building design process Fragmented market structure Landlord/tenant split and misplaced incentives Administrative and regulatory barriers (e.g. in the incorporation of distributed generation technologies) Imperfect information
Behavioral and organizational non-optimalities	Behavioral characteristics of individuals and organizational characteristics of companies that hinder energy efficiency technologies and practices	Tendency to ignore small opportunities for energy conservation Organizational failures (e.g. internal split incentives) Non-payment and electricity theft Tradition, behavior, lack of awareness, and lifestyle Corruption Lack of enforcement/implementation/monitoring

Source: Carbon Trust, 2005.

lack of or limited access to financing, cost of capital, and energy subsidies that do not allow for estimating the real energy cost savings (IPCC, 2007; de T'Serclaes, 2007).

Table 10.12 illustrates that transaction costs, high discount rates, and lack of real-time pricing are studied to the largest extent in the financial group of barriers. Based on the information available, the financial barriers are concluded to be very strong. For example, the transaction costs of energy efficiency and renewable energy projects in the buildings sector reach as high as 20% of investment costs.

Higher investment costs of efficiency technologies require significant investment in research as well as government programs to push market development further and faster. Since energy codes are relatively new in several developing nations, green products and services including insulation, CFLs and T5 lamps, efficient glass, and efficient HVAC systems – required by buildings to comply with some code requirements – are not readily and abundantly available or competitively priced.

Consolidation of the majority of global production in a handful of multinational manufacturers for each product type creates an opportunity for a well-designed set of policies, including voluntary labels, mandatory performance regulations, and financial incentives, to rapidly increase the production of energy efficient products. The energy efficiency standards and the Energy Star program for some products provide examples where most production has shifted toward higher efficiency, having several sequential updates for some products. Updating those policies and continued R&D are necessary to maintain the long-term trend in decreasing energy use and costs per product to support sustainable development in the building sector. For building retrofits, accessible mechanisms for providing information and capital are needed.

10.5.4 Market Failures

In traditional economic analysis, market failures refer to flaws in the ways that markets operate in practice compared to theoretically perfect markets. They are violations of one or more of the neoclassical economic assumptions that define an ideal market, such as rational behavior, costless transactions, and perfect information (Brown, 2001; Jaffe and Stavins, 1994). These failures are caused by misplaced incentives; administrative and regulatory barriers; imperfect information; unpriced environmental, health, social, and other external costs and benefits; fragmented market structure; and limitations of the typical building design and construction process. Decisions that result in the energy performance of buildings are fragmented, being made by building owners, architects, craftsmen, and occupants. Their motivations and opportunities for efficiency differ. Recently, social science gained new insights concerning positive drivers for efficiency and renewables – e.g., social prestige using the image of being socially responsible or "green," education, and social networking.

As Table 10.12 shows, the impacts of market failures are rarely measured, with the exception of misplaced incentives which have been covered recently by a few publications. Meier and Eide (2007) estimated that up to 100% of energy services and up to 80% of primary energy use of buildings are affected by misplaced incentives. If the barrier is removed, the energy savings, for instance for space and water heating, may reach 50–75%. The impact of administrative and regulatory barriers is difficult to measure, but researchers attest that there are a number of distorting policies against installing distributed generation (see, e.g., Brown, 2001). Methodologies exist for monetizing externalities, but final consumers have little information or incentive to undertake the level of effort required to include external costs and benefits in their decisions.

Table 10.12 | Selected barriers to GHG mitigation in the buildings sector and methodologies for their quantification in the literature.

Barriers	Case study	Methodologies used for quantification	Quantified impact of the barriers	References
Financial (including hidden) costs and benefits				
Transaction costs	World, India (IN), Sweden (SE), United Kingdom (UK), United States (US)	Empirical survey Literature review Interviews Review of official documentation Bottom-up estimates	SE: ≤ 20.5% and ≤ 14.4% of project costs for energy efficiency and renewable energy projects in buildings respectively IN: about 100 US$/t CO_2 in CDM projects targeting EE in housing sector World: ≤ 100 €/tCO_2 for CDM on EE in buildings World: €30,000 – €100,000 per JI/CDM project (almost for any project size) UK: 30% and 10% of investment costs for cavity wall insulation and CFL respectively US: information cost, vendor information cost, and consumer preferences add US$10, US$5, and US$5 respectively to the CFL price that was US$10/piece	Krey, 2005; UNFCCC, 2002;Michaelowa and Lotzo, 2005;Mundaca, 2007; Sathaye and Murtishaw, 2004; UNIDO, 2003
Limited access to capital	Australia (AU)	Case study	AU: In the case of energy efficient mortgage, stretching the credit ratio[a] by 2% allows a further 5% of loan application to succeed. However, as financial institutions frequently avoid softening the loan conditions, it may be concluded that these 5% loan application fail postponing or cancelling the intended energy efficiency projects	de T'serclaes, 2007
Lack of real-time pricing	US, Japan (JP), Canada (CA), Netherlands (NL)	Case study on residential energy demand feedback devices Questionnaires Statistical analysis (Chi-square test)	USA: energy savings from providing real-time energy feedback is about 10–15% JP: feedback devices caused a 18% electricity saving and 9% gas saving CA: feedback displays produced electricity savings of 13% NL: daily feedback led to a 10% reduction in household gas consumption USA, CA: electricity savings totaled to 5% in Quebec and to 7% in California	Parker et al, 2006; Ueno et al, 2006; Dobson and Griffin, 1992; Van Houwelingen and Van Raaij, 1989; Hutton et al, 1986.
Behavioral: high discount rates (DR)	US, NL	Present value analyses Option value method Capital asset pricing model (CAPM) Discrete choice models Econometric models	US: DR is 5.1% – 89%/yr. in an inverse order of household income; average 20%/yr. US: DR < 15% a choice is a CFL; > 15% – an incandescent; = 15%/yr – indifferent US: DR = 21%/yr. at electricity price of 4.8 ¢/kWh; 27–28%/yr. – 6 ¢/kWh, and 30%/yr. – 7.5 ¢/kWh NL: high-efficiency durables to be bought at DR 0% – 1.4%/yr. or, in a more myopic case, 8% – 101.3%/yr; low-efficiency durables – at DR 1.4% – 8%/yr. or, in a more myopic case,> 101.3%/yr.[b]	Hausman, 1979; Thompson, 1997; Sanstad et al, 1995; Kooreman and Steerneman, 1998

Market failures

Market fragmentation	US	Empirical study	US: Uncaptured social returns from R&D are ≥ 50% than estimated private returns because of the fragmentation within the sector	CEA, 1995
Uncertainty in future prices	US	Economic model of irreversible investment	USA: Investment in energy efficiency under uncertainty falls from 25%/yr. to ~ 1%/yr. and is less than 5% after 20 years	Hassett and Metcalf, 1993
Misplaced incentives (principal agent problem)	US, Norway (NO), NL, AU, JP	Survey Authors' estimation Top-down analysis Case study	NO: the problem in the commercial offices affects around 80% of the energy use; energy use is ca 50 kWh/m2 higher in leased office space; the potential energy savings are 3.24 to 5.4 PJ/yr., or 15% of the total energy use in the Norwegian commercial sector US: savings from 2003 sales of water heaters would amount to 9.6 PJ/yr. of final energy and 12.6 PJ/yr. (about 7%) of primary energyuseif PA problems were solved NL: after removing the barrier, ca 20% of houses can be additionally insulated with an energy saving of 50%-75%/house for space heating (2–4 PJ/yr.) NL: The energy use affected by problems is more than 24.5 PJ per year, which accounts for over 40% of total energy use in commercial offices JP: energy use in commercial offices affected by the problem is 0–1.5% of total national electricity consumption	Meier and Eide, 2007;Ecofys, 2001;OECD/IEA, 2007

Behavioral and organizational non-optimality

Lack of understanding	US	Case study, regression analysis, survey	US: energy savings amounted to 10% due to campaign (ca $130,000/yr.)	McMakin et al., 2002

[a] The ratio, expressed as a percentage, which results when a borrower's monthly payment obligation on long-term debts is divided by his or her net income or gross monthly income (Business Dictionary Online, 2008)

[b] Difference: more myopic consumers are more reluctant to buy EE goods (this is true for both high- and low-efficient durables). On the other hand, consumers are more myopic in case of low-efficiency durables.

Box 10.5 | Design Challenges of Efficient Buildings

In developing countries located in tropical regions, most of the residential buildings and several commercial building types are still designed as unconditioned spaces. Thus, the construction methods do not address issues of proper weather stripping, infiltration control or use of appropriate building materials, such as insulation or doubling glazing. However, changing climate and lifestyles have triggered retrofit of such buildings with window/packaged air conditioners. This leads to inefficient energy use.

Designing efficient buildings typically leads to a reduction in cooling loads and lighting loads and thus reduced sizes for installed system, with an imminent reduction of capital expenditure on systems. Typically a consultant's remuneration is a percentage of the capital cost, and thus a consultant may not be positively inclined to reduce the capital cost of a project, if there is no provision for added incentive for doing so. Designers and contractors are often slow to adopt energy efficient designs due to inertia or lack of training.

10.5.5   Behavioral and Organizational Non-optimalities

There are different behavioral and cognitive characteristics of individuals and organizational characteristics of companies that hinder energy efficiency technologies and practices in buildings. Perhaps the strongest barrier in this category is a major lack of technical, economic, and general knowledge related to low-energy buildings (see Box 10.5). This knowledge gap exists not only among building designers and architects, but also among politicians, investors, tenants, and consumers.

Beginning with the design side, there is a lack of knowledge among designers of how to best incorporate efficiency practices into building design to meet or exceed building code requirements. This is due to the novelty of energy efficiency concepts in buildings, especially in developing countries until recently. Furthermore, the capability of designers and architects to perform energy simulations to quantify the potential savings from energy efficiency is also very limited in developing countries.

Even once some efficient buildings have been designed effectively, the building construction industry is not generally prepared to apply these measures practically on site and remains largely unaware of the environmental impacts of its operations. Actual energy performance is generally not reported back to designers, builders, or contractors and they are not held accountable for inefficient designs or poor construction practices.

There are very few university programs in architecture and engineering in which curricula include energy efficiency issues in buildings. A report in the United States on workforce education and training identified 43 (of 492) higher education and/or training programs that meet minimum criteria of a specific emphasis on energy efficiency (Goldman et al., 2009). Due to the fact that courses on energy efficiency topics are not always compulsory, the number of graduates who have studied these subjects is even lower.

Insufficient education for building professionals causes the problem of ineffective communication with specialists in related fields and lower production in the construction process. Gallaher et al. (2004) estimated that the costs caused by ineffective communication between different stakeholders in the US building industry, including architects, engineers, general constructors, suppliers, owners, etc., is about US$15.8 billion. Improvement of building professionals' communication skills and ability to use modern software and electronic systems for the exchange of information can reduce such costs. In addition to the education of architects and effective communication of opportunities to consumers, an adequate trained workforce is needed to install and maintain efficient buildings and equipment. The training of a modest workforce (about 24,000 jobs) to commission non-residential buildings is expected to yield significant benefits, including US$30 billion in energy savings by year 2030 and annual greenhouse gas emissions reductions of about 340 megatonnes of carbon dioxide (Mills, 2009). Studies of energy service companies in the United States estimate the 2008 workforce to be about 120,000 person/yr equivalents (PYE), or equivalent to about 400,000 employees, and it is expected to grow by a factor of two to four by 2020, if they can overcome a number of key challenges (Goldman et al., 2010).

For designers and architects to shift their design and building practices, there needs to be a demand for efficient buildings from investors, developers, owners, and building occupants. This demand is currently low in developing countries due to several factors. There is a lack of knowledge about green investments and returns on efficient buildings. The value of energy efficient designs are underestimated by appraisers and reduced energy bills are not generally considered by mortgage lenders. Most consumers are also unaware of the comparative costs of the future operation and maintenance of buildings, and therefore they do not take building efficiency into consideration when making purchasing decisions. Because of this, most builders do not consider the future costs of operating the building during design and construction. Finally, there is a lack of awareness of the financial, environmental, and health benefits of operating buildings efficiently.

Also, many consumers are still unaware of the availability of green products and energy efficiency labeled products. With attention primarily on purchase costs and without adequate consideration of the lifecycle costs and benefits of efficient products, the perception of higher upfront cost limits purchases of efficient products. An owner who wants to improve his or her house may find it difficult or impossible to get a proper offer from a construction firm on how to refurbish a building with a focus on the whole house and its energy efficiency.

Decisions about installing original or replacement appliances, lighting, and equipment (e.g., heating and cooling, water heating) in buildings face similar barriers. The situation is improved by the existence of energy labels, and more so by minimum energy performance standards and building codes that mandate cost-effective efficiency levels.

10.5.6 Barriers Related to Energy Efficiency Options in Buildings in Developing Countries

Table 10.12 presents an assessment of impacts of different groups of barriers on the scope or the costs of energy conservation potential. Analysis of Table 10.12 finds that the research has extensively covered only a few barriers, namely transaction costs, lack of real-time pricing, and the principal-agent problem, and only in developed countries. Table 10.12 also illustrates that studies use different indicators to measure the impacts of different barriers, making it difficult to compare these impacts or to estimate the overall aggregated effect of the barriers on energy conservation potential in the built environment and its costs and benefits. Therefore, unification of the methods is important to have a comprehensive analysis. This section attempts to look at the barriers through a regional lens.

Major barriers to energy efficiency improvements in the building sector in developing countries include lack of awareness of the importance of, and the potential for, energy efficiency improvements, lack of financing, lack of qualified personnel, and insufficient energy service levels (Ürge-Vorsatz and Koeppel, 2007). Also, negative experiences with energy efficient equipment such as in the case of some low cost CFLs that fail prematurely, can pose barriers. The biggest building market – single-family homes – is often unorganized and outside the control of local authorities. When homes are part of the informal sector, building codes or standards are not applied. In Brazil, for example, 75% of the residential sector falls under the informal category.

Subsidized energy prices are another strong barrier in many developing countries. However, these subsidies enable access to minimal energy service levels for certain population groups, which means that removing subsidies may be socially difficult and undesirable. In these cases energy efficiency programs may be especially important because improved efficiency can either reduce the need for public subsidies or enable elevated service levels and the more effective use of subsidies. In countries or regions with a lack of access to reliable energy supply, such as parts of Africa, the priority of governments may be to improve access to energy for inhabitants rather than to improve

energy efficiency. In such cases, renewable energy projects and rural electrification often play a more important role for governments than energy efficiency. A scenario for implementing cost-effective efficiency for electricity in India, primarily through end-use technologies in buildings, is expected to reduce government payments of energy subsidies, to involve lower capital costs than the costs of new supply, to eliminate the chronic electricity supply shortage in a few years' time, and to improve the national economy by increasing the availability of electricity for commercial and industrial enterprises (Sathaye and Gupta, 2010).

The increased influence of western architecture, such as glass-dominant structures for commercial use, is very common in India. Being in primarily a cooling-dominant climate, this often leads to large cooling loads and hence increased energy demand. Also in developing countries, the regulatory frameworks for the implementation of energy efficiency in buildings are often inadequate. In India, for example, while there are regulations – such as environmental clearance of large construction projects by the state or central environment departments or ministries – implementation and monitoring mechanisms are inadequate.

Lack of knowledge among architects and system providers to incorporate energy efficiency is another major barrier. Energy efficiency in buildings is not taught as a part of the curriculum in most schools of architecture. Another key barrier is inadequate availability of products and services related to energy efficient buildings, which often leads to a monopoly of a few providers and thus higher costs. The absence of suitable financial products, such as low interest loans for energy efficient buildings and robust energy performance contract mechanisms to offset and/or build confidence in incremental costs, is a major barrier that hinders the penetration of energy efficiency in buildings.

10.6 Pathways for the Transition: Scenario Analyses on the Role Of Buildings in a Sustainable Energy Transition

10.6.1 Key Messages

This section presents scenarios for future regional and global energy use in the buildings sector that meet the multiple objectives outlined in the GEA. The energy demand scenarios developed here served as input to and have been harmonized with the main assumptions in the GEA transition pathways presented in Chapter 17. The scenarios demonstrate that reducing approximately 46% of final thermal energy use in buildings is possible by 2050, as compared to 2005. This is achievable through the proliferation of existing best practices in building design, construction, and operation, as well as accelerated state-of-the-art retrofits. This is attainable in concert with increases in thermal comfort, the elimination of energy poverty, economic development, and living space increases in some regions. Realization of this potential requires undiscounted cumulative investments of approximately US$14.2 trillion (or US$18.6 trillion without technology learning) by 2050. At the same time, these costs will be substantially recovered by the approximately US$58 trillion in undiscounted energy cost savings during the same period. However, scenarios also show a great risk of lock-in effect – about 80% of thermal energy savings can be locked in the global building sector if suboptimal solutions continue to be pursued.

New appliances, IT, and other electricity using equipment also have significant potential for energy use reduction – up to 65% by 2020 in relation to the baseline – due to the worldwide utilization of present and foreseen cutting-edge technologies.

10.6.2   Description of the GEA Building Thermal Energy Use Model

Providing thermal comfort in buildings contributes significantly to global energy use, yet this energy end-use sector is also the most poorly understood one. Little data and detailed information exist related to the heating and cooling of our buildings. Consequently, few global models exist. The model in this report is a newly constructed one prepared for the GEA pathway analysis, but is built on earlier results as well as present state-of-the-art work in progress, using a novel approach.

The energy demand scenarios developed here served as input to and have been harmonized with the main assumptions in the GEA transition pathways presented in Chapter 17.

The building thermal energy use model constructed for the GEA is novel in its method, as compared to earlier global world energy analyses. It reflects a new emerging paradigm that builds on an emerging approach to building energy transformation: one that takes advantage of the fact that buildings are complex systems rather than sums of components. This holistic approach is based on a performance-oriented concept of building energy use, as opposed to a component-oriented approach. It also focuses on providing energy services rather than energy per se. Applying this approach to building energy saving potential assessment typically results in much higher energy saving potentials than earlier approaches, which do not integrate systemic opportunities or opportunities emerging through focusing on the provision of energy services. However, this approach is consistent with the empirically observed opportunities presented earlier in this report for various technologies and know-how, based on the savings that are possible by treating buildings as systems rather than as sums of separate components. Electricity use by appliances (except cooling appliances) and other plug loads is treated separately and does not require the consideration of system level savings opportunities. These energy uses (in contrast to cooling), are more complex from modeling purposes and do require the consideration of system level opportunities; thus, they have been modeled using conventional approaches in a separate module.

The following subsections describe the models of thermal energy use and electric plug loads and their results. The driving questions were:

* How large of a role can buildings play in an energy transition for sustainability?

* How far can buildings take us in mitigating climate change and addressing other energy-related challenges outlined in this report?

The scenarios presented here analyze pathways in which energy efficiency in buildings is pushed toward the state-of-the-art, but do not extend to assessing building-integrated sustainable energy generation options such as renewables, or to assessing the role of lifestyle/behavioral changes, due to time and other constraints.

They represent feasible deployment potentials (i.e., techno-economic potentials that also consider deployment constraints), assuming a very strong supporting policy framework globally. It is important to note, however, that the building scenarios share the fundamental philosophy of the GEA in that they presume the increase in the thermal energy service to satisfactory levels to all populations worldwide, i.e., they assume that fuel poverty is eliminated and sufficient thermally conditioned minimal living space is provided for all by the end of the modeling period. Originating from the nature of these end-uses, the timeframe for these projections is different. In contrast to other GEA pathways, the building scenarios only extend until 2050 since, due to the shorter lifetime and high changeability of the equipment covered, any projections beyond the midterm become extremely speculative and thus lack robustness.

10.6.2.1   Model Design and Novelty

Prior models of building energy use or mitigation opportunities have focused on individual building components or the equipment used for heating and cooling, or other end-uses and alternatives to these that can save energy. One example of a model mastering this approach is the Bottom Up Energy Analysis System (BUENAS) of the Lawrence Berkeley National Laboratory, focusing on appliance efficiency. The model projects increases in appliance – space heating and cooling systems included efficiency on total final energy use. This model was used for the plug load part of the GEA pathways. (The BUENAS model is described in more detail in Section 10.6.3.)

A new thinking has been emerging recently in building energy science and analysis that represents a shift of paradigm – a system-based, performance-based, holistic approach. This replaces the component-based, piecemeal approaches of earlier efforts. The new approach recognizes that buildings are more complex systems than just the sum of their components, and that there are many synergistic opportunities and trade-offs, too. It also recognizes that the same levels of energy performance can be obtained through different pathways – i.e., different packages of energy-efficiency measures, which gives optimal freedom for the constructors and designer to reduce energy use in a particular set of circumstances (Laustsen, 2008).[23] This new thinking is reflected in performance-based building energy regulations – i.e., that specify building codes based on energy use per square meter useful space, or other similar complex systemic performance indicators, rather than those regulating individual building components.

Following this paradigm change, a number of countries and jurisdictions have been revising their building energy codes based on new performance-based approaches (Hui, 2002). These include building regulations in the United States, Canada, the European Union and its member states (i.e., the Energy Performance of Buildings Directive (EPBD)), and Singapore (Hui, 2002).

At the same time, building energy and climate scenario modeling related to buildings has not yet reflected this paradigm change. The GEA building pathway assessment is among the first models using the performance-based approach, and has been developed in close cooperation with the few other ongoing efforts using a similar logic for global building energy modeling (Harvey, 2010; Laustsen, forthcoming).

Another novel aspect of the present model is in that it focuses on providing energy services rather than energy per se. This is reflected in the fact that the model's end goal is to provide adequate thermal comfort for living and commercial floor spaces needed by the population, and first examines options how this can be provided with the least energy input. As a result, architectural and engineering solutions that maximize the thermal performance of buildings are emphasized, often significantly reducing, and sometimes eliminating, the heating and/ or cooling load that needs to be met by energy input, even before a technological solution needs to be applied. Therefore, a focus on energy services rather than energy allows for many non-energy solutions or a larger portfolio of innovative options to reducing energy use, and thus unlocks much larger mitigation options and energy saving potential.

10.6.2.2   Modeling Logic, Structure, and Main Assumptions

As described earlier, the GEA model is grounded in a performance-based logic that considers buildings as entire complex systems and not the sum of their components. Specifically, buildings energy use is not modeled based on individual energy efficiency measures, but are computed based on marker exemplary buildings with measured, documented energy performance levels and associated investment costs. A fundamental thesis of the model is that building energy performance depends less on precise degree days (cooling and heating) and technical

efficiencies of individual devices than on state-of-the-art design, construction, and operation know-how and technology packages, as well as main climate types. The total energy requirement for thermal loads is derived as the product of energy intensity and floor area, summed over all building types, vintages, regions, and climate zones. Thus, key model inputs include floor space developments and specific energy demand values for existing and replicable, economically feasible, exemplary buildings in each region and each climate zone.

The logic of the model leaves it to the creativity of the architect and energy engineers to decide how – through which technologies or design and operational measures – the state-of-the-art performance level is exactly achieved. It assumes that once the selected type of exemplary buildings have demonstrated the feasibility of a certain ambitious level of energy performance and the promise of economic viability in the respective climate and building type, such levels are broadly attainable in that particular climate zone, building type, and vintage. The model then presumes that such state-of-the-art construction and renovation becomes the standard, e.g., through strictly enforced building codes.

Each time a new building is constructed or reaches its retrofit cycle, it is assumed to reach state-of-the-art specific energy demand levels for its category and climate type, after a certain transition period. The transition period is allowed so that markets and industries have ramp-up time for large-scale deployment of exemplary building construction technologies, materials, and know-how, as well as allowing time for the needed ambitious enabling policies to be enacted and the necessary supporting institutional framework to be introduced. The model assumes ten years for this transition period, which is shown by recent literature likely to be a very conservative assumption (Ürge-Vorsatz et al., 2010; in preparation).

Separate final energy intensity levels are specified for different building types (single family (detached or attached), multifamily (four or more levels, terraced, etc.), and commercial and public buildings) in four different climate zones (warm moderate, cold moderate, tropical, and arid) in each of 11 GEA regions (North America (NAM), Western Europe (WEU), Pacific OECD (PAO), Eastern Europe (EEU), Former Soviet Union (FSU), Centrally Planned Asia (CPA), South Asia (SAS), Other Pacific Asia (PAS), Middle East and North Africa (MEA),Latin America and the Caribbean (LAC), and sub-Saharan Africa (AFR)).

The specified energy intensity levels for advanced new buildings and retrofits are based on demonstrated energy and financial performance results in each region, but energy values are adjusted upward to allow for difficulties in achieving the best-observed performance in all cases.

The model distinguishes three different categories of buildings: all three categories of buildings exist in all eleven GEA regions of the world and are then split by four climate types (for regions, climates, and the floor area model see the sections below). The model uses business-as-usual construction and demolition rates. However, as became clear in the first

---

23   Note that with greater flexibility for designers, the performance-based standards require using computer-based models and a deeper understanding of the building principles (Laustsen, 2008). However, there are developments in this area as well, and European Committee for Standardization (CEN) and the International Organization for Standardization (ISO) are developing international standards founded on performance-based approaches (Laustsen, 2008).

modeling runs, retrofit dynamics fundamentally determine the attainability of ambitious sustainability goals in the mid-century, so the model assumes an approximately doubled retrofit rate (i.e., 3% as opposed to the average 1.5%). This requires policy intervention. Most of the model's non-building input data, such as on GDP and population growth, urbanization, and other key macroeconomic parameters are derived from the main GEA pathway work, and as such is consistent with its assumptions on GDP and population growth, urbanization, and other key macroeconomic parameters. These are explained in Chapter 17.

The model estimates the additional investment costs needed for energy efficient construction and retrofits assumed and an estimation of the resulting energy cost savings. The cost values are calculated based on marginal expenditures as compared to standard construction and retrofit investments, while overall energy cost savings are based on comparison with a business-as-usual scenario, taking into account policies in place or in the pipeline. The overall model logic diagram is shown in Figure 10.A.4 in the GEA Chapter 10 Online Appendix.

10.6.2.3   Description of the Analyzed Scenarios

Two scenarios have been elaborated in the presented global building energy use model: state-of-the-art and sub-optimal efficiency scenarios, which are described below.

State-of-the-art Efficiency Scenario
This scenario demonstrates how far today's state-of-the-art construction and retrofit know-how and technologies can take us in meeting the GEA objectives as far as the provision of thermal comfort in buildings is considered, were they to become standard practice after a transition period. These standards are applied to all buildings in their respective categories as they are retrofitted or constructed during the modeling period, except for the small share of heritage buildings where lower efficiency levels can be achieved in renovation.

Sub-optimal Efficiency Scenario
The rationale for this scenario is to illustrate the potential lock-in effect in building infrastructure that can be caused by accelerated major policy efforts (such as the ones currently implemented by many governments and international organizations for climate change mitigation)which do not mandate sufficiently ambitious performance levels. Specifically, the scenario assumes the same accelerated renovation rates as the state-of-the-art scenario, to reflect that many countries recognize the importance of energy-efficient retrofits and energy-efficient building codes, but these accelerated retrofits and advanced new buildings are still built and renovated to far less efficient levels than are achievable according to the state-of-art scenario; thus they are referred to as "suboptimal" levels.

New buildings in this scenario are assumed to be built to the building codes for the region. Only in Western Europe (WEU) is it assumed that

highly efficient buildings are being built in relatively large numbers, but the maximum fraction of these advanced building is only 5% in this scenario. Renovations are carried out to achieve approximately 35% energy savings from the stock average, as opposed to the state-of-the-art savings, which can be as high as 95% in some climate and building types, as demonstrated by best practices.

10.6.2.4   Main Assumptions and Input Data

The model's main assumptions and data sources are briefly described in the GEA Chapter 10 Online Appendix. For a more detailed description, see Ürge-Vorsatz and Petrichenko et al. (in preparation).

The world's building stock is broken into the same eleven regions as are used elsewhere in the GEA (such as Chapter 17), which are presented in the technical appendix. These are: North America (NAM), Western Europe (WEU), Pacific OECD (PAO), Eastern Europe (EEU), Former Soviet Union (FSU), Centrally Planned Asia (CPA), South Asia (SAS), Other Pacific Asia (PAS), Middle East and North Africa (MEA),Latin America and the Caribbean (LAC), and sub-Saharan Africa (AFR).

Specific Energy Demand Assumptions
Table 10.13 shows the specific energy demand figures that are used as an input to the model. It is important to note that it has been extremely challenging to arrive at these figures. Necessary statistics are rarely available, and even if they are, they contain different groupings, and thus during the regrouping new assumptions needed to be introduced. For the majority of the regions, however, experience transfer and extrapolations, combined with interviews and expert judgments, had to be applied to derive these figures. New and retrofit data represent advanced standard practice today – i.e., reflecting new building codes and relatively ambitious energy retrofits taking place as a result of support programs. Advanced new and advanced retrofit data are based on exemplary buildings for the respective climate zones, assuming that these can become standard practice from a technological and economic perspective, after sufficient learning and deployment phases, and under mature market conditions, but with some allowance for lesser performance when scaled up to the entire building stock. Figure 10.5, shown earlier, maps these values weighted by respective floor areas.

10.6.2.5   Results of the World Building Thermal Energy use
Scenario Analysis

As Figure 10.1 demonstrates, if today's existing regional best practices in building construction and retrofit proliferate and become the standard, approximately 46% of global building heating and cooling final energy use can be saved by 2050 as compared to 2005 levels. This is in spite of the approximately 126% increase in floor area during the period and a significant increase in comfort and energy service levels arising

Table 10.13 | Specific heating and cooling final energy use (kWh/m²/yr) figures assumed in the scenarios for different building types, vintages, and regions and climate zones.[24]

Region	Climate Type	Single Family					Multi-Family					Commercial and Public				
		Existing	New	Adv New	Retrofit	Adv Retrofit	Existing	New	Adv New	Retrofit	Adv Retrofit	Existing	New	Adv New	Retrofit	Adv Retrofit
NAM	Warm Mod.	150	65	15	105	20	170	65	10	119	15	220	65	15	154	17
	Cold Mod.	191	65	20	134	30	200	65	15	140	20	340	65	15	238	20
	Tropical	75	65	17	53	25	75	65	17	53	25	131	65	25	92	30
	Arid	87	65	12	61	20	87	65	12	61	20	114	65	20	80	25
WEU	Warm Mod.	160	50	12	112	15	155	50	10	109	15	130	50	10	91	17
	Cold Mod.	261	50	14	183	20	225	50	14	158	20	209	50	14	146	20
PAO	Warm Mod.	100	55	15	70	20	95	60	10	67	15	90	66	15	63	17
	Cold Mod.	150	65	20	105	30	130	80	15	91	20	90	66	15	63	20
	Tropical	65	55	17	46	25	63	55	17	44	25	131	65	25	92	30
	Arid	155	65	12	109	20	155	60	12	109	20	114	65	20	80	25
EEU	Warm Mod.	240	145	14	168	15	205	120	10	144	15	180	120	10	126	17
	Cold Mod.	280	123	20	196	20	245	150	15	172	20	280	111	14	196	20
FSU	Warm Mod.	240	150	15	168	25	205	130	15	144	20	180	120	10	126	17
	Cold Mod.	280	180	20	196	20	246	150	20	172	25	353	150	14	247	20
	Arid	210	100	12	147	20	210	100	15	147	20	210	65	18	147	25
CPA	Warm Mod.	65	42	15	46	20	65	42	10	46	15	96	62	15	67	17
	Cold Mod.	140	91	20	98	30	120	78	15	84	20	150	98	15	105	20
	Tropical	60	39	17	42	25	55	36	17	39	25	96	62	25	67	30
	Arid	70	46	12	49	20	55	36	12	39	20	96	62	20	67	25

Continued next page →

24 NAM has the same energy demand for multi-family and single-family buildings due to the aggregation of the residential sector in US EIA's RECS data by climate but not by climate and type of dwelling.

Table 10.13 | (cont.)

Region	Climate Type	Single Family					Multi-Family					Commercial and Public				
		Existing	New	Adv New	Retrofit	Adv Retrofit	Existing	New	Adv New	Retrofit	Adv Retrofit	Existing	New	Adv New	Retrofit	Adv Retrofit
SAS	Warm Mod.	65	42	15	46	20	65	42	10	46	15	96	55	15	75	17
	Tropical	35	23	17	25	25	35	23	17	25	25	96	65	25	75	30
	Arid	35	23	12	25	20	35	23	12	25	20	96	65	18	75	18
PAS	Warm Mod.	65	42	15	46	20	65	42	10	46	15	96	55	15	75	17
	Tropical	35	23	17	25	25	35	23	17	25	25	96	65	25	75	30
MEA	Arid	87	50	12	50	20	62	60	12	60	20	62	65	20	75	25
LAC	Warm Mod.	81	50	15	50	20	81	60	10	60	15	91	55	15	55	17
	Cold Mod.	196	50	20	50	30	170	60	15	60	20	209	65	15	65	20
	Tropical	63	50	17	50	25	63	55	17	55	25	131	65	25	65	30
	Arid	87	50	12	50	20	155	60	12	60	20	114	65	20	65	25
AFR	Warm Mod.	120	50	15	50	20	100	60	10	60	15	100	55	15	55	17
	Tropical	63	50	17	50	25	63	55	17	55	25	65	65	25	65	30
	Arid	87	50	12	50	20	62	60	12	60	20	62	65	20	65	25

Note: Sources of data are diverse, ranging from national statistics through literature review to personal interviews and expert judgments, largely by the author team of this chapter, for regions without documented data.

Source: US DOE, 2009a; 2009b; US EIA, 2003a; 2005; BMVBS, 1993; ICC, 2007; Feist et al., 2001; Schnieders, 2003; Mitthone, 2010.

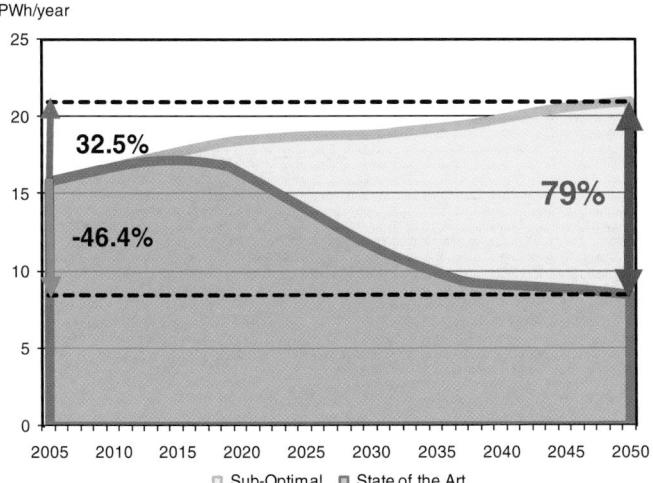

PWh/year

Figure 10.21 | World space heating and cooling final energy use, 2005–2050, suboptimal and state of the art efficiency scenarios.

from a general improvement in affluence. The savings correspond to a drop from 15.8 Petawatt-hour (PWh)/yr to 8.5 PWh/yr in final heating and cooling energy use during the period. At the same time, fuel poverty is eliminated as a basic assumption, in accordance with the GEA multiple objectives.

At the same time, if similarly broad and strong efforts are invested in proliferating building codes and accelerating energy-efficient retrofits worldwide, but only suboptimal performance levels are mandated instead of the state-of-the-art ones that have been demonstrated to be feasible and economic in the particular region, global building heating and cooling final energy use will *increase* by 33% by 2050 as compared to 2005 instead of decreasing (Figure 10.21). Since buildings are constructed or renovated for very long periods, this represents a significant lock-in – 79% of 2005 total global heating and cooling final energy demand in this case – as it is not feasible or is extremely uneconomic to capture the remaining energy savings opportunities outside of renovation and construction cycles. The lock-in problem is described in more detail below.

In the state-of-the-art scenario, most regions are able to decrease final thermal energy use in buildings, with the largest drop in OECD countries (73%), followed by emerging economies (66%) (see Figure 10.22 for the five aggregate GEA regions). Even in Asia, the final energy use decreases, after an initial increase, ending 16.5% lower than in 2005. Regions in which the increase in conditioned floorspace and thermal comfort levels exceed efficiency gains are Latin America and the Caribbean, as well as the Middle East and Africa, with 15% and 71.5% increases, respectively.

The Significance of the Lock-in Effect

The model demonstrates the major risk of the lock-in effect in the building infrastructure. If present standards prevail for new construction,

combined with suboptimal efficiency levels[25] for renovation, 79% of 2005 final heating and cooling energy use will be locked in by 2050, even with accelerated renovation rates. This will result in 21 PWh/yr consumption in the sub-optimal scenario – a 32.5% increase as opposed to a consumption rate of 8.5 PWh/yr in the state-of-the-art scenario.

There has not been an extensive discussion in the literature of the lock-in risk in the buildings sector. The GEA scenarios show, for the first time, the significance of strong policies that are insufficiently ambitious in efficiency targets – ones that prevail today in many developed countries. While from merely an energy savings perspective, the lock-in effect is less problematic since energy saving targets may be reached at a later stage, i.e., in the next renovation or construction cycles although some potentials will never be possible to unlock, which is more due to building structures related to urban design, plot sizing, and orientation, etc. From the climate change perspective, it is essential that buildings deliver greater energy savings in the midterm, such as 2050. Since this chapter shows that buildings are one of the lowest cost options to reach GEA objectives, including climate change, locked-in potential in buildings means that other options will need to replace building-related measures for reaching very ambitious midterm climate change goals. This may be problematic, because building-related measures come with a wide range of multiple benefits, as later sections here and Chapter 17 show, which may not be present in the case of the replacement mitigation measures at comparable costs. Also, there may not be alternative measures of such magnitude at similar cost levels.

The architecture of the suboptimal scenario is based on present efforts taking place in countries, jurisdictions, and institutions strongly committed to solving the climate problem. Many countries and multilateral international foundations and institutions recognize the importance of the building sector, and have passed improved building codes or encouraged high-efficiency or even zero-energy buildings and facilitated an acceleration of energy efficiency retrofit activities. However, in few of these cases are energy efficiency levels close to what is achievable by the state-of-the-art scenario, especially for retrofits. Therefore, the suboptimal scenario already depicts a world in which strong efforts are devoted to solving the building energy problem, and thus shows the danger with which even a well-intended path may be associated.

The lock-in problem originates from the fact that if suboptimal performance levels become the standard in new buildings or retrofits, it can either be impossible or extremely uneconomic to go back for the potentially remaining measures for many decades to come, or in some cases, for the entire remaining lifetime of the building. For instance, lower performance levels originating from suboptimal land use planning and constraints related to plot and building orientation can never be corrected in the building's life or longer. If, during a refurbishment or

---

25 Suboptimal renovation levels are determined for each region, climate, and building type, as shown in Table 10.2, but are typically in the 35% energy savings range where no other data were found.

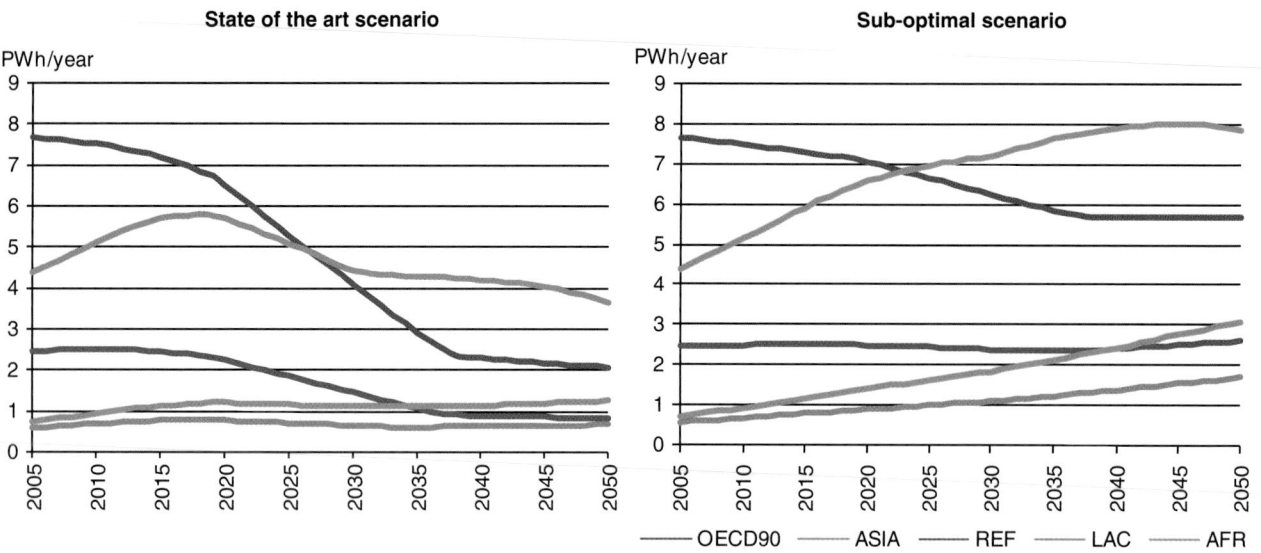

Figure 10.22 | Space heating and cooling final energy use in the five aggregate GEA regions, 2005–2050, in state-of-the-art and sub-optimal scenarios for the five regions: heating and cooling final energy use development for scenarios (i.e., assuming sub-optimal renovation and construction energy performance levels).

new construction, a holistic optimization is not followed, later installation of even the highest efficiency equipment or building materials will not be able to capture the savings otherwise attainable in a comprehensive refurbishment. For instance, heat losses and gains will still occur through other, non-optimized building parts. Finally, each retrofit is associated with significant transaction costs and inconveniences, including finding contractors, planning, preparing contracts, perhaps obtaining the financing, putting up scaffolding or other construction support structures, painting and finishing surfaces after it is done, etc. Thus, in subsequent "top-up" retrofits, energy savings are smaller and costs higher, with fixed costs comparable to those for a comprehensive, deep retrofit. As a result, going back for non-captured savings after suboptimal retrofits or new construction is typically so expensive on a specific cost, such as cost/tCO$_2$ saved, basis that other mitigation or sustainability measures will likely become much more attractive, whereas this is not the case if they are originally part of an integrated, deep design retrofit or construction.

Figure 10.22 shows large increases in energy demand in the suboptimal scenario for the regions of the developing world, where the ASIA region (Asia, excluding the OECD90 countries[26]) has the most pronounced increase. The results for the OECD90 countries show that there is still a decrease in total building energy demand in the sub-optimal scenario due to already gradually strengthening new building codes, less dynamically growing floor space, and actions taken toward efficient retrofits. Conversely, in ASIA there is a major increase in energy demand due to presently unsaturated thermal energy service levels and partially less ambitious building codes.

The lock-in risk is high in all regions, in the range of 40–200% of 2050 state-of-the-art energy use, but its composition is different. The relative importance of renovation and new construction is different in each region. Figures 10.23–10.26 show state-of-the-art and sub-optimal renovation scenarios in the eleven GEA regions broken down by vintage (age) and efficiency level. In OECD, the difference in energy use is primarily due to differences in renovation efficiency levels (see Figure 10.23), while in Asia it is almost entirely due to sub-optimal performance standards in the new building stock (Figure 10.24). In regions of the world that are highly developed or had built the majority of their building stock up to 100 years ago, there is great potential to incur a future energy penalty due to the renovation lock-in effect. This occurs when there is a large part of the building stock that has been built to lower energy standards in the past and is not scheduled for demolition or a deep energy retrofit in the near term. This problem is exacerbated by the fact that buildings have a very long service life, over 150 years in some parts of Europe, and will continue to have the same energy demand until they are appropriately renovated. This points to the importance of different priorities in building-related policymaking in the different regions. In historic regions, ambitious renovation policy – consisting of accelerated renovation dynamics emphasizing state-of-the-art energy performance levels – is important, while in dynamically developing regions new building codes are paramount for achieving a low building energy future. Figure 10.23 demonstrates that retrofits are also important in the more dynamically developing regions.

The Rebound Effect and the Elimination of Fuel Poverty

The logic of the model is that when a new building is constructed it applies the average specific energy use of the new stock – either low or high efficiency – that is either based on state-of-the-art case studies or building codes, i.e., assuming full thermal comfort in the entire building. As a result,

---

26   OECD90 comprises countries that were members of OECD in 1990.

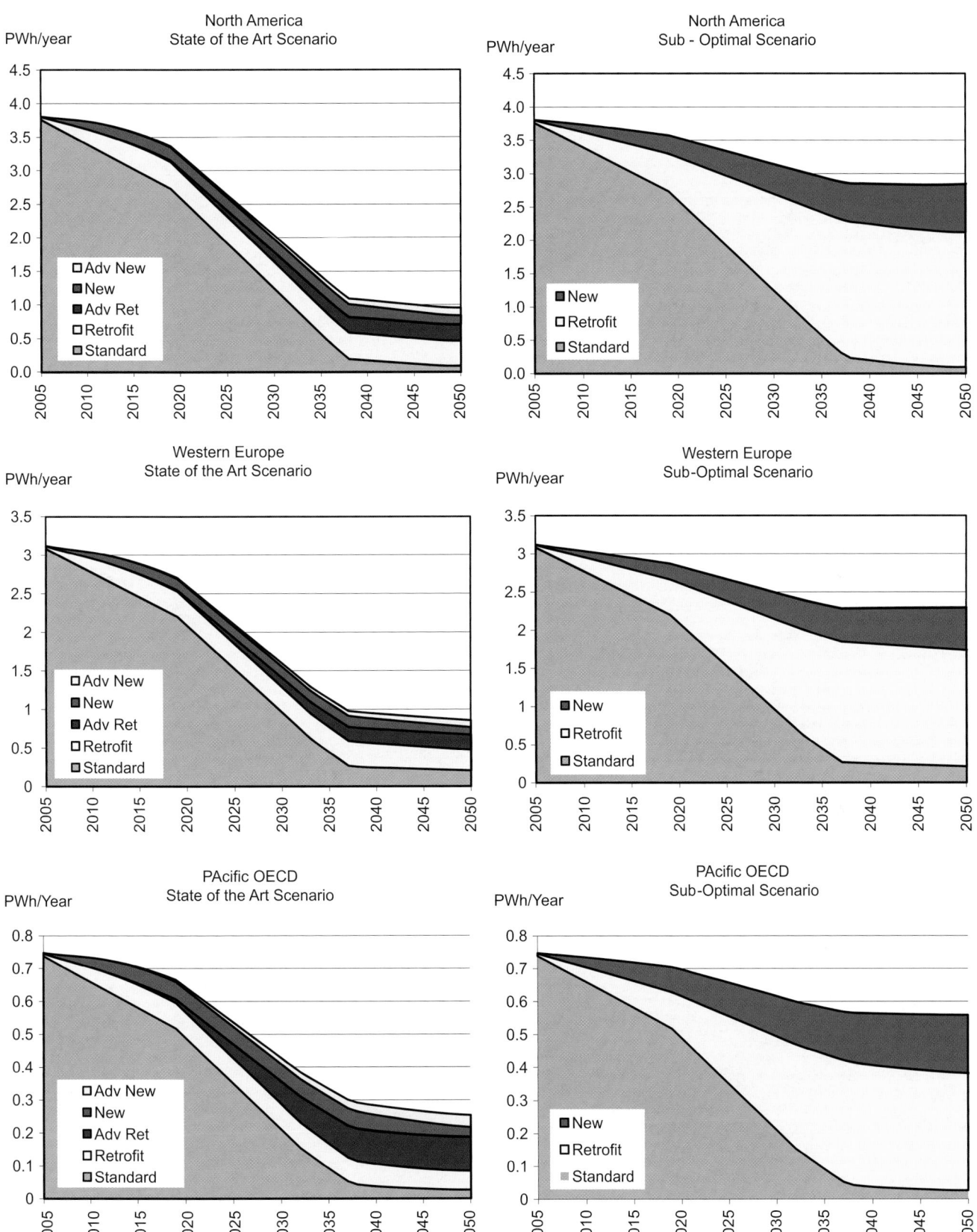

Figure 10.23 | Comparison of the two GEA building scenarios for OECD90 regions, final thermal energy use by construction type.

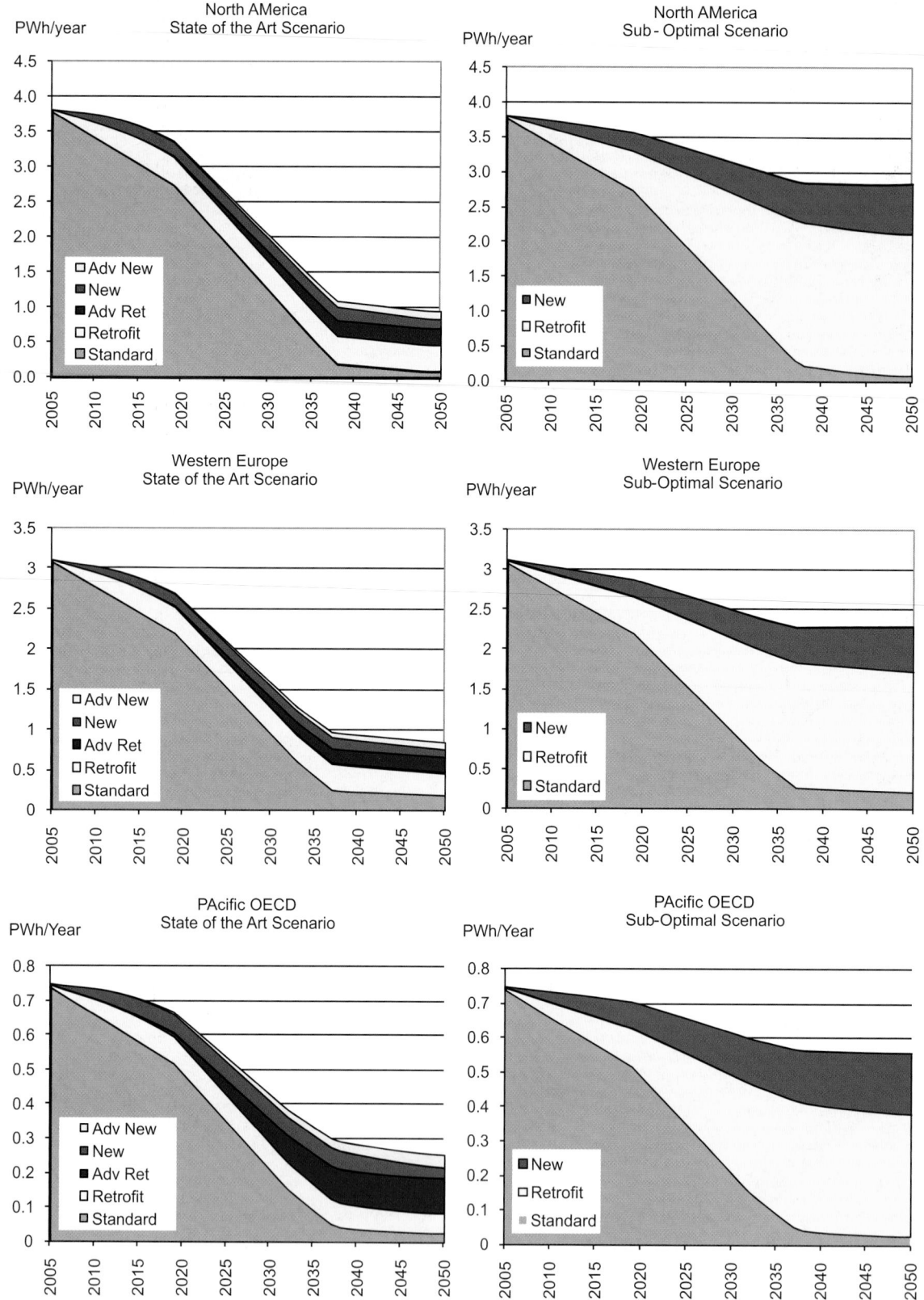

Figure 10.24 | Comparison of the two GEA building scenarios for ASIA Regions, final thermal energy use by construction type.

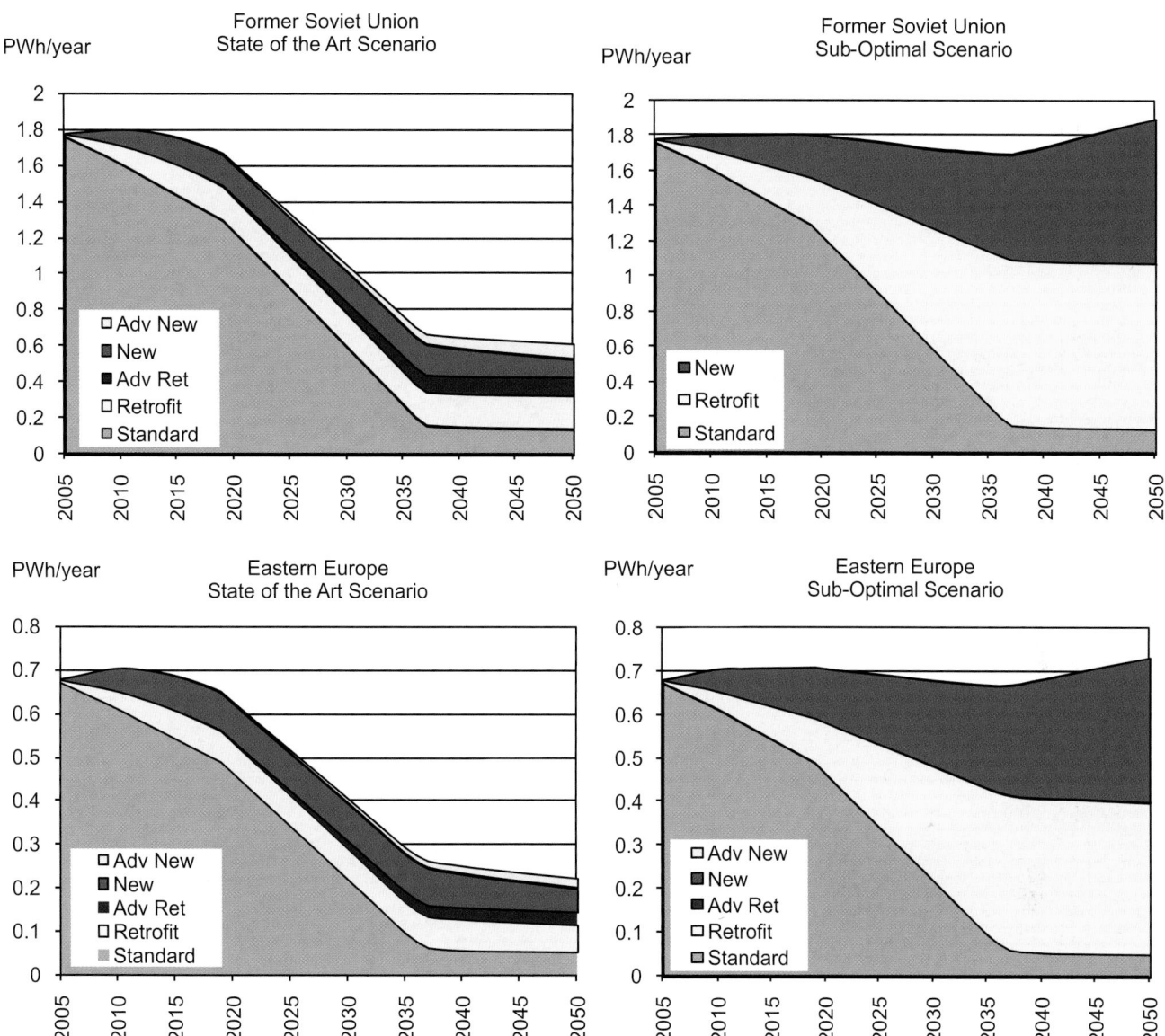

Figure 10.25 | Comparison of the two GEA building scenarios for REF regions, final thermal energy use by construction type.

fuel/energy poverty is fully eliminated by the time the entire building stock has been either replaced or renovated, i.e., before 2050 through assuming full thermal comfort to everyone. This results in significant thermal comfort increases, especially in regions having unsaturated thermal comfort levels – i.e., spaces not heated or cooled to medically acceptable levels – in all of our scenarios. As a result, it is possible that the state-of-the-art new building is actually more energy intensive than the inefficient existing ones. Traditionally, this is referred to as the rebound effect; it is important to note that the direct rebound effect has been fully considered in the scenarios. It is not possible to have more thermal comfort, since heating or cooling beyond the comfortable levels will not result in increases in well-being but rather compromises it. However, this is not considered an undesirable effect, but rather, one of the primary goals of the scenarios and an integral part of the GEA approach: to provide adequate energy service levels to those presently suffering from energy poverty or other limitations due to inadequate thermal energy services.

Investment Costs and Energy Cost Savings

Implementation of the state-of-the-art-scenario worldwide requires approximately US$_{2005}$\$14.2 trillion of cumulative undiscounted investments until 2050, if a 60% cost learning[27] is assumed for new technologies and know-how. This value is US$_{2005}$\$18.6 trillion without any cost learning. In contrast, these investments result in a US$_{2005}$\$57.9 trillion cumulative undiscounted energy cost saving for the same period. Figure 10.27 shows these results on cumulative investments and energy cost savings by 2050 for the different GEA regions of the world.

The cost values are calculated based on marginal expenditures as compared to standard construction and retrofit investments, and energy

---

27 Approximately 60% reduction in costs of low-energy buildings construction and renovation by 2030 as a result of large market size, technology learning, and optimization.

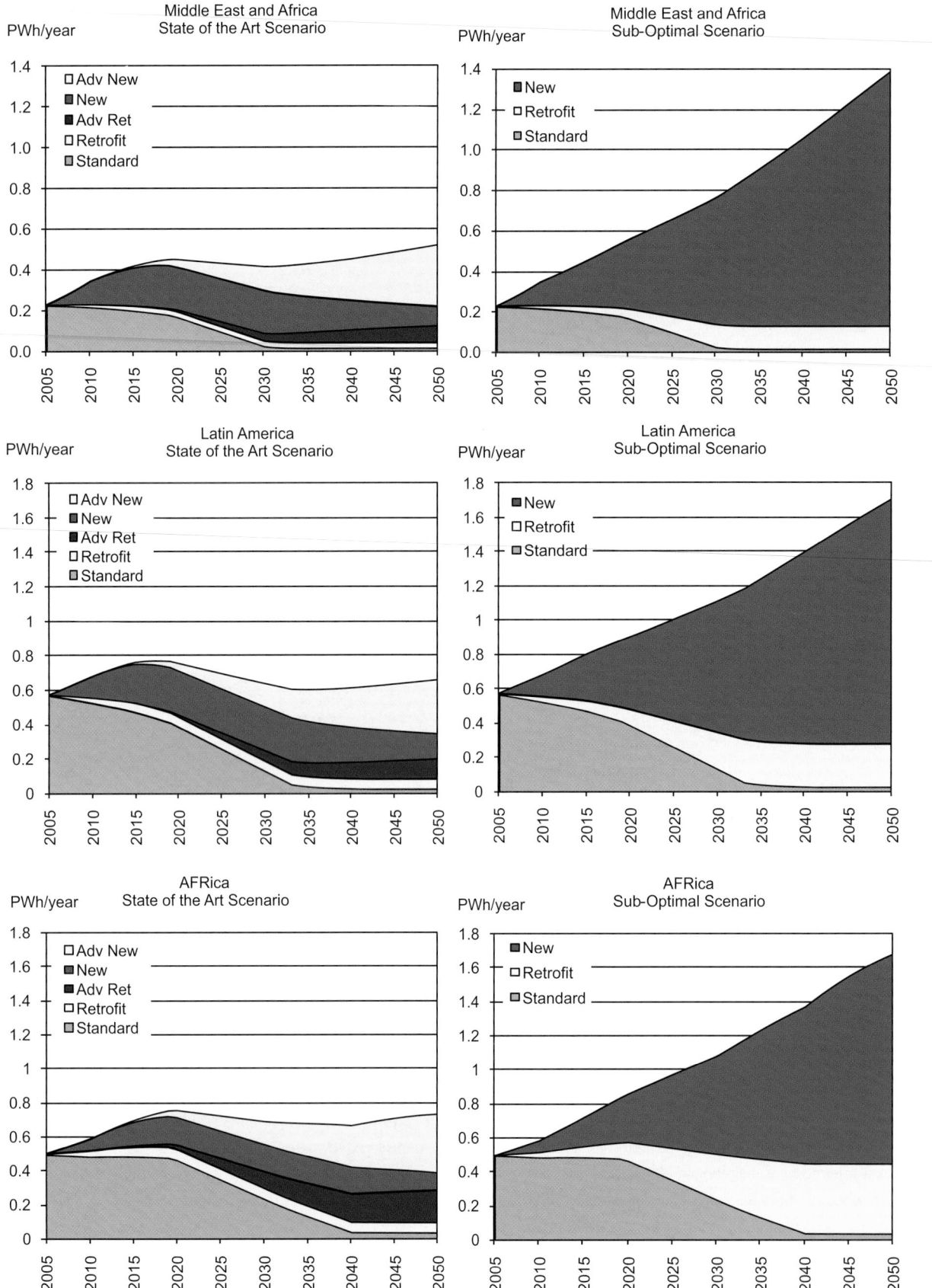

Figure 10.26 | Comparison of the two GEA building scenarios for MEA, LAC, and AFR regions, final thermal energy use by construction type.

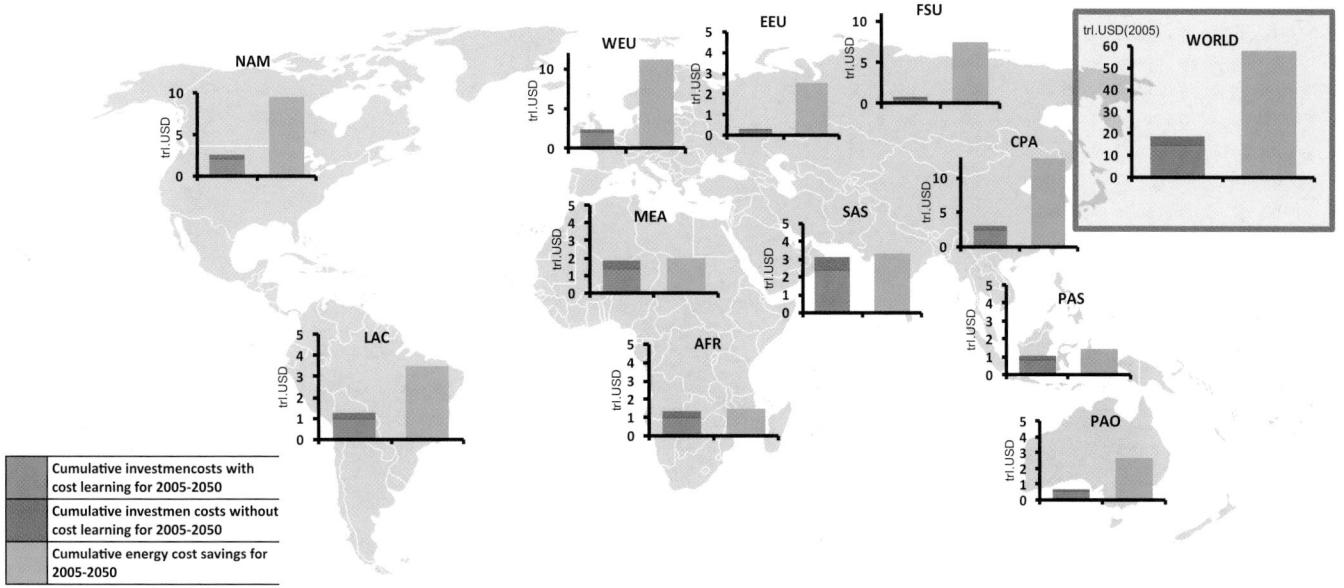

Figure 10.27 | Cumulative undiscounted investment costs and energy cost savings in different regions and worldwide, US$_{2005}$$ trillion.

cost savings as compared to a business-as-usual scenario, taking into account policies in place or in the pipeline. Energy prices and projections are based on IEA statistics and US EIA data (2010). An update of the calculations and further details of the methodology and assumptions is documented in (Ürge-Vorsatz and Petrichenko et al., in preparation).

The results presented in Figure 10.27 are comparable to other existing estimations. For example, in Laustsen (forthcoming) cumulative additional incremental investments for the period 2010–2030 in a case with considerable reduction in final energy use for heating, cooling and hot water are about US$_{2005}$$16 trillion. Cumulative investment costs for the realization of the Blue Scenario in IEA (2010a) – which assumes maximizing the deployment of energy-efficient technologies, achieving substantial renovation of three-quarters of the OECD building stock by 2050, and ensuring the widespread deployment of new technologies – are around US$12 trillion for the period 2007–2050.

Comparison of the Results with Other Existing Models
In order to verify the findings of the model, the authors have calibrated global estimations of energy use in the building sector of a few landmark and reliable recent studies. The results of the comparison are presented in detail in Ürge-Vorsatz and Petrichenko et al. (in preparation), and summarized in Figure 10.28, as well as in Table 10.14. However, it is important to note a few major points that limit how far the messages of such comparison can be interpreted. First, the GEA buildings model specifically covers the final energy use for space heating and cooling, while most other models include other end-uses – e.g., Laustsen (forthcoming) also considers domestic hot water (IEA, 2006; IIASA, 2007; IEA, 2008a; IEA, 2010a; IEA, 2010b). Moreover, methodologies, assumptions, and metrics differ among the analyzed studies. Thus, precise comparison among the models is not possible. However, the general trends can be captured. Figure 10.28 illustrates all the scenarios analyzed and Table 10.14 presents the percentage of change in thermal energy use from 2005 to 2050.

Therefore, while numbers should not be precisely compared, the figures illustrate well that most major building energy use scenarios reinforce the achievability of this sector's significant energy use reduction potential with ambitious policies, and its substantial growth without strong efforts (except WEO10 New Policy and WEO06 ALT).

The GEA Chapter 10 Online Appendix discusses, in detail, how these scenarios relate to the main GEA scenarios described in Chapter 17, and further discuss the limitations of the model.

10.6.2.6   Conclusions from the GEA Building Thermal Energy Scenarios

GEA building scenario analysis has demonstrated that building thermal energy use can significantly decrease by the mid-century despite expected growth in living space, well-being, and energy service levels, and despite the GEA assumption that fuel poverty is fully eliminated in two decades. Assuming that today's state-of-the-art becomes standard practice in new construction and retrofit, world space heating and cooling energy use can decline by 46% in 2050 from 2005 levels, in spite of the approximately 126% growth in global floor space, elimination of fuel poverty, and significant increases in thermal comfort levels. The implementation of such a scenario requires an approximately US$14.2 trillion undiscounted investment (US$18.6 trillion without technology learning), but results in US$58 trillion savings in undiscounted energy expenditures. However, while this scenario is achievable at net profit, it does require significant policy effort.

At the same time, the analysis has also demonstrated the significant risk of the lock-in effect. If policies are implemented to reduce buildings energy use, such as building codes and support to accelerate energy-efficient renovation, but these do not mandate the state-of-the-art,

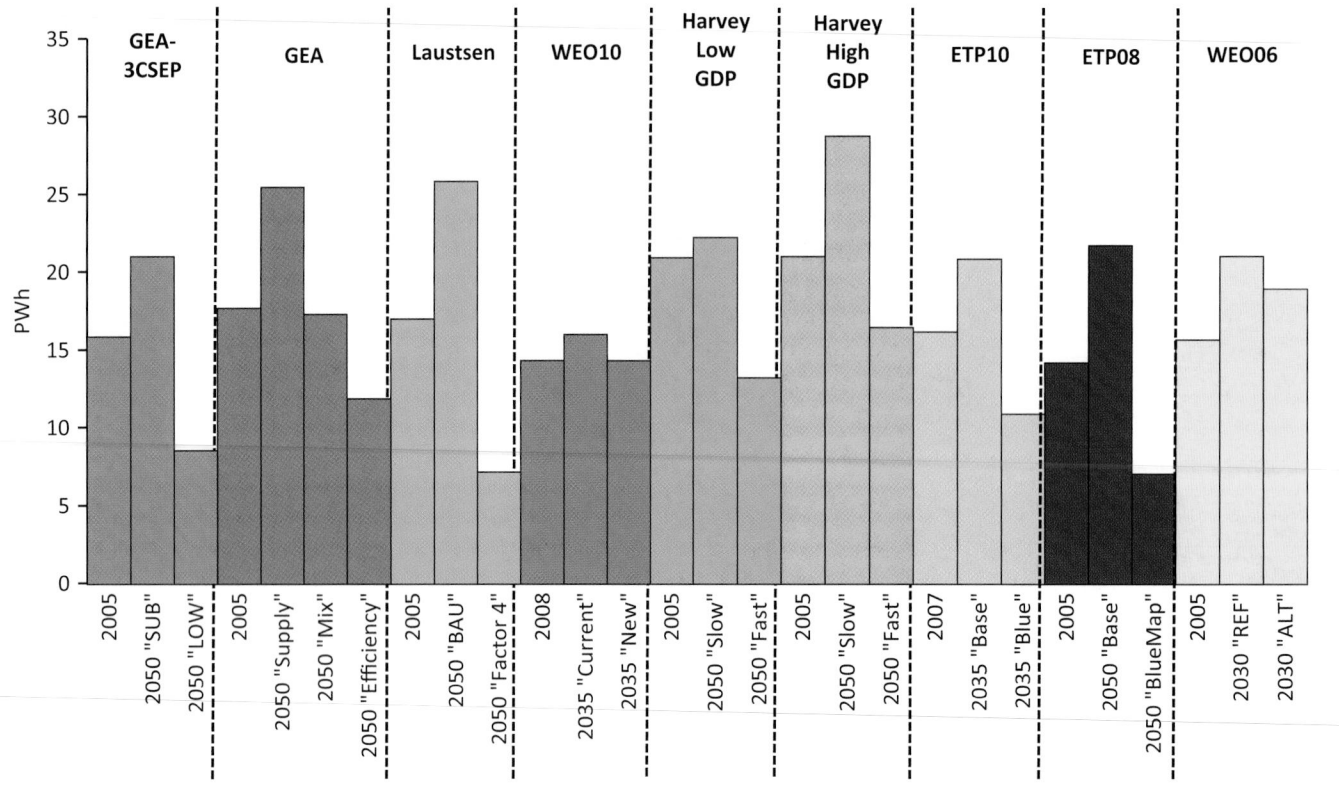

Figure 10.28 | Thermal energy use in the global building sector for different scenarios projected by different models. Source: for WEO06 ALT and WEO06 REF – IEA, 2006; for GEA Efficiency, GEA Mix, and GEA Supply – IIASA, 2007; for ETP08 ACT Map, ETP08 BlueMao, and ETP08 Base – IEA, 2008a; for Harvey Low Fast, Harvey High Fast, Harvey Low Slow, and Harvey High Slow – Harvey, 2010; for ETP10 Blue and ETP 10 Base – IEA, 2010a; for Laustsen Factor 4 and Laustsen BAU – Laustsen, forthcoming; for WEO10 New WEO 10 Current Policy – IEA, 2010b.

Note: Bars should not be compared precisely due to the reasons described in the text.

but rather only suboptimal efficiency levels, there will be substantial energy penalties; thus, energy use is locked-in for many decades due to the very long lifetime and renovation cycle of buildings. As a result, energy use increases by 32.5% by 2050 instead of a 46% decline as compared to 2005, resulting in a 79% lock-in by 2050 as expressed in terms of 2005 world building thermal energy use. Therefore, it is essential that building-related energy policies do not compromise target performance levels, and are most ambitious from as early as possible.

The state-of-the-art scenario, while extremely ambitious, can be achieved through a combination of policy instruments, of which building codes and equipment energy performance standards are the pillars. While state-of-the-art new construction is often a little more expensive than conventional new-built, it can sometimes be less expensive, and deep retrofits do incur substantial capital investments. Although these are investments that pay back well within the remaining life of the building, financing is key for renovation. Section 10.8 is devoted to exploring the policy space and the menu and effectiveness of different policy options that can take us to this more sustainable building energy future, and which are necessary for the implementation of the state-of-the-art scenario.

10.6.3   Description of Appliance Energy Scenarios

If the world embarks upon an ambitious strategy to improve the energy performance of its buildings, currently already pursued in several regions, the building energy demand toward the middle of the century may start to be dominated by electric appliances and other plug-in loads. Therefore it is paramount to also investigate the possibilities of energy savings through improved efficiency in energy-using equipment. The present section describes the potential of energy savings if very aggressive energy efficiency programs would be applied to the equipment in buildings in different world regions.

The description of the Bottom-Up Energy Analysis System (BUENAS) model, developed by LBNL, and its adoption to the GEA scenario exercise is included in the GEA Chapter 10 Online Appendix.

10.6.3.1   BUENAS Model Results

In the long term, that is, beyond 2030 and up to 2050, the efficiency improvements to the stock initiated in 2010 and enhanced in 2020

Table 10.14 | Relative change in building thermal energy use from buildings for different global scenarios.

Scenario	2005–2030	2008–2035	2005–2050
GEA-3CSEP LOW	**−26%**	**−39%**	**−46%**
GEA Efficiency	−25%		−33%
Laustsen Factor 4	−30%		−58%
WEO10 450		−8%	
WEO10 New Policy		12%	
Harvey LowFast	−9%		−37%
Harvey HighFast	−1%		−21%
ETP10 Blue		−33%	−33%
ETP08 ACT Map			−16%
ETP08 Blue Map			−51%
WEO06 ALT	21%		
GEA-3CSEP SUB	**19%**	**18%**	**32%**
GEA Mix	−2%		−2%
GEA Supply	28%		44%
Laustsen BAU	31%		52%
WEO10 Current Policy		23%	
Harvey LowSlow	23%		6%
Harvey HighSlow	40%		37%
ETP10 Base		29%	29%
ETP08 Base			53%
WEO06 REF	35%		

will have permeated the stock almost completely due to the replacement of old appliances with more efficient ones. Therefore, relative energy use of the stock after 2030 will largely scale according to the efficiency level in 2020 relative to the baseline. In order to avoid dependence on assumptions of market-driven efficiency improvements, BUENAS calculates energy savings demand versus a frozen efficiency case rather than a business-as-usual scenario. Frozen efficiency electricity demand, absolute savings, and percent savings are shown in Table 10.15.

The results show that electricity consumption for appliances studied, which is estimated at 1582 terawatt hours (TWh) in 2005, will double by 2025, and will nearly double again to 5696 TWh in 2030 in the absence of significant efficiency improvement. Much of the growth in appliance use during this period will occur in developing countries, especially in ASIA. By 2050, the OECD90 countries will still have the highest consumption for most appliances, but will be surpassed by ASIA for refrigerators, fans, and televisions, appliances that will be present in nearly all Asian households by that time.

Electricity savings will vary between appliances and regions. In absolute terms, savings in 2050 will be largest for refrigerators, with 1171 TWh, followed by standby power with 961 TWh and televisions with 842 TWh. In relative terms, standby power offers the largest opportunity for savings, with the assumption of the technical capability

to reduce standby nearly to zero (0.1 W) per appliance. The total savings for all appliances is 3718 TWh, or 65% of the demand. The savings for all other appliances is 50% or greater, with the exception of washing machines in Latin America and the Caribbean. In that case, efficiency gains are expected to be largely offset by increases in capacity and market shifts from semi- to fully-automatic washing machines.

Figure 10.29 shows a graphical representation of the efficiency scenario. In this picture, appliance efficiency is currently rising rapidly, but growth will be slowed somewhat by standards in 2010, then more dramatically curtailed in 2020. From 2020 to about 2030, growth will actually be negative, and by 2030 total consumption will be roughly equivalent to 2005 levels. After that, however, the demand begins to grow again, although at a much lower rate than base case demand.

In conclusion, the scenario shows that a very significant reduction in appliance electricity demand is possible, given the current state of technologies. It is unlikely, however, that technologies common on today's market will result in energy demand that is only a small fraction of today's consumption. For that to happen, new, very high efficiency technologies must be developed, marketed, and adopted on a wider scale. Such a high tech scenario, which at this point would be somewhat speculative, is an interesting topic for further study.

Table 10.15 | Appliance electricity demand and savings in 2025 and 2050.

Appliance	Region	Demand (TWh)			Savings (TWh)		Percent Savings	
		2005	2025	2050	2025	2050	2025	2050
Refrigeration	ASIA	226	511	867	201	592	39%	68%
	OECD90	194	267	357	61	206	23%	58%
	REF	76	79	79	22	51	28%	65%
	LAC	58	100	150	40	105	40%	70%
	MEA	32	98	336	43	216	44%	64%
	Total	586	1055	1788	368	1171	35%	65%
Fan	ASIA	56	107	153	33	77	31%	50%
	OECD90	32	42	53	12	27	30%	50%
	REF	12	12	12	4	6	28%	50%
	LAC	16	25	36	8	18	30%	50%
	MEA	15	38	107	12	54	33%	50%
	Total	131	224	362	69	181	31%	50%
Washing Machine	ASIA	22	92	155	43	102	47%	66%
	OECD90	171	245	348	65	257	27%	74%
	REF	33	34	34	11	26	32%	75%
	LAC	9	16	24	3	4	18%	16%
	MEA	8	27	120	10	65	38%	54%
	Total	243	414	682	132	455	32%	67%
Television	ASIA	122	414	834	170	417	41%	50%
	OECD90	91	215	335	88	167	41%	50%
	REF	26	46	53	18	27	40%	50%
	LAC	24	71	124	29	62	41%	50%
	MEA	16	82	339	35	169	42%	50%
	Total	279	828	1685	341	842	41%	50%
Standby	ASIA	62	139	565	101	553	73%	98%
	OECD90	134	194	276	135	271	70%	98%
	REF	10	13	18	9	17	69%	98%
	LAC	15	26	46	18	45	71%	98%
	MEA	15	32	76	23	75	72%	98%
	Total	236	403	980	287	961	71%	98%
Oven	OECD90	96	133	181	49	98	37%	54%
	REF	10	16	18	6	10	37%	54%
	Total	106	148	200	54	108	37%	54%
GRAND TOTAL		1582	3073	5696	1251	3718	41%	65%

10.7 Co-benefits Related to Energy Use Reduction in Buildings

10.7.1 Key Messages

A future involving highly energy-efficient buildings also results in significant associated benefits – typically with monetizable benefits at least twice the operating cost savings, in addition to non-quantifiable or non-monetizable benefits now and avoided impacts of climatic change in the future. Multiple benefits beyond climate change mitigation

include: improvements in energy security and sovereignty; net job creation; elimination or reduction in indoor air pollution-related mortality and morbidity; other health benefits; alleviation of energy poverty and improvement of social welfare; new business opportunities, mostly at the local level; stimulation of higher skill levels in building professions and trades; improved values for real estate and enhanced ability to rent; and increased comfort, well-being and productivity.

A survey of quantitative evaluations of such multiple benefits shows that even single energy efficiency initiatives in buildings in individual

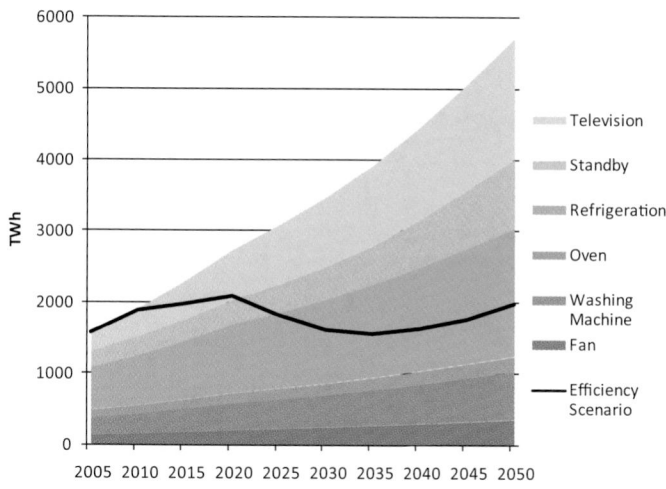

Figure 10.29 | Appliance electricity demand in base case and high efficiency scenario.

countries and regions alone have resulted in benefits with value in the range of billions of dollars annually for single benefits, such as health improvement-related productivity gains and cost aversions. Due to their significance, co-benefits can often present attractive entry points to policymaking, and point to the crucial importance of policy integration.

10.7.2   The Importance of Non-energy Benefits as Entry Points to Policymaking and Decision Making

Traditionally, energy cost savings have been regarded as a key rationale for implementing energy efficiency measures and environmental benefits for introducing policies to promote them. However, only a limited portion of such opportunities has been taken up by individuals as a result of the cost saving motivation and few policies implemented. Energy efficiency being the key strategy in climate change mitigation has provided a new rationale, giving rise to further policies that foster efficiency in developed countries. However, strong policy commitment to climate change mitigation only exists in a few countries and even there energy efficiency policies still fall very short of capturing the potential for cost-effective efficiency.

This section demonstrates that co-benefits are very significant in the building sector and offer new entry points into policy- and decision making. In jurisdictions where environmental benefits do not play a strong role in public policy, other benefits, such as poverty alleviation, employment creation, or improved energy security may be important enough to motivate such policies. For private decision makers, for whom energy cost savings are not sufficient to take steps, other benefits, such as improved comfort and health or corporate productivity gains, can unlock action.

While a single benefit, such as climate change mitigation, energy cost saving, or energy security gains, may not be sufficient to motivate action to capture saving potentials or to fare positive in a cost-benefit

calculation, considering multiple benefits together may help the total benefits significantly outweigh the costs, or be sufficient motivation to enable action. However, this is only possible if public policymaking frameworks are adequately integrated to allow benefits in such otherwise disjointed areas to be combined. Private decision makers can take advantage of multiple benefits in decision making if tools are available that facilitate complex, integrated assessments of costs and benefits.

This section focuses on demonstrating the significance of co-benefits to sustainable energy action in buildings in several areas to illustrate that they are sufficient enough for offering alternative avenues for decision- and policymaking.

10.7.3   Typology of Benefits Of Energy Efficiency And Building-Integrated Renewable Energy

The co-benefits – often also called non-energy benefits – of energy efficiency and distributed energy generation in buildings are numerous. Many of the existing studies do not give an explicit classification of these benefits in the building sector. One classification is proposed by Skumatz and Dickerson (1997). They group non-energy benefits depending on the recipient: as (1) energy efficiency program participants, e.g., increased comfort, improved health; (2) society, e.g., cleaner outdoor air, employment creation, lower energy prices; or (3) a utility, e.g., lower bad debt write-off, decreased transmission costs. In addition, there are a few classifications of benefits of energy efficiency and GHG emission mitigation in general that might be applied to the benefits of energy-using sectors. For instance, Davis et al. (2000) suggest three categories of co-benefits: health, ecological, and economic and the IPCC (2007) lists co-benefits in the buildings sector in a similar way adding improved social welfare and poverty alleviation. Table 10.16 classifies co-benefits of energy efficiency in the buildings sector synthesizing these approaches. Table 10.16 also contains indicators showing how specific benefits can be measured. Once quantified, and with certain caution, many of these figures could be monetized and integrated into cost-benefit assessments of energy efficiency and saving actions. The following sections elaborate on a selection of co-benefits, avoiding repetitions of previous assessments.

10.7.4   Health Effects

The existing links between public health and the use of energy at home have been explored and quantified following two main directions: the health impacts of indoor and outdoor (regional) pollution, and the health impacts of inadequate access to energy, mostly heating in regions with a cold season. That way, it is likely that the most important health non-energy benefit of providing more energy-efficient solutions in buildings is the large number of lives that could be potentially saved through the provision of safe, clean, and energy-efficient cooking plus heating and lighting equipment in developing countries for population segments not

Table 10.16 | Typology of benefits of energy efficiency and distributed energy use in the buildings sector and selected indicators for their potential quantification.

Category	Non-energy benefits	Examples of indicators and concepts for its quantification
Health effects	Reduced mortality	Mortality risk (acute and chronic), Years of Life Lost (YLL), Loss of Life Expectancy (LLE), Value of a Life Year (VOLY), Value of a Statistical Life (VSL), hedonic wages.
	Reduced morbidity and other negative physiological effects	Avoided hospital admissions, Restricted Activity Days (RAD), Years Lived with Disability (YLD), Disability Adjusted Life Years (DALY), Quality Adjusted Life Years (QALY), Cost of Illness (COI: direct medical costs plus cost to the society from lost earnings), productivity loss and lower learning performance.
Ecological effects	Reduced impacts on ecosystems and crops	Acidification, eutrophication, exposure to tropospheric ozone, atmospheric deposition of pollutants, critical loads, number of protected hectares, Potentially Disappeared Fraction (PDF)
	Construction and demolition waste reduction	Percentage of reduction in construction and demolition wastes.
	Lower water consumption and sewage production	Percentage of reduction in water consumption and sewage production,
Economic effects	Lower energy prices[1]	Inverse price elasticity of supply.
	Employment creation[2]	Employments per unit of investment, multiplier effect, working age population relying on unemployment benefits,
	New business opportunities	New market niches
	Rate subsidies avoided[3]	Decrease in the number of subsidized units of energy sold, percentage of the energy price subsidized,
	Enhanced value of the buildings capital stock.	Higher resale and rental prices.
	Energy security	Reduced dependence on imported energy,.
	Improved productivity	Drop in absenteeism rates, reductions in voluntary terminations, GDP/income/profit generated.
Service provision benefits	Transmission and distribution loss reduction	Value of eliminated energy losses.
	Fewer emergency service calls	Saving staff time and resources necessary for attending the calls.
	Utilities' insurance savings[4]	Decrease in the insurance costs of utility companies.
	Lower bad debt write-off[5]	Decrease in the average size of bad debt written off, decline in the number of such accounts.
Social effects	Fuel poverty alleviation	Reduced expenditures on fuel and electricity; reduced fuel / electricity households debt; reduced excess winter deaths.
	Increased comfort	Mean household temperature, reduction of outdoor noise infiltration (dB).
	Safety increase (fewer fires)	Reduced number of fires and fire calls.
	Increased awareness	(Conscious) reductions in energy use, higher demand for energy efficiency measures.

Notes:
1. However, it is unlikely that initiatives at local, regional or even national scales bring sufficient reductions in the overall energy demand to affect prices set internationally.
2. To be incorporated as a benefit into cost-benefit analysis, only net employment creation can be accounted for.
3. Rate subsidies can be defined as lower, subsidized rates provided by utilities for their low-income customers (Schweitzer and Tonn 2002).
4. Reducing gas leaks and repair of faulty appliances (as a part of weatherization programs) decreases the insurance costs of utility companies.
5. Writing off the portion of a bad debt which is not paid by customers to the utilities (Schweitzer and Tonn 2002).

having access to clean energy sources. The health benefits of avoiding the inefficient indoor burning of traditional biomass and other solid fuels is thoroughly discussed in Chapter 4; it could translate into avoiding up to 1.5 million deaths/yr by 2030 (IEA, 2010d).

Providing access to cleaner fuels like LPG and more efficient stoves with enhanced ventilation systems would substantially improve in-house air quality and reduce household fuel collection and cooking time (Hutton et al., 2007). Furthermore, indoor air pollution is also a health concern in developed countries because of problems related to inadequate ventilation (Kats, 2005) and the so-called Sick Building Syndrome (SBS), discussed earlier in the chapter, which result in low work and learning performance

and loss of productivity (WHO, 2000). Improving building ventilation and insulation allows the control of air exchange rates and reduces indoor air pollution and outdoor noise infiltration (Jakob, 2006). This provides opportunities for enhancing public health protection through retrofitting and weatherization programs.. In particular, state-of-the-art buildings eliminate, or significantly reduce, the health effects of indoor radon, dampness and mold, house dust mites, Volatile Organic Compounds (VOCs), NO_2, and secondhand tobacco smoke (see Chapter 4).

The buildings sector will also play a role in reducing mortality and morbidity if energy-efficiency and -saving programs are able to reduce outdoor air pollution, thus lowering end-use energy demand, especially when emissions arise from the direct combustion of fossil fuels like coal and oil for heating and cooking (see Chapter 4). Many studies (Aunan et al., 2000; Samet et al., 2000; DEFRA, 2004; Rypdal et al., 2007; van Vuuren et al., 2008) have translated the reduction of human mortality and morbidity stemming from outdoor air quality improvements and climate policies into monetary terms. In the European Union, large research initiatives such as ExternE (Bickel and Friedrich, 2005), NEEDS (Ricci, 2009), and CASES (FEEM, 2008) have identified and put an economic value on the negative health-related externalities of the life cycle of energy provision that can also be avoided through energy demand reductions.

Additionally, connections have been established between cold and damp houses and excess winter deaths, respiratory illness, asthma, and impaired mental health (Morrison and Shortt, 2008). Therefore, improving building capital stocks for fuel poverty alleviation is also expected to generate physical health (Clinch and Healy, 2001) and psychosocial and mental health benefits related to warmer indoor environments, as discussed in Chapter 4. Such weatherization programs are especially beneficial in countries with poor housing conditions, where the problem of fuel poverty is especially acute (Clinch and Healy, 1999).

10.7.5   Ecological Effects

Reducing energy use in buildings results in a lower concentration of outdoor air pollutants (NO_x, NH_3, SO_2, VOC, or PM) that damage ecosystems and crops, which will then be better protected against acidification and eutrophication and less exposed to elevated ground level tropospheric ozone concentrations (EEA, 2006; van Vuuren et al., 2008). This includes the reductions in of various negative externalities like the impacts of acid rain and ozone in forests and the effects of acidification on recreational fisheries, as well as a reduction in noise pollution, visual amenity disruption, and major accident risks (European Commission, 1995). A growing body of literature on the economic value of ecosystem services (Costanza et al., 1997; Torras, 2000; CBD, 2001; Hein et al., 2005; Nahuelhual et al., 2007) provides a basis to establish a connection between energy efficiency and saving actions and enhanced ecosystem services provision.

Retrofitting and weatherization programs also extend the lifetime of buildings and increase resource use efficiency. For instance, Kats (2005) and SBTF (2001) estimated that building green and efficient houses could reduce construction and demolition wastes over 50%, and up to a maximum of 99%, as compared to an average practice. In the United States, Schweitzer and Tonn (2002) found significant reductions in water consumption and sewage production over the lifetime of energy-efficiency measures, i.e., low-flow showerhead and faucet aerators. Since building construction, operation, and decommissioning are energy-using activities, as is water provision and treatment

(see Section 10.1.4), longer building lifetimes and lower resource consumption will bring about reduced amounts of embodied energy and GHG emissions.

10.7.6   Economic Effects

Creating employment and enhancing the overall productivity of the economy are broad macroeconomic goals that energy efficiency and saving programs can help to achieve. In underdeveloped, often rural regions the lack of modern energy is a primary cause of poverty. Improving energy efficiency would result in better energy security and less dependence on imported energy sources (IEA, 2004; Behrens and Egenhofer, 2007). Also other co-benefits, such as the increased value of real estate and lower energy prices, have welfare implications for households.

Characteristics of buildings and indoor environments significantly influence rates of communicable respiratory illness, allergy and asthma symptoms, sick building symptoms, and worker performance (Fisk, 2000). Related to that, productivity increase at the micro level has been documented and may reach about 6–16% in efficient buildings (Lovins, 2005), which translates into direct financial benefits: a 1% increase in productivity, ~5 minutes/employee/day, is calculated at US$600–700/employee/yr or US$3/ft2/yr in the United States (Kats, 2003).

Research also indicated that investing in energy efficiency renovation in the buildings sector has positive net employment effects once job losses in energy supply sectors are accounted for. Given the distributed nature of direct, indirect, and induced employment effects, additional jobs are expected to be geographically widespread (for Hungary, see Ürge-Vorsatz et al., 2010). The experience has pointed at spatial overlaps of fuel poverty and high unemployment. Promoting energy-efficient renovations in fuel poverty affected areas will benefit fuel-poor households by also providing additional income-earning opportunities (ACE 2000).

10.7.7   Service Provision Benefits

Improvement of energy efficiency and emission reduction in the buildings sector may result in higher quality provision of a number of energy-related services. This category of co-benefits include, inter alia, transmission and distribution (T&D) loss reduction, fewer emergency service calls, utilities' insurance savings, and lower bad debt write-off (Schweitzer and Tonn, 2002; Stoecklein and Skumatz, 2007). Even though these are mostly related to the functioning of utility companies, they can well translate into economic benefits, i.e., positive welfare changes, as long as similar comfort and service provision levels are achieved with fewer resources.

10.7.8   Social Effects

Energy efficiency in the buildings sector can also contribute to tackling social issues, such as poverty and fuel poverty. High-efficiency retrofitting of the existing building stock or the construction of near-zero-energy new buildings, can alleviate, and in some cases fully eliminate, fuel poverty. This, in turn, saves large amounts of public funds that are being spent on relief for those in fuel or energy poverty.

In general, improved domestic, corporate, and public energy efficiency lowers energy expenditures, therefore leaving higher disposable incomes for bill payers, and thereby improving social welfare.

In addition to immediate social co-benefits – including higher thermal comfort levels, noise protection, and improved indoor air quality – there is an increase in safety – namely fire prevention, as well as a number of long-term social benefits. The conveniences of education and health have far-reaching societal consequences on the development of equity, citizens' rights, and gender and child protection.

10.7.9   Worldwide Review of Studies Quantifying the Impact of Benefits Related to Energy Savings in the Built Environment

When societal interests are considered, many of the identified co-benefits related to improved building energy efficiency should be included in economic cost-benefit assessments that support decision-making processes and to determine whether certain measures or actions are justified on a societal basis or not. Similarly, non-energy benefits, especially those obtained at micro (household or firm) level, are important determinants of private decision making. At the same time, there are a limited number of potential studies or other cost-benefit assessments related to energy efficiency and GHG emission mitigation strategies that incorporate such benefits into the analysis.

Table 10.17 reviews the literature that is available in the public domain that has quantified non-energy benefits in the building sector. Typically these studies quantify physical impacts of energy conservation or GHG mitigation and monetize them. The survey revealed that different types of co-benefits have been examined to different extents. For instance, the effects of reducing outdoor air pollution – e.g., avoided morbidity and mortality – and productivity gains have been intensively studied. The authors were unable to locate research on the quantification of co-benefits such as new business opportunities and costs avoided due to increased awareness.

Most studies focus only on a few regions. The United States is subject of many studies, followed by only a few countries of the European Union. No studies were found that aggregated the quantified co-benefits, especially at regional or national levels. A global aggregation would be especially challenging, because ideally such an effort applies a uniform

methodology and approach which has not yet been possible for potential assessments either.

10.8   Sector-Specific Policies to Foster Sustainable Energy Solutions in Buildings

10.8.1   Key Messages

- A wide range of policies has been demonstrated that are successful and cost-effective in reducing energy use in buildings. These include stringent and well enforced building and appliance standards, codes, and labeling, applying also to retrofits.

- Urgent introduction of strong building codes mandating near-zero-energy performance levels, progressively improving appliance standards, as well as strong promotion of state-of-the-art efficiency levels in accelerated retrofits of the existing building stock are crucial. Particular attention should also be paid to addressing non-compliance related to building codes. However, net-zero-energy building mandates may be not feasible for every type of building and in all regions, and in many cases their economics are unfavorable compared to high-efficiency buildings. Policy instruments to encourage deep retrofits should be implemented, including performance standards, performance contracting, energy audits and incentive mechanisms.

- Appropriate energy pricing is fundamental for promoting energy efficiency in buildings. Taxation provides an impetus for a more rational use of energy sources, but especially in poor regions or population segments, subsidies of highly efficient capital stock can be more effective and acceptable.

- Awareness campaigns, education, and the provision of more detailed and direct information, including smart metering, enhance social and behavioral changes. Combining regulation, incentives and information measures has the highest potential to increase energy efficiency in buildings.

10.8.2   Overall Presentation and Comparison of the Policy Instruments

The previous sections, in addition to earlier work, have demonstrated that there is a very broad spectrum of technologies and know-how that can save significant amounts of energy in buildings without compromising the level of energy services provided, often at net societal benefits rather than costs (Levine et al. 2007; Ürge-Vorsatz et al., 2007). Much of this potential though has not been captured due to the especially strong and diverse barriers that prevail in this sector. However, many of these barriers can be removed or lowered by appropriate policies, programs, and measures.

Table 10.17 | Benefits of GHG mitigation in the buildings sector and methodologies for their assessment in the worldwide literature.

Co-benefits	Country/ region	Impact of GHG emissions or energy demand reduction		References
		Physical indicator	Monetary indicator	
Health effects				
Morbidity and other negative physiological effects reduction	USA, New Zealand, Denmark	• USA: Improved indoor air quality reduced asthma occasions by 38.5%; a 25% reduction in asthma incidence in an average 900-student school translates into 20 fewer children/yr. with asthma. • Denmark: The improvement of thermal air quality led to better concentration of 15% of respondents and a 34% decrease in the number of respondents with sick building syndrome.	• USA: Ventilation increased may result in net savings of US$_{2005}$527/employee-yr. that represent the productivity lost on a national scale of US$_{2005}$30 billion/yr. • USA: Better ventilation and indoor air quality could reduce influenza and cold by 9–20% in population (ca. 16–37 million fewer cases) that translates into annual savings of US$_{2005}$7.5–17.4 billion.	National Medical Expenditure Survey, 1999; Kats, 2005 Milton et al., 2000; Schweitzer andTonn, 2002; Stoecklein andSkumatz, 2007; Fisk, 1999
Mortality reduction	Hungary; USA, Ireland, Norway, worldwide	• USA: Every 10 g/m^3 increase in ambient particulate matter (the day before deaths occur) brings a 0.5% increase in the overall mortality. • Ireland, Norway: The share of excess winter mortality attributable to poor thermal housing standards is 50% for cardiovascular disease and 57% for respiratory disease. • Worldwide: more than 3 billion people cook with solid fuels on open fires or traditional stoves, resulting in more than 1.5 million deaths annually and a multitude of negative economic and environmental impacts.	• Hungary: The energy saving program resulted in the total health benefit of US$_{2005}$854 million/yr. due to a decrease of chronic respiratory diseases and premature mortality. • reland: the expected health-benefits (mortality and morbidity reduction) of a proposed domestic energy efficiency program amount to US$_{2005}$1.88 billion (at the 5% Irish government's discount rate) over 20 years. • Worldwide: the global annual economic benefits (incl. less expenditure on health care, health-related productivity gains, fuel collection and cooking time savings, and environmental impacts) of halving the population currently lacking access to cleaner fuels (LPG) would amount to US$_{2005}$91 billion at a net intervention cost (incl. fuel, stove, and program costs, from which monetary fuel cost savings are subtracted) of US$_{2005}$13 billion. Improving stoves would generate US$_{2005}$105 billion in economic benefits at a negative net cost of US$_{2005}$34 billion.	Aunan et al., 2000; Samet et al., 2000; Clinch and Healy, 1999; Clinch and Healy, 2001; Hutton et al., 2007
Ecological co-benefits				
Reduced impacts on ecosystems and crops	Europe	• The application of the Kyoto protocol in Europe is expected to reduce from 16.1% (93.4 million ha in 1990) to 1.5% (8.7 million ha in 2010) the share of ecosystems unprotected to acidification. The figures for areas with excess deposition of nutrient nitrogen are, respectively, 30.5% (166 million ha in 1990) and 18.8% (103 million ha in 2010).		Van Vuuren et al., 2008
Construction and demolition waste benefits reduction	USA	• The Construction and demolition diversion rates are 50–75% lower in green buildings (with the maximum of 99% in some projects) as compared to an average practice		SBTF, 2001; Kats, 2005

Continued next page →

Table 10.17 | (cont.)

Co-benefits	Country/ region	Impact of GHG emissions or energy demand reduction		References
		Physical indicator	Monetary indicator	
Lower water consumption and sewage production	USA		• The NPV of reduction in waste water and sewage over the measure lifetime is US$3.4–657/household.	Schweitzer and Tonn, 2002
Economic co-benefits and ancillary financial impacts				
Lower energy prices	USA	• A 1% decrease of the national natural gas demand through energy efficiency and renewable energy measures leads to a long-term wellhead price reduction of 0.8% – 2% or 0.75% – 2.5%;		Wiser et al., 2005; Platts Res. &Consult, 2004
Employment creation	USA, EU	• USA: The US Department of Energy's Office of Energy Efficiency and Renewable Energy (EERE) macroeconomic (input-output) analysis of its energy efficiency residential and commercial buildings program found a potential to increase employment by up to 446,000 jobs by 2030. EU: the SAVE program, based on data for 9 EU countries in the 1990s, estimated that investing in energy efficiency in residential and commercial/public buildings generated between 4.0 and 12.3 jobs per million US$_{2005}$.		Scott et al., 2008; Wade et al., 2000
Rate subsidies avoided	USA	• The NPV of avoided rate-subsidies over the lifetime of the measures is US$6 – 70/household.		Schweitzer and Tonn, 2002
Enhanced value of the buildings capital stock.	Switzerland, the Netherlands	• An economic analysis using the hedonic pricing approach shows a valuation of energy efficient windows of 2–3.5% of the selling price of existing single-family houses and reveals that new single-family houses certified with the 'Minergie' Label yield higher selling prices by almost 9% (with a standard error of about 5%) • A hedonic value analysis of the Dutch housing sector, where an early, voluntary adoption of the EU EPBD energy labeling system is in place, found out a premium in the sale value of energy-efficient properties, with a 2.8% higher transaction price in houses with an A, B, or C certificate.		Borsani and Salvi, 2003; Brounen and Kok, 2010.
Energy security	USA	• The Oak Ridge National Laboratory estimated the energy security benefits of reduced US oil use at $13.58: $8.90 coming from monopsony component plus $4.68 related to macroeconomic disruption / adjustment costs. This approach estimates the incremental benefits to society not reflected in the market price of oil, in US$_{2004}$per barrel, of reducing US imports.Omitted from this "oil premium" calculation are environmental costs and possible non-economic or unquantifiable effects, such as effects on foreign policy flexibility or military policy		Leiby, 2007
Improved productivity	USA	• In efficient buildings labor productivity rises by about 6–16%; students' test scores shows ~20–26% faster learning in schools well day lighted; retail sales can rise 40% in well day lighted stores	• Estimated potential annual savings and productivity gains of better indoor environments related with building energy efficiency in the US$_{2005}$\$7.5–17.4 billion (reduced respiratory diseases); US$_{2005}$\$1.2–5 billion (reduced allergies and asthma); US$_{2005}$\$12.5–37.3 billion (reduced sick building syndrome symptoms); and US$_{2005}$\$25–200 billion (direct improvements in worker performance unrelated to health)	Lovins, 2005; Fisk, 2000; Kats, 2003;

Service provision benefits

T&D loss reduction	USA	• The NPV over the lifetime of the measures installed ranges US$33–80/household.	Schweitzer andTonn, 2002
Fewer emergency gas service calls	USA	• The NPV of fewer emergency gas service calls over the measure lifetime is US$39–201/household.	Schweitzer andTonn, 2002
Utilities' insurance savings	USA	• The NPV of utilities insurance cost reduction over the lifetime of the measures is US$0–2/household	Schweitzer andTonn, 2002
Decreased number of bill-related calls	New Zealand	• Bill-related calls became less frequent after the implementation of weatherization program, which amounted savings of US$$_{2005}$$24.6 /household-yr, that is 7% of the total saved energy costs	StoeckleinandSkumatz, 2007
Lower bad debt write-off	USA	• The NPV of lower bad debt write-off over the lifetime of the measures is US$15–3462 /household.	Schweitzer andTonn, 2002

Social co-benefits

Fuel poverty alleviation	UK	• UK: Energy efficiency schemes applied to 6 million households in January-December 2003 resulted in an average benefit of US$_{2005}$$19.9/household-yr.	DEFRA, 2005
Increased comfort,	Ireland, New Zealand, Switzerland	• Ireland: the mean household temperature once completed the proposed domestic energy efficiency program increases from 14 to 17.7 °C. • Switzerland: old windows reduce the level of external noise in the interior of the building by about 20–25 dB, whereas new ones achieve 33–35 dB. 38–40 dB reductions are also possible if asymmetric triple glazing is applied	• Ireland: The total comfort benefits of a proposed domestic energy efficiency program amount to US$_{2005}$$748 million discounted at 5% (Irish government's discount rate) over 20 years; • New Zealand: Comfort (including noise reduction) benefits gained after the implementation of weatherization program amount to US$_{2005}$$130.3/household-yr, equivalent to 43% of the saved energy costs.	StoeckleinandSkumatz, 2007; Jakob, 2006; Clinch and Healy, 2001
Safety increase (fewer fires)	USA	• The NPV over the lifetime of the measures installed is US$0–555/household.	Schweitzer andTonn, 2002

Note: (i) conversion of monetary units to US$_{2005}$$ was done whenever possible by applying Consumer Price Index (CPI) and exchange rates of the corresponding years or periods and nations.

725

A great variety of policy instruments have been implemented worldwide to promote energy efficiency in the buildings sector. There is no single policy instrument that can capture the entire energy saving potential. Due to the especially diverse and strong barriers in this sector, it requires an equally diverse portfolio of policy instruments for effective and far-reaching energy conservation. Policy instruments for promoting energy efficiency in the buildings sector can be classified in five major categories, namely:

- *Control and regulatory mechanisms*, which are institutional rules aimed at directly influencing the energy performance of buildings and/or energy equipment used in buildings. Policy instruments classified in this group include appliance standards, building codes, procurement regulations, energy efficiency obligations, quotas, etc.

- *Regulatory informative instruments*, which are also institutional schemes that aim to inform energy users about energy efficiency. These comprise mandatory labeling and certification programs, mandatory audit programs, utility demand-side management programs, etc.

- *Economic and market based instruments*, such as energy performance contracting[28], cooperative technology procurement, energy efficiency certificate schemes, the Kyoto flexible mechanisms, regional carbon trading platforms and carbon offset programs, etc. These are directly or indirectly aimed at steering economic actors toward improved energy efficiency.

- *Fiscal instruments*, which usually correct energy prices through either the implementation of taxes, tax exemptions or reductions, public benefit charges, the removal of fossil fuel subsidies, etc., or by providing financial support, e.g., capital subsidies, grants, subsidized loans, rebates, property-assessed clean energy (PACE)[29] financing, etc., if first cost-related barriers are to be addressed.

- *Support and information programs and voluntary action*. Instruments classified under this group are very diverse and comprise voluntary certification and labeling programs, voluntary agreements, public leadership programs, awareness-raising, education and information campaigns, detailed billing and disclosure programs, the establishment

Table 10.18 | Indicators of Cost-effectiveness of policy instruments based on best practice cases.

	High	Medium	Low
cost of energy conserved US$_{2005}$$/kWh	<0	0 – 0.01	>0.01
benefit-cost ratio (B/C)	>1	0.8 – 1	<0.8

(and sufficient funding) of energy agencies, awards and competitions, personalized advice, training of building professionals, etc.

In this section, the final effects of key categories of policy instruments used in the building sector to improve energy efficiency are reviewed and evaluated. In total, over 80 studies, review articles, and other relevant publications were identified from over 50 countries, covering each inhabited continent. Experts agree that even a brilliantly designed policy tool may lose its value if inadequately enforced. However, the greatest attention is usually given to policy design, whereas implementation and enforcement processes are often neglected (Khan et al., 2007). Due to this reason, special attention was given to limitations and success factors.

Recent reviews (Khan et al., 2007; Novikova, 2010) examined the outcomes of policies. These are: (a) the degree to which a policy tool achieves the target, often referred to as policy effectiveness; (b) the extent to which a tool has made a difference compared to the situation without it, referred to as net impact of the policy tool; and (c) the relationships between the net impact and spending required, referred to as cost-effectiveness. What follows is summarized from perspectives of effectiveness for energy saving and cost-effectiveness of policy tools. Also, the best attempt has been made to identify limitation and success factors.

Since studies reviewed used different methodologies for evaluation of policy effectiveness, these estimates were converted to a uniform format. For each implemented policy case study, the amount of energy saved as a result of the policy instrument in question was determined, both in absolute and relative terms – i.e., compared to a logical baseline, such as total national energy or electricity consumption in the particular sector and/or end-use. However, this was often not possible due to lack of data. Furthermore, the comparability of these estimates with other cases or policy instruments is in many cases very limited. Thus, the effectiveness of the various policy instruments examined was evaluated in a qualitative way, by assigning grades of "low," "medium," and "high" based on energy saving figures, but taking into account the overall applicability and potential of the instrument. The effectiveness of policies working in limited end-use categories was balanced with those affecting most end-uses; if the instrument works in a narrow energy end-use but can achieve an important reduction in that category, such as appliance standards, it could qualify as "high." (See Table 10.18.)

The cost-effectiveness of the policy instruments examined was also evaluated with qualitative grades, which are based on best practice cases and on the approximate ranges presented in Table 10.18.

---

28 Energy performance contracting is not a policy tool *per se*, but a business model that delivers a similar impact on transformation of the market toward higher energy efficiency as policy tools. Due to this reason, energy performance contracting is often added to policy tools.

29 The PACE model is a relatively new financing structure that enables local governments to raise money through the issuance of bonds or other sources of capital to fund energy efficiency and renewable energy projects, thus lowering the up-front cash payment for property owners. The financing is repaid over a set number of years through a special tax or assessment only on those property owners who voluntarily choose to attach the cost of their energy improvements to their property tax bill. The PACE approach attaches the obligation to repay the cost of improvements to the property, not the individual borrower, creating a way to pay for the improvements if the property is sold.

The comparative assessment of policy instruments presented in Table 10.19 reveals significant differences between them, especially concerning cost-effectiveness. The governmental costs of policy tools in the sample varied widely: figures ranged between -0.13 US$_{2005}$/kWh (i.e., a significant net benefit) and 0.11 US$_{2005}$/kWh. Despite the fact that economic performance of the policy instruments examined is presented in a variety of forms – i.e., economic cost or benefit per unit energy saved, benefit-cost ratio, total amount of estimated savings, etc. – making comparative evaluation extremely difficult, it can be generally stated that appliance standards, energy efficiency obligations and quotas, utility demand-side management (DSM) programs, and tax exemptions were found to be the most cost-effective policy tools in the sample, all achieving significant energy savings at negative costs in several applications. Regarding effectiveness for energy saving overall, appliance standards, building codes, labeling, utility DSM programs, and tax exemptions achieved the highest savings in the sample. More specifically, the implementation of building codes and tax exemptions (investment tax credits) policies in the United States are the two single policy instruments in the sample that have resulted in the maximum absolute energy savings, amounting to 174 TWh/yr each.

When comparing the five different categories of measures, the collected case studies indicate that regulatory and control measures are probably the most effective in terms of energy savings as well as the most cost-effective category, at least in developed countries. They all achieved ratings of high or medium according to both criteria, i.e., effectiveness and cost-effectiveness. Measures that can be designed both as voluntary and as mandatory, such as labeling or energy efficient public procurement policies, have been revealed as more effective when they are mandatory. The Mesures d'Utilisation Rationnelle de l'Energie (MURE) database (MURE, 2007; 2008), which collects and evaluates ex-post estimates of energy savings delivered by policies in the European Union member states and their neighbors, confirms the findings above.

The effectiveness of economic instruments varies, but some of them, such as energy performance contracting (EPC) and cooperative procurement, are promising. Project-based instruments that require credits for savings, e.g., white certificates, may have limited effectiveness due to the complex nature of buildings and resulting high transaction costs, the many efficiency upgrades, and the small project size, if complex monitoring and evaluation are required, but can otherwise be highly cost effective (Eyre et al., 2009).

Fiscal instruments also vary considerably in effectiveness and have numerous success conditions. For instance, in the short run, instruments that increase the energy price such as taxation are often less effective than fiscal incentives for capital investment in energy efficiency, such as tax exemptions, loans, and subsidies, due to the limited energy price elasticity in buildings – i.e., the percentage change in energy demand associated with each 1% change in price. Financing grants and rebates are especially needed in both developed and developing countries, particularly for low-income households, because the first cost barrier often

prevents energy efficiency improvements. In general, tax exemptions were found to be the most effective tool in the category of fiscal instruments, while subsidies, grants, and rebates can also achieve high savings, but are usually costly to society.

Voluntary instruments vary in effectiveness that depends, for example, on the demand for energy-efficient products in the case of voluntary labeling and on whether the companies take voluntary commitments seriously. Though they have often failed to reach their goals, they can be a good starting point for countries that are just introducing building energy efficiency policies or when mandatory measures are not possible. Private sector commitments may be more effective where there is the clear prospect of regulation as an alternative, i.e., where they are in the context of negotiated agreements rather than purely voluntary. Finally, information instruments can be effective, but have to be specifically tailored to the target group.

Identification of the most cost-effective instruments was much more difficult because for some instruments, no quantitative information could be found. In the assessed sample, appliance standards, mandatory audits, utility demand-side management programs, mandatory labeling, energy efficiency obligations, energy performance contracting, cooperative procurement, and tax exemptions seem to be the most cost-effective policy measures. Thus, the category of regulatory and control instruments is apparently also the most cost-effective one, in contrast to a generally prevailing expectation that economic instruments are the most cost-effective (IPCC, 1995). These findings are partly confirmed by the MURE database. Such results are specific to the building sector, and might be explained by considering which barriers specific policy instruments address and the low sensitivity to prices of most non-intensive energy users.

Table 10.20 summarizes the major barriers and corresponding potential policy instruments to overcome them.

10.8.3   Combinations or Packages of Policy Instruments

Every policy measure is tailored to overcome one or a few market barriers, but none can address all the barriers and all targets and target groups. In addition, most instruments achieve higher savings if they operate in combination with other tools, and often these impacts are synergistic, i.e., the impact of the two is larger than the sum of the individual expected impacts (IEA, 2005b). Figure 10.A.8 in the GEA Chapter 10 Online Appendix diagrams the combined effect of the three policy instruments: appliance standards, labeling, and financial incentives.

A number of combinations of policy instruments are possible, as illustrated in Table 10.21. Usually, combining sticks (regulations), and carrots (incentives), with tambourines (measures to attract attention such as information or public leadership programs) has the highest potential to increase energy efficiency (Warren, 2007) by addressing a number of barriers.

Table 10.19 | Comparative assessment of all policy instruments from a governmental standpoint.

Policyinstrument	Country/ regions examples	Effective-ness	Energy reductions for selected best practices	Cost-effectiveness	Cost of energy reduction for selected best practices in US$_{2005}$$	Special conditions for success, major strengths and limitations, co-benefits	References
Control and regulatory mechanisms – normative instruments							
Appliance standards	EU, USA, JPN, AUS, BRA, CHN, MAR, DEU, NLD	High	DEU: app. 214 GWh/yr USA: 129 TWh in 2000 = 2.5% of electricity use, 6% = 252 TWh in 2020 NLD (coupled with rebate): 0.06 TWh/yr. in 1995–2004	High / Medium	AUS: -0.13$/kWh EU: -0.08$/kWh USA: -0.05$/kWh MAR: 0.009$/kWh NLD: 0.07–0.11 $/kWh	Factors for success: periodic update of standards, independent control, information, communication and education	IEA 2005b, Schlo-mann et al. 2001, Gillingham et al. 2004, ECS 2002, WEC 2004, Australian GHG office 2005, IEA 2003, Fridley and Lin 2004, Khan et al.2007
Building codes	SGP, PHL, HKG, EGY, USA, UK, NLD, CHN, EU	High	HKG: 1% of total el. saved USA: 174 TWh in 2000 EU: up to 60% energy savings for new dwellings CHN: 15–20% of bdg. energy saved in urban regions NLD: 0.2 TWh/yr. or 0.1% of the sectoral reference energy use in 1996–2004	Medium	NLD: 0.02–0.05$/kWh for society, -0.09 – -0.002 $/kWh for end-user	No incentive to improve beyond target. Only effective if enforced.	WEC 2001, Lee and Yik 2004, Schaefer et al. 2000, Joosen et al. 2004, Geller et al. 2006, ECCP 2001, IEA 2005a, DEFRA 2006c, Khan et al 2007
Procurement regulations	USA, EU, CHN, MEX, KOR, JPN	High	MEX: 4 cities saved > 5,000 MWh in 1 year CHN: 4.6 TWh expected EU: 52–115 TWh potential EU: Energy+ programme for cold appliances 0.3 TWh/yr. in 1999–2004 USA: 18–61.6 TWh by 2010	Medium	MEX: $1million in equipment purchases saves $ 726,000/yr. in electricity costs EU: 0.007$/kWh EU: Energy + programme for cold appliances 0.001$/kWh	Factors for success: Enabling legislation, energy efficiency labelling and testing. Energy efficiency specifications need to be ambitious.	Borg et al. 2003, Harris et al. 2005, Van Wie McGrory et al. 2006, Gillingham et al. 2006, Khan et al 2007
Energy efficiency obligations and quotas	UK, BEL, FRA, ITA, DNK, IRL	High	UK: In 2002–2005: 2.5 TWh/yr. or 0.5% of sectoral reference energy use; 2010 savings: 6.1 TWh of electricity, 7.4 TWh of natural gas BEL: 0.36 TWh/yr. or 0.2% of sectoral reference energy in 2003–2004	High	Flanders: -0.04$/kWh for households, -0.014$/kWh for other sectors UK: -0.02$/kWh UK: 0.00004 $/kWh BEL: 0.0001 $/KWh	Continuous improvements necessary: new energy efficiency measures, savings target change, short term incentives to transform markets etc.	UK government 2006, Sorell 2003, Lees 2008, Collys 2005, Bertoldi and Rezessy 2006, DEFRA 2006c, Khan et al 2007

Continued next page →

Regulatory/informative instruments

Instrument	Countries	Rating	Energy savings	Rating	Cost	Effectiveness	References
Mandatory labelling and certification programs	USA, CAN, AUS, JPN, MEX, CHN, CRI, EU, ZAF	High	AUS: 6.3 TWh savings 1992–2000, 102 TWh 2000–2015; ZAF: 623 GWh/yr; DNK: 8 TWh	High	AUS: -0.02$/kWh	Effectiveness can be boosted by combination with other instrument, and regular updates.	WEC 2001, OPET network 2004, HoltandHarrington 2003, IEA 2003
Mandatory audit programs	USA, FRA, NZL, EGY, AUS, CZE	High but variable	USA: Weatherisation pro-gram: 22% saved in weatherized households after audits (30% according to IEA)	Medium/ High	USAWeatherisation program: BC-ratio: 2.4	Most effective if combined with other measures such as financial incentives, regular updates, stakeholder involvement in supervisory systems.	WEC 2001, IEA 2005b
Utility demand-side management programs	USA, CHE, DNK, NLD, DEU, AUT, THA, JAM	High	USA: 54.7 TWh in 2004; JAM: 13 GWh/year, 4.9% less electricity use; DNK: 179 GWh; THA: 5.2 % of annual electricity sales 1996–2006	High / Medium	EU: -0.07$/kWh; DNK: ca.-0.06$/kWh; USA: average costs app. -0.05$/kWh; THA: 0.01$/kWh	More cost-effectivein the commercial sector than in residences, success factors: combination with regulatory incentives, adaptation to local needs & market research, clear objectives.	IEA 2005b, Kushler et al. 2004, Evander et al. 2004, Mills 1991, Parfomak and Lave 1996, Gillingham et al. 2006

Economic and market-based instruments

Instrument	Countries	Rating	Energy savings	Rating	Cost	Effectiveness	References
Energy performance contracting/ ESCO support	DEU, AUT, FRA, SWE, FIN, USA, JPN, HUN, CHN	High	FRA, SWE, USA, FIN: 20–40% of buildings energy saved; EU: 104–144 TWh by 2010; USA: 6.3 TWh/yr; CHN: 44 TWh	Medium/High	EU: mostly at no cost, rest at <0.008$/kWh; USA: Public sector: B/C ratio 1.6, Private sector: 2.1	Strength: no need for public spending or market intervention, co-benefit of improved competitiveness.	ECCP 2003, OPET network 2004, Singer 2002, IEA 2003, WEC 2004, Goldman et al 2005, Evander et al. 2004
Cooperative technology procurement	DEU, ITA, SVK, UK, SWE, AUT, IRL, USA, JPN	Medium/ High	USA: 15–56 TWh/yr estimated by 2010; USA: 192,750 MWh saved in refrigerator procurement; DEU: German telecom company: up to 60% energy savings for specific units	High/ Medium	USA: -0.07$/kWh; SWE: 0.12$/kWh (BELOK-program)	Combination with standards and labelling, choose products with technical and market potential.	Oak Ridge National Lab 2001, Le Fur B. 2002, Borg et al. 2003, Nilsson 2006, Gillingham et al. 2006
Energy efficiency certificate schemes	ITA, FRA	Medium	ITA: 3.25TWh in 2006, 9.1 TWh by 2009 expected	Medium	FRA: max. 0.01$/kWh estimated	No long-term experience yet. Transaction costs can be high. Adv. institutional structures needed. Profound interactions with existing policies. Benefits for employment.	OPET network 2004, Bertoldi/Rezessy 2006, Lees 2006, Defra 2006c, IEA 2006, Beccis 2006

Table 10.19 | (cont.)

Policyinstrument	Country/ regions examples	Effective-ness	Energy reductions for selected best practices	Cost-effectiveness	Cost of energy reduction for selected best practices in US$_{2005}$$	Special conditions for success, major strengths and limitations, co-benefits	References
Fiscal instruments and incentives							
Taxation (on household fuels)	NOR, DEU UK, NLD, DNK, CHE, SWE	Low/ Me-dium	DEU: household energy usereduced by 0.9 % in 2003: app. 10 TWh NOR: 0.1–0.5% 1987–1991 NLD: 1.1–1.6 TWh in 2000 SWE: 5% 1991–2005, 48 TWh	Low		Effect depends on price elasticity. Revenues can be earmarked for further energy efficiency improvements. More effective when combined with other tools.	WEC 2001,Kohlhaas 2005, Larsen and Nesbakken 1997, MURE 2007, Brink and Erlandsson 2004
Tax exemptions/ reductions	USA, FRA, NLD, KOR	High	USA: 174 TWh in 2006 FRA: 25 TWh NLD¹: 0.4 TWh/yr. or 0.3% of sectoral reference energy in 1997–2004	High / Medium	USA: B/C ratio Com-mercial buildings: 5.4 New homes: 1.6 NLD: 0.02$/kWh	If properly structured, stimulate introduction of highly efficient equipment and new buildings.	Quinlan et al. 2001, Geller and Attali 2005, MURE 2007, Khan et al 2007
Public benefit charges	BEL, DNK, FRA, NLD, USA, BRA	Medium	USA: 0.1–0.8% of total el. sales saved each year, 2.8 TWh/yr savings in 12 states NLD: 7.4TWh in 1996 BRA: 1.95 TWh	Medium	USA:0.03–0.05$/kWh	Success factors: Independent administration of funds, involve-ment of all stakeholders, regular evaluation/ monitoring& feed-back, simple and clear program design, multi-year programs	Western Regional Air Partnership 2000, Kushler et al. 2004, Lopes et al. 2000
Capital subsidies, grants, sub-sidised loans	JPN, SVN, NLD, DEU, CHE, USA, CHN, UK, ROU, DNK, BRA	High/ Medium	SVN: up to 24% energy savings for buildings BRA: 5.2 TWh UK: 13 GWh/yr ROU: 414 GWh DEU: 0.7–1.0 TWh/yr. or 0.1% of sectoral reference energy in 1996–2004	Estimates vary from Low to High	DNK: −0.004$/kWh UK: 0.01$/kWh for government, −0.05$/kWh net NLD: 0.02–0.05$/kWh DEU: 0.02–0.04$/kWh	Positive for low-income households, risk of free-riders, may induce pioneering investments.	ECS 2002, Martin Y. 1998, Schaefer et al. 2000, Geller et al. 2006, Joosen et.al. 2004, Shorrock 2001, Berry and Schweitzer 2003, Khan et al 2007
Support, information and voluntary action							
Voluntary certification and labelling	DEU, CHE, USA, THA, BRA, FRA	Medium/ High	BRA: 5.3 TWh in 1998 USA: 53 TWh in 2004, 3558 TWh in total by 2012 THA: 311 MWh	High	USA: −0.03$/kWh BRA: $20 million saved	Effective with fiscal incentives, voluntary agreements and regulations, adaptation to local market is important.	OPET 2004, Geller et al. 2006,WEC 2001, Webber et al. 2003, US EPA 2003

Instrument	Countries		Energy savings		Cost	Conditions for success	References
Voluntary and negotiated agreements	Mainly Western Europe, JPN, USA, DEN	Medium/ High	USA: app. 275 TWh in 2000 EU: 100 GWh/yr (300 buildings) UK: app. 41 TWh in 2006 DEN (coupled with subsidies): 0.3 TWh/yr. or 0.5% of sectoral reference energy in 1996–2003	Medium	SWE: 0.02$/kWh DEN: 0.01$/kWh	Can be effective when regulations are difficult to enforce. Effective if combined with fiscal incentives, and threat of regulation. Inclusion of most important manufacturers, and all stakeholders, clear targets, effective monitoring important.	Geller et al. 2006, Cottrell 2004, Gillingham et al. 2006, Bertoldi et al. 2005, Bertoldi and Rezessy 2007, DEFRA 2007a, Khan et al 2007
Public leadership programs	NZL, MEX, USA, PHL, ARG, BRA, ECU, ZAF, DEU, GHA	Medium/High	DEU: 25% public sector energy savings in 15 years USA: 4.8 GWh/yr USA: 0.4% of sectoral reference energy in 1985–2004 BRA: 140 GWh/yr GHA: 27 MWh (14% of baseline) MEX: 200 GWh/yr(13% of baseline)	Medium	USA: DOE/FEMP estimates $4 savings for every $1 invested USA: 0.04 $/kWh EU: $13.5 billion savings by 2020 ZAF: 0.07$/kWh BRA: -0.08$/kWh	Can be used to demonstrate new technologies and practices. Man-datory programs have higher potential than voluntary ones. Clearly state, communicate and monitor, adequate funding and staff, involve building managers and experts.	Borg et al. 2003 &2006, Harris et al. 2005, Van Wie McGrory et al. 2006, OPET 2004, Van Wie McGrory et al. 2002, Khan et al 2007
Awareness raising, education, information campaigns	DNK, USA, UK, FRA, CAN, BRA, JPN, SWE, ARG	Low/ Medium	UK: app. 50 GWh annually ARG: 25% in 2004/05, 4.1 TWh FRA: 1 GWh/yr BRA: 69 GWh/yr, 203–381 TWh/yr with voluntary labeling 1986–2005 SWE: 48 GWh/yr	Medium/ High	BRA: -0.002$/kWh UK: -0.02$/kWh (households), -0.02$/kWh(business) SWE: 0.02$/kWh	More applicable in residential sector than commercial. Deliver understandable message and adapt to local audience.	Bender et al. 2004, Dias et al. 2004, IEA 2005b, Darby 2006b, Ueno et al. 2006, Defra 2006c, Lutzenhiser 1993, Savola, 2007
Detailed billing and disclosure programs	Ontario, ITA, SWE, FIN, JPN, NOR, AUS, USA, CAN, UK	Medium	Max.20% energy savings in households concerned, usually app. 5–10% savings UK: 3% NOR: 8–10 % USA: 2.1%	High	NOR: -0.04$/kWh USA: -0.07$/kWh	Success conditions: combination with other measures and periodic evaluation. Comparability with other households is positive.	Crossleyet al., 2000, Darby 2000, RobertsandBaker 2003, Energywatch 2005, Wilhite and Ling 1995, Allcot 2010.

Country name abbreviations: DZA – Algeria, ARG – Argentina, AUS – Australia, AUT – Austria, BEL – Belgium, BRA – Brazil, USA/Cal – California, CAN – Canada, BC/CAN – British Columbia, CEE – Central and Eastern Europe, CHN – China, CRI – Costa Rica, CZE – Czech Republic, DEU – Germany, DNK – Denmark, ECU – Ecuador, EGY – Egypt, EU – European Union, FIN – Finland, FRA – France, GHA – Ghana, HKG – Hong Kong, HUN – Hungary, IRL – Ireland, ITA – Italy, JAM – Jamaica, JPN – Japan, KOR – Korea (South), MAR – Morocco, MEX – Mexico, NLD – Netherlands, NOR – Norway, NZL – New Zealand, PHL – Philippines, POL – Poland, ROU – Romania, ZAF – South Africa, SGP – Singapore, SVK – Slovakia, SVN – Slovenia, CHE – Switzerland, SWE – Sweden, THA – Thailand, UK – United Kingdom (Great Britain and Northern Ireland), USA – United States.

[1] The tool covers the commerce and the industry and the results are for both sectors.

731

Table 10.20 | Barriers to energy efficiency and policy instruments as remedies.

Barrier category	Examples of barriers	Instrument category	Policy instruments as Remedies
Economic barriers	Higher up-front costs for more efficient equipment / buildings; lack of access to financing; energy subsidies; lack of internalization of environmental, health, and other external costs	Regulatory-normative/ regulatory-informative	Appliance standards, building codes, energy efficiency obligations, requiring the use of a technology or a building with specific standards. Procurement regulations ensuring the purchase of energy efficient equipment. DSM programs.
		Economic instruments	Under EPC, ESCOs finance energy efficiency improvements (addressing the issues of high front-up costs as well the lack of access to financing) and are paid from the energy cost reductions achieved. Cooperative procurement, which is used by big buyers to specify the energy efficient standards for the equipment they use. Energy efficiency certificates. Other risk-sharing/financing instruments like co-operation with banks/public-private partnerships, guarantee schemes.
		Fiscal instruments	Tax exemptions, subsidies/rebates/grants to reduce up-front costs. Taxation and/or public benefit charges to increase energy prices and to internalize externalities.
Hidden costs/benefits	Costs and risks due to potential incompatibilities, transaction costs, etc.	Regulatory-normative	Appliance standards and building codes, requiring the use of particularly technologies or solutions. Labelling for informing the public about product standards, etc., resulting in reducing transaction costs for seeking information.
		Economic instruments	ESCOs can undertake a significant part of these hidden costs and risks.
		Support action	Public leadership programs to demonstrate new technologies and to gain experience from their implementation. Information and advice programs.
Market imperfections	Limitations of the typical building design process; fragmented market structure; landlord/tenant split and misplaced incentives; administrative and regulatory barriers (e.g., in the incorporation of distributed generation technologies)	Regulatory-normative/ regulatory/informative	Appliance standards, building codes, energy efficiency obligations, imposing specific standards to buildings/equipment. A holistic approach engaging engineers, architects, etc., from the early stages of the design process. Mandatory labelling for informing the public about product standards. Procurement regulations and DSM programs. Regular monitoring and evaluation of programs and implemented policies.
		Economic instruments	ESCOs can handle more effectively the administrative barriers for implementing energy conservation in buildings. Energy efficiency certificates and Kyoto Flexibility mechanisms can create incentives for implementing energy conservations measures.
		Fiscal instruments	Taxation and public benefit charges giving the right price signals. Tax exemptions, subsidies/rebates/grants to accelerate first-movers.
		Support, information, voluntary action	Public leadership programs, awareness raising campaigns. Education programmes for builders, architects, engineers, etc.
Cultural/ behavioral barriers	Tendency to ignore small opportunities for energy conservation, organizational failures (e.g., internal split incentives); tradition, behavior, lack of awareness and lifestyle; non-payment and electricity theft.	Support, information, voluntary action	Awareness raising through information campaigns. Detailed billing programmes, providing information to the final consumers as regards the most energy-requiring uses and equipment. Public leadership programs to demonstrate new technologies and practices. Voluntary agreements between governmental bodies and a business or organization, which undertakes the responsibility to increase its energy efficiency. Voluntary labelling programmes.
		Fiscal instruments	Taxation and/or public benefit charges to increase energy prices and to give the right signal in the market.
		Regulatory-normative	Appliance and equipment standards as well as other normative measures are a good way to overcome the tendency to ignore small opportunities (cf. phasing out of incandescent bulbs).
Information barriers	Imperfect information.	Support, information, voluntary action	Awareness raising through information campaigns. Detailed billing programmes, providing information to the final consumers as regards the most energy-requiring uses/equipment. Public leadership programs to demonstrate new technologies and practices. Voluntary labelling programmes.
		Regulatory-informative	Mandatory labelling for informing the public about product standards. Procurement regulations ensuring the purchase of energy efficient equipment. DSM programs. Mandatory audits.

Source: IPCC, 2007; Carbon Trust, 2005; Ürge-Vorsatz et al., 2007.

Table 10.21 | Characteristic examples of possible policy instrument packages and examples of commonly applied combinations.

Measure	Regulatory Instruments	Information Instruments	Financial /Fiscal Incentives	Voluntary Agreements
Regulatory instruments	Building codes and standards for building equipment	Standards and information programs	Building codes and subsidies	Voluntary agreements with a threat of regulation
Information instruments	Appliance standards and labelling	Labelling, campaigns, and retailer training	Labelling and subsidies	Voluntary MEPS and labelling
Financial/Fiscal Incentives	Appliance standards and subsidies	Energy audits and subsidies Labelling and tax exemptions	Taxes and subsidies	Technology procurement and subsidies
Voluntary Agreements	Voluntary agreements with a threat of regulation	Industrial agreements and energy audits	Industrial agreements and tax exemptions	

10.8.3.1   Standards, Labeling, and Financial Incentives

There are several effective policy options available to accelerate the transformation of appliance markets toward higher efficiency, including MEPS (Minimal Energy Performance Standards), voluntary or mandatory consumer information labels, and publicity or rebate programs sponsored by utilities or government agencies. Appliance efficiency standard and labeling programs have by now become a core part of energy efficiency programs in many countries.

Among the oldest and most comprehensive programs are the US federal MEPS program, the comparative labeling program implemented by the European Union, and the Energy Star endorsement label program, which is a program of the United States but has become widely recognized internationally. Minimum performance standards are used in the European Union, as well as in developing countries such as China, Tunisia, and Thailand. Appliance standards have been perhaps the most successful policy for improving energy efficiency, in part because they capture 100% of the market in a few years' time (US DOE, 2008) and they remove key market barriers – such as lack of interest, incentives, etc. – and transaction costs.

For office equipment, voluntary information programs have induced manufacturers to improve efficiency (US EPA, 2007). The year 2007 brought new Energy Star specifications for office and imaging equipment. In addition to reducing power use of the products themselves, the new specifications also set additional requirements for accessories. If an imaging product is sold with an external power adapter, cordless handset, or digital front-end, these accessories must meet Energy Star External Power Supply, telephony, or computer specifications. These requirements ensure that the Energy Star represents only the market's most energy-efficient products. Energy Star-qualified office and imaging products use 30–75% less electricity than standard equipment.

Due largely to the example set by the success of programs in the United States, the European Union, and Japan, there has been a proliferation of similar programs throughout the world in the past two decades. The number of programs exceeds 60 (Wiel and McMahon, 2005), and they cover dozens of different residential, commercial, and industrial products.

Appliance standards are often combined with labeling and rebates in order to give incentives for investments beyond the level required by the minimum energy efficiency standard. McNeil et al. (2008) demonstrate the importance of implementing both energy efficiency labeling and standards, showing that enforcing such policies on a global scale would lead to worldwide savings of 1113 TWh/yr of electricity and 327 TWh/yr of fuels by 2020, and 3385 TWh/yr of electricity and 928 TWh/yr of fuels by 2030. In addition, rebates for the most energy-efficient products encourage consumers to buy these, which reinforces and sustains market transformation.

The Japanese Top Runner approach is another unique and successful method to improve the energy efficiency of appliances (Murakoshi et al., 2005). In the Top Runner approach, government sets target energy efficiency values and years for appliances, including scope, based on the highest energy efficiency products on the market, and encourages manufacturers to make products better than this target energy efficiency value. Energy efficiency values and indicator labels are voluntarily displayed in catalogs and other advertising and publicity material so that consumers can consider energy efficiency when making purchases. In addition, the Top Runner program sets fleet standards for appliances. This system of voluntary agreements between the Japanese government and manufacturers has been highly effective, leading Japan's appliance market to be among the most efficient in the world.

Building codes can also be combined successfully with voluntary or mandatory certification of buildings (IEA, 2010c) such as through rating systems like the British BREEAM,[30] the Japanese CASBEE,[31] and the

---

30   BREEAM (BRE Environmental Assessment Method) is a widely used environmental assessment method for buildings. It was developed by BRE (Building Research) Trust Companies and covers a wide range of building types (www.breeam.org/).

31   CASBEE (Comprehensive Assessment System for Built Environment Efficiency) is an assessment tool based on the environmental performance of buildings or urban area. CASBEE evaluates a building from the two viewpoints of environmental quality and performance (Q=quality) and environmental load on the external environment (L=load) and defines a new comprehensive assessment indicator, the Building Environmental Efficiency (BEE), by Q/L (www.ibec.or.jp/CASBEE/english/index.htm).

American LEED system.[32] The European Union's EPBD[33] is actually an example of combining codes with certification. A number of developing countries, such as China, intend to introduce building rating schemes to complement building codes (Ürge-Vorsatz and Koeppel, 2007). In the United States, mandatory energy efficiency regulations are sometimes coupled with voluntary labels, such as ENERGY STAR, and tax credits to manufacturers and to consumers. This combination eliminates the least efficient products, while compensating manufacturers for some of the increased production costs both through tax credits and through premiums charged for ENERGY STAR designs.

10.8.3.2 Regulatory and Information Programs

Regulatory policy instruments are usually effective, but lack of enforcement can be a barrier and the rebound effect may result in some benefits being taken as increased service rather than reduced consumption. Awareness might improve compliance and help overcome the rebound effect in more affluent population groups where energy service levels are not constrained.

10.8.3.3 Public Leadership Programs and Energy Performance Contracting (EPC)

By improving its own energy efficiency, the public sector can not only save costs, but also demonstrate to the private sector the potential and feasibility of energy efficiency improvements and trigger market transformation. EPC in the public sector is especially advantageous, as the budget of many public administrations is limited. Executive orders that oblige public authorities to reduce energy use by 30% and the federal energy management program in the United States, as well as the Energy Saving Partnership in Berlin, Germany, have boosted the energy service company (ESCO) industry (Ürge-Vorsatz and Koeppel, 2007). However, significant barriers still hamper EPC in the public sector in developing countries.

10.8.3.4 Financial Incentives and Labeling

In order for financial incentives such as loans, subsidies, and tax credits to be most effective, the labeling of energy efficient products is necessary, which ensures that only the most efficient categories of equipment are financially supported (Menanteau, 2007). On the other hand, labeling, particularly voluntary labeling alone, might not be effective (Menanteau, 2007), because if the premium labeled products are substantially more expensive, that discourages especially low-income

households from purchasing them. This implies that governments might consider incentive schemes for companies that undertake labeling.

10.8.4 Policy Instruments Addressing Selected Barriers and Aspects Toward Improved Energy Efficiency in Buildings

10.8.4.1 Policies for Retrofit

While the majority of broad policy approaches – information, labeling, standards, incentives, etc. – are in principle as applicable to building retrofit as new buildings, there are clearly important differences in the policy measures required. In most countries, due to the long lifetime of buildings replacement rates of the building stock are low, and therefore retrofit will be essential to achieving rapid progress in energy efficiency.

Securing low-energy retrofit activity raises some different issues. This is partly because codes and standards are expected to deliver much more in new construction. Retrofitting existing buildings is a discretionary investment – no action is an option, and often an easier option. Building owners and occupiers therefore need to be persuaded not only of the merits of energy investment, but to finance it and bear whatever disruption it entails. Incentives may therefore need to be higher than for new buildings. This sub-section therefore focuses upon the differences in policy mix that may be required for building retrofit.

The incentive policies reviewed above are particularly relevant to retrofitting policies. In some cases, e.g., in the United Kingdom, there is support for energy efficiency measures in low-income households from government-funded programs. More commonly energy efficiency retrofits are supported through a variety of fiscal measures, including income tax credits – e.g., in the United States and France – and low rates of relevant sales taxes – e.g., reduced rate value added tax (VAT) in some European Union countries. In some cases, these formed part of green stimulus packages in 2009.

Programs of financial support for building energy efficiency through energy utility regulation are increasingly common and growing in size. First known as Demand Side Planning or Integrated Resource Planning and used in the United States in the 1980s in vertically integrated monopoly utilities, they have now been adapted to a range of regulatory environments. In principle, any category of energy efficiency may be addressed in this way, but in practice building retrofits predominate. In some states of the United States and some European Union countries, the United Kingdom, Italy and France, program savings approach or exceed 1% of regulated energy use (e.g., York, 2008; Eyre et al., 2009).

Performance contracting by energy service companies is widely used in many countries for the retrofitting of commercial buildings. Subsidized energy advice has been used widely to increase information on energy

---

32 LEED (The Leadership in Energy and Environmental Design) is a Green Building Rating System developed by the United States Green Building Council and providing a suite of standards for environmentally sustainable construction (www.usgbc.org/).

33 EPBD (EU Directive on the Energy Performance of Buildings) is a range of provisions aimed at improving energy performance of residential and non-residential buildings, both newly built and already existing (www.buildingsplatform.org/cms/).

efficiency opportunities in the existing building stock. These have been linked to incentive schemes, which may be in the form of grants, loans, tax incentives, or energy company incentive payments.

An audit is generally recognized as the precursor to effective retrofit investment. The logic of this is reflected in policies that increasingly require mandatory labeling and audit and certification of buildings, e.g., European Performance of Buildings Directive. The first aim of such labeling is to provide consumer information, but labeling schemes are also essential as a tool in incentive or regulatory policies that are linked to building performance.

Although codes and standards are best known for use in new buildings, they are increasingly used in retrofit. This can be to require performance standards at the time of major refurbishment, and this applies in the European Union to buildings through the Energy Performance of Buildings Directive. Standards are also applied to individual components in retrofit, e.g., heating, glazing, air conditioning, and this is very cost effective (e.g., DEFRA, 2007b). Germany has extended this principle to both fabric elements and whole building performance at the point of major refurbishment (Dilmetz, 2009). However, to date there has been very limited use of standards for whole building performance to require refurbishment, for example at the point of sale or rent. At present, actual regulation of this type has probably been confined to a few parts of the United States – San Francisco, Berkeley, Davis, Burlington, Ann Arbor, and the state of Wisconsin (CLG, 2010). The adoption of mandatory labeling and retrofit codes (e.g., in Europe) in principle provides a basis for wider use of this approach.

It is increasingly recognized that substantial retrofit will be required if older buildings are to reach the energy efficiency standards implied by the ambitious targets of many governments. Traditional incentive mechanisms that support individual components will not be sufficient to deliver the major changes in fabric, airtightness, and heating and cooling system efficiency that are required. Policies that support very substantial improvement are beginning to be explored. For example, in Germany there are 100% low interest loans up to Euro 50,000 (approximately US$_{2005}$\$59,000) for CO_2 rehabilitation of buildings, supporting very low energy refurbishment (Schonborn, 2008). In Berkeley, California, to encourage energy-efficient renovations of residential and commercial buildings, the municipality provides funds that are to be repaid within 20 years through property taxes (Fuller et al., 2009). Since the 2009 financial crisis, with lending significantly reduced, schemes of this type are of increasing importance for supporting retrofits. However, policies of this type are on hold in the United States because of concerns that they infringe on contractual obligations to mortgage lenders.

Policies for retrofit to low energy standards also need to deal with practical complexities inherent in the diversity of current buildings, many already significantly altered since original construction and often poorly built, maintained, and documented. Retrofit has to deliver the range of outcomes defined by building owners or managers and provide the

services expected by future occupants, when known. In general, these clients will have little energy knowledge and very little insight into the challenge of low energy retrofits. Moreover, in the case of minor changes and even major ones to small buildings, e.g., single-family dwellings, retrofit may frequently be done without the oversight of architects or energy services professionals. In this environment, delivering the very low levels of air infiltration and thermal bridging implied by passive building standards involves some practical challenges for the retrofit process. At the very least, the widespread adoption of low energy retrofit programs will also require a substantial program of training in the building sector (Killip, 2008).

10.8.4.2   Policies Addressing Non-compliance Related to Building Codes

Building codes have served as a major policy tool for reducing energy required, especially for building services such as heating, cooling, water heating, and lighting (Listokin et al., 2004). Existing practices in several countries show there are mainly two types of building codes: (1) overall performance-based codes requiring that a building's predicted energy demand or energy cost (usually determined through an energy modeling software), is equal to or lower than a baseline target that has been specified by the code; and (2) prescriptive codes, which set separate energy performance levels for major envelope and equipment components. A combination of an overall performance requirement with some component performance targets (e.g., wall insulation) is also possible.

Computer simulation tools have existed since the 1970s to calculate the energy performance of buildings based upon their design, and the results have been validated – or improved – by comparison with measured energy use. The full potential for energy savings from building codes has not been achieved, in part because compliance and enforcement are not complete. The range of experiences is broad, from jurisdictions where even the structural integrity of buildings may be compromised – sometimes revealed when earthquakes cause widespread damage in a region – to buildings failing to meet fire codes, to buildings meeting some but not all requirements, to those meeting all requirements.

Good statistics are lacking to quantify the problem globally. In many parts of the world, there are no legal requirements for building code enforcement. Even in some of the most prosperous regions, the level of resources available for enforcement, such as the number of local code enforcement officials or information and tools available to them, is often inadequate. In some developed areas, code compliance has been shown to be about 50% (Usibelli, 1997). A recent California study found rates of non-compliance by measure ranging from 28–73% for residential and 44–100% for non-residential. In both cases, the lowest non-compliance was for lighting, and the highest was for ducts (Sami Khawaja et al., 2008).

While performance measurement, by means of commissioning or research on occupied buildings, may be necessary to rigorously compare actual performance with the energy performance projected by computer models using design parameters, simpler inspections are sufficient to detect significant non-compliance. Failure to comply can occur because actual construction practice deviated from design, or inferior materials or equipment were substituted, or installation was flawed or incomplete. Studies have suggested ways to improve building regulations (for a recent European example, see Garcia Casals, 2006).

Aiming at enhancing the effectiveness of building codes, there has been much discussion recently about a new energy code compliance framework based on actual post-construction energy performance outcomes of buildings, called outcome-based building codes (Hewitt et al., 2010). In other words, owners would have the flexibility to pursue whichever retrofit strategies they deem appropriate to their individual buildings, but would be required to actually achieve a pre-negotiated performance target, demonstrated through mandatory annual reporting of energy use. Although no current codes regulate actual energy use, outcome-based codes may be a very critical tool toward deep energy savings in existing buildings. As existing buildings do have an energy performance history, benchmarking plays a vital role in establishing an outcome-based compliance path in energy codes (Denniston et al., 2010). An outcome-based performance path may be based on either requiring certain percent improvement in energy performance, assuming old energy performance needs are known, or absolute performance goals, given sufficient data about building performance needs in order to establish those goals are available.

Generally, successful building energy codes will likely require: (1) clear, consistent code documentation, that is as simple as possible; (2) providing sufficient information, training, and motivation to practitioners; (3) adequate local resources to check compliance, keeping in mind that the sheer size of the construction industry argues for simpler methods and sampling using statistical methods, rather than rigorous checking of every building; (4) feedback to architects, designers, builders, contractors, and consumers to identify good and bad practice; (5) penalties for consistent bad practice; and (6) performance-based rating systems to close the loop from actual performance back to design and construction.

10.8.4.3   Policies Addressing Professional, Social, and Behavioral Opportunities and Challenges

A substantial part of the huge energy conservation potential in buildings requires the effective engagement of human dimensions of providing, adopting, and using energy efficient technologies and buildings. These are set out in Section 10.4.8 above. In many cases, these factors constitute significant barriers for capturing even low cost, energy conservation potential. To overcome these barriers, appropriately designed policies or portfolios of policies can be implemented that are briefly described below.

The adoption of behavior to use less energy by building occupants can be enhanced through information, advice, and educational programs and the provision of more detailed and direct feedback on energy use by end-use and energy expenditures. It has been noted that, in many cases, people underestimate the amount of energy they use for specific energy uses. To this end, detailed billing and disclosure programs have been estimated to potentially save up to 20% of energy (Darby, 2000) and are mostly cost-effective (Ürge-Vorsatz et al., 2007).

However, better feedback to final consumers on energy use may increase costs and reduce revenues for the energy supplier. The policy implication is that regulation is likely to be needed to set a minimum standard for the quality and frequency of energy use information. This could include frequency of meter reading and billing and requirements for comparison with historical data and consumption of similar users. The advent of smart meters provides new opportunities to enhance information across the energy supply chain, including to final users. Again, this needs a clear regulatory framework to be successful, including timescales for deployment and standards for inter-operability, consumer information, and data transfer.

Carefully designed and targeted awareness-raising programs to secure specific outcomes can have a place in policy packages to reduce energy use. The Japanese "Cool Biz" campaign is a good example (see Box 10.3). However, there is limited evidence that general exhortation, e.g., advertising campaigns, has a significant lasting impact. Energy saving information needs to be clear and relevant and provided at the point of key decisions; product labeling is the best example of this. Advice on energy saving opportunities is generally most effective when it is specific to personal circumstances, especially when provided by trusted sources independently of energy companies and as part of policy packages including incentives and other support to act. Consumers are unlikely to seek this out, or pay for it at cost, and therefore it needs to be provided as a public or community service, for example by local government or specialist energy agencies.

Energy pricing may also be a useful tool for influencing energy use behavior in buildings. Furthermore, even if residential energy price elasticities are relatively low in high income countries, appropriate energy pricing systems, e.g., with the adoption of staggered rates, may provide impetus for a more rational use of energy sources in buildings (Reiss and White, 2008; National Action Plan for Energy Efficiency, 2009).

Finally, non-technological options to reduce energy use in new buildings related to architectural decisions at an early design stage, e.g., building shape, form, orientation, size, etc. (see Section 10.4.2), can be supported through appropriately designed building codes, regulations on urban density, building heights, and the mix of land uses, etc. (WBCSD, 2008). These mechanisms already exist in many countries. However, their effective implementation and particularly their revision to incorporate energy efficiency aspects remain a challenge in most developing and in several developed countries.

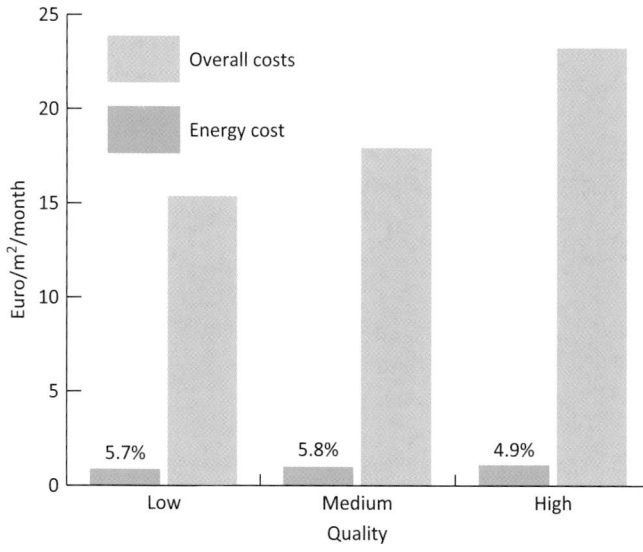

Figure 10.30 | Average values of total costs by quality of fittings. Source: Jones Lang LaSalle, 2006; based on 397 buildings with six million m² in 2006.

Most policies to address human dimensions of energy use in buildings are designed to improve energy efficiency, rather than to bring about more fundamental lifestyle changes. However, broader policy frameworks do have implications for culture and lifestyle, although these raise complex and controversial issues. Greater attention to these may be required in the future and should be the subject of further research. These broader lifestyle issues are addressed in Chapter 21.

10.8.4.4   Energy Cost Information and Analysis

Energy is typically a small proportion of total operating costs for buildings. For example, in a high quality office building in Germany, heating and electricity made up less than 5% of the total running cost of the building – about 1.1 out of every €23.3 (1.4 out of US$_{2005}$$29.8) spent (see Figure 10.30).

Energy costs are more significant for direct investors. Energy efficiency is part of the due diligence for the procurement of new properties by those who will own and operate them. For other investors, energy costs are not important. There is emerging evidence that an energy-efficient building can command a premium. One US study (McGraw-Hill Construction, 2006) found that professionals expect greener buildings to achieve an average increase in value of 7.5% over comparable standard buildings, together with a 6.6% improved return on investment. Average rents were expected to be 3% higher.

The insurance industry faces substantial risks from climate change. Based on the consideration of those risks, many insurance companies are providing incentives toward more energy efficient, and climate neutral, buildings. Green insurance is not very common yet, and only exists in developed countries, but is gaining increasing significance.

While energy costs are a relatively small part of total occupancy costs, they can still be a significant factor in motivating energy efficiency action. But profitable opportunities for energy savings are often overlooked because of inadequate cost information. Despite real estate managers' stated interest in energy efficiency, a study in 2006 found that only two-thirds of companies tracked energy data and only 60% tracked energy costs (WBCSD, 2007).

In the United States, only 30% of real estate managers or facilities managers claimed to have included energy efficiency requirements into requests for proposals. Despite these findings, the study surprisingly suggested that energy costs are the most important driver for energy efficiency, both currently and in the future. Energy managers and investment decision makers need to develop a common methodology and language for valuing energy efficiency projects in a similar manner to other investments (Jackson, 2008). With such a risk analysis framework in place, energy efficiency experts and investment decision makers could exchange the information they need to expand investment into energy efficient buildings projects. Accurate and robust analysis demands a high level of understanding of the physical aspects of energy efficiency, which enables physical performance data to be translated into the language of investment. However, while there is a general recognition that energy efficiency practices and products are becoming more widespread in the market place, there are limited data on how these factors impact the value of buildings. The financial effectiveness of capital improvements that target energy demand reduction is usually assessed in terms of simple payback times and does not typically reflect a property investor's valuation methods.

10.8.4.5   New Business Models – ESCOs

Appropriate commercial relationships can increase the focus on energy costs by altering commercial relationships, removing the split incentives problem and introducing more effective incentives for reducing energy use and costs.

An energy performance contract is an arrangement between a property owner and an ESCO that covers both the financing and management of energy-related costs. It involves a variety of mechanisms to help property owners use the knowledge of energy professionals to reduce energy costs. While the effectiveness of ESCOs is well documented, miscalculation of financial and technical risks has caused many failures of these firms.

ESCOs generally act as project developers, installers, and operators over a seven–ten year time period. They assume the technical and performance risk associated with the project. The services offered are bundled into the project's cost and are repaid through operational savings generated, with the ESCO's profit coming from a proportion of cost savings or a fixed fee based on projected energy savings.

As an additional service in most contracts, the ESCO provides any specialized training needed so that the customer's maintenance staff can take over at the end of the contract period. ESCOs have placed great emphasis on measurement and verification and have led the way to verify, rather than estimate, energy savings. One of the most accurate means of measurement is the relatively new practice of metering, which is the direct tracking of energy savings according to sanctioned engineering protocols.

10.8.4.6   New Financial Instruments

Ways to shift the financial equation in favor of energy-efficient investment include reducing the first cost or increasing the savings in the early years.[34] One widely recognized way of increasing potential savings is to increase the cost of energy. These are useful mechanisms across the broader economy. However, WBCSD modeling shows they are likely to have a limited impact on building energy investment decisions if set at a level that is acceptable politically and economically. Even a relatively high carbon price, for example US$60/t$CO_2$, does not add enough to the energy cost to make energy savings sufficiently attractive.

Potential savings can be increased through commercial means. In some countries, utility-charging practices may encourage waste because of discounts for higher use – the unit rate typically declines above specified consumption levels. Reversing this practice would increase the cost of energy at higher consumption levels. This is already the case in Japan, where the first 120kWh of monthly electricity consumption are charged at JPY17.87/kWh (US¢18), increasing to JPY22.86 (US¢23) up to 300kWh and JPY24.13 (US¢24) above that level. A high feed-in tariff for renewable energy supplied to the grid may encourage investment in on-site renewable generation, as is already the case in countries like Germany and France.

WBCSD/EEB modeling has clearly shown that many potentially attractive energy investments do not meet the short-term financial return criteria of businesses, investors, and individuals. While significant savings are possible with relatively modest investment premiums, a first-cost-sensitive buyer will never adopt transformative solutions.

Solutions are to attract new sources of funding, learning from best practice and experience with business models such as ESCOs. Several opportunities are available to open up finance for energy investment:

- Pay as you save: the first cost is financed in full or in part by an energy utility, which recoups the outlay through regular surcharges on the monthly bill; these surcharges attach to the house, not the specific customer.

- ESCOs: utilities or other providers contract to achieve specified energy performance for a commercial building and share the savings with the owner.

- Energy performance contracting: schemes enabling energy services companies or other players to offer innovative contracts guaranteeing the level of services and the energy savings to the customer, as above.

- Locally available loans: local authorities provide loans to finance energy investment, and repayments are made through an addition to the property tax charge.

- Cross-subsidized mortgages for energy-efficient buildings: higher rates for low-efficiency buildings and lower rates for efficient ones.

- Energy efficiency investment funds: capitalizing on the lower risk of mortgage lending on low-energy housing, funds to provide such investment could be attractive to socially responsible investment funds.

10.8.4.7   Toward Low-energy and Zero-energy Buildings

Recently, several countries and organizations adopted ambitious goals with far-reaching implications regarding energy use in buildings. In the United Kingdom, the government is already committed to all new housing being carbon neutral from 2016. In the United States, a number of zero-energy initiatives have been adopted at the state level. For example, in California the zero-energy target has been specified for all new residential buildings by 2020 and for all new commercial buildings by 2030. Analogous targets have been adopted by several other countries, such as Denmark, France, the Netherlands, etc. Table 10.22 reviews these ambitious targets and the related policies that have been adopted by key countries to substantially reduce energy use in new buildings.

Also, the recast of the European Union's EPBD is stimulating Member States to develop frameworks – national plans, etc. – for higher market uptake of low- or zero-energy and carbon buildings. To this end, Member States shall set specific targets for 2020 with respect to the penetration of these buildings in relation to the total number of buildings and the total useful floor area.

The WBCSD in the context of the Energy Efficiency in Buildings project has adopted a vision in which all buildings in the world will consume zero net energy by 2050. In addition, the buildings must also be aesthetically pleasing and meet other sustainability criteria, especially for air quality, water use, and economic viability (WBCSD, 2008). Also, Architecture 2030, an independent nonprofit organization, proposed the 2030 Challenge in 2006, tasking the global architecture and building community to adopt a plan for making new buildings and major renovations carbon neutral by 2030. The plan sets an immediate energy

---

34 Increasing access to capital requires a very different set of tools, such as preferential rate loans, risk-guarantee mechanisms, or others.

Table 10.22 | Examples of countries and regions which already adopted ambitious targets toward low-energy and zero-energy buildings.

Country / Region	Nature of target	Explanation of the target	Year of target	Source	Legal status of target
UK	Zero carbon homes[1]	All new houses built in the UK must be carbon neutral by 2016. Intermediate targets: carbon performance of new houses to be improved by 25% by 2010 and 44% by 2013, as compared to 2006 Building Regulations.	2016 (with intermediate targets for 2010 and 2013)	Department for Communities and Local Government (2007)	Government plan to be cast into law over time according to the interim targets, until then not legally binding
California	Zero-net-energy buildings[2]	All new residential construction will be zero-net-energy by 2020, while all new commercial construction will be zero-net-energy by 2030.	2020 (for all new residential buildings) 2030 (for all new commercial buildings)	California Public Utilities Commission (CPUC), California Energy Efficiency Strategic Plan (2008)	Strategic goal, not legally binding
Denmark	Continuous reduction of energy use in new buildings	Maximum energy demand in new buildings must be, compared to the requirements[3] of 2006: – 25% lower by 2010 – 50% lower by 2015 – 75% lower by 2020	2010, 2015, 2020	Engelund Thomsen (2008), Danish Government (2009)	Agreement among all major parties of Danish parliament in 2008, confirmed by government strategy in 2009, but not yet cast into law
Austria	Passive house standard[4]	Social housing subsidies[5] shall only be granted for passive buildings from 2015.	2015	Engelund Thomsen et al. (2008), Federal Ministry of Agriculture, Forestry, Environment and Water Management (2007)	Goal of federal government, but not binding for federal states in charge of implementation
France	Energy positive buildings[6]	All new buildings must be energy positive from end of 2020. Intermediate target: maximum primary energy demand of 50 kWh/m² for new tertiary buildings from end of 2010 and all new buildings from end of 2012.	2020 (with intermediate target for 2010–2012)	Assemblée Nationale (2008)	Legislative process for binding law ongoing as of May 2009
Germany	Climate-neutral buildings	All new buildings should use zero amount of primary energy by 2020. Existing buildings should undergo efficiency renovation in 2020–2050 to achieve a 80% reduction in their primary energy demand.	2020 for new buildings; 2050 for existing buildings	Ministry of Economics and Technology and Ministry of Environment, Nature Conservation, and Nuclear Safety (2010)	The targets set in the national Energy Concept. The supporting Laws, regulations, and other implementation documents are under discussion and development.
Hungary	Zero carbon residential and public buildings	Granting building permits for residential and public buildings only if they are carbon neutral.	2025	Ministry of Environment and Water (2008)	Strategic goal, not legally binding
The Netherlands	Energy-neutral buildings[7]	All new buildings shall be energy neutral from 2020. The following intermediate targets apply: – 2011: 25% improved energy performance[8] compared to regulations of 2007 – 2015: 50% improved energy performance compared to regulations of 2007	2020 (with intermediate targets for 2011 and 2015)	Engelund Thomsen et al. (2008), Ministry of Economic Affairs (2007), Ministry of Housing, Spatial Planning and the Environment (2007)	Government target, not yet cast into law

1 Zero carbon means that a home should be zero carbon (net over the year) for all energy use in the home. This would include energy use from cooking, washing and electronic entertainment appliances as well as space heating, cooling, ventilation, lighting and hot water.

2 CPUC defines zero-net-energy building as a building with a net energy use of zero over a typical year, i.e. the amount of energy provided by on-site renewable energy sources is equal to the amount of energy used by the building.

3 The minimum requirements for energy efficiency in buildings in Denmark as of April 1, 2006, are defined as follows: the total demand of supplied energy to a building to deal with heat losses, ventilation, cooling and domestic hot water must not exceed 70 + 2200/A kWh/m²/a for residential and 95 + 2200/A kWh/m²/a for offices, schools, institutions and other buildings (A is the heated gross floor area).

4 Passive house standard is defined in Austria as annual heating energy use ≤15 kWh/m², where the area refers to net area in Austria in general, but to the useful area in the federal state of Styria and to heated area in the federal state of Tirol.

5 The share of social housing is high in Austria, this measure is therefore expected to have a great impact on market development for low-energy buildings.

6 Energy positive buildings are defined as buildings with a lower consumption of primary energy than the amount of energy they produce themselves from renewable sources.

7 Energy-neutral buildings are defined as not requiring any fossil fuel.

8 The Netherlands use an Energy Performance Coefficient to measure energy performance. In 2007, its maximum value for new residential buildings was 0.8. This will be lowered to 0.6 from 2011 and to 0.4 from 2015, which roughly corresponds to passive house standard for heating.

reduction target of 50% for all new buildings and major renovations compared to the regional or national average of each building type, with a goal of increasing this target by 10% every five years, reaching 100% by 2030. These goals have been adopted by several organizations, including the American Institute of Architects, the United States Green Building Council, the Environmental Protection Agency, and ASHRAE (Arens, 2008).

However, as mentioned earlier, zero-energy buildings may not be feasible for every type of building and in all regions, and in many cases their economics are unfavorable compared to high-efficiency buildings. They also may not be the most environmentally sustainable solution, as compared to very high-efficiency buildings supplied with partially on-site renewable and zero-carbon grid energy. In fact, universal net zero-energy mandates may even have negative environmental consequences. First, they may encourage urban sprawl, since net zero-energy buildings are easier – or even technologically only feasible – to implement in low-density, low-rise developments, but these increase transportation energy use. Second, if limited budgets of investors and building owners are forced to be spent on relatively more expensive PV panels, larger opportunities are foregone that would have been unlocked if the same investments had been made into efficiency measures that might have eliminated the need to produce more energy.

Nevertheless, in cases where zero-energy and very low-energy buildings are economically and environmentally justified, their promotion requires the implementation of a portfolio of policies that will motivate the several stakeholders and the billions of building owners and occupants.

The adoption of stricter energy performance requirements is of particular importance. As an example, in the United Kingdom the tightening of local planning regulations and building codes will be the basic policy instruments to achieve the carbon neutral target for all new buildings by 2016. Also, the adoption of stricter appliance and equipment standards that are periodically updated will decrease the total electricity loads that should be compensated by on-site generation.

The implementation of economic incentives in the form of subsidies, tax exemptions, etc., either for entire buildings, or for specific equipment components, will improve the economic performance of low-energy and zero-energy buildings and will accelerate the integration of renewable or other efficient technologies in their design. As an example, in France, new buildings respecting certain environmental criteria can be exempted from property taxes for 15–30 years (European Commission, 2009). There is some evidence that economic incentive programs are critical for preparing the market to adopt zero-energy goals in the present (premature) phase (Arens, 2008).

The adoption of preferably mandatory energy performance certification schemes and labeling programs will ensure compliance with energy code requirements and achievement of the targets set, thus enhancing market value. In fact, development of the Passivhaus scheme in Germany

and the MINERGIE standards in Switzerland was instrumental in the strong development of low-energy buildings over the past decade.

Lack of awareness and technical capabilities for constructing low-energy and zero-energy buildings among practicing architects, engineers, interior designers, and professionals in the building industry may be a major impediment to their development and the fulfillment of the ambitious goals set. Consequently, the development of capacity building programs to expand know-how is of particular importance. More specifically, there is a significant need in most countries to create comprehensive, integrated programs at universities and other educational establishments to train current and future building professionals in the design and construction of low-energy buildings. Finally, the development of pilot projects is helpful to demonstrate the effectiveness and the everyday functionality of the new buildings. In this context, the role of the public sector as an early adopter is crucial.

10.8.5   Energy conservation versus the rebound effect

Energy efficiency measures often have other, unintended effects on society. These ripple effects include rebound effects, co-benefits, and trade-offs with other resource use and pollution, and spillovers (Hertwich, 2005a). The rebound effect is defined first in Chapter 2, and discussed in detail in Chapter 22. Here it is reviewed with regard to building energy efficiency programs.

There have been a large number of empirical and modeling studies addressing the rebound effect (Greening et al., 2000; Schipper and Grubb, 2000; Hertwich, 2005a; Sorrell, 2007). Empirical studies of the micro-rebound effect consistently find that simple engineering-economic estimates of energy savings from an energy efficiency measure overestimate the savings by 0–30% (Table 10.23), with some outliers in the case of energy poverty (Roy, 2000). When people cannot afford to heat or cool their homes, but efficiency measures make these energy services affordable, these efficiency measures can in fact increase energy use. In all cases, the "rebound" implies that consumers enjoy a higher level of energy service as a result of increased efficiency.

With regard to macro effects, the main concern is that increased efficiency leads to the substitution of energy services for other inputs to production such as labor, capital, and land, alleviating resource constraints and thus enabling economic growth. Empirical studies show that the substitution effect does not lead to an increase in energy use for the same basket of goods (Schipper and Grubb, 2000). As energy is usually a small portion of the overall costs of a product or service, changes in the energy content of the product or service are unlikely to have a substantial effect on demand. It can be observed, however, that energy productivity has increased more slowly than labor productivity. Overall, the input of physical work has increased in lock-step with economic output (Ayres and Warr, 2005), suggesting that increased efficiency of converting primary energy to physical work may have contributed

Table 10.23 | Rebound effect for energy efficiency measures for different energy services.

Energy service	Rebound	Region/study
Residential space heating	10–30% 10–30% (1.4–60%) 20–30%	Review USA (Greening et al., 2000) Review OECD (Sorrell, 2007) Austria (Haas and Biermayr, 2000)
Residential space cooling	0–50% 1–36% 57–70%	Review USA (Greening et al., 2000). OECD (Sorrell, 2007) S.Korea (Jin, 2007)
Appliances	0 0	Austria (Haas, Biermayr et al., 1998) USA (Greening et al., 2000).
Lighting	5–12% 200%	USA (Greening et al., 2000) India (Roy, 2000)
Automotive transport	0 10–30% 5–26% 10–30% (5–87%) 57–67%	Switzerland (de Haan et al., 2006) USA (Greening et al., 2000) USA (Small and Van Dender, 2005) Review global (Sorrell, 2007) Germany (Frondel et al., 2008)

substantially to economic growth. Increased input of energy services in agriculture was essential both for raising yields and for freeing labor for work in industry (Erb et al., 2008). Empirical evidence for the importance of energy efficiency for economic growth, however, remains contested (Schipper and Grubb, 2000). The debate about how much energy efficiency increases economic growth and thus demand for energy services has yet to be resolved.

Taxes on energy or energy-related pollution or a cap and trade approach have been identified as an effective way to counteract both micro-rebound and macro-rebound effects because these effects are envisaged to operate through a price mechanism. The rebound effect, while limiting the effectiveness of efficiency measures in reducing overall demand, provides additional economic welfare arguments for energy efficiency as necessary for overcoming energy poverty and as being potentially beneficial for economic growth.

10.8.6   Focus on Developing Countries

10.8.6.1   Existing Policy Instruments in Developing Countries

A recent review of building energy codes (Janda, 2009) found that at least 61 countries had energy codes for buildings in 2008 of which the majority (40) were mandatory. The majority of these are developed countries, but several developing countries are included in this list: Mexico, Singapore, Vietnam, the United Arab Emirates, Jamaica, Jordan, Kazakhstan, Kuwait, Cote d'Ivoire, Thailand, India, China, Chile, South Africa, Indonesia, Malaysia, the Philippines, Lebanon, Tunisia,

and Pakistan. Several more plan to introduce building energy codes, including Morocco, Brazil, and Colombia (Janda, 2009), as well as Kenya and Uganda, often supported by international organizations. The most commonly applied measures are voluntary and mandatory labeling, appliance standards, building codes, public leadership programs, DSM programs, subsidies, grants and rebates, awareness-raising campaigns, and mandatory audits. However, only very few evaluations of instruments in these countries are available.

10.8.6.2   Enabling Factors: High Energy Price Levels and Energy Shortages

Increasing energy prices are often considered the most important precondition for improved energy efficiency in developing countries (Levine et al., 2007). Low, subsidized energy prices in many developing countries imply very long payback periods for energy efficiency investments, which renders such projects unprofitable. The differences in energy prices explain why certain governments in the Mediterranean region, such as Tunisia and Morocco, are interested in energy efficiency while others, especially oil-producing countries such as Algeria, are not or are less interested. Revenues from lower energy price subsidies can be rechanneled into rebates for energy efficient programs, loans, and special assistance for low-income households to increase their energy efficiency and thereby reduce energy costs. Since policymakers often consider energy efficiency a lower priority than more vital economic goals, such as poverty alleviation or increased employment, it is essential that non-energy benefits are well mapped, quantified, and well understood by policymakers.

10.8.6.3   Need For Technical Assistance and Training

Sustainable construction know-how needs to be introduced into the base curriculum of architects and other construction-related professions all over the world. This is even more important in developing countries because of the often much more dynamic new construction rates. As the training of a countries' own nationals will take some time, technical assistance through international organizations can bridge this gap for a period. Even in Tunisia – which is often considered a best practice developing country due to its successful energy efficiency policy in the buildings sector (Ürge-Vorsatz and Koeppel, 2007) – representatives of the energy efficiency agency request technical assistance for the development of thermal building standards due to a lack of national expertise in this area (Ürge-Vorsatz and Koeppel, 2007).

10.8.6.4   Need For Demonstration Projects and Information

In addition to the lack of information and awareness, there are also human barriers – for instance, a lack of trust. Trust and awareness can be raised through pilot projects or demonstration projects in the public sector. Demonstration programs at all levels, from the capital city to villages, such as the Green Buildings for Africa program in South Africa, prove the advantages of energy efficiency to every citizen irrespective of education level.

10.8.6.5 Need For Financial Assistance or Funding Mechanisms

The higher first cost of energy efficient technologies may still hamper penetration in developing countries, especially if the technologies must be imported. For example, high-performance glass is an expensive proposition in many countries, including India. Though it is proven to save energy, the higher first cost of about six-seven times that of the conventional single glazing systems deters many users from investing in it. However, as the market grows, the industry is expected to achieve economies of scale and thus the cost of such technologies should be reduced. Tax breaks can provide an initial push toward energy efficient technologies. Other financial mechanisms such as low interest loans from banks for energy efficient and renewable energy technologies are important, too. For example, in India, interest subsidy is available to buy solar water heating systems.

Especially poorer consumers need investment support or affordable loans, governmental funding, or ESCO financing (Deringer et al., 2004). For example, in July 2008, the World Bank approved a US$15 million grant to Argentina in order to support energy efficient projects and develop incentives for reducing demand on fossil fuels (WB, 2008). Developing countries can raise money on their own through public benefit charges or taxes. For instance, Brazil has obliged utilities to spend 1% of annual revenues on end-use energy efficiency improvements and on R&D. In 2007, India introduced prepayment electricity metering in public buildings and for private consumers to ensure bill payment (IEA Online Database). In Thailand, the government has raised funds through a petrol tax since 1992 (Brulez and Rauch, 1998). The tax revenues are collected in a fund and are now used to support energy efficiency projects. It is important that such funds are managed by independent agencies or institutions to avoid political influence.

10.8.6.6 The Role of Regulatory Measures

Many developing countries, such as Malaysia, Brazil, Morocco, and partly Thailand, first introduced voluntary standards or voluntary labeling for appliances or buildings which are, however, often less effective than mandatory ones. Mandatory audits for public buildings and commercial sector buildings above a certain annual demand are a frequently used instrument, applied, for example, in Tunisia and Thailand. However, compliance is often difficult to achieve. In order to ensure enforcement, special efforts are necessary, such as combining regulatory measures with incentives like subsidies or awards.

10.8.6.7 Importance of Monitoring and Evaluation

While many strategies to encourage or mandate energy efficient buildings or energy efficiency get introduced by government, many fail due to the lack of proper implementation and monitoring mechanisms. There is a major gap between political statements and actual action or changes in building design and construction. For instance, in India, while the government has initiatives to encourage integration of the Energy Conservation Building Code with the National Building Code for the uniform and larger adoption of the energy code, there is currently no concrete plan for the implementation of the code, or for monitoring and verification. Incentives, both financial and symbolic, are crucial for the wider adoption of these programs.

In many countries, baseline data on energy demand are missing. This is problematic, since measuring the success of implemented policy instruments requires knowledge of the baseline consumption. Regular monitoring and evaluation of programs are necessary in order to adapt the program, if possible, to changing circumstances and maximize its outcome. Evaluation studies quantifying energy savings are needed to determine cost-effectiveness and make necessary program adjustments (Jannuzzi, 2005).

10.8.6.8 Role of Institutionalization

Developing countries with successful energy efficiency policies have usually started with the adoption of an Energy Efficiency law or an Energy Efficiency Strategy, as is the case in Thailand, South Africa, and Tunisia. For example, the Tunisian National Agency for Energy Management is one of the main drivers behind the country's currently successful energy efficiency programs. Numerous Arab states are currently introducing such agencies, often with external assistance. The agency can be established as a nonprofit foundation, which provides flexibility in hiring and contracting (Szklo and Geller, 2006). The aim of this institutionalization is to get energy efficiency recognized as a priority among government officials, as well as among utilities and other stakeholders. Furthermore, in universities, the establishment of energy management curricula can contribute to knowledge dissemination and the training of professionals. These professionals can then become competent staff members of the mentioned institutions.

10.8.6.9 Importance of Adaptation to Local Circumstances

Finally, although best practices and experiences can be shared and regional cooperation is useful, the success of programs depends, among other factors, on adaptation to the local economic, political, social, and cultural context. Many programs have already failed because they copied programs from other countries without taking into account differences in culture, political systems, or other areas (Ürge-Vorsatz and Koeppel, 2007). Therefore, a thorough assessment of the local social, economic, political, and cultural fabric as it affects the operation of the policy instrument is important before decisions are taken. In large countries, the design of energy efficiency programs is most effective if adjusted to different regional contexts and institutional realities, such as the frailty of the regulatory certainty or the tendency of high contract failures. Moreover, the specificities of emerging economies – the social, legal, or economic context – can result in an increasing difficulty for customers in accessing capital. As it is, the higher rates of contractual failures in India or China, for instance, have resulted in investors finding it even more difficult to invest in energy efficiency projects (Taylor et al., 2008).

10.8.7   Implications of Broader Policies on Energy Efficiency in Buildings

10.8.7.1   Liberalization and Restructuring of Electricity Markets

Substantial literature exists on the impacts, both actual and estimated, of the liberalization and resulting restructuring of electricity markets on energy use in final demand sectors and particularly in residential and commercial buildings.

A number of authors (Burtraw et al., 2000; Sondreal et al., 2001; Sevi, 2004; Pollitt, 2008) suggest that electricity restructuring results in lower average prices for electricity, particularly in cases in which the regulated utilities were relatively inefficient. They point out that lower prices are likely to generate higher demand from consumers. However, as household sector long run price elasticity is relatively low, energy market liberalization may have only a small effect on demand even if price reductions are quite substantial. Following liberalization, prices will not necessarily fall in all areas, as price changes may start from different initial levels. If the local regulated utility is a low cost supplier of electricity compared to its neighbors, then prices in the local area could actually rise under competition (Palmer, 1999; Sevi, 2004). Furthermore, the experience with retail competition in Massachusetts and Pennsylvania shows that large industrial and some medium-size customers are the likely beneficiaries of lower prices, while the average price to residential and small commercial customers is likely to rise over time (Sverrisson et al., 2003).

The restructuring of electricity markets is also expected to produce more widespread use of time-differentiated pricing of electricity. This form of pricing will lead to a shifting of demand from peak to off-peak periods and will encourage building occupants to improve their energy-using behavior.

Several studies also indicate that the trend in new utility-funded DSM programs has been downward in the United States due to the deregulation of electricity markets (Eikeland, 1998; Palmer, 1999; Dubash, 2003; Sverrisson et al., 2003; Sevi, 2004; Blumstein et al., 2005) unless there is a regulatory environment that decouples sales of electricity

from profits (WBCSD, 2008). For instance, in the United States, many state restructuring laws and federal restructuring bills also include a mechanism for funding DSM initiatives, e.g., through an electricity surcharge, which does not discriminate among electricity suppliers and could result in some energy savings. The Consortium for Energy Efficiency reports a doubling of total DSM expenditures by 88% of the utilities in the United States from 2006 through 2009 (Nevius et al., 2010). In many European countries, energy market reform has been accompanied by the introduction of a formal regulatory system for the first time. In some cases the opportunity has been taken to introduce energy saving obligations on energy companies, using "white certificates" or related mechanisms. In these countries, the introduction of energy efficiency obligations in conjunction with market restructuring has resulted in significantly increased activity on energy efficiency by energy companies (Pavan, 2008; Lees, 2008; Eyre et al., 2009).

The impact of energy market restructuring on energy efficiency is therefore dependent on the prior conditions and details of policy. The restructuring of electricity markets may also lead to increased penetration of distributed generation, both in industrialized and developing countries. More specifically, in a region with relatively high electricity prices stemming, for example, from expensive past investments, a customer can avoid these expenses by operating a distributed generator (e.g., PV units, small wind mills, etc.).

In developing countries, different patterns of current energy use and the relative unavailability of electricity distribution infrastructure will likely lead to very different effects of electricity restructuring than are likely in the developed world. The removal of subsidies that many developing nations provide to energy use, as well as the diversification in the level of service and pricing to reflect local actual costs, could lead to higher electricity prices in many cases, enhancing the attractiveness of investments in rural electrification (Burtraw et al., 2000; Nagayama, 2008). However, without an explicit effort, energy markets restructuring will result in decreased electricity accessibility to the poorer segments of the population (Dubash, 2003).

10.8.7.2   Energy Taxation and High Energy Prices

Inevitably, higher end-use energy prices, while not addressing all the market failures in the buildings sector, increase the energy conservation potential. However, the effect of price increases on energy demand depends on how sensitive demand is to energy price fluctuations. As already pointed out, household sector long-term price elasticity is relatively low, at least in developed economies, indicating that variability in energy prices may have only a small effect on demand even if price increases are quite substantial. For example, in the Netherlands, short run price elasticity for electricity in households was estimated to be between 0 and -0.25, whereas the long run elasticity was estimated to be -0.3 to -0.45 (Berkhout et al., 2000). This is mainly attributable to the fact that in richer countries, energy prices are often a relatively insignificant cost component and therefore receive inadequate attention from building occupiers and owners (WBCSD, 2008).

There is some evidence that energy use behavior may be seriously affected when a significant part of available income is spent on energy. Energy tends to be used carefully in developing countries, and this is also true in richer countries during the last five years due to very high increases in international fuel prices. In the United States, high energy prices over the last five years have stimulated energy saving initiatives. In 2007, two-thirds of United States homebuilders were planning to build green in 15% of their projects, citing customer concern about energy costs as the main reason (Kelleher, 2006).

The increase of end-use energy prices through the imposition of energy taxes at some point in the energy supply chain may provide a second means to energy efficiency through the investment of the tax revenues – or at least part of them – in energy conservation related activities, such as mandatory DSM measures, subsidy schemes, green funds, or other mechanisms.

10.9   Gaps in Knowledge

While buildings are ubiquitous and some aspects are well researched and documented, such as the engineering aspects, there is surprisingly little understanding about their energy use and thus how problems related to their energy demand can be mitigated – both at the micro and macro levels. This section identifies a selection of important gaps in this knowledge, as considered important by the authors of this chapter.

Perhaps the most glaring problem with the knowledge is the shortage of related data and information. Little data exist on how energy in buildings is concretely used and how it is broken down by end-uses, building types, technologies, or other variables. Knowledge gaps exist about detailed energy use by energy service. Another major knowledge gap pertains to region-specific costs of new buildings in relation to their energy performance and region-specific costs of retrofits of existing buildings in relation to the savings in energy use achieved. Sufficient knowledge is also lacking about best practices, that is, the most sustainable means for providing energy services in each developmental, cultural, and individual situation. The interaction between life cycle energy use, environmental impacts, and cost aspects are also not well studied. Most of the published literature on life cycle assessments of different building types is very recent and the field is not yet very mature. Most engineering, environmental, and economic assessments related to sustainability options for buildings rely on their direct energy use, costs, and emissions. However, considering the entire life cycle would probably affect the validity of many decisions and policies, as there are often trade-offs.

The situation regarding energy policies in developing countries clearly requires further research. Only very few ex-post policy impact evaluation studies are currently available and even fewer include quantitative data on effectiveness, cost-effectiveness. The cumulative effect of policy packages, incremental and double counting effects of policy tools, as well as synergy effects, are poorly understood. While the area of co-benefits or ancillary benefits of energy efficiency in the buildings sector also needs to be further explored, even less is known about ancillary costs, which are seldom mentioned in literature. This relates to the better researched field of the obstacles or barriers to the deployment of energy efficiency in the sector.

10.10   Novelties in GEA's Global Building Energy Assessment

There have been several assessments completed recently on building-related opportunities for climate change mitigation or sustainable energy by various organizations. There are several new elements in this chapter as compared to these earlier assessments. These include:

- An energy service-centered approach: the discussion and consideration of opportunities recognizes that energy services are needed rather than energy per se, allowing for a broader spectrum of more innovative alternatives and solutions.

- Life cycle energy and emissions versus only operational ones: considering life cycle energy use and emissions when possible rather than just the operational energy and emissions in buildings, and recognizing the trade-offs. Novel policy recommendations originate from applying such a perspective.

- Applying a holistic, performance-based approach that recognizes that buildings are complex, integrated systems rather than sums of individual components.

- A novel global building energy use model using a performance-centered logic; presenting building thermal en pathways until 2050.

- The importance of the lock-in effect in the building sector has been shown for the first time, as well as detailed quantification of it.

- Large database on quantified or monetized co-benefits, illustrating the large orders of magnitude of such benefits.

- A detailed assessment of non-technological opportunities and challenges, emphasizing human dimensions.

References

ACE, 2000: *Energy Efficiency and Jobs. UK Issues and Case Studies*. Report prepared by the Association for the Conservation of Energy (ACE) for the Energy Saving Trust, London, UK.

ACEEE, 2010: *Panel 10 – Workforce Training. Proceedings of the ACEEE Summer Study on Energy Efficiency in Buildings*. American Council for an Energy-Efficient Economy (ACEEE), Asilomar, CA.

ADEME, 2009: *Économies d'énergie, Faisons vite ça chauffe*. Agence de l'Environnement et de la Maîtrise de l'Energie (ADEME), Paris, France. www.faisonsvite.fr/ (accessed March 15, 2011).

Aebischer, B., M. Jakob, G. Catenazzi, and G. Henderson, 2007: Impact of climate change on thermal comfort, heating and cooling energy demand in Europe. In *Proceedings of the European Council for an Energy Efficient Economy (ECEEE) Summer Study 2007*. Colle sur Loup, France.

AEC, 2002: *Energy Design Brief: Building Integrated Photovoltaics. Architectural Energy Corporation (AEC)*, Boulder, CO. www.archenergy.com (accessed December 16, 2010).

Ahrestani, S., 2010: *A multi-criteria site selection model for green roofs in the Sydney CBD: A GIS approach*. Institute of Environmental Studies and Faculty of the Built Environment, University of New South Wales, Australia.

Akbari, H., A. H. Rosenfeld, and H. Taha, 1990: *Cooling urban heat islands. Proceedings of the Fourth Urban Forestry Conference*. P. D. Rodbell (ed.) St. Louis, MO.

Akbari, H., S. Menon, and A. Rosenfeld, 2008: Global cooling: Increasing worldwide urban albedos to offset CO_2. *Climatic Change*, **95**(3–4): 275–286.

Allcott, H., 2010: *Social Norms and Energy Conservation*. Working Paper, June 9, 2010, Massachusetts Institute of Technology (MIT) and New York University (NYU).

Alliance to Save Energy, 2008: *India: Manual for the Development of Municipal Energy Efficiency Projects*. The Alliance to Save Energy, Washington, DC.

Amstalden, R. W., M. Kost, C. Nathani, and D. M. Imboden, 2007: Economic potential of energy-efficient retrofitting in the Swiss residential building sector: the effects of policy instruments and energy price expectations. *Energy Policy*, **35**(3): 1819–1829.

Arens, E., 2008: *Getting to zero: How CBE Industry Partners are meeting net-zero energy goals*. Centerline: Newsletter of the Center for the Built Environment at the University of California, Berkeley, CA, USA.

Argiriou, A. A., S. P. Lykoudis, C. A. Balaras, and D. N. Asimakopoulos, 2004: Experimental study of a earth-to-air heat exchanger coupled to a photovoltaic system. *Journal of Solar Energy Engineering*, **126**(1): 620–625.

Armstrong, P. R., S. B. Leeb, and L. K. Norford, 2006: Control with building mass-Part II: Simulation. *ASHRAE Transactions*, **112**(1): 462–473.

Atif, M. R. and A. D. Galasiu, 2003: Energy performance of daylight-linked automatic lighting control systems in large atrium spaces: report on two field-monitored case studies. *Energy and Buildings*, **35**(5): 441–461.

Aunan, K., H. Aaheim, and H. Seip, 2000: Reduced damage to health and environment from energy saving in Hungary. In *Proceedings of the IPCC Workshop on Assessing the Ancillary Benefits and Costs of Greenhouse Gas Mitigation Strategies*. Washington, DC, USA.

Audenaert, A., S. H. de Cleyn and B. Vankerckhove, 2008: Economic analysis of passive houses and low-energy houses compared with standard houses. *Energy Policy*, **36**(1): 47–5.

Australian Greenhouse Office, 2005: *When you keep measuring it, you know even more about it! Projected impacts 2005–2020*. National Appliance and Equipment Program, Department of the Environment and Heritage, Australian Greenhouse Office, Commonwealth of Australia.

Ayres, R. U. and B. Warr, 2005: Accounting for growth: the role of physical work. *Structural Change and Economic Dynamics*, **16**(2): 181–209.

Baker, N. and K. Steemers, 1999: *Energy and Environment in Architecture: A Technical Design Guide*. E and FN Spon, London, UK.

Balaras, C. A., 2001: Energy retrofit of a neoclassic office building: social aspects and lessons learned. *ASHRAE Transactions*, **107**(1): 191–197.

Banks, N., 2008: *Implementation of Energy Performance Certificates in the Domestic Sector*. UK Energy Research Centre (UKERC), Oxford, UK.

Bryan Christie, 2006: In *Tools for better living: Smoke-free*, Barkeman, E., Fortune. CNN Money.com. http://money.cnn.com/popups/2006/fortune/better_living/5.html (accessed December 31, 2010).

Barry, J.A., 2007: *Watergy: Energy and Water Efficiency in Municipal Water Supply and Wastewater Treatment. Cost-Effective Savings of Water and Energy*. The Alliance to Save Energy, Washington, D.C.

Batlles, F. J., S. Rosiek, I. Muñoz, and A. R. Fernández-Alba, 2010: Environmental assessment of the CIESOL solar building after two years operation. *Environmental Science and Technology*, **44**(9): 3587–3593.

Beccis, F., 2006: *Energy efficiency targets using tradable certificates: microeconomic aspects and lessons from Italy. Proceedings of the biennial international workshop – advances in energy studies. "Perspectives on Energy Future,"* September 14, Portovenere, Italy.

Behrens, A. and C. Egenhofer, 2007: *An initial assessment of policy options to reduce – and secure against – the costs of energy insecurity*. Cost Assessment for Sustainable Energy Market. Centre for European Policy Studies, Unit for Climate Change and Energy, Brussels, Belgium.

Bell, M. and R. Lowe, 2000: Building regulation and sustainable housing. Part 2: technical issues. *Structural Survey*, **18**(2): 77–88.

Bender, S., M. Moezzi, M. Gossard, and L. Lutzenhiser, 2004: Using mass media to influence energy consumption behavior: California's 2001 flex your power campaign as a case study. In *Proceedings of the American Council for an Energy Efficient Economy (ACEEE)*, Summer Study, 2004. California Energy Commission, Sacramento, CA.

Berkhout, P. H. G., J. C. Muskens, and J. W. Welthuijsen, 2000: Defining the rebound effect. *Energy Policy*, **28**(6–7): 425–432.

BERR, 2001: *UK Fuel Poverty Strategy*. Report prepared by the Department for Business Enterprise and Regulatory Reform (BERR), London, UK.

Berrueta, V., R. Edwards, and O. R. Masera, 2008: Energy performance of wood-burning cookstoves in Michoacan, Mexico. *Renewable Energy*, **33**(5): 859–870.

Berry, L. and M. Schweitzer, 2003: *Metaevaluation of National Weatherization Assistance: Program Based on State Studies 1993–2002*. Oak Ridge National Laboratory report prepared for the US Department of Energy (US DOE), Oak Ridge, TN, USA.

Bertoldi, P. and C. Ciugudeanu, 2005: *Five-Year Assessment of the European Green Light Programme*. Presentation at the 6th International Conference on Energy Efficient Lighting, 9–11 May, 2005, Shanghai, China.

Bertoldi, P. and S. Rezessy, 2006: *Tradable certificates for energy savings (White Certificates)*. Report prepared by the Joint Research Center of the European Commission (ISPRA), Rome, Italy.

Bertoldi, P. and S. Rezessy, 2007: Voluntary agreements for energy efficiency: review and results of European experiences. *Energy and Environment,* **18**(1):37–73.

Bhattacharya, S. C. and M. A. Leon, 2004: *Prospects for biomass gasifiers for cooking applications in Asia.* Energy Field of Study, Asian Institute of Technology, Bangkok, Thailand.

Bickel, P. and R. Friedrich, 2005: *ExternE. Externalities of energy.Methodology 2005 update.*Institut für Energiewirtschaft und Rationelle Energieanwendung (IER), Universität Stuttgart, Germany. Directorate-General for Research, Sustainable Energy Systems, European Commission (EC), Luxembourg.

BioLite, 2009: *Thermoelectrics in biomass stove systems.* Presented at the "Cookstove Workshop," November 18, Asian Institute of Technology, Bangkok, Thailand.

BioLite, 2011: *Introducing the BioLite HomeStove.* http://www.biolitestove.com/HomeStove.html (accessed 19 September 2011).

Blengini, G. A., and T. Di Carlo, 2010: Energy-saving policies and low-energy residential buildings: an LCA case study to support decision makers in Piedmont (Italy). *International Journal of Life Cycle Assessment,* 15(7): 652–665.

Blondeau, P., M. Sperandio, and F. Allard, 1997: Night ventilation for building cooling in summer. *Solar Energy,* **61**(5): 327–335.

Blumstein C., C. Goldman, and G. Barbose, 2005: Who should administer energy-efficiency programs? *Energy Policy,* **33**(8): 1053–1067.

Boardman, B., S. Darby, G. Killip, M. Hinnells, C. N. Jardine, J. Palmer, and G. Sinden, 2005: *40% House.* Environmental Change Institute, University of Oxford, UK.

Boardman, B., 2010: *Fixing Fuel Poverty. Challenges and Solutions.* Earthscan, London, UK.

Bodart, M. and A. De Herde, 2002: Global energy savings in office buildings by the use of daylighting. *Energy and Buildings,* **34**(5): 421–429.

Bohne, R. A., H. Brattebo, and H. Bergsdal, 2008: Dynamic eco-efficiency projections for construction and demolition waste recycling strategies at the city level. *Journal of Industrial Ecology,* **12**(1): 52–68.

Boonstra, C., I. Thijssen, and R. Vollebregt, 1997: *Glazed Balconies in Building Renovation.* James and James, London, UK.

Borg, N., Y. Blume, S. Thomas, W. Irrek, H. Faninger-Lund, P. Lund, and A. Pindar, 2003: *Harnessing the Power of the Public Purse.* Final report from the European PROST study on energy efficiency in the public sector, Stockholm, Sweden.

Borg, N., Y. Blume, S. Thomas, W. Irrek, H. Faninger-Lund, P. Lund, and A. Pindar, 2006: Release the power of the public purse. *Energy Policy,* **34**(2): 238–250.

Borjesson, P. and L. Gustavsson, 2000: Greenhouse gas balances in building construction: wood versus concrete from life-cycle and forest land-use perspectives. *Energy Policy,* **28**(9): 575–588.

Bornehag, C. G., J. Sundell, C. J. Weschler, T. Sigsgaard, B. Lundgren, M. Hasselgren, and L. Hägerhed-Engman, 2004: The association between asthma and allergic symptoms in children and phthalates in house dust: a nested case control study. *Environmental Health Perspectives,* **112**(14): 1393–1397.

Bornstein, R. and Q. Lin, 1999: *Urban Convergence Zone Influences on Convective Storms and Air Quality. Proceedings of the WMO International Conference, Congress of Biometeorology and International Conference on Urban Climate,* World Meteorology Organisation (WMO), Sydney, Australia.

Borsani, C. and M. Salvi, 2003: *Analysebericht zum Minergie-Standard.* Memorandum of Züricher Kantonalbank (ZKB) to CEPE-Team. Zurich, Switzerland.

Bottrill, C., 2007: *Internet Tools for Behaviour Change. Proceedings of the European Council for an Energy Efficient Economy (ECEEE),* June 4–9, Cote d'Azur, France.

Bourassa, N., P. Haves, and J. Huang, 2002: A Computer Simulation Appraisal of Non-residential Low Energy Cooling Systems in California. In *Proceedings of the 2002 American Council for an Energy Efficient Economy (ACEEE) Summer Study on Energy Efficiency in Buildings,* Vol. 3, Washington, DC, USA.

Bourbia, F. and H. B. Awbi, 2004: Building cluster and shading in urban canyon for hot dry climate, part 1: air and surface temperature measurements. *Renewable Energy,* **29**(2): 249–262.

Brandemuehl, M. and J. Braun, 1999: *The Impact of Demand-Controlled and Economizer Ventilation Strategies on Energy Use in Buildings.* American Society of Heating, Refrigerating and Air-Conditioning Engineers Inc., Atlanta, GA, USA.

BRE, 2003: *Energy Consumption Guide 19. Energy use in Offices.* Report prepared by the Government's Energy Efficiency Best Practice Programme (BRE), Watford, UK.

Bressand, F., N. Zhou, and J. Lin, 2007: Energy use in commercial buildings in China current situation and future scenarios. In *Proceedings of the European Council for an Energy Efficient Economy(ECEEE) 2007 Summer Study – Saving Energy Just Do it*! La Colle sur Loup, France, pp.1065–1071.

Brink, A. and M. Erlandsson, 2004: *Energiskatternas effekt på energianvändningen 1991–2001.* Rapport till Energimyndigheten. [In Swedish: Energy taxes effect on energy consumption 1991–2001. Report to the Swedish Energy Agency], Stockholm, Sweden.

Brophy, V., J.P. Clinch, F. J. Convery, J. D. Healy, C. King, and J. O. Lewis, 1999: *Homes for the 21st Century – the Costs and Benefits of Comfortable Housing for Ireland.* Report prepared for Energy Action Ltd. by the Energy Research Group and Environmental Institute, University College Dublin, Dublin, Ireland.

Brounen, D. and N. Kok, 2010: *On the Economics of EU Energy Labels in the Housing Market.* RICS Research, Erasmus University and Maastricht University, the Netherlands.

Brown, M., 2001: Market failures and barriers as a basis for clear energy policies. *Energy Policy,* **29**(14): 1197–1207.

Brown, M., J. Chandler, M.V. Lapsa, and B. K. Sovacol, 2008: *Lock-in: Barriers to Deploying Climate Change Mitigation Technologies,* Oak Ridge National Laboratory ORNL/TM-2007–124 (revised November, 2008), Oak Ridge, TN.

Brown, M., J. Chandler, M. V. Lapsa, and M. Ally, 2009: *Making Buildings Part of the Climate Solution: Policy Options to Promote Energy Efficiency.* Oak Ridge National Laboratory, TN.

Brulez, D. and R. Rauch, 1998: *Energy Conservation Legislation in Thailand: Concepts, Procedures and Challenges.* In *Compendium on Energy Conservation Legislation in Countries of the Asia and Pacific Region.* United Nations Economic and Social Commission for Asia and the Pacific (UNESCAP), Bangkok, Thailand.

Brunklaus, B., C. Thormark, and H. Baumann, 2010: Illustrating limitations of energy studies of buildings with LCA and actor analysis. *Building Research and Information,* **38**(3): 265–279.

Bureau of Energy Efficiency, 2009:*Official website of the Bureau of Energy Efficiency (BEE).* Ministry of Power, Government of India, New Delhi.

Burgess, J., 2003: Sustainable consumption: is it really achievable? *Consumer Policy Review,* **13**(3): 78–84.

Burtraw, D., K. Palmer, and M. Heintzelman, 2000: *Electricity restructuring: Consequences and Opportunities for the Environment, Resources for the Future,* Discussion paper 00–39, Washington, D.C.

Busch, J.F., 1992: A tale of two populations: thermal comfort in air-conditioned and naturally ventilated offices in Thailand. *Energy and Buildings,* **18**(3–4):235–249.

Buzar, S., 2007: *Energy Poverty in Eastern Europe: Hidden Geographies of Deprivation*. Ashgate Publishing Group, Aldershot, UK.

Çakmanus, I., 2007: Renovation of existing office buildings in regard to energy economy: an example from Ankara, Turkey. *Building and Environment, 42*(3): 1348–1357.

Carbon Trust, 2005: *The UK Climate Change Programme: Potential Evolution for Business and the Public Sector*. Technical report prepared by Carbon Trust, London, UK.

Cartalsi C., A. Synodinou, M. Proedrou, A. Tsangrassoulis and M. Santamouris, 2001: Modifications in energy demand in urban areas as a result of climate change: an assessment for the southeast Mediterranean region. *Energy Conversion and Management, 42*(14) pp.1647–1656.

CEA, 1995: *Supporting Research and Development to Promote Economic Growth: The Federal Government's Role*. Council of Economic Advisers (CEA), Washington, D.C.

Cebon, P. B., 1992: Twixt cup and lip: organizational behavior, technical prediction, and conservation practice. *Energy Policy, 20*(9): 802–814.

Cedar, J., 2009: *Thermoelectrics in Biomass Stove Systems*. Presented at "Cookstove Workshop." Asian Institute of Technology, November 18, Bangkok, Thailand.

CERES, 2009: *From Risk to Opportunity: Insurer responses to Climate Change*, Report prepared by CERES, 2009.

Chakravarty, S., A. Chikkatur, H. de Coninck, S. Pacala, R. Socolow, and M. Tavoni, 2009: Sharing global CO_2 emission reductions among one billion high emitters. In *Proceedings of the National Academy of Sciences of the United States of America*, 106(29): pp.11884–11888.

Chapman, R., P. Howden-Chapman, H. Viggers, D. O'Dea, and M. Kennedy, 2009: Retrofitting houses with insulation: a cost–benefit analysis of a randomised community trial. *Journal of Epidemiology and Community Health, 63*(4): 271–277.

Claridge, D. E., M. Liu, S. Deng, W. D. Turner, J. S. Haberl, S. U. Lee, M. Abbas, and H. Bruner, 2001: Cutting heating and cooling use almost in half without capital expenditure in a previously retrofit building. In *Proceedings of the European Council for an Energy Efficient Economy (ECEEE), Summer 2001, 4*: 74–85.

CLG, 2010: *Residential Energy Conservation Ordinance Factsheet*. Prepared by Climate Leaders Group (CLG).

Clinch, J. P. and J. D. Healy, 1999: *Housing Standards and Excess Winter Mortality in Ireland*. Environmental Studies Research Series Working Paper 99/02, Department of Environmental Studies, University College Dublin, Ireland.

Clinch, J. P. and J. D. Healy, 2001: Cost-benefit analysis of domestic energy efficiency. *Energy Policy, 29*(2): 113–124.

Cooremans, C., 2007: Strategic Fit of Energy Efficiency (Strategic and cultural dimensions of energy-efficiency investments).In *Proceedings of the European Council for an Energy-Efficient Economy (ECEEE) Summer Study*, June 4–9, 2007, la Colle Sur Loop, France.

Costanza, R., R. d'Arge, R. de Groot, S. Farber, M. Grasso, B. Hannon, K. Limburg, S. Naeem, R. V. O'Neill, J. Paruelo, R. G. Raskin, P. Sutton, and M. van den Belt, 1997: The value of the world's ecosystem services and natural capital. *Nature, 387*: 253–260.

Cottrell, J., 2004: *Ecotaxes in Germany and the United Kingdom – A Business View*. Green Budget Germany Conference Report, Berlin, Germany.

Crawford, R. H. and G. J. Treloar, 2004: Net energy analysis of solar and conventional domestic hot water systems in Melbourne, Australia. *Solar Energy, 76*(1–3): 159–163.

Crossley, D., M. Maloney, and G. Watt, 2000: *Developing Mechanisms for Promoting Demand-side Management and Energy Efficiency in Changing Electricity Businesses*. Research report 3, task 6 of the International Energy Agency (IEA) Demand-side Management Programme, Energy Futures Australia Pty Ltd., New South Wales, Australia.

Cullen, J. M. and J. M. Allwood, 2009: The role of washing machines in life cycle assessment studies. *Journal of Industrial Ecology, 13*(1): 27–37.

da Graça, G. C., Q. Chen, L. R.Glicksman and L.K. Norfold, 2002: Simulation of wind-driven ventilative cooling systems for an apartment building in Beijing and Shanghai. *Energy and Buildings, 34*:1–11.

Darby, S., 2000: *Making it obvious: designing feedback into energy consumption*. In Effective Advice. Boardman and S. Darby (eds.), Appendix 4, Environmental Change Institute, University of Oxford, UK.

Darby, S., 2003: Making Sense of Energy Advice. In *Proceedings of the European Council for an Energy Efficient Economy (ECEEE) Summer Study*. Stockholm, Sweden.

Darby, S., 2006a: *The Effectiveness of Feedback on Energy Consumption*. A review for the Department for Environment Food and Rural Affairs (DEFRA) of the literature on metering, billing and direct displays. Environmental Change Institute, University of Oxford, UK.

Darby, S., 2006b: Social learning, household energy practice and public policy: lessons from an energy-conscious village. *Energy Policy, 34*: 2929–2940.

Darby, S., 2008: Energy feedback in buildings: improving the infrastructure for demand reduction. *Building Research and Information, 36*(5): 499–508.

Dascalaki, E. and M. Santamouris, 2002: On the potential of retrofitting scenarios for offices. *Building and Environment, 37*(6): 557–567.

Davie, G. S., M. G. Baker, S. Hales and J. B. Carlin, 2007: *Trends and determinants of excess winter mortality in New Zealand: 1980 to 2000*. BMC *Public Health, 7*:263.

Davis, D. L., A. Krupnick, and G. McGlynn, 2000: Ancillary benefits and cost of greenhouse gas mitigation: an overview. In *Proceedings of Workshop on Assessing the Ancillary Benefits and Costs of Greenhouse Gas Mitigation Strategies*, March 27–29, 2000, Washington, DC, USA.

de Dear, R. J. and G. S. Brager, 1998: Developing an adaptive model of thermal comfort and preference. *ASHRAE Transactions, 104*(1A):145–167.

de Dear, R. J. and G. S. Brager, 2002: Thermal comfort in naturally ventilated buildings: revisions to ASHRAE Standard 55. *Energy and Buildings, 34*(6):549–561.

de Garbino, J. P., 2004: *Children's health and the environment. A global perspective. A resource manual for the health sector*. World Health Organization (WHO), Geneva, Switzerland.

de Haan, P., M. G. Mueller, and A. Peters, 2006: Does the hybrid Toyota Prius lead to rebound effects? Analysis of size and number of cars previously owned by Swiss Prius buyers. *Ecological Economics, 58*(3): 592–605.

de T'Serclaes, P., 2007: *Financing Energy Efficient Home. Existing Policy Responses to Financial Barriers*. Organization for Economic Co-operation and Development/International Energy Agency (OECD/IEA), Paris, France.

DECC, 2009: *Annual Report on Fuel Poverty Statistics 2009*. Department of Energy and Climate Change (DECC), London, UK.

DeCanio, S. J. and W. E. Watkins, 1998: Investment In Energy Efficiency: Do the Characteristics of Firms Matter? *The Review of Economics and Statistics,* **80**(1): 95–107.

DEFRA, 2004: *Valuation of Health Benefits Associated with Reductions in Air Pollution.* Department for Environment, Food and Rural Affairs (DEFRA), London, UK.

DEFRA, 2005: *Fuel Poverty Monitoring: Energy Company Schemes 2003.* Annex to UK fuel Poverty Strategy, 3rd Annual Progress Report 2005,Department for Environment, Food and Rural Affairs (DEFRA), London, UK.

DEFRA, 2006b: *Review and Development of Carbon Abatement Curves for Available Technologies as Part of the Energy Efficiency Innovation Review.* Department for Environment, Food and Rural Affairs (DEFRA), ENVIROS Consulting, Ltd., London, UK.

DEFRA, 2006c: *Synthesis of Climate Change Policy Evaluations.* Department for Environment, Food and Rural Affairs (DEFRA), London, UK.

DEFRA, 2007a: Climate Change Agreements: Results of the Third Target Period Assessment. Department for Environment, Food and Rural Affairs (DEFRA), London, UK.

DEFRA, 2007b: *Energy Efficiency Action Plan.* Report prepared by the Department for Environment, Food and Rural Affairs (DEFRA), London, UK.

DEFRA/BERR, 2008: *The UK Fuel Poverty Strategy 6th Annual Progress Report.* Department for Environment, Food and Rural Affairs (DEFRA) / Department of Business Enterprise & Regulatory Reform (BERR), London, UK.

DEG, 2004: *Development of an Improved Two-Stage Evaporative Cooling System.* Davis Energy Group, California Energy Commission Report P500–04–016, St. Davis, CA.

Demirbilek, F. N., U. G. Yalçiner, M. N. Inanici, A. Ecevit, and O. S. Demirbilek, 2000: Energy conscious dwelling design for Ankara. *Building and Environment,* **35**: 33–40.

Den Baars, S., 2008: Energy efficient white LEDs for sustainable solid-state lighting, In *Physics of Sustainable Energy, Using Energy Efficiently and Producing it Renewably,* D. Hafemeister, B. G. Levi, M. D. Levine and P. Schwartz (eds), American Institute of Physics, Melville, NY.

Denniston, S., L. Dunn, J. Antonoff, and R. DiNola, 2010: Toward a future model energy code for existing and historic buildings. In *Proceedings of the American Council for an Energy-Efficient Economy (ACEEE) Summer Study on Energy Efficiency in Buildings,* Pacific Grove, CA.

Deringer, J., M. Iyer, and Yu Joe Huang, 2004: Transferred Just on Paper? Why Doesn't the Reality of Transferring/Adapting Energy Efficiency Codes and Standards Come Close to the Potential? In *Proceedings of the 2004 American Council for an Energy-Efficient Economy (ACEEE) Summer Study on Energy Efficiency in Buildings.* August 2004, Pacific Grove, CA.

DEWHA, 2008: *Energy Use in the Australian Residential Sector 1986–2020. Part 1.* Report prepared by Energy Efficient Strategies, Department of Environment, Water, Heritage and the Arts (DEWHA), Canberra, Australia.

Dharmadhikari, S., D. Pons, and F. Principaud, 2000: Contribution of stratified thermal storage to cost-effective trigeneration project. *ASHRAE Transactions,* **106**(2): 912–919.

Dias, R., C. Mattos, and J. Balesieri, 2004: Energy education: breaking up the rational energy use barriers. *Energy Policy,* **32**(11): 1339–1347.

Dietz, T., G. T. Gardner, J. Gilligan, P.C. Stern, and M.P. Vandenbergh, 2009: Household actions can provide a behavioral wedge to rapidly reduce US carbon emissions. In *Proceedings of the National Academy of Sciences of the United States of America (PNAS),* **106**(44): 18452–18456.

Dilmetz, K., 2009: *Energy Efficiency for Buildings.* Presentation for German American Chamber of Commerce of the Southern United States, Inc. (GACC South), Houston, TX.

Dobson, J. and J. Griffin, 1992: Conservation effect of immediate electricity cost feedback on residential consumption behavior. In *Proceedings of American Council for an Energy-Efficient Economy (ACEEE) 1992 Summer Study on Energy Efficiency in Buildings*10, Pacific Grove, CA, pp.33–35.

Dodoo, A., L. Gustavsson, and R. Sathre, 2010: Life cycle primary energy implication of retrofitting a wood-framed apartment building to passive house standard. *Resources, Conservation and Recycling,* **54**(12): 1152–1160.

DSDNI, 2008: *Ending fuel Poverty: A Strategy for Northern Ireland.* Department for Social Development in Northern Ireland (DSDNI), Belfast, Northern Ireland.

Dubash, N. K., 2003: Revisiting electricity reform: the case for a sustainable development approach. *Utilities Policy,* **11**: 143–154.

Durkin, T., 2006: Boiler system efficiencies. *ASHRAE,* **48**(51–57): 42–43.

Dutta, S., I. H. Rehman, P. Malhotra, and V. P. Ramana, 1997: *Biogas, the Indian NGO Experience.* A FPRO-CHF Network Program, New Delhi, India.

EC, 1995: *ExternE – Externalities of Energy.* Vol. 2: Methodology, EUR 16521. European Commission (EC), Office for Official Publications of the European Communities, Luxembourg.

EC, 2009: *Low Energy Buildings in Europe: Current State of Play, Definitions and Best Practice.* Info note, European Commission (EC), Brussels, Belgium.

ECCJ (Energy Conservation Center, Japan), 2008: *Energy saving appliance catalog.* http://www.eccj.or.jp/index.html.

ECCP, 2001: *Long Report.* European Climate Change Program (ECCP), Brussels, Belgium.

ECCP, 2003: *Can We Meet Our Kyoto Targets?*2nd European Climate Change Programme Progress Report, European Climate Change Program (ECCP), Brussels, Belgium.

Ecofys, 2001: *Economic Evaluation of Sectoral Emission Reduction Objectives for Climate Change. Top-down Analysis of Greenhouse Gas Emission Reduction Possibilities in the EU.* Contribution to a study for DG Environment, European Commission by Ecofys Energy and Environment, AEA Technology Environment and National Technical University of Athens, Greece.

ECS, 2002: *Fiscal Policies for Improving Energy Efficiency. Taxation, Grants and Subsidies.* Energy Charter Secretariat (ECS), Brussels, Belgium.

EEA, 2006: *Air Quality and Ancillary Benefits of Climate Change Policies.* European Environment Agency (EEA), Technical Report 4/2006, Copenhagen, Denmark.

Efficiency Vermont, 2009: *For My Home.* Vermont Energy Investment Corporation: Burlington, VT. www.efficiencyvermont.com/for_my_home.aspx (accessed March 15, 2011).

Eichholtz, P., N. Kok, and J. Quigley, 2009: *Why Do Companies Rent Green? Real Property and Corporate Social Responsibility.* Energy Policy and Economics, University of California Energy Institute, Berkeley, CA, USA.

Eicker, U., M. Huber, P. Seeberger, and C. Vorschulze, 2006: Limits and potentials of office building climatisation with ambient air. *Energy and Buildings,* **38**: 574–581.

Eiffert, P., 2003: *Guidelines for the Economic Evaluation of Building-Integrated Photovoltaic Power Systems.* International Energy Agency PVPS

Task 7: Photovoltaic Power Systems in the Built Environment, Technical Report NREL/TP-550–31977, National Renewable Energy Laboratory, Golden, CO.

Eikeland, P. O., 1998: Electricity market liberalization and environmental performance: Norway and the UK. *Energy Policy*, 26(12): 91–927.

Enerdata, 2009: Global Energy Intelligence. Database: www.enerdata.net (accessed 1 September, 2009)

Evander, A., G. Sieböck, and L. Neij, 2004: *Diffusion and Development of New Energy Technologies: Lessons Learned in View of Renewable Energy and Energy Efficiency End-use Projects in Developing Countries*. Report 2004: 2, International Institute for Industrial Environmental Economics, Lund University, Sweden.

Energy Data and Modeling Center, 2008: *Handbook of Energy and Economic Statistics in Japan*. The Energy Conservation Centre, Japan.

Energywatch, 2005: *Get Smart: Bring Meters into the 21st Century*. Energywatch, July 2005. www.energywatch.org.uk/uploads/Smart_meters.pdf (accessed May 30, 2011).

EST, 2009: *Easy Ways to Stop Wasting Energy*. Energy Saving Trust (EST): London, UK. www.energysavingtrust.org.uk/Easy-ways-to-stop-wasting-energy/Stop-wasting-energy-and-cut-your-bills/Tips-to-help-you-stop-wasting-energy/Getting-started (accessed March 15, 2011).

Erb, K. H., S. Gingrich, F. Krausmann, and H. Haberl, 2008: Industrialization, fossil fuels, and the transformation of land use. *Journal of Industrial Ecology*, 12(5–6): 686–703.

Eurowinter Group, 1997: Cold exposure and winter mortality from ischaemic heart disease, cerebrovascular disease, respiratory disease, and all causes in warm and cold regions of Europe. *The Lancet*, 349: 1341–1346.

Eyre, N., M. Pavan, and L. Bodineau, 2009: Energy company obligations to save energy in Italy, France and the UK: what have we learnt? In *Proceedings of the 2009 European Council for an Energy Efficient Economy (ECEEE) Summer Study*. La Colle sur Loupe, France.

FEEM, 2008: *Social Costs of Electricity Generation and Policy Design*. In *Proceedings of the final conference of the CASES Project*, Fondazione Enni Enrico Matei (FEEM), September 29–30, 2008, Milan, Italy.

Feist, W., 1996: Life-cycle Energy Analysis: Comparison of Low-energy House, Passive House, Self-sufficient House. In *Proceedings of the International Symposium of CIB W67*. Vienna, Austria, pp.183–190.

Feist, W., J. Schnieders, V. Dorer, and A. Haas, 2005: Re-inventing air heating: convenient and comfortable within the frame of the Passive House concept. *Energy and Buildings*, 37: 1186–1203.

Feist, W., 2007: *Tasks – challenges – perspectives*. Conference Proceedings, 11th International Passive House Conference, Bregenz, Passive House Institute, Darmstadt, Germany, pp. 383–392.

Fisher, D., F. Schmid, and A. J. Spata, 1999: Estimating the energy-saving benefit of reduced-flow and/or multi-speed commercial kitchen ventilation systems. *ASHRAE Transactions*, 105(1): 1138–1151.

Fisk, B., 1999: *Indoor Air Quality Handbook*, McGraw Hill, New York, NY.

Fisk, W. J., 2000: Health and productivity gains from better indoor environments and their relationship with building energy efficiency. *Annual Review of Energy and the Environment*, 25(1): 537–566.

Florides, G. A., S. A. Tassou, S. A. Kalogirou, and L. C. Wrobel, 2002: Measures used to lower building energy consumption and their cost effectiveness. *Applied Energy*, 73: 299–328.

Ford, B., N. Patel, P. Zaveri, and M. Hewitt, 1998: Cooling without air conditioning: The Torrent Research Centre, Ahmedabad, India. *Renewable Energy*, 15: 177–182.

Francisco, P. W., L. Palmiter, and B. Davis, 1998: Modeling the thermal distribution efficiency of ducts: comparisons to measured results. *Energy and Buildings*, 28: 287–297.

Frank, T., 2005: Climate change impacts on building heating and cooling energy demand in Switzerland. *Energy and Buildings*, 37: 1175–1185.

Fridley, D. and J. Lin, 2004: *Potential Carbon Impact of Promoting Energy Star in China and Other Countries*. Lawrence Berkeley National Laboratory (LBNL), Berkeley, USA.

Friel, S. 2007. Housing, fuel poverty and health: A Pan-European analysis (book review). *Health Sociology Review*, 16(2): 195–196.

Frondel, M., J. Peters, and C. Vance, 2008: Identifying the rebound: evidence from a German household panel. *Energy Journal*, 29(4): 145–163.

Fujino, J., G. Hibino, T. Ehara, Y. Matsuoka, M. Yuzuru, K. Toshihiko, and M. Kainuma, 2008: Back-casting analysis for 70% emission reduction in Japan by 2050. *Climate Policy*, 8: S108-S124.

Fülöp, O., 2009: Ösztönzött pazarlás – Lakossági energiaárak állami támogatása 2003–2009. Összefoglaló elemzés. Energia Klub, Budapest, Hungary.

Fuller, Merrian C., S. Compagni Portis, and D. M. Kammen, 2009: Toward a low-carbon economy: municipal financing for energy efficiency and solar power. *Environment Science and Policy for Sustainable Development*. January-February 2009. www.environmentmagazine.org/Archives/Back%20Issues/January-February%202009/FullerPortisKammen-full.html (accessed May 31, 2011).

Gadgil, A., S. Al-Beaini, M. Benhabib, S. Engelage, and A. Langton, 2007: *Domestic Solar Water Heater for Developing Countries*. Lawrence Berkeley National Laboratory (LBNL), Berkeley, CA, USA.

Gallaher, M.P., A. C. O'Connor, J. L. Dettbarn, Jr., and L. T. Gilday, 2004: *Cost Analysis of Inadequate Interoperability in the U.S. Capital Facilities Industry*. US Department of Commerce Technology Administration National Institute of Standards and Technology, Gaithersburg, MD, USA.

Gamble, D., B. Dean, D. Meisegeier, and J. Hall, 2004: Building a path towards zero energy homes with energy efficient upgrades. In *Proceedings of the 2004 American Council for an Energy Efficient Economy (ACEEE) Summer Study on Energy Efficiency in Buildings*. Vol. 1, Washington, DC, USA. pp.95–106.

Garbesi, K. and L.B. Desroches, 2010: *Max Tech and Beyond: Maximizing Appliance and Equipment Efficiency by Design*. Lawrence Berkeley National Laboratory, Berkeley, CA.

Garcia Casals, X., 2006: Analysis of building energy regulation and certification in Europe: Their role, limitations and differences. *Energy and Buildings*, 38(5): 381–392.

García-Frapolli E., A. Schilmann, V. Berrueta, H. Riojas-Rodríguez, R. Edwards, M. Johnson, A. Guevara-Sanginés, C. Armendariz, and O. Masera, 2010: Beyond fuelwood savings: valuing the economic benefits of introducing improved biomass cookstoves in the Purépecha region of Mexico. *Ecological Economics*, 69(12): 2598–2605.

Geller, H. and S. Attali, 2005: *The Experience with Energy Efficiency Policies and Programmes in IEA Countries: Learning from the Critics*. International Energy Agency (IEA) Information Paper, Paris, France.

Geller, H., P. Harrington, A. H. Rosenfeld, S. Tanishima, and F. Unander, 2006: Policies for increasing energy efficiencies. 30 years of experience in OECD-countries. *Energy Policy*, 34(5): 556–573.

Genchi, Y., Y. Okano, and T. Ihara, 2007: *Environmental impact assessment of Urban Heat Island phenomena based on Endpoint-type LCIA Methodology: Part 4 Evaluation of environmental impacts on sleep disorder caused by urban heat island in Tokyo* (in Japanese). Summaries of Technical Papers of Annual Meeting Architectural Institute of Japan, Environmental Engineering I, 771–772.

Gillingham, K., Newell, R., and K. Palmer, 2004: The effectiveness and cost of energy efficiency programmes. *Resources. Resources for the Future* (RFF) Technical paper, Washington, DC, USA.

Gillingham, K., R. Newell, and K. Palmer, 2006: Energy efficiency policies: a retrospective examination. *Annual Review of Environmental Resources*, 31: 161–92.

Goldman, C., N. Hopper, and J. Osborn, 2005: Review of the US ESCO industry market trends: an empirical analysis of project data. *Energy Policy*, 33(3): 387–405.

Graham, P., 2005: *Design for Adaptability: An introduction to the principles and basic strategies*. The Environment Design Guide GEN 66, Royal Australian Institute of Architects, Canberra, Australia.

Gram-Hanssen, K. 2004. Domestic electricity consumption: consumers and appliances. In *The Ecological Economics of Consumption*. L. A. Reisch and I. Røpke (eds), Edward Elgar Publishing, London, UK, pp.132–149.

Green, M. A., 2006: Consolidation of thin-film photovoltaic technology: the coming decade of opportunity. *Progress in Photovoltaics: Research and Applications*, 14: 383–392.

Greening, L., D. Green, and C. Difiglio, 2000: Energy efficiency and consumption – the rebound effect – a survey. *Energy Policy*, 28(6–7): 389–401.

Gustavsson, L. and R. Sathre, 2006: Variability in energy and carbon dioxide balances of wood and concrete building materials. *Building and Environment*, 41(7): 940–951.

Gutschner, M. and Task-7 Members, 2001: *Potential for Building Integrated Photovoltaics*. International Energy Agency (IEA) of the Organisation for Economic Cooperation and Development (OECD), Paris, France.

Guy, S. and E. Shove, 2000: *A Sociology of Energy, Buildings, and the Environment*. Routledge, London, UK.

Haas, R., P. Biermayr, J. Zoechling, and H. Auer, 1998: Impacts on electricity consumption of household appliances in Austria: a comparison of time series and cross-section analyses. *Energy Policy*, 26(13): 1031–1040.

Haas, R. and P. Biermayr, 2000: The rebound effect for space heating – Empirical evidence from Austria. *Energy Policy*, 28(6–7): 403–410.

Halozan, H. and R. Rieberer, 1997: Air heating systems for low-energy buildings. *IEA Heat Pump Centre Newsletter*, 15(4): 21–22.

Haller, A., E. Schweizer, P. O. Braun, and K. Voss, 1997: *Transparent Insulation in Building Renovation*. James and James, London, UK.

Hamada, Y., M. Nakamura, K. Ochifuji, S. Yokoyama, and K. Nagano, 2003: Development of a database of low energy homes around the world and analysis of their trends. *Renewable Energy*, 28: 321–328.

Haney, A. B., T. Jamasb, and M. G. Pollitt, 2008: *Smart Metering and Electricity Demand: Technology, Economics and International Experience*. ESRC Electricity Policy Research Group and Faculty of Economics, University of Cambridge, UK.

Harrigan, M. and E. Curley, 2010: *Global Environmental Education, Lacking Energy*. Proceedings of American Council for an Energy-Efficient Economy (ACEEE) 2010 Summer Study on Building, Pacific Grove, CA, USA.

Harris, J., B. Aebischer, J. Glickman, G. Magnin, A. Maier, and J. Vigand, 2005: Public sector leadership: transforming the market for efficient products and services. In *Proceedings of the European Council for an Energy Efficient Economy (ECEEE) Summer Study* 2005. ADEME Editions, Paris, France.

Hart & Cooley, Inc., 2009. Composition and Properties of Air. *TechTalk*, 1(2), September-October.

Hartway, R., S. Price, and C. K. Woo, 1999: Smart meter, customer choice and profitable time-of-use rate option. *Energy*, 24(10): 895–903.

Harvey, L.D.D., 2006: *A Handbook on Low-Energy Buildings and District-Energy Systems*. Earthscan, London, UK.

Harvey, L. D. D. and M. Siddall, 2008: Advanced glazing systems and the economics of comfort. *Green Building Magazine*, Spring 08: 30–35.

Harvey, L. D. D., 2009: Reducing energy use in the buildings sector: measures, costs, and examples. *Special Issue: Energy Efficiency: How Far Does it Get Us in Controlling Climate Change?*, 2(2): 139–163.

Harvey, L. D. D., 2010: *Energy and the New Reality, Volume 1: Energy Efficiency and the Demand for Energy Services*. Earthscan, London, UK.

Hassett, K. and G. Metcalf. (1993). Energy conservation investments: do consumers discount the future correctly? *Energy Policy*, 21(6), 710–716.

Hastings, R. and M. Wall, 2007a: *Sustainable Solar Housing, Volume 1: Strategies and Solutions*. Earthscan, London, UK.

Hastings, R. and M. Wall, 2007b: *Sustainable Solar Housing, Volume 2: Exemplary Buildings and Technologies*. Earthscan, London, UK.

Haus der Zukunft, 2006: *The very first reconstruction in Austria of a one-unit house to passive house standard (Model project in Pettenbach/Upper Austria)*. www.hausderzukunft.at/results.html/id3955 (accessed 30 August, 2011).

Haus der Zukunft, 2007: *PASSIVe house renovation, Makartstrasse, Linz*. www.hausderzukunft.at/results.html/id3951 (accessed 30 August, 2011).

Hausman, J. A., 1979: Individual discount rates and the purchase and utilization of energy-using durables. *Bell Journal of Economics*, 10(1): 33.

Healy, J. D., 2004: *Housing, Fuel Poverty, and Health: a Pan-European Analysis*. Ashgate Publishing, Aldershot, UK.

Healy, J. D. and J. P. Clinch, 2002: Fuel poverty, thermal comfort and occupancy: results of a national household-survey in Ireland. *Applied Energy*, 73(3–4): 329–343.

Hegarty, D., 2006: Satisfying a burning need. *Philips Research Password*, 28(October): (28–31)

Hegedus, S., 2006: Thin film solar modules: The low cost, high throughput and versatile alternative to Si wafers. *Progress in Photovoltaics. Research and Applications*, 14: 393–411.

Heijne, C., 2003: *Energy Education Hitting Home*. Centre for Sustainable Energy, Bristol, UK.

Hein, L., K.V. Koppen, R.S. de Groot, and E.C. van Ierland, 2005: Spatial scales, stakeholders and the valuation of ecosystem services. *Ecological economics*, 57(2): 209–228.

Henderson, G., 2005: *Home air conditioning in Europe – how much energy would we use if we became more like American households?* Proceedings of the European Council for an Energy Efficient Economy (ECEEE), June 2005, Mandelieu, France.

Hermelink, A., 2006: A retrofit for sustainability: meeting occupants' needs within environmental limits. In *Proceedings of the American Council for an Energy-Efficient Economy (ACEEE) Summer Study on Energy Efficiency in Buildings*. Pacific Grove, CA, USA.

Hernandez, P. and P. Kenny, 2010: From net energy to zero energy buildings: defining life cycle zero energy buildings (LC-ZEB). *Energy and Buildings*, 42(6): 815–821.

Hertwich, E. G., 2005a: Consumption and the rebound effect: An industrial ecology perspective. *Journal of Industrial Ecology*, **9**(1–2): 85–98.

Hertwich, E. G., 2005b: Lifecycle approaches to sustainable consumption: a critical review. *Environmental Science and Technology*, **39**(13): 4673–4684.

Hertwich, E. G, 2011: The life cycle environmental impacts of consumption. *Economic Systems Research*, **23**(1): 27–48.

Hestnes, A. G. and N. U. Kofoed, 1997: *OFFICE: Passive Retrofitting of Office Buildings to Improve their Energy Performance and Indoor Environment*. JOULE Programme: JOR3-CT96–0034, European Commission Directorate General for Science Research and Development, Brussels, Belgium.

Hestnes, A. G. and N. U. Kofoed, 2002: Effective retrofitting scenarios for energy efficiency and comfort: results of the design and evaluation activities within the OFFICE project. *Building and Environment*, **37**(6): 569–574.

Hewitt, D., M. Frankel, and D. Cohan, 2010: The future of energy codes. In *Proceedings of the American Council for an Energy-Efficient Economy (ACEEE) 2010 Summer Study on Energy Efficiency in Buildings*. Pacific Grove, CA, USA.

Hoffman, A. J. and R. Henn, 2008: Overcoming the social and psychological barriers to green building. *Organization & Environment*, **21**(4): 390–419.

Holt, S. and L. Harrington, 2003: Lessons learnt from Australia's standards and labelling programme. In *Proceedings of the European Council for an Energy Efficient Economy (ECEEE)2003 Summer Study*. Saint Raphaël, France.

Holton, J. K. and P. E. Rittelmann, 2002: Base loads (lighting, appliances, DHW) and the high performance house. *ASHRAE Transactions*, **108**(Part 1): 232–242.

Howe, M., D. Holland, and A. Livchak, 2003: Displacement ventilation: Smart way to deal with increased heat gains in the telecommunication equipment room. *ASHRAE Transactions*, **109**(Part 1): 323–327.

Howieson, S. G., 2005: Multiple deprivation and excess winter deaths in Scotland. *The Journal of the Royal Society for the Promotion of Health*, **125**(1): 18–22.

Hughes, P. J. and J. A. Shonder, 1998: *The Evaluation of a 4000-Home Geothermal Heat Pump Retrofit at Fort Polk, Louisiana*, ORNL/CON-460, Oak Ridge National Laboratory, TN, USA.

Hui, S. C. M., 2002: *Using performance-based approach in building energy standards and codes. In Proceedings of the Chonqing-Hong Kong Joint Symposium 2002*.8–10 July 2002, Chongqing, China, pp.A52–61.

Humm, O., 2000: Ecology and economy when retrofitting apartment buildings. *IEA Heat Pump Centre Newsletter*, **15**(4): 17–18.

Hunt, A., Watkiss, P., 2011: Climate Change Impacts and Adaptation in Cities: A Review of the Literature. *Climatic Change*, **104** (1): 13–49

Hutton, R. B., G. A. Mauser, P. Filiatrault, and O. T. Ahtola, 1986: Effects of cost-related feedback on consumer knowledge and consumption behaviour: a field experimental approach. *Journal of Consumer Research*, **13**(3): 327–336.

Hutton, G., E. Rehfuess, and F. Tediosi, 2007: Evaluation of the costs and benefits of interventions to reduce indoor air pollution. *Energy for Sustainable Development*,11(4): 34–43.

Hutson, S. S., N. L. Barber, J. F. Kenny, K. S. Linsey, D. S. Lumia and M. A. Maupin, 2004: *Estimated Use of water in the United States in 2000*. USGS Circular 1268, United States Geological Survey (USGS), Reston, VA, USA.

IBM, 2007:*Ontario Energy Board Smart Price Pilot Final Report*, IBM Global Business Services and eMeter Strategic Consulting, Ontario Energy Board, Toronto, ON, Canada.

IEA, 1999: *District Cooling, Balancing the Production and Demand in CHP*. Netherlands Agency for Energy and Environment, International Energy Agency (IEA), Sittard, the Netherlands.

IEA, 2003: *Cool Appliances, Policy Strategies for Energy Efficient Homes*. International Energy Agency (IEA) of the Organisation for Economic Cooperation and Development (OECD), Paris, France.

IEA, 2004: *Energy Balances of non-OECD Countries 2001–2002*.International Energy Agency (IEA) of the Organisation for Economic Cooperation and Development (OECD), Paris, France.

IEA, 2005a: *Evaluating Energy Efficiency Policy Measures & DSM Programmes, Volume 1 Evaluation Guidebook*. International Energy Agency (IEA) of the Organisation for Economic Cooperation and Development (OECD), Paris, France.

IEA, 2005b: *Saving Electricity in a Hurry – Dealing with Temporary Shortfalls in Electricity Supplies*. International Energy Agency International Energy Agency (IEA) of the Organisation for Economic Cooperation and Development (OECD), Paris, France.

IEA, 2006: *World Energy Outlook 2006*. International Energy Agency (IEA) of the Organisation for Economic Cooperation and Development (OECD), Paris, France.

IEA, 2007a: *Mind the Gap- Quantifying Principal-Agent Problems in Energy Efficiency*. International Energy Agency (IEA) of the Organisation for Economic Cooperation and Development (OECD), Paris, France.

IEA, 2007b: *Energy Use in the New Millennium. Trends in IEA countries. In Support of the G8 Plan of Action*. International Energy Agency (IEA) of the Organisation for Economic Cooperation and Development (OECD), Paris, France.

IEA, 2008a: *Energy Technology Perspectives, 2008: Scenarios & Strategies to 2050: in Support of the G8 Plan of Action*. International Energy Agency (IEA) of the Organisation for Economic Cooperation and Development (OECD), Paris, France.

IEA, 2008b: *World Energy Outlook 2008*. International Energy Agency (IEA) of the Organisation for Economic Cooperation and Development (OECD), Paris, France.

IEA, 2008c: *Trends in Photovoltaic Applications, Survey Report of Selected IEA Countries Between 1992 and 2007*. Report IEA-PVPS T1–17,International Energy Agency (IEA) of the Organisation for Economic Cooperation and Development (OECD)

IEA. 2010a: *Energy Technology Perspective. Scenarios & Strategies to 2050*. International Energy Agency (IEA). Paris, France.

IEA, 2010b: *World Energy Outlook 2010*. International Energy Agency (IEA), Paris, France.

IEA, 2010c: *Energy Performance Certification of Buildings*. International Energy Agency (IEA), Paris, France.

IEA, 2010d: *Energy Poverty. How to make energy access universal?* Special early excerpt of the World Energy Outlook 2010 for the UN General Assembly on the Millennium Development Goals. International Energy Agency (IEA), United Nations Development Programme (UNDP) and United Nations Industrial Development Organization (UNIDO), Paris, France.

IEA Online Statistics, 2007: Statistics and Balances. International Energy Agency (IEA) of the Organisation for Economic Cooperation and Development (OECD), Paris, France, www.iea.org/stats/index.asp (accessed 28 April 2011).

IIASA, 2007: *GGI Scenario Database*. International Institute for Applied System Analysis (IIASA), Laxenburg, Austria. www.iiasa.ac.at/Research/GGI/DB/ (accessed April 2011).

Indiastat, 2010: *Figures at All-India, State level. Revealing India statistically*. Indiastat.com. www.indiastat.com/power/26/consumptionandsale/70/stats.aspx (accessed April 2011).

Interface Engineering, 2005: *Engineering a Sustainable World: Design Process and Engineering Innovations for the Center for Health and Healing at the Oregon Health and Science University, River Campus*. Interview by Maria Sharmina, 2009. www.interface-engineering.com.

IPCC, 1996: *Climate Change 1995. IPCC Second Assessment*. Report of the Intergovernmental Panel on Climate Change (IPCC), Cambridge University Press, Cambridge, UK and New York, NY, USA.

IPCC, 2007: *Climate Change 2007 – Mitigation*. Contribution of Working Group III to the Fourth Assessment Report of the Intergovernmental Panel on Climate Change, B. Metz, O.R. Davidson, P.R. Bosch, R. Dave, L.A. Meyer (eds), Cambridge University Press, Cambridge, United Kingdom and New York, NY, USA.

Itard, L. and G. Klunder, 2007: Comparing environmental impacts of renovated housing stock with new construction. *Building Research and Information, 35*(3): 252–267.

Jaboyedoff, P., C. A. Roulet, V. Dorer, A. Weber, and A. Pfeiffer, 2004: Energy in air-handling units – results of the AIRLESS European project. *Energy and Buildings, 36*(4): 391–399.

Jackson, J., 2008: *Energy Budgets at Risk (EBaR): A Risk Management Approach to Energy Purchase and Efficiency Choices*. Wiley Finance, Hoboken, NJ, USA.

Jackson, T., 2009: Prosperity without growth-Economics for a finite planet, Earthscan, Oxford.

Jaffe, A. B. and R. N. Stavins, 1994: The energy-efficiency gap. *Energy Policy, 22*(10): 804–810.

Jain, M., V. Gaba, and L. Srivastava, 2007: *Managing Power Demand – A Case Study of Residential Sector in Delhi*. The Energy & Resources Institute (TERI), Teri Press, New Delhi, India.

Jakob, M., 2006: Marginal costs and co-benefits of energy efficiency investments – the case of the Swiss residential sector. *Energy Policy, 34*(2): 172–187.

Janda, K. B., 1998: *Building Change: Effects of Professional Culture and Organizational Context on Energy Efficiency Adoption in Buildings*. Energy and Resources Group, University of California at Berkeley, Berkeley, CA, USA.

Janda, K. B., 1999: *Re-Inscribing Design Work: Architects, Engineers, and Efficiency Advocates. Proceedings of the European Council for an Energy-Efficient Economy (ECEEE) Summer Study*. June 1–6, 2009, Mandelieu, France.

Janda, K. B., C. Payne, R. Kunkle, and L. Lutzenhiser, 2002: What organizations did (and didn't) do: three factors that shaped conservation responses to California's 2001 'crisis'. In Proceedings of the American Council for an Energy-Efficient Economy (ACEEE) 2002 Summer Study, Asilomar, CA, USA.

Janda, K. B., 2008: *Implications of Ownership in the U.S. and U.K.: An Exploration of Energy Star Buildings & the Energy Efficiency Accreditation Scheme. Proceedings of the American Council for an Energy-Efficient Economy (ACEEE) Summer Study on Energy Efficiency in Buildings*, August 17–22, 2008, Asilomar, CA, USA.

Janda, K.B., 2009: Worldwide status of energy standards for buildings: a 2009 update. In *Proceedings of the European Council for an Energy Efficient Economy (ECEEE) Summer Study*, June 1–6, 2009, Cote d'Azur, France.

Jannuzzi, G. M., 2005: Energy Efficiency and R&D Activities in Brazil: Experiences from the Wirecharge Mechanism (1998–2004). In *Developing Financial Intermediation Mechanisms for Energy Projects in Brazil, China and India*. Prepared by Econergy International Corporation, Angra dos Reis, Brazil.

Jennings, J. D., F. M. Rubinstein, D. DiBartolomeo, and S. L. Blanc, 2000: *Comparison of Control Options in Private Offices in an Advanced Lighting Controls Testbed*. LBNL-43096 REV, Lawrence Berkeley National Laboratory, Berkeley, CA, USA.

Jin, S. H, 2007: The effectiveness of energy efficiency improvement in a developing country: Rebound effect of residential electricity use in South Korea. *Energy Policy, 35*(11): 5622–5629.

Jochem E., T. Barker, S. Scrieciu, W. Schade, N. Helfrich, O. Edenhofer, N. Bauer, S. Marchand, J. Neuhaus, S. Mima, P. Criqui, J. Morel, B. Chateau, A. Kitous, G. J. Nabuurs, M. J. Schelhaas, T. Groen, L. Riffeser, F. Reitze, E. Jochem, G. Catenazzi, M. Jakob, B. Aebischer, W. Eichhammer, A. Held, M. Ragwitz, U. Reiter, and H. Turton, 2009: *Adaptation and Mitigation Strategies: Supporting European Climate Policy*. Report D-M1.2 of the reference scenario for Europe. Fraunhofer ISI, Karlsruhe, Germany.

John, M., M. Smith, and S. Korosec, 2005: *Integrated Energy Policy Report*. California Energy Commission, Sacramento, CA.

Johnson, M., R. Edwards, C. Alatorre, and O. Masera, 2008: In-field greenhouse gas emissions from cookstoves in rural Mexican households. *Atmospheric Environment, 42*(2008): 1206–1222.

Jones Lang LaSalle/CREIS, 2006: *OSCAR – Office Service Charge Analysis Report. Based on 397 buildings with 6 million m2 in 2006*.Jones Lang LaSalle. Chicago, IL, USA.

Joosen, S., M. Harmelink, and K. Blok, 2004: *Evaluation Climate Policy in the Built Environment 1995–2002*. Ecofys report prepared for the Dutch Ministry of Housing, Spatial Planning and the Environment, Utrecht, the Netherlands.

Jyukankyo Research Institute and Japan Centre for Climate Change Action, 2007: *Fact sheet of energy saving home electrical appliances II. Energy saving of home electrical appliances (1) Air conditioner (in Japanese)*. Jyukankyo Research Institute and the Japan Center for Climate Change Action, Tokyo, Japan.

Kats, G. 2003: *The Costs and Financial Benefits of Green Buildings*. A report prepared for California's Sustainable Building Task Force, Sacramento, CA, USA.

Kats, G., L. Alevantis, A. Berman, E. Mills, and J. Perlman, 2003: *The Costs and Financial Benefits of Green Buildings*. A report to California's Sustainable Building Task Force, Sacramento, CA, USA.

Kats, G., 2005: *National Review of Green Schools: Costs, Benefits, and Implications for Massachusetts*. A report for the Massachusetts Technology Collaborative, Boston, MA.

Katsev, R. D. and T. R. Johnson, 1987: *Promoting Energy Conservation: An Analysis of Behavioral Research*. Westview Special Studies in Natural Resources and Energy Management, Westview Press, Boulder, CO, USA.

Kelleher, E., 2006: *US Homebuilders Go Green*. Clean Energy Solutions, US Department of State, Bureau of International Information Programs, Washington, DC, USA.

Kempton, W., 1992: Psychological research for new energy problems. *American Psychologist, 47*(10): 1213.

Keshner, M. S. and R. Arya, 2004: *Study of Potential Cost Reductions Resulting from Super-Largescale Manufacturing of PV Modules*. Final Subcontract Report, NREL/SR-520–56846. Washington, DC, USA.

Khan, J., M. Harmelink, R. Harmsen, W. Irrek, and N. Labanca, 2007: From Theory Based Policy Evaluation to SMART Policy Design. Summary Report of the AID-EE Project. *Project executed within the framework of the Energy Intelligence for Europe program*, EIE-2003–114, Brussels, Belgium.

Kikegawa, Y., Y. Genchi, H. Kondo, and K. Hanaki, 2006: Impacts of city-block-scale countermeasures against urban heat-island phenomena upon a building's energy-consumption for air-conditioning. *Applied Energy, 83*(6): 649–668.

Killip, G., 2008: *Building a Greener Britain: Transforming the UK's Existing Housing Stock*. Environmental Change Institute, Oxford, UK.

King, A. D. (ed.), 1980: *Buildings and Society*. Routledge and Kegan Paul, London, UK.

Kinney, S. and M. A. Piette, 2002: *Development of a California Building Energy Benchmarking Database*. LBNL 50676, Lawrence Berkeley Laboratory Report, Berkeley, CA, USA.

Koch-Nielsen, H., 2002: *Stay Cool: A Design Guide for the Built Environment in Hot Climates*. James and James, London, UK.

Koebel, C. T., 2008: Innovation in Homebuilding and the Future of Housing. *Journal of the American Planning Association*, **74**(1): 45–58.

Kohlhaas, M., 2005: *Gesamtwirtschaftliche Effekte der ökologischen Steuerreform*. (Macroeconomic effects of the ecological tax reform), DIW Berlin, Abteliung Energie, Verkehr, Umwelt, Berlin, Germany.

Kok, N., P. Eichholtz, R. Bauer, and P. Peneda, 2010: *Environmental Performance: A Global Perspective on Commercial Real Estate*. European Centre for Corporate Engagement, Maastricht University School of Business and Economics, Maastricht, the Netherlands.

Kolokotroni, M., 2001: Night ventilation cooling of office buildings: parametric analyses of conceptual energy impacts. *ASHRAE Transactions*, **107**(1): 479–489.

Kooreman, P. and T. Steerneman, 1998: A note energy efficiency investments of an expected cost minimizer. *Resource and Energy Economics*, **20**: 373–381.

Kosmo, M., 1989: Commercial energy subsidies in developing countries: opportunity for reform. *Energy Policy*, **17**(3): 244–253.

Krapmeier, H. and E. Drössler (eds.), 2001: CEPHEUS: *Living Comfort Without Heating*. Springer-Verlag, Vienna, Austria.

Krey, M., 2005: Transaction costs of unilateral CDM projects in India – results from an empirical survey. *Energy Policy*, **33**(18): 2385–2397.

Kushler, M, D. Work, and P. Witte, 2004: *Five Years In: An Examination of the First Half Decade of Public Benefits Energy Efficiency Policy*. American Council for an Energy Efficient Economy (ACEEE). Report U041, Washington, D.C., USA.

Larsen, B. M. and R. Nesbakken, 1997: Norwegian emissions of CO2 1987–1994. *Environmental and Resource Economics*, **9**(3): 275–90.

Larsson, N., 2001: Canadian green building strategies. In *Proceedings of the 18th International Conference on Passive and Low Energy Architecture*. 7–9 November, Florainopolis, Brazil, pp.17–25.

Laustsen, J., 2008: *Energy Efficiency Requirements in Building Codes, Energy Efficiency Policies for New Buildings*. IEA information paper. International Energy Agency (IEA) of the Organisation for Economic Cooperation and Development (OECD), Paris, France.

Laustsen, J., forthcoming: *Reducing Energy in Buildings with a Factor 4. An IEA Factor Four Strategy for Buildings*. International Energy Agency (IEA) of the Organisation for Economic Cooperation and Development (OECD), Paris, France.

Le Fur, B., 2002: *Panorama des dispositifs d'économie d'énergie en place dans les pays de L'Union Européenne*. Club d'Amélioration de l'Habitat, Paris, France.

Leckner, M. and R. Zmeureanu, 2011: Life cycle cost and energy analysis of a Net Zero Energy House with solar combisystem. *Applied Energy*, **88**(1): 232–241.

Lee, K. H. and R. K. Strand, 2008: The cooling and heating potential of an earth tube system in buildings. *Energy and Buildings*, **40**(4): 486–494.

Lee, W. L. and F. W. Yik, 2004: Regulatory and voluntary approaches for enhancing building energy efficiency. *Progress in Energy and Combustion Science*, **30**(2004): 477–499.

Lees, E., 2006: *Evaluation of the Energy Efficiency Commitment 2002–05*. Oxon, London.

Lees, E. W., 2008: *Evaluation of the Energy Efficiency Commitment 2005–08*. Report to the Department of Energy and Climate Change. Eoin Lees Energy, Oxford, UK.

Lehtomäki, A., S. Huttunen, and J. A. Rintala, 2007: Laboratory investigations on co-digestion of energy crops and crop residues with cow manure for methane production: Effect of crop to manure ratio. *Resources, Conservation and Recycling*, **51**(3): 591–609.

Lenzen, M. and G. Treloar, 2002: Embodied energy in buildings: wood versus concrete – reply to Borjesson and Gustavsson. *Energy Policy*, **30**(3): 249–255.

Lenzen, M., M. Wier, C. Cohen, H. Hayami, S. Pachauri, and R. Schaeffer, 2006: A comparative multivariate analysis of household energy requirements in Australia, Brazil, Denmark, India and Japan. *Energy*, **31**(2–3): 181–207.

Levine, M., D. Ürge-Vorsatz, K. Blok, L. Geng, D. Harvey, S. Lang, G. Levermore, A. Mongameli Mehlwana, S. Mirasgedis, A. Novikova, J. Rilling, and H. Yoshino, 2007: Residential and commercial buildings. In *Climate Change 2007: Mitigation*. Contribution of Working Group III to the Fourth Assessment Report of the Intergovernmental Panel on Climate Change (IPCC) B. Metz, O.R. Davidson, P.R. Bosch, R. Dave, L.A. Meyer (eds.), Cambridge University Press, Cambridge, United Kingdom and New York, NY, USA.

Lewis, M., 2004: Integrated design for sustainable buildings. Building for the future. A Supplement to *ASHRAE Journals*, **46**(9): 2–3.

Li, D. H. W. and J. C. Lam, 2003: An investigation of daylighting performance and energy saving in a daylight corridor. *Energy and Buildings*, **35**: 365–373.

Li, J.-F., O. W. H. Wai, Y. S. Li, J.M. Zhan, Y. A. Ho, J. Li, and E. Lam, 2010: Effect of green roof on ambient CO2 concentration. *Building and Environment*, **45**: 2644–2651.

Liddell, C. and C. Morris, 2010: Fuel poverty and human health: A review of recent evidence. *Energy Policy*, **38**(6): 2987–2997.

Lin, B. and Z Jiang, 2011: Estimates of energy subsidies in China and impact of energy subsidy reform. *Energy Economics*, **33**(2): 273–283.

Lin, Z. and S. Deng, 2004: A study on the characteristics of nighttime bedroom cooling load in tropics and subtropics. *Building and Environment*, **39**(9):1101–1114.

Listokin, D. and D. Hattis, 2004: *Building Codes and Housing*. Rutgers University, NJ, USA and Building Technology Inc., MA, USA.

Liu, M. and D. E. Claridge, 1999: Converting dual-duct constant-volume systems to variable-volume systems without retrofitting the terminal boxes. *ASHRAE Transactions*, **105**(Part 1): 66–70.

Liu, M., D. E. Claridge and W. D. Turner, 2003: Continuous commissioning of building energy systems. *Journal of Solar Energy Engineering*, **125**(3): 275–281.

Lloyd, B., 2006: Fuel poverty in New Zealand. *Social Policy Journal of New Zealand*, **27**: 142–155.

Lopes, C., L. Pagliano and S. Thomas, 2000: Conciliating the economic interest of energy companies with demand-side management – A review of funding mechanisms in a changing market. In *Proceedings of the UIE International Conference – Electricity for a Sustainable Urban Development*, Lisbon, Portugal.

Lorenzioni, I., S. Nicholson-Cole, and L. Whitmarsh, 2007: Barriers perceived to engaging the UK public with climate change and their policy implications. *Global Environmental Change,* **17**(3–4): 445–459.

Loudermilk, K. J., 1999: Underfloor air distribution solutions for open office applications. *ASHRAE Transactions,* **105**(Part 1): 605–613.

Lovins, A. B., 2005: *Energy End-Use Efficiency. A Part of its 2005–06 Study, Transitions to Sustainable Energy Systems*. Inter Academy Council, Amsterdam, the Netherlands.

Lu, L. and H. X. Yang, 2010: Environmental payback time analysis of a roof-mounted building-integrated photovoltaic (BIPV) system in Hong Kong. *Applied Energy,* **87**(12): 3625–3631.

Lutzenhiser, L., 1993: Social and behavioral aspects of energy use. *Annual Review of Energy and the Environment,* **18**: 247–289.

Lutzenhiser, L., 1994: Innovation and organizational networks. *Energy Policy,* **22**(10): 867–876.

Lutzenhiser, L., K. B. Janda, R. Kunkle, and C. Payne, 2002: *Understanding the Response of Commercial and Institutional Organizations to the California Energy Crisis California Energy Commission*. Sacramento, CA.

MacAvoy, H., 1997: *All-Ireland Policy Paper on Fuel Poverty and Health*. Institute of Public Health in Ireland, Dublin, Ireland.

Manisha, J., V. Gaba and L. Srivastava, 2007: *Managing Power Demand – A case study of the residential sector in Delhi*. The Energy and Resources Institute (TERI), New Delhi, India.

Mansur, E. T., R. Mendelsohn, and W. Morrison, 2005: *A Discrete-Continuous Choice Model of Climate Change Impacts on Energy*. Columbia University, New York, NY, USA.

Marchand, C., M. H. Laurent, R. Rezakhanlou, and Y. Bamberger, 2008: Le bâtiment sans énergie fossile? Les bâtiments pourront-ils se passer des énergies fossiles en France à l'horizon 2050. *Futuribles,* **343**(July): 79–100.

Martens, R., B. Bass, and S. S. Alcazar, 2008: Roof–envelope ratio impact on green roof energy performance. *Urban Ecosystems,* **11**(4): 399–408.

Martinsen D, V. Krey, and P. Markewitz, 2007: Implications of high energy prices for energy system and emissions – the response from an energy model for Germany. *Energy Policy,* **35**(9): 4504–4515.

Masera, O. R., B. D. Saatkamp, and D. M. Kammen, 2000: From linear fuel switching to multiple cooking strategies: a critique and alternative to the energy ladder model for rural households. *World Development,* **28**(12): 2083–2103.

Masera, O., R. Edwards, C. Armendáriz, V. Berrueta, M. Johnson, L. Rojas, and H. Riojas-Rodríguez, 2007: Impact of "Patsari" improved cookstoves on indoor air quality in Michoacan, Mexico. *Energy For Sustainable Development,* **11**(2): 45–56.

McDonell, G., 2003: Displacement ventilation. *The Canadian Architect,* **48**(4): 32–33.

McDonell, G., 2004: *Personal communication. Omicron Consulting,* Vancouver, Canada, December, 2004.

McGraw-Hill Construction, 2006: *Green Building SmartMarket Report*. McGraw-Hill Construction, New York, NY, USA.

McMakin, A. H., E. L. Malone, and R. E. Lundgren, 2002: *Motivating Residents to Conserve Energy without Financial Incentives,* **34**(6): 848–863.

McNeil, M. and V. Letschert, 2008: *Future Air Conditioning Energy Consumption in Developing Countries and what can be done about it: The potential of Efficiency in the Residential Sector*. Lawrence Berkeley National Laboratory, Berkeley, CA, USA.

McNeil, M.A., V.E. Letschert, and S. de la Rue du Can, 2008: *Global Potential of Energy Efficiency Standards and Labeling Programs*. LBNL-760E, Lawrence Berkeley National Laboratories, Berkeley, CA, USA.

McNeil, M. A. and V. E. Letschert, 2010: Modeling diffusion of electrical appliances in the residential sector. *Energy and Buildings,* **42**(6):783–790.

Meier, A. and A. Eide, 2007: How many people actually see the price signal? Quantifying market failures in the end-use of energy. In *Proceedings of the European Council for an Energy Efficient Economy (ECEEE) Summer Study,* 2007, Stockholm, Sweden.

Melichar, J., M. Havranek, V. Maca, M. Scasny, and M. Kudelko, 2004: *Implementation of ExternE Methodology in Eastern Europe*. Final Report on Work Package 7, Contract No. ENG1-CT-2002–00609, Extension of Accounting Framework and Policy Applications, The European Commission (EC), Brussels, Belgium.

Menanteau, P., 2007: *Policy Measures to Support Solar Water Heating: Information, Incentives and Regulations*. Technical Report prepared for the World Energy Council (WEC), London, UK and Agence de l'Environnement et de la Maîtrise de l'Energie (ADEME), Paris, France.

Meyers, S., J. E. McMahon, and B. Atkinson, 2008: *Realized and Projected Impacts of U.S. Energy Efficiency Standards for Residential and Commercial Appliances –* LBNL-63017. Lawrence Berkeley National Laboratory, Berkeley, CA, USA.

Michaelowa, A. and F. Lotzo, 2005: Transaction costs, institutional rigidities and the size of the clean development mechanism. *Energy Policy,* **33**(4): 511–523.

Mills, E., 1991: Evaluation of European lighting programmes: utilities finance energy efficiency. *Energy Policy,* 19 (3): 266–278.

Mills, E., 2009: Building Commissioning: *A Golden Opportunity for Reducing Energy Costs and Greenhouse Gas Emissions*. LBNL report prepared for California Energy Commission Public Interest Energy Research (PIER), Lawrence Berkeley National Laboratory, Berkeley, CA, USA.

Milne, G. and B. Boardman, 2000: Making cold homes warmer: the effect of energy efficiency improvements in low-income homes. A report to the Energy Action Grants Agency Charitable Trust. *Energy Policy,* **28**(6–7): 411–424.

Milton D., P.M. Glencross, and M.D. Walters, 2000: Risk of sick leave associated with outdoor ventilation level, humidification, and building related complaints. *Indoor Air,* **10**(4): 212–21.

Mirasgedis, S., Y. Sarafidis, E. Georgopoulou, K. Kotroni, K. Lagouvardos, and D.P. Lalas, 2007: Modeling framework for estimating impacts of climate change on electricity demand at regional level: case of Greece. *Energy Conversion and Management,* **48**(5): 1737–1750.

MOE, 2006: Ministry of Environment, Government of Japan, press release on 11/10/2006 (in Japanese). www.env.go.jp/press/press.php?serial=7690 (accessed December 31, 2010).

Monahan, J. and J. C. Powell, 2011: An embodied carbon and energy analysis of modern methods of construction in housing: a case study using a lifecycle assessment framework. *Energy and Buildings,* **43**(1): 179–188.

Morch, A. Z., J. Parsons, and J. C. P. Kester, 2007: *Smart Electricity Metering as an Energy Efficiency Instrument: Comparative Analyses of Regulation and Market Conditions in Europe*. In European Council for an Energy Efficient Economy (ECEEE) Summer Study, Stockholm, Sweden.

Morrison, C., and N. Shortt, 2008: Fuel poverty in Scotland: refining spatial resolution in the Scottish Fuel Poverty Indicator using a GIS-based multiple risk index. *Health & Place,* **14**(4): 702–717.

Mundaca, L., 2007: Transaction costs of tradable white certificate schemes: the energy efficiency commitment as case study. *Energy Policy*, **35**(8): 4340–4354.

Murakoshi, C., H. Nakagami, M. Tsuruda, and N. Edamura, 2005: *New Challenges of Japanese Energy Efficiency Program by Top Runner Approach*. Proceedings of the European Council for an Energy Efficient Economy (ECEEE) 2005 Summer Study, Stockholm, Sweden.

MURE, 2007: *Mesure d'Utilisation Rationnelle de l'Energie (MURE) Database*. www.isis-it.com/mure/ (accessed December 31, 2010).

MURE, 2008: *Mesure d'Utilisation Rationnelle de l'Energie (MURE) Database*. www.isis-it.com/mure/ (accessed December 31, 2010).

Nagayama, H., 2008: Electric Power Sector Reform Liberalization Models and Electric Power Prices in Developing Countries: An Empirical Analysis Using International Panel Data. *Energy Economics*, **31**(3): 463–472.

Nahuelhual, L., P. Donoso, A. Ñara, D. Nuñez, C. Oyárzun, and E. Neira, 2007: Valuing ecosystem services of Chilean temperate forests. *Environment, Development and Sustainability*, **9**(4):481–499.

Narumi, T., 2007: *Impact of the Temperature Change upon the Human Health in Osaka Prefecture* (in Japanese). Summaries of Technical Papers of Annual Meeting. Architectural Institute of Japan. *Environmental Engineering* I: 763–764.

Nassen, J., J. Holmberg, A. Wadeskog, and M. Nyman, 2007: Direct and indirect energy use and carbon emissions in the production phase of buildings: an input-output analysis. *Energy*, **32**(9): 1593–1602.

National Action Plan for Energy Efficiency, 2009: *Customer Incentives for Energy Efficiency Through Electric and Natural Gas Rate Design*. Prepared by William Prindle, ICF International, Inc.

National Medical Expenditure Survey, 1999: The economic burden of asthma in US children. *Journal of Allergy, Clinical Immunology*, **104**(5): 957–963.

National Research Council, 2009: *America's Energy Future: Technology and Transformation*. National Research Council (NRC) of the National Academies Press, Washington, DC, USA.

NETL, 2008: *National Energy Technology Laboratory Advance Metering Infrastructure*, Office of Electricity Delivery and Energy Reliability, U.S. Department of Energy (US DOE), Washington, DC, USA.

Nevius, M, R. Eldridge, and J. Krouk 2010: *The State of the Efficiency Program Industry Budgets, Expenditures, and Impacts 2009*. Consortium of Energy Efficiency (CEE), Boston, MA, USA.

Nilsson, I., 2006: *Evaluation of BELOK-procurement group for commercial buildings (Sweden)*, EIE-2003–114, Active Implementation for the European Directive on Energy Efficiency (AID-EE), Intelligent Energy-Europe, Brussels, Belgium.

Niu, J. L., L. Z. Zhang, and H. G. Zuo, 2002: Energy savings potential of chilled-ceiling combined with desiccant cooling in hot and humid climates. *Energy and Buildings*, **34**(5): 487–495.

Nordman, B. and K. Christensen, 2010: The next step in reducing IT energy use. Green IT column. *IEEE Computer*, **43**(1): 91–93.

Novikova, A., 2010: *Methodologies for Assessment of Building's Energy Efficiency and Conservation: A Policy-Maker View*. DIW Berlin, Working Paper. German Institute for Economic Research (DIW-Berlin). Berlin, Germany.

Oak Ridge National Laboratory, 2001: *Improving the methods used to evaluate voluntary energy efficiency*. Programs Report, US Department of Energy (US DOE) and US Environmental Protection Agency (US EPA), Washington, DC, USA.

Oke, T.R., 1982: The energetic basis of the urban heat island. *Quarterly Journal of the Royal Meteorological Society*, **108**(455): 1–25.

OPET Network, 2004: *OPET Building European Research Project*: final reports.

Osawa, H. and M. Hayashi, 2009: Present status of the indoor chemical pollution in Japanese houses based on the nationwide field survey from 2000 to 2005. *Building and Environment*, **44**(7): 1330–1336.

Pachauri, S. and D. Spreng, 2002: Direct and indirect energy requirements of households in India. *Energy Policy*, **30**(6): 511–523.

Pachauri, S. and L. W. Jiang, 2008: The household energy transition in India and China. *Energy Policy*, **36**(11): 4022–4035.

Palmer, K., 1999: *Electricity Restructuring: Shortcut or Detour on the Road to Achieving Greenhouse Gas Reductions?* Resources for the Future Climate Issue Brief 18, Resources for the Future, Washington, DC, USA.

Parfomak, P.W. and L.B. Lave, 1996: How many kilowatts are in a negawatt?: verifying ex post estimates of utility conservation impacts at the regional level. *The Energy Journal*, **17**(4): 59–88.

Parker, D. S., J.R. Sherwin, J.K. Sonne, S.F. Barkaszi, D.B. Floyd, and C. R. Withers, 1998: *Measured energy savings of a comprehensive retrofit in an existing Florida residence*. In *Proceedings of the 1998 American Council for an Energy Efficient Economy (ACEEE) Summer Study on Energy Efficiency in Buildings*, vol. 1, Washington, DC, USA pp. 235–251.

Parker, D. S., J. K. Sonne, and J. R. Sherwin, 2002: Comparative Evaluation of the Impact of Roofing Systems on Residential Cooling Energy Demand in Florida. In *Proceedings of the 2002 American Council for an Energy Efficient Economy (ACEEE) Summer Study on Energy Efficiency in Buildings*, Vol.1., Washington, D.C., pp.219–234.

Parker, D., D. Hoak, A. Meier, and R. Brown, 2006: *How Much Energy Are We Using? Potential of Residential Energy Demand Feedback Devices*. In *Proceedings of 2006 Summer Study on Energy Efficiency in buildings*. Florida Solar Energy Center. Lawrence Berkley National Laboratory, Berkeley, CA, USA.

Passivehaus Institut, 2009: Passive House Institute US. www.passivehouse.us/passiveHouse/PHIUSHome.html (accessed August 2, 2010).

Pavan, M., 2008: Not Just Energy Savings: Emerging Regulatory Challenges from the Implementation of Tradable White Certificates. In *Proceedings of the 2008 American Council for an Energy Efficiency Economy (ACEEE) Summer Study on Energy Efficiency in Buildings*, August, 2008, Washington, DC.

CBD, 2001: *The value of ecosystems. Convention on Biological Diversity*, CBD Technical Series No. 4, Convention on Biological Diversity (CBD), Montreal, Canada.

Peters, G. P., C. L. Weber, D. Guan, and K. Hubacek, 2007: China's growing CO_2 emissions-a race between increasing consumption and efficiency gains. *Environmental Science & Technology*, **41**(17): 5939–5944.

Petersdorff, C., T. Boermans, J. Harnisch, O. Stobbe, S. Ullrich, and S. Wartmann, 2005a: *Cost-Effective Climate Protection in the EU Building Stock*. Report established by ECOFYS, Utrect, the Netherlands, for the European Insulation Manufacturers Association (EURIMA), Brussels, Belgium.

Petersdorff, C., T. Boermans, S. Joosen, I. Kolacz, B. Jakubowska, M. Scharte, O. Stobbe, and J. Harnisch, 2005b: *Cost-Effective Climate Protection in the Building Stock of the New EU Member States: Beyond the EU Energy Performance of Buildings Directive*. Report established by ECOFYS, Utrect, the Netherlands, for the European Insulation Manufacturers Association (EURIMA), Brussels, Belgium,

Piette, M. A., S. K. Kinney, and P. Haves, 2001: Analysis of an information monitoring and diagnostic system to improve building operations. *Energy and Buildings,* **33**: 783–791.

Platts Research and Consulting, 2004: *Hedging Energy Price Risk with Renewables and Energy Efficiency*, New York, NY, USA.

Pollitt, 2008: Electricity reform in Argentina: lessons for developing countries. *Energy Economics,* **30**(4): 1536–1567.

Porta-Gándara, M. A., E. Rubio, and J. L. Fernández, 2002: Economic feasibility of passive ambient comfort in Baja California dwellings. *Building and Environment,* **37**(10): 993–1001.

Power, M., 2006: Fuel poverty in the USA: the overview and the outlook. *Energy Action,* **98**.

Quinlan, P., H. Geller, and S. Nadel, 2001: *Tax incentives for innovative energy-efficient technologies (Updated)*. Report Number E013, American Council for an Energy Efficient Economy (ACEEE), Washington, DC, USA.

Rajvanshi, A. K., S. M. Patil, and B. Mendonca, 2007: Low concentration ethanol stove for rural areas of India. *Energy for Sustainable Development,* **11**(1):63–68.

Ramesh, T., R. Prakash, and K. K. Shukla, 2010: Life cycle energy analysis of buildings: an overview. *Energy and Buildings,* **42**(10): 1592–1600.

Rapoport, A., 1969: *House Form and Culture*. Prentice-Hall, Englewood Cliffs, N.J.

Reinhart, C. F., 2002: *Effects of Interior Design on the Daylight Availability in Open Plan Offices*. In *Proceedings of the 2002 American Council for an Energy Efficient Economy (ACEEE) Summer Study on Energy Efficiency in Buildings*, vol. 3., Washington, DC, USA, pp. 309–322.

Reiss, P. and M. White, 2008: What changes energy consumption? Prices and public pressures. *The RAND Journal of Economics,* **39**(3): 636–663.

Ricci, A., 2009: *Policy use of the needs results*. New Energy Externalities Developments for Sustainability (NEEDS), Rome, Italy.

Rivière, A, L. Grignon-Masse, S. Legendre, D. Marchio, G. Nermond, S. Rahim, P. Riviere, P. Andre, L. Detroux, J. Lebrun, J. L'hoest, V. Teodorose, J.L. Alexandre, E. SA, G. Benke, T. Bogner, A. Conroy, R. Hitchin, c. Pout, W. Thorpe and S. Karatasou, 2008: Preparatory study on the environmental performance of residential room conditioning appliances (airco and ventilation), Tasks 1 to 5. Draft reports as of July 2008, *Directorate-General for Energy, European Commission*, Brussels, Belgium.

Roberts, S., 2008: Energy, equity and the future of the fuel poor. *Energy Policy,* **36**(12): 4471–4474.

Roberts, S. and W. Baker, 2003: *Towards Effective Energy Information: Improving Consumer Feedback on Energy Consumption*. A report to Ofgem. Center for Sustainable Energy, Bristol, UK.

Romieu, I., H. Riojas-Rodriguez, A. T. Marrón-Mares, A. Schilmann, R. Perez-Padilla, and O. Masera, 2009: Improved biomass stove intervention in rural Mexico: Impact on the respiratory health of women. *American Journal of Respiratory and Critical Care Medicine,* **180**(7): 649–656.

Rosenfeld, A. H., H. Akbari, J. J. Romm, and M. Pomerantz, 1998: Cool communities: strategies for heat island mitigation and smog reduction. *Energy and Buildings,* **28**(1): 51–62.

Roth, K. W., F. Goldstein, and J. Kleinman, 2002: *Energy Consumption by Office and Telecommunications Equipment in Commercial Buildings*. Volume 1: Energy Consumption Baseline. Arthur D. Little Inc., Cambridge, MA, USA.

Roth, K.W., D. Westphalen, and J. Brodrick, 2003: Saving energy with building commissioning. *ASHRAE Journal,* **45**(11):65–66.

Roudil, N., 2007: *Artisans et énergies renouvelables: une chaîne d'acteurs au cœur d'une situation d'innovation*. Les Annales de la Recherche Urbaine. Ministère de l'Ecologie, du Développement et de l'Aménagement Durables (MEDAD), **103**:101–111, Paris, France.

Roux, C., 2010: *The Life Cycle Performance of Energy Using Household Products*. Department of Energy and Process Engineering, Norwegian University of Science and Technology, Trondheim, Norway.

Roy, J., 2000: The rebound effect: some empirical evidence from India. *Energy Policy,* **28**(6–7): 433–438.

Roy, A. N., A. R. Mahmood, O. Baslev-Olesen, S. Lojuntin, C. K. Tang, and K. S. Kannan, 2005: *Low Energy Office Building in Putrajaya, Malaysia. Case studies and innovations. In Proceedings of Conference on Sustainable Building South Asia*.11–13, Kuala Lumpur, Malaysia, pp.223–230.

Rubel, F. and M. Kottek, 2010: Observed and projected climate shifts 1901–2100 depicted by world maps of the Köppen-Geiger climate classification. *Meteorologische Zeitschrift,* **19**:135–141.

Rubinstein, F. and S. Johnson, 1998: *Advanced Lighting Program Development*. Final Report, Lawrence Berkeley National Laboratory, Berkeley, CA, USA.

Rubinstein, F., H. J. Jennings, and D. Avery, 1998: *Preliminary Results from an Advanced Lighting Controls Testbed*. LBNL-41633, Lawrence Berkeley National Laboratory, Berkeley, CA, USA.

Rudd, A., P. Kerrigan, Jr., and K. Ueno, 2004: What will it take to reduce total residential source energy use by up to 60%? In *Proceedings of the 2004 American Council for an Energy Efficient Economy (ACEEE) Summer Study on Energy Efficiency in Buildings*.Vol.1, Washington, DC, USA, pp.293–305.

Rypdal, K., N. Rive, S. Aström, N. Karvosenoja, K. Aunan, J. L. Bak, K. Kupiainen, and J. Kukkonen, 2007: Nordic air quality co-benefits from European post-2012 climate policies. *Energy Policy,* **35**(12)6309–6322.

Sadler, R., 2002: *The Benefits of Energy Advice*. Energy Efficiency Partnership for Homes (EEPH), London, UK.

Sailor, D.J. 2008: A green roof model for building energy simulation programs. *Energy and Buildings,* **40**: 1466–1478.

Saiz, S., Kennedy, C., Bass, B., Pressnail, K., 2006: Comparative Life Cycle Assessment of Standard and Green Roofs. *Environmental Science and Technology,* **40**: 4312–4316.

Samet, J., S. Zeger, F. Dominici, F. Curriero, I. Coursac, D.W. Dockery, J. Schwartz, and A. Zanobetti, 2000: The national morbidity, mortality, and air pollution study – part II: morbidity and mortality from air pollution in the United States. *Research Report Health Effects Institute,* **94**(Part 2):5–70.

Sami Khawaja, M. et al., 2008: Statewide Codes and Standards Market Adoption and Noncompliance Rates. Final Report CPUC Program, No. 1134–04, SCE0224.01, San Francisco, CA, USA.

Sanstad, A. H., C. Blumstein, and S. Stoft, 1995: How high are option values in energy-efficiency investments? *Energy Policy,* **23**(9): 739–744.

Santamouris, M., 2001: *Energy and Climate in the Urban Built Environment*, James and James, London, UK.

Santamouris, M., Pavlou, C, Doukas, P., Mihalakakou, G., Synnefa, A., Hatzibiros, A., Patargias, P., 2007: Investigation and analysing the energy and environmental performance of an experimental green roof system installed in a nursery school building in Athens, Greece. *Energy,* **32**: 1781–1788.

Sartori, I. and A. G. Hestnes, 2007: Energy use in the life cycle of conventional and low-energy buildings: A review article. *Energy & Buildings,* **39**(3): 249–257.

Sathaye, J. and A. Gupta, 2010: *Eliminating Electricity Deficit through Energy Efficiency in India: An Evaluation of Aggregate Economic and Carbon Benefits*, LBNL-3381E, Lawrence Berkeley National Laboratory, Berkeley, CA, USA.

Savola, H., 2007: International Institute for Industrial Environmental Economics (IIIEE), Lund, Sweden. Email communication, contribution by filling out the questionnaire for several evaluation studies in Swedish, sent on May 7, 2007.

SBTF, 2001: *Building Better Buildings: A Blueprint for Sustainable State Facilities*. California State and Consumer Services Agency and Sustainable Building Task Force (SBTF), Sacramento, CA, USA.

Schaefer, C., C. Weber, H. Voss-Uhlenbrock, A. Schuler, F. Oosterhuis, E. Nieuwlaar, R. Angioletti, E. Kjellsson, S. Leth-Petersen, M. Togeby, and J. Munksgaard, 2000: *Effective Policy Instruments for Energy Efficiency in Residential Space Heating – an International Empirical analysis*. University of Stuttgart, Institute for Energy, Stuttgart, Germany.

Schiellerup, P. and J. Gwilliam, 2009: Social production of desirable space: an exploration of the practice and role of property agents in the UK commercial property market. *Environment and Planning C: Government and Policy*, **27**(5): 801–814.

Schipper, L. and M. Grubb, 2000: On the rebound? Feedback between energy intensities and energy uses in IEA countries. *Energy Policy*, **28**(6–7): 367–388.

Schlomann, B., W. Eichhammer, and E. Gruber, 2001: Labelling of electrical appliances – an evaluation of the Energy Labelling Ordinance in Germany and resulting recommendations for energy efficiency policy. In *Proceedings of the European Council for an Energy Efficient Economy (ECEEE) Summer Study*.11–16 June 2001, Mandelieu, France.

Schmidt, T., D. Mangold and H. Muller-Steinhagen, 2004: Central solar heating plants with seasonal storage in Germany. *Solar Energy*, **76**(1–3):165–174.

Schnieders, J. and A. Hermelink, 2006: CEPHEUS results: Measurements and occupants satisfaction provide evidence for Passive Houses being an option for sustainable building. *Energy Policy*, **34**(2): 151–171.

Schonborn, M., 2008: *Energy efficiency for residential buildings*. Presentation at London School of Economics, December, 2008. London, UK.

Schweitzer, M. and B. Tonn, 2002: Nonenergy benefits from the weatherization assistance program: A summary of findings from the recent literature. Prepared for US Department of Energy, Oak Ridge National Laboratory, Oak Ridge, TN, USA.

Scott, S., 1996: *Social welfare fuel allowances…to heat the sky?* Working Paper No. 74. The Economic and Social Research Institute (ESRI), Dublin, Ireland.

Scott, S., S. Lyons, C. Keane, D. McCarthy, and R.S.J. Tol, 2008: *Fuel Poverty in Ireland: Extent, Affected Groups and Policy Issues*. Working Paper No. 262. Prepared by the Economic and Social Research Institute (ESRI), Dublin, Ireland.

Sekhar, S. C. and K. J. Phua, 2003: Integrated retrofitting strategy for enhanced energy efficiency in a tropical building. *ASHRAE Transactions*, **109**(1): 202–214.

Seppanen, O. A., W. J. Fisk, and M. J. Mendell, 1999: Association of ventilation rates and CO2 concentrations with health and other human responses in commercial and institutional buildings. *Indoor Air*, **9**(4): 226–52.

Sevi, B., 2004: *Consequences of electricity restructuring on the environment: a survey*.Technical Paper. Centre de Recherche en Economie et Droit de l'Energie (CREDEN), Montpellier Cedex, France.

Shepherd, D. and A. Zacharakis, 2001:The venture capitalist-entrepreneur relationship: control, trust and confidence in co-operative behaviour. *Venture Capital: An International Journal of Entrepreneurial Finance*, **3**(2): 129–149.

Shorrock, L., 2001: Assessing the effect of grants for home energy efficiency improvements. In *Proceedings of the European Council for an Energy Efficient Economy (ECEEE) Summer Study*.11–16 June 2001, Mandelieu, France.

Singer, T., 2002: *IEA DSM Task X- Performance Contracting- Country Report: United States*. International Energy Agency (IEA), Paris, France.

Skumatz, L.A. and C. A. Dickerson, 1997: Recognizing All Program Benefits: Estimating the Non-Energy Benefits of PG and E's Venture Partner Pilot Program (VPP).In *Energy Evaluation Conference*, Chicago, IL, USA.

Small, K. and K. Van Dender, 2005: *A Study to Evaluate the Effect of Reduced Greenhouse Gas Emissions on Vehicle Miles Traveled*. Sacramento: State of California Air Resources Board.

Smeds, J. and M. Wall, 2007: Enhanced energy conservation in houses through high performance design. *Energy and Buildings*, **39**: 273–278.

Smith, K.R., R. Edwards, K. Shields, R. Bailis, K. Dutta, C. Chengappa, O. Masera, and V. Berrueta, 2007: Monitoring and evaluation of improved biomass cookstove programs for indoor air quality and stove performance: conclusions from the household energy and health project. *Energy for Sustainable Development*, **11**(2): 5–18.

Smith-Sivertsen, T., E. Díaz, D. Pope, R.T. Lie, A. Díaz, J. McCracken, P. Bakke, B. Arana, K. R. Smith, and N. Bruce, 2010:Effect of reducing indoor air pollution on women's respiratory symptoms and lung function: the RESPIRE Randomized Trial, Guatemala. *Am. J. Epidemiol*, **170**(2): 211–220.

Socolow, R. H., 1978: *Saving energy in the home: Princeton's experiments at Twin Rivers*. Ballinger Press, Cambridge, MA, USA.

SOM, 2010: *Lever House – Curtain Wall Replacement*. Skidmore, Owings & Merrill LLP, New York, NY, USA. www.som.com/content.cfm/lever_house_curtain_wall_replacement (accessed 22 September, 2010).

Sondreal, E., M. Jones, and G. Groenewold, 2001: Tides and trends in the world's electric power industry. *The Electricity Journal*, **14**(1): 61–79.

Sorrell S., E. O'Malley, J. Schleich, and S. Scott, 2004: *The Economics of Energy Efficiency*. Edward Elgar Pub, Cheltenham, UK.

Sorrell, S., 2007: *The Rebound Effect: An Assessment of the Evidence for Economy-Wide Energy Savings from Improved Energy Efficiency*. UK Energy Research Centre (UK ERC), London, UK.

Spala, A., Bagiorgas, H.S., Assimakopoulos, M.N., Kalavrouziotis, J. Matthopoulos, D., Mihalakakou, G., 2008: On the green roof system. Selection, state of the art and energy potential investigation of a system installed in an office building in Athens, Greece, *Renewable Energy*, **33**: 173–177.

Springer, D., G. Loisos, and L. Rainer, 2000: Non-compressor cooling alternatives for reducing residential peak load. In *Proceedings of the 2000 American Council for an Energy Efficient Economy (ACEEE)Summer Study on Energy Efficiency in Buildings*.Vol.1., Washington, DC, pp.319–330.

Stern, P.C., 1999: Information, incentives and pro-environmental consumer behaviour. *Journal of Consumer Policy*, **22** (4): 461–478.

Stetiu, C., and H. E. Feustel, 1999: Energy and peak power savings potential of radiant cooling systems in US commercial buildings. *Energy and Buildings* **30**(2): 127–138.

Still, D., 2009: *New Biomass Stoves and Carbon Credits*. Aprovecho Research Center, Cottage Grove, OR, USA.

Summerfield, A.J., R.J. Lowe, and T. Oreszczyn, 2010: Two models for benchmarking UK domestic delivered energy. *Building Research &Information***38** (1): 12–24.

Stoecklein, A. and L. Skumatz, 2007: Zero and low energy homes in New Zealand: the value of non-energy benefits and their use in attracting homeowners. In *Proceedings for the European Council for an Energy Efficient Economy (ECEEE) Summer Study*.4–9 June, La Colle sur Loup, France.

Sverrisson, F., J. Li, M. Kittell, and E. Williams, 2003: *Electricity restructuring and the environment: lessons learned in the United States*, Center for Clean Air Policy, Washington, DC, USA.

Swanson, R. M., 2006: A vision for crystalline silicon photovoltaics. *Progress in Photovoltaics: Research and Applications*, 14(5): 443–453.

Szklo, A. and H. Geller, 2006: Policy Options for Sustainable Energy Development. In *Brazil: A Country Profile on Sustainable Energy Development*. International Atomic Energy Agency (IAEA), Vienna, Austria.

Takebayashi, H., and M. Moriyama, 2007: Policy Options for Sustainable Energy Development. *Building and Environment*, 42(8): 2971–2979.

Taylor, R., C. Govindarajalu, J. Levin, A. S. Meyer, and W. A. Ward, 2008: *Financing Energy Efficiency: Lessons from Brazil, China, India, and Beyond*. World Bank, Washington, DC, USA.

Tenorio, R., 2007: Enabling the hybrid use of air conditioning: a prototype on sustainable housing in tropical regions. *Building and Environment*, 42(2): 605–613.

Thompson, P., 1997: Evaluating energy efficiency investments: accounting for risk in the discounting process. *Energy Policy*, 25(12): 989–996.

Thiers, S., and B. Peuportier, 2008: Thermal and environmental assessment of a passive building equipped with an earth-to-air heat exchanger in France. *Solar Energy*, 82(9): 820–831.

Tirado Herrero, S. and D. Ürge-Vorsatz, 2010: *Fuel poverty in Hungary. A first assessment*. Center for Climate Change and Sustainable Energy Policy (3CSEP), Central European University, Budapest, Hungary.

Tirado Herrero, S., in preparation: *Fuel poverty in Hungary: Assessing the co-benefits of residential energy efficiency*. PhD dissertation. Center for Climate Change and Sustainable Energy Policy (3CSEP), Central European University, Budapest, Hungary

Tochihara, Y., 1999: Bathing in Japan: A review. *Journal of Human-Environment System*, 3(1): 27–34.

Torcellini, P. A., R. Judkoff, and S. J. Hayter, 2002: Zion National Park Visitor Center: Significant energy savings achieved through a whole-building design process. In *Proceedings of the 2002 American Council for an Energy Efficient Economy (ACEEE) Summer Study on Energy Efficiency in Buildings*, Vol. 3., Washington, D.C., pp.361–372.

Torcellini, P. A. and D. B. Crawley, 2006: Understanding zero-energy buildings. *ASHRAE Journal*, 48(9): 62–69.

Torras, M., 2000: The total economic value of Amazonian deforestation, 1978–1993. *Ecological Economics*, 33(2): 283–297.

Treloar, G., R. Fay, P. E. D. Love, and U. Iyer-Raniga, 2000: Analysing the life-cycle energy of an Australian residential building and its householders. *Building Research and Information*, 28(3): 184–195.

Ueno, T., F. Sano, O. Saeki, and K. Tsuji, 2006: Effectiveness of an energy-consumption information system on energy savings in residential houses based on monitored data. *Applied Energy*, 83(2): 166–183.

UKERC, 2009: Making the *Transition to a Secure and Low-Carbon Energy System*: Synthesis Report prepared by the UK Energy Research Centre (UKERC), London, UK.

UNIDO, 2003: *Bundling small-scale energy efficiency projects*: Issue Paper, United Nations Industrial Development Organization (UNIDO), Vienna, Austria.

UNEP, 2009: *Common Carbon Metric for Measuring Energy Use and Reporting Greenhouse Gas Emissions from Building Operations*, United Nations Environment Programme (UNEP), Sustainable Buildings and Climate Initiative (SCBI), Paris, France.

UNFCCC, 2002: AIJ project reports to the United Nations Framework Convention on Climate Change (UNFCCC), http://unfccc.int/program/coop/aij/aijproj.html. (accessed 10 February, 2010).

Unruh, G. C., 2000: Understanding carbon lock-in. *Energy Policy*, 28(12): 817–830.

Ürge-Vorsatz, D. and S. Koeppel, 2007: *An assessment of Energy Service Companies worldwide*. Report submitted to the World Energy Council (WEC), London, UK.

Ürge-Vorsatz, D., S. Koeppel, and S. Mirasgedis, 2007: An appraisal of policy instruments for reducing buildings' CO2 emissions. *Building Research and Information*, 35(4): 458–477.

Ürge-Vorsatz, D., A. Telegdy, S. Fegyverneki, D. Arena, S. Tirado Herrero, and A.C. Butcher, 2010: *Employment Impacts of a Large-Scale Deep Building Energy Retrofit Programme in Hungary*. Prepared by the Center for Climate Change and Sustainable Energy Policy (3CSEP) of Central European University on behalf of the European Climate Foundation. Budapest, Hungary.

Ürge-Vorsatz, D., K. Petrichenko, A. Butcher, and M. Sharmina, in preparation: *World Building Energy Use and Scenario Modeling*. Prepared for United Nations Environmental Programme, the Sustainable Buildings and Climate Initiative, Paris, France and the International Institute for Applied Systems Analysis (IIASA), Laxenburg, Austria

Ürge-Vorsatz, D., E. Wojcik-Gront, S. Tirado Herrero, P. Foley et.al. In preparation: *Employment Impacts of a Large Scale Deep Building Energy Retrofit Programme in Poland*. Prepared by the Center for Climate Change and Sustainable Energy Policy (3CSEP) of Central European University, Budapest, in collaboration with FEWE (Polish Foundation for Energy Efficiency) for the European Climate Foundation, the Hague, the Netherlands.

US Census, 2000: *Population*. www.allcountries.org/uscensus/2_population.html (accessed December 31, 2010).

US Census Bureau, 2008: *Annual Estimates of the Resident Population for the United States, Regions, States, and Puerto Rico: April 1, 2000 to July 1, 2008*. www.census.gov/popest/states/tables/NST-EST2008–01.xls.

US DOE, 2008: *2008 Buildings Energy Data Book*. Report prepared for the Buildings Technologies Program, Office of Energy Efficiency and Renewable Energy, United States Department of Energy by D and R International, Ltd., Washington, D.C.

US DOE, 2010: *Summary of Gaps and Barriers for Implementing Residential Buildings Energy Efficiency Strategies*. Report prepared by the National Renewable Energy Laboratory for the United States Department of Energy (US DOE) Building Technologies, Washington D.C.

US EIA, 2005: *Residential Energy Consumption Survey*. U.S. Energy Information Administration (US EIA), Washington, DC, USA. http://www.eia.doe.gov/emeu/recs/contents.html (accessed 2009/2010).

US EIA, 2008: *Annual Energy Review. Energy consumption by Sector 1949–2008*. United States Energy Information Administration (US EIA), Washington, DC, USA.

US EIA, 2010: *Annual Energy Outlook*. United States Energy Information Administration (US EIA), Report #: DOE/EIA-0383(2010), Washington, DC, USA.

US EIA, 2011: *State Energy Consumption Database*, June 2011 for 1980–2009. U.S. Energy Information Administration (US EIA), Washington, DC, USA.

US EIA, 2012: *Annual Energy Outlook 2012 Early Release*, Jan. 2012, Summary Reference Case. U.S. Energy Information Administration (US EIA), Washington, DC, USA.

US EPA, 1991: *Indoor Air Facts No. 4 (revised) Sick Building Syndrome*. Air and Radiation Research and Development (6609J). United States Environmental Protection Agency (US EPA), Washington, DC, USA.

US EPA, 2003: Energy Star: *The Power to Protect the Environment Through Energy Efficiency*. United States Environmental Protection Agency (US EPA), EPA 430-R-03–008. Washington, DC, USA.

US EPA, 2007: Energy Star: *New Specifications for Many Office Products in 2007*. United States Environmental Protection Agency (US EPA), Washington, DC, USA. www.energystar.gov/index.cfm?c=ofc_Equip.pr_office_Equipment. (Accessed 12 May, 2010).

US EPA, 2011: Fans, Ceilings for Consumers. URL: www.energystar.gov/index.cfm?fuseaction=find_a_product.showProductGroup8pgw_code=CF. (Accessed 5 January, 2011)

Usibelli, T., 1997: *The Washington State Energy Code: The Role of Evaluation in Washington State's Non-Residential Energy Code. Washington State University Cooperative Extension Energy Program*, WSU/EEP 97–007, Pullman, WA, USA.

van Houwelingen, J. and F.W. van Raaij, 1989: The effect of goal setting and daily electronic feedback on in-home energy use. *Journal of Consumer Research*, **16**(1): 98–105.

van Vuuren, D.P., J. Cofala, H. E. Eerens, R. Oostenrijk, C. Heyes, Z. Klimont, M. G. J. den Elzen, and M. Amann, 2008: Exploring the ancillary benefits of the Kyoto Protocol for air pollution in Europe. *Energy Policy*, **34**(2006): 444–460.

van Wie McGrory, L., J. Harris, M. Breceda Lapeyre, S. Campbell, S. Constantine, M. della Cava, J. G. Martínez, S. Meyer, and A. M. Romo, 2002: Market Leadership by Example: Government Sector Energy Efficiency in Developing Countries. In *Proceedings of the 2002 American Council for an Energy Efficient Economy (ACEEE) Summer Study*. Report LBNL-5098, Asilomar, CA, USA.

Vasile, C. F., 1997: *Residential Waste Water Heat Recovery System: GFX*. Center for the Analysis and Dissemination of Demonstrated Energy Technologies (CADDET) newsletter, December 4, 1997.

Venkatataraman, C., and S. Maithel, 2007: *Advanced Biomass Energy Technology for cooking*. DACE Workshop, October 12–13, 2007, Mumbai, India.

Viridén, K., T. Ammann, P. Hartmann, and H. Huber, 2003: *P+D – Projekt Passivhaus im Umbau* (in German), Viriden + Partner, Zürich, Switzerland

Vojdani, A., 2008: Smart Integration. *Power and Energy Magazine, IEEE*6 (6): 71–79.

Voss, K., 2000: Solar energy in building renovation – results and experience of international demonstration buildings. *Energy and Buildings*, **32**(3): 291–302.

Voss, K., S. Herkel, J. Pfafferott, G. Löhnert, and A. Wagner, 2007: Energy efficient office buildings with passive cooling – results and experiences from a research and demonstration programme. *Solar Energy*, **81**(3): 424–434.

Wade, J., W. Wiltshire, and I.Y. Scrase, 2000: *National and Local Employment Impacts of Energy Efficiency Investment Programmes*. Vol.1: Summary Report. SAVE contract XVII/4.1031/D/97–032, Association for the Conservation of Energy, London, UK.

Wagner, A., S. Herkel, G. Löhnert, and K. Voss, 2004: *Energy efficiency in commercial buildings: Experiences and results from the German funding program SolarBau*. Presented at EuroSolar 2004, Freiburg, Germany.

Walker, C. E., L. R. Glicksman and L.K. Norford, 2007: Tale of two low-energy designs: Comparison of mechanically and naturally ventilated office buildings in temperate climates. *ASHRAE* Transactions, **113** (Part 1): 36–50.

Wang, U., 2008: *Building a Cool World, With New Roofs*. Greentechmedia. www.greentechmedia.com/articles/read/building-a-cool-world-with-new-roofs-5438/ (accessed June 27, 2010).

Warren, A., 2007: Taxation. *Presentation at the Energy Efficiency in Buildings (EEB) Forum. Driving Investments for Energy Efficiency in Buildings*. July 5, Brussels, Belgium.

WB, 2008: *Global Environment Facility and World Bank Donate US$15 Million to Support Energy Efficiency In Argentina*, Press Release No: 2008/212/LCR, World Bank, Washington, DC, USA.

WBCSD, 2007: *Energy Efficiency in Buildings: Business Realities and Opportunities*. Report prepared by the World Business Council for Sustainable Development (WBCSD), Geneva, Switzerland.

WBCSD, 2008: *Energy Efficiency in Buildings: Business Realities and Opportunities*. Report prepared by the World Business Council for Sustainable Development (WBCSD), Geneva, Switzerland.

WBCSD, 2009: *Energy Efficiency in Buildings – Transforming the Market*. Report prepared by the World Business Council for Sustainable Development (WBCSD), Geneva, Switzerland.

WEA, 2000: *World Energy Assessment Energy and the Challenge of Sustainability*. United Nations Development Programme (UNDP), *United Nations Department of Economic and Social Affairs (UNDESA)*, New York, NY, USA, and World Energy Council (WEC), London, UK.

WEA, 2004: *World Energy Assessment, Energy and the Challenge of Sustainability, Overview2004 Update*. United Nations Development Programme (UNDP), United Nations Department of Economic and Social Affairs (UNDESA), New York, NY, USA, and World Energy Council (WEC), London, UK.

Webber, C., R. Brown, M. McWhinney, and J. Koomey, 2003: *2002 Status Report: Savings Estimates for the ENERGY STAR Voluntary Labeling Program*. Lawrence Berkeley National Laboratory, LBNL-51319, Berkeley, CA, USA.

Weber, C. L., 2009: *Energy Cost of Living Calculation*, personal communication, January 6, 2009.

WEC, 2001: *Energy Efficiency Policies and Indicators*. World Energy Council (WEC), London, UK.

WEC, 2004: *Energy Efficiency: A Worldwide review. Indicators, Policies, Evaluation*. World Energy Council (WEC), London, UK.

Wei, Y. M., L. C. Liu, Y. Fan, and G. Wu, 2007: The impact of lifestyle on energy use and CO2 emission: An empirical analysis of China's residents. *Energy Policy*, **35**(1): 247–257.

Western Regional Air Partnership, 2000: Air Pollution Prevention Meeting Summary. In *Air Pollution Prevention Forum Meeting*, May 31-June 1, San Francisco, CA, USA.

WHO, 2000: *Development of World Health Organisation (WHO) Guidelines for Indoor Air Quality*. Report on a Working Group Meeting, October 23–24, 2006, Bonn, Germany.

WHO, 2007: *Housing, energy and thermal comfort. A review of 10 countries within the WHO European Region*. World Health Organization (WHO).Regional Office for Europe, Copenhagen, Denmark

WHO, 2008: World Health Organization (WHO) Statistics 2008 – quote on COPD's www.who.int/respiratory/copd/World_Health_Statistics_2008/en/ (accessed February 25, 2010).

WHO, 2009: The Energy Access Situation in Developing Countries. *A Review focusing on the Least Developing Countries and Sub-Saharan Africa*. United Nations Development Programme (UNDP), New York, NY, USA, and World Health Organisation (WHO), Geneva, Switzerland.

Wilhite, H., H. Nakagami, T. Masuda, Y. Yamaga, and H. Haned, 1996: A cross-cultural analysis of household energy use behaviour in Japan and Norway. *Energy Policy,* **24**(9): 795–803.

Wilhite, H., and R. Ling, 1995: Measured energy savings from a more informative energy bill. *Energy and Buildings,* **22**(2): 145–155.

Williams, E., 2004: Energy intensity of computer manufacturing: Hybrid assessment combining process and economic input-output methods. *Environmental Science and Technology,* **38**(22): 6166–6174.

Wilson, A. and J. Boehland, 2005: Small is beautiful: US house size, resource use, and the environment. *Journal of Industrial Ecology,* **9**(1–2):277–287.

Winner, L., 1977: *Autonomous Technology.* MIT Press, Boston, MA.

Winther, B. N. and A. G. Hestnes, 1999: Solar versus green: the analysis of a Norwegian row house. *Solar Energy,* **66**(6): 387–393.

Wiser, R., M. Bolinger, and M. St. Clair, 2005: *Easing the Natural Gas Crisis: Reducing Natural Gas Prices through Increased Deployment of Renewable Energy and Energy Efficiency.* Testimony prepared for a hearing on Power Generation Resource Inventive & Diversity Standards, Senate Committee on Energy and Natural Resources, Lawrence Berkeley National Laboratory, Berkeley, CA, USA.

Withers, C. R. and J. B. Cummings, 1998: Ventilation, humidity, and energy impacts of uncontrolled airflow in a light commercial building. *ASHRAE Transactions,* **104**(2): 733–742.

Wong, N.H., Tay, S.F., Wong, R., Ong, C.L. Sia, A., 2003: Life cycle cost analyses of rooftop gardens in Singapore, *Building and Environment,* **38**: 499–509.

Wood, G. and M. Newborough, 2003: Dynamic energy-consumption indicators for domestic appliances: environment, behaviour and design. *Energy and Buildings,* **35**(8): 821–841.

Yanbing, K. and W. Qingpeng, 2005: Analyses of the impacts of building energy efficiency policies and technical improvements on China's future energy demand. *International Journal of Global Energy Issues,* **24**(3–4): 280–299.

Yang, J., Q. Yu, and P. Gong, 2008: Quantifying air pollution removal by green roofs in Chicago. *Atmospheric Environment,* **42**(31): 7266–7273.

Yeang, K., 1994: *Bioclimatic Skyscrapers.* Ellipsis, London, UK.

Yonehara, T., 1998: Ventilation windows and automatic blinds help to control heat and lighting. *CADDET Energy Efficiency Newsletter,* December 1998: 9–11.

York, D., 2008: What's working well: lessons from a national review of exemplary energy efficiency programs. In *Proceedings of the American Council for an Energy Efficient Economy (ACEEE),* Asilomar, CA, USA.

Yoshino, H., F. Hasegawa, H. Arai, K. Iwasaki, S. Akabayashi, and M. Kikuta, 1985: Investigation of the relationship between the indoor thermal environment of houses in the winter and cerebral vascular accident (CVA). *Japanese Journal of Public Health,* **32**(4).

Zhang, S., X. Yang, Y. Jiang, and Q. Wei, 2010: Comparative analysis of energy use in China building sector: current status, existing problems and solutions. *Frontiers of Energy and Power Engineering in China,* **4**(1).

Zhen, B., L. Shanhou, and Z. Weifeng, 2005: Energy efficient techniques and simulation of energy consumption for the Shanghai ecological building. In *Proceedings 2005 World Sustainable Building Conference.* September 27–29, Tokyo, Japan, pp.1073–1078.

Zhivov, A. M., and A. A. Rymkevich, 1998: Comparison of heating and cooling energy consumption by HVAC system with mixing and displacement air distribution for a restaurant dining area in different climates. *ASHRAE Transactions,* **104**(part 2): 473–484.

Zhou, N., M. A. McNeil, D. Fridley, L. Jiang, L. Price, S. de la Rue du Can, J. Sathaye, and M. Levine, 2007: *Energy Use in China: Sectoral Trends and Future Outlook.* Lawrence Berkeley National Laboratory Report LBNL-61904, Berkeley, CA, USA.

Zweibel, K., 2005: *The Terawatt Challenge for Thin-Film PV.* NREL Technical Report, NREL/TP-520–38350, National Renewable Energy Laboratory (NREL), Golden, CO, USA.

11

Renewable Energy

Convening Lead Author (CLA)
Wim Turkenburg (Utrecht University, the Netherlands)

Lead Authors (LA)
Doug J. Arent (National Renewable Energy laboratory, USA)
Ruggero Bertani (Enel Green Power S.p.A., Italy)
Andre Faaij (Utrecht University, the Netherlands)
Maureen Hand (National Renewable Energy Laboratory, USA)
Wolfram Krewitt[†] (German Air and Space Agency)
Eric D. Larson (Princeton University and Climate Central, USA)
John Lund (Geo-Heat Center, Oregon Institute of Technology, USA)
Mark Mehos (National Renewable Energy Laboratory, USA)
Tim Merrigan (National Renewable Energy Laboratory, USA)
Catherine Mitchell (University of Exeter, UK)
José Roberto Moreira (Biomass Users Network, Brazil)
Wim Sinke (Energy Research Centre of the Netherlands)
Virginia Sonntag-O'Brien (REN21, France)
Bob Thresher (National Renewable Energy Laboratory, USA)
Wilfried van Sark (Utrecht University, the Netherlands)
Eric Usher (United Nations Environment Programme)

Contributing Authors (CA)
Dan Bilello (National Renewable Energy Laboratory, USA)
Helena Chum (National Renewable Energy Laboratory, USA)
Diana Kraft (REN21, France)
Philippe Lempp (German Development Ministry)
Jeff Logan (National Renewable Energy Laboratory, USA)
Lau Saili (International Hydropower Association, UK)
Niels B. Schulz (International Institute for Applied systems Analysis, Austria and Imperial College London, UK)
Aaron Smith (National Renewable Energy Laboratory, USA)
Richard Taylor (International Hydropower Association, UK)
Craig Turchi (National Renewable Energy Laboratory, USA)

Review Editor
Jürgen Schmid (Fraunhofer Institute for Wind Energy and Energy System Technology, Germany)

Contents

Executive Summary

Renewable energy sources – including biomass, geothermal, ocean, solar, and wind energy, as well as hydropower – have a huge potential to provide energy services for the world. The renewable energy resource base is sufficient to meet several times the present world energy demand and potentially even 10 to 100 times this demand. This chapter includes an in-depth examination of technologies to convert these renewable energy sources to energy carriers that can be used to fulfill our energy needs, including their installed capacity, the amount of energy carriers they produced in 2009, the current state of market and technology development, their economic and financial feasibility in 2009 and in the near future, as well as major issues they may face relative to their sustainability or implementation.

Present uses of renewable energy

Since 1990 the energy provided from renewable sources worldwide has risen at an average rate of nearly 2% a year, but in recent years this rate has increased to about 5% annually (see Figure 11.1.) As a result, the global contribution of renewables has increased from about 74 EJ in 2005 to about 89 EJ in 2009 and represents now 17% of global primary energy supply (528 EJ, see Figure 11.2). Most of this renewable energy comes from the traditional use of biomass (about 39 EJ) and larger-scale hydropower (about 30 EJ), while other renewable technologies provided about 20 EJ.

As summarized in Table 11.1 many renewable technologies have experienced high annual growth rates – some (biofuels, wind, solar electricity, solar thermal, and geothermal heat) even experiencing double-digit growth rates globally over the past five years – and now represent an economy with more than US$_{2005}$$230 billion of investment annually. With cumulative installed contributions to the power, fuel, and thermal heat markets growing rapidly, turnkey costs reflect not only the capital intensity and most often zero or low fuel costs of these solutions, but also the technology and scale advancements of the past decades. The levelized costs of energy, particularly for the more mature renewable technologies, offer competitively priced solutions in some markets but are still comparatively expensive in others under current economic pricing schemes.

In 2009, the contribution of renewable energy technologies to the world's electricity generation was roughly 3800 TWh, equivalent to about 19% of global electricity consumption. Renewable power capacity additions now represent more than one-third of all global power capacity additions.

Note that in the figures presented in Table 11.1, the contribution of renewables to the primary energy supply based on the substitution calculation method is presented. Using this method, non-traditional renewables contributed 50 EJ in 2009. Following other calculation methods – see Chapter 1 and Table 11.2 – the total result for 2009 would be different: 28.5 EJ when using the physical content calculation method and 26 EJ when using the direct equivalent calculation method.

Potential and obstacles for renewable energy technologies

The potential of renewables to provide all the energy services needed is huge, as described in Chapter 7 and in the Special Report on Renewable Energy Sources and Climate Change Mitigation of the Intergovernmental Panel on Climate Change, published in 2011 (IPCC, 2011). Further developing and exploiting renewable energy sources using modern conversion technologies would enhance the world's energy security, reduce the long-term price of fuels from conventional sources, and conserve reserves of fossil fuels, saving them for other applications and for future generations. It would also reduce pollution and avoid safety risks from conventional sources, while offering an opportunity to reduce greenhouse gas emissions to levels that will stabilize greenhouse gases in the atmosphere, as agreed upon globally. It could also reduce dependence on imported fuels, minimize conflicts related to the mining

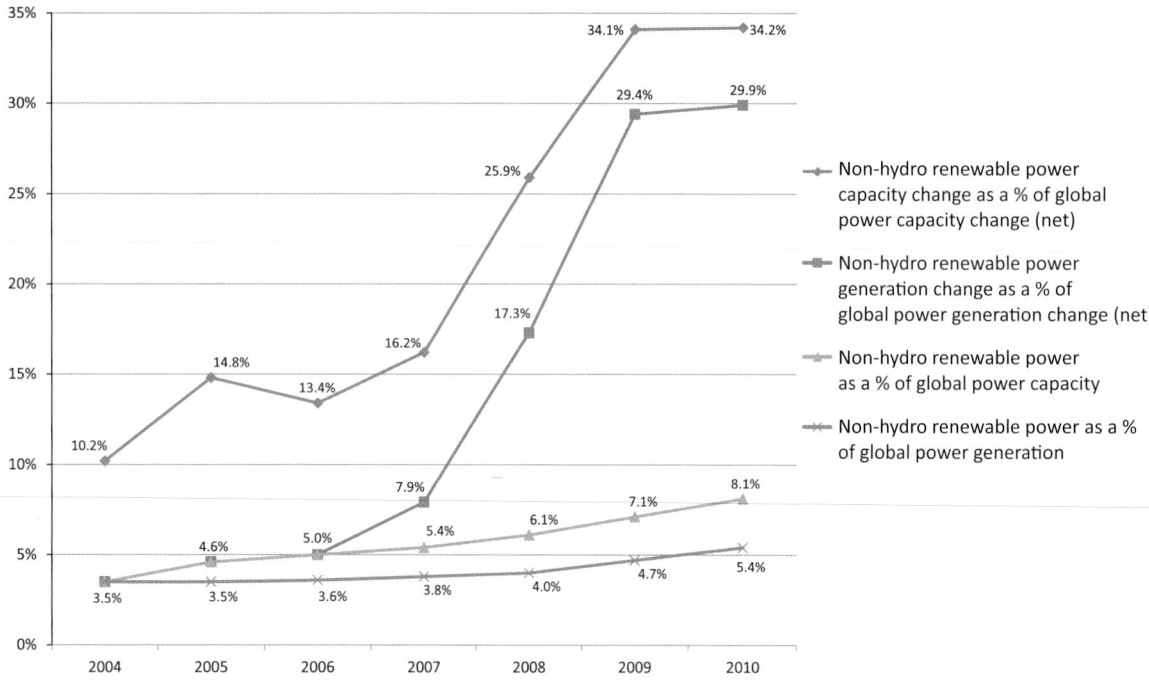

Figure 11.1 | The generating power and capacity of non-hydropower renewable energy sources as a proportion of global power generation and capacity in the period 2004–2010. Source: UNEP and BNEF, 2011.

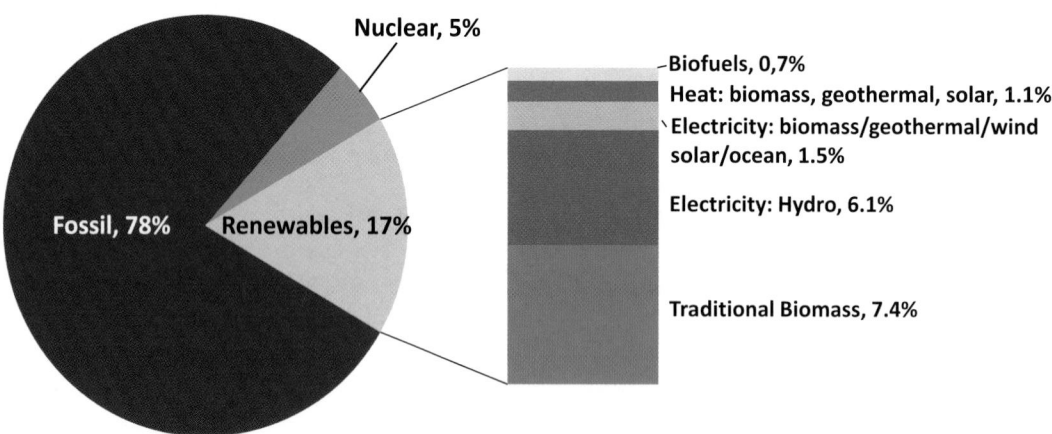

Figure 11.2 | Renewable share of primary energy supply in 2009 (528 EJ). Source: fossil and nuclear fuel use: IEA 2011; renewables: this Chapter.

and use of limited available natural resources, and spur economic development, creating new jobs and regional employment.

But using energy from renewable sources also faces a number of challenges because of their often low spatial energy intensity (J/m²) or energy density (J/m³) compared with most fossil fuel and nuclear energy sources, their generally capital-intensive installation costs, their sometimes higher-than-desirable operational costs, and a variety of environmental and

Table 11.1 | Status of Renewable Energy Technologies as of 2009 (all financial figures are in US$_{2005}$).

Technology	Installed capacity increase in past five years (percent per year)	Operating capacity end 2009	Capacity factor (percent)	Secondary energy supply in 2009	Primary energy supply in 2009 (EJ/yr) based on the substitution calculation method	Turnkey investment costs ($/kW of output)	Current energy cost of new systems	Potential future energy cost
Biomass energy								
Electricity	6	54 GW$_e$	51[a]	~ 240 TWh$_e$	3.3	430–6200	2–22¢/kWh$_e$	2–22¢/kWh$_e$
Bioethanol	20	95 bln liter	80[a]	~ 76 bln liter	2.7	200–660	11–45 $/GJ	6–30 $/GJ
Biodiesel	50	24 bln liter	71[a]	~ 17 bln liter	0.9	170–325	10–27 $/GJ	12–25 $/GJ
Heat / CHP	~ 3	~ 270 GW$_{th}$	25–80	~ 4.2 EJ	5.2	170–1000	6–12¢/kWh$_{th}$	6–12¢/kWh$_{th}$
Hydroelectricity								
Total capacity	3	~ 950 GW$_e$	30–80	~ 3100 TWh$_e$	32	1000–3000	1½–12¢/kWh$_e$	1½–10¢/kWh$_e$
Smaller scale plants (<10 MW)	~ 9	~ 60 GW$_e$	30–80	~ 210 TWh$_e$	2.2	1300–5000	1½–20¢/kWh$_e$	1½–20¢/kWh$_e$
Geothermal energy								
Electricity	4	~ 10 GW$_e$	70–90	~ 67 TWh$_e$	0.7	2000–4000	3–9¢/kWh$_e$	3–9¢/kWh$_e$
Direct-use of heat	12	~ 49 GW$_{th}$	20–50	~ 120 TWh$_{th}$	0.5	500–4200	2–19¢/kWh$_{th}$	2–19¢/kWh$_{th}$
Wind electricity								
Onshore	27	~ 160 GW$_e$	20–35	~ 350 TWh$_e$	3.6	1200–2100	4–15¢/kWh$_e$	3–15¢/kWh$_e$
Offshore	28	~ 2 GW$_e$	35–45	~ 7 TWh$_e$	0.07	3000–6000	7–25¢/kWh$_e$	5–15¢/kWh$_e$
Solar Photo-Voltaics								
Electricity	45	~ 24 GW$_e$	9–27	~ 32 TWh$_e$	0.33	3500–5000	15–70¢/kWh$_e$	3–13¢/kWh$_e$
Solar thermal electricity (CSP)								
Without heat storage	15	0.8 GW$_e$	30–40	~ 2 TWh$_e$	0.02	4500–7000	10–30¢/kWh$_e$	5–15¢/kWh$_e$
With 12h heat storage	-	-	50–65	-	-	8000–10,000	11–26¢/kWh$_e$	5–15¢/kWh$_e$
Low- temperature solar thermal energy								
Low-temperature heat	19	~ 180 GW$_{th}$	5–12	~ 120 TWh$_{th}$	0.55	150–2200	3–60¢/kWh$_{th}$	3–30¢/kWh$_{th}$
Ocean energy								
Tidal head energy	0	~ 0.3 GW$_e$	25–30	~ 0.5 TWh$_e$	0.005	4000–6000	10–31¢/kWh$_e$	9–30¢/kWh$_e$
Current energy	-	exp. phase	40–70	negligible	-	5000–14,000	9–38¢/kWh$_e$	5–20¢/kWh$_e$
Wave energy	-	exp. phase	25	negligible	-	6000–16,000	15–85¢/kWh$_e$	8–30¢/kWh$_e$
OTEC	-	exp. phase	70	negligible	-	6000–12,000	8–23¢/kWh$_e$	6–20¢/kWh$_e$
Salinity gradient energy	-	R&D phase	80–90	-	-	-	-	-

[a] industry-wide average figure; on plant level the CF may vary considerably.

social concerns related to their development. An additional important issue is the intermittent character of wind, solar, and several ocean energy, requiring backup system investments or other innovations to secure a reliable energy supply.

System integration of renewable energy technologies

System integration studies show no intrinsic ceiling to the share of renewables in local, regional, or global energy supplies, depending on the resource base and energy demand. Intelligent control systems, supported by appropriate

Table 11.2 | Contribution of modern renewables to primary energy supply in 2009 using three calculation methods: the substitution method (with GEA conversion efficiencies), the physical content method, and the direct equivalent method.

Technology	Primary energy supply in 2009 (EJ/yr) using the substitution method	Primary energy supply in 2009 (EJ/yr) using the physical content method	Primary energy supply in 2009 (EJ/yr) using the direct equivalent method
Biomass energy	12.1	12.1	12.1
Hydroelectricity			
Total hydropower capacity	32	11.2	11.2
Smaller scale plants (<10 MW)	2.2	0.76	0.76
Geothermal energy	1.2	3.3	0.67
Wind electricity	3.7	1.3	1.3
Solar PV electricity	0.33	0.12	0.12
Solar thermal electricity (CSP)	0.02	0.05	0.007
Low-temperature solar thermal energy	0.5	0.43	0.43
Ocean energy	0.005	0.002	0.002
Total supply	49.9	28.5	25.9

energy storage systems and energy transport infrastructure, will help renewable energy meet the energy demands of different sectors. However, the variability of wind, solar, and several ocean energy resources can create technical or cost barriers to their integration with the power grid at high levels of penetration (20% or above).

To reduce or overcome these barriers, the main approaches in the electricity sector involve: drawing power from geographically larger areas to better balance electricity demands and supplies; improving network infrastructures; increasing the transmission capacity, including the creation of so-called supergrids for long-distance power transmission; developing the Smart Grid further; applying enhanced techniques to forecast intermittent energy supplies hours and days ahead with high accuracy; increasing the flexibility of conventional generation units (including dispatchable renewables) to respond to load changes; using demand-side measures to shift loads; curtailing instantaneous renewable supplies when necessary to guarantee the reliability of power supplies; and further developing and implementing energy storage techniques.

Financial and investment issues

Annual investments in new renewables have increased tremendously, from less than US_{2005}\$2 billion in 1990 to about US_{2005}\$191 billion in 2010, excluding investments in larger-scale hydropower. Including large hydropower, investments were about US_{2005}\$230 billion in 2010. The quickest growth in sustainable energy financing has come from three sectors of the financial community that had previously shown little interest: venture capital and private equity investors, who provide the risk capital needed for technological innovation and commercialization; public capital markets, which mobilize the resources needed to take companies and projects to scale; and investment banks, which help raise capital and arrange mergers and acquisitions. The engagement of these new financial actors has signaled a broader scale-up in asset financing of the investment in actual generating plant capacity.

In the period 2004–2009, the annual growth rates in renewable energy investments were 32% for financing technology commercialization, 45% for financing the construction of projects, and 85% for financing equipment manufacturing and scale-up. This fast-tracking of alternative energy technologies into the commercial mainstream is beginning to change the energy paradigm, although much more growth will be needed before renewables can become the world's primary source of energy.

Policy instruments and measures

A key issue is how to accelerate the deployment of renewables so that deep penetration of these technologies into the energy system can be achieved quickly. Renewable energy technologies face a number of factors that make it harder for them to compete based solely on costs: their capital intensity, scale, and resource risk; their discounted value to traditional utility operators; their real or perceived technology risk; the absence of full-cost accounting for environmental impacts on a level playing field; generous subsidies to the use of conventional energy sources; and a lack of recognition of their long-term value for reducing utilities' exposure to variable fuel costs. These issues have a negative impact for investors by extending the time frame for returns or by increasing risk and the expected rate of return. The risks can be reduced by using public-sector financial or market instruments such as guarantees in terms of market access, market size, and price security.

A wide range of regulatory, fiscal, and voluntary policies have been introduced globally to promote renewable energy – whether renewable electricity, renewable heat, or renewable fuels. These serve a range of technology-specific objectives, including innovation, early-stage development and commercialization, manufacturing, and market deployment, as well as wider political goals such as creating new manufacturing bases for a technology, local and global environmental stewardship, and economic prosperity. These policies all help to reduce risk and encourage renewable energy development, and they are generally used in combination. Integrating renewable energy into the conventional energy system will require a portfolio approach that addresses key issues such as comprehensive and comparable cost-benefit analysis of all energy options, provision of stable and predictable policy environments, and removal of market barriers and competing subsidies for fossil fuels, thus increasing the probability for successful innovation and commercialization, provided that the policies complement one another.

Of the market pull policies, two are most common. Feed-in tariffs (FITs) ensure that renewable energy systems can connect and supply their power to the grid and offer a set price for renewable energy supplies. Policies known as quota or obligation mechanisms (also referred to as Renewable Portfolio Standards, Renewable Electricity Standards, or Renewable Fuel Standards) set an obligation to buy but not necessarily an obligation on price. So far, FITs have been used for electricity only, although some countries are now considering how to provide them for heat. Quotas have so far been used for electricity, heat, and transport. Biofuel mandates are now common globally.

The rapid expansion in renewables, which has largely taken place in only a few countries, has usually been supported by incentives or driven by quota requirements. The FITs used in the majority of European Union countries, China, and elsewhere have been exceptionally successful. The number of states, provinces, and countries that have introduced policies to promote renewable energy doubled in the period 2004–2009.

Future contribution of renewable energy

Many studies have been done on the potential of renewable energy in the remainder of the twenty-first century. Most of them indicate that the contribution of these sources, excluding traditional and non-commercial uses, could increase from today's 10% of world energy supply to 15–30% in 2030, to 20–75% in 2050, and to 30–95% in 2100, depending on assumptions made about economic growth, the volume of investments in energy efficiency and energy technology development, policies and measures to stimulate the deployment of different technologies, and public acceptance of these technologies. Some studies suggest that by 2050, renewable sources could provide 75–95% of the world's energy, or even all of it, if there is enough societal and political will to choose a path of clean energy development that focuses mainly on renewable sources; on new energy transmission, distribution, and storage systems; and on strong improvements in energy efficiency. There is, however, no consensus on whether such a deep penetration of renewables can be achieved in practice within the indicated time frames because of physical limits on the rate at which new technologies can be deployed, the need to design targeted policies to accelerate the deployment of specific technologies, and the difficulty of curtailing energy use through actions on the demand side.

11.1   Introduction

This chapter presents an in-depth examination of major renewable energy technologies, including their installed capacity and energy supply in 2009, the current state of market and technology development, their economic and financial feasibility in 2009 and in the near future, as well as major issues they may face relative to their sustainability or implementation. Renewable energy sources have been important for humankind since the beginning of civilization. For centuries, biomass has been used for heating, cooking, steam generation, and power production; solar energy has been used for heating and drying; geothermal energy has been used for hot water supplies; hydropower, for movement; and wind energy, for pumping and irrigation. For many decades renewable energy sources have also been used to produce electricity or other modern energy carriers.

Renewable energy can be defined as "any form of energy from solar, geophysical, or biological sources that is replenished by natural processes at a rate that equals or exceeds its rate of use" (Verbruggen et al., 2011). In a broad sense, the term renewable energy refers to biomass energy, hydro energy, solar energy, wind energy, geothermal energy, and ocean energy (tidal, wave, current, ocean thermal, and osmotic energy). In the literature (see, e.g., UNDP et al., 2000) the term "new renewable" is also used, referring to modern technologies and approaches to convert energy from renewable sources to energy carriers people can use, taking into account sustainability requirements. In general, "new renewable" includes modern biomass energy conversion technologies, geothermal heat and electricity production, smaller-scale use of hydropower, low- and high-temperature heat production from solar energy, wind conversion machines (wind turbines), solar photovoltaic and solar thermal electricity production, and the use of ocean energy (UNDP et al., 2000; REN21, 2005; Johansson et al., 2006).

11.1.1   Renewable Energy Sources and Potential Energy Supplies

The energy of renewable sources originates from solar radiation (solar energy and its derivatives), geothermal heat (from the interior of the earth), and gravitational energy (mainly from the moon). The energy flow from these sources on Earth is abundant. Tapping just a small fraction of it would in theory be enough to deliver all energy services humanity needs.

A major energy source on Earth is solar energy. As indicated in Figure 11.3, the amount of solar energy directly available for energy conversion in principle is at least more than 1000 times the primary energy use of humankind at present (about 528 exajoules (EJ) in 2009). And in theory, the potential of geothermal energy and ocean energy is also impressive. In practice, however, only a fraction of these potentials can be exploited. Nevertheless, this chapter describes how this fraction can be equivalent to perhaps several times present global energy use.

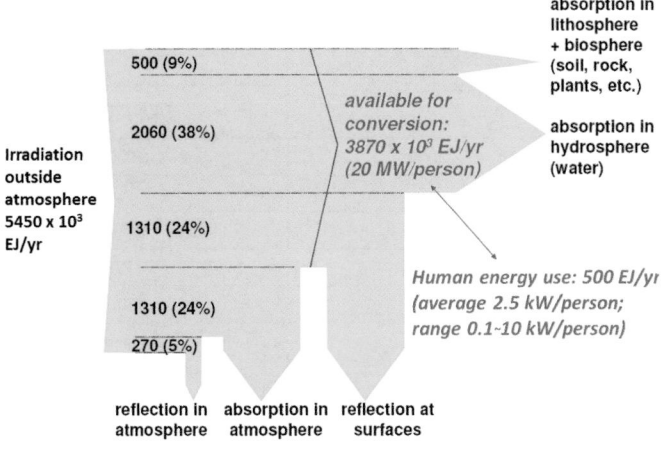

Figure 11.3 | The solar resource and its flows on Earth. The amount of solar energy available on Earth (estimated at 3.9 million EJ/yr) is many times the present human energy use (~528 EJ in 2009).

Table 11.3 presents a summary of the theoretical and technical potential of renewable sources to contribute to the production of energy carriers, based on the *World Energy Assessment* (WEA) published in 2000 (UNDP et al., 2000), on the Special Report on Renewable Energy Sources and Climate Change Mitigation of the Intergovernmental Panel on Climate Change (IPCC, 2011) and on Chapter 7. This chapter focuses on the techno-economic potential of renewables in the longer term, which is a portion of the technical potential, taking into account socioeconomic and environmental constraints. In the literature, "market potential" and "implementation potential" of renewables are also sometimes distinguished, showing a further reduction of the potential of renewables as socioeconomic barriers are taken into account and, in the case of "implementation potential," as policies to promote the deployment of renewables as well as public attitudes are considered.

11.1.2   Renewable Energy Conversion Technologies

A wide variety of technologies are available or under development to provide affordable, reliable, and sustainable energy services from renewables (see Table 11.4), but the stage of development and the competitiveness of various technologies differ greatly. In addition, performance and competitiveness are determined by local aspects such as resource availability, technological infrastructure, socioeconomic conditions, policy measures, and the cost of other energy options.

All renewable energy sources can be converted to electricity. Only a few of them can be used to produce solid, liquid, or gaseous fuel directly, as well as heat. Some of the major sources are intermittent, e.g., solar and wind energy, which can create challenges in adopting these sources while maintaining the reliability of the overall energy supply, depending on how widely they are used.

Table 11.3 | Overview of the technical potential of renewable energy sources in EJ/yr.

Renewable Energy Source	Technical Potential (EJ/yr)		
	WEA (2000)	Chum et al. (2011)	GEA (2012)
Geothermal energy[a,b]	5000	130–1420	810–1545
Direct solar energy[c,a]	> 1575	1575–50,000	62,000–280,000
Wind energy[d,a]	640	85–580	1250–2250
Biomass energy – land-based[e,a]	> 276	50–500	160–270
Hydropower[a]	50	50–52	50–60
Ocean energy[f,a]	not estimated	7–331	3240–10,500
TOTAL	> 7600	1900–52,800	76,500–294,500

[a] Figures from WEA and GEA are based mainly on their own assessment, whereas those from Chum et al. are the result of a review of available literature.

[b] The WEA estimates relate to the amount of energy stored underground; the GEA estimates, to the annual terrestrial heat flow.

[c] Differences in outcome between WEA and GEA are due to different approaches to calculating the potentials.

[d] One reason for the increase of the technical potential between WEA and GEA is the inclusion of offshore wind energy.

[e] Excluding marine biomass energy; the difference in theoretical potential between WEA and GEA can be explained by different calculation methods; restrictions related to sustainability criteria are the main reason for the decrease in estimated technical potential between WEA and GEA.

[f] The indicated technical potential comes mainly from ocean thermal energy conversion; the GEA numbers refer to the potential before conversion, the IPCC SRREN numbers to the potential after conversion.

11.1.3 Advantages and Disadvantages of Renewables

Developing and exploiting renewable energy sources using modern conversion technologies can be highly responsive to national and international policy goals formulated because of environmental, social, and economic opportunities, objectives, and concerns:

- Diversifying energy carriers for the production fuels, electricity, and heat; enhancing energy security; and reducing the long-run price of fuels from conventional sources;

- Reducing pollution, environmental emissions, and safety risks from conventional energy sources that damage human health, natural systems, crops, and materials;

- Mitigating greenhouse gas emissions down to levels that can be sustained;

- Improving access to clean energy sources and conversion technologies, thereby helping to meet the Millennium Development Goals (MDGs) while taking advantage of the local availability of renewables;

- Reducing dependence on and minimizing spending on imported fuels;

- Reducing conflicts related to the mining and use of limited available natural resources, as most renewable energy sources are well distributed;

- Spurring economic development, creating new jobs and local employment, especially in rural areas, as most renewable energy technologies can be applied in small-, medium-, and large-scale systems in distributed and centralized application;

- Balancing the use of fossil fuels, saving them for other applications and for future generations.

Making use of renewable energy sources also has some disadvantages and drawbacks, part of which are intrinsic and part of which are due to the status of technology development:

- The spatial energy intensity (J/m²) or density (J/m³) of renewable energy sources is often low compared with most fossil fuel and nuclear energy sources. Consequently, space is needed where these renewable sources are converted to allow them to deliver most – and finally, perhaps all – of the energy needed. But this may create competition with other claims and requirements for the use of land, including food production, the protection of ecosystems, and biodiversity conservation. Strategies to mitigate these concerns include multifunctional land use, technologies with high conversion efficiencies, and the combination of renewables with measures to improve energy efficiency.

- Although the energy from renewable sources is most often available for free (which reduces vulnerabilities associated with the price volatility of fossil fuels), renewable energy conversion technologies are often quite capital-intensive: operating costs (fuel costs) are replaced by initial costs (installed capital costs). This can make renewables less attractive, especially when high discount rates are

Table 11.4 | Modern renewable energy conversion technologies.

Technology	Energy Product	Status
Biomass energy		
Combustion (domestic scale)	Heat (cooking, space heating)	Widely applied; improvement of technologies
Combustion (industrial scale)	Heat, steam, electricity, CHP (combined heat and power)	Widely applied; potential for improvement
Gasification/power production	Electricity, heat, CHP	Demonstration phase; large-scale deployment of small units in certain countries
Gasification/fuel production	Hydrocarbons, Methanol, H2	Development and demonstration phase
Hydrolysis and fermentation	Ethanol	Commercially applied for sugar/starch crops; production of fuels from lignocellulose under development
Pyrolysis/production liquid fuels	Bio-oils	Pilot and demonstration phase
Pyrolysis/production solid fuels	Charcoal	Widely applied commercially
Extraction	Bio-diesel	Commercially applied
Digestion	Biogas	Commercially applied
Marine biomass production	Fuels	R&D phase
Artificial photosynthesis	H2 or other fuels	Fundamental and applied research phase
Hydropower		
Mini-hydro	Movement	Remotely applied; well-known technology
Small and larger-scale hydropower	Electricity	Commercially applied
Geothermal energy		
Power production	Electricity	Commercially applied locally
Direct heating	Heat, steam	Commercially applied locally
Heat pumps	Heat	Increasingly applied
Wind energy		
Small wind machines	Movement, electricity	Water pumping / battery charging
Onshore wind turbines	Electricity	Widely applied commercially
Offshore wind turbines	Electricity	Demonstrated; initial deployment phase
Solar energy		
Passive solar energy use	Heat, cold, light, ventilation	Demonstrations and application, combined with energy-efficient buildings
Low-temperature solar energy use	Heat (water and space heating, cooking, drying) and cold	Solar collectors commercially applied; solar drying and cooking locally applied
Photovoltaic solar energy conversion	Electricity	Widely applied, remote and grid-connected; high learning rate
Concentrated solar power	Heat, steam, electricity	Demonstrated; initial deployment phase
Ocean energy		
Tidal head energy	Electricity	Applied locally; well-known technology
Wave energy	Electricity	Some experience; research, development, and demonstration (RD&D) phase
Tidal and ocean current energy	Electricity	Some experience; RD&D phase
Ocean thermal energy conversion	Heat, cold, electricity	Some experience; some application of cold use; other technologies mainly in RD&D phase
Salinity gradient / osmotic energy	Electricity	RD&D phase

applied, depending on the level of investments needed as well as governmental interventions.

- The levelized cost of energy (LCOE) from renewables is often not yet competitive in the (distorted) marketplace, especially in grid-connected applications. This may change, however, as renewable energy costs are brought down through technological learning while penetrating markets. Also, using conventional energy sources will become more expensive due to resource depletion and policies to internalize external costs.

- The exploitation of renewable energy sources may entail environmental and social concerns, as experienced with, for instance, electricity production from hydropower and wind energy and the use of biomass resources.

- The intermittent character of the production of energy carriers from wind, solar, and wave energy may set specific requirements for the total energy system to achieve a reliable energy supply. It may require methods to predict renewable energy supplies many hours ahead, management of energy demands, availability of backup power, development of storage options, and enhanced flexibility of energy systems.

As most of these issues have to be solved in an acceptable manner within a few decades (see Cluster I, Chapters 2–6 of this assessment), this chapter also discusses policies and measures to create an environment that will enable deep penetration of renewable energy technologies into existing and new energy systems in the period 2020–2050.

11.1.4   Contribution of Renewables to Global Energy and Electricity Supply

In 2009, renewable energy, including traditional use of biomass, contributed about 89 EJ (17%) to the world's primary energy use, mostly

Table 11.5 | Contribution of renewables to global primary energy supply in 1998, 2005, and 2009, calculated using the substitution method (see Chapter 1).

Category	Renewable Energy Supply		
	1998 (WEA)	2005 (IPCC 2007)	2009 (this chapter)
Traditional biomass[a]	38 EJ	37 EJ	39 EJ
Larger-scale hydropower	26 EJ	24 EJ[b]	30 EJ
New renewables			
- Modern biomass	7.0 EJ	9.0 EJ	12.1 EJ
- Geothermal energy	0.6 EJ	1.9 EJ	1.2 EJ
- Wind energy	0.2 EJ	0.9 EJ	3.7 EJ
- Smaller-scale hydropower	0.9 EJ	0.8 EJ	2.2 EJ
- Low temp. solar energy	0.05 EJ	0.2 EJ	0.55 EJ
- Solar PV	0.005 EJ	0.2 EJ	0.33 EJ
- Conc. Solar Power / STE	0.01 EJ	0.03 E	0.02 EJ
- Ocean energy	0.006 EJ	0.02 EJ	0.005 EJ
	9 EJ	13 EJ	20 EJ
TOTAL	73 EJ	74 EJ	89 EJ

[a] After correction for differences between calculation methods applied.

[b] It is unclear why in 2005 the energy produced is below the WEA figure for 1998, whereas installed capacity increased substantially (see also Table 11.6).

through traditional biomass (about 39 EJ or 7%)[1] and larger-scale hydropower (about 30 EJ or nearly 6%), whereas the share of new renewables was 20 EJ (about 4%).[2] The contribution of renewables to electricity production in 2009 was around 3800 terawatt-hours (TWh), equivalent to about 19% of global electricity consumption, with new renewables accounting for about 900 TWh (4.5%). Table 11.5 presents an overview of the share of renewables in global energy use in 1998, 2005, and 2009. Note that these figures originate from different sources and that they represent savings on fossil fuel consumption as explained in Chapter 1 and footnote 2.

Since 1990 the use of energy from renewable sources has increased about nearly 2%/yr. As total primary energy use has increased at about the same rate, the relative contribution remained almost constant (IEA, 2009), but it is increasing the last few years. Investments in new renewables have increased tremendously, from less than US$_{2005}$2 billion in 1990 to about US$_{2005}$191 billion in 2010, excluding investments in larger-scale hydropower. Including larger-scale hydropower, investments in 2010 were about US$_{2005}$230 billion (UNEP and BNEF, 2011; see Section 11.11).

To illustrate some impacts, Table 11.6 shows the increase in installed power generating capacity, heat and hot water production, and the

production of liquid fuels between 1998 and the years 2008, 2009, and 2010 based on data published by UNDP et al. (2000) and by REN21 (2009; 2010; 2011). Throughout this chapter, an in-depth analysis of the status and contribution of each technology is made, sometimes providing slightly different numbers.

Many of the developments indicated in Table 11.6 are taking place in western industrial countries and in countries like China, Brazil, and India. However, impressive developments are also found in remote areas of developing countries (REN21, 2009). At present, nearly 600 million households depend on traditional biomass (fuelwood, dung, agricultural residues) for cooking and heating. Improving the efficiency of biomass stoves can save 10–50% of biomass consumption, strongly improve indoor air quality, and reduce time spent on collection of the biomass. More than 220 million improved stoves are now found around the world, mainly in China (180 million) and India (34 million), whereas in Africa this number is over 8 million (REN21, 2009).

In rural areas, small-scale thermal biomass gasification is also applied, especially in India and China, where the gas is used for cooking, milling, drying, and electricity generation, for example. About 25 million households worldwide, especially in China (20 million) and India (3 million), receive energy for lighting and cooking from biogas produced in household-scale plants (REN21, 2009). By 2007 more than 2.5 million households were receiving electricity, mainly for lighting and communication, from solar home systems. On the order of 1 million small-scale windmills are in use for mechanical water pumping, mainly in Argentina and South Africa. In remote areas and on islands worldwide, tens of thousands of mini-grids exist, often powered by hybrid electricity supply systems using renewables like solar PV, wind, biomass, and mini-hydro combined with batteries or backup power provided by a diesel generator (REN21, 2009). More than half of the global supply of low-temperature solar heat is found in China (see Section 11.8).

Many studies have been done on the potential contribution of renewable energy to the global energy supply in the twenty-first century. Figure 11.4 show the results for the BLUE Map scenario of the International Energy Agency (IEA), whereas Figure 11.5 also shows results for the baseline scenario of IEA and for other energy sources (IEA, 2010a). Most studies indicate that the contribution of renewables to future energy supply – excluding traditional and non-commercial uses – could increase from the present figure of about 10% up to 15–30% in 2030, 20–76% in 2050, and 30–95% in 2100, depending on assumptions made about economic growth, investments in energy efficiency and energy technology development, policies and measures to stimulate the deployment of different technologies, and public acceptance of these technologies. (See, e.g., UNDP et al., 2000; IPCC, 2000, 2007a, 2011; WGBU, 2003; Johansson et al., 2006; IEA, 2008a; 2008b; 2009a; 2010a; 2010b; Shell, 2008; 2011; Greenpeace and EREC, 2007; 2010; Krewitt et al. 2009; NEAA, 2009; ECF, 2010; WWF and Ecofys, 2011; See also Chapter 17.)

---

1 The contribution of traditional biomass in 2008 and 2009 can be estimated at 39 ±10 EJ; see Section 11.2.

2 Different studies present different numbers for the contribution of renewables to primary energy supply. This is partly due to uncertainty about the contribution of traditional biomass energy use. The main reason, however, is the calculation method applied when converting generated heat and electricity from renewables to primary energy; see Chapter 1 and table 11.2. GEA applies the substitution method for non-combustible fuels assuming a conversion efficiency of 35% in electricity production and 85% for heat production.

Table 11.6 | Installed capacity to generate electricity from renewable energy sources (gigawatts, GW) as well as amount of process heat (GW$_{th}$) and liquid fuels (billion liters per year) produced in 1998, 2008, 2009 and 2010

POWER GENERATION[a]	Existing end 1998 (UNDP et al., 2000)	Existing end 2008 (REN21, 2009)	Existing end 2009 (REN21, 2010)	Existing end 2010 (REN21, 2011)
Hydropower	665 GW	945 GW	980 GW	1010 GW
Biomass power	40 GW	52 GW	54 GW	62 GW
Wind power	10 GW	121 GW	159 GW	198 GW
Geothermal power	8 GW	10 GW	11 GW	11 GW
Solar PV (grid connected)	0.2 GW	13 GW	21 GW	40 GW
Concentrated Solar Power	0.3 GW	0.5 GW	0.6 GW	1.1 GW
Ocean power	0.3 GW	0.3 GW	0.3 GW	0.3 GW
HOT WATER / HEATING				
Biomass heating	~ 200 GW$_{th}$	~ 250 GW$_{th}$	~ 270 GW$_{th}$	~ 280 GW$_{th}$
Solar collectors	~ 18 GW$_{th}$[b]	145 GW$_{th}$	180 GW$_{th}$	185 GW$_{th}$
Geothermal heating	11 GW$_{th}$	~ 45 GW$_{th}$	~ 51 GW$_{th}$	~ 51 GW$_{th}$
LIQUID FUELS				
Bio-ethanol production	18 bln liters/yr	67 bln liters/yr	76 bln liters/yr	86 bln liters/yr
Bio-diesel production	negligible	12 bln liters/yr	17 bln liters/yr	19 bln liters/yr

[a] For comparison: total installed electric power capacity in 2008 was 4700 GW.

[b] As indicated in the WEA update in 2004 (UNDP et al., 2004), this figure is probably too low.

Source: UNDP et al., 2000; REN21, 2009; 2010; 2011.

Some studies suggest that 75–95% or even 100% contributions from renewables can be achieved by 2050 regionally (in the European Union (EU), for instance) as well as globally if there is enough societal and political will to choose an energy development focusing mainly on renewables and new energy transmission, distribution, and storage systems, along with strong improvements in energy efficiency (Greenpeace and EREC, 2007; 2010; Krewitt et al, 2009; ECF, 2010; WWF and Ecofys, 2011). Figures 11.6 and 11.7 show results of two of these studies.

It should be noted, however, that there is no consensus about whether deep penetration of renewables up to the level indicated in these studies can be achieved in practice within the time frames indicated in the Figures because of physical limits to the rate at which new technologies can be deployed, the need to design policies targeted at specific technologies to accelerate deployment, and actions required on demand side to (strongly) increase energy efficiencies and curtail energy use (Kramer and Haigh, 2009).

The remainder of this chapter investigates the status and potential development of renewable energy technologies as well as their market development, costs, environmental performance, deployment barriers, and incentives to overcome the barriers. The cross-cutting issues discussed in the final sections are the integration of renewables into reliable and secure energy systems, developments in renewable energy investments, and policies and measures to enhance the development and application of renewables.

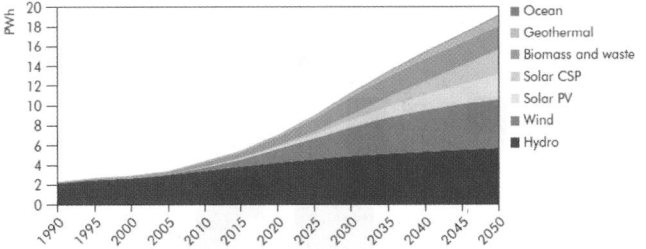

Figure 11.4 | IEA ETP-2010 BLUE Map scenario: Historical and projected contribution of renewables to global electricity generation. Source: IEA, 2010a. ©OECD/International Energy Agency 2010.

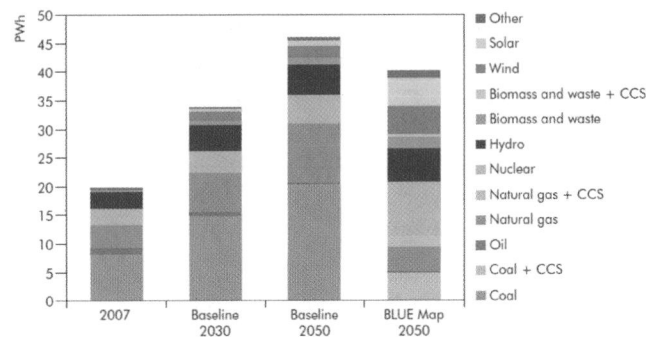

Figure 11.5 | The contribution of different energy sources to the global electricity supply in the Baseline and the BLUE Map scenario developed by IEA. Source: IEA, 2010a. ©OECD/International Energy Agency 2010.

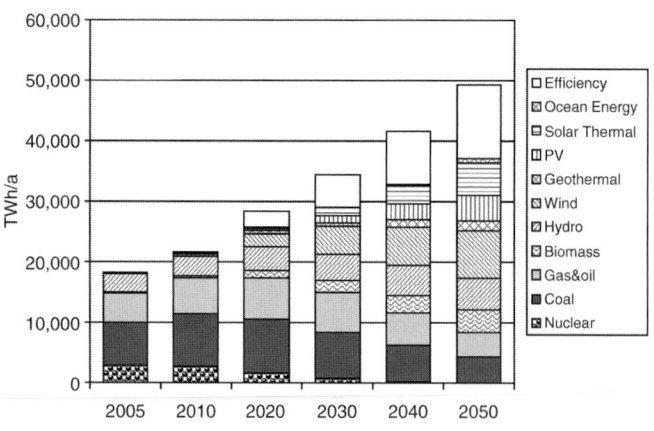

Figure 11.6 | Global electricity generation in the Energy [R]evolution scenario ("efficiency" = savings compared with Reference scenario) Source: based on Greenpeace and EREC, 2007; Krewitt et al., 2009b.

11.2   Biomass Energy

11.2.1   Introduction

Globally, photosynthesis stores energy in biomass at a rate about seven times the current 500 EJ/yr rate of total global energy use. Less than 2% of this biomass is used for human energy consumption today. Biomass resources are diverse, and the global consumption volume of the largest category of these resources (fuelwood in developing countries) rivals that of industrial roundwood (see Figure 11.8). The geographic distribution of biomass resources is uniform relative to that of most fossil fuels (see Chapter 7).

When exploited sustainably, biomass can be converted to modern energy carriers that are clean, are convenient to use, and have little or no associated greenhouse gas (GHG) emissions on a life-cycle basis. Various conversion technologies are available or under development. Sustainable bioenergy has the potential to make large contributions to rural and economic development, to enhance energy security, and to reduce environmental impacts (Chum et al., 2011).

Projections from the IEA, among others, and many national targets count on increasing biomass production and use going forward. Chapters 7 and 20 discuss factors relating to sustainable feedstock production and estimate the quantities of supplies that might be available to 2050. This chapter focuses on summarizing the status and future prospects of modern biomass energy technologies.

11.2.2   Potential of Bioenergy

Biomass accounts today for about 10% (51 EJ) of global primary energy consumption (IEA, 2010a; see also Table 11.5). Its largest, mostly traditional use is found in developing countries. Dominating the traditional use is firewood for cooking and heating. On the other hand, an

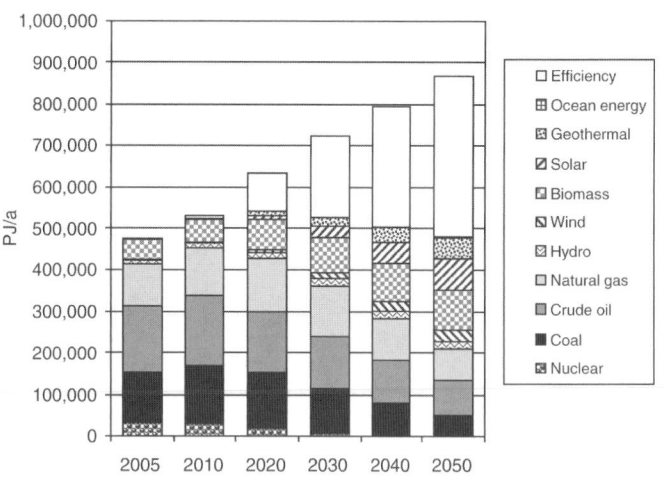

Figure 11.7 | Primary energy use in the Energy [R]evolution scenario ("efficiency" = savings compared with Reference scenario). Source: based on Greenpeace and EREC, 2007; Krewitt et al., 2009b.

estimated 12.1 EJ of primary biomass energy was converted to modern energy carriers in 2009: an estimated 3.3 EJ were converted to 241 TWh of electricity (Electricité de France and Observ'ER, 2010) (more than 1% of all global power generation); about 3.6 EJ were converted to 2.3 EJ of biofuels (REN21, 2010) (about 2% of all global transportation energy) from primarily corn, rape, soy, and sugarcane; and the remainder (about 5.2 EJ) was converted to 4.2 EJ heat, including heat from combined heat and power (CHP) systems.[3]

The analysis in Chapter 7 indicates projected sustainable supplies by region of five categories of biomass supplies – energy crops, forestry residues, crop residues, municipal solid wastes (MSW), and animal wastes – in 2050. Combining these estimates with estimates of prospective efficiencies of biomass electricity generation or of so-called second-generation biofuels production described in this Chapter enables order of magnitude estimates to be made for the maximum potential supplies of low-GHG electricity or liquid transportation fuels in 2050.

Table 11.7 shows that if all sustainably producible biomass supplies were to be converted to electricity, the total electricity generation could approach today's global level of electricity generation from all sources.

---

3 The estimate of 12.1 EJ/yr biomass energy converted to modern energy carriers assumes an average biomass power generating efficiency of 26.3% (IEA, 2010a). Also, the 2.3 EJ of ethanol plus biodiesel produced in 2009 (REN21, 2010) is assumed to have been made from primary biomass with an average energy efficiency of about 60% (Chum et al., 2011). To estimate biomass use for heat, it was conservatively assumed that all biomass use reported for the Organisation for Economic Co-operation and Development by the IEA (2010a), other than biomass used for power generation and for biofuels production, was used for heat or combined heat/power. In addition, Brazil's use of biomass in the industrial sector (as reported by the IEA) is assumed to be for modern heating (e.g., bagasse CHP in the sugarcane industry, or black liquor CHP in the pulp and paper industry). All other non-power/non-biofuel biomass in the world, as reported by the IEA, is assumed to be used in traditional fashion. The IEA estimates are for 2008. It is assumed that there was no change in heat use of biomass between 2008 and 2009. The efficiency of converting primary biomass energy into heat is assumed to be 80% (Chum et al., 2011).

(a)

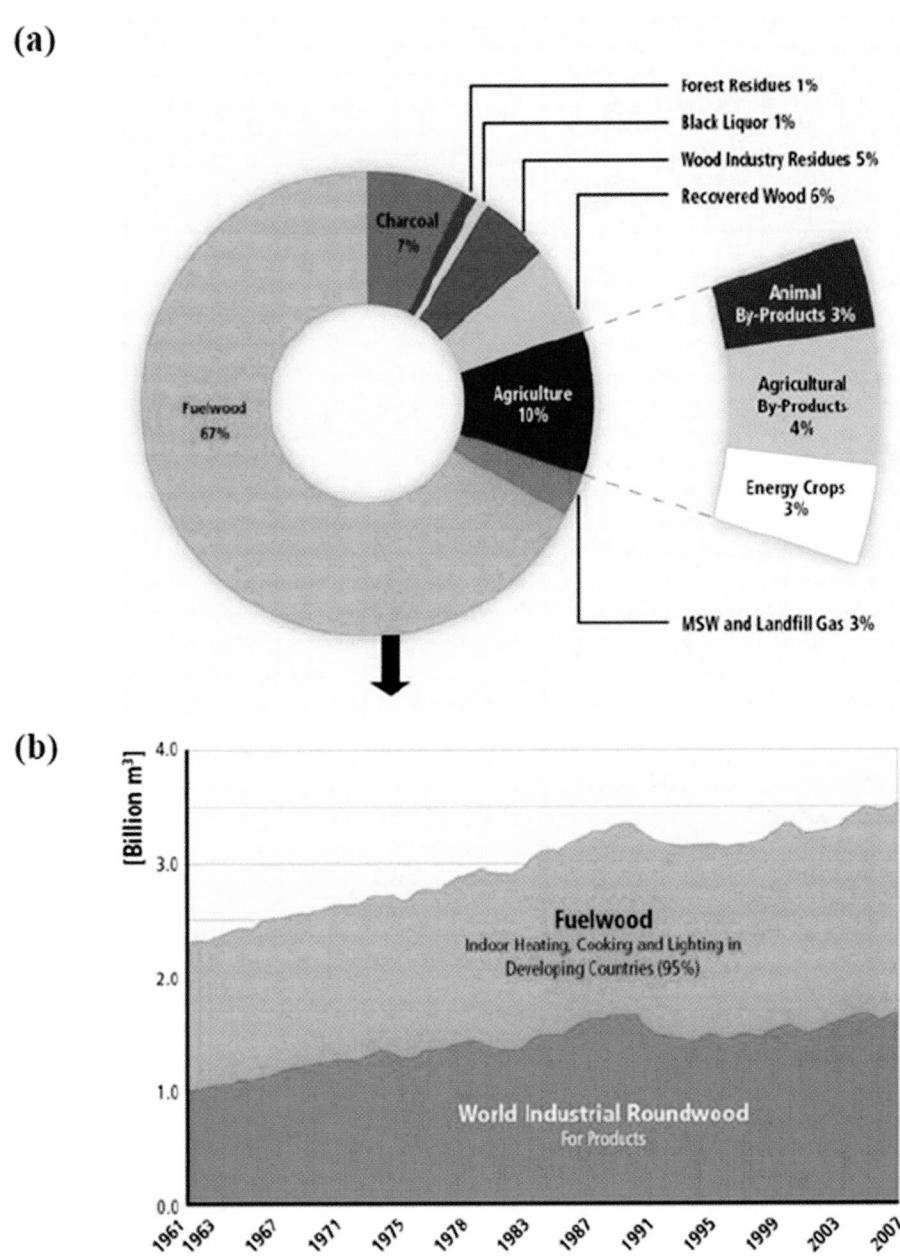

(b)

Figure 11.8 | (a) Shares of primary biomass sources in global primary energy use, (b) Fuelwood used in developing countries and world industrial roundwood production levels. Source: Chum et al., 2011.

Alternatively, if all of the biomass were to be converted to second-generation biofuels, the amount of fuel would be comparable to 70–85% of all petroleum-based transportation fuel use in the world at present. If the biomass were converted to transportation fuels in coal and biomass to liquids (CBTL) systems – that is, those that co-process some coal and capture and store by-product CO_2 (as described briefly in this chapter and more extensively in Chapter 12) – twice as much low GHG-emitting liquid transportation fuel could be produced as the level of petroleum transportation fuels consumed today.

It should be noted that in this assessment the potential to produce transportation fuels from algae was not taken into account because it

is more speculative at this time. With the large number of diverse algal species in the world, upper range productivity potentials of up to several hundred EJ for microalgae and up to several thousand EJ for macroalgae have been reported (Sheehan et al., 1998; Florentinus, 2008; van Iersel et al., 2009; Chum et al., 2011). No reliable estimates are available yet on the techno-economic potential of energy from algae.

11.2.3   Market Developments

Nearly 80% of the total biomass use for energy today occurs in rural areas of developing countries (Chum et al., 2011), and biomass as a

Table 11.7 | Order of magnitude estimate of the technical potential of biomass supply for low-carbon electricity or liquid transportation fuels in 2050 taking into account sustainability requirements.

GEA Region[d]	2050 Potential Biomass Supply[a] (EJ/yr, higher heating value basis)						Electricity (TWh/yr)[b]		Second-Generation Transport Fuels (EJ/yr LHV)[c]		
	Crop residues	Animal waste	MSW	Forest residues	Energy crops	Totals	Small scale	Large scale	Low	High	High CBTL[e]
CAN	0.6	0.47	0.12	2	2.45	5.6	441	587	1.7	2.2	5.3
CHN	7.81	4.99	1.54	2.5	6.30	23.1	1,616	2,155	6.7	8.6	19.1
EAF	0.91	1.67	0.19	0	3.27	6.0	399	532	1.6	2.0	4.6
EEU	0.66	0.55	0.21	1.5	1.71	4.6	352	469	1.4	1.8	4.2
FSU	1.92	1.65	0.41	3	5.85	12.8	966	1,288	3.9	5.0	11.5
IND	6.75	6.17	0.77	0	1.66	15.3	893	1,191	3.6	4.6	9.9
JPN	0.14	0.17	0.31	1	0.37	2.0	155	207	0.6	0.8	1.8
LAC	11.28	7.9	1.75	3	22.67	46.6	3,390	4,519	14.1	18.0	40.4
MEE	0.95	0.64	0.46	0	0.59	2.6	180	240	0.8	1.0	2.1
NAF	0.93	1.33	0.2	0	1.06	3.5	211	281	0.8	1.1	2.3
OCN	0.67	1.65	0.08	0.5	5.07	8.0	561	748	2.2	2.9	6.5
OEA	0.66	0.46	0.13	0	3.16	4.4	339	452	1.4	1.8	4.1
OSA	1.95	1.46	0.33	0	0.38	4.1	252	336	1.1	1.3	2.9
PAS	4.52	1.35	1.09	0.5	4.28	11.7	894	1,192	3.9	5.0	11.0
SAF	1.63	1.14	0.27	0	4.99	8.0	598	797	2.5	3.2	7.2
USA	3.28	3.03	1.33	7	11.09	25.7	1,955	2,606	7.8	10.0	23.2
WCA	2.22	1.51	0.55	0.5	8.27	13.1	993	1,324	4.2	5.3	12.0
WEU	2.56	3.09	1.28	5.5	4.98	17.4	1,258	1,677	4.9	6.3	14.7
TOTALS	49	39	11	27	88	215	15,453	20,603	63	81	183

World production from all sources in 2008 → Electricity[f]: 20,183 — Transport Fuels[g]: 95

[a] Average of high and low estimates for 2050 of potential sustainable biomass energy production (from Chapter 7).

[b] Assuming any available biomass is used only for electricity generation. All categories of biomass resource shown here except animal manures are assumed to be converted to electricity at the same efficiency. For small-scale electricity systems, the assumed conversion efficiency is 30% (gross heating value or HHV basis), a value that can be achieved today with biomass gasifier internal combustion engine systems. For large-scale systems, an efficiency of 40% (HHV) is assumed. This level is likely to be achievable with future gasifier gas turbine combined cycle systems. Animal wastes are assumed to be converted at 25% efficiency to biogas prior to conversion to electricity at the 30% and 40% efficiencies.

[c] Animal wastes are neglected for the transport fuel calculations. Forest residues (assumed HHV of 20 GJ/dry t) are assumed to yield 5.9 and 7.6 GJ$_{LHV}$ per dry t biomass for the low and high estimates. Crop residues and municipal solid waste (assumed to have HHV of 18 GJ/dry t) are assumed to yield 7.3 and 9.3 GJ$_{LHV}$ per dry t biomass for the low and high estimates. Liquid fuel yield from energy crops is assumed to be the average of yields from forest residues and crop residues. The "High with CBTL" case assumes 19.5 GJ$_{LHV}$ liquid yield per dry tonne of biomass for each feedstock

[d] The global regions are defined in Annex II of this Assessment.

[e] CBTL means co-processing coal-and-biomass-to-liquid (CBTL) fuels.

[f] Source: US DOE, 2010a.

[g] Source: IEA, 2010b.

whole accounts for 70% or more of total primary energy use in many of these countries (Karakezi et al., 2004).

Rural biomass use consists in general of charcoal, wood, agricultural residues, and manure used primarily by direct combustion for cooking, lighting, and space heating, with attendant negative health and socio-economic impacts that affect primarily the poor (see Chapters 2 and 4). In Asia, China's biomass consumption represented about 10% of total energy (2007), while in India and some other countries the average was

about 25%. Latin American biomass use was about 20% of primary energy, while in Africa the average approached 50% (IEA, 2009a).

Biomass is used in industrial countries differently, typically first being converted into clean modern energy carriers (electricity, process heat, transport fuels). On average, biomass accounts for 3–4% of total energy use in these countries, although in countries with supportive policies (Sweden, Finland, Austria, and others), the contribution reaches 15–20% (REN21, 2011).

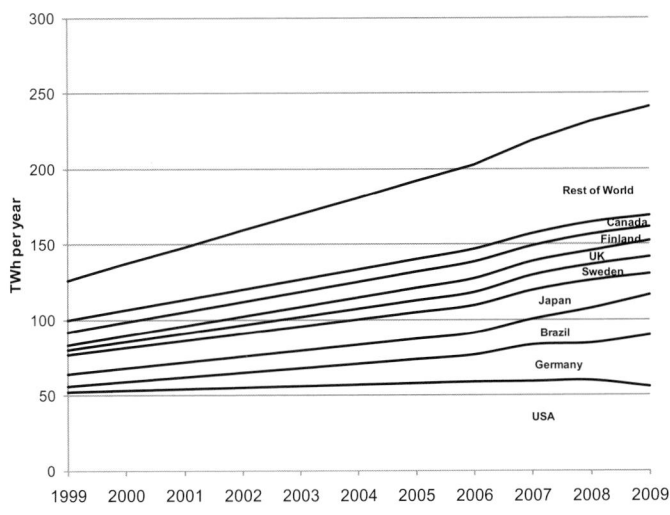

Figure 11.9 | Electricity generation from biomass in selected countries, 2007. Source: based on Electricité de France and Observ'ER, 2010.

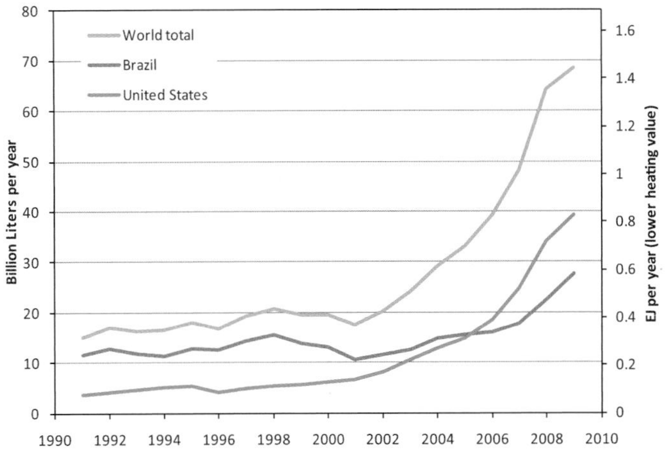

Figure 11.10 | Ethanol production from sugar and starch crops. Source: RFA, undated; UNICA, undated; BP, 2009.

Most electricity generation from biomass occurs in Organisation for Economic Co-operation and Development (OECD) countries, with Brazil being the leading producer outside the OECD (see Figure 11.9). Production of transportation fuels (biofuels) based on agricultural crops (principally corn, soy, rape, and sugarcane) has been growing faster than biopower generation. Figure 11.10 shows the recent rapid growth in ethanol production. Biodiesel production in the United States (from soybeans) and in Europe (from rapeseed) similarly has been growing rapidly in recent years. International trade of biofuels has also grown in recent years, with around 10% of all biofuels traded internationally. Similarly, a third of biomass pellets produced for energy are traded internationally (Junginger et al., 2011).

Combined heat and power systems in the pulp and paper industry accounted for most of the estimated 8500 MW of installed biomass-electric generating capacity in the United States by the end of 2009 (REN, 2010). Biomass CHP plants, with heat used for district heating, are found in Sweden, Finland, Austria, and elsewhere. In developing countries, biomass power generation is found most notably at sugarcane processing factories, where residue from juice extraction is the fuel. Globally, there is an estimated 4700 MW of sugarcane biomass power generating capacity.[4]

Co-combustion in existing coal-fired power plants is an important and growing conversion route for biomass in many EU countries (Spain, Germany, the Netherlands, and others).

Biogas production globally was an estimated 1 EJ in 2009, based on the following analysis. In China, 18 million small-scale anaerobic biomass waste digesters were installed in 2005, having increased an average of

16% per year starting in 2000 (Chen et al., 2010). Assuming an average per-digester production of about 400 m³ of biogas per year (22 MJ/m³ energy content) and continued 16% per year growth in installations from 2005–2009, the total biogas production (from 32.8 million digesters) in 2009 would have been 0.29 EJ. India is estimated to have several million digesters installed, and there are several hundred thousand in other developing countries (REN21, 2010). The number of digesters in industrial countries is much lower, but unit sizes are generally larger – located at large livestock processing facilities (stockyards), municipal sewage treatment plants, and landfills. Biogas production from all sources in the EU totaled 0.35 EJ in 2009, up 4.1% over 2008 (EurObserv'ER, 2010). Germany accounted for 50% of the production and the United Kingdom accounted for 21%. In the United States, biogas production in 2008 was an estimated 0.28 EJ, having increased at 4.3% per year from 2004 (US DOE, 2010a). Assuming continued growth at this rate, US production in 2009 was 0.29 EJ.

Large-scale partial-combustion biomass gasifier systems are not commercially operating today, but as a result of considerable research, development, and pilot-scale demonstration work during the past 25 years (Engstrom et al., 1981; Strom et al., 1984; Kosowski et al., 1984; Evans et al., 1987; Lau et al., 2003) these technologies are nearly commercially ready. The "downstream" components in systems for making liquid fuels from syngas derived via gasification are fully commercial in many cases. (See Chapter 12.)

The only operating commercial demonstration plant for second-generation ethanol (made from non-edible biomass) started in 2004 in Canada, using separate saccharification and fermentation to produce about 3 million liter/yr from wheat straw. Additional commercial-scale demonstration plants are under construction or in planning, largely in the United States (IEA Bioenergy Task 39, undated; US DOE, 2010c), where significant government incentives are available.

4   Based on the update of a survey carried out by The World Alliance for Decentralized Energy (WADE) in 2005, reported in Survey of Energy Resources (WEC, 2007).

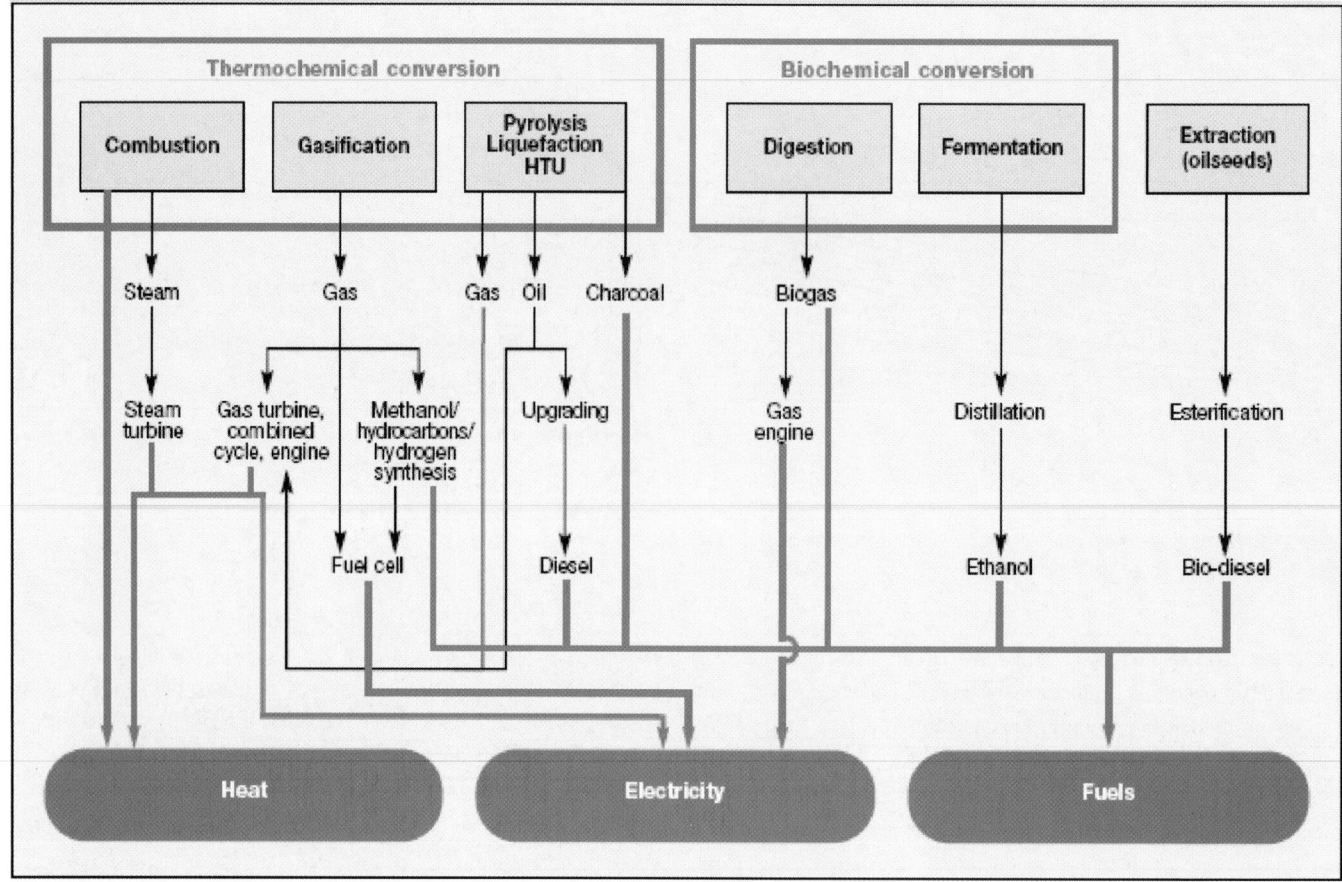

Figure 11.11 | Main conversion routes for biomass to secondary energy carriers. Source: UNDP et al., 2000.

11.2.4 Biomass Energy Conversion Technology Development

A wide variety of technologies are used today to convert raw biomass into modern energy carriers. Figure 11.11 is a simplified road map of the main options. Not shown in this figure but discussed in this section are biomass conversion systems that would include capture and storage of by-product CO_2 to achieve negative greenhouse gas emissions.

Figure 11.12 summarizes the development status of a range of technologies for producing heat and power from biomass, ranging from anaerobic digestion, co-firing, gasification, and combustion of biomass to the production of heat. The development status of different biomass densification techniques is also shown.

11.2.4.1 Electricity and Combined Heat/Power from Biomass

The predominant technology for generating megawatt levels of electricity from biomass today is the steam-Rankine cycle. Efficiencies are modest – often under 20%. To improve efficiency, most steam cycle

power plants are located at industrial sites, where they are configured for combined heat/power production. Low efficiencies, together with relatively high capital costs, explain the reliance of most existing biomass power plants on captive, low-value biomass (primarily residues of agro- and forest-product-industry operations).

Biomass is co-fired in some existing coal-fired power plants. Benefits of this approach include typically high overall efficiency (~40%) because of the large scale of the plants and low investment costs. Aside from GHG emission reductions, co-firing also leads to lower sulfur and other emissions (Faaij, 2006). Generally, relatively low co-firing shares can be deployed with very limited consequences for boiler performance and maintenance.

Many EU coal plants are now equipped with some co-firing capacity. The interest in co-firing higher shares (for instance, up to 40%) is rising. At such high levels, good technical performance (of feeding lines and boiler, for example) is more challenging. Development efforts are focusing on such issues (van Loo and Koppjan, 2002). Power plants capable of co-firing various biomass types with natural gas or with coal exist in Denmark, where alkali-rich straw is a common fuel. Increased corrosion and slagging are common alkali-related problems

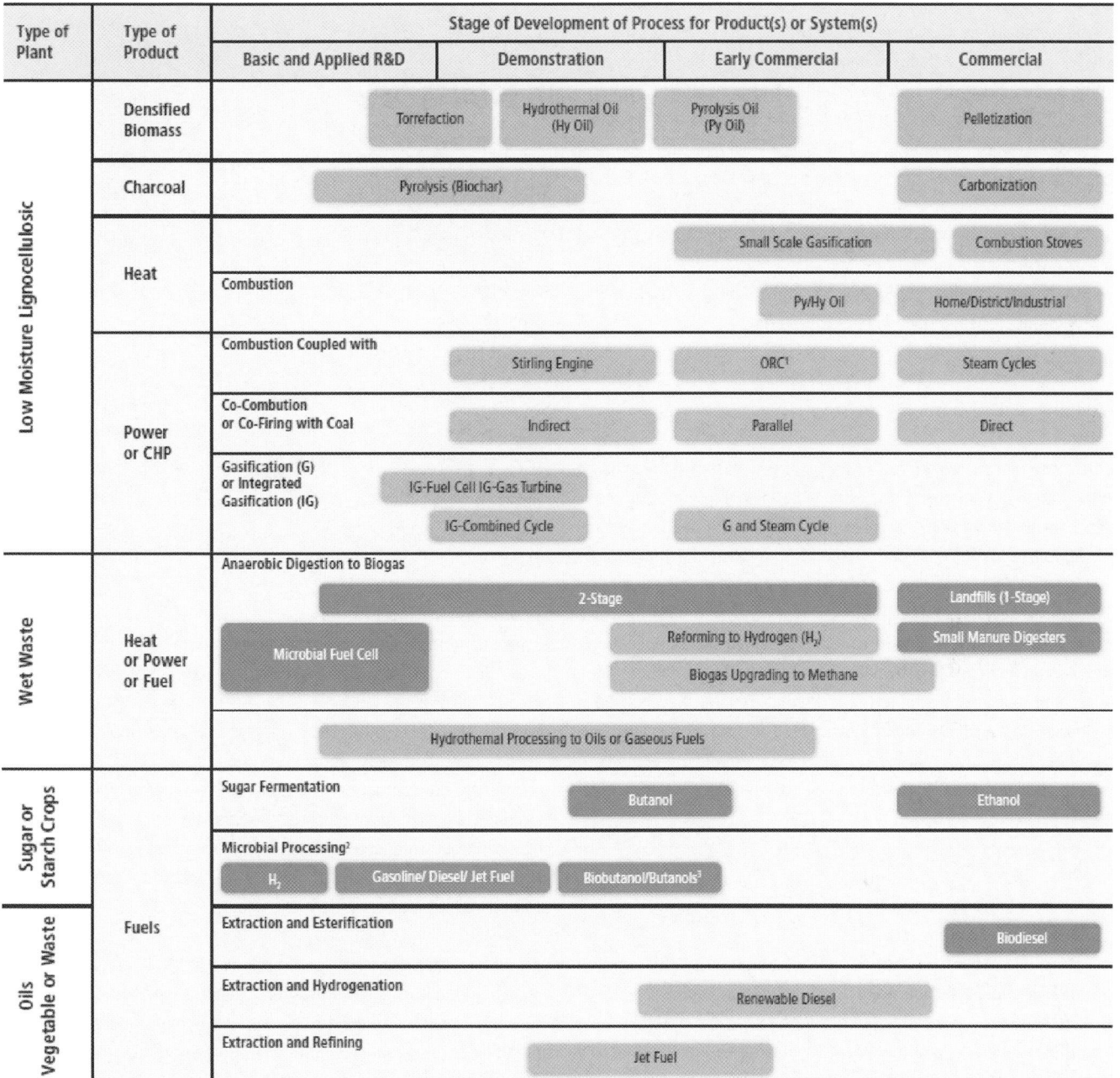

Figure 11.12 | Examples of stages of development of bioenergy: thermochemical (yellow), biochemical (blue), and chemical routes (red) for heat, power, and liquid and gaseous fuels from solid lignocellulosic and wet waste biomass streams, sugars from sugarcane or starch crops, and vegetable oils. Source: Chum et al., 2011.

Notes: 1. ORC: Organic Rankine Cycle. 2. Genetically engineered yeasts or bacteria to make, for instance, isobutanol (or hydrocarbons) developed either with tools of synthetic biology or through metabolic engineering. 3. Several four-carbon alcohols are possible and isobutanol is a key chemical building block for gasoline, diesel, kerosene and jet fuel and other products.

in conventional combustion systems. In multi-fuel systems, however, these problems can be largely circumvented by using straw to raise low temperature steam, which is then superheated using heat from fossil fuel combustion (Nikolaisen et al., 1998). Finally, there is also growing interest in co-feeding of biomass with fossil fuels in gasification-based systems, as briefly discussed later in this chapter and more extensively in Chapter 12.

Advanced biomass power technologies have been the focus of considerable research, development, and demonstration work over the past 25 years to improve conversion efficiency and reduce power generation costs. Gasification of biomass (producing a gas rich in hydrogen and carbon monoxide) coupled with gas engines (Bolhar-Nordenkampf et al., 2003; Salo, 2009) or with gas turbine combined cycles (Larson et al., 2001; Williams, 2004; Sipila and Rossi, 2002; Sydkraft et al.,

1998, 2001; DeLong, 1995; Lau, 2005) have gotten the most attention. Projected conversion efficiencies for Biomass Gasification Combined Cycle systems at a scale of tens to hundreds of MWs exceed 40% (higher heating value basis). At 5–10 MWs, Biomass Gasifier Engine (BGE) efficiencies of 28–30% (lower heating value, or LHV basis) are expected, with total efficiency reaching 87% (LHV basis) in combined heat/power mode (Salo, 2009). Key technical challenges for gasification-based systems are feeding of biomass, especially against pressure (Kurkela, 2008), and cracking of heavy hydrocarbons (tars and oils) that form at typical biomass gasification temperatures (Bergman et al., 2002; Kurkela, 2008; Swedish Energy Agency, 2008; Dayton, 2002; Pfeifer and Hofbauer, 2008). Progress is being made in both these areas.

11.2.4.2   Fuel Gas from Biomass

Biomass can be converted into several types of fuel gases. These gases can be used to generate heat or electricity but also to produce synthetic natural gas or liquid fuels. Anaerobic digestion is by far the most widely used option today for gas production from biomass. The term anaerobic digestion commonly refers to low-temperature biological conversion, with the resulting product (biogas) typically being 60% methane and 40% CO_2. Its use is limited to relatively small-scale applications. Animal and human wastes, sewage sludge, crop residues, carbon-laden industrial-processing by-products, and landfill material have all been widely used as feedstocks. High-moisture feedstocks are especially suitable. Anaerobic digestion has important direct non-energy benefits: it produces concentrated nitrogen fertilizer and neutralizes environmental waste.

11.2.4.3   Liquid Transportation Fuels from Biomass

A popular (non-technical) classification for liquid fuels made from biomass includes first-, second-, and third-generation biofuels. The different generations are distinguished primarily by the feedstocks from which they are derived and the extent to which they are commercially developed.

First-generation biofuels

First-generation biofuels are the only ones being used in significant commercial quantities today. These are made from sugars, grains, or seeds – that is, from only a specific (usually edible) portion of a crop. Relatively simple processing steps convert these feedstocks into fuels (see Figure 11.13), which provides for relatively low processing costs.

Representative fuel yields for the most common first-generation biofuels range from about 2 GJ ethanol per tonne of sugarbeet or sugarcane (LHV) up to about 15 GJ biodiesel per tonne of rapeseed. In terms of land utilization efficiency, the values range from an average of about 20 GJ/ha/yr for US soy biodiesel to about 160 GJ/ha/yr for Malaysian palm biodiesel (see Table 11.8).

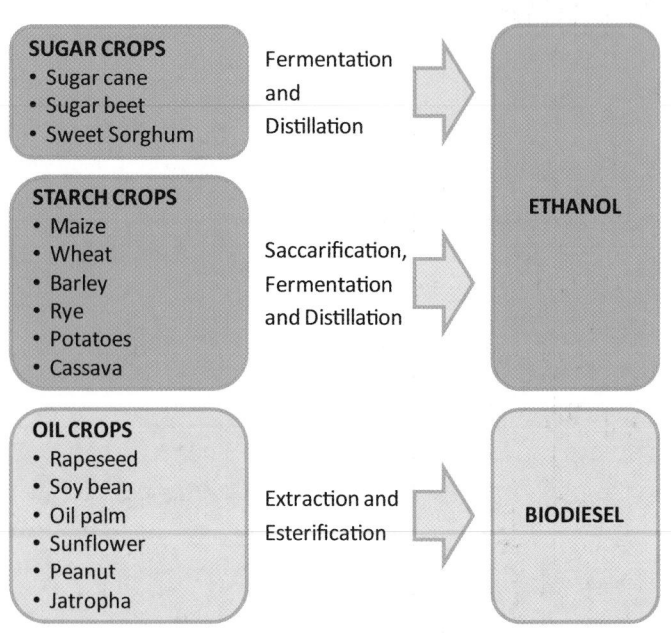

Figure 11.13 | Main feedstocks and conversion steps for first-generation biofuels.

Second-generation biofuels

Second-generation biofuels are those made from land-based non-edible lignocellulosic biomass, either residues of food crop production (such as corn stalks or rice husks) or whole-plant biomass (such as grasses or trees) (UNCTAD, 2008). There are a variety of technology routes for second-generation fuels production (Huber et al., 2006). Among these, second-generation ethanol or butanol can be made via biochemical processing, while most other second-generation fuels are made via thermochemical processing (see Figure 11.14), in some cases using processing steps that are nearly identical to those that would be used to produce synthetic fuels from fossil fuels (see Chapter 12).

Second-generation biochemically produced alcohol fuels are produced via pre-treatment, saccharification, fermentation, and distillation. Pre-treatment is designed to help separate cellulose, hemicelluloses, and lignin so that cellulose and hemicellulose can be broken down by enzyme-catalyzed or acid-catalyzed hydrolysis (water addition) into their constituent simple sugars. The use of acid hydrolysis was practiced commercially as long ago as the 1930s for ethanol production, but involves high capital and operating costs and low yields. Processes using enzymes for hydrolysis promise more competitive ethanol (UNCTAD, 2008).

A variety of process designs have been proposed for the production of second-generation ethanol, including separate enzyme-catalyzed hydrolysis (or saccharification) and fermentation steps, simultaneous saccharification and fermentation in a single reactor (Aden et al., 2002; Jeffries, 2006), and consolidated bioprocessing that incorporates enzyme production (from biomass) with saccharification and fermentation (Zhang and Lynd, 2005). Less work has been done on butanol, but similar processing ideas as for ethanol can be envisioned (UNCTAD, 2008).

Table 11.8 | Representative average yields of first-generation biofuels.

	Fuel yield from grain /seed[a]		Seed/Grain Yield[b]	Fuel per ha/yr	
	Liters/tonne	GJ/tonne	tonne/ha/yr	Liter/ha/yr	GJ/ha/yr
Bioethanol					
Corn (US)	413	8.7	10.2	4220	89
Sugarbeet (US)	100	2.1	60.6	6056	128
Sugarcane (Brazil)	95	1.8	70	5950	126
Biodiesel					
Soybean (US)	204	7.0	2.8	575	20
Rapeseed (EU)	441	15.2	3.6	1601	55
Oil palm (Malaysia)	230	7.9	20.6	4738	163

[a] Fuel yields from corn, sugarbeets, and soybean are estimated averages for 2012 in the United States (FAPRI, undated). Fuel yield from rapeseed is an estimate for the United Kingdom (ESRU, undated). Oil palm estimates are from FAO (2008). Fuel yield of ethanol from sugarcane is the current average in the Center-South region of Brazil in liters per tonne of sugarcane stalk (Hassuani, 2009). The lower heating values for ethanol and biodiesel are 21.1 MJ/liter and 34.5 MJ/liter, respectively.

[b] Corn grain, sugarbeet, and soybean yields are average 2008 yields in the United States (ERS, undated). Rapeseed yield is a representative 2008 average for the European Union (FAS, undated). Sugarcane yield is an estimate of the current Brazilian average in tonnes of cane/ha/yr (Matsuoka et al., 2009). Oil palm estimates are from FAO (2008).

Another biological approach to second-generation biofuels is "synthetic biology," which involves engineering microbes to produce specifically desired fuels, especially hydrocarbon fuels that are "drop-in" replacements for petroleum diesel and gasoline. The work that has been done thus far has been targeting the development of microbes that "eat" sugar molecules and excrete diesel-like fuel (Lee et al., 2008).

Thermochemical (gasification-based) processing of biomass can produce Fisher-Tropsch Liquid fuels, dimethyl ether, gasoline made via methanol, methanol, ethanol, and other fuels.

Gasification takes place in oxygen or via indirect heating (to avoid diluting the resulting gas with nitrogen). The resulting synthesis gas (syngas) is cleaned of contaminants, and in some cases the composition of the gas is adjusted to prepare it for further downstream processing (UNCTAD, 2008). Carbon dioxide is a diluent in the syngas and so is removed to facilitate downstream reactions and reduce equipment sizes. The major components of the clean, concentrated syngas are carbon monoxide (CO) and hydrogen (H_2), usually with a small amount of methane that is unavoidably formed at typical gasification temperatures (<1000°C).

Further gas processing can follow different routes. In one class of processes the syngas is passed over a catalyst that promotes reactions between the CO and H_2 to form liquid fuels molecules. The design of the catalyst determines what fuel is produced. Not all of the syngas passing over the catalyst will be converted to fuel. The unconverted syngas can be burned to make electricity for some or all of the power needed to run the facility and in some cases to export electricity to the grid (UNCTAD, 2008).

In a second class of processes, the CO_2-free syngas goes to a reactor containing specially designed micro-organisms that ferment the CO and H_2 into ethanol or butanol (Spath and Dayton, 2003). This combined thermochemical/biochemical route to a pure alcohol, if it can be made commercially viable, would enable the lignin in the biomass feedstock, as well as the hemicellulose and cellulose, to be converted to fuel (UNCTAD, 2008), unlike the case for purely biochemical "cellulosic ethanol," where the lignin is unfermentable.

Second-generation biofuel yields can be high. Table 11.9 gives estimated yields per tonne of dry biomass processed for several second-generation biofuels. Also shown are estimates of yields per unit land area in the United States on two qualities of land: one suitable for agriculture and one not well suited for it. On good-quality lands, yields for second-generation biofuels are likely to exceed by a considerable margin the yields of first generation-biofuels given in Table 11.8. Yields would be higher still on good-quality lands in tropical climates, due to the longer growing season.

Direct liquefaction is one additional class of second-generation conversion that deserves mention. Fast pyrolysis (Bridgwater and Peacocke, 2000) and hydrothermal upgrading (for moist biomass) (Goudriaan and Naber, 2008) represent different variants of liquefaction. Different reaction pressures and temperatures, catalysts, and rates of heating in the absence of oxygen produce different liquid, gas, and solid output compositions (Huber et al., 2006). The liquid product is typically a "biocrude oil" that is denser than water, highly acidic, odoriferous, and carcinogenic. It requires refining into finished transportation fuels. So-called green diesel can be produced from such oils using catalytic processing that resembles conventional petroleum refining (McCall et al., 2008).

A final comment is warranted regarding the concept of biorefining because of the potential it has for optimizing economics of biofuels production. Biorefining is analogous to current petroleum refining, which leads to an array of products, including liquid fuels, other energy products, and chemicals (Kamm et al., 2006). Although the biofuel and

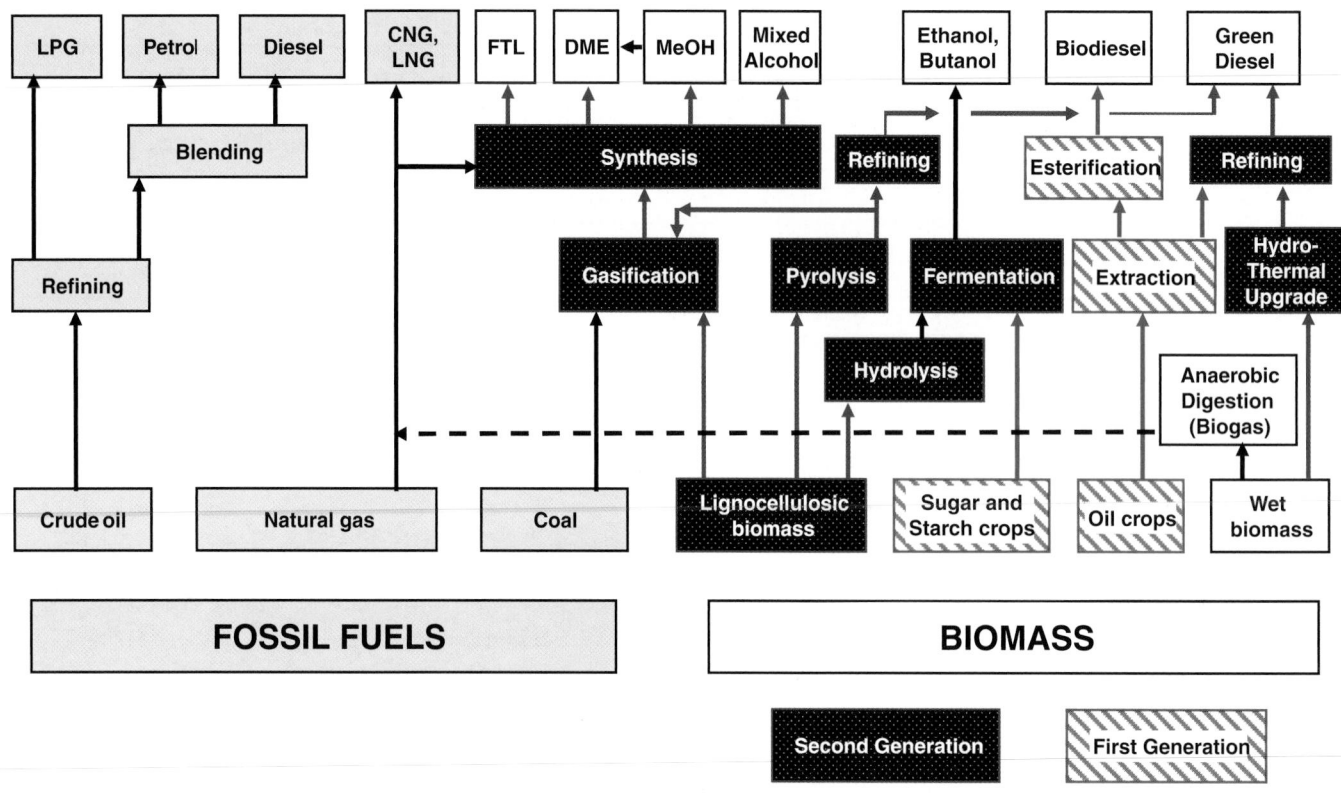

Figure 11.14 | Production paths to liquid fuels from biomass and, for comparison, from fossil fuels.

associated co-products market are not fully developed, first-generation operations that focus on single products (such as ethanol or biodiesel) are regarded as a starting point in the development of sustainable biorefineries. New sugarcane processing facilities are examples of commercial biorefineries, producing both ethanol and exported electricity (EPE, 2008). Advanced or second generation biorefineries would be based on more sustainably derived biomass feedstocks. These biorefineries aim to optimize the use of biomass and resources in general (including water and nutrients), while mitigating GHG emissions (Chum et al., 2011).

Third-generation biofuels

Third-generation biofuels are those that require considerable research and/or technology advances before they can begin to approach commercial viability. Fuels in this category include biological solar hydrogen production and photo-electrochemical fuel production (Aartsma et al., 2008), as well as algae-derived fuels (Sheehan et al., 1998a; Brennan and Owende, 2010). Such options have the potential to more easily meet sustainability constraints, especially land requirements, in part via substantially higher efficiencies of solar energy conversion than for second-generation biofuels. (See Figure 11.15.)

Among third-generation fuels, algae-derived fuels are attracting the most attention today. Algae production does not require the use of good-quality land (though water needs can be substantial). With the

large number of diverse algal species in the world, upper range productivity potentials of up to several hundred EJ for microalgae and up to several thousand EJ for macroalgae have been reported (Sheehan et al., 1998a; Florentinus, 2008; van Iersel et al., 2009; Chum et al., 2011). No reliable estimates are available on the techno-economic potential of energy from algae.

Aquatic phototrophic organisms in the world's ocean (halophytes) produce annually 350–500 billion tonnes of biomass and include microalgae (such as Chlorella and Spirulina), macroalgae (seaweeds), and cyanobacteria (also called "blue-green algae") (Garrison, 2008). Microalgae such as Schizochytrium and Nannochloropsis can accumulate lipids, from which diesel-like oils can be extracted at reportedly greater than 50% of their dry cell weight (Chum et al., 2011). The US Department of Energy (US DOE) operated a significant algae energy research and development (R&D) program from the late 1970s until the mid-1990s, in which considerable progress was made in understanding the biochemistry of different microalgaes and their potential as a source of lipids. (Still, fewer than 100 microalgae species have been tested or used industrially out of about 100,000 known species.) A realistic yield of unrefined oil from algae with a 50% oil content located on the equator has been estimated to be 40,000–50,000 lit/ha/yr, and a mere 10% yield from this oil could match palm oil productivities of about 4700 lit/ha/yr (Weyer et al., 2009). Uncultivated macro-algae can reach yields that are higher than those of sugarcane (per unit area)

Table 11.9 | Estimated yields for second-generation biofuels with current and projected future technology levels (note: none of these systems are currently commercial).

Technology	Fuel yield per tonne dry biomass		Fuel yields from dedicated bioenergy US land, with projected 2020 biomass yields[f]	
	Liters/tonne	GJ_{LHV}/tonne	Low-quality land GJ_{LHV}/ha/yr	High-quality land GJ_{LHV}/ha/yr
Ethanol via enzymes[a]				
Current (woody biomass)	279	5.9	66	139
Best future (woody biomass)	362	7.6	86	180
Current (low lignin biomass)	346	7.3	82	172
Likely future (low lignin biomass)	396	8.3	94	197
Best future (low lignin biomass)	441	9.3	105	219
Ethanol via syngas fermentation[b]				
Future low	120	2.5	28	60
Future high	160	3.4	38	80
Fischer-Tropsch diesel/gasoline				
Current low[c]	75	2.6	29	61
Current high[c]	200	6.9	77	162
Future[d]	236	8.1	91	191
Future, biomass/coal co-processing w/CCS[e]	568	19.5	219	460

[a] Source: NAS, 2009. Biochemical processing of plantation-grown short-rotation poplar. "Current" means currently understood technology.

[b] Source: Sims et al., 2010, citing Putsche, 1999. Details of the feedstock and process designs considered are not given. The process likely begins with thermochemical gasification followed by fermentation of the syngas to ethanol.

[c] Source: Sims et al., 2010. Assumptions behind these estimates are not provided, but they appear to refer to only the diesel fuel that would be produced from a biomass-to-liquids (BTL) system (ignoring naphtha and other co-products).

[d] Source: Liu et al., 2011. Simulation results for a BTL system using herbaceous feedstock and producing both diesel and gasoline. Gasoline is produced by refining the naphtha fraction of the raw Fischer-Tropsch product. See Chapter 12 for addition discussion of BTL options.

[e] These estimates are for the CBTL-OTA-CCS (CO2 capture and storage) process configuration described in Chapter 12.

[f] Average yield of 11.2 dry t/ha/yr on lower quality lands, such as Conservation Reserve Program lands, and 23.6 t/ha/yr on lands with moist, fertile soils (NAS, 2009).

(Zemke-White and Ohno, 1999). Photosynthetic cyanobacteria can produce fuels like hydrogen directly.

Aside from questions about economic viability (Pienkos, 2009), fuels from algae grown by fertilizing with CO_2-rich flue gases from fossil fuel power plants, as is being proposed by many start-up companies today, would be far from "carbon-neutral" since the carbon in the algae originates from fossil sources and since considerable energy inputs are required for algae growing, harvesting, and oil extraction systems (Lardon et al., 2009; Cazzola, 2009).

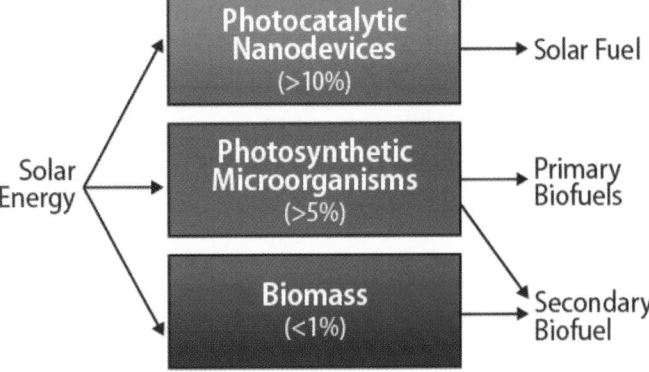

Figure 11.15 | Third-generation biofuels, with the potential for higher solar energy conversion efficiencies than first- or second-generation biofuels. Source: Aartsma et al., 2008.

11.2.5   Economic and Financial Aspects

Figure 11.16 presents projected ranges of commercial production costs for biomass-to-power and biomass-fired CHP plants (Bauen et al., 2009). Not all of the technologies included there are commercially mature, but the figure illustrates that there can be a wide range in costs for any technology, depending on feedstock cost, annual operating hours, and byproduct credit for heat in the case of CHP. A self-consistent set of electricity

generating cost estimates for several commercially mature technologies is shown in Table 11.10. Analogous estimates for commercial biomass-fired heat production technologies are provided in Table 11.11.

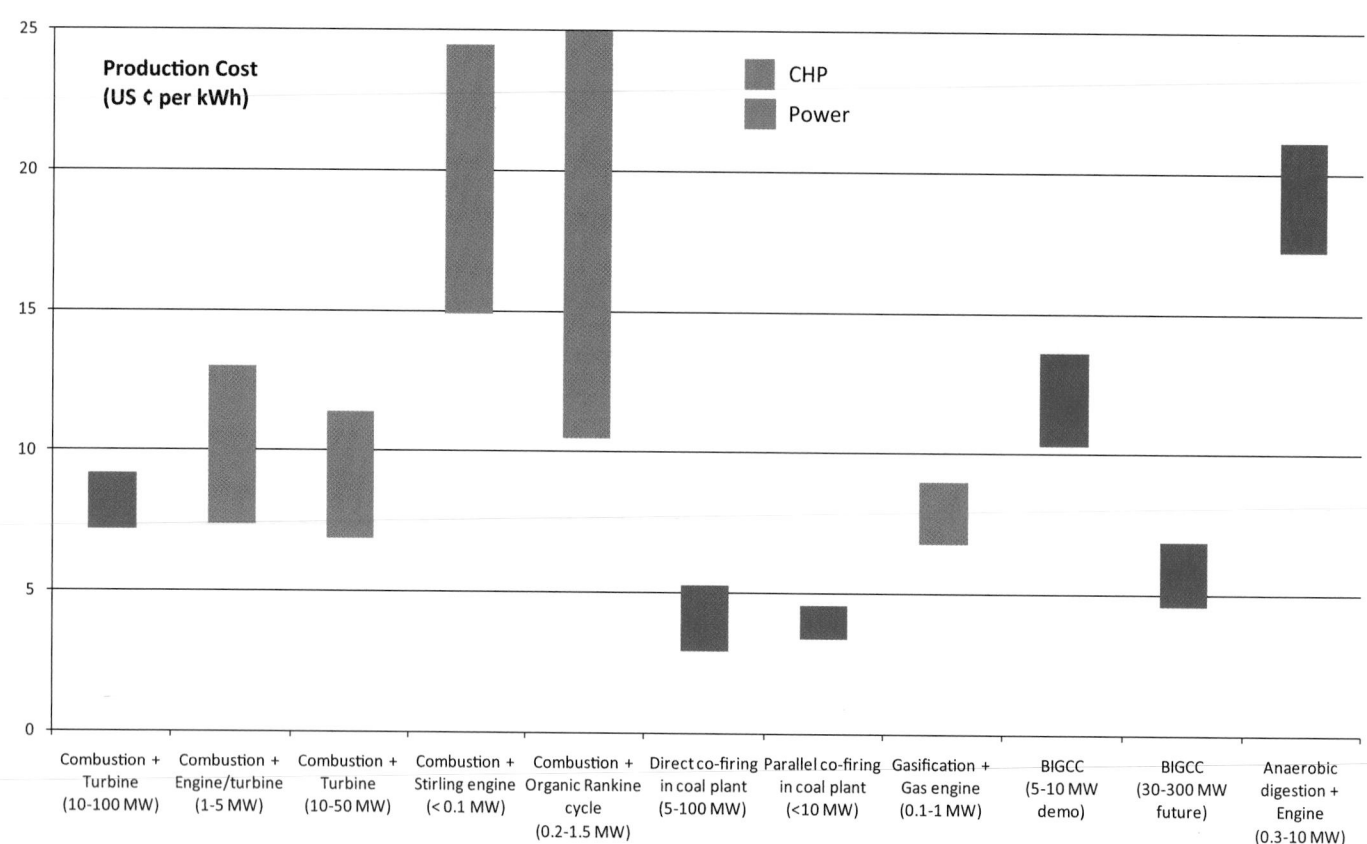

Figure 11.16 | Projected production costs for biomass to power and to CHP. Anaerobic digestion can also be operated in CHP mode; production cost will reduce by 60–80% (depending on technology and plant size) if "free" biomass feedstock is used, such as MSW, manure, or wastewater. Source: based on Bauen et al., 2009.

Table 11.10 | Estimated levelized cost of electricity for commercial biomass-fired electricity generating technologies. Capital investment, operating and maintenance costs, conversion efficiencies, and feedstock costs are from Annex III in IPCC (2011).

	Capacity range, MW_e	Capital investment, $US\$/kW_e$	Annual capacity factor, %	Conversion efficiency, % (HHV)	US$_{2005}$¢/kWh$_e$			
					5% Discount rate		10% Discount rate	
					Biomass Price, $/GJ (HHV)			
					1.25	5	1.25	5
Stoker boiler/ steam turbine	25–100	2600	75	27	6.1	11.1	7.6	12.6
		4000			7.9	12.9	10.1	15.1
Co-feed/co-fire with coal	20–100	430	75	36	2.1	5.8	2.3	6.0
		500			2.2	5.9	2.5	6.2
Small steam CHP[a]	2.5–10	4100	62	18	7.3	14.8	10.2	17.7
		6200			10.5	18.0	14.7	22.2

[a] Includes a byproduct heat credit of US¢5.4/kWhe.

For most first-generation biofuels, fluctuating crop prices during the past decade created large swings in production costs. Even in periods with low feedstock costs, however, essentially all first-generation biofuels (with one important exception) historically have been unable to compete on cost with prevailing petroleum-derived fuels. Table 11.12

provides a self-consistent comparison of first-generation biofuel production costs.

Ethanol made from sugarcane in Brazil is the lone first-generation fuel that has been able to compete with gasoline at prevailing oil prices

Table 11.11 | Estimated levelized cost of heat from biomass CHP. Capital investment, operating and maintenance costs, conversion efficiencies, and feedstock costs are from Annex III in IPCC (2011).

					US$_{2005}$/GJ			
					5% Discount rate		10% Discount rate	
					Feedstock Price, US$/GJ (HHV)			
	Capacity range, MW$_{th}$	Capital investment, US$/kW$_{th}$	Annual capacity factor, %	Conversion efficiency, % (HHV)	3.7	6.2	3.7	6.2
Biomass steam turbine CHP	12–14	370	69	25	16	26	17	27
		1000			19	29	20	30
					3.7	6.2	3.7	6.2
Biogas CHP	0.5–5	170	80	25	18	28	18	28
		1000			21	31	22	32

Table 11.12 | Estimated levelized cost of fuel for commercial first-generation biofuels. Capital investment, operating and maintenance costs, conversion efficiencies, and feedstock costs are from Annex III in IPCC (2011).

					US$_{2005}$/GJ (HHV)			
					5% Discount rate		10% Discount rate	
	Capacity range, kW$_{th}$	Capital investment, US$/kW$_{th}$	Annual capacity factor, %	Conversion efficiency, % (HHV)	Feedstock Price, US$/GJ (HHV)			
					2.1	6.5	2.1	6.5
Brazil, cane ethanol	170–1000	200	50	17	15.0	40.9	17.7	39.2
		660			17.4	43.2	18.9	44.8
					4.2	10	4.2	10
USA, corn ethanol	140–550	168	95	54	10.8	21.5	11.0	21.7
		253			11.0	21.7	11.3	22.0
					9.7	24	9.7	24
USA, soy biodiesel	44–440	168	95	103	12.7	26.6	12.9	26.8
		316			13.1	27.0	13.5	27.4
					7	18	7	18
Brazil, soy biodiesel	44–440	168	95	103	9.7	20.4	10.0	20.6
		326			10.2	20.8	10.6	21.3

[a] Byproduct credits are included in each case. See Annex III of Chum et al., 2011 for details.

in the recent past. The Brazilian industry, launched in the 1970s, was subsidized (in decreasing amounts with time) for nearly 30 years. Long-term sugarcane breeding programs, together with research and development on agronomic and distillery practices, have led to significant reductions in sugarcane production costs over time in Brazil,[5] to the

point where Brazilian ethanol competes today without subsidy with petroleum-derived gasoline (see Figure 11.17). Also, essentially all of the energy required for processing the cane into ethanol is derived from bagasse, the fibrous residue of juice extraction, so producers incur little or no purchased-energy costs, and net greenhouse gas emissions are modest.

The reliability of cost estimates for second-generation biofuels is difficult to assess due to the pre-commercial nature of technologies and the uncertainty of future feedstock costs, among other variables. With presently understood (but not yet commercially demonstrated) systems, ethanol is estimated to be competitive with oil-derived gasoline when

5   Increased sugarcane yields achieved by concerted research and development have been the main driving force behind cost reductions: sucrose content and thus ethanol yield have been increased; ratoon harvesting (multiple harvests from one planting) has been extended; efficiency of manual harvesting has improved; mechanical harvesting has increasingly displaced manual harvesting; and the use of larger transport trucks has further reduced feedstock costs. Residues used for electricity production, along with electricity production efficiency, have also been increasing.

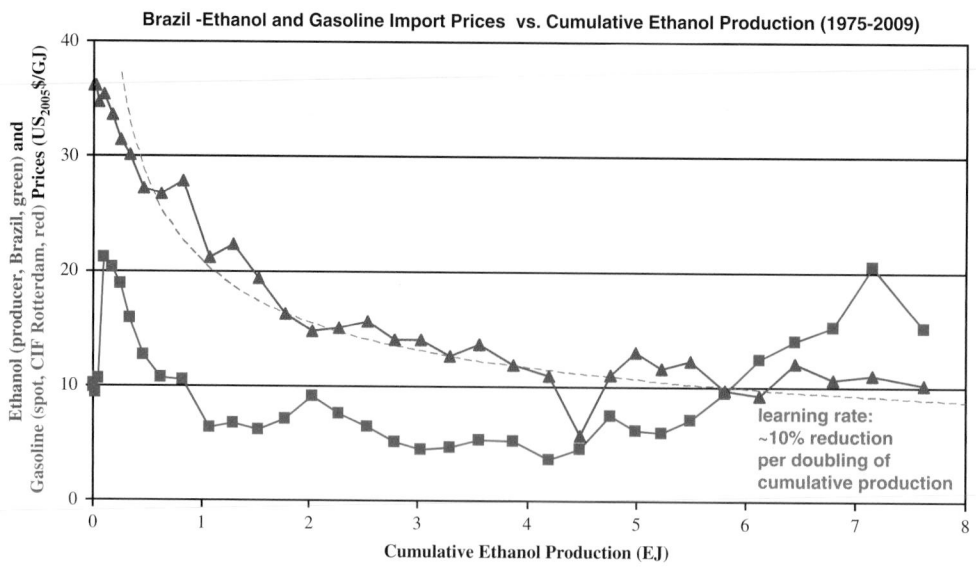

Figure 11.17 | Selling price for ethanol in Brazil versus cumulative production, compared with Rotterdam price for gasoline. Source: Mytelka and de Sousa Jr., 2011; see also van den Wall Bake et al., 2009.

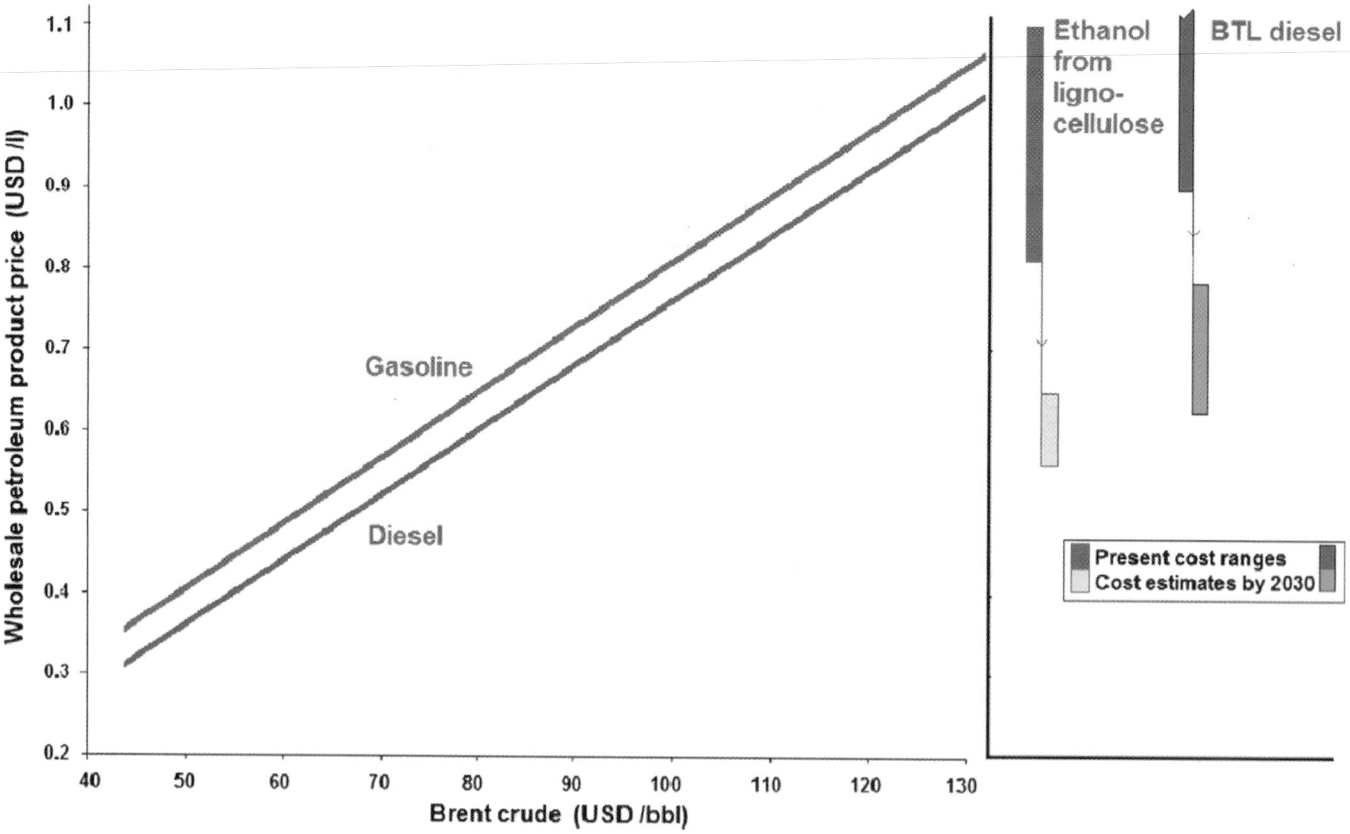

Figure 11.18 | Range of production cost estimates for second-generation biofuels (in US$_{2006}$ per liter of gasoline equivalent). The 2030 estimates assume significant investment in RD&D. Source: Sims et al., 2010.

the crude oil price is in the range of US$100–140/bbl with the range for BTL (biomass-to-Fischer-Tropsch-liquid) being higher than this. (See Figure 11.18.) With significant investments in RD&D efforts over the next 20 years, the oil price at which ethanol competes with gasoline may fall to US$60–80/bbl (see, e.g., Hamelinck and Faaij, 2006) and to US$80–100/bbl in the case of BTL. Sims et al. (2010) suggest that greater cost reductions may be anticipated for ethanol than for BTL because many of the components of BTL systems are already commercially mature (in other applications). Table 11.13 provides another perspective on projected costs for liquid biofuels in the 2020–2030 timeframe.

An important consideration is how the cost and performance of advanced biomass conversion technologies can be expected to change as commercial experience is gained. Bioenergy systems show technological learning and related cost reductions with progress ratios (a measure of the rate of reduction in cost with increasing cumulative production) comparable to those of other renewable energy technologies (Chum et al., 2011). This applies to cropping systems, supply systems, and logistics (as clearly observed in Scandinavia) and in conversion (ethanol production,

power generation, biogas, and biodiesel). There has been clear technology learning for several important bioenergy systems (see Table 11.14), but with the exception of ethanol from sugarcane in Brazil or systems with unusually low biomass feedstock costs, most still require subsidies to be competitive.

In the 2030 timeframe, the performance of existing bioenergy technologies can be improved considerably, while new technologies offer the prospect of more-efficient and competitive deployment (Chum et al., 2011).

11.2.6 Sustainability Issues

Fossil fuel inputs are significant for producing most first-generation biofuels, as shown in Figure 11.18, with the notable exception of ethanol from sugarcane in Brazil. For the other biofuels in Figure 11.19, two parameters are especially important: the amount of fossil energy consumed that is assigned to non-biofuel co-products of the process and the energy source used to provide the process heat needed at the conversion facility.

The case of ethanol made from wheat grain illustrates the significance of these parameters. For each of the wheat ethanol cases shown in Figure 11.19, the low and high estimates correspond, respectively, to use of the co-product (distillers dried grains, or DDGs) as an energy source at the facility (reducing the need for external energy inputs) or

Table 11.13 | Projected biofuel production costs in the period 2020–2030, in US_{2005}/GJ.

Biofuel	Projected Production Costs
Sugarcane ethanol, Brazil[a]	9–10
Corn ethanol, USA[a]	16
Rapeseed biodiesel[a]	25–37
Lignocelulose sugar-based biofuels[b]	6–30
Lignocelulose syngas-based biofuels[b]	12–25
Lignocelolose pyrolysis-based biofuels[b]	14–24
Gaseous biofuels[b]	6–12

[a] Numbers from Chum et al., 2011 (Table 2.7).

[b] Numbers from Chum et al., 2011 (Table 2.18).

Table 11.14 | Overview of experience curves for biomass energy technologies and energy carriers. Cost/price data from various sources.*

System	Learning Rate (%)	Time frame	Region	N	R²
Feedstock production					
Sugarcane (tonnes sugarcane per ha/yr)[a]	32 ± 1	1975–2005	Brazil	2.9	0.81
Corn (tonnes corn per ha/yr)[b]	45 ± 1.5	1975–2005	USA	1.6	0.87
Logistics chains					
Forest wood chips (Sweden)[c]	12–15	1975–2003	Sweden/Finland	9	0.87–0.93
Conversion investment & operation and maintenance (O&M) costs					
CHP plants (€/kWe)[c]	19–25	1983–2002	Sweden	2.3	0.17–0.18
Biogas plants (€/m3biogas/day)[d]	12	1984–1998		6	0.69
Ethanol from sugarcane[a]	19 ± 0.5	1975–2003	Brazil	4.6	0.80
Ethanol from corn (only O&M costs)[b]	13 ± 0.15	1983–2005	USA	6.4	0.88
Final energy carriers					
Ethanol from sugarcane[e]	7 / 29	1980–1985	Brazil	~6.1	n.a.
Ethanol from sugarcane[a]	20 ± 0.5	1975–2003	Brazil	4.6	0.84
Ethanol from corn[b]	18 ± 0.2	1983–2005	USA	7.2	0.96
Electricity from biomass CHP[d]	8–9	1990–2002	Sweden	~9	0.85–0.88
Electricity from biomass[f]	15	Unknown		n.a.	n.a.
Biogas[d]	0–15	1984–2001	Denmark	~10	0.97

* The learning rate is defined as the percentage reduction in cost for each doubling in cumulative production. N is the number of doublings observed in cumulative production, and R² is the regression correlation coefficient for the data.

References: [a]van den Wall Bake et al., 2009; [b]Hettinga et al., 2009; [c]Junginger et al., 2005; [d]Junginger et al., 2006; [e]Goldemberg et al., 2004; [f]IEA, 2000.

Source: Chum et al., 2011.

First Generation Biofuels

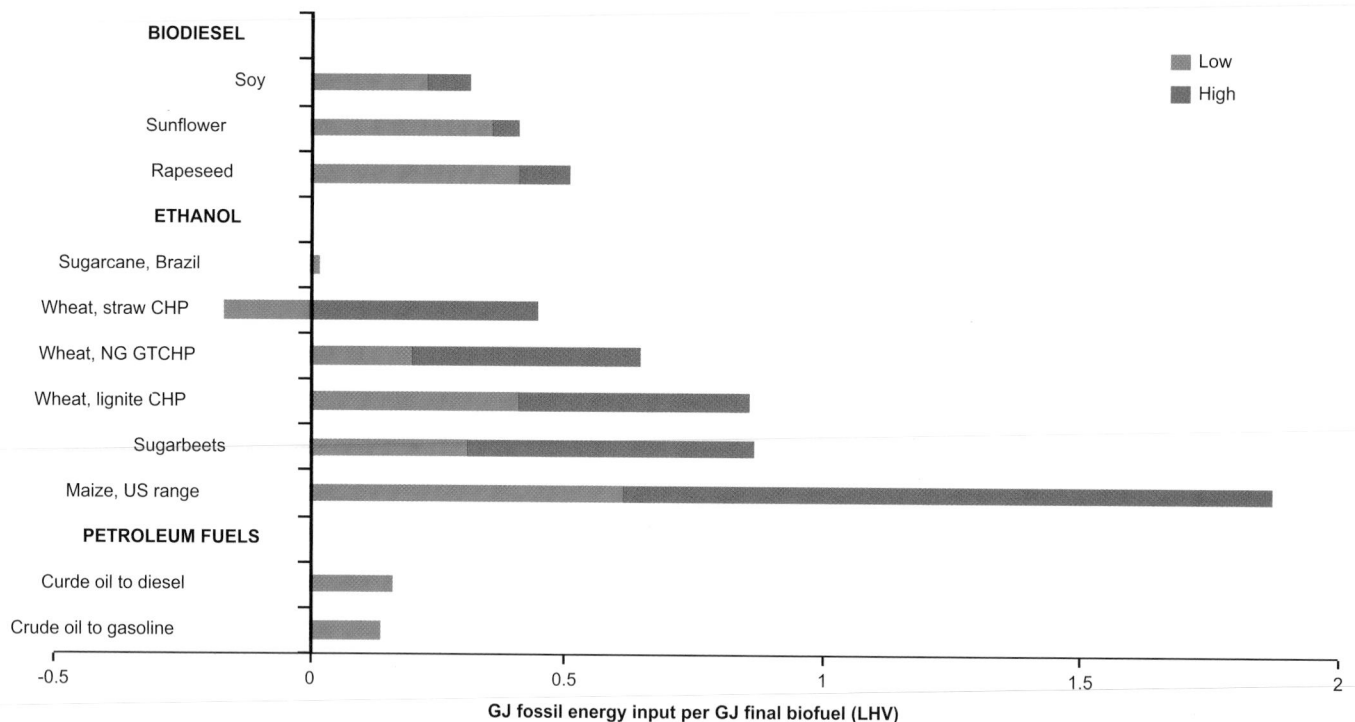

Figure 11.19 | Lifecycle fossil energy use estimates in production of several first generation biofuels and two petroleum fuels. Source: based on CONCAWE et al., 2008; Pradhan et al., 2009; Sheehan et al., 1998b; Farrell et al., 2006.

as an animal feed. For the latter case, the animal feed co-product was assigned fossil fuel emissions equivalent to the fossil energy that would have been consumed to produce animal feed from traditional sources (CONCAWE et al., 2007). The significance of external fuel used for processing energy is evident also in the wheat ethanol cases: fossil energy use is highest when lignite is used, somewhat lower when natural gas is used, and lowest when wheat straw (a non-fossil fuel) is used. In fact, when wheat straw and DDGs are used for fuel onsite, the facility is able to produce an excess of electricity that can be exported to replace electricity that would otherwise have been produced elsewhere from fossil fuels, resulting in net negative fossil fuel consumption.

Biofuels are often valued for the enhanced security of liquid fuel supplies that they provide. Thus petroleum use in the production of biofuels is an important subset of total fossil fuel use. For essentially all biofuels, petroleum use is relatively low. Estimates vary, depending on specific assumptions. In the case of corn ethanol in the United States, estimates of about 0.03–0.3 MJ of petroleum input per MJ of ethanol produced are indicated in a summary by Farrell et al. (2006).

Estimates of life-cycle GHG emissions associated with the production of first-generation biofuels span a wide range, depending on the assumptions used. As with the petroleum estimates, a particularly important assumption relates to how GHG emissions are allocated among co-products. Figure 11.20 shows ranges depending on co-product assumptions for several first-generation biofuels.

It is important to note that the results shown in Figure 11.20 exclude any considerations of GHG emissions arising from land use changes associated with production of the biofuels. Land use change emissions can result when land is converted from an existing use to production of biomass for energy (so-called direct land use change). Emissions may also result from land use changes in one region of the world as a result of establishing biomass energy production in another region (indirect land use changes). Chapter 20 discusses land use change impacts of biomass production.

The high yields for second-generation biofuels in term of GJ/ha/y (see Table 11.9) are achieved with relatively less need for fossil fuel inputs than for first-generation fuels. This is in part because well-designed second-generation conversion systems will use generally unconvertable portions of the incoming biomass as fuel to generate the energy needed to run the conversion facility. For example, lignin is envisioned to be used for this purpose in many biochemical ethanol production systems. The larger scale envisioned for second-generation biofuel production plants also enables more-energy-efficient processing.

The low fossil energy requirements result in relatively low life-cycle GHG emissions. Figure 11.21 shows estimates of emissions relative to petroleum-derived gasoline for a range of second-generation fuels (excluding any emissions associated with direct or indirect land use change). Especially notable are the highly negative emissions for systems that use CO_2 capture and storage (CCS, see Chapter 13) as part of the process.

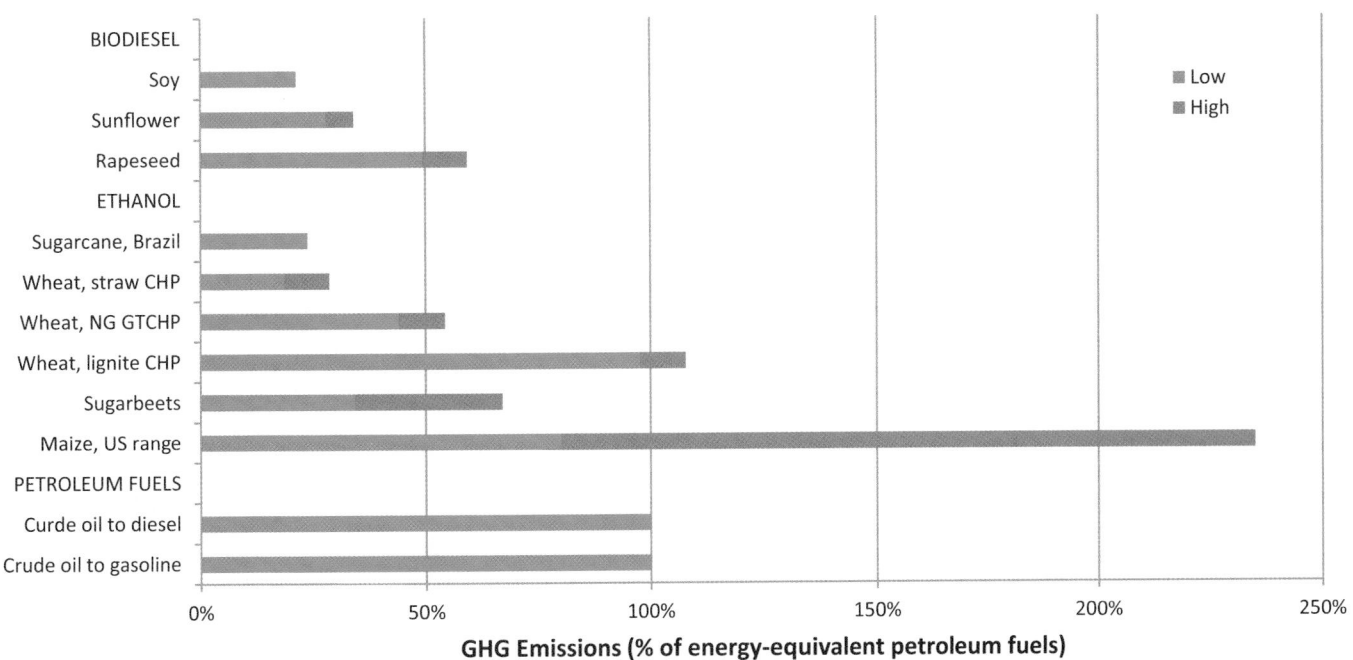

Figure 11.20 | Estimated life-cycle GHG emissions for first-generation biofuels, excluding any impacts of land use change. Source: based on CONCAWE et al., 2008; Pradhan et al., 2009; Sheehan et al., 1998b; Farrell et al., 2006; Macedo et al., 2008. The low and high values for maize ethanol are based on analysis by Shapouri et al. (2002) and Pimental and Patzek (2005), respectively, as reported by Farrell et al. (2006).

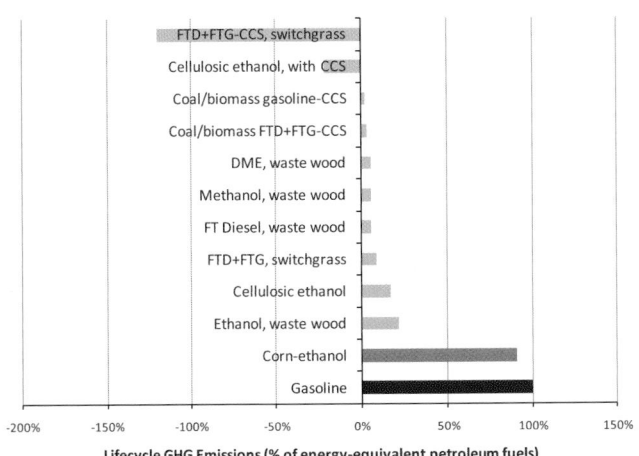

Figure 11.21 | Estimated life-cycle GHG emissions for second generation-biofuels, excluding any impacts of land use change. Estimates for fuels from waste wood are from CONCAWE et al. (2008). The corn-ethanol estimate represents the low estimate from Figure 11.20. The cellulosic ethanol estimates are based on Figure 12.22 in Chapter 12. The ethanol case with CCS assumes capture and storage of fermentation CO_2. Others are based on Table 12.15 in Chapter 12. FTD + FTG refers to synthetic diesel and gasoline made via Fischer-Tropsch synthesis. DME is dimethyl ether.

One additional interesting second-generation thermochemical route to fuels production is co-processing of biomass with coal at a single facility. Fuels such as Fisher-Tropsch Liquids, methanol to gasoline, and others can be made purely from coal in much the same manner as they would be made from biomass, but doing so leads to fuel cycle GHG emissions that could be about double those of the petroleum

fuels displaced. If CCS were included as part of the facility design, net GHG emissions would be about the same as for petroleum fuels (see Chapter 12). By using CCS and co-processing an appropriate amount of biomass with the coal (so-called CBTL systems), net GHG emissions can be reduced to zero or less, since capture of photosynthetic carbon from the biomass provides negative GHG emissions to off-set unavoidable positive emissions with coal (van Vliet et al., 2009; Larson et al., 2010).

Careful and detailed life-cycle assessments are needed to understand the gains to be made with different algal biofuel systems and thereby help guide future developments toward sustainable systems (IEA, 2010c). Kreutz (2011) has estimated the potential for GHG emissions reduction via reuse of coal power plant CO_2 to fertilize microalgae grown in ponds and converted to biodiesel. His analysis shows that with this approach the carbon mitigation potential is not sufficient to achieve deep GHG emissions reductions. As Kreutz states, "[u]sing the carbon twice fails to meet the objective of deep GHG emission reductions."

For bioelectricity, the life-cycle emissions are generally low, between about 10–50 g CO_2-eq/kWh, but also extremes like −1350 and +350 gCO_2-eq/kWh can be obtained depending on the technology and assumptions used (Chum et al., 2011). Negative emissions are achieved when biomass use is combined with CCS, allowing removal of CO_2 from the atmosphere by storing it underground.

11.2.7   Implementation Issues

A wide variety of existing and potential factors hinder the further deployment of bioenergy, related to supplies, technologies, and markets. Some essential supply side concerns relate to risks of biological production and the availability of residues. El Niño, drought, floods, fire, pests, and insect attacks affect production of biomass as well as of food. Some of these risks (fire, pests, insect attacks) can be reduced through proper management, but they cannot be eliminated. In general, diversity is the best mechanism to minimize these risks.

Concerning technologies, the main issues to be dealt with in an economic and environmentally sound manner are robustness of biomass conversion technologies to feedstock variability, the handling and storage of biomass feedstock, the commercialization of technologies with improved economics at small scale, the handling of co-products (e.g., ash, digestate) containing contaminants, and the need for flue gas cleansing meeting stringent limits on toxic emissions (e.g., NO_x, CO, particulates).

In terms of markets, the main risks and barriers are related to feedstock availability and price (representing 50–90% of the production costs of bioenergy), the competitiveness of different biomass conversion routes, the need for clarity and foresight in regulatory aspects such as planning regulation and emission standards, an unstable and unsupportive policy environment, and the interaction with other sectors – such as food and forestry – and the policies that affect them.

In the past few decades, bioenergy developments have been promoted and supported by governments of many countries through a wide variety of policy instruments. Typical examples for liquid biofuels include the Proálcool program launched in Brazil in 1975 to reduce dependence on imported oil; the Common Agricultural Policy in the EU, including production quotas for oilseed food crops as well as exemptions from certain taxes; and the US support included in several farm bills and state and federal incentives for ethanol production (Worldwatch Institute, 2006). Subsidies in one form or another to encourage consumption of first-generation biofuels amounted in 2006 to over US$6 billion in the United States and nearly US$5 billion in the European Union (FAO, 2008).

Successful policies to promote biomass for heat have focused on more centralized applications for heat or combined heat and power production in district heating and industry (Bauen et al., 2009). In the power sector, feed-in-tariffs have gradually become the most popular incentive. In addition, policies such as fiscal measures and soft loans have been supportive (Global Bioenergy Partnership, 2007). Quota systems have so far been less successful (van der Linden et al., 2005). Priority grid access for renewables is applied in most countries where bioenergy technologies have been successfully deployed (Sawin, 2004a).

As discussed by Chum et al. (2011), the main drivers behind governmental support for the bioenergy sector have been concerns about energy security and climate change as well as the desire to support the farm sector through increased demand for agricultural products (FAO, 2008). An estimated 69 countries had one or several biomass support policies in place in 2009 (REN21, 2010).

Concerns about the effects of these policies on food prices, questions about the GHG emission savings of biofuels, and doubts about the environmental sustainability of bioenergy have also seen countries rethinking their policies and targets for biofuels blending (IEA, 2009a). In addition, uncoordinated targets for renewables and biofuels without an overall biomass strategy may enhance competition for biomass between the power and transportation sectors (Bringezu et al., 2009). Some national targets will require increased imports, thus contributing to competition for biomass globally. An overall strategy would have to consider all types of use, especially for food and non-fiber biomass (Chum et al., 2011; see also Chapter 20).

11.3   Hydropower

11.3.1   Introduction

Hydropower is a form of renewable energy derived from moving water. It has been applied to generate mechanical power, at watermills for example, for several centuries, and has been used to generate electric power for more than 100 years. Hydropower projects may be usefully categorized by the way they harness water to generate power:

- Hydrokinetic – a project that places in a watercourse devices capable of generating electrical power from the flow of water;

- Run of river – a project that uses a watercourse to pass water through a power plant, with limited storage;

- Reservoir – a project that impounds a watercourse for storage, forming a reservoir, for release through a power plant;

- Pumped storage – a project that pumps water from a lower level to a reservoir at a higher elevation for storage in a cyclical fashion, for release through a power plant at times of high demand.

It should be noted that water is also used for power generation through other means, such as ocean energy (see Section 11.9), and water is used as a medium to drive turbines in thermal power stations or to produce hydrogen.

Hydropower plants are able to operate in isolation, but most of them are connected to a transmission network. The maximum output of individual units ranges from 0.1 kW (models) to 852 MW (Three Gorges power station, China). Three Gorges is nearing its full capacity of 22.5 GW and is producing around 84 TWh/yr. By comparison, Itaipu power station, on the border between Brazil and Paraguay, generated a record of 95 TWh in 2008, with an installed capacity of 14 GW. This reflects the different operating regimes between the two stations, with Three Gorges fulfilling flood-control and navigational functions in addition to the generation of power. It is common for hydropower facilities, especially of the reservoir type, to serve multiple purposes.

11.3.2   Potential of Hydropower

The theoretical, technical, and economic potentials of hydropower for electricity production were presented in the *World Energy Assessment* (UNDP et al., 2000). An update of those figures is presented in Chapter 7. The main results are as follows:

- The gross theoretical potential for electricity production can be estimated at 40–55 x 10^3 TWh/yr (about 150–200 EJ$_e$/yr).

- The technically feasible potential can be estimated at 14–17 x 10^3 TWh/yr (about 50–60 EJ$_e$/yr).

- The potential used at present is about 3.0–3.2 x 10^3 TWh/yr (about 11 EJ$_e$/yr).

- The unexploited economic potential with production costs between US¢2–8/kWh can be estimated at about 5 x 10^3 TWh/yr (about 18 EJ$_e$/yr); at production costs ranging from US¢2–20/kWh, this figure would be about 11–14 x 10^3 TWh/yr (40–50 EJ$_e$/yr).

Climate change may have an impact on the potential of hydropower and the availability of hydro capacity in particular river basins, but it is unlikely to affect the global totals (see Section 11.3.3).

Recent energy scenario studies suggest that in 2050 the potential used could have increased to 18–35 EJ$_e$/yr (Greenpeace and EREC, 2007; IEA, 2008b; 2009a). This figure translates into savings on primary energy use through conventional power generation of 36–70 EJ/yr in 2050, assuming a significant improvement in the average conversion efficiency of fossil-fueled power plants from about 35% to 50% in the period 2010–2050.

11.3.3   Market Developments

In 2008 and 2009, hydropower provided about 3100±100 TWh, nearly 16% of the world's electricity generation, which was more than 80% of renewable energy-sourced electricity generation (US EIA, 2011; BP, 2011). This figure translates to fossil fuel savings in conventional power production equivalent to about 32 EJ/yr (assuming a conversion efficiency of 35%) or about 6% of global primary energy

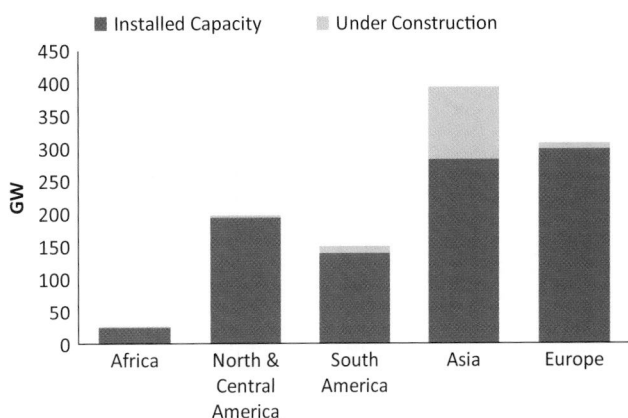

Figure 11.22 | Global installed hydropower capacity, by region, in 2009, including capacity under construction. Source: based on IHA, 2010a.

use. Hydropower is currently generating electricity in some 160 countries, using more than 11,000 stations with around 27,000 generating units (Taylor, 2010).

11.3.3.1   Installed Capacity and Capacity under Construction

Global installed capacity estimates for 2007 range from 860 to 950 GW (IHA, 2010a). According to REN21, global installed capacity can be estimated at about 920 GW in 2007, 950 GW in 2008, 980 GW in 2009, and 1010 GW in 2010 (REN21, 2009; 2010; 2011). Included in these figures are hydropower projects with less than 10 MW installed capacity (about 60 GW in 2009). A regional breakdown of installed capacity and capacity under construction for 2009 is presented in Figure 11.22. In 2009, 32% of the global capacity was installed in Asia, 29% in Europe, 20% in North and Central America, 16% in South America, and 3% in Africa.

According to regular surveys of hydro equipment suppliers by the International Hydropower Association (IHA), 2007 and 2008 were record years in the history of hydropower deployment. From 2005–2009 approximately 135 GW of additional capacity were commissioned, an average of 27 GW/yr, indicating a growth rate of 3% a year. It would appear realistic to assume at least this rate of development will continue into the foreseeable future. For smaller-scale hydro (<10 MW), the growth rate is estimated at about 9% a year.

11.3.3.2   Regional Developments

Hydropower is undergoing rapid development in Asia and Latin America in line with the remarkable growth of these regions. Aside from the recent economic transformations, other factors contributing to growth there include using hydropower to further stimulate economic growth, improve energy and water security, and mitigate and adapt to climate

Table 11.15 | Estimated changes in hydropower generation due to climate change by 2050.

Region	Installed capacity in 2005 [GW]	Generated electricity in 2005 [TWh/yr]	Change by 2050 [TWh/yr]
Africa	22	90	0.0
Asia	246	996	2.7
Europe	177	517	−0.8
North America	161	655	0.3
South America	119	661	0.3
Oceania	13	40	0.0
TOTAL	**737**	**2931**	**2.5**

Source: Hamududu and Killingtveit, 2010.

change. Despite high levels of existing deployment in North America and Europe, these regions also continue to show sizable growth. These developments are in sharp contrast to sub-Saharan Africa, a region with proportionally the lowest deployment to potential (less than 10% of its technical potential). Governance, institutional capacity, and financing rather than the lack of available resource remain major impediments to hydropower development in this region.

11.3.3.3 Potential Impacts of Climate Change on Hydropower

Table 11.15 presents the potential impacts of climate change on hydropower. Projected power generation in 2050 is compared with 2005 based on 12 different climate models. These suggest that losses in some areas will be offset by gains in others, resulting in a largely unchanged or slightly improved global picture of hydropower availability in 2050.

Further work is required to ascertain the impacts of climate change on existing hydropower infrastructure and future development. At present these are uncertain, but it seems clear that climate change will alter the hydrologic cycle at the river basin level. Although this does not change the amount of water in the global hydrologic cycle, changes, some of them significant, are anticipated in the spatial and temporal facets of precipitation and glacial discharge, which will vary from river basin to river basin. The changes this may bring to flows in some catchments could affect hydropower availability, particularly if reservoir storage is managed poorly or if there is limited storage capacity (run of river). Overall, it may be that the positive and negative impacts of climate change balance each other out if the value of water storage is appreciated in forward planning, e.g., by increasing capacity or adding spillways.

Pressure on scarce water resources coupled with a changing hydrologic cycle also increases the importance of hydropower's water storage capabilities. The impetus for developing reservoir projects grows if it is appreciated that water storage also provides energy storage and ancillary service capabilities.

11.3.4 Hydropower Technology Developments

Hydroelectricity generation is generally regarded as a proven, mature technology. For example, conventional turbines have reached 96% efficiency. Yet advances in the technology are still being made.

11.3.4.1 Technology Improvements

Inside the powerhouse, improvements include abrasion-resistant turbines, variable speed technology, and fish-friendly equipment. Also beyond conventional project types, developments in ultra-low head technology and hydrokinetic turbines show promise, especially for existing non-hydropower facilities, such as water reservoirs, weirs, barrages, canals, and falls. Outside the powerhouse, improvements include tunnel boring by machine, roller compacted concrete dams, and use of geomembranes, allowing potential civil works cost savings.

Further efficiencies are expected to be made through modernization and up-rating of aging hydropower stations. Most of the current generation capacity will require refurbishment within the next 30 years. (On average, major refurbishment of generation equipment and machinery is required every 40 years or so.) Consequently, these are ongoing benefits.

11.3.4.2 System Integration

The role that hydro will play in future energy networks is also changing. Most of the early hydropower projects were developed to provide continuous supply (base load) to the power system. This pattern will continue in countries where hydropower provides a significant share of power generation. As other electricity generation technologies have developed, however, the role of hydropower has evolved to encompass a supporting service. Given its unique abilities to store energy and move quickly to full capacity from standstill, it has assumed importance in peak loading when demand requires it.

As other renewable energy use expands on national grids, these abilities on the part of hydropower will assume greater importance: some renewable energies tend to supply electricity on a variable basis (solar photovoltaics, for instance, and wind). By matching these with hydropower, synergies develop from hydro's capacity to supply power on demand, which allows renewable variability to be balanced, as well as matching supply with demand.

The variable nature of some renewables, as well as the costs associated with matching output at times of reduced demand with fossil fuels, geothermal, and nuclear thermal generation options (resulting in these being kept running through periods of low demand), means that there is often excess power in a grid in times of low usage. This has created an increasingly important role for pumped storage hydro through the recycling of stored water.

11.3.4.3   Pumped Storage

Pumped storage facilities make use of the energy of water pumped from a lower level to a reservoir at a higher elevation. The water is brought up to the reservoir when demand for power is low (and when there is excess capacity in the generation system, resulting in cheaper electricity), and it is stored for use to meet peaks in demand. It thus provides peak load capabilities, as well as being available to deal with the intermittency issues surrounding other renewable energies. In 2008, pumped storage capacity in operation worldwide has been estimated to be 127 GW (Ingram, 2009).

It is anticipated that the market for pumped storage will increase by 60% by 2014 (Ingram, 2009). However, as pumped storage is a net user of electricity (about 20% of the energy is lost in the cycle of pumping and generating), its viability depends on clear and predictable differentials in price between periods of low and peak demand.

11.3.5   Economic and Financial Aspects

Hydropower projects are very site-specific. This makes it difficult to predict the amount of engineering required, and therefore the final cost, until investigations are well advanced. In complicated cases, this can lead to sizable cost overruns during construction.

Projects typically have a high upfront capital cost and risk profile. But operation and maintenance costs are very low. (See Table 11.16.) The long life of projects, electro-mechanical equipment about 40 years and civil works about 80–100 years, means that once capital expenditure is amortized, electricity can be produced very economically. This gives hydropower projects an economic life cycle that is quite different from other energy options.

Development costs on hydropower plants may range from US$1000~5000/kW installed capacity. O&M costs can be estimated at US¢0.3–1/kWh. The capacity factor of hydropower plants may be between 30–80%, depending, among other factors, on the characteristics of the energy system within which the hydro power plant operates. Assuming an economic lifetime of the system of 40–80 years and a discount rate ranging from 5–10%, the levelized cost of energy may be between US¢1.5–12/kWh for larger-scale systems (>100 MW) and between US¢1.5–20/kWh for smaller-scale systems (<10 MW). (See Table 11.17.) Because the plant is usually sited far from the point of electricity use, investments may also be required for transmission, perhaps adding another US¢1/kWh.

While these characteristics make many hydropower projects economically viable from a public sector perspective, they do not necessarily translate into financial viability for the private sector. Figure 11.23 presents a range of electric utility and project experiences, comparing average energy tariffs required to make private hydropower financially viable with the average generation cost of most electric utilities. The figure shows a high hydropower tariff in the first 10–20 years because

Table 11.16 | Costs of Hydropower Development.

Project size (MW)	Development cost (US$ million/MW)	Operational cost (US$/MWh)
< 10	1 to > 5	3 to 10
10 to 100	1 to 3	3 to 7
> 100	1 to 2.5	3 to 7

Source: Taylor, 2008.

of the relatively high investment costs, but a low tariff when return of investments has been achieved.

The main challenges for hydropower are therefore reducing risk and raising investor confidence, especially prior to project permitting. These challenges are compounded in the least developed countries, where international public financing (multilateral or regional development banks, for example, or bilateral development assistance) continues to play a strong role.

Given the capital cost and risk profile, there have been relatively few successful independent power producer projects above 100 MW (Trouille and Head, 2008). Most projects have required a substantial amount of state involvement to get off the ground. However, there are increasing pressures on the public sector to provide a favorable environment to attract private-sector capital. This is becoming important for hydropower developments in emerging markets. Brazil, for example, is actively using public-private partnerships to drive the development of its hydro capacity (Ray, 2009).

Financing, rather than resource availability, along with governance and institutional capacity, continue to inhibit the growth of hydropower in least developed countries. Given the role that hydropower can play not only in providing energy and water services but also in contributing to the sustainable development of these countries, it is vital that financing, governance, and institutional support mechanisms are adapted.

11.3.6   Sustainability Issues

The development of a hydropower project, whatever its scale, generates a variety of positive and negative effects and involves working with all stakeholders. With over 100 years of development experience, the effects of projects have been relatively well documented and studied. (See Box 11.1.) Modern construction of hydropower plants tries to include in the system design several approaches that minimize social and ecological impacts. Some of the most important impacts are changes in habitat, fish stocks and other species, sedimentation, water quality, and downstream flow regimes. Hydropower reservoirs may also create opportunities for ecological services, tourism, fisheries, irrigation, and secured water supply.

Of note are advances made in proactive avoidance and reduction of negative effects prior to construction of hydropower projects, particularly in the past 20 years in response to changing societal values, increased public scrutiny, and evolving environmental and social

Table 11.17 | Cost of electricity as a function of capacity factor, turnkey investment costs, discount rate, and O&M costs. The O&M costs are assumed as US¢0.3–0.7/kWh for larger-scale systems and as US¢0.3–1.0/kWh for smaller systems. The economic lifetime of a hydro system is assumed to be 40–80 years.

Capacity factor	Turnkey investment costs per kWe installed (*plant > 100 MW*)	Discount rate	
		5%	10%
	US$	US¢/kWh	
80%	1000	1.4–1.8	2.0–2.4
	3000	3.5–3.9	5.3–5.7
30%	1000	2.5–2.9	4.2–4.6
	3000	6.9–7.3	11.9–12.3

Capacity factor	Turnkey investment costs per kWe installed (*plant < 10 MW*)	Discount rate	
		5%	10%
	US$	US¢/kWh	
80%	1300	1.4–2.1	2.3–2.9
	5000	4.4–5.1	7.6–8.3
30%	1300	3.2–3.9	5.3–6.0
	5000	11.3–12.0	19.7–20.4

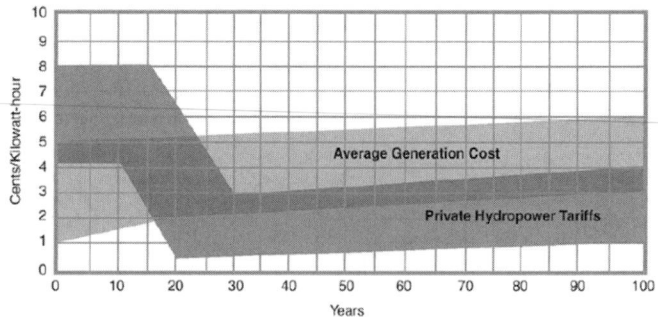

Figure 11.23 | Hydropower tariff versus average electricity generation costs in US$_{2008}$¢/kWh. Source: Trouille and Head, 2008.

standards. Managing the positive and negative environmental, social, and economic effects of a given project at local, national, regional, and even international levels remains complex.

11.3.6.1 Sustainability Assessment Protocol

The hydropower sector has been actively addressing issues of sustainability for more than a decade through engagement with key stakeholders. This has led the International Hydropower Association to develop Sustainability Guidelines (2004) and a Sustainability Assessment Protocol (2006) to guide and measure the sustainability performance of a hydropower project from conception to operation. The Hydropower Sustainability Assessment Forum in 2008 was formed as a direct result of these efforts. The Forum contains key stakeholders: representatives of industrial and developing countries, environmental and social NGOs, commercial and development banks, and the hydropower industry. A 2010 Forum-reviewed version of the Protocol adopted by IHA has now moved into the implementation phase. It measures performance against relevant criteria over four stages of a project. (See Figure 11.24.)

The consensus among key stakeholders on how to measure project sustainability paves the way for improved sustainability performance and more informed decision-making on hydropower development, particularly in the policy and financial arenas.

11.3.6.2 GHG Status of Reservoir Hydropower

As a renewable energy source, hydropower is recognized as being both clean and a low-carbon technology. A life-cycle assessment (LCA)

published by the International Energy Agency in 2000 suggests that GHG emissions could range from 2–48 gCO_2-eq/kWh for reservoir type systems and from 1–18 gCO_2-eq/kWh for run-of-river systems (IEA, 2000). An overview published by the IPCC concludes that "the majority of lifecycle GHG emissions estimates for hydropower cluster between about 4–14 gCO_2-eq/kWh, but under certain scenarios there is the potential for much larger quantities of GHG emissions" (Kumar et al., 2011).

In conclusion, there is uncertainty and no consensus yet on these figures among experts (or, separate from the numbers, on LCA methodology for hydropower). In some circumstances freshwater reservoirs can produce highly elevated GHG emissions, a phenomenon that is the subject of ongoing research. Specific problems have included a lack of standard measurement techniques, limited reliable information from a sufficient variety of sources, and the lack of standard tools for assessing net GHG exchanges (Goldenfum, 2011).

The lack of scientific consensus has impeded progress in decision-making on carbon accounting and in carbon markets. For example, guidance is needed to support national GHG inventories, to develop methods (measurement and predictive modeling) of establishing the GHG footprint of new reservoirs (hydro, multipurpose, and non-hydro alike), and to quantify more precisely the carbon offsets of hydropower projects for GHG emissions trading. These circumstances led to the establishment of the UNESCO/IHA GHG Status of Freshwater Reservoirs Research Project in 2008 (Goldenfum, 2011). The project's main goals are:

- Developing, through a consensus-based, scientific approach, detailed measurement guidance for net GHG assessment

- Promoting scientifically rigorous field measurement campaigns and the evaluation of net emissions from a representative set of freshwater reservoirs throughout the world

- Building a standardized, credible set of data from these representative reservoirs

Box 11.1 | Sustainable Hydropower in Nepal

Source: IHA, 2005.

The Andhikhola Hydel & Rural Electrification Hydro Scheme in western Nepal was built in 1991 with technical and financial assistance from the Norwegian Development Agency.

A concrete gravity diversion weir on the Andhikhola River diverts water through a 1.3 km long tunnel and a 234 m vertical drop shaft. The 5 MW powerhouse is currently equipped with secondhand turbines previously used in Norway. The tunnel system also diverts water for irrigation. The opportunity was taken during construction to build Nepalese experience in tunneling technology as well as other areas of technical capability. Various elements of the scheme have demonstrated exceptional innovation to fit with aspects of capacity building, resource availability, and the remoteness of the site.

Some 100,000 people in the region now enjoy the benefits of an electricity supply for the first time. With the available power, more than 200 small enterprises have been established, creating employment for hundreds of people.

- Developing predictive modeling tools to assess the GHG status of unmonitored reservoirs and potential sites for new reservoirs

- Developing guidance and assessment tools for mitigation of GHG emissions at sites vulnerable to high net emissions.

The project has benefited from collaboration among some 160 researchers, scientists, and professionals working in the field, representing more than 100 institutions. In July 2010 the project met its first goal by publishing *GHG Measurement Guidelines for Freshwater Reservoirs*, which represents the state of the art in measurement guidance for net GHG assessment of reservoirs (Goldenfum, 2011).

11.3.7   Implementation Issues

Financing is by far the biggest obstacle for scaling up hydropower development. Barriers are especially high in the poorest countries. It is thus important for the public and private sectors to work together to reduce risk profiles and unlock finance.

A substantial opportunity to adapt is provided by the economic dimension of climate change policy. The important role that climate change financing can play is indicated by the Clean Development Mechanism (CDM): hydropower accounts for 30% of all registered CDM projects and 48% of all registered renewable energy projects (UNEP Risø, 2011). The additional income generated for projects is generally at levels below 5% of a project's internal rate of return, a stream of revenue not normally categorized as influential. Despite this, such revenues have strong impacts on overall project feasibility. An important factor is that this revenue will be in "hard currency" and can mitigate exchange rate risks for the developer. Further climate change policy and financing innovations to target the sustainable development of developing countries may increase the appetite for investment. For example, the addition of adaptation-driven multilateral and bilateral finance looks likely to emerge as a significant lever for further hydropower development.

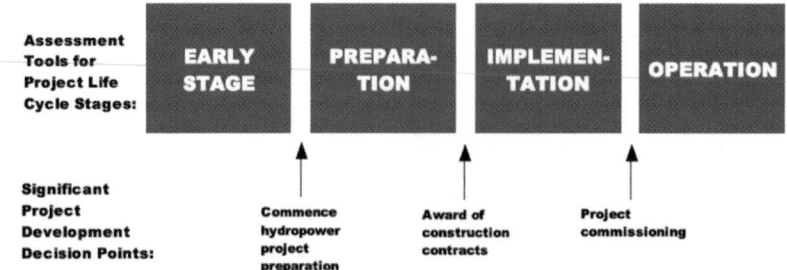

ES - Early Stage	P - Preparation	I - Implementation	O - Operation
ES-1 Demonstrated Need	P-1 Communications & Consultation	I-1 Communications & Consultation	O-1 Communications & Consultation
ES-2 Options Assessment	P-2 Governance	I-2 Governance	O-2 Governance
ES-3 Policies & Plans	P-3 Demonstrated Need & Strategic Fit		
ES-4 Political Risks	P-4 Siting & Design		
ES-5 Institutional Capacity	P-5 Environmental & Social Impact Assessment & Mgmt	I-3 Environmental & Social Issues Mgmt	O-3 Environmental & Social Issues Mgmt
ES-6 Technical Issues & Risks	P-6 Integrated Project Management	I-4 Integrated Project Management	
ES-7 Social Issues & Risks	P-7 Hydrological Resource		O-4 Hydrological Resource
ES-8 Environmental Issues & Risks			O-5 Asset Reliability & Efficiency
ES-9 Economic & Financial Issues & Risks	P-8 Infrastructure Safety	I-5 Infrastructure Safety	O-6 Infrastructure Safety
	P-9 Financial Viability	I-6 Financial Viability	O-7 Financial Viability
	P-10 Project Benefits	I-7 Project Benefits	O-8 Project Benefits
	P-11 Economic Viability		
	P-12 Procurement	I-8 Procurement	
	P-13 Project Affected Communities & Livelihoods	I-9 Project Affected Communities & Livelihoods	O-9 Project Affected Communities & Livelihoods
	P-14 Resettlement	I-10 Resettlement	O-10 Resettlement
	P-15 Indigenous Peoples	I-11 Indigenous Peoples	O-11 Indigenous Peoples
	P-16 Labour & Working Conditions	I-12 Labour & Working Conditions	O-12 Labour & Working Conditions
	P-17 Cultural Heritage	I-13 Cultural Heritage	O-13 Cultural Heritage
	P-18 Public Health	I-14 Public Health	O-14 Public Health
	P-19 Biodiversity & Invasive Species	I-15 Biodiversity & Invasive Species	O-15 Biodiversity & Invasive Species
	P-20 Erosion & Sedimentation	I-16 Erosion & Sedimentation	O-16 Erosion & Sedimentation
	P-21 Water Quality	I-17 Water Quality	O-17 Water Quality
		I-18 Waste, Noise & Air Quality	
	P-22 Reservoir Planning	I-19 Reservoir Preparation & Filling	O-18 Reservoir Management
	P-23 Downstream Flow Regimes	I-20 Downstream Flow Regimes	O-19 Downstream Flow Regime

Figure 11.24 | IHA Sustainability Assessment Protocol assessment tools and major decision points, with topics by section. Source: based on IHA, 2010b.

The sustainability performance of hydropower projects remains an ongoing challenge, but this is becoming more manageable with the efforts the sector continues to make at national, regional, and international levels, such as the Hydropower Sustainability Assessment Protocol.

The construction and operation of hydropower plants in transboundary river basins requires international cooperation to avoid conflicts over water – particularly as a changing hydrological cycle brought about by climate change is likely to increase pressure on water resources in some river basins.

As hydropower can cover base load and peak load demands, its integration into transmission systems can balance the output from variable systems such as wind and solar photovoltaics and can increase the economic value of produced electricity (US DOE, 2004).

In rural areas, smaller-scale hydropower may be used alone or connected to a mini-grid to provide electricity or mechanical power for local industrial, agricultural, and domestic uses. Depending on local circumstances, it can play a major role in rural electrification, as has been demonstrated in China and other developing countries. Although

Table 11.18 | Total geothermal use in 2009.

Use	Installed capacity [GW]	Electricity production / direct use [TWh/yr]	Capacity factor	Number of countries reporting
Electric Power	10.7	67.3	0.72	24
Direct Use	48.5	117.8	0.28	78

Source: Lund et al., 2010; Bertani, 2010.

Table 11.19 | Geothermal energy use in 2009 by continent.

Region	Electric Power			Direct Use		
	Installed capacity	Electricity production	Number of countries reporting	Installed capacity	Energy use	Number of countries reporting
Africa	1.6%	2.1%	2	0.3%	0.7%	7
Americas	42.6%	39.0%	6	30.1%	19.0%	15
Asia	34.9%	35.1%	6	28.7%	34.9%	16
Europe	14.5%	16.2%	7	40.0%	43.1%	37
Oceania	6.4%	6.7%	3	0.9%	2.3%	3

Source: Lund and Bertani, 2010.

support of these smaller-scale/rural, often financially challenged projects is desirable, the categorization of hydropower by size in policy and markets in an unsophisticated fashion can distort development outcomes.

11.4 Geothermal Energy

11.4.1 Introduction

Geothermal energy has been used for thousands of years for washing, bathing, and cooking. But it was only in the twentieth century that geothermal energy was harnessed on a large scale for space heating, electricity production, and industrial use. The first large municipal district heating service started in Iceland in the 1930s, and it now provides geothermal heat to about 99% of the 200,000 residents of Reykjavik.

The use of geothermal energy has increased rapidly since the 1970s. In the first decade of the new century, the globally installed direct use capacity tripled from 15 to nearly 50 GW_{th} whereas the installed capacity for electricity production increased from 8.0 to 10.7 GW_e.

11.4.2 Potential of Geothermal Energy

At present, geothermal energy is used by 78 countries for heating purposes (called "direct use") and by 24 countries for electricity production (Lund et al., 2010; Bertani, 2010). Table 11.18 presents the worldwide geothermal electric and direct-use capacity as well as the generated amount of heat and electricity in 2009, based on 70 country papers at the World Geothermal Congress 2010.

The figures for electric power capacity (MW_e) and annual generation values (GWh) appear to be fairly accurate. The direct-use figures are less reliable due to reporting errors and lack of data from some countries. Table 11.19 reports on geothermal energy use by continent.

The flow of heat from Earth's interior to its surface is 1400–1500 EJ/yr, with about 315 EJ/yr onshore (Stefansson, 2005; see also Chapter 7). The upper limit for the technical potential to use energy from geothermal resources is estimated at 50–60 TW_{th} (Stefansson, 2005). Part of these resources can be used to produce 1–2 TW_e electricity. Most resources, however, are suitable for direct use only, giving access to 22–44 TW_{th}. The technical potential for electricity production can be enhanced if the heat from hot dry rocks can be exploited using enhanced geothermal systems (EGS) technology. This may enlarge the technical potential for electricity production by about a factor of 5–10 (see, e.g., Tester et al., 2006). This may result in a total technical potential of about 700 EJ/yr; the economic potential in 2050 might be as high as 75 EJ/yr (see Chapter 7). The conversion efficiency for electricity generation is around 10% and for direct use nearly 100%.

11.4.3 Market Developments

In 1975 only 10 countries reported electrical production from or direct use of geothermal energy. By now the figure is 78 countries. At least another 10 countries are actively exploring for geothermal resources and are expected to be online by 2015 (Bertani, 2010; Lund and Bertani, 2010).

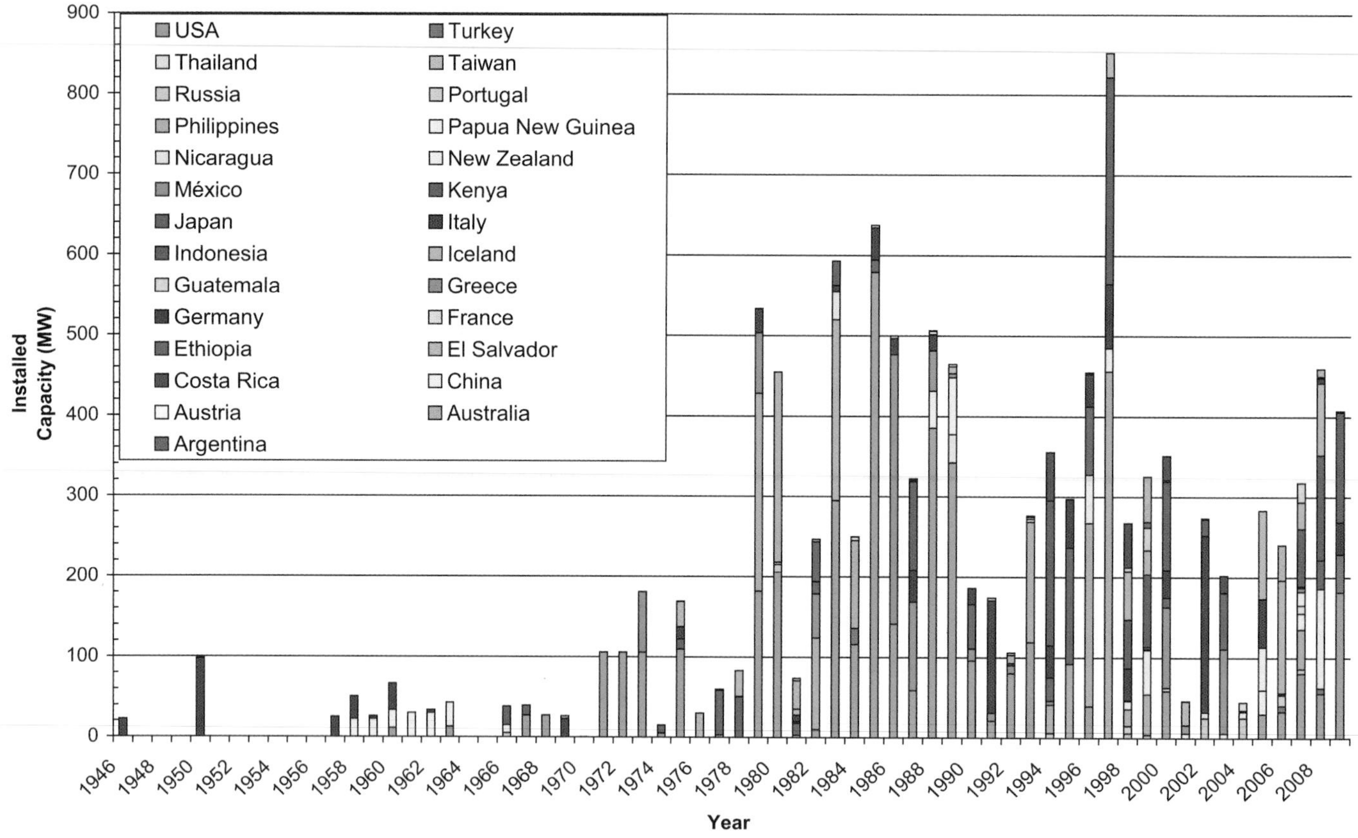

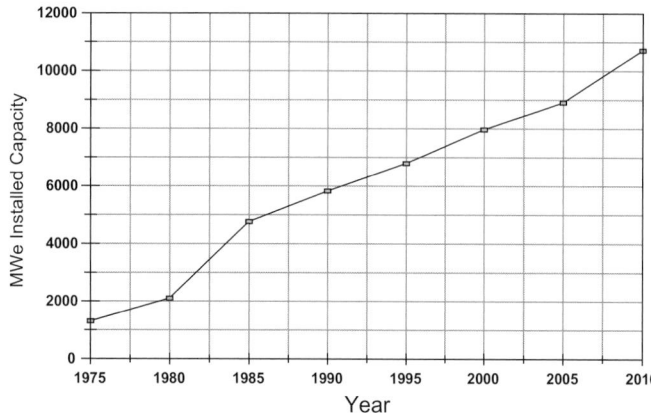

Figure 11.25 | Installed geothermal electricity capacity, 1946–2010, by country and in total. Source: data from Bertani, 2010.

Figure 11.26 | Growth of the globally installed capacity of geothermal power production. Source: Bertani, 2010.

11.4.3.1   Geothermal Electricity Production

Use of high-temperature geothermal energy for electric power production started experimentally at Larderello, Italy, in 1904; the first commercial plant (250 kW$_e$) became available in 1913 and was connected to the electricity grid. In 1958, the Wairakei A station of 69 MWe came online in New Zealand. This was the first "wet steam" plant in the world. It was followed by plants using "dry steam" at Pathè, Mexico, in 1959 (3.5 MW$_e$); at The Geysers in the United States in 1960 (12 MW$_e$);

and at Matsukawa in Japan in 1966 (23 MW$_e$). The first low-temperature plant using a binary (organic Rankine) cycle plant was opened in 1967 at Paratunka, Kamchatka, in Siberia (680 kW$_e$).

Figure 11.25 shows the annual installed capacity in the 27 countries that had initiated geothermal power production by 2010, starting in 1946. The worldwide installed capacity since 1975 is presented in Figure 11.26. The average growth rate between 1975 and 2010 has been 6.5% per year. Since 2005, major increases have occurred in El Salvador, Iceland, Indonesia, New Zealand, Turkey, and the United States.

The top 10 countries in terms of installed capacity of geothermal power plants in 2009 were the United States, The Philippines, Indonesia, Mexico, Italy, New Zealand, Iceland, Japan, El Salvador, and Kenya or Costa Rica. (See Figure 11.27.) If the criterion were the percentage contribution of geothermal plants to the total generating capacity of the country or region, however, the top 10 would be Lihir Island (Papua New Guinea), Tibet, San Miguel Island (Azores), El Salvador, Tuscany (Italy), Iceland, Kenya, the Philippines, Nicaragua, and Guadeloupe (Caribbean).

11.4.3.2   Direct Use of Geothermal Energy

Since 1975 the installed capacity of geothermal direct use has increased at an average rate of about 10% annually (see Figure 11.28). In the

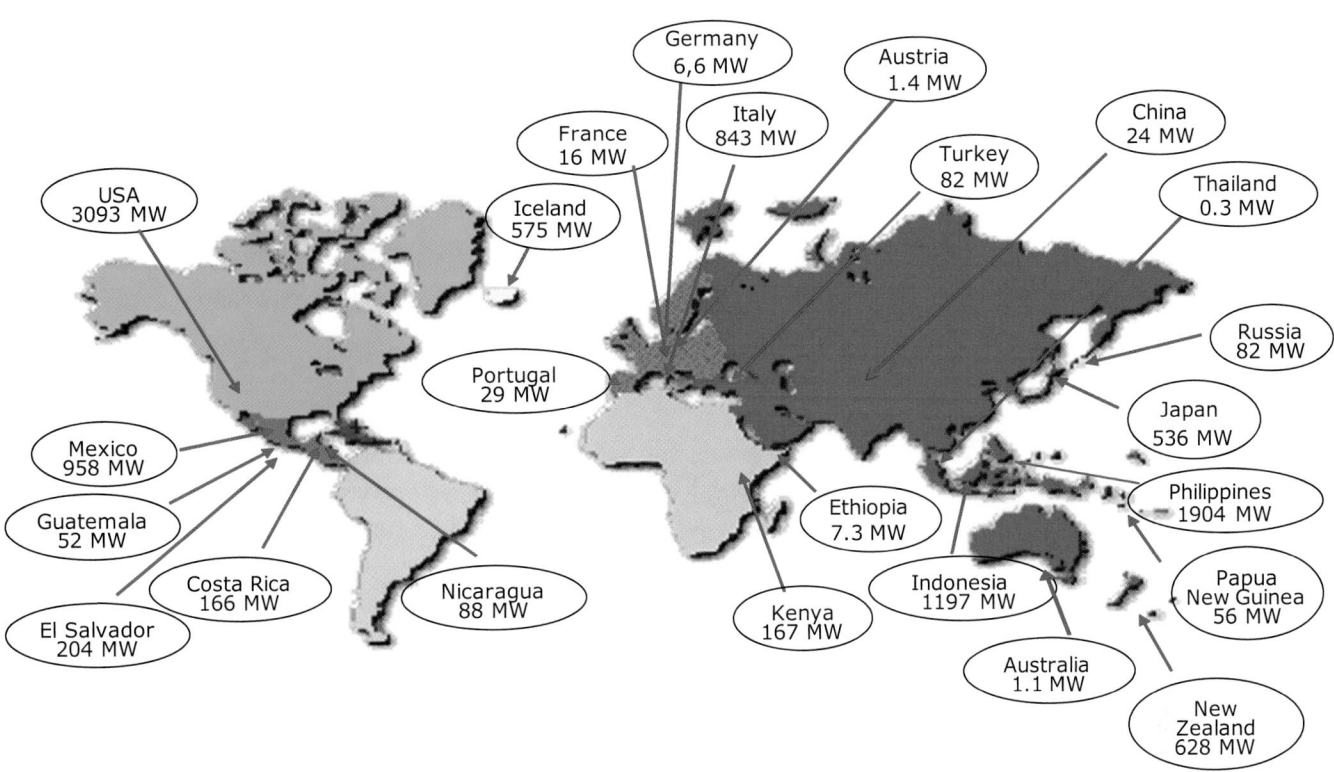

Figure 11.27 | Mapping of the 10.7 GW installed geothermal electric capacity in 2009. Source: data from Bertani, 2010.

period 2005–2009, heat production for direct use grew nearly 12% a year. Ground-source heat pumps (GHPs) alone have increased heat production about 20% per year in this period. This is due in part to the ability of geothermal heat pumps to use groundwater or ground-coupled temperatures anywhere in the world.

The annual savings achieved by geothermal energy use amounted to 36 million tons of fuel oil and 32 million tonnes of carbon emissions compared with power production from fuel oil (Lund et al., 2010).

In 2009, the top five countries with the largest installed direct-use capacity were the United States, China, Sweden, Norway, and Germany, accounting for about 63% of the total capacity worldwide. The five countries in 2009 with the largest direct use were China, the United States, Sweden, Turkey, and Japan, accounting for 55% of the world geothermal energy use.

Looking at the data in terms of the country's land area or population, however, the smaller countries dominate. In terms of GJ/km^2, the top five countries in 2009 were the Netherlands, Switzerland, Iceland, Norway, and Sweden. In terms of GJ/capita, the ranking was Iceland, Norway, Sweden, Denmark, and Switzerland. In Iceland, geothermal energy saves about US$100 million in imported oil (Lund et al., 2010).

The largest increases in direct use in terms of GJ/yr over the past five years were achieved in the United Kingdom, the Netherlands, South Korea, Norway, and Ireland, while the largest increases in installed

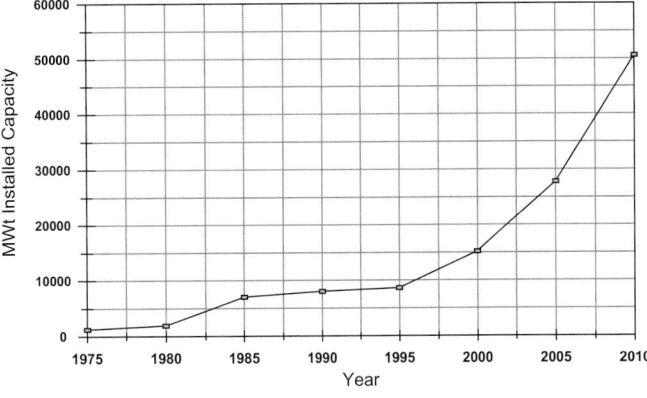

Figure 11.28 | Worldwide growth of installed capacity of geothermal direct use. Source: Lund et al., 2010.

capacity (MW$_{th}$) occurred in the United Kingdom, South Korea, Ireland, Spain, and the Netherlands, mainly due to an increased use of geothermal heat pumps.

In 1990 there were 14 countries reporting an installed capacity of over 100 MW$_{th}$. By 2000 this number had increased to 23, by 2005 it was 33, and by 2009 it reached 36. (See Box 11.2 for more details on some of these countries.)

Table 11.20 provides data on direct geothermal energy use in terms of installed capacity and thermal energy utilization for several recent years.

Box 11.2 | National Developments in Geothermal Direct Use as of 2009

- Iceland: Geothermal direct use meets 89% of the country's space heating needs.

- Japan: Over 2000 resorts, over 5000 public bath houses, and over 15,000 hotels visited by 15 million guests per year use natural hot springs.

- Tunisia: Geothermal heating of greenhouses has increased from 100 ha to 194 ha over the last five years.

- Turkey: Geothermal space heating has increased 40% in the five years, supplying 201,000 equivalent residences; 30% of the country will be heated with geothermal energy in the future.

- Switzerland: The country has installed 60,000 geothermal heat pumps; this is $1/km^2$. Also, 2000 km of boreholes were drilled in 2009. Drain water from tunnels is used to heat nearby villages. Several geothermal projects have been developed to melt snow and ice on roads.

- United States: The country has installed 1 million geothermal heat pumps, mainly in the midwestern and eastern states, with about 12% annual growth. Around 100,000 to 120,000 units are installed per year.

Table 11.20 | Various geothermal direct-use categories worldwide, 1995–2009.

Direct-Use Category	Installed capacity [GW$_{th}$]				Thermal energy utilization [PJ/year]			
	1995	2000	2005	2009	1995	2000	2005	2009
Geothermal Heat Pumps	1.85	5.28	15.38	33.13	14.6	23.3	87.5	200.2
Space Heating	2.58	3.26	4.37	5.40	38.2	42.9	55.3	63.0
Greenhouse Heating	1.09	1.25	1.40	1.54	15.7	17.9	20.7	23.3
Aquaculture Pond Heating	1.10	0.61	0.62	0.65	13.5	11.7	11.0	11.5
Agricultural Drying	0.07	0.07	0.16	0.13	1.1	1.0	2.0	1.6
Industrial Uses	0.54	0.47	0.48	0.53	10.1	10.2	10.9	11.7
Bathing and Swimming	1.09	3.96	5.40	6.70	15.7	79.5	83.0	109.4
Cooling / Snow Melting	0.12	0.11	0.37	0.37	1.1	1.1	2.0	2.1
Others	0.24	0.14	0.09	0.04	2.2	3.0	1.0	1.0
Total	8.66	15.15	28.27	48.49	112.4	190.7	273.4	423.8

Source: Lund et al., 2010.

The most recent use of low-grade geothermal energy is in the form of ground-source heat pumps that use the natural temperature of the Earth (between 5° and 30°C) to produce both heating and cooling with a limited amount of electric energy input. The first commercial building installation using a groundwater heat pump took place in Portland, Oregon, in 1946. Europe began using the technology around 1970. In 2009 heat pumps were the largest portion of the installed direct-use capacity (70%) and contributed 49% to the direct use of geothermal energy. The actual number of installed GHP units is around three million. Units are found in 43 countries, although they are mainly in the United States, Canada, and Europe (Lund et al., 2010).

11.4.3.3 Development of Enhanced Geothermal Systems

While conventional geothermal resources cover a wide range of uses for power production and direct uses in profitable conditions, a large part of the scientific and industrial community has been involved for more than 20 years in promoting enhanced geothermal systems (Ledru et al., 2007; Fridleifsson et al., 2008; Tester et al., 2006). The principle of EGS is simple: in the deep subsurface where temperatures are high enough for power generation (above 150°C), an extended fracture network is created or enlarged to act as new pathways. Water from the deep wells, or cold water from the surface, is circulated through this deep reservoir using injection and production wells and is then recovered as steam or hot water (Fridleifsson et al, 2008). These wells and further surface installations complete the circulation system. After using energy for power generation, the fluid can be cascaded for direct-use applications such as for district heating. EGS plants, once operational, can be expected to have great environmental benefits (such as zero CO_2 emissions).

The enhancement challenge is based on several conventional methods for exploring, developing, and exploiting geothermal resources that are not economically viable yet. The original idea calls for general applicability, since the temperature increases with depth everywhere. But a number of basic problems still need to be solved; mainly, techniques need to be developed for creating, characterizing, and operating the deep fracture systems. Some environmental issues, like the chance of triggering seismicity and the availability of surface water, also need careful assessment and management (Bertani, 2009).

Targeted EGS demonstrations are under way in several places: Australia can claim a large-scale activity through a number of stock market-registered enterprises. A real boom can be observed: 19 companies are active in 140 leases (a total of 67,000 km^2 in four states), with an investment volume of US$650 million. The project developers plan to establish the first power plants (with a few MWe of capacity) in the coming years (Beardsmore, 2007). The EU project EGS Pilot Plant in Soultz-sous-Forêts, France, started in 1987 and has installed a power plant of 1.5 MW$_e$ to use the enhanced fracture permeability at 200°C. In Landau, Germany, the first EGS-plant with a capacity of 2.5–2.9 MW$_e$ went into operation in fall 2007 (Baumgärtner et al., 2007). Another approach is being made for deep sediments in the in-situ geothermal laboratory in Groß Schönebeck, Germany, using two research wells (Huenges et al., 2007).

One of the main future demonstration goals in EGS will be to see whether and how power plant size could be upscaled to several tens of MW$_e$. In the United States, the potential for EGS power generation is estimated at 1250 GW$_e$ (based on a conservative 2% recovery factor). Assuming a capacity factor of 90%, this would correspond to 35.4 EJ/yr of electricity production (Tester et al., 2006).

11.4.3.4   New Developments: Drilling for Higher Temperatures

Production wells in high-temperature fields are commonly 1.5–2.5 km deep, and the production temperature is 250–340°C. The energy output from individual wells is highly variable, depending on the flow rate and the heat content of the fluid, but it is commonly in the range 5–10 MW$_e$ and rarely over 15 MW$_e$ per well.

It is well known from research on eroded high-temperature fields that much higher temperatures are found in the roots of the high-temperature systems. The International Iceland Deep Drilling Project is a long-term program to harness deep unconventional geothermal resources (Fridleifsson et al., 2007). Its aim is to produce electricity from natural supercritical hydrous fluids (at high temperature and pressure above the critical point where there is a phase change) from drillable depths. Producing supercritical fluids will require drilling wells that produce temperatures of 450–600°C. The current plan is to drill and test at least three 3.5–5 km deep boreholes in Iceland within the next few years. A deep well producing 0.67 m^3/sec steam (~2400 m^3/h) from a reservoir with a temperature significantly above 450°C could yield enough steam to generate 40–50 MW of electric power. This exceeds by an order of magnitude the power typically obtained from conventional geothermal wells (Fridleifsson et al. 2007).

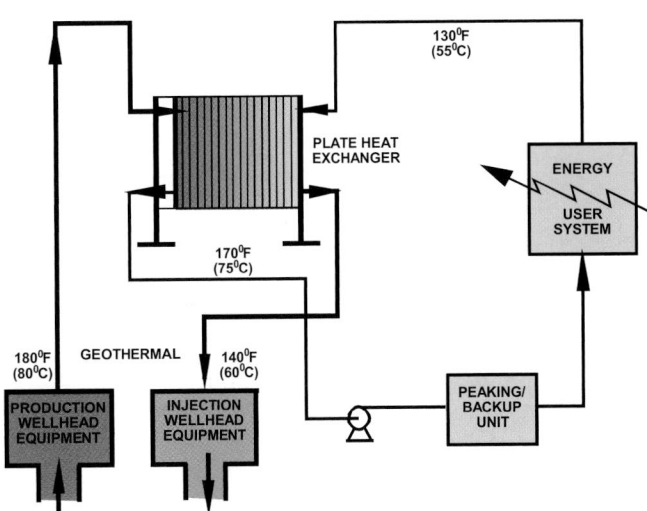

Figure 11.29 | Typical direct use geothermal heating system configuration. Source: based on Geo-Heat Center, 1998.

11.4.4   Geothermal Energy Technology Developments

11.4.4.1   Direct Use

The main advantage of using geothermal energy for direct use projects in the low- to intermediate-temperature range is that these resources are more widespread and exist in at least 80 countries at economic drilling depths. In addition, there are no conversion efficiency losses, and projects can use conventional water-well drilling and off-the-shelf heating and cooling equipment. Most projects can be online in less than a year. Projects can be on a small scale such as for an individual home, single greenhouse, or aquaculture pond, but they can also be a large-scale commercial operation such as for district heating/cooling, food and lumber drying, and mineral ore extraction.

It is often necessary to isolate the geothermal fluid from the user side to prevent corrosion and scaling. Care must be taken to prevent oxygen from entering the system (geothermal water normally is oxygen-free), and dissolved gases and minerals such a boron and arsenic must be removed or isolated, as they are harmful to plants and animals. Hydrogen sulfide, even in low concentrations, will cause problems with copper and solder and is harmful to humans. On the other hand, carbon dioxide, which often occurs in geothermal water, can be extracted and used for carbonated beverages or to enhance growth in greenhouses. The typical equipment for a direct-use system is illustrated in Figure 11.29; it includes downhole and circulation pumps, heat exchangers (normally the plate type), transmission and distribution lines (normally insulated pipes), heat extraction equipment, peaking or back-up plants (usually fossil-fuel-fired), and fluid disposal systems (injection wells). Geothermal energy can usually meet 80–90% of the annual heating or cooling demand, yet it is only sized for 50% of the peak load (Lund, 2005).

A well-known major example of geothermal direct-use is the district heating system in Reykjavik, Iceland; in 1930, some official buildings and about 70 private houses received hot water from geothermal wells close to old thermal springs. The results were so encouraging that other geothermal fields were explored near the city. Now 52 wells produce 2400 liters/s of water at a temperature of 62–132°C. Later the municipal district heating agency added a high-temperature field about 27 km away. Today the geothermal water from the wells flows through pipelines to six large reservoir tanks and then to six storage tanks in downtown Reykjavik that hold 24 million liters. Nine pumping stations distribute the water to consumers.

Reykjavik Energy uses either a single or a double distribution system. In the double system, the used geothermal water from radiators runs back from the consumer to the pumping stations. There it is mixed with hotter geothermal water, which serves to cool the water to the proper 80°C before it is recirculated. In the single system, the backflow drains directly into the sewer system (Gunnlaugsson and Gíslason, 2003). The system has over 2000 km of pipelines and an installed capacity of 830 MW$_{th}$. A fossil fuel peaking station is used to increase the fluid temperature in extremely cold weather (Lund, 2005; Ragnarsson, 2010).

11.4.4.2   Geothermal Heat Pumps

Geothermal (ground-source) heat pumps use the relatively constant temperature of Earth to provide heating, cooling, and domestic hot water for homes, schools, government, and commercial buildings. A small amount of electricity input is required to run a compressor (approximately 25% of a normal baseboard electric heating system). However, the energy output is about four times the energy input in electricity form, described as the coefficient of performance (COP) of 4.0. These "machines" cause heat to flow "uphill" from a lower to a higher temperature location. "Pump" is used to describe the work done. The temperature difference is called "lift." The greater the lift, the greater the energy input required. The technology is not new, as Lord Kelvin developed the concept in 1852. GHPs gained commercial popularity in the 1960s and 1970s (Lund et al., 2003).

GHPs come in two basic configurations: ground-coupled (closed loop) systems, which are installed in the ground, and groundwater (open loop) systems, which are installed in wells and lakes. The type chosen depends on the soil and rock type at the installation, the land available, and whether a water well can be drilled economically or is already on-site. In the ground-coupled system, a closed loop of pipe, placed either horizontally (1–2 m deep) or vertically (50–70 m deep) is placed in the ground, and a water-antifreeze solution is circulated through the plastic pipes (high-density polyethylene) to either collect heat from the ground in the winter or reject heat to the ground in the summer (Rafferty, 2008). The open loop system uses groundwater or lake water directly in the heat exchanger and then discharges it into another well, into a stream or lake, or on the ground (say, for irrigation), depending on local laws.

The efficiency of GHP units is described by the coefficient of performance in the heating mode and by the energy efficiency ratio in the cooling

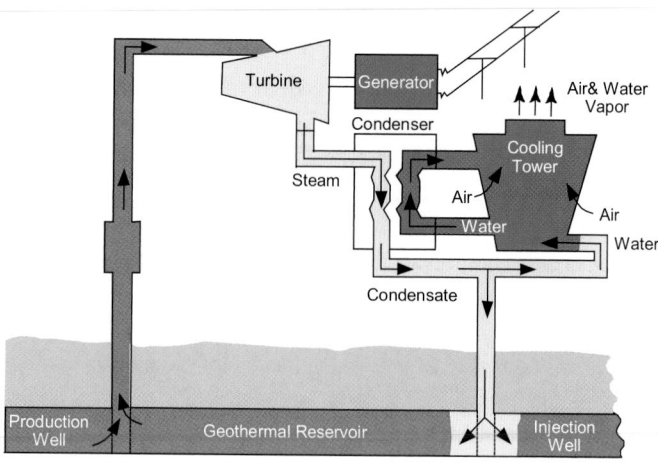

Figure 11.30 | Steam plant using a vapor or dry steam dominated geothermal resource. Source: modified from Lund, 2007.

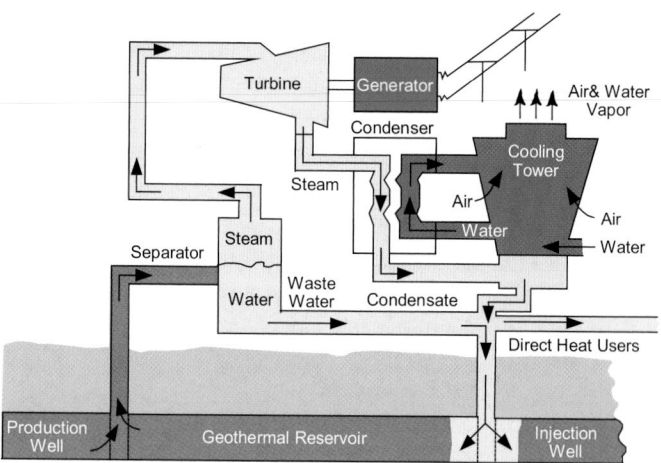

Figure 11.31 | Single-stage flash steam plant using a water-dominated geothermal resource with a separator to produce steam. Source: modified from Lund, 2007.

mode (COP$_h$ and COP$_c$, respectively, in Europe), which is the ratio of the output thermal energy divided by the input energy (electricity for the compressor). The higher the number, the better the efficiency. The ratio varies from three to six with present equipment. In comparison, an air-source heat pump has a COP of around two and depends on backup electrical energy to meet peak heating and cooling requirements (Lund et al., 2003; Curtis et al., 2005; Bertani, 2010).

11.4.4.3   Electric Power Generation

Geothermal power is generated by using steam or a hydrocarbon vapor to turn a turbine-generator set to produce electricity. A vapor-dominated (dry steam) resource (see Figure 11.30) can be used directly, but a hot water resource (see Figure 11.31) needs to be flashed by reducing the pressure to produce steam, normally in the 15–20% range. Some plants use double and triple flash to improve efficiency (IEA, 2010a). In the

Table 11.21 | Average capacity and electricity produced per plant category and the share of each category.

Type of plant	Average capacity per unit (MW)	Average electricity production per unit (GWh/yr)	Share in number of plants	Share in total capacity (MWe)	Share in electricity produced (GWh)
Binary plant	5	27	44%	4%	4%
Back Pressure plant	6	50	5%	5%	6%
Single Flash plant	31	199	27%	25%	26%
Double Flash plant	34	236	12%	28%	30%
Dry Steam plant	46	260	12%	38%	34%

Source: Bertani, 2010.

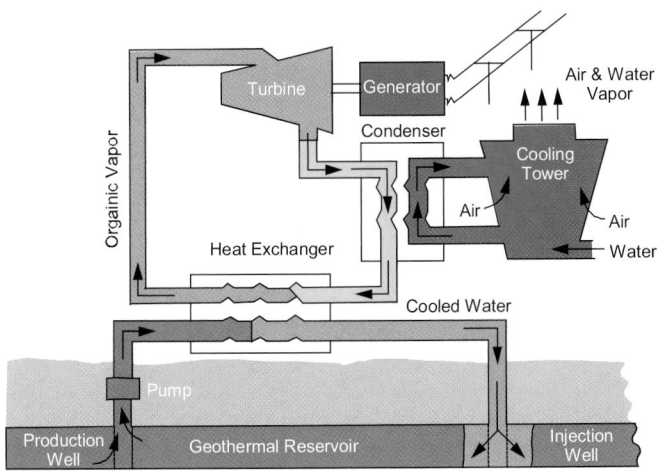

Figure 11.32 | Binary power or organic Rankine cycle plant using a low temperature geothermal resource and a secondary fluid of a low boiling-point hydrocarbon. Source: modified from Lund, 2007.

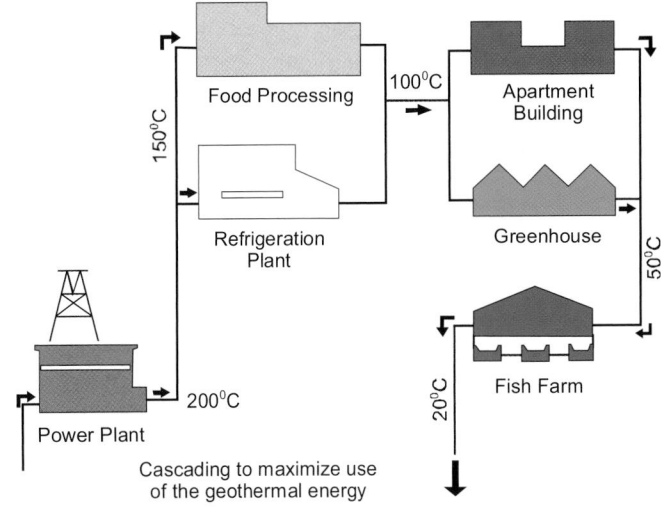

Figure 11.33 | Cascading the use of a geothermal resource for multiple applications. Source: Geo-Heat Center, 2003.

case of low-temperature resources, generally below 180°C, a secondary low boiling point fluid (typically a hydrocarbon) is needed to generate the vapor, resulting in a binary or organic Rankine cycle plant. (See Figure 11.32.)

Usually a wet or dry cooling tower is used to condense the vapor after it leaves the turbine to maximize the temperature and pressure drop between the incoming and outgoing vapor and thus increase the efficiency of the operation. Dry cooling is often used in arid areas where water resources are limited. Air cooling normally has lower efficiencies in summer months, when air temperatures are high and humidity is low.

The standard plant classifications are used here: *binary*, *back pressure*, *single flash*, *double flash*, and *dry steam* plant. Back pressure units (noncondensing) are used where condensing water is not available or based on economics to get the unit on line sooner. Table 11.21 indicates the share of each category in the total installed capacity, the electricity produced annually, and the total number of units. Average values per unit for installed capacity and annually produced electricity are also given. Hybrid plants using more than one form of energy are excluded from this overview, but their share is at present almost zero.

As indicated, there are three major families of plants: *small plants* (binary and back pressure plants) of about 5 MW per unit, *medium plants* (single and double flash plants) of around 30 MW per unit, and *big plants* (dry steam plants) of around 45 MW per unit or larger.

More recently, the use of combined heat and power plants has made low-temperature resources and deep drilling more economical. District heating using the spent water from a binary power plant can make a marginal project economical, as demonstrated at Neustadt-Glewe, Landau, and Bad Urach in Germany and at Bad Blumau in Austria. This was also found for high-temperature combined heat and power plants in Iceland (Geo-Heat Center, 2005). Options for cascading are shown in Figure 11.33, where the geothermal fluid is used for a number of applications at progressively lower temperatures to maximize energy use.

11.4.5   Economic and Financial Aspects

The economics of electricity production are influenced by drilling costs and resource development. The typical capital expenditure quota is 40% for reservoir and 60% for plant. The productivity of electricity per well is a function of the thermodynamic characteristics of the reservoir fluid (phase and temperature). The higher the energy content of the reservoir fluid, the lower is the number of required wells; as a consequence, the reservoir typical capital expenditure quota is reduced. Single geothermal wells can produce from 1 to 5 MW_e, but sometimes even as high as 30 MW_e.

The cost of geothermal projects and the production of energy carriers vary considerably from site to site and from region to region, depending mainly on the depth, quality, quantity, and location of the resource. For any geothermal project, the costs can be divided into land acquisition or leasing; resource exploration and characterization; drilling and reservoir development; gathering and transmission pipelines; plant design and construction; energy or product transmission to consumers; operation and maintenance; the cost of financing, debt, and royalty payments; and costs related to permitting, legal, and institutional issues.

In this section, the levelized cost of energy is calculated for a number of applications, assuming a discount rate of 5–10%.

11.4.5.1   Electric Power Projects

Taking into account cost increases during 2006–2008, a new 50 MWe greenfield project costs in the range of US$2000–4000 per installed kW_e (see, e.g., Geothermal Task Force Report, 2006; Bromley et al., 2010). Thus, a 50 MWe plant is estimated to cost on average US$150 million. Of this, drilling would average US$1500/$kW_e$ or about US$2200/m for a 3 MW, 2000m deep well. Average well cost are estimated to vary from US$2–5 million, though it can approach US$9 million (Kagel, 2006).

A typical cost breakdown for a geothermal power project in the United States is as follows: exploration 5%, confirmation 5%, permitting 1%, drilling 23%, stream gathering 7%, power plant 54%, and transmission 4% (Hance, 2005). O&M costs can be estimated at 1.8–2.6US¢/kWh with an average of 2.2US¢/kWh in the United States (Owens, 2002; Hance 2005), but at 1.0–1.4US¢/kWh in New Zealand (Barnett and Quinlivan, 2009). The calculations here use an average value ranging from US_{2005}¢1.5–2.2/kWh.

At present, the capacity factor of geothermal power plants is worldwide on average 71–72% (Bertani, 2010). But new plants can achieve more than 90%. A value ranging from 70% to 90% is used here. The economic lifetime of the system is assumed to be 30 years. Based on these figures, an LCOE ranging from about US_{2005}¢3–9/kWh is found. (See Table 11.22)

In the United States, a federal production tax credit of US_{2005}¢1.8–2.1/kWh would just about offset the average O&M cost, increasing revenues and shortening the payback period. FITs, such as those provided in Germany, would also increase the income from the plant.

Smaller plants, around 5 MWe, are estimated to cost 20% more per installed kWe, while binary plants of the 5 MWe size cost approximately

Table 11.22 | Cost of electricity as a function of capacity factor, turnkey investment costs, discount rate, and O&M costs. The O&M costs are assumed at ¢1.5–2.2/kWh. The assumed lifetime is 30 years.

Capacity factor	Turnkey investment costs/kWe	Discount rate	
		5%	10%
	US$	US_{2005}¢/kWh	
90%	2000	3.1–3.8	4.2–4.9
	4000	4.8–5.5	6.9–7.6
70%	2000	3.6–4.3	5.0–5.7
	4000	5.7–6.4	8.3–9.0

30% more. Larger plants in the 100 MWe range can cost 10% less per installed kWe (Al-Dabbas, 2009). Water cooling versus air cooling is estimated to cost 15% more.

11.4.5.2   Direct-use Projects

Direct-use project costs have a wide range, depending on the specific use, the temperature and flow rate required, the associated O&M and labor costs, and the income from the product produced. In addition, new construction usually costs less than retrofitting older structures. Here, estimates are made for the following: individual space heat for a residence, a greenhouse project, district heating, and an industrial application. Well drilling and casing costs would vary from US$150–300/m for depths up to 500 m. Drilling cost will increase with depth and can approach US$500/m, but the cost can be highly non-linear with depth (Chad et al., 2006). These values are based on projects in the United States, and they can vary for other locations, depending on resource temperature and flow, labor and materials costs, and rig availability.

Individual space heating for a building of 200 square meters may have a load of 43MJ_{th}/h, requiring a generating capacity of 12 kW_{th}. Depending on the well depth and temperature of the resource, the system could cost US$10,000–25,000 in addition to the cost of land. This could result in total investment costs of about US$1600–4200/$kW_{th}$. The capacity factor is about 30%. Assuming an economic lifetime of 30 years and O&M costs of US¢2/kWh_{th}, the LCOE ranges between US¢6–19/kWh_{th}

Greenhouses are covered facilities costing approximately US$150/$m^2$. Thus, a commercial facility of 2.0 ha outdoors would cost US$3 million. The peak heating requirement is about 1.0 MJ/m^2. Thus an installed capacity of 5.6 MW_{th} is needed. The geothermal system may cost US$500–1000 per installed kW_{th}. With a capacity factor of 50%, this set-up would result in a geothermal heat use of 24.3 million kWh/yr. Pumping costs and other O&M for the geothermal system may be approximately US¢2/kWh. The economic lifetime is assumed to be 20 years. The LCOE ranges between US¢3–5/kWh_{th}.

District heating may be defined as the heating of two or more structures from a central heat source. Heat may be provided in the form of either steam or hot water and may be used to meet process, space, or domestic hot water requirements. The heat is distributed through a network of insulated pipes consisting of delivery and return mains. Thermal load density (heating load per unit of land area) is critical to the feasibility of district heating, as the distribution network may be the largest single capital expense, at approximately 35–75% of the entire project cost. Thus, high-rise buildings downtown are better candidates than single family residential areas. Generally, a thermal load density above 1.2 GJ/hr/ha or a favorability ratio of 2.5 GJ/ha/yr is recommended for district heating projects. Often fossil fuel peaking is used to meet the coldest period, rather than drilling additional wells or pumping more fluid, improving the efficiency and economics of the total system (Bloomquist et al., 1987). Biomass could also be used effectively for peaking.

One example cost for a district heating projects is found in Germany (Reif, 2008), where two geothermal wells are drilled to 3200m to provide a capacity of 35 MW$_{th}$, and 66 GWh$_{th}$ a year of heat to customers (load factor of 0.22). The total cost of the project, including a fossil fuel peak heating load plant, was US$58.5 million broken down into the following components: 22.6% drilling, 1.7% pumps and accessories, 4.6% geothermal station and equipment, 1.7% peak-load heating plant, 42.4% distribution network, 14.4% service connections, 11.7% heat-transfer stations, and 0.9% land.

A smaller example is found in Elko, Nevada, in the United States (Bloomquist, 2004). The Elko Heat Company system was built in 1989 for US$1.4 million, which at today's costs would be approximately US$5 million; 15% was for resource assessment, 15% for drilling the production well (disposal is to a local river), 29% for the distribution system, 26% for retrofitting customer heating systems, and the remaining 15% for contract services and materials. The estimated capacity of the system is 3.8 MW$_{th}$ and the energy provided to customers is 6.5 GWh$_{th}$ a year. So the capacity factor is nearly 20%. In 2001, the annual operating revenue was US$184,270. The operating expenses were US$47,840, the maintenance US$19,105, and contract services and materials cost US$22,135. Translated to 2005 prices, this suggests O&M costs of US¢1.5/kWh$_{th}$.

From these two examples, investment costs ranging from US$1300–1700/kW$_{th}$ can be derived. However, lower figures are also possible, down to US$600/kW$_{th}$ (Lund et al., 2009). Assuming an economic lifetime of 30 years, an LCOE ranging from US$_{2005}$¢4–12/kWh$_{th}$ can be calculated.

Industrial applications are more difficult to quantify, as they vary widely depending on the energy requirements and product produced (Goldstein et al., 2011). These plants normally require higher temperatures and often compete with power plant use. However, they do have a high load factor of 0.40–0.70, which improves the economics.

One recent study looked at an onion drying facility in Oregon (Geo-Heat Center, 2006). A single-line dryer handling 4500 kg/h of fresh onions

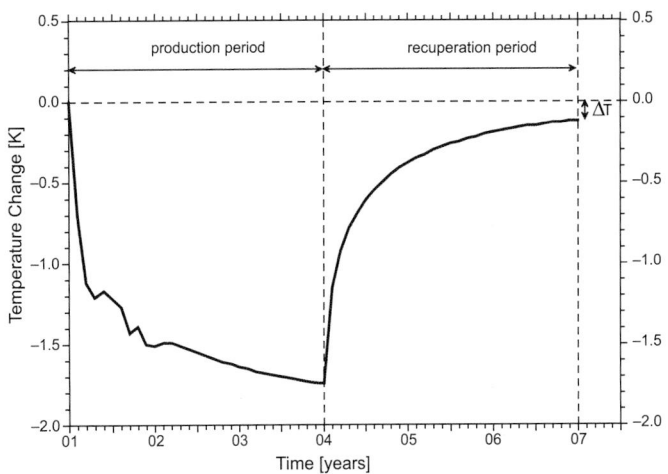

Figure 11.34 | Calculated ground temperature change at a depth of 50 m and at a distance of 1m from a 105 m long borehole heat exchanger over a production period and a recuperation period of 30 years each. Source: Eugster and Rybach, 2000.

for 24 hr/day over a season of 250 days (0.68 load factor) produces 900 kg/h of dried product. About 35 GJ/h (about 10 MWth) of direct use is required and 210 TJ would be the annual energy use, requiring a resource of at least 120°C at 57 liters per sec flow. The total cost for this facility would be US$13.3 million. The geothermal system including the wells may add US$3.6 million (US$360/kW$_{th}$). Assuming an economic lifetime of 20 years and US¢2/kWh$_{th}$ O&M costs, the LCOE would be US¢2.5–2.7/kWh$_{th}$. As in practice higher investment costs are also found (up to US$1000/kW$_{th}$), an LCOE ranging from about US$_{2005}$¢2–4/kWh$_{th}$ is calculated.

11.4.6   Sustainability Issues

Geothermal energy is generally classified as a renewable resource, where "renewable" describes a characteristic of the resource: the energy removed from the resource is continuously replaced by more energy on time scales similar to those required for energy removal (Stefansson, 2000). Consequently, geothermal energy use is not a "mining" process. Whether the exploitation of a geothermal source can be classified as "sustainable" depends on issues like the durability of energy supplies and environmental concerns.

11.4.6.1   Durability of Geothermal Energy Supplies

It appears natural to define the term "sustainable production" as production that can be maintained for a very long time. In Iceland, a reference period for a production well of 100–300 years has been proposed (Axelsson et al., 2005), while in New Zealand, production for more than 100 years is used as a criterion (Bromley et al., 2006). Much longer time scales, such as those comparable to the lifetimes of geothermal resources, are considered unrealistic in view of human endeavors. As geothermal heat is coming from the internal part of Earth and is related to natural decay processes of radioactive isotopes, the flow of energy to Earth's surface of 1400–1500 EJ a year will continue for many million years.

The production of geothermal fluid/heat continuously creates a hydraulic/heat sink in the reservoir. This leads to pressure and temperature gradients, which in turn – after termination of production – generate fluid/heat inflow to re-establish the pre-production state. The regeneration of geothermal resources occurs over various time scales, depending on the type and size of the production system, the rate of extraction, and the attributes of the resource. Time scales for re-establishing the pre-production state following the cessation of production have been determined using numerical model simulations (for details, see Rybach and Mongillo, 2006; Axelsson et al., 2005). The results show that after production stops, recovery driven by natural forces like pressure and temperature gradients begins. This can be illustrated by the production and recuperation periods presented in Figure 11.34. The recovery typically shows an asymptotic behavior and theoretically takes an infinite amount of time to reach its original state. However, practical replenishment (for instance, 95%) will occur much earlier, generally on time scales of the same order as the lifetime of the geothermal production systems (Axelsson et al., 2002).

Examples of long-term production and use from high-temperature geothermal fields include Larderello in Italy for over 100 years (see, e.g., Cappetti, 2009) and The Geysers in northern California for almost 50 years – both of which generate electrical energy. In recent years, however, both fields have experienced reduced production – mainly due to not injecting all the spent fluid from the plants. Another example of what appears to be a sustainable use of a low-temperature geothermal field is the Reykir field (Mosfellssveit), which has been used for district heating of Reykjavik (Iceland) since 1943 (Gunnlaugsson and Gíslason, 2003).

11.4.6.2   Environmental Aspects

Geothermal fluids contain a variable quantity of gas (largely nitrogen) and carbon dioxide, plus some hydrogen sulfide and smaller proportions of ammonia, mercury, radon, and boron. The amounts depend on the geological conditions of different fields. Most of the chemicals are concentrated in the disposal water that is routinely reinjected into drill holes and thus not released into the environment. The concentration of the gases is usually not harmful, and they can be vented to the atmosphere. However, the technology for removing harmful non-condensable gases does exist, and these systems are installed at most geothermal power plants. Removing the hydrogen sulfide released from geothermal power plants is a requirement in, for example, the United States and Italy.

CO_2 emission from electricity production using high-temperature geothermal fields in the world is variable, but it is much lower than that for fossil fuel plants. Bertani and Thain (2002) reported on CO_2 emission data obtained in 2001 from 85 geothermal power plants operating

in 11 countries. These plants had an operating capacity of about 6650 MWe, which constituted 85% of the world geothermal power plant capacity at the time. The collected data showed a wide spread – from 4 g/kWh to 740 g/kWh – with the weighted average being 122 g/kWh. This compares well with the US value reported by Bloomfield et al. (2003) of 91 g CO_2/kWh. A recent comprehensive literature review by IPCC of life cycle assessments for geothermal power plants, however, concluded that "lifecycle GHG emissions are less than 50 gCO_2-eq/kWh for flash steam plants and less than 80 gCO_2-eq/kWh for projected EGS plants" (Goldstein et al., 2011).

Where there is a high natural release of CO_2 from the geothermal fields prior to development, geothermal power development may also decrease this natural emission, as happened, for example, at the Larderello field in Italy. The GHG emissions from low-temperature geothermal resources are normally only a fraction of those from the high-temperature fields used for electricity production. The gas content of low-temperature water is in many cases minute, as in Reykjavik, where the CO_2 content is lower than that of the cold groundwater. In sedimentary basins, such as the Paris basin, the gas content may cause scaling if it is released. In such cases the geothermal fluid is kept under pressure within a closed circuit (the geothermal doublet) and reinjected into the reservoir without any de-gassing taking place.

No systematic collection has been made of data about GHG emissions from geothermal district heating systems. The CO_2 emissions from low-temperature geothermal water can be less than 1 gCO_2/kWh$_{th}$ depending on the carbonate content of the water. As an example, for Reykjavik District Heating the emissions from low-temperature areas are about 0.00005 gCO_2/kWh$_{th}$. Data from geothermal district heating systems in China (Beijing, Tianjin, and Xianyang) are limited but also indicate emissions of less than 1 gCO_2/kWh$_{th}$ (Gunnlaugsson, 2007). The district heating system in Klamath Falls, Oregon, has about zero emissions, as all the geothermal water is used and reinjected in a closed system. Life-cycle analyses, taking into account indirect emissions, show GHG emissions ranging from 14–58 gCO_2-eq/kWh$_{th}$ (Kaltschmidt, 2000).

The GHG emission rates of geothermal heat pumps depend on the energy efficiency of the equipment as well as the fuel mix and the efficiency of electricity generation. In most cases, heat pumps are driven by auxiliary electric power, so the CO_2 emissions depend on the energy source for electricity generation. The average CO_2 emission associated with generation of electricity in Europe in 2005 has been estimated to be about 0.55 kgCO_2/kWh. With proper system design, the electrically driven geothermal heat pump reduces the CO_2 emissions of an oil-fired boiler by 45% and those of a natural-gas-fired boiler by 33% (ISEO, 2010). Based on life-cycle analysis, Kaltschmitt (2000) found emission rates for GHPs ranging from 180–200 gCO_2-eq/kWh$_{th}$. If the electricity that drives the heat pump is produced from a renewable energy source, the emission rate will be much smaller. The total CO_2 emission reduction potential of heat pumps has been estimated to be 1.2 Gt/yr, or about 6% of global emissions (ISEO, 2010).

There are also local environmental impacts related to land and water use and the operation of an energy system. In addition, there could be some specific phenomena such as discharges of gas other than CO_2. But these can be reduced strongly where gas injection is used and nearly eliminated when binary geothermal plants are installed for power generation. Since most direct-use projects use only hot water and the spent fluid injected, the polluting emissions are nearly eliminated.

The exploration of geothermal resources can also have impacts on outstanding natural features and landscapes. And geothermal systems using wells may have an impact on seismicity and ground vibrations and contribute to risks such as hydrothermal steam eruptions. Proper design and management as well as monitoring and control are needed to avoid or mitigate these risks. Recent exploratory work on EGS has produced small earthquakes (up to 3.4 on the Richter scale) that have caused local concerns. The mitigation of these earthquakes is currently under investigation.

11.4.7   Implementation Issues

The technology to produce electricity and generate heat for direct use is mature and can be applied cost effectively depending on local circumstances. New technology developments may enhance access to geothermal energy use. Getting access to geothermal energy fields may also require the construction of transmission capacity, adding to the total costs of energy supply. Other implementation issues could include siting and permitting delays, high capital costs, the concerns of local populations, and public perceptions and support, including lack of knowledge about benefits.

Geothermal power plants are operated in base load. Geothermal energy may fit well with the use of other renewable resources such as wind, solar, biomass, and hydro, offering the potential of a reliable and secured supply of heat and electricity. Given the barriers and constraints, enhanced deployment of geothermal energy use requires government policies, regulations, and initiatives, including incentives such as a FIT for geothermal pricing, as demonstrated by a number of countries (see, e.g., Rybach, 2010; REN21, 2010). Enhanced deployment also requires education, including training and outreach, along with improvements of the technology.

11.5   Wind Energy

11.5.1   Introduction

Wind energy has been used for thousands of years in a variety of applications, but it was largely overshadowed by other fuels for much of this time for a variety of technical, social, and economic reasons. The oil crises of the 1970s, however, renewed interest in wind energy technology for grid-connected electricity production, water pumping, and power supply in remote areas (WEC, 1994; UNDP et al., 2000). This section focuses on utility-scale, grid-connected wind technology deployed either on land or offshore.

Wind power capacity grew to meet nearly 2% of global electricity demand in 2009. Onshore wind is currently one of the most economical renewable energy generation technologies. In areas with good wind resources, generating electricity with wind turbines is already competitive. As a result, these installations have grown rapidly. Offshore wind projects are almost twice as capital-intensive as their land-based counterparts, but in Europe and Asia some countries have set aggressive goals for their deployment. The experience gained through this is expected to reduce costs and improve performance.

11.5.2   Potential of Wind Energy

Wind energy is broadly available but diffuse. There is a vast global wind resource that could be tapped (see Chapter 7) and provide carbon-free electricity. Figure 11.35 illustrates the global land-based wind resource at 50-m above ground level.

The *technical potential* of wind energy to fulfill energy needs is very large. Estimates range from about 20,000 TWh/yr (onshore only) to 125,000 TWh/yr (onshore and near-shore). The range suggests that wind could supply in principle anywhere from one to six times the 2009 global electricity production of about 20,000 TWh (IEA 2010b; BP, 2011). Although wind resource quality varies around the globe, there is sufficient potential in most regions to support high levels of wind energy generation. Wind resources are not a global barrier to expansion of the use of this technology in the coming decades.

It has been noted that local and global climate change could affect wind resources (IPCC, 2007b), although research in this area is nascent (Wiser et al., 2011). Climate change could have impacts on wind patterns locally, but it is unlikely to be a large enough magnitude to change the global technical potential of wind energy greatly (IPCC 2011).

The *economic potential* of harnessing wind energy is defined by capital investment, corresponding annual power production, competitiveness with other energy technologies, and policy measures. The *realizable (or implementation) potential* also depends on aspects such as access of a particular site to electricity markets through transmission lines; rules favorable to variable generation technologies; sufficient mitigation of visual, acoustic, and wildlife impacts; public acceptance; and maintenance costs. For a given site, the economic potential depends on the annual power production from wind technology, requiring knowledge of, for example, average wind speeds at heights above ground corresponding to wind turbine hubs, the statistical distribution of wind speeds throughout the year, turbulence intensity, and the impact of terrain features near the plant. Continued development of calculation models and increased availability of observational wind speed data are critical to optimizing the annual power production of

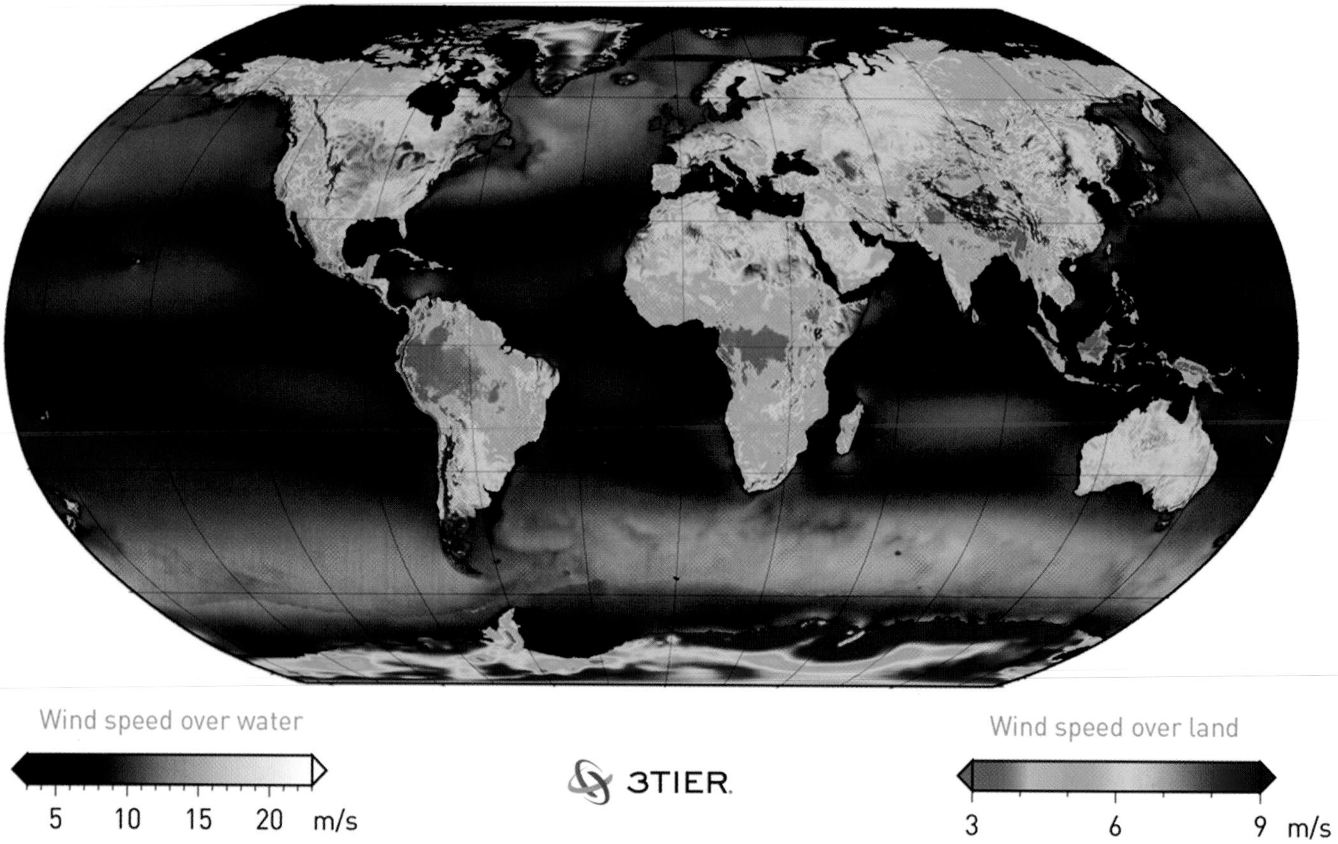

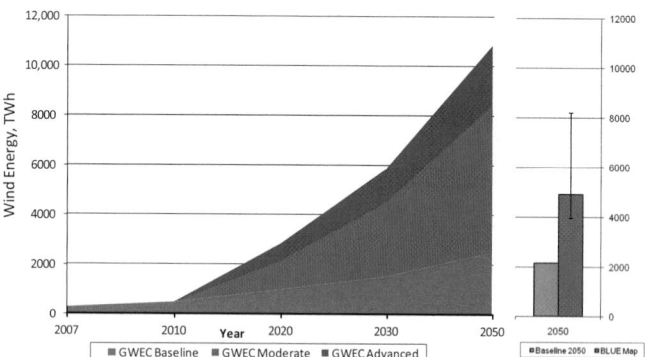

Wind speed over water   🔵 3TIER.   Wind speed over land

5   10   15   20   m/s   3   6   9   m/s

Figure 11.35 | Global wind resource. The red color represents the strongest wind and the blue color denotes the weakest wind resource. Source: 3 Tier, 2011. (c) 2011 3TIER, Inc.

wind turbines and assessing the economic potential of wind resources (IEA, 2010a).

Numerous studies and approaches have been used to develop scenarios for economic deployment of wind energy. The Global Wind Energy Council (GWEC) uses a simulation model to create scenarios for wind technology through 2050 based on different levels of political support (GWEC and Greenpeace, 2010a). GWEC poses three scenarios: Reference, Moderate, and Advanced. The Reference scenario is based on the IEA's *World Energy Outlook* 2009. The Moderate scenario accounts for all policy measures to support renewable energy either enacted or in planning stages. The Advanced scenario represents the extent to which the wind industry could grow in a base case "wind energy vision."

A second estimate, developed by the IEA in *Energy Technology Perspectives* 2010 (ETP), uses an optimization model to create global energy scenarios based on different levels of commitment to carbon emission reduction (IEA, 2010a). It poses two scenarios, Baseline and BLUE Map. The Baseline scenario is based on the *World Energy Outlook* 2009 extrapolated to 2050 and assumes that no new policies are intro-

Figure 11.36 | Wind energy generation estimates from two studies (left panel: GWEC and Greenpeace, 2010a; right panel: IEA, 2010a).

duced. The BLUE Map scenario represents a reduction of CO_2 emissions to 50% below 2005 levels by 2050.

Figure 11.36 shows the estimated wind energy generation in 2050 for these two studies. The GWEC scenario projections include both onshore and offshore wind but do not estimate the proportion from each. The error bar on the Blue Map scenario (4900 TWh/yr in 2050) represents the variation in wind energy generation estimates based on sensitivity runs.

These two studies suggest significant generation potential for wind technology, up to 10,000 TWh annually in the GWEC Advanced scenario. Wind generation as a percentage of total global electricity demand varies from 10% to 26% for the non-reference / non-baseline scenarios, depending on assumptions that influence projections of total electricity demand and the role of competing low-carbon generation technologies. These studies use different approaches to reach internally consistent scenarios. The GWEC simulation model focuses on wind industry growth potential based on historical expansion levels and considering growth rates, turbine rating and capacity factor, capital costs, and progress ratios. The IEA-ETP simulation model assesses wind technology contribution to projected electricity sector expansion relative to other generation technologies under various policy influences.

A third approach by Hoogwijk et al. (2004) explores the economic potential of onshore and offshore wind generation based on geographic location, exclusion areas for environmental and competing uses, wind speed estimates, wind technology generation estimates, and costs. This study finds that wind technology could generate an amount roughly equal to 2001 world electricity consumption, 16,000 TWh, at a cost of up to US$_{2004}$¢7/kWh.

Although all these studies use very different approaches, they demonstrate that economically feasible wind energy generation potential is significant. The cost of wind generation relative to other generation technologies becomes the ultimate driver.

11.5.3   Market Developments

The wind industry has enjoyed sustained global growth since 1990. Almost 160 GW of wind generation capacity were installed through 2009, generating nearly 350 TWh electricity annually. At the end of 2010, almost 200 GW was operating worldwide. In the period 2004–2009, cumulative installed capacity grew at an annual average rate of approximately 27%.

Europe is the global leader in terms of installed wind capacity (76 MW in 2009) but cumulative installations in Asian and American markets grew rapidly between 2008 and 2009 at 68% and 39% respectively. (See Figure 11.37.)

European countries such as Denmark, Germany, and Spain led growth in the late 1990s. The United States and China became the fastest-growing markets from 2005 to 2010. India has moved into the top 10 wind markets. By the end of 2010, the highest installed wind capacity was found in China (45 GW), followed by the United States (40 GW), Germany (27 MW), Spain (20 GW), and India (13 GW) (GWEC, 2011).

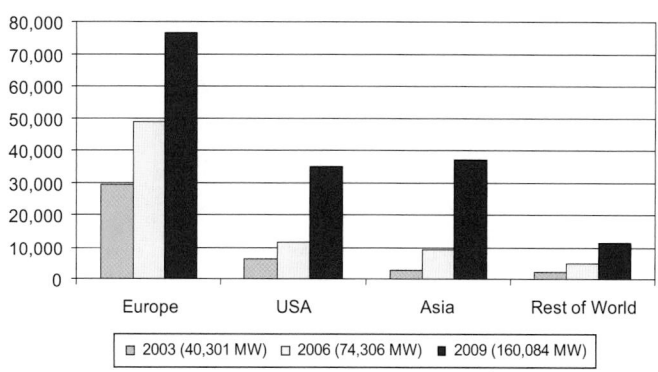

Figure 11.37 | Cumulative installed capacity in Europe, United States, Asia, and the rest of the world, 2009. Source: BTM Consult, 2010; AWEA project database for US capacity.

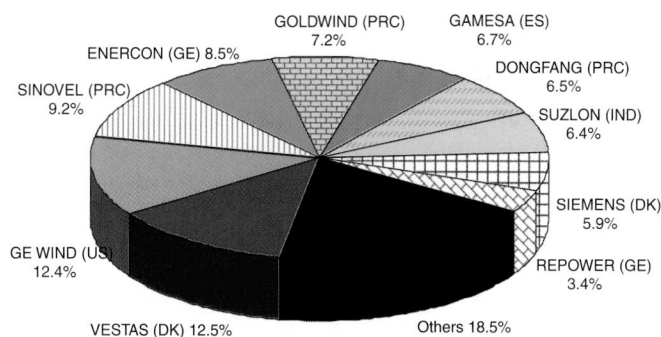

Figure 11.38 | Top 10 wind turbine suppliers, 2009. Source: based on BTM Consult, 2010.

The wind industry remains highly concentrated: just six countries are home to the top 10 wind turbine suppliers in 2009 (see Figure 11.38) – all from Europe, North America, or Asia.

While the industry was previously dominated by small independent project developers, electric utilities and large independent power producers are increasingly investing in wind projects. This is leading to increasing globalization and competitiveness in the wind turbine supply chain (BTM Consult, 2010). In 2009, investments in wind power installations totaled nearly US$_{2009}$$52 billion; direct employment in the wind energy sector has been roughly estimated at 600,000 jobs (GWEC and Greenpeace, 2010a).

In Europe and the United States, wind has become a major source of capacity additions to the electric sector. Between 2000 and 2009, wind capacity additions were second only to natural gas and were ahead of coal. In 2009, 39% of all capacity additions in both the United States and the European Union came from wind power. (See Figure 11.39.)

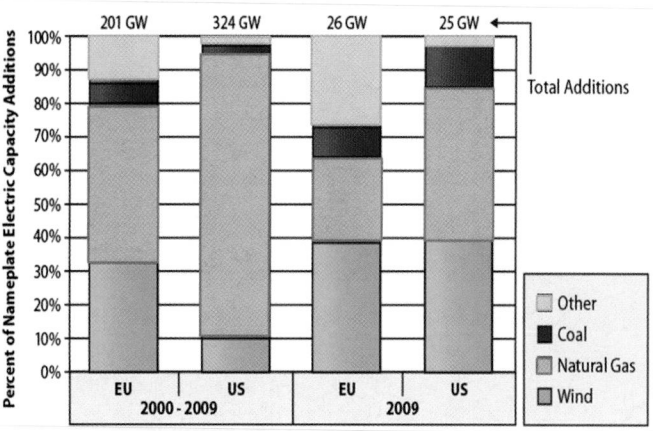

Figure 11.39 | Relative contribution of generation types to capacity additions in the United States and the European Union, 2000–2009 and 2009. Source: Wiser and Bolinger, 2010; EWEA, 2010.

By the end of 2009, the wind industry had developed 38 offshore projects in nine European countries, bringing total capacity to above 2000 MW. Annual installations grew to 577 MW in 2009, up 54% from 2008 (EWEA, 2010). The European Wind Energy Association (EWEA) estimates that the 2056 MW of capacity installed at year-end 2009 will generate more than 7 TWh of electricity annually. The development of offshore wind projects has so far been concentrated in Europe, but 2010 saw the first Chinese project commissioned and the final permitting of the Cape Wind project in the United States (Musial and Ram, 2010).

11.5.3.1   Role of Wind Turbine Standards

The development of a suite of international standards for wind turbines has been a major contributor to the evolving market for wind turbine technology over the past 15 years. The International Electrotechnical Commission (IEC) is a global organization that prepares and publishes international standards for electrical, electronic, and related technologies (IEC, 2010). These standards serve as the basis for national standards and as a reference when drafting international tenders and contracts. The standards cover wind turbine systems and subsystems, such as mechanical and internal electrical systems, support structures, and control and protection systems.

11.5.3.2   The Near-term Market Future

Near-term forecasts indicate continued rapid growth of the global wind industry from the current level of 160 GW. The Global Wind Energy Council forecasts a cumulative installed capacity (onshore and offshore) of 409 GW by 2015, meeting about 4% of global electricity demand (up from 2% in 2009) (GWEC and Greenpeace, 2010b). This equates to an investment of more than USUS_{2005}\$225 billion. BTM Consult reaches a similar but slightly more aggressive forecast for the period, predicting that wind will achieve 447 GW by 2014 (BTM Consult, 2010).

Prospects for growth in the offshore wind market are considerable. EWEA (2010) foresees European offshore wind installations growing to 150 GW by 2030, depending on the policies implemented. The US DOE and the US Department of the Interior have adopted a strategy to realize 54 GW of offshore wind in the United States by 2030 (US DOE, 2011a). In Europe at the end of 2009, some 3500 MW of offshore wind was under construction, 16,000 MW had received final permits, and 100,000 MW had been proposed (EWEA, 2010).

11.5.3.3   Small Wind Turbines

There is also a market for small-scale wind turbines with capacities in the hundreds of watts up to 100 kW range. In rural areas they are used for battery charging and in stand-alone hybrid electricity systems. It has been estimated that annual global installations of small wind turbines may approach 40 MW (AWEA, 2009). In addition, probably more than 1 million water-pumping wind turbines (wind pumps) manufactured in developing countries supply water to livestock (UNDP et al., 2000).

11.5.4   Wind Energy Technology Development

11.5.4.1   Onshore Wind Turbine Technology: Current Status

Commercial-scale wind turbines have largely coalesced around a horizontal-axis design with three pitch-regulated blades that capture the wind resource upwind of the tower and have a diameter of 60–120m. The rotor and blades are attached to a hub and main shaft through which mechanical energy is transferred to a gearbox (depending on the design) and finally to a variable speed generator, where the energy is converted to electricity. These components are contained in a housing made of fiberglass, called a nacelle, which protects components from the elements. The nacelle is mounted on a 50–120m tower to allow the rotor to capture higher-quality wind resources than found near the ground.

Wind turbines are typically grouped together into wind power plants (wind projects, wind farms) for the commercial production of electricity (as opposed to community or residential electricity production). The electricity generated is aggregated at an on-site substation and then exported to the utility system grid. Modern onshore wind projects typically range in size from 25 to 400 MW, though the largest plant currently operating has more than 800 MW of capacity.

During the past 25 years, average wind turbine ratings have grown almost linearly since the introduction of 50 kW turbines in the early 1980s. Current commercial machines are rated at 1.5–3 MW for land-based turbines, offshore turbines as large as 5 MW are being deployed, and larger machines are on the drawing boards of several manufactures. Wind turbine designers in the last two decades have continually predicted that the current generation of turbines had grown as large as

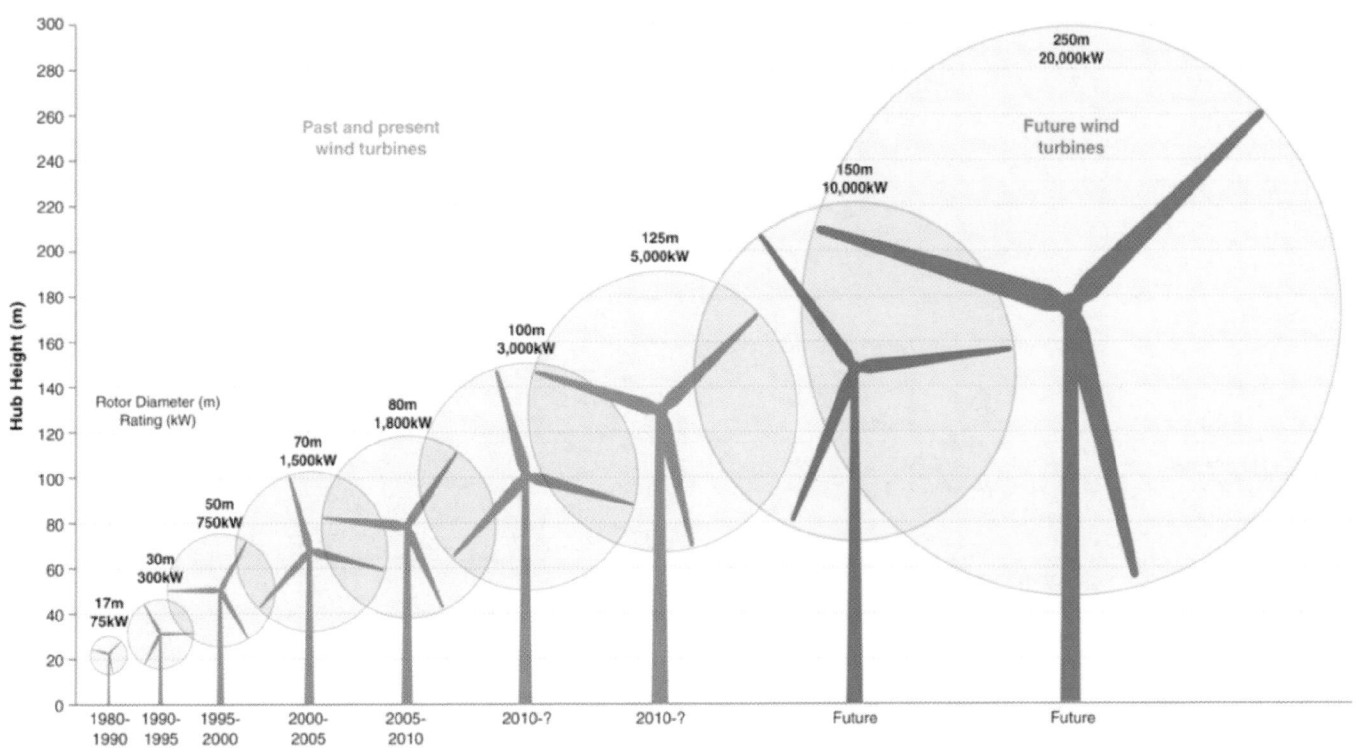

Figure 11.40 | Development and size growth (rotor diameter) of wind turbines since 1980. Source: courtesy of NREL.

they would ever be. But with each new generation of turbines, size has increased linearly and resulted in a reduction in life-cycle cost of energy. This impressive evolution of wind turbine technology is illustrated in Figure 11.40.

11.5.4.2   Potential Future Land-based Turbine Technology Improvements

The long-term drive to develop larger turbines stems from a desire to increase power production (by tapping higher-quality wind resources with larger rotors and towers), reduce investment costs per unit of capacity, and reduce operation and maintenance costs per unit of capacity. Increased size also leads to the consideration of wind turbines as energy plants – which is important to large-scale implementation – and to the introduction of more sophisticated technologies such as smart-blade designs.

There are constraints to this continued growth, however (see, e.g., Thresher et al., 2008a). In general, it costs more to build a larger turbine. The primary argument for a size limit for wind turbines is based on the "square-cube law." Roughly stated, it says that "as a wind turbine rotor increases in size, its energy output increases with the rotor-swept area (the diameter squared), while the volume of material required, and therefore its mass and cost, increases as the cube of the diameter." Consequently, at some size the cost for a larger turbine will grow faster than the resulting energy output and revenue, making scaling unattractive on economic terms. Engineers have successfully skirted this law by optimizing design

characteristics for a given turbine size and removing material or by using material more efficiently to trim weight and cost (Burton et al., 2001).

Constraints in transporting very large blades, towers, and nacelles over land also can pose limiting factors to wind-turbine growth. An additional constraint is the cost and availability of cranes that are capable of lifting components into place. If designers are unable to address these problems, future land-based turbines are likely to be redesigns of 2–5MW turbines, with an emphasis on greater efficiency and reliability. Transportation to offshore sites is not limited by size, and offshore wind energy is a major driver of the development of larger turbines.

Both the European Technology Platform for Wind Energy (TPWind, 2008) and the US DOE (WindPACT, 1999; US DOE, 2008a) have identified a broad array of wind energy R&D activities that have the potential to improve the cost and performance of wind technology significantly. Potential improvements are summarized in Table 11.23, which also shows the manufacturing learning-curve effect generated by several doublings of turbine manufacturing output expected. The impact on capital cost reduction is assumed to range from zero in a worst-case scenario to the historic level in a best-case scenario, with the most likely outcome halfway in between. The most probable scenario is a sizable (~ 45%) increase in capacity factor with a modest (~ 10%) drop in capital cost (Thresher et al., 2008a).

Thus although no "big breakthrough" is on the horizon for land-based wind technology, many evolutionary R&D steps can cumulatively bring

Table 11.23 | Areas of potential wind energy technology improvements.[a]

Technical Area	Potential Advances	Increments from Baseline (Best/Expected/Least)	
		Annual Electricity Production per kW (%)	Turbine Investment Cost per kW (%)
Advanced Tower Concepts	* Taller towers in difficult locations * New materials and/or processes * Advanced structures/foundations * Self-erecting, initial or for service	+11/+11/+11	+8/+12/+20
Advanced (Enlarged) Rotors	* Advanced materials * Improved structural-aero design * Active controls * Passive controls * Higher tip speed/lower acoustics	+35/+25/+10	-6/-3/+3
Reduced Energy Losses and Improved Availability	* Reduced blade soiling losses * Damage-tolerant sensors * Robust control systems * Prognostic maintenance	+7/+5/0	0/0/0
Advanced Drive Trains (Gearboxes and Generators and Power Electronics)	* Fewer gear stages or direct drive * Medium/low-speed generators * Distributed gearbox topologies * Permanent-magnet generators * Medium-voltage equipment * Advanced gear tooth profiles * New circuit topologies * New semiconductor devices * New materials (GaAs, SiC)	+8/+4/0	-11/-6/+1
Manufacturing Learning	* Sustained, incremental design and process improvements * Large-scale manufacturing * Reduced design loads	0/0/0	-27/-13/-3
Totals		+61/+45/+21	-36/-10/+21

[a] The baseline for these estimates was a 2002 turbine system in the United States. There have already been sizable improvements in capacity factor since 2002, from just over 30% to almost 35%, while investment costs have increased due to large increases in commodity costs in conjunction with a drop in the value of the US dollar. Therefore, working from a 2008 baseline, a more-modest increase in capacity factor could be expected, but the 10% investment cost reduction is still quite possible, if not conservative, particularly from the higher 2008 starting point. The table does not consider any changes in the overall wind turbine design concept (e.g., two-bladed turbines).

Source: US DOE, 2008a.

about a 30–40% improvement in the cost effectiveness of this wind technology over the next two decades.

11.5.4.3   Offshore Wind Turbine Technology: Current Status

The typical offshore wind turbine to date has essentially been a marinized version of the standard land-based turbine installed in shallow water, with some system redesigns to account for ocean conditions (Musial and Ram, 2010). Modifications include upgrades to the support structure to address added loading from waves, pressurized nacelles and environmental controls to prevent corrosive sea air from degrading critical drive train and electrical components, and personnel access platforms to facilitate maintenance and provide emergency shelter (Thresher et al., 2008b). Offshore turbines must also have corrosion protection systems at the sea interface and high-grade marine coatings on most exterior components. For marine navigational safety, turbine arrays are equipped with warning lights, vivid markers on tower bases, and fog signals. To minimize expensive servicing, offshore turbines can be equipped with, for example, enhanced condition monitoring systems, automatic bearing lubrication systems, and on-board service cranes – all of which exceed the standard for land-based designs (Thresher et al., 2008b).

Today's offshore turbines range in capacity from 2 to 5 MW and typically are represented by architectures that include a three-bladed horizontal-axis upwind rotor, nominally 80–126m in diameter. The drivetrain topology consists of a modular three-stage hybrid planetary-helical gearbox that steps up to generator speeds between 1000 and 1800 rpm, which is generally run with variable speed torque control, although there is some evidence that direct drive generators might offer a smaller, lighter, cheaper, and more reliable alternative. Tower heights offshore are generally less than those of land-based turbines because wind-shear profiles are less steep, tempering the energy capture gains sought with increased elevation. Lower offshore tower heights also reduce the potential for overturning – an important consideration for floating platforms (Thresher et al., 2008b).

The offshore foundation systems differ substantially from land-based turbines. Offshore wind turbines installed to date have used three main foundation designs: monopile, gravity base, and multipile. Monopile and multipile foundation technologies require specialized installation vessels to drive piles into the seabed. Gravity bases can simply be towed to the site, where they are filled with ballast and sunk carefully onto each site. Once the foundation is prepared, the turbine is installed using a specialized crane ship or barge. Mobilization of the infrastructure and logistical support for a large offshore wind farm is a significant portion of the cost (Thresher et al., 2008b).

In 2009, some 88% of offshore turbines were installed on monopiles, 8% on gravity bases, and 3% on multipiles (EWEA, 2009). The choice of foundation technology is largely governed by project economics. Monopiles and gravity bases are the most cost-effective technologies for turbines rated lower than 5 MW and installed in shallow water (less than 30m). It is expected that it will be cost-prohibitive to use these technologies with larger turbines or transitional depths (30–60m). So the use of multipiles is expected to expand considerably.

11.5.4.4   Potential Future Offshore Turbine Technology Improvements

Three pathways for offshore technology represent progressive levels of complexity and development that will lead to cost reductions and greater deployment potential (Thresher et al., 2008b). The first path is to lower costs and remove deployment barriers for *shallow-water technology* in water depths of 0–30m, where technology has already been deployed and proven. The second path is *transitional depth technology*, which is needed for depths where current technology no longer works, up to the point where floating systems are more economical. This technology deals mostly with substructures that will be adapted from existing offshore oil and gas practices. The third path is to develop technology for *deep-water* depths of 60–900m. This could use floating systems, which will require a higher level of R&D to optimize turbines that are lightweight and can survive additional tower motion on anchored, buoyant platforms. Deep-water designs would open up major areas to wind energy development where the turbines would not be visible from shore and where competition with other human activities would be minimal. Such platforms would allow mass production of all system components, introducing a major new opportunity for cost reduction. At this time, all three of these development pathways are being explored.

The European *UpWind* research project (Hjuler Jensen, 2007) envisions offshore wind turbines growing in scale to 8–10 MW and having rotor diameters greater than 120m, which is a challenge to design, build, install, and operate at sea. The project, established to address this multitude of engineering challenges, addresses these technical areas: aerodynamics and aero-elasticity, rotor structure, and materials; foundations and support structure; controls systems; remote sensing; condition monitoring; flow; electrical grid; and management. The *UpWind* project is also designed to address systems integration topics like integrated system design, standards, metrology (measurement), training

and education, innovative rotor systems, electricity transmission and conversion, smart rotor blades, and system up-scaling.

In general, offshore wind technology is expected to gain improvements similar to those envisioned for land-based wind turbines (as outlined in Table 11.23) but at a much larger machine scale and in a more hostile operating environment. Clearly the researchers must design larger and lighter rotors.

11.5.4.5   Future Underlying Science Challenges

The very significant wind energy technology improvements and related cost reductions currently achieved have been enabled by the application of improved engineering analysis and design techniques and by testing each new wind turbine component and system. However, wind energy technology has matured to a point where it will be difficult to sustain this rapid rate of improvement without a major advance in understanding of the basic physical processes underlying wind energy science and engineering. There are fundamental knowledge barriers to further progress in virtually all aspects of wind energy engineering: scientists' understanding of atmospheric flows, unsteady aerodynamics and stall, turbine dynamics and stability, and turbine wake flows and related array effects. Even climate effects might be caused locally by the large-scale use of wind energy, both onshore and offshore (Thresher et al, 2008a).

Research in these focus areas has developed in relative isolation from the others. Continued progress in wind energy technology will require interdisciplinary reunification, especially with the atmospheric sciences, to exploit previously untapped synergies. Also, experiments and observations need to be applied in a coordinated fashion with computation and theory. The use of high-penetration wind energy deployment requires an unprecedented ability to characterize the operation of large wind turbines deployed in gigawatt-scale wind plants (Thresher et al, 2008a).

11.5.5   Economic and Financial Aspects

11.5.5.1   Present Wind Energy Costs

Capital costs for wind projects have declined dramatically since the 1980s. Figure 11.41 shows historic installed capital costs for wind projects in the United States and Denmark from 1982 to 2009. Improved design methods and experience with installations as well as upscaling of turbine technology contributed to the decreases observed until 2002/2003. Since 2004, however, capital costs have increased, driven by turbine performance improvements, rising commodity prices, currency fluctuations, high demand for turbines, supply chain constraints, higher labor costs, and increased margins for original equipment manufactures, developers, and component suppliers. Milborrow (2010) reports that the global average capital costs for onshore wind projects installed in 2009 ranged from US\$1400–2100/kW, with an average of US_{2005}\$1750/kW.

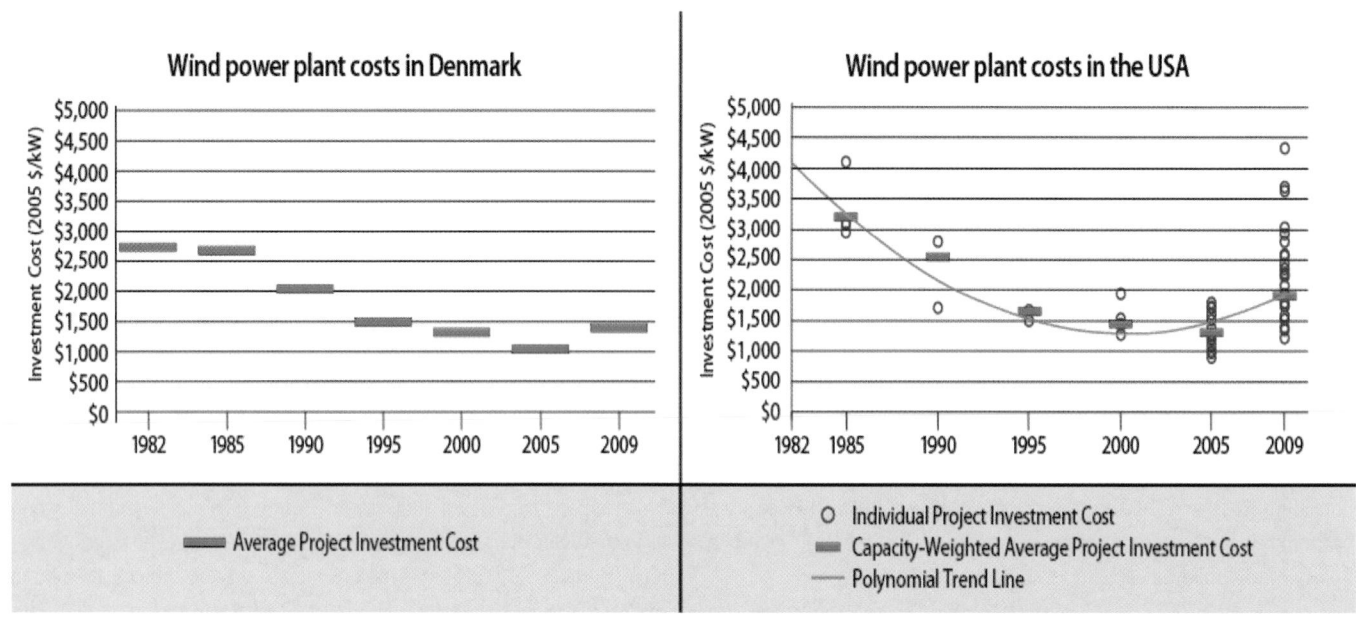

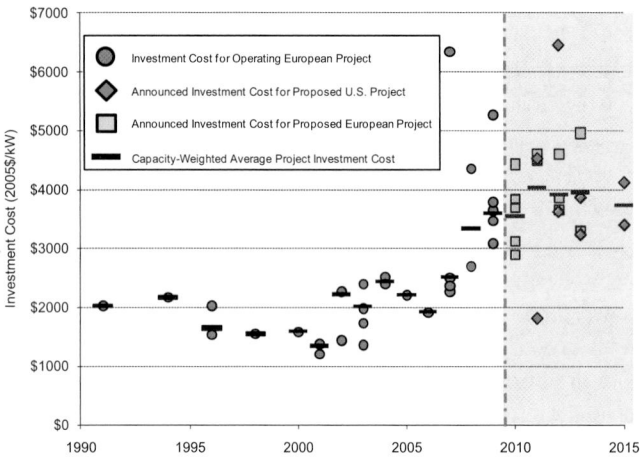

Figure 11.42 | Capital costs for installed and announced offshore wind projects. Source: Musial and Ram, 2010. Courtesy of NREL.

Wiser and Bolinger (2010) report a slightly higher average capital cost in the United States of US$1900/kW. Capital costs for Chinese projects are substantially lower, with an average in the range of US$1000–1300/kW as a result of low-cost turbines supplied by Chinese manufacturers (Li and Ma, 2009; Li, 2010). Some of the cost pressures (supply chain, commodity prices) appear to be easing, leading to expectations that capital costs will decrease over time (Wiser and Bolinger, 2010).

Capital costs for offshore projects are far less certain due to the relative immaturity of the technology. Figure 11.42 shows the capital costs for installed and announced offshore wind projects from 1990 to 2015. Offshore wind project capital costs rose from about US$1300/kW to current levels of US$3000–6000/kW, driven by similar factors as land-based wind capital costs.

But these increases were exacerbated by sector-specific factors, including a lack of competition in the market for offshore wind turbines (two manufactures had a 90% market share in 2009), limited availability of specialized installation vessels, complexity of siting projects in deeper water and farther from shore, and increasingly robust turbine designs necessitated by a better understanding of technical risks (Musial and Ram, 2010). While the near-term trend for offshore wind capital costs is uncertain, there is a positive sign in that the supply chain is becoming more competitive. In 2010, some 21 manufactures announced 29 different wind turbine models for the offshore market (EWEA, 2011).

Power prices for onshore wind energy have risen considerably since the low reached in 2002 and 2003. Wiser and Bolinger (2010) estimate that prices for electricity generated by onshore wind projects built in 2009 range from about US¢4–8/kWh, with a capacity-weighted average of about US¢6/kWh. (The price for projects in the United States reflects the value of state and federal incentives available to wind projects, such as the federal Production Tax Credit, worth US$20/MWh, and the Investment Tax Credit/Cash Grant, worth 30% of installed capital cost.) This is double the average price of projects built in 2002 and 2003 and 20% higher than the average price of projects built in 2008. The authors attribute increasing prices to elevated capital costs and lower project capacity factors. EWEA (2009) reports unsubsidized levelized costs of energy ranging from US¢6–13/kWh.

Offshore wind power prices are much less well understood than land-based prices due to a lack of available data and the substantial spike in capital costs between 2004 and 2009. EWEA (2009) calculated unsubsidized LCOE for 10 European projects installed between 2001 and 2008, resulting in a range of US$_{2005}$¢7–11/kWh. Mott MacDonald (2010) estimates that the LCOE for offshore wind projects installed between

Table 11.24 | Cost of electricity as a function of capacity factor, turnkey investment costs, discount rate, and O&M costs. The O&M costs are assumed at US¢1–2/kWh onshore and US¢2–4/kWh offshore. The economic lifetime is 30 years.

Capacity factor	Turnkey investment costs per kWe (*onshore*)	Discount rate	
		5%	10%
	US$_{2005}$$	US$_{2005}$¢/kWh	
35%	1200	3.5–4.5	5.1–6.1
	2100	5.5–6.5	8.3–9.3
20%	1200	5.4–6.4	8.2–9.2
	2100	8.7–9.7	13.5–14.5

Capacity factor	Turnkey investment costs per kWe (*offshore*)	Discount rate	
		5%	10%
	US$_{2005}$$	US$_{2005}$¢/kWh	
45%	3000	7.0–9.0	10.1–12.1
	6000	11.8–13.8	18.1–20.1
35%	3000	8.4–10.4	12.4–14.4
	6000	14.7–16.7	22.8–24.8

2010 and 2020 will range from US¢15–26/kWh. These estimates of LCOE differ considerably, based on assumptions about the capital cost of offshore wind projects: EWEA uses a range of US$1500–3300/kW, while Mott Mac Donald uses US$4356–6825/kWh.

The performance of wind power plants varies with the wind resource of the site, the technology deployed, and maintenance. Globally, individual project capacity factors range from a low of about 20% to a high of 50%. Average capacity factors have been estimated for Germany at about 21% (BTM Consult, 2010), China at 23% (Li, 2010), India at 20% (Goyal, 2010), and the United States at 30% (Wiser and Bolinger, 2010). Offshore turbines installed to date have had a more narrow range of capacity factors, from 35–45% (Lemming et al., 2009), but in the United Kingdom initially also about 30% was found (UKERC, 2010).

The O&M costs for wind projects onshore are estimated to range at present from about US¢1–2/kWh, while those for offshore plants are about US¢2–4/kWh (EWEA, 2009; Lemming et al., 2009; IEA, 2009b, 2010a; Milborrow, 2010; UKERC, 2010; Wiser and Bolinger, 2010).

From these figures it can be calculated, assuming an economic lifetime of the investment of 30 years, that the electricity production costs (the LCOE) range from about US¢4–15US¢/kWh for onshore wind and from US¢7–25/kWh for offshore wind. (See Table 11.24.) Assuming a lifetime of 20 years, the lowest figure increases by about US¢1/kWh.

11.5.5.2   Estimates of Future Wind Energy Costs

Estimating future wind technology costs is difficult, and various approaches have been used. Potential future cost reductions have been shown through "bottom-up" engineering models through reduced capital cost and increased energy capture of –10% and 45%, respectively. Another methodology for the calculation of future cost reduction potential is learning curve analysis, which calculates a learning rate defined as the cost reduction that occurs with each doubling of deployment.

Learning rates for wind technology summarized for IEA over different time periods range from 8–17% for capital costs and from 18–32% for electricity production costs (IEA, 2010a). The IEA study assumed a learning rate of 7% would result in cost-competitive onshore wind technology by 2020–2025, representing a capital cost reduction of 25%. A learning rate of 9% for offshore wind technology would reach a cost-competitive level by 2030–2035, representing a capital cost reduction of 38%. The IEA study assumed that global average capacity factors would increase from current averages of 20% offshore and 38% onshore to 30% and 40% respectively.

11.5.5.3   System Integration Consideration and Costs

Electric systems have historically linked large, centralized generators to consumers. Wind resources are dispersed and often located some distance from electricity customers. Integration of large amounts of wind-generated electricity will depend on the degree to which electric networks and markets evolve to accommodate variable dispersed generation technologies. Power systems are designed to handle variability, and wind generation adds to that. Experience has shown a number of ways to mitigate this: aggregation of wind generation over large geographic areas to reduce the system-level variability; forecasting methodologies implemented in control rooms to reduce operational impacts and costs; and more flexibility in the system to increase the potential for integrating wind energy. Methods to increase flexibility include additional flexible capacity in the generation mix (such as gas turbines, pumped hydro storage, or other storage technologies), increases in the size of balancing areas, trades closer to real-time (short gate-closure), and encouragement of demand-side flexibility. (See also Section 11.10).

Costs associated with managing wind variability are small at low penetrations. As wind penetration increases, these increase. Additional system balancing costs vary widely across markets, depending on plant mix and fuel costs among other factors. Recent estimates of balancing costs range from US¢0.1–0.5/kWh for wind energy penetrations between 10% and 20% (IEA-Wind, 2010). Expansion of the transmission system to remote wind resources and reinforcing the grid will also add costs. In the United States, the transmission cost to achieve 20% of projected US electricity demand from wind generation in 2030 was estimated to add US$150–290/kW to the investment cost of wind plants (US DOE, 2008a). Estimates by system operators in Germany, the United Kingdom, the Netherlands, and Portugal are of approximately US$70–170/kW of wind capacity to reach penetration levels between 15–25% (IEA-Wind, 2010).

System-level costs related to balancing or transmission expansion are borne by different entities based on the operation practices of a particular electric system. As electric systems expand to meet growing demand, including the characteristics of variable generation in the planning and design stages will enable higher penetrations of wind and other renewable technologies.

11.5.6   Sustainability Issues

This section covers sustainability issues outlined by the National Research Council of the National Academies of Sciences in the United States in *Environmental Impacts of Wind-Energy Projects* (NRC, 2007).

There are a number of life-cycle analysis studies of wind turbines. The most extensive is the ExternE study on the externalities of all energy sources. There is general agreement that a modern wind turbine system has an energy payback period of less than half a year (Wiser et al., 2011). Calculated life-cycle GHG emissions for wind range generally from about 8–20 gCO_2-eq/kWh (ExternE, 1995; Tester et al., 2005; Wiser et al., 2011). It should, however, be noted that substantial penetrations of wind energy into the grid will enhance the need for backup power and/or storage and reduce the conversion efficiency of backup plants, adding to these emissions (Gross and Hepponstall, 2008).

In terms of human health and well-being, concerns are related to noise and shadow flicker caused by the turbine blades passing through the sun and creating a flickering light shadow effect that is visually annoying. The loudest noises at a wind farm are generally emitted during construction and are similar for all construction projects. Operating wind turbines are now designed to meet a maximum noise output according to the IEC Standards (IEC, 2010). A shadow flicker analysis for close neighbors is done by the developers during the layout of the wind farm. At that time turbines can be moved to eliminate the possibility of turbine rotor shadows impinging on any close residences.

Aesthetic impacts are generally handled using techniques like an opinion survey of local inhabitants and organizations that are found in or that use the area, to determine the least visually intrusive layout for the wind farm and thereby gain general public acceptance. Responsible wind developers have established methods for gaining acceptance, and they attempt to make the projects a real economic benefit to the local residents and to minimize the visual impacts (see also Wolsink, 2007).

Regarding cultural impacts, the concern is to avoid possible intrusion on important historical, sacred, or archeological sites and to preclude possible litigation.

The economic and fiscal impacts on the local inhabitants are generally felt to be beneficial. Jobs are created locally, and the tax base is often increased along with the monetary benefits associated with direct payments to landowners who allow turbines on their property.

Turbines are sited with an eye to any possible interference with microwave communications systems and electromagnetic interference. Television interference has not been much of a problem since the widespread use of cable and dish satellite TV broadcasting. Interference between wind facilities and radar is real. If the radar unit cannot perform its intended functions, the effects must be mitigated. This can involve blocking the interference using a software fix or moving the turbines. Sometimes the radar unit must be moved or additional radar units added to the system (NWCC, 2006).

The environmental issues for wind technology concern the impacts of facilities on wildlife and wildlife habitat, including wildlife fatalities and habitat loss and modification, as well as animal displacement and fragmentation both for onshore and offshore wind facilities. This impact on wildlife can be direct (for example, fatalities or reduced reproduction) or indirect (such as habitat loss or behavioral displacement). The largest impacts to date are on flying animals, such as birds and bats. At the current level of deployment, land-based wind energy does not appear to have a significant impact on bird fatalities compared with other sources of fatalities, such as collisions with buildings and communication towers as well as predation by cats.

The general conclusion from site monitoring in the United States indicates that the average fatality rate is approximately three birds per MW annually, with about 70% of the fatalities being passerines (song birds) (Wildlife Society, 2007; NRC, 2007). Bat fatalities have been reported at land-based facilities either anecdotally or by post-construction monitoring. The highest fatalities have been recorded at wind farms located on ridges in eastern deciduous forests of the United States. Recent reports showing greater bat fatalities in the open prairie regions of southern Alberta, Canada, and in mixed agricultural and forested lands in New York suggest that the impacts on bats could be higher than currently assumed. Bats are long-lived and have low reproductive rates, so their ability to recover from population declines is limited. This increases the risk of local population extinctions and the loss of species biodiversity (Wildlife Society, 2007). Similar observations are found in Europe. In general, the variability of fatalities between facilities is quite high.

Commercial offshore wind energy facilities require a relatively large expanse of seabed floor for foundations and related structures to fix (or anchor) the structures and for deployment of interconnection cabling. These facilities will cause a degree of physical disturbance to the sea and surrounding seabed, with possible ecological responses (Gill, 2005). Therefore ecologists have expressed concerns about the impact on benthic communities, fisheries resources, marine mammals, sea turtles, and birds. The available studies on possible impacts are, however, limited (Minerals Management Service, 2007). In spite of the limited data available from existing studies, current results show limited negative direct impacts.

As with onshore wind farms, the collision of birds and bats with offshore wind turbine rotors is a major environmental and public concern. Radar studies at offshore Nysted wind farm in Denmark indicate that the

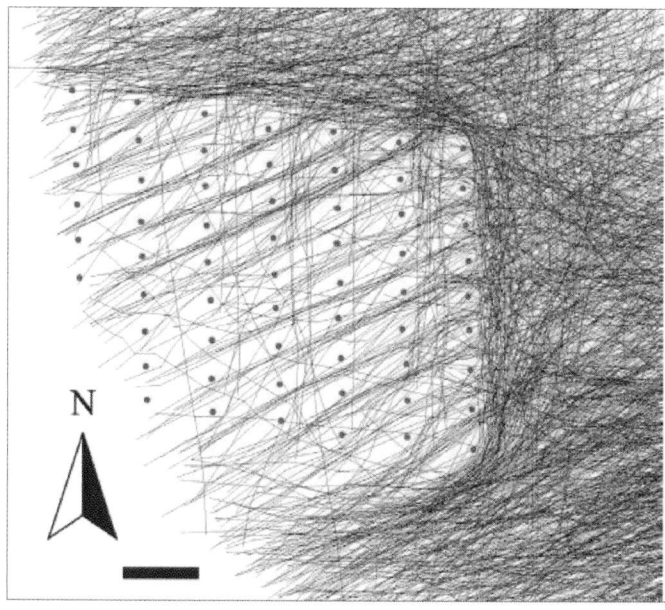

Figure 11.43 | Westerly orientated flight trajectories during the initial operation of the wind farm at Nysted off the southern coast of Denmark. Black lines show the trajectories of water bird flocks and the red dots are the wind turbine locations. The black bar is 1000 m. Source: Desholm and Kahlert, 2005.

diurnal percentage of flocks entering the wind farm area decrease significantly (by a factor of 4.5) from preconstruction to operational conditions (Desholm and Kahlert, 2005). At night about 14% of a flock entered the area of operating turbines, but only 6.5% of those flew within 50m of an operating turbine. During the day these numbers declined to 12.3% and 4.5% respectively. Radar trajectories of water bird flocks flying through the wind farm area are shown in Figure 11.43. It illustrates that water birds avoid wind farms, which reduces the risk of collisions.

11.5.7   Implementation Issues

11.5.7.1   Regional Deployment

The deployment of wind technology has historically been concentrated in Europe, North America, India, and China. High penetration scenarios such as GWEC's advanced scenario of about 2300 GW by 2030 require more diverse geographic deployment. In the advanced scenario, GWEC projects that regions other than Europe, North America, and China will account for 30% of cumulative wind capacity compared with 17% in the reference scenario (GWEC and Greenpeace, 2010a). Enabling the spread of the technology to other regions poses challenges and opportunities for the global industry.

11.5.7.2   Supply Chain Issues

Due to rapid increases in annual installations, from 2005 to 2008 the wind industry suffered acute shortages in the supply of gearboxes,

bearings, and skilled labor. In addition, the offshore industry supply chain experienced a shortage in specialized ports and installation vessels. These factors increased delivery time for turbines and installation time for offshore projects, and they generally drove costs upward. The industry has largely addressed these issues through investment in new manufacturing capacity, the globalization of the supply chain, and the widespread expansion of firms supplying the market. The supply chain does not appear to offer any insurmountable barriers despite the high level of annual installations required in high penetration scenarios.

11.5.7.3   Economics

Deployment will depend on the economic attractiveness of wind technologies balanced against competing technologies under prevailing market conditions. Onshore wind is competitive with conventional generation technologies in areas with strong wind resources. The cost of generating offshore wind power far exceeds market prices for electricity in most regions. While costs are expected to be reduced, near-term deployment of offshore wind technology will remain highly dependent upon policy. This presents a substantial barrier to achieving the high penetration scenarios that suggest that 18–32% of total capacity installed by 2050 may come from offshore (Lemming et al., 2009; IEA, 2010a).

11.5.7.4   Transmission and Integration

Studies have shown that electric systems can integrate 20% wind generation with relatively modest integration costs and without encountering insurmountable technical barriers (US DOE, 2008a; European Commission, 2010; IEA-Wind, 2010). System impacts for deployment above 20% are less well understood but could be addressed through structural changes in the electric market, including better forecasting techniques, increasing fidelity of dispatching procedures, flexible deployment of other generating plants, demand response measures, increased international coordination and interconnection (like the Supergrid and other proposals in Europe (FOSG, 2010; EEGI, 2010)), "smart grids," deployment of storage technologies (including electric transport system), and wind curtailment. (See Section 11.10.) The costs of integration and of maintaining electric system reliability will increase with higher levels of penetration and at some point are likely to constrain further deployment on economic terms. Substantial investments in transmission will be required to deliver power from both onshore and offshore wind plants to load centers. New transmission is required in even the low to moderate penetration scenarios and will limit wind energy deployment if not built.

11.5.7.5   Social and Environmental Concerns

Concerns about the social and environmental impacts of wind power plants – including bird and bat collision fatalities, habitat and ecosystem modifications, visibility, acoustics, competing uses, and radar interference – could limit the deployment of wind technology. These concerns are likely to become more acute as wind technology reaches higher levels of penetration. The wind industry can limit concerns over social and environmental impacts by seeking to understand the nature and magnitude of these impacts, developing mitigation strategies, and educating the public on the results of these efforts (Wolsink, 2007; Minerals Management Survey, 2007). Advanced deployment scenarios might require that regulators adopt a more streamlined permitting process for wind projects.

11.5.7.6   Policies

In the last two decades, an increasing number of countries have developed and implemented policy measures to promote renewables, including wind (REN21, 2010). Financial support for R&D, establishment of generation or capacity targets based on Renewable Portfolio Standards (RPS) or Obligations, and FITs have been used successfully. Additional measures sometimes include tax incentives, regulation of GHG emissions, bidding overseen by government, appropriate administrative procedures for wind farm planning, priority access to transmission grids, and transmission grid expansion. (See section 11.12.)

11.6   Photovoltaic Solar Energy

11.6.1   Introduction

Photovoltaic solar energy is the direct conversion of sunlight into electricity. The basic building block of a PV system is the solar module, which consists of a number of solar cells. Solar cells and modules come in many different forms that vary greatly in performance and degree of maturity. Applications range from consumer products (milliwatts) and small-scale systems for rural use (tens or hundreds of watts) to building-integrated systems (kilowatts) and large-scale power plants (megawatts and soon up to gigawatts). The PV market – and hence, the PV industry – is developing rapidly as a result of market support programs in a number of countries. So far, self-sustained markets are modest in size, but this may well change during this decade.

11.6.2   Potential of PV Solar Energy

The theoretical potential of solar energy is huge, as expressed in the popular statement that the amount of sunlight hitting Earth in one hour equals the total annual primary energy use worldwide. (See also Chapter 7.) However, this impressive fact has little practical significance. Therefore, it is important to consider the technical and realizable potentials.

11.6.2.1   Technical and Economic Potentials

The technical potential of solar energy is estimated to be between 1600 and 50,000 exajoules (EJ) a year (UNDP et al., 2000), with the wide range expressing the very different assumptions made in the analyses. Higher figures are mentioned too, however (see Chapter 7; also Figure 11.3). Even though the lower limit exceeds the current and estimated future worldwide energy use, there is no consensus on the economic potential of solar energy in general and of PV in particular because of the many technical, economic, and societal aspects at play.

11.6.2.2   PV Roadmaps and Scenarios

At the end of 2009, the installed solar PV capacity was 24–25 gigawatts peak (GWp), consisting of 21 GWp of grid-connected systems and 3–4 GWp of off-grid systems (REN21, 2010; IEA-PVPS, 2010; Jäger-Waldau, 2010). This generated 30–35 terawatt-hours (TWh) a year.[6] In 2010, another 16–17 GWp were added (versus about 7 MW in 2009), bringing the total installed capacity worldwide to roughly 40 GWp, producing some 50 TWh a year (EPIA, 2011).

In the 2010 IEA *PV Technology Roadmap*, the share of PV in global electricity consumption in 2050 is estimated at 11% (4600 TWh/yr) and the installed capacity at 3.2 TWp. Higher projections of the installed capacities can also be found. In *Solar Generation 6*, Greenpeace and the European Photovoltaic Industry Association project up to 4.7 TWp (6800 TWh) for 2050 (Greenpeace and EPIA, 2010). A study on very large-scale use of PV and concentrating solar power (in combination with compressed air energy storage) in the United States suggests that the combined contribution to electricity consumption could be as high as 69% in 2050; this would correspond to 35% of primary energy use, a share that could increase to 90% in 2100 (Fthenakis et al., 2009).

Komoto et al. (2009) prepared a detailed analysis of the potential of large-scale PV systems in sunbelt (desert) regions of the world. For six major regions, they determined a total potential of 465 TWp, allowing 750,000 TWh of solar electricity generation per year (2700 EJ$_e$/yr). In an ambitious scenario, the total installed PV capacity could be 10 TWp in 2050 and 133 TWp in 2100, of which 2 TWp and then 67 TWp would be in the form of very large-scale systems (the remainder would be urban and rural systems).

11.6.2.3   Potential of PV in the Built Environment

Because PV is a highly modular technology and does not involve moving parts, it can be integrated into buildings (roofs and facades) and infrastructure objects such as noise barriers, railways, and roads. This makes PV a suitable technology for use in urban and industrial areas. A number of studies have shown that the potential expressed as the fraction of

---

6   Because about half of the installations are still in Germany, which has relatively low insolation, the ratio between electricity generated and power installed will increase as the market share of "sunny" countries increases (from 1.25 to at least 1.5 kWh/Wp per year).

solar electricity from roofs and facades to the total electricity consumption per country varies from 20% to 60%, mainly depending on the electricity consumption per capita and the insolation level (IEA-PVPS, 2002; Lehmann et al., 2003). This assumes a rather conservative 10% energy conversion efficiency of the PV system. Future generations of PV modules may well cover a significantly higher fraction of the demand.

11.6.3   Market Developments

The market for PV systems can be divided into two main categories and several subcategories:

- *Grid-connected PV systems* – can be building-integrated and building-adapted systems (distributed systems), ground-based systems (power plants), and others (such as systems on sound barriers).

- *Off-grid / stand-alone PV systems* – can be solar cells integrated in consumer products, professional systems (e.g., telecom), rural PV systems, mini-grid systems, and others.

The market can also be divided according to the type of ownership, such as households, housing corporations, industries, companies, utilities, and institutional investors.

11.6.3.1   PV Market Deployment

Figure 11.44 shows the evolution of grid-connected and off-grid PV systems from 1995 to 2009. The total cumulative installed capacity has increased by about 30% per year on average. For grid-connected systems alone, the average growth was 50% a year.

Table 11.25 presents the cumulative installed PV capacity by country at the end of 2010, showing the leading positions of Germany (50%), Spain (11%) and Japan (10%) (see also REN21, 2011). On a per capita basis, Italy also belongs to the lead. These leading positions have been achieved in various ways (see IEA-PVPS, 2010; 2011). Japan has been a pioneer in market development and saw steady market growth until 2005, after which the market stabilized for a few years as a result of changes in market incentives; in 2009, Japan caught up again. Spain only recently joined the leaders, with an explosive (and unsustainable) growth of the Spanish market by 500% in 2008; but in 2009, the market fell back to pre-2006 levels due to drastically reduced market incentives. Germany is the only country with a substantial and steadily growing market through 2010.

The global PV market depends at present on market support policies in a very small number of countries. For sustainable and yet rapid growth, it is essential that the global market relies on more countries. Thus it is noteworthy that the group of countries with significant PV markets is growing, with Italy, France, and the United States as important examples. A

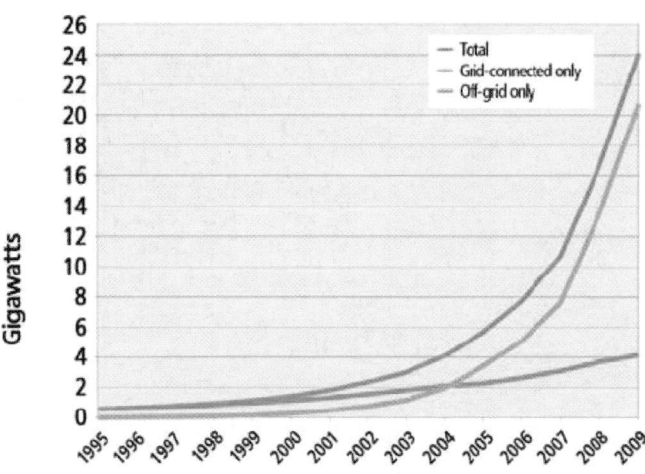

Figure 11.44 | Cumulative installed PV capacity, 1995–2009. In 2010 the total installed capacity was 40 GW (REN21, 2011). Source: REN21, 2010.

key question is how fast new markets will develop in, for example, China and India and countries in sunbelt regions, such as northern Africa.

Although currently the global market for PV systems depends heavily on various market support schemes, this situation is expected to change rapidly as system prices – and therefore, electricity generation costs – continue to decrease. In its *PV Roadmap*, the IEA distinguishes three levels of competitiveness (IEA, 2010). In the first level, the current situation, PV is only competitive in selected applications and regions of the world. In the second level, PV generation costs are lower than retail electricity prices. This is expected to happen between 2012 and 2030 almost everywhere. In the third level, PV electricity can compete with wholesale prices of electricity and, after that, with bulk power generation costs in a number of markets. This is expected to occur from 2020 onward. This scenario agrees with other major studies, such as the *Implementation Plan* of the European Photovoltaic Technology Platform (EPTP, 2009) and the Japanese *Roadmap PV2030+* of New Energy and Industrial Technology Development Organization (NEDO) (NEDO, 2009; Aratani, 2009).

11.6.3.2   Development of PV Industry

Figure 11.45 shows the growth of PV cell and module production in different regions. Manufacturers are located throughout the world, with recent rapid industry expansion in Asia. Table 11.26 shows the number of jobs in the PV sector, including R&D, as of the end of 2009. While the total number of PV-related jobs there is close to 200,000, China, Taiwan, and other Asian countries are not included in this list. So the real total is much higher (especially industry-related jobs, not deployment-related jobs).

The 2010 global turnover of the PV sector is roughly estimated to be US$_{2010}$\$100 billion. This number includes the value of turnkey system installations as well as investments in new production capacity and R&D. Although turnkey system prices are decreasing, market expansion

Table 11.25 | Cumulative installed PV capacity by country and application type, by the end of 2010. (Only the countries participating in the International Energy Agency-Photovoltaic Power Systems Programme (IEA-PVPS) are shown)

Country	Cummulative off-grid PV capacity (MW)		Cummulative grid-connected PV capacity (MW)		Cumulative installed PV power (MW)	Cumulative installed per capita (W/capita)	PV power installed during 2010 (MW)	Grid connected PV power installed during 2010 (MW)
	Domestic	non-domestic	distributed	centralized				
Australia	44.2	43.6	479.3	3.8	570	25.19	383.3	379.5
Austria		3.8	91.7		95.5	11.36	42.9	42.7
Canada	22.9	37.2	37.7	193.3	291.1	8.43	196.6	171.7
Switzerland		4.2	104.1	2.6	110.9	14.1	37.3	37.1
China					800	0.6	500	
Germany		50	17320		17370	212.47	7411	7406
Denmark	0.2	0.5	6.4	0	7.1	1.28	2.4	2.3
Spain					3915	84.83	392	389
France		29.8	830.3	194.2	1054.3	16.02	719	716
United Kingdom					69.8	1.07	40.1	
Israel	3	0.3	66.6	-	69.9	9.02	45	45
Italy	4	9	1532	1957	3502	57.76	2321	2321
Japan	3.4	95.4	3496	23.3	3618.1	28.28	991	986.8
Korea	1	5	131.3	518.3	655.6	13.38	131.2	131.2
Mexico	19.1	6.3	4.2	1	30.6	0.27	5.6	3.9
Malaysia		11		1.6	12.6	0.46	1.5	0.5
Netherlands					88	5.27	21	0.1
Norway	8.4	0.5	0.2	0	9.1	1.83	0.4	0.1
Portugal		3.1	33.1	94.6	130.8	12.3	28.6	28.5
Sweden	4.9	0.8	5.4	0.3	11.4	1.21	2.7	2.1
Turkey	1.2	4.2	0.6	0	6	0.08	1	0.1
USA		440	1727	367	2534	8.13	918	887
Estimated totals for IEA PVPS Countries (MW)	980		33973		34953		14192	14098

Source: IEA-PVPS, 2011.

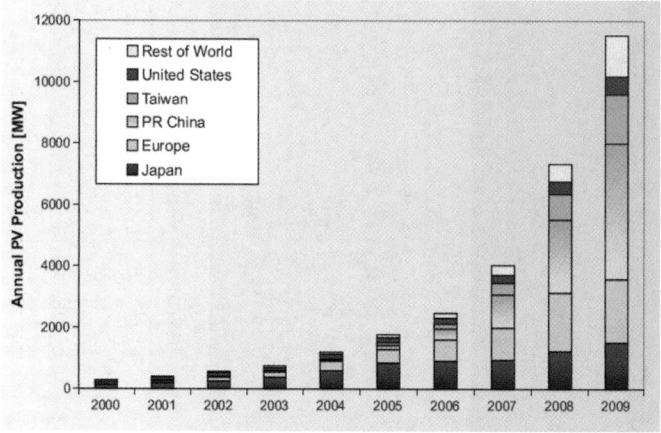

Figure 11.45 | Annual solar cell/module production by region. Source: Jäger-Waldau, 2010.

is so rapid that the related turnover is increasing rapidly. Investments in expansion of manufacturing capacity increase even more rapidly than market volumes, in view of expected PV market growth.

11.6.4   PV Technology Development

The discovery of the photovoltaic effect is usually attributed to Edmond Becquerel (Becquerel, 1839). Practical applications for power generation, however, have only come within reach after the successful development in the early 1950s of methods for well-controlled processing of semiconductor silicon, the material that most solar cells are still made of. The research group at Bell Laboratories in the United States played a key role in this development (Perlin, 1999). Many people immediately saw the great potential of PV for large-scale use, but the number and size of applications remained very modest until the 1980s. An exception was the use of PV to power satellites, which began successfully in 1958 and has remained their standard power source.

11.6.4.1   Basic Principles of Operation

The photovoltaic effect is based on a two-step process (see Figure 11.46), summarized by Sinke (2009) as follows:

- The absorption of light (consisting of light particles, or photons) in a suitable (usually semiconductor) material, by which negatively charged electrons are excited and mobilized. The excited electrons leave behind positively charged "missing electrons," called holes, which can also move through the material.

- The spatial separation (collection) of generated electrons and holes at a selective interface, which leads to a buildup of negative charge on one side of the interface and positive charge on the other side. As a result of this charge separation, a voltage (an electrical potential difference)

builds up over the interface. In most solar cells, the selective interface (junction) is formed by stacking two different semiconductor layers—either different forms of the same semiconductor (in homojunction cells) or two different semiconductors (in heterojunction cells).

The key feature of a semiconductor junction is that it has a built-in electric field, which pushes/pulls electrons to one side and holes to the other side. When the two sides of the junction are contacted and an electrical circuit is formed, a current can flow (i.e., electrons can flow from one side of the device to the other). The combination of a voltage and a current represents electric power. When the solar cell is illuminated, electrons and holes are generated and collected continuously and the cell can thus generate power.

11.6.4.2   Features of Sunlight and Solar Cell Efficiency Limits

The total annual amount of solar energy per unit area (the insolation) varies over Earth's surface and roughly ranges from 700 kWh/m^2 in polar regions to 2800 kWh/m^2 in selected dry desert areas for horizontal planes (Šúri, 2006).[7] When comparing the insolation on optimally inclined surfaces, the range becomes smaller, with the lower values increasing to roughly 900 kWh/m^2. In other words, the global range of electricity production potentials per m^2 of fixed, optimally inclined surface area roughly spans a factor of three.

The maximum intensity of sunlight is about 1 kW/m^2 everywhere. The differences in insolation primarily result from varying time fractions with low light levels (seasonal, but also daily variations). Whereas daily variations are inherent to the use of sunlight, the magnitude of seasonal variations in the daily (or weekly or monthly) amount of solar energy received may have important implications for system design and implementation.

Sunlight consists of a wide range of colors (from infrared to ultraviolet) and corresponding photon energies that make up the solar spectrum. The shape of the spectrum and the total intensity of the light depend on the position of the observer with respect to the sun and on atmospheric conditions. When the sun is exactly overhead and the sky is clear, the spectrum is "Air Mass 1," which means that the sunlight has passed through Earth's atmosphere in the shortest possible path: it has crossed "1 air mass."[8] Upon passing through the atmosphere, some light is absorbed and some light is scattered, leading to characteristic features in the spectrum shape. When the sun is incident at another angle, the

---

7   High-quality, extensive information on insolation in Europe and Africa can be found at sunbird.jrc.it/pvgis/, which also gives links to databases covering other regions of the world. An excellent source of US information is rredc.nrel.gov/solar/old_data/ nsrdb/redbook/atlas/.

8   Actually, definitions are much stricter than described here; see, e.g., rredc.nrel.gov/ solar/spectra.

Table 11.26 | PV-related labor places, selected countries, 2009.

Country	R&D, manufacturing, and deployment labor places	Country	R&D, manufacturing, and deployment labor places
Australia	5,300	Mexico	119
Austria	2,870	Malaysia	3,172
Canada	2,700	Norway	1,485
Denmark	350	Sweden	630
France	8,470	South Korea	6,500
Germany	65,000	Switzerland	8,100
Great Britain	1,171	Turkey	300
Italy	8,250	USA	46,000
Japan	26,700		

Source: IEA-PVPS, 2010.

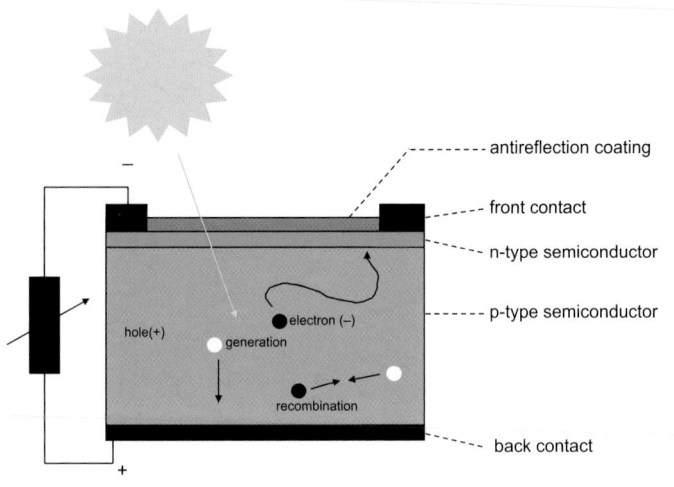

Figure 11.46 | Schematic cross-section of a solar cell.

Air Mass number changes accordingly. For measurement purposes, Air Mass 1.5 is often taken as a reference, where the path of light is at an angle through 1.5 times the thickness of the atmosphere.

The efficiency (η) of solar cells is defined as the maximum power output of the cell ($P_{max,electrical}$) divided by the power input in the form of light (P_{light}). For reasons of easy and fair comparison, efficiency values are normally defined and measured under Standard Test Conditions (STC): Air Mass 1.5 spectrum (AM1.5), 1 kW/m² light intensity, and 25°C operating temperature, although other standardized conditions are also useful and necessary for certain types of cells and modules, especially concentrators.

The rated power of PV cells, modules, and systems is expressed in terms of watt-peak (Wp), which is the power produced at STC. The efficiency of cells, and hence, of modules and systems, depends in part on operating conditions. Therefore, the average efficiency over a year differs from the STC efficiency. In most cases, the average value is lower. In particular, higher operating temperatures, lower light intensities, and different light spectra lead to deviations from STC efficiency. To calculate the electricity yield, the rated power has to be multiplied by the equivalent number of hours of full sun (e.g., per year) and corrected for the effects of non-STC conditions.

Semiconductors can be characterized by their light absorption behavior. Each semiconductor has a specific threshold energy above which photons can be absorbed and below which the material is basically transparent. This threshold energy is the so-called bandgap energy, the minimum energy it takes to free an electron from its original position in the crystal lattice of the material.[9] The bandgap energy varies

substantially from one semiconductor to another and determines which part of the solar spectrum can be absorbed. Low-bandgap materials absorb more of the solar spectrum than high-bandgap materials and can thus, in principle, generate more current. However, the output voltage of a solar cell is also related to the bandgap: the higher the bandgap, the higher the voltage (at least for ideal devices). This implies that to achieve maximum efficiency, it is important to choose an optimum bandgap.

In any cell made of a single semiconductor (i.e., one bandgap), there are large spectral losses. Part of the spectrum cannot be absorbed at all and the rest can only be partially used. The combined spectral losses add to more than 50% even for the optimum bandgap and perfect material. In addition to the spectral losses, any cell suffers from some fundamental losses – for example, those related to electrons and holes recombining. For cells made from one type of material and operating under natural, unconcentrated sunlight, the efficiency is therefore limited to a maximum of about 30%. The best single-material solar cells made so far have an efficiency of 25–26% (Green et al., 2010), indicating that these devices are already close to perfect. Commercial solar cells and modules have significantly lower efficiencies, as discussed in the next sections.

Two common strategies to move the efficiency beyond the limit are to use multiple materials (bandgaps) for light absorption and to use concentrated sunlight. By stacking solar cells with different absorption characteristics, it is possible to achieve better coverage of the solar spectrum and to reduce the spectral losses. These are called multijunction, multigap, multilayer, or tandem solar cells. By using concentrated sunlight (up to 1000x), the effects of recombination of electrons and holes can be reduced and the output voltage can be increased. This requires a dedicated device design, because the currents generated by concentrated sunlight are large, and significant amounts of heat need to be extracted. The combination of these two approaches pushes the fundamental efficiency limit up to 75% (Green, 2003).

---

9   The bandgap energy and also photon energies are usually expressed in electron-volts (eV)—the energy an electron takes up when it passes through a 1-volt (V) potential difference: 1.6×10^{-19} joule (J). The eV is a convenient unit in relation to the behavior of single electrons, such as light absorption. Bandgap energies are typically between a few tenths of an eV and a few eV. Solar photon energies are in the same range.

Table 11.27 | Overview of PV technologies, PV conversion concepts, and PV efficiency boosters.

State of Development	Category	Technology Type
Commercial	Flat-plate	Wafer-based crystalline silicon (mono-crystalline, cast multi-crystalline, ribbon)
		Thin-film silicon (amorphous, nano- and microcrystalline)
		Thin-film cadmium telluride (CdTe)
		Thin-film copper-indium/gallium-diselenide/sulfide (CIGSS)
	Concentrator	Silicon-based
		Compound (III-V) semiconductor-based
Emerging (typically advanced laboratory or pilot production)	Flat-plate	Polymer cells and modules
		Dye-sensitized cells and modules
		Alternative forms of inorganic thin films (e.g., printed CIGSS) and hybrid materials
	Concentrator (low concentration)	Luminescent concentrators using silicon or compound semiconductor cells
Novel concepts (laboratory only – research or proof-of-principle phase)	Not yet known	Intermediate-band semiconductors ("intrinsic multijunction materials")
		Spectrum converters ("external efficiency boosters")
		Various electronic and optical applications of quantum dots (e.g., "all-silicon tandems")
		Plasmonic structures for light management
		Hot-carrier devices

Source: EPTP, 2007.

There is growing interest in achieving such efficiencies through "third-generation photovoltaics." Some of these concepts offer fundamental efficiency limits of 75–85% (Green, 2003); however, they are all in a very early stage of development.

11.6.4.3   Practical Solar Cells and Modules

Solar modules are commonly divided into two categories:

- Flat-plate modules, in which the active cell area is roughly equal to the light-harvesting area.

- Concentrator modules, in which a small-area solar cell is illuminated by sunlight collected on a (much) larger area. These modules have a lens or mirror to focus sunlight onto the solar cell.

Concentrator modules need to track the sun because the light must come from a well-defined angle to produce a high-quality focus on the cell. For similar reasons, concentrator modules only use the direct (not the diffuse) part of the sunlight. Therefore, they work best in regions where the fraction of direct radiation is high, typically countries with an insolation of 1500 kWh/m^2/yr or more. Low-concentration-factor modules that do not require tracking have also been developed; "flat-plate concentrators" or "luminescent concentrators" fall in this category (see, e.g., Currie et al., 2008).

For flat-plate modules, it has been calculated that the performance gain of two-axis tracking compared with using an optimal but fixed orientation ranges from 10% to more than 40%, depending on the geographical location (Huld et al., 2008; Komoto et al., 2009).

It is also common to categorize cell and module technologies according to the active material(s) – the semiconductors – used for the solar cells. At the highest level, the technologies can be divided into wafer-based technologies and thin-film technologies.

Although the term "wafer-based" usually refers to flat-plate technologies, concentrator cells may also be (and usually are) wafer-based. The wafer may be silicon, germanium, or gallium arsenide, although the latter two are only for concentrator applications. In the discussion here, "wafer-based" is used only regarding silicon technology. The individual cells produced from wafers are electrically connected before or on encapsulation in a module.

In the case of thin-film technologies, the cells are deposited on a substrate (glass or metal foil) or a superstrate (glass) in the form of a very thin layer. Typical thicknesses are on the order of 1 micrometer (10^{-6} meter).

In the framework of a Strategic Research Agenda (EPTP, 2007), the European Photovoltaic Technology Platform prepared an overview of the present state of the art and expected future developments in cell and module technologies. The technology categories and types are summarized in Table 11.27.

Figure 11.47 presents the historic development of record laboratory cell efficiencies for a selection of technologies and concepts mentioned in Table 11.27. The figure shows gradually increasing efficiencies for most technologies, with occasionally more rapid increases and a few cases of saturation (indicated by termination of trend lines).

While Figure 11.47 shows record efficiencies for small-area laboratory cells (typically 1 cm^2), efficiencies for large-area commercial module

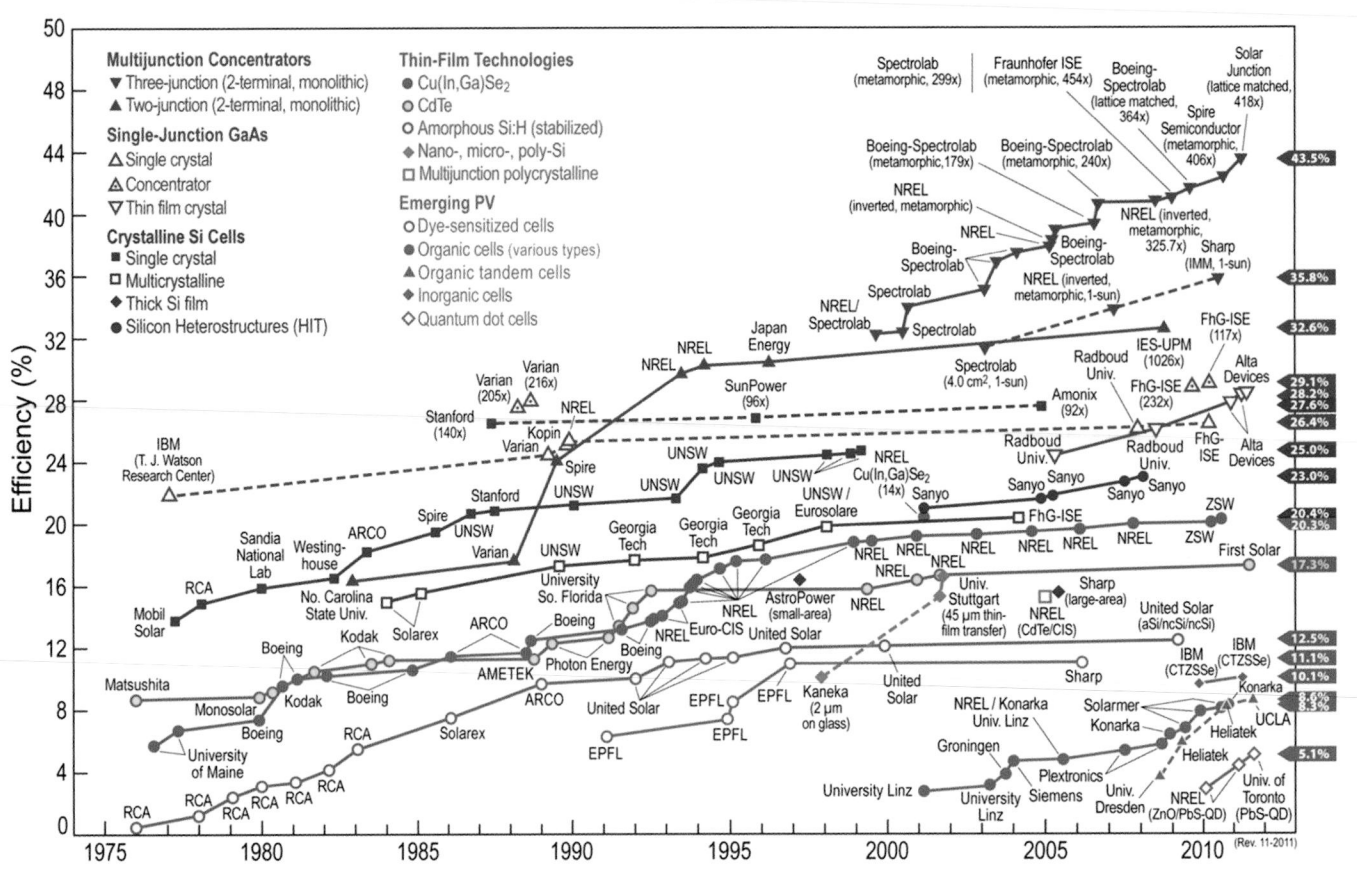

Figure 11.47 | Trends in conversion efficiencies for various laboratory solar cell technologies. Source: Kazmerski, 2011. Courtesy of NREL.

efficiencies (typically 1 m²) are significantly lower. Efficiencies of currently available most commercial modules and the possible efficiency evolution of improved, as well as new, technologies are presented in Figure 11.48. The figure shows a gradual but robust increase for all existing technologies and the emergence of new technologies (aimed at very high efficiencies or at very low manufacturing costs) in the course of this decade.

11.6.4.4   PV Systems and Systems Terminology

Complete PV systems consist of modules (also referred to as panels) that contain solar cells, and the so-called balance-of-systems (BOS). The BOS mainly comprises electronic components, cabling, support structures, and, if applicable, electricity storage, optics and sun trackers, and/or site preparation. The BOS costs also include labor costs for turnkey installation.

PV systems are often divided into grid-connected systems, whether integrated or ground-based, and stand-alone (or autonomous) systems. PV systems feeding into or connected to a mini-grid fall in between these categories.

Systems consist of modules that are electrically connected in series and/or in parallel. Modules connected in series are called a string. A number of strings connected in parallel are called an array or sub-array. A number of arrays that function together are called a system or subsystem. To be able to quantify the performance of grid-connected PV systems and to allow comparisons with other electricity-generating technologies, the following terms and definitions are helpful:

- *System power:* The nominal (nameplate, rated) power of PV cells, modules, and systems is expressed in watt-peak – the power produced under STC. The power of complete systems is often simply expressed as the sum of the powers of the individual modules that make up the systems, although the actual direct-current (DC) power of many modules connected in series and in parallel will never equal the sum of the individual powers.

- *Performance ratio:* The (dimensionless) Performance Ratio (PR) of a PV system is defined as the average alternating-current (AC) system efficiency divided by the STC module efficiency. The PR is usually taken as the average over a year. In the PR, the effects on efficiency of very different factors are taken together. Most important are module mismatch, cabling and inverter (DC/AC conversion) losses, and

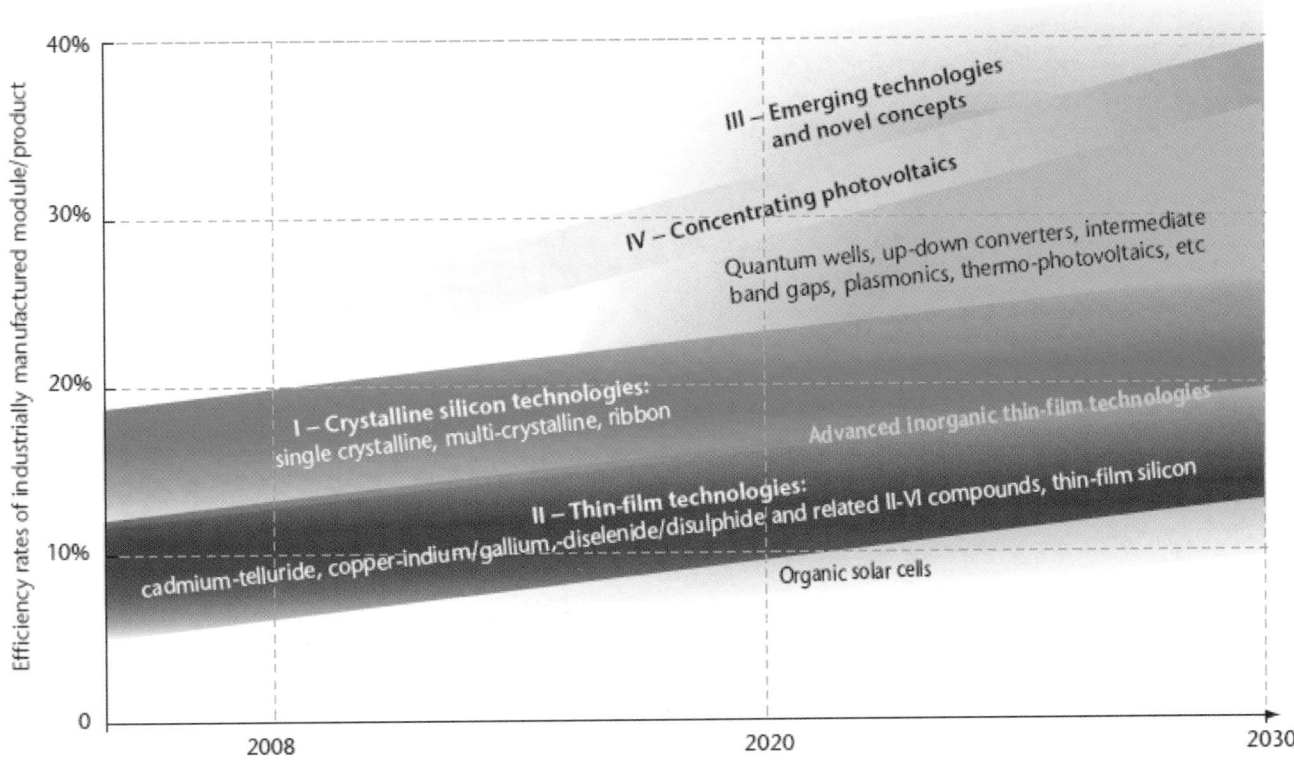

Figure 11.48 | State of the art and possible development of commercial PV module efficiencies. Source: IEA, 2010d. ©OECD/International Energy Agency 2011.

cell/module operating temperatures, as well as light intensities, light spectra, and angles of incidence that deviate from standard conditions. Also, system outages, if they occur, influence the PR. Typical PR values for recent well-functioning systems are in the range of 0.70~0.85 (IEA-PVPS, 2007). With improvements, the PR may have a modest upward potential to 0.9 ultimately.

- *PV system performance in practice:* Extensive data on the performance of PV systems in different countries are collected in the framework of IEA-PVPS Task 2 (IEA-PVPS, 2007; Nordmann, 2008). From 1991 to 2005, the typical system PR improved significantly – from 0.64 to 0.74 – as a result of, among other reasons, higher inverter efficiencies and fewer and shorter system outages and more accurate module rating. This increase does not appear to be saturated yet. In 2005, the best systems had a PR in the range of 0.80 to 0.85.

- *Capacity factor:* The (dimensionless) Capacity Factor (CF) of a PV system is determined by the insolation (in the plane of the modules) at the location of the system and the system PR. It is defined as (N × PR) / 8760, in which 8760 is the number of hours in a year and N is the equivalent number of full-sun hours in a year. N is equivalent to the insolation (in kWh/m²/yr) divided by the intensity of full sun (1 kW/m²). It is noted that the numeric value of N is equal to the insolation, because on Earth, "full sun" can defined as 1 kW/m². Taking the most relevant range of insolation values (in the plane of the modules) as 1000–2500 kWh/m²/yr and the range of PRs as

0.75–0.85, the global range of CFs becomes 0.08–0.21, with upward potential to 0.23. It is important to note that these values refer to systems without sun tracking. If tracking is applied, the CF of systems in high-insolation regions may ultimately reach values around 0.30. The CF is a particularly useful parameter when comparing different electricity generation technologies.

- *Specific electricity yield:* The specific final AC electricity Yield (Y_f) of a system is defined as the annual AC electricity output of the system (P × N × PR) divided by the system power (P). The global range of Y_f values is 750 to 2100 kWh/kWp/yr, with upward potential to 2250 kWh/kWp/yr, excluding yield gains by sun tracking. With tracking, however, which is especially likely to be applied for all ground-based desert systems, Y_f could go up to almost 3000 kWh/kWp/yr.

Stand-alone systems (Luque, 2003) come in a wide variety of types and sizes (powers), ranging from mini systems integrated in consumer products with a typical power well below 1 Wp, through solar home systems for use in rural areas with a power in the order of 100 Wp, to larger systems for industrial use and village electrification above 1 kWp. They may also be combined with other electricity generators such as wind turbines in a hybrid system.

Although the market share of stand-alone systems is small and decreasing, their value for the user is often very high. This is because these systems are generally the user's only source of electricity, and the alternatives are either more expensive or less convenient. It is difficult to define a simple set of performance indicators as in the case of grid-connected systems. The performance of stand-alone systems, however, can also be measured in terms of their availability – that is, the fraction of time they are able to supply the electricity needed by the user.

11.6.5 Economic and Financial Aspects

Although PV modules are the most visible part of PV systems, the economics of PV obviously depends on the price and performance of complete, turnkey systems. The price of a turnkey PV system is the sum of the price of the modules and of the BOS; it also contains labor costs, such as those related to engineering and installation.

11.6.5.1 Module Price Development

The price evolution of PV modules can be described well by a price-experience or "learning" curve, in which the average selling price (ASP) is plotted on a double-logarithmic scale as a function of cumulative production (van Sark et al., 2008; Hoffmann et al., 2009). Figure 11.49 presents data for 1980–2008. The straight line indicates the price decrease as a fixed percentage for each doubling of the cumulative production (or shipments). The progress ratio is defined as 100% minus this percentage.

The figure shows that the ASP of crystalline silicon modules (the main technology in this period) decreased by about 20% for each doubling of the cumulatively shipped volume. The curve also shows the effect on prices of the temporary silicon feedstock shortage in the early 2000s, due to rapid market expansion. By 2010, prices had largely returned to the historic trend line, as the shortage had been solved. By December 2009, prices were falling below $2/Wp in some instances (REN21, 2010). Note that the curve shows prices, not costs. The black dot in the figure shows that the price level for thin-film modules was comparable to that of crystalline silicon modules at its lowest point, albeit at a much lower cumulative production volume. This underscores the strength of thin-film technologies as low-cost options.

Although it is not possible to simply extrapolate the curve to higher cumulative production volumes (and thus, implicitly, into the future), the potential for further cost reductions by technology development and economies of scale is still substantial for both crystalline silicon (Sinke et al., 2009; Swanson, 2006) and thin-film technologies (Hoffmann et al., 2009).

11.6.5.2 Value of Higher Module Efficiency

Turnkey system prices of installed systems are a better indicator of the competitive position of PV than module prices as such. Because the BOS component of system prices consists of a power-related part (such as

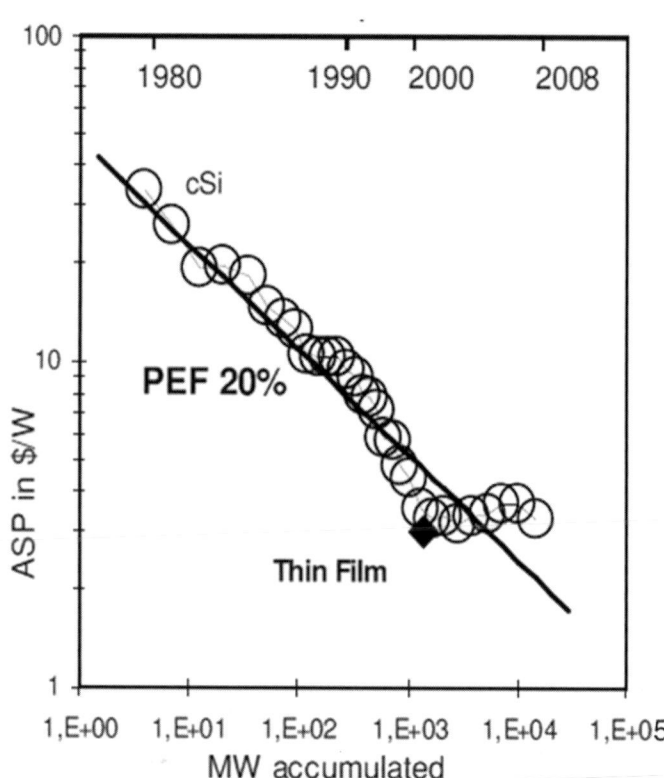

Figure 11.49 | PV module experience curves showing the average selling price (ASP) of modules in US_{2008}$ as a function of the cumulative module production. Note that the Progress Ration (or PEF: Price Experience Factor) is about 20%. See text for further explanation of the curve. Source: Hoffmann et al., 2009.

the inverter) and an area-related part (such as the mounting or support structure), module efficiency has an influence on system price (US$/Wp). This could be turned around to say that for an equal turnkey system price, higher efficiency modules are generally allowed to be somewhat more expensive than lower efficiency modules. For example, assume an area-related BOS price of US$100/m² and a power-related BOS price of US$0.50/Wp. Then a turnkey system price of US$3/Wp may allow a module price of US$1.50/Wp when the conversion efficiency of the module would be 10%, but US$2/Wp when the module efficiency would be 20%. In this specific example, the "value" of 20% over 10% efficiency is thus US$0.5/Wp.

11.6.5.3 PV System Price Development

System prices vary much more than module prices do because of the wide variety of system types and sizes, country-to-country differences in experience and installation practice, and other factors. For systems that can be compared, a European study has indicated that the BOS part of system prices may follow an experience curve with a progress ratio similar to that of modules, although the uncertainties are much larger (Schaeffer et al., 2004). Moreover, it is uncertain whether this trend can be maintained over the long term, because drastic possibilities for price reduction are perhaps less obvious for BOS than for modules. Therefore

Table 11.28 | Turnkey investment costs of PV systems and corresponding (rounded) levelized electricity generation costs for 2009, 2020, and 2050.

Typical Irradiation on Fixed Optimally Oriented Plane (kWh/m²/yr)	Capacity Factor (and Corresponding Annual Yield in kWh/kWp)	Typical Turn-key Investment Costs (US$_{2005}$/kWp)	Cost of Electricity (¢/kWh)	
			Discount rate	
			5%	10%
Current (2009)				
1000	9% (790)	4500 (typical range 3500–5000)	46.1	68.5
2000 / 1500 (without / with sun tracking)	18% (1580)		23.1	34.2
2300 (with sun tracking)	27% (2370)		15.4	22.8
2020				
1000	9% (790)	2000 (possible range < 1500–2500)	19.0	29.4
2000 / 1500 (without / with sun tracking)	18% (1580)		9.5	14.7
2300 (with sun tracking)	27% (2370)		6.3	9.8
Long-Term (2050)				
1000	9% (790)	900 (possible range ~ 700–1200)	8.6	13.2
2000 / 1500 (without / with sun tracking)	18% (1580)		4.3	6.6
2300 (with sun tracking)	27% (2370)		2.9	4.4

it has been argued that the progress ratio for BOS on the longer term may be 85% or 90%, rather than 80% as for modules.

The evolution of system prices is monitored by IEA-PVPS (see, e.g., IEA-PVPS, 2010). The 2009 turnkey system prices of larger grid-connected systems were roughly in the range of US$_{2009}$3.5–7.5/Wp in major markets – with the lower end of the range being characteristic for systems installed in countries with a well-developed and competitive market. For off-grid systems this range was US$_{2009}$7–22/Wp (IEA-PVPS, 2010).

Turnkey system price can be translated into electricity generation costs using the levelized cost of energy method. Although the calculation involved is rather straightforward (see, e.g., EPTP, 2010), it requires assumptions and estimates of parameters that may not be made in a straightforward manner. More precisely, in addition to the turn-key investment price, values are needed for the O&M costs (normally expressed as a percentage of the investment per year), system economic lifetime (depreciation/amortization period), cost of capital, and specific electricity yield (kWh per year per watt-peak of system power). However, it is difficult to "standardize" the parameters used in the calculation because they may vary substantially depending on, among other items, the type of technology and system, geographical location, and type of ownership.

Table 11.27 gives 2009 and (target) 2020 and long-term turnkey investment costs and corresponding levelized electricity generation costs.

Indicative turnkey investment costs (in mature markets), O&M costs, and depreciation times are taken from the IEA-PVPS *Trends Report* (IEA-PVPS, 2010) and IEA *PV Technology Roadmap* (IEA, 2010d). It is noted, however, that more aggressive cost reduction targets can also be found in the literature – see, for instance, the US DOE's SunShot Initiative, which mentions a target of US$1/Wp to be reached even before 2020 (US DOE, 2011b).

In Table 11.28, the capacity factor ranges indicated roughly correspond to low-, medium-, and high-insolation regions of the world, assuming a performance ratio of 80% and sun tracking for high-insolation regions. Gains of sun tracking are indicative and vary per region (see section 11.6.4.4 and Huld et al., 2008). In the calculations, annual O&M costs were assumed as 1% of the investment costs. The system lifetime assumed is 25 years in 2009, 30 years in 2020, and 40 years in 2050. Note that the table is primarily meant to show trends and typical numbers; in practice, other figures are found (IEA-PVPS, 2010).

11.6.5.4 Grid and Investment Parity

PV will gradually reach various levels of competitiveness (IEA, 2010d; Greenpeace and EPIA, 2008; Breyer, 2010). (See Figure 11.50.) For grid-connected systems, which compete with electricity from the grid, these levels are usually referred to as "grid parity" (e.g., with consumer

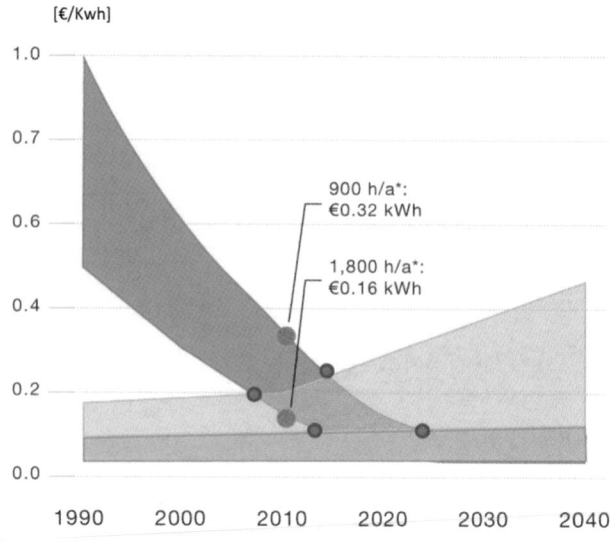

[€/Kwh]

900 h/a*:
€0.32 kWh

1,800 h/a*:
€0.16 kWh

■ PV GENERATION COST AT LOWEST PRICE
□ UTILITY PEAK POWER
■ UTILITY BULK POWER

Figure 11.50 | Schematic representation of grid parity points (in blue, see text). *h/a: Hours of sun per annum, 900 h/a corresponds to northern countries of Europe. 1800 h/a corresponds to southern countries of Europe. Source: European Photovoltaic Industry Association/Greenpeace International, 2011.

prices and wholesale prices respectively). The concept of "grid parity" is based on a comparison of LCOE and electricity prices in a specific year. The concept of grid parity does not take into account the expected increase of conventional electricity prices; therefore "investment parity" or "dynamic grid parity" has been proposed (EPIA, 2009) as a more useful concept in relation to decision-making. In this case, the life-cycle costs of PV electricity generation are compared with the anticipated total value of avoided electricity purchase and revenues of sales over the relevant period (such as 15, 20, or 25 years).

Note that in the direct comparison between PV generation costs and electricity prices, other costs and benefits are not taken into account. Specifically, the comparison does not take into account costs for grid transport, backup power, or storage (even though these need not be attributed exclusively to PV). Nor does it take into account benefits such as avoided carbon dioxide emissions and (passive or active) grid support. Therefore, the point of grid parity is useful only as a rough indicator of the competitive position of PV.

11.6.6   Sustainability Issues

PV technology may inherently be renewable, but it is not automatically sustainable. The sustainability of PV systems depends on, among other considerations, equivalent CO_2 emissions, the use of hazardous and non-Earth-abundant elements and materials, and possibilities

for recycling. The equivalent CO_2 emissions are related to the energy needed for manufacturing and installation (often expressed in terms of the energy payback time) and the fuel mix for generation of electricity used in PV manufacturing.

To date, most attention has focused on the energy payback time of PV systems and the equivalent CO_2 emissions of PV electricity generation. Figure 11.51 shows the energy payback time of PV technologies (including recycling) when applied in reasonably sunny regions (in-plane irradiation of 1700 kWh/m²/yr) (De Wild-Scholten, 2010). Energy payback times are all well below 2 years and even below 1.5 years for thin-film technologies, which is to be compared with a service lifetime of PV systems of 25 years or more. The energy payback times are expected to decrease further substantially as the technology develops (EPTP, 2009).

The equivalent CO_2 emissions per kWh produced with PV electricity are 16 gCO₂-eq/kWh at an in-plane insolation of 1700 kWh/m²/yr for the best current PV systems (de Wild-Scholten, 2010). Based on a review of life-cycle analysis studies, Arvizu et al. (2011) conclude that "the majority of lifecycle GHG emission estimates cluster between about 30–80 gCO₂-eq/kWh." It is expected that this value may be reduced to 10 g or less, along with shortening of the energy payback time and transformations in the energy system, decreasing the indirect emissions (Reich et al., 2011). These values refer to equivalent emissions directly related to the PV plants. High degrees of penetration of PV necessitate modifications on a system level (e.g., adding backup power, small- or large-scale storage, smart grids) to accommodate the power generated by PV. These modifications may lead to increased – or, in specific cases, decreased – overall emissions.

Recently, the rapidly increasing production volumes and high prices of a wide range of materials have drawn attention to the use of non-Earth-abundant elements such as silver, indium, and tellurium in PV cells and modules. Multi-gigawatt-scale or even terawatt-scale manufacturing of specific PV technologies is only possible at very low cost when materials constraint can be avoided. This has led to research on alternatives for active and passive parts of PV cells and modules. Examples are copper- and carbon-based conductors instead of silver in wafer-silicon cells and zinc-tin instead of indium in the light absorber of CIGS modules.

For many years, there has been a debate about the use of hazardous materials in PV modules, particularly Cadmium telluride (CdTe). CdTe is the biggest thin-film PV technology in the market today, and take-back and recycling systems have been developed and implemented. Moreover, CdTe is a very stable compound. Therefore, it has been argued that CdTe can be used in a safe and sustainable way (Raugei, 2010).

Companies in the PV sector joined forces in 2007 and founded the PV Cycle Association. It aims to "implement the photovoltaic industry's commitment to set up a voluntary take back and recycling program for

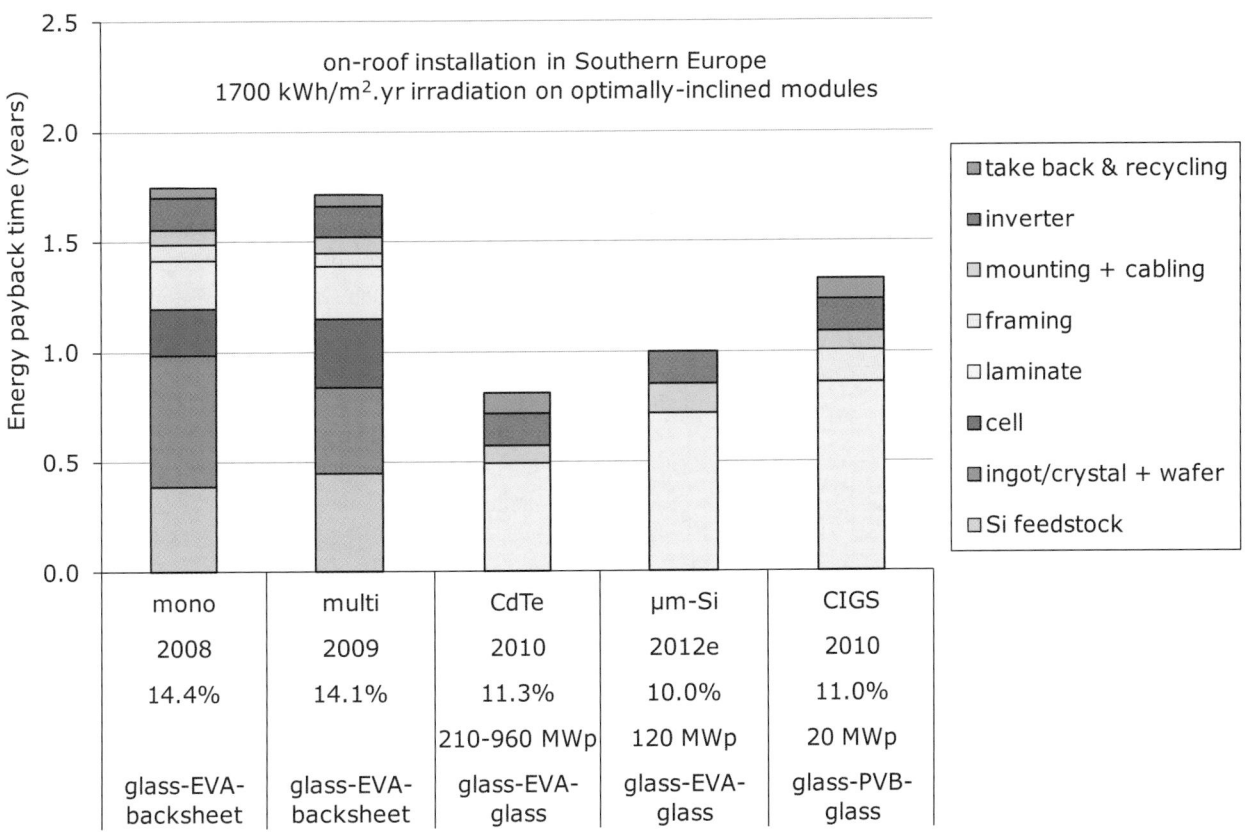

Figure 11.51 | Energy payback time (EPBT) of on-roof PV systems for an in-plane irradiation of 1700 kWh/m²/yr. Source: de Wild-Scholten, 2011.

end-of-life-modules and to take responsibility for PV modules throughout their entire value chain" (PV Cycle, undated). In 2010, there were about 40 full members.

11.6.7   Implementation Issues

The main issues to be addressed in relation to very large-scale deployment of PV are the cost of electricity generation and grid integration. In the long term, materials availability also needs to be considered. In addition, a range of less fundamental issues to be addressed include the development of dedicated products for building integration and standardization.

The cost of electricity is generally considered to be the key driver for PV deployment, but grid integration is certainly a key enabler (FVEE, 2010). Grid integration is not a problem today, with the exception of some local constraints. However, ambitious deployment scenarios may bring PV into the regime where grid adaptations are necessary (Braun, 2009). The framework of the Grand Solar Plan (Fthenakis et al., 2009) in the United States illustrates how very large volumes may be integrated into the electricity system by choosing a portfolio approach—in particular, by combining PV with concentrating solar power and compressed air energy storage.

Large-scale deployment of PV, and thus high levels of penetration into the grids, requires sufficient grid flexibility to enable integration of the varying output of PV systems and to provide the backup power needed when the sun does not shine. This can be achieved by introducing a proper portfolio approach – that is, combining PV with other generators such as wind energy, biomass, natural-gas-fired plants, and hydropower – and demand-side management (FVEE, 2010). In addition, storage will become important on small and large scales (including, for instance, the capacity available in electric vehicles) and for different time scales (day-night; summer-winter).

11.7   High-Temperature Solar Thermal Energy

11.7.1   Introduction

High-temperature solar thermal technologies, also referred to as concentrating solar thermal, use mirrors that reflect and concentrate sunlight onto receivers. The receivers convert the solar energy to thermal energy, which is used in a steam turbine or heat engine to drive an electric generator. These concentrating solar power (CSP) systems might also allow the production of chemical fuels for transportation, storage, and industrial processes (Meier and Steinfeld, 2010). CSP systems perform best in regions having a high direct-normal component of solar radiation.

Table 11.29 | Potential growth of cumulative CSP capacities until 2050

Name of Scenario and Year	Installed CSP Capacity [GW$_e$]				
	2000	**2015**	**2020**	**2030**	**2050**
CSP Global Outlook 2009 (Reference scenario)	0.35	4	7	13	18
CSP Global Outlook 2009 (Moderate Scenario)	0.35	24	69	231	831
CSP Global Outlook 2009 (Advanced Scenario)	0.35	29	84	342	1524
IEA Reference Scenario (2008)	0.35	N.A.	N.A.	< 10	N.A. (but competetive)
IEA ACT Map (2008)	0.35	N.A.	N.A.	250	380
IEA BLUE Map (2008)	0.35	N.A.	N.A.	250	630

Source: IEA, 2008; Greenpeace et al., 2009.

The primary advantage of CSP is its potential for integrating thermal storage, which would allow dispatchable power generation. Dispatchable CSP electricity can have a higher value than variable renewable energy technologies (Madaeni et al., 2011). The four main types of CSP systems are parabolic trough, power tower, dish/engine, and linear Fresnel reflector. Systems that use concentrators to focus sunlight onto high-efficiency solar cells for utility-scale electricity generation are discussed in section 11.6.

11.7.2 Potential of High-Temperature Solar Thermal Energy

CSP requires significant levels of direct-normal irradiance, which generally occurs in semiarid areas between 15° and 40° north or south latitude. Closer to the equator the humidity is generally too high, and at higher-latitude regions there is usually too much cloud cover. A threshold of 1800–2000 kWh/m²/yr is often considered suitable for CSP development. The regions with the best resource potential are the Mediterranean, North Africa, Middle East, South Africa, portions of southern Asia, Australia, Chile, the southwestern United States, and Mexico.

By the end of 2010, more than 1.1 GW of grid-connected CSP plants were installed worldwide, generating 2.9 TWh of electricity a year (REN21, 2011). The installed capacity in 2009 was 610 MW, generating 1.6 TWh a year.

The *Global CSP Outlook 2009* – developed jointly by SolarPACES, the European Solar Thermal Electricity Association, and Greenpeace International – used an advanced industry development scenario, with high levels of energy efficiency, to project future electricity demands. The study estimates that CSP could meet up to 7% of the world's power needs by 2030 and 25% by 2050. The global CSP capacity for 2050 could be 1500 GW with annual energy output of 7800 TWh. More moderate

assumptions for future market development put combined solar power capacity at around 830 GW by 2050, with annual deployments of 41 GW. This would meet 3.0–3.6% of global electricity demand in 2030 and 8.5–11.8% in 2050 (Greenpeace et al., 2009).

A recent analysis of CSP potential in the United States projects capacity of 11,000 GW in Arizona, California, Colorado, Nevada, New Mexico, Texas, and Utah by 2030 (Mehos et al., 2009). Another study projects up to 30 GW of parabolic trough systems with thermal storage could be deployed in this region by 2030 (Blair et al., 2006). Capacity in the Middle East and North Africa is projected to be 390 GW by 2050 by the German Aerospace Center. That study concludes that with the high capacity factors resulting from integration of CSP with thermal storage, this capacity could provide about half this region's electricity (German Aerospace Institute, 2005). The trends and potential for CSP capacities from recent studies are shown in Tables 11.29 and 11.30.

In 2010, IEA published a technology roadmap for CSP, which foresees a potential generation of 4000 TWh in 2050, contributing 10% to global electricity production (IEA, 2010e). CSP electricity production and consumption predicted in the roadmap is shown in Figure 11.52. The roadmap authors expect North America to be the largest producing region, followed by Africa, India, and the Middle East. Africa would be by far the largest exporter of electricity, and Europe the largest importer. The Middle East and North Africa considered together, however, would produce almost as much electricity as the United States and Mexico (IEA, 2010e).

11.7.3 Market Developments

The number of CSP plants built or planned each year has increased since 2004. The size of new plants has increased from several megawatts to hundreds of megawatts. By the close of 2010, CSP generation was able to meet intermediate-load demand, particularly in plants that have thermal energy storage. As the amount of storage increases, CSP could become cost-competitive in the base-load electricity market if the generation cost of conventional fossil base-load technologies includes a carbon cost.

Between 1985 and 1991, about 354 MW of solar parabolic trough technology were deployed in southern California, and most of these plants are still in commercial operation. CSP technology is most economically viable in large-scale installations. In the 1990s, world energy prices dropped and remained relatively low. Low prices and the lack of incentives discouraged additional large installations.

The emerging demand for cuts in GHG emissions, as well as the need to decrease dependence on fossil fuels, may improve the market outlook for CSP. Worldwide, interest in CSP is increasing in the United States, Spain, and the Middle East-North Africa. Figure 11.53 shows the development of CSP from 1985 to 2008 and the estimated CSP project pipeline, by country, for 2009–2014.

Table 11.30 | Potential growth of electricity generated by CSP until 2050.

Name of Scenario and Year	Electricity production by CSP [TWh/yr]				
	2000	2010	2020	2030	2050
IEA Reference Scenario (2008)	N.A.	N.A.	N.A.	< 15	25
IEA ACT Map (2008)	N.A.	N.A.	N.A.	625	890
IEA BLUE Map (2008)	N.A.	N.A.	N.A.	810	2080
Shell (Scramble)	N.A.	N.A.	110	1450	5220
Shell (Blueprints)	N.A.	56	390	1220	4110
Greenpeace Reference Scenario (2010)	0.63	5	38	121	254
Greenpeace Revolution Scenario (2010)	0.63	9	321	1447	5917
Greenpeace Advanced Scenario (2010)	0.63	9	689	2734	9012

Source: IEA, 2008; Shell, 2008; Greenpeace and EREC, 2007.

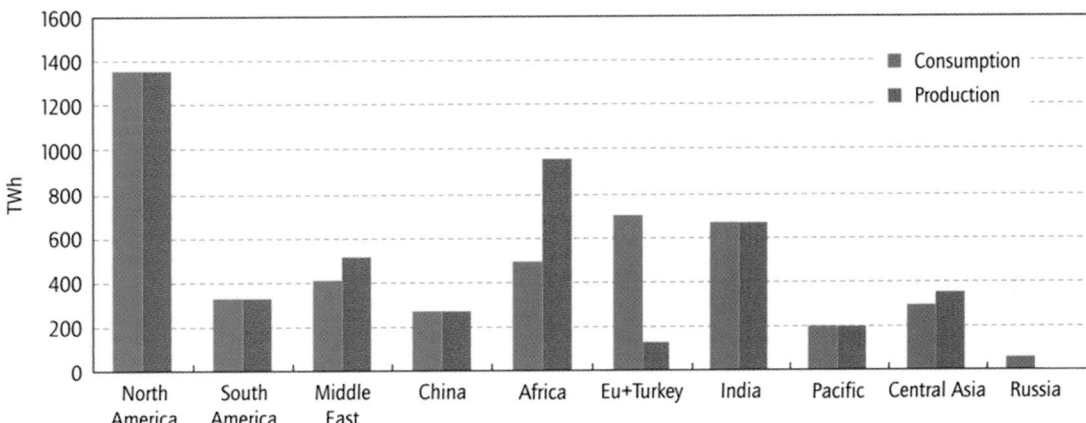

Figure 11.52 | Possible production and consumption of CSP electricity in 2050. Source: IEA, 2010e. ©OECD/International Energy Agency 2010.

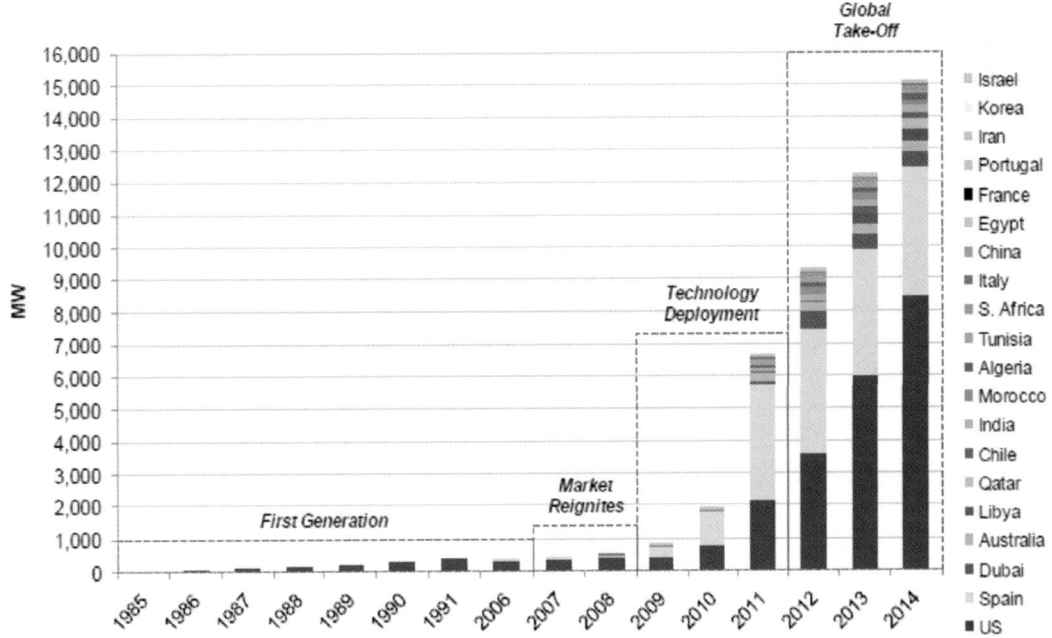

Figure 11.53 | CSP development, 1985–2008, project pipeline by country, 2009–2014. Source: IHS EER, 2009.

Table 11.31 | CSP plants in operation, under construction, and under development in the United States, end of 2010.

Technology	Capacity Operational		Capacity Under Construction		Capacity Under Development	
	MWe	# projects	MWe	# projects	MWe	# projects
Trough	496	14	5	1	5406	19
Tower	5	1	392	3	2545	10
Dish/engine	2	1			964	5
Linear Fresnel	5	1	2	1	231	3
Total	508	17	399	5	9146	37

Source: EIA, 2011.

Table 11.32 | CSP plants in operation, under construction, and under development in Europe, Asia, and North Africa, end of 2010.

Technology	Capacity Operational		Capacity Under Construction		Capacity Under Development	
	MWe	# projects	MWe	# projects	MWe	# projects
Trough	617	11	810	15	2911	60
Tower	34	4	2214	5	250	6
Dish/engine			986	2	181	9
Linear Fresnel	4	2	33	2	190	7
Total	655	17	4043	24	3532	82

Source: GTM Research, 2011.

At the end of 2010, most of the 1.1 GW of grid-connected CSP plants worldwide used parabolic trough technology (about 39 MW used power-tower technology). Tables 11.31 and 11.32 provide details about these plants in the United States, Europe, Asia, and North Africa. The tables also show that more than 100 CSP projects were in the planning phase at the close of 2010. Contracts specify when the projects must start delivering electricity between 2010 and 2014.

In the United States, 500 MW of CSP were in operation at the end of 2010, with more than 9 GW on hold with signed power purchase agreements. A 64 MW parabolic trough plant came online in 2007 in Nevada, and a 75 MW plant began operating in 2010 in Florida. A 5 MW power tower and a 5 MW Linear Fresnel project came online in 2009 in California. The US federal solar investment tax credit, which allows developers and utilities to offset 30% of plant costs, was extended through the end of 2016. Companies may use a 30% federal grant in lieu of the tax credit and can also apply for a federally backed loan guarantee. These policies are expected to stimulate plans for and construction of CSP plants in the United States in the next several years.

In Europe, Asia, and North Africa, about 655 MW were in operation at the end of 2010, and more than 7 GW of CSP capacity were under construction or are being developed over the next several years. In Spain, more than 400 MW of commercial CSP projects were operational, and about 2000 MW have provisional registration. Spanish Royal Decree 436/2004 (dated 12 March 2004) guaranteed a FIT of €0.27/kWh for 25 years. In the *Plan de Energías Renovables en España* (2005 to 2010), a total

capacity of 500 MW was foreseen. Royal Decree 661/2007 raised the cap to 500 MW and allowed the tariff to float, adjusting with the market price. In response, developers in Spain have identified projects well beyond the 500 MW cap, and the government must reassess the number of projects to support. Recent projects in Spain include the 11 MW PS10 power tower plant in 2007, the 20 MW PS20 tower in 2009, and the 50 MW parabolic trough system, with 7.5 hours of storage, in 2008. Several additional plants came online in 2009 and 2010.

Since 2000, the CSP industry grew from a negligible activity to one that has more than 1400 MW of generation commissioned or under construction in 2009 worldwide. In 2006, two or three companies were in a position to build for commercial-scale plants. In 2009, more than 10 companies were active in building or preparing for such plants. These companies range from large organizations with international construction and project management expertise who have acquired rights to specific technologies to start-up companies using their own technology (IHS EER, 2009; Arvizu et al, 2011).

The CSP industry provides employment in countries with growing capacities. About 50–60% of the cost of CSP projects is for equipment leading to manufacturing jobs (Stoddard et al., 2006). In the United States, the National Renewable Energy Laboratory (NREL) estimates that realizing 4000 MW of CSP would require domestic investment of US$13 billion. Such an investment would create 145,000 jobs in construction and engineering and 3000 direct permanent jobs for operation and maintenance (Stoddard et al., 2006). Similar numbers would be likely in other regions of the world.

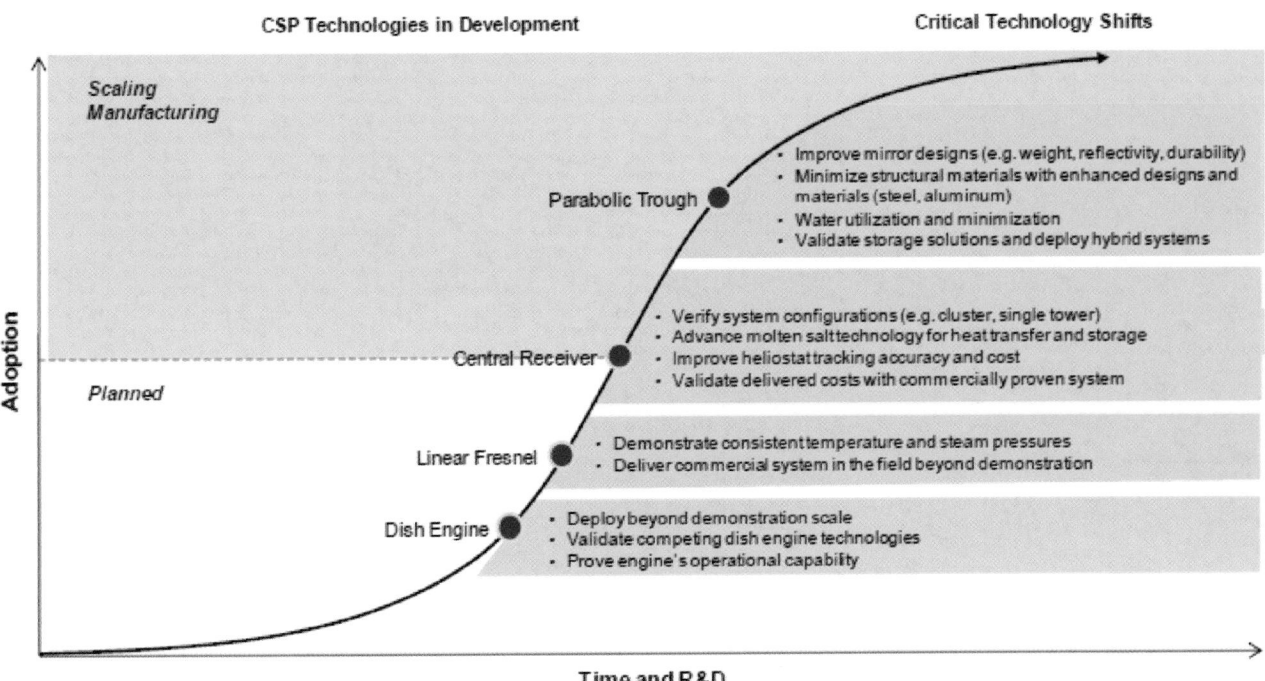

Figure 11.54 | CSP technology curve and evolutionary changes. Source: IHS EER, 2009.

11.7.4   CSP Technology Development

Four main CSP technologies, with different levels of maturity, are discussed in this section (Sargent and Lundy, 2009). (See Figure 11.54.)

11.7.4.1   Status of CSP technologies

A *parabolic trough system* has long parallel rows of trough-like reflectors – typically, glass mirrors. Controllers make the reflectors follow the sun as it moves from east to west by rotating them along their axes. Each trough focuses the sun's energy on a thermal receiver or heat-collection element located along its focal line. A heat-transfer fluid – typically oil at temperatures as high as 390°C – is circulated through the heat-collection element. Then the hot fluid is pumped to a central power-block area, where it passes through a series of heat exchangers. Superheated steam is created at 370°C and 100 bar. Once past the steam generator, parabolic trough plants behave just like conventional steam plants. The steam is used to drive a conventional steam-Rankine turbine generator, and the plant can store heat or use additional heat from fossil fuels to generate electricity when the sun is not shining. The annual solar-to-electric efficiency of parabolic trough plants is approximately 15%, a figure that includes losses due to non-normal solar incidence, part-load operation, and plant availability throughout the year.

The first commercial trough/oil CSP plants were installed and commissioned between 1985 and 1991 in the United States. Nine plants with a combined 354 MW capacity were built by Luz, and they continue

operation with new owners. In 2007, the next commercial plant was built and owned by Acciona. This 64 MW Nevada Solar One uses aluminum rather than steel troughs.

With the increasing interest in CSP construction, strong competition is emerging among companies in the supply chain for components and construction. However, in 2010 only Schott and Solel were capable of supplying several 100 MW/yr of the large evacuated tubes (heat-collection elements) designed specifically for use in trough/oil systems for power generation (Arvizu et al., 2011). The trough concentrator requires know-how in both structure and thermally sagged glass mirrors. Companies are offering new trough designs and considering alternatives to conventional rear-silvered glass (such as new polymer-based reflective films). But the essential technology remains unchanged (Arvizu et al., 2011).

Commercial systems today are limited by the maximum operating temperature of the heat-transfer fluid – synthetic oil with a maximum temperature of 390°C. Direct steam generation in troughs may allow trough systems to operate at higher temperatures, and this concept is being demonstrated. In other designs, molten salt has the potential advantage of operating at higher temperatures than steam systems and allows for integration with direct two-tank salt storage systems similar to those used for molten-salt tower configurations. In 2010, the Italian utility ENEA began operating a small prototype molten salt trough plant in Sicily. A disadvantage of this concept is the potential for the salt freezing in the solar field and the need to design a system to recover from such an event.

Figure 11.55 | PS10 and PS20 power towers in Spain (left) and a power tower solar thermal power plant in the California desert (right). Source: Left photo courtesy of Abengoa Solar.

In a *power tower system* (also called a central-receiver system), a field of two-axis tracking mirrors, called heliostats, reflect solar energy onto a receiver mounted on a central tower. (See Figure 11.55.) To maintain the beam of concentrated sunlight on the receiver at all times, each heliostat must track a position in the sky that is midway between the receiver and the sun. A heat-transfer fluid heated in the receiver is used to generate steam that runs a conventional turbine to generate electricity. Some power towers use water and steam directly as the heat transfer fluid. Central-receiver systems typically operate at higher temperatures than parabolic troughs, with superheated steam temperatures of 550°C for proposed steam and molten-salt systems. As with parabolic troughs, central-receiver systems can be integrated with thermal storage; the amount of storage available depends on the heat transfer fluid used in the receiver. Direct-steam receivers offer limited storage capacities, typically less than one hour, due to the high costs associated with storing high-temperature steam.

Power tower systems based on molten-salt receivers integrated with thermal storage have been demonstrated on the pilot scale. Because molten-salt receivers have superior heat-transfer and energy-storage capabilities, annual efficiencies are projected to be higher than for oil-based parabolic trough systems. In early 2011, a commercial unit was under construction in Spain, and several commercial units were under development in the United States.

Power tower systems are just entering the market on a commercial scale, and this should open the way for new industry participants and a diversity of system designs. Primary design choices include the heliostat (1m² to >100m²), receiver (cavity or external), and heat transfer fluid type (steam, molten salt, or air). The high concentration of solar energy associated with tower systems allows heat transfer fluid operating temperatures as high as 1000°C, as for example in the air heat-transfer fluids for solar Brayton cycles. So far, there is little consensus on the best

approach for achieving low cost, high performance, and high market value in power tower systems.

A *dish/engine system* tracks the sun and focuses solar energy into a cavity receiver; the receiver absorbs the energy and transfers it to the heat engine/generator that generates electric power. Dish/engine systems have demonstrated peak efficiencies greater than 30%, and the projected annual conversion efficiency is 24% (Arvizu et al., 2011). Effort is going into developing a commercial product using Stirling engines as the power-conversion device, although Brayton engines are also an alternative. The technology may be able to take advantage of existing know-how such as on the Stirling engine mass-produced through the automotive industry.

A *linear Fresnel reflector (LFR) system* uses a series of flat or shallow-curvature mirrors to focus light onto a linear receiver located at the focal point of the mirror array. Linear Fresnel systems could have lower capital cost than systems with parabolic mirrors because the mirrors are flat and located close to the ground. But they have also lower operating efficiencies than parabolic trough systems (Häberle et al., 2002). Because no large-scale commercial Fresnel-based systems are in operation at present, it is unclear whether the lower upfront capital cost will offset the efficiency problem.

Research on Linear Fresnel reflector systems has used steam as the heat-transfer fluid. A significant disadvantage of steam-based systems is their current incompatibility with long-term (>1 hour) thermal storage. The US DOE (US DOE 2008b) is supporting development of a linear Fresnel reflector system that uses molten salt as a heat-transfer fluid. As with parabolic troughs, the freezing of the salt in the field is still a primary concern. However, unlike parabolic trough systems, the linear Fresnel reflector receiver is stationary, so engineering freeze protection should be much more straightforward.

Figure 11.56 | Two-tank molten-salt indirect storage system at the 50 MW Andasol 1 parabolic trough plant in Spain. Source: Photo courtesy of Solar Millennium.

11.7.4.2   Thermal Storage

Thermal storage is an important attribute of CSP. Even 30 minutes to 1 hour of full-load storage can reduce the impact of thermal transients (such as clouds) on the plant and of electrical transients to the grid. Thermal storage can result in significantly higher energy and capacity value over equivalent systems without storage.

Spain is the first country to incorporate thermal storage into commercial installations. Parabolic trough plants such as Andasol 1 in Spain have been designed for 7.5 hours of full-load storage. This allows operation well into the evening, when peak demand can occur and tariffs are high. Several parabolic trough plants in Spain use molten-salt storage. (See Figure 11.56) Depending on the electricity market, storage can be increased up to 16 hours to allow 24-hour-a-day electricity generation.

Power towers operate at high temperatures and can charge and store molten salt more efficiently and less expensively than other CSP systems. The 17 MW GEMASOLAR power tower being developed in Spain is designed to operate 6500 hours a year – a 74% capacity factor. A 100 MW plant in Nevada in the southwest United States with 10 hours of molten-salt thermal storage has a signed power purchase agreement.

With thermal storage, the heat from the solar field is stored prior to reaching the turbine. Storage media include molten salt, steam accumulators (for short-term storage only), solid ceramic particles, high-temperature phase-change materials, graphite, and high-temperature concrete (Gil et al., 2010; Medrano et al., 2010). Figure 11.57 shows that thermal storage allows the CSP plant to dispatch power to meet demand after sunset.

Significant R&D is under way to develop thermal storage technologies that are compatible with the steam-based heat-transfer fluids for power tower, parabolic trough, and linear Fresnel systems. Storage systems using phase-change materials are more thermodynamically compatible with the latent energy associated with evaporation and condensation of steam. The intrinsic nature of phase-change systems results in decreased storage volumes, thereby reducing materials costs.

11.7.4.3   Power Cycles

In general, thermodynamic cycles will perform more efficiently at higher temperatures. The solar collectors providing the thermal energy must perform efficiently at these higher temperatures. Development to optimize the linkage between solar collectors and higher-temperature thermodynamic cycles is under way (Arvizu et al., 2011). The most commonly used power block is the steam turbine (Rankine cycle), which is most efficient and cost-effective in large capacities. Parabolic trough plants using oil as the heat-transfer fluid limit steam-turbine temperatures to 370°C and turbine cycle efficiencies to around 37%. This leads to design-point solar-to-electric efficiencies of 24% and annual average efficiency of 15% (Arvizu et al., 2011). To increase efficiency, alternatives to using oil as the heat transfer fluid – such as producing steam directly in the receiver or using molten salts – are being developed for parabolic troughs. Power towers and dishes/engine systems can reach the upper limits of existing fluids (around 600°C for current molten salts) for advanced steam-turbine cycles. Power towers can also provide the temperatures needed for higher-efficiency cycles (Arvizu et al., 2011).

11.7.4.4   Solar Thermal Hydrogen Production

The global use of hydrogen, mainly as feedstock in industrial processes, was estimated to be around 5 EJ/yr in 2004 (IEA, 2005). In 2050, in a carbon-constrained world, the demand for hydrogen as an energy carrier might be as large as 11–44 EJ/yr (IEA, 2008b; European Commission, 2006). It would be used in the transport sector and for stationary applications. High-temperature solar thermal energy can be used in several ways to produce the hydrogen (see, for example, Pregger et al., 2009).

One commercially available route is splitting water molecules into hydrogen and oxygen using electricity from a CSP plant: $2H_2O \rightarrow 2H_2 + O_2$. Another approach, solar thermal high-temperature electrolysis, splits water into pure oxygen and hydrogen at about 700–1000°C. Splitting water is possible using very high temperature heat (2300–2600°C) produced by solar power towers. Yet another option is cracking methane using solar heat: $CH_4 \rightarrow C + 2H_2$. Solar heat (1200–2000°C) can also be used to assist in steam reforming methane: $CH_4 + H_2O \rightarrow CO + 3H_2$, followed by a water-gas shift reaction: $CO + H_2O \rightarrow CO_2 + H_2$. The hydrogen can be blended with natural gas in existing pipelines and distribution networks. Capture and storage of the CO_2 (see Chapter 13) can be applied to largely prevent its emission into the atmosphere. According to the IEA-CSP roadmap, this technology will be used in the Middle East, Central Asia, and the US Southwest starting in 2030 (IEA, 2010e).

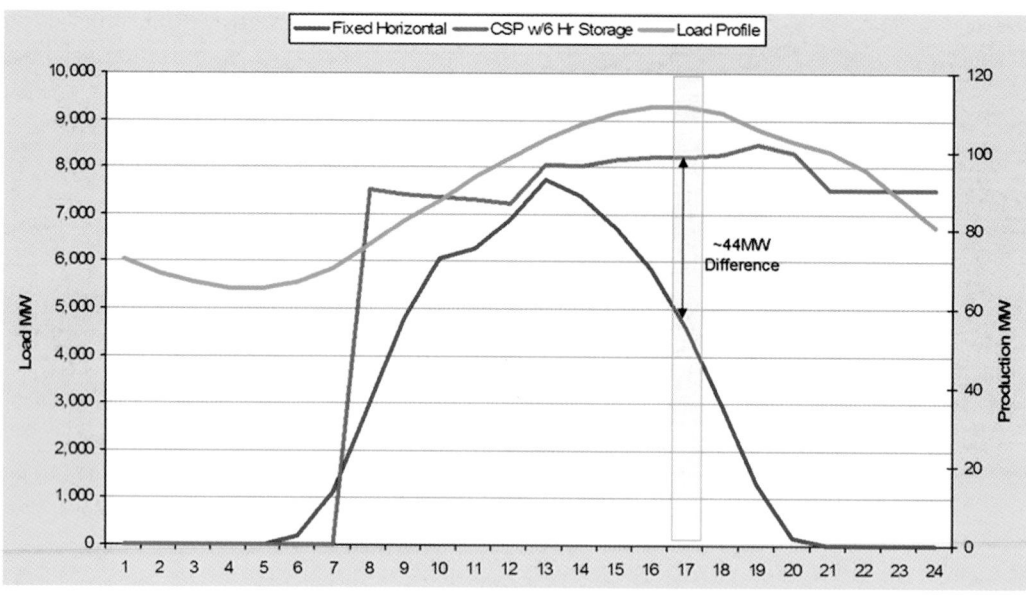

Figure 11.57 | Customer load profile (green) and power production profile of a CSP system with storage (red) and a non-tracking PV system (blue) over 24 hours. Source: Courtesy of Arizona Public Service.

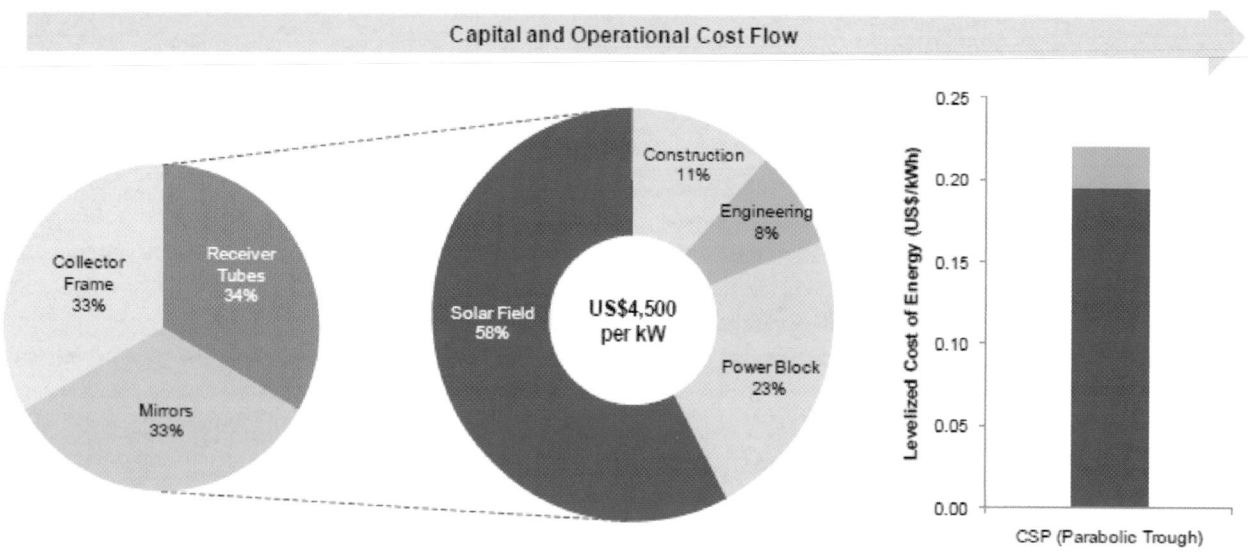

Figure 11.58 | Capital cost breakdown for parabolic trough plant. Source: IHS EER, 2009.

Thermo-chemical production of hydrogen is in the early development stages. An example is the Solzinc process, developed in Israel, using solar heat at a temperature above 1200°C to decompose zinc oxide. The zinc is subsequently combined with water and transformed back to zinc oxide, producing hydrogen: $Zn + H_2O \rightarrow ZnO + H_2$.

Because of the high costs involved and the current high conversion losses (cumulative up to 80–90%), R&D is needed to substantially improve the performance and competitiveness of solar hydrogen systems. Hydrogen production using fossil fuels is expected to remain the cheapest hydrogen source until at least 2030 (Pregger et al., 2009).

11.7.5   Economic and Financial Aspects

CSP must compete with electricity generation rates in utility markets rather than end-use electricity rates in residential or commercial markets. The levelized cost of energy for a CSP plant is calculated using the plant's upfront capital costs and its projected operating and maintenance costs, annual generation of electricity, finances, and lifetime.

In 2007, the installed cost of the 64 MW Nevada Solar One plant was US$260 million. The installed costs for parabolic trough CSP plants were US$4200/kW for a 100 MW system without thermal storage and

US$4900/kW for the same system with six hours of thermal storage. O&M fees for the system have been estimated at US¢1–3/kWh or more (US DOE, 2010a; IEA, 2010e).

An analysis by Lazard (2009) of CSP capital costs shows a range from US$4500–6300/kW, with parabolic troughs representing the low end and power towers representing the high end. In the IEA's technology roadmap (IEA, 2010e), the estimated capital costs range from US$4200–8400/kW. A breakdown of the estimated capital cost is presented in Figure 11.58. Land costs generally add less than 3%.

In 2010, the cost of installing a 100 MW CSP parabolic trough plant was estimated at US$4900/kW for a system without thermal storage and US$8400/kW for the same system with six hours of thermal storage (Turchi, 2010). The increased cost projected for current plants is primarily due to an increase in raw material costs (such as steel, concrete, and sodium/potassium nitrate salts) that occurred prior to the worldwide economic downturn at the end of 2007. Since 2008, costs have remained relatively steady, although a global recovery would likely put upward pressure on material costs.

The levelized cost of energy can be calculated from the estimated installation and O&M costs. The LCOE presented here assumes little or no storage, a capacity factor of 30–40%, a discount rate of 5% and 10%, an economic lifetime of 30 years, and average O&M costs of US¢2/kWh. The LCOE ranges from US¢10–20/kWh if total investment costs were US$4500/kW and from US¢15–30/kWh for investments of US$7000/kW. (See Table 11.33.) Similar LCOE figures were found for systems with 12-hour heat storage.

A recent analysis supported by the US DOE projected cost reductions for parabolic troughs based on advances in collector design, economies of scale for larger plants, and the use of molten salt as the heat transfer fluid. The capital cost for a trough system with 12 hours of storage is projected to be US$6500/kW in 2020. The same study projected a capital cost for an advanced tower system with 12 hours of storage to be US$5900/kW in 2025 (Turchi, 2010). The lowest energy cost in the long-term may be US¢5–7/kWh. This would result from technical learning, material improvements, and increased system performance (see, for example, IEA, 2010e).

11.7.6   Sustainability Issues

The environmental impacts of CSP considered here include land use, water use, and environmental emissions, and they depend on the technology used. Solar thermal power plants require relatively large areas of land – up to 8 km^2 for a 250 MW plant with six hours of storage. The acreage is often in desert regions, where plant construction and operation may affect sensitive habitats (Pregger et al., 2009).

CSP plants use a continuous supply of water for steam generation, cooling, and cleaning the solar mirrors. For Rankine cycle systems, a water

Table 11.33 | Cost of electricity as a function of capacity factor, turnkey investment costs, discount rate, and O&M costs. The O&M costs are assumed to be US¢2/kWh; the lifetime is assumed to be 30 years.

Capacity factor	Present turnkey investment costs per kW_e (*without storage*)	Discount rate	
		5%	10%
	US$_{2005}$$	US¢/kWh	
40%	4500	10.3	15.6
	7000	15.0	23.2
30%	4500	13.1	20.2
	7000	19.3	30.2

Capacity factor	Future turnkey investment costs per kW_e (*with 12 hour storage*)	Discount rate	
		5%	10%
	US$_{2005}$$	US¢/kWh	
65%	8000	11.2	16.9
	10,000	13.4	20.6
50%	8000	13.9	21.3
	10,000	16.8	26.2

source for cooling is desirable to achieve higher turbine efficiencies. Where water is limited, air cooling or a combination of wet/dry hybrid cooling can eliminate up to 90% of the water usage (Kutscher, 2009). Such approaches for steam-generating CSP plants reduce electricity produced by 2–10%, depending on geographic location, electricity pricing, and water costs. Dry cooling performs least efficiently during the summer months, when solar energy is most abundant and when the plants should have the greatest output to meet the higher electricity demand (WorleyParsons Group, 2009).

Most environmental emissions from CSP plants occur during the manufacturing of plant components that are produced with fossil fuels. Compared with the life-cycle emissions of fossil-fueled power plants, CSP power plants generate significantly lower levels of greenhouse gases and other emissions (Pehnt, 2006). Studies conducted by NREL concluded that a 4000 MW solar power plant could offset 300 tons of nitrogen oxide, 180 tons of carbon monoxide, and 7.6 million tons of carbon dioxide (Western Governors' Association, 2006; also cf. Greenpeace et al., 2009). Another study by NREL provides a life-cycle assessment indicating that a reference 100 MW parabolic trough plant with six hours of storage would generate GHG emissions estimated at 26 gCO_2-eq/kWh (Burkhardt et al., 2010). In life-cycle studies, several assumptions have to be made, such as the fuel mix in the power sector. Consequently, somewhat lower figures can be found in the literature. The indicated reference plant would also cumulatively demand 0.43 MJ/kWh of energy and consume 4.7 L/kWh of water.

According to the IPCC Special Report on *Renewable Energy Sources and Climate Change Mitigation*, most estimates of life-cycle GHG emissions fall between 14–32 gCO_2-eq/kWh for parabolic trough, power tower, dish/engine, and linear Fresnel reflector systems, whereas the energy payback time of CSP systems can be as low as five months (Arvizu et al., 2011). According to the *CSP Global Outlook 2009* (Greenpeace et al., 2009) results presented in Table 11.29, the moderate scenario will provide annual savings of 148 million metric tonnes of CO_2 in 2020, rising to 2.1 gigatonnes (Gt) annually by 2050. The cumulative savings would account for about 0.6 $GtCO_2$ by 2020 and 28 Gt by 2050.

11.7.7   Implementation Issues

The barriers and challenges to implementation include issues related to cost, land access, water use, transmission of electricity, system integration, and policy development.

11.7.7.1   Land Access

Land generally represents a minor portion of the cost of the whole CSP plant. However, aesthetic and environmental issues may cause a "not in my backyard" response to news of potential CSP development. When seeking to acquire private or public lands, developers may encounter difficulties such as high costs and permitting delays.

A 100 MW CSP plant would require 200–400 hectares, depending on the technology and the degree of storage integrated into the plant. The land should ideally have less than a 1% grade and no more than a 2% grade, particularly for parabolic trough and linear Fresnel reflector systems. Land for development must be near transmission lines and roads and should not be in an environmentally sensitive area. Although the mirror area covers only 25–35% of the land, the arid nature of a solar plant site will mean it is not suitable for other agricultural pursuits (Arvizu et al., 2011).

11.7.7.2   Transmission

Utility-scale CSP plants (50–300 MW_e) must be linked to the transmission network, so developing the grid infrastructure is critical to the widespread implementation of CSP.

North European countries are studying the installation of long transmission lines to get power from CSP plants in Southern Europe and North Africa. The DESERTEC Foundation has proposed using solar thermal power plants throughout the Sahara Desert to send power to Europe via a super grid running from Iceland to the Arabian Peninsula and from the Baltic Sea to the west coast of Africa (DESERTEC Foundation, 2011). (See Figure 11.59.)

In the United States, the Energy Policy Act of 2005 directed DOE to analyze the state of transmission capacity across the country and to identify areas requiring improvements. In the southwestern and western United States, many power lines operate at or near capacity, and bringing the power from remote locations to cities can be difficult. After conducting several studies to determine the impact of renewable on the US power transmission system, DOE concluded it was in the national interest to create an energy highway to allow power to travel more easily from the West Coast to the East Coast. In October 2007, DOE designated two national transmission corridors as the first step to a national power transmission system. The solar industry petitioned the Federal Energy Regulatory Commission (FERC) to clear any transmission bottlenecks to "greening the grid." In late July 2008, FERC granted the California Independent System Operator Corporation (a nonprofit organization charged with managing the flow of electricity in California's wholesale power grid) the ability to open the grid to renewable energy in California.

11.7.7.3   System Integration

CSP can be combined with fossil fuel or biomass plants in so-called integrated solar combined cycle plants to conserve fuel at relatively low cost. These plants are being built in Algeria, Australia, Egypt, Iran, Italy, and the United States (IEA, 2010e). Solar-augment of existing fossil power plants offers a lower-cost and lower-risk alternative to stand-alone CSP plant construction. A recent study found the potential in the southern half of the United States for over 11 GW_e of parabolic trough and over 21 GW_e of power tower capacity that could be added to coal-fired and natural gas combined-cycle plants whether existing, under construction, or planned (Turchi et al., 2011).

Combined with storage, CSP can enhance the reliability of power production and even offer base-load capacity. Consequently, CSP can contribute to grid flexibility and accommodate a larger share of variable energy sources in electricity systems. Losses in thermal storage cycles are much smaller than in other existing electricity storage technologies, such as pumped hydro and batteries (IEA, 2010e).

11.7.7.4   Policies

A number of countries are subsidizing R&D to increase the performance of CSP and reduce costs. Policy measures such as tax credits, emission trading schemes, and feed-in-tariffs help to create markets for CSP and achieve cost competitiveness. In recent years Spain has been the most active market in CSP development as a result of Royal Decrees enacted in 2004 and 2007 offering long-term and profitable FITs for solar thermal electricity. By contrast, the US market is driven primarily by Renewable Portfolio Standards, which require utilities to purchase a specified fraction of electricity generation from renewable energy facilities – sometimes with a specific "set-aside" requiring generation from solar. Combined with attractive federal tax incentives for solar installations, the United States represents a burgeoning near-term CSP market (see also Section 11.12).

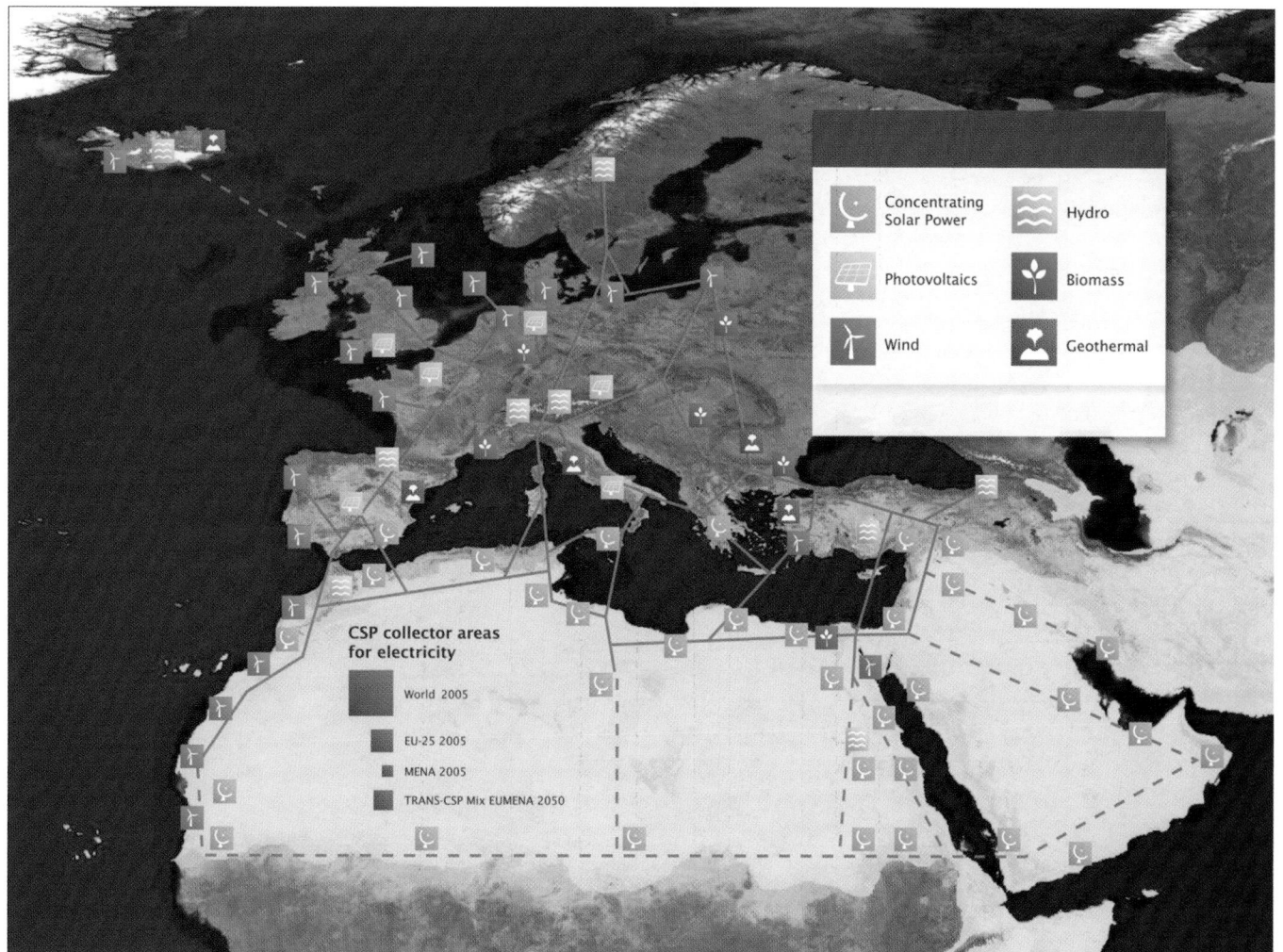

Figure 11.59 | Super grid envisioned by DESERTEC to bring power from the Middle East and North Africa to population centers throughout the region and in Europe. Source: DESERTEC Foundation, 2010.

11.8 Low-Temperature Solar Energy

11.8.1 Introduction

Low-temperature solar energy technologies, with operating temperatures up to 100° C, are perhaps the simplest way to use solar resources. These systems can be active or passive. In active conversion systems, heat from a solar collector is transported to the end process by a heat transfer system. In passive systems, no active components are needed to use the solar resource for heating or lighting (UNDP et al., 2000). This section is focused mainly on active systems that convert sunlight to thermal energy for water heating, space heating, space cooling, cooking, and crop drying.

11.8.2 Potential of Low-Temperature Solar Energy

Solar thermal energy use varies greatly by country and region depending on the maturity of the market, policy incentives, and available solar resource. The total installed capacity has been estimated at 152 GW_{th} in 2008 and 180 GW_{th} in 2009 (Weiss and Mauthner, 2010; REN21, 2010). Table 11.34 provides a breakdown for different countries for 2008. China has the most capacity, with 87.5 GW_{th}, followed by Europe (28.5 GW_{th}), the United States and Canada (15.1 GW_{th}), and Japan (4.4 GW_{th}). The energy yield from solar collectors in 2008 worldwide was about 110 TWh_{th} (395 PJ) (Weiss and Mauthner, 2010), and in 2009 it was about 130 TWh (470 PJ) – saving about 0.5 EJ of primary fossil fuel consumption.

Figure 11.60 shows solar heating capacity in 2008 and the types of collectors used in the 10 leading countries. China, the world leader in total capacity, uses more evacuated tube liquid collectors than any other country. The United States uses a high percentage of unglazed liquid collectors (for solar pool heating). Australia also uses many unglazed collectors, whereas glazed liquid collectors are the leading technology in other countries.

IEA developed an energy scenario in which energy-related CO_2 emissions in 2050 are 50% below today's level. In this scenario, called BLUE MAP, world solar thermal capacity growth is 8%/yr (IEA, 2010a). This trend

Table 11.34 | Total low-temperature solar heating capacity in operation, end of 2008.

Region[a]	Water collector[b] (GW_th)			Air collector[c] (GW_th)		Total (GW_th)
	Unglazed	Glazed	Evacuated tube	Unglazed	Glazed	
NAM	12.9	1.5	0.4	0.1	0.1	15.1
- United States	12.4	1.5	0.4	<0.05	0.1	14.4
- Canada	0.5	<0.05	<0.05	0.1	<0.05	0.7
WEU	1.6	24.4	1.1	0.6	<0.05	27.8
- Germany	0.5	6.5	0.7	-	<0.05	7.8
- Turkey	-	7.4	-	-	-	7.4
- Austria	0.4	2.3	<0.05	-	-	2.8
- Greece	-	2.7	-	-	-	2.7
- France	<0.05	1.2	<0.05	-	-	1.3
- Spain	<0.05	1.0	<0.05	-	-	1.1
- Switzerland	0.1	0.4	<0.05	0.6	-	1.1
- Cyprus	-	0.6	<0.05	-	-	0.6
- Netherlands	0.3	0.2	-	-	-	0.5
EEU	<0.05	0.6	<0.05	<0.05	<0.05	0.7
FSU	-	<0.05	-	-	-	<0.05
PAO	2.9	5.5	0.1	-	0.3	8.7
- Japan	-	4.0	0.1	-	0.3	4.4
- Australia	2.9	1.4	<0.05	-	-	4.3
CPA	-	7.2	80.3	-	-	87.5
- China	-	7.2	80.3	-	-	87.5
SAS	-	1.8	<0.05	-	<0.05	1.8
- India	-	1.8	<0.05	-	<0.05	1.8
PAS	<0.05	2.2	<0.05	-	-	2.2
- Taiwan	<0.05	1.2	<0.05	-	-	1.2
- Korea	-	1.0	-	-	-	1.0
MEA	<0.05	3.3	0.2	<0.05	-	3.5
- Israel	<0.05	2.6	-	<0.05	-	2.7
- Jordan	-	0.4	0.2	-	-	0.6
LAC	0.9	2.8	-	-	-	3.7
- Brazil	0.6	2.4	-	-	-	3.0
- Mexico	0.3	0.4	-	-	-	0.7
AFR	0.5	0.2	<0.05	-	-	0.7
- South Africa	0.5	0.2	<0.05	-	-	0.7
Total	18.8	49.5	82.3	0.7	0.5	151.9

[a] Includes only countries with installed capacity of at least 0.5 GWth.

[b] If no data given, no reliable database is available.

[c] Unglazed air collector in Switzerland is a simple site-built system for drying hay.

Source: based on data from Weiss and Mauthner, 2010.

would result in about 4000 GW_th installed capacity by 2050. Assuming an average capacity factor of 8%, solar thermal systems would yield 2800 TWh_th a year (10 EJ a year). Since 2004, the actual increase in capacity has been 19%/yr.

In the United States alone, the technical potential of solar water heating (SWH) has been estimated at 300 TWh_th (about 1 EJ) of primary energy savings per year, equivalent to an annual CO_2 emissions reduction of 50–75 million metric tons. For US consumers, this could save more than US$8 billion per year in retail energy costs. Natural gas is used to heat a large fraction of hot water in the United States. It is used directly in gas water heaters or indirectly in electric water heaters,

where the electricity is generated using natural gas as the marginal fuel (Denholm, 2007).

In Europe, the European Solar Thermal Technology Platform formulated a target that in the long term 50% of the heating demand should be covered by solar thermal, while accounting for 100% of the heating and cooling demands in new buildings (ESTTP, 2008).

Low-temperature solar heat is also considered an excellent option for crop drying. Its low, even temperatures do not harm delicate foods and are as effective as fired heating methods. Solar crop drying can be used for coffee, tea, beans, rice, fruit, cocoa, spices, rubber, and timber.

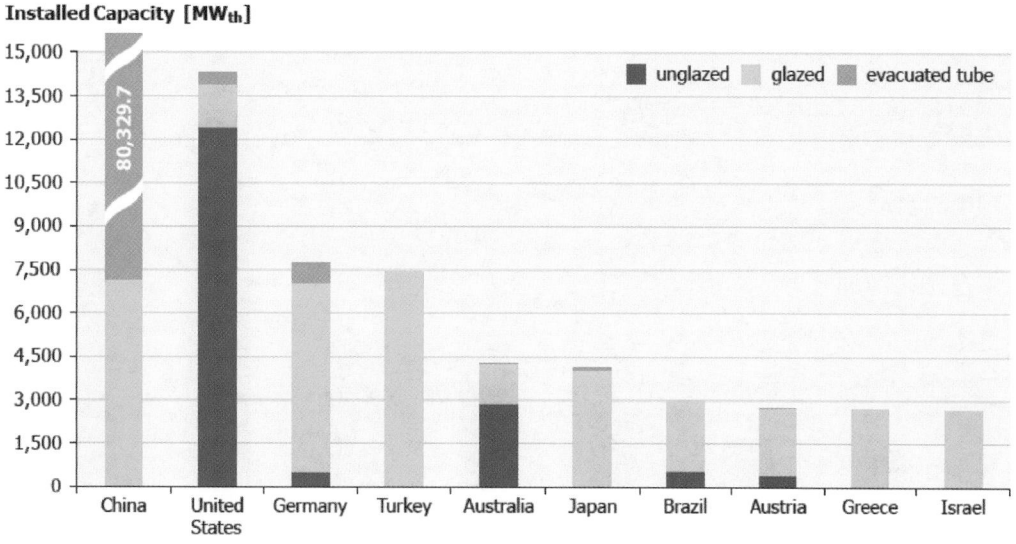

Figure 11.60 | Total capacity of solar collectors, 2008, in top 10 countries. Source: Weiss and Mauthner, 2010.

Expanding its use could displace between 300 and 900 PJ annually. This estimate is based on displacing fuel-fired dryers for crops operating at temperatures below 50°C. The use of solar energy for these markets is largely undeveloped (IEA-SHC, 2009).

In 2004, the organization Solar Cookers International ranked India, China, Pakistan, Ethiopia, and Nigeria as countries with the highest potential for solar cooking. The study considered annual solar radiation, the percentage of the country with forests, estimated populations in 2020, and the estimated share of the population within each country with both good solar insolation and fuel scarcity. In India, 80 concentrating systems of different capacities cover 25,000 m² of dish area. The world's largest system, at Shirdi, cooks food for 20,000 people every day (Solar Cookers International, 2009).

11.8.3   Market Developments

Since the early 1990s the solar thermal energy market has been growing. Installed capacity of flat plate and evacuated collectors increased by about a factor of four from 2000 to 2008. (See Figure 11.61.) China is the world's largest market today.

In 2008, 29 GW$_{th}$ (41.5 million m²) of solar collectors were installed worldwide. The average annual increase between 2004 and 2008 was 19%, but between 2007 and 2008 the increase was 43%. Some markets for glazed collectors (flat-plate and evacuated tube collectors) in Europe had growth of 62% in one year. In the United States and Canada, growth was nearly 42%, and in Australia and New Zealand, nearly 40%. New installations in China increased in 2008 by about 35%.

By the end of 2008, nearly 217 million m² of collector area were in operation in the 53 countries tracked by the IEA (Weiss and Mauthner, 2010).

These countries represent about 61% of the world's population. The installed capacity in these countries represents an estimated 85–90% of the solar thermal market worldwide. Note that 217 million m² is equivalent to the installed capacity of 152 GW$_{th}$ already mentioned, using an average conversion factor of 0.7 kW$_{th}$/m².

In China, Europe, and Japan, flat-plate and evacuated tube collectors are used mainly to provide hot water and space heating, while in the United States and Canada, swimming pool heating with unglazed polymer collectors is the dominant application.

Solar water heating offers the least-cost option in certain locales. This is the case in China, which is experiencing a boom in sales of these systems.

Figure 11.62 shows the regional distribution of glazed flat-plate and evacuated tube liquid low-temperature solar energy systems at the end of 2008. Most systems produce domestic hot water for single and multi-family houses. Only in Europe do combination systems for solar space heating and water heating account for a measureable number of low-temperature solar energy applications.

Although growing fast, especially in Europe, the market for solar cooling systems is still small. The IEA Solar Heating and Cooling Program has identified over 450 solar cooling systems in Europe and 30 in North America (Wiemken, 2009).

The most glazed flat-plate and evacuated tube collectors are found in Cyprus (527 kW$_{th}$ per 1000 inhabitants), followed by Israel (371), Austria (285), Greece (253), Barbedos (203) and Jordan (102) (Weiss und Mauthner, 2010). Currently, glazed flat-plate collector manufactures are producing about 27 million m² of solar collectors per year (Epp, 2009).

Annually Installed Capacity [MW$_{th}$/a]

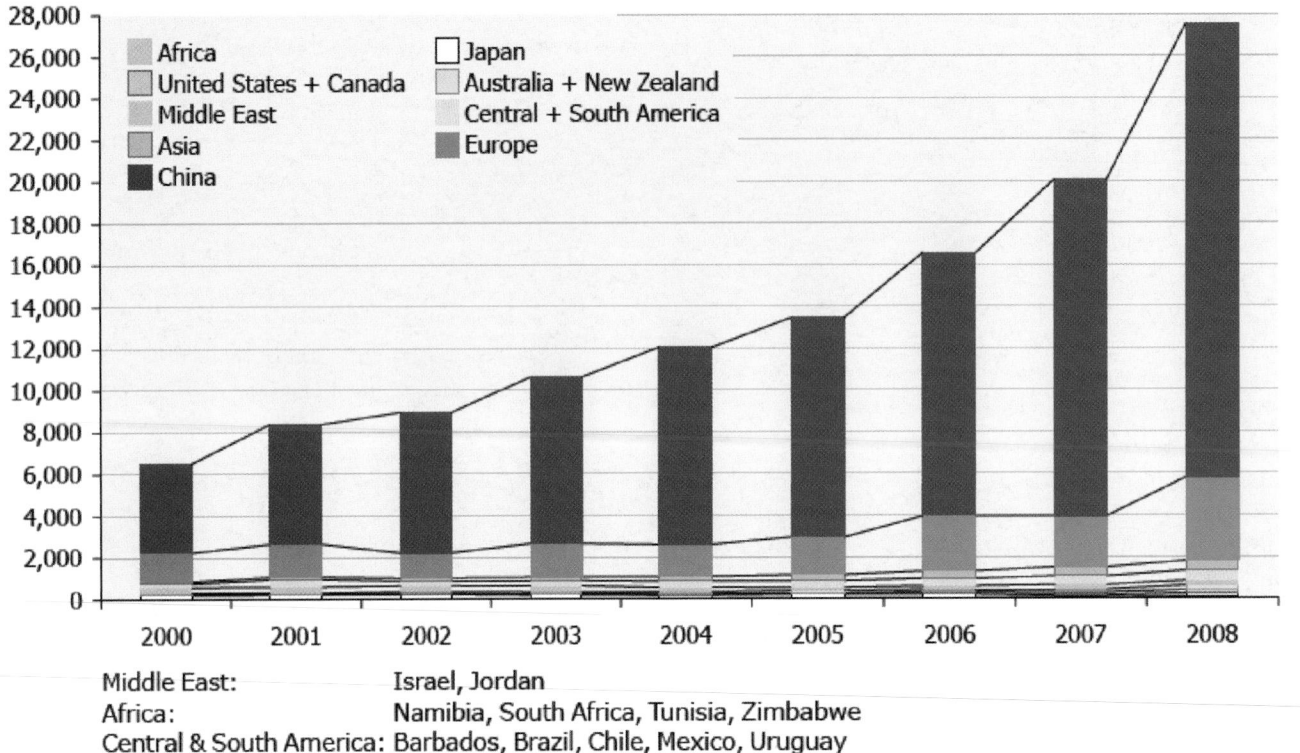

Middle East: Israel, Jordan
Africa: Namibia, South Africa, Tunisia, Zimbabwe
Central & South America: Barbados, Brazil, Chile, Mexico, Uruguay
Asia: India, South Korea, Taiwan, Thailand
Europe: EU 27, Albania, Macedonia, Norway, Overseas Dep. of France, Switzerland, Turkey

Figure 11.61 | Annual installed capacity of flat-plate and evacuated tube collectors, 2000–2008, total and per region. Source: Weiss and Mauthner, 2010.

The IEA Solar Heating and Cooling Programme has identified promising applications and sectors for solar heat in industrial processes as follows (Vannoni et al., 2008):

- cleaning, primarily in food processes, but also for process equipment and metal treatment plants (galvanizing, anodizing, and painting);

- commercial laundries;

- car washes;

- drying requirements after cleaning in both the food and chemical industries;

- pasteurization and sterilization for the food and biochemistry sectors; and

- preheating of boiler feed water.

Local sourcing, local jobs, and local sales are hallmarks of low-temperature solar technologies. According to detailed country reports for 2007, production, installation, and maintenance of solar thermal plants created 200,000 jobs worldwide (Weiss and Mauthner, 2010).

11.8.4 Low-temperature Solar Energy Technology Development

The working temperature ranges for the active solar thermal technologies used for water heating, space heating, space cooling, pool heating, crop drying, and cooking and the typical types of solar collectors used in these application are shown in Figure 11.63 (For concentrating technologies to generate high-temperature heat, see section 11.7.)

Most active solar energy technologies have four basic components:

- Solar thermal collector(s) – flat-plate and evacuated tube collectors are the most typical

- Storage system – in order to meet the thermal energy demand when solar radiation is not available

- Heat transfer system – piping and valves for liquids and ducts and dampers for air; pumps, fans, and heat exchangers, if necessary

- Control system – to manage the collection, storage, and distribution of thermal energy.

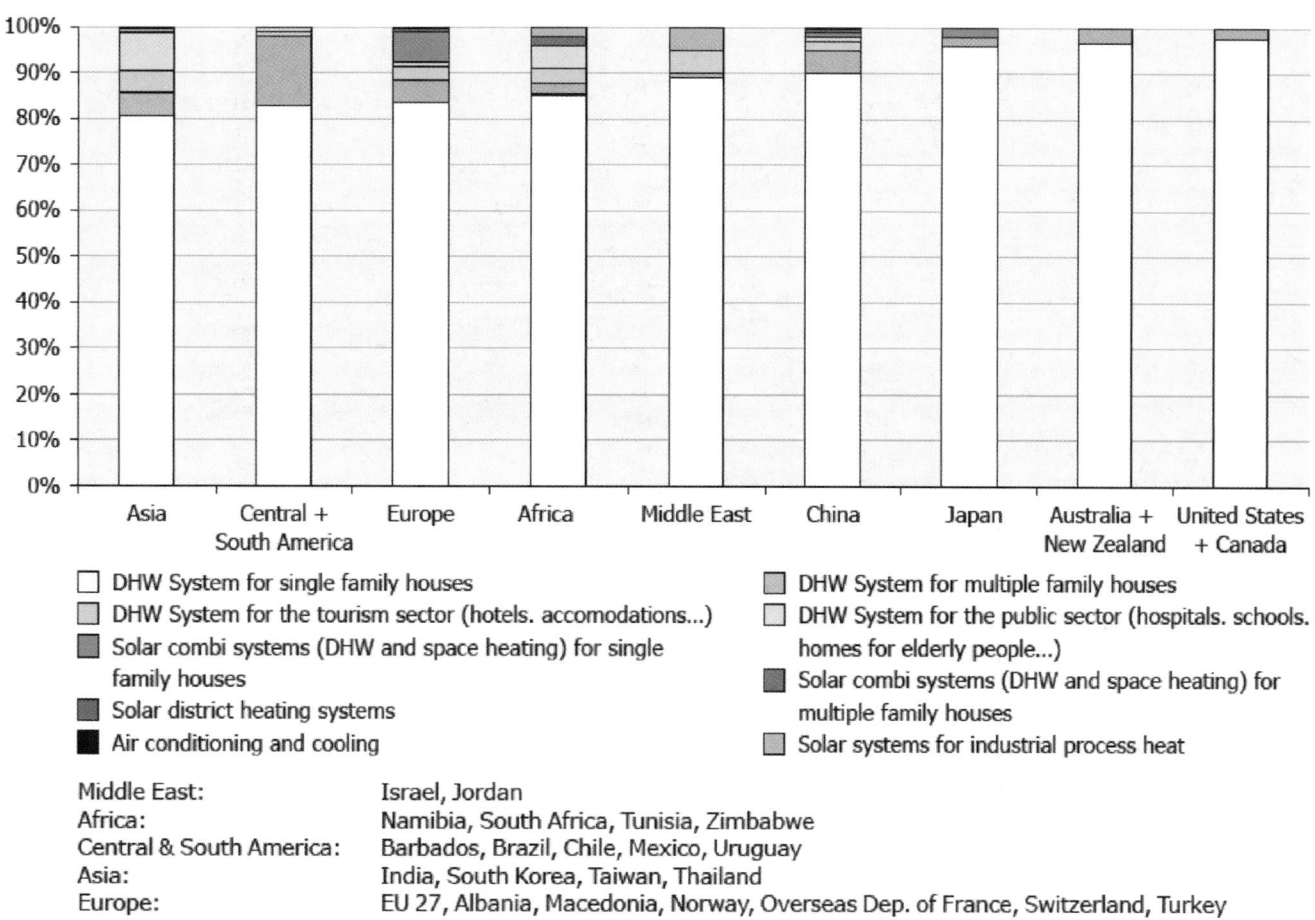

Middle East: Israel, Jordan
Africa: Namibia, South Africa, Tunisia, Zimbabwe
Central & South America: Barbados, Brazil, Chile, Mexico, Uruguay
Asia: India, South Korea, Taiwan, Thailand
Europe: EU 27, Albania, Macedonia, Norway, Overseas Dep. of France, Switzerland, Turkey
DHW: Domestic Hot Water

Figure 11.62 | Applications of glazed and evacuated tube collectors, by region, 2008. Source: Weiss and Mauthner, 2010.

11.8.4.1   Solar Water Heating

Solar water heating (SWH) is the most widely used application of low-temperature solar heat. Conventional collectors use either flat-plate and evacuated tube approaches. A flat-plate collector is an insulated metal box with a dark, heat-absorbing metal plate inside and a cover of glass or plastic. Sunlight passes through the cover and heats up the dark absorber plate. The heat is transferred to a liquid (usually water or propylene glycol) flowing through pipes inside the absorber plate. In areas with freezing temperatures, liquid collectors must contain anti-freeze or have a system to drain the water when the temperature drops. Evacuated tube systems have a row of glass tubes that contain small metal pipes with heat transfer fluid that act as heat absorbers. This type of collector has higher temperature differences between the ambient air and the collector fluid, leading to higher operating efficiency than flat-plate collectors at cold outdoor temperatures.

Active SWH systems use a circulating pump, sometimes powered by a small solar electric panel, to circulate fluid through the heating system. Passive systems rely on water pressure, the buoyancy of warm liquids,

and gravity to move the heat-transfer fluid through the system. (See Figure 11.64.) Tables 11.35 and 11.36 show characteristics of SWH systems and combination solar space heating and water heating systems for single-family and multi-family residences in three regions of the world.

SWH technologies have improved significantly in the last 20 years. Areas that can be improved further include the following:

- Increase durability and reliability while reducing costs. Identify and develop low-cost polymer materials, predict degradation from optical and mechanical processes, develop protective coatings, and improve active system components such as electronic sensors and controls.

- Improve freeze protection. Expand the geographic range of SWH markets from systems primarily made of low-cost polymers.

- Standardize SWH system components. Develop easy-to-assemble systems that incorporate standardized, packaged sets of subsystems and components (pumps, valves, controls, and tanks).

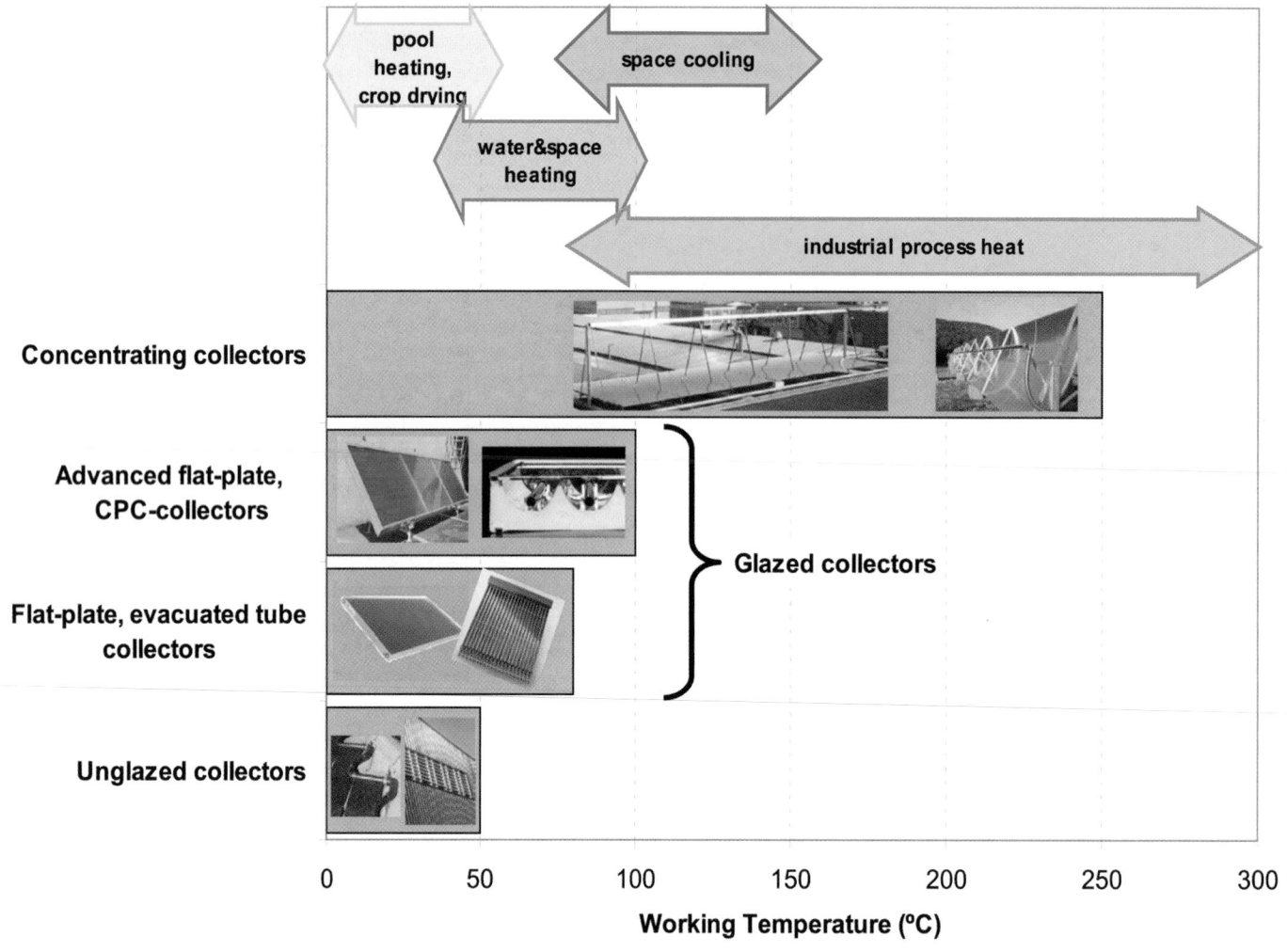

Figure 11.63 | Active solar thermal technologies, collectors, and working temperature ranges. Source: IEA-CERT, 2011. Courtesy of ORNL.

- Develop "combination" technologies and integrate SWH into buildings. Combine solar thermal technologies with other water-heating and building-system technologies.

11.8.4.2 Large Solar Water Heating Systems

Solar thermal systems can provide heat and hot water for direct use. They can also provide preheated water to boilers that generate steam. Large water heating systems can be used by hotels, hospitals, homes for the elderly, and public institutions such as correctional facilities. Other markets for large water heating systems are fertilizer and chemical factories, textile mills, dairies, and food-processing units. Large (MW-scale) solar thermal systems in the low temperature range are used for district heating and for cooling and low/medium temperature process heating.

11.8.4.3 Solar Space Heating

Active solar space heating systems for residential and commercial buildings use a solar collector to heat liquid or air. Thermal energy is transferred directly to an interior space or to a storage area for later use. Liquid-based heating systems are used when storage is desired. Solar-heated air is typically used for ventilation air heating. For example, the transpired solar collector draws air through the perforations of a solar absorber, warming the air in the process. This heated air is used directly in the building, or it may serve as pre-warmed air for a conventional heating/ventilation system.

Space heating often uses liquid-based systems that heat ordinary water or an antifreeze solution such as glycol, depending on the climate. The hot liquid may be used in a fan coil, a hydronic system, or a radiant floor system. R&D activities are under way to improve the performance, cost, and reliability of solar collectors and associated thermal-storage systems.

Solar-assisted heat pump systems are being installed in Europe. Four main components interact in these combined solar and heat pump systems:

- solar collectors: glazed, evacuated, or unglazed;
- a heat pump: air source, water source, or ground source;

Table 11.35 | Characteristics of a typical single-family solar water heating system and combination solar space and water heating system, 2007.

	OECD Europe	OECD North America	OECD Pacific
Water heating: typical size (kW$_{th}$)	2.8–4.2	2.6–4.2	2.1–4.2
Water heating: useful energy: (GJ/system/yr)	4.8–8.0	9.7–12.4	6.5–10.3
Combi systems: typical size (kW$_{th}$)	8.4–10.5	8.4–10.5	7.0–10.0
Combi systems: useful energy (GJ/system/yr)	16.1–18.5	19.8–29.2	17.2–24.5
Installed cost: new build (US$/kW$_{th}$)	1140–1340	1200–2100	1100–2140
Installed cost: retrofit (US$/kW$_{th}$)	1530–1730	1530–2100	1300–2200

Source: data from IEA-CERT, 2011.

Table 11.36 | Characteristics of a typical multi-family solar water heating system and combination solar space and water heating system, 2007.

	OECD Europe	OECD North America	OECD Pacific
Water heating: typical size (kW$_{th}$)	35	35	35
Water heating: useful energy (GJ/system/yr)	60–77	82–122	86
Combi systems: typical size (kW$_{th}$)	70–130	70–105	70
Combi systems: useful energy (GJ/system/yr)	134–230	165–365	172
Installed cost: new build (US$/kW$_{th}$)	950–1050	950–1050	1100–1850
Installed cost: retrofit (US$/kW$_{th}$)	1140–1340	1140–1340	1850–2050

Source: data from IEA-CERT, 2011.

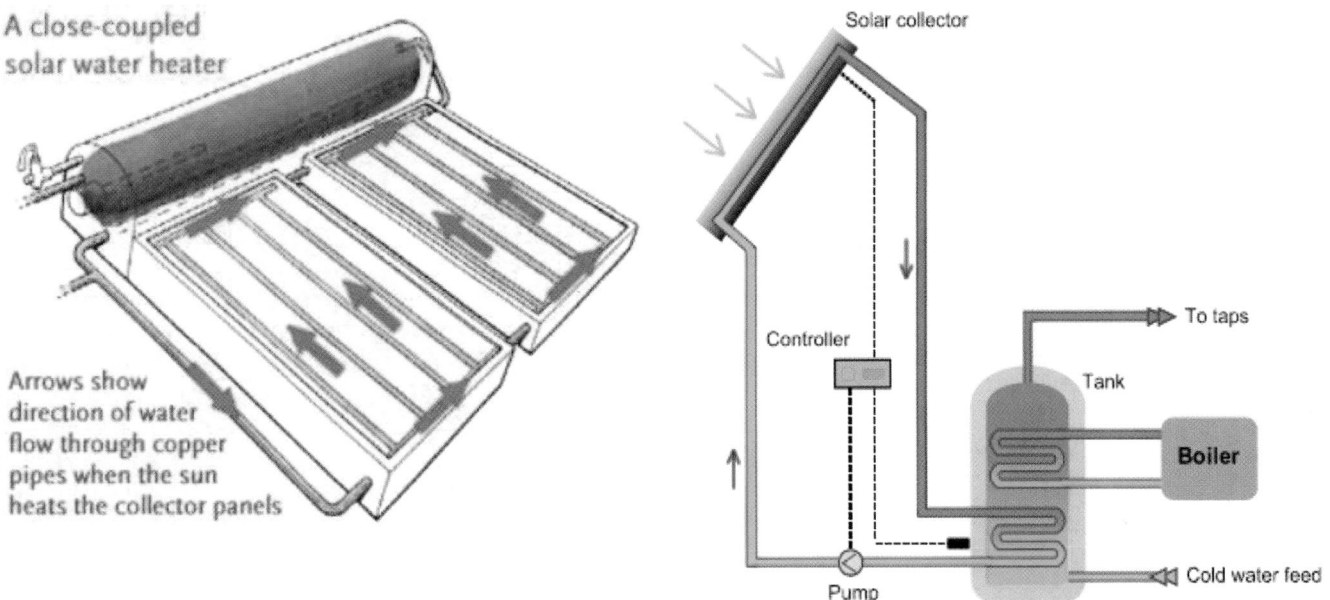

Figure 11.64 | A passive thermo syphon SWH system (left) and an active forced circulation SWH system (right).

- storage: a water tank, the building structure, the ground, phase change materials, or thermo chemical heat storage (IEA-CERT, 2011); and

- controls that determine how the other components interact.

11.8.4.4   Solar Space Cooling

Space cooling is the newest building application of solar thermal energy. Solar cooling systems can reduce summer peak demand on electricity grids. In addition to the four basic components of a typical solar thermal system, a space cooling system also requires a heat rejection system such as a wet cooling tower or a dry condenser.

Thermally driven cooling uses either a closed cycle for sensible cooling or an open sorption cycle for latent cooling. Closed cycles include absorption cooling, adsorption cooling, and ejector cooling. Absorption cooling systems use working fluids such as water paired with ammonia or solutions of certain salts such as lithium bromide. Closed adsorption cooling systems use solids such as silica gel paired with water. Absorption chillers that use thermal energy from natural gas or other fossil fuels provided the first space cooling before the advent of electric vapor compression cooling equipment. However, smaller-scale absorption chillers designed to operate at temperatures more suitable for solar thermal systems are now under development in Europe and China.

In hot, humid climates, removing moisture from the conditioned air is a key factor in space cooling. In an open sorption cooling cycle, solar thermal energy can be used in the regeneration phase to dry either solid or liquid desiccant material that has absorbed moisture. Depending on the desiccant material, flat-plate or evacuated tube collectors can be used to provide the regeneration energy. Liquid desiccant systems that work well below 80°C are under development.

11.8.4.5   Solar Pool Heating

Solar pool heating for homes is the largest use of solar thermal energy in the United States. Hotels, municipal governments, and other commercial customers are also starting to adopt this technology. Solar pool heating systems use the pool-filtration system to pump water from the pool to a solar collector. The sun heats the water as it flows through the collector, and the heated water is returned directly to the pool. Solar pool heating collectors are usually unglazed, operate at a slightly warmer temperature than the surrounding air, and are in general made from polymers.

11.8.4.6   Solar Cooking

Solar cookers can cook food, boil water, and pasteurize food or water. Simple solar cookers use diffuse sunlight. Even a simple insulated box with a transparent lid can produce temperatures in the range of 50–100°C. More complex designs use reflectors in dish or trough concentrators to produce temperatures up to 300°C using direct sunlight. In countries where firewood is traditionally used for cooking, solar cookers can slow deforestation and decrease indoor and outdoor air pollution from wood smoke.

11.8.4.7   Solar Crop Drying

Drying agricultural products uses large quantities of low temperature heat usually supplied by burning firewood or fossil fuels, such as diesel and propane. In many cases, air-based solar collectors could provide the needed heat. In Finland, Norway, and Switzerland, this technology is used to dry hay.

The IEA Solar Heating and Cooling Programme (IEA-SHC, 2009) sponsored a project in India to demonstrate a solar air preheating system to help dry coir pith, a byproduct manufactured from coconut shells marketed as an absorbent substitute for potting soil. The coir pith is pressed to reduce moisture content and then final drying takes place in a fluidized bed burner. Solar energy is used to preheat air to the burner. Perforated transpired collector panels on the roof draw air through the small openings heating it to 20°C above ambient. The air is then ducted to the burner, where the temperature is raised to the required 105°C to feed the dryers. After three years of operation, the solar heating system displaced 14% of the heating fuel, which resulted in a two-year payback.

11.8.4.8   Passive Solar Energy

Applying passive solar design principles to new buildings can reduce energy demands for heating, cooling, ventilation, and lighting. A successful solar building design integrates individual building components to achieve optimal energy performance. Computer simulations can help designers understand and quantify the interactions among the various components and systems. It has been estimated that 13% of the heat demand of buildings could be met by passive solar energy use. For optimized buildings, this figure could reach 30% without major investments (Brouwer and Bosselaar, 1999).

The key to passive solar design is the building envelope – the interface between the interior of the building and the outdoor environment. Energy pathways through the building envelope include the roof, walls, windows, air infiltrations, thermal storage, and insulation. Numerous advances are being made within these various components (Kutscher, 2007; Walker et al., 2003):

* New roofing materials with pigments can reflect more heat than conventional materials. Preventing heat from entering a building through the roof can help to reduce the amount of energy needed to cool the interior space.

* New wall designs can help control heat loss by reducing the amount of framing and optimizing insulating materials, such as structural insulated panels and insulated concrete forms. When retrofitting existing buildings, new insulating fabrics can be hung or applied to interior walls to control indoor temperatures.

* High-quality windows address the three main energy paths: radiant energy, heat conduction through the frame, and air leakage around the window's components. Low-emissivity window coatings increase the window's R-value by reducing the flow of infrared energy out of the building; other low-emissivity coatings can block infrared energy from entering the window to reduce the cooling load.

* Passive daylighting is combined with very efficient lighting systems to meet additional lighting needs. In daylight design, windows provide adequate interior illumination while minimizing glare and controlling interior temperatures. Building designs can also include clerestory windows, skylights, light tubes, and light shelves to bring light into the deeper recesses of buildings. Energy use can be offset both directly by replacing artificial lighting or indirectly by reducing cooling loads.

* Sunshine can heat a space passively through direct solar gain, where the sun shines into a building and warms materials in the space such as bricks, concrete, or adobe. These materials store thermal energy as they are heated during the day and release their heat to warm the space at night. A Trombe wall, separated from the outdoors by a glass wall, is designed to absorb solar heat and release it into the interior

Table 11.37 | Cost of energy for a single-family solar thermal system as a function of capacity factor, turnkey investment costs, and discount rate. The O&M costs are assumed at US¢2/kWh for a capacity factor of 10–12% and US¢4/kWh for 5–6%. Lifetime is 20 years.

Capacity factor	Turnkey investment costs per kWth (*water heating*)	Discount rate	
		5%	10%
	US$_{2005}$$	US¢/kWh	
12%	1100	10.4	14.2
	2200	18.7	26.5
6%	1100	20.8	28.5
	2200	37.5	53.0
Capacity factor	Turnkey investment costs per kWth (*space and water heating*)	Discount rate	
		5%	10%
	US$_{2005}$$	US¢/kWh	
10%	1100	12.0	16.7
	2200	22.1	31.4
5%	1100	24.1	33.4
	2200	44.2	62.8

of a building. The future may see special phase-change materials used for thermal storage and molecular or nanocomposite materials. Ventilation takes place during periods of cool outside temperatures.

11.8.5   Economic and Financial Aspects

Active low-temperature solar technologies require an auxiliary energy source. In solar water heating systems, 20–70% of the demand is met by solar, and in combined solar water and space heating systems, the figure is 20–60% (IEA-CERT, 2011). However, when thermal energy storage is included, this percentage will increase.

Solar has a first cost to the user that must be amortized over the years of service. Based on the data in Tables 11.35 and 11.36, the levelized cost of energy for these systems can be calculated. This calculation assumes discount rates of 5% and 10%, an economic system lifetime of 20 years, and O&M costs ranging from US¢1–4/kWh$_{th}$, depending on the system and the capacity factor achieved. (See Tables 11.37 and 11.38) The energy costs range from US¢10–63/kWh$_{th}$ for single-family systems and from US¢8–55/kWh$_{th}$ for multi-family systems. In the long term, further improvements in the performance and investment costs may reduce the energy cost by a factor of two (ESTTP, 2008). Solar thermal systems in developing countries cost US¢2–3/kWh$_{th}$ because they are simple thermosiphon systems and usually without freeze-protection (IEA-CERT, 2011).

11.8.6   Sustainability Issues

Low-temperature solar energy technologies have the potential to displace burning wood and fossil fuels. Solar water heating and solar

cooking technologies can conserve local biomass and reduce the burden of collecting firewood. They can purify water and mitigate the health impacts of cooking with wood, which creates smoky indoor environments. They can reduce costly diesel fuel use and pollution of the local environment. Low-cost solar crop drying added to the local infrastructure can produce feed for livestock and bring more protein into the diet.

Solar technologies emit no CO_2 while operating. The CO_2 emissions attributed to these technologies come mostly from the energy expended in producing the raw materials, fabricating the products, and transporting and installing them. Displacing combustion with solar thermal heat can offset considerable amounts of CO_2 (Arvizu, 2008), as the energy payback time of a thermosiphon type solar collector and storage system could be 1.3–4.0 years (Harvey, 2006). Worldwide, the energy yield of solar collectors in 2008 roughly translated to an oil equivalent of over 12 million metric tons and an annual CO_2 emissions reduction of 39 million metric tons (Weiss and Mauthner, 2010). Under the IEA BLUE Map scenario, the total installed worldwide solar thermal capacity in 2050 could reduce CO_2 emissions by roughly 450 million metric tons (IEA-CERT, 2011).

11.8.7   Implementation Issues

Barriers to the widespread use of solar thermal systems include initial costs, financing problems, policy uncertainty, and consumer and institutional ignorance (IEA-CERT, 2011). Key barriers to wider use of solar crop drying are lack of information about the cost-effectiveness of these systems, about their technical details, and about practical experience (IEA-SHC, 2009).

Table 11.38 | Cost of energy for a multi-family solar thermal system as a function of capacity factor, turnkey investment costs, and discount rate. The O&M costs are assumed at US¢1/kWh for a capacity factor of 10–12% and US¢2/kWh for 5–6%. Lifetime is 20 years.

Capacity factor	Turnkey investment costs per kWth (*water heating*)	Discount rate	
		5%	10%
	US$_{2005}$$	US¢/kWh	
12%	950	8.2	11.6
	2000	16.2	23.3
6%	950	16.5	23.2
	2000	32.5	46.6

Capacity factor	Turnkey investment costs per kWth (*space and water heating*)	Discount rate	
		5%	10%
	US$_{2005}$$	US¢/kWh	
10%	950	9.7	13.7
	2000	19.3	27.3
5%	950	19.4	27.4
	2000	38.5	55.4

Incentive programs and policies that include environmental mandates can help reduce economic and other barriers. These actions may be taken at the local, regional, national, or international level – from a desire for a city to be "green," to a state or province mandating air quality levels, to a country striving to meet the guidelines of the Kyoto Protocol (Arvizu, 2008). Tax credits for companies and consumers, a carbon tax on fossil fuel use, and renewable energy portfolio mandates designed for utilities are a few examples of policies and practices that encourage use of solar technologies (Johansson and Turkenburg, 2004; REN21, 2010). (See also section 11.12.) Legislation can be very effective. For example, in 1980 the Israeli government required solar water heaters in all new homes (except tall buildings with insufficient roof area). In Israel, 85% of households now use solar thermal systems. Spain also introduced a national solar thermal obligation for new buildings in 2006 (ESTIF, 2007).

In the (draft) *Roadmap for Energy Efficiency in Buildings*, focusing on heating and cooling (IEA-CERT, 2011), the following areas are identified for policy support:

- Increased technology R&D, significant demonstration programs, and development beyond the present best available technologies;

- Improved information for consumers and robust metrics for analyzing the energy savings, CO_2 emission savings, and financial benefits of heating and cooling technologies;

- Market transformation policies to overcome the current low uptake of many low/zero carbon heating and cooling technologies;

- International cooperation to maximize the benefits of policy intervention as well as the transfer of technical knowledge between countries;

- The creation of strong partnerships for the promotion of efficient and low/zero carbon heating and cooling technologies.

11.9   Ocean Energy

11.9.1   Introduction

In principle, the oceans represent one of the largest renewable energy resources on earth. Energy is stored in the seas as kinetic energy from the motion of waves and currents and as thermal energy from the sun. Energy can be extracted from the mixing of fresh and salt water ("salinity gradient energy") where rivers run into the sea. Ocean water can be used for cooling purposes, and oceans can be used to produce marine biomass for energy services (as discussed in section 11.2).

Most ocean energy resources are diffuse and far from where they are needed. Yet energy from the oceans can be captured for practical use (Turkenburg et al., 2000). Tidal energy can be extracted where extreme tidal ranges or currents exist. Wave energy can be exploited where a higher-than-average wave climate occurs. Ocean thermal energy conversion (OTEC) can be achieved where the temperature difference between surface waters and water near the seabed is sufficient. Using ocean water to cool power plants or buildings is feasible when the distance between source and end user is short. These requirements tend to limit the use of the ocean resource to certain areas of coastline where the resource and a market for the energy are within reach.

Many techniques have been proposed to exploit this resource to generate electricity, produce fuels, provide cooling, and make potable water. Virtually all of them, however, are still at the early development or demonstration stage.

11.9.2 Potential of Ocean Energy Technologies

The main ocean energy resources considered here are tidal head energy, tidal current energy, ocean current energy, wave energy, ocean thermal energy, and salinity gradient energy.

Tidal head energy: Tidal head (or barrage) energy makes use of the potential energy from the difference in height between high and low tides. The output of a tidal head plant is entirely predictable. The largest tidal heads can be found in estuaries in Canada (Bay of Fundy), the United Kingdom (Severn Estuary), France (Baie du Mont Saint Michel), and India (Gulf of Cambay). The global potential of tidal head energy is estimated at 3 TW, with about 1 TW being available at comparable shallow waters (Charlier et al, 1993). The capacity factor for these plants varies from 22–29%. Consequently, the theoretical potential of tidal head energy can be estimated at about 2500 TWh/yr (9 EJ$_e$/yr). The technical potential is estimated at about 1,000 TWh/yr (3.6 EJ$_e$/yr) (see Chapter 7).

Tidal current and ocean current energy. Tidal current energy makes use of the kinetic energy of water in a tide. Ocean currents are generated by winds and by differences in water temperature and salinity. In most places, the movements of seawater are too slow to permit practical energy exploitation. Water velocity can be increased by a reduction in cross-section of the flow area. This happens in straits, around the ends of headlands, in estuaries, and in other narrowing topographical features (Turkenburg et al., 2000). As with wind energy, a cube law governs instantaneous power to fluid velocity. An average peak marine current of 2.5 mps (5 knots), which is not unusual, translates to a power flux of about 8 kW/m^2. Potential sites must have flows exceeding 1.5 mps for a reasonable period (Fraenkel, 1999; IT Power, 1996).

Estimates of the tidal current energy resource have been made for the European Union, Asia, the United Kingdom, and Canada (CEC, 1996; CEC, 1998; UK-DTI, 2004; Cornett, 2006). Sites have also been identified in Japan, New Zealand, and South America. In Europe, about 100 locations have been identified where power production from tidal current energy may become attractive, offering a potential electricity supply of 48 TWh/yr (compared with European electricity demand of about 5000 TWh/yr in 2010). Countries with promising sites include France, Greece, Ireland, Italy, and the United Kingdom (CEC, 1996). In China, 14 GW of tidal current power installations might be possible (Wang and Lu, 2009). This would translate to 50 TWh/yr, assuming a high capacity factor of about 40% (Harvey, 2010).

Assessments for open ocean current flows have also been made (see, e.g., Leaman et al., 1987). Potential locations have been identified in South Africa, East Asia, Australia, and North America, including the Florida current of the Gulf Stream, 15–30 km off the coast. Ocean currents offer the potential of a relatively high capacity factor of 70–80% (Boud, 2002). The theoretical potential of ocean current energy has been estimated at about 6000 TWh/yr (Sörensen and Weinstein, 2008). The technical potential could be 1000–2000 TWh/yr and maybe more, depending on assumptions made.

Wave energy: Wave energy is generated by friction transfers from wind energy passing over the ocean. The energy density of waves is substantial: a wave 1.5 m high has a capacity of about 10 kW/m of wave front. The greatest wave power densities occur off northwestern Europe, western Canada and Alaska, and southern South America and Australia (Harvey, 2010). Energy fluxes of up to 75 kW/m have been found off the coast of Ireland and Scotland (Clément et al., 2002).

The global theoretical potential of wave energy has been estimated at 8000–80,000 TWh/yr (Harvey, 2010). Another study has proposed approximately 30,000 TWh/yr (108 EJ$_e$/yr) (Mørk et al., 2010). In Chapter 7, the technical potential is estimated at about 20,000 TWh/yr (72 EJ$_e$/yr).

The wave energy potential that can be harnessed in practice has been estimated at 260 TWh/yr in the United States and at 120–200 TWh/yr in Europe (Boud, 2002; Harvey, 2010). Wave energy is variable in time, resulting in capacity factors of 25–35%, depending on the location. However, the available wave energy can be predicted accurately 24–48 hours in advance (Harvey, 2010).

Ocean thermal energy: Using some form of heat engine to exploit natural temperature differences in the ocean has been discussed and investigated for many decades (Boyle, 1996). To deliver a feasible system, technically and economically, as large a temperature difference as possible is required. A temperature difference of about 20°C is required for OTEC. A few tropical regions with very deep water (a depth of 1 km or so) have this characteristic. In a recent assessment, a global technical potential of about 40,000 TWh/yr (144 EJ$_e$/yr) was calculated (Nihous, 2007). However, in another study the technical potential was assessed at 10,000 TWh/yr (Harvey, 2010). An even earlier assessment presented an estimate of 30,000–90,000 TWh/yr (108–324 EJ$_e$/yr) (Charlier and Justus, 1999).

Deep water from the ocean can also be used to cool buildings in coastal areas, especially in the tropics (see, e.g., Makai Ocean Engineering, 2010). The technology is called seawater air conditioning (SWAC). No estimates could be found about the technical potential of this technology.

Salinity gradient energy: Mixing salt water and fresh water releases latent heat that can be converted to work when a river runs into the sea (Pattle, 1954). According to Norman (1976), cited by Post (2009), "the tremendous energy flux available in the natural salination of fresh water is graphically illustrated if one imagines that every stream and river in the world is terminated at its mouth by a waterfall 225 meter high," which is the potential energy equivalent of the latent heat from mixing fresh water with seawater. In a recent study, the theoretical potential of this technology was calculated at 1.7 TW/yr, or about 50 EJ/yr (Post, 2009). For most continents, the theoretical potential is about 300 GW (nearly 10 EJ$_e$/yr); only Europe (94 GW, or 3 EJ$_e$/yr) and Australia (30 GW, or 1 EJ$_e$/yr) have significantly lower potentials.

The global technical potential of salinity gradient energy can be estimated at 1 TW (Post, 2009). Assuming a capacity factor of 80–90%, this would translate to about 7500 TWh/yr (27 EJ$_e$/yr). In another publication the technical potential was estimated at about 150–170 GW$_e$ (van den Ende and Groeman, 2007); this capacity could generate about 1200 TWh/y (4.3 EJ$_e$/yr). However, citing Statkraft of Norway, Criscione (2010) suggests that by 2030 the deployment of this energy source could already be 1600–1700 TWh/yr.

11.9.3   Market Developments

The world is only beginning deployment of ocean energy conversion systems. Not many systems are operational and most activities are undertaken by research institutes and small and medium-size enterprises, mainly in Canada, Norway, the United Kingdom, and the United States (Kahn et al., 2009; Harvey, 2010). Other countries involved include Australia, China, Denmark, France, India, Ireland, Japan, the Netherlands, and Sweden. Most of these countries participate in the IEA Ocean Energy Systems Implementing Agreement (IEA-OES-IA).

The 2010 annual report of IEA-OES-IA (2011) indicates that at the end of that year the installed power, reported by member countries, was 2 MW for wave energy, 4 MW of tidal stream, and 259 MW of tidal barrage energy. Consequently, the total installed capacity for ocean energy in 2009 and 2010 was less than 300 MW, mainly tidal head capacity, generating about 0.5 TWh/yr.

Tidal head energy: Tidal head energy has been exploited on a small scale for centuries using water mills. In France, the 240 MW La Rance scheme, built in the 1960s, is a large modern version. It uses bulb turbines like those applied in river-based hydro-electric projects. These turbines can pump to increase storage and produce about 0.5 TWh/yr (Andre, 1976; Kerr, 2007). A 254 MW plant at Lake Sihwa on the western coast of South Korea was expected to become operational in 2011. In addition, a handful of smaller plants have been built in Canada, China, and Russia (Cavanagh et al., 1993; Strange et al., 1994; Kerr, 2007). About 43 GW of tidal range capacity is under investigation, with an estimated electricity production potential of 63 TWh/yr (Kerr, 2007).

Tidal and ocean current energy: Large-scale energy generation from currents requires a totally submerged turbine, which may look like a wind turbine, adapted for harsh marine conditions. It can have a horizontal or a vertical axis. About a dozen devices are in the RD&D phase. A 1.2 MW grid-connected system, called SeaGen, has been built by Marine Current Turbines and has been in operation in Northern Ireland since 2008. It is accredited as an official UK power station. Marine Current Turbines's earlier Seaflow 200 kW prototype operated from 2003 to 2009 in the Bristol Channel at the Devon coast (Willis, 2010). Other devices include the Hammerfest Strøm 300 kW turbine in Norway, installed in 2004; the TGL 500 kW device of Rolls Royce in the United Kingdom, installed in 2010; and the Verdant Energy array of 6 x 35 kW (small) turbines in East River, New York, that operated between 2005 and 2008.

Wave energy: The development of wave energy through 2007 was supported by Australia, Canada, Chile, China, India, Ireland, Japan, New Zealand, Norway, Portugal, Russia, Sweden, the United Kingdom, and the United States. In the last 5–10 years, the wave energy converters introduced by several companies include the Limpet (operational in Scotland since 2000) of Voith Hydro Wavegen, the PowerBuoy of OPT, the Oyster of Aquamarine Power, the Wave Energy Converter of Pelamis Wave Power, the Archimedes Waveswing of AWS Ocean Energy, and the Wave Energy Converters of Oceanlinx.

Ocean thermal energy: The first OTEC plant (using a low-pressure turbine) was a 22 kW device built in 1940 in Cuba. Some later attempts off the coast of Brazil and the Ivory Coast were initiated but failed. OTEC technology has been developed in the United States since 1974 (Hawaii) and in Japan since 2004. In the United States, prototype plants with rated capacities of 50 kW up to 1 MW were tested around 1980. A 210 kW system was tested from 1993 to 1998. Several small-scale power plants, including a 30 kW hybrid OTEC prototype plant, have been built in Japan. Unfortunately, Pacific Ocean storms knocked out a number of pioneering installations. OTEC systems have a very long pipe extending to deep water, making the system vulnerable to damage from rough seas. This vulnerability and the high system costs may explain why in 2006 only one ongoing project could be identified. It was in operation near the coast of India and developed by the Institute of Ocean Energy in Japan. Almost all the major US OTEC experiments have taken place at the Natural Energy Laboratory of Hawaii Authority (NELHA). In 2010, a 1.2 MW demonstration plant was under construction (Harvey, 2010). About one-third of the capacity will be used to run the pumps and the system, which would result in a net production of 800 kilowatts.

In Hawaii, cold seawater is also being used to air-condition buildings. This technology is applied to cool buildings of NELHA, which saves the laboratory nearly $4000 per month in electricity costs. The system requires much less maintenance than traditional compressor systems (State of Hawaii, 2010). Similar projects have been developed or are being developed in Sweden and elsewhere (IEA-DHC, 2002; IEA, 2009c; Makai Ocean Engineering, 2010).

Salinity gradient energy: Two technologies to extract energy from mixing fresh water and salt water, using quite different physical principles, are in the prototype stage. The Pressure Retarded Osmosis technology has been under investigation by Statkraft of Norway since 1997 (Scråmestø et al., 2009). The Reversed Electro Dialysis technology has been investigated by Wetsus Institute of the Netherlands since about 2004 (Post, 2009).

Ocean energy in future energy scenario studies: Because of the early stage of development, ocean energy is projected to have a minor role in most energy scenarios for 2030–2050. Potential contributions in 2050 range from 25–600 TWh/yr. The European Renewable Energy Council and Greenpeace International, however, project a figure of 1900 TWh in 2050 (Krewitt et al., 2009; EREC & Greenpeace, 2010; IEA 2010a).

11.9.4 Ocean Energy Technology Developments

A recent study identified 135 ocean energy technologies that have been under development since the 1990s (Khan and Bhuyan, 2009). Within constrained funding environments, and due to the difficult operating environment and resource diversity, maturity of various ocean energy technologies has been relatively slow. (See Figure 11.65.)

Tidal head energy technology: Tidal head energy conversion technology uses similar technology to conventional low-head hydro-electric systems and therefore is considered mature. At present it is used in estuaries where a basin is created by means of a barrage. (See Figure 11.66) The high tide fills the basin and the impounded water is held behind the barrage until the receding tide creates a suitable head. The water is then released through the hydroelectric converter until the rising tide reduces the head to a minimum operating point. This sequence is repeated with the tides (Strange et al., 1994). In this approach, called "ebb generation," power production on both ebb and flood is also possible. To obtain continuity of supply, plant configurations and operating routines of greater complexity are needed, including linked and paired basins (Strange et al., 1994). Single-basin plants deliver one or two intermittent pulses of energy per tide. The pulses recur at a period of 12 hours and 25 minutes. Consequently, the capacity (or load) factor of the plant is limited to 25–35%. Recent studies also investigated options to produce electricity using artificial basins offshore (called tidal lagoons). Increased output can be obtained by using the turbines as pumps using external power to increase the water level and therefore the generating head (Strange et al., 1994). Turbine and generator system configurations include the bulb, the Straflow, and the tubular turbine (Harvey, 2010).

Tidal and ocean current energy technology: Marine current energy machines work very much like underwater wind turbines; the physical principles of kinetic energy conversion are directly analogous in moving water or air. Currents are complicated by short-term variations in both velocity and direction, caused by turbulence and velocity shear effects. In particular, surface currents move much faster than the currents deeper down, to the extent that the majority of the energy in most locations with strong currents is in the upper half of the water column. Full understanding is needed about how variation in velocity combined

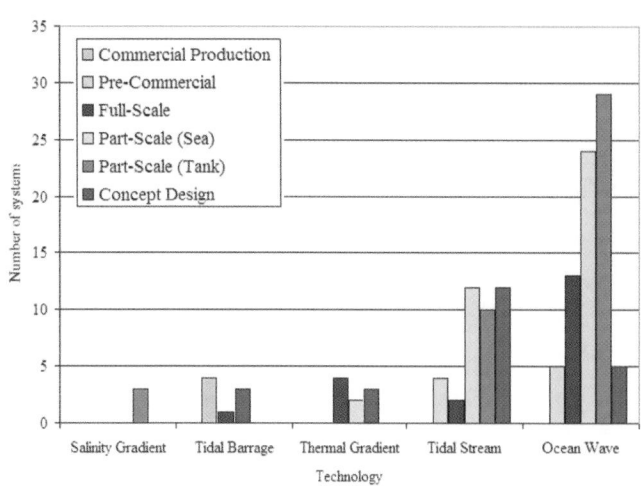

Figure 11.65 | Number of ocean energy conversion schemes and their maturity. Source: Kahn and Bhuyan, 2009.

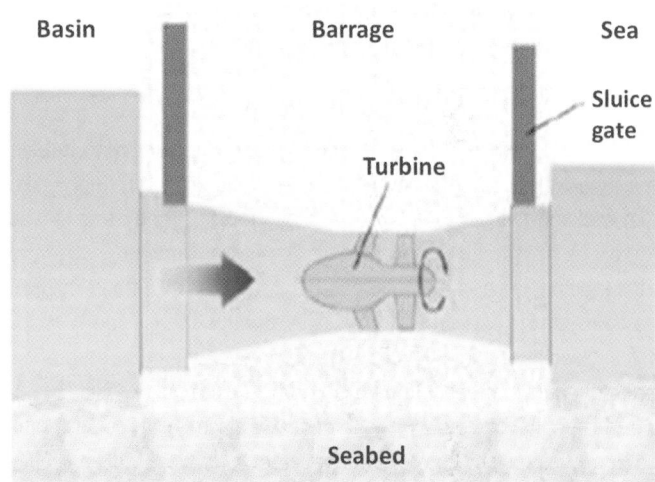

Figure 11.66 | Tidal barrage type of energy device. Source: Sørensen and Weinstein, 2008.

with waves from any direction will affect structural strength and power-conversion mechanisms (Salter et al., 2006).

The density of water is some 850 times higher than that of air at normal atmospheric pressure, so a water current turbine can be considerably smaller than an equivalent powered wind turbine. When applied to convert tidal current energy, which is a bi-directional flow, the turbine should be able to respond to reversing flow directions. This is not an issue when converting energy from ocean currents, which are generally unidirectional.

The biggest problem with underwater kinetic energy conversion systems stems from the very large forces generated by taking momentum from such a dense medium. Although the literature tends to focus on rotor

Figure 11.67 | Three technologies to convert ocean current energy: reciprocating hydrofoil (left), horizontal axis turbine (middle), vertical axis turbine (right). Source: DP Energy, 2010.

concepts, the overwhelming problem in engineering a large system to extract kinetic energy from moving water, whether it is waves or currents, is the strength needed both to react to very large forces and to be anchored securely to the seabed. Typically a tidal turbine experiences thrust forces amounting to around 100 tonnes per MW with a rated current of about 2.5m/s (typical), whereas a wind turbine will experience about one third of this force per MW.

Tidal turbine developers are following in the footsteps of wind turbine developers 30 years ago by experimenting with a variety of rotor types: axial-flow turbines, cross-flow turbines, and reciprocating devices. (See Figure 11.67.) When using axial-flow or cross-flow turbines, horizontal axis as well as vertical axis configurations can be applied and shrouds can be used to increase the flow velocity through the turbine. A reciprocating device uses the flow of water to produce the lift or drag of an oscillating part transverse to the flow direction. The mechanical energy from this oscillation can be used to generate electricity. It seems probable that the pros and cons of these different rotor types will be similar in water to those in air, notably that the open axial flow rotor is more efficient, more controllable, and more cost-effective than any other. This is why virtually 100% of commercial wind turbines (and the majority of low-head hydro turbines too) use pitch regulated axial flow rotors.

Although in an early stage of development, experiments have been conducted with prototypes in the open sea in the United Kingdom. At least one system (Marine Current Turbines's 1.2 MW SeaGen) has reached the commercial stage, is an accredited UK power station, and will be replicated in larger numbers in the near future.

Wave energy technology: At least 50 wave-energy devices are under development (US DOE, 2009; Khan and Bhuyan, 2009). One group is devices operating on the shoreline, near-shore, or offshore. Another group is installations located at the water surface versus installations located at a depth of about 5–15 m (Sørensen and Weinstein, 2008). In a third group, technologies are divided according to the mechanism applied: oscillating water column devices, surge devices, heaving float devices, heaving and pitching float devices, and pitching devices.

(See Figure 11.68.) Each mechanism reacts to the waves in a different way. For example, surge devices often have a flap that is pushed back and forth by the waves; heaving float devices move up and down as the wave passes and activate either a pump or a generator to extract energy; pitching devices rock usually fore and aft and generally react one floatation device against another, often compressing hydraulic fluid in the process to drive a generator.

To increase confidence in wave energy, demonstrations are needed that wave devices can survive the worst climate (wind, temperature) and ocean conditions (saltwater corrosion, fouling, heavy storms, seawater pollution) and perform as expected. Less expensive installation methods are needed, along with test platforms so that large numbers of components and subassemblies can be operated in parallel.

Ocean thermal energy technology: The three main concepts of OTEC are all based on the Rankine (steam/vapor) cycle: an open cycle system, a closed cycle system, and a combination of these two. The open cycle concept quickly evaporates warm seawater to vapor (at reduced pressure) and draws it though a turbine by condensing the vapor with cold seawater. The closed cycle system uses warm seawater to boil a low-temperature fluid, such as ammonia or propane. (See Figure 11.69.) That vapor is then drawn through a turbine by being condensed in a heat exchanger with cold seawater and recycled back to the boiler by a pump (Turkenburg et al., 2000). Closed cycle systems offer in principle a better conversion efficiency and a smaller turbine. The hybrid system flash evaporates warm seawater, and the generated steam is used as a heat source for a closed cycle.

Offshore OTEC is technically difficult because of the need to pipe large volumes of water from the sea bed to a floating system and the need for huge areas of heat exchanger. Other operational challenges include storm resistance, corrosion, maintenance of vacuum, and fouling. Transmitting power from a device floating in deep water to the shore is also an issue, because of the costs involved. As a result, OTEC technology has yet to mature to commercial and economic viability.

Seawater air-conditioning systems can be applied to produce chilled water. Such systems pipe cold water from the deep ocean to the shore. The cold (1–7°C) water, which is denser than warm water, is pumped from a depth of 600–1000 meters, and the return system water is 8–12°C (IEA-DHC, 2002; IEA, 2009c). Using a heat exchanger, the cold can be made available to air condition hotels, offices, and other buildings, reducing the energy use for air conditioning by 90% (Van Wijk, 2010). The cold can also be used to condense water vapor for irrigation purposes and to produce potable water. The capacity of a SWAC system can be many MW. Commercial production is feasible, especially in regions where a high capacity factor can be achieved.

Salinity gradient energy technology: In a Pressure Retarded Osmosis system – also called osmotic power system – a semi-permeable membrane is placed between a salt solution (seawater) and fresh water (Post,

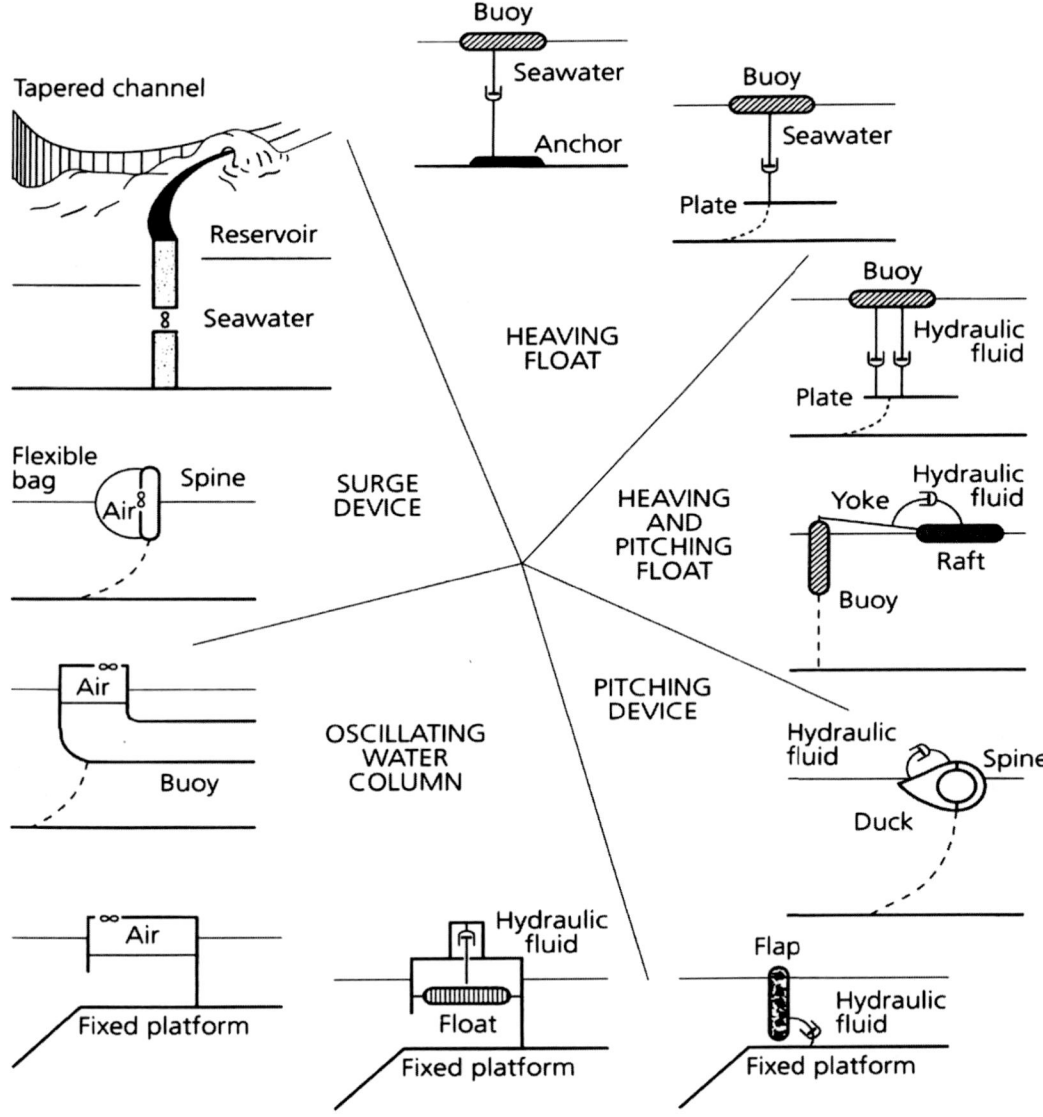

Figure 11.68 | Schematic representation of five mechanisms and 11 devices to use wave energy. Source: Cavanagh et al.,1993.

2009). The membrane allows the solvent (the water) to permeate, but it acts as a barrier to the dissolved salts. The chemical potential difference between the solutions causes transport of water from the diluted salt solution to the more concentrated salt solution. This transport results in a pressurization of the volume of transported water, which can be used to generate electricity in a turbine. The great difference in osmotic pressure (equal to a 240 m head) between fresh water and seawater is used. A first pilot plant based on this technology was commissioned by Statkraft in Norway in 2009 (Post, 2009).

A Reversed Electro Dialysis system is essentially a salt battery. A number of cation and anion-exchange membranes are stacked in an alternating pattern between a cathode and an anode to create an electrical battery. The compartments between the membranes are alternately filled with a concentrated salt solution and a diluted salt solution. The salinity gradient results in a potential difference (80 millivolt for sea water and

river water) over each membrane. The potential difference between the outer compartments of the membrane stack is the sum of the potential difference over each membrane. First prototypes of this technology have been tested at the Wetsus Institute in the Netherlands (Post, 2009).

These technologies are still at an early stage of experimentation, and at present it is unclear whether they can be made sufficiently efficient to become commercially and economically viable.

11.9.5   Economic and Financial Aspects

Most ocean energy technologies are in the RD&D phase. Their development is driven by governmental R&D funding and policy objectives. Consequently, only preliminary cost figures can be presented, with a high degree of uncertainty.

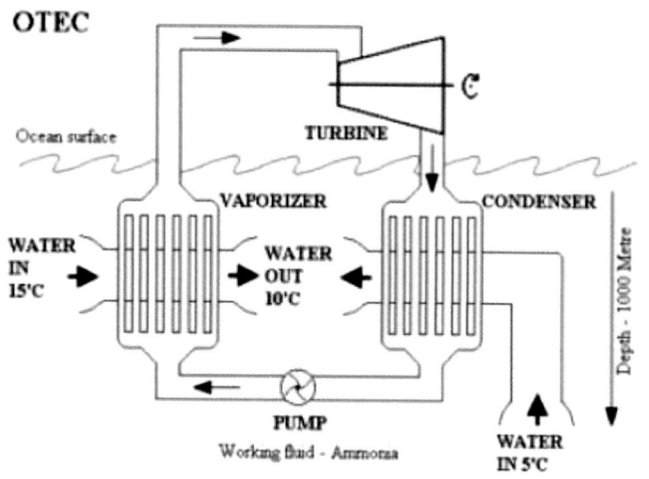

OTEC

Figure 11.69 | A closed cycle OTEC power plant. Source: Sørensen and Weinstein, 2008.

When calculating the cost per kWh, the assumptions are a discount rate ranging from 5–10%, annual O&M costs equivalent to 1% of the initial investment costs, and an economic lifetime of the system ranging from 30–40 years (depending on the technology involved). It should be noted that if the technical lifetime of a system is much longer than the economic lifetime – as is the case with tidal barrage systems – the kWh production costs will in general be substantially lower in the long term. Due to anticipated technical learning, it can be expected that future electricity production costs will come down by a factor of two or more for most of these technologies. An overview of the kWh cost calculations is presented in Table 11.39.

It must be noted that corrosion issues and fouling can have a large impact on O&M costs and the capacity factor of the conversion system. Seawater is bad for electronic circuitry as well as for other parts of the system. Failure of the simplest, cheapest components can cause outage for long periods of time because bad weather, for example, can hinder and delay even minor repairs.

Tidal head electricity: Big barrage systems, have been estimated to have investment costs of US$_{2005}$$4000–6000/kW (Turkenburg et al., 2000; IEA-ETSAP, 2010) and a capacity factor of 25–30%, with resulting production costs of electricity of US¢10–31/kWh. The 254 MW Sihwa scheme under construction in South Korea is the only other large tidal barrage to be installed after the French La Rance scheme of the 1960s. The system is based on an existing barrage built primarily for flood control purposes, which thereby reduced its capital costs to the marginal cost of adding turbines and generating plant. Economic estimates, however, are not available. The Severn Barrage in the United Kingdom, which was assessed in 2010 for the UK government, would have been rated at 7000 MW and cost over US$30 billion; it was rejected for governmental support primarily because it was evaluated to be uneconomical.

Tidal and ocean current electricity: Tidal and marine turbines are modular devices that may have individual power outputs in the 1–5 MW range and that are designed for build-out into arrays or "farms." Therefore they can be built up incrementally into large projects eventually, allowing quicker start to the repayments compared with big barrage systems.

It is estimated that the initial investments for open ocean current energy systems may range from US$_{2005}$$7000–14,000/kW. Assuming a capacity factor of these systems of 70%, the initial kWh production costs would be US¢9–27/kWh. For tidal current systems, the initial investment costs will probably be lower, such as US$5000–10,000/kW. However, the capacity factor will also be less: maybe about 35%. This would result in a kWh cost of US¢12–38/kWh. It should be noted that higher initial investment costs are mentioned in the literature than assumed here, up to US$14,000/kW (Callaghan, 2006; IEA-ETSAP, 2010).

Wave power electricity: Wave power devices tend to be heavier than wind ones, because for a given power output they need to handle much larger forces, demanding more structural material. This gives wave energy a disadvantage if cost predictions are based on weight. Plants can run steadily for days, even weeks at a time. But because energy comes in pulses at twice the wave frequency, and wave amplitudes follow the Gaussian distribution, there are large power peaks – 10 or more times the mean values. Assuming present investment costs ranging from US$_{2005}$$6000–16,000/kW (Turkenburg et al., 2000; Callaghan, 2006; IEA-ETSAP, 2010) and a capacity factor of 25–35%, the initial electricity production cost could be US¢15–85/kWh. It is projected that energy generated by wave energy installations can reach an average price below about US¢13/kWh in the longer term (Sørensen and Weinstein, 2008). Harvey (2010) gives a longer term figure of US¢10–14/kWh.

OTEC electricity: Investment costs of initial commercial plants have been estimated at US$6000–12,000/kW (see, e.g., Vega, 2002; Cooper et al., 2009). Assuming a capacity factor of 70%, this would result in production costs of US¢8–23/kWh.

Salient gradient electricity: No cost estimates have been completed for this technology, and thus no figures for the investment and kWh costs can be presented. It is estimated that the investment costs per kW will be substantially higher than for other mature renewable energy technologies, but this is compensated by the expected high capacity factor (80–90%).

It should be noted that the cost figures presented in Table 11.39 are higher than expected about 10 years ago, when the World Energy Assessment was published (Turkenburg et al., 2000). The main reasons are a steep increase in the overall investment costs and lack of technological learning, reflecting the relatively immature nature of these technologies.

Table 11.39 | Production cost of electricity as a function of capacity factor, turnkey investment costs, and discount rate. For all technologies, the annual O&M costs are assumed to be equivalent to 1% of the turnkey investment costs. The economic lifetime is assumed to be 40 years for tidal range, and 30 years for other systems.

Tidal head electricity (economic lifetime 40 years)			
Capacity factor	Turnkey investment costs (US$_{2005}$)	Discount rate	
		5%	10%
30%	$4000/kW	10.4¢/kWh	17.0¢/kWh
	$6000/kW	15.5¢/kWh	25.6¢/kWh
25%	$4000/kW	12.4¢/kWh	20.5¢/kWh
	$6000/kW	18.6¢/kWh	30.7¢/kWh

Tidal current electricity (economic lifetime 30 years)			
Capacity factor	Turnkey investment costs (US$_{2005}$)	Discount rate	
		5%	10%
35%	$5000/kW	12.3¢/kWh	19.0¢/kWh
	$10,000/kW	24.5¢/kWh	37.8¢/kWh

Ocean current electricity (economic lifetime 30 years)			
Capacity factor	Turnkey investment costs (US$_{2005}$)	Discount rate	
		5%	10%
70%	$7000/kW	8.6¢/kWh	13.2¢/kWh
	$14,000/kW	17.1¢/kWh	26.5¢/kWh

Wave electricity (economic lifetime 30 years)			
Capacity factor	Turnkey investment costs (US$_{2005}$)	Discount rate	
		5%	10%
35%	$6,000/kW	14.6¢/kWh	22.7¢/kWh
	$16,000/kW	39.1¢/kWh	60.5¢/kWh
25%	$6000/kW	20.5¢/kWh	31.8¢/kWh
	$16,000/kW	54.8¢/kWh	84.7¢/kWh

OTEC electricity (economic lifetime 30 years)			
Capacity factor	Turnkey investment costs (US$_{2005}$)	Discount rate	
		5%	10%
70%	$6000/kW	7.9¢/kWh	11.3¢/kWh
	$12,000/kW	14.6¢/kWh	22.8¢/kWh

11.9.6 Sustainability Issues

All ocean energy technologies are able to support local energy supplies with almost no environmental emissions to the atmosphere during operation. Consequently, using ocean energy would contribute to reducing GHG emissions as well as acidification and other environmental problems associated with conventional energy use. Further development and deployment of these technologies can also contribute to regional and local economic development and employment. In coastal regions, the marine energy industry could create 10–20 jobs/MW (Sørensen and Weinstein, 2008).

Life-cycle assessments of tidal and wave energy conversion systems indicate GHG emissions ranging from about 15–50 gCO$_2$-eq/kWh and an energy payback time ranging from about 1–2.5 years (Raventós et al., 2010).

Locally ocean energy systems could also have some negative impacts that should be minimized. Tidal head energy systems, for example, when located in estuaries, will have an impact on currents and on sediment transport and deposits. Current subjects of investigation include the impacts of construction of the barrage on local biodiversity, of a large human-made seawater lake behind the barrage, and of offshore tidal lagoon systems on fishing, fish, bird breeding, and feeding. Wave energy systems will have a visual impact when large arrays of floating devices are installed near shore.

Current energy devices are expected to have limited environmental impact. Underwater noise and vibrations can be a concern as well as impacts on fishing activities. To avoid accidents with surface vessels, these devices must be installed deep enough (resulting, however, in a lower energy yield) or be made to be surface-piercing and hence visible to shipping. For OTEC devices that use closed-circuit hydraulics, spills of working fluids or leakage is a subject for investigation.

The corrosive effects of seawater as well as debris and fouling are having an impact on the durability and performance of ocean energy systems and are issues of concern.

11.9.7 Implementation Issues

A major issue is the nascent stage of development of most ocean energy technologies, the high investments costs per kW, and the need to reduce these costs through R&D and learning by doing. Therefore government support is needed to bring promising technologies from the R&D phase to the stage of prototype and pilot plant developments. A drawback then could be the large diversity of technologies investigated. A barrier for early deployment can also be the remote location of most ocean energy systems, increasing installation costs but also the need to develop or adjust infrastructures to connect these systems to the grid. Yet some synergies may be found when also deploying wind turbines offshore. And the high predictability of energy supplies from these sources reduces system integration challenges. An advantage of tidal barrage systems can be the possible integration with energy storage (pumped hydro) to enlarge the flexibility in electricity supplies.

At a country level, high-quality resource assessment and inclusion of these technologies into energy development and planning portfolios is needed. Of the few countries that have completed these initial steps, the United Kingdom has set a target of 2 GW ocean capacity installed by 2020 (UKERC, 2008). Following this model, a strategy to commercialize and deploy ocean energy technologies should cover in principle all aspects of the innovation chain for emerging technologies, ranging from R&D and capacity building to, for example, the manufacturing of systems, the development of standards, and financial arrangements to facilitate high-volume deployment.

11.10 System Integration

11.10.1 Introduction

At present, only a handful of countries have non-traditional renewable penetration above 10% (with a few above 30%). A number of studies

suggest that this figure could steeply increase in the coming decades (see, e.g., Krewitt et al., 2009b; IEA, 2010a; ECF, 2010; Sims et al., 2011; see also Chapter 17). Because of the partly fluctuating nature of some renewable energy sources, ranging from minutes to annual seasons, a key challenge will be to match load and supply of energy properly.

Experience demonstrates that large shares of variable renewables are possible if appropriate measures are taken to support system integration. The common and historical use of hydroelectricity – in some countries providing more than 50% of the overall electricity supply – is an excellent example of how annual seasonal variation can be successfully managed. In 2007, about 20% of the Danish electricity demand was met by wind energy (IEA, 2008a), with instantaneous penetration levels exceeding 100% of electricity demand in the west. Only a few years earlier, such an achievement was considered impossible (IEA, 2008c).

The current energy supply system is capable of dealing with significant variability in demand by combining "base load," "middle load," and "peak load" power plants. With a growing contribution from renewable energies, the concept of base versus peak load power plants will become increasingly obsolete. The goal will be to guarantee firm capacity and supplies from a variety of energy technologies and sources such that a reliable and environmentally sound supply is achieved at competitive costs. Innovative ways to pool various decentralized renewable energy sources – such as energy storage options, advanced infrastructures (like "smart grids," "virtual power plants," and "microgrids"), and long-distance energy transport between supply and demand clusters (FOSG, 2010; ENTSO-E, 2010) – are required to integrate especially variable renewable energy sources into low-carbon energy supply systems (see, e.g., ECF, 2010; Sims et al., 2011).

A distinction can be made between renewables being dispatchable (having a high potential to deliver when needed) and variable (having a low potential to deliver on each moment). Hydropower, biomass energy, geothermal energy, and ocean energy options like OTEC are dispatchable renewables. Solar energy, wind energy, and ocean energy options like tidal range are variable renewables.

The integration of variable renewables into energy systems, which is most challenging, is the focus of this section, especially their integration into larger-scale electricity and thermal heating and cooling systems. Some aspects of integrating biofuels in liquid fuel transportation systems are discussed in Section 11.2. Integrating energy systems is discussed in Chapter 15.

11.10.2 Integration of Renewables into Electricity Supply Systems

Normally, electricity systems consist of a number of power plants (units) that generate electricity to fulfill load demands and a transmission and distribution system that can deliver the generated electricity to customers. Two concepts are of special importance: unit commitment and load dispatch. When a unit is running, it is committed. It does not need to be generating power, but all the implications of its start-up have been accepted by the system operator. After carrying out the commitment procedure, the operator has to organize the dispatch of the load to all the running units. These procedures are essentially optimization problems. Solving these problems can be difficult, especially when high levels of renewable energy sources and energy storage are incorporated (Van Wijk and Turkenburg, 1992). The unit commitment problem can be defined as "commit the units in an order such that total fuel and start-up costs of the system are minimized;" the load dispatch problem is to "dispatch the load and the spinning reserve to all committed units, such that the total fuel costs are minimized." Spinning reserve has to be available to cover power shortages caused by sudden outages of units or sudden sharp increases in the load (Van Wijk and Turkenburg, 1992).

Penetration of dispatchable renewables can take advantage of complementarity along the different periods of the year. In particular, biomass is harvested during part of the year to take advantage of its energy content and the better logistic facility. Harvesting time usually is coupled with bioenergy production, at least to reduce storage feedstock costs. On the other hand, hydroelectricity availability peaks during the rainy season and quite often at this time water supply exceeds demand or generation installed capacity. In some regions the harvest period for bioenergy feedstock occurs mainly during the low rain period, providing synergism of significant economic value, mainly when bioelectricity is one of the final energy forms obtained from biomass.

Deep penetration of variable renewables reduces the consumption of conventional fuels and the associated production of waste and emissions to the atmosphere. It also reduces somewhat the need for conventional power plants, because of the capacity credit of wind turbines and solar systems. However, it may also force dispatchable plants to contribute more to the spinning reserve and to operate in partial load, reducing conversion efficiencies and increasing fuel consumption. Enough backup power should be available to guarantee security of supply when solar or wind energy is not available. And there should be enough flexibility in the system to meet the variations in load and in availability of wind and solar resources. Finally, the transmission and distribution system should be able to cope with increases in distributed electricity generation associated with the use of renewables.

11.10.2.1 Distributed Electricity Generation

In conventional grid structures, electricity is fed into the grid at high voltage levels by relatively few large power stations and is brought to the customer via several intermediate grid levels. As generation becomes more widely distributed, the number of electricity sources increases and the direction of flow can be reversed. The distribution grids assume the function of transporting electricity in different (bidirectional) directions

and become service providers between generators and consumers. Central power stations will continue to exist, but in addition there will be a large number of smaller, distributed systems. This change requires investments in many high-voltage distribution lines when the supply comes from MW-size power plants and in coordination of the operation of a large number of systems in the electricity distribution and transportation networks, facilitated by adequate information and communication technologies (Degner et al., 2006). Current grids are not designed to allow major amounts of electricity to be fed into the distribution grid. However, several solutions are under development to solve this, including converter technologies that also provide ancillary services to the grid and control schemes that enable a high penetration of distributed generation. These approaches are integral to the development of so-called Smart Grids.

As defined by the European Technology Platform for Electricity Networks of the Future (also called Smart Grids ETP), "a Smart Grid is an electricity network that can intelligently integrate the actions of all users connected to it—generators, consumers and those that do both—in order to efficiently deliver sustainable, economic and secure electricity supplies." A Smart Grid "employs innovative products and services together with intelligent monitoring, control, communication, and self-healing technologies to better facilitate the connection and operation of generators of all sizes and technologies, to allow consumers to play a part in optimizing the operation of the system, to provide consumers with greater information and choice of supply, to significantly reduce the environmental impact of the whole electricity supply system, and to deliver enhanced levels of reliability and security of supply" (Smart Grids ETP, 2010).

Quality of supply, power quality, grid control and stability, safety and protection, and standardization are considered the main challenges for a major deployment of decentralized generation. Scheepers et al. (2007) listed a number of ways to address these challenges:

- Active management of distribution systems to increase the accommodation of distributed generation (DG); typical examples are voltage control in rural systems and fault level control in urban systems through network switching;

- Development of intelligent networks where technological innovations on power equipment and information and communication technologies are combined to allow a more efficient use of distribution network capacities;

- Further development of the microgrid concept, based on the assumption that large numbers of micro-generators, connected to the network, can be used to reduce the requirement for transmission and high-voltage distribution assets;

- Pooling small (renewable energy) generators to create a virtual plant, either for the purpose of trading electrical energy or to provide system support services.

Microgrids are low-voltage or medium-voltage distribution systems with distributed generators, storage devices, and controllable loads. They are operated connected to the main power network or isolated from the main grid (Hatziargyriou, et al., 2007; Hatziargyriou, 2008; Microgrids, 2011). From the grid's point of view, a microgrid can be regarded as a controlled entity that can be operated as a single aggregated load or generator and, given attractive remuneration, as a small source of power.

Many system operators regard distributed generation as an additional complexity and thus fear additional costs (Scheepers et al., 2007). Nevertheless, distributed generation also offers potential benefits to electric system planning and operation. (See Table 11.40). However, the regulatory context is of key importance for deployment of distributed generation (Coll-Mayor et al., 2007).

11.10.2.2 Forecasting the Short-term Availability of Wind and Solar Energy

The availability of renewable energy resources, especially wind and solar, can vary dramatically (Holltinen et al., 2007). Technical options like load management, energy storage, and improved interconnection of grids help match supply and demand, but good information on the availability of renewable energy sources in space and time is required to apply these options optimally.

The challenge with variable renewable energy at present "is not so much its variability, but rather its predictability" (IEA, 2008). To facilitate system integration and to reduce related costs, it is necessary to predict the available variable power hours or days ahead as accurately as possible. Sophisticated wind forecasting tools are available today, and interest in forecasting solar radiation is rising in parallel to the market uptake of PV systems and concentrating solar thermal power plants. But predictability is not the only issue to be solved. Reliability is also important. Long periods of low wind or sun, even if predicted accurately, require the power system to meet demand by running other capacity or importing electricity from elsewhere, which imposes additional cost.

There are two main approaches for wind power prediction (Jursa and Rohrig, 2008). One uses physical models of wind farms to determine the relationship between weather data from a numerical weather prediction model and the power output of the wind farm. The other is a mathematical modeling approach, in which statistical or artificial intelligence methods, such as neural networks, are used to model the relationship between weather prediction data and power output from historical data sets. This is currently used by several transmission system operators in three European countries. Artificial neural networks provide a time series of the instantaneous and expected wind power for a control zone for up to 96 hours in advance (Rohrig and Lange, 2008). Use of this system in Germany to make predictions a day ahead resulted in errors of

Table 11.40 | Potential benefits of distributed generation.

Potential Benefit	Explanation
System reliability	DG can add to supply diversity. A distributed network of smaller sources may also provide a greater level of adequacy than a centralized system with fewer large sources. Reliability can also be enhanced if islanding is planned in case of faults.
Reduction in peak power requirements	Reduction in peak load can displace or defer capital investment, as the need for more expensive power plants is reduced.
Ancillary services	DG can provide ancillary services, particularly those that are needed locally, such as reactive power, voltage control, and local black start.
Power quality	DG can address power quality problems, particularly when the systems involve energy storage, power electronics, and power conditioning equipment. However, large-scale introduction of DG may also lead to instability of the voltage profile.
T&D capacity	DG can reduce transmission and distribution requirements and can contribute to reduced grid losses. On-site production of energy carriers can result in savings in T&D costs.

Source: based on US DOE, 2007; Pepermans et al., 2005; Rawson, 2004.

around 5% (root mean square error, in percent of the installed capacity) for the entire German system. In addition, short-term (15 minutes to 8 hours ahead) high-resolution forecasts of wind power generation are generated, taking into account data from online wind power measurements of representative sites. A number of short-term prediction tools, both physical and statistical, are currently in use in Denmark, Germany, United States, Spain, Ireland, and Greece (Costa et al., 2008).

The difference between the forecasted and instantaneous wind power production can be minimized with advanced control strategies of wind farms. Pooling of geographically distributed large wind farms in a portfolio of one supplier allows optimized control of variable generation. This minimizes imbalance penalties.

Solar forecasting is not yet as mature as wind forecasting, but several models are under development. In general there are two approaches: short-term forecasting (a few hours ahead) and medium-term forecasting (up to two days ahead) (Heinemann et al., 2006). Short-term forecasting can be based on satellite data about solar irradiation and an algorithm calculating the motion of clouds. This approach improves considerably the prediction accuracy with respect to assuming persistency (that is, irradiation at hour h is identical to that at hour h-1). When forecasting one hour ahead, a relative root mean square error of about 30% can be obtained for an area of 31x45 km^2, while for a six-hour forecast this error increases to 40–50% (Heinemann et al., 2006). Medium-term forecasting can be performed using weather forecasts, with a three-hour temporal and a 0.25°x0.25° spatial resolution. Lorenz et al. (2009) have shown that the relative root mean square error could be 36% for a single location in Germany for a one-day-ahead forecast. Taking the whole of Germany, this error would reduce to 13%.

Table 11.41 | Storage options in electricity systems.

Scale	Response time	Typical discharge times	Storage technologies	Applications
Very large scale	> 15 min	days to weeks	Hydrogen storage systems	Reserve power compensating for long-lasting unavailability of wind energy
Large scale	< 15 min	hours to days	Compressed air storage Hydrogen storage systems Pumped hydro	Secondary reserve Minute reserve Load leveling
Medium scale	1–30 sec[a] 15 min[b]	minutes to hours	Batteries (Li-ion, lead-acid, NiCd) High-temperature batteries Zinc-bromine batteries Redox-flow batteries	Load leveling Peak shaving

[a] Primary reserve.

[b] Secondary and minute reserve.

Source: Kleimaier et al., 2008; Leonard et al., 2008; Weimes, 2009.

11.10.2.3 Energy Storage Options

Energy storage in an electricity system allows generation to be decoupled from demand. Storage can also improve the economic efficiency and use of the entire system (Ummels et al., 2008). The broad range of available electricity storage technologies differ with respect to capacity, duty cycle, response time to full power, and load following capability.

Managing seasonal variation on electricity supply from hydro sources uses well-known storage procedures based in large water reservoirs able to keep above the average river flows due to either the rainy season or ice melting annual period, mainly in countries/regions with large penetration of this renewable source. Table 11.41 provides a summary of storage technologies appropriate for short periods compared with annual seasons and their typical applications.

Recent developments in electricity storage (see, e.g., Chen et al., 2009; Ibrahim et al., 2008; Meiwes, 2009) include the following:

- Adiabatic compressed air energy storage (CAES) incorporates compression heat into the expansion process and thus does not need additional fuel. The efficiency for adiabatic CAES is up to 70%.

- Lithium-ion batteries, with an efficiency of 90–95%, have become an important storage technology in portable applications.

- Redox-flow batteries allow the electrolyte to be stored in large tanks, making redox-flow batteries well-suited for large-scale applications. The system efficiency is in the range of 60–75%.

Hydrogen is considered a suitable storage option for fluctuating renewable electricity. The achievable energy density of compressed hydrogen is more than one order of magnitude higher than the one of compressed air. The storage of compressed hydrogen in salt caverns is currently the only technology with a technical potential for single storage systems in the 100 GWh range (Kleimaier et al., 2008) and for long (seasonal) storage periods. Different electrolyser technologies are under development to produce hydrogen from electricity. The most efficient conversion back to electricity can be achieved in combined-cycle power plants or in fuel cells. Round-trip efficiencies are expected to be in the range of 35–40% (Kleimaier et al., 2008), but a demonstration project in Norway shows 10–20% (Ulleberg et al., 2010).

The economic performance of a storage system depends on operating conditions and system costs, which can vary depending on required volume, conversion and generation capacities, and response times.

Figure 11.70 shows life-cycle costs for two storage applications: longer-term storage (500 MW, 100 GWh, 200 hour full load, ~1.5 cycle/month) and load-leveling (1 GW, 8 GWh, 8 hour full load, 1 cycle/day) (Kleimaier et al., 2008; Meiwes, 2009). For each storage option, Figure 11.70 shows a range of costs that spans from a low value representing "achievable costs expected within 10 years" to a high value of present (2008) costs. The life-cycle costs have been calculated assuming 8% interest rate and investment cost for the storage system, including necessary auxiliary units, power converters, and interfaces to the grid (Leonard et al., 2008). Component replacement has also been included, as well as the electricity price needed to cover losses. In both storage applications, pumped hydro systems are reported to be the most economic option, assuming appropriate geographic conditions. For long-term storage, hydrogen benefits from low volume-related costs due to its high energy density compared with adiabatic CAES. For load leveling, however, compressed air appears viable.

Battery technologies also can be used for load leveling, but costs are in the range of US_{2008}¢6.5–10/kWh up to more than US_{2008}¢30/kWh, depending on the technology used, and thus are well above the costs shown in Figure 11.70.

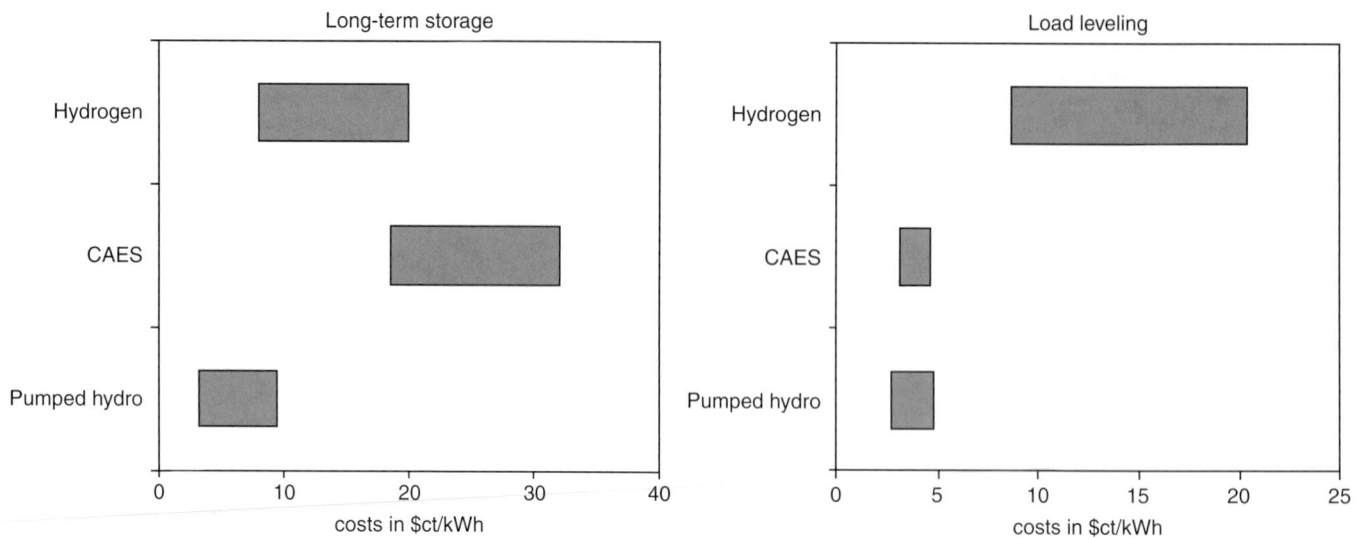

Figure 11.70 | Electricity storage costs for different storage options (US$_{2008}$ ¢t/kWh). Source: Kleimaier et al., 2008; Leonard et al., 2008; Meiwes, 2009.

11.10.2.4   Vehicle-to-Grid Options

Because of the partly complementary nature of the electric power system and the light vehicle fleet, "vehicle-to-grid" (V2G) technologies may improve the ability of the electricity grid and the transportation system to integrate larger proportions of fluctuating renewables. The basic concept of vehicle-to-grid power is that electric vehicles (EVs) – including battery electric vehicles, fuel cell vehicles, and plug-in hybrids – can not only be charged by the grid but also can provide electricity and other services to the grid while parked. Each vehicle must have a connection to the grid for electrical energy flow, options for communication with the grid operators, and on-board control and metering devices. Figure 11.71 schematically illustrates the integration of electric vehicles and solar and wind technologies into an electricity supply system (Kempton and Tomić, 2005a).

Lund and Kempton (2008) show that wind electricity shares of more than 50% in the Danish power system lead to substantial excess wind power production, which reduces the revenues from wind electricity generation. The introduction of electric vehicles can reduce excess production. It can also improve the ability of power systems to integrate large proportions of wind power.

While the capital costs of power plants are high and thus motivate high use, personal vehicles are cheap per unit of power and are used only about 5% of the time for transportation, making them potentially available 95% of time for a secondary function (Kempton and Tomić, 2005a). Because vehicles have limited storage capacity and higher per kWh cost of energy than power plants, the generation of bulk power will not be the main business case for V2G concepts. V2G systems can, however, be competitive in providing spinning reserve and regulation capacity (Kempton and Tomić, 2005b; Dallinger, 2009).

11.10.2.5   Long-distance Electricity Transmission

While energy storage decouples supply and demand in time, transmission can decouple supply and demand in space. Transmission allows the pooling of different renewable energy sources, even on a transcontinental level (DESERTEC Foundation, 2011). Transmission distances have been limited because conventional alternating current transmission technology is suited to transmitting electricity only about 500 km. However, newer high voltage direct current (HVDC) technology can be used to link areas with large renewable resources to demand centers, thus facilitating the provision of dispatchable renewable bulk electricity over a larger distance (see Figures 11.71 and 11.72).

One of the advantages of HVDC is the low cost – in the range of US$_{2005}$¢0.7–2/kWh (for 700 TWh, 150 GW, 3000 km) – for transmission of very high power over very long distances (Asplund, 2007; Trieb and Müller-Steinhagen, 2007). The total losses for point-to-point transfer of power over 3000 km may be on the order of 5%. The loss rate increases with each injection/extraction point along the line (Klimstra and Hotakainen, 2011). Today's HVDC schemes have a maximum power of 3000 MW and a transmission distance of around 1000 km. Unlike with alternating current cables, there are no physical limitations on the distance or power level for underground or submarine HVDC cables. HVDC can be used to transfer power from wind parks offshore and to strengthen the electricity grid in areas where overhead lines are not feasible. Maximizing implementation of HVDC, however, will require international coordination and cooperation on a master plan for developing and optimizing super-grids for long-distance transmission. Technical limitations exist that need to be addressed, as well as some regulatory obstacles and economic concerns (Van Hertem and Ghandhari, 2010).

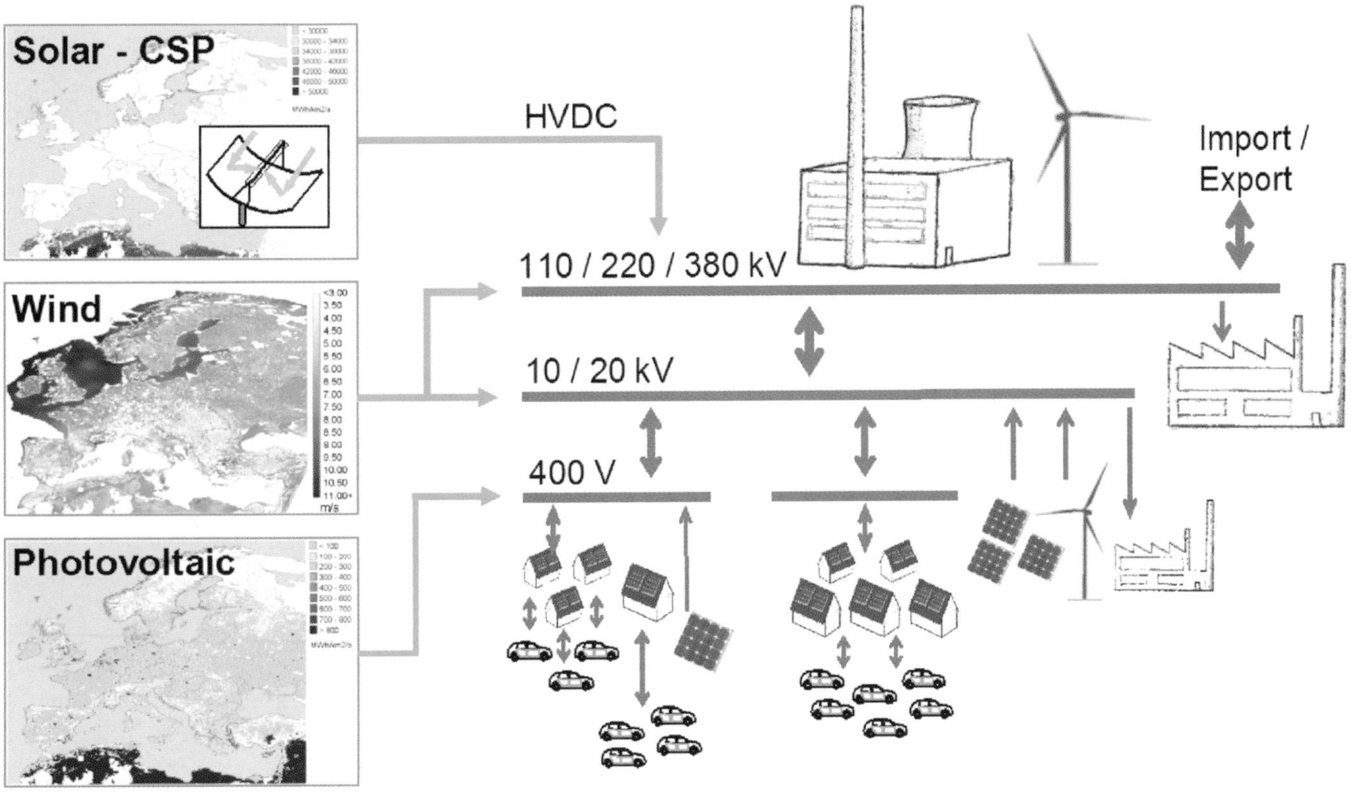

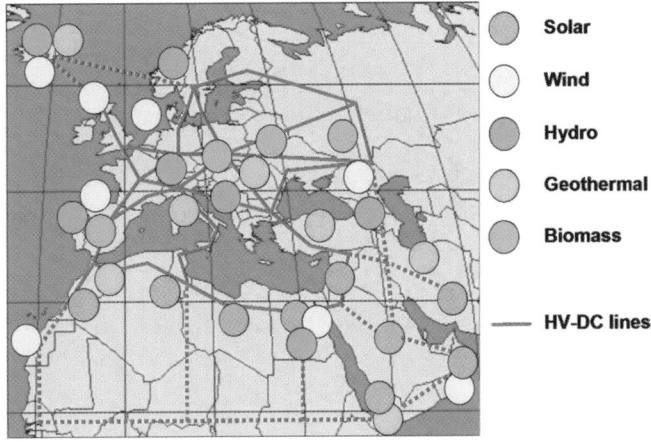

Figure 11.71 | Integration of renewables and electric vehicles into a supply system. Source: Sketch provided by Yvonne Scholz, German Aerospace Centre, 2011.

11.10.2.6 German Renewable Energy Combined Power Plant Demonstration Project

Following a 2006 German energy summit, leading German manufacturers of renewable energy technologies announced that they would prove the feasibility of a secure and constant provision of power using renewable energies only. The goal of the project was to demonstrate that the demand for electric power can be met fully by combining different forms of renewable electricity generation.

The so-called renewable energy Combined Power Plant (Combi-Plant) links and controls 36 wind (12.6 MW), solar PV (5.5 MW), biogas CHP (4 MW), and pumped hydro (1 MW) installations throughout Germany (Mackensen et al., 2008). (See Figure 11.73.) The plant is scaled to meet 1/10,000 of the electricity demand in Germany. This corresponds to the annual electricity requirements of a small town with around 12,000 households.

A central control unit is the core of the Combi-Plant. A control algorithm determines the optimum energy mix at any given point in time. Biogas and hydropower are used to balance fluctuations in wind and solar resources.

The Combi-Plant has been in operation since July 2007, and several months of real world operation were used to calibrate tools that control

Figure 11.72 | Concept of a HVDC-based transcontinental "super-grid." Source: modified from Trieb and Müller-Steinhagen, 2007.

the plant. Figure 11.74 shows the simulated generation mix based on electricity demand data for 2006. Mackensen et al. (2008) conclude that the Combi-Plant is able to meet electricity demand entirely from renewable energy sources. The use of biogas, in particular, plays a central role in controlling the Combi-Plant by covering peak loads and balancing the natural fluctuations in wind and solar energy (Mackensen et al., 2008).

Combined Power Plant

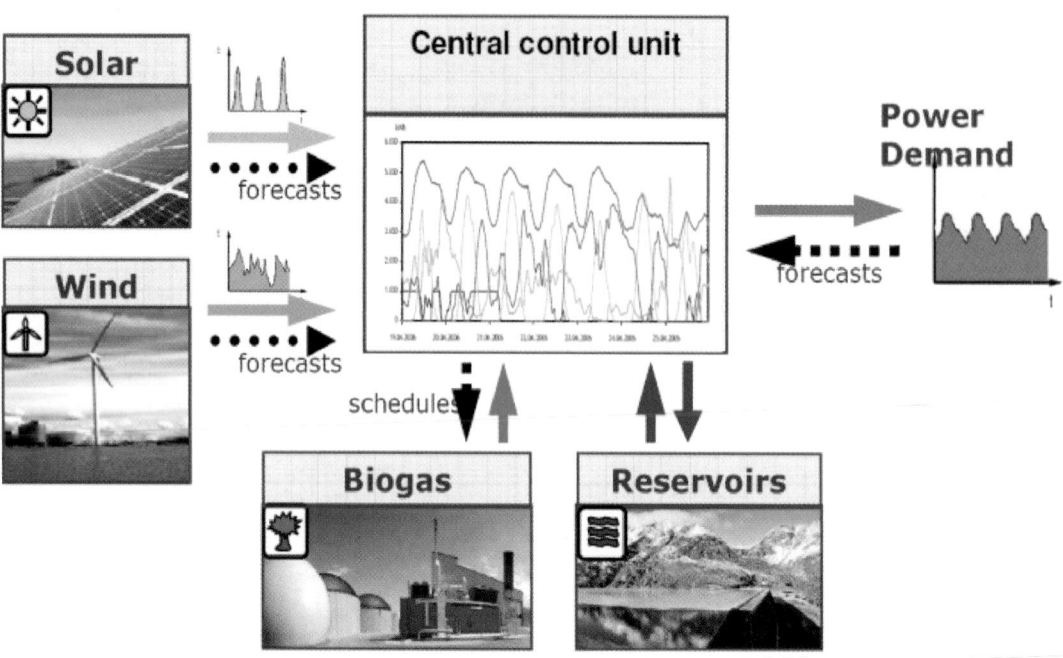

Figure 11.73 | The Renewable Combi-Plant. Source: Mackensen et al., 2008.

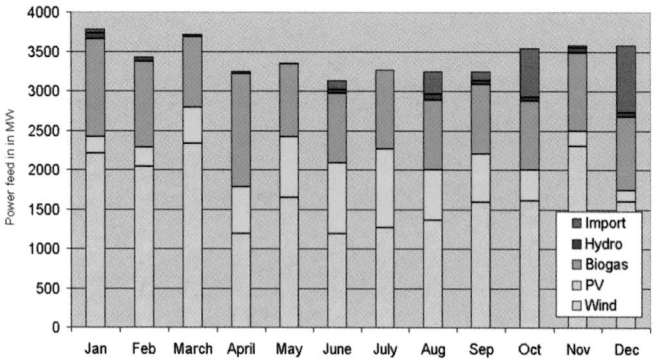

Figure 11.74 | Renewable energy mix of the Combi-Plant: simulation for 2006. Source: Mackensen et al., 2008.

11.10.3   Renewables in Heating and Cooling

Heating and cooling demands from the industrial, commercial, and domestic sectors constitute around 40–50% of global final energy demand (IEA, 2007). Although there is a large potential for renewables in this sector, the use of renewables for heating and cooling generally receives less attention than electricity generation or the production of biofuels for transportation.

Solar thermal, biomass energy, and geothermal heat can provide various categories and scales of heating and cooling services. (See Table 11.42.)

A key challenge is to match seasonal variations in demand and supply. The increasing use of renewable energies for heating and cooling – but also the intensified use of waste heat from industrial processes, the expansion of combined heat and power generation, and the growing demand for air conditioning – calls for more sophisticated interaction between energy demand and supply options.

A fully autonomous energy supply for buildings using locally available sources is feasible. In most regions, however, an independent off-grid supply system causes oversizing of the capacity to produce energy carriers from renewables, in order to overcome the seasonal mismatch of demand and supply. In areas with an existing infrastructure, bringing together a variety of demand and supply options may facilitate the cost-effective integration of renewables, as complementary supply and demand patterns can help reduce system requirements.

The design, costs, and operation strategies of integrated renewable heating and cooling systems depend on local conditions – that is, the availability of renewable resources and the demand for heating and cooling services as a function of time.

11.10.3.1   Renewables in District Heating and Cooling Systems

The potential of renewable heating can be fully exploited primarily in settlement areas having a district heating system, in which centralized

Table 11.42 | Suitability of renewable energy resources for various categories and scales of heating and cooling.

Sector	Solar thermal	Solid biomass	Biogas	Biomass from waste	Shallow geothermal	Deep geothermal
Dwellings	✓	✓	(✓)		✓	
Settlements (district heat)	✓	✓	✓	✓	✓	✓
Commerce and services	✓	✓	✓		✓	
Agriculture	✓	✓	✓	✓	✓	✓
Industry	✓	✓	✓	✓		✓

Source: IEA, 2007.

plants create and distribute heat to residential and commercial customers. These systems are considered more efficient than distributed systems, such as boiler systems. Temperatures provided by them are typically 80–90°C, with about 45–60°C in the return system.

Figure 11.75 shows that world district heating supplies are about 10–11 EJ a year. District heating meets much of the heat demand of the residential, service, and other sectors in Iceland (94%), Russia (63%), Sweden (55%), Lithuania (50%), Finland (49%), and Poland (47%) (Euroheat & Power, 2007). The energy source can be fossil fuels, waste heat, biomass, geothermal energy, and (partly) solar thermal energy.

In the case of deep geothermal heat, a large demand is required to compensate for high drilling costs. In most cases, such a demand is only present in district heating networks. Selling waste heat might be crucial for the economic performance of geothermal electricity production, and that is possible only if large consumers or networks of consumers are available. It should be noted, however, that heating and cooling of buildings can also be achieved by applying heat pumps and using near-surface geothermal energy (see Section 11.4). Moreover, improving the building envelope to reduce heating and cooling needs will influence the competitiveness of district heating systems.

As with deep geothermal heat sources, the scale provided by district heating systems is a prerequisite for the economical integration of large-scale solar thermal systems with seasonal storage. In this case, large scale assumes solar meets 50% or more of the total heat demand. The energy gained by the solar collectors can be delivered via a collecting network to the heating central (see, e.g., Bodmann et al., 2005). From there, heat either can be supplied directly to the consumer or, in summer, stored for use in autumn and winter. A gas or biomass boiler can cover the remaining heat demand. A key component in such a system is the seasonal heat storage.

A district cooling system distributes cold from a central source to multiple buildings through a network of underground pipes. The distribution medium in this case is usually chilled water, which is typically generated by compressor-driven chillers, absorption chillers, or other sources like ambient cooling or "free cooling" from deep lakes, rivers, aquifers, or oceans. The temperature of the supply system is normally 1–7°C, while

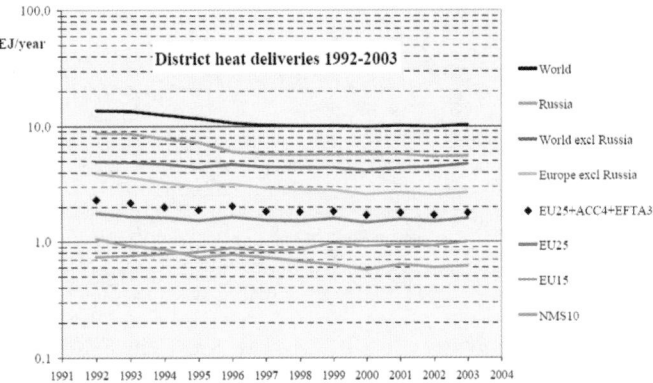

Figure 11.75 | Development of district heat in various parts of the world, 1992–2003. Source: Ecoheatcool, 2005–2006.

the return system is at 8–12°C (IEA-DHC, 2002; IEA, 2009). Cold water from the sea or a lake is applied in a number of projects in Sweden and other countries (see Section 11.9).

11.10.3.2   Thermal Energy Storage Options

The capacity of thermal storage systems ranges from a few kWh up to a GWh, the storage time from minutes to months, and the temperature from –20°C up to 1000°C. Capacity depends in part on the storage materials – solid, water, oil, salt, air – each of which has its own mechanism to store thermal energy. In household applications, water is almost exclusively the medium to store heat.

Storage systems may also rely on sorptive heat storage or latent heat storage (using so-called phase change materials). Both of these options allow thermal storage for an almost unlimited period of time but are at present in early development (Sharma et al., 2009).

The development of very large systems for seasonal heat storage has shown considerable progress in the last few decades. Various demonstration plants have been built. Four different storage types have been developed (Bauer et al., 2007, 2010; Mangold, 2007):

- *Hot water heat storage* has the widest range of possibilities, as it is independent of geological conditions and can be used in small

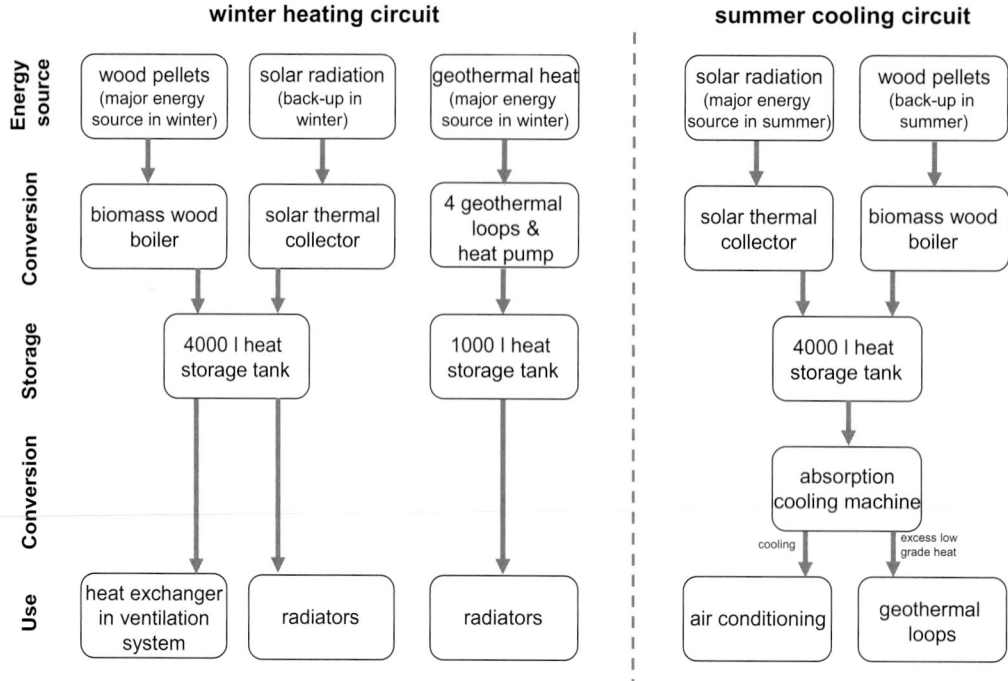

winter heating circuit   **summer cooling circuit**

Figure 11.76 | Renewable heating and cooling system in the Renewable Energy House, Brussels, Belgium. Source: adopted from EREC, 2008.

amounts, such as heat storage for a period of days. It consists of a water-filled containment of steel-enforced concrete, which is partly submerged into the ground.

- *Gravel/water heat storage* consists of a pit sealed with a water-proof synthetic foil, filled by a storage medium of gravel and water.

- Using *borehole thermal energy storage*, heat is conducted via U-tube probes directly into water-saturated soil. These polybutane tubes are inserted into bore holes with a diameter of 10–20cm, which are 20–100m deep. The operational behavior is slow, as heat transport within the store occurs mainly by conduction.

- *Aquifer heat storage* uses natural layers of underground water to store heat. Water is taken out of the aquifer through well boreholes, heated, and then pumped back into the store through another borehole.

The suitability of a storage system depends on local geological and hydro-geological conditions.

Both ice and chilled water storage are used in district cooling plants. Chilled water storage systems are generally limited to a temperature of 4°C due to density considerations. Ice-based storage systems can achieve temperatures of 0.5–1°C. The cool storage is most commonly sized to shift part of the cooling load, which allows the chillers to be sized closer to the average than the peak load (IEA-DHC, 2002). An alternative can be to store cold underground or in aquifers. In Europe, this technology is most widely applied in Sweden (Norden, 2006).

11.10.3.3   Two System Integration Case Studies

The Renewable Energy House in central Brussels is an office building with meeting facilities of approximately 2.800 m². The plan for refurbishing the 140-year-old building was designed to reduce the annual energy use for heating, ventilation, and air conditioning by 50% compared with a reference building and to cover energy demand for heating and cooling with 100% renewable energy sources (EREC, 2008). Key elements of the system, illustrated in Figure 11.76, are two biomass wood pellet boilers (85 kW + 15 kW), 60 m² solar thermal collectors (30 m² evacuated tube collectors, 30 m² flat plate collectors), four geothermal energy loops (115m deep) exploited by a 24 kW ground source heat pump in winter and used as a "cooling tower" by the thermally driven cooling machine in summer, and a thermally driven absorption cooling machine (35 kW cooling capacity at 7–12°C).

In winter, the heating system mainly relies on the biomass pellet boilers and the geothermal system. The solar system and the biomass boilers heat the same storage tank, while the geothermal system operates on a separate circuit. The core of the cooling system is the thermally driven absorption cooling machine, which is powered from relatively low temperature solar heat (85°C) and a small amount of electrical power for the control and pumping circuits. Because solar radiation and cooling demands coincide, the solar thermal system provides most of the heat required for cooling. The solar system is backed up on cloudy days by the biomass boiler. The geothermal borehole loops absorb the low-grade excess heat from the cooling machine, thus serving as a seasonal heat storage system in the winter.

In Crailsheim, Germany, a former military area is currently being transferred into a new residential area in which more than half of the total heat demand will be covered by solar energy. A prerequisite for achieving such high solar shares is the use of a seasonal heat storage facility. The new residential area consists of former military barracks, a school and a sports hall equipped with 700 m² of solar collectors, and new single-family buildings. The residential area is separated from a commercial area by a noise protection wall, which houses the main part of the solar collectors. The first phase of the project focuses on 260 housing units with an expected total annual heat demand of 4100 MWh. The total solar collector area is 7300 m². A borehole thermal energy storage system with 80 boreholes at a depth of 55m provides seasonal storage. In a second phase, the residential area is extended by 210 housing units. The total collector area will then be around 10,000 m², and the seasonal storage system will consist of 160 boreholes. The solar system is separated into diurnal and seasonal parts. Solar heat costs are estimated to be around $US_{2005}¢15/kWh$ (Mangold and Schmidt, 2006; Mangold 2007).

11.10.4 The Way Forward

Various studies on the integration of renewables into electricity grids focus on the challenges it poses and inform strategies on how to proceed. A number of studies have assessed the impacts of up to 35% renewables in the western and eastern grid areas of the United States (EnerNex, 2010; IEEE, 2009; Piwko et al., 2010). The main findings were that renewable energy represents a near-term, leveragable opportunity, provided that issues like siting, access to transmission, and systems operations are well addressed (Piwko et al., 2010). Renewable energy penetration on the order of 30–35% (30% wind, 5% solar) is operationally feasible, assuming significant changes to current operating practice (NERC, 2008; Arent, 2010). NERC (2008) made several specific recommendations:

- Diversify supply technologies across a large geographical region to leverage resource diversity.

- Use advanced control technology to address ramping, supply surplus, and voltage control.

- Ensure access to and the installation of new transmission lines.

- Add flexible resources, such as demand response, V2G systems, and storage capacity.

- Improve the measurement and forecasting of variable generation.

- Use more-comprehensive system-level planning.

- Enlarge balancing areas to increase access to larger pools of generation and demand options.

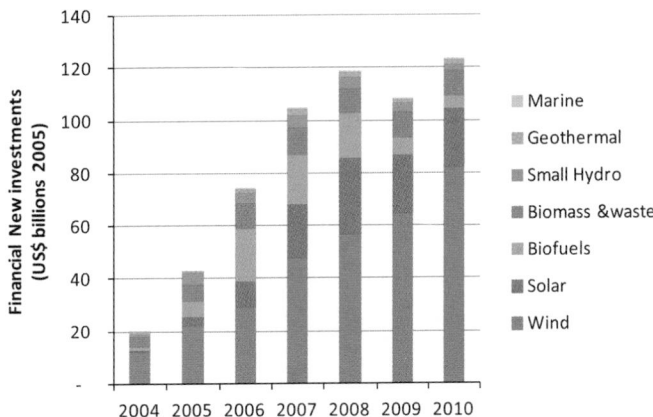

Figure 11.77 | Financial New Investments in New Renewable Energy, 2004–2010 (US₂₀₀₅$ bn). Note that investments in larger-scale hydropower (>50MW) are excluded. Source: UNEP and BNEF, 2011.

Recent investigations are focusing on system-level solutions, in which information technology-enabled power management, advanced forecasting, adaptive and shiftable loads, and advances in energy storage go hand in hand in moving toward power systems with a larger share, possibly a majority, of renewable generation (Denholm et al., 2010; US DOE, 2010b; Krewitt et al., 2009a; Sterner, 2009).

Another important development is the European Electricity Grid Initiative, which is aiming in vision and strategy to enable high penetration rates of renewables (EEGI, 2010) through the integration of national networks into a market-based pan-European network.

11.11 Financial and Investment Issues

11.11.1 Introduction

This section provides information on developments in financing renewable energy for the period 2004–2010. A breakdown of global transactions in non-hydro renewables in 2010 is also shown. The section then describes trends in public policies and public finance mechanisms that aim to stimulate private investments in renewables. From the results, it is clear that – in terms of investments – renewables have become a mainstream energy option.

11.11.2 Commercial Financing

The renewable energy sector has mostly seen increasing levels of financing in the past 10 years. Figure 11.77 shows the trend for financial new investments in new renewables (excluding large hydropower, governmental and corporate R&D, small projects, and solar water heaters) for the period 2004–2010, with a breakdown for different technologies. Between 2004 and 2010 financial new investment increased six-fold to

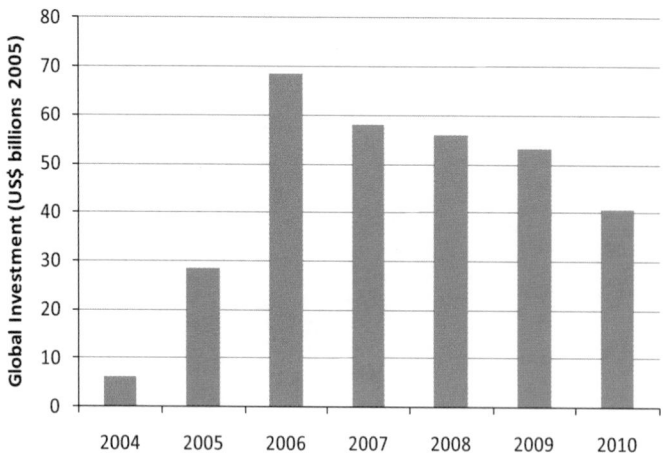

Figure 11.78 | Investments in larger-scale hydropower, 2004–2010 (US$_{2005}$ bn). Source: based on International Journal of Hydropower and Dams Annual Directory, 2004; 2005; 2006; 2007; 2008; 2009, estimating US$1500/MW average cost of new plant and UNEP and BNEF (2011) for the 2010 value.

US$_{2005}$$123 billion, indicating a compound annual growth rate of 36% (UNEP and BNEF, 2011).[10]

Figure 11.78 shows similar investment data for larger-scale hydroelectricity generating capacity for the period 2004–2010. Reporting investment figures in hydropower on a year by year basis is not easy due to the relatively long building time of hydro developments. Generating capacity can also be added at established dams, and progress on both activities is rarely reported annually.

Government and corporate R&D and small projects totaled US$_{2005}$$59 billion in 2010, while investments in solar water heater were estimated at US$_{2005}$$9 billion (UNEP and BNEF, 2011). Including estimated new investments in larger-scale hydropower of US$_{2005}$$40 billion, the total new investment in renewables in 2010 was about US$_{2005}$$230 billion.

A regional breakdown of financial new investments in renewables for the period 2004–2010 is presented in Figure 11.79. It shows strong increases in total investments in Asia and Oceania but stagnation in 2008–2010 in North America, Europe, and South America.

Figure 11.80 provides a breakdown of the total investment of US$_{2005}$$182 billion in 2010 by different financing type: venture capital investments ($2 billion), corporate and governmental R&D support ($3 billion and $5 billion, respectively), private equity investments ($3 billion), public equity investments ($13 billion), asset finance ($110 billion), and investment in small distributed capacity ($52 billion).

Different forms of financing are used for technology development and commercialization, equipment manufacture and scale-up, project construction, and re-financing and sale of companies (mergers and acquisitions, or M&A). This last category is not shown in the Figures since it does not represent new investment in the sector but rather a recycling of funds between early investors and those that later buy them out. The trends in financing along this continuum represent successive steps in the innovation process and provide indicators of the sector's current and expected growth as follows (see also Mitchell et al, 2011):

• Trends in technology investment indicate the mid- to long-term expectations for the sector – investments are being made in new technology developments that will only begin to pay off several years down the road.

• Trends in manufacturing investment indicate near-term expectations for the sector – essentially, that market demand will be sufficient to justify new manufacturing capacity.

• Trends in new project investment indicate current sector activity – that is, the number of new renewable energy plants being constructed.

• Trends in industry mergers and acquisitions indicate the overall maturity of the sector, since increasing refinancing activity over time shows that larger, more conventional investors are entering the sector, buying up successful early investments from first movers.

Each of these trends is discussed in the following sections.

Table 11.43 provides information about the variety of financing types, arranged by phase of technology development. An important catalyst that will influence these trends is the price of fossil fuels, particularly oil and gas. As "easy oil" becomes scarce, the dependence on fossil fuel imports increases and environmental emissions decrease; renewable energy investments in all four dimensions are also expected to increase (see, e.g., REN21, 2010). However, the availability and pricing of natural gas present competitive challenges to the attractiveness of investing in renewable energy technologies.

11.11.2.1 Financing Technology Development and Commercialization

While governments fund most of the basic R&D, and large corporations fund applied or "lab-bench" R&D, venture capitalists begin to play a role once technologies are ready to move from the lab-bench to the early market deployment phase (Mitchell et al., 2011). According to Moore and Wüstenhagen (2004), venture capitalists have initially been slow to pick up on the emerging opportunities in the energy technology sector, with renewable energies accounting for only 1–3% of venture capital investment in most countries in the early 2000s. Since 2002, however, venture

---

10 For conversion of current value US$ to US$_{2005}$$ throughout this section the international time series of USDA www.ers.usda.gov/Data/Macroeconomics/ was used (downloaded in August 2011). In US$_{2010}$$ Financial new investment was $143bn, plus $68bn in Governmental and corporate R&D and Small projects and an estimated 10bn$ decentralized and small scale investment in solar water heater.

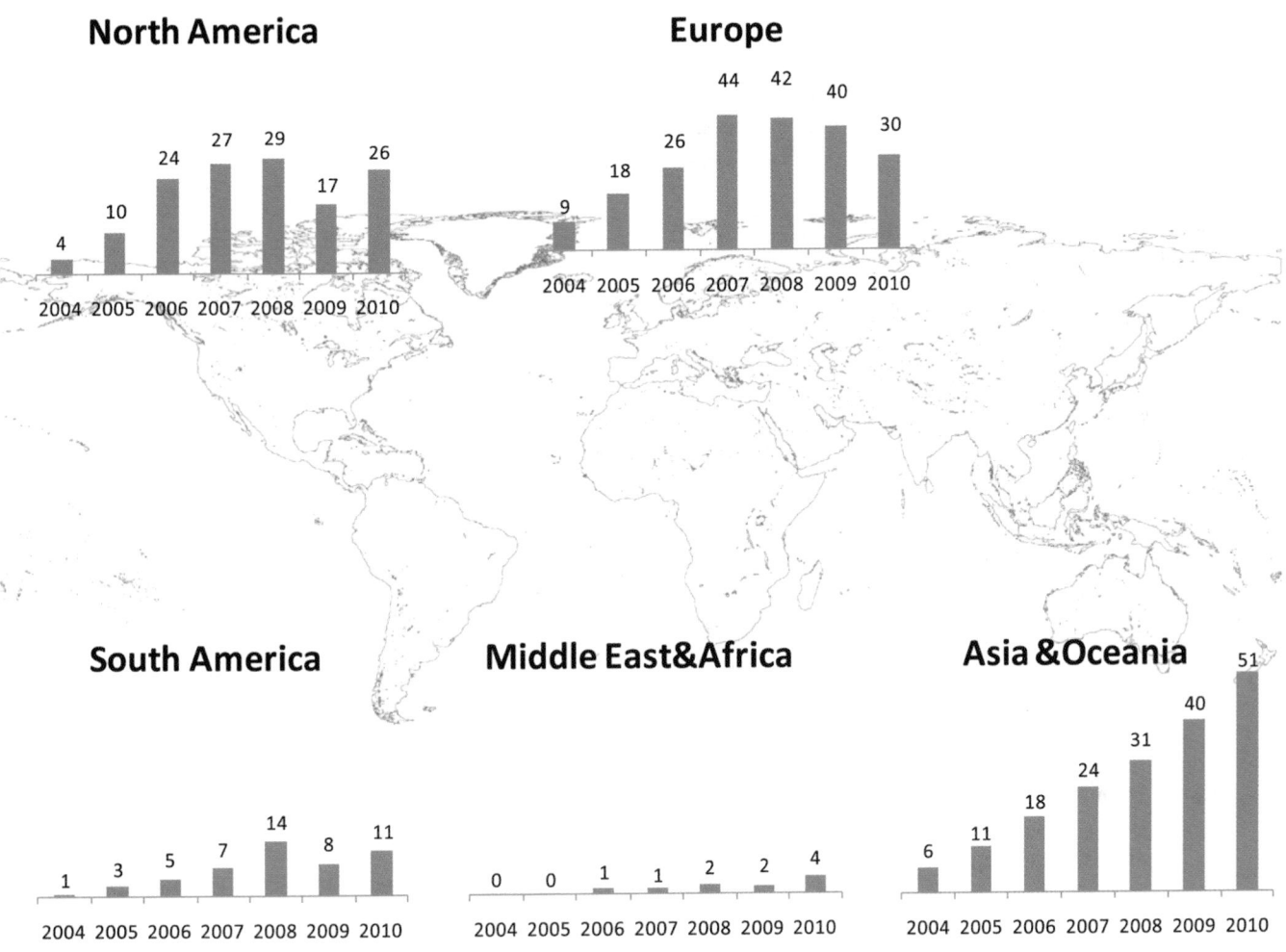

Figure 11.79 | New financial investments in renewable energy, by region, 2004–2010 (US$_{2005}bn). New investment volume adjusts for reinvested equity; total values include estimates for undisclosed deals. This comparison does not include small-scale distributed energy projects or larger-scale hydropower investments. Source: UNEP and BNEF, 2011.

capital investment in RE technology firms has increased markedly. Venture capital into RE companies grew from about US$_{2005}$400 million in 2004 to US$_{2005}$2.2 billion in 2010, representing a compound annual growth rate of 33% (UNEP and BNEF, 2011). This growth trend in technology investment has been an indicator that the finance community expects continued growth in the RE sector. Downturns such as those experienced in 2008/2009 may slow or reverse the trend in the short term, but in the longer term an increasing engagement of financial investors is foreseen in RE technology development (UNEP and BNEF, 2010; 2011).

11.11.2.2   Financing Equipment Manufacturing and Scale-up

Once a technology has passed the demonstration phase, the capital needed to set up manufacturing facilities will usually come initially from private equity investors (those financing un-listed companies) and subsequently from public equity investors buying shares of companies listed on public stock markets (Mitchell et al., 2011). Private and public equity investment in RE grew from about US$1 billion in 2004 to US$16 billion in 2010, representing a compound annual growth rate of 68% (UNEP and BNEF, 2011).

As discussed in Mitchell et al. (2011), in 2008 and 2009 the trading prices of shares in publicly listed companies on the global stock markets in general dropped sharply from the peaks of 2007, but RE shares fared worse due to the initial energy price collapse and the fact that investors shunned stocks with any sort of technology or execution risk, particularly those with high capital requirements (UNEP and BNEF, 2010). By 2010, even with the overall financing picture for renewables much improved, the prices of publicly traded stocks remained depressed largely due to concerns over changes in subsidy regimes and the lower profit margins that manufacturers were earning in many countries (UNEP and BNEF, 2010).

11.11.2.3   Financing Project Construction

Financing construction of renewable energy generating facilities involves a mix of equity investment from the owners and loans from the banks

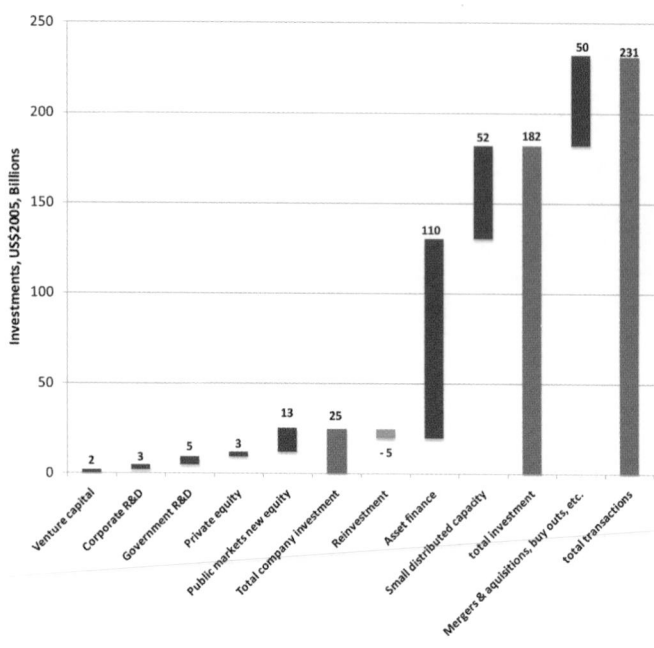

Figure 11.80 | Global Transactions in Renewable Energy in 2010 (US$_{2005}$ bn). Includes all investments as well as acquisition. Source: UNEP and BNEF, 2011.

Table 11.43 | Overview of financing types arranged by phase of technology development.

Development Phase	Financing Types
R&D	Public and corporate R&D support to (further) develop technology is provided through a range of funding instruments.
Technology Commercialization	Venture capital is a type of private equity capital typically provided for early-stage, high-potential technology companies in the interest of generating a return on investment through a trade sale of the company or an eventual listing on a public stock exchange.
Equipment Manufacturing and Scale-up	Private equity investment is capital provided by investors and funds directly into private companies often for setting up a manufacturing operation or other business activity. (This can also apply to project construction.)
	Public equity investment is capital provided by investors into publicly listed companies most commonly for expanding manufacturing operations or other business activities, or to construct projects. (This can also apply to project construction.)
Project Construction	Asset finance is a consolidated term that describes all money invested in generation projects, whether from internal company balance sheets, from debt finance, or from equity finance.
	Project finance is debt obligations (loans) provided by banks to distinct, single-purpose companies, whose energy sales are usually guaranteed by power off-take agreements. Often known as off-balance sheet or non-recourse finance, since the financiers rely mostly on the certainty of project cash flows to pay back the loan.
	Corporate finance is debt obligations provided by banks to companies using "on-balance sheet" assets as collateral. Most mature companies have access to corporate finance but have constraints on their debt ratio and therefore must rationalize each additional loan with other capital needs.
	Bonds are debt obligations issued by corporations directly to the capital markets to raise financing for expanding a business or to finance one or several projects.
Small-scale Technology Deployment	Small and medium-size enterprise finance is realized in different forms – such as consumer loans, micro-finance, and leasing – and is generally provided to help companies set up the required sales and service infrastructure.
Carbon Mitigation	Carbon finance can be in the form of loans or investment obtained from some banks or investors in return for future carbon revenue streams. Examples include the CDM and JI under the Kyoto Protocol.
Refinancing and Sale of Companies	Mergers & acquisitions involve the sale and refinancing of existing companies and projects by new corporate buyers.

Source: adapted from Sonntag-O'Brien and Usher, 2004.

(private debt) or capital markets (public debt raised through bond offerings). The share of equity and debt in a project typically ranges from 20/80 to 50/50, depending on the project context and the overall market conditions. Both types of finance are combined into the term "asset finance," which represents all forms of financing secured for renewable energy projects (Mitchell et al., 2011).

Asset financing to this sector grew from US$_{2005}$$18 billion in 2004 to US$_{2005}$$110 billion in 2010, representing a compound annual growth rate of 35% (UNEP and BNEF, 2011). In recent years capital flows available to RE projects have become more mainstream and have broadened, meaning that the industry has access to a far wider range of financial sources and products than it did around 2004/2005 (UNEP and BNEF, 2011). For instance the largest component of total renewable energy capital flows today is through project finance investment (Deutsche Bank, 2010a,b), an approach that mobilizes large flows of private-sector investment in infrastructure (Mitchell et al., 2011).

11.11.2.4 Financing Small-scale Technology Deployment

Consumer loans, micro-finance, and leasing are some of the instruments that banks offer to households and other end-users to finance the purchase of small-scale technologies. (See Box 11.3) But most investment in such systems comes from the end-user themselves, usually through purchases made on a cash basis. UNEP and BNEF (2011) estimates that US$_{2005}$$52 billion was invested in 2010 in small-scale RE projects ("small distributed capacity"), up from US$_{2005}$$9 billion in 2004, representing a compound growth rate of 34%.

An interesting development, especially with decentralized solar technologies, is the advent of new small-scale financing approaches. In industrial countries, solar equipment manufacturers in the United States have led the way, realizing that they could help overcome capital-cost barriers by acting as financial intermediaries. One of the main financing tools used is the third-party power purchase agreement, which by some estimates drove 60% of the solar capacity installed in California in 2007 (Deutsche Bank, 2010a). Under a third-party power purchase agreement, a third party designs, builds, owns, operates, and maintains

Box 11.3 | End-user Financing for Small-scale Renewables

Many developing countries have rural electrification programs today, and an increasing number of these rely on renewables and distributed financing models to provide access in off-grid areas. Besides electrification, many other clean energy systems and services are being installed with a range of end-user finance approaches.

For example, in Tunisia UNEP has jointly run the Prosol Solar Water Heating Programme with the energy agency and electric utility, helping realize over 95,000 household installations through bank financing made via customer utility bills. UNEP has also run a loan program with two of India's largest banking groups, Canara Bank and Syndicate Bank, helping to kick-start the consumer credit market there for solar home system financing. More than 19,000 homes where financed over three years, and the market continues to grow, with other banks now beginning to lend. Today UNEP has 20 such small-scale end-user finance programs operating in the developing world.

Source: Haselip et al; 2011

the solar systems and sells back solar-generated electricity to the end-user. This model removes the burden of significant upfront costs from the end-user and also allows the solar contractor, who has significantly greater expertise than the end-user, to assume the responsibility for system installation and maintenance (UNEP, 2008).

11.11.2.5   Financing Carbon Mitigation

The carbon markets include a range of instruments used to monetize the CO_2 offset value of climate mitigation projects. According to the World Bank (2009), the primary carbon markets associated with actual emission reductions – the CDM, joint implementation (JI), and voluntary transactions – decreased from US$8.2 billion in 2007 to US$7.2 billion in 2008.

According to the UNEP Risø CDM Pipeline Analysis (UNEP Risø, 2011), renewable energy projects now account for the majority of CDM projects, with 60% of all validated and registered CDM projects, 35% of expected certified emissions reductions (CERs) by 2012, and 13% of CERs issued to date. The low share of CERs issued is mostly due to the very large industrial gas mitigation projects that have been small in number but quick to build, accounting for 75% of CERs issued to date.

The Risø CDM Pipeline Analysis has also calculated the total underlying investment associated with building the proposed 4968 carbon mitigation projects that have reached at least the CDM validation stage in 2010. Of the US$60 billion of total projected investment, US$39 billion (65%) is for renewable energy projects.

11.11.2.6   Refinancing and the Sale of Companies

In 2010, about US$2005$50 billion worth of mergers and acquisitions took place involving the refinancing and sale of renewable energy companies

and projects, up from about US$2005$6.5 billion in 2004, for 41% compound annual growth (UNEP and BNEF, 2011). M&A transactions usually involve the sale of generating assets or project pipelines or of companies that develop or manufacture technologies and services. Increasing M&A activity in the short term is a sign of industry consolidation, as larger companies buy out smaller, less well capitalized competitors. In the longer term, growing M&A activity indicates the increasing mainstreaming of the sector, as larger entrants prefer to buy their way in rather than developing businesses from the ground up (Mitchell et al., 2011).

11.11.3   Linking Policy and Investments

Policies and their design play an important role in improving the economics of renewable energy systems, and as such they can be central to catalyzing private finance and influencing longer-term investment flows. Stern (2009) has proposed that governments have a role to play in reducing the cost of capital and improving access to it by mitigating the key risks associated with renewable energy investments, particularly non-commercial risks that cannot be directly controlled by the private sector. FITs, for instance, have been found to be particularly effective for mobilizing commercial investment (REN21, 2010). They enhance the possibility of project financing, as the guaranteed cash flow increases the willingness of banks to lend money (Deutsche Bank, 2010a; Couture 2010).

Private-sector investment decisions are underpinned by an assessment of risk and return. Financiers want to make a return proportional to the risk they undertake; the greater the risk, the higher the expected return (Justice, 2009). Expectations about the level of risk that will be taken, and the returns required, vary with different financial institutions. A policy framework to induce investment will need to be designed to reduce risks and enable attractive returns and to be stable over a time frame relevant to the investment. To be fully effective, or "investment grade," a policy needs to cover all the factors that are relevant to a particular investment or project (Hamilton, 2009).

11.11.4   Public Financing

The most important role of governments may be to create an enabling environment (a "policy framework") to achieve accepted policy objectives. For many countries, however, creating enabling policies on their own may be insufficient to mobilize the many types of investment needed to move a technology from innovation to deployment to market transformation. Public funding and related interventions may therefore be needed to bring down market barriers, bridge gaps, and share risks with the private sector (UNEP, 2008a).

Public finance mechanisms (PFMs) have a twofold objective: to directly mobilize or leverage commercial investment into renewable energy projects and to indirectly create scaled-up and commercially sustainable markets for these technologies. To make the best use of public funding, it is essential that both these outcomes are sought when designing and implementing such mechanisms. For example, direct short-term benefits should not create market distortions that indirectly hinder the growth of sustainable long-term markets (UNEP, 2008a).

There is a growing body of experience with the use of PFMs for promoting investments in renewable energy deployment. Their role is to help commercial financiers act within a national policy framework, filling gaps and sharing risks where the private sector is initially unwilling or unable to act on its own (UNEP, 2008a).

According to UNEP (2008a), the following are the most common types of PFMs used to mobilize investment in climate change mitigation broadly, all of which are relevant to the renewable energy sector:

- credit lines to local commercial financial institutions for providing debt financing (loans) to projects;

- guarantees to share with local commercial financial institutions and sponsors the risks of providing financing to projects and companies;

- debt financing provided directly to projects;

- loan softening programs, to mobilize domestic sources of capital;

- financing of private equity funds that invest risk capital in companies and projects;

- equity financing provided directly to companies and projects; and

- grants and contingent grants to share elevated project development, transaction, or capital costs.

Using these mechanisms to facilitate access to finance is necessary, but access alone is seldom sufficient for getting clean energy or other low-carbon technologies deployed and scaled up (UNEP, 2008a). Successful mechanisms typically combine access to finance with technical assistance programs designed to help prepare projects for investment and build capacity.

Many finance facilities were created but did not disburse funds because they failed to find and generate sufficient demand for the financing (UNEP, 2008a). Successful public funding mechanisms actively reach back into the project development cycle to find and prepare projects for investment; that is, they work on both the supply and the demand sides of the financing equation. Strategies to generate a flow of well-prepared projects for financing can involve partnerships with many market actors, such as utilities, equipment suppliers and project developers, end-user associations, and government authorities (UNEP, 2008a).

11.11.5   Renewables as a Mainstream Energy Option

Whether renewables can now be considered a mainstream energy option can be best seen by examining the power sector (Usher, 2008). In total, and excluding large-scale hydropower (above 50 MW), about US_{2005}\$170 billion was invested in new renewable power generation plants in 2010, compared with about US_{2005}\$200 billion for fossil-fueled power plants (UNEP and BNEF, 2011). This is about the same as more than half of Spain's total electricity generating capacity. So this is not only about success in Germany, Spain, Denmark, and a few other countries; it is becoming a global phenomenon.

As shown in Figure 11.1, new renewables still represented only about 8% of global power capacity and about 4.5% of global electricity generation in 2010. According to UNEP and BNEF (2011), new renewables accounted for about one-third (about 60 GW) of total global electricity capacity installed in 2010, whereas all renewables accounted for about 47% (84 GW) of net power additions worldwide. REN21 (2011), however, estimates the net power additions worldwide at 194 GW in 2010, with renewables accounting for approximately half (about 96 GW).

More will be needed to meet the challenges ahead. In August 2007, the Secretariat of the United Nations Framework Convention on Climate Change published *Investment and Financial Flows to Address Climate Change*, which estimated that US\$200–210 billion in additional investment will be required annually by 2030 to meet global GHG emissions reduction targets (UNFCCC, 2007). The technical paper concluded that although the lion's share of this investment will need to come from the private sector, substantial public funding will be required to mobilize and leverage the needed private capital (UNEP, 2008a).

11.12 Policy Instruments and Measures

11.12.1 Introduction

Increased use of renewable energy technologies (RETs) can address a broad range of goals, including energy security, economic development, and emission reductions. For this to occur, policy measures have been used to overcome the barriers within the current energy system that prevent wider uptake of renewables. In addition to continued R&D aimed at performance improvements, cost reductions, and system integration, a key issue is how to accelerate the deployment of renewable energy so that deep penetration of renewables into the energy system can be achieved quickly.

This section considers some of the key market failures in developing and deploying renewable energy, the policies in place to overcome them, the need to encourage technological and social innovation, and the appropriate frameworks for investment. It also explores positive links between renewable energy and other policies, such as climate change policies and water policies. (Chapters 22 to 25 provide more in-depth analysis of energy policies in general.)

11.12.2 Need for Policy Intervention and Policy Frameworks

The overall supply and use of energy within the global marketplace does not currently reflect the importance of energy's role in relation to economic, social, and environmental goals. Brown and Chandler (2008) evaluated more than 300 policies and measures grouped into nine categories of "deployment activities": tax policy and other financial incentives; technology demonstrations; codes and standards; coalitions and partnerships; international cooperation; market conditioning, including government procurement; education, labeling, and information dissemination; legislative act of regulation; and risk mitigation. The comprehensive taxonomy underscores how many different policies can affect the innovation value chain related to the reduction of GHGs (and many specific to renewable energy) and the importance of choosing policies that will have the intended impacts (Mitchell et al., 2011).

Given the enormous size and momentum of the current global energy system, in combination with capital intensity and technology lock-in, new technologies such as renewables face significant market barriers. To address these, policy measures should ideally enable not just a level playing field where renewables can compete fairly with other forms of energy; they should also support RE development so that they can overcome the additional hurdles to their deployment that result from inertia to change the incumbent socioeconomic-technical system (see, e.g., Hughes, 1983).

11.12.2.1 Market Failures

While competitive markets operate effectively for many goods and services, a number of failures need to be addressed in relation to energy. A central concern is the way that markets currently favor conventional forms of energy by not fully incorporating the externalities they are responsible for and by continuing to subsidize them, thereby making it harder to incorporate new technologies, new entrants, and new services in the energy system. This both distorts the market and creates barriers for RE (Johansson and Turkenburg, 2004; Mitchell, 2008).

The negative externalities of using fossil fuels include impacts on the environment, health, safety, and security (Johansson and Turkenburg, 2004; Bosello et al., 2006). Similarly, the potential benefits that renewables can offer – such as increased energy security, access to energy, reduced economic impact volatility, climate change mitigation, and new manufacturing and employment opportunities – are also often not accounted for when evaluating the return on investment.

These issues are exacerbated by ongoing subsidies for fossil fuels. Global fossil fuel subsidy levels have been estimated at between US_{2005}170–317 billion – and even as high as US_{2005}516 billion (IEA, 2010f), compared with US_{2005}15.5 billion for renewable energy (UNEP and BNEF, 2008).

11.12.2.2 Failures in the Innovation System

Innovation is central to the development and deployment of new and existing energy technologies. Innovation can be considered as a chain in which technologies move through a number of stages from basic R&D to full commercialization, as well as encouragement for social innovation. (Social innovation concerns the ability of people and institutions to be able to change the way they do things so as to adapt and support the emergence and deployment of RE technologies; Kok et al., 2002). Technologically, this comprises both a supply push from the R&D side of the chain and a demand pull from the market as a technology approaches commercialization.

As technologies advance through these stages, there are both performance improvements and cost reductions (Grubb, 2004). At each stage of the chain, policy instruments may enable (or deter) the appropriate funding to be available – through either public or private means. Figure 11.81 provides an overview of the innovation chain. It shows the need for a seamless linking of policies to provide basic and applied R&D funds early on through to market expansion-type policies later on.

As discussed in Chapter 24, however, innovation is a complex, non-linear process that involves dynamic feedback among the actors, organizations, and networks that include market mechanisms and the flow of knowledge (see also Foxon et al., 2007). As Grubb (2004) describes, the playing field is not level in terms of the introduction of new technologies

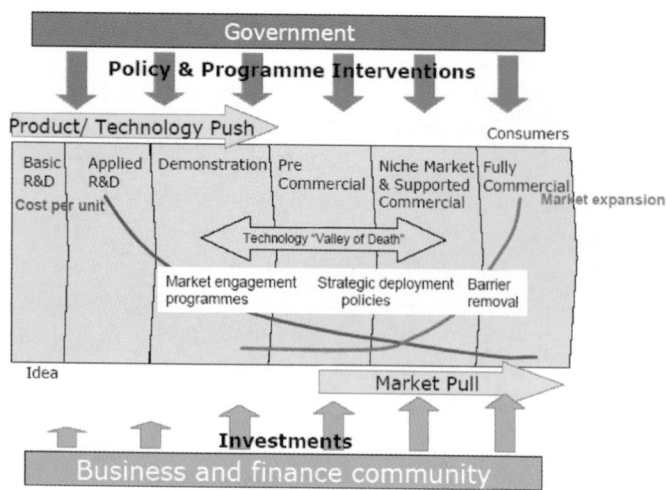

Figure 11.81 | Overview of Innovation System. Source: Grubb, 2004.

due to the nature of large energy systems, which are capital-intensive with long life spans that therefore do not encourage innovation and rapid technology turnover.

Incumbent energy technologies are also mature in terms of their cost and the infrastructure, skills, and knowledge that accompany them (Johansson and Turkenburg, 2004). This results in "lock-in" of technologies and institutions, which favors the energy incumbents and creates a range of barriers for RETs. Some technologies can become stuck along the way. This occurs from a range of risks that relate to the technology itself not performing, market uncertainty, concerns of regulation, and the issues with the system itself, such as lock-in (Turkenburg, 2002). These risks and uncertainties can make it difficult to secure the necessary investment to get a technology through the technology "valley of death."

Policies to address these risks and market failures can also encourage a balanced approach to funding across the innovation chain: one that focuses on research, development, and demonstration as well as deployment phases and that helps reduce the risk that technologies will become stuck (Turkenburg, 2002; Watson, 2008; Foxon et al., 2007).

11.12.2.3   Failures Related to Investment

It is clear that RETs face a number of factors that make it harder for them to compete based solely on costs:

- capital intensity, scale, and resource risk;

- a discounted value to traditional utility operators;

- real or perceived technology risk;

- the absence of full-cost accounting for environmental impacts on a level playing field; and

- a lack of recognition of the long-term value owing to reduced variable fuel cost exposure.

These issues have a negative impact on attractiveness for investors by extending the time frame for returns or by increasing risk and expected rate of return. These risks can be reduced by using public-sector financial or market instruments such as guarantees in terms of market access, market size, and price security (Johansson and Turkenburg, 2004; Mitchell, 2008; Baker et al., 2009).

Support for technological innovation can help create a positive environment for development, demonstration, and commercial deployment investment (Foxon et al., 2007). In contrast, there are many examples of the impact of inconsistent and unsustained policies on renewable energy in terms of creating uncertainty, damaging commercial interest, preventing manufacturing, and creating boom-and-bust markets (Sawin, 2004b). The result is market instability, an increasing perception of risk, and a loss of investor confidence, which can affect development of individual technologies.

11.12.3   Policy Approaches for Renewable Energy

A wide range of regulatory, fiscal, and voluntary policies have been introduced globally to promote renewable energy. These serve a range of technology-specific objectives, including innovation, early-stage development and commercialization, manufacturing, and market deployment, as well as wider political goals such as new manufacturing bases for a technology, local and global environmental stewardship, and economic prosperity. These policies all help to reduce risk and encourage RET development and are generally used in combination. (See Table 11.44).

Using a portfolio of policies helps to control total costs and increase successful innovation and commercialization, providing the policies complement each other (Foxon et al., 2007). The means of using financial instruments are described in section 11.11. To get renewable technologies through the "valley of death," it is important to note that (IEA, 2008b):

- market growth results from the use of combinations of policies;

- long-term, predictable policies are important;

- multi-level involvement and support from national to local actors is important; and

- each policy mechanism evolves as experience with it increases.

Policy approaches for RE intend to address the innovation chain both technologically and socially, to pull technologies to the marketplace and

Table 11.44 | Summary of renewable energy policies.

Policy	Definition	End-use Sector		
		Electricity	Heat/Cooling	Transport
Regulatory Policies				
Targets	A voluntary or mandated amount of RE, usually a percentage of total energy supply	X	X	X
Access-related Policies				
Net metering	Allows a two-way flow of electricity between generator and distribution company and also payment for the electricity supplied to the grid	X		
Priority access to network	Allows RE supplies unhindered access to network for remuneration	X	X	
Priority dispatch	Ensures RE is integrated into energy system before supplies from other sources	X	X	
Quota-driven Policies				
Obligation, mandates, Renewable Portfolio Standards	Set a minimum percentage of energy to be provided by RE sources	X	X	X
Tendering/bidding	Public authorities organize tenders for a given quota of RE supplies and ensure payment	X		
Tradable certificates	A tool for trading and meeting RE obligations	X	X	
Price-driven Policies				
Feed-in tariff (FIT)	Guarantees RE supplies with priority access, dispatch, and a fixed price per unit payment (sometimes declining) delivered for a fixed number of years	X	X	X
Premium payment	Guarantees RE supplies an additional payment on top of their energy market price or end-use value	X	X	
Quality-driven Policies				
Green energy purchasing		X	X	
Green labeling	Usually government-sponsored labeling that guarantees that energy products meet certain criteria to facilitate voluntary green energy purchasing	X	X	X
Fiscal Policies				
Accelerated depreciation	Allows for reduction in tax burden	X	X	X
Investment grants, subsidies, and rebates	One-time direct payments usually from government but also from other actors, such as utilities	X	X	X
Renewable energy conversion payments	Direct payment by government per unit of energy extracted from RE sources	X	X	
Investment tax credit	Provides investor/owner with an annual tax credit related to investment amount	X	X	X
Other Public Policies				
Research and development	Funds for early innovation	X	X	X
Public procurement	Public entities preferentially purchasing RE or RE equipment	X	X	X
Information Dissemination and Capacity Building	Communications campaigns, training, and certification	X	X	X

commercialize them, and to improve the financial attractiveness and investment opportunities of RE.

Of the market pull policies, two are most common: policies that set a price to be paid for RE and that ensure connection to the grid and offtake (i.e., FITs) and policies that set an obligation to buy but not necessarily an obligation on price (often known as a quota or obligation mechanism but also referred to as a Renewable Portfolio Standard or Renewable Fuel Standard (RFS) and sometimes as a Renewable Electricity Standard). So far, FITs have been for electricity only, although some countries, such as the United Kingdom, are now considering how to provide them for heat (DECC, 2009). Quotas have so far been used for electricity, heat, and transport. Biofuel mandates are now a common occurrence globally.

Table 11.45 indicates the number of countries that have introduced policies to promote renewable, which is increasing nearly across the board.

Table 11.45 | Recent development of the implementation of Renewable Energy policy instruments.

Policy instrument	End 2005	End 2007	End 2009	End 2010
Countries with policy targets	55	66	80	96
States / provinces / countries with FITs	41	56	80	87
States / provinces / countries with RPS policies	38	44	51	63
States / provinces / countries with biofuels mandates	38	53	64	60

Source: REN21, 2008; 2010; 2011.

11.12.3.1 Innovation-driven

R&D is the policy instrument intended to meet the early requirements of the innovation chain. However, support needs to be balanced across this innovation chain (IEA, 2008b) because learning through doing or utilization can focus R&D beneficially (Neij, 2008; Junginger et al., 2010).

Research and development is generally targeted to both increase performance and reduce costs of newer technologies. New forms of collaboration between researchers, such as joint road mapping (NEDO, 2009) or the Scholarships for Excellence program (BIS, 2010), may also lead to increased and targeted innovation. Investments in energy R&D are low within total national R&D budgets (UNEP, 2008b), and only a small part of those budgets is directed toward renewables (IEA, 2010g).

Changing energy behaviors is not simple, but it is critical to reduce overall energy use and enable the widespread adoption and application of renewable energy. There is often a gap between support for RE in opinion polls and buying, investing, or enabling RE projects to gain planning permission (Devine-Wright, 2005). Increasingly, policies are being implemented that are intended to support RE but also to change people's attitudes and behaviors. For example, the United Kingdom has a Low Carbon Challenge program that has supported 25 communities implementing RE but that also has used the process to educate communities and individuals about reducing GHG emissions (DECC, 2010).

11.12.3.2 Price-driven

The most widely used price-based policy is the FIT. Other RE price policies are premium payments, which are often set as a percentage or an amount over an electricity price (for example, Spain offers the choice between an amount over the retail price (Lucas Porta, 2009). While beneficial, premium payment structures are more risky because the price of electricity moves up and down, so investors may not be sure about the payment they will receive.

FITs guarantee a price for the sale of qualified (such as renewably generated) electricity and allow the supply of investment to determine the resulting volume or capacity. FIT rules differ by country, but a "standard" electricity FIT includes an obligation to connect the renewable project to the grid, its priority dispatch (see section 11.10), and the purchase of any

electricity generated at a fixed minimum price (which is generally above the market level for a set time period). In addition, the payment usually declines over time according to a rate known at the time of initial contracting, to take account of technological learning and maturity (Couture et al., 2010). This approach minimizes the risk of investment and, if designed appropriately, should provide attractive risk-adjusted returns on investment.

FITs were first introduced in the United States in the late 1970s and since then have become the most widely adopted class of promotional policy globally. Much of the growth in their use has been within the last decade as some countries enact new policies and others adjust and refine existing tariffs. Their popularity is linked to their ability (if well designed and implemented) to offer sufficient rates of return over the economic life of the technology at low risk (Klessman et al., 2010). To date, FITs have had a large impact on the development and deployment of wind, PV, biomass, and small hydro (REN21, 2010). The transaction costs involved with FITs are lower than those of quotas (Mitchell et al., 2006). This means that the number and diversity of actors taking up the opportunity of an FIT are far greater than with quota mechanisms, which in turn has a number of other positive benefits, such as new market entrants and reduced excess profits (Jacobsson et al., 2009). However, the payment and cost structure for FITs, as exemplified by recent restructuring in Germany and Spain, have raised concerns over FIT design and best practices (Kreycik et al., 2011, Coture et al., 2010).

11.12.3.3 Quantity-driven

Renewable Portfolio Standards set an obligation for renewable capacity or volume such as a percent of electricity or share of fuels, and the market determines the price at which this is achieved. This obligation often increases over time toward a set total or date, and the requirement to meet it can be placed on producers, distributors, or consumers (Sawin, 2004a,b). Most RPS policies require a renewable power share of 5–20% by a set date, generally 10–20 years ahead. Such targets translate into expected future investments, but the means of achieving it are often flexible (REN21, 2008).

For electricity generation, quota systems can be based on obligation or certificate mechanisms in which investors, generators, or utilities have to achieve specific targets and are penalized if they do not. Credits are created through the generation of renewable electricity such as green certificates, green labels, or renewable energy credits. To meet the obligation, it is possible to generate renewable electricity, enter into a contract with others to do so, or buy a valid credit in the marketplace. Quotas or obligations can be more complicated to administer than FITs (Mitchell et al., 2006). Moreover, like the UK mechanism, they can do little to minimize the risk of investment if their obligation is only for purchasing rather than including a minimum price or a minimum contract length. But they have the benefit of letting government know the maximum cost per year (Mitchell et al., 2006).

An alternative approach within quotas can involve tendering systems in which both the total share of electricity and a maximum price per kilowatt-hour are set. Developers then submit prices for contracts in a one-off bidding round to compete for funds and contracts (Sawin, 2004).

11.12.3.4   Quality-driven

Quality-driven RE policies aim to ensure energy that says it is renewable actually is RE (and produced sustainably), in the hope that this will encourage greater voluntary RE purchasing. For example, green labeling ensures that the consumer buys RE that is generated and verified as original source. The European Union's requirement for a guarantee of origin on all RE generated within Europe enables trading but also ensures that each unit of energy can only be used once. The US voluntary green power market has similar quality assurance standards and has been growing strongly for many years (Bird et al., 2009).

Voluntary green energy purchasing can include utility pricing programs, retail sales in liberalized markets, and voluntary trading of renewable energy certificates (used in some countries within quota-based systems). Within these schemes, consumers pay a premium price for the green power, which helps to encourage investment in RETs, although the level of premiums may be declining. It is estimated that in 2009 there were over 6 million green power customers in Australia, Canada, Europe, Japan, and the United States (REN21, 2010).

11.12.3.5   Financial Incentives

A range of financial incentives are also in use globally to support RETs, such as policies that use taxes, rebates, grants, payments, and subsidies. (See also section 11.11.) They are used to help "level the playing field" either by lowering the cost of RETs relative to conventional energies or by increasing the value of the RE sold to make it easier for the technologies to compete in the market (Sawin, 2004). They can be linked to a price per kilowatt-hour or to an installed capacity.

Tax incentives are commonly used to provide financial support, through either a production tax credit or an investment tax credit, both of which encourage the investment and build-out of a RET. Production tax credits in the United States provide a financial benefit based on the generated amount of energy from renewable energy sources, which helps reduce the payback period and therefore increase the rate of return, making investment more attractive. As the reward is based in the amount of energy generated, these credits can encourage technologies and installations that are reliable and continuously improving. Other forms of tax relief can reduce the cost of investing in renewables, such as accelerated depreciation, value added tax exemptions, or the removal of import duties (Sawin, 2004).

In 2010, at least 52 countries had some sort of direct capital investment policy in place (REN21, 2011), from national to local levels, offering typically 30–50% of installed cost (mostly for smaller-scale distributed systems such as solar PV). These have been tied to certain technologies in many countries, with PV in particular receiving widespread support that has led to its rapid market growth; in other countries financial support has been offered for a wider range of renewables (REN21, 2010). There is debate about whether a payment should be made upfront on the "capital" outlay or on the output (the energy). Early US experiences shaped global RE policy perceptions that payment on outputs was more efficient (Karnoe, 1993). Yet both instruments, or options among them, may be used. In some cases grants may also be used, either when tax-related policies are deemed ineffective (Bolinger et al., 2009) or when the goal is to encourage small-scale investors (DECC, 2009).

Another means of providing increased revenue (and thus a better rate of return) to investors is net metering, which allows grid-connected generators to sell excess power into the grid, with an obligation for utilities to purchase the excess. (See, for example, US DOE, 2011c). The reimbursement or rebate may be based on retail or wholesale prices and may include a time-of-day value as well. Net metering laws now exist in at least 14 countries and nearly all US states (REN21, 2011).

11.12.4   Other Enabling RE Policies

11.12.4.1   Targets

Targets provide a marker by which all actors involved can assess the situation and act in their own best interests. Establishment of a goal or target may increase investor and developer confidence and may also provide a basis for possible future policy or regulation direction, but it does not set a legal obligation.

Targets have been steadily developing over the last three decades, but the rate of growth and their geographic spread have increased significantly only recently. At the end of 2010, an estimated 96 countries worldwide had policy targets for renewable energy (see Table 11.45), with further targets existing in individual states, provinces, municipalities, and individual cities.

The approach taken for renewable energy targets varies by country and can include a percent share of electricity production, a share of total primary/final energy supply, an installed capacity target, or total amounts of energy extracted from renewable sources. In addition, the time frame for targets varies from 2010–2012 to 2020–2025 and beyond. Some countries have technology-specific goals, and many also have targets for biofuels or blending requirements.

In the European Union, policies adopted in late 2008 as part of the energy and climate package included agreement on a 20% final energy from renewables target and a 10% target for transport fuels, with sustainability criteria. The legally binding targets are spread across all 27 member states and build on earlier EU-wide targets for 2010 (European Parliament, 2009). China has adopted a national target of 15% of primary energy from renewable by 2020, while India has a range of long-term goals to 2032 that include 15% of power capacity and 10% of transport fuels from renewable energy (REN21, 2010).

11.12.4.2   Economic Regulation

As already highlighted, fossil fuels and nuclear power continue to receive significant subsidies globally, and there can be hidden barriers to RE in economic regulatory approaches. For example, obligations to purchase RE, public investment in distribution networks, and access rules to transmission networks may all favor incumbents and create a market barrier to new entrants (Baker et al., 2009). Exemptions from risks or liabilities and the structure of energy markets may also unfairly penalize intermittent forms of generation (Sawin, 2004; Mitchell, 2008; Morthorst et al., 2007). Removal of these barriers addresses the cost competitiveness of renewables against conventional energies and it may enable the energy system to be more efficient and secure (Gottstein, 2010). For example, the United States has recently developed transmission policies to take account of the difficulties in predicting output for some renewables, giving them a fairer chance in a system designed for conventional generation (REN21, 2008). Allowing electricity markets to include wind forecasts that are nearer market closure is also helpful (Jónsson et al., 2010).

11.12.4.3   Planning and Permitting

Almost all surface areas of the globe are now influenced by ownership, conservation, traditional use, or commercial interests. As a result, the growing deployment of RE technologies may create tensions. Policymakers therefore attempt to balance a planning regime that broadly supports RE use and deployment while at the same time establishing processes that ensure public insight and environmental protection.

Obtaining planning permission for an RE (or any other) project is a social process when different actors and stakeholders are able to become involved and share their views. There is clearly a positive side to this (Ellis et al., 2009). On the other hand, lengthy permitting processes, high application costs, lack of local or regional capacity to

deal with RE applications, and (sometimes) local opposition can make the permitting process expensive and time-consuming (Breuker and Wolsink, 2007).

Social acceptance and a commitment to effective planning and permitting help reduce risk and the cost of deployment of RE solutions. Codifying the framework into a set of legal, formal rules and procedures to address the differences and mediate conflicting interests and values is a long-term transition process; the planning and policy process is only one part of that (Agterbosch et al., 2009). Streamlining application processes, adopting benefit-sharing schemes, simplifying legal documents, and so forth are all helpful to permitting for RE (Ellis et al., 2009). However, speeding up the planning process may mean that local participation is inherently reduced.

11.12.5   Heating and Cooling and Transport

11.12.5.1   RE Heating and Cooling

Relative to electricity or liquids for RE fuel, renewable heat has had very little policy support, and RE for cooling has been the subject of even fewer mechanisms. The use of renewable heat is widespread. (See sections 11.2, 11.4, and 11.8.) Examples include solar water heating in China, geothermal heating in Iceland, and biomass CHP in Sweden, although these have resulted from a confluence of factors rather than a specific RE support policy. Policy mechanisms in place are similar to those for electricity, although they have different names: bonus mechanisms (similar to the FIT for electricity) in the United Kingdom; a "use" obligation, where building regulations can compel the adoption of renewable heat technologies (in Germany); standards and building regulations to ensure a minimum quality of hardware and installation alongside the "use" obligations; and fiscal instruments such as tax credits, capital grants, and soft loans (Seyboth et al., 2008).

11.12.5.2   RE Transport

Currently, 95% of the world's transport relies on petroleum, and there is evidence of this increasing annually (IPCC, 2007a; IEA, 2010b). Considerable attention is being given to increasing the share of transport services provided by renewables directly through liquid and gaseous fuels and indirectly via electricity.

Again, policies for renewable transport are similar to those for RE electricity and heat. Policies in Europe and North America that promote the use of RE in transport applications include renewable/low carbon fuel standards, tax incentives, R&D, RFSs, GHG emission standards, preferential government purchasing, and regulation standards and licenses for production and sale of renewable energy carriers (Altenburg et al., 2008; Felix-Saul, 2008).

Care is needed to ensure that the increased use of biofuels (and biomass more generally) does not have negative sustainability impacts and that bio-resource policies account for the wider context of energy policy, as biofuel production may reduce options for other forms of biomass, for food production, and for other uses or destinations. (See also Chapter 20.)

11.12.6   Links with Other Policies

Supporting renewable energy may have other benefits, such as climate change mitigation, poverty eradication, rural development, protection of water resources, energy security, infrastructure development, employment, and demand reduction through energy efficiency.

11.12.6.1   GHG Reduction and Local Environmental Improvements

One of the most obvious and strategic policy links is with climate change, as the supply and use of energy is the main driver of GHG emissions globally (IPCC, 2007b; IPCC, 2011). As a result, there are significant RE policies around the globe to reduce GHG emissions (REN21, 2010). In addition, there are other environmental benefits in relation to air quality and pollution. This includes the reduction of particulates, low-level ozone, sulfur, and nitrogen oxides everywhere (IPCC, 2007a). One exception to this general assumption, however, is the traditional use of biomass, particularly in developing countries, where poor combustion for heat and cooking lowers indoor air quality and has serious health impacts. In these circumstances, access to better technology and cleaner fuels, including fossil fuels, can dramatically improve air quality and health. (See Chapter 4.)

11.12.6.2   Poverty Eradication

The opportunities for renewable energy to help with poverty eradication and the broader Millennium Development Goals are significant, as energy services have been shown to be essential to achieving these. This requires policies to increase both the quality and the quantity of affordable energy services in the world's poorest countries (Modi et al., 2006). The UN has emphasized the close links between energy use and the eight MDGs (UN, 2009). The IEA estimates that 22% of the global population was without access to electricity in 2008 – around 1.5 billion people, 85% of whom live in rural areas. The organization estimated that this figure had fallen by 500 million since 2002. The extent to which this related to the deployment of renewables is less clear. But several programs have focused on this; for example, China has introduced an RE program, including the Riding the Wind Plan and the Brightness Project on solar energy to provide electricity to people in remote areas not been served by small hydropower efforts (World Bank, 2010; see also Chapter 19).

11.12.6.3   Invigorating Agriculture and Rural Areas

Efforts are increasingly being made to replace fossil fuels with biofuels and with biomass from agriculture in order to reduce GHG emissions (IPCC, 2007a) and fossil fuel imports. This can include the use of crops, crop residues, wood, and animal wastes, and additional resources are available from forestry and waste streams. Linking agricultural and renewable policies more closely could help maximize this resource potential while supporting rural development, creating jobs, increasing local energy availability, and yielding other social and environmental benefits (WIREC, 2008). But a range of issues need to be addressed to ensure that these benefits do not have negative impacts on food production, biodiversity, sustainable management of soil and water, and other sustainability criteria. (See Chapter 20.) Increasing the scale of bioenergy deployment globally will also need to take careful account of the socioeconomic-political conditions within a country as well as local resources and assets (WIREC, 2008).

11.12.6.4   Provision and Protection of Water Resources

There are a number of direct links between energy and water policy, including the possible massive reduction of water needs for thermal power plant cooling and the use of RETs to replace expensive diesel for water pumping, desalination, and purification (Bates et al., 2008). These applications, often at small scale, can directly support the MDGs, as they can provide local reliable and clean water while reducing the time it takes to gather water in many developing countries (UN, 2009). Water is also a key global resource for generating power, particularly the large-scale hydro that provides a large amount of electricity in developing and industrial countries. At the same time, biomass-based renewables and concentrated solar power in desert areas may introduce new demands for water. (See Chapter 20.) The substitution of RE for other energy sources may also have a beneficial impact on water usage and quality, such as mining for coal.

11.12.6.5   Energy Security

Energy security policies, while not uniformly framed across countries, have begun to link the development and support of RETs to the reduction in risk of supply and economic volatility of fossil fuels (see, e.g., University of Exeter and University of Sussex, undated). RE can improve the security of energy supply in a variety of ways, including reducing dependence on imported fuels, helping to diversify supply, enhancing the national balance of trade, and reducing vulnerability to price fluctuations (Mitchell et al., 2011).

These various benefits are driving a number of governments to adopt policies to promote RE. In the United States, for example, development and extension of the national Renewable Fuel Standard were framed within the context of reducing imported oil (Arent et al., 2009). In Japan, with few domestic energy resources, solar energy allows the domestic generation of electricity, with associated policies to support R&D and market deployment. For the last 30 years Brazil has promoted ethanol from sugarcane as an alternative to fossil transport fuels in order to decrease dependency on imported fuels (Solomon et al., 2007). China established its 2005 Renewable Energy Law in part to diversify energy supplies and safeguard energy security. A number of municipalities and local communities are also adopting RE plans to become more energy self-sufficient (St. Denis and Parker, 2009; Mitchell et al., 2011).

11.12.6.6   Infrastructure Development

As discussed in section 11.10, expanded use of renewables will require infrastructure developments that allow distribution of energy for heat, power, and transport applications. Heat tends to be used close to the point of production, often through local heat distribution networks. Changes to transport infrastructure depend on the type of renewable energy that may be exploited, but a significant switch to hydrogen or battery vehicles, for example, may require considerable new up-front planning and infrastructure (Lund and Clark, 2008). An integrated, holistic planning approach may also benefit changes in electricity infrastructure to accommodate more renewables (Baker et al., 2009; Arent, 2010).

The existing infrastructure in many countries was originally designed around large fossil fuel and nuclear generating plants. As such, the operational mindset may be focused on the original intent, so quick change can be difficult. To date, this has not been a major problem in many countries, as grid infrastructure has been designed to meet peak load requirements, meaning most countries have been able to accommodate increasing amounts of renewables while maintaining supply security and reliability. This is likely to become an increasing challenge, however, as the contribution from renewables increases, since some (like wind and solar) are variable resources.

Needed changes can be achieved through planned upgrades and pre-investment in preparation for future renewables or through making the grid infrastructure more active. Linked to the latter approach is the increasing interest in more-intelligent smart or micro grids, demand response, storage, and load shifting. These are based on clusters of connected, distributed generation that are collectively controlled to manage output (Battaglini et al., 2009). Both the European Union and the United States are developing such grids to improve reliability and lower costs (Coll-Mayor et al., 2007). Another approach being considered is "supergrids," which are based on large-scale transmission of renewable electricity over very long distances (Battaglini et al., 2009; FOSG, 2010; DESERTEC Foundation, 2011). This can include connections between existing national grids to balance power, as well as the much more strategic construction of new high-voltage distribution lines to bring in larger areas of supply potential. There are also opportunities to combine approaches to create a "supersmart" grid (Battaglini et al., 2009).

11.12.6.7   Improving Economic Development and Employment

The global financial crisis in 2008 spawned an unprecedented policy response, totaling US_{2005}2.65 trillion in stimulus spending by March 2009 (UNEP, 2009). The UNEP 2010 Sustainable Energy Finance Initiative report stated that 9% of the US_{2005}165 billion in global green stimulus packages had been spent by the end of 2009, with greater shares expected in 2010 and 2011(UNEP and BNEF, 2010). Broadly, these "global green stimulus packages" refer to direct and indirect expenditures on a broad suite of energy solutions, including RETs, energy efficiency, advanced materials, and "smart grids."

There are considerable employment opportunities within the renewable energy sector. UNEP (2009) estimated that approximately 2.3 million people found employment in renewables in recent years; REN21 reports that around 3.5 million direct jobs had been created by 2010 (REN21, 2011). UNEP also stated that projected investments to 2030 – not taking into account the full impacts of the financial and economic crisis – could result in at least 20 million additional jobs globally. These figures mask the complexity involved, as any major international switch to renewables is likely to be accompanied by job losses in the fossil-fuel-based energy sectors.

Frankhauser et al. (2008) provide a useful analysis of employment over time. They highlight the work of Kammen et al. (2006) that compared 13 studies within the European Union and the United States. The more labor-intensive nature of new renewables resulted in net increases in employment over the short term. In the medium term there are economy-wide effects that reinforce overall employment gains. Research in Germany from BMU (2006) showed that just over 50% of 157,000 renewable energy-related jobs were directly related to manufacture and operation, with the rest in related industries. The biggest effects will be felt in the long term, as technical change and innovation create a dynamic impact that results in job creation, productivity improvement, and growth. Although jobs are lost in conventional energy industries, the authors suggest that overall employment may increase on a net level. A more recent EU report (EU, 2009) generally supports this. Additional research is required, however, to better capture the employment and macroeconomic impacts of renewable energy at the local, regional, and global level.

Relative to conventional energy, the renewable market is still small, with a few countries accounting for the bulk of installations. For example, the top five countries for wind power have 72% of global installed capacity. Many of these top countries also have the manufacturing capabilities for RETs and their associated jobs. Some have made specific policy decisions to link renewables to their national economic strategies. Germany's support for renewable energy, for instance, resulted in a competitive export industry (Frankhauser et al., 2008), enabling it to obtain a global share of the market, particularly in wind and PV (UNEP, 2008b). This is true for other countries too: China has become a major global player in many RETs in a short time and is capturing a growing part of the global market for, among other technologies, solar hot water. However, this pattern also suggests that although competitive advantages can be gained by being an early mover, over time such advantages may balance out as other countries – currently China and India – develop their own manufacturing and export capabilities for RETs (UNEP, 2008b).

11.12.7 Development Cooperation

Technology transfer between countries is a prominent area of discussion in international meetings and summits (Brewer, 2008), but it has been a sticking point at many negotiations between industrial and developed countries (Ockwell et al., 2008). As this topic is discussed in Chapters 22–25, this section is limited to a discussion of cooperation on renewable energy.

11.12.7.1 National Systems of Innovation

National Systems of Innovation play an important role in technology development and its potential for wider distribution within the market. (See Chapter 24.) Foxen et al. (2005) describe the concept of national systems in innovation used by the OECD, which characterizes the innovation system in terms of complex flows of knowledge, influence, and market transactions between a wide range of actors and institutions. These processes vary between countries, but they set the framework for innovation at a national level and help shape the process of technology development.

In the case of renewables, innovation is influenced by national policy intervention, targets, and wider policies relating to R&D and infrastructure development. This should result in a reduction in the cost of RETs and should increase their commercial uptake (Ockwell et al., 2008), assuming that the policies are well designed and implemented. National systems of innovation are therefore important in helping to develop new, commercial renewables for deployment within both industrial and developing countries.

11.12.7.2 Capacity Building

A central part of cooperation between nations involves building capacity within developing countries. (See Chapter 25.) As Ockwell et al. (2008) highlight, the transfer of technology in itself may not have a sustained impact on the uptake and development of low-carbon technologies unless it is accompanied by a transfer of knowledge and expertise, such as information on installing, operating, and repairing the equipment. These broader educational aspects help increase the capacity of companies and therefore the likelihood of effective deployment.

Article 4.5 of the UN Framework Convention on Climate Change (UNFCCC) obliges Annex I countries to ensure the availability of affordable clean technologies to non-Annex I countries. Parties to the convention have agreed on a technology transfer framework (UNFCCC, 2007) that includes five areas for action: technology needs and needs assessments, technology information, enabling environments, capacity building, and mechanisms for technology transfer.

As discussed in Mitchell et al. (2011), perhaps the most important insight in the evolution of technology and innovation in the past 30 years is the recognition that technology transfer is not an end in itself but a means to achieving a greater strategy of technological capacity building (Mytelka, 2007). Technology transfer mostly takes place between firms via the market – through the use of products or services that incorporate a specific technology or through licensing the capability to produce such products, either by an indigenous firm or through a joint venture arrangement or foreign direct investment (Kim, 1991, 1997; UNCTAD, 2010). Its sustainability relies crucially on the successful learning of one party from another and the effective application of that information and knowledge in generating marketable products and services.

11.12.7.3 Impact of the Clean Development Mechanism

The CDM enables Annex I countries to support the development of projects to reduce GHG emissions within developing countries. As of August 2011, a total of 3392 projects were registered, and more than 50% of these were for renewable projects (cdm.unfccc.int). In assessing the contribution of CDM to technology transfer, Schneider et al. (2008) suggest that although the program was not designed for this, it does contribute to the process, as developing countries can gain access to technologies that may not have been available previously. The literature, they suggest, shows that the CDM contributes to the transfer of equipment and capacity building by lowering several existing barriers and by increasing the quality of transfers. They suggest that it is currently the strongest mechanism that the UNFCCC has for technology transfer of RE, although its effectiveness varies considerably with geography, technology, and project size.

11.12.7.4 Role of International Institutions, Arrangements, and Partnerships

Given the nature of development cooperation today and the potential benefits that renewables can offer, a number of initiatives are in place to encourage the transfer of technologies. This includes the work of specific bodies that support countries on climate change, development, and sustainability, including bodies like the UN Development Programme and the UN Industrial Development Organization.

There is an ongoing interest in technology transfer under the mechanisms agreed as part of the UNFCCC:

- Article 4 includes promotion and cooperation on development, application, and diffusion, including the transfer of technologies that help to mitigate emissions. It also calls for practical steps that promote, facilitate and finance the transfer of environmentally sound technologies.

- The Marrakesh Accords include agreement on the framework of the five themes for technology transfer activities. A summary on the actions being taken under these themes, including the UN bodies involved, shows how many of these can link to energy and renewables (UN, 2008).

- The seventh Conference of the Parties also established the Expert Group on Technology Transfer to analyze and identify ways to facilitate and advance technology transfer.

- The Bali Road Map agreed to at the thirteenth Conference of Parties to the UN-FCCC called for enhanced action on technology development and transfer to support action on climate mitigation and adaptation.

Additional initiatives include the Global Environment Facility (GEF) that supports developing countries. GEF is managed by the World Bank and aims to make the economics of low-impact technologies, such as renewables, more favorable. The 2006–2010 fund amounts to US$_{2005}$$2.67 billion, of which nearly 12% is earmarked for renewable energy (REN21, 2007; 2008; 2010). However, irregular voluntary funding to GEF has created major obstacles to its effective functioning (REN21, 2007). Schneider et al. (2008) suggest that the investment that takes place through the CDM is more significant for technology transfer than the GEF.

The IEA Technology (or Implementing) Agreements encourage cooperation between member and non-member governments and organizations to meet the challenges of energy security, competitiveness, and climate change through technological solutions. They provide a legal contract with standard rules and regulations for a range of technologies that allow the pooling of resources, research, deployment, and development. Currently the renewable-related agreements include bioenergy, geothermal, hydrogen, hydropower, ocean energy systems, PV, solar heating and cooling, concentrating solar power, wind, and RET deployment. Examples of agreements that can support technology transfer include the Networks of Expertise in Energy Technology that works to foster better international cooperation, particularly with non-IEA countries, and the Climate Technology Initiative, which aims to foster international cooperation to accelerate development and diffusion of environmentally sound technologies and practices (IEA, 2009d).

Other examples of international partnerships that include agreements and principles on energy, renewables, and technology transfer, include the following:

- Asia-Pacific Partnership on Clean Development and Climate Change, which includes cooperation on technology transfer and development.

- Johannesburg Renewable Energy Coalition, which focuses on political initiatives to help promote renewable energy at national, regional, and international levels.

- Mediterranean Renewable Energy Program, which includes the objectives to provide modern energy services to rural populations and contribute to climate change mitigation by increasing the share of renewable energy technologies in the region.

- New Partnership for Africa's Development, which includes objectives to tackle poverty and place African countries on a path of sustainable growth and development.

- Small Island Developing States, which aims to support the sustainable development of these counties, including initiatives on renewable energy and climate change.

11.12.7.5 Role of Dedicated Renewable Energy Partnerships

A wide range of renewable energy partnerships also have a role within development cooperation and the wider support of renewable energy. As Suding and Lempp (2007) describe, these include federations, business associations and societies for renewables and specific technologies that include conventional organizations and structures, and numerous much more diverse partnerships and networks. They help to bring like-minded partners together to pool skills and resources to work toward common goals. Examples of these in terms of development cooperation include:

- Global Network on Energy for Sustainable Development

- Global Bioenergy Partnership

- Global Village Energy Partnership

- International Science Panel on Renewable Energies

- International Solar Energy Society

- International Renewable Energy Agency (IRENA)

- Renewable Energy and Energy Efficiency Partnership

- Renewable Energy Policy Network for the 21st Century

- World Council for Renewable Energy.

Of these, IRENA has the most ambitious goals. It was founded in 2009, and as of August 2011, a total of 148 countries and the European Union had signed the agency's statute. IRENA is to provide advice and support to governments on RE policy, capacity building, and technology transfer.

Table 11.46 | Elements in the UN-FCCC/Kyoto Protocol relevant to Renewable Energy.

Element	Current impact on RE	Potential future impact on RE
Long-term goal (UNFCCC, esp. Article 2)	Limited, as not broken down into technology-specific goals	If defined as GHG emissions or concentration, only an indirect signal. If long-term goals were established for global technology shares, then their effect on RE could be significant.
Emission reduction targets (Kyoto Protocol Annex B)	Fair impact, but countries aim to reach the short-term targets with today's low-cost mitigation options, often not RE	Future emission targets will provide an indirect long-term signal. Depends on the stringency of required reductions and the number of countries participating.
Joint Implementation (Kyoto Protocol Article 6)	Limited, as JI volume is small	Limited, as large JI volumes are unlikely.
Clean Development Mechanism (Kyoto Protocol Article 12)	Fair impact, but other reduction options are often more cost-effective	Growing fast, but the number of host countries will decrease. The level is driven by stringency of emission reduction targets of industrial countries. Additionality criterion may be an obstacle to comprehensive national frameworks for RE.
Technology transfer and financial mechanisms (Kyoto Protocol, esp. Article 11; UNFCCC, esp. Article 4	Limited, as the funds are small	Larger only if there is an automatic flow of resources into the funds.

Source: REN21, 2007.

It is to improve the flow of financing and to collaborate with existing RE institutions. Its goal is to increase the share of RE worldwide.

11.12.8   International Policy Initiatives to Stimulate Renewable Energy

11.12.8.1   UN Programs and Initiatives

International policy processes from the UN include programs, summits, and initiatives that link directly and indirectly to renewable energy. By providing an arena for countries and other stakeholders, the UN processes are important for working toward common goals or agreements that translate into national renewable policies (Suding and Lempp, 2007). UN World Summits, such as the 1992 UN Conference on Environment and Development in Rio, initiated many of the processes that continue today. The transition to different energy sources and the promotion of renewable energy were included within Agenda 21 that was approved at that conference. Governments in Rio also adopted the UN-FCCC, approved development of the GEF, and created the Commission on Sustainable Development.

The IPCC was created in 1988 to assess the science, impacts, and possible responses to climate change. To date the IPCC has produced four comprehensive assessments and a number of special and technical reports on climate change; the fifth assessment report is being prepared, and a Special Report on Renewable Energy Sources and Climate Change Mitigation was published in 2011.

The UN-FCCC was the political response to concerns raised by the IPCC. Several of the elements of the UN-FCCC are relevant to renewable energy, including the binding targets agreed under the Kyoto Protocol. (See Table 11.46.)

11.12.8.2   The Group of 8 (G8)

Renewable energy has been featured in several G8 summits:

- The 2005 Gleneagles Summit included a focus on the urgent need for action on climate change. The Gleneagles Plan of Action included a statement to continue development and commercialization of renewable energy, building on the commitments made at a renewables conference in Bonn. The meeting also pledged to work with the IEA on integrating renewables into electricity grids.

- The Russian summit in 2006 agreed to the St. Petersburg Plan of Action on Global Energy Security, which included recognition of the role of renewable energy in creating a secure energy mix.

- The Heiligendamm Summit in Germany in 2007 included a declaration on energy cooperation, in which renewable energy was specified.

- The 2008 G8 Summit in Japan included a reaffirmation of the aim to tackle climate change through the UN-FCCC process. There was a recognition of the important role that renewables have in reducing emissions and improving energy security and commitments to increase investment in R&D.

11.12.8.3   International Renewable Energy Action Plans and Declarations

In addition to the UN world summits, a set of Action Plans and Declarations have been agreed to in order to create additional momentum to advance renewable energy policies and technologies. Following an initial meeting in Bonn, these have been organized and monitored by REN21:

- Bonn 2004 – Numerous Action Plans were developed to maintain momentum developed at the 2002 World Summit on Sustainable Development in Johannesburg; several key outcomes were adopted:

  - an International Action Program that included around 200 actions and commitments from a wide range of stakeholders to develop renewable energy (follow-up on these in 2006 by REN21 suggested 79% were being implemented, resulting in significant annual carbon savings); and

  - a political declaration to create and work within a global policy network, which led to the creation of REN21.

- Beijing 2005 – The Beijing Declaration signed by 78 countries reaffirmed the commitments made under previous UN summits to "substantially increase – with a sense of urgency – the global share of renewable energy in the total energy supply." This included recognition of the need for finance and investment in renewables and a call for greater international cooperation for capacity building in developing countries.

- Washington 2008 – The Washington International Action Program collected pledges on specific and measurable renewable initiatives, including policy commitments, targets, and programs from a wide range of stakeholders. Progress on these pledges will be monitored by REN21 (WIREC, 2008).

- Delhi 2010 – The Delhi International Action Program was announced to encourage governments, international organizations, private companies, industry associations, and civil society organizations to take voluntary action to scale up renewable energy within their jurisdictions or spheres of responsibility.

References

3 Tier, 2011: *Renwable Energy Risk Analysis.* http://www.3tier.com/en/ (accessed 12 May 2011).

Aartsma, T., E-M. Aro, J. Barber, D. Bassani, R. Cogdell, T. Flüeler, H. de Groot, A. Holzwarth, O. Kruse, and A. W. Rutherford, 2008: *Harnessing Solar Energy for the Production of Clean Fuel*, European Science Foundation Policy Briefing 34. Strasbourg, France.

Aden, A., M. Ruth, K. Ibsen, J. Jechura, K. Neeves, J. Sheehan, B. Wallace, L. Montague, A. Slayton, and J. Lukas, 2002: *Lignocellulosic Biomass to Ethanol Process Design and Economics Utilizing Co-current Dilute Acid Prehydrolysis and Enzymatic Hydrolysis for Corn Stover.* National Renewable Energy Laboratory, Golden, CO, USA.

Agterbosch, S., R. M. Meertens, and W. J. V. Vermeulen, 2009: The relative importance of social and institutional conditions in the planning of wind projects. *Renewable and Sustainable Energy Reviews* **13**(2): 393–405.

Al-Dabbas, M. A. A., 2009. The Economical, Environmental and Technological Evaluation of Using Geothermal Energy. *European Journal of Scientific Research* **38**(4): 626–642.

Altenburg, T., H. Schmitz, and A. Stamm, 2008: Breakthrough? China's and India's transition from production to innovation. *World Development* **36**(2): 325–344.

Andre, H., 1976: Operating experience with bulb units at the Rance tidal power plant and other French hydropower sites. *IEEE Transactions on Power Apparatus and Systems* **95**: 1038–1044.

Aratani, F., 2009: *Accelerated and Extended Japanese PV Technology Roadmap PV2030+*. New Energy and Industrial Technology Development Organization and Ministry of Economy, Trade and Industry, Tokyo.

Arent, D., F. Verrastro, E. Peterson, and J. Bovair, 2009: *Alternative Fuel and Vehicle Technology, Challenges and Opportunities*. Center for Strategic and International Studies, Washington, DC, USA.

Arent, D. J., 2010: The role of renewable energy technologies in limiting climate change. *The Bridge* **40**: 31–39.

Arvizu, D., 2008: Potential role and contribution of direct solar energy to the mitigation of climate change. In *Proceedings Intergovernmental Panel on Climate Change (IPCC) Scoping Meeting on Renewable Energy Sources*, 20–25 January, Lübeck, Germany, pp.33–58.

Arvizu, D., P. Balaya, L. Cabeza, T. Hollands, A. Jäger-Waldau, M. Kondo, C. Konseibo, V. Meleshko, W. Stein, Y. Tamaura, H. Xu, and R. Zilles, 2011: Direct Solar Energy. In *IPCC Special Report on Renewable Energy Sources and Climate Change Mitigation*, O. Edenhofer et al. (eds.), Cambridge University Press, Cambridge, UK and New York, NY, USA.

Asplund, G., 2007: *Transmission – The Missing Links towards a Sustainable Climate*. ABB Power systems, Ludvika, Sweden.

AWEA, 2009: *AWEA Small Wind Turbine Global Market Study: Year Ending 2008*. American Wind Energy Association, Washington, DC, USA.

Axelsson, G., V. Stefansson, and G. Björnsson, 2005: Sustainable utilization of geothermal resources for 100–300 years. In *Proceedings of World Geothermal Congress 2005*, 24–29 April, Antalya, Turkey.

Axelsson, G., V. Stefansson, and Y. Xu, 2002: Sustainable management of geothermal resources. In *Proceedings of 2002 Beijing International Geothermal Symposium*. Liu Jiurong (ed.), 29–31 October, Beijing, China, pp.277–283.

Baker, P., C. Mitchell, and B. Woodman, 2009: *The Extent to Which Economic Regulation Undermines the Move to a Sustainable Electricity System*. UK Energy Research Centre, London, UK.

Barnett, P., and P. Quinlivan, 2009: *Assessment of Current Costs of Geothermal Power Generation in New Zealand (2007 Basis)*, SKM report for New Zealand Geothermal Association.

Bates, B. C., Z. W. Kundzewicz, S. Wu, and J.P. Palutikof (eds.), 2008: *Climate Change and Water. Technical Paper of the Intergovernmental Panel on Climate Change*. IPCC Secretariat, Geneva, Switzerland.

Battaglini, A., J. Lilliestam, A. Haas, and A. Patt, 2009: Development of SuperSmart Grids for a more efficient utilisation of electricity from renewable sources. *Journal of Cleaner Production* **17**: 911–918.

Bauen, A., B. Göran, M. Junginger, M. Londo, and F. Vuille, 2009: *Bioenergy – A Sustainable and Reliable Energy Source*. IEA Bioenergy, Rotorua, New Zealand.

Bauer, D., W. Heidemann, and H. Müller-Steinhagen, 2007: Central solar heating plants with seasonal heat storage. In *Proceedings CISBAT 2007, International Conference On Renewable in a changing climate – Innovation in the Built Environment*, 4–5 September, Lausanne, Switzerland.

Bauer, D., R. Marx, J. Nußbicker-Lux, F. Ochs, W. Heidemann, and H. Müller-Steinhagen, 2010: German central solar heating plants with seasonal heat storage. *Solar Energy* **84**(4): 612–623.

Baumgärtner, J., H. Menzel, and P. Hauffe, 2007: The Geox GmbH project in Landau – The first geothermal power project in Palatinate / Upper Rhine Valley. In *Proceedings of First European Geothermal Review, Geothermal Energy for Electric Power Production*, 29–31 October, Mainz, Germany.

Beardsmore, G., 2007: The burgeoning Australian geothermal industry. *Quarterly Bulletin* **28**(3): 20–26.

Becquerel, E., 1839: Mémoire sur les Effets Électriques Produits sous l'Influence des Rayons Solaires. *Comptes Rendues* **6**: 561–567.

Bergman, P. C. A., S. V. B. van Paasen, and H. Boerrigter, 2002: The novel "OLGA" technology for complete tar removal from biomass producer gas. *Pyrolysis and Gasification of Biomass and Wastes Expert Meeting*, 30 September-1 October, Strasbourg, France.

Bertani, R., 2009: *Long-Term Projections of Geothermal-Electric Development in the World*. ENEL Green Power, Rome, Italy.

Bertani, R., 2010: Geothermal Power Generation in the World: 2005–2010 Update Report. In *Proceedings of World Geothermal Congress 2010*, 25–30 April, Bali, Indonesia.

Bertani, R., and I. Thain, 2002: Geothermal power generating plant CO_2 emission survey. *IGA News* **49**: 1–3.

Bird, L., C. Creycik, and B. Friedman, 2009: *Green Power Marketing in the United States: A Status Report*. National Renewable Energy Laboratory, Golden, CO, USA.

BIS, 2010: *UK-China Scholarships for Excellence*. UK Department for Business Innovation & Skills (BIS). www.bis.gov.uk/policies/higher-education/international-education/china/scholarships-for-excellence (accessed 20 April 2010).

Blair, N., W. Short, M. Mehos, and D. Heimiller, 2006: Concentrating solar deployment systems (CSDS) – A new model for estimating U.S concentrating solar power market potential. *ASES Solar 2006 Conference*, 8–13 June.

Bloomfield, K. K., J. N. Moore, and R. N. Nelson, 2003: Geothermal energy reduces greenhouse gases. *Bulletin* **32**: 77–79.

Bloomquist, R. G., 2004: Elko Heat Company district heating system – A case study. *Quarterly Bulletin* 25(2): 7–10.

Bloomquist, R. G., J. T. Nimmons, and K. Rafferty, 1987: *District Heating Development Guide – Legal, Institutional and Marketing Issues, Vol. 1.* Washington State Energy Office, Olympia, WA.

BMU, 2006: *Erneuerbare Energien: Arbeitsplatzeffekte, Wirkungen des Ausbaus erneuerbarer Energien auf den deutschen Arbeitsmarkt.* Federal Environment Ministry (BMU), Berlin, Germany.

Bodmann, M., D. Mangold, J. Nußbicker, S. Raab, A. Schenke, and T. Schmidt, 2005: *Solar unterstützte Nahwärme und Langzeit-Wärmespeicher (Februar 2003 bis Mai 2005).* Forschungsbericht zum BMWA/BMU-Vorhaben, Stuttgart, Germany.

Bolhar-Nordenkampf, M., R. Rauch, and H. Hofbauer, 2003: Biomass CHP plant Güssing – Using gasification for power generation. In *Proceedings of the 2nd RCETCE Conference*, 12–14 February, Phuket, Thailand, pp.567–572.

Bolinger, M., R. Wiser, K. Cory, and T. James, 2009: *PTC, ITC, or Cash Grant? An Analysis of the Choice Facing Renewable Power Projects in the United States.* National Renewable Energy Laboratory, Golden, CO, USA.

Bosello, F., R. Roson, and R. S. J. Tol, 2006: Economy-wide estimates of the implications of climate change: Human health. *Ecological Economics* 58(3): 579–591.

Boud, R., 2002: Status and Research and Development Priorities 2003 – Wave and Marine Current Energy. IEA Implementing Agreement on Ocean Energy Systems. www.iea-oceans.org (accessed 2 June 2010).

Boyle G. (ed.), 1996: *Renewable Energy, Power for a Sustainable Future.* Oxford University Press, Oxford, UK.

BP, 2009: *Statistical Review of World Energy 2009.* BP, London, UK. www.bp.com/sectiongenericarticle.do?categoryId=9023791&contentId=7044194 (accessed 19 March 2011).

BP, 2011: *Statistical Review of World Energy June 2011.* BP, London, UK. www.bp.com/statisticalreview (accessed 19 March 2011).

Braun, M., 2009: *Grid Integration of Gigawatts of Photovoltaics. CrystalClear Final Event*, 29 May, Munich, Germany.

Brennan, L., and P. Owende, 2010: Biofuels from microalgae – A review of technologies for production, processing, and extractions of biofuels and co-products. *Renewable and Sustainable Energy Reviews* 14(2): 557–577.

Breuker, S., and M. Wolsink, 2007: Wind power implementation in changing institutional landscapes: An international comparison. *Energy Policy* 35: 2737–2750.

Brewer, T. L., 2008: International Energy Technology Transfers for Climate Change Mitigation. Paper prepared for *CESifo Summer Institute Workshop Europe and Global Environmental Issues*, Venice, Italy, 14–15 July 2008.

Breyer, Ch., Gerlach, A., 2010: Global overview on grid parity event dynamics. In *Proceedings of the 25th European PV Solar Energy Conference and Exhibition*, 6–10 September, Valencia, Spain, pp.5283–5304.

Bridgwater, A. V., and G. V. C. Peacocke, 2000: Fast pyrolysis processes for biomass. *Renewable and Sustainable Energy Reviews* 4(1): 1–73.

Bringezu, S., H. Schütz, M. O'Brien, L. Kauppi, R. W. Howarth, and J. McNeely, 2009: *Towards Sustainable Production and Use of Resources: Assessing Biofuels.* United Nations Environment Programme, Nairobi, Kenya.

Bromley, C. J., M. A. Mongillo, B. Goldstein, G. Hiriart, R. Bertani, E. Huenges, H. Maraoka, A. Ragnarsson, J. Tester, and V. Zui, 2010: IPCC renewable energy report – The potential contribution of geothermal energy to climate change mitigation. In *Proceedings of World Geothermal Congress 2010*, 25–30 April, Bali, Indonesia.

Brouwer, L. C., and L. Bosselaar, 1999: Policy for stimulating passive solar energy in the Netherlands. In *Proceedings of the Northsun 1999 Conference*, Edmonton, Canada.

Brown, M. A., and S. J. Chandler, 2008: Governing confusion: How statutes, fiscal policy, and regulations impede clean energy technologies. *Stanford Law and Policy Review* 19(3): 472–509.

BTM Consult, 2010: *International Wind Energy Development: World Market Update (2009).* Ringkobing, Denmark.

Burkhardt, J., G. Heath, and C. Turch, 2010: Life Cycle Assessment of a Model Parabolic Trough Concentrating Solar Power Plant with Thermal Energy Storage. In *Proceedings of ASME 2010 4th International Conference on Energy Sustainability*, 17–22 May, Phoenix, Arizona, USA.

Burton T., D. Sharpe, N. Jenkins, and E. Bossanyi, 2001: *Wind Energy: Handbook.* J. Wiley, New York, NY, USA.

Callaghan, J., 2006: *Future Marine Energy; Results of the Marine Energy Challenge – Cost competitiveness and growth of wave and tidal stream energy.* Research Report, Carbon Trust, London, UK

Cappetti, G. 2009: *Larderello: a case history of production sustainability.* CEGL Workshop on Geothermal Energy: Opportunities and Challenges, Pomarance, Tuscane, Italy, 3–4 September 2009.

Cavanagh, J. E., J. H. Clarke, and R. Price, 1993: Ocean Energy Systems. In *Renewable Energy: Sources for Fuels and Electricity*. T. B. Johansson et al. (eds.), Island Press, Washington, pp.513–547.

Cazzola, P., 2009: *Algae for biofuel production: process description, life cycle assessment and some information on cost.* IEA Bioenergy Executive Committee 64, Liege, Belgium, 1 October.

CEC, 1996: *Wave Energy Project Results: The Exploitation of Tidal Marine Currents.* Commission of the European Communities (CEC), DGXII, Brussels, Belgium.

CEC, 1998: *Promotion of New Energy Sources in the Zhejiang Province, China – Final Report.* Program SYNERGY, Commission of the European Communities (CEC), DGXVII, Brussels, Belgium.

Chad, A., J. W. Tester, B. J. Anderson, S. Petty, and B. Livesay, 2006: A comparison of geothermal with oil and gas well drilling costs. *Thirty-first Workshop on Geothermal Reservoir Engineering, January 30-February 1*, Stanford University, Stanford, CA, USA.

Charlier, R. H. and J. R. Justus, 1993, *Ocean Energies: Environmental, Economic and Technological Aspects of Alternative Power Sources.* Elsevier Oceanography Series, Elsevier, Amsterdam, Netherlands.

Chen, H., T. N. Cong, W. Yang, C. Tan, Y. Li, and Y. Ding, 2009: Progress in electrical energy storage system: a critical review. *Progress in Natural Science* 19: 291–312.

Chen, Y., G. Yang, S. Sweeney, and Y. Feng, 2010: Household biogas use in rural China: A study of opportunities and constraints, *Renewable and Sustainable Energy Reviews* 14(1): 545–549.

Chum, H., A. Faaij, J. Moreira, G. Berndes, P. Dhamija, H. Dong, B. Gabrielle, A. Goss-Eng, W. Lucht, M. Mapako, O. Masera, T. McIntyre, T. Minowa, and K. Pingoud, 2011: Bioenergy. In *IPCC Special Report on Renewable Energy Sources and Climate Change Mitigation*. O. Edenhofer et al. (eds), Cambridge University Press, Cambridge, UK and New York, NY, USA.

Clément, A., P. McCullen, A. Falcão, A. Fiorentino, F. Gardner, K. Hammarlund, G. Lemonis, T. Lewis, K. Nielsen, S. Petroncini, M.T. Pontes, P. Schild, B.O. Sjöström, H. C. Sørensen, and T. Thorpe, 2002: Wave Energy in Europe:

current status and perspectives. *Renewable and Sustainable Energy Reviews* **6**(5): 405–431.

Coll-Mayor, D., M. Paget, and E. Lightner, 2007: Future intelligent power grids: Analysis of the vision on the European Union and the United States. *Energy Policy* **35**: 2453–2465.

CONCAWE, EUCAR, and JRC, 2007: *Well-to-Wheels Analysis of Future Automotive Fuels and Powertrains in the European Context: Well-to-Tank Report*, version 2c. Conservation of Clean Air and Water in Europe (CONCAWE), European Council for Automotive R&D (EUCAR), and Joint Research Centre, European Commission (JRC).

CONCAWE, EUCAR, and JRC, 2008: *Well-to-Wheels Analysis of Future Automotive Fuels and Powertrains in the European Context*, Well-to-tank report version 3.0, Appendix 2: Description and detailed energy and GHG balance of individual pathways. Conservation of Clean Air and Water in Europe (CONCAWE), European Council for Automotive R&D (EUCAR), and Joint Research Centre, European Commission (JRC).

Cooper, D. J., L. J. Meyer, and R. J. Varley, 2009: OTEC Commercialization Challenges. *In Proceedings Offshore Technology Conference*, 4–7 May 2009, Houston, Texas, USA.

Cornett, A. M., 2006: *Inventory of Canada's Marine Renewable Energy Resources*. Canadian Hydraulics Centre, Natural Research Council Canada, Ottawa, Canada, CHC-TR-041.

Costa, A., A. Crespo, J. Navarro, G. Lizcano, H. Madsen, and E. Feitosa, 2008: A review on the young history of the wind power short-term prediction. *Renewable and Sustainable Energy Reviews* **12**: 1725–1744.

Couture, T. D., K. Cory, C. Kreycik, E. Williams, 2010: Policymaker's Guide to Feed-in Tariff Policy Design, NREL Report No. TP-6A2–44849, 144pp.

Criscione, V., 2010: Norway's pioneering role in osmotic power. *Nortrade.com* 16 September 2010.

Currie, M. J., J. K. Mapel, T. D. Heidel, S. Goffri, and M. A. Baldo, 2008: High-efficiency organic solar concentrators for photovoltaics. *Science* **321**: 226–228.

Curtis, R., J. Lund, B. Sanner, L. Rybach, and G. Hellström, 2005: Ground source heat pumps – Geothermal energy for anyone, anywhere: Current worldwide activity. In *Proceedings of World Geothermal Congress 2005*, 24–29 April 2005, Antalya, Turkey.

Dallinger, D. 2009: Analysis of possible vehicle-to-grid benefits to battery vehicle owners. *Technical Conference on Advanced Battery Technologies for Automobiles and their Electric Power Grid Integration*, 20–21 January. Haus der Technik, Essen, Germany.

Dayton, D., 2002: *A Review of the Literature on Catalytic Biomass Tar Destruction*. Milestone Completion Report. National Renewable Energy Laboratory, Golden, CO, USA.

de Wild-Scholten, M., 2010: *Life Cycle Assessment of Photovoltaics: from cradle to cradle*, First International Conference on PV Module Recycling, 26 January 2010, Berlin, Germany. www.epia.org/events/past-events/epia-events-within-the-last-year/1st-international-conference-on-pv-module-recycling.html (accessed 6 June 2011).

de Wild-Scholten, M., 2011: *Environmental profile of PV mass production: globalization*, 26th European Photovoltaic Solar Energy Conference, 5–9 September 2011, Hamburg, Germany.

DECC, 2009: *The UK Renewable Energy Strategy*. UK Department of Energy & Climate Change (DECC), London, UK.

DECC, 2010: *Low Carbon Communities Challenge*. UK Department of Energy & Climate Change (DECC), London, UK.

Degner, T., J. Schmid, and P. Strauss (eds.), 2006: *DISPOWER-Distributed Generation with High Penetration of Renewable Energy Sources*. Final Public Report, ISET 5–2006, Kassel, Germany.

DeLong, M. M., 1995: *Economic Development Through Biomass System Integration: Summary Report*. National Renewable Energy Laboratory, Golden, CO, USA.

Denholm, P., 2007: *The Technical Potential of Solar Water Heating to Reduce Fossil Fuel Use and Greenhouse Gas Emissions in the United States*. National Renewable Energy Laboratory, Golden, CO, USA.

Denholm, P., E. Ela, B. Kirby, and M. Milligan, 2010: *The Role of Energy Storage with Renewable Electricity Generation*. National Renewable Energy Laboratory, Golden, CO, USA.

DESERTEC Foundation, 2010: DESERTEC in EU-MENA. Available at www.desertec.org/press/#c1413 (accessed 14 November 2011).

DESERTEC Foundation, 2011: *Red Paper: An Overview of the DESERTEC Concept*. Hamburg, Germany.

Desholm, M., and J. Kahlert, 2005: Avian collision risk at an offshore wind farm. *Biology Letters* **1**(3): 296–298. (p297, Figure 1."The westerly oriented flight trajectories during the initial operation of the wind turbines. Black lines indicate migrating waterbird flocks, red dots the wind turbines" used as Figure 11.43).

Deutsche Bank, 2010a: *GET FiT Program – Global Energy Transfer Feed-In Tariffs for Developing Countries*. New York, NY, USA.

Deutsche Bank, 2010b: *Investing in Climate Change 2010 – A Strategic Asset Allocation Perspective*. New York, NY, USA.

Devine-Wright, P, 2005: Beyond Nimbyism: Towards an integrated framework for understanding public perceptions of wind energy. *Wind Energy* **8**(2): 125–139.

DP Energy, 2010: Tidal Energy. www.dpenergy.com/information/tidal.html (accessed 06 Dec 2011).

ECF, 2010: *Roadmap 2050 – A Practical Guide to a Prosperous, Low Carbon Europe*. European Climate Foundation (ECF), The Hague, Netherlands. www.roadmap2050.eu/ (accessed 17 May 2011).

Ecoheatcool, 2005–2006: *Possibilities with More District Heating in Europe*. Ecoheatcool Work package 4, Final Report. Ecoheatcool and Ecoheat & Power, Brussels, Belgium.

EEGI, 2010: *The European Electricity Grid Initiative – Roadmap 2010–2018 and Detailed Implementation Plan 2010–12*. European Network of Transmission System Operators for Electricity (ENTSO-E) and European Distribution System Operators for Smart Grids (EDSO-SG). European Electricity Grid Initiative (EEGI), Brussels, Belgium.

Electricité de France and Observ'ER, 2010: *Worldwide Electricity Production from Renewable Energy Sources, 12th inventory*. Paris, France. energies-renouvelables.org/observ-er/html/inventaire/Eng/methode.asp (accessed 6 June 2010).

Ellis, G., R. Cowell, C. Warren, P. Strachan, J. Szarka, R. Hadwin, P. Miner, M. Wolsink, and A. Nadaï, 2009: Wind Power: Is There A 'Planning Problem'? *Journal of Planning Theory and Practice* **10**(4): 521–547.

EnerNex, 2010: *Eastern Wind Integration and Transmission Study*. Prepared for the National Renewable Energy Laboratory, EnerNex Corporation, Golden, CO, USA.

Engstrom, S., N. Lindman, E. Rensfelt, and L. Waldheim, 1981: A new synthesis gas process for biomass and peat. *Energy from Biomass and Wastes V.* Institute of Gas Technology, Chicago, IL, USA.

ENTSO-E, 2010: *Ten-Year Network Development Plan 2010–2020*. European Network of Transmission System Operators for Electricity (ENTSO-E), Brussels, Belgium.

EPE, 2008: *PLANO DECENAL DE EXPANSÃO DE ENERGIA 2008–2017*. Prepared for the Ministry of Mines and Energy, Secretariat of Planning and Energy Development, Empresa de Planejamento Energético (EPE), Brasilia, Brazil.

EPIA, 2009: *SET for 2020 (Executive Summary).* European Photovoltaic Industry Association (EPIA), Brussels, Belgium.

EPIA, 2011: *Global Market Outlook for Photovoltaics until 2015*, European Photovoltaic Industry Association (EPIA), Brussels, Belgium.

Epp, B., 2009: World map of flat plate collector manufacturers 2009, In *Sun & Wind Energy* December, Solrico, Bielefeld, Germany.

EPTP, 2007: *A Strategic Research Agenda for Photovoltaic Solar Energy Technology*. European Photovoltaic Technology Platform (EPTP), Office for Official Publications of the European Communities, Luxembourg.

EPTP, 2009: *Today's Actions for Tomorrow's PV Technology; An Implementation Plan for the Strategic Research Agenda of the European Photovoltaic Technology Platform*. European Photovoltaic Technology Platform (EPTP), Office for Official Publications of the European Communities, Luxembourg.

EPTP, 2010: *Calculation tool*. European Photovoltaic Technology Platform (EPTP), www.eupvplatform.org/pv-development/tools.html (accessed 18 May 2011).

EREC, 2008: *The Renewable Energy House – Europe's Headquarters for Renewable Energy*. European Renewable Energy Council (EREC), Brussels, Belgium.

EREC and Greenpeace, 2010: *Energy [R]evolution – a Sustainable World Energy Outlook*. European Renewable Energy Council (EREC), Brussels, Belgium and Greenpeace International, Amsterdam, Netherlands.

ERS, undated: *Crops, U.S. Department of Agriculture*. Economic Research Service (ERS), Washington, DC. www.ers.usda.gov/Browse/view.aspx?subject=Crops (accessed 15 October 2010).

ESRU, undated: *Biofuels, University of Strathclyde*. Energy Systems Research Unit (ESRU), Glasgow, Scotland. www.esru.strath.ac.uk/EandE/Web_sites/02–03/biofuels/quant_biodiesel.htm (accessed 7 July 2010).

ESTIF, 2007: *Solar Thermal Markets in Europe 2006*. European Solar Thermal Industry Federation (ESTIF), Brussels, Belgium.

ESTTP, 2008: *Solar Heating and Cooling for a Sustainable Energy Future in Europe – Strategic Research Agenda*. European Solar Thermal Technology Platform (ESTTP), Brussels, Belgium.

EU, 2009: *The Impact of RE policy on economic growth and employment in the EU*. European Union (EU), Brussels, Belgium.

Eugster, W. J. and L. Rybach, 2000: Sustainable production from borehole heat exchanger systems. In *Proceedings of World Geothermal Congress 2000*, 28 May–10 June, Kyushu-Tohoku, Japan, pp.825–830.

EurObserv'ER, 2010: Solid Biomass Barometer. *Systemes Solaires, le Journal des Energies Renouvelables* 200 (November).

Euroheat & Power, 2007: *District Heating and Cooling – 2007 Statistics.* www.euroheat.org/Statistics-69.aspx (accessed 23 May 2010).

European Commission, 2006: *World Energy Technology Outlook – 2050*. Brussels, Belgium.

European Commission, 2010: *European Wind Integration Study*. Sixth Framework Programme, Brussels, Belgium.

European Parliament, 2009: *Directive of the European Parliament and the Council on the promotion of the use of energy from renewable sources amending and subsequently repealing*, Directives 2001/77/EC and 2003/30/EC. Brussels, Belgium.

European Photovoltaic Industry Association/Greenpeace International, 2011: *Solar Generation 6*. Brussels, Belgium.

Evans, R. J., R. A. Knight, M. Onischak, and S. P. Babu, 1987: Process performance and environmental assessment of the renugas process. In *Energy from Biomass and Wastes X*, D. L. Klass (ed.), Elsevier Applied Science, London, UK and Institute of Gas Technology, Chicago, IL, USA, pp.677–694.

EWEA, 2009: *The Economics of Wind Energy*. European Wind Energy Association (EWEA), Brussels, Belgium.

EWEA, 2010: *European Offshore Wind Industry – Key Trends and Statistics*. European Wind Energy Association (EWEA), Brussels, Belgium.

EWEA, 2011: *European Offshore Wind Industry – Key Trends and Statistics 2010*. European Wind Energy Association (EWEA), Brussels, Belgium.

ExternE, 1995: *Externalities of Energy – Volume 6, Wind and Hydro*. European Commission, Directorate General XII, Science, Research and Development, Brussels, Belgium.

Faaij, A., 2006 : Modern biomass conversion technologies. *Mitigation and Adaptation Strategies for Global Change* 11(2): 335–367.

FAO, 2008: *The State of Food and Agriculture: Biofuels Prospects, Risks, and Opportunities*. Food and Agriculture Organization (FAO), Rome, Italy.

FAPRI, undated: *Biofuel Conversion Factors*. Food and Agricultural Policy Research Institute (FAPRI), Ames, IA, USA.

Farrell, A. E., R. J. Plevin, B. T. Turner, A. D. Jones, M. O'Hare, and D. M. Kammen, 2006: Ethanol Can Contribute to Energy and Environmental Goals. *Science* 311(5760): 506–508.

FAS, undated: *EU Rapeseed Yield*. Foreign Agricultural Service (FAS), US Department of Agriculture, Washington, DC, USA. www.pecad.fas.usda.gov/highlights/2007/05/EU_21May07/EURapeseedYield.htm (accessed 17 February 2011).

Felix-Soul, R, 2008: Assessing the impact of Mexico's Biofuels Law. *Biomass Power and Thermal*. June 2008.

Florentinus, A., 2009: *Worldwide potential of aquatic biomass: algae as the new sustainable bio-energy resource*. World Bio Energy – Clean Vehicles and Fuels, Stockholm, Sweden.

Florentinus, A., C. Hamelinck, S. de Lint, and S. van Iersel, 2008: *Worldwide Potential of Aquatic Biomass*, Ecofys, Utrecht, The Netherlands.

FOSG, 2010: *Position paper on the EC Communication for a European Infrastructure Package*. Friends of the Supergrid (FOSG), Brussels, Belgium.

Foxon, T., R. Gross, P. Heptonstall, P. Pearson, and D. Anderson, 2007: *Energy Technology Innovation: A Systems Perspective – Report for the Garnaut Climate Change Review*. Sustainability Research Institute, University of Leeds and ICEPT Centre for Environmental Policy, Imperial College, London, UK.

Fraenkel, P. L., 1999: New Developments in Tidal and Wavepower Technologies. In *Proceedings of the Silver Jubilee Conference 'Towards a Renewable Future,'* Solar Energy Society, 13–15 May, 1989, Brighton, UK.

Frankhauser, S., F. Sehlleier, and N. Stern, 2008: Climate change, innovation and jobs. *Climate Policy* 8: 421–429.

Fridleifsson, G. O., A. Albertsson, B. Stefansson, E. Gunnlaugsson, and H. Adalsteinsson, 2007: Deep unconventional geothermal resources: A major opportunity to harness new sources of sustainable energy. In *Proceedings of 20th World Energy Conference*, 12–15 November, Rome, Italy.

Fridleifsson, I. B., R. Bertani, E. Huenges, J. W. Lund, A. Ragnarsson, and L. Rybach, 2008: The possible role and contribution of geothermal energy to the mitigation of climate change. *IPCC Geothermal Report*, presented at Potsdam, Germany, 11 February, 2008.

Fthenakis, V., J. E. Mason, and K. Zweibel, 2009: The technical, geographical, and economic feasibility for solar energy to supply the energy needs of the US. *Energy Policy* **37**(2): 387–399.

FVEE, 2010: *Energiekonzept: Eine Vision für ein nachhaltiges Energiekonzept auf Basis von Energieeffizienz und 100% erneuerbaren Energien (A Vision for a Sustainable Energy Concept Based on Energy Efficiency and Renewable Energies)*. ForschungsVerbund Erneuerbare Energien (FVEE), Berlin, Germany.

Garrison, T., 2008: *Essentials of Oceanography*. Brooks Cole, Pacific Grove, CA, USA.

Geo-Heat Center, 1998: *Geo-Heat Center Quarterly Bulletin* **19**(1): March.

Geo-Heat Center, 2003: *Geo-Heat Center Quarterly Bulletin* **24**(3): September.

Geo-Heat Center, 2005: Combined geothermal heat & power plants, *Geo-Heat Center Quarterly Bulletin* **26**(2).

Geo-Heat Center, 2006: *Feasibility Study for the Direct Use of Geothermal Energy for Onion Dehydration in Vale/Ontario Area, Oregon*. Oregon Institute of Technology, Klamath Falls, OR, USA.

Geothermal Task Force Report, 2006: Clean and Diversified Energy Initiative, Western Governors' Association, Geothermal Energy Association, Washington, DC.

German Aerospace Institute, 2005: Concentrating Solar Power for the Mediterranean Region (MED-CSP). www.dlr.de/tt/desktopdefault.aspx/tabid-2885/4422_read-6575/ (accessed 9 March 2011).

Gil, A., M. Medrano, I. Martorell, A. Lázaro, P. Dolado, B. Zalba, and L. F. Cabeza, 2010: State of the art on high-temperature thermal energy storage for power generation. Part 1 – Concepts, materials and modellization. *Renewable and Sustainable Energy Reviews* **14**: 31–55.

Gill, A. B., 2005: Offshore renewable energy: Ecological implications of generating electricity in the coastal zone. *Journal of Applied Ecology* **42**: 605–615.

Global Bioenergy Partnership, 2007: *A Review of the Current State of Bioenergy Development in G8+5 Countries*. Food and Agriculture Organization (FAO), Rome, Italy.

Goldemberg, J., S.T. Coelho, P. M. Nastari, and L. O., 2004: Ethanol learning curve – the Brazilian experience. *Biomass and Bioenergy* **26**: 301–304.

Goldenfum, J. A., 2011: Tool of the trade: details on the UNESCO/IHA GHG Measurement Guidelines for Fresh Water Reservoirs, *International Water Power & Dam Construction* January 2011. www.waterpowermagazine.com/storyprint.asp?sc=2058692 (accessed 7 February 2011).

Goldstein, B., G. Hiriart, R. Bertani, C. Bromley, L. Gutiérrez-Negrín, E. Huenges, H. Muraoka, A. Ragnarsson, J. Tester, V. Zui, 2011: Geothermal Energy. In *IPCC Special Report on Renewable Energy Sources and Climate Change Mitigation*, O. Edenhofer et al. (eds.), Cambridge University Press, Cambridge, UK and New York, NY, USA.

Gottstein, M. and L. Schwartz, 2010: *The Role of Forward Capacity Markets in Increasing Demand-side and Other Low-carbon Resources: Experience and Prospects*. Regulatory Assistance Project, Montpelier, VT, USA.

Goudriaan, F., and J. E. Naber, 2008: HTU diesel: From wet waste streams. In *Symposium New Biofuels*, May 2008, Berlin, Germany.

Goyal, M., 2010: Repowering – The next big thing in India. *Renewable and Sustainable Energy Reviews* **14**: 1400–1409.

Green, M.A., 2003: *Third Generation Photovoltaics; Advanced Solar Energy Conversion*. Springer-Verlag, Berlin, Germany.

Green, M. A., K. Emery, Y. Hishikawa, and W. Warta, 2010: Solar cell efficiency tables (version 37). *Progress in Photovoltaics: Research Applications* **19**: 84–92.

Greenpeace and EPIA, 2008: *Solar Generation V.* Greenpeace International, Amsterdam, Netherlands and European Photovoltaic Industry Association (EPIA), Brussels, Belgium.

Greenpeace and EPIA, 2010: *Solar Generation 6 (Executive Summary)*. Greenpeace International, Amsterdam, Netherlands and European Photovoltaic Industry Association (EPIA), Brussels, Belgium.

Greenpeace and EREC, 2007: *Energy [R]evolution – A Sustainable World Energy Outlook*. Greenpeace International, Amsterdam, Netherlands and European Renewable Energy Council (EREC), Brussels, Belgium.

Greenpeace and EREC, 2010: *Energy [R]evolution – Towards a Fully Renewable Energy Supply in the EU-27*. Greenpeace International, Amsterdam, Netherlands and European Renewable Energy Council (EREC), Brussels, Belgium.

Greenpeace, ESTELA, and SolarPACES, 2009: *Concentrating Solar Power: Global Outlook 2009 – Why Renewable Energy Is Hot*. Greenpeace International, Amsterdam, Netherlands, European Solar Thermal Electricity Association (ESTELA), Brussels, Belgium, and Solar PACES, Tabernas, Spain.

Gross, R. and P. Hepponstall, 2008: The costs and impacts of intermittency: an ongoing debate. *Energy Policy* **36**: 4005–4007.

Grubb, M., 2004: Technological innovation and climate change policy: An overview of issues and options. *Keio Economic Studies* **41**(2): 103–132.

GTM Research, 2011: *Concentrating Solar Power 2011: Technology, Costs and Markets*. www.gtmresearch.com/report/concentrating-solar-power-2011-technology-costs-and-markets (accessed 9 November 2010).

Gunnlaugsson, E. and G. Gíslason, 2003: Reykjavik energy – District heating in Reykjavik and electrical production using geothermal energy. In *Lectures on the Sustainable Use and Operating Policy for Geothermal Resources*. I. B. Fridleifsson and M. V. Gunnarsson (eds.), United Nations University Geothermal Training Programme, pp.67–78.

Gunnlaugsson, E., 2007: *Personal Communication, Reykjavik District Heat System*, Reykjavik, Iceland.

GWEC and Greenpeace, 2010a: *Global Wind Energy Outlook 2010*. Global Wind Energy Council (GWEC), Brussels, Belgium and Greenpeace International, Amsterdam, Netherlands.

GWEC and Greenpeace, 2010b: *Global Wind 2009 Report*. Global Wind Energy Council (GWEC), Brussels, Belgium and Greenpeace International, Amsterdam, Netherlands.

Häberle, A., C. Zahler, H. Lerchenmüller, M. Mertins, C. Wittwer, F. Trieb, and J. Dersch, 2002: The Solarmundo line focussing Fresnel collector. Optical and thermal performance and cost calculations, In *Proceedings of the 11th International Solar Paces Conference*, September 2002, Zürich, Switzerland.

Hamelinck, C. N., and A. P. C. Faaij, 2006: Production of advanced biofuels. *International Sugar Journal* **108**(1287): 168–175.

Hamilton, K., 2009: *Unlocking Finance for Clean Energy: The Need for 'Investment Grade' Policy*. Chatham House, London, UK.

Hamududu, B. and Å. Killingtveit, 2010: Estimating Effects of Climate Change on Global Hydropower Production, In *Hydropower10: 6th International Conference on Hydropower*, Tromsø, Norway, 1–3 February 2010.

Hance, C. N., 2005: *Factors Affecting Costs of Geothermal Power Development*. Geothermal Energy Association, Washington, DC.

Harvey, L. D., 2006: *A handbook on Low-Energy Building and District-Energy Systems: Fundamentals, Techniques and Examples*. Earthscan, Sterling, VA, USA.

Harvey, L. D., 2010: *Carbon-Free Energy Supply*. Earthscan, London, UK, pp. 311–324.

Haselip, J., Nygaard, I., Hansen, U., Ackom, E. (eds.), 2010: Diffusion of renewable energy technologies: case studies of enabling frameworks in developing countries. Technology Transfer Perspectives Series, UNEP Riso Centre, Denmark, pp 3–32.

Hassuani, S. J., 2009: *Personal communication*. Sugarcane Technology Center, Piracicaba, Brazil, 18 December.

Hatziargyriou, N, H. Asano, R. Iravani, and C. Marnay, 2007: Microgrids. *IEEE Power & Energy Magazine* **5**(4): 78–94.

Hatziargyriou, N., 2008: Microgrids, the key to unlock distributed energy resources? *IEEE Power & Energy Magazine* **6**(3): 26–30.

Heinemann, D., E. Lorenz, and M. Girodo, 2006: Forecasting of solar radiation. In *Solar Energy Resource Management for Electricity Generation From Local Level to Global Scale*. E. D. Dunlop, L. Wald, and M. Šúri (eds.), Nova Science Publishers, Hauppage, NY, pp.83–94.

Hettinga, W. G., H. M. Junginger, S. C. Dekker, M. Hoogwijk, A. J. McAloon, and K. B. Hicks, 2009: Understanding the reductions in US corn ethanol production costs: An experience curve approach. *Energy Policy* **37**(1): 190–203.

Hjuler Jensen, P., 2007: UpWind: Wind energy research project under the 6th Framework Programme. In *Proceedings of International Energy Conference 2007 on Energy Solutions for Sustainable Development*, 22–24 May, Risø, Denmark. L. Sønderberg Petersen and H. Larsen (eds.), Risø National Laboratory, pp.135–139.

Hoffmann, W., S. Wieder, and T. Pellkofer, 2009: Differentiated price experience curves as evaluation tool for judging the further development of crystalline silicon and thin film solar electricity. In *Proceedings of the 24th European PV Solar Energy Conference*, 21–25 September 2009, Hamburg, pp.4387–4394.

Holttinen, H., B. Lemström, P. Meibom, H. Bindner, A. Orths, F. Van Hulle, C. Ensslin, A. Tiedemann, L. Hofmann, W. Winter, A. Tuohy, M. O'Malley, P. Smith, J. Pierik, J. O. Tande, A. Estanqueiro, E. Gomez, L. Söder, G. Strbac, A. Shakoor, J. C. Smith, P. Parsons, M. Milligan, and Y. Wan, 2007: *Design and Operation of Power Systems with Large Amounts of Wind Power – State-of-the-Art Report*. VTT Technical Research Centre of Finland, Espoo, Finland.

Hoogwijk, M., B. de Vries, and W. Turkenburg, 2004: Assessment of the global and regional geographical, technical, and economic potential of onshore wind energy, *Energy Economics* **26**: 889–919.

Huber, G. W., S. Iborra, and A. Corma, 2006: Synthesis of transportation fuels from biomass: Chemistry, catalysts, and engineering. *Chemical Reviews* **106**: 4044–4098.

Huenges, D., L. Moeck, A. Saadat, W. Brand, A. Schulz, H. Holl, D. Bruhn, G. Zimmerman, G. Blöcher, and L. Wohlgemuth, 2007: Geothermal research well in a deep sedimentary reservoir. *Geothermal Resources Council Bulletin* 36. Geothermal Resources council (GRC), Davis, CA, USA.

Hughes, T., 1983: *Networks of Power: Electrification in Western Society 1880–1930*. John Hopkins University Press, Baltimore, MD, USA.

Huld, T., M. Šúri, and E. D. Dunlop, 2008: Comparison of potential solar electricity output from fixed-inclined and two-axis tracking photovoltaic modules in Europe. *Progress in Photovoltaics: Research Applications* **16**: 47–59.

Ibrahim, H., A. Ilinca, and J. Perron, 2008: Energy storage systems – characteristics and comparisons. *Renewable and Sustainable Energy Reviews* **12**: 1221–1250.

IEA, 2000: *Hydropower and the Environment: Present Context and Guidelines for Future Action. Vol. II: Main Report*, Chapter 3: "Comparative Environmental Analysis of Power Generation Options." International Energy Agency (IEA), Paris, France.

IEA, 2005: *Prospects for Hydrogen and Fuel Cells*. International Energy Agency (IEA), Paris, France.

IEA, 2007: *Renewables for Heating and Cooling*. International Energy Agency (IEA), Paris, France.

IEA, 2008a: *World Energy Outlook 2008*. International Energy Agency (IEA), Paris, France.

IEA, 2008b: *Energy Technology Perspectives 2008 – Scenarios and Strategies to 2050*. International Energy Agency (IEA), Paris, France.

IEA, 2008c: *Empowering Variable Renewables – Options for Flexible Electricity Systems*. International Energy Agency (IEA), Paris, France.

IEA, 2008d: *Deploying Renewables: Principles for Effective Policies*. International Energy Agency (IEA), Paris, France.

IEA, 2009a: *World Energy Outlook 2009*. International Energy Agency (IEA), Paris, France.

IEA, 2009b: *Technology Roadmap – Wind Energy*. International Energy Agency (IEA), Paris, France.

IEA, 2009c: *Cities, Towns & Renewable Energy – Yes in My Front Yard*. International Energy Agency (IEA), Paris, France.

IEA, 2009d: *Multilateral Technology Initiatives (Implementing Agreements)*. International Energy Agency (IEA), Paris, France. www.iea.org/Textbase/techno/index.asp (accessed 22 April 2009).

IEA, 2010a: *Energy Technology Perspectives 2010 – Scenarios and Strategies to 2050*. International Energy Agency (IEA), Paris, France.

IEA, 2010b: *World Energy Outlook 2010*. International Energy Agency (IEA), Paris, France.

IEA, 2010c: *Algae – The Future for Bioenergy?* International Energy Agency (IEA), Paris, France.

IEA, 2010d: *Technology Roadmap: Solar Photovoltaic Energy*. International Energy Agency (IEA), Paris, France.

IEA, 2010e: *Technology Roadmap – Concentrating Solar Power*. International Energy Agency (IEA), Paris, France.

IEA, 2010f: *Energy Subsidies: Getting the Prices Right*. International Energy Agency (IEA), Paris, France.

IEA, 2010g: *Key World Energy Statistics*. International Energy Agency (IEA), Paris, France.

IEA, 2011: *Energy Balances of Non-OECD Countries 2011*. International Energy Agency (IEA), Paris, France.

IEA Bioenergy Task 39, undated: *Commercializing 1st and 2nd Generation Liquid Fuels from Biomass*. biofuels.abc-energy.at/demoplants/projects/mapindex (accessed 2 February 2011).

IEA-CERT, 2011: *Energy Efficiency in Buildings – Heating and Cooling: Technology Roadmap (Draft 28 January 2011)*. International Energy Agency (IEA)-Committee on Energy Research and Technology (CERT), Paris, France.

IEA-DHC, 2002: *District Heating and Cooling Connection Handbook*. IEA District Heating and Cooling (IEA-DHC), Paris, France.

IEA-ETSAP, 2010: Marine Energy. *Technology Brief* 13. IEA Energy Technology Systems Analysis Program (IEA-ETSAP), Paris, France.

IEA-OES-IA, 2011: *Implementing Agreement on Ocean Energy Systems – Annual Report 2010*. IEA Ocean Energy Systems Implementing Agreement OES-IA (IEA-OES-IA), Wave Energy Centre, Lisbon, Portugal.

IEA-PVPS, 2002: *Potential for Building Integrated Photovoltaics*. International Energy Agency-Photovoltaic Power Systems Programme (IEA-PVPS), Paris, France.

IEA-PVPS, 2007: *Cost and Performance Trends in Grid-connected Photovoltaic Systems and Case Studies*. International Energy Agency-Photovoltaic Power Systems Programme (IEA-PVPS), Paris, France.

IEA-PVPS, 2010: *Trends in Photovoltaic Applications 1992–2009*. International Energy Agency-Photovoltaic Power Systems Programme (IEA-PVPS), Paris, France.

IEA-PVPS, 2011: *Trends in Photovoltaic Applications 1992–2010*. International Energy Agency-Photovoltaic Power Systems Programme (IEA-PVPS), Paris, France.

IEA-SHC, 2009: *Solar Crop Drying*. International Energy Agency – Solar Heating & Cooling Programme (IEA-SHC). www.iea-shc.org/task29/index.html (accessed 21 January 2009).

IEA-Wind, 2010: *IEA Wind Energy Annual Report 2009*. Executive Committee for the Implementing Agreement for Co-operation in the Research, Development and Deployment of Wind Energy Systems of the International Energy Agency, Paris, France.

IEC, 2010: *Various Electrical Standards*. International Electrotechnical Commission (IEC). www.iec.ch/ (accessed 12 May 2010).

IEEE, 2009: Wind and the grid: The challenges of wind integration. Institute of Electrical and Electronics Engineers (IEEE). *IEEE Power and Energy Magazine* **7**.

IHA, 2005: Andhikhola Rural Development. International Hydropower Association (IHA). www.hydropower.org/iha_blue_planet_prize/information.html (accessed 3 April 2010).

IHA, 2010a: *2010 Activity Report: Status of the Hydropower Sector*. International Hydropower Association (IHA), London, UK. p.5.

IHA, 2010b: *Background Document – Hydropower Sustainability Assessment Protocol 2010*. International Hydropower Association (IHA), London, UK. pp.7, 9.

IHS EER, 2009: *Global Concentrated Solar Power Markets and Strategies, 2009–2020*. IHS Emerging Energy Research (EER), Cambridge, MA, USA and Barcelona, Spain.

Ingram, E. A., 2009: Pumped storage – development activity snapshots. *Hydro Review Worldwide* **17**(6): 13–25.

International Journal of Hydropower and Dams, Annual Directory 2004, *Aqua-Media International*, Surrey, UK.

International Journal of Hydropower and Dams, Annual Directory 2005, *Aqua-Media International*, Surrey, UK.

International Journal of Hydropower and Dams, Annual Directory 2006, *Aqua-Media International*, Surrey, UK.

International Journal of Hydropower and Dams, Annual Directory 2007, *Aqua-Media International*, Surrey, UK.

International Journal of Hydropower and Dams, Annual Directory 2008, *Aqua-Media International*, Surrey, UK.

International Journal of Hydropower and Dams, Annual Directory 2009, *Aqua-Media International*, Surrey, UK.

IPCC, 2000: *Special Report on Emission Scenarios*. Intergovernmental Panel on Climate Change (IPCC), Cambridge University Press, Cambridge, UK.

IPCC, 2007a: *Climate Change 2007: Mitigation of Climate Change, Contribution of Working Group III to the Fourth Assessment Report of the Intergovernmental Panel on Climate Change*. Intergovernmental Panel on Climate Change (IPCC), Cambridge University Press, Cambridge, UK.

IPCC, 2007b: *Fourth Assessment Report of the Intergovernmental Panel on Climate Change*. Intergovernmental Panel on Climate Change (IPCC), Cambridge University Press, Cambridge, UK.

IPCC, 2007c: *Climate Change 2007: The Physical Science Basis*. Contribution of Working Group I to the Fourth Assessment Report, Intergovernmental Panel on Climate Change (IPCC), Cambridge University Press, Cambridge, UK.

IPCC, 2011: *Special Report on Renewable Energy Sources and Climate Change Mitigation* O. Edenhofer et al. (eds.), Intergovernmental Panel on Climate Change (IPCC), Cambridge University Press, Cambridge, UK, and New York, NY, USA.

ISEO, 2010: *Environmental Benefits of Heat Pumps*. International Sustainable Energy Organisation for Renewable Energy and Energy Efficiency (ISEO), Geneva.

IT Power Ltd., 1996: *The exploitation of Tidal and Marine Currents*. Tecnomare SpA, Report EUR 16683 EN, Office for Official Publications of the European Communities, Luxembourg.

Jacobsson, S., A. Bergek, D. Finon, V. Lauber, C. Mitchell, D. Toke, and A. Verbruggen, 2009: EU Renewable Energy Support Policy: Faith or Facts? *Energy Policy* 37(6): 2143–2146.

Jäger-Waldau, A., 2010: *PV Status Report 2010*. Office for Official Publications of the European Communities, Luxembourg.

Jeffries, T. W., 2006: Engineering yeasts for xylose metabolism. *Current Opinion in Biotechnology* 17(3): 320–326.

Johansson, T. B., K. McCormick, L. Neij, and W. C. Turkenburg, 2006: The potentials of Renewable Energy. In *Renewable Energy: A Global Review of Technologies, Policies and Markets*. D. Assmann et al. (eds.), Earthscan, London, pp.15–47.

Johansson, T. B., and W. C. Turkenburg, 2004: Policies for renewable energy in the European Union and its member states: An overview. *Energy for Sustainable Development* 8(1): 5–24.

Jónsson, T., P. Pinson, and H. Madsen, 2010: On the market power of wind energy forecasts. *Energy Economics* 32(2): 313–320.

Junginger, M., A. Faaij, R. Björheden, and W. C. Turkenburg, 2005: Technological learning and cost reductions in wood fuel supply chains in Sweden. *Biomass and Bioenergy* 29(6): 399–418.

Junginger, M., E. de Visser, K. Hjort-Gregersen, J. Koornneef, R. Raven, A. Faaij, and W. Turkenburg, 2006: Technological learning in bioenergy systems. *Energy Policy* 34(18): 4024–4041.

Junginger, M., W. van Sark, and A. Faaij (eds.), 2010: *Technological Learning in the Energy Sector – Lessons for Policy, Industry and Science*, Edward Edgar, Cheltenham, UK.

Junginger, M., J. van Dam, S. Zarrilli, F. A. Mohammed, D. Marchal, and A. Faaij, 2011: Opportunities and barriers for international bioenergy trade. *Energy Policy*, 39(4): 2028–2042.

Jursa, R. and K. Rohrig, 2008: Short-term wind power forecasting using evolutionary algorithms for the automated specification of artificial intelligence models. *International Journal of Forecasting* 24(4): 694–709.

Justice, S., 2009: *Private Financing of Renewable Energy: A Guide for Policymakers*. Chatham House, UN Environment Programme, and Bloomberg New Energy Finance. London, UK.

Kagel, A., 2006: *A Handbook on the Externalities, Employment and Economics of Geothermal Energy*. Geothermal Energy Association, Washington, DC.

Kaltschmitt, M., 2000: Environmental effects of heat provision from geothermal energy in comparison to other resources of energy. In *Proceedings of World Geothermal Congress 2000*, 28 May–10 June, Kyushu-Tohoku, Japan, pp.803–808.

Kamm, B., P. R. Gruber, and M. Kamm, 2006: *Biorefineries – Industrial Processes and Products: Status Quo and Future Directions*, Volumes 1 and 2. WILEY-VCH, Weinheim, Germany.

Kammen, D., K. Kapadia, and M. Fripp, 2006: *Putting Renewables to Work: How Many Jobs Can the Clean Energy Industry Generate?* Renewable and Appropriate Energy Laboratory, University of California, Berkeley, April 2004 (corrected January 2006).

Karakezi, S., K. Lata, and S. T. Coelho, 2004: *Traditional Biomass Energy: Improving its Use and Moving to Modern Energy Use*. Thematic background paper for International Conference for Renewable Energies, Bonn, Germany.

Karnoe, P., 1993: *Approaches to Innovation in Modern Wind Energy Technology: Technology Practices, Science, Engineers and Craft Traditions*, Centre for Economic Policy Research, Stanford, CA, USA.

Kazmerski, L., 2010: *Compilation of Best Research Solar Cell Efficiencies (Revision Sept. 2010)*. National Renewable Energy Laboratory, Golden, CO, USA.

Kazmerski, L., 2011: *Best Research Solar Cell Efficiencies, 1976–2011*, compiled by L. L. Kazmerski, National Renewable Energy Laboratory, Golden, CO, USA.

Kempton, W., and J. Tomić, 2005a: Vehicle-to-grid power fundamentals: Calculating capacity and net revenue. *Journal of Power Sources* **144**: 268–279.

Kempton, W., and J. Tomić, 2005b: Vehicle-to-grid power implementation: From stabilizing the grid to supporting large-scale renewable energy. *Journal of Power Sources* **144**: 280–294.

Kerr, D., 2007: Marine Energy. *Phil. Trans. R. Soc. A* **365**: 971–992.

Khan, J., and G. S. Bhuyan, 2009: *Ocean Energy: Global Technology Development Status*, Report prepared by Powertech Labs for the IEA-OES.

Kim, L., 1991: Pros and cons of international technology transfer: A developing country view. In *Technology Transfer in International Business*. Oxford University Press, New York, NY, USA.

Kim, L., 1997: *Imitation to Innovation: The Dynamics of Korea's Technological Learning*. Harvard Business School Press, Boston, MA, USA.

Kleimaier, M., U. Buenger, F. Crotogino, C. Gatzen, W. Glaunsinger, S. Huebner, M. Koenemund, H. Landinger, T. Lebioda, W. Leonhard, D. U. Sauer, H. Weber, A. Wenzel, E. Wolf, W. Woyke, and S. Zunft, 2008: *Energy storage for improved operation of future energy supply systems*. Paper presented at CIGRE 42nd Biennial Session, Paris, France, 24–28 August.

Klessmann, C., P. Lamers, M. Ragwitz, and G. Resch, 2010: Design options for cooperation mechanisms under the new European renewable energy directive. *Energy Policy* **38**(8): 4679–4691.

Klimstra, J., and M. Hotakainen, 2011: *Smart Power Generation – The Future of Electricity Production*. Avain Publishers, Helsinki, Finland.

Kok, M., W. Vermeulen, A. Faaij, and D. de Jager (eds.), 2002: *Global Warming and Social Innovation*. Earthscan, London, UK.

Komoto, K., M. Ito, P. van der Vleuten, D. Faiman, and K. Kurokawa (eds.), 2009: *Energy from the Desert: Very Large Scale Photovoltaic Systems – Socioeconomic, Financial, Technical and Environmental Aspects*. Earthscan, London, UK.

Kosowski, G. M., M. Onischak, and S. P. Babu, 1984: Development of biomass gasification to produce substitute fuels. In *Proceedings of the 16th Biomass Thermochemical Conversion Contractors' Meeting*, 8–9 May, Pacific Northwest Laboratory, Richland, WA, USA, pp.39–59.

Kramer, G. J. and M. Haigh, 2009: No quick switch to low-carbon energy, *Nature* **462**: 568–569.

Kreutz, T. G., 2011: Prospects for producing low carbon transportation fuels from captured CO_2 in a climate constrained world. In *Proceedings of the 10th International Greenhouse Gas Control Technologies Conference (GHGT-10)*, September, Amsterdam, Netherlands.

Krewitt, W., K. Nienhaus, C. Klessmann, C. Capone, E. Sticker, W. Graus, M. Hoogwijk, N. Supersberger, U. Von Winterfeld, and S. Samadi, 2009a: *Roles and Potential of Renewable Energy and Energy Efficiency for Global Energy Supply*. Federal Environment Agency (Umweltbundesamt), Germany.

Krewitt, W., S. Teske, S. Simon, T. Pregger, W. Graus, E. Blomen, S. Schmid, and O. Schäfer, 2009b: Energy [r]evolution 2008 – A sustainable world energy perspective. *Energy Policy* **37**: 5764–5775.

Kreycik, C., T. D. Couture, K. S. Cory, 2011: Innovative Feed-In Tariff Designs that Limit Policy Costs, NREL Report No. TP-6A20–50225, 48 pp.

Kumar, A., T. Schei, A. Ahenkorah, R. Caceres Rodriquez, J.-M. Devernay, M. Freitas, D. Hall, Å. Killingtveit, Z. Liu, 2011: Hydropower. In *IPCC Special Report on Renewable Energy Sources and Climate Change Mitigation*, O. Edenhofer et al. (eds.), Cambridge University Press, Cambridge, UK, and New York, NY, USA.

Kurkela, E., 2008: *Biomass gasification technologies for advanced power systems and synfuels: Status and present R&D activities at VTT*. National Flame Days, Tampere, Finland, 23 January.

Kutscher, C. F. (ed.), 2007: *Tackling Climate Change in the U.S.—Potential Carbon Emission Reductions from Energy Efficiency and Renewable Energy by 2030*. American Solar Energy Society, Boulder, CO, USA.

Kutscher, C., 2009: *Concentrating Solar Power Commercial Application Study: Reducing Water Consumption of Concentrating Solar Power Electricity Generation*, Report to Congress, US Department of Energy, Washington, DC.

Lardon, L., A. Helia, B. Sialve, J-P. Steyer, and O. Bernard, 2009: Life-cycle assessment of biodiesel production from microalgae. *Environmental Science and Technology* **43**(17): 6475–6481.

Larson, E. D., R. H. Williams, and M. R. L. V. Leal, 2001: A review of biomass integrated-gasifier/gas turbine combined cycle technology and its application in sugarcane industries, with an analysis for Cuba. *Energy for Sustainable Development* **5**(1): 54–76.

Larson, E. D., G. Fiorese, G. Liu, R. H. Williams, T. G. Kreutz, and S. Consonni, 2010: Co-production of decarbonized synfuels and electricity from coal and biomass with CO_2 capture and storage: An Illinois case study. *Energy and Environmental Science* **3**(1): 28–42.

Lau, F. S., D. A. Bowen, R. Dihu, S. Doong, E. E. Hughes, R. Remick, R. Slimane, S. Q. Turn, and R. Zabransky, 2003: *Techno-economic Analysis of Hydrogen Production by Gasification of Biomass*. Final technical report for the period 15 September 2001–14 September 2002, US Department of Energy, Gas Technology Institute, Des Plaines, IL, USA.

Lau, F., 2005: *Integrated gasification combined cycles and other advanced concepts for biomass power generation*. 2nd Annual California Biomass Collaborative Forum, Sacramento, California, USA, 1 March.

Lazard, 2009: *Levelized Cost of Energy Analysis – Version 3.0 (February 2009)*. Lazard, New York, NY, USA.

Leaman, K. D., R. Molinari, and P. S. Vertes, 1987: Structure and Variability of the Florida Current at 27°N. *Journal of Physical Oceanography* **17**(5): 565–583.

Ledru, P., D. Bruhn, P. Calcagno, A. Genter, E. Huenges, M. Kaltschmitt, C. Karytsas, T. Kohl, L. Le Bel, A. Lokhorst, A. Manzella, and S. Thorhalsson,

2007: Enhanced geothermal innovative network for Europe: The state-of-the-art. *Geothermal Resources Council Bulletin*.

Lee, S. K., H. Chou, T. S. Ham, T. S. Lee, and J. D. Keasling, 2008: Metabolic engineering of microorganisms for biofuels production: from bugs to synthetic biology to fuels. *Current Opinion in Biotechnology* 19: 556–563.

Lehmann, H., and S. Peter, 2003: *Assessment of Roof & Façade Potentials for Solar Use in Europe*. Institute for Sustainable Solutions and Innovations, Aachen, Germany.

Lemming J. K., P. E. Morthorst, N. E. Clausen, and P. Hjuler Jensen, 2009: *Contribution to the Chapter on Wind Power in Energy Technology Perspectives 2008, IEA*. Risø National Laboratory, Roskilde, Denmark.

Leonard, W., U. Buenger, F. Crotogino, C. Gatzen, W. Glaunsinger, S. Huebner, M. Kleimaier, M. Koenemund, H. Landinger, T. Lebioda, D. U. Sauer, H. Weber, A. Wenzel, W. Wolf, W. Woyke, and S. Zunft, 2008: *Energy Storage in Power Supply Systems with a High Share of Renewable Energy Sources, Significance, State of the Art, Need for Action*. ETG Energy Storage Task Force, VDE Association for Electrical, Electronic, and Information Technologies, Frankfurt am Main, Germany.

Li, J., 2010: Decarbonizing power generation in China – is the answer blowing in the wind? *Renewable and Sustainable Energy Reviews* 14: 1154–1171.

Li, J., and L. Ma, 2009: *Background Paper: Chinese Renewables Status Report*. REN21, Paris, France.

Liu, G., E. D. Larson, R. H. Williams, T. G. Kreutz, and X. Guo, 2011: Making Fischer-Tropsch fuels and electricity from coal and biomass: Performance and cost analysis. *Energy & Fuels* 25(1): 415–437.

Lorenz, E., J. Hurka, D. Heinemann, and H.-G. Beyer, 2009: Irradiance Forecasting for the Power Prediction of Grid-Connected Photovoltaic Systems. *IEEE Journal of Selected Topics in Applied Earth Observations and Remote Sensing* 2: 2–10.

Lorenz, E., J. Hurka, D. Heinemann, and H.-G. Beyer*Los Angeles Times*, 2009: China, green? In the case of solar water heating, yes. 6 September.

Lucas Porta, H., 2009: *International Feed-in Cooperation – Mitigation through Renewable*, IDAE, Spain, presented at Side Event, Conference of the Parties, Copenhagen, Denmark.

Lund, H., and W. Clark, 2008: Sustainable energy and transportation systems introduction and overview. *Utilities Policy* 16(2): 59–62.

Lund, H., and W. Kempton, 2008: Integration of renewable energy into the transport and electricity sectors through V2G. *Energy Policy* 36: 3578–3587.

Lund, J. W., 2007: Characteristics, Development and Utilization of Geothermal Resources. *GHC Bulletin*, Geo-Heat Center, Oregon Institute of Technology, June 2007.

Lund, J. W. (ed.), 2005: Hitaveita Reykjavikur and the Nesjavellir geothermal co-generation power plant. *Quarterly Bulletin* 26(3): 19–24.

Lund, J. W. and T. Boyd, 2009: Oregon Institute of Technology geothermal uses and projects – Past, present and future. In *Proceedings of the Thirty-Fourth Workshop on Geothermal Reservoir Engineering*, 9–11 February, Stanford University, Stanford, CA, USA.

Lund, J. W., B. Sanner, L. Rybach, R. Curtis, and G. Hellström, 2003: Ground-source heat pumps – A world overview. *Renewable Energy World* 6(4): 218–227.

Lund, J. W., D. H. Freeston, and T. L. Boyd, 2010: Direct utilization of geothermal energy 2010 worldwide review. In *Proceedings of World Geothermal Congress 2010*, 25–30 April, Bali, Indonesia.

Lund, J. W., and R. Bertani, 2010: Worldwide Geothermal Utilization 2010, In *Proceedings of the Geothermal Resources Council Annual Meeting*, Davis, CA, USA (CD-ROM).

Luque, A., and S. Hegedus, 2003: *Handbook of Photovoltaic Science and Engineering*. John Wiley & Sons, Ltd., Chichester, UK.

Macedo, I. C., J. E. A. Seabra, and E. A. R. Silva, 2008: Greenhouse gases emissions in the production and use of ethanol from sugarcane in Brazil: The 2005/2006 averages and a prediction for 2020. *Biomass and Bioenergy* 32(7): 582–595.

Mackensen, R., K. Rohrig, and H. Emanuel, 2008: *Das regenerative Kombikraftwerk*. Abschlussbericht. ISET Kassel, Schmack Biogas AG.

Madaeni, S., R. Sioshansi, and P. Denholm, 2011: How thermal energy storage enhances the economic viability of concentrating solar power, submitted to *Proceedings of the IEEE*.

Makai Ocean Engineering, 2010: *SWAC – SeaWater Air Conditioning*, www.makai.com/e-swac.htm (accessed 7 November 2010).

Mangold, D., and T. Schmidt, 2006: The new central solar heating plants with seasonal storage in Germany, In *Proceedings of Eurosun 2006*, June 27–30, Glasgow, UK.

Mangold, D., 2007: Seasonal storage – A German success story. *Sun & Wind Energy* 1(2007): 48–58.

Matsuoka, S., J. Ferro, and P. Arruda, 2009: The Brazilian experience of sugarcane ethanol industry. *In Vitro Cellular & Developmental Biology – Plant* 45: 372–381.

McCall, M. J., A. Anumakonda, A. Bhattacharyya, and J. Kocal, 2008: Feed-flexible processing of oil-rich crops to jet fuel. In *2008 AIChE Spring National Meeting*, New Orleans, LA, USA, 6–10 April.

Medrano, M., A. Gil, I. Martorell, X. Potau, and L. F. Cabeza, 2010: State of the art on high-temperature thermal energy storage for power generation. Part 2 – Case studies. *Renewable and Sustainable Energy Reviews* 14: 56–72.

Mehos, M., D. Kabel, and P. Smithers, 2009: Planting the seed – Greening the grid with concentrating solar power. *IEEE Power & Energy* 7(3): 55–62.

Meier, A., and A. Steinfeld, 2010: Solar thermal production of fuels. *Advances in Science and Technology* 74: 303–312.

Meiwes, H., 2009: Technical and Economic Assessment of Storage Technologies for Power-Supply Grids, *Acta Polytechnica* 49(2–3): 34–30.

Microgrids, 2011: The EU projects 'Microgrids' and 'More Microgrids,' European Research Project Cluster "Integration of RES + DG." Brussels, Belgium. www.microgrids.eu (accessed 8 May 2010).

Milborrow, D., 2010: Annual power cost comparison: what a difference a year can make. *Windpower Monthly* 26: 41–47.

Minerals Management Service, 2007: *Worldwide Synthesis and Analysis of Existing Information Regarding Environmental Effects of Alternative Energy on the Outer Continental Shelf*. US Department of the Interior, Report MMS 2007–038. www.mms.gov/offshore/AlternativeEnergy/Studies.htm (accessed 8 November 2010).

Mitchell, C., 2008: *The Political Economy of Sustainable Energy*. Palgrave MacMillan, Basingstoke, UK.

Mitchell, C., D. Bauknecht, and P. M. Connor, 2006: Effectiveness through risk reduction: A comparison of the renewable obligation in England and Wales and the feed-in system in Germany, *Energy Policy* 34(3): 297–305.

Mitchell, C., J. Sawin, G.R. Pokharel, D. Kammen, Z. Wang, S. Fifita, M. Jaccard, O. Langniss, H. Lucas, A. Nadai, R. Trujillo Blanco, E. Usher, A. Verbruggen, R. Wüstenhagen, and K. Yamaguchi, 2011: Policy, Financing and Implementation. In *IPCC Special Report on Renewable Energy Sources and Climate Change Mitigation*, O. Edenhofer et al. (eds.), Cambridge University Press, Cambridge, UK, and New York, NY, USA.

Modi, V., S. McDade, D. Lallement, and J. Saghir, 2006: *Energy and the Millennium Development Goals*. Energy Sector Management Assistance Programme, United Nations Development Programme, UN Millennium Project, and World Bank, New York, NY, USA.

Moore, B., and R. Wüstenhagen, 2004: Innovative and sustainable energy technologies: The role of venture capital. *Business Strategy and the Environment* **13**: 235–245.

Mørk, G., S. Barstow, A. Kabuth, and M.T. Pontes, 2010: Assessing the global wave energy potential. In *Proceedings of the OMAE-2010*, 6–11 June 2010, Shanghai, China.

Morthorst, P. E. (ed.), 2007: Detailed Investigation of Electricity Market Rules. In *Further Developing Europe's Power Market for Large Scale Integration of Wind Power*. Tradewind Project, Intelligent Energy-Europe, Brussels, Belgium.

Mott MacDonald, 2010: *U.K. Electricity Generation Costs Update*. Commissioned by the U.K. Department of Energy and Climate Change, London, UK.

Musial, W. and B. Ram, 2010: *Large-Scale Offshore Wind Power in the United States – Assessment of Opportunities and Barriers*. National Renewable Energy Laboratory, Golden, CO, USA.

Mytelka, L. and P. Teixeira de Sousa, Jr., forthcoming: Ethanol in Brazil. In *Energy Technology Innovation: Learning from Success and Failure: 20 Case Studies of Energy Technology Innovation*. A. Grubler and C. Wilson (eds.), Cambridge University Press, Forthcoming.

NAS, 2009: *Liquid Transportation Fuels from Coal and Biomass: Technological Status, Costs, and Environmental Impacts*. America's Energy Future Panel on Alternative Liquid Transportation Fuels, National Academy of Sciences (NAS), Washington, DC, USA.

NEAA, 2009: *Meeting the 2 °C Target – From Climate Objective to Emission Reduction Measures*. Netherlands Environmental Assessment Agency (NEAA), Bilthoven, Netherlands.

NEDO, 2009: *Outline of the Roadmap PV2030+*. New Energy and Industrial Technology Development Organization (NEDO), Tokyo, Japan.

Neij, L., 2008: Cost development of future technologies for power generation – A study based on experience curves and complementary bottom-up assessments. *Energy Policy* **36**(6): 2200–2211.

NERC, 2008: *Accommodating High Levels of Variable Generation*. North American Electric Reliability Corporation (NERC), Princeton, NJ, USA.

Nielson, P., J. K. Lemming, P. E. Monthorst, H. Lawetz, E. A. James-Smith, N. E. Clausen, S. Strøm, J. Larsen, N. C. Bang, and H. H. Lindboe, 2010: *The Economics of Wind Turbines*. EMD International, Aalborg, Denmark.

Nihous, G. C., 2007: A Preliminary Assessment of Ocean Thermal Energy Conversion Resources. *Journal of Energy Resources Technology* **129**: 10–17.

Nikolaisen, L., C. Nielsen, M. G. Larsen, V. Nielsen, U. Zielke, J. K. Kristensen, and B. Holm-Christensen, 1998: *Straw for Energy Production*. Centre for Biomass Technology, Soeborg, Denmark.

Norden, 2006: *Underground Cold Storage in Buildings Can Save Energy*. Nordic Energy Research (Norden), Oslo, Norway.

Nordmann, T. and L. Clavadetscher, 2008: Reliability of grid-connected photovoltaic systems – the learning curve in yield and system cost. In *Proceedings of 23rd EU PV Solar Energy Conference*, Valencia, Spain, pp.3217–3221.

Norman, R. S., 1976: Water Salination – Source of Energy. *Science* **186**(4161): 350–352.

NRC, 2007: *Environmental Assessment of Wind-Energy Projects*. National Research Council (NRC), National Academy Press, Washington, DC.

NWCC, 2006: *Issue Forum Brief: Wind Power and Radar*. National Wind Coordinating Collaborative (NWCC), Washington, DC, USA.

Ockwell, D. G., J. Watson, G. MacKerron, P. Pal, and F. Yamin, 2008: Key policy considerations for facilitating low carbon technology transfer to developing countries. *Energy Policy* **36**(11): 4104–4155.

Owens, B., 2002: *An Economic Evaluation of a Geothermal Production Tax Credit*. Technical Report, National Renewable Energy Laboratory, Golden, CO, USA.

Pattle, R. E., 1954: Production of Electric Power by Mixing Fresh and Salt Water in the Hydroelectric Pile. *Nature* **174**(4431): 660.

Pehnt, M., 2006: Dynamic life-cycle assessment (LCA) of renewable energy technologies. *Renewable Energy* **31**: 55–71.

Pepermans, G., J. Driesen, D. Haeseldonckx, R. Belmans, and W. D'haeseleer, 2005: Distributed generation: definition, benefits and issues. *Energy Policy* **33**: 787–798.

Perlin, J., 1999: *From Space to Earth: The Story of Solar Electricity*. Aatec publications, Ann Arbor, MI, USA.

Pfeifer, C., and H. Hofbauer, 2008: Development of catalytic tar decomposition downstream from a dual fluidized bed biomass steam gasifier. *Powder Technology* **180**(1–2): 9–16.

Pienkos, P. T., 2009: Algal biofuels: Ponds and promises. In *13th Annual Symposium on Industrial and Fermentation Microbiology*, Lacrosse, WI, USA, 1 May.

Pimentel, D. and T. Patzek, 2005: Ethanol Production Using Corn, Switchgrass, and Wood; Biodiesel Production Using Soybean and Sunflower. *Natural Resources Research* **14**(1): 65–76.

Piwko, R., D. Clark, L. Freeman, G. Jordan, and N. Miller, 2010: *Western Wind and Solar Integration Study: Executive Summary*. National Renewable Energy Laboratory, Golden, CO, USA.

Post, J. W., 2009: *Blue Energy – Electricity Production from Salinity Gradients by Reverse Electro Dialysis*. PhD Thesis, Wageningen University, Netherlands.

Pradhan, A., D. S. Shrestha, A. McAloon, W. Yee, M. Haas, J. A. Duffield, and H. Shapouri, 2009: *Energy Lifecycle Assessment of Soybean Biodiesel*. Agricultural Economics Report 845, U.S. Department of Agriculture, Washington, DC, USA.

Pregger, T., D. Graf, W. Krewitt, C. Sattler, M. Roeb, and S. Moller, 2009: Prospects of solar thermal hydrogen production processes. *International Journal of Hydrogen Energy* **34**(10): 4256–4267.

Putsche, V., 1999: *Complete Process and Economic Model of Syngas Fermentation to Ethanol*. C Milestone Completion Report, National Renewable Energy Laboratory, Golden, CO, USA.

PV Cycle, undated: *Making Photovoltaics "Double Green."* www.pvcycle.org/index.php?id=4 (accessed 9 April 2011).

Rafferty, K., 2008: *An Information Survival Kit for the Prospective Geothermal Heat Pump Owner*. HeatSpring Energy, Cambridge, MA.

Ragnarsson, A., 2010: Geothermal development in Iceland 2005–2009. In *Proceedings of World Geothermal Congress 2010*, 25–30 April, Bali, Indonesia.

Raugei M., and V. M. Fthenakis, 2010: Cadmium flows and emissions from CdTe PV: Future expectations. *Energy Policy* **38**: 5223–5228.

Raventós, A., T. Simas, A. Moura, G. Harrison, C. Thomson, J.-D. Dhedin, 2010: *Life Cycle Assessment for Marine Renewables*. Wave Energy Centre, The University of Edinburgh, UK and European Development Fund (EDF), Commission of the European Communities, Brussels, Belgium.

Rawson, M., 2004: *Distributed Generation Costs and Benefits Issue Paper*. Staff paper 500–04–048. Public Interest Energy Research, California Energy Commission, Sacramento, CA, USA.

Ray, R. W., 2009: A review of the hot hydro market in Latin America. *Hydro Review Worldwide* 17(6): 26.

Reich, N. H., E. A. Alsema, W. G. J. H. M. van Sark, and W. C. Turkenburg, 2011: Greenhouse gas emissions associated with photovoltaic electricity from crystalline silicon modules under various energy supply options. *Progress in Photovoltaics: Research Applications* 19(3): 603–613.

Reif, T., 2008: *Economic Aspects of Geothermal District Heating and Power Generation. German Experience Transferable?* Presented at Innovaatilised Lahendused Energeetikas: Maasoojusenergia at Tallinn University of Technology, Tallinn, Estonia, 17 April 2009.

REN21, 2005: *Renewables Global Status Report – 2005*. Paris, France.

REN21, 2007: *Renewable Energy and the Climate Change Regime – Considerations from REN21 ahead of Bali COP13*. REN21 Secretariat, Paris, France.

REN21, 2008: *Renewables 2007 Global Status Report*. REN21 Secretariat and Worldwatch Institute. Paris, France and Washington, DC, USA.

REN21, 2009: *Renewables Global Status Report – 2009 Update*. REN21 Secretariat, Paris, France.

REN21, 2010: *Renewables 2010 – Global Status Report*. REN21 Secretariat, Paris. France.

REN21, 2011: *Renewables 2011 – Global Status Report*. REN21 Secretariat, Paris. France.

RFA, undated: *Ethanol Industry Statistics*. U.S. Renewable Fuel Association (RFA). www.ethanolrfa.org/pages/statistics (accessed 13 June 2010).

Rohrig, K., and B. Lange, 2008: Improving security of power system operation applying DG production forecasting tools. *IEEE PES General Meeting 2008*, Pittsburgh, PA, USA.

Rybach, L., 2010: Legal and regulatory environment favourable for geothermal development investors. In *Proceedings of the World Geothermal Congress 2010*, Bali, Indonesia, 25–30 April 2010.

Rybach, L., and M. Mongillo, 2006: Geothermal sustainability – A review with identified research needs. *Geothermal Resources Council Transactions*, 30: 1083–1090.

Salo, K., 2009: *Applications of bubbling fluidized bed gasification*. Centro de Investigaciones Energéticas Medioambientales y Tecnológicas (CIEMAT), Madrid, Spain, 12–13 November.

Salter, S, K. MacGregor, and C. Jones, 2006: *Scottish Energy Review – Scotland's Opportunity, Scotland's Challenge*, SNP, Scotland.

Sargent & Lundy, LLC, 2009: *Assessment of Parabolic Trough, Power Tower, and Dish Solar Technology Cost and Performance Forecast 2008*. Chicago, IL, USA.

Sawin, J., 2004a: *Mainstreaming Renewable Energy in the 21st Century*. Worldwatch Institute, Washington, DC, USA.

Sawin, J., 2004b : *National Policy Instruments – Policy Lessons for the Advancement & Diffusion of Renewable Energy Technologies Around the World*. Thematic Background Paper for Conference for Renewable Energies, Bonn, Germany.

Schaeffer, G. J., E. Alsema, A. Seebregts, L. W. M. Beurskens, H. H. C. de Moor, W. G. J. H. M. van Sark, M. Durstewitz, M. Perrin, P. Boulanger, H. Laukamp, and C. Zuccaro, 2004: *Learning from the Sun – Analysis of the Use of Experience Curves for Energy Policy Purposes: The Case of Photovoltaic Power. Final report of the Photex project*. ECN-C – 04–035, Energy Research Centre of the Netherlands, Petten, Netherlands.

Scheepers, M., D. Bauknecht, J. Jansen, J. de Joode, T. Gómez, D. Pudjianto, S. Ropenus, and G. Strbac, 2007: *Regulatory Improvements for Effective Integration of Distributed Generation into Electricity Distribution Networks – Summary of the DG-GRID project results*. Energy Research Centre of the Netherlands, Petten, Netherlands.

Schneider, M., A. Holzer, and V.H. Hoffmann, 2008: Understanding the CDM's contribution to technology transfer. *Energy Policy* 36(8): 2920–2928.

Scråmestø O.S., S.E. Skilhagen, and W.K. Nielsen, 2009: Power production based on Osmotic Pressure, *Waterpower* 16 (July): 10pp.

SEIA, 2011: *Utility-Scale Solar Projects in the United States Operational, Under Construction, and Under Development – Updated February 8, 2011*. Solar Energy Industries Association (SEIA), Washington, DC, and Cambridge, MA.

Seyboth, K., L. Beurskens, O. Langniss, and R. E. H. Sims, 2008: Recognising the potential for renewable energy heating and cooling. *Energy Policy* 36(7): 2460–2463.

Shapouri, H., J. A. Duffield, and M. Wang, 2002: *The Energy Balance of Corn-Ethanol, An Update*. Agricultural Economic Report 813. Office of Energy Policy and New Uses, U.S. Department of Agriculture, Washington, DC, USA.

Sharma, A., V. V. Tyagi, C. R. Chen, and D. Buddhi, 2009: Review on Thermal Energy Storage with Phase Change Materials and Application. *Renewable and Sustainable Energy Reviews* 13: 318–345.

Sheehan, J., T. Dunahay, J. Benemann, and P. Roessler, 1998a: *A Look Back at the U.S. Department of Energy's Aquatic Species Program – Biodiesel from Algae*. National Renewable Energy Laboratory, Golden, CO, USA.

Sheehan, J., V. Camobreco, J. Duffield, M, Graboski, and H. Shapouri, 1998b: *Life Cycle Inventory of Biodiesel and Petroleum Diesel for Use in an Urban Bus*. National Renewable Energy Laboratory, Golden, CO, USA.

Shell, 2008: *Shell Energy Scenarios to 2050*. Shell International BV, The Hague, Netherlands.

Shell, 2011: *Signals and Signposts – Shell Energy Scenarios to 2050*. Shell International BV, The Hague, Netherlands.

Sims, R. E. H., W. Mabee, J. N. Saddler, and M. Taylor, 2010: An Overview of Second Generation Biofuel Technologies. *Bioresource Technology* 101: 1570–1580.

Sims, R., P. Mercado, W. Krewitt, G. Bhuyan, D. Flynn, H. Holttinen, G. Jannuzzi, S. Khennas, Y. Liu, M. O'Malley, L. J. Nilsson, J. Ogden, K. Ogimoto, H. Outhred, Ø. Ulleberg, F. Van Hulle, 2011: Integration of Renewable Energy into Present and Future Energy Systems. In *IPCC Special Report on Renewable Energy Sources and Climate Change Mitigation*, O. Edenhofer et al. (eds.), Cambridge University Press, Cambridge, UK and New York, NY, USA.

Sinke, W. C., W. van Hooff, G. Coletti, B. Ehlen, G. Hahn, S. Reber, G. Beaucarne, J. John, E. van Kerschaver, M. de Wild-Scholten, and A. Metz, 2009: Wafer-based crystalline silicon PV modules at 1 € per watt-peak: Final results from the CrystalClear Integrated Project. In *Proceedings of the 24th European PV Solar Energy Conference*, Hamburg, 21–25 September, pp.845–856.

Sinke, W. C., 2009: Design Guidelines – Basics of Photovoltaics. In *Photovoltaics in the Urban Environment*. B. Gaiddon, H. Kaan and D. Munro (eds.), Earthscan, London, UK.

Sipila, K. and M. Rossi (eds.), 2002: Power Production from Waste and Biomass IV: Advanced Concepts and Technologies. In *Proceedings of VTT Symposium 222*, Technical Research Center of Finland, Espoo, Finland.

Smart Grids ETP, 2010: *Strategic Deployment Document for Europe's Electricity Networks of the Future*. European Technology Platform for the Electricity

Networks of the Future (ETP), Brussels, Belgium. www.smartgrids.eu (accessed 20 February 2011).

Solar Cookers International, 2009: http://www.solarcookers.org (accessed 21 January 2009 and 18 February 2011).

Solomon, B. D., J. R. Barnes, and K. E. Halvorsen, 2007: Grain and cellulosic ethanol: History, economics and energy policy. *Biomass and Bioenergy* **31**(6): 416–425.

Sonntag-O'Brien, V. and E. Usher, 2004: *Mobilising Finance for Renewable Energies*. Secretariat of the International Conference for Renewable Energies, Bonn, Germany.

Sørensen, H. C., and A. Weinstein, 2008: Ocean Energy – Position Paper for IPCC. In *Proceedings IPCC Scoping Conference on Renewable Energy*, Lübeck, Germany, 20–25 January, 8pp.

Spath, P. L., and D. C. Dayton, 2003: *Preliminary Screening Technical and Economic Assessment of Synthesis Gas to Fuels and Chemicals with Emphasis on the Potential for Biomass-derived Syngas*. National Renewable Energy Laboratory, Golden, CO, USA.

St. Denis, G., and P. Parker, 2009: Community energy planning in Canada: The role of renewable energy. *Renewable and Sustainable Energy Reviews* **13**(8): 2088–2095.

State of Hawaii, 2010: *Ocean Thermal Energy*. hawaii.gov/dbedt/info/energy/renewable/otec (accessed 15 January 2010).

Stefansson, V., 2000: The renewability of geothermal energy. In *Proceedings of the World Geothermal Congress 2000*, Kyushu-Tohoku, Japan, pp.883–888.

Stefansson, V., 2005: World geothermal assessment. In *Proceedings of the World Geothermal Congress 2005*, Antalya, Turkey, 24–29 April 2005.

Stern, N., 2009: *Meeting the Climate Challenge: Using Public Funds to Leverage Private Investment in Developing Countries*. Grantham Institute for Climate Change and the Environment, London School of Economics. London, UK.

Sterner, M., 2009: *Bioenergy and Renewable Power Methane in Integrated 100% Renewable Energy Systems*. Kassel University Press, Kassel, Germany.

Stoddard, L., J. Abiecunas, and R. O'Connell, 2006: *Economic, Energy, and Environmental Benefits of Concentrating Solar Power in California*. National Renewable Energy Laboratory, Golden, CO, USA.

Strange, D. L. P., T. P. Tung, G. W. Mills, A. Bartle, K. Goldsmith, F. Jenkin, L. P. Mikhailov, and A. A. Zolotov, 1994: Ocean Energy. In *New Renewable Energy Resources*. E. P. Volkov et al. (eds.), World Energy Council, Kogan Page Limited, London, UK, pp.321–358.

Strom, E., L. Liinanki, K. Sjostrom, E. Rensfelt, L. Waldheim, and W. Blackadder, 1984: Gasification of biomass in the MINO-process. *Bioenergy 84, Vol. III (Biomass Conversion)*, H. Egneus and A. Ellegard (eds.), Elsevier Applied Science Publishers, London, UK, pp.57–64.

Suding, P. and P. Lempp, 2007: The multifaceted institutional landscape and processes of international renewable energy policy. *IAEE Energy Forum*, **Second Quarter**: 4–9.

Šúri, M., 2006: Solar resource data and tools for an assessment of photovoltaic systems. In *PV Status Report 2006*. A. Jäger-Waldau (ed.), Office for Official Publications of the European Communities, Luxembourg.

Swanson, R. M., 2006: A vision for crystalline silicon photovoltaics. *Progress in Photovoltaics: Research Applications* **14**(5): 443–453.

Swedish Energy Agency, 2008: *Chrisgas: Fuels from Biomass*. Intermediate Project Report. www.chrisgas.com (accessed 18 May 2010).

Sydkraft, Elforsk, and Nutek, 1998: *Varnamo Demonstration Plant: Construction and Commissioning, 1991–1996*. Skogs Satteri AB, Trelleborg, Sweden.

Sydkraft, Elforsk, and Nutek, 2001: *Varnamo Demonstration Plant: The Demonstration Program, 1996–2000*. Berlings Skogs, Trelleborg, Sweden.

Taylor, R. M., 2008: The possible role and contribution of hydropower to the mitigation of climate change. *IPCC Scoping Meeting on Renewable Energy Sources Proceedings*, Lubeck, Germany, p.85.

Taylor, R. M., 2010: *Hydropower*. In Survey of Energy Resources. World Energy Council, London, UK, pp.287–290.

Tester, J. W., E. M. Drake, M. J. Driscoll, M. W. Golay, and W. A. Peters, 2005: *Sustainable Energy: Choosing Among Options*. The MIT Press, Cambridge, MA, USA.

Tester, J., and Panel Members, 2006: *The Future of Geothermal Energy – Impact of Enhanced Geothermal Systems (EGS) on the United States in the 21st Century*. Massachusetts Institute of Technology, Cambridge, MA, USA.

Thresher, R., M. Robinson, and P. Veers, 2008a: *Wind energy status and future wind engineering challenges*. National Renewable Energy Laboratory, Golden, CO, USA.

Thresher, R., M. Robinson, and P. Veers, 2008b: *Wind Energy Technology: Current Status and R&D Future*. National Renewable Energy Laboratory, Golden, CO, USA.

Tomić, J. and W. Kempton, 2007: Using fleets of electric-drive vehicles for grid support. *Journal of Power Sources* **168**: 459–468.

TPWind, 2008: *Strategic Research Agenda—Market Deployment Strategy from 2008 to 2030, Synopsis—Preliminary Discussion Document*, European Technology Platform for Wind Energy, TPWind Secretariat Brussels, Belgium.

Trieb, F. and H. Müller-Steinhagen, 2007: Europe-Middle East-North Africa cooperation for sustainable electricity and water. *Sustainability Science* **2**: 205–219.

Trouille, B. and C. R. Head, 2008: Introduction. *Hydro Finance Handbook*, HCI Publications, Kansas City, MO, USA, pp.3–4.

Turchi, C., 2010: *Parabolic Trough Reference Plant for Cost Modeling with the Solar Advisor Model (SAM)*, Technical Report. National Renewable Energy Laboratory, Golden, CO, USA.

Turchi, C., N. Langle, R. Bedilion, and C. Libby, 2011: *Solar-Augment Potential of U.S. Fossil-Fired Power Plants*, Technical Report. National Renewable Energy Laboratory, Golden, CO, USA.

Turkenburg, W. C., 2002: The innovation chain: Policies to promote energy innovations. In *Energy for Sustainable Development: A Policy Agenda*. T. B. Johansson and J. Goldemberg (eds.), UN Development Programme, New York, NY, USA, pp.137–172.

Turkenburg, W. C., J. Beurskens, A. Faaij, P. Fraenkel, I. Fridleifsson, E. Lysen, D. Mills, J. R. Moreira, L. J. Nilsson, A. Schaap, and W. C. Sinke, 2000: Renewable Energy Technologies. In *World Energy Assessment – Energy and the Challenge of Sustainability*. J. Goldemberg (ed.), UNDP and UN-DESA, New York, NY, USA, WEC, London, UK, pp.219–272.

UK-DTI, 2004 : *The World Offshore Renewable Energy Report 2004–2008*. UK Department of Trade and Industry (UK-DTI), London, UK.

UKERC, 2008: *UKERC Marine (Wave and Tidal Current) Renewable Energy Technology Roadmap*, UK Energy Research Centre (UKERC), London, UK and University of Edinburgh, Edinburgh, UK.

UKERC, 2010: *Great Expectations: The Costs of Offshore Wind in UK Wales – Understanding the Past and Projecting the Future*. UK Energy Research Centre (UKERC), London, UK.

Ulleberg, Ø., T. Nakken, and A. Eté, 2010: The wind/hydrogen demonstration system at Utsira in Norway: Evaluation of system performance using operational data and updated hydrogen energy system modeling tools. *International Journal of Hydrogen Energy* 35(5): 1841–1852.

Ummels, B. C., E. Pelgrum, and W. L. Kling, 2008: Integration of large-scale wind power and use of energy storage in the Netherlands. *IET Renewable Power Generation* 2(1): 34–46.

UN, 2008: *Acting on Climate Change: the UN System Delivering as One*. United Nations (UN), New York, NY, USA.

UN, 2009: *The Millennium Development Goals Report*. United Nations (UN), New York, NY, USA.

UNCTAD, 2008: *Biofuel Production Technologies: status, prospects and implications for trade and development*, prepared by E. D. Larson. United Nations Conference on Trade and Development (UNCTAD), Geneva, Switzerland.

UNCTAD, 2010: *World Investment Report*. United Nations Conference on Trade and Development (UNCTAD), Geneva, Switzerland.

UNDP, UN-DESA, and WEC, 2000: *World Energy Assessment – Energy and the Challenge of Sustainability*. United Nations Development Programme (UNDP) and United Nations Department of Economic and Social Affairs (UN-DESA), NY, New York, USA, and World Energy Council, London, UK.

UNDP, UN-DESA, and WEC, 2004: *World Energy Assessment – Energy and the Challenge of Sustainability: 2004 update*. United Nations Development Programme (UNDP) and United Nations Department of Economic and Social Affairs (UN-DESA), NY, New York, USA, and World Energy Council, London, UK.

UNEP, 2008a: *Public Finance Mechanisms to Mobilise Investment in Climate Change Mitigation*. United Nations Environment Programme (UNEP), Nairobi, Kenya.

UNEP, 2008b: *Green Jobs: Towards Decent Work in a Sustainable, Low-Carbon World*. United Nations Environment Programme (UNEP), Nairobi, Kenya.

UNEP, 2009: *Global Green New Deal – A Policy Brief*. United Nations Environment Programme (UNEP), Nairobi, Kenya.

UNEP, Undated: *Solar Water Heating Loan Facility in Tunisia and Indian Solar Loan Programme*. Energy Branch, Division of Technology, Industry, and Economics, United Nations Environment Programme (UNEP), Nairobi, Kenya. www.unep.fr/energy/activities/medrep/tunisia.htm and www.uneptie.org/energy/activities/islp (accessed 18 November 2010).

UNEP and BNEF, 2010: *Global Trends in Sustainable Energy Investment 2010: Analysis of Trends and Issues in the Financing of Renewable Energy and Energy Efficiency*. United Nations Environment Programme (UNEP), Nairobi, Kenya and Bloomberg New Energy Finance (BNEF), London, UK.

UNEP and BNEF, 2011: *Global Trends in Renewable Energy Investment 2011: Analysis of Trends and Issues in the Financing of Renewable Energy*. United Nations Environment Programme (UNEP), Nairobi, Kenya and Bloomberg New Energy Finance (BNEF), London, UK.

UNEP Risø, 2008: *Year end snapshot of the CDM*. CDM Pipeline. UNEP Risø Centre on Energy, Climate and Sustainable Development (URC), Roskilde, Denmark.

UNEP Risø, 2011: CDM/JI Pipeline Analysis and Database. UNEP Risø Centre on Energy, Climate and Sustainable Development (URC), Roskilde, Denmark. cdmpipeline.org/ (accessed 4 March 2011).

UNFCCC, 2007: *Investment and Financial Flows Relevant to the Development of an Effective and Appropriate International Response to Climate Change*. United Nations Framework Convention on Climate Change (UNFCCC), Bonn, Germany.

UNICA, undated: *Quotes and Stats*. Brazilian Sugarcane Industry Association (UNICA), english.unica.com.br/dadosCotacao/estatistica (accessed 8 August 2009).

University of Exeter and University of Sussex, undated: *Welcome to Energy Security in a Multipolar World*. www.exeter.ac.uk/energysecurity (accessed 8 November 2010).

US DOE, 2004: *Annual Energy Review 2003*. US Department of Energy (US DOE), Washington, DC, USA.

US DOE, 2007: *The potential Benefits of Distributed Generation and Rate-released Issues That May Impede their Expansion*. US Department of Energy (US DOE), Washington, DC, USA.

US DOE, 2008a: *20% Wind Energy by 2030: Increasing Wind Energy's Contribution to the U.S. Electricity Supply*. US Department of Energy (US DOE), Washington, DC, USA.

US DOE, 2008b: *Solar Energy Technologies Program Multi-Year Program Plan 2008–2012*. US Department of Energy (US DOE), Washington, DC, USA.

US DOE, 2009: *Marine & Hydrokinetic Technologies*. Office of Energy Efficiency and Renewable Energy, US Department of Energy (US DOE), Washington, DC, USA.

US DOE, 2010a: *Renewable Energy Annual, 2008 Edition*. Energy Information Administration, US Department of Energy (US DOE), Washington, DC, USA.

US DOE, 2010b: *Renewable Energy Futures Study*. US Department of Energy (US DOE), Washington, DC, USA.

US DOE, 2010c: *Biofuels, Biopower, and Bioproducts: Integrated Biorefineries*. Office of Energy Efficiency and Renewable Energy, US Department of Energy (US DOE), Washington, DC, USA.

US DOE, 2011a: *A National Offshore Wind Strategy: Creating an Offshore Wind Energy Industry in the United States*. Office of Energy Efficiency and Renewable Energy, US Department of Energy (US DOE) and the Bureau of Ocean Energy Management, Regulation, and Enforcement, US Department of the Interior, Washington, DC, USA.

US DOE, 2011b: *SunShot Initiative*. US Department of Energy (US DOE), Washington, DC, USA.

US DOE, 2011c: *Metering and Rate Arrangements for Grid-Connected Systems*. US Department of Energy (US DOE), Washington, DC, USA. www.energysavers.gov/your_home/electricity/index.cfm/mytopic=10600 (accessed 14 May 2011).

US EIA, 2011: *International Energy Statistics*. US Energy Information Administration (US EIA). tonto.eia.doe.gov/cfapps/ipdbproject/IEDIndex3.cfm# (accessed 22 March 2011).

Usher, E., 2008: Decarbonizing Energy: Are Financial Markets Taking the Lead? *Cogeneration & On Site Power Production*. 1 July. www.cospp.com/articles/print/volume-9/issue-4/perspective/decarbonizing-energy-are-financial-markets-taking-the-lead.html (accessed 9 February 2010).

Van den Ende, K., and F. Groeman, 2007: *Blue Energy*. KEMA, Arnhem, Netherlands. www.leonardo-energy.org/drupal/book/export/html/2243 (accessed 19 February 2010).

van den Wall Bake, J. D., M. Junginger, A. Faaij, T. Poot, and A. Walter, 2009: Explaining the experience curve: Cost reductions of Brazilian ethanol from sugarcane. *Biomass and Bioenergy* 33: 644–658.

van der Linden, N. C., M. A. Uyterlinde, C. Vrolijk, L. J. Nilsson, J. Khan, K. Åstrand, K. Erisson, and R. Wiser, 2005: *Review of International Experience with Renewable Energy Obligation Support Mechanisms*. Energy Research Centre of the Netherlands, Petten, Netherlands.

Van Hertem, D., and M. Ghandhari, 2010: Multi-terminal VSC HVDC for the European supergrid: Obstacles. *Renewable and Sustainable Energy Reviews* 14(9): 3156–3163.

Van Iersel, S., L. Gamba, A. Rossi, S. Alberici, B. Dehue, J. van de Staaij, and A. Flammini, 2009: *Algae-Based Biofuels: A Review of Challenges and Opportunities for Developing Countries*. Food and Agriculture Organization (FAO), Rome, Italy.

van Loo, S. and J. Koppjan (eds.), 2002: *Handbook Biomass Combustion and Co-firing*. Twente University Press, Enschede, Netherlands.

van Sark, W. G. J. H. M., E. A. Alsema, H. M. Junginger, H. H. C. de Moor, and G. J. Schaeffer, 2008: Accuracy of progress ratios determined from experience curves: the case of crystalline silicon photovoltaic module technology development. *Progress in Photovoltaics: Research Applications* 16(5): 441–453.

van Vliet, O. P. R., A. P. C. Faaij, and W. C. Turkenburg, 2009: Fischer-Tropsch diesel production in a well-to-wheel perspective: A carbon, energy flow and cost analysis. *Energy Conversion and Management* 54(4): 855–876.

van Wijk, A. J. M., and W. C. Turkenburg, 1992: Costs avoided by the use of wind energy in the Netherlands. *Electric Power Systems Research* 23: 201–216.

Van Wijk, A., 2010: *Hoe kook ik een ei – een frisse kijk op duurzame energie voor iedereen*. MGMC, the Netherlands, pp.39–41.

Vannoni, C., R. Battisti, and S. Drigo, 2008: *Potential for Solar Heat in Industrial Processes*. Solar Heating & Cooling Programme, International Energy Agency, Paris, France.

Vega, L. A., 2002: Ocean Thermal Energy Conversion Primer. *Marine Technology Society Journal* 6(4): 25–35.

Verbruggen, A., W. Moomaw, J. Nyboer, 2011: Annex I: Glossary, Acronyms, Chemical Symbols and Prefixes. In: *IPCC Special Report on Renewable Energy Sources and Climate Change Mitigation*. O. Edenhofer et al. (eds.), Cambridge University Press, Cambridge, UK, and New York, NY, USA.

Walker, A., D. Renne, S. Bilo, C. Kutscher, J. Burch, D. Balcomb, R. Judkoff, C. Warner, R.J. King, and P. Eiffert, 2003: Advances in solar buildings. *Transactions of the ASME* 125: 236–244.

Wang C. K., and W. Lu, 2009: *Analysis Methods and Reserves Evaluation of Ocean Energy Resources*. Ocean Publication, Beijing, China.

Watson, J., 2008: *Setting Priorities in Energy Innovation Policy: Lessons for the UK*. Discussion paper 2008–08. Belfer Center for Science and International Affairs, Cambridge, MA, USA.

WBGU, 2003: *World in Transition – Towards Sustainable Energy Systems*. German Advisory Council on Global Change (WGBU), Earthscan, London, UK.

WEC, 1994: *New Renewable Energy Resources: A Guide to the Future*. World Energy Council (WEC), Kogan Page, London, UK.

WEC, 2010: *2010 Survey of Energy Resources*. World Energy Council (WEC), London, UK.

Weiss, W. and F. Mauthner, 2010: *Solar Heat Worldwide – Markets and Contribution to the Energy Supply 2008*, 2010 edition. Solar Heating & Cooling Programme, International Energy Agency (IEA-SHC), Paris, France.

Western Governors' Association, 2006: *Solar Task Force Report*. Clean and Diversified Energy Initiative, Washington, DC, USA.

Weyer, K. M., D. R. Bush, A. Darzins, and B. D. Wilson, 2010: Theoretical Maximum Algal Oil Production. *BioEnergy Research* 3: 204–213.

Wiemken, E., 2009: *Market Review and Analysis of Small and Medium Sized Solar Air Conditioning Applications: Survey of Available Technical Solutions and Successful Running Systems – Cross Country Analysis*. Fraunhofer ISE, Munich, Germany.

Wildlife Society, 2007: *Impacts of Wind Energy Facilities on Wildlife and Wildlife Habitat*, Technical Review 07–2. Bethesda, MD, USA.

Williams, R. B., 2004: *Technology Assessment for Biomass Power Generation*. Draft final report. Department of Biological and Agricultural Engineering, University of California, Davis, CA, USA.

Willis, M., I. Masters, S. Thomas, R. Gallie, J. Loman, A. Cook, R. Ahmadian, R. Falconer, Binliang Lin, Guanghai Gao, M. Cross, N. Croft, A. Williams, M. Muhasilovic, I. Horsfall, R. Fidler, C. Wooldridge, I. Fryett, P. Evans, T. O'Doherty, D. O'Doherty, and A. Mason-Jones, 2010: Tidal turbine deployment in the Bristol Channel: a case study. *Energy* 163(3): 93–105.

WindPACT, 1999: *The Wind Partnerships for Advanced Component Technology: Various Projects on Advanced Wind Technology*. www.nrel.gov/wind/advanced_technology.html (accessed 2 July 2010).

WIREC, 2008: *Conference Report 2008*. Washington International Renewable Energy Conference (WIREC), Washington, DC, USA.

Wiser, R. and M. Bolinger, 2010: *2009 Wind Technologies Market Report*. Energy Efficiency and Renewable Energy, US Department of Energy (US DOE), Washington, DC, USA.

Wiser, R., Z. Yang, M. Hand, O. Hohmeyer, D. Infield, P.H.Jensen, V. Nikolaev, M. O'Malley, G. Sinden, and A. Zervos, 2011: Wind Energy. In *IPCC Special Report on Renewable Energy Sources and Climate Change Mitigation*. O. Edenhofer et al. (eds.), Cambridge University Press, Cambridge, UK and New York, NY, USA.

Wolsink, M., 2007: Planning of renewables schemes: Deliberative and fair decision-making on landscape issues instead of reproachful accusations of non-cooperation. *Energy Policy* 35: 2692–2704.

World Bank, 2009: *State and Trends of the Carbon Market 2009*. Washington, DC, USA.

World Bank, 2010: *Annual Report 2010*. Washington, DC, USA.

Worldwatch Institute, 2006: *Biofuels for Transportation – Global Potential and Implications for Sustainable Agriculture and Energy in the 21st Century*. Washington, DC, USA.

WorleyParsons Group, 2009: *Dry Cooling Option: Addendum to CSP Parabolic Trough Plant Cost Assessment*. National Renewable Energy Laboratory, Golden, CO, USA.

WWF and Ecofys, 2011: *The Energy Report – 100% Renewables by 2050*. Gland, Switzerland.

Zemke-White, L., and M. Ohno, 1999: World seaweed utilisation: An end-of-century summary. *Journal of Applied Phycology* 11: 369–376.

Zhang, Y-H. P., and L. R. Lynd, 2005: Cellulose utilization by clostridium thermo-cellum: bioenergetics and hydrolysis product assimilation. *Proceedings of the National Academy of Sciences* 102(20): 7321–7325.

12

Fossil Energy

Convening Lead Authors (CLA)
Eric D. Larson (Princeton University and Climate Central, USA)
Zheng Li (Tsinghua University, China)

Lead Author (LA)
Robert H. Williams (Princeton University, USA)

Contributing Authors (CA)
Theo H. Fleisch (BP America (retired), USA)
Guangjian Liu (North China Electric Power University)
George L. Nicolaides (Wildcat Venture Management, USA)
Xiangkun Ren (Shenhua Coal Liquefaction Research Center, China)

Review Editor
Peter McCabe (Commonwealth Scientific and Industrial Research Organization, Australia)

Contents

Executive Summary

Analysis in Chapter 12 shows that a radical transformation of the fossil energy landscape is feasible for simultaneously meeting the multiple sustainability goals of wider access to modern energy carriers, reduced air pollution health risks, enhanced energy security, and major greenhouse gas (GHG) emissions reductions.

Fossil fuels will dominate energy use for decades to come. Two findings apply to developing and industrialized countries alike. First, fossil fuels must be used judiciously – by designing energy systems for which the quality of energy *supply* is well matched to the quality of energy *service* required, and by exploiting other opportunities for realizing high efficiencies. Second, continued use of coal and other fossil fuels in a carbon-constrained world requires that carbon capture and storage (CCS) becomes a major carbon mitigation activity.

Since developing and industrialized countries have different energy priorities, strategies for fossil energy development will be different between these regions in the short term, but must converge in the long term. The focus in developing countries should be on increasing access to modern and clean energy carriers, building new manufacturing and energy infrastructures that anticipate the evolution to low carbon energy systems, and exploiting the rapid growth in these infrastructures to facilitate introduction of the advanced energy technologies needed to meet sustainability goals. Rapidly growing economies are good theaters for innovation. In industrialized countries, where energy infrastructures are largely already in place, a high priority should be overhauling existing coal power plant sites to add additional capabilities (such as coproduction of power and fuels) and CCS. (Simply switching from coal to natural gas power generation without CCS will not achieve the ultimately needed deep carbon emission reductions.)

Analysis in Chapter 12 highlights the essential technology-related requirements for a radical transformation of the fossil energy landscape: (i) continued enhancement of unit energy conversion efficiencies, (ii) successful commercial deployment of carbon capture and storage, (iii) co-utilization of fossil and renewable energy in the same facilities, and (iv) efficient coproduction of multiple energy carriers at the same facilities.

Among the fossil fuel-using technologies described in this chapter, only coproduction strategies using some biomass with the fossil fuel and with CCS have characteristics such that they can simultaneously address all four of the major energy-related societal challenges identified by GEA, as shown in Figure 12.1. It is plausible that these technologies could begin to be deployed in the relatively near term (2015–2020) because nearly all of the technology components of such systems are already in commercial use. Hydrogen made from fossil fuels with CCS is an energy option in the long term, but infrastructure challenges associated with hydrogen distribution and end use (especially for mobile applications) amplify the magnitude of the fossil energy challenge and are likely to limit hydrogen as an option in the near term. Other energy options may emerge in the post-2050 timeframe, and some ideas are touched upon briefly in this chapter.

The energy performance, cost, and GHG emissions of many of the power generation and coproduction technologies described in this chapter are summarized in Table 12.1 and Table 12.2. (Similar metrics for hydrogen production from fossil fuels and for smaller scale coproduction systems that coprocess biomass and coal or biomass and natural gas can be found in the main body of this chapter.)

Table 12.2 includes coal-biomass coprocessing systems with CCS that provide liquid fuels and electricity via coproduction. These technologies are attractive both as repowering and repurposing options for existing coal power plant sites and for greenfield projects. The economics of such systems depends on the greenhouse gas emissions price and the oil price, as discussed quantitatively in this chapter.

Clear benefits of this coproduction approach include:

- greatly reduced carbon emissions for electricity and transportation fuels;

- enhanced energy supply security;

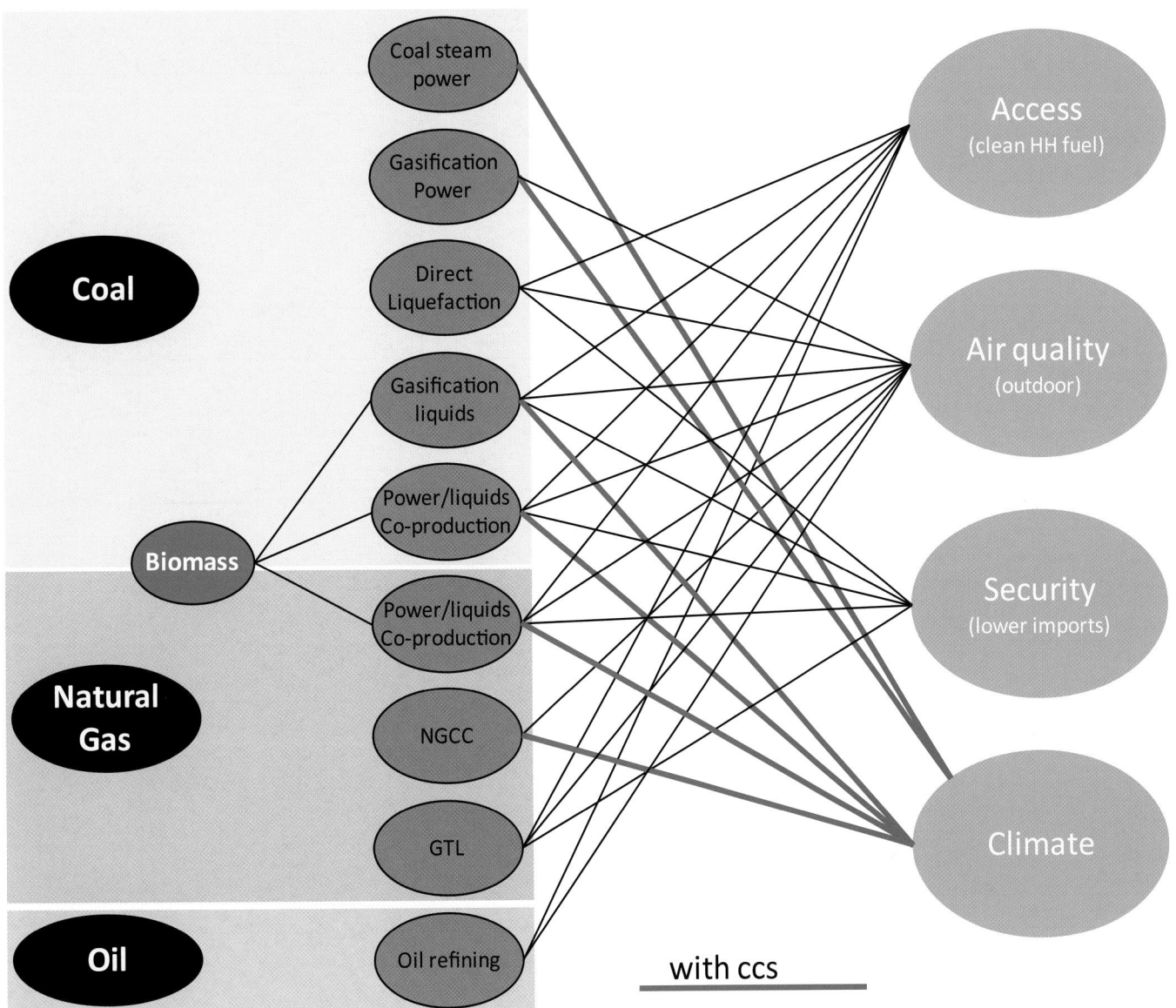

Figure 12.1 | Commercial or near-commercial fossil energy technologies discussed in this chapter and their suitability for addressing four major energy-related challenges. Among the listed technology options, only coproduction systems (that include coprocessing of some biomass) are capable of addressing all major challenges.

- provision of transportation fuels that are less polluting in terms of conventional air pollutants than petroleum-derived fuels;

- provision of super-clean synthetic cooking fuels such as liquefied petroleum gas (LPG) and dimethyl ether (DME) as alternatives to cooking with biomass and coal, which is critically important for developing countries; and

- greatly reduced severe health damage costs due to particulate matter ($PM_{2.5}$) air pollution from conventional coal power plants.

Coprocessing biomass with coal in these systems requires half or less biomass to provide low-carbon transport fuels as required for advanced fuels made only from biomass, such as cellulosic ethanol.

Table 12.1 | Performance and cost estimates in US$_{2007}$$ from this chapter for power generating technologies under US conditions. (To express costs in other-year US$, use the Chemical Engineering Plant Cost Index as shown in Figure 12.4 and discussed in text accompanying it.)

	Installed cost (US$/kW$_e$)	Capacity (MW$_e$)	Levelized electricity cost[a] (US$/ MWh)	Plant inputs (MW HHV)			Life cycle GHG[c] kgCO$_2$-eq per MWh
				Coal	Biomass[b]	Natural gas	
WITHOUT CO$_2$ capture and storage							
Sub-critical pulverized coal	1598	550	62	1496			896
Super-critical pulverized coal	1625	550	61	1405			831
Coal-IGCC (GEE radiant)	1865	640	68	1673			833
Coal-IGCC (Conoco-Phillips)	1788	623	65	1586			823
Coal-IGCC (Shell)	2076	636	72	1546			787
Coal-IGCC (GEE quench)	1901	528	69	1405			833
Biomass IGCC (Carbona)	2008	317	92		699		25
NGCC (F class GT)	572	560	51			1102	421
WITH CO$_2$ capture and storage							
Sub-critical pulverized coal	2987	550	114	2211			187
Super-critical pulverized coal	2961	546	111	2004			171
Coal-IGCC (GEE radiant)	2466	556	92	1709			138
Coal-IGCC (Conoco-Phillips)	2508	518	94	1634			162
Coal-IGCC (Shell)	2755	517	101	1616			136
Coal-IGCC (GEE quench)	2677	435	100	1405			126
Biomass IGCC (Carbona)	2779	259	129		699		−776
NGCC (F class GT)	1209	482	77			1102	110

a Assuming capacity factor of 0.85. Prices assumed (US$/GJ$_{HHV}$) for coal, biomass, and natural gas are US$2.04, US$5, and US$5.11, respectively.
b As-received biomass moisture content is 15% by weight.
c Includes GHG emissions associated with feedstock production and delivery to the power plant.

Coproduction also represents a promising approach for gaining early market experience with CCS, because CO$_2$ capture is easier in coproduction than for stand-alone power plants. In the near term, coproduction could serve as a bridge to enabling CCS as a routine activity for biomass energy, with corresponding negative greenhouse gas emissions, in the post-2030 era. Analysis in this chapter shows that this could plausibly become a major industrial activity under a carbon mitigation policy for economically poor but biomass-rich regions, where it could make clean cooking fuels widely available and affordable in the regions while making major contributions to decarbonization of the transport sector worldwide.

No technological breakthroughs are needed to get started with coproduction strategies, but there are formidable institutional hurdles created by the need to manage two disparate feedstocks (coal and biomass) and provide

Table 12.2 | Summary of performance and cost estimates in US$_{2007}$ from this chapter for alternative coproduction systems. Electricity is coproduced (to greater or lesser degree) with liquid fuels in all of these systems. FTL refers to Fischer-Tropsch liquids. MTG stands for methanol-to-gasoline. (To express costs in other-year US$, use the Chemical Engineering Plant Cost Index as shown in Figure 12.4 and discussed in text accompanying it.)

	Installed cost US$/(bbl$_{eq}$/d)[a]	Capacity (bbl$_{eq}$/d)[a]	O&M 10^{6}US$/yr	Plant Inputs[b]		Plant Outputs[b]				GHG Emission Index (GHGI)[d]
				Coal MW$_{HHV}$	Biomass[c] MW$_{HHV}$	Synthetic Diesel MW$_{LHV}$	Synthetic Gasoline MW$_{LHV}$	Synthetic LPG MW$_{LHV}$	Electricity MW$_e$	
WITHOUT CO$_2$ capture and storage										
Coal FTL	97,033	50,000	194	7559	9	2006	1153		404	1.71
Coal MTG	80,757	50,000	162	6549	0		2913	309	126	1.76
Coal FTL/ Power	122,958	35,706	176	7559		1431	825		1260	1.31
Coal MTG/ Power	126,167	32,579	164	6549			1898	202	959	1.37
Biomass FTL	160,189	4521	29		661	182	104		42	0.063
Biomass MTG	171,520	4630	32		661		270	28	32	0.066
WITH CO$_2$ capture and storage										
Coal FTL	98,372	50,000	197	7559		2006	1153		295	0.89
Coal MTG	82,099	50,000	164	6549			2913	309	36	0.97
Coal FTL/ Power	128,093	35,706	183	7559		1431	825		1058	0.70
Coal MTG/ Power	132,293	32,579	172	6549			1898	202	760	0.56
Biomass FTL	162,927	4521	29		661	181	105		31	−0.95
Biomass MTG	174,131	4630	32		661		270	28	20	−1.07
C+B FTL	139,091	9845	55	804	661	395	227		53	0.029
C+B MTG	129,200	10,476	54	781	661		610	69	11	0.018
C+B FTL/ Power	177,526	8036	57	1011	661	322	186		257	0.093
C+B MTG/ Power	180,110	11,582	83	1651	661		675	68	292	0.089

a bbl$_{eq}$/d is energy-equivalent barrels (LHV basis) per day of petroleum-derived fuels that could be replaced by the synthetic liquids.
b LHV is lower heating value and HHV is higher heating value.
c As-received biomass moisture content is 15% by weight.
d GHGI = system wide life cycle GHG emissions for production and consumption of the energy products divided by emissions from a reference system producing the same amount of liquid fuels and electricity. Here the reference system consists of equivalent crude oil-derived liquid fuels plus electricity from a stand-alone new supercritical pulverized coal power plant venting CO$_2$. See Table 12.15, note (c) for additional details.

simultaneously three products (liquid fuels, electricity, and CO$_2$) serving three different commodity markets. Creative public policies can help overcome these and other hurdles. Most importantly:

- Policy is urgently needed that sets a price on greenhouse gas emissions high enough to motivate CCS as a commercial activity.

- Stricter limits on air pollution are needed, especially from existing coal power plants and from indoor direct combustion sources. For the latter, policies should be designed to induce a shift, especially among the poor, from cooking by direct combustion of biomass or coal to using clean fluid fuels. Added costs that would result from stricter air pollution limits can be justified on the basis of the large reductions in public health damage costs that would follow.

- Incentives are urgently needed that specifically target integrated CCS demonstration projects at megascale. To minimize spending on such incentives, governments should aim to pursue projects from which maximum learning is derived per dollar spent. This would include multilateral financial support for these demonstrations, since all countries needing CCS technologies will benefit from these early projects if the learning is well documented and shared.

- Policies are needed that support early deployment of promising new technologies and systems at commercial scale, such as coproduction with CCS. Without incentives for first-of-a-kind projects that offer major public benefits, promising new technologies will enter the market slowly or not at all. Incentives should include ones that encourage new inter-industry partnerships where needed. It is desirable that policy instruments specify performance goals rather than specific technologies, and maximize use of market forces in meeting the goals.

- CO_2 storage prospects are not well known in many countries where sorely-needed clean liquid cooking fuels could be produced from coal or biomass while storing byproduct CO_2 underground. This is especially true for many biomass-rich but coal-poor countries. Detailed assessments of CO_2 storage prospects are needed on a reservoir-by-reservoir basis in these countries. Financial support for these assessments from the international community would be appropriate.

- International collaboration is needed to speed up the needed global energy transformation, including assistance from industrialized to developing countries for technological and institutional capacity development.

- New public policies are needed to facilitate industrial collaborations between companies producing transportation fuels, electricity, and clean cooking fuels and to encourage coprocessing of coal and biomass in regions having significant supplies of both (e.g., United States and China). It is desirable that policy instruments specify performance rather than technology and maximize use of market forces in meeting performance goals.

12.1   Introduction

In 2009, the world used 11,164 million tonnes of oil equivalent (Mtoe) or 469 exajoules (EJ) of commercial energy in total, nearly 90% of which was from fossil sources (BP, 2010). Due to advantages in cost, technological maturity and established industry and infrastructure, fossil energies are very likely to remain as a major component of world energy supply for several decades (especially coal-based power generation and liquid hydrocarbon fuels for transport), even as the world increasingly transitions to renewable energy technologies. At the same time, as discussed in earlier chapters, the world today faces four major challenges stemming from fossil energy use: a widespread lack of access to affordable modern energy carriers (Chapter 2), climate change (Chapter 3), air pollution (Chapter 4), and energy insecurity (Chapter 5). Given that continued use of fossil fuels is likely for at least the next several decades, how can they be used to address effectively these four challenges? This question frames the content of this chapter. Figure 12.2 shows the broad filtering criteria applied to focus the discussion in this chapter.

A technology "missing" from Figure 12.2 is combined heat and power (CHP). Large energy and environmental benefits can be achieved by replacing separate stand-alone power and heat production systems with CHP. Carbon emission reductions can be especially significant when stand-alone coal-fired systems are replaced with natural gas fired

CHP systems (Krause et al., 1994). We do not discuss CHP in this chapter in large part because the analysis presented on this topic in the World Energy Assessment (Williams, 2000) is still relevant today.

Hydrogen as a vehicle fuel is also not analyzed in this chapter. Technologies for fossil fuel conversion to hydrogen are described, but because hydrogen distribution and end-use infrastructural challenges associated with using it in vehicles likely would require at least several decades to overcome, the emphasis on transportation fuels in this chapter is on liquid fuels that can be made from hydrocarbon (fossil or biomass) resources.

In Chapter 12, power generation technologies are discussed in Section 12.2 with an emphasis on their ability to reduce carbon emissions. Section 12.3 discusses the possibilities for carbon mitigation in conventional petroleum refineries. Section 12.4 discusses alternative transportation fuel technologies that can ease energy security tension and also help reduce carbon emissions from transportation. Section 12.5 discusses roles of non-petroleum feedstocks for production of clean household fuels that can help to address the problem of the widespread lack of access to modern energy carriers. In Section 12.6, strategies for coproduction of electricity and fuels are discussed. These offer the prospects for comprehensive solutions to using fossil fuels efficiently, economically, and with low environmental impacts, both in retrofitting

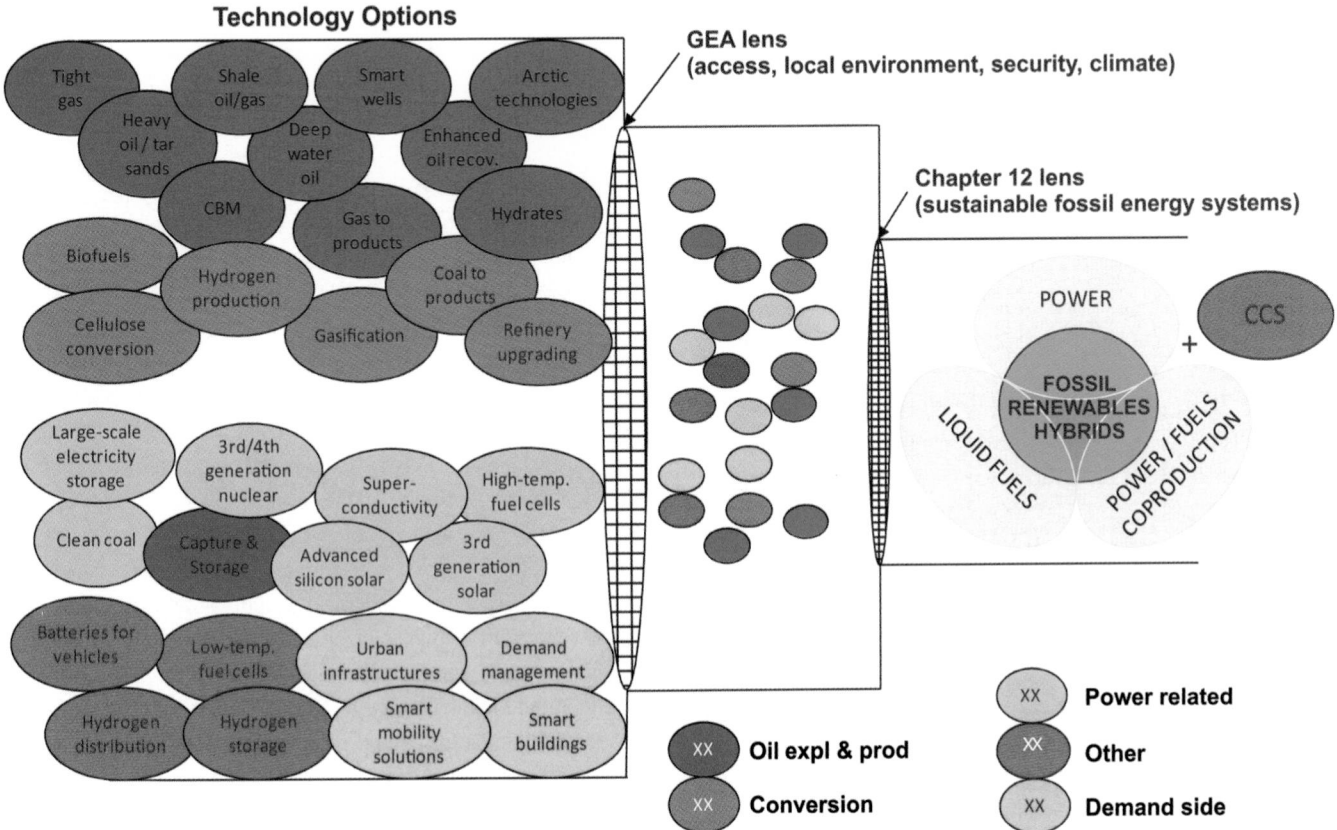

Figure 12.2 | Technology filters for this chapter.

existing energy facilities and in new installations needed to meet growing energy demands. Section 12.7 touches briefly on long-term technology options. Section 12.8 steps back from technology to consider strategic and policy issues.

12.2   Fossil Energy Technologies for Power Generation

For the foreseeable future, electricity will be one of the major energy carriers used by society. The problem lies in the large amount of fossil energy (and emissions) associated with electricity generation today to meet global demands.

Fossil fuels are the predominant primary energy at present in the world, accounting for nearly 90% of commercial energy use (BP, 2010). They are also the dominant fuel for power generation: producing about two thirds of our electricity today and projected to provide a similar fraction in 2035 (IEA, 2010). Today, fossil fuels are the most mature and economic source for power generation. However, they also account for most local conventional pollution and global carbon dioxide emissions. The future of fossil energy power generation in a carbon-constrained world depends on a compromise between growth in electricity demand and reduction in carbon dioxide emissions.

This section focuses on comparing the energy, environmental, and economic performance of fossil energy power generation technologies including coal-steam power, integrated coal gasification combined cycle (IGCC) and natural gas combined cycle (NGCC) with and without carbon capture and storage (CCS). The section also touches on other issues related to sustainable development, including water usage, health damage from environmental pollution, and co-utilization of coal with biomass, reflecting the intention of this section to explore alternatives for using fossil energy wisely and with lower carbon emissions.

12.2.1   Overview of Global Electricity Demand and Supply

The International Energy Agency (IEA) describes current and projected future electricity demand and supply (IEA, 2010). Here we cite relevant numbers from the IEA to give a general overview of fossil energy power generation.

12.2.1.1   Current Electricity Demand and Supply

Electricity is the "blood" of modern society that supports human prosperity. Electricity demand has invariably increased along with economic growth. In 2008, world end-use energy utilization was 8423 Mtoe (354 EJ) (IEA, 2010), including coal, oil, natural gas, electricity, heat, biomass, and waste. Oil usage ranked first with a fraction of 42%. Electricity was second, at 17%. This indicates the importance of electricity in modern society.

In 2008, 16,814 terawatt hours (TWh) of electricity was consumed in end uses globally but with a regional imbalance. Member countries of the Organisation for Economic Co-operation and Development (OECD), with 18% of world population,[1] consumed 9244 TWh (55% of the total). Non-OECD countries, with 82% of world population, consumed 7570 TWh (45%). Per capita electricity demand in non-OECD countries was about 1300 kilowatt hours (kWh) per person, or only one sixth of that in OECD countries. Today, some 1.4 billion people still have no electricity supply, some 85% of them in rural areas (IEA, 2010). No access to electricity means not only energy deprivation, but also a lower capacity for economic growth, which has deep and long-term impacts.

The largest electricity consumption, 3814 TWh (23% of the world total) among OECD countries is by the United States. Among developing countries, China has the largest electricity consumption, 2884 TWh (15% of world total). Since China has about 20% of the world's population, per capita electricity consumption there is still lower than the world average.

The two countries also lead in power production. In 2008, the total electricity generation globally was 20,183 TWh, of which 53% was in OECD countries. The United States and China were the largest power-generating countries in OECD and in the developing world, respectively, accounting for 22% and 17% of global power production. Power generation in Africa was only 621 TWh, or 3% of the global total.

Total global power generating capacity was 4719 gigawatts (GW) in 2008. The primary energy sources used for power generation by percentage were coal (41%), oil (5%), natural gas (21%), nuclear (14%), hydro (16%), and other renewable energy (3%). The share of fossil energy power capacity (coal, oil, and gas) is 67%.

The primary energy used for power generation differs by geographical region, depending on resource endowments as well as the state of economic and technological development. In general, the share of nuclear power is much higher in OECD countries (26% in 2008) than in non-OECD countries (5% in 2008). This may be due to the advantage of OECD countries in mastering nuclear technology as well as their economic power. Hydropower depends on resource availability. The share of hydropower is 40% in Latin America, for example. In countries with abundant coal, coal-steam power is the low-cost choice. In China, India and the United States, respectively, coal power accounts for about 79%, 69%, and 49% of power generation. Natural gas is the best feedstock among fossil fuels for power generation in the sense of energy efficiency and environmental pollution. However, its application depends either on resources or on economic power. This resource is available in Russia, for example, with 48% of its power from natural gas, and in the Middle East, with 58%. The impact of economic power is evident in the use of natural gas by OECD countries, with 22% of their power from natural gas, in contrast to non-OECD Asia, with 10%.

---

[1]   The total population of OECD countries was 1.18 billion in 2007. The total population in the world was 6.6 billion.

In China, because natural gas is rare and valuable, only 1% of power generation is from natural gas.

Globally, power generation is one of the major sources of CO_2 emissions, accounting for 11.9 gigatonne (Gt) in 2008, or 41% of the world's total fossil fuel CO_2 emissions. In many developing countries, however, conventional environmental pollution is considered a more urgent issue than CO_2 emissions because of the damage to the environment and human health it is causing today. China's emissions of sulfur oxides and nitrogen oxides (SO_x and NO_x) and dust were 25.9 million tonnes (Mt), 15.2 Mt, and 10.9 Mt in 2006. Contributions from power generation were 45%, 41%, and 29%, respectively (State Environmental Protection Administration of China, 2007). India's emissions of sulfur dioxide (SO_2), NO_x, and particulate matter less than 2.5 micrometers in diameter ($PM_{2.5}$) were 6.7 Mt, 4.1 Mt, and 4.7 Mt in 2005. The reduction of conventional pollutants is imperative and urgent for both countries.

Public health costs of air pollution (especially SO_x, NO_x, and PM) from fossil fuel power generation are discussed later in this section. A large amount of new power generation infrastructure is being established daily in developing countries, especially in countries such as China and India that are undergoing rapid industrialization and urbanization. It is of great importance for the sustainable development of societies like China's and India's to find ways to simultaneously address conventional pollutant emissions and carbon emissions.

12.2.1.2   Future Electricity Demand and Supply Expansion

In the "current policies" scenario of the World Energy Outlook 2010 (IEA, 2010), world electricity demand is projected to increase 95% from 2008 to 2035, reaching 32,919 TWh. Every region of the world is projected to increase, though at different rates. The increase in non-OECD consumption accounts for 82% of the total projected global increase. Even in this case, annual per capita electricity consumption in non-OECD countries only reaches 4600 kWh, about half the average for OECD countries (9200 kWh). In Africa, annual per capita electricity consumption is projected to increase only modestly, to 700 kWh per person.

In this scenario, which assumes a future world with essentially no price on carbon emissions, China is projected to have the largest increase in both total electricity consumption and annual per capita consumption among developing countries. China's electricity consumption is projected to be 9420 TWh in 2035, surpassing the United States as the world's largest consumer. The large projected increase in annual per capita consumption in China brings it to 6400 kWh, a level slightly under 70% of the OECD average.

In the current policies scenario, world power production is projected to increase by 90% from 2008 to 2035, reaching 38,423 TWh. Likewise, total installed power generation capacity is projected to be 8875 GW,

representing an 88% increase compared to 2008. The share of electricity by source is projected to change only modestly from 2008 to 2035: coal from 41% to 43%, oil from 5% to 2%, natural gas from 22% to 21%, nuclear from 14% to 11%, hydropower from 16% to 13%, and renewable energy (other than hydro) from 3% to 10%.

Correspondingly, global CO_2 emissions from fossil fuels are projected to increase by 46%, going from 29,260 Mt in 2008 to 42,589 Mt in 2035. CO_2 emissions from power generation are projected to increase by 59% to 18,931 Mt. Together, China and India account for 75% of the total projected global increase in power sector CO_2 emissions from 2008 to 2035. Power sector emissions in 2035 are projected to be 7130 Mt in China and 2068 Mt in India. Overall, emissions from coal power top the global list. Emissions from coal power are projected to be 14,403 Mt, accounting for 76% of all power generation emissions and one third of total global fossil fuel emissions.

12.2.2   Steam Electric Power Generation Using Pulverized Coal

12.2.2.1   Process description

At present, coal-steam power based on the Rankine cycle is the most commonly applied power generation technology worldwide. Utility coal boilers are generally divided into pulverized coal (PC) and circulating fluidized bed (CFB) units, which describes the method of combustion in the furnace. Since it is the predominant form of coal-steam power generation, PC combustion is the main focus of the discussion here. Some basic information about CFB is provided in Box 12.1, including coal-biomass cofiring, an option that is attracting increased attention with the drive toward greater use of renewable energy.

Figure 12.3 shows the process of a typical pulverized coal combustion power plant. Major equipment includes the boiler, steam turbine, and electric generator. The system can be simply described by following the fluid loops inside the process.

- Water and steam loop: This is the working fluid in the power plant. Cold water from the condenser is boiled and converted into superheated steam and sent to the steam turbine where it expands to generate mechanical rotating power. This mechanical power drives an electric generator to generate electricity, which is then transformed into high voltage and sent into the electricity grid. The steam exhausted from the turbine is cooled in the condenser and then sent back to the boiler to repeat the cycle.

- Air-flue gas loop: Air from the atmosphere is pumped into the boiler furnace to provide combustion air to burn the coal, the hot combustion products from which heat the water-steam loop. After releasing heat, the flue gas first goes through a selective catalytic reduction unit to get rid of NO_x, then to an electrostatic precipitator to reduce

Box 12.1 | Other Combustion Options

Circulating fluidized bed (CFB) combustion power generation technology was originally developed as a low-cost approach to sulfur control and to facilitate the use of low-quality coals. Currently, the main objective is to facilitate the use of low-quality coals.

In fluidized bed coal combustion, coal is crushed into pieces smaller than 10 mm and mixed with a large amount of fluidized bed material to burn inside the furnace. The typical bed temperature is 850°C, an appropriate level to minimize formation of thermal NO_x. Limestone is fed into the furnace along with coal in order to absorb sulfur dioxide formed during combustion. Up to 90% or more of the sulfur can be removed by simply adding limestone. This makes CFB a favorite lower-cost clean coal technology in developing countries. In China, for example, more than 2000 CFB boilers are in operation and the largest unit capacity is 300 MW. Units up to 600 MW (supercritical CFB boilers) are under development.

The energy efficiency of CFB units in general scores moderately lower than their PC counterparts for the same steam parameters due to slightly higher parasitic power consumption. With regard to carbon emissions, CFB units are somewhat less competitive because limestone used for in-situ desulphurization emits CO_2. Capital costs are slightly higher than for PC plants because of the requirements for more auxiliary facilities and anti-erosion refractory.

Cofiring coal and biomass provides a flexible method for using biomass, the supply of which may fluctuate seasonally. Cofiring coal and biomass reduces net carbon emissions compared to pure coal burning. Cofiring can also increase the efficiency of biomass use compared to a small scale power plant fired purely by biomass. Cofiring with coal is an efficient and effective way to use biomass and also to offset carbon emissions from coal power generation, at least until large-scale biomass use for fuel production becomes a reality. As for capital cost, there is one plant in China that has retrofitted an existing 140 MW PC boiler to utilize up to 20% (heat) biomass. The results also would be indicative for a CFB boiler. The incremental cost corresponding to 20% biomass power is much lower than the specific initial capital investment of a new biomass-fired power plant – and its efficiency is much higher.

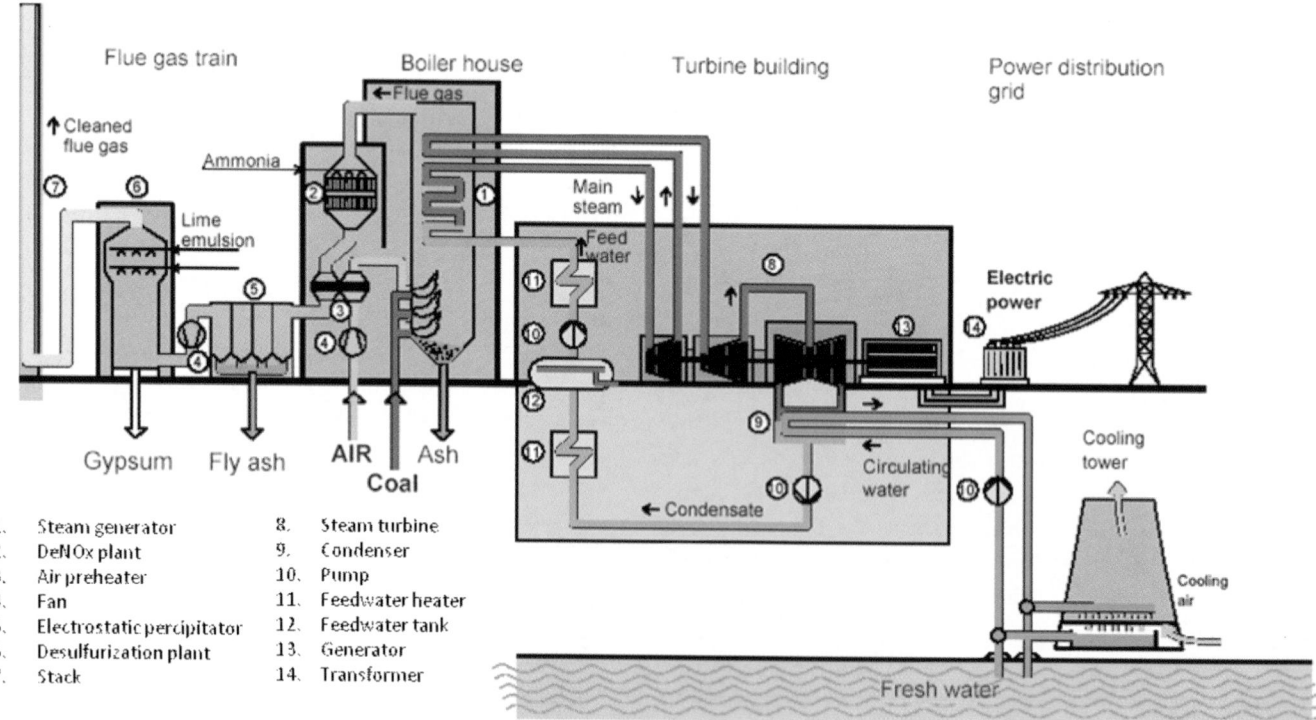

1. Steam generator
2. DeNOx plant
3. Air preheater
4. Fan
5. Electrostatic precipitator
6. Desulfurization plant
7. Stack
8. Steam turbine
9. Condenser
10. Pump
11. Feedwater heater
12. Feedwater tank
13. Generator
14. Transformer

Figure 12.3 | Pulverized coal combustion power plant. Source: Termuehlen and Empsperger, 2003.

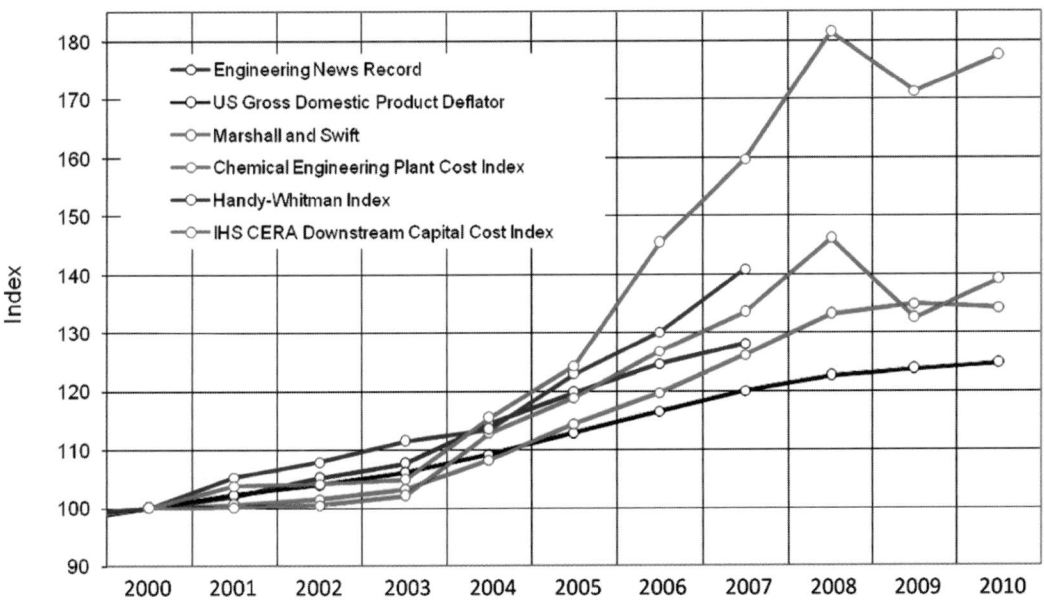

Figure 12.4 | Alternative cost indices, each normalized to 100 in year 2000. The Chemical Engineering Plant Cost Index is the most appropriate one to use for many of the energy supply systems discussed in this chapter. Source: updated from Kreutz et al., 2008.

dust, and then through flue gas desulphurization to get rid of SO_x. The cleaned flue gas returns to the atmosphere via the stack.

- Cooling water loop: Low temperature cooling water is sent to the condenser to condense the exhausted steam at the outlet of the steam turbine. Normally, the lower the temperature of the cooling water, the higher the efficiency of the power plant. Cooling water is supplied either from a water pond within the power plant or from the river or sea. In the first case, water is recycled and a cooling tower is required (as shown in Figure 12.3). Evaporation in the cooling tower causes a transfer of cooling water to the atmosphere, comprising the major water use for a power plant.

12.2.2.2   Efficiency and Steam Conditions

The net efficiency of a power plant is defined as the amount of electric power sent to the grid divided by the total energy input as coal. The efficiency depends mainly on the temperature and pressure of the steam as it enters the steam turbine. Coal-steam power plants can be classified as subcritical, supercritical, and ultra-supercritical, depending on the conditions of the steam entering the turbine. Steam parameters and typical corresponding efficiency levels are shown in Table 12.3.

To further improve the efficiency of pulverized coal power plants, development efforts are ongoing on advanced supercritical pulverized coal designs, with steam operating temperatures of 700°C. The main hurdles are related to development of high-temperature materials. The efficiency target for such plants is as high as 50% (Quinkertz, 2010). Gains in efficiency translate to reduced CO_2 emission/MWh. It is conceivable that this technology may become available before carbon capture and storage can be widely applied, because it represents an incremental improvement on a technology with which there is already considerable commercial experience. Thus, this technology may serve to help reduce carbon emissions from coal-fired power generation in the near term.

12.2.2.3   Construction Costs

There were sharp increases in construction costs for new power plants in the middle part of the past decade, particularly in OECD countries. The increases have been especially marked in the United States as evidenced by the much higher rate of increase than GDP of various cost indices relevant to the energy and power sectors (Figure 12.4). Table 12.4 provides a view from one power plant equipment manufacturer of the factors influencing cost trends. The main factors contributing to the increasing

Table 12.3 | Classification of coal-steam power plants according to steam conditions.

Classification of power plant	Temperature (C)	Pressure (MPa)	Efficiency (%)	Typical unit capacity (MW)
Subcritical	~540	16.7	38	300–600
Supercritical	~560	~25	40–42	600
Ultra-supercritical	>560	>25	42–45	600–1000

Table 12.4 | One power plant vendor's view of factors influencing power plant pricing.

	Price trend since 2003
Civil Engineering	
Construction Materials (cement, steel, etc.)	↗
Labour costs (local, international)	↗
Power Plant	
Main mechanical components (boiler, generator, turbine)	↑ (approx. +270%)
Other mechanical equipment (e.g. piping)	↑ (+150%)
Electrical assembly and wiring:	↑
-cables	(+150%)
-transformers	(+60%–90%)
Engineering and plant start up	↗
Additional factors	
Transportation and logistic costs	↗
Power plant demand	↗
Cost of capital and inflation	→

Note: Vertical arrow (↑) = large increase; slanted arrow (↗) = moderate increase; horizontal arrow (→) =no increase

Source: IEA, 2008a.

costs in OECD countries were (i) substantial increases in raw materials costs, e.g., a quadrupling in the cost of iron ore and a doubling in the price of steel between 2003 and 2008 (IEA, 2008a), (ii) higher demand for materials generally, (iii) higher crude oil prices, (iv) increases in labor costs due to shortages of engineering, construction, and procurement personnel, and (v) a weakening US dollar during the decade.

Cost escalations for other parts of the world may be different. As an example, Table 12.5 summarizes the actual capital investment for coal and natural gas power in China. The investment levels were quite stable during 2004–2007. The main reason for this is that the major component technologies are manufactured locally.

In this chapter, unless otherwise indicated, we have expressed costs in US$_{2007}$. We have used the Chemical Engineering Plant Cost Index (CEPCI) to adjust costs from other year values. As Figure 12.4 shows, using the CEPCI and 2007 as the reference year captures most of the escalation observed in the decade: the index for 2007 is approximately the same as for 2009 (i.e., no relative price increases) and only modestly lower than the preliminary value reported for 2010.[2]

12.2.2.4   Carbon Capture from Coal Steam Power Plant

Two approaches can be considered for capture of CO_2 from a steam electric plant. One way is to capture CO_2 from the flue gases of a conventional

plant (referred to as post-combustion capture). A second alternative is to use oxygen rather than air for the combustion ("oxyfuel" combustion). The advantage of the latter over post-combustion is the greater ease of CO_2 separation from the flue gases, which is accomplished by condensing out water from the combustion product mix of mainly CO_2 and water vapor. Offsetting this advantage are the significant added costs for an air separation plant and the fundamental power plant redesign required. (See Chapter 13 for additional discussion of this technology.)

Representative mass and energy balances of subcritical and supercritical coal power plants with and without post-combustion carbon capture are shown in Figures 12.5 and 12.6. Their energy, economic and environmental performances are shown in Table 12.6.[3, 4] The important messages revealed by comparing power plants without and with CCS are:

- The energy efficiency of both subcritical and supercritical power plants decrease by ~12 percentage points with the addition of carbon capture and storage. This is mainly due to steam consumption for regeneration of the solvent used to capture the CO_2 and mechanical power consumption for CO_2 compression.

- Water consumption more than doubles in both cases due to the large amount of heat required for regenerating the solvent used to capture the CO_2.

- The cost of electricity will nearly double (Table 12.6).

12.2.3   Gasification-based Power Generation

12.2.3.1   Technology

The first demonstration of gasification-based power generation at a significant scale dates to the early 1970s in Europe (163 MW plant in Lünen, Germany (Morehead and Hannemann, 2005)) and the mid-1980s in the United States (100 MWe Cool Water project (Alpert et al., 1987)). Since this time there has been considerable development of the technology, several commercial demonstration projects, and a few fully-commercial implementations in applications with low-cost waste feedstocks such

---

2   The GEA technical guidelines provide methodological assistance for readers who want to convert these numbers to alternative year prices, e.g. the 2005 numbers used in most other chapters.

3   In this table, and as much as possible elsewhere in this chapter, we show both lower heating value (LHV) and higher heating value (HHV) for fuels. We have chosen not to use only LHV (the convention in much of Europe) for several reasons: (i) LHV can be an ambiguous value in the case of biomass (and we include some analysis of biomass/coal systems in this chapter); (ii) the US convention for energy prices is HHV; (iii) using LHV implies that the latent heat of condensation from combustion of a fuel is not recoverable. In fact, it can, and is, recovered in some circumstances; and (iv) use of HHV in this chapter motivates a search for opportunities to capture this latent heat in other applications – and is thus a good index for a carbon-constrained world.

4   These performance and cost estimates (and some subsequent estimates in this chapter) are based on the work of Woods et al. (2007). At approximately the time the writing of this chapter was completed, an updating of the work of Woods et al. was published (Haslbeck et al., 2010). A comparison of estimates in the original study and the revised study reveals only modest differences in plant efficiencies and installed capital costs. For example, total installed plant costs ($/kW) in the revised study (in US$_{2007}$$) are lower by 1.5–5% for pulverized-coal plants and for natural gas combined cycle plants. Unit costs are 9–16% lower for coal integrated gasification combined cycle (CIGCC) plants.

Table 12.5 | Capital costs (nominal RMB/kW)[a] for thermal power plants in China.

Type / capacity	2004	2005	2006	2007
	Cost (2004 RMB/kW)	Cost (2005 RMB/kW)	Cost (2006 RMB/kW)	Cost (2007 RMB/kW)
Subcritical PC / 2x300 MW	4853	4596	4292	4401
Supercritical PC / 2x600 MW	4074	3919	3608	3643
Ultra supercritical PC / 2x1000 MW	4128	3924	3604	3724
NGCC / 2x300 MW, GE, 9F	3106	3060	3039	3155
NGCC / 2x180 MW, GE, 9E	3137	2946	2912	2998

a For reference, the exchange rate in mid-2008 was US$1 = RMB6.9.

Source: EPPDI and CPECG, 2008.

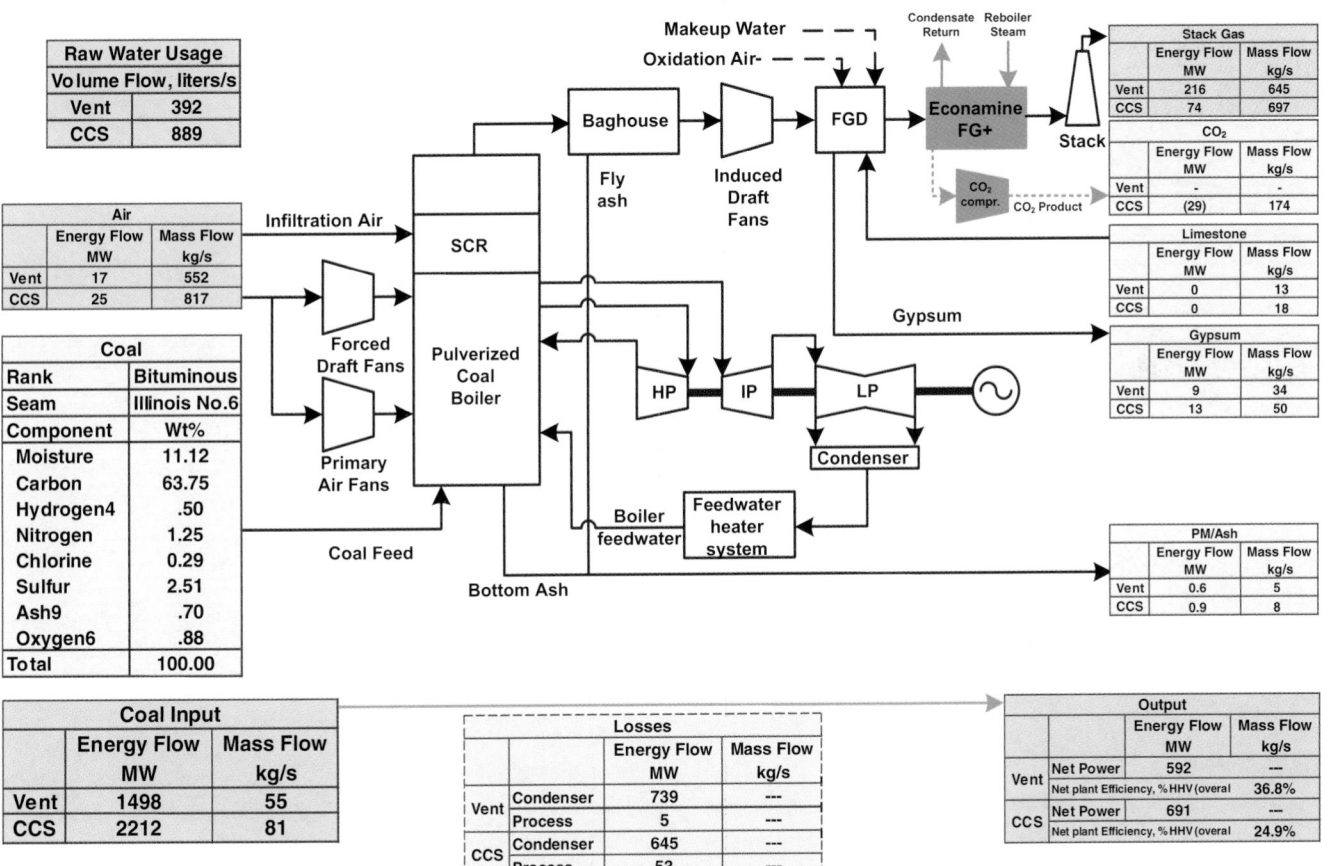

Figure 12.5 | Mass and energy balance of a typical subcritical power plant without and with carbon capture ('Vent' and 'CCS,' respectively). In the case with CCS, the net power output accounts for CO_2 compression to 153 bar, which is assumed to be sufficient to transport and inject the CO_2 into an underground storage reservoir. Source: based on Woods et al., 2007.

as petroleum residuals. Gasification-based power generation with fossil fuels, including coal and petroleum residuals, is thus commercial technology, but without widespread operating experience. A distinguishing feature of integrated gasification combined cycle (IGCC) technology is the very low emissions of conventional pollutants that can be achieved, especially SO_2 and particulates. Interest in IGCC has grown recently in part because of the prospectively lower cost of producing low-carbon electricity from coal compared to the cost with pulverized coal (PC) combustion technologies with post-combustion CO_2 capture.

Figure 12.7 is a simplified representation of a coal-IGCC system (CIGCC). Coal, water (or steam), and pure oxygen[5] are fed to a

5   The oxygen for gasification is produced in a dedicated air separation unit (ASU), with some or all of the air feed to the ASU coming from the compressor of the gas turbine. When 100% of the air used in the ASU originates from the gas turbine, the plant is regarded as fully integrated (and the name integrated gasification combined cycle derives from this feature) and has the highest efficiency in theory. In practice, however, full integration has proved operationally difficult. Most IGCC facilities today consider a maximum of 50% of the ASU air requirement being provided by the gas turbine.

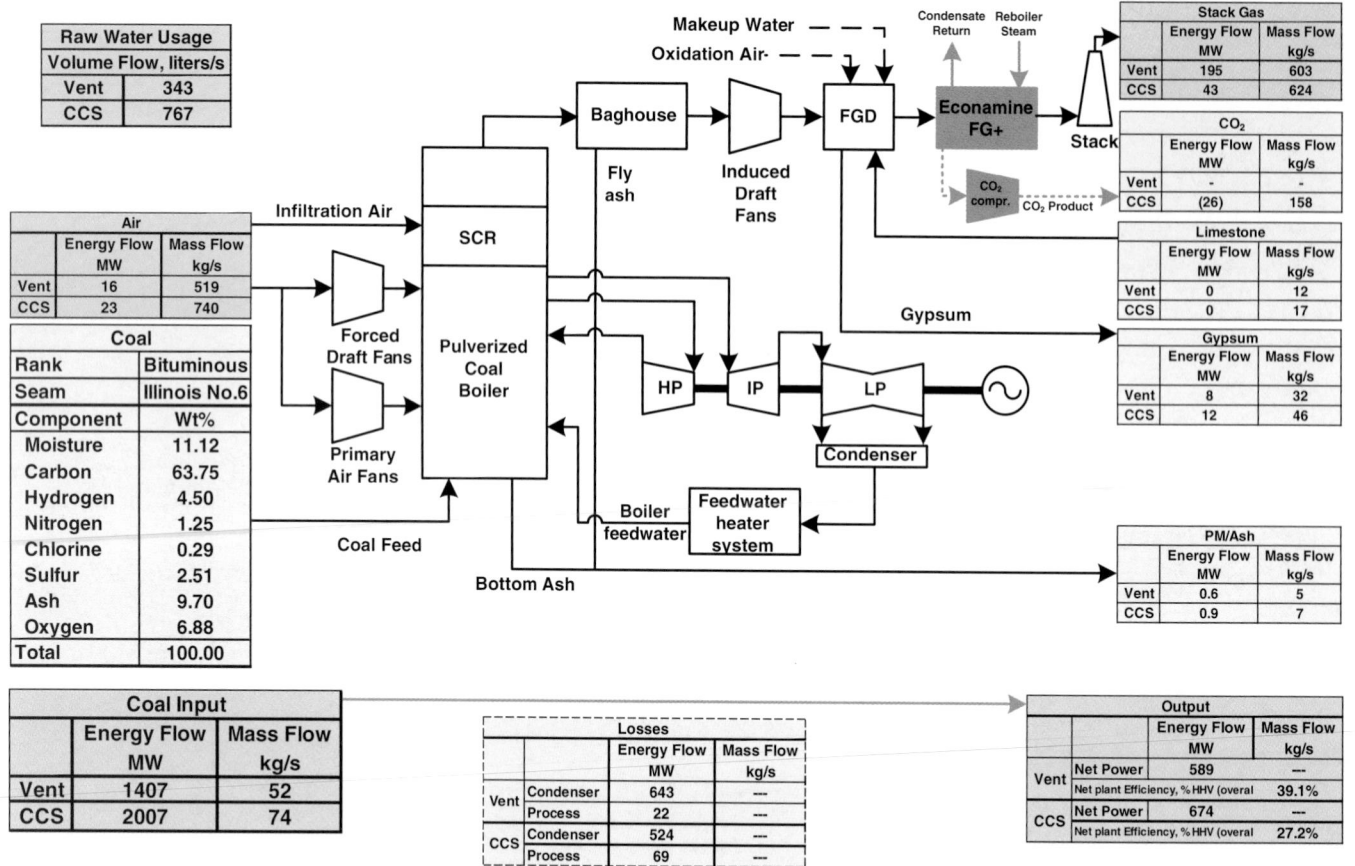

Figure 12.6 | Mass and energy balance of a typical supercritical power plant without and with carbon capture ('Vent' and 'CCS,' respectively). In the case with CCS, the net power output accounts for CO_2 compression to 153 bar, which is assumed to be sufficient to transport and inject the CO_2 into an underground storage reservoir. Source: based on Woods et al., 2007.

pressurized gasifier in which partial oxidation reactions occur, along with some water gas shift reaction. The resultant syngas, composed mainly of CO and H_2, is cooled and cleaned, including removal of essentially all particulate matter, which might otherwise create operating difficulties in downstream processes. Sulfur compounds in the syngas (primarily H_2S, with some COS (carbonyl sulfide)) are then removed, e.g., using the Selexol® physical absorption process, to ensure that air emissions limits are met. The syngas is then fed to a gas turbine designed to operate on a CO and H_2 mixture. The turbine generates power, and its high temperature exhaust passes to a heat recovery steam generator (HRSG) to generate superheated steam. The steam drives a steam turbine bottoming cycle to generate additional power.

If CO_2 is to be captured for storage, three additional steps are required (orange shading in Figure 12.7). First, the composition of the cleaned syngas is adjusted following gasification via the water gas shift reaction to primarily H_2 and CO_2. This is followed by capture of both H_2S and CO_2 in the acid gas removal step, and then by compression of the captured CO_2 to an elevated pressure for pipeline transport to a storage site for underground injection. Typically, CO_2 that would otherwise have been emitted to the atmosphere can be reduced by 90% using these additional steps.

Several different coal gasifier designs are offered by vendors. Two key features that differentiate designs are the type of feeding system (dry coal feed or coal-water slurry feed) and the syngas cooling strategy (radiant cooling or quench cooling). Dry-feed gasifiers are able to utilize a range of coal types, including lower-rank coals (lignite, sub-bituminous). Slurry-feed gasifiers cannot utilize lower-rank coals because the oxygen requirements for reaching the high temperature needed for effective gasification become prohibitive. Dry-feed gasifiers have higher efficiencies than slurry-feed gasifiers when both are operating on bituminous coal. On the other hand, slurry-feed gasifiers can operate at considerably higher pressures than dry-feed gasifiers, which can be advantageous in some circumstances. The use of radiant syngas cooling generally provides for higher overall plant efficiency than quench cooling, but also requires greater capital investment. For IGCC-CCS systems, the efficiency advantage with radiant cooling is offset to a significant degree by the added benefit with quench cooling of saturating the syngas with moisture, which avoids the need to raise steam for injection to the water gas shift (WGS) reactor.

Table 12.6 | Performance and cost estimates for PC power plants burning bituminous coal.

	Subcritical steam cycle		Supercritical steam cycle	
CO₂ vented or captured/stored →	Vent	CCS	Vent	CCS
As-received coal input, t/day	4,768	7,044	4,477	6,386
Coal input, MW HHV	1,496	2,211	1,405	2,004
Coal input, MW LHV	1,427	2,108	1,340	1,911
Gross electricity production, MW	583.3	679.9	580.3	663.4
On-site power consumption, MW	32.9	130.3	30.1	117.5
Net power generation, MW	550.4	549.6	550.2	546
Net generating efficiency (% HHV)	36.8	24.9	39.2	27.2
Life cycle GHG emissions,[a] kgCO₂eq/MWh$_{net}$	896	187	831	171
CO₂ emissions, kgCO₂/MWh$_{net}$	855	126	792	115
SO₂ emissions, gSO₂/MWh$_{net}$	357	negligible	335	negligible
NO$_x$ emissions, gNO$_x$/MWh$_{net}$	295	436	277	398
PM emissions, gPM/MWh$_{net}$	55	81	51	74
Hg emissions, gHg/MWh$_{net}$	0.005	0.007	0.005	0.007
Installed capital cost (US siting, US$_{2007}$$)				
Total plant cost (TPC), million US$_{2007}$$	880	1,642	894	1,617
Specific TCP, US$_{2007}$$/kWnet	1,598	2,987	1,625	2,961
Levelized generating cost, US$_{2007}$$/MWhb				
Capital (at 14.38% capital charge rate)	33.1	61.8	33.6	61.3
O&M (at 4% of TPC per yr)	8.6	16.0	8.7	15.9
Coal (at 2.04 US$/GJ, HHV)	19.9	29.5	18.7	26.9
CO₂ transportation and storage	0	7.0	0	6.4
Levelized cost of electricity, US$_{2007}$$/MWh	61.6	114.4	61.1	110.5

a Life cycle GHG emissions include CO₂ emissions at the plant plus emissions (expressed as CO₂-eq) that occur during coal mining and transportation.

b The following assumptions, representing US financial conditions, are adopted for all levelized cost calculations: coal price (US$/GJ$_{HHV}$) is 2.04; annual operating and maintenance (O&M) costs are 4% of total plant cost; the annual capital charge rate on TPI is 14.38%; TPI is total plant investment (TPC plus interest during construction); the capacity factor for power-only plants is 85%; the capacity factor for plants (described later) that produce liquid fuels or hydrogen (with or without coproduct electricity) is 90%. Also, costs for pipeline transportation and underground injection of compressed (150 bar) CO₂ are estimated based on work by Ogden, 2003; Ogden, 2004, as discussed in Liu et al., 2011a.

Source: Woods et al., 2007.

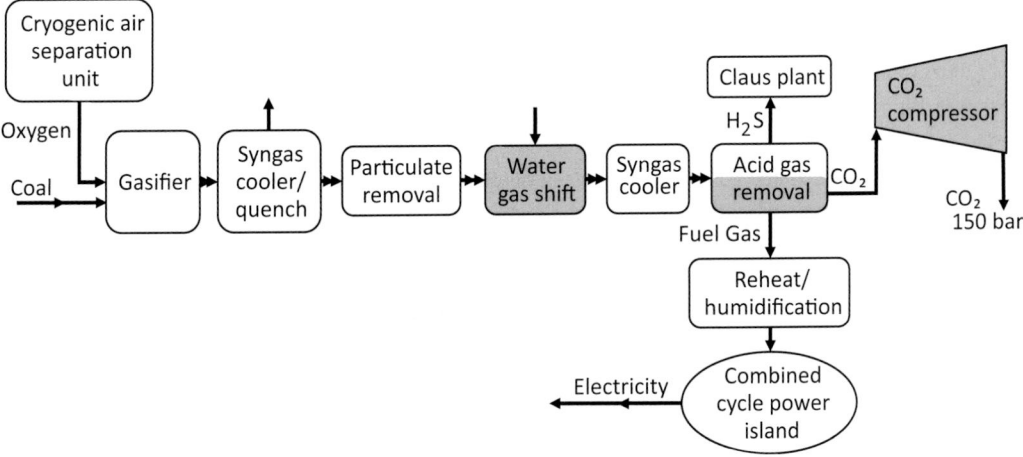

Figure 12.7 | Generic representation of coal-IGCC technology. Shaded blocks indicate units added for carbon capture and storage (CCS). When CCS is utilized, the acid gas removal unit is re-designed to capture CO₂ in addition to H₂S.

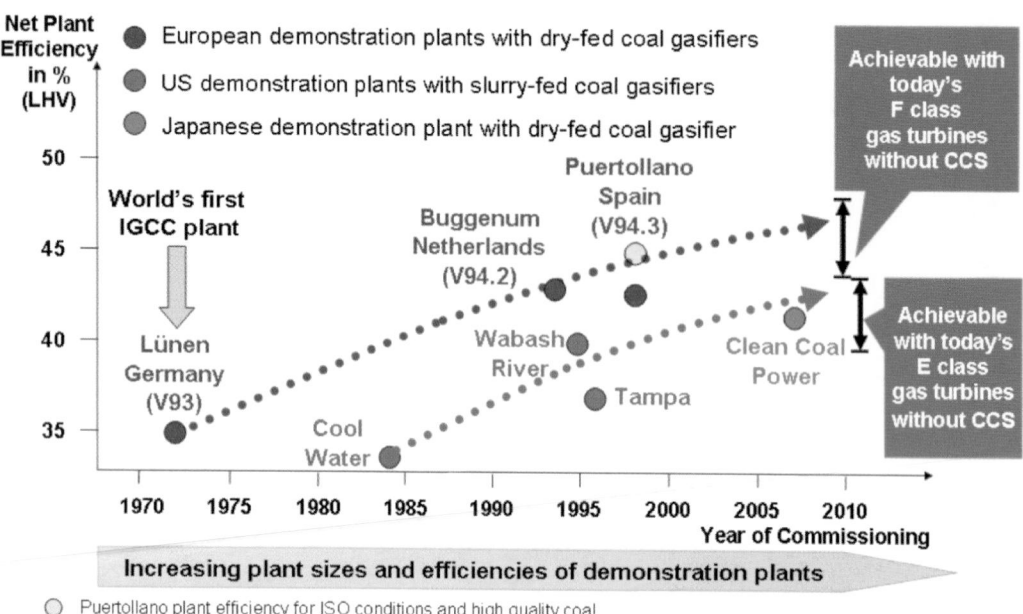

Net Plant Efficiency in % (LHV)

- European demonstration plants with dry-fed coal gasifiers
- US demonstration plants with slurry-fed coal gasifiers
- Japanese demonstration plant with dry-fed coal gasifier

○ Puertollano plant efficiency for ISO conditions and high quality coal

Figure 12.8 | Development of IGCC net plant efficiencies for coal-based plants without CCS. Source: Karg, 2009.

12.2.3.2 IGCC Energy and Environmental Performance

Figure 12.8 illustrates the rising trend in plant efficiencies for IGCC systems since the first plant was installed in the early 1970s. The figure also illustrates the generally higher efficiencies achievable using dry-feed gasifiers. There is a future potential for raising IGCC efficiencies through various technological means, e.g., by adapting the latest generation of gas turbines to operate on syngas, by adopting more efficient air separation technologies that are under development today, and others.

A detailed material and energy balance developed by engineers at the US National Energy Technology Laboratory (Woods et al., 2007) is shown in Figure 12.9 for a coal-IGCC system using a slurry-feed gasifier and venting CO_2 to the atmosphere. Figure 12.9 also shows a balance for a similar IGCC design, but with CO_2 capture and compression. Table 12.7 summarizes the performance of these two designs and shows estimates also from Woods et al. (2007) for two alternative IGCC designs using different gasifier technologies. With venting of CO_2, the electricity generating efficiencies for these three designs range from 38–41% (HHV basis). Capturing 90% of the generated CO_2 and compressing it for storage incurs a six to nine percentage point efficiency penalty depending on the design (or a 16–22% reduction in electricity output per unit of coal input). Emissions of conventional pollutants are low, especially SO_2 and PM, regardless of whether CO_2 is vented or captured.

Gasification-based power generation can also be utilized with biomass feedstocks. The right two columns in Table 12.7 show performance estimates for biomass-IGCC (BIGCC) systems made for this chapter by a team at the Princeton Environmental Institute (PEI) at Princeton

University. For consistency of comparisons, PEI estimates for two CIGCC systems are also shown.[6] Feedstock collection and delivery logistics for biomass use will generally limit BIGCC plant sizes. The BIGCC systems shown in Table 12.7 are designed for a biomass input capacity of about 3000 t/day dry biomass fed to a pressurized oxygen-blown fluidized-bed gasifier (such as the one originally developed at the Gas Technology Institute (GTI) in the 1980s, the license for which is now owned by Carbona).

With venting of CO_2, the BIGCC system efficiency exceeds that of a CIGCC system by several percentage points due primarily to three factors: reduced onsite electricity use for supplying oxygen (since biomass itself contains some oxygen and the gasification temperature with biomass is much lower than with coal), no requirement for sulfur removal (since most biomass has negligibly low sulfur content), and less nitrogen injection in the gas turbine for NO_x control due to the higher methane content of biomass-derived syngas (which yields lower combustion flame temperatures compared with coal-derived syngas).

A key benefit of utilizing biomass is the negative greenhouse gas emissions that can be achieved. Even with CCS, a CIGCC will still release some CO_2 to the atmosphere. When biomass is used, the CO_2 that is stored underground originated in the atmosphere, so net life cycle GHG emissions (including emissions associated with growing the biomass) are negative (Table 12.7).

---

6   The Princeton CIGCC estimates, which are based on designs with slurry-feed gasification and syngas quench cooling (rather than radiant cooling), are consistent with the estimates of the National Energy Technology Laboratory, US Department of Energy.

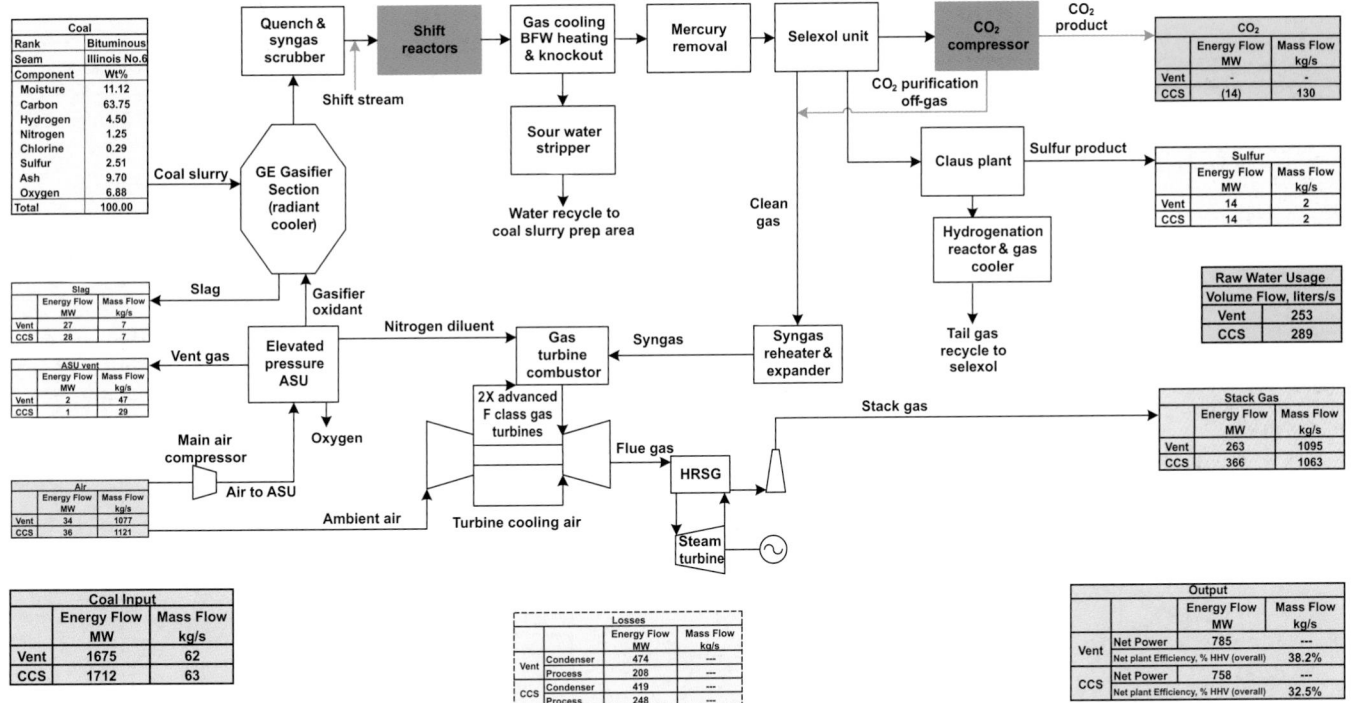

Figure 12.9 | Mass/energy balance for a coal-IGCC plant using slurry-feed gasifier, radiant syngas cooling without and with CO₂ capture ('Vent' and 'CCS,' respectively). In the CCS case, the net power output accounts for CO₂ compression to 153 bar, which is assumed to be sufficient to transport and inject the CO₂ into an underground storage reservoir. Source: based on Woods et al., 2007.

12.2.3.3   IGCC Economic Performance

The lack of a clear cost advantage for IGCC technology is an important reason that the technology has not yet seen widespread commercial deployment, despite having been first demonstrated commercially in the 1970s. While IGCC systems have clear environmental benefits over combustion systems for coal power generation, the least costly IGCC electricity (under US conditions, Table 12.7) is not less costly than electricity from a new supercritical-steam pulverized coal (PC) plant when CO₂ capture/storage is not considered (Table 12.6).

For CIGCC systems with CCS, specific capital costs (US$/kW) increase by 30–40% compared with IGCCs that vent CO₂ (Table 12.7), with correspondingly higher levelized costs of electricity generation. But the levelized cost of electricity (LCOE) generation for a CIGCC-CCS system is considerably lower than for a PC-CCS system, because supercritical pulverized coal plants require even higher incremental capital investment for CCS and sacrifice more efficiency points in the process. As shown in Figure 12.10, the LCOE for CIGCC-CCS reaches the LCOE for a pulverized coal plant with CO₂ venting (PC-V plant) at a much lower GHG emission price (US$55/tCO₂-eq) compared to that for the PC-CCS plant (US$75/tCO₂-eq). In the BIGCC-CCS case, although the LCOE at US$0/tCO₂-eq is much greater than for all the other options shown, its LCOE reaches that for a PC-V plant at a still lower GHG emission price (US$42/tCO₂-eq) because of the strong downward slope of the LCOE curve that arises from the negative GHG emissions for this option.

12.2.4   Natural Gas Combined Cycle Technology

Natural gas-fired combined cycles are among the cleanest and most efficient fossil fuel based power generating technologies. A natural gas combined cycle (NGCC) consists of a Brayton (gas turbine) cycle, the exhaust heat from which passes through a heat recovery steam generator to raise steam for a Rankine (steam turbine) bottoming cycle. The mass and energy balance shown in Figure 12.11 is for a NGCC with an overall efficiency of 50.8% on a HHV basis. With state of the art gas turbine technologies, NGCCs can reach efficiencies up to about 55% (HHV basis).

Since the hydrogen/carbon ratio of methane is relatively high, the volumetric CO₂ concentration in the flue gas of a NGCC is low, typically about 5%. The low partial pressure of CO₂ in flue gases requires use of an amine solvent for post-combustion CO₂ capture (Figure 12.11). (See also Chapter 13.) Regenerating the solvent requires considerable energy, leading to high efficiency penalties for NGCC systems with CCS. In the example case shown in Figure 12.11, the efficiency penalty is about seven percentage points. More detailed comparison data are shown in Table 12.8.

To reduce the energy penalty for CO₂ capture, some companies are developing CO₂ recycling processes that re-circulate part of the exhaust gas back to the inlet of the gas turbine compressor, resulting in a higher CO₂ concentration in the flue gases.

Table 12.7 | Estimates for coal or biomass fueled IGCC performance and costs in US$_{2007}$.

Source of estimate>>>	US National Energy Technology Laboratory (NETL)[a]						Princeton Environmental Inst. (PEI)[b]			
Feedstock>>>	Bituminous Coal						Bituminous Coal		Switchgrass biomass	
Gasifier technology >>>	GE Energy (slurry)		Conoco-Phillips (slurry)		Shell (dry)		GEE-Quench		GTI fluid bed	
CO$_2$ vented or captured>>>	Vent	CCS	Vent	CCS	Vent	CCS	Vent	CCS	Vent	CCS
Coal input rate										
As-received, metric t/day	5,330	5,447	5054	5206	4927	5151	4,477	4,477	0	0
Coal, MW HHV	1673	1709	1586	1634	1546	1616	1,405	1,405	0	0
Biomass input rate										
As-received, metric t/day	0	0	0	0	0	0	0	0	3,792	3,792
Biomass, MW HHV	0	0	0	0	0	0	0	0	699	699
Electricity output										
Gross production, MW	770	745	743	694	748	694	594	582	348	306
On-site consumption, MW	130	189	119	176	112	176	67	146	31	47
Net exports, MW	640	556	623	518	636	517	528	435	317	259
Efficiency (HHV)	38.3%	32.5%	39.3%	31.7%	41.1%	32.0%	37.5%	31.0%	45.3%	37.0%
Pollutants, grams per MWh										
SO$_2$	51.4	45.7	49.1	41.7	39.4	50.9	not estimated			
NO$_x$	222	223	234	243	220	236				
Particulate matter	28.9	34.1	28.1	34.6	26.7	34.7				
Carbon dioxide										
CO$_2$ vented, tonne/hr	508	51	489	60	477	46	418	34	224	15
CO$_2$ stored, tonne/hr	0	469	0	444	0	453	0	385	0	209
LC GHG emis., kgCO$_2$eq/ MWh	833	138	823	162	787	136	833	126	25	-776
Economics and metrics[c]										
Total plant cost (TPC), 10^{6}US$_{2007}$\$	1194	1370	1115	1300	1320	1424	1003	1166	636	720
Specific TCP, US$_{2007}$\$/kWe	1865	2466	1788	2508	2076	2755	1901	2677	2008	2779
Levelized cost of electricity (US$_{2007}$\$/MWh)[c]										
Capital (at 14.38% of TPI)	38.6	51	37.2	51.9	43.0	57.0	39.3	55.4	41.6	57.5
O&M (at 4% of TPC per yr)	10	13.2	9.6	13.5	11.2	14.8	10.2	14.4	10.8	14.9
Coal (at 2.04 US\$/GJ, HHV)	19.1	22.5	18.6	23.1	17.8	22.9	19.5	23.6	0	0
Biomass(at 5 US\$/GJ, HHV)	0	0	0	0	0	0	0	0	39.7	48.6
CO$_2$ transportation & storage	0	5.6	0	5.9	0	5.9	0.0	6.2	0.0	7.6
Cost of electricity, US2007\$/MWh	68	92	65	94	72	101	69	100	92	129

a Performance and capital costs from Woods et al., 2007. Capital costs escalated to January-mid US$_{2007}$\$ costs using the ratio of the 2007 average chemical engineering plant cost index to the January 2007 value (1.03). Levelized electricity costs calculated assuming financing parameters described in note (c) below.

b These are previously unpublished estimates that have been made for this chapter using the methodology and assumptions consistent with the work of Liu et al., 2011a. For cases utilizing biomass, the moisture content of the delivered biomass is 15%. No separate biomass drying step is included.

c See note (b) of Table 12.6 for financial parameter assumptions.

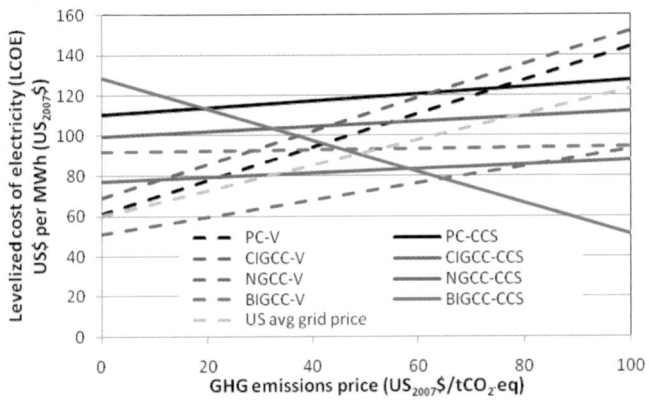

Figure 12.10 | Estimated levelized cost of electricity generation (LCOE) for alternative technologies as a function of GHG emissions price. (See Table 12.6, Table 12.7, and Table 12.8 for detailed cost assumptions.) The CIGCC and BIGCC systems shown here correspond to the PEI performance and cost estimates shown in Table 12.7.) Also shown is the 2007 average price paid to electricity generators in the United States (at zero GHG price) and how this price would change with GHG emissions price if the emissions are assumed to be the US grid-average for 2007 (636 kgCO$_2$-eq/MWh).

The minimum GHG emissions price needed to motivate adding CCS to a NGCC plant (US$83/tCO$_2$-eq) is even higher than for a PC plant, as a result of the much lower concentration of CO$_2$ in the flue gas and the smaller amount of CO$_2$ emitted per kWh of electricity. However, for the assumed fuel prices in Figure 12.10, the LCOE for NGCC-CCS at this GHG emissions price is lower than for any of the coal options considered.

12.2.5 Health Damage Costs of Power Plants

Air pollution (especially SO$_x$, NO$_x$, and particulate matter) from fossil fuel power generation imposes public health costs that are not reflected in power generating costs discussed to this point in this chapter. The health damage costs vary with the power generating fuel and technology, as well as with income level and demographics of the exposed populations (NRC, 2010). There are inherent and large uncertainties in estimating health damage costs, but a preponderance of published studies suggests that costs are significant in many cases, especially with coal-fired power generation

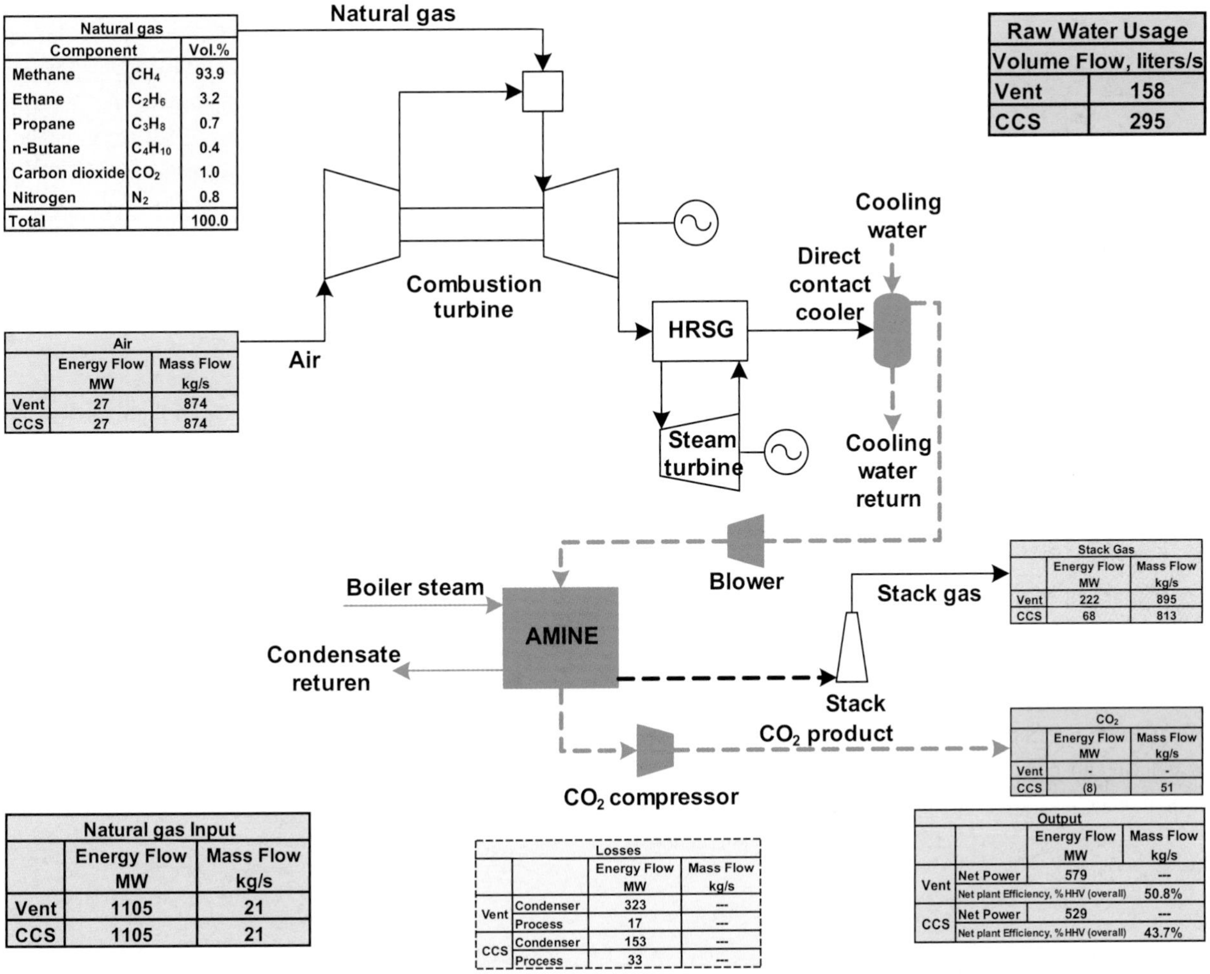

Figure 12.11 | Mass and energy balance of NGCC without CO$_2$ capture (when shaded components are excluded). Including shaded components, the schematic represents NGCC with post-combustion CO$_2$ capture. For the case with CCS, the net power output accounts for CO$_2$ compression to 153 bar, which is assumed to be sufficient to transport and inject the CO$_2$ into an underground storage reservoir. Source: based on Woods et al., 2007.

Table 12.8 | NGCC cost, performance, and environmental profile.

Advanced "F Class" gas turbine	NGCC (NETL)	
CO$_2$ vented or captured→	V	CCS
Natural gas input rate		
t/day	1,798	1,798
HHV, MW	1,102	1,102
Electricity output		
Gross production, MW	570.2	520.1
On-site consumption, MW	9.8	38.2
Net exports, MW	560.4	481.9
Efficiency (HHV)	50.8%	43.7%
Pollutants, grams/ MWh$_{net}$		
SO$_2$	-	-
NO$_x$	27.7	32.3
Carbon dioxide		
CO$_2$ vented, t/hr	203.3	20.3
CO$_2$ stored, t/hr	0.0	183.0
LC GHG emis., kg CO$_2$-eq/MWh$_{net}$	420.8	109.6
Capital costs		
Total plant cost (TPC), million US$_{2007}$\$	321	583
Specific TCP, US$_{2007}$\$/kWe	572	1,209
Levelized cost of electricity (US$_{2007}$\$/MWh)[a]		
Installed capital (at 14.38% of TPI)	11.8	25.0
O&M (at 4% of TPC per yr)	3.1	6.5
NG (at US\$5.11/GJ$_{HHV}$)	36.2	42.1
CO$_2$ transportation and storage	0.0	3.5
Total cost of electricity, US2007\$/MWh	51.1	77.1

a See note (b) of Table 12.6 for financial parameter assumptions.

Source: Woods et al., 2007.

located near population centers. We discuss here estimates of costs of damages due to air pollution in the United States, China, and Europe.

The environmental and cost impacts of air-pollution control strategies are explicitly considered in the future-pathways analysis presented in Chapter 17. Details of the pollutant emissions projections are discussed in Section 17.5.2, and Section 17.7 concludes that stringent policies for climate mitigation to decarbonize the energy system may bring the co-benefit of reducing air pollution control costs by up to 75% globally, with attendant reductions in health impacts and related damage costs of the type we discuss here. Also Chapter 4 addresses the interaction of energy and health in detail.

12.2.5.1   United States

Since publication of the pioneering Harvard School of Public Health study (Pope III et al., 1995), it has been known that the most significant

health hazards from air pollution are associated with small "PM$_{2.5}$" particles (particulate matter with particle diameters less than 2.5 microns). Inhaled, these small particles become lodged in and build up in the alveoli of the lungs and give rise to significant life-shortening. PM$_{2.5}$ particles are emitted directly by power plants and other combustion systems and also formed in the atmosphere from gaseous primary emissions of SO$_2$ and NO$_x$.

One study (Abt Associates Inc., 2000) estimated that over 30,000 premature deaths each year were attributable to fine particle pollution from power plants near the turn of the century in the United States. An analysis by the United States Environmental Protection Agency (US EPA, 1997) estimated an average life-shortening of 14 years for a person dying prematurely as a result of exposure to fine particle air pollution in the United States.

A recent Harvard School of Public Health study (Levy et al., 2009) has systematically assessed the public health damage costs for air pollution from 407 coal power plants in the United States that account for over 90% of US power plant SO$_2$ and direct PM$_{2.5}$ emissions and over 80% of NO$_x$ emissions. The most significant health damages are the result of premature deaths arising from PM$_{2.5}$ air pollution. The authors of this report estimated median costs of emissions (US\$/tonne) based on estimated dollar values of the health damages, as well as values at the 5th and 95th percentiles for uncertainty. In the left-most column in Table 12.9, the median estimates of health damages (US\$/tonne) from this study[7] are applied to average emission rates for US existing coal power plants in 2007. This table shows that SO$_2$ emissions account for over 70% of the total health damage costs.

At the median level of health damage cost estimates, the specific cost of health damages (US\$87/MWh) is 2.5 times the direct cost of electricity from a fully-depreciated coal plant (US\$35/MWh).[8] At the 5th and 95th percentiles of uncertainty (see Table 12.9, note d) the public health damage costs would amount to 1.3 times and 4.5 times the private cost of generation for written-off plants.

Coal power plant owners would like to keep old written-off units running as long as possible because they are so profitable – the average US selling price for electricity is about twice the generation cost for a written-off plant. However, these plants are profitable only in terms of private costs – not total societal costs. Taking into account

---

7   Another major study for the United States (NRC, 2010) that used a similar methodology as Levy et al. (2009) found considerably lower damage costs, highlighting the inherent uncertainties involved in externality costing. Nevertheless, the two studies can be reconciled largely by the fact that the National Research Council study uses a concentration-response function for premature mortality based on Pope et al. (2002), rather than on the more recent work by Schwartz et al. (2008).

8   This is based on average conditions for US coal power plants in 2009: HHV efficiency of 32.6% and an average coal price for power generators of \$2.08/GJ (EIA, 2010b) and an operation and maintenance cost of \$11.93/MWh for existing coal power plants as estimated in Simbeck and Roekpooritat (2009).

Table 12.9 | Estimated health damage costs (US_{2007}\$) from air pollution for alternative power plants.

	US average coal plant (2009)[a]	Subcritical PC[b]		Supercritical PC[b]		CIGCC[b]		NGCC[b]		XTL-OT-CCS[c]
		Vent	CCS	Vent	CCS	Vent	CCS	Vent	CCS	
Emission rates, kg/MWh										
SO_2	2.97	0.357	0.0	0.335	0.0	0.051	0.046	0.0	0.0	0.0
NO_x	1.70	0.295	0.436	0.277	0.398	0.222	0.223	0.028	0.032	0.164
$PM_{2.5}$	0.196	0.034	0.050	0.032	0.046	0.018	0.021	0.0	0.0	0.016
Health damage costs[d], US_{2007}\$ MWh										
SO_2	62.2	7.5	0.0	7.0	0.0	1.1	1.0	0.0	0.0	0.0
NO_x	9.0	1.6	2.3	1.5	2.1	1.2	1.2	0.2	0.2	0.9
$PM_{2.5}$	15.6	2.7	4.0	2.5	3.6	1.4	1.7	0	0	1.2
Total	87	12	6.3	11	5.7	3.7	3.8	0.2	0.2	2.1
Total relative to average US electricity generation price in 2009	149%	20.1%	10.8%	18.9%	9.8%	6.3%	6.5%	0.3%	0.3%	3.6%
HHV efficiency of power plant	32.6%	36.8%	24.9%	39.2%	27.2%	38.3%	32.5%	50.8%	43.7%	–

a   In 2009 total emissions of SO_2 and NO_x from coal power plants in the US power sector were 5.19 and 1.81 Mt, respectively (US EIA, 2010a) when generation from coal in the power sector totaled 1749 million MWh. For $PM_{2.5}$ it is assumed that in 2009 the $PM_{2.5}$ emission rate is based on the median 1999 estimate of 0.0413 lb per million BTU (17.8 gm/GJ) of coal input for the 407 coal plants investigated in Levy et al. (2009).

b   Emission rates for SO_2, NO_x, and PM from these new plants are from Woods et al. (2007). For PC plants the emission rates are based on BACT (best available control technologies), exceeding NPPS (new source performance standards) requirements. For CIGCC plants, the emission rates are based on the Electric Power Research Institute's Coal Fleet User Design Basis Specification. $PM_{2.5}$ emissions are assumed to be 0.62 x PM emissions, following Dockery and Pope (1994).

c   For XTL-OT-CCS plants emission rates for NO_x and $PM_{2.5}$ are assumed to be the same per MWh of gross gas turbine output as for CIGCC-CCS plants, but the SO_2 emission rate is assumed to be zero because protecting synthesis catalysts requires reducing the sulfurous compounds in syngas essentially to zero.

d   For 407 US coal power plants (which accounted in 1999 for over 90% of US power plant emissions of SO_2 and $PM_{2.5}$ and over 80% of US power plant emissions of NO_x) median estimates of health damage costs were found to be US\$20.9, US\$5.3, and US\$79.3 per kg for SO_2, NO_x, and $PM_{2.5}$, respectively, in 1999 (Levy et al., 2009). These valuations were assumed for all the alternative power plants in this table. At the 5th and 95th percentiles for uncertainty, the damage costs (in US_{2007}\$) were estimated to be, respectively, US\$11.0 and US\$35.3 per kg for SO_2, US\$2.0 and US\$9.4 per kg for NO_x, and US\$45.2 and US\$198.4/kg for $PM_{2.5}$. At the 5th and 95th percentiles for uncertainty, total health damage costs for US coal-fired power plants are US\$45.0/MWh and US\$159.5/MWh, respectively.

health damages from air pollution and considering the low estimate of health damage costs presented in Table 12.9 (5th percentile of uncertainty (see Table 12.9, note d)) the total cost of generation (private cost for a written-off plant + air pollution damage cost) would be 37% higher than the US average electricity generation price in 2009 (US\$58/MWh). This is the case without any cost assigned to greenhouse gas emissions.

Estimates of health damage costs from air pollution are also indicated for alternative new power plants in Table 12.9, based on emission rate estimates from Tables 12.7 and 12.8. This table shows that damage costs are extremely low for NGCC plants, still significant for new subcritical or supercritical PC-V plants, but quite low for all the plants with CCS. The XTL-CCS plants in the table refer to plants that co-produce electricity and Fischer-Tropsch Liquids (FTL) transportation fuels from X, where X is coal, biomass, or coal+biomass (as will be described in detail in later sections of this chapter). The XTL-CCS plants are characterized by damage costs that are half or less than the damage costs for all the coal stand-alone power plants with CCS. The emission for the PC-CCS plants are low because the SO_2 level in the flue gas has to be reduced to extremely low levels upstream of the

amine scrubber to prevent amine solvent degradation. Similarly, in the XTL-CCS case, protecting downstream process catalysts requires removal of sulfur.

12.2.5.2   China

Studies of health damage costs due to air pollution are more limited for China, and studies specific to pollution from electric power generation are even scarcer. Studies that have estimated public health costs of air pollution in China include those published by the World Bank (World Bank, 1997; World Bank and SEPA, 2007) and jointly by Harvard and Tsinghua Universities (Ho and Nielsen, 2007; Ho and Jorgenson, 2008).

The Harvard/Tsinghua study (Ho and Jorgenson, 2008) is the only one that has attempted to estimate health damage costs due to power generation in China. It estimated total air pollution damage costs (from all activities) in China using a willingness-to-pay approach (Hammitt and Zhou, 2006). The estimated total cost of premature deaths, chronic bronchitis, and asthma attacks was RMB_{2002}213 billion, or about

Table 12.10 | Estimated health damage costs in billion RMB$_{2003}$ (billion US$)[a] due to urban outdoor air pollution in China in 2003 (WB and SEPA, 2007).

	Excess deaths	Chronic bronchitis	Direct hospital costs	Indirect hospital costs	Total	% GDP
Adjusted human capital approach						
95th percentile	178.7 (25.5)	47.7 (6.8)	4.82 (0.69)	0.670 (0.096)	231.8 (33.1)	1.8
Average	110.9 (15.8)	42.5 (6.1)	3.41 (0.49)	0.470 (0.067)	157.3 (22.5)	1.2
5th percentile	35.8 (5.1)	36.9 (5.3)	1.88 (0.27)	0.264 (0.038)	74.9 (10.7)	0.57
Willingness-to-pay approach						
95th percentile	641.1 (91.6)	136.7 (19.5)	4.82 (0.69)	0.670 (0.096)	783.3 (111.9)	4.9
Average	394.0 (56.3)	122.1 (17.4	3.41 (0.49)	0.470 (0.067)	519.9 (74.3)	3.3
5th percentile	135.6 (19.4)	106.2 (15.2)	1.88 (0.27)	0.264 (0.038)	243.9 (34.8)	1.6

a Estimated costs in 2003 Yuan RMB have been converted to US$ using a nominal exchange rate of 7 RMB/US$.

US$30 billion at a nominal exchange rate of 7 RMB/US$. The study also estimated that 28% of the total damage cost, or about RMB60 billion (US$8.6 billion), was attributable to electric power generation. Assuming that most of the health damage was due to coal-fired generation, which accounted for about 81% of power generated in 2002 (1338 TWh) (National Bureau of Statistics China, 2007), the estimated health damage cost was about RMB0.044/kWh (US$0.006/kWh) of coal fired generation.

The Harvard/Tsinghua estimate of total health damage cost due to air pollution is on the low end of costs estimated in the joint study by the World Bank and China State Environmental Protection Administration. That study found the average health damage costs in 2003 to be equivalent to 1.2–3.3% of GDP (RMB157–520 billion, or US$23–76 billion, assuming 7 RMB/US$). The low estimate was based on a human-capital estimation approach and the high figure was based on a willingness-to-pay approach. Uncertainty ranges in the damage costs were also estimated (Table 12.10). If 28% of total damage costs are attributed to electric power generation (as indicated by the Harvard/Tsinghua study), and coal power generation is assumed to contribute most of these damages, then the range of damage costs per kWh of coal-fired power (1580 TWh in 2003) based on Table 12.10 is RMB0.047–0.14 / kWh (US$0.007–0.02/kWh).[9]

Taken together, the World Bank/SEPA and the Harvard/Tsinghua studies suggest that health damage costs from electric power generation in the middle of this decade were equivalent to between 10–40% of the cost of new coal power generation in China today. This is far lower than damage costs estimated for the United States, and is explained in part by the large difference in per capita income between the two

countries, which reduces individuals' willingness to pay to avoid health damages, as well as lower medical costs in China and fewer health impacts being included in the damage estimates for China as compared to the US estimates (US-China Joint Economic Research Group, 2007).

As per capita incomes rise in China, the willingness to pay to avoid health damages will rise proportionately; China's GDP is expected to more than quadruple between 2005 and 2030 (IEA, 2007). At the same time coal power generation is expected to increase substantially, and pollution controls are expected to tighten significantly in this period. Also, the geographical distribution of population in relation to power plants may be significantly different in 2030 from that at present. Studies taking into account trends for these several factors are needed to understand better prospective health damage costs per MWh of coal power generation in the 2030 time frame.

12.2.5.3   Europe

The ExternE Study in Europe was carried out over a period of more than 15 years beginning in the early 1990s. It involved a comprehensive and detailed assessment of air pollution and global warming damage costs from the lifecycle of energy use in Europe, estimated on the basis of the principle of willingness to pay to avoid damages (Rabl and Spadaro, 2000; Friedrich, 2005). Estimates were published in the late 1990s for 14 countries (CIEMAT, 1999) and additional estimates were published in 2004 for three countries in Eastern Europe (Melichar et al., 2004a). The estimates correspond to the technologies and demographics in these countries in the mid-to-late 1990s. A reassessment of damage costs today would yield different results, but it is difficult to guess whether damage costs would be lower, due to improved pollution controls, or higher, due to higher population densities and higher per capita incomes. (The latter translates into higher levels of willingness to pay to avoid damages.)

The original ExternE estimates indicate that health damage costs from conventional power generation in Europe are not inconsequential. Table 12.11

---

9   Another study, involving some of the same authors as the Harvard/Tsinghua study, estimated health damage costs from SO_2 emissions in the power sector to be RMB6555/t (US-China Joint Economic Research Group, 2007). The average SO_2 emissions rate from coal-fired power plants in China in 2005 was 5.4 g/kWh (based on total SO_2 emission from power sector of 11.12 million tonnes (Gao et al., 2009) and 2,047 TWh of thermal power generation (National Bureau of Statistics China, 2007). This gives a much higher damage cost estimate of RMB0.36/kWh (US$0.051/ kWh using a nominal exchange rate of 7 RMB/US$).

Table 12.11 | Estimated health damage costs of power generation in 17 European countries, as of mid/late 1990s.

Country[a]	Health damage costs (US_{2007}\$ per MWh)							
	COAL				NATURAL GAS			
	Power Plant Emissions		Other fuel cycle	Total	Power Plant Emissions		Other fuel cycle	Total
	Mortality	Morbidity			Mortality	Morbidity		
Austria					2	1	0	**3**
Belgium	25	4	0	**29**				
Czech Rep.[b]	9	4	5	**17**				
Finland	3	1	1	**5**				
France	65	13	0	**79**	15	3	0	**18**
Germany	17	2	2	**21**	4	1	2	**7**
Greece					3	1	0	**4**
Holland	11	2	0	**13**	3	1	0	**4**
Hungary[b]	76	37	1	**114**	6	3	0	**9**
Ireland	47	9	0	**57**				
Italy					9	2	0	**11**
Norway					0	0	0	**0**
Poland[b]	30	19	6	**56**				
Portugal	26	5	11	**42**	0	0	0	**1**
Spain	35	6	1	**42**	5	1	0	**6**
Sweden	1	0	2	**3**				
UK	32	6	0	**38**	4	1	0	**5**

a Except for Czech Republic, Hungary, and Poland, original estimates given by CIEMAT, 1999 in 1995 ECU/MWh have been converted to US_{2007}\$/MWh by assuming 1.25 US\$/ECU and using the ratio of 2007:1995 US GDP deflators (1.3).
b Original estimates given by Melichar et al., 2004b in 2002 Euro/MWh have been converted to US_{2007}\$/MWh by assuming 1.25 US\$/Euro and using the ratio of 2007:2002 US GDP deflators (1.15).

shows these estimates by country for coal and for natural gas-fired power generation. Damages with coal are considerably higher than with natural gas. Also, there is a wide range in estimates for coal-fired generation damages from one country to another. Costs in Finland and Sweden are under US\$5/MWh, most likely reflecting the low density of populated areas, the high level of pollution control technologies, and the modest amount of coal-fired power generation in these countries. The majority of countries in Western Europe have damage costs in the range of US\$40–80/MWh – comparable to the cost of electricity production for a new coal-fired power plant. At the extreme high end is Hungary at US\$114/MWh.

12.3 Technological Changes at Refineries and Opportunities for CCS

12.3.1 Context

12.3.1.1 Demand for Liquid Fuels

Liquid fuels are the world's single largest source of energy and are expected by many to remain so for decades to come. The United States Energy Information Administration (US EIA) (2010b) forecasts an increase in world liquid fuel consumption from 86 million bbl/day in 2007 (192 EJ/yr) to 111 million bbl/day in 2035 (248 EJ/yr) (Figure 12.12). The US EIA projects this increase to occur primarily in non-OECD Asia (a 93% increase for that region), followed by Central and South America, and the Middle East. Liquid fuel share of world energy use drops from 35% in 2007 to 30% in 2035, still remaining the single largest global source of energy. The vast majority of these liquids (96% in 2007 and 80–91% projected for 2035) are from conventional petroleum, with the rest coming from unconventional sources, including heavy oil and tar-sand bitumen (processed in refineries and tar sand upgraders), biofuels, coal-to-liquids and gas-to-liquids plants (Figure 12.13). Biofuels, such as renewable diesel, may be coprocessed in oil refineries but are unlikely to amount to more than 1% of total refinery throughput by 2035. Transportation accounted for nearly half of all liquid consumption in 2007, with industry accounting for about a third. The transportation fraction is projected to reach 60% by 2035, with industry falling to 29% (Figure 12.14).

These projections are consistent with trends over the past decade. From 1998 to 2008, petroleum consumption grew by 15%, with Central and

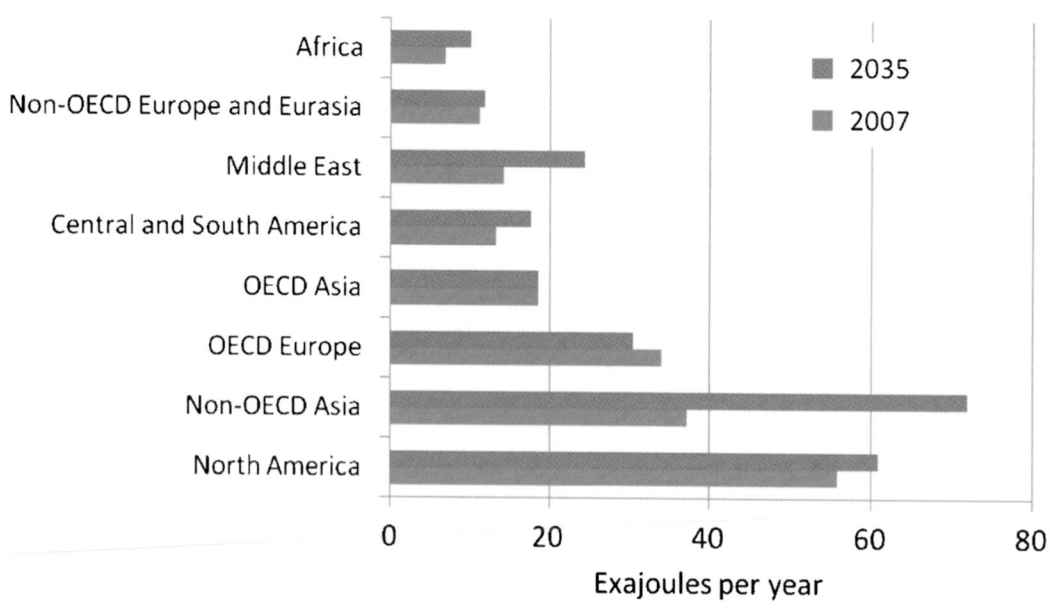

Figure 12.12 | World liquids consumption by region. Source: based on US EIA, 2010a.

South America, China, India, and the Middle East accounting for 72% of the increase (BP, 2009). The US EIA projection is also largely consistent with projections in *The Outlook for Energy: A View to 2030* (ExxonMobil, 2009). ExxonMobil has forecast world liquid fuel demand of 108 million bbl/day by 2030 (241 EJ/yr), with transportation fuel demand of 62 million bbl/day (138 EJ/yr). They expect 94% of the latter demand to be met from oil, 5% from biofuels, and 1% from natural gas. In this *Outlook*, US demand for transportation fuels peaks around 2015 and declines by 10% to 13 million bbl/day by 2030 (29 EJ/yr), the European Union demand stays flat at about 9 million bbl/day (20 EJ/yr), and China's demand triples to 8 million bbl/day (18 EJ/yr).

The IEA (2009a) forecasts global oil demand of 105 million bbl/day in 2030 (234 EJ/yr) with 97% of the increase from transportation fuels and all growth in the non-OECD countries. Their 450 Scenario (an aggressive timetable of actions to limit atmospheric greenhouse gas concentration to 450 ppm CO_2-eq) projects oil demand lower by 12 million bbl/day (27 EJ/yr), but still higher than current demand.

Forecasts of biofuel use are in the range of 3–6 million bbl/day by 2030 (6.7–13 EJ/yr). ExxonMobil expects about 3 million bbl/day (6.7 EJ/yr) with 30% from US corn ethanol, 30% from Brazil sugar cane and the remaining from biodiesel worldwide. The US EIA (2009b) forecasts 5.9 million bbl/day (13 EJ/yr), assuming that cellulosic technology will be significant from 2012.

12.3.1.2   Crude Oil Supply

World oil reserves have increased steadily over the past several decades, despite periodic predictions to the contrary. They have increased from 998 billion barrels (bbl) in 1988 (6088 EJ) to 1069 billion bbl in 1998 (6521 EJ) to 1258 billion bbl (7674 EJ) at the end of 2008 (BP, 2009). Current proven reserves are sufficient for the next 40 years and it is widely assumed that additional reserves are yet to be discovered. (See Chapter 7 for additional discussion of oil reserves.) There is a significant mismatch, however, between the location of reserves and the centers of demand for liquid fuels. This mismatch is likely to become more pronounced. For example, regions where 75% of fuel demand is located hold only 9.5% of the current reserves (Table 12.12). The extensive need to import crude oil that major fuel consuming countries face will continue, with its concomitant supply security concerns. It should be noted, however, that, since the first short oil embargo and the Iranian revolution in the mid- and late seventies, and despite several wars and conflicts in oil-producing regions, there has been no world supply disruption that would justify uneconomic or environmentally unsound measures and investment for the sake of "security." Indeed, the world's fastest-growing oil importer, China, has successfully embarked on a course of securing supplies all over the world through alliances and outright acquisition of reserves.

Crude oil is primarily categorized and priced by its gravity and sulfur content with heavier, more sour crudes being the cheapest and most difficult to process into usable, lighter fuels. About 20% of world reserves are classified as "sweet," i.e., light, low sulfur crudes, 65% are "light or medium sour," and 15% "heavy sour." This mix will change slowly over the next couple of decades, probably with more sour and heavy sour crudes in production. Any new refining capacity being built now will readily accommodate anticipated crude mix changes.

Estimates for reserves of heavy and extra heavy oil, bitumen from oil sands, and shale oil vary depending on technology and economic feasibility assumptions but are generally considered to be approximately 3735 EJ and resources in place to be approximately 56,000 EJ (Table 7.8), with only a minor fraction produced to date.

12.3.1.3 Oil Refining Capacity

Global refining capacity (Table 12.13) increased by 11% in the last decade with 69% of the increase occurring in China, India and the Middle East and 15% in the United States. These numbers do not include the new 580,000 BD refinery added to the Reliance complex at Jamnagar (India) that started up in 2009 and makes Jamnagar one of the largest and most modern refinery sites in the world. Nor do the numbers include the 240,000 BD Fujian refinery expansion that also started in 2009 in China. Future capacity additions will generally follow the liquid

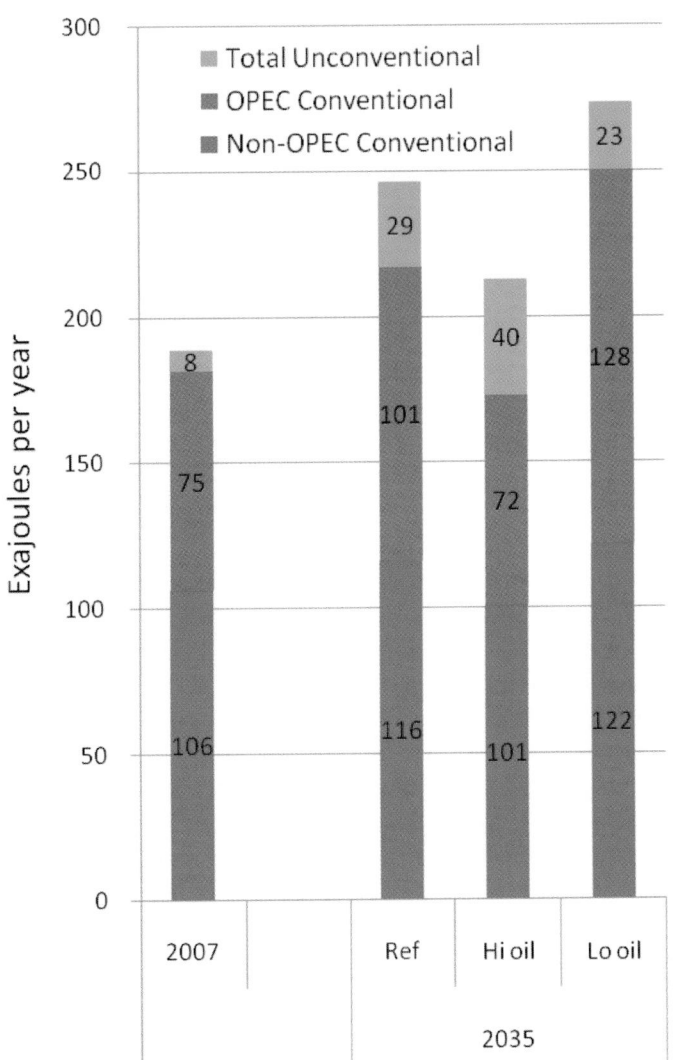

Figure 12.13 | Projections of world liquids supply in 2035 for three alternative oil price projections. Source: based on US EIA, 2010a.

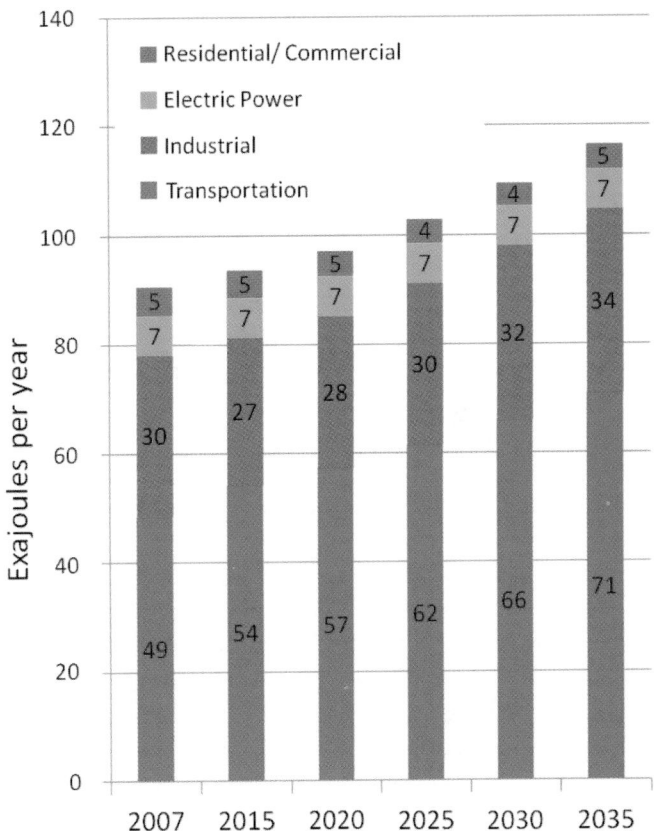

Figure 12.14 | Projections of world liquids consumption by sector. Source: based on US EIA, 2010a.

Table 12.12 | Oil reserves and consumption by region in 2008.

	Reserves 10^9 bbl (EJ)	Consumption 10^3 bbl/d (EJ/yr)	% of Reserves	% of Consumption
World	1,258 (7,674)	84,455 (188)	100	100
United States	30.5 (186)	19,419 (43)	2.4	23.0
European Union	6.3 (38.4)	14,765 (33)	0.5	17.5
China	15.5 (94.6)	7,999 (18)	1.2	9.5
Middle East	754 (4,599)	6,423 (14)	59.9	7.6
Asia Pacific (incl.China)	42 (256)	25,339 (57)	3.3	30.0
North America (incl.US)	70.9 (432)	23,753 (53)	5.6	28.1
N. Am., Asia Pac., & EU	119.2 (727)	63,857 (142)	9.5	75.6

Source: BP, 2009.

Table 12.13 | Refining capacity 1988–2008, 10³ bbl/day.

	1998	2008	Capacity increase	% increase	% of world capacity increase
World	79,699	88,627	8,928	11	100
United States	16,261	17,621	1,360	8.4	15
European Union	15,262	15,788	526	3.4	6
Central and South America	6,114	6,588	474	7.8	5
Middle East	6,202	7,592	1,390	22	16
China	4,592	7,732	3,140	68	35
India	1,356	2,992	1,636	121	18

Source: BP, 2009.

fuels demand with most new capacity expected in non-OECD Asia. In the United States and the European Union, some capacity will be shut down for economic reasons (e.g., in 2010 Valero announced the permanent shutdown of its Delaware refinery, or 7% of its capacity, and BP announced plans to sell its Texas City, Texas, and Carson, California, refineries) while some revamps and expansions will occur in the United States for processing heavy Canadian crude. China is expected to add around 3 million bbl/day of capacity by 2015 while shutting some of the smaller, less efficient refineries that have a combined capacity of between 1 million bbl/day and 1.5 million bbl/day.

12.3.2 Refining Technology and Economics

Refineries are complex processing plants that can generally handle a wide range of crude oil feedstocks and produce hundreds of products of varying specifications. No two refineries are alike. Each one has been designed for a particular market and presumed crude slate and then modified over time as product specifications, environmental regulations, and crude supply economics evolve. The complexity of a particular refinery results from the trade-off between the costs of investment and operating vs the capability to process heavier, less expensive crudes and produce lighter, higher-value transportation fuels or petrochemicals.

The major conversion processes in refining have not changed fundamentally for about six decades, though there have been steady, significant improvements and occasional breakthroughs in product yields, product quality, energy efficiency, and environmental performance. A brief description of refining technology as well as a detailed analysis of life cycle GHG emissions of petroleum-based fuels has been given by the National Energy Technology Laboratory, US Department of Energy (NETL, 2008).

Briefly, crude oil is heated in fired furnaces and separated into different boiling point fractions in atmospheric and vacuum distillation columns. The simplest refining configuration today is a hydroskimming/topping refinery processing a light sweet crude. It includes only a reformer and a distillate hydrotreater in addition to distillation. The reformer increases

the octane of gasoline-range boiling material (naphtha) by dehydrogenating naphthenes into aromatics, isomerizing linear paraffins into branched ones, and converting paraffins into aromatics over a fixed-bed catalyst at moderate pressure. Hydrogen is produced as a by-product. The hydrotreater removes sulfur in a high pressure hydrogen atmosphere by converting it to hydrogen sulfide, also over a fixed bed catalyst. Such a refinery would produce LPG, gasoline, kerosene/jet fuel, diesel/heating oil, and a large amount (30–35%) of low-value heavy fuels.

The next step would be to add a fluid catalytic cracker (FCC) to convert some of these heavy fuels, products from the vacuum distillation column, into diesel, gasoline and lighter gases. The FCC is a complex, atmospheric pressure process where oil vapor is contacted with a fluidized circulating catalyst. Coke is deposited and then burned off continuously from the catalyst in a regeneration step. The FCC is a major source of refinery carbon dioxide and other emissions. In modern refineries, a great deal of equipment is required to reduce sulfur/nitrogen oxides and catalyst particulate emissions from FCC units. An FCC feed hydrotreater, operating at higher pressure than the distillate hydrotreater, would desulfurize the FCC feed and improve the FCC light product yields by saturating aromatic rings. Such a medium conversion refinery would process higher-sulfur crudes and create higher value products than the hydroskimming refinery.

The next step would be to add a coker, a thermal cracking process that can handle the heaviest residue from vacuum distillation and convert it to diesel, gasoline, lighter gases and solid petroleum coke. Such a high conversion refinery could handle the heaviest crudes in its feed and create the maximum value-added between feed and products. In the early 1980s some complex catalytic units were built to hydrotreat heavy residues but the simpler coking process has been the economic choice worldwide since then.

Another major conversion process is hydrocracking, which handles feed similar to the FCC but cracks it in a very high pressure hydrogen atmosphere over a fixed catalyst bed. A refinery can have either of the two catalytic cracking processes or both. Hydrocracking capacity worldwide is about one third that of FCC. Hydrocracking is preferable when a higher diesel/gasoline ratio is needed and is particularly suited for making very-low-sulfur products. Other refinery processes include (i) naphtha hydrotreating, necessary to treat the reformer feed or for gasoline fractions to meet current low-sulfur specifications, (ii) alkylation that converts FCC gases into high-octane gasoline material, (iii) isomerization, a reforming process that increases the octane of light naphtha, (iv) visbreaking, another thermal process for heavy residues, and (v) a multitude of treating processes to remove impurities and contaminants from various intermediate and product streams.

As refinery complexity and feedstock flexibility increases, more severe hydrotreating for distillates and FCC feed may require more hydrogen than is produced by the reformer; a hydrocracker will definitely require another source of hydrogen. A refinery then needs a hydrogen plant or needs to import hydrogen from an external facility. Most refineries use natural gas and refinery off-gas to produce hydrogen through steam reforming. There are several options and technological opportunities in this area. In 1999, BOC Gases (now a part of The Linde Group) and BP Amoco developed a partial oxidation scheme using the Texaco gasification process to supply hydrogen to a new refinery hydrocracker in Bulwer Island, Australia, and to recover pure carbon dioxide (Ramprasad et al., 1999). Depending on the investment cost and the availability or cost of natural gas, refineries can also use gasification of residues or even of petroleum coke. The Shell Group's Pernis refinery in the Netherlands, one of its largest facilities at 400,000 bbl/day, chose a 1650 t/day heavy residue gasification route to produce hydrogen for a new hydrocracker (Shell, 2009).

12.3.2.1   Heavy Oils Technology

The production of heavy oils and bitumen from tar sands generally requires use of steam in one of several techniques, e.g. cyclic steam stimulation, steam assisted gravity drainage, horizontal cyclic steam, or some combination of these (Anonymous, 2008). If heavy oil is to be transported to a distant refining center, lighter oil must be used as diluent to allow flow in a pipeline. Alternatively, such oil can be partially processed where it is produced and a "syncrude" product then shipped to conventional refineries. The processing of very heavy oil or tar sands occurs through application of essentially conventional refining technology. Coking is the major conversion process, followed by severe hydrotreating of the ensuing products before further processing. As crude oil prices rose in the 2004–2007 period, a great deal of investment was initiated to upgrade Canadian tar sands and to process Canadian heavy oil in northern US refineries.

12.3.2.2   Refining Economics

Refining is a capital-intensive and competitive industry that has not been particularly profitable in the long term, except for short periods and in specific "niche" markets. The most recent example is the "Golden Age" of refining in the United States, lasting less than four years, which was caused by a steep global demand increase and unexpected shutdown of refining capacity. Refining margins are determined by market forces and are affected primarily by factors such as the balance between product demand and available refinery capacity, the differential price between heavy and light crudes, and the global balance of fuel oil. (High differentials and low fuel oil prices tend to cause refineries with simpler "topping" configurations to cut production as their variable margins reach break-even and to favor refineries with cracking capacity that can convert the heavier oils.)

As with other commodities, periodic overbuilding of capacity has caused margins to be cyclical and weather or operational variations in available capacity have caused volatility. New capacity costs have historically been around US$10,000/bbl/day of capacity but have risen to US$20,000–30,000 in recent years. Generally, margins have not justified new capacity construction in OECD countries. Government policies or incentives, niche supply or market conditions, or particular synergies with other petrochemical or power projects have supported new capacity additions, primarily in Asia. With the Golden Age in the past, non-mandatory investments in refining will be difficult to justify in OECD countries in the future.

12.3.3   Greenhouse Gas Emissions

12.3.3.1   Extent of Refinery Emissions in the Life Cycle of Petroleum-based Fuels

Emissions of carbon dioxide from stationary fossil fuel sources account for 60% of total global fossil fuel emissions. Of the major stationary sources, refineries account for 6%, cement manufacture for 7%, and power generation for 78% (CO_2 Capture Project, 2009). A detailed analysis of life cycle GHG emissions of petroleum-based fuels has been conducted by NETL (2008; 2009). The scope of this work is extensive and includes analysis of the impact of different crude sources on the GHG emissions of transportation fuels sold in the United States. A summary of some of their results is given in Figure 12.15. Not surprisingly, a key conclusion is that the extent of GHG emissions that is attributed to refinery processing itself is small (6–10%), compared to the ultimate use of the fuels (80–83%). NETL (2008) conclude that the best way to reduce transportation life cycle GHG emissions is by lower overall use of fuels through more efficient modes of transportation and vehicle efficiency. They estimate, as an example, that seven miles per gallon improvement in light vehicle efficiency would be equivalent to eliminating all oil production, transportation, and processing "well-to-tank" emissions. They consider refineries already efficient and cite ongoing continuous improvement programs in the US industry. Although their study is centered on US transportation fuels, this conclusion is quite general. New capacity in Asia is being built to very high standards of efficiency and environmental performance and, as smaller, less efficient refineries are phased out for economic as well as environmental reasons, global refinery emissions will decrease on a continuing basis.

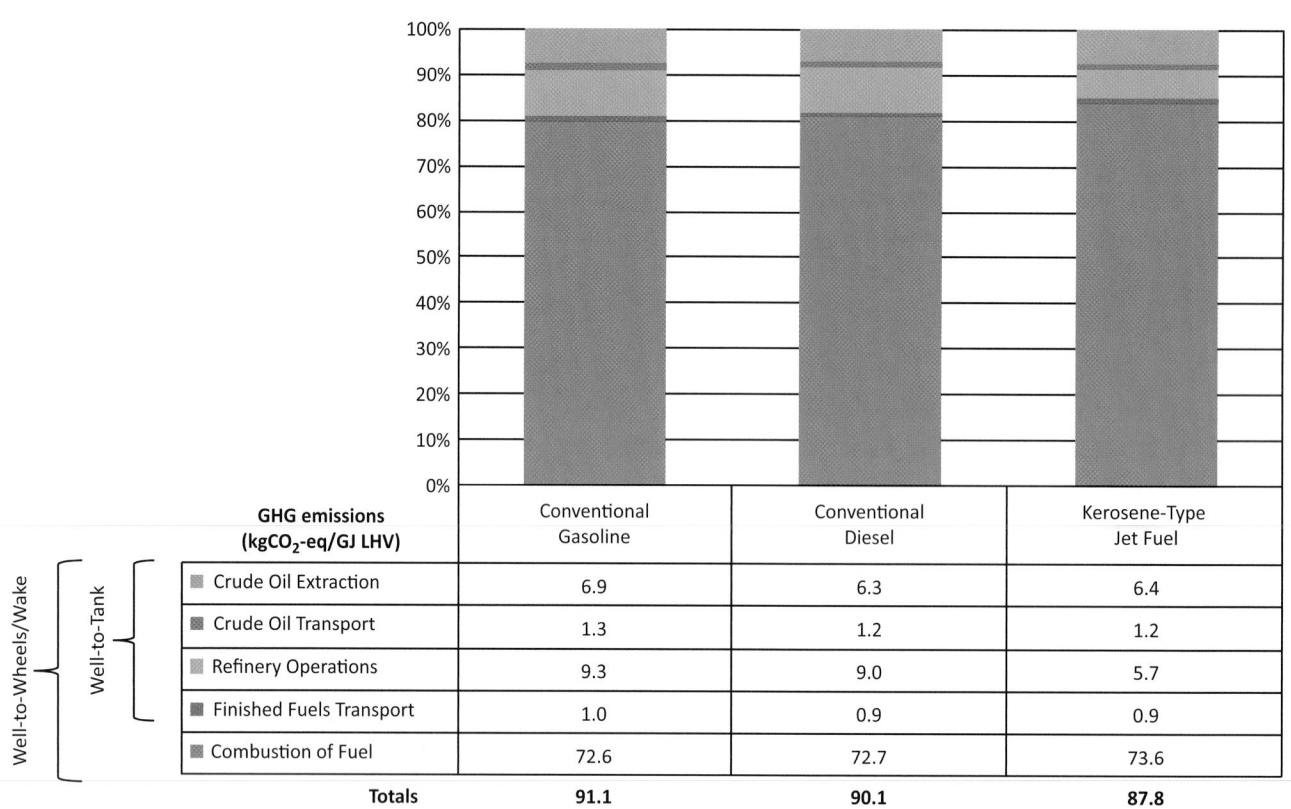

GHG emissions (kgCO$_2$-eq/GJ$_{LHV}$)	Conventional Gasoline	Conventional Diesel	Kerosene-Type Jet Fuel
Crude Oil Extraction	6.9	6.3	6.4
Crude Oil Transport	1.3	1.2	1.2
Refinery Operations	9.3	9.0	5.7
Finished Fuels Transport	1.0	0.9	0.9
Combustion of Fuel	72.6	72.7	73.6
Totals	91.1	90.1	87.8

Figure 12.15 | Baseline life cycle emissions of CO$_2$-eq for fuels derived from petroleum and sold or distributed in the United States in 2005. Source: based on NETL, 2009.

Emissions from the production and transportation of heavy oils are higher by 9.5–14 kgCO$_2$/GJ$_{LHV}$ diesel (NETL, 2009), than those from more conventional crudes, which vary by comparable amounts from each other. An analysis of the GHG emissions associated with the production and transportation of different crudes and heavy oils must be conducted in terms of global impact. There is little overall benefit, for example, if the use of particular crude or heavy oil is limited in the United States through a carbon tax or other regulatory measures only to route this crude to an alternative refining center elsewhere in the world. Opportunities for reducing flaring during production, which are higher for some crudes, no doubt will be pursued. The upgrading of heavy oil lends itself to carbon capture and sequestration. Refineries constantly optimize their crude mix on a global basis based on the demand of their market, their configuration, and the cost of crudes relative to their value for a specific refinery. If a globally consistent and well-designed cap-and-trade system emerges, the cost associated with specific feedstocks, including conventional crudes, heavy crudes, or tar sands, should lead to the optimal use of all global resources.

12.3.3.2   Opportunities for Carbon Capture from Refinery Emissions

Carbon dioxide capture at a refinery has been studied by the CO$_2$ Capture Project, a partnership of seven major energy companies working closely with government organizations, academic bodies, and global research institutes (CO$_2$ Capture Project, 2009). The major sources of emissions considered are boilers and fired heaters, fluid catalytic cracker units, and hydrogen production operations. The FCC unit has its own intrinsic combustion emissions due to coke burning during catalyst regeneration. Emissions from other major processes, such as crude distillation, hydrotreating, and hydrocracking, are generated by fired heaters used in the process. Fired heaters can be treated as a group and are amenable to common analysis, optimization, consolidation, and carbon capture approaches. Three CO$_2$ capture approaches are being considered:

- post-combustion: absorption of flue gas CO$_2$ from major sources into MEA (monoethanolamine) solution;

- oxy-firing: use of oxygen instead of air in combustion to produce a high concentration of CO$_2$ in the flue gases thus facilitating capture; and

- pre-combustion: gasification to produce a hydrogen-rich fuel and separate CO$_2$ prior to combustion.

Two recent studies associated with CO$_2$ Capture Project have considered opportunities for carbon capture in refineries. Shell (van Straelen et al., 2009) has evaluated the feasibility of post-combustion capture at a world-scale refinery. They identified scale (1–2 MtCO$_2$/yr) and CO$_2$ concentration (above about 8%) to be major factors in the cost-effectiveness of refinery carbon capture. They concluded that the best opportunities lie in capturing the emissions from hydrogen production (5–20% of refinery emissions) and large flue gas stacks (another 30% of emissions), finding smaller sources unlikely to be practical. They estimated that such carbon capture would be economical at carbon trading "3–4 times higher than the current." Shell is also studying opportunities for oxy-firing burners and operating the FCC regeneration on oxygen, so as to facilitate CO_2 capture from flue gases. Petrobras and Lummus (de Mello et al., 2009) have studied post-combustion and oxy-firing the FCC and concluded that oxy-firing is feasible and more cost-effective although it would entail higher investment. They did not show absolute cost estimates for using such technology.

The FCC is a major source of refinery emissions. This process has been dominant globally and especially in China for a number of reasons. It was the choice of Reliance for the new refinery in Jamnagar, Gujarat, India. For new capacity, understanding the impact of high carbon costs on the choice of catalytic cracking technology would be worthwhile, especially if demand for diesel continues to increase, as it has in Europe over the last decade.

12.3.4   Conclusions

Current refining technology is likely to continue as the major source of liquid fuels for the next few decades. The global refining industry will grow with some capacity shifting from the OECD, especially the United States, to non-OECD Asian countries, primarily China. Significant technological or regulatory breakthroughs would be needed to change this course, given the scale, technological maturity, and economic fundamentals of this industry. Nevertheless, material opportunities for continuous improvement to decrease refinery GHG emissions and for carbon capture exist and should be developed further.

12.4   Transportation Fuels from Non-petroleum Feedstocks

Fossil fuels other than petroleum can be converted to transportation fuels. Technologies are available and commercially operated today for converting natural gas, coal, or biomass into liquids that closely resemble diesel and gasoline derived from crude oil. These processes involve converting the gas, coal, or biomass first into a synthesis gas (primarily CO and H_2) that is then reacted over specialized catalysts to form liquid hydrocarbons. The most widely used synthesis process for making transportation fuels today is Fischer-Tropsch (F-T) synthesis. The crude Fischer-Tropsch liquid product (FTL) can be refined into middle distillate fuels (65–85% of the final liquid volume produced), naphtha or gasoline (15–25% of the final product), and heavy waxes or lubricating oils (0–30% of the final product) (Fleisch et al., 2002). Oxygenated fuels, such as methanol, higher alcohols (C3 to C8, propanol to octanol), or dimethyl ether (DME), can also be produced from synthesis gas

using different catalysts, and methanol can be further processed catalytically into synthetic gasoline and synthetic liquefied petroleum gas (LPG). There are other technologies, not yet commercialized, that pursue conversion through other intermediates than synthesis gas such as acetylene (Biello, 2009) or methanesulfonic acid (MSA) (Richards, 2005). Technology is also available for direct liquefaction of coal to produce a mix of products, including synthetic gasoline, diesel, LPG, and jet fuel, as well as a variety of chemicals.

12.4.1   Gas to Liquids

It is estimated that nearly 40% (or 71 trillion cubic meters) of the world's current natural gas reserves are "remote" or "stranded," defined as too far from the market place for economic delivery via pipelines. Natural gas resources in Australia, Trinidad and Tobago, and Qatar are good examples. In these instances, gas monetization is sought via liquefying the gas for shipping. The siting of facilities to re-gasify liquefied natural gas is now very challenging in many countries so, increasingly, gas monetization is being pursued via conversion to liquid fuels and chemicals. The high oil prices of recent years are providing a significant impetus for these so-called gas-to-liquids (GTL) efforts. Another driver for the advancement of GTL is the reduction or elimination of the flaring of associated gas. Rather than flaring gas associated with crude oil production, it could be converted via GTL technology into a synthetic crude oil and blended with the produced crude oil for shipping to refineries. The estimated 425 million cubic meters of gas flared daily if converted via GTL technology could produce some 1.5 million bbl/day (3.3 EJ/yr) of additional liquid hydrocarbon products.

Billions of dollars have been invested in GTL technology development over the last several decades by major oil companies and several technology companies. The investment has been primarily in technology for Fischer-Tropsch conversion. (The term GTL in popular usage usually refers to Fischer-Tropsch systems.) A lower level of investment has gone into technologies for producing other liquids, including methanol, higher alcohols, DME, and synthetic gasoline (via methanol). At current rates of commercial deployment, GTL systems could be displacing as much as 1 million bbl/day (2.2 EJ/yr) of petroleum-derived transportation fuels by 2020.

12.4.1.1   Fischer-Tropsch Technology

Transportation fuels are produced by Fischer-Tropsch processing, with chemical feedstocks as byproducts. F-T systems consist of three major integrated sub-systems: (i) syngas production, (ii) syngas conversion to a syncrude, and (iii) syncrude upgrading to finished products.

First, syngas (consisting predominantly of CO and H_2) is produced from methane (CH_4) by reforming using one of a variety of proven technologies. For smaller plants, steam methane reforming (SMR) is the technology of choice where natural gas is reacted with steam at high temperatures and pressures in an externally-heated multi-tubular reactor over a nickel-based catalyst. At plant sizes over about 10,000 bbl/day, autothermal reforming (ATR) or partial oxidation reforming (POX) have become the norm. (The larger size justifies the costs of a dedicated onsite air separation unit for oxygen production.) In these systems, part of the natural gas is combusted with oxygen and the hot combustion products (CO_2 and H_2O) react with additional natural gas over a nickel-based catalyst to form syngas. Most of the heat requirements of the endothermic reforming reaction are met by the heat from the combustion process.

Following reforming, syngas is converted over a catalyst into predominantly long chain paraffinic hydrocarbons with water as a side product. The reaction is quite exothermic and great care must be taken to remove the heat from the reaction vessels. Cobalt-based catalysts have become the norm because of their much higher activity and selectivity for producing the desired paraffins compared with traditional iron-based catalysts. The former are optimized to produce a heavy wax intermediate and reduce the formation of lighter material (C1 to C5). The CO_2 yield from the F-T reactor is less than 1%. The wax is hydrotreated with additional hydrogen in the "upgrading" section to make primarily diesel and naphtha as well as other high value materials such as lube oils, jet fuel, and detergents. The iron-based catalysts produce a broader product portfolio including branched hydrocarbons (gasoline), olefins, and oxygenates along with the paraffins. They still find applications in coal-to-liquids because of the water gas shift activity of these catalysts, which produces additional hydrogen from CO through reaction with water. Coal-to-liquids are discussed further in Section 12.6.

Upgrading of F-T syncrude to finished products, the third and final step in a GTL system, is a well-proven conventional refining step. Paraffinic hydrocarbon chains are hydrocracked into shorter chains, predominantly in the diesel range (C_8 to C_{13} hydrocarbons). A small desirable degree of isomerisation can be achieved to provide proper diesel blending properties.

The basic catalytic reaction for syngas conversion was discovered in the 1920s by Franz Fischer and Hans Tropsch, two German scientists. It was sparingly used after its discovery, but there has been a resurgence of interest over the last 30 years supported by billions of dollars of research and development investments. A number of technologies that use proprietary catalysts have been developed by Sasol, Shell, ExxonMobil and other companies:

- Sasol, a South African company, has almost 50 years of experience with F-T technology, mostly in connection with coal conversion. Current production at their Sasolburg and Secunda facilities in South Africa is about 135,000 bbl/day (from coal). Cumulative production of F-T fuels using Sasol technology exceeds one billion barrels. The Sasol process has undergone significant advancements over the years from the original fixed bed technology (Arge process) to a circulating bed process (Advanced Synthol Process). The Petroleum Oil and Gas Corporation of South Africa (PetroSA), the state-owned GTL producer, started up its

Mossgas plant in 1993 based on Sasol technology and now produces 22,500 bbl/day of finished products. The latest Sasol technology development is its Slurry Phase process. The process was commercialized in the Oryx GTL facility in Ras Laffan, Qatar, in 2006, which produces about 34,000 bbl/day and cost nearly US$1 billion to build. The process involves bubbling hot syngas through a slurry consisting of catalyst particles suspended in liquid hydrocarbon products. The capacity of a single Sasol slurry reactor is 17,500 bbl/day. In partnership with Chevron and the Nigerian National Petroleum Corporation, Sasol is providing technology, engineering services and engineering support for the construction of a carbon-copy of the Oryx facility on a site adjacent to the Escravos River in Nigeria. The Escravos GTL plant will convert about 8.5 million cubic meters per day of currently-flared gas into 22,300 bbl/day of diesel, 10,800 bbl/day of naphtha and 1000 bbl/day of LPG. The project was hit by global construction cost escalations starting in 2007 and may end up costing more than five times the same-sized Oryx plant, some US$5.9 billion.

- Shell is the other leader in commercial gas-to-liquids experience as a result of its GTL plant in Bintulu, Malaysia (the Shell Middle Distillate Synthesis plant). The Shell technology is a tubular fixed bed reactor containing a proprietary cobalt-based catalyst. The Bintulu plant was designed to convert 2.8 million m³/day of gas into 12,500 bbl/day of GTL products. It started operation in May 1993. In 1997, an explosion in the air separation unit damaged the plant. The plant was rebuilt and production restarted in mid-2000 and has been operating since that time at full capacity. The Bintulu plant was the inspiration for Shell's Pearl gas-to-liquids project in Ras Laffan, Qatar. Pearl is designed to produce 140,000 bbl/day in four trains of 35,000 bbl/day each. The trains are planned to come online between 2011 and 2014. The overall project cost announced in 2001 of US$5 billion has ballooned to allegedly US$20 billion (nominal estimate, circa 2008), mainly due to escalating construction costs. However, the oil price has risen fourfold as well, making all products more valuable and compensating for the higher capital costs.

- ExxonMobil's Advanced Gas Conversion for the 21st Century (AGC-21) is a Fischer-Tropsch hydrocarbon synthesis process that converts syngas to heavy hydrocarbons over a cobalt-based catalyst suspended in a slurry. The AGC-21 hydrocarbon synthesis technology is protected by about 1200 patents and has not yet been operated commercially. In the last decade, ExxonMobil explored several commercialization options. It advanced a 160,000 bbl/day project in Qatar, but abandoned the project (and perhaps the technology) in 2006.

- BP has been involved with GTL technology since the 1980s. It operated a GTL test facility in Nikiski, Alaska, from 2002–2009 and continues to pursue commercial applications. The test facility converted about three million cubic feet of natural gas into an estimated 300 barrels of synthetic crude a day. The BP technology uses a cobalt-based catalyst in a tubular fixed bed reactor. BP is licensing the technology and is pursuing projects around the globe.

- Conoco was an aggressive player in GTL starting in 1998, pursuing a catalytic partial oxidation process for syngas production and a slurry phase reactor with cobalt-based catalyst for F-T conversion. A 400 bbl/day pilot plant in Ponca City, Oklahoma, was approved in 2001, and a large commercial project in Qatar was being pursued. However, the technology was abandoned shortly after Conoco merged with Phillips in 2002.

- Rentech is focused on the development of an iron-based catalyst in a slurry-phase process to be able to utilize syngas derived from not only natural gas but also solid or liquid hydrocarbon feedstocks. The Rentech F-T process was verified in a 235 bbl/day facility in Pueblo, Colorado, in 1993. Rentech is licensing their technology and pursing numerous projects with different feedstocks.

- Syntroleum's GTL process has been under development since the 1980s. It involves the use of a cobalt-based catalyst in a slurry to convert syngas produced from a proprietary air-fed autothermal reformer. A 70 bbl/day demonstration facility was successfully operated, but no commercial project has advanced.

- World GTL in partnership with Petrotrin, the Petroleum Company of Trinidad and Tobago, is building a small GTL plant using a refurbished and re-engineered methanol plant. They have developed their own cobalt based, multi-tubular F-T technology. There have been significant cost overruns and the project went into receivership in 2010. Nevertheless, the 2,250 bbl/day plant is expected to come on line in 2012, and World GTL is targeting small gas fields and associated gas with its low-cost modular technology.

- There are a number of related advanced technology developments underway. Petrobras is piloting a micro-reactor based GTL system developed by Velocys and CompactGTL (CompactGTL, 2010). Micro-reactors are low-cost modular systems where the reactions and the heat exchange are conducted in small channels of about 1mm diameter. These systems have a much smaller footprint than conventional systems and could be deployed on platforms or floating vessels to convert off-shore associated gas. Another major development underway for more than 15 years is the use of ionic membranes to separate oxygen from air and to react the oxygen with natural gas to syngas. A smaller footprint and cost reduction of 25% or more over conventional reformers have been touted by the lead developer, Air Products (Air Products, 2008).

Energy and Environmental Performance of F-T Systems

The energy efficiency of a large modern commercial GTL plant today (see Figure 12.16) is about 60–65% on a LHV basis (53–58% HHV basis). Its carbon efficiency (fraction of carbon input as natural gas that appears in the liquid products) will reach 75–80%. For comparison, the methanol process is more efficient than GTL with today's advanced technologies

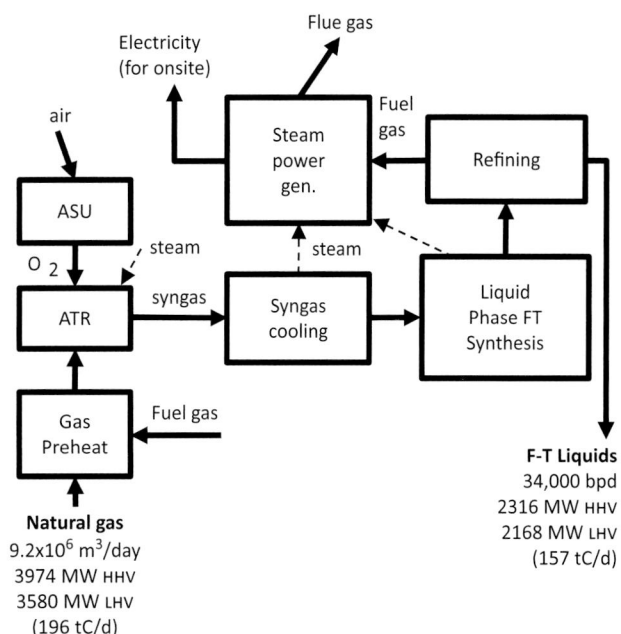

Figure 12.16 | Overall energy and carbon balance for a modern two-train GTL facility. Source: based on Simbeck and Wilhelm, 2007.

already having energy and carbon efficiencies of about 70% and 85%, respectively.

Two processes basically determine the overall efficiency of a GTL plant: (1) the inherent water production in the conversion of methane to higher hydrocarbons (typically one barrel of water will be produced for every barrel of hydrocarbon liquid) and (2) process fuel losses, which refers to the combustion of some of the feed methane (or fuel gas derived therefrom) to provide the heat needed for reforming. From a theoretical perspective, the overall methane-to-liquids reaction can be represented as:

$$(12+x)CH_4 + (5.5+2x)O_2 \Rightarrow C_{12}H_{26} + (11+2x)H_2O + x(CO_2) \qquad (1)$$

The maximum efficiency will be for the case where x is zero (no net CO_2 produced). This theoretical maximum is 78% (LHV). The remaining 22% of the input energy went into making water. At this maximum thermal efficiency, the methane carbon conversion into liquid products would be 100%. For a carbon efficiency of 75% (approximately the level with current technology), x is about three. Technology advances are projected to raise energy and carbon efficiencies to as high as 73% and 90%, respectively, within the next decade (Fleisch et al., 2003). This would make the carbon efficiency of the GTL process comparable to that of petroleum refining.

For a large, modern GTL facility that vents CO_2, the life cycle GHG emissions associated with the F-T liquid products amounts to some 101 kgCO$_2$-eq/GJ$_{LHV}$ (where GJ stands for gigajoule) or about 10% above GHG emissions associated with an equivalent amount of crude-oil derived products.[10, 11] The prospects for reducing these emissions are limited, since only a relatively small amount of CO_2 is available for capture. If available CO_2 were to be captured at a GTL facility today, life cycle GHG emissions would be about 10% below emissions from an equivalent amount of crude-oil derived products. The expectation is that further advances in GTL technology will reduce the amount of CO_2 available for capture, which makes this figure a lower bound on potential GHG emissions of GTL fuels.

Prospective F-T Economics

The gap is widening between the value of natural gas and that of liquid transportation fuels such as diesel and gasoline, influenced by oil prices consistently above US$60/bbl and increasing bullishness regarding future natural gas supplies (see Section 12.7.2.2). In the largest market, the United States, gasoline and diesel sold for two to four times Henry Hub natural gas prices on an energy equivalent basis in recent times (Henry Hub is the pricing point for natural gas futures contracts traded on the New York Mercantile Exchange). During most of the 1980s and 1990s, gas hovered around US$2/GJ while diesel and gasoline averaged US$4–6/GJ. More recently, the gap has widened because of plentiful gas supplies and the capping of gas prices by coal in the competition for power generation. In 2009, we have seen gas prices below US$3/GJ while low sulphur diesel was at US$14/GJ and above. These are strong economic drivers for GTL especially with commercially proven technologies. It can be shown that at oil prices of US$60/bbl, a delivered gas price of about US$8/GJ is needed to achieve a similar net present value for GTL and LNG (liquefied natural gas) projects. At higher oil prices and/or lower gas prices, GTL is economically advantaged over LNG while at lower oil prices and/or higher gas prices LNG is the economic choice.

The economics for an integrated upstream gas field and GTL plant will typically be more attractive than those for a stand-alone GTL plant. Shell's 140,000 bbl/day Pearl project is an example of an integrated project. It includes gas production platforms, gas processing plants and the GTL facilities. The economic returns from the condensates and LPG in the natural gas are high, putting less demand on the return from the GTL project.

Capital costs of GTL Fischer-Tropsch projects dropped from about US_{2005}\$80,000 per bbl/day (US_{1991}\$60,000) of installed capacity in 1991 for Mossgas to just below US_{2005}\$29,000 per bbl/day (US_{2006}\$30,000) installed capacity for Oryx in 2006. These costs made GTL economic with oil at US$20/bbl. Capital costs have more than doubled in the last four years, however, due to sky rocketing oil prices and the corresponding

increases in project costs in the oil and gas industry. Many GTL projects have been delayed for a "cooling down" period. However, the first train of the Shell Pearl project started up in mid-2011, to be followed by the other three trains through 2014. This project may prove the techno-economic attractiveness of large scale GTL once and for all and set the stage for widespread global applications with ongoing improvements. Meanwhile, research and development is continuing and will also contribute to reducing capital costs as it has in the past.

Thus, GTL has the potential to become a prominent part of the international energy business. Continued high oil prices and the widening gap between gas prices and oil prices will favor GTL projects. Most importantly, the concerns of technology risks will wane with commercial plants operating safely and reliably around the world. The problem of technology access will disappear: an increasing number of licensing opportunities and patents are expiring.

12.4.1.2   Methanol to Gasoline

Production of methanol from natural gas is a well-established technology (Cheng and Kung, 1994; Olah et al., 2009). Methanol consumption globally exceeds 40 Mt/yr primarily in chemical processing. Use of methanol as a vehicle fuel attracts considerable interest in some provinces of China today (Dolan, 2008), as it did in the 1970s and 1980s in the United States, where interest has since faded largely as a result of health and environmental concerns from the use of methanol-derived MTBE (methyl tert-butyl ether) as an additive to gasoline. There is renewed interest in higher alcohols for both gasoline and diesel blending (IGP, 2010). These alcohols overcome some disadvantages of methanol and ethanol and have been shown to increase engine efficiencies and lower tailpipe emissions (Yacoub et al., 1997). Furthermore, there is now growing interest in China, the United States and elsewhere in the production of synthetic gasoline from synthesis gas via a methanol intermediate. This so-called methanol-to-gasoline process is the subject of this section.

The first step is methanol production by reforming of natural gas into synthesis gas (CO and H_2), which is then converted over a catalyst to methanol. The two most common reforming technologies, SMR and ATR, have been described in Section 12.4.1.1. A syngas H_2/CO ratio of about two leaving the reformer will optimize methanol synthesis yields, but reforming typically yields a syngas with H_2/CO higher than two. One option for reducing the H_2/CO ratio is to feed recycled CO_2 to the reformer. Katofsky (Katofsky, 1993) showed that this has the benefit of increasing methanol yield and overall process efficiency by several percentage points.

Two companies currently offer technology for synthesis of gasoline from methanol. A key distinction between the technologies is that the technology offered by Haldor Topsoe utilizes an initial single-step conversion of syngas into DME/methanol, followed by conversion to gasoline

---

10   This estimate includes 9.10 $kgCO_2$-eq/GJ_{LHV} associated with extraction, preprocessing, and transportation of the feedstock natural gas (based on the GREET model (Argonne National Laboratory, 2008)).

11   If natural gas that would otherwise have been flared (a relatively small potential resource for GTL) were the feedstock, then the net life cycle GHG emissions associated with these GTL fuels would be considerably less than for equivalent petroleum-derived fuels, since the GTL fuels would be displacing the petroleum-derived fuels (FWI, 2004) while eliminating the emissions from the flared gas.

in a separate reactor (Nielsen, 2009). The ExxonMobil process involves methanol production from syngas followed by partial conversion of methanol to DME in a separate reactor, followed by conversion of the DME/methanol mixture into gasoline in a third reactor (Tabak et al., 2009). In either case, most of the DME/methanol mixture that flows to the gasoline synthesis reactor is converted in a single pass over the catalyst. Some propane/butane (LPG) and a small amount of methane are coproducts.

Because the H/C ratio of methane is higher than that of the final gasoline product, no significant CO_2 by-product stream is available for capture and storage, giving natural gas-to-MTG systems a carbon footprint not substantially different from petroleum-derived gasoline.

12.4.2   Direct Coal Liquefaction

Direct coal liquefaction refers to a process in which pulverized coal reacts with a hydrogen-donor solvent over a catalyst, causing hydrocracking of the coal into liquids. Direct liquefaction is distinct from indirect liquefaction. As discussed in Section 12.4.3, indirect liquefaction utilizes coal gasification followed by a separate catalytic step to convert the gasified product into liquid fuels. (Fischer-Tropsch fuels are perhaps the best known of the different fuels that can be produced via indirect liquefaction.)

Direct coal liquefaction will produce not only gasoline, diesel, LPG, and jet fuel, but also a variety of chemicals such as benzene, toluene, xylene, and other raw olefins to further produce ethylene and propylene. The reaction can be summarized as follows:

$$nC + (n+1)H_2 \rightarrow C_nH_{2n+2} \tag{2}$$

The first generation of direct coal liquefaction technology was developed during World War II. Germany was the first country to realize the industrialization of direct coal liquefaction. Twelve direct coal liquefaction plants were built and the total capacity exceeded 4 Mt/yr. However, the first generation technology was limited by its harsh reaction condition (Pressure: around 70 MPa) and costly catalysts. After World War II, all the plants were shut down due to technical defects and economic disadvantage driven by the development of cheap oil. The 1970s oil crisis brought attention back to direct coal liquefaction. Many countries developed a wide range of modern coal liquefaction technologies (Pressure: 10~30 MPa and less expensive catalysts): American H-Coal and Hydrocarbon Technology, Inc. (HTI), the German Integrated Gross Oil Refining (IGOR) technology, the Japanese NEDOL technology, and others. These technologies completed 350~600 t/day pilot tests but did not achieve large scale industrialization. Table 12.14 lists known technologies developed to at least the 50 t/day coal input scale. The main differences between these processes lie in catalysts, reactors, hydrogen donor solvents, and system design.

Table 12.14 | World testing of coal liquefaction technologies.

Single stage processes	Two stage processes
• Kohleoel (Ruhrkohle, Germany) • NEDOL (NEDO, Japan) • H-Coal (HRI, USA) • Exxon Donor Solvent (EDS) (Exxon, USA) • SRC-I and II (Gulf Oil, USA) • Imhausen high-pressure (Germany) • Conoco zinc chloride (Conoco, USA)	• Catalytic Two-Stage Liquefaction (CTSL) (USDOE and HRI, now HTI, USA) • Liquid Solvent Extraction (LSE) (British Coal Corp., UK) • Brown Coal Liquefaction (BCL) (NEDO, Japan) • Consol Synthetic Fuel (CSF) (Consolidation Coal Co, USA) • Lummus ITSL (Lummus Crest, USA) • Chevron Coal Liquefaction (CCLP) (Chevron, USA) • Kerr-McGee ITSL (Kerr-McGee, USA) • Mitsubishi Solvolysis (Mitsubishi Heavy Industries, Japan) • Pyrosol (Saarbergwerke, Germany) • Amoco CC-TSL (Amoco, USA) • Supercritical Gas Extraction (SGE) (British Coal Corp., UK)

Source: DTI, 1999.

The most active country today in direct liquefaction is China, which became a net oil importing country in 1993. Since then, the fraction of oil imported has been increasing, and it reached about 50% by the end of 2008. Concern about energy security gives impetus for making oil from coal in China. Direct coal liquefaction became a major candidate because it is believed to be more energy efficient and less water consuming than many alternatives.

In 2001, China's Shenhua Group, the largest coal company in the world, started development of a demonstration project to produce one Mt/yr of liquids from coal by direct liquefaction. In early 2009, the facility conducted the first test run which lasted about 300 hours and then shut down as planned. This project is ongoing and is expected to provide improved understanding about potential energy and economic performances of this technology (see Box 12.2).

12.4.2.1   Process Description

Direct coal liquefaction includes a catalytic liquefaction step in the presence of hydrogen followed by solid-liquid separation and upgrading. Pulverized coal is blended with a solvent and the catalyst, together making a coal slurry. In the liquefaction unit, weak-bond breaking is achieved, leading to free radical hydrogenation. Then, the inorganic minerals and un-reacted coal are removed by a series of solid-liquid separation processes such as vacuum distillation, filtration, extraction, and sedimentation. Finally, a catalytic hydrogenation process is required to increase the hydrogen-to-carbon ratio in the liquid product and remove impure elements.

Generally speaking, apart from anthracite, all other types of coal can be liquefied to some extent. In rough terms, the difficulty of liquefaction increases with the age of the coal: peat < young lignite < lignite < high-volatile bituminous coal < low-volatile bituminous coal. In addition, excessive ash content also has a negative impact on coal liquefaction.

Box 12.2 | Shenhua Direct Coal Liquefaction Demonstration Project

China's Shenhua Group's direct coal liquefaction demonstration project in Ordos, Inner Mongolia Autonomous Region, China, is designed to produce 1 Mt/yr of oil. The Chinese government approved the project in 2001, construction started in August 2004, and the first testing run succeeded in early 2009.

Shenhua developed its own synthetic catalyst to lower costs. Hydrogen donor solvent cycle is used to mitigate the slurry properties. A slurry bed compulsory intra-circulation reactor is used to improve the capacity of reaction. A mix of advanced and mature unit technologies reduces project risks.

The process flow of Shenhua direct coal liquefaction project is shown in Figure 12.17. At present, 1.08 Mt of liquefied oil can be produced with the input of 4.10 Mt of coal, of which 1.32 Mt is used for hydrogen production and 0.53 Mt for industrial boiler firing. Of the liquefied oil, 70% is diesel and 20% is naphtha. The naphtha products, with a high content of aromatics, are very good reforming materials. The water consumption in this project is about 7 to 8 tonnes of water per tonne of product.

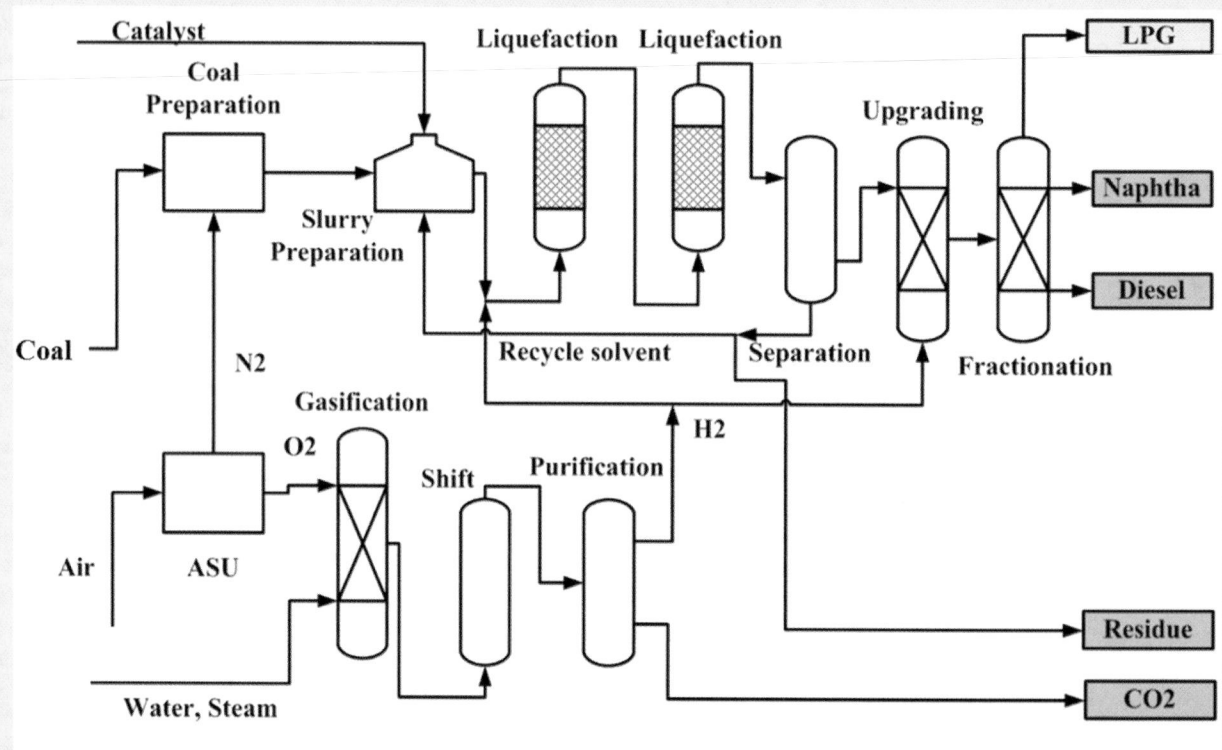

Figure 12.17 | Shenhua direct coal liquefaction demonstration project process flow.

The liquefied oil from direct coal liquefaction, including naphtha, diesel, and liquefied heavy oil, retains some characteristics of raw coal, such as a high content of aromatic hydrocarbons and hetero-atoms (e.g., Oxygen, Nitrogen). A process of hydrogenation is carried out to upgrade the quality of the liquefied oil, which causes the refining costs to be much higher than for conventional petroleum refining.

12.4.2.2   Resource Consumption and Economic Performance

Compared with indirect coal liquefaction and other coal conversion technologies, the efficiency of direct coal liquefaction is higher, up to 60%. Shenhua estimates that the coal consumption of a direct coal liquefaction project is about 0.061 tonne of coal equivalent (tce) per

GJ of liquid product, compared with 0.075 tce/GJ for indirect coal liquefaction.

Water consumption is always a big concern for coal conversion. Compared to other conversion technologies, direct coal liquefaction appears to be relatively lower in specific water consumption at about 8 t water per metric tonne of product (t/t), compared with about 11–12 t/t for indirect coal liquefaction. Water is mainly consumed in the gasification unit for providing hydrogen for hydrocracking and the hydrogenation upgrading processes.

The investment of a direct coal liquefaction plant is still uncertain. The first-of-a-kind plant may cost US$100,000 to US$150,000 per bbl/day output capacity. This number can be decreased with further technology development and improvements in engineering.

12.4.2.3   Carbon Dioxide Emissions

In a coal liquefaction plant producing one Mt/yr of liquids, the carbon dioxide emissions will amount to more than 3.6 Mt/yr, excluding emissions from combustion of the liquid products. The plant CO_2 emissions are predominantly (about 80% of emissions) from the hydrogen production unit. The CO_2 leaves this unit in high concentration (87–99%), which can facilitate low-cost capture of CO_2 for underground storage. (Other CO_2-containing emission streams, e.g., flue gas from the flare system, have relatively low concentrations of CO_2, making capture for storage more difficult.) About one-third of carbon input as coal is available for capture as a relatively pure stream from the H_2 production plant. In this respect, direct coal liquefaction has some similarity to gasification-based coal-to-fuels and coal-to-chemicals processes, since a pure stream of CO_2 is available (just as there is in gasification-based processes). For additional discussion of direct versus indirect liquefaction, see Williams and Larson (2003) and Lepinski et al. (2009).

12.4.3   Gasification-based Liquid Fuels from Coal (and/or Biomass)

There is growing interest in making synthetic fuels from coal – known as coal-to-liquid (CTL) fuels – in light of coal's relatively low prices and the abundance of coal in China, the United States, and other countries that are not politically volatile. Much of this attention has been focused on so-called indirect liquefaction to produce Fischer-Tropsch liquids (Bechtel Corporation et al., 2003; van Bibber et al., 2007; Bartis et al., 2008; AEFP, 2009). Synthetic gasoline (made via a methanol intermediate, see Section 12.4.1.2) is also beginning to attract interest (AEFP, 2009).

Coal can do much to improve energy security if it is used to make liquid fuels. Moreover, these synfuels would be cleaner than the crude oil products they would displace (having essentially zero sulfur and other contaminants and low aromatics content). Also, if produced using modern entrained flow gasifiers, the air pollutant emissions from the production facility would be extremely low. When synthetic fuels are made from coal without capture and storage of by-product CO_2, however, net fuel-cycle GHG emissions are about double those from petroleum fuels they would displace (AEFP, 2009). And even with carbon capture and storage (CCS), the net GHG emission rate would be only about the same as the crude oil products displaced.

One approach to reducing GHG emissions below petroleum-fuel levels is to exploit negative GHG emissions opportunities to offset the emissions. One important opportunity is synthetic fuels production from biomass with CCS (Larson et al., 2006). An intrinsic feature of synthetic fuels production from coal or biomass is the production of a by-product stream of pure CO_2, accounting for about one half of the carbon in the feedstock. If this CO_2 can be captured and stored via CCS while producing synthetic fuels from sustainably grown biomass, the biofuels produced would be characterized by strong negative GHG emissions, because of the geological storage of photosynthetic CO_2. However, sustainably-recovered biomass is expensive, and the size of the biomass-to-liquid (BTL) facilities will be limited by the quantities of biomass that can be gathered in a single location – which implies high specific capital costs for BTL.

The shortcomings of the BTL with CCS option could be overcome to an extent by coprocessing biomass with coal in the same facility. The economies of scale inherent in coal conversion could thereby be exploited and the average feedstock cost would be lower than for a pure BTL plant. And if CCS were carried out at the facility, the negative CO_2 emissions associated with the biomass could offset the unavoidable positive emissions with coal (Figure 12.18), leading to liquid fuels with low, zero, or negative fuel-cycle emissions depending on the relative amounts of coal and biomass input (AEFP, 2009; Larson et al., 2010; Liu et al., 2011a). Interest in the CBTL-CCS concept is growing in the United States (Tarka et al., 2009).

The equipment for gasification-based production of liquid fuels from coal and biomass are commercial or nearly-commercial in all cases. Coal gasifiers are commercially available and deployable today, with more than 420 gasifiers already in commercial use in some 140 facilities worldwide (AEFP, 2009). The technology for cogasifying biomass and coal is close to being ready for commercial deployment; the commercial Buggenum IGCC facility in the Netherlands has been cogasifying coal and some biomass in a coal gasifier since 2006 (van Haperen and de Kler, 2007). Stand-alone biomass gasification technology is an estimated five to eight years from being ready for commercial-scale deployment (AEFP, 2009). Technologies for converting syngas into Fischer-Tropsch diesel and gasoline are in commercial use today. Those for making synthetic gasoline via methanol can be considered commercially deployable (AEFP, 2009), with technology offered by vendors such as Haldor-Topsoe (Nielsen, 2009) and ExxonMobil (Tabak et al., 2009), and projects in development (Doyle, 2008; ExxonMobil, 2009). See Section 12.4.1.2.

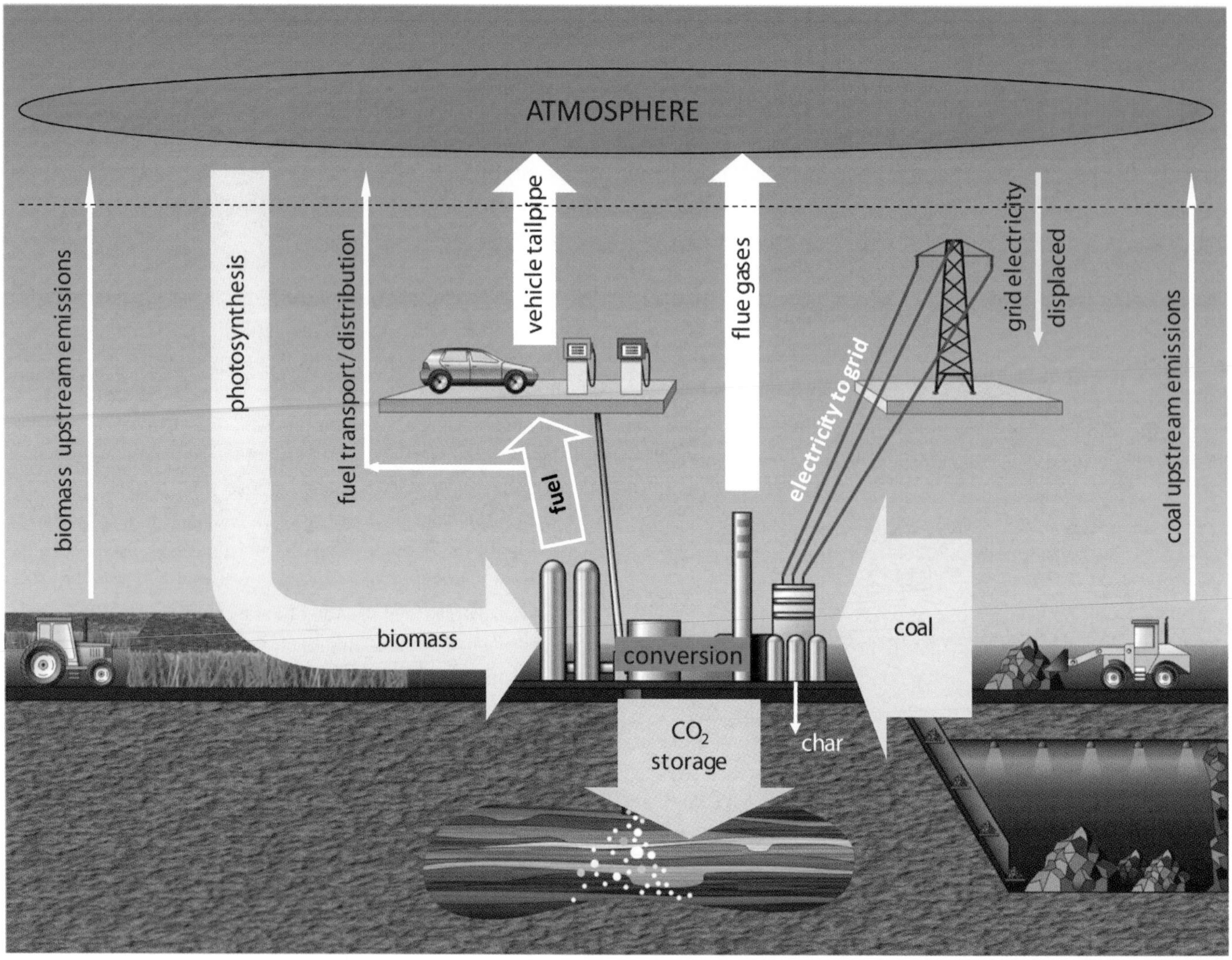

![Carbon flows diagram showing atmosphere, biomass, coal, conversion, CO₂ storage]

Figure 12.18 | Carbon flows for conversion of coal and biomass to liquid fuels and electricity. When biomass is approximately 30% of the feedstock input (on a higher heating value basis), the net fuel cycle GHG emissions associated with the produced liquid fuels and electricity would be less than 10% of the emissions for the displaced fossil fuels. Source: Larson et al., 2010.

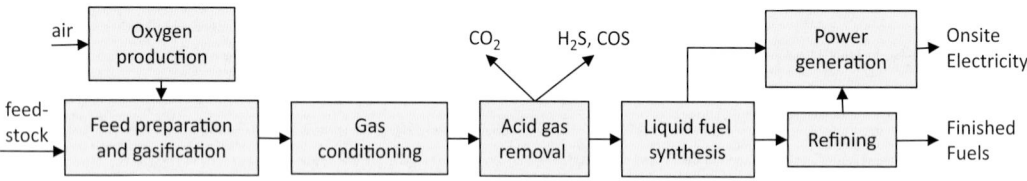

Figure 12.19 | Production of liquid fuels from coal and/or biomass feedstocks.

12.4.3.1   Process Descriptions

Figure 12.19 illustrates generically gasification-based production of liquid fuels from coal or biomass. The feedstock is gasified in oxygen and steam, with subsequent gas conditioning that includes cleaning of the raw synthesis gas (syngas) and in some cases adjusting the composition of the syngas using a water gas shift reaction to achieve the requisite H_2:CO ratio for downstream catalyst-assisted synthesis into liquids. Prior to synthesis, CO_2 and sulfur compounds are removed in

the acid gas removal step to increase the effectiveness and reduce the required size of downstream equipment, as well as avoid sulfur poisoning of catalysts. The CO_2 may be vented (-V) or captured and stored underground (-CCS). The liquid fuel synthesis island is designed with recycle (RC) of unconverted syngas to maximize liquids production. A purge stream from the recycle loop, together with light gases generated in the refining area, provide fuel for power generation, which primarily goes to meet onsite needs. (Alternatively, as discussed in Section 12.6, if none – or only some – of the unconverted syngas is recycled, the

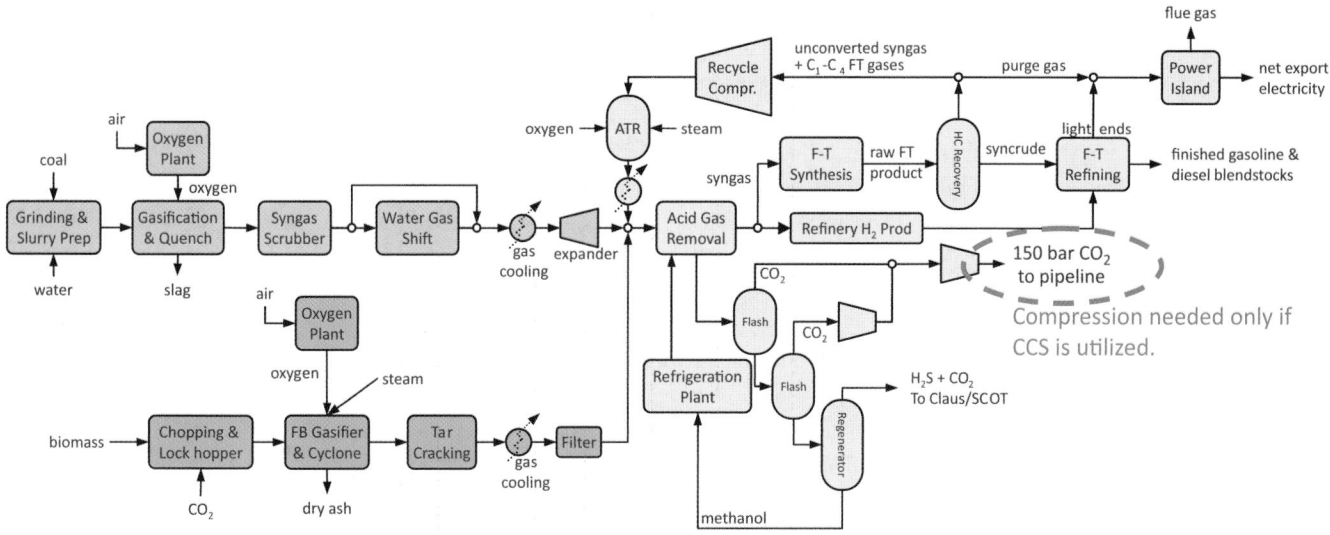

Figure 12.20 | Alternative process configurations for maximizing production of FTL from coal (CTL-RC-CCS, blue-shaded components) or from biomass (BTL-RC-CCS, green-shaded components) with capture and storage of carbon. Yellow-shaded components are common to both CTL and BTL systems.

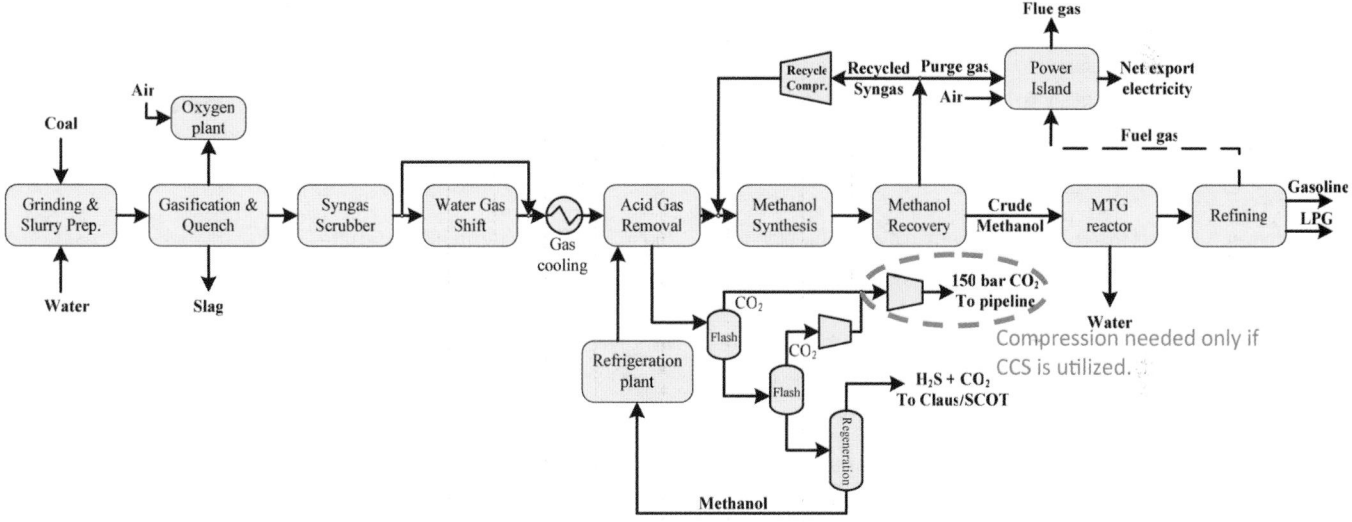

Figure 12.21 | Plant configurations for maximizing gasoline output with only coal as input feedstock.

unconverted syngas can be used to increase electricity generation and provide an exportable electricity co-product.)

More detailed process depictions of coal conversion into F-T fuels are shown in Figure 12.20 (blue and yellow shading). Figure 12.21 shows coal conversion to synthetic gasoline. In these particular plant designs, a slurry of bituminous coal and water is fed into a pressurized gasifier along with oxygen from a dedicated onsite cryogenic air separation unit. The syngas leaving the gasifier is wet-scrubbed before some of it enters a water gas shift reactor. The fraction of the syngas sent to the WGS reactor is adjusted to ensure an appropriate H_2:CO ratio for the later synthesis reactor. Following the WGS reactor, the syngas is cooled in preparation for acid gas removal and, in the case of F-T production, expanded to near the pressure required

for later synthesis. CO_2 and H_2S are then removed by the acid gas removal system. Several acid gas removal technologies are commercially available (see Chapter 13). A physical absorption process using chilled methanol (e.g., Rectisol®) is indicated in Figures 12.20 and 12.21. The captured H_2S is processed via a Claus/SCOT system into elemental sulfur. The captured CO_2 can be either vented to the atmosphere or compressed to a supercritical state for pipeline delivery and underground injection for storage. Following acid gas removal, the syngas is delivered to the synthesis reactor.

In the case of F-T synthesis (Figure 12.20) most of the syngas passes to the synthesis reactor, where it reacts over a catalyst to produce a mixture of olefinic and paraffinic hydrocarbons. When an iron-based synthesis catalyst is used, the syngas H_2:CO ratio entering the reactor is about one, and the raw liquid product is a mix of hydrocarbons with a wide range of carbon numbers. With a cobalt-based catalyst, the ratio is about two, and the liquid product is largely heavy hydrocarbons (waxes). After synthesis, unconverted syngas and light gases generated during synthesis are recycled back to the synthesis reactor to increase F-T liquids production. (In some situations, separation for sale of the light components that constitute LPG may be economic.) The recycled gases undergo ATR to convert light hydrocarbons, resulting in a stream returning to the synthesis reactor that consists primarily of CO, H_2, and CO_2. A small portion of the recycle gases are drawn off as a purge stream to avoid the buildup of inert gases in the recycle loop. The purge gases are mixed with light hydrocarbons produced in the F-T refining area, and the mixture of gases is burned in the boiler of a steam turbine power system. Electricity production is primarily to meet onsite needs.

The final step in FTL production is refining of the raw liquid product (mostly a crude diesel and naphtha). Hydrogen addition is typically required in this step. In the design shown in Figure 12.20, some syngas is diverted upstream of synthesis for use in producing the requisite hydrogen. A minimum refining step is required to stabilize the liquid products if they are to be shipped to a conventional refinery for further refining. This minimum step includes hydrogenation of the naphtha and diesel range hydrocarbons and hydrocracking of the heavier hydrocarbons (waxes). Alternatively, further refining to finished diesel and gasoline blendstocks can be done onsite. Reforming of the naphtha fraction is required in this case to produce an acceptable gasoline blendstock (Liu et al., 2011a; Guo et al., 2011). The naphtha could be sold instead as a chemical feedstock, thereby avoiding added cost and energy expenditures for catalytic reforming, but chemical markets for naphtha are relatively small (compared to gasoline markets), so this would be a limited option if FTL production were to become widespread.

In the particular process shown in Figure 12.21 for making synthetic gasoline via methanol (MTG) from syngas, the first step following acid gas removal is methanol synthesis, for which the requisite syngas H_2:CO ratio is about two and optimal pressure is 50–100 bar (considerably higher than for F-T synthesis). Most of the unconverted syngas is recycled to increase methanol production, with a small purge stream extracted to avoid building up inert gases. (No reformer is required in the recycle loop as in the FTL design, since there is no significant hydrocarbon content in this stream.) The crude liquid methanol is vaporized and sent to a DME reactor, where it is catalytically converted to an equilibrium mixture of methanol, dimethyl ether and water. The mixture flows to the MTG reactor, where most of the gas is converted in a single pass over a catalyst into gasoline-range hydrocarbons. Some propane/butane (LPG) and a small amount of methane are coproducts. The liquid products are sent for fractionation and finishing (primarily durene treatment), resulting in finished gasoline and LPG products. The purge gas from the methanol synthesis area and light gases evolved in the MTG area fuel the power island, where a steam cycle generates all the electricity needed to run the facility plus a modest amount of export electricity.

Process designs for making FTL or MTG from biomass feedstocks would be similar to those described above for coal conversion. Figure 12.20 shows a configuration for FTL from biomass (BTL) (green plus yellow shading). There are important differences from the coal design (also shown in Figure 12.20): (i) a pressurized fluidized-bed gasifier, which avoids the energy-intensive grinding of biomass that is required with an entrained flow gasifier and enables ash to be removed as a dry material that might be returned to the field for its inorganic nutrient content; (ii) a tar cracking step following gasification to convert into light gases the heavy hydrocarbons that form at typical biomass gasification temperatures (which are lower than coal gasification temperatures) and that would otherwise condense and cause operating difficulties downstream; (iii) reforming of the recycle stream in the MTG design due in part to the higher methane production from biomass gasification compared to coal gasification.

As noted earlier, shortcomings of the BTL option include high feedstock costs and steep scale economies for the plant capital cost. These challenges can be mitigated by coprocessing some biomass with coal (CBTL) (Blades et al., 2008; Tarka et al., 2009; Liu et al., 2011a). Figure 12.20 (all colors of shading) shows a detailed design for a coal/biomass coprocessing system to make FTL fuels (CBTL) utilizing separate coal and biomass gasifiers. With different downstream processing steps, synthetic gasoline could similarly be produced from coal and biomass (CBTG).

12.4.3.2   Performance Estimates

Results from a set of detailed and self-consistent designs and performance simulations of coal and/or biomass conversion to FTL and MTG transportation fuels are presented in Table 12.15. (Illinois #6 bituminous is the coal type, and switchgrass is the biomass type.) The simulations utilize design assumptions for each unit operation (gasification, water gas shift, acid gas removal, FTL synthesis, MTG synthesis, etc.) that are consistent with performance that has been demonstrated in existing commercial applications for all except biomass gasification/tar cracking. For the latter, design assumptions are based on pilot-plant performance. The greenhouse gas emissions estimates include net life cycle emissions for synfuel production and consumption, including emissions associated with activities upstream of the conversion plant, such as coal mining and biomass growing, harvesting, and transportation, as well as emissions downstream of the plant, including transport of the liquid products to refueling stations and combustion of the fuels in vehicles. The process of making synthetic gasoline has some efficiency benefit. For systems using only coal as feedstock and producing liquid fuels at a rate of 50,000 petroleum-fuel-equivalent barrels per day of liquids, Table 12.15 indicates that when making synthetic gasoline (MTG) more of the input coal is converted to liquid fuel (and less to electricity) than for the FTL designs considered here. The result is an overall efficiency advantage of about 5 percentage points for MTG due to the intrinsically higher thermodynamic efficiency of converting syngas to liquids compared to converting it to electricity. There is only

Table 12.15 | Performance estimates for conversion of coal, biomass, and coal + biomass to FTL or MTG.

Process configuration >>>	CTL-RC-V	CTL-RC-CCS	CTG-RC-V	CTG-RC-CCS	BTL-RC-V	BTL-RC-CCS	BTG-RC-V	BTG-RC-CCS	CBTL-RC-CCS	CBTG-RC-CCS
Coal input rate										
As-received, metric t/day	24,087	24,087	20,869	20,869	–	–	0	0	2562	2489
Coal, MW HHV	7559	7559	6549	6,549	–	–	0	0	804	781
Biomass input rate										
As-received metric t/day	0	0	0	0	3581	3581	3581	3581	3581	3581
Biomass, MW HHV	0	0	0	0	661	661	661	661	661	661
% biomass HHV basis	0	0	0	0	100	100	100	100	45	46
Liquid production capacities										
LPG, MW LHV	–	–	309	309	–	–	28	28	–	69
Diesel and/or Gasoline, MW LHV[a]	3159	3159	2913	2913	286	286	270	270	622	610
bbl/day crude oil products displaced (excl. LPG)	50,000	50,000	50,000	50,000	4521	4521	4630	4630	9845	10476
Electricity										
Gross production, MW	849	849	545	545	77	77	78	78	157	145
On-site consumption, MW	445	555	419	509	35	46	46	58	104	134
Net exports, MW	404	295	126	36	42	31	32	20	53	11
ENERGY RATIOS										
Liquid fuels out (HHV)/Energy in (HHV basis)	45.0%	45.0%	52.8%	52.8%	46.5%	46.5%	48.4%	48.4%	45.7%	50.2%
Net electricity/Energy in (HHV)	5.3%	3.9%	1.9%	0.6%	6.4%	4.7%	4.9%	3.1%	3.6%	0.7%
Total products (HHV)+ electricity/Energy in (HHV)	50.3%	48.8%	54.7%	53.4%	52.9%	51.2%	53.3%	51.5%	49.3%	51.0%
CARBON ACCOUNTING										
C input as feedstock, kgC/sec	178	178	154	154	17	17	17	17	35	35
C stored as CO_2, % of feedstock C	0	52	0	49	0	56	0	60	54	54
C in char (unburned), % of feedstock C	4.0	4.0	4.0	4.0	3.0	3.0	3.0	3.0	3.5	3.5
C vented to atmosphere, % of feedstock C	51.6	10.3	56.9	8.2	63.9	8.2	63.4	3.8	9.0	6.7
C in liquid fuels, % of feedstock C	34.1	34.1	39.1	39.1	33.1	33.1	33.7	33.7	33.7	36.1
C stored, 10^6 tCO_2/yr (at 90% capacity)	0	9.54	0	7.80	0	0.96	0	1.03	1.98	1.95
Lifecycle GHG emissions, kgCO_2-eq/GJ liquid fuels LHV[b]	207	101	195	100	7.9	–110	8.5	–125	3.2	1.9
GHGI[c]	1.71	0.89	1.76	0.97	0.063	– 0.95	0.066	– 1.07	0.029	0.018

a Finished diesel and gasoline for FTL cases (63.5% and 36.5% on LHV basis). Finished gasoline in the MTG case.

b If all emissions are charged to gasoline and/or diesel fuels.

c GHGI, the greenhouse gas emissions index, is the system wide life cycle GHG emissions for production and consumption of the energy products relative to emissions from a reference system producing the same amount of liquid fuels and electricity. For FTL systems, the reference liquid fuels are a mix of gasoline and diesel from crude oil for which the average GHG emission rate is 91.6 kgCO_2-eq/GJ$_{LHV}$. For MTG systems the reference liquid fuels are gasoline and LPG having life cycle GHG emission rates of 91.3 kgCO_2-eq/GJ$_{LHV}$ and 86.0 kgCO_2-eq/GJ$_{LHV}$, respectively. In all cases the reference system electricity is assumed to be from a new supercritical pulverized coal power plant for which the average GHG emissions rate is 830.5 kgCO_2-eq/MWh$_e$.

Source: Liu et al., 2011a; Liu et al., forthcoming.

about 1.5 percentage points penalty in efficiency (with either MTG or FTL) when CCS is added. The penalty is primarily due to the added electricity needed onsite to compress the captured CO_2 for transport and injection underground.

For systems described in Table 12.15 that use biomass it is assumed that the total biomass input is 1 Mt/yr (dry basis), a practical limit on the biomass delivery rate. For pure biomass systems (see Figure 12.20), this implies a liquid fuel production capacity of 4500–4600 bbl/day. The overall efficiency for the biomass FTL design is modestly higher than that for the coal-FTL plant. For the MTG plant, however, the efficiency is lower than for the coal-only design due to the syngas compression required in the biomass design to raise the pressure of the syngas to that needed for methanol synthesis. (The gasification pressure is higher for coal than for biomass, so no syngas compressor is required in the coal-to-gasoline designs.) The need for reforming in the recycle loop also contributes to the reduced efficiency for biomass conversion to MTG relative to coal.

The carbon mitigation performance of alternative options is indicated by a greenhouse gas emissions index (GHGI), the system wide life cycle GHG emissions relative to emissions from a reference system producing the same amount of fuels and electricity. It is assumed that the reference system consists of equivalent crude oil-derived liquid fuels and electricity from a new stand-alone supercritical pulverized coal power plant venting CO_2. The GHGI for each option is listed at the bottom of Table 12.15 and

summarized in Figure 12.22. For coal conversion to FTL or MTG without capture of CO_2, GHGI is 1.7 to 1.8 and GHGI is 0.9 to 1.0 with CCS. For biomass conversion, fuel cycle GHG emissions are < 0.07 when CO_2 is vented and strongly negative when CO_2 is captured and stored. For the designs coprocessing coal and biomass (see Figure 12.20), a mix (about 55% coal and 45% biomass, HHV basis) is chosen so that GHGI < 0.1. This GHGI constraint and the assumption of a biomass processing rate of 1 Mt/yr dry biomass imply that the liquid fuel production capacity is about 10,000 bbl/day in the CBTL and CBTG designs.

Biomass is a relatively scarce resource and the only carbon-bearing renewable energy source. Thus, the effectiveness of its use is an important consideration. Figure 12.23 shows one measure of effectiveness: liters of low/zero-GHG gasoline-equivalent FTL or MTG fuel produced per dry tonne of biomass consumed. Shown for comparison is an estimate for future cellulosic ethanol (EtOH) made (without and with CCS) from switchgrass biomass via enzymatic hydrolysis. Not surprisingly, with the FTL and MTG systems that coprocess coal and biomass, the total liquid fuel produced per unit of biomass input is high (because of the coal coprocessing). In the case of pure biomass FTL and MTG and EtOH systems with CCS, it is assumed that the negative GHG emissions provide "room in the atmosphere" for using some conventional crude oil-derived fuels while maintaining overall zero net GHG emissions. The total low-C liquid fuel yields for the cellulosic ethanol production options would be comparable to the biomass-only FTL and MTG systems that vent CO_2 but less than half the yields of low carbon fuels realized for all FTL and MTG systems with CCS.

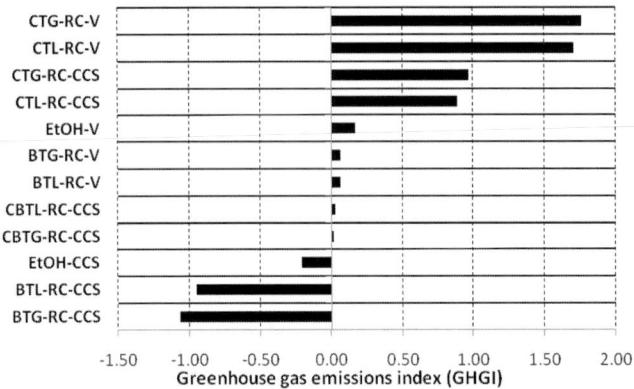

Figure 12.22 | Values of GHGI for the synthetic fuel options described in Table 12.15, along with the GHGI for future cellulosic-biomass ethanol technologies. See Table 12.15, note (c) for definition of GHGI. For details on the cellulosic ethanol options see Box 12.3 and Liu et al., 2011a.

12.4.3.3   Cost Estimates

For each of the plant designs described in Table 12.15, cost estimates are given in Table 12.16. Costs are reported here in US$_{2007}$\$ as discussed in Section 12.2.2.3.

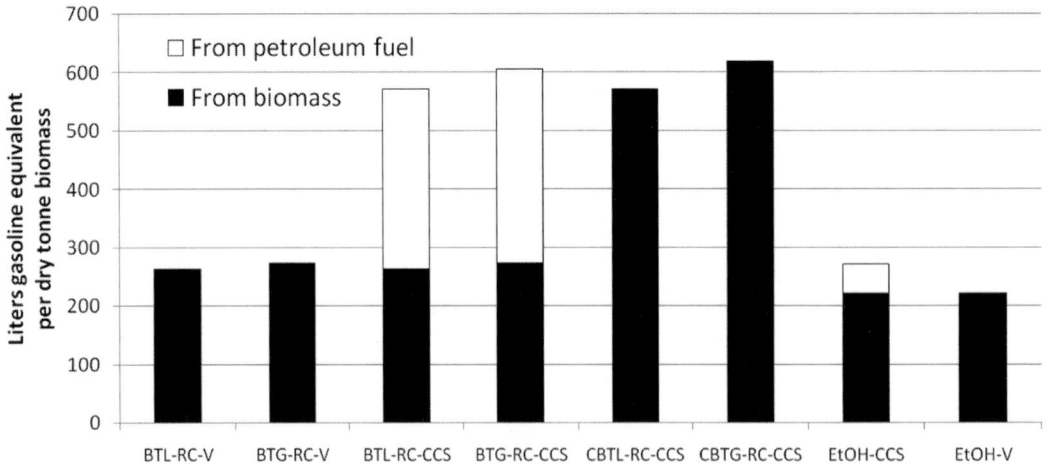

Figure 12.23 | Yields of low/zero net GHG emitting liquid fuels from biomass, liters of gasoline equivalent per dry tonne. For the biomass-only designs that incorporate CCS (BTL-RC-CCS, BTG-RC-CCS, and EtOH-CCS), life cycle GHG emissions are negative, leaving "room in the atmosphere" for some crude oil-derived fuel that can be used while keeping zero net GHG emissions for the biomass + crude oil-derived liquid fuels. Based on Liu et al., 2011a. See also Box 12.3 for a discussion of the cellulosic ethanol options.

Table 12.16 | Capital cost and production cost estimates (US$_{2007}$$) for conversion of coal, biomass, and coal+biomass to FTL or MTG.

Process configuration >>>	CTL-RC-V	CTL-RC-CCS	CTG-RC-V	CTG-RC-CCS	BTL-RC-V	BTL-RC-CCS	BTG-RC-V	BTG-RC-CCS	CBTL-RC-CCS	CBTG-RC-CCS
Coal input rate, MW HHV	7559	7559	6549	6549	–	–	0	0	804	781
Biomass input rate, MW HHV	0	0	0	0	661	661	661	661	661	661
Diesel and/or gasoline production, MW LHV	3159	3159	2913	2913	286	286	270	270	622	610
bbl/day crude oil products displaced, excl LPG	50,000	50,000	50,000	50,000	4521	4521	4630	4630	9845	10,476
Co-product LPG, MW LHV	–	–	309	309	–	–	28	28	–	69
Net export to grid, MW	404	295	126	36	42	31	32	20	53	11
Plant capital costs, million US2007$										
Air separation+ O$_2$ and N$_2$ compression	808	808	645	645	100	100	109	109	208	211
Biomass handling, gasification, and gas cleanup	0	0	0	0	336	336	340	340	335	347
Coal handling, gasification, and quench	1468	1468	1301	1301	0	0	0	0	226	189
water gas shift, acid gas removal, Claus/SCOT	849	849	589	598	59	59	89	89	158	162
CO$_2$ compression	0	67	0	59	2	14	2	14	22	22
F-T synthesis & refining or methanol synthesis	882	882	506	506	137	137	89	89	244	147
Naptha upgrading or MTG synthesis & refining	86	86	526	526	21	21	80	80	33	141
Power island topping cycle	35	27	0	0	0	0	0	0	7	0
Heat recovery and steam cycle	723	0	470	470	69	69	86	86	136	135
Total plant cost (TPC), million US$_{2007}$$	4852	4919	4038	4105	724	737	794	806	1369	1354
Specific TPC, US$_{2007}$$/bbl/day	97,033	98,372	80,757	82,099	160,189	162,927	171,520	174,131	139,091	129,200
Liquids production cost, US$_{2007}$$/GJ$_{LHV}$ (with zero GHG emissions price)[a]										
Capital charges (at 14.38% of Total Plant Inv., TPI)	8.34	8.45	7.52	7.65	13.77	14.00	15.97	16.22	11.95	12.03
O&M charges (at 4% of TPC/year)	2.16	2.19	1.95	1.98	3.57	3.63	4.15	4.21	3.10	3.12
Coal (at 2.04 US$/GJ$_{HHV}$, 55 US$/t, as-received)	4.87	4.87	4.58	4.58	0.00	0.00	0.00	0.00	2.63	2.60
Biomass (at 5 US$/GJ$_{HHV}$,94US$/t, dry)	0	0	0	0	11.56	11.56	12.24	12.24	5.31	5.41
CO$_2$ emissions charge	0	0	0	0	0	0	0	0	0	0
CO$_2$ disposal charges	0	0.52	0	0.46	0	1.38	0	1.41	0.94	0.90
Coproduct electricity (at 60 US$/MWh)	−2.13	−1.56	−0.72	−0.21	−2.46	−1.81	−1.98	−1.26	−1.42	−0.29
Co-product LPG revenue (at 100 US$/bbl)	–	–	−2.19	−2.19	–	–	−2.16	−2.16	–	−2.16
Total, US$_{2007}$$/GJ LHV	**13.24**	**14.48**	**11.14**	**12.27**	**26.44**	**28.76**	**28.22**	**30.67**	**22.51**	**21.62**
Total, US$_{2007}$$/gallon gasoline equivalent	1.59	1.74	1.33	1.47	3.17	3.45	3.38	3.68	2.70	2.59
Breakeven oil price, US$_{2007}$$/bbl[b]	61	67	48	53	133	145	126	137	111	96
Cost of avoided CO$_2$, US$_{2007}$$/t	–	12.4	–	12.8	–	20.9	–	19.3	16.9	29.4

a See note (b) of Table 12.6 for financial parameter assumptions.

b The breakeven oil price (BEOP) is calculated assuming the LPG co-product is sold for its wholesale price assuming the crude oil price is US$100/bbl. The wholesale price of LPG is determined as a function of crude oil price from a regression correlation of wholesale propane prices and refiner crude oil acquisition costs in the United States (propane (US$/bbl) = 0.7212 * Crude acquisition cost (US$/bbl) + 5.2468). See Kreutz et al., 2008 for additional discussion of the BEOP calculation.

Source: Liu et al., 2011a; Liu et al., forthcoming.

Specific capital costs for coal conversion without CCS range from US$81,000 per bbl/day to US$98,000 per bbl/day. Adding CCS involves a relatively small cost increment, since the primary additional cost is equipment for CO$_2$ compression. Specific costs for biomass conversion are considerably higher due largely to the much smaller scale of the conversion facility. Systems coprocessing coal and biomass are larger in scale than biomass-only systems, but smaller than the coal-only systems, which largely accounts for the intermediate level of specific capital costs.

Table 12.16 also shows both the levelized cost of fuel (LCOF) production and the crude oil price at which the FTL and MTG fuels would compete with petroleum-derived fuels when the price of GHG emissions is zero. For the coal-only plants, capital charges are the most significant production cost component, while for biomass-only facilities capital and feedstock are of comparable importance. For systems that vent CO$_2$ the breakeven oil price (BEOP) is in the range US$48–61/bbl for coal plants and US$126–133/bbl for biomass plants. For systems with CCS the BEOP is in the range

US$53–67/bbl for coal plants, US$137–145/bbl for biomass plants, and US$96–111/bbl for plants that coprocess coal and biomass.

As shown in Figure 12.24, when non-zero GHG emissions prices are considered, the relative economics of alternative process designs can be considerably different from those in Table 12.16. At GHG emission prices above a modest US$10–20/tCO$_2$-eq the –CCS variant of each option realizes a lower BEOP than the –V variant because the cost of CO$_2$ capture is low as a result of the production of a by-product stream of pure CO$_2$ as an intrinsic part of the design of gasification-based liquid fuels production. At GHG emission prices above US$65–75/tCO$_2$-eq the biomass –CCS option realizes a lower BEOP than the corresponding coal –CCS option (with more than 10 times the output capacity), as a result of the strong negative GHG emission rates for the biomass with –CCS options.

12.4.4 Hydrogen from Non-petroleum Feedstocks for Transportation

Hydrogen production from fossil fuels, the subject of this section, is well-established commercially in petroleum refining, ammonia production, and other industries where hydrogen is needed as a chemical feedstock.

H$_2$ produced electrolytically from non-carbon energy sources (wind, solar) is more costly than projected costs of making H$_2$ with ultra-low GHG emissions from coal or natural gas with CCS (Williams, 2002). The higher cost is largely because electricity purchases account for the largest share of total H$_2$ production costs with electrolysis. Wind and solar electricity are still more expensive than fossil fuel electricity today, despite reductions in costs for wind or solar electricity and escalations in

costs for construction of fossil energy conversion facilities. A key point is that electrolytic H$_2$ could plausibly be competitive as a transportation fuel (via use in fuel cell vehicles) only if offpeak/low-cost electricity (regardless of source) is used to make H$_2$. But only a tiny fraction of transportation fuel demand (which globally is 1.6 times electricity generation) could be satisfied with H$_2$ from offpeak electricity (a small fraction of total electricity).

Hydrogen is not used as a transportation fuel today, but its attractions in this application include the potential for low emissions of local pollutants and of greenhouse gases, as well as the energy security benefits arising from being able to shift transportation from oil dependence. Such attractions have made research on distribution and end-use systems for H$_2$ as a transportation fuel the focus of important government research and development programs in the United States, China, and elsewhere beginning in the 1990s. However, these R&D efforts have brought recognition that there are still major challenges to be overcome in H$_2$ delivery infrastructure and end use before it can become a real option for the transportation sector (CASFHPU, 2004; Agnolucci, 2007; CARNFCHT, 2008). Here we limit our discussion to a review of technologies for hydrogen production.

Globally, natural gas is the most commonly used feedstock for hydrogen production (Consonni and Vigano, 2005; Rostrup-Nielsen, 2005), but hydrogen can also be made from coal (Chiesa et al., 2005; Kreutz et al., 2005), as is the predominant commercial practice in China. Hydrogen can also be produced from biomass in systems closely resembling those for coal conversion (Hamelinck and Faaij, 2002; Lau et al., 2002).

Figure 12.25 shows a simplified block flow diagram for hydrogen production from coal. Gasification technologies for production of CO and H$_2$-rich synthesis gas from coal and biomass have been discussed in Section

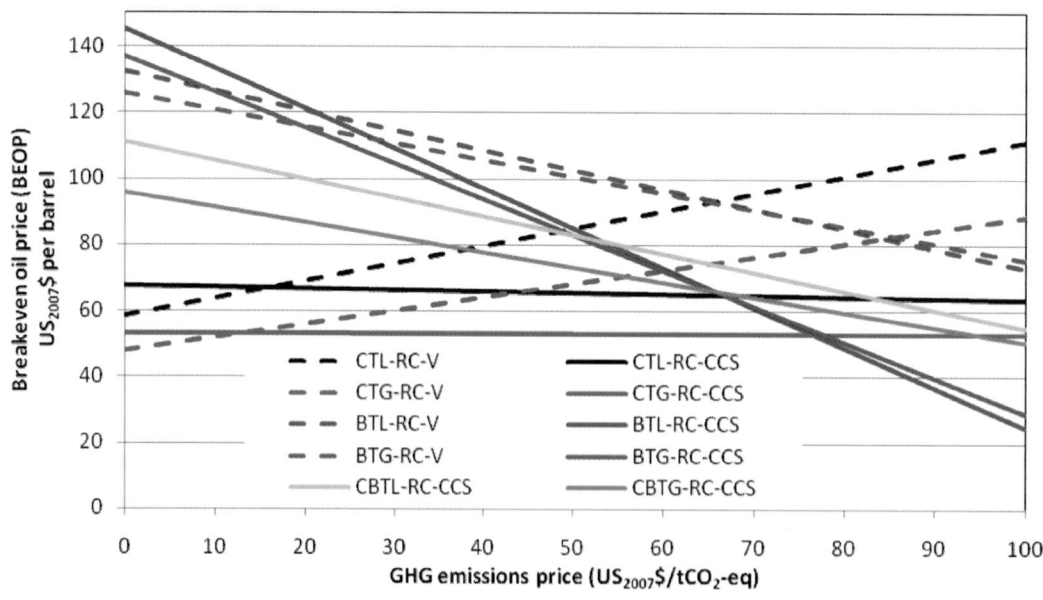

Figure 12.24 | Breakeven oil prices (US$_{2007}$$) as a function of GHG emission price for coal, biomass, and coal/biomass conversion to FTL or MTG. (See Table 12.6, note (b) for financial parameter assumptions. Also, electricity sales are assumed at the US average grid price plotted in Figure 12.10.)

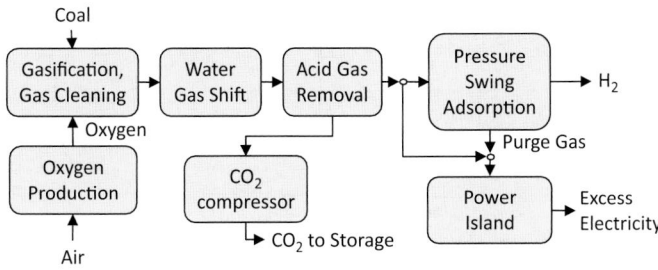

Figure 12.25 | Simplified process diagram for H_2 production from coal with CO_2 capture.

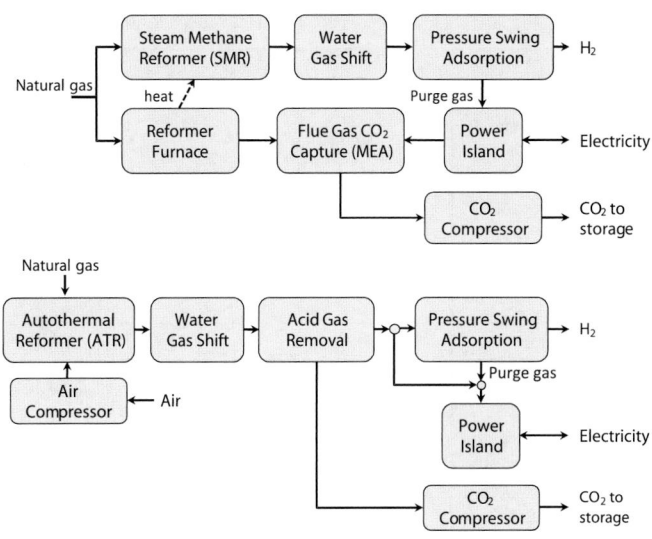

Figure 12.26 | Simplified process diagrams for H_2 production from natural gas via steam-methane reforming (top) or via autothermal reforming (bottom). When CO_2 is captured for storage (as shown), import of electricity may be required to meet on-site needs. Without CO_2 capture, there can be net electricity exports to the grid from the facility.

12.4.3.1. Gasification is followed by a water gas shift, after which sulfur species are removed to prevent poisoning of downstream catalysts. This presents a natural point for also removing CO_2 with little added capital cost. Acid gas removal using a physical solvent (e.g., methanol, as in the Rectisol process) would typically be used because of the elevated pressure of the gas mixture containing the acid gases. The captured sulfur is recovered from the solvent and converted, for sale or disposal, to elemental sulfur using the Claus/SCOT process. The captured CO_2 can be compressed for pipeline transport to an underground storage site. Following acid gas removal, the remaining hydrogen-rich stream is concentrated using pressure swing adsorption (PSA) technology to purity as high as 99.999%. Purge gases from the PSA, supplemented as needed by some syngas bypassing the PSA, are burned in the power island to provide the electricity needs of the plant and additional exportable power.

For hydrogen production from natural gas (Figure 12.26), syngas can be produced using any of several different reforming technologies (Rostrup-Nielsen and Rostrup-Nielsen, 2002). The two most common, SMR and ATR, have been discussed in Section 12.4.1.1. With natural gas conversion based on SMR (Figure 12.26, top), sulfur removal is not needed following the water gas shift, since any sulfur present in the methane input to the plant has been removed prior to the SMR to avoid poisoning the SMR catalyst. If CO_2 is to be captured for storage, this would most effectively be accomplished using an acid gas removal system on the flue gases from the power island and the reformer furnace. This design enables maximum CO_2 capture, because all flue gases are collected before CO_2 removal. However, the low-pressure of the CO_2 in the flue gases requires using a chemical solvent (such as MEA) to effectively capture the CO_2. The work to subsequently compress the captured CO_2 for pipeline transport will be higher than in the systems producing hydrogen using ATR, for which the CO_2 is available at a higher starting pressure. For a system using ATR (Figure 12.26, bottom), capture of CO_2 would take place after the water gas shift using a physical solvent, such as Rectisol.

Table 12.17 summarizes comparative mass and energy balance simulation results developed for this chapter for hydrogen production. It is based on detailed system designs represented in simplified form in Figures 12.25 and 12.26. For designs that capture CO_2, the overall energy efficiency of coproducing hydrogen and electricity is 67% with coal and 74% with natural gas. The efficiency penalty for systems that

capture CO_2 compared with systems that vent the CO_2 is largest for the natural gas case using SMR (Figure 12.26, upper). This is due to the substantial heat required to liberate the dissolved CO_2 from the solvent used to capture it. The heat is provided as steam extracted from the power island, which reduces the on-site power generation significantly. Natural gas conversion using the ATR also requires some net import of electricity due to the substantial power requirements for air separation and CO_2 compression.

Capital and operating cost estimates for the systems described in Table 12.17 are provided in Table 12.18 using the same framework and component capital cost database as for systems described in Sections 12.4.3 and 12.6. Hydrogen production costs using natural gas, despite the considerably lower capital cost intensity compared with coal designs, are nevertheless higher than for coal due to the much higher assumed feedstock prices. Production costs with natural gas are also higher in the CCS cases due to the need to purchase electricity to operate the facility rather than selling excess electricity to the grid.

The final row of Table 12.18 shows the avoided cost of CO_2 emissions when CCS is considered. The avoided cost for coal is modest (US$11/$tCO_2$-eq) because capturing the CO_2 at these plants involves only adding a CO_2 compressor. The situation is similar for the natural gas system using ATR, but the avoided CO_2 cost is higher than for a coal-CCS design of comparable scale because the added cost includes that for acid-gas capture in addition to a CO_2 compressor. (The cost for acid gas removal is modest because only half as much CO_2 must be captured as in the coal case due to the lower carbon intensity of natural gas.) In the SMR design for natural gas, costs for the added equipment to capture dilute CO_2 from flue gases and compress it are charged against the CO_2, leading to relatively higher avoided costs.

Table 12.17 | Mass and energy balances for hydrogen production for different feedstocks.

Feedstock >>>	Coal[a]		N. gas – SMR[b]		N. gas – ATR[b]	
CO₂ vented or captured >>>	Vent	CCS	Vent	CCS	Vent	CCS
Power island technology >>>	Combined Cycle		Steam Rankine Cycle			
Coal input, as-received t/day	12,615	12,615				
Coal input, MW HHV	3,817	3,817				
Biomass input, as-received t/day						
Biomass input, MW HHV						
Natural gas input, t/day			5,561	5,561	5,519	5,519
Natural gas input, MW HHV			3,335	3,335	3,310	3,310
Hydrogen Production, MW LHV	2,083	2,083	2,083	2,083	2,083	2,083
Hydrogen Production, MW HHV	2,461	2,461	2,461	2,461	2,462	2,462
Electricity						
Gross production, MW	424.8	424.8	217.5	85.6	262.0	262.0
On-site consumption, MW	272.4	349.2	27.2	103.4	220.4	274.3
Net export to grid, MW	152.4	75.6	190.3	-17.8	41.6	-12.3
ENERGY RATIOS (HHV basis)						
H₂ out / energy in	64.5%	64.5%	73.8%	73.8%	74.4%	74.4%
net electricity / energy in	4.0%	2.0%	5.7%	-1.4%	1.3%	-0.4%
H₂ + electricity / energy in	68.5%	66.5%	79.5%	72.4%	75.6%	74.0%
CARBON ACCOUNTING						
C input as feedstock, kgC/sec	**89.5**	**89.5**	**47.7**	**47.7**	**47.3**	**47.3**
C stored as CO₂, % of feedstock C	0.0	91.2	0.0	90.0	0.0	82.8
C in char (unburned), % of feedstock C	0.8	0.8	0.0	0.0	0.0	0.0
C vented to atmosphere, % of feedstock C	99.2	8.0	100	9.9	100	17.1
C stored, MtCO₂/yr (90% capacity factor)	0.0	8.5	0.0	4.5	0.0	4.1
Lifecycle GHG emissions						
Net emissions, kgCO₂-eq/ GJ H₂ LHV	163	19.1	84.3	10.8	83.6	16.1
GHGI[c]	1.74	0.22	0.86	0.14	1.03	0.21

a Results based on performance simulations of Chiesa et al., 2005.

b Results based on performance simulations of Zhang, 2005.

c GHGI, the greenhouse gas emissions index, is the system wide life cycle GHG emissions for production of H₂ and electricity relative to emissions from a reference system producing the same amount of H₂ and electricity. The reference system consists of large-scale centralized H₂ production by steam reforming of natural gas, with lifecycle emissions of 9.22 kgCO₂-eq/kgH₂ (NRC, 2004), plus electricity from a supercritical pulverized coal power plant with GHG emissions rate of 830.5 kgCO₂-eq/MWhe.

d In the GHGI calculation, net electricity consumed in these process designs is charged 830.5 kgCO₂-eq/MWhe. All other designs in this table have net exports of electricity.

Table 12.18 | Cost estimates (US$_{2007}$$) for hydrogen production systems described by Table 12.17.

Feedstock >>>	Coal		N. gas – SMR		N. gas – ATR	
CO_2 vented or captured >>>	Vent	CCS	Vent	CCS	Vent	CCS
Coal input rate, MW HHV	3817	3817				
Biomass input rate, MW HHV						
Natural gas input rate, MW HHV			3335	3335	3310	3310
hydrogen production rate, MW LHV	2083	2083	2083	2083	2083	2083
Net export to grid, MW	152.4	75.6	190.3	-17.8	41.6	-12.3
Plant capital costs, million US$_{2007}$$ [a]						
Air separation unit + O_2, N_2, air compressor	404	404	0.0	0.0	79	79
Biomass handling, gasification, gas cleanup	0.0	0.0	0.0	0.0	0.0	0.0
Coal handling, gasification, and quench	791	791	0.0	0.0	0.0	0.0
Reforming (SMR or ATR)	0.0	0.0	737	737	244	244
WGS, acid gas removal, Claus/SCOT[b]	581	581	161	161	222	264
MEA system for SMR case (CO_2 removal)	0.0	0.0	0.0	135	0.0	0.0
CO_2 compression	9.3	62	0.0	48	0.0	38
PSA section (including H_2 compression)	83	83	34	34	76	58
Power island topping cycle	62	62	0.0	0.0	0.0	0.0
Heat recovery and steam cycle	139	139	110	79	222	237
Total plant cost (TPC), million US$_{2007}$$	2067	2120	1042	1194	844	920
US$_{2007}$$/kW$_{HHV}$ of input feedstock	542	555	312	358	255	278
Levelized hydrogen cost with no carbon emission price, US$_{2007}$$/GJ LHV[c]						
Capital charges	5.4	5.5	2.7	3.1	2.2	2.4
O&M charges	1.4	1.4	0.7	0.8	0.6	0.6
Coal (at 2.04 US$/GJ, HHV; 55 US$/t, as-rec'd)	3.7	3.7	0.0	0.0	0.0	0.0
Biomass (at 5 US$/GJ, HHV; 93.7 US$/t, dry)	0.0	0.0	0.0	0.0	0.0	0.0
NG (at 5.11 US$/GJ, HHV)	0.0	0.0	8.2	8.2	8.1	8.1
CO_2 transportation and storage	0.0	0.7	0.0	0.4	0.0	0.4
Electricity sales or purchase (at 60 US$/MWh)	-1.2	-0.6	-1.5	0.1	-0.3	0.1
Total hydrogen cost, US$_{2007}$$/GJ LHV	9.3	10.8	10.1	12.7	10.6	11.7
Cost of avoided CO_2, US$_{2007}$$/tCO$_2$ed	–	11	–	47	–	17

a Component costs are based on Liu et al., 2011a, except for: SMR (Simbeck, 2004); ATR (Simbeck and Wilhelm, 2007); MEA system (Kreutz et al., 2005; Woods et al., 2007).

b In the case of N.gas SMR with CCS, only the WGS cost is included in this line since acid gas removal is done via MEA (separate cost line) and no sulfur treatment via Claus/SCOT is needed.

c See note (b) of Table 12.6 for financial parameter assumptions.

d Levelized H_2 production cost for system with CCS minus that for system without CCS, divided by the difference in system-wide life cycle emissions of CO_2-eq/GJ$_{LHV}$ of H_2 (given in Table 12.17).

Figure 12.27 plots hydrogen production costs as a function of the price of greenhouse gas emissions. Coal provides the least costly low-GHG hydrogen (US$10–12/GJ$_{LHV}$) in the system with CCS for a GHG emission price above US$10/tCO$_2$-eq.

An important final comparison is between the costs of hydrogen production with CCS and those for low-GHG liquid fuels. For GHG emission prices from US$0–100/tCO$_2$-eq, the lowest production costs for low-GHG liquid fuels are with systems that coproduce liquids and power (see Section 12.6). Liquid fuel costs from such systems are $US15–17/GJ$_{LHV}$ depending on the GHG emissions price (see Figure 12.36). This is higher than the US$10–12/GJ$_{LHV}$ estimated for hydrogen (Figure 12.27). But use of hydrogen for transportation will require new delivery and refueling infrastructures, as well as new vehicle technologies, unlike for petroleum-like liquid fuels.

Cost estimates in the literature for new infrastructure in the United States delivering H$_2$ from centralized production facilities to vehicle fuel tanks are much higher than for new infrastructure for liquid fuels. Published hydrogen infrastructure cost estimates include US$_{2007}$$14–16/GJ$_{LHV}$ (Ogden et al., 2004), US$_{2007}$$9–20/GJ$_{LHV}$ (Mintz et al., 2002), and US$_{2007}$$44/GJ$_{LHV}$ (Simbeck, 2003). Compared to these costs, the new-infrastructure costs for delivery of liquid fuels, particularly petroleum-like fuels, would be small. Moreover, in the industrialized countries, the investment in a full liquid fuel infrastructure has already been made. Thus, when production and delivery infrastructure costs are considered together, delivered hydrogen costs would be significantly higher than delivered costs for liquid fuels.

Beyond questions around fuel delivery infrastructure, it remains uncertain when the cost of hydrogen fuel cell vehicles, which have been the focus of considerable development efforts over the past two decades, can be reduced to competitive levels. One study (IEA, 2009a) estimates that in the near term the GHG emissions price needed to induce by market forces a shift to hybrid fuel cell vehicles is more than US$1000/tCO$_2$-eq when the crude oil price is US$60/bbl and almost US$800/t with US$120/bbl of crude oil. The major technical challenges for fuel cell vehicles are difficulties of onboard H$_2$ storage due to its low volumetric energy density and high projected costs for fuel cell engines (in part due to platinum requirements). These challenges may not be insurmountable, but overcoming them will require time and sustained large government investments in R&D (CARNFCHT, 2008).[12] Thus, it is likely to be many decades before H$_2$ fuel cell vehicles will be in a position to make significant contributions toward reducing GHG emissions.

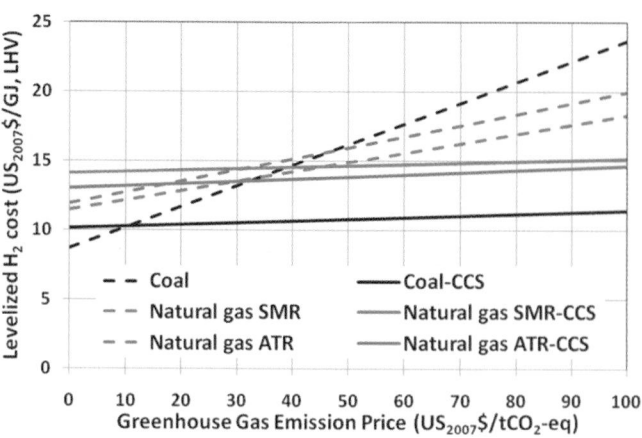

Figure 12.27 | Levelized hydrogen production costs as a function of GHG emissions price. (See Table 12.6, note (b) for financial parameter assumptions. Also, electricity sales are assumed at the US average grid price plotted in Figure 12.10.) Source: based on Simbeck, 2004; Simbeck and Wilhelm, 2007; Liu et al., 2011a.

12.5 Clean Household Fuels Derived from Non-petroleum Feedstocks

Studies have shown that human welfare, as measured by the Human Development Index (HDI), increases with diminishing returns as the level of modern energy services provided increases (WEA, 2004). The HDI increase is especially large for provision of the first increments of modern energy carriers to satisfy basic needs such as cooking and heating, for which demand is very inelastic (cooking and boiling water are essential for survival). There is wide recognition of the importance of the role that electricity plays in economic development and the fact that more than a billion people do not have access even to the minimal amounts of electricity required to satisfy basic needs. Similarly, it is recognized that there are nearly three billion people who cook their food today using traditional open fires inside their homes, suffering considerable health damages in the process (see Hutton et al., 2006 and Chapters 4, 17, and 19).

Fluid hydrocarbon fuels offer a much cleaner means of providing cooking services than solid fuels (Smith, 2002). Moreover, fluid fuels enable much higher efficiency (Dutt and Ravindranath, 1993) and controllability than cooking with solid fuels.[13] Historical real data confirm the gains from efficiency and controllability (Figure 12.28). Such gains along with consideration that a shift to cooking with clean fuels leads to demands on a global basis that are relatively small compared to energy demands for other purposes, means that a relatively small amount of fluid fuel would be sufficient to replace all current solid fuel used for cooking.

---

12 Waning enthusiasm for addressing these challenges was evident in the Obama Administration's proposed 2010 Department of Energy (DOE) budget, which included a cut in the federal hydrogen fuel cell research and deployment budget by more than two thirds, eliminating funds for the H$_2$ fuel cell vehicle program and market transformation programs.

13 Consider LPG stoves. Not only are they much more energy-efficient than biomass stoves, but also an LPG stove can be instantly turned on and off with the demand for cooking services, whereas a biomass stove must be started up long before cooking begins and continue burning long after cooking stops. Of course, the continued burning of biomass after a meal is cooked in the evening may often be for lighting. This implies that, if LPG is to be substituted for biomass for cooking, this cooking fuel switch should often be accompanied by the introduction of alternative means for lighting.

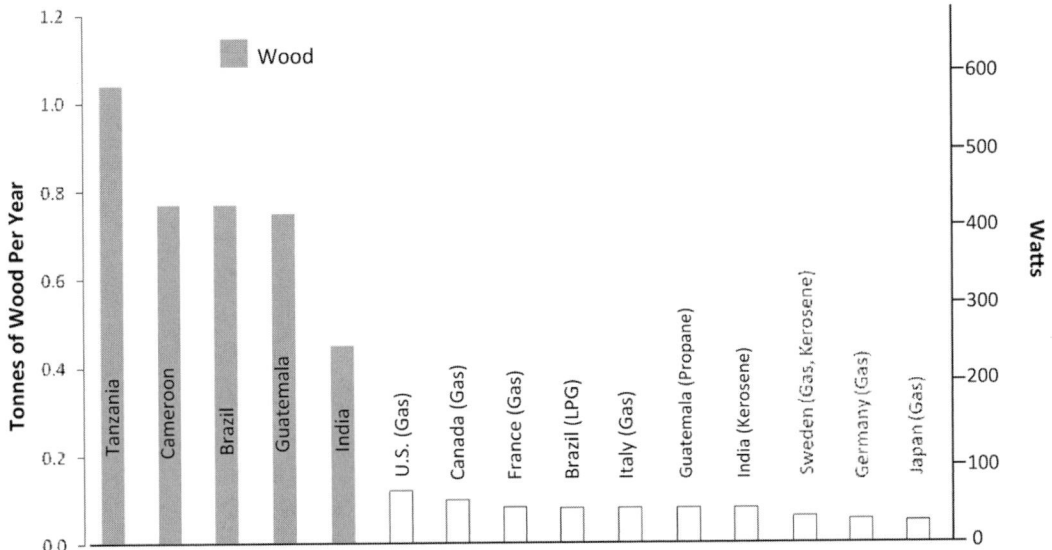

Figure 12.28 | Per capita energy use rate for cooking in the early 1980s. For both wood stoves and stoves burning high-quality energy carriers, the per capita energy use rate is in Watts; for wood fuel, the rate is also in t/yr of dry wood. (Assuming 1 tonne = 18 GJ, 1 t/yr = 570 Watts). Source: Goldemberg et al., 1985.

For perspective, suppose that all three billion people currently using solid fuels for cooking were instead to use liquefied petroleum gas (LPG) at an estimated average needed rate of 25 kg/capita/yr (36 W/capita).[14] The total annual requirement would be 3.4 EJ/yr, or 1.3% of current total global oil and gas consumption.

A shift from solid to fluid fuels for cooking would yield substantial public benefits in terms of improved public health and time saved not spent collecting solid fuels, which would make time available to pursue educational and other opportunities (see Chapter 4). Another benefit would be reduced deforestation, to the extent that some biomass collected for cooking is removed unsustainably from forests.

Such considerations have led to proposals for concerted global efforts to replace solid cooking fuels with clean fluid fuels worldwide (Goldemberg et al., 2004; IEA, 2006; WHO, 2006; GACS, 2011).

Concerns about rising costs of and overdependence on petroleum imports have created interest in alternatives to petroleum-derived fluid cooking fuels such as kerosene and LPG. This section discusses production systems for expanding clean fluid cooking fuel supplies. It considers the use of synthetic fluid fuels (synthetic LPG and DME) derived via gasification of coal and/or biomass, without and with carbon capture and storage. Coal and biomass are the most widely available feedstocks in regions where solid fuels are now used for cooking. DME as a cooking

fuel could also be made from natural gas (Naqvi, 2002). The growing optimism that shale gas might prove to be widely available (see Section 12.7.2.2) suggests also giving close attention to DME derived from natural gas for cooking, though this topic is not covered here.

12.5.1 Dimethyl Ether from Coal

Dimethyl ether is a colorless gas at ambient temperature and pressure, with a slight ethereal odor. It requires mild pressurization, similar to that required for LPG, to be stored as a liquid. It burns with a clean blue flame over a wide range of air/fuel ratios. It can be used as a diesel engine fuel (Semelsberger et al., 2006) or blended with LPG for use as a household or commercial sector fuel. In the latter application, the focus of the discussion here, the DME can be blended up to about 25% by volume without the need to change end-use combustion equipment. Table 12.19 compares some physical properties of DME with those of the two main constituents of LPG.

Until recently, DME was used primarily as an aerosol propellant in hair sprays and other personal care products and was produced globally at a rate of about 150,000 t/yr (Naqvi, 2002). This production level has increased dramatically in the past few years, with the added DME being used primarily as an LPG supplement for household use. The increase has been most substantial in China, where an estimated total DME production capacity of nearly 14 Mt/yr from coal have recently commenced production or construction, and a comparable amount of additional capacity is at various planning, feasibility, or engineering stages (Zheng et al., 2010).

Production of DME from synthesis gas is similar in many respects to synthesis of methanol, a well-established commercial process. In fact, a key step in the synthesis process is catalytic synthesis of methanol, followed

---

14 In a spreadsheet accompanying the World Health Organization paper (Hutton et al., 2006) that was made available to the authors, estimates were developed by this WHO group of the LPG that would be required to replace direct use of biomass for cooking in regions throughout the world. Worldwide the average per capita amount of LPG needed annually was estimated to be 24.6 kg, but this rate varies from an average of 14.3 kg for Africa, to 25.6 kg for the region including India, to 29.3 kg for the region including China, to 47.0 kg for the region including Brazil.

Table 12.19 | Physical properties of DME, propane, and butane.

	DME	Propane	Butane
Boiling point (°C)	−24.9	−42.1	−0.5
Vapor pressure ant 20°C (bar)	5.1	8.4	2.1
Liquid density at 20°C (kg/m³)	668	501	610
Lower heating value (MJ/kg)	28.4	46.4	45.7
Auto-ignition temperature at 1 atm pressure (°C)	235 – 350	470	365
Flammability limits in air (vol %)	3.4 – 17	2.1 – 9.4	1.9 – 8.4

Source: Larson and Yang, 2004.

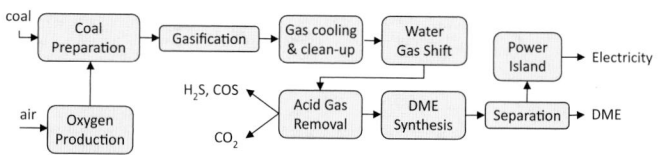

Figure 12.29 | Process steps for DME production from coal.

by dehydration of methanol to form DME: $2CH_3OH \leftrightarrow CH_3OCH_3 + H_2O$. Methanol can be synthesized at one plant location and transported to another location where dehydration can be done. This is how most DME has historically been produced. Alternatively, methanol and dehydration plants can be integrated at a single facility. Moreover, technology is now available for a single-step synthesis of DME from syngas: some methanol synthesis catalyst and some dehydration catalyst are used together in the same reactor so that methanol is dehydrated as it forms. The single-step synthesis chemistry can be represented as follows: $3CO + 3H_2 \leftrightarrow CH_3OCH_3 + CO_2$. This approach gives rise to higher yields than the traditional two-step process and so is likely to be the technology of choice in the future.

Figure 12.29 illustrates one possible process arrangement for converting coal to DME. Coal is gasified in oxygen to produce a raw syngas that is cooled and cleaned before having its H_2:CO ratio adjusted in a water-gas-shift reactor (using a sulfur-tolerant catalyst) to an optimum value for subsequent catalytic synthesis of DME. The synthesis step can use a recycle of unconverted gas to increase DME production or a single-pass of the gas through the synthesis reactor ("once-through" design). The unconverted gas can be burned in a gas turbine/steam turbine combined cycle to generate coproduct electricity. Removal of CO_2 is an essential part of the DME production process, since excess CO_2 reduces the efficiency of the downstream synthesis step and would also necessitate larger downstream equipment. The captured CO_2 can be released to the atmosphere or compressed for pipeline transport and underground storage.

Celik et al. (2004), building on work by Larson and Ren (2003), present detailed process designs and production costs for large-scale single-step "once-through" DME synthesis systems starting with coal as the feedstock. Designs with and without CCS were analyzed (Table 12.20). For the design labeled "UCAP" (shorthand for "upstream CO_2 capture"), CO_2 for storage is captured upstream of the DME synthesis area (as depicted in Figure 12.29). Only about 30% of the carbon in the coal feedstock is captured in this case. Nearly 70% of the carbon is released to the atmosphere as flue gas from the power island and when the product DME is burned. For the systems labeled DCAP (shorthand for "downstream plus upstream CO_2 capture"), additional CO_2 is removed by subjecting the unconverted syngas after synthesis to varying levels

of water gas shift before the power island. In the case with the highest amount of CO_2 capture (DCAP-3), nearly 80% of the carbon in the coal is captured.

The GHGI, defined in Table 12.20 note (b), without CCS is 1.3. The UCAP design reduces this to 0.94. With more aggressive CCS (DCAP designs), greenhouse gas emissions can be reduced to less than 50% of the reference system emissions. The cost of avoided CO_2 emissions relative to the VENT design are modest for the UCAP case (last row, Table 12.20), since the cost of capture is an intrinsic part of the DME production process regardless of whether CO_2 storage is contemplated. Costs of avoided CO_2 emission are higher for the DCAP designs because the cost for the additional CO_2 capture equipment is fully charged to CO_2 capture.

The added capture equipment also leads to higher DME production costs. For US conditions, with the financial assumptions noted in Table 12.20 (note d), the VENT design produces DME at an estimated cost of US$423/tLPG-eq, corresponding to a breakeven crude oil price (BEOP) of about US$40/bbl. The BEOP is only slightly higher for the UCAP case, but significantly higher for the DCAP cases, reaching US$81/bbl for the DCAP-3 design. Larson and Yang (2004) estimate that costs of DME production in China today might be 15% lower than these estimates for US conditions.

12.5.2   Synthetic LPG from Coal and/or Biomass

Clean cooking fuels can also be produced from coal and/or biomass via the F-T or methanol-to-gasoline processes described in Section 12.4.3. The designs discussed there are for production of synthetic transportation fuels as the primary products, but significant quantities of C3 and C4 hydrocarbons are produced as intermediate or final products. In either system, these lighter hydrocarbons can be separated from the heavier transportation fuel products for sale as synthetic LPG.

For the design of the F-T systems described earlier, it was assumed that the light hydrocarbon fraction was consumed internally as a component of the fuel gas for the power island. Alternatively, the lighter fraction could have been separated as an additional coproduct. The raw synthesis product can contain as much as 20% by weight of C3 and C4 compounds that constitute synthetic LPG.

Table 12.20 | Performance and cost (US$_{2007}$$) estimates for DME production from coal with different levels of CCS. "UCAP" refers to designs with CO_2 removal upstream of DME synthesis (as in Figure 12.29). "DCAP" refers to designs that additionally capture some CO_2 downstream of synthesis. Three alternative downstream capture designs are considered. See Celik et al. (2004) for details.

	No CCS	With varying levels of CO_2 capture			
	Vent	UCAP	DCAP-1	DCAP-2	DCAP-3
Coal input, MW LHV	2203	2203	2203	2203	2203
DME output, MW LHV	600	600	600	600	600
Gross power production, MW	628	628	589	590	586
Net power export, MW	490	469	367	365	353
Fraction of coal LHV converted to DME	0.272	0.272	0.272	0.272	0.272
Fraction of coal LHV converted to net power	0.223	0.213	0.167	0.166	0.160
Total efficiency (LHV basis)	49.5	48.5	43.9	43.8	43.2
Plant carbon balance, tC/hr					
Input as coal	199.7	199.7	199.7	199.7	199.7
Buried as char	2.0	2.0	2.0	2.0	2.0
Captured as CO_2	0	57.5	143.5	147.6	156.4
Total C captured, % of coal C	0	30%	73%	75%	79%
System life cycle GHG emissions[a]					
kgCO_2-eq/GJ$_{DME}$ LHV	349	251	105	98	84
GHGI[b]	1.27	0.94	0.46	0.43	0.38
Costs					
Overnight Installed Capital (million US$_{2007}$$)[c]	1,306	1,198	1,269	1,308	1,317
Levelized DME production cost (US$_{2007}$$/GJ$_{LHV}$)[d]					
Capital, US$/GJ$_{LHV}$	13.85	12.70	13.46	13.87	13.96
O&M, US$/GJ$_{LHV}$	3.45	3.17	3.35	3.46	3.48
Coal, US$/GJ$_{LHV}$	7.49	7.49	7.49	7.49	7.49
CO_2 transport and storage, US$/GJ$_{LHV}$	0	1.56	3.72	3.84	4.05
Electricity sales, US$/GJ$_{LHV}$	−13.61	−13.03	−10.19	−10.14	−9.80
Total (US$_{2007}$$/GJ$_{LHV}$)	**11.18**	**11.89**	**17.83**	**18.52**	**19.18**
Total (US$_{2007}$$/tDME)	317	337	506	526	544
Total (US$_{2007}$$/tLPG-eq)	514	546	820	852	882
Breakeven crude oil price (US$_{2007}$$/bbl)[e]	50	54	85	88	92
Cost of CO_2 captured (US$_{2007}$$/tCO_2 avoided)[f]		7.3	27	29	30

a Including emissions associated with coal mining and delivery (1.024 kgC-eq/GJ$_{COAL,LHV}$), emissions at the conversion plant, and emissions from combustion of the DME.

b GHGI, the greenhouse gas emissions index, is the system wide life cycle GHG emissions for production of DME and electricity relative to emissions from a reference system. The reference system consists of LPG from conventional sources, with estimated lifecycle emissions 86 kgCO_2/GJ$_{LHV}$, plus electricity from a supercritical pulverized coal power plant with GHG emissions rate of 830.5 kgCO_2-eq/MWhe.

c Converted from US$_{2003}$$ in Celik et al. (2004) to US$_{2007}$$ using the Chemical Engineering Plant Cost Index.

d Assuming 7% interest during construction, 15% per year capital charge, 80% capacity factor, annual O&M cost of 4% of overnight capital cost, coal cost of US$2.04/GJ, electricity revenue of US$60/MWh (the 2007 US average generator sale price), and CO_2 transport and storage cost of US$15/t$CO_2$, and zero GHG emissions price.

e A linear regression of monthly wholesale propane price and refiner acquisition cost for crude oil in the United States for the period October 1990 to March 2009 US EIA, 2009b gives the following correlation: Propane price (US$/gallon) = 0.0168 * (US$/bbl$_{oil}$) + 0.133 [R^2 = 0.92]. Assuming the propane density given in Table 12.19, the breakeven oil price in US$/bbl is (US$/t$_{propane}$ − 70.13) / 8.87. We assume this correlation holds equally for LPG.

f This is the difference in US$/GJ$_{LHV}$ levelized cost of DME production with CO_2 capture and the VENT design divided by the difference in system-wide life cycle emissions of CO_2-eq/GJ$_{LHV}$ of DME.

Table 12.21 | Key features of alternative process designs for producing synthetic gasoline and LPG from coal and/or biomass.

Process description[a]	Output capacities		LPG output (10⁶ kg/y)	TPC (millionUS₂₀₀₇$)	Biomass fraction (HHV)	Biomass input (10⁶ dt/y)	CO₂ storage rate (10⁶ t/y)	CCS primary energy penalty (%)	GHGI[b]
	Gasoline (bbl/day)	Electricity (MW$_e$)							
CTG-PB-V	32,579	959	125	4,110	0	0	0	–	1.37
CTG-PB-CCS	32,579	760	125	4,310	0	0	10.2	9.8	0.56
CBTG1-PB-CCS	32,579	782	123	4,526	0.1	0.99	10.3	9.5	0.40
CBTG-PB-CCS	11,582	292	42.2	2,086	0.29	1.0	3.7	8.9	0.098
BTG-RC-V	2,315	16.0	8.72	475	1.0	0.5	0	–	0.066
BTG-RC-CCS	2,315	10.2	8.72	482	1.0	0.5	0.513	5.7	– 1.07

a CTG = coal to gasoline+LPG; CBTG = coal+biomass to gasoline+LPG; CBTG1 = coal+biomass to gasoline+LPG with reduced biomass fraction; BTG = biomass to gasoline+LPG; PB = partial bypass of syngas around synthesis island for use in power island; RC = recycle of unconverted syngas to maximize liquids production; V = venting of CO₂; CCS = carbon capture and storage.

b GHGI, the greenhouse gas emissions index, is the system wide life cycle GHG emissions for production and consumption of the energy products relative to emissions from a reference system producing the same amount of liquid fuels and electricity. The reference system consists of electricity from a stand-alone new supercritical pulverized coal power plant venting CO₂ plus equivalent crude oil-derived liquid fuels. For details, see Table 12.15, note (c).

For the MTG systems described in Section 12.4.3, a synthetic LPG coproduct is produced, equivalent to about 10% of the synthetic gasoline output (LHV energy basis, see Table 12.15). Systems such as this one might contribute to addressing the challenge of providing universal access to clean cooking fuels, as discussed in the next section.

12.5.3 Co-providing synthetic cooking and transport fuels in the context of a carbon mitigation policy

Increasing conventional LPG use to meet the basic cooking fuel needs of those currently cooking with solid fuels would make a relatively small total energy impact and GHG emission impact. As such, one might argue that meeting the critical energy needs of the energy poor should not be constrained by a requirement that access to energy be provided in a manner consistent with simultaneously mitigating the climate change impacts of the cooking fuel consumed.

In some cases, however, a strong carbon mitigation policy may actually improve the prospects for providing clean energy to satisfy basic cooking needs. This judgment is illustrated here by considering the technology (Table 12.21) and economics (Figure 12.30) of six MTG process designs. Figure 12.30 shows that for five of these systems the economics improve with GHG price. These five systems (four of which involve CCS[15] and four of which involve biomass) all offer substantial reductions in GHG emission rates relative to the crude oil derived products (CODP) and PC-V coal electricity displaced (see final column in Table 12.21). The two pure biomass designs in this table have been

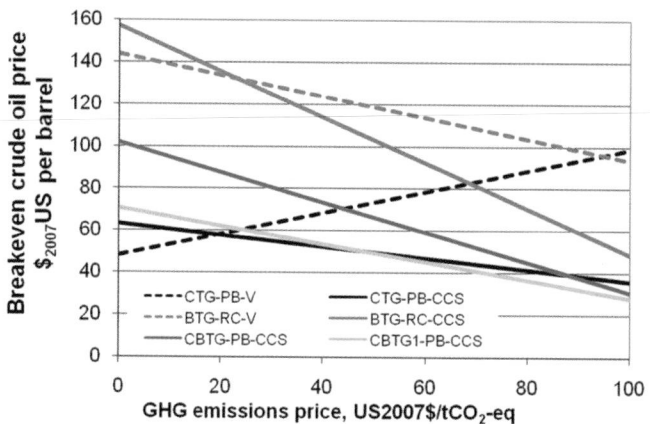

Figure 12.30 | Breakeven crude oil price (BEOP) as a function of the GHG emissions price for the MTG options described in Table 12.21. (See Table 12.6, note (b) and Table 12.16, note (b) for financial parameter and other assumptions. Electricity sales are assumed at the US average grid price plotted in Figure 12.10.)

discussed in Section 12.4.3. The other four designs, which involve the generation of electricity as a major co-product, are described in Section 12.6.

Key observations can be made about the relative economics of the six MTG options:

- CTG-PB-CCS is a coal-only option offering a 44% reduction in system-wide GHG emissions relative to the CODP and PC-V electricity displaced. It offers less costly gasoline than the GHG-emissions-intensive CTG-PB-V for GHG emissions prices greater than US$20/tCO₂-eq.

- CTG-PB-CCS and CBTG1-PB-CCS (an option with biomass accounting for 10% of input energy offering a 60% reduction in system-wide GHG

---

15 For all CCS options it is assumed that the captured CO₂ is compressed to 150 bar, transported 100 km, and stored 2 km underground in a deep saline formation via wells for which the maximum injectivity is 2500 tonnes per well per day.

GHG emissions) are both competitive as gasoline providers when crude oil is US$70/bbl at zero GHG emissions price.

- CBTG1-PB-CCS offers less costly gasoline than both CTG-PB-V and CTG-PB-CCS for GHG emissions prices greater than US$50/tCO$_2$-eq.

- CBTG-PB-CCS is an option with biomass accounting for 30% of input energy offering a 90% reduction in system-wide GHG emissions. It could provide gasoline at the least cost of all the options shown at GHG emissions prices greater than about US$100/tCO$_2$-eq.

- BTG-RC-CCS is an option for which biomass accounts for 100% of input energy offering strongly negative GHG emissions. It could provide less costly gasoline than gasoline derived from US$90/bbl of crude oil at GHG emissions prices greater than US$62/tCO$_2$-eq.

- BTG-RC-V can provide less costly gasoline than gasoline derived from US$90/bbl of crude oil at GHG emissions prices greater than US$106/tCO$_2$-eq.

- BTG-RC-CCS can provide less costly gasoline than BTG-RC-V at GHG emissions prices greater than US$23/tCO$_2$-eq.[16]

Analysis in the *World Energy Outlook* 2009 (IEA, 2009b) is helpful in understanding the relative competiveness of alternative technologies in Table 12.21 in a world with high oil prices and policies that constrain GHG emissions. The *Outlook* considers the current situation regarding carbon trading and analyzes prices for crude oil and for GHG emissions for a world on a path toward ultimate stabilization of GHG concentrations in the atmosphere at 450 ppmv. The IEA estimates for its 450 ppmv stabilization scenario that: (i) the world oil price would be stable during 2020–2030 at US$90/bbl (considerably lower than the oil price in this period for the IEA Reference Scenario), (ii) if there were separate emissions trading regimes for OECD+ countries and for other major economies (China, Russia, Brazil, South Africa, Middle East), the GHG emissions price would rise from US$50/t to US$110/t in OECD+ countries between 2020–2030 and would reach US$65/t in other major economies by 2030; and (iii) if there were a single global carbon market for emissions trading, the global GHG emissions price would be US$70/t by 2030.

Thus under such carbon policy constraints, well before 2030, the processes of CTG-PB-CCS and CBTG1-PB-CCS would become cost competitive in both OECD+ and Other Major Economies. Also, CBTG-PB-CCS technologies would be cost-competitive in the post-2030 time frame

in OECD+ countries. Early experience, prior to 2030, with widespread deployment of CTG-PB-CCS and CBTG1-PB-CCS technologies in coal-rich regions such as the United States and China would establish in the market all the technological components needed for subsequent deployment of CBTG-PB-CCS in coal-rich regions and BTG-RC-CCS in coal-poor but biomass-rich regions. The economic prospects for BTG-RC-V are not auspicious, although the BTG-RC-CCS option is prospectively economically viable in coal-poor but biomass-rich regions in the period after 2030 where there is adequate CO$_2$ storage capacity and emissions trading opportunities. For example, when the crude oil price is US$90/bbl, when coal coprocessing is not a realistic option, and the emissions trading price is US$70/t the BTG-RC-CCS option would be highly competitive.

12.5.3.1   Two Thought Experiments

Two thought experiments are presented here, one for China (a coal-rich region) and one for Africa (a biomass-rich but coal-poor region). In this experiment, both regions are under a carbon policy constraint. The discussion focuses on the strategic linkages between providing fuels/electricity for transportation on the one hand, versus providing LPG to satisfy basic human needs for cooking on the other hand.

China

Why China? China is a good candidate for early deployment of CTG-PB-CCS and CBTG1-PB-CCS technologies to make simultaneously liquid fuels for transportation and LPG for cooking. There are many reasons:

- China is a coal-rich country, accounting for more than 70% of coal use by developing countries in 2007 (IEA, 2009b).

- China has a strong coal gasification-based chemical process industry (making gasoline is very much like making chemicals via coal gasification).

- China already has experience with CTG technology. A demonstration plant producing 2600 barrels of gasoline per day came online in 2009, built by Uhde for the Shanxi Jincheng Anthracite Coal Mining Co. Ltd. This plant uses the ExxonMobil methanol-to-gasoline process and is coupled to a fluidized bed hard coal gasifier and a plant for making methanol from coal via gasification.

- Potential demand for clean cooking fuels in China is huge. Hundreds of millions of Chinese still cook with solid fuels.

- As a large food producer, China has substantial crop residue resources that could be used for energy purposes.[17]

---

16   This breakeven GHG emissions price for shifting from the V to the CCS variant of the BTG-RC option is higher than the US$20/tCO$_2$-eq indicated for this pair of options in Figure 12.24 because in the present case the assumed biomass input rate is half as large.

17   Prospective crop residue supplies in China have been estimated by Li et al. (1998), who estimate that crop residue supplies potentially available for energy applications in China in 2010 were 376 million dry t/yr.

- China is a good candidate for pursuing synthetic transportation fuels derived from secure domestic coal and biomass supplies because its domestic oil supplies are scarce and its demand for liquid fuels for transportation is rapidly growing.

- The coproduction approach to CCS, which offers low energy and water penalties and low capture costs, may be perceived as an attractive approach for reducing coal-related GHG emissions in China.

The thought experiment: To illustrate the possibilities for early action for coal-based technologies, suppose that in China in the period prior to 2030 enough MTG plants are built to provide LPG sufficient to meet the needs of the 1.06 billion people that were cooking with solid fuels as of 2001.[18]

Assumptions: It is assumed that during this period the carbon policy in China becomes sufficiently stringent to warrant deployment of both CTG-PB-CCS and CBTG1-PB-CCS systems. Furthermore, it is assumed that a mix of these two technologies is deployed such that on average gasoline, LPG, and electricity are provided at one half the GHG emission rate of the CODP and PC-V electricity displaced. This implies that 62% of the capacity would be CTG-PB-CCS plants and 38% would be CBTG1-PB-CCS plants (so that, on average, biomass accounts for 3.8% of primary energy input).

Findings: The system-wide features of this combination of plants would yield 100% satisfaction of the need for clean cooking fuels and a 50% reduction in GHG emissions relative to the energy products displaced. In addition, the features would be the following:

- the required investment (Total Plant Cost (TPC)) would be US$1.1 trillion (US$_{2007}$$);

- synthetic gasoline would be produced at a rate of 13.5 EJ/yr (322 Mtoe/yr or 66% of projected transportation energy demand in China in 2030 (IEA, 2009b);

- electricity would be produced at a rate of 1516 million MWh/yr (67% of projected increase in coal electricity generation in China, 2015–2030 (IEA, 2009b);

- the biomass required is 95 Mt/yr dry biomass or 25% of prospective crop residue supplies available for energy in China;[17]

- CO_2 would be stored in deep geological formations at a rate of 2.57 Gt/yr.

This coproduction approach also would offer significant advantages relative to provision of the same energy products via use of CTG-RC-CCS (a

coal-only design that maximizes gasoline output) plants plus CIGCC-CCS plants to provide the electricity needed in excess of what can be provided by the CTG-RC-CCS plants. Relative to the case with production in separate facilities, the coproduction with coal/biomass coprocessing approach would involve comparable total investment, total primary input, and CO_2 storage requirements, but would generate 15% less GHG emissions. Moreover, when evaluating the coproduction system as an electricity generator, the levelized cost of electricity at crude oil and GHG emissions prices of US$90/bbl and US$65/tCO_2-eq, respectively, would be only 14% as large as for a 2028 MW$_e$ CIGCC-CCS plant having the same primary energy input as the average coproduction unit that provides 768 MW$_e$ of electricity. The reasons for the outstanding economic performance of coproduction plants evaluated as electricity generators compared to stand-alone power plants are discussed in Section 12.6.3.

Africa

Why Africa? The region of Africa is a good candidate for deployment of BTG-RC-CCS technology in the 2030+ time frame. There are many reasons:

- Much of Africa is biomass-rich but coal-poor (except for Botswana and South Africa).

- Much of Africa is economically poor and in need of industrial development such as that which BTG-RC-CCS technology could help provide.

- Africa has a huge population in need of clean cooking fuels, with some 710 million people (Hutton et al., 2006) dependent on solid fuels for cooking.[19]

- Much of Africa must spend precious export earnings on fuels for transportation as well as on LPG for cooking and thus stands to benefit economically from having a domestic BTG-RC-CCS synfuels industry.

- Preliminary indications are that there might be significant CO_2 storage opportunities in Africa (see Chapter 13).

- Although at low GHG emissions prices it would make more economic sense for biomass-rich/coal-poor countries to import coal and make gasoline by coprocessing coal and biomass, biomass to gasoline plants would become more cost competitive at high GHG emissions prices.[20] So, if there were a reasonable expectation of such high GHG emissions prices in the future, perhaps before 2050, such

---

18 Following the findings of Hutton et al., (2006) it is assumed that the average per capita LPG requirement for China is 29.3 kg/yr (43 Watts).

19 In 2006 biomass used for cooking in the developing world (assumed to be total biomass use for energy, minus biomass used for power generation, minus industrial use of biomass, and minus biomass used to make transport fuels) was 229 Mtoe (9.6 EJ) in Africa, 224 Mtoe (9.4 EJ) in China, 131 Mtoe (5.5 EJ) in India, 142 Mtoe (6.0 EJ) in other non-OECD Asia, and 32 Mtoe (1.3 EJ) in Latin America (IEA, 2008b). Thus Africa accounted for about 30% of global biomass use for cooking in 2006.

20 If the competition were between CBTG-PB-CCS and BTG-RC-CCS plants each consuming 0.5 million dry tonnes of biomass annually, the BTG-RC-CCS option would provide gasoline at a lower levelized cost of fuel when the GHG emissions price is greater than US$108/t$CO_2$-eq.

countries might be reluctant to make such coal-related infrastructure investments.

The thought experiment: A thought experiment is presented that envisions widespread deployment of BTG-RC-CCS technology in Africa in the 2030+ time frame.

Assumptions: The following conditions are assumed:

- Before 2030 all the needed technological components are established in the market somewhere in the world via earlier widespread deployment of CTG-PB-CCS and CBTG1-PB-CCS technologies (e.g., as described in the previous China thought experiment).

- The GHG emissions price is high enough (greater than US$62/tCO$_2$-eq) that BTG-RC-CCS is able to sell gasoline that is competitive with gasoline derived from US$90/bbl of crude oil (see Figure 12.30).

- There are concerted multilateral activities prior to 2030 aimed at: (i) identifying geological CO$_2$ storage opportunities in Africa's biomass-rich regions; (ii) identifying prospective biomass supplies that can be provided on a sustainable basis (avoiding supplies that involve deforestation, destruction of soil carbon stores, and competition with food production); (iii) building in currently economically poor, biomass-rich regions the physical infrastructures and human capacity needed to support rapid development of a BTG-RC-CCS industry.

- BTG-RC-CCS plants are deployed at modest scales to keep biomass supply logistics and CO$_2$ infrastructure challenges from being too daunting—producing gasoline at a scale ~2300 bbl/day, processing only 0.5 Mt/yr of dry biomass, and storing underground only 0.5 Mt/yr of CO$_2$, see Table 12.21. To get a sense of the scale of the activities, consider that Campbell et al. (2008) have estimated for tropical regions that yields for growing mixed prairie grasses on abandoned cropland in tropical regions are 7–20 dry t/yr-hectare. For these yields the amount of land required to serve a single biofuels plant is 250–714 km^2. The average biomass transport distance is 30–50 km if the biomass is available on 20% of the land around the plant and 40–70 km if the biomass is available on only 10% of the land.

- For the thought experiment it is assumed that the amount of biomass available for prospective BTG-RC-CCS plants is the same as the actual estimated amount of biomass that was used for cooking in 2006 (229 Mtoe/yr = 9.59 EJ/yr).[19] If this much biomass were grown as an energy crop at a yield of 7–20 dry t/hectare-yr, some 27–78 million hectares would be required for all of Africa. To put this into perspective, Cai et al. (2009) estimates that worldwide the amount of land available for growing biomass for energy on both abandoned and/or degraded cropland and grassland, savanna, and shrubland with marginal productivity suitable for use with low-input

high diversity prairie grasses as energy crops is 1343 million ha of which one third to one half (450–670 million hectares) is in Africa.[21]

Making cooking with LPG affordable: The analysis below explores prospects for earning income that current users of biomass for cooking might pursue. They could sell their biomass to "biomass-supply-logistics" agents, who would in turn make the biomass available to operators of BTG-RC-CCS plants. And they could generate thereby enough revenue to cover both purchases of LPG for cooking and the annualized cost of purchasing an LPG stove and storage canisters.

Compare the energy requirements for cooking with these fuels. The current rate of use of biomass for cooking by 710 million people in Africa (229 Mtoe/yr, 9.6 EJ/yr),[19] corresponds to a wood consumption rate of 0.7 dry tonnes per capita/yr = 428 Watts/capita. According to Hutton et al. (2006) the average per capita consumption rate of LPG for cooking in Africa as a substitute for wood would be 14.4 kg/yr or 21 Watts. Both of these cooking rates are consistent with historical rates for cooking with wood and fluid fuels (Figure 12.28).

Energy requirements for cooking via LPG are only ~5% of the energy requirements for cooking via the direct burning of biomass. This makes LPG prospectively affordable even for very poor households. The following illustrative calculation suggests how LPG might be made affordable: Suppose that the crude oil and GHG emissions prices are US$90/bbl and US$70/tCO$_2$-eq, respectively, so that synthetic gasoline produced with carbon capture and storage would be cost-competitive with gasoline from crude oil (Figure 12.30). At these crude oil and GHG emissions prices, the estimated average retail LPG price would be US$1.39/kg in rural Africa.[22] Thus the total average cost of LPG for a family of 4.4 (the average household size in Africa) would be:

$$4.4 \times \left(\frac{US\$1.38}{kg} \right) \times \frac{14.3 \dfrac{kg}{capita}}{yr} = \frac{US\$87}{yr} \qquad (3)$$

The total cost of cooking also includes capital expenses for the stove and storage canisters, which are estimated to be about US$50 (IEA, 2006) or an annualized cost of US$8/yr,[23],[24] so that the total annualized

---

21   Of the global total Cai et al. (2009) estimate that 256–463 million hectares is abandoned and/or degraded cropland. For comparison, Campbell et al. (2008) estimate that globally the amount of land available on abandoned cropland is 385–472 million hectares.

22   Based on a wholesale LPG price of $1.07/kg for $90/bbl + a $0.31/kg markup to retail for rural Africa, as estimated in Hutton et al. (2006).

23   Assuming a 10% discount rate and a 10-year system life, the capital recovery factor is 16.3%/yr for the capital equipment.

24   Some sort of microfinancing program may be needed to overcome the expenditure "lumpiness" hurdle of the stove/canisters investment, but otherwise the required investment would not appear to be a show-stopper for poor households.

cost is US$95/yr per average African household. The price P_W (in US$/t) at which a family would have to sell biomass currently used for cooking to a biomass-supply-logistics agent of a synfuel producer in order to be able to afford the LPG without an additional income stream is given by:

$$4.4 \times \left(0.70 \frac{\text{tonnes} \frac{\text{wood}}{\text{capita}}}{\text{yr}} \right) \times P_W = \frac{\text{US}\$95}{\text{yr}} \Rightarrow P_W = \text{US}\$31 \text{ per tonne} \quad (4)$$

This is a plausible selling price because it is only 30% of the assumed price of wood delivered to the conversion facility[25] – allowing a margin of revenue to the biomass-supply-logistics agent that is plausibly enough to pay for profitably getting the biomass purchased from cooking fuel consumers to the synfuel producer. (Detailed biomass supply logistics analysis is needed to ascertain the validity of this very preliminary judgment.)

Implications of the thought experiment for Africa: The biomass now used for cooking in Africa could support the operation of 1023 BTG-RC-CCS plants like those described in Table 12.21. If such a shift in biomass use were made, there would be Africa-wide implications:

- a total investment requirement for BTG-RC-CCS plants of US$528 billion;

- enough LPG to meet the cooking fuel needs of 94%[26] of the population currently cooking with biomass;

- 2.28 million bbl/day average gasoline output, equivalent to 90% of Africa's transportation fuel demand for 2030 as projected by the IEA (IEA, 2010) or 11% of world gasoline output of refineries in 2007, some 21.3 million bbl/day (47 EJ/yr) (US EIA, 2009b);

- 88 million MWh/yr of electricity, equivalent to 7% of Africa's electricity generation in 2030 as projected by the IEA; and

- annual storage in geological formations of 562 MtCO$_2$/yr.

Toward a business plan for BTG-RC-CCS technology deployment in Africa: This thought experiment suggests that widespread deployment of BTG-RC-CCS technology in Africa might not only go a long way

towards meeting Africa's transportation fuel needs but also might help to catalyze widespread use of LPG for cooking, even among very poor households.

But much new thinking is needed about business strategies and public policies required to convert this thought experiment into a plausible energy projection for Africa. A list of proposed public policies for both the China thought experiment and the Africa thought experiment is presented in Section 12.5.3.2.

Although articulating appropriate business plans is beyond the scope of the current analysis, we conclude the economic analysis with a suggestion for one possible element of such business plans: involvement of industrial firms that produce and/or use crude oil-derived transportation fuel as investors in BTG-RC-CCS systems.

The reason for this suggestion is that such firms may be interested in procuring credits for the strong negative GHG emissions characterizing BTG-RC-CCS systems to offset emissions from the crude oil-derived products they produce or consume. The production of each barrel of gasoline by a BTG-RC-CCS plant provides enough negative GHG emissions to offset the emissions of 1.37 barrels of crude oil-derived gasoline.[27] Assuming that crude oil and GHG emissions prices are US$90/bbl and US$70/t, respectively, the annual cost of purchasing credits from a single BTG-RC-CCS plant would be US$34 million. The present worth of purchasing such credits over the life of the plant would be US$357 million,[28] which is about three fourth of the investment cost (TPC) for a BTG-RC-CCS plant (Table 12.21).

The industrial firms that are producers and/or users of crude oil-derived transportation fuel could either try to buy emissions credits from BTG-RC-CCS plant owners in an emissions trading market or they could instead invest in BTG-RC-CCS plants. In the former case, they would risk not being able get the full amount of credits they are seeking to obtain, while in the latter case they would have guaranteed access to these credits.

---

25 For all the systems presented in Figure 12.30 and involving biomass it is assumed that the delivered price of biomass at the conversion plant is $5.0GJ (HHV), which corresponds to a wood price of $103/dry tonne.

26 The cooking fuel production could be increased to 100% of the current cooking fuel needs by reducing the gasoline output of the BTG-RC-CCS plants in favor of producing some DME as a coproduct and blending this with LPG for use as a cooking fuel. So doing would be straightforward because the first step in the production of gasoline from methanol (CH_3OH) is methanol dehydration, which produces DME (CH_3OCH_3) and water. See additional discussion of DME in Section 12.5.1.

27 Alternatively, offsets might be sought from plants that make electricity via gasification of biomass with CCS, in which case a much larger fraction of the C in the biomass can be stored underground. If the negative emissions from power plants are used to offset emissions from crude oil-derived gasoline, a comparable amount of "effective" zero GHG emitting liquid fuels would be provided via the biomass to power with CO_2 capture and storage route as via the BTG-RC-CCS route (1.04 and 0.94 GJ of zero net GHG-emitting gasoline is provided per GJ of dry biomass input in the BTG-RC-CCS and power-only cases, respectively). In the power-only case, the gasoline provided is 100% crude oil-derived gasoline offsets; in the BTG-RC-CCS case, 42% of the gasoline is actually produced and 58% is in the form of crude oil-derived offsets. Thus if a biomass-rich country seeks to reduce oil imports as well as mitigate climate change, it is likely to choose the synfuel option over the power option. Another consideration that might tip the balance in favor of the synfuel option is that CO_2 storage capacity is a non-renewable resource, and the power option stores 1.6 times as much CO_2 per tonne of biomass as the synfuel option.

28 Assuming a 7% discount rate and a 20-year economic life of a BTG-RC-CCS plant.

12.5.3.2   Public Policy Issues

Despite the huge public benefits and seemingly attractive economics indicated by the China and Africa thought experiments, converting these thought experiments into projects will require new public policy initiatives. There would be many hurdles to overcome, including issues related to the viability of CO_2 storage at gigascale, CO_2 storage potential, sustainable biomass production potential, commercialization of large biomass gasifiers, new industrial collaborations, the rural lighting challenge, and physical infrastructure and human capacity for a BTG-RC-CCS industry. These public policy issues are outlined here.

Viability of CO_2 storage at gigascale: There is widespread belief in the scientific community that CCS is a viable carbon mitigation option at scales storing billions of tonnes of CO_2 annually worldwide (IPCC, 2005 and Chapter 13). Demonstration projects are needed, however, to prove and gain a high degree of confidence in CCS viability and also to provide a solid scientific and engineering basis for widespread deployment of CCS technologies post-2020. Commercial-scale integrated CCS demonstration projects worldwide are needed during the coming decade, with emphasis on CO_2 storage in deep saline formations, which account for most of the geological storage opportunity. Such projects are needed for several reasons. They could address scientific questions that can only be answered in projects that inject and store CO_2 at rates comparable to those for commercial projects. They could demonstrate to the satisfaction of a wide range of stakeholder groups that CCS is a viable major option to be included in the portfolio of carbon mitigation options. And they could provide the experience base needed for formulating practicable regulations governing CO_2 storage.

An international political framework for early CCS action has already been established. In July 2008, an agreement was reached by the G8 countries at the G8 Summit in Japan that 20 large-scale fully integrated CCS demonstration projects worldwide would be deployed by the middle of the next decade, with the aim of establishing the basis for broad commercial deployment of CCS technologies after 2020. In July 2009, the leaders of the G8 countries re-iterated their call for the projects, and in February 2010 US President Obama issued a Presidential Memorandum calling for five to ten commercial scale CCS demonstration projects to be up and running in the United States by 2016.

Much if not all of the incremental cost of CCS for the 20 projects called for by the G8 will probably have to be paid for by governments (individually or collectively) because of the likelihood that carbon prices will be lower initially than what will be needed to make pursuit of CCS a profitable activity for private companies.

If governments will have to pay for the incremental CCS cost, they will want to pursue projects in which they can maximize the learning about the gigascale prospects of CCS per dollar spent.

An important consideration is that coproduction systems based on coal or coal + biomass generate, as a natural part of the process of their manufacture, relatively pure streams of CO_2 for which the incremental cost of CO_2 capture is low. Accordingly, coproduction facilities (such as the CTG-PB-CCS and CBTG1-PB-CCS described in Table 12.21) built in coal-rich countries should be considered seriously as candidates for some of the needed CCS early action projects and supported financially jointly by the governments of several coal-intensive energy economies.

CO_2 storage potential: CO_2 storage prospects are not well known in countries where clean cooking fuels are sorely needed and where attractive economics for providing these fuels to poor households can plausibly be realized in conjunction with the building of synfuel plants with CCS. "Bottom-up" assessments of storage prospects, including the construction of supply curves (storage capacity in tonnes vs cost in US\$/t), are needed on a reservoir-by-reservoir basis. These assessments should be carried out in each of the major regions requiring clean cooking fuels – with financial support from the international community.

Sustainable biomass production potential: Assessments should be carried out in biomass-rich regions to determine the prospects for biomass production for energy. Such production should be on a sustainable basis in which conflicts with food production, adverse indirect land-use impacts, and biodiversity conflicts are minimized. Emphasis should be on agricultural residues, forest residues (including mill residues, logging residues, diseased tree removals, fuel treatment thinnings, and productivity enhancement thinnings), and the growing of dedicated energy crops on abandoned croplands and other degraded lands. The growing of bioenergy crops on marginal lands should be done in ways that enhance the wellbeing of poor indigenous populations currently use such lands for their livelihoods. One way of expanding biomass production on marginal lands without forcing off the land local populations would be to encourage the local populations to grow biomass for energy by creating corporate smallholder partnerships that establish agreements for industries to purchase biomass from smallholders. Outgrower schemes such as these have been common for some time in agriculture; smallholders are now playing an increasingly important role in the establishment and management of planted forests (Cushion et al., 2010).

These assessments should be carried out in each of the major regions requiring clean cooking fuels – with financial support from the international community.

Commercialization of large biomass gasifiers: Successful demonstration projects have been carried out for biomass gasifiers at small scales (processing tens of MW of biomass). But there are no commercial biomass gasifiers capable of processing 300–600 MW of biomass – the scales for the conversion systems described here. Policies are needed to encourage commercialization of biomass gasifiers suitable for coupling to synthetic fuel production units at these scales. Such commercialization efforts should be carried out in parallel with early deployment of CBTG1-PB-CCS systems (designed for ~10% biomass or less) that could involve instead co-gasification of biomass and coal in suitable coal gasifiers.

New industrial collaborations: New public policies are needed to facilitate industrial collaborations between companies producing transportation fuels, electricity, and clean cooking fuels and to encourage coprocessing of coal and biomass in regions having significant supplies of both (e.g., United States and China). It would be desirable to identify policy instruments that specify performance rather than technology and maximize use of market forces in meeting performance goals. Promising approaches along these lines include mandating a Low Carbon Fuel Standard (as in California); a low carbon standard for coal electricity, perhaps modeled after Renewable Portfolio Standards or green certificate markets; and a Universal Clean Cooking Fuel Standard in regions requiring major infusions of clean cooking fuels, perhaps modeled after the "obligation to serve" mandates of the rural electrification programs introduced in the United States in the 1930s. The latter could plausibly facilitate the formation of strategic industrial alliances that would be capable of guaranteeing universal access to clean cooking fuels without major subsidy.

Rural lighting challenge: Policies aimed at inducing a shift from biomass to clean cooking fuels should be complemented by policies to promote universal access to modern lighting technologies (see also Chapter 23).

Physical infrastructure and human capacity for a BTG-RC-CCS industry: Official development assistance (ODA) should be expanded for economically poor but biomass-rich and coal-poor regions. The increment should be directed to developing the physical infrastructures and human capacities needed to build and manage large BTG-RC-CCS industries. This additional ODA should aim for established infrastructures and capacities *before* GHG emission prices are high enough to launch BTG-RC-CCS technologies in the market.

12.6 Coproduction of Liquid Fuels and Electricity from Non-petroleum Feedstocks

Coproduction can enhance cost-competitiveness. The discussion of gasification-based synfuels in Section 12.4.3 was focused on recycle (RC) systems designed to maximize liquid fuel output. In RC systems, syngas unconverted in a single pass through the synthesis reactor is recycled to the reactor to maximize synfuel yield. A major study of alternative system configurations for making FTL from coal, from biomass, and from coal + biomass (Liu et al., 2011a) found that FTL can often be produced more cost-effectively in so-called "once-through" (OT) systems, in which syngas unconverted in a single pass is burned to make coproduct electricity in a gas turbine/steam turbine combined cycle power plant (See also Larson et al., 2010). OT configurations are worthwhile exploring from a synfuels production perspective for the slurry-phase synthesis reactors investigated by Liu et al. (2011a) because these reactors can yield high

one-pass conversion of $CO+H_2$ to liquids – much higher than is feasible with gas-phase reactors. A major study of alternative system configurations for making gasoline via the methanol-to-gasoline process from coal, from biomass, and from coal + biomass (Liu et al., forthcoming) also found that gasoline can often be produced more cost-effectively via systems that produce electricity as a major coproduct.

Here the merits of coproduction for both FTL and MTG systems are discussed for the six systems described in Table 12.22 that involve coal (both –V and –CCS configurations) and coal + biomass (only –CCS configurations). Since detailed descriptions of RC variants of FTL and MTG systems were already made in Section 12.4.3, here only the major differences introduced by the coproduction designs are discussed.

Figure 12.31 shows several OT configurations for coal and coal/biomass FTL systems. When iron-based synthesis catalysts are used (as in the analysis in Liu et al., 2011a), CO_2 removal from synthesis gas in –CCS configurations is required downstream as well as upstream of synthesis because of the water-gas-shift activity of iron-based catalysts, and downstream CO_2 removal accounts for more than half of the total CO_2 removed. Also, a gas turbine/steam turbine combined cycle rather than a steam turbine is used to generate electricity on the power island.

The MTG designs that coproduce electricity and gasoline use a bypass of some of the syngas around the methanol synthesis reactor to directly feed a gas turbine/steam turbine combined cycle power plant in so-called partial bypass (PB) configurations (Figure 12.32).[29] An additional water gas shift reactor is introduced to convert the bypass syngas to mainly CO_2 and H_2, from which the CO_2 is captured for storage, and H_2 is the main constituent of the fuel gas delivered to the combined cycle power plant.

12.6.1 Performance Estimates

Greenhouse gas emissions are the focus of the discussion of performance estimates here. To facilitate comparisons, the coal-only plants (Figure 12.31, Table 12.22) have the same levels of coal inputs as the corresponding RC plants described in Table 12.15 that provided 50,000 bbl/day of crude oil products displaced for both FTL and MTG. As shown in Table 12.22, the OT and PB system configurations produce less liquid fuel and much more electricity from this coal than the corresponding RC systems. As in the RC cases, an important metric for comparing emissions for different system configurations is the GHGI: the ratio of the fuel cycle wide emissions of greenhouse gases from these systems are compared to emissions from a reference system.

---

29 A partial-bypass, rather than a "once through" (OT), process design is utilized for the MTG systems here because the assumed methanol synthesis reactor is a gas-phase design, for which the per-pass conversion of syngas is small, making it ill-suited for use in an OT process configuration. A liquid-phase methanol synthesis reactor would have much higher single-pass conversion and as a result might be better-suited for an OT application.

Table 12.22 | Performance estimates for converting coal and coal + biomass to electricity + FTL or MTG.

Process configuration >>>	CTL-OT-V	CTL-OT-CCS	CTG-PB-V	CTG-PB-CCS	CBTL-OT-CCS	CBTG-PB-CCS
Coal input, as received metric t/day	24,087	24,087	20,869	20,869	3220	5260
Coal input, MW HHV	7559	7559	6549	6549	1011	1651
Biomass input, as received metric t/day	0	0	0	0	3,581	3,581
Biomass input, MW HHV	0	0	0	0	661	661
Production of all liquids, MW LHV	2256	2256	2100	2100	508	743
Production of LPG, MW LHV	-	-	202	202	-	68
Production, bbl/day COPD[a], excl. LPG	35,706	35,706	32,579	32,579	8,036	11,582
Gross electricity production, MW	1661	1653	1417	1369	384	521
On-site electricity consumption, MW	401	595	459	609	127	229
Net electricity exports to grid, MW	1260	1058	959	760	257	292
Energy ratios						
Liquids out/energy in (HHVs)	32.1%	32.1%	34.4%	34.4%	32.5%	34.5%
Net electricity/energy in (HHV)	16.7%	14.0%	14.6%	11.6%	15.4%	12.6%
(Liquids+electricity)/energy in (HHVs)	48.8%	46.1%	49.1%	46.0%	48.1%	47.2%
Carbon accounting						
C input as feedstock, kgC/second	178	178	154	154	40	55
C stored as CO_2, % of feedstock C	0.0%	52.2%	0.0%	63.7%	54.0%	64.9%
C in char (unburned), % of feedstock C	4.0%	4.0%	4.0%	4.0%	3.6%	3.7%
C vented to atm., % of feedstock C	71.6%	19.5%	70.5%	6.8%	18.2%	6.3%
C in liquid fuels, % of feedstock C	24.4%	24.4%	25.5%	25.5%	24.2%	25.1%
C stored, 10^6 tCO_2/yr (at 90% cap factor)	0	9.6	0	10.2	2.3	3.7
Full fuel cycle GHG emissions (incl. from feedstock supply and conversion, fuels distribution and use)						
$kgCO_2$-eq/GJ liquid fuel LHV	289.6	139.1	298.3	108.9	19.4	19.7
tCO_2-eq/hr	2352	1129	2038	744	35	48
Reference system emissions, tCO_2-eq/hr	1790	1623	1483	1318	381	538
GHG emissions index – GHGI[b]	1.31	0.70	1.37	0.56	0.093	0.089

a COPD = crude oil products displaced.
b GHGI, the greenhouse gas emissions index, is the system wide life cycle GHG emissions for production and consumption of the energy products relative to emissions from a reference system producing the same amount of liquid fuels and electricity. For details, see Table 12.15, note (c).

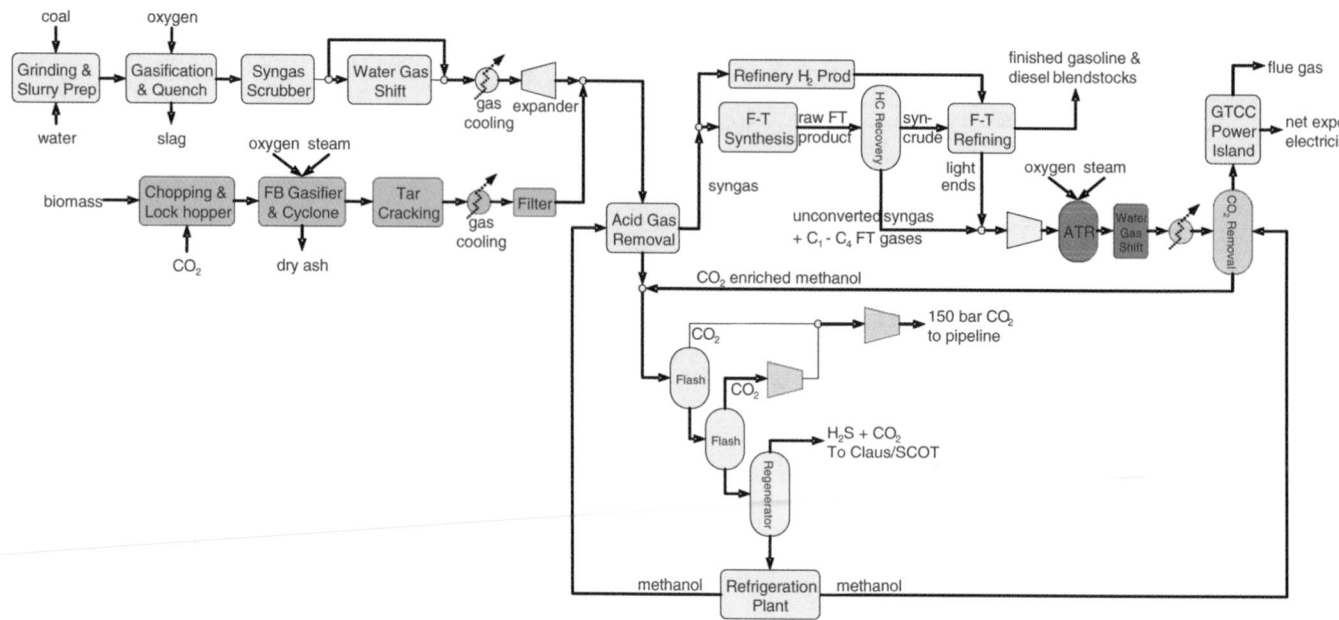

Figure 12.31 | CTL-OT-V system (including only yellow-shaded components) that provides F-T liquids + electricity from coal via gasification while venting the captured CO_2 coproduct. Upstream of the synthesis reactor CO_2 accounting for 25% of the C in the coal is captured along with H_2S using Rectisol. The H_2S is converted to elemental sulfur in a Claus plant and the CO_2 is vented.

With the addition of the blue-shaded components, the system (CTL-OT-CCS) includes capture of CO_2 both upstream of the synthesis reactor (accounting for 25% of C in the coal) and downstream of it (accounting for 27% of C in the coal). The CO_2 is compressed to 150 bar and delivered to a pipeline for transport to a geological storage site.

With the further addition of the green-shaded components, the system (CBTL-OT-CCS) co-processes biomass with coal.

And finally, the further addition of the red components creates a design (CBTL-OTA-CCS) that includes autothermal reforming (ATR) of the C_1-C_4 components in the fuel stream to the power island, followed by water gas shifting. The autothermal reforming (ATR) plus WGS creates more CO_2 for capture downstream of synthesis than is the case with the CBTL-OT-CCS system.

Both the CBTL-OT-CCS and CBTL-OTA-CCS systems have been designed with enough biomass to realize GHGI < 0.10. CBTL-OT-CCS, storing as CO_2 54% of the feedstock C, requires coprocessing 40% biomass, whereas CBTL-OTA-CCS, storing as CO_2 65% of the feedstock C, requires coprocessing only 29% biomass to realize essentially the same GHGI value.

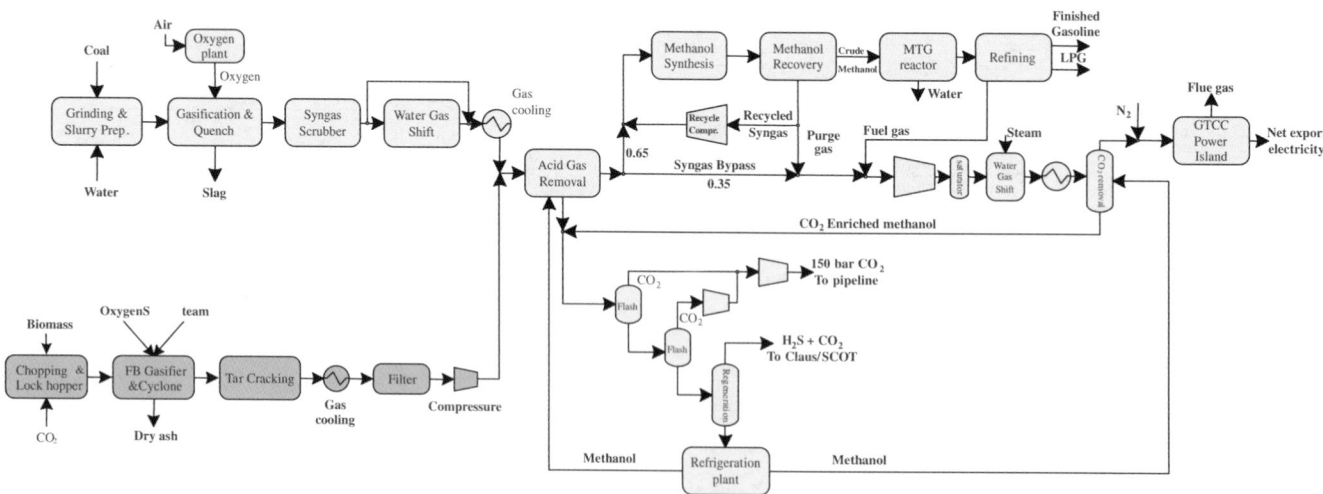

Figure 12.32 | CTG-PB-CCS system (yellow-shaded components) that provides gasoline and LPG + electricity from coal. H_2S is converted to elemental sulfur, and CO_2 is captured both upstream and downstream of synthesis (accounting for 64% of C in the coal), compressed to 150 bar, and delivered to a pipeline for transport to a geological storage site. With the addition of the green-shaded components, the system (CBTG-PB-CCS) produces gasoline and LPG + electricity from coal and biomass. CO_2 accounting for 65% of C in the feedstocks is captured and sent to geological storage.

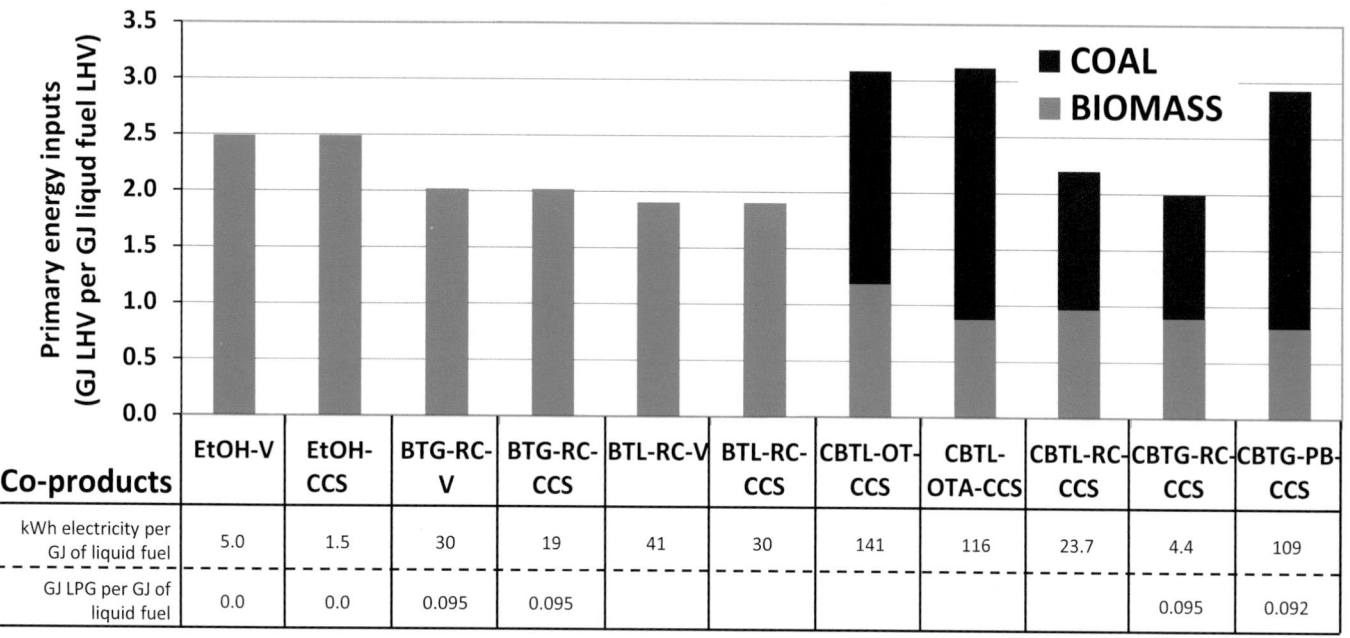

Figure 12.33 | Biomass (and coal) required to produce alternative fuels having low/zero fuel-cycle-wide GHG emissions. For this graph, all primary energy input is allocated to liquid fuels even though all systems also provide electricity. For details on the cellulosic ethanol options (EtOH-V and EtOH-CCS) see Box 12.3. Source: based on Liu et al., 2011a.

For the coal-only designs, the GHGI is 1.3–1.4 for the systems with venting of CO_2, indicating that emissions would be considerably higher than for the reference system. Adding carbon capture and storage reduces emissions relative to the reference system by 30–40% (GHGI = 0.56–0.70).

The two coal + biomass systems described in Table 12.22 are each designed for a total biomass consumption of 1 Mt/yr dry biomass, together with an amount of coal that results in a GHGI of 0.1 or less, i.e., 10% of the emissions of the reference system or less. For the FTL systems, 40% of the input feedstock energy is biomass. It is 29% for the MTG system. Unlike systems that produce liquid fuels using only biomass as a feedstock, the energy content in the near-zero GHG liquid fuels produced from coal/biomass systems is comparable to or greater than the energy contained in the input biomass (see Figure 12.33). The reason is that a large percentage of the energy input for these coal/biomass coprocessing options is provided by coal, as shown in Figure 12.33.

12.6.2   Cost estimates

When coproduction systems are evaluated from a fuels perspective, the LCOF is given by:

$$LCOF = \frac{\text{levelized system cost}\frac{US\$}{yr} - \text{levelized value of electricity coproduct}\frac{US\$}{yr}}{\text{levelized fuel production rate}\frac{GJ}{yr}} \quad (5)$$

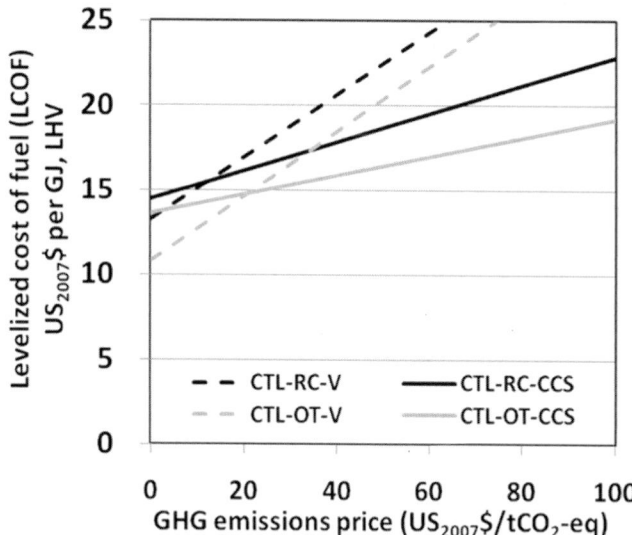

Figure 12.34 | Comparison of CTL production costs (US_{2007}\$): OT vs RC options, both with CO_2 vented (–V) and with carbon capture and storage (–CCS). (See Table 12.6, note (b) for financial parameter assumptions. Also, electricity sales are assumed at the US average grid price plotted in Figure 12.10.) Source: based on Liu et al., 2011a.

In Liu et al. (2011a) and in Liu et al. (forthcoming) it is assumed that the electricity coproduct is valued at the US average grid generation price in 2007 (US\$60/MWh) augmented by the value of the average grid GHG emission rate in that year (see Figure 12.10).

Figures 12.34 and 12.35 present LCOFs vs GHG emissions price for the coal only – OT and –PB systems, respectively, as well as the corresponding LCOFs for the RC alternatives.

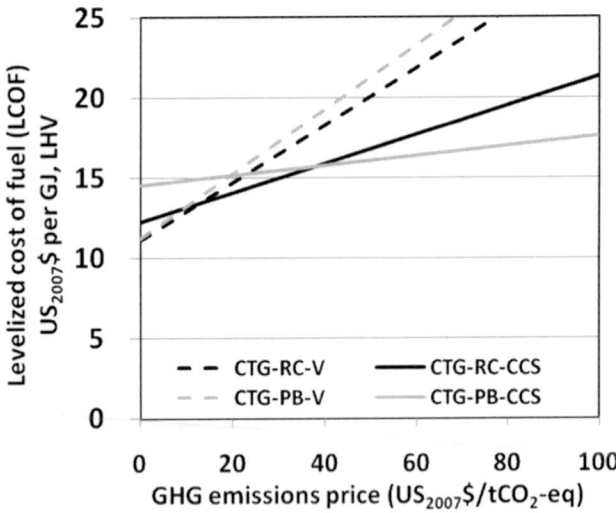

Figure 12.35 | Comparison of CTG production costs (US$_{2007}$): PB vs RC options, both with CO_2 vented (–V) and with carbon capture and storage (–CCS). Same financial parameter and electricity price assumptions as Figure 12.34. Source: based on Liu et al., forthcoming.

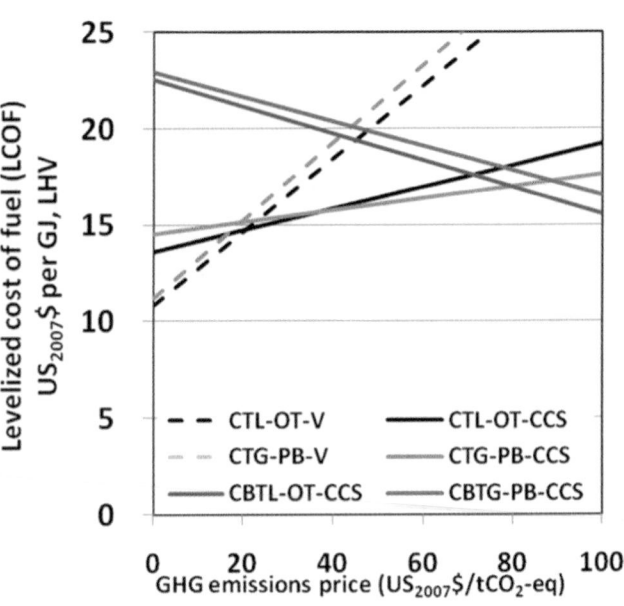

Figure 12.36 | Liquid fuel production costs (US$_{2007}$) for alternative FTL-OT and MTG-PB options, involving coal and coal + biomass as feedstocks. Same financial parameter and electricity price assumptions as Figure 12.34. Source: based on Liu et al., 2011a.

Figure 12.34 shows that the LCOF is 18% less for the CTL-OT-V case than for the CTL-RC-V case when the GHG emissions price is zero. This economic advantage of the OT option can be attributed in large part to the high marginal efficiency of power generation (Liu et al., 2011a): for OT and RC plants having the same FTL outputs, the difference in electricity outputs divided by the extra coal required for the OT case is 45% for the –V case and 39% for the –CCS case (HHV basis). In both instances these marginal efficiencies for power generation for OT options are much higher than for stand-alone power plants, e.g., 37.5% and 31.0% for CIGCC in –V and –CCS configurations, respectively (Table 12.7, PEI cases). For CTG systems, there are also gains in marginal efficiency, but the gains are not as pronounced as for CTL systems: the marginal efficiencies of incremental power generation are 38% for the –V case and 32% for the –CCS case.

For CTG systems (Figure 12.35) the PB option offers no cost advantage relative to the RC case when CO_2 is vented, but for GHG emissions prices greater than US$40/t there is a significant production cost advantage for systems with CCS.

One reason why the economic benefit of coproduction is less for MTG systems than for FTL systems is that in the CTL case a shift from an RC to an OT configuration involves elimination of an energy- and capital-intensive autothermal reformer, whereas neither of the MTG recycle systems use autothermal reformers. Also, MTG systems need an extra water gas shift reactor to decarbonize the syngas stream going to the power plant, which is not needed for FTL systems.

OT systems for FTL and PB systems for MTG also offer less costly approaches to decarbonizing power than stand-alone power systems. To see this consider the cost of GHG emissions avoided (US$/tCO$_2$-eq),

which is the GHG emissions price at which costs are equal for the –V and –CCS system configurations, i.e., the minimum GHG emissions price needed to induce CCS by market forces. This minimum GHG emissions price for both FTL and MTG systems is ~US$20/tCO$_2$-eq (Figures 12.34 and 12.35) compared to US$45/t or more for stand-alone power systems (Figure 12.10). The reason for this large difference is that for synfuels production systems a substantial fraction of the carbon in the feedstock has to be removed from the shifted syngas upstream of synthesis as a natural part of the part of the process of making synthetic fuels. Moreover, in the FTL cases removal of the extra CO_2 downstream of synthesis is not very costly because no extra water gas shift reactors are required.

Figure 12.36 shows the LCOFs for all six OT and PB systems considered as a function of the GHG emissions price, based on the performance information presented in Table 12.22 and the additional economic information presented in Table 12.23. An important feature of the curves presented in Figure 12.36, is the rapid rate of decline in LCOF with GHG emissions price for both the CBTL-OT-CCS system and the CBTG-PB-CCS system – in stark contrast to the flat LCOF curves for the corresponding RC systems presented in Section 12.4.3. This rapid rate of cost reduction reflects the rising value of decarbonized electricity with GHG emissions price (Figure 12.10). As a result of this sharp decline the LCOFs at US$100/tCO$_2$-eq are US$15.6/GJ$_{LHV}$ (US$1.9/gallon of gasoline equivalent) for FTL via CBTL-OT-CCS (73% of the LCOF for FTL via CBTL-RC-CCS) and US$16.6/GJ$_{LHV}$ (US$2.0/gallon of gasoline equivalent) for gasoline via CBTG-PB-CCS (80% of the LCOF for gasoline via CBTG-RC-CCS). In both instances the liquid fuels are characterized by near-zero net GHG emissions, and large quantities of decarbonized coproduct power are provided.

Table 12.23 | Capital cost and production cost estimates (US$_{2007}$$) for coal and coal + biomass to electricity + FTL or MTG.

CO$_2$ vented or captured >>>	CTL-OT-V	CTL-OT-CCS	CTG-PB-V	CTG-PB-CCS	CBTL-OT-CCS	CBTG-PB-CCS
Coal input rate, MW HHV	7559	7559	6549	6549	1011	1651
Biomass input rate, MW HHV	0	0	0	0	661	661
Liquids production rate, MW LHV	2256	2256	2100	2100	508	743
LPG production rate, MW LHV	0	0	202	202	0	68
bbl/day crude oil products displaced, excl. LPG	35,706	35,706	32,579	32,579	8036	11,582
Plant capital costs, million US$_{2007}$$						
Air separation unit (ASU) + O$_2$ & N$_2$ compressors	711	742	681	675	217	270
Biomass handling, gasification & gas cleanup	0	0	0	0	335	353
Coal handling, gasification & quench	1468	1468	1301	1301	263	402
All water gas shift, acid gas removal, Claus/SCOT	636	727	598	705	151	324
CO$_2$ compression	0	60	0	70	24	34
F-T synthesis & refining or Methanol synthesis	519	519	347	347	171	155
Naphtha upgrading or MTG synthesis & refining	71	71	359	359	29	190
Power island topping cycle	272	280	227	231	80	141
Heat recovery and steam cycle	713	708	597	622	155	219
Total plant cost (TPC), million US$_{2007}$$	4390	4574	4110	4310	1427	2086
Specific TPC, US$_{2007}$$/bbl/day	122,958	128,093	126,154	132,293	177,526	180,110
Levelized liquid fuel cost (US$_{2007}$$/GJ$_{LHV}$) at US$0/tCO2[a]						
Capital charges	10.57	11.01	11.75	12.32	15.26	16.77
O&M charges	2.74	2.86	3.05	3.20	3.96	4.35
Coal (at US$2.04/GJ$_{HHV}$; US$55/t, as received)	6.82	6.82	7.02	7.02	4.05	4.98
Biomass (at US$5/GJ$_{HHV}$; US$94/t, dry)	0.00	0.00	0.00	0.00	6.51	4.89
CO$_2$ transport and storage	0.00	0.73	0.00	0.86	1.22	1.22
Co-product electricity revenue (at US$60/MWh)	-9.31	-7.82	-8.42	-6.67	-8.44	-7.22
Co-product LPG revenue (for US$100/bbl crude oil)	-	-	-2.20	-2.20	-	-2.09
Total liquid fuel cost, US2007$/GJLHV	**10.8**	**13.6**	**11.2**	**14.5**	**22.6**	**22.9**
Total liquid fuel cost, US$_{2007}$$/gallon gasoline eq (US$/gge)	1.3	1.6	1.3	1.7	2.7	2.8
Breakeven crude oil price, US$_{2007}$$/bbl[b]	47	63	48	63	112	102
Cost of CO$_2$ emissions avoided, US$_{2007}$$/tCO$_2$	-	21	-	20	24	21

a See note (b) of Table 12.6 for financial parameter assumptions.

b See note (b) of Table 12.16 for discussion of the breakeven oil price calculation for MTG cases.

12.6.3 Coproduction Economics from the Electricity Production Perspective

The economics discussed in the prior section focused on estimating the cost of making liquid fuels via coproduction systems given assumed prices for the electricity coproducts. Alternatively, electricity production costs can be estimated based on assumed liquid fuel selling prices.[30] This alternative approach for evaluating the economics of

---

30 Here it is assumed that the synfuels are sold at the wholesale (refinery-gate) prices for the crude oil products displaced, including the valuation of the fuel cycle-wide GHG emissions for these products. The relevant GHG emissions for the crude oil products displaced are given in Table 12.15, note (c).

Table 12.24 | Alternative systems with 550 MW$_e$ of electric power capacity.

Technology	Primary energy input, MW (HHV)	Biomass input, kt/y (%, HHV basis)	FTL output, barrels /day (kt/year)	CO_2 stored, Mt/yr (% of feedstock C stored)	TPC, million US$_{2007}$$	GHGI[a]
Sup PC-V	1410	0	0	0	894	1.00
CIGCC-CCS	1780	0	0	3.62 (88)	1460	0.15
CTL-OT-V	3300	0	15,600 (658)	0	2200	1.31
CTL-OT-CCS	3930	0	18,600 (783)	5.01 (52)	2570	0.70
CBTL1-OT-CCS	3810	710 (12)	18,100 (763)	4.96 (53)	2630	0.50
CBTL1-OTA-CCS	4820	370 (5)	22,800 (963)	7.58 (64)	3300	0.50

a GHGI = system wide life cycle GHG emissions for production and consumption of the energy products divided by emissions from a reference system producing the same amount of liquid fuels and electricity. Here the reference system consists of electricity from a stand-alone new supercritical pulverized coal power plant venting CO_2 and equivalent crude oil-derived liquid fuels. For details, see Table 12.15, note (c).

coproduction is of interest for three important reasons: (i) the incremental cost for carbon capture and storage is much less than for stand-alone power plants because, as already noted in Section 12.6.2, much CO_2 must be removed from syngas prior to synthesis as a natural part of fuels manufacture, so that capture costs are low; (ii) the economics of power generation are attractive at current and prospective high oil prices, resulting in a huge credit against the cost of electricity generation; and (iii) these systems can defend high capacity factors and force down capacity factors of competing technologies in economic dispatch competition. Coproduction is considered here for systems making Fischer-Tropsch liquids. First, we consider coproduction in the context of an evaluation of alternative options for new coal-using power plants (Section 12.6.3.1), with potential applications in China (Section 12.6.3.2). Second, we consider coproduction as an option for repowering old coal power plant sites (Section 12.6.3.3), with potential applications in the United States (Section 12.6.3.4).

12.6.3.1   XTL vs Stand-alone Power Options for New Plant Construction

Table 12.24 lists key system characteristics for four coproduction options and two conventional stand-alone power options – each of which is designed to have 550 MW$_e$ of electric generating capacity. The reference technology is a new supercritical coal power plant that vents CO_2 (sup PC-V), the least costly option for new stand-alone coal power plants in the absence of a carbon constraint (see Section 12.2.2). Also listed is a coal integrated gasification combined cycle plant with carbon capture and storage (CIGCC-CCS), currently the least costly CCS option for new stand-alone bituminous coal power plants (Section 12.2.3). The CTL-OT options are identical to those considered in the previous two sub-sections except that the scales have been adjusted to 550 MW$_e$ of electric capacity.[31] Two options are considered that involve coprocessing just enough biomass to reduce the system GHGI to 0.5. Biomass

accounts for 12% of energy for one of these options, CBTL1-OT-CCS, which involves capturing only the naturally concentrated streams of CO_2 in the syngas[32] (accounting for 53% of carbon in the feedstock). Biomass accounts for 5% of energy input for the other option, CBTL1-OTA-CCS, which involves more aggressive CO_2 capture[33] (accounting for 64% of carbon in the feedstock). Thus the same GHG emissions mitigation level is realized in one case by using more biomass and in the other by capturing more CO_2.

For each of these options, curves for the levelized cost of electricity (LCOE) vs GHG emissions price is shown in Figure 12.37 for a levelized crude oil price of US$90/bbl, for US economic conditions. This is the oil price for the period 2020–2030 as projected by the IEA for a future in which the global community is then on a course aimed at stabilizing the atmospheric GHG concentration at 450 ppmv (IEA, 2009b).

Several notable observations can be made about the curves in Figure 12.37:

- A GHG emissions price of only ~US$10/t is needed to induce a transition from CTL-OT-V to CTL-OT-CCS. In contrast a GHG emissions price of more than US$50/t is needed to induce a transition in stand-alone power generation from sup PC-V to CIGCC-CCS.

- The LCOE curves for CBTL1-OT-CCS and CBTL1-OTA-CCS lie nearly atop each other, meaning the extra capital cost for the more aggressive capture option is largely compensated by the lower fuel cost.[34]

---

31 In adjusting plant scales it is assumed that system conversion efficiencies do not change, but capital cost scale economy effects are taken into account.

32 Both upstream and downstream of synthesis.

33 In the CBTL1-OT-CCS case, C1 to C4 gases in syngas downstream of synthesis are burned in the gas turbine combustor, thereby generating CO_2 that is vented to the atmosphere on the power island. In the CBTL1-OTA-CCS case an autothermal reformer and a water-gas-shift reactor are inserted downstream of synthesis to convert most of these C1 to C4 gases to mainly CO_2 and H_2 (see Figure 12.31). The extra CO_2 thus created is captured and stored. In this case the gas burned to make electricity is mostly H_2. For additional details regarding this "autothermal reforming" option for once-through (OT) systems, see Liu et al., 2011a.

34 The assumed (HHV) fuel prices are $2.04/GJ for coal and $5.0/GJ for biomass.

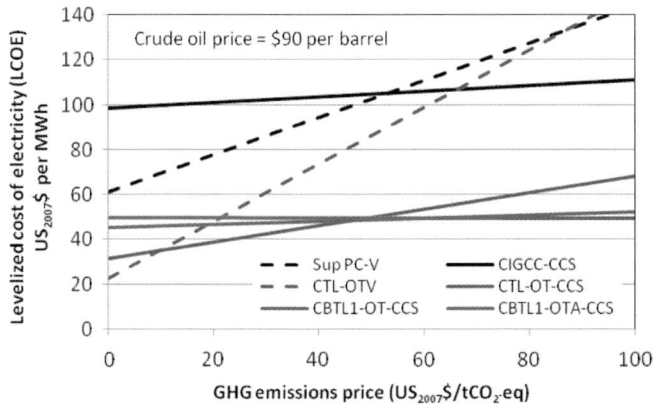

Figure 12.37 | Levelized cost of electricity (LCOE) for crude oil at $90/bbl for the electric power options described in Table 12.24. (See Table 12.6, note (b) for financial parameter values assumed.) The IEA (2009b) projects that if the global community were to pursue a GHG emissions trajectory consistent with stabilizing the atmospheric GHG concentration at 450 ppmv the world oil price would be flat at US$90/bbl, 2020–2030, considerably lower than its projection of the world oil price for its Reference Case (BAU) scenario.

- Although not shown in the figure, the LCOE for coproduction systems declines sharply with the crude oil price: for each US$10/bbl increase in the crude oil price the LCOE at a GHG emissions price of US$0/t is reduced by US$12/MWh for CTL-OT-V and by US$17/MWh for CBTL1-OTA-CCS.

- The LCOE for CBTL1-OTA-CCS is lower than for CTL-OT-CCS when the GHG emission price is US$57/tCO₂-eq.

The LCOE values represented by these curves are for design capacity factors of 85% for power only systems and 90% for coproduction systems. In a market economy, however, capacity factors are determined not by the design engineers but rather by economic dispatch competition. Once a power plant is built, its capital cost (*a sunk cost*) is not taken into account in the determination of the merit order dispatch on an electric grid. The dispatcher operating the grid determines, on the basis of competitive bids, the order in which plants are dispatched to meet electricity demand – with those plants being dispatched first that offer to sell electricity at the lowest price. It is worthwhile for a power generator to bid to sell electricity as long as the revenue it gets from power (in US$/MWh) is not less than its short run marginal cost of producing electricity (SRMC, which excludes sunk costs). This minimum acceptable selling price for the generator is called the minimum dispatch cost (MDC).

As power generators, the XTL-OT plants considered here were designed as "must-run" baseload units. Their MDC depends on the oil price: oil revenues reduce the revenue required from electricity. The MDC is very low for coproduction systems at sufficiently high oil prices. In the CBTL1-OTA-CCS case, with zero GHG emissions price the MDC is the same as for sup PC-V if the crude oil price is US$37/bbl, and the MDC falls to zero when the crude oil price is US$51/bbl. Thus coproduction will be able to defend high capacity factors in economic dispatch competition

and force down the capacity factors of competing technologies as their deployment on the electric grid increases.

12.6.3.2   Thought Experiment for China

Coproduction technologies with CCS can be effective in addressing simultaneously climate change and energy security challenges. To illustrate these benefits of coproduction strategies, a thought experiment is constructed in which it is imagined that all new power generation in China over the period 2016–2030 is shifted from the business-as-usual path of sup PC-V to a CCS path – considering for CCS, in turn, the CIGCC-CCS, CTL-OT-CCS, and CBTL1-OTA-CCS options listed in Table 12.24. Although none of these alternatives to sup PC-V have been built, all system components for these three options are either commercial or commercially ready, so it would be technically feasible to deploy each of these options commercially beginning later in this decade.[35] The aim of the thought experiment is to gain a better understanding of the strategic implications of pursuing coproduction options for power generation.

There are several reasons for focusing on China for this power-sector oriented thought experiment that are similar to the rationale for the China clean cooking fuel thought experiment discussed in Section 12.5.3.1:

- China has powerful reasons for becoming a world leader in identifying and pursuing low-cost approaches to decarbonizing coal energy conversion such as via deployment of coproduction systems with CCS, as suggested by Figure 12.37.[36] China is the world's largest consumer of coal, the most carbon-intensive fossil fuel, accounting for 48% of world coal consumption in 2010. And the expectation is that under business-as-usual conditions coal demand will continue to grow in China (IEA, 2009b).

- China has a strong incentive to explore alternatives to oil imports. It has evolved from being a net oil exporter in the early 1990s to being the world's second largest consumer of imported oil in 2010, with the expectation of continued rapid oil import growth.

- China has more experience with coal gasification technology than any other country. Essentially all of this experience is in China's chemical process industry, which is largely based on coal because of the scarcity of China's oil and gas resources. There are nearly 400 operating and planned coal-based chemical plants in China using

---

35 The modeling reported above for the CBTL1-OTA-CCS option (coprocessing 5% biomass) is for a system with separate gasifiers for coal and biomass. For systems deployed in this decade, it is more likely that a single suitable coal gasifier that can cogasify modest quantities of biomass with coal would be deployed instead.

36 Of course the LCOEs shown in this figure are only suggestive for China, because they were estimated for the United States, not Chinese, economic conditions. Moreover, the LCOEs for first of a kind plants would be higher than those shown even in the US case, because the indicated costs are for N[th] of a kind plants.

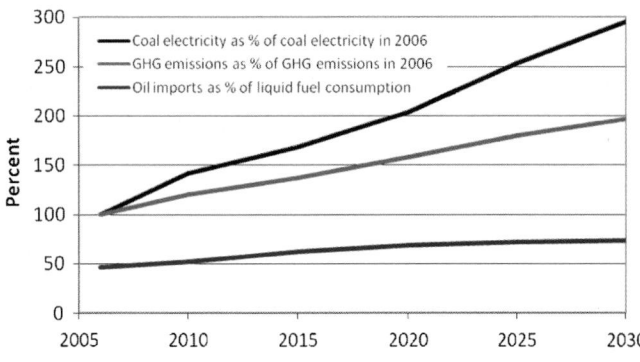

Figure 12.38 | Business-as-usual (BAU) projection by US Energy Information Administration of China's energy future, 2006–2030. Source: US EIA, 2009a.

coal gasification (Zheng et al., 2010); most are ammonia plants but there are also methanol plants, other chemical plants, and a few synthetic fuels plants. Evolving from the manufacture of chemicals and synthetic fuels in such a relatively mature gasification-based industry to the coproduction of liquid transportation fuels and electricity represents a relatively modest step forward from a technological perspective.

- China has one of the world's fastest-growing economies, and rapidly growing economies are the most favorable theatres for technological innovation.

The starting point for the thought experiment is the US Energy Information Administration's Reference Scenario for China (US EIA, 2009a), in which it is projected that between 2006 and 2030 coal power generation will almost triple, total national GHG emissions will almost double, and oil imports will increase 3.4 times from 42% to 75% of liquid fuel consumption (Figure 12.38).

Implications of the thought experiment for carbon mitigation are indicated in Figure 12.39, which shows that in 2030, China's GHG emissions would be less than for the business-as-usual case by the amounts 1.3 GtCO$_2$-eq/yr, 1.9 GtCO$_2$-eq/yr, and 2.3 GtCO$_2$-eq/yr for the CTL-OT-CCS, CIGCC-CCS, and CBTL1-OTA-CCS trajectories, respectively. That the CBTL1-OTA-CCS path leads to 20% more GHG emissions mitigation than the CIGCC-CCS path even though GHGI = 0.5 for CBTL1-OTA-CCS compared to 0.15 for CIGCC-CCS arises because the coproduction option reduces emissions for liquid fuels as well as for electricity.

Implications of the thought experiment for China's oil imports are indicated in Figure 12.40. For the business-as-usual and CIGCC-CCS trajectories, oil imports grow from 162 to 551 Mt/yr (3.3 to 11.2 million bbl/day) between 2006 and 2030 but are reduced by 2030 to + 33 and − 86 Mt/yr (+ 0.68 and − 1.74 million bbl/day) for the CTL-OT-CCS and CBTL1-OTA-CCS trajectories, respectively.

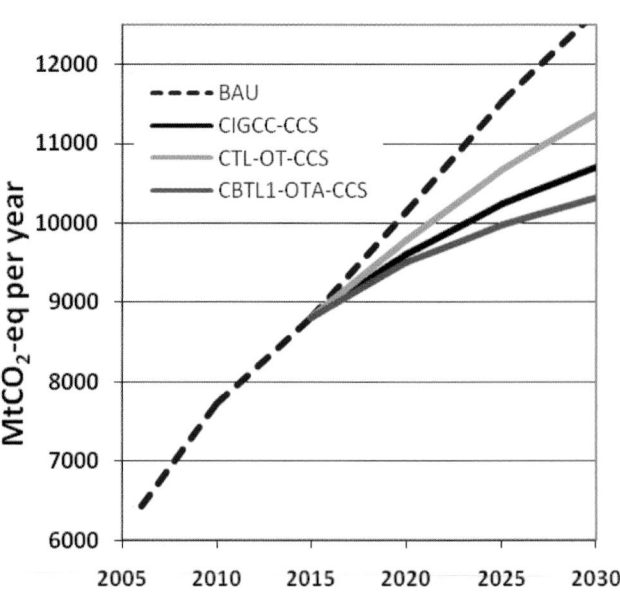

Figure 12.39 | GHG emissions for China: for the BAU scenario based on US EIA (2009a) and for the three thought experiment variants discussed in the main text.

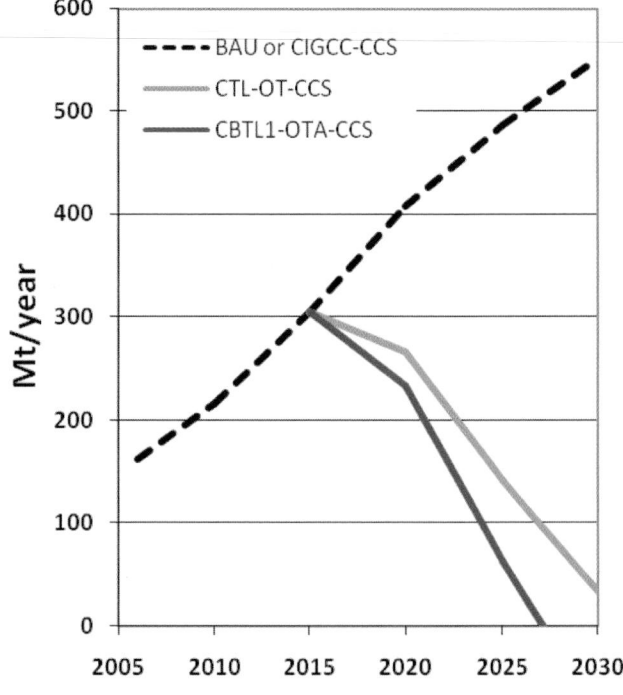

Figure 12.40 | Oil imports for China: for the BAU scenario based on US EIA (2009a) and for the three thought experiment variants discussed in the main text.

In 2030 the biomass required for the CBTL1-OTA-CCS trajectory would be 235 Mt/yr – some 62% of the crop residue supplies potentially available for energy applications in China in 2010 (Li et al., 1998).

The coproduction trajectories might well require a substantial increase in China's coal imports. However, coal imports are likely to pose less of an energy security challenge than oil imports, and the reduced oil import bill is likely to more than compensate for the increased cost of coal imports.[37]

Of course, the economics of coproduction relative to conventional power must be evaluated under China's economic conditions, and the coal/biomass supply implications of the thought experiment must be well understood before one can have a high level of confidence that this would be a cost-effective and otherwise attractive energy strategy for China. But the thought experiment suggests that investigating these issues would be worthwhile.

12.6.3.3   XTL-OT Systems as Repowering Options for Existing Coal Power Plant Sites

Decarbonization will be needed not only for new coal power plants but also for existing power plant sites. In industrialized countries such as the United States, the focus will be on existing power plant sites because overall electricity demand growth is expected to be slow. In this section, five systems providing Fischer-Tropsch liquid transportation fuels as coproducts of electricity and three stand-alone power systems are considered as repowering decarbonization options for sites of existing written-off pulverized coal plants that vent CO_2 (WO PC-V). The coproduction options include both coal-only systems and systems that coprocess coal and biomass. Each of these repowering options is compared to a CCS retrofit of the WO PC-V plant (PC-CCS retrofit), based on analyses in Williams et al. (2011) and Liu et al. (2011a). Key system characteristics for these ten systems are presented in Table 12.25. The alternative systems are compared with regard to GHG mitigation performance, CO_2 storage requirements, energy penalties for CCS, site water requirements, and economics. However, the emphasis is on GHG emissions mitigation performance and economics.

A narrow definition of repowering is scrapping an existing power plant but keeping the site and its infrastructure for use by a new facility. Not all sites can accommodate repowering. There has to be enough space to accommodate all equipment associated with repowering, there have to be suitable CO_2 storage opportunities, and (for cases in which biomass is coprocessed with coal) biomass supplies have to be available. However, the definition of repowering is broadened somewhat to include also the option of abandoning the site entirely and rebuilding at a greenfield site if the targeted site is unsuitable. The economics change only relatively modestly in a shift from building a new plant at an existing site to building a new plant at a greenfield site. The economic benefits associated with saving the infrastructure are not taken into account.

For each of the ten options in Table 12.25 the LCOE vs GHG emissions price is shown in Figure 12.41 for US economic conditions. For the options involving CCS, it is assumed that the cost of CO_2 transport and storage is US$15/t for all options – higher than would be typical for new plant construction. For coproduction systems, the LCOE is evaluated for a levelized crude oil price of US$90/bbl, the oil price for the period 2020–2030 as projected in *World Energy Outlook* 2009 (IEA, 2009b) for a future in which the global community is then on a course aimed at stabilizing the atmospheric GHG concentration at 450 ppmv.[38] As in the case of the larger-scale coproduction systems discussed in Section 12.6.3.1, LCOEs are very sensitive to the crude oil price.[39] To assess the risk from an oil price collapse, the breakeven crude oil price for the five coproduction repowering systems is shown as a function of GHG emissions price in Figure 12.42.[40]

Particular attention is given to the five options for which GHGI is less than 0.20, in light of the fact that political leaders of many industrialized countries are targeting emissions reductions of 80% or more for their countries by mid-century.

Three of the coproduction systems were evaluated earlier from a fuels production perspective, though in some cases at different plant scales: CTL-OT-V, CTL-OT-CCS, and CBTL-OT-CCS (Figure 12.31). One of the coproduction systems not discussed earlier is CBTL-OTA-CCS (also depicted in Figure 12.31), which has a GHGI below 0.1, as does CBTL-OT-CCS (0.077 vs 0.083, see Table 12.25) but realizes the low emission rate via more aggressive CO_2 capture (65% vs 54% of the C in the feedstock is captured) along with a lower biomass input percentage (29% vs 40%). The other new coproduction option considered is CBTL2-OT-CCS for which a GHGI = 0.5 is realized with about 9% biomass input. This option is included because systems involving such a modest percentage of biomass in the feed could be ready for deployment in commercial-scale demonstration projects in this decade.

---

37   The extra coal required for the CBTL1-OTA-CCS option amounts to 2.1 tonnes of coal equivalent (tce) per tce of Fischer-Tropsch liquid transportation fuels produced. The IEA projects for its Reference Scenario that the coal import price for OECD countries will be $100/t in 2020 and $110/t in 2030 (IEA, 2009b). For China, the cost of oil imports avoided for $90/bbl of crude oil would be 1.7 times the increased cost of coal imports if coal imported into China were to cost $110/t.

38   For this evaluation it is assumed that: i) the synthetic liquid fuel coproducts are sold at the wholesale (refinery-gate) prices of the crude oil-derived products displaced (7.90 and 8.51¢ per liter refining markups for diesel and gasoline, respectively), and ii) selling prices increase with GHG emissions price by an amount equal to the fuel-cycle-wide GHG emission rates for these crude oil-derived products times the emissions price.

39   At a GHG emissions price of $0/t, the LCOE increases by an amount ranging from $12/MWh (for CTL-OT-V) to $16/MWh (for CBTL-OTA-CCS) for each $10/bbl reduction in the levelized crude oil price.

40   The breakeven crude oil price is a metric for evaluating the economics of coproduction systems from a synfuel producer's perspective. It is the crude oil price at which the levelized costs of manufacturing the synfuel coproducts equal the wholesale (refinery-gate) prices of the crude oil products displaced. In this calculation it is assumed that the electricity coproduct is sold at the average selling price for US electricity generation in 2007 ($60/MWh) augmented by the value of the US average electric grid emission rate in 2007 (638 kgCO_2-eq/MWh). Figure 12.10 shows this electricity price vs GHG emissions price.

Table 12.25 | CCS retrofit and repowering options for sites of written-off PC-V plants.

| | Capacities[d] | | | | Energy penalty for CCS[e] (%) | Site raw H_2O use at full output,[f] liters/s (% of WO PC-V) | GHGI[g] | CO_2 stored, Mt/yr (% of input C stored) | TPC[h], million US$_{2007}$\$ |
| | Inputs | | Outputs | | | | | | |
	Fuel, MW, HHV	Biomass, Mt/yr (%, HHV basis)	Electricity, MW$_e$	FTL bbl/day (MW, LHV)					
WO PC-V[a]	1613	0	543	0	–	310 (100)	1.00	0	0
PC-CCS retrofit[a]	1613	0	398	0	36.4	413 (133)	0.19	3.48 (90)	426
Repowering options									
CIGCC-CCS[b]	1613	0	500	0	21.1	294 (95)	0.13	3.29 (88)	1369
CTL-OT-V[b]	1613	0	269	7,619 (481)	–	232 (75)	1.17	0	1235
CTL-OT-CCS[b]	1613	0	226	7,619 (481)	8.6	245 (79)	0.63	2.06 (52)	1280
CBTL2-OT-CCS[b]	1694	0.23 (8.8)	242	8,036 (508)	8.6	253 (82)	0.50	2.19 (53)	1348
CBTL-OT-CCS[b]	1671	1.0 (40)	257	8,036 (508)	8.5	237 (76)	0.083	2.26 (54)	1427
CBTL-OTA-CCS[b]	2272	1.0 (29)	287	10,861 (686)	17.1	271 (87)	0.077	3.71 (65)	1784
NGCC-V[c]	1102	0	560	0	–	113 (37)	0.42	0	321
NGCC-CCS[c]	1102	0	482	0	16.3	171 (55)	0.11	1.36 (90)	583

a System characteristics as developed in Simbeck and Roekpooritat, 2009.

b Source: Williams et al., 2011 and Liu et al., 2011a.

c System characteristics as developed by Woods et al., 2007.

d Capacities for systems using coal were determined by the following algorithms: i) Coal input rates cannot exceed the coal input rate of the WO PC-V plant displaced; ii) biomass input rates cannot exceed 1.0 Mt/year; iii) CBTL2-OT-CCS has the same FTL output capacity as CBTL-OT-CCS.

e For power-only systems, penalty = 100 x $(\eta_v / \eta_c - 1)$, where η_v and η_c are HHV plant efficiencies for –V and –CCS options, respectively. For XTL-OT-CCS options, penalty = 100 x (extra coal energy required via CIGCC-CCS to make up for lost power in shifting from –V → –CCS)/(coal energy use by XTL-OT-V).

f Raw water usage = consumption − water recycled; estimated by authors for gasification energy systems; based on Woods et al., 2007 for combustion energy systems, as discussed in the main text.

g GHGI = greenhouse gas emissions index = system wide life cycle GHG emissions for production and consumption of the energy products relative to emissions from a reference system producing the same amount of power and liquid fuels. In this instance the reference system consists of electricity from a WO PC-V power plant (for which the GHG emission rate is 998.8 kgCO_2-eq/MWh$_e$) and equivalent crude oil-derived liquid fuels (for which the GHG emission rate is 91.6 kgCO_2-eq/GJ).

h This is the total plant cost (TPC), or "overnight capital cost" (which excludes interest during construction).

A CIGCC system is included as a repowering option because, among currently available technologies, this option offers the lowest LCOE among stand-alone coal power systems in new construction applications (see Section 12.2.3). The NGCC is also given close attention in light of recent bullishness about US gas supplies (MIT, 2010).

Simbeck and Roekpooritat (2009) analyzed and estimated LCOEs for several CCS retrofit options based on near-term technologies and showed that a simple CCS retrofit (without plant modification) based on an amine post-combustion scrubber offers the lowest LCOE for a retrofit, although the alternatives would be less energy-intensive. The least-costly CCS retrofit option identified in Simbeck and Roekpooritat is adopted with some of the cost assumptions adjusted to enable self-consistent comparisons in the analytical framework of Williams et al. (2011) and Liu et al. (2011a).

The findings of this analysis for various alternatives to a WO PC-V plant are as follows:

PC-CCS retrofit: The CCS retrofit option is widely thought to be the preferred option if coal use is to persist under decarbonization. This option is by far the least disruptive of the status quo. Its main attraction is that it requires the least capital investment among decarbonization options that involve coal use (see Table 12.25). However, this option involves a huge 36% energy penalty for CCS mainly because of the large amount of heat required to regenerate the amine solvent after it has absorbed the CO_2 from flue gases in which its partial pressure is only 0.14 atmospheres. This high energy penalty gives rise to a large (~33%) increase in the water requirements for the site, which will often greatly limit the viability of this option. The high energy penalty also implies a high minimum GHG emissions price ~US\$70/tCO_2-eq to induce via market forces a shift from WO PC-V to PC-CCS retrofit (see Figure 12.41).[41]

---

41 The point at which the PC-CCS retrofit and WO PC-V curves cross determines the minimum GHG emissions price needed to induce CCS via market forces.

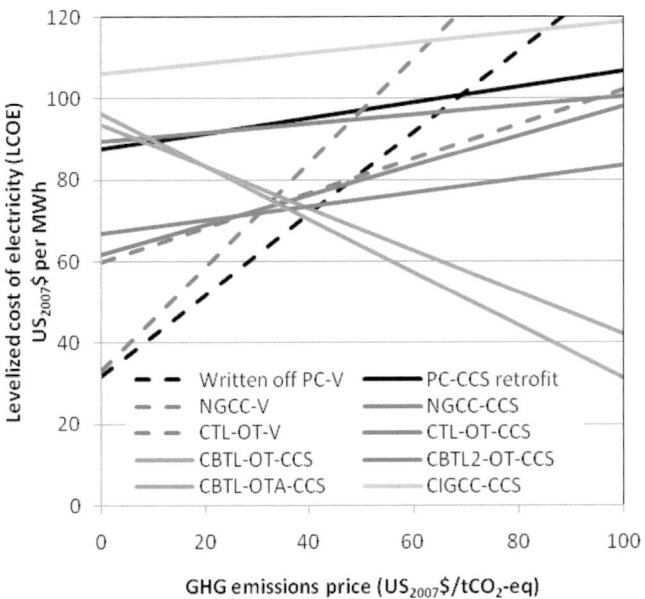

Figure 12.41 | Levelized cost of electricity (LCOE) vs GHG emissions price under US conditions for the alternative power options in Table 12.25 when the levelized crude oil price is US$90/bbl. (See Table 12.6, note (b) for financial parameter values assumed.) Source: based on Liu et al., 2011a; Williams et al., 2011.

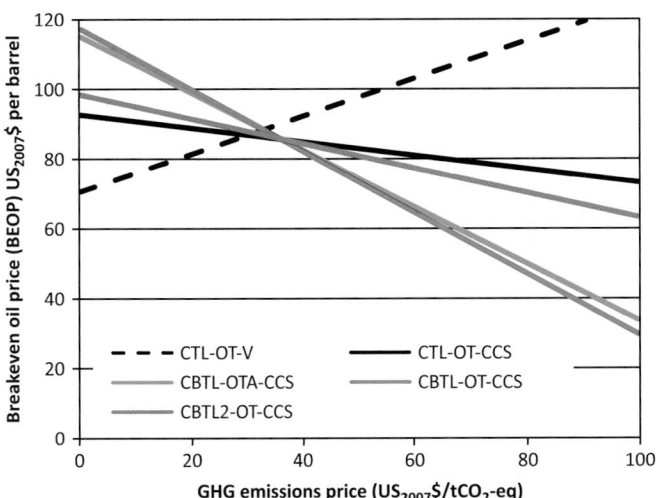

Figure 12.42 | Breakeven crude oil price (BEOP) vs GHG emissions price for the alternative coproduction systems in Table 12.25. (See Table 12.6, note (b) for financial parameter values assumed.) Source: based on Liu et al., 2011a; Williams et al., 2011.

Repowering via CIGCC-CCS: With pre-combustion capture, the energy penalty for CCS and water requirements are both reduced more than 40% relative to the PC-CCS retrofit (see Table 12.25). In part this is because the CO_2 is captured at high partial pressure from shifted syngas instead of from flue gases. But also the low water requirement reflects the fact that only ~1/3 of the power output of the combined cycle power plant is from the steam turbine, which requires the

consumption of water for condenser cooling purposes (the gas turbine topping cycle does not), and cooling water dominates water requirements in all cases. These technical advantages do not imply a lower LCOE for the CIGCC-CCS repowering option compared to the PC-CCS retrofit via post-combustion capture (see Figure 12.41), in contrast to the situation for new construction. This is largely because in this application the capital investment is ~three times that for the PC-CCS retrofit (see Table 12.25) whereas in new construction, where CIGCC-CCS does offer a lower LCOE, the capital cost for CIGCC-CCS is likely to be less than for a new pulverized coal plant with post-combustion capture (see Figure 12.10).

Repowering via NGCC: The natural gas combined cycle power plant venting CO_2 (NCCC-V) is the least capital-intensive and least water-intensive of the options shown in Table 12.25. But its carbon-mitigation potential (GHGI = 0.42) is far less than what is likely to be required by mid-century to realize the US goal of more than an 80% reduction in the nation's total GHG emissions. This implies that, if NGCC-V is deployed in the near term as a repowering option at WO PC-V plant sites, the NGCC-V would have to be replaced by NGCC-CCS (as a retrofit or via repowering option) at some point during this half-century. As shown by Figure 12.41 this option would be roughly competitive with a PC-CCS retrofit, and the energy penalty for CCS and water requirements would be less than half of those for a PC-CCS retrofit.[42] One serious economic challenge posed by the NGCC-CCS option is that the minimum GHG emissions price needed to induce CCS by market forces is nearly US$100/t (see Figure 12.41). Another relates to the prospect that it will be very difficult for NGCC-CCS to defend the high assumed 85% capacity factor in economic dispatch competition if there is much coproduction capacity on the electric grid because of the technology's high minimum dispatch cost compared to coproduction options (see Figure 12.43).[43] Still another problem posed by the NGCC-CCS option is that prospective natural gas supplies in the United States are likely to fall far short of what would be required to meet fully the decarbonization challenge posed by existing coal power plants, as discussed in Section 12.6.3.4.

Repowering via XTL-OT technologies: Though not a decarbonization option (GHGI = 1.18), CTL-OT-V is included here as a repowering option because it offers the least costly electricity at low GHG emissions prices.[44] Its inclusion highlights the GHG emissions prices needed to induce a transition to coproduction systems offering significant reductions in GHG emissions. Notably, for all coproduction options the energy

---

42 This may seem surprising because, like the PC-CCS retrofit, CCS involves post-combustion capture and the CO_2 partial pressure in flue gases is much lower (0.04 atmospheres). The modest penalty arises in this case because the capture rate is only 0.38 t/MWh compared to 1.17 t/MWh for PC-CCS retrofit – reflecting the much lower carbon intensity of natural gas compared to coal.

43 See discussion of dispatch competition in Section 12.6.3.1.

44 At $0/t its LCOE is the same as for WO PC-V.

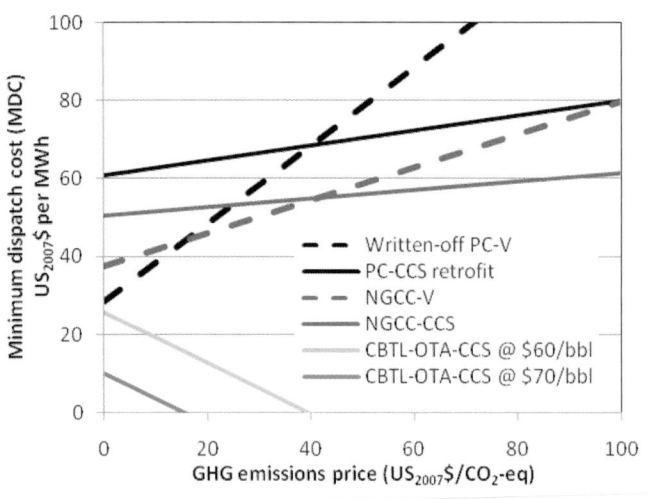

Figure 12.43 | Minimum dispatch cost (MDC) vs GHG emissions price for the alternative coproduction systems listed in Table 12.25. Source: based on Liu et al., 2011a; Williams et al., 2011.

penalty for CCS is one quarter to one half of that for the PC-CCS retrofit, and in all cases water requirements are less than two-thirds as much as for the PC-CCS retrofit. The CCS energy penalty is small largely because a substantial amount of CO_2 must be removed from syngas upstream of synthesis even in the absence of a carbon policy.[45] Water requirements are low for three reasons: (i) the low energy penalty for CO_2 capture; (ii) a significant part of the power output is from the gas turbine part of the power cycle, which does not require water for cooling (as in the NGCC and CIGCC cases); (iii) two thirds or more of the net energy output is in the form of liquid fuels, the production of which requires much less water than does electricity generation.

A notable feature of Figure 12.41 is that the coproduction options with CCS that do not offer deep reductions in emissions (CTL-OT-CCS (GHGI = 0.63) and CBTL1-OT-CCS (GHGI = 0.50)) are never the least costly electricity generation options. Rather, WO PC-V is the least costly option until the GHG emissions price reaches ~US$40/t, above which CBTL-OTA-CCS (GHGI = 0.077) is the least costly option.

It may seem counterintuitive that the more capital-intensive CBTL-OTA-CCS, which involves "aggressive" CO_2 capture, offers a lower LCOE than CBTL-OT-CCS at high GHG emissions prices. There are three reasons for this. First, the average feedstock price for CBTL-OTA-CCS (with 29% biomass) is 10% less than for CBTL-OT-CCS (with 40% biomass). Second, the capital intensity (in US$/kW$_e$) of CBTL-OTA-CCS falls from being 18% higher than for CBTL-OT-CCS when their FTL output capacities are the same to 12% higher when the CBTL-OTA-CCS output capacity is increased to the point where the biomass consumption rates are the same (1 Mt/yr) for these options. The net effect of the lower feedstock price and the higher capital intensity is that the LCOE is only 3% higher

for CBTL-OTA-CCS than for CBTL-OT-CCS at US$0/t. Third, the more rapid rate of decline of LCOE with GHG emissions price for CBTL-OTA-CCS arises because its FTL/electricity output ratio (and thus the credit for GHG emissions avoided by displacing crude oil derived products) is 21% higher for CBTL-OTA-CCS than for CBTL-OT-CCS.

A single XTL-OT plant requires an investment of US$1.2–1.8 billion (see Table 12.25). Investors will worry about the risk of oil price collapse, which is a very real concern because marginal oil production costs are lower than US$90/bbl, the assumed oil price for the LCOE calculations presented in Figure 12.41. In particular, successful market establishment of coproduction technologies could plausibly drive down oil prices.

How might investors be protected against the risk of oil price collapse? To help address this question, Figure 12.42 was constructed to show the breakeven crude oil prices for the alternative coproduction technologies. This figure shows that those who invest in XTL-OT-CCS technologies that coprocess biomass at high rates (29–40%) would be protected against falling oil prices down to less than US$60/bbl if the average GHG emissions price were US$69/t, the minimum GHG emissions price needed to make a PC-CCS retrofit competitive with WO PC-V. In contrast, the XTL-OT-CCS options that coprocess no or only modest amounts of biomass offer only modest protection against the risk of oil price collapse. Thus, the combination of a strong carbon mitigation policy and a high rate of biomass coprocessing is key to simultaneously realizing deep reductions in GHG emissions and enhancing transportation fuel security.

Finally, an attractive attribute of XTL-OT systems is their low minimum dispatch costs (MDCs), which would enable them not only to defend high capacity factors in economic dispatch competition but also would make it possible for them to drive down the capacity factors of competing technologies as XTL-OT market penetration expands. Figure 12.43 illustrates the point for CBTL-OTA-CCS. At crude oil prices of US$37/bbl and US$57/bbl, respectively, this system would have MDCs that are the same as for PC-CCS retrofits and WO PC-V plants when the GHG emissions price is $0/tCO$_2$eq. For US$74/bbl crude oil the MDC = US$0/MWh for this system.

12.6.3.4 Repowering Thought Experiment for the United States

The outstanding economics at sufficiently high oil and GHG emissions prices of coproduction systems coprocessing coal and substantial amounts of biomass suggest that it would be worthwhile exploring the repowering thought experiment for the United States that follows.

The point of departure for the thought experiment is an assumption that the United States enacts public policy mandating that all coal-fired power plants be retrofitted or repowered at a linear rate over the period 2016–2050, until all existing coal power plant capacity is replaced. Since the projected level of coal capacity for 2015 is 322 GW$_e$ (US EIA, 2010b), the coal power retrofit/repowering rate would be 9.2 GW$_e$/yr during

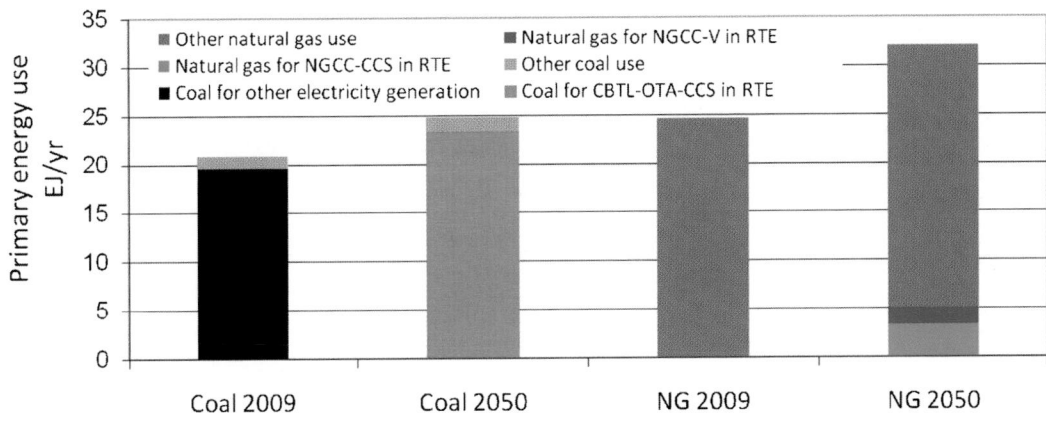

Figure 12.44 | Coal and natural gas demand implications of US repowering thought experiment (RTE).

2016–2050. It is further assumed that market forces are effective in deploying the least-costly options and that crude oil and GHG emissions prices are sufficiently high to make CBTL-OTA-CCS technology the least costly power generation option, and to make both NGCC-V and NGCC-CCS technologies less costly options than both the WO PC-V and PC-CCS retrofit options. (One set of prices for which these conditions would be met is US$90/bbl of crude oil and a GHG emissions price greater than US$50/t, see Figure 12.41.) Also it is assumed that under repowering conditions a site would not produce any electricity for a period of five years, which is the time required to bulldoze the existing site and build a new repowering unit there. Under these conditions, it would be cost-effective to retire WO PC-V plants, no PC-CCS retrofits would be deployed, and the following would be a cost-effective scenario:

- During 2016–2020, greenfield NGCC-V plants come on line at a rate of 9.2 GW_e/yr to exactly compensate for coal power plant retirements.

- During 2021–2050, each year 9.2 GW_e of coal power capacity is replaced by 4.9 GW_e of CBTL-OTA-CCS repowering capacity + 4.3 GW_e of NGCC-CCS makeup capacity built at greenfield sites to compensate for coal plant retirements.

The following are the results of this repowering thought experiment, assuming for simplicity that over the period to 2050 the US energy economy is frozen at the level of energy activities of 2015 as projected in US EIA (2010b) for all energy sectors except coal power:

- GHG emissions for 1810 TWh of electricity (all coal electricity in 2015) would be reduced in 2050 by 1.61 GtCO$_2$-eq/yr (85% reduction).

- GHG emissions for 5.4 million barrels/day of gasoline equivalent liquid transportation fuels would be reduced in 2050 by 0.84 GtCO$_2$-eq/yr (92% reduction), bringing the total GHG emissions avoided in 2050 to 2.45 GtCO$_2$-eq/yr.

- Replacement electricity generation in 2050 would be made up of 64% CBTL-OTA-CCS, 22% NGCC-CCS, and 14% NGCC-V.

- The CO$_2$ storage requirements in 2050 would be 2.04 GtCO$_2$ (93% via CBTL-OTA-CCS and 7% via NGCC-CCS).

- The biomass required for CBTL-OTA-CCS plants in 2050 would be 508 Mt/yr.

- US coal and natural gas use in 2050 would be, respectively, 19% and 31% higher than in 2009 (see Figure 12.44).

- The total investment required for the repowering thought experiment through 2050 is US$1.1 trillion (see Figure 12.45).

The results of the repowering thought experiment have some important strategic implications:

- The low C transportation fuels produced would be at about the level of transportation fuels produced from US domestic crude oil in 2009.[46]

- The total CO$_2$ storage requirements in 2050 would be about the same if PC-CCS retrofits were pursued[47] instead of this repowering strategy, even though the retrofit strategy would involving decarbonising only coal power – a startling result that reflects largely the difference in energy penalties for CCS (36% for PC-CCS retrofits vs 17% and 16% for CBTL-OTA-CCS and NGCC-CCS, respectively, see Table 12.25).

- The potentially available biomass resources converted using CBTL-OTA-CCS systems could produce much more low-GHG liquid fuel than if converted to cellulosic ethanol, because producing 1 GJ of

---

46 Crude oil was produced in the United States in 2009 at a rate of 5.36 million barrels per day, from which 5.18 million bbl/day of gasoline equivalent transportation fuels (gasoline, diesel, and jet fuel) were produced.

47 In the CCS retrofit strategy 73% of the system capacity and 97% of the system generation in 2050 are provided by PC-CCS retrofit plants, while 27% of system capacity and 3% of generation are provided by NGCC-CCS. The CO$_2$ storage rate in 2050 is 2.08 GtCO$_2$, 99% is via PC-CCS retrofit plants.

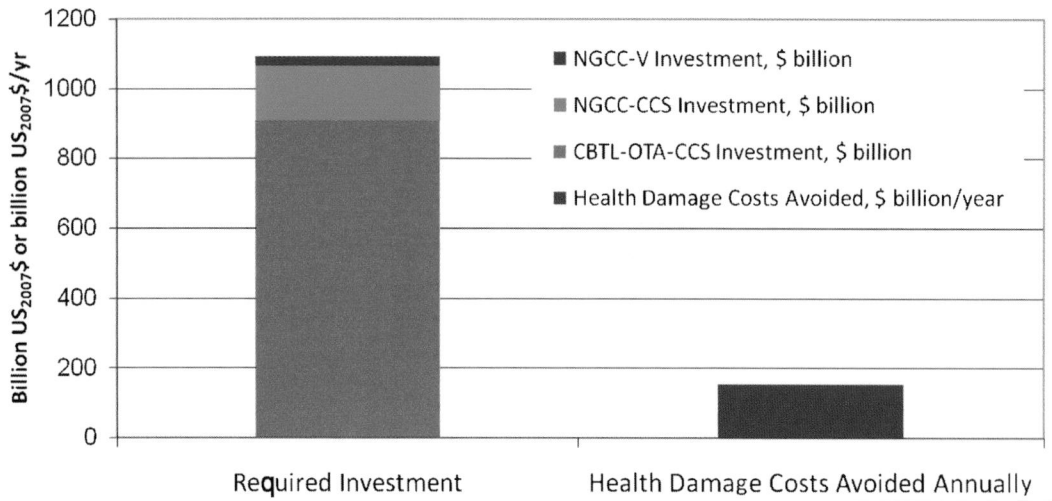

Figure 12.45 | The simple payback on the total new plant investment, 2016–2050, for the RTE would be about seven years if the sole benefit considered is the reduced public health damage costs from reduced $PM_{2.5}$ air pollution. (See Table 12.9.)

Fischer-Tropsch transportation fuels via CBTL-OTA-CCS requires only 36% as much biomass as producing 1 GJ of cellulosic ethanol (see Figure 12.33). If the same amount of synthetic low-carbon liquid fuel as would be produced in 2050 in the repowering thought experiment were instead produced as the energy equivalent amount of cellulosic ethanol, some 1400 Mt/yr of biomass would be required (see Section 12.6.4). In 2005 it was widely believed that the United States has 1300 Mt/yr of prospective biomass supplies for energy. Concerns about food/fuel competition, indirect land-use impacts, and biodiversity loss have reduced the estimate of sustainable US biomass production to ~500 Mt/y (the level for the repowering thought experiment in 2050) if growing dedicated energy crops on good cropland is not allowed – according to the *America's Energy Future* Study of the US National Research Council (AEFP, 2009).

- The incremental natural gas supply required for the RTE beyond that for 2009 is 7.5 EJ/yr by 2050 and 5.9 EJ/yr for the intermediate year 2035. The latter is much less than the 9.4 EJ/yr of incremental shale gas which the Energy Information Administration estimates will be available by 2035 (AEFP, 2009). However, US EIA also projects substantial reductions for other gas supplies in the period to 2035 (see Figure 12.46), so that the net incremental gas supply in 2035 relative to 2009 is only 3.1 EJ/yr, which is enough to satisfy only 52% of the natural gas requirements[48] for the RTE by 2035, assuming all the incremental gas is dedicated to power generation.

- Alternative renewable electricity supplies would have to grow four times between 2009 and 2035 if all of net incremental gas supply projected for 2035 were used for power generation and the short-

fall in electricity generation were provided entirely instead by these renewable sources.

- The simple payback on the investment would be ~seven years (see Figure 12.45) if the only benefit considered for this repowering strategy were the reduced health damage costs from $PM_{2.5}$ air pollution (see Table 12.9).

12.6.4 Alternative Approaches for Low GHG-emitting Liquid Fuels Using Biomass

The analysis in the preceding section indicates that coproduction systems with CCS that coprocess ~30% biomass would be very competitive as repowering options for sites of old coal power plants at high oil and GHG emissions prices, while simultaneously providing synthetic liquid fuels with near zero net GHG emissions. But biomass is a scare resource. How does this biomass use option compare to other options for providing low-GHG-emitting transport fuels? This question must be addressed to understand better the extent to which biomass coprocessing with coal and CCS for coproduction systems should be given priority in public policies relating to future uses of biomass for energy.

To address this question, the CBTL-OTA-CCS option for which 29% of the feed (HHV basis) is biomass is compared to alternative ways to use biomass to provide low-C fuels. Attention is focused on six alternatives. One alternative (BIGCC-CCS) was discussed in Section 12.2.3. The negative emissions with BIGCC-CCS provide "room in the atmosphere" for emission from some petroleum-derived fuels. Three alternatives (CBTL-RC-CCS, BTL-RC-V, BTL-RC-CCS) were discussed in Section12.4.3. Two new systems not yet considered, are also introduced as alternatives: (i) a future cellulosic ethanol option (EtOH-V), which is currently a high priority for biofuels development in the United States; and (ii) a –CCS variant of this (EtOH-CCS), see Box 12.3. In total, seven options are

---

48 Since the natural gas demand for NGCC-V systems deployed during 2016–2020 would have to be satisfied first in the RTE, only 15% of the natural gas needed for NGCC-CCS systems by 2035 would be available from net incremental natural gas supplies.

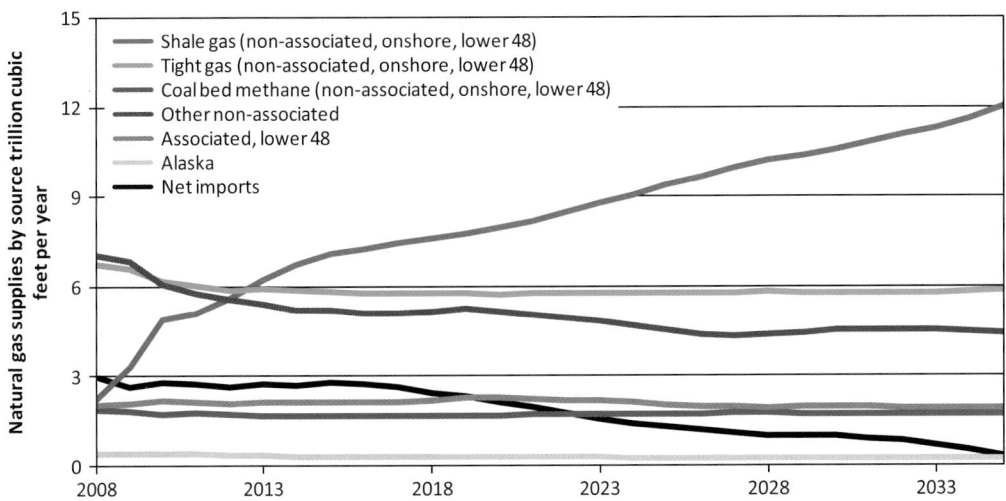

Figure 12.46 | Outlook for natural gas supplies in the United States. To convert cubic feet to cubic meters, divide by 35.3. Source: modified from US EIA, 2010a.

Box 12.3 | Prospective Cellulosic Ethanol Technologies

EtOH-V: The main cellulosic ethanol option considered here is the future switchgrass-to-ethanol option analyzed in a study by America's Energy Future Panel (AEFP, 2009), for which the estimated yield is 80 gallons of ethanol (EtOH) per dry short ton (334 L/t). For reference, the yield from switchgrass with currently understood technology is ~69 gallons per dry short ton (288 L/t). That study estimated a capital cost of US$_{2007}$\$156 million for a plant producing 40 million gallons (156 million liters) of ethanol annually (an output capacity of 1941 bbl/day, or 309,000 L/day, of gasoline equivalent).

EtOH-CCS: It is feasible to carry out CCS for EtOH production if there are adequate CO_2 storage opportunities nearby because in the fermenter 1 mol of CO_2 is generated in a pure stream for each mol of EtOH) produced. The capital cost penalty for this CCS option is modest (see Table 12.26).

presented in Table 12.26. In all cases it is assumed that the feedstock is switchgrass consumed at a rate of about 0.5 Mt/yr dry biomass – the reference scale assumed for future cellulosic ethanol plants (EtOH-V) assessed in a recent United States National Research Council report (AEFP, 2009).

Six of the plants listed in Table 12.26 involve CCS, but the CO_2 capture rates vary by a factor of 17 from the lowest (EtOH-CCS) to the highest (CBTL-OTA-CCS). To illustrate the impact of capture rate on system economics it is assumed in all CCS cases that the CO_2 is transported 100 km and stored in a saline formation 2 km underground, and that the maximum CO_2 injectivity is 2500 t/day per well. The CO_2 transport and storage cost model cited by Liu et al. (2011a) is used to estimate CO_2 transport and storage costs (in US$_{2007}$\$/t). Alternative biomass use options are compared with respect to five metrics. Three of these are defined and presented for the alternative options in Table 12.26: a biomass input index (BII), a greenhouse gas emissions index (GHGI), and a zero-emission fuels index (ZEFI). In addition, Figure 12.47 presents the LCOF for the six options that provide fuels,

and Figure 12.48 presents the real internal rate of return (IRRE) for four options.

BII: This index (defined in Table 12.26, note (c)) indicates that the coprocessing options (CBTL-RC-CCS and CBTL-OTA-CCS) require much less biomass to make low-C liquid fuels than the biofuel options (e.g., 36–40% as much as is required for the EtOH options).

GHGI: All options have outstanding GHG emissions mitigation performance, with GHGI less than 0.20. Four options have negative emission rates that could offset emissions from other fossil energy systems.

ZEFI: This performance index (defined in Table 12.26, note (e)) highlights the strategic importance of storing underground a large fraction of the biomass feedstock carbon not contained in the energy products, thereby exploiting the negative GHG emissions benefit of photosynthetic CO_2 storage. It quantifies simultaneously the carbon mitigation and liquid fuel insecurity mitigation potentials of an option. The higher the value of ZEFI, the better the option. Four of the options store enough photosynthetic

Table 12.26 | Alternative low carbon fuel options – each consuming 457,000 dt/yr biomass[a].

| Plant name | Capacities | | | | BII[c] | GHGI[d] | ZEFI[e] | CO_2 stored[f], Mt/y (% of feedstock C stored) | TPC[g], million US$_{2007}$ |
| | Inputs | | Outputs | | | | | | |
	Fuel, MW, HHV	% biomass, HHV basis	Electricity, MW$_e$	Liquid fuel, barrels/day of gasoline equivalent (MW, LHV)					
EtOH-V	302	100	2.03	1,941 (113)	2.49	0.17	0.33	0	156
EtOH-CCS	302	100	0.62	1,941 (113)	2.49	−0.21	0.49	0.11 (15)	158
BTL-RC-V	302	100	19.3	2,241 (131)	2.15	0.063	0.42	0	408
BTL-RC-CCS	302	100	14.2	2,241 (131)	2.15	−0.95	1.02	0.44 (56)	416
CBTL-RC-CCS	670	45	24.3	4,882 (284)	0.99	0.029	0.97	0.91 (54)	733
CBTL-OTA-CCS	1041	29	131	5,395 (314)	0.90	0.086	0.92	1.70 (65)	939
BIGCC-CCS[b]	320	100	118	0	–	−0.93	0.99	0.71 (90)	398

a Based on Liu et al., 2011a. See also Box 12.3.
b Based on unpublished modeling carried out by the authors of Liu et al., 2011a in a manner that is consistent with the modeling of the systems presented here that make liquid
   fuels + electricity, see Table 12.7 (in which the BIGCC-CCS case here corresponds to the switchgrass case in the PEI section of the table).
c BII (biomass input index) ≡ Biomass energy (in GJ LHV dry) required to produce 1 GJ of liquid fuel (LHV).
d GHGI is the system-wide life cycle GHG emissions relative to emissions from a reference system producing the same amount of power and fuels. The reference system consists of
   equivalent crude oil-derived liquid fuels and electricity from a stand-alone new supercritical pulverized coal power plant venting CO_2. For details, see Table 12.15, note (c).
e ZEFI (Zero-emissions fuels index) ≡ Amount (in GJ) of equivalent zero GHG-emitting liquid fuel provided per GJ of biomass input.
f It is assumed that captured CO_2 is compressed to 150 atmospheres, transported via pipeline 100 km to a storage site, and injected into a deep saline formation 2.5 km
   underground.
g TPC (total plant cost) ≡ "overnight capital cost" (which excludes interest during construction).

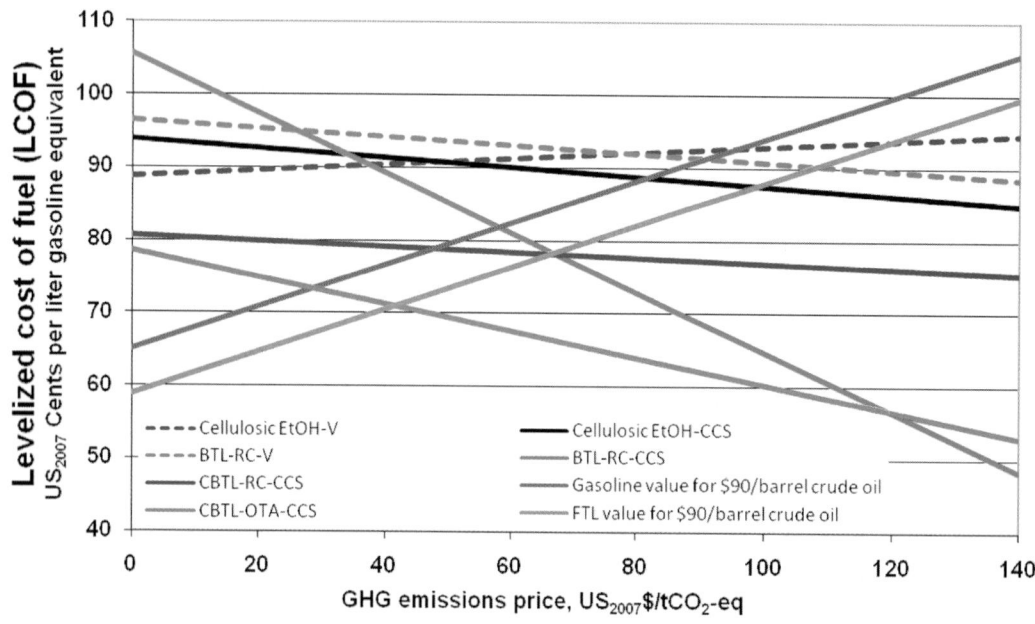

Figure 12.47 | LCOF vs GHG emissions price for the six alternative low-carbon fuel options listed in Table 12.26. Also shown are values of petroleum-derived gasoline and an
average for gasoline plus diesel corresponding to the proportions of these fuels found in the FTL flues in the CBTL and BTL cases (see Table 12.15, note (a)). (See Table 12.6,
note (b) for financial parameter assumptions. Also, electricity sales are assumed at the US average grid price plotted in Figure 12.10.) Based on Liu et al., 2011a.

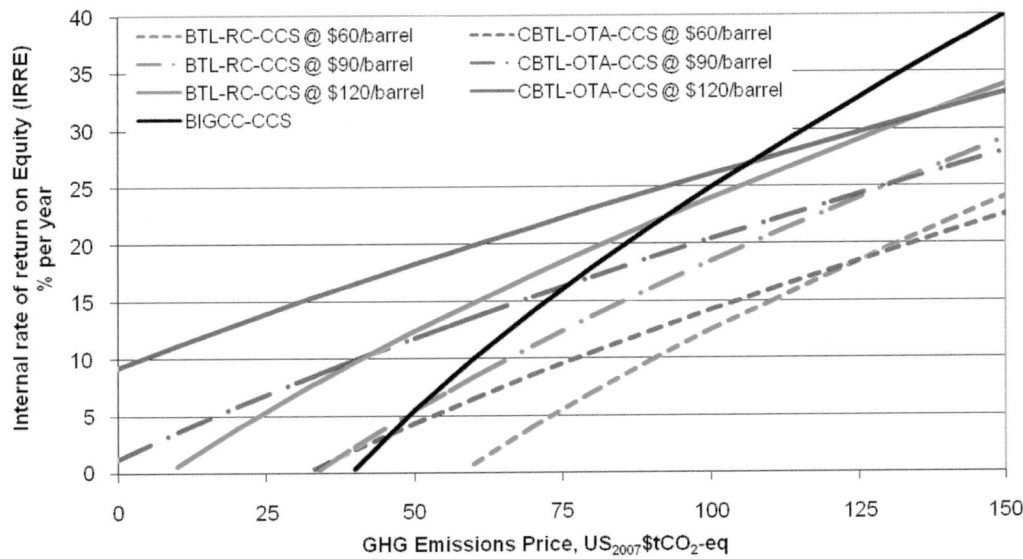

Figure 12.48 | Real internal rate of return on equity (IRRE) for BIGCC-CCS (see Table 12.26) and for the FTL options with the lowest LCOF (see Figure 12.47) for three levelized crude oil prices. Electricity sales are assumed at the US average grid price plotted in Figure 12.10. Based on Liu et al., 2011a.

CO_2 to convert (in effect) 92% or more of the biomass energy into useful zero GHG emitting transportation fuel energy. Although the BIGCC-CCS option does not provide liquid transportation fuels, the negative GHG emissions of this option enables it to offset emissions from crude oil-derived products; assuming the crude oil product displaced is gasoline, the ZEFIs for this option is high (0.99). The –V options perform poorly in terms of this metric. Notably, EtOH-CCS also has a relatively low ZEFI, reflecting the fact that only 15% of the C in the feedstock is stored underground so that the potential photosynthetic storage benefit is modest.

LCOF: Figure 12.47 shows that at all GHG emissions prices below $120/tCO_2eq CBTL-OTA-CCS offers the least costly liquid fuel, becoming competitive with the crude oil products displaced when the GHG emissions price exceeds US$40/tCO_2-eq and the crude oil price is US$90/bbl. In contrast EtOH-V requires a GHG emissions price of more than US$95/t to be competitive at this crude oil price. Notably, pursuing CCS does not markedly improve the economics of cellulosic ethanol at high GHG emissions prices – reflecting the relatively modest photosynthetic CO_2 benefit for this option (storing only 15% of the feedstock C).

Although at US$0/t, BTL-RC-CCS is the most costly transportation fuel option, its LCOF declines rapidly with GHG emissions price as a result of its strong negative GHG rate (GHGI = –0.95). This technology cannot compete with CBTL-OTA-CCS until the GHG emissions price exceeds US$120/t. However, in regions where coal coprocessing is not a realistic option, BTL-RC-CCS would become competitive with crude oil derived products for GHG emissions prices > US$67/t when the oil price is US$90/bbl. As noted earlier, *World Energy Outlook 2009* estimates that if the world community were on a path to stabilizing the atmospheric concentration of GHG gases at 450 ppmv, the world oil price would be stable during 2020–2030 at US$90/bbl, and if, in addition, there were a single global carbon trading price, that price would be about US$70/t by 2030 (IEA, 2009b). Thus it is plausible that in the post-2030 time period BTL-RC-CCS would be cost-competitive in biomass-rich but coal-poor regions if the world community were on this carbon mitigation path.

IRRE: The real IRRE is a good metric for comparing the economics of power-only options with liquid-fuel options. Figure 12.48 shows the IRRE for the BIGCC-CCS option and the two most promising liquid fuel options analyzed here (CBTL-OTA-CCS and BTL-RC-CCS) for three crude oil prices: US$60, US$90, and US$120/bbl.[49] Several conclusions can be drawn from these curves. First, at very high GHG emissions prices, BIGCC-CCS has the highest IRRE. Second, at all three oil prices, CBTL-OTA-CCS has the highest IRRE at low GHG emission prices, and break-even with BIGCC-CCS occurs at a GHG emissions price that increases with the crude oil price: US$46/t at US$60/bbl, US$76/t at US$90/bbl, and US$107/t at US$120/bbl. Third, BTL-RC-CCS will become more profitable than CBTL-OTA-CCS at GHG emissions prices in the range US$125–130/t.

12.6.5   Coal/biomass Coprocessing with CCS as a Bridge to CCS for Biofuels

Establishing in the market over the next 20 years technologies involving coal/biomass coprocessing with CCS for the coproduction of liquid fuels and electricity could plausibly serve as a bridge to widespread biofuels production with CCS in the post-2030 time frame. This is suggested by the analyses of Sections 12.6.3 and 12.6.4 considered together. This is

---

49   For these options it is assumed that the debt/equity ratio is 45/55 and that electricity is sold at the US average grid price indicated in Figure 12.10. For other financing assumptions relating to these IRRE calculations see Liu et al., 2011a.

an important consideration because biomass supplies in coal-rich countries are relatively limited but are abundant in many coal-poor regions of the world. In coal-poor regions, the economics of liquid fuel production via thermochemical conversion of biomass with CCS become economically attractive relative to crude oil-derived products at prospective GHG emissions prices for the post-2030 period if the global community were to pursue stabilization of the GHG concentration in the atmosphere at 450 ppmv. Moreover, the capital needed to build these plants should become available in a world with carbon trading and high GHG emissions prices. This is likely because not only would these plants be very competitive in providing liquid fuels, but they also would generate huge revenue streams from the sale of carbon credits.[50]

12.6.6 Global Coal/biomass Thought Experiment for Transportation

A global coal/biomass thought experiment for transportation is presented to illustrate the carbon mitigation potential for transportation by 2050 via widespread use of low-carbon fuels provided via coal and/or biomass with CCS in many regions. In coal-poor regions this would be accomplished via BTL-RC-CCS, and in coal-rich regions via CBTL-OTA-CCS. This thought experiment is set in the context of a world in which the growing of dedicated energy crops on cropland is off-limits (Tilman et al., 2009) because of concerns about impacts on food prices (Rosegrant, 2008) and land-use effects (Fargione et al., 2008; Searchinger et al., 2008).

It is assumed for the supply side of this thought experiment that the only biomass supplies are residues[51] and mixed prairie grasses grown with minimal inputs on abandoned agricultural lands. The estimated total global biomass supply under these conditions is approximately 6 Gt/yr dry biomass[52] – all of which is assumed to be used in the thought experiment. It is assumed that CBTL-OTA-CCS plants use 1 Gt/yr of biomass worldwide and that the rest of the biomass is used in BTL-RC-CCS plants located in coal-poor regions. Many of these plants would be located in developing regions such as sub-Saharan Africa and Latin America.

---

50 For example, at a GHG emissions trading price of $70/tCO$_2$-eq, the present worth of the stream of revenues (over the assumed 20-year economic life of the plant) from the sale of carbon emissions credits for the BTL-RC-CCS plant described in Table 12.26 would be equivalent to 73% of the TPC for the plant.

51 Agricultural residues, forest residues, forest thinnings to reduce forest fire risk and enhance productivity for commercial species, urban wood wastes, municipal solid wastes.

52 In energy terms, the residue supply is assumed for 2050 to be 75 EJ/yr (the global residue supply for energy assumed for 2050 in IEA, 2008b). The assumed potential supply of biomass grown on abandoned agricultural lands is 32 EJ per year (average yield of 4.3 dry tonnes/yr on 429 million hectares worldwide) based on Campbell et al. (2008). The total biomass supply, 107 EJ/yr, is 2.3 times total biomass use for energy in 2005. For comparison, the global biomass supply for 2050 in the IEA Blue Map scenario is 150 EJ/yr (which includes biomass grown for energy on cropland), and the maximum biomass resource availability for 2050 is estimated in Chapter 7 to be still higher than this. For perspective, the total assumed global biomass supply (107 EJ/yr) is comparable to total US primary energy use in 2006 (97 EJ/yr).

Dedicating the entire global biomass supply to BTL and CBTL plants is clearly an unrealistic assumption. However, as the preceding analysis has shown, these applications are likely to be hard to beat in terms of both the carbon mitigation per tonne of scarce biomass and economics in a world of high GHG emissions prices and high oil prices. Moreover, such a starkly defined global coal/biomass thought experiment helps to clarify the tradeoffs associated with the allocation of scarce biomass resources.

The demand for global transportation energy in 2050 in the thought experiment is assumed to be essentially that presented for the IEA's BLUE Map transportation scenario (IEA, 2009a), an update for the transport sector of its earlier projection to 2050 (IEA, 2008b).[53] A general characterization of the BLUE Map scenario is that it envisages technological change bringing emissions to 50% of the 2005 level by 2050. This includes a strong emphasis on improving energy efficiency but excludes lifestyle changes and modal shifts. It is appropriate to assume an energy-efficient future as a context for exploring the prospects for carbon mitigation via biomass and coal with CCS because, typically, energy efficiency improvement represents "the low-hanging fruit" in carbon mitigation (e.g., see Chapter 10) and thus should be given priority. Transportation is also one of the main branching points of different pathways considered in Chapter 17; Figure 17.2 shows possible final energy shares for transportation up to 2100 for different transportation system strategies (discussed further in Chapter 17, Section 17.3.3.4).

The challenge for global transportation is to reduce GHG emissions at midcentury by a factor of four relative to the IEA Baseline Projection in order to reduce global emissions by 50% relative to emissions in 2005. As shown in Figure 12.49, this baseline projection, which involves nearly a doubling of both energy demand and GHG emissions relative to 2005, illustrates the dimensions of this challenge.

In our global coal/biomass thought experiment, while transportation demand in 2050 is essentially that in the BLUE Map scenario, we modify that scenario by excluding some technologies. We are working from the assumption that no plug-in hybrid vehicles, no all-electric vehicles, and no H$_2$ fuel cell vehicles are included in the light-duty vehicle (LDV) mix in the period to 2050. This is in contrast to the actual BLUE Map scenario, which assumes a high penetration of these technologies in the mix by midcentury.[54] Obviously, this exclusion is not a realistic assumption, because these technologies are being heavily promoted by governments. The main reason for the exclusion is to bring to the attention of policymakers that, as shown in this chapter, deep reductions can be plausibly realized (e.g., via pursuit of CCS for coal and biomass) at attractive production costs with essentially no costly changes in transportation fuel infrastructures and with evolutionary rather than revolutionary changes

---

53 For the IEA Baseline Projection the total number of light-duty vehicles (LDVs) increases three times to 2.15 billion, 2005–2050, air travel increases 3.9 times, and truck freight increases 1.9 times.

54 In 2050 the LDV fleet in the IEA BLUE Map Scenario (IEA, 2009a) includes 37% plug-in hybrids, 21% all-electric vehicles, and 25% H$_2$ hybrid fuel cell vehicles.

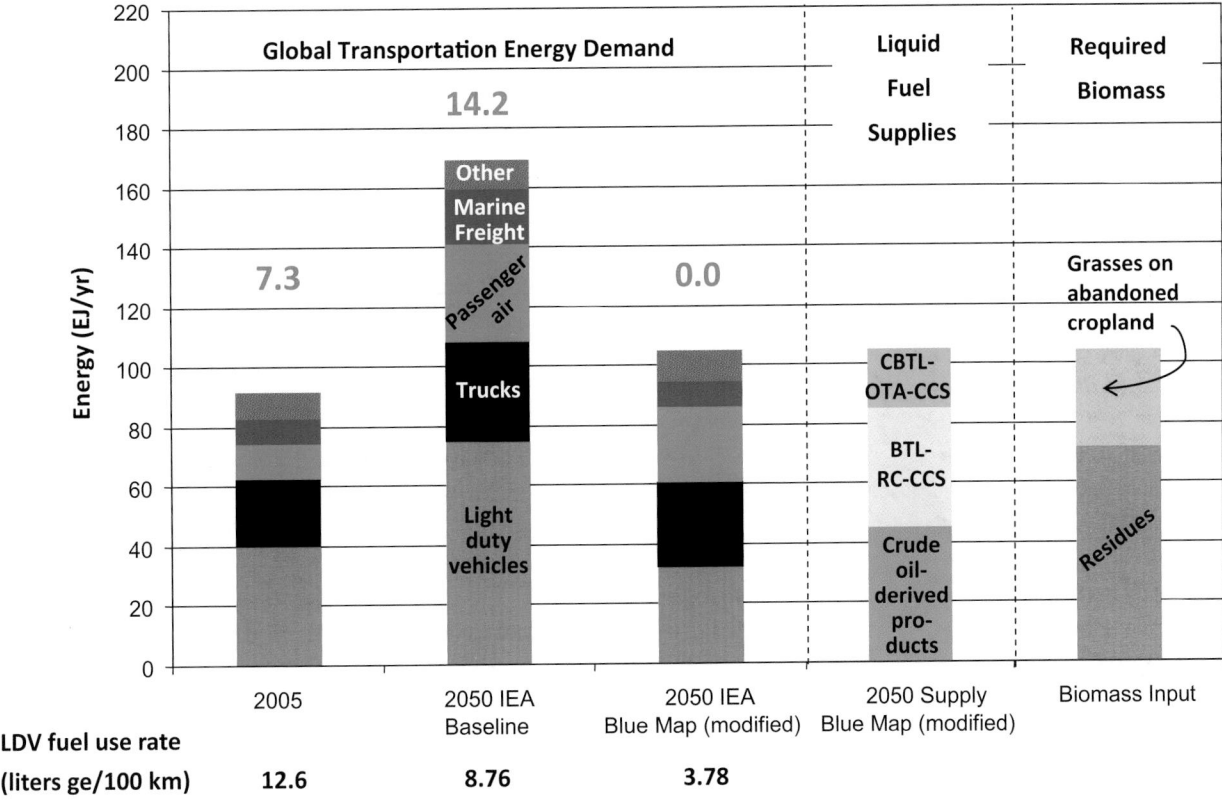

Figure 12.49 | Energy and GHG emissions for global coal/biomass thought experiment and for IEA 2050 Baseline (business-as-usual) scenario. Numbers at tops of bars are total GHG emissions in GtCO$_2$-eq/yr. LDV fuel use rate is in liters of gasoline equivalent per 100 km. Source: see discussion in text.

in automotive engine technologies. In particular, energy-efficient LDVs[55] (e.g., advanced gasoline and diesel hybrid electric vehicles) are emphasized in our modified BLUE Map scenario. Our hope is that this modified scenario will inspire policymakers to take a more balanced approach toward decarbonizing the transportation sector and to create policies that enable the low carbon transportation technology options that have been the focus of this chapter to compete alongside the technologies that are intensely promoted in current policies.

The resulting liquid fuel demand and supply for global transportation under the modified BLUE Map scenario are shown as the 3rd and 4th bars of Figure 12.49. Total net GHG emissions are zero for global transportation in 2050 under the conditions of the global coal/biomass thought experiment. The liquid fuel supply would be made up of 19% FTL via CBTL-OTA-CCS, 38% FTL via BTL-RC-CCS, and 43% crude oil–derived products. The level of crude oil-derived products used in transportation

is 51% of the amount of crude oil derived products used for transportation worldwide in 2005. The projected GHG emissions are zero despite the large share of crude oil-derived products in the mix because it is assumed for the thought experiment that the negative emissions from BTL-RC-CCS systems are used to offset positive GHG emissions from both crude oil-derived products and CBTL-OTA-CCS systems.

In 2050 coal use for the modified BLUE Map scenario (with the global coal/biomass thought experiment) is 1.15 times global coal use in 2005, compared to 0.78 times for the IEA BLUE Map scenario and 3.04 times for the IEA Baseline Scenario.

Under the conditions of the thought experiment, CO$_2$ storage would be carried out in 2050 at rates of 3.7 Gt/yr for CBTL-OTA-CCS plants and 4.8 Gt/yr for BTL-RC-CCS plants. This finding highlights the importance of ascertaining the extent to which there are good geological CO$_2$ storage opportunities in biomass-rich but coal-poor regions, where consideration of CCS as a carbon mitigation option has probably not been given much thought.

Is it plausible that about ten thousand BTL-RC-CCS plants[56] might be up and running by 2050 (each processing 0.5 Mt of biomass annually)? If one imagines starting with 20 such plants by 2030 the annual average

---

55 In the IEA (2009a) scenarios, average fuel economies for new LDVs in 2050 are 5.9 liters of gasoline equivalent per 100 km (l$_{ge}$/100 km) or 40 miles per gallon of gasoline equivalent (mpg$_{ge}$) for the Baseline Scenario and 2.9 l$_{ge}$/100 km (81 mpg$_{ge}$) for the BLUE Map scenario. The average fuel economy assumed for the entire LDV stock in 2050 in the modified BLUE Map scenario is 2.9 l$_{ge}$/100 km (62 mpg$_{ge}$), which is consistent with the IEA assumptions for the average fuel economy for new LDVs in the actual BLUE Map scenario. There are reasonable prospects that this average fuel economy for the LDV stock could be realized without use of plug-in hybrids or fuel cell vehicles. Kromer and Heywood (2007) have estimated that by 2030 a mid-sized conventional hybrid-electric car (essentially a 2030 version of a Prius) could have a fuel economy of 3.1 l$_{ge}$/100 km (76 mpg$_{ge}$).

56 Each of which provides 2,200 bbl/day of gasoline equivalent FTL + 14 MW$_e$ of electricity and requires an investment of US$420 million (see Table 12.26).

growth rate in plant deployment, 2030–2050, would have to be 37%/yr. This extraordinarily high growth rate would not take place without supportive public policy, but it is not inconceivable. In its heyday, nuclear power grew worldwide at a sustained average growth rate of 37%/yr from 1957–1977.

Much of this growth would take place in now-impoverished developing countries, where this technology would represent a major opportunity for industrial growth. But many such countries currently have neither the needed physical infrastructures (e.g., roads, railroads, pipelines, port facilities for managing exports, etc.) nor the human capacity to manage such industrial growth. The technology is not likely to be cost competitive until about 2030 even under strong carbon-mitigation policies worldwide, however, so there is a strategic opportunity to build those needed capacities in the interim – if the global community comes to think that there is a pressing need for this industry.

Finally, if the goal for decarbonizing global transportation were to reduce by 2050 GHG emissions for transportation worldwide not to zero but rather to half the 2005 level (in line with current long-term policy goals for carbon mitigation), the global biomass requirements for 2050 drop to a level that is only 38% higher than global biomass use in 2005 – so that considerable biomass would be available for purposes other than to satisfy the global coal/biomass thought experiment needs. Moreover, crude oil-derived products (in an amount equivalent to 69% of the level for transportation in 2005) would account for 60% of total transportation energy in 2050.

12.7 Long-term Considerations

The preceding analyses in Chapter 12 focused mainly on technologies that are near at hand. It is beyond the scope of Chapter 12 to present a comprehensive review of advanced fossil energy conversion technologies. Rather, the focus in this short section is on some advanced technologies that might be especially helpful in evolving to promising longer-term future low-GHG emitting fossil-fuel-based electricity and synfuels technologies and strategies.

12.7.1 Low-carbon Electricity Generation Technologies in the Long-term

For systems that provide only electricity, a brief review is presented of prospects for substantial cost reduction in systems with CCS. The review is restricted to gasification-based systems for three reasons. First, with current bituminous coal technologies, CIGCC systems with pre-combustion CO_2 capture outperform PC systems with post-combustion capture – as shown in Section 12.2.3. Second, synfuel/electricity coproduction systems that require gasification have astonishingly good economic prospects even with current technologies, as discussed in Section 12.6.3. Reason number three has to do with the fact that coal gasification technologies are at a much earlier stage in their technological evolution than

are coal combustion technologies. Over the longer term, it is plausible that conversion efficiencies will be higher and costs for generating coal electricity with CCS will be lower with advanced gasification technologies than the costs presented in detail in earlier sections of Chapter 12.

A recent report from the US National Energy Technology Laboratory gives a sense of the possibilities: Gray et al. (2009) describe two alternative technological paths for gasification energy to provide power from coal with CCS: (i) an evolutionary change path that involves a series of incremental changes in CIGCC technology and (ii) a revolutionary change path that involves a shift from combustion energy conversion to electrochemical energy conversion via use of solid oxide fuel cells (SOFCs). Here key findings of Gray et al. (2009) are sketched out to highlight the importance of a major research and development effort in this area.

The evolutionary change approach for CIGCC involves the following incremental changes over time:

- a continuation of the historical trend toward improved gas turbine performance as a result of increasing firing temperatures and higher pressure ratios, but with emphasis in a carbon-constrained world on gas turbines burning hydrogen;

- introduction of a dry coal feed pump to reduce the energy penalty associated with evaporating water via a coal/water slurry feed system and to provide a simplified alternative to the lockhopper dry feed systems currently used;

- warm acid gas removal with CO_2 capture in Selexol instead of the current cold CO_2 capture in Selexol, to greatly reduce the energy penalty otherwise required for Selexol regeneration;

- warm acid gas removal with H_2 and CO_2 separation via a hydrogen separating membrane that delivers hydrogen in a pressurized N_2/H_2 mixture to the gas turbine combustor while enabling CO_2 recovery as a separate stream at elevated pressure, thereby reducing the CO_2 compressor load;

- use of an ion transport membrane[57] instead of a cryogenic air separation unit to provide O_2 for the gasifier – the main impact of which would be to reduce the capital cost for O_2 production;

- increased system reliability, availability, and maintainability – resulting in an increased system capacity factor.

Gray et al. (2009) carry out a systems analysis suggesting that success in all of these areas over time could plausibly lead, for US conditions, to

---

57 A non-porous ceramic membrane through which O_2 ions flow at high temperature (800–900°C) and are converted to O_2 molecules on the permeate side of the membrane, while electrons flow countercurrent to the retentate side of the membrane.

an eventual IGCC-CCS efficiency ~40% (HHV basis) and a capital cost (TPC) less than US$1700/kW$_e$ – values very similar to the corresponding values for new supercritical steam-electric plants in the US today (the PC-V option described in Section 12.2.2).

The revolutionary SOFC approach involves coupling a high-efficiency SOFC operating at pressure to a catalytic gasifier that produces methane-rich syngas, such as the gasifier being developed by Great Point Energy (a company in the United States) for making substitute natural gas.[58] As the syngas passes through the anode of the SOFC the methane is reformed with steam to form CO and H$_2$, and these gases in turn react with steam and oxygen to form CO$_2$ and H$_2$O while providing electric power for the external circuit. Part of the heat released in these reactions is manifest as a rise in the syngas temperature from 600°C to 900°C as it passes through the anode. The oxygen is provided as O$_2^-$ ions (undiluted with nitrogen) via transport through the fuel cell's non-porous ceramic membrane electrolyte[59] from the hot air streaming through the cathode. The benefits of this approach to power generation (Grol et al., 2007) include the following:

- Making electricity directly from chemical energy without first having to burn the fuel to make heat gives rise to higher energy conversion efficiency than can be realized via fuel-powered heat engines.

- The lower gasification temperature of the catalytic gasifier means that less energy is needed to heat the gas to its operating temperature, so that the gasification efficiency is higher.

- Another benefit is energy and cost savings, given the fact that with a sufficiently high fraction of methane in the gasifier-generated syngas, the methanation reaction exotherm can provide enough heat for gasification – which implies significant energy and cost savings associated with air separation that would otherwise be needed to provide oxygen for gasification.

- The SOFC produces steam when H$_2$ in the syngas and pure O$_2$ react on the anode; the steam reacts with methane in the syngas producing CO and H$_2$; the CO reacts with more steam to produce H$_2$ and CO$_2$, so that what ultimately leaves the system is a stream of CO$_2$ and H$_2$O – from which CO$_2$ is easily separated for piping to an underground storage reservoir.

- The strong endotherm of the methane steam reforming reaction on the anode absorbs a considerable amount of heat released from the oxidation of H$_2$ to produce water, thereby reducing the amount

of air required to cool the SOFC – one manifestation of which is a reduction of the amount of air compression that would otherwise be required to provide this cooling air.

- If the SOFC operates at an elevated pressure, the subsequent CO$_2$ compression required for pipeline transport to suitable storage sites would be less than if the SOFC operates at atmospheric pressure.

- The steam reforming of methane on the anode eliminates the need for a separate steam reformer when methane-rich syngas is the feed.

- The "shifting" of CO to CO$_2$ by reaction on the anode with water, producing H$_2$, eliminates the need to shift CO to CO$_2$ in a separate "water-gas-shift" reactor.

Gray et al. (2009) have estimated prospective efficiency, capital cost, and levelized electricity production costs for such SOFCs and compared them to the same quantities that would arise with the evolutionary strategy. They found that the system efficiency would be 40% higher (~56%), the specific capital cost (US$/kW$_e$) would be ~5% higher and the levelized cost of electricity would be about the same – although perhaps requiring a longer period of time to achieve such estimated performance and cost values. Many of the technological advances analyzed would be applicable to systems that coproduce synthetic fuels and electricity, as described for current technologies in Section 12.6. These findings suggest that both the evolutionary and the revolutionary approaches warrant comparable levels of research and development support. It does not necessarily follow, however, that these two approaches would offer comparable overall benefits when the technologies involved are fully developed.

12.7.2 Technologies and Strategies for Low-carbon Liquid Fuels in the Long-term

Maintaining the option of sustaining liquid hydrocarbon fuels as the primary energy carriers for the transportation sector even under a severe carbon policy constraint is extraordinarily important in light of the ease of transporting, storing, and using these fuels compared to the alternatives such as H$_2$ and electricity (e.g., see near the end of Section 12.4.4 for some discussion of the challenges for evolving a ground transportation system based on hydrogen fuel cell vehicles).

In the earlier Chapter 12 analysis of opportunities for producing synthetic fuels characterized by low greenhouse gas emission rates, some of the most important findings were the strategic importance of:

- CCS for biomass as a carbon mitigation strategy for a carbon-constrained world;

- the opportunity to get started with CCS for biomass via coal/biomass coprocessing; and

---

58 Based on the original Exxon process, Great Point Energy's catalytic gasifier is a relatively low-temperature gasifier that involves the production of methane as a considerable fraction of the syngas output as a result of a potassium-catalyzed methanation reactions.

59 The SOFC electrolyte permits the flow of O$_2$ ions (like the ceramic non-porous ion transport membrane mentioned above for O$_2$ production) but does not permit electrons to flow in the reverse direction.

- electricity + synfuels coproduction w/CCS as a critical path for coprocessing in realizing simultaneously:

  - Low costs for low-C electricity;

  - Low costs for low-C synfuels;

  - Energy insecurity mitigation in a manner consistent with realization of carbon-mitigation objectives.

Chapter 12 points out that first generation technologies based on these novel concepts can be deployed in the market during the current decade. Much can be done to improve such systems using advanced technologies. Although a systematic review of advanced concepts is beyond the scope of the present analysis, what is discussed here is of possible strategic importance:

- a new approach to chemical synthesis in the context of a carbon constrained world;

- shale gas for a carbon-constrained world; and

- extension of the coproduction concept from the present focus on coal/biomass coprocessing to include as well natural gas/biomass coprocessing.

12.7.2.1   Synthesis chemistry for a carbon-constrained world

The chemical process industry and the embryonic synfuels industry are based on making synthesis gas (mainly CO and H_2) from a fossil fuel and/or biomass and then combining the synthesis gas molecules CO and H_2 to make chemicals and/or fuels. A recent paper (Hildebrandt et al., 2009a) points out that there are potentially significant gains in energy conversion efficiency to be exploited by shifting both the gasification process from the production of mainly CO and H_2 to the production of instead mainly H_2 and CO_2 and basing the subsequent chemical synthesis on combining H_2 and CO_2 instead of H_2 and CO. Applying the idea to the production of F-T liquids from coal via gasification, those authors point to an 18% theoretical potential reduction in the amount of net work required for the overall process and a 15% reduction in the amount of coproduct CO_2 generated in the process.[60]

This idea might be helpful in the context of a carbon mitigation policy, for evolving from the present situation where energy is based mainly on fossil fuels, to an energy future in the very long term when synthetic liquid hydrocarbon fuels might be made mainly by combining H_2 derived from water using a non-carbon primary energy source[61] such as solar energy or thermonuclear fusion, and CO_2 extracted directly from the air[62] to make liquid fuels – a process that might start with methanol production:

$$CO_2 + 3\,H_2 \rightarrow CH_3OH + H_2O, \qquad (6)$$

followed by conversion of the methanol to hydrocarbon liquid fuels such as gasoline via the already commercial MTG processes (see Sections 12.4.1.2 and 12.4.3.1) or via future processes that would produce middle distillates as well as gasoline (Keil, 1999).

Though seemingly different in the case of methanol production from the present approach that involves making methanol from syngas via:

$$CO + 2H_2 \rightarrow CH_3OH, \qquad (7)$$

it is now well known (Hansen, 1997) that what really happens in making methanol from syngas is that CO first reacts with H_2O to form CO_2 and H_2 via the water gas shift reaction followed by the hydrogenation of CO_2 on the catalyst surface. Moreover, making methanol from CO_2 and H_2 is actually not a novel approach.[63]

At present H_2 produced from non-carbon energy sources is far more costly than making H_2 with ultra-low GHG emissions from coal or natural gas with CCS (Williams, 2002), and direct extraction of CO_2 from the air is far from being economic.[64] Nevertheless, the concept of making chemicals and synfuels from $CO_2 + H_2$ could potentially be of strategic importance even in the near term in fossil fuel and fossil/biomass systems that make synfuels (or synfuels + electricity) and produce CO_2 as a major coproduct. Under a serious carbon policy constraint, the excess CO_2 has to be captured and stored underground in any case; a key insight implicit in the Hildebrandt analysis (Hildebrandt et al., 2009a; Hildebrandt et al., 2009b) is that there may be strategic advantages of using this readily available CO_2 to improve overall energy system performance before delivering the excess to underground storage.

This discussion ends with a word of caution. The importance of potentially improving system efficiencies by exploring gasification and synthesis

---

60   An e-letter comment on the Hildebrandt et al. (2009a) paper by Desmond and Gibson (2009) points out that at the high gasification temperature considered it is not practical to design a gasifier that produces mainly H_2 and CO_2 unless an enormous amount of water is added to promote the water-gas-shift reaction, but the energy penalty of heating that water would eliminate potential energy savings. Hildebrandt et al. (2009b) responded that it would be practical to produce mainly H_2 and CO_2 if one considers not a single piece of equipment but rather a gasifier followed by a water-gas-shift reactor operated at a much lower temperature than the gasifier. But if that were done it is not clear how one could realize energy savings for the gasification part of the system, although there would still be significant savings by basing synthesis on $CO_2 + H_2$. Moreover, the GHG emissions associated with the water heating referred to in the Desmond and Gibson (2009) criticism could be mitigated by heating the water with solar energy instead of burning fossil fuel.

---

61   See, for example, Lewis and Nocera (2006) and OSL (2003).

62   See, for example, Section 2.1.4: Carbon dioxide capture from ambient air, in The Royal Society (2009).

63   Olah et al. (2009) point out that the very first methanol plants operating in the 1920s and 1930s were commonly using CO_2 and H_2 from other processes to make methanol.

64   In a recent report of the American Physical Society, it is estimated that with current technology direct air capture of CO_2 would cost $600 or more per tonne of CO_2, under optimistic assumptions (APS, 2011).

based on CO_2 chemistry should not be construed as suggesting that CO_2 reuse to make synthetic fuels is a viable alternative to CCS.

There is widespread interest in CO_2 reuse for making industrial products as an alternative to CCS. For example, the US Department of Energy has dedicated more than US\$107 million to the exploration of twelve approaches to reuse (DOE, 2010). Of these concepts, the most numerous and globally significant are systems that use microalgae to capture CO_2 from power plant flue gas and convert it (via sunlight, water, and nutrients) into natural oils that are readily processed into liquid transportation fuels such as biodiesel. The concept is tantalizing when one considers, as an example, that the United States in 2008 consumed fossil fuel carbon in the amounts of 500 Mt in the form of transportation fuels while simultaneously emitting 540 Mt from pulverized coal steam electric power plants. If the CO_2 in flue gases were captured and reused by being converted into carbonaceous liquid fuels using solar energy or another non-carbon energy source to provide the needed process energy, enormous quantities of transportation fuels could be provided without increasing GHG emissions. But this example contains the essence of the shortcomings of reuse to make transportation fuels: If, hypothetically, 100% of the CO_2 from flue gases of US coal power plants were converted via carbon-free energy sources into transportation fuels, the carbon emissions for the system of making both transportation fuels and electricity from coal would be reduced by only 50% from the level of the current system that makes electricity from coal with CO_2 venting and transportation fuels from crude oil.

Kreutz (2011) examines two cases to estimate the potential for GHG emissions reduction via reuse of coal power plant CO_2 to make liquid transportation fuels: the growing of algae in algal ponds to make biodiesel and the use of concentrated sunlight to reduce CO_2 to CO and O_2 and/or H_2O to H_2 and O_2 to make liquid fuels via Fischer-Tropsch synthesis. He shows that the carbon mitigation potential of reuse strategies to make synthetic transportation fuels is at best very modest in a world where the carbon price is high enough to induce decarbonization of coal power plants either by shifting to a carbon-free or low-carbon power source or by making CCS more economical than venting CO_2 (in which case the CO_2 would be captured and available mainly in pipelines to underground storage sites so that its diversion from this purpose would be equivalent to extracting CO_2 from underground).[65] According to Kreutz: "Using the carbon twice fails to meet the objective of deep GHG emission reductions across the entire energy economy; only one sector (either power or transportation) – but not both – can claim the benefit of carbon neutrality." Thus CO_2 reuse strategies to make synthetic fuels would not enable the deep (over 80%) reductions in GHG emissions that leaders of industrialized countries are targeting for their countries by midcentury.

12.7.2.2   Implications of Abundant and Ubiquitous Shale Gas

It is truly remarkable how fast views of the fossil fuel energy future change. In its 2003 report to the US Secretary of Energy, the National Petroleum Council (NPC, 2003) predicted that future North American natural gas supplies would be flat or declining. In sharp contrast, there is currently much bullishness about the future prospects of shale gas, largely as a result of applications of new effective, economic hydraulic fracturing and horizontal drilling technologies.[66] It is now thought that natural gas extracted from shale in sedimentary basins might turn out to be abundant and ubiquitous, with reasonable production costs. Although most empirical data are from the United States, there is much optimism that the judgments that are now being formed about US shale gas prospects might be more or less valid for sedimentary basins throughout the world (see Chapter 7).

If this bullishness about shale gas proves to be sound, and growing environmental concerns associated with extraction of shale gas (Rahm, 2011) can be addressed, it has potentially far-reaching implications for the future of fossil fuel energy in a carbon-constrained world.

To begin, CCS would need to be pursued for NG energy conversion systems based on shale gas in order to meet carbon mitigation obligations over the longer term. Consider first use of shale gas for making electricity in NGCC-CCS plants. One advantage offered relative to CIGCC-CCS plants is that less than half as much CO_2 would have to be stored per MWh. Moreover, the capital investment required to generate electricity via NGCC-CCS is likely to be less than half that for a CIGCC-CCS plant having the same output capacity, and the LCOE for such a plant is likely to be less than for a CIGCC-CCS plant at the GHG emissions price needed to make a CIGCC-CCS plant cost competitive with a PC-V plant for natural gas prices up to four times the coal price.[67] The downside for the NGCC-CCS system is that the GHG emissions price required to induce a shift from NGCC-V to NGCC-CCS is very high (see Figure 12.10).

12.7.2.3   Gas/biomass to Liquids and Electricity with CCS

Should shale gas prove to be as abundant and ubiquitous as some believe, it will be of interest to consider the merits of using shale gas for liquid fuels production. The high breakeven GHG emissions price required to induce CCS for gas-based power generation could be greatly reduced while simultaneously making it possible to realize deep reductions in GHG emissions for synthetic liquid fuels if natural gas were coprocessed

---

65 This is in contrast to the case at low carbon prices, when using CO_2 from flue gases is equivalent to extracting CO_2 from the atmosphere, because a profit-motivated power generator would rather vent the CO_2 to the atmosphere and pay the emissions fine than invest in CCS.

66 There are fundamental differences in the production of gas from shale and from other unconventional sources such as tight sands (Frantz Jr. and Jochen, 2005). The latter yield a tremendous amount of gas for the first few months, but then production declines significantly and often becomes uneconomic after a relatively short time. In contrast, shale gas wells do not come on as strong as gas from tight sands but, once production stabilizes, the wells will produce consistently for 30 years or more.

67 For a coal price of US\$2.0/GJ.

with biomass to make liquid fuels + electricity with CCS – analogous to the CBTL-OT-CCS and CBTL-OTA-CCS systems discussed in Section 12.6.

Liu et al. (2011b) present analysis of systems coproducing FTL and electricity from natural gas and from natural gas + biomass (switchgrass). Two of the systems analyzed are described in Table 12.27. The GBTL-OT-CCS option (see Figure 12.50), which provides 300 MW$_e$ of net electricity + 9750 bbl/day of FTL transportation fuels, was designed to coprocess enough biomass (34%) to reduce GHGI to 0.10. The output capacities of the system were determined by the design criterion that the system consumes 1 Mt/year dry biomass. Its system features are compared to the same system without CCS (GBTL-OT-V) and two coal based systems that also coprocess 1 Mt of biomass annually: CBTL-OTA-CCS, which coprocesses 29% biomass to realize GHGI = 0.0855 (discussed in Section 12.6.3.3 (see Figure 12.31)), and a variant without CCS.

Notable differences between the natural gas-based and coal-based coproduction systems can be gleaned from Table 12.27:

- The natural gas-based systems convert 34.5% of the feedstock C to FTL compared to 24.2% for the coal-based systems – a consequence of the much lower H/C ratio for coal compared to natural gas (0.8 vs 4.0);

- The FTL output amounts to 34.4% of input energy for natural gas-based systems compared to 32.5% for coal-based systems, showing that the low rate of carbon conversion in the coal case is largely compensated for by the water gas shift reaction, which entails a relatively minor energy penalty;

- The CO_2 storage rate for GBTL-OT-CCS is only half that for CBTL-OTA-CCS, reflecting both the lower H/C ratio for coal and the fact that the latter option vents as CO_2 only 6.6% of the C in the feedstock compared to 12.4% in the GBTL-OT-CCS; and

- The GBTL-OT-CCS option is much less capital intensive than the CBTL-OTA-CCS option.

Figure 12.51 presents the levelized cost of electricity (LCOE) vs GHG emissions price for these four options along with the LCOEs for four stand-alone power plants (discussed in Section 12.2.2), assuming for all cases that: (i) the levelized fuel prices are US$2.04/GJ$_{HHV}$, US$5.1/GJ$_{HHV}$, and US$5.0/GJ$_{HHV}$ for coal, natural gas, and biomass, respectively (see note (b) of Table 12.6 for additional financial parameter assumptions), (ii) a US$90/bbl crude oil price, and (iii) the captured CO_2 that is pressurized to 150 atmospheres is transported via pipeline 100 km and stored in a deep saline formation 2 km below ground with a maximum injectivity per well of 2500 t/day. The important results from this set of curves are:

- Compared to other stand-alone power plants, the GBTL-OT-CCS option has outstanding performance under a carbon policy constraint,

having breakeven GHG emission prices of: US$25/t in competing with sup PC-V, US$28/t in competing with NGCC-CCS, and US$47/t in competing with NGCC-V; moreover, GBTL-OT-CCS would be able to compete with CIGCC-CCS even at $0/t. For comparison the breakeven GHG emissions price for NGCC-CCS competing with NGCC-V is about US$83/t.

- At the assumed reference feedstock prices GBTL-OT-CCS is not competitive with CBTL-OTA-CCS.

- However, the relative competitiveness between the natural gas-based and coal-based coproduction systems depends sensitively on the relative coal and natural gas prices. For example, at a GHG emissions price of US$0/t the LCOEs for these two coproduction options with CCS would be equal with a modest decrease of 8% in the assumed natural gas price (to US$4.7/GJ$_{HHV}$) and a modest increase of 8% in the assumed coal price (to US$2.19/GJ$_{HHV}$).

Finally, a global gas/biomass thought experiment for transportation is constructed in the same spirit as the global coal/biomass thought experiment for transportation presented in Section 12.6.6 to illustrate how zero net GHG emissions for transportation in 2050 might be realized via widespread deployment of a combination of GBTL-OT-CCS + BTL-RC-CCS systems instead of CBTL-OTA-CCS + BTL-RC-CCS systems, assuming the same demand levels as in the global coal/biomass thought experiment presented in Figure 12.49.

A comparison of some attributes of the two global thought experiments is presented in Table 12.28. This table shows that overall production rates for FTL fuels, the amounts of crude oil products used, the total GHG emissions avoided in displacing conventional fossil fuels, and electricity generation rates are comparable for the two thought experiments. The amount of coal used in the coal/biomass thought experiment (which produces a larger fraction of FTL via coproduction) is 25% more than the amount of gas used in the gas/biomass thought experiment. The amount of CO_2 storage required with natural gas is only 80% of that for coal, reflecting in large part the lower carbon content of natural gas compared to coal.

The relative roles of coal-based and natural-gas based systems will vary from region to region depending on relative coal and natural gas prices. Despite the bullishness about shale gas in the United States, total US domestic gas supplies are not likely to be adequate to support a substantial role for GBTL-OT-CCS. This conclusion is suggested by the analysis in Section 12.6.3.4. However, the finding presented in Table 12.28 that total worldwide natural gas use in the modified BLUE Map scenario (which includes the natural gas used for the global gas/biomass thought experiment) is only 0.81 times the projected worldwide natural gas use in the IEA Baseline Scenario – suggests that at the global level gas supplies might be adequate for realization of the global gas/biomass thought experiment. In any case, the potential benefits of the gas/biomass coproduction option as indicated in

Table 12.27 | Performance and capital cost (US$_{2007}$\$) estimates for coproduction systems with biomass/fossil fuel coprocessing.

	CBTL-OT-V	CBTL-OTA-CCS	GBTL-OT-V	GBTL-OT-CCS
Input capacities				
Coal, as-received mt/day (MW$_{HHV}$)	5150 (1616)	5150 (1616)		
Natural gas, mt/day (MW$_{HHV}$)			2084 (1278)	2084 (1278)
Biomass, as-received mt/day (MW$_{HHV}$)	3581(660.5)	3581(660.5)	3581(660.5)	3581(660.5)
Biomass % of total input, HHV basis	29.0%	29.0%	34.1%	34.1%
Output capacities				
Synthetic diesel + gasoline, MW LHV (bbl/d crude oil products displaced)	687 (10,882)	687 (10,882)	619 (9,752)	619 (9,752)
Diesel fraction of FTL (LHV basis)	0.634	0.634	0.669	0.669
Gasoline fraction of FTL (LHV basis)	0.366	0.366	0.331	0.331
Gross electricity production, MW	521.1	465.9	484.0	430.5
Net electricity exports, MW	407.8	287.4	384.5	300.2
FTL/Electricity output ratio	1.69	2.39	1.61	2.06
ENERGY RATIOS (HHV basis)				
Liquid fuels out /Energy in	32.5%	32.5%	34.4%	34.4%
Net electricity/Energy in	17.9%	12.6%	19.8%	15.5%
Fuels + electricity/Energy in	50.4%	45.1%	54.2%	49.9%
C input as feedstock, kgC/sec	54.5	54.5	34.4	34.4
C stored as CO_2, % of feedstock C	0	65.5	0	51.7
C in unburned char, % of feedstock C	3.7	3.7	1.4	1.4
C vented, % of feedstock C	45.0	6.6	64.0	12.4
C in FTL, % of feedstock C	24.2	24.2	34.5	34.5
C stored, MtCO$_2$/yr[a]	0	3.71	0.0	1.85
GHGI[b]	0.903	0.0855	0.539	0.105
Plant capital costs, million US$_{2007}$\$				
Air separation unit + O_2/N_2 compressors	243	255	217	216
Biomass handling and gasification	335	335	295	295
Coal handling and gasification	396	396	-	-
Syngas cleanup	180	215	153	197
NG handling	-	-	4.0	4.0
CO_2 compression	1.6	33	0	21
F-T synthesis & refining	208	208	204	203
Naphtha upgrading	35	35	31	31
Autothermal reformer	0	55	87	87
Power island gas turbine	98	86	92	94
Heat recovery and steam cycle	224	168	154	166
Total plant cost (TPC), million US$_{2007}$\$	1,720	1,786	1,236	1,313
Specific capital cost, US$_{2007}$\$/kW$_e$	4,218	6,215	3,213	4,375

a With plant operating at 90% capacity factor.
b GHGI is the system-wide life cycle GHG emissions relative to emissions from a reference system producing the same amount of power and fuels. The reference system consists of electricity from a stand-alone new supercritical pulverized coal power plant venting CO_2 plus equivalent crude oil-derived liquid fuels. See Table 12.15, note (c), for details.

Table 12.28 suggest the importance of exploring further gas/biomass coproduction systems – which might offer a key to exploiting abundant and ubiquitous shale gas in a carbon-friendly manner. However, environmental concerns about "fracking" technologies used for shale gas recovery with respect to protection of water (Osborn et al., 2011) and soil (Jones, 2011), as well as concerns about life cycle GHG balances for shale gas (Howarth et al., 2011), need to be addressed effectively.

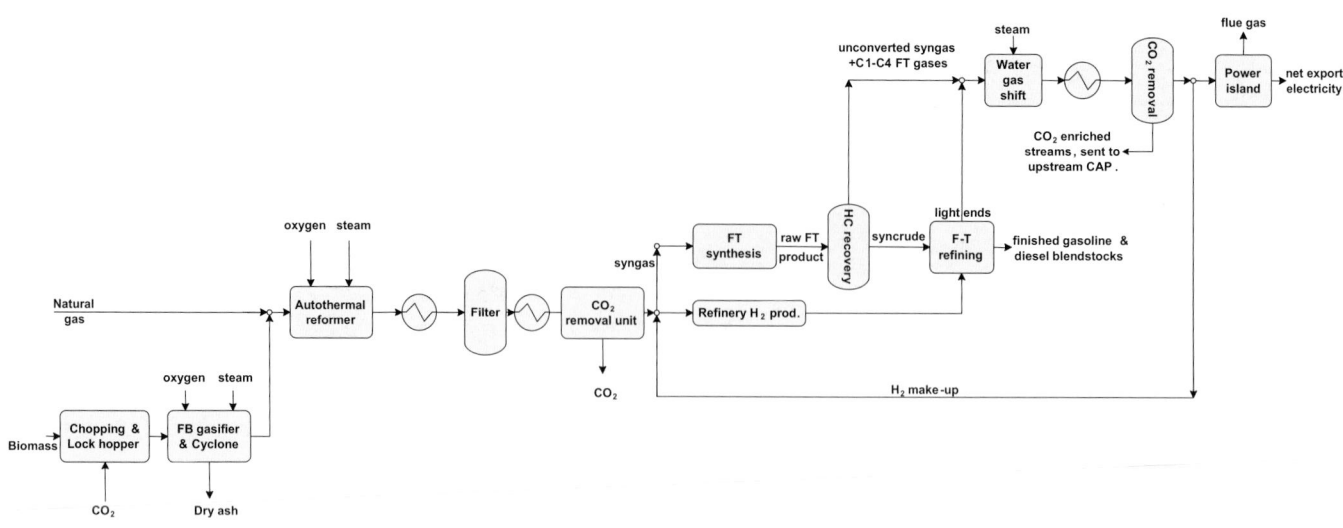

Figure 12.50 | GBTL-OT-CCS system that provides electricity + FTL (via use of a cobalt catalyst) from natural gas and biomass. Natural gas is converted to syngas in an autothermal reformer. Syngas is generated from biomass in a fluidized bed gasifier. The autothermal reformer is also used to crack (i.e., eliminate) the tars from the biomass-derived syngas. 52% of C in the feedstocks is captured, compressed to 150 bar, and sent via pipeline to a geological storage site. The system was designed with enough biomass coprocessing (34%) to realize GHGI = 0.10. Source: based on Liu et al., 2011b.

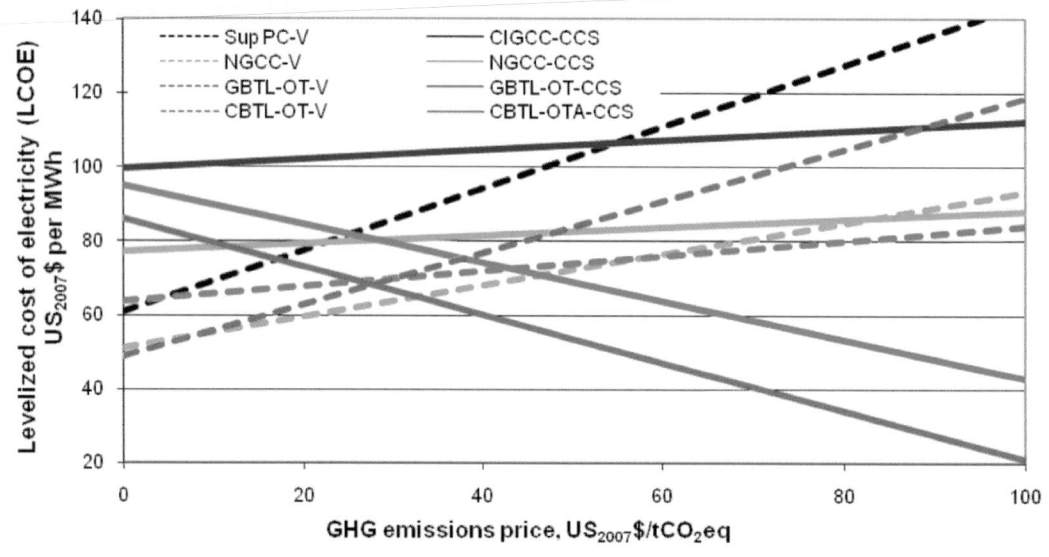

Figure 12.51 | Levelized cost of electricity vs GHG emissions price for the coproduction technologies in Table 12.27 and stand-alone power options – assuming a US$90/bbl of crude oil price and reference case coal and natural gas pricing (US$2.04/GJ and US$5.1/GJ, respectively). (See Table 12.6, note (b) for financial parameter assumptions.) Source: based on Liu et al., 2011b.

12.8 Strategies and Policies for Radically Transforming the Fossil Energy Landscape

The analysis in this chapter shows that a radical transformation of the fossil energy landscape is feasible for simultaneously meeting the multiple societal objectives of wider access to modern energy carriers, reduced air pollution health risks, enhanced energy security, and major GHG emissions reductions.

12.8.1 Strategies for a Radical Transformation

Developing countries have quite different economic circumstances at present and hence different energy priorities from industrialized countries. The strategies for fossil energy development in developing and industrialized countries will, accordingly, be different in the short term, but necessarily must be convergent in the long term.

Table 12.28 | Thought experiments with zero GHG emissions for global transportation in 2050.

	Global C/B TE	Global G/B TE
Decarbonized synfuels produced for transportation in 2050, EJ/yr LHV	59.8	59.0
FTL via coproduction, %	32.7	29.8
FTL via BTL-RC-CCS, %	67.3	70.2
Decarbonized net electricity generation in 2050, million MWh/yr	3,484	3,621
Crude oil-derived products used in transportation in 2050, EJ/yr LHV	45.0	45.2
Fossil fuel feedstock for making FTL in thought experiment, EJ/yr HHV	45.7	36.3
Biomass for making FTL in thought experiment, EJ/yr HHV (a)	111.7	114.4
Fossil fuel feedstock (coal or NG) needed for Modified Blue Map Scenario relative to use of that feedstock in IEA Blue Map Scenario in 2050	1.46	1.26
Fossil fuel feedstock (coal or NG) needed for Modified Blue Map Scenario relative to use of that feedstock in IEA Baseline Scenario in 2050	0.38	0.81
Fossil fuel feedstock (coal or NG) needed for Modified Blue Map Scenario relative to use of that feedstock in 2005	1.15	1.59
Overall GHG emissions avoided in displacing conventional energy, GtCO$_2$-eq/yr	12.8	12.6
CO$_2$ storage rate, GtCO$_2$/yr	8.47	6.75

a Estimates of sustainable biomass resource potentials are subject to discussion (e.g., see Chapters 7, 20).

The key for developing countries is to lay a sound foundation for further evolution to low carbon energy systems. In high economic growth developing countries like China and India, reducing conventional pollution is still the most urgent energy challenge. The priority of economic development of these societies tends to lead to increasing emissions of CO$_2$. Accordingly, carbon emissions for such countries will increase in the near term as they complete their industrialization and will subsequently decrease under appropriate sustainable development policies. In the poorest developing countries, economic poverty and associated lack of access to modern forms of energy are the major problems, and growing conventional pollution is a risk as these countries develop. In order to minimize overall carbon emissions from the use of fossil fuels in developing countries, efforts should concentrate on building manufacturing and infrastructure for the most efficient use of hydrocarbon energy resources and flexibility for incorporation of low carbon energy and feedstocks into the energy mix. Technology leapfrogging in energy-using technologies is another essential component of radical transformation strategies.

Although industrialized countries have made major advances in controlling conventional pollution, much more needs to be done in this regard—especially for old coal power plants. Moreover, their energy systems have not been designed for minimal CO$_2$ emissions. At the same time, the energy infrastructure is well established and much of

it is several decades old and ready for replacement or upgrade. Thus, the emphasis going forward in industrialized countries should be on replacing or upgrading obsolete infrastructure, e.g., via repowering sites of old fossil fuel power plants with technologies offering additional capabilities (such as coproduction of electricity and fuels, as discussed in this chapter) and pursuing CCS retrofits.

While different strategies are needed regionally, this chapter has identified four key technology-related requirements as essential for transforming the fossil energy landscape: (i) continued enhancement of energy conversion efficiencies, (ii) carbon capture and storage (CCS), (iii) co-utilization of fossil fuel and biomass in the same facilities, and (iv) coproduction of multiple energy carriers at the same facilities.

Fossil fuel/biomass coprocessing with CCS to coproduce clean liquid fuels and electricity with low greenhouse gas emissions is a technically and economically feasible strategy that can make significant contributions in addressing all the major challenges posed by present energy systems, although additional strategies are also needed to solve all the problems posed by fossil fuels in the electric power and transportation sectors. Moreover, coproduction via fossil fuel/biomass coprocessing with CCS also represents a promising approach for launching CCS technologies in the market early on (facilitating as a result of early experience CCS for power-only systems in both greenfield and brownfield applications). Also, it offers a promising route for gaining early experience using lignocellulosic biomass to make liquid fuels thermochemically at attractive costs. And it can serve as a bridge to enabling CCS as a routine activity for biomass energy (with corresponding negative greenhouse gas emissions) in the post-2030 era. The latter could plausibly become a major industrial development opportunity for economically poor and fossil fuel-poor but biomass-rich regions, while making major contributions to decarbonization of the transport sector worldwide.

No technological breakthroughs are needed to get started with coproduction technologies and strategies that involve fossil fuel/biomass coprocessing with CCS. Most of the technical challenges are engineering issues best addressed via commercial-scale experience (learning by doing).

Despite the multiple attractions of coproduction systems, the concept faces formidable institutional hurdles because of complexities at the systems level. Success requires managing two disparate feedstocks (a fossil fuel and biomass) to provide simultaneously three commodity products (liquid fuels, electricity, and CO$_2$) that would serve three very different markets.

With fossil fuel/biomass coproduction systems, fossil fuel consumers for power generation would have to become consumers of biomass as well. There has been some experience cofiring existing coal power plants with biomass but that has been at biomass levels far more modest than the levels envisioned for the coproduction technologies described here.

Moreover, there is little reason to expect that electricity generators will want to take on the risks of producing and marketing liquid fuels. The oil industry, which could do that, has shown little interest in producing either synfuels from coal or electricity for sale to the electric grid, which offers much lower profit margins than what the oil industry has been accustomed to in exploration and production (E & P).

Yet these obstacles might be surmountable. Coal power generators are coming to realize they will have to embrace CCS in order to survive in a carbon-constrained world and will have to give priority in industrialized countries to CCS for existing coal power plant sites. Coal suppliers who are eager to see coal use expand into synfuels production are coming to realize that this cannot be done without both CCS and coprocessing of biomass under a stringent carbon policy constraint. Biomass suppliers might become interested in the concept once they recognize the potential for greater profits selling biomass to operators of such systems rather than to producers of "pure" biofuels, such as cellulosic ethanol. The coproduction concepts described in this chapter offer the potential for meeting these obligations in a more profitable manner than other routes. And some multinational oil companies might come to embrace synfuels production both because they are running out of E & P investment opportunities as a result of being denied adequate access to the oil resources controlled by national oil companies, and because at high oil and GHG emissions prices, profit margins with coproduction systems can approach historical norms for E & P investments.

Strategic corporate alliances involving partnerships across industries could facilitate deployment of coproduction systems. Encouraging power industry/oil industry partnerships is likely to be especially important. Power companies could, of course, manage the power side of the business but may be uncomfortable with gasification technology – which doesn't resemble the "boiler technology" that they are accustomed to managing. Moreover, the power industry has no experience with either CO_2 capture or management of the geological structures needed for CO_2 storage. The oil industry could, of course, manage liquid fuels production and marketing quite well. It has extensive experience with CO_2 capture (at refineries and chemical process plants for purposes unrelated to CO_2 storage), with CO_2 transport (for enhanced oil recovery), with management of geological structures similar to those that will be used for CO_2 storage, and with gasification of petroleum residuals at both refineries and chemical process plants that could facilitate their becoming comfortable with coal and biomass gasification. Moreover, oil companies that also sell natural gas might become interested in natural gas/biomass coprocessing systems that make fuels and electricity with CCS.

12.8.2   New Policies for a Radical Transformation

A radical transformation of the fossil energy landscape is unlikely to occur without new facilitating public policies. Needed critical broad cross-cutting policies include:

- Carbon mitigation policies in the near term for which implicit GHG emission prices are high enough to make CCS competitive as a routine commercial activity, and which are implemented soon in the context of the expressed G8 support for a global effort to limit the increase in average global temperature due to climate change to 2°C above the pre-industrial level. This goal is written into the Copenhagen Accord and reconfirmed during COP16 in Cancun. The G8 statement in July 2009 noted that this formidable goal will require a global reduction in emissions by at least 50% by 2050 (relative to the 1990 level), and they supported the goal for industrialized countries to achieve 80% reductions by 2050 to help meet the global 50% reduction.

- New air pollution control policies are needed, with highest priority given to pollution from existing coal power plants in industrialized countries and indoor air pollution in developing countries. The analysis in Chapter 12 shows the value of crafting new air pollution control policies and GHG mitigation policies together because of the helpful synergisms that would arise with "co-control" approaches.

- Technological innovation policies in support of the plausible radical transformation described in this chapter should include both support for R&D on promising technological options and incentives for early deployment, including incentives that would encourage needed new inter-industry partnerships. First-of-a-kind projects are always risky, both technologically and financially. Despite the attractive economic features of coprocessing, coproduction, and other technologies discussed in this chapter, without incentives for first-of-a-kind and early deployment projects that offer major public benefits, such technologies will enter the market only slowly or not at all.

High priority should be given to encouraging early CCS action, especially for coal, because if coal is to be widely used in a future carbon-constrained world (via coproduction and coprocessing with biomass or via any other means), so doing will be viable only if the option is available to safely store CO_2 in geological media. For geologic storage, there appear to be neither technical issues that cannot be managed nor economic show-stoppers (see Chapter 13). Nevertheless, there is a pressing need to carry out a significant number of integrated commercial-scale CCS projects worldwide (each involving the annual storage of at least one $MtCO_2$/yr) with an emphasis on storage in deep saline formations, to: (i) address scientific and engineering issues at scale for storage; (ii) provide a sound empirical basis for developing the regulatory and political framework for site selection, permitting, operation, monitoring, and closure of storage sites for routine commercial CCS projects; and (iii) satisfy a wide range of stakeholders as to the viability of CCS as a "gigascale" carbon mitigation option.

The international political framework for early CCS action has already been established. As noted earlier, the G8 Summit in 2008 produced an agreement to sponsor at least twenty commercial-scale, fully integrated CCS demonstration projects worldwide that would be committed by 2010 with the aim of establishing the basis for broad commercial deployment of CCS technologies after 2020. The G8 reiterated its call for these CCS projects in 2009. In August 2010, a high-level task force in the U.S. created by President Obama issued its report on a comprehensive CCS strategy for the United States that includes bringing five to ten commercial demonstration projects online by 2016.

Much if not all of the incremental cost of CCS for the 20 projects called for by the G8 will probably have to be paid for by governments (individually or collectively). It is likely that carbon prices will be lower initially than what will be needed to make pursuit of CCS a profitable activity for private companies in many, if not most, parts of the world. To minimize spending, governments should aim to pursue projects in which they can maximize the learning about the gigascale prospects of CCS per dollar spent. In this context, the coproduction systems described in this chapter are good candidates for CCS early action projects because they have much lower capture costs than power only systems. If these projects are allowed to compete for government awards for CCS demonstrations, the cost to government might be significantly less than for CCS demonstration projects based on power-only systems. Accordingly, coproduction facilities coprocessing fossil fuel and biomass with CCS, especially those built in coal-rich countries, should be considered seriously as candidates for some of the needed CCS early action projects and supported financially jointly by the governments of several coal-intensive energy economies.

Since it will require at least a decade of demonstration efforts before CCS systems can begin to be routinely deployed, the deployment of CCS systems in conjunction with fossil energy conversion would need to proceed very rapidly thereafter to achieve deep reductions in global GHG emissions from fossil fuel burning by 2050.

Analysis in this chapter indicates that biomass conversion with CCS could play an essential role in enabling continued fossil fuel use. This would be accomplished by off-setting GHG emissions from fossil fuels via the strongly negative emissions of biomass/CCS systems. Two new initiatives needed in this area are:

- New policies should encourage demonstration and commercialization of large biomass gasifiers (300–600 MW biomass input) because these represent a key technology for economically attractive biomass/CCS systems. To date, biomass gasifiers have only been successfully demonstrated at the scale of tens of MW biomass input.

- Detailed assessments should be carried out of the prospects for biomass production for energy on a sustainable basis in ways that minimize conflicts with food production, adverse indirect land-use impacts, and biodiversity. Emphasis should be on agricultural

residues, forest residues (including mill residues, logging residues, diseased tree removals, fuel treatment thinnings, and productivity enhancement thinnings), and the growing of dedicated energy crops on abandoned cropland and other degraded lands. These assessments should be carried out in each of the major world regions.

New policies are also needed to promote universal access to clean cooking fuels (as discussed in detail in Chapter 19, Section 19.2.2) derived from coal and/or biomass systems with CCS (Section 12.5):

- As a basis for new policy, CO_2 storage assessments should be carried out in each of the major regions requiring clean cooking fuels – with financial support from the international community. CO_2 storage prospects are not well known in countries where clean cooking fuels are sorely needed. This is especially true in biomass-rich but coal-poor regions – where consideration of CCS as a carbon mitigation option has probably not been given much thought. Accordingly, "bottom-up" assessments of storage prospects, including the construction of supply curves (storage capacity in tonnes vs cost in US$/t), are needed on a reservoir-by-reservoir basis.

- For regions where there are good prospects for biomass energy with CCS, official development assistance should be expanded to help develop the physical infrastructures and human capacities needed to build and manage large industries based on modern biomass conversion technologies with CCS. This additional ODA should aim to establish this industry so that it can take off as GHG emission prices approach levels high enough that market forces are sufficient to support commercial biomass/CCS activities.

Finally, new public policies are needed to facilitate in the near term industrial collaborations between companies that would produce simultaneously fuels (clean cooking fuels as well as transportation fuels) and electricity from fossil fuels and biomass in regions having significant supplies of both. It would be desirable to identify policy instruments that specify performance rather than technology and maximize use of market forces in meeting performance goals. Promising approaches along these lines include: (i) mandating a Low Carbon Standard for Fuels (Andress et al., 2010; Sperling and Yeh, 2010), (ii) a low carbon standard for electricity generation, perhaps modeled after Renewable Portfolio Standards or green certificate markets; (iii) feed-in tariffs for environmentally-qualified electricity, and (iv) a Universal Clean Cooking Fuel Standard in regions requiring major infusions of clean cooking fuels, perhaps modeled after the "obligation to serve" mandates of the rural electrification programs introduced in the United States in the 1930s. The latter could plausibly facilitate the formation of strategic industrial alliances that would be capable of guaranteeing universal access to clean cooking fuels without major subsidy.

References

Abt Associates Inc., 2000: *The Particulate-Related Health Benefits of Reducing Power Plant Emissions.* Prepared for Clean Air Task Force, Abt Associates Inc. with ICF Consulting and E.H. Pechan Associates, Inc., Bethesda, MD.

AEFP, 2009: *Liquid Transportation Fuels from Coal and Biomass: Technological Status, Costs, and Environmental Impacts.* America's Energy Future Panel (AEFP) on Alternative Liquid Transportation Fuels, National Academy of Sciences, Washington, DC.

Agnolucci, P., 2007: Hydrogen Infrastructure for the Transport Sector. *International Journal of Hydrogen Energy*, **32**(15):3526–3544.

Air Products, 2008: *Air Separation Technology – Ion Transport Membrane (ITM).* Brochure, Air Products and Chemicals, Inc., Allentown, PA.

Alpert, S. B., S. S. Penner and D. F. Wiesenhahn, 1987: Gasification for Electricity Generation. *Energy, The International Journal*, **12**(8–9):639–646.

Andress, D., T. D. Nguyen and S. Das, 2010: Low-carbon fuel standard—Status and analytic issues. *Energy Policy*, **38**(1):580–591.

Anonymous, 2008: *Unconventional Oil: A Supplement to E&P.* Houston, T.

APS, 2011: *Direct Air Capture of CO2 wih Chemicals: A Technology Assessment of the APS Panel on Public Affairs*, American Physical Society, Washington, DC, April.

Argonne National Laboratory, 2008: *GREET Model*. Available at greet.es.anl.gov/ (accessed 30 December, 2010).

Bartis, J. T., F. Camm and D. S. Ortiz, 2008: *Producing Liquid Fuels from Coal: Prospects and Policy Issues*. Project Air Force and Infrastructure, Safety, and Environment, Prepared for the United States Air Force and the National Energy Technology Laboratory (NETL), United States Department of Energy, Santa Monica, CA, USA.

Bechtel Corporation, Global Energy Inc. and Nexant Inc., 2003: *Gasification Plant Cost and Performance Optimization*. Contract No. DE-AC26–99FT40342, Prepared for the National Energy Technology Laboratory (NETL), United States Department of Energy, Houston, TX and San Francisco, CA.

Biello, D., 2009: *Flame Off! Turning Natural Gas Pollution into Gasoline*. Available at www.scientificamerican.com/article.cfm?id=turning-natural-gas-pollution-into-gasoline (accessed 30 April, 2010).

Blades, T., D. Henson, C. Peters, I. Martinalbo and Z. Yu, 2008: Reducing the Carbon footprint with combined Carbo-V® & CCG® technologies. *Gasification Technologies Conference*, October 2008, Washington, DC.

BP, 2009: *BP Statistical Review of World Energy: June 2009.* BP plc., London, UK.

BP, 2010: *BP Statistical Review of World Energy: June 2010.* BP plc., London, UK.

Cai, X., D. Wang and X. Zhang, 2009: *The Impact of Global Trade in Biofuels on Water Scarcity and Food Security in the World*. Available at www.energybioscinesinstitute.org/index.php?option=com_content&task=view&id=145&Itemid=20 (accessed 30 December, 2010).

Campbell, J. E., D. B. Lobell, R. C. Genova and C. B. Field, 2008: The Global Potential of Bioenergy on Abandoned Agriculture Lands. *Environmental science and technology.*, **42**(15):5791–5794.

CARNFCHT, 2008: *Transitions to Alternative Transportation Technologies – A Focus on Hydrogen*. Committee on Assessment of Resource Needs for Fuel Cell and Hydrogen Technologies (CARNFCHT), The National Academies, Washington, DC.

CASFHPU, 2004: *The Hydrogen Economy: Opportunities, Costs, Barriers, and R&D Needs.* Committee on Alternatives and Strategies for Future Hydrogen Production and Use (CASFHPU), The National Academies, Washington, DC.

Celik, F., E. D. Larson and R. H. Williams, 2004: Transportation Fuel from Coal with Low CO_2 Emissions. *Proceedings of the 7th International Conference on Greenhouse Gas Control Technologies*, M. Wilson, E. S. Rubin, D. W. Keith, C. F. Gilboy, T. Morris, K. Thambimuthu and J. Gale, (eds.), Vancouver, BC.

Cheng, W. H. and H. H. Kung, 1994: *Methanol Production and Use.* Marcel-Dekker, New York, USA.

Chiesa, P., S. Consonni, T. Kreutz and W. Robert, 2005: Co-production of Hydrogen, Electricity and CO2 from Coal with Commercially Ready Technology. Part A: Performance and Emissions. *International Journal of Hydrogen Energy*, **30**(7):747–767.

CIEMAT, 1999: *ExternE: Externalities of Energy*. Directorate-General XII, Vol XX: National Implementation, Prepared for The European Commission by Centro de Investigaciones Energéticas, Medioambientales y Tecnológicas (CIEMAT), Brussels.

CO_2 Capture Project, 2009: Carbon Dioxide Capture for Storage in Deep Geologic Formations – Results from the CO_2 Capture Project. In *Advances in CO_2 Capture and Storage Technology Results (2004–2009)*. L. I. Eide, (ed.), CPL Press, Speen, Newbury, UK, Vol. Volume 3.

CompactGTL, 2010: *The Modular Gas Solution*. Available at www.compactgtl.com/ (accessed

Consonni, S. and F. Vigano, 2005: Decarbonized Hydrogen and Electricity from Natural Gas. *International Journal of Hydrogen Energy*, **30**(7):701–718.

Cushion, E., A. Whiteman and G. Dieterle, 2010: *Bioenergy Development: Issues and Impacts for Poverty and Natural Resource Management.* Agriculture and Rural Development, The World Bank, Washington, DC.

de Mello, L. F., R. D. M. Pimenta, G. T. Moure, O. R. C. Pravia, L. Gearhart, P. B. Milios and T. Melien, 2009: A Technical and Economical Evaluation of CO2 Capture from FCC Units. *Energy Procedia*, **1**(1):117–124.

Desmond, M. J. and V. C. Gibson, 2009: Is There a Lower Work Pathway for Fuel Formation? *Science Online*, **E-Letter** (30 Sep 2009).

Dockery, D. W. and C. A. Pope, 1994: Acute Respiratory Effects of Particulate Air Pollution. *Annual Review of Public Health*, **15**(1):107–132.

DOE, 2010: *Innovative Concepts for Beneficial Reuse of Carbon Dioxide*. Available at fossil.energy.gov/recovery/projects/beneficial_reuse.html (accessed 10 July, 2010).

Dolan, G., 2008: Methanol Fuels: The Time has Come. *The 17th International Symposium on Alcohol Fuels*, October 2008, Taiyuan, China.

Doyle, J., 2008: Coal to Liquids: Transport Fuel For a Supply Constrained World Oil Market. *Gasification Technologies Conference*, October 2008, Washington, DC.

DTI, 1999: *Technology Status Report 010: Coal Liquefaction*. DTI/pub URN 99/1120, UK Department of Trade and Industry, London, UK.

Dutt, G. S. and N. H. Ravindranath, 1993: Bienergy: Direct Applications in Cooking. In *Renewable Energy: Sources of Fuels and Electricity*. T. Johansson, B, H. Kelly, A. Reddy, K. N. and R. H. Williams, (eds.), Island Press, Washington, DC pp.653–699.

EPPDI and CPECG, 2008: *2007 Thermal Power Project Construction Cost Handbook*. Electric Power Planning and Design Institute (EEPDI) and China Power Engineering Consulting Group (CPECG), Beijing, China.

ExxonMobil, 2009: *Outlook for Energy: A View to 2030*. ExxonMobil, Irving, TX.

Fargione, J., J. Hill, D. Tilman, S. Polasky and P. Hawthorne, 2008: Land Clearing and the Biofuel Carbon Debt. *Science*, **319**(5867):1235–1238.

Fleisch, T., R. Sills, M. Briscoe and J. F. Freide, 2003: GTL-FT in the Emerging Gas Economy. *Petroleum Economist*,(Special issue):39–41.

Fleisch, T. H., R. A. Sills and M. D. Briscoe, 2002: 2002 – Emergence of the Gas-to-Liquids Industry: a Review of Global GTL Developments. *Journal of Natural Gas Chemistry*, **11**(2002):1–14.

Frantz Jr, J. H. and V. Jochen, 2005: *Shale gas: When Your Gas Reservoir is Unconventional Do is our Solution*. White Paper, Schlumberger, Houston, TX.

Friedrich, R., 2005: ExternE: Methodology and Results. *Meeting External Costs of Energy and their Internalisation in Europe, a Dialogue with Industry, NGO, and Policy-makers*, The European Commission (EC), Brussels.

FWI, 2004: *Gas to Liquids Life Cycle Assessment*. Synthesis Report, Prepared by Five Winds International (FWI) for ConocoPhillips Company, Sasol Chevron Consulting Ltd. and Shell International Gas Ltd.

GACS, 2011: *The Global Alliance for Clean Cookstoves*. Available at cleancookstoves. org/ (accessed 11 October, 2010).

Gao, C., H. Yin, N. Ai and Z. Huang, 2009: Historical Analysis of SO2 Pollution Control Policies in China. *Environmental Management*, **43**(3):447–457.

Goldemberg, J., T. B. Johansson, A. K. N. Reddy and H. Williams Robert, 1985: Basic needs and much more with one Kilowatt per capita. *Ambio*, **14**(4–5):190–200.

Goldemberg, J., T. B. Johansson, A. K. N. Reddy and R. H. Williams, 2004: A Global Clean Cooking Fuel Initiative. *Energy for sustainable development: the journal of the International Energy Initiative*, **8**(3):5–12.

Gray, D., J. Plunkett, S. Salerno, C. White and G. Tomlinson, 2009: *Current and Future Technologies for Gasification-Based Power Generation: Volume 2: A Pathway Study Focused on Carbon Capture Advanced Power Systems R&D Using Bituminous Coal*. Report DOE/NETL-2009/1389, National Energy Technology Laboratory, US Department of Energy.

Grol, E., J. DiPietro, J. H. J. S. Thijssen, W. Surdoval and H. Quedenfeld, 2007: *The Benefits of SOFC for Coal-Based Power Generation*. National Energy Technology Laboratory (NETL), Department of Energy, Washington, DC.

Guo, X., Liu, G., and Larson, E.D., 2011: "High octane gasoline production by upgrading low-temperature Fischer-Tropsch syncrude," *Industrial & Engineering Chemistry Research*, **50**(16): 9743–9747, 11 July 2011.

Hamelinck, C. N. and A. P. C. Faaij, 2002: Future Prospects for Production of Methanol and Hydrogen from Biomass. *Journal of Power Sources*, **111**(1):1–22.

Hammitt, J. K. and Y. Zhou, 2006: The Economic Value of Air-Pollution-Related Health Risks in China: A Contingent Valuation Study. *Environmental and Resource Economics*, **33**(3):399–423.

Hansen, J. B., 1997: *Methanol Synthesis*. Handbook of Heterogeneous Catalysis. G. Ertl, H. Knözinger and J. Weitkamp. Weinham, Germany, Wiley-VCH. Volume 4, 1856

Haslbeck, J. L., N. J. Kuehn, E. G. Lewis, L. L. Pinkerton, J. Simpson, M. J. Turner, E. Varghese and M. C. Woods, 2010: *Cost and Performance Baseline for Fossil Energy Plants: Volume 1: Bituminous Coal and Natural Gas to Electricity*. Report

DOE/NETL-2010/1397, Original Issue Date, May 2007, Revision 2, November 2010, National Energy Technology Laboratory (NETL), Department of Energy (DOE), Washington, DC.

Hildebrandt, D., D. Glasser, B. Hausberger, B. Patel and B. J. Glasser, 2009a: Producing Transportation Fuels with Less Work. *Science*, **323**(5922):1680–1682.
    2009b: Response to M. J. Desmond and V. C. Gibson's E-Letter. *Science Online*, **E-Letter** (30 Sep 2009:).

Ho, M. S. and C. P. Nielsen, (eds.), 2007: *Clearing the Air: the Health and Economic Damages of Air Pollution in China*. MIT press, Cambridge, MA.

Ho, M. S. and D. W. Jorgenson, 2008: Greening China: Market-based Policies for Air-pollution Control. *Harvard Magazine*, September-October 2008, pp. 32–37.

Howarth, R., R. Santoro and A. Ingraffea, 2011: Methane and the green-house-gas footprint of natural gas from shale formations. *Climatic Change*, **106**(4):679–690.

Hutton, G., E. Rehfuess, F. Tediosi and S. Weiss, 2006: *Evaluation of the Costs and Benefits of Household Energy and Health Interventions at Global and Regional Levels*. World Health Organization (WHO), Geneva, Switzerland.

IEA, 2006: *World Energy Outlook 2006*. International Energy Agency (IEA) of the Organization for Economic Co-operation and Development (OECD), Paris.

IEA, 2007: *World Energy Outlook 2007: China and India Insights*. International Energy Agency (IEA) of the Organization for Economic Co-operation and Development, Paris.

IEA, 2008a: *World Energy Outlook 2008*. International Energy Agency (IEA) of the Organization for Economic Co-operation and Development (OECD), Paris.

IEA, 2008b: *Energy Technology Perspectives 2008: Energy Technology Perspectives to 2050*. International Energy Agency (IEA) of the Organization for Economic Co-operation and Development (OECD), Paris.

IEA, 2009a: *Transport, Energy and CO2: Moving Toward Sustainability*. International Energy Agency (IEA) of the Organization for Economic Co-operation and Development (OECD), Paris.

IEA, 2009b: *World Energy Outlook 2009*. International Energy Agency (IEA) of the Organization for Economic Co-operation and Development (OECD), Paris.

IEA, 2010: *World Energy Outlook 2010*. International Energy Agency (IEA) of the Organization for Economic Co-operation and Development (OECD), Paris.

IGP, 2010: *Higher Alcohols*. Available at www.igp.co/IGP_higheralcohols.htm (accessed 30 December, 2010).

IPCC, 2005: *Carbon Dioxide Capture and Storage*. Special Report to Working Group III of the IPCC, B. Metz, O. Davidson, H. de Coninck, M. Loos and L. Meyer, (eds.), Intergovernmental Panel on Climate Change (IPCC), New York.

Jones, N., 2011: *United States investigates fracking safety*. Available at www.nature. com/news/2011/110512/full/news.2011.282.html (accessed 15 May, 2011).

Karg, J., 2009: IGCC experience and further developments to meet CCS market needs. *Coal-Gen Europe Conference & Exhibition*, Katowice, Poland.

Katofsky, R. E., 1993: *The Production of Fluid Fuels from Biomass*. Master's Thesis, Center for Energy and Environmental Studies, Princeton University, Princeton, NJ.

Keil, F. J., 1999: Methanol-to-hydrocarbons: Process Technology. *Microporous and Mesoporous Materials*, **29**(1–2):49–66.

Krause, F., J. Koomey, H. Becht, D. Olivier, G. Onufrio and P. Radanne, 1994: *Fossil Generation:The Cost and Potential of Conventional and Low-Carbon – Electricity Options in Western Europe*. International Project for Sustainable Energy Paths, El Cerrito, CA.

Kreutz, T., R. Williams, S. Consonni and P. Chiesa, 2005: Co-production of Hydrogen, Electricity and CO2 from Coal with Commercially Ready Technology. Part B: Economic Analysis. *International Journal of Hydrogen Energy*, **30**(7):769–784.

Kreutz, T. G., E. D. Larson, G. Liu and R. H. Williams, 2008: Fischer-Tropsch Fuels from Coal and Biomass. *Proceedings of the 25th Annual International Pittsburgh Coal Conference*, Pittsburgh, Pennsylvania, USA.

Kreutz, T. G., 2011: Prospects for Producing Low Carbon Transportation Fuels from Captured CO_2 in a Climate Constrained World. *Proceedings of the 10th International Greenhouse Gas Control Technologies Conference*, Amsterdam, The Netherlands.

Kromer, M. A. and J. B. Heywood, 2007: *Electric Powertrains: Opportunities and Challenges in the U.S. Light-Duty Vehicle Fleet*. Publication No. LFEE 2007–03 RP, Laboratory for Energy and the Environment, Massachusetts Institute of Technology (MIT), Cambridge, MA.

Larson, E. D. and T. Ren, 2003: Synthetic Fuel Production by Indirect Coal Liquefaction. *Energy for Sustainable Development*, **7**(4):79–102.

Larson, E. D. and H. Yang, 2004: Dimethyl Ether from Coal as a Household Cooking Fuel in China. *Energy for sustainable development: the journal of the International Energy Initiative*, **8**(3):115–126.

Larson, E. D., R. H. Williams and H. Jin, 2006: *Fuels and Electricity from Biomass with CO_2 Capture and Storage*. Proceedings of the International Conference on Greenhouse Gas Control Technologies. Trondheim, Norway.

Larson, E. D., Jin, H., and Celik, F. E., 2009: "Large-scale gasification-based coproduction of fuels and electricity from switchgrass," *Biofuels, Biorprod. Bioref*. **3**:174–194.

Larson, E. D., G. Fiorese, G. Liu, R. H. Williams, T. G. Kreutz and S. Consonni, 2010: Co-production of Decarbonized Synfuels and Electricity from Coal + Biomass with CO2 Capture and Storage: an Illinois Case Study. *Energy & Environmental Science*, **3**(1):28–42.

Lau, F. S., D. A. Bowen, R. Dihu, S. Doong, E. E. Hughes, R. Remick, R. Slimane, S. Q. Turn and R. Zabransky, 2002: *Techno-Economic Analysis of Hydrogen Production by Gasification of Biomass*. Final Technical Report for the Period September 15, 2001 to September 14, 2002, Report prepeared by Gas Technology Institute for United States Department of Energy, Des Plaines, IL.

Lepinski, J., T. Lee and S. Tam, 2009: Recent Advances in Direct Coal Liquefaction Technology. *The American Chemical Society National Meeting*, March 26, Salt Lake City, Utah.

Levy, J. I., L. K. Baxter and J. Schwartz, 2009: Uncertainty and Variability in Health-Related Damages from Coal-Fired Power Plants in the United States. *Risk Analysis*, **29**(7):1000–1014.

Lewis, N. S. and D. G. Nocera, 2006: Powering the Planet: Chemical Challenges in Solar Energy Utilization. *Proceedings of the National Academy of Sciences*, **103**(43):15729–15735.

Li, J., J. Bai and R. Overend, (eds.), 1998: *Assessment of Biomass Resource Availability in China*. MOA/DOE Project Expert Team. China Environmental Science Press, Beijing, China.

Liu, G., E. D. Larson, R. H. Williams, T. G. Kreutz and X. Guo, 2011a: Making Fischer–Tropsch Fuels and Electricity from Coal and Biomass: Performance and Cost Analysis. *Energy & Fuels*, **25**(1):415–437.

Liu, G., R. H. Williams, E. D. Larson and T. G. Kreutz, 2011b: Design/economics of Low-carbon Power Generation from Natural Gas and Biomass with Synthetic Fuels Co-production. *10th International Greenhouse Gas Control Technologies Conference*, Amsterdam, Netherlands.

Liu, G., E. D. Larson, R. H. Williams, J. Katzer and T. G. Kreutz, forthcoming: Gasoline from Coal and Biomass with CO_2 Capture and Storage. Manuscript in Preparation. Princeton Environmental Institute, Princeton University.

Melichar, J., M. Havranek, V. Maca, M. Scasny and M. Kudelko, 2004a: *Implementation of ExternE Methodology in Eastern Europe*. Final Report on Work Package 7, Contract No. ENG1-CT-2002–00609, Extension of Accounting Framework and Policy Applications, The European Commission (EC), Brussels.

Melichar, 2004b: *Implementation of ExternE Methodology in Eastern Europe*. Final Report on Work Package 7, under ExternE Extension of Accounting Framework and Policy Applications.

Mintz, M., S. Folga, J. Molburg and J. Gillette, 2002: *Cost of Some Hydrogen Fuel Infrastructure Options*. Presentation to Transportation Research Board, Transportation Technology Research and Development Center, Argonne National Laboratory,, Argonne, IL.

Termuehlen, H. and W. Empsperger, 2003: *Clean and Efficient Coal Fired Power Plants*. ASME, New York, NY.

MIT, 2010: *The Future of Natural Gas*. An Interdisciplinary MIT Study. Interim Report, MIT Energy Initiative, Massachusetts Institute of Technology, Cambridge, MA, USA.

Morehead, H. and F. Hannemann, 2005: Siemens Technology Improvements Enhance IGCC Plant Economics. *Gasification Technologies Conference 2005*, October 12, San Francisco, CA.

Naqvi, S., 2002: *Dimethyl Ether as Alternate Fuel*. Report 245, Process Economics Program, SRI Consulting, Menlo Park, CA.

National Bureau of Statistics China, 2007: *China Energy Statistical Yearbook 2007*. China Statistics Press, Beijing, China.

NETL, 2008: *Development of Baseline Data and Analysis of Life Cycle Greenhouse Gas Emissions of Petroleum-Based Fuels*. Report DOE/NETL-2009/1346, National Energy Technology Laboratory (NETL), U.S. Department of Energy.

NETL, 2009: *Consideration of Crude Oil Source in Evaluating Transportation Fuel GHG Emissions*. Report DOE/NETL-2009/1360, National Energy Technology Laboratory (NETL), Department of Energy, Washington, DC.

Nielsen, P. E. H., 2009: From Coal Gas to Liquid Product: the Topsoe TIGAS Technology. *3rd International Freiberg Conference on IGCC & XtL Technologies*, May 18–21, Dresden, Germany.

NPC, 2003: *Balancing Natural Gas Policy – Fueling the Demands of a Growing Economy*. Volume I. Summary of Findings and Recommendations, National Petroleum Council (NPC), Washington, DC.

NRC, 2004: *The Hydrogen Economy: Opportunities, Costs, Barriers and R&D Needs*, The National Research Council (NRC) of the National Academies, The National Academies Press, Washington, DC.

NRC, 2010: *Hidden Costs of Energy: Unpriced Consequences of Energy Production and Use*. The National Research Council (NRC) of the National Academies, The National Academies Press, Washington, DC.

Ogden, J. M., 2003: *Modeling infrastructure for a fossil hydrogen energy system with CO2 sequestration*. Proceedings of the 6th International Conference on Greenhouse Gas Control Technologies. J. Gale and Y. Kaya. Oxford 1069–1074

Ogden, 2004: *Conceptual Design of Optimized Fossil Energy Systems with Capture and Sequestration of Carbon Dioxide*. Research Report UCD-ITS-RR-04–34, Institute of Transportation Studies, University of California, Davis, USA.

Ogden, J. M., H. Williams Robert and D. Larson Eric, 2004: Societal Lifecycle Costs of Cars with Alternative Fuels/engines. *Energy Policy*, **32**(1):7–27.

Olah, G. A., A. Goeppert and G. K. Surya Prakash, 2009: *Beyond Oil and Gas: The Methanol Economy*. Wiley-VCH, Weinheim, Germany.

Osborn, S. G., A. Vengosh, N. R. Warner and R. B. Jackson, 2011: Methane contamination of drinking water accompanying gas-well drilling and hydraulic fracturing. *Proceedings of the National Academy of Sciences*, **108**(20):8172–8176.

OSL, 2003: *Basic Research Needs for the Hydrogen Economy*. Report on the Basic Energy Sciences Workshop on Hydrogen Production, Storage and Use, Office of Science Laboratory (OSL), United States Department of Energy, Washington, DC.

Pope III, C. A., M. J. Thun, M. M. Namboodiri, D. W. Dockery, J. S. Evans, F. E. Speizer and C. W. Heath Jr, 1995: Particulate Air Pollution as a Predictor of Mortality in a Prospective Study of U.S. Adults. *American Journal of Respiratory and Critical Care Medicine*, **151**(3):669–674.

Pope III, C. A., R. T. Burnett, M. J. Thun, E. E. Calle, D. Krewski, K. Ito and G. D. Thruston, 2002: Lung Cancer, Cardiopulmonary Mortality and Long-Term Exposure to Fine Particulate Air Pollution. *JAMA: Journal of the American Medical Association*, **287**(9):1132–1141.

Quinkertz, R., 2010: Efficiency improvements in coal-fired power plants. *IEA Workshop on efficiency and clean coal technologis*, International Energy Agency (IEA) of the Organisation for Economic Development and Cooperation, Moscow, Russia.

Rabl, A. and J. V. Spadaro, 2000: Public Health Impact of Air Pollution and Implications for the Energy System. *Annual Review of Energy & the Environment*, **25**(1):601–628.

Rahm, D., 2011: Regulating hydraulic fracturing in shale gas plays: The case of Texas. *Energy Policy*, **39**(5):2974–2981.

Ramprasad, R., T. Vakil, J. Falsetti and M. Islam, 1999: Competitiveness of Gasification at the Bulwer Island, Australia Refinery. *1999 Gasification Technologies Conference*, October 17–20, San Francisco, CA.

Richards, A. K., 2005: *Anhydrous Processing of Methane into Methane-Sulfonic Acid, Methanol, and Other Compounds*. PCT/US2004/019977, WO/2005/069751, Patent.

Rosegrant, M. W., 2008: *Biofuels and Grain Prices: Impacts and Policy Responses*. Testimony for the U.S. Senate Committee on Homeland Security and Governmental Affairs, International Food Policy Research Institute, Washington DC.

Rostrup-Nielsen, J. R. and T. Rostrup-Nielsen, 2002: Large-Scale Hydrogen Production. *CATTECH*, **6**(4):150–159.

Rostrup-Nielsen, T., 2005: Manufacture of Hydrogen. *Catalysis Today*, **106**(1–4):293–296.

Schwartz, J., B. Coull, F. Laden and L. Ryan, 2008: The Effect of Dose and Timing of Dose on the Association between Airborne Particles and Survival. *Environmental Health Perspectives*, **116**(1):64–69.

Searchinger, T., R. Heimlich, R. A. Houghton, F. Dong, A. Elobeid, J. Fabiosa, S. Tokgoz, D. Hayes and T.-H. Yu, 2008: Use of U.S. Croplands for Biofuels Increases Greenhouse Gases through Emissions from Land-Use Change. *Science*, **319**(5867):1238–1240.

Semelsberger, T. A., R. L. Borup and H. L. Greene, 2006: Dimethyl Ether (DME) as an Alternative Fuel. *Journal of Power Sources*, **156**(2):497–511.

Shell, 2009: *Pernis refinery*. Available at www.shell.com/home/content/globalsolutions/about_global_solutions/key_projects/pernis/ (accessed 31 December, 2010).

Simbeck, D., 2003: *Biggest Challenge for the Hydrogen Economy Hydrogen Production & Infrastructure Costs*. Long-Term Technology Pathways to Stabilization of Greenhouse Gas Concentrations, Presentation to the Aspen Global Change Institute, Aspen, CO.

Simbeck, D. and W. Roekpooritat, 2009: Near-Term Technologies for Retrofit CO_2 Capture and Storage of Existing Coal-fired Power Plants in the United States. *MIT Coal Retrofit Symposium*, Cambridge, MA.

Simbeck, D. R., 2004: Hydrogen Costs with CO_2 Capture. *7th International Conference on Greenhouse Gas Control Technologies (GHGT-7)*, September 6–10, Vancouver, British Columbia, Canada.

Simbeck, D. R. and D. J. Wilhelm, 2007: *Assessment of Co-Production of Transportation Fuels and Electricity*. Report 1014802, Electric Power Research Institute (EPRI), Palo Alto, CA.

Smith, K. R., 2002: In Praise of Petroleum? *Science*, **298**(5600):1847–1848.

Sperling, D. and S. Yeh, 2010: Toward a global low carbon fuel standard. *Transport Policy*, **17**(1):47–49.

State Environmental Protection Administration of China, 2007: *Annual Statistic Report on Environment of China 2006*. China Environmental Science Press, Beijing, China.

Tabak, S., M. Hindman, A. Brandl and M. Hienritz-Adrian, 2009: An Alternative Route for Coal To Liquid Fuel: the ExxonMobil Methanol to Gasoline (MTG) Process. *World CTL Conference*, March 2009, Washington, DC.

Tarka, T. J., J. G. Wimer, P. C. Balash, T. J. Skone, K. C. Kern, M. C. Vargas, B. D. Morreale, C. W. White III and D. Gray, 2009: *Affordable, Low-Carbon Diesel Fuel from Domestic Coal and Biomass*. Report DOE/NETL-2009/1349, National Energy Technology Laboratory (NETL), United States Department of Energy.

The Royal Society, 2009: *Geoengineering the Climate: Science, Governance and Uncertainty*. Available at royalsociety.org/Geoengineering-the-climate/ (accessed 10 October, 2010).

Tilman, D., R. Socolow, A. Foley Jonathan, J. Hill, E. Larson, L. Lynd, S. Pacala, J. Reilly, T. Searchinger, C. Somerville and R. Williams, 2009: Beneficial Biofuels-The Food, Energy, and Environment Trilemma. *Science*, **325**(5938):270–271.

US-China Joint Economic Research Group, 2007: *US-China Joint Economic Study: Economic Analyses of Energy Saving and Pollution Abatement Policies for the Electric Power Sectors of China and the United States*. Summary for Policymakers, US-China Strategic Economic Dialogue, State Environmental Protection Administration (SEPA) of the People's Republic of China and the United States Environmental Protection Agency (EPA), Beijing, China and Washington, DC.

US EIA, 2009a: *Annual Energy Outlook 2009: With Projections to 2030*. Report DOE/EIA-0383(2009), United States Energy Information Administration (US EIA), Washington, DC, USA.

US EIA, 2009b: *Weekly U.S. Propane Wholesale/Resale Price*. Available at www.eia.gov/dnav/pet/hist/LeafHandler.ashx?n=PET&s=W_EPLLPA_PWR_NUS_DPG&f=W (accessed 30 December, 2010).

US EIA, 2010a: *Annual Energy Outlook 2011*. AEO2011 Early Release Overview, Report Number: DOE/EIA-0383ER(2011), US Energy Information Administration (US EIA), Washington, DC, USA.

US EIA, 2010b: *International Energy Outlook 2010*. Report DOE/EIA-0484(2010), United States Energy Information Administration (US EIA), Washington, DC.

US EPA, 1997: *The Benefits and Costs of the Clean Air Act, 1970 to 1990*. Prepared for the United States Congress, United States Environmental Protection Agency (US EPA), Washington, DC.

van Bibber, L., E. Shuster, J. Haslbeck, M. Rutkowski, S. Olson and S. Kramer, 2007: *Technical and Economic Assessment of Small-Scale Fischer-Tropsch Liquids Facilities.* Report DOE/NETL-2007/1253, National Energy Technology Laboratory (NETL), United States Department of Energy.

van Haperen, R. and R. de Kler, 2007: Nuon Magnum. *Gasification Technologies Conference*, October 15, San Francisco, CA.

van Straelen, J., F. Geuzebroek, N. Goodchild, G. Protopapas and L. Mahony, 2009: CO2 capture for refineries, a practical approach. *Energy Procedia*, **1**(1):179–185.

WB and SEPA, 2007: *Cost of Pollution in China: Economic Estimates of Physical Damages.* The World Bank and the State Environmental Protection Administration P. R. China, Washington, D.C., USA.

WEA, (ed.) 2004: *World Energy Assessment (WEA): Overview 2004 Update.* United Nations Development Programme (UNDP), United Nations Department of Economic and Social Affairs (UNDESA), World Energy Council (WEC), New York.

WHO, 2006: *Fuel for Life: Household Energy and Health.* World Health Organization, Geneva.

Williams, R. H., 2000: Advanced Energy Supply Technologies. In *World Energy Assessment: Energy and the Challenge of Sustainability*, United Nations Development Programme, New York, NY pp.273–329.

Williams, 2002: Decarbonized Fossil Energy Carriers and their Energy Technology Competitors. *Proceedings of the Workshop on Carbon Dioxide Capture and Storage*, Regina, Canada.

Williams, R. H. and E. D. Larson, 2003: A Comparison of Direct and Indirect Liquefaction Technologies for Making Fluid Fuels from Coal. *Energy for Sustainable Development*, **7**(4):103–129.

Williams, R. H., G. Liu, T. G. Kreutz and E. D. Larson, 2011: Alternatives for Decarbonizing Existing USA Coal Power Plant Sites. *Proceedings of the 10th International Greenhouse Gas Control Technologies Conference*, Amsterdam, The Netherlands.

Woods, M. C., P. J. Capicotto, J. L. Haslbeck, N. J. Kuehn, M. Matuszewski, L. L. Pinkerton, M. D. Rutkowski, R. L. Schoff and V. Vaysman, 2007: *Cost and Performance Baseline for Fossil Energy Plants: Volume 1: Bituminous Coal and Natural Gas to Electricity.* Report DOE/NETL-2007/1281, National Energy Technology Laboratory (NETL), Department of Energy (DOE), Washington, DC.

World Bank, 1997: *Clear Water, Blue Skies.* China 2020: China's Environment in the New Century, T. Johnson, F. Liu and R. S. Newfarmer, (eds.), The World Bank, Wahington, DC.

World Bank and SEPA, 2007: *Cost of Pollution in China: Economic Estimates of Physical Damages.* The World Bank and the State Environmental Protection Administration (SEPA) of the People Republic of China, Washington, DC and Beijing, China.

Yacoub, Y., R. Bata, M. Gautam and D. Martin, 1997: *The Performance Characteristics of C1-C5 Alcohol – Gasoline Blends with Matched Oxygen Content in a Single Cylinder SI Engine.* Department of Mechanical and Aerospace Engineering, West Virginia University, Morgantown, WV.

Zhang, B., 2005: *Carbon Dioxide Mitigation in Polygeneration Energy Systems.* Thermal Engineering Department, Tsinghua University, Beijing.

Zheng, Z., E. D. Larson, Z. Li, G. Liu and R. H. Williams, 2010: Near-term Mega-scale CO2 Capture and Storage Demonstration Opportunities in China. *Energy & Environmental Science*, **3**(9):1153–1169.

13

Carbon Capture and Storage

Convening Lead Author (CLA)
Sally M. Benson (Stanford University, USA)

Lead Authors (LA)
Kamel Bennaceur (Schlumberger, France)
Peter Cook (Cooperative Research Centre for Greenhouse Gas Technologies, Australia)
John Davison (IEA Greenhouse Gas R&D Programme, UK)
Heleen de Coninck (Energy research Centre of the Netherlands)
Karim Farhat (Stanford University, USA)
Andrea Ramirez (Utrecht University, the Netherlands)
Dale Simbeck (SFA Pacific, USA)
Terry Surles (Desert Research Institute, USA)
Preeti Verma (The Climate Group, India)
Iain Wright (BP, UK)

Review Editor
John Ahearne (Sigma Xi, USA)

Contents

Executive Summary

Emissions of carbon dioxide, the most important long-lived anthropogenic greenhouse gas, can be reduced by Carbon Capture and Storage (CCS). CCS involves the integration of four elements: CO_2 capture, compression of the CO_2 from a gas to a liquid or a denser gas, transportation of pressurized CO_2 from the point of capture to the storage location, and isolation from the atmosphere by storage in deep underground rock formations. Considering full life-cycle emissions, CCS technology can reduce 65–85% of CO_2 emissions from fossil fuel combustion from stationary sources, although greater reductions may be possible if low emission technologies are applied to activities beyond the plant boundary, such as fuel transportation.

CCS is applicable to many stationary CO_2 sources, including the power generation, refining, building materials, and the industrial sector. The recent emphasis on the use of CCS primarily to reduce emissions from coal-fired electricity production is too narrow a vision for CCS.

Interest in CCS is growing rapidly around the world. Over the past decade there has been a remarkable increase in interest and investment in CCS. Whereas a decade ago, there was only one operating CCS project and little industry or government investment in R&D, and no financial incentives to promote CCS. In 2010, numerous projects of various sizes are active, including at least five large-scale full CCS projects. In 2015, it is expected that 15 large-scale, full-chain CCS projects will be running. Governments and industry have committed over USD 26 billion for R&D, scale-up and deployment.

The technology for CCS is available today, but significant improvements are needed to support widespread deployment. Technology advances are needed primarily to reduce the cost of capture and increase confidence in storage security. Demonstration projects are needed to address issues of process integration between CO_2 capture and product generation, for instance in power, cement and steel production, obtain cost and performance data, and for industry where capture is more mature to gain needed operational experience. Large-scale storage projects in saline aquifers are needed to address issues of site characterization and site selection, capacity assessment, risk management and monitoring.

Successful experiences from five ongoing projects demonstrate that, at least on this limited scale, CCS can be safe and effective for reducing emissions. Five commercial-scale CCS projects are operational today with over 35 million tonnes of CO_2 captured and stored since 1996. Observations from commercial storage projects, commercial enhanced oil recovery projects, engineered and natural analogues as well as theoretical considerations, models, and laboratory experiments suggest that appropriately selected and managed geological storage reservoirs are very likely to retain nearly all the injected CO_2 for very long times, more than long enough to provide benefits for the intended purpose of CCS.

Significant scale-up compared to existing CCS activities will be needed to achieve large reductions in CO_2 emissions. A 5- to 10-fold scale-up in the size of individual projects is needed to capture and store emissions from a typical coal-fired power plant (500 to 1000 MW). A thousand fold scale-up in size of today's CCS enterprise would be needed to reduce emissions by billions of tonnes per year (Gt/yr).

The technical potential of CCS on a global level is promising, but on a regional level is differentiated. The primary technical limitation for CCS is storage capacity. Much more work needs to be done to realistically assess storage capacity on a worldwide, regional basis and sub-regional basis.

Worldwide storage capacity estimation is improving but more experience is needed. Estimates for oil and gas reservoirs are about 1000 $GtCO_2$, saline aquifers are estimated to have a capacity ranging from about 4000 to 23,000 $GtCO_2$. However, there is still considerable debate about how much storage capacity actually exists, particularly in saline aquifers. Research, geological assessments and, most importantly, commercial-scale demonstration projects will be needed to improve confidence in capacity estimates.

Costs and energy requirements for capture are high. Estimated costs for CCS vary widely, depending on the application (e.g. gas clean-up vs. electricity generation), the type of fuel, capture technology, and assumptions about the baseline technology. For example, with today's technology, CCS would increase cost of generating electricity by 50–100%. In this case, capital costs and parasitic energy requirements of 15–30% are the major cost drivers. Research is underway to lower costs and energy requirements. Early demonstration projects are likely to cost more.

The combination of high cost and low or absent incentives for large-scale deployment are a major factor limiting the widespread use of CCS. Due to high costs, CCS will not take place without strong incentives to limit CO_2 emissions. Certainty about the policy and regulatory regimes will be crucial for obtaining access to capital to build these multi-billion dollar projects.

Environmental risks of CCS appear manageable, but regulations are needed. Regulation needs to ensure due diligence over the lifecycle of the project, but should, most importantly, also govern site selection, operating guidelines, monitoring and closure of a storage facility.

Experience so far has shown that local resistance to CO_2 storage projects may appear and can lead to cancellation of planned CCS projects. Inhabitants of the areas around geological storage sites often have concerns about the safety and effectiveness of CCS. More CCS projects are needed to establish a convincing safety record. Early engagement of communities in project design and site selection as well as credible communication can help ease resistance. Environmental organisations sometimes see CCS as a distraction from a sustainable energy future.

Social, economic, policy and political factors may limit deployment of CCS if not adequately addressed. Critical issues include ownership of underground pore space (primarily an issue in the US); long-term liability and stewardship; GHG accounting approaches and **ve**rification; and regulatory oversight regimes. Governments and the private sector are making significant progress on all of these issues. Government support to lower barriers for early deployments is needed to encourage private sector adoption. Developing countries will need support for technology access, lowering the cost of CCS, developing workforce capacity and training regulators for permitting, monitoring and oversight.

CCS combined with biomass can lead to negative emissions. Such technologies are likely to be needed to achieve atmospheric stabilization of CO_2 and may provide an additional incentive for CCS adoption.

13.1 Introduction: The Need for Carbon Capture and Storage

13.1.1 Introduction to Carbon Capture and Storage (CCS)

In 2008 fossil fuels provided over 85% of our energy supply and emitted over 30 Gt (billion tonnes) of carbon dioxide (CO_2) into the atmosphere. Stabilizing greenhouse gas (GHG) concentrations in the atmosphere at levels that avoid dangerous interference with the climate system will require reducing emissions by an estimated 50–80% by 2050 (IPCC, 2007). Fossil fuel use continues to grow worldwide, especially in countries with rapidly developing economies. Heavy reliance on fossil fuels for all aspects of our energy system makes the transformation to a sustainable future with lower GHG emissions very challenging. The principal benefit of CCS is that it reduces emissions from fossil fuel use, especially from power generation and industrial processes, thus enabling reducing or slowing growth of emissions while other lower GHG emission energy technologies mature and deploy more widely. In addition, over the longer term, CCS could be used to reduce emissions from sources that are difficult to eliminate in any other way, such as energy-intensive industrial processes, natural gas cleanup, hydrogen production, fossil fuel refining, petrochemical industries, and steel and cement manufacturing. The availability of scalable CCS technology by 2020 to 2030 would be most beneficial to lessen the disruption of this transformation by providing low-emission energy services from fossil fuels while alternatives are still developed and scaled-up to meet current and growing energy demands.

For heavily coal-dependent and coal-rich counties such Australia, Canada, China, India, the United States, and Russia, it will be difficult to provide adequate energy supplies while rapidly reducing emissions if CCS is not possible. Figure 13.1 illustrates the current reliance on coal and rate of capacity growth for new coal-fired plants for several of the world's largest economies. Among these heavily coal-reliant economies, those with the most rapid economic growth continue to install new plants at a rapid rate. Large new coal plants will each emit over 100 $MtCO_2$ over their 30- to 40-year lifetime, unless they can be retrofit with capture in the future.

CCS involves the integration of four elements: CO_2 capture, compression of the CO_2 from a gas to a liquid or a denser gas, transportation of pressurized CO_2 from the point of capture to the storage location, and isolation from the atmosphere by storage (see Figure 13.2). Technologies are available to carry out each of these elements, but today, implementation of CCS is challenging because the cost for capture is high, integration of CCS with electricity production or industrial processes is not demonstrated at a large scale, and more confidence is needed that storage can be safe and effective over time periods of 1000 years or longer. While in principle CCS could be deployed on a much wider basis today, the challenge of doing so should not be underestimated. Integrating CCS into existing power generation

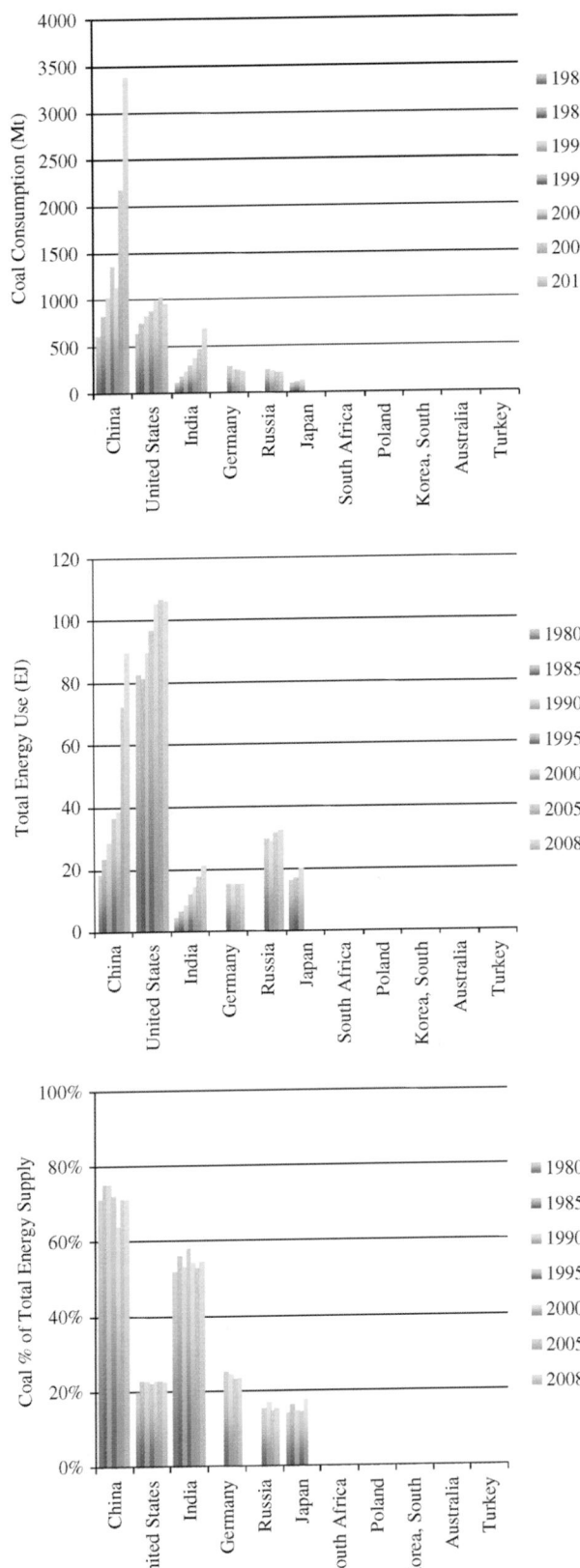

Figure 13.1 | Analysis of trends in the 11 largest coal consuming countries: (a) coal consumption from 1980 to 2010 (Mt), (b) total energy use (Exajoules), (c) percentage of total energy from coal.

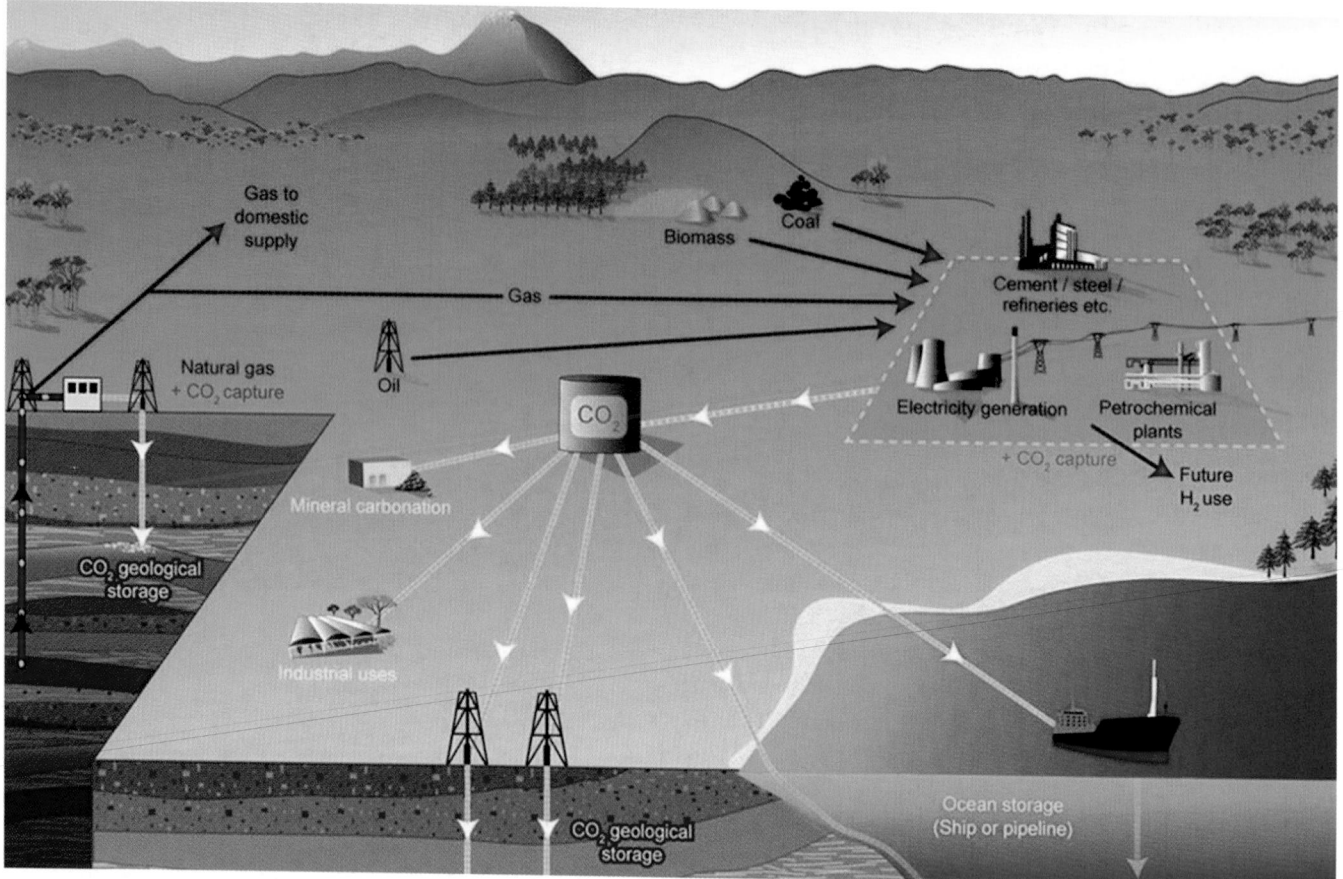

Figure 13.2 | Schematic of a fossil fuel-based energy system using carbon capture and storage to reduce CO_2 emissions. The red lines in the figure illustrate the extraction of oil, gas, and coal from natural resources and transport to facilities used for power generation, refining, and industrial applications. Carbon dioxide capture is integrated with the facilities for burning fossil fuels. After the CO_2 is captured, it is compressed to a dense gas, or more typically a liquid. From the compressor, CO_2 is put into pipelines for transport as shown by the blue lines. The CO_2 is transported to the storage sites, whereas shown, it could be injected into deep underground geological formations, converted to minerals, used to increase biological productivity in greenhouse and algae, or perhaps injected into the deep ocean. With today's technology, storage in deep geological formations is the most advanced of these storage options and could be applied on the scale needed to significantly reduce emissions. Source: IPCC, 2005.

facilities and other industrial operations that demand highly reliable performance is fraught with technological challenges, on top of the large capital investment and significant operating costs required for CO_2 capture. Moreover, with today's capture technology, from 10 to 30% of the output of the power plant may be consumed by the capture unit, depending on the capture technology, the vintage and type of power plant, and the degree of systems integration. Advances not only in capture technology, but ongoing improvements in the efficiency of power generation will be needed to offset the energy penalty for CO_2 capture.

An enormous effort is now devoted to advancing this technology with over 234 active or planned projects (GCCSI, 2011a). While many of these are small-scale pilot projects, 77 are large-scale integrated projects in various stages of the asset life cycle, nine large-scale projects are operating with two more under construction, and an additional 65 potential projects are at various stages of planning (GCCSI, 2011a). The vast majority of these projects have not yet made a final investment decision to go ahead with the construction and operation phases of these projects. Consequently, it is far too early to

tell what will be the outcome of the large amount of activity in this area. As of April 2010, this represented US$26 billion of government investment.

Today, five CCS projects capture 0.5–2 $MtCO_2$/yr from industrial sources and store it in deep geological formations and have been operating for years to a decade or more, demonstrating that, at least on this limited scale, CCS can be safe and effective for reducing emissions. For four of these projects, CO_2 is captured from natural gas cleanup operations while in the fifth, CO_2 is captured from a coal to synthetic natural gas plant. Relevant experience from nearly 40 years of CO_2 enhanced oil recovery (EOR) also shows that CO_2 can safely be injected underground. While most of the CO_2 used for EOR is from natural CO_2 reservoirs, a small fraction is produced from industrial sources such as natural gas cleanup, hydrogen production, and ammonia production. A five- to ten-fold scale-up in the size of individual projects would be needed to capture and store emissions from a single large coal-fired power plant (e.g. current CCS projects store about 1 Mt/yr, while a 500–1000 MW power plant with capture would need to store from 5–10 $MtCO_2$/yr). Globally, a thousand-fold scale-up from current CCS operations would be needed to achieve a contribution to emissions reductions on the order of 20% of current fossil fuel derived emissions over the next century (IEA, 2008a).

13.1.2   The Potential Role of CCS in Climate Change Mitigation

The goal of CCS is to reduce CO_2 emissions from large stationary sources such as power generation, natural gas processing, hydrogen (H_2) production from coal or gas, cement manufacturing, or steel making. Considering full life cycle emissions, using CCS technology on an individual facility can reduce about 65–85% of CO_2 emissions from fossil fuel, although greater reductions may be possible if low emission technologies are applied to activities beyond the plant boundary, such as fuel transportation. In reality, the optimal degree of emission reduction will depend on the tradeoffs between the amount of emission reduction and the cost of capture and age of the facility on which it is deployed. Partial capture may in some cases be more advantageous than striving for the largest emissions reductions possible from a particular facility. For example, a low cost and widely deployed retrofit technology that captures 50% of the emissions from existing coal-fired power plants in China may be one of the most cost effective ways for near-term emission reductions. On the other hand, for a newly built power plant with integrated capture, emissions reductions of 90% may be preferable.

Worldwide assessments suggest that under a range of stabilization scenarios, the contribution of CCS is anticipated to be about 20% of needed emission reductions over the next century, on par with the contributions from renewable energy supplies and end-use efficiency gains (IPCC, 2005; 2007; IEA, 2008c). While the early focus of CCS has been on reducing emissions from coal and natural gas electricity generation, it is estimated that by 2050 about 2–3 $GtCO_2$/yr from industrial sources will need to be captured and stored (IEA, 2009). In the future, CCS may also contribute significantly to emission reductions from the transportation sector via hydrogen production and use for light- and heavy-duty vehicles, electrification of vehicles, production of synthetic fuels using captured CO_2, or the manufacturing of other products such as cement or polymers from captured CO_2 (see also Chapter 12). The importance of CCS will vary by region, depending on the mix of primary energy supply and storage options. Key determinants of the extent to which CCS is likely to be deployed include:

- capacity for storage in appropriate sites in suitable geological formations;

- policy frameworks to encourage emissions reductions, likely to include incentives for early deployments for capacity building;

- lack of lower cost opportunities for reducing emissions such as the increased use of renewable energy resources or nuclear power;

- pace of technological progress to lower the cost and increase confidence in the safety and permanence of geological storage;

- interest in widespread deployment of CO_2-EOR;

- public acceptance of CCS in the local communities where projects are deployed; and

- access to the large capital investment needed for CCS projects, on the order of several billion US$ for a 500 MW power generation plant, particularly for first-of-a-kind facilities (Al-Juaied and Whitmore, 2009; US DOE, 2010).

Today it is difficult to predict the extent to which CCS will be deployed, given the rapidly evolving energy technology and policy landscape.

In the long run, CCS combined with biomass gasification/combustion could be used to create net negative emissions. Net negative emissions are possible when CO_2 is captured from the atmosphere by plants and then used as an energy source with CCS. However, the extent to which this is possible will depend on a number of factors, which are described below (see Section 12.2.3.2 for additional discussion).

Co-location of Geological Storage Opportunities, Energy Services Demands, and Woody Biomass Resources: Rock formations suitable for geological storage are not distributed equally around the world. Additionally, pipeline transportation of CO_2 over distances longer than several hundred kilometers can become costly and impractical, especially in urbanized areas. A quantitative assessment of the co-location of biomass resources, sequestration capacity, and electricity/heat demand is needed to provide a realistic assessment of negative emissions potential.

Flue Gas Composition from Biomass Combustion: Technology options, energy requirements, and cost for capture depend on flue gas composition, which could range from about 15% CO_2 for fully combusted biomass to less than 5% for partially oxidized biomass. Alternatively, combustion in pure oxygen could be used to achieve very high concentrations of CO_2 that would need little post-combustion processing, but this would come at the cost of reduced efficiency caused by the energy needed to separate oxygen from air. A systematic study to evaluate technology options, energy requirements, and cost of capture of CO_2 from woody biomass is needed – with an overlay of the practicality of various technology options depending on the scale of the woody biomass resource, technology access in different parts of the world, and desired balance between CO_2 and biochar production.

Scalability of Negative Emissions from Capture and Storage of Woody Biomass: Today's paradigm for the capture and storage of CO_2 emissions is based on deployment at very large central station facilities (greater than 1 $MtCO_2$/yr emissions). CCS is a complex technological endeavor embedded in a policy framework regulated by numerous local, national, and potentially international laws. Depending on the size and continued availability of these biomass resources, this large-scale approach to deployment using stationary facilities may not be appropriate. The technological and socioeconomic requirements for scaling down this technology and operating storage facilities in a far more distributed manner must be assessed.

13.1.3   Consequences of Excluding CCS from the Mitigation Portfolio

The consequences of excluding CCS from the emission reduction options fall into three broad categories.

First, the cost of the overall emission reduction actions will be greater if CCS is excluded. Integrated assessment studies indicate that the overall cost of cutting emissions will be significantly higher if CCS technology is not included in the portfolio of emissions reductions opportunities (e.g., EPRI, 2008; IEA, 2008c).

Second, sufficiently large emissions reductions may not be possible without CCS. Maintaining a reliable supply of electricity that meets needs for both base-load and peak-load power may not be possible in the 2050 timeframe without CCS. Renewable energy sources are intermittent and of limited geographic distribution. Firming these supplies either by having large-scale energy storage or an extremely extensive electrical grid and needed storage may not be practical in the 2050 timeframe. Additionally, while rapid advances in demand-side energy management are promising, the extent to which these can be aggregated and used to balance supply and demand remain unproven. Nuclear power may provide more of the energy services currently provided by fossil fuels, but to displace them entirely may require an unrealistically rapid scale-up of the nuclear power industry. Additionally, nuclear power plants are not well suited to providing the variable, low-load factor generation currently provided by fossil fuel plants.

Third, political support for reducing emissions may not be sufficient if CCS is excluded from the portfolio of emission reduction options. In many parts of the world, including China, North America, and the Middle East, coal and natural gas provide the major source of electricity. Emission mitigation strategies that do not enable continued use of these domestic and plentiful resources are likely to meet with insurmountable political opposition. Furthermore, some geographic areas lack alternative options for providing reasonably priced electricity with lower emissions. Hence, CCS would be favored under these conditions. Excluding CCS as an option would hinder international negotiations seeking global solutions to the climate change problem.

13.1.4   Key Conditions for CCS to Contribute to Climate Change Mitigation

There are a number of conditions that must be met if CCS is to contribute to climate change mitigation at a significant scale. First, if CCS is implemented as a major part of this transformation, large volumes of CO_2 will need to be captured and stored, on the order of several Gt/yr, with cumulative totals over the next century on the order of hundreds to thousands of Gt. The pathways described in Chapter 17 indicate that a maximum of 400 Gt of CO_2 storage will be needed over the range of decarbonization pathways examined. In addition, these pathways indicate this will require large storage capacities broadly distributed over the globe within several hundred kilometers of large emission sources. Second, unless the technology reaches maturity sometime in the next several decades and wide-scale deployment begins shortly thereafter, other options for reducing emissions may emerge and displace CCS in the carbon mitigation portfolio. Technical maturity will require demonstrating that:

- CCS is a cost-competitive way to reduce emissions;

- storage sites can be selected effectively and managed during and after operations are complete;

- populations living near storage sites feel safe and that storage does not unreasonably compromise property values;

- policies are established to reduce emissions;

- regulations are developed and effectively implemented for managing a CCS project over its entire life cycle, including the post-injection period where long-term stewardship of stored CO_2 is required; and

- methods are available for integrating power generation with CCS in the evolving electricity generation, transmission, and distribution system.

All of these issues are addressed in the following sections.

13.2   Carbon Dioxide Capture and Compression Technology

The objective of this section is to present a brief status of CO_2 capture (separation) and compression technology, focusing on the updates since the *IPCC Special Report on Carbon Dioxide Capture and Storage* (IPCC, 2005). This includes CO_2 capture for energy-intensive industrial applications and electricity production with fossil fuel and biomass feedstocks. The focus is on the new developments in the three CO_2 capture technology process options (pre-combustion, post-combustion, and oxyfuel combustion), as well as CO_2 compression technology.

Costs of CCS are addressed by first developing CO_2 avoidance costs (see Section 13.5.2.1 for a precise definition of CO_2 avoidance cost) for both new and existing coal power plant baselines. The CO_2 avoidance costs are then applied as a CO_2 tax to determine the "triple-point" where product cost (i.e., electricity) is the same for coal with and without CCS and natural gas combined cycle (NGCC) without CCS based on varying the price of natural gas (NG). Learning curves and prospects for technology improvements to reduce CCS costs are also addressed.

In the seven years since the *IPCC Special Report on Carbon Dioxide Capture and Storage* there have been significant technology developments in CO_2 capture and transport. This brief section is an update of only the CO_2 capture technology and costs. A brief description of next-generation capture research and development is also provided in Section 13.3.2.9.

13.2.1 Applications and Feedstocks for CO₂ Capture

The greatest focus of CCS is on coal-based electric power generation. This focus is logical because close to 40% of total human-made fossil fuel CO_2 emissions are from this one application and feedstock. Coal-based power plants also dominate the lists as the biggest stationary CO_2 point sources. Nevertheless, there are other applications and feedstocks for CCS. For example, all of the large commercial-scale CCS projects currently in operation are for industrial applications such as natural gas clean-up, biofuels production, and production of synthetic natural gas from coal. Only one of the five commercial scale projects operating today uses coal.

13.2.1.1 Industrial CO₂ Capture

Examples of CCS projects of approximately 1 Mt/yr or greater of CO_2 storage that have been successfully operated for over five years (listed from the oldest) include:

- Statoil's Sliepner offshore gas platform in the North Sea between Scotland and Norway, with CO_2 from associated natural gas purification injected into an undersea deep saline formation;

- Dakota Gasification in North Dakota, United States, with brown coal (lignite) gasification via synthetic natural gas (SNG) purification into a CO_2 pipeline to Weyburn, Saskatchewan, Canada, for use in enhanced oil recovery (CO_2-EOR);

- ExxonMobil's LaBarge facility in Wyoming, United States, with CO_2 from natural gas purification into a CO_2 pipeline for CO_2-EOR in the Powder River Basin of Wyoming;

- BP's In Salah facility in Algeria, with CO_2 from natural gas purification with injection into the same formation, but at a distance from the production well; and

- Statoil's Snøhvit Project, which sequesters CO_2 from a liquefied natural gas facility in a saline formation underneath the Barents Sea.

Most of these successful commercial-scale CCS projects have the economic advantage of producing a large pure CO_2 vent from raw natural gas purification. Natural gas produced from these fields contains from about 2% to over 15% CO_2 or more. Carbon dioxide removal or capture is required to meet natural gas pipeline energy content specifications, regardless of the CO_2 mitigation issue. CO_2-EOR projects have the added economic advantage of a small market value credit for the CO_2 after its compression and CO_2 pipeline delivery expenses to nearby oil fields (EPRI, 1999). Thus, the overall costs of these CCS projects have significantly lower CCS costs than those discussed later in this section.

There has been some debate about whether CO_2 used for EOR should be considered as CO_2 storage. While the amount remaining underground varies from place to place and well to well, about 50% of the injected

CO_2 never returns to the surface (Stevens et al., 2003). Moreover, in almost all cases, the CO_2 produced with the oil is separated and injected back into the reservoir, primarily because it is a valuable commodity and avoids the need to purchase more CO_2. Beyond this, some argue that the CO_2 emitted into the atmosphere when the oil is used negates the benefits of storage. However, at least on the margin, if this CO_2 was not geologically stored in EOR, the alternative replacement oil from other sources would not have reduced the CO_2 in the atmosphere at all. In the United States, where most of the CO_2-EOR is carried out, CO_2 utilization credits for EOR are only about US$10/t (or US$0.53 per thousand standard cubic foot (scf) in common EOR terms). That is significantly less than the total cost of CCS unless the CO_2 is being captured regardless of the CO_2 mitigation issue.

There are several important industrial applications that produce large, nearly pure CO_2 vents. These include raw natural gas and synthesis gas (H_2 and CO from gasification) purification, plus the production of high-value products like ammonia and sometimes other synthesis gas products such as hydrogen, synthetic natural gas (SNG), coal-to-liquids, or methanol. It should be noted that most hydrogen is made from natural gas via steam methane reforming (SMR), which normally does not produce a pure CO_2 stream (Shah et al., 2006). However, hydrogen made from heavy oil, petcoke, or coal via gasification does produce large, pure CO_2 streams.

Industries with moderately large point sources of CO_2 include cement kilns, iron/steel making, oil refining, and bulk chemicals. Large industrial CO_2 point sources have typically less than 1 $MtCO_2$/yr. These CO_2 sources for CCS are relatively small compared to the big coal power plants. Nevertheless, the pure industrial CO_2 vent sources for CCS should not be overlooked due to their lower CCS costs. Nations, such as China with over 65 coal gasification plants and 20 GWt of synthesis gas capacity for ammonia, methanol, and hydrogen, have these pure CO_2 vents for lower cost CCS (Simbeck, 2009a).

13.2.1.2 Electric Power Generation

As already discussed, electric power generation and large industrial operations are the focus of CCS efforts due to their large overall CO_2 emissions and large point sources. Typical central coal power plants (resulting in 40% of total worldwide fossil fuel CO_2 emissions) emit 0.8–1.0 tCO_2/TWh of net electricity generated. A 1000 MWe coal power plant at a 75% annual load factor emits about 6 $MtCO_2$/yr. An equivalent baseload 1000 MWe NGCC power plant emits about 3 $MtCO_2$/yr. There are very few industrial point sources that match the size of fossil fuel power plants.

Fossil fuel power plants in wealthy nations are the most susceptible to CO_2 reduction mandates and thus CCS may be strategic to their future CO_2 mitigation options. Furthermore, unlike other energy-intensive industries, with few exceptions, power plants cannot move to nations with fewer CO_2 restrictions. CO_2 taxes on energy-intensive industries are only effective if the imported products are taxed based on their manufacturing and shipping direct and indirect CO_2 emissions (Davis and Caldeira, 2010). Otherwise, the movement of industries to countries with fewer CO_2 restrictions would hurt the economies of countries with high CO_2 restrictions while failing to reduce total world CO_2 emissions.

13.2.1.3   CCS Feedstocks

Feedstocks for CCS reflect the industry, applications, and fuel availability. More importantly, CCS also requires large point sources located near good geologic CO_2 storage sites. This key issue favors large coal uses for CCS, which are dominated by central coal-fired power plants but also large coal-based integrated steel mills and cement kilns.

Most industries are natural gas-based, except for nations where it is not readily available, such as China, where coal is widely used by energy-intensive industries. Industrial natural gas applications tend to be relatively small CO_2 point sources, thus generally not attractive for potential CCS. Industrial exceptions are the large oil refineries, cement kilns, and iron/steel making and ethanol fermentation facilities.

Oil refinery CO_2 emissions are generally from processing raw crude oil into usable oil products. The heavier the crude oil feed and the lighter the products, the higher the CO_2 emissions. If available, oil refineries prefer using natural gas for hydrogen production and fuel gas, thus reducing the need for more expensive oil for their internal use. CO_2 emissions from the process of oil refining are only about 9% of the carbon in the raw crude oil feed. Also, the extensive use of residual oil coking for heavy oil conversion avoids a significant amount of CO_2 emissions by oil refiners. Most CO_2 emissions from crude oil are emitted by the end user of the oil refinery products, both premium distillate fuels and low-value petcoke. Emissions from the end user of petroleum products are widely distributed and are often from small and mobile sources such as cars, trucks, and airplanes, which are, of course, difficult to capture.

Cement kilns are large CO_2 producers due to both the fuel requirement and the conversion of limestone to lime. It should be noted that cement kilns have the highest CO_2 emissions per unit value (or per million dollars) of product. Cement kilns are also ideal for the utilization of waste fuels. If waste fuels are not available, coal or petcoke are commonly used.

Integrated steel mills are traditionally coal-based and make mostly virgin iron from iron ore via coke ovens and blast furnaces. This should not be confused with mini-mills recycling scrap steel via electric furnaces. Integrated steel mills have very high CO_2 emissions per unit value of product. Mini-mills have much lower overall energy use, and most of their CO_2 emissions are indirect, e.g., located at the electric generation facilities.

A potentially interesting and often overlooked CCS CO_2 source could be the high purity CO_2 vent of ethanol fermentation from biomass (US DOE,

2010). Ethanol plants are extensive and growing for gasoline replacement in the United States and Brazil due to government mandates and subsidies. Currently, US ethanol production is about 54.27 billion liters/yr (13.3 billion gallons/year) and almost all of it comes from corn fermentation. The high purity CO_2 vented from corn ethanol fermentation in the United States alone is about 30 Mt/yr. However, a challenge is the small size of the CO_2 ethanol fermentation vents. With about 150 ethanol plants operating in the United States, the average point source is about 0.2 $MtCO_2$/yr per plant. Nevertheless, many of these US corn ethanol facilities are concentrated in the corn belt of the upper Midwest, from Indiana to eastern South Dakota, with the major capacity located in Iowa, Illinois, and Nebraska.

Electric power generation is traditionally focused on coal feedstocks for "baseload" (60–80% annual load factor) power generation due to its low and stable fuel costs (US EIA, 2009). Natural gas is more commonly utilized for cycling and peaking power generation unless the local grid does not have enough coal or nuclear available for baseload. The demand for natural gas-based cycling and peaking generation increases with mandates for more intermittent wind and solar renewable power. Therefore, CCS for electric power generation logically focuses first on coal-based power. However, this could change in the future due to the heightened interest in natural gas-based power. If stable long-term supplies of lower natural gas prices are available, and gas is used in high-efficiency cogeneration (COGEN) or baseload power with high annual load factors, natural gas power generation may become the focus of CCS.

The focus on large coal power plants for CCS creates an interesting longer-term opportunity for biomass, as discussed in Section 13.1.2 and Chapter 12. The CCS infrastructure would likely begin with coal due to its low delivered fuel cost. However, as CO_2 avoidance values grow, the economics begin to favor blending waste biomass (whenever available) with coal to increase CO_2 reductions using biomass with CCS.

13.2.2 CO_2 Capture Status

CCS has several important decision steps and process choices. The following discussion simply and briefly explains these choices so the CO_2 capture and compression technology status can be discussed.

13.2.3   The Basics of CO_2 Capture

CCS requires large CO_2 stationary point sources within reasonable distances of good geologic storage locations. CO_2 can be from an existing source considering CO_2 capture retrofit or rebuild or from a potential new CO_2 source. It should be noted that a new CO_2 source adding CCS would only avoid increasing CO_2 emissions unless it were based on biomass or if it displaced an existing source without CCS. CO_2 emissions reductions based on fossil fuels require existing CO_2 sources to have retrofits, rebuilds, or new unit replacements, all with CCS.

Existing CO_2 sources adding CCS are subject to net capacity and efficiency losses (Simbeck and Roekpooritat, 2009a; 2009b). It may be useful to also consider rebuilding existing CO_2 sources for CCS to avoid efficiency losses, particularly in the case of older, inefficient coal plants. Rebuilding old power plants increases the total capital but can also avoid the net capacity and efficiency losses of adding CCS (relative to the older existing power plant). Existing CO_2 sources may also consider moving new replacement units with CCS to be co-located with CO_2 storage sites, thus avoiding CO_2 pipeline costs and permitting issues at the expense of permitting and the capital outlay of an entire new replacement plant.

The first step of CO_2 concentration, recovery, or capture to a high purity CO_2 stream is the most expensive and has the most options. This step is normally subdivided into three general process options (IPCC, 2005):

- pre-combustion,
- post-combustion, and
- oxyfuel combustion.

A schematic illustrating the major process steps and material flows for each of these options is provided in Figure 13.3. The choice of CO_2 capture process is complicated by the many different process technologies being developed as well as their state of their development.

13.2.3.1 CO_2 Capture and Storage Ready

The concept of making plants "CO_2 capture and storage ready" is to enable plants to be retrofitted with CCS when the necessary economic and policy drivers are in place, which in most countries currently are not. The term "CO_2 capture and storage ready" means different things to different people. The Global Carbon Capture and Storage Institute (GCCSI), in collaboration with the International Energy Agency (IEA) and Carbon Sequestration Leadership Forum (CSLF), has produced a definition of CCS ready (GCCSI, 2010), which states that in order for a facility to be considered capture and storage ready the project developer should:

- carry out a site-specific study in sufficient detail to ensure that the facility is technically capable of being fully retrofitted with CO_2 capture using one or more choices of technology that are proven or whose performance can be reliably estimated as being suitable;

- demonstrate that retrofitted capture equipment can be connected to the existing equipment effectively and without an excessive outage period and there will be sufficient space to construct and safely operate additional capture and compression facilities;

- identify realistic pipeline or other routes to storage of CO_2;

- identify one or more potential storage areas that have been appropriately assessed, and found suitable for safe geological storage of the projected full lifetime volumes and rates of captured CO_2;

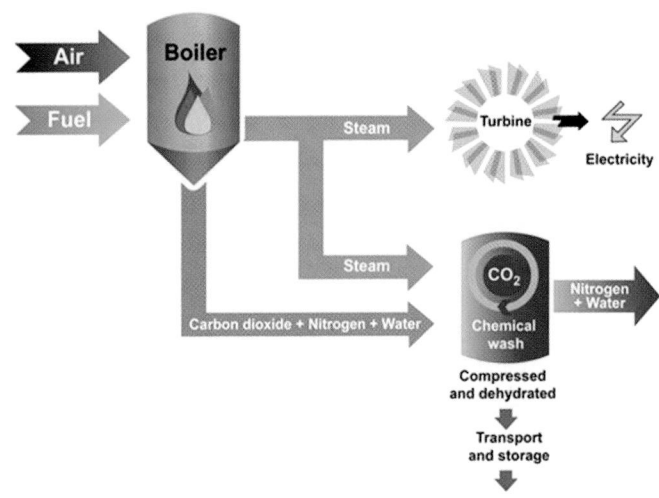

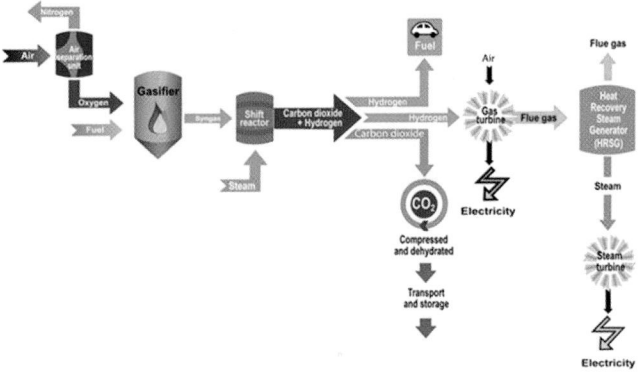

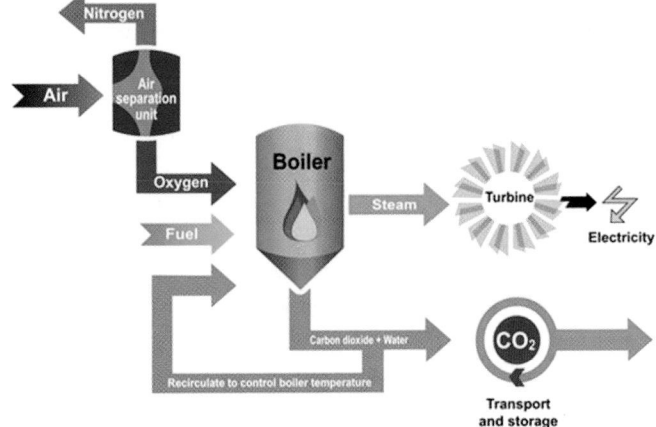

Figure 13.3a-c | Schematic showing the major steps and materials flows associated with the (a) post-combustion capture, (b) pre-combustion capture, and (c) oxy-combustion. Source: courtesy of ZEP, 2011.

- identify other known factors that could prevent the installation and operation of CCS and identify credible ways in which they could be overcome;

- estimate the likely costs of retrofitting CCS;

- engage in appropriate public engagement and consideration of health, safety, and environmental issues; and

- review CCS status and report on it periodically.

A much more dubious meaning of "CO_2 capture ready" is that processes or plants simply claim that space has been left for the addition of retrofitting CO_2 capture at a later date. This usually occurs for new natural gas or coal power plants. In these cases, there is no large pure CO_2 vent, just a normal natural gas or coal power plant with some extra space. However, once the investment is made, the economics of CCS greatly change. Sunk capital cost (paid-off or not) power plants that are considering adding CCS require much higher CO_2 avoidance costs in order to economically justify adding retrofit CCS than if the investment had been made before the original large power plant was built. This is shown in the later cost sections.

Additionally, the concept of "CCS ready" applies to processes or plants removing (capturing) CO_2 in their normal operation and venting it as a large, high-purity CO_2 stream. These sources provide good opportunities for low-cost CCS when compression and storage are added to the existing operations. Examples include most industrial synthesis gas gasification plants, some natural gas-based hydrogen plants, and some natural gas purification plants that remove high concentrations of CO_2.

13.2.4   Tracking Real and Proposed CCS Developments

There are numerous groups and organizations developing and promoting potential CCS technologies as well as bench scale research and development (R&D), pilot, demonstration, and commercial-scale projects. These include governments (at all levels), national laboratories, academics, and private organizations ranging for one-person government grant recipients to venture capital start-ups to major international corporations. A recent report by GCCSI lists hundreds of organizations working on CCS (GCCSI, 2009a).

Keeping up-to-date with the many developments and promotions of CCS projects is very time-consuming. Nevertheless, there are some excellent websites and organizations that are tracking the real and numerous proposed CCS projects. Some of these organizations are also attempting to document worldwide CO_2 point sources for potential CCS, as well as potential geologic CO_2 storage locations.

Some of the organizations tracking real and proposed CCS developments include:

- Bellona: www.bellona.org/ccs
- IEA GHG R&D: www.ieaghg.org

- MIT: sequestration.mit.edu/index.html
- NETL: www.netl.doe.gov/technologies/carbon_seq/global/database/index.html
- Scottish CCS: www.geos.ed.ac.uk/sccs/

In addition to the numerous government organizations funding CCS technologies and R&D, there are some important joint ventures and institutes actively involved in CCS:

- CO_2 Capture Project (CCP): www.co2captureproject.org
- Cooperative Research Centre for Greenhouse Gas Technologies (CO2CRC): www.co2crc.com.au
- Electric Power Research Institute: www.epri.com
- Global CCS Institute (GCCSI): www.globalccsinstitute.com
- Global Climate & Energy Project (GCEP): gcep.stanford.edu
- Zero Emissions Platform: www.zeroemissionsplatform.eu

13.2.4.1   Pre-combustion – Industrial

Pre-combustion CO_2 capture has the most commercial experience. However, all of this large-scale use has been in industrial applications, and not in electric power generation. All five of the large successful CCS projects discussed in the application subsection involve pre-combustion CO_2 capture in industrial oil and gas applications.

There are natural gas purification- and gasification-based synthesis gas (H_2 + CO) purification plants that remove (or capture) CO_2 at amounts greater than 1 Mt/yr. However, that high purity CO_2 is normally just vented. The CO_2 is simply removed to meet product gas (usually natural gas or H_2) specifications. The feed gas is at high pressure without any oxygen, making the CO_2 capture relatively easy. This CO_2 removal or capture is usually accomplished by scrubbing the high-pressure feed gas and then, with low-pressure steam use, stripping the CO_2 from highly loaded tertiary amine chemical solvents, or just high-pressure drop flashing the CO_2 from highly loaded physical solvents for even lower energy requirements.

Of the 50 GWt (synthesis gas) capacity of operating gasification plants around the world, most include large CO_2 capture. The exception is the 8 GWt of synthesis gas converted into 4 GWe of IGCC electric generation. This industrial CO_2 capture is due to the raw synthesis gas having a lower H_2/CO ratio, whereas the higher value uses are for a high H_2/CO ratio such as for hydrogen, ammonia, and methanol production. Therefore, H_2O is added to the CO-rich synthesis gas that reacts over catalysis into H_2 and CO_2 and the CO_2 is then removed. Examples of industrial gasification plants with pure CO_2 vents with CCS include the Dakota Gasification SNG plant in the United States, the Shell Oil Pernis refinery in the Netherlands, and the Shenhua Group coal liquefaction plant in China (CO_2 from the Shell coal gasifiers supplying the hydrogen). The Shenhua Group CCS plant is a smaller first stage demonstration of only 0.1 $MtCO_2$/yr, and it is still under construction. However, it will be the first CCS plant operating in China in 2011, and it may be expanded to a 3 Mt/yr CCS operation.

There has also been prior extensive experience with the combustion of H_2-rich fuel gas in oil refineries when there was excess hydrogen used as a fuel in refinery gas. However, this experience is for older and smaller gas turbines operating at relatively low firing temperatures, and it is only useful today for COGEN when the low gas turbine firing temperature does not hurt the overall efficiency.

Pre-combustion industrial CCS development tends to get much less publicity than electric generation. Nevertheless, industrial CCS costs can be much lower if the industrial process makes a large, pure CO_2 vent regardless of the CO_2 mitigation issue. This is commonly the case for the production of hydrogen (via gasification), ammonia, synthesis gas (for chemicals), and natural gas (when significant amounts of CO_2 are in the raw natural gas).

There is increasing interest in developing hydrogen units for oil refineries via steam-methane reforming (SMR) of natural gas that produce a pure CO_2 vent, as is already the case when making hydrogen via the gasification of heavy feedstocks. Most modern SMR units do not make a pure CO_2 vent because of the use of high-pressure swing absorbers (PSA). With the improvement of natural gas supplies and increase in hydrogen demand due to upcoming heavy fuel oil upgrading and desulfurization mandates for maritime fuel, an SMR unit with a pure CO_2 vent could be an excellent opportunity for low-cost CCS.

Research on vacuum pressure swing absorbers (VPSA), along with staged PSAs by the major industrial gas companies for integration into SMR to produce a pure CO_2 vent, is underway. Interestingly, this is already commercially applicable but not for SMR hydrogen units. A Linde VPSA system is used for synthesis gas recycle purification as part of a Corex/Midrex iron making unit in South Africa (Linde, 2009). However, VPSAs would have about half of the total SMR CO_2 emissions unless a portion of the product H_2 is used to fire the reforming furnace, which would greatly increase the overall cost. There are also commercially proven heat exchange reformers by several vendors that recover all of the CO_2 as a pure vent in making H_2 from natural gas. However, these designs require oxygen, whereas traditional SMR does not.

ExxonMobil is developing a cryogenic process for removing CO_2 from high pressure raw natural gas that is high in CO_2. It is called the Controlled Freeze Zone process and it is being demonstrated in a US$100 million unit at its LaBarge NG facility in Wyoming, United States (Mart, 2009). One of the advantages of this process is the production of liquid CO_2 at pressure, thus greatly reducing the CO_2 compression costs for CCS.

Due to high CO_2 emissions from iron/steel production, there is increased interest in advanced iron making processes that produce CO_2 in a pure stream for potential CCS. An option is to convert the air-blown blast furnaces to an oxygen-blown operation (plus adding CO_2/CO top gas for recycle). More advanced options are the utilization, integration, or advanced versions of synthesis gas-based direct reduced iron (DRI) processes, such as Corex or Midrex (ULCOS, 2008). The synthesis gas

conversion pre-pass over the iron ore is low, and thus the top gas is recycled after removal of H_2O and CO_2 (products of iron oxide reduction to iron). As mentioned above, this is already commercially done with VPSA on a Corex/Midrex plant in South Africa. In addition, several coal-based DRI steel mills are under construction in India. The mills are based on separate coal gasification that will increase the H_2/CO ratio before feeding into the Midrex DRI units; thus, a second large pure CO_2 vent is produced (Jindal, 2009).

The ammonia solvent CCS process is usually associated with post-combustion CO_2 capture. However, it has some of the same benefits for use in pre-combustion CO_2 capture of stripping at pressure to reduce CO_2 compression costs. It has the added benefits of high solvent loading, fewer side reactions, no refrigeration, and fewer ammonia leakage issues when operated on high-pressure synthesis gas. Testing is currently being done at SRI International for application with SNG for coal with CCS.

13.2.4.2   Pre-combustion – Electric Generation (See also Chapter 12)

Pre-combustion CO_2 capture in electric generation is mostly focused on coal via integrated gasification combined cycle (IGCC) and not for NGCC. NGCC CO_2 capture is more focused on post-combustion CCS, as this avoids major modification to the standard NGCC design. Plus, the CO_2 capture can be by-passed when needed for additional peaking power.

Even before the CO_2 capture issue, the electric power industry had been slow to accept the IGCC option relative to conventional direct combustion pulverized coal (PC) boiler steam systems. This was due to the high capital cost, low availability plus complex chemical processing, and the integration requirements of IGCC. There is also an ongoing debate about the lack of expertise in the electric power industry because gasification is a very complex chemical process. There was a similar slow acceptance and learning curve by electric generators of wet limestone scrubber flue gas desulfurization and selective catalysis reduction nitrogen oxide (NO_x) controls, which are very simple chemical processes.

The addition of CO_2 capture to pre-combustion IGCC has some challenges. First, the hot raw synthesis gas cooling heat recovery is lost due to the large excess steam (H_2O) requirements of the CO conversion to H_2 and CO_2. More important are the special challenges of firing H_2-rich gas in high temperature gas turbines. This is generally due to the negative impact of the much higher heat transfer coefficient of H_2O vapor (after combustion of the H_2-rich fuel) on the critical hot gas path metal parts associated with the hottest initial stages of the gas turbine exhaust gas expansion. The options for addressing this issue include: reducing the firing temperature (lower capacity and efficiency) or shorter hot gas path metal parts life (i.e., higher maintenance), adding large amounts of high pressure nitrogen to the H_2 fuel (reducing net efficiency or most expensive air-blown gasification), or developing a special and challenging gas turbine pre-mix air and H_2 combustor. Reducing the firing temperature would be the worst choice, unless utilized in the industrial COGEN of combined heat and power (CHP).

Perhaps the biggest challenge facing pre-combustion CCS is the high capital cost of the few IGCC projects actually being built at this time. For example, the latest capital cost of Duke Energy's Edwardsport, Indiana IGCC project (currently under construction) has increased to US$3.3 billion for 618 MWe (net), or US$5593/kW, with no capture and storage. However, this very high capital cost does include "as spent" current and future dollars, some infrastructure for a potential second unit, allowance for funds during construction, and other owners' costs. While this high unit capital cost is reducing interest in IGCC, at the same time it is going unnoticed that the few conventional coal boiler power plants under construction in the United States have unit capital costs almost as high. A good example is Duke Energy's Cliffside project in North Carolina, at a cost of about US$3000/kW with no capture or storage. The few US nuclear plants currently under construction have also increased to significantly higher unit capital costs.

The positive activity in pre-combustion CO_2 capture for electricity is associated with increasing competition and innovative designs. This is likely a result of the many studies showing the incremental costs plus net power and efficiency losses of converting IGCC to CCS as being lower than for post- and oxyfuel combustion (US DOE, 2007; EPRI, 2008). However, the issue is higher power plant costs and the lower availability of IGCC compared to more commercially proven PC boilers before considering CCS.

There are more than 10 IGCC with CCS projects being promoted based on at least six different coal gasification technologies. Also, many pre-combustion power projects are being proposed by independent power producers (IPP), some having innovative designs based on "polygeneration" versus just stand-alone central power plant IGCC. The use of high H_2/CO ratio synthesis gas with CCS for various high-value products such as hydrogen, SNG, methanol, olefins, and plastics, as well as COGEN CHP, is involved. As previously explained, COGEN avoids the added challenges and costs of high-temperature gas turbine-fired H_2-rich fuel gas while also greatly increasing overall efficiency and reducing water demands (Simbeck, 2009a; 2009b). The combination of this added competition, especially from the more aggressive and innovative IPPs, will reduce overall costs and improve performance. In fact, the key potential advantage of pre-combustion CO_2 capture for electricity is that hydrogen has more uses and flexibility than just steam from boilers.

13.2.4.3   Post-combustion – Industrial

Globally, there are over 2500 industrial sources with CO_2 emissions over 0.1 Mt/yr. These emissions sources tend to be smaller than for coal-fired power generating stations, averaging about 1 Mt/yr (IPCC, 2005).

Carbon dioxide concentrations in flue gas vary widely, from 7–10% for gas-fired boilers to 14–33% for cement kilns (IPCC, 2005). Post-combustion CCS for industry has been successfully demonstrated at a relatively small scale. The largest operating unit is only about 0.1 Mt/yr. There are few commercial post-combustion CO_2 capture (without special circumstances) units but MHI (Mitsubishi Heavy Industries) has for example built around 10 of the largest units, which handle flue gas from NG-SMR ammonia plant reforming furnaces to capture CO_2 to make urea from the ammonia.

The low pressure and low CO_2 concentration in flue gas after combustion requires large absorbers, high CO_2 removal solvent circulation rates, and high stripping steam use. The presence of oxygen in flue gas also eliminates many potential solvents and leads to a small amount of waste from the reaction with the solvents. However, for boiler systems it is possible to add flue gas catalysis combustion via the use of a small amount of NG to consume the few percent of oxygen in the flue gas and then use better solvents. This was commercially done in the early days of CO_2 EOR before large CO_2 pipeline sources of cheaper CO_2 became available.

Perhaps the greatest attribute of post-combustion CO_2 capture is the ability to retrofit any flue gas steam with little impact on the original process other than the added utility demands (which reduce net capacity and efficiency) and space requirements. The added utility demand, and consequently added energy demand, can be met by adding a small boiler with a steam turbine or gas turbine in a COGEN system for the CO_2 compressor power and low pressure CO_2 stripping steam. The CO_2 scrubber could be located in the base of a new wet stack, as has been done for a retrofit flue gas desulfurization (FGD) unit with space limitations. Back-end CO_2 capture also means the CO_2 capture process can be by-passed at any time, thus assuring higher availability than pre-combustion or oxyfuel combustion and the ability to meet any peak demands if CO_2 emissions are allowed.

Most of the effort on post-combustion CO_2 capture is focused on electric power generation applications. One exception is Statoil's current demonstration of chilled ammonia post-combustion CO_2 capture on flue gas from the fluid catalytic cracker (FCC) at its Mongstad refinery in Norway (Alstom, 2007).

13.2.4.4   Post-combustion – Electric Generation

Post-combustion CO_2 capture for electric generation has gained greater interest in the last five years. This increasing post-combustion CO_2 capture interest is due to many factors, including the following:

* a decreased interest in pre-combustion CO_2 capture, likely due to the high capital costs and slow commercial acceptance of IGCC (with or without CCS). Current reported capital costs for IGCC plants average about US$6400/kW normalized to a 460 MW plant with 90% capture (Al-Juaied and Whitmore, 2009);

- the number and scale of emissions from existing and planned PC power plants;

- improved designs for post-combustion CO_2 capture with more vendor competition and choices of chemical solvents;

- minimal impact to the traditional NGCC or PC power plant process other than the large need for low pressure steam for CO_2 stripping and for CO_2 compressor power;

- the ability to easily by-pass the back-end flue gas scrubber process when problems with the CO_2 system occur or when there is a need for additional peaking power; and

- lower total capital outlay (not to be confused with CO_2 avoidance costs) and ease of retrofit to the existing power plant, except for accounting for the moderately high net capacity and efficiency losses plus additional space requirement.

As already discussed, most interest in natural gas-based electric generation CCS is with post-combustion CO_2 capture for the CO_2 emission avoidance reasons. Lower natural gas prices and improved supplies are making NGCC more competitive with coal-based electric generation for baseload power. However, NGCC without CCS replacing coal is generally more economical than coal with CCS until there are very high CO_2 avoidance values or high NG prices. Also, with high natural gas prices, CCS economics can favor coal with CCS over natural gas with CCS.

There are at least five quality vendors offering improved amine CO_2 capture systems via advanced amines and/or innovative integration. There are also at least two quality vendors claiming even better performance based on less developed ammonia solvent in the place of amines (Black, 2006; McLarnon and Duncan, 2008).

The advantages of ammonia over amine CO_2 capture are higher solvent loading (units of CO_2 captured per unit of total solution recirculation), significantly lower stripping steam requirements, and sizable CO_2 compression power plus capital savings. However, ammonia to ammonium bicarbonate flue gas chemical scrubbing system is still in the early stages of demonstration. Nevertheless, ammonia for both pre-combustion and especially post-combustion CCS is likely the most significant development in CO_2 capture technology in the last five years.

The current small (0.1 MtCO_2/yr) "Chilled Ammonia" demonstration plants on coal boilers in North America and on NGCC and oil refinery FCC fuel gas in Europe are being followed closely (Alstom, 2009; Spitznogle, 2009). These demonstration plants should resolve concerns about potential ammonia leakage in the stack gas, solid ammonia bicarbonate buildups, the impact of higher quality steam to strip the CO_2 at pressure, expensive refrigeration, and extensive heat exchange requirements.

The Powerspan ammonia CO_2 capture process claims it can avoid the ammonia leakage concern and refrigeration requirements. However, they have only begun testing a very small CO_2 pilot plant. Powerspan

also has the added challenge of integrating this process to its only demonstrated SO_2 and NO_x control process that utilizes high-cost ammonia for conversion into lower grade and less utilized ammonia sulfate fertilizer that may saturate local markets.

There is increasing fundamental R&D interest in more advanced chemical absorption and adsorption of CO_2, especially with solid sorbents and strong bases that could adsorb CO_2 more effectively (Rhudy, 2009). However, these ideas are at the very early stage of development and not close to the minimal small pilot plant stage or successful development. Solid sorbents range from non-volatile amine polymers to sodium carbonate to metal organics to lithium silicates. Adsorption and regeneration configurations, energy requirements, and sweep gas issues are yet to be resolved. Also under consideration is oxyfuel combustion for the regeneration to make the higher purity CO_2 stream. CO_2 flue gas capture into advanced cement and building materials are discussed separately in Section 13.2.4.7 due to their close association with reducing the large cement making CO_2 emissions.

13.2.4.5   Oxyfuel Combustion – Industrial

Oxyfuel combustion replaces air with oxygen, thus generating a CO_2-rich flue gas. Oxyfuel CO_2 capture is the least developed of the three CO_2 capture processes. However, it continues to gain interest and development. This is likely due to its potential advantage of greatly simplifying the overall CO_2 capture process plus avoiding most of the chemical processing associated with pre-combustion and post-combustion capture. Oxyfuel combustion also has the interesting potential to increase existing process efficiency in retrofit applications. This is due to its much lower gas volume of oxygen with recycle CO_2-rich flue gas replacing high N_2-content air with a much higher total gas volume per tonne of oxygen. However, the challenge here is the massive oxygen requirement at about 60% more oxygen needed than for pre-combustion CO_2 capture. The benefits of higher combustion efficiency with oxygen combustion is lost once the large energy demands of producing oxygen plus smaller energy demands of CO_2 recycle are considered. Also, oxygen from external sources or from air separation units using grid electricity will usually have large indirect or off-site CO_2 emissions.

The oxyfuel CO_2 capture is being seriously considered for CCS demonstrations by the CO_2 Capture Project (CCP) for oil refinery-fired heaters and FCCs as well as steam boilers. Petrobras is already testing oxyfuel in its FCC pilot plant in Brazil and Suncor is working with the provincial government of Alberta (Canada) on a potential oxyfuel retrofit of an existing NG-fired field steam boiler for steam generation in oil sands production via steam-assisted gravity drainage (de Mello et al., 2008).

Advanced oxygen production looping cycles are being developed at a small scale by numerous groups and companies. This involves the oxidation and reduction cycling of metals on ceramic particles at moderate temperatures. The high solid flow recycle rates per tonne of oxygen and contacting requirements of solids and gases generally favor the use of modified circulating fluid beds with long-term particle attrition being a key issue. Looping cycle oxyfuel CCS is being developed for both natural gas and coal as well as at atmospheric pressure and pressurized. Oxygen looping with natural gas is easier, as it avoids the added challenge of solids and coal ash. Oxygen looping cycles also apply to pre-combustion CCS while reducing the oxygen demands by about 60%.

13.2.4.6   Oxyfuel Combustion – Electric Generation

Despite being the least developed of the CO_2 capture options, oxyfuel combustion is perhaps the CO_2 capture option that the traditional electric power industry likes best. This is most likely due to the lack of chemical processing compared to the simple chemical processing of post-combustion and especially the complex chemical processing of pre-combustion CO_2 capture. There is hope that raw CO_2 from the oxygen combustion flue gas might be directly compressed for CO_2 storage, thus avoiding additional SO_2 and NO_x controls as well as 100% CO_2 capture and no stack or emissions. Oxyfuel combustion at moderate pressures could also noticeably reduce the CO_2 compression capital costs and power requirements relative to the added power/costs of making or compressing the oxygen at moderate pressure.

Oxyfuel combustion has several challenges, such as the large net capacity and efficiency losses associated with making the massive amounts of oxygen, which are 60% larger than required for pre-combustion. There are also challenging CO_2 compression, pipeline, and injection issues on purity relative to the raw CO_2 coming from oxyfuel combustion. Specifically, trace amounts of SO_2 and NO_x are likely to react and even a 1% residual amount of oxygen (from oxyfuel combustion) is generally not acceptable in a traditional CO_2 pipeline (White et al., 2008; Darde et al., 2008), so CO_2 purification using low temperature separation of other means will be required. Finally, N_2 that remains in the oxygen from the air separation requires higher pressure CO_2 to avoid a two-phase flow and can increase the injection well back-pressure.

Oxyfuel combustion for electric generation CCS has been under development by Vattenfall for its large Eastern European brown coal power plants for many years. There is now a small demonstration scale unit of 0.1 $MtCO_2$/yr at its large brown coal power plant at Schwarze Pumpe in Germany. FutureGen 2.0 in the United States will demonstrate retrofit of an old coal plant with oxyfuel combustion.

All of the major air separation vendors are working with most of the major boiler vendors on oxyfuel combustion. Larger oxyfuel combustion CO_2 capture demonstration projects are being promoted for several power plants. The projects are still smaller in size, and there are fewer proposed projects than for pre-combustion and significantly less than for post-combustion. A large oxyfuel demonstration proposed in Western Canada was abandoned due to its high capital costs. Most proposals are for PC boilers, both new and retrofits, where the mass of recycled

CO_2-rich flue gas mixed with oxygen matches the mass flow and heat exchanges similar to conventional air and coal combustion boilers.

Praxair and Foster Wheeler have proposed a new circulating fluidized bed combustion boiler steam cycle oxyfuel CCS demonstration power plant in Western New York, United States. Circulating fluidized bed combustion boilers have the advantage over PC boilers of a much greater feed fuel flexibility to co-process larger amounts of waste biomass for double CO_2 reductions in CCS.

Oxyfuel combustion for CO_2 capture has revitalized renewed R&D in the magnetohydrodynamics (MHD) process for the advanced electric generation process from fossil fuels. The original MHD development suffered a big loss in performance because of the N_2 in its air combustion.

Most of the interest in oxyfuel for NG-based power generation CCS has been based on a modified steam turbine used like a gas turbine without the big air compressor. Large amounts of water injected with NG-oxygen combustion generate a high pressure of mostly hot steam working in the fluid for expansion as in the modified steam turbine. This produces a high steam temperature with direct NG-oxygen combustion plus high temperature double reheats with smaller oxyfuel combustors and then steam condensing. This is the basis for the Clean Energy System under development in both North America and Europe. It is also being proposed for integration with coal gasification.

For some, the potential of 100% CO_2 capture, avoiding separate SO_2 and NO_x capture, and totally avoiding a stack via oxyfuel CCS, is a great incentive to investigate sequestering "dirty CO_2." The only significant impurity in large commercial CO_2 operations is the H_2S from pre-combustion CCS. Specifically, Dakota Gasification's CCS project in Weyburn, Canada leaves all of the coal gasification H_2S in the CO_2. This is also common in EOR field CO_2 recycles as the oil production often has associated gas with H_2S. However, transporting H_2S-rich CO_2 gas through long pipelines carries greater risks and costs, thus negating some of the benefits of this approach.

Researchers are investigating innovative ways to integrate, convert, and/or process dirty CO_2. The trade-off of CO_2 compressors and pipelines versus varying the CO_2 purity has not been fully addressed. Traditional commercial EOR operations pipeline CO_2 purity specifications (except for the water content) have been developed based on the natural CO_2 dome sources, not CCS. Other options for dirty CO_2 include the reaction and recovery of SO_2 and NO_x before or during CO_2 compression and low temperature catalysis combustion with a small amount of natural gas to react all of the residual oxygen in the raw oxyfuel combustion flue gas.

13.2.4.7 CO_2 Capture via Advanced Building Materials

A growing area of interest is the development of advanced building materials that utilize CO_2, perhaps directly from fossil fuel stacks. This can be a "win-win" situation if CO_2 captured materials can be effectively utilized in place of traditional cement, thereby replacing limestone-based cement kilns, which are the second largest source of human-made stationary CO_2 emissions (after power plants).

Most of the focus appears to be based on the utilization of naturally occurring magnesium silicates to replace limestone in cement making. The magnesium silicate reacts to MgO by adding CO_2 to make cement that thereby reduces the CO_2 that is generated from current limestone cement kilns. The challenges are many, including traditional building material standards that have been based on chemical composition instead of performance required for the building material application. There are at least five developers of this general type of process. Somewhat related applications are the use of carbon-rich building materials to replace limestone-based cement and coal-based steel. Possibilities are a more carbon-rich asphalt replacing cement for road construction and a carbon fiber or CO-based hard plastic replacing steel.

All building material applications for CO_2 reduction have the challenge of the much larger mass of CO_2 emissions versus the order of magnitude smaller building material markets. In addition, large amounts of naturally occurring oxides of silica, alumina, and iron aggregate added to building materials are used because they are inexpensive and also because they have no impact on CO_2 emissions.

13.2.4.8 CO_2 Capture from the Air

A small group of mostly advanced research physicists have been promoting the idea of CO_2 capture from the air for some time (APS, 2011). If practical, this would be an elegant solution, as it would completely separate the CO_2 capture locations from the CO_2 emission sources and would indirectly solve the CO_2 reduction issue for mobile transportation fuels from mostly oil.

However, practical and economically competitive CO_2 capture from air is a major challenge due to its very low pressure and concentration in air of only about 390 ppmv of CO_2. The ultra-low CO_2 partial pressure in air (at just 0.0004 atmospheric pressure) would likely require strong bases to capture most of that CO_2 from the air as well as a very large adsorber or absorber contactor. This also could mean large energy requirements to regenerate the CO_2 from the strong basic sorbent into a high purity CO_2 stream for compression. Conversely, if a practical CO_2 capture system from the air is possible, it could be much cheaper with better performance if first used at higher pressure and much higher concentration CO_2 sources. CO_2 capture from the air is in the very early stages of development. A recently completed study suggests that direct air capture using chemicals will cost more than $600/t$CO_2$ and will therefore not be competitive with capture from higher concentration sources (APS, 2011).

Basic R&D is also being done using biological processes for CO_2 capture from air. This could also be applied to higher concentration and

pressure CO_2 sources (like fossil power plants). However, like microalgae for CCS, the annual load factors of CO_2 capture could be very low if it only works during warm, sunny daylight hours. The economics of CCS generally favor processes with high annual load factors due to the high capital costs.

13.2.4.9   Prospects for Advanced Capture Technology and Capture Research

Over the past five years or so, there has been a growing interest by governments, a new community of academic scientists, and industry in developing new materials and separations techniques that could dramatically improve the efficiency and lower the cost of CO_2 capture. While these new approaches are a long way from commercial implementation, in principle, they hold significant promise. Among these new methods are (e.g., US DOE, 2010):

- metal-organic frameworks, a new class of nano-structured hybrid materials with exceptionally high surface area that can improve the efficiency of absorption of traditional and novel organic solvents;

- ionic liquids with higher absorption rates and comparatively smaller energy penalties for regenerating the solvent;

- Si-based solvents requiring much less water for CO_2 capture;

- biologically motivated approaches that utilize nature-inspired catalysts such as carbonic anhydrase to capture and convert gaseous CO_2 to liquid or solid forms;

- membranes coated with CO_2 affinity materials to improve the selectivity and permeance to CO_2;

- hydrogen-conducting membranes for pre-combustion capture;

- catalytic membranes to simultaneously separate CO_2 and carry out the water-gas shift reaction;

- oxygen separation membranes to lower the cost of oxygen production; and

- solid adsorbents such as activated carbon, carbon nanotubes, or other nano-structured solids.

As these new approaches reach maturity, they will need to compete based on cost and performance with existing approaches for CO_2 capture as described above. These existing technologies are, of course, expected to improve as they become more widely deployed. For this reason, it is difficult to anticipate which of the existing and new technologies will emerge as market leaders.

13.2.5 CO_2 Compression

CO_2 compressors are commercially well proven from their use in EOR over the last 30 years. In 2010, 50 MtCO_2 were compressed and transported through over 4800 km of CO_2 pipelines. Nevertheless, there are some development opportunities in this area. The impact on overall CCS cost and performance improvements, however, is likely much smaller than in the development of better CO_2 capture processes.

Current CO_2 compressors have high capital costs of over US$1000/kW, and the power requirements are significant at 100–150 kWe per t/hr of CO_2 (McCollum and Ogden, 2006). The best and easiest improvement is doing the CO_2 capture at high pressure. An operating pressure of 3–5 atmospheres will begin to significantly reduce CO_2 compression costs.

Development is occurring on an advanced CO_2 compressor via shock wave compression. This might reduce the total installed unit capital cost ($/kWe) by a large amount, as this type of compressor may require only two stages of compression compared to current CO_2 compressors that require 8–12 stages plus more intercoolers. The developer, Ramgen Power Systems, is working with Dresser-Rand on commercialization (see Ramgen, 2011).

As discussed in the oxyfuel CO_2 capture subsection, there is ongoing development for advanced compressors and pipeline standards for handling dirty CO_2. The raw flue gas from oxygen combustion emits mostly CO_2, but also significant amounts of H_2O, N_2, and O_2, as well as lesser amounts of SO_2 and NO_x. H_2O and O_2 may require removal or conversion, respectively. Current CO_2 compressors may unintentionally convert the SO_2 and NO_x into ammonia sulfate and plug the compressor.

13.2.6 CO_2 Capture Costs (See also Chapter 12)

There are numerous reports documenting current and future estimates for the cost of capture for power generation and industrial emission sources (e.g., IPCC, 2005; MIT, 2007; Al-Juaried and Whitmore, 2009; US DOE, 2010). Costs of capture for the nth-of-a-kind plant range from about US$30–100/tCO$_2$ avoided. First-of-a-kind plants are expected to cost significantly more with estimates in the range of US$100–150/tCO$_2$ avoided. Costs of capture for first generation CCS power plants available in the early 2020s are estimated to be about $45/tCO$_2$ avoided for coal and $115/tCO$_2$ avoided for natural gas (ZEP, 2011). Estimates vary widely, in part because they use a variety of different assumptions about baseline technology, capture technology, discount rates, material and labor inflation, regional indices, first-of-a-kind plants versus n^{th}-of-a-kind plants, etc. We do not repeat these here, but provide an overview that describes not only the complexity of these estimates, but the underlying drivers as well. Examples providing costs are provided for a variety of illustrative scenarios. CO_2 capture cost estimating for CCS is extremely challenging for several reasons, including:

- Costs are application specific. Therefore, while it is necessary to analyze the CCS costs for coal-based power plants, the use of natural gas or biomass with CCS for a double CO_2 reduction needs to be considered, as well as other applications especially those for industrial applications with existing large, pure CO_2 vents, regardless of the CO_2 mitigation issue.

- Costs will also be site specific, depending on labor rates, material costs, construction codes, safety codes, local construction constraints, etc.

- Comparison of costs with and without capture needs to also consider baseline costs in a broad context. Fuel type, fuel switching, and future fuel price impacts – especially as a carbon-constrained world develops – will have a significant effect on costs. Coal energy prices are currently low and stable and may even go down slightly as a carbon-constrained world develops and coal use declines. However, natural gas energy prices are significantly higher than coal and are highly volatile with few, if any, long-term fixed price contracts, except for take-or-pay Liquefied Natural Gas (LNG) contracts. Furthermore, natural gas prices will almost certainly go up as a carbon-constrained world develops. Natural gas supplies will be stressed as natural gas begins replacing some of the large coal use applications long before economics warrant serious consideration of coal or natural gas with CCS. Also at high natural gas prices, the cost of NGCC with CCS quickly becomes higher than coal with CCS.

- Costs can be assessed based on a variety of metrics, such as the increase in product cost (i.e., $/MWh for electricity) with CCS versus without CCS or the CO_2 costs of CCS as CO_2 capture costs or CO_2 avoidance costs.

- Existing CO_2 sources considering retrofit CCS have much different costs than those for a proposed new CO_2 source considering CCS. There are also issues that a new CO_2 source with CCS would only minimize CO_2 emissions growth unless it replaces an old existing CO_2 source that is shut down. Also to be considered is the large impact of CO_2 avoidance costs based on the baseline fuel with coal having about twice the baseline CO_2 emissions as natural gas.

- Basic capital cost estimates have changed due to the large run-up in construction costs (materials, labor shortages, and equipment) from 2005–2008, but since then there has been moderate construction cost decline. There is also the issue of site-specific cost factors such as contingencies, capitalization of allocation of funds during construction (common for regulated US electric utilities), and cost of capital or project capital return rates due to mainly the debt/equity ratio, but also due to local taxes, depreciation rates, and the design basis (e.g., extreme weather conditions or earthquake rating). For example, recent all-in capital costs estimates by SaskPower for the commercial-scale oxyfuel and post-combustion capture demonstrations for coal boiler power plants are over CAN$12,000/kW (US$12,453/kW).

- CCS costs are traditionally estimated based on the assumption of proven commercial operating experience, sometimes called the "n^{th}" plant design, which signifies that the cost estimate is not for the 1st, 2nd, or 3rd commercial unit. In other words, there are much lower cost estimates than for the first large-scale demonstration or first commercial plant using a specific CCS technology because it has already progressed down the learning curve.

- Estimating cost for developmental technology without any large-scale operating demonstration plant experience is difficult. In this case, it is common to excessively overestimate performance and underestimate costs.

This section attempts to reduce these complexities to simple economics. More importantly, the economics are presented in a logical and transparent way with an emphasis on the key variables and relative consistency. Learning curves and impact of the state of development are also addressed, as the less developed CCS systems are more likely to have their performance overestimated and costs grossly underestimated.

13.2.6.1　What Does CO_2 Avoidance Cost Mean?

Most economics of CCS are stated in terms of CO_2 avoidance cost. Therefore, it is essential to fully understand what the CO_2 avoidance cost matrix means as well as what the key inputs, sensitivities, and other options are to CCS.

It is also essential to understand the key difference between CO_2 capture and CO_2 avoidance costs, especially with CCS. This is because CCS usually reduces net capacity and efficiency. CO_2 capture cost is calculated based on the CO_2 captured per unit net product. Thus, larger the efficiency losses can erroneously appear to lower CO_2 capture costs. However, CO_2 avoidance is calculated based on the CO_2 reduction or avoidance to the atmosphere per unit net product. Thus, the larger the efficiency losses, the higher the CO_2 avoidance costs, especially when the fuel costs are high. The goal of CCS is reducing CO_2 emissions into the atmosphere; thus, it is almost always best to work with and estimate just CO_2 avoidance costs.

It is typically assumed that CO_2 avoidance cost is the likely minimal CO_2 tax required for a major human-made CO_2 emissions source to start seriously considering CCS. Using a coal-fired power plant as a simple example, the CO_2 avoidance cost is the $/tCO_2$ emissions tax at which the $/MWh electric "loaded" (capital charges, fuel, and operations and maintenance) price is the same as paying the CO_2 tax or adding CCS to avoid paying most of the CO_2 tax. In reality, the CO_2 tax (or CO_2 avoidance costs) must be even higher to justify the added capital and much higher risks of adding CCS versus simply paying the CO_2 tax.

The formula for CO_2 avoidance cost is relatively simple. As shown below, the formula estimates the product costs (in $/MWh) and CO_2 emissions per unit of product (tCO_2/MWh) for a traditional plant (called "b" for baseline case) and then estimates the higher product costs but with lower CO_2 emissions (called "c" for carbon reduction cases). The lower CO_2 emissions case can simply be a higher efficiency or lower carbon fuel with or without CCS. The CCS added option is the most common comparison option due to the larger potential CO_2 reduction. Nevertheless, conversion without CCS to higher efficiency or lower carbon fuel switching cannot be overlooked or ignored, due to its higher efficiency and much lower capital and avoiding CO_2 storage liability risks.

The formula is as follows:

$$\$/tCO_2 \text{ avoidance cost} = (\$/MWh_c - \$/MWh_b) \div (tCO_2/MWh_b - tCO_2/MWh_c) \qquad (1)$$

This simple formula makes it easy to understand the conditions where the CO_2 avoidance cost estimates can be high or low. Low CO_2 avoidance costs occur when small increases in power costs give large CO_2 reductions. High CO_2 avoidance costs occur when there are large increases in power costs and small reductions in CO_2 emissions. The following three simple examples show the fluctuation of CO_2 avoidance costs depending on the baseline and choice of CO_2 reduction option. All three examples are for a coal power plant baseline and assume baseload (high annual load factor) power prices at the plant gate, without added transmission or distribution costs.

- First, consider the CO_2 avoidance cost of an old, dirty, and inefficient, but paid-off coal power plant that is replaced with a new, clean, state-of-the-art coal plant without CCS. The electricity price increase would be about US$40/MWh_c, assuming US$80/MWh_c for the new, efficient coal power (with all of the capital charges) up from only US$40/MWh_b for the old paid-off baseline plant (with no capital charges and just operating costs). The CO_2 reduction would be only about 0.2 t/MWh, assuming 1.0 t/MWh_b for the old subcritical coal plant baseline, and reduced to 0.8 t/MWh_c for the new efficiency supercritical plant. Thus, the CO_2 avoidance cost would be very high at US$200/tCO_2 based on a US$40/MWh power increase divided by 0.2 t/MWh of CO_2 reduction.

- Second, consider the CO_2 avoidance cost for a proposed new coal power plant contemplating the addition of CCS. The electricity price would increase about US$40/MWh_c, assuming US$120/MWh_c for the new coal power with CCS, up from only about US$80/MWh_b for the normal new state-of-the-art coal unit. The CO_2 reduction would be large at about 0.7 t/MWh, assuming 0.8 t/MWh_b for the new state-of-the-art coal plant baseline reduced to only 0.1 t/MWh for the similar new coal unit but now having CCS. Thus, the CO_2 avoidance cost would be US$57/tCO_2 based on a US$40/MWh power increase divided by 0.7 t/MWh of CO_2 reduction.

- Third, consider the CO_2 avoidance for replacing the old coal power plant in the first example with the new coal plant with CCS in the second example. The CO_2 avoidance would now be US$89/tCO_2 based on a US$80/MWh power increase divided by 0.9 t/MWh of CO_2 reduction.

In general, CO_2 avoidance costs are usually high if the baseline is a paid-off existing facility with low cost fuel or if the CO_2 reduction is relatively low. Conversely, CO_2 avoidance costs are the lowest when the baseline is for a new proposed CO_2 source and the CO_2 reduction are high.

A CO_2 avoidance cost involving natural gas can be more complex due to the large potential variation of prices and the relatively low CO_2 emissions of natural gas. Over the long term, as a carbon-constrained world develops, natural gas prices will likely increase and supplies may be stressed as coal is replaced with natural gas long before it becomes economical to consider CCS.

13.2.7 Triple CCS Point Economics

Due to the above complexity of CCS CO_2 avoidance costs with coal versus a similar scenario with natural gas, it is useful to develop "triple point economics" from a coal baseline that also includes natural gas. This would avoid the suggestion that a CO_2 tax or CO_2 avoidance cost where coal with CCS could become competitive while ignoring the simple use of natural gas without CCS to replace coal. Three simple steps define the CO_2 tax triple point economics where the power price is the same for natural gas without CCS versus coal with or without CCS.[1]

The following examples of triple point economics are estimated for two recent projects. As discussed, the baseline choice of existing versus proposed new CO_2 sources greatly impact the economics of coal-based CCS and natural gas price where it would be less expensive to simply replace coal with natural gas and avoid CCS.

13.2.7.1 Triple Point CCS Economics for an Old Existing Coal Power Plant Baseline

Existing coal-fired power plants represent most of the current large CO_2 point sources that are good prospects for CCS. For example, in the United States there are over 300 GW of existing coal power plants with a MWe-weighted average age of almost 40 years old. Recent estimates for this option have been made assuming US$2.00/MMBtu of coal and all costs in constant US$_{2008}$$ (Simbeck and Roekpooritat, 2009a; 2009b). The lowest CO_2 avoided cost for continued coal use was a simple add-on retrofit post-combustion CCS to the old existing boiler at a CO_2 avoided cost of US$74/MtCO_2$. This reduced the old existing power plant net capacity by about 27% and dropped the net efficiency (HHV) from 34% to only 25%. A rebuilt new supercritical steam coal plant with CCS would avoid the net capacity loss and decrease most of the net efficiency loss

of CCS relative to the old subcritical steam coal power plant without CCS (US DOE, 2010). The new rebuilt coal units with CCS had higher CO_2 avoidance costs, however, due to the significantly higher total capital investment associated with the new power plant.

Existing paid-off power plant baselines generally force higher CO_2 avoidance costs due to the inexpensive baseline power costs. In this case, the existing steam coal power cost was estimated at only US$37/MWh (plant gate). At this relatively high US$74/tCO_2$ avoided, the triple point electricity price was US$108/MWh. At this price for power generation, by converting to natural gas via repowering with a new NGCC unit without CCS, the same cost per tonne of CO_2 avoided could be obtained at US$8.31/MMBtu for natural gas. Consequently, repowering an aging steam coal power plant with new NGCC without CCS could be an attractive option if the natural gas industry and electric power generators become willing to agree to amenable long-term supply/price contracts.

13.2.7.2 Triple Point CCS Economics for a New Proposed Coal Power Plant Baseline

The more traditional CCS economics are based on proposed new coal and NG power plants. However, this would only minimize CO_2 emissions growth from the new power plant capacity addition unless an equivalent-size old fossil fuel power plant is shut down and replaced by the new power plant with CCS.

This scenario was evaluated in 2009. The lowest CO_2 avoided cost for a new coal power plant with CCS was the pre-combustion option at a CO_2 avoided cost of US$48/MtCO_2$. However, the post-combustion and oxyfuel combustion CCS option costs and performance were relatively close. In comparison to the new supercritical steam coal plant baseline, the addition of CCS reduced the power plant net capacity by about 20% and dropped its net efficiency (HHV) from 39% to 32%.

As previously explained, setting a proposed new fossil power plant as the baseline reduces the CO_2 avoidance costs, as the new power plant baseline will have relative power costs even before the additional cost of CCS. At a moderate US$48/tCO_2$ tax, the triple point electricity price would be US$113/MWh for the new coal unit with or without CCS or, if converting to a new NGCC without CCS, it would be US$11.15/MMBtu for natural gas.

Both the new and old coal power plant baseline estimates required about the same US$110/MWh power price for CCS. However, the electric costs increased by about 300% for the CCS addition to the old paid-off power plants, whereas the electric costs only increased by about 50% for CCS added to a new proposed power plant. The CO_2 avoidance cost and natural gas alternative prices (without CCS at the given CO_2 tax) were also much different for the two coal power plant baselines. Existing coal power plants would find it cheaper and a much lower risk to simply pay the CO_2 tax than to significantly reduce CO_2 emissions until the CO_2 tax is very high. Another option is to mandate

---

1   Start with estimating CO_2 avoidance costs for a coal plant baseline without CCS versus adding CCS. This is the most logical baseline due to the high CO_2 emissions per unit of coal energy, which is the dominant use in large central power plants having large point sources and a likely stable coal price even as a carbon-constrained world develops. The only major issue would be if the baseline is an existing PC power plant or a new state-of-the-art PC power plant. The CCS case can be retrofitted, rebuilt, or a new plant, and can include any or all of the CCS process options (pre-, post-, or oxyfuel combustion). Next, enter the CO_2 avoidance cost estimated in Step 1 for the same coal baseline estimate and CCS option with the lowest cost. The power cost should now be the same for the coal baseline without CCS (paying the CO_2 tax) and the best coal CCS option (avoiding most of the CO_2 tax). Finally, the third key alternative of natural gas without CCS is simply replacing coal. The natural gas replacement option is estimated by varying the natural gas price until this option (without CCS) has the same electricity price as the Step 2 coal cases (with and without CCS).

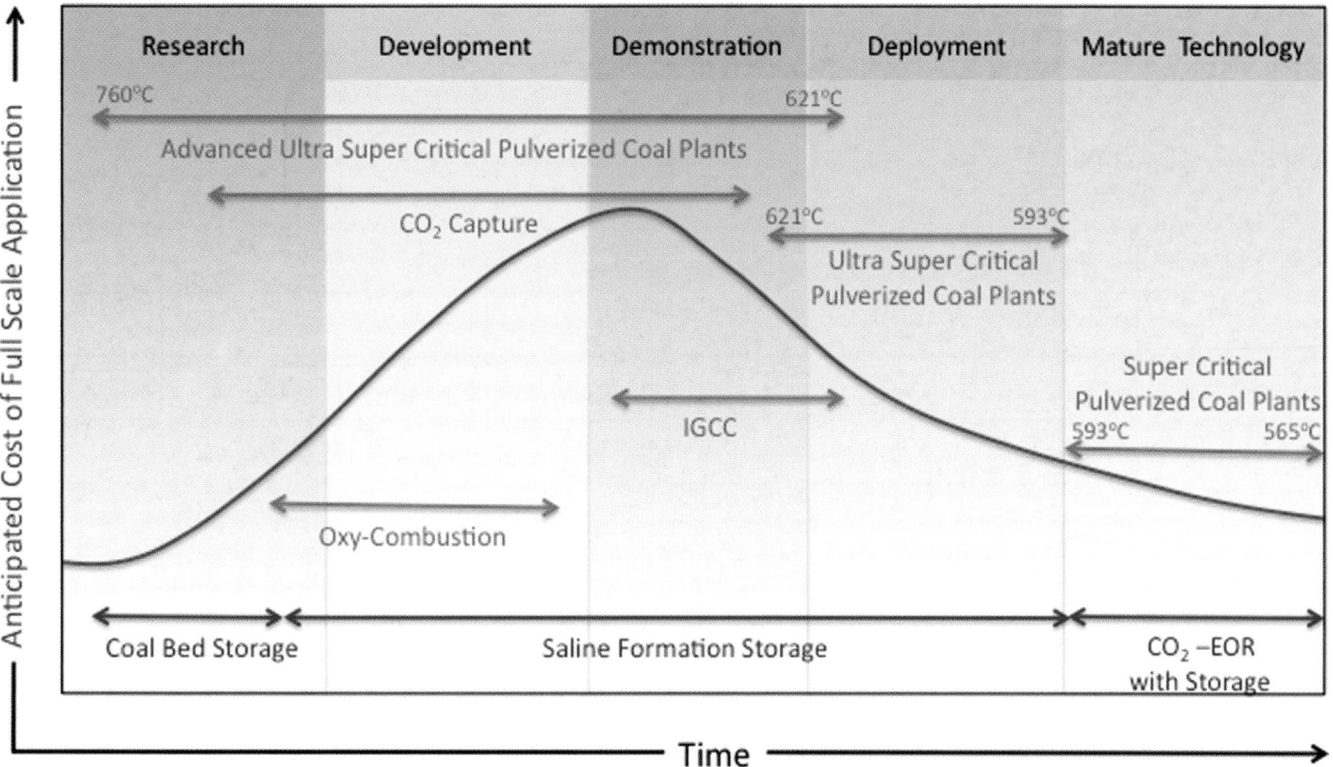

Figure 13.4 | Utility and petroleum industry perspective on CCS and advanced power generation technology for coal. Source: modified from Booras, 2009.

that old coal power plants shut down after 45 years of service, as is currently being proposed in Canada. In the meantime, ongoing talk about CO_2 mitigation encourages the life extension of existing old coal power plants with high emissions. However, life-extending old coal units maximize CO_2 emissions before potential CO_2 reduction mandates take place and increase the quantity of inexpensive CO_2 cap and trade allocations that existing coal unit owners will likely receive. This also keeps electric power prices relatively low in nations such as the United States, which have large fleets of aging coal power plants.

13.2.7.3   CCS Cost Estimating Learning Curves

There are very real learning curves associated with the increasing use of specific technologies in industries where the unit costs go down as the total installed capacity increases. This is sometimes referred to as "learning by doing." In the electric generation industry, these cost improvement "learning curves" are well documented for the growth in the use of flue gas desulfurization (FGD) SO_2 controls and selective catalytic reduction (SCR) NO_x controls (Yeh and Rubin, 2007). However, this is not always the case, as exemplified by the large cost increases for nuclear power plants (Grubler, 2010).

Learning curves are sometimes applied to CCS technologies. Only time will tell the accuracy of the estimates. Nevertheless, it is essential to realize that learning curves almost always start to reduce unit costs after the respective

technology is established in commercial service and the total commercial capacity is growing at a high rate with time. From pilot to demonstration to the first commercial unit, the unit costs almost always increase before decreasing via the learning curve. This is illustrated in Figure 13.4, which shows cost versus stated development for coal technologies and CCS options (Booras, 2009). In the early stages of development, advanced technologies often overestimate performance and underestimate costs. The less developed the CCS technology, the more uncertain the estimates.

All CO_2 mitigation options performance and cost estimates are highly questionable until successfully demonstrated in large commercial-scale operations. The time and cost to progress from research to development is long and expensive for advanced energy and environmental technologies. It is even longer and more expensive for the electric power industry. This is because highly regulated industries do not reward risks, and traditional electric power generation, especially for CCS, is more economical in very large and expensive power plants. Innovative policies and incentives will be required to change this.

13.2.8   Air Pollution from Power Plants with CCS

Contrary to CO_2, where emission levels are mainly dependent on the fuel type, non-CO_2 emissions are mainly influenced by parameters related to specific conditions such as fuel composition, type of technology, combustion, operating and maintenance conditions, size and age,

Table 13.1 | Average minimum and maximum emission factors for energy conversion concepts with and without CO_2 capture as reported in the literature. The ranges, in brackets, report the minimum and maximum values found in 171 studies.

Capture Technology	Conversion Technology	CO_2 g/kWh	NO$_x$	SO$_2$	NH3	VOC	PM
			mg/kWh				
No-capture	IGCC	766 (694–833)	229 (90–580)	64 (40–141)	n.a.	n.a.	28 (27–29)
	NGCC	370 (344–379)	168 (90–262)	n.a.	n.a.	n.a.	n.a.
	PC	826 (706–1004)	374 (159–620)	414 (100–1280)	7 (3–10)	10 (9–11)	39 (7–51)
Oxyfuel combustion	GC	10 (0–60)	n.a.	n.a.	n.a.	n.a.	n.a.
	NGCC	8 (0–12)	0	n.a.	n.a.	n.a.	n.a.
	PC	47 (0–147)	172 (0–390)	25 (0–98)	n.a	n.a.	3 (0–10)
Post-combustion	NGCC	55 (40–66)	188 (110–275)	n.a.	6 (2–19)	n.a.	n.a.
	PC	143 (59–369)	537 (205–770)	9 (1–13)	209 (187–230)	n.a.	52 (9–74)
Pre-combustion	GC	21 (0–42)	n.a.	n.a.	n.a.	n.a.	n.a.
	IGCC	97 (71–152)	209 (100–550)	28 (10–51)	n.a.	n.a.	34 (34–35)

n.a.: no data available. The emissions factors are based on various fuels and power plant configuration and performance. Post-combustion capture includes capture with amine-based solvents and chilled ammonia.

Source: Koornneef et al., 2010.

and emission control policy (EMEP, 2009). Table 13.1 shows an overview of emission factors reported in the literature for power plants with and without CO_2 capture. General trends can be summarized as follows.

13.2.8.1 Sulfur Dioxide – SO$_2$

SO$_2$ emissions are expected to be very low for power plants with CO_2 capture. This will mainly be due to plants using fossil fuels with very low sulfur content (e.g., NGCC or GCs), a limitation in SO$_2$ content in the flue gas to avoid solvent degradation (post-combustion with amine-based solvents), or to a high level of recovery of sulfur compounds from the syngas in IGCC plants.

13.2.8.2 Nitrogen Oxides – NO$_x$

In post-combustion capture concepts, NO$_2$ needs to be removed since it can react with amine-based solvents and cause degradation. NO$_2$ accounts, however, for only ~10% of the NO$_x$, making the net impact of its removal limited. A major factor will be the increased use of fuel due to the efficiency penalty. Such an increase may result in higher levels of NO$_x$ (/kWh) for the whole power sector (Tzimas et al., 2007; Koornneef

et al., 2010). Studies in the literature of pre-combustion concepts report both lower and higher NO$_x$ values, compared to similar plants without CO_2 capture (e.g., IEA GHG, 2006; Davison, 2007; US DOE/NETL, 2007a). The levels of NO$_x$ reported are dependent on the performance of the gas turbines and the efficiency penalties. NO$_x$ emissions in oxyfuel concepts are expected to be very low, due to a reduction of fuel NO$_x$ and inhibition of thermal NO$_x$ (Croiset and Thambimuthu, 2001; Tan et al., 2006) and removal of NO$_x$ during CO_2 purification (White et al., 2008).

13.2.8.3 Ammonia – NH$_3$

Changes in NH$_3$ levels are only relevant for post-combustion capture. Formation of NH$_3$ is the result of reactions between flue gases and monoethanolamine (MEA)-based solvents as a consequence of oxidation or elevated temperatures.

13.2.8.4 Particulate Matter – PM

The emission of particulate matter from natural gas-fired cycles in general can be considered negligible. In the case of coal-fired plants, significant reductions are expected for the post-combustion capture process since low levels of PM are necessary to assure a stable capture process. However, in terms of emissions/kWh, there may be an overall increase in PM emissions as a result of increased use of fossil fuel due to the efficiency penalty. PM in IGCC plant cycles without CO_2 capture is already low due to high removal efficiencies in the gas cleaning section. Preliminary estimates indicate that the application of pre-combustion capture may lower further PM emissions (US DOE/NETL, 2007a). Also, for coal-fired oxyfuel concepts, PM emissions are estimated in the literature to be lower per kWh, compared to conventional pulverized coal-fired power plants.

13.2.8.5   Volatile Organic Carbon – NMVOC

Quantitative estimates for the impact of CO_2 capture on the level of NMVOC emissions are absent in the (open) literature. It is therefore not possible to indicate whether and to what extent CO_2 capture technology will affect the net level of these emissions.

13.2.9   Water Consumption

Water consumption will increase as a consequence of additional fuel use to make up for the energy penalty and the demand of the CO_2 capture system itself. For instance, coal-fired power plants with post-combustion capture (MEA-based) have large cooling water make up requirements, while increased water demand in IGCCs with pre-combustion capture is mainly driven by the increased cooling load required to further cool the syngas and steam for the water gas shift reactor and the increase auxiliary load (US DOE/NETL, 2009a). The impact on water use appears larger for PCs and NGCCs than for IGCCs and oxyfuel. Ranges reported in the literature indicate relative increases in water use, compared to similar plants without capture, between 50–95% for PC with post-combustion capture, 30–50% for IGCC with pre-combustion capture, and 30–35% for oxyfuel with CO_2 removal (US DOE/NETL, 2007a; US DOE/NETL, 2007b; RECCS, 2008). Water use could be reduced by using dry cooling instead of wet cooling, even to levels below those of wet cooled plants without CCS. However, operating power for a given heat load in a dry cool-based system can be four to six times larger than that for an optimized wet system (Maulbetsch, 2002) and this could result in an increase in the energy penalty in the power plant.

13.2.9.1   Waste and By-products from CCS

As a consequence of increased fuel consumption, it is expected that waste streams such as bottom ash, fly ash, boiler slag, and reclaimer waste will increase with the implementation of CO_2 capture. Relative increases reported in the literature for power plants are between 20–40% (bottom ash/fly ash); 18–29% (slag); 18–25% (sulfur); and 31–47% (gypsum) (US DOE/NETL, 2007a; Davison, 2007; Rubin et al., 2007). Depending on market conditions, gypsum (from PC plants) and sulfur (from IGCC plants) may be considered sub-products instead of waste.

Reclaimer waste generated in post-combustion capture processes may become a point of concern. This waste, containing heat stable salts, heavy metal corrosion inhibitors, and absorption solvents among others, is considered hazardous. Amounts of reclaimer waste for post-combustion with amine-based solvents are reported in the range of 1.6–4.5 g/kWh (Rubin et al., 2007; Koornneff et al., 2008; Schreiber et al., 2009; Korre et al., 2010). This amount could be even higher, depending on the rate of solvent slip stream. Thitakamol et al. (2007) have indicated that an increase of 0.5–2% in the slip stream could result in a factor four increase in the amount of reclaimer waste.

13.2.10   Capture Conclusions

To make a large contribution to CO_2 emissions reductions, capture technologies are needed both for new sources and existing sources. The same technology is unlikely to be optimal for both situations. Four approaches are available for capture from the power and industrial sources. Three of them – the so-called post-combustion capture, pre-combustion capture, and oxy-combustion capture approaches – produce CO_2 gas with a relatively small concentration of contaminants. This needs to be compressed for transport to a storage location. All have been separately demonstrated; some are used routinely today for other applications, but little experience is available for integration and optimization with power production or most industrial applications. The fourth approach, mineralization, produces solids whose composition depends on the specific process. Mineralization is much less mature than the other capture technologies. Mineralization of CO_2 is an attractive option because the end product is a carbonate rock, thus avoiding the need to store large amounts of CO_2. However, mineralization requires large volumes of rocks to provide a source of magnesium (or other cations) and is more expensive and energy intensive than other options. If a use for these minerals is not found, they will become a waste that must be managed.

Today's capture and compression technology has a parasitic energy requirement of 15–30% of the electricity generated from combustion sources. This increases the cost of capture, the use of valuable energy resources, and the upstream impacts of energy supply. Progress is being made on reducing electricity requirements for capture and compression, but more progress is needed to make CCS less energy intensive.

Impacts of CO_2 capture on other air pollutants are highly dependent in the capture approach used. SO_2 emissions are expected to be very low for power plants with CO_2 capture. In the post-combustion concepts, NO_x emissions are believed to be largely unaffected by the capture process while for oxyfuel capture these emissions are expected to be very low. For pre-combustion, consensus on the possible effects seems to be absent. Only post-combustion capture NH_3 emissions are estimated to significantly increase (with more than a factor of 20 for conventional amine solvents). This aspect will have to be addressed by the facilities and, while technology already exists which could abate these emissions, it will have an impact on the investment costs. PM emissions are expected to increase per kWh as a result of the efficiency penalty. There is little experience in the power generation sector with operating capture facilities. Capture facilities are large chemical processing plants. Operating these facilities requires different expertise than operating steam boilers and turbine generators. Building the expertise and experience to integrate power production and CCS to provide electricity with the high reliability expected today will require significant capacity building in the power generation sector.

The capture of CO_2 directly from air has also been proposed. Air capture is appealing for a number of reasons, namely: capture facilities could be co-located with ideal storage locations; emissions from nonpoint sources could be captured; and emissions from noncompliant parties could be offset. However, the low concentration of CO_2 in the air makes capture with today's technology expensive and energy intensive. The high cost of direct air capture compared to capture from concentrated point sources makes it unlikely that air capture will play a role in CO_2 emission reduction until far into the future.

The cost of capture is challenging to determine, but is estimated to range from US$50 to over US$150/tCO_2 avoided, depending on the type of fuel, capture technology, assumptions about the baseline technology, and whether it is the first plant or the nth of a kind. Costs of capture for first generation CCS power plants available in the early 2020s are estimated to be about $45/$tCO_2$ avoided for coal and $115/$tCO_2$ avoided for natural gas. The combined cost of capture and compression would increase the cost of power production by about 50% to over 100%. The costs of early demonstration projects are likely to be significantly higher. Worldwide, large investments in improving capture technologies are being made by governments and the private sector. It remains to be seen whether these investments are sufficiently large to quickly reduce costs and energy requirements for capture and compression, particularly those investments in fundamental research, which could produce technological breakthroughs with dramatic improvements.

13.3   Carbon Dioxide Transport

13.3.1   Introduction

One common requirement for all CCS schemes is the need to transport CO_2 from the capture to storage sites. CO_2 can be transported by land via pipelines, motor carriers, or railway, or by water via ships or barges. Each of the individual possibilities can be considered mature, but their integration at large-scale will be complex; in fact, the scale and the speed at which transport networks will be needed are regarded as the main challenges for CO_2 transport (IPCC, 2005; IEA, 2008a).

Experiences in CO_2 transport by truck or ship are mainly found in the food and brewery industry. CO_2 is generally transported as a compressed liquid (e.g., -50°C, 0.7–0.8 MPa). The transport of CO_2 by ship or train will require the development of loading and unloading infrastructures and temporary CO_2 storage in steel tanks or rock caverns, which make the options costly. R&D is being carried out to increase the cost-effectiveness of the option, for instance by using integrated tug barges instead of ships. Although this concept has not been demonstrated, it could theoretically eliminate the need for intermediate storage, reducing cost and logistic complexity (Haugen et al., 2009). Though transport of CO_2 via ships or trains is not regarded as the preferred option for large-scale systems (e.g., IPCC 2005; MIT, 2009), it may prove to be cost effective during a transition phase (when pipeline networks are not yet available) and on a case-specific basis (specific combinations of pipeline, ships, and/or trucks).

CO_2 transport by pipeline is currently considered the most mature transport option. There are currently some 5800 km of CO_2 pipelines in operation in the United States (Parfomak and Folger, 2008). These pipelines transport CO_2 in the liquid or dense phase (CO_2 pressure above 7.38 MPa), are land-based and sectioned (typically less than 30 km), are generally made out of carbon steel, transport relatively clean CO_2, do not use internal coating, and mainly transport CO_2 for EOR purposes (Oosterkamp, 2008). The IEA estimates that about 150,000 km of dedicated CO_2 pipelines will be needed in the European Union. In the United States, up to 37,000 km of CO_2 pipelines will be needed between 2010 and 2050 (Dooley et al., 2009). Given its importance, the focus of this section is on pipeline transport.

Gas compression is well developed around the globe and uses mature technologies, as described in Section 13.2.4. CO_2 needs to be compressed generally from atmospheric pressure, at which point it exist as a gas, up to a pressure suitable for pipeline pressure in either the liquid or "dense" phase regions, depending on the temperature. In the gas phase, a compressor is required, while in the liquid/dense phase, a pump can be used to boost the pressure. It is generally assumed that the cut-off pressure for switching from compressor to pump is 7.38 MPa (McCollum and Ogden, 2006). At this pressure and at temperatures lower than 20°C, CO_2 would have a density of between 800–1200 kg/m^3. At these conditions, larger mass per unit volume can be transported, since CO_2 would behave as a liquid and have liquid-like density while having the compressibility and viscosity of CO_2 in the gas phase. As an example, a 30-inch pipeline could transport about 5 MtCO_2/yr in gas phase. In the liquid or dense phase, the same pipeline could transport about 20 Mt/yr.

13.3.2   Transportation Operational Issues

13.3.2.1   Pressure Drop

As CO_2 flows through a pipeline, there is frictional loss. For single phase flow, the pressure drop (ΔP) depends on the pipe's inner diameter, CO_2 flow velocity, viscosity, and density, and the pipe's roughness factor. At constant temperature, pressure drop in a typical 20-inch pipeline transporting CO_2 in dense phase is about 30 kPa/km (Essandoh-Yeddu and

Gulen, 2008). In order to maintain the inlet pressure to the pipeline, it is necessary either to increase the pipeline inlet pressure to levels that secure that after losses CO_2 would still have at least 7.38 MPa or to install boosters every 100–200 km to make up the pressure losses.

13.3.2.2   Corrosion

CO_2 and components such as SO_2, NO, and H_2S may form acid compounds in the presence of water that are highly corrosive. Furthermore, H_2S could react with carbon steel to form thin films of iron sulphide (FeS). FeS may dislodge at times and coat the inside surface, decreasing heat transfer efficiency and potentially causing operational problems in the compression units. Control of water content by dehydration is therefore essential for safe and cost-effective pipeline operation. Other possibilities to minimize corrosion are material selection (e.g., stainless steel instead of carbon steel), use of corrosion inhibitors, use of protective coating, and cathodic protection. These possibilities will increase pipeline costs.

13.3.2.3   Hydrate Formation

In the presence of water, CO_2 and H_2S may form hydrate compounds, which can block the line and plug and damage equipment. Hydrate formation can largely be stopped by drying the CO_2 and removing the "free water" that is present. Maximum allowable water content recommended in the literature is in the range of 50–500 ppm (Visser et al., 2007).

13.3.2.4   Operating Temperatures

CO_2 pipelines' operating temperatures are generally dictated by the temperature of the surrounding soil or water. In northern latitudes, the soil temperature could reach below zero in winter and to 6–8°C in summer, while in tropical locations, soil temperatures of 20°C are common. At the discharge of compression stations, after-cooling of compressed CO_2 may be required to ensure that the temperature of CO_2 does not exceed the allowable limits for either the pipeline coating or the flange temperature (McCoy, 2008). CO_2 cools dramatically during decompression, so pressure and temperature must be controlled during routine maintenance (Gale and Davison, 2003).

13.3.2.5   Impurities

Depending on the source of the flue gas and the type of CO_2 capture process, CO_2 streams may contain trace concentrations such as H_2S, SO_2, NO_x, O_2, CO, HF, Hg, N_2, and Ar. These impurities might have an impact on (Visser et al., 2007; McCoy, 2008; Oosterkamp, 2008):

- *the physical state of the CO_2 stream*, affecting the operation of the compressors, pipelines and storage fields;

- *CO_2 compressibility*, which is non-linear in the range of pressures common for pipeline transport and is highly sensitive to any impurities such as H_2S or CH_4. Changes in compressibility may result in the reduction of the CO_2 volume that can be transported;

- *CO_2 density*, since impurities such as N_2, O_2, H_2, Ar, and CH_4 reduce the density of the flow, which in turn reduces the CO_2 volume that can be transported;

- *pipeline integrity*, for instance, by producing hydrogen of sulphide-induced stress cracking;

- *safe exposure limits* required for the pipeline; and

- *minimum miscibility pressure* of the CO_2 in oil, which could affect the possibilities to use CO_2 for EOR.

13.3.3   Cost of Transportation

CO_2 transport costs are a function of pipeline length, diameter, material, route of the pipeline, and safety requirements, among other things. Figure 13.5 shows relations between capital costs, levelized cost, and CO_2 flows outlined in the literature.

The transport of CO_2 by pipeline benefits from economies of scale: i.e., average costs decrease as scale increases. The amount of fluid that can flow through a pipeline increases non-linearly with diameter, so larger diameters are preferred to smaller ones. Pressure drop, ΔP, is inversely related to the diameter. Lower pressure drops translate to lower costs, since costs to compress or pressurize a fluid are linearly related to ΔP. Finally, the marginal cost of constructing a pipeline decreases with diameter (Bielicki, 2008; MIT, 2009). Economies of scale regarding the transport of large amounts of CO_2 by pipeline indicate that it may be more efficient to encourage hub-and-spoke transport systems than point-to-point systems. Returns to scale are, however, not constant; i.e., they do not continually increase as the system expands (Figure 13.5b). The point at which economies of scale are reached is case dependent. Some studies found that economies of scale are reached with annual CO_2 flow rates in excess of 60 Mt/yr (e.g., van den Broek et al., 2010), while other studies report lower values, e.g., 10 Mt/yr (Heddle et al., 2003). The increase in costs with scale is partly due to the increase in material costs. McCoy (2008) reports that doubling pipeline diameter would result in a three-fold material costs increase. This is a concern for most investors, due to the high price fluctuations of steel in recent years (Parformak and Folger, 2008). For example, in late 2001, steel prices were about US$600/t. By late 2007, it was US$1400/t. Studies in the literature have also shown a high variability in pipeline costs depending on the length assumed (e.g., McCollum and Ogden, 2006; Perfomak and Folger 2008). In this regard, CO_2 pipeline costs may present the costs component in integrated CCS schemes with the greatest potential variability (IPCC 2005; MIT, 2007).

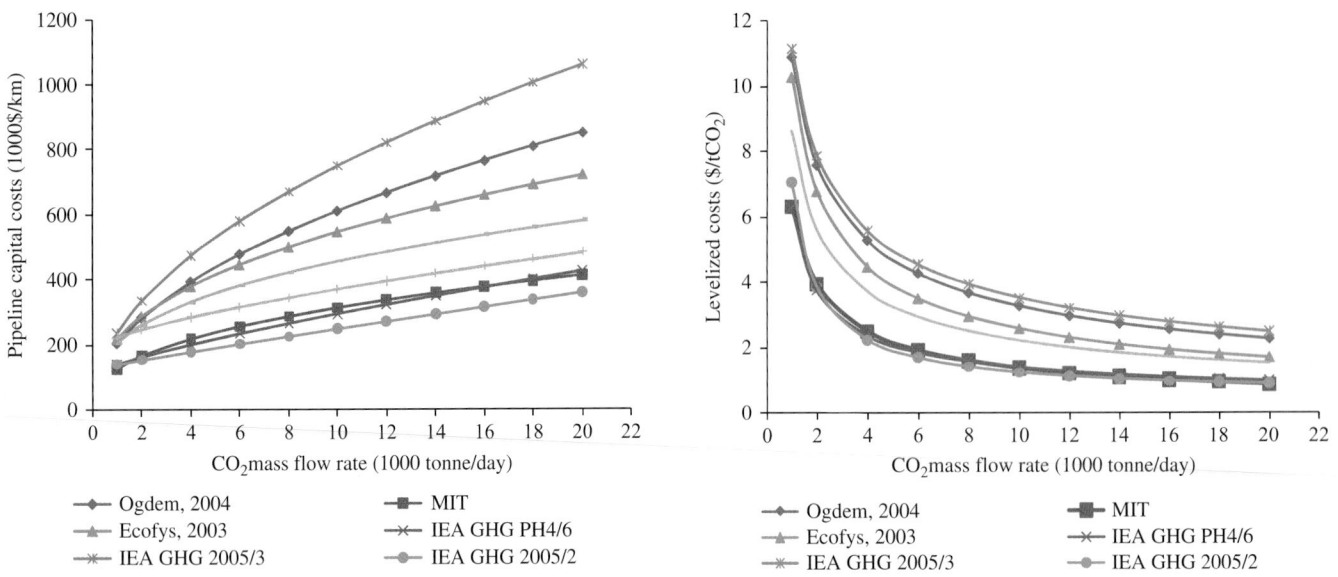

Figure 13.5 | Comparison of pipeline costs (a) and levelized costs (b) for a pipeline with a length of 100 km. Costs are reported in US$_{2005}$. Source: McCollum and Ogden, 2006.

13.3.4   Transportation Safety

Several studies indicate that CO_2 transport by pipeline does not pose a higher risk than is already tolerated for transporting hydrocarbons (e.g., Heinrich et al., 2004; Hooper et al., 2005; Damen et al., 2006). For instance, cumulative failure rates reported in the literature for CO_2 pipelines range from 0.7–6.1 per 10,000 km/yr, which is in the same range of failure rates reported for hydrocarbon pipelines (Koornneef et al., 2009). Currently there are no minimum standards for pipeline quality CO_2 for CCS purpose. Such standards could depart from those already used in existing CO_2 pipelines. However, since most of those pipelines (e.g., EOR in the United States) predominantly run through sparsely populated areas, the impact of a potential incident is limited. Existing standards and risk assessment models will have to be modified to take into account the vicinity to densely populated areas. Additional measures can include (Skovholt, 1993; Koornneef et al., 2009):

- appropriate monitoring facilities and safety systems, e.g., sectioning valves to reduce the quantity of CO_2 that could leak out and shorter distances between valves near populated areas;

- safety zones on both sides of the pipeline. Ranges recommended in the literature vary between less than 1 m and 7 km;

- increased pipe wall thickness near populated areas; and

- protection from damage (e.g., burying the line).

13.3.5   Conclusions

From a technical point of view, transport of CO_2 by pipeline and ship has successfully been conducted for many years, indicating that issues

associated with the design and operation of CO_2 transport can be addressed. Nevertheless, the transport of CO_2 for geological storage also has challenges that need to be addressed. Factors of concern for the appropriate and safe management of CO_2 pipelines include water content, hydrate formation, corrosion, two-phase flow, and toxic components. Benefits of scale indicate the development of pipeline networks as an optimal strategy for the medium and long term. The challenge remains with the scale and speed at which transport networks will need to be constructed, often through densely populated areas. Further development and validation of models that deal with the operation of CO_2 pipeline networks and the management of external safety is therefore necessary.

13.4   Carbon Dioxide Storage Science and Technology

13.4.1   Introduction

Carbon dioxide storage in deep geological formations has emerged over the past 15 years as a feasible component in the portfolio of options for reducing GHG emissions. Five commercial projects now operating provide valuable experience for assessing the efficacy of carbon storage. These projects, in addition to more than 100 CO_2-EOR located primarily in North America, provide a growing experience base for assessing the potential for geological storage. However, if CCS is implemented on the scale needed for large reductions in CO_2 emissions, one Gt or more of CO_2 will be sequestered annually – a 250-fold increase over the amount sequestered annually today. Effectively sequestering these large volumes requires building a strong scientific foundation for predicting the coupled hydrological-geochemical-geomechanical processes that govern the long-term fate of CO_2 in the subsurface. In addition, methods to characterize and select storage sites, subsurface engineering to optimize performance and cost, safe operations, monitoring technology, remediation methods, regulatory oversight, and country-specific institutional approaches for managing long-term liability will all be needed.

13.4.1.1 Options for the Storage of CO_2

Geological formations are not the only option considered for CO_2 storage. Beginning in the early 1990s, there was a great deal of interest in storing CO_2 in the ocean. Two different approaches were pursued: biological sequestration via ocean fertilization and direct injection of a concentrated stream of CO_2.

Scientific experiments have been conducted to evaluate whether adding iron to the ocean could increase biological productivity, thus increasing the rate of ocean uptake CO_2. In 2001, the Southern Ocean Iron Experiment was conducted in the southern Pacific (Buesseler et al., 2004). Results from this and similar experiments showed rapid increases in biological productivity, but many questions remain regarding long-term ecosystem impacts and the effectiveness of this technique for lowering atmospheric CO_2 concentrations. Consequently, at the present time, ocean fertilization is not under serious consideration for large-scale CCS.

CO_2 can also be injected into the mid-depth ocean (1000–3000 m deep), which enables storage for hundreds to thousands of years before returning to the atmosphere via ocean circulation. The injected CO_2 would dissolve and be transported with ocean currents. Alternatively, it could be injected near the ocean bottom, to create stationary "pools" of CO_2. The potential capacity for ocean storage is large – on the order of a trillion tonnes of CO_2 (IPCC, 2005). Little ocean storage research is still being actively pursued. Concerns about unknown biological impacts, high costs, impermanence of ocean storage, and concerns regarding public acceptance have decreased interest and investment in this technology over the past five years.

Hybrid storage schemes that rely on a combination of ocean storage and geological storage have also been proposed recently (Schrag, 2009). For a sufficiently cold ocean, at water depths of greater than 3000 m CO_2 transitions from being lighter than water to heavier than water. Under these conditions, CO_2 would remain on the ocean bottom; however, over time the CO_2 would dissolve into the ocean water, leading to ocean acidification and gradual release back into the atmosphere. To prevent dissolution of CO_2, it is proposed to inject the CO_2 under a thin layer of ocean bottom sediments, thus combining aspects of geological and ocean storage. This relatively new idea is the subject of research and is much less well developed than conventional geological storage.

It has also been suggested that CO_2 storage could be combined with subsea production of methane hydrates (Ohgaki et al., 1996). As methane is being released from the hydrate structure, CO_2 could replace it. In principle not only would this provide a secure storage option, but it would increase hydrate production as well. This idea is in the early stages of conceptual development.

Due to its comparative maturity, geological storage will be the focus of the remainder of this section.

13.4.2 Geological Storage in Deep Underground Formations

13.4.2.1 Technology Description

Under a thin veneer of soils or sediments, the earth's crust is made up primarily of three types of rocks: igneous rocks, formed by cooling magma from either volcanic eruptions or magmatic intrusions far beneath the land surface; sedimentary rocks, formed as thick accumulations of sand, clay, salts, and carbonates over millions of years; and metamorphic rocks of either origin that have undergone deep burial with accompanying pressure and thermal alteration.

To date, sedimentary rocks, located in so-called sedimentary basins, have been the primary focus for geological storage of CO_2 because the storage on geological timescales has already been proven through the presence of oil and gas accumulations in them. Sedimentary basins underlie much of the continents and are co-located with many major large CO_2 emission sources (see Figure 13.6). Until recently, storage in sedimentary basins has been the exclusive focus of geological storage technology and capacity assessments. However, in the past five years there has been a significant effort to understand the potential of volcanic rocks, primarily basalt, for the storage of CO_2 (McGrail et al., 2006; Kumar et al., 2007). Experiments testing the feasibility of storage in basalt are underway in Iceland, India, and two locations in the United States. Motivation for evaluating storage in basalt is two-fold: first, some countries with large CO_2 emissions, such as India, Brazil, and the United States, are underlain primarily by basaltic rocks, and second, it is hypothesized that a large fraction of the stored CO_2 would be converted to stable minerals, assuring permanent storage.

Sedimentary basins often contain many thousand meters of sediments where the tiny pore spaces (10^{-3}–10^2 µm) in the rocks are filled with saltwater (saline aquifers) and where oil and gas reservoirs are found. An example of a cross section through a sedimentary basin is shown in Figure 13.7. Sedimentary basins consist of many layers of sand, silt, clay, carbonate, and evaporite (rock formations composed of salt deposited from evaporating water). The sand layers provide storage space for oil, water, and natural gas. The silt, clay, and evaporite layers provide the seal that can trap these fluids underground for millions of years and longer. Geologic storage of CO_2 would take place deep in sedimentary basins trapped below silt and clay layers, much in the same way that oil and natural gas are trapped today (Holloway, 1996; Gunter et al., 2004). Within sedimentary basins, possible storage formations include oil reservoirs, gas reservoirs, saline aquifers, and even coalbeds (see Figure 13.8).

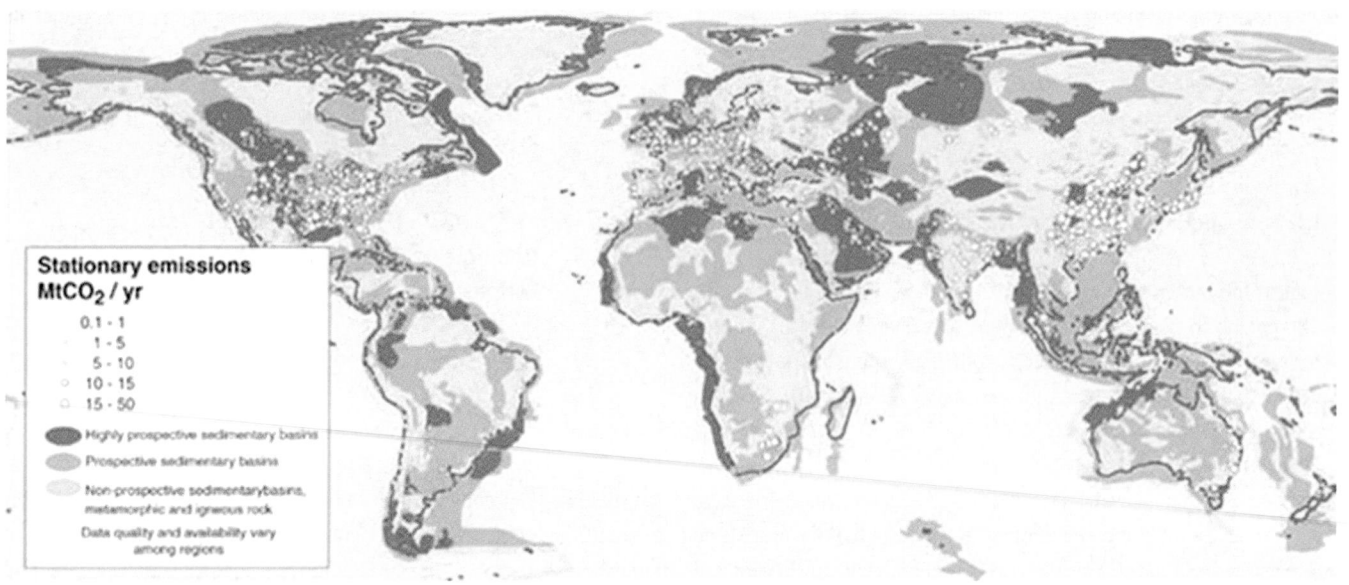

Figure 13.6 | Geographical relationship between CO_2 emission sources and prospective geological storage sites. The dots indicate CO_2 emission sources of 0.1–50 $MtCO_2$/yr. Prospectivity is a qualitative assessment of the likelihood that a suitable storage location is present in a given area based on the available information. This figure should be taken as a guide only, because it is based on partial data, the quality of which may vary from region to region, and which may change over time and with new information. Source: IPCC, 2005.

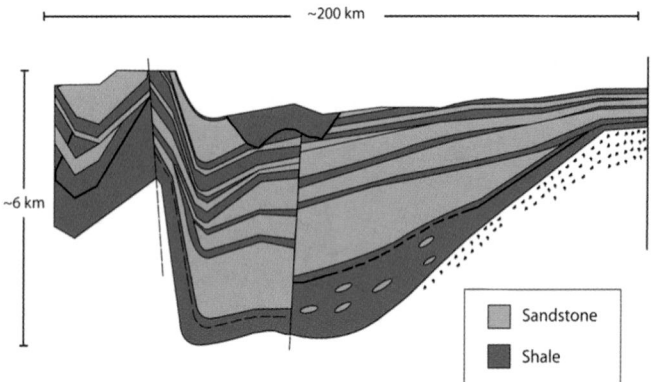

Figure 13.7 | Cross-section of the San Joaquin Valley, California, showing alternating layers of sand and shale. Sand layers provide storage reservoirs; shale layers provide seals.

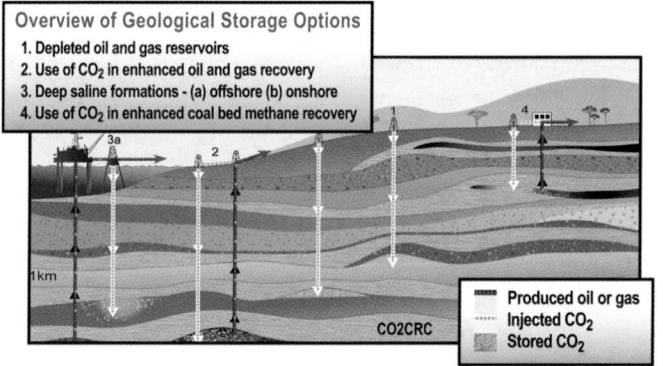

Figure 13.8 | Options for geological storage within a sedimentary basin, including oil and gas formations, saline aquifers, and coalbeds. Source: modified from IPCC, 2005.

The presence of an overlying, thick, and continuous layer of shale, silt, clay, or evaporite is the single most important feature of a geologic formation that is suitable for geological storage of CO_2. These fine-textured rocks physically prevent the upward migration of CO_2 by a combination of viscous and capillary forces. Oil and gas reservoirs are found under such fine-textured rocks and the mere presence of the oil and gas demonstrates the presence of a suitable reservoir seal. In saline aquifers, where the pore space is initially filled with water, after the CO_2 has been underground for hundreds to thousands of years, chemical reactions will dissolve some or all of the CO_2 in the saltwater and eventually some fraction of the CO_2 will be converted to carbonate minerals, thus

becoming a part of the rock itself (Gunter et al., 2004). These so-called secondary trapping mechanisms that continue to increase storage security as time goes on have been the subject of significant scientific research over the past five years, with hundreds of relevant publications (Benson and Cole, 2008).

Capillary trapping, sometimes referred to as residual gas trapping, occurs primarily after injection stops and water begins to imbibe into the CO_2 plume. The mechanism immobilizes CO_2, slowing migration towards the surface as the trailing edge of the plume is immobilized. This mechanism is particularly important for storage in dipping formations that do not have structural closure. Studies by Ide et al. (2007) and Hesse et al. (2008) suggest that eventually all the CO_2 in a plume can be immobilized by this mechanism and develop analytical approaches for predicting how quickly this happens and how far the leading edge of the plume moves before it is immobilized.

The dissolution of CO_2 (and other flue gas contaminants) in saline aquifers can lead to the mechanism known as "solubility trapping." The solubility depends on several factors, most notably pressure, temperature, and salinity of the brine (see e.g., Spycher et al., 2003; Lagneau et al., 2005; Oldenburg, 2005; Koschel et al., 2006). For typical storage conditions, the solubility of CO_2 in brine ranges from about 2–5% by mass. Bench-scale experiments have shown that the dissolution of CO_2 is rapid at high pressure when the water and CO_2 share the same pore space (Czernichowski-Lauriol et al., 1996). However, in a real injection system, dissolution of CO_2 may be limited due to the variability of the contact area between the CO_2 and the fluid phase. The principal benefit of solubility trapping is that once the CO_2 is dissolved, it decreases the amount of CO_2 subject to the buoyant forces that drive it upwards. The amount of solubility trapping is expected to increase over periods of hundreds to thousands of years because the density of CO_2-saturated brine is several percentage points higher than the *in situ* brine, leading to convectively enhanced mixing and dissolution.

The third type of secondary trapping is known as "mineral trapping," which occurs when acidic brines enriched in dissolved CO_2 react directly or indirectly with minerals in the geologic formation, leading to the precipitation of stable secondary carbonate minerals (Gunter et al., 2004). This mechanism is potentially attractive because it could immobilize CO_2 permanently. However, a significant degree of mineral trapping could take thousands of years due to sluggish rates of silicate mineral dissolution and carbonate mineral precipitation, so the overall impact may not be realized until far into the future. Moreover, the amount of mineral trapping depends heavily on the mineralogical makeup of the storage reservoir rock. Rocks with large fractions of feldspar minerals are expected to have a significant amount of mineral trapping, while quartz-dominated reservoirs may have little to no mineral trapping.

Schematics illustrating the relative contribution of each of these mechanisms and the consequent increase in storage security over time are shown in Figures 13.9 and 13.10. The range of trapping contributions from each of these processes is highly site specific.

One of the key questions for geologic storage is: how long will the CO_2 remain trapped underground? There are a number of lines of evidence that suggest that for well-selected and -managed storage formations, retention rates will be very high and more than sufficient for the purpose of avoiding CO_2 emissions into the atmosphere (IPCC, 2005). Specifically:

- Natural oil, gas, and CO_2 reservoirs demonstrate that buoyant fluids such as CO_2 can be trapped underground for millions of years.

- Industrial analogues such as natural gas storage, CO_2-EOR, acid gas injection, and liquid-waste-disposal operations have developed methods for injecting and storing fluids without compromising the integrity of the caprock or the storage formation.

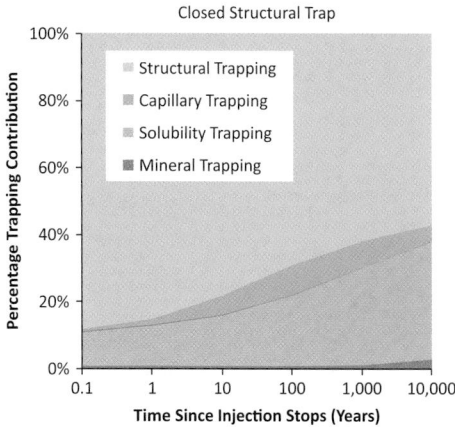

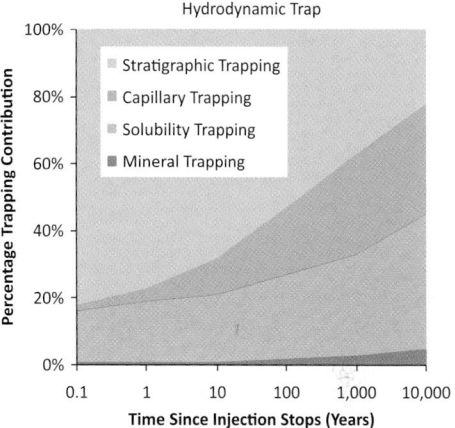

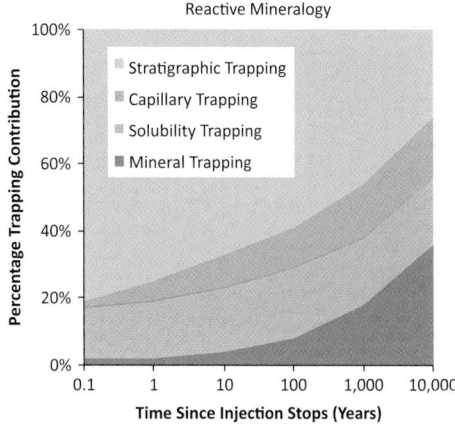

Figure 13.9 | Conceptual schematic showing trapping mechanisms and their evolution over a 10 thousand year period, as expressed as a percentage of the total trapping contribution (modified from IPCC, 2005). The relative importance of each trapping mechanism will be different depending on the attributes of the formation. For example, in a closed structural trap, which provides excellent containment beneath a dome-shaped seal, the secondary trapping mechanisms are comparatively small. In a formation with a dipping seal where the CO_2 moves upgradient due to buoyancy effects, the CO_2 will dissolve more quickly and a large fraction be subject to capillary trapping. For geological formations with a large fraction of reactive minerals such as feldspar or olivine, a significant fraction of the CO_2 may be converted to minerals.

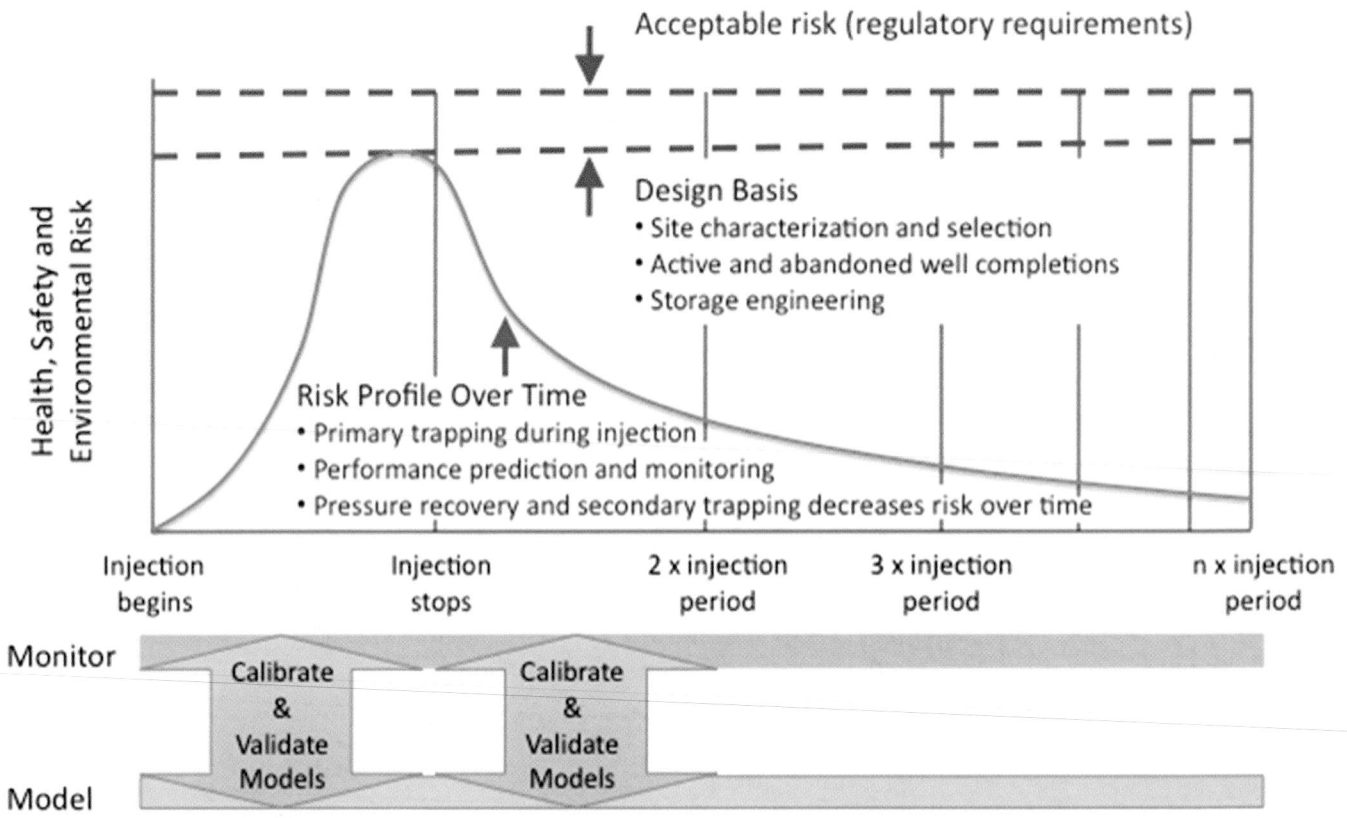

Figure 13.10 | Conceptual schematic illustrating the anticipated magnitude of health, safety and environmental risks over the lifetime of a typical geological storage project. Performance specifications, or acceptable risks, will be set by regulatory authorities. Projects will be designed to conform to regulatory requirements – or even lower depending in the design specifications. Primary risk management tools include site characterization and selection, identification and assessment of abandoned wells (including potential remediation), and storage engineering to ensure CO_2 containment and management of injection pressures. Actual risks will change over time, with growing risks during the early stages of the project, as CO_2 is first injected into the storage reservoir. Eventually, information gained from the combination of performance modeling and acquisition of monitoring data will provide assurance that the project is conforming to the design specifications – or remediation measures will be taken to address unforeseen risks. After injection stops, the pressure in the storage reservoir will begin to decrease, lessening the risk of CO_2 leakage or brine migration. Over time, as indicated in Figures 13.9a-c, secondary trapping mechanisms will further reduce the risks of health, safety and environmental impacts.

- Multiple processes contribute to long-term retention of CO_2, including physical trapping beneath low permeability rocks, dissolution of CO_2 in brine, capillary trapping of CO_2, adsorption on coal, and mineral trapping. Together, these trapping mechanisms increase the security of storage over time, thus further diminishing the possibility of potential leakage and surface release.

- Experiences with projects having large amounts of monitoring data, such as the Sleipner Project in the North Sea and the Weyburn Project in Saskatchewan, Canada, have demonstrated a high degree of containment.

The technology for storing CO_2 in deep underground formations is adapted from oil and gas exploration and production technology. For example, technologies to drill and monitor wells that can safely inject CO_2 into the storage formation are available. Methods to characterize sites are fairly well developed. Models are available to predict where the CO_2 moves when it is pumped underground, although more work

is needed to further develop and test these models, particularly over the long timeframes and large spatial scales envisioned for CO_2 storage. Monitoring of the subsurface movement of CO_2 is currently being successfully conducted at several sites, although again, more work is needed to refine and test monitoring methods.

13.4.2.2   Existing and Planned CO_2 Storage Projects

Maps showing the location of current and planned CO_2 storage projects are shown in Figures 13.11a and 13.11b. Many of these are pilot tests or small-scale demonstrations. Today, each of five commercial projects store from about 1–3 $MtCO_2$/yr in deep geological formations at Sleipner, Weyburn, In Salah, Powder River Basin, Wyoming and Snøhvit (Torp and Gale, 2003; Riddiford et al., 2003; Moberg et al., 2003). The Sleipner, In Salah, and Snøhvit projects were designed with CCS as their primary purpose. The Weyburn and Wyoming projects designed initially as an enhanced oil recovery project but has evolved into a project that combines enhanced oil recovery with CO_2 storage. Today,

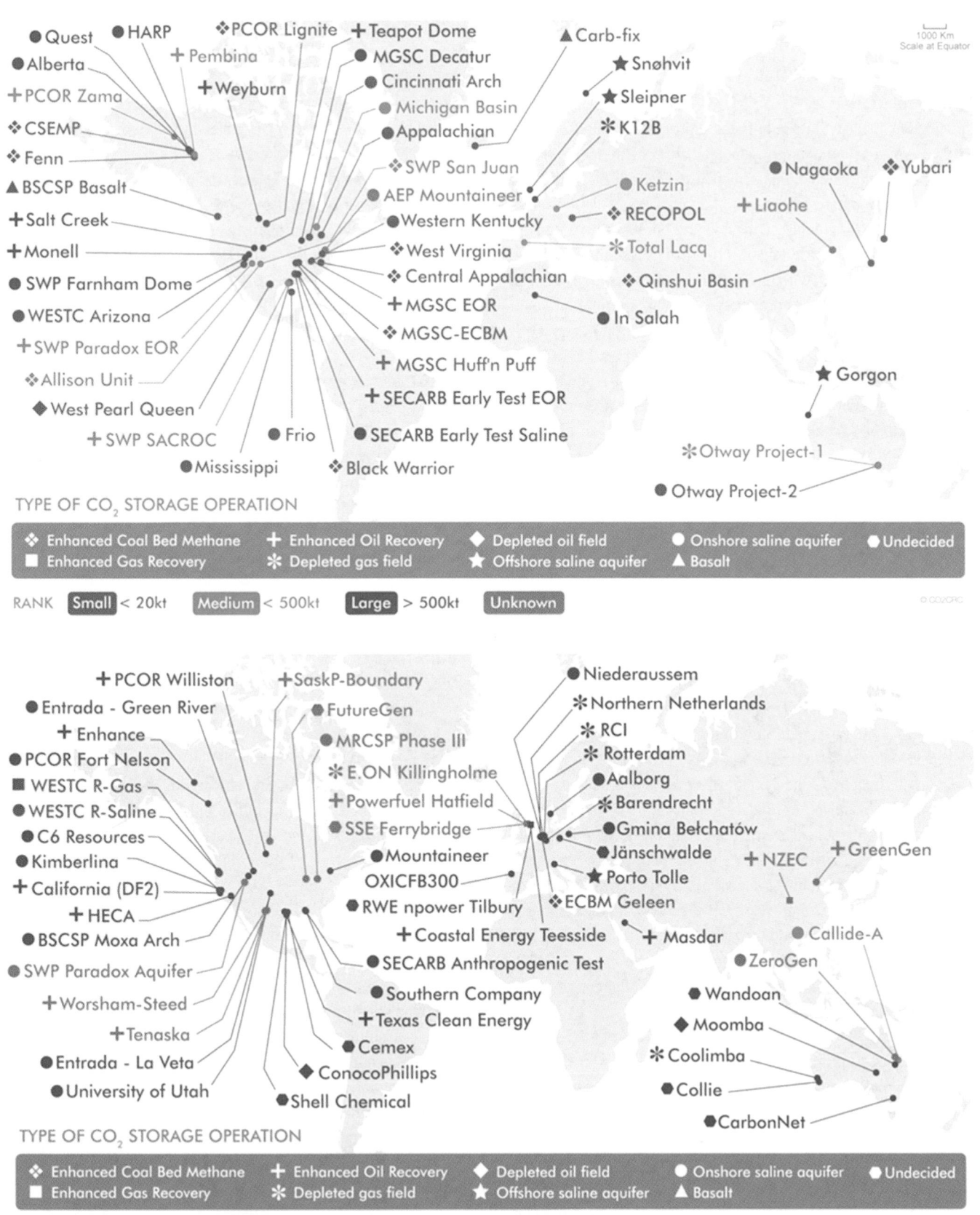

Figure 13.11 | (a) Location of existing CCS projects, including commercial projects, pilot tests, and demonstration projects. (b) Location of planned CCS projects, including commercial projects, pilot tests and demonstration projects (ES1). Source: courtesy of CO2CRC, 2011.

over 30 years of cumulative experience have been gained from these projects.

Sleipner Project, Norway

Beginning in 1996, Statoil initiated storage of approximately 1 $MtCO_2$/yr in a saline aquifer underneath the North Sea. Injection is carried through a single well screened near the base of the 200 m thick Utsira sandstone, at a depth of about 1 km beneath the sea level. The formation is highly permeable and porous, with a number of thin layers of shale distributed over the thickness of the formation. Injection has taken place with few difficulties. A comprehensive seismic monitoring program has provided invaluable information about the movement of CO_2 in the subsurface, demonstrating that CO_2 is securely trapped beneath a thick shale bed (Arts et al., 2010). After 16 years of injection, the plume covers an extent of about 15 km^2 and is moving slowly below the base of the seal. Numerous studies have compared predictions from computer models to observed migration, with reasonably good results as long as the geological complexity of the storage reservoir is included in the model (Bickle et al., 2007). Overall, the Sleipner Saline Aquifer Storage Project has done a great deal to generate confidence in the geological storage of CO_2.

Weyburn Project, Canada

The Weyburn Project in Saskatchewan, Canada combines CO_2-EOR and storage in a carbonate oil reservoir located north of the United States-Canada border. Beginning in 2000, PanCanadian Oil Company, and later Encana, initiated injection of a total of 2–3 $MtCO_2$/yr containing about 2% H_2S from the Dakota Gasification Plant into an array of wells designed to optimize oil recovery. The IEA Greenhouse Gas Program coordinates an extensive monitoring effort, including seismic imaging, geochemical sampling of produced brine, and soil gas sampling. The oil field is situated under a thick seal of anhydrite (salt formation), which is expected to provide an excellent seal. Results from the monitoring program document the movement of CO_2 in the reservoir (White el al., 2009), geochemical interactions between CO_2 and the carbonate rocks (Emberly et al., 2005), and the lack of detectable changes in soil gas composition associated with injection (Wilson and Monea, 2004). Overall, the Weyburn Project demonstrates effective retention of CO_2 in the oil field.

In Salah Project, Algeria

The In Salah Project in Algeria, initiated by BP, Sonatrech, and Statoil, is the most technologically complex storage project undertaken to date. CO_2 separated from natural gas is pumped back into the flanks of the gas reservoir from which the CO_2 is produced. Due to the low permeability of the reservoir rocks, three horizontal wells (with open intervals of 1000–1500 m) are used to inject CO_2 into the 20-meter thick reservoir at a depth of about 1800 m. Since 2004, about 0.7 $MtCO_2$/yr have been injected into the reservoir. In 2008, satellite-based land surface deformation monitoring enabled mapping of the region where pressure buildup was occurring, as documented

by uplift on the order of 2 cm/yr. This discovery, confirmed now by many research groups, provides a new method for monitoring CO_2 storage projects (Vasco et al., 2008). These observations, combined with the detection of a small leak up an exploratory well, suggested that the flow of CO_2 is in part controlled by fractures in the reservoir (Iding and Ringrose, 2010). After the leak was detected, injection in the closest well was stopped while the exploratory well was plugged with cement and abandoned properly. Injection was started again after the well was plugged. The In Salah Project provides an interesting case study regarding the ability to detect and stop leaks, as well as illustrating how subsurface complexity controls CO_2 movement in the storage reservoir. Seismic data has been collected and is now being interpreted with regard to CO_2 movement.

Snøhvit Project, Norway

The Snøhvit Project, located in the Barents Sea north of Norway, is the most recently initiated project, with injection starting in 2008. Approximately 0.7 $MtCO_2$/yr is injected into a sub-sea saline aquifer situated underneath the producing gas reservoir. The storage facility is associated with a large liquefied natural gas project. The injectivity was initially lower than expected. The large pressure buildup leading to low injectivity was attributed to compartmentalization of the storage reservoir and a remedy has been implemented (Eiken et al., 2011).

CO_2 EOR Projects

Vast experience pumping CO_2 into oil reservoirs also comes from nearly 30 years of CO_2-EOR, with nearly 50 Mt injected in 2010. About 100 projects are underway worldwide, with the vast majority in North America. When CO_2 is pumped into an oil reservoir, it mixes with the oil, lowering the viscosity and density the oil. Under optimal conditions, oil and CO_2 are miscible, which results in the efficient displacement of oil from the pore spaces in the rock. An estimated increase in oil recovery of 10–15% of the initial volume of oil-in-place is expected for successful CO_2-EOR projects. Not all of the injected CO_2 stays underground, as 30–60% is typically produced back with the oil. On the surface, the produced CO_2 is separated from the oil and re-injected into the reservoir. If CO_2 is left in the reservoir after oil production stops, most of the CO_2 injected over the project lifetime remains stored underground. The majority of CO_2-EOR projects today uses CO_2 from naturally occurring CO_2 reservoirs. The high cost and limited availability of CO_2 has restricted the deployment of CO_2-EOR to those areas with favorable geological conditions and a readily available source of CO_2. A few projects use CO_2 captured from industrial sources, notably the Weyburn Project discussed above and the Salt Creek Project in Wyoming, which is injecting several million tonnes per year.

The recent high oil prices have spurred interest in expanding the application of CO_2-EOR. This possibility, together with prospects for obtaining tradable credits for storing CO_2, has attracted considerable interest by the oil and gas industry.

Other Planned Projects, Pilot Tests, and Small-Scale Demonstrations

As shown in Figure 13.11, there are more than 280 pilot tests, demonstration projects, and commercial projects that are underway or in the planning stages. Most of these are small-scale pilot tests, supported by government-industry partnerships and designed primarily to gain experience with site selection and monitoring and to increase our understanding of the behavior of CO_2 in the subsurface. Launching a commercial scale storage project takes many years and in some cases a decade or more, as a result of the need to characterize the site, obtain injection permits, secure financing, drill wells, and manage all of the other components of these large-scale engineering endeavors. Several large projects, notably the Gorgon Project in Australia (3–4 $MtCO_2$/yr), are under construction and are expected to begin operations in the next five years.

13.4.3 Storage Capacity Estimation

The distribution and capacity of storage reservoirs are two of the most important parameters with regard to the widespread deployment of CCS technology. If storage capacity is not sufficient or it is not co-located with large emission sources, CCS is unlikely to play a major role in reducing CO_2 emissions. The IPCC (2005) gathered worldwide capacity estimates and concluded that there is sufficient capacity to meet at least one hundred years (technical potential ranging from 200–2000 Gt) of needed storage capacity, but that the distribution was variable and not all nations or regions had equally good potential for storage. The storage capacity in oil and gas formations ranged from 675–900 Gt, in coalbeds from 3–200 Gt, and in saline aquifers from 1000–10,000 Gt (IPCC, 2005).

In 2005, the quality of information about storage capacity was highly variable, with only a few regions having undertaken systematic capacity assessments. Moreover, the IPCC report also highlighted the uncertainties with respect to the storage capacity of saline aquifers. No generally accepted methodology for saline aquifer capacity assessment was available (Bachu et al., 2007). Since that time, many nations have undertaken systematic capacity assessments, using more clearly documented methods (e.g., US DOE, 2010). Specifically, the majority of saline aquifer capacity assessments are carried out by identifying locations where the following criteria are met:

- deep sedimentary basins where permeable strata are present;
- depths are greater than 800–1000 m;
- seals are present above the storage reservoir;
- salinity exceeds specific standards (e.g. 10,000 ppm in the United States); and
- access is not otherwise precluded due to other beneficial uses (e.g., military bases, parks, etc.)

After these screening criteria are applied, the mass of CO_2 that can be stored is calculated by equations of the following type:

$$Capacity = \rho_{CO_2} \cdot A \cdot T \cdot \Phi \cdot E$$
$$where:$$
$$\rho_{CO_2} = density\ of\ CO_2\ (kg/m^3)$$
$$A = Area\ (m^2)$$
$$T = net\ sand\ thickness\ (m)$$
$$\Phi = porosity$$
$$E = Efficiency\ Factor \tag{2}$$

The efficiency factor typically ranges from 0.01–0.05, or from 1–5% of the pore space of the rocks (NETL, 2010a). The efficiency factor is estimated based on the numerical simulation of a wide range of storage scenarios. Variations on these kinds of volumetric capacity estimation methods that account for a number of other factors are also being developed and applied. These factors include considering those areas where there is a structural closure that limits the migration of CO_2 (European method; USGS, 2010); multiphase flow dynamics (Juanes et al., 2010); and the amount of CO_2 dissolved in the saline aquifer. Capacity estimates using these different methods can vary significantly, particularly when accounting for only those regions with structural closure. Consequently, while much progress has been made with regard to capacity estimation, many uncertainties remain.

One of the largest questions regarding storage capacity is the extent to which saline aquifers without structural closure can be used for storage. Structural closure refers to the presence of a typically domed-shaped seal where buoyant fluids can accumulate. Arguments in favor of using these formations without closure rely on scientific theories indicating that CO_2 will be immobilized by a combination of capillary trapping, solubility trapping, and mineral trapping (IPCC, 2005). These processes gradually reduce the fraction of CO_2 that is free to continue moving after injection has stopped (Ide et al., 2007; Hesse et al., 2008; MacMinn and Juanes, 2009). In theory, once CO_2 stops moving, structural closure is not needed to securely trap CO_2 over geological time periods. Arguments against using portions of saline aquifers without structural closure focus on uncertainties about the extent and timing of these secondary trapping mechanisms. If these processes do not effectively immobilize the CO_2 quickly enough, CO_2 could migrate into drinking water aquifers or be released to the atmosphere. Additional understanding of these trapping processes through a combination of theoretical, laboratory, and field research is needed. The monitoring of existing projects described in Section 13.4.2.2 will also provide crucial insights into the long-term mobility of CO_2 in the subsurface.

Over the past several years, another issue has arisen with regard to capacity estimation. Several authors have suggested that pressure buildup in the storage reservoir will limit storage capacity (van der Meer and van Wees, 2006; Economides and Ehlig-Economides, 2009; Birkholzer and Zhou, 2009). For small- to moderate-size reservoirs that are completely sealed above, below, and on all sides, pressure buildups can be very large if even a small fraction of the pore space (~1% or more) is filled with CO_2, resulting from the extremely low compressibility of water. Overly large pressure buildup can lead to hydraulic fracturing of the reservoir and seal, thus limiting the maximum pressure buildup in the storage reservoir. For completely sealed reservoirs, pressure buildup would indeed limit storage capacity. The importance of pressure buildup constraints on capacity will depend on the degree to which completely sealed reservoirs are used for storage and how many such reservoirs exist. Most reservoirs are not completely sealed and only a top seal is absolutely essential and, perhaps, desirable. Additionally, several authors have also shown it is possible to extract water while storing CO_2 to control the extent of pressure buildup (e.g., Guénan and Rohmer, 2011). Consequently, the extent to which pressure buildup will limit storage capacity remains to be seen. Importantly, pressure limits on storage capacity should not be confused with pressure limits on injectivity, which is a very real constraint on the rate that CO_2 can be injected into a well.

13.4.3.1 Regional and Global Geological Storage Capacity Estimates

Table 13.2 summarizes the results from regional and national capacity assessments. Capacity estimates for oil and gas formations range from 996–1150 $GtCO_2$, coalbeds from 93–150 $GtCO_2$, and saline aquifers from 3963–23,171 $GtCO_2$. Current estimates suggest that global capacity will be at the high end of the range estimated in IPCC (2005).

Table 13.2 | Compilation of current storage capacity estimates. (Note this is not a complete accounting because capacity assessments have not been completed in many areas of the world. Additionally, common methods are not used and significant uncertainties in capacity assessment methodology remain. Different methodologies and assumptions have been used and comparisons between capacities for different regions should be viewed with caution.)

Region	Estimated Storage Capacity ($GtCO_2$)				Source	Note
	Depleted Oil and Gas Reservoirs	Saline Formations	Coal Seams	TOTAL		
North America	143	1653–20,213	60–117	1856–20,473	1	
Latin America	89	30.3	2	NA	14	a
Brazil	NA	2000	0.2	2000.2	2	
Australia	19.6	28.1	11.3	59	3, 4	b
Japan	0	1.9–146	0.1	2–146.1	5, 6, 14	
Centrally Planned Asia and China (CPA)	9.7–21	110–360	10	1445 -3080	7, 8, 9, 10, 17	c
Other Pacific Asia (PAS)	56–188	NA	NA	56–188	11, 12	d
South Asia (SAS)	6.5–7.4	NA	0.36–0.39	6.86–7.79	12	e
Former Soviet Union (FSU)	177	NA	NA	177	13	f
Sub-Saharan Africa	36.6	34.6	7.6	48.3	14	g
Middle East and North Africa	439.5	9.7	0	449.2	14	
Europe	20.22–30	95.72–350	1.08–1.5	117–381	15, 16	h
World	996 – 1150	3963 – 23,171	93 – 150			i

(a) All countries in Central and South America, including Brazil.

(b) The storage capacities are only in Australia. There is no storage capacity assessment in New Zealand.

(c) The numbers correspond to the storage capacity in China only. The numbers are not additive since they come from different references.

(d) The numbers correspond to the storage capacity in Indonesia only. No information was available for other countries.

(e) The numbers correspond to the storage capacity in India, Pakistan, and Bangladesh. The same report states that significant storage capacity exists in saline formations, but it has not been quantified yet.

(f) Does not include Estonia, Latvia, and Lithuania, even though they are former Soviet states. The capacity in the aforementioned states is considered in the European Union region.

(g) Includes Eastern, Western, and Southern Africa (not Northern Africa)

(h) Not all European countries are included in the Geocapacity Study. The countries that are NOT included in the Geocapacity study (specifically Finland, Ireland, Portugal, Switzerland, and Sweden) were individually studied, and no information was found about their CO_2 storage capacity.

(i) Equal to the sum of the values in Table 13.2 for oil and gas field, coalbeds and saline formations.

Source: (1) Carbon Sequestration Atlas of the United States and Canada, Second Edition, 2010 (2) Ketzer et al., 2007 (3) Carbon Storage Taskforce, 2009 (4) Bradshaw et al., 2004 (5) Nakanishi et al., 2009 (6) Takahashi et al., 2009 (7) PetroChina Company Limited, 2007 (8) Wang, L., 2010 (9) Lou, 2008 (10) APEC, 2005 (11) Indonesia CCS Study Working Group, 2009 (12) IEA GHG, 2008 (13) Zakharova, 2005 (14) Hendriks et al., 2004 (15) EU Geocapacity, 2008 (16) Anthonsen et al., 2009 (17) Dahowski et al., 2009.

13.4.4 Risks of Geological Storage of CO_2

Risks of CO_2 storage are typically separated into two broad categories: 1) risks associated with the release of CO_2 back into the atmosphere, and 2) health, safety, and environmental risks associated with the local impacts of the storage operations and potential leakage out of the storage reservoir.

The consequences of releases of CO_2 back into the atmosphere are that CCS would be less effective as a mitigation measure than anticipated and there would be financial liabilities associated buying credits or otherwise assuming responsibility for those emissions. Ultimately, these kinds of risks are associated with the duration and security of storage, which are addressed in Section 13.4.4.1. A particular complexity of CCS is that these financial liabilities are perceived to persist over periods of up to several hundred years or longer, but, as discussed in Section 13.4.4.1, the risk of releases to the atmosphere are greatest during the period of CO_2 injection, which for any particular project are limited to decades. Additionally, as discussed in Section 13.4.5, legal and administrative mechanisms for managing long term liabilities, well beyond the period of operation and post-injection assurance monitoring are being developed by governments worldwide.

The local health, safety, and environmental concerns of a CO_2 storage project are similar to those typically associated with producing oil and gas fields, such as road traffic, noise, habitat fragmentation, and infrequent uncontrolled releases from wells. In addition, if CO_2 or brine leaks out of the storage reservoir it may affect groundwater quality, and result in locally hazardous concentrations of CO_2 in the air and microseismicity if injection pressures are very high. The local health, safety, and environmental concerns for CO_2 storage projects are discussed in Section 13.4.4.2.

Because of the operational and technological similarities, the probability and consequences of risks from geological storage are generally assumed to be similar to those of existing activities such as oil and gas production, natural gas storage, and acid gas injection (IPCC, 2005). The risks are managed on a routine basis through a combination of operational controls, management oversight, monitoring, maintenance, regulatory oversight, and insurance. Similar practices are needed to manage the risks of geological storage. In fact, each of the five active projects described in Section 13.4.1 have successfully managed these risks.

Based on analogous experience from the natural gas storage industry, the three largest sources of risk for geological storage projects are inadequate site selection due to inadequate seal characterization, leakage up active or abandoned wells, and leakage through undetected faults or fractures in the storage reservoir seal.

With experience, site selection for the natural gas storage industry has improved to the point where these risks are relatively small. For natural gas storage projects, if a sufficiently thick seal with structural closure, low permeability, and high capillary entry pressure covers the entire extent of the storage project, the risks of leakage are small. However, in comparison, a typical underground plume of CO_2 may grow to extend over hundreds of km², far larger than the size of a typical natural gas storage project; this suggests that time and experience will be needed to perfect the site selection process.

Leakage up active and abandoned wells remains the largest concern from natural gas storage projects and indeed has resulted in a small amount of leakage from the In Salah Project (Mathleson et al., 2011). The potential for wellbore leakage from CO_2 storage reservoirs has been the subject of a large amount of experimental and theoretical research over the past five years (Gasda et al., 2004; Carey et al., 2007; Nordbotten et al., 2009; Duguid and Scherer, 2009). Clearly, if CO_2 encounters a wellbore that is not cemented to the rock, CO_2 can migrate up the annulus and, if left unchecked, can reach the surface. More problematical is the influence of CO_2 on a cemented well. Laboratory and theoretical studies suggest that chemical reactions with water and CO_2 can degrade cement, creating leakage pathways through a largely intact cement seal. On the other hand, several wellbores in active CO_2 reservoirs have been tested and suggest that cement degradation is minimal. Additional studies are needed to resolve this issue.

Leakage through undetected faults and fracture is another source of leakage risk. Usually a combination of well drilling, logging, pressure testing, and seismic imaging is used to assess the quality of the reservoir and seal. Under certain circumstances, particularly where the offset across a fracture or fault is small, a fault may go undetected. Site characterization methods are getting better all the time, decreasing the risk that this could occur. Nevertheless, this risk is usually minimized by avoiding storage projects in areas with known faulting or at least with no active faults. Over time, as the best storage sites are used, greater efforts will be needed to assure that undetected faults and fractures do not compromise the storage site.

13.4.4.1 Duration and Security of Storage

Based on our best scientific understanding of the processes controlling CO_2 migration in the subsurface, CO_2 should remain securely stored in the subsurface for geological time periods (thousands of years and longer) if the following conditions are met:

- The seal has a low enough permeability and high enough capillary pressure to prevent migration into the seal.
- CO_2 cannot migrate around the edge of the seal or through breaches in the seal caused by leaking wells, faults, or fractures.
- The injection pressure is low enough to avoid fracturing the seal.

While in principle these are straightforward enough to understand, the bigger challenge is to identify sites that meet these conditions. Subsurface geology is by nature complex and geological storage sites are by necessity large, with CO_2 plumes potentially covering hundreds of km². For this reason, it is not easy to provide an unqualified answer about the duration and security of storage. The IPCC (2005) dealt with this issue by making qualified statements about the retention of CO_2:

> "Observations from engineered and natural analogues as well as models suggest that the fraction retained in appropriately selected and managed geological reservoirs is very likely to exceed 99% over 100 years and is likely to exceed 99% over 1000 years" (IPCC, 2005).

This statement was unsatisfying to many people for several reasons. First, the statement was based on expert opinion as opposed to actual experience. Second, the quantitative measures appear to be somewhat arbitrary, using terms like "likely" and "very likely," which perhaps imply a greater degree of precision than is possible. Finally, the statement is somewhat circular, implying that a good outcome will be achieved if the "the job is done right" but without prescribing exactly how to do the job right. The question then becomes, what have we learned in the intervening five years that could shed further light on the issue of storage permanence, especially considering the tremendous amount of research carried out over the past five years, numerous pilot tests, and demonstration and commercial projects?

From the five existing commercial scale projects, available information indicates that overall, they have performed as expected, with little or no leakage, and leakage that did occur was caused by abandoned wells. Once identified, the leaking wells can be sealed. This suggests that it is possible to find sites that meet the conditions needed for secure storage. Monitoring methods have shown to be effective for tracking CO_2 and the pressure buildup caused by CO_2 injection. Calibrated models also reasonably replicate CO_2 plume movement. All of these build confidence in the security of storage and our ability to "do the job right." But, it is still early in the lifetime of the projects and more will be learned as time goes on.

From the vast amount of research and pilot projects, much has been learned about how CO_2 migrates through the subsurface and how CO_2 geochemically interacts with rocks and cement. Understanding and confidence are growing in the secondary trapping mechanisms' ability to immobilize CO_2 and increase storage security in the post-injection period. Many new and improved methods for monitoring are being developed and tested, improving our ability to detect even small amounts of leakage. Methodologies for characterizing sites have been proposed and are being tested in numerous demonstration projects around the world (WRI, 2009, NETL, 2010b, DNV, 2010). Leakage up abandoned wellbore remains the biggest vulnerability for otherwise high-quality sites.

In light of the progress made over the past five years, what new can be said about storage duration and security? Observations from *commercial storage projects*, engineered and natural analogues, as well as *theoretical considerations*, models, and *laboratory experiments*, suggest

that appropriately selected and managed geological storage *reservoirs are very likely to retain nearly all the injected CO_2 for very long times, more than long enough to provide benefits for the intended purpose of CCS.*

13.4.4.2 Local Health, Safety, and Environmental Concerns for CO_2 Storage

CO_2 is used in a wide variety of industries, from chemical manufacture to beverage carbonation and brewing, from enhanced oil recovery to refrigeration, and from fire suppression to inert atmosphere food preservation. Because of its extensive use and production, the hazards of CO_2 are well known and routinely managed. Engineering and procedural controls are well established for dealing with the hazards of compressed and cryogenic CO_2.

While CO_2 is generally regarded as a non-toxic inert gas, exposure to concentrations in excess of several percent can lead to adverse consequences. In particular, since CO_2 is denser than air, hazardous situations arise when large amounts of CO_2 accumulate in low-lying, confined, or poorly ventilated spaces. While the chances of this occurring are very low, if a large amount of injected CO_2 were to escape from a storage site, it could present risks to health and the local environment. However, hazardous conditions would only persist several hundred meters from the site of the release, even for the largest possible leakage rates (Aines et al., 2009). Such releases could be associated with surface facilities, injection wells, or leakage from the storage formation itself. There may be small-scale diffuse leaks or leaks concentrated near the injection facilities. Leakage, if unchecked, could harm groundwater and ecosystems. Persistent leaks could suppress respiration in the root zone or result in soil acidification and eventually lead to tree-kills such as those associated with soil gas concentrations in the range of 20–30% CO_2 which have been observed at Mammoth Mountain, California, where volcanic out gassing of CO_2 has been occurring for several decades (Martini and Silver, 2002).

CO_2 storage projects are designed with the goal of avoiding these potential hazards. It is important, however, to have a comprehensive understanding of what could go wrong. Potential local health, safety, and environmental concerns from geological storage include the following. This list is approximately ordered with the largest to smallest risks, based on analogous experience from existing commercial operations:

- Occupational risks associated with well field operations. These risks are well understood both in terms of the nature of the risks and the frequency of occurrence. Overall, the injury rate for oil and gas field projects in the United States are lower than for many industries, but the severity is on the higher side, caused mainly by vehicle accidents (US Bureau of Labor Statistics (BLS, 2011), on-line database). For example, from the

---

2   The TRC is defined as the number of recordable injury and illness cases per 100 full-time workers.

period 2004–2007, the TRC (Total Recordable Case Rate) for the oil and gas production workers ranged from 1.4–1.7, as compared to 4.3–5.4 for all or the goods-producing industries.[2] For the same period, injuries or illnesses involving days away from work were 0.4–1.0 for oil and gas workers compared to 1.2–1.5 for all goods-producing industries.

- While CO_2 storage projects are expected to take place in storage reservoirs far deeper than drinking water aquifers, if CO_2 or brine leaks out of the storage reservoir it could migrate into drinking water aquifers. Risks of groundwater contamination include CO_2 intrusion into a fresh water aquifer, secondary contamination resulting from geochemical reactions between CO_2 and the aquifer rocks, from displaced saline brines, or in the case of storage in oil and gas fields (and potentially coalbeds), organic contaminants transported with the CO_2. In addition, risks will be site specific, depending on the hydrogeological environment and mineralogy of the rocks (Wilkin and DiGiulio, 2010). Of particular emphasis is the potential for brine migration even if no CO_2 leaks from the storage reservoir, which could increase the salinity and potentially the trace element content of drinking water sources (Nicot, 2008; Birkholzer et al., 2009). A significant concern is the large region over which the pressure will increase in the storage formation, a region far larger than the extent of the CO_2 plume. If abandoned wells or faults provide a conduit to short-circuit the seal, brine could invade a drinking water aquifer and degrade water quality. However, an extensive field study above one of the largest CO_2-EOR projects did not find any evidence of groundwater degradation due to CO_2 intrusion or brine migration (Romanak et al., 2010).

- Resource damage to nearby oil and gas fields or coalbeds, due to unwanted CO_2 migration into nearby mineral resources. The probability of these risks is low unless the storage reservoir is close to an oil or gas field.

- Ecosystem impacts in the event that CO_2 is released into soil, wetlands, or surface waters. If CO_2 is released to the surface, some local impacts to ecosystem productivity and function would be expected.

- Public safety risks from exposure to elevated CO_2 concentrations if CO_2 is released at the surface. The risks to the public will always be small compared to worker risk. Additionally, around the world there are many known sites of CO_2 releases into the atmosphere that pose little to no hazard, such as the Crystal Geyser site in Utah (Aines et al., 2009) and many sites in Europe (Beaubien et al., 2004; Voltattorni et al., 2006). However, if CO_2 leaks to the surface and accumulates in a low-lying area, exposure to even fatal concentrations of CO_2 is possible such as occurred at Mammoth Mountain, CA (Martini, and Silver, 2002). Risks from areas of known leakage can and are controlled by limiting access to the hazardous areas through signage and fences (Beaubien et al., 2004).

- Structural damage associated with land surface deformation or microseismicity. These risks are relatively well understood and should be

managed by controlling the injection pressure to avoid unacceptable levels of microseimicity or land surface deformation. High-resolution remote sensing maps depicting subsurface faults, fractures, and fracture density patterns of geological storage sites may be prepared to understand the structure and associated risks. Regulatory oversight is needed to ensure the project is managed with proper controls.

Importantly, the five existing projects have experienced none of these environmental impacts. Even though an abandoned well that intersected the storage reservoir at In Salah was found to be leaking through a routine monitoring program, no health safety or environmental impacts resulted from this event (Mathleson et al., 2011).

Extensive industrial experience with injection of CO_2 and gases in general indicates that risks from geologic storage facilities are manageable by using standard engineering controls and procedures. Loss of well control, probably the highest risk event that could occur in a CO_2 storage project, is infrequent. For example, one study focused on the largest oil and gas producing region in California indicated an overall rate of one blowout/12,000 wells/yr (Jordan and Benson, 2008). Furthermore, this study showed that property damage and human health impacts were small from blowouts that did occur (Jordan and Benson, 2008). Regulatory oversight and institutional controls further enhance the safety of these operations, and ensure that the site selection and monitoring strategy are robust. Employed on a scale comparable to existing industrial analogues, the risks associated with CCS are comparable to those of today's oil and gas operations. Eventually, if CCS were to be deployed on the grand-scale needed to significantly reduce CO_2 emissions (billions of tonnes annually), the scale of operations would increase to become as large as or larger than existing oil and gas operations (Burruss, 2004). In this eventuality, experience gained in the early years of CCS would be critical for assessing and managing the risks of the very large-scale geological storage projects.

13.4.4.3 Monitoring and Risk Management

In the 2005 *IPCC Special Report on Carbon Dioxide Capture and Storage*, the authors concluded that "the local health, safety and environment risks of geological storage would be comparable to risks of current activities such as natural gas storage, EOR, and deep underground disposal of acid gas." To achieve this high level of performance, a comprehensive approach is needed to manage environmental, health, and safety risks throughout the life cycle of a storage project, from site selection past the operational lifetime of the storage project. Monitoring must play a key role to observe the behavior of the injected CO_2, calibrate and validate predictive models, and provide any early warning that leakage may be imminent. In the event of imminent or actual leakage, remediation measures such as plugging abandoned wells will be needed. A regulatory oversight capacity is needed to ensure due diligence for siting, engineering, operating, monitoring, and remediating storage projects. Finally, private and/or public sector mechanisms are needed to ensure financial responsibility for any short- and long-term liabilities created by the project.

Every storage project is likely to use a combination of monitoring techniques (e.g., geophysics, hydrology, geochemistry) that will, at a minimum, track migration of the CO_2 plume, detect leakage out of the storage reservoir, monitor injection rates and pressure, and detect microseismic activity. Technology for monitoring geologic storage of CO_2 is available from a variety of other applications, including the oil and gas industry, natural gas storage, disposal of liquid and hazardous waste in deep geologic formations, groundwater monitoring, food preservation and beverage industries, fire suppression, and ecosystem research. Many of these techniques have also been demonstrated at the five existing storage projects and many smaller-scale pilot tests around the world (e.g., Arts et al., 2004; Hovorka et al., 2006). Specific regulatory requirements for monitoring have yet to be established. The principle methods of monitoring are described below.

Geophysical Monitoring

Several geophysical monitoring methods can be used to monitor the location of the CO_2 plume. Seismic imaging can detect changes in compressional wave velocity and attenuation caused by the presence of separate phase CO_2. Electromagnetic imaging can detect decreases in electrical conductivity caused by the presence of CO_2 in the pore spaces of the rock. Gravity measurements are sensitive to the decrease in bulk density of the rock caused by the presence of CO_2. Seismic methods for monitoring have been used successfully at Sleipner, Weyburn, the Frio Brine Pilot, and the Otway Basin Pilot Project, as well as others (Arts et al., 2009; Hovorka et al., 2006; White, 2009; Pevzner et al., 2010).

Geochemical Monitoring

Two types of geochemical measurements can be deployed to monitor CO_2 injection. The first involves the use of direct techniques including measurements of brine chemistry and introduced or natural tracers in samples obtained from injection horizons in observation wells. The second involves monitoring the near-surface for possible CO_2 leakage in the immediate vicinity of injection and observation wells, as well as from soils and shallow wells within the injection area.

To date, geochemical methods have been used primarily for the pilot scale tests because of the important insights they provide about the geochemical interactions between CO_2 and the storage reservoir rocks (Kharaka et al., 2006a; 2006b). They have also been used extensively at the Weyburn Project (Emberly et al., 2005). The simplest monitoring systems include pH, alkalinity, and gas-compositions. Of these, pH is probably the most diagnostic indicator of brine-CO_2 interactions and typically exhibits marked decreases that correlate closely with the breakthrough of CO_2 to monitoring wells. Major, minor, trace elemental chemistry and stable isotope geochemistry are used to assess the extent of water-CO_2-rock interactions. Enrichments of constituents such as Fe, Mn, and Sr compared to pre-injection fluid concentrations have been shown to be indicative of mineral dissolution reactions occurring

at depth during brine-CO_2-rock interactions (Emberly et al., 2005; Kharaka et al., 2006a; 2006b).

Monitoring of surface fluxes can also directly detect and measure leakage. Surface CO_2 fluxes may be measured directly with eddy covariance towers, flux accumulation chambers, and by techniques such as a field-portable, high-resolution infrared (IR) gas analyzers (Klusman, 2003; Miles et al., 2005; Lewicki et al., 2007; Lewicki et al., 2009; Spangler et al., 2010). A great deal of progress has been made to quantify detection levels, compare various surface monitoring approaches, and increase the number of options for monitoring CO_2 leakage at the surface (Spangler et al., 2010; Krevor et al., 2010).

13.4.5   Legal and Regulatory Issues for CO_2 Storage

There are a number of legal and regulatory issues that must be addressed before widespread deployment of CCS is adopted (IEA, 2007). If these issues are not addressed, they could become impediments to the deployment of CCS.

13.4.5.1   CCS Policy Context Related to Other Environmental Laws

From a regulatory and institutional perspective, any laws, regulations, and institutional changes that may affect CCS must be viewed within a larger regulatory framework. While not the emphasis here, it is important to note these other regulatory and legal requirements may impact CCS. For example, in the United States, these include, but are not limited to: Source Emission Regulations and Emerging Requirements; Clean Air Act after *Mass v. USEPA*; Tailoring Rule; USEPA policy guidance on BACT; USEPA GHG Reporting Rules; USEPA Conditional Exemption for CCS Under RCRA; DOT Regulations that May Impact Carbon Dioxide Pipelines; Safety (Department of Transportation); Siting (primarily state-based permitting); Rate Regulation; Safe Drinking Water Act under USEPA; and USEPA Underground Injection Control (UIC) Class II, V, and VI Rule (publication of Class VI rule will be specific to CCS). Solutions to the current institutional and legal issues facing CCS cannot be developed without considering the myriad of related laws and regulations and their intent.

13.4.5.2   CCS Regulatory and Legal Issues and their Current Status

The most pressing issues facing CCS include long-term stewardship and related liability concerns; pore space (subsurface) ownership; monitoring, measurement, and verification protocols for accounting purposes and health, safety, and environmental oversight; and regulatory oversight. These issues are discussed in turn below.

Long-Term Stewardship and Liability

Liability associated with long-term management and/or ownership of a carbon storage facility is of great concern to the private sector. Two types of liabilities can be distinguished: the health and safety related liability of releases of CO_2 on the long term potentially harming people or ecosystems, and the climate-related liability of releases of CO_2 on the long term, contributing to climate change. For most phases of any geologic storage project, the risks and attendant insurability of the project phases are well known. However, although many models and analyses conclude that the long-term probability, post-closure, of a significant release of gas is low and major existing projects have proven to be very effective in storing CO_2, the long timeframes associated with storage extend beyond existing institutional experience.

The rationale for a government role in addressing (and possibly indemnifying) long-term liability is due to the belief that CCS is in the public interest and that long-term liability issues should not, at this early stage in the development of the industry, be a barrier to further development. For example, in the case of the FutureGen project in the US, the acceptance of long-term liability became a competitive tool for Illinois and Texas and was deemed beneficial to those competing states.

CO_2 must remain underground for a long period – thousands of years and longer – to meet the goals of atmospheric climate change amelioration. This is well beyond the historic lifespan of companies and most governments. This will require institutional, administrative, and regulatory approaches for long-term stewardship to protect the public and to properly assess the efficacy of the removal of CO_2 from the atmosphere. As a result, the major barrier for industry and its supporting financial community to undertaking CCS projects is the undefined and open-ended liability for any CCS project. Any organization accepting liability will likely (without the development of institutional initiatives) be held responsible for the expenses of continuing monitoring and verification activities, any mitigation or remediation required, and compensation for any damages if leakage occurs.

One option is to create an industry-supported fund managed by the government. Such an approach is under discussion in the United States. For example, bills have been introduced in the US Congress that would establish a carbon storage stewardship trust fund financed by fees from operators to ensure compensation for potential damages. Long-term liability schemes have been adopted for other industries, including bond provisions by the UIC program, trust accounts funded through fees to operators that are administered by state or industry organizations such as the Acute Orphan Well Account, the Price-Anderson indemnity program that pools risk for the nuclear industry, and the National Flood Insurance Program.

The most common option is for government agencies to take on the long-term responsibility for CCS sites. This approach has been agreed to in the member states of the European Union, Australia and Norway. In addition, some states in the United States have adopted legislation to accept limited liability, but there has been little consistency in the time frames or agreement as to where the liability should ultimately reside. States, including oil-producing states like Texas and Louisiana and coal-producing states like Wyoming and Montana, have enacted

laws relating to CCS development. There are some common elements in these statutes:

- state policy declaration that CO_2 is a valuable commodity and that CO_2 storage provides a public benefit by reducing GHG emissions;

- a fee-based structure to cover the state's responsibility for administering long-term monitoring and oversight of CO_2 injection and storage;

- post-closure monitoring by the drilling or reservoir operator for a period of 10 years or longer;

- a certificate of completion to be issued by a designated state or federal agency, following permanent closure; and

- in some cases, a transfer of the state's responsibility for long-term (post-closure) monitoring and verification to the federal government after a designated period of years (e.g., 10 years or longer).

There are examples of governmental support for accepting CCS liability for the long-term in other countries. These include the Norwegian government acceptance of long-term liability from Statoil for the Sleipner project. Australian federal and state governments jointly accepted long-term liability for the Gorgon facility. In 2009, the European Parliament issued Directive 2009/31/EC on the geological storage of CO_2. Provisions in this directive are similar to those issued by other governments; particularly that financial security must be established for the operations and an anticipated post-injection phase of a minimum of 30 years. Liability may be transferred to a "competent authority" after a minimum of 20 years. "Competent authority," however, is not defined.

In summary, liability and the related long-term stewardship issues are potentially the most significant impediments to creating a global CCS industry. While some governments have taken a number of tentative steps toward solving this problem, it is currently unclear how the issue will be successfully resolved.

13.4.5.3   Pore Space (Subsurface) Ownership Issue Status

Specifically in the United States, there are a number of legal issues surrounding pore space ownership. In some other parts of the world, this is not an issue as the deep pore space is not owned by private individuals but by government.

There are a number of legal issues surrounding pore space ownership in the United States. While there is some interest in applying the same subsurface rules currently used for CO_2 ownership when it is considered a commercially-recoverable resource, most pore space laws and regulations are much more complicated in terms of federal or state ownership, mineral rights, and other subsurface regulation (CIEE, 2010). These rules and requirements can become complicated, but the key issues can be simply stated as:

- Who owns the pore space and can authorize CO_2 storage in them?

- How are ownership rights between surface and subsurface resolved?

- When does plume migration constitute trespass?

- How can rights to pore space be aggregated to enable large-scale projects, particularly in saline aquifers?

The answer to these questions varies from country to country, as discussed in Section 13.6, "Regional Outlook." However, regardless of the approach taken by a country, these issues need to be resolved before large-scale deployment of CCS is likely. In some regions like Europe and China, the State is the owner of the pore space. In this case, the resolution of these issues is comparably simple. These laws and requirements are either poorly developed or not developed at all in emerging economies, as will be addressed in the following section. In other regions, like North America, the situation is quite different.

In the United States, there is a body of statutory and case law that provides guidelines to resolving any issues associated with injecting CO_2 into deep subsurface saline formations. Original case law suggested that mineral rights ownership was to "the center of the earth." However, recent case law suggests that the answer is not as clear. This is based on a number of CCS-analog court cases that have legal, as well as scientific and technological, precedent. This includes natural gas storage, hazardous waste injection (under UIC regulations), fresh water storage in aquifers, and EOR. Much of current case law addresses "trespass," or the migration of materials from injection sites to the subsurface areas whose rights are owned by someone else. In some cases, under common tort law, the verdict has been that these trespasses interfered with possible future use of the subsurface formation. Thus, fines and penalties were assessed. In other instances, seemingly for the same type of trespass, it was judicially determined that there were no damages to the subsurface formation and, therefore, no fines or damages were assessed. For cases involving EOR or fresh water storage, these are deemed to be in the public interest and no fines were assessed. In cases where there are no mineral rights owners for the pore space, ownership devolves to the state. It is clear that new, more straightforward, laws will need to be enacted to clarify ownership, operation in the public interest, and legal recourse for CCS. Some states, such as Montana, Wyoming, and North Dakota, have taken the initiative to develop laws that clarify pore space ownership and liability with respect to CCS. In particular, these laws must clarify CCS's impact on saline formations and how the state will chose to regulate these resources and operations.

13.4.5.4   Monitoring, Verification, and Accounting (MVA) Protocols

There has been considerable discussion of Monitoring, Verification, and Accounting (MVA) at international level and national levels. This work is driven by three major public concerns. First, monitoring needs to be established to ensure public health and safety are maintained and not compromised by CO_2 releases. Second, national emissions are reported using uniform and documented protocols. The IPCC (2006) developed guidelines for inventory verification that included how to treat CCS projects. Third, as laws emerge mandating emission reductions (e.g., carbon taxes, cap-and-trade legislation, and performance standards), there will be verification and accounting protocols that determine that the volume of CO_2 that an organization is taking credit for is, in fact, staying underground.

Many international and national agencies have focused on this issue and a considerable body of information has been developed. A recent summary of these activities has been developed as part of a supporting background document for the Carbon Capture and Storage Blue Ribbon Panel in California (CIEE, 2010). However, these findings and data need to be translated into laws. The legal requirements are to ensure that human health and safety and other living systems (flora and fauna) will be protected. Thus, MVA protocols will have two purposes. The primary purpose will be measurement to allow for the proper crediting of CO_2 that is sequestered. However, monitoring protocols must also be developed to ensure that public health, safety, and the environment are properly protected. This will lead to emissions requirements for CCS projects.

Most importantly, there must be protocols developed that will allow for crediting organizations that sequester CO_2. These protocols will have value to the organizations, as they can take credit for CO_2 removed from emission streams as a means for capturing financial credits, either through cap-and-trade rules (similar to current sulfur dioxide rules) or minimizing taxes under a tax system. Until CO_2 is regulated and treated as a commodity, there will be no reason to inject it into saline formations. To treat it as a commodity, rules and regulations need to be developed under international MVA guidelines that allow for the crediting of CO_2 removal from emission streams. Similarly, when CCS is deployed under the clean development mechanism (CDM), MVA protocols will be needed to ensure that expected emission reductions are in fact achieved.

13.4.5.5   Regulatory Oversight

All of the requirements for regulation will eventually fall on regulatory agencies for enforcement and verification. There are several issues facing these agencies at this time.

Laws covering CCS implementation on a large commercial scale are under development worldwide. In the United States, the most important law appears to be the Underground Injection Control (UIC) Law. The US Environmental Protection Agency has developed regulations under the UIC for underground injection of CO_2 for the purposes of long-term storage. However, additional focused legislation to appropriately regulate and validate CCS activities is likely to be required.

The second issue facing any developer of CCS projects is the overlapping authorities for many of the phases of operations that define a CCS project (Tsang et al., 2002). For example, a thorough analysis of either the lack of jurisdiction or the existence of overlapping authorities has been done in California (CIEE, 2010). As stated above, many of the rules and regulations that CCS projects must operate under have not been promulgated. However, as noted in Section 13.1, many projects will face multiple jurisdictions and legal requirements, each of which will have different metrics for permitting. The nature of this issue could cause a developer to not move forward on a project, simply because the timeline associated with obtaining all regulatory approvals from multiple state and federal agencies is too long from a financing perspective.

An additional issue is the capability of regulatory agencies to meet the needs of new types of industrial projects such as CCS. Most agencies are significantly understaffed. Further, the rapid scientific and technical changes require agency staff to be better equipped to address new regulatory requirements. In fact, most agencies are becoming less adept at keeping staff that can stay abreast of these changes. In countries without existing regulatory authorities, the situation is even more challenging, as the regulatory infrastructure will need to be established. These issues are addressed in the "Regional Outlook," Section 13.6.

An additional issue related to those described earlier in this section is the need to develop regulations and institutional requirements that can address some of the issues described in the previous subsections. Continued lack of clarity on pore space ownership and trespass, post-closure liability, the extent of long-term stewardship, and MMV precision and accuracy requirements will preclude any large-scale development until resolved.

13.4.5.6   Summary of Regulatory and Legal Issues

Currently, many international, federal, and state legislative bodies understand the nature of these issues and are trying to develop legislation that can allow CCS to move forward. It will be critical over the early part of this decade to see if institutional and regulatory solutions can be developed to address legal and regulatory issues. Failure to do so could have a bigger impact on CCS than issues of carbon dioxide capture costs and the low, but uncertain, risks associated with CCS.

13.4.6   Cost of Storage

Costs for geological CO_2 storage consist of four elements: site characterization costs; project capital costs (e.g., costs for surface equipment for each well, cost of drilling, costs of additional CO_2 compression if required); operating and maintenance (O&M) costs; and monitoring, verification, and closure costs. Site characterization costs are lower for storage in oil and gas formations (compared to saline formations and deep coal seams), since the main characterization of these types of fields has already occurred during their exploration. Site characterization costs will be dependent on the area of review (aerial extent of plume spread in CO_2), which will be determined by regulatory regimes in place (Rubin et al., 2007). Drilling costs are mainly dependent on the number of wells (including water production wells if necessary), the injectivity of the field, and the allowed overpressure. O&M costs for CO_2 injection are assumed to be comparable to the costs of water injection for secondary oil recovery (Bock et al., 2003; Rubin et al., 2007).

Storage costs vary depending on numerous factors, including type of reservoir, existing information/infrastructure for the site, onshore versus offshore storage, extent of monitoring, and regional factors. Cost estimates found in the literature are limited to capital and operational costs, and do not include potential costs associated with long-term liability. Table 13.3 shows an overview of these costs by field type. A first assessment of global investment needed for CO_2 storage alone, under a stabilization scenario of 450 ppmv, indicate a range between US$_{2009}$\$0.8–5.6 billion in 2020, and US$_{2009}$\$88–650 billion in 2050 (IEA, 2009).

Table 13.3 | Overview of indicative geological storage costs published in the literature. The figures represent average data; when available, ranges are provided between brackets. Data is presented in US$_{2008}$\$/tCO_2 stored.

	Ecless et al., 2009	Heddle et al., 2003	Hendriks et al., 2004	IEA, 2008	IPCC, 2005	Kober and Blesl, 2010	McKinsey, 2008	Ramírez et al., 2010
Depleted gas and oil fields		(2–24)[a]	(1–11)	(10–25)	(1–14)	(1–21)		(1–20)
-onshore			3 (1–5)		(1–14)	5 (1–21)	6	(1–9)
-offshore			8 (5–11)		(4–9)	6 (4–12)	16	(4–20)
Aquifer	(2–7)	(1–15)	(3–14)	(10–20)	(0.2–33)	(2–35)		>25
-onshore			4 (3–8)		(0.2–7)	3 (2–15)	7	
-offshore			9 (7–14)		(1–33)	6 (3–35)	18	

[a] average value for depleted gas fields is \$6/tCO_2 and for depleted oil fields \$5/tCO_2

Estimates of storage costs derived from current commercial-scale projects are in the order of US$11–17/tCO$_2$ (Sleipner); US$20/tCO$_2$ (Weyburn) and US$6/tCO$_2$ (In Salah) (ITFCCS, 2010). CO$_2$ storage costs for EOR projects could be offset by the revenues provided by the additional gas or oil produced. In these cases, storage costs are determined by the marginal value of oil, the underlying production costs, and the potential cost of supply of CO$_2$. Currently, the largest single cost component in EOR projects is the purchase of CO$_2$ (IPCC, 2005; Kuuskraa and Ferguson, 2008). CO$_2$ purchase prices found in the literature are in the order of US$38–49/tCO$_2$. The large variability in the economics of EOR makes it difficult to provide representative levelized annual costs of CO$_2$ stored, and therefore large cost ranges are reported in the literature. For instance, Haddle et al. (2003) reported a range of −91–74 US$_{2008}$/tCO$_2$, while Hendriks et al. (2004) reported a range of −12–24 US$_{2008}$/tCO$_2$.

13.4.7   CO$_2$ Storage Conclusions

Storage in deep underground formations is the most mature storage option because it uses technology developed from over a century of oil and gas exploration and production. Ocean storage has also been studied; however, due to lack of permanence and potential environmental impacts, ocean storage is not being considered at this time. Stimulating the primary productivity of the oceans by adding trace nutrients (e.g., iron), so-called ocean fertilization, has also been proposed as a means of extracting CO$_2$ from the atmosphere. The long-term effectiveness of this method of removing CO$_2$ from the atmosphere is uncertain. Impacts to ocean ecosystems are unknown and potentially large.

Cumulative learning from over 30 years of experience is available from five active storage projects. Deep underground formations suitable for CO$_2$ storage are located in sedimentary basins and include depleted or depleting oil and gas fields, saline aquifers, and deep, unminable coalbeds. Other options for geological storage, such as storage in volcanic rocks, which rely on in situ mineralization for long-term storage, may be developed over time. Rapid in situ mineralization, a comparatively new idea, is the subject of fundamental research investigations that will take many years to mature. Ocean bottom sediments in very deep and cold water may provide an attractive option for emission sources near coasts or where other options are not available. However, this approach has not been tested and considerable research and demonstration will be needed before this option becomes available.

The permanence of storage in deep geological formations in sedimentary basins will depend on a number of factors, the most important one being the presence of a high-quality seal at the top of the storage reservoir. A very high degree of permanence (defined as retention of greater than 99% of the injected CO$_2$ over a period of 1000 years) is likely for sites that have good seals, are characterized and selected carefully, are operated to stay within a safety envelop

to protect the seal and prevent well leakage, and are routinely monitored and regulated to ensure due diligence in all of the aforementioned activities.

The biggest health, safety, and environmental risks from geological storage are groundwater pollution, short-duration well failures, and leakage through undetected faults or fractures in the reservoir seal. These risks should be manageable, as they are similar to the risks that are routinely managed in the oil and gas industry. Some failures are, however, to be expected, especially in the early projects as experience in geological storage is growing. Worst-case environment, health, and safety impacts are likely to be less than the worst case accidents that occur in the oil and gas exploration and production industry because CO$_2$ is not explosive and it does not accumulate in ecosystems like oil does.

The cost of storage is highly variable, depending on the location of the project, the number of wells needed, whether the project is onshore or offshore, and if the project is in a hydrocarbon reservoir or saline aquifer. Costs from three of the existing projects are in the range of US$6–20/tCO$_2$.

13.5   Energy Systems Synergies and Tradeoffs Influencing CCS Deployment

13.5.1   Life Cycle Analysis of CCS

Life Cycle Assessment (LCA) is one of the most frequently used tools for evaluating the potential environmental impact of products and materials, since it allows upstream and downstream elements to be included in the analysis. For CCS, this implies that besides CO$_2$ capture, CO$_2$ transport, and CO$_2$ storage, the acquisition of the fuel (e.g., coal mining), its transport, and the materials and energy required for the capital equipment and the infrastructure throughout the complete chain are also included.

The most common categories examined in LCA-CCS studies are global warming potential (GWP), increase energy use, acidification potential, and eutrophication. Impact categories such as land use, habitat alteration, impacts on biodiversity, human toxicity and ecotoxicity, and photo-oxidant formation are not addressed by most studies. This, however, is not a unique feature for LCAs of CCS systems. It has already been reported that these impact categories are not (formally) addressed by most published LCAs and due to significant data gaps (e.g., Finnveden, 2000).

A comparison of key results recently published for different CCS technologies is shown in Table 13.4. The results indicate that CCS decreases the global warming potential (GWP) of fossil fuel-fired power plants by 65–85% depending on the technology selected. Indirect emissions are responsible for between 15–46% of the GWP and are mainly originated

Table 13.4 | Key results of selected LCAs for CCS chains.

Author	Type of technology			1. Key results			
	CO$_2$ Capture	Trans.	Storage	GWP	Primary energy demand	Acidification	Euthrophication
				gCO$_2$-eq/ kWh	MJ/kWh	gSO$_2$-eq/kWh	gPO$_4{}^3$-eq/kWh
Koornneef et al., 2008[1]	PC-MEA	pipeline	CO$_2$ storage in gas field	1092 (PC) 837 (PC BAT) 243 (PC MEA)		2.76 (PC) 1.44 (PC BAT) 2.10 (PC MEA)	0.29 (PC) 0.16 (PC BAT) 0.29 (PC MEA)
Korre et al., 2010[2]	PC-MEA PC- K$^+$/PZ PC-KS-1	pipeline	CO$_2$ storage in aquifer	846 (PC) 179 (PC MEA) 160 (PC K$^+$/PZ) 152 (PC KS-1)		0.39(PC) 0.47(PC MEA) 0.31(PC K$^+$/PZ) 0.36 (PC KS-1)	0.04 (PC) 0.06 (PC MEA) 0.04 (PC K$^+$/PZ) 0.05 (PC KS-1)
Pehnt and Henkel, 2009	PC-MEA; IGCC-selexol; Oxyfuel	pipeline	CO$_2$ storage in gas field	940 (PC) 190 (PC MEA) 900 (IGCC) 150 (IGCC CCS) 145 (Oxy CCS)	8.0 (PC) 13.5 (PC MEA) 7.8 (IGCC) 9.5 (IGCC CCS) 11.0 (Oxy CCS)	0.65 (PC) 0.58 (PC MEA) 0.25 (IGCC) 0.35 (IGCC CCS) 0.13 (Oxy CCS)	0.05 (PC) 0.09 (PC MEA) 0.03 (IGCC) 0.04 (IGCC CCS) 0.01 (Oxy CCS)
RECCS, 2008	PC (MEA), Oxyfuel; Gas CC (MEA); IGCC (Rectisol)	pipeline	CO$_2$ storage in aquifer	792 (PC) 262 (PC MEA) 176 (oxy CCS) 396 (gas CC) 132 (gas CC MEA) 774 (IGCC) 244 (IGGC CCS)	7.7 (PC) 9.9 (PC MEA) 10.4 (oxy CCS) 7.0 (gas CC) 8.4 (gas CC MEA) 7.7 (IGCC) 9.5 (IGGC CCS)	0.85 (PC) 0.77 (PC MEA) 0.19 (oxy CCS) 0.56 (gas CC); 0.69 (gas CC MEA); 0.64 (IGCC); 0.84 (IGGC+CCS)	0.07 (PC) 0.08 (PC MEA) 0.10 (oxy CCS) 0.05 (gas CC); 0.07 (gas CC MEA) 0.05 (IGCC) 0.06 (IGGC CCS)
Schreiber et al., 2009	PC-MEA			852 (PC) 212 (PC MEA retrofit) 179 (PC MEA Greenfield)	10.0 (PC) 14.3 (PC MEA retrofit) 12.1 (PC MEA Greenfield)	1.2 (PC) 1.3 (PC MEA retrofit) 1.1 (PC MEA Greenfield)	0.140 (PC) 0.245 (PC MEA retrofit) 0.198 (PC-MEA Greenfield)

1: In this study the authors compare an average PC plant of 2005 with a PC using best available technologies (BAT) and a PC with post combustion capture.
2: In this study, two other solvents were analyzed besides MEA: a hindered amine (KS-1) and a promoted potassium carbonate (K$^+$/PZ).

in the fuel supply chain. Contributions due to capital equipment and infrastructure are estimated to be below 5%. Due to the energy penalty induced by CO$_2$ capture, there is a relative increase in total primary energy demand in the order of 20–40%, with up and down streams accounting for 15–30% of this increase. Impacts on the acidification and eutrophication potentials differ and are highly dependent on the type of technology examined. Gas and oxyfuel concepts tend to show better performances than coal-based systems, especially post-combustion concepts. However, the type of solvent selected appears as the main determinant on the environmental performance of post-combustion capture.

13.5.2   Potential Impacts of CCS on Reducing Vehicle Emissions (See also Chapter 12.4)

Several studies have been undertaken on well-to-wheels emissions of GHGs from vehicles (see Chapter 12). The IEA Greenhouse Gas R&D Programme undertook a study to assess the addition of CCS to transport fuel production, which was based on a well-to-wheels study undertaken by JEC, CONCAWE, and EUCAR (IEA GHG, 2005). The costs and CO$_2$-eq emissions of various transport fuel options are summarized in Table 13.5. The base case for calculation of the costs of GHG abatement is a typical European car, such as a Volkswagen Golf, with a gasoline port injection spark ignition engine (PISI). A study looking at just hydrogen options undertaken by the European HySociety project (Mulder et al., 2007) resulted in somewhat more optimistic results for CCS.

Production of Fischer-Tropsch (F-T) diesel and dimethyl ether (DME) from natural gas with CCS results in modest (22–31%) reductions in emissions compared to the gasoline base case. The percentage emissions reduction from the use of hydrogen produced from fossil fuels with CCS is 60–78%, depending on whether the fuel is coal or natural and whether fuel cells or an IC engine is used. Without CCS, some of the hydrogen production options would have emissions greater than the reference case. The fossil-fuel electricity cases with CCS result in the greatest emissions reduction (79–88%). The biomass-based cases result in negative net emissions, assuming the biomass is produced from sustainable sources, but the results should be viewed cautiously due to a

Table 13.5 | Potential impacts of CCS on reducing vehicle emissions

Pathway, Fuel	Vehicle	Emissions gCO$_2$-eq/km (mean)	Avoided cost relative to the gasoline base case €/tCO$_2$ avoided
Gasoline (base case)	PISI	165	0
Diesel (conventional)	DICI	159	-
CNG	PISI hybrid	109	281
DME	DICI hybrid	139	1076
CNG with CCS	PISI hybrid	101	247
FT diesel with CCS	DICI hybrid	129	1089
DME with CCS	DICI hybrid	114	762
Hydrogen: coal with CCS	ICE	66	479
Hydrogen: natural gas with CCS	ICE	49	296
Hydrogen: biomass with CCS	ICE	-261	109
Hydrogen: coal with CCS	Fuel cell	37	510
Hydrogen: natural gas	Fuel cell	93	827
Hydrogen: natural gas with CCS	Fuel cell	37	420
Hydrogen: biomass with CCS	Fuel cell	-147	189
Electricity: natural gas with CCS	Electric	20	805
Electricity: coal with CCS	Electric	34	918
Electricity: biomass with CCS	Electric	-118	468

PISI: Port injection spark ignition engine (Gasoline)

DICI : Direct injection compression ignition engine (Diesel)

CNG: Compressed Natural Gas

FT: Fischer-Tropsch

DME: Di-Methyl Ether

ICE: Internal Combustion Engine

whole range of factors, from indirect land use change to differing engine efficiencies with biofuels.

The costs of CO$_2$ abatement are in all cases substantially greater than the costs of abatement by use of CCS in fossil fuel power generation. For the cases where there are comparable data, the use of CCS reduces the overall costs per tCO$_2$ abated. The costs depend strongly on plant construction costs and fuel prices, which have been highly volatile in recent times, and in some cases the costs of technologies that have not yet been widely used, such as fuel cells and hydrogen storage in vehicles.

13.6   Regional Outlook for CCS

13.6.1   Methodology

This section discusses in four parts the activities in various regions in the field of CCS. Technical potential and source-sink matching comprise the presence of CO$_2$ sources amenable to capture, what is known about the storage capacity, and its proximity to CO$_2$ sources.

The local aspects of economic viability of CCS in the region will be discussed under the section on factors influencing regional CCS costs. The economic activities that relate to CCS, such as oil and gas operations, related human capacity and education, and CCS-related R&D in the region, are addressed in the third part. Fourth, the developments related to legislation and policy are discussed. Last, there is a general discussion of politics and, where data is available, public perception.

13.6.2   Europe

13.6.2.1   Technical Potential and Source-Sink Matching

From 2006–2009, a Europe-wide consortium of research institutes conducted a geological storage capacity estimate and mapped CO$_2$ point sources and prospective storage reservoirs over the 27 Member States, Norway, and several states in former Yugoslavia and the Balkans (Geocapacity, 2009). An overall estimate of prospective reservoirs yielded 360 GtCO$_2$ storage capacity, the majority in saline aquifers, and more than half in saline aquifers off the coast of Norway. Much of this storage capacity may not have trapping structures, making storage integrity less certain (see the discussion in Section 13.4.2.2). A more conservative estimate yields a storage potential of 96 $GtCO_2$ in saline aquifers, around 20 Gt in hydrocarbon reservoirs and another one Gt in coalbeds all over Europe.

In 2008, the European Union emitted 4.9 $GtCO_2$-eq, of which around 2.6 $GtCO_2$ are stationary CO_2 sources that could be amenable to capture, and 1.3 $GtCO_2$ is electricity production. The sources of CO_2 are widely distributed over Europe, although there is a concentration in the German Ruhr area, the Netherlands, the United Kingdom, and southern Poland (Geocapacity, 2009).

13.6.2.2   Factors Influencing Regional CCS Costs

Costs of CCS are built from capture and compression, transport, and storage costs. Capture and compression costs can be affected by the type of CO_2 sources – if there is a significant amount of high-purity, storage-ready CO_2 sources, capture costs can be kept low. If transport distances can be kept short and in sparsely populated, flat areas, this positively affects the transport costs. And if storage can be done in EOR, or in onshore fields where injection facilities are still usable, this reduces CCS costs as well. Other indirect factors that impact the siting, and therefore the costs, of CCS operations include population density and public perception, types of industrial activities, and access to resources.

In terms of the cost curve for CCS in Europe, there are a number of relevant factors. First, much of the storage potential in Europe seems to be located offshore and in saline aquifers. Therefore, storage costs may be higher, as saline aquifers are often not characterized well and offshore drilling adds to the costs. Second, the population and industrial density as well as the landscape in much of Europe make transport a relatively costly matter. Third, with relatively high consumer electricity prices, the cost increase of CCS compared to current electricity prices are relatively low compared to other countries: incremental cost of CCS for a typical residential consumer is estimated to be about 20%.

13.6.2.3   Technical and Human Capacity

Most countries in Europe are developed countries with a highly educated workforce. Several European countries have a significant oil and gas industry, with ensuing human and technical capacity on the underground.

Many universities and research institutes in Europe conduct research and education related to various technical, economic, and social aspects of CCS. Although the capacity in Europe is considered high, in order to enable broad application of CCS, a larger base of CCS experts and practitioners will need to be developed over the years to come.

In terms of R&D, the European Union indicates that through 2009, €115 million was spent on CCS-related R&D activities through its major program.[3] In addition, member states invest significant sums in CCS R&D; the Netherlands, for example, has since 2003 spent and planned roughly €50 million on CCS research, which was matched by industry.[4] The Zero Emissions Platform (ZEP), a European platform for stakeholders of CCS, indicates that its industry members' investment in CCS-related R&D have amounted €635 million over 2003–2008.[5]

13.6.2.4   CCS Legal and Policy Initiatives

In December 2008, the European Parliament passed the "Climate and Energy package," a collection of new EU Directives including a commitment for the European Union to, by 2020, reduce GHG emission by 20% compared to 1990, use 20% renewable energy, and improve energy efficiency by 2% annually. Part of the package was legislation to enable CCS in the European Union. It consisted of a "Directive on geological CO_2 storage" and a series of modifications to other Directives that took away obstacles to CCS in, for instance, directives on landfills and on large combustion plants.[6] The geological storage directive outlines the legal conditions that any CO_2 storage project has to meet. The provisions are further detailed and implemented by the Member States.

The Climate and Energy package also modified the directive regulating the EU Emissions Trading Scheme (ETS), providing a carbon price incentive for CCS projects. However, due to various factors, carbon prices, steady at around US$_{2010}$\$14.64 halfway through 2010, remained too low to enable CCS (Sijm, 2009). In addition to the inclusion in the ETS, the European Union decided to reserve 300 million emission allowances, corresponding to the value of 300 million not emitted tCO2, from the "New Entrants Reserve" for demonstration of innovative energy technologies, which is widely considered to go mainly to CCS demonstration. In 2009, the European Union decided to grant around €1 billion to six demonstration projects as part of its economic recovery package in response to the 2009 economic crisis. Later, it was decided in addition to the economic recovery package to make available 300 million emission allowances from the New Entrants Reserve in the EU Emissions Trading Scheme for demonstration of innovative renewable and CCS projects. At a price of roughly 10 EUR/tCO2, this amounts to another EUR 3 billion in total. Combined, the EU hopes to enable up to 12 large-scale demonstrations of CCS by 2015. Most of these demonstrations are expected to be in the power sector, but some are intended to take place in industry.

3   CCS EII Implementation Plan 2010–2012. ec.europa.eu/energy/technology/ initiatives/doc/ccs_implementation_plan_final.pdf.

4   www.co2-cato.nl/cato-2/program-overview.

5   CCS EII Implementation Plan, 2010. See ec.europa.eu/energy/technology/ initiatives/doc/ccs_implementation_plan_final.pdf. It is unclear what is included in this number; it may include investments of European companies in CCS projects elsewhere, for instance BP's investment in the In Salah project.

6   Directive 2009/31/EC on the geological storage of carbon dioxide: eur-lex.europa. eu/LexUriServ/LexUriServ.do?uri=OJ:L:2009:140:0114:0135:EN:PDF.

13.6.2.5   Discussion of CCS in the European Union

CCS is considered a political reality in the European Union. Many policy documents indicate the importance of the option to achieving the European Union's ambitious climate goals and steps have been undertaken to stimulate demonstration of CCS, establish a legal framework, and develop human and business capacity and knowledge. On the other hand, CCS in Europe experiences some of the strongest resistance. Although several environmental organizations support or tolerate CCS, others are vehemently opposed. In addition, companies and governments in Denmark, Germany, and the Netherlands have experienced strong public opposition to CCS demonstration projects by local inhabitants resulting in the cancellation of several projects.

CCS is not expected to become a reality all over the continent. R&D and policy activities are concentrated in a few active countries, including Norway, the Netherlands, Germany, Denmark, and the United Kingdom. Those countries are characterized by a strong fossil fuel tradition, high industrialization levels, a coastline on the prospective North Sea, a relatively high per capita national income, and a relatively well-developed environmental awareness. In addition, countries like France, Poland, Italy, and Spain have shown interest and have worked on the option. But CCS is not considered to be an option in countries with low storage potential, a small fossil fuel industry, and low incomes.

13.6.3   North America

13.6.3.1   Technical Potential and Source-Sink Matching

In 2008, the CO_2 emissions in the United States and Canada reached 5921.2 and 574 Mt, respectively. The Regional Carbon Sequestration Partnerships (RCSP), initiated by the US Department of Energy (US DOE), have been leading the efforts to determine the most suitable technologies, regulations, and infrastructure needed for CCS in different areas in North America. The *2010 Carbon Sequestration Atlas of the United States and Canada,* released by the US DOE and National Energy Technology Lab (NETL, 2010a), provides information about the currently estimated CO_2 emissions in 42 US states and four Canadian provinces, divided into the seven main partnership regions. The Atlas documents more than 4700 stationary CO_2 sources with total annual emissions of over 3200 Mt. Most of these emissions come from power generation (83%), followed by refining and chemical processes (6%), industrial processes (4%), petroleum and natural gas processing (3%), cement production (3%), and ethanol plants (1%) (NETL, 2010a).

Significant storage capacity has been documented in depleted and depleting oil and gas reservoirs, saline aquifers, and unminable coal seams throughout the seven partnership regions. The RCSPs document the location of almost 143 Gt of geologic CO_2 storage potential in 9667 oil and gas reservoirs distributed over 27 American states and three Canadian provinces. Similarly, the CO_2 geologic storage potential in unminable coal seams is estimated at 187–217 Gt, distributed over 24 states and three provinces. Finally, 3600–13,000 Gt is the estimated CO_2 storage potential in saline formations. The capacity estimation range in the case of coal seams and saline aquifers is primarily due to the uncertainty inherent in the calculation methodology (NETL, 2010a).

The National Carbon Explorer (NATCARB) is another initiative by the US DOE that aims to link CO_2 geological storage and emissions sites databases across several regional centers. NATCARB interactive maps show close proximity between CO_2 sources and sinks throughout North America where many CO_2 sources are located on or near CO_2 storage sites. Nevertheless, exceptions exist and source-sink matching is not very good, including parts of the US upper mid-west and northeast.

13.6.3.2   Technological Maturity

North America is among the leading regions in CCS activities. Today, many pilot projects exist, and commercial-scale projects are in the early stages of planning, the majority of which are in the United States. This includes pre-combustion, post-combustion, and oxy-combustion capture facilities, as well as geologic storage projects at the R&D, demonstration, and industrial/commercial scales (BRGM, 2009). In addition to CCS-related endeavors, a well-developed CO_2 infrastructure exists in North America, which facilitates the early deployment and testing of CCS. In this regard, EOR through CO_2 injection (CO_2-EOR) is implemented widely in the United States and Canada. The 2008 World EOR Survey issued by the Oil and Gas Journal lists 100 ongoing CO_2-EOR projects in the United States and seven projects in Canada (Koottungal, 2008).

13.6.3.3   Factors Influencing Regional CCS Costs

As mentioned before, the geographical distribution of sources and sinks plays a major role in determining the economic competitiveness and feasibility of CCS deployment. To start, most of the currently implemented and proposed CCS-related projects in North America are onshore, and, as noted previously, many CO_2 point sources are located either on or close to potential geologic storage formations. Both factors favor economic implementation of CCS. Equally important, both the United States and Canada have conducted several studies to investigate the effect of population density, environmental factors, and topography on the cost of CCS implementation. In 2010, NETL published *Site Screening, Selection, and Initial Characterization for Storage of CO_2 in Deep Geologic Formations*, which aims to provide a framework and an overview of processes for selecting suitable sites for geologic storage. The proposed framework discusses several aspects related to site screening, site selection, and initial characterization. Among other important topics, the document investigates the effect of protected and sensitive areas (wetlands, source water protection areas, protected areas, and protected species); population centers; existing resources development; and pipeline right-of-way on the feasibility of CCS deployment (NETL, 2010b). Similarly, *Carbon*

Dioxide Capture and Storage: A Compendium of Canada's Participation (Natural Resources Canada, 2006) lists several studies and R&D projects that aim to optimize the economic aspects of CO_2 transportation network and infrastructure design, including: *Optimisation of integrated CO_2 capture, transportation, and storage in Canada (Waterloo),* and *Integrated economic model for CO_2 capture and storage (ARC and others).*

In 2009, more than 80% of US energy use was fossil fuel based, and almost 60% was from oil and coal (US EIA, 2010a). Similarly, more than 77% of Canada's energy use in 2008 was fossil fuel based. As of 2010, Canada's oil reserves are the second largest in the world after Saudi Arabia, and a significant portion of these reserves is unconventional CO_2-intensive oil shale (US EIA, 2009). As such, both countries are highly dependent on fossil fuels, which, in the absence of earlier, cheaper, low-carbon alternatives, improves the economic attractiveness of CCS as an effective climate change mitigation option.

13.6.3.4   Human Capacity, Research and Development

Several governmental agencies, academic institutes, and industrial/commercial businesses in the United States and Canada are actively involved in CCS and are conducting CCS-related R&D locally and internationally. In the United States, the Department of Energy, specifically the Office of Fossil Energy (FE) and NETL, are guiding local and international research initiatives on CO_2 geologic sequestration. FE/NETL is directly supporting more than 75 research projects across the United States and internationally. The CCS R&D network database of the Global CCS Institute (GCCSI) shows that a total of 209 organizations in the United States are involved in CCS R&D, the majority of which is related to CCS technologies (GCCSI, 2011b). Canada is actively involved in CCS R&D through 25 CCS research centers (including eight universities with substantial engagement); 23 companies that are developing, testing, using, or analyzing the effects of CCS technologies; and 13 governmental programs supporting CCS projects. In that regard, 59 CCS-related projects in Canada are led by universities, 42 are led by government research agencies (including provincial research organizations), 23 are led by industry (a category that includes any for-profit company) and two are led by nongovernment organizations (Natural Resources Canada, 2006). Similar to the United States, the CCS R&D network database of the GCCSI shows that a total of 58 organizations in Canada are involved in CCS R&D, the majority of which is related to CCS technologies (GCCSI, 2011b).

In addition, the United States and Canada take part in international collaboration on CCS. Both countries are members of the Carbon Sequestration Leadership Forum (CSLF), GCCSI, and the *IEA Greenhouse Gas* R&D Programme (IEAGHG). Both countries are also engaged in bilateral and multilateral agreements promoting CCS, including the US-China Fossil Energy Cooperation Protocol; the National Energy Technology Laboratory-Korea Institute of Energy Research Memorandum of Understanding (Smouse, 2007) the CO_2 Capture Project; and Weyburn-Midale Monitoring Organizations (Natural Resources Canada, 2006). All

aforementioned research endeavors, as well as the pilot CCS projects and the well-developed industrial and energy infrastructure discussed in the section above, contribute significantly to building a skilled human capital that is competently educated about CCS and would be expected to play a major role in advancing this technology. In that regard, both the United States and Canada are also investing in developing, supporting, and initiating educational opportunities (conferences, workshops, publications) that guide various stakeholders on how to engage the public and raise awareness and support for CCS-related activities. (Natural Resources Canada, 2006; The ECO ENERGY Carbon Capture and Storage Task Force, 2008; US DOE/NETL, 2009b).

13.6.3.5   CCS Legal and Policy Initiatives

Several regulatory frameworks contribute to the deployment of CCS. Regulations governing the deployment of CCS-related activities have been proposed and are currently considered in both the United States and Canada. In July 2008, the US EPA published the Federal Requirements under the Underground Injection Control (UIC) Program for Carbon Dioxide (CO_2) Geologic Sequestration (GS) Wells Proposed Rule. This proposal, which applies to owners or operators of CO_2 injection wells, establishes a new class of wells, as well as minimum technical criteria for the geologic site characterization, fluid movement, area of review and corrective action, well construction, operation, mechanical integrity testing, monitoring, well plugging, post-injection site care, and site closure, to ultimately protect underground sources of drinking water (USEPA, 2010). More recently, in March 2010, a bill titled the Carbon Capture and Sequestration Deployment Act of 2010 was introduced in the US Congress. The bill aims to "provide financial incentives and a regulatory framework to facilitate the development and early deployment of CCS technologies, and for other purposes. (IIIth US Congress, 2010)" On the other hand, even though no comprehensive national legislation specifically addressing CCS has been proposed yet in Canada, several of the Canadian CCS activities are adequately regulated under the oil and gas legislation (IEA, 2010b). In 2009, the Government of Alberta introduced the Carbon Capture and Storage Funding Act, which aims to "encourage and expedite the design, construction and operation of carbon capture and storage projects in Alberta." (Province of Alberta, 2009a).

In March 2012 the EPA issued the first Clean Air Act performance standard for carbon pollution from future power plants (epa.gov/carbonpollutionstandanrd). The Standard limits emissions to 1000 lbs CO_2/MWh (454.4 kg CO_2/MWh), which is about one-half of the typical emissions from a coal-fired power plant and about equal to typical emissions from a combined cycle natural gas power plant. The Standard applies to all power plants that will be constructed beginning 12 months hence. In the short term this will encourage the deployment of natural gas power generation over coal-fired generation. Over the long run, if natural gas power generation over coal-fired generation. Over the long run, if natural gas prices rise above the level where coal generation plus 50% CCS is economically competitive, the Standard will encourage the deployment of CCS.

These CCS-specific legal frameworks are part of both countries' efforts to manage carbon emissions, and in their national policies that promote CCS as a major contributor to climate change mitigation. While there is no nationwide commitment to reduce emissions in the United States, both the American Power Act (2010) and the American Clean Energy and Security Act (2009) propose forming a nationwide cap-and-trade program and creating other incentives and standards for increasing energy efficiency and low-carbon energy use (US EPA, 2010). In Canada, even though no national legislative framework addressing CO_2 emissions reduction has been officially proposed, get several provinces have adopted legislation that tax or cap CO_2 emissions. In 2003, Alberta introduced the Climate Change and Emissions Management Act, requiring companies to reduce the intensity of their GHG emissions by 12% and providing funding for CCS-relates activities (Province of Alberta, 2009b). Ontario introduced regulations for a cap and trade system in 2009. Similarly, in 2009, Quebec introduced a bill that provides a framework for a provincial GHG emissions cap-and-trade scheme (Quebec National Assembly, 2009), and British Columbia introduced a carbon tax in 2008 (Legislative Assembly of British Columbia, 2008).

Another important legislative aspect that affects the deployment of CCS is subsurface resources ownership and mineral rights. In the United States, no national legislation exists to regulate subsurface and pore space ownership, especially for the purpose of CO_2 geologic storage. Nevertheless, taking into account that property rights are generally regulated at the state rather than the federal level, some states have taken the initiative to address the issue of pore space ownership. In 2008, Wyoming enacted legislation that gives the ownership of subsurface pore space in the state to the owners of the surface (Duncan et al., 2009). In Canada, no specific legislation regulates subsurface rights and pore space ownership. Nevertheless, the Canadian Mineral Resources Act gives the Crown (and thus the federal government) ownership over "all minerals existing or which may be found within, upon or under lands in the province. (Government Canada, 2009)."

Equally important, both the United States and Canada have devoted large amounts of federal funds to promote the deployment of CCS. In 2009, the United States announced US$2.4 billion funding for CCS projects (USDOE, 2008). Similarly, Canada committed to more than $US_{2009}$$48 million funding for CCS R&D initiatives and around $US_{2009}$$2 billion for large-scale demonstrations. Here is it important to note that the Government of Alberta announced a $US_{2010}$$1.9 billion fund to encourage the large-scale deployment of CCS (Mourits, 2009). Along with budget allocations, both countries have taken the effort to raise public awareness and analyze the community's perspectives on CCS through focus groups and interviews. In the US, some studies found that factors such as past experience with government, existing low socioeconomic status, desire for compensation, and/or perceived benefit to the community contribute to shaping, along with the risks associated with CCS technology, the public opinion on CCS deployment (Bradbury et al., 2009). In Canada, several studies have been conducted and several initiatives have been taken to enhance public engagement in discussions about CCS deployment in Canada (Natural Resources Canada, 2006).

13.6.4   China

13.6.4.1   Technical Potential and Source-Sink Matching

In 2007, China ranked first in global CO_2 emissions with 6538 Mt (IEA, 2008a). Electricity and heat production, as well as industrial manufacturing, contribute to the majority of these emissions, followed by residential use, transportation, and others. In 2000, the IEA estimated China's CO_2 emissions from stationary sources to be almost 3000 Mt, projected to increase up to 4600 Mt in 2010. The power sector contributes more than 90% of stationary CO_2 emissions, and 73% of these emissions come from coal-fired power stations. In 2005, the six largest CO_2 point sources existed in China, all of which are power stations emitting a total of 227 $MtCO_2$ annually. Other CO_2 point sources include iron, steel, ammonia, cement, ethylene, and ethylene oxide facilities as well as refineries. Until 2005, 77 pure CO_2 sources existed in China, forming attractive early opportunities for low-cost CO_2 capture (APEC, 2005).

Significant storage capacity has been documented in depleted and depleting oil and gas reservoirs, saline aquifers, and unminable coal seams. The *CO_2 Storage Prospectivity of Selected Sedimentary Basins in the Region of China and South East Asia* report, issued by Asia Pacific Economic Cooperation in 2005, identifies several potential storage basins with high, intermediate, and low prospectivity (APEC, 2005). An overall storage capacity of 1445 $GtCO_2$ has been reported in China by APEC, and more recent investigations indicate the capacity may be as high as 3080 $GtCO_2$ (Dahowski et al., 2009). In more detailed studies, depleted and depleting oil and gas fields storage potential has been estimated to be 9.7–21 Gt, which makes it a short-term sink for a small fraction of China's CO_2 emissions from stationary sources (approximately 3–7 years of storage capacity based on 2007 emissions level) (Wang, 2010). Geological storage potential in unmineable coal seams is estimated to be approximately 10 $GtCO_2$, including 4 $GtCO_2$ in Ordos, 2 $GtCO_2$ in Turpankumul, and more than 1 $GtCO_2$ in Dzungaria (Wang, 2010). In fact, one study reports that CO_2 storage capacity by enhanced coalbed methane (ECBM) can be up to 12.08 $GtCO_2$ (Petra China Company Limited, 2007). Finally, 110–360 $GtCO_2$ is the estimated CO_2 storage potential in saline formations (Wang, 2010; 240, 2008). A more recent assessment suggests there is a capacity of about 3080 $GtCO_2$ in China, with all but 780 $GtCO_2$ available on shore (Dahowski et al., 2009).

The aforementioned APEC report shows that most large CO_2 point sources (10–55Mt/yr) are located either on or close to high- and intermediate-prospectivity storage basins. Thus, the distribution of major CO_2 stationary point sources and analyzed storage basins throughout China shows reasonable proximity between sources and sinks and some potential for source-sink matching.

13.6.4.2   Technological Maturity

China continues to develop the technological infrastructure required for CCS deployment. The NETL reports eight CCS projects in the country, many of which are currently active. This includes pre-combustion capture, EOR, and ECBM, at the R&D, demonstration, and industrial/commercial scales (Wang, 2010; BRGN, 2009). In this regard, CO_2-EOR is being actively developed, and more than 10 pilot projects have been implemented. In addition to increasing oil production, some of these projects aim to test permanent storage of injected CO_2, such as the Research on Exploitation of Natural Gas with Higher CO_2 Concentration, CO_2 Storage and Comprehensive Utilization of Resources in Jilin Oil Field project, which was initiated in August 2007. This rapid development of pilot projects enhances China's capability of early deployment and testing of CCS (Luo, 2008).

According to the US Energy Information Administration (US EIA), China's total oil and gas production in 2007 reached 192.2 t/yr (4.0 million bbl/day) and 69.2 billion m³/yr (2,446 bcf), respectively. Since 1987, oil production and consumption in China increased by almost 25% and 350%, respectively, and both gas production and consumption increased by almost 500% (US EIA, 2010b). As such, it would be reasonable to argue that China's experience in oil and gas exploration and production and rapidly developing energy infrastructure help in the early deployment of CCS.

13.6.4.3   Factors Influencing Regional CCS Costs

As mentioned before, many of the currently proposed CO_2 storage sinks in China are onshore, and many CO_2 point sources are located either on or close to potential geologic storage formations (APEC, 2005). Both factors favor economic implementation of CCS. Nevertheless, even though several CCS pilot projects are currently implemented, the design and layout of the potential CO_2 infrastructure and transportation network in the country are not fully investigated yet. Accordingly, little information exists on the effect of population density, landscape topography, and environmental factors on the construction of the CCS infrastructure.

Still, China's significant dependence on coal in fueling its economy and energy sector favors the country's adoption of CCS as an economically attractive option to reduce CO_2 emissions. According to the IEA *World Energy Outlook* in 2007, coal accounts for approximately two-thirds of China's energy needs, 80% of its electricity fuel mix, 50% of its industrial fuel use, and 60% of its chemical fuel use. In addition, oil and coal account for 81.6% of the country's overall energy demand in 2005; renewable energy (including nuclear) accounts for less that 14% of the energy demand. In terms of future projections of energy use, China generated around 622 GW of electricity in 2006, with additional installed capacity of 100 GW since 2005. Over 90% of this capacity increase was coal-fired. By 2030, the electricity generation capacity is expected to increase up to 1755 GW, Taking into account that China's domestic coal reserves are the second largest in the world, and given the coal's relatively low price, some studies suggest that even with strong policy incentives for energy efficiency, renewables, and other low carbon technologies, coal will remain a major contributor to China's energy fuel mix for the coming two decades. Thus, CCS is expected to be a major, option for reducing China's CO_2 emissions as part of its efforts to combat climate change (FIndlay et al., 2009).

13.6.4.4   Human Capacity, Research and Development

Several Chinese governmental agencies, academic institutes, and industrial/commercial businesses are actively involved in many aspects of CCS both locally and internationally. CCS-related research initiatives have been authorized, initiated, and supported by governmental agencies, including: Research for Utilizing Greenhouse Gas as Resource in EOR and Storing It Underground (Project 973), supported by the Major State Basic Research Development Program of the People's Republic of China, and Carbon Capture and Storage Techniques (Project 863), supported by the National High Technology Research and Development Program of China (Wang 2010). In addition, many aspects of the CCS technology are investigated by Chinese academic institutions, including: CO_2 absorption, CO_2 adsorption, membrane separation, membrane adsorption, CO_2 storage, and others (Wang, 2010). China National Offshore Oil Corporation and PetroChina are also active participants in CCS research (Wang 2010). China National Offshore Oil Corporation supports a project investigating the utilization techniques of CO_2 on large-scale. PetroChina is involved in several projects promoting CCS, including: the aforementioned Project 973, Pilot Test of CO_2 EOR and Storage in Jilin Oil Field, and Research on Phase Theory of Multiphase and Multi-component during CO_2 Flooding Process (Shen and Jiang, n.d.) In this regard, the CCS R&D network database of GCCSI shows that a total of 18 organizations in China are involved in CCS R&D, the majority of which is related to CCS technologies (GCCSI, 2011b).

In addition to the local efforts, China is actively engaging in international collaboration on CCS initiatives. For example, China is a member of the Carbon Sequestration Leadership Forum (CSLF), GCCSI, and the Asia Pacific Partnership on Clean Development and Climate Change. In addition, China is collaborating with the European Union through the Cooperation Action within CCS China-EU, Geo-Capacity, and Support to Regulatory Activities for Carbon Capture and Storage. Similar collaboration efforts exist between China and the United Kingdom through Near Zero Emissions Coal, Australia through the China-Australia Geological Storage, and the United States through some projects supported by the US EPA (Findlay et al., 2009). All aforementioned research efforts, as well as the vigorously developing energy sector, discussed in the section above, contribute to building a competently educated human capital that would be expected to play a major role in advancing CCS development. In that regard, China have both hosted and participated in several conferences and workshops that educate and train on CCS topics and guide various stakeholders on how to engage the public in CCS activities.

13.6.4.5   CCS Legal and Policy Initiatives

Although no specific mandates currently exist to regulate CCS activities in the country, China's national policies promote CCS as a major climate change mitigation option. *China's National Climate Change Programme* report, issued in 2007 by the National Development and Research Commission, states that strengthening the development of advanced and suitable technologies such as carbon dioxide capture, utilization, and storage is one key area to combat climate change (CNDRC, 2007). However, China Roundtable Meeting Notes from the IEA CCS Roadmap shows that the CCS profile in the Chinese climate and energy policy is currently low, due to its high cost and perceived status as an emerging option for GHG mitigation that requires more R&D (OECD/IEA, 2009). In terms of subsurface resources, pore-space ownership, and mineral rights, the Chinese constitution and the Mineral Resources Law of China explicitly state that the "mineral resources are owned by the state." The State Council exercises the state ownership over the mineral resources. In 1996, the National Mineral Resources Commission was established to strengthen the central government's control over mineral resources and protect its mineral rights (Chinese Government, 2003).

Another important aspect affecting the deployment of CCS is governmental funding. Currently, only a modest funding is provided by the Chinese Ministry of Science and Technology on CCS R&D. The government is not providing support to CCS due to its high cost and energy penalty. In that regard, some suggest that it would be valuable to develop a roadmap for China's development of CCS, which would particularly focus on early opportunities for profitable CO_2-EOR and explain the costs and benefits of a progressive expansion of the Chinese CCS industry from early demonstration. In addition, the Chinese government can be expected to increase expenditure on CCS if international funding resources become available (OECD/IEA, 2009). Still, money alone cannot ensure a wide deployment of CCS; public support is an essential element too. Some studies have been conducted to raise public awareness and analyze the community's perspectives on carbon capture and storage. The results showed that the awareness of CCS was low among the surveyed public in China, compared to other clean and renewable energy options. Further analysis of the results shows that the community's understanding of the characteristics, risks, and potential regulations of CCS are all important in predicting and promoting CCS public acceptance (DUAN, 2010).

13.6.5 Sub-Saharan Africa

13.6.5.1 Technical Potential and Source-Sink Matching

The short-term technical potential for CCS is generally considered low as the amount of CO_2 point sources amenable to capture are limited. With the exception of South Africa, many sub-Saharan countries lack large, energy-intensive industries, coal- or gas-fired power production and other stationary CO_2 sources. Most countries have very low or even negative GHG emissions. According to the 2006 version of the IEA GHG database on CO_2 point sources, sub-Saharan Africa has a total of 155 point sources, 79 of which are power and 76 cement, iron and steel, ethylene and ammonia, amounting to some 241 $MtCO_2$ emissions from power and some 46 $MtCO_2$ industrial emissions (IEA, 2008c). Of both, around half of the emissions originates in South Africa.

In the future, CO_2 emissions in Africa are projected to rise. IEA (2009) projects CCS activities in Africa from around 2030 onwards.

Several African countries have an abundance of coal and plan to use it.

Geological storage potential for CO_2 in sub-Saharan Africa is generally not known, as no comprehensive assessments have been carried out. However, various countries have known sedimentary basins or have done underground exploration for other economic reasons, such as hydrocarbon exploration or searching for underground drinkable water reservoirs in arid areas. For instance, it is likely that oil-exporting countries along the west coast of Africa, such as Nigeria, Angola, and Gabon, have areas of high geological storage prospectivity (IPCC, 2005) but also a country like Mozambique with oil and gas exploration (ECN, 2010).

South Africa has done a focused geological storage atlas. Although the atlas was not public at the time, some early results can be conveyed (Cloete, 2010). South Africa does not appear to have an abundance of suitable sedimentary basins, and early results suggest that the coal fields in the northeast (northern Karoo), where most of the large-scale emission sources are, also do not appear prospective.

13.6.5.2 Factors Influencing Regional CCS Costs

Literature around costs of CCS in sub-Saharan Africa is sparse. What is remarkable is the presence, however, of the largest single high-purity source of CO_2 in the world: around 30 $MtCO_2$ from a Sasol-operated coal gasification plant in South Africa (Cloete, 2010; Surridge, 2010). This could be an early opportunity if suitable storage capacity can be found nearby.

In oil-producing countries, in the longer term, potential for EOR might exist but such suggestions are highly hypothetical at the moment and not supported by literature.

13.6.5.3 Technical and Human Capacity

South Africa has an unofficial target of realizing one demonstration plant for CCS by 2020 and developing the technical and human capacity for this demonstration along the way. The main instrument for developing human capacity is the South African Centre of Excellence on CCS, founded in 2009 during a conference on CCS, which was also attended by representatives from surrounding countries. The Centre, however, still needs to gain critical mass. In addition, there is capacity on CCS within South African companies such as Sasol and Eskom. Additionally, South African industry representatives support the CO2CRC in Australia.

In Africa broadly, a few capacity building activities have taken place. In 2007, two regional workshops were held on CCS and CDM, one aimed at West Africa in Dakar, Senegal and the second focusing on Southern Africa in Gaborone, Botswana (ECN, 2010). In 2010, country workshops were held in Botswana, Mozambique, and Namibia. There is no R&D on CCS taking place in sub-Saharan Africa.

13.6.5.4   CCS Legal and Policy Initiatives

There are no sub-Saharan countries with any developments toward legislation on CCS. The only potential incentive for CCS might have come through the CDM. Botswana and South Africa are the only sub-Saharan countries that have mentioned CCS in their submissions to the UNFCCC for the Copenhagen Accord. The World Bank is interested in providing financing for feasibility studies on CCS in Botswana. Capacity for regulatory permitting of CCS projects in sub-Saharan Africa is currently largely absent, and apart from capacity building efforts mentioned above, little is being undertaken now to change this.

13.6.5.5   Discussion of CCS in Sub-Saharan Africa

CCS in sub-Saharan Africa is an option for the longer term, given the presence of coal in southern Africa that is projected to be used to fulfill the continent's growing energy and development needs. Before CCS can be realized, human and technical capacity needs to increase, regulation and policy need to be developed, and better insight in geological storage capacity is needed. South Africa is most progressed with a demonstration plant anticipated for 2020 and a pathway toward it. Also in South Africa, short-term, low-cost CCS potential may exist in the form a large coal gasification plant if suitable storage reservoirs can be found within a reasonable distance.

13.6.6   Former Soviet Union

13.6.6.1   Technical Potential and Source-Sink Matching

CO_2 emissions related to fuel combustion in the Former Soviet Union (FSU) region decreased by 34% from 1990 to 2008, reaching 2427 Mt, while population grew by 8%. Russia's emissions declined by 27% to 1594 Mt and Ukraine by as much as 55% to 310 Mt. Energy supply decreased by 27% over the same period in the FSU, with a first period of decline from 1990 to 2000, followed by an expansion from 2000 onwards. Electricity and heat production contributed in 2008 to 50% of the emissions from fuel combustion, followed by manufacturing (17%) and transport (14%). Gas represented 50% of the emissions, declining by only 7%, to be compared with coal (-42%) and oil (-57%). Given the abundance of natural gas resources in the region, applications of CCS would be mostly limited to storage from fuel transformation and natural gas-fueled power plants. A number of gas fields in the Caspian region have relatively high CO_2 and/or H_2S content, and some of the field developments have been delayed due to the issue of sour gas handling.

An assessment of sources and sinks matching in the Baltic States has been made within the EU-funded Geocapacity and CO2NET East projects (Shogenova et al., 2008). There are 24 large sources of CO_2 emissions (greater than 0.1 Mt/yr), with Estonia's upstream production (associated with the extraction and use of oil shales) amongst the largest. While

there are no suitable storage areas in Estonia, several prospective formations exist in the Baltic sedimentary basin, with solubility trapping capacities in the range of 13 Gt.

Russia is the country with the single highest storage potential after the United States, with more than 2000 Gt; the capacity of depleted oil and gas fields in the Western Siberian Basin alone is in the order of 150–200 Gt. However, given the distribution of CO_2 emissions (mostly located in the European part of Russia) and the potential storage sites, distances for pipeline transport of 2000–4000 km have to be considered, significantly increasing the cost for the CCS chain.

13.6.6.2   Technological Maturity

The Russian oil and gas and petrochemical industries have a long experience with CO_2 capture, transport, and storage. Use of CO_2 from anthropogenic sources has been investigated since the early 1980s in Russia (Kuvshinov, 2006). Large-scale pilot tests have been carried out to inject CO_2 and other flue gas for enhanced oil recovery. Areas that have promising source and sink matching include the onshore Black Sea area (oil fields near Krasnadar), the Baskortostan (near Ufa), Tatarstan (near Samara), and Perm oil fields. Enhanced coalbed methane potential also exists in the coal fields in the southern part of Russia, but given the current levels of oversupply in gas markets, the application of technology in those prospects is unlikely.

13.6.6.3   Factors Influencing Regional CCS Costs

The cost of CO_2 transport and subsurface injection is well-documented in the FSU, given the volume of oil and gas operations in the region.

13.6.6.4   Human Capacity, Research and Development

In addition to the EU FP6-funded projects for CCS potential in the Baltic States, a call for proposals for clean coal applications has been made in 2010 by the European Commission to explore the potential for some of the Caspian states, including Kazakhstan. In Russia, several technological institutes related to oil and gas production have in-house capabilities for CO_2-EOR, as well as subsurface knowledge. The oilfield service sector has also developed the capabilities to use the building blocks of site assessment for CO_2 storage.

13.6.6.5   CCS Legal and Policy Initiatives

Emission levels decreased significantly during the economic collapse that followed the separation of the FSU states. Therefore, there are no incentives for Russia to use CCS, except if appropriate international mechanisms allowed a monetization of storage. While the current efforts are directed toward gas flaring reduction as a priority, political willingness exists for promoting CCS. The 2006 G8 Summit in St. Petersburg included a joint statement from the Russian Academies of Science promoting research, development, and demonstration in the areas of carbon dioxide storage for energy sustainability.

13.6.7   Australia

13.6.7.1   Technical Potential and Source Sink Matching

CO_2 emissions related to fuel combustion in Australia increased by 56% from 1990 to 2009, reaching 417 Mt, while population grew by 25% (EIA on-line data base). The single largest contributor is stationary sources (particularly coal-fired power station), with 2006 total stationary emissions around 280 $MtCO_2$.

Like most countries, Australia is pursuing a broad portfolio approach to GHG mitigation, although it currently excludes nuclear power from the portfolio. It has in place a Mandatory Renewable Energy Target (MRET) of 20% of electricity from renewables by 2020 and a longer-term goal to reduce emissions by 60% from 2000 levels by 2050. It also has a range of mitigation options under consideration, such as biomass, enhancement of soil carbon, fuel switching (coal to natural gas), and greater energy efficiency. Important though all these measures are, the reality is that at present the rate at which they are being implemented is not keeping pace with the rise in demand for electricity, with most of that increase in demand being met from increased use of coal.

Australia has abundant gas reserves, but currently coal-fired power generation is cheaper than gas in the absence of a price on carbon. In 2011 the Australian governments announced a carbon tax for the 500 top emitters of US_{2011}\$24. However, a carbon price in the range US\$20–30/tCO_2 is unlikely to produce a marked change in consumer behavior or significant deployment of CCS in Australia in the short term. At the same time, there is widespread recognition that given Australia's high level of economic dependency on inexpensive coal for power generation (and its coal exports), it potentially has more to gain than most countries from the successful deployment of CCS. In the short term there is scope for replacing some coal-based power generation with gas, with a commensurate decrease in GHG emissions, and almost certainly this will happen to some extent. In addition, given Australia's move to 20% renewable power, there is likely to be increased use of gas to handle the intermittency of wind power. Gas substitution may provide some reprieve from ever-increasing emissions in the next decade, but in the longer term gas, like coal, becomes part of the greenhouse problem, as gas-related CO_2 emissions rise. There may also be some increase in the use of biomass, particularly for COGEN. However, whether it is coal-, gas-, or even biomass-based power generation, the likelihood is that CCS will play a future role. With this in mind, Australia has taken a number of recent CCS initiatives, including an update of its storage potential through the Carbon Storage Taskforce (CST, 2009).

The major sources of CO_2 are located primarily in eastern Australia, where the major population centers are located, although it is anticipated that the emissions will grow significantly in northwestern Australia as new LNG production comes on stream along with other industrial developments. The Carbon Storage Taskforce (CST, 2009) has identified 10 concentrations of stationary emitters, with sources sufficiently close together that a "hub approach" to CCS may be feasible. There is, or will be, some storage potential in depleted oil and gas fields, estimated by the Taskforce (CST, 2009) at 16.5 $GtCO_2$, most of which is offshore. The depleted fields of the Bowen-Surat Basin in southeast Queensland are the most prospective onshore, with the Gippsland Basin the most prospective offshore region. However, many of the fields are still some years away from being depleted, and therefore overall, the potential for storage in depleted oil or gas fields is quite modest in the short to medium term. CO_2-EOR is regarded as having only very limited potential because of the light nature of Australian crude oil.

The major CO_2 storage opportunities for Australia lie in saline aquifers. The Taskforce (CST, 2009) has determined the technical storage capacity (at 90% confidence level) as 10.9–87.5 $GtCO_2$ for eastern Australia, using storage efficiencies of 0.5% and 4%, respectively, and for western Australia the capacities are 12.3 $GtCO_2$ (0.5% efficiency) and 98.5 $GtCO_2$ (4% efficiency). Total storage capacity is estimated at between 33 and 226 $GtCO_2$, suggesting adequate storage capacity for at least this century and possibly for several centuries, assuming a storage rate of 2–300 $MtCO_2$/yr. It is important to treat these figures with caution, as knowledge of many saline formation systems is still quite limited. Nonetheless, the values do provide confidence that Australia has sufficient storage potential to meet its CCS needs for many years to come.

13.6.7.2   Factors Influencing Regional CCS Costs

Most of Australia is quite sparsely populated with widely dispersed CO_2 sources, which adds significantly to the cost of CCS infrastructure. However, as pointed out earlier, there are a number of emission hubs, where a coordinated approach could be taken to CCS in order to bring down costs. The most extensively investigated hub is that of the Latrobe Valley in eastern Victoria, where brown coal-fired power stations collectively produce one of the largest regional sources of CO_2 in the country. The Gippsland Basin, located only 150 km away, is an excellent prospect for large-scale CO_2 storage with a technical storage capacity of 4–5 $GtCO_2$ at the P90 level, assuming a conservative storage efficiency of 0.5%. Therefore, it is highly likely that a Latrobe-Gippsland source-sink could function very effectively for the remainder of this century. South Central Queensland may also offer similar opportunities, using the Bowen-Surat Basin. Other source-sink matches are also possible with some of the natural gas/LNG-related sources offering effective CCS options for the future. Some of the other emission nodes are less favorably located for nearby large-scale storage, with transport distances in excess of 500 km being contemplated in some instances, such as New

South Wales. Therefore, in some areas, the need for large-scale long distance pipelines will add considerably to the cost of CCS. Those costs will be increased by the need for recompression over long distances with substantial power requirements.

The possible impact of CO_2 storage on other resources could also add to the cost of CCS and result in delays in implementation. For example, while the Latrobe-Gippsland source-sink hub is the highest ranked in terms of storage capacity and cost, it is also a major oil and gas producing basin and will be for quite some years to come. Careful management of CO_2 storage would obviously be necessary to ensure that there is no adverse impact on oil or gas production. In the case of some of the Queensland storage options, the major concern is to ensure that any CO_2 storage does not adversely impact on the significant fresh water resources of the Great Artesian Basin.

The Taskforce (CST, 2009) summarizes the main factors affecting the economics of CO_2 storage as "location (the distance from the CO_2 source to the storage location determines pipeline costs), reservoir depth (influencing well costs) and injectivity parameters (notably permeability and differential pressure, which determine the number of wells needed)." As a result of these factors, CCS costs vary considerably throughout Australia, ranging (for transport plus storage only) from less than US$10/t$CO_2$ avoided for the Victorian Latrobe-Gippsland option to more than US$100/t$CO_2$ avoided for the proposed New South Wales Sydney-Cooper proposal.

The further challenge to deployment in Australia and elsewhere is to establish the appropriate business model for taking the hub concept forward. Several models are under consideration at the moment, including as a government-owned utility, as a public-private partnership, and as a privately-owned utility. There is general agreement on the need for any pipeline network to have excess capacity to handle future growth in emissions, but at the same time, there are no agreed mechanisms to pay for that upfront investment to provide the excess capacity. Finally, in many parts of Australia, as in many parts of the world, there is a lack of knowledge of the deep geology of many sedimentary basins, which in turn makes it difficult to provide a confident assessment of storage capacity. Paucity of geological information represents a significant investment risk in many areas. The Taskforce (CST, 2009) has proposed a national program of precompetitive storage assessment costing in excess of US$250 million, and there is no question that such a program would be extremely useful, but for the present the program remains unfunded.

On a more positive note, while knowledge of CCS in the Australian community is limited, there have been extensive programs to inform the wider community on CCS through the activities of state and federal governments, CO2CRC, and CSIRO. As a consequence, there is no entrenched community opposition to CCS. Indeed, Australia has had an operational storage project, the CO2CRC Otway Project, underway since March 2008, which has been able to proceed with a significant level of local support and which has received national media coverage. Therefore, while every CCS project will need to involve the community in a careful and considered manner, there are no obvious showstoppers at this time in terms of community concerns regarding the technology.

13.6.7.3 Human Capacity, Research and Development

Australia has a strongly developed skills base in the resource sector, particularly in the geosciences, materials science, and engineering, all of which are skills directly relevant to CCS.

Research and training in CCS commenced in Australia in 1998 through the Petroleum Cooperative Research Centre and its Geodisc Program and greatly expanded from 2003 onwards through the successor Cooperative Research Centre for Greenhouse Gas Technologies (CO2CRC). Through this Centre, there is an unusually broad and active program of collaborative research involving most of the major universities as well as CSIRO and Geoscience Australia. The Centre also works closely with overseas research bodies. Through CO2CRC and its collaborating universities, a large number of people have been trained in CCS. More recently a number of the bodies within CO2CRC have also developed their own specific research activities, but the Centre is still the main research body that brings CCS research together. CO2CRC is currently funded to 2015 and, with a combination of industry and government funding and in-kind contributions from research providers, has an annual budget of US$20–25 million for its research into CO_2 capture transport and storage.

CSIRO, Australia's main research body, has a CCS program focused in particular on post-combustion capture of CO_2 with several pilot projects underway. It is also engaged in research into CO_2 storage, although most of that is through its collaboration with CO2CRC.

Geoscience Australia (the Federal Geological Survey), together with a number of the State Surveys, have a national program underway to assess storage capacity and develop infrastructure plans.

Australia has been active in the development of CCS training and capability in a number of countries, particularly in East Asia, where a number of courses have been delivered by CO2CRC, GA, and GCCSI.

At present, the Government of Australia has chosen to use direct financial assistance to take CCS forward, and the states have a similar though more modest approach. At the state level, the planning approval process has also acted as a way of ensuring that projects consider CCS as a mitigation option. However, up to now, the primary focus of the government, and to some extent industry, has been to provide funding through a range of CCS initiatives that will run until 2015 or beyond.

In late 2009, the federal government announced the formation of GCCSI with an initial annual contribution of $US_{2005}$$ 65.5 million ($A_{2009}$$ 100 million) from Australia for four years and the hope that other countries might contribute at some stage in the future. The GCCSI is not involved in research and is focused on global deployment of CCS and the aim of the G20 countries to have 20 large-scale CCS projects by 2020.

The Australian National Low Emissions Coal Research and Development is a recent joint government-coal industry initiative. $US_{2010}\$245$ million will be made available over five years to 2015–2016 to support low emission coal technology demonstrations in Australia and to understand and address the deployment risks that early demonstration projects may face.

The National Low Emission Coal Council (now called the National CCS Council) was established in 2008 to provide guidance to the government on the overall direction of CCS development and deployment in Australia. The initial focus on coal had the effect of leaving the oil and gas industry largely out of the deliberations, despite the importance of that sector to future CCS deployment. This has now been addressed and in future initiatives will deal more broadly with CCS. The related Carbon Storage Taskforce has involved the oil and gas industry and has now produced a valuable report for a national carbon mapping and infrastructure plan, which, if implemented, will produce a major national assessment of opportunities for geological storage.

The federal government's CCS Flagships Program, should help to accelerate the deployment of large-scale integrated CCS projects in Australia through federal grants totaling $US_{2005}\$$ 1.0 billion ($A_{2009}\$$ 1.3 billion). Along with matching industry and state funding, it is expected that this will fund two large-scale CCS projects. Bids are currently being assessed from project proponents. Along with the Flagships Program, the Australian Government has established a companion Education Infrastructure Fund, totaling approximately $US_{2005}\$$ 130 million ($A_{2009}\$$ 100 million), to support research partnerships between the Flagships and research institutions, with CSIRO, and CO2CRC the designated Lead Research Organizations.

13.6.7.4   CCS Legal and Regulatory Initiatives

Australia has established an offshore regulatory regime for the geological storage of CO_2 for offshore areas under federal control. Victoria has its own onshore CCS legislation in place and several other states are in the process of defining onshore regulations for CO_2 storage. Western Australia has to date taken a project-specific approach to regulation to enable it to permit CO_2 storage under Barrow Island as part of the Gorgon LNG Project. The issue of long-term liability has proved complex in Australia, as it has with most jurisdictions. However, an agreed process is now in place to transfer offshore storage liability to the federal government at the conclusion of offshore injection and once a number of other performance criteria have been met. The onshore approach to long-term liability (under State jurisdiction) is less clear and varies from state to state. In Victoria, a regulatory regime under

the R&D provisions of the EPA enabled the CO2CRC Otway Project to go ahead, but under the more recent CCS-specific provisions of that state's GHG legislation, it would be much more difficult to take such a research project forward. Therefore, one of the clear lessons from the Australian experience is to ensure that regulations take full account of the research needs of CCS. Nonetheless, overall Australia is probably more advanced than most other countries in the development of CCS legislation.

13.6.7.5   Australian CCS projects

A combination of various state and federal financial initiatives, coupled with general recognition of the importance of CCS to the Australian economy and the Australian resource and power industry, has resulted in a number of planned or proposed projects, which are summarized below and in Figure 13.11b.

The Callide Oxyfuel Project in Queensland involves oxy fuel conversion of an existing 30MW unit at Callide A (currently underway) with power generation and capture of CO_2 commencing in 2013. A second stage of the project may involve the injection and storage of captured CO_2 into a saline aquifer or depleted oil/gas fields over about three years, commencing in 2013, but as yet a suitable site has not been identified. The cost estimate for the project is approximately $US_{2010}\$218$ million. The project involves CS Energy, IHI, Schlumberger, Mitsui, J-Power, and Xstrata, with extra funding from the Australian Coal Association and the Australian and Queensland governments.

The CarbonNet Project in Victoria, a CCS Flagship Project proposal, is for the development of the infrastructure for a storage and transport hub in the Latrobe Valley. It is coordinated by the state of Victoria. It aims to collect, transport, and store 3–5 $MtCO_2$/yr from Latrobe Valley industry, including coal-fired power plants (and may involve both pre- and post-combustion capture), with storage in the Gippsland Basin.

The Collie South West Hub Project in Western Australia is a CCS Flagship Project proposal coordinated by the state of Western Australia. It aims to store up to three $MtCO_2$/yr captured from industrial and power plants located southwest of Perth. At the present time, the biggest challenge the project faces is identification of a suitable storage site.

The CO2CRC Otway Project in Victoria is Australia's only operational storage demonstration project. Injection of CO_2 from a nearby gas well, initially into a depleted gas field at a depth of two km, began in April 2008 with injection of 65,000 tCO_2-rich gas to date. A major program of monitoring and verification has been implemented. A new well was drilled in early 2010 and a new phase of injection has been initiated. The A\$60 million project, which is supported by 15 companies and seven government agencies, involves researchers from Australia, New Zealand, Canada, Korea, and the United States. Partners include major gas, coal, and power companies, research organizations and governments.

Additional financial support is provided by the Australian Government, the Victorian Government, and the US DOE through Lawrence Berkeley National Laboratory.

The Coolimba Project of Aviva Corporation Ltd is an early stage proposal in Western Australia for 2x200 MW coal-fired base-load power stations with the plant built ready for conversion to CO_2 capture. Sequestration sites have been sought for the storage of about 3 $MtCO_2$/yr, but the project is presently suspended.

The Gorgon Project of Chevron (operator), Shell, and Exxon is at the early construction stage. It involves a major storage project linked to the Gorgon LNG Project. The separated CO_2 will be injected under Barrow Island at a depth of about 2.3 km, with injection of 3–4 $MtCO_2$/yr. A total of 125 Mt will be injected over the life of the project. A data well has been drilled, and a major study of the subsurface is underway. All government approvals have been granted, and the final investment decision for the project to proceed has been made. A number of contracts have been awarded, and construction is underway. The storage component of the project will cost in excess of A\$1 billion.

Galilee Power has an early stage proposal for a new 900 MW coal-fired power station incorporating CCS with storage of captured CO_2 in the Galilee Basin. Prefeasibility studies are underway, but activity is quite limited at the present time.

A post-combustion capture plant is operating at International Power's Hazelwood Power Station in Victoria. The solvent capture plant began operation in 2009 and is capturing and chemically sequestering CO_2 at a nominal rate of 10,000 tCO_2/yr. This project is partly funded by the Australian and the Victorian Governments.

The Latrobe Valley Post Combustion Capture Project (LVPCC) in Victoria is developing technologies for post-combustion capture from coal-fired power stations in the Latrobe Valley. The LVPCC involves International Power, Loy Yang Power, CO2CRC, and CSIRO and is partly funded by the Victorian Government. It comprises work at the Hazelwood Power Station by CO2CRC and by CSIRO at Loy Yang.

The H3 Capture Project, Hazelwood, Victoria, led by CO2CRC, is based at International Power's Hazelwood plant and exploits synergies with the Hazelwood Capture Project. A range of solvents and different process configurations are being tested using the solvent post-combustion capture plant. In addition, post-combustion techniques using adsorbent and membrane technologies are being developed, using two purpose-built rigs.

The Loy Yang Project in Victoria involved a CSIRO mobile pilot post-combustion capture facility. This has begun operation at Loy Yang Power Station and is capturing around 1000 tCO_2/yr. The facility investigated a range of solvent technology for CO_2 capture and has now concluded the research.

At the CO2CRC/HRL Mulgrave Capture Project in Victoria, CO_2 emissions were captured from HRL's research gasifier at Mulgrave in a pilot-scale pre-combustion project. The capture technologies were evaluated to identify which are the most cost-effective for use in a coal gasification power plant. Partners included CO2CRC and HRL, with funding from the Victorian Government. The research has now been concluded.

The Munmorah PCC Project in New South Wales has investigated the post-combustion capture (PCC) ammonia absorption process, and the ability to adapt it to suit Australian conditions. Tests to capture up to 3000 tCO_2 have been successfully completed. Partners involved in this project are Delta Electricity, CSIRO, and the ACA. The pilot plant has now been relocated to another site and a larger-scale demonstration project, incorporating geological storage, is under consideration.

CSIRO and Tarong Energy have installed a post-combustion capture pilot plant using an amine-based solvent at Tarong Power Station near Kingaroy, Queensland. The pilot plant will capture 1500 tCO_2/yr over a two-year research program. Construction has commenced.

Stanwell and Xstrata Coal are proposing the Wandoan Project, Queensland as a CCS Flagship Proposal. Identification of suitable storage sites in the Surat Basin of Queensland is being undertaken by the consortium.

13.6.8   Latin America

13.6.8.1   Technical Potential and Source-Sink Matching

CO_2 emissions related to fuel combustion in Latin America increased by 70% from 1990 to 2009, reaching 1212 Mt, while population grew by 32% (EIA on-line data base). Latin America includes South America, including Brazil, and Central America, including Mexico and the Caribbean. The region is a major oil producer; Mexico and Venezuela jointly produced 7.2 % of worldwide crude oil in 2009 (IEA, 2010a), and Brazil is up and coming as an oil producer and exporter. EOR is clearly seen as a possibility. Almost all countries in Latin America have CO_2 sources from cement production, refineries, and power, although much of the power in the region originates from renewables, in particular hydropower. Several of the larger countries have steel and ammonia plants and ethylene production. As companies are starting to explore EOR in the region, they are looking for high-purity CO_2 sources. In Mexico, an ammonia plant has been explored, and in Brazil, biomass production is being explored.

Except for Brazil, no comprehensive CO_2 storage atlas or assessment has been done in Latin America, although in Mexico and Argentina several reservoirs have been explored for EOR or storage. In Brazil, the CARBMAP project (Rockett et al., 2011) has done a rough assessment of sources, pipeline corridors, and reservoirs.

An additional possibility in the Latin American region is the combination of biomass and CCS (Möllersten et al., 2003). According to the IEA GHG (2008) database on CO_2 emissions, the CO_2 emissions from ethanol and biomass production are around 72 MtCO_2/yr, all of them in Brazil. These sources are high-purity and therefore amenable to capture, but also relatively small-scale at ca. 100,000–150,000 tCO_2 per source per year. Most of the sources are in the São Paulo region, relatively close to the ocean shore, although it is unclear at this point whether there might actually be storage potential close by. The Global Environment Facility may fund a small-scale bioethanol CCS project in Brazil.

13.6.8.2   Factors Influencing Regional CCS Costs

Distances are large and currently it is unclear whether there is storage potential and where it is. In Mexico and particularly Brazil, much of the EOR potential is likely to be offshore, negatively affecting costs.

13.6.8.3   Technical and Human Capacity

CCS does not have a long history in the Latin American region, but capacity has increased over the past five years. Countries like Mexico and Venezuela, with a significant oil industry, have some embedded capacity. In Brazil, a dedicated CCS center was established, the Brazilian Carbon Storage Research Center at the Pontifical Catholic University in Porto Alegre. Research is also done at the Sindicato da Industria Carbonifera de Santa Catarina (SIECESC). The Brazilian Carbon Storage Research Center's capabilities are clear from publications in peer-reviewed journals (e.g., Ketzer et al., 2007). The Center is partly funded by Petrobras, which is building up capacity internally for CCS and for CO_2-EOR.

13.6.8.4   CCS Legal and Policy Initiatives

There are no known legal or policy initiatives in Latin America. The Global CCS Institute commissioned a study, which examined current initiatives in Brazil (GCCSI, 2009b). This study reported that in Brazil, there is existing legislation for pipelines, environment impacts, and mining that would apply to CCS operations, but no major barriers were identified. Similar conclusions were drawn for Mexico (GCCSI, 2009c). In terms of policy incentives, the CDM may be a driver for low-cost opportunities provided the conditions in the latest UNFCCC decisions can be fulfilled.

13.6.8.5   Discussion of CCS in Latin America

CCS plays a minor role in the climate and energy debates in most Latin American countries. It is only seriously considered in Argentina, Brazil, and Mexico. In those countries, the first demonstrations are likely to be from industrial CO_2 sources rather than power, as much of the power in the region originates from renewables and there is some potential

for low-cost capture opportunities. Although in private and academic sectors capabilities are being developed, legal and regulatory frameworks are still absent in all countries. It appears that either EOR or international instruments, potentially CDM, with which the region has had some success, will be the main drivers for CCS in the coming years.

13.6.9   Middle East and North Africa

13.6.9.1   Technical Potential and Source-Sink Matching

CO_2 emissions related to fuel combustion in the Middle East and North Africa (MENA) region increased by 128% from 1990 to 2007, reaching 1770 Mt, significantly outpacing population growth over the same period (+44%). Over 95% of those emissions were related to the use of oil and gas, with oil-based transport and gas-based power generation having the largest growth over the last two decades. Given the abundance of natural gas resources in the region (and its lack of coal resources), applications of CCS would be mostly limited to storage from fuel transformation, and natural gas-fueled power plants. A number of gas fields in the MENA region have relatively high CO_2 and/or H_2S content, and some of the field developments have been delayed by the issue of sour gas handling. In North Africa, the main potential for capture is in Algeria, Libya, and to a lesser degree, Tunisia. Both in Salah Gas and Gassi Touil projects in Algeria have a CO_2 content as high as 10% with nearby storage reservoirs. In Libya most of the potential is from offshore fields with a potential of use of CO_2 for EOR, while in Tunisia the largest gas field in the country (Miskar) has nearly 13% CO_2 content. In the Middle East 60% of the proven gas reserves have more than 100 ppm of H_2S and/or 2% CO_2 (IEA, 2008b). Other opportunities for capture in the areas are in fuel transformation, particularly gas-to-liquids, as well as in the growth of gas-fired power plants, and in the developing petrochemical sector.

While no detailed study of storage potential has been made in the area, the global assessment performed points to a highly favorable sedimentary environment for the MENA region. The Middle East represents the largest future potential for storage in depleted fields with the five biggest sites having a combined 180 Gt of capacity. With the caveat that deep saline aquifers are poorly understood, combined storage capacity ranges for the MENA region are estimated at (IEA, 2008a) 200–1200 Gt for oil and gas fields and 50–550 Gt for saline formations.

The region has the highest potential incremental recovery from CO_2-EOR, with estimates of additional volumes of oil ranging from 80–120 billion barrels (IEA, 2008a). Given the lack of availability of CO_2 and the incremental cost, attempts to develop this tertiary method in the region are still limited. In 2009 Saudi Arabia announced plans for a CO_2-pilot project in a waterflood of the Arab-D reservoir (Ghawar field) that could be started in 2013 with the injection of 0.8 MtCO_2/yr, but stressed that the country did not need EOR on a large scale at this point. In the United Arab Emirates, a pilot project (the first in the Middle East) was started by

Abu Dhabi Company for Onshore Oil Operations (ADCO) at the end of 2009 for the injection of CO_2 in the Northeast Bab's Rumaitha carbonate reservoir, while a study was launched in 2010 to use CO_2 for EOR in the Lower Zakum oil field in Abu Dhabi.

13.6.9.2   Technological Maturity

Efforts to reduce the emission of GHGs from upstream operations have been ongoing for the last two decades, including a significant reduction in gas flaring. The MENA region has the first large-scale CCS projects outside of the Organisation for Economic Co-operation and Development with the BP-Sonatrach-Statoil In-Salah Gas project (see Section 13.4.2.2.3). The region also benefits from the experience gained in both surface and subsurface processes from the oil and gas industry. Turkey had one of the earliest CO_2-EOR project in the heavy oil field of Bat Raman. In 2009 the Abu Dhabi Future Energy Company confirmed plans for a major initiative to reduce emissions from the Emirates by half using CCS. The first phase of the project would involve the capture of up to 5 $MtCO_2$/yr from three sources (a gas-fired power plant, a steel mill in Mussafah, and an aluminum smelter at Taweelah). The plan also includes the development of a specific pipeline network and the injection in Abu Dhabi National Oil Company's oilfields.

13.6.9.3   Factors Influencing Regional CCS Costs

The cost of CO_2 transport and subsurface injection is well-documented given the volume of oil and gas operations in the region. The feasibility and optimization of Water Alternating Gas processes in enhanced oil recovery still requires detailed evaluation, as the conditions of EOR operations in North America are significantly different: well spacing, reservoir heterogeneity, and thickness, along with crude gravity, all play an important role in the incremental oil recovery with CO_2. Also the development of interstate pipeline networks would allow further cost optimization. The MENA region has ample and widely distributed storage capacity, which would allow matching sources and sinks relatively easily. What remains to be determined is the potential for the region to host the emissions from nearby European sources.

13.6.9.4   Human Capacity, Research and Development

In November 2007, the Organization of Petroleum Exporting Countries (OPEC) announced pledges for a US$750 million fund to develop clean energy technologies, in particular CCS, with the participation of Saudi Arabia, Kuwait, Qatar, and the United Arab Emirates. Several initiatives have been started in the region to develop technological capabilities, including the Masdar project, and the recently created Qatar Carbonates and Carbon Storage Research Centre. Many international workshops have been convened in the region to increase awareness and assess which areas of research are most appropriate in the Middle East context. Efforts to promote technology transfer in the region have been led by the Society of Petroleum Engineers and other professional societies, along with OPEC and national organizations.

13.6.9.5   CCS Legal and Policy Initiatives

Saudi Arabia and other OPEC countries have successfully negotiated for the inclusion of CCS as a Clean Development Mechanism, allowing developed countries to offset their emissions. While awareness in the region about carbon abatement rationales and options is generally low, there is also the concern by governments to prevent a strong curbing of hydrocarbon use, which may impact the region's economic growth. Therefore, options such as fuel switching for natural gas or CCS rank amongst the highest.

13.6.10   Japan

13.6.10.1   Technical Potential and Source-Sink Matching

In 2007 Japan emitted almost 1214 $MtCO_2$. The industrial sector, led by iron and steel manufacturing, contributes to the majority of the emissions (451 Mt), followed by transportation (257 Mt), commercial sector (215 Mt), residential sector (201 Mt), and energy conversion sector (90 Mt) (Aoshima, 2009). Thermal power stations, iron and steel manufacturing plants, and cement plants are the three major stationary sources assessed for CO_2 capture (Nakanishi et al., 2009).

A storage capacity of 146.1 $GtCO_2$ has been reported in Japan (Nakanishi et al., 2009; Takahashi et al., 2009). A total of 27 potential storage areas are being investigated throughout the country, four of which correspond to regions with large emission sources: Tokyo Bay, Ise Bay, the Osaka Bay area, and northern Kyushu (Nakanishi et al., 2009). Studies identify 18 saline aquifers that can be suitable for CO_2 with storage capacity ranging between 0.01–7 Gt per saline aquifer. However, the accuracy of evaluating storage volume and effectiveness varies significantly among the identified saline formations (Ogawa et al., 2009). Existing oil and gas reservoirs and formations investigated by exploration wells and seismic surveys are estimated to have a storage capacity of approximately 36.2 Gt. Other formations, investigated by seismic surveys only, are expected to store around 109.9 Gt (Nakanishi et al., 2009; Takahashi et al., 2009). The spatial distribution of major CO_2 stationary point sources and analyzed storage basins throughout Japan shows close proximity between sources and sinks and some potential for source-sink matching (Takahashi et al., 2009).

13.6.10.2   Technological Maturity

Japan continues to develop the technological infrastructure required for CCS deployment. Eight CCS projects are completed and/or currently active. This includes post-combustion capture, ECBM, and storage projects, at the R&D, demonstration, and industrial/commercial-scales (BRGM, 2009; Lund et al., 2008). Post-combustion capture has been implemented in Japan since 1991 to capture CO_2 emissions from Nanko power facility in Osaka. Recently, several post-combustion capture projects have been implemented in coal and gas power plants and chemical facilities. Between 2000 and 2007, the Nagaoka project involved a successful storage of 10,400 tCO_2 in a 1000 m deep saline aquifer. Also, in 2002, the first ECBM project was launched in the country with the aim of injecting CO_2 in 6–9 m thick coalbed 900 m underground in Yubari, Hokkaido (Lund et al., 2008). This rapid development of CCS projects improves Japan's experience in CCS and enhances its capability of early deployment and testing of CCS.

According to the US Energy Information Administration (US EIA), Japan has little domestic oil and natural gas reserves, which makes it the second-largest net importer of crude oil and largest net importer of liquefied natural gas in the world. In 2007 the country's total consumption of oil reached almost 5 million bbl/day (249 Mt/yr), compared to 130,000 bbl/day (6.47 Mt/yr) of production. Similarly, the total consumption of natural gas in the same year was about 100 billion m³ (3.5 trillion cubic feet), only 1 billion m³ (32 billion cubic feet) of which is produced domestically. Nevertheless, Japan has developed significant experience in the oil and gas industry by leading in technology and equipment development, establishing a robust refining infrastructure, and participating in exploration and production projects overseas. Today, Japan's refining capacity is the second-largest in the Asia-Pacific region, and almost 15% of the country's oil imports come from Japanese-owned concessions around the world, primarily in the Middle East (US EIA, 2010). As such, it would be reasonable to argue that Japan's experience in oil and gas exploration and production enhances the country's capability of developing CCS infrastructure both locally and internationally.

13.6.10.3   Factors Influencing Regional CCS Costs

As mentioned before, many of the currently proposed CO_2 storage sinks in Japan are onshore, and most CO_2 point sources are located either on or close to potential geologic storage formations (Nakanishi et al., 2009). Today, even though several CCS pilot projects are currently implemented, the design and layout of the potential CO_2 infrastructure and transportation network in the country has not been fully investigated (Nakanishi et al., 2009). Nevertheless, the safety aspects of CCS projects, as well as the effect of topography and various environmental factors on the cost and effectiveness of CCS implementation, have been recently addressed in a study by the Carbon Dioxide Capture and Storage Study Group within the Industrial Science and Technology Policy and Environment Bureau at the Japanese Ministry of Economy, Trade and Industry. In addition to detailing the geological requirements for safe storage of CO_2 at a large demonstration-scale, the study suggests a CO_2 transportation standard, a regulatory framework for assessing

operations' safety, a methodology for environmental impact assessment of CCS-related activities, and a list of monitoring techniques to track CO_2 before and during injection for safe operation of a CCS demonstration project (Japan CCS Study Group, 2009).

According to the US EIA, around 83% of the total energy use in Japan in 2005 was met by coal, oil, or natural gas; renewable energy accounts for about 4% of the energy demand. Electricity production is estimated to increase from almost 245 GW in 2007 to around 305 GW in 2030, 40% of which will be generated from fossil fuels (compared to 59% in 2007) (US EIA, 2010). Thus, even with strong policy initiatives and major technological developments that incentivize for the implementation of energy efficiency, renewable energy, and other low carbon technologies, fossil fuels will remain a major contributor to Japan's energy fuel mix for the coming two decades. Although not the only option, CCS is expected to be a major, economically feasible tool to reduce Japan's CO_2 emissions as part of its efforts to combat climate change (Ministry of Economy, Trade and Industry of Japan, 2010).

13.6.10.4   Human Capacity, Research and Development

Japanese governmental agencies, academic institutes, and private industrial/commercial businesses are actively involved in CCS-related endeavors both locally and internationally. At the governmental level, CCS-related research initiatives have been initiated and supported by governmental agencies within the Ministry of Economy, Trade and Industry, Ministry of Science and Education, Ministry of Environment, and the Prime Minister Cabinet Office, with plans to reduce 100 $MtCO_2$ emissions through CCS by 2020 (Lund et al., 2008). Two major government initiatives are the CO_2 Storage Research Group within the Research Institute of Innovative Technology for the Earth under the Ministry for Environment, Trade and Industry (see RITE, 2011) and the New Energy and Industrial Technology Development Organization (Lund et al., 2008). The private sector is also promoting CCS initiatives either individually or in collaboration with the Japanese Government. In 2008 the Japanese CCS Company Ltd was launched by 29 power and energy firms in Japan to advise the government on the feasibility CCS implementation. Electric utilities, industrial firms, and private Japanese international companies have also funded research projects to develop advanced capture technologies and investigate ocean sequestration (Lund et al., 2008). In this regard, the CCS R&D network database of the Global CCS Institute shows that a total of 32 organizations in Japan are involved in CCS R&D, the majority of which is related to CCS technologies (GCCSI, 2011b).

In addition to the local efforts, Japan is actively engaged in international collaboration on CCS initiatives. Japan is a member of CSLF, GCCSI, and IEA CCS roadmap. In addition, Japan is collaborating with China on an EOR project that involves capturing 1–3 $MtCO_2$ from the Harbin thermal power plant in Heilungkiang Province to be injected and stored in Daqing Oilfield, China's largest oil field. The Yantani

IGCC project is another example of collaboration between both countries where Japan is primarily involved through Mitsubishi Heavy Industry. In addition to China, Japan is collaborating with Vietnam on the White Tiger CCS Project, Malaysia on Bintulu CCS Project, Australia on CS Energy Oxy-Fuel Project, and United Arab Emirates on JODCO EOR project. All aforementioned initiatives contribute significantly to building a competently educated human capital that would be expected to play a major role in advancing CCS technology (Lund et al., 2008).

13.6.10.5   CCS Legal and Policy Initiatives

Although there is currently no specific regulations, for carbon capture and storage activities in the country, Japan's national policies promote CCS as major climate change mitigation option. The *Cool Earth-Innovative Energy Technology Program* report, issued in March 2008 by the Ministry of Environment, Trade and Industry, identifies CCS as one of the major technologies that need to be focused on and developed in order to achieve substantial reductions in CO_2 emissions by 2050. The report includes CCS on the top of 21 priority-technologies to support. The program lays out a multiphase plan to concurrently accelerate technological enhancement, reduce cost, implement large-scale projects, maintain strong international collaboration, and draft domestic regulations and laws between 2020 and 2050 (Ministry of Economy, Trade and Industry, Japan, 2008).

Another important aspect affecting the deployment of CCS is the sufficiency of governmental funding. Currently, modest funding is provided by the Japanese Ministry of Economy, Trade and Industry to the Office of Environmental Affairs whose R&D budget was 5.66 billion Yen (~US$50 million) in 2006 and 4.3 billion Yen (~US$37 million) in 2007, with main focus on two areas: ocean/saline aquifer storage and clean coal technologies. Another US$10 million was allocated for international coal utilization projects by the New Energy and Industrial Technology Development Organization (Lund et al., 2008).

Studies show that the Japanese population is still not very well informed about CCS as an option to mitigate climate change: almost 68.8% of the population is not familiar with CCS, compared to less than 5% for solar or wind energy. Public media seems to have great impact on people's awareness, as 72% of the people learn about CCS from either television programs or newspapers. In that regard, four major factors are found to shape people's perception about CCS: concern about risk and leakage, understanding of effectiveness of CCS, responsibility for mitigation of CO_2, and concern about use of fossil fuel. Surveying the Japanese public opinions shows that people have a positive attitude toward the implementation of CCS, but they become less supportive when asked about CCS implementation in specific applications and locations. As such, it is unlikely that the Japanese public would voice strong opposition against CCS implementation in the future, but educating people about the CCS technologies and

associated risks would play a major role in increasing public acceptability (Itaoka et. al., 2011).

13.6.11   Conclusions from the Regional Outlook

Technical potential of CCS on a global level is promising but on a regional level is differentiated. Much more work is required to realistically assess storage capacity on a worldwide, regional, and subregional basis. The information available to date indicates that there is sufficient worldwide capacity for storing at least 100 years of emissions from stationary sources in deep geological formations. On a regional basis, the distribution of deep geological formations is highly variable. In general, those regions with large fossil fuel resources, particularly oil and gas, have the largest storage potential. Some regions, including the Former Soviet Union, Northern Europe, North America, and the Middle East, have large storage capacity. Other areas, notably India, and parts of Southeast Asia, seem to have more limited geological storage capacity.

How and when CCS becomes an economically feasible option depends on the matching over time between CO_2 sources and sinks in specific regions. Besides the technical potential, the matching will depend on factors such as the electricity mix, the regulation of industry, national targets for CO_2 abatement, national and regional targets for renewables, local policies regarding nuclear power, and the existence of economic incentives. CCS comes at a cost and therefore requires appropriate incentives. Many institutional factors influence the degree to which CCS is implemented, including environmental regulations, mineral and property rights, carbon credits for CCS, international trading of carbon credits, and resolution of long-term liability issues.

13.7   Public Perception and Acceptance of CCS

The deployment of CCS technology relies on a myriad of interactions among technologies, markets, institutions, policies, regulations, and society (IPCC, 2005). Confidence about the technical feasibility of CCS has been growing, yet like any technology its deployment will be influenced by many social factors. Experiences with other technologies suggest that if public acceptance of a technology is lacking, large-scale global deployment is inhibited (Renn et al., 1995; Cormick, 2002; Kalaitzandonakes et al., 2005). Likewise, public perception and support will be critical if CCS technology is to achieve its potential as a GHG mitigation strategy (IEA, 2008a). As the number of CCS-related projects grows, and the potential for CCS technology to contribute to reducing CO_2 emissions becomes more prominent in societal debates, issues related to the public perception and social acceptance of this technology are becoming more salient. Public engagement and communication on CCS is therefore an increasingly important issue as large-scale deployment advances. This section outlines key public perception issues based on scholarly research and real world CCS projects. Additionally it presents lessons learned from public engagement in the proposed and existing CCS projects.

13.7.1   What is Public Perception in the CCS Context?

13.7.1.1   Definition of 'Public'- What Does Public Constitute in the Context of CCS?

In the context of CCS, the term 'public' encompasses diverse subgroups or 'publics' – referring to general public, globally or within the vicinity of a project. The public also includes other segments of society such as policymakers, regulators, industry, academia, NGOs, and media, which helps the general public or communities form an opinion.

There are different dimensions of how public perception is formed and influenced –narrowly, at the project level, and more broadly, at national and global levels. At the project level, public perception and support for CCS technology will be influenced by the regulatory processes that are in place to ensure public participation in the decision-making. The direct public engagement at the project level is likely to happen through a variety of mechanisms during the different stages of a project, depending on the legal requirements or social norms in the country where the project is planned. For example, a common mechanism for public engagement includes public hearings as a part of the project approval process. These are the more concrete interactions with the public, focusing at a specific project level.

The policy debate surrounding CCS technology is the other dimension that will define broader public perception of the technology. The public is generally represented by and organized in non-governmental organizations, such as environmental NGOs, community groups and the like. How governments and other stakeholders perceive CCS, as part of their national climate change mitigation strategies, and how they frame national and international policies, will also affect the public's perception and support of the technology.

13.7.1.2   Introduction to Research Methods and Issues around Public Perception of CCS

Public perception of CCS technology has been assessed through numerous research and case studies, conducted across the world at national as well as project-specific levels. These research studies have relied on different research methods such as survey instruments, focus groups, mental model approach, factor analysis, information choice questionnaires, and discrete choice analysis to identify public perceptions of CCS at the broader societal level. Most of the studies use one or a combination of three methods: a written or digital survey method, focus or discussion groups, and experiments. In surveys, experimental surveys and experiments respondents are often given written information to read. In focus or discussion groups participants are often informed by researchers, experts or handouts, or a combination thereof (Huijts, 2003; Curry

et al., 2004; 2006; Palmgren et al., 2004; Shackley et al., 2005; Sharp et al., 2006; Reiner et al., 2006; Best-Waldhober et al., 2006, Ha-Duong et al., 2009; Itaoka et al., 2004; 2006; 2009). More recently, the public has also been engaged at the specific project level (see case studies summarized in Table 13.8), although the motivations behind engaging the public living in the vicinity of the projects are slightly different than broad national-level social science-based research. The project-level public engagement aims for making better project decisions, supporting the successful design and implementation of the proposed project and, in some cases meeting the regulatory requirements.

13.7.2   Public Perception of CCS Technology

13.7.2.1   A Summary and Analysis of Research Conducted over the Past Few Years to Assess Public Perception of CCS across the World

The numbers of research studies to assess public perception of the technology around the world have grown over the past five years (for a review, see Reiner, 2008; Ashworth et al., 2007). At the broad societal level, these studies show consistency in some areas and differences in others. Both across jurisdictions and over time, all studies show that the vast majority of the public is not aware of CCS and even fewer understand the technology, its risks and benefits (Johnsson et al., 2009; Reiner et al., 2006). Differences, however, generally arise in the level of public understanding and support for CCS across countries (Reiner et al., 2006).

13.7.3   An Overview of CCS Public Perception Survey Research

Most of the studies to assess public perception of CCS have been conducted in industrialized countries. Studies to evaluate public acceptance and support of CCS in the major coal-dependent emerging economies are practically nonexistent. This can be attributed to very few to almost no CCS projects in these countries due to limited R&D, lack of government support for projects, and in those areas where projects exist, limited regulatory and institutional capacity to engage the general public.

The studies to assess public perception of CCS around the world and their findings are summarized below:

* In the Netherlands a study conducted in 2003 (Huijts, 2003) indicated that residents not living above a likely CO_2 storage site had neutral to positive attitudes about CCS but were less supportive to its development in their community or nearby, suggesting "Not In My Back Yard" (NIMBY) characteristics of public perception of CCS.

* The survey conducted by MIT to assess general public perception of CCS in the United States (Curry et al., 2004; 2007) confirms very low public awareness of CCS. The internet-based survey among samples of 1200 respondents conducted in 2003 and 2006 reveals that more

than 90% of the respondents had never heard of CCS and the results were largely consistent during both survey periods. Support for CCS is also linked to public attitudes about fossil fuel alternatives such as renewable energy, efficiency measures, and demand reduction.

- A 2004 study in the United States by Carnegie Mellon University (Palmgren et al., 2004) found that people were significantly less willing to pay for CCS than for any other major option to reduce CO_2 emissions, including new nuclear power plants.

- Studies in the United Kingdom (Shackley et al., 2005; 2007) found "slight support" for CCS in concept but also a belief that as a stand-alone policy "CCS might delay more far-reaching and necessary long-term changes in society's use of energy."

- A 2006 survey of 900 respondents in Australia (Miller et al., 2007) found that although most respondents believe it is very important to reduce greenhouse gas emissions at a national level, many are "neutral" toward CCS as a strategy. This study found that approximately 40% of the public believes CCS would be "a quick fix that would not solve the greenhouse gas problem." A less representative but more recent Australian study confirmed those average results (Ashworth et al., 2009).

- A 2006 study in Canada (Sharp et al., 2006) found low awareness of CCS in a survey among 1972 respondents in Canada. Although between 10% of respondents in Alberta and Saskatchewan and 15% of respondents in the rest of Canada said they had heard of CCS, very few were able to correctly identify the problem that will be addressed by CCS.

- Reiner et al. compared the awareness of CCS in the United States, United Kingdom, Sweden, and Japan (Reiner et al., 2006). The four samples in this study were mostly drawn at random from the countries' adult population and consisted of at least 742 respondents per sample. The study found low awareness of CCS in all four countries, ranging from 22% of respondents confirming they had heard or read about CCS in Japan to as little as 4% in the United States.

- A recent survey of 1076 respondents in France revealed that a vast majority of respondents were not strictly opposed of CCS but were more suspicious than supportive. The study also showed that a vast majority of respondents had not heard of CCS and only 6% of a representative sample was able to define the term (Ha Duong et al., 2009).

- A recent study in Switzerland (Wallquist et al., 2009) investigating lay people's concepts of CO_2 and CCS showed that some people were worried that CO_2 might cause cancer or even that CO_2 leaking from storage might cause DNA changes. These kinds of misconceptions are not likely to be anticipated by experts and therefore less likely to be investigated in polls or surveys or addressed in information about CCS technology.

- Other surveys in the Netherlands conducted in 2006, 2007, and 2008 (Best-Waldhober et al., 2006; 2008; Lambrichs, 2008) showed similar results. Depending on the kind of CCS technology that the three samples, a total of 918 respondents, were asked about,

between 51.2–91.4% of respondents stated they were unaware of the technology.

- The survey research conducted in Japan (Itaoka et al., 2004; 2006; 2009; Tokushige et al., 2007) suggests very limited understanding of CCS technology among the respondents. The effectiveness of CCS (i.e., its effect on CO_2 emission reduction) and perception of potential benefits of CCS were the two most influential factors for public acceptance of CCS.

Some researchers have considered traditional questionnaire methodologies less suitable for examining public perceptions of CCS because with limited or no knowledge about an issue there is a risk of producing variable responses or in some instances "pseudo opinions" (Best-Waldhober et al., 2009). Malone et al. (2009) further state that, "because of the inherent difficulty of providing information in an unbiased way, surveys may be compromised at the outset if they seek to educate." Providing respondents with elaborate, understandable, recent, accurate, and balanced information on CCS technologies as well as their consequences and context is difficult and highly time- and resource-consuming. Most studies can only partially overcome these issues and are thereby unable to rule out susceptibility to bias. Therefore, the opinions found in these studies should be interpreted with care.

13.7.3.1   The Role of Information

Studies found that in some cases, when information was provided solely as part of a questionnaire, individuals were more negative toward the technology. Conversely, when information was provided with increasing depth and interactivity, individual attitudes toward CCS tended to be more positive. Some studies have tried to address this by providing respondents with information about CCS to see how informed and uninformed perceptions of CCS vary among the general public. In the United States, a survey of over 100 respondents shows that exposure to information about CCS technology increased the public level of understanding as well as support for advancing the technology (Stephens et al., 2009). Another study conducted in the Netherlands tested traditional survey methodology on a sample of 327 respondents and information choice questionnaires on 995 respondents (where information about CCS is provided to the respondents) and reported that the public is more supportive of the technology after processing relevant information (Best-Waldhober, 2009). Similar results were obtained from Japan, where a survey administered on a sample of 1006 adults in Tokyo revealed that additional information, prior to the survey, led to increased support for the technology (Itaoka, 2006; 2009).

13.7.3.2   Which Actors Are Trusted?

Research by Ashworth et al. (2006), Mors (2009), and Terwel (2009) shows that the relative trust individuals place in the information source is a key factor influencing acceptance of CCS. People put more

Table 13.6 | Strengths and weaknesses of existing public perception research.

Strengths of existing public perception research	• Successful identification of public's concerns and issues about CCS • A range of methods have been tested to assess public perception of CCS in different countries, which has not only provided insights into how public perceives CCS but also shed light on advantages and disadvantages of using different research methods • The results of public perception research has provided some guidance to policy makers and project developers • Comparison beginning to occur across countries to identify cultural similarities and differences toward CCS • Meaningful dialogue is more successful in engaging public–some activities have ventured beyond survey research
Weaknesses of existing public perception research	• Large focus on survey research in comparison with meaningful engagement and communication about CCS • Many activities directed at students, and as a subset of lay public, their views are not necessarily representative of the society • Little emphasis on evaluating perception of public toward CCS in developing countries • Lack of coordination among researchers in conducting public perception research • Limited investment and funding for CCS public communication research

Source: Ashworth et al., 2007; Reiner, 2008.

trust in environmental NGOs than in industrial organizations or the government. The previous experience with the organizations or the actors involved, concerns over accountability, and openness can also play important roles in shaping public trust (Reiner, 2008). The communication strategies by untrustworthy stakeholders may result in negative public attitudes toward CCS. Additionally, communication about CCS may result in more positive perceptions when stakeholders work together to provide information to the public rather than as separate "stand alone" organizations. The role of the CCS community, as an epistemic community, is also thought to play a role in the perception of the technology (Stephens et al., 2011).

13.7.3.3   Analysis of Existing Public Perception Research

The conclusions that can be drawn from the CCS public perception research studies depend on the kind and quality of the information as well as the method used. Methods using some kind of discussion group often have the advantage of giving insight into the perceptions lay people have of CCS technology, which might be very different from the perceptions of experts. The downside is that these kinds of methods are expensive and time-consuming, which often leads to only a few small discussion or focus groups being used, resulting in conclusions that cannot be generalized to the larger population (Shackley et al., 2005). Surveys with representative samples do not have this disadvantage. However, surveys often use a restricted set of questions leading to a restricted set of possible answers. This restriction might lead to missing certain issues that the experts or researchers had not thought of but that may be important to the public (Curry et al., 2006; Best-Waldhober, 2006; Itaoka et al., 2006; Malone et al., 2009).

Some studies have analyzed the existing CCS public perception research and communication activities, illustrated in Table 13.6 (Ashworth et al., 2007; Reiner, 2008). They conclude that little work has been done to inform the general public about CCS and the majority of activities have been surveys used to inform research, policy, and environmental NGO communities. Beyond the survey there has been very little communication activity targeted at the general public, and as a result overall public awareness of CCS is still low.

The total investment in communication of CCS also remains significantly lower compared to the allocated budgets of the CCS technological research and development programs. Limited budgets have adversely impacted the scope and methodology employed for public perception research. On the positive side, the existing research provides some useful insights into similarities and differences on how the public perceives CCS across nations, globally, and locally. The studies also provide an understanding of issues of public concerns regarding CCS.

13.7.3.4   Public Perception of CCS at the Global Level

Globally, the understanding and support for CCS among the public remains low. The public opinion on CCS varies based on socioeconomic status, level of education, culture, and professional backgrounds. The results of several studies suggest that people with a higher education background have more positive attitudes toward the technology. Conversely, those of lower socioeconomic background and education levels tend to be more skeptical (Bradbury et al., 2009). Likewise, the energy industry seems to be more supportive of CCS whereas some environmental advocacy organizations and local public interest groups have come out openly against CCS. Other major environmental NGOs favor CCS along with a broad portfolio of GHG mitigation options. The lower understanding of CCS is attributed to limited engagement and lack of open and honest communication with the public about the technology. In some cases, the lack of awareness regarding CCS is also tied to the public's limited understanding of and belief in climate change (Parfomak, 2008). Some studies tried to place CCS in a broader context by investigating whether the public considers CO_2 emissions as a problem to be solved. Giving people several societal issues to rank, Curry et al. (2007) report that protecting the environment was the eleventh highest ranked priority for Americans (out of 18 issues), while Palmgren et al. (2004) respondents ranked "reducing climate change" as the lowest social priority of the 15 choices provided them. Although general awareness regarding climate change and its implications are reported to have been increasing worldwide, there are some parts of society that still do not believe or give credibility to anthropogenic climate change as a pressing global problem.

13.7.3.5   Public Perception of CCS at the Local Level

When it comes to siting a CCS project in their vicinity, the public tends to become more reluctant and often negative in their support for the technology. Research study in Japan and the Netherlands reported that the public seems to be supportive of CCS but expressed NIMBY concerns when asked if they would support a CCS project in their community (Itaoka et al., 2004; 2006; Huijts, 2003). The existing research and experience with engaging the public in the vicinity of the proposed CCS projects demonstrate that at the local level, public support or opposition to CCS is motivated by both concerns as well as benefits of the technology. For example, some communities perceive CCS projects as economic opportunities and hence support the project in their vicinity (Hund et al., 2009), while others may focus upon the unfair distribution of hazards and hence oppose a project in their vicinity (Bradbury et al., 2009; Simpson, 2009) (see case studies Table 13.8).

13.7.3.6   What Are the Public's Main Concerns Regarding CCS?

The public's concerns about CCS stem in part from the lack of understanding of the role of technology as a climate change mitigation option and its potential environmental health and socioeconomic risks.

At the specific project level, the public is concerned about potential impacts of siting a CCS project in their community. A recent study across the three regional carbon sequestration partnerships in the United States (Bradbury et al., 2009) identified a range of social concerns regarding CCS among local communities (see Table 13.7). In addition to fear of underground CO_2 storage risks, the public expressed lack of trust in government authority and the private sector as a source of information and raised concern about the fairness of CCS implementation procedures. The public was also concerned about being neglected or ignored if the project turned out to be more harmful than expected. Concerns also varied based on the potential impacts of a proposed CCS project at the individual level; thus local landowners in the vicinity of proposed projects are concerned about the effect of conducting seismic surveys on their property, the effect of potential CO_2 seepage, and other issues.

13.7.3.7   Factors Affecting Public Perception of CCS

Public opinion and support for CCS is influenced by several factors such as cultural, educational, and socioeconomic background; past experience with other energy infrastructure and development projects; perceived risks and benefits of CCS technology; and influence by media and other stakeholders such as academia, NGOs, and industry.

Table 13.7 | An overview of public concerns regarding CCS technology.

Public concerns about CCS	• Safety risks of CO_2 leak • Contamination of ground water • Upfront impacts of increased coal mining • Effect on local environment including plants and animals near site • Assumption that CO_2 is explosive or poisonous • Effect of conducting seismic survey on their property General concerns: • Availability of enough storage sites • Long-term viability and who is liable for stored CO_2 • Lack of infrastructure to support large-scale deployment • Diversion of interest and funding from alternative energy • Lack of clarity on issues such as pore space ownership • Cost-economic concerns • Risk of unknown technology
CCS benefits as cited by public	• It could provide a good bridge to the future • If successful, can avoid large quantities of CO_2 from getting into atmosphere • Allows continued use of fossil fuels which provides economic advantage for some countries or regions • Helps clean up coal-based power plants • Allows emissions to be reduced without changing lifestyle too much

Source: Ashworth et al., 2006; Bradbury et al., 2009.

13.7.3.8   Stakeholder Perception of CCS Technology

Besides understanding the general public's view of CCS, some studies have also assessed other stakeholders' perceptions of the technology. Usually those stakeholders are grouped in environmental NGOs, academia, government, and industry. However, there can be more specific groupings, such as environmental justice groups in the United States, the financing industry, or different ministries in countries that affect the overall government position. Surveys have been used to assess the views and concerns of stakeholders on CCS, but study of position papers can also lead to insights.

A global study (IEA Working Party on Fossil Fuels, 2007) evaluated stakeholder perceptions toward CCS across North America, Australia, the European Union, New Zealand, Japan, China, South Africa, and India and found relatively low levels of awareness of and support for CCS in developing countries compared to industrialized countries. The study also highlighted concerns and issues around CCS such as costs, lack of policy incentives and regulatory frameworks, and local risks of safety. It also found a growing interest in CCS among stakeholder groups in most regions. These findings are roughly consistent with the conclusions of a study specifically on stakeholder perceptions in Germany (Fischedick et al., 2008) with the exception of public acceptance: 65% of respondents in the German study flagged that as "very important," but it was not mentioned as a survey outcome in the IEA Working Party study.

A research study in Europe evaluated the perspectives of different stakeholder groups on CCS in more detail (Shackley et al., 2008) through a survey conducted among 512 respondents from most European countries.

The sample was not representative as the number of respondents was much greater from countries with an active CCS community, such as Norway, than from countries with low levels of CCS engagement, such as Hungary. The survey results showed that majority of the sample was moderately supportive of CCS and believed that it had a role to play in their own country's plans to mitigate carbon emissions. Safety risks were seen as a major concern by environmental NGOs, but much less so by other stakeholders.

An NGO survey was also done in the United States, where perceptions of national-level NGOs were inventoried through semistructured interviews supplemented by content analysis of their documents (Wong-Parodi et al., 2008). It was found that while all NGOs are committed to combating climate change, their views on CCS as a mitigation strategy vary considerably. Some NGOs, such as Greenpeace (2008), oppose CCS, arguing that it will become available too late, that costs and energy use are forbiddingly high, and that liability issues cannot be feasibly addressed. Other NGOs, particularly US environmental NGOs such as the Natural Resources Defense Council and the Environmental Defense Fund, but also the Norwegian Bellona Foundation (Stangeland et al., 2006), are generally favorable or even very supportive of CCS, although they generally warn of the risks.

In developing countries, the situation differs. A study to assess stakeholder perceptions of CCS in India (Shackley and Verma, 2008) revealed very low government support for CCS, whereas industry was fairly enthusiastic about demonstrating the technology. The study showed very limited interest and engagement by the NGO community. A multistakeholder survey of 700 participants in Brazil (Cunha et al., 2006) reported high support for technology from government and academia. Brazilian NGOs were less supportive of the technology, and the lay public is generally not aware of CCS. A specific case of acceptance of CCS is the support for the technology for inclusion in the Kyoto Protocol's Clean Development Mechanism. Stakeholder and government perceptions varied greatly on that matter, citing concerns over immaturity of the technology and lack of sustainable development benefits (de Coninck, 2008). For example, although Brazil is not opposing CCS in general, it opposed including CCS in the CDM.

Environmental NGOs often indicate the possibility of diverging attention and resources away from renewable energy and energy efficiency toward CCS as a major issue, as this would delay a fully sustainable energy system (Greenpeace, 2008). In the European Union study (Shackley et al., 2008) this was also evaluated, resulting in 51% of the respondents thinking there would be no such negative impacts on improving energy efficiency and reducing energy demand; 44% thought there would be some effect; very few thought such effects would be large. The study clearly indicated that NGO respondents were much more concerned about the implications for renewable energy than other stakeholders.

A relatively new area of study is how the CCS expert community itself generates and assesses information and communicates with the lay public. The CCS community is growing rapidly, exemplified by increasing attendance of conferences, expanding specific academic journals, and a long list of companies joining organizations like the Global CCS Institute. The internal communication on CCS, within the community, generally has a positive tone and is rarely critical of the technology, or of advocacy work done by CCS experts. This signals a community that is conveying a strong and coherent message on CCS to policymakers but that also, potentially, has a complacent attitude toward the technology itself, which is strengthened by the group process in the community. This signals a community that is conveying a strong and coherent message on CCS to policymakers but that also, potentially, has a complacent attitude toward the technology itself, which is strengthened by the group process in the community (Coninck, 2010; Stephens et al., 2011).

13.7.4   Implications of Public Perception for CCS

13.7.4.1   Implications of Not Acting on Public Perception of CCS

In the last few years, evidence has arisen on public perception of CCS projects in communities which provides information about what might happen if governments and project developers failed to act on public perception. The project portfolio shows mixed results. Some CCS demonstration projects were well received by the community and have been able to move forward, while others encountered local public opposition, which in several cases contributed to cancellation or stalling of the project. The case studies provide insight into what affects a community's and individual's views on CCS and on CCS projects and what the crucial factors are that may lead to the community being receptive or rejecting of a project.

A selection of the most notable and best-documented CCS demonstration projects and project plans is given in Table 13.8. It also reviews motivations behind public support or opposition.

13.7.4.2   Lessons on Public Engagement and Communication

The existing experience engaging the public indicates that at the local level, public acceptance and support for CCS is largely driven by the potential socioeconomic benefits of the project activity. In cases where the public is not supportive of a project, there is public concern about potential environmental, health, and economic impacts of deploying CCS in their vicinity. Clearly, when it comes to actually siting a project in a community, potential risks of CCS dominate over the public's concern about the potential impacts of global warming. Some studies have analyzed the existing public engagement and communication activities around CCS broadly as well as at specific project levels (Reiner, 2008; Ashworth et al., 2006; 2007; Simpson and Ashworth, 2009; Hund et al., 2009) and provide some useful recommendations for public engagement on CCS. In addition, the experience engaging with communities in the proposed and existing CCS projects (for example Table 13.8) also provide some useful lessons on public engagement, as summarized below:

Table 13.8 | A summary of some proposed CCS projects and motivation for public support or opposition.

Project	Location	Public Support	Motivations behind Public Perception
FutureGen	Illinois, USA	Yes	Industry and Government funded first-of-a-kind-in- the-world project 'cool factor', public recognition of socioeconomic benefits of the project to the community (Bielicki, 2008; Hund and Judd, 2008; Hund and Greenberg, 2010; Greenberg, 2009)
Wallula Energy Resource Center	Washington, USA	Yes	Focus on win-win attributes of the projects such as potential social and economic benefits (Hund, 2009)
Jamestown	New York, USA	No	Concerns regarding unproven geologic storage capacity, support for renewable rather then coal-fired power plant; community believes coal-fired plant will be environmentally and economically unsustainable (Simpson, 2009)
Carson	California, USA	No	Communities past experience with industries in the area and perceived environmental concerns (Stephens et al., 2009)
Barendrecht	Netherlands, EU	No	Concerns regarding added burden of another industrial facility on the environment, style of communication with public–lack of transparent and honest engagement with communities (Brunsting and Mikunda, 2010)
Vattenfall	Germany, EU	No	Concerns regarding health and environmental risks of potential CO_2 leakage, lack of large scale demonstration, diversion of funding and interest away from alternative energy, extra burden on consumers (Slavin and Jha, 2009)

- It is important to communicate with the public about CCS technology in the context of climate change mitigation and present it as a tool in a portfolio of options including renewables, energy efficiency, and demand reduction (Hund et al., 2009; Bielicki and Stephens, 2008; Simpson, 2009; Best-Waldhober, 2009; Itaoka et al., 2009).

- Honest, transparent, and clear communication is critical in establishing trust with the public. In addition, when providing information it is always better to respond to the questions and concerns expressed by communities (Simpson and Ashworth, 2009; Hund et al., 2009; Bielicki and Stephens, 2008; Simpson, 2009; Brunsting and Mikunda, 2010; and Greenberg, 2009).

- CCS is a new technology with little operational experience. Additional field tests and a demonstrated ability to mitigate risks should they arise will be necessary to improve the public's perception of risk from CCS technologies (Singleton et al., 2009). In addressing public concern about a proposed CCS project, it is important to communicate openly about potential or perceived risks and present a mitigation plan in cases where necessary to make the public aware of measures that can be taken to tackle those risks (Slovic et al., 1993).

- Engaging the public early on in the vicinity of a proposed CCS project is crucial as opinions may be slow or difficult to change once formed (Hund and Judd, 2008; Hund, 2009; de Coninck and Feenstra, 2009; Bielicki and Stephens, 2008).

- Concentric communication and engagement with the public involving thought leaders and influence groups such as local NGOs and media, in addition to general members of public, is important for successful public engagement around a proposed project (Hund and Judd, 2008; Hund, 2009; Greenberg, 2009).

- There is no one-size-fits-all approach; different publics will require different engagement and communication strategies (Bielicki and Stephens, 2008).

13.8   Summary and Conclusions

Over the past decade, there has been a remarkable increase in interest and investment in CCS. A decade ago, there was only one operating project, little corporate or government investment in R&D, and no financial incentives to promote CCS. Today there are over 234 projects of various sizes and stages of development. Many companies have significant investments in technology development, and governments around the world have committed billions of dollars for R&D, scale-up, and deployment. The coming decade will be critical in the technology development and the ultimate role this option plays in reducing GHG emissions. While the outlook is quite promising, there are a number of economic, scientific, and social challenges ahead.

CCS involves the integration of four elements: CO_2 capture, compression of, transportation to the storage location, and isolation from the atmosphere by pumping it into appropriate saline formations, oil and gas reservoirs, and coalbeds with effective seals to keep it safely and securely trapped underground. Storage in other rock types such as basalts, oil and gas shales, and subsea bed sediments may also be possible, but much less is known about their potential. Component technologies are in different stages of development, some fully mature, such as compression, and some such as storage in saline formations, in the early stages of demonstration.

Three approaches are available for capture from the power and industrial sources that produce CO_2 gas with relatively small concentration of contaminants: post-combustion capture, pre-combustion capture, and oxy-combustion capture. All have been demonstrated; some are used routinely today for other applications, but little experience is available for integration and optimization with power production or most industrial applications. Considering full life cycle emissions, CCS technology can reduce up to about 65–85% of CO_2 emissions from fossil fuel combustion from stationary sources. CCS is applicable to the power generation and the industrial sectors (see also Chapter 12). In the future, CCS may contribute significantly to the transportation sector via hydrogen production and/or electrification of light duty vehicle and public transport, or through reducing emissions from biofuel production processes (see Chapter 17 for analyses on different pathways).

Successful experiences from five ongoing projects demonstrate that, at least on this limited scale, CCS can be safe and effective for reducing emissions. Moreover, relevant experience from nearly 40 years, currently at the rate of 45 $MtCO_2$/yr for enhanced oil recovery, also shows that CO_2 can safely be pumped and retained underground. Our best understanding of storage security can be summarized as follows:

> Observations from **commercial storage projects**, engineered and natural analogues as well as **theoretical considerations**, models, and **laboratory and field experiments** suggest that appropriately selected and managed geological storage **reservoirs are very likely to retain nearly all the injected CO_2 for very long times, more than long enough to provide benefits for the intended purpose of CCS.**

A five- to ten-fold scale-up in the size of individual projects operating today would be needed to capture, transport, and store emissions from a large (500–1000 MW) coal-fired power plant. A thousand fold scale-up in size of today's CCS enterprise would be needed to reduce emissions by billions of tonnes per year (Gt/yr). Herein lies one of the major challenges for CCS, which raises a number of issues. Specifically, is there sufficient capacity to store these large quantities of CO_2? What will this cost? And finally, what are the institutional, economic, and technical constraints for implementing CCS on this scale? Regional factors are likely to play a major role in the extent and timing of CCS implementation. Furthermore, the scale and speed at which CO_2 transport networks will need to be built, often through densely populated areas, will pose a challenge. Although successful experiences with CO_2 transport in the past indicate that issues associated with the design and operation of such networks can be addressed, aspects such as impurities, fluctuating demand, and securing stable operational conditions (e.g., avoiding two-phase flow) need to be further addressed.

On a regional basis, storage capacity and the quality of information is highly variable. The Former Soviet Union, Northern Europe, North America, and the Middle East appear to have the largest storage capacity. Other areas, notably Japan, India, and parts of Southeast Asia appear to have more limited geological storage capacity. The IPCC 2005 Special Report concluded there is sufficient capacity to store 100 years of emissions at the high end of the technical potential (2000 Gt). Recently, more detailed capacity estimates suggest that capacity will be on the high end of the IPCC range, but there is still considerable debate about how much storage capacity actually exists. The debate is particularly significant about saline formations, which are believed to have the greatest capacity. Research, geological assessments, and most importantly commercial-scale demonstration projects, will be needed to improve confidence in and reliability of capacity estimates.

Costs of capture for first generation CCS power plants available in the early 2020s are estimated to be about \$45/tCO_2 avoided for coal and \$115/tCO_2 avoided for natural gas. Estimates vary widely depending on whether the plant is the first or n^{th}-of-a-kind, the type of fuel, capture technology, and assumptions about the baseline technology. Capital costs and parasitic energy requirements of 15–30% are the major cost drivers. Research is underway to lower costs and energy requirements. Early demonstration projects are likely to cost more. Due to high costs, CCS will not take place without strong incentives to limit CO_2 emissions. Access to capital for large-scale deployment will be a major factor limiting the widespread use of CCS, particularly if the policy regime for emission reductions and regulatory requirements for storage – especially long-term liability for stored CO_2 – remains uncertain. Estimated costs of storage range from US\$2–35/tCO_2 and experience from operating projects fall in the middle of this range of US\$6–20/tCO_2. Transportation costs are highly site specific, depending on the transport distance and size of the pipeline. Assuming a 500 MW coal plant emitting about 4 $MtCO_2$/yr and an onshore pipeline transport over a distance of 100 km, costs for transport are estimated to range from US\$1.25–3.5 tCO_2. Overall costs of CCS including capture, transportation and storage are estimated to range from US\$50–70/tCO_2 avoided for coal based electricity generation. This could increase the cost of generating electricity by an estimated 50–100%.

The environmental risks of CCS appear to be manageable, but regulations are needed to ensure due diligence over the life cycle of the project, most importantly: siting decisions, operating guidelines, and monitoring and closure of a storage facility. Many members of the public have concerns about the safety and effectiveness of CCS. More CCS projects and education are needed to establish a convincing safety record.

Social, economic, policy, and political factors may limit deployment of CCS if not adequately addressed. Critical issues include ownership of underground pore space, long-term liability and stewardship, GHG accounting approaches and verification, and regulatory oversight regimes. Significant progress is being made by governments and the private sector on all of these issues. Government support to lower barriers for early deployments is needed to encourage private sector adoption. Developing countries will need support for technology access, lowering the cost of CCS, and developing workforce and regulatory capacity for permitting, monitoring, and oversight. CCS, combined with biomass gasification, can lead to net removal of CO_2 from the atmosphere, which is likely to be needed to achieve atmospheric stabilization of CO_2 and may provide an additional incentive for CCS adoption.

Finally, public support for CCS is crucial for large-scale deployment. Today, the public remains largely unaware of the purpose and nature of CCS technology. Much work remains in this area.

References

111th US Congress, 2010: *S.3589 – Carbon Capture and Sequestration Deployment Act of 2010.* United States Congress, Washington, DC USA.

Aines, R. D., Leach, M. J., Weisgraber, T.H., Simpson, M. D, Friedmann S. J., and Bruton, C. J., 2009: Quantifying the potential exposure hazard due to energetic releases of CO_2 from a failed sequestration well. *Energy Procedia,* **1**(2009): 2421–2429.

Al-Juaied, M. and A. Whitmore, 2009: Realistic Costs of Carbon Capture. Discussion Paper 2008–09, July 2009. Energy Technology Innovation Research Group, Belfer Center for Science and International Affairs, Harvard Kennedy School, Cambridge, MA.

Alstom, 2007: "Alstom and Statoil to jointly develop project for chilled ammonia-based CO_2 capture for natural gas in Norway," Alstom Press Release, June 21, 2007.

Anthonsen, K., T. Vangkilde-Pedersen, and L. H. Nielsen, 2009: Estimates of CO_2 storage capacity in Europe. Geological survey of Denmark and Greenland. www.co2captureandstorage.info/docs/Copenhagen/EU_Storage_capacity_KLA.pdf. Also, IOP Conf. Series: Earth and Environmental Science 6 (2009) 172006 doi:10.1088/1755-1307/6/7/172006.

Aoshima, M., 2009: Comparative Analysis on Japan's CO_2 Emissions Computation – Gaps between the Japanese GHG Inventory Report and IEA Statistics and Their Factor Decomposition. Statistics Information Group, Energy Data and Modeling Center, IEEJ: February 2009.

APEC (Asia Pacific Economic Cooperation), 2005: *CO_2 Storage Prospectivity of Selected Sedimentary Basins in the Region of China and South East Asia.* publications.apec.org/publication-detail.php?pub_id=393.

APS (American Physical Society), 2011: *Direct Air Capture of CO_2 with Chemicals.* A Technology Assessment for the APS Panel on Public Affairs. American Physical Society.

Arts, R., G. Williams, N. Delepine, V. Chlochard, K. Labat, S. Sturton, M.-L. Buddensiek, M. Dillen, M. Nickel, and A. L. Lima, 2010: Quantitative analysis of time-lapse seismic monitoring data at the Sleipner CO_2 storage operation. *The Leading Edge,* **29**(2): 170.

Ashworth, P., A. Pisarski, and A. Littleboy, 2006: *Understanding and Incorporating Stakeholder Perspectives to Low Emission Technologies in Queensland.* Research report prepared for Commonwealth Scientific and Industrial Research Organisation (CSIRO), Centre for Low Emission Technology, Brisbane, Australia.

Ashworth, P., N. Boughen, M. Mayhew, and F. Millar, 2007: An Integrated Roadmap of Communication Activities around Carbon Capture and Storage in Australia and beyond. Report No. 2007/975, CEM, CSIRO, Pullenvale. www.cslforum.org/aboutus/australia.html.

Ashworth, P., S. Carr-Cornish, N. Boughen, and K. Thambimuthu, 2009: Engaging the public on Carbon Dioxide Capture and Storage: Does a large group process work? *Energy Procedia,* **1**(1): 4765–4773.

Bachu, S., D. Bonijoly, J. Bradshaw, R. Burruss, S. Holloway, N. P. Chistensen, and O. M. Mathiassen, 2007: CO_2 Storage capacity estimation: methodology and gaps. *International Journal of Greenhouse Gas Control*, **1**(4): 430–443.

Beaubien, S.E., S. Lombardi, G. Ciotoli, A. Annunziatellis., G. Hatziyannis, A. Metaxas, and J. M. Pearce, 2004: Potential hazards of CO_2 leakage in storage systems: learning from natural systems. Proceedings, Seventh Internal Conference on Greenhouse Gas Control, Vancouver, BC.

Benson, S. M. and D. R. Cole, 2008: CO_2 Sequestration in Deep Sedimentary Formations, *ELEMENTS*, Vol, 4, pp. 325–331, DOI: 10.2113/gselements.4.5.325.

Best-Waldhober, M., D. Daamen, and A. Faaij, 2006: *Public Perceptions and Preferences Regarding Large Scale Implementation of Six CO_2 Capture and Storage Technologies: Well-informed and Well-considered Opinions versus Uninformed Pseudo-opinions of the Dutch public*. Report commissioned by NWO/SenterNovem Project "Transition to sustainable use of fossil fuel," Centre for Energy and Environmental Studies, Faculty of Social Sciences, Leiden University, & Department of Science, Technology and Society, Copernicus Institute, Utrecht University.

Best-Waldhober, M., D. Daamen, C. Hendriks, E. de Visser, A. Ramirez, and A. Faaij, 2008: *How the Dutch Evaluate CCS Options in Comparison with Other CO_2 Mitigation Options.* Results of a Nationwide Information-Choice Questionnaire Survey. Leiden University, Leiden.

Best-Waldhober, M., D. Daamen, and A. Faaij, 2009: Informed and uninformed public opinions on CO_2 capture and storage technologies in the Netherlands. *International Journal of Greenhouse Gas Control,* **3**(3): 322–332.

Bickle, M., A. Chadwick, H. E. Huppert, M. Hallworth, and S. Lyle, 2007: Modelling carbon dioxide accumulation at Sleipner: Implications for underground carbon storage. *Earth and Planetary Science Letters*, **255**(1–2): 164–176.

Bielicki, J., 2008: *Returns to scale in carbon capture and storage infrastructure and deployment*. Discussion Paper 2008–04, Harvard Kennedy School, Cambridge, MA.

Bielicki, J. and J. C. Stephens, 2008: *Public Perception of Carbon Capture and Storage Technology Workshop Report.* Energy Technology Innovation Policy Group Workshop Series, Harvard Kennedy School, Cambridge MA.

Birkholzer, J. T. and Q. Zhou, 2009: Basin-scale hydrogeologic impacts of CO_2 storage: capacity and regulatory implications. *International Journal of Greenhouse Gas Control*, **3**(6): 745–756.

Black, S., 2006: Chilled Ammonia Scrubber for CO_2 Capture. Presentation at the MIT Sequestration Forum, Alstom, Cambridge, MA.

BLS, 2011. Industries at a Glance. www.bls.gov/iag/tgs/iag_index_naics.htm.

Bock, B., R. Rhudy, H. Herzog, M. Klett, J. Davison, D. De La Torre, and D. Simbeck, 2003: *Economic Evaluation of CO_2 Storage and Sink Enhancement Options.* Final Technical Report submitted by TVA Public Power Institute, Muscle Shoals, AL.

Booras, G., 2009: Economic assessment of advanced coal-based power plants with CO_2 capture. Presentation for MIT Carbon Sequestration Forum IX – Advancing CO_2 Capture, Cambridge, MA.

Bradbury, J., I. Ray, T. Peterson, S. Wade, G. Wong-Parodi and A. Feldpausch, 2009: The Role of Social Factors in Shaping Public Perceptions of CCS: Results of Multi-State Focus Group Interviews in the U.S. *Energy Procedia*, **1**(1): 4665–4672.

Bradshaw, J., G. Allinson, B. E. Bradshaw, et al., 2004: Australia's CO_2 geological storage potential and matching of emission sources to potential sinks. *Energy,* **29**: 1623–1631.

BRGM, 2009: CO_2 Capture and Storage Projects Around the World. Presented at the 3rd international symposium organised by ADEME, BRGM and IFP on the Capture and geological storage of CO_2, "Accelerating deployment."

Brunsting, S. and T. Mikunda, 2010: *Case study on the Barendrecht project*, NEARCO2 Final Report, IEEP, Brussels, Belgium.

Buesseler, K. O., J. E. Andrews, S. M. Pike, and M. A. Charette, 2004: The effects of iron fertilization on carbon sequestration in the Southern Ocean. *Science*, 304: 414–417.

Burruss, R., 2004: Geologic storage of carbon dioxide in the next 50 years: an energy resource perspective. Proceedings, *Pew Center 10–50 Workshop*, Washington, D.C.

California Institute for Energy and Environment (CIEE), 2010: Background Report for the California Carbon Capture and Storage Review Panel, Berkeley, CA.

Carbon Sequestration Atlas of the United States and Canada, Second Edition, 2010: U.S. Department of Energy, Office of Fossil Energy, National Energy Technology Laboratory.

Carey, J. W., M. Wigand, S. Chipera, G. WoldeGabriel, R. Pawar, P. Lichtner, S. Wehner, M. Raines, and G. D. Guthrie., Jr., 2007: Analysis and performance of oil well cement with 30 years of CO_2 exposure from the SACROC Unit, West Texas, USA. *International Journal of Greenhouse Gas Control*, 1: 75–85.

Chinese Government, 2003: *China's Policy on Mineral Resources.* Chinese Government's Official Web Portal, English.gov.cn/official/2005-07/28/content_17963.htm (accessed 1 October, 2010).

Cloete, M., 2010. CO_2 Atlas Assessment of Geological Storage Potential. Presentation for Workshop CCS in South Africa. Council for Geoscience.

CNDRC, 2007: *China's National Climate Change Programme.* China National Development and Reform Commission (CNDRC), People's Republic of China, Beijing, China.

de Coninck, H., 2008: Trojan horse or horn of plenty? Reflections on CCS and the CDM. *Energy Policy, 36*(3): 929–936.

de Coninck, H. and C. F. J. (Ynke) Feenstra, 2009: *CCS and Community Engagement: Guidelines for Community Engagement in CCS Capture, Transportation and Storage Projects*. Case Study #1: Barendrecht CCS Project – Barendrecht (The Netherlands). World Resources Institute.

de Coninck, H., 2010. Advocacy for carbon capture and storage could arouse distrust. Nature 463, p. 293.

Duan, H., 2010: The public perspective of carbon capture and storage for CO_2 emission reductions in China. *Energy Policy*, 38(9): 5281–5289.

Cormick, C., 2002: Australian Attitudes to GM Foods and Crops. *Pesticide Outlook, 13*(6): 261–263.

Croiset, E. and K. Thambimathu, 2001: Nox and SO_2 from O_2/CO_2 recycle coal combustion. *Fuel, 80*(14): 2117.

CST (Carbon Storage Taskforce), 2009: Australia's Potential for the Geological Storage of CO_2.

Cunha, C., S. Paulo, Z. Santos Estevão dos Márcia, and M. Aurélio, 2006: *Public Perception on CCS in Brazil*. Petrobras Research Center/Ecoplan Institute/Ecoar Institute.

Curry, T., D. Reiner, S. Ansolabehere, and H. Herzog, 2004: How aware is the public of carbon capture and storage? In *Proceedings of 7th International Conference on Greenhouse Gas Control Technologies*. Volume 1: Peer-Reviewed Papers and Plenary Presentations, IEA Greenhouse Gas Programme, Cheltenham, UK.

Curry, T. E., S. Ansolabehere, and H. J. Herzog, 2007: *A Survey of Public Attitudes towards Climate Change and Climate Change Mitigation Technologies in the United States: Analyses of 2006 Results*. MIT LFEE 2007-01 WP.

Czernichowski-Lauriol, I., B. Sanjuan, C. Rochelle, K. Bateman, J. Pearce, and P. Blackwell, 1996: The underground disposal of carbon dioxide. *Inorganic Geochemistry, 2*(7): 183–276.

Dahowski, R. T., X. Li, C. L. Davidson, N. Wei, and J. J. Dooley, 2009: *Regional Opportunities for Carbon Dioxide Capture and Storage in China: A Comprehensive CO_2 Storage Cost Curve and Analysis of the Potential for Large Scale Deployment of CCS in the People's Republic of China*. Pacific Northwest National Laboratory. Richland, WA.

Damen, K., A. Faaij, and W. Turkenburg, 2006: Health, safety and environmental risks of underground CO_2 storage – overview of mechanisms and current knowledge. *Climate Change, 74*(1): 289–318.

Darde et al., 2008: *Air separation and flue gas compression and purification units for oxy-coal combustion systems*. Presented at the 9th Greenhouse Gas Technologies Conference (GHGT-9) by Air Liquide, Washington, D.C., *Energy Procedia, 1*: 1035–1042.

Davis, S. J. and K. Caldeira, 2010: Consumption-based accounting of CO_2 emissions, Stanford University, CA. www.pnas.org/cgi/doi/10.1073/pnas.0906974107.

Davison, J., 2007: Performance and costs of power plants with capture and storage of CO_2. *Energy, 32* (7): 1163.

DNV, 2010. CO2QUALSTORE Guideline for Selection and Qualification of Sites and Projects for Geological Storage of CO_2. DNV Report No.: 2009–1425.

Dooley, J. J., R. T. Dahowski, and T. Davidson, 2009: Comparing existing pipeline networks with the potential scale of future U.S. CO_2 pipeline networks. *Energy Procedia, 1*: 1595–1602.

Duguid, A. and G. W. Scherer, 2009: Degradation of Oilwell Cement Due to Exposure to Carbonated Brine, *Int. J. Greenhouse Gas Control*, 4(2010): 546–560.

Duncan, I. J., S. Anderson and J.-P. Nicot, 2009: Pore space ownership issues for CO_2 sequestration in the US. *Energy Procedia, 1*(1): 4427–4431.

ECN, 2010: *CCS in Southern Africa. An assessment of the rationale, possibilities and capacity needs to enable CO_2 capture and storage in Botswana, Mozambique and Namibia*. ECN Report ECN-E--10-065, Petten, Netherlands.

Economides, M. and C. Ehlig-Economides, 2009: *Sequestering CO_2 in a closed Underground Volume*. SPE-124430. Presented at the 2009 SPE Annual Conference and Technical Exhibition, New Orlean, LA.

Eiken, O., P. Ringrose, C. Mermanrud, B. Nazarian, T. Torp and L. Hoier, 2011: Lessons learned from 14 years of CCS operations. *Energy Procedia, 4*: 5541–5548.

Emberly, S., I. Hutcheon, N. Shevalier, K. Durocher, B. Mayer. W. D. Gunter, and E. H. Perkins, 2005: Monitoring of fluid-rock interaction and CO_2 storage through produced fluid sampling at the Weyburn CO_2-injection enhanced oil recovery site Saskatchewan, Canada. *Applied Geochemistry, 20*: 1131–1157.

EMEP (European Monitoring and Evaluation Programme), 2009: *EMEP/EEA Air Pollutant Emission Inventory Guidebook*, TFEIP.

EPRI (Electric Power Research Institute), 1999: *Enhanced Oil Recovery Scoping Study*. TR-113836. EPRI, Palo Alto, CA.

EPRI, 2008: *The Power to Reduce CO_2 Emissions: the Full Portfolio: 2008 Economic Sensitivity Studies.* EPRI, Palo Alto, CA.

Essandoh-Yeddu, J. and G. Gulen, 2008: *Economic modeling of carbon dioxide integrated transport network for enhanced oil recovery and geologic sequestration in the Texas Gulf Coast region.* Presented at the 9th International Conference on Greenhouse Gas Control Technologies, Washington, D.C.

EU Geocapacity, 2008: Assessing European Capacity for Geological Storage of Carbon Dioxide, Publishable Final Activity Report.

Findlay, M., N. Mabey, R. Marsh, S. Ng and S. Tomlinson, 2009: *Carbon Capture and Storage in China.* Briefing Paper, E3G Report for Germanwatch, Germanwatch, Bonn, Germany.

Fischedick, M., K. Pietzner, N. Supersberger, A. Eskena, W. Kuckshinrichs, P. Zapp, J. Linßen, D. Schumann, P. Radgen, C. Cremer, E. Gruber, N. Schnepf, A. Roser, and F. Idrissova, 2008: *Stakeholder acceptance of CCS in Germany.* Paper presented at GHGT9, Washington, D.C.

Gale, J. and J. Davison, 2003: *Transmission of CO_2 – Safety and Economic Considerations*. Proceedings GHGT-6, Kyoto, Japan.

Gasda, S. E., S. Bachu, and M. A. Celia, 2004: The potential for CO_2 leakage from storage sites in geological media: analysis of well distribution in mature sedimentary basins. *Environmental Geology,* **46**(6–7): 707–720.

GCCSI (Global Carbon Capture and Storage Institute), 2009a: *Strategic Analysis of the Global Status of Carbon Capture and Storage. Report 4: Existing Carbon Capture and Storage Research and Development Networks around the World.* www.globalccsinstitute.com/resources/publications/strategic-analysis-global-status-carbon-capture-storage.

GCCSI, 2009b: *(Strategic Analysis of the Global Status of Carbon Capture and Storage. Report 3) Country Study Brazil*. www.globalccsinstitute.com/resources/publications/strategic-analysis-global-status-ccs-country-study-brazil.

GCCSI, 2009c: *Strategic Analysis of the Global Status of Carbon Capture and Storage. Report 3: Country Study Mexico.* www.globalccsinstitute.com/resources/publications/strategic-analysis-global-status-ccs-country-study-mexico (accessed January 2011).

GCCSI, 2010: *(CCS Ready Policy) Considerations and Recommended Practices for Policymakers.* Prepared for The Global Carbon Capture and Storage Institute, IFC International, Fairfax, VA.

GCCSI, 2011a: *The Global Status of CCS, 2010.* www.globalccsinstitute.com/resources/publications/global-status-ccs-2010.

GCCSI, 2011b: *The Global CCS Institute CCS R&D network database – Strategic Analysis of Global Status of CCS.* www.globalccsinstitute.com/resources/data.

Geocapacity, 2009: *Assessing European Capacity for Geological Storage of Carbon Dioxide.* Final Report of Geocapacity Project.

Government Canada, 2009: *Mineral Resources Act.* Chapter M-7, Legislative Counsel Office, Ottawa, Canada.

Greenberg, S. 2009: personal communication with the authors on September 4, 2009.

Greenpeace, 2008: *False hope: Why Carbon Capture and Storage Won't Save the Climate.* Greenpeace International, Amsterdam, Netherlands.

Grubler, A., 2010. The cost of the French nuclear scale-up: A case of negative learning by doing. *Energy Policy,* **38**(9): 5174–5188.

Gunter, W. D., S. Bachu, and S. Benson, 2004: The role of hydrogeological and geochemical trapping in sedimentary basins for secure geological storage for carbon dioxide. In *Geological Storage of Carbon Dioxide: Technology*. S. Baines and R. H. Worden (eds.), Special Publication of Geological Society, Special Publication 233, London, United Kingdom, pp.129–145.

Ha-Duong, M., A. Nadaï, and A. S. Campos, 2009: A survey on the public perceptions of CCS in France. *Energy Procedia,* **1**: 4757–4765.

Haugen H. A., N. Eldrup, C. Bernstone, S. Liljemark, H. Pettersson, M. Noer, J. Holland, P. A. Nilsson, G. Hegerland, and J. O. Pande, 2009: *Options for Transporting CO_2 from Coal fired Power Plants*. Case Denmark. GHGT-9, Washington, D.C. *Energy Procedia,* **1**: 1665–1672.

Heddle G., H. Herzog, M. Klett, 2003: *The Economics of CO_2 Storage*. MIT LFEE 2003–003 RP, Laboratory for Energy and the Environment, Cambridge, MA.

Heinrich J., H. Herzog, and D. Reiner, 2004: *Environmental Assessment of Geologic Storage of CO_2*. MIT LFEE Report, Cambridge, MA.

Hendriks, C., W. Graus, and F. van Bergen, 2004: *Global carbon dioxide storage potential and costs.* Prepared by order of the Rijksinstituut voor Volksgezondheid en Milie, Utrecht, The Netherlands.

Hesse, M., F. M. Orr, and H. Tchelepi, 2008: Gravity currents with residual trapping. *Journal of Fluid Mechanics*, **611**: 35–60.

Holloway, S. (ed.), 1996: *The Underground Disposal of Carbon Dioxide*. Final report of Joule 2 Project No. CT92–0031. British Geological Survey, Keyworth, Nottingham, United Kingdom, pp.355.

Hooper B., L. C. Murray, and C. Gibson-Poole, 2005: Latrobe Valley CO_2 Storage Assessment, CO2CRC, Melbourne, Australia.

Hovorka, S. D., S. M. Benson, C. Doughty, B. M. Freifeld, S. Sakurai, T. M. Daley, Y. K. Kharaka, M. H. Holtz, R. C. Trautz, H. S. Nance, L. R. Myer, and K. G. Knauss, 2006: Measuring permanence of CO_2 storage in saline formations: the Frio experiment. *Environmental Geosciences*, **13**(2): 105–121.

Huijts, N., 2003: *Public Perception of Carbon Dioxide Storage.* Master's Thesis. Eindhoven University of Technology, Netherlands.

Hund, G., 2009: *Case Study on Wallula*, CCS project.

Hund, G., and K. Judd, 2008: Stakeholder acceptance issues concerning advanced coal-power plants with carbon capture and storage: lessons learned from FutureGen. In *Proceedings of the 9th International Conference on Greenhouse Gas Control Technologies (GHGT-9).* November 16–20, 2008, Washington, D.C., USA.

Hund, G. and Greenberg S. 2010: FutureGen Case Study. A collaboration between the Pacific Northwest National Laboratory and the Illinois State Geological Survey – Advanced Energy Technology Initiative in fulfillment of Task 1 for CSIRO on behalf of the Global CCS Institute: International Comparison of Public Outreach Practices Associated with Large Scale CCS Projects. pp36.

Ide, S. T., K. Jessen, and F. M. Orr, Jr., 2007: Storage of CO_2 in saline aquifers: effects of gravity, viscous, and capillary forces on amount and timing of trapping. *Journal of Greenhouse Gas Control,* **1**: 481–491.

Iding, M. and P. Ringrose, 2010: Evaluating the impact of fractures on the performance of the In Salah CO_2 storage site. *International Journal of Greenhouse Gas Control,* **4**(2): 242–248.

IEA (International Energy Agency), 2007: *Legal Aspects of CO_2 Storage: Update and Recommendations*, IEA/OECD, Paris.

IEA, 2008a: *CO_2 Capture and Storage: A Key Abatement Option*, IEA/OECD, Paris, France.

IEA, 2008b: *World Energy Outlook*, IEA/OECD, Paris, France.

IEA, 2008c: Energy Technology Perspectives, Scenarios and Strategies to 2050. www.iea.org/publications/free_new_Desc.asp?PUBS_ID=2012.

IEA, 2009: *CCS Roadmap.* International Energy Agency (IEA) of the Organisation for Economic Cooperation and Development (OECD), Paris, France.

IEA, 2009b: *Technology Roadmap on Carbon Capture and Storage*, IEA, Paris, France.

IEA, 2010: *Key World Energy Statistics*. IEA, Paris, France.

IEA, 2010b: *Carbon Capture and Storage: Legal and Regulatory Review.* Edition 1, International Energy Agency (IEA) of the Organisation for Economic Cooperation and Development, Paris, France.

IEA GHG (IEA Greenhouse Gas R&D Programme), 2005: *Low Greenhouse Gas Emission Transport Fuels – The Impact of CO_2 Capture and Storage on Selected Pathways*. Report No. 2005/10, IEA GHG, Cheltenham, UK.

IEA GHG, 2006: *Environmental impact of solvent scrubbing of CO_2.* Report No. 2006/14, IEA GHG, Cheltenham, UK.

IEA GHG, 2008: *A Regional Assessment of the CO_2 Storage Potential of the Indian Subcontinent.* Report No. 2008/2, IEA GHG, Cheltenham, UK.

IEA Working Party on Fossil Fuels, 2007: *CO_2 Capture Project, commissioned by IEA Working Party on Fossil Fuels. Public Perception of Carbon Dioxide Capture and Storage: Prioritised Assessment of Issues and Concerns*. March 2007.

Indonesia CCS Study Working Group, 2009: *Understanding Carbon Capture and Storage Potential in Indonesia.* APEC/MEMR/IEA Joint Workshop, Jakarta, Indonesia.

IPCC (Intergovernmental Panel on Climate Change), 2005: *IPCC Special Report: Carbon Dioxide Capture and Storage*, Cambridge University Press, Cambridge, UK.

IPCC, 2007: *IPCC Fourth Assessment Report, Working Group III. Mitigation.* Cambridge University Press, Cambridge, UK.

Itaoka, K., A. Saito, and M. Akai, 2004: Public acceptance of CO_2 capture and storage technology : A survey of public opinion to explore influential factors. E.S. Rubin, D.W. Keith, and C.F. Gilboy, (eds.), *Proceedings of 7th International Conference on Greenhouse Gas Control Technologies*, Volume 1: Peer-reviewed Papers and Plenary Presentations, IEA Greenhouse Gas Program, Cheltenham, UK.

Itaoka, K., A. Saito, and M. Akai, 2006: *A path analysis for public survey data on social acceptance of CO_2 capture and storage technology.* Presented at the 8th International Conference on Greenhouse Gas Control Technologies (GHGT-8), Trondheim, Norway.

Itaoka, K., Y. Okuda, A. Saito, and M. Akai, 2009: Influential information and factors for social acceptance of CCS: the 2nd round survey of public opinion in Japan. *Energy Procedia*, **1**: 4803–4810.

Itaoka, K., A. Saito, and M. Akai, 2011. A study on roles of public survey and focus groups to assess public opinions for CCS implementation, *Energy Procedia*, **4**: 6330–6337.

Japan CCS Study Group, 2009: For safe operation of a CCS demonstration project. Ministry of Economy, Trade and Industry of Japan. Tokyo, Japan.

Jindal, 2009: "Jindal Steel & Power Limited Announces New Coal Gasification Midrex Plant in India," Jindal Steel News Release, December 20, 2009.

Johnsson F., D. Reiner, K. Itaoka, and H. Herzog, 2009: Stakeholder attitudes on carbon capture and storage – an international comparison. *International Journal of Greenhouse Gas Control*, **1**(1): 4819–4826.

Jordan, P. and S. M. Benson, 2008: Well Blowout Rates and Consequences in California Oil and Gas District 4 from 1991 to 2005: Implications for Geological Storage of Carbon Dioxide, *Environmental Geology*, **57**(5): 1103–1123.

Juanes, R., C. W. MacMinn, and M. L. Szulczewski, 2010: The footprint of the CO_2 plume during carbon dioxide storage in saline aquifers: storage efficiency for capillary trapping at the basin scale. *Transport in Porous Media*, **82**(1): 19–30.

Kalaitzandonakes, N., L. A. Marks, and S. S. Vickner, 2005: Sentiments and acts towards genetically modified foods. *International Journal of Biotechnology*, **7**(1–3): 161–177.

Ketzer, M., J. A. Villwock, G. Caporale, et al., 2007: Opportunities for CO_2 Capture and Geological Storage in Brazil: The CARBMAP Project. Sixth Annual Conference on Carbon Capture and Sequestration, Pittsburgh, Pennsylvania.

Kharaka, Y. K., D. R. Cole, S. D. Hovorka, W. D. Gunter, K. G. Knauss, and B. M. Freifield, 2006a: Gas-water-rock interactions in the Frio formation following CO_2 injection: implications for the storage of greenhouse gases in sedimentary basins. *Geology*, **34**: 577–580.

Kharaka, Y. K., D. R. Cole, J. J. Thordsen, E. Kakouros, and H. S. Nance, 2006b: Gas-water-rock interactions in sedimentary basins: CO_2 sequestration in the Frio Formation, Texas, USA. *Journal of Geochemistry and Exploration*, **89**: 183–186.

Klusman, R.W., 2003: Rate measurements and detection of gas seepage to the atmosphere from enhanced oil recovery/sequestration project, Rangely, Colorado, USA. *Applied Geochemistry*, **18**: 1825–1838.

Koornneef, J., T. van Keulen, A. Faaij, and W. Turkenburg, 2008: Life cycle assessment of a pulverized coal power plant with post-combustion capture, transport and storage of CO_2. *International Journal of Greenhouse Gas Control*, **2**(4): 448–467.

Koornneef, J., M. Spruijt, M. Molag, A. Ramírez, W. Turkenburg, and A. Faaij, 2009: Quantitative risk assessment of CO_2 transport by pipelines – A review of uncertainties and their impacts. *Journal of Hazardous Materials*, **177**(1–3): 12–27.

Koornneef, J., A. Ramírez, T. van Harmelen, A. van Horssen, W. Turkenburg, and A. Faaij, 2010: The impact of CO_2 capture in the power and heat sector on the emission of SO_2, NOx, particular matter, volatile organic compounds and NH3 in the European Union. *Atmospheric Environment*, **44**: 1369.

Koottungal, L., 2008: Worldwide EOR Survey. *Oil and Gas Journal*, **1**: 47–59.

Korre, A., Z. Nie, and S. Durucan, 2010: Life cycle modelling of fossil fuel power generation with post-combustion CO_2 capture. *International Journal of Greenhouse Gas Control*, **4**: 289–300.

Koschel, D., J. Y. Coxam, L. Rodier, and V. Majer, 2006: Enthalpy and solubility data of CO_2 in water and NaCl (aq) at conditions of interest for geological sequestration. *Fluid Phase Equilibria*, **247**: 107–120.

Krevor, S. C., J-C. Perrin, A. Esposito, C. Rella, and S. M. Benson, 2010: Rapid detection and characterization of surface CO_2 leakage through the realtime measurement of ^{13}C signatures in CO_2 flux from the ground. *International Journal of Greenhouse Gas Control*, **4**(5): 811–815.

Kumar, B., S. N. Charan, R. Menon, and S. Panicker, 2007: *Geological CO_2 Sequestration in Basalt Formations of Western India – A Feasibility Study*. IWCCS-07, NGRI, Hyderabad, India.

Kuuskraa, V. and V. Fergeson, 2008: Storing CO_2 with enhanced oil recovery. National Energy Technology Laboratory. U.S. Department of Energy. Washington, D.C., USA.

Kuvshinov, V. A., 2006: The use of CO_2 and combustion gases for Enhanced Oil Recovery in Russia. *NATO Science Series*, **65**: 271–275.

Lagneau, V., A. Pipart, and H. Catalette, 2005: Reactive transport modeling of CO_2 sequestration in deep saline aquifers. *Oil & Gas Science and Technology*, **60**: 231–247.

Lambrichs, C., 2008: *Pseudo-opinions on Carbon Capture and Storage Technology: When and Why are Pseudo-opinions Unstable?* Master's Thesis. Leiden University, Leiden, the Netherlands.

Legislative Assembly of British Columbia, 2008: *Bill 37 – Carbon Tax Act.* 2008 Legislative Session: 4th Session, 38th Parliament, Legislative Assembly of British Columbia, Victoria, BC, Canada.

Le Guénan, T. and J. Rohmer, 2011: Corrective measures based on pressure control strategies for CO_2 geological storage in deep aquifers. *International Journal of Greenhouse Gas Control Technology*, **34** (2010): 571–578.

Lewicki, J. L., G. E. Hilley, T. Tosha, R. Aoyagi, K. Yamamoto, and S. M. Benson, 2007: Dynamic coupling of volcanic CO_2 flow and wind at the Horseshoe Lake tree kill, Mammoth Mountain, California. *Geophysical Research Letters*, **34**.

Lewicki, J. L., G. E. Hilley, M. L. Fischer, L. Pan, C. M. Oldenburg, L. Dobeck, and L. Spangler, 2009: Eddy covariance observations of surface leakage during shallow subsurface CO_2 releases. *Journal of Geophysical Research – Atmospheres*, **114**.

Linde, A. G., 2009: PSA – Pressure Swing Adsorption Plants. www.linde.com.

Lou, Z., 2008: Status of CCS in China. In *2nd US-China Symposium on CO_2 Emission Control Science & Technology.* Hangzhou, China.

Lund P., A. Prashad, P. Holtedahl, and T. Völker, 2008: International CCS (Carbon Capture and Storage) Technology Survey. Innovation Norway.

Luo, Z., 2008: Status of CCS in China. *2nd U.S.-China Symposium on CO_2 Emission Control Science & Technology*, Hangzhou, China.

MacMinn, C. W. and R. Juanes, 2009: Computational post-injection spreading and trapping of CO_2 in saline aquifers: impact of the plume shape at the end of injection. *Geosciences*, **13**:483–491.

Malone, E. L., J. A. Bradbury, and J. J. Dooley, 2009: Keeping CCS stakeholder involvement in perspective. *Energy Procedia*, **1**: 4789–4794.

Mart, C. J., 2009: *Controlled Freeze Zone™ Technology – An Integrated Solution for Processing Sour Natural Gas*. Presentation for the Global Climate & Energy Project, Stanford University.

Martini, B. and E. Silver, 2002: The evolution and present state of tree-kills on Mammoth Mountain, California: tracking volcanogenic CO_2 and its lethal effects. In *Proceedings of the 2002 AVIRIS Airborne Geoscience Workshop*, Jet Propulsion Laboratory, California Institute of Technology, Pasadena, CA.

Maulbetsch, J. S., 2002: *Comparison of Alternate Cooling Technologies for California Power Plants: Economic, Environmental and Other Tradeoffs*. PIER/EPRI Technical Report. Prepared for the Electric Power Research Institute and the California Energy Commission. Sacramento, CA.

McCollum, D. and J. Ogden, 2006: Techno-Economic Models for Carbon Dioxide Compression and Transport and Storage & Correlations for Estimating Carbon Dioxide Density and Viscosity. Institute of Transportation Studies, University of California, Davis, Research Report UCD-ITS-RR-06–14.

McCoy, S. 2008: *The Economics of CO_2 Transport by Pipeline and Storage in Saline Aquifers and Oil Reservoirs*. PhD thesis. Carniege Mellon University, Pittsburg, PA.

McGrail, B. P., H. T. Schaef, A. M. Ho, Y.J. Chien, J. J. Dooley, and C. L. Davidson, 2006: Potential for carbon dioxide sequestration in flood basalts. *Journal of Geophysical Research*, **111**.

McLarnon, C. R. and J. L. Duncan, 2008: *Testing of Ammonia Based CO_2 Capture with Multi-Pollutant Control Technology*. Paper presented at GHGT-9 by Powerspan, Washington, D.C.

de Mello, L., L. Gearhart, P.B. Milios, and T. Melien, 2008: *A Technical and Economical Evaluation of CO_2 Capture from FCC Units*. Presented at the 9th Greenhouse Gas Technologies Conference (GHGT-9), Washington, D.C.

Miles, N., K. Davis, and J. Wyngaard, 2005: Detecting leakage from CO_2 reservoirs using micrometeorical methods, Carbon Dioxide Capture for Storage in Deep Geologic Formations. Results from the CO_2 Capture Program, v. 2: Geologic Storage of Carbon Dioxide with Monitoring and Verification. Elsevier Science, London, pp.1031–1044.

Miller, E., L. Bell, and L. Buys, 2007: Public understanding of carbon sequestration in Australia: socio-demographic predictors of knowledge, engagement and trust. *Australian Journal of Emerging Technologies and Society*, **5**(1), 15–33.

Ministry of Economy, Trade and Industry, Japan, 2008: *Cool Earth-Innovative Energy Technology Program*. Tokyo, Japan.

Ministry of Economy, Trade and Industry of Japan, 2010: The Strategic Energy Plan of Japan. Tokyo, Japan.

MIT, 2007. The Future of Coal. Massachusetts Institute of Technology. web.mit.edu/coal/.

MIT, 2009. Carbon capture and sequestration technologies program. Carbon Management GIS: CO_2 pipeline Transport Costs estimation. Contract DE-FC26–02NT41622.

Moberg, R., D.B. Stewart, and D. Stachniak, 2003: The IEA Weyburn CO_2 Monitoring and Storage Project. In *Proceedings of the Sixth International Conference on Greenhouse Gas Control Technologies (GHGT-6)*, J.Gale, and Y.Kaya (eds.), 1–4 October 2002, Kyoto, Japan, pp.219–224.

Möllersten, K., J. Yan, and J. R. Moreira, 2003: Potential markets niches for biomass supply with CO_2 capture and storage – Opportunities for energy supply with negative CO_2 emissions. *Biomass and Bioenergy*, **25**: 273–285.

Mors, E., 2009: *Dealing with Information about Complex Issues: The Role of Source Perceptions*. PhD Dissertation. Leiden University, Leiden, Netherlands.

Mourits, F., 2009: Overview of Canadian Policy and Project Activities in Carbon Capture and Storage (CCS). *2009 APEC Clean Fossil Energy Technical and Policy Seminar*, Incheon, Republic of Korea.

Mulder, G., J. Hetland, and G. Lenaers, 2007: Towards a sustainable hydrogen economy: hydrogen pathways and infrastructure. *International Journal of Hydrogen Energy*, **32**: 1324–1331.

Nakanishi, S., Y. Mizuno, T. Okumura, H. Miida, T. Shidahara, and S. Hiramatsu, 2009: Methodology of CO_2 aquifer storage capacity assessment in Japan and overview of the project. *Energy Procedia*, **1**(1): 2639–2646.

Natural Resources Canada, 2006: *Carbon Dioxide Capture and Storage: A Compendium of Canada's Participation*. www.nrcan.gc.ca/com/resoress/publications/carbone/carbone-eng.php#dowtel.

NETL (National Energy Technology Laboratory), 2010a. Carbon Sequestration: 2010 Carbon Sequestration Atlas of the United States and Canada – Third Edition (Atlas III). U.S. Department of Energy, Washington, D.C.

NETL, 2010b. Site Screening, Selection, and Initial Characterization for Storage of CO_2 in Deep Geologic Formations, DOE/NETL-401/090808. U.S. Department of Energy, Washington, D.C.

Nicot, J. P., 2008: Evaluation of large-scale CO_2 storage on fresh-water sections of aquifers: an example from the Texas Gulf Coast Basin. *International Journal of Greenhouse Gas Control*, **2**(4): 582–593.

Nordbotten, J. M., D. Kavetski, M. A. Celia, and S. Bachu, 2009: A semi-analytical model estimating leakage associated with CO_2 storage in large-scale multi-layered geological systems with multiple leaky wells. *Environmental Science & Technology*, **43**(3), 743–749.

Ogawa, T., T. Shidaharab, S. Nakanishi, T. Yamamoto, K. Yoneyama, T. Okamura, and T. Hashimoto, 2009: Storage capacity assessment in Japan: comparative evaluation of CO_2 aquifer storage capacities across regions. *Energy Procedia*, **1**: 2685–2692.

Ohgaki, K., K. Takano, H. Sangawa, T. Matsubara, and S. Nakano, 1996: Methane hydrate exploitation by carbon dioxide from gas hydrates – phase equilibria for CO_2-CH_4 mixed hydrate systems. *Journal of Chemical Engineering of Japan*, **29**: 478–483.

Oldenburg, C., 2005. Migration mechanisms and potential impacts of CO_2 leakage and seepage. *In Carbon Capture and Sequestration – Integrating Technology, Monitoring, Regulation*. E. Wilson and D. Gerard (eds.), Blackwell Publishing, pp.127–146.

Oosterkamp, R., 2008: *State-of-the-art Overview of CO_2 pipeline transport with relevance to offshore pipelines*. Haugesund: R&D Foundation Polytec.

Palmgren, C. R., G. Morgan, W. Bruine de Bruin, and D. W. Keith, 2004: Initial public perceptions of deep geological and oceanic disposal of carbon dioxide. *Environmental Science &Technology*, **38**(24): 6441–6450.

Parfomak, P.W., 2008: *Community Acceptance of Carbon Capture and Sequestration Infrastructure: Siting Challenges*. A CRS Report to Congress, July 29, 2008.

Parfomak, P. and P. Folger, 2008. *Pipelines for Carbon Dioxide Control: Network Needs and Costs Uncertainties*. CRS Report for Congress.

Pehnt, M., and J. Henkel, 2009: Life cycle assessment of carbon dioxide capture and storage from lignite power plants. *International Journal of Greenhouse Gas Control*, **3**(1): 49–66.

PetroChina Company Limited, 2007: *CCS Activities and Developments in China*. Beijing, China.

Pevzner, R., V. A. Shulakova, M. Kepic, and M. Urosevic, 2010. Repeatability analysis of land time-lapse seismic data: CO2CRC Otway pilot project case study. *Geophysical Prospecting*, **59**(1): 66–77.

Province of Alberta, 2009a: *Carbon Capture and Storage Funding Act*. Alberta Queen's Printer, Government of Alberta, Edmonton, Alberta.

Province of Alberta, 2009b: *Climate Change and Emissions Management Act*. SA 2003, c C-16.7, Alberta Queen's Printer, Government of Alberta, Edmonton, Canada.

Quebec National Assembly, 2009: *An Act to amend the Environment Quality Act and other legislative provisions in relation to climate change*. Bill 42, Chapter 33, First Session, 39th Legislature, Québec Official Publisher, Québec City, QC, Canada.

Ramgen, 2011: Ramgen Power Systems, Ramgen Document 0800–00153. www.ramgen.com.

RECCS, 2008: Ecological, economic and structural comparison of renewable energy technologies with carbon capture and storage – an integrated approach. Federal Ministry for the Environment, Nature Conservation and Nuclear Safety (BMU).

Reiner, D., 2008: *A looming rethorical gap: a survey of public communication activities for carbon dioxide capture and storage Technologies*. Electricity Policy Research Group Work Papers No 0801, University of Cambridge, Cambridge, UK.

Reiner, D., T. Curry, M. de Figueredo, H. Herzog, S. Ansolabehere, K. Itaoka, M. Akai, F. Johnsson, and M. Odenberger, 2006: *An international comparison of public attitudes towards carbon capture and storage technologies*. Paper presented at 8th International Conference on Greenhouse Gas Control Technologies (GHGT-8), Trondheim, Norway.

Renn, O., T. Webler, and P. Wiedeman (eds.), 1995: *Fairness and competence in citizen participation: evaluating models for environmental discourse*. Risk, Governance and Society Series, vol. 10, Kluwer Academic Publishers, Dordrecht.

Rhudy, D., 2009: *CO_2 Capture Primer and Industry/EPRI Initiatives*. Presentation for SECARB Annual Meeting, March 4, 2009, Atlanta, Georgia.

RITE, 2011: Research Institute of Innovative Technology for the Earth. www.rite.or.jp/index_E.html.

Riddiford, F. A., A. Tourqui, C. D. Bishop, B. Taylor and M. Smith, 2003: A cleaner development: The In Salah Gas Project, Algeria. In *Proceedings of the 6th International Conference on Greenhouse Gas Control Technologies (GHGT-6)*. J. Gale and Y. Kaya (eds.), 1–4 October 2002, Kyoto, Japan, Vol. I, pp.601–606.

Rockett, G.C., C. X. Machado, J.M.M. Ketzer and C.I. Centeno, 2011. The CARBMAP project: Matching CO_2 sources and geological sinks in Brazil using geographic information system, *Energy Procedia*, **4**: 2764–2771.

Romanak K. D., R. Smyth, C-Y Yang, and S. Hovorka, 2010: *Detection of anthropogenic CO_2 in dilute groundwater: field observations and geochemical modeling of the Dockum aquifer at the SACROC oilfield, West Texas, USA*. GCCC Digital Publication Series #10–06.

Rubin, E., C. Chen, and A. Rao, 2007: Cost and performance of fuel power plants with CO_2 capture and storage. *Energy Policy*, **35**, 4444–4454.

Schrag, D. P., 2009: Storage of carbon dioxide in offshore sediments. *Science*, **325**: 1658–1659.

Schreiber, A., P. Zapp, and W. Kucksinrichs, 2009: Environmental assessment of German electricity generation from coal-fired power plants with amine-based carbon capture. *International Journal of Life Cycle Assessment*, **14**: 547–559.

Shackley S, C. McLachlan, and C. Gough, 2005: The public perception of carbon dioxide capture and storage in the UK: Results from focus groups and a survey. *Climate Policy* 4: 377–398.

Shackley, S., H. Waterman, P. Godfroij, D. Reiner, J. Anderson, K. Draxlbauer, and T. Flach, 2007: Stakeholder perceptions of CO_2 capture and storage in Europe: Results from a survey, *Energy Policy*, **35**(10): 5091–5108.

Shackley, S. and P. Verma, 2008: Tackling CO_2 reduction in India through the use of CO_2 capture and storage (CCS): Prospects and challenges. *Energy Policy*, **36**(9): 3554–3561.

Shackley, S., D. Reiner, P. Upham, H. de Coninck, G. Sigurthorsson, and J. Anderson, 2008: The acceptability of CO_2 capture and storage (CCS) in Europe: An assessment of the key determining factors. Part 2: The social acceptability of CCS and the wider impacts and repercussions of its implementation. *International Journal on Greenhouse Gas Control*, **3**: 344–356.

Shah, M., D. Bonaquist, and J. Shine, 2006: "Attributes of Existing Industrial Sources Part 2. Steam Methane Reformer Plants," A Part of the CEED CO_2 Shortcourse #13 entitled "CO_2 Sourcing for Enhanced Oil Recovery," Midland, Texas, December 6, 2006.

Sharp, J., Jaccard, M., Keith, D., 2006. *Public attitudes toward geological disposal of carbon dioxide in Canada*. Report No. 384. Simon Fraser University, Burnaby, BC, Canada.

Shen, P. and H. Jiang, n.d.: *China utilization of greenhouse gas as resource in EOR and the storage of it underground*. Research Institute of Petroleum Exploration & Development, PetroChina, Beijing, China.

Shogenova, A., Sliaupa, S., Shogenov, K., Sliaupiene, R., Pomeranceva, R., Vaher, R., Uibu, M., Kuusik, R., 2008: Possibilities for geological storage and mineral trapping of industrial CO_2 emissions in the Baltic Region. In *9th*

International Conference on Greenhouse Gas Control Technologies, Washington, 16–20 November 2008.

Sijm, J. P. M., 2009: Tradable carbon allowances: the experience of the EU and lessons learned. ECN report ECN-E-09-078, November 2009.

Simbeck, D., 2009a: *Insights on Current Gasification Issues for CO_2 Capture and Storage.* Presented at the Seventh Annual EOR Carbon Management Workshop, 7 December 2009, Houston, Texas.

Simbeck, D., 2009b: Trends and Outlook for Gasification and Coking Technologies. Presentation for the ERTC Coking and Gasification Conference, 20–22 April 2009, Budapest, Hungary.

Simbeck, D. and W. Roekpooritat, 2009a: *Near-term technologies for retrofit CO_2 capture and storage of existing coal-fired power plants in the United States.* MIT Coal Retrofit Symposium, 23 March 2009. "Energy Technology Perspectives—2010: Scenarios & Strategies to 2050," International Energy Agency, OECD/IEA, 2010.

Simbeck, D. and W. Roekpooritat, 2009b: *Near-Term Technologies for Retrofit CO_2 Capture and Storage of Existing Coal-Fired Power Plants in the United States.* White paper for the MIT Coal Retrofit Symposium, March 2009. web.mit.edu/mitei/ research/reports.html.

Simpson, P. and P. Ashworth, 2009: ZeroGen new generation power – a framework for engaging stakeholders. *Energy Procedia,* 1: 4697–4705.

Simpson, W., 2009: Personal communication with author on August 8, 2009.

Singleton, G., H. Herzog, and S. Ansolabehere, 2009: Public risk perspectives on the geologic storage of carbon dioxide. *Energy Procedia,* 1: 100–107.

Skovholt, O., 1993: CO_2 transportation system. *Energy Conversion Management,* 34(9–11): 1095–1103.

Slavin, T. and A. Jha, 2009: Not under our backyard, say Germans, in blow to CO_2 plans. *Guardian U.K.* www.guardian.co.uk/environment/2009/jul/29/germany-carbon-capture (accessed February 2011).

Slovic, P., B. Fischhoff, and S. Lichtenstein, 1993: Perceived risk, dread, and benefits. *Risk Analysis,* 13(3): 259–264.

Smouse, S. M., 2007: *International Cooperation on CCS Technology.* Expert Group Meeting: Carbon Capture and Storage and Sustainable Development, Division for Sustainable Development, UN Department of Economic and Social Affairs (UN DESA), New York, NY, USA.

Spangler, L. H., L. M. Dobeck, K. S. Repasky, et.al., 2010: A shallow subsurface controlled release facility in Bozeman, MT, USA, for testing near surface CO_2 detection techniques and transport models. *Environmental Earth Sciences,* 60: 227–239.

Spitznogle, G. O., 2009: *AEP CCS Project Update – Mountaineer Plant – New Haven, WV.* Presentation for the West Virginia 2009 Energy Summit, 8 December 2009, Roanoke, West Virginia.

Spycher, N., K. Pruess, and J. Ennis-King, 2003: CO_2-H_2O mixtures in geological sequestration of CO_2, I. Assessment and calculation of mutual solubility from 12 to 100°C and up to 600 bar. *Geochimica Et Cosmochimica Acta,* 67: 3015–3031.

Stangeland, A., B. Kristiansen, and A. Solli, 2006: How to close the gap between global energy demand and renewable energy production. Bellona Paper, The Bellona Foundation, Oslo, Norway.

Stephens, Jennie C., Anders Hansson, Yue Liu, Heleen de Coninck, Shalini Vajjhala, 2011. Characterizing the international carbon capture and storage community. Global Environmental Change 21 (2011) 379–390.

Stephens, J. C., J. Bielicki, and G. M.Rand, 2009: Learning about carbon capture and storage: Changing stakeholderperceptions with expert information. *Energy Procedia,* 2009: 4655–4663.

Stevens, S. H., C. Fox, T. White, S. Melzer and C. Byrer, 2003: Production operations at natural CO_2 Fields: Technologies for geologic sequestration. In *Proceedings of the 6th International Conference on Greenhouse Gas Control Technologies (GHGT- 6).* J. Gale and Y. Kaya (eds.), 1–4 October 2002, Kyoto, Japan, Pergamon, Vol. I, pp.429–433.

Surridge, T., 2010. Carbon capture and storage. www.ccs-africa.org/fileadmin/ccs-africa/user/docs/Gabarone_9_4/01_Surridge_-_CCS_-_Botswana_-_8–9Apr10.pdf.

Takahashi, T., T. Ohsumi, K. Nakayama, K. Koide, and H. Miida, 2009. Estimation of CO_2 Aquifer Storage Potential in Japan. *Energy Procedia,* 1(1): 2631–2638.

Tan Y., E. Croiset, M. Douglas, and K. Thambimuthu, 2006: Combustion characteristics of coal in a mixture of oxygen and recycled flue gas. *Fuel,* 85(4): 507.

Terwel, B., 2009: Origins and Consequences of Public Trust: Towards and Understanding of Public Acceptance of Carbon Dioxide Capture and Storage. Dissertation. Leiden University, Leiden, Netherlands.

The ecoENERGY Carbon Capture and Storage Task Force, 2008: *Canada's Fossil Energy Future – The Way Forward on Carbon Capture and Storage.* National Resources Canada, Ottawa, Canada.

Thitakamol, B., A. Veawab, A. Aroonwilas, 2007. Environmental impacts of absorption-based CO_2 capture unit for post-combustion treatment of flue gas from coal-fired power plant. *International Jouranl of Greenhouse Gas Control,* 1(1): 318–343.

Tokushige, K., K. Akimoto, and T. Tomoda, T., 2007: Public perceptions on the acceptance of geological storage of carbon dioxide and information influencing the acceptance. *International Journal of Greenhouse Gas Control,* 1: 101–112.

Torp, T. A. and J. Gale, 2003: Demonstrating storage of CO_2 in geological reservoirs: the Sleipner and SACS projects. In *Proceedings of the 6th International Conference on Greenhouse Gas Control Technologies (GHGT-6),* J. Gale and Y. Kaya (eds.), 1–4 October 2002, Kyoto, Japan, Pergamon, Amsterdam, Vol. I, pp.311–316.

Tsang, C. F., S. M. Benson, B. Kobelski, and R. Smith, 2002: Scientific Considerations Related to Regulation Development for CO_2 Sequestration in Brine Formations. *Environmental Geology,* 42(2–3): 275–281.

Tzimas, E., A. Mercier, C. Cormos, and S. Peteves, 2007: Trade-off in emission of acid gas pollutants and of carbon dioxide in fossil fuel power plants with carbon capture. *Energy Policy,* 35(8): 3991.

US DOE (US Department of Energy), 2007: *Cost and Performance Baseline for Fossil Energy Plants, Volume 1: Bituminous Coal and Natural Gas to Electricity,* National Energy Technology Laboratory, DOE/NETL-2007/1281.

US DOE, 2009: *Secretary Chu Announces $2.4 Billion in Funding for Carbon Capture and Storage Projects.* Fossil Energy Techline, US Department of Energy (US DOE), http://www.fossil.energy.gov/news/techlines/2009/09029-DOE_Announces_Stimulus_Funding.html (accessed 1 May, 2010).

US DOE, 2010: DOE/NETL Advanced Carbon Dioxide Capture Technology R&D Program: Technology Update.

US DOE/NETL, 2009a: *Estimating Freshwater Needs to Meet Future Thermoelectric Generation Requirements, 2009 Update.* National Energy Technology Laboratory, US Department of Energy, Washington, D.C.

US DOE/NETL, 2009b: *Public Outreach and Education for Carbon Storage Projects.* DOE/NETL-2009/1391, US Department of Energy (US DOE) and the National Energy Technology Laboratory (NETL), Washingyon, DC, USA.

US DOE/NETL, 2007a: *Cost and Performance Baseline for Fossil Energy Plants, Volume 1: Bituminous Coal and Natural Gas to Electricity.* National Energy Technology Laboratory, US Department of Energy, Washington, D.C.

US EIA, 2009: Electric Power Industry 2009: Year in Review. *Energy,* 1–12. www.eia.doe.gov/cneaf/electricity/epa/epa_sum.html.

US EIA, 2010a: U.S. Energy Consumption by Fuel (1980–2035), www.eia.doe.gov/oiaf/forecasting.html.

US EIA, 2010b: *China Country Brief Analysis.* US Energy Information Administration (US EIA), http://205.254.135.7/countries/country-data.cfm?fips-CH#cde (accessed 1 May, 2010).

US EPA, 2009: *EPA Analysis of the American Clean Energy and Security Act of 2009 H. R. 2454.* Office of Atmospheric Programs, US Environmental Protection Agency (US EPA), Washington, DC, USA.

US EPA, 2010: *Underground Injection Control (UIC) Program Requirements for Geologic Sequestration of Carbon Dioxide Final Rule.* Office of Water, US Environmental Protection Agency (US EPA), Washington, DC, USA.

US EPA, 2010: *EPA Analysis of the American Power Act in the 111th Congress.* Office of Atmospheric Programs, US Environmental Protection Agency (US EPA), Washington, DC, USA.

van den Broek, M., A. Ramirez, H. Groenenberg, F. Neele, P. Viebah, W. Turkenburg, and A. Faaij, 2010: Feasibility of storing CO_2 in the Utsira formation as part of a long term Dutch CCS strategy, an evaluation based on GIS/MARKAL tool box. *International Journal of Greenhouse Gas Control,* **4**: 351–366.

van der Meer, B. and L. G. H. van Wees, 2006: *Limitations to Storage Pressure in Finite Saline Aquifers and the Effects of CO_2 Solubility on Storage Pressure,* SPE 103342, presented at the SPE Annual Conference and Technical Exhibition, CA.

Vasco, D. W., A. Ferretti, and N. Fabrizio, 2008: Reservoir monitoring and characterization using satellite geodetic data: Interferometric Synthetic Aperture Radar observations from the Krechba field, Algeria. *Geophysics,* **73**: WA113-WA122.

de Visser, E., C. Hendriks, G. de Koeijer, S. Liljemark, M. Barrio, A. Austegard, and A. Brown, 2007: DYNAMIS CO_2 quality recommendations. www.dynamis-hypogen.com.

Voltattorni, N., G. Caramanna, D. Cinti D, G. Galli G, L. Pizzino L, and F. Quattrocchi, 2006: Study of natural emissions in different Italian geological scenarios: a refinement of natural hazard and risk assessment. Advances in the geological storage of carbon dioxide, NATO Science Series, Springer 2006; 175–190.

Wallquist, L., V. H. M. Visschers, and M. Siegrist, 2009: Lay concepts on CCS deployment in Switzerland based on qualitative interviews. *International Journal of Greenhouse Gas Control,* **3**(5): 652–657.

Wang, L., 2010: *Research and Application of CCS in China.* Presented at 5th IEA GHG Risk Assessment Network Workshop. May 17–19, 2010. International Performance Assessment Centre for Geologic Storage of CO_2 (IPAC-CO_2). www.ipac-co2.com/images/stories/files/presentation/NCEPU-IEA-GHG-Lidong-Wang.pdf.

White, D., 2009: Monitoring CO_2 storage with EOR at the Weyburn-Midale Field. *The Leading Edge,* **28**(7): 838–842.

White, V., L. Torrente-Murciano, D. Chadwick, and D. Sturgeon, 2008: *Purification of Oxyfuel-Derived CO_2.* Presented at the *9th Greenhouse Gas Technologies Conference (GHGT-9).* www.sciencedirect.com.

Wilkin, R. T., and D. C. DiGiulio, 2010: Geochemical Impacts to Groundwater from Geologic Carbon Sequestration: Controls on pH and Inorganic Carbon Concentrations from Reaction Path and Kinetic Modeling, *Environment, Science and Technology,* **44**(12): 4821–4827.

Wong-Parodi, G., I. Ray and A. Farrell, 2008: Environmental non-government organizations' perceptions of geologic sequestration. *Environmental Research Letters,* **3**. www.iop.org/EJ/abstract/1748-9326/3/2/024007.

WRI (World Resources Institute), 2009. CCS Guidelines: Guidelines for Carbon Dioxide Capture, Transport, and Storage. Washington, D.C.

Yeh, S. and E. Rubin, 2007: *Incorporating Technological Learning in the Coal Utility Environmental Cost (CUECost) Model: Estimating the Future Cost Trends of SO_2, NO_x, and Mercury Control Technologies.* Prepared for ARCADIS Geraghty & Miller, Inc., Research Triangle Park, North Carolina, February 20, 2007.

Zakharova, E., 2005: *Some aspects of application of CO_2 capture and storage technologies for CO_2 emission reduction in the Russian electric power production sector.* Greenhouse Gas Cnt. Tech.

ZEP, 2011. The Cost of Capture, Transport and Storage. www.globalccsinstitute.com/resources/publications/costs-co2-capture-transport-and-storage.

14

Nuclear Energy

Convening Lead Author (CLA)
Frank von Hippel (Princeton University, USA)

Lead Authors (LA)
Matthew Bunn (Harvard University, USA)
Anatoli Diakov (Moscow Institute of Physics and Technology, Russia)
Ming Ding (Delft University, the Netherlands)
Robert Goldston (Princeton Plasma Physics Laboratory, USA)
Tadahiro Katsuta (Meiji University, Japan)
M.V. Ramana (Princeton University, USA)
Tatsujiro Suzuki (Tokyo University, Japan)
Suyuan Yu (Tsinghua University, China)

Contributing Authors (CA)
Charles McCombie (Independent Consultant, USA)

Review Editor
John Ahearne (Sigma Xi, USA)

Contents

Executive Summary

In the 1970s, nuclear energy was expected to quickly become the dominant generator of electrical power. Its fuel costs are remarkably low because a million times more energy is released per unit weight by fission than by combustion. But safety requires redundant cooling and control systems, massive leak-tight containment structures, very conservative seismic design, and extremely stringent quality control. As a result, the capital costs of nuclear power plants at least, in Western Europe and North America, proved to be quite high and nuclear power did not become the dominant generator of electrical power.

The routine health risks and greenhouse gas emissions from fission power are small relative to those associated with coal, but there are also high-consequence risks: nuclear weapons proliferation and the possibility of overheated fuel releasing massive quantities of fission products to the environment. The public is sensitive to these risks. The 1979 Three Mile Island and 1986 Chernobyl accidents, along with the high capital costs, ended the rapid growth of global nuclear power capacity (Figures 14.1 and 14.2). After these accidents, the industry improved its overall safety culture, particularly with regard to operator training. This chapter was completed before the large releases or radioactivity from the Fukushima Daichi nuclear power plant that began in March 2011. That event has resulted in reviews of the adequacy of nuclear power safety design and regulation worldwide and, in some countries, a reconsideration of plans for new reactors and/or reactor operating license extensions.

Today, China has 24 GW$_e$ of nuclear capacity under construction (IAEA-PRIS, September 28, 2010) and much more planned. But, Germany has decided to phase out nuclear power; and nuclear power elsewhere in Western Europe and North America, which together account for 63% of current global capacity, is being dogged again by high capital costs and it is not yet clear that new construction will offset the losses due to the retirement of old capacity. Cost escalation is better contained in East Asia, where the International Atomic Energy Agency (IAEA) expects 44–68% of global nuclear capacity expansion by 2030 to occur in China. In Japan, however, following the Fukushima accident, the government has decided to reduce the country's dependence on nuclear power and the debate is ongoing whether to phase it out entirely. Even for its high-nuclear-growth projection which assumes a doubling of current generating capacity by 2030, the IAEA acknowledges that nuclear power's current 14% share of global electric power generation will not increase.

An important societal debate is still ongoing. Do the potential environmental benefits from low-carbon nuclear power outweigh the risks inherent in the technology? These risks occur in reactor operation and possibly in disposal facilities,

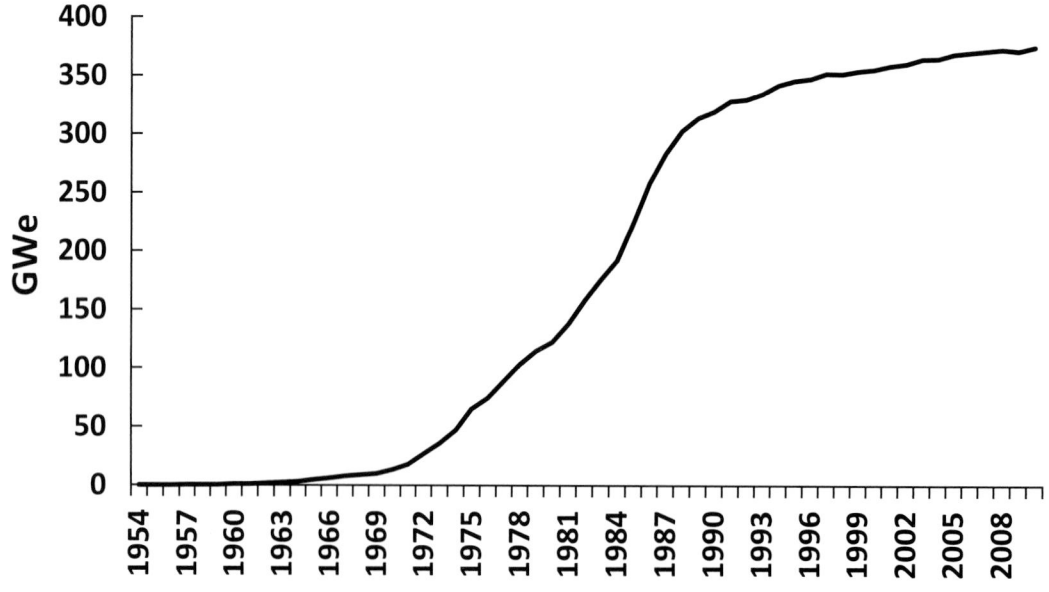

Figure 14.1 | Growth of global nuclear power capacity (GW$_e$). Source: data from IAEA-Pris, 9 January 2010.

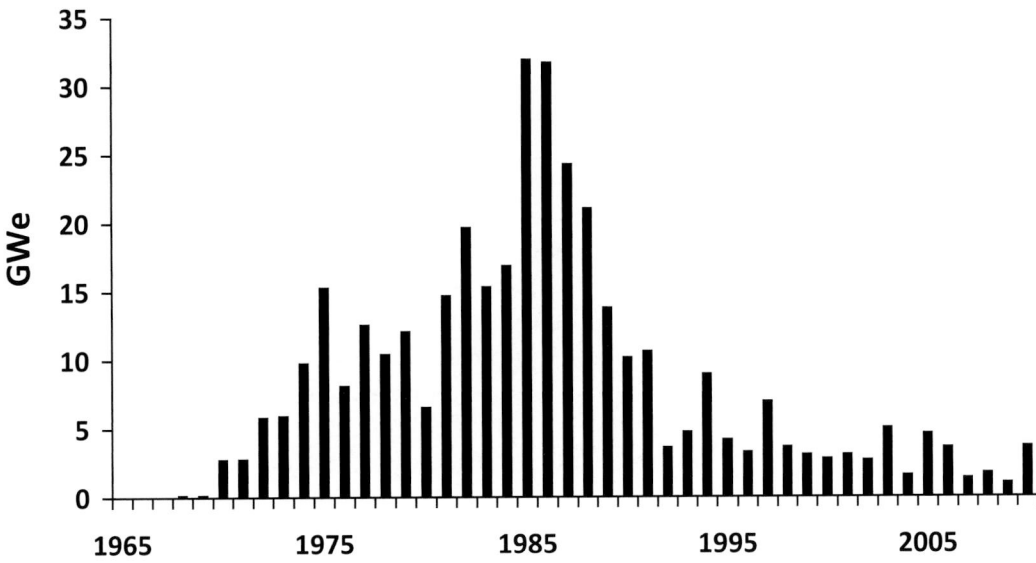

Figure 14.2 | Global nuclear capacity installed by year (GW$_e$). Source: data from IAEA-Pris, 9 January 2010.

but, in the view of the authors of this chapter, the most important risk from nuclear power is that its technology or materials may be used to make nuclear weapons. Of the 30 nations that have nuclear power today, seven are nuclear weapon states,[1] and most of the non-weapon states have had their non-weapon status stabilized either by being part of the European Union, the North Atlantic Treaty Organization (NATO) or otherwise being under the security umbrella of the United States, or by having been part of the Soviet Union or the Warsaw Pact in the past. The non-weapon states with the weakest security ties to the United States and Soviet Union – Argentina, Brazil, South Africa and Sweden – for a time used their nuclear power programs as covers for nuclear weapon programs. The majority of the countries that have expressed an interest in acquiring their first nuclear power plants (see Introduction, Table 14.2) are similarly not tied to constraining alliances such as NATO and the former Warsaw Pact, and some may have mixed motives for their interest in acquiring nuclear technology. That nuclear weapons may spread with nuclear power technology is therefore a danger that must be taken seriously.

The dominant type of nuclear power reactor in operation today, the light-water reactor (LWR), is relatively proliferation resistant when operated on a "once-through" fuel cycle. It is fueled with low-enriched uranium (LEU), which cannot be used to make nuclear weapons without further enrichment. Its spent fuel contains about 1% plutonium but it is mixed with highly radioactive fission products that make it inaccessible except by "reprocessing" with remotely controlled apparatus behind thick radiation shielding. Given the availability of low-cost uranium and the possibility to dispose spent fuel as waste there is no compelling economic or waste-management reason today to separate out this plutonium.

Much of the leadership of the global nuclear energy establishment, including in France, India, Japan, and Russia, however, continue to promote the uranium conservation and waste-reduction benefits of recovering plutonium from the spent fuel and recycling it. These arguments provided cover for India's nuclear weapon program, which used plutonium produced using a research reactor supplied under the international "Atoms for Peace" program to make its first nuclear explosion in 1974, and also for the weapons dimensions of at least six other nuclear programs.[2] Even when done for

---

1   In historical order: the United States, Russia, the United Kingdom, France, China, India and Pakistan. Israel and North Korea have nuclear weapons but do not have nuclear power plants.

2   Argentina, Brazil, South Korea, Pakistan, Sweden, and Taiwan. Fortunately, all but Pakistan have abandoned their nuclear weapons programs.

peaceful purposes, plutonium recycling is destabilizing because it dramatically reduces the time required for a country to implement a decision to acquire nuclear weapons.

The other route to nuclear weapons involves the enrichment of uranium to a level above 20% uranium-235 (typically to more than 90%). Historically, acquiring this capability required a massive investment in a gaseous diffusion plant, with thousands of stages of compression of an ever-smaller stream of corrosive uranium hexafluoride gas through porous barriers. Today, however, the dominant enrichment technology is the gas centrifuge, which, as Brazil, India, Iran, and Pakistan have demonstrated, can be deployed in affordable plants that can begin operating on an even smaller scale than that required to fuel a single gigawatt-scale LWR. Unfortunately, such plants can easily be used or reconfigured to produce weapon-grade uranium, and a plant sized to fuel a 1-gigawatt electric (GW_e) LWR could produce enough material for 25 nuclear weapons per year. Today, members of the nonproliferation community are devoting much of their attention to preventing the spread of small national centrifuge enrichment plants.

The final issue that contributes to the uncertainty of the future of nuclear energy is the persistent opposition from a significant portion of the public. As memories of the accidents at Three Mile Island and Chernobyl faded and concerns about the consequences of climate change increased, the trend was toward public opinion that was more favorable. The Fukushima accident has revived concerns about reactor safety, however. Public concern about radioactive waste and opposition to the siting of central spent-fuel storage sites have also helped keep reprocessing plants alive as alternative destinations for spent fuel, despite their poor economics and the proliferation dangers they pose. Of the countries that are most advanced in siting repositories, Finland and Sweden do not reprocess and France does. The radiological hazards from properly designed deep underground waste repositories are small in comparison with those of a Chernobyl-scale release to the atmosphere from a nuclear power plant accident. Perhaps it is due to their recognition of this fact that the communities that have agreed to host radioactive waste repositories already host nuclear power plants.

In the 1970s, nuclear power proponents expected that by 2010 nuclear power would produce perhaps 80–90% of all electrical energy globally (US AEC, 1974). Today, the official high-growth projection of the Organisation for Economic Co-operation and Development Nuclear Energy Agency estimates that nuclear power plants will generate about 20% of all electrical energy in 2050 (NEA, 2008a). Thus, nuclear power could make a significant contribution to the global electricity supply. At the other extreme, it could be phased out, especially if another accident or terrorist incident causes a Chernobyl-scale release of radioactivity. If the spread of nuclear energy cannot be decoupled to a much larger extent from the spread of nuclear weapons, for example, by ending reprocessing and shifting from national to multinational enrichment, it should be considered a last resort energy option.

14.1   Introduction

Fission energy is released when the nucleus of a very heavy atom such as uranium-235 or plutonium-239 splits into two. Fission is induced by the absorption of a neutron and releases typically two or three neutrons. If there is a sufficient concentration and mass of fissile material, i.e., a "critical mass," a sustained fission chain reaction can occur.

Most current-generation fission reactors are "slow-neutron" reactors.[3] The fast neutrons emitted by fission are slowed by multiple collisions with the nuclei of a "moderating" material before they cause additional fissions (Figure 14.3). Because the probability that fissile nuclei will capture neutrons increases greatly at low neutron velocities, this makes it possible to sustain a chain reaction in a mixture in which the fissile atoms are quite dilute. Indeed, the first reactors were fueled by natural uranium in which uranium-235 constituted only one out of 140 uranium atoms but captured about half of the slow neutrons. The remaining atoms in natural uranium are virtually all non-fissile uranium-238, which captures most of the slow neutrons not absorbed by uranium-235 and is thereby converted into chain-reacting plutonium-239.

Fission power is climate friendly. The emissions of greenhouse gases per kilowatt-hour (kWh) from fission power on a life-cycle basis are on the order of a few percent of those from fossil-fueled power plants.[4] A nuclear capacity of 500–700 GW_e, i.e., 1.3–1.8 times current global nuclear capacity, could forestall the annual release of 10^9 tonnes of carbon to the atmosphere if used to replace coal-fired power plants that do not sequester their carbon dioxide emissions.[5] This would be about one-eighth of the global amount of carbon released into the atmosphere from fossil fuel use and cement production in 2005 (IPCC-PSB, 2007: 139) and 5–12% of the releases projected for 2030 in the full range of IPCC scenarios (IPCC-SRES, 2000: Figure 5–2). The other routine occupational and environmental impacts of nuclear power plants per kWh are relatively low compared with those of fossil power (see Chapter 4). But the potential for catastrophic releases of radioactivity makes the reputation of the global nuclear industry vulnerable to unsafe practices in any country. It is therefore critical to maintain

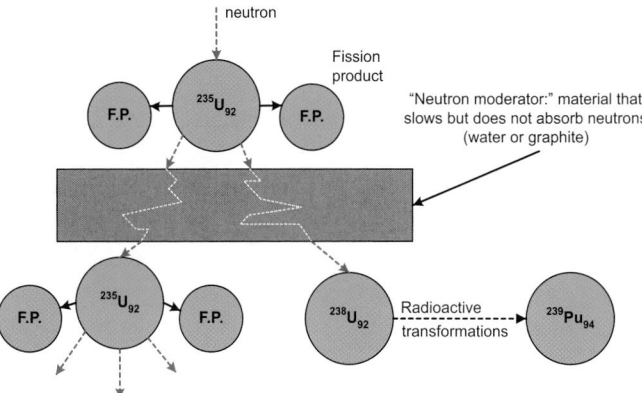

Figure 14.3 | A fission chain reaction in a slow-neutron reactor. Each fission splits a nucleus of a fissile atom (shown here as U-235) into two unequal medium-weight nuclei (fission products, F.P.) and also produces typically two or three neutrons that can go on and cause further fissions. In almost all of today's power reactors, these neutrons are slowed by collisions with the nuclei in "moderating" material (typically with hydrogen nuclei in water) which increases the probability that they will cause fissions. In a reactor operating in steady state, one fission causes on average one fission. The extra neutrons are mostly absorbed by U-238, converting it into U-239, which decays by electron (and antineutrino) emission into neptunium-239 (which has 93 protons) and then into plutonium-239 (94 protons). Plutonium-239 is itself a fissile isotope and contributes increasingly to the chain reaction in the reactor core as the U-235 is depleted and the plutonium concentration builds up.

high safety and security standards in design, construction, and operation everywhere.

Although relatively little nuclear capacity has been added in recent years (Figure 14.2), the average capacity factors of nuclear power plants have increased steadily to about 80%.[6] Between 1988 and 2001, they therefore maintained their share of generated electricity at about 17%, before dropping to about 14% by 2009 (IAEA, 2010d: Tables 3 and 4).[7] For nuclear energy to maintain its share of the global electrical power market there will have to be a dramatic increase in nuclear capacity construction – especially as most existing nuclear capacity will have to be replaced during the period 2010–2050.[8]

Given the uncertain capital costs of nuclear power plants, the risks associated with uncertain demand growth projections in North America and Europe, and the possibility of catastrophic accidents, private capital is unlikely to be available to fund nuclear power plant construction without government guarantees. Such support is available, however, in the form

---

3   "Slow" here is used as a relative term. Ultimately, neutrons are slowed down so that they have the kinetic energy associated with the atoms in the material through which they are passing. At this velocity, they are often called "thermal" neutrons. The velocity associated with the temperature of the water in a light-water reactor (about 300°C) is about 4 km/s, more than ten times the speed of sound in the atmosphere.

4   Emissions from coal-fired power plants are about 1000 g CO_2/kWh. Emissions estimates for nuclear power plants range from 1.4 to 200 gCO_2/kWh (Sovacool, 2008a). Eliminating incomplete estimates and estimates associated with extremely low grades of uranium ore, and reducing two inexplicably high estimates for the emissions associated with decommissioning, we obtain a range of 38±27 g CO_2/kWh. The uranium ore being mined today typically has concentrations of 0.1% uranium and above (van Leeuwen, 2008, Figure D-3).

5   Assuming that 25.8 kg of carbon are released to the atmosphere per 10^9 J of energy from bituminous coal (World Energy Assessment, 2000, box D.1), an efficiency range for new coal power plants of 35–50% and an average nuclear-power plant capacity factor of 90% (478.5–683.6 GWe). See also Pacala and Socolow (2004).

6   There are substantial variations in these capacity factors, even among those states with the largest nuclear programs. For the period 2007–2009, the average capacity factors for the United States and France were 91 and 95%, respectively, and in Japan and Russia 63% and 81%, respectively (IAEA, PRIS, 11 Nov. 2010).

7   The average capacity factor ("unit capability factor" in the IAEA's terminology) is the ratio of the average output of a power plant as a percentage of its full generating capacity. The "up-rating" of nuclear power plants, i.e. operating them at higher power than their original design capacity, has also contributed to a lesser degree to increasing the number of kilowatt hours being generated by nuclear power during the period when few new plants were being built.

8   Assuming operating lifetimes of 40–60 years (see Figure 14.2).

of direct government funding, loan guarantees, or guaranteed payback of investments through government-regulated markets for electric power.

In China, the construction of nuclear power plants by state-owned companies is centrally approved in the five-year plans of China's National Development and Reform Commission, and investors receive tax incentives and low-interest loans. In France, two large government-owned companies – AREVA, which sells reactors and fuel-cycle services, and Électricité de France, the national utility – are partnering to finance and build reactors in other countries. In Russia, Rosatom, the government-owned company that builds and operates reactors and supplies fuel-cycle services, is using government funding and its own income to invest in a major expansion of both domestic capacity and sales overseas. In India, the national government is financing the construction of nuclear power reactors. In the United States, the US Congress passed a major package of incentives and loan guarantees in 2005 to restart reactor orders after a hiatus of three decades. Some state regulators in the United States also are allowing utilities to charge their customers for the costs of construction before their reactors start generating power.

All this government support will certainly result in the construction of some nuclear power reactors. Whether the new construction will be significant on a global scale remains to be seen. The IAEA believes that, with high growth rates for electric power consumption and favorable public policies, both electric power demand and nuclear power production could approximately double by 2030, with nuclear power slightly increasing its current 14% share of the global market for electric power to 16% by 2030. The IAEA's low-growth scenario shows global nuclear power capacity increasing by 47% by 2030 but its market share staying constant (IAEA, 2010d).

Much of the continuing political support for nuclear power stems from the large government nuclear research and development (R&D) establishments in the nuclear weapon states. The first power reactors in the former Soviet Union, the United Kingdom, and France were derivatives of their natural-uranium-fueled, graphite-moderated, plutonium production reactors.[9] Canada developed a natural-uranium-fueled heavy-water-moderated reactor and exported it to other countries interested in independence from foreign suppliers of enrichment services, most notably in India. Today's most successful power reactor, the low-enriched uranium-fueled light-water reactor (LWR), stems from the compact water-cooled reactors developed for submarine propulsion.

Most of the initial R&D relevant to nuclear power technology was therefore paid for by government military nuclear budgets. Later, separate

---

9   Graphite and heavy water are used in reactors fueled with natural uranium to "moderate" (slow down) the typically two or three neutrons emitted by fissions to speeds where a large fraction of them will be absorbed by the U-235 and continue the chain reaction. The probability that the graphite or heavy water will itself absorb the neutrons is relatively low. The probability of neutron absorption is higher in the ordinary water used in light-water cooled reactors, which is why LWRs require enriched uranium fuel.

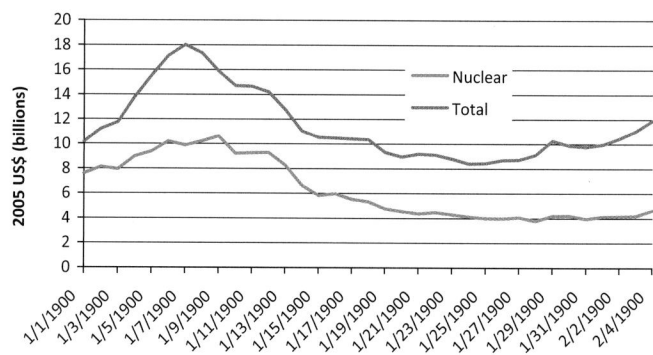

Figure 14.4 | Government energy RD&D expenditures in the OECD countries, 1974–2006. Nuclear includes both fission and fusion. Source: based on IEA, 2008.

civilian nuclear energy R&D programs developed but they continued to have privileged access to national treasuries relative to other energy R&D programs. Over the past three decades they have received more than 50% of government expenditures on energy research, development and demonstration (RD&D) projects (Figure 14.4). This bias toward nuclear energy spread to the national R&D establishments of some non-weapon states, most notably that of Japan.

Perhaps the most important question with regard to fission power is whether it can grow and spread without spreading nuclear weapons. Various efforts have been undertaken in the past to contain the spread of national spent-fuel reprocessing and uranium enrichment plants that remain the keys to producing nuclear weapon materials. There is reluctance in have-not countries, however, to forgo the option of acquiring a national enrichment plant. And, even though reprocessing and plutonium recycling are uneconomic and have few environmental benefits when used with current nuclear power reactors, many local communities have opposed plans to expand spent-fuel storage facilities at reactor sites or to host central interim storage facilities. This opposition has sustained reprocessing in Japan as an alternative destination for spent fuel, and has helped foster a revival of interest in reprocessing in the United States and South Korea.

14.1.1   Global and Regional Nuclear Capacity

Today, 29 countries plus Taiwan have operating nuclear power plants: 18 in Europe (including Armenia, Russia, and Ukraine), five in Asia (plus Taiwan), five in the Americas, and one in Africa (South Africa) (see Figure 14.5).

At the end of 2009, global nuclear generating capacity was 372 GW$_e$, of which 46% was in Europe (including Eastern Europe, Armenia, Ukraine and Russia), 30% in North America, and 21% in the Far East. The rest of the world – Africa, Latin America, the Middle East and South Asia – accounted for only 3% of this capacity (Table 14.1), but contributed a large part of the controversy over the proliferation implications.

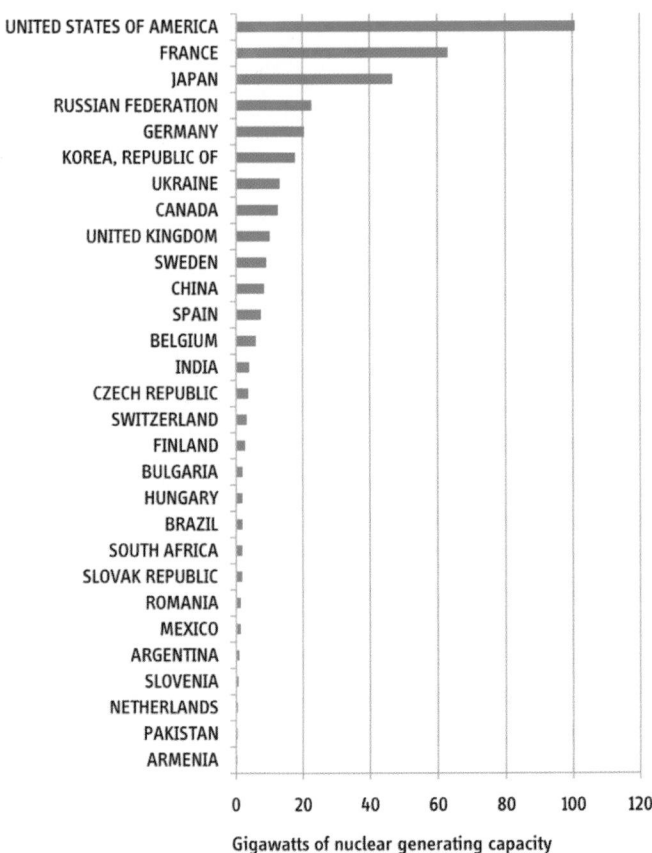

Figure 14.5 | Capacities of nuclear power reactors in operation by country, end of 2009. Source: data from IAEA-PRIS, January 9, 2010.

14.1.2   Projections for Expansion

The IAEA makes annual projections of global nuclear growth. Between 1985 and 1995, even the low projections were higher than what was actually built by 2000 and 2005 (IAEA, 2007a, figures 29, 30). There

Table 14.1 | Global distribution of nuclear power generating capacity, end of 2009.

Region	Nuclear generating capacity (GW$_e$)
North America	113.3
Western Europe	122.7
Pacific OECD Far East	46.8
Eastern and Central Europe	11.2
Former Soviet Union	36.4
Centrally Planned Asia and China	8.4
South Asia	4.4
Other Pacific Asia	22.7
Middle East and North Africa	0.0
Latin America	4.1
Sub-Saharan Africa	1.8
World Total	371.9

Source: IAEA, 2010d.

were few new orders for nuclear power plants and many orders were either cancelled or delayed because of falling electric power consumption growth rates and licensing and construction delays. As a result, projections of growth declined through 2000, and the low projections in 2000 even showed future declines in global nuclear generating capacity as the number of old plants being retired exceeded new builds.

New orders resumed in 2005, however, and most power reactor licenses in the United States are being extended to allow operation for up to 60 years. The projections, therefore, began increasing again. The IAEA's 2010 projection was for a net increase in global nuclear generating capacity of 174–431 GW$_e$ by 2030 (IAEA, 2010d). The high end of the range corresponds to more than a doubling of the 2009 global nuclear capacity and assumes an average net annual addition of new capacity of 25 GW$_e$/yr between 2020 and 2030. This corresponds to a growth rate that was only achieved in the past during the late 1980s (Figure 14.2).[10]

Much of the projected increase would be in the Far East (119–189 GW$_e$) and Eastern Europe (including Russia) (36–63 GW$_e$), reflecting in particular the ambitious plans of China and Russia (see the country studies in Section 14.3). North America is also projected for an increase (15–53 GW$_e$). In the low projection, nuclear capacity in Western Europe declines by 37 GW$_e$, while in the high projection, it increases by 35 GW$_e$. The other world regions – Latin America, Africa, the Middle East, South Asia, Southeast Asia, and the Pacific – together are projected to add 38–79 GW$_e$.

The IAEA's low and high growth projections for nuclear power are associated respectively with 2.1 and 3.1% average annual growth rates in global electric power production between 2009 and 2030 (IAEA, 2010d). For comparison, between 1996 and 2006, global electricity consumption increased at an average annual rate of 3.3% (US EIA, 2008b). Given the likelihood of an increase in electricity prices associated with a shift away from fossil-fuel-based generating capacity, global electricity consumption growth rates could decline below the range assumed by the IAEA. On the other hand, if a significant fraction of automobile transport shifts to electric cars or plug-in hybrids, that could help offset the price effect.[11]

Reflecting the revived interest in nuclear power, as of 2010, 61 countries had requested advice from the IAEA about acquiring their first nuclear

---

10 According to the press release accompanying the 2009 IAEA projection, "The low projection ... assumes that ... there are few changes in the laws and regulations affecting nuclear power... The high projection assumes ... that recent rates of economic growth and electricity demand, especially in the Far East, continue. It also assumes that national policies to reduce greenhouse gas emissions are strengthened, which makes electricity generation from low-carbon technologies, like nuclear power and renewables, more attractive."

11 The current global population of automobiles is about 700 million (Transportation Energy Databook, 2009, Table 3.1). If they travel 15,000 km each on average, that would total about 10^{13} automobile-km/yr. Assuming that 0.2 kWh would be required per km (Electric Auto Association Europe, 2008) about 2×10^{12} kWh/yr would be required, equivalent to the output of about 250 GWe of generating capacity operating at an average capacity of 90%.

power plants (IAEA, 2010e). Excluding Iran, whose first nuclear power plant is virtually complete, and including Israel, which is the only country that is on a similar list developed by the Organisation for Economic Co-operation and Development (OECD) Nuclear Energy Agency but not on the IAEA list (NEA, 2008a, Table 2.1), the 61 countries are listed in Table 14.2.[12] In 2009, one of them, the United Arab Emirates (UAE), contracted with South Korea to build four 1.4-GW$_e$ LWRs (Reuters, December 27, 2009).

Table 14.2 | Countries that have recently expressed an interest in acquiring a first nuclear power plant (IAEA, 2010e); their GDPs (World Bank, 2009) and rough equivalent generating capacities in 2006, 2007 or 2008 (US CIA, 2010, for kWh generated); and those that pass a screening test for GDPs greater than US$50 billion/year and electricity consumption roughly equivalent to an output of 5 GW$_e$.

Country	2009 GDP(billion 2005 US$)	Estimated grid capacity (2006, 2007 or 2008) (kWh/5000 h = GW$_e$)	GDP >US$50 billion/yr and estimated grid capacity >5 GW$_e$
Albania	12	0.5	
Algeria	156	6	×
Bahrain	20	2	
Bangladesh	72	5	
Belarus	55	6	×
Benin	6	0.02	
Bolivia	15	1	
Cameroon	22	1	
Chile	156	12	×
Columbia	221	10	×
Cote d'Ivoire	21	1	
Croatia	63	2	
Dominican Rep.	42	2	
Egypt	148	24	×
El Salvador	20	1	
Estonia	21	2	
Ethiopia	24	0.7	
Georgia	12	2	
Ghana	26	1	
Greece	319	12	×
Haiti	6	0.1	
Indonesia	465	27	×

Country	2009 GDP(billion 2005 US$)	Estimated grid capacity (2006, 2007 or 2008) (kWh/5000 h = GW$_e$)	GDP >US$50 billion/yr and estimated grid capacity >5 GW$_e$
Israel	184	10	×
Italy	2090	55	×
Jamaica	13	1	
Jordan	21	2	
Kazakhstan	121	14	×
Kenya	27	1	
Kuwait	135	9	×
Latvia	31	1	
Libya	85	5	×
Madagascar	9	0.2	
Malawi	4	0.1	
Malaysia	202	21	×
Mongolia	5	1	
Morocco	81	4	
Myanmar (Burma)		1	
Namibia	8	0.3	
Niger	5	4	
Nigeria	188	4	
Oman	55	3	
Peru	117	5	×
Philippines	152	11	×
Poland	481	29	×
Qatar	101	3	
Saudi Arabia	432	30	×
Senegal	12	0.4	
Singapore	176	8	×
Sri Lanka	37	2	
Sudan	52	1	
Syria	50	7	×
Tanzania	19	0.2	
Thailand	248	27	×
Tunisia	37	3	
Turkey	664	36	×
UAE	243	14	×
Uganda	13	0.4	
Uruguay	28	2	
Venezuela	283	23	×
Vietnam	74	13	×
Yemen	24	1	

---

12 A 2008 report by the OECD's NEA lists 25 countries, five of which have "planned or approved projects" and 20 of which have "proposed or intended" projects. "Planned or approved projects": Bangladesh, Belarus, Indonesia, Iran, Turkey and Vietnam. "Proposed or intended projects": Bahrain, Bangladesh, Egypt, Georgia, Ghana, Israel, Kazakhstan, Kuwait, Libya, Malaysia, Namibia, Nigeria, Oman, Philippines, Qatar, Saudi Arabia, Thailand, Uganda, UAE, and Yemen. More detailed information on how serious the interest is in many of these countries can be found in the Survey of Emerging Nuclear Energy States by the Centre for International Governance Innovation (CIGI, 2010).

The widespread interest in nuclear power reflects a broadly shared perception of the need to shift away from fossil fuels because of concerns about climate change. In some countries, nuclear power also is seen as a way to reduce the dependence on imported fuels. There is a concern, however, that a small fraction of countries are also interested in moving toward a nuclear weapon option. Currently, there is special concern that the nuclear weapon option inherent in Iran's uranium enrichment program may stimulate efforts by some of its neighbors to pursue their own nuclear weapon options. In the UAE–US Agreement for Peaceful Nuclear Cooperation, the UAE agreed to forgo the acquisition of uranium enrichment or spent-fuel reprocessing technologies (UAE–US, 2009) but other countries in the region have been unwilling to give up these rights.

Table 14.2 lists the 2007 gross domestic product (GDP) for the 61 countries and the generating capacity in GW$_e$ that would have been required, at a 60% capacity factor, to produce the electric energy that they generated in 2006, 2007, or 2008. In terms of GDP, some countries, such as Mongolia (2009 GDP, US$5 billion) are so poor that it is difficult to understand how they could pay back the US$_{2005}$4 billion cost of a standard 1-GW$_e$ nuclear power plant (World Bank, 2009). The World Bank and Asian Development Bank do not provide loans for the purchase of nuclear power reactors (Schneider et al., 2009: 55). Until recently, South Africa (2009 GDP, US$_{2005}$252 billion) had plans to add 20 GW$_e$ of nuclear capacity by 2025. In December 2008, however, the South African government announced the cancellation of its request for tenders for the first 4 GW$_e$ of capacity because "it is not affordable at this present juncture" (*WNN*, December 5, 2008). It seems unlikely that a country with an annual GDP of less than US$50 billion could afford a US$4 billion nuclear power plant. This situation could be eased if nuclear power plant exporters were willing to provide low-cost loans.

A second issue is that the capacity of many countries' grids may not be large enough to accommodate a standard 1-GW$_e$ nuclear power plant. The IAEA recommends that a single nuclear power reactor not constitute more than 5–10% of the generating capacity on a grid (IAEA, 2007b: 39; 2010f, para. 59).

Of the 61 countries interested in acquiring a first nuclear power plant, listed in Table 14.2, only 24 pass both a US$50 billion annual GDP and a 5-GW$_e$ grid capacity screening requirement (indicated with crosses, ×). Even though the threshold size for a grid required to support a nuclear power reactor has been reduced to 5 GW$_e$, to allow for the possibility of a doubling of the grid capacity before the first nuclear power plant comes online, the grid requirement appears to be the most stringent. It is mitigated, however, in regions where there is a strong supranational grid. There is an existing grid connecting Malaysia, Singapore, and Thailand, for example, and a proposed West Africa grid that would include Ghana and Nigeria. The grid constraint could also be eased by the use of nuclear power plants with lower generating capacities.

Beyond 2030. The OECD's Nuclear Energy Agency (NEA) has made high, low and phase-out projections of global nuclear capacity beyond 2030

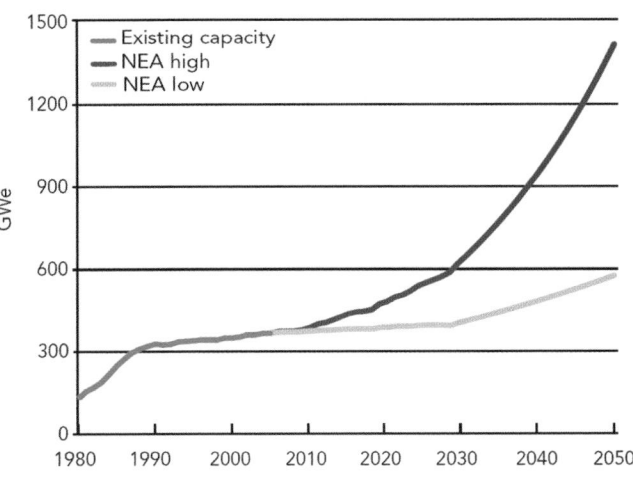

Figure 14.6 | Two Nuclear Energy Agency growth scenarios for nuclear power to 2050. Phase-out scenarios were also considered. Source: NEA, 2008a, Figure 3.11.

(Figure 14.6, phase-out scenarios not shown). They are rather arbitrary, but the high projection reflects a judgment as to the maximum credible rate at which nuclear power could be expanded worldwide. In the high scenario, it is assumed that, after the industry tools up over the next two decades, it will be able to bring online an average of more than 40 GW$_e$/yr of nuclear capacity during 2030–2050, slightly more than twice the rate of buildup between 1972 and 1987. Even so, the NEA high scenario has nuclear power generating only 22% of global electric energy in 2050 – up from 14% in 2009 (NEA, 2008a: 105).[13] In 2010, the IAEA produced almost identical projections for 2050 (IAEA, 2010d).

14.1.3   Fuel cycles

Nuclear fuel is derived from natural uranium. For use in the dominant reactor type, the LWR,[14] the chain-reacting uranium-235 is enriched from its natural level of 0.7% to between 4% and 5%. Uranium in this enrichment range is called low-enriched uranium (LEU).[15] The remainder of the uranium is almost entirely non-chain-reacting uranium-238.

The fuel resides in the reactor core for a few years until most of the uranium-235 has been fissioned. About 2% of the uranium-238 is converted by neutron absorption into plutonium. About half of this plutonium is also fissioned, so that, at the time of discharge, plutonium constitutes about 1% of the heavy elements (uranium plus reactor-produced transuranic elements) in the fuel (Figure 14.7).

---

13   At the country level, China and India have more ambitious plans than the capacity assumed in the NEA high scenario, which assumes approximately 120 GWe in China and 90 GWe in India in 2050 (see country studies below).

14   The core of an LWR is cooled by ordinary "light" water. The hydrogen nuclei in the water also slow down or "moderate" the neutrons in the chain reaction.

15   The IAEA defines uranium enriched to less than 20% in uranium-235 as LEU (IAEA, 2001). Uranium enriched to 20% or higher is called highly enriched uranium (HEU) and is considered weapon useable.

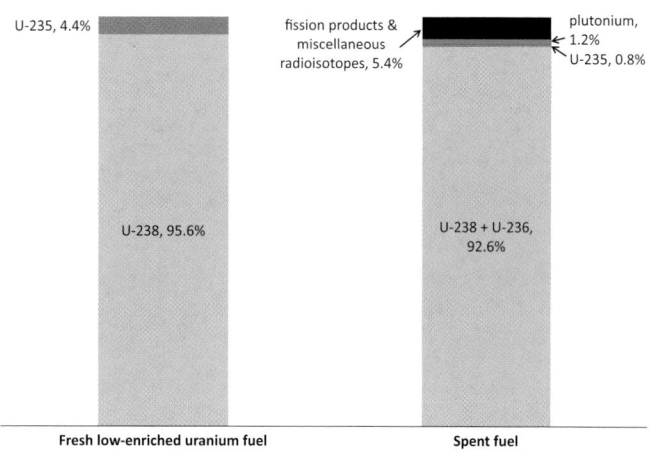

Figure 14.7 | Example of the composition of fresh and spent LWR fuel. Fresh fuel used in standard LWRs is "low-enriched" in uranium-235 (in this case 4.4% U-235) when it is put into the reactor core. Three to five years later, when the fuel is "spent," most of the U-235 has been fissioned, and some has been converted to U-236 by neutron absorption without fission. About 2% of the non-chain-reacting U-238 has been converted to plutonium and heavier "transuranic" isotopes, but more than half of the plutonium has been fissioned. Adapted from: NEA, 1989, Table 9, assuming 53 MW-days/kgU energy release.

The fission products in "spent" fuel are highly radioactive and generate so much heat that the fuel must be water cooled in deep pools for several years. After this period, the fuel can be placed in air-cooled, radiation-shielded dry casks for either transport or storage (Alvarez et al., 2003). After about a century, the radiation level from a spent fuel assembly will drop to a level below that considered "self-protecting" by the IAEA.[16]

14.1.3.1   Uranium enrichment

There are two technologies in commercial use for enriching uranium: diffusion and centrifugation of uranium hexafluoride (UF_6) gas. A third technology, based on the selective ionization of UF_6 molecules containing uranium-235 with finely tuned lasers, may soon be commercialized.

Gaseous diffusion. Gaseous diffusion was the first uranium enrichment technology to be used on a large scale. It was originally used in the United States, Russia, the United Kingdom, France, and China to produce highly enriched uranium (HEU) for weapons. Because of the compression work involved, gaseous diffusion enrichment is very energy intensive and, because of the thousands of stages involved, economies of scale resulted in enormous plants. A 10-million separative work unit (SWU)[17] gaseous diffusion plant that can supply enrichment services for 65 GW_e of LWR capacity requires about 3 GW_e of electrical power to operate at

full capacity (Zhang and von Hippel, 2000, endnote 8). Because of its energy inefficiency, gaseous diffusion is used today only in one plant each by the United States and France, and these are being replaced.

Gas centrifuge. Gas centrifuge enrichment technology is more than ten times more energy efficient than gaseous diffusion, and was first deployed in the early 1960s in the Soviet Union and in the 1970s in Western Europe. UF_6 gas is spun at high speed in a vertical cylinder. Because the molecules containing uranium-238 atoms are slightly heavier, they concentrate near the wall. The fraction of the gas nearest the cylinder's axis is thereby slightly enriched in uranium-235 and can be skimmed off.

In a gas centrifuge system, the separation factor of a single stage is 1.3 to 1.7 (Glaser, 2008, Table 2).[18] For a separation factor of 1.5, only 15 stages are required to produce low-enriched uranium, and 40 stages to produce weapon-grade uranium. With a smaller number of stages, it is possible to build small plants, and more countries have been able to acquire them – in some cases for weapons purposes (Table 14.3).

Laser enrichment. Laser enrichment technology has been a candidate to compete with centrifuge enrichment since the 1980s, but was unsuccessful due to unresolved technical difficulties. The difficulties may now have been overcome, however, and, in 2008, a joint subsidiary of GE and Hitachi, later joined by the Canadian uranium producer, Cameco, was formed to commercialize in the United States a laser-enrichment technique developed by the Silex Company of Australia (GE-Hitachi, 2008).

If all the planned capacity shown in Table 14.3 is built, the global enrichment capacity in around 2017 will be about 70 million SWU/yr, enough to support at least 500 GW_e of LWR capacity.

14.1.3.2   Spent-fuel reprocessing

In France, India, Japan, and the United Kingdom, most spent fuel is shipped to "reprocessing plants" where it is dissolved using equipment operated remotely behind thick radiation shielding, and the uranium and plutonium are separated from the fission products. In France, the recovered plutonium is mixed back with uranium (about 7–8% plutonium) to make "mixed oxide" (MOX) fuel for LWRs (Figure 14.8). Since the plutonium from about 7 tonnes of spent LEU fuel is required to make about 1 tonne of MOX fuel (NEA, 1989), plutonium recycling can reduce EU fuel requirements by approximately 15%. This is currently being done in France. Japan has begun to do the same but its program has been delayed by mistakes and public opposition for about a decade (CNIC, November/December 2008). Russia and China reprocess only a small fraction of their spent fuel. Some of the uranium recovered by

---

16   The IAEA's self-protection standard is more than 1 Gray (100 rad) per hour at a distance of 1 meter (IAEA, 1999: Infcirc-225, Rev. 4, p.11, footnote).

17   SWU is a measure of the output of a uranium enrichment plant. Production of 1 kg of uranium containing 5% U-235 from natural uranium containing 0.72% U-235 with 0.3% U-235 remaining in the depleted uranium requires 7.2 SWUs.

18   The definition of the separation factor is $(e^p(1 − e^t))/((1 − e^p)e^t)$, where e^p and e^t are respectively the fractional amounts of uranium-235 of the product and depleted "tails" from a single stage of enrichment. For low enrichment, the separation factor can be approximated as e^p/e^t.

Table 14.3 | Centrifuge and laser-enrichment plants, operating, under construction and planned (including planned expansions). All plants, other than the proposed GLE laser-enrichment plant in the United States, are gas centrifuge plants.

Country	Plant (year for projected growth) (reference)	Capacity (10⁶ SWU/yr)		
		In operation	Under construction	Planned
Brazil	Resende (2015)	0.12 → 0.2		
China	Shaanxi (IBR, 2008)	1.0 → 1.5		
	Lanzhou II	0.5		
France	George Besse II (AREVA, 2008)		7.5	
Germany	Gronau (URENCO, 2007)	1.8 → 4.5		
India	Rattehalli (military)	0.004–0.01		
Iran	Natanz	0.005 → 0.125		
Japan	Rokkasho (2017) (JNFL, 2007)	→ 1.5		
Netherlands	Almelo (URENCO, 2007)	3.6		
Pakistan	Kahuta (military)	0.015–0.02		
	Chak Jhumra, Faisalabad			0.15
Russia (IBR, 2004)	Novouralsk, Sverdlovsk region (2011)	→ 13.9		
	Zeleznogorsk, Krasnoyarsk region (2011)	→ 8.3		
	Angarsk, Irkutsk region (2015) (WNN, June 25, 2007)	2.5		
	Seversk, Tomsk region (2011)	→ 4.1		
United Kingdom	Capenhurst (URENCO, 2007)	4.2		
United States	URENCO, NM (URENCO, 2008)		5.9	
	AREVA, Idaho			3
	USEC, Portsmouth, Ohio			3.5
	GLE, NC (laser) (GE-Hitachi, 2008)			3.5–6
	Total	→ 45	13.4	10.15–12.65

Sources: Unless stated otherwise, IPFM, 2008, table 4.2.

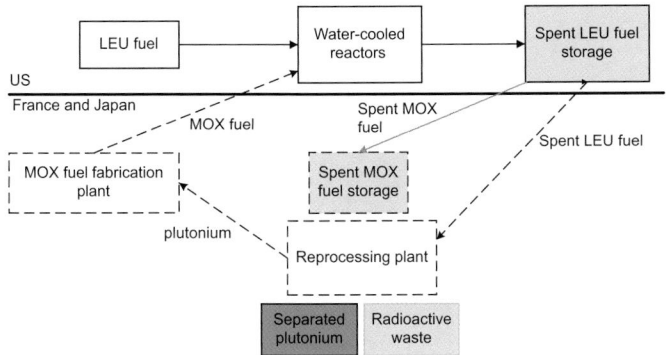

Figure 14.8 | Current spent-fuel disposal strategies. The "once-through" fuel cycle as currently practiced in the United States and many other countries is shown above the horizontal line; LEU fuel is irradiated in a LWR and then stored. The reprocessing and recycling system that is in operation in France and is planned in Japan is shown below the horizontal line. It currently involves the separation of the plutonium for recycling once in MOX fuel. The spent MOX fuel is then stored. Because of the high cost of reprocessing, it is much more expensive to produce MOX fuel than LEU fuel, and most countries have decided that it is not worthwhile. MOX fuel reduces the reactivity safety margins of LWRs somewhat. Except for LWRs specially designed for MOX fuel, the fraction of the core that is made up of MOX fuel is therefore limited to about one third.

reprocessing is also recycled (IAEA, 2007e). If it were all recycled, it could reduce the demand for natural uranium by almost an additional 10%. Similar reductions in uranium demand could also be achieved by reducing the depleted uranium assay for enrichment from the typical value of 0.3% to lower values.

A study for the French Prime Minister in 2000 estimated that reprocessing and plutonium recycling increase the cost of nuclear power by about 0.2 US¢/kWh (Charpin et al., 2000, converting 5 French francs = US$_{2005}$1). A study for the Japan Atomic Energy Commission in 2004 found the cost increase due to reprocessing in Japan to be about three times higher, at ¥0.6/kWh (about US$_{2005}$¢0.6/kWh) (CNIC, 2004b).

14.1.4 Current Reactor Technology

The dominant power reactor technology today is the LWR, which accounts for 89% of global operating nuclear power capacity (IAEA-PRIS, January

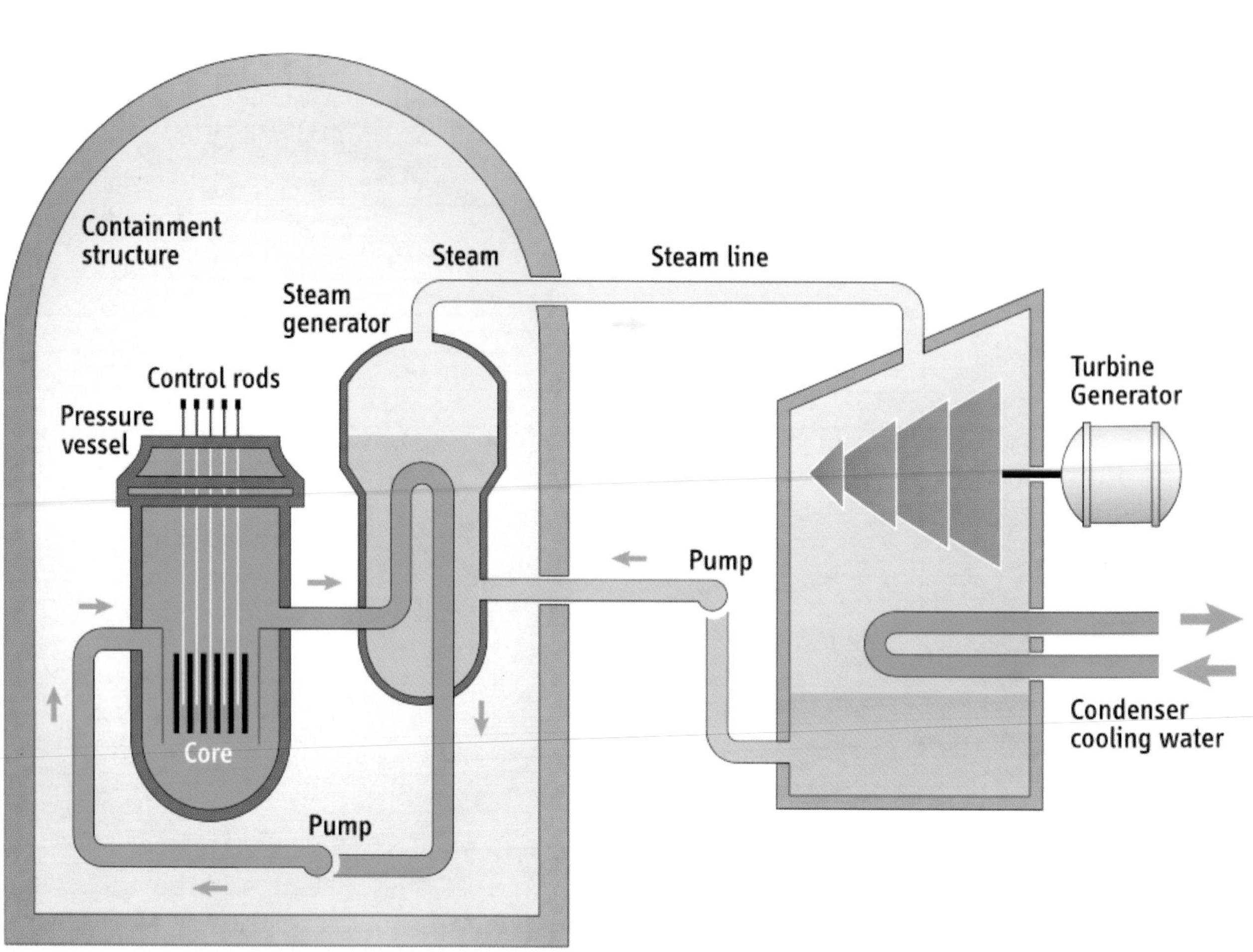

Figure 14.9 | Pressurized light-water reactor. Source: adapted from US Nuclear Regulatory Commission, 2010b.

9, 2009).[19] The fuel is in the form of cylindrical uranium oxide pellets about 1 cm in diameter stacked inside long, thin, sealed zirconium alloy tubes. This fuel is immersed in pressurized water that both slows down ("moderates") the neutrons in the chain reaction as they travel from rod to rod (Figure 14.3) and removes the fission heat from the fuel.

There are two basic types of LWR. In a pressurized water reactor (PWR), the superheated water is not allowed to boil but rather transfers its heat to secondary water that boils in a "steam generator," and the steam then drives a turbo-generator (Figure 14.9). In a boiling water reactor (BWR), the water boils in the reactor and the high-pressure steam goes directly to the turbine.

14.2   The Costs of Nuclear Power

Capital cost. The cost of nuclear power is determined primarily by the capital cost of the plant. For LWRs ordered today, this capital cost

is both uncertain and in flux. Based on recent orders worldwide, the median capital costs are around $US_{2005}\$4000/kW_e$, and projected total generating costs are in the region of US\$0.08/kWh (MIT, 2009). The capital costs, however, can vary by $\pm$US\$2000/kW_e.

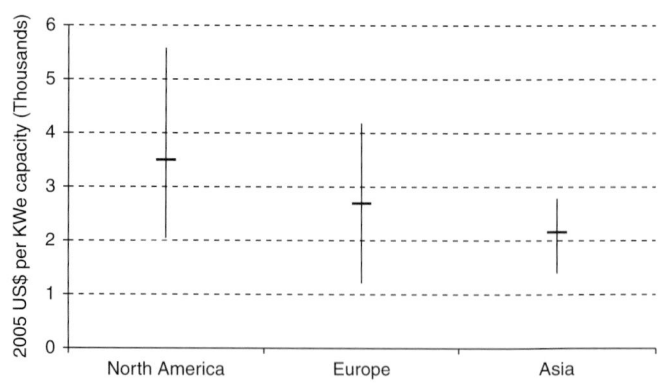

Figure 14.10 | Ranges of 2007–2008 "overnight" construction costs (excluding interest during construction) for plants in North America, Europe (including Russia), and Asia. Source: IAEA, 2009a.

---

19 Some 6% of global nuclear-power generating capacity is heavy-water-moderated reactors; 5% is graphite-moderated reactors (both water and gas cooled), and two liquid-sodium-cooled fast-neutron breeder reactors constitute 0.2%.

Figure 14.10 shows the results of a compilation of "overnight" costs (most of them are estimated) of the construction of 1 kW_e of nuclear generating capacity for plants in North America, Europe, and Asia in 2007–2008. "Overnight" costs exclude the interest during construction. Over a 4–10 year construction period, a 10% annual interest rate would increase the capital cost by 28–75%.

The large range of costs can be attributed to a number of factors, including:

- Whether costs such as site costs and transmission connections are included.

- Biases in the estimates, depending upon the institutional interests of the estimator (e.g., a vendor estimating low in order to obtain a contract to build a reactor, or a utility estimating high because it wishes to obtain a larger loan guarantee).

- Whether the estimate is for the first or second reactor at a particular site (follow-on reactors at the same site should be less costly to build).

- Assumptions about escalating material costs relative to general inflation.

There is a high level of uncertainty about estimates of nuclear power plant construction costs in North America since no new construction has been launched since 1978. Many projects have been announced but cancellations, postponements and cost increases are announced monthly. In Europe, the high end of the cost estimates are based primarily on the only two units that are under construction by AREVA in Finland and France (see Section 14.3.7). The low end reflects costs quoted for Russian-built units. The IAEA lists 15 nuclear power reactors as under construction by Rosatom in Russia and in Eastern Europe. Ten of these units have, however, been nominally under construction since the 1980s (Schneider et al., 2009). About half of the 52 power reactors listed by the IAEA as under construction today are in East Asia (25 units in China, South Korea, Japan, and Taiwan). Today, nuclear power plant construction costs are the lowest in these countries.

The fact that the estimated capital costs of nuclear power plants are higher in North America and Western Europe, where the nuclear power plant industry is being restarted with new designs, suggests that costs should come down as more plants are built. This is not certain, however. Figure 14.11 shows that during the late 1970s and 1980s, when France and the United States brought most of their current nuclear power plants online, the capital costs of LWR construction in the two countries (measured in constant French francs and US$, respectively) actually increased. There are several reasons why cost savings were not realized from industrial learning:

- Much of the construction of nuclear power plants is on-site, and locally hired workers tend not to benefit from experiences at other sites.

- The lack of standardization. In the United States, designs were customized, while in France, they were standardized across the country but new models were introduced as the program developed.

- Quality standards in the nuclear industry are necessarily very high, and mistakes often require defective work to be torn out and done again.

Table 14.4 | Civilian spent-fuel reprocessing plants.

Country	Reprocessing plant (Reference)	Level of activity[a] (tonnes of spent fuel per year)
China	Yumenzhen (LWR) (Nuclear Fuel, 7 April 2008)	50 → 100 (design)
France	La Hague, UP1 + UP2 (LWR) (AREVA-EDF, 2008)	1050 (domestic use)[b]
India	Tarapur and Kalpakkam (natural-U-fueled HWRs)[c]	100 + 100 (Mian et al., 2006)
Japan	Rokkasho (LWR)	Not operating, 800 (design)
Russia	Ozersk, RT-1 (LWRs, BN-600, isotope production, naval & research reactor fuel)	50[d]
UK	B-205 (Magnox)[e] and Thorp (LWR), Sellafield	B-205 to be shut down; future of Thorp uncertain.

a For data on design throughput, see IAEA, 2008a, Annex I.

b The reprocessing of foreign fuel once constituted about 50% of the reprocessing activity at La Hague, but Germany and Japan, the largest foreign customers, as well as Switzerland and Belgium, decided not to renew their contracts. Thus almost all the spent fuel now reprocessed at La Hague is domestic fuel (Schneider and Marignac, 2008).

c HWR = heavy-water moderated reactor. In heavy water, ordinary hydrogen is replaced by heavy hydrogen (deuterium) in which the atomic nucleus contains a neutron as well as a proton.

d Anatoli Diakov, personal communication, 13 October 2009.

e Magnox reactors are natural-uranium-fueled, graphite-moderated, gas-cooled reactors. Their phase-out is to be completed in 2012.

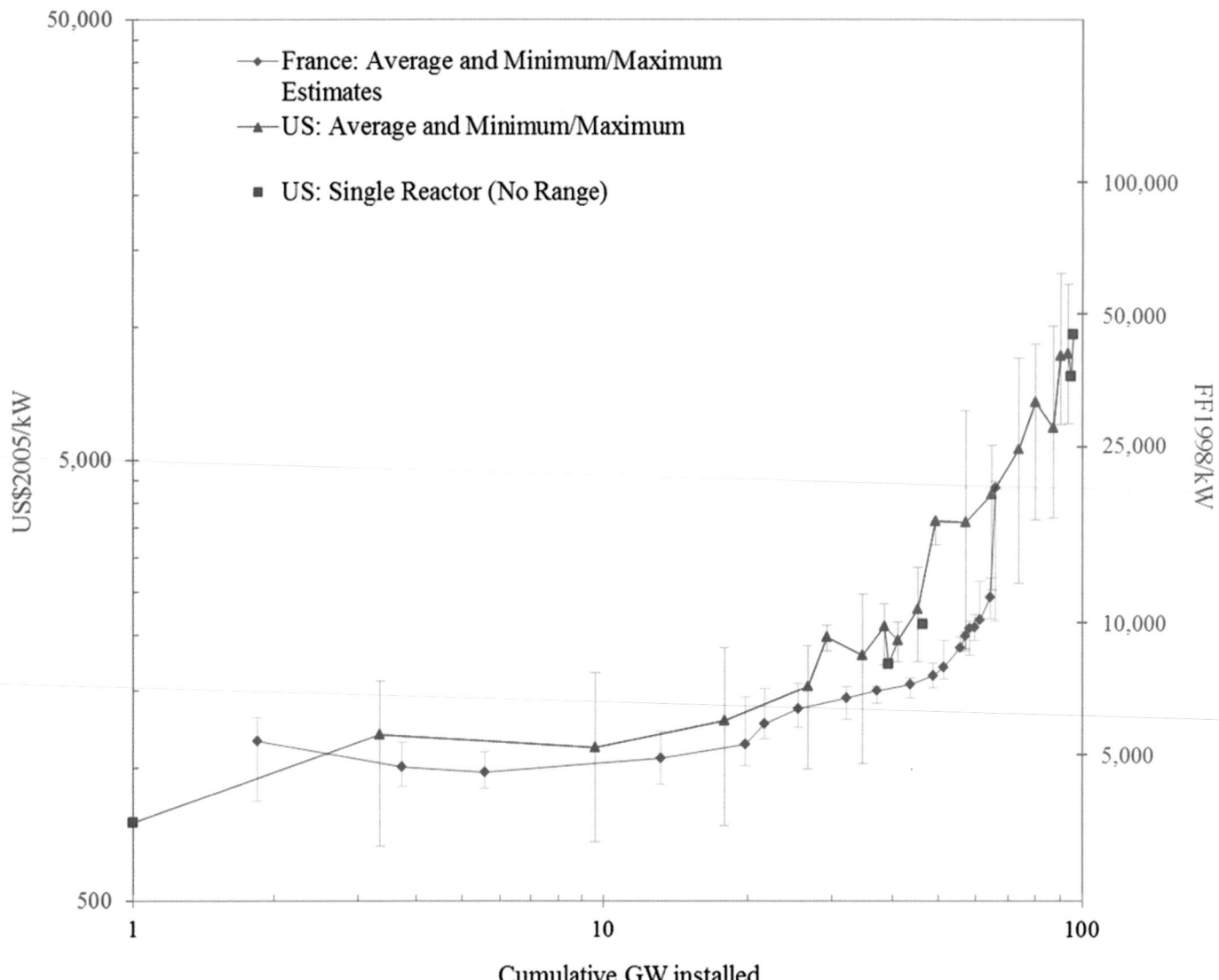

Figure 14.11 | French and US nuclear reactor construction costs (including interest during construction), by completion year, average for all reactors completed in a given year and minimum/maximum (in US2005 per kW installed capacity). In the United States, after the 1979 accident at Three Mile Island, there were prolonged delays in licensing plants under construction. The original French cost data (in French Francs 1998) are also shown for comparison. Data sources: US: Koomey and Hultman, 2007; all other data: Grubler, 2010.

- Regulatory requirements become more stringent as the understanding of potential design problems improves.

- Delays in completion result in extra interest charges.

In France, the increase in nuclear power plant construction costs was roughly the same as that of construction costs in general. In the United States, this was true until the 1979 accident at Three Mile Island, after which there were prolonged delays in licensing plants under construction. In France, the duration of construction also increased because of a shift to higher-power reactors with larger components and changes in component designs (Hultman et al., 2007; Grubler, 2010).

Other costs. The costs of decommissioning nuclear reactors are higher than those for other infrastructure because of the neutron-induced radioactivity of the primary pressure vessel and its internal components. The World Nuclear Association cites, without sources, decommissioning costs of US$190–520/kW$_e$ for water-cooled reactors (WNA, 2007, converted to US2005), while the IAEA reports that the costs for decommissioning the 0.4 GW$_e$ LWRs built by the former Soviet Union ranged from US2005\$600–3800/kW$_e$ (IAEA, 2002).[20] Operating and maintenance costs are about US2005\$0.02/kWh (Harding, 2007). For a "once-through" fuel cycle, fuel

---

20 The large cost range reflects in part the different approaches to decommissioning in different countries. Finland, which had the least costly project, disposed of the waste on site. Germany had the most costly project. Costs in the East European countries fell between these extremes.

costs are only about US$0.007/kWh[21] and interim spent-fuel storage costs are about US$0.00035/kWh.[22] Spent-fuel reprocessing costs in Japan are estimated at ¥0.63/kWh (US$_{2005}$0.006/kWh; JAEC, 2004).

Because of their high capital costs and relatively low operating costs, nuclear power plants are ordinarily operated in a "baseload" fashion, i.e., at full power whenever they are not down for refueling, inspection, maintenance, and/or repair.[23]

14.2.1   Government Subsidies[24]

Subsidies for nuclear power include government-funded research, development and demonstration projects; limitations on liability for catastrophic accidents; low-cost and guaranteed loans; and guarantees of private investments.

Research, development and demonstration. Between 1974 and 2007, the nuclear weapon states and Japan made huge national investments in fission-energy RD&D. In addition to investments in the development of enrichment and reactors for the production of nuclear weapon materials, and the development of LWRs for naval propulsion,[25] Japan, the United States, France, the United Kingdom, and Germany spent an estimated US$_{2005}$156 billion on civilian fission-energy R&D (IEA R&D Statistics, 2009). This corresponds to about US$700/kW$_e$ of nuclear capacity in these countries at the end of 2009 (IAEA, 2010d).

Loan and export guarantees. Some governments – notably, France, Japan, Russia, and the United States – are supporting their nuclear power industries with loan and export guarantees. These subsidies are critical because the repayment of the capital cost of nuclear power plants largely determines the cost of the power, and loan guarantees allow the purchasers of reactors to obtain the lowest possible interest rates. Loan guarantees also make it possible to finance a larger fraction of the cost of a plant with debt. This is an advantage because, even on low-risk projects, investors require about twice the rate of return on their capital as banks charge in interest on loans (IAEA, 2008b; US NAS, 1996: 427–428).

Guarantees are especially important for nuclear power plants, which are considered risky investments. In 2003, the US Congressional Budget Office estimated, based on historical data, that the risk of default on guaranteed loans for nuclear power plants "to be very high – well above 50%" (US CBO, 2003). The US Energy Policy Act of 2005 (Title XVII) provides government loan guarantees up to 80% of the project costs. Congress authorized up to US$18.5 billion for this purpose in the Consolidated Appropriations Act of 2008 and, in its budget proposal for fiscal year 2011, the Obama Administration recommended increasing the limit to US$54.5 billion (US DOE, 2010b: 259). One US utility has estimated that a loan guarantee would reduce the costs of generating electricity from its proposed nuclear power plant by 40% (Schneider el al., 2009: 79).

Limitations on liability for catastrophic accidents. In many countries, nuclear power plants are government owned and the governments would decide after the fact on how much restitution to make for the consequences in the event of a catastrophic accident. In the United States, the government began encouraging private investment in nuclear power in 1954 but found the private market unwilling to invest without its liability being limited. Liability limitation was granted in the Price–Anderson Act of 1957, which has been modified and extended four times, most recently in the 2005 Energy Policy Act. In the current version, plant owners are required to obtain the maximum amount of liability insurance available from private insurers (US$300 million in 2004). Beyond, that, if an accident occurs, each owner is required to contribute up to US$96 million (to be adjusted for inflation) to a pool of about US$10 billion per incident to help cover damages. Beyond that, the government would be responsible (Hore-Lacy et al., 2008). Estimates of the cost savings to utilities from such limitations on liability are highly uncertain (Heyes and Liston-Heyes, 1998; Heyes, 2002).

14.3   Country Studies

This section provides brief case studies of the current nuclear expansion programs of China, India, Japan, South Korea, Russia, the United States, and Western Europe.

14.3.1   China[26]

China's engagement with nuclear power began in 1970. The technology has been drawn from France, Canada, and Russia, with local

---

21 Assuming a cost of US$150/kg of natural uranium, US$150/SWU, US$11/kg to convert natural uranium to UF$_6$ or back, and US$300/kg of uranium for fuel fabrication (von Hippel, 2008b, 2005 US$).

22 Assuming a capital cost for interim dry-cask storage of US$150/kg of spent fuel (Alvarez et al., 2003).

23 Nuclear power reactors constitute such a large fraction of France's generating capacity that they operate in load-following mode.

24 Energy subsidies are discussed more generally in Chapter 6.

25 The United States spent about US$130 billion on the production of plutonium and HEU between 1948 and 1966 and US$51 billion on naval nuclear propulsion between 1948 and 1996 (Schwartz, 1998: 65, 143).

26 Professor Yu Suyuan (Tsinghua University, China) and Dr. Ming Ding (Delft University, the Netherlands), lead authors. The sources for this section include the China National Development and Reform Commission, *State Mid–Long Term Development Plan for Nuclear Power Plants (2005–2020)*, October 2007; and *China Nuclear Power*, Vol. 1, Nos. 1–4, 2008. See also *Nuclear Power in China* (WNA, 2010b) and "China's nuclear industry at a turning point" (Kubota, 2009). We would like to thank Dr Yun Zhou, currently with Harvard University's Managing the Atom Project, for sharing with us a draft of her working paper, *China's Nuclear Energy Policy: Expansion and Security Implications*.

development based largely on French designs. The latest technology has been acquired from the United States (the AP1000 reactor) and France (the European Pressurized Reactor (EPR)). As of the end of 2009, China had 9 GW_e of operating nuclear capacity under two companies: China National Nuclear Corporation (CNNC) and China Guangdong Nuclear Power Holding Company (CGNPC). CNNC is state owned, has a major R&D capability and provides architect-engineer services to CGNPC. Because of the huge planned expansion of China's nuclear generating capacity, additional power companies are co-investing in nuclear power plants but are not building or operating the plants themselves.

Prior to 2000, China was in an exploratory mode with regard to nuclear technology. Over a period of about 15 years, it built an indigenous 0.3-GW_e PWR, ordered two GW_e-scale reactors each from Canada (heavy-water reactors), France, and Russia (PWRs) and built two PWRs in a China–France joint venture. Recently, however, the Chinese government has committed itself to a large-scale PWR construction program that emphasizes initially an indigenized version of the French PWR, the CPR1000. Domestic production of pressure vessels for the Westinghouse AP1000 reactor has begun and it is proposed to develop an indigenized 1.4 GW_e version, the CAP1400 (Kubota, 2009).

In March 2008, China's newly formed State Energy Bureau set a target for 2020 of 5% of electricity from nuclear power, requiring at least 50 GW_e to be in operation by then. In June 2008, the China Electrical Council projected 60 GW_e of nuclear capacity by 2020. The total capacity of the nuclear power units under construction as of the end of 2010 was 25 GW_e (IAEA-PRIS, October 12, 2010).[27]

In May 2007, China's National Development and Reform Commission announced that its target for nuclear generating capacity for 2030 was 120–160 GW_e, which corresponds to an average rate of construction of 5–7 GW_e/yr. Sites have been nominated for a potential total nuclear generating capacity of about 155 GW_e. Various tax incentives have been provided for the construction of nuclear power plants.[28]

This projected growth of China's nuclear capacity cannot be dismissed. However, while China has realized extraordinary growth rates in other areas of infrastructure, the rapid expansion of the nuclear industry faces multiple challenges, in particular the lack of trained personnel. The universities are not producing sufficient nuclear engineers,

so engineers with other backgrounds have been recruited and given one year of training to familiarize them with nuclear technology. Construction company staff with experience building coal-fired power plants are also being trained to build nuclear power plants, starting with the non-nuclear buildings and turbo-generators. The capabilities of China's nuclear regulatory agency, the National Nuclear Safety Administration, will also have to be strengthened. Finally, China's nuclear operators will have to develop a safety culture, including information sharing among plants with regard to safety-related incidents. Li Ganjie, director of China's National Nuclear Safety Administration, has warned that, "if we are not fully aware of the sector's over-rapid expansion, it will threaten construction quality and operation safety of nuclear power plants" (*New York Times*, December 16, 2009). He has also indicated that China's nuclear industry faces challenges on all fronts (Kubota, 2009):

- shortages of trained personnel;
- an inadequate foundation in R&D;
- lack of manufacturing and installation capabilities;
- inadequate management;
- weak safety oversight; and
- insufficient dialogue with the concerned public.

China's ambition to generate nuclear energy on a large scale has attracted the country's largest heavy engineering enterprises to develop the capacity to manufacture nuclear power plant equipment. Much of the equipment used in nuclear power plants, including steam generators, main pumps, and high-pressure piping, can be manufactured in China. The China First Heavy Industries Corporation has also developed the capability to produce pressure vessels for GW_e-class pressurized water reactors.

Uranium supply. China's known uranium resource at a recovery cost of less than US$130/kg (US$_{2005}$$122/kg) is 70,000 tonnes, but estimates of undiscovered resources in favorable areas exceed 1 million tonnes (NEA, 2008c: 155, Table 2). Domestic production of 840 tonnes/yr provides about half of China's current requirements, and the remainder is reportedly imported from Kazakhstan, Russia, and Namibia. In 2006, China signed a deal with Australia, the world's leading uranium mining country, to buy up to 20,000 tonne/yr, enough to supply about 100 GW_e of LWR capacity (BBC, April 3, 2006).

Fuel cycle – front end. China's original enrichment plants used gaseous-diffusion technology but these have been replaced with gas centrifuge plants imported from Russia under agreements made in the mid-1990s between the Tenex subsidiary of Rosatom and the China Nuclear Energy Industry Corporation. The agreements have resulted in the construction of 1.5 million SWU/yr enrichment capacity at two sites in China, based on Russian sixth-generation centrifuges, and work to expand this capacity by an additional 0.5 million SWU/yr is underway.

---

27 In 2008, China began construction of 6 GWe of capacity; in 2009, 9 GWe; and as of October, 6 GWe in 2010. China's most recently completed nuclear power plants, Tianwan 1 and 2, took eight and seven years to build, respectively.

28 These tax incentives for the construction of nuclear power plants include: 1) a rebate of 75% on the value-added tax during the first five years of operation, decreasing to 70% in the subsequent five years and 55% for the third five-year period; 2) a waiver of tariffs on imports of nuclear energy equipment and materials that cannot be produced domestically; 3) a rebate on taxes on land associated with nuclear power plants; and 4) a 15% income tax rate with a reduced tax base and possible tax waiver (Dr Yun Zhou, personal communication, April 22, 2009).

This is enough for about 17 GW$_e$ of LWR capacity. Reportedly, additional capacity expansion is planned (Kubota, 2009). China also buys enrichment services abroad. It has contracted URENCO, an international enrichment company, to supply 30% of the enrichment for the two 0.944-GW$_e$ LWRs at Daya Bay. Tenex, Russia's nuclear materials exporting company, has agreed to supply 6 million SWU in LEU to China between 2010 and 2021.

Spent fuel and reprocessing. In 1987, China announced at an IAEA conference that it was pursuing a "closed" fuel cycle, i.e., one in which plutonium would be recycled from spent to fresh fuel. Accordingly, the CNNC has drafted a state regulation requiring the reprocessing of power reactor spent fuel. Construction of a centralized spent-fuel storage facility for a pilot reprocessing plant at the Lanzhou Nuclear Fuel Complex near Yumenzhen, in Gansu province, in western China, began in 1994. The initial storage capacity is 550 tonnes, which could be doubled. The pilot PUREX reprocessing plant was opened in 2006 with a capacity of 50 tonnes/yr, which also could be doubled.

In November 2007, AREVA and CNNC signed an agreement to assess the feasibility of building commercial-scale reprocessing and MOX fuel fabrication plants in China, at an estimated cost of €20 billion (US$_{2005}$$25 billion). In mid-2008, the CNNC stated that an 800 tonnes/yr reprocessing plant would start operations in 2025, probably in Gansu province. High-level reprocessing wastes would be vitrified, encapsulated and put into a geological repository 500 m below the surface. Site selection for a repository is focused on six candidate locations and is to be completed by 2020. An underground research laboratory would then operate for 20 years and actual disposal would begin in 2050.

14.3.2   India[29]

India's nuclear power program dates back to the late 1940s. Thanks to decades of sustained government support, the Department of Atomic Energy (DAE) has developed expertise and facilities that cover the entire nuclear fuel cycle, from uranium mining to the reprocessing of spent nuclear fuel (Sundaram el al., 1998).

Most of India's current power reactor capacity is based on 0.22 GW$_e$ heavy-water reactors (HWRs), modified versions of the CANDU reactors that India imported from Canada before its 1974 nuclear test resulted in a cutoff of its nuclear imports. Two Russian 1-GWe LWRs are under construction at Koodankulam at the southern tip of India, and there are plans to build two to four more such reactors at the same site over the next decade. A 0.5-GW$_e$ prototype fast breeder reactor, to be fueled with MOX (plutonium–uranium) fuel, is under construction at Kalpakkam on the southeast coast.

The DAE's program is still based on the three-stage strategy first announced in 1954 by Homi Bhabha, the founder of India's nuclear program (Bhabha and Prasad, 1958):

1. Heavy-water-moderated reactors are fueled with natural uranium, and the spent fuel is reprocessed to recover the produced plutonium.

2. The separated plutonium is to be used to provide startup cores for fast-neutron plutonium-breeder reactors. These breeder reactors would produce more plutonium than they consumed, and the excess would be used to provide startup fuel for additional breeder reactors.

3. After a large enough fleet of breeder reactors has been established, thorium is to be substituted for uranium in the fast-breeder reactor blankets to produce fissile uranium-233. The bred uranium-233 is to be used to fuel the fast-neutron reactors, which would operate with a lower breeding ratio – but still above a self-sustaining level – using India's abundant thorium resources as their ultimate fuel.

DAE planners have a history of making optimistic projections for the growth of nuclear power in India. In 1962, Bhabha predicted that India would have 20–25 GW$_e$ of installed heavy-water and breeder-reactor capacity by 1987 (Hart, 1983: 61). This was subsequently replaced by the goal of 43.5 GW$_e$ of nuclear capacity by 2000 (Sethna, 1972). At the end of 2009, however, India's nuclear capacity amounted to just 4.5 GW$_e$, about 3% of the country's total electric power generating capacity (IAEA-PRIS, January 11, 2010).

Prior to the 2008 lifting of the Nuclear Suppliers Group (NSG) ban on uranium and nuclear technology trade with India, the DAE projected a nuclear generating capacity of 20 GW$_e$ by the year 2020 and 275 GW$_e$ by the year 2052 (Grover and Chandra, 2006). Since the NSG waiver, there have been even higher predictions of 40 GW$_e$ by 2020 and 470 GW$_e$ by 2050 (*Financial Express*, October 14, 2008; India, Ministry of Power, 2008). The US Energy Information Administration (US EIA) reference case projection is for India's nuclear power capacity to grow to 20 GW$_e$ by 2030 (US EIA, 2008c).

Uranium constraint. India has known resources of about 60,000 tonnes of low-cost uranium (NEA, 2008c: 207), sufficient for a 40-year-lifetime supply for only about 10 GW$_e$ of HWR capacity. India's relatively small resource base of uranium has been the primary justification for the DAE's plans to focus on breeder reactors designed to have a very high breeding ratio for plutonium.[30] This justification has not

---

29 Dr. M.V. Ramana of Princeton University, lead author.

30 The breeding ratio is increased by eliminating material that could slow down the neutrons. For this reason, all of India's fast breeder reactors after 2020 are to be fueled with metal fuel rather than the higher-melting-point oxide fuel that has been used in demonstration reactors in other countries (Grover and Chandra, 2006).

yet changed despite the recent lifting of the NSG ban on natural and enriched uranium exports to India.

The plutonium supply constraint. The rate at which India can build up its breeder capacity is limited by the rate at which it can produce excess plutonium for the initial cores. The DAE assumes a starting capacity of 6 GW_e of high-breeding-ratio, metal-fueled breeder reactors in 2022. This would require about 22 tonnes of fissile plutonium for startup fuel. Because of the limited rate of plutonium production by India's heavy-water reactors, the DAE's stock of fissile plutonium is unlikely to exceed this amount by 2022.[31] Of this inventory, the DAE plans, however, to use at least 15 tonnes for startup fuel (including the first two fuel reloads) for the four oxide-fuel-based breeder reactors with a low breeding ratio that are to be an intermediate step toward the more advanced metal-fueled breeder reactors. The remaining plutonium will therefore be sufficient only to start about 1 GW_e of metal-fueled breeder reactor capacity by 2022.

The DAE's projected growth rates after 2022 are also unachievable. Even with a fuel residence time of two years inside the reactor and an optimistic out-of-reactor time of only two years to cool the spent fuel, reprocess it, and fabricate the extracted plutonium into new fuel, it would take four years for a given batch of plutonium loaded into a breeder reactor to become available for recycling with some extra bred plutonium that could be used as startup fuel for new breeder reactors. A careful calculation finds that the resulting plutonium growth rate is only 17–40% of the DAE's estimates, depending on whether realistic or optimistic assumptions are used for various parameters (Ramana and Suchitra, 2009). Unless India's nuclear establishment shifts its focus away from breeder reactors, nuclear power is unlikely to contribute significantly to electricity generation in India for the next several decades.

Cost of breeder electricity. Even if the capital costs of breeder and heavy water reactors were the same, electricity from the breeders would be more expensive because of their high fuel-cycle costs. The cost of electric power generated from India's first commercial-scale breeder reactor will be at least 80% higher than from heavy water reactors, mostly because of the high costs associated with reprocessing and fabricating plutonium-containing fuel (Suchitra and Ramana, 2011). Breeders are competitive with heavy water reactors fueled with natural uranium and operating on a once-through fuel cycle only for uranium at prices

well above US$1000/kg. In recent decades, the average price of uranium has been around US$50/kg.

14.3.3   Japan[32]

Japan has the world's third largest nuclear generating capacity. It is one of the few non-weapon states with an enrichment program and the only one that reprocesses spent fuel.

Nuclear generation capacity and projections. As of the end of 2009, Japan had 53 operational commercial LWRs with a generating capacity of 47 GW_e. Two LWRs (2.7 GW_e) are under construction[33] and 10 (13.6 GW_e) were to be commissioned by 2020. Some of Japan's older reactors are being decommissioned, while others are proposed to have their licenses extended.[34] Prior to the 2011 Tōhoku earthquake and tsunami, a total of 66 LWRs (65.1 GW_e) were expected to be operating in 2020 according to the plans of the electric utilities. These plans now seem likely to be scrapped, at least for the near term. Indeed, Prime Minister Kan, before stepping down, proposed phasing out nuclear power entirely.

Uranium enrichment capacity. Japan Nuclear Fuel Limited (JNFL), the nuclear fuel cycle subsidiary of Japan's nuclear utilities, started operating the country's first commercial enrichment plant with a capacity of 150,000 SWU/yr in 1992. Its capacity was increased every year by one module with a capacity of 150,000 SWU/yr until it reached a nominal capacity of 1.05 million SWU/yr in January 2009. All seven modules have been permanently shut down due to technical troubles, however. Cumulatively, only about 1 million SWUs have been produced.[35]

JNFL launched the development of a more advanced type of replacement centrifuge in 2000. Prototypes began testing with UF_6 gas in 2007. JNFL plans to introduce the new centrifuges into commercial operation starting in 2010 and hopes to achieve an enrichment capacity of 1.5 million SWU/yr by around 2020 (JNFL, 2007).

Spent-fuel management. The local governments hosting Japan's power reactors are generally opposed to the expansion of on-site storage. Japan's policy is therefore to reprocess its spent fuel. In the late

---

31 The fissile isotopes of plutonium that chain-react with slow neutrons are plutonium-239 and plutonium-241. The spent fuel of India's pressurized heavy-water reactors contains about 2.6 kg of fissile plutonium per tonne. The amount of plutonium that India can separate is limited by the capacity of its reprocessing plants. It is estimated that India will have separated about 10 tonnes of fissile plutonium by 2018. That is the earliest that India could bring online significantly more reprocessing capacity. Assuming that its reprocessing capacity increases tenfold to 2000 tonne/yr thereafter, and operates at 80% capacity, it is estimated that India could separate out anather 13 tonnes of fissile plutonium during 2019–2021 (Ramana and Suchitra, 2009).

32 Dr Tatsujiro Suzuki, Tokyo University (now vice chair of Japan's Atomic Energy Commission) and Professor Tadahiro Katsuta, Meiji University, Tokyo, lead authors.

33 Two 1.37-GWe advanced boiling water reactors, Shimane-3 and Ohma, are scheduled to go into operation in 2011 and 2012, respectively.

34 On February 17, 2009, Japan Atomic Power published its plan to extend the operation of Tsuruga-1, a 0.357-GWe boiling water reactor, commissioned in 1966, for another 20 years.

35 Some 1599 tonnes of uranium have been enriched. Assuming 4.4% enrichment with 0.3% uranium-235 in depleted uranium, this would correspond to 1.2 million SWU.

1970s, Japan contracted to ship 5500 tonnes of spent fuel to France and the United Kingdom for reprocessing (Albright et al., 1997, Tables 6.4 and 6.5). Japan subsequently decided to build a domestic reprocessing plant at Rokkasho, at the northern tip of the main island, with a design capacity of 800 tonne/yr of uranium throughput. Commercial operation of the plant was to begin in 2003 but, due to various technical problems, has repeatedly been postponed, most recently until late 2012 (*Japan Times*, 2010).

As a result of the many years of delays in starting up the Rokkasho reprocessing plant, Japan has a developing shortage of spent-fuel storage pool capacity. As of September 2007, 12,140 tonnes of spent nuclear fuel were being stored at nuclear power plant sites (JAEC, 2008). Utilities were installing new racks for denser storage in the pools and transferring spent fuel from one pool to another within the same site. A 3000-tonne capacity storage pool at the Rokkasho reprocessing plant started accepting spent nuclear fuel in 2000, but was almost full (2817 tonnes of spent fuel) as of the end of 2008 (JAEC, 2008). The first away-from-reactor interim storage facility at Mutsu City in Aomori prefecture near the reprocessing plant is under construction and is planned to start operation in 2012. Ultimately, it is to have 5000 tonnes of spent-fuel storage capacity. Japan will need about five interim storage facilities of this scale by 2050, even if the reprocessing plant operates as planned (Katsuta and Suzuki, 2006; Japan-METI, 2008).

The reprocessing of spent fuel that Japan sent to France has been completed. Its reprocessing contract with the United Kingdom, which was to have been completed in 2003, has been delayed by problems at the UK reprocessing plant (Forwood, 2008). As a result of its foreign reprocessing program and domestic pilot program, as of the end of 2008, Japan had about 47 tonnes of separated plutonium, most of it in France and the United Kingdom (IAEA, 2009c, Japan).[36] Japan's second commercial reprocessing plant, originally due to begin operating in 2010, is now not scheduled to become operational before 2040.

MOX fuel program. As a result of various scandals and public opposition, Japan's plan to partially fuel 16–18 LWR power plants with MOX (plutonium–uranium) fuels made from Japanese plutonium in Europe by 2010 has been delayed (CNIC, 2008). The first MOX fuel was finally loaded into a Japanese LWR (Genkai 3) in October 2009 (*WNN*, November 5, 2009).

JNFL plans to produce MOX fuel for Japan's LWRs from plutonium separated in Japan. A commercial MOX fuel fabrication plant is to be built next to the Rokkasho reprocessing plant with a maximum capacity matched to that of the reprocessing plant, about 10 tonnes of plutonium mixed with 120 tonnes of uranium to make 130 tonne/yr of MOX fuel. Construction was to have started in October 2007, but the plant is still

in the pre-construction licensing phase. According to JNFL, commercial operation will start in 2015 (JNFL, 2010).

Breeder reactor R&D. Japan's first experimental breeder was the Joyo (140 megawatt thermal (MW$_t$), no electricity generation), which has operated about 27% of the time since it achieved first criticality in 1977. Japan's prototype 280 megawatt-electric (MW$_e$) fast breeder reactor, Monju, suffered a sodium leak and fire in 1995 after its first three months of operation. After repairs and many delays, it finally restarted 15 years later, in May 2010, only to be shut down indefinitely again due to a refueling accident in August 2010. In 2006, the Nuclear Energy Subcommittee of the Ministry of Economy, Trade and Industry's advisory committee published a long-term program under which a follow-on demonstration breeder reactor would be built by 2025. Commercialization of breeder reactors, the original justification for Japan's reprocessing program, has slipped by 80 years from 1970 to 2050 (Japan-METI, 2006; Suzuki, 2010).

High-level radioactive waste disposal. In May 2000, Japan passed a "Law Concerning the Final Disposal of Specific Radioactive Waste" that outlines legal responsibilities, cost sharing, and site selection processes. A voluntary site selection process started in December 2002. Thus far, only one application (Toyo Town) has been received and officially accepted, but it was subsequently withdrawn due to local public opposition.

Budget. Japan accounts for almost half of the total nuclear energy R&D carried out in the OECD countries. In 2007, Japan devoted ¥261 billion (US$_{2005}$2 billion) to nuclear energy R&D, about 65% of its budget for energy R&D (IEA, R&D Statistics).

14.3.4   South Korea[37]

The Republic of Korea (ROK) has the world's fifth largest nuclear generating capacity; 20 units with a capacity of 17.7 GW$_e$, with eight units (9.6 GW$_e$) under construction as of the end of 2009 and four more (5.6 GW$_e$) planned for completion by 2021. Except for four heavy-water reactors, all of these are PWRs, located at four sites. The new reactors are all of Korean design with all the major components, including pressure vessels, produced in Korea. South Korea has been actively trying to export its reactors and, at the end of 2009, obtained a US$20 billion contract from the United Arab Emirates for four 1.4-GW$_e$ reactors to be completed by 2020, and another US$20 billion contract to jointly operate them for 60 years (Reuters, December 27, 2009).

South Korea has the world's largest nuclear power program without a national enrichment or reprocessing facility. This reflects the desires of its close ally, the United States, and also the 1992 Joint Declaration with North Korea on the Denuclearization of the Korean Peninsula, under which the two countries agreed not to acquire enrichment or reprocessing facilities. North Korea violated this agreement and there is

---

36 Since 2007, Japan's government has reported to the IAEA only the quantity of "fissile" isotopes (plutonium-239 and plutonium-241) held abroad. Based on a comparison of the declarations of 2005 and 2006 with 2007, these numbers must be multiplied by approximately 1.5 to obtain the tonnage of total plutonium.

37 This section is mostly based on von Hippel (2010).

resentment within South Korea's nuclear establishment that the United States acquiesced to Japan acquiring these technologies and not South Korea. Following North Korea's nuclear test in May 2009, there were calls from South Korea's opposition party for "nuclear sovereignty," i.e. that South Korea should have the same rights as Japan.

As in Japan, the spent-fuel pools at South Korea's older reactors are filling up and local governments are resisting the construction of more on-site storage. As a solution to the problem, South Korea's Korea Atomic Energy Research Institute (KAERI) proposes a form of reprocessing, "pyroprocessing,"[38] the electro-refining of the fuel in molten salt and the use of liquid-sodium-cooled fast-neutron reactors to fission the recovered plutonium and minor transuranic elements. This vision is supported by one of the ROK's R&D ministries, the Ministry of Education, Science and Technology, but not the Ministry of Knowledge and Economy, which is worried about the cost. The G.W. Bush Administration was also interested in pyroprocessing and supported joint R&D between the US Department of Energy's (US DOE) nuclear laboratories and KAERI.

South Korea, like most other countries with nuclear power programs, is having difficulty siting radioactive waste repositories. It succeeded in siting an underground low- and intermediate-level radioactive waste repository at one of its reactor sites in exchange for US$300 million plus US$600 per waste drum for up to 800,000 waste drums for the local government and a commitment by the government-owned utility, Korea Hydro and Nuclear Power, to move its headquarters and staff from Seoul to a small city near the site (Park Seong-won et al., 2010). This is still a small cost, however, compared with Japan's ¥1 trillion (US$10 billion) programmed payments to Aomori prefecture for accepting the Rokkasho reprocessing plant (Takubo, 2008), and the ¥11 trillion (US$_{2005}$94 billion) estimated cost of building, operating and decommissioning the Rokkasho reprocessing plant (CNIC, 2004a).

14.3.5   Russia[39]

Russia has the world's fourth largest nuclear generating capacity, 22.7 GW_e, provided by 16 PWRs, 11 graphite-moderated, water-cooled RBMK-1000 (Chernobyl-type) reactors and the BN-600 sodium-cooled fast breeder prototype reactor.[40] The expansion of this capacity and foreign sales of Russian-designed reactors and fuel-cycle services have become a key economic goal of the Russian government. In April 2007, then

President Vladimir Putin consolidated Russia's civilian nuclear activities into one giant state-owned company, Atomenergoprom, under another state-owned company, Rosatom, which also operates Russia's military nuclear programs (*Moscow Times*, May 2, 2007).[41]

Seven 1-GW_e PWRs and a 0.8-GW_e demonstration breeder reactor are currently under construction (IAEA-PRIS, October 8, 2010).[42] According to the Russian government's 2008 long-term plan (Russia Federal Target Program, 2008), starting in 2009, construction was to be initiated each year on two new 1.2-GW_e PWRs. By the end of 2015, 11 new nuclear power units were to be put into operation, and the construction of a further 10 initiated (see Figure 14.12). In late 2009, the schedule slipped, and only eight of the PWRs are expected to be completed by the end of 2015 (WNA, 2010a).[43] Beyond 2015, Rosatom is supposed to find its own funding.[44]

Outside Russia, Rosatom has under construction two VVER-1000 PWRs[45] at the Koodankulam nuclear power plant in India and one at the Bushehr plant in Iran (Rosatom, September 26, 2008). Russia has agreements, but not in most cases binding contracts, to construct: in India, eight more VVER-1200s (four at Koodankulam and four in West Bengal); in China, two more units at the Tianwan nuclear power plant, where two VVER-1000s are already operating, and two BN-800 breeder reactors; in Bulgaria, two VVER-1000 units; in Turkey, four VVER-1000 units; in Armenia, one VVER-1000 unit; and in Ukraine, two VVER-1200 units.[46]

Uranium supply. Russia's confirmed uranium reserves could support about 100 GW_e of LWR capacity for 45 years.[47] All uranium mining activity has been consolidated within Rosatom under Atomredmetzoloto JSC

---

38   In pyroprocessing, spent fuel is dissolved in molten salt and the transuranics are separated electrochemically.

39   Professor Anatoli Diakov, Moscow Institute of Physics and Technology, lead author. This section provides an update of a more detailed chapter in the *Global Fissile Material Report* 2007 (IPFM, 2007). www.fissilematerials.org.

40   The numerical suffixes indicate the approximate gross electric power generating capacity in MW_e. The net capacity is typically about 10% lower (see IAEA-PRIS).

41   See Rosatom's English-language website: www.rosatom.ru/en.

42   One graphite-moderated reactor, Kursk-5, has also been listed as under construction since 1985. Barge-mounted reactors are discussed separately below.

43   The reactors under construction at the end of 2010: two VVER-1200 light-water reactors at Rostov nuclear power plant (NPP), two VVER-1200 units at Leningrad-2 NPP, two VVER-1200 units at Novovoronezh-2 NPP, one VVER-1000 unit at Kalinin NPP, and one BN-800 unit at Beloyarskaya NPP (IAEA-PRIS). Rosatom's website also lists two VVER-1200 units under construction at the Baltiyskaya NPP (www.rosenergoatom.ru/rus/development).

44   After 2015, Rosatom plans to build two VVER-1200 units/yr, costing a combined US$4.6 billion (2005 US$). According to Figure 14.12, in 2015, Atomenergoprom would have a capacity of 33 GWe. Assuming that these reactors operate at an 80% capacity factor, they would generate 0.23 trillion kWh/yr. Atomenergoprom therefore would have to be raising capital investments of US$0.02/kWh generated.

45   VVER: Vodo-Vodyanoi energetichesky reactor, the Russian version of the pressurized water reactor.

46   See www.nuclear.ru/rus/press/nuclearenergy/2117970/ (in Russian).

47   Operating at a 90% capacity factor, a 1-GWe VVER would require 140–200 tonne/yr of natural uranium, assuming 0.1–0.3% uranium-235 in the depleted uranium. In 45 years, 100 GWe of VVER capacity would, therefore, require 630,000–900,000 tonnes of natural uranium. Russia is reported to have 545,600 tonnes of uranium

(ARMZ). ARMZ plans a major expansion of its uranium mining, including joint ventures in Kazakhstan. If these plans are realized, in 2025, ARMZ will be mining enough uranium to support about 100 GW$_e$ of LWR capacity. ARMZ is also planning joint ventures with companies in Canada, Armenia, Mongolia, and the Republic of South Africa (Atomenergoprom, 2008).

Reprocessing. Russia has a small reprocessing plant (Mayak) near Chelyabinsk in the Urals, where it reprocesses the spent fuel of the last six of its first-generation VVER-400 reactors along with its naval and research-reactor spent fuel. Russia has declared to the IAEA that, as of the end of 2008, about 45 tonnes of civilian separated plutonium were stored at its reprocessing plants (IAEA, 2009c, Russia).[48] Rosatom is funding R&D related to the possibility of building a pilot reprocessing plant on the site of the never-completed RT-2 reprocessing plant at the Krasnoyarsk Mining and Chemical Combine in Zheleznogorsk (Russia Federal Target Program, 2006).[49]

Breeder reactors. As shown in Figure 14.12, Russia's nuclear establishment, like India's, is still planning on the near-term, large-scale commercialization of plutonium breeder reactors. One 0.8-GW$_e$ BN-800 liquid-sodium-cooled reactor, on which construction began in 1985, is now scheduled to be put into operation in 2014 (AtomInfo, 2009). China is considering ordering two units (WNA, 2010a). A 1.8-GW$_e$ BN-1800 is being designed to be deployed in the 2020s. In January 2010, the Russian government approved a 10-year, 110 billion ruble (US$3.6 billion) federal target program for the development of fast-neutron reactors and their fuel cycle (Rosatom, 2010).

MOX (plutonium–uranium) fuel for Russia's first plutonium breeder reactors could be fabricated using Russia's separated civilian plutonium. In addition, Russia has committed to dispose of 34 tonnes of excess weapon-grade plutonium in parallel to US use of an equal amount of excess weapon-grade plutonium in MOX LWR fuel. Russia plans to dispose of its excess weapon-grade plutonium in breeder-reactor fuel. Until Russia builds a MOX fuel fabrication pilot plant, however, the BN-800

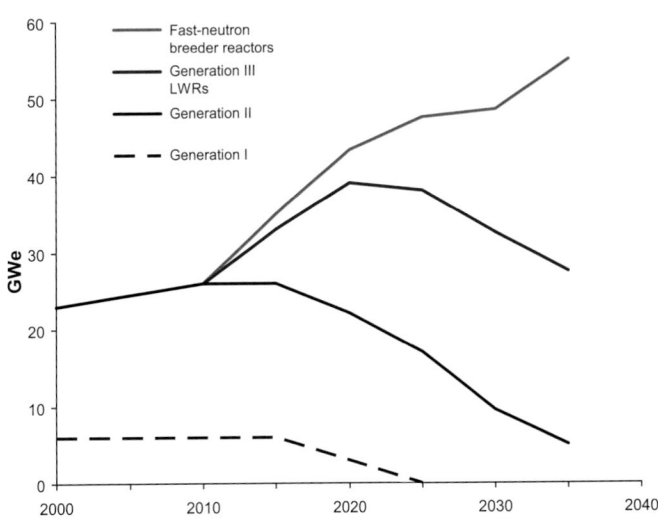

Figure 14.12 | Plans for Russian nuclear power expansion as of 2007. Source: Adapted from Saraev, 2007.

will be fueled with HEU enriched to slightly above 20%, as the BN-600 has been since 1980 (Nigmatulin and Kozyrev, 2008).

14.3.6   United States

The United States has the world's largest nuclear-generating capacity: 104 power reactors with a net generating capacity of 100.6 GW$_e$ as of the end of 2009 (IAEA-PRIS). The construction of all of these reactors began before 1978, more than three decades ago (US EIA, 2008d). As of the end of 2009, US utilities had applied for 18 combined construction and operating licenses for 28 reactors with a total capacity of 37 GW$_e$ (US NRC, 2010a), but only 12 of these applications were active,[50] and only five had signed engineering, procurement and construction contracts with reactor vendors (for nine reactors) (WNA, 2010c). The IAEA still listed only one reactor under construction in the United States; the Tennessee Valley Authority's Watts Bar II, a reactor on which construction began in 1973 and was suspended in 1988 when it was about 80% complete because of a reduction in the growth rate of US electric power demand (Reuters, August 1, 2007).

Renewed US interest in nuclear power reflects in part concerns about the future cost of natural gas, following a temporary tripling in real wellhead prices between 1998 to 2008 (US EIA, 2010)[51] and a move away from coal-fired power plants in anticipation of policies aimed at reducing greenhouse gas emissions. It also reflects government incentives. In the

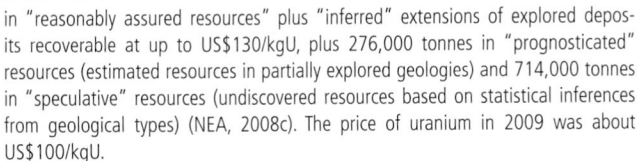

---

in "reasonably assured resources" plus "inferred" extensions of explored deposits recoverable at up to US$130/kgU, plus 276,000 tonnes in "prognosticated" resources (estimated resources in partially explored geologies) and 714,000 tonnes in "speculative" resources (undiscovered resources based on statistical inferences from geological types) (NEA, 2008c). The price of uranium in 2009 was about US$100/kgU.

48 It is possible that this includes some of the separated plutonium stored at the Seversk and Zhelznogorsk reprocessing plants. These plants produced weapon-grade plutonium for Russia's weapons program but, according to the 1997 Russia–US "Agreement … Concerning Cooperation Regarding Plutonium Production Reactors," any plutonium separated at these plants after January 1, 1997 will not be used for weapons purposes (Annex III, Subsidiary Arrangement B, Article II).

49 According to Task 30 of the 2006 Federal Target Program, sources other than the federal budget (presumably Rosatom) are to supply 1.617 billion rubles (US$65 million at US$24.6/ruble) through 2015 for R&D in support of a pilot reprocessing plant at the Krasnoyarsk Mining and Chemical Combine.

50 The applications for Callaway Unit 2, Grand Gulf Unit 3, River Bend Unit 3, Victoria County Station Units 1 and 2 are shown as suspended, and that for Nine Mile Point Unit 3 has been inactive since 2008. The review of the application for Turkey Points Units 6 and 7 has not yet begun (see application review schedule for each project at US NRC, 2010a).

51 Prices dropped by a half between 2008 and 2009, however, as gas from hydrofracturing shale began to enter the market in large quantities (US EIA, 2010).

Energy Policy Act of 2005, the US government created major incentives to investors to commit quickly to build new nuclear power plants. These included (Title XVII) government loan guarantees equal to up to 80% of project costs. Congress authorized up to US$18.5 billion for this purpose in the Consolidated Appropriations Act of 2008. In June 2008, the US DOE solicited requests for the loan guarantees. The response was applications for US$122 billion in guarantees to cover 65% of the cost of 21 reactors with a total generating capacity of 28.8 GW$_e$ (US DOE, 2008a). In its budget proposal for fiscal year 2011, the Obama Administration proposed to increase the funding available for nuclear loan guarantees to US$54.5 billion (US DOE, 2010b: 259). The four companies that were on the short-list for the first tranche of loan guarantees were also among the five companies that signed engineering, procurement and construction contracts for new nuclear power plants (Bloomberg, December 17, 2009). (The fifth was responding to state-level incentives; see below). In January 2010, one utility received a loan guarantee for US$8.3 billion (US$_{2005}$$7.6 billion) that reportedly would cover up to 70% of the project costs for two 1.1-GW$_e$ reactors, amounting to at least US$5.4 billion/GW$_e$ (US$_{2005}$$4.9 billion/GW$_e$). Additional loan guarantees might be provided by the Japanese government because the reactors will be built by Westinghouse, which is now a subsidiary of Toshiba (*New York Times*, February 17, 2010).

The Energy Policy Act also allows for up to US$2 billion to compensate companies building the first six nuclear power reactors for regulatory delays in the startup process. Finally, Title XIII provides for a production tax credit of US$0.018/kWh, up to a total of US$6 billion, for power produced by 6 GW$_e$ of advanced nuclear power capacity during the first eight years of operation.[52]

State-level policy is also important. Out of the 50 states, 36 regulate the investments of utilities in the generation and transmission of electric power.[53] Under these regulations, if a state regulatory authority authorizes investment in the construction of a power plant, the investor is allowed to charge customers for the cost of building and operating that power plant, plus a guaranteed rate of return.

In Florida, the Public Service Commission has gone further and permits investors to start charging the customers even before a nuclear power plant is under construction. If, for some reason, the plant is never completed, the owners of the reactors still will be entitled to recoup "prudent" costs from their customers.[54] The only utility that has signed

engineering, procurement and construction contracts for new nuclear power plants without the expectation of a loan guarantee is Progress Energy, a Florida utility that is the beneficiary of such a ruling. Georgia has a similar policy (*WNN*, 18 March 2009). Plans for a plant in Missouri were shelved after the utility proposing it was unable to obtain the repeal of a state law banning charges for construction work in progress (*Fuel Cycle Week*, April 24, 2009).

In late 2009, the US EIA projected that only 8.4 GW$_e$ of new nuclear electric generating capacity will actually come online in the United States by 2035 (US EIA, 2009). After 2030, US nuclear power plants will reach 60 years at an average rate of about 5 GW$_e$/yr. US power reactors will have to be licensed to operate for more than 60 years – a possibility that is already being discussed – or the rate of reactor construction will have to increase greatly if US nuclear capacity is not to decline (US EIA, 2008a).

14.3.7   Western Europe

At the end of 2009, nuclear capacity in Western Europe totaled 122 GW$_e$, with two new 1.6-GW$_e$ units under construction in Finland and France by the French company AREVA Nuclear Power (IAEA-PRIS).

In seven of the nine WEU countries with operating nuclear reactors, the youngest reactor was built in the 1980s (or earlier, in the case of the Netherlands). The United Kingdom completed one reactor in the 1990s. France completed an average of one power reactor per year during the 1990s, but none since then.[55] Due to retirements, Western Europe's nuclear generating capacity has declined by about 4 GW$_e$ since 2000 (IAEA, 2007a; 2010d). Unless the reactor licenses are extended or the rate of construction picks up, Western Europe's nuclear capacity will continue to decline during the next few decades.

France accounts for a little more than half of Western Europe's nuclear capacity (63.3 GW$_e$) and for one of the two new units under construction (IAEA, 2010d, Table 1). The equivalent of about 76% of France's electricity is generated by nuclear power (IAEA, 2010a). France is a major net exporter of electric power (Schneider et al., 2009: 101). France's national utility, Électricité de France, is also considering investing in nuclear power plants in China, the United Kingdom and the United States, and possibly also in Italy and South Africa (*WNN*, December 4, 2008). Both of the new 1.6-GW$_e$ EPR reactors being built by AREVA are suffering from serious delays and cost overruns, however. As of the end of 2008, the EPR under construction at Flamanville,

---

52   The US Congressional Budget Office puts the limit at US$7.5 billion (US CBO, 2008).

53   Some 14 states deregulated electric power production (Connecticut, Delaware, Illinois, Maryland, Maine, Massachusetts, Michigan, New Jersey, New Hampshire, New York, Ohio, Pennsylvania, Rhode Island, Texas). Eight have re-regulated (Arkansas, Arizona, California, Montana, Nevada, New Mexico, Oregon, Virginia). The remaining 28 states never deregulated (US EIA, 2008e).

54   In late 2008, the Florida Public Services Commission authorized two utilities to charge their customers US$0.6 billion during 2009 for pre-construction expenses they expected to incur for four nuclear power reactors that they hoped to build (Florida Public Service Commission, 2007/8). One of the utilities, Florida Public and

---

Light, estimated that the cost for completing the building of two Westinghouse AP1000s would be US$5780–8071/kW$_e$ (*Nucleonics Week*, February 21, 2008).

55   The last reactor came on line in Belgium in 1985, Finland in 1980, France in 1999, Germany in 1989, the Netherlands in 1973, Spain in 1988, Sweden in 1985, Switzerland in 1984, and the United Kingdom in 1995 (IAEA-PRIS).

France, was expected to cost €4 billion (US$_{2005}$5.8 billion) or US$3600/kW$_e$ (*WNN*, December 4, 2008). In early 2009, the Finland's Olkiluoto EPR was 18 months behind schedule and expected to cost close to €5 billion (US$_{2005}$7.2 billion) (*Nucleonics Week*, March 5, 2009). AREVA NP's client in Finland, TVO, was suing for compensation of €2.4 billion (US$_{2005}$3.5 billion) for power replacement and other losses due to the delay. AREVA, for its part, accused TVO of having slowed down the licensing procedure more than necessary and filed an arbitration case with the International Chamber of Commerce for about €1 billion in compensation (*Nucleonics Week*, March 19, 2009).

In both cases, a large part of the problem seems to be inadequately trained workers and poor quality control leading to the rejection of completed work by safety inspectors (*New York Times*, 29 May 2009). To some extent, these problems reflect a loss of expertise in the nuclear industry that might be overcome if the number of orders increases to the point where crews can move from one project to another at a nearby location in the same country. At the moment, this condition is met only in China and South Korea.

Of the remaining eight West European countries with operating nuclear power plants, two have laws mandating a phase-out: Belgium (with 54% of electric power generated by nuclear plants in 2008) and Germany (28%) (IAEA, 2010a). Spain's current government favors a nuclear energy phase-out (WNA, 2010g). The Netherlands is currently considering the construction of new nuclear power plants (WNA, 2010f). Despite its experience with AREVA, Finland is considering buying a second new nuclear power reactor from another vendor (*WNN*, February 5, 2009). Sweden had a phase-out law but, in 2010, decided to allow replacement of its current units as they are retired (WNA, 2010e).

The United Kingdom has the third largest nuclear capacity in Western Europe (10 GW$_e$).[56] The last of its first-generation "Magnox" graphite-moderated, gas-cooled nuclear power plants are to be shut down in 2012 (UK NDA, 2010). Its advanced gas reactors (AGRs) have a design life of 35 years, which would have them all shut down by 2024.[57] In early 2008, the UK government, which has an ambitious plan to reduce carbon dioxide emissions, came out in support of the building of new nuclear power plants, but declared that it would not subsidize their construction (*Times*, January 11, 2008). In September 2008, France's government-owned utility, Électricité de France, bought the UK nuclear utility, British Energy, for US$23 billion with the intention of building four new 1.6-GW$_e$ EPR LWRs on the AGR sites (*International Herald Tribune*, September 24, 2008). Other companies are also considering building new nuclear reactors in the United Kingdom (WNA, 2010d).

14.4 Advanced Reactor Technology

The major reactor vendors have developed and are licensing and selling advanced (generation III+) LWRs (see, e.g., US NRC, 2010a). With the renewed interest in nuclear power, however, there has also been renewed interest in exploring alternatives to the LWR. This section first briefly discusses advanced LWRs, and then considers two types of alternative reactors – fast-neutron and slow-neutron[58] – and finally, small and transportable reactors, which may be of interest to countries or regions with small power grids.

14.4.1 Generation III+ Light-water Reactors

After the 1979 accident at Three Mile Island, there were few reactor orders. The reactor vendors that survived used this period of slow business to develop and license evolutionary designs of LWRs intended to be both safer and less costly per unit output. The resulting so-called Generation III+ LWR designs and their instrumentation have been simplified and standardized, and some contain "passive" safety systems that operate automatically even if electrical power to the control system and pumps is lost. In the Westinghouse-Toshiba AP1000, for example, valves are designed to open automatically when the level of water in the reactor falls below a certain level. Emergency cooling water then is driven into the reactor, initially with steam and nitrogen pressure. After the reactor vessel is depressurized, water flows in from elevated tanks without pumping and, after evaporating, is condensed at the top of the containment building and returns to the tanks to flow into the reactor again (Westinghouse, 2009). These systems are calculated to reduce considerably the probability of a core meltdown accident.

Despite their inherent reliability, the pressures generated by gravity-driven passive systems are modest in comparison with those produced by pumps. Their performance is therefore less certain in a situation where hot fuel can generate steam backpressure. Also, by giving credit to the passive systems for reducing the probability of a core meltdown, the US Nuclear Regulatory Commission (US NRC), whose regulations are usually treated as world standards, has reduced the reliability requirements on the active backup systems by not requiring them to be safety grade, and has allowed less robust containments. As a result, the net effect on overall safety of installing the passive systems is uncertain (Lyman, 2008).

---

56 Germany has the second largest capacity, with 20 GWe.

57 They may have to be shut down even earlier (*New Scientist*, March 25, 2004).

58 In slow-neutron reactors, fission neutrons are "moderated" or slowed down by collisions with the nuclei of light elements. In contrast, neutrons lose relatively little energy to the recoil of the heavier nuclei of the liquid metals used to cool fast-neutron-reactor fuel. A primary advantage of slow-neutron reactors is that they can sustain a chain reaction in low-enriched or even, in some cases, in natural uranium. A primary advantage of fast-neutron reactors is that, when fueled with plutonium, they can be designed to breed more plutonium than they consume.

14.4.2   Fast-neutron Reactors for Breeding and Burning

Although there is no reason to expect the dominance of LWRs to end in the foreseeable future, the major government nuclear-energy R&D establishments continue to develop potential successors. The OECD's Generation IV (Gen IV) International Forum coordinates research on six reactor types (Gen IV, 2008) and the IAEA-based International Project on Innovative Nuclear Reactors and Fuel Cycles focuses on methodologies and generic technical challenges (IAEA, 2010b).

The most attention – but relatively little funding – is going to the liquid-sodium-cooled "fast-neutron" reactor, the reactor type in which the nuclear R&D establishments invested their greatest efforts in the 1960s and 1970s (Figure 14.13). Fast-neutron reactors fueled with plutonium can be designed to produce more plutonium from uranium-238 than they consume. This makes uranium-238, which constitutes 99.3% of natural uranium, the ultimate fuel of fast-neutron breeder reactors.

Plutonium breeder reactors were pursued when it appeared that resources of high-grade uranium ore were very limited and global nuclear power capacity was expected to increase by 2010 to several thousand GW_e (IPFM, 2010). But nuclear capacity plateaued and high-grade uranium ore proved to be much more abundant than previously believed. Enough low-cost uranium has been found to sustain 500 GW_e of capacity for 50 years (NEA, 2008c)[59] and much more probably remains to be discovered (see discussion below). The contribution of the cost of uranium to the cost of power at the cutoff grade in these estimates (uranium recoverable at a cost of US$130/kg or less) would only be about 0.3¢/kWh.

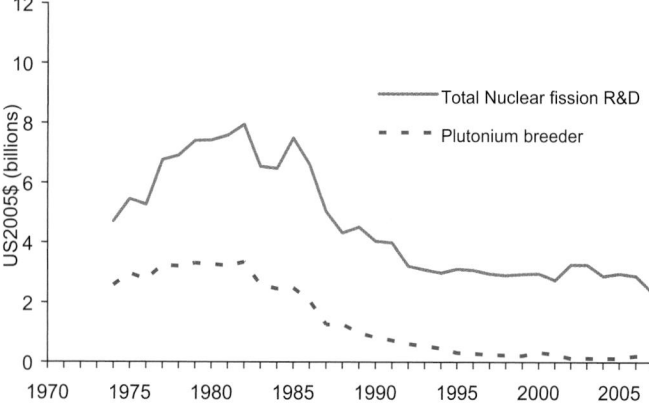

Figure 14.13 | Total fission and breeder research, development and demonstration (RD&D) funding in the OECD countries with substantial breeder programs, 1974–2007 (Belgium, France, Germany, Italy, Japan, Netherlands, UK, and United States). The breeder share is understated because France, which had the world's largest breeder program in the 1980s and 1990s, did not report most of its breeder RD&D activities as such. Source: data from IEA, R&D Statistics, 2009.

Fast-neutron reactors cannot be cooled by water, because (as occurs in the collisions of billiard balls) neutrons are drastically slowed down by a relatively small number of collisions with the light nuclei of the hydrogen atoms in the water. Liquid metal therefore is used because (as with the collision of a golf ball with a boulder) the heavy nuclei of the atoms take away little energy in a collision. As already noted, most development has focused on reactors cooled by molten sodium. Because sodium burns on contact with air or water, however, sodium-cooled reactors have proved to be much more costly and difficult to operate than water-cooled reactors, and only a few experimental and "demonstration" reactors have been built with government support. Japan's 0.28-GW_e Monju fast breeder demonstration reactor, which began operating in 1995, shut down a few months later as a result of a sodium fire, and only restarted briefly 15 years later in May 2010. The largest demonstration liquid-sodium-cooled reactor built to date, France's 1.2-GW_e Superphénix, spent so much time in repair that it had an average capacity factor of only 7% over its operating life (1985 to 1996). Russia's BN-600 is the exception. Despite 15 sodium fires in 23 years, it has been kept online with an average capacity factor of about 74% (Oshkanov et al., 2004).[60]

There are various ideas for reducing the cost of fast-neutron reactors, including the use of alternative coolants such as molten lead. Helium is also being considered, but has the safety disadvantage that it has little heat capacity if there is a loss of pumping power.

Breeder reactors and uranium resources. Superphénix cost about three times as much as an LWR of the same capacity.[61] If the capital cost of a commercialized fast-neutron reactor were higher than that of an LWR by only US$1000/$kW_e$, it would require the cost of uranium to rise to about US$1200/kg for the uranium savings from a breeder reactor to offset its extra capital cost (Bunn et al., 2005, Figure 3; MIT, 2003; IPFM, 2010). This is about 10 times the cost of uranium in early 2009.

Most uranium exploration has focused on deposits with recovery costs of less than US$80/kg and finds are reported only when the recovery costs would be less than US$130/kg (NEA, 2008c). These resources range from 0.03–20% uranium (IAEA, 2009b). The concentration at which the amount of electric energy extractable from the uranium in 1 tonne of ore with a once-through fuel cycle would equal the amount of energy extracted from 1 tonne of coal is approximately 0.005–0.02%.[62]

---

60 The BN-600 is actually fueled by HEU because its core would not be stable with plutonium fuel. It is therefore not a breeder reactor.

61 The capital cost of the 1.2-GWe Superphénix was FF 34.4 billion (about US$8 billion in 2005 US$) according to France's public accounting tribunal, the Cour des Comptes (*Nucleonics Week*, October 17, 1996).

62 At 0.005% uranium, 1 tonne of ore would contain 50 g of natural uranium. That would produce 6.25 g of 4.4% enriched uranium (assuming 0.2% uranium-235 is left in the depleted uranium). For a burnup of 53 MWt-days/kg, the amount of fission energy released from the 6.25 g of LEU would be 28×10^9 J, about the amount of energy released from the combustion of 1 tonne of coal. Van Leeuwen has the breakeven level at 0.02%, in large part because he uses 50% gross thermal efficiency of the coal plant versus 32% for an LWR and a uranium recovery factor of 50% (Van Leeuwen, 2008).

---

59 Between 160 and 200 tonnes of natural uranium are required annually to fuel a 1-GWe LWR operating at a capacity factor of 90%, assuming 0.2–0.3% uranium-235 remaining in the depleted uranium from enrichment. With a global LWR capacity of 500 GWe, 10 million tonnes of uranium would last 100–125 years.

Despite consumption and inflation, known uranium resources with recovery costs less than US$40/kg continue to increase from year to year (NEA, 2008c, Table 1). Resources also are likely to go up rapidly at higher recovery costs (Deffeyes and MacGregor, 1980). Figure 14.14 compares the known conventional resources of uranium reported by the OECD's Nuclear Energy Agency (NEA, 2008c) with crustal abundance models and the estimated cost of recovering uranium from seawater (Schneider and Sailor, 2008):

- The cost of recovery used to translate estimated crustal abundance to cost in Schneider and Sailor's "conservative" and "optimistic" crustal models is assumed to be simply proportional to the amount of rock that must be mined, crushed, and leached to recover 1 kg of uranium. Thus, for ore with half the concentration of uranium, it would cost twice as much to recover 1 kg of uranium.

- The Kim & Edwards cost curve assumes that the recovery cost per kilogram increases somewhat more rapidly than inversely with declining ore concentration. Thus, for example, the cost of recovering a kg of uranium from an ore grade with one-tenth the concentration would be 19 times as high.

- The ocean contains about 4.5 billion tonnes of uranium but at a very low concentration of 3.3 parts per billion by weight. The estimate shown in Figure 14.14 that uranium would be recoverable from seawater at a cost of US$200/kg was developed by the US Department of Energy Generation IV Fuel Cycle Cross Cut Group (see also Tamada et al., 2006).

All the curves in Figure 14.14 are constrained to agree on uranium resources at a uranium recovery cost of about US$40/kg.

If the crustal model approach is correct, 20–60 million tonnes of uranium should be recoverable at a cost of less than US$130/kg. Some 25 million

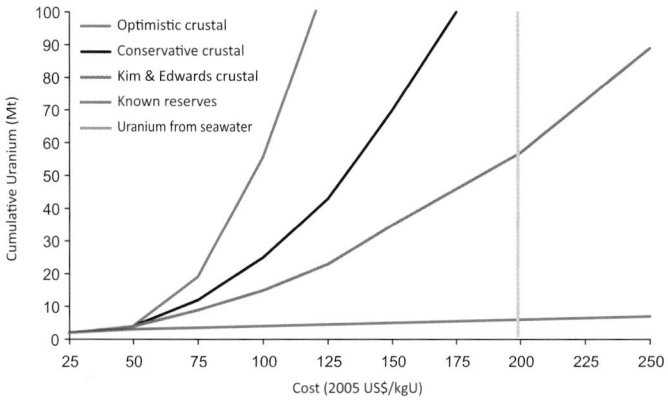

Figure 14.14 | Uranium resource availability. Known conventional reserves as reported in the NEA–IAEA "Red Book" [NEA, 2008c] compared with three geological estimates of crustal abundance as a function of recovery cost and an estimated cost of US$200/kg for recovery of uranium from seawater. adapted from: Schneider and Sailor, 2008.

tonnes of uranium would be required to sustain a once-through LWR economy until the year 2100 if global nuclear capacity increased linearly from 2020 to approximately 4000 GW_e. An LWR capacity of 4000 GW_e would require about 6.4 million tonnes of uranium per decade.[63] Thus breeder reactors would be unlikely to be competitive until well beyond the end of the century, even if global nuclear capacity climbs into the thousands of GW_e. A recent MIT study has come to a similar conclusion (MIT, 2010).

It would be much more useful to determine whether the crustal model is approximately correct and to refine the technology and cost estimates for recovering uranium from seawater than to embark on the promotion of a hugely expensive proliferative technology involving the separation and recycling of plutonium because of probably unfounded fears of uranium shortages.

Fast-neutron "burner" reactors and the spent-fuel problem. In 2006, the US DOE proposed to design and build fast-neutron reactors as "burner" rather than breeder reactors (US DOE, 2006). This was because of the presence of long-lived transuranic isotopes (plutonium, neptunium, americium, and curium) in spent nuclear fuel and public concerns about uncertainties about the performance of geological repositories over a time scale of hundreds of thousands of years. Unlike LWRs, fast-neutron reactors can fission all the long-lived transuranic isotopes in spent LWR fuel relatively efficiently.

This was not a new proposal. In 1992, the US DOE had asked the US National Academy of Sciences (US NAS) to study proposals to reduce the longevity of the radioactive waste problem through separation and transmutation of long-lived radioisotopes. The resulting study (US NAS, 1996) was quite skeptical. It concluded that:

- "Although a significant fraction (90–99%) of many of the most troublesome isotopes could be transmuted (to shorter-lived or stable isotopes) this reduction of key isotopes is not complete enough to eliminate all the process streams containing HLW (high-level radioactive waste)… Transmutation thus, would have little effect on the need for the first repository…"

- "It would take about two centuries of operating time to reduce the inventory of the residual (transuranics) to about 1% of the inventory of the reference LWR once-through fuel cycle…"

- "Estimates of changes in dose (from nuclear power and radioactive waste) are small… Taken alone, none of the dose reductions seem large enough to warrant the expense and additional operational risk of transmutation…."

- "The excess cost of (a separation and transmutation) disposal system over the once-through disposal of the 62,000 (tons heavy metal

---

63   Assuming 160 tonnes of natural uranium per GWe-yr, i.e., a depleted uranium assay of 0.2%.

in) LWR spent fuel (the approximate legislated limit on what could be stored in Yucca Mountain before a second US repository came into operation) is uncertain but is likely to be no less than US$50 billion and easily could be over US$100 billion ($US_{2005}$$120 billion ($US_{2005}$$0.3–0.6+¢/kWh), not including the extra cost of the fast-neutron reactors or other transmutation systems)."

- "The committee concluded that the once-through fuel cycle should not be abandoned… (T)his has the advantage of preserving the option to retrieve energy resources from the wastes for an extended period of time. This can be achieved by adopting a strategy that will not eliminate access to the nuclear fuel component for a reasonable period of time, say about 100 years, or by preserving easy access to the repository for a prescribed period of time, or by extending the operating period of the repository… A reason for supporting continued use of the once-through fuel cycle is that it is more economical under current conditions."

- "Widespread implementation of (separation and transmutation) systems could raise concerns of international proliferation risks…"

The last comment relates to the fact that, as currently envisioned, fast-neutron reactors only achieve the benefits of uranium savings and more complete fissioning of transuranics with a "closed" fuel cycle (one in which spent fuel is processed and the plutonium and other transuranic elements are recycled repeatedly in new fuel). Since the transuranic elements could be used to make nuclear weapons, this creates a proliferation risk. We discuss this issue below.

In 2007, the US Office of Management and Budget requested that the National Academies of Sciences review US DOE's nuclear energy R&D program. The response was even more unequivocal; "domestic waste management, security and fuel supply needs are not adequate to justify early deployment of commercial-scale reprocessing and fast reactor facilities" (US NAS, 2007: S-8).

14.4.3   Thermal (slow)-neutron Reactors

On its list of reactor types of interest, the Generation IV collaboration has three types of thermal-neutron reactors: supercritical water-cooled reactors, very high-temperature gas-cooled reactors, and molten-salt reactors. The supercritical reactor would allow LWRs to take advantage of the increased efficiency of the conversion of heat into electricity at higher coolant temperatures. Some fossil fuel plants already operate at supercritical temperatures.

Very high-temperature gas-cooled reactor designs, with coolant temperatures up to 950°C, are being examined in the United States, primarily as a way to produce hydrogen by heat-driven instead of electricity-driven chemical reactions. The "Nuclear Hydrogen" project was launched in the

Energy Policy Act of 2005. The US DOE and the US Nuclear Regulatory Commission have defined a joint research program to provide the analytical tools to license such a reactor (US DOE, 2008d). Otherwise, R&D in this area has been confined primarily to the development of thermochemical processes (US DOE, 2008c). Charles Forsberg, a nuclear engineer at the Massachusetts Institute of Technology, has suggested that a more important use of high-temperature gas-cooled reactors that could produce heat with a temperature of 700°C would be to replace fossil fuels in providing process heat for oil refineries and for extracting liquid fuels from oil shales and tar sands (*Oil and Gas Journal*, August 11, 2008).

Finally, the molten-salt reactor would have its fuel dissolved in molten salt. The heat would be extracted by pumping the salt through a heat exchanger and the fission products could be removed and new fuel added by chemically processing a side stream. The problem with this design, however, is the complexity of operating a reactor with an integrated small reprocessing plant (Gen IV, 2008).

14.4.4   Low-power and Transportable Reactors

The IAEA recommends that a single nuclear power reactor should not constitute more than 5–10% of the generating capacity on a grid (IAEA, 2007b). As discussed previously (Table 14.2), gigawatt-scale reactors therefore require large grids. Smaller reactors could provide an alternative for countries and regions with small grids and expensive power. Proponents also claim that economies of scale in factory production could make them less costly than today's reactors, which involve costly field construction. A few relatively low-power LWR designs are available (WNA, 2008).

The low-power (~0.2 GW_e) graphite-moderated high-temperature gas (helium)-cooled reactor has been under development since the 1970s and continues to be the most plausible, relatively safe alternative to the LWR in the near term. It is being actively investigated in China, Japan, Russia, South Africa and the United States.

Some small reactor designs emphasize long core life. The tradeoff is that the initial core would be more costly per unit output (IAEA, 2005a, Table 5; IAEA, 2007c; US NRC, 2009).

One transportable nuclear power plant is under construction, a barge carrying twin 0.035-GW_e reactors based on a design used in some of Russia's nuclear-powered icebreakers. These reactors would be refueled after four years of operation. The first floating power plant is being constructed in St Petersburg with completion projected for 2011 (Bellona, May 18, 2009; IAEA, 2005a, Annex 6.5; Greencross, 2004). In 2006, Rosatom was planning to complete seven floating nuclear power plants by 2015 (Rosatom, 2006a, b). It is believed that, because of Russia's interest in exporting these reactors, they will be fueled with LEU rather than the weapon-usable highly enriched uranium used in Russia's submarines and nuclear-powered icebreakers (Sokov, 2006).

14.4.5   Safer Reactor Designs

Operational safety has been improved and, as discussed above, the new Generation III+ light-water reactor designs have passive safety features that might make emergency cooling independent of the availability of power for days. Nevertheless, the probability of terrorists attempting to cause Chernobyl-scale releases of radioactivity appears to be greater today than it was in the 1980s. This is a major reason why attention should be devoted to less vulnerable designs as well as to improved physical security.

In the case of the high-temperature gas-cooled reactor, attempts to give it inherent safety have been made by putting the uranium into small particles, encapsulated in layers of pyrolytic carbon and silicon carbide, in order to contain the fission gases. The chain reaction would shut down because of negative temperature feedback effects on the reactivity and the reactor would eventually reach thermal equilibrium by radiating away to the cooler wall of the containment the heat generated by the declining radioactivity of the fission products in the fuel. If the reactor power is low enough (less than 0.3 GW$_e$), the peak temperature could be kept below the failure temperature of the particles (Labar, 2002). Oxidation of the graphite moderator by penetrating air could provide an additional source of heat, but one analysis has found that, even with a break in the largest coolant pipe, the rate of air inflow would be limited to a level where graphite oxidation would not drive the core temperature significantly higher (Ball et al., 2006).

A recent report on the operational history of Germany's 46 MW$_t$ gas-cooled AVR (Arbeitsgemeinschaft Versuchsreaktor) pebble-bed reactor, which operated between 1967–1988, has called into question the adequacy of its safety design. It was revealed that the reactor had suffered a serious leakage of fission products into the helium coolant. One possible reason was "inadmissible high core temperatures … more than 200 K higher than calculated." Another was that cesium-137 (30-year half-life), the most dangerous radioisotope released by the Chernobyl accident, appears to have diffused through intact particle coatings. It was therefore concluded that the reactor would require a leak-proof containment similar to that required for modern LWRs, which would erase a major cost saving. Additional safety issues were noted for designs such as the AVR, in which water ingress into the graphite was possible. "Thus a safe and reliable AVR operation at high coolant temperatures (does) not conform with reality" (Moormann, 2008, abstract).

With questions about the safety of what has been claimed by General Atomics for decades to be an "inherently-safe" reactor design (General Atomics, 2010), it would be useful to launch a new R&D program to consider the possibilities for a truly inherently safe, reliable, and economic design.

14.5   Once-through versus Plutonium Recycling

Today, five weapon states (China, France, India, Russia, and the United Kingdom) plus Japan reprocess at least some of their spent fuel. The Netherlands has contracted with France to have the spent fuel from its single reactor reprocessed (van der Zwann, 2008). Of the reprocessing states, France, India, and Japan currently plan to reprocess most of their spent fuel. The United Kingdom is expected to end reprocessing when it has fulfilled its existing contracts (*Nuclear Fuel*, June 18, 2007, July 28, 2008a). Russia reprocesses only the spent fuel from its first-generation VVER-440 LWRs and its BN-600 demonstration fast-breeder reactor, with a combined capacity of 3 GW$_e$ (IAEA-PRIS). It also receives and mostly stores spent fuel from Bulgaria and Ukraine. China has built a pilot reprocessing plant.

Of the remaining 21 countries with nuclear energy programs, 11 have not reprocessed their spent fuel[64] and 10 that shipped their spent fuel to France, Russia, or the United Kingdom for reprocessing in the past, have not renewed their contracts.[65] All 21 have decided on interim storage. As a result, measured in terms of fission energy released, worldwide, about one-third of spent fuel is reprocessed today (Table 14.5). The percentages shown for some countries in the first column reflect various limitations on the fraction of the spent fuel reprocessed.

There are two primary reasons why almost all customer countries have stopped shipping their spent fuel abroad for reprocessing:

1.   reprocessing and plutonium recycling are much more costly than spent-fuel storage, and

2.   countries providing reprocessing services are requiring (or, in the case of Russia, keeping the option to require) their foreign customers to take back the high-level waste from reprocessing.

Thus, foreign reprocessing simply converts, at considerable cost, a politically difficult spent-fuel disposal problem into a politically difficult spent MOX fuel and high-level waste disposal problem. It is politically attractive only for an interim period because it buys a decade or so of respite.

Only Russia has routinely kept its customers' separated plutonium. Among France's and the United Kingdom's former reprocessing customers, Belgium, Germany, Japan, and Switzerland have been recycling their separated plutonium in MOX (plutonium–uranium) fuel in the LWRs that produced it. France is doing the same. India and Russia plan to use their separated plutonium for startup cores for plutonium breeder reactors. France has kept Spain's separated plutonium[66] and presumably will do the same for Italy. It has included that offer in proposed reprocessing contracts to other countries such as South Korea (*Nuclear Fuel*, 13 July 2009).

---

64   Ignoring US reprocessing prior to 1973.

65   In 2008, Italy contracted with France to have reprocessed 235 tonnes of irradiated fuel from reactors that were shut down after the 1986 Chernobyl accident.

66   The reprocessed fuel was from the Vandellós-1 reactor, a graphite-moderated, gas-cooled reactor that operated from 1972 to 1990 (WISE, 1999).

Table 14.5 | Civilian spent-fuel reprocessing by country

Countries that reprocess spent fuel (GWe)		Customer countries that have quit or are planning to quit reprocessing (GWe)		Countries that have not reprocessed (GWe)	
Bulgaria (in Russia)	1.9	Armenia (in Russia)	0.4	Argentina	0.9
China (30%)	10.0	Belgium (in France)	5.9	Brazil	1.9
France (80%)	63.3	Czech Republic (in Russia)	3.7	Canada	12.6
India (~50%)	4.2	Finland (in Russia)	2.7	Mexico	1.3
Japan (90% planned)	46.8	Germany (in France/UK)	20.5	Pakistan	0.4
Netherlands (in France)	0.5	Hungary (in Russia)	1.9	Romania	1.3
Russia (15%)	22.7	Slovak Republic (in Russia)	1.8	Slovenia	0.7
UK (ending?)	10.1	Spain (in France/UK)	7.5	South Africa	1.8
Ukraine (in Russia, ~50%)	13.1	Sweden (in France/UK)	9.3	South Korea	18.7
		Switzerland (in France/UK)	3.2	Taiwan, China	4.9
				US (since 1972)	100.6
Total	172.6	Total	56.9	Total	145.2

The United Kingdom is not recycling its own separated plutonium and as yet has no disposal plans. By the time its domestic spent-fuel reprocessing contracts are fulfilled, the UK stockpile of separated plutonium will amount to about 100 tonnes – enough for more than 10,000 nuclear explosives. The UK Nuclear Decommissioning Authority is now examining disposal options (UK NDA, 2009). The storage of separated plutonium and reprocessing waste is significantly more expensive than storage of unreprocessed spent fuel. In addition, after several years in storage, americium-241, a decay product of plutonium-241 (14-year half-life), builds up in plutonium and has to be separated before fuel fabrication.[67]

Where plutonium is being recycled in LWR MOX, it is only being recycled once. Irradiation results in a net reduction of the plutonium in the MOX fuel by about one-third, but also results in a shift of the isotopic mix in the plutonium toward the even isotopes (plutonium-238, plutonium-240 and plutonium-242) that are less easily fissioned in slow-neutron reactors (NEA, 1989, Table 12B). With repeated recycling in "non-fertile" fuel with LWRs, i.e., without uranium-238 in which neutron capture produces more plutonium, it would be possible eventually to completely fission plutonium and the other transuranics except for reprocessing and fabrication losses. It would require shielded fuel fabrication, however, and long intervals (20 years) are recommended between cycles to allow radioactive decay to offset the steady buildup of neutron-emitting curium and californium isotopes. Achieving a significant reduction in the global inventory of transuranic elements would therefore take centuries (Shwageraus et al., 2005).

In its Global Nuclear Energy Partnership (GNEP) initiative, the G.W. Bush Administration proposed that "fuel-cycle countries" would supply fresh fuel and take back and reprocess spent fuel from countries

with reactors but no enrichment or reprocessing facilities. The "fuel-cycle countries" would recycle the separated transuranic elements domestically in fast-neutron reactors and dispose of the reprocessing waste in domestic geological repositories. Although about US$100 billion has so far been spent worldwide in efforts to commercialize fast-neutron reactors, no country has yet succeeded. Nor has any country yet been willing to volunteer to take other countries' radioactive waste. The US Congress became skeptical about GNEP; the Obama Administration cancelled the proposal to build a reprocessing plant; and the US reprocessing program has returned to a focus on R&D, and in particular on the feasibility of developing more economic and proliferation-resistant methods of recycling plutonium and other transuranic elements. An evaluation of the resistance of the alternative recycling technologies proposed thus far against national proliferation has been discouraging, however (Bari et al., 2009). Attention has therefore been turning to "breed and burn" concepts in which plutonium is bred and burned in place without separation from fission products (Finck, 2010).

14.5.1   Radioactive Waste[68]

Geological disposal is very widely accepted in the nuclear community as technically feasible and adequately safe (NEA, 2008b). Absolute proof that there will be no significant releases over 100,000 years or more as a result of natural processes or human intrusion is impossible, however. In the United States, Congress mandated in the 1987 amendments to the Nuclear Waste Policy Act that a site characterization program for a geological repository for spent power-reactor fuel be carried out only at Yucca Mountain, Nevada, and, if justified by the results of that program, a repository should be built and licensed by 1998. More than US$10 billion have been spent on the project and an application for a license was

---

67   Current MOX fuel fabrication plants cannot process LWR plutonium after the americium has built up for more than 3–5 years.

68   Charles McCombie, executive director, Association for Regional and International Underground Storage, adviser.

submitted in 2008 (US DOE, 2008b). In this sense, this repository was the most advanced in the world. It may never be completed, however, because of fierce political and legal opposition from the state of Nevada, now supported by President Obama, who has proposed to cancel the repository project and has established a "Blue Ribbon Commission" to study alternatives (US DOE, 2010a).

Other countries have encountered similar opposition from potential host communities for geological repositories. This is resulting in the abandonment of centralized "decide–announce–defend" siting approach in favor of a more consultative approach with possible host communities (Isaacs, 2006).

Finland and Sweden have adopted the consultative approach and, until its recent site selection, Sweden actually had two communities with nuclear power plants competing to host its repository (*WNN*, 3 June 2009). In Finland, the construction of an underground test facility that is expected to become a spent-fuel repository is underway next to a nuclear power plant, following acceptance by the local community and formal approvals granted by the regulator and the parliament (McCombie and Chapman, 2008). More recently, local governments in Spain have competed to provide the site for a national radioactive waste repository. Here again, the finalists are communities that already host nuclear power plants (Deutsche Welle, 2010).

In the design envisioned for the Finnish and Swedish repositories, the spent fuel is to be encapsulated in a 5-cm thick copper cask and then embedded in bentonite clay, which swells when it is wet. Recently a technical challenge has emerged to the assumed durability of the cask (Hultquist et al., 2009). Whether this will derail progress toward the repositories remains to be seen.

The fact that communities that already have nuclear facilities appear to be more willing to host radioactive waste repositories suggests that they may have a different assessment of both the risks and benefits than do communities without nuclear facilities. This certainly makes sense on an objective basis since, as the Chernobyl accident showed, the potential scale of radioactive contamination of the surface from an operating nuclear facility dwarfs any potential surface contamination from an underground facility. Also, if no off-site destination can be found for a nuclear power plant's spent fuel, putting the spent fuel underground nearby would reduce the long-term risk to the local community.

Given the already large number of relatively small national nuclear energy programs, there is interest in regional radioactive waste repositories in Europe and East Asia, although no country has yet expressed interest in hosting one. In the past, Russia has taken spent fuel back from Eastern Europe and Ukraine. There is still interest in Russia's nuclear establishment in doing so. Disposing of foreign spent fuel is seen as potentially profitable and the plutonium in the spent fuel is seen as a future energy resource. Much of Russia's public disagrees, however,

and, for now, the leadership of Rosatom is not pushing to import foreign spent fuel other than Russian-origin fuel from power reactors exported by the Soviet Union or Russia.

In Europe, the European Commission has encouraged projects aimed at developing shared repositories for its smaller member states (ERDO, 2010). There should be economies of scale in the construction of repositories. A theoretical exploration, based on an identification of fixed and variable costs in the cost models developed by the Swedish, Finnish, and Swiss repository projects, finds savings of 5–10% from building one repository instead of two, each with half the capacity. It estimates 60% savings if 14 European countries with small nuclear energy programs share a single repository, but notes that 60% of those savings would result from the countries sharing repository R&D costs (Chapman et al., 2008).

Economies of scale may not be realized in the real world, however. The estimated cost of the large US geological repository proposed for Yucca Mountain was as high as or higher per tonne of spent fuel than the costs of the smaller disposal projects being developed in Europe. In 2008, the estimated cost of the US repository, not including transportation costs, was US$76.8 billion [US$_{2005}$$72 billion] for the equivalent of 122,100 tonnes of spent fuel,[69] or about US$_{2005}$$590/kg (US DOE, 2008b). For comparison, the estimated costs for disposing of 9500 tonnes of spent fuel in Sweden was about US$_{2005}$$650/kg;[70] for 5600 tonnes in Switzerland, about US$_{2005}$$500/kg;[71] and for 5800 tonnes in Finland, about US$_{2005}$$365/kg.[72]

Since implementing geological repositories is politically difficult and not technically urgent, and interim dry-cask storage is inexpensive and relatively safe, it is not surprising that interim storage at nuclear power plants has become the de facto spent-fuel management strategy in the United States, Germany and a number of other countries. It also avoids the risks of dispersal of radioactive waste while it is in liquid form at the reprocessing plant.[73]

Interim storage is not immune to controversy, however, because of concerns that interim may become permanent. Indeed, with a few

---

69 This total would include 109,300 tonnes of spent civilian fuel. The remainder would be "defense nuclear wastes," including solidified high-level waste from US plutonium production for weapons and naval-reactor spent fuel.

70 46.5 billion Swedish krona (SEK), assuming SEK7.5 (2003) per 2005 US$ (SKB, 2003).

71 Assuming that 2065 packages of spent fuel and 720 packages of vitrified high-level waste are equivalent to 5570 tonnes of spent fuel, and that 4.4 billion Swiss francs (2001 SFR) equal US$2.8 billion (2005 US$) (Chapman et al., 2008, Appendix).

72 Assuming that 2899 spent-fuel containers hold 5800 tonnes of spent fuel and that €2.54 billion (2005 €) equal US$2.12 billion (2005 US$) (Chapman et al., 2008, Appendix).

73 Both France and the UK have accumulated years of production of high-level liquid waste at their reprocessing plants because of technical problems with the vitrification (glassification) process. This waste contains on the order of 100 times the amount of cesium-137 (30 year half-life) that was released in the Chernobyl accident.

exceptions, local governments in Japan and South Korea have vetoed the construction of additional on-site interim storage. This is one of the reasons for the persistence of reprocessing in Japan (Katsuta and Suzuki, 2006) and the interest in reprocessing in South Korea (von Hippel, 2010).

The cost of dry-cask interim spent-fuel storage is relatively low (US$_{2005}$\$100–200/kg, or 0.02–0.05¢/kWh) and keeps open all future options, including deep underground disposal and reprocessing/recycling. It is relatively safe because the fuel is typically about 20 years or more old and the heat generated by the radioactivity has declined to less than 2 kW$_t$/tonne (Alvarez et al., 2003: Figure 5).[74] Ten tonnes in a typical 100-tonne cask therefore generate less heat than an ordinary automobile engine and only passive air-cooling is required. The temperature of the fuel in the cask remains well below the fuel operating temperature in a reactor and its zirconium-alloy fuel rod cladding is expected to remain intact indefinitely. In Germany, Switzerland and Japan, the casks are stored inside thick-walled buildings. In the United States, they are stored outside (Figure 14.15). It is possible to puncture such casks with a missile tipped with a shaped charge but, based on an experiment with simulated fuel, it was concluded that, for a single puncture, only a few parts per million of the cesium-137 in the cask would be released. Even a fractional release 100 times larger would still be negligible on the scale of the Chernobyl accident, where the equivalent

of the amount of cesium-137 in approximately three casks of spent fuel was released (Alvarez et al., 2003).

14.6   Risks from Large-scale Releases of Radioactivity to the Atmosphere

The most serious release of radioactivity to the environment from a nuclear power plant accident occurred at Chernobyl, on the border between Belarus and Ukraine, in late April and early May 1986. It was caused by an accidental supercriticality that produced a power spike that ruptured the cooling tubes, followed by a steam explosion as the water contacted the hot graphite in the core, and finally a graphite fire after the core was opened to the air.

The physical consequences of the Chernobyl accident included the following:[75]

- The deaths of 28 emergency workers from radiation illness within weeks.

- Exposure to high radiation fields of 600,000 civilian and military "liquidators" who were involved in the emergency decontamination of the reactor, the reactor site and nearby roads, and in the construction of the temporary "sarcophagus" over the reactor.[76]

Figure 14.15 | Dry cask storage of older spent fuel at a US nuclear power plant. Each year, a 1000-MW$_e$ light-water reactor discharges spent fuel that originally contained about 20 tonne of uranium. Each cask typically holds about half that amount and costs US$_{2005}$\$1–2 million. Reprocessing spent fuel costs about ten times as much. Source: Connecticut Yankee, 2008.

---

74 Spent fuel can be placed in dry-cask storage as soon as three years after discharge, when the decay heat is about 6 kW$_t$/tonne.

75 Unless otherwise stated, the source of information in this section is UNSCEAR, 2000, vol. II, Annex J.

76 The dose limit was 0.25 sieverts (Sv). The average recorded doses were 0.17 Sv in 1986, 0.13 Sv in 1987, 0.03 Sv in 1988 and 0.015 Sv in 1989 (*ibid.*, p. 470).

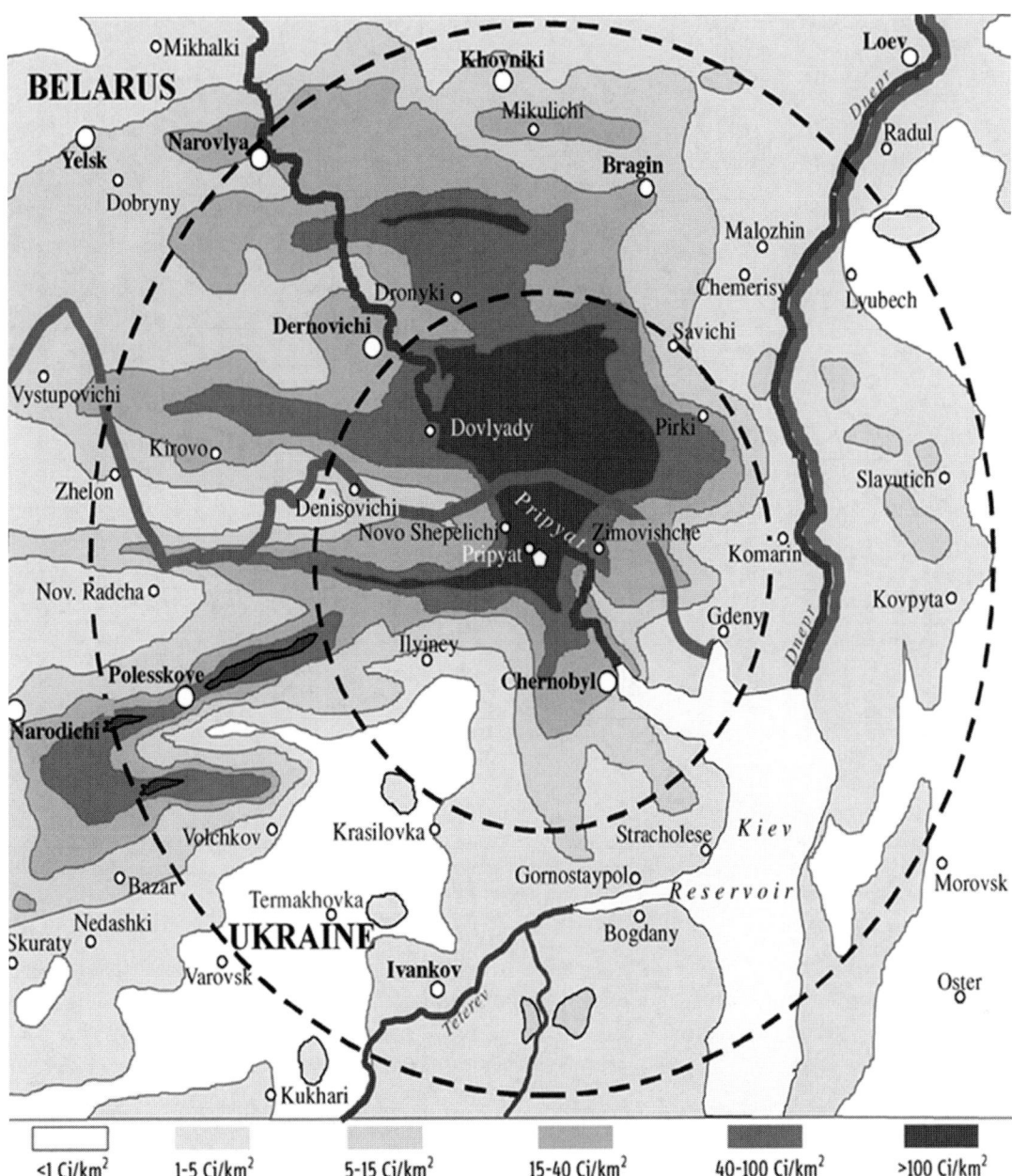

Figure 14.16 | Cesium-137 contamination levels around the Chernobyl reactor measured during the decade after the accident. Source: UNSCEAR, 2000, vol. II, Annex J, Figure VII. About 3000 km² contaminated to greater than 1.48 MBq/m² (40 Curies (Ci) cesium-137 per km²) are still officially evacuated. Areas with contamination levels between 0.56 and 1.48 MBq/m² (15 and 40 Ci/km²) were designated as areas of strict radiation control, requiring decontamination and control of intake of locally grown food. As a result, the annual accident-related effective dose in these areas has been kept below a level of about 5 mSv/yr – approximately twice the global average natural background rate (UNSCEAR, 2000, vol. 1, Table 1; Annex, para. 108). A population subject to such a dose rate indefinitely would incur an extra cancer risk of about 3.5%, about half of which would be fatal (US NAS, 2006: 4, 8).

- Radioactive contamination of an area of about 3000 km² by cesium-137, the 30-year half-life gamma emitter, to levels that resulted in the long-term evacuation of residents (Figure 14.16).[77]

- A still-growing epidemic of thyroid cancer among people in the region who received large doses from ingested and inhaled radioactive iodine (see Figure 14.17).

- Other radiogenic cancers are suspected but undetectable in a much larger background of cancers due to other causes.[78] One recent theoretical estimate, based on dose estimates and dose–risk coefficients derived from Hiroshima and Nagasaki survivors, is typical: 4000 extra cancer deaths among the 600,000 Chernobyl liquidators, 5000 among the 6 million people living in "contaminated areas" (above 37 kBq/m² of cesium-137), and about 7000 in the population

---

77   The area within a 30-km radius of the reactor was evacuated, as well as some heavily contaminated villages outside this zone.

78   Almost 30% of deaths in developed countries are from cancer (American Cancer Society, 2007).

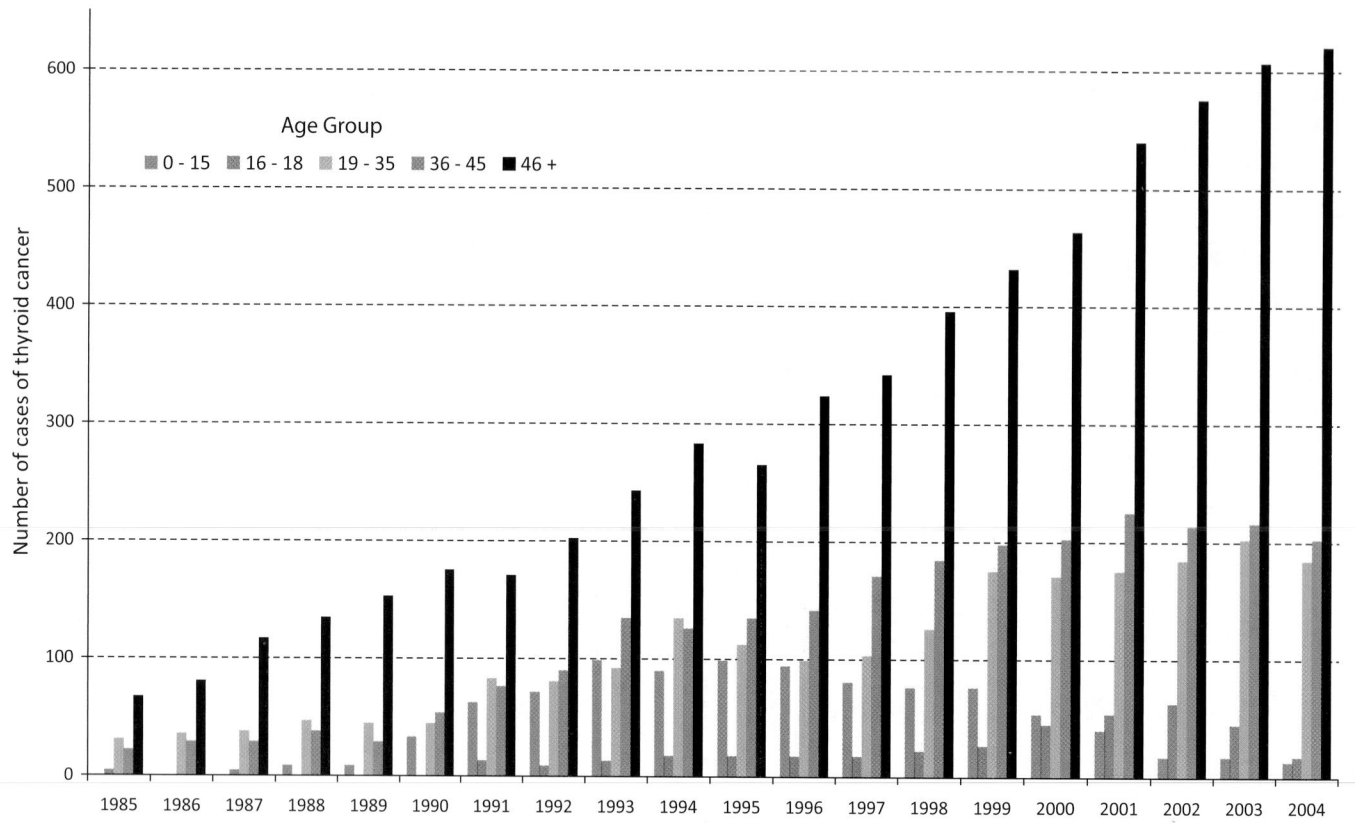

Figure 14.17 | Epidemic of thyroid cancer in Belarus following the 1986 Chernobyl accident. Annual number of cases of thyroid cancer grouped by age at the time of diagnosis. Note that the incidence of thyroid cancer in the youngest group (0–15 years) returned to low levels in 2002 when all the exposed population had graduated into older groups. In the younger groups the death rate due to thyroid cancer has been low. This does not appear to be true, however, for the increasing number of cases among those aged 46 and over. In 2004, this group would have been over 27 at the time of the Chernobyl accident. Source: Bespalchuk et al., 2007.

of 500 million of the rest of Europe who were subjected to lower doses. The total number of extra cancer deaths over the expected lifetime of the exposed population was estimated at 6000–38,000 (95% confidence level) (Cardis et al., 2006).[79]

Averaged over the approximately 10,000 GW$_e$-yrs of nuclear reactor capacity accumulated as of the end of 2008, an estimated 16,000 deaths from the Chernobyl accident amount to less than two cancer deaths per GW$_e$-yr. This is a rather modest level compared with the occupational and air-pollution deaths associated with coal-fired power plants. Perhaps the greatest harm from the Chernobyl release, however, has been the social and psychological trauma to the approximately 200,000 people who were permanently evacuated from their homes, and the millions of people now living in dread of the long-term consequences of their radiation exposure (UNSCAER, 2000, Appendix J, II.B&V.D).

Estimates of the economic cost of the Chernobyl accident range from US$6.7 billion (Sovacool, 2008b) to US$235 and US$148 billion by the

governments of Belarus and Ukraine, respectively (years unspecified). In Belarus, the costs of dealing with Chernobyl amounted to 20% of the national budget in 1992, falling to 5% in 2001. These costs were paid for in part by a special tax of 18% on all wages paid by non-agricultural firms in 1994 (UNDP, 2002, sections 5.04ff). Estimates of the potential costs due to the evacuation of the population and the loss of assets due to contamination by hypothetical spent-fuel pool fires at a range of US sites also run to hundreds of billions of dollars (Beyea et al., 2004).[80]

The Chernobyl accident occurred in a reactor type that is now being phased out. The Three Mile Island accident in 1979, where the reactor core partially melted but there was not a major release of radioactivity from the containment occurred in a pressurized water reactor and the three core meltdowns at the Fukushima Daiichi nuclear power plant, which released on the order of one tenth as much radioactivity to the atmosphere as the Chernobyl accident, were in boiling water reactors (von Hippel, 2011). In the Chernobyl and Three Mile Island accidents, the operators' lack of understanding of what was happening was a key factor. Since that time, operator training has been greatly improved with the use of simulators. A wide range of other steps have been taken to

79 An estimated figure of 4000 cancer deaths from the IAEA's 2005 Chernobyl Forum is often quoted in rebuttal to higher estimates, but the Chernobyl Forum estimate was limited only to the projected cancer deaths from doses in the most contaminated areas of Belarus, Russia, and Ukraine (IAEA, 2005c, Table 5.13).

80 The range of releases of cesium-137 considered was 130–1300 PBq (3.5–35 megacuries, MCi), 1.5–15 times the estimated release from Chernobyl.

improve safety culture, learn lessons from safety incidents, share best practices, and review safety-related aspects of the design and operation of individual plants. There were about 1500 GW$_e$-yrs of nuclear power before the Chernobyl accident, and about 8500 after than until the March 2011 Fukushima Daichi accident.[81] That accident raised concerns about the possible release of radioactivity from the spent fuel pools. The dangers of loss of water from spent fuel pools and possible remedial actions had been the subject of debate in the United States (Alvarez, 2003; US NAS, 2006b).

In 2002, at the Davis-Besse nuclear power plant (Ohio, United States), it was discovered that leaking boric acid had eaten almost through a reactor pressure vessel head before it was discovered, despite the presence of iron oxide in the air and dried boric acid deposits on the outside of the vessel. The incident was a potent reminder that nuclear safety requires constant vigilance (US GAO, 2004).

Major efforts will also be necessary to ensure that countries building nuclear power plants for the first time, or those rapidly expanding their reactor fleet, as in China and India, put effective safety measures in place. These measures include instilling a strong safety culture and granting independent regulators the power, resources, and expertise they need to do their jobs.

Given the steps that have been taken in recent decades to improve safety, the probability of a catastrophic release occurring purely by accident may be lower than the probability of such a release occurring as a result of malevolent action. Yet there is far less focused attention today to reactor security than to reactor safety. The possibility of terrorism puts an even greater premium on trying to design reactors that are more inherently safe than they were in the past.

In many cases, design for safety and design for security are complementary. Ensuring that redundant control systems cannot all be disabled by one fire or one explosive charge, for example, is important for both safety and security. Protecting against terrorism, however, also requires effective physical protection measures, designed to ensure that major nuclear facilities are adequately protected against attack by small groups on the ground or from the air.

14.7　Nuclear Weapon Proliferation

Nuclear weapon acquisition was the first priority of the United States and most other early national nuclear programs. Civilian nuclear energy programs contributed to some later nuclear weapon programs as a vehicle for acquisition of technology and building infrastructure and expertise for parallel nuclear weapon programs. Indeed, all of the

countries outside the two former Cold War blocs[82] that have acquired nuclear power have done so in the context of nuclear weapon programs. Fortunately, most of these countries abandoned the weapon dimensions of their nuclear programs.[83]

Will it be possible to extend nuclear power to tens more countries without spreading the bomb along with it? This will depend on both technological and institutional choices.

14.7.1　The Nonproliferation Regime

Today, an extensive regime to limit the spread of nuclear weapons is in place, with the Treaty on the Non-Proliferation of Nuclear Weapons (NPT) as its cornerstone. Only three countries have not joined the NPT – India, Israel and Pakistan – and only North Korea has withdrawn. While this regime has been highly successful, it is now under substantial stress.

A wide range of proposals to strengthen the nonproliferation regime and reduce the potential proliferation impact of nuclear power have been put forward. International support for these measures will require the nuclear-weapon states – especially Russia and the United States – to live up to their end of the NPT bargain and drastically reduce the numbers, roles, and readiness of their nuclear weapons (see, e.g., WMD Commission, 2006; ICNND, 2009; Perkovich et al., 2005).

Under the NPT, all non-nuclear-weapon states commit not to acquire nuclear explosives and to accept IAEA inspections of all their nuclear activities to assure that they are peaceful. The traditional safeguards agreement negotiated to fulfill this NPT requirement focuses primarily on accountancy and containment and surveillance to provide "timely detection" of the diversion of "significant quantities" of uranium and plutonium (IAEA, 1972). The IAEA adopted the recommendation of its Standing Advisory Group on Safeguards Implementation (SAGSI) that a "significant quantity" of nuclear material – the amount required to make a first nuclear weapon, taking into account likely losses in processing – should be taken as 8 kg of plutonium or uranium-233, or 25 kg of uranium-235 contained in HEU (IAEA, 2001: 23). For practical reasons, however, the IAEA set its timeliness objective for detection of the diversion of a significant quantity of material at one month – longer than recommended by SAGSI.[84]

More fundamentally, at a large reprocessing plant such as Japan's Rokkasho, which is designed to separate 8 tonnes of plutonium (1000

81　This does not mean that there had not been worrisome incidents at many nuclear power plants (Kastchiev et al., 2007).

82　The regions included in the Cold War blocs were North America/Western Europe/Japan, and the former Soviet Union/Eastern Europe, and China. Countries with nuclear power programs outside these blocs include Argentina, Brazil, India, Pakistan, South Africa, South Korea, and Taiwan.

83　Including Argentina, Brazil, South Africa, South Korea, and Taiwan.

84　SAGSI estimated the times required to convert various types of nuclear material into nuclear-weapon components as one week for plutonium or HEU metal (IAEA, 2001: 22).

significant quantities) per year, measurement uncertainties make it impossible to verify that one significant quantity has not been diverted – especially in the case of small diversions occurring over an extended period. Critics therefore argue that safeguards at large bulk-processing facilities are ineffective (Sokolski, 2008). IAEA experts respond by arguing that a wide range of containment and surveillance measures implemented throughout the plants provide substantial (though unquantifiable) additional confidence that no material has been diverted.

The IAEA inspections in Iraq after the 1991 Persian Gulf War dramatically demonstrated that the focus of IAEA safeguards at the time were too narrow. Iraq had mounted a program to produce highly enriched uranium for nuclear weapons, largely at undeclared facilities that were therefore not under safeguards. In response to this wakeup call, member states of the IAEA agreed to take a series of steps to extend the reach of safeguards. Some of these required the negotiation of an "Additional Protocol" to complement the traditional safeguards agreement.

The Additional Protocol (IAEA, 1998) requires states to provide the IAEA with more information and access to a broader range of sites, in particular relevant facilities with technology and equipment that could contribute significantly to a capacity to produce plutonium or HEU. The IAEA has been integrating this information with open-source data (including commercial satellite photographs) intelligence provided by member states, and information from its own inspection activities, into an overall picture of the nuclear activities of each state. This so-called "state-level approach" makes it possible for the Agency to raise questions and focus resources on questionable activities (Cooley, 2003).[85]

Export controls are another critical element of the nonproliferation regime. The NPT requires that states only export nuclear materials or technologies for producing them to non-weapon states if they will be under safeguards. The Zangger Committee was established under the NPT to define what specific items should be controlled to fulfill this requirement. After India's nuclear detonation in 1974, the major suppliers established a separate NSG under which each participant makes a political commitment to follow much more restrictive export guidelines.

There is an ongoing struggle, however, between those states that are attempting to slow the spread of sensitive technologies and those trying to acquire them. After the 1991 Persian Gulf War, it was discovered that Iraq had succeeded in illicitly importing a wide range of controlled items for its

nuclear weapons program from companies in many countries (Fitzpatrick, 2007). This provoked many countries to strengthen their nuclear export control systems. In 1992, the NSG supplemented its rules with restrictions on exports of "dual-use technologies" and called for states to adopt "catch-all" provisions covering any technology that an exporter suspected was going to an entity involved in proliferation activities.

Nevertheless, in 2003, it was revealed that a global black-market nuclear technology network led by Pakistan's A. Q. Khan had been marketing centrifuge technology and even nuclear-weapon designs, and had been operating in some 20 countries for more than 20 years. These revelations made it clear that far more needs to be done to control the spread of the most sensitive nuclear technologies (Fitzpatrick, 2007).

14.7.2 Controlling Enrichment and Reprocessing Technologies

The most important potential proliferation impact of the civilian nuclear energy system is through the spread of what the 1946 Acheson–Lilienthal Report called the "dangerous" nuclear technologies for uranium enrichment and the chemical "reprocessing" of spent fuel to recover plutonium (Acheson–Lilienthal, 1946). Concern about the spread of these technologies declined during the late 1950s and 1960s, when the United States and former Soviet Union promoted competitive "Atoms for Peace" programs. But India's use of US-supplied reprocessing technology to separate plutonium for its 1974 "peaceful" nuclear explosion convinced the US government to stop promoting reprocessing both at home and abroad, and to organize the Nuclear Suppliers Group as a forum in which it could be agreed that sales of reprocessing and enrichment technology would no longer be used as "sweeteners" in the international competition for sales of nuclear power plants.

Today, there is a similarly catalytic international crisis over Iran's insistence on its "inalienable right," under Article IV.1 of the NPT, to build a national uranium-enrichment plant.[86] The nominal purpose of the plant is to produce LEU for Iran's future nuclear power plants, although Iran has acknowledged that its centrifuge program began in 1985, a time when it had no plans for nuclear power plants and was locked in a war with Iraq, which was using chemical weapons and was suspected of seeking nuclear weapons. The plant could potentially be converted to the production of HEU for nuclear weapons or provide a civilian cover for a parallel clandestine enrichment program.

---

85 An inadvertent testimonial to the effectiveness of these new approaches was provided by Hassan Rohani, then Iran's nuclear negotiator and secretary of Iran's Security Council, in a speech to the Supreme Council of the Cultural Revolution in 2005. Rohani complained that, as a result of the IAEA finding a dissertation and a journal article that mentioned certain covert nuclear activities, "the IAEA was fully informed about most of the cases we thought were unknown to them" (Rohani, 2005).

86 Article IV.1 of the NPT reads: "Nothing in this Treaty shall be interpreted as affecting the inalienable right of all the Parties to the Treaty to develop, research, production and use of nuclear energy for peaceful purposes without discrimination and in conformity with articles I and II of this Treaty." The debate over Iran's program has concerned whether its intentions are peaceful. The IAEA has found that Iran repeatedly failed to comply with its safeguards obligations and the UN Security Council has legally obligated Iran to suspend all its enrichment activities and make its nuclear program fully transparent to the IAEA. However, Iran has refused to comply.

The Acheson–Lilienthal report proposed that enrichment and reprocessing be allowed only at plants owned by an international "Atomic Development Authority." Attenuated versions of this idea were discussed during the 1970s and early 1980s, including the idea of multinationally controlled reactor parks in which spent fuel could be reprocessed and the recovered plutonium recycled into on-site reactors (see, for example, Chayes and Lewis, 1977; SIPRI, 1980). In 2003, former IAEA Director-General Mohammed ElBaradei proposed another look at multinational control (ElBaradei, 2003) and subsequently initiated a high-level study of multilateral approaches to the fuel cycle (IAEA, 2005b).

In 2004, President G.W. Bush proposed that reprocessing and enrichment plants not be built outside of countries already operating full-scale plants, i.e., the nuclear-weapon states, Western Europe and Japan. A number of leading non-weapon states firmly rejected such a two-class solution,[87] however, and the Bush Administration proposed to include states with pilot-scale facilities.[88] Currently, efforts are underway to give countries such as Iran greater confidence in foreign sources of enrichment services as an alternative to building their own enrichment plants. This includes an IAEA-controlled bank of LEU as a last resort.[89] Over the longer term, ElBaradei has argued, "the ultimate goal … should be to bring the entire fuel cycle, including waste disposal, under multinational control, so that no one country has the exclusive capability to produce the material for nuclear weapons" (ElBaradei, 2008).

Enrichment. The fundamental issue with enrichment is that the same technology that can be used to produce LEU for civilian fuel can produce material for nuclear weapons. Indeed, most of the enrichment work required to produce 90% enriched HEU for weapons has already been done in enriching material to 4% for reactor fuel.[90] Gas-centrifuge cascades, now the dominant technology for producing LEU, can be used or relatively quickly reconfigured to produce weapon-grade uranium (Glaser, 2008).

With regard to the spread of enrichment technology, there are two contradictory trends:

- URENCO, an international enrichment group, is expanding its enrichment plants in Germany, the Netherlands, and the UK, and is building large new enrichment plants in France and the United States while Russia is doing the same in China.

- Small national enrichment plants are being built in Brazil and Iran and are being proposed in Argentina and South Africa (Nuclear Fuel, August 25, 2008).

Japan is an intermediate case. It has for a long time had a medium-sized enrichment plant that has not been economically competitive and whose centrifuges have mostly failed, but plans to rebuild its enrichment capacity on the same scale (JNFL, 2007).

The small national enrichment plants in Brazil and Iran have different histories. Brazil's program grew out of its navy's ambition to build nuclear-powered submarines. The primary public rationale for Iran's enrichment plant has been to provide it with fuel security for its nuclear power plants.

Iran currently has only one nuclear power plant, whose fuel is being supplied by Russia. Iran has announced, however, an ambitious program for bringing 20 GW_e of nuclear capacity online by 2025 (NEA, 2008c). Given its bad relationship with the United States, and its earlier history of being refused enrichment services by the European Gaseous Diffusion Uranium Enrichment Consortium (EURODIF),[91] Iran states that it is unwilling to depend upon other countries for enrichment services. Its limited resources of natural uranium, however, would require its proposed large nuclear program to depend upon imported uranium.[92] It would therefore have to stockpile imported natural uranium to protect itself against uranium supply disruptions. If so, why not stockpile imported LEU to protect itself from disruptions of uranium enrichment services as well (von Hippel, 2008b)?

Argentina's interest in enrichment goes back to the military nuclear program that it abandoned in tandem with Brazil in 1990. South Africa's interest in enrichment similarly goes back to its nuclear-weapon program, which it ended in 1991. Canada's largest uranium company, Cameco, was interested in adding value to its exports by acquiring

---

87 Including Argentina, Brazil, Canada, and South Africa.

88 Including Argentina and Brazil, according to a statement by Richard Stratford, Director of the US Department of State's Office of Nuclear Energy Affairs at the Carnegie Endowment International Nonproliferation Conference in June 2004.

89 American billionaire Warren Buffett offered US$50 million through the US-based Nuclear Threat Initiative toward establishing an IAEA fuel bank, and the US government, European Union, United Arab Emirates, and Norway have pledged contributions. The United States and Russia are also establishing supplementary reserves of LEU on their territories (in the US case, to be produced by blending down excess HEU), upon which countries in good standing with regard to their nonproliferation commitments will be able to draw (US NAS-RAS, 2008).

90 One way to understand this is to note that, by the time 4% enrichment has been achieved, the uranium-235 has been separated from over 80% of the uranium-238 in natural uranium. A separative work unit (SWU or, more precisely, a kg-SWU) is a measure of the amount of work done in isotope separation. To extract 1 kg of uranium-235 from natural uranium, which contains about 0.7% uranium-235, and concentrate it to 90% enriched "weapon-grade" uranium, leaving 0.3% uranium-235 in the depleted uranium, would require about 200 SWUs. About two-thirds of that separative work would be required to concentrate the same quantity of uranium-235 to 4.5% enrichment.

91 Iran still owns 10% of EURODIF via a 40% interest in Sofidif (Société franco–iranienne pour l'enrichissement de l'uranium par diffusion gazeuse), which holds 25% of EURODIF SA. The company has not paid out dividends to Iran since various restrictions were imposed on Iran following its non-compliance with the UN Security Council order of 31 July 2006.

92 Iran also does not currently have the technology to fabricate fuel for light-water reactors.

an enrichment plant. In 2008, however, after URENCO refused to sell it a gas-centrifuge enrichment plant, Cameco bought a 24% share of a laser-enrichment company whose plant is to be built in the United States (GE-Hitachi, 2008).[93]

Multilateral arrangements. As noted above, the controversy over Iran's uranium enrichment program has revived the idea of non-national – this time multinational – ownership of fuel-cycle facilities. In 2004, IAEA Director-General ElBaradei created an expert group to study the multinational option. In its report, the expert group noted that four multinationally owned enrichment plants already exist – the EURODIF plant in France and the three URENCO plants in Germany, the Netherlands, and the United Kingdom (IAEA, 2005b).

In the case of EURODIF, France built and operated a large gas-diffusion enrichment plant in which other countries (Italy, Spain, Belgium, and Iran) invested, in exchange for rights to a share of the enrichment work. Iran loaned the consortium US$1 billion for the construction of the plant and prepaid US$0.18 billion for future enrichment services. After Iran's 1979 revolution, it temporarily lost interest in nuclear power and requested its money back. After a protracted process, it did get back its US$1 billion plus interest in 1991. When it requested delivery of the enrichment services for which it also had paid, however, France's position was that the contract had expired. Iran views this refusal as proof of the unreliability of outside nuclear supplies and uses the EURODIF episode to argue that it requires its own enrichment plant (Meier, 2006).

Recently, Russia, in an arrangement very similar to EURODIF, created an International Uranium Enrichment Center (IUEC) at Angarsk as a commercial open joint stock company. The IUEC will buy enrichment services from the Angarsk enrichment plant and, perhaps in the future, a share in the plant itself. Holders of IUEC stock will have a guaranteed supply of enriched uranium and/or a share in the profits. Russia will continue to manage the enrichment plant and have sole access to its technology.

Thus far, Kazakhstan has committed to buy 10% of the IUEC (*Nuclear Fuel*, September 24, 2007).[94] Ukraine (*Nuclear Fuel*, November 30, 2009) and Armenia (*Nuclear Fuel*, April 20, 2009) also are expected to become partners. Russia offered Iran a share as an alternative to Iran building its own enrichment plant, but the offer was declined.

This arrangement has been at least partially successful as a nonproliferation initiative, however, in that it has apparently convinced the partner countries that they do not need to have their own national enrichment plants. But it appears that the operation of the plant will be no more transparent to the investors than to non-owner customers. In a non-weapon state such as Iran, therefore, this form of multinational ownership would not provide an additional level of nonproliferation assurance beyond that provided by IAEA inspections.

URENCO provides another model for multinational arrangements. Each of the original partner countries (Germany, the Netherlands, and the United Kingdom) had its own technology R&D team and enrichment plant. Obviously, the joint management and sharing of technology within URENCO provides greater transparency among the partners. In the past, however, the consortium has not maintained effective control of the technology. URENCO subcontractors were the source of the technology that A. Q. Khan used to build Pakistan's enrichment complex and to export centrifuge enrichment technology to Iran, Libya, North Korea, and perhaps other countries. Iraq similarly acquired centrifuge technology through German companies that were supplying URENCO with centrifuge components (Kehoe, 2002).

More recently, URENCO has expanded its business through a joint subsidiary, Enrichment Technology Company (ETC), to provide centrifuges and design services for enrichment plants in France and the United States. France's nuclear services provider, AREVA, has purchased a 50% share of ETC, but without access to the technology. The centrifuges are being built in ETC facilities in Germany and the Netherlands, and are assembled into cascades by ETC employees in France and the United States (ETC, 2008). Russia has similarly built enrichment plants in China (*Nuclear Fuel*, December 19, 2005). The centrifuges are described as "black boxes" as far as the host country is concerned, though regulators in the countries where these plants operate inevitably have to understand some aspects of the technology to be able to confirm its safety. Since France, the United States and China are all weapon states, however, URENCO and Russia have not yet faced the full challenge of protecting their technologies in a non-weapon state.

Canada's uranium company Cameco has been refused a black-box enrichment plant by URENCO and it appears that the United States will not allow export of a laser-enrichment plant to Canada because of doubts about the feasibility of operating this technology in black-box mode (*Nuclear Fuel*, August 25, 2008). URENCO has rejected a proposal to resolve the international crisis over Iran's enrichment program by putting it under multinational control and replacing Iran's centrifuges with black-boxed URENCO centrifuges (*Nuclear Fuel*, July 30, 2007). Such an arrangement would not likely be of interest to Iran either.

Former IAEA Director-General ElBaradei has proposed that all future enrichment and reprocessing facilities should be under some form of multinational or international control. The nonproliferation advantages and disadvantages of such approaches have been discussed (US NAS-RAS, 2008; Thomson and Forden, 2006). If a plant were owned by several countries, or by an international institution, with the plant location

---

93 The three largest suppliers of uranium are Kazakhstan, Canada, and Australia. Kazakhstan has become a partner in Russia's Angarsk enrichment facility. Australia has supported enrichment R&D but is not currently actively pursuing the idea of building its own enrichment plant.

94 In a separate arrangement, Kazakhstan and Russia have agreed to make equal investments in a new 5-million SWU enrichment plant adjoining the existing Angarsk facility (*Nuclear Fuel*, 28 July 2008b).

designated as extra-territorial – as are embassies and the laboratory of the European Organization for Nuclear Research (CERN) in Switzerland – this would pose a somewhat higher political barrier to the host state seizing the plant to use it for weapons purposes, as this would require expropriating the property of other states or a multinational organization. If the full-time operating staff of such a plant included multinational personnel, this would provide greater transparency into plant operations than IAEA inspections do, and relationships among the foreign and host-state personnel might provide greater insight into whether some of the host state experts were disappearing to work on a covert facility. On the other hand, any multinational approach would have to pay extremely careful attention to technology protection. Access to sensitive technologies should be limited to staff from countries that already possess such technology, with appropriate clearance and screening.

Reprocessing plants. With regard to reprocessing, the simplest alternative would be to forgo the practice. As practiced today, reprocessing and plutonium recycling are not economic and do not significantly simplify spent-fuel disposal (von Hippel, 2007, 2008a; Schneider and Marignac, 2008; Forwood, 2008). Reprocessing costs about ten times as much as interim storage of spent fuel in dry casks, and recycling plutonium in LWRs once, as is the current practice, does not significantly reduce its long-term radiological hazard. Most countries are abandoning reprocessing (see Table 14.5).

An exception is Japan, where it is politically unacceptable to allow spent fuel to accumulate at nuclear power plants and prefectures have been reluctant to host centralized spent-fuel storage facilities.[95] A reprocessing plant, with large tax payments to the local town and prefecture, turned out to be more attractive and is being used to provide a centralized interim destination for Japan's spent fuel and also for high-level waste being returned from the reprocessing of Japanese spent fuel in France and the United Kingdom (Katsuta and Suzuki, 2006). Japan's nuclear establishment also argues that eventually, if fast-neutron plutonium breeder reactors are introduced, plutonium recycling could make Japan independent of uranium imports.

Japan's reprocessing plant, when operating at its design capacity of 800 tonne/yr of spent fuel, will separate about 8000 kg/yr of plutonium. The first-generation Nagasaki bomb contained 6 kg of weapon-grade plutonium metal (almost pure plutonium-239), which would be roughly equivalent, in terms of critical masses, to 8 kg of power reactor-grade plutonium (Kang and von Hippel, 2005, Table 1).

A shift to more "proliferation-resistant" reprocessing technologies was proposed by the G.W. Bush Administration in 2003 (US DOE, 2003). An evaluation of the added proliferation resistance of the proposed

---

95 At the end of 2008, however, Chubu Electric Power Company proposed to build a dry-cask storage facility with a capacity of 700 tonnes of spent fuel, in connection with a proposal to build a new 1.4 GWe reactor to replace two old reactors with a comparable generating capacity (CNIC, 2009).

technologies found, however, that it was not significant (see, e.g., Collins, 2005; Hill, 2005; Kang and von Hippel, 2005). Ultimately, the Administration proposed to deploy a reprocessing plant very little different from those in France and Japan. It insisted that pure plutonium would not be separated, i.e., that it would be mixed with uranium. Since it is not difficult to separate plutonium from uranium, however, this would be of only modest significance.

14.7.3   Risk of Nuclear-explosive Terrorism

In addition to the problem of proliferation of nuclear weapons to more nations, there is also the risk that terrorists could acquire and detonate a nuclear explosive (Bunn, 2010). Repeated studies by the US and other governments have concluded that, if a well-organized and well financed terrorist group acquired plutonium or HEU, it might well be able to make at least a crude nuclear explosive. Attempts by groups such as al-Qaeda and the Japanese cult Aum Shinrikyo to acquire nuclear weapons or the materials needed to make them, and to recruit nuclear experts, have demonstrated that the danger is more than theoretical. A number of cases of theft and smuggling of at least small quantities of plutonium and HEU have already occurred (Zaitseva, 2007).

Neither HEU nor separated plutonium are present when current-generation nuclear power plants operate on a once-through fuel cycle. The fresh fuel is made from LEU, which cannot support an explosive nuclear chain reaction without further enrichment – a challenge that is beyond plausible terrorist capabilities in the near term – and it would be very difficult for terrorists to steal the intensely radioactive spent-fuel assemblies and separate out plutonium for use in a nuclear weapon. For decades, however, there have been concerns that fuel cycles involving plutonium separation and recycling might significantly increase the risk of nuclear theft and terrorism (Willrich and Taylor, 1974; Mark et al., 1987).

Weapon-usability of power-reactor plutonium. The Acheson–Lilienthal report contained a misunderstanding concerning the weapon-usability of power-reactor plutonium. It stated that both "U-235 and plutonium can be de-natured" for weapon use (Acheson–Lilienthal, 1946: 30). That is correct for uranium-235. When diluted with uranium-238 to less than 6% concentration, uranium-235 cannot sustain an explosive chain reaction. Indeed, when the percentage is less than 20%, the fast critical mass is considered too large for fabrication of a practical nuclear weapon (IAEA, 2001, Table II). This is the basis for the belief that LEU, defined as containing less than 20% uranium-235, is not directly weapon-usable.

The authors of the Acheson–Lilienthal report apparently believed that the isotope plutonium-240 could be used to denature plutonium for weapons. Plutonium-240 fissions spontaneously and therefore generates neutrons continually at a low rate. In the Nagasaki weapon implosion design, these neutrons could start the fission chain reaction before the optimal time for maximum yield. In the Manhattan Project, great efforts therefore were made to keep the percentage of plutonium-240 below a few percent. LWR plutonium contains about 25% plutonium-240 (NEA, 1989, Table 9). In the Nagasaki design, this could have reduced the yield from 20,000 tonnes of chemical explosive equivalent to as low as 1000 tonnes (Oppenheimer, 1945; Mark, 1993). Such an explosion would still be devastating, however. The radius of total destruction, which was 1.6 km at Hiroshima, would still be 0.7 km for a one-kilotonne (1 kt) explosion.[96] For more advanced designs, such as those in the arsenals of the NPT weapon states, there might be no significant reduction in yield (US DOE, 1997: 38–39).

Today, therefore, any mix of plutonium isotopes containing less than 80% plutonium-238 is considered weapon usable (IAEA, 2001, Table II).[97] Since the amount of plutonium-238 in the world is only 1–3% as large as the total amount of plutonium (NEA, 1989: Table 9), it would be impractical to attempt to use it to denature a significant fraction of the world's plutonium.

14.8   Institutional Requirements[98]

Because of the safety, security, and proliferation risks it poses, the use of nuclear energy requires worldwide vigilance. Each nation operating nuclear facilities is responsible for their safety and security. But all states have an interest in making sure that other states fulfill these responsibilities, creating a need for international institutions. For nuclear power to grow enough to make a significant contribution to mitigating global climate change, stronger institutions at both the national and international levels will be required.

In the decades since the Chernobyl accident, many countries have strengthened their safety practices and regulations substantially, but there is clearly more to be done. The Fukushima accident has provoked a global discussion concerning what national and international institutions and approaches need to be changed. Even in the United States, which has some of the world's most stringent nuclear safety regulations and more reactor-years of operating experience than any other country, both internal and external critics continue to argue that the Nuclear Regulatory Commission too often subordinates enforcement to the industry's cost concerns (US NRC, 2002; UCS, 2007). Countries building nuclear power plants for the first time will need to build up adequate groups of trained personnel, put in place effective nuclear regulatory structures, and forge nuclear safety cultures (IAEA, 2007b; Acton and Bowen, 2008). Countries such as India and China, which are rapidly expanding their civilian nuclear infrastructures, will have to take

care that the expansion does not outpace the growth of capabilities to provide expert personnel to build, operate and regulate these facilities.

The development of regulatory requirements for securing nuclear facilities against sabotage and the theft of fissile and radioactive material is at a much earlier stage. Some countries still have no regulations specifying what insider and outsider threats should be defended against, and some do not require armed guards even to protect weapon-usable nuclear material from theft. Substantial steps are needed worldwide to reduce vulnerability (Bunn, 2010).

National institutions also play a critical role in nonproliferation. Foreign ministries, export controls and intelligence agencies all have key roles. And IAEA safeguards cannot function without each state having an effective state system of accounting and control.

14.8.1   International Institutions

International institutions promoting safety, security, and nonproliferation include not only the IAEA but also industry organizations such as the World Association of Nuclear Operators (WANO), the Western European Nuclear Regulators Association and professional associations such as the Institute for Nuclear Materials Management.

Safety. The IAEA's International Nuclear Safety Group (INSAG) has produced a diagram (Figure 14.18) showing the international organizations, networks, and activities to promote nuclear power plant safety that have grown up in the two decades since the Chernobyl accident.[99] Ultimately,

Figure 14.18 | The global nuclear safety regime. Source: IAEA-INSAG, 2006, Fig. 1.

---

96   The radius of blast destruction is proportional to the one-third power of the yield (Glasstone and Dolan, 1977, equation 3.61.1).

97   Plutonium-238 is relatively short-lived (88-year half-life) and generates a great deal of decay heat (0.56 kW$_t$/kg). It therefore would be difficult to fabricate into a nuclear explosive. It has been exempted from safeguards because it is used as a heat source for applications such as space probes to the outer planets, but the exemption has been drawn as narrowly as possible.

98   Professor Matthew Bunn, Kennedy School of Government, Harvard University, USA, Lead Author.

however, decisions on nuclear safety measures are still left to each state. The Convention on Nuclear Safety, for example, does not set specific safety standards and reporting on safety problems is entirely voluntary.[100]

The IAEA plays a critical role by publishing standards, guides and recommendations, and organizing discussions of critical issues and best practices. It manages an incident reporting system with the OECD Nuclear Energy Agency that collects and assesses information on operating experience and safety-related incidents. It also organizes in-depth, three-week safety reviews of facilities by an international team of safety experts. In those cases where a follow-up mission has been performed, the IAEA has found that sites either have implemented or are implementing some 95% of the teams' recommendations (IAEA, 2007d).

IAEA safety peer reviews occur, however, only when a member state asks to be reviewed, and only a minority of the world's power reactors have ever undergone such a review.[101] In 2008, a "Commission of Eminent Persons" appointed by Director-General ElBaradei recommended that states "enter into binding agreements to adhere to effective global safety standards and to be subject to international nuclear safety peer reviews" (IAEA-CEP, 2008).

WANO, an industry group established after the Chernobyl accident, is another key international nuclear safety institution.[102] WANO is divided into four regional groups with headquarters in Atlanta, Moscow, Tokyo, and Paris. A reactor's affiliations with one or more of these headquarters is determined by a combination of its location and reactor type. All operators of nuclear power reactors worldwide are participants in WANO and accept international peer reviews as a condition of membership. WANO also manages a system for reporting incidents and operating experiences, and helps organize exchanges of best practices. The reactor vendors also play a key role helping countries to put effective regulations and operating practices into place. Also both the G8 countries and the European Union have pursued extensive nuclear safety assistance programs, especially in former Soviet-bloc countries. Finally, there are also several international groupings of nuclear regulators (IAEA, 2007d).

Despite all of these efforts, INSAG has reported important weaknesses in international information sharing and actions. For example, the fact that WANO maintains confidentiality can delay national regulators

becoming aware of the incidents being reported (IAEA-INSAG 2006) and some types of incidents continue to recur. In 2006, Luc Mampaey, then WANO managing director, complained that some utilities were not reporting at all (*Nucleonics Week*, September 27, 2007).

Security. Most countries shroud their security practices in secrecy, and international institutions for promoting nuclear security are therefore substantially weaker than those for nuclear safety. The Conventions on the Physical Protection of Nuclear Material and Facilities, and on the Suppression of Acts of Nuclear Terrorism do not set specific standards for how secure nuclear materials or facilities should be, and include no mechanisms for verifying that states are complying with their commitments. The IAEA has published physical protection recommendations, but they too are vague. They call for having a fence with intrusion detectors around significant stocks of plutonium or HEU, for example, but say nothing about standards of effectiveness.

UN Security Council Resolution 1540 legally obligates all UN member states to provide "appropriate effective" security and accounting for any nuclear weapons or related materials they may have. A common interpretation of what key elements are required for a nuclear security and accounting system to be considered "appropriate" and "effective" therefore could provide the basis for a legally binding global nuclear security standard (Bunn, 2008).

Since the mid-1990s, bilateral and multilateral assistance programs have played a critical role in improving nuclear security. The United States in particular has invested billions of dollars in programs designed to help former Soviet-bloc countries install and operate improved security and accounting systems at sites with significant quantities of plutonium and HEU. It has also mounted a global program outside Russia to convert research reactors to use LEU rather than HEU (Bunn, 2010). Less attention has been devoted, however, to protecting nuclear power plants, fuel cycle facilities and nuclear shipments against terrorist actions. In 2008, the World Institute of Nuclear Security (WINS) was established, modeled in part on WANO. It is designed to provide a confidential forum for nuclear security managers around the world to exchange best practices and discuss issues they have confronted in the hope of improving nuclear security practices worldwide (Howsley, 2008).[103]

In April 2010, leaders from 47 countries gathered in Washington D.C. for the first nuclear security summit and endorsed the objective of securing all vulnerable nuclear material worldwide within four years. They agreed on a broad communique and modestly more specific work plan, and some countries made important national commitments, such as Ukraine's commitment to eliminate all the HEU on its soil by 2012. A second nuclear security summit was scheduled in Seoul for March 2012. This summit process has elevated nuclear security to the top levels of government.

---

99 For a useful overview of international activities related to nuclear safety, see IAEA (2007d).

100 For a critique and a suggestion of a more robust approach, see (Barkenbus and Forsberg, 1995).

101 The first-ever IAEA Operation Safety Review Team review in Russia was held in 2005 at the Volgodonsk plant. The plant prepared for months and received a very positive review (M. Lipar, Head, Operational Safety Section, IAEA, personal communication, April 2008).

102 The US Institute of Nuclear Power Operations was formed to play a similar function at the national level after the 1979 Three Mile Island accident near Harrisburg, Pennsylvania.

---

103 WINS' first director is Roger Howsley, previously head of security, safeguards and international affairs at British Nuclear Fuel Services Limited: www.wins.org.

Nonproliferation. The 1968 NPT is the foundation for all international efforts to stem the spread of nuclear weapons and has been highly successful. The nonproliferation regime is now under stress, however. Iran's refusal to comply with the UN Security Council's demand that it suspend its enrichment program, combined with North Korea having become the first state ever to withdraw from the NPT and manufacture nuclear weapons, have raised concerns about the ability of the international community to enforce compliance. In addition, the treaty's legitimacy has been undercut by the perception that the NPT nuclear-weapon states have not lived up to their obligation under Article VI of the NPT "to pursue negotiations in good faith on … nuclear disarmament."[104] Many non-weapon states also see efforts by the United States and some other states to prevent the spread of national enrichment and reprocessing plants as undermining the treaty's Article IV guarantee of the "inalienable right of all the Parties to the Treaty to develop, research, production and use of nuclear energy for peaceful purposes and without discrimination…"

Of the institutions established to implement the nonproliferation regime, the IAEA is the most important. IAEA safeguards play a critical role in verifying the peaceful use of nuclear energy around the world. The IAEA faces important constraints in access to sites, information, resources and technology, however, as well as challenges in balancing its efforts to maintain essential positive relationships with states with an appropriate investigatory attitude.

The Additional Protocol to the NPT is a major advance with regard to access to sites and information, but many issues remain. First, more than a decade after its adoption, there are a number of non-weapon states with significant nuclear activities or ambitions that have not acceded to the Additional Protocol.[105] Also, despite its expansion beyond the traditional focus on nuclear materials, the Additional Protocol focuses primarily on the IAEA's rights to inspect sites with technologies related to the production of nuclear materials. As a result, when the IAEA wanted to investigate a site in Iran where implosion experiments related to nuclear-weapon design allegedly had taken place, there were no undisputed legal grounds for doing so.[106] Pierre Goldschmidt, former IAEA Deputy Director-General for safeguards, has suggested that the UN Security Council pass a resolution that would require any state found to be in violation of its safeguards agreements to provide access beyond

that required by the Additional Protocol and to allow IAEA inspectors to interview, in private, key scientists and other participants in nuclear programs (Goldschmidt, 2008). The UN Security Council has, in fact, demanded that Iran provide such a level of transparency.[107]

With respect to resources, the IAEA's regular budget for implementing nuclear safeguards worldwide in 2007 is only US$100 million, or about 0.004¢/ kWh generated by the world's nuclear power plants (IAEA, 2010d). In the context of renewed hiring in the nuclear industry, the Agency also has increasing difficulty recruiting and even retaining nuclear experts. This is especially serious, given that, in 2008, roughly half of all senior IAEA inspectors and managers were within five years of the agency's mandatory retirement age (IAEA-CEP, 2008). The IAEA also does not have the resources to do its own R&D to develop new safeguards technologies. It depends on support programs from member states.[108]

The IAEA also plays a major promotional role by helping states acquire and apply nuclear technology for research, medical and agricultural purposes. Overall, by informal agreement among the member states, the IAEA budget for promoting and assisting with nuclear energy and other applications of nuclear technology is kept at about the same size as the budget for safeguards.[109]

Despite a call from former IAEA Director-General ElBaradei for negotiation of a universal nuclear export control regime, no progress has been made in that direction. The Nuclear Suppliers Group has tried to fill this space but faces ongoing challenges to its legitimacy because it is a self-selected group. Also, the decision to exempt India from the NSG requirement of membership in the NPT has strengthened the impression that economically powerful countries do not have to comply with the rules. The NSG has traditionally operated by consensus but, as more and more states have joined, consensus on strengthening its rules has become more and more difficult to achieve. Most NSG participants, for example, strongly support making the Additional Protocol a condition for nuclear exports from NSG states, but Brazil (which has not accepted the Protocol) has resisted.

---

104 The entire article, whose interpretation has been clarified by a legal opinion (International Court of Justice, 1996) and subsequent commitments by the weapon states at the 1995 and 2000 NPT Review Conferences (UN, NPT, 2000), reads as follows: "Each of the Parties to the Treaty undertakes to pursue negotiations in good faith on effective measures relating to cessation of the nuclear arms race at an early date and to nuclear disarmament, and on a Treaty on general and complete disarmament under strict and effective control."

105 Including Algeria, Argentina, Brazil, Egypt, Iran (signed but not ratified), Iraq (signed but not ratified), Mexico (signed but not ratified), Syria, United Arab Emirates (signed but not ratified), Venezuela and Vietnam (signed but not ratified) (IAEA, 2010c).

106 The IAEA asked Iran to voluntarily accept a visit to that site, which Iran eventually did.

107 In its Resolution 1803 of 3 March 2008, the UN Security Council ordered Iran to "take the steps required by the IAEA Board of Governors in its resolution GOV/2006/14, which are essential to build confidence in the exclusively peaceful purpose of its nuclear programme and to resolve outstanding questions." IAEA Board of Governors resolution GOV/2006/14 calls on Iran to "implement transparency measures, as requested by the Director General, including in GOV/2005/67, which extend beyond the formal requirements of the Safeguards Agreement and Additional Protocol, and include such access to individuals, documentation relating to procurement, dual use equipment, certain military-owned workshops and research and development as the Agency may request in support of its ongoing investigations."

108 In recent years, however, the IAEA has established an expanded effort to identify new technological approaches to address some of its key safeguards needs, and to work with member states to develop and deploy them (Khlebnikov et al., 2007).

109 In 2007, the total IAEA budget was US$268 million. Excluding central management and information services, it was US$199 million, of which US$103 million went to safeguards, US$71 million to nuclear energy and development and US$22 million to nuclear safety and security. An additional US$37 million of extra-budgetary funds were contributed by interested countries – mostly for the safeguards, safety and security programs (IAEA, 2008c).

14.9   Public Acceptance[110]

Historically, fission power has inspired more public opposition than any other energy source, except possibly hydropower in India and a few other countries. According to a survey of public opinion in 18 countries done for the IAEA in 2005, on average 62% of the respondents did not want existing nuclear power plants to be shut down, although almost the same percentage opposed building new ones (Figure 14.19). In Western Europe and the United States, a majority of the population has consistently opposed the construction of new nuclear reactors since the early 1980s (Rosa and Dunlap, 1994; Bolsen and Cook, 2008; EC, 2007). As memories of the accidents at Three Mile Island (1979) and Chernobyl (1986) faded and concerns about the consequences of global warming increased, however, the trend prior to the 11 March 2011 accident at Japan's Fukushima Daiichi nuclear power plant was toward more pro-nuclear public opinion, and government policies were moving even more rapidly in that direction. Public opposition also diminishes when new nuclear power plants are built at existing sites, as is often the case. In China, where many new sites are being established, public opposition was relatively limited, at least prior to the Fukushima Daiichi accident (WISE, 2007).

Individuals oppose nuclear power for different reasons. Some feel that the technology is too expensive, while others are concerned about the fact that nuclear energy technologies can be used to produce nuclear weapons. The main source of opposition, however, is the public concern

about the production of radioactive waste and the potential for high-impact accidents.

This public perception of risk has been something of a puzzle to many technical experts, since they do not view the risk to the public from nuclear power plants as especially high.[111] Technical experts often assess risk probabilistically through injuries and deaths per GW_e-yr of nuclear energy generated. On this scale, nuclear power does not seem particularly dangerous.

Public perceptions are the result of various psychological, social, and cultural processes, however, that can heighten or attenuate risk signals (Kasperson et al., 1988). Typically, attenuation occurs with everyday hazards such as indoor radon, smoking, and driving. In the case of nuclear power, in contrast, the risks are often amplified. Indeed, some scholars studying public perceptions have argued that nuclear energy is "subject to severe stigmatization" (Gregory et al., 1995).

Faced with public antipathy, the nuclear industry and some governments have tried to persuade the public to see risk the way experts see it, such as through campaigns pointing out that the annual risk from living near a nuclear power plant is equivalent to the risk of riding an extra three miles in an automobile (Slovic, 1996). But such comparisons do not address the aspects of the risk that people believe to be important, and often produce more anger than enlightenment.

The assumption that public opposition results from ignorance may not be correct (US OTA, 1984). One analysis of the debate over the risks from the Diablo Canyon nuclear plant in California found that,

> "proponents and opponents were equally knowledgeable about nuclear power factual information, but those who supported nuclear energy expressed more trust in the credibility of information received from government and industry officials and were more trusting that the officials would protect the public" (Levi and Holder, 1988).

Indeed, many studies reveal a widespread belief that the institutions that manage nuclear power are untrustworthy as sources of information (Wynne, 1992). A 2001 survey by the European Commission found that only 12% of Europeans trusted the nuclear industry (EC, 2008). Both trust and distrust tend to reinforce and perpetuate themselves (Slovic, 1993). Today, however, concerns about nuclear power are confronted by another major concern: the consequences of global climate change. The nuclear industry, some independent scientists, and some governments are increasingly reframing the debate as one about whether public fears

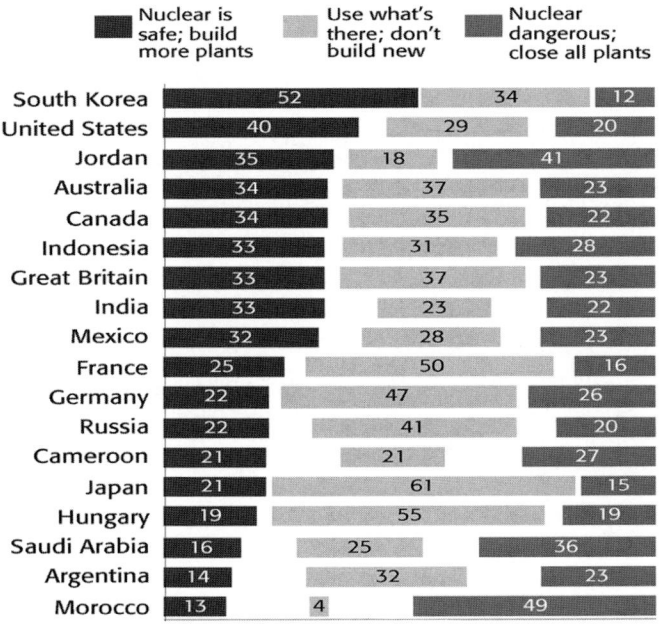

Figure 14.19 | Attitudes to nuclear power by country. The white spaces represent "don't know," "none of the above," "other" or "no answer." Source: Globscan, 2005.

---

110   Dr. M. V. Ramana, Princeton University, USA, lead author.

111   Enrico Fermi, who designed the first chain reacting "pile," apparently anticipated the public's concerns, however, well before the US nuclear power program was launched. He is reported to have commented (as paraphrased by Alvin Weinberg) that it "is not certain that the public will accept an energy source that produces vast amounts of radioactivity as well as fissile material that might be diverted by terrorists" (Weinberg, 1994).

about nuclear power have to be subordinated to the need to limit climate change. In a 2005 survey, this argument resulted in a 10% increase in public support for building new nuclear power plants (Globescan, 2005).

In the United Kingdom, the debate has been particularly intense because of the national commitment to reduce carbon dioxide emissions, the declining supplies of North Sea gas, and the retirement of the country's first-generation Magnox reactors. The government, the nuclear industry, major scientific leaders and professional societies have all been promoting a "new build" of nuclear capacity. One study used a survey and focus groups, to evaluate the impact on public attitudes of a reframing of the issue of nuclear power around the need to reduce carbon emissions, and found "reluctant acceptance" (Bickerstaff et al., 2008). The study found, however, that radioactive waste was regarded with even greater dread than climate change, and there was great mistrust of the competence of the nuclear power establishment and the government to manage nuclear power safely. The respondents were also concerned about the possibility of terrorist attacks on nuclear facilities. If it were feasible to make a more rapid shift to renewable sources of energy, that would attract greater support. In the United States, a national poll in 2008 found that 42% of the population supported an increased commitment to nuclear power, compared with 93% for solar, 90% for wind, 52% for natural gas, 33% for coal, and 22% for oil (Greenberg, 2009).

14.10   Policy Recommendations

Throughout its history, the debate on nuclear power has focused on two questions: is it necessary, and to what extent can the dangers that it poses be reduced?

The first question can only be answered in the larger context of an examination of the alternatives, the rates at which they can be deployed, and their costs, both economic and external. This is the subject of several chapters in this volume, especially Chapter 17 (Global and Regional Scenarios). If public attitudes are to be respected, however, nuclear power should be introduced or expanded in a country only after a comparative assessment, with public review and participation, of alternative means of matching energy supply and demand.

With regard to the second question, the greatest dangers that need to be minimized are Chernobyl-scale releases of radioactivity into the environment and the possibility that nuclear power facilitates the proliferation of nuclear weapons and nuclear terrorism. There are a number of initiatives that could help to reduce but not eliminate both of these risks.

14.10.1   Reducing the Risk of Catastrophic Releases of Radioactivity

The light-water reactors (LWRs) that dominate nuclear power today, and will continue to do so for the foreseeable future, were originally developed for naval propulsion. The primary design consideration was therefore that they be compact. When they were adapted for use as power reactors, that constraint was loosened and redundant emergency cooling systems and a containment building were added. These additions helped, but as the recent Fukushima Daiichi accident showed, a core meltdown accident with a large release of radioactivity to the human environment is still possible. Indeed, the successful containment of the 1979 Three Mile Island melt-down was to some extent a matter of luck. Some US containment buildings would not have been able to withstand the pressure increase from the hydrogen burn that occurred during the Three Mile Island accident, while others would be over-pressured by the carbon dioxide that would be released if a molten core began to eat its way through the concrete floor of the containment (Beyea and von Hippel, 1982).

The owners of LWRs have made significant improvements in operator training since the Three Mile Island accident and the "Generation III" LWRs that are being introduced today have significant improvements in safety design. The pressure to increase safety has not been so effective in the regulatory area, however. In 2002, after the US NRC acceded to the operator's insistent demands and allowed the Davis-Besse nuclear reactor to continue to operate in what was later established to be an extremely dangerous condition, the US NRC Inspector General commented that "NRC appears to have informally established an unreasonably high burden of requiring absolute proof of a safety problem, versus lack of reasonable assurance of maintaining public health and safety…" (US NRC, 2002).

And what of the alternatives to the conventional LWR that are currently being examined in the Generation IV (Gen IV) reactor R&D effort? Although safety is a desideratum, relative safety does not appear to be a criterion for selecting among the different types of reactors under consideration. Rather, safety studies are being pursued with the objective of making each existing design type as safe and licensable as practicable (Gen IV, 2008: 54). In any case, Gen IV industry representatives have repeatedly stated that, whatever the design type, commercial operation is still decades away.

An effort should therefore be mounted, with a higher priority than the Gen IV efforts, to design a reactor for safety, including associated spent-fuel storage, and then to see how the design could be optimized economically, rather than the other way around.

Given license extensions to 60 years and perhaps beyond,[112] many of the existing plants are likely to be operating for a very long time. In the aftermath of the Fukushima Daiichi accident, national regulations and international standards should be tightened, to ensure that nuclear facilities are prepared for the full spectrum of foreseeable earthquakes, floods, blackouts, and other disasters – and for terrorist attacks of the

---

112 In its 2009 projections for US nuclear capacity, the US Energy Information Administration assumed that the licenses of US nuclear power plants will be extended to 80 years (US EIA, 2009).

scale and sophistication of the September 2001 aircraft hijackings. Reactor operators should be required to protect against spent fuel burning if a spent fuel pool loses water and fuel that is sufficiently cool should be moved into safer dry casks. They also should be required to retrofit reactor containments with robust filtered vents that could greatly reduce the amount of radioactivity released in a severe accident (Beyea and von Hippel, 1982; Schlueter and Schmitz, 1990). Operators and nearby agencies such as police and fire departments must have effective emergency plans in place and regularly exercise them, so they all know what to do in the event of a crisis. The role and capabilities of the IAEA and WANO also should be significantly strengthened, including expanded and more transparent peer reviews.

14.10.2   Increasing Proliferation Resistance

There are three obvious steps by which the proliferation resistance of civilian nuclear energy could be increased:

1.   phase out reprocessing as quickly as possible,

2.   place enrichment plants under multinational ownership and management, and restrict them to politically stable regions, and

3.   establish regional spent-fuel storage and repositories.

1. Phase out reprocessing as quickly as possible. In the fuel cycle of an LWR fueled with LEU, weapon-usable material becomes directly accessible only as a result of spent-fuel reprocessing, i.e., the separation of plutonium (possibly mixed with other transuranic elements) from the intensely gamma-emitting fission products that, 10 or more years after discharge, are dominated by cesium-137, with a half-life of 30 years. As discussed above, there is general agreement that, for the foreseeable future, fuel cycles involving reprocessing and plutonium recycling will not be economically competitive with "once-through" fuel cycles in which the spent fuel is stored. Proliferation resistance is therefore aligned with economics in this case, and there would only be economic benefits from phasing out reprocessing. In some countries, such as Japan and South Korea, political obstacles would have to be overcome to the extended interim storage of spent fuel on nuclear power plant sites or at a central site.

2. Place enrichment plants under multinational ownership and management and restrict them to politically stable regions. Multinational ownership and management, if well designed, could make it more difficult for a host government to convert an enrichment plant to the production of HEU for weapons or to divert expertise and components to the construction of a clandestine national enrichment plant. Multinational ownership does not necessarily mean ownership by multiple governments. It could include ownership by companies that are owned by or answerable to multiple governments. An arrangement intermediate between that of URENCO, which involves technology sharing, and that of EURODIF

and Angarsk, in which non-host countries are passive investors, might be optimal. This could include multinational teams of operators in the enrichment plant control room. Indeed, the black-box model that has been adopted by URENCO in France and the United States, and by Russia in China, might be near the correct balance. In this arrangement, management of the plant can be shared, making operations, but not the technology, more transparent among the partners.

In order to make this approach politically feasible, it will probably be necessary to convert existing national facilities in the weapon states into multinational facilities. This could be done without significantly disrupting existing commercial arrangements, for example, if companies from other countries bought shares of ownership in existing facilities, and ultimately began participating in the staffing of those facilities. It will also be desirable to agree that new facilities not be built until a minimal level of contracted demand exists to make it economically viable (at least the equivalent of 10 GW$_e$ of LWR capacity, corresponding to an enrichment capacity greater than 1 million SWU/yr).

It would be desirable that such facilities be built in politically stable regions. If, as in the case of Iran, neighboring countries feel that their security would be threatened by a proposed enrichment plant, that perception should be regarded as a major argument against the facility.

Finally, it would be important to have arrangements to reduce the danger that new enrichment plants will be used to justify the host countries mastering enrichment technology. Today, URENCO's ETC and Russia's Rosatom make the most cost-effective gas centrifuges.

3. Establish regional or international spent-fuel storage facilities and repositories. In the not too distant future, perhaps 50 countries, many with only a few nuclear power reactors, will be accumulating spent fuel. This may create the danger that countries will begin to "reprocess" their spent fuel as a "solution" to their spent-fuel problem as is being urged in South Korea today. Also, after about a century, the radiation field around spent fuel declines to a level where it is no longer considered self-protecting and "quick-and-dirty" reprocessing would become easier.

Interim spent-fuel storage facilities and geological repositories should be established in countries willing to host them for a price lower than the cost of reprocessing and national disposal of the associated reprocessing wastes. This puts quite a high ceiling on the price and, given the minimal risks involved from a well-designed spent-fuel storage facilities and repositories, could make hosting them economically attractive. The designs of the spent fuel storage facilities and repositories should be subject to international standards and oversight in order to ensure that the import of spent fuel does not create environmental hazards in countries with less well developed regulatory infrastructures. The repositories should also be designed to be retrievable for a period of a century or so in order to keep options open for alternative disposal strategies.

14.11   Fusion Power[113]

Nuclear fusion may be an alternative to a longer-term commitment to fission energy. In nuclear fusion, energy is produced by fusing together the nuclei of deuterium and tritium, the two heavy isotopes of hydrogen, to form helium and a neutron.[114] Effectively unlimited quantities of the primary fuels, deuterium and lithium (from which tritium is produced), are easily available. (The quantity of uranium is also effectively unlimited, when used in fast breeder reactors.) Due to the low fuel inventory and the high heat capacity of a fusion reactor, an explosive runaway reaction or a meltdown of a fusion energy system are not possible. Radioactive waste products from fusion decay with half-lives of decades. The proliferation risk from fusion is greatly reduced since the introduction of "fertile" materials from which fissile materials such as plutonium could be produced could be made easily detectable in a pure fusion system under safeguards. In a "breakout" scenario, fissile material would not become available until after significant operation of the fusion power plant, and could be prevented by international action.

Current fusion energy research is focused on the confinement of hot ionized gas, called a plasma, in a toroidal (doughnut-shaped) magnetic field. Substantial progress has been made in developing a quantitative understanding of the physical processes that determine the behavior of fusion plasmas. Laboratory experiments have now produced about 10 MW_t of heat from fusion for about one second. The international thermonuclear experimental reactor (ITER) project, under construction in France by a consortium that includes China, Europe, India, Japan, Russia, South Korea, and the United States, is expected to produce 300–500 MW_t of heat from fusion for hundreds of seconds. It is designed to be able to produce about 100 MW_t averaged over a period of weeks.

An alternative route to fusion power production is to heat a small pellet of fuel to high temperature so rapidly that it burns and produces significant fusion energy before it disassembles. The disassembly time is set by the mechanical inertia of the fuel, so this approach is called "inertial fusion energy." It is being pursued today predominantly for its value in providing understanding of the physics of nuclear weapons, but it presents an alternative approach to commercial fusion energy as well. The National Ignition Facility (NIF) has recently begun operation at the Lawrence Livermore National Laboratory in California.

ITER and NIF are fusion research facilities at the scale of fusion power plants. They are first-of-a-kind facilities, and have proven to be expensive, more so than originally planned. Critics of fusion tend to focus on specific technological issues such as the production

of tritium fuel or the development of neutron-resistant materials (Moyer, 2010), for which solutions are under development (Hazeltine et al., 2010). There is an appropriate overall concern, however, that fusion power plants will be large and complex high-tech facilities, and as a result their economic practicality cannot be assured despite favorable projections (Maisonnier et al., 2005; Najmabadi et al., 2006). Very considerable R&D is required to move from scientific feasibility to technological feasibility to practical demonstration (FESAC, 2003; US BPO, 2009). In parallel with ITER, supporting research is planned in each of the national fusion R&D programs to facilitate progress toward higher power and continuous operation, and to qualify advanced materials and components to withstand the heat, particle and neutron fluxes from fusion plasmas. Some national programs anticipate demonstration fusion power plants in the 2035–2040 timeframe, with commercialization starting in mid-century.

While fusion will not provide a solution to reducing carbon emissions in the near term, energy needs will continue to grow in the second half of the 21st century, and carbon emissions will need to continue to decline. Fusion energy, if it is indeed developed by mid-century, could in principle obviate the need for reprocessing and fission breeder reactors, which carry with them significant safety and severe proliferation risks.

14.11.1   Resources

The current focus is on producing fusion energy from the two heavy isotopes of hydrogen, deuterium (D) and tritium (T).[115] Tritium is not available in significant quantities in nature, because it has a half-life of only 12.3 years. However the D–T reaction produces a neutron:

$$D + T \rightarrow {}^4He + n + 17.6 \text{ MeV}$$

This neutron, multiplied moderately through (n, 2n) reactions with beryllium or lead can be used to produce tritium through the reaction:

$$n + {}^6Li \rightarrow T + {}^4He + 4.8 \text{ MeV}$$

Thus the basic fuels for D–T fusion energy are deuterium and lithium-6. Deuterium is present in water at 154 atoms per million hydrogen atoms, and lithium-6 comprises about 7.5% of natural lithium. Other materials used in the construction of fusion systems are not considered to limit the expansion of fusion power.

---

113   Robert Goldston, Princeton University, USA, lead author.

114   The nucleus of deuterium contains a proton and a neutron and that of tritium a proton and two neutrons, as distinct from an ordinary hydrogen atom, whose nucleus contains no neutrons.

---

115   The power density achievable for a given plasma pressure is about 35 times higher for the D–T reaction than for any other potential fusion fuel system. This ratio is appropriate for comparison with the D–D reaction, assuming that all T and ${}^3He$ produced in the D–D reaction is subsequently burned as well.

Three gigawatt-years (GW-yr) of thermal energy from fusion, including the 4.8 MeV produced in generating the needed tritium, requires the burning of 90 kg of deuterium and 265 kg of lithium-6. Therefore, the natural resources required for a year's operation of a 1 GW$_e$ fusion power plant would be the deuterium in 5000 tonnes of natural water and the lithium-6 in 4 tonnes of natural lithium. Deuterium is already being separated from water on an industrial scale to produce the moderator for heavy-water reactors. Assuming US$_{2005}$$300–600/kg for heavy water (Miller, 2001; Ramana, 2007) and ignoring the relatively minor cost of recovering deuterium from heavy water by electrolysis, the cost would be US$3000/kg of deuterium or US$0.003/kWh of electric energy generated from fusion, assuming an efficiency of one-third in the conversion of thermal to electrical energy. At the year-2000 price of about US$20/kg (US$_{2005}$$23/kg), world resources of lithium are estimated at more than 12 million tonnes (USGS, 2007; Fasel and Tran, 2005). This corresponds to a cost of about US$300/kg of lithium-6, which, at that price, would contribute only about 0.001¢/kWh to the price of fusion-generated electricity. Increasing lithium-6 enrichment to the levels proposed for fusion reactors (up to ~80%) would cost approximately US$_{2005}$$3000/kg of lithium-6, or 0.01¢/kWh (Rhinehammer and Wittenberg, 1978).

Lithium reserves of 12 million tonnes would allow the full-power operation of 2000 1- GW$_e$ power plants for 1500 years. It should be economically possible to extract an additional 200 billion tonnes of lithium from seawater for this purpose.[116] The reserves of deuterium in seawater are effectively unlimited.

14.11.2   Technologies

14.11.2.1   Magnetic confinement fusion

The primary approach to developing fusion energy is to use strong magnetic fields to confine a very hot, circa 100 million °C, ionized deuterium–tritium gas, or plasma. Ions and electrons spiral along magnetic fields and magnetic field lines can be contained within a closed volume, specifically, a torus. The hot plasma can therefore be suspended in a vacuum while it is at fusion temperature. The main toroidal magnetic field (i.e., going around the torus the long way) is provided by toroidal field (TF) coils (see Figure 14.20). Shielding these coils from the fusion neutrons are the first-wall components that face the hot plasma and lithium-bearing blankets in which the neutrons are absorbed and the tritium fuel is produced. Some of the heat energy from the plasma flows in the form of energetic ions, directed by magnetic field lines into localized "divertor" regions at the top and/or bottom of the vacuum chamber.

The overall D-shaped cross-section of the plasma is produced by the interplay between the magnetic field produced by a strong electrical current that flows within the plasma around the torus (in the tokamak configuration shown in Figure 14.20) and the magnetic fields produced by currents flowing in poloidal field coils. (These coils produce magnetic fields with no component in the toroidal direction; their fields face only the short way around the torus.) The toroidal current is initiated by a pulsed central solenoidal magnet that creates a toroidally directed electric field and is sustained by external means such as radio-frequency waves. D–T fuel is injected through ports in the form of frozen pellets. Coolant is supplied and tritium is removed through pipes connected to the blankets.

14.11.2.2   Inertial confinement fusion

An inertial fusion energy system would consist of four major components:

- A large laser, ion beam, or other means to provide megajoules (1 megajoule = 1 MW-s) of energy to a millimeter-sized target in about 10 ns, in order to compress and heat the fusion target to high temperature before it disassembles.

- A factory to produce precision cryogenic D–T targets at low cost, at a rate of 5–15 per second.

- A system to inject a new target accurately and rapidly into the target chamber 70–200 ms after each explosion.

- A target chamber capable of withstanding repetitive explosions (and being cleared rapidly thereafter), converting the fusion power to heat, and with a lithium blanket to capture the neutrons and produce replacement tritium.

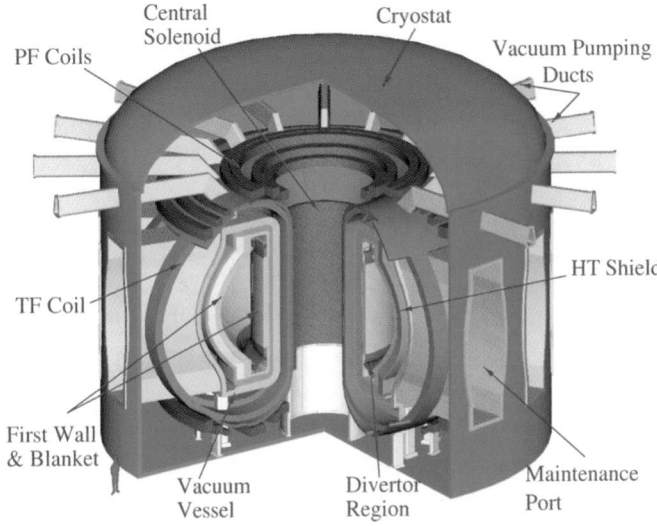

Figure 14.20 | Schematic of the ARIES Advanced Tokamak fusion power core. Source: Najmabadi et al., 2006.

---

116   Seawater contains 0.18 ppm lithium by weight; the concentration of uranium by weight is 56 times lower, while the requirement for power production in LWRs is 50 times higher.

14.11.3   Comparison of Fusion and Fission Power

14.11.3.1   Safety

A magnetic fusion reactor cannot undergo the equivalent of a prompt criticality accident, such as occurred at Chernobyl. The time constant for a thermal excursion is set at about 5 s by the ratio of the heat content of the fuel to the heating rate of the plasma by the 3.5 MeV helium nuclei from fusion. This time constant is much longer than the approximately10 ms characteristic time constant of a prompt-critical thermal neutron fission reactor. Were the plasma density and fusion rate to rise on this a 5-second scale, the limits to the pressure of the plasma, set by the underlying plasma physics, would rapidly extinguish the reaction. Most likely, before this occurred the increased heat flux to the plasma-facing components would erode their surfaces with a resulting influx of impurities into the plasma that would also quench the reaction. The most energetic physically possible excursion could damage internal components, but could not massively breach the vessel containing the fuel, as happened at Chernobyl.

Like fission reactors, fusion reactors could have loss-of-coolant or loss-of-coolant-flow accidents. Even after the fusion reaction stopped, the decay heat released by radioactive isotopes created by transmutation of the materials of internal components would continue to heat the reactor's structure. The structure of a fusion reactor is more massive than that of a fission reactor core, however, and with appropriate choice of reactor materials, the radioactive decay heat source would be reduced to the point where, even with no active cooling, the heat could be conducted and radiated away from the hottest components as fast as it was generated without a risk of breaching the vessel containing the fuel (Petti et al., 2006).

Fusion systems have significant radioactive inventories because of the large flux of energetic fusion neutrons through the reactor structure facing the plasma, which transmutes non-radioactive atoms into radioactive species. There is also a significant inventory of tritium within the vacuum vessel. While the physical mechanisms that could massively breach the vessel of a fission system do not exist in a fusion system, it is still important to determine the biological hazard potential of all components of the reactor, and to examine how much radioactivity in the structure could be mobilized and released into the atmosphere if the largest credible leak in the vessel were sustained. Figure 14.21 compares the amount of air that would be required to dilute the amount of tritium in a fusion reactor to permissible concentrations with that required to dilute the amount of radioactive iodine in a fission reactor.

In one study of a hypothetical reactor, designed with a silicon carbide structure to reduce the inventory of long-lived activation products, it was found that the hazard was dominated by tritium absorbed into the surface of the plasma chamber, activation products in the molten lead coolant, and activation products in structural tungsten.[117] If the full

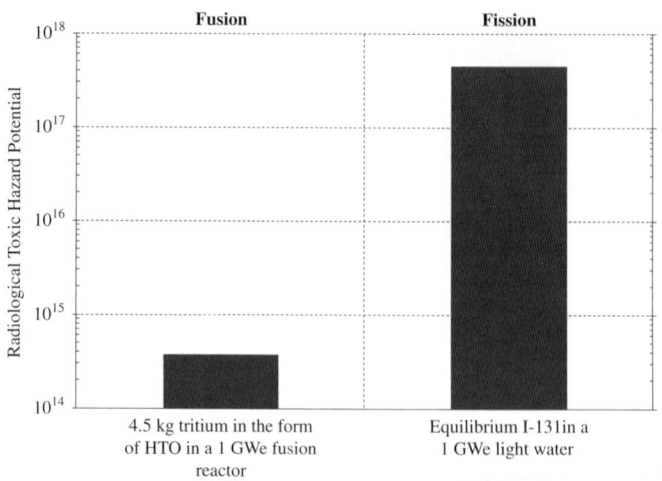

Figure 14.21 | Comparison of biological hazards from potential releases of tritium and iodine-131 from nuclear fusion and fission reactors, respectively, measured in terms of the volume of air required to reduce their concentrations to maximum allowed levels. Source: Kikuchi and Inoue, 2002.

"mobilizable inventory" were released during average weather conditions, doses of more than 1 sievert (Sv) could result at the site boundary 1 km away, and 10 Sv for worst-case meteorology. A dose of a few sieverts within a period of a few weeks could be lethal. Such a release would require a massive breach of the containment vessel, however, which the authors of the study found physically not credible. They found that, for the worst credible leak, the reactor structure and containment building would reduce the releases to a level where the site boundary dose would be below the 0.01 Sv threshold at which evacuation would be required (Petti et al., 2006).

In summary, nuclear fusion has the safety advantages over fission that the primary fission accident initiators of a prompt criticality event (Chernobyl) or meltdown (Three Mile Island, Fukushima-Daichi) are absent. Furthermore, the biological hazard potential of the mobilizable inventory is far below that of a fission reactor. The consequences of an accident are thus reduced to levels dramatically below those of worst-case fission accidents by the choices of structural materials and coolants. Calculations of reductions in releases due to the performance of the reactor containment and other measures below the level required for an evacuation plan would nonetheless likely be controversial and require rigorous regulatory review.

14.11.3.2   Proliferation

The proliferation risk from fusion energy systems is much smaller than that from fission systems because no chain-reacting fissile material

---

117 The most important neutron-activation products in lead are polonium-210 (138-day half-life) and mercury-203 (47 days). The most important activation products

in tungsten are: rhenium-184 (38 days), rhenium-184m (165 days), rhenium-186 (4 days), and rhenium-188 (17 hours); and tungsten-181 (121 days), tungsten-185 (75 days), and tantalum-182 (114 days) and tungsten-187 (24 hours) (Petti et al., 2006).

or fertile material that could be converted to fissile materials by neutron absorption need to be present in a fusion power plant at any time (Glaser and Goldston, 2011b). Thus, while an inspection regime would be required, it would only need to detect the presence or absence of heavy fissile or fertile isotopes.[118] Fairly simple detection schemes could very sensitively detect small quantities of fertile materials or fission products in a fusion system whose neutrons are being used to breed fissile material.

One could be concerned about a "breakout" scenario for fusion, in which a nation expelled IAEA inspectors or unplugged remote monitoring devices and then reconfigured a fusion power system to breed fissile material. There is a very important qualitative difference between fission and pure fusion in the breakout scenario, however. In the fission case, such as occurred in North Korea, the reactor owner already has in hand spent and cooled fuel containing plutonium at the moment that the inspectors are expelled. There is nothing that can be done, short of military invasion, to prevent the separation of this plutonium for use in nuclear weapons. Bombing a fission reactor and its spent fuel storage facilities risks spreading radioactivity, or, if highly controlled, leaves the plutonium available to be mined from the rubble. Neither bombing nor invasion was considered practical in the case of North Korea, nor would they likely be considered practical in many other cases.[119]

By contrast, in the case of a fusion power plant, at the time of "breakout," no fissile material is yet in hand, so the challenge becomes the prevention of the operation of a reconfigured version of the power plant, capable of fissile material production. A fusion power plant could be rendered inoperable quickly and easily by a conventional cruise-missile strike on any of a number of support facilities: cryogenic systems, power conversion systems, or even cooling systems, without risk of a significant release of radioactivity.[120]

A fusion power plant consumes – and so must replace – about 130 kg of tritium per full-power year. Access to grams of tritium to "boost" the yield of nuclear weapons would be harder to prevent if many fusion power systems were in operation around the world. Tritium is not now controlled under the Non-Proliferation Treaty, whose focus is on preventing the diversion of fissile materials for use in weapons. Consideration should be given to strengthening international controls over tritium (Kalinowski and Colschen, 1995).

One might also be concerned that the science associated with inertial confinement fusion would be proliferated along with inertial fusion energy systems, and that key tests relevant to nuclear weapons could be undertaken on these facilities, as they will be at the NIF in California. Classified radiation-hydrodynamics codes are used to predict the performance of nuclear weapons. Their equivalents in the non-classified world could be calibrated against inertial fusion experiments, and could become widely available. A second concern is that if plutonium or other nuclear materials were used in the target of an inertial fusion R&D facility, critical information could be gained about relevant equations of state. The design of first-generation fission weapons, however, does not require experimentally validated advanced design codes, nor information on the equation of state of materials under such extreme conditions. In principle, these could help in the design of advanced weapons, but further review is required (Goldston and Glaser, 2011a).

Some researchers have considered "hybridizing" fusion and fission; these ideas have recently been reviewed and evaluated (Freidberg and Finck, 2010). In principle, the neutrons from fusion can be used for three purposes related to fission power:

- multiplying the 20-MeV energy output from each fusion reaction by using fusion neutrons to induce fission reactions (200 MeV each) in a sub-critical fission blanket surrounding the fusion system;

- breeding fuel for fission systems by transmuting uranium-238 or thorium-232 to plutonium-239 or uranium-233, respectively; and/or

- using the energetic neutrons from fusion to "burn" plutonium-239 and other transuranics recovered from the reprocessed spent fuel of fission power plants.

The advantages are that the fusion system could operate at lower gain and possibly lower neutron wall loading, and the fission system could operate sub-critically, reducing some of the accident potential compared with fast-neutron breeders. It also appears, however, that fusion-fission reactors would combine the majority of the scientific and technological development issues of the fusion systems with the majority of the proliferation risks of fission systems with reprocessing.

It should be noted that even if fusion comes into large-scale use after mid-century, plutonium and other transuranics will remain from LWRs and will constitute a potential proliferation risk. In case 3, where fusion-fission hybrids are used to burn transuranics, possibly more efficiently than "burner" fast reactors, one would have to balance the diversion

---

118 Uranium-238 and thorium-232 are both termed fertile isotopes because the addition of a neutron followed by radioactive decay produces artificial chain-reacting or fissile isotopes: plutonium-239 and uranium-233, respectively.

119 Attacks on reactors have happened. Iran and Israel both bombed Iraq's Osiraq reactor, in 1980 and 1981 respectively, while it was under construction, because they thought that Iraq would try to use it or its fuel to make nuclear weapons. Similarly, Iraq repeatedly bombed Iran's Bushehr reactor while it was under construction during the period 1984–88. The Unites States was on the verge of bombing North Korea's power reactor at Yongbyon in 1994 because it was being used to make plutonium, but was deterred in part because of concern about the international reaction to the first attack against an operating reactor and the associated radiation release.

120 See, for example, the layout of the ITER complex, www.iter.org/gallery/com_image_download.

risks from reprocessing this material multiple times as it is consumed, versus those of placing it in monitored underground repositories.

14.11.3.3   Radioactive waste

Fusion radioactive waste, unlike that from fission, does not originate from the burning of the fuel, but rather from neutron irradiation and consequent activation of the structural and blanket materials that face the reacting fuel. In typical designs, the first 20 cm or so of the chamber wall would need to be replaced approximately every four years. Neutron activation products can have half-lives of millennia, just like long-lived transuranic elements and fission products.[121] Power plant studies indicate, however, that structural materials can be selected that reduce the radiological hazard from the fusion reactor waste to as low as one-hundred thousandth that of fission-reactor waste (Fetter, 1987) or even one-millionth, as shown in a more recent Japanese study (Kikuchi and Inoue, 2002; see Figure 14.22). The concentration of the radioactivity would be low enough so that the material could be classified as Class C low-level radioactive waste and shallow burial (less than 30 m deep) would be permitted by US regulatory standards (Henderson et al., 2000; US Federal Code of Regulations, Part 61.7(5)).[122] It has also been proposed that the waste from fusion systems could be stored on site, and/or that a significant fraction could be recycled. Nevertheless, despite the potential for greatly reduced waste production, fusion power could still face significant public concerns about the disposal of its wastes.

14.11.3.4   Cost

At this point in the R&D process, it is difficult to project the costs of constructing and operating a fusion power plant. Based on simple consideration of the mass of the major technical components of a 1- GW_e fusion system relative to that of a fission system, and the complex materials required for some components, it appears unlikely that, without accounting for externalities, electricity produced by fusion systems would be less expensive than that from light-water reactors. Studies of magnetic fusion-power systems have been undertaken in the United States (Najmabadi et al., 2006) and in Europe (Maisonnier et al., 2005). Assuming success with the R&D issues discussed below, they conclude that the cost of electricity from a 1- GW_e magnetic fusion power plant would be in the range of US_{2005}\$0.05–0.13/kWh, for a tenth-of-a-kind power plant. The cost of electricity is estimated to be reduced by 20% for 1.5- GW_e plants, due to economies of scale. These estimates should, however, be treated with extreme caution, due to the distance of extrapolation.

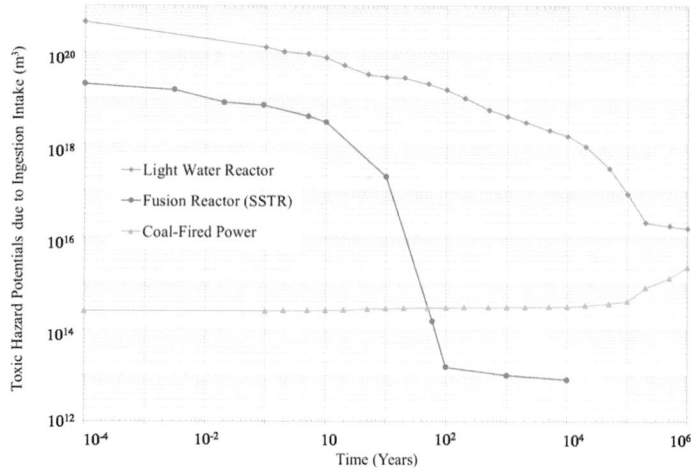

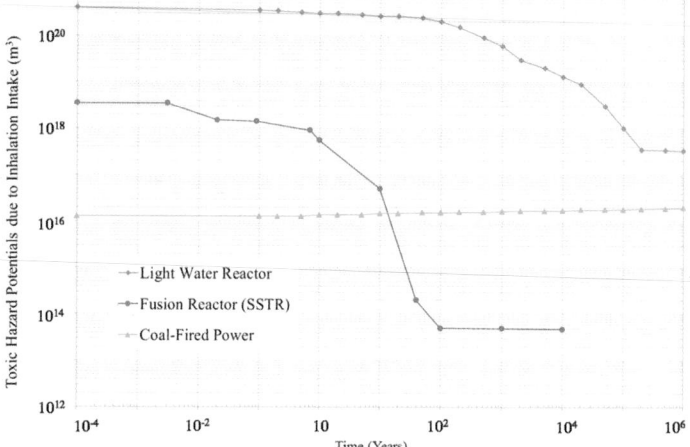

Figure 14.22 | Comparison of radioactivity hazard potential from lifetime operation of fission and fusion reactors. Based on the tritium inventory and the calculated lifetime production of radioactive transmutation products by the Japanese-designed Steady State Tokamak Reactor fusion reactor compared with that of the spent fuel discharged over the lifetime of a light-water reactor, and the uranium and decay products in the ash from a coal-fired power plant accumulated over its operating lifetime. Source: Kikuchi and Inoue, 2002.

Integrated cost projections have not been undertaken recently for inertial fusion energy, but the extrapolation distance to a cost estimate for inertial fusion energy is even greater with respect to the repetition rate of the driver beams, the lifetime required for high-power final optics for systems involving laser-driven pellet implosion, and the required cost reduction for target fabrication.

14.11.4   R&D Status

14.11.4.1   Magnetic confinement fusion

The basic concept behind magnetic fusion is that strong magnetic fields are used to contain a plasma, so that it can be heated to high temperature and can burn in a sustained manner. The two central scientific challenges are:

1. thermally insulating the plasma sufficiently well so that the fusion process itself can provide most of the needed plasma heating,

---

121   Important long-lived activation products in fusion-reactor studies are: carbon-14 (half-life: 5700 years), aluminum-26 (710,000 years), and three activation products of molybdenum, molybdenum-93 (3500 years), niobium-94 (24,000 years), and technicium-99 (213,000 years) (Henderson et al., 2000).

122   US Code of Federal Regulations, Part 61.55 gives the limits in Ci/m³ for a small number of relevant nuclides for waste to qualify as Class C waste: carbon-14 (8); nickel-59 (220), and niobium-94 (0.2).

allowing a high power gain (energy output divided by energy input) to be sustained; and

2. containing a high enough pressure plasma that it can provide sufficient fusion power density to justify the cost of the magnetic "bottle."

Very substantial scientific progress has been made in addressing both of these challenges. The basic mechanisms that allow heat to escape across magnetic fields have been identified and modeled computationally.

While issues remain for scientific confirmation, the overall experimental picture is consistent with the presence of fine-scale turbulence driven largely by the gradient in the temperature between the center of the hot plasma and its cooler edge. This results in an overall energy confinement time (energy stored in the plasma divided by power required to heat it) that scales consistently across the many experimental devices that have been operated around the world (Figure 14.23). Scaling from these experiments gives a projection that the international ITER project will have a gain of 10, meaning that 10 times more fusion power will be produced than the heat input from microwaves or other inputs from outside of the plasma required to sustain it at fusion temperature. Since 20% of the heat from fusion stays within the plasma in the form of energetic helium nuclei, this means that two-thirds of the power heating the plasma will come from the fusion reactions themselves. Demonstrating this self-sustaining plasma heating is the primary scientific goal of ITER. A magnetic fusion power plant will require a gain of about 25.

Substantial progress has also been made in identifying the limits to the plasma pressure that can be contained in a magnetically confined fusion plasma. Indeed, these pressure limits, as determined by limits of the ratio of plasma kinetic pressure to the pressure of the magnetic field, are now accurately predicted on the basis of theory. Since the fusion rate is approximately proportional to the square of the plasma pressure, this sets the power production capability of fusion systems. ITER is predicted to be able to produce at least 500 MW_t of fusion power. Fusion power production multiplied by the pulse length gives the energy released per pulse from fusion systems. In the 1970s, the toroidal magnetic confinement configuration called the "tokamak" achieved fusion power production of 1/10 of 1 W for one-hundredth of a second. ITER, also configured as a tokamak, is planned to operate for at least 300–500 s at a gain of at least 10, with a goal of effectively steady-state operation at gain of 5. Because ITER will produce significant power levels from fusion for significant periods of time, many of the technologies for ITER are similar to those that would be used in a fusion power plant. Thus the mission of ITER is to "demonstrate the scientific and technological feasibility of fusion energy for peaceful purposes."

14.11.4.2   Inertial confinement fusion

The concept underlying inertial fusion is that a small pellet of D–T fuel is compressed to very high density, but mostly at low temperature. A few percent of the fuel is heated to fusion temperature, however, and, as it burns, it "ignites" a fraction of the remaining fuel, which burns as well, igniting more fuel and ultimately providing adequate gain for net power production. The key recent scientific advances have been in the development of fully three-dimensional codes that can predict the evolution of the fundamentally unstable compression process, as well as the unstable burn process. These calculations define the requirements on the manufacturing precision required for the spherical fusion targets, and on the timing and uniformity of the laser or other beams used to compress and heat the targets. Furthermore, new ideas are being developed on means to heat the "hot spot" that initiates the burn, for example using special very short-pulse lasers (called "fast ignition") or carefully timed shocks. These may allow higher gain or lower laser driver energy for fusion energy systems.

A second issue, particular to inertial fusion driven by lasers, is laser–plasma interaction. The very high power laser light can interact with the plasma it produces in the vicinity of the target, with the result that the laser beam is steered away from the target, and/or energetic electrons are produced that heat the target and impede implosion. This is an active topic of research at the National Ignition Facility.

In inertial fusion, gain is defined as fusion yield divided by the laser light energy input. It is reduced by the relatively low efficiency (~20%) of conversion of laser light to X-rays which actually impinge on the pellet and implode it, in the geometry used at the NIF. To set a clear goal, a US NAS (1997) report defined ignition at the NIF as gain of unity. The total fusion energy released per pulse at the NIF will, at gain of unity, be 1–2 MW-s, 100,000 less than anticipated in ITER. The pulse repetition rate at high gain will be of the order of a few per day, as compared with ~50/day in ITER.

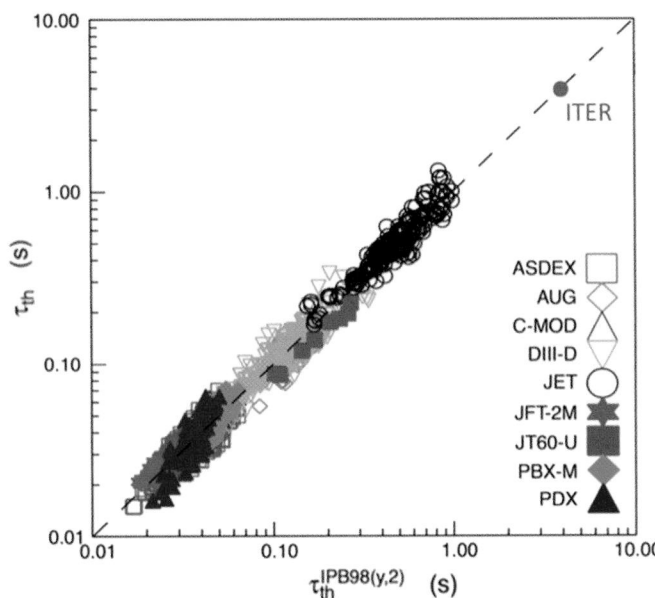

Figure 14.23 | Experimental confinement time of thermal energy vs. regression fit [IPB98(y,2)] to experimental results from nine tokamak experiments world-wide. Source: Shimada, 2000

While magnetic fusion needs a gain of about 25 for a practical power plant, the low overall efficiency of the pulsed system in laser-driven inertial fusion requires a gain of about 150. The NIF will perform pellet shots at a rate of a few per day, as compared with a fusion power system that would need to perform shots at a rate of order 5–15 per second. Since the technologies used at the NIF are quite different from those to be used in a fusion power plant, it is viewed as demonstrating the scientific feasibility of inertial fusion, but not its technological feasibility.

14.11.5   R&D Needs

14.11.5.1   Magnetic confinement fusion

The key R&D needs for magnetic fusion (FESAC, 2007) are in support of:

1. higher power gain than ITER (25 versus 10) at higher total power output (~2500 MW$_t$ versus 500 MW$_t$) in fully continuous operation;

2. efficient techniques to handle the power and ion flux from such a plasma; and

3. efficient techniques to handle the neutron flux from such a plasma while producing the needed tritium fuel.

Major new experimental facilities in China, South Korea, Japan, and Europe are now under construction or are beginning operation to investigate advanced operating modes and plasma configurations to address the first issue. These are superconducting tokamaks and "stellarators," comparable to the size of current experiments, but capable of very long pulse operation. To avoid irradiation problems, these experiments are designed to use hydrogen and deuterium "fuel" for physics studies, not DT for power production. Stellarators are more complex to construct than tokamaks, having a cross-section that rotates and distorts around the torus. They have the advantage, however, that they do not require external energy inputs to sustain internal plasma current and so can operate more efficiently in steady state than tokamaks. Results from these experiments are anticipated in parallel with ITER operations, and ITER itself will explore higher-performance operating modes once its basic goals are achieved.

In the Unites States, experiments with D–D fuel are also being considered that could cost-effectively address the issue of how best to handle the power and ion flux from the plasma (OFES, 2009).

The European Union and Japan are working together on the design and engineering validation of a facility to address the issue of qualifying materials to handle the high fluence (time-integrated neutron flux) of very high-energy neutrons from fusion plasmas. This would be based on an intense deuterium ion beam penetrating a liquid lithium target to produce energetic neutrons. Material tests are to be done with displacements per atom and volumetric helium production expected for fusion power plant plasma-facing structures. Promising experimental results have already been achieved with nano-composited ferritic alloys, tested using ion beams to simulate energetic neutrons.

ITER will provide a test bed for tritium breeding modules at relatively low neutron fluence. Integrated testing in a pilot plant or demonstration power reactor would be needed before commercialization could be undertaken.

14.11.5.2   Inertial confinement fusion

The key R&D needs for inertial fusion are:

1. higher gain than the NIF's base mode of operation (150 versus 1);

2. cost-effective, repetitively pulsed driver systems (about 5–15 per second, versus a few per day for the NIF), perhaps using beams of heavy ions that can be produced with smaller energy conversion losses than laser light;

3. cost-effective, very-high-throughput, precision D–T target manufacture;

4. techniques to place targets in the fusion chamber with high accuracy and speed; and

5. target chambers and final beam focusing systems that can handle repeated 400 MJ explosions (the equivalent of 100 kg of TNT) while providing the required precisely controlled environment quickly after each explosion.

Figure 14.24 | Laser beam lines at the National Ignition Facility at Livermore National Laboratory, California. Source: FESAC, 2003.

The National Ignition Facility itself will likely investigate higher-gain operation, as will a fast ignition experiment, called FIREX, in Japan. The Omega-EP experiment at the University of Rochester in the United States will also study fast ignition. Small efforts are underway in the United States, Japan and Europe to investigate other technological issues, such as the development of ion beams as a more efficient alternative to lasers (allowing lower pellet gain). But the main purpose of the large inertial fusion studies in the United States and France (Laser Mega Joule facility) has been to support nuclear-weapon science.

14.11.6   Possible Deployment Scenarios

Progress toward fusion energy depends on funding as well as scientific and technological success. In the United States, the Magnetic Fusion Engineering Act of 1980 projected a demonstration power plant by 2000, and authorized funding for magnetic fusion of US_{2005}\$34 billion. The actual level of funding appropriated was about one-third of this.

A study was undertaken (FESAC, 2003) to determine the program that would be required to bring on line a fusion demonstration power plant within 35 years. The scientific and technological issues listed above were considered in detail, and it was estimated that the United States could be one of several leaders worldwide in the development of fusion if an investment of about US_{2005}\$27 billion were made over the 35-year period. This would include aggressive R&D in both magnetic and inertial fusion, until a down-selection before construction of major new DT facilities.

If a number of successful demonstration fusion power plants were brought online around the world in the time frame of 2035–2040, one could imagine the commercial construction of fusion power plants beginning in 2050. Whether this is practicable will depend on whether the significant R&D issues discussed above can be resolved. Fission power, during its period of growth in 1975–1990, increased its share of electricity production at a rate of 1.2%/yr. If we posit fusion energy growth at 0.9%/yr of total electricity production, it could achieve about a 30% share of world electricity production by 2100. Assuming world electricity consumption of 12 TW_e in 2100, this would correspond to the construction of an average of 50 1.5- GW_e power plants per year, worldwide.

14.11.7   Policy Recommendations

Fusion is clearly attractive from the points of view of waste, safety and proliferation, particularly when compared with fission based on reprocessing and breeding. At the same time, it should be recognized that success with the technical development of fusion energy is not assured, and that the cost of fusion energy is difficult to project. Each of the major candidates for baseload electric power in the next decades – coal with carbon sequestration, major expansion of fission, and renewables with large-scale energy storage and transportation – faces serious challenges, however. Thus the availability of fusion energy could be important for stabilizing CO_2 levels in the atmosphere as energy demand continues to grow in the latter half of the 21st century (Goldston, 2011).

Further opportunities for international collaboration/coordination should be pursued. As the goal is to determine whether commercialization of fusion can be achieved by mid-century, however, it will be prudent to continue to pursue the development of fusion energy science and technology in the world's domestic fusion programs, in parallel with international participation in the ITER project and other projects, perhaps a neutron irradiation facility.

Taking at face value the development plan articulated in the United States, and perhaps multiplying by four to obtain a global figure, an investment in the range of US\$100 billion over the next 30 years would be required to bring fusion online on the timescale discussed above. This corresponds to a worldwide investment rate of US\$3.3 billion/yr, comparable to the investment rate in fusion R&D in the 1970s. Thus, once ITER construction is well under way, the world level of investment in fusion research will need to approach US\$3 billion/yr to sustain healthy domestic research efforts as well as construct ITER. If the yearly level of investment were to remain at about that level after completion of ITER, this could provide the needed resources for the R&D and demonstration projects needed to support the beginning of the commercialization of fusion by mid-century. Because of the large size of the energy market, investments of this scale in energy R&D can be justified on a purely economic basis (Goldston, 2006).

In the opinion of the present authors, diverting current fusion research towards fission–fusion hybrids could be counter-productive. The destruction of wastes from fission might be an application of fusion energy, but the primary attraction of fusion is its high level of safety, low level of waste, and, especially, its low proliferation risk when compared with fission. A fusion-fission hybrid likely would not have those advantages.

References

Acheson–Lilienthal, 1946: C.I. Barnard, J.R. Oppenheimer, C.A. Thomas, H.A. Winne, and D.E. Lilienthal (Chairman), *A Report on the International Control of Atomic Energy*, prepared for the Secretary of State's Committee on Atomic Energy, Washington, DC. www.fissilematerials.org/ipfm/site_down/ach46.pdf

Acton, J. and W. Bowen, 2008: Nurturing nuclear neophytes, *Bulletin of the Atomic Scientists*, September/October, p. 27.

Albright, D., F. Berkhout, and W. Walker, 1997: *Plutonium and Highly Enriched Uranium 1996* Oxford University Press.

Alvarez, R., J. Beyea, K. Janberg, J. Kang, E. Lyman, A. MacFarlane, G. Thompson, and F. von Hippel, 2003: Reducing the hazards from stored spent power-reactor fuel in the United States, *Science & Global Security* 11: 1.

American Cancer Society, 2007: *Global Cancer Facts and Figures, 2007*.

AREVA, 2008: *Georges Besse II: A new era for enrichment*, www.areva-nc.com/areva-nc/liblocal/docs/download/Eurodif_Presentation_projet_G_Besse_II_en.pdf (accessed December 25, 2008).

AREVA-EDF, 2008: AREVA and EDF create long-term used fuel management partnership, press release, December 19, 2008.

ARIES studies: ARIES Program public information site, www-ferp.ucsd.edu/ARIES/DOCS/final-report.shtml

Atomenergoprom, 2008a: "International Activities," www.atomenergoprom.ru/en/produce/foreign (accessed 29 December, 2008).

Atominfo, 2009: Start of BN-800 will be delayed up to 2014 (translation), April 15, 2009, www.atominfo.ru/news/air6318.htm (in Russian).

Ball, S., M. Richards and S. Shepelev, 2006: Sensitivity Studies of Air Ingress Accidents in Modular HTGRs, *HTR 2006*, Johannesburg, RSA, October 1–4, 2006.

Bari, R., L-Y Cheng, J. Phllips, J. Pilat, G. Rochau, I. Therios, R. Wigeland, E. Wonder, and M. Zentner, 2009: *Proliferation Risk Reduction Study of Alternative Spent Fuel Processing*, BNL-90264–2009-CP, Brookhaven National Laboratory.

Barkenbus, J. N. and C. Forsberg, 1995: Internationalizing nuclear safety: the pursuit of collective responsibility, *Annual Review of Energy and Environment* 20: 179–212.

BBC, April 3, 2006: China to buy Australian uranium, http://news.bbc.co.uk/2/hi/asia-pacific/4871000.stm

Bellona, May 18, 2009: Russia begins building its first floating nuke plant amid pep talk for Medvedev from Kiriyenko, http://bellona.org/articles/articles_2009/fnpp_starts

Bespalchuk, P. I., Y. E. Demidchik, E. P. Demidchik, Z. E. Gedrevich, A. P. Dubovskaya, V. A. Saenko, and S. Yamashita, 2007: *Thyroid Cancer in Belarus after Chernobyl*, International Congress Series 1299: 27–31.

Beyea, J. and F. von Hippel, 1982: Containment of a reactor meltdown, *Bulletin of the Atomic Scientists*, August/September, p.52.

Beyea, J., E. Lyman and F. von Hippel, 2004: Damages from a major release of ^{137}Cs into the atmosphere of the United States, *Science & Global Security* 12: 125.

Bhabha, H. J. and N. B. Prasad, 1958: A study of the contribution of atomic energy to a power programme in India, *Proceedings of the Second UN International Conference on the Peaceful Uses of Atomic Energy*, Geneva, pp. 89–101.

Bickerstaff, K., I. Lorenzoni, N.F. Pidgeon, W. Poortinga, and P. Simmons, 2008: Reframing nuclear power in the UK energy debate: Nuclear power, climate change mitigation and radioactive waste, *Public Understanding of Science* 17: 145–169.

Bloomberg, December 17, 2009: *Southern said to be in line for U.S. loan guarantee (Update2)*, www.bloomberg.com/apps/news?pid=20601103&sid=aaId1 p6YD01Q.

Bolsen, T. and F. L. Cook, 2008: Public opinion on energy policy: 1974–2006, *Public Opinion Quarterly* 72(2): 364–388.

Bunn, M. 2008: *'Appropriate Effective' Nuclear Security and Accounting – What is It?* Presentation to 'Appropriate Effective' Material Accounting and Physical Protection, Joint Global Initiative/ UNSCR 1540 Workshop, Nashville, Tennessee, 18 July 2008. http://belfercenter.ksg.harvard.edu/files/bunn-1540-appropriate-effective50.pdf

Bunn, M., 2010: *Securing the Bomb 2010: Security All Nuclear Material in Four Years* (Cambridge, MA amd Washington DC: Project on Managing the Atom, Harvard University, and Nuclear Threat Initiative), www.nti.org/securingthebomb.

Bunn, M., S. Fetter, J. Holdren and B. van der Zwaan, 2005: The economics of reprocessing versus direct disposal of spent nuclear fuel, *Nuclear Technology* 150: 209.

Cardis, E. et al., 2006: Estimates of the cancer burden in europe from radioactive fallout from the Chernobyl Accident, *International Journal of Cancer* 119(6): 1224–35.

Chapman, N., C. McCombie, and P. Richardson, 2008: *Strategic Action Plan for Implementation of European Regional Repositories: Stage 2, Economic Aspects of Regional Repositories*, April 2008. ftp://ftp.cordis.europa.eu/pub/fp6-euratom/docs/sapierr-2–3-economic-aspects-of-regional-repositories_en.pdf

Charpin, J.M., B. Dessus and R. Pellat, 2000: *Report to the Prime Minister [of France]: Economic Forecast Study of the Nuclear Power Option*, Tables on pp. 43, 56, 214, 215.

Chayes, A. and W.B. Lewis (eds), 1977: *International Arrangements for Nuclear Fuel Reprocessing*. Ballinger, Cambridge, MA.

China National Development and Reform Commission, 2007: *State Mid-Long Term Development Plan for Nuclear Power Plants (2005–2020)*, October.

CIGI, 2010: *Survey of Emerging Nuclear Energy*. Centre for International Governance Innovation, *States*, www.cigionline.org/senes.

CNIC, 2004a: Japanese nuclear industry's back end cost, *Nuke Info Tokyo*, No. 98, November 2003/ February 2004, p.10. Citizen's Nuclear Information Center.

CNIC, 2004b: Long-term Nuclear Program Planning Committee publishes costs of nuclear fuel cycle, Citizens Nuclear Information Center, *Nuke Info Tokyo*, No.103, November/December 2004. Citizen's Nuclear Information Center.

CNIC, 2008: Opposition to dangerous MOX fuel, *Nuke Info Tokyo*, November/December 2008. Citizen's Nuclear Information Center.

CNIC, 2009: Replacement of Hamoaka Reactors 1 and 2, *Nuke Info Tokyo*, January/February 2009, p. 3. Citizen's Nuclear Information Center.

Collins, E.D., 2005: *Oak Ridge: Closing the fuel cycle can extend the lifetime of the high-level-waste repository*, American Nuclear Society 2005 Winter Meeting, Washington, DC.

Connecticut Yankee, 2008: *Dry Fuel Storage*. www.connyankee.com/html/fuel_storage1.html (accessed 16 December 2008).

Cooley, J., 2003: Integrated nuclear safeguards: genesis and evolution, *Verification Yearbook 2003* (Vertic), www.vertic.org/assets/YB03/VY03_Cooley.pdf

Deffeyes, K.S. and I.D. MacGregor, 1980: World uranium resources, *Scientific American* January: 50–60.

Deutsche Welle, 2010: Spanish villages vie for nuclear waste dump, March 12, 2010.

EC, 2007: *Gallup: Attitudes on Issues Related to EU Energy Policy*, Eurobarometer 206a, European Commission, April 2007.

EC, 2008: *Attitudes toward Radioactive Waste*, Eurobarometer 297, European Commission, June 2008.

ElBaradei, M., 2003: *Towards a safer world*, *The Economist*, October 16, 2003.

ElBaradei, M., 2008: *Reviving Nuclear Disarmament,* address to the Conference on Achieving the Vision of a World Free of Nuclear Weapons, Oslo, Norway, February 2008. www.iaea.org/NewsCenter/Statements/2008/ebsp2008n002.html

Electric Auto Association Europe, 2008: *CO$_2$-emissions of Electric Autos.* http://eaaeurope.org/emissions_of_Eas.html (accessed December 16, 2008).

ERDO, 2010: *Working on a shared solution for radioactive waste*, www.erdo-wg.eu/ERDO-WG_website/Home.html

ETC, 2008: *Enrichment Technology, Geographical Operations,* Enrichment Technology Corporation. www.enritec.com/FullStory.aspx?m=106.

Fasel, D. and M.Q. Tran, 2005: Availability of lithium in the context of future D–T fusion reactors, *Fusion Engineering and Design* 75–79: 1163.

FESAC, 2003: Goldston, R. et al., *A Plan for the Development of Fusion Energy*, DOE/SC-0074, Fusion Energy Sciences Advisory Committee (FESAC). www.ofes.fusion.doe.gov/More_HTML/FESAC/DevReport.pdf

FESAC, 2007: *Priorities, Gaps, and Opportunities: Towards a Long-Range Strategic Plan for Magnetic Fusion Energy*, Fusion Energy Sciences Advisory Committee (FESAC). www.ofes.fusion.doe.gov/fesac.shtml

Fetter, S., 1987: The radiological hazards of magnetic fusion reactors, *Fusion Technology* **11**: 400.

***Financial Express*, October 14, 2008**: "35 pc power from n-plants by 2050: Kakodkar."

Finck, P., 2010: *Initial Assessment of Fuel Cycles, Idaho National Laboratory, August 17, 2010.*

Fitzpatrick, M. (ed), 2007: *Nuclear Black Markets: Pakistan, A.Q. Khan, and the Rise of Proliferation Networks: A Net Assessment*. International Institute for Strategic Studies, London.

Florida Public Service Commission, 2007/8: "Florida Public Service Commission rule to encourage nuclear power development," February 13, 2007; "FPSC Approves Need for Two New Nuclear Units at Progress Levy County Facility," July 15, 2008; FPSC Approves Need for Two New Nuclear Units at FPL Turkey Point Facility," March 18, 2008; "PSC Votes on Nuclear Cost Recovery for Florida Power & Light and Progress Energy Projects," October 14, 2008.

Freidberg, J. and P. Finck, 2010: *Research Needs for Fusion-Fission Hybrids.* Report of the Research Needs Workshop, September 30–October 2, 2009, Gaithersburg, MD, chair J. Freidberg (MIT), vice-chair P. Finck (Idaho National Laboratory). web.mit.edu/fusion-fission/

Forwood, M., 2008: *The Legacy of Reprocessing in the United Kingdom*. International Panel on Fissile Materials.

***Fuel Cycle Week*, April 24, 2009**: J. Mazer, "Without CWIP, AmerenUE shelves Callaway-2."

GE-Hitachi, 2008: "Global laser enrichment: Uranium enrichment using advanced laser technology"; "Cameco joins GE and Hitachi in global laser enrichment venture," GE Press releases, June 20, 2008.

General Atomics, 2010: *GT-MHR: Inherently Safe Nuclear Power for the 21st Century*, http://gt-mhr.ga.com (accessed January 12, 2010).

Gen IV, 2008: *GenIV International Forum: 2007 Annual Report*.

Glaser, A., 2008: Characteristics of the gas centrifuge for uranium enrichment and their relevance for nuclear weapon proliferation, *Science & Global Security* **16**: 1.

Glaser, A. and Goldston, R.J., 2011a: Inertial Confinement fusion energy R&D and nuclear proliferation: The need for direct and transparent review, Bulletin of the Atomic Scientists, **67**(3) 59

Glaser, A. and Goldston, R.J., 2011b, Proliferation Risks of Magnetic Fusion Energy: Clandestine Production, Covert Production and Breakout: Submitted to Nuclear Fusion

Glasstone, S. and P. J. Dolan (eds.), 1977: *The Effects of Nuclear Weapons*, 3rd edn, DOD and DOE, Washington, DC.

Globescan, 2005: *Global Public Opinion on Nuclear Issues and the IAEA: Final report from 18 countries*, prepared for the IAEA, www.iaea.org/Publications/Reports/gponI_report2005.pdf

Goldschmidt, P., 2008: *IAEA Safeguards: Dealing Preventively with Non-Compliance*, Carnegie Endowment for International Peace.

Goldston, R.J., 2006: Is Fusion Research Worth It?, *2006 IAEA Fusion Energy Conference*, http://www-pub.iaea.org/MTCD/Meetings/FEC2006/se_p2–1.pdf

Goldston, R.J., 2011: Climate Change, Nuclear Power and Nuclear Proliferation: Magnitude Matters, Science and Global Security, **19**: 130.

Greenberg, M., 2009: Energy sources, public policy, and public preferences: Analysis of US national and site-specific data, *Energy Policy* 37: 3242.

Greencross, 2004: *Floating Nuclear Power Plants in Russia: A Threat to the Arctic, World Oceans and the Non-proliferation Treaty* (Greencross, Russia), www.greencross.ch/pdf/gc_fnpp_book.pdf

Gregory, R., J. Flynn, and P. Slovic, 1995: Technological stigma, *American Scientist*. **83**(3): 220–223.

Grover, R. B. and S. Chandra, 2006: Scenario for growth of electricity in India, *Energy Policy* 34(17): 2834–2847.

Grubler, A. 2010, "The costs of the French nuclear scale-up: A case of negative learning by doing," *Energy Policy*, **38**, p. 5174.

Harding, J., 2007: Economics of nuclear power and proliferation risks in a carbon-constrained world, *Electricity Journal*, **20**(10): 65.

Hart, D., 1983: *Nuclear Power in India: A Comparative Analysis*, George Allen & Unwin, London.

Hazeltine, R., Porkolab, M., Prager, S., and Stambaugh, R., 2010. Letter to the Editor, *Scientific American*, **July 2010**.

Henderson, D. et al., 2000: Activation, decay heat, and waste disposal analyses for the ARIES-At power plant, *Proceedings of the 14th ANS Topical Meeting on Technology of Fusion Energy*, October 2000, Park City, Utah. www-ferp.ucsd.edu/LIB/REPORT/CONF/ANS00/henderson.pdf

Heyes, A., 2002: Determining the price of Price–Anderson, *Regulation*, winter 2002–2003: 26.

Heyes, A. and C. Liston-Heyes, 1998: Subsidy to nuclear power through Price–Anderson liability limit: Comment, *Contemporary Economic Policy*, **16**: 122–124.

Hill, R. N. 2005: *Advanced Fuel Cycle Systems: Recycle/Refabrication Technology Status,* presentation, Argonne National Laboratory, 7 Sept. 2005.

Hore-Lacy, I. *et al.*, 2009: Price–Anderson Act of 1957, United States, *Encyclopedia of Earth,* World Nuclear Association, www.eoearth.org/article/Price-Anderson_Act_of_1957,_United_States (accessed 6 December 2009).

Howsley, R., 2008: *WINS – From Concept to Reality*, plenary address, 49th Annual Meeting of the Institute for Nuclear Materials Management, Nashville, TN, July 13–17.

Hultman N., J. Koomey and D. Kammen, 2007: What history can teach us about the future costs of U.S. nuclear power, *Environmental Science & Technology*, April, p. 2088.

Hultquist, G. *et al.*, 2009: Water corrodes copper, *Catalysis Letters*, July 28, 2009.

IAEA, 1972: *The Structure and Content of Agreements between the Agency and States Required in Connection with the Treaty on the Non-Proliferation of Nuclear Weapons*. Information Circular 153, INFCIRC/153, corr. www.iaea.org/Publications/Documents/Infcircs/

IAEA, 1998: *Model Protocol Additional to the Agreements between States and the (IAEA) for the Application of Safeguards.* Information Circular 540, INFCIRC/540, corr. 1.

IAEA, 1999: *The Physical Protection of Nuclear Material and Nuclear Facilities*. Information Circular INFCIRC/225, Rev. 4.

IAEA, 2001: *IAEA Safeguards Glossary*, 2001 edition.

IAEA, 2002: *Decommissioning costs of WWER-440 nuclear power plants,* Interim report. IAEA-TECDOC-1322, 2002.

IAEA, 2005a: *Innovative Small and Medium Sized Reactors: Design features, safety approaches and R&D trends*. IAEA-TECDOC-1451, 2005.

IAEA, 2005b: *Multinational Approaches to the Nuclear Fuel Cycle: Expert Group Report Submitted to the Director General of the International Atomic Energy Agency.* IAEA, INFCIRC/640, 2005.

IAEA, 2005c: *Environmental Consequences of the Chernobyl Accident and their Remediation: Twenty Years of Experience: Report of the Chernobyl Forum Expert Group.* www.who.int/ionizing_radiation/chernobyl/IAEA_Pub1239_web%5b1%5d.pdf

IAEA, 2007a: *Energy, Electricity and Nuclear Power: Developments and Projections*, p.47.

IAEA, 2007b: *Milestones in the Development of a National Infrastructure for Nuclear Power*.

IAEA, 2007c: *Status of Small Reactor Designs without On-site Refueling*.

IAEA, 2007d: *Nuclear Safety Review for 2007*.

IAEA, 2007e: *Management of Reprocessed Uranium: Current Status and Future Prospects,* IAEA-TECDOC-1529.

IAEA, 2008a: *Spent Fuel Reprocessing Options,* IAEA-TECDOC-1587.

IAEA, 2008b: *Financing of New Nuclear Power Plants*.

IAEA, 2008c: *IAEA Annual Report 2007*, Tables A1 and A2.

IAEA, 2009a: *Nuclear Technology Review 2009,* Report by the Director General to the IAEA General Conference, GC(53)INF/3, July 31, 2009, Fig. A-1; see also slight changes in *Nuclear Technology Review 2010*, Fig. A-3.

IAEA, 2009b: *World Distribution of Uranium Deposits (UDEPO) with Uranium Deposit Classification*, IAEA-TECDOC-1629.

IAEA, 2009c: *Reports to the IAEA concerning Policies and Stocks of Civilian Plutonium*, Information Circular 549, INFCIRC-549.

IAEA, 2010a: *Nuclear Energy Share in Generation, 2008,* www.iaea.org/cgi-bin/db.page.pl/pris.nucshare.htm (accessed January 16, 2010).

IAEA, 2010b: *INPRO – International Project on Innovative Nuclear Reactors and Fuel Cycles,* www.iaea.org/OurWork/ST/NE/NENP/NPTDS/Projects/INPRO/index.html

IAEA, 2010c: *Strengthened Safeguards System: Status of Additional Protocols,* www.iaea.org/OurWork/SV/Safeguards/sg_protocol.html (accessed January 16, 2010).

IAEA, 2010d: *Energy, Electricity and Nuclear Power Estimates for the Period up to 2050*.

IAEA, 2010e: H. H. Rogner, IAEA, personal communication, April 13, 2010.

IAEA, 2010f: *International Status and Prospects of Nuclear Power,* Report by the Director General to the IAEA Board of Governors and General Conference, GOV/INF/2010/12-GC(54)/INF/5, September 2, 2010.

IAEA-CEP, 2008: *Reinforcing the Global Nuclear Order for Peace and Prosperity: The Role of the IAEA to 2020 and Beyond*, IAEA Commission of Eminent Persons, www.iaea.org/NewsCenter/News/PDF/2020report0508.pdf

IAEA, INSAG, 2006: *Strengthening the Global Nuclear Safety Regime*, International Nuclear Safety Group, INSAG-21. www-pub.iaea.org/MTCD/publications/PDF/Pub1277_web.pdf

IAEA-PRIS, various years: IAEA Power Reactor Information System. www.iaea.org/programmes/a2/

IBR, 2004: *Russian Uranium Enrichment Industry, State and Prospects*, Annual Report 2004, Table A.7.1. International Business Relations, LLC.

IBR, 2008: *Perspectives for Phase 4 of the Enrichment Plant in China,* International Business Relations, LLC.

ICNND, 2009: *Eliminating Nuclear Threats*, International Commission on Nuclear Non-Proliferation and Disarmament, www.icnnd.org/reference/reports/ent/index.html

IEA, 2008: *Energy Technology Perspectives, 2008*. International Energy Agency.

IEA R&D Statistics, various years: International Energy Agency. www.iea.org/Textbase/stats/rd.asp

India, Ministry of Power, 2008: Speech of Sushilkumar Shinde on U.S.–India Business Council's "Green India" Summit at Washington, Press Information Bureau, Government of India, October 16.

International Court of Justice, 1996: *Advisory Opinion on the Legality of the Threat or Use of Nuclear Weapons.* www.prop1.org/2000/icjop1.htm

International Herald Tribune, 2008: "Électricité de France agrees to pay $23 billion for British Energy," September 24. www.iht.com/articles/2008/09/24/business/edf.php

IPFM, 2007: *Global Fissile Material Report 2007*. International Panel on Fissile Materials. www.fissilematerials.org/ipfm/site_down/gfmr07.pdf

IPFM, 2008: *Global Fissile Material Report 2008,* International Panel on Fissile Materials. www.fissilematerials.org/ipfm/site_down/gfmr08.pdf

IPFM, 2010: T. B. Cochran, H. A. Feiveson, W. Patterson, G. Pshakin, M. V. Ramana, M. Schneider, T. Suzuki and F. von Hippel, *Fast Breeder Reactor Programs: History and Status*, International Panel on Fissile Materials. www.fissilematerials.org/ipfm/site_down/rr08.pdf

IPCC-PSB, 2007: *Climate Change 2007: The Physical Science Basis.* Contribution of Working Group I to the Fourth Assessment Report of the Intergovernmental Panel on Climate Change, S. Solomon, D. Qin, M. Manning, Z. Chen, M. Marquis, K.B. Averyt, M. Tignor and H.L. Miller (eds), Cambridge University Press.

IPCC-SRES, 2000: *Special Report on Emissions Scenarios*. A special report of Working Group III of the Intergovernmental Panel on Climate Change, N. Nakicenovic et al. (eds), Cambridge University Press.

Isaacs, T., 2006: Radwaste management: going underground, *Nuclear Engineering International Magazine*, January 12, 2006.

JAEC, 2004: *New Nuclear Policy Planning Council Interim Report Concerning the Nuclear Fuel Cycle*, Japan Atomic Energy Commission, November 12, 2004, Table 6, http://cnic.jp/english/topics/policy/chokei/disposalcost.html#table6 (accessed December 21, 2008).

JAEC, 2008: White Paper on Nuclear Energy 2007, Japan Atomic Energy Commission, March 2008, p.103 (in Japanese), www.aec.go.jp/jicst/NC/about/hakusho/hakusho2007/3.pdf

Japan-METI, 2006: "Main Points and Policy Package" in *Japan's Nuclear Energy National Plan*, Ministry of Economy, Trade and Industry. www.enecho.meti.go.jp/english/report/rikkokugaiyou.pdf

Japan-METI, 2008: Ministry of Economy, Trade and Industry, *Outlook for Long-Term Energy Supply and Demand by 2030*, Ministry of Economy, Trade and Industry. www.enecho.meti.go.jp/topics/080523.htm (in Japanese).

Japan Times, 2010: "Rokkasho nuke plant delayed two more years," September 11, 2010.

JNFL, 2007: *Start of the Centrifuge Cascade Test using Uranium Hexafluoride*, Japan Nuclear Fuel Limited. www.jnfl.co.jp/english/press_release/presse2007/20071112–1.html

JNFL, 2010: '*MOX Fuel*', Plutonium Utilization for Light Water Reactors, Japan Nuclear Fuel Limited. www.jnfl.co.jp/english/mox.html (accessed January 16, 2010).

Kalinowski, M.B. and Colschen, L.C., 1995: International Control of Tritium to Prevent Horizontal Proliferation and to foster Nuclear Disarmament, *Science and Global Security*, **5**:131

Kang, J. and F. von Hippel, 2005: Limited proliferation-resistance benefits from recycling unseparated transuranics and lanthanides from light-water reactor fuel, *Science & Global Security* **13**: 169.

Kasperson R. E., O. Renn, P. Slovic, H. S. Brown, J. Emel, R. Goble, J. X. Kasperson, and S. Ratick, 1988: The social amplification of risk: a conceptual framework, *Risk Analysis* **8**(2): 177–187.

Kastchiev, G., W. Kromp, S. Kurth, D. Lochbaum, E. Lyman, M. Sailer, and M. Schneider (coordinator), 2007: *Residual Risk: An Account of Events in Nuclear Power Plants since the Chernobyl Accident in 1986*. www.greens-efa.org/cms/topics/dokbin/181/181995.pdf

Katsuta T. and T. Suzuki, 2006: *Japan's Spent Fuel and Plutonium Management Challenges* International Panel on Fissile Materials, 2006.

Kehoe, R.B., 2002: *The Enriching Troika: A History of URENCO to the Year 2000*. Urenco.

Khlebnikov, N., D. Parise, and J. Whichello, 2007: Novel technologies for the detection of undeclared nuclear activities, in *Addressing Verification Challenges, Proc. International Safeguards Symposium, Vienna, October 16–20, 2006* (IAEA-CN-148/32), www.npec-web.org/Essays/20070301-IAEA-NovelTechnologiesProject.pdf

Kikuchi, M. and K. Inoue, 2002: *Role of Fusion Energy for the 21 Century Energy Market and Development Strategy with International Thermonuclear Experimental Reactor*, World Energy Council.

Koomey, J. and Hultman, N.E., 2007: A reactor-level analysis of busbar costs for US nuclear plants, 1970–2005, *Energy Policy*, **35**: 5630–5642.

Kubota, H., 2009: China's nuclear industry at a turning point, *E-journal of Advanced Maintenance* **1**(3). www.jsm.or.jp/ejam/Vol.1.No.3/GA/6/article.html

Labar, M. P., 2002: *The Gas Turbine – Modular Helium Reactor: A promising option for near term deployment*, General Atomics, GA-A23952.

Levi, D. and E. Holder, 1988: Psychological factors in the nuclear power controversy, *Political Psychology*, **9**(3): 445–57.

Lyman, E. 2008: Can nuclear plants be safer? *Bulletin of the Atomic Scientists*, September/October 2008.

Maisonnier, D., et al, 2005: The European power plant conceptual study, *Fusion Engineering and Design* 75–19: 1173.

Mark, J. C. 1993: Explosive properties of reactor-grade plutonium, *Science & Global Security*, **4**: 111.

Mark, J. C., T. Taylor, E. Eyster, W. Maraman, and J. Wechsler, 1987: Can terrorists build nuclear weapons? In P. Leventhal and Y. Alexander (eds), *Preventing Nuclear Terrorism*, Lexington Books, p. 55.

McCombie, C. and N. Chapman, 2008: *A Nuclear Renaissance without Disposal?* Presentation at a *Conference on International High-level Radioactive Waste Management*, Las Vegas, Nevada, September 2008.

Meier, O., 2006: Iran and foreign enrichment: A troubled model, *Arms Control Today*, January/February 2006, www.armscontrol.org/act/2006_01–02/JANFEB-iranenrich.

Mian, Z., A.H. Nayyar, R. Rajaraman, and M.V. Ramana, 2006: *Fissile Materials in South Asia: The Implications of the U.S.-India Nuclear Deal* (International Panel on Fissile Materials).

Miller, A. I., 2001: Heavy water: A manufacturer's guide for the hydrogen century, *Canadian Nuclear Society Bulletin* 22(1).

MIT, 2003: *The Future of Nuclear Power*, http://web.mit.edu/nuclearpower/

MIT, 2009: Yangbo Du and John E. Parsons, *Update on the Cost of Nuclear Power*, Center for Energy and Environmental Policy Research, Massachusetts Institute of Technology.

MIT, 2010: I. A. Matthews and M. J. Driscoll, *A Probabilistic Projection of Longterm Uranium Resource Costs*. MIT Report, MIT-NFC-TR-119.

Moorman, R., 2008: *A Safety Re-evaluation of the AVR Pebble Bed Reactor Operation and its Consequences for Future HTR Concepts*. Jülich Forschungszentrum, Jül-4275.

Moscow Times, May 2, 2007: "Putin orders atomic energy holding," p. 5; see also G. Mukhatzhanova, "Russian nuclear industry reforms: Consolidation and expansion," Monterey Institute for International Studies, Monterey Center for Nonproliferation Studies, May 22, 2007. http://cns.miis.edu/stories/070522.htm

Moyer, M., 2010: Fusion's false dawn, *Scientific American*, February, p. 52. www.scientificamerican.com/article.cfm?id=fusions-false-dawn

Najmabadi, F. and the ARIES Team, 2006: The ARIES-AT advanced tokamak, Advanced technology fusion power plant, *Fusion Engineering and Design* 80: 3. www.nap.edu/catalog.php?record_id=12477.

NEA, 1989: *Plutonium Fuel: An Assessment*, OECD Nuclear Energy Agency.

NEA, 2008a: *Nuclear Energy Outlook*, NEA 6348, OECD Nuclear Energy Agency.

NEA, 2008b, *Moving Forward with Geological Disposal of Radioactive Waste*, OECD Nuclear Energy Agency. www.nea.fr/html/rwm/reports/2008/nea6433-statement.pdf

NEA, 2008c: *Uranium 2007: Resources, Production and Demand*, OECD Nuclear Energy Agency.

New Scientist, March 25, 2004: R. Edwards, "Cracks may force shutdown of UK reactors," www.newscientist.com/article/dn7171

New York Times, May 29, 2009: J. Kanter, "Not so fast, nukes: Cost overruns plague a new breed of reactor.

New York Times, **December 16, 2009**: K. Bradsher, "Nuclear power expansion in China stirs concerns."

New York Times, **February 17, 2010**: "DOE delivers its first, long-awaited nuclear loan guarantee."

Nigmatulin, B. and M. Kozyrev, 2008: *Russian Nuclear Energy Power: The Time of Missed Opportunities*, May 6, 2008. www.proatom.ru/modules.php?name=News&file=article&sid=1334 (in Russian).

Nuclear Fuel, **December 19, 2005**: M. Hibbs and A. MacLachlan, "French centrifuge plant will be 'black box' equipped with TC-21 machine."

Nuclear Fuel, **June 18, 2007**: P. Marshall and D. Horner, "Thorp gears up for restart as reprocessing position clarified."

Nuclear Fuel, **July 30, 2007**: M. Hibbs, "Urenco 'not enthusiastic' about plan to set up gas centrifuge plant in Iran."

Nuclear Fuel, **September 24, 2007**: A. MacLachlan, "Russia's Angarsk international enrichment center open for business."

Nuclear Fuel, **July 28, 2008**: P. Marshall, "UK cleanup authority tackling issue of plutonium stockpiles."

Nuclear Fuel, **August 25, 2008**: M. Hibbs and D. Horner, "Canada's enrichment prospects uncertain."

Nuclear Fuel, **April 20, 2009**: A. MacLachlan, "Russia said to woo India as partner in Angarsk IUEC."

Nuclear Fuel, **July 13, 2009**: M. Hibbs, "Reprocessing costs might exceed KHNP's [Korea Hydro and Nuclear Power Company's] spent fuel management fees."

Nuclear Fuel, **November 30, 2009**, A. McLachlan, "Agreements lay foundation for Ukrainian fuel cycle."

Nucleonics Week, **October 17, 1996**: "Accounting Panel Pegs Superphénix (construction plus decommissioning) Cost at FF 60-billion to 2000."

Nucleonics Week, **September 27, 2007**: A. MacLachlan, "WANO warns safety lapse anywhere could halt 'nuclear renaissance'."

Nucleonics Week, **February 21, 2008**: P. Radtke, "FPL says cost of new reactors at Turkey Point could top $24 billion."

Nucleonics Week, **March 5, 2009**: A. MacLachlan, "Areva reveals 47% cost overrun on contract for Olkiluoto-3."

Nucleonics Week, **March 19, 2009**: A. Sains, "Okiluoto-3 arbitration could last 'several years', TVO says."

OFES, 2009: *Research Needs for Magnetic Fusion Energy*. Office of Fusion Energy Sciences. www.burningplasma.org/renew.html

Oil and Gas Journal, **August 11, 2008**: "Nuclear heat advances oil shale refining in situ."

Oppenheimer, J.R., 1945: Letter to General L. Groves, July 1945, quoted by A. Wohlstetter, *Foreign Policy* 25: 160.

Oshkanov, N.N., M.V. Bakanov, and O.A. Potapov, 2004: Experience in operating the BN-600 unit at the Belyiyar nuclear power plant, *Atomic Energy* **96**(5): 315.

Pacala, S. and R. Socolow, 2004: Stabilization wedges: solving the climate problem for the next 50 years with current technologies, *Science* **305**(5686): 968–972.

Park Seong-won, M. A. Pomper, and L. Scheinman, 2010: *The Domestic and International Politics of Spent Nuclear Fuel in South Korea: Are We Approaching Meltdown?* Korea Economic Institute.

Perkovich, G., D. Choubey, R. Gottemoeller, J. T. Mathews, and S. Squassoni, 2005: *Universal Compliance: A Strategy for Nuclear Security*, Carnegie Endowment for International Peace, www.carnegieendowment.org/files/univ_comp_reportcard_final.pdf

Petti, D.A., et al., 2006: ARIES-AT safety design and analysis, *Fusion Engineering and Design* 80: 111.

Ramana, M. V., 2007: Heavy subsidies in heavy water: Economics of nuclear power in India, *Economic and Political Weekly*, August 25, p. 3483. www.cised.org/wp-content/uploads/heavy-subsidies-in-heavy-water-epw.pdf

Ramana, M. V. and J. Y. Suchitra, 2009: Slow and stunted: plutonium accounting and the growth of fast breeder reactors in India, *Energy Policy* 37: 5028–5036.

Reuters, August 1, 2007: "TVA to complete Tennessee Watts Bar 2 reactor."

Reuters, December 27, 2009: "South Korea wins landmark Gulf nuclear power deal," www.reuters.com/article/idUSLDE5BQ05O20091227

Rhinehammer T. B and L. J., Wittenberg, 1978: *Evaluation of Fuel Resources and Requirements for the Magnetic Fusion Energy Program*, MLM-2419, Mound Facility, Ohio.

Rohani, H., 2005: Secretary of the Supreme National Security Council, "Beyond the Challenges Facing Iran and the IAEA Concerning the Nuclear Dossier," Text of speech to the Supreme Cultural Revolution Council (place and date not given): Tehran Rahbord, Journal of the Center for Strategic Research, September 30, 2005, pp. 7–38 (in Farsi) FBIS-IAP20060113336001.

Rosa, E. A. and R. E. Dunlap, 1994: Nuclear power: three decades of public opinion, *Public Opinion Quarterly* 58(2): 295–324.

Rosatom, 2006a: "The first offshore nuclear heat and electrical power plant of small capacity is planned to operate in October 2010 in Severodvinsk (Artkhangelsk district)." Press release, December 15, 2006.

Rosatom, 2006b: "The project for the construction of floating nuclear thermal power plants (FNTPP) was discussed during a conference on Dec. 20." Press release, December 20, 2006.

Rosatom, September 26, 2008: V. Karogodin, "Russia will construct 12 nuclear power units in other countries by year 2020," www.rosatom.ru/news/11994_26.09.2008 (in Russian).

Rosatom, 2010: "The Russian Government approved FTP 'Nuclear Power Technologies of New Generation in 2010–2015 and until 2020'." Press release, January 22, 2010. www.rosatom.ru/en/about/press_centre/news_ROSATOM/index.php?id4=16272.

Russia Federal Target Program, 2006: Development of the Nuclear Power Industry Complex of Russia for 2007–10 and Further to 2015, Decree #605, approved October 6, 2006.

Russia Federal Target Program, 2008: On a Program for the State Nuclear Corporation Rosatom Development in 2009 thru 2015, Decree #705, approved September 20, 2008.

Russia-US Agreement, 1997: Agreement between the Government of the (U.S.) and the Government of (Russia) Concerning Cooperation Regarding Plutonium Production Reactors. www.ransac.org/new-web-site/related/agree/bilat/coreconv.html

Saraev, O., 2007: Rosatom: *Prospects of Establishing a New Technology Platform for Nuclear Industry Development in Russia*, International Congress on Advances in Nuclear Power Plants, Nice, May 13–18, 2007.

Schlueter, R.O. and R.P. Schmitz, 1990: Filtered vented containments, *Nuclear Engineering and Design*, **120**: pp. 93–103.

Schneider, E. and W. Sailor, 2008: Long-term uranium supply estimates, *Nuclear Technology* **162**: 379.

Schneider, M. and Y. Marignac, 2008: *Spent Nuclear Fuel Reprocessing in France*. International Panel on Fissile Materials.

Schneider, M., S. Thomas, A. Froggatt, and D. Koplow, 2009: *World Nuclear Industry Status Report 2009*. German Federal Ministry of Environment, Nature Conservation and Reactor Safety.

Schwartz, S. I., 1998: *Atomic Audit: The Costs and Consequences of U.S. Nuclear Weapons since 1940.* Brookings Institution Press, Washington, DC.

Sethna, H., 1972: India, past achievements and future promises, *IAEA Bulletin*, 1**4**(6): 36–44.

Shwageraus, E., P. Hejzlar, and M. S. Kazimi, 2005: A combined nonfertile and UO₂ PWR fuel assembly for actinide waste minimization, *Nuclear Technology*, 149: 281.

Shimada, 2000, M. Shimada, D.J. Campbell, M. Wakatani, H. Ninomiya, N. Ivanov, V. Mukhovatov and the ITER Joint Central Team and Home Teams, IAEA Fusion Energy Conference, 2000, ITERP/05

SIPRI, 1980: *Internationalization to Prevent the Spread of Nuclear Weapons.* Stockholm International Peace Research Institute (SIPRI), Taylor & Francis, London.

SKB, 2003: *Plan 2003. Costs for Management of the Radioactive Waste Products from Nuclear Power Stations.* Technical Report TR-03–11. www.skb.se/upload/publications/pdf/TR-03–11webb.pdf

Slovic, P., 1993: Perceived risk, trust and democracy, *Risk Analysis*, **13**(6): 675–82.

Slovic, P., 1994: Perception of risk and the future of nuclear power, *Physics and Society*, **23**(1). www.aps.org/units/fps/newsletters/1994/january/ajan94.html

Slovic, P., 1996: Perception of risk from radiation, *Radiation Protection Dosimetry* **68**(3/4): 165–180.

Sokolski, H. (ed), 2008: *Falling Behind: International Scrutiny of the Peaceful Atom*. U.S. Army Strategic Studies Institute, Carlisle, PA. www.npec-web.org/files/20080327-FallingBehind.pdf

Sokov, N., 2006: Construction of Russia's first floating nuclear power plant raises potential nonproliferation issues, Opportunities, *WMD Insights,* September 2006. www.wmdinsights.com/I8/I8_R5_ConstructionofRussia.htm.

Sovacool, B. 2008a: Valuing the greenhouse gas emissions from nuclear power: A critical survey, *Energy Policy* **36**: 2950–63.

Sovacool, B., 2008b: The costs of failure: A preliminary assessment of major energy accidents, *Energy Policy* **36**: 1802.

Suchitra, J.Y., and M.V. Ramana, 2011: The costs of power: Plutonium and the economics of India's prototype fast breeder reactor, *International Journal of Global Energy Issues* **35** (1–23).

Sundaram, C. V., L. V. Krishnan, and T. S. Iyengar, 1998: *Atomic Energy in India: 50 Years*, Department of Atomic Energy, Government of India, Mumbai.

Suzuki, T., 2010: Japan's plutonium breeder reactor and its fuel cycle: fading into the future?, in *Fast Breeder Reactor Programs: History and Status*. International Panel on Fissile Materials, 2010.

Takubo, M., 2008: Wake up, stop dreaming: reassessing Japan's processing program, *Nonproliferation Review*, **15**(1).

Tamada M., et al., 2006: Cost estimation of uranium recovery from seawater with system of braid type adsorbent, *Nippon Geshiryoku Gakkai Wabun Ronbunshi*, **5**(4): 358–363 (in Japanese).

Thomson, J. and G. Forden, 2006: *Iran as a Pioneer Case for Multilateral Nuclear Arrangements*, Science, Technology and Global Security Working Group, MIT, Cambridge, MA.

The Times, January 11, 2008: Race for nuclear Britain begins. www.timesonline.co.uk/tol/news/politics/article3168648.ece

Transportation Energy Data Book, 2009, Oak Ridge National Laboratory, ORNL-6984.

UAE–US, 2009: Agreement for Cooperation Between the Government of the United Arab Emirates and the Government of the United States of America Concerning Peaceful Uses of Nuclear Energy, www.npec-web.org/us-uae/20090115-UsUae-Revised123Agreement.pdf

UCS, 2007: L. Gronlund, D. Lochbaum, and E. Lyman, *Nuclear Power in a Warming World: Assessing the Risks, Addressing the Challenges*, Union of Concerned Scientists, December 2007.

UK NDA, 2009: *NDA Plutonium Topic Strategy: Current Position*. UK Nuclear Decommissioning Authority.

UK NDA, 2010: *Magnox Operating Program (MOP8)*, including addendum.

UNDP, 2002: *The Human Consequences of the Chernobyl Nuclear Accident: A Strategy for Recovery*, February 6, 2002, http://chernobyl.undp.org/english/docs/strategy_for_recovery.pdf

UN, NPT, 2000: *2000 Review Conference of the Parties to the Treaty on the Non-Proliferation of Nuclear Weapons, Final Document*, NPT/CONF.2000/28 (Parts I and II).

UNSCEAR, 2000: *Sources and Effects of Ionizing Radiation*. UN Scientific Committee on the Effects of Atomic Radiation.

URENCO, 2007: *Building for the Future: Annual Report and Accounts, 2007*, www.urenco.com/Content/117/Reports.aspx

URENCO, 2008: *National Enrichment Facility Expansion*, December 12, 2008. www.urenco.com/content/169/National-Enrichment-Facility-expansion-.aspx.

US AEC, 1974: *Draft Environmental Statement on the Liquid Metal Fast Breeder Reactor,* WASH-1535. U.S. Atomic Energy Commission.

US BPO, 2009: *Research Needs for Magnetic Fusion Energy Science,* Final Workshop Report, US Burning Plasma Organization.

US CBO, 2003: US Congressional Budget Office Cost Estimate, *Energy Policy Act of 2003.* www.cbo.gov/doc.cfm?index=4206&type=0.

US CBO, 2008: *Nuclear Power's Role in Generating Electricity.* US Congressional Budget Office. www.cbo.gov/ftpdocs/91xx/doc9133/05–02-Nuclear.pdf

US CIA, 2010: *World Fact Book*, Country Comparison, Electricity Production, Central Intelligence Agency, www.cia.gov/library/publications/the-world-factbook/rankorder/2038rank.html (accessed January 15, 2010).

US Code of Federal Regulations, Part 61. *Licensing Requirements for Land Disposal of Radioactive Waste*, www.nrc.gov/reading-rm/doc-collections/cfr/part061/full-text.html

US DOE, 1997: *Nonproliferation and Arms Control Assessment of Weapon-usable Fissile Material Storage and Excess Plutonium Disposition Alternatives*, DOE/NN-0007, US Department of Energy.

US DOE, 2003: *U.S. Department of Energy Report to Congress, Advanced Fuel Cycle Initiative: The Future Path for Advanced Spent Fuel Treatment and Transmutation Research.* US Department of Energy.

US DOE, 2006: *Report to Congress: Spent Nuclear Fuel Recycling Program Plan*. US Department of Energy.

US DOE, 2008a: *Requests for Nuclear Loan Guarantees far Exceed Funds*, US Department of Energy, October 15, 2008, http://apps1.eere.energy.gov/news/news_detail.cfm?news_id=12041 (accessed December 19, 2008).

US DOE, 2008b: *Analysis of the Total System Life Cycle Cost of the Civilian Radioactive Waste Management Program, Fiscal Year 2007*, DOE/RW-0591, US Department of Energy.

US DOE, 2008c: *U.S. Department of Energy, Fiscal Year 2009 Congressional Budget Request*, US Department of Energy, "Nuclear Hydrogen Initiative", Vol. 3a, p. 611.

US DOE, 2008d: *Next Generation Nuclear Plant Licensing Strategy: A Report to Congress.* US Department of Energy. www.ne.doe.gov/pdfFiles/NGNP_report-toCongress.pdf

US DOE, 2010a: "Secretary Chu Announces Blue Ribbon Commission on America's Nuclear Future," US Department of Energy press release, January 29, 2010, www.energy.gov/news/8584.htm

US DOE, 2010b: *US Department of Energy, FY 2011 Congressional Budget Request*, vol. 2. US Department of Energy. www.cfo.doe.gov/budget/11budget/Content/Volume%202.pdf US EIA, 2008a: *Nuclear Power: 12 Percent of America's Capacity, 20 Percent of the Electricity*, US Energy Information Administration, www.eia.doe.gov/cneaf/nuclear/page/analysis/solution4.pdf

US EIA, 2008b: *International Electricity Consumption*, US Energy Information Administration, www.eia.doe.gov/emeu/international/electricityconsumption.html.

US EIA, 2008c: *Mid-Term Prospects for Nuclear Electricity Generation in China, India, and the United States.* US Energy Information Administration. www.eia.doe.gov/oiaf/ieo/negen.html (accessed December 16, 2008).

US EIA, 2008d: *U.S. Nuclear Reactor List*, US Energy Information Administration. www.eia.doe.gov/cneaf/nuclear/page/nuc_reactors/operational.xls

US EIA, 2008e: *Electricity Restructuring by State*, US Energy Information Administration. www.eia.doe.gov/cneaf/electricity/page/restructuring/restructure_Elect.html (accessed December 19, 2008).

US EIA, 2009: *Annual Energy Outlook, Early Release Overview*, Report No. DOE/EIA-0383(2009), US Energy Information Administration, December 14, 2009. www.eia.doe.gov/oiaf/aeo/overview.html – elecgen.

US EIA, 2010: *Natural Gas Wellhead, City Gate, and Imports Prices, 1949–2009*, US Energy Information Administration. www.eia.doe.gov/emeu/aer/txt/ptb0607.html

US GAO, 2004: *Nuclear Regulation: NRC Needs to More Aggressively and Comprehensively Resolve Issues Related to the Davis-Besse Nuclear Power Plant's Shutdown*. GAO-04-415. US Government Accountability Office, Washington, DC, May 2004.

USGS, 2007: *Lithium*, US Geological Survey, www.minerals.usgs.gov/minerals/pubs/commodity/lithium/lithimcs07.pdf

US NAS, 1996: *Nuclear Wastes: Technologies for Separations and Transmutation.* National Academy of Sciences, National Academy Press, Washington, DC.

US NAS, 1997: *Review of the Department of Energy's Inertial Confinement Fusion Program: The National Ignition Facility.* National Academy of Sciences, National Academy Press, Washington, DC.

US NAS, 2006a: *Health Risks from Exposure to Low Levels of Ionizing Radiation.* National Academy of Sciences, National Academy Press, Washington, DC.

US NAS, 2006b: *Safety and Security of Commercial Spent Nuclear Fuel Storage.* National Academy of Sciences, National Academy Press, Washington, DC.

US NAS, 2007: *Review of DOE's Nuclear Energy Research and Development Program.* National Academy of Sciences, National Academy Press, Washington, DC.

US NAS-RAS, 2008: *Internationalization of the Nuclear Fuel Cycle: Goals, Strategies, and Challenges*. National Academy Press, Washington, DC.

US NRC, 2002: *NRC's Regulation of Davis-Besse regarding Damage to the Reactor Vessel Head.* Inspector General Report on Case No. 02–03S, US Nuclear Regulatory Commission, p. 23. www.nrc.gov/reading-rm/doc-collections/insp-gen/2003/02–03s.pdf.

US NRC, 2009: *Backgrounder on New Nuclear Plant Designs*, US Nuclear Regulatory Commission. www.nrc.gov/reading-rm/doc-collections/fact-sheets/new-nuc-plant-des-bg.html (accessed January 16, 2010).

US NRC, 2010a: *Combined License Applications for New Reactors*, US Nuclear Regulatory Commission. www.nrc.gov/reactors/new-reactors/col.html (accessed January 11, 2010).

US NRC, 2010b: *The Pressurized Water Reactor*, US Nuclear Regulatory Commission, www.nrc.gov/reading-rm/basic-ref/students/animated-pwr.html (accessed March 25, 2010).

US OTA, 1984: *Nuclear Power in an Age of Uncertainty*. US Congressional Office of Technology Assessment.

van der Zwaan, R., 2008: Nuclear waste repository case studies: The Netherlands, *Bulletin of the Atomic Scientists,* May 19, 2008.

van Leeuwen, J.-W. S., 2008: *Nuclear Power-The Energy Balance*, www.storms-mith.nl

von Hippel, F., 2007: *Spent Fuel Management in the United States: The Illogic of Reprocessing*, International Panel on Fissile Materials.

von Hippel, F., 2008a: Rethinking nuclear fuel recycling, *Scientific American*, May, p. 88.

von Hippel, F., 2008b: National fuel stockpiles: An alternative to a proliferation of national enrichment plants? *Arms Control Today* 20. www.armscontrol.org/act/2008_09/vonhippel

von Hippel, F., 2010: South Korean reprocessing: An unnecessary threat to the non-proliferation regime, *Arms Control Today* 22. www.armscontrol.org/act/2010_03/VonHippel

von Hippel, F., 2011: The radiological and pschological consequences of the Fukushima Daiichi accident, *Bulletin of the Atomic Scientists*, Septermber 2011.

Weinberg, A. M., 1994: From technological fixer to think-tanker, *Annual Reviews of Energy and the Environment* 19: 15–36.

Westinghouse, 2009: AP1000 Documentation, www.ukap1000application.com/AP1000Documentation.aspx.

Willrich, M. and T. B. Taylor, 1974: *Nuclear Theft: Risks and Safeguards*: Ford Foundation Energy Policy Project.

WISE, 1999: The curious case of Vandellós-1, World Information Service on Nuclear Energy, Plutonium Investigation **16**: 4.

WISE, 2007: China's emerging anti-nuclear movement, World Information Service on Nuclear Energy, *Nuclear Monitor* **663**: 1, November 29.

WMD Commission, 2006: *Weapons of Terror: Freeing the World of Nuclear, Biological and Chemical Arms.* International Weapons of Mass Destruction Commission. www.wmdcommission.org

WNA, 2007: *Decommissioning Nuclear Facilities*, information paper, World Nuclear Association, www.world-nuclear.org/info/inf19.html.

WNA, 2008: *Small Nuclear Power Reactors*, World Nuclear Association, www.world-nuclear.org/info/inf33.html.

WNA, 2010a: *Nuclear Power in Russia*, World Nuclear Association, www.world-nuclear.org/info/inf45.html (accessed January 11, 2010).

WNA, 2010b: *Nuclear Power in China*, World Nuclear Association, www.world-nuclear.org/info/inf63.html (accessed January 11, 2010).

WNA, 2010c: *Nuclear Power in the USA*, World Nuclear Association, www.world-nuclear.org/info/inf41.html

WNA, 2010d, *Nuclear Power in the United Kingdom*, World Nuclear Association, www.world-nuclear.org/info/inf84.html (accessed January 16, 2010).

WNA, 2010e, *Nuclear Power in Swede*n, www.world-nuclear.org/info/inf42.html (accessed October 12, 2010).

WNA, 2010f, *Nuclear Power in Netherlands*, World Nuclear Association, www.world-nuclear.org/info/inf107.html (accessed October 24, 2010).

WNA, 2010g: *Nuclear Power in Spain*, World Nuclear Association, www.world-nuclear.org/info/inf85.html (accessed October 24, 2010).

WNN, June 25, 2007: "Enrichment capacity at Angarsk to be boosted," *World Nuclear News.*

WNN, December 4, 2008: "EDF plans for future nuclear growth," *World Nuclear News.* www.world-nuclear-news.org/C-EdF_plans_for_future_nuclear_growth-0412084.html (accessed December 20, 2008).

WNN, December 5, 2008: "Eskom shelves new nuclear project," *World Nuclear News*. www.world-nuclear-news.org/NN-Eskom_shelves_new_nuclear_project-0512084.html

WNN, February 5, 2009: "Fortum submits application for Loviisa 3," *World Nuclear News.*

WNN, March 18, 2009: "Georgia PSC approves new Vogtle units." *World Nuclear News.*

WNN, June 3, 2009: "Forsmark for Swedish nuclear waste." *World Nuclear News.*

WNN, November 5, 2009: "Japan starts using MOX fuel." *World Nuclear News.*

World Bank, 2009: *Gross Domestic Product 2009*. World Development Indicators database, World Bank, December 15, 2010. http://siteresources.worldbank.org/datastatistics/resources/gdp.pdf

World Energy Assessment, 2000: *Energy and the Challenge of Sustainability.* UN Development Programme.

Wynne, B., 1992: Risk and social learning: reification to engagement, in S. Krimsky and D. Golding (eds), *Social Theories of Risk*, Praeger, Westport, CT, pp.275–297.

Zaitseva, L., 2007: Illicit trafficking in radioactive materials, *Nuclear Black Markets*, International Institute for Security Studies, pp. 119–138.

Zhang, H. and F. von Hippel, 2000: Using commercial imaging satellites to detect the operation of plutonium-production reactors and gaseous-diffusion plants, *Science & Global Security* **8**: 219.

15

Energy Supply Systems

Convening Lead Authors (CLA)
Robert N. Schock (World Energy Council, UK and Center for Global Security Research, USA)
Ralph Sims (Massey University, New Zealand)

Lead Authors (LA)
Stan Bull (National Renewable Energy Laboratory, USA)
Hans Larsen (Technical University of Denmark)
Vladimir Likhachev (Russian Academy of Sciences)
Koji Nagano (Central Research Institute of Electric Power Industry, Japan)
Hans Nilsson (FourFact, Sweden)
Seppo Vuori (VTT Technical Research Centre of Finland)
Kurt Yeager (Electric Power Research Institute and Galvin Electricity Initiative, USA)
Li Zhou (Tsinghua University, China)

Contributing Author (CA)
Xiliang Zhang (Tsinghua University, China)

Review Editor
John Weyant (Stanford University, USA)

Contents

Executive Summary

A sustainable future depends on more efficient use of the Earth's abundant energy resources in order to meet the rapidly increasing demand for energy services as well as to provide broader access to everyone. In 2005 the overall efficiency of the energy system from primary energy to useful energy was only about 34%. Owing to diverse geographic inequities in both sources and people, supply cannot always meet the demand where needed. Energy pathways from source through conversion, transmission, storage, and distribution to end-users are complicated and presently consist of numerous discrete pathways that differ widely for each energy source and carrier. These include solid fuels, liquid fuels, gaseous fuels (including hydrogen), electricity and heat. Aging equipment, congested networks, and extreme demands complicate this picture in many countries of the Organisation for Economic Co-operation and Development (OECD). Development of new infrastructure in both non-OECD and OECD countries will lock-in future dependence on conventional or non-conventional energy sources. This chapter aims to assist decision-makers by providing up-to-date knowledge on the full range of energy pathways, their management, and operation. Energy systems to achieve a sustainable future should be made much more flexible in order to deal with societal needs and the probable deployment of technologies not yet commercially available (such as smart appliances, electric vehicles, fuel cells, and carbon capture and storage). Technology and policy solutions are available for supporting more energy for sustainable development, but in order to meet the transition necessary to avoid unacceptable events such as social unrest and/or climate change driven temperature rise, they should be put in place rapidly, and done in concert with each other.

Major energy supply pathways[1] today include oil, natural gas, and coal fossil fuel sources, nuclear energy, and renewable energy converted to energy carriers mostly as liquids, electricity and/or heat, and then used in households, for transport, and by industry. In future, larger contributions can be expected from purpose-grown energy crops and waste streams, wind, geothermal, small hydro and solar energy, as well as natural gas and, at least in some places, nuclear power. Electricity, and perhaps hydrogen, are expected to play increasing roles in the overall energy system, and perhaps with a trend away from centralized energy systems to decentralized and distributed generation. Increased use of electricity is key to greater energy access and more energy for sustainable development across the world, and to increasing energy efficiency throughout the energy supply-chain. Electricity is distinguished by its capacity to transform a broad array of raw energy resources efficiently and precisely into useful energy services (such as building thermal comfort and street lighting, motor-drives for industrial applications and mobility, heating and cooling), irrespective of scale. Local, state and national governments will continue to place a high priority on supplying heating, cooling, transport and particularly electricity to their citizens.

The goal for the energy system should be to make the transition from analogue to modern digital technologies having optimized interconnected networks using up-to-date information and communications technology and linking energy carrier, transmission and distribution sub-systems. In planning, constructing and operating such an energy system, redundancy, robustness and flexibility are critical. Integrated energy storage is an area where technology lags and needs intense development if systems with optimum overall efficiency gains are to be attained.

In most cases, the private sector will lead in developing and deploying the most effective energy system approaches, but will need to work closely with a stable governance policy framework. Success will depend on rapid creation and implementation of robust global-scale public-private partnerships that will, in turn, depend on unprecedented levels of integration and cooperation between the partners. This must be accompanied by increased investment in, and development of cleaner energy-sector technologies together with their wide deployment and efficient management and operation.

---

1   Quantities of energy carriers required vary according to the energy end use efficiency level achieved, see Chapters 8–10 and 17.

15.1   Overview

Chapter 15 reviews recent literature on the movement and conversion of energy from its source to end-use. The various conversion steps involve "energy carriers[2]," which may include a secondary (converted) form of energy, such as electricity, heat, or hydrogen, or a primary energy source itself, such as natural gas, wind, or biomass (Figure 15.1). For all the known sources of energy displayed in the lower left part of the Figure, from crude oil to nuclear, as one moves across the horizontal line, the intersections denote conversions of the energy source to a form in which energy is moved to a point of use. The size of the dot at the intersection denotes which conversions are currently major uses of sources for an energy carrier. Moving from left to right, the first two lines represent solid fuels (coal, peat or woody biomass); the next two vertical lines represent liquid fuels (gasoline, diesel, jet fuel, biofuels); the following two lines represent gaseous fuels (natural gas, biomethane, propane or hydrogen); and the two lines, farthest to the right, represent electricity and heat forms of energy carriers. Energy end-uses are shown at the top.

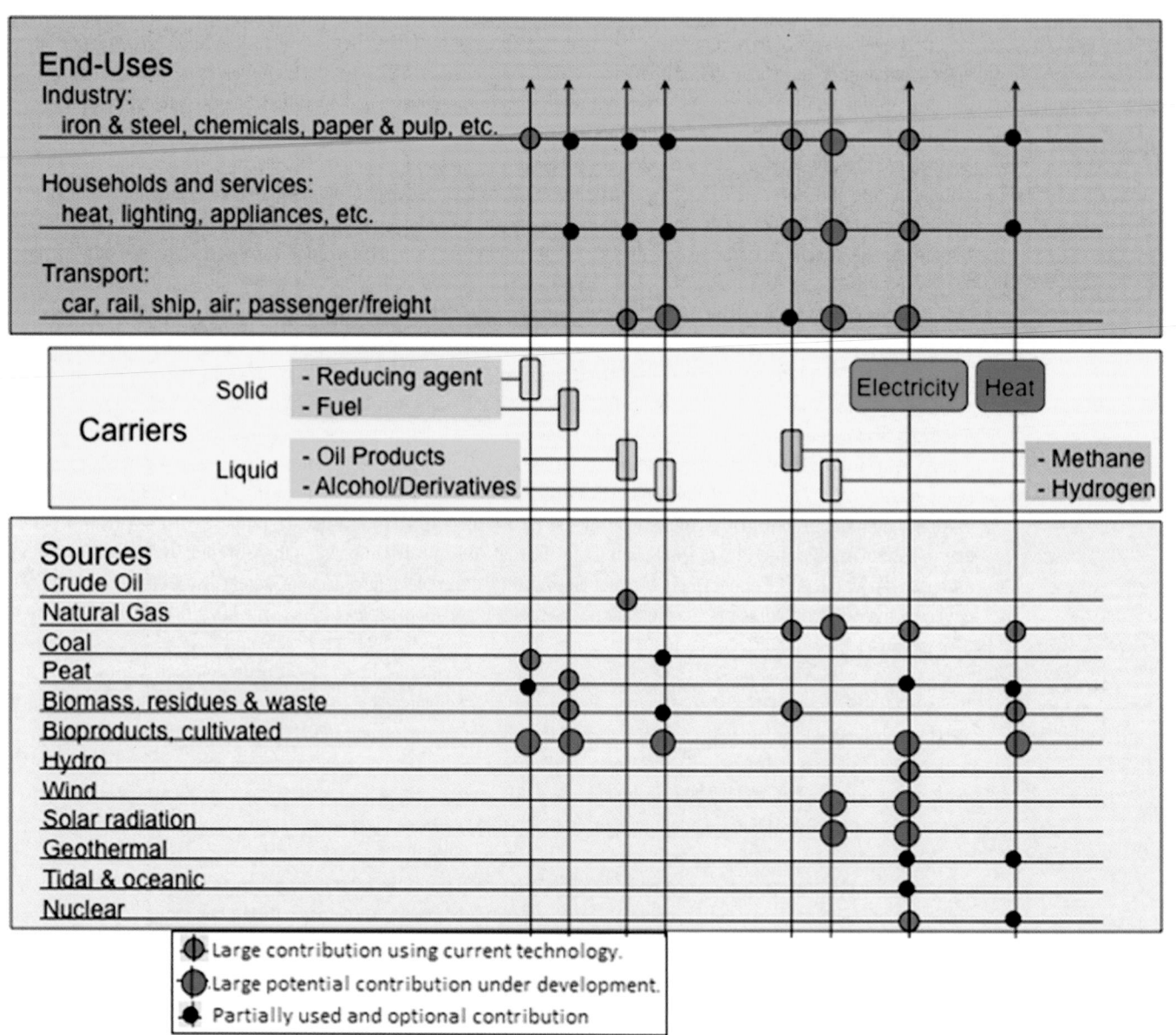

**Figure 15.1 | Relationships from energy sources through conversion to energy carriers and end-uses, showing intersections where current technologies make large contributions, where the intersection is partially used and/or optional and where underdevelopment and a large future contribution to conversion and distribution is expected.

2   An energy carrier is a substance or phenomenon that has the ability to produce mechanical work or heat, or to operate chemical or physical processes (from ISO 13600).

The largest dots indicate the major uses of energy carriers for generalized areas of end-use, e.g., vehicles, homes, industry.

Primary and secondary energy can be converted to provide energy services in the forms of heating, cooling, lighting, entertainment, motor engines, and mobility (transport). This Chapter assesses the interrelationships between energy sources, conversion technologies, and energy carriers (referred to as "secondary energy" in Chapter 1), in terms of both their distribution and delivery to end-users today, and as innovative possibilities for the future. It presents potential pathways to a sustainable future. Energy sectors differ between regions, between major economies, and between developing and developed countries. The chapter describes varying approaches toward the transitions necessary for the creation of sustainable development, while noting how and why supportive energy policies that work successfully in one region may fail in another.

The focus is on energy carriers and the distribution systems for these carriers, including, but not limited to, electricity, solid fuels, liquid fuels, gaseous fuels, and heat. The chapter examines technologies and their potential development for these conversion and distribution steps. It gives more space to electricity because of the advanced state of electricity production and distribution networks, their ubiquity in modern economies, and the future potential for electricity to power heat pumps and electric vehicles, thereby displacing some current fossil fuel demand in the heat and transport sectors. Other energy carrier and distribution systems can learn a great deal from the electricity supply and distribution network.

Research, development, and demonstration (RD&D) in this area of energy supply and operation is assessed in terms of expenditures, both current and likely future costs, along with an examination of where funds may have to come from and where investments will need to be made (see Chapter 24 for a comprehensive discussion of RD&D). Critical questions include whether future competition will or should be between energy sources alone, between end-uses, between sources and end-uses, between integrated systems, or some combination of these. In this Chapter, current policies are assessed for effectiveness, and then related to more effective action in the future (see also Chapters 7–16 as well as Chapter 22).

15.2   Background

Throughout history, mankind's ability to live in harmony with the natural environment and its resources has depended on the availability of energy. Only through broader and more efficient access to the world's diverse energy resources, made possible by continuous advances in science and technology, can a sustainable global energy future be achieved. The overall efficiency of the energy supply chain (extraction, conversion, delivery, and use) has reached only about 34% (Figure 1.5). Major opportunities for improving efficiency exist at every link in this chain.

For millennia, human societies have obtained useful energy for heating and cooking by the combustion of "traditional" solid fuels, such as wood, animal dung, whale blubber, and other forms of biomass. In addition, manual labor and animal power depended on chemical stores of energy in the form of food and animal feed. Coal became prominent during the mid-19th century replacing much human and animal power as the northern hemisphere industrialized. The liquid-fuel era began in the late 1800s as the first oil fields were developed. It was not until after the middle of the 20th century that oil began to dominate the global primary energy mix. Natural gas was also consumed to a limited degree over this period, but real growth in demand only occurred in the past two to three decades as gaseous fuels began to replace liquid fuels. This on-going trend from solid to liquid to gaseous fuels continues to be driven by cost and convenience.

Between 1850 and the early 1900s, as human and animal power was replaced, primary global energy demand grew from around 17 EJ/yr (0.4 Gtoe/yr) sourced from traditional biomass and coal, to around 42 EJ/yr, as oil and gas now made up around 25% of the mix. Throughout the 20th century, the world's population grew, people in developed countries became more mobile, and more countries became industrialized. As a result, primary energy demand continued to increase, growing by over 10 times that original figure (see Chapter 1, Figure 1–3). By 2008, it was about 500 EJ/yr.

There are major differences between nations in energy use when considered on a per capita basis. Residents of the least developed countries, such as Eritrea, Haiti, and Senegal, use around 12 GJ/person each year on average, whereas people living in the wealthiest regions of the world (such as the Middle East and the United States) use 30–40 times more per person. The global average annual energy use is currently around 70 GJ/capita (average rate of 2.3 kW). China is 25% below average, but rising. Overall, each year, the billion wealthiest people on the planet use more than 25 times more primary energy than the poorest two billion who still rely on traditional biomass fuels to provide their very limited energy services.

Energy use per unit of gross domestic product (GDP) is another indicator of regional differences, as are greenhouse-gas emissions per unit of GDP. Countries in the Organization for Economic Cooperation and Development (OECD) generally have lower greenhouse gas emission intensities (around 0.5–0.7 $kgCO_2$-eq/GDP US$PPP[3]) than transition and developing countries, which can have intensities up to three times higher due to the higher use of fossil fuels and less efficient energy conversion equipment including boilers, vehicles, and power plants (IPCC, 2007).

The critical requirement for transforming energy into refined carriers (electricity or heat) from the locations where it is converted into useful work and services is a system to transfer the energy (so that it is readily available where and when it is needed). Any leakage or loss during the

---

3   PPP = purchasing power parity.

transmission and distribution process will result in increased cost and emissions per unit of energy for final use. Electricity, oil and natural gas transmission networks cover hundreds and even thousands of kilometers, linking energy conversion centers and energy users. These networks have been developed to meet the needs of large facilities, such as central electrical generation plants or large manufacturing plants, often located at some remote distance from demand. For electricity systems, transmission and distribution networks account for 54% of the global capital assets (IEA, 2008a).

Energy systems have four overall requirements: they should meet demand fluctuations, they should adapt to supply fluctuations, they should operate efficiently, and they must abide by regulations. Energy transfer may often, but not always, involve storing the energy in some form that can be utilized at a later time. This infrastructure demands storage, transmission, and distribution elements, all of which are connected with an energy carrier. Both energy carriers and storage systems contain energy that can be converted to useful energy for use at another time or in a different physical location.

Storage decouples energy supply and energy demand, thereby giving greater system flexibility, essential for both technical and economic reasons. Energy may be stored in chemical, mechanical, kinetic, electromagnetic, or thermal forms. Fossil fuels, such as coal and natural gas, already contain stores of energy, whereas electricity, gasoline, ethanol, and hydrogen contain energy derived from some other source. Some forms of energy are easily stored in their existing states: solid fuels require little in the form of containment, whereas liquid fuels require a sealed container. No technology advances are needed, as current methods are sufficient. However, gaseous fuels are not so easily stored for all intended applications, and so new storage technologies should better manage their inherent low density. Devices exist for converting electricity into energy forms that can be more easily stored using advances in technology (including pumped hydro, compressed-air, compressed natural gas, batteries, flywheels, super-capacitors, and superconducting magnetic systems). Such storage is useful for enabling more cost-effective and efficient, and therefore expanded, use of electricity, particularly in small autonomous systems and in systems with high penetration levels of variable renewable energy (such as wind and solar photovoltaics). Heat is generally stored simply by heating a thermal mass (e.g., bricks, stones, concrete, and water) to an elevated temperature and then extracting the heat for later use. Technology developments could lead to much better use of heat energy. The status and potential of storage technologies will be described later in this Chapter, along with the corresponding energy carrier.

The concept of an energy pathway from the energy source to its end-use, with storage, transmission, and distribution in between, sounds simple. However, in actual applications, it may not be this simple three-step bridge, but may comprise storage-transmission-distribution-storage, or some other combination. Moreover, in the latter example, the two storage steps may be substantially different. It is therefore essential to take an overall integrated system approach – or energy pathway – to a life-cycle-based, benefit-cost analysis of energy-supply operations. The other key element that requires robust development is the infrastructure for each of the energy carriers or combinations thereof.

In the case of electricity, it is widely recognized that aging infrastructure needs to be modernized, but that alone is insufficient. A "grid of the future" or "smart grid" could be developed to accommodate energy sources in a more optimal way. In the case of hydrogen, for example, it is not a modification of the infrastructure that is needed, but rather an entirely new infrastructure specifically tailored for a hydrogen carrier and storage system.

Aging equipment, congested networks, and extreme peak-load demands contribute to system losses and low reliability, especially for electricity systems in developing countries that often require substantial upgrades. Existing infrastructure also needs to be modernized to improve security, to add information and controls, and to reduce emissions. Future infrastructure and control systems will become more complex in order to be able to handle higher and more variable loads; to recognize and dispatch energy at distribution and end-use sites; and to integrate variable and decentralized sources with higher load flows, frequency oscillations, and quality (e.g., electric voltage or gas pressure) without reducing system performance. In general, energy supply systems need to have more flexibility to accommodate higher penetrations of variable renewable energy. These problems are exacerbated by different degrees of decentralization and differing strengths of distribution systems, so that one solution cannot be applicable to all.

Superconducting cables, sensors, and rapid response controls that could help reduce electricity costs and line losses are already available or close to commercial development. System management can also be improved by incorporating devices that will help efficiency, for example, by rapidly routing flows on the grid, and introducing advanced pricing methods (such as time-based pricing, including, but not limited to, time-of-use (TOU) pricing, critical peak pricing, real-time pricing, and peak load reduction credits). The aim is to avoid peaks, as well as to take advantage of less expensive, off-peak electricity. Energy security challenges from technical failures, theft, physical threats, and geopolitical actions will become more important. Although co-utilization by various end-uses (e.g., plug-in hybrid vehicles) is introduced in this report, it will be developed in Chapter 16, with particular discussion of the concept and its related economic, social and environmental consequences.

In planning, constructing, and operating energy systems, we should not simply focus on a cost optimized system, but also include a sufficient level of risk preparedness, which may be implemented in terms of redundancy, robustness, or flexibility. These aspects are important because energy is essential for human society, which is now both complex and interdependent, with connections between states and regions forged by a highly globalized market economy. A single incident in an energy system can have enormous negative effects, as we have seen in recent years in several large power-transmission network failures. Those blackouts vividly illustrated how deeply dependent the system is on reliable networks, rather than on traditional one-way routes from suppliers to end-users.

Also, cultural differences, as well as physical differences, mean that local/regional/national energy systems, although they may converge on a common set of best-available technologies, are likely to evolve in different ways that are locally suitable and affordable. This means that they may eventually vary considerably from each other, at least within the time horizon of this assessment.

15.3 Sources

15.3.1 Overview

In terms of final consumer energy use (including heat, electricity, direct combustion of oil, coal, gas and biomass, and other renewable energy products), industry uses about 27%, transportation 28%, residential and commercial buildings 36%, and feedstock 9% (see also Chapter 1). Electricity generation accounts for over a third of the world's primary energy demand, with an average conversion efficiency of just over 40% (ranging from up to 90% for large hydro to less than 15% for some older coal-fired power stations).

Of primary energy use, 40% results in useful heat (IEA, 2008a). Where feasible, using "waste" heat from thermal and geothermal power stations, industrial process heat, district-heating schemes, etc. makes combined heat and power (CHP) systems much more efficient. Almost half of all electricity generated is used in buildings (households, commercial services, and the public sector) and about one-third by industry.

15.3.2 Oil and Gas

Known reserves of oil and natural gas (including unconventional sources) will meet global demand for many decades at current rates of use (WEC, 2010; Moniz et al., 2010). The uses of oil and gas are partly interchangeable (for boiler fuels, light vehicle fuels, power generation), and this compatibility largely explains their correlation in global price fluctuations. The International Energy Agency's *World Energy Outlook 2008* analyzed ultimately recoverable reserves in detail and concluded that remaining proven oil reserves are equal to over 40 years of production at current rates of consumption, while proven gas reserves are equal to over 60 years (IEA, 2008a; see also Chapter 7).

Besides conventional fossil-fuel reserves, unconventional sources of liquid fuels are available from oil shale, oil sands, coal-to-liquids, gas-to-liquids, heavy oils, and biofuels. Unconventional gaseous fuels include coal-bed methane, shale gas, methane hydrates, and clathrates. These are very large in quantity (see Chapter 7) but challenging to extract and refine, and the environmental impacts are many. Therefore, unconventional liquid and gaseous fuels may cost more.

15.3.3 Coal and Peat

From 2003 to 2008, coal was the fastest growing fossil fuel (EIA, 2009) in spite of its relatively high greenhouse gas and local air pollution emissions. Coal is the world's most abundant fossil fuel, and continues to be a vital resource in many countries (see Chapters 12 and 13 for detail). In 2009, it accounted for around 29% of total world energy use (about 153 EJ), primarily in the electricity and industrial sectors. Global proven recoverable reserves of coal are at least 20,000 EJ and an estimated additional possible resource of more than 400,000 EJ for all types. Coal resources have also been discovered on the Antarctic continent and in newly explored parts of Australia, but it is difficult to estimate the exact amounts. For details about the distribution of coal resources, see Chapter 7. In general, the northern hemisphere has more coal than the southern, especially concentrated in temperate and sub-boreal regions. Together, coal resources represent stores of over 12,100 GtC. Consumption in 2009 introduced approximately 14.7 $GtCO_2$/yr into the atmosphere.

Peat (partially decayed plant matter together with minerals) has been used as a heating fuel for thousands of years, particularly in northern Europe. In Finland, it still provides 7% of electricity and 19% of district heating (IPCC, 2007), but globally it is a small resource.

15.3.4 Renewable Energy[4]

Renewable energy, by definition, is essentially time-limitless energy stemming from various energy forms. At present, enough renewable energy flows are captured and converted to provide around 18% of global electricity (mainly from hydroelectric power) and around 4–5% of the global heat market (from solar thermal, geothermal, and modern bioenergy). In addition, traditional biomass for cooking and heating, which is mainly used in rural areas, accounts for around 10% of global primary energy use (IEA, 2008a). Liquid biofuels provide around 1.5% of the world's transport fuels. Detailed analyses of the potentials for each renewable energy resource category are given in Chapter 11 (and in IPCC, 2011) – but major uncertainties must be acknowledged, particularly for biomass and ocean energy.

Most energy scenarios assume rapid growth in renewable energy for the next few decades. For example, the IEA, in its "Blue-MAP" scenario (IEA, 2008a) suggests that to achieve 450 ppm CO_2-eq stabilization of atmospheric greenhouse gases, in tandem with nuclear power and thermal power generation plus carbon capture and storage (CCS), electricity generation from hydro, wind, solar and geothermal would need to rise to around 35% of global total power generation by 2030 and to almost

---

4   See also Chapter 11.

half of total generation by 2050. In GEA pathways, which achieve limiting climate change to less than 2°C compared with preindustrial times, the share of renewable energy of total primary energy ranges from 28–74% in 2050 (Table 17.12). This is supported by recent analysis in the IPCC renewable energy special report (IPCC, 2011). Increased heat demand could be met in part by solar (mainly for water heating and passive-solar building designs), geothermal (from direct heating and ground source heat pumps), and modern biomass (including from crop and forest residues, landfill gas, biogas, and CHP). Assessing the future potential for transport fuels based on renewable energy generated electricity, hydrogen and biofuels is very difficult for light-duty vehicles: the rate of introduction of electric cars, plug-in hybrids, fuel-cell vehicles, high-speed electric trains, behavioral changes, etc., are uncertain, as are the likely improvements to efficiencies of conventional vehicle engines and drive trains.

Increased renewable energy growth is just one component required if the necessary transition of the energy system to combat climate change is to be achieved (together with energy efficiency, nuclear, and CCS, see Chapter 17). Continued extraction and use of fossil fuels without CCS would be unsustainable due to the resulting carbon emissions. The IEA has analyzed the transition needed to achieve a 50% reduction of greenhouse-gas emissions from current levels by 2050 (roughly equivalent to a 2°C temperature rise limit) and has identified this huge challenge (IEA, 2007; see also Chapter 3).

15.3.5   Nuclear Power[5]

In the long term, the potential of nuclear power depends on available uranium resources. Reserve and resource estimates of uranium vary with the assumptions for its use (see Table 7.22, Chapter 7). Used in typical light-water reactors (LWRs), identified reserves and resources of 5.5 Mt of uranium, at prices up to US$130/kg, correspond to about 3200 EJ of primary energy and are sufficient for about 100 years at the 2006 level of consumption (Chapter 7). The total conventional proven (identified) and probable (yet undiscovered) uranium resources are about 16 Mt (9300 EJ). Unconventional uranium resources are contained in phosphate minerals (7–22 Mt), which are recoverable for between US$60–100/kg (NEA and IAEA, 2004, IPCC, 2007, see also Chapter 3). Together the total conventional resources and uranium in phosphate minerals could amount to 38 Mt (22,000 EJ, using current LWR technology) and would last for about 700 years at the 2006 consumption level. Furthermore, huge total amounts of uranium are contained in seawater, albeit in low concentrations, resulting in a resource base about a hundred times larger.

When used in current reactor designs with a "once-through" fuel cycle, only a small percentage of the energy content is utilized from the fissile

isotope ^{235}U (0.7% in natural uranium). Around 60% of the present mining of uranium takes place in three countries: Canada 23%, Australia 21%, and Kazakhstan 16% (NEA and IAEA, 2010).

During recent years, the price of natural uranium has been rather volatile, and in mid-2007 the price peaked at US$297/kg (US$_{2005}$$280/kg). Since then, the price has declined to a level of US$110/kg (US$_{2005}$$104/kg), which is roughly at the level of the long-term average price, as expressed in constant US$_{2007}$$.

Nuclear fuels could also be made from thorium. There are identified and probable resources of about 6 Mt (NEA and IAEA, 2010); India, in particular, has large reserves. Thorium-based reactors appear capable of at least doubling the effective resource base, but the technology is yet to be developed in order to determine its commercial feasibility (IAEA, 2005). The thorium (or ^{232}U) fuel cycle is more proliferation-resistant than the uranium (^{235}U) cycle because fissionable ^{233}U is produced instead of plutonium, and emits high-energy photons, making the material difficult to handle.

15.4   Solid Fuel Applications (including Coal, Biomass, Uranium, and Thorium)

15.4.1   Overview

15.4.1.1   Coal

Total recoverable reserves of coal around the world are estimated at over 20,000 EJ thermal energy (Table 7.18, Chapter 7) – thus reflecting a reserves-to-production ratio of 137 (EIA, 2009). Historical estimates of recoverable coal reserves, although relatively stable, have declined gradually from 1145 Gt in 1991 to 1083 Gt in 2000 to 929 Gt in 2006 (see Chapter 7).

15.4.1.2   Biomass

Biomass is the major source of food and fiber for both people and animals. It is also used for heating, generating electricity, liquid fuels (Section 15.5), structural materials, and chemicals. Sources include forest and agricultural residues, livestock wastes, energy plantations and dedicated crops, and the organic part of waste streams such as municipal solid waste, biogas, and landfill gas. Uses of biomass for heating purposes range from the more traditional, such as cooking (with fuel wood, dung, and crop residues), to more modern applications, such as conversion into industrial process heat and district heating plants (Figure 15.2, Biomass currently provides about 50 EJ/yr of primary energy; see also Chapters 1 and 11).

Energy carriers for biomass include solid fuels such as chips, pellets, briquettes, and logs, as well as liquid biofuels and gases (see Figure 15.1).

---

5   See also Chapter 14.

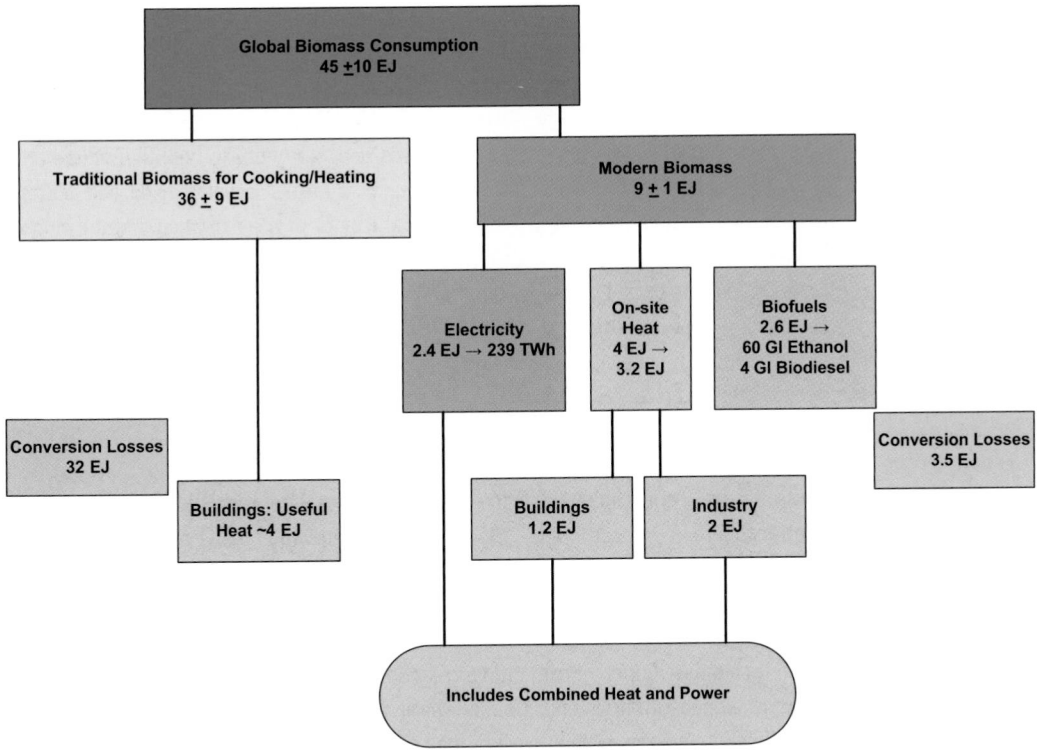

Figure 15.2 | Breakdown of global biomass consumption by type. Source: IEA, 2008a.

Before conversion, biomass feedstocks usually have a lower specific-energy than fossil fuels, and much lower specific-energy than nuclear fuel. Thus, the cost of accumulation, storage, transport, and handling raises the cost-per-unit of energy. In cases where the material is already accumulated for other reasons, such as residues from wood or sugarcane processing, these costs can be drastically reduced when the bioenergy is used on site.

15.4.1.3 Uranium[6]

Nuclear reactors reflect three basic fuel cycles. Conventional thermal reactors operate in a "once-through" mode, where spent fuel is sent directly to disposal. Thermal reactors can operate in a "closed" fuel cycle, where waste products are separated from unused and recycled fissionable material. Fast reactors use reprocessed fuel in a closed cycle, which dramatically increases the nuclear fuel supply. Each of these nuclear fuel cycles has its advantages and disadvantages in addressing the four core issues related to the use of nuclear energy (cost, safety, waste, and nuclear-weapons proliferation). The development choice depends on the priorities placed on each of these issues, and how soon the deployment decision must be made. Even the most advanced closed-cycle reactor designs could be ready for large-scale commercial deployment by 2030 if the necessary development investment is promptly committed (see also Chapter 24).

With fast reactors in a closed fuel cycle, reprocessing of spent fuel and extraction of the un-utilized uranium and plutonium, reserves of natural uranium may be extended to several thousand years at current consumption levels. In the recycle option, fast-neutron spectrum reactors utilize depleted uranium, and only plutonium is assumed to be recycled, so that the uranium resource efficiency is increased by a factor of 30 (Pauluis and Van den Durpel, 2001). As a result, the estimated availability of the total conventional uranium resources corresponds to about 240,000 EJ of primary energy. If advanced breeder reactors are introduced in the future to efficiently utilize recycled or depleted uranium and all actinides, the resource utilization efficiency will be further improved by an additional factor of up to eight (NEA, 2006; NEA and IAEA, 2008), depending on the detailed design of these reactors.

15.4.1.4 Thorium

Further technical development is underway, predominantly in India (WNN, 2009). A government committee in Norway studying the opportunities for using thorium as an energy source concluded that current knowledge of thorium-based electricity generation and related geology is not sufficient to assess the potential value for Norway of a thorium-based system for long-term electricity supply. However, the committee recommended that the thorium option be kept open since it may offer the potential to complement the use of uranium, and so strengthen the long-term sustainability of nuclear energy. It is also worth bearing in mind that the thorium

---

6   See also Chapter 14.

cycle leaves only half the amount of long-lived radioactive waste per unit of energy compared with mainstream LWRs (WNN, 2009).

15.4.2   Conversion

Coal is unevenly distributed and abundant, and can be converted to liquids, gases, heat, and electricity, although more intense use demands viable CCS technologies if greenhouse-gas emissions are to be limited. Bioenergy can be used in similar conversion technologies to those used for coal, as outlined below. Although the conversion efficiencies are lower for bioenergy, atmospheric CO_2 concentrations can be reduced, assuming all the biomass used is regrown. Moreover, the process of biomass pyrolysis can produce bio-char as a co-product – and this can be incorporated in the soil as a long-term carbon sink (Lehmann et al., 2006), thereby removing carbon from the atmosphere.

Most coal-fired, electricity-generating plants are a conventional subcritical, pulverized-fuel design, with typical conversion efficiencies of about 35% for the more modern units, although lower for biomass. Supercritical steam plants are in commercial use in many developed countries and are being installed in greater numbers in developing countries such as China. Current supercritical technologies employ steam temperatures of up to 600°C and pressures of 280 bar, delivering fuel to electricity at cycle efficiencies of about 42%. Conversion efficiencies of almost 50% are possible in the best supercritical plants, but these are more costly. Improved efficiencies have reduced the amount of waste heat and CO_2 that would otherwise have been emitted per unit of electricity generation (see also Chapters 12 and 13).

Technologies have changed little since those reported in the IPCC's 4th Assessment Report (IPCC, 2007). Supercritical plants are now built to an international standard. An Australian Commonwealth Scientific and Industrial Research Organization (CSIRO) project is underway to investigate the production of ultra-clean coal that reduces ash below 0.25%, and sulfur to low levels. With combined-cycle direct-fired turbines, it can reduce greenhouse-gas emissions by 24% per kWh, compared with conventional coal-fired power stations. Gasifying coal before conversion to heat reduces the emissions of sulfur, nitrogen oxides, and mercury, resulting in a much cleaner fuel, while reducing the cost of capturing CO_2 emissions from the flue gas. Continued development of conventional combustion integrated gasification combined cycle (IGCC) systems is expected to further reduce emissions (IPCC, 2007).

Coal-to-liquid (CTL) is well-understood and regaining interest, but will increase greenhouse-gas emissions significantly without CCS. Liquefaction can be performed by direct solvent extraction, and hydrogenation results in liquid production at up to 67% conversion efficiency. Indirect gasification can also produce liquids by employing the Fischer-Tropsch (FT) process, producing synthetic diesel fuel (80%) plus naphtha (20%) at 37–50% thermal efficiency. Lower-quality coals reduce thermal efficiency, whereas co-production with electricity and heat increases it,

thereby reducing costs by around 10% (IPCC, 2007; see also Chapters 12 and 13).

The Fischer-Tropsch process is performed in three main types of reactors: fluidized-bed, slurry-phase bed, and fixed-bed. Many companies are competing to further develop fixed bed and slurry reactors. The conversion efficiency of the fixed bed and the slurry bed is almost the same. The main advantages of slurry-bed reactors are lower operating and reactor construction costs. Fixed-bed reactors are easily scaled up, which means it is easier to realize large capacity. Wax/catalyst separation is easier and costs less than for fluidized- and slurry-bed reactors, which also have high capital costs and difficult process controls (Dry, 2002).

The first coal-to-methanol technology was commercialized by BASF in 1923. A major improvement was achieved in the 1960s by producing a sulfur-free synthesis gas that enabled the British ICI company to use the more active Cu/ZnO catalyst. This resulted in a significant reduction of the compression and heat exchange duty in the recycle loop. The lower reaction temperature also improved the selectivity by virtually suppressing the co-production of light hydrocarbons. Besides reducing the consumption of synthesis gas by a few percent, it also saved 5–10% cooling duty by avoiding the heat released by the side reactions (Lange, 2001). Nowadays, there are several kinds of methanol-composing technologies, and the technology is very mature. In the modern methanol-synthesizing production technology, the low-pressure methanol synthesis process is more widely used than the high- or mid-pressure processes because of the energy saving. ICI and the German Lurgi Company provide this technology for 70% of production facilities (Lou et al., 2006).

Dimethyl ether (DME), a hydrocarbon fuel that can be used to displace liquid petroleum gas (LPG) in vehicles, cooking stoves etc., can be synthesized from coal or biomass by methanol dehydration or direct synthesis. Methanol dehydration is called the "two-step DME process," while direct synthesis is the "one-step process." The two-step process offers some advantages, such as low temperature, high conversion ratio, and high selectivity, but the costs are too high and the process also has serious environmental problems, so it has been gradually phased out.

The one-step DME process uses two kinds of reactors: a gas-phase fixed bed and a slurry-phase bed. The slurry reactor offers a number of advantages: for example, although the DME synthesis reaction is highly exothermic and makes removing the heat of reaction difficult, in a slurry reactor, the heat of reaction is quickly absorbed by a solvent with a large heat capacity and high heat conductivity, and the temperature within the reaction vessel is easily controlled. In addition, there are fewer restrictions on the shape and strength of the catalyst in the slurry bed than in the fixed bed. Generally, the slurry bed could use coal-based syngas, which has a high DME productivity ratio, high selectivity, low energy use, and industrial scalability, with obvious economic and social benefits (Adachi et al., 2000; Lou and Wang, 2006).

Solid biomass is similar to coal in that it can be readily converted into liquids, gases, heat, or electricity. Many conversion technologies are mature for producing bioenergy carriers, while others are under development. The use of biomass products such as sugarcane bagasse, bark, or sawdust used for CHP in industrial, residential, and commercial uses is widespread globally and expanding. Co-firing of biomass with existing fossil-fuel conversion technologies, particularly coal, to produce heat or electricity is a rapidly evolving field. Conversion to liquid biofuels has many routes with various conversion efficiencies, typically around 30–40% (Sims et al., 2010).

The energy converted from uranium or thorium is essentially all used to generate electricity, but with considerable residual heat (see Chapter 14), and thus the energy converted is usually only about 35% using current technology.

15.4.3   Transmission and Distribution

Solid fuels can be costly to transport by road, rail, or ship but less so if the energy density is high. Movement over long distances is usually not cost effective for biomass due to its aforementioned low energy density. It is thus usually utilized near its place of production and harvest (e.g., wood process residues, sugarcane bagasse, cereal straw, energy crops). Wood pellets are an exception as they have a higher energy density (16–18 GJ/t) and about 50 PJ/yr (0.05 EJ) are traded internationally (Zwart, 2010).

Coal, with its higher energy density (20–30 GJ/t) saw about 23 EJ of primary (thermal) energy transported between countries in 2006 (EIA, 2011), representing about 17% of world coal production. There is active movement of uranium ore and enriched uranium (much to Europe), since more than half is currently mined in Australia, Kazakhstan, and Canada (see Section 15.3.5) and of those only Canada now utilizes uranium energy. This is cost-effective owing to the very high energy density of uranium (80 TJ/kg in LWR fuel), one kg being equivalent, on average, to 3000 tonnes of coal.

An innovative technology is being developed to transport coal through a water-filled pipeline over long distances. This "coal log pipeline[7]," technology has low water requirements and costs comparable to existing railways. The coal-to-water-mass ratio is three to one (Marrero, 2006).

Biomass is normally used locally, but transport for sale elsewhere of commercial firewood, wood chips, pellets, and liquid biofuels by truck, train, or boat is commonplace. The energy density is generally too low to pay for long-distance transportation, meaning conversion usually takes place near the source, although millions of tonnes of wood pellets are being shipped annually from the west coast of Canada to Europe. Bioenergy gases can be upgraded and transported in trucks or in natural gas pipelines as blends.

---

7   Coal is pressed into logs and then floated through the pipeline.

15.4.4   Storage

Storage of energy along any of the paths in Figure 15.1 is optional and may be desirable (but is not shown in the figure). Coal is often transported long distances and stockpiled at various points between the source and the user. Thus, coal piles come in many shapes and sizes, from the huge multi-line longitudinal piles frequently found at ports, to ring blending beds at large power plants, to simple conical or irregular piles common at industrial plants. Although many of the same issues that apply to most other bulk materials are encountered when storing coal, the risk of spontaneous combustion makes it a special case. The US National Fire Protection Association (NFPA), in its publications NFPA 850 and 120, identifies the hazards associated with storing and handling of coal and the recommendations against these hazards. The NFPA recommends that storage structures be made of non-combustible materials, and designed to minimize the surface area on which dust can settle. This includes the installation of cladding to reach underneath the building's structural elements. However, because of coal's propensity to heat spontaneously, ignition sources are almost impossible to eliminate, and any enclosed area where loose dust accumulates is at great risk. Furthermore, even a small conflagration can result in a catastrophic secondary explosion if the small event releases a much larger dust cloud (Geometrica, 2006).

If the biomass is seasonally produced, storage is essential if the feedstock is to be supplied to a bioenergy, biogas or biofuel conversion plant all year round. Using a mix of feedstocks in a plant could reduce the storage requirements. Wet feedstocks can be stored as silage but some energy content loss will occur naturally over time, as is also the case for stored dry feedstocks such as straw or woodchips but at a lower rate. Biomass storage can also lead to possible spontaneous combustion, for example bagasse. Liquid biofuels can deteriorate in energy properties over time since they have a biological base, and their cold properties can be problematic for winter and aviation use. Biomass is usually used relatively soon after harvest, but when needed, storage is not an issue for solid or liquid biomass resources. Current trends indicate that investments in technical processes for biomass conversion technologies can pay dividends in reduced costs of utilization (IPCC, 2007). Developing a bioenergy plant can be challenging, particularly in gaining resource consent and securing long-term biomass supplies (IEA, 2007). Current concerns relating to providing sustainable biomass supplies, with minimal impacts from land-use change, related greenhouse-gas emissions, or water supplies, are being debated by several organizations, including the Global Bioenergy Partnership (GBEP, 2011).

15.4.5   Policy and Investment Considerations

15.4.5.1   Raw Materials

As international coal prices spiked in 2008, US coal prices tripled (Nelder, 2008). As Asian production and consumption of coal fell, local suppliers there suffered. The loss of China's coal exports indirectly translated to higher demand for North American coal, driving its price upwards. Reduced output from foreign producers, along with record prices, has meant that more of US coal has been sent to buyers in Europe and in Asia than was the case in 2007.

Biomass resource costs delivered to the processing plant vary with source and location. Processing residues produced and used on site can be around US$1–2/GJ, or even negative, where for the "waste" biomass disposal costs have been avoided by using the resource. Collection and transport of residues from forests and crops are around US$4–6/GJ depending on transport distance, whereas biomass produced from dedicated energy crops can be double this or higher. Transporting biomass materials between locations can spread diseases, pests, and weeds, so some border controls may be advisable. However, such contamination risks can be avoided if sufficient heat for sterilization is generated when the biomass is processed into pellets, bio-oil, liquid fuels, etc. before it is transported. Policies to support biofuel production and use also need to take account of the fact that significant subsidies already exist for the fossil fuel industries (IEA, 2008a), and that sustainable land use practices, including carbon emissions from changes in land use, need careful assessment (Fritsche et al., 2010).

15.4.5.2 Final Product

Production costs of CTL appear competitive when crude oil is around US$35–45/bbl, assuming a coal price of US$1/GJ. Converting lignite at US$0.50/GJ close to the mine could compete with production costs of about US$30/bbl. The CTL process is less sensitive to feedstock prices than the gas-to-liquid (GTL) process, but the capital costs are much higher. An 80,000 bbl/day CTL installation would cost about US$5 billion and would need at least 2–4 Gt of coal reserves available to be viable (IPCC, 2007).

Biomass resources converted to liquids or gases often require government subsidies to make them cost-competitive. Exceptions to this are sugarcane-bagasse-ethanol, produced in Brazil, landfill gas from MSW, and biogas from sewage treatment systems using anaerobic digestion.

15.4.5.3 Policy Considerations

There are significant opportunities for technology to improve the efficiency and environmental performance of coal use. The key is to refine rather than burn coal. Coal refining depends on the conversion of coal under reducing conditions into a synthesis gas composed primarily of methane and carbon monoxide (see Section 15.8.2). This concentrated synthesis gas is purified of contaminants, and can be used either as a clean fuel for relatively high efficiency combustion turbine/combined-cycle power generation, or as feedstock for synthetic petroleum, diesel, aviation fuel, and chemical production, as well as for hydrogen. These coal refineries would

most efficiently operate as flexible, around-the-clock facilities. Their synthesizing gas production would be selectively routed either to electricity generation or liquid fuel/chemical production, as market demands dictate. The necessary technology is well developed and has been demonstrated to be economically competitive under a variety of conditions worldwide.

The future investments needed by the whole coal industry are rather difficult to predict. However, it is possible to estimate them for some application technologies, especially for the main equipment components. This is also true of heat and power from bioenergy. The general flow of CTL technologies is described in Figure 15.3 (see also extensive discussion in Chapter 12).

Static system investment is calculated by adding up all the separate equipment costs. Each budgetary investment is estimated by plant cost indices and an exponential coefficient method. The overnight cost C, of a component having size S, is related to the cost Co, of a single unit of a reference component of similar size.

$$C = \sum C_i = \sum [n \times Co_i \times (S_i / So_i)^f]\qquad(1)$$

where n = domestic factors and f = scale factor.

Table 15.1 gives the investment costs for the main equipment used in CTL technologies as one example. Parameters used for estimating overnight capital costs (including installation, balance-of-plant, general facilities, engineering, overhead and contingencies) are in $US_{2002}\$^{8}$.

Nuclear power has been a technologically dependable choice for filling the gap between reducing dependence on fossil fuels and the deployment of renewable energy. Recent developments in Japan have caused this belief to be reassessed. About 85% of the world's nuclear power generation today is by reactors derived from designs originally developed for naval use. Today, so-called "third-generation" reactor designs include greater design standardization, longer (60 years) operating life, reduced core melt potential, and higher fuel burn-up efficiency for less fuel use and waste. However, expansion will depend on resolving to public satisfaction several currently perceived limiting issues, including cost, waste, accident safety and proliferation.

Over the coming decades, two additional issues must be resolved if nuclear power is to fulfill its service capability. These are the quality and availability of waste repository space, and uranium resource availability. "Fourth-generation" nuclear reactor systems are intended to shape this more robust global future for nuclear energy. For example, the Generation IV International Forum, an international government and industry task force, selected six such advanced systems.

---

8 To convert $US_{2002}\$$ into $US_{2005}\$$, multiply $US_{2002}\$$ value by 1.085.

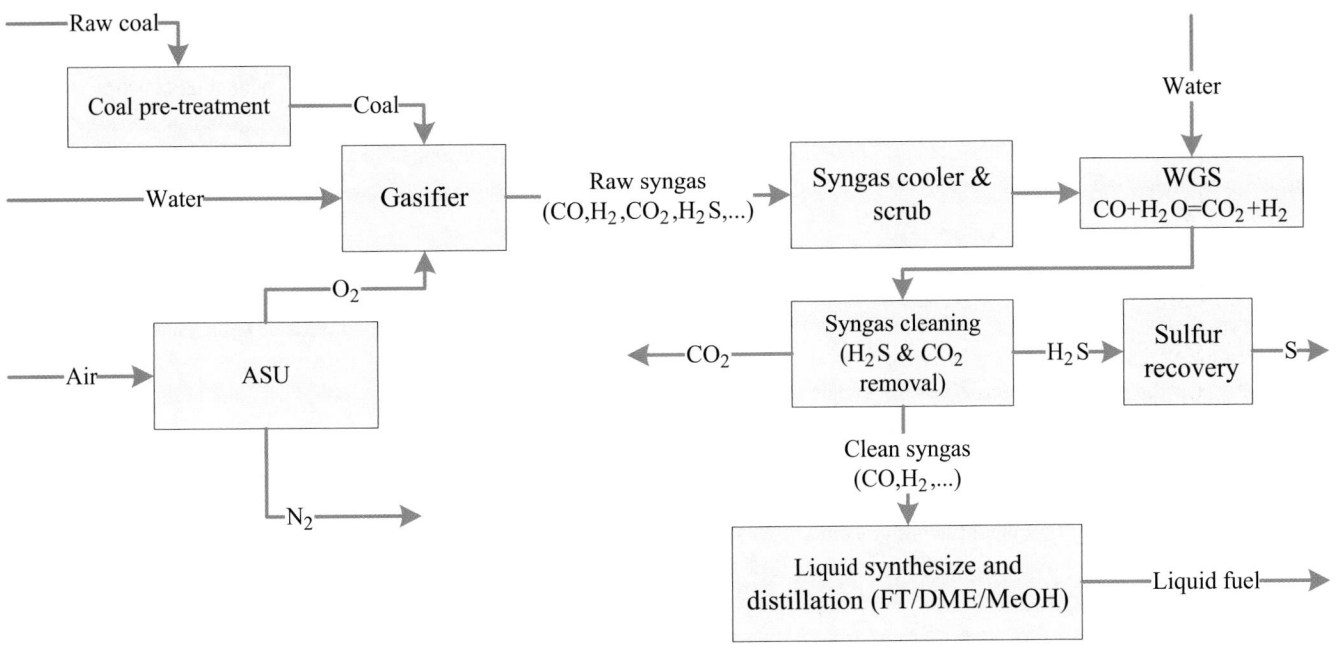

Figure 15.3 | The general flow of coal-to-liquid technologies.

Table 15.1 | Investment (US$_{2002}$$) of main equipment used in CTL technologies.

Equipment or process	Scaling parameter	So	Unit of So	Co (M$)	f
Coal storage, prep, handling	Raw Coal feed	27.4	kg/s	29.10	0.67
Gasifier + syngas cooler & scrub	MAF[9] coal input (LHV)	697	MW	144.30	0.67
Air separation unit (O_2 at 1.05 bar)	Pure O_2 produced	21.28	kg/s	40.40	0.5
O_2 compression (from 1.05 bar)	Compression power	10.0	MW	6.30	0.67
N_2 compressor (GT NO_x control)	Compression power	10.0	MW	4.7	0.67
WGS reactors, heat exchangers	MAF coal input (LHV)	1377	MW	39.80	0.67
Selexol H_2S removal & stripping	Sulfur input	81	t/d	33.6	0.67
Selexol CO_2 absorption, stripping	Pure CO_2 captured	2064.4	mol/s	32.80	0.67
Sulfur recovery (Claus, SCOT)	Sulfur input	29.3	mol/s	22.90	0.67
Fischer Tropsch reactor	FT liquid produced	100	MW	73.5	1
DME recycling synthesize and distillation	Amount of feed gas (H/C=1)	8680	mol/s	87.37	0.65
MeOH recycling synthesize and distillation	Amount of feed gas (H/C=2)	10810	mol/s	81.77	0.65

Source: Tijmensen, 2000; Chiesa, 2003; Kreutz et al., 2005; Zhou et al., 2009.

These are: gas-cooled fast reactors, lead-cooled fast reactors, sodium-cooled fast reactors, molten salt reactors, supercritical water-cooled reactors, and very high-temperature gas reactors. Beyond these nuclear fission systems lies the relatively unlimited, but also unproven, potential of nuclear fusion-based energy. For a more comprehensive discussion, see Chapter 14.

15.5 Liquid Fuel Applications (including Petroleum and Biofuels)

15.5.1 Overview

Most of the world's liquid fuels today are refined from conventional crude oil (petroleum) and used mainly in the transport sector, but liquid energy carriers made from unconventional deposits (heavy oil, oil sands, oil shale, coal-to-liquids, processing of various biomass feedstocks)

---

9   Moisture and ash free coal.

are increasingly being made available in the marketplace where their promise is to reduce the very high dependence on conventional petroleum. Resulting GHG emissions (kgCO$_2$-eq/km traveled) are higher for some alternatives and claims of reduced GHGs from using biofuels are being carefully scrutinized since land use changes are often involved. The utilization of liquid fuels is shown schematically in Figure 15.1.

15.5.1.1   Petroleum

Of the world's total final energy demand, 44% comes from petroleum (Chapter 1). Most of the production comes from the Middle East (30%) and Russia (13%), while almost 50% of consumption is in the United States and Europe (BP, 2010). This results in a significant movement of oil from source to consumers, at a rate of over 50 Mbbl/day. Conventional oil resources are estimated at two trillion barrels, or enough to supply the world for over 60 years at present rates of consumption. However, consumption rates are expected to increase over time. About three times as much oil resources are estimated from unconventional reserves, but these are likely to cost more to extract, even with new technology.

15.5.1.2   Liquid Biofuels

Raw biomass materials can be broadly classified into wet or dry resources as well as liquids or solids. Solid dry sources are best used for thermo-chemical processes, and wet liquids and gases for biochemical conversion. A number of conversion processes are commercially available to produce both bioethanol and biodiesel (triglyceride esters) as vehicle fuels. Other processes still under development include hydrothermal gasification, enzymatic hydrolysis of lignocellulosic feedstocks, biodiesel hydrogenation and the production and use of algal oils.

15.5.2   Conversion

Petroleum is converted in a refinery to consumer products such as gasoline, lubricating oil, or asphalt. Global refinery capacity utilization in 2008 was between 70 and 90% (BP, 2010). Because of demand centers, almost 28% of refinery throughputs are in Europe, whereas the United States accounts for almost 20%. Both these percentages are decreasing, but Asia Pacific countries account for 28% of throughput and this figure is rising. Most recent refinery additions have been in Asia, particularly in China, to meet this increasing demand for petroleum products.

Solid or liquid biomass feedstocks can be converted using numerous technologies to provide more convenient energy carriers in the form of liquid fuels, such as methanol, ethanol, biodiesel, and bio-oil. Biomass feedstocks with a high moisture-content are usually preferred for anaerobic digestion, pyrolysis, or biofuel production, although there is continuing interest in producing advanced biofuels from lignocellulosic feedstocks (Sims et al., 2010). The energy inputs, related carbon savings, and water

demands needed for biofuel vary with the process. High-energy input/output ratios (for example, corn ethanol processing when using coal-fired power and heat) can lead to limited GHG mitigation benefits. Stringent planning regulations, feedstock supply security and sustainable production can constrain biofuel plant developments (IEA, 2007).

15.5.3   Transmission and Distribution

Petroleum is moved, either as liquid crude oil or as refined products, predominantly by pipeline, or, in the case of intercontinental transmission, in oil tankers. The petroleum that moves from the Middle East is mostly by ship and from the former Soviet Union is mostly by pipeline. Current oil markets and trade are shown in Figure 15.4.

Liquid biofuels ranging from raw vegetable oils to highly refined liquid transport fuel blends can be transported in similar fashion, usually by road tanker or ship. Pipelines are also being planned for bioethanol. In Brazil for example, the company Uniduto Logistica has begun construction of a 600 km, US$800 million pipeline from São Paolo to the port of Santos, being the first phase of a major network. In the United States, Magellan Midstream Partners and the biofuel producer Poet are together undertaking a feasibility study of a 2700 km pipeline from northwest Iowa to the New York harbor (IEA, 2008a). At a cost of US$3.5 billion, it would collect ethanol produced in Iowa, South Dakota, Minnesota, Illinois, Indiana, and Ohio. Technical challenges include the corrosive nature of ethanol, its affinity for absorbing moisture, and how it might react with other products and substances within the pipeline. A dedicated ethanol line is therefore preferable, but at a relatively high cost per volume carried if the pipeline is not used to its full capacity.

15.5.4   Storage

Storage of vegetable oils and biofuels over time is problematic due to the biological origin of these materials, as well as their being hazardous and flammable. In general, the standard storage and handling procedures used for petroleum diesel can be used for biodiesel. Ideally, a clean, dark environment free of moisture is required, using storage tanks made from aluminum, steel, polyethylene, or polypropylene. Biodiesel corrodes copper, brass, lead, tin, and zinc, and therefore these metals should be avoided in all components.

For ethanol, some components and equipment used for storing and dispensing of conventional gasoline, as well as for internal combustion engines, do not have adequate compatibility with ethanol or gasoline-ethanol blends above E10. Hence, a range of national and state regulations exists (Wisconsin Department of Commerce, 2005).

Metals such as zinc, brass, or aluminum, commonly found in conventional fuel storage and dispensing systems, are incompatible with ethanol. In addition, components made from natural rubber, polyurethane,

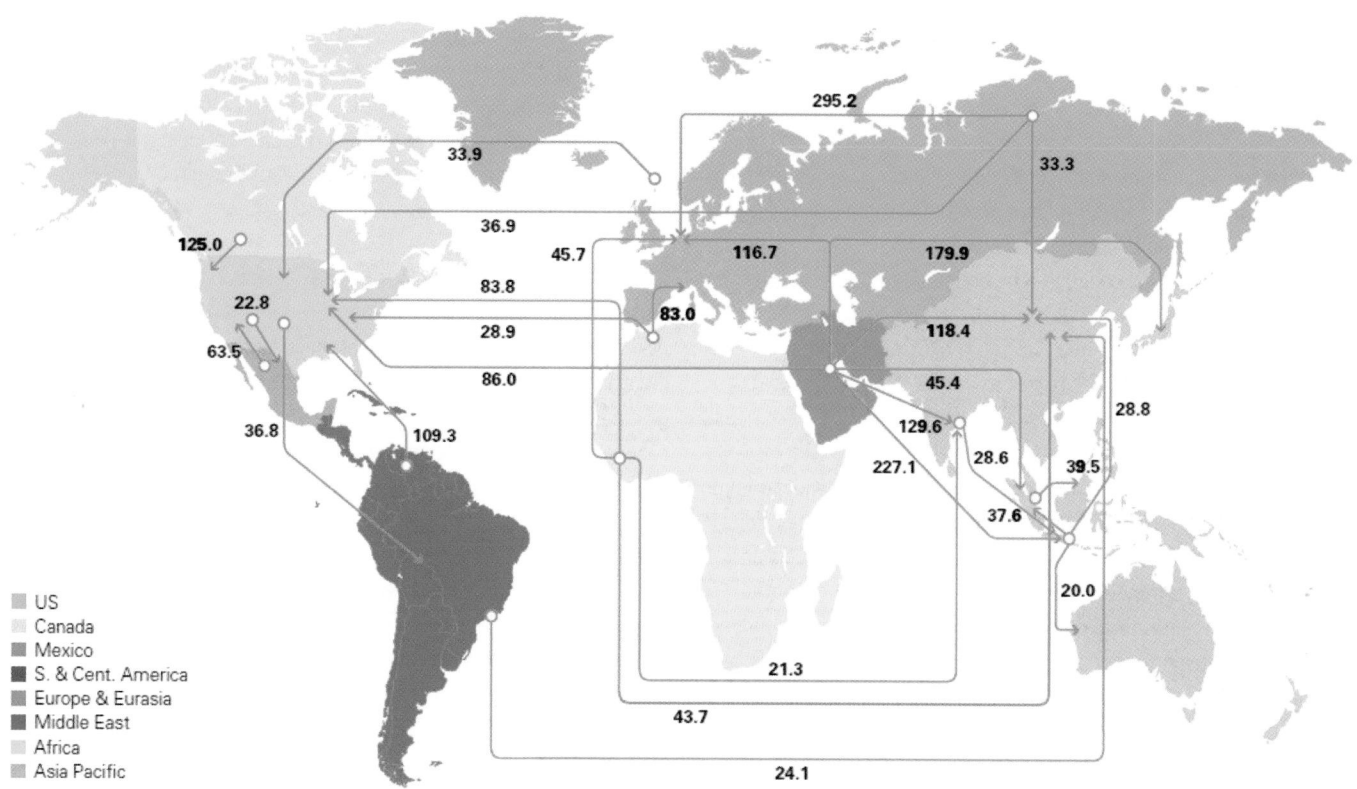

Figure 15.4 | Current oil markets and trade. Source: BP, 2011, data based on Cedigaz and GIIGNL.

cork, certain adhesives, elastomers, and polymers used in flex piping, bushings, gaskets, meters, and filters may also degrade. If water is present, it can cause premature failure of components such as probes, line leak detectors, submersible pumps, fuel dispensers, piping, hoses, and nozzles. Being miscible with ethanol, it can result in phase separation of the blend. Storage in single-walled tanks can leak and condense, so lined- or double-walled underground storage tanks at service stations are usually necessary.

Emissions of liquid effluents or leachates from biofuel plants could affect watercourses, groundwater, and soils, and so need treatment before disposal to land or water. Policy drivers for biofuels are increasing and now exist in around 80 countries and states (REN21, 2011).

15.6 Gaseous Fuel Applications (Including Natural Gas and Biogas)

15.6.1 Overview

Natural gas is the cleanest burning of all fossil fuels, but known reserves are somewhat limited. However, recent successes in extracting gas from deep shale deposits using horizontal drilling techniques, especially in

North America, are releasing more resources worldwide. Estimates of future natural gas availability vary widely from hundreds to thousands of years. Such estimates depend on assumptions about the future rate of use, and the development of technology to drill for gas in more difficult geographical conditions. The utilization of gaseous fuels is shown schematically in Figure 15.1. Methane can also be extracted from coal, peat, and oil shale as well as from organic materials, such as landfill gas or biogas. If these sources can be successfully exploited, the world's methane supply will be extended another 500 or more years (see Chapter 7).

In 2007, the world's proven reserves of natural gas were nearly 177 trillion cubic meters (Tm³), of which 45 Tm³, or 25%, were in Russia (Chapter 7). Following the depletion of cheap local gas, natural gas markets will require larger investments to transport gas from more remote areas. A growing number of market participants, each with different institutional frameworks, leads to higher uncertainties and risks.

Exploration for gas is currently at a significantly lower degree of completeness than for oil, with the exception of certain regions, such as North America. Thus, significant portions of the world's conventional gas resources are yet to be discovered. Future gas exploration is anticipated in the Arctic domains, in fold-belt provinces, and in deep sedimentary basins. Results from recent hydrocarbon exploration, combined with new technological progress, suggest a high potential for future discoveries. Even though exploration in geological fold belts has contributed to past success, it has been focused on shallow-water locations that are the easiest to identify. Deepwater exploration has not been undertaken on any sizeable scale, leaving potential for future major discoveries.

The Arctic basins present a very high potential for gaseous hydrocarbons, since the offshore area is underexplored. On-shore resources, even though more developed, also present significant potential. The remaining exploration potential could, in a favorable scenario, be of the same order as the resources already discovered.

15.6.2 Conversion

As production from the most developed gas fields enters the depletion stage, accelerating depletion rates and falling rig efficiency rates indicate that the gas resource base could be reaching maturity, though annual world production continues to increase (see Figure 7.11, Chapter 7). In the past, natural gas recovered while extracting petroleum could not be profitably transported for sale and was simply flared at the oil field. This wasteful practice is now illegal in many countries. Oil and gas companies now recognize that revenue from the gas can often be achieved by conversion to liquefied natural gas (LNG), compressed natural gas (CNG), or other energy carriers easier to transport.

Producing biomethane from landfill gas collection and from anaerobic digestion of farm and food-processing wastes and dedicated green crops involve mature technologies, with the gas usually being used for heat and power at the local and community scale. After cleanup and compression, the gas can be used to power vehicles, if the engines are designed or converted as if running on CNG. DME produced from coal or biomass can be used in similar fashion to displace liquefied petroleum gas.

Synthesis gas can be produced from biomass (as well as from coal) via gasification, but cleaning the gas and removing tars remains a challenge in some gasifier designs at both small (100 kW) and medium (20 MW) scales. Hydrogen can also be produced from biomass or fossil fuels as well as through water electrolysis. Direct conversion through reduction of water at high temperatures is also possible but a more costly option. Storage of small hydrogen molecules remains an issue, but their use through direct combustion or in fuel cells is reasonably well understood.

15.6.3 Transmission and Distribution

The major difficulty in transmitting natural gas and biomethane is their low density. Natural gas pipelines are presently impractical for transmission across oceans. The development of new technologies for gas

processing and transport, especially over greater distances (Figure 15.5), to a certain extent remove "regional borders." Biomethane can be produced at high enough quality so that it can be injected into natural gas pipelines. The volume of the world's LNG trade, including intercontinental and spot markets, has been growing at 6–7% annually, faster than the growth of pipeline gas exports at 2–2.6% annually. This growth is expected to continue due to technological advances in gas liquefaction and LNG transport and utilization. This means that national and regional markets are no longer isolated from each other.

Capital intensity of LNG production has halved in the past decade, with high-capacity tankers now costing 50–60% less than 10 years ago. Additionally, demand for LNG has been enhanced by the importers' drive to diversify supplies and the overall liberalization of the natural gas market. The pool of LNG exporters is currently limited to Indonesia, Algeria, Malaysia, Qatar, Trinidad and Tobago, Nigeria, and Australia, with Russia joining Atlantic and Pacific markets.

With 15 nations accounting for 84% of the world's gas production, access to natural gas has become a significant factor in international economics and politics, and control over international gas pipelines is a major strategic factor. Natural gas is transmitted through various high-pressure pipelines, forming national and international gas-transmission networks connected to medium- and lower-pressure pipelines, operated by distribution companies that eventually reach end-users. Gas distribution systems are the piping networks that deliver the gas to buildings and businesses, downstream of a city's gate stations. The main problem in distribution systems is ensuring adequate maintenance and monitoring of the safety of these piping networks.

LNG ship carriers transport natural gas across oceans, while tank trucks can carry LNG or CNG over shorter distances. They may transport natural gas directly to end-users, or to distribution points such as pipelines.

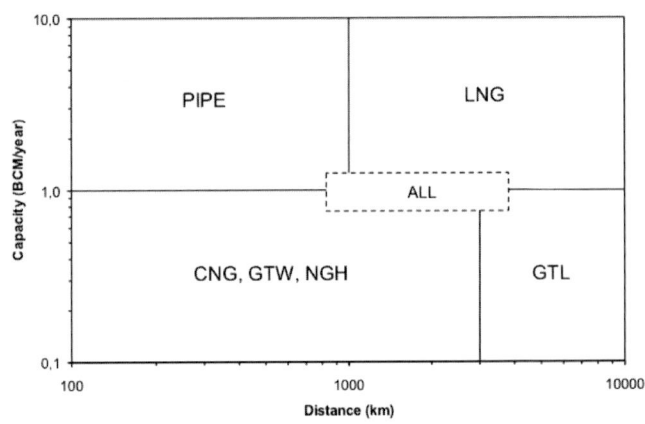

Figure 15.5 | Capacity versus distance for forms of natural gas transport. Pipe is gas pipeline, LNG: liquefied natural gas, CNG: compressed natural gas, GTW: gas to wire (electricity conversion in place), NGH: natural gas hydrate, and GTL: gas-to-liquids. Source: Gudmundsson and Mork, 2001.

Additional facilities incur additional costs – for liquefaction or compression at the production point, and then gasification or decompression at the end-use facility or for a pipeline.

The limited number of countries with a meaningful gas-pipeline export capacity includes Canada (92 Gm³ in 2009), Russia (176 Gm³), Norway (96 Gm³), the Netherlands (50 Gm³), Algeria (31 Gm³), and the United Kingdom (12 Gm³) (BP, 2010; see Figure 15.6).

Any future system development requires the continued creation of a gas infrastructure. This includes:

- ensuring that the growth of LNG production, transportation, and utilization facilities is faster than consumption, thus raising the LNG share of global gas consumption;

- building intercontinental gas pipelines that, by 2020, should include the Europe-Russia-Central Asia-North Africa network, Russia-Central Asia-China-Asia-Pacific network, a network covering much of Latin America, and the Australia part of an Oceania-Southeast Asia network;

- shaping a single system of well-structured technological management standards for gas-transport facilities to promptly reroute gas, including routine and emergency reserves, into regions in need, thus rendering the entire gas transport system more stable and reliable.

During the gas liquefaction process, impurities are removed, and heavy hydrocarbons separated. The material is cooled down until the gas liquefies and is then stored. The construction of a liquefaction plant is a critical factor in the LNG chain, taking the largest portion of overall costs. The power required for operation of a liquefaction plant can be up to 900 MW, as large as a thermal power station. Therefore, the costs of running the liquefaction plant represent a substantial part of the overall costs of the LNG chain, and depend on the energy efficiency of the plant. For example, the use of large axial compressors can reduce energy use by 15% compared with centrifugal compressors.

Significant cost reductions can also be obtained through economies of scale and more advanced technologies to generate electricity. In older plants, the power to drive the compressors is produced with steam turbines (with a thermodynamic efficiency at lower heating value (LHV) of about 30–35%), while more modern units use combined-cycle gas

Major trade movements

Trade flows worldwide (billion cubic metres)

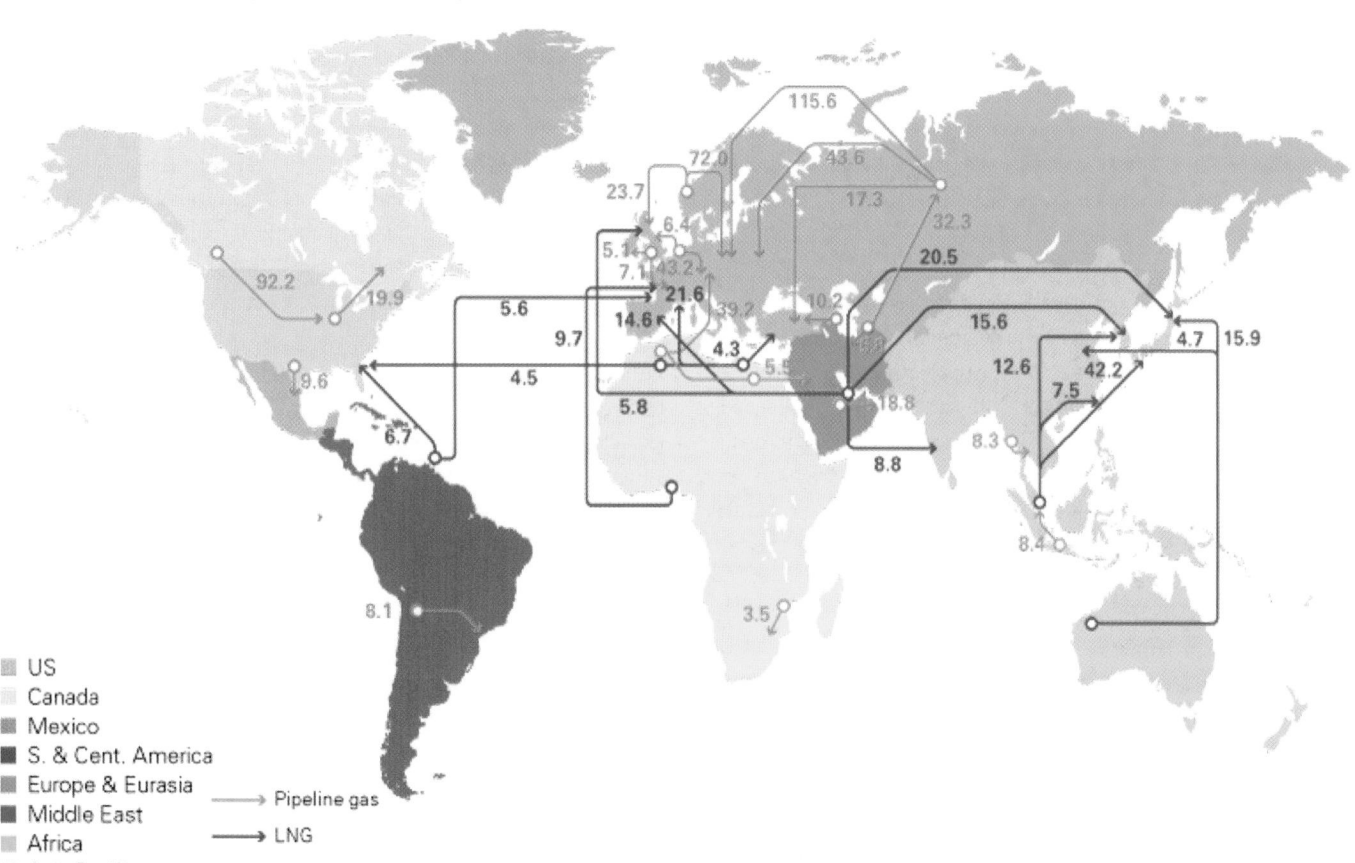

Figure 15.6 | Current gas markets with transport by pipeline or as LNG. Source: BP, 2010, data based on Cedigaz and GIIGNL.

turbines (CCGT). Natural gas fuels the turbine, and the heat generated produces steam, which in turn drives a steam turbine. The overall conversion efficiency of CCGT can be around 50–60%.

While no floating liquefaction plants have yet been constructed, the concept and design is currently being developed. Such floating plants could exploit remote or stranded gas fields that are not economical for fixed platforms. Shorter construction times and the fact that they can be reused for other sites allow the costs of such plants to be spread over several projects. Floating liquefaction plants can also minimize environmental impacts on coasts. The main challenge they present is the carrying out of liquefaction, storage, and loading on a mobile platform, under changing weather and sea conditions.

Due to public opposition to LNG terminal construction and safety issues associated with the operation, two new concepts for a coastal reception terminal have been developed. A structure resting on the seabed is constructed at depths ranging from 15–25 m, in a similar fashion to building offshore concrete production platforms. A floating storage and regasification unit (FSRU) is modeled on the floating production, storage, and off-loading unit from the upstream oil industry. FSRUs range from converted LNG tankers to terminals planned and built specifically for the purpose. Specially designed regasification vessels transport LNG, and then vaporize it to send it through a sub-sea pipeline at the destination port.

Shipping efficiency is essential to reduce transportation costs. A typical carrier can transport 145,000–155,000 m^3 of LNG, which will become about 89–95 Mm^3 standard of natural gas when vaporized. LNG carriers are similar in size to aircraft carriers, but significantly smaller than very large crude oil tankers. Because LNG ships are extremely capital intensive, they cannot afford to have idle time.

LNG vessels usually do not have a liquefaction facility on board and use boil-off gas for propulsion at a maximum of about 0.15% of cargo volume a day. Because the gas cargo supplements the fuel oil, LNG tankers arrive at the destination port with less LNG than was loaded at the liquefaction plant. LNG tankers mainly use steam turbines as their propulsion system. Although the fuel efficiency is low, they can easily be adapted to use the boil-off gas because LNG carriers also normally retain a small percentage of the cargo (the "heel") to cool the tanks down to the required temperatures before reloading.

The regasification terminal is usually the least capital-intensive link in the LNG supply chain and several systems are in operation today. Key selection factors are operational costs (fuel consumption, maintenance), environmental costs, rates of emissions, and availability of the equipment suppliers on the market. Climate and geographical constraints are also important factors.

Gas quality and interchangeability are important issues because the gas regasified from LNG goes into the pipeline system. Also, the two fastest growing importers, the United Kingdom and the United States, require gas caloric specifications that are lower than other traditional gas importers, such as Japan.

In June 2008, 80 LNG liquefaction trains at 19 sites in 15 countries had a total liquefaction capacity of 194 Mt/yr. By August 2008, 65 LNG regasification facilities in 19 countries had a total capacity of 438 Mt/yr (599 Gm^3/yr), and a total storage capacity of 28 Mm^3 of LNG (equivalent to 17.2 Gm^3 of natural gas).

Zero CO_2 emissions for LNG terminals can be achieved by avoiding fuel consumption for regasification. Using open rack vaporizers has been the most common option for many years. Air heat exchangers are also used when climatic conditions are favorable.

Energy integration with other industries saves energy and costs. These include: "cold/hot energy" exchanges at the power plant; direct use of cold energy for industrial processes, air liquefaction, and cryogenic processes; and power production (direct expansion, Rankine cycle).

As the major gas consumption centers become more dependent on gas imports, LNG trade will play an increasing role as a transmitter of price signals between the regional markets of Europe, Asia Pacific, and North America. However, significant differences in the import structures among these markets are likely to remain until 2030 and beyond.

As Europe's indigenous production declines, the resulting deficits will be supplied by both imports of LNG and pipeline gas. North America will predominantly rely on LNG to replace its falling production, and in the Asia Pacific region, rising demand will be met by increases in indigenous production along with LNG imports (but with increased intra-regional trade). The recent findings of large amounts of shale gas will affect this. LNG will fill the supply-demand gaps in all three main gas consumption regions, but the trade of pipeline gas will remain focused on Europe.

Although most natural gas imports to Europe will still be via pipeline under contracts until 2025 and beyond, several new LNG suppliers are expected to emerge. The ability of LNG suppliers to compete in downstream markets is due to its cost-competitiveness with pipeline gas.

Historically, rising gas prices and technological developments have resulted in substantial cost reductions in the LNG business and have helped improve the competitiveness of LNG compared with pipeline gas. Between 1990 and 2000, liquefaction costs fell by 25% to 35% and shipping costs by 20% to 30%. However, more recently, this trend was reversed, with strong increases in steel prices and even stronger increases in the cost of constructing liquefaction plants.

Comparisons between the costs of gas pipeline transport onshore and offshore with the costs of LNG transport have been published since the late 1970s, when LNG transport first arose for the export of Algerian gas. A comparison included in the European Commission's Energy

Sector Inquiry report (European Commission, 2007) was based on varying throughputs of 10, 25, and 40 Gm^3/yr; and for pipeline gas, (including capital expenditure related to laying pipelines on land and building gas compressor stations); and an LNG tanker of 135,000 m^3.

This analysis between pipeline transport and LNG showed break-even gas-transport distances of 3000 km and 6500 km for projects of 10 Gm^3/yr and 25 Gm^3/yr, respectively, with shorter break-even distances for more difficult terrain, and substantially shorter break-even distances for offshore pipelines.

Recent studies suggest that a single LNG train is cheaper than all other options when the distance is over 4500 km. With further cost reductions, the break-even distance will become even shorter in the future. However, it must be kept in mind that choices between LNG and pipeline transportation are rather the exception. North African and maybe Nigerian gas to Europe as well as Middle Eastern gas to Europe are the main cases where a choice is feasible, possibly together with some gas sent from the Gulf to Pakistan and India. Even in these cases, the distances are not the same, due to differences between the shipping route and the pipeline routing onshore. For Japan and North America, LNG remains the major gas import option. Possible pipeline projects for Japan would have to originate from regions that do not supply LNG. Similarly, most gas from Russia is located in the middle of the Eurasian continent and inevitably has to be transported by pipeline.

Gas transport costs for internationally traded gas can be critical. Pricing in a tight market is based on a netback value either derived from the replacement value of gas, or alternatively from the market price of gas in a deep and liquid market. Transport costs are eventually deducted from the revenue of the exporter. For an exporter, the main question is whether or not the net present value of the compensation it receives for the depletion of its resources is attractive. The transport costs are only one element among many for the resource owner to consider when making a decision on whether, how, and where to market its gas.

15.6.4   Storage

Gas is stored during periods of low demand, and withdrawn during periods of peak demand. It is also used for a variety of secondary purposes, including:

- balancing flow in pipelines;
- leveling production over periods of fluctuating demand;
- insuring against unforeseen events; and
- meeting regulatory obligations regarding reliability.

The capacity of underground gas storage (UGS) systems among countries (Table 15.2) is highest in the United States, followed by Russia, Ukraine, and Germany. Russia's working gas volumes include long-term strategic reserves.

Table 15.2 | Number of underground storage facilities and working gas volumes (Mm^3) in 2004/2005 by nation.

Nations	No. of UGS Facilities	Total Installed Working Gas volume of UGS Facilities
		10^6 m^3
USA	385	100,846
Russia*	22	93,533
Ukraine	13	31,880
Germany	42	19,179
Italy	10	17,415
Canada	49	14,820
France	15	11,643
Netherlands	3	5,000
Uzbekistan	3	4,600
Kazakhstan	3	4,203
Hungary	5	3,610
United Kingdom	4	3,267
Czech Republic	8	2,891
Austria	4	2,820
Latvia	1	2,300
Romania	5	2,300
Slovakia	2	2,198
Spain	2	1,981
Poland	6	1,556
Azerbaijan	2	1,350
Australia	4	934
Denmark	2	820
Belarus	2	750
China	1	600
Croatia	1	558
Belgium	1	550
Japan	4	542
Bulgaria	1	500
Ireland	1	210
Argentina	2	200
Armenia	1	110
Kyrgyzstan	1	60
Sweden	1	9
Total	**606**	**333,235**

*including long-term strategic reserves.

Source: IGU, 2006.

The working gas volume of regional UGS facilities in operation in 2004/2005 was 333 Gm^3 (IGU, 2006). The most important type of gas storage is in depleted gas reservoirs, aquifer reservoirs, or salt cavern reservoirs. Each possesses distinct physical and economic characteristics that govern its suitability for a given application. Most of the working-gas volume is installed in former oil and gas fields (Figure 15.7), followed by aquifer structures, and caverns in salt. Abandoned mines and rock caverns are of little relevance on a world scale.

Depleted gas reservoirs are the most common form of storage. They are generally the cheapest to develop, operate, and maintain. Obviously, location is a very important factor in determining whether or not a depleted gas field makes an economically viable storage facility. Aquifer reservoirs can be used for natural-gas storage in some cases. Usually these facilities are operated on a single annual cycle as with depleted gas reservoirs. Salt formations are well suited to natural-gas storage. Salt caverns allow very little of the injected natural gas to escape from storage unless specifically extracted. A cavern is leached in the salt deposit.

15.6.5 Policy and Investment Considerations

Gas demand is still growing rapidly. It already provides 22% of all primary energy (compared with only 10% in 1960). According to IEA estimates, its share may reach 25% before 2030. For the foreseeable future, natural gas will continue to be used primarily for direct residential and commercial heating and cooking, as well as for electric power generation, and industrial heat processes. Power generation will continue to drive the gas market according to long-term forecasts. At the same time, new uses for gaseous fuels in new sectors (such as transport) could be achieved.

Methane is a diverse and flexible fuel which, when combusted, provides both heat and/or electricity. It can also be used in the transport sector in the form of CNG, LNG, or compressed or liquefied biomethane. The use of CNG as a transport fuel is not new, but could continue to grow (Figure 15.8), although improvements in electric vehicles may dampen any dramatic increase in CNG-powered engines. The relatively favorable environmental characteristics of combusting natural gas compared with coal and oil will certainly help it hold its position at the forefront of fossil-fuel consumption.

In countries with well-developed domestic gas markets, liberalization has privatized unbundled segments enabling consumers to choose suppliers through non-discriminatory, third-party access to gas transport systems. These policies have led to the prompt development of spot gas trading. Thus, terms of contracts have been gradually reduced, and their linkage to other energy carriers such as oil, has been replaced with a linkage to gas, with spot-market prices. The result for gas-supplying countries is negative, as it is now more difficult to guarantee returns on large-scale investments within the framework of extremely volatile spot prices. There is a real danger that this situation will destroy the financing tools for large capital-intensive projects that have developed in the last decades on the basis of long-term contracts. Nor does this situation encourage the huge investments necessary for infrastructure and production development. Participants in the gas market simply delay their investment decisions because of high uncertainty. The question of timely investments in gas project development has become more important than ever.

With institutional frameworks changing rapidly in gas-producing, -transit, and -consuming countries, there is a need for additional guarantees of contract fulfillment. The mutual distrust of many market participants makes them protect their traditional territories, leading to energy nationalism and protectionism. Competition between national energy companies (representing mainly gas-producing countries), and international energy companies (representing gas-consuming countries) are increasing.

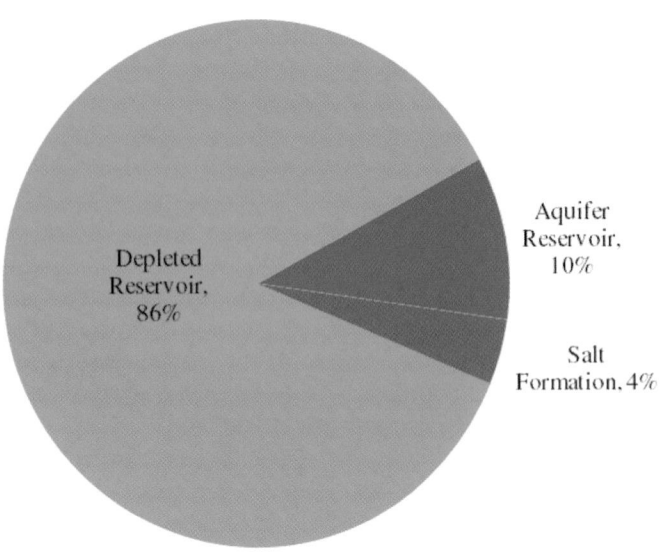

Figure 15.7 | Working gas capacity distribution by underground storage types in 2005. Source: adapted from EIA, 2006.

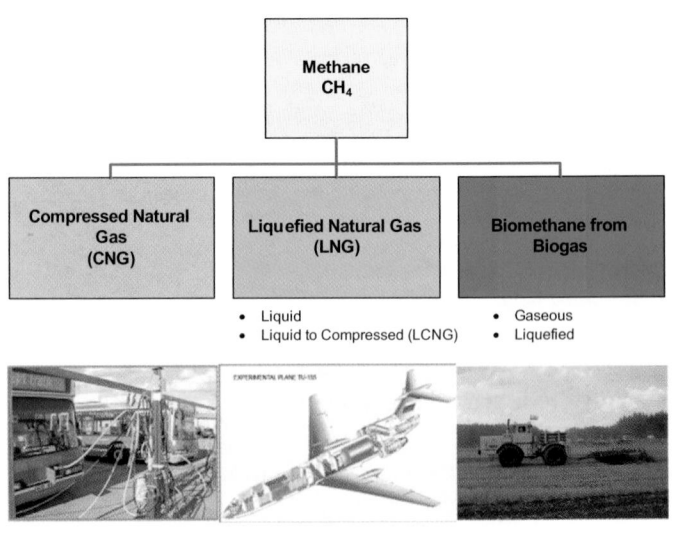

Figure 15.8 | Methane as a diverse fuel. Sources: adapted from Matic, 2006.

Gas pricing is still inconsistent: regional prices usually follow global oil prices with a slight delay. The market is currently divided into three main areas – North America, Europe, and the Asia Pacific – where different infrastructures lead to different prices, supply scheduling, and pipeline/LNG supply percentages. In Asia Pacific, demand is met mainly by LNG, with a premium over European and American prices. Future liberalization of gas markets and streamlining of trading procedures will probably eliminate the premium. LNG tankers are helping to bring regional prices down.

Solid or liquid biomass feedstocks can be converted using numerous technologies to provide more convenient energy carriers in the form of gaseous fuels, such as biogas from anaerobic digestion, landfill gas, synthesis gas from gasification, or hydrogen (see Chapter 11). Biomethane (i.e., scrubbed biogas) can be used directly for a range of local applications or injected into natural gas pipelines. The total market potential of biogas by 2020 has been estimated to be nearly 10 times the current biogas production of over 90 PJ/yr (IEA, 2007). Beneficial environmental impacts from the use of some biomass feedstocks include improving sewage treatment before discharging effluent and sludge to waterways or oceans, avoiding methane emissions from landfills, and reducing odors from direct application of animal wastes to land by first processing in a biogas plant.

Recovery of methane from modern biogas plants, following anaerobic digestion of the wet biomass feedstock, such as animal (Figure 15.9), human, food, and organic wastes and green crops, has increased in recent years. More than 4500 installations (including sewage gas and landfill gas recovery plants) were operating in Europe alone in 2002. The moisture content of slurries and wet biomass feedstocks used with anaerobic digestion plants is usually less important than for combustion of solid biomass, because the feedstock is usually not transported long distances and will not deteriorate in the short time between processing. A lower total solids content of the feedstock liquid can have an adverse effect on the biogas plant efficiency, along with the additional costs of having to store larger feedstock volumes before processing takes place.

Biogas can be directly combusted to produce heat. It can also be fed into natural gas grids or distributed to filling stations for dedicated or dual gas-fueled vehicle engines, or for power generation. For these applications, the biogas first requires scrubbing to remove any corrosive hydrogen sulfide, and carbon dioxide where gas-storage volumes are limited resulting in fairly pure biomethane gas.

Because biomass tends to have relatively low-energy density (whether as a solid, liquid, or gaseous fuel), and is organic, the storage of large volumes can be more costly than equivalent fossil fuels. For example, biogas needs either large plastic or steel storage tanks, or to be compressed and stored in cylinders, both expensive options. Therefore, matching the biogas production rate to the energy demand is the recommended approach to avoid storage costs.

Figure 15.9 | Anaerobic digestion plant and biogas storage tank on a 4000-sow pig farming in South England using two gas engines (housed in the closed shed) to provide electricity for the farm as well as low-grade heat for drying Lucerne horse feed as an ancillary farm operation. Source: courtesy of Ralph Sims.

Gasification of dry, solid biomass produces synthesis (producer) gas consisting mainly of CO and H_2. Development of efficient biomass integrated gasification combined-cycle (BIGCC) systems is nearing commercial realization, but the challenges of gas cleanup, to remove tars and condensates, remain. Several pilot and demonstration projects have been evaluated with varying degrees of success. The gas produced can be used in gas turbines, gas engines, or as feedstock for a range of liquid biofuels based on the Fischer-Tropsch process. Gas transport and storage are not usually problematic, because the gas can be converted onsite to produce heat, power, or liquid fuels as it is produced. Co-firing of biomass with coal in a gasification plant can provide increased efficiency and carbon mitigation, especially if linked with CCS, as can co-firing for heat and power generation. Torrefaction of biomass (bio-coal) using microwaves is reaching commercialization and could be co-fired (Rotawave, 2010). For an extensive discussion see Chapters 11 and 12.

15.7   Heating and Cooling

District heating systems transfer and distribute heat from one or more heating or CHP plants to individual homes, institutions, and industrial consumers, primarily in urban areas. District-heating systems are based on mature technology with a high degree of reliability. The infrastructure to deliver heat is costly, but is cost-effective when utilized efficiently. Furthermore, efficiency depends on climate and whether the infrastructure already exists. For instance, the centralized systems in the largest communities in Finland have district heating in more than 90% of homes. More than 75% of district heating in Finland is produced by CHP. In Sweden, district heating is supplied by a combination of CHP and large heat pumps. In 2008, 50 TWh district heating was produced, out of which 5.5 TWh was supplied by heat pumps and the majority of the rest (71%) from CHP fuelled mainly with woody biomass.

Major improvements in overall efficiency through CHP largely offset losses in transporting heat. In the case of heat applications in industrial processes, CHP loses less heat, as the heat transportation distances are short. Greenhouse-gas emissions are reduced because carbon-neutral wood residues and black liquor from pulp production are usually used to fuel CHP plants. In Japan, there are many cases where heating and cooling systems using heat pumps and/or induced heating have higher overall efficiency and economy than a conventional steam-based system.

Insulation material used for heat transfer pipes is now being improved continually, allowing heat to be distributed over long distances with relatively lower losses. This will facilitate using surplus heat from power plants. Further improvements in the performance of district-heating systems could develop low-temperature systems with reduced losses and more efficient operating conditions for CHP plants.

Efficiency improvements in end-use are a challenge to the district-heating sector. New low-energy houses need very little heat and only at selected times, such as cold winter nights (Larsen and Petersen, 2005). Such houses might be self-sufficient with hot water and will be unreliable consumers. In Europe, many of the district-heating systems are old and need upgrades and maintenance. General development in the electricity sector toward distributed generation works in two ways for the co-generated heat. On the one hand, generated heat is produced closer to the consumer, thereby reducing loss; on the other hand, the system gets more dispersed with smaller producers. As in the electricity sector, there is a compelling need to make the distribution system more intelligent, with closer links between producers and end-users. This is of utmost importance for the future development of sustainable heat-supply systems. Another step towards increased sustainability would be to combine district-heating systems with solar heating and heat storage, optimizing the total system.

Heating of water and building space using solar-thermal systems, ground-source heat pumps, as well as bioenergy (pellets or firewood) at the domestic and small business scale have rapidly growing markets (IEA, 2007). This is due to technological improvements and more competition, and brings the added benefit of increased reductions in greenhouse-gas emissions. Heating demands can be reduced by energy efficiency measures such as insulation, temperature adjustment, behavioral changes, etc. There is also increasing use of heat pumps, air-to-air, and air-to-water.

District-cooling systems are analogous to district-heating systems. The cooling medium is normally water from lakes and the ocean extracted at a temperature of 6–10°C, with a temperature rise of about 10°C at the consumer end before its return to the source. A water-glycol or ice-slurry-water mixture is also used. Cooling energy is produced by compressors using electricity or by absorption machines using district heating from a CHP plant or solar energy, usually at the domestic scale.

The district-heating system might drive cooling systems during the summer. Heat-driven cooling machines might also offset part of the electricity for air-conditioning, thus reducing the peak load of electricity in hotter areas of the world. Solar irradiation energy reaches its highest level around early afternoon, as building-space cooling demands peak. More direct district cooling would require the circulation of chilled water to end-users. Presently, Helsinki and Turku in Finland are among a number of cities that integrate district cooling and heating. District cooling also makes it possible to prolong the utilization period of the maximum load in CHP plants.

15.8   Electricity

15.8.1   Overview

Electricity is the key to greater energy access and more energy for sustainable development across the world, and to increasing energy efficiency throughout the energy supply-chain. Figure 15.1 vividly shows the many sources used to generate electricity and its almost universal use. By mid-century, at least two-thirds of the world's population will be concentrated in urban locations. This further underscores the importance of electricity as the primary carrier of energy from relatively remote production locations to these ever-growing urban population centers (Planck Foundation, 2009).

Electricity's capacity to transform the broad array of raw energy resources most efficiently and precisely into useful goods and services, irrespective of scale, distinguishes it from all other energy forms. Electricity also enables technological innovation and productivity growth – the lifeblood of a modern society. Electricity is indeed the equal-opportunity conversion, delivery, and end-use vehicle for all the world's energy resources (EPRI, 2003b; Yeager, 2007).

The profound impact of electricity on economic development and quality of life indicates that governments around the world will continue to place a very high priority on supplying electrical service to their citizens. Today, the annual individual consumption of electricity ranges from zero to well over 10,000 kWh. The empirical dividing line between advanced and developing economies is about 2000 kWh of annual electricity consumption per person. The resulting "electrification gap" effectively excludes nearly *half* the world's population from the potential benefits of a global economy.

Electrification offers the developing world an opportunity to "leapfrog" over the earlier, energy-intensive development of the West. For example, by 2050, if electricity provided even 40% of the world's total energy, as opposed to the 20% put forward in business-as-usual projections, improvements in global sustainability would be profound. For example, relative to business-as-usual, there would likely be at least a two-thirds reduction in global carbon emissions, made possible through higher efficiency and low-carbon fuels, and the same reduction in global oil consumption, through the use of electric vehicles. Equally important is the potential for at least a 50% increase in developing the world's economic output (IAC, 2007).

At the same time, the global electricity supply system is currently undergoing fundamental changes in its infrastructure, associated not just with the rapidly increasing amounts of renewable energy, but also with the development of new production and end-use technologies. One change is an increase in the large number of distributed production units that are significantly smaller than traditional thermal power plants. This development will include low-voltage connections from micro-distributed generation/CHP plants in individual households. Another important development is active control of this low-voltage demand, introducing a new method of providing flexibility in power balancing.

Parallel with this development is the increased use of information and communications technologies (ICT). The communications capabilities of electric devices are expanding rapidly, while also becoming less expensive. This introduces two-way communication with end-users, and is therefore an important enabling technology for future power systems. Advances in measurement technology and computational methods, e.g., for predicting weather, energy demand, and prices, create new ways to control the entire power system (Larsen and Petersen, 2009).

15.8.2   Conversion

Today's electricity supply worldwide overwhelmingly depends on centralized sources. These electricity-generating plants are designed and built to take maximum advantage of economies of scale and relatively low-cost fossil fuel (primarily coal) and hydropower. As has been the case with power delivery, short-term economics have restricted the application of innovative technology to improve the efficiency, security, and environmental performance of these centralized plants. The United States, for example, depends on thousands of centralized power plants, whose average age is approaching 40 years, because they still produce electricity at significantly lower economic cost than any newer alternative, even though the resulting environmental cost can be high.

This structure and approach has made it very difficult to achieve sustainable global energy for development. The primary issue restricting progress is not technology per se, but rather the disruptive impact new technology may have on the long-established and deeply embedded status quo. In the face of rapidly growing quantity and quality demands, it is crucial that this institutional and cultural inertia is promptly resolved so that the significant opportunities currently available to improve electricity generation are realized (Cicchetti and Long, 2003).

Increasing prices for petroleum and natural gas, along with concerns about CO_2 emissions, have contributed to increased interest in advanced coal-based power generation, including supercritical pulverized coal, circulating fluidized-bed combustion, and integrated gasification combined-cycle (IGCC) coal plants. With extensive positive experience in Europe, Japan, and Korea over the last decade, the superior efficiency and resulting improved environmental performance of supercritical plants makes them the generally preferred coal-fired power-generation choice today. Over the long run, the key to meeting the challenges of using coal to generate electricity is to refine rather than burn. Coal refining depends on the conversion of coal under reducing conditions into a synthesis gas composed principally of methane and carbon monoxide. This concentrated synthetic gas can be purified of contaminants and used, either as a clean fuel for high-efficiency combustion turbine/combined-cycle electricity generation, or as a feedstock for synthetic petroleum and chemical production (see Chapters 12 and 13). This technology is well developed and has been shown to be economically competitive with US$50/bbl, without consideration of the sizeable energy security and carbon-capture value that such plants provide (National Academy of Sciences, 2009). Ultimately, it may be feasible to extend coal refining through carbon capture and more advanced technologies, such as carbon reduction and increased nuclear and renewable energy, in order to achieve nearly zero-emission power generation. Demand side management (DSM) is today an increasingly cost and environmentally effective alternative to building new fossil fuel-fired peak generation capacity. Strategically for example, as discussed in the Appendix 15.A and Chapters 8–10, DSM through community aggregation and the decoupling of profits from electricity sales volume, also has the potential to become a much larger contributor and to effectively become an "Energy Efficiency Power Station."

The technology gap is also evident in nuclear power, preventing any realistically achievable strategy for the use of clean energy. A variety of advanced nuclear reactor, power-generation system designs are in various stages of development to address the four core issues of cost, safety, waste, and nuclear-weapon proliferation (see Section 15.3.5 and Chapter 14).

Renewable energy is also, potentially, a very important way of meeting global electrification needs in both rural and urban situations, depending on whether or not current barriers to large-scale deployment can be overcome. Despite recent gains and over one third of new power generation investments, renewable energy resources still only supply around 18% of the world's electricity, mostly from hydropower. If renewable energy is to play a greater role in the power mix, it must be reliable for the electricity system operator to use. The many options of renewable energy are technologically unrelated to each other, and, in terms of integration into present systems, each option has its own challenges. The implications for the electricity carrier and storage infrastructure are especially significant and urgent. Renewable energy power-plant capabilities, grid-planning and operation, energy and power management, and energy markets are each important to the large-scale integration of renewable energy at the system level (National Academy of Sciences, 2010). For further discussion on renewable energy, see Chapter 11.

Large proportions of wind power and other variable renewable energy generation make constantly maintaining a supply-demand balance an even larger challenge. In such cases, the system's flexibility in generation, demand management, and intra-area transmission may need to be increased (IEA, 2009b). The layout and basic structure of the grid, as well as operational practices, need to be adapted to manage the presence of large amounts of variable supply. An energy system with large-scale integration of renewable energy, particularly wind power, is expected to meet the same requirements for the security of supply and economic efficiency as the current energy conversion and delivery systems, while delivering better environmental performance, especially with regard to CO_2 emissions and lessening dependence on fossil fuels.

Arguably the most effective mechanism to rapidly grow the renewable energy supply is locally distributed power generation. Policymakers have also set other objectives that are most effectively addressed by distributed generation (DG), since it can maximize efficiency and minimize the need for new large-scale generation plants, and upgrading of the transmission and distribution infrastructure. DG includes the incorporation of combined heat, power, and cooling for end-users, who are then able to more effectively contribute to maintaining supply-demand balance (Figure 15.10).

Isolated DG systems may be connected to the primary grid, or they may operate independently, as microgrids, or confined within a building. They are generally not centrally controlled, and with few exceptions at the present time, they cannot be switched on and off according to the needs of the grid – unless they incorporate energy conversion and storage. These current realizations of a possible "DG future" are attracting considerable interest (IEA, 2007).

The eventual goal of DG is to "reinvent" the grid itself. Instead of electricity being produced in large central plants and transmitted in one direction, DG will provide end-users with the following benefits:

- a degree of energy independence;
- opportunities for local control to improve the security of supply;
- financial optimization with energy markets;
- equal or better power quality; and
- a cleaner environment.

The perceived benefits of distributed-electricity generation systems include increased reliability of service, improved power quality, the ability to defer investment on extending the grid, and greater energy efficiency through better use of waste heat. Of these drivers, reliability of service is the most important and is linked with energy security. More and more consumers need uninterrupted electric power, yet many existing grids cannot operate without occasional blackouts. Using DG as a backup power source can largely eliminate these blackouts. Finally, the redundancy offered by a DG network, with its intertwined multitude of generators, converters, and connections, will certainly enhance the security of the power system.

To provide a reliable electricity supply, DG based on a high proportion of renewable energy will depend on a number of support technologies. These will include energy storage and load management to deal with variable power from renewable energy sources such as wind turbines.

Once the concept is fully developed, DG, in addition to the obvious benefit of providing a cleaner environment, has the advantage that it is easy to add generating capacity as required, using local energy resources. The cost of such expansion is predictable over the life cycle of the generating plant, regardless of price fluctuations and shortages that may affect some fossil fuels and/or uranium in the future.

As the technology and benefits associated with it become more widely available, DG will contribute to electrification in the developing world. In this bottom-up approach, electrification can occur at the village scale, using local, renewable energy-powered "microgrids". The fundamental concept of a microgrid can be summarized as an integrated energy system having multiple distributed generation sources and multiple electrical loads, operating either in connection to, or separate from, the existing bulk power grid. A microgrid is thus a small-scale version of the electricity grid that the majority of electricity consumers rely on for power service today. Perhaps the most compelling feature of a microgrid is the ability to separate and isolate itself – known as "islanding" – from the bulk power distribution system during brownouts or blackouts. These local power cooperatives provide consumers with direct access and market transparency that traditional top-down, centralized, electricity-supply systems lack. These microgrids also offer basic building blocks for developing world electricity system expansion in the most sustainable manner. As the number of end-users (and their electricity needs) steadily grows, local microgrids can be interconnected into a regional grid combining the best of both distributed and centralized electricity supplies (Energy Business Reports, 2008).

A global future with energy for sustainable development will depend on a wise combination of both centralized and distributed generation in a system that best captures the advantages of both. The optimal balance will depend on local circumstances and the employment to best advantage of state-of-the-art technological advancements throughout the supply chain. Only in this manner can all the world's essential energy resources be made efficiently and cleanly available to serve the global population within this century.

15.8.3   Transmission and Distribution

The delivery of electricity depends on a system of overhead wires and underground cables, collectively called circuits or grids, which connect electrical loads with diverse sources of electric-power generation. This delivery and distribution system begins at the buss bar located at the generation plant, and extends to end-user meters. Today's delivery systems are largely based on technology developed in the first half of the 20th century. The strain on this aging analog, electromechanically controlled system is evident as it tries to keep pace with the precise power requirements of a digital economy – there is increasing circuit congestion, and the need for better system security. This translates into a large and growing gap between the performance capability of today's outmoded bulk-electricity delivery systems and the needs and expectations of end-users.

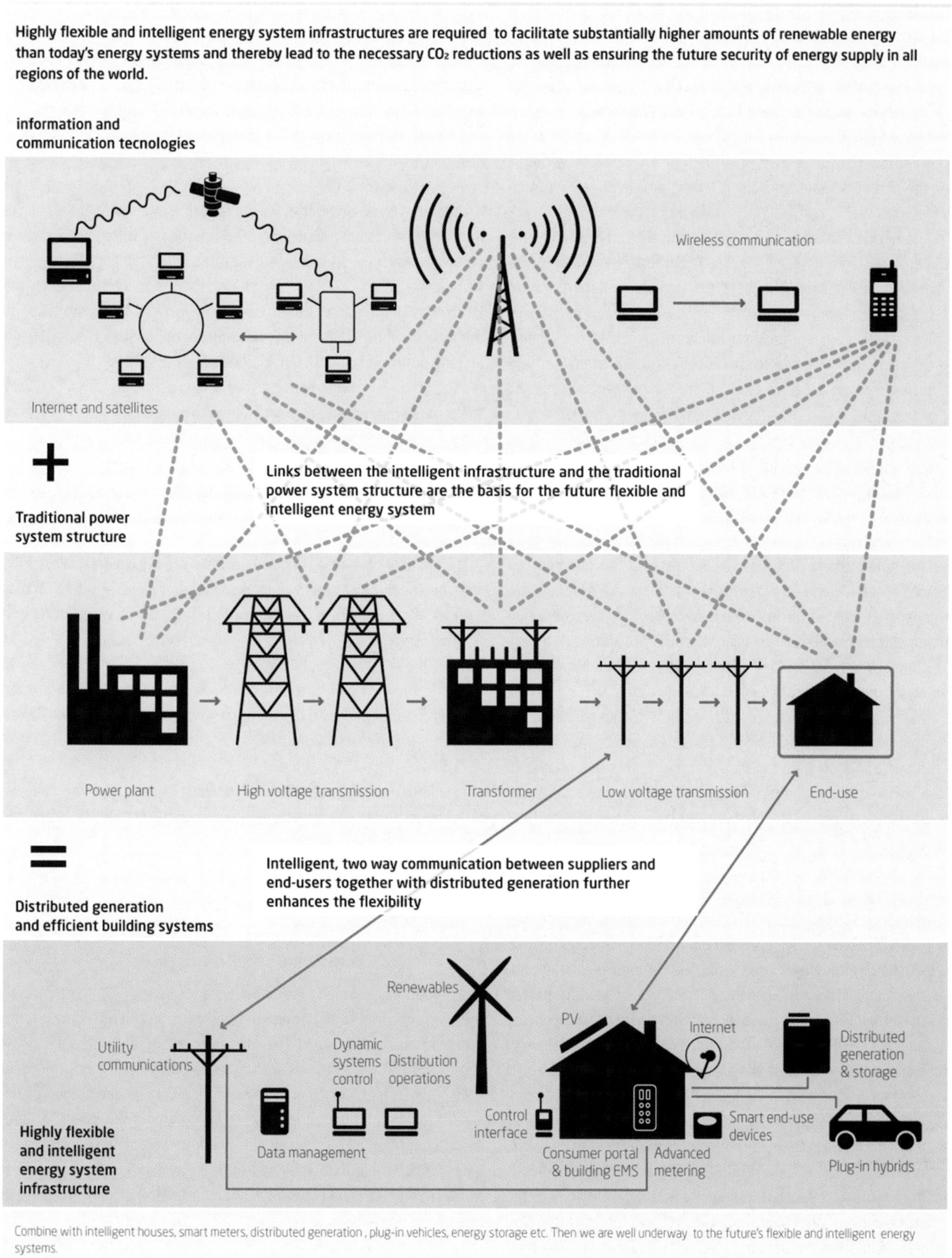

Highly flexible and intelligent energy system infrastructures are required to facilitate substantially higher amounts of renewable energy than today's energy systems and thereby lead to the necessary CO_2 reductions as well as ensuring the future security of energy supply in all regions of the world.

information and communication tecnologies

Wireless communication

Internet and satellites

+

Traditional power system structure

Links between the intelligent infrastructure and the traditional power system structure are the basis for the future flexible and intelligent energy system

| Power plant | High voltage transmission | Transformer | Low voltage transmission | End-use |

=

Distributed generation and efficient building systems

Intelligent, two way communication between suppliers and end-users together with distributed generation further enhances the flexibility

Renewables

PV

Internet

Utility communications

Dynamic systems control

Distribution operations

Distributed generation & storage

Control interface

Highly flexible and intelligent energy system infrastructure

Data management

Consumer portal & building EMS

Advanced metering

Smart end-use devices

Plug-in hybrids

Combine with intelligent houses, smart meters, distributed generation, plug-in vehicles, energy storage etc. Then we are well underway to the future's flexible and intelligent energy systems.

Figure 15.10 | Overview of an intelligent energy system infrastructure noting flexibility and distributed generation. Source: Larsen and Petersen, 2009.

The ultimate force pulling the transmission and distribution system into the 21st century is rapidly advancing technologies: specifically, intelligent electronic technologies that enable ever-broader consumer involvement in defining and controlling electricity-based needs. Historically, issues of reliability, security, quality, and availability have been measured and dealt with in a fragmented manner. In the future, these issues must become a highly integrated set of design and operating criteria, meeting the requirements of consumers (Nahigian, 2008).

Today's alternating current (AC) electricity delivery systems have two basic dimensions: high-voltage, long-distance transmission, and lower-voltage, local distribution. Throughout this process, voltage is raised or lowered by transformers to meet particular circumstances. Various techniques can increase the quantity of electricity carried on existing corridors. New conductors with carbon-fiber cores, for example, have higher current-carrying capacity. Because of their greater strength and lighter weight, they sag less at the line temperatures associated with high power-flow rates, and can operate continuously at temperatures above 100°C. High Voltage Direct Current (HVDC) Light, an underground cable that moves huge amounts of power with very low losses over thousands of km, is also viewed favorably is under consideration for the European super grid (Galvin and Yeager, 2009). By doubling voltage, for example, one increases the power capacity of a transmission line by a factor of four. As load growth increases and the use of renewable power generation located far from the load center becomes more globally dominant, there will be increasing demand for HVDC technology. HVDC technology may be the only effective means of increasing power flow on an existing AC transmission corridor (Gellings, 2011). Electric power companies, government agencies, and industry are collaborating on "high-temperature" (liquid nitrogen temperature), superconducting direct current (DC) cables. As the technical and cost issues related to these are resolved, they may triple the electricity-carrying capacity of today's conventional conductors. Current technical issues include fault current susceptibility, reliability, and efficient use of cryogenics.

Renewable energy on a large scale represents a paradigm shift for the transmission and distribution system: the pacing issue is their natural variability. Because electricity supply and demand must be in constant balance, the carrier system or "grid" must be flexible, able to accommodate the sudden loss of a generation resource or an unexpected increase in demand. For small amounts of non-hydro renewable energy on a system, this can be accommodated reasonably well. However, with continued growth in such renewable energy, particularly wind, the goal to accommodate a significant fraction, 20% or more, requires new solutions. This issue has been underscored by the experience of Vattenfall, which controls northeast Germany's electricity transmission network, where the world's greatest concentration of wind-energy generation is located. In order to accommodate the unpredictable availability of the wind resource, conventional power plants have been forced to cycle on and off inefficiently, and the company has had to make emergency electricity purchases at high prices. Improved forecasting would help mitigate the problem. So would designing the grid to be more flexible (IEA,

2008a), integrating a wider portfolio of generation types (Awerbuch, 2006), and making loads responsive and active participants in power system operations. Building-integrated solar PV systems that compete with retail power prices can also have variability issues, whereas concentrating solar power (CSP) systems, combined with thermal storage, can help to overcome variability during short periods of cloud cover or darkness (UNEP, 2007).

An additional consideration is the demand that renewable energy places on the carrier system for ancillary services, such as voltage and reactive power, which must be managed to ensure the stability of the carrier system. The European Wind Integration Study, produced in 2007 by the European Transmission System Operators, focused on these issues as the amount of installed European wind capacity increased from 41 GW in 2005 to an expected 67 GW in 2008 (ETSO, 2007).

Other major upgrades to the electricity carrier system are also necessary to meet ever-growing demands, and ensure sustainability, including:

- extremely reliable delivery of high quality, "digital-grade" power, needed by a growing number of end-uses (EPRI, 2003a);

- availability of a wide range of "always-on, price smart" electricity-related consumer and business services that stimulate the economy and offer consumers greater control over their energy usage and expenses;

- a transmission and distribution infrastructure confidently protected from natural and man-made threats, which can be quickly restored in the event of an interruption;

- minimized environmental and societal impacts through the use of much more energy-efficient equipment, distributed renewable energy resources, and CHP; and

- improved economic productivity and growth, with decreased electricity intensity.

These demands, coupled with climate change concerns and increasingly involved consumers, are pushing the transmission and distribution infrastructure and its operators toward a technology and business model commonly known as "the smart grid revolution." The terms "intelligent grid," "smart grid," or "digital energy" may best be understood as the overlaying of a unified digital electronic communications and control system on the entire electricity delivery infrastructure. The goal is to provide the right information to the right entity (end-use devices, transmission and distribution controls, and consumers) at the right time to enable the right action to be taken – all at the speed of light.

The smart grid's transmission and distribution system (Figure 15.11) will constantly fine-tune itself to achieve and maintain an optimal state of

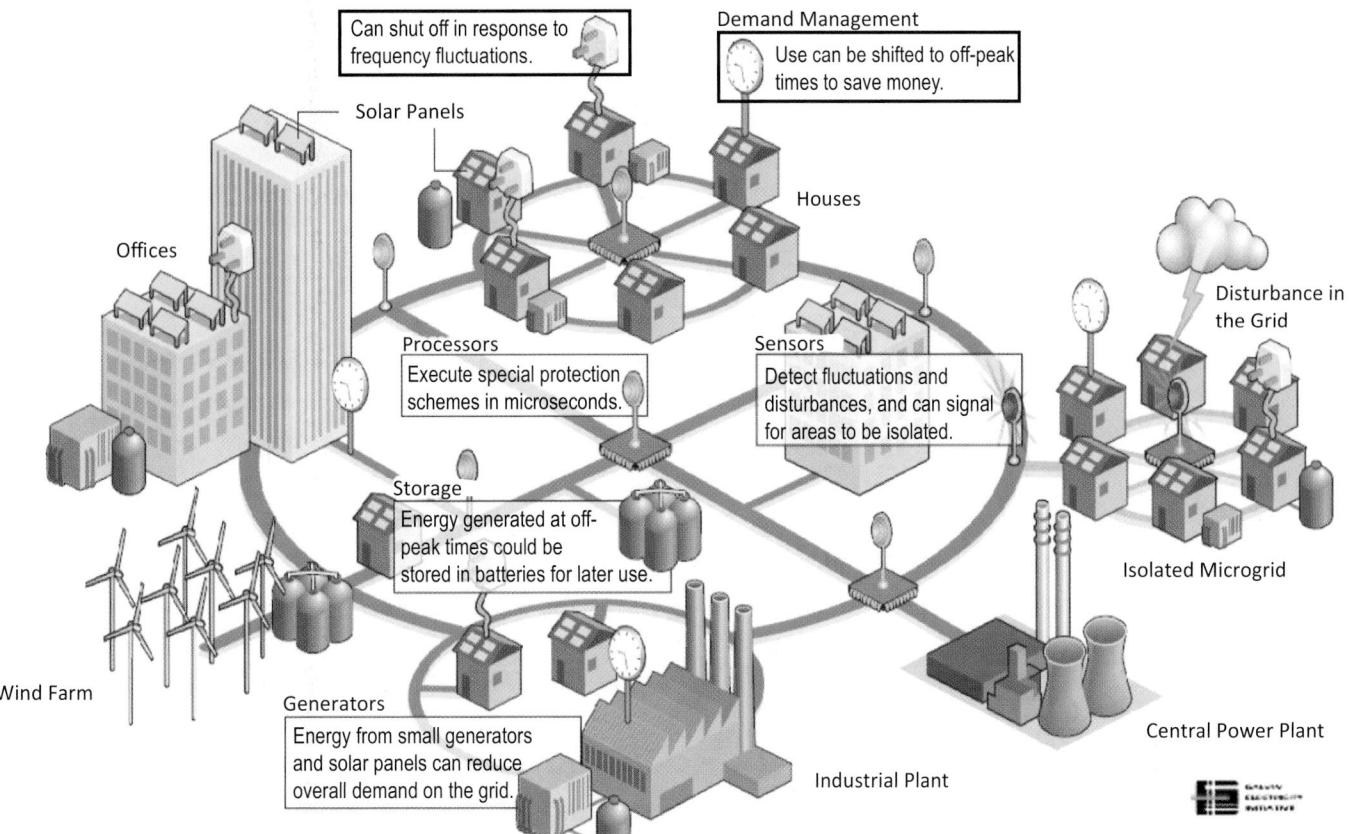

Figure 15.11 | An example of a Smart Grid, a network of integrated microgrids that can monitor and heal itself. Source: Amin, 2008.

operation, while monitoring for potential problems that could interrupt service. When a potential problem is detected, appropriate corrective action will automatically be taken. This includes "intelligent islanding," which can instantaneously separate the transmission and distribution system into self-sustaining parts, so as to maintain electricity supply under all conditions. The result is a system that optimally balances electricity supply and delivery, minimizes losses, is self-healing, and enables next-generation electricity supply, energy efficiency, and demand-response. First and foremost, it opens the door to a strong demand-side response to price signals, thereby reducing the need for peak generation, improving the use of capital, and affording better asset management throughout the entire energy chain – all with positive repercussions for the environment. This smart delivery system can also pave the way to more rapid introduction of new, more energy-efficient, end-use technology.

Perhaps the most important ultimate attribute of a smart electricity-delivery system is its capacity to assimilate significant quantities of variable renewable energy. As the National Research Council has recently concluded, the current, electromechanically controlled, electricity delivery system generally cannot reliably transport more than single-digit percentages of non-hydro, variable renewable energy without requiring self-defeating quantities of backup electricity generation or bulk-storage capacity, which is limited today to pumped hydropower (National Academy of Sciences, 2009). A green electricity system must

indeed be a smart electricity system and will incorporate the following basic technological modernization improvements (NETL, 2007):

- Transitioning the electricity transmission and distribution system from what is now typically a radial design to a true network will ensure absolute connectivity from electricity generation sources to end-uses.

- Converting the transmission and distribution system from an electromechanical to a comprehensive, digitally monitored, electronically controlled network will enable continuous, instantaneous two-way communication between electricity users and suppliers, so that all consumers have the ability to move from passive to active marketplace participation.

- Incorporating locally distributed energy and CHP resources will translate into much higher efficiency and reliability, and much greater use of variable renewable energy.

All the elements of the smart grid are linked and in constant communication to optimize performance. It is indeed a supreme irony that computers, sensors, and computational ability have transformed every major industry except electric power-generation, whose product is the lifeblood of the modern global economy.

Several specific technology-related capabilities are pacing the implementation and full value operation of smart electricity-delivery systems (EPRI, 2010):

- Operating models and system impact algorithms. Simulation tools address the transient behavior of the modern grid incorporating renewable energy. These include improved operator visualization techniques, new training methodologies, and advanced simulation tools. Advanced simulations provide an accurate and complete understanding of grid behavior, as well as assisting system planners in designing reliable power-delivery systems.

- Nanotechnology sensors. Fundamental to system reliability is the comprehensive incorporation of sensors that instantaneously track conditions of the system. It is therefore surprising that nano-sensors based on magneto-resistance (the change of resistance in the presence of a magnetic field) are still not widely used in electricity carrier systems. By comparison, 250 million such sensors are produced for the hard disk industry, at a cost of a few dollars per sensor.

- Phasor measurement units (PMUs). PMUs are ideal for monitoring and controlling the dynamic performance of a carrier system in measurement systems for wide geographic areas. By 2012, for example, China plans to have PMUs throughout its electricity carrier grids, including all 500 kV substations and all power plants of 300 MW and above. All PMUs are connected via private network, and performance signals are received within 40 milliseconds on average.

- Standards and protocols for system interoperability. The lack of consistent, open standards for comprehensive system communication among all automatic components, including distributed energy resources, leads to market fragmentation, a major deterrent to smart-grid progress. The IEC 6/850 standard for substations is a valuable first step, but it falls short of addressing the interoperability requirements raised by the modernization of the entire electricity delivery system.

- Cyber-security solutions. As the control system becomes more automated, cyber security becomes a pacing issue. Although numerous security products exist on the market today, each is designed to address a specific security concern and there is no "one-size-fits-all" solution. The security for this carrier control system must provide in-depth defense, by seamlessly integrating an array of technologies to ensure that the carrier is controlled safely, securely, and meets all reliability expectations at all times.

An additional advantage provided by smart-delivery systems is greater use of direct current. DC has several advantages for distribution networks. DC-distributed links, for example, can supply power directly to digital devices at the customer's site, and connect renewable energy resources, without the need for costly and inefficient individual DC-to-AC convertors. These DC-distributed links also increase reliability by reducing the spread of disturbances from one customer to another, while enabling each facility to operate independently using distributed generation and storage. Because of its cost-saving advantages, DC distribution is becoming more widely used, especially in high-technology facilities such as data centers. Several Japanese suppliers are actively marketing DC systems for broad commercial and residential applications. Portable computers, televisions, and game consoles are all powered by DC electricity. Even washing machines, air conditioners, and fluorescent lamps convert AC into DC, and then convert it back to high-frequency AC with inverters.

Today's bulk-electricity supply and delivery systems were typically designed and built with the primary objective of keeping pace with the rapid growth in demand associated with initial electrification. Those days are long over throughout the developed world, yet little has been done in most countries to update either the infrastructure or the business incentives to focus on efficiency and quality. The electricity meter, for example, still holds retail consumers hostage to an electricity supplier monopoly over which they have essentially no market leverage. Technology is available to break down this iron curtain meter, just as the internet transformed communications. And, like the internet, consumers, suppliers, and society alike will benefit from a differentiated, service-based market business model. The incentives should be to add maximum value to each electron, not to maximize the quantity of bulk electricity sold to captive consumers (Yeager, 2008).

One of the quickest and most effective ways to achieve the greatest consumer and community benefits from the smart grid is through the microgrid – a small-scale, power-supply network, designed to provide power to a small community or a few buildings. Modern microgrids, utilizing smart grid technology, are an emerging distribution configuration. They offer significant economic and environmental benefits compared with the alternative of simply expanding the legacy electricity transmission and distribution systems. The benefits of microgrids include autonomy, stability, compatibility, flexibility, scalability, efficiency and economics. Microgrids can provide a superior match between electricity generation and load. Thus, they have a low impact on the existing electricity delivery system, despite incorporating significant amounts of variable renewable energy resources, such as photovoltaics (PVs). Because small electricity generators are also close to the users in microgrids, any waste heat can be easily recovered, and total energy efficiencies in excess of 80% can be achieved. This compares very favorably with the 30% or less efficiency for a typical electricity-generation and delivery system today (Energy Business Reports, 2008).

A key feature of the microgrid is the seamless interconnection of supply and demand. Particularly important here is the incorporation of buildings as suppliers, as well as users, of electricity. The sustainability implications here are profound because buildings are typically the largest users of electricity on a national scale, and thus also account for a large fraction of carbon emissions. Embedding intelligence into the physical fabric of each building allows all functions, appliances, and energy sources to "communicate" with one another, automatically coordinating their activities for the greatest efficiency and economy. The result makes zero net-energy buildings a realistic goal in most regions of the world (Cheung and Wilshire, 2010; see also Chapter 10).

It is also becoming feasible to incorporate mobile electricity storage in the form of plug-in electric vehicles. Microgrids will enable an electricity delivery network strategy, where transportation can become an efficient energy-supply source, particularly during high-cost, peak-demand periods. The essential enabling policy step to realize the full benefits of DG, microgrids and plug-in electric vehicles is time-of-use electricity pricing, which encourages consumers to be efficient and to take advantage of resulting savings.

In effect, the microgrid is a local electricity refinery, raising the reliability, efficiency, and quality of the electricity supply. Although a variety of microgrid ownership and operation alternatives are possible, microgrids intrinsically lend themselves to local "co-operative" ventures, where consumers are also suppliers. Metering and billing arrangements are agreed on locally to reflect market needs within each microgrid. Microgrids can also take full advantage of currently commercial energy-storage technology, so as to accommodate rapid fluctuations in electricity demand or supply.

Microgrid development in the United States has taken a somewhat different path from similar efforts in Japan and Europe. In the United States, the emphasis has been on compensating for the relatively poor reliability and power quality provided by the nation's bulk-electricity supply system. This is in contrast to Europe and Japan, where incorporating clean, distributed electricity generation has been emphasized. In other countries DG and microgrids are seen as a means of avoiding the costly upgrades of transmission lines that are nearing capacity limits due to increasing local demand. Microgrid development efforts in the United States have been led by the Consortium for Electricity Reliability Solutions (CERTS) and, more recently, the California Energy Commission (2008) and the Galvin Electricity Initiative (2010). Prototype microgrid demonstrations range from the Fort Bragg military base in North Carolina to the Illinois Institute of Technology campus in Chicago. General Electric is also developing a Microgrid Energy Management framework to provide a unified control, protection, and energy management platform.

In the EU, several major research efforts have been devoted to DG microgrids. Demonstration sites include Kathnos Island, Greece, and Mannheim-Wallstadt, Germany. Japan is also a leader in microgrid demonstration projects that generally emphasize ensuring that the variable power of distributed renewable energy does not degrade Japan's outstanding electricity reliability and power quality.

In terms of the growing international concern regarding energy security and sustainability, modern microgrids are also attractive. In addition to providing a cost-effective approach to the modernization of transmission and distribution systems, they can be dispersed through the electricity

delivery system close to consumers, where they can protect against the failure of a large power plant or bulk-electricity transmission facility. They also diversify the energy resource base by incorporating distributed renewable energy resources. However, clear policy and associated regulatory instruments are needed to capture the benefits of microgrids and to integrate them into existing electricity distribution networks. With a very high share of variable renewable energy sources, power balancing can become a huge challenge. Such a power-supply system would require the use of all potential balancing measures, including new transmission lines between regions, new flexible generating plants, demand-side management, energy storage, advanced weather forecasting, clear rules from the system operator, and the use of existing distributed resources within the system.

End-users have the potential to contribute to system balancing. Several types of demand, notably electric heating and cooling systems, can be operated in a flexible manner that responds to signals from the power company. As the number of small-scale DG units increases, many of these can also be used in ways that help balance the system. Future storage technologies, such as electric vehicle batteries, also have the potential to act as flexible balancing measures.

Power systems are currently undergoing some fundamental changes in structure and operation. These changes are associated not only with the rapidly increasing amounts of renewable energy being connected to the system, but also with the development of new types of production and end-use technologies. One such change is a general increase in the number of distributed production units that are smaller in scale than traditional thermal power plants. This development will, in the future, include low-voltage connections from micro-CHP plants in individual households. On the low-voltage side, another important trend is the active control of demand, which introduces a new way to provide some of the necessary flexibility in power balancing (UNEP, 2007).

In parallel with this development is the increased use of Information and Communications Technologies (ICT). The communications capabilities of electric devices are expanding rapidly while also becoming cheaper. This enables a power system which incorporates two-way communication with end-users, and is therefore one of the most important enabling technologies for future power systems. Advances in measuring technology and advanced computational methods, e.g., for predicting weather, energy demand, and price, create new ways to control the entire power system (Larsen and Petersen, 2009).

Few countries have a comprehensive and coordinated set of policies and incentives to promote micro-grids. Subsidies for renewable distributed generation, utility revenue decoupling, time-of-use pricing, and independent zero-energy districts all act as enablers for microgrid development.

Distributed generation comprises the cornerstone of the new microgrid paradigm. While CHP units will certainly play a role, the escalation of renewable distributed generation will be the focal point. The microgrid development model allows the ability to maximize value from private customer investments. By far the largest and most important renewable distributed generation technology is distributed photovoltaics. Germany in particular has set the most aggressive goal, aiming to generate 25% of its total electricity capacity from solar photovoltaics. This level of penetration will require microgrid-like controls and management. Small wind turbines are also becoming a more important energy asset enabled by microgrids. Micro-scale energy-storage systems, particularly batteries, complement intermittent renewable energy sources and are also considered a key immediate driver of microgrids. These storage systems offer clear economic benefits for renewable distributed generation. When the sun is shining or the wind is blowing, a storage system builds up the electricity reserve for sale or dispatch.

Since the interest in smart-grid technologies is in dramatic ascent, microgrids are attracting considerable attention from both policy and investment sectors. While much of Europe and Japan have upgraded their grid infrastructure, these countries have traditionally emphasized more aggressive policy support for the distributed generation technologies that are driving the acceleration of microgrid developments. That said, Japan and Korea, two leaders for microgrid technology, enjoy power surpluses and therefore have limited need for microgrids today. However, within the next five years, the new generation of microgrid configurations is expected to be fully commercial, setting the stage for major global commercialization progress. Elsewhere around the world, Denmark and China are leaders in the microgrid market. An important economic advantage of microgrids is related to the payback period. Estimates of paybacks for microgrids range from two to five years. Renewable energy microgrids also offer increased co-benefits such as improved health, employment, rural development, system reliability and avoided transmission infrastructure, compared to traditional utility power, with non-renewable microgrids offering a benefit rate of three-to-five-fold over today's basic grid services (Lovins and Cohen, 2010).

15.8.4 Storage

Unlike other energy forms, electricity cannot be easily stored in large quantities where hydro capacity is not available. Without storage, there is little flexibility in managing electricity production and delivery, or in seamlessly linking electricity supply and end-use. Likewise, variable renewable energy resources are constrained in their ability to enter the bulk-electricity market and keep pace with continuously changing demands. Advanced storage technologies promise to change the nature of electricity markets by providing much greater operational and financial flexibility, and enabling carrier systems to resolve system transients and bottlenecks. The critical pacing issue is the accelerated development and implementation of more cost-effective, higher-capacity storage options, including batteries, flywheels, supercapacitors, hydrogen, and superconducting systems. Today, most of these bulk-storage options are relatively unproven, and their value

proposition is complex and poorly understood (National Academy of Sciences, 2009).

Currently, the only battery commercially available for large-energy storage is the lead-acid battery. Advanced batteries being developed for large-scale applications, include nickel cadmium, sodium sulfur, sodium nickel chloride, lithium ion, and zinc bromine. Although considerable progress is being made, these generally remain too expensive for large-scale applications. The first of these advanced batteries is anticipated to be fully commercialized in the very near future, and significant cost reductions are expected as modular design and factory assembly become the norm, and production volumes increase substantially. Particularly notable from a strategic perspective are the lithium-ion battery (for electric vehicles) and the sodium sulfur battery. Japan has demonstrated a total of more than 20 MW of energy storage using sodium sulfur batteries. In August 2003, the world's largest battery energy-storage system began operating in Fairbanks, Alaska, using nickel-cadmium cells to generate 40 MW for up to seven minutes at an installed cost of US$35 million.

Over the long term, energy storage can potentially resolve the intermittency challenge of renewable energy. The types of storage under consideration range from pumped hydro and compressed air energy storage to advanced, stationary batteries. A notable example of wind-hydro synergy is Denmark's grid: this can absorb a great deal of wind because of its strong electricity transmission ties to the hydroelectric systems in Norway, Germany, and Sweden. In the absence of storage, the alternative today is to locate conventional fossil-fired power generation near to the variable renewable energy resource. This, of course, diminishes the environmental value of the renewable energy resource and raises the competitive hurdle.

Energy carriers such as hydrogen and ethanol may become important in interconnection and the storage of energy from renewable energy sources, and they have the potential to provide the interface for renewable energy sources to mobile users.

Hydrogen is an important implicit dimension of the ultimate "electrified world" scenario. The electro-hydrogen economy refers to hydrogen being generated from low-cost, clean, off-peak electricity generation, and stored and delivered as feedstock to fuel transportation, electricity generation, and industrial processes.

15.8.5 Policy and Investment Considerations

The transition to a greater use of locally distributed generation will increasingly make the local microgrid the design and operational focus of electricity supply systems, and the entire enterprise must be redesigned accordingly. The 20th century principle that an electric utility is a vertically integrated, natural monopoly has been slowly evolving under the pressure of technological progress and deregulation. Unfortunately, much of the world is still laboring under an electricity regulatory structure. This maintains an obsolete business model with incentives to simply produce and deliver more commodity electricity, rather than compensating suppliers based on the efficiency and reliability of their electricity services.

Market reform for electricity supply and delivery is a relatively new initiative now being applied in a variety of situations throughout the developed world. As a result, sufficient experience has been gained to begin to evaluate its advantages and disadvantages as a tool in achieving universal, sustainable, global electrification. The potential of decentralized competitive markets to match supply and demand most efficiently is a well-proven principle at the heart of the developed world's commercial economic system; based on that success, it is rapidly becoming the standard for economic systems around the world.

Nonetheless, the application of market-based liberalization to the electricity sector over the past decade has had mixed results. Particularly with regard to developing regions of the world, it is important to consider the lessons learned and the possible causes of this checkered performance. Does this simply reflect a learning curve, or are there some basic limitations in applying market reforms to the electricity sector? The reforms in place typically bear little resemblance to the theoretical, market-oriented ideal. Certainly, the critical values of electricity to economic prosperity and societal welfare make it unusually susceptible to political manipulation, irrespective of governmental structure or the level of economic development (Smil, 2005).

What then can be concluded about the role of market reform in facilitating global electrification? Because surplus electricity supply is, by definition, not a circumstance facing the developing world, it is doubtful that market reform as it has been widely applied in much of the developed world, will stimulate investment on the scale required. Certainly, its initial applications have failed. Market reform is primarily a tool dependent on the more fundamental conditions of national performance. These include governmental and institutional stability, the rule of law, and sound fiscal policies. In the absence of these conditions, no amount of market reform is likely to make a difference. Where these conditions do exist, market reform can be valuable if applied to achieve its intended purpose – sustainable economic development and efficiency.

One interesting variation on market reform is the local market paradigm emerging in many rural regions of the developing world, where electrification is being initiated from the bottom-up, rather than the top-down, through village-scale microgrids under local management and control. These local cooperatives provide the direct consumer access and market transparency that top-down systems generally lack and – by virtue of their politically based institutions and commodity business culture – have great difficulty in achieving. The process of local rural electrification, whereby these distributed "PC-like" microgrids seek to network with each other to expand their capability, is also a fertile context for extending the locally established market culture. All other factors being equal, perhaps the most important advantage in building markets from the

ground up, relative to reforming the status quo, is the more encouraging environment they may produce for both innovation and investment.

The lessons learned about what works and what does not show that neither the central planning approach of the 1950s and 1960s, nor the minimal government, free-market approach broadly advocated over the last 20 years will necessarily be successful. The most effective approaches to electrification will be led by the private sector in most cases, but with a stable governance framework that facilitates physical infrastructure and human capital investments, along with the social cohesion necessary for economic development and poverty reduction. Institutional development has too often been neglected in past policy discussions, but it is essential to achieve energy access and sustained poverty reduction. Successful development of this new paradigm will therefore depend on a robust, strategic public-private partnership for global development (Yeager, 2007).

Global electrification-related policies must become better at anticipating and synthesizing the revolutionary, disruptive changes underway in demographics, technology, communications, and commerce, which are currently shaping global politics, and energy and overall security in the 21st century. Because of these changes, people's aspirations – to exercise their free will and transform their lives – are rising in all corners of the globe. Policymakers must also recognize that progress is not automatic. The course of this century will be fraught with great risks. These must be wisely managed as we aim for the unambiguous goal of enabling universal wellbeing, a goal that must also be sustainable within our evolving civilization. A policy of universal electrification is the necessary energy foundation on which to resolve the pacing global "trilemma" of people, poverty, and pollution, and most importantly, to avoid failed aspirations that create intolerable levels of frustration, despair, and anger (see also Chapter 23). Failure is not an option and hope is not a strategy for addressing this global survival challenge (Patterson, 2009).

Impatience for immediate results that maximize short-term commercial returns is a lever of vulnerability impeding global electrification and modernization. Reviving the commitment to strategic investment in human and capital infrastructure is essential to sustainable global development and wellbeing, and the public-private partnerships and policies needed to realize this survival goal must be given renewed priority. Integral to this priority is the urgent need to boost development and investment in advanced energy technologies. The lag time between research and large-scale commercial deployment is sobering, yet funding for energy R&D continues to decline.

> "Civilizations decline when they stop the application of surplus to new ways of doing things. In modern terms we say the rate of investment decreases. This happens because the social groups controlling the surplus have a vested interest in using it for non-productive but ego-satisfying purposes ... which distribute the surpluses to consumption but do not provide more effective methods of production" (Huntington, 1998).

15.9 Hydrogen

15.9.1 Overview

Hydrogen is a substance that may be used as an energy carrier and as a storage medium. Although hydrogen is the most abundant element in the universe, it does not exist in pure concentrations of its elemental form or as molecular hydrogen on earth. Hydrogen is abundant in the molecular structure of coal, petroleum, natural gas, biomass, and water and can be produced from these materials. Hence, hydrogen is not a source of energy, but energy is required to separate hydrogen from any of these hydrogen-containing materials. In principle, the energy source for the production of hydrogen may be any of the fossil fuels, nuclear, or renewable energy forms, although using nuclear or renewable energy enables a straightforward route to low-carbon and sustainable hydrogen, although likely at some additional cost.

Hydrogen as a complimentary energy medium offers clear benefits to the world's future energy economy. However, the potential for the hydrogen revolution still remains a speculative vision. Not the least of the challenges is the high energy-losses involved in producing hydrogen, which means that this system demands plentiful, low-cost electricity. The long-term goal is to achieve an emission-free energy system. Since the synthetic hydrogen energy carrier cannot be cleaner or more efficient than the energy from which it is produced, this means renewable or nuclear energy. High temperature gas-cooled reactors with the ability to operate at core outlet temperatures of 850°C or above could, for example, be used to drive a variety of thermal processes for hydrogen as well as electricity production. Fuel cells are projected as the primary end-user for hydrogen with water as the byproduct. A mega-infrastructure in which liquid hydrogen pipelines also act as the cryogenic vehicle for the superconducting, zero-resistance transmission of electricity is certainly an exciting opportunity for the second half of the 21st century.

Hydrogen can be used in the same way as other gaseous fuels in combustion, or in an engine for conventional power generation, such as automobile engines and power-plant turbines. The current use of hydrogen engines is a relatively well-developed technology in the United States, with the space shuttle and unmanned engines. Vehicles with hydrogen-fueled internal combustion engines are now in the demonstration phase. However, the highest value use of hydrogen is in fuel cells, which are relatively cleaner and more efficient than conventional technologies, including onsite generation for individual homes and for businesses. Hydrogen in a fuel cell produces electricity and water with zero greenhouse-gas emissions, and efficiencies greater than when hydrogen is used in an engine.

Fuel cells are in various stages of development and efficiencies are generally in the 40 to 50% range. Phosphoric-acid fuel cells are the most developed fuel cells for commercial use. Many stationary units have been installed to support the grid with reliable backup power, and mobile units are powering buses and other large vehicles. Solid oxide and molten carbonate fuel cells operate at high temperatures and are therefore best in generating electricity in stationary, combined-cycle applications, and applications in which the excess heat can be used for cogeneration of electricity or as process heat. They are also a good fit for portable power and transportation applications, especially large trucks. Alkaline fuel cells have been used in military applications and space missions and are currently being tested for transportation applications. Polymer electrolyte membrane (PEM) fuel cells operate at low temperatures and are being demonstrated in transportation, stationary, and portable applications. Interest in PEM fuel cells has experienced a large upsurge over the past few years, and most major automobile manufacturers are developing fuel cell cars at the demonstration stage.

Nearly all major automobile manufacturers have a hydrogen fuel vehicle program, with various targets for demonstration. The early fuel cell demonstrations will consist of fleets of between 10 to 150 vehicles. To limit initial capital investment, these early vehicles were deployed in fleets with a centralized or shared refueling infrastructure. Hydrogen-fueled internal combustion engine vehicles are generally viewed as a near-term, lower-cost option, which can foster the development of a hydrogen infrastructure.

Stationary power production includes backup power units, grid management, power for remote locations, stand-alone power plants, DG, and cogeneration (see, for example, USDOE, 2002 and USDOE, 2007). The industry for commercial fuel cells for stationary applications is still in its infancy. Most existing fuel-cell systems are being used in commercial settings and operate on hydrogen reformed onsite from natural gas. Extensive use in this setting would require widespread availability of hydrogen.

The development of hydrogen as an energy carrier and storage media requires the evolution of technology for a new infrastructure consisting of three system elements: (1) conversion; (2) transmission and distribution; and (3) storage (see also Chapters 12 and 13).

15.9.2 Conversion

Hydrogen can be produced using any of the fossil fuels, nuclear power, or renewable energy, and be derived from any of the feedstock materials coal, petroleum, natural gas, biomass, or water. Most hydrogen produced today in the United States is used for chemical production, petroleum refining, metal treating, and electrical applications. Thus, hydrogen is primarily used as a feedstock, intermediate chemical, or as a specialty chemical. Steam reforming of natural gas accounts for 95% of the hydrogen produced in the United States. The least costly route to produce hydrogen is steam reforming of natural gas and is carried out in large, centralized facilities with carbon dioxide as the principle byproduct (USDOE, 2002; 2007).

Challenges include:

- high hydrogen conversion costs,
- low demand inhibiting development of production, and
- current technologies are not optimized for making hydrogen and produce large quantities of carbon dioxide.

Research, development, and demonstrations are needed to provide innovative methods of economically producing hydrogen. There is a portfolio of existing commercial processes such as steam reforming, multi-fuel gasifiers, and electrolyzers, and on the development of advanced techniques such as biomass pyrolysis and nuclear water splitting, photoelectrochemical electrolysis, and biological methods which are all promising options for the future. Renewable energy pathways offer a variety of carbon-free hydrogen that are generally longer-term but promising options (Figure 15.12).

The paths forward are different for each of the hydrogen technology areas, but recommendations for policy actions include maintaining a stable research and development funding base, providing incentives for pilot and demonstration scale tests and operating facilities, and providing market incentives as the technologies show promise.

15.9.3   Transmission and Distribution

A key element of the hydrogen-energy infrastructure is the delivery system that transports hydrogen from the point of conversion to end-use. For current applications, pipeline or trucks hauling high-pressure cylinders or cryogenic tankers are used. Rail cars or barges ship a small amount. Pipelines are currently limited to a few areas of the United States and other OECD nations where large refineries and chemical plants are concentrated and are owned and operated by merchant hydrogen producers.

The vision for hydrogen transmission and distribution is a network or infrastructure to accommodate both centralized and decentralized

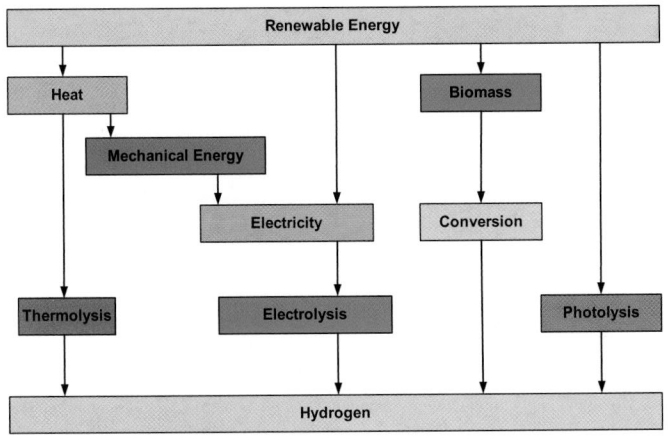

Figure 15.12 | Renewable paths to hydrogen.

production facilities. Pipelines could transport hydrogen to high demand areas, and trucks and rail cars will distribute hydrogen to rural and other low demand areas. Onsite hydrogen conversion and distribution facilities could be built where demand is high enough to sustain operations.

The path forward is to enhance current transmission and distribution systems for safe and affordable operations through policy actions such as increasing research and development on delivery systems, developing a cost analysis of hydrogen transmission and distribution systems, and providing incentives and funding for hydrogen delivery demonstrations, such as fueling stations and power parks, in order to evaluate the hydrogen infrastructure components.

15.9.4   Storage

Storage issues cut across the conversion, transmission and distribution, and energy end-use pathways for the use of hydrogen as an energy carrier. Vehicle applications drive the development of safe, space-efficient, and cost-effective storage systems. However, other applications will benefit from all the technological advances made for these on-vehicle hydrogen storage systems (US DOE, 2002; 2007).

Hydrogen can be stored as a gas, liquid, or chemical compound. Technologies that are currently available permit the physical storage, delivery, and end-use conversion of gaseous or liquid hydrogen in tanks and pipeline systems. Compressed gaseous hydrogen storage is a mature technology and adequate vehicle range can be achieved, but with a significant volume and weight penalty. Liquid hydrogen requires less storage volume than gas, but requires cryogenic temperatures and associated insulated containers. The liquefaction of hydrogen is an energy-intensive process and results in energy losses up to approximately one-third of the energy content of the hydrogen. Other technologies in the early stage of development are reversible metal hydrides or absorption on carbon structures. Hydrogen is absorbed and released from these materials by relatively modest changes in temperature or pressure. Chemical hydrides are also being investigated as a hydrogen-storage alternative.

The vision for hydrogen storage is to achieve relatively lightweight, low cost, and low volume hydrogen-storage devices to meet the hydrogen-economy needs. Compressed hydrogen storage is a mature technology and adequate range can be achieved; however, an even better storage method would be desirable. Several automobile manufacturers are considering liquid-hydrogen storage because of its good volumetric storage efficiency. However, the special handling requirements, long-term storage losses, and cryogenic liquefaction energy requirements currently detract from its commercial viability. Metal hydrides offer the advantage of lower-pressure storage, comfortable shapes, and reasonable volumetric-storage efficiency.

Before hydrogen can become an acceptable energy option for the consumer the technology must be made transparent to end-use, similar to today's experience with internal-combustion, gasoline-powered vehicles. Specific challenges include:

- research and development investment levels that are currently insufficient;

- costs of the round-trip, chemical hydride formation process that are too high;

- weight penalties are too high and thermal-management issues for metal hydrides are unresolved; and

- carbon materials adsorption and release of hydrogen that do not have practical processes developed.

15.9.5   Policy and Investment Considerations

Policy actions needed for the development of fuel cells include research and development to lower costs and enhance manufacturing capabilities, support for industry to develop profitable business models for distributed power systems and optimizing fuel cell designs for mobile and stationary applications, and incentives for demonstration facilities to provide information on operating performance. Policy actions for the development of hydrogen storage include increased research and development

in hydrogen storage, supporting the development of innovative and high-risk technologies, and providing incentives and funding for a large-scale production process for the most promising hydrogen-storage materials.

15.10   Summary and Conclusions

Cumulative investments are needed throughout the energy sector in order to be able to meet the growth in energy demand in a sustainable manner. From 2007–2030, these investments, mainly in the power, oil, and gas sectors, have been estimated by the IEA to be over US$26 trillion (Figure 15.13). Over 65% of these investments are projected to be in non-OECD countries (IEA, 2008a). Delays in investments may result from the financial crisis that began in 2008, and there will be changes in these investments if an international agreement on greenhouse-gas emissions should come into force.

To stabilize atmospheric greenhouse-gas emissions at around 550 ppm-eq in the long term, and aim for a global temperature rise of around 3°C, would require an additional investment of US$4.1 trillion to deploy and improve existing technologies. This equates to around 0.24% of annual world GDP. Cumulative savings from concurrent energy-efficiency measures could yield US$7 trillion (IEA 2008a).

To stabilize at 450 ppm CO_2-eq will require new technologies to be developed (IEA, 2009a), as well as rapid deployment of both new and

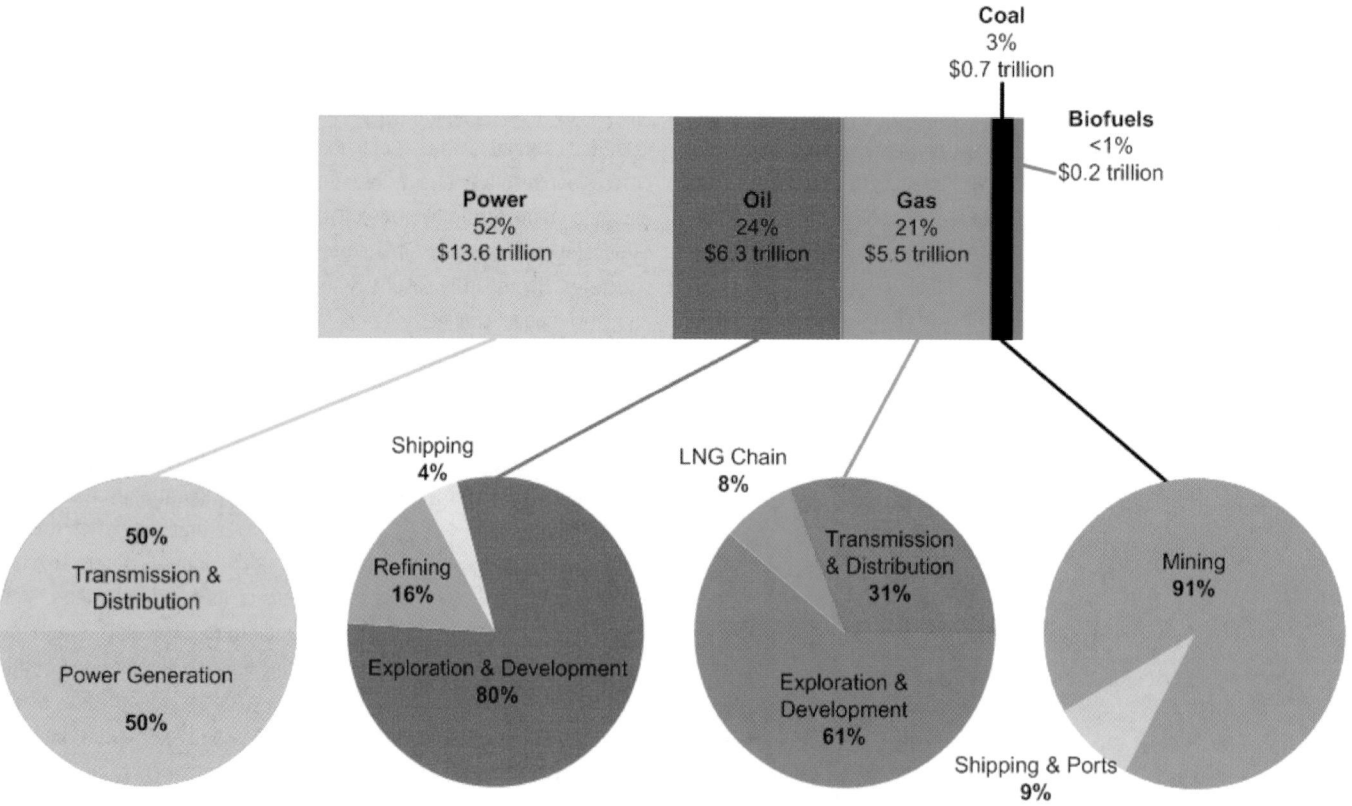

Figure 15.13 | Cumulative world investment in the energy supply sector and infrastructure under business-as-usual to meet growing energy demand and replace old plant stock at end-of life. Source: data from IEA, 2008a.

existing technologies (see Figure 15.10). An additional investment of US$2.7 trillion in more energy-efficient equipment, buildings, and appliances, together with US$2.4 trillion in low-carbon, power-generation capacity would be needed, in total equating to around 0.55% of annual world GDP by 2030. Cumulative savings would again help offset the investment costs. The GEA pathways estimates required investment needed to achieve less than 2°C compared with preindustrial times to be between US$_{2005}$68–87 trillion or US$_{2005}$1.7–2.2 trillion/yr or about 1.8–2.3% of global GDP (see Figure 17.26, Chapter 17).

Different models and pathways all show the importance of investment in renewable and end-use technologies to reach climate targets. The IEA and GEA analyses show that investments in power-generation energy technologies are necessary for all fuels, but with less emphasis on coal and gas (other than CCS) and more on renewable and nuclear (IEA, 2009a). The same investment strategy applies to end-use technologies, such as heating/cooling and conversion for transport fuels, biofuels, or hydrogen fuel cells. GEA pathways show in addition that the transformation toward the targets would be more cost-effective under pathways that focus on efficiency improvements.

15.11   Insights

In the 21st century, with a population approaching 10 billion people, it is essential that no energy resource be overlooked. By mid-century, at least two-thirds of the world's population is projected to be concentrated in urban locations (see Chapter 18). This underscores the importance of considering all sources of energy and their conversion, transmission, and perhaps most importantly, energy end-use efficiency improvements. With many and varied energy sources available, it is end-use utilizations of energy that have the most potential to achieve energy and energy-system sustainability.

In addition to energy sources, sustainable methods for conversion to carriers, along with transmission and distribution systems that can be employed for the most important end-uses, are crucial. This places particular emphasis on energy carriers to move energy from relatively remote production locations to growing urban population centers. Natural gas, as a lower-carbon fossil fuel and energy carrier, could in some situations be a bridge to the low-carbon future.

Electricity and hydrogen can be efficiently used and are critical to the availability of affordable and acceptable energy. Markets and the scale of international carbon trading will determine how much of each technology is employed in each geographic region, when this happens, and how rapidly. Heat is important to industrial processes, as well as for residential and commercial uses, and should be employed wherever feasible.

Providing integrated storage systems for this energy is essential for making optimal use of the energy sources and carriers that exist. If affordable,

these energy-storage systems are key to achieving a sustainable system that is also economic and secure, and under minimum stress.

The entire energy grid for each system, from energy source through conversion to end-use must be optimized into smart grids. These will use the latest ICT in order to transition to a digital system that communicates between source and end-use continuously, employing all energy carriers and their transmission and distribution systems, heat, electricity, and hydrogen in a coordinated fashion. In this way, the operation of the entire energy system can be optimized to take advantage of the qualities, efficiencies, and productivity of each component. The production of hydrogen from surplus electricity in off-peak periods, and its conversion back to electricity for lighting or mobility, is just one example; all combinations become possible.

The capacity for competitive markets to match supply and demand most efficiently is at the heart of the developed world's economic system – and means that this model is rapidly becoming the standard for economic systems around the world. These markets will also most likely determine the future of decentralized grids worldwide.

In countries with well-developed energy markets, liberalization has often been imposed, along with privatization, unbundling, and enabling consumers to choose energy suppliers. This has led to spot-market energy trading, with the result that long-term contracts are less often the norm. The result for energy supplying countries is generally negative, as it becomes more difficult to ensure returns on large-scale investments. There is thus a threat of destroying the financing tools for large capital-intensive projects in the critical energy-supplying countries (see also Chapter 22).

15.11.1   Conclusions

The private sector will lead in developing and deploying most of the effective approaches, but will need a stable governance framework, facilitation of physical infrastructure, capital investments, and the social cohesion necessary for economic development and poverty reduction, while protecting public health and the environment. Success depends on the implementation of robust, global public-private partnerships that can achieve unprecedented cooperation and integration between governments, between businesses, and between governments and businesses. To have an effect on the changing and growing energy sector, this must happen rapidly.

In order to improve current energy systems, and their capacity to meet rapidly changing needs, it is essential to boost development and investment in advanced energy-sector technologies and integrated systems, from the energy source, through conversion and transmission, to distribution and end-use. The lag time between research and large-scale commercial deployment is long, and funding for energy RD&D continues to decline worldwide. This trend should be reversed with enhanced cooperation between private businesses, between public agencies, as well as between the public and private sectors (see also Chapter 24).

Appendix A: Demand Side Management

The term 'demand-side management' (DSM) was once coined as the utility alternative to supply-side "overspending" in energy systems. The demand-side "negawatt hour" (nWh) was made the conceptual alternative to the supply-side megawatt hour (MWh). However, DSM is not only a utility-tool, but also covers the more general large-scale deployment by use of programs to attract and raise customer interest in which utilities may be a part. Implementation of DSM is also evolving due to both new technological possibilities and requirements regarding energy security and environmental sustainability of systems.

15A.1 One System, Different Measures and Different Actors[10]

- The problem: optimization of a system that is operated by many actors, not all of whom consider (or recognize) energy as an issue.

- The solution: finding a way to deploy (mainly existing) end-use technology on a larger scale on the demand side, in order to obtain the cheapest solution for the energy services (light, climate, motive power).

Industry could be assumed to be more knowledgeable about the energy issues than, for example, residential customers. However, in industry, a great proportion of the energy used even in the energy-intensive branches is used for "trivial purposes" in motors, lighting, ventilation systems, small tools, etc. It is seldom considered and calculated in the way that is done for process-equipment.

The measures to manage demand are basically two: either (1) mandate that something should be done or (2) make use of the market and the economic instruments (Table 15A.1). Mandating is typically used to give explicit information or explicit tasks about certain technologies and certain actors that should be activated, whereas the market acceptance is used when the object cannot be easily identified, but the performance characteristic can be well defined.

The technical problem and challenge: there is a need to change the load shape (peaks and valleys) and to change the load level (conservation and growth). Whereas in the old days the objectives were formulated from the utilities need (and wish) to get a more flat and predictable load curve, the task today is more to serve societal needs and customers. The task is to keep the energy system working and to prevent blackouts and to shift from carbon-fat to carbon-lean systems, as illustrated in Figures 15A.1 and 15A.2.

15A.2 Energy Efficiency Supply Curves[11]

Since improvements in energy demand work on the energy system the same way as do additions in supply, it is necessary to compare demand-side reductions in the same terms as supply-side extensions. A revived technique is to use energy efficiency supply curves, i.e., curves that show the marginal costs for energy efficiency measures (McKinsey &

Table 15A.1 | Approaches, types of measures, suitability and examples of demand- side management to improve energy efficiency.

Approach	Type		Example of Measure	Suitability for Industry
Mandated	Standards		1. Minimum performance (MEPS)	Small motors, Lighting
			2. Top-runner standard	c.f. Energy Star computers
	"Agreed Actions"		3. Voluntary Agreements	For branches
			4. (Technology) Procurements	LCC Procurement guidelines
	Delegated Actions	By actor	5. Regional bodies	Chambers of commerce
			6. Municipalities	–
		By Means	7. Commitments	For SMEs
			8. Certificates	Quota obligations
Market Acceptance	Price-responsive customers		9. Taxes; Tax reduction	Combined with audits and agreements
			10. Price elasticity (Demand Response)	Common and neutral for all types
	Non-price responsive Customers	"Commoditizing" energy efficiency	11. Energy Services (ESCO) and Performance Contracting (EPC)	Outsourcing and facility management
			12. Labels	–

---

10 See also Chapter 8.

11 Early examples are presented in Fickett et al., 1990. The technique is also called Energy Conservation Supply Curves (see Worrell et al., 2003). Here the term is "energy efficiency supply curves" and is used to stress that the energy services (output) are at least similar to that of the system that uses more energy.

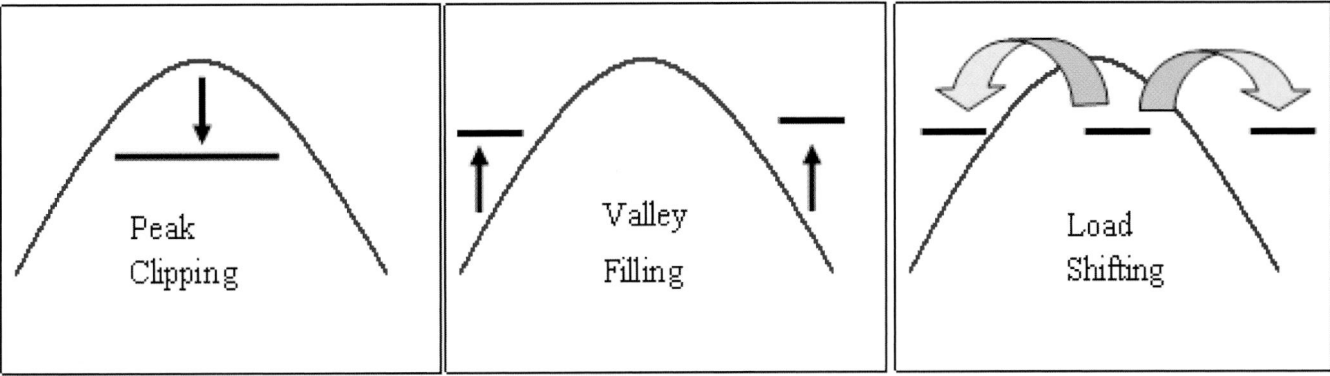

Figure 15A.1 | Load shape changes. Source: Gellings, 1982.

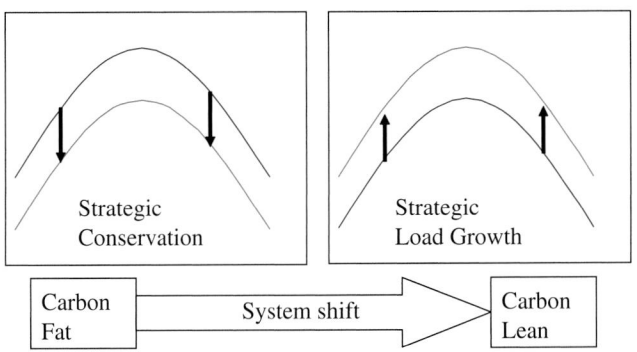

Figure 15A.2 | Load level changes. Source: Gellings, 1982.

Company, 2010). Energy Efficiency Cost (EEC) is calculated as the full cost of the alternatives minus the full cost of the reference solution divided by the annual energy savings, where:

- Full cost includes annualized capital costs over the improvement expected life-time (LCC)

- The reference solution is often the installed one and does not carry further capital costs

Changes in operating and maintenance costs are included in full costs, which also includes the cost for energy supply of each of the alternatives. Transaction costs are not included. Subsidies are generally not included, but could be added, depending on the purpose of the calculation.

Changes in productivity can be added in this calculation. This normally means a further reduction of the cost, since the basic assumption is that the pure energy services should be at least equal in the comparison.

With this formulation all economically attractive alternatives have negative costs.

The critique comprises two elements:

- Even when a calculation can show good reasons, from the point of view of the economic life-cycle, to undertake energy efficiency measures, it can still make sense for a single company to hesitate. The company may have severe difficulties and need to overlook the longer life-time of the alternative. On the other hand, they can make severe losses paying for wasted energy if they are too strict in their requirements for pay-back.

- Transaction costs and also costs for delivery of energy-efficiency measures can be quickly and substantially lowered by the "market learning process", when customers begin to demand the alternative. A programmed DSM-activity can be very useful in this process.

DSM, as a form for programmed large-scale deployment, is a useful measure to both lower the transaction costs and to release the market learning that facilitates the dissemination of cost-efficient technologies to the markets. The profitable efficiency improvements (having negative costs) are often in the range of 20–50% of the energy use. The major part of these will remain locked in without DSM-programs.

15A.3   DSM Formats

The implementation and utilization of DSM will be different in different parts of the world. In some places governments will be the operational agent, in others utilities – in some cases regulated by governments and in others operation in the free market. In any case, it is important that governments set the rules, enforce them. Rules must also not be changed very often so as to attract capital investments.

An excellent overview of the implementation of DSM, with examples from different regions and from the perspective of various government, business and consumer interests, along with costs and benefits of various approaches, can be found in IEA (2008b).

References

Adachi, Y., M. Komoto, I. Watanabe, Y. Ohno and K. Fujimoto, 2000: Effective Utilization of Remote Coal through Dimethyl Ether Synthesis. *Fuel*, **79**(3–4): 229–234.

Amin, M., 2008: Interview with Massoud, Amin, "Upgrading the Grid". *Nature*, **454**: 570–573.

Awerbuch, S., 2006: Portfolio-Based Electricity Generation Planning: Policy Implications for Renewables and Energy Security. *Mitigation and Adaptation Strategies for Global Change*, **11**(3): 693–710.

BP, 2010: *Statistical Review of World Energy 2010*. BP plc., London, UK.

BP, 2011: *Statistical Review of World Energy 2011*. BP plc., London, UK.

Cheung, A. and M. Wilshire, 2010: *White Paper from Consortium on Digital Energy 2009–2010*. Bloomberg New Energy Finance, London, UK.

Chiesa, P., 2003: *Co-Production of Hydrogen, Electricity and Co$_2$ from Coal Using Commercially Ready Technology*. Second Annual Conference on Carbon Sequestration: Developing & Validating the Technology Base to Reduce Carbon Intensity, Alexandria, VA, USA.

Chuang, A., J. Hughes, M. McGranaghan and X. Mamo, 2008: *Integrating New and Emerging Technologies into the California Smart Grid Infrastructure*. CEC-500-2008-047, Electric Power Research Institute (EPRI) and the California Energy Commission, Palo Alto and Sacramento, CA, USA.

Cicchetti, C. J. and C. M. Long, 2003: *Restructuring Electricity Markets: A World Perspective Post-California and Enron. Visions Communications*, New York, USA.

Dry, M. E., 2002: High Quality Diesel Via the Fischer–Tropsch Process – a Review. *Journal of Chemical Technology & Biotechnology*, **77**(1): 43–50.

EIA, 2006: *U.S. Underground Natural Gas Storage Developments: 1998–2005*. Office of Oil and Gas, Energy Information Administration, Washington, D.C., USA.

EIA, 2009: *International Energy Outlook 2009*. Report No. DOE/EIA-0484(2009), United States Energy Information Administration (EIA), Washington, D.C., USA.

EIA, 2011: *International Coal Imports and Exports*. www.eia.doe.gov/emeu/international/coaltrade.html (accessed 28 February, 2011).

Energy Business Reports, 2008: *Microgrids Market Potential. Energy Business Reports*, Anthem, AZ, USA.

EPRI, 2003a: *Electricity Technology Roadmap: Power Delivery and Markeys*. Product No. 1009321, Electric Power Research Institute, Palo Alto, CA, USA.

EPRI, 2003b: *Electricity Technology Roadmap: Meeting the Critical Challenges of the 21st Century – 2003 Summary and Synthesis*. Product No. 1010929, Electric Power Research Institute (EPRI), Palo Alto, CA, USA.

EPRI, 2010: *Distributed Energy Storage Systems Testing and Evaluation 2010 Interim Results*. Product No. 1020079, Electric Power Research Institute, Palo Alto, CA, USA.

ETSO, 2007: *The European Wind Integration Study (Ewis) Towards a Successful Integration of Wind Power into European Electricity Grids, Phase 1*. Final Report, European Transmission System Operators (ETSO), Brussels, Belgium.

European Commission, 2007: *Energy Sector Inquiry*. DG Competition Report, SEC(2006) 1724, Energy, Basic Industries, Chemicals and Pharmaceuticals, Brussels, Belgium.

Fickett, A. P., C. W. Gellings and A. B. Lovins, 1990: Efficient Use of Electricity. *Scientific American*, **263**(3): 64–68, 71–74.

Fritsche, U. R., R. E. H. Sims and A. Monti, 2010: Direct and Indirect Land-Use Competition Issues for Energy Crops and Their Sustainable Production – an Overview. *Biofuels, Bioproducts and Biorefining*, **4**(6): 692–704.

Galvin, R. and K. Yeager with J. Stuller, 2009: *Perfect Power*, McGraw-Hill, New York, 37–38.

GBEP, 2011: *Global Bioenergy Partnership*. Rome, Italy. http://www.globalbioenergy.org/.

Gellings, C. W., 1982: *Load Shape and Level Changes*. In H. Nilsson, Demand Side Management (DSM) – A Renewed Tool for Sustainable Development in the 21st Century (2007), FourFact, Stockholm, Sweden.

Gellings, C., 2011: *Estimating the Costs and Benefits of the Smart Grid: A Preliminary Estimate of the Investment Requirements and the Resultant Benefits of a Fully Functioning Smart Grid*, Electric Power Research Institute Technical Report, Product ID 1022519, Palo Alto, CA, USA, March 2011.

Geometrica, 2006: *Geodesic Domes for Storing Coal, Petcoke and Other Combustible Bulk Materials*. www.coalstorage.com/ (accessed 28 February, 2011).

Gudmundsson, J. S. and M. Mork, 2001: *Stranded Gas to Hydrate for Storage and Transport. International Gas Research Conference*, Amsterdam, the Netherlands.

Huntington, S. P., 1998: *The Clash of Civilizations and the Remaking of World Order. Simon & Schuster*, New York, USA.

IAC, 2007: *Lighting the Way – toward a Sustainable Energy Future. InterAcademy Council (IAC)*, Amsterdam, the Netherlands.

IAEA, 2005: *Thorium Fuel Cycle – Potential Benefits and Challenges*. IAEA-TECDOC-1350, International Atomic Energy Agency (IAEA), Vienna, Austria.

IEA, 2007: *Renewables for Heating and Cooling – Untapped Potential*. International Energy Agency (IEA) of the Organisation for Economic Cooperation and Development (OECD), Paris, France.

IEA, 2008a: *World Energy Outlook 2008*. International Energy Agency (IEA) of the Organisation for Economic Cooperation and Development (OECD), Paris, France.

IEA, 2008b: *Implementing Agreement on Demand-Side Management Technologies and Programmes*. International Energy Agency (IEA) of the Organization for Economic Cooperation and Development (OECD), Paris, France.

IEA, 2008c: *Empowering Variable Renewables – Options for Flexible Electricity Systems*. International Energy Agency (IEA) of the Organisation for Economic Co-Operation and Development (OECD), Paris, France.

IEA, 2009a: *World Energy Outlook 2009*. International Energy Agency (IEA) of the Organisation for Economic Co-Operation and Development (OECD), Paris, France.

IEA, 2009b: *From 1st to 2nd Generation Biofuel Technologies – an Overview of Current Industry and Rd&D Activities*. R. E. H. Sims, M. Taylor, J. N. Saddler and W. Mabee, International Energy Agency (IEA) of the Organisation for Economic Cooperation and Development (OECD), Paris, France.

IGU, 2006: *Underground Storage of Gas*. Report of Working Committee 2, Triennium 2003–2006, S. Khan, International Gas Union (IGU), Oslo, Norway.

IPCC, 2007: *Mitigation of Climate Change*. Contribution of working group III to the Fourth Assessment report of the Intergovernmental Panel on Climate Change (IPCC), B. Metz, O. R. Davidson, P. R. Bosch, R. Dave and L. A. Meyer, Cambridge University Press, Cambridge, UK.

IPCC, 2011: *Special Report on Renewable Energy and Climate Change Mitigation. Intergovernmental Panel on Climate Change (IPCC) and Cambridge University Press*, New York, USA and Cambridge, UK.

Kreutz, T., R. Williams, S. Consonni and P. Chiesa, 2005: Co-Production of Hydrogen, Electricity and Co2 from Coal with Commercially Ready Technology. Part B: Economic Analysis. *International Journal of Hydrogen Energy*, **30**(7): 769–784.

Lange, J.-P., 2001: Methanol Synthesis: A Short Review of Technology Improvements. *Catalysis Today*, **64**(1–2): 3–8.

Larsen, H. and L. P. Petersen, Eds. 2005: *The Future Energy System – Distributed Production and Use*. Risø Energy Report 4. Risø National Laboratory for Sustainable Energy, Technical University of Denmark, Roskilde, Denmark.

Larsen, H. and L. P. Petersen, 2009: *The Intelligent Energy System Infrastructure for the Future. Risø Energy Report 8, Risø National Laboratory for Sustainable Energy, Technical University of Denmark*, Roskilde, Denmark.

Lehmann, J., J. Gaunt and M. Rondon, 2006: Bio-Char Sequestration in Terrestrial Ecosystems – a Review. *Mitigation and Adaptation Strategies for Global Change*, **11**(2): 395–419.

Lou, L. W. and B. Wang, 2006: Current Situation of Dme Synthesis Technology. *Guizhou Chemical Industry*, **31**(4): 9–15.

Lou, R., R.-l. Xu and S.-l. Lou, 2006: Application of Methanol Synthesis Reactor to Large-Scale Plants. *Applied Chemical Industry*, **35**: 260–268.

Lovins, A. B. and B. Cohen, 2010: *Renewables, Micropower, and the Transforming Electricity Landscape*. Report No. U10–08, Rocky Mountain Institute, Snowmass, CO, USA.

Marrero, T. R., 2006: *Long-Distance Transport of Coal by Coal Log Pipeline*. Capsule Pipeline Research Center, University of Missouri, Columbia, MO, USA.

Matic, D., 2006: *Global Opportunities for Natural Gas as a Transportation Fuel for Today and Tomorrow. IGU Study Group – Natural Gas for Vehicles, World Gas Conference* International Gas Union (IGU), Amsterdam, the Netherlands.

McKinsey & Company, 2010: *Greenhouse Gas Abatement Cost Curves*. www.mckinsey.com/clientservice/sustainability/costcurves.asp (accessed 23 March, 2011).

Moniz, E. J., H. D. Jacoby and A. J. M. Meggs, 2010: *The Future of Natural Gas: An Interdisciplinary Mit Study. Energy Initiative, Massachusetts Institute of Technology*, Cambridge, MA, USA.

Nahigian, K. R., 2008: *The Smart Alternative: Securing and Strengthening Our Nation's Vulnerable Electric Grid*. Advancing the Reform Agenda, Reform Brief, The Reform Institute, Alexandria, VA.

National Academy of Sciences, 2009: *America's Energy Future: Technology and Transformation, the National Academies Energy Summit Report*. National Academies Press, Washington, D.C., USA.

National Academy of Sciences, 2010: *Electricity from Renewable Resources: Status, Prospects, and Impediments*. National Academies Press, Washington, D.C., USA.

NEA and IAEA, 2004: *Uranium 2003: Resources, Production and Demand*. NEA Report No. 5291, Nuclear Energy Agency (NEA) of the Organisation for Economic Cooperation and Development (OECD) and the International Atomic Energy Agency (IAEA), Paris, France.

NEA, 2006: *Advanced Nuclear Fuel Cycles and Radioactive Waste Management Nuclear Energy Agency (NEA) of the Organisation for Economic Cooperation and Development (OECD)*, Paris, France.

NEA and IAEA, 2008: *Uranium 2007: Resources, Production and Demand. Nuclear Energy Agency (NEA) of the Organisation for Economic Cooperation and Development (OECD) and the International Atomic Energy Agency (IAEA)*, Paris, France.

NEA and IAEA, 2010: *Uranium 2009: Resources, Production and Demand. Nuclear Energy Agency (NEA) of the Organisation for Economic Cooperation and Development (OECD) and the International Atomic Energy Agency (IAEA)*, Paris, France.

Nelder, C., 2008: *How to Play the China Coal Crisis*. www.energyandcapital.com/articles/coal-investment-stocks/747 (accessed 30 December, 2010).

NETL, 2007: *Barriers to Achieving the Modern Grid*. The NETL Modern Grid Initiative – Powering our 21st-Century Economy, United States National Energy Technology Laboratory (NETL) for the United States Department of Energy (DOE), Washington, D.C., USA.

Patterson, W., 2009: *Keeping the Lights On. Earthscan*, London, UK.

Pauluis, G. and L. Van den Durpel, 2001: *Trends in the Nuclear Fuel Cycle – Economic, Environmental and Social Aspects*. Facts and Opinions, NEA News No. 19.2, Nuclear Energy Agency (NEA) of the Organisation for Economic Cooperation and Development (OECD), Paris, France.

Planck Foundation, 2009: *Situation 2009*. Global Resources Analysis Planck Foundation, Washington, D.C., USA.

Rotawave, 2010: *"Intelligent" Energy System Set to Revolutionise the World's Biomass Fuel Supply Market*. News Release, Ref: 0110, Rotawave Ltd, Aberdeen, UK.

Sims, R. E. H., W. Mabee, J. N. Saddler and M. Taylor, 2010: An Overview of Second Generation Biofuel Technologies. *Bioresource Technology*, **101**(6): 1570–1580.

Smil, V., 2005: *Energy at the Crossroads: Global Perspectives and Uncertainties. MIT Press*, Cambridge, MA, USA.

Tijmensen, M. J. A., 2000: *The Production of Fischer-Tropsch Liquids and Power through Biomass Gasification*. NWS-E-2000-29, Science Technology Society, Utrecht University, Utrecht.

UNEP, 2007: *Buildings and Climate Change – Status, Challenges, and Opportunities. United Nations Environment Programme*, New York, USA.

US DOE, 2002: *National Hydrogen Energy Roadmap*. United States Department of Energy (US DOE), Washington, DC, USA.

US DOE, 2007: *Hydrogen, Fuel Cells & Infrastrucutre Technologies Program, Multi-Year Research, Development and Demonstration Plan*. www.hydrogen.energy.gov (accessed 15 December, 2010).

WEC, 2010: *Survey of Energy Resources: Focus on Shale Gas. World Energy Council (WEC)*, London, UK.

Wisconsin Department of Commerce, 2005: *Ethanol Motor Fuel Storage Overview*. Reivsed Program Letter, Environmental & Regulatory Services, Wisconsin Department of Commerce, Madison, WI.

WNN, 2009: *Thorium-Fuelled Exports Coming from India*. WNN (World Nuclear News). London, UK.

Worrell, E., J. A. Laitner, M. Ruth and H. Finman, 2003: Productivity Benefits of Industrial Energy Efficiency Measures. *Energy*, **28**(11): 1081–1098.

Yeager, K., 2008: Striving for Power Perfection. *IEEE Power and Energy Magazine*, **6**(6): 28–35.

Yeager, K. E., 2007: *Electricity Enterprise: U.S., Prospects*. Encyclopedia of Energy Engineering and Technology. Taylor & Francis, London, UK.

Zhou, L., S. Hu, D. Chen, Y. Li, B. Zhu and Y. Jin, 2009: Study on Systems Based on Coal and Natural Gas for Producing Dimethyl Ether. *Industrial & Engineering Chemistry Research*, **48**(8): 4101–4108.

Zwart, R., 2010: *Wood Pellets Markets and International Trade*. Reference Document for the 2nd Biomass Power and Trade Conference, Centre for Management and Technology (CMT), Rotterdam.

16

Transitions in Energy Systems

Convening Lead Author (CLA)
Anand Patwardhan (Indian Institute of Technology-Bombay)

Lead Authors (LA)
Ines Azevedo (Carnegie Mellon University, USA)
Tira Foran (Commonwealth Scientific Industrial Research Organisation, Australia)
Mahesh Patankar (Independent Energy Sector Consultant)
Anand Rao (Indian Institute of Technology-Bombay)
Rob Raven (Eindhoven University of Technology, the Netherlands)
Constantine Samaras (Rand Corporation, USA)
Adrian Smith (University of Sussex, UK)
Geert Verbong (Eindhoven University of Technology, the Netherlands)
Rahul Walawalkar (Customized Energy Solutions, India)

Contributing Authors (CA)
Riddhi Panse (Indian Institute of Technology-Bombay)
Saumya Ranjan (Indian Institute of Technology-Bombay)
Neha Umarji (Indian Institute of Technology-Bombay)

Review Editor
John Weyant (Stanford University, USA)

Contents

Executive Summary

This chapter examines the theme of transitions in energy systems. It assesses the literature that explores the genesis, growth, and management of transitions. This literature provides a multi-level framework for large-scale, transformative change in technology systems, involving a hierarchy of changes from experiments to niches to technology regimes.

The chapter also covers specific innovation systems and experiments in the energy sector that may have the potential for larger impact and could lead to new niches or technology regimes. These experiments include technology-driven innovations in generation and end-use; system-level innovations that could reconfigure existing systems; and business model innovations centered on energy service delivery. Experiments in generation include hybrid systems, where multiple primary energy sources help address issues such as intermittency. Experiments in end-use include technology options for the simultaneous delivery of multiple energy services, or energy and non-energy services. System-level experiments include innovations in storage, distributed generation, and the facilitation of energy efficiency by effectively monetizing savings in energy use.

In some of these experiments, technology can lead to changing relationships between actors or changing roles for actors; for example, the process of consumers becoming producers is seen in small-scale biogas projects. These changing relationships present both challenges and opportunities for influencing the transition process. The chapter also discusses policy and institutional issues that affect transitions. Finally, it is seen that although technological research, development, and innovation are important, a wide-scale, equitable, and accessible transformation to energy systems for sustainable development needs to be tackled as a socio-political issue.

16.1   Introduction

The energy sector is evolving rapidly. This evolution is shaped by the convergence of several factors, including the realization that energy is a key enabler for achieving development goals, such as the Millennium Development Goals (MDGs). Chapters 2 and 6 of this report review the link between economic growth and energy use, while the effects of energy system operation for human health and the local & global environment are reviewed in Chapters 3 and 4, respectively.

This evolution simultaneously poses a challenge and an opportunity. The challenge is to manage multiple goals and objectives, while the opportunity is to use this evolution to move society towards an energy future that promotes sustainability. The challenge is complicated by the fact that, in many countries, existing energy policy institutions were not designed for this task. Many countries have energy institutions designed for the liberalization and privatization of their energy systems in the 1990s. The challenge of investing and developing energy systems for sustainable development may require new institutional innovations (Helm, 2005; Smith, 2009).

Accomplishing the transition in energy systems will require an accelerating pace of technological change throughout the developed and developing world. Energy policies will need to facilitate this acceleration, while at the same time avoiding lock in to sub-optimal solutions. The variety of energy resources as well as end-use and supply-side technologies that may be available are assessed in earlier chapters of this assessment (Chapters 7–15). While these chapters have assessed the state of the energy system as it exists today, Chapter 17 describes possible pathways by which energy systems could be modified to simultaneously achieve multiple objectives, including environmental sustainability. In many ways, these pathways require transformative change in the energy system.

Transformative change and transitions therefore form the focus of this chapter, and the literature assessed is intended to reflect some of our learning and experience with large-scale transitions in energy systems and in other large, complex socio-economic systems; to provide insights into how transitions may be studied; and, more importantly, how they can be managed. We also review recent experiments and innovations in energy systems, which, if scaled up appropriately, could offer significant opportunities for transforming the energy system.

These experiments include approaches for generating electricity by combining different primary energy sources and approaches for the simultaneous delivery of multiple energy services. An emphasis on providing the energy services while reducing energy wastage (energy efficiency), demand-side management, and a supportive regulatory and policy environment is also encouraging innovations in business models, with energy service companies being a prime example. Finally, in many parts of the developing world, the challenge of achieving energy access goals more rapidly is leading to efforts aimed at supporting a new generation of energy-sector entrepreneurs. Taken individually, each of these trends might be seen as an isolated example, with little significance for the large-scale context. But taken together, and governed effectively, they offer a possibility of creating new transformation pathways for energy systems. In this way, the chapter serves to connect the assessment of the expectations from, and the current state of, the energy system (Clusters I and II) with the pathways to energy systems for sustainable development (Clusters III and IV).

The remainder of this chapter is structured as follows. Section 16.1 reviews literature on transitions and transition management that underpins the importance of considering ancillary and complementary factors while examining policy-driven or autonomous change. Several examples of experiments that may have the potential to influence future energy transitions are grouped into two broad categories. The first section (16.2) includes technology-driven experiments in electricity generation and end-use. Then Section 16.3 considers innovative systems experiments in overall energy system optimization, configuration, and operation. The final section (16.4) looks at a number of policy and institutional issues that may positively or negatively influence energy system transitions.

16.1.1   Transitions and Transition Management

A detailed historical perspective on the long-term evolution of energy systems is given in Chapter 1, including the way in which the industrial revolution changed the fabric of human society and our natural environment. Chapter 1 also discusses the characteristics of historical energy transitions, such as the change in resource use from traditional biomass to coal, oil, and gas. These long-term transformations were driven by both endogenous factors, such as the emergence of electricity after the invention of the dynamo by Michael Faraday, and by exogenous factors, such as the 1973 Oil Crisis. Endogenous technological change is a key element of these long-term transitions. A combination of historical analysis and new modeling techniques can help in identifying predictable patterns of technological change and innovation (Grubler et al., 1999) and thus improve the treatment of technology in models used to project trends in energy and economic and environmental futures.

16.1.1.1   Determinants and Drivers of Transitions

Transformations in energy systems are long-term change processes (decadal or longer) in technology, the economy, institutions, ecology, culture, behavior, and belief systems. They typically cover all aspects of energy systems, including resource extraction, conversion, and end-use. While these transitions are not deterministically caused by technology, technological change is a useful entry point through which long-term transitions may be analyzed and explained. In relation to the transport sector, for example, transitions have been explained as changes in the mode of transport, such as from sailing ships to steamships (Geels, 2002) and from horses to motorcars (Grubler et al., 1999).

A recent approach to understanding transitions is the "multi-level perspective" (Rip and Kemp, 1998; Geels, 2002; Geels and Schot, 2007; Markard and Truffer, 2008). Combining insights from evolutionary theory and the sociology of technology, this perspective conceptualizes major

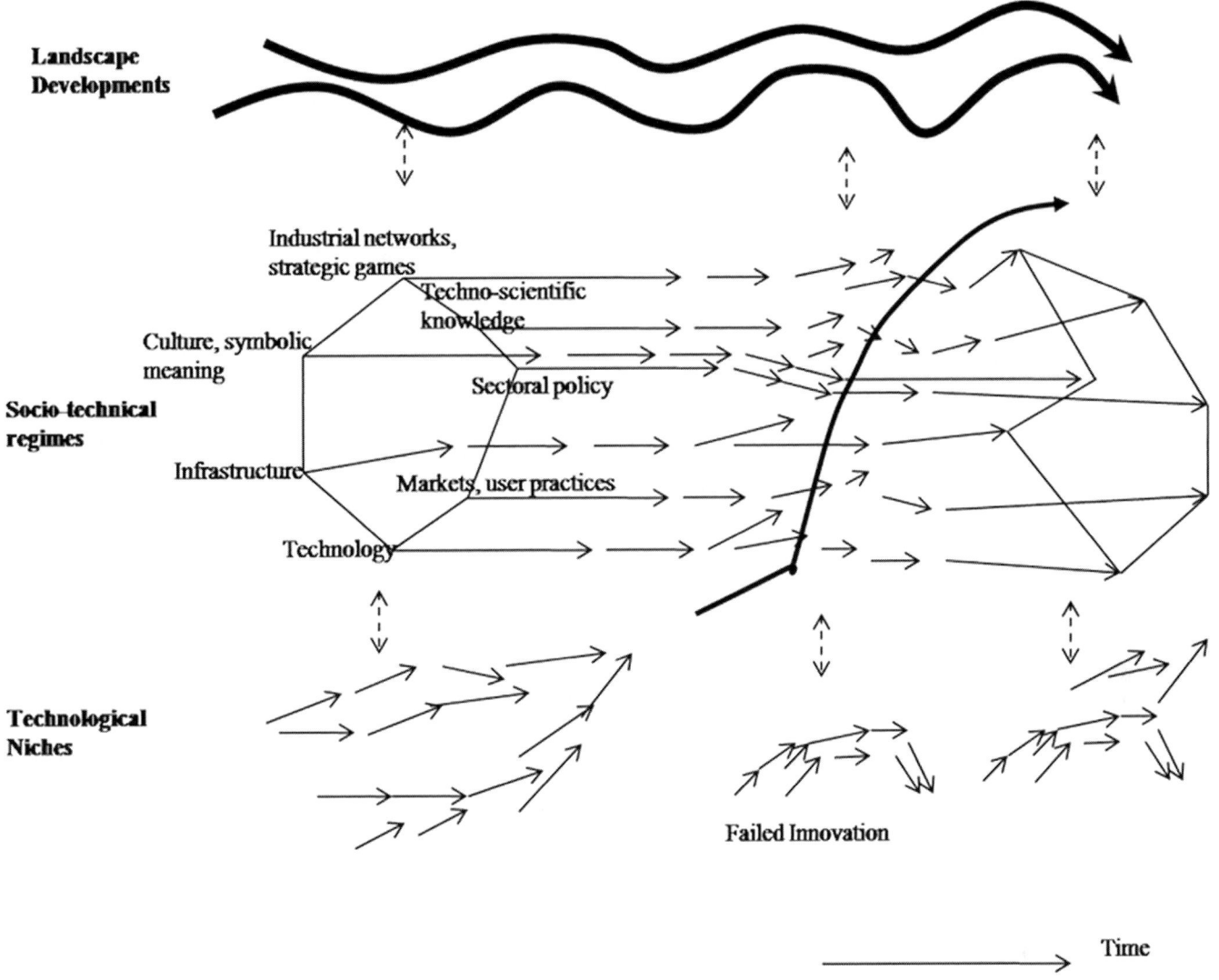

Figure 16.1 | A dynamic representation of the multi-level perspective on transitions. This perspective distinguishes between the micro-level of niches, the meso-level of socio-technical regimes, and the macro-level of landscapes. Innovations and experiments can only break through when there is sufficient pressure on the socio-technical regime. The small arrows at the bottom represent the niche innovations and experiments, which join together, become powerful (as in the thicker arrows), start influencing the socio-technical regimes, leading to major changes in technology, market, user practices, etc., and eventually become part of the landscape. Source: Geels, 2002.

transformative change as the product of interrelated processes at the three levels of niche, regime, and landscape. The framework emphasizes the incremental nature of innovation in socio-technical regimes. A niche is a network of similar projects carried out by innovating actors who seek to challenge the incumbent and dominant socio-technical practice (regime), such as a distributed generation using renewable energy versus a regime of centralized electricity generation. Regimes and niches develop in the context of a socio-technical landscape, which consists of both hard geographical features, such as resource availability and infrastructure, and "soft" elements, such as political conditions, societal trends, and economic fluctuations. The socio-technical landscape provides the exogenous environment for regime change and is a source of major selection pressures on prevailing regimes. Transitions, i.e., shifts from one stable socio-technical regime to another, occur when regimes are destabilized through landscape pressures, which in turn provide breakthrough opportunities for niche innovations. Figure 16.1 visualizes the multi-level model.

Similar frameworks have been developed by other authors. For example, Van Dijk (2000) defines the technological regime as a particular combination of opportunity, appropriateness, cumulativeness, conditions, and properties of the knowledge base that are common to specific activities of innovation and production and are shared by a population of firms undertaking those activities. Stier (1983) and Kathuria (1998) have used examples from product/process and system diffusion to describe changes in technologies and technology systems. Technology products or processes form part of technology systems, which in turn form part of technology regimes.

The transitions literature has advanced from an initial one-pathway conceptualization of transitions, in which niches grow and change regimes, to distinguishing between pathways, depending on different configurations and multi-level interactions (Smith et al., 2005; Geels and Schot, 2007), as summarized in Box 16.1.

Box 16.1 | Pathways that Represent and Characterize Transitions

1. Reorientation/transformation

This pathway is characterized by external pressure (from the landscape level or outsider social groups) and gradual adjustment and reorientation of existing regimes. Although external pressures create windows of opportunity for wider change, niche innovations are insufficiently developed to take advantage of them. Change is therefore enacted primarily by regime actors, who reorient existing development trajectories. Outside criticism from social movements and public opinion is important because it creates pressure on regime actors, especially when they spill over towards stricter environmental policies and changes in consumer preferences. Although regime actors respond to these pressures, the changes in their search heuristics, guiding principles, and research and development investments are modest. The result is a gradual change of direction in regime trajectories. New regimes thus grow from older ones through cumulative adjustments and reorientations. Regime actors survive, although some change may occur in social networks. Furthermore, regime actors may import external knowledge if the distance with regime knowledge is not too large. Such symbiotic niche innovations add to the regime, rather than disrupting the basic architecture.

2. Reconfiguration

In this pathway, niche innovations are further developed when regimes face problems and external landscape pressures. In response, the regime adopts certain niche innovations into the system as add-ons or component substitutions. This leads to a gradual reconfiguration of the basic architecture and changes in some guiding principles, beliefs, and practices. In the reconfiguration pathway, the new regime also grows out of the old regime. But it differs from the transformation pathway insofar as the cumulative adoption of new components changes the regime's basic architecture substantially. The main interaction is between regime actors and niche actors, who develop and supply the new components and technologies. This is therefore a more radical transition than that of the transformation pathway.

3. Technological substitution

Here, landscape pressures produce problems and tensions in regimes, which create windows of opportunity for niche innovations. Niche innovations can use these windows when they have stabilized and gathered momentum. Diffusion of these new technologies usually takes the form of niche-accumulation, with innovations entering increasingly bigger markets and eventually replacing the existing regime. In this pathway newcomers (niche actors) compete with incumbent regime actors.

4. De-alignment and re-alignment

Major landscape changes lead to huge problems in the regime. The regime experiences major internal problems, collapses, erodes, and de-aligns. Regime actors lose faith in the system's future. The regime's destabilization creates uncertainty about dimensions on which to optimize innovation efforts. The sustained period of uncertainty is characterized by the coexistence of multiple niche innovations and widespread experimentation. Eventually, one option becomes dominant, leading to a major restructuring of the system (new actors, guiding principles, beliefs, and practices).

Source: Smith et al., 2005; Geels and Schot, 2007.

The multi-level perspective on transitions is still under active development and has raised some criticisms. In their review of current transitions research, Genus and Coles (2008) have identified limitations of this perspective that need to be addressed to improve the understanding of processes of innovation affecting the transformation. Few case studies presented in the literature systematically identify or analyze the meso-level socio-technical regimes said to be central to stability and change in socio-technical systems, not least with respect to the rules and routines acknowledged as key to the activities of groups in those regimes. There has also been a tendency to focus on successful technologies and methodological issues concerning the multi-level perspective's functionality, and the poor record of historical case studies appears to have been undervalued. Moreover, there is a danger that some ideas implicit in this treatment of the perspective can seep into the policymaking domain, so that the reality of a neat, mechanistic model of transition could become the dominant interpretation of the perspective. Markard and Truffer (2008) suggest that the multi-level perspective approach is largely confined to the niche level in its analysis of emerging technologies. Frameworks for the analysis of transitions need to:

- consider more explicitly the innovation processes perceived at the micro-level of organizations (strategies, agency);

- take account of mutual dependencies between actors and institutions;

- develop consistent performance comparisons to recommend how to support the development of particular innovations; and

- facilitate systematic identification and assessment of the broad range of factors (e.g., events, developments, institutional effects, actor behavior) that influence innovation processes.

Fouquet (2010) attempts to analyze the role of economic drivers in past energy transitions in order to identify commonalities that may be useful in anticipating future transformations. The analysis proposes that the same micro-economic drivers that were important in the past, such as better and cheaper services, will be relevant in the future. In terms of policy findings, governments and regulatory agencies will have to protect niches rather than rely solely on consumers' willingness to pay for niche technologies. This would involve incentivizing the provision of ever cheaper services and highly valued additional attributes, and minimizing the problem of negative aspects of new technologies. The study further suggests that the periods of niche development, refinement, and large-scale adoption will probably spread over decades.

16.1.2   Managing Transitions

The emerging understanding of transitions and transformations have raised the possibility that they may be actively influenced or managed through policy and other interventions. For example, strategic niche management (SNM) conceptualizes innovation projects with novel technologies as sustainability experiments taking place in real-life contexts (Hoogma et al., 2002; Raven, 2005; Schot and Geels, 2008). SNM also emphasizes the critical role of social network dynamics, articulating expectations, learning processes, and clever niche regime interfaces. The perspective has been applied to a large number of case studies in the field of sustainability, such as sustainable transport (Hoogma et al., 2002), photovoltaic cells (Van Mierlo, 2002), sustainable food supply chains (Wiskerke, 2003; Smith, 2006), bioenergy technologies (Raven, 2005), biofuels (Van Eijck and Romijn, 2008; Van der Laak et al., 2007), wind turbines (Kemp et al.,

2001; Healey, 2008), the hydrogen economy (Agnolucci and Ekins, 2007), and low-energy housing (Lovell, 2007; Smith, 2007).

Another part of the literature deals with specific social groups as niches for sustainable technologies, such as Non Governmental Organizations (NGOs), civil society, and social entrepreneurs (Verheul and Vergragt, 1995; Seyfang and Smith, 2007; Witkamp et al., 2011). More fundamentally, Hegger et al. (2007) and Monaghan (2009) suggest that rather than emphasizing technology, niche management could include experimentation with concepts and guiding principles. Ieromonachou et al. (2004) propose an adapted version of strategic niche management based on policy niches, rather than technological niches, to investigate urban road pricing schemes.

A similar method of conceptualizing the role of experiments is the Bounded Socio-Technical Experiment, in which the mapping and monitoring of the learning process can be conceptualized into four levels (Brown and Vergragt, 2008):

- problem solving using pre-determined objectives;

- problem definition in relation to the particular technology-societal problem coupling;

- dominant interpretive frames; and

- worldview, which denotes deeply held values with regard to the preferred social order.

A detailed case study by Brown and Vergragt (2008) of a zero-fossil-fuel residential building led them to conclude (though by no means generic) that the learning involved in bringing about system changes takes place at both the individual and team levels, that it mainly involves changes in problem definition, and that the process involved is just as important as the product obtained. However, a study of several green niches by Smith (2007) found that different actors drew different lessons from projects in niches, depending upon their institutional position and interests. Learning is important, but so too is translating lessons between actors operating in very different contexts. Low-carbon construction techniques that work for green builders may be less amenable to mainstream volume house-builders, for instance, who adopt only elements of the greener techniques. Policy has to support more effective translation processes between niche actors and those in the incumbent regime. Smith et al. (2005) have downplayed the role of niches as instruments in transition governance and introduced the concept of transition contexts to better understand how regimes might adapt to changing selection pressures, such as emerging sustainability preferences in society. They argue that governance must be understood as articulating selection pressures and coordinating and providing resources for adaptation.

The management of transformations could be considered as the transformation of complex adaptive systems (Kemp et al., 2007; Loorbach,

2007), where governance is a process of envisioning sustainable futures in a participatory manner, with front-runners in policy, industry, and society.

Another approach examines the functions of technological innovation systems. This approach conceptualizes transformative change towards sustainability as a process of fulfilling a set of different system functions that are developing and interacting over time. Change occurs through motors that produce either virtuous or vicious cycles, causing system growth or system collapse (Suurs, 2009). Governance advice following an analysis of the functioning of a specific technological innovation system often entails arguments to support certain system functions that are preventing further growth and breakthrough.

Since technology is an important element of transitions, the planning and coordination of research and development efforts may be relevant for guiding transitions. Shafiei et al. (2009) have developed a mathematical and conceptual model of energy research and development resource allocation with an explicit perspective of developing countries (in their case, Iran), which has been linked to a bottom-up energy-systems model and can help in determining the optimal allocation of research and development resources. A similar effort was produced for the United States energy system through a recent project involving several US national laboratories. This model, an open source, long-range Stochastic Lite Building Module, which is part of a larger economic model for the US energy system (the Stochastic Energy Modeling System), estimates the impact of different policies and consumer behavior on the market penetration of low-carbon building technologies (Stadler et al., 2009). The tool can be used, among other things, to simulate the impact of different levels of research and development funding for energy-related technologies. Recent work from the Lawrence Berkeley National Laboratory using the Stochastic Energy Modeling System assesses the impact of research and development levels for photovoltaic technology and solid-state lighting (Stadler et al., 2009). These models can also be helpful in getting at least an approximation of the expected duration of a major energy transition.

As the variety of technology options and combinations increases, there is a greater need for assessment, evaluation, and planning tools. Azevedo (2009) has developed a flexible and transparent, user-friendly tool that assesses the most cost effective way to provide residential end-use energy services (heating, cooling, clothes washer, dishwashers, hot water, cooking, clothes dryer, refrigerator, freezers, and lighting) from different fuels (natural gas, electricity, kerosene, wood, geothermal, coal, solar, distillate, and LPG). With this tool, known as the Regional Residential Energy Efficiency Model, simulations can be performed in any number of the following US census division regions: New England, Middle Atlantic, East North Central, West North Central, South Atlantic, East South Central, West South Central, Mountain and Pacific. The model considers different energy and climate policy constraints.

Transition processes have been studied in the context of national policies. Marinova and Balaguer (2009) present a comparative study of photovoltaic industry growth in three countries – Australia, Germany, and Japan. While Australia has seen successes in sophisticated photovoltaic off-grid systems, it has not been able to achieve great success in the development of on-grid systems, due to the lack of market development and also because of a shortage of funding available to smaller or newer research groups emerging in the photovoltaic domain. In contrast, the growth in Germany and Japan has been more successful, because development took place in a more business-like and industrial environment. In Germany, the strong industrial environment in sophisticated manufacturing allowed the vertical integration of the photovoltaic industry from feedstock materials to module fabrication. In Japan, meanwhile, the existing and huge electronics industry served as a natural base for solar cell technology. In both these countries, the photovoltaic industry was directed towards on-grid applications and was well nurtured by subsidies that helped create markets and mainstream the technologies. These studies suggest the importance of considering energy, industrial, and technology policies in an integrated manner.

In the developing world, an important transition is the replacement of traditional non-commercial energy carriers by modern energy sources. A recent analysis of the household energy transition in India and China reported that, despite sharp differences in the absolute amount of energy used and penetration of electric infrastructure, the overall trends in energy use and the factors influencing a transition to modern energy in both nations are similar (Pachauri and Jiang, 2008). Compared with rural households, urban households in both nations consume a disproportionately large share of commercial energy and are much further along in the transformation to modern energy. However, total energy use in rural households exceeds that of urban households because of a continued dependence on inefficient solid fuels, which contributes to over 85% of rural household energy needs in both countries. It has been found that, in addition to urbanization, key drivers of the transition in both nations include income, energy prices, energy access, and local fuel availability. Similar trends have also been observed in other parts of the developing world.

The study of transitions is an active area of research, and may provide many useful insights into the way that emerging trends and options for meeting energy-related goals might be facilitated and accelerated into larger-scale system transitions. This will require careful identification, documentation, and analysis of experiments that may have a potential to enhance sustainability, and that may be replicable and scalable. Crucially, research also shows that opportunities for these experiments to influence wider changes may depend upon processes that destabilize the incumbent energy regime and its institutions, and it may be necessary to consider the creation of sustainable alternatives in tandem with tensions within prevailing energy systems (Smith, 2005; Smith et al., 2005; Geels and Schot, 2007).

16.2   Technology-Driven Innovations Experiments

Experiments play an important role in long-term, large-scale transitions, particularly where there is a supportive policy environment that allows for promising experiments to evolve into niches that could challenge incumbent regimes. The challenges of access, cost, security, environmental, and health concerns are pushing for new ideas and technologies to emerge in the energy sector, creating a range of possibilities. This section assesses emerging experiments that are largely driven by technology innovation. Three categories of experiments are considered. The first consists of experiments that combine multiple primary energy sources for electricity generation, to overcome problems such as intermittency, and to increase reliability and availability. The second considers multiple energy end uses that may be addressed together, while the third category addresses experiments in which energy and non-energy services are provided simultaneously.

16.2.1   Combining Different Primary Energy Sources

Renewable energy technologies are A key element of all energy system evolution pathways that meet the GEA normative goals for sustainability (see Chapter 17). However, renewable energy sources, such as biomass, wind, and solar, vary greatly in intensity and availability, both diurnally and seasonally. This intermittency is an important constraint to their use. Performance may be improved by connecting non-correlated, multiple sources together, to create a hybrid system. A hybrid system typically consists of two or more energy sources, together with power conditioning equipment and optional energy storage. There is increasing experience with the design and operation of such hybrid systems. Mohamed et al. (2004) have demonstrated that hybrid energy systems can significantly reduce the total lifecycle cost of stand-alone power supplies in many situations, whilst providing a more reliable supply of electricity through the combination of energy sources. An example of hybrids that combine multiple renewable energy technologies are solar-wind LED-based lighting systems that are being used extensively in China for street lighting (Liu and Wang, 2009).

Sometimes, for reasons of resource availability, landscape, climate, or investment difficulties, it is not possible or feasible to harness more than one source of renewable energy. In such situations, conventional fossil fuels could be used to supplement the electricity generated through renewable options. For example, hybrid dual-fuel engine pumps using biogas obtained from community biogas plants are often the least expensive option for pumping irrigation water. The unit cost is further reduced for large-capacity biogas plants, reflecting the effect of an economy of scale in their capital cost, in comparison to pumps operating on-grid electricity, wind power, solar photovoltaic, and stand-alone biogas plants (Purohit, 2007). Similarly, Shaahid and El-Amin (2009) have discussed the techno-economic feasibility of using a hybrid solar photovoltaic-diesel battery power generation system for off-grid supply to Saudi Arabia's remote, unelectrified villages. Another approach to issues of variability and intermittency is the addition of storage or coproduction of another energy carrier, such as hydrogen. Energy storage technologies and related issues are covered in Section 16.3.1, while Chapter 15 examines the generation, storage, and use of hydrogen as a future energy carrier.

Hybridization may offer not only greater reliability, but also optimize the utilization of various energy resources. To maximize revenue from such systems, the optimum configuration will depend on resource availability, the capital and fuel input prices, and the prices of electricity generation from different sources (Dufo-López et al., 2009).

One interesting review by Paska et al. (2009) lists various hybrid combinations that are commercially available (see also Chapter 11), in development, or, at least, plausible (Figure 16.2). Potentially, hybrid energy systems are proving capable of adding reliability and quality to the existing power generation system, particularly as renewable energy systems seem to offer distinct, technology-specific advantages. This approach can be extremely useful in addressing region-specific power issues.

It is clear that most technologies mentioned above have been demonstrated successfully at the research and development stage, and even in small-scale application. To achieve an economy of scale, appropriate policy and enabling environment interventions may be needed.

Research into hybrid systems has focused mostly on the actual design and operation of generation technologies. However, constructing even a pilot plant or innovative, experimental system to evaluate the economic feasibility for each new location where a hybrid or renewable energy system is being proposed can be very cumbersome and capital intensive and can lead to disinterest among decision makers. In such situations, the use of mathematical modeling and simulation techniques may be an attractive alternative for assessing techno-economic feasibility. For example, Deshmukh and Deshmukh (2008) have produced mathematical models of hybrid solar and wind energy systems, including a diesel generator and a battery.

16.2.2   Addressing Multiple End-uses

The idea of hybridization could also be used to address multiple energy end-uses. The most common example is the use of heat produced during electricity generation for thermal applications or the use of waste heat for electricity production, i.e., cogeneration, or combined heat and power (CHP).

Low temperature heat produced by heat engines has very few applications, because most applications require heat at high temperature.

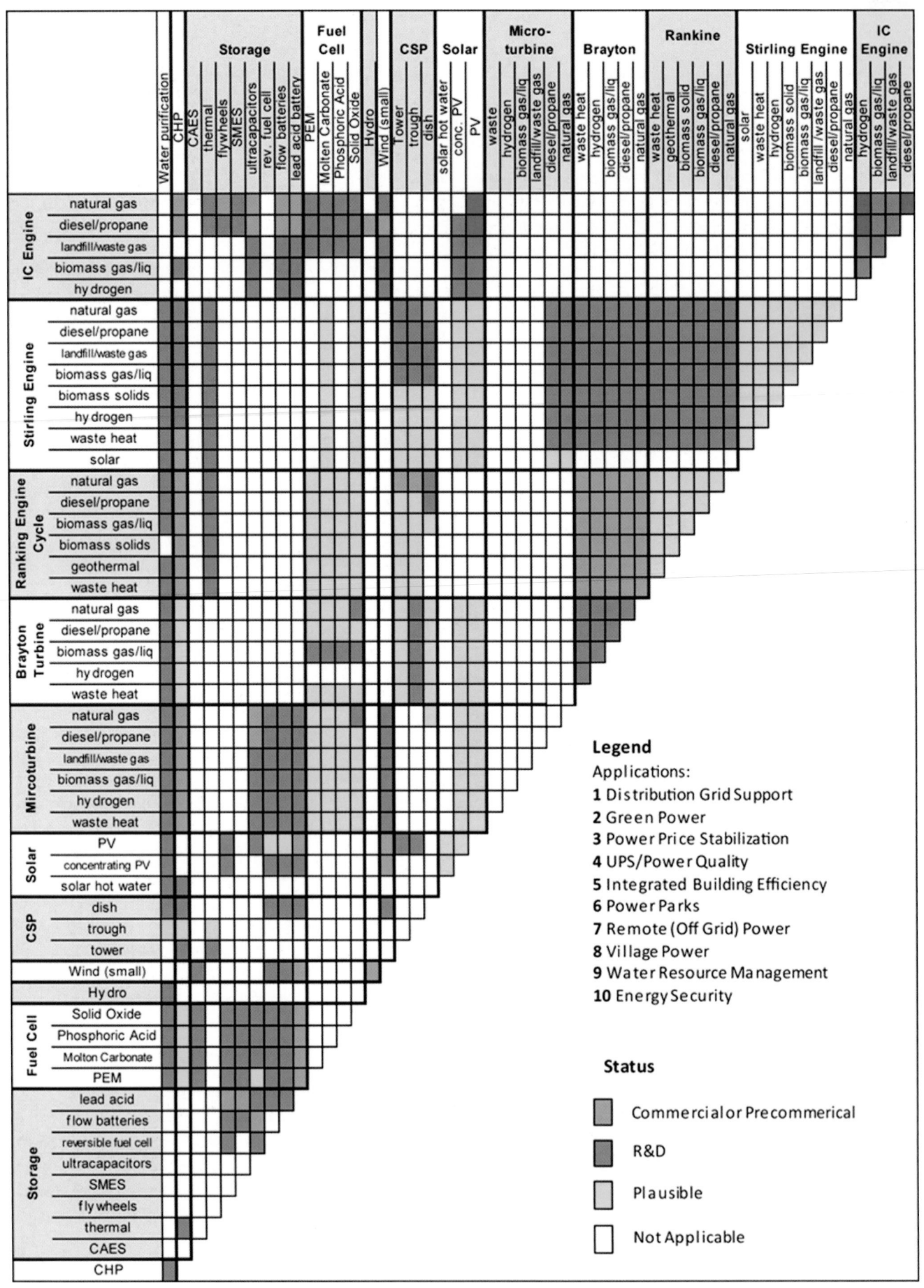

Figure 16.2 | Matrix showing overview of commercialized, research and development stage, and plausible combinations of hybrid technologies, including storage. Source: Paska et al., 2009.

However, if this heat can be harnessed to produce electricity on-site, which is technically possible, a CHP system is obtained. This category of micro-CHP system is generating much interest (Laughton, 1996; Hinnells, 2008). Such systems can operate in homes or small commercial buildings and are driven by heat-demand, delivering electricity as the by-product. Micro-CHPs have the added advantage of being able to run on renewable energy sources, such as wood, biomass, and solar thermal systems, as well as on fossil fuels. Gas turbines are used in most small systems because of their high efficiency, small size, clean combustion, durability, and low maintenance requirements. Natural gas is normally used to provide heat in a boiler or furnace by combustion, and the waste heat (usually from the waste gas stream) is used to generate electricity from a dynamo (generator). Marbe et al. (2004) show the economic advantage of cooperation between an industry and an energy company when a biofuel CHP plant delivers both process stream and district heating. District heating systems transfer and distribute heat from one or more centralized heating plants or CHP plants to individual homes, institutions, and industrial consumers for space heating and industrial processes. In Europe many heating systems are old and need upgrades and maintenance. The general development in the electricity sector towards distributed generation may have two advantages for the cogenerated heat. On the one hand, the generated heat is produced closer to the consumer, thereby reducing the losses. On the other hand, the system gets more dispersed with widespread smaller producers. As in the electricity sector, there will be a strong future need for making the distribution system more intelligent, with closer links between producer and end-user. One further development of district heating systems toward higher levels of sustainability is to combine the systems with solar heating systems and storage facilities, allowing a higher degree of overall optimization of the total system.

The district heating system might, in summer, be used for heat-driven cooling systems. Heat-driven cooling machines might also be used to offset part of the need for electricity for air-conditioning and thus reduce the peak load of electricity in the world's hot regions. A more direct way of district cooling is to develop specially piped chilled water circulating inside the cooling systems to end-use customers. Presently, the integration of district cooling and heating systems is used in the Finnish cities of Helsinki, Turku, and Lahti (Finnish Energy Industries, 2004). District cooling also makes it possible to prolong the utilization period of the maximum load in CHP plants.

Another interesting approach for the simultaneous delivery of heat and electricity is the use of fuel cells based on natural gas, where household needs for electricity and space heating could be met simultaneously. Brown et al (2007) describe micro-CHP using fuel cells in Japan. Simultaneous production of heat and electricity by hybrid solar photovoltaic and thermal systems could reach higher rates of solar energy conversion than stand-alone solar photovoltaic systems. Evaluation of relative payback time periods for both the systems demonstrate that the hybrid systems are cheaper (Kalogirou and Tripanagnostopoulos, 2006). End-use hybrids could also include systems where two different thermal end-uses are provided simultaneously without the intermediate

production of electricity. For example, Rane and Dasgupta (2003) report a Multi-utility Vapor Compression System that simultaneously produces hot and cold water streams for different end-use applications, such as bathing and air-conditioning.

16.2.3 Delivering Energy and Non-energy Services

New opportunities may arise from the possibility of combining the delivery of energy services with non-energy services. One example that has attracted much recent attention is that of plug-in hybrid vehicles (PHEV), which create interesting possibilities around the combination of transport and electricity. Other examples include the use of waste heat for desalination of water for domestic and industrial use, the integration of waste management with electricity production, and more generally, the concept of energyplexes, where multiple energy carriers and other byproducts may be produced simultaneously.

The PHEV contains an internal combustion engine as well as batteries that can be recharged by connecting a plug to an electrical power source. It therefore shares characteristics with both traditional hybrid electric vehicles (HEVs), having an electric motor and an internal combustion engine, and battery electric vehicles (BEVs), also having a plug to connect to the electric grid. Depending on the vehicle control design, a PHEV can be driven solely on electricity until the storage battery charge is depleted, after which the vehicle can be operated as a traditional HEV powered primarily from petroleum (Bradley and Frank, 2009). This gives users the advantages of driving on electricity as with a BEV, without the limitation of a shorter range.

Benefits of the PHEV include reduced carbon emissions, economic operating costs, and more efficient use of existing electric system capacity (EPRI, 2007; Shiau et al., 2009). Even when the production of the vehicle's storage battery is included, the life cycle amount in gCO_2-eq/km of a PHEV are below that of a conventional hybrid vehicle, when the PHEV is charged with electricity that has about 650–750 gCO_2-eq/kWh (Samaras and Meisterling, 2008). Other benefits include fewer fill ups at service stations, the convenience of home recharging, opportunities to provide emergency backup power in the home, and vehicle to grid applications. Disadvantages include a battery pack with increased size and cost, the requirement of public and home electrical outlets, potentially increased peak electrical loads, and increased pollution in some areas from increased electricity production from coal, although overall pollution declines (Simpson, 2006; EPRI, 2007). The availability of a distributed system of PHEVs may create the opportunity for large-scale storage and integration of variable renewable energy in the future (Kempton and Tomic, 2005).

The user cost, fuel use, and greenhouse gas emissions of PHEVs also depend critically on the size of the battery and the distance traveled. Further efficiency can be added to PHEVs by using newer batteries, such as closed-system regenerative fuel cells, instead of the regular battery. The regenerative fuel cell uses the conversion of water into its

Box 16.2 | The Future of Solid-state Lighting for General Illumination

The International Energy Agency has estimated that worldwide lighting is responsible for emissions of approximately 1900 $MtCO_2$/year, with 80% of these emissions being associated with electricity generation, and 20% from the 1% of global lighting that is produced by the direct combustion of paraffin and oil lamps used (IEA, 2006). Dramatically improved lighting system efficiency, together with electrification, could make a large contribution to controlling global CO_2 emissions. The most widely used technology, the incandescent bulb, converts only between 1% and 5% of the electricity into usable light. To address and overcome such inefficiency and concerns for the environment, energy security, and affordability, legislatures and regulators in the European Union (EU), Australia, Brazil, Canada, Ireland, Italy, New Zealand, the United States, and Venezuela have all recently moved to implement a mandatory phase-out of most standard incandescent bulbs over the coming decade. Most remaining countries in the EU are likely to adopt similar policies. Solid-state lighting uses technology based on a light-emitting diode, which has the potential to achieve such objectives.

Monochromatic LEDs have seen a very sharp increase in efficacy over a very short period of time when compared with other lighting technologies. Today, red and green LED efficacies are as good as, or better than, fluorescent and high intensity discharge technologies. Commercialized white solid-state lighting is expected to reach those levels in just the next few years, and still is far from reaching theoretical limits that have already constrained future improvements in incandescent and fluorescent lamps.

Solid-state lighting shows great promise as a source of efficient, affordable, color-balanced white light. Azevedo et al. (2009) have shown that, assuming market discount rates, white solid-state lighting already has a lower annualized cost than incandescent bulbs. The annualized cost for white solid-state lighting will be lower than that of the most efficient fluorescent bulbs by the end of this decade. However, much of the literature suggests that households do not make their decisions in terms of simple expected economic value. Assuming higher discount rates on consumer decision-making will delay the adoption of solid-state lighting technology. Successful consumer adoption will depend on, among other things, the availability of solid-state lighting products providing high color quality, high efficiency, and a low up-front cost.

Sources: IEA, 2006; Azevedo et al., 2009.

components, hydrogen and oxygen gases, whose recombination (again, to produce water) generates energy. The gasoline consumption of the vehicle can be reduced by over 80%, whilst reducing the operating cost of the vehicle by about US$200/year by 2010 and US$500/year if less expensive air-cooled engines are used in plug-in fuel cell hybrid electric vehicles (PFCHEV) (Suppes, 2006). A further advantage is found in solar chargers to recharge these plug-in hybrid vehicle systems during the day, when most people are at work (Birnie, 2009). Parking lots can thus become large solar recharging stations for commuters.

The use of low-energy lighting can also provide co-benefits (Box 16.2).

Another interesting example is the use of industrial waste heat, generated in a power plant or a chemical plant, to desalinize water from the sea and make it potable. Desalinization is a highly energy-intensive process, where seawater is distilled by heat from coal- or oil-fired boilers. Multi-stage flash distillation currently accounts for over 85% of the world's desalinized water, while the rest uses mostly membrane-based reverse osmosis process (Water Technology, 2009). The heat currently generated by fossil fuels dedicated solely to desalinization units can alternatively come from waste heat from power plants. Chacartegui et al. (2009) examine one example. Their research, taking a thermal power plant in Spain as a case study, shows that desalinization is actually the

most efficient and most economically and technically feasible option of cogeneration that can be obtained from a power plant. This is because the waste heat from the power plant is at a temperature that is too low for most other options, such as residential and district heating/cooling and oil-refining. The desalinization unit can be installed with minor configuration changes, such as increasing the plant's condenser pressure and installing a heat exchanger at the top of the stack.

Mahmoudi et al. (2008) have assessed the feasibility of using hybrid (wind and solar) energy conversion systems to meet the energy needs to power a seawater greenhouse.[1] Examination of the data collected confirms that it is technically feasible to take advantage of renewable energy to run the seawater greenhouse and produce freshwater without the back-up support of fossil fuel energy sources. An attempt to scale-up these concepts in practice is the Sahara Forest Project, a collaborative effort of the Jordanian and Norwegian governments. The project, targeted to commence in 2015, attempts to simultaneously provide water for a greenhouse for growing crops, desalination for drinking water, and solar thermal for electricity

---

1   Seawater Greenhouse: The seawater greenhouse is a method of cultivation adopted in arid coastal areas that provides desalination, cooling, and humidification in an integrated system. Its purpose is to provide a sustainable means of agriculture where scarcity of fresh-water and expense of desalination make agriculture unviable (Davies et. al, 2004).

production and is a good example of an integrated system that attempts to generate energy and non-energy co-benefits.

Waste-to-energy projects are gaining popularity in the developing as well as the developed world. These experiments and niches offer the opportunity of generating electricity and heat as well as assisting in waste management, thus generating non-energy related co-benefits. A recent experiment in India illustrates the interaction of social, technological, economic, and policy-related aspects around electricity generation using cattle-waste as a primary fuel (Patankar et al., 2010). This experiment represents local technology adaptation of bio-methanation and electricity generation technologies prevalent in other applications, and independent plant management as success parameters to ensure waste collection and energy throughput. In addition to electricity generation, the experiment contributes to waste management and greenhouse gas mitigation.

Solid waste management offers opportunities to simultaneously meet waste management, climate mitigation, and electricity generation objectives. Landfills are the primary mode of disposal for municipal solid waste in developing countries. Methane emissions from landfills, occurring as a result of the natural decomposition of waste, are estimated to account for between 3% and 19% of anthropogenic emissions globally (Mor et al., 2006). The landfill gas generated can be extracted from the landfills through a gas recovery system. Once collected, this gas can be utilized for electricity generation or used as an alternative vehicle or industrial fuel such as in Sweden (Lantz et al., 2007). Municipal solid waste could also be converted into refuse-derived fuel, for use in electricity production or other industrial applications as a substitute for coal or traditional biomass. This conversion helps in higher and constant heating value; homogeneity of physical-chemical composition; ease of storage, handling, and transportation; lower pollutant emissions; and a reduced excess air requirement during combustion. While the production and sale of refuse-derived fuel alone might not be economically viable, there are benefits associated with regard to waste management and the reduction of pressure on landfills. The economics may improve further if the refuse-derived fuel is used for downstream electricity production, as that offers additional opportunities for incentivization (Caputo and Pelagagge, 2002a; 2002b).

Another promising alternative is the development and deployment of integrated energy conversion and end-use systems. In energy conversion, these integrated systems are also known as energyplexes. They are highly efficient and incorporate advanced technologies that may have fuel flexibility and allow for various combinations of electricity, liquid fuels, hydrogen, chemicals, and/or heat. Yamashita and Barreto (2005) have explored three co-production strategies based on coal gasification, namely hydrogen and electricity, Fischer-Tropsch (F-T) liquids and electricity, and methanol and electricity. Using these examples, they have highlighted the important role that integrated energy systems and enabling poly-generation strategies may play in long-term global energy supply systems. Specifically, emphasis has been laid on the role of synthesis gas (or syngas) as a key energy carrier for a multi-fuel, multi-

product system based on carbonaceous feedstocks. Since syngas can be obtained not only from natural gas but also from solid energy carriers such as coal and biomass, this allows for its conversion into higher quality, cleaner, and more flexible energy carriers. Because syngas production systems are similar, or at least compatible to some extent, this could facilitate the introduction of multi-fuel systems, provided that technical issues over the quality and variety of feedstocks can be overcome. Hu et al. (2011) have attempted a techno-economic evaluation of coal-based polygeneration systems. China is abundant in coal reserves but deficient in oil and gas, which poses a challenge of utilizing the fossil fuels efficiently and cleanly. Polygeneration system technologies for chemical and power coproduction are therefore becoming more important. When compared with conventional synthetic fuel production systems and power generation technologies, with system integration, the polygeneration technology can achieve a trade-off between primary installed capital cost and fuel savings. That can effectively reduce the cost penalty for CO_2 avoidance. The polygeneration system, without shift reaction and the adoption of a partial recycle scheme, can achieve the optimal primary cost saving of over 10%. Another novel technique proposes the synthetic utilization of coal and natural gas through a combined cycle process of dual fuel-reforming. In this technique, the thermal heat released from burning coal is absorbed by the reforming reaction and transformed into chemical energy of the syngas (Han et al., 2007). Since the chemical energy of natural gas and coal is used more efficiently, the net thermal efficiency of coal to power can reach from 43.5% to 46.3%, which is almost equal to the efficiency of the integrated gasification combined cycle. Sensitivity analysis shows that the performance of the new cycle can be improved further to about 50%. Further, the net specific investments of coal to power, and the operating cost of the new combined cycle, are also lower than the integrated gasification combined cycle.

16.3 Experiments and Niches in System Configuration and Operation

Technological change and innovation is also creating opportunities for improvements in overall system operation. For example, the smart grid is an umbrella term used for a range of technologies to enhance system operation and performance and perhaps transform the grid from a centrally controlled entity to several self-controlled sub-networks (Hamidi et al., 2010). The array of technologies observed in today's smart grid ranges from smart metering to simulating business models with variations in configuration and operation. The improvements in computer communications and networks enhance the ability to optimize generation, demand management, and integration of renewable energy resources (Rahman, 2009).

Amongst the range of experiments related to system configuration and operation, four emerging ideas seem to have significant transformational potential for the future. These include energy storage, distributed generation, micro-grids, and business models around energy service delivery.

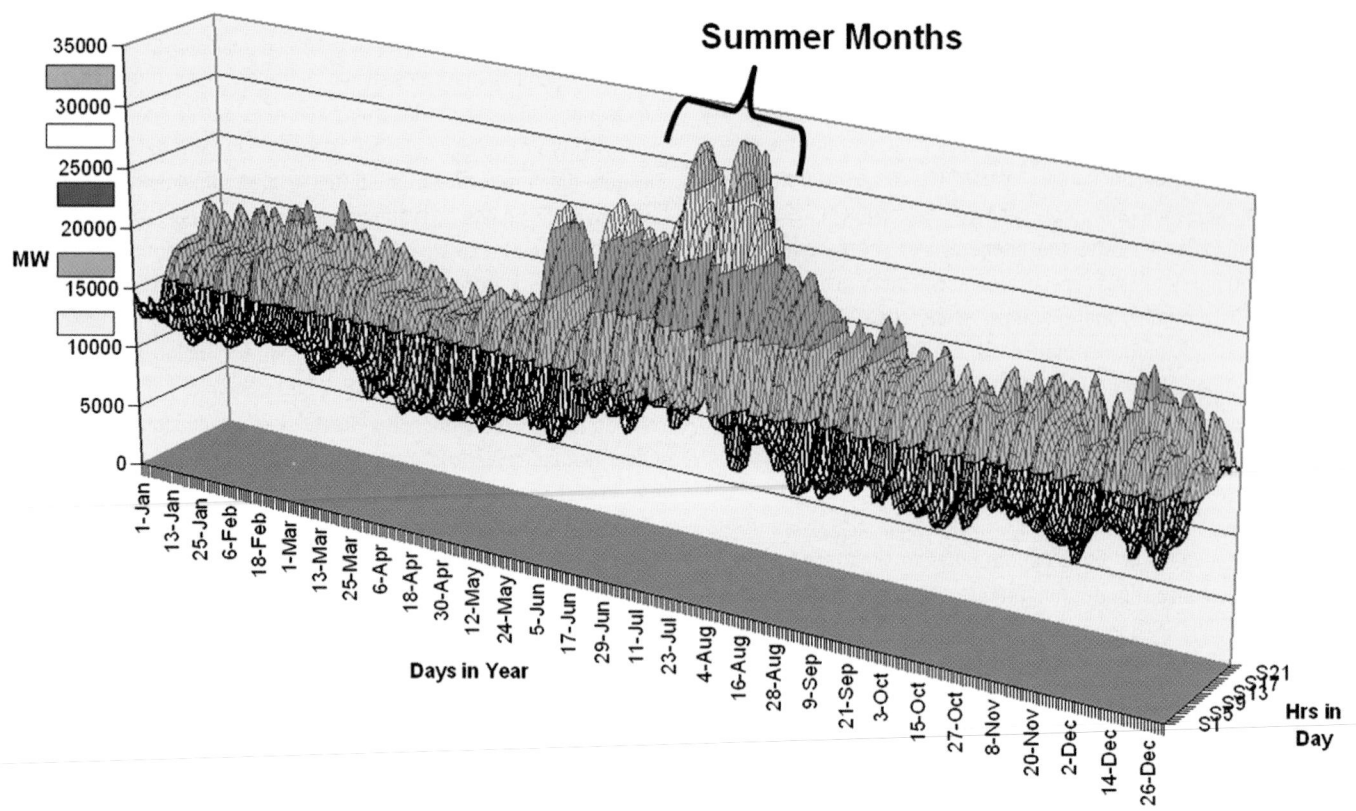

Figure 16.3 | Variability in daily system load curves for NY (using 2006 load data from NYISO). Source: Walawalkar, 2008.

16.3.1   Energy Storage

Although the present electric grid operates effectively without storage, cost-effective ways of storing electrical energy can help make the grid more efficient and reliable. The amount of load in most electrical grids changes from hour to hour and from day to day (Figure 16.3). In most regions, system operators typically try to meet this demand by using least-cost economic dispatch, based on available generation and transmission. Electric energy storage (EES) can be used to accumulate excess electricity generated at off-peak hours and discharge it at peak hours, thus reducing the need for peak generation and reducing the strain on transmission networks. EES can also provide critically important ancillary services such as grid frequency regulation, voltage support, and operating reserves, thereby enhancing grid stability and reliability.

Unlike other energy forms, electricity cannot be easily stored in large quantities. Without storage, there is little flexibility in managing electricity production and delivery. Likewise, intermittent renewable energy resources such as solar and wind are constrained in their ability to practically enter the bulk electricity market and keep pace with its continuously changing demands (see Chapter 11). Advanced storage technologies promise to change the nature of electricity markets by providing much greater operational and financial flexibility and enabling carriers to resolve system transients and bottlenecks. The accelerated development and implementation of more cost-effective, higher capacity storage

options, including batteries, flywheels, super-capacitors, hydrogen, and super-conducting systems, is the critical issue underlying further diffusion. Today, most of these bulk storage options are relatively unproven, and their value proposition is complex and poorly understood.

For example, the lead acid battery is the most used for large energy storage as well as all types of stationary applications (Perrin et al., 2005). However, advanced batteries being developed for large- scale applications include nickel cadmium, sodium sulfur, sodium nickel chloride, lithium ion, and zinc bromine. Advanced batteries are moving towards commercialization[2], although they remain too expensive for large-scale electricity carrier and storage system application. However, significant cost reductions are expected as modular design and factory assembly become the norm and production volumes increase substantially. Particularly notable from a strategic perspective is the lithium ion battery, being targeted for electric vehicle applications, and the sodium sulfur battery, with the latter being used in Japan for more than 200 MW of energy storage installations (Baker, 2008). The total worldwide market for energy storage products was estimated at around US$26 billion in 2005 (Frost and Sullivan, 2006). About 59% of this market came from

---

2   A recent commercial-scale deployment of Na-S batteries is a project in Texas, US, where a 4 MW Na-S was energized to the grid in March 2010, with the objective of providing transmission backup in the event of a transmission line outage, along with additional benefits of improvement in power quality. For more information, see ETT, 2011.

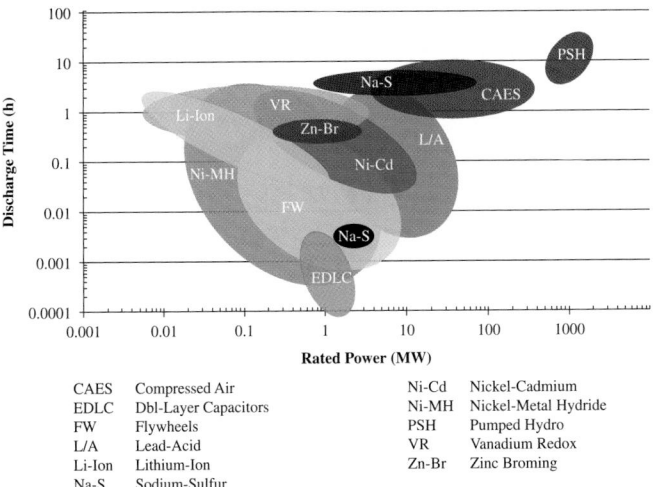

Figure 16.4 | Energy storage system ratings (rated power vs. discharge time) for installed systems (November 2008). Source: Roberts, 2009. © Energy Storage Association.

CAES	Compressed Air	Ni-Cd	Nickel-Cadmium
EDLC	Dbl-Layer Capacitors	Ni-MH	Nickel-Metal Hydride
FW	Flywheels	PSH	Pumped Hydro
L/A	Lead-Acid	VR	Vanadium Redox
Li-Ion	Lithium-Ion	Zn-Br	Zinc Broming
Na-S	Sodium-Sulfur		

sales of automotive starting batteries. Another 27% of the total market was concentrated in portable electronics, while motive power (batteries for, e.g., forklifts, golf carts, and electric vehicles) constitutes about 6.6%. Only about 6% of the market, or around US$1.5 billion, came from stationary power products (Frost and Sullivan, 2006). Most of these sales were for backup power applications, such as UPS systems, substation and generating plant backup batteries, and telecom backup batteries. The market for energy storage in utility applications, other than their use as backup batteries, is relatively small. EES technologies can be grouped as electrochemical and non-electrochemical EES technologies. The most common EES technologies are:

- Electrochemical EES
- Lead Acid Battery
- Sodium-Sulfur Battery
- Flow Batteries
- Vanadium Redox Battery
- Zinc Bromine Battery
- Nickel Cadmium Battery
- Nickel Metal Hydride Battery
- Lithium Ion Battery
- Non-Electrochemical EES
- Pumped Storage Hydroelectric
- Compressed Air Energy Storage
- Flywheel
- Ultra-Capacitor
- Superconducting Magnetic Energy Storage (SMES)

The EES technologies listed above and shown in Figure 16.4 are described in detail in EPRI (2003; 2004) and Gyuk et al. (2005). In general, large-scale applications of energy storage have been limited to the utility industry. Utility-scale projects based on storage technologies other than pumped hydroelectric storage have been built, though they have

not become common. Existing US facilities include one Compressed Air Energy Storage system, several plants based on lead-acid batteries, one based on nickel-cadmium batteries, and recent installations of sodium sulfur batteries. There are also recent installations of lithium ion batteries and flywheels for frequency regulation in regions where there are spate markets for frequency regulation through Independent System Operators/Regional Transmission Organizations (ISO/RTOs). In all, some 2.5% of the total electric power delivered in the US passes through energy storage, largely pumped hydroelectric. The percentages are somewhat larger in Europe and Japan, at 10% and 15%, respectively (EPRI, 2003). The most widespread form of electricity storage used for utility scale applications is pumped hydroelectric storage, with over 90 GW of installed capacity throughout the world. According to the US Energy Information Administration, the United States had 150 pumped storage plants with a total nameplate capacity of 19.5 GW in 2005 (US EIA, 2005).

There are technical as well as market barriers for the wide-scale integration of energy storage for wholesale market applications. At present, most energy storage technologies have higher capital costs than peaking power alternatives, such as gas turbines. While capital costs are falling somewhat due to technology improvements, significant manufacturing economies of scale have not yet been realized (EPRI, 2003; 2004).

In the longer term, the introduction of energy storage can potentially resolve the renewable energy intermittency challenge. The type of storage under consideration ranges from advanced stationary batteries to other emerging technologies, such as electrochemical capacitors, thermal storage using ice, pumped hydro technology, and compressed air energy storage (Roberts, 2009). The advanced compressed air energy storage system essentially uses the concept of compressing air and storing it during off-peak hours when energy prices are very low and running a turbine with the released air to produce electricity when needed. Because the compressed air is blended with the input fuel in the turbine, it produces much more power than conventional stand-alone gas turbines. Such systems have already been installed in places like Handorf, Germany, and McIntosh, AL, United States. Locations considered ideal for these systems are underground geological formations, such as mines, salt caverns, or depleted gas wells. The largest of these, with a capacity of 800 MW, was proposed to be installed in the United States as of August 2009 (Roberts, 2009). While these systems have mostly been built on the principle of using energy from a conventional electricity grid, they can also be designed as a hybrid to use renewables like wind energy (from turbines) to compress the air.

16.3.2   Decentralization of Power: Distributed Generation

The provision of electricity remains a major challenge in many developing countries. The typical policy response has been to mount aggressive and large-scale rural electrification schemes. Advances in technology and the possibilities provided by hybrid systems may make it more attractive technically and economically to provide these modern energy services through distributed, off-grid generation options. Recent studies have tried to make explicit the trade-offs and choices between these two modes for electrification and energy service delivery.

Distributed generation is the concept of having decentralized, small generating units that may or may not be connected to the main grid, but are usually located close to the point of end-use (Bayod-Rújula, 2009). Distributed systems may be connected to the grid, or they may operate independently. They are generally not centrally controlled and, with few exceptions at the present time, are not dispatchable; that is, they cannot be switched on and off according to the needs of the grid unless they incorporate suitable grid support technologies, such as energy conversion and storage (Larsen and Peterson, 2005).

Distributed generation provides consumers with opportunities to gain benefits, such as a degree of energy independence, opportunities for local control to improve security of supply, financial optimization, equal or better power quality, and a cleaner environment. It also offers a possibility of including millions of small suppliers, providing a high proportion of renewable energy sources. With distributed generation, homes and businesses could produce their own electric power and heat using various technologies or a mixture thereof. Excess power would be sold to the grid, and consumers could obtain from the grid any power that they did not generate themselves.

Distributed generation is often described as 'integrated' and 'decentralized' to distinguish it from traditional centralized systems, in which power is generated at large, centrally located, and centrally controlled plants. Electricity is presently the dominating medium of exchange in distributed generation, although heat from, for example, combined heat and power (CHP) plants also plays an important role in district heating and other kinds of energy supply. In the long run, energy carriers such as hydrogen and biofuels are expected to contribute very significantly to the energy exchange in distributed generation. Regardless of size, fuel, or technology, the central issue in distributed generation is the large-scale integration of decentralized energy resources.

The generating technologies and energy sources behind distributed generation include wind, solar, micro-hydro, biomass, geothermal energy, and wave or tidal power. These share several attributes:

- energy resources that are geographically dispersed and unevenly distributed;

- energy resources that are typically intermittent, varying by the hour or the season, and not available on demand (some technologies, such as fuel cells and micro turbines, are not intermittent);

- energy that is typically produced as electricity, rather than some other energy carrier that is easier to store; exceptions include biomass and hydropower based on reservoirs; and

- in electricity-driven technologies, heat may be a by-product; in the same way, electricity may be a by-product in heat-driven technologies.

These characteristics, and the inflexibility of most renewable energy technologies, pose important constraints on and requirements to energy systems based on distributed generation. As the share of renewables grows, energy systems (and their necessary support technologies) will face increasing challenges to their operation and future development, affecting both the supply and demand side. Information and Control Technology will be very important to the successful integration of these renewable energy options.

The perceived benefits of distributed power systems include increased reliability of service, improved power quality, the ability to defer investment in extending the grid, and greater energy efficiency, e.g., through better use of waste heat. Of these drivers, reliability of service is one of the most important. Increasingly, consumers need uninterrupted power supplies, yet many existing grids cannot operate without occasional blackouts. Finally, the redundancy offered by a distributed generation network, with its intertwined multitude of generators, converters, and connections, will certainly enhance the security of the power system to deal with acts of terrorism.

Distributed generation based on a high proportion of renewable energy will depend on a number of support technologies that include energy storage and load management to deal with the fluctuating power and intermittency from renewable sources such as wind turbines. Once the concept is fully developed, distributed generation, in addition to the obvious benefit of providing a cleaner environment, has the advantage in that it is easy to add generating capacity as required, using local energy resources. The cost of such expansion is predictable over the life cycle of the generating plant, regardless of price fluctuations and shortages that may affect fossil fuels in the future.

Of the various barriers to a robust implementation of distributed power, key factors include the need for reliable interconnections between distributed generation systems and the grid, and the quality assurance that includes the testing and certification of distributed generation systems and components. Developing standards, procedures, and techniques for the design, testing, and certification of distributed generation systems is relatively new, but it can draw on experiences developed in the wind industry over the last two decades for both systems and projects. Other pacing issues include the lack of experience in understanding how various distributed generation systems interact when a variety of designs are broadly deployed, and the increasing complexity of physical and cyber security issues as the number of electric system participants grows. Proper availability and provision of skills for the maintenance and repair of such systems is also an issue.

New electricity pricing and electricity supplier business structures will be needed in many jurisdictions to remove the reduced bulk electricity sales

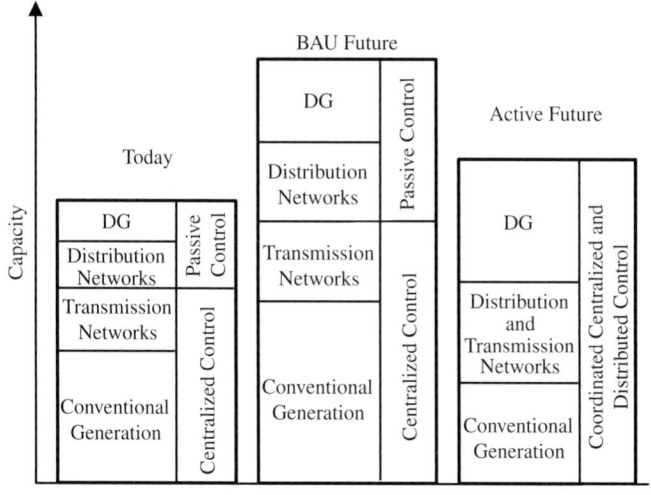

Figure 16.5 | Passive vs. active management of distribution networks. Note: BAU = Business As Usual. Source: Cossent et al., 2009.

disincentives that monopoly electricity suppliers now experience when they encounter user-sited distributed generation. These can be replaced with incentives that encourage distributed generation investments that achieve widely accepted societal benefits such as carbon reduction. The use of a more distributed power generation system will be an important element in protecting consumers against power interruptions and blackouts, whether these are caused by technical faults, natural disasters, or terrorism. Figure 16.5 proposes a Coordinated Centralized Control Mechanism for better control of grids that use distributed generation (Cossent et al., 2009).

Distributed generation appears to be a more promising approach where grid connectivity does not exist. In such regions, distributed generation offers an alternative to grid extension. One such case study identifies niche areas in India where renewable energy-based decentralized generation options can be financially more attractive than grid extension for providing electricity (Nouni et al., 2008). They estimated the cost of delivering electricity in remote areas, considering the cost of electricity generation, and also the cost of its transmission and distribution in the country, where about 70% of the population live in rural areas (see Chapter 24).

Nouni et al. (2008) also analyze the relative contributions of different factors to the levelized unit cost of electricity. At any location receiving electricity through the grid, the delivered cost of electricity consists of three components: (a) cost of generating electricity at the bus bar of the centralized plant; (b) cost of transmitting electricity through the transmission network; and (c) cost of distribution. For electricity generated from coal thermal power plants, the delivered cost of electricity in remote areas located in the distance range of 5–25 km was found to vary from INR[3] 3.18–231.14/kWh

(US$_{2005}$0.073–5.32/kWh), depending on the peak electrical load up to 100 kW and on the load factor (Nouni et al., 2008). The paper concluded that all renewable energy-based decentralized electricity supply options (such as micro-hydro, dual-fuel-biomass gasifier systems, small wind electric generators, and photovoltaics) could be more financially attractive than grid extension in providing access to electricity in remote villages.

A similar study in Cameroon (Nfah and Ngundam, 2009) uses software simulation to examine the breakeven distances for grid extension compared to distributed generation using hybrid systems. Pico-hydro and photovoltaic hybrid systems incorporating a biogas generator were simulated for remote villages in Cameroon based on a load of 73 kWh/day and 8.3 kW peak. For a single wire grid extension cost of €5000/km (US$_{2005}$6329/km), operation and maintenance costs of €125/km (US$_{2005}$158/km) per annum, and a grid power price of €0.10/kWh, the breakeven grid extension distances were 12.9 km for pico-hydro/biogas/battery systems and 15.2 km for photovoltaic, biogas, and battery systems, respectively.[4]

By giving more emphasis to distributed generation, the United Kingdom can potentially reduce its energy losses in the transmission and distribution system of electricity. Including the waste heat generated during electricity production, this currently amounts to about 65% of the energy input (Allen et al., 2008). However, technical barriers still cause significant impediments to their growth. These include grid integration, planning permission, and licensing issues, because the main grid is designed for large-capacity centralized power generation (Sauter and Watson, 2007). Financial and economic barriers also exist, notably including the poor performance of the Low Carbon Buildings Programme grant given directly to home owners to encourage distributed generation using renewables. For instance, according to the UK Department of Trade and Industry, a domestic solar photovoltaic installation in 2006 could cost up to £10,400 (US$_{2005}$19,259),[5] and it could supply about 51% of the average annual electricity demand of the house. However, the maximum available Low Carbon Buildings Programme grant for this is £2500 (US$_{2005}$4630), which is too little to make the distributed generation system an economic option (Allen et al., 2008). Wolfe (2008) has also described the Low Carbon Buildings Programme grants as too limited in nature to cause a significant surge in distributed generation. A general lack of monitoring and information on distributed generation also prevents a faster growth of successful commercialization of these new systems.

Nevertheless, many European Union countries have managed to achieve appreciable distributed generation contributions using renewable energies in their electricity supply. Figure 16.6 shows the percentage

---

3   US$1.00 = INR45.80 (October 20, 2006)

4   US$1.00 = €0.71 (July 01, 2009)

5   US$1.00 = £0.54 (July 01, 2006)

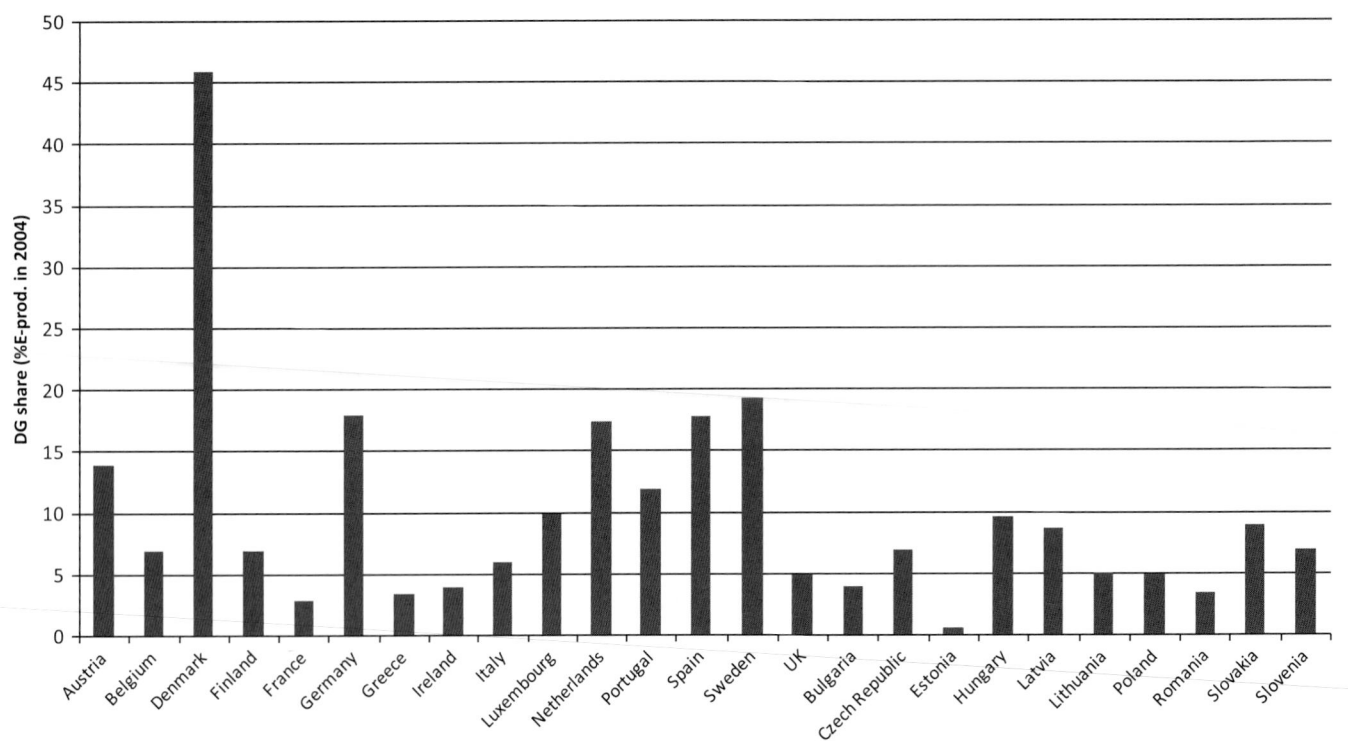

Figure 16.6 | Distributed generation shares in total electricity production in 25 EU countries (2004). Source: Cossent et al., 2009.

contribution of distributed generation in each EU member's total electricity generation in 2004 (Cossent et al., 2009). Denmark scores the highest, with over 45% of its electricity coming from distributed generation resources. It is interesting to note that historical circumstances, next to favorable policies, have played a critical role in the success of distributed generation in Denmark, particularly the role of the cooperative movement in the Danish power industry (van der Vleuten and Raven, 2006).

16.3.3 Aggregating the Distributed Energy Resources (DERs): The Micro-grid

Distributed generation has many benefits and, although it is commercially stable, its growth has been rather slow. DERs can be much more economic if they are realized as a micro-grid. A micro-grid is systems architecture: an aggregate of DERs in a small power network that can provide electric power and often heating and cooling to a small group of closely located customers (King, 2006). This combines the advantages of DERs with those of the micro-grid, the latter being aggregated demand, increased reliability, and an ability to share the capital costs.

Intelligently interconnected buildings and distributed electricity generation are the most critical elements of the micro-grid architecture. It is also becoming feasible to incorporate mobile electricity storage in the form of plug-in electric vehicles (PHEVs). Micro-grids will enable an

electricity delivery network strategy where transportation can become an efficient energy supply source, particularly during high cost, peak demand periods. The essential enabling policy step to realizing the full benefits of distributed generation, micro-grids, and PHEVs is time-of-use electricity pricing, which incentivizes consumers to be efficient and to take advantage of the resulting savings.

The interface between the micro-grids and the wider electricity distribution network is through a well-defined and controlled interface utilizing smart grid technologies. The micro-grid is designed and operated to best serve the needs of its customers, ensuring a precise quantity and quality of electricity supply that most efficiently and cost-effectively meets the needs of those customers at all times. In effect, the micro-grid acts as a local electricity refinery, raising the reliability, efficiency, and quality of the bulk electricity supply system.

Although there are a variety of potential micro-grid ownership and operation alternatives, micro-grids intrinsically lend themselves to local cooperative ventures where consumers may also be suppliers (Sauter and Watson, 2007). Metering and billing arrangements are agreed upon locally to reflect the market needs for power within each micro-grid. Micro-grids can also take full advantage of current commercial energy storage technology to accommodate rapid fluctuations in electricity demand or supply.

Micro-grid development in the United States has taken a somewhat different path than similar efforts in Japan and Europe. In the United

States, the emphasis has been on compensating for the relatively poor reliability and power quality provided by the nation's bulk electricity supply system. This is in contrast to Europe and Japan, where the emphasis has been on incorporating clean distributed electricity generation. Micro-grid development efforts in the United States have been led by the Consortium for Electricity Reliability Solutions and, more recently, by the California Energy Commission and the Galvin Electricity Initiative. Prototype micro-grid demonstrations include the Fort Bragg military base in North Carolina, the Beach Cities Micro-grid project in San Diego, CA, and the Illinois Institute of Technology campus in Chicago (GalvinPower, 2010). General Electric is also developing a Micro-grid Energy Management framework to provide a unified control, protection, and energy management platform. Unfortunately, micro-grids are still not defined in most US states' regulatory laws and are usually subject to case-by-case interpretation of regulatory laws for the main power grids, their suppliers, and their customers (King, 2006). Steps such as formalizing the definition and legal rights of micro-grids, adopting formalized procedures for their interconnection with the main grid, and formalizing the responsibilities of the micro-grid owner would greatly help to encourage this new power architecture, along with the great scope for innovation and creativity that it potentially has.

In the European Union, several major research efforts have been devoted to distributed generation micro-grids. These include a consortium led by the National Technical University of Athens, along with utilities, equipment suppliers, and research teams from many EU countries. Demonstration sites include Kythnos Island, Greece, and Mannheim-Wallstadt, Germany. Japan is also a leader in micro-grid demonstration projects. These generally serve the purpose of ensuring that the fluctuating power of distributed renewable energy resources does not degrade Japan's outstanding electricity reliability and power quality. These projects are demonstrating the technical feasibility of micro-grids incorporating renewable energy, while providing multiple levels of power quality and reliability with constant bulk electricity delivery system inflows.

16.3.4 Business Models for Energy Service Delivery

Optimization across demand and supply has created the possibility of new business models centered on the energy service companies (ESCOs). ESCOs are private or public companies that can provide the technical, commercial, and financial services needed for energy efficiency projects. ESCOs may assume a variety of risks, including project performance risk (technical risks associated with the project), arrange financing for the

project, and, depending on their reach and agreement with the client, may also take customer credit risk (financial risks). This is done through a performance contract between the ESCO and customer. Two popular models of ESCO operation are:

- *Shared saving contracts*: the ESCO finances the project either from its own fund or by borrowing from a third party. Thus, the ESCO assumes the performance as well as the credit risk.

- *Guaranteed savings contracts*: the customer finances the project by borrowing funds from a third party. Finance is usually arranged by an ESCO, but the loan contract is between the bank and the customer. The ESCO takes only the performance risk by guaranteeing the savings.

In general, the services provided in the two models above are characterized in Table 16.1.

ESCO models may be evaluated with regard to their business sustainability and viability, attractiveness to entrepreneurs, technology linkages and requirements, and the enabling environment required. At the level of firms or in the public systems (buildings, water pumping, streetlighting), the role of ESCOs has proven to be successful at various levels. ESCOs provide multiple services related to the diagnostics of existing energy-use patterns and the implementation of identified technical interventions, including technical, operation, and maintenance services.

Different levels of success characterize the maturity of the ESCO industry in North America and Europe, compared to the evolving ESCO industry in Asia and Africa. ESCOs are still at a nascent stage in developing countries. Painuly et al. (2003) describe a number of barriers to ESCOs in developing countries, including market barriers, institutional barriers, financing barriers, and knowledge and information barriers. They suggest that market development through the active involvement of governments as a customer, information provider, and policymaker is required to promote ESCOs. Local financing markets need to be boosted through the development of special energy efficiency financing windows in appropriate financial institutions such as banks. Skills for energy efficiency project appraisal and for the development of financial products to execute energy efficiency projects also need to be developed. In some cases, specialized energy funds and guarantee funds may be needed to kick-start the investment in energy efficiency projects.

ESCOs have a much longer history and a richer knowledge base in developed countries. Vine (2005) reports that, based on an international

Table 16.1 | Two models of energy service companies.

	Energy auditing	Project structuring	Project financing	Implementation	M&V	Stake in perpetual savings
Guaranteed savings	Yes	Yes	No	Yes	Yes	Limited
Shared savings	Yes	Yes	Yes	Yes	Yes	High

survey, the total amount of ESCO activity outside the United States in 2001 was between US$560 million and US$620 million. This is approximately half to one-third of the ESCO revenues in the United States for 2001.

Goldman et al. (2005) present a detailed empirical analysis of project data and market trends in the United States. They estimate that industry investment for energy efficiency related services reached US$2 billion in 2000, following a decade of strong growth. ESCO activity is concentrated in states with high economic activity and strong policy support. Typical projects save 150–200 MJ/m^2/year and are cost-effective, with median benefit to cost ratios of 1:6 and 2:1 for institutional and private sector projects, respectively. The median simple payback time is seven years among institutional customers; three years is typical in the private sector. They conclude that appropriate policy support, both financial and non-financial, can jump-start a viable private sector energy efficiency services industry that targets large institutional and commercial or industrial customers.

The ESCO market is quite diverse. Historically ESCOs were distinguished from other energy efficiency providers by their use of performance contracting as a core business activity. Increasingly, however, ESCOs are moving away from performance contracting, instead installing energy efficiency projects on a design/build or fee-for-service basis and offering additional services such as energy consulting and information services. Other ESCOs also provide performance-based services beyond the traditional shared savings and

guaranteed savings mechanisms, such as build/own/operate contracts for major energy facilities at customer sites. Some ESCOs have pursued new business opportunities in restructured electricity and natural gas markets, combining commodity procurement with energy price risk management and energy efficiency services in a single, bundled product.

Although the ESCO concept has been working well globally, there are challenges facing ESCOs in developing countries, including:

- the creditworthiness (technical and financial) of ESCOs;

- the availability and acceptability of well-drafted energy service contracts and/or agreements;

- arranging finances for the implementation of projects (both equity and loans) for successful demonstration projects;

- performance guarantees: users want the ESCO to indemnify them if promised energy savings do not materialize;

- a lack of minimum performance standards; and

- a lack of clear measurement and verification protocols.

Box 16.3 describes a specific ESCO case study from India (Kalra and Shekhar, 2006).

Box 16.3 | Case Study of an ESCO with Nasik Municipal Corporation, India

A specific example of an ESCO in India is the experience of Nasik Municipal Corporation (NMC). In 2003, ICICI Bank financed an ESCO project implemented, following a tender process, by Sahastratronic Controls Private Limited (SCPL) for NMC in the city of Nasik, Maharashtra. The NMC project, based on shared savings, had SCPL as the borrower work to upgrade the existing street lighting facilities on NMC's premises. The contract was for SCPL to supply 460 energy-saving devices, which represented almost half of NMC's total requirement. Under the Energy Services Agreement, SCPL was required to invest in establishing energy efficiency measures, namely Street Light Controllers, including capital assets, and maintain the same for five years.

ICICI Bank provided financial assistance of INR8.3 million (US$_{2005}$$190,900) in two installments (INR4.5 million and INR3.8 million [US$_{2005}$$103,500 and US$_{2005}$$87,400]) to meet part of the project's cost, from a INR20.0 million (US$_{2005}$$460,000) Line of Credit facility sanctioned to the ESCO through USAID's ECO Programme. Repayments were secured by a direct payment mechanism to NMC through an escrow arrangement. Energy audit studies had estimated an energy savings potential of at least INR7.5 million per year (US$_{2005}$$172,500). SCPL had guaranteed a minimum 25% in energy savings. A large part of the saving was supposed to be shared with SCPL as compensation for their establishment and maintenance of the energy saving devices. The first phase, which consisted of 361 panels with a total load of 4000 kVA, was commissioned in December 2004. Subsequent to the successful implementation of this project, NMC awarded the second contract to the ESCO for 125 panels for which ICICI Bank sanctioned a term loan of INR3.8 million (US$_{2005}$$87,400).

NMC draws about 5000 kW of energy per hour in a 12-hour day, throughout the year for its street lighting application. This amounts to an energy bill of about INR5.5 million per month (US$_{2005}$$126,500) payable to the Maharashtra State Electricity Board (at a tariff of INR3 per kWh), for an annual expenditure of around INR65 million (US$_{2005}$$1,495,000) on street lighting alone. The SCPL area had an actual load of 2700 kW. On this basis, the monthly energy bill payable to the Maharashtra State Electricity Board was about INR35 million (US$_{2005}$$805,000). On the basis of 25% assured savings, the savings for this area was estimated to be 0.25 x 35.0 = INR8.8 million (US$_{2005}$$202,400). In line with this, the NMC project has realized energy savings of about INR11 million (US$_{2005}$$253,000) on a total load of 4000 kVA in one year, with a capital investment of INR12 million (US$_{2005}$$276,000). The average savings per year have worked out at 31% with peak savings on some sub-sections as high as 44%. SCPL has so far received INR7.8 million (US$_{2005}$$179,400) as its share of savings, while NMC's share was INR3.2 million (US$_{2005}$$73,600).

Impact

The project was the first of its kind in India. It has served as a model for other ULBs, municipalities, and government organizations to follow. Successful execution of the project and timely repayments to the bank has not only convinced other banks to lend to similar ESCOs but has also encouraged other municipal corporations to undertake similar street lighting projects.

16.4   Policy and Institutional Issues

The previous sections (Section 16.3 and 16.4) enumerate several emerging experiments both with regard to technological innovations as well as changes in systemic configuration and business models. Many of the experiments can become niches, provided they are suitably nurtured and supported. One of the key factors influencing the scaling up of niches to larger regimes is the existing policy environment. Energy transitions are as much affected by policy and institutional issues as they are by technological and systemic ones. Subsequent chapters (see Chapters 22 – 25) in this Assessment cover energy policies and related policies for innovation and capacity-building. This section focuses on the role of policies in facilitating and guiding transitions.

16.4.1   Role of Policy in Transition Management

To appreciate the role of policy in energy transitions, it is important to note that only very few historical regime transitions were explicitly directed by collective, socially deliberated, long-term goals like sustainability (Smith and Stirling, 2010). Nevertheless, there have been historical studies that trace the emergence of new regimes back to originating niches, and they do inspire ideas for sustainability transitions. Transition management focuses on facilitating an evolution of sustainable regimes from green niches. This should include helping green niches by putting incumbent regimes under significant pressure to reach sustainability targets through policy building, and thereby favoring the environment for these niches, although these are often underplayed in transition policy recommendations (Smith and Stirling, 2010).

Niches, by their very nature, provide protective settings that are less susceptible to prevailing market pressures. Radical innovations pertaining to those niches that carry systemic implications typically need this kind of space to develop, improve, and get much needed support. Transition management puts this niche-based, evolutionary view of change within an iterative, four-stage cyclical governance framework (Smith and Stirling, 2010):

1. *Problem structuring and goal envisioning*: In this step, the development of a shared vision for attaining sustainability goals by multi-stakeholder transition arenas is often facilitated by a government department. Practical visions are formed from the sustainability goals by scenario-building techniques, and these visions then provide a basis and a direction for subsequent government policies.

2. *Transformation pathways and experiments*: Pathways toward transition visions are identified by the participants using backcasting methods, which provide a framework for the development of the niche experiments. Successful pre-development of those niches is followed by a period of take-off and acceleration, before culminating in stabilization within a more sustainable regime.

3. *Learning and adaptation*: This step provides the essential links between long-term goals, socio-technical pathways, and short-term actions in niche experiments. Lessons are learnt for both the improvement of the niche and the institutional reforms. Understanding the institutional constraints and opportunities for the sustainable practices is central to the niche experiments.

4. *Institutionalization*: Institutionalization is often given the least consideration in transition management literature, despite the fact that it is the most important factor in transition management. Institutionalization involves the mobilization of serious selection pressures against the incumbent regime, and the redirection of vast institutional, economic, and political commitments into promising niches along certain feasible pathways. This is the point at which serious commitments are needed, to the extent that the incumbent regime suffers and is undermined as a result. This step is difficult politically as well as economically.

A promising approach for transformative change appears to be strategic niche management, which suggests that such transformative

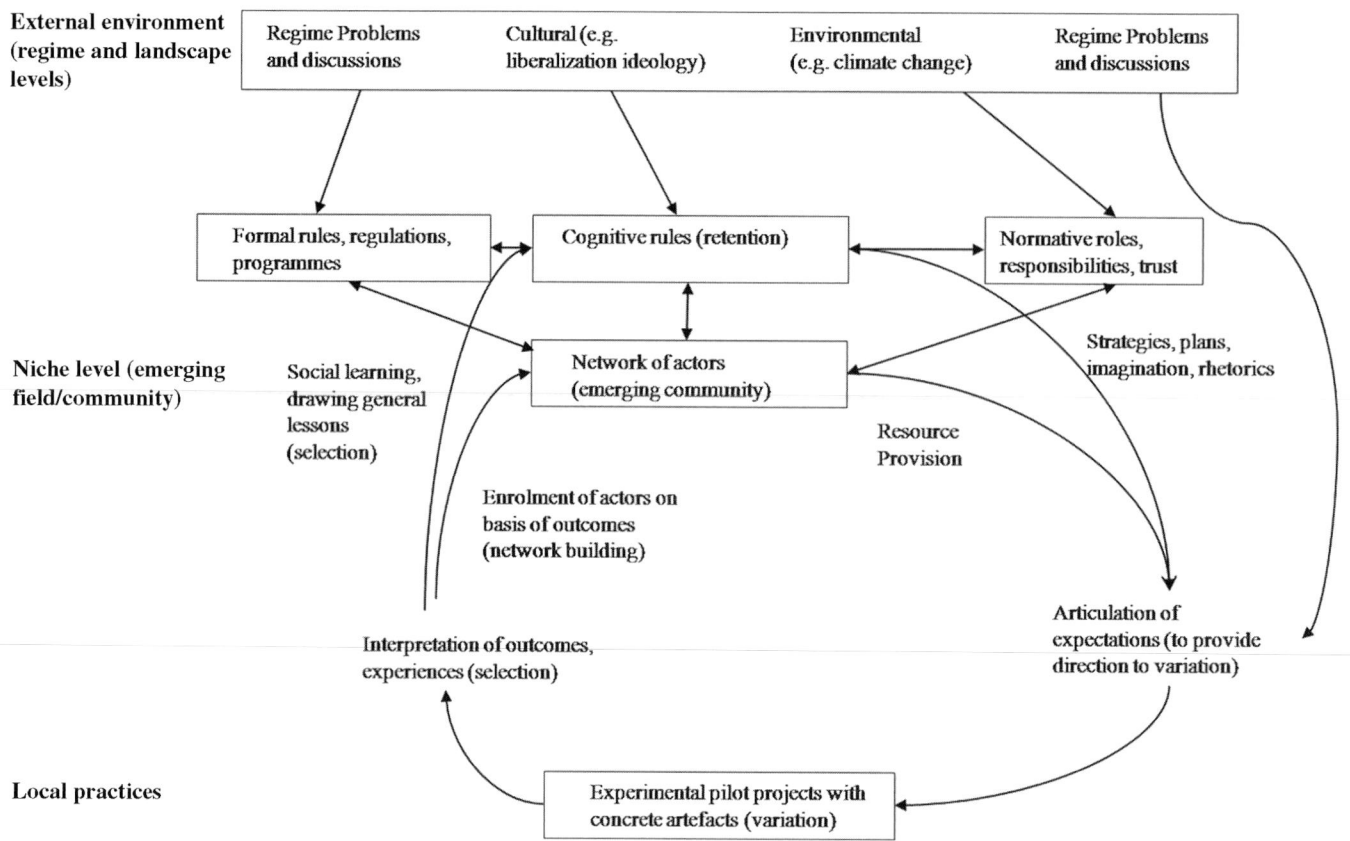

Figure 16.7 | Dynamics in socio-cognitive technology evolution. Source: Raven and Geels, 2010.

change could be facilitated by the creation of technological niches, i.e., protected spaces that allow experimentation with the co-evolution of technology, user practices, and regulatory structures (Geels and Schot, 2007). While the understanding of the evolution of a technological niche to a regime shifts, the following three internal niche processes have been distinguished (Elzen et al., 1996):

- The articulation of expectations and visions provides direction to the learning processes.

- The building of social networks becomes important for developing a constituency behind the new technology and facilitating interactions with relevant stakeholders.

- To facilitate alignments between variation and selection and create "configurations that work," the projects and experiments in technological niches should provide learning processes on dimensions such as:

  - Technical aspects and design specifications
  - Market and user preferences
  - Cultural and symbolic meaning
  - Infrastructure and maintenance network
  - Industry and production networks

- Regulation and government policy
- Societal and environmental effects

A further broader, more recent view of the role of niches in regime shift has led to the principle of multi-level perspective, which distinguishes three analytical levels. The niches form the micro-level, where the novelties emerge; the socio-technical regime forms the meso-level, which accounts for the stability of existing large-scale systems; and the socio-technical landscape depicts the macro-level, an exogenous environment beyond the direct influence of the niche and regime actors (Figure 16.7).

The manner in which transitions occur in the real world is a function of the dynamics between the three levels, as well as of internal niche interactions. The attributes that influence the transitions change depending upon the niche's setup and location. The following sections attempt to analyze transitions in practice which have occurred at different settings and scale.

16.4.2　Creating Policy Environment for Transitions

Overall, transition management incorporates goal-directing processes into socio-technical transformations. This approach is attracting increasing interest in many countries' environmental policies, such as in the Netherlands and Finland. The European Commission is funding research into transition foresight under its seventh framework program. Think tanks in the United Kingdom and elsewhere are advocating the approach. The Austrian Program on Technologies for Sustainable Development promotes the so-called Energy Regions, which can be considered niche spaces for experimentation within a larger systemic perspective (Smith and Stirling, 2010).

The Swiss Energy program offers an interesting case study. It has been highlighted by Geels et al. (2008) as a systems-transformation approach to energy policy for sustainability. Switzerland's energy policy is characterized by strong vertical policy linkages, including both top-down initiatives and bottom-up engagement. The top-down initiatives include the Federal Energy Act 1998 and the CO_2 Act 2000, which encourage environmentally friendly energy sources and the reduction of CO_2 emissions. The bottom-up engagement is facilitated by the Swiss tradition of decentralization of power, so that the cantons (the provincial units) have a great say in decisions pertaining to energy matters and in their implementation. Furthermore, there is an emphasis on using voluntary measures to enhance cooperation between the state and industry. Lastly, the Swiss tradition of direct democracy empowers people to influence the local, cantonal, and federal policies by referendums and initiatives. Because of these vertical linkages, the Swiss energy policy is characterized by a combination of regulatory instruments and process-based implementation. The Swiss Energy program has focused on five key areas: modernization of buildings, renewable energy, energy-efficient appliances, efficient use of waste heat, and energy-efficient, low-emission transportation. Overall, Geels et al. (2008) believe that the Swiss Energy program should be considered a success, because without it, Swiss CO_2 emissions would have been about 2.8 $MtCO_2$ higher than present day levels, and fossil fuel consumption, about 8% higher. The key message here is how the strong vertical policy linkages have been utilized as an instrument to bring about a significant transition.

Nykvist and Whitmarsh (2008) provide a comparative analysis of sustainable transport-related policies in two countries, the United Kingdom and Sweden. In Sweden particularly, where the major historical focus of transport policy was on safety, a remarkable change is seen in the way it has embraced sustainability goals. Swedish policies, such as a law in operation since 2006 that requires filling stations above a certain size to supply biofuels, highlight the government's efforts to reduce dependence on traditional fuels. In the United Kingdom, there is graduated vehicle excise duty and a company car tax, which reduces the tax in line with carbon emissions. The governments of both countries have also been exploring the options of creating a modal shift from cars to public transport systems and car pools or car clubs. A decrease in the registration for driving licenses can also be possibly linked to an overall societal change of moving from private to public modes of transport. The authors note that there are several reasons for the emergence of such policies. Key amongst the determining factors are landscape pressures – including environmental (e.g., climate change), economic (e.g., oil prices, automotive and Information and Communication Technoloy

markets), and cultural (value/behavior change) factors – which in turn have an impact on policies at the national and European level. At the niche level, there is interest in exploiting opportunities arising from these trends, e.g., among agricultural and emerging biofuel industries. Networks are starting to emerge in some areas (e.g., biofuels), supported by a favorable policy system. Regime actors are also beginning to respond to landscape pressures and exploit technological opportunities and new markets (e.g., biofuels, HEVs, and small, fuel-efficient cars).

Van der Laak et al. (2007) have studied three recent experiments on biofuels in the Netherlands in order to develop policy guidelines. The strategic niche management approach has been used to explain the success and failures of the three projects, and thereby develop key policy pointers. The guidelines have been categorized into three broad areas: shaping expectations, network building, and learning processes. Their analysis suggests that sustainable technologies often lack a clear advantage for individual entrepreneurs or users (e.g., financial gains), because advantages are at the wider level of society (e.g., reduction of greenhouse gases). Nevertheless, in particular situations (niche markets, local benefits, and ideological arguments) there might be local entrepreneurs willing to nurture and develop a technology that is potentially sustainable. These bottom-up initiatives are important, because they can trigger a process of increasing societal awareness, enable the emergence of local demand, and benefit from wider diffusion. Also, additional (unanticipated) advantages may prove to be viable. Issues such as new job creation, reinforcing local economies, and visual improvements to the countryside may render first generation biofuels socially and economically attractive, despite their limited potential for reducing greenhouse gas emissions.

16.4.3   Examples of Sectoral Transitions

16.4.3.1   Rural Energy Sector

Rural communities across the globe rely on cheap and inefficient fuels, such as traditional biomass. With rapid population growth and the absence of reliable alternatives to traditional fuels, these communities face a grave situation. Renewable energy options or other modern energy technologies are almost non-existent, either due to their high cost implications or because they are unreliable. Field experience has shown that energy, such as electricity, should be employed in enterprises that employ local people or add value to local resources (Bastakoti, 2003).

The energy sector and, in particular, energy service delivery could be an important source of new economic opportunities, including livelihood options. Many hybrid systems described in this chapter offer opportunities for new enterprises based around more sustainable energy solutions, and create the possibility of achieving energy, developmental, and economic goals. Innovation support programs of national governments and multilateral institutions may be used to target energy related sectors. Decentralization of power can be a possible solution to the problem of introducing clean energy technologies to rural areas of developing nations. Distributed generation provides consumers with opportunities to gain a number of benefits, such as a degree of energy independence, opportunities for local control to improve security of supply, financial optimization, equal or better power quality, and a cleaner environment.

The *Uttam Urja* project was developed by TERI (2009) as a field project for dissemination of photovoltaic lighting technologies in rural India. The project developed a novel business model that promoted the delivery and management of energy services through the development of a local entrepreneurial chain. By customizing the financial assistance, providing affordable quality products teamed with good after-sales service through local energy service networks, this model has been able to function and sustain effectively without any foreign aid (financial/technical) (Rehman et al., 2010).

Another example is the African Rural Energy Enterprise Development (AREED) which demonstrates a partnership based model that accomplishes multiple objectives: new venture creation, support to small and medium enterprises, and energy service delivery (AREED, 2009; see also Chapter 25). An entrepreneur and an energy business are at the heart of the REED (the UNEP Rural Energy Enterprise Development) model, which seeks to tap and support the entrepreneurial spirit and help create sustainable energy businesses through appropriate financing, business development, technology support, and enabling government policies. Following the successful examples of AREED model, the REED approach has been applied in Yunnan province and neighboring areas of western China. China Rural Energy Enterprise Development (CREED) functions through coupling Enterprise Development Services (EDS) with closely targeted start-up financial for entrepreneurs, enabling them to deliver cleaner and higher quality energy services through new business ventures. CREED also offers support for consumer credit and income-generation loans through Green Village Credit. It creates strong links with local government agencies, NGOs and financial institutions active in the areas of energy, environmental protection, consumer credit, and income generation. Through these links, CREED aims to influence broader energy and development shifts underway in China, and redirect existing sources of finance and support to sustainable energy activities.

In general, the rural poor are willing and able to pay for energy services if they have appropriate financing options and are able to meet the first costs of access and/or of the appliance (Rehman et al., 2010).

16.4.3.2   Transport Sector

Delhi and Mumbai are the economic hubs of urban India. As a result of fast-paced economic development, the cities have seen a substantial increase in transport infrastructure and services. The existing transport technological regime, characterized by the use of low-efficiency fuel, was pressurized by public interest groups to improve local air quality. This triggered the development of institutions involved in clean technologies in the transport sector. These institutions analyzed various technological

innovations to achieve the goal of better air quality, and consequently improved health conditions.

An increased awareness and concern over the deteriorating air quality was demonstrated through Public Interest Litigations and NGO movements. This resulted in actions taken at the policy level to address the issue. The transport authorizing bodies issued specific directives which resulted in phasing out older vehicles that depended on inefficient fuel. These directives were further supported through court interventions. These legal directives acted as a market pull for technology developers and service providers.

The transition in the transport sector for the two cities is characterized by both top-down initiatives and bottom-up engagement. The top-down approach includes the transport authority directives and court judgments. The bottom-up approach involves NGOs and citizen groups, who expressed their concerns through Public Interest Litigations, plus the interest shown by public and private organizations to develop ancillary infrastructure to support the transition. This is an example of an established technology regime that is supported by infrastructure, finance, and user interest, with clearly defined roles for policymakers and regulators.

The co-benefits associated with the change in the technological regime provided a major impetus to the entire transition phase. The concerns of citizens led technology and service providers, policymakers, and the judiciary to mobilize the implementation process rapidly. Although fuel switching in the transport sector is a micro-change, it has triggered the use of natural gas in the industrial and domestic sectors. Therefore, through the application of a supporting policy structure, a micro-change can be successfully converted into a large-scale technological regime (Patankar and Patwardhan, 2006).

16.4.3.3   Small and Medium Enterprises Sector

For governments and manufacturing companies, global warming, rising energy prices, and increasing customer awareness have pushed energy efficient manufacturing to the top of the agenda (Bunse et al., 2010). Energy management is the judicious and effective use of energy to maximize profits, to enhance competition through organizational measures and optimization of energy efficiency in the process. Realizing the importance of energy efficiency, many developing countries have initiated policy and regulatory mechanisms focusing exclusively on conservation of energy. These policies are a mix of both obligating and voluntary approaches (Lindhqvist, 2001).

Lack of financial and technical assistance is the most prominent hindrance in the adoption of energy efficient technologies in small to medium enterprises (SMEs). Most of the existing conservation policies provide some form of incentives to the existing industries, to motivate them to adapt efficient and clean technologies. This may be in the form of soft loans and/or tax incentives. Long-term, low interest funding from international institutions like the World Bank and GEF can be utilized by countries to establish a separate dedicated financial institution with equity participation from the government to address the financial issues. One such example is the Pollution Control and Abatement Fund, a US$5 million fund established in Sri Lanka to provide financial assistance to financially viable industrial enterprises towards waste minimization, resource recovery, pollution control, and abatement.

The scheme has two components: (a) Technical assistance and (b) Credit component. The loan can be obtained from any of the six participating credit institutions. Loan disbursement is affected only after obtaining the Environmental Protection Licence. Under the Technical Assistance component, reimbursement up to 75% of cost towards cost of consultancy services for the investigation of waste minimization, preparation of designs, selection, supervision, installation, and operation of the equipment is affected. Under the Credit Component, finance up to a maximum amount of US$128,000 per industry at zero real rate of interest is provided. Maximum repayment period will be seven years including a maximum of one year of grace period. Security needed for the loan is a mortgage over the project assets. For projects that involve investment for modernization entailing a financial return in addition to the desired environmental effects, a loan amount of 50% of such costs would be provided and for all other cases it could be 100%. This loan could be used for purchase of equipment or phasing out of hazardous substances. Over 75 industries have benefited from this scheme (Thiruchelvam et al, 2003).

Demonstration projects in energy intensive sub-sectors undertaken as a part of government led initiatives can also act as an example of successful implementation of EE activities.

For example in India, The Energy Research Institute (TERI) has been involved in improvising the foundry segment of the SMEs. Energy efficient divided blast cupola (DBC), developed by the British Cast Iron Research Association (BCIRA), has been promoted by TERI. The implementation of these DBCs has improvised the profitability through reduced fuel costs. Further, it also delivers molten metal at higher temperatures and substantially reduces other (silicon and manganese) losses, thereby further reducing production costs (Patel et al., 2009).

For SMEs, energy is usually a small portion of the total production cost and therefore receives relatively little attention. Barriers to the introduction of energy management in SMEs are lack of innovation, information availability, and expertise of entrepreneurs. Further absence or lack of data on benchmarks, good practices, and standards makes comparative analysis difficult (Kannan and Boie, 2003).

References

Agnolucci, P. and P. Ekins, 2007: Technological transitions and strategic niche management: The case of the hydrogen economy. *International Journal of Environmental Technology and Management*, **7**(5/6): 644–671.

Allen, S. R., G. P. Hammond and M. C. McManus, 2008: Prospects for and barriers to domestic micro-generation: A United Kingdom perspective. *Applied Energy*, **85**: 528–544.

AREED, 2009: *Examples of AREED Enterprises*. African Rural Energy Enterprise Development (AREED). www.areed.org/projects/index_projects.htm (accessed September 5, 2009).

Azevedo, I. M. L., 2009: *Energy Efficiency in the U.S. Residential Sector: An Engineering and Economic Assessment of Opportunities for Large Energy Savings and Greenhouse Gas Emissions Reductions*. Diss., Department of Engineering and Public Policy, Carnegie Mellon University, Pittsburgh, PA.

Azevedo, I. L., G. Morgan and F. Morgan, 2009: The transition to solid state lighting. *The Proceedings of the IEEE*, **97**(3): 481–510.

Baker, J., 2008: New technology and possible advances in energy storage. *Energy Policy*, **36**(12): 4368–4373.

Bastakoti, B. P., 2003: Rural electrification and efforts to create enterprises for the effective use of power. *Applied Energy*, **76**(1–3): 145–155.

Bayod-Rújula, A. A., 2009: Future development of the electricity systems with distributed generation. *Energy*, **34**: 377–383.

Birnie, D. P., 2009: Solar-to-vehicle (S2V) systems for powering commuters of the future. *Journal of Power Sources*, **186**: 539–542.

Bradley, T. H. and A. A. Frank, 2009: Design, demonstrations and sustainability impact assessments for plug-in hybrid electric vehicles. *Renewable and Sustainable Energy Reviews*, **13**(1): 115–128.

Brown, J., Hendry, C.N., Harborne, P., 2007: An emerging market in fuel cells? Residential combined heat and power in four countries. *Energy Policy*, **35**: 2173–2186.

Brown, H. S. and P. J. Vergragt, 2008: Bounded socio-technical experiments as agents of systemic change: The case of a zero-energy residential building. *Technological Forecasting & Social Change*, **75**: 107–130.

Bunse, K., M. Vodicka, P. Schönsleben, M. Brülhart and F. O. Ernst, 2010: Integrating energy efficiency performance in production management: Gap analysis between industrial needs and scientific literature. *Journal of Cleaner Production*, **19**(6–7): 667–679.

Caputo, A. C. and P. M. Pelagagge, 2002a: RDF production plants I: Design and costs. *Applied Thermal Engineering*, **22**: 423–437.

Caputo, A. C.and P. M. Pelagagge, 2002b: RDF production plants II: Economics and profitability. *Applied Thermal Engineering*, **22**: 439–448.

Chacartegui, R., D. Sánchez, N. di Gregorio, F. J. Jiménez-Espadafor, A. Muñoz and T. Sánchez, 2009: Feasibility analysis of a MED desalination plant in a combined cycle based cogeneration facility. *Applied Thermal Engineering*, **29**(2–3): 412–417.

Clery, D., 2011: Greenhouse-power plant hybrid set to make Jordan's desert bloom. *Science Magazine*, **331**(6014): 136.

Cossent, R., T. Gómez and P. Frías, 2009: Towards a future with large penetration of distributed generation: Is the current regulation of electricity distribution ready? Regulatory recommendations under a European perspective. *Energy Policy*, **37**(3): 1145–1155.

Davies, P., Turner, K., Paton, C. (2004): *Potential of the seawater greenhouse in middle eastern climates*. In the *Proceedings of International Engineering Conference*, Jordan: 523–540.

Deshmukh, M. K. and S. S. Deshmukh, 2008: Modeling of hybrid renewable energy systems. *Renewable and Sustainable Energy Reviews*, **12**: 235–249.

Dufo-López, R., J. L. Bernal-Agustín and F. Mendoza, 2009: Design and economical analysis of hybrid PV-wind systems connected to the grid for the intermittent production of hydrogen. *Energy Policy*, **37**(8): 3082–3095.

Elzen, B., B. Enserink and W. A. Smit, 1996: Socio-technical networks: How a technology studies approach may help to solve problems related to technical change. *Social Studies of Science*, **26**(1): 95–141.

EPRI, 2003: *Handbook of Energy Storage for Transmission and Distribution Applications*. Electric Power Research Institute (EPRI), Paolo Alto, CA.

EPRI, 2004: *Handbook of Energy Storage for Grid-Connected Wind Generation Applications*. Electric Power Research Institute (EPRI), Paolo Alto, CA.

EPRI, 2007: *Environmental Assessment of Plug-in Hybrid Electric Vehicles, Volume 1: Nationwide Greenhouse Gas Emissions*, and *Volume 2: United States Air Quality Analysis Based on AEO-2006 Assumptions for 2030*. Electric Power Research Institute (EPRI), Paolo Alto, CA.

ETT, 2011: *Electric Transmission Texas Brings Largest Utility-scale Battery in the United States to One of Oldest Cities in Texas*. Electronic Transmission Texas (ETT), Austin, TX. www.ettexas.com/projects/presnas.asp (accessed October 9, 2010).

Finnish Energy Industries, 2004: *Energy – A Key for Competitiveness in Finland*. Report by the Finnish Energy Sector for *Government Analysis: Finland in the Global Economy*, Helsinki, Finland.

Fouquet, R., 2010: The slow search for solutions: Lessons from historical energy transitions by sector and service. *Energy Policy*, **38**: 6586–6596.

Frost and Sullivan, 2006: Today's rechargeable battery industry: Current state and recent innovations. *Battery Power Online Magazine*, April 2006.

GalvinPower, 2010: *What are some examples of smart microgrids?* Galvin Electricity Initiative. www.galvinpower.org/smart-microgrids/smart-microgrid-examples (accessed October 9, 2010).

Geels, F. W., 2002: Technological transitions as evolutionary reconfiguration processes: A multilevel perspective and a case-study. *Research Policy*, **31**: 1257–1274.

Geels, F. W. and J. W. Schot, 2007: Typology of sociotechnical transition pathways. *Research Policy*, **36**: 399–417.

Geels, F. W., M. Eames, F. Steward and A. Monaghan, 2008: *The Feasibility of Systems Thinking in Sustainable Consumption and Production Policy*. Report prepared for the Department for Environment, Food and Rural Affairs (DEFRA), Brunel University, London.

Genus, A. and A.-M. Coles, 2008: Rethinking the multi-level perspective of technological transitions. *Research Policy*, **39**(7): 1426–1445.

Goldman, C. A., N. C. Hopper and J. G. Osborn, 2005: Review of US ESCO industry market trends: An empirical analysis of project data. *Energy Policy*, **33**(3): 387–405.

Grubler, A., N. Nakicenovik and D. Victor, 1999: Dynamics of energy technologies and global change. *Energy Policy*, **27**: 247–280.

Gyuk, I., P. Kulkarni, J. H. Sayer, J. D. Boyes, G. P. Corey and G. H. Peek, 2005: The United States of storage – electric energy storage. *IEEE Power & Energy Magazine*, **3**(2): 31–39.

Hamidi, V., K. S. Smith and R. C. Wilson, 2010: *Smart Grid Technology Review within the Distribution and Transmission Sector*. In *Innovative Smart Grid Technologies Conference Europe (IGST Europe)*, IEEE, Gothenburg.

Han, W., J. Hongguang and W. Xu, 2007: A novel combined cycle with synthetic utilization of coal and natural gas. *Energy*, **32**: 1334–1342.

Healey, G. P., 2008: *Fostering Technologies for Sustainability: Improving Strategic Niche Management as a Guide for Action Using a Case Study of Wind Power in Australia*. Diss., MRIT University, Melbourne, Australia.

Hegger, D. L. T., J. van Vliet and B. J. M. van Vliet, 2007: Niche management and its contribution to regime change: The case of innovation in sanitation. *Technology Analysis and Strategic Management*, **19**(6): 729–746.

Helm, D., 2005: The assessment: The new energy paradigm. *Oxford Review of Economic Policy*, **21**(1): 1–18.

Hinnells, M., 2008: Technologies to achieve demand reduction and microgeneration in buildings. *Energy Policy*, **36**: 4427–4433.

Hoogma, R., R. Kemp, J. Schot and B. Truffer, 2002: *Experimenting for Sustainable Transport: The Approach of Strategic Niche Management*. Routledge, London and New York.

Hu, L., J. Hongguang, G. Lin and H. Wei, 2011: Techno-economic evaluation of coal-based polygeneration systems of synthetic fuel and power with CO2 recovery. *Energy Conservation and Management*, **52**: 274–283.

IEA, 2005: *World Energy Outlook 2005*. International Energy Agency (IEA) of the Organization for Economic Cooperation and Development (OECD), Paris, France. www.iea.org/weo/2005.asp (accessed September 10, 2010).

IEA, 2006: *Light's Labour's Lost: Policies for Energy-efficient Lighting, in Support of the G8 Plan of Action*. International Energy Agency (IEA) of the Organization for Economic Cooperation and Development (OECD), Paris, France.

Ieromonachou, P., S. Potter and M. Enoch, 2004: Adapting strategic niche management for evaluating radical transport policies – The case of the Durham Road access charging scheme. *International Journal of Transport Management*, **2**(2): 75–87.

Kalogirou, S. A. and Y. Tripanagnostopoulos, 2006: Hybrid PV/T solar systems for domestic hot water and electricity production. *Energy Conversion and Management*, **47**: 3368–3382.

Kalra, P. K. and R. Shekhar, 2006: *Urban Energy Management: India Infrastructure Report*. 3iNetwork, Oxford University Press, Oxford, UK.

Kannan, R. and W. Boie, 2003: Energy management practices in SME-case study of a bakery in Germany. *Energy Conversion and Management*, **44**: 945–959.

Kathuria, V., 1998: Technology transfer and spillovers for Indian manufacturing firms. *Development Policy Review*, **16**(1): 73–91.

Kemp, R., A. Rip and J. W. Schot, 2001: Constructing transition paths through the management of niches. In *Path Dependence and Creation*, R. Garud and P. Karnoe (eds.), Lawrence Erlbaum, New Jersey, NJ.

Kemp, R., D. Loorbach and J. Rotmans, 2007: Transition management as a model for managing processes of co-evolution towards sustainable development. *International Journal of Sustainable Development and World Ecology*, **14**(1): 78–91.

Kempton, W. and J. Tomic, 2005: Vehicle-to-grid power implementation: From stabilizing the grid to supporting large-scale renewable energy. *Journal of Power Sources*, **144**(1): 280–294.

King, D. E., 2006: *Electric Power Micro-grids: Opportunities and Challenges for an Emerging Distributed Energy Architecture*. Diss., Engineering and Public Policy, Carnegie Mellon University, Pittsburg, PA.

Lantz, M., Svensson, M., Björnsson, L., Börjesson, P., 2007: *The prospects for an expansion of biogas systems in Sweden: Incentives, barriers and potentials*. In *Energy Policy*, **35**(3): 1830–1843.

Larsen, H., Petersen, L. S., 2005: *The Future Energy System: Distributed Production and Use*. Risø Energy Report 4, Risø National Laboratory. Risø, Denmark.

Laughton, M., 1996: Combined heat and power: Executive summary. *Applied Energy*, **53**: 227–233.

Leach, G. and M. Gowen, 1987: *Household Energy Handbook*. World Bank, Washington, DC.

Lindhqvist, T., 2001: *Cleaner production: government policies and strategies*. In *Journal of Cleaner Production*, **24**(1–2): 41–45.

Liu, L. and Z. Wang, 2009: The development and application practice of wind-solar energy hybrid generation systems in China. *Renewable and Sustainable Energy Reviews*, **13**: 1504–1512.

Loorbach, D., 2007: *Transition Management: New Mode of Governance for Sustainable Development*. International Books, Utrecht, Netherlands.

Lovell, H., 2007: The governance of innovation in socio-technical systems: The difficulties of strategic niche management in practice. *Science and Public Policy*, **34**(1): 35–44.

Mahmoudi, H., S. A. Abdul-Wahab, M. F. A. Goosen, S. S. Sablani, J. Perret, A. Ouagued and N. Spahis, 2008: Weather data and analysis of hybrid photovoltaic-wind power generation systems adapted to a seawater greenhouse desalination unit designed for arid coastal countries. *Desalination*, **222**: 119–127.

Marbe, Å., S. Harvey and T. Berntsson, 2004: Biofuel gasification combined heat and power – new implementation opportunities resulting from combined supply of process steam and district heating. *Energy*, **29**(8): 1117–1137.

Marinova, D. and A. Balaguer, 2009: Transformation in the photovoltaics industry in Australia, Germany and Japan: Comparison of actors, knowledge, institutions and markets. *Renewable Energy*, **34**: 461–464.

Markard, J. and B. Truffer, 2008: Technological innovation systems and the multi-level perspective: Towards an integrated framework. *Research Policy*, **37**: 596–615.

Mohamed, E. S. and G. Papadakis, 2004: Design, simulation and economic analysis of a stand-alone reverse osmosis desalination unit powered by wind turbines and photovoltaics. *Desalination*, **164**(1): 87–97.

Monaghan, A., 2009: Conceptual niche management of grassroots innovation for sustainability: The case of body disposal practices in the UK. *Technological Forecasting and Social Change*, **76**: 1026–1043.

Mor, S., K. Ravindra, A. De Visscher, R. P. Dahiya and A. Chandra, 2006: Municipal solid waste characterization and its assessment for potential methane generation: A case study. *Science of the Total Environment*, **371**(1–3): 1–10.

Nfah, E. M. and J. M. Ngundam, 2009: Feasibility of pico-hydro and photovoltaic hybrid power systems for remote villages in Cameroon. *Renewable Energy*, **34**(6): 1445–1450.

Nouni, M. R., S. C. Mullick and T. C. Kandpal, 2008: Providing electricity access to remote areas in India: Niche areas for decentralized electricity supply. *Renewable Energy*, **34**(2): 430–434.

Nykvist, B. and L. Whitmarsh, 2008: A multi-level analysis of sustainable mobility transitions: Niche development in the UK and Sweden. *Technological Forecasting and Social Change*, **75**: 1373–1387.

Pachauri, S. and L. Jiang, 2008: The household energy transition in India and China. *Energy Policy*, **36**: 4022–4035.

Painuly, J. P., H. Park, M.-K. Lee and J. Noh, 2003: Promoting energy efficiency financing and ESCOs in developing countries: mechanisms and barriers. *Journal of Cleaner Production*, **11**(6): 659–665.

Paska, J., P. Biczel and M. Klos, 2009: Hybrid power systems – An effective way of utilizing primary energy sources. *Renewable Energy*, **34**: 2414–2421.

Patankar, M. and A. Patwardhan, 2006: The switch to CNG in two urban areas in India: How was this achieved? In *The Business of Sustainable Mobility: From Vision to Reality*. P. Nieuwenhuis, P. Vergragt and P. E. Wells (eds.), Greenleaf Publishing Ltd., pp.171–186.

Patankar, M., A. Patwardhan and G. Verbong, 2010: A promising niche: waste to energy project in Indian dairy sector. *Environmental Science and Policy*, **13**: 282–290.

Patel, M. H., P. Pal and A. Nath, 2009: *Savings from Divided Blast Cupola: A Case-Study of Successful Implementation at a Foundry Unit at Rajkot (Gujarat).*

Perrin, M., Saint-Drenan, Y.M., Mattera, F., Malbranche, P., 2005: *Lead-acid batteries in stationary applications: competitors and new markets for large penetration of renewable energies.* In *Journal of Power Sources*, **144**(2): 402–410.

Purohit, P., 2007: Financial evaluation of renewable energy technologies for irrigation water pumping in India. *Energy Policy*, **35**: 3134–3144.

Rahman, S., 2009: Smart grid expectations [in my view]. *IEEE Energy and Power Magazine*, **7**(5): 88, 84. ieeexplore.ieee.ord/stamp.jsp?tp=&arnumber=520843009 (accessed January 10, 2011).

Rane, M. V. and A. Dasgupta, 2003: *International Patent PCT/IN2003/000408, for Multiutility Vapor Compression System.*

Raven, R. P. J. M., 2005: *Strategic Niche Management for Biomass: A Comparative Study on the Experimental Introduction of Bioenergy Technologies in the Netherlands and Denmark*. Eindhoven University Press, Eindhoven, Netherlands.

Rehman, I. H., K. Abhishek, R. Raven, D. Singh, J. Tiwari, R. Jha, P. K. Sinha and A. Mizra, 2010: Rural energy transitions in developing countries: A case of the *Uttam Urja* initiative in India. *Environmental Science and Policy*, **13**: 303–311.

Rip, A. and R. Kemp, 1998: Technological change. In *Human Choice and Climate Change*, vol. 2. S. Rayner and E. L. Malone (eds.), Battelle Press, Columbus, OH, pp.327–399.

Roberts, B., 2009: Performance, purpose and promise of different storage technologies. *IEEE Power and Energy Magazine*, (July-August **2009**): 33–41.

Samaras, C. and K. Meisterling, 2008: Life cycle assessment of greenhouse gas emissions from plug-in hybrid vehicles: Implications for policy. *Environmental Science and Technology*, **42** (9): 3170–3176.

Sauter, R. and J. Watson, 2007: Micro-generation: a disruptive innovation for the UK energy system? In *Governing Technology for Sustainability*. J. Murphy (ed.), Earthscan, London.

Schot, J. W. and F. W. Geels, 2008: Strategic niche management and sustainable innovation journeys: theory, findings, research agenda, and policy. *Technology Analysis and Strategic Management*, **20**(5): 537–554.

Seyfang, G. and A. Smith, 2007: Grassroots innovations for sustainable development: Towards a new research and policy agenda. *Environmental Politics*, **16**(4): 584–603.

Shaahid, S. M. and I. El-Amin, 2009: Techno-economic evaluation of off-grid hybrid photovoltaic-diesel-battery power systems for rural electrification in Saudi Arabia – A way forward for sustainable development. *Renewable and Sustainable Energy Reviews*, **13**(3): 625–633.

Shafiei, E., Y. Saboohi and M. B. Ghofrani, 2009: Impact of innovation programs on development of energy system: Case of Iranian electricity-supply system. *Energy Policy*, **37**(6): 2221–2230.

Shiau, C.-S. N., C. Samaras, R. Hauffe and J. J. Michalek, 2009: Impact of battery weight and charging patterns on the economic and environmental benefits of plug-in hybrid vehicles. *Energy Policy*, **37**(7): 2653–2663.

Simpson, A., 2006: *Cost-benefit Analysis of Plug-in Hybrid Electric Vehicle Technology.* Conference Paper NREL/CP-540–40485.

Smith, A., 2005: The alternative technology movement: An analysis of its framing and negotiation of technology development. *Human Ecology Review*, Special Issue on Nature, Science and Social Movements, **12**(2): 106–119.

Smith, A., 2006: Green niches in sustainable development: The case of organic food. *Environment and Planning,* Government and Policy, **24**: 439–458.

Smith, A., 2007: Translating sustainabilities between green niches and socio-technical regimes. *Technology Analysis and Strategic Management*, **19**(4): 427–450.

Smith, A., 2009: Energy governance: The challenges of sustainability. In *Energy for the Future: A New Agenda*. J. I. Scrase and G. Mackerron (eds.), Palgrave, London, pp.54–75.

Smith, A. and A. Stirling, 2010: The politics of social-ecological resilience and sustainable socio-technical transitions. *Ecology and Society*, **15**(1): 11.

Smith, A., A. Stirling and F. Berkhout, 2005: The governance of sustainable socio-technical transitions. *Research Policy*, **34**(10): 1491–1510.

Stadler, M., C. Marnay, I. L. Azevedo, R. Komiyama and J. Lai, 2009: The open source stochastic building simulation tool SLBM and its capabilities to capture uncertainty of policymaking in the U.S. building sector. Presented at the *32nd IAEE International Conference on Energy, Economy, Environment: The Global View*, San Francisco, CA, US on June 21–24, 2009. www.osti.gov/bridge/purl.cover.jsp;jsessionid=40A18DAB 655DE10339ADDED847495A71?purl=/957409-hMtopY/ (accessed April 9, 2009).

Stier, J. C., 1983: Technological substitution in the united states pulp and paper industry: The sulfate pulping process. *Technological Forecasting and Social Change*, **23**: 237–245.

Suppes, G. J., 2006: Roles of plug-in hybrid electric vehicles in the transition to hydrogen economy. *International Journal of Hydrogen Energy*, **31**: 353–360.

Suurs, R. A. A., 2009: *Motors of sustainable innovation- Towards a theory on the dynamics of technological innovation systems*. Ph D Thesis, Utrecht University, Netherlands

TERI, 2009: *The Energy and Resources Institute (TERI), TERI Energy Data Directory and Year Book*, New Delhi, India.

Thiruchelvam, M., S. Kumar and C. Visvanathan, 2003: *Policy options to promote energy efficient and environmentally sound technologies in small and medium-scale industries. Energy Policy*, **31**(2003): 977–987.

US EIA, 2005: *Electric Power Annual Report with Data for 2005: Existing Capacity by Energy*. United States Energy Information Administration (US EIA).

van der Laak, W. W. M., R. P. J. M. Raven and G. P. J. Verbong, 2007: Strategic niche management for biofuels: Analyzing past experiments for developing new biofuels policies. *Energy Policy*, **35**(6): 3213–3225.

van der Vleuten, E. and R. P. J. M. Raven, 2006: Lock-in and change: Distributed generation in Denmark in a long-term perspective. *Energy Policy*, **34**(18): 3739–3748.

van Dijk, M., 2000: Technological regimes and industrial dynamics: The evidence from Dutch manufacturing. *Industrial and Corporate Change*, **9**: 173–94.

van Eijck, J. and H. Romijn, 2007: Prospects for Jatropha Biofuels in Tanzania: An analysis with strategic niche management. *Energy Policy*, **36**(1): 311–325.

van Mierlo, B., 2002: *Kiem van Maatschappelijke verandering*. Diss. (in Dutch), Aksant, University of Amsterdam, the Netherlands.

Verheul, H., and Ph. J. Vergragt, 1995: Social experiments in the development of environmental technology: a bottom-up perspective. *Technology Analysis and Strategic Management*, **7**(3):315–26.

Vine, E., 2005: An international survey of the energy service company (ESCO) industry. *Energy Policy*, **33**: 691–704.

Walawalkar, R., 2008 : *Emerging Electrical Energy Storage Applications for Public Power Utilities in Competitive Electricity markets*. Report presented to American Public Power Association under the Demonstration of Energy Efficiency Developments Scholarship.

Wallmark, C. and P. Alvfors, 2003: Technical design and economic evaluation of a stand-alone PEFC system for buildings in Sweden. *Journal of Power Sources*, **118**(1–2): 358–366.

Water Technology, 2009: *Shoaiba Desalinization Plant at Saudi Arabia.* www.water-technology.net/projects/shuaiba/ (accessed April 18, 2009).

Wiskerke, J. S. C., 2003: On promising niches and constraining socio-technical regimes: The case of Dutch wheat and bread. *Environment and Planning*, **35**: 429–448.

Witkamp, M. J., L. M. M. Royakkers and R. P. M. Raven, 2011: From cowboys to diplomats: challenges for social entrepreneurship in The Netherlands. *Voluntas: international journal of voluntary and nonprofit organizations*, **22**(2): 283–310.

Wolfe, P., 2008: The implications of an increasingly decentralized energy system. *Energy Policy*, **36**: 4509–4513.

Yamashita, K. and L. Barreto, 2005: Energyplexes for the 21st century: Coal gasification for co-producing hydrogen, electricity and liquid fuels. *Energy*, **30**: 2453–2473.

3 Cluster 17–21

17

Energy Pathways for Sustainable Development

Convening Lead Author (CLA)
Keywan Riahi (International Institute for Applied Systems Analysis, Austria)

Lead Authors (LA)
Frank Dentener (Joint Research Center, Italy)
Dolf Gielen (United Nations Industrial Development Organization)
Arnulf Grubler (International Institute for Applied Systems Analysis, Austria and Yale University, USA)
Jessica Jewell (Central European University, Hungary)
Zbigniew Klimont (International Institute for Applied Systems Analysis, Austria)
Volker Krey (International Institute for Applied Systems Analysis, Austria)
David McCollum (University of California, Davis, USA)
Shonali Pachauri (International Institute for Applied Systems Analysis, Austria)
Shilpa Rao (International Institute for Applied Systems Analysis, Austria)
Bas van Ruijven (PBL, Netherlands Environmental Assessment Agency)
Detlef P. van Vuuren (PBL, Netherlands Environmental Assessment Agency)
Charlie Wilson (Tyndall Centre for Climate Change Research, UK)

Contributing Authors (CA)
Morna Isaac (PBL, Netherlands Environmental Assessment Agency)
Mark Jaccard (Simon Fraser University, Canada)
Shigeki Kobayashi (Toyota Central R&D Laboratories, Japan)
Peter Kolp (International Institute for Applied Systems Analysis, Austria)
Eric D. Larson (Princeton University and Climate Central, USA)
Yu Nagai (Vienna University of Technology, Austria)
Pallav Purohit (International Institute for Applied Systems Analysis, Austria)
Jules Schers (PBL, Netherlands Environmental Assessment Agency)
Diana Ürge-Vorsatz (Central European University, Hungary)
Rita van Dingenen (Joint Research Center, Italy)
Oscar van Vliet (International Institute for Applied Systems Analysis, Austria)

Review Editor
Granger Morgan (Carnegie Mellon University, USA)

Contents

Executive Summary

Chapter 17 explores possible transformational pathways of the future global energy system with the overarching aim of assessing the technological feasibility as well as the economic implications of meeting a range of sustainability objectives simultaneously. As such, it aims at the integration across objectives, and thus goes beyond earlier assessments of the future energy system that have mostly focused on either specific topics or single objectives. Specifically, the chapter assesses technical measures, policies, and related costs and benefits for meeting the objectives that were identified in Chapters 2 to 6, including:

- providing almost universal access to affordable clean cooking and electricity for the poor[1];
- limiting air pollution and health damages from energy use;
- improving energy security throughout the world; and
- limiting climate change.

The assessment of future energy pathways in this chapter shows that it is technically possible to achieve improved energy access, air quality, and energy security simultaneously while avoiding dangerous climate change. In fact, a number of alternative combinations of resources, technologies, and policies are found capable of attaining these objectives. From a large ensemble of possible transformations, three distinct groups of pathways (GEA-Supply, GEA-Mix, and GEA-Efficiency) have been identified and analyzed. Within each group, one pathway has been selected as "illustrative" in order to represent alternative evolutions of the energy system toward sustainable development. The pathway groups, together with the illustrative cases, depict salient branching points for policy implementation and highlight different degrees of freedom and different routes to the sustainability objectives. The characteristics of the pathways thus differ significantly from each other, depending on the choices made about technologies, infrastructures, behaviors, and lifestyles, as well as on future priorities with respect to the portfolio of supply- and demand-side policies. These choices, in turn, have broad implications for issues of technological availability and scale-up, institutional and capacity requirements, and financing needs.

The analysis in this chapter shows that achieving all the objectives simultaneously remains an extremely ambitious task. Although a successful transformation is found to be technically possible, it will require the rapid introduction of policies and fundamental political changes toward concerted and coordinated efforts to integrate global concerns, such as climate change, into local and national policy priorities (such as health and pollution, energy access, and energy security). An integrated policy design will thus be necessary in order to identify cost-effective "win-win" solutions that can deliver on multiple objectives simultaneously.

The transition can be achieved from different levels of energy demand as well as through alternative combinations of energy resources. An in-depth modeling sensitivity analysis shows, however, that efficiency improvements throughout the energy system are the most important options to achieve the energy transformation toward a more sustainable future. Under assumptions of high energy efficiency (the GEA-Efficiency pathways), it is feasible to achieve the transformation under any of the analyzed supply-side portfolio restrictions. This includes in particular the feasibility of the transformation in absence of carbon dioxide (CO_2) capture and storage in combination with the phase-out of nuclear as well as cases without bioenergy with carbon capture and storage, or without relying on carbon sink management. Under the contrary assumption of high energy demand (the GEA-Supply pathways), however, the rapid and simultaneous growth of many advanced technologies is required. For instance, with high energy demand the sustainability targets remain out of reach if the supply of intermittent renewables or carbon capture and storage (CCS) is restricted, thus making these two "options" in effect mandatory in the absence of important improvements on the demand side. Assuming a nuclear phaseout, on the other hand, was found compatible with the transformation also at high energy demand.

---

1   The target is "almost universal access" because reaching the remotest rural populations is exceedingly expensive.

Despite the flexibility and choices available to direct the energy system transformation, a large number of robust and nondiscretionary components of an energy transition would need to begin being implemented now. These are referred to in the chapter as necessary conditions, summarizing the commonalities across all pathways to achieve the objectives. They include the following:

- Future improvements of at least the historical rate of change in the energy intensity of the economy, to reduce the risk that the sustainability objectives become unreachable. Further improvements in energy intensity, entailing aggressive efforts to improve end-use efficiency, increase the flexibility of supply and the overall cost-effectiveness of the energy system transformation.

- A broad portfolio of supply-side options, focusing on low-carbon energy from non-combustible renewables, bioenergy, nuclear energy, and CCS, achieving low-carbon shares in primary energy of at least 60–80% by 2050. These include:
    - strong growth in renewable energy beginning immediately and reaching 165–650 exajoules (EJ) of primary energy by 2050;

    - an increasing requirement for storage technologies to support system integration of intermittent wind and solar energy;

    - growth in bioenergy in the medium term to 80–140 EJ by 2050 (including extensive use of agricultural residues and second-generation bioenergy to mitigate adverse impacts on land use and food production);

    - nuclear energy plays an important role in the supply-side portfolio in some transition pathways, but the assessment of pathways with "restricted" portfolios suggests that it is also feasible to phase out nuclear and still meet the sustainability targets; and

    - fossil CCS as an optional bridging or transitional technology in the medium term, and increasing the contribution of biomass with CCS in the long term, unless energy demand is high, in which case cumulative storage of up to 250 gigatons of carbon dioxide ($GtCO_2$) by 2050 would be needed in order to limit global average temperature change to below 2°C.
- Aggressive decarbonization in the electricity sector, reaching low-carbon shares of 75% to almost 100% by 2050; phase-out of conventional coal power (i.e., without CCS); natural gas power could act as a bridging or transitional technology in the short to medium term.

- Enhancements of the transportation sector through electrification or the introduction of hydrogen vehicles to improve end-use efficiency, increase the flexibility of supply, and improve the overall cost-effectiveness of the energy system transformation.

- A peak in oil use in the transportation sector by 2030, followed by a phase-out over the medium term; a strong growth of liquid biofuels in the short to medium term, after which the mix of liquid and gaseous fuels depends on transportation system choices and technological breakthroughs.

- Availability of energy resources (fossil and non-fossil) does not limit deployment on an aggregated global scale but may pose important constraints regionally, particularly in Asia, where energy demand is expected to grow rapidly.

The analysis of the GEA pathways shows, similarly to earlier assessments, that the transformation of the energy system would require dedicated efforts to increase global energy-related investments to between US$1.7 trillion and US$2.2 trillion annually, compared with about US$1.3 trillion in annual investment today. Out of this total, about US$300 to US$550 billion of efficiency-related investments are required on the demand-side of the pathways. This includes only the efficiency-increasing part of the investment to improve energy intensity beyond historical improvement rates. The full demand-side investments into all energy components of appliances might thus be significantly higher. Total investments into energy supply and efficiency-related investments at the demand-side correspond in sum to a small fraction (about 2%) of global gross domestic product (GDP). Future transitions with a focus on energy efficiency achieve the targets at more modest cost and thus represent the lower bound of the investment range.

Meeting the sustainability objectives will require the further tightening of present and planned legislation and the introduction of new policies:

- *Universal access to electricity and clean cooking* requires the rapid shift from the use of traditional biomass to cleaner fuels and/or clean cooking technologies. This is feasible over the next 20 years, provided that sufficient financial resources are made available for investments on the order of US$36 billion to US$41 billion/year (half of it in Africa).

- *Pollution control measures* across all sectors need to be tightened beyond those in present and planned legislation so that the majority of the world population is meeting the World Health Organization (WHO) air quality guideline (annual PM2.5 (particulate matter less than 2.5 μg in size) concentration < 10 μg/m^3 by 2030), while remaining populations are staying well within the WHO Tier I-III levels (15–35 μg/m^3 by 2030). Estimated global costs to meet the air pollution target are about US$200 billion to US$350 billion/year to 2030 (about 10–20% of energy costs). This estimate accounts for ancillary benefits of stringent climate change mitigation policies that reduce overall pollution control costs by about 50–65%.

- *Limiting global temperature change to less than 2°C over preindustrial levels* (with a probability of $> 50\%$) is achieved through rapid reductions of global CO_2 emissions from the energy sector, which peak around 2020 and decline thereafter to 30–70% below 2000 emissions levels in 2050, reaching finally almost zero or even negative CO_2 emissions in the second half of the century.

- *Enhanced energy security* for regions can be achieved by increasing the use of domestic energy sources and by increasing the diversity and resilience of energy systems. A focus on energy efficiency improvement and renewable deployment increases the share of domestic (national or regional) supply in primary energy by a factor of 2 and thus significantly decreases import dependency. At the same time, the share of oil in global energy trade is reduced from the present 75% to under 40% and no other fuel assumes a similarly dominant position in the future.

Achieving society's near-term pollution reduction and health objectives is greatly furthered by climate change mitigation, and similarly, stringent climate policy can help further the energy security goals of individual countries. The simultaneous achievement of climate change mitigation, energy security, and air pollution control comes thus at a significantly reduced total energy cost when the multiple economic benefits of each are properly accounted for. This concerns:

- the added costs of future air pollution control measures at the global level, which can be cut significantly (by up to US$500 billion annually to 2030) in the case of stringent climate policy;

- energy security costs, which can be substantially decreased under increasingly stringent levels of decarbonization, approaching almost zero for very stringent climate policies and translating to an annual cost savings of about US$130 billion annually in 2030; and

- subsidies of carbon-intensive oil products and coal amount at present to about US$132 billion to US$240 billion/year. Rapid decarbonization of the energy system reduces the need for these subsidies by about US$70 billion to US$130 billion/year by 2050.

The transformation toward the sustainability objectives offers multiple benefits that cannot be assigned monetary values at a detailed level, but are nevertheless important to account for. The following are some important nonpecuniary benefits of the transformation:

- Universal access to electricity and clean cooking increases the productivity of the poorest people and thus contributes to overall well-being and more equitable economic growth. In addition, such access results in significant health benefits of more than 24 million disability-adjusted life years (DALYs) saved in 2030.

- Stringent pollution control policies to meet the WHO air quality guidelines for the majority of the world population result in health benefits on the order of 20 million DALYs saved in 2030.

- Limiting climate change to less than 2°C compared with preindustrial times reduces the risks of a number of different types of climate impacts, summarized by five main reasons for concern: the risk to unique or threatened systems; the risk of more frequent episodes of extreme weather; an inequitable distribution of impacts (given that some regions, countries, and populations may face greater harm from climate change); large aggregate damages (assessing comprehensive measures of impacts through efforts to aggregate into a single metric, such as monetary damages); and the risk of large-scale discontinuities (e.g., tipping points associated with very large impacts, such as the deglaciation of the West Antarctic or the Greenland ice sheet).

- Rapid decarbonization and thus stronger reliance on efficiency improvements and low-carbon energy (e.g., renewables) may create new job opportunities, thus providing additional economic benefits.

17.1 Introduction

17.1.1 Scenarios and Energy Transformations

Chapter 17 represents an integrative module of the Global Energy Assessment (GEA). It builds on Clusters I and II of this report to shed light on the question of how future energy systems can address multiple challenges and sustainability goals, ranging from issues of energy access to climate change mitigation. Specifically, the analysis of integrative future energy pathways presented in this chapter aims at illustrating how the energy system components, technologies, and resources described in Cluster II can be combined to address the challenges and realize the sustainability goals identified in Cluster I. The resulting energy transitions achieve multiple goals simultaneously and include various combinations of policy measures and instruments as well as lifestyle and value changes. The results of this scenario analysis thus prepare the ground for Cluster IV, which assesses policy packages and institutional and governance changes for realizing the different sustainable futures.

The two main objectives of developing the transformational pathways are, first, to provide a quantitative and qualitative framework for the identification of policies and measures for a transition toward an energy system that supports sustainable development, and second, to facilitate the integration of diverse energy issues and consistency across the different chapters of the GEA.

The existing literature contains a large number of scenarios, following different traditions in scenario design, development process and objectives. Broadly, one can distinguish between scenarios along "qualitative versus quantitative" lines or along "normative versus descriptive" lines. Whereas quantitative scenarios provide detailed numerical information about underlying processes and dynamics, qualitative scenarios aim at a textual and narrative description of how the future might unfold, thus providing an overarching story (see, e.g., Schwartz, 1991). A few scenario exercises have combined the two traditions and developed quantitative scenarios with so-called underlying storylines (among the first of which were those in the IPCC (Intergovernmental Panel on Climate Change) Special Report on Emissions Scenarios (SRES) (Nakicenovic and Swart, 2000) and the scenarios developed by the Millennium Ecosystem Assessment (Carpenter and Pingali, 2005). Descriptive scenarios usually aim at exploring a wide scenario space and thus improving our understanding of future uncertainties (given the variation of underlying assumptions about driving forces). Prominent examples include the IPCC reference greenhouse gas (GHG) emissions scenarios, such as the IS92 (Leggett et al., 1992; Pepper et al., 1992) and the SRES. Normative scenarios, by contrast, explore the underlying dynamics of change in order to achieve specific desirable outcomes or targets, usually assuming the deployment of a certain set of measures or policies. Consequently, normative scenarios usually do not aim at exploring the whole uncertainty space of possible future developments, but rather focus on the main characteristics of the transition that are considered necessary to achieve specific objectives.

Although various combinations of the above scenario designs are possible, a descriptive or a qualitative scenario design would not be sufficient to address the main aim of the GEA scenario analysis, which is to identify specific measures and policies that would enable the transformation of the energy system. Instead the GEA adopts a combination of a normative and a quantitative scenario approach, whereby specific targets for various energy objectives are defined and formal modeling approaches are used to quantify how, over what time frame, and at what costs those objectives can be achieved.

The GEA energy transition pathways presented in this chapter are designed to describe transformative changes toward a more sustainable future. A specific feature of these pathways is that they simultaneously achieve normative goals related to all major energy challenges, including the environmental impacts of energy conversion and use, energy security, and how to provide access to clean and affordable energy services for growing populations and higher standards of living (particularly for the world's poorest 3–4 billion people). Emphasis is given to the identification of potential synergies, or in other words, of integrated solutions and "win-win" strategies in addressing multiple energy objectives at the same time. One possible way of understanding the GEA pathways is to regard them as alternative interpretations of one overarching GEA scenario in which the energy system is transformed under normative, sustainable goals. The pathways highlight different degrees of freedom and routes to these goals.

17.1.2 Roadmap of the Chapter

The chapter is structured as follows (Figure 17.1). First, the GEA scenario logic and taxonomy are introduced, followed by assumptions about the main sustainability objectives and targets as defined by various chapters of Cluster I of the report. Next, the main characteristics of

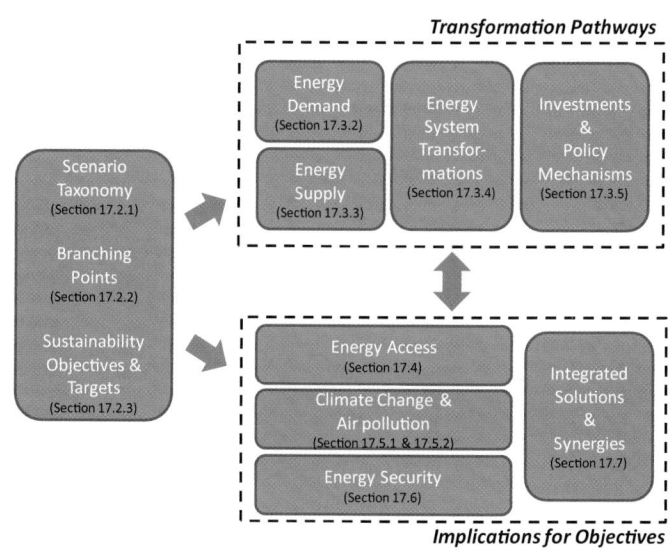

Figure 17.1 | Roadmap of Chapter 17.

Box 17.1 | Definitions of Key Terms Used in Chapter 17

GEA scenario	An overarching storyline of energy system transformation to meet normative sustainability objectives
Pathways	Qualitative and quantitative descriptions of demand- and supply-side energy system transformations falling within the overarching GEA scenario
Pathway groups	Groups of pathways distinguished by their level of energy demand and used as an organizing framework for the modeling of specific supply-side pathways
GEA-Efficiency	Pathway emphasizing demand-side and efficiency improvements
GEA-Supply	Pathway emphasizing the supply-side transformation at relatively high energy demand
GEA-Mix	Pathway emphasizing regional diversity at an intermediate level of demand between GEA-Efficiency and GEA-Supply
Illustrative pathway	A single pathway selected from one of the three pathway groups to illustrate in more depth the similarities and differences between pathways and to explore further implications
Branching points	Substantive alternatives or "choices" causing a divergence of pathways and contrasting characteristics:
	level of demand (low, intermediate, or high)
	transportation system transformation (conventional or advanced)
	portfolio of supply-side options (full or restricted)
Counterfactual	Hypothetical no-policy baseline describing the evolution of the energy system in absence of any transformational policies for the demand- or supply-side of the energy system

the energy transformation, including demand-side efficiency enhancements, supply-side transitions, technology deployment, and investment needs are analyzed. In the first instance, a wide range of possible transformation pathways and associated uncertainties are identified and explored. From the resulting ensemble of pathways, three illustrative cases are selected to represent salient differences in choices of how to meet the sustainability objectives. After addressing the issue of how the transition can be achieved, the chapter moves to the individual objectives and elaborates on what can be improved through which measures. Specific attention is given to identifying cost-effective policy portfolios for addressing energy access, environment (climate and pollution), and energy security objectives. The chapter concludes with a synthesis of how multiple sustainability indicators can be reached simultaneously. Box 17.1 sets out the key terms to be explained and used throughout the chapter.

17.2   GEA Scenario Logic

17.2.1   Scenario Taxonomy

The GEA comprises essentially a single normative scenario of the sustainability transition. Within this single scenario, alternative pathways are developed that describe transformations toward normative

objectives related to the environmental impacts of energy conversion and use, energy security, and energy access. All pathways fulfill these objectives by reaching specific and clear targets. For example, they all limit the future global mean temperature increase to not more than 2°C above preindustrial levels, and they all lead to almost universal access to clean energy services throughout the world by 2030. Another feature common to all pathways is that all economic and demographic changes within them are consistent with the GEA's aspirational goals with respect to sustainable development.

Achieving all these goals simultaneously is an enormous challenge that requires substantial effort and fundamental change in the energy system. Although the direction of change in the GEA is clearly defined by the sustainability objectives, the specific characteristics of the transition pathways may differ significantly and will depend on choices about technologies, infrastructures, behaviors, and lifestyles, as well as future priorities with respect to the portfolio of supply- and demand-side policies. These choices, in turn, have broad implications for issues of technological availability and scale-up, institutional and capacity requirements, and financing needs.

A fundamental assumption underlying the pathways is that the coordination required to reach the multiple objectives simultaneously can be achieved. The pathways thus illustrate the extent of coordination that is necessary and the benefits of policy integration across local and global concerns. By doing so, they inform decision making about the impacts of successful policy implementation. They do not, however, aim at developing recommendations of how the favorable political environment that is also necessary for successful policy coordination and implementation should be achieved (see Chapters 22 and 24).

The main aim of the GEA pathways is thus to provide a better understanding of what combination of measures, over which time frames and at what costs, is needed to deliver the necessary solutions. Although some combination of both supply- and demand-side measures is needed to transform the energy system, emphasis on one side or the other constitutes an important point of divergence between different policy choices that may drive the energy system in alternative directions. Thus, a critical factor is to what extent demand-side efficiency measures, together with lifestyle and behavioral changes, can reduce the amount of energy used for mobility, housing, and industrial services, and thus help fulfill the GEA's aspirational goals across virtually the whole range of sustainability objectives. If energy demand is low, any of a number of alternative supply-side configurations might be able to fulfill the goals. By contrast, a lower emphasis on reducing energy demand will require a much more rapid expansion of a broader portfolio of supply-side options. Hence, the successful implementation of demand-side policies increases the flexibility of supply-side options, and, vice versa, more rapid transformation of the supply side increases flexibility on the demand side.

Figure 17.2 illustrates this concept, which is the logical basis of the overarching GEA scenario and of the different GEA pathways. Three GEA pathway groups, labeled GEA-Efficiency, GEA-Mix, and GEA-Supply, are constructed to represent different emphases in terms of demand-side and supply-side changes. Each group varies in particular with respect to assumptions about the comprehensiveness of demand-side policies to enhance efficiency, leading to pathways of comparatively low energy demand (GEA-Efficiency), intermediate demand (GEA-Mix), and high demand (GEA-Supply). Within each group, a range of alternative pathways for the supply-side transformation are explored. These include a large diversity of supply portfolios in the GEA-Efficiency group of pathways, exploring, for example, the implications of the transformation with limited contributions of either nuclear, carbon capture and storage (CCS), or renewable technologies. In contrast, the GEA-Supply pathways involve much less flexibility with respect to supply-side measures, as most options need to expand pervasively and successfully, given the assumed high level of demand. By the same token, the GEA-Supply pathways show the most flexibility on the demand side of the energy system, requiring, for example, a much less pervasive introduction of efficiency measures to reduce energy demand for services. The pathways thus explore not only alternative combinations of supply- and demand-side policy portfolios, but also different choices with respect to overall strategy and level of implementation. In this context, the GEA-Mix pathways explore the degrees of freedom offered by more diverse energy systems, from resource extraction to services delivered to end users. The emphasis

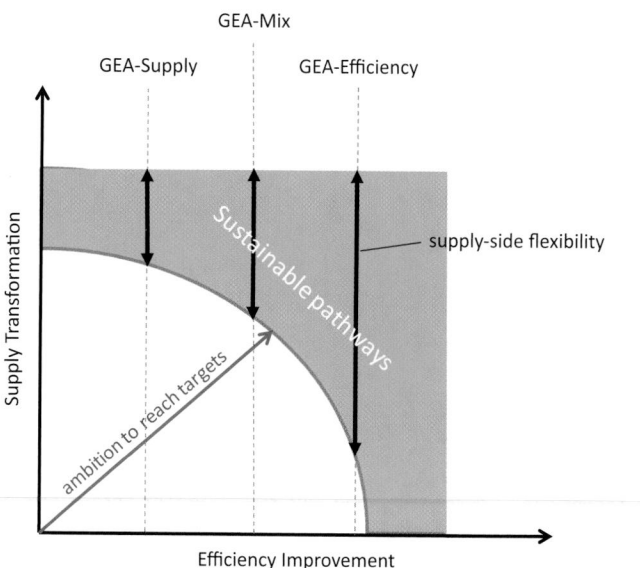

Figure 17.2 | Schematic illustration of the GEA pathways. The different pathways (GEA-Supply, GEA-Mix, and GEA-Efficiency) explore alternative combinations of efficiency improvements and supply-side transformations to achieve ambitious targets for sustainable development. The ambitiousness of the targets defines the feasibility frontier for the combinations of supply and efficiency measures. High levels of efficiency improvements, as depicted by the GEA-Efficiency pathways, increase the supply-side flexibility to reach the targets, and vice versa, the rapid transformation of supply increases the flexibility for the required efficiency improvements to limit energy demand.

of GEA-Mix is on the diversity of the energy supply mix, to enhance the system's resilience against innovation failures or technology shocks. This emphasis also implies that the GEA-Mix group of pathways is not necessarily intermediate between the other two groups in terms of other salient scenario characteristics (e.g., the required policy portfolio,[2] costs, fuel choices, or deployment of individual technologies).

17.2.2 Branching Points and the GEA Pathways

Many alternative GEA pathways fulfill the normative objectives set out for the global energy system. Moving from these objectives to a specific pathway entails three critical choices or "branching points." The first branching point involves a choice among alternative levels of energy demand and efficiency improvements, leading to distinct pathway groups of low, high, and intermediate demand (GEA-Efficiency, GEA-Supply, and GEA-Mix, respectively).

Another branching point explores alternative transformations on the supply side with the main aim of testing the flexibility of different supply-side configurations to fulfill the GEA sustainability objectives, given the levels of energy demand resulting from the choice at the first branching point. One aim was specifically to use the GEA Integrated Assessment Models to explore whether any of the supply options were mandatory.

---

2   The emphasis on policy in GEA-Mix is on developing and maintaining a diversity of demand- and supply-side options through a diversity of policy choices.

To do this, constraints were set on the portfolio of supply-side options by prohibiting or limiting the availability of specific technologies, including nuclear, CCS, biomass, and other renewables.

A third branching point, whose importance was revealed by this supply-side analysis, concerns changes in the transportation system. A "conventional" transportation system relying on liquid fuels has substantively different implications for supply flexibility than an "advanced" system dominated by electric or hydrogen-powered vehicles. Although any major transformation in an end-use sector that entails fuel switching will impact the energy supply, the magnitude of the impact of such a transformation in the transportation system alone warranted its inclusion as an explicit branching point.

The sequencing of these branching points is important and reflects a central tenet of an integrated, systemic approach to efficient environmental design that is equally applicable to wider technological systems (von Weizsäcker et al., 1997). In the context of energy systems, it is the demand for energy services such as mobility, heating, and industrial processes that drives the system. Hence, systems design should begin with the demand for energy services, emphasizing efficiency improvements and other means of reducing demand. This "sizes" the overall system and forms the basis for exploring supply-side options to meet this demand.

Whereas the first branching point thus addresses the main question of which level of resources needs to be mobilized in order to make the provision of energy services more efficient as well as reduce overall demand for those services, the other two branching points address issues of technological risk and uncertainty related to potential barriers to the deployment of specific supply technologies, which would hinder their adoption at full scale. These barriers might include, for example, the high investment requirements of a hydrogen distribution and refueling infrastructure, system constraints on the scale-up of specific technologies (e.g., integrating large amounts of power from intermittent renewable energy sources into electricity grids), potential public opposition (e.g., to the widespread deployment of CCS or nuclear power), and other specific risks of individual technologies (e.g., proliferation in the case of nuclear).

The branching points also depict irreversibilities, "lock-ins," and path dependencies within the system, reflecting the fact that once technological change is initiated in a particular direction, it becomes increasingly difficult to change its course. A prominent historical example of lock-in is the success of the internal combustion engine; in the same way, the two branches for the transportation sector – toward either electric or hydrogen-powered vehicles or clean liquid fuels – depict two alternative and not easily reversible directions of technological change for the future.

These branching points generate a wide range of alternative GEA pathways exploring different interactions between possible energy

Table 17.1 | Branching points and GEA pathways.

Branching point 1: *What is the level of energy demand?*	Branching point 2: *What are the dominant transportation fuels and technologies?*	Branching point 3: *How diverse is the portfolio of supply-side options?*
GEA-Efficiency (low demand) GEA-Supply (high demand) GEA-Mix (intermediate demand)	Conventional (liquid fuels) Advanced (electricity, hydrogen)	Full portfolio (all options) Restricted portfolio (excludes or limits particular options):[1] No CCS No BioCCS No sinks No nuclear No nuclear and no CCS Limited renewables Limited biomass Limited biomass and renewables Limited biomass, no BioCCS, no sinks

1 For further details and rationales of specific restrictions, see Table 17.9 and Section 17.3.3.5.

demand- and supply-side changes; these are summarized in Table 17.1 and illustrated in Figure 17.3. The first branching point, as already noted, leads to three pathway groups of low, high, and intermediate demand (GEA-Efficiency, GEA-Supply, and GEA-Mix, respectively). The other two branching points, relating to the transportation system and supply-side flexibility, in reality do not occur in a neat sequence, but rather are elaborated through an iterative process of pathway modeling and analysis. However, it is convenient to present them sequentially so that the pathways can be more easily understood; doing so in no way affects the underlying scenario logic. Thus, the second branching point, relating to the transportation system, gives rise to two scenarios, labeled Conventional Transportation and Advanced Transportation, in each of the three pathway groups. Conventional Transportation refers to the continuation of a predominantly liquid-based transportation system, whereas Advanced Transportation requires either fundamental changes in infrastructures (in the case of high penetration of electric vehicles) or major breakthroughs in transportation technology (e.g., in hydrogen fuel cells). The third branching point, relating to supply-side flexibility, then generates 10 alternative pathways in each of these six scenarios, giving a total of 60 alternative GEA pathways. Of these, 19 were rejected as they failed to fulfill the GEA objectives. That is, no feasible solution could be found within these pathways that would meet the "stringent" sustainability objectives described in Section 17.7. The issue of feasibility is discussed further in Section 17.3 and summarized in Section 17.3.6.

17.2.3 Energy Goals and Targets of the Sustainability Transition

There is a large body of literature on different types of objectives for sustainable development that addresses the environmental and social,

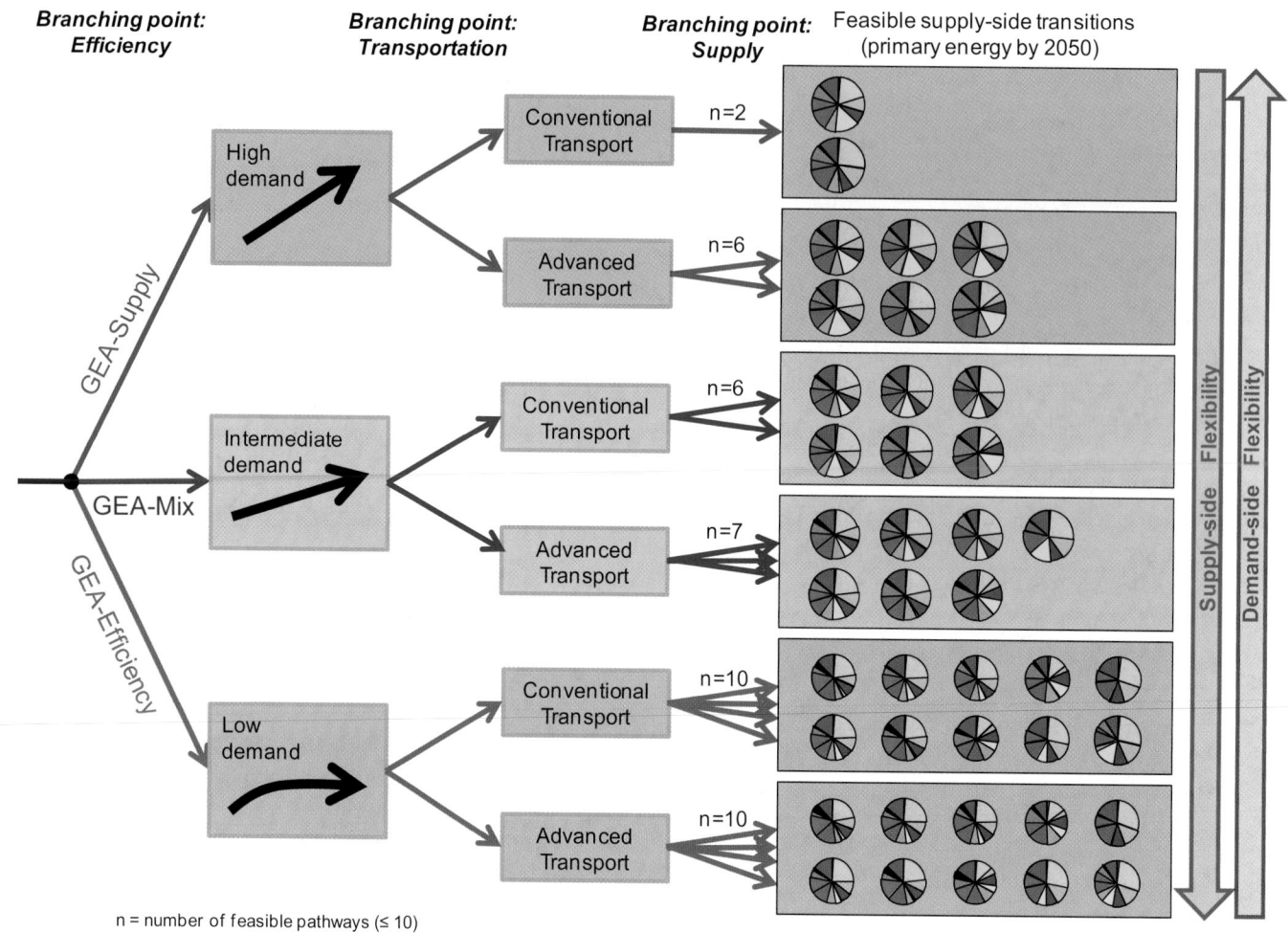

Figure 17.3 | Schematic illustration of the GEA pathways and the three branching points. The scenario setup features alternative choices for the combination of demand-side efficiency improvements and supply-side transformations, describing alternative policy emphases that would enable the transformation of the energy system. The pie charts represent primary energy portfolios of feasible transformation pathways under different branching point assumptions.

as well as the economic, dimensions of sustainability (Hirschberg et al., 2007; Vera and Langlois, 2007). This section does not intend to be comprehensive with respect to all these dimensions but instead focuses on the main energy challenges, and thus on selected objectives that are directly or indirectly affected by energy use. The targets identified here thus refer only to the necessary changes in local and global energy systems; much more is required in other sectors of societies for overall sustainability to be realized.

The definition of the targets builds upon the assessment of the objectives presented in the chapters of Cluster I of this report. Their selection has, to the extent possible, been guided by agreements and aspirations expressed by the international community or by United Nations actions and resolutions.

The targets are of central importance, since they define the ambitiousness and the magnitude and pace of the required transformation. The targets are thus major drivers of the pathways, defining the policy

stringency and portfolio of measures to respond to the energy challenges (see Sections 17.3 to 17.7). The model-based assessment in this chapter focuses predominantly on the technological feasibility, required policies, and associated costs and benefits of reaching the targets. The political feasibility of the assessed pathways will depend, in addition, on whether international and regional agreements for the implementation of the policies are put in place (see Chapters 22 and 26).

Table 17.2 summarizes the main target levels. These are used in the analysis of pathways to sustainability as the main boundary conditions or formal constraints in the integrated assessment modeling frameworks MESSAGE and IMAGE (see Box 17.2). The targets are defined in quantitative terms and prescribe a specific time schedule for meeting certain goals. They cover goals for all four principal energy challenges: energy access, air pollution and health, climate change, and energy security. In addition to these goals, the GEA also adopted adequate energy services to support economic growth as a normative goal (see Chapter 6).

Table 17.2 | Targets for the four main energy challenges and key characteristics of the corresponding transition pathways. In addition to these targets, the GEA also adopted adequate energy services to support economic growth as a normative goal (see Chapter 6).

Objective/Goal	Target and timeline	Pathway characteristics	Further details
Improve energy access	Almost universal access to electricity and clean cooking fuels by 2030 (see also Chapters 2 and 19)	Diffusion of clean and efficient cooking fuels and appliances. Extension of both high-voltage electricity grids and decentralized microgrids. Increased financial assistance from industrialized countries to support clean energy infrastructure.	Section 17.4
Reduce air pollution to improve human health	Achieve global compliance with WHO air quality guidelines (annual PM2.5 concentration < 10 µg/m³) for the majority of the world population, and the remaining populations staying well within the WHO Tier I-III levels (15–35 µg/m³) by 2030 (see also Chapters 3 and 4)	Tightening of air pollution legislation across all energy sectors (e.g., vehicles, shipping, power generation, industrial processes). Decarbonization to support pollutant emissions controls. Fuel switching from traditional biomass to modern energy forms for cooking in developing countries.	Section 17.5.2, 17.7
Avoid dangerous climate change	Limit global average temperature change to 2°C above preindustrial levels with a likelihood > 50% (see also Chapter 3)	Widespread diffusion of zero- and low-carbon energy supply technologies, with substantial reductions in energy intensity. Global energy-related CO_2 emissions peak by 2020, are reduced to 30–70% of 2000 levels by 2050, and approach almost zero or negative levels in the very long term. Globally comprehensive mitigation efforts covering all major emitters. Financial transfers from industrialized countries to support decarbonization.	Section 17.5.1, 17.7
Improve energy security	Limit energy trade; increase diversity and resilience of energy supply (both by 2050; see also Chapter 5)	Increase in domestic energy supply options (e.g., renewables to provide 30–75% of primary energy by 2050), and reduction of the share of oil in global energy trade from the present 75% to under 40% (and no other fuel assumes a similarly dominant position). Increase in diversity of energy supply as well as all endues sectors and regions by 2050. Infrastructure expansion and upgrades to support interconnections and backup, including increased capacity reserves, stockpiles, and energy storage technologies.	Section 17.6, 17.7

Note: For further details see Section 17.3.

Box 17.2 | Scenario Development Process

The GEA scenarios were developed in parallel by two integrated assessment modeling frameworks and through an iterative and participatory process so as to achieve integration across various chapters of the GEA. Figure 17.4 illustrates the scenario development process, showing the flow of information from individual chapters to the scenario development team and the iterations across various knowledge clusters.

Important inputs to the GEA scenarios include quantitative technoeconomic information such as technology costs, energy resources, and potentials provided by other GEA clusters. In addition, a series of workshops and a scenario questionnaire were prepared by the GEA writing team and external experts to solicit input for defining the main characteristics of the GEA scenario taxonomy and the set of objectives for a sustainable energy system with specific targets and timelines. These inputs are used by two modeling frameworks for the development of the GEA pathways:

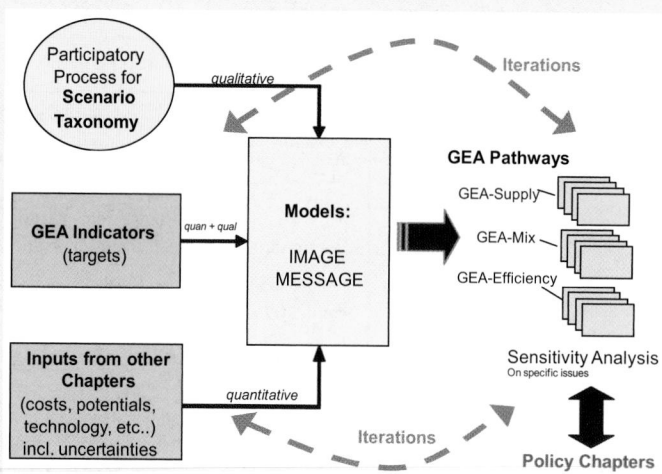

Figure 17.4 | The GEA scenario development process.

MESSAGE (Model for Energy Supply Strategy Alternatives and their General Environmental Impact) is a systems engineering optimization model used for medium- to long-term energy system planning, energy policy analysis, and scenario development (Messner and Strubegger, 1995; Riahi et al., 2007). The model provides a framework for representing an energy system with all its interdependencies from resource extraction, imports and exports, conversion, transport, and distribution to the provision of energy end-use services such as light, space heating and cooling, industrial production processes, and transportation. The framework covers all GHG-emitting sectors, including agriculture, forestry, energy, and industrial sources, for a full basket of greenhouse gases and other radiatively active gases: CO_2, methane, nitrous oxide, nitrogen oxides, volatile organic compounds, carbon monoxide, sulfur dioxide, black carbon and organic carbon, tetrafluoromethane, hexafluoroethane, various hydrofluorocarbons (HFC125, HFC134a, HFC143a, HFC227ea, HFC245ca), and sulfur hexafluoride. MESSAGE is used in conjunction with MAGICC (Model for Greenhouse Gas Induced Climate Change) version 5.3 (Wigley and Raper, 2001) for calculating internally consistent scenarios for atmospheric concentrations, radiative forcing, annual-mean global surface air temperature, and global-mean sea level implications.

IMAGE is an integrated assessment modeling framework consisting of a set of linked and integrated models (Bouwman et al., 2006). Together the framework describes important elements in the long-term dynamics of global environmental change, such as air pollution, climate change, and land use change. Important subcomponents of the model are the global energy model TIMER, the land use and land cover submodels of IMAGE, the detailed description of the carbon cycle, and the MAGICC 6.0 (Meinshausen et al., 2009) model that is included as the climate model within IMAGE. The model focuses on several dynamic relationships within the energy system, such as inertia, learning-by-doing, depletion, and trade among the different regions. Technological choices are made on the basis of relative costs (using multinomial logit equations). The land cover submodels in the earth system simulate the change in land use and land cover at a resolution of 0.5×0.5 degrees (driven by demands for food, timber and biofuels, and changes in climate). The earth system also includes a natural vegetation model to compute changes in vegetation in response to climate change feedbacks from changes in temperature, precipitation, and atmospheric CO_2 concentrations.

Both models use a set of harmonized assumptions about future drivers of change (including targets) to generate the GEA pathways. Many of these drivers are specified externally to the modeling frameworks and were provided by other chapters of the GEA report (Table 17.3). The pathways thus also aim to integrate information (e.g., on resources, technologies, costs) provided elsewhere in the GEA.

Although the models were applied to develop the three illustrative pathways within each of the pathways groups, the assessment relies on the strengths of the individual models with respect to specific sensitivity analysis. For instance, both models explored the sensitivity of the results with respect to energy access; however, the assessment mostly builds upon simulations from IMAGE for detailed land use projections and mainly uses the MESSAGE model to explore supply-side flexibility and to calculate pollutant emissions. The atmospheric chemistry and dispersion modeling for the assessment of health impacts from air pollution were conducted with the TM5 model hosted

Table 17.3 | Model structure and assumptions used to generate GEA pathways.

Examples of externally specified or harmonized variables across models	Constraints on model outputs or "boundary conditions" for least-cost model solutions	Examples of internally generated or "endogenous" model outcomes
Population growth	Energy access target	Diffusion of supply-side technology options and their shares in primary energy
Reference economic growth	Environmental impact targets	Demand-side portfolios and fuel consumption
Reference energy intensity improvements	Energy security targets	Price-induced changes in energy demand
Resource availability and costs		Changes in land use and land cover
Technology availability and costs		Exposure to pollutant emissions
		Energy system investments
		Costs of alternative policy packages for energy access, environment, and security
		Costs of emissions reductions
		Carbon price

at the Joint Research Centre of the European Commission (Dentener et al., 2006; Stevenson, 2006; Kinne et al., 2006; Textor et al., 2007; Bergamaschi et al., 2007). As with any model-based assessment, any specific conclusions are conditional on the applied methods and assumptions.

Detailed scenario data for the individual GEA pathways are publicly available in the GEA database at www.iiasa.ac.at/web-apps/ene/geadb. The GEA database provides interactive features for data visualization and a user interface for the download of scenario information in different formats.

Because the GEA objectives are strongly normative, the targets are all designed to be ambitious. The elaborated GEA pathways suggest that all the targets can be reached, if appropriate policies are introduced and energy investments are scaled up considerably. Table 17.2 lists some general characteristics of the GEA pathways as influenced by each of the objectives.

The target of ensuring *almost universal access to electricity and clean cooking by 2030*[3] is driven by the current reliance of a large fraction of the population in developing countries on traditional biomass to satisfy basic energy needs. Their lack of access to electricity and to affordable and clean fuels for cooking has vast impacts on human health, productivity, and land conservation. Section 17.4 presents a comprehensive analysis of the combinations of policies that can achieve the GEA goal of universal access by 2030. Specific focus is given to microcredits or grants to finance appliances as well as subsidies to improve the affordability of clean fuels for cooking. In addition, the same section assesses the need for infrastructure investments for transmissions

and distribution networks to connect the rural poor to the grid (see also Chapters 2 and 19).

The target of *reducing air pollution in compliance with WHO (World Health Organization) air quality guidelines*[4] *by 2030* is explored in depth in Section 17.5.2 through a bottom-up, technology-based assessment of main measures across main pollutant emissions sources and sectors. Many countries around the world have adopted antipollution legislation and have specific plans for further implementation of legislation in the short term. As the analysis in Section 17.5.2 indicates, however, current legislative plans in the aggregate are not sufficient to achieve the GEA target. Hence, a major focus of that section is on identifying specific policy levers for individual sectors and regions, and the associated costs, to deliver further improvements consistent with the overall objective (see also Chapter 4).

---

3   The target is "almost universal access" because reaching the remotest rural populations is exceedingly expensive.

4   The WHO air quality guidelines are given for an annual PM2.5 concentration < 10 µg/m^3. In the GEA pathways the majority of the population meets this guideline by 2030, while the remaining populations stay well within the WHO Tier I-III levels of 15–35 µg/m^3.

With respect to climate change, the GEA adopts the target of *limiting global average temperature change to 2°C above preindustrial levels with a likelihood of more than 50%*.[5] This target is consistent with various scientific assessments of the increasing risk of climate-related impacts above that threshold (Smith et al., 2009) as well as with EU and UN policy recommendations (European Commission, 2007; United Nations Conference of the Parties, 2009). Such a global target calls for globally comprehensive and stringent GHG emissions reductions. Section 17.5.1 analyzes the required emissions pathways, measures to reduce emissions, costs, and equity implications of the transition (see also Chapter 3).

Last but not least, the GEA objective of improving energy security is achieved partly as a convenient co-benefit of decarbonization, which is illustrated in the GEA transition through two related objectives on *limiting energy trade across major importing regions* and *increasing the diversity and resilience of energy supply*. Although many different types of energy security indicators are summarized in the literature (e.g., Jansen et al., 2004; Scheepers et al., 2007; Kruyt et al., 2009; Sovacool, 2009; Sovacool and Brown, 2010), the GEA uses a relatively simple dual taxonomy to define security: sovereignty of the energy system based on the degree of energy trade, and resilience based on the degree of diversity of types of energy sources. The sovereignty dimension is incorporated by limiting energy trade as a fraction of total primary energy at a regional scale (discussed in Section 17.7). Although the resilience dimension is not a direct limitation in the GEA pathways, the analysis in Section 17.6 shows that diversity increases in all energy subsystems (total primary energy supply, fuel supply for end uses, and regional mixes). Section 17.6 elaborates on these indicators and on the different strategies to improve energy security and their implications for the transition (see also Chapter 5).

Without policies to enable the sustainability transformation, the energy system would continue its heavy reliance on fossil fuels. This is illustrated by the hypothetical no-policy baseline (counterfactual) of the GEA, which describes the evolution of the energy system in absence of any transformational policies to meet the GEA objectives. In the GEA counterfactual fossil fuels more than double their contribution by 2050 (reaching about 900 EJ). As a consequence greenhouse gas emissions would continue to grow at present rates for many decades to come, leading to an average global mean temperature change of about 5°C in the long term. Increasing use of fossil fuels would also increase import dependency and worsen energy security, particularly in resource poor regions in Asia. Lack of incentives to strengthen policies to control the emissions of air pollutants would result in an increase of outdoor air-

pollution induced health impacts from 23 million disability-adjusted life years (DALYs) lost globally in 2005 to more than 33–40 million by 2030. In addition, the lack of financing for clean cooking fuels and electricity for the poor would leave the energy access problem unresolved, leading to health impacts from household fuel pollution of about 40 million DALYs by 2030.

Changing the energy system to support sustainable development requires thus dedicated policies so that all the GEA goals are met concurrently. Hence, a major focus of the assessment is to explore integrated and holistic solutions that take into account potential trade-offs and help to identify *synergies* from achieving all the different objectives simultaneously. These are discussed in detail in Section 17.7.

17.3   The GEA Energy Transition Pathways

This section describes the main underlying dynamics and transformational changes featured on both the demand and the supply side of the energy system. The pathways are described initially in a disaggregated way, separating out macro drivers, demand-side improvements, and supply-side transformations. Then the pathways are reintegrated using three illustrative pathways to provide comprehensive storylines of what the energy system transformation might look like if the overarching GEA scenario is to be fulfilled. Once these "what" questions are answered, the chapter turns to questions of "how." The section that follows sketches out an answer to the question of how such a transformation might come about, pointing the way to more detailed analysis later in this chapter as well as in the remainder of this report.

This part of the chapter is organized as follows. Section 17.3.1 describes the main socioeconomic and demographic trends common to all the GEA pathways. Section 17.3.2 covers changes in energy intensity and final energy demand and draws together evidence from other parts of the GEA on the potential for efficiency improvements in different end-use sectors. Based on this demand-side analysis, three groups of pathways are set up corresponding to low, high, and intermediate levels of demand: these are the GEA-Efficiency, GEA-Supply, and GEA-Mix pathways, respectively. Section 17.3.3 turns to the supply side of the energy system. The lowest-cost portfolio of supply-side transformations (assuming the full availability of all advanced future technologies on a large scale) is described, followed by an analysis of the importance of fuel and technology transformations in the transportation sector. This leads into a broader analysis of flexibility in supply-side portfolios and the potential for specific supply-side options to be either limited or omitted completely. Section 17.3.4 integrates the analysis of macro drivers, efficiency improvements, and supply transformations to present the GEA pathways in an integrated form. Initially, three illustrative pathways are explored in depth to establish key characteristics, similarities, and differences. Then the full diversity of pathways is compared and contrasted, with particular emphasis on regional-level analysis and on the implications for land and food supply,

5   The likelihood of 50% refers to physical climate change uncertainties, including climate sensitivity, aerosol forcing, and ocean diffusivity. It thus depicts the chances that a specific GHG pathway would stay below the 2°C temperature target. The likelihood does not imply any probability of the political implementation of the targets, nor does it correspond to the likelihood of specific technologies becoming available in the future.

given bioenergy's potential contribution to the transformation. Section 17.3.5 is concerned with how the pathways might be implemented. Two critical issues are addressed: costs and investments, and policies. Because the overarching GEA scenario is strongly normative, all the pathways analyzed within this scenario require strong interventions to induce and direct the energy system transformation.

17.3.1   Economic Growth and Demographic Change

The GEA pathways share a common median demographic projection whereby the global population increases from almost 7 billion today to about 9 billion by the 2050s before declining toward the end of the century (UN DESA, 2009). Figure 17.5 illustrates this population projection in the context of the full range of global demographic developments from a very low to an improbably high number of people by 2100. The median development path is a challenging one, as the global population will be aging rapidly through the century and concentrating ever more in urban areas.

The GEA pathways also share a median economic development path, expressed in terms of world GDP that allows for significant development in the 50 or so poorest countries in the world, while at the same time reflecting increased resource productivity and demand growth in the richest countries, dampened by changing consumption patterns and lifestyles. This GDP development path builds on the updated IPCC B2 scenario projection by Riahi et al., (2007); for details see also the GEA database at www.iiasa.ac.at/web-apps/ene/geadb. Main changes include updates of short-term trends and revisions of regional projections consistent with the sustainability objectives of the GEA. The economic projection used in all the GEA pathways is illustrated in Figure 17.6, which also shows the full range of economic trajectories for the global energy scenarios in the literature (Nakicenovic et al., 2006).

The socioeconomic development pathway is chosen to be consistent with global aspirations toward a sustainable future while also attaining this goal with a high degree of confidence. Global real per capita income in the GEA pathways grows at an annual average rate of 2% over the next 50 years, but with significant differences in the pace of development across regions. Today's developing and emerging economies continue to grow at a relatively rapid pace, with their combined economic output surpassing that of the industrialized world by around 2040 (see inset in Figure 17.6). This pathway is also consistent with other central projections in the literature (Nakicenovic et al., 2006) and hence provides a good reference point for placing the GEA energy pathways within a comparative context.

17.3.2   Energy Demand and Services

The adequate provision of energy services is a prerequisite for human well-being and productivity, and ultimately it is the demand for these

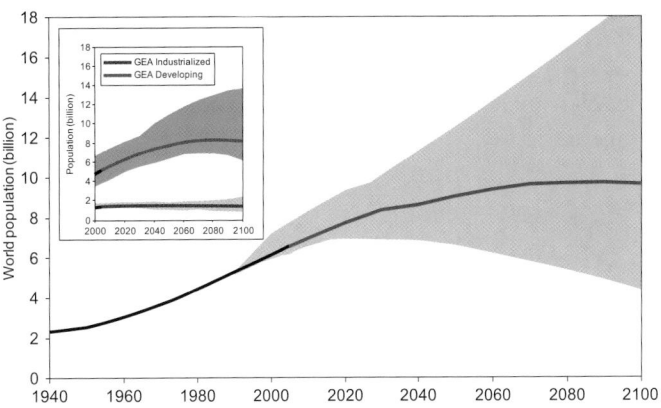

Figure 17.5 | Global population projections. The line indicates the median GEA development pathway and the fan indicates the range of population projections from the literature (Nakicenovic et al., 2006). The insert shows the median projections and ranges for the industrialized and developing regions separately.

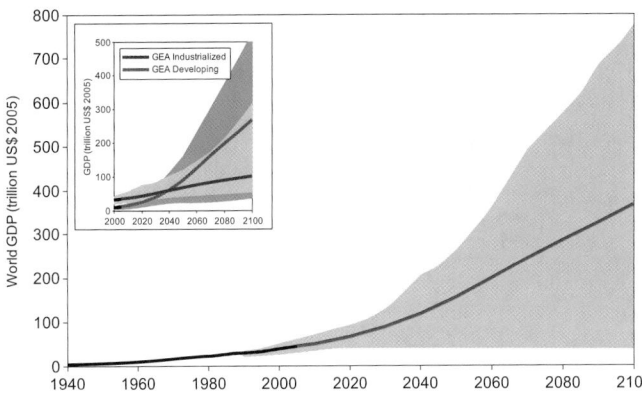

Figure 17.6 | Global economic development projections. The line indicates the median GEA development pathway and the fan indicates the range of economic projections from the literature (Nakicenovic et al., 2006). The insert shows the median projections and ranges for the industrialized and developing regions separately.

services that drives the energy system and its continuing expansion. Increasing affluence has historically been one of the major drivers of energy demand, and both the quantity and the quality of energy services determine in turn the magnitude of environmental and social impacts associated with the energy system. It is these impacts that are addressed by the normative objectives enshrined in the overarching GEA scenario.

Energy services are typically provided by end-use technologies, which convert energy from a particular form (biomass, petroleum, natural gas, electricity, and so forth) into services useful to a final consumer (heating and cooking, mobility, industrial processing, entertainment, and others). Consequently, end-use technologies and the efficiency with which they convert energy into useful services are inseparably connected with the levels and types of energy services demand. As a result, one can identify three broad and interrelated approaches to tackling demand-side challenges in the energy system:

- *improve technological efficiency*, e.g., increase vehicle fuel efficiency;

- *change the structure of energy services demand*, e.g., substitute physical mobility with "virtual" mobility enabled by electronic communications; and

- *reduce the level of energy services demand*, e.g., reduce travel needs by living closer to work or amenities.

Although all three of these approaches are explored in the GEA pathways as means of reducing final demand for energy, the emphasis throughout this section is on efficiency improvements. As a means for potentially decoupling energy demand from economic growth, energy efficiency represents a central lever for policy to target. Moreover, efficiency contributes to all the sustainability objectives. The degree to which efficiency improvements can limit energy demand growth is – by design – one of the main distinguishing characteristics of the GEA pathways. It should be noted, however, that efficiency improvements can be offset by both rebound effects and scale effects (Greening et al., 2000; Birol and Keppler, 2000; Hanley et al., 2009). Rebound effects describe an increase in demand for energy services as improvements in efficiency lower their effective cost. These effects can be direct (the savings from greater efficiency are spent on the same energy service), indirect (the savings are spent on a different energy service), or economy-wide (the savings contribute to economic and income growth, which increases demand). Rebound effects can be mitigated by price and other policies, which are discussed further in Section 17.3.5. Scale effects describe an increase in demand for energy services due to rising population or to rising economic output. Both rebound and scale effects make it important to consider the other approaches to demand-side transformation described above. Hence, both the structure and the level of energy services demand are also important parts of the GEA pathways described in this section.

The rest of this section is organized as follows. First, the headline trends in each group of pathways are discussed, covering the efficiency of the economy as a whole as well as on a per capita basis. Second, the GEA-Efficiency group of pathways is explored in more depth, sector by sector, drawing on material from the corresponding chapters of this report. Third, similarities and differences in the structure of energy demand (e.g., its distribution between end-use sectors) are considered.

17.3.2.1　　Energy Intensity Improvements

Energy intensity is energy used per unit of output, typically expressed in megajoules per US dollar (MJ/US$) of GDP or value added. Energy intensity metrics are widely used to represent the overall energy productivity of an economy or sector. The final energy intensity of the global economy has fallen historically at a rate of about 1.2%/year since the early 1970s. However, some regions have experienced substantially

more rapid reductions over certain periods. For example, China's energy intensity declined at a rate of about 4%/year between 1990 and 2000 (followed by a slower decline in the subsequent period). The causes of the energy intensity declines are many. They include, first, technological improvements in individual energy end-use appliances and technologies combined with substitution among fuels, such as the replacement of fuelwood with electricity or liquefied petroleum gas (LPG) for cooking. They also include changing patterns of energy end use; urbanization, which is characterized by generally higher system efficiencies: changes in the structure of the economy, including shifts toward higher shares of the less energy-intensive services sector; and finally, changing lifestyles, which affect both the type and the level of energy services demanded. Although not every such change has resulted in declining energy intensities in the past, taken together the overall trend is persistent and pervasive (Nakicenovic et al., 1998).

Energy intensity improvements can continue for a long time to come. Despite the energy efficiency and intensity improvements that have already been implemented to date, the efficiency of the energy system remains far from the theoretical potential. Although the full realization of this potential may never be possible, many estimates indicate that energy intensity reductions of a factor of 10 or more may be possible in the very long run (see Nakicenovic et al., 1993; Gilli et al., 1995; Nakicenovic et al., 1996).

The degree of energy intensity improvement is a crucial uncertainty for the future. All three groups of GEA pathways depict energy intensity futures that are driven by policies to improve energy efficiency, leading to global energy intensity improvement rates at or above historical experience. This is partly a result of the increasing importance of some low-income regions with relatively high rates of intensity improvement, but it is also partially due to the assumed move away from inefficient traditional fuels in the developing world. Energy intensity improvements thus vary significantly at the regional level, with some regions also developing more slowly than the historical rate, particularly in the GEA-Supply and -Mix pathways. The resulting global average reduction in energy intensity varies across the GEA pathways between about 1.5% and 2.2% annually to 2050. The lower end of the range is slightly faster than the historical experience, whereas the higher end is roughly double that and corresponds to a reduction in energy intensity of 60% by 2050. Cumulatively, these intensity improvements lead to substantial differences in per capita energy demand across the three pathway groups (see Figure 17.7).

Studies have shown that it is possible to improve energy intensity radically through a combination of behavioral changes and the rapid introduction of stringent efficiency regulations, technology standards, and environmental externality pricing, which mitigates rebound effects (see also Chapters 8, 9, and 10). The group of GEA-Efficiency pathways depicts such a development with a radical departure from historical trends. This group of pathways thus deliberately explores the consequences of demand-side interventions that lead to substantial declines

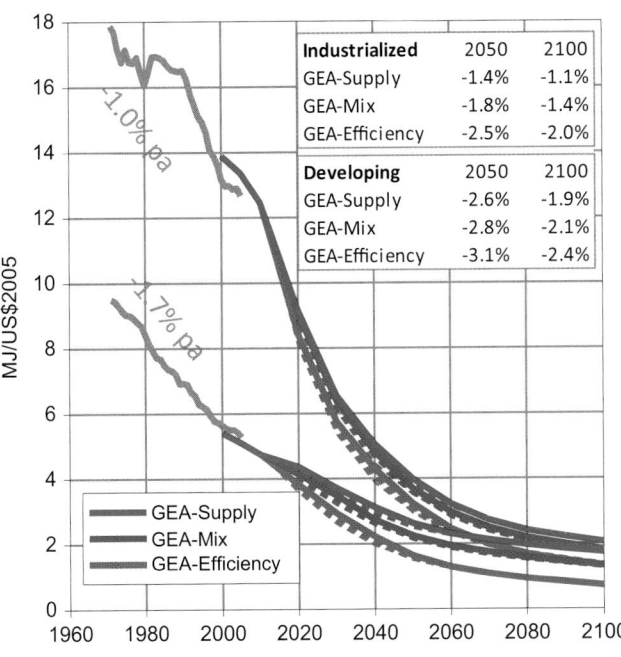
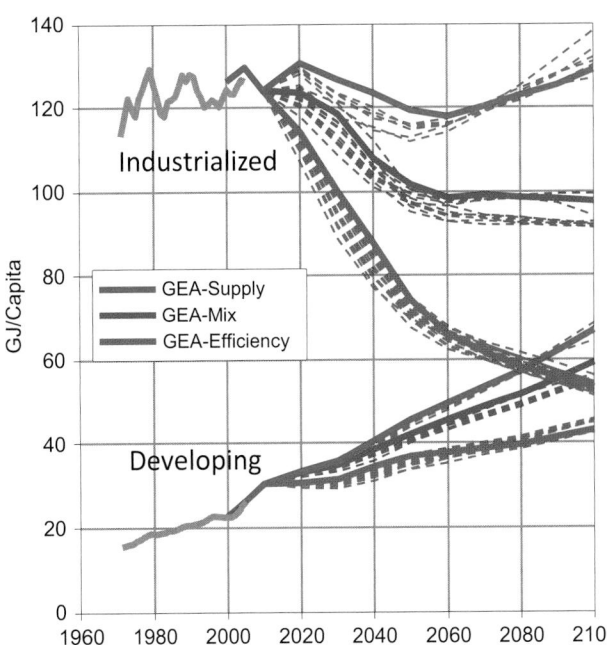

Figure 17.7 | Historical and projected energy intensity (left panel) and per capita final energy use (right panel) in the developing world and the industrialized world. Solid lines denote the illustrative GEA pathways within each of the three pathway groups; dashed lines show changes in energy intensity due to supply-side variations (discussed further in Section 17.3.3). The inset in the left panel shows rates of improvement in energy intensity (calculated using GDP at market exchange rates) between 2005 and 2050 and between 2005 and 2100. Source of historical data: IEA, 2010.

in per capita energy use in the industrialized world of about 45% in 2050 (from 130 GJ per capita in 2005 to about 75 GJ by 2050). Energy intensity rates in the developing world under the GEA-Efficiency pathways decline at 3.1%/year between now and 2050, and then slow down corresponding to an average of 2.4%/year over the course of the century. These rates are also considerably higher than historical experience. Given expected economic growth in the developing world, however, per capita energy demand continues to increase over the course of the century, although at a considerably slower pace than in the other GEA pathways groups (0.75%/year in the GEA-Efficiency pathways compared with 1.3%/year in the GEA-Supply pathways to 2050).The magnitude and pace of these efficiency improvements in the GEA-Efficiency pathways will undoubtedly require concerted and dedicated demand-side policies and measures. These are discussed in general terms through the remainder of this section and in more detail in Section 17.3.5.

As noted, the GEA-Efficiency pathways group depicts the upper bound of potential efficiency improvements and thus the lower bound of energy demand in the GEA pathways. The GEA-Supply pathways group depicts the opposite, that is, the lower bound of potential efficiency improvements giving rise to an upper bound of energy demand across all GEA pathways. The GEA-Supply pathways thus place much less emphasis on efficiency and other demand-side measures, focusing instead on supply-side transformations, which are discussed further in Section 17.3.3. In the GEA-Supply pathways, the long-term improvement rate in global energy intensity over the course of the century is slightly above the historical record of 1.2%/year. Over the medium term to 2050, however, both

developing and industrialized regions experience intensity improvements about 40% higher than in the past (1.4%/year and 2.6%/year compared with 1%/year and 1.7%/year in the past, respectively). As a result, per capita energy use in the industrialized world stays at roughly 2005 levels, while per capita demand in the developing world catches up, increasing by almost a factor of 2 in the long term (Figure 17.7). The GEA-Mix pathways group is characterized by intermediate efficiency improvements, giving rise to energy intensities both economy-wide and per capita that lie between the aggressive GEA-Efficiency pathways and the less prescriptive demand-side trends of the GEA-Supply pathways.

17.3.2.2 Sectoral Measures to Improve Energy Efficiency

Introduction

Increasing affluence typically results in additional demand for energy. However, per capita energy use today varies widely even between countries at comparable income levels (see Table 17.4). The reasons include differences in the type and amount of energy services demanded, in the efficiency of end-use technologies, and in the way these services and these technologies form part of broader structural patterns of behavior and lifestyle.

The use of energy for mobility provides a prominent example of these differences in energy use across countries. The average North American consumes about 54 GJ annually traveling by car, compared with about half of this amount in the other member countries of the Organisation

Table 17.4 | Final energy use and income per capita for Industrialized and Developing Regions, 2005 Actual and 2050 under GEA-Efficiency and GEA-Supply.

	2005		2050			
			Industrialized[1]		Developing[2]	
	Industrialized[1]	Developing[2]	GEA-Efficiency	GEA-Supply	GEA-Efficiency	GEA-Supply
GDP per capita (2005 US$ at market exchange rates)	3487–40,050	671–4905	24,446–52,535	24,446–52,535	6029–19,829	6029–19,829
Total final energy (GJ per capita)	73–219	7–46	62–98	104–156	28–50	32–71

1 Aggregated ranges of five GEA regions representing the industrialized world: North America, Western Europe, Pacific OECD, Eastern Europe, and Former Soviet Union. For full regional definitions see the electronic appendix to this chapter.

2 Aggregated ranges of six GEA regions representing the developing world: Centrally Planned Asia and China, South Asia, Pacific Asia, Middle East and North Africa, sub-Saharan Africa, and Latin America. For regional definitions see the electronic appendix to this chapter.

for Economic Co-operation and Development (OECD). Three main factors, in addition to the slightly higher per capita income in North America, explain the difference: the lower fuel economy of the typical individual vehicle (3 MJ/km in North America versus 2.6 MJ/km in the other OECD countries), longer distances traveled (as a result of both preferences and structural characteristics of urban form and land area), and more individualized use of cars (average occupancy is about 1.3 passengers/vehicle in North America compared with up to 1.5 passengers/vehicle in Eastern Europe, for example). This illustrates well the combined effect of efficiency and of behavior and lifestyles (levels and types of energy service demanded) on fuel consumption (see also Chapter 9).

Similar differences in per capita energy use can be found between other regions of the world as well as for other sectors, such as residential and industry. Large-scale improvements in the energy intensity of an economy therefore require a portfolio of measures that stimulate the adoption of highly efficient end-use technologies, complemented by policies to promote changes in energy services demand through behavioral and lifestyle shifts. In addition, structural changes in the economy play an important role.

The overarching finding from the sectoral analysis is that the rapid energy intensity improvements depicted by the GEA-Efficiency group of pathways are feasible with currently available technologies. The necessary magnitude of change, however, requires a fundamental shift in the way energy is used across all major sectors of the economy. The following sections summarize the nature of these shifts and the policies that might drive them in the GEA-Efficiency pathways.

The Residential and Commercial Sector in the GEA-Efficiency Pathways

In the residential sector, economic growth is expected to further increase the floor areas of dwellings by increasing living standards, particularly in developing countries. This will result in additional energy demand for space heating and cooling. As noted in Chapter 10, however, the potential for efficiency improvements in the use of energy for this purpose is

vast. In the GEA-Efficiency group of pathways, a large fraction of this potential is successfully tapped. Policies to improve thermal insulation as well as retrofits to advanced building types (passive house standards or lower) lead to improvements in energy use per unit of floor area by a factor of 4 in the industrialized world, from about 400–900 MJ/m^2 down to 100–230 MJ/m^2 by 2050 (Table 17.5). Improvement rates are similar in the developing world, on the order of a factor of 2 to 3.

The potential efficiency gains from buildings in terms of energy use avoided are among the highest across all end-use sectors. Achieving these gains requires the rapid introduction of strict building codes and retrofit standards for almost the complete global building stock. The rate of retrofit would need to increase to about 3% annually to 2050, about three times the historical rate.

In the GEA-Efficiency pathways, demand for energy from centralized sources and grids is further reduced by the adoption of technologies that enable space heating and cooling with net zero use of centralized energy. These include solar water heating, solar heating, air-source or ground-source heat pumps powered by solar photovoltaics, and biomass-based heating. Combined with efficiency improvements to building shells, these technologies would significantly reduce the need for centralized solutions for thermal comfort; centralized energy infrastructure would largely provide the additional energy required for lighting, cooking, and appliances.

Per capita electricity use in the residential and commercial sector is expected to grow significantly because of rising incomes and the adoption of modern household appliances and other electric devices. This trend is particularly pronounced in the developing world. Despite high efficiency standards, electricity use in the developing world increases in the GEA-Efficiency pathways group by a factor of 3 to 8 by 2050 (Table 17.5). The increase is more modest in the lower-income countries of the industrialized world, whereas in the higher-income countries of North America and Western Europe, per capita electricity use peaks and then declines toward 2050 to levels below that of 2005. Although overall demand for electricity continues to increase in the residential sector,

Table 17.5 | Energy service indicators for the residential and commercial sector in Industrialized and Developing Regions, 2005 Actual and 2050 under GEA-Efficiency pathways.

	2005, actual		2050, GEA-Efficiency	
	Industrialized[1]	Developing	Industrialized	Developing
GDP per capita (2005 US$ at market exchange rates)	3487–40,050	671–4905	24,446–52,535	6029–19,829
Floor area (m² per capita)[2]	26–55	9–32	48–58	19–52
Share of buildings with advanced technology (%)				
Single-family	<1	<1	74–86	75–83
Multifamily	<1	<1	74–79	82–91
Commercial and public	<1	<1	81–92	85–96
Heating demand (MJ/m²)				
Single-family	443–875	112–241	101–230	50–79
Multifamily	443–781	112–277	104–198	47–68
Commercial and public	475–914	173–371	101–166	54–97
Residential and commercial electricity demand (GJ per capita)[3]	11–45	1–6	22–33	8–15

1 Industrialized and developing regions are defined as in Table 17.4.

2 Includes public and commercial buildings based on the bottom-up analysis in Chapter 10.

3 Includes electric cooling and heating as well as lighting and appliances.

Table 17.6 | Energy service indicators for the transportation sector in Industrialized and Developing Regions, 2005 Actual and 2050 under GEA-Efficiency pathways.

	2005, actual		2050, GEA-Efficiency	
	Industrialized[1]	Developing	Industrialized	Developing
GDP per capita (2005 US$ at market exchange rates)	3487–40,050	671–4905	24,446–52,535	6029–19,829
Passenger-kilometers per capita[2]	14,293	2499	15,925	3892
Car	8778	404	6539	1009
Bus and train	2855	1461	3334	1368
Aviation	2274	198	5579	795
Other[3]	386	437	473	720
No. of light-duty vehicles per capita	0.46	0.03	0.52	0.11
Fuel use for mobility (GJ per capita)	30.8	2.4	24.0	4.6
Freight-kilometers per capita (thousands)	8219	1059	15,969	2774
Truck	4544	606	6925	1370
Rail	3675	453	9044	1404
Fuel use for freight (GJ per capita)	13.0	2.4	12.4	2.8

1 Industrialized and developing regions are defined as in Table 17.4.

2 Estimates from Chapter 9. Because of differences between the regional definitions used in that chapter and those used for the GEA scenarios, transport indicators are given as regional averages of the whole developing and industrialized world only.

3 Includes two- and three-wheeled vehicles.

efficiency improvements significantly slow this growth. As a result, per capita consumption across all income groups is about 25–50% lower in the GEA-Efficiency pathways group than it would be without a concerted emphasis on the demand-side transformation.

The Transportation Sector in the GEA-Efficiency Pathways

The slow growth of energy demand in the transportation sector in the GEA-Efficiency pathways results in part from efficiency improvements in the vehicle fleet, but also from structural shifts toward public transport (including rail and bus) and limits to car ownership, with implications for behavior and lifestyle (see also Chapter 9). In the GEA-Efficiency pathways group, about half of the overall improvement in energy intensity by 2050 comes about through technical efficiency improvements across all modes of passenger transportation. The compound global effect of

these efficiency gains reduces fuel consumption from about 1.7 MJ/km in 2005 to 1.3 MJ/km by 2050. Gains are largest for vehicles, with some significant differences across world regions (the range is from 1.9 to 0.9 MJ/km). The other half of the overall intensity improvement is achieved by reducing demand for mobility as an energy service (e.g., by substituting travel with teleconferencing) and shifting demand for mobility to public transportation (e.g., trains and buses). Large differences in modal split across countries already exist world-wide. Although demand is thus significantly lower in relative terms in the GEA-Efficiency pathways than in the GEA-Supply pathways, in absolute terms mobility continues to increase.

In the industrialized world, the proportion of total mobility (expressed in passenger-kilometers) provided by cars declines from about 60% in

2005 to 40% in the GEA-Efficiency pathways (Table 17.6). Trends are different in the developing world, where a large fraction of the population already relies on public transportation. Increasing affluence will make cars more affordable and thus increase reliance on individual mobility. As a result, car ownership in the developing world is expected to increase by almost a factor of 5 even in the GEA-Efficiency pathways (from 2 to 11 cars per 100 people by 2050). Although this is a considerable increase, the expected growth in the absence of any policies to support public transportation and limit car ownership would be some 30% higher still. Also, despite this large increase, transportation by bus and train in 2050 in the GEA-Efficiency pathways covers a much larger fraction of total passenger transport demand in the developing world than in the industrialized world (35% versus 20%; see Table 17.6).

In addition to individual mobility, freight transport continues to be a strong driver of energy demand in the transportation sector. An important feature of the GEA-Efficiency pathways group is therefore the switch toward higher shares of railway transportation (Table 17.6) combined with improvement in the overall efficiency of freight transportation by about a factor of 2 by 2050, from 1.3 MJ/t-km (tonne-kilometers) on average in 2005 down to 0.7 MJ/t-km in 2050. In the industrialized world this leads to relatively constant per capita energy use for freight transportation despite the near doubling of transport volume from 8,200 t-km per capita to about 16,000 t-km per capita by 2050. Although efficiency gains are of a similar order of magnitude in the developing world, increases in freight demand more than offset those gains, leading to an increase in per capita energy use for freight by about 20% to 2050. In absolute terms, however, by 2050 energy demand for this purpose in developing countries remains considerably below that of today's industrialized countries.

The Industry Sector in the GEA-Efficiency Pathways

In the GEA-Efficiency pathways, energy efficiency in the industrial sector improves by about 1.5%/year, resulting in an overall demand of about 200 EJ in 2050. This is around 20% below what it would be in the absence of a concerted approach to demand-side transformation, and it equates to a 50% reduction in the overall energy intensity of industrial production (see Table 17.7 for related data in per capita terms).

The demand-side emphasis of the GEA-Efficiency pathways features a number of different measures in the industrial sector to improve energy efficiency, promote structural change, and optimize industrial systems design to reduce energy demand. These measures can be broadly split into the following categories:

- widespread adoption of best available technology for new investments;

- retrofit of existing plants to improve energy efficiency;

- optimization of energy and material flows through systems design, quality improvements, lifecycle product design, and enhanced recycling; and

- further electrification and a switch to renewable energy.

Table 17.7 | Energy service indicators for the industry sector in Industrialized and Developing Regions, 2005 Actual and 2050 under GEA-Efficiency pathways.

	2005, actual		2050, GEA-Efficiency	
	Industrialized[1]	Developing	Industrialized	Developing
GDP per capita (2005 US$ at market exchange rates)	3486–40,054	671–4905	24,446–52,535	5661–19,829
Final energy intensity (MJ/ dollar of GDP)	1.2–10.7	3.0–9.8	0.7–1.3	0.9–2.5
Final energy (GJ per capita)	26–65	3–17	33–46	15–26
Process heat (all thermal)	15–28	2–11	12–17	8–13
Feedstock	6–23	0.3–6	6–14	1–7
Other (nonthermal, e.g., electric)	4–15	1–4	12–16	5–9

1 Industrialized and developing regions are defined as in Table 17.4.

The adoption of best available technology for industrial processes can yield an efficiency improvement of around 15% (IEA, 2007; Saygin et al., 2010). More systemic approaches to optimizing the use of combined heat and power, pumps, fans, compressed air and steam systems, and so on can yield another 15% (IEA, 2007; Price and McKane, 2009). Further reductions in energy intensity in the industrial sector can be achieved through the optimization of material flows and the widespread adoption of new high-efficiency technologies currently at niche scales (WBCSD/IEA, 2009). Moreover, a switch to 25% renewable energy throughout the manufacturing industry yields a 10% "efficiency" gain through electrification and reduced used of fossil resources, although this is balanced by a similar loss from widespread adoption of CCS (see Chapter 8). The efficiency potentials of the five most energy intensive industrial subsectors (iron and steel making, chemicals and petrochemicals, cement making, pulp and paper, and aluminum), which account for about two-thirds of industrial energy use, are discussed in more detail in the online electronic appendix to this chapter, as well as in Chapter 8.

Energy Efficiency by Sector in the GEA-Supply Pathways

The sectoral analysis above provides some specific detail as to how the fundamental demand-side transformation represented by the GEA-Efficiency pathways can be achieved. Central to this effort is the rapid and pervasive introduction of energy efficiency measures throughout the world. However, technical measures alone will not be sufficient. They need to be complemented by measures to both shift and limit the underlying demand for energy services, build institutional capacity (see Chapter 25), remove market and nonmarket barriers to increased energy efficiency (see Chapters 22 and 26), and mobilize the substantial investment needed (see Chapter 6). These policy and investment needs are discussed further in Section 17.3.5

but are emphasized here as an integral and essential feature of the GEA-Efficiency group of pathways.

The GEA-Supply pathways, in contrast, represent the extent of potential demand-side transformation without this concerted policy and investment emphasis, and without many of the specific efficiency measures described in the sectoral analysis above. However, the GEA-Supply pathways group is not simply a business-as-usual continuation of historical trends. All the GEA-Supply pathways fulfill the sustainability objectives of the overarching GEA scenario set out in Section 17.2.3. As noted, these pathways are implemented through the achievement of highly ambitious targets relating to energy access, environmental impacts, and energy security. Reaching these targets requires a raft of policy and other initiatives, discussed in detail in Sections 17.4–17.7 of this chapter, that lead to a transformation of the global energy system. Whereas the GEA-Efficiency pathways emphasize the transformative potential on the demand side, the GEA-Supply pathways concentrate on supply-side measures. However, the latter also impact energy demand, albeit indirectly. To take one simple example, a carbon tax implemented to reduce the share of fossil fuels in electricity generation might indirectly raise the cost of final energy and so reduce demand. The more general point is that the level of energy demand and energy intensity improvements in the GEA-Supply pathways shown in Figure 17.7 falls well below the upper bound of demand projections found in the scenario literature to represent business as usual. Compared with, for example, the extensive scenario database of the IPCC Fourth Assessment Report (IPCC AR4) (Fisher et al., 2007), energy intensity improvement rates in the GEA-Supply pathways correspond roughly to an intermediate demand projection close to the median of the scenario distribution by 2050. Compared with the upper 90th percentile of the full scenario set reviewed in the IPCC AR4, the GEA-Supply pathways achieve more than double the intensity improvements by 2050 (1.5%/year compared with 0.6%/year in the scenarios reviewed by the IPCC).

Table 17.8 summarizes indicators of per capita energy use for the GEA-Efficiency and GEA-Supply groups of pathways in all end-use sectors: residential and commercial, transportation, and industrial. Projections for 2050 are compared with efficiencies of energy use as of 2005. The table combines information from the detailed bottom-up technology assessments of each sector in Chapters 8–10 with information from the global GEA pathways discussed in this chapter.

In the residential and commercial sector, floor area as an underlying determinant of energy services demand is the same in both pathway groups. However, the penetration of advanced buildings combining major efficiency improvements with decentralized energy technologies is minimal in the GEA-Supply pathways. Resulting heat demand per unit of floor space is consequently a factor of 2 to 3 higher, with per capita consumption around double that in the GEA-Efficiency pathways in both industrialized and developing countries.

In the transportation sector, total passenger demand for mobility is around 20–30% higher in the GEA-Supply pathways, because the measures described above to limit and shift services demand in the context of the GEA-Efficiency pathways are not implemented. Car ownership is also around a factor of 7 higher than current levels, compared with the factor of 5 increase in the GEA-Efficiency pathways. Although levels of car ownership increase proportionally more in developing countries, they remain at far higher absolute levels in the industrialized world. Meanwhile, although overall demand for freight mobility is similar, the GEA-Supply pathways have a higher proportion of freight moving by road than rail, the opposite of the case in the GEA-Efficiency pathways. Together with lower efficiency gains in the vehicle fleet, this results in 15–20% higher per capita fuel use. Demand for aviation is assumed to be the same in both the GEA-Supply and the GEA-Efficiency pathways.

In the industrial sector, the absence of major new efficiency policies in the GEA-Supply pathways results in energy demand more than doubling, to 260 EJ, in 2050 compared with 210 EJ in the GEA-Efficiency pathways. Energy intensity does improve, by about 1%/year, but this is lower than the 1.5%/year improvements achievable from the concerted demand-side transformation that occurs in the GEA-Efficiency pathways.

These large differences in efficiency improvements between the GEA-Supply and the GEA-Efficiency groups of pathways have major implications for the required transformation of the supply side of the energy system. The higher level of demand in the GEA-Supply pathways means fewer degrees of freedom in terms of supply options. This interdependency between the demand- and the supply-side features of the pathways is discussed in detail in Section 17.3.4.

17.3.2.3   The Structure of Final Energy Demand

Despite the large differences in efficiency improvements between the GEA-Efficiency and the GEA-Supply groups of pathways described in the previous section, certain structural characteristics of final energy use are remarkably consistent across all GEA pathways, including the group of GEA-Mix pathways. These relate to the provision of energy services using higher-quality forms of energy. All GEA pathways depict a demand-side transformation toward ever more flexible, more convenient, and cleaner forms of energy at the point of final use (see also Nakicenovic et al., 1998). Figure 17.8 presents the same findings graphically by distinguishing the share of final energy provided by solid, liquid, and grid-based or on-site-generated forms of energy. The low variation across the three groups of GEA pathways is shown by the limited extent of cross-hatching or overlap.

A pervasive characteristic of all the GEA pathways is the continuing shift from energy used in its original, often solid form, exemplified by

Table 17.8 | Energy services indicators inIndustrialized and Developing Regions, 2005 Actual and 2050 under GEA-Efficiency and GEA-Supply pathways.

	2005		2050			
			Industrialized		Developing	
	Industrialized[1]	Developing	GEA-Efficiency	GEA-Supply	GEA-Efficiency	GEA-Supply
GDP per capita (2005 US$ at market exchange rates)	3487–40,050	671–4905	24,446–52,535	24,446–52,535	6029–19,829	6029–19,829
Total final energy (GJ per capita)	73–219	7–46	62–98	104–156	28–50	32–71
Residential and commercial						
Floor area (m² per capita)[2]	26–55	9–32	48–58	48–58	19–52	19–52
Share of buildings with advanced technology (%)						
Single-family	<1	<1	74–86	0–0	75–83	0–0
Multifamily	<1	<1	74–79	0–1	82–91	0–0
Commercial and public	<1	<1	81–92	0–3	85–96	0–0
Heating demand (MJ/m²)						
Single-family	443–875	112–241	101–230	324–619	50–79	90–187
Multifamily	443–781	112–277	104–198	266–518	47–68	90–209
Commercial and public	475–914	173–371	101–166	288–475	54–97	205–230
Electricity demand (GJ per capita)[3]	11–45	1–6	22–33	35–46	8–15	10–20
Transportation						
Passenger-kilometers per capita[4]	14,293	2499	15,925	20,302	3892	4632
Car	8778	404	6539	11,045	1009	1775
Bus and train	2855	1461	3334	3205	1368	1342
axiatim	2274	198	5579	5579	795	795
other	386	437	473	473	720	720
No. of light-duty vehicles per capita	0.46	0.03	0.52	0.64	0.11	0.14
Fuel use for mobility (GJ per capita)	30.8	2.4	24.0	33.3	4.6	5.9
Freight-kilometers per capita	8219	1059	15,969	16,209	2774	2856
Truck	4544	606	6925	9032	1370	1786
Rail	3675	453	9044	7177	1404	1070
Fuel use for freight (GJ per capita)	13.0	2.4	12.4	14.3	2.8	3.3
Industry						
Final energy (GJ per capita)	26–65	3–17	33–46	42–63	15–26	17–33
Process heat (all thermal)	15–28	2–11	12–17	16–24	8–13	9–16
Feedstock	6–23	0.3–6	6–14	9–20	1–7	1–11
Other (nonthermal, e.g., electric)	4–15	1–4	12–16	17–19	5–9	5–12

1 Industrialized and developing regions are defined as in Table 17.4.

2 Includes public and commercial buildings. Based on the bottom-up analysis in Chapter 10.

3 Includes electric cooling and heating as well as lighting and appliances.

4 Estimates from Chapter 9. Because of differences between the regional definitions used in that chapter and those used for the GEA scenarios, transport indicators are given as regional averages of the whole developing and industrialized world only.

the traditional direct uses of coal and biomass, to more sophisticated systems of energy conversion and delivery. All the GEA pathways include the phase-out of traditional biomass in the residential and commercial sector by 2030, due to dedicated energy access policies in the developing world (see Section 17.4). In addition, stringent climate policies (see Section 17.5.1) lead to the phase-out of the direct use of coal in both the industrial and the residential sectors by 2050. The share of solid fuel shown in Figure 17.8 after 2050 is predominantly biomass in the industrial sector, as a substitute for coal in industrial

processes where carbonaceous fuels are required (e.g., iron and ore reduction processes).[6]

A second major transformation is the increasing degree to which energy is delivered by dedicated transport infrastructures, such as pipelines and networks. This enables similar end-use patterns across regions, as end

---

6   In addition, the possibility of direct reduction with hydrogen is considered by the pathways as a long-term option for the substitution of coal in these processes.

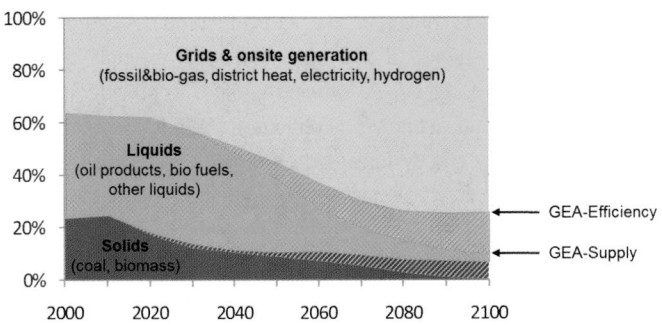

Figure 17.8 | World final energy shares of solid fuels, liquid fuels, and grids and on-site generation. Overlapping hatched areas indicate variations across the GEA-Efficiency and the GEA-Supply pathways. The GEA-Mix pathways lie in between the two others.

uses can be linked to fundamentally different and potentially distant primary energy supplies (see Section 17.3.3 for more details). An additional and related transformation, represented particularly in the GEA-Efficiency pathways, is the increasing development of on-site generation of both heat and electricity by renewable energy technologies.

These transformations in the use of final energy also involve changes in the economic structures that underpin the pathways. Continued industrialization of the developing world and increasing demand for industrial goods as incomes rise, result in comparatively larger shares of industrial energy demand in the future. This trend is most pronounced in the GEA-Efficiency group of pathways because aggressive efficiency programs in the residential and transportation sectors limit their shares of final energy demand, as shown in Figure 17.9.

In sum, all the GEA pathways share some key features in terms of the structure of final energy demand: a shift away from traditional solid forms of energy; an increase in modern, cleaner, grid-delivered and on-site-generated forms of energy; and a rising share of industrial sector energy use as incomes rise. There is, however, one major point of difference among the GEA pathways. This relates to the nature of transformation in the transportation sector. The environmental and energy security objectives of the overarching GEA scenario necessitate a lower reliance on oil in all GEA pathways. By 2050, oil use in the transportation sector is reduced by 35–50% from 2005 levels. However, the substitution away from oil branches into alternative transportation systems, as described in Section 17.2.1. Broadly speaking, these alternatives can be described as "conventional" and "advanced" transportation.

A future conventional transportation system would rely predominantly on liquid fuels (including some oil), biofuels, liquefied natural gas, and potentially the direct use of biogas and natural gas. This represents the least discontinuity from current trends in terms of both end-use technologies and fuel supply and distribution infrastructure. In contrast, an advanced transportation system involves a more fundamental transformation, requiring largely new infrastructure systems in the case of hydrogen fuel cell vehicles, or new uses for existing infrastructure in the case of plug-in hybrids or fully electric vehicles.

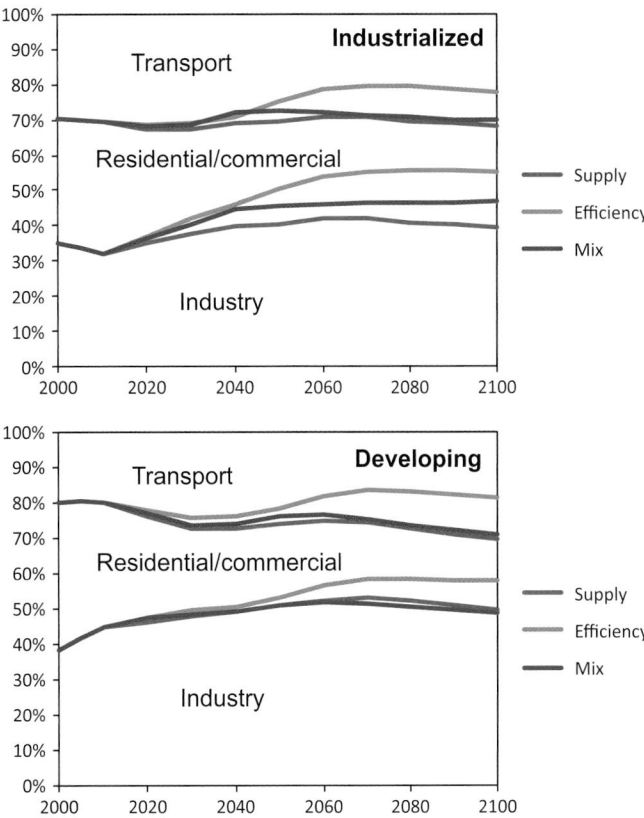

Figure 17.9 | Shares of final energy by sector in the three pathway groups in industrialized and developing countries.

Both these alternatives are feasible within all three of the GEA pathway groups. Their outcome is essentially one of choice and direction rather than necessity. This is why they are explicitly included as a branching point in the scenario taxonomy. The importance of this branching point stems from the magnitude of the implications on both the demand and the supply sides. On the demand side, for example, vehicle technologies would follow very different innovation and development paths, either reducing the costs and improving the reliability of fuel cells in the hydrogen-based Advanced Transportation system, or improving the flexibility of engines to use both biofuels and fossil-derived liquids in the Conventional Transportation system. However, the impacts of the transportation branching point are felt particularly strongly on the supply side, as they potentially reduce the flexibility of supply-side portfolios given the GEA sustainability objectives. Similarly, any changes on the supply side can have major implications for the choice of demand-side technologies. These interdependencies are discussed further in the next section, which turns to consider the supply side of the energy system transformations represented in the GEA pathways.

17.3.3   Energy Supply

17.3.3.1   Introduction

The level of energy demand determines the flexibility of supply-side portfolios. This is particularly the case given the ambitiousness of the

GEA sustainability objectives and the energy transition required to reach the associated targets. Across all the GEA pathways, energy access objectives constrain the use of traditional fuels in developing countries; energy security objectives limit the amount of energy trade and foster the increasing diversity of energy supply; climate change objectives constrain the use of carbon-intensive energy forms in electricity generation; and so on. Within these already tight constraints, low energy demand allows a greater number of viable options for energy supply, whereas high energy demand reduces the choices available and makes it more difficult to limit or omit specific supply options. Similarly, the future fuel needs of the transportation sector have a further impact on the extent to which supply-side portfolios can be varied in response to political, resource, land, or other requirements.

Having established and analyzed the different levels of demand in the GEA-Efficiency, GEA-Supply, and GEA-Mix groups of pathways, this section explores variability in the corresponding supply-side transformations, including transmission and distribution infrastructure.[7] The aim is twofold: to enrich the storylines represented by the GEA pathways in fulfilling the sustainability objectives, and to identify both the necessities and the choices available on the supply side of the energy system if these sustainability objectives are to be met.

This section is organized using the branching point approach set out in Section 17.2.2. First, the principal options on the supply side are set out, covering both energy forms and relevant technologies. Second, the viable portfolios of supply-side options across the GEA-Efficiency, GEA-Supply, and GEA-Mix groups of pathways are assessed, given the different levels of energy demand in each group – the first branching point. Third, the effect of alternative transportation system transformations, either Conventional or Advanced, is explored in terms of supply-side portfolio flexibility – the second branching point. Fourth, the potential for further limiting or omitting specific options from the supply-side portfolio is assessed through an extensive sensitivity analysis – the third branching point. Through this process, the three groups of GEA pathways become first 6 and then 60, although ultimately they are reduced to the 41 that are feasible. The feasibility analysis indicates how important certain supply-side options are for a the energy transition. However, the results should not be mistaken for predictions. Rather, they can be interpreted as an assessment of the necessary technological changes, exploring the "option values" of different technology clusters that might, for example, guide future investment decisions.

17.3.3.2   Supply-Side Options and Portfolios

There is a large portfolio of options on the supply side to provide the energy needed to meet the demand for energy services. These options comprise different forms of energy and their attendant conversion

technologies: crude oil converted into petroleum products by refineries to provide transportation fuel and thus mobility, for example, or wind energy converted into electricity by wind turbines to provide lighting. Throughout this section, supply-side options are distinguished in terms of both the primary energy form (e.g., bioenergy, coal, solar energy) and the conversion or processing technology (e.g., biomass-to-liquids, biomass power generation, coal power generation with CCS, solar photovoltaic). Each option also has implications for the transmission and distribution infrastructure: rigs, pipelines, and filling stations in the case of the energy conversion chain from crude oil to mobility; electricity grids and transformer stations in the case of conversion from wind energy to lighting. These, too, will be considered here.

In scenario studies (as well as historically), there are also a wide range of factors that shape and constrain the shares of final energy provided by different supply-side options. The clearest determining factors relate to cost, efficiency, and other performance attributes. Availability, based on the underlying resource potential, is another factor, although a detailed assessment of the bioenergy and other renewable, fossil, and uranium resources in the context of the GEA pathways show that none of these supply-side options face an absolute resource constraint at the global level (see Box 17.3 and the electronic appendix to this chapter on resource potentials). At the regional level, however, some resource categories could become scarce.

In the context of sustainability assessments like the GEA, other factors also come into play, including environmental impacts (e.g., air pollution, GHG emissions), social impacts (e.g., electrification and clean cooking), and geopolitical considerations (e.g., energy security). The supply-side options used in all the GEA pathways must allow the GEA sustainability objectives to be fulfilled within the timelines and to the extent set out by their associated targets (see Section 17.2.3). This means that certain supply-side options are preferred over other options in the energy transition toward its objectives. For instance, technologies like bioenergy and other renewable energy sources, nuclear energy, and CCS have the potential to help meet the climate target. CCS can also be used in combination with bioenergy (BioCCS) to produce net negative carbon dioxide (CO_2) emissions (see Chapter 11, Section 11.3 for more details about the technology). This is another potentially important option in the context of climate stabilization objectives.

An alternative option to reduce CO_2 emissions is carbon sink enhancement through afforestation. As both BioCCS and carbon sinks can significantly affect the magnitude and timing of emissions reductions on the energy supply side, both are included in this analysis.

Cost, performance, resource availability, and sustainability criteria are not the only factors influencing the projected success of these supply-side options. Some options require advanced technological knowledge, which is not universally available (and which has contributed to historical differences in primary energy supply patterns at the country or regional level). Other options face barriers to a rapid scaling up (Wilson,

---

7   The supply-side flexibility analysis relies primarily on the MESSAGE modeling framework, but the findings have been checked for consistency with the pathways generated with the IMAGE model.

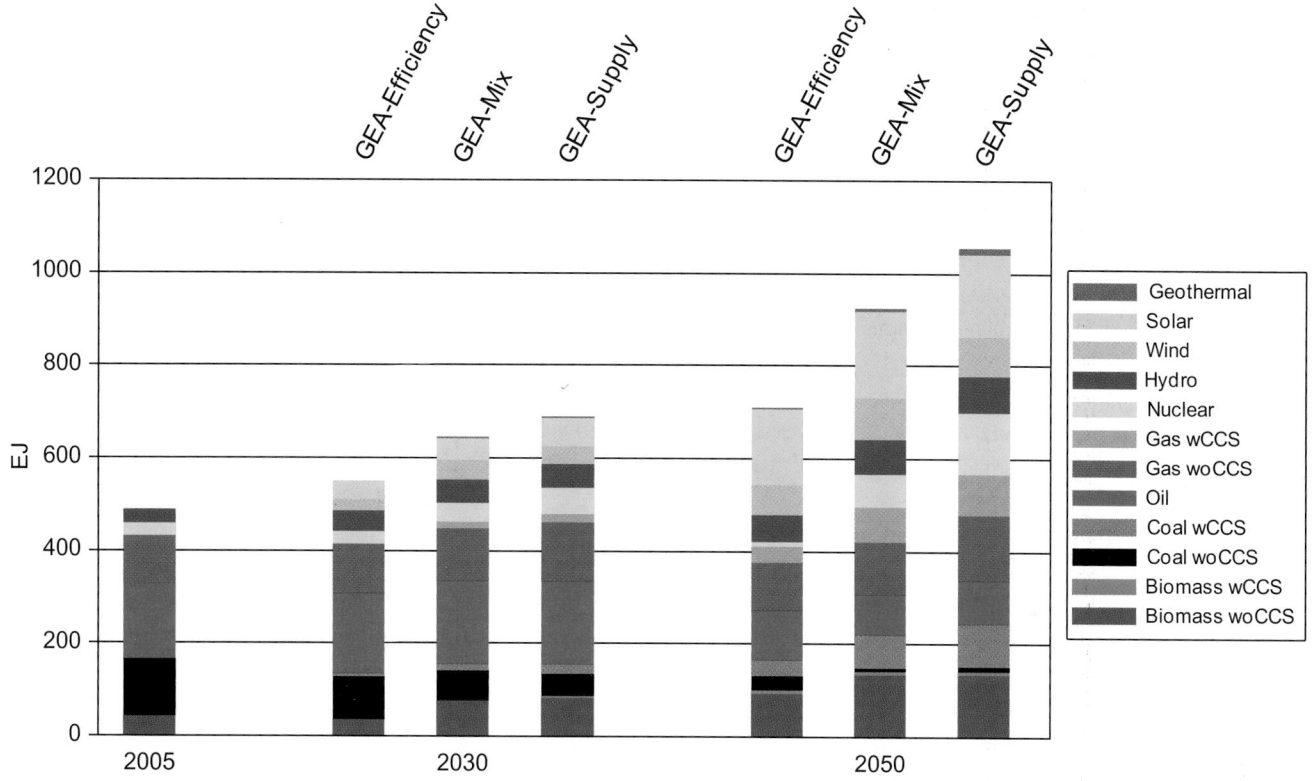

Figure 17.10 | Composition of global primary energy supply in 2005, 2030, and 2050 across pathway groups under an unrestricted supply portfolio and Conventional Transportation setup.

2009). Integrating high proportions (e.g., 20% or more) of intermittent energy sources such as wind or solar in electricity grids is an example. Still other options face issues of public acceptance. Nuclear energy in some countries is an obvious example, but some forms of renewable energy, such as large-scale hydropower, bioenergy, on-shore wind, and CCS, are others. Some options, such as nuclear, entail also other societal risks due to accidents or proliferation of fissile material for weapons use. The overall conclusion of the GEA assessment on nuclear is thus that until the proliferation potential is better controlled and safer reactor designs are available, countries (especially those considering building their first nuclear power plants) should consider other climate-friendly power supply options first (for a further discussion, see Chapter 14). Finally, the requirements of some supply-side options in terms of new physical infrastructure and distribution systems are highly capital intensive but face initially low overall demand, and thus are often unattractive to both private investors and resource-limited public investors.

In light of all these potential issues, the approach taken here begins by elaborating the broadest possible decision space, or range of possibilities, in terms of supply-side portfolios in each of the three GEA pathway groups (GEA-Efficiency, GEA-Mix, and GEA-Supply). First, the full range of supply-side options is considered, subject to cost, performance, and system integration constraints but always respecting the overarching need to comply with the GEA targets. As noted earlier, the level of demand has a significant impact on supply-side flexibility: the greater efficiency improvements

and reductions in energy services demand of the GEA-Efficiency group of pathways leave more options open on the supply side. Next, this maximal decision space is reduced in stepwise fashion. The impacts of major changes in the transportation system on this unrestricted supply portfolio are explored. Finally, the impacts of specific restrictions or omissions of particular supply-side options are considered, to reflect the sensitivities or concerns surrounding their widespread deployment. These restricted supply portfolios, as well as the transportation analysis, provide a broad sensitivity analysis around the unrestricted supply portfolio, illustrating which options are "musts" and which others are choices.

17.3.3.3　Supply-Side Options under Different Levels of Demand

Figure 17.10 summarizes the result of the unrestricted supply portfolio analysis for the three levels of energy demand represented in the GEA-Efficiency, GEA-Mix, and GEA-Supply pathway groups. It compares the primary energy supply mix in 2030 and 2050 under each of the three groups with that in 2005.[8] Each future primary energy supply mix depicted can be interpreted as the least-cost portfolio subject to

---

8　A well-recognized problem with reporting primary energy supply is how to include noncombustible energy forms (e.g., nonbiomass renewables and nuclear energy). Here the substitution method is used to back-calculate primary energy by assigning a 35% efficiency for electricity generation from noncombustible sources and an 85% efficiency for heat generation (see Chapter 1 for details).

the cost and performance characteristics of the different supply-side options and the need to fulfill the GEA objectives with respect to access, environment, and security.

The most striking difference across pathway groups is in the total demand provided by the energy system. In terms of supply-side options, the figure also shows the breadth of the supply portfolio needed to meet the GEA sustainability objectives: most if not all options contribute across all three pathway groups. Nuclear energy makes a greater proportional contribution in the GEA-Supply pathways group than in the other groups, which has less flexibility in terms of portfolio restrictions.[9] Conversely, as will be explored further below, the GEA-Efficiency pathways group can tolerate the restriction or even omission of various individual supply-side options.

An equally important difference across pathway groups is the varying degree of urgency for change on the supply side in the medium term. With the ambitious effort on the demand side in the GEA-Efficiency pathway, the change from current supply-side structures can be less rapid. In 2030, with the exception of wind and solar (which grow considerably in absolute terms), the primary energy supply mix in the GEA-Efficiency pathway is only modestly different from that of today. In contrast, the GEA-Mix and, in particular, the GEA-Supply pathways require more radical changes in energy supply. This includes a more rapid scaling up of all renewable supply options, and CCS, which by 2030 needs to remove up to 10% of CO_2 emissions from fossil fuel combustion in the GEA-Supply pathways, increasing to about 50% by 2050. The same is true for nuclear energy, which in the GEA-Efficiency pathway (with an unrestricted supply portfolio) continues to contribute about the same amount of energy as today or less through 2050, whereas in the GEA-Mix and GEA-Supply pathways a two- to fivefold increase up to 2050 is observed. For pathways with a nuclear phase-out see sections 17.3.3.5 and 17.3.4.

17.3.3.4   Supply-Side Options under Different Transportation Systems

As noted in Section 17.3.2, the structure of the transportation sector decisively influences the feasibility of supply-side portfolios. Therefore, the analysis below distinguishes between two sets of assumptions about the transportation sector transition, labeled Advanced Transportation and Conventional Transportation. The Advanced Transportation setup is characterized by a transition to electricity or hydrogen, or both, as main transportation fuels in the medium to long term. By 2050 these two fuels would have to deliver between roughly 20% and more than 60% of the transportation sector's final energy, depending strongly on

overall transportation demand. This implies a massive buildup of new infrastructure over the coming decades. Whereas such a transition could proceed more gradually in the case of electrification, the transition to hydrogen is more challenging, because bulky investments in a new distribution infrastructure would need to be made. On the other hand, hydrogen would be more compatible with the existing refueling infrastructure and business model, which might have to change significantly in a largely electrified transport sector (Andersen et al., 2009). In contrast, the Conventional Transportation story would stay mostly within current modes of operation, largely relying on liquid fuels and, in some regions, on gas. Still, a growing share of electricity would also be needed in this conventional world, reflecting a combination of a modal shift toward public transportation and some electrification of at least short-distance individual transport.

Two different interpretations of the Advanced Transportation setup are realized in the GEA scenario analysis: an electric route and a hydrogen route. These have in common that numerous additional energy sources (e.g., nonbiomass renewable energy, nuclear energy) become available to the transportation sector on a large scale.[10] The electric route leads to a substitution process, dominated by electric vehicles and plug-in hybrids in combination with biofuels. The alternative route, hydrogen, explores a transition toward a long-term transportation sector that is dominated by hydrogen fuel cell vehicles after 2050. In contrast, the Conventional Transportation setup tends to follow a regionally more diversified path, depicting the coevolution of a wide portfolio of fuels and technologies with similar shares, including hybrid vehicles, flexible cars using biofuels in conjunction with fossil liquids from natural gas (in combination with CCS to reduce carbon emissions), and direct use of biogas and natural gas. These alternative transportation sector configurations also have important implications for the required technological innovation and improvements in vehicle engines. R&D and deployment incentives are needed in the Advanced Transportation setup to reduce costs and improve the reliability of either fuel cells or the next generation of batteries. The transition under the Conventional Transportation setup relies more heavily on advanced and more flexible designs of internal combustion engines.

As illustrated by Figure 17.11, the differences between the alternative transportation sector assumptions tend to play out more severely in the GEA-Mix and particularly the GEA-Supply groups of pathways, simply because demand is significantly higher by 2030 and still higher by 2050. Total primary energy supply is lower in an Advanced Transportation world than in a Conventional Transportation world, because the well-to-wheel efficiency of the electric and hydrogen routes is generally higher than that of the liquid route (van Vliet et al., 2010; 2011). A more subtle difference concerns the higher uptake of the available bioenergy potential under the Conventional Transportation option by 2030 across the three groups of

---

9   The main reasons for the high nuclear contribution in the GEA-Supply pathways is the high demand of energy, which reduces the flexibility of supply (see Section 17.3.3.5) and thus results in comparatively higher prices for energy. The higher energy prices, in turn, increase the demand for more costly energy options, such as nuclear. Different phase-out pathways for nuclear, however, show that the transformation toward the sustainability objectives are in principle technically possible also in the GEA-Supply pathways (Section 17.3.3.5).

10   The Advanced Transportation sector in the GEA-Supply pathways relies largely on hydrogen, whereas in the GEA-Efficiency pathways the electric route is chosen. The GEA-Mix pathways rely to a greater extent on electricity in their MESSAGE interpretation and more on hydrogen in the IMAGE interpretation.

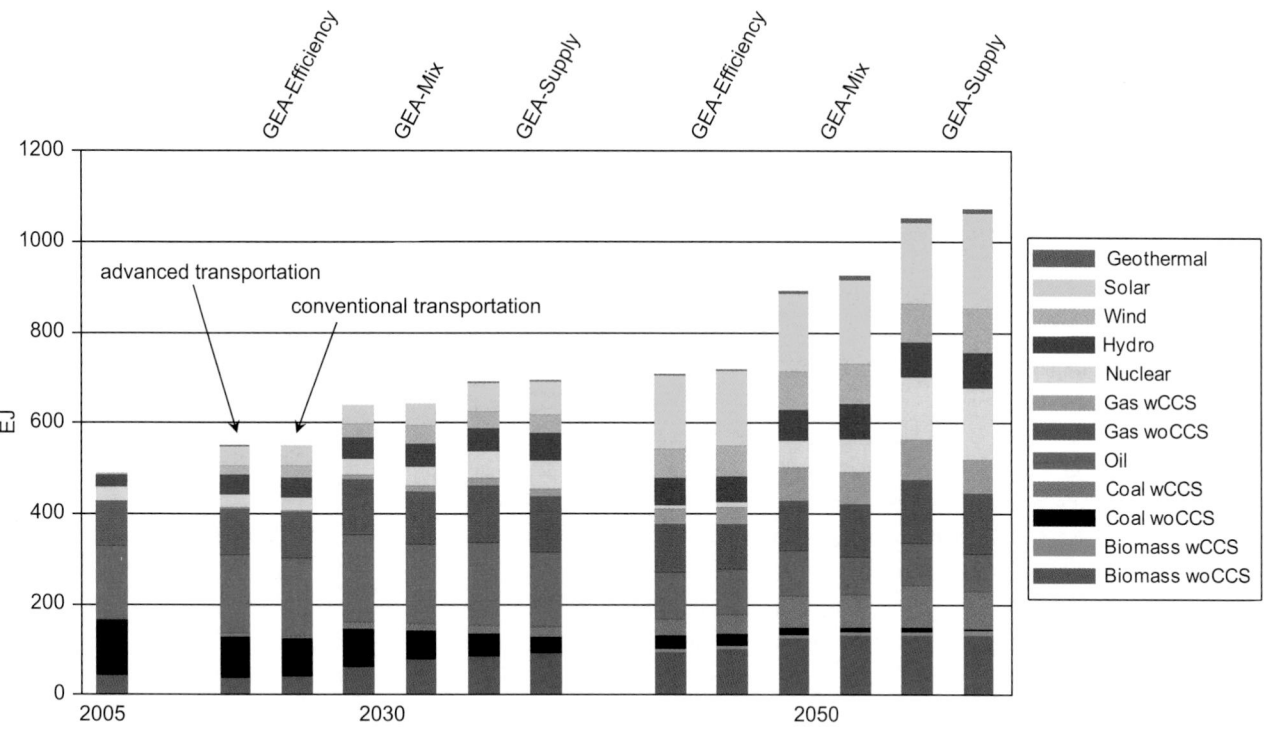

Figure 17.11 | Composition of global primary energy supply in 2005, 2030, and 2050 across pathway groups under an unrestricted supply portfolio and alternative transportation sector assumptions.

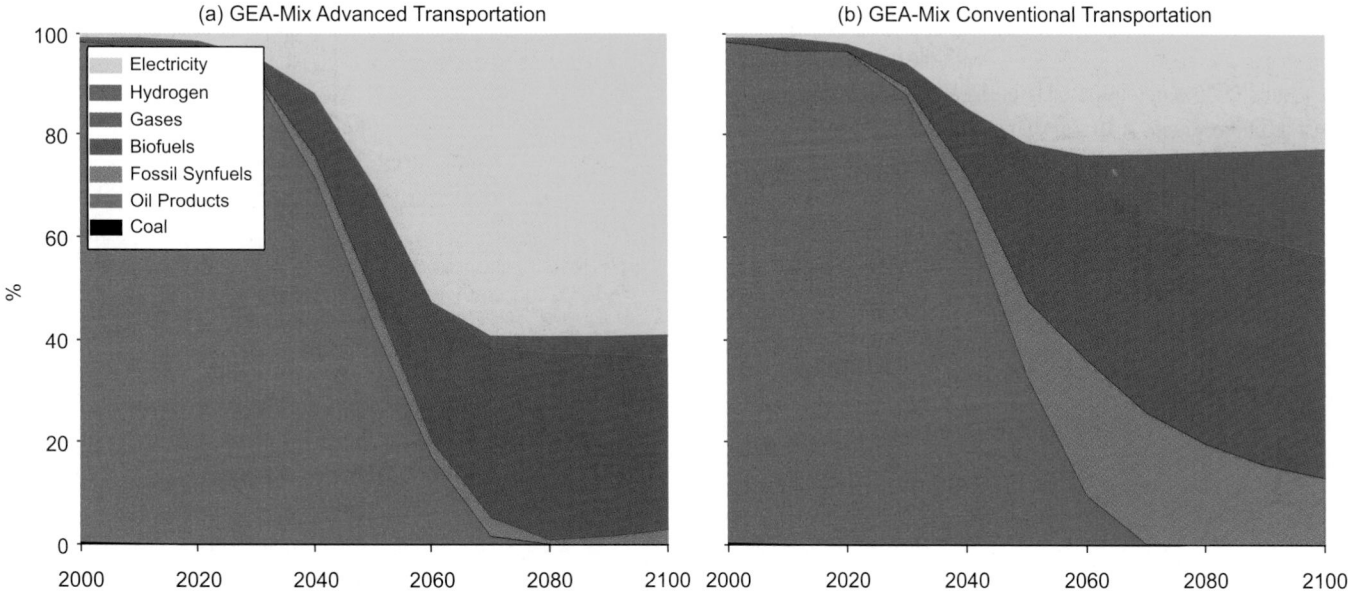

Figure 17.12 | Development of global final energy fuel shares in the transportation sector under Advanced and Conventional assumptions for the GEA-Mix pathways group.

pathways. The different final energy patterns of the two transportation sector configurations are shown in Figure 17.12 for the GEA-Mix pathway.

Given the ambitious goals of the energy transition, this difference in the transportation sector has profound implications for the supply-side

choices, particularly for the GEA-Mix and GEA-Supply groups of pathways. The Advanced Transportation setup generally opens additional supply routes for transportation fuels such as electricity and hydrogen from non-biomass renewables or nuclear energy, whereas Conventional Transportation offers fewer routes, to a large extent relying on fossil fuels and bioenergy and leading to reduced flexibility on the supply side. The limited potential of sustainable bioenergy is the main determinant of the reduced number of feasible transition pathways under the Conventional Transportation assumption (see Section 17.3.4.3 for details), because bioenergy is one of the few remaining ways to reduce GHG emissions in the transportation sector and in selected other parts of the energy system (e.g., bioenergy feedstocks for nonenergy use; see Dornburg and Faaij, 2005).

17.3.3.5   Supply-Side Options under Different Portfolio Restrictions

Portfolio Restrictions as a Sensitivity Analysis

The analysis presented here relies on a set of "restricted portfolio" pathways in which selected supply-side options are either limited or excluded completely, in order to focus on overall questions of feasibility and on economic and resource implications. These pathways should therefore be interpreted as sensitivity analyses around the central case of the full or "unrestricted" portfolio for each of the three GEA pathway groups.[11] An important assumption underlying this restricted portfolio analysis is that the level of energy demand in each group of pathways is fixed.[12]

In total, nine different restricted supply portfolios are explored for each of the six possible combinations of GEA-Efficiency, GEA-Mix, and GEA-Supply pathway groups and two transportation system transformations (Conventional and Advanced). Together with the unrestricted portfolios, this results in 60 different possible pathways (3 levels of demand × 2 transportation systems × 10 supply portfolios).

Issues, concerns, and potential constraints facing different supply-side options were the basis for the choice of restricted portfolios analyzed. Six supply-side options were either limited or excluded, either in isolation or in combination with other options.[13] These options and the corresponding restricted portfolio pathways are shown in Table 17.9. Also presented are summaries of the rationales for including these particular restricted portfolios.

---

11   Similar analyses, mostly in the context of climate change mitigation, have been done in the ADAM (Edenhofer et al., 2010b) and RECIPE (Edenhofer et al., 2010b) modeling comparison studies and in several individual publications (Krey and Riahi, 2009).

12   Generally, excluding options from the supply portfolio leads to increased prices of energy services, because more expensive supply options have to be utilized. In this situation, standard economic theory suggests a price-induced demand response, which in this analysis is not considered, because by design the supply-side flexibility is investigated given a fixed level of demand for energy services.

13   An additional potential option would have been to limit the availability of unconventional fossil resources. This has not been implemented, however, since the stringent climate target allows unconventional resources no significant role in the pathways (even if no restriction is assumed).

Table 17.9 | Overview of restricted supply portfolios.

Supply-side option	Main rationales for restriction	Restricted portfolio pathways	Description
CO$_2$ capture and storage (CCS)	Storage availability Social acceptability Infrastructure requirements Environmental risks	No CCS[1]	CCS excluded
Bioenergy with CCS (BioCCS)	See entries for CCS and bioenergy	No BioCCS[1]	Bioenergy used only for co-firing in fossil CCS facilities (no dedicated BioCCS facilities)
Carbon sinks (afforestation)	Resource availability Land use impacts Political acceptability	No sinks	No additional afforestation beyond baseline assumption of no net global deforestation from 2070 onward
Bioenergy	Resource availability Land use impacts Food security risks Environmental risks	Limited bioenergy	Bioenergy potential reduced to 50% of central estimate to reflect potential implementation issues for sustainable bioenergy
Nuclear energy	Environmental risks Social acceptability Proliferation risk	No nuclear[1]	No new nuclear power plants built after 2020, leading to full phase-out after 2060 (assuming 40-year plant lifetime)
Renewable energy	Systems integration	Limited renewables	Intermittent renewables (wind, solar) restricted to 20% of final electricity consumption
Combinations[2]		Limited bioenergy + Limited renewables No nuclear[1] + No CCS[1] No BioCCS[1] + No sinks + Limited bioenergy	

1   Option was fully excluded from the portfolio; for other options the restriction was implemented in terms of limited potentials.

2   See individual options for rationales and descriptions.

The results of the portfolio analyses are shown in Figure 17.13 for the 3 × 2 × 10 matrix of pathways. The 19 blank columns, each marked with an X, show those pathways that were not feasible given the portfolio restrictions.

A headline conclusion of the portfolio analysis is that the low level of energy demand in the GEA-Efficiency group of pathways makes it possible to reach the sustainability objectives in the absence of both nuclear energy and CCS. For the intermediate and high levels of energy demand under the GEA-Mix and GEA-Supply pathways, respectively, excluding either nuclear or CCS is typically possible, but in the high-demand case this requires transforming the transportation system away from liquid fuels. In the context of climate change mitigation, only a limited number of studies (e.g., Krewitt et al., 2009; Teske et al., 2010; Delucchi and Jacobson, 2011; Føyn et al., 2011; Jacobson

Box 17.3 | Resource Potentials

Integrated Assessment Models like IMAGE and MESSAGE typically do not consider the full technical potential of energy resources but include additional criteria, such as sustainability or economic criteria, which are not fully captured within the models but which lead to a significant reduction of the technical potential. For the GEA pathways, the ranges of these deployment potentials are summarized in Table 17.10. The resource assumptions for all sources are within the ranges of resource uncertainties assessed in Chapter 7 (see also the electronic appendix to this chapter for more details).

Table 17.10 | Fossil fuel resources and renewable energy potentials.

Energy source	GEA pathways: reserves and resources (ZJ)[1]	Chapter 7 Reserves (ZJ)	Chapter 7 Resources (ZJ)
Fossil fuels			
Coal	259 – 376	17.3 –21.0	291 –435
Conventional oil	9.8– 11.1	4.0 –7.6	4.2 –6.2
Unconventional oil	8.9 – 23.0	3.8– 5.6	11.3– 14.9[2]
Conventional gas	11.6 –16.8	5.0 –7.1	7.2 –8.9
Unconventional gas	23.0– 96.4	20.1 – 67.1	40.2 – 122
	Deployment potential in 2050 (EJ/year)	Technical potential (EJ/year)	
Renewables			
Bioenergy	145–170	160–270	
Hydro	18.7–28	50–60	
Wind	170–344	1250–2250	
Solar photovoltaic	1650–1741	62,000–280,000	
CSP	990[3]	810–1400	
Geothermal	23[4]		

Notes: The deployment potentials for noncombustible renewable energy sources in the pathways are specified in terms of the electricity or heat that can be produced by specific technologies (secondary energy perspective). By contrast, technical potentials from Chapter 7 refer to the flow of energy that could become available as inputs for technology conversion (e.g., the technical potential for wind is given as the kinetic energy available for wind power generation, whereas the deployment potential as reported in this chapter gives the electricity that can be generated by wind turbines). In addition to the renewable energy potentials stated in this table, technology diffusion and systems integration constraints may apply in the pathways and prevent the potentials from being fully utilized. Note that elsewhere in Chapter 17 the substitution method is used to report primary energy from non-combustible sources and therefore primary energy numbers can exceed those reported as deployment potential in this table.

1 One zettajoule (ZJ) equals 1000 EJ, or 10^{21} joules.

2 Estimates for unconventional oil that are not separated into reserves and resources reach significantly higher values than reported in this summary from Chapter 7 (see Table 7.8a/b).

3 The potential is from MESSAGE as the IMAGE modeling framework does not include CSP.

4 Geothermal energy is exogenously determined in the IMAGE scenarios; therefore no deployment potential can be specified.

and Delucchi, 2011) have looked at ambitious climate stabilization scenarios that exclude nuclear and CCS completely from the supply-side portfolio, thus relying exclusively on a combination of renewable energy and energy efficiency.

Another important insight from the portfolio analysis is that the low energy demand in the GEA-Efficiency pathways also enables an energy transition with limited contributions from bioenergy, without BioCCS and without relying on carbon sink management. All of these land use-related supply options have potentially adverse impacts and are controversial in the literature (see Section 17.3.4.7 and the electronic appendix to this chapter).

Furthermore, how the transportation sector is configured has profound implications for supply-side flexibility. Under the Advanced Transportation setup, the GEA-Supply group of pathways is still feasible if either BioCCS, carbon sink enhancement, nuclear energy, full bioenergy supply, or the large-scale deployment of other renewable energy is not considered an option. Under the Conventional Transportation setup, only nuclear energy can be excluded to keep the GEA sustainability targets within reach.

The situation is somewhat improved in the GEA-Mix group of pathways, where the Conventional Transportation setup still allows for the same choices as the Advanced Transportation setup under high energy demand. CCS turns out to be a crucial technology under these conditions, because

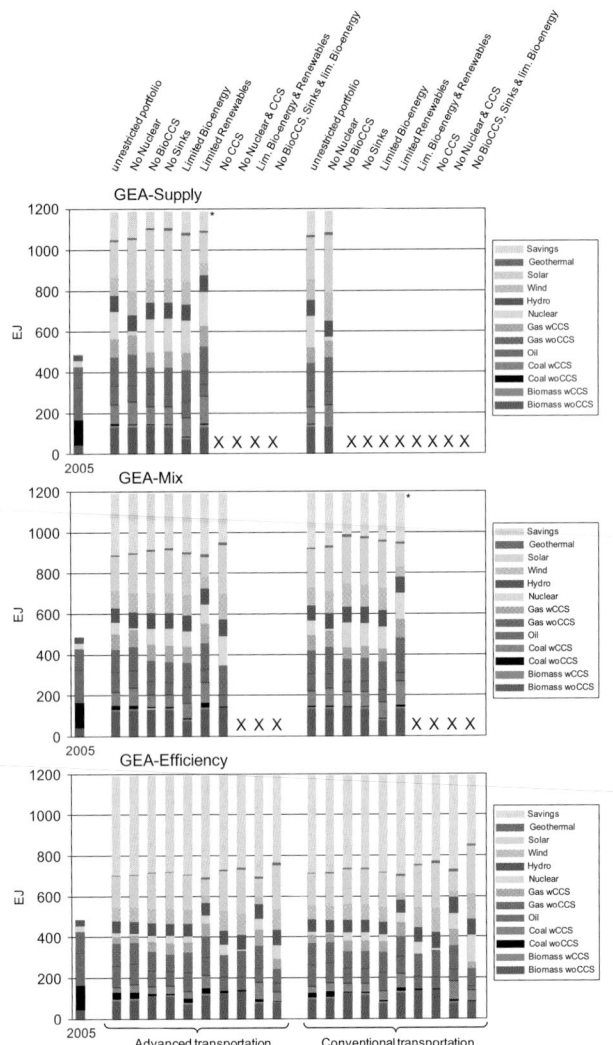

Figure 17.13 | Composition of global primary energy supply in 2005 and 2050 for the three pathway groups under alternative transportation sector assumptions and supply portfolios. See Table 17.9 for details of the restricted supply portfolios. Xs indicate infeasible pathways. Pathways marked with an asterisk are bordering on infeasibility, with carbon prices slightly higher than defined in the note below. Energy savings are calculated by comparison with a hypothetical case with energy intensity improvements compatible with historical trends and no additional climate and energy policies (see also Section 17.3.2.1).

Note: Feasibility is technically defined here as the inability of the supply side to deliver the (fixed) useful energy demand. As uncertainties grow over time, modest undersupply of service demands of up to 5% beyond 2050 is still interpreted as feasible. As in other studies (e.g., Clarke et al., 2009), this concept is operationalized by declaring pathways with carbon prices higher than US$1000/tCO$_2$ in 2012 (as a result of so-called demand backstop penetration) infeasible. This relaxation may lead to limited comparability of economic indicators (e.g., energy-related investments) across the restricted portfolio pathways after 2050.

in the absence of a major transition to electricity or hydrogen, or both, biofuels are the only alternative available to decarbonize the sector. As in the GEA-Supply groups of pathways, the limited sustainable bioenergy potential is a constraining factor, and CCS is important to remove the carbon from bioenergy feedstocks that does not end up in the liquid biofuel itself.

In contrast, the strong focus on energy efficiency, to some extent combined with lifestyle changes (transportation demand is most affected;

see Section 17.3.2 on energy demand), in the GEA-Efficiency group of pathways allows the greatest flexibility on the supply side, essentially independent of the transportation sector assumptions – at least for the portfolio of options examined within this analysis.

Restrictions on CO$_2$ Capture and Storage
Background to Restrictions – Although CCS has not been deployed in energy applications beyond the demonstration level, the scenario literature indicates that it could play an important role as a bridge or transitional technology under stringent climate targets (Edenhofer et al., 2009; Krey and Riahi, 2009; Luderer et al., 2009; Edenhofer et al., 2010a).[14] However, the use of CCS raises various issues. First, the deployment potential is limited by storage capacity at both the global and the regional scale (see Section 17.3.4.4 on regional analysis). Second, even relatively low leakage rates (e.g., between 0.1% and 1%) can compromise CCS as a climate stabilization option (van der Zwaan and Smekens, 2009). Third, upstream emissions from fossil fuel extraction (including of non-CO$_2$ gases) reduce the comparative effectiveness of fossil CCS as a low-carbon supply-side option. Fourth, in contrast to other low-carbon options, CCS is a single-purpose technology with limited or no ancillary benefits, and it imposes an energy penalty, thus increasing resource consumption (see also Section 17.7 on multiple benefits).

Results of Restrictions – Figure 17.13 shows the importance of CCS as a transitional supply-side option. For all pathways with high levels of energy demand (the GEA-Supply pathways group), including those with Advanced Transportation, the No CCS restricted supply portfolio is not feasible. The No CCS restriction is feasible for the GEA-Mix pathways group (with intermediate levels of energy demand), but only if the transportation system does move away from the currently dominant liquid fuels.

In other words, the supply-side analysis shows that CCS is a necessary supply option if the level of demand and the associated energy intensity improvements remain on current trajectories, and if the transportation sector remains more or less compatible with existing infrastructures and business models. These infrastructures can continue to serve a large fraction of energy demand, and the required decarbonization of the energy system is accomplished using CCS in an "end of the pipe" mode.

Restrictions on BioCCS and Sinks (Negative GHG Emissions Options)
Background to Restrictions – Supply-side options that can produce considerable amounts of net negative GHG emissions include carbon sink enhancement and BioCCS (bioenergy in combination with CCS). Capture of CO$_2$ from the atmosphere (CO$_2$ air capture, artificial trees) is another option but is not addressed here (for details see, e.g., Baciocchi et al., 2006; Keith et al., 2006; Zeman, 2007). Although the current emphasis with CCS is on fossil fuel conversion technologies, large-scale biomass cofiring (practiced today in coal-fired power plants typically at the level of a few percent; see, e.g., De and Assadi, 2009) could become an attractive

---

14 CCS can also become economically feasible in non-intervention scenarios, because of the benefits of the captured CO$_2$ for use in enhanced oil and gas recovery, for example. However, the potential for these applications is limited, and therefore CCS deployment in baseline scenarios is typically constrained to a relatively low level compared with climate stabilization scenarios.

supply-side option, helping to reduce residual emissions from coal CCS toward zero without having to increase the capture rate to 100%, which currently appears to be economically unattractive. Once infrastructure has been built to transport CO_2 from its place of origin (e.g., a power plant or cement production plant) to a suitable storage location, the building of smaller-scale BioCCS plants may also become an option.

Results of Restrictions – As can be seen from Figure 17.13, the No BioCCS and No Sinks restricted portfolio pathways produce fairly similar primary energy supply mixes by 2050, in particular within the GEA-Efficiency and GEA-Mix pathways groups. In both cases, cumulative negative carbon fluxes are similar. In contrast to BioCCS and atmospheric CO_2 capture, carbon sink enhancement is a supply option that is available immediately, but large annual reductions in carbon emissions result only with relatively long lag times. Although their deployment occurs at scale only after 2050, the availability of net negative GHG emissions options is crucial for near-term targets, as they reduce the need for immediate GHG mitigation while still allowing the 2°C climate stabilization target to be reached by 2100.[15] So, in contrast to the other supply-side options considered here, these negative net emissions options affect the supply portfolio throughout the time period of the analysis.

The No BioCCS and No Sinks restrictions require a more aggressive deployment of other low-carbon options in the first half of the century, in particular nuclear energy and renewables (Figure 17.13). In combination, however, with the additional restriction of Limited Bioenergy, this can only be achieved at low levels of energy demand (the GEA-Efficiency pathways).

Restrictions on Bioenergy

Background to Restrictions – Bioenergy could play an important role in future energy systems and can contribute to multiple objectives, such as increased energy security (e.g., Brazilian ethanol program; (Goldemberg, 2007)) and GHG mitigation. Biofuels in the transportation sector represent a relatively low cost alternative to fossil fuels that (unlike hydrogen, for example) also requires less additional infrastructure. Even in scenarios that aim to limit bioenergy use, it might still be attractive to use biofuels for specific transport modes (e.g., air traffic, shipping, or long-distance freight (LBST, 2008). In the power sector, bioenergy also represents a relatively low-cost option to reduce GHG emissions. Unlike other renewable energies, it does not pose intermittency challenges. In ambitious climate stabilization scenarios, bioenergy is particularly attractive because of the possible combination with CCS (see Section 17.3.3.5). Despite these potential benefits, however, if bioenergy production is not implemented in a sustainable way, multiple adverse effects can occur, including competition with food production, net increases in GHG emissions from deforestation, and biodiversity losses (Rajagopal and Zilberman, 2007; Wise et al., 2009); see also the electronic appendix to this chapter). Estimates

of bioenergy available in 2050 range as high as approximately 360 EJ, a sevenfold increase over current use (Dornburg et al., 2010). Under stricter sustainability criteria and less favorable assumptions about water scarcity and yield improvements, however, the estimated potential can be as low as 60–70 EJ in 2050 (van Vuuren et al., 2010).

Results of Restrictions – Limiting bioenergy in the supply portfolio is feasible in all but the GEA-Supply (high demand) pathways group with Conventional Transportation (see Figure 17.13). Limited bioenergy implies about 80 EJ/year of sustainable bioenergy, including agricultural residues, by 2050, and less than 125 EJ/year by 2100. The 2050 level would mean less than a doubling of total bioenergy use compared with today, taking into account a substitution of traditional biomass for clean forms of bioenergy as required by energy access objectives.

Restrictions on Nuclear Energy

Background to Restrictions – Although nuclear energy can potentially contribute to energy security objectives (by reducing the imported share of primary energy) as well as to climate stabilization and air pollution objectives, it is a controversial supply-side option for various reasons, including the unresolved problem of long-term waste disposal, the risk of catastrophic accidents and the associated liabilities, and the possible proliferation of weapons-grade fissile material (see also the electronic appendix to this chapter). An additional concern at present is the imbalance of R&D portfolios in favor of nuclear energy, leading to a diversion of government resources from other important options. Compared with actual nuclear generation capacity, R&D spending is, for instance, among the highest levels of government support across all supply-side options (Grubler and Riahi, 2010). The overall conclusion of the GEA assessment on nuclear is thus that until the proliferation potential is better controlled and safer reactor designs are available, countries (especially those considering building their first nuclear power plants) should consider other climate-friendly power supply options first. For a further discussion of risks and concerns, see Chapter 14.[16] The extent to which nuclear energy can contribute to the GEA objectives is, given the risks and concerns described above, thus very uncertain at the global scale, and even more so at the regional or country level.

Results of Restrictions – The No Nuclear restricted portfolio pathway is feasible under all levels of demand and transportation system alternatives. Although Figure 17.13 shows a continued contribution of nuclear energy to the supply mix in 2050 in these pathways, after 2060 this has largely been reduced to zero as the last plants built in 2020 under the No Nuclear restrictions are retired. Most importantly, nuclear energy can in general be seen as a choice rather than a necessity. In other words, alternatives can substitute nuclear energy at a global scale without endangering a successful energy transition, which also implies that

---

15   The potential for future BioCCS to allow a postponement of more costly supply-side options (in present value terms) comes at a price in the form of a reduced likelihood of staying below a certain temperature threshold (see Section 17.5.1 on climate change for details).

16   It should be noted that following historical experience (see Grubler, 2010), all pathways assume that the cost of nuclear power is at best stable or even increasing further over time. In response to addressing proliferation risks, the analysis is restricted to light water reactors, i.e., a plutonium fuel cycle including fast breeder reactors is not considered to be an option, which can have implications for uranium resource availability at some point (see electronic appendix to this chapter).

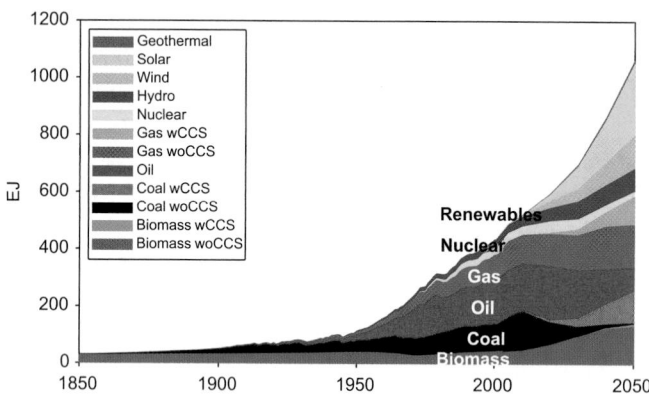

Figure 17.14 | Development of primary energy in the GEA-Supply pathway with a nuclear phaseout shortly after 2050.

different attitudes toward nuclear energy at the national or the regional level can be facilitated. Hence, the phaseout of nuclear is even possible under the relatively high energy demand projection of the GEA-Supply group of pathways (see Figure 17.13 and Figure 17.14). The phase-out of nuclear might lead, however, to somewhat higher costs (see, e.g., investment needs of restricted portfolios in Section 17.3.5, and IEA, 2010).

Restrictions on Other Renewable Energy

Background to Restrictions – Like demand reductions, renewable energy can contribute to all the energy system objectives, including energy access (by decentralizing supply), climate stabilization, and energy security (by reducing energy imports and diversifying supply). Although the global renewable energy resource base (technical potential) vastly exceeds projected future primary energy use (see Box 17.3), harvesting renewable resources may be limited by factors including land competition and systems integration. The latter issue is discussed further in Chapter 15, Section 15.8 and in a number of wind energy integration studies, which conclude that penetrations of wind-generated electricity of up to – and, in a limited number of cases, exceeding – 20% are technically feasible, but not without challenges (see, e.g., Gross et al., 2007; Smith et al., 2007; Holttinen et al., 2009; Milligan et al., 2009). In addition, institutional challenges to the expansion of transmission, including cost allocation and siting, can be substantial (Benjamin, 2007; Vajjhala and Fischbeck, 2007; Swider et al., 2008). In interpreting the results of this restriction, it needs to be taken into account that 20% of electricity from wind and solar would be a substantial increase from today's level, which essentially only Denmark has reached.

Results of Restrictions – The Limited Renewables portfolio restriction reduces the contribution of wind and solar to electricity generation, and thus to the primary energy supply mix (Figure 17.13). The effect is somewhat less pronounced for solar energy, because solar thermal (e.g., solar rooftop collectors) and concentrating solar power (CSP) with thermal storage at the power plant level are affected less or not at all by systems integration issues. Restrictions on wind are intermediate in their impact, whereas the reduction of output is strongest for solar photovoltaic, as the curtailment of electricity generation from this source at peak hours

(around noon) can occur at a relatively low installed capacity compared with the system's overall size.[17]

A related and counterbalancing effect of the pathways that assume limited potential of intermittent renewables is that dispatchable renewables (hydro, geothermal, and bioenergy) and other low-carbon electricity generation sources (nuclear and fossil CCS) are deployed to a larger extent.[18]

17.3.4 The GEA Pathways: Energy System Transformations

17.3.4.1 Introduction

The pathways analysis presented in the previous sections focused on several different dimensions of a successful energy transformation toward a more sustainable future. These dimensions correspond to the branching points of the scenario design process described in Section 17.2.2. This analysis led to a total of 60 pathways, of which 41 were feasible alternative representations of an energy transition that fulfills the GEA objectives. The approach taken here is to use a selection of three illustrative GEA pathways, interpreted by two different modeling frameworks – IMAGE and MESSAGE – to integrate the demand-side analysis of Section 17.3.2 with the supply-side analysis of Section 17.3.3. Narrowing the task down to these three pathways clarifies the important points of similarity and difference within the broader range of 41 pathways. Having firmly established the characteristics of these three illustrative pathways, the section turns in more detail to questions of supply-side flexibility and to regionally disaggregated findings.

17.3.4.2 Three Illustrative GEA Pathways

The three illustrative GEA pathways were selected to combine specific choices at the different branching points. They are *not* intended to be central cases, most likely cases, or even representative cases. Rather, their intent is to illustrate the types of options available throughout the energy system and to explore critical choices in terms of the magnitude and direction of policies, investments, and broader efforts to transform the energy system.[19]

---

17 The lower penetration of intermittent renewables in the Limited Renewables portfolio pathways also leads to reduced demand for technologies to address load management issues (such as backup generation capacities, hydrogen electrolyzers, pumped hydro, or compressed air storage).

18 There appear to be no insurmountable technical barriers to high penetration rates of renewables if ancillary changes to system management and infrastructure are implemented appropriately (see, e.g., the high penetration rates of renewables in the GEA pathways with unrestricted renewables).

19 The overarching criterion for the selection of illustrative cases is that they reflect the main characteristics of each of the groups of GEA pathways (Efficiency, Mix, Supply). In addition, the selection of illustrative cases is guided by the ranges of different energy options in the peer-reviewed scenario literature (eg the recent assessment by

The main distinguishing dimensions of the three illustrative GEA pathways are as follows:

- *Demand versus supply focus.* While the assessment shows that a combination of supply- and demand-side measures is needed to transform the energy system, emphasis on either side is an important point of divergence between different policy choices. A critical factor is thus the extent to which changes in demand for energy services, together with demand-side efficiency measures, can reduce the amount of energy to provide mobility, housing, and industrial services and thus help meet the goals across virtually the whole range of sustainability objectives. This dimension is one of the main distinguishing characteristics, and it motivates the naming of the three illustrative pathways.

- *Global dominance of certain energy options versus regional and technological diversity.* Once technological change is initiated in a particular direction, it becomes increasingly difficult to change its course. Whether the transformation of the future energy system follows a globally more uniform or a diverse path has thus important implications, given irreversibility, "lock-in," and path dependency of the system. GEA-Efficiency and GEA-Supply depict worlds with global dominance of certain demand/supply options, while GEA-Mix is characterized by higher levels of regional diversity.

- *Incremental versus radical new solutions.* Given the ambitious sustainability objectives, transformational changes need to be introduced very rapidly across all GEA pathways. For instance, all pathways feature decreasing shares of presently dominant fossil fuels. The pathways differ, however, with respect to the emergence of new solutions. Some rely more heavily on today's advanced options and infrastructures

(such as biofuels in GEA-Mix), while others depict futures with radically new solutions (such as hydrogen in GEA-Supply).

In terms of the supply mix that provides for overall levels of energy demand across the three illustrative pathways, Figure 17.15 shows the ongoing transition from primarily fossil fuels, which have dominated since the late 19th century (see Nakicenovic et al., 1998) to the low-carbon options – nuclear energy, renewable energies, and fossil energy with CCS – of the second half of the 21st century.[20] The two alternative interpretations of these illustrative transition pathways have been harmonized to a great extent. They start from identical socio-economic assumptions, but the development of final energy is very close as a result of largely harmonized service demand levels (see Box 17.2). However, despite sharing many characteristics, there are three dimensions in which the interpretations are quite different, and this allows for important insights:

- The counterfactuals (or business-as-usual scenarios) underlying these successful transition pathways are considerably different, which materializes in different levels of energy savings across the MESSAGE and IMAGE interpretation (as shown in Figure 17.15). While the MESSAGE counterfactual is more optimistic about the future availability of hydrocarbons and, consequently, many sectors will continue to rely on this option, the IMAGE counterfactual features a transition to coal and nuclear energy becoming the dominant primary energy sources and inducing a shift to synthetic fuels in, for example, the transport sector. Both share, however, similar characteristics when it comes to environmental problems, as well as an inability to successfully address energy access and security issues. This illustrates that a successful transition can be achieved from very different starting points.

- While the demand-side changes, in particular the role of energy efficiency, have been largely harmonized, the supply-side portfolios are different across the two interpretations of the illustrative GEA pathways. The MESSAGE interpretations rely to a much greater extent on renewable energy, while in IMAGE, fossil CCS tends to be of greater importance, particularly over the second half of the century.[21] The role of nuclear energy is very similar in the GEA-Efficiency and GEA-Supply pathways in both frameworks while in the GEA-Mix pathway, MESSAGE foresees a significantly larger role for nuclear after 2050 than IMAGE.

---

the IPCC (2011)). Each of the illustrative cases represents a distinctly different pathway, and together they illustrate the different types of transformations that were explored in more detail through sensitivity analysis of the wider set of 41 feasible pathways. (1) The group of GEA-Efficiency pathways is characterized by the highest degree of "supply-side flexibility". This is illustrated by sensitivity cases with respect to supply-side technology restrictions, all of which were found technologically possible under the GEA-Efficiency conditions. To illustrate this high degree of supply-side flexibility and the feasibility to reach the GEA objectives under even restricted portfolios, a pathway with a restriction for nuclear was selected as illustrative for the GEA-Efficiency group of pathways. (2) The group of GEA-Mix pathways is intermediate with respect to many pathway characteristics (eg, energy demand and the required pace for the upscaling of supply-side options). A major characteristic of the Mix pathways, which distinguished this group from the others, is the diversity of options that contribute toward reaching the GEA objectives. This diversity is most pronounced in the GEA-Mix pathway that assumes restrictions for the electrification of transport, and thus require a variety of "conventional transport" options to satisfy the increasing demand for mobility. Hence, this pathway was selected as the illustrative case for the group of GEA-Mix pathways. (3) The GEA sustainability goals become out of reach in most of the GEA-Supply sensitivity cases with restricted supply-side options. The GEA-Supply pathways are thus characterized by the least supply-side flexibility, requiring the more rapid and pervasive upscaling of supply-side options compared to the other groups of pathways. To illustrate the limited supply-side flexibility and the need of rapid upscaling of supply-side options, an unrestricted portfolio case that allows all options to penetrate the market was selected as the illustrative pathways for GEA-Supply.

---

20 The reporting convention for primary energy in the GEA follows the substitution method. It cannot be overemphasized that for transition pathways towards a sustainable energy future that imply fundamental change in the energy system, the accounting method can introduce a significant perception bias. A more detailed discussion of this issue can be found in Chapter 1.

21 It should be noted that the representation of solar energy in the IMAGE modeling framework is restricted to solar photovoltaic, which leads to significantly lower solar energy contributions in the IMAGE interpretation of the GEA pathways. For other forms of renewable energy, the differences between the IMAGE and MESSAGE interpretations are much less pronounced.

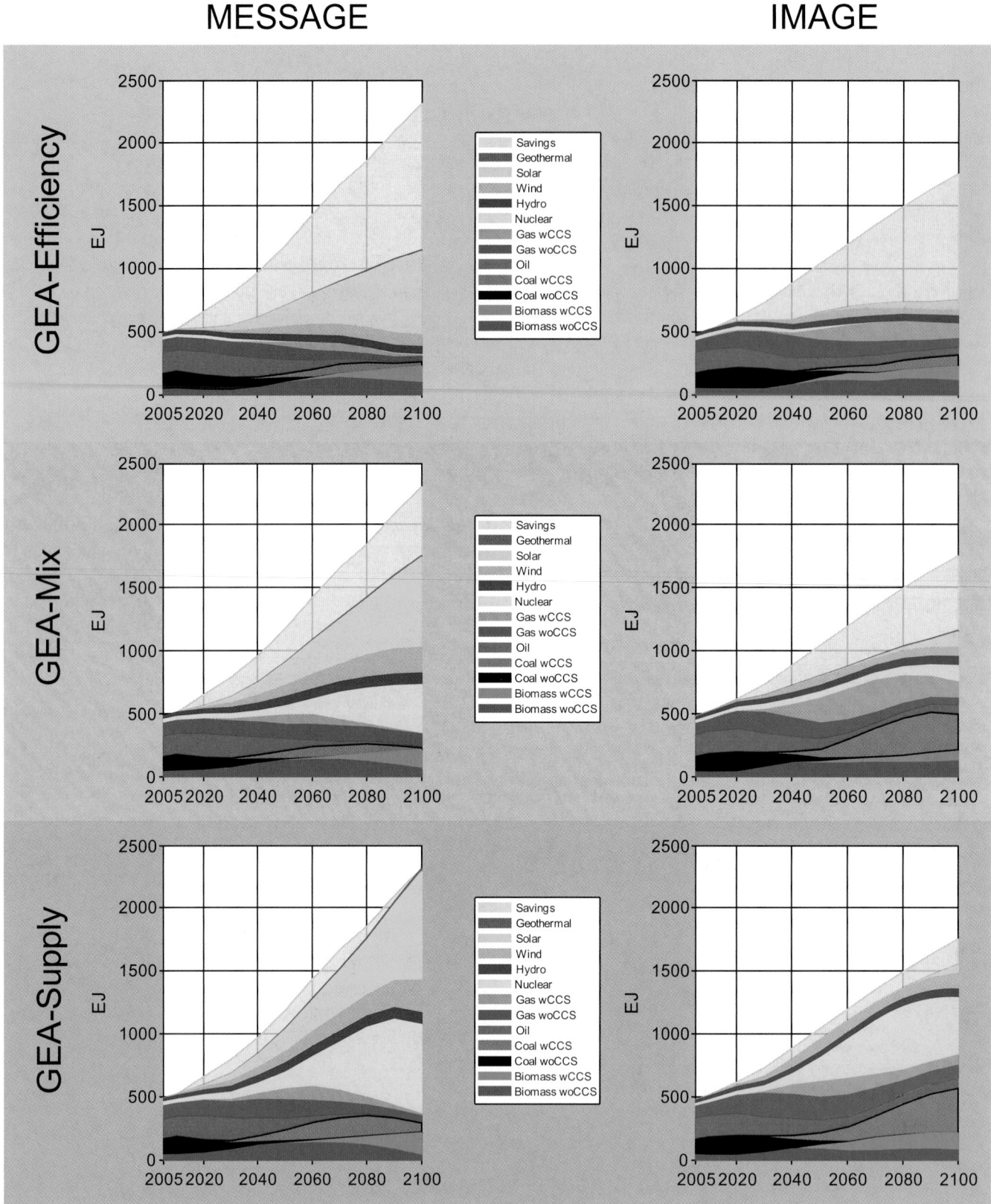

Figure 17.15 | Development of primary energy supply in the three illustrative GEA pathways in both their MESSAGE (left-hand side) and IMAGE (right-hand side) interpretations.

Notes: Energy savings were calculated compared to hypothetical cases without climate or any other energy policies – the so-called counterfactuals – and are roughly compatible with historical energy intensity improvements.

- The level of final energy use is very similar in the respective IMAGE and MESSAGE interpretations of a particular GEA pathway, but due to the primary energy reporting convention adopted in this report – the substitution method (see Chapter 1 for details) – the resulting primary energy supply mix appears very different across the two interpretations.

Specific characteristics of the illustrative GEA pathways are summarized below, followed by a more detailed discussion of the commonalties and choices across all pathways in the next section.

- The illustrative *GEA-Efficiency* pathway features a strong emphasis on efficiency and a heavy reliance on renewable energy, with a share of between 50–90% of primary energy in the long term (2100). Nuclear is assumed to be phased out over the lifetime of existing capacities to reduce the proliferation risk, and CCS provides an optional bridge for the medium-term transition toward a renewables-based energy system. The strong emphasis on efficiency reduces the growth of energy demand over the course of the century by about a factor of 2 (see comparison of the GEA-Efficiency and GEA-Supply pathways in Figure 17.15). This corresponds to roughly doubling the energy intensity improvement rate over historical experience and requires policies in the buildings sector to improve by a factor of 4 by 2050 (a global retrofit rate of 3%/year); in the industry sector, it requires the rapid adoption of best available technology, retrofitting of existing plants, enhanced recycling, and lifecycle product design; in the transportation sector, it requires reducing energy demand through both aggressive efficiency standards (in both freight and passenger transport) and behavioral and lifestyle changes (a switch to public transport and the reduction of demand for private mobility). Table 17.11 recalls the relative contributions of different demand-side transformations in the illustrative GEA-Efficiency pathway, whose absence in the GEA-Supply pathways explains the higher primary energy supply trend. Although efficiency improvements make the largest contributions toward the lower levels of demand in the GEA-Efficiency pathways, these are not viable without complementary policies and measures (including direct and indirect energy pricing) to mitigate demand for energy services.

- The illustrative *GEA-Supply* pathway has a major focus on the rapid upscaling of all supply-side options. The more modest emphasis on energy efficiency and conservation leads to energy intensity improvement rates slightly above historical experience. Massive upscaling of R&D and deployment investments leads to the emergence of new infrastructures and fuels (such as hydrogen vehicles in the transportation sector). Fossil CCS becomes an essential building block to decarbonize fossil fuels, and new nuclear power plants gain a significant market share after 2030.[22] A prerequisite of the

Table 17.11 | Approaches to reducing energy demand in the GEA-Efficiency pathways.

Approach	Relative contribution to reducing energy demand		
	Residential and commercial sector	Transport sector	Industrial sector
Improvements in technological efficiency	High	High	High
Structural changes in the type of energy services demanded	Low	Medium	Low
Reductions in the level of energy services demanded	Low	Medium	Low

GEA-Supply pathway is thus that the associated proliferation risks of weapons-grade fissile material are addressed successfully through, e.g., internationalization of the nuclear fuel cycle. Renewable energy contributes significantly in the long term, with bioenergy playing a more important role in the developing world while wind deployment is occuring in about half of today's OECD countries. The role of other renewable energy sources is more heterogeneous in the two different interpretations by MESSAGE and IMAGE.

- The illustrative *GEA-Mix* pathway is intermediate with respect to many scenario characteristics, such as the focus on efficiency and the pace of supply transformation (i.e., upscaling of advanced and clean supply-side technologies). The main emphasis is on diversity of the energy supply mix, thus enhancing system resilience against innovation failures or technology shocks. Large differences in regional implementation strategies reflect local choices and resource endowments. Whereas the absolute level of nuclear energy deployment is different between the MESSAGE and IMAGE interpretations, the regional focus of its deployment is Asia in both frameworks where more than two thirds of new capacity is commissioned. In the MESSAGE interpretation of the pathway, this results in the co-evolution of multiple fuels, particularly in the transport sector where, for example, second generation biofuels (in Latin America, Former Soviet Union, North America, and Pacific OECD), liquid fossil fuels with CCS (China and the Middle East), and electricity (which gains importance globally, although the generation portfolio differs significantly across regions)

---

22 In the illustrative GEA-Supply case nuclear energy grows to the level of 1850 GWe installed capacity, which is above the IAEA high projection of 1228 GWe, and the present (May 2011) level of 366 GWe of mostly relatively aged capacities. This

level of expansion might be increasingly difficult to achieve in the aftermath of the Fukushima accident. In the past, nuclear growth has been below the IAEA's high projection – and, until 2000, even below its low projections. The range of nuclear in the GEA pathways is similar to the ranges found in other studies in the peer-reviewed scenario literature. For example, the IPCC Report on Renewable Energy (Fischedick et al, 2011) has assessed the recent literature on mitigation scenarios. The nuclear deployment of the underlying scenarios was collated by Clarke and Krey (personal communication, 2012). An analysis of the underlying data shows that 25% of the 137 mitigation scenarios assessed by the IPCC Report on Renewable Energy have higher nuclear deployment by 2050 than the GEA-Supply pathway. The lower bound of the GEA is representative of other low nuclear pathways from the literature that explore GHG mitigation in case of a nuclear phase-out.

gain importance in different regions. The same interpretation of the pathway using the IMAGE model shows less transport fuel diversification and feeds into a hydrogen-dominated transportation sector much as in the GEA-Supply storyline. The IMAGE interpretation, in turn, shows a stronger diversification of electricity generation and hydrogen production, particularly in the long term.

17.3.4.3   Commonalities and Choices across the GEA Pathways

Introduction

Despite major differences in the levels of energy demand and in the nature of transportation system transformation, the three illustrative GEA pathways share certain supply-side characteristics. All show a decarbonization of the supply mix away from conventional fossil fuels, particularly coal. All show an ever-increasing share of energy services demand being met by renewable energy, particularly by the end of the century. All show a substitution of traditional biomass for clean forms of bioenergy.

These commonalities are pervasive features of all the transition pathways for the global energy system toward the sustainability objectives. They can be interpreted as "musts" – that is, required elements of the supply-side transformation if the access, environmental, and security objectives are to be fulfilled.

In other areas, the three illustrative pathways have major points of difference. The most obvious, and the most influential, is the level of energy demand, which distinguishes the pathways by design. Emphasis on demand-side transformation varies massively between the GEA-Efficiency pathways and the GEA-Supply pathways at the extremes. Another point of difference, again by design, is the nature of transformation in the transportation sector, either Conventional, with an ongoing reliance on liquid fuels, or Advanced, a more radical departure from historical trends and existing infrastructures and technologies.

These points of difference are analogous to broad, systemic choices about how and where to direct attention, investment, and policies in order to transform the energy system. None of the outcomes of these choices precludes a transition pathway that fulfills the GEA sustainability objectives: all are therefore feasible within these normative bounds. However, the interdependencies within the energy system mean that choices made in one part of the system have potentially major enabling or constraining effects elsewhere. This was most clearly demonstrated in the comparison of the illustrative GEA-Efficiency and GEA-Supply pathways as to their degree of supply-side flexibility.

This section moves from an analysis of the three illustrative GEA pathways using both IMAGE and MESSAGE to the full suite of pathways, including those 41 feasible pathways based on additional MESSAGE analysis, to establish further commonalities and choices. These are organized in six related sections: the first three relate to the major

energy conversion chains of electricity generation, the "other" conversion sector (mainly liquid and gaseous fuels), and the upstream sector that supplies fuels; the other three relate to supply-side decarbonization, looking at low-carbon energy in general and CCS and bioenergy in particular.

Commonalities and Choices: Electricity Generation

The commonalities in primary energy supply with respect to decarbonization also hold for electricity generation. First, across all GEA pathways, the fossil electricity generation technologies that dominate today's system are on the retreat. Second, low-carbon alternatives, taken together, grow across the board. As a group of supply-side options, these low-carbon alternatives include fossil CCS, nuclear power, and renewable electricity generation.

Beyond these commonalities (which are largely intuitive given the challenge of climate stabilization), electricity generation has three further common features across all GEA pathways. First, conventional coal power generation has to decrease very soon, and by 2030 should not supply more electricity than today. This implies that new construction of coal power plants without CCS must stop, and that some existing plants will have to be retired prematurely or, if possible, retrofitted with CCS (see Chapter 12, Section 12.6.3 for a discussion of CCS retrofitting).

Second, gas power generation, mostly in combined-cycle configurations but also as gas turbines for load balancing, sees considerable growth until around 2030 and only thereafter faces a decline. This is also a commonality across most GEA pathways. In other words, whereas coal power without CCS must phase out rapidly, gas power can be considered a short- to medium-term bridge or transitional technology until longer-term options become more available at scale.

Third, renewable power technologies show significant increases compared with the role they play in electricity generation today. This is clearly visible in Figure 17.16. This finding is consistent with a recent large-scale analysis of renewable energy in long-term transition scenarios by Krey and Clarke (2011). The GEA pathways analysis did not consider scenarios that exclude renewables completely, but limitations of various options were explored. Under these constraints, relatively mature technologies such as hydropower and onshore wind experience strong growth to 2030 and to 2050, with limited variability between pathway groups. Solar photovoltaic and CSP[23] are more variable and show stronger deployment after 2030, although by 2020 the average deployment shows a multifold increase compared with today's levels. Biomass and geothermal electricity generation show much lower deployment levels on average compared with these other renewable technologies.

Three important points relating to these common trends in renewable power should be noted. First, even the lower end of market volumes

---

23 Note that IMAGE does not include CSP and therefore the above statement only refers to the MESSAGE interpretation of the pathways.

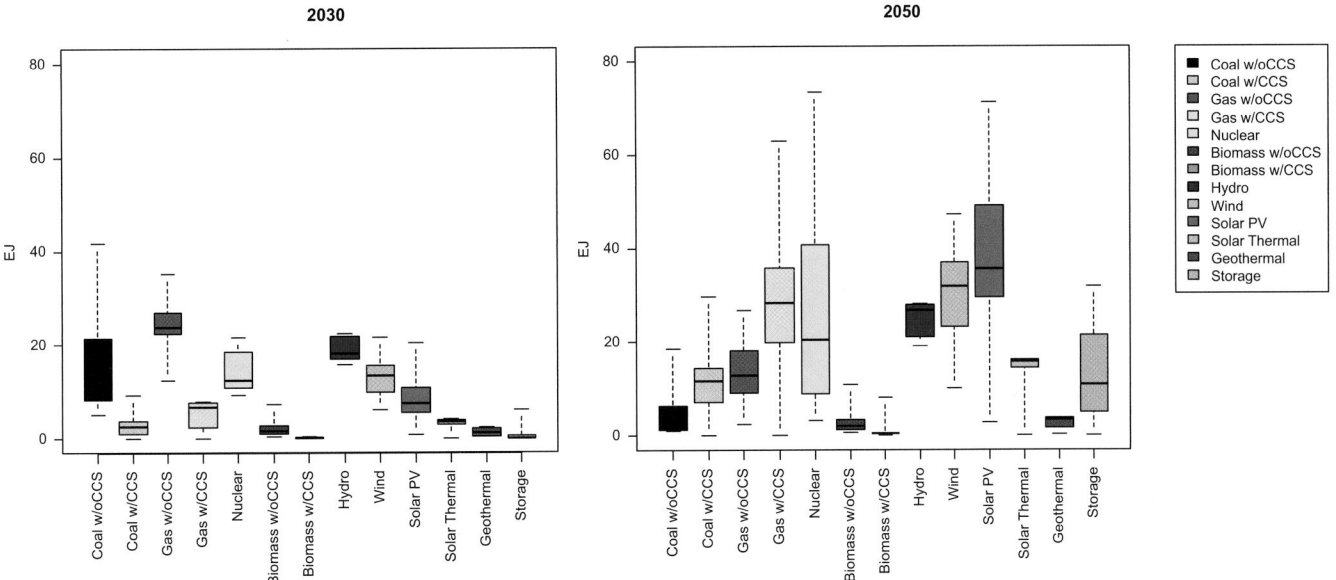

Figure 17.16 | Electricity supplied by different generation technologies in 2030 and 2050 in the GEA pathways. The boxes represent interquartile (25th-75th percentile) ranges, and the horizontal lines within boxes represent medians across all feasible GEA pathways from MESSAGE and IMAGE. Error bars indicate the full range across all feasible pathways.

shown for 2020 will require effective and stable policy frameworks, extending those that are increasingly observed in the current situation. Second, deployment levels vary greatly by region according to local resource potentials. This is discussed further in Section 17.3.4.4. Third, increasing requirements of storage technologies are characteristic for pathways with very high deployments of intermittent sources (wind, solar photovoltaic, and to a lesser extent CSP). This trend is more pronounced in the medium and long term,, when renewable energy become the dominant source of energy supply in some of the regions. Storage can be supplemented with demand-side management and/or so-called smart-grids. These requirements also depend on the availability of negative carbon options (BioCCS and carbon sinks). If negative emissions technologies do not become available on large scale, more rapid early action, including larger contributions of wind and solar photovoltaic are needed. The same is true if nuclear power is excluded, with the slack taken up by intermittent renewable sources again. Like the renewables themselves, storage technologies are diverse, unevenly distributed (e.g., pumped storage hydropower), and in some cases less mature, more costly, and more dependent on new infrastructures (e.g., hydrogen electrolyzers and fuel cells) and business models.

Fossil CCS provides a bridge or transitional option for the power sector. However, in contrast to conventional gas power generation, this is not common to all pathways, as the most efficient pathways do not necessarily include CCS. The most attractive option to combine with CCS in power generation is natural gas, with its cleaner fuel supply chains, lower upstream GHG emissions, higher conversion efficiencies, and significantly lower capital intensity. However, this is not entirely consistent with current R&D activities, which are focused on coal power generation with CCS. The focus on coal is, in turn, driven by the relative cost and abundance of the resource, as well as concerns over dependence on imported gas. The

global preference for gas with CCS is particularly weak in coal-rich regions such as China. Although bioenergy with CCS plays an increasingly important role under more stringent climate stabilization targets, the deployment focus can be in either the electricity or the synthetic fuels sector, depending on the overall system configuration (on advanced designs of BioCCS technologies see also Chapter 11, Section 11.3).

The major choice in terms of supply-side flexibility to emerge from the pathways analysis relates to nuclear power. Nuclear energy can become one of the central sources of electricity generation by 2050, and it is among the supply-side options with the highest deployment across all GEA pathways. Such a development can only materialize if effective technological, institutional, governance, and legal frameworks are introduced to avoid present risks of nuclear energy, including in particular the risk of proliferation. It is thus important to emphasize that in all pathways nuclear power can also be fully phased out after 2060, with no new plants built after 2020. The global "choice" of excluding nuclear power from the supply mix has implications for energy costs, as do any of the other restricted portfolio options (as, by definition, the unrestricted portfolio has the lowest cost). This is discussed further in Section 17.3.5.

Some of the trends observed between 2030 and 2050 in Figure 17.16 continue into the second half of the century, namely, considerable growth in electricity generation from solar and wind energy as well as nuclear power. On the other hand, fossil fuel-generated electricity with CCS tends to decline again toward the end of the century, because of non-negligible GHG emissions in the entire fuel supply chain.

Commonalities and Choices: Liquid and Gaseous Fuels
Today the processing of liquid and gaseous fuels is a very different energy conversion chain than the generation of electricity from primary

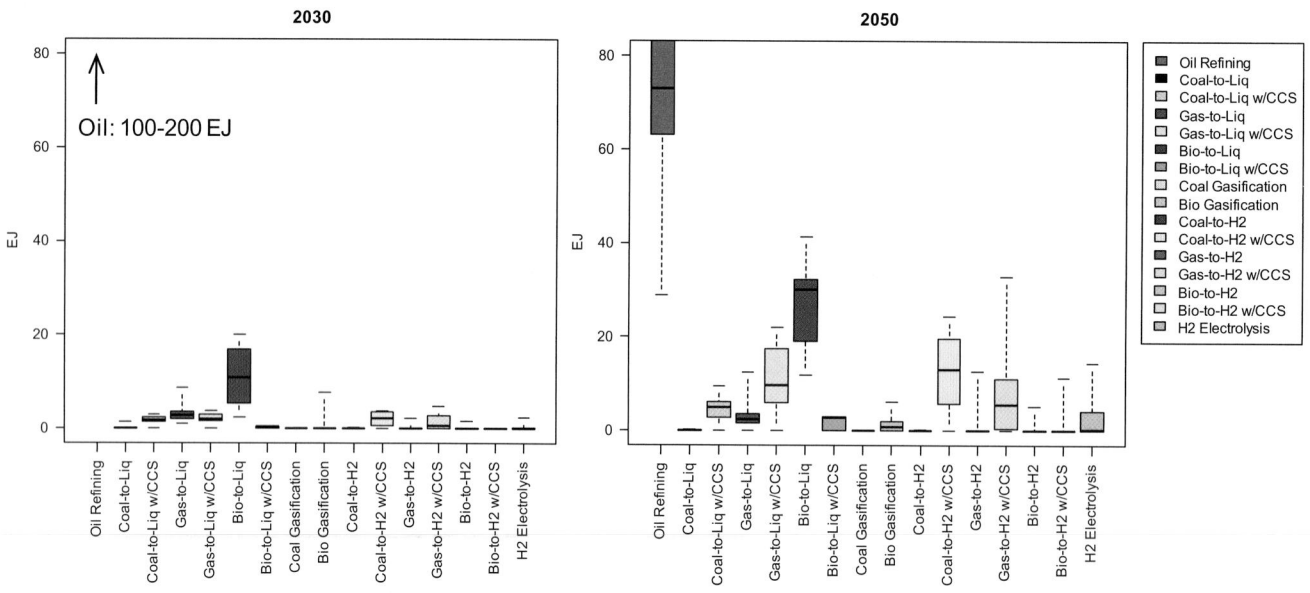

Figure 17.17 | Deployment of liquid and gaseous fuel production technologies in 2030 and 2050 in the GEA pathways. Boxes represent interquartile (25th-75th percentile) ranges, and horizontal lines within boxes represent medians of all feasible GEA pathways from MESSAGE and IMAGE. Error bars indicate the full range across all feasible pathways.

energy sources, as it is almost entirely dominated by oil products. Only about 1.3% of liquid fuel production – about 2.2 EJ out of a total of about 165 EJ in 2007 (IEA, 2009b) – originates from biofuels. In addition, this minor biofuel contribution is regionally very heterogeneous and is essentially dominated by Brazil, the United States, and the European Union, all of which have specific biofuel policies in place. As a result, in the left-hand panel of Figure 17.17, which shows the situation by 2030, the contribution of oil refining is still outside the range of the figure (oil products contribute between 100 and 200 EJ in 2030), while all other technologies contribute less than 20 EJ, even in the most extreme case. The first major commonality, therefore, is the short-to-medium term dominance of oil in the production of liquid and gaseous fuels as energy carriers destined primarily for the transportation sector.

The second commonality is that in the medium term, the biofuel contribution grows substantially. This occurs in regions that already have supportive policies in place, as well as in regions with advantageous conditions for biofeedstock production (e.g., sub-Saharan Africa and Australia/New Zealand). By 2030, the range will be somewhere between today's level and approximately 20 EJ, which corresponds to almost a 10-fold increase over 20 years. The higher end of this range is driven by transportation sector assumptions: the higher deployment range typically comes from scenario variations with a Conventional Transportation setup. Nevertheless, even in the transformation process under an Advanced Transportation setup, biofuels play an important transitional role, and in the very long term (beyond 2050), liquid biofuels may still have an important role in, for example, aviation and heavy freight transport. Once second-generation biofuel technologies become available at a larger scale, sometime between 2020 and 2030, a continued diversification of biofuel production is foreseen.

Unlike liquid biofuels, the potential substitution of oil products with synthetic fuels from natural gas (gas-to-liquid conversion) is a choice rather than a common feature across all pathways. However, this depends on related choices made with respect to CCS.[24] Coal-to-liquid conversion, on the other hand, plays a less important role at the global scale even with CCS, except in regions with abundant coal resources.

In terms of gaseous fuel production, biomass gasification will be limited even in 2050 (depending on the choices made with respect to bioenergy production), although it could be readily integrated into existing natural gas infrastructures. The major choices with respect to gaseous fuels concern hydrogen; but again, these depend on the choices made with respect to Conventional versus Advanced Transportation systems and, within the latter, whether electricity or hydrogen is the preferred route (although hydrogen can also supply some industrial applications). If the build-up of a hydrogen-only infrastructure turns out to be too ambitious, the injection of hydrogen into the gas grid is a favorable (relatively low-cost) option that helps reduce direct CO_2 emissions in the end-use sectors (Riahi and Roehrl, 2000; Midilli et al., 2005; NATURALHY, 2010). If it is derived from fossil fuels, however, hydrogen is an attractive option only in combination with CCS. In the longer term its predominant source would be nuclear or renewable energy. In the latter case, hydrogen electrolyzers would offer the opportunity to deal with the intermittency of wind and solar electricity and thus serve as a storage technology (Sherif et al., 2005; Yang, 2008).

---

24   CCS is usually a low cost add-on for gas-to-liquid and coal-to-liquid technologies because one of the predominant processes, Fischer-Tropsch synthesis, produces a concentrated CO_2 stream that can be captured at little extra cost (see Chapter 12).

Commonalities and Choices: Fossil Resource Extraction

Currently the upstream sector is dominated by fossil fuel extraction, most importantly oil but increasingly natural gas and coal, which has seen considerable expansion over the last several years, largely driven by significant price increases for hydrocarbons. Compared with these three groups of fossil fuels, renewable energy sources, such as purpose-grown bioenergy as well as agricultural residues and byproducts from other uses of biomass feedstock (e.g., in paper and pulp production), are small today.

However, it needs to be emphasized that the sustainability goals built into the GEA scenario, in particular the climate target, put limitations on the use of fossil fuels. This implies that resource limitations for fossil fuels are generally not a concern for the GEA pathways (Figure 17.18). Therefore, the pathways in general show peak oil and gas behavior, however, not because of the assumed physical scarcity of hydrocarbons, but because of the limited carbon emission budgets under, for example, the 2°C target (see also Verbruggen and Al Marchohi, 2010).

The extraction of conventional hydrocarbons lies within a smaller range than that of the unconventional categories,[25] largely because they still play an important role during the energy transition over the coming decades. The largest part of the ranges in oil and gas extraction shown in Figure 17.18 is due to variations in the deployment of unconventional resources, which tend to play a significant role only under specific conditions because of their relatively more energy- and emissions-intensive extraction processes. Unconventional oil plays a limited role in the GEA-Efficiency group of pathways and in the unrestricted portfolio pathways with the Advanced Transportation option of both the GEA-Mix and the GEA-Supply pathway groups, because here the transition away from oil to other fuels is permissible at a steadier pace (see Section 17.3.3.3). In contrast, unconventional gas extraction is most relevant in the GEA-Mix and GEA-Supply groups of pathways under the No Nuclear and Limited Renewables pathways, where CCS is elevated in importance compared with the unrestricted supply portfolio.

Coal extraction declines significantly over the next couple of decades across almost all transition pathways. However, after 2030, when CCS could become available at a larger scale, two distinct developments are possible, leading to a very wide range of possible levels of coal extraction by 2050. If CCS is excluded as a supply-side option, which is a possibility under the GEA-Mix and GEA-Efficiency groups of pathways, then coal extraction has to almost completely disappear by the middle of the century. On the other hand, if CCS can be successfully deployed at scale, a revival of coal extraction, reaching current levels and even going beyond, is an option. The absolute level depends on overall demand: in general, coal extraction is highest in the GEA-Supply group of pathways, followed by the GEA-Mix and the GEA-Efficiency groups.

Commonalities and Choices: Low-Carbon Energy Shares

Figure 17.19 shows low-carbon energy shares for total primary energy supply, electricity generation, and final energy demand (overall and in the transportation sector) in 2020, 2030, and 2050 for the different energy demand levels of the three GEA pathway groups. At the primary energy level and in electricity generation, nuclear energy, renewables, and fossil with CCS are counted as low-carbon energy. At the final energy level, fuels without direct CO_2 emissions (i.e., electricity, district heat, and hydrogen) as well as solid biomass and biofuels are counted as low-carbon energy.[26]

One important finding shown clearly by Figure 17.19 is that the low-carbon energy share is consistently lowest in the GEA-Efficiency group of pathways. From a climate perspective, what matters is not the share of GHG-emitting fuels but the absolute amount of emissions. Therefore, at a lower level of total energy demand, the share of GHG-emitting fuels can obviously be higher (see Section 17.5.1 on climate change).

A consequence is that the required speed of decarbonization is very ambitious under the GEA-Supply pathways with high energy demand; up to a doubling of the low-carbon energy share is needed within the decade to 2020. In the GEA-Efficiency pathways, a less aggressive decarbonization of global primary energy supply is permissible. Across all pathways, 60–90% of primary energy supply has to come from low-carbon sources by 2050.[27]

By far the most complete decarbonization has to be achieved in electricity generation; the threshold is in the range of 40–60% globally by 2020, starting from today's share of around 35%, largely due to nuclear and hydropower. By 2050 almost full decarbonization of electricity generation (80–100%) is required. This implies the need for a continued expansion of renewable electricity generation and of nuclear power, or a rapid commercial deployment of CCS, or both (see the discussion in Section 17.3.4.3). Financing for these mostly very capital intensive technologies (Section 17.3.5) and technology transfer mechanisms to enable deployment in developing countries (Section 17.5.1.4) are major challenges.

At the final energy level, the transition appears less challenging, at least in the short term. However, traditional biomass is included as a low-carbon energy source, accounting for some 30 EJ or about 10% of final energy

---

25 The definition of conventional and unconventional hydrocarbon resources used here follows Rogner (1997). For a comprehensive discussion of the difficulties in the distinction of conventional and unconventional hydrocarbons see Chapter 7, Section 7.2.

26 Note that the latter definition implies the assumption that the generation of electricity, heat, and hydrogen is decarbonized to a large extent, although in a transition period (in particular 2020–2030) significant carbon emissions in the conversion sector may be implied. In addition, it is noted that biomass and liquid biofuels cause direct CO_2 emissions, but if derived from sustainably grown biomass their lifecycle carbon emissions can become close to zero or even negative (in combination with CCS).

27 These values are comparable with those from a study conducted by O'Neill et al. (2010b) that systematically analyzes mid-century targets for keeping long-term climate stabilization options open. O'Neill et al. (2010b) identify low-carbon primary energy shares above about 65% by 2050 to achieve a similar probability (>60% based on uniform prior climate sensitivity probability density function by Forest et al. (2002); see Section 17.5.1 on climate change for details on staying below 2°C warming by 2100 compared with preindustrial levels.

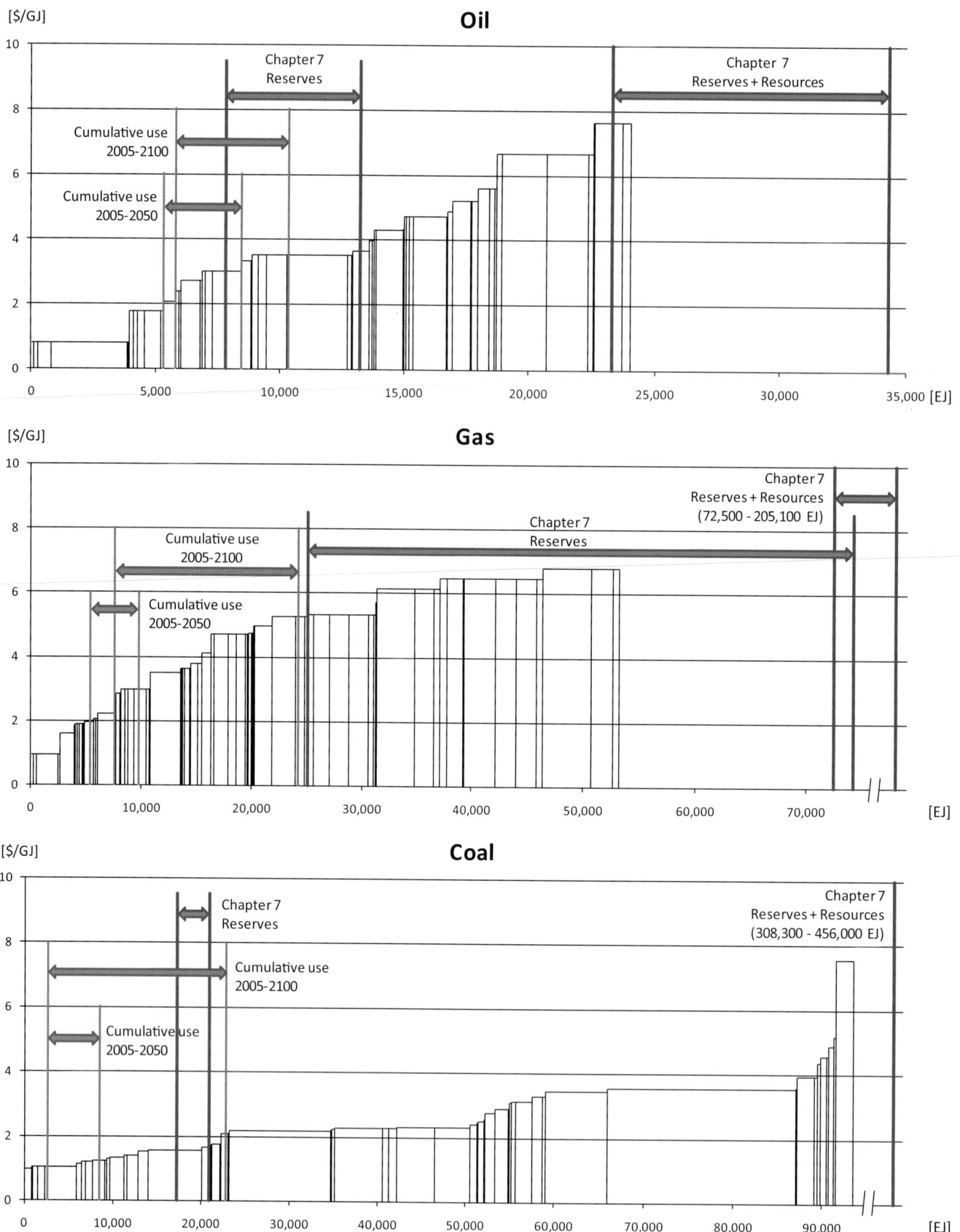

Figure 17.18 | Cumulative global resource supply curves for oil, gas, and coal in the MESSAGE model. The double-headed arrows show the range of cumulative fuel extraction across all feasible GEA pathways from MESSAGE and IMAGE between 2005 and 2050 and between 2005 and 2100. Vertical lines show the central reserve and resource estimates from Chapter 7.

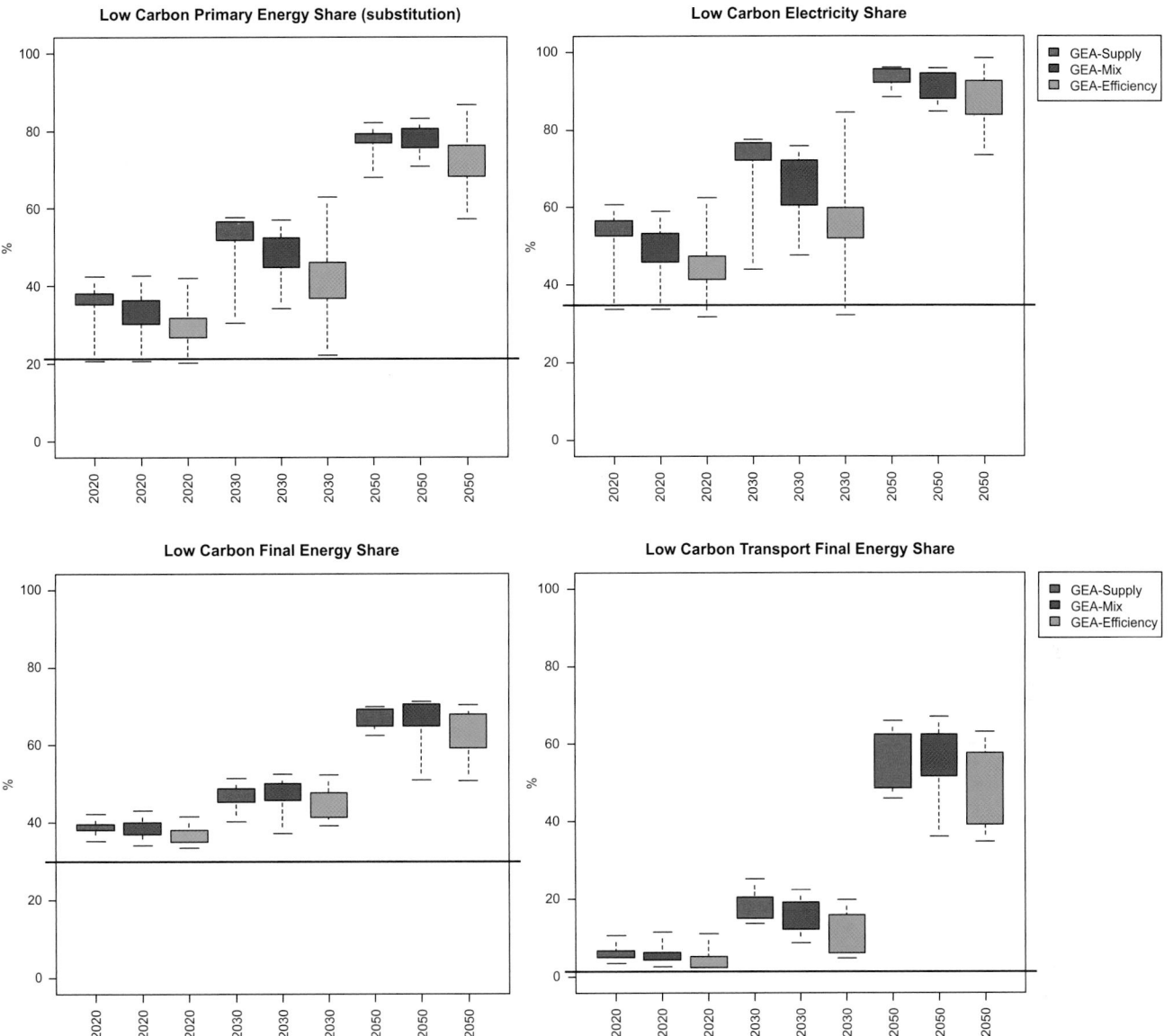

Figure 17.19 | Low-carbon shares in primary energy supply, electricity generation, total final energy use, and transportation sector final energy use by 2020, 2030, and 2050 across GEA pathways from MESSAGE and IMAGE. Boxes represent the interquartile (25th–75th percentile) range and error bars the full range across all feasible pathways in the indicated group. The black horizontal lines indicate low-carbon shares in 2005.

use in 2005. The energy access objective requires that traditional biomass be replaced with cleaner energy forms; in this light, the decarbonization challenge with respect to final energy demand becomes clearer.

In the transportation sector, the threshold level of a less than 10% low-carbon energy share to be reached by 2020 appears relatively modest compared with the low-carbon fuel shares in the other sectors, although it has to be kept in mind that the starting point is close to zero at present. Also, this required development is opposite to the historically observed downward trend in the share of public and rail-bound freight transport (see Chapter 18, Section 18.6.3). By 2050, low-carbon energy shares in transport have to reach a range of 35–75%, depending on the demand level.

Commonalities and Choices: CO$_2$ Capture and Storage

In those pathways that do not exclude CCS as an option, considerable amounts of CO$_2$ would need to be stored between 2020 and 2030, quickly rising to 2050, by which time cumulative storage needs to be no less than 55 GtCO$_2$ and closer to 250 GtCO$_2$. The bulk of this would come from fossil CCS (see above); bioenergy in combination with CCS takes off only around 2040 but increases its contribution in the latter half of the century.

By mid-century, cumulative CO$_2$ storage is strongly determined by the demand level, with relatively little overlap across the high, intermediate, and low demand levels that are assumed in the GEA-Supply, GEA-Mix, and GEA-Efficiency pathway groups, respectively (Figure 17.20). This observation accords well with the fact that under low demand, the need

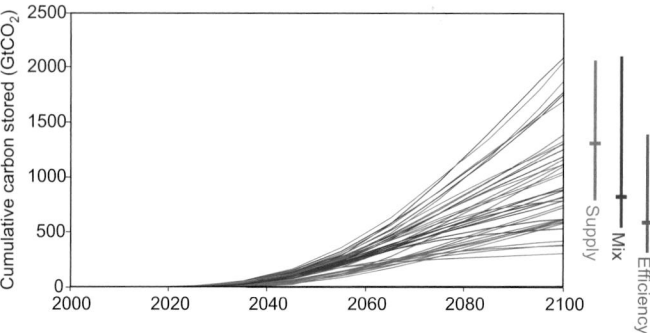

Figure 17.20 | Cumulative global CO_2 storage in the GEA pathways between 2020 and 2100. The vertical bars to the right of the chart show the range of cumulative storage needed within each group of pathways, with the median pathways marked as horizontal bars.

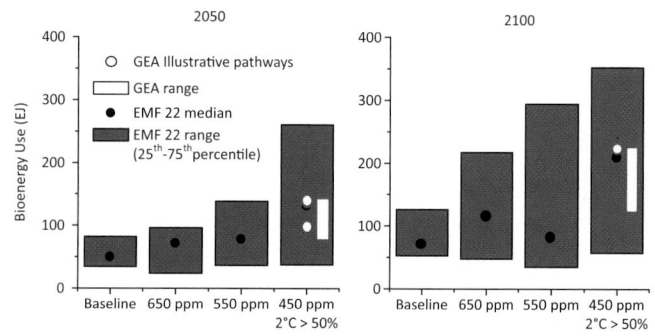

Figure 17.21 | Bioenergy use in 2050 and 2100 for alternative CO_2-equivalent atmospheric concentration targets in scenarios submitted to Energy Modeling Forum 22 (Clarke et al., 2009). Concentration targets (horizontal axis) are in parts per million. White squares and circles indicate the relative positions of the three GEA illustrative pathways. Fewer than three white squares and circles indicate that the pathways overlap.

for CCS is even independent of the transportation sector setup; that is, CCS is a choice, not a "must."

The above figure for global CO_2 storage by 2100 is compatible with the best estimate of 2000 $GtCO_2$ storage capacity, as presented in the IPCC Special Report on CCS (IPCC, 2005), although a few cases come close to this estimate: the GEA-Supply and GEA-Mix pathways with limited renewable energy and those without nuclear energy and the IMAGE interpretations of these two pathways that generally rely to a greater extent on CCS than their MESSAGE counterparts. In these cases, the demands placed on CCS would rely on the more optimistic storage estimates – including saline aquifers, in particular – to materialize, which is supported by more recent studies (see Chapter 13, Section 13.4.3 on CO_2 storage estimates). The high levels of storage and the continued growth toward the end of the century are mostly due to bioenergy with CCS; fossil CCS is generally a bridge technology for application within the medium term, although in the absence of other mitigation options it may have to serve throughout the entire century.[28]

Commonalities and Choices: Bioenergy and Land

The range of bioenergy used in the GEA pathways reaches up to almost 150 EJ by 2050 and 225 EJ by 2100 (in the IMAGE and MESSAGE models). A considerable part of this energy is assumed to be supplied from residues, thus conforming to sustainability criteria. Figure 17.21 shows the use of bioenergy in the three illustrative GEA pathways, set within the broader context of bioenergy estimates from a range of modeling studies exploring increasingly stringent climate change stabilization targets. As is immediately evident, all three illustrative pathways show a similarly high use of bioenergy on the order of 150 EJ in 2050, rising to around 225 EJ in 2100 unless bioenergy use is further restricted.

The main commonality, therefore, with respect to bioenergy is its increasing role in the global energy system. Most studies using energy scenarios under stringent climate policy targets show that it is attractive to use bioenergy as a feedstock to create biofuels, electric power, or hydrogen, possibly in combination with CCS. Bioenergy might also be attractive as a feedstock for the production of materials. The impacts of this increased bioenergy production depend on several factors. First and foremost, first-generation bioenergy production routes may lead to extensive land use (either directly or indirectly) and are therefore likely to have negative impacts on biodiversity and food security. These impacts are expected to be considerably less for second-generation biofuels and electricity generation feedstocks (Dornburg and Faaij, 2005). However, second-generation bioenergy could also lead to significant GHG emissions. In the longer term, biomass can also play an essential role in achieving low GHG concentration targets by making negative emissions a possibility if combined with CCS. Nevertheless, the potential impacts of bioenergy on other policy and sustainability objectives imply that additional policies and strict monitoring of bioenergy and its land use implications will be necessary.

A key factor associated with bioenergy expansion relates to the availability of land.[29] As the GEA pathways use similar levels of bioenergy (see Figure 17.21), the illustrative GEA-Mix pathway is used here to indicate relevant trends in land use and food production.

As shown in Figure 17.22, grassland use (pastures) in the GEA-Mix pathway is more or less stable. This is the result of two opposing trends: an increase in meat consumption and an intensification of animal husbandry. The use of cropland for food production in the GEA-Mix pathway is expected to increase slowly up to 2050 and then stabilize. The land use trends for food production in the GEA pathways are therefore within, but somewhat on the low side of, the range of projections in the literature, which typically show a 0–40% increase (see, for instance, van Vuuren et al., 2009).

---

28 Even these high levels will need to be phased out some time during the first half of the 22nd century to avoid atmospheric CO_2 concentrations falling below pre-industrial levels, which appears to be neither desirable nor justifiable based on the argument that dangerous anthropogenic interference with the climate system should be avoided. In addition, within this time frame storage capacity could become a problem even at the global level.

29 The IMAGE modeling framework was used to assess the land use implications of the GEA pathways.

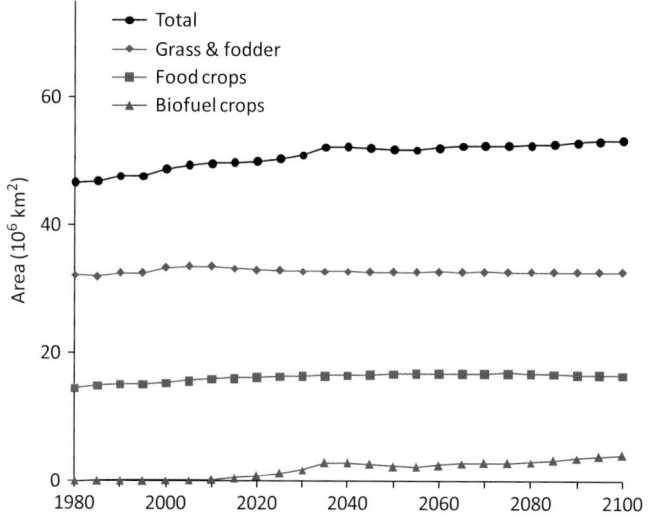

Figure 17.22 | Land use in the illustrative GEA-Mix pathway.

The increase in modern bioenergy use in the GEA scenarios implies an increase in land use for bioenergy production. The exact amount of land used depends strongly on assumed yield increases (see the electronic appendix to this chapter) and the types of bioenergy that are used. Still, as most other studies (e.g., van Vuuren et al., 2010) have found, bioenergy remains a relatively small category compared with other forms of land use. At the same time, the increase in land for bioenergy production in the GEA pathways is about equal to the total increase in agricultural land use. This implies that increases in global land use will lead to some further biodiversity loss and land scarcity. In the context of the GEA sustainability objectives, it will thus be particularly important that policies be put in place that can avoid a strongly adverse impact on crop prices (or the risk of such impacts in a situation of sharply rising energy prices).

17.3.4.4   Regional Analysis of the GEA Pathways

Thus far, the discussion of the GEA pathways has focused predominantly on the transition at the global scale, with regional issues raised in only a few cases. The following sections present explicit regional detail. The starting point of the analysis is energy demand, which starts from a very heterogeneous basis and is tightly linked to economic development in the different regions. On the supply side, three different topics with important regional implications are highlighted: renewable energy deployment, bioenergy use and related land use issues, and CCS, which links fossil energy resources and geological CO_2 storage potential.

This regional analysis is limited in scope and usually remains at the level of illustrative examples rather than being fully comprehensive. Full regional detail of the quantitative GEA scenario analysis is, however, available in the web-based GEA scenario database at www.iiasa.ac.at/web-apps/ene/geadb (see also Box 17.2 on scenario development).

Regional Analysis of Energy Demand

A core concern of the GEA transition pathways is to explore strategies to overcome the current, extremely inequitable distribution of incomes and the associated lack of access to clean and efficient energy services worldwide, while at the same time improving the environmental performance of energy end use and supply. This illustrates the comprehensive nature of the GEA transition pathways under the three main dimensions of sustainability: economic, social, and environmental.

This discussion illustrates some generic patterns in the regional development of energy demand in the GEA transition pathways. In an ideal case, such an analysis would first of all concentrate on actual levels of energy services provided to people, as the ultimate driver of energy systems. However, the diversity of energy end uses and their correspondingly different metrics (thermal comfort, distances traveled, industrial output, etc.) do not allow their aggregation into a commensurable uniform metric. Therefore, this section focuses on final energy demand, which has the advantage of being well covered by current statistics and available modeling and scenario methodologies.

Figure 17.23 summarizes per capita final energy demand as of 2005 as well as the situation in 2050 for the three illustrative GEA transition pathways. Final energy demand is disaggregated by major end-use application type: residential and commercial, industrial, and transportation. For the scenario projections to 2050, an additional efficiency (improvement) potential (grey shaded areas in the bottom panel of Figure 17.23) is shown, corresponding to the difference in final energy use between the lowest energy demand scenario, that of the illustrative GEA-Efficiency pathway, and the more supply-side-focused illustrative GEA-Supply pathway. All three GEA transition pathways achieve comparable levels of energy services provision. Their different levels of final energy use do not represent a difference in energy demand proper, but rather a difference in the efficiency with which comparable levels of energy demand can be provided. These levels of efficiency of energy end use across pathways can be compared with current levels as ranging from efficient (GEA-Supply) to extremely efficient (GEA-Efficiency). For greater legibility, Figure 17.23 summarizes the scenarios for only six of the 11 GEA world regions.

The current disparities in income and energy use across low-income (Africa and South Asia), middle-income (Latin America), and high-income countries (Western Europe and North America) are immediately apparent, as are important differences in the structure of final energy use (a very low industry share in Africa versus a high share in China and Centrally Planned Asia) and the much greater importance of transport energy use in North America than in Latin America. The figure also demonstrates the importance of energy end-use efficiency: Western Europe and North America currently enjoy comparable levels of income, but Western Europe uses only about half as much final energy per capita.

The world in 2050 will look decidedly different from today when viewed through the lens of the GEA transition pathways. Current distinctions between low- and higher-income countries will be largely obsolete, as

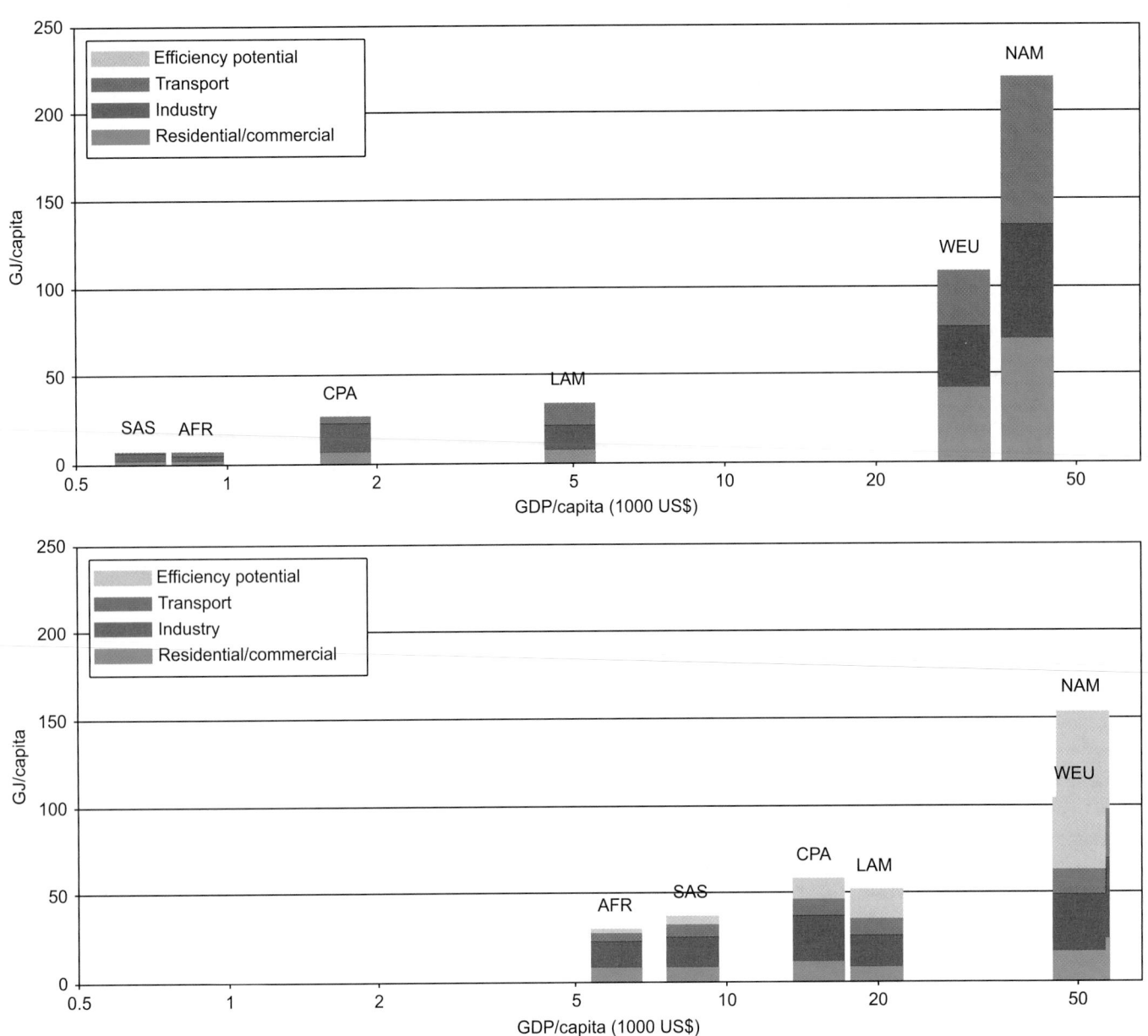

Figure 17.23 | Per capita final energy use and income for selected world regions in 2005 (top panel) and in the illustrative GEA-Efficiency pathway in 2050 (bottom panel). The efficiency potential is calculated from the difference in final energy use between the illustrative GEA-Efficiency and the GEA-Supply pathway. Income is in thousands of 2005 US dollars.

even the regions with the lowest per capita incomes today (sub-Saharan Africa and South Asia) will have advanced to lower-middle-income levels (annual per capita incomes in 2005 dollars of US$5000–10,000), while other developing regions (Centrally Planned Asia and China, Latin America and the Caribbean) will have attained middle-class incomes and lifestyles (US$15,000–20,000) characteristic of the affluent OECD countries in the 1990s. The GEA transition pathways thus describe a pattern of conditional convergence in incomes. This tendency is even more pronounced when one considers that corresponding final energy use per capita remains below roughly 50 GJ per capita across all currently developing regions and transition pathways, because of significant

improvements in energy end-use efficiency that combine appropriate policies at the local to regional scale with globalized availability of energy-efficient technologies and devices. Nor will the potential for improvement have been exhausted even by 2050, as indicated by the scenario differences between the GEA-Efficiency and the GEA-Supply pathways.

Undeniably, the biggest transition challenge will face today's high-income OECD countries, above all in North America. The much-discussed "factor 4" (see von Weizsäcker et al., 1997) – a simultaneous doubling of income and halving of energy use – characterizes the GEA-Efficiency

Table 17.12 | Ranges of renewable energy deployments across GEA Pathways, by region, 2050 (in Exajoules unless otherwise stated).

Region	Bioenergy	Hydropower	Wind	Solar[1]	Geothermal	All renewables	All renewables as % of total
Sub-Saharan Africa	8.8–40.5	2.0–5.5	0.0–19.6	0.5–25.5	0.0–0.3	11.4–91.4	31–94
Centrally Planned Asia and China	6.9–24.7	9.7–10.3	3.7–8.8	0.9–40.1	0.0–0.3	21.2–84.2	24–50
Eastern Europe	1.3–2.8	0.8–1.0	0.7–5.0	0.2–6.1	0.0–0.3	2.9–15.3	23–85
Former Soviet Union	2.9–10.1	2.7–15.8	1.4–7.4	0.3–9.7	0.0–1.0	7.4–43.9	25–93
Latin America and the Caribbean	10.5–22.5	10.7–17.6	3.6–12.4	0.5–21.8	0.0–1.8	25.3–76.1	40–100
Middle East and North Africa	1.2–5.1	0.8–1.2	1.3–8.7	0.5–15.8	0.0–0.3	3.8–31.1	17–40
North America	10.0–21.5	7.2–7.9	2.6–36.7	1.2–41.6	0.0–3.4	21–111	38–89
Pacific OECD	3.4–11.3	1.4–1.7	0.6–4.9	0.2–5.4	0.1–0.8	5.7–24	26–89
Pacific Asia	5.0–11.9	1.9–7.2	1.0–2.0	0.4–14.5	0.2–1.3	8.6–36.9	15–63
South Asia	5.2–20.8	3.5–4.3	1.1–6.7	1.0–79.0	0.0–0.2	10.7–111	21–65
Western Europe	3.9–11.0	5.7–7.6	3.0–30.2	0.7–28.9	0.1–2.1	13.4–79.8	34–83
World	78.3–139	49.9–80.1	28.5–134	7–285	0.6–11.9	164–651	28–74

Note: Ranges include restricted supply portfolios and are calculated as primary energy supply using the substitution method.

1 The representation of solar energy in the IMAGE modeling framework is restricted to solar photovoltaic, which leads to significantly lower solar energy contributions in the IMAGE interpretation of the GEA pathways that typically mark the lower end of solar deployment in this table.

pathway for this region. Although challenging, such a major transition both is technologically feasible by 2050 (see Section 17.3.2.1), considering historical rates of capital turnover, and entails significant benefits in terms of increasing energy security, improving local and regional environments, and avoiding damages from climate change. Compared with these benefits, the required adjustments in energy end-use patterns and lifestyles (toward more energy-efficient, more compact vehicles and greater use of high-quality public transit) will be either modest or nonexistent (e.g., in the case of zero-energy homes that do not compromise living space and comfort).

The biggest challenge revealed by the transition pathways faces not consumers, who can confidently expect expanded and improved levels of energy services in the future, but rather entrepreneurs and policymakers. They need to embrace decidedly different views from those widely held today, focusing on energy services provision rather than mistakenly viewing technology- and policy-dependent levels of primary energy use as immutable, given consumer demand. Policymakers must also embark on different policies that combine both carrots and sticks in order to include stricter building, appliance, and vehicle efficiency standards and changes in relative prices through taxes, subsidies, feed-in tariffs, and other measures. This would open up new business opportunities (e.g., for energy services companies), thereby creating new markets (e.g., for efficiency technologies) and leveraging the power of market forces to meet social concerns and public policy choices.

Regional Analysis of Renewable Energy Sources

Renewable energy sources play an important role in essentially all GEA transition pathways, as discussed at length in Section 17.3.3. Even under the Limited Bioenergy and Limited Renewables pathways (see Table 17.9), renewable energy sources reach some 40% of total global primary energy supply by 2050. Regionally, however, the contribution of renewables varies considerably, as Table 17.12 illustrates for 2050. Several reasons deserve mentioning in this context. First, the resource supply curves (i.e., technical potential as well as resource quality) for the various renewable energy sources differ significantly across regions (see Box 17.3 and the electronic appendix to this chapter). Second, the availability of low-carbon supply-side alternatives (nuclear energy and fossil CCS), which ultimately determine the economic potential for renewable energy sources, is also strongly region dependent. Third, the tradability of renewable energies or of secondary energy carriers derived from them is very heterogeneous. Whereas liquid biofuels are easy to trade and can even rely on existing infrastructures, the scope for trading electricity (e.g., from wind, solar photovoltaic and CSP, and hydropower) at the global scale is much more limited, and for heat (e.g., solar thermal, geothermal), trade is not an option at all. This generally leads to higher exploitation rates of bioenergy potentials than of other renewables. For example, sub-Saharan Africa and Latin America, with the largest sustainable bioenergy potentials, export significant quantities of liquid biofuels starting after 2020 across almost all GEA pathways.

Beyond bioenergy, the relative contribution of the other renewable energy sources varies considerably. In regions with advantageous wind conditions, such as North America and Europe, wind power becomes the largest or second-largest source in terms of secondary energy provided. In most other regions, by 2050 solar energy can become the dominant renewable energy source.[30] Hydropower continues to provide a sizeable share in North America, Latin America, Europe, the Former Soviet Union countries, Centrally Planned Asia and China, and Pacific Asia. Geothermal energy deployment is less pronounced globally and most relevant in North America, Western Europe, Latin America and the Caribbean, and Pacific Asia, in that order (see Chapter 7, Section 7.4.5).

The largest variation in deployment across GEA pathways naturally occurs for those renewable energy sources that are explicitly constrained in the restricted supply portfolio analysis: bioenergy, wind, and solar. For hydropower, deployment ranges tend to be narrower, with a few exceptions. The role of renewable energies generally varies greatly across regions. However, two points are worth noting. First, sub-Saharan Africa and Latin America have the highest renewables deployments by 2050, which, even at the lower end, means that about 40% of primary energy supply comes from renewables; this rises to more than 90% if other supply options are not available. These high shares are related to the bioenergy potentials in these regions not being solely exploited for domestic use, but also being converted to liquid biofuels for export. Second, all of the Asian regions lie toward the lower end of the spectrum, with renewable shares of typically less than 50% by 2050. This is primarily due to those regions' high population density and to potential land use and other conflicts that limit, for example, their sustainable bioenergy or wind energy potential.

Regional Analysis of Bioenergy and Land Use

The use of bioenergy is expected to lead to an expansion of agricultural land. As Figure 17.24 (which is meant only for illustration purposes) suggests, this expansion may occur in many parts of the world. In today's high-income regions such as Europe, North America, and Russia, the GEA-Mix pathway shows a small decrease in agricultural area leading to land abandonment. This decrease in land use results from a stabilizing population, further increases in yields, and increasing food imports. The pathway thus shows that some land could be abandoned for agricultural purposes, to be compensated for by the expansion of agricultural land in low-income areas. These abandoned agricultural areas are used in the GEA-Mix scenario for bioenergy production, which also provides alternative rural income.

Land use for production is also expected to expand in low-income regions such as sub-Saharan Africa, Latin America, and Southeast Asia.

Here, an important driver is relatively low production costs. Again, this opens up routes for bioenergy production. Figure 17.24 shows important bioenergy production areas developing in Latin America and sub-Saharan Africa (which also retain vast forest areas). The trends in the other GEA pathways are broadly comparable to those shown here and are therefore not discussed in detail.

Regional Analysis of CCS

As discussed in previous sections, the role of CCS is very heterogeneous in the electricity and liquid and gaseous fuel sectors as well as across the globe. Regionally, the situation depends on available alternatives such as nonbiomass renewable energy sources, but it depends even more so on the resource basis and costs of fossil fuels and bioenergy. As a result, the amount of CO_2 captured and stored, or shipped to appropriate storage elsewhere, also varies widely. Figure 17.25 shows regional cumulative CO_2 storage over the century in the three illustrative GEA pathways in their IMAGE and MESSAGE interpretations.

The regions with the highest storage volumes are those with large coal resources and correspondingly high utilization of coal with CCS (Centrally Planned Asia and China), large bioenergy potential (sub-Saharan Africa), a combination of the two (North America) or a lack of alternatives (South Asia). Regional CO_2 storage potentials reported in the literature (see Chapter 13, Section 13.4.3; IPCC, 2005; and Hendriks et al., 2004) indicate that in some regions, storage beyond the best estimate may be needed under the higher CCS deployment levels derived in the present analysis.[31] Particularly difficult is the situation in South Asia, where even the comparatively modest need for CO_2 storage under conditions of low demand comes close to the higher estimate by Hendriks et al., (2004). It must be acknowledged that the global best estimate of Hendriks et al. (2004), about 1660 $GtCO_2$, is almost 20% lower than the best estimate of the IPCC Special Report on CCS (IPCC, 2005) published shortly afterward. The overview of CO_2 storage estimates presented in Chapter 13, Section 13.4.3 indicates that uncertainties of storage estimates in saline aquifers – potentially the largest storage option by far, with up to more than 20,000 $GtCO_2$ at the global scale – are very large and that reliable estimates at the regional level are often missing.

17.3.5   Energy Investments

Having presented the main transformational changes of the GEA pathways, the discussion moves now to financial resources, and specifically to the energy investments that need to be mobilized to transform the system.

An important characteristic of the energy sector is its long-lived capital stock, with lifetimes for infrastructure and energy conversion facilities of

---

30 Note that this only holds for the MESSAGE interpretation of the GEA pathways because the representation of solar energy in the IMAGE modeling framework is restricted to solar photovoltaic, which leads to significantly lower solar energy contributions in the IMAGE interpretation of the GEA pathways.

31 Because of the large uncertainties in the regional CO_2 storage potentials, no explicit limitations were imposed in the scenario generation process with the MESSAGE model, whereas IMAGE explicitly includes limitations for CO_2 storage based. This section presents an ex post comparison of storage needs in the GEA pathways with ranges from the literature undertaken here.

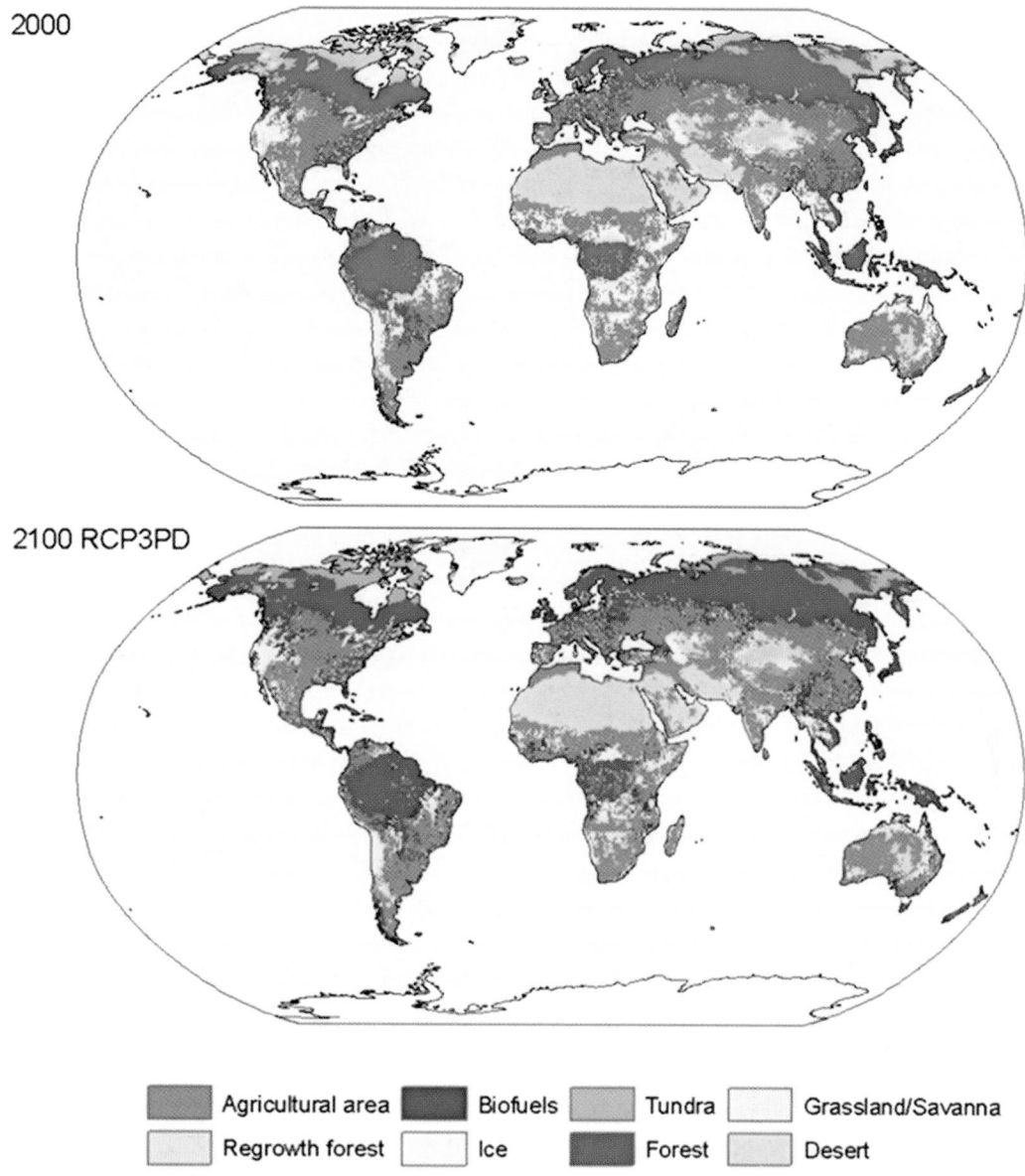

Figure 17.24 | Land uses in 2000 and in the illustrative GEA-Mix pathway in 2100.

30–60 years and sometimes longer. This longevity translates into high inertia in energy supply systems, which impedes rapid transformation. The energy investment decisions of the next several years are thus of central importance, since they will have long-lasting implications and will critically shape the direction of the energy transition path for many years to come.

17.3.5.1 Present Energy Investments

To put energy sector investment into context, it is helpful to first compare current worldwide energy investment with overall economic activity. Following a detailed, bottom-up cost calculation for the entire energy sector, from resource extraction (e.g., coal mining, oil wells) through development and production to delivery and transmission, as well as accounting for historical capacity extensions (and replacement schedules), the present study estimates total global supply-side investment in 2010 at about

US$960 billion.[32] This corresponds to about 2% of global GDP that, while a relatively small share, varies greatly among countries at different stages of economic development. At 3.5% of GDP on average, energy investments are a much larger part of the economy in the developing world than in the industrialized world, where they average 1.3% of GDP.

Understanding the order of magnitude of demand-side investments is of critical importance, particularly because the lifetimes of end-use technologies can be considerably shorter than those on the supply side.

---

32 The calculations of present and future investments rely on estimates from the systems engineering MESSAGE model, which includes a detailed vintage structure and information on the development pathways of historical capacities. All monetary values are given in 2005 US dollars at market exchange rates unless stated otherwise.

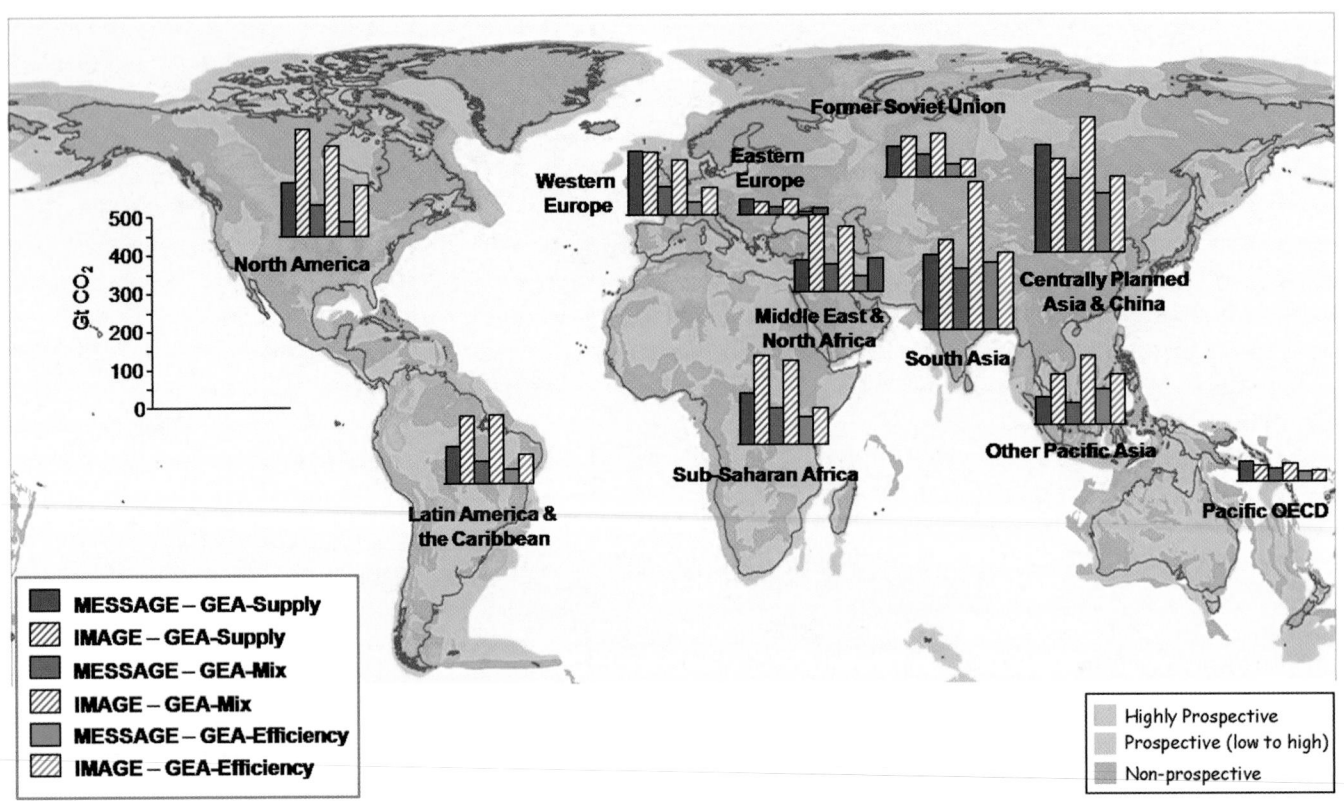

Figure 17.25 | Cumulative CO$_2$ storage by region until 2100 across the illustrative GEA pathways from MESSAGE and IMAGE. The figure also shows prospective areas in sedimentary basins where suitable saline formations, oil or gas fields, or coal beds may be found (prosperity map from Bradshaw and Dance, 2005).

Demand-side investments might thus play an important role in achieving pervasive and rapid improvements in the system. Following the analysis of Chapter 24, around US$300 billion is additionally invested in energy components at the service level, such as engines in cars, boilers in building heating systems, and compressors, fans, and heating elements in large household appliances. Demand-side investments are, however, subject to considerable uncertainty due to a lack of reliable statistics and difficulties in clearly defining what constitutes a purely energy-related investment. Chapter 24 thus reports a relatively wide range of energy component investments on the demand side of about US$100 billion to US$700 billion. In addition, accounting for the full cost of demand-side energy technologies (not only the energy components) would increase investment (but also uncertainty) by about an order of magnitude, to about US$1700 billion (with a range of US$1000 billion to US$3500 billion; see Chapter 24 for more details).

Uncertainties are considerably smaller for total supply-side investment. The estimates presented here are, for instance, similar to those of the International Energy Agency (IEA, 2009b).[33] There is nonetheless some

uncertainty about investment in specific technologies, such as nuclear power. The estimates used for this study include about US$5 billion of investment into approximately 2 GW of new nuclear capacity additions worldwide. In addition, proportional investment in ongoing construction of about 43 GW capacity and investments in fuel processing and lifetime extensions are taken into account. These categories are subject to relatively more uncertainty, but they account for the bulk of total investment in nuclear by up to US$40 billion.

Figure 17.26 summarizes present investment for individual supply-side sectors. Investments are most capital intensive in the power sector, which includes generation, transmission, and distribution. This sector thus accounts for about 42% of total investment, with generation (US$270 billion) accounting for about the same share as transmission or distribution (US$260 billion). The remaining supply-side investment is dominated by the fossil fuels upstream sector: US$130 billion for natural gas, US$210 billion for oil, and US$33 billion for coal.[34] As mentioned above, the uncertainties are particularly large for demand-side investments, which account for at least 24% of total investment (if only energy components are considered).

---

33 Unfortunately, the IEA does not report all investment categories for the base year but focuses rather on cumulative numbers to 2030. The present analysis thus reconstructed the IEA base-year numbers for individual categories using activity numbers as proxies. Note also that the IEA investments are reported for 2008, whereas those reported here are for 2010. Hence, some of the difference might be due to the different base years.

34 Upstream investments include investment in extraction as well as transportation and distribution and upstream conversion facilities (such as LNG terminals and refineries). They exclude, however, investment for fossil fuel exploration (on the order of about US$50 billion).

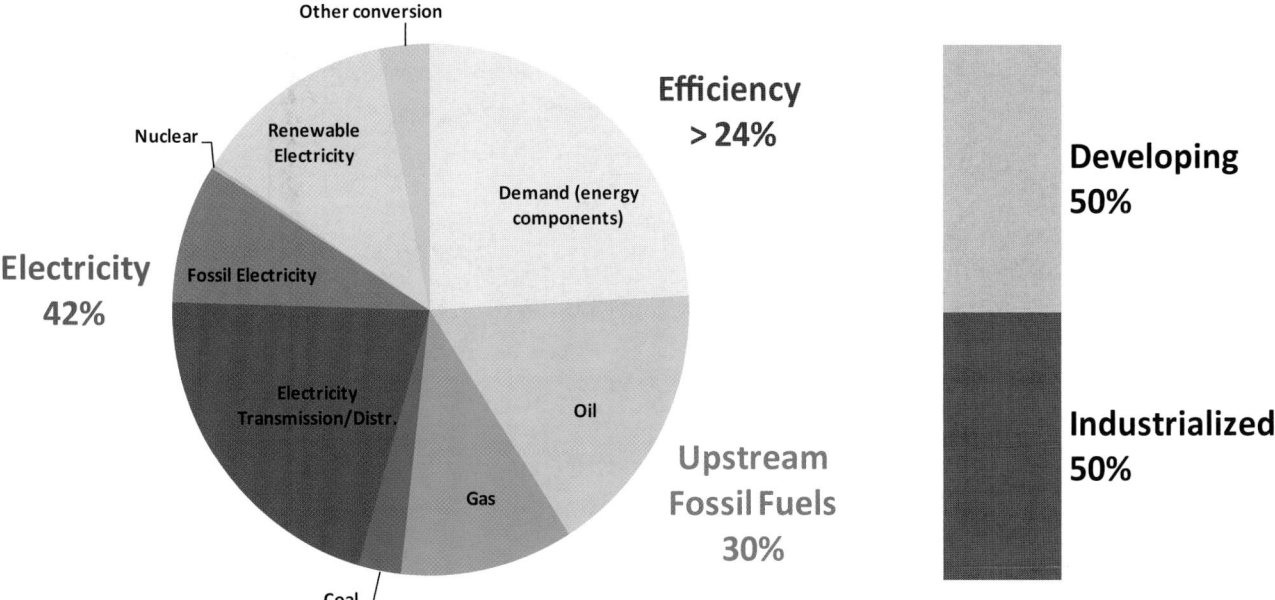

Figure 17.26 | Composition of energy investment in 2010. Total supply-side investment, excluding investment in fossil fuel exploration, is about US$960 billion. In addition >300 billion investments are made into energy components at the demand level.

The composition of investment has been especially dynamic in the past few years. Renewable energy investment, in particular, grew at an unprecedented rate of more than 50% annually between 2004 and 2008, reaching US$83 billion in the latter year, and is presently about US$190 billion (of which US$160 billion goes into power generation). By comparison, investment in fossil power generation in 2010 was about US$110 billion.

17.3.5.2   Future Investment Needs for Transformational Change

Investments in energy supply and demand will be critical for achieving virtually all energy objectives. Figure 17.27 shows the cumulative investment projection up to 2050 for each of the three illustrative GEA pathways. The figure indicates that achieving the GEA climate targets (Section 17.5.1) while also improving energy security (Section 17.6) and access and reducing pollution (Section 17.4) will require a scaling up of investment by almost a factor of 2 compared with today.[35] This corresponds to average annual investment globally of between US$1.7 trillion and US$2.2 trillion, or about 1.8–2.3% of global GDP.

In addition to the need to scale up investment, all the GEA transformational pathways depict significant changes in the structure of the investment portfolio. On the supply side, the transformation of the system is achieved through pronounced shifts of investment away from the upstream fossil fuel sector to downstream electricity generation and transmission. Consequently, the share of upstream fossil fuel-related

supply-side investment in total investment decreases from 30% at present to about 12–23% by 2050. At the same time, electricity investment increases its share on average from about 55% to up to 68% by 2050.

Among all supply-side options, the largest increase in investment needs is for renewable power generation, ranging from US$160 billion/year in pathways with restricted renewables penetration to US$800 billion/year in pathways without CCS and nuclear power (compared with US$160 billion/year in 2010). Another priority for future investment is in building electricity transmission and distribution systems with sufficient operation and capacity reserves to increase reliability, as well as in power storage to allow the integration of intermittent renewables. Global average electricity grid investment (including storage) by 2050 thus increases to about US$310 billion to US$500 billion/year across the GEA pathways, compared with US$260 billion in 2010.

As discussed in the previous section, nuclear power and CCS play a prominent role in some of the GEA-Mix and GEA-Supply pathways, but the full portfolio also includes transformations excluding these options. The uncertainty ranges of these options are thus relatively wide. Investment in CCS ranges from zero to about US$65 billion/year, and investment in nuclear is between US$5 billion and US$210 billion/year. As Figure 17.28 indicates, the higher-bound estimates correspond to pathways in each GEA group that assume limited potential for other technologies.

Investment requirements for each pathway are the result of detailed, bottom-up cost calculations. Each technology of the energy system is characterized by a set of technical and economic parameters, one of which is investment cost measured in US dollars per kilowatt of installed capacity (US$/kW). In the long term, specific investment costs are not static. Innovation and technological learning tend to lower such costs,

---

35   Note that future demand-side investments of the pathways consider only efficiency-related investments at the margin. Comparable global investments into efficiency improvements for the year 2010 are not available. Hence, for 2010 investments for the demand-side consider the full investments into energy components. Future investment needs compared to the year 2010 might thus be an underestimate.

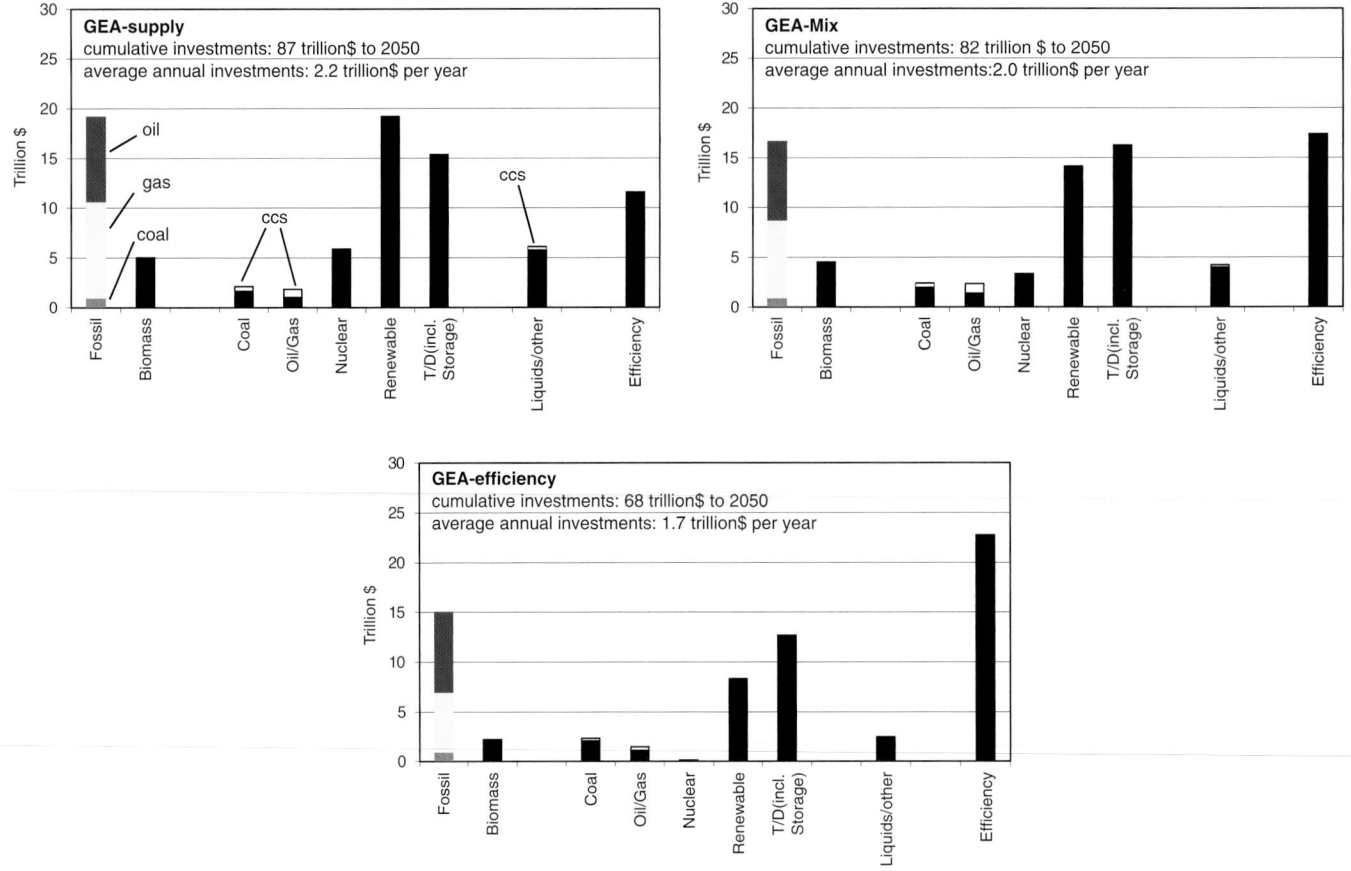

Figure 17.27 | Cumulative energy investment over 2010–2050 for the three illustrative GEA pathways. Within the "efficiency" category, only investments dedicated to improving efficiency are considered. Total demand-side investments for all energy components and appliances would be an order of magnitude larger.

and future energy sector investment requirements will depend greatly on the degree to which innovation and learning improve specific investment costs, efficiencies, emissions, and other performance characteristics (Nakicenovic et al., 1998; Roehrl and Riahi, 2000). Environmental regulation and resource depletion, on the other hand, tend to increase specific investment costs. In the past, innovation has more than compensated for depletion, and often for environmental regulation as well. The extent to which this trend continues in the future varies across pathways. The ranges of specific investment costs assumed for several key energy technologies are presented in the electronic appendix to this chapter, as well as in the GEA database at www.iiasa.ac.at/web-apps/ene/geadb/.

Generally, the present analysis suggests that the transition pathways that focus on energy efficiency achieve the targets at more modest cost and thus represent the lower bound of the investment range (Figures 17.27 and 17.28). One reason for this is the multiple benefits of efficiency measures (and behavioral and lifestyle changes) that limit energy demand and thus contribute to meeting virtually all energy objectives. By contrast, many supply-side measures, such as end-of-pipe pollution control, help improve the sustainability of the system with respect to one objective (local air pollution control) but do not necessarily contribute to others (e.g., climate change mitigation). The other reason why the efficiency pathways depict more modest costs has to do with the nonlinearity of the aggregate supply cost curve: the lower the demand, the less the need to deploy supply-side options with higher marginal costs.

Achieving high levels of efficiency enhancement is not, however, a free lunch. In the GEA-Efficiency pathways, about one-third of overall investment is efficiency related (Figures 17.27 and 17.28). Efficiency investment is calculated using a top-down methodology and thus includes investments on the margin only. In other words, only the efficiency-increasing part of an investment that directly contributes to improving energy intensity compared with a counterfactual (baseline) is accounted for. We thus do not consider the full demand-side investments in end-use devices.[36] Considering the latter would increase overall investment considerably (see the previous section and Chapter 24), but would not change the main conclusion with respect to the economic effectiveness of the efficiency measures.[37]

---

36 The baseline assumes the continuation of energy intensity improvement at historical rates.

37 Efficiency investments are calculated compared with a hypothetical case where the decline in the energy intensity of demand follows globally the historical trend of about 1%/year. For the accounting of macroeconomic feedbacks and price elasticity effects, the present analysis uses a macroeconomic equilibrium model (MACRO) linked to the systems engineering model MESSAGE from which are derived internally consistent energy intensity improvement rates for the alternative pathways (Messner and Schrattenholzer, 2000). Efficiency investments are then computed by assuming that, in equilibrium, the marginal investment to reduce demand would equal the marginal investment in supply. Efficiency investments thus include only investments that have been made to enhance the efficiency of demand in order to offset supply-side investments. Calculated efficiency investment thus does not represent all demand-side investments, including, for example, the component costs of appliances, which would be an order of magnitude larger (see Chapter 24).

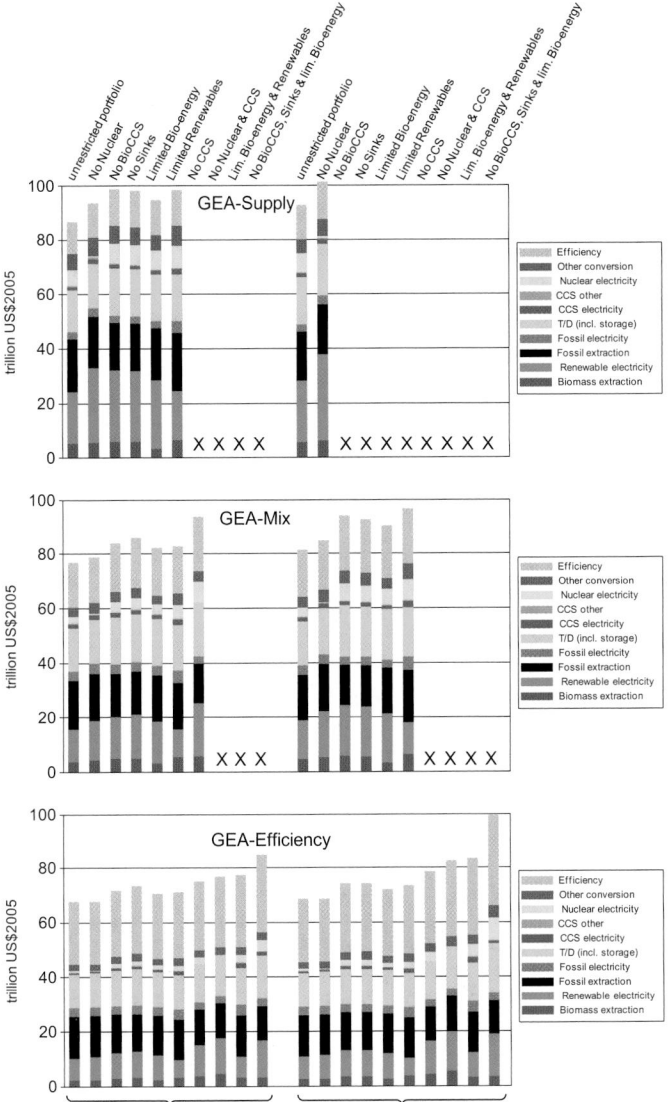

Figure 17.28 | Cumulative energy investment over 2010–2050 for all GEA pathways. Within the "efficiency" category, only dedicated investments to improve efficiency are considered. Total demand-side investments for all demand-side energy components and appliances would be an order of magnitude larger.

As in earlier analysis of stringent climate change mitigation scenarios (e.g., Fisher et al., 2007), the present study finds that the effect of the different investment patterns on the macroeconomy is relatively small. Compared to the counterfactual without policy interventions to achieve the GEA objectives, the projected loss to consumption by 2050 ranges from 0.6% for the GEA-Efficiency pathways to 1.4% for the GEA-Mix pathways and up to about 2.0% for the GEA-Supply pathways. This should be compared with 200% growth in overall consumption over the same period.[38]

---

38   Note that macroeconomic losses are indicative and do not, for example, include costs of overcoming policy barriers, effects of efficiency improvements, reduced losses from air pollution and climate change mitigation, and benefits of improved energy security.

17.3.5.3   Policies to Mobilize Financial Resources

Although the GEA pathways reveal considerable uncertainty about future needs for investment in specific technology options, they clearly illustrate that present investment in energy is neither sufficient nor compatible in structure with a sustainable investment portfolio. Mobilizing the required financial resources for the transformation will thus be a major challenge.

Increasing investment in the energy system as depicted by the GEA pathways requires the careful consideration of a wide portfolio of policies in order to create the necessary financial incentives. The portfolio needs to include regulations and technology standards in sectors with relatively low price elasticity, in combination with externality pricing, in order to avoid rebound effects, as well as targeted subsidies to promote specific "no-regrets" options while addressing affordability. In addition, attention must be given to building an enabling technical, institutional, legal, and financial environment to complement traditional deployment policies (particularly in the developing world).

Table 17.13 identifies effective combinations of policies for specific technology options (see also Chapters 22 and 26) and puts these in the context of the required future investment needs. In addition, the costs and policies for different technology options are compared with those for promoting energy access (see Section 17.4 for further details). Different types of technologies and objectives will require different combinations of policy mechanisms to attract the necessary investment. Table 17.13 thus distinguishes among various mechanisms: "essential" policy mechanisms are those that must be included for a specific option to achieve the rapid energy system transformation; "desired" policy mechanisms are those that would help but are not a necessary condition; "uncertain" policy mechanisms are those where the outcome will depend on the policy emphasis and thus might favor or disfavor a specific option; and "complement" policies are those that are inadequate on their own but could complement other essential policies.

As the table illustrates, future investment needs are comparatively modest for some objectives, such as access, but a variety of different policy mechanisms including subsidies, regulation, and capacity building need to be in place. Regulation and standards are also essential for almost all the other options; externality pricing (e.g., a carbon tax to promote the diffusion of renewables, CCS, or efficiency) might also be necessary for capital-intensive technologies to achieve rapid deployment. Capital requirements for energy infrastructure are among the highest of the options listed in Table 17.13. Thus, high priority needs to be given to future policies (including regulations) to address security and reliability aspects of the energy infrastructure. In addition, subsidies will need to ensure that customers can afford the reliability levels they value. For a more detailed discussion of implementation and policy issues, see Chapters 22 and 26.

Table 17.13 | Energy investments needed to achieve GEA sustainability objectives and illustrative policy mechanisms for mobilizing financial resources.

	Investment (billions of US$/year)		Policy mechanisms			
	2010	2010–2050	Regulation, standards	Externality pricing	Carefully designed subsidies	Capacity building
Efficiency	n.a.[1]	290–800[2]	Essential (elimination of less efficient technologies every few years)	Essential (cannot achieve dramatic efficiency gains without prices that reflect full costs)	Complement (ineffective without price regulation, multiple instruments possible)[3]	Essential (expertise needed for new technologies)
Nuclear	5–40[4]	15–210	Essential (waste disposal regulation and, of fuel cycle, to prevent proliferation)	Uncertain (GHG pricing helps nuclear but prices reflecting nuclear risks would hurt)	Uncertain (has been important in the past, but with GHG pricing perhaps not needed)	Desired (need to correct the loss of expertise of recent decades)[5]
Renewables	190	260–1010	Complement (renewable portfolio standards can complement GHG pricing)	Essential (GHG pricing is key to rapid development of renewables)	Complement (feed-in tariff and tax credits for R&D or production can complement GHG pricing)	Essential (expertise needed for new technologies)
CCS	<1	0–64	Essential (CCS requirement for all new coal plants and phase-in with existing)	Essential (GHG pricing is essential, but even this is unlikely to suffice in near term)	Complement (would help with first plants while GHG price is still low)	Desired (expertise needed for new technologies)[5]
Infrastructure[6]	260	310–500	Essential (security regulation critical for some aspects of reliability)	Uncertain (neutral effect)	Essential (customers must pay for reliability levels they value)	Essential (expertise needed for new technologies)
Access[7]	n.a.	36–41	Essential (ensure standardization but must not hinder development)	Uncertain (could reduce access by increasing costs of fossil fuel products)	Essential (grants for grid, microfinancing for appliances, subsidies for cooking fuels)	Essential (create enabling environment: technical, legal, institutional, financial)

1 Global investments into efficiency improvements for the year 2010 are not available. Note, however, that the best-guess estimate from Chapter 24 for investments into energy components of demand-side devices is by comparison about 300$ billion per year. This includes, for example, investments into the engines in cars, boilers in building heating systems, and compressors, fans, and heating elements in large household appliances. Uncertainty range is between US$100 billion and US$700 billion annually for investments in components. Accounting for the full investment costs of end-use devices would increase demand-side investments by about an order of magnitude (see Chapter 24 for details).

2 Estimate includes efficiency investments at the margin only and is thus an underestimate compared with demand-side investments into energy components given for 2010 (see note 1).

3 Efficiency improvements typically require a basket of financing tools in addition to subsidies, including, for example, low- or no-interest loans or, in general, access to capital and financing, guarantee funds, third-party financing, pay-as-you-save schemes, or feebates as well as information and educational instruments such as labeling, disclosure and certification mandates and programs, training and education, and information campaigns.

4 Lower-bound estimate includes only traditional deployment investments in about 2 GW capacity additions in 2010. Upper-bound estimate includes, in addition, investments for plants under construction, fuel reprocessing, and estimated costs for capacity lifetime extensions.

5 Note the large range of required investments for CCS and nuclear in 2010–2050. Depending on the social and political acceptability of these options, capacity building may become essential for achieving the high estimate of future investments.

6 Overall electricity grid investments, including investments for operations and capacity reserves, back-up capacity, and power storage.

7 Annual costs for almost universal access by 2030 (including electricity grid connections and fuel subsidies for clean cooking fuels).

17.3.6   Key Features of the Energy Transition

Fulfilling the GEA objectives is an extremely ambitious task, but it is technically possible. The full suite of GEA pathways, grouped according to the aggressiveness with which energy demand can be reduced,

show the potential role for a range of energy conversion chains, from primary energy sources to conversion technologies to end-use technologies. Although there are a number of choices available to direct the energy system transformation, there is also a large number of givens – nonnegotiable, nondiscretionary components of an energy transition

Table 17.14 | Targets for the four main energy challenges and illustrative examples of policies and investments quantified by the GEA pathways. In addition to these targets, the GEA also adopted adequate energy services to support economic growth as a normative goal (see Chapter 6).

Objective/Goal	Target and timeline	Pathway characteristics	Examples of policies and investments	Further details
Improve energy access	Universal access to electricity and clean cooking by 2030	Diffusion of clean and efficient cooking appliances Extension of both high-voltage electricity grids and decentralized microgrids Increased financial assistance from industrialized countries to support clean energy infrastructure	Microcredits and grants for low-emission biomass and LPG stoves in combination with LPG and kerosene fuel subsidies for low-income populations *Estimated cost to provide clean cooking: US$17billion to US$22 billion per year to 2030* Grants for high-voltage grid extensions and decentralized microgrids *Estimated cost to provide rural grid connections: US$18.4 billion to US$19 billion per year to 2030*	Section 17.4
Reduce air pollution and improve human health	Achieve global compliance with WHO air quality guidelines (PM2.5 concentration < 10 μg/m³) for the majority of the world population, and the remaining populations staying well within the WHO Tier I-III levels (15–35 μg/m³) by 2030	Tightening of technology standards across transportation and industrial sectors (e.g., vehicles, shipping, power generation, industrial processes) Combined emissions pricing and quantity caps (with trading) Fuel switching from traditional biomass to modern energy forms for cooking in developing countries	Vehicles: Euro 3–4 standards for vehicles in developing countries by 2030 (e.g., -60% NO_x, PM reductions by 2030) Shipping: Revised MARPOL Annex VI and NO_x Technical Code 2008 (-80% SO_x, NO_x reductions by 2030) Industry/power: rapid desulfurization, de-NO_x, and PM control around the world by 2030 *Estimated cost to meet air pollution targets: US$200 billion to US$350 billion/year in 2030 (about 12% of energy costs); co-benefits of stringent climate mitigation policies reduce overall pollution control costs by about 50–65%*	Section 17.5.2 Section 17.7
Avoid dangerous climate change	Limit global average temperature change to 2°C above preindustrial levels with a likelihood >50% by 2100	Widespread diffusion of zero- and low-carbon energy supply technologies, with substantial reductions in energy intensity Energy-related CO_2 emissions peak by 2020 and are reduced to 30–70% by 2050 from 2000 levels Globally comprehensive mitigation efforts covering all major emitters Financial transfers from industrialized countries to support decarbonization	Combination of cap-and-trade and carbon taxes (with initial carbon price >US$30/t$CO_2$, increasing over time) *Upscaling of investments into low-carbon technologies and efficiency measures to >US$600 billion/year to 2050* Additional financial transfers to developing countries of about 3–12% of total energy systems costs to 2050, depending on the domestic commitment of industrialized countries	Section 17.5.1
Improve energy security	Limit energy trade; increase diversity and resilience of energy supply (both by 2050)	Increase in domestic energy supply options (e.g., renewables to provide 30–75% of primary energy by 2050), and reduction of the share of oil in global energy trade. Increase in diversity of energy supply as well as end-use sectors and regions by 2050. Infrastructure expansion and upgrades to support interconnections and backup, including increased capacity reserves, stockpiles, and energy storage technologies.	Public procurement strategies and regulations to support local supplies (e.g., renewable obligations) Interconnection and back-up agreements between energy network operators Stockpiling of critical energy resources for coordinated release during acute market shortages *Estimated cost of infrastructure upgrades for the electricity grid: >US$310 billion/year by 2050, co-benefits of stringent climate mitigation policies reduce overall security costs (import dependency and diversity) by more than 75%.*	Section 17.6 Section 17.7
Further details	Section 17.2.3	Section 17.3.4	Section17.3.5 (overview) and Sections 17.4–17.6 (details)	

that must begin immediately. These commonalities across all pathways (previously referred to as "musts" or necessities) are summarized here:

- improvements to at least the historical rate of energy intensity reduction (more rapid improvements in energy intensity, and thus aggressive efforts to improve end-use efficiency, would increase the flexibility of supply as well as the overall cost-effectiveness of the energy system transformation);

- a rapid shift from traditional biomass to widely accessible, clean, flexible energy forms;

- important regional constraints on availability of energy resources, although such constraints do not limit deployment on an aggregated global scale;

- a broad portfolio of supply-side options focusing on low-carbon energy from renewables, bioenergy, nuclear, and CCS and including:

  - strong growth in renewable energy beginning immediately, and a rising requirement for storage technologies to support the integration of intermittent wind and solar power into electrical grids;

  - strong bioenergy growth in the medium term, with extensive use of agricultural residues and nonagricultural feedstocks (second-generation bioenergy) to mitigate adverse impacts on land use and food production;

  - nuclear energy as an important part of the supply-side portfolio in many transition pathways, although it is also feasible to phase out nuclear energy completely; and

  - CCS as an optional bridging or transitional technology in the medium term unless energy demand is high, in which case CCS becomes necessary.

  - aggressive decarbonization in the electricity sector (especially in the high-demand case), a rapid phase-out of conventional (i.e., without CCS) coal power, and natural gas power as a bridging or transitional technology in the short to medium term;

- at least some electrification of the transportation sector, even in a conventional liquid fuels-based system;

- continued dominance of oil among liquid and gaseous fuels into and beyond the medium term, strong growth in liquid biofuels in the medium term, and thereafter the mix of liquid and gaseous fuels depends on transportation system choices and technological breakthroughs;

- substantial increases in investment on both the demand and the supply side (including energy infrastructure); and

- concerted and aggressive policies to support energy system transformation, including strong regulation and standards and externality pricing.

The storylines of the required energy system transformations that are quantified and elaborated on in the GEA pathways are far richer than these commonalities suggest. Nevertheless, this collation of all the required features of an energy system transformation describes the trunk off of which the many choices and possibilities branch.

Many of these choices are strongly influenced by one or more of the GEA objectives with respect to energy access, air pollution, climate change, and energy security. These are the subject of the second half of this chapter. Table 17.14 provides a link from Sections 2 and 3 on the GEA pathways to Sections 4–7 on the GEA objectives. Some of the main characteristics of the pathways are summarized in the context of each objective. More detailed policy and investment requirements are then given to illustrate how these pathways might be driven, and are explored at length in Sections 4–7.

17.4 Access to Modern Energy Carriers and Cleaner Cooking

This section builds on issues highlighted in Chapters 2 and 19 concerning the need for and benefits of providing universal access to clean cooking and electricity by 2030. This section discusses possible future scenarios for improving access to clean cooking and electricity to meet household energy needs in developing countries. All GEA scenarios are consistent with meeting a target of almost universal access by 2030.[39] The section starts with the GEA-Mix pathway and provides a detailed breakdown of specific access policies and their impacts toward reaching the target for the period 2005–2030. The detailed access modeling presented here focuses on three key regions where lack of access is currently the most acute – sub-Saharan Africa,[40] South Asia, and Pacific Asia – and for which disaggregate data are available on energy choices and use in the household sector. The detailed results from these regions are used to inform the estimation of costs and impacts of alternative policies to improve access to clean cooking and electricity. The section distinguishes between, on the one hand, access to clean fuels and stoves for cooking and, on the other, access to electricity for lighting and appliances. Electricity, even when available, is rarely used for cooking in most developing country households. Therefore, access to modern fuels is as important as access to electricity, if not more so, for meeting the thermal energy needs of most households.

---

39 The target is "almost universal access" because reaching the remotest rural populations is exceedingly expensive.

40 While Sudan is not included in the sub-Saharan Africa region in GEA, it is included in this region for the access analysis because Sudan has severe issues with energy access.

17.4.1   Access to Clean Cooking

There is enormous diversity in the types and amounts of fuels used for cooking in households in developing countries. The starting point for this analysis is data on existing energy choices and demands to meet cooking energy needs in each of the three regions considered. The estimates of energy choices and demand are based on bottom-up estimates using detailed household survey data for key nations in each of the regions (see Ekholm et al., 2010, and Pachauri et al., forthcoming, for details regarding data sources and methods). Most rural and low-income urban households in developing nations still depend predominantly on biomass to meet their cooking energy needs. For the base year 2005, the total quantity of final energy used for cooking in households for the three regions depicted in Figure 17.29 amounted to 15.8 EJ, of which 13.6 EJ was from biomass (including charcoal). This estimate differs substantially from that of the IEA for total residential sector biomass consumption: about 18.5 EJ for the same three regions in 2005. There are several reasons for this difference. Apart from the large uncertainties associated with biomass demand estimates globally, the IEA estimates are generally higher than most national estimates. This study bases its estimates of biomass demand on bottom-up estimates from national household surveys and corrects these for differences in biomass consumption patterns across nations within regions. The resulting estimates are then further compared with national estimates of biomass consumption, wherever available, and scaled up to derive the regional estimates of consumption. As can be seen from Figure 17.29, in rural sub-Saharan Africa and South Asia, the share of biomass (including charcoal) in total final cooking energy was as high as 97–98% in 2005. Among households in rural Pacific Asia, this share was about 60%. In urban centers of South and Pacific Asia, a larger share of kerosene and LPG is used for cooking. However, even in urban sub-Saharan Africa, about 87% of total final energy used for cooking is biomass (again including charcoal).

The GEA access scenarios also estimate the numbers of people dependent on biomass and other solid fuels. Since this study considers the total population dependent on these fuels and not only the share of the population that uses them as their primary source of cooking energy, our estimates tend to be slightly higher than other global estimates from the United Nations Development Programme, WHO, and the IEA (IEA, 2006; UNDP and WHO, 2009; IEA, 2010). This study finds that including only populations that report biomass or other solid fuels to be their primary source of cooking energy tends to underestimate the total population consuming solid fuels. Often populations that use solid fuels as a supplementary fuel actually consume a significant amount of these fuels and meet a large proportion of their total cooking energy needs from them.

Past efforts to model residential sector energy demand for cooking have been limited, particularly in developing countries. The reason is that empirical data for the least developed countries and regions are sorely lacking. Even in emerging nations, finding reliable data for the household sector is a huge challenge. Given the heterogeneity of fuel choices and demand in the household sector, data at an aggregate scale is insufficient for such analysis. Besides the lack of data, uncertainties concerning socioeconomic and demographic trends in these countries add to the challenge of energy demand modeling. Other difficulties with modeling energy demand and choices in developing countries have to do with the special circumstances and conditions in these nations. These have been discussed in detail by Pandey (2002), Pachauri (2007), and van Ruijven et al. (2008).

The GEA access scenarios for residential cooking energy employ the MESSAGE-Access modeling framework (see Ekholm et al., 2010, and Pachauri et al., forthcoming, for details of the model). The model has several novel features that capture some of the special circumstances prevailing in developing countries. Demand is disaggregated both by rural and urban region and for heterogeneous income or expenditure groups. Data from detailed household surveys for key nations in each region are used to calibrate the model.

Various scenarios simulating different combinations of policy packages are modeled within the MESSAGE-Access framework to determine their impact on access to cooking fuels in these regions. Although the specific choice of fuels and cooking technologies will certainly need to be context specific, for the GEA access scenarios this study considers a final transition to LPG as the fuel of choice for cooking for those who have access to and can afford it. This should not in any way be interpreted as an endorsement of LPG as the best of the available choices. Clearly, other alternative cooking fuels, such as biogas, natural gas, and other emerging sources such as ethanol gel and dimethyl ether, in combination with different stove technologies, might be better suited to certain regions or nations. In some regions, there might even be a transition to electricity for cooking. However, in order to quantify the costs and impacts of alternative policies, this study uses LPG as a proxy for all clean cooking fuels.

The main policies considered to encourage a more rapid transition away from solid fuels for cooking include fuel subsidies, to reduce the cost of cleaner fuels, and grants or microlending, to make access to credit

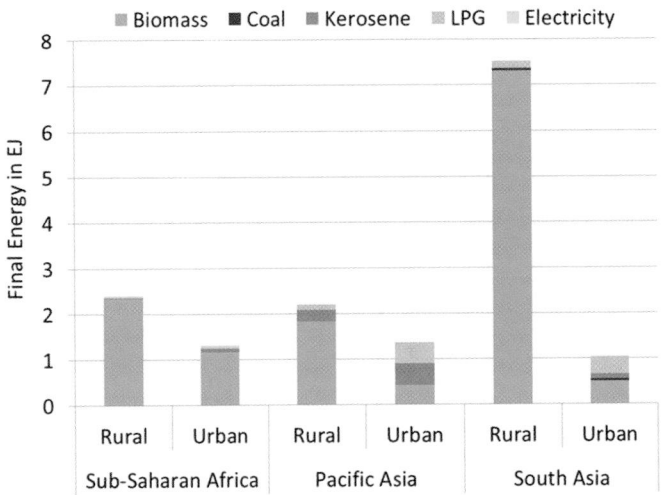

Figure 17.29 | Composition of final energy use for cooking in rural and urban households in three developing regions in 2005.

easier and lower households' cost of borrowing. This makes it cheaper and easier for households to purchase both the fuel and the end-use equipment (cook stoves). Purchasing the stoves that use cleaner fuels often involves a capital outlay beyond the reach of poor and rural households, which often have irregular cash inflows. Policy packages that combine different levels of subsidies with microfinance options are also modeled.

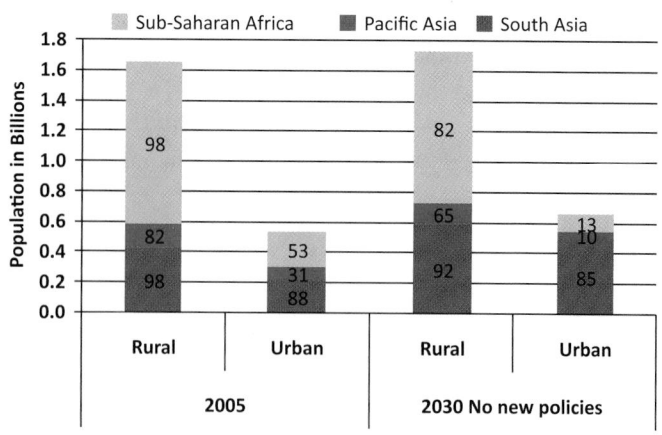

Figure 17.30 | Population relying on solid fuels in three developing regions in 2005 and in 2030 in the absence of new policies. Numbers on the bars are percentages of the total population in the indicated region and year.

17.4.1.1 Populations Dependent on Solid Fuels for Cooking

The GEA cooking fuel access scenarios project that the total population dependent on solid fuels for cooking will rise from 2.2 billion to 2.4 billion in South Asia, Pacific Asia, and sub-Saharan Africa between 2005 and 2030, in the absence of new policies to improve access. The population dependent on solid fuels is projected to decline marginally in South Asia and more significantly in Pacific Asia, whereas in sub-Saharan Africa the numbers rise during this period. In all regions the percentage of the population dependent on solid fuels decreases between 2005 and 2030. This decrease is significantly more rapid in urban centers than in rural regions (Figure 17.30). However, in urban sub-Saharan Africa, population growth is projected to outstrip the decrease in the percentage of population dependent on solid fuels, so that the total population dependent on these fuels continues to rise. These projections are based on outputs of the MESSAGE-Access model that account for changes in income level and distribution, urbanization, and population growth and for the consequent impact of these factors on the transition in cooking energy choices.

The impact of the alternative policy packages considered on the numbers of people dependent on solid fuels varies across the different regions from slight to dramatic. Figure 17.31 depicts the impact of the policies on the number of people dependent on solid fuels for each region and for the urban and rural sectors separately. A subsidy policy that reduces the price of clean fuels by 20% below existing prices in each region would reduce the number of people dependent on solid fuels in all three regions from 2.4 billion, in the case with no new policies, to 1.9 billion. A policy that provides cheaper microfinance options for upfront costs and the purchase of end-use equipment would also reduce that number to 1.9 billion. In estimating the effect of the microfinance policy, it is assumed that the interest charged on loans is 15%/year. This is at the low end of the range estimated by Robinson (1996) for interest rates on loans by microfinance institutions to the poor in developing countries, and much lower than the internal discount rate of poor households in these nations. The scenarios that combine a fuel subsidy with microfinance are more effective in all regions in accelerating a shift away from solid fuels than either a subsidy-only policy or providing microfinance alone, as Figure 17.31 also shows. However, even the policy scenario that combines a subsidy of 50% on the existing price with microfinance leaves about 500 million people, virtually all of them in sub-Saharan Africa, reliant on solid fuels in 2030.

17.4.1.2 Costs of Policies to Reduce Dependence on Solid Fuels

The GEA access scenarios quantify the costs of reducing dependence on solid fuels for several of the different policy packages considered. The net present value of the costs is estimated for each policy scenario and compared with the impact of the policy in reducing the number of people dependent on solid fuels to determine the relative effectiveness of each scheme. The cost of microfinance schemes is estimated to be zero for governments, as it is assumed that microfinance companies are able to cover the costs of their operations through the interest payments they receive. If however, the capital costs of new stoves are met through some form of public grants, these obviously represent a police cost. Although the objective of all access policies is to accelerate the transition away from the use of solid fuels to modern forms of liquid or gaseous cooking fuels, not all policies are able to achieve this equally. For those households that remain dependent on solid fuels, an estimate was made of what it would cost to provide them with improved cook stoves. Chapter 19 provides information on a range of improved cook stove technologies developed around the globe. These vary tremendously in design, sophistication, cost, emissions, and performance. However, it is assumed that, given the rapid improvements in stove technology, future deployment of such stoves will meet a minimum standard in terms of both efficiency and emissions as defined in Chapter 2. Table 17.15 provides a breakdown of costs by region and type cumulatively between 2010 and 2030. Figure 17.32 relates the cost per person gaining access per year to the number of people gaining access up until 2030, to provide an indication of the effectiveness of alternative policies in providing improved access to clean fuels.

The costs of policies aimed at encouraging a more rapid transition to the use of clean cooking fuels depend on the combination of the policy instruments deployed and the extent of subsidy, as shown in Table 17.15. Even a low-cost policy of providing easier access to credit through microfinance institutions is projected to substantially reduce

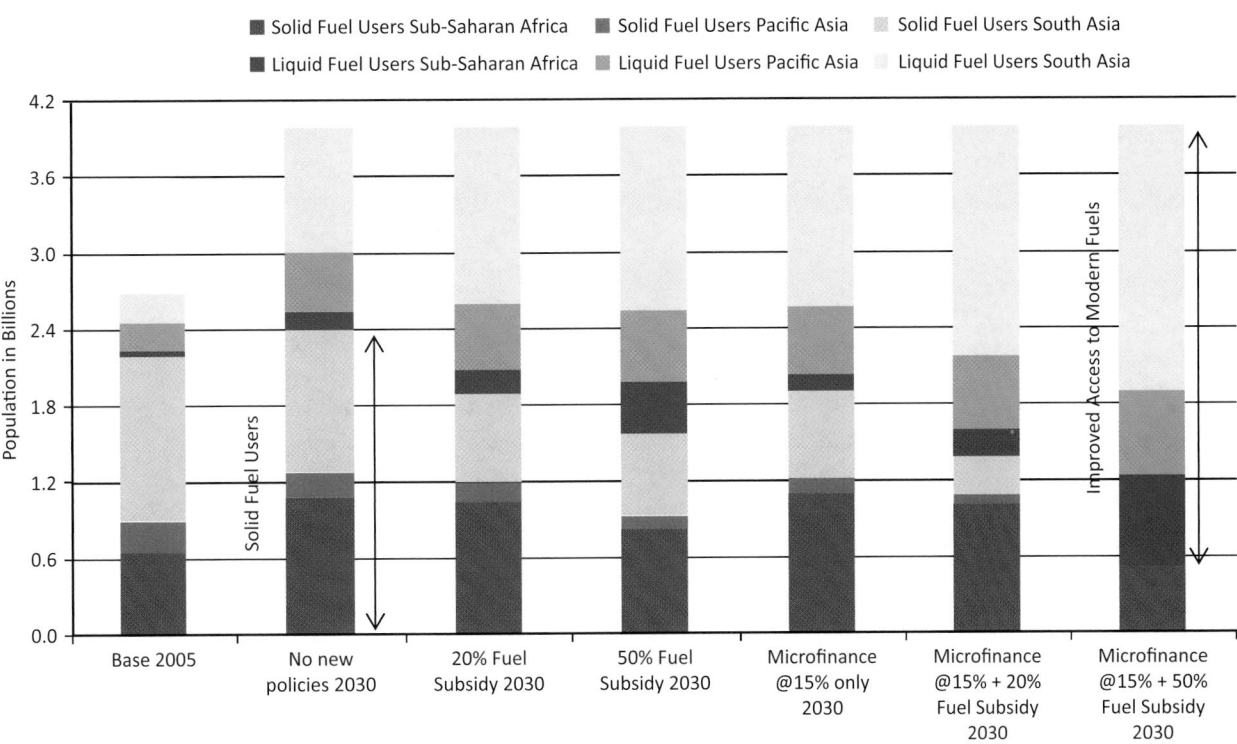

Figure 17.31 | Impact of alternative policy scenarios on access to clean cooking fuels in three developing regions. Subsidies are relative to consumer price levels and are additional to existing subsidies.

dependence on solid fuels among urban populations and the richer rural households in South Asia and Pacific Asia by 2030. However, a policy that promotes microfinance alone leaves about 1.4 billion people still dependent on solid fuels in 2030. Such a policy, if combined with a

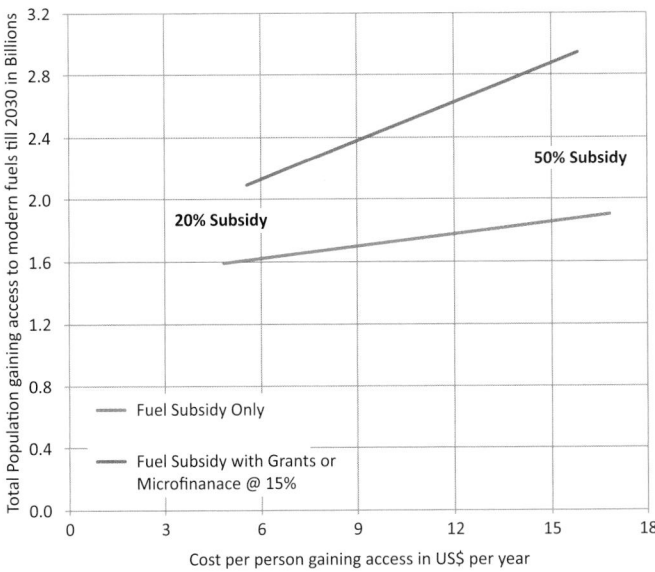

Figure 17.32 | Costs of alternative policy packages and their effectiveness in reducing dependence on solid fuels in three developing regions (sub-Saharan Africa, Pacific Asia, and South Asia) between 2010 and 2030.

massive scale-up of improved cook stoves that are more efficient and less polluting than conventional stoves, along with better ventilation in cooking areas, might be a cost-effective interim solution for many rural households for whom a shift away from biomass may be out of reach in the short term. The financial requirements of such a policy are the lowest among the entire combination of policy scenarios assessed: on the order of US$1.1 billion to US$1.6 billion/year until 2030. A more stringent access target would require a combination of grants or microfinance for the purchase of stoves with a fuel subsidy and would cost considerably more, about US$15.8 billion to US$17.0 billion/year until 2030. The wide range of uncertainty in the cost estimates reflects the high capital costs associated with the use of cleaner fuels. The lower range of the cost estimates assumes that the entire cost of the stoves is met through cheap loans provided by microfinance institutions that are able to recover their costs. The total spending required to meet an access target for clean cooking fuels and stoves would be higher if stove costs have to be funded through public grants.

Clearly, the choice of policies, the stringency of the targets, and the exact combination of clean fuels and end-use stove technologies promoted are likely to be specific to each country or region. However, the analysis presented here is indicative of the range of costs of different combinations of policies and their effectiveness in achieving different access targets. What is clear from this analysis is that, although fuel subsidies are necessary to increase access for the poorest households and regions, subsidies alone are likely to be less effective in accelerating a transition

Table 17.15 | Cumulative financing required to provide access to clean cooking fuels and devices in developing Africa and Asia, 2010–2030 (in billions of US$).

Policy intervention	Region	Fuel subsidy	New LPG stoves	Improved biomass cook stoves	Total, all three regions
20% fuel subsidy	Sub-Saharan Africa	7.54	0.43	8.98	59.6–67.2
	Pacific Asia	3.47	0.75	2.93	
	South Asia	27.56	6.41	9.11	
50% fuel subsidy	Sub-Saharan Africa	91.71	3.60	6.93	202.2–214.3
	Pacific Asia	10.42	0.95	3.01	
	South Asia	81.49	7.55	8.60	
Microfinance only[1]	Sub-Saharan Africa	0.00	2.19	9.66	21.6–31.2
	Pacific Asia	0.00	0.87	3.05	
	South Asia	0.00	6.54	8.92	
Microfinance + 20% fuel subsidy	Sub-Saharan Africa	9.04	0.89	8.72	85.0–100.0
	Pacific Asia	5.35	1.28	2.43	
	South Asia	50.87	12.88	8.56	
Microfinance + 50% fuel subsidy	Sub-Saharan Africa	130.67	6.52	5.20	315.2–339.4
	Pacific Asia	16.72	1.71	2.60	
	South Asia	152.65	15.97	7.36	

[1] It is assumed that no public costs are associated with microlending and that microfinance institutions are able to recover their full costs from the interest charged. However, these can be considered costs if purchase of the stoves is financed from public grants.

to the use of clean fuels for cooking than a policy that combines subsidies with improved access to credit through microfinance institutions. Such a policy would make it easier for households to cover the capital costs associated with a switch to cleaner fuels.

Fuel subsidies are considered controversial in many nations, and many developing countries already have generous subsidies on kerosene and LPG. Although such subsidies may be justified on social grounds, they have often resulted in market distortions, been appropriated largely by richer consumers, and led to poor economic returns for energy suppliers and distributors. Leakages to the black market from existing subsidies to households for kerosene in India have been estimated to be as high as 44% (Planning Commission, 2006). However, "smart" and targeted subsidy schemes and lifeline tariffs for poor customers can be designed and have proved successful, as in the case of the Bolsa Familia program in Brazil, which couples assistance to low-income families for the purchase of LPG fuel with mandatory child school attendance. Removing subsidies in a phased manner once incomes reach a level where households have the ability to pay can be challenging for governments, but this can be achieved if coupled with increased social spending in other areas. In other countries, such as many in sub-Saharan Africa, low-income

consumers face prices for modern fuels that are at times even higher than the competitive market prices of these fuels in Europe. This difference in prices often arises from nonmarket factors such as weak institutions and safety and stability concerns, which urgently need to be addressed. For the design and implementation of more targeted subsidy schemes and removal of nonmarket barriers, additional enabling conditions will need to be created in these nations. This will require additional capacity building to strengthen the administration of governance systems and local institutions. Chapter 25, especially, addresses the issue of capacity building and concludes that good governance in the energy sector is especially critical for attracting investments in needed infrastructural development in the least developed and emerging nations, which face the greatest challenge to expanding energy access.

17.4.2   Access to Electricity

Improving access to electricity requires accelerating the pace of electrification in the least developed countries and regions. Decisions about setting targets for grid expansion are generally made by national governments or regional bodies. However, the literature shows that public or private utilities generally bear the financial responsibility for these programs as the executors of these decisions (Zomers, 2001; Kemmler, 2007; World Bank, 2008). In the best case, decisions about where to expand electrification are grounded in standards or criteria for electrification. In general, such criteria support electrification in places where it is cheapest. Thus, utilities often select projects that require the least infrastructure investment relative to demand. Villages or communities that are closest to existing grids, that have the highest population density, or where economic activity is greatest are generally connected to the grid first. Social criteria, including preferential selection of the poorest households or more remote rural regions, also influence the decision for grid expansion in some nations, but less so because these regions are not the logical choice from an economic perspective for electric utilities or developing country governments. In general, one can expect that electrification will proceed most rapidly where the costs are lowest.

In many countries, households also have to pay a connection fee and have to make their own decision about whether or not to get an electricity connection. Factors that influence whether households opt for grid connection are the amount of the connection fee, whether payments can be spread over time, and the household's understanding of the fees, tariffs, subsidies, and billing (Zomers, 2001; Gaunt, 2005; World Bank, 2008).

This section analyzes electrification using two separate model frameworks. Within the MESSAGE-Access and IMAGE models, rural electrification and grid infrastructure expansion are modeled in slightly different ways. As a starting point, both models take existing levels of electrification by nation, or by subpopulation within a nation, to calibrate the base year. For the purposes of quantification, two alternative levels of demand are assumed for household consumption within both models, corresponding to different electricity service levels:

- Low demand or minimal access: each household has one conventional light bulb (40W), and one out of three households has a television set (60W); on the assumption that these are used for three hours a day, this amounts to approximately 65 kWh/household/year.

- High demand or sustainable universal access: consumption is assumed to be 250W for four hours per day for lighting and other applications, as in the Tanzanian reference study of Modi et al. (2005), amounting to 420 kWh/household/year.

Electrification is defined differently in the two models. Within the IMAGE model, access is defined as connection to the grid. Thus, once the grid has been extended to reach a certain region, all households in that region are considered connected. The MESSAGE-Access model defines electrification in terms of whether a household's electricity demand exceeds an amount considered the basic minimum required to meet household needs: 65 kWh/year in the low-demand case and 420 kWh/year in the high-demand case.

Future rates of electrification in both models are driven by future income growth. However, within the IMAGE framework, in the base case a regression model is developed first by regressing national electrification levels on GDP per capita (in US dollars at purchasing power parity) in order to project what the future electrification level by region will be, based on future income (see the appendix for details of the methodology). Within the MESSAGE-Access model, future electrification in the base case is determined by income growth and distribution across rural and urban income groups. Thus, the MESSAGE-Access model incorporates a greater degree of heterogeneity on the demand side. By contrast, the IMAGE model takes into account a much finer degree of spatial resolution on the supply side, determining investment needs for rural electrification at the level of 0.5 × 0.5 degree grid cells. Each cell is considered to have either complete access to electricity or no access at all. Within a world region, grid cells are electrified over time, starting with those with the lowest levelized transmission and distribution costs.

17.4.2.1   Rural Populations with Access to Electricity

Figure 17.33 shows rates of access by region in the base year and projections to 2030 for both models in two separate cases: one with no additional new policies or resources for improving the rate of electrification, and another with universal access. Differences are observed across the two models both in base-year rural access and in progress with electrification across time for the two scenarios. Rural electrification levels differ in the base year across the two models in part because of differences in regional definitions in the IMAGE and MESSAGE models. For sub-Saharan Africa and South Asia, differences in regional composition across the two models are minor, and thus rural electrification levels are fairly similar. However, the regional definition for the Pacific Asia region differs significantly across the two models, as does the base-year electrification level. Differences in the sources of data used in the two models

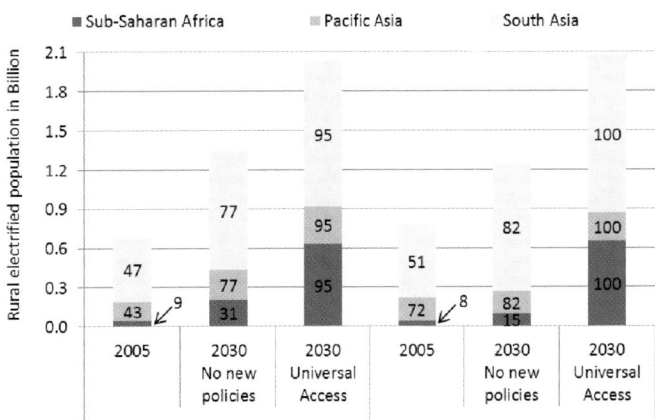

Figure 17.33 | Rural population with electricity in three developing regions in 2005 and in 2030 under business-as-usual and universal access scenarios. Numbers on the bars are percentages of the total rural population in the indicated region and year. Rural electrification level in the MESSAGE model refers to the 420 kW/household scenario.

also account for part of the variation in base-year electrification levels. Thus, for instance, in the MESSAGE-Access model, base-year electrification levels for Pacific Asia are based in large part on bottom-up estimates of access levels across rural income quintile groups in Indonesia. In contrast, in the IMAGE model, electrification levels in Pacific Asia are determined by estimates of rural electrification levels published in IEA (2006) and UNDP and WHO (2009). Progress with rural electrification in the two scenarios differs across the two models because of differences in base-year electrification levels and in methodology.

In sub-Saharan Africa, rural electrification in 2005 covers less than 10% of households in both models. Following a trend with increasing GDP per capita, in the no new policies scenario, this is projected to increase to 31% in 2030 according to the IMAGE model, but only to 15% according to the MESSAGE-Access model. In South Asia, the projected increase under the no new policies case is the largest, from 47% in 2005 to 77% in 2030 according to the IMAGE projections, and from 51% to 82% according to the MESSAGE-Access model. Thus, the shortfall with respect to universal access in 2030 is largest in sub-Saharan Africa. In other developing regions such as Latin America, rural electrification levels are already relatively high and are expected to reach over 90% by 2030 under the no new policies projections. For this reason, rural electrification is not modeled here for regions other than South Asia, Pacific Asia, and sub-Saharan Africa, which remain the regions where the gap between the no new policies scenario and the universal electrification scenario remains widest.

17.4.2.2   Investments for Improving Access to Electricity

The amount of investment required to increase electrification levels depends on the assumptions made about the costs of transmission and distribution, but also on population density. Costs rise as required capacity expands to meet rural household electricity demand. For example, regions with relatively high population density, such as South Asia, have lower costs per unit of capacity than less densely populated regions such

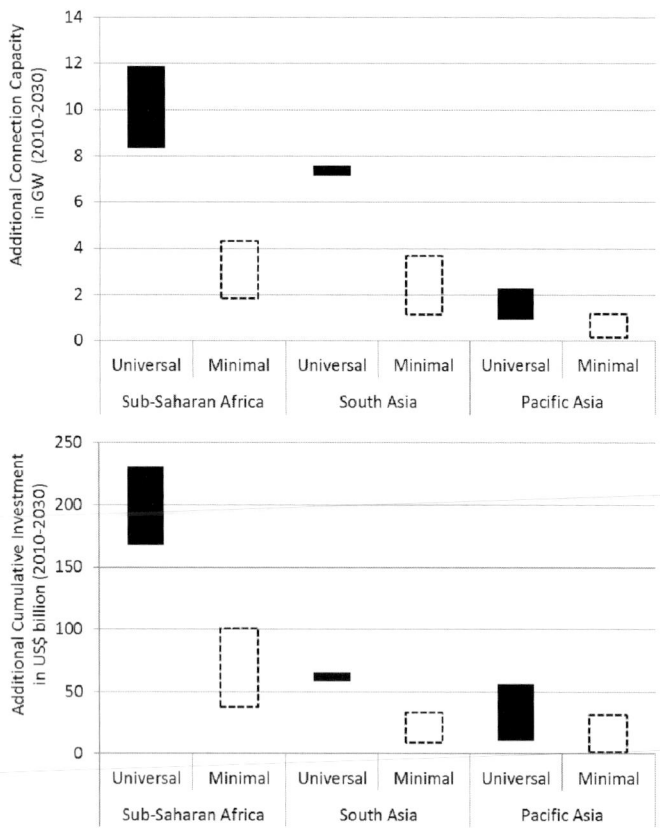

Figure 17.34 | Additional connection capacity and cumulative investment required to achieve almost universal rural electrification in three developing regions. Additional connection capacity equates to additional generation capacity of approximately 2 GW for the low case and 22 GW for the high case in the MESSAGE-Access model.

as sub-Saharan Africa. Levelized transmission and distribution costs for electrification within both models are determined in the following way. Within the IMAGE model, costs are based on the required increase in capacity and the distance over which transmission is required. Thus, costs are determined by spatial factors such as population density and distance from an existing electricity network at the grid cell level. In the MESSAGE-Access model, a simple three-step, region-specific technology cost curve differentiated by grid capacity is used to estimate costs. Given the least-cost optimization approach of the MESSAGE-Access model, the low-cost grid technology deploys first and the high-cost grid technology next as electricity demand increases. Both models estimate the costs of almost universal power supply in a given region through grid connection by estimating the total cost of extending transmission and distribution to all populated parts of the region. Decentralized technology alternatives such as mini-grids, off-grid, or stand-alone options might be more economic in some circumstances, but the present analysis does not include these.

Figure 17.34 shows, for each region, the additional connection capacity and total cumulative investment needed until 2030 to achieve rural electrification and compares results across the two models. The largest investment needs, not surprisingly, are in sub-Saharan Africa, where cumulative investment to achieve universal access amounts to an additional US$230

billion between 2010 and 2030. In the low demand scenario, where minimal access is assumed, the cost is significantly lower – about US$37 billion in the case of sub-Saharan Africa. However, this may not be considered sufficient and sustainable electricity access in the longer term. In general, the range in estimates depicted in the figure reflects the difference between the results from the two alternative models used. In Pacific Asia and South Asia, the majority of investment takes place in the no new policies case, so that the additional investment needed is relatively lower. This implies that additional investment for universal access in the three regions of sub-Saharan Africa, South Asia, and Pacific Asia is estimated at about US$300 billion cumulatively between 2010 and 2030.

17.4.3   Impacts of Access Policies on Energy Demand and Greenhouse Gas Emissions

The impacts of alternative policies for improving access to electricity and clean fuels for cooking are relatively modest in comparison with changes in demand in other sectors. As seen in Figure 17.35, compared with the base year 2005, energy demand in 2030 is projected to almost double in the case where no new access policies are implemented, from 17.7 EJ to 33.2 EJ, with most of this rise accounted for by additional LPG demand for cooking and kerosene and electricity for lighting and appliances. In an access scenario with no fuel subsidy but easier access to credit through microfinance and minimal electricity access, total energy demand in the low electricity demand case in 2030 is lower than in the no new policies case, but LPG and electricity demand are higher. In this scenario, in addition to improved microfinance, if it is assumed that all households dependent on solid fuels are provided with improved biomass stoves that double the efficiency of combustion, then biomass demand in this scenario could be cut in half, from 10 EJ, as shown in Figure 17.35, to about 5 EJ. Finally, in the case where a 50% fuel subsidy is combined with improved microfinance and high electricity demand with universal access, total energy demand actually drops to 16.8 EJ. This is explained by a rapid shift away from biomass to more efficient LPG for cooking and a substitution away from kerosene to electricity for lighting. Total LPG demand in this scenario is projected to rise from 1.1 EJ in the base year to 9.4 EJ in 2030; biomass demand declines from 13.4 EJ to 1.7 EJ over the same period. This increase in LPG demand over the entire projection period for the three developing regions amounts to less than half of energy use in 2005 in the Western European transportation sector alone. Electricity demand rises in this scenario from 1.7 EJ in 2005 to 5.7 EJ by 2030, displacing about 6.6 EJ of kerosene.

The changes in final energy demand due to various access policies also have implications for GHG emissions. Figure 17.35 presents the impacts of various access policies on total GHG emissions relative to the base year of 2005. The grey columns (scale on the right axis) depict total emissions, assuming that all biomass consumption is sustainably harvested, and the error bars indicate emissions in the case where 20% of biomass consumption is assumed to be harvested unsustainably. Without any access policy, total GHG emissions increase by 65%, to 4.7 gigatonnes

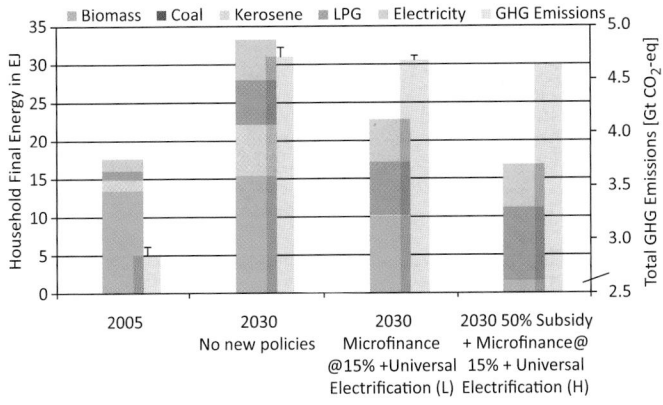

Figure 17.35 | Household final energy demand and total GHG emissions in three developing regions (sub-Saharan Africa, South Asia, and Pacific Asia) in 2005 and in 2030 under alternative access policy scenarios. GHG emissions include those on both the supply and the demand side. Error bars represent additional emissions on the assumption that 20% of biomass consumption in households is not sustainably harvested.

of carbon dioxide-equivalent ($GtCO_2$-eq.) in 2030 compared with 2.9 $GtCO_2$-eq. in 2005. As a consequence of implementing access polices, GHG emissions decline marginally, having a negligible impact overall.

17.4.4 Summary of the Costs and Impacts of Access Policies

The previous subsections have highlighted the level of existing access to both clean cooking and electricity in developing countries, the policies and measures required to accelerate access, and the relative costs and effectiveness of these policies and measures in achieving access goals and targets. Detailed assessments and scenarios were constructed for the three major regions of the world where the lack of access is most acute, namely, South Asia, Pacific Asia, and sub-Saharan Africa. These regions account for over 85% of the total global population without access to electricity and over 70% of the global population still dependent on solid fuels. Extrapolating the cost estimates for these three key regions to arrive at a global estimate of the costs of access policies suggests that between US$36 billion and US$41 billion will need to be spent annually until 2030 to ensure that almost universal access to clean cooking and electricity is achieved. For the high end of the estimate, about half of this amount will need to be spent on improving access to electricity and the rest on improving access to clean cooking. The largest share of this spending (more than a third of the total cost to achieve clean cooking access and two-thirds of the electrification bill) will need to occur in sub-Saharan Africa. The wide range in estimated costs is a consequence of whether the cost of stoves (LPG and improved biomass stoves) is included in the estimates or assumed to be provided through microfinance instruments that recover these costs. However, even the high end of this estimate is less than 5% of global energy sector investment today.

Spending on policies and measures to achieve access goals by 2030 will improve the welfare of those benefiting in several ways. Health impacts

from improved household air quality are quantified in Section 17.5.2.3. Access policies will result in averting between 0.6 million and 1.8 million premature deaths, on average, every year until 2030, or a savings of over 24 million DALYs annually. Additional benefits that are likely to be substantial include time savings for women and children and the potential for improved livelihood opportunities.

17.5 Energy and the Environment

17.5.1 Climate Change

The ultimate goal of international climate change policy, as stated in Article 2 of the United Nations Framework Convention on Climate Change, is to "avoid dangerous anthropogenic interference with the climate system." This goal has motivated a wide array of analyses of potentially dangerous climate change impacts and of mitigation strategies that might limit GHG concentrations or global average temperature increases. (For an overview see, for example, Smith et al., 2009, or the report by IPCC AR4 Working Group II, IPCC, 2007). Political attention has increasingly focused on limiting global average warming to 2°C above preindustrial levels, as reflected most recently in the acknowledgment by the Copenhagen Accord of the scientific basis for such a limit (O'Neill et al, 2010b).

The 2°C limit on warming has also been adopted by the GEA as one of the main sustainability objectives. This target is one of the fundamental drivers of the demand- and supply-side transformations portrayed in Sections 17.3.2 and 17.3.3, respectively. The sequel of this section will focus on the consequences of the transformation for the required reductions of GHG emissions, the pace at which the energy system will need to decarbonize, associated costs, and finally, some potential implications with respect to the regional equity of the solutions.

17.5.1.1 Probability of Staying below 2°C Temperature Change

The relationship among future GHG emissions, resulting changes in GHG concentrations in the atmosphere, and the ultimate effect in terms of temperature change is subject to large uncertainty. Major reasons for this uncertainty include the limited present understanding of important carbon cycle feedbacks and, in particular, the uncertainty surrounding the so-called climate sensitivity, defined as the increase in global mean temperature resulting from a doubling of the GHG concentration in the atmosphere.

Implications of this uncertainty are manifold. First, climate change needs to be seen within the context of an adaptive risk management problem. That is, the risks of exceeding future thresholds for specific impacts need to be viewed in the context of measures undertaken today and in the future to reduce those risks, and the costs of those measures. Second, targets such as the 2°C limit need to be studied in a probabilistic context.

In other words, one has to define the likelihood with which a certain temperature target can be achieved to properly define the objective.

The GEA pathways aim at an ambitious target that maximizes the chances of keeping the global temperature increase below 2°C, while at the same time providing sufficient flexibility in the system to allow for multiple pathways to reach the target. Setting an ambitious target is important for limiting the risk of dangerous interference with the climate system with high likelihood. Flexibility of solutions is central for identifying decarbonization strategies that are robust against multiple uncertainties due, for example, to potential technological failure and the associated risks (see also the discussion about flexibility in Section 17.3.3). An extensive sensitivity analysis was therefore conducted to assess the "maximum" likelihoods under a range of assumptions for the stringency of emissions reductions. (For an illustration of likelihood estimates of different emissions pathways, see, for example, Figure 17.53.) Probabilistic assessment of the relationship between GHG emissions and global temperature change has been studied by den Elzen and van Vuuren, (2007), Keppo et al. (2007), Meinshausen (2006), Meinshausen et al. (2009) and O'Neill et al. (2010a). Like these earlier studies, the present analysis finds that under very stringent emissions reductions, the 2°C target can be achieved with a likelihood exceeding 50% (maximum likelihoods found in the analysis were around 67%).[41]

Exact numerical values for the likelihood of meeting the 2°C target differ slightly across the individual GEA pathways. In principle, however, all GEA-Efficiency, GEA-Mix, and GEA-Supply pathways stay below the 2°C target with a probability between 50% and 67%.

17.5.1.2   GEA Emissions Pathways

The target of limiting temperature change to 2°C with a probability above 50% translates into very stringent emissions reductions, comparable to the lowest emissions scenarios that have been developed so far with integrated assessment models. This section focuses on CO_2 emissions, as these make up the largest share of greenhouse gas emissions from energy and industry by far. For non-CO_2 emissions of the GEA pathways, see the online GEA database: www.iiasa.ac.at/web-apps/ene/geadb.

Figure 17.36 compares the total global CO_2 emissions pathways of the GEA with selected scenarios from the literature, including the most stringent climate change mitigation scenarios assessed by the IPCC AR4 (category I, Fisher et al., 2007) as well as high-emissions scenarios assuming no interventions or climate policies in the future (Nakicenovic and Swart, 2000). As the figure illustrates, total CO_2 emissions (from land use, energy, and industry) in the GEA pathways follow a trajectory comparable to those of the most stringent IPCC scenarios. In these

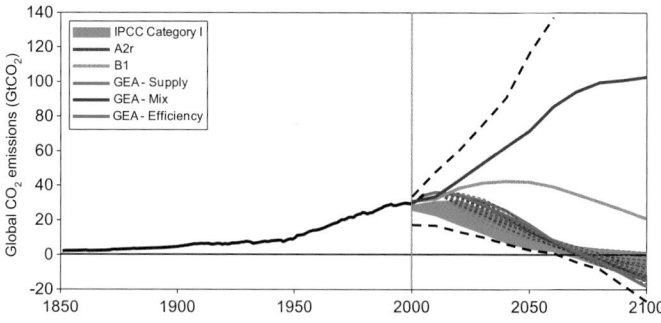

Figure 17.36 | Actual global CO_2 emissions, 1850–2000, and projections of the GEA pathways for 2000–2100. Solid lines in red, blue and green denote emissions under the three illustrative GEA pathways, and dashed lines those for individual pathways in the full set. Shaded area indicates the 90th percentile range of emissions under the most stringent mitigation scenarios of the IPCC AR4 (category I). Brown and grey lines of the A2r and B1 scenarios show the approximate range of nonintervention scenarios in the literature (Nakicenovic and Swart, 2000), assuming no implementation of climate policies.

low-emissions pathways, emissions may continue to increase for a very short period but have to peak and decline rapidly thereafter to reach zero to negative emissions in the long term.

The low-emissions pathways of the GEA and the IPCC category I scenarios are compatible with long-term atmospheric CO_2 concentrations below 400 parts per million (ppm). In fact, most of the GEA pathways reduce CO_2 concentrations to around today's concentration of about 390 ppm.[42] These low concentrations are the result of achieving globally negative emissions due to enhancements of the terrestrial sink potential (e.g., afforestation and reforestation) in combination with BioCCS in the late 21st century. Further details on emissions mitigation options are provided below. Accounting for the direct and indirect effects of non-CO_2 GHG emissions and other radiatively active substances results in long-term concentration levels under the GEA scenarios of 440–450 ppm CO_2-equivalent.

The CO_2 emissions of the GEA pathways are driven by stringent GHG mitigation policies to reduce emissions intensities across all sectors and sources (see Section 17.5.1.3). The magnitude of the challenge is huge, as Figure 17.36 illustrates by comparing the GEA emissions pathways with scenarios without any future climate change mitigation policies. Although, again, emissions in the absence of climate policies are subject to relatively large uncertainties, the GEA pathways depict reductions of about 70–85% by 2050 compared with scenarios without any policy interference.

Arguably, a more informative indicator of the necessary emissions reductions is obtained by comparing future emissions with today's levels. For this purpose, Figure 17.37 considers CO_2 emissions from energy and industrial sources only. The corresponding emissions profiles of the GEA pathways feature three major characteristics for the short, medium, and long term:

---

41   For the estimation of likelihoods of temperature outcomes, the probability distribution of the climate sensitivity of Forest et al. (2002) was used. The methodology is described in detail in O'Neill et al. (2010a) and Keppo et al. (2007).

42   As reported by the Mauna Loa observatory (www.esrl.noaa.gov/gmd/ccgg/ trends/#mlo).

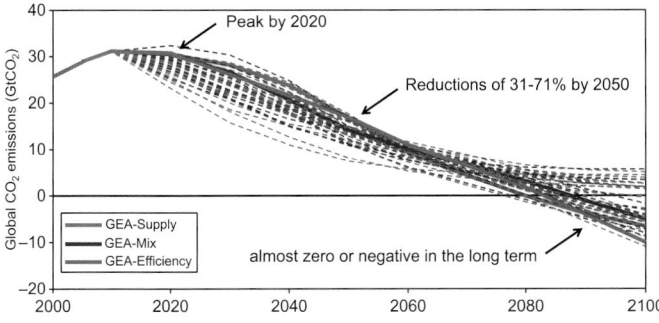

Figure 17.37 | Projected global CO_2 emissions from energy and industry in the GEA pathways. Solid lines denote emissions under the three illustrative GEA pathways, and dashed lines those for individual pathways in the full set.

- rapid introduction of climate change mitigation measures over the next decade to stop emissions growth, resulting in a peaking of emissions by about 2020;

- further strengthening of climate policies over the medium term to achieve CO_2 emissions reductions of about 30–70% by 2050 compared with 2000; and

- net negative emissions by the end of the century in the majority of pathways (particularly those that allow for BioCCS).

The relatively wide range of emissions reductions by 2050 reflects uncertainties with respect to emissions reduction potentials in the long and the short term. It is derived from the comprehensive sensitivity analysis across the transformation pathways and reflects choices as well as uncertainties with respect to policy implementation and technological development on both the demand and supply sides of the energy system (see Sections 17.3.2 and 17.3.3). Generally, pathways that have restricted supply-side portfolios (e.g., limited potential for renewables, or no CCS) require more rapid emissions reductions early in the century, to compensate for the loss of mitigation potential in the long term. For example, in the absence of bioenergy and CCS, emissions from the energy sector cannot become negative in the long term and thus need to be reduced comparatively more early in the century. A later subsection revisits the issue of how technology assumptions may influence the emissions profile.

The stringency of the emissions reductions becomes apparent when reviewing the cumulative emissions budgets of the GEA pathways. Given the cumulative nature of climate change, aggregate emissions over the full century represent one of the central boundary conditions for staying below the 2°C target. In the GEA pathways, the allowable emissions budget is on average around 1180 $GtCO_2$ between 2010 and 2100 (full range is 940–1460 $GtCO_2$). At today's rate of emissions, this "headroom" would be spent on average in about 38 years (full range between 30–45 years). With continuing growth in emissions in the absence of any new climate policies, the headroom would shrink further to about 27 years (full range between 22–32 years) before the overall objective for the full century would become out of reach.

Table 17.16 summarizes the characteristics of the GEA pathways and compares them with the lowest emissions scenarios assessed by the IPCC (category I). In addition to the IPCC scenarios, this analysis considers the three main recent studies that have looked into the relationship between short- and medium-term emissions characteristics of a wide set of scenarios. Van Vuuren and Riahi (2011) have conducted a survey of recent scenarios and updated the IPCC assessment with a wider set of new scenarios published since that assessment (collated from different sources). In addition, the results of studies by den Elzen and van Vuuren (2007) and O'Neill et al. (2010b) are shown, since they explicitly analyze short-term emissions reductions in the context of long-term temperature and GHG concentration targets.

The results across the studies are relatively similar, and all studies suggest the need for very ambitious short-term emissions reductions if CO_2 concentrations are to be kept below 400 ppm (corresponding to the 2°C target with a likelihood exceeding 50%). However, both the most recent studies and the GEA pathways indicate that there might be slightly greater flexibility for emissions reductions than indicated by the IPCC assessment. As noted by van Vuuren and Riahi (2011), a main reason for this difference is that a large number of new scenario studies have been published since the IPCC AR4 (IPCC, 2007), especially for very low long-term concentration levels. For instance, global emissions peak around 2020 in the GEA pathways as well as in the recent literature, which is around five years later than reported by the IPCC. Similarly, 2050 emission reductions in the least reduction scenarios are about 30% in the GEA pathways, compared to 50% at the time of the IPCC assessment. Studies that explicitly explored emissions thresholds that, if surpassed, would make the lowest long-term targets infeasible suggest even less stringent emissions reductions (O'Neill, 2010b). This latter conclusion depends, among other things, on assumptions about the future availability of technology and the feasibility of negative emissions in the second half of the century, which is reviewed next.

Impact of Technology Assumptions on Required Short-Term Emissions Reductions

As indicated earlier, the trajectory of emissions in the GEA pathways depends strongly on assumptions about technologies, the portfolio of abatement options considered, and their potentials. Crucial technological options include energy efficiency-enhancing technologies, renewables, CCS, and nuclear energy, as well as technologies that would allow for negative emissions later in the century, such as carbon plantations and BioCCS. For a discussion of the deployment of these options and how they shape the energy transformation, see Sections 17.3.2 and 17.3.3.

The full set of GEA pathways explores alternative combinations of the above options, including pathways with restricted supply-side portfolios (Section 17.3.3.5). These restrictions have significant implications for the short-term emissions pathway. Generally, pathways that assume limits on the potential of individual options in the long term require stronger short-term emissions reductions in order to stay within the cumulative emissions budget (dictated by the stringent climate change

Table 17.16 | Emissions trends in the GEA pathways and in the literature.

Study	Year of peak emissions	Emissions reduction in 2050 from 2000 level (%)	No. of scenarios	Cumulative emissions (GtCO$_2$)[4]	
				2000–2050	2000–2100
Van Vuuren and Riahi (2011)	Before 2020	-85 to -40	27	807–1357	807–1522
IPCC (2007, category I)[1]	2000–2015	-85 to -50	6	n/a	n/a
O'Neill et al. (2010b)[2]	Before 2030	-85 to -15	9	1393–1760	770–1503
Den Elzen and van Vuuren (2007)	Before 2020	-65 to -40	12	1144–1320	1364–1723
GEA (illustrative pathways)[3]	Before 2020	-45 to -35	3	1290–1350	1490–1520
GEA (full set)	Before 2020	-70 to -30	41	980–1400	1230–1540

1 IPCC AR4 ranges refer to the 90th percentile of the scenario distribution.

2 Includes scenarios down to 415 ppm CO$_2$-eq. by the end of the century.

3 Ranges across the three illustrative GEA pathways for GEA-Supply, GEA-Mix, and GEA-Efficiency.

4 CO$_2$ emissions from fossil energy and industry.

n/a, not available.

Source: den Elzen and van Vuuren, 2007; van Vuuren and Riahi, 2011; IPCC, 2007; O'Neill et al., 2010b.

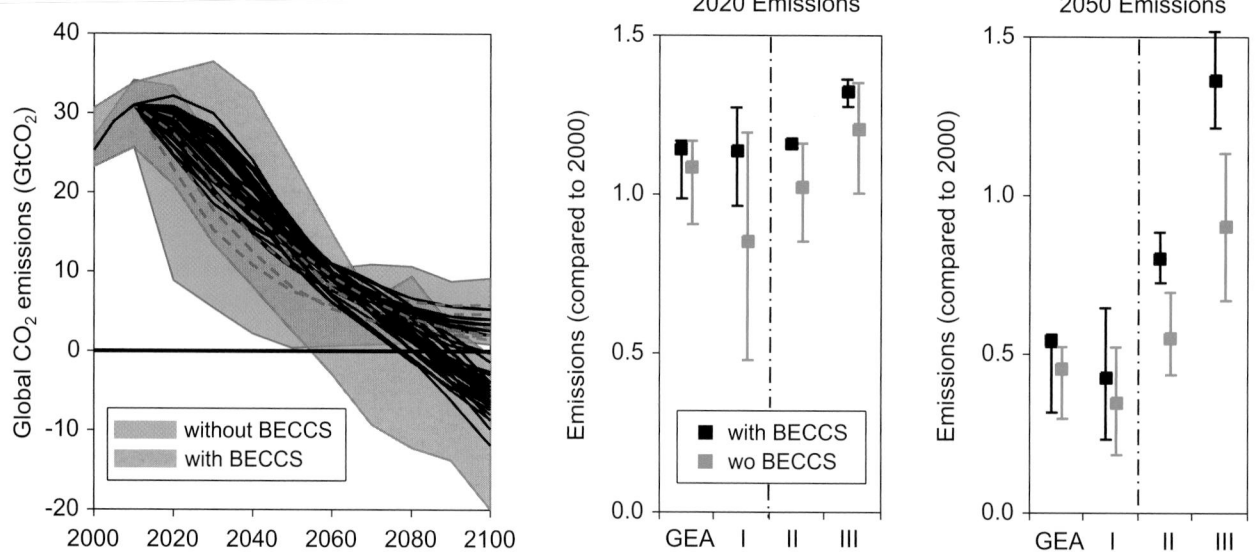

Figure 17.38 | Future CO$_2$ emissions with and without BioCCS. Left panel shows projected CO$_2$ emissions in the GEA pathways with and without BioCCS. Shaded areas show the corresponding ranges of the lowest (category I) mitigation scenarios in the literature. Middle and right panels compare average emissions in the GEA pathways by 2020 and 2050 with estimates from the scenario literature (van Vuuren and Riahi, 2011), with and without BioCCS. Error bands indicate 15–85% percentile range across the scenarios. "I" summarizes scenarios from the literature corresponding to similar targets as the GEA (category I of the IPCC, <400 ppm CO$_2$); "II" corresponds to scenarios of category II (400–440 ppm CO$_2$), and "III" corresponds to scenarios of category III (440–480 ppm CO$_2$).

objective). Although this is the case for all restricted pathways and technology combinations that were analyzed, Figure 17.38 shows the order of magnitude of this effect by using BioCCS as an illustrative example. The figure compares results of the scenario survey of van Vuuren and Riahi (2011) with the GEA pathways both for cases with BioCCS and for cases assuming that BioCCS does not become available in the future.

In general, pathways that include BioCCS allow for more modest emissions reductions in 2020 and 2050. Despite the fact that

BioCCS is rather a long-term option (see Section 17.3.3.5), the differences across pathways with respect to emissions are already relatively large by 2020 (Figure 17.38). From a systems perspective, the results thus also illustrate the path dependency of the energy system and the importance of long-term planning for short-term decisions. In addition, this finding highlights the importance of the branching point concept and the restricted portfolio analysis of the GEA for deriving robust policy conclusions for the short term (see next section).

Comparison with Present Pledges

Having reviewed the emissions under the GEA pathways, this section turns to how they compare with present plans for GHG emissions reductions.

Various countries have made commitments to mitigation actions in the context of the Copenhagen Accord. The compound effect of these pledges on global GHG emissions is subject to uncertainty. Estimates differ between studies that have collated individual country pledges and translated them into global emissions levels due to different assumptions about, for example, the business-as-usual scenario, national actions, the use of offsets included in other countries' targets, particular emissions categories, and the role of land use change (UNEP, 2010). Rogelj et al. (2010), for example, estimate that the present pledges are likely to lead to global emissions of 47.9–53.6 $GtCO_2$-eq. by 2020, and UNEP (2010) estimates a range between 48.8–51.2 $GtCO_2$-eq. by that year.

Figure 17.39 compares the range of emissions expected to result from the pledges by 2020 with the emissions reductions under the GEA pathways. As the figure illustrates, even the most optimistic assumptions about future implementation of pledges lead to emissions levels at around the upper bound of the GEA pathways. Present commitments are therefore not sufficient and thus inconsistent with the vast majority of the GEA pathways, which aim at limiting global temperature increase to 2°C compared with preindustrial times (with a likelihood of above 50%).

The gap between the present pledges and the GEA pathways ranges between none (a slight overlap of around 2 $GtCO_2$-eq.) to as large as

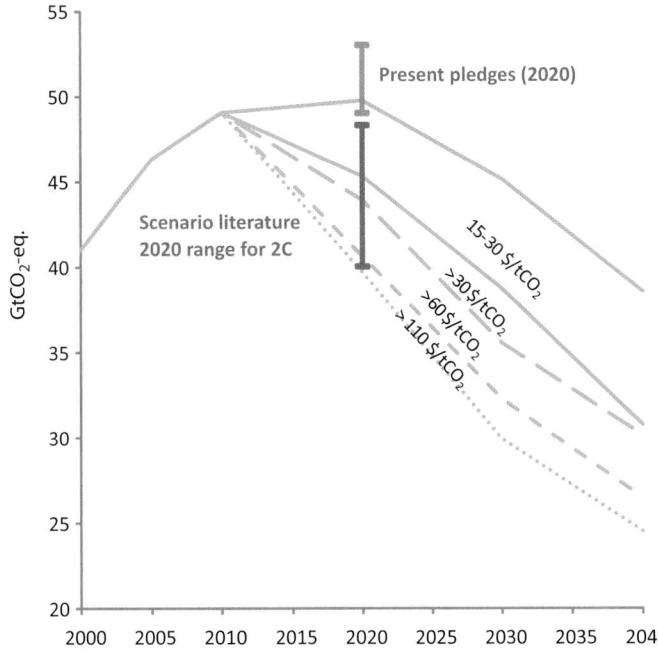

Figure 17.39 | Short-term GHG emissions in the GEA pathways compared with the range of emissions resulting from present pledges by 2020. The lower error bar gives the 90th percentile range of 2020 emissions from low-mitigation scenarios in the literature (van Vuuren and Riahi, 2011). The GEA pathways are grouped according to ranges of CO_2 prices by 2020.

11 $GtCO_2$-eq. The pathways with no gap combine the most optimistic assumptions about the emissions reductions resulting from present pledges with the highest emissions estimate from all 41 feasible GEA pathways in 2020. However, as discussed earlier, the GEA pathways with the highest emissions in the short term coincide with those cases that employ the most optimistic assumptions about the future availability of technology, and in which the full portfolio of all mitigation options can expand pervasively and successfully. Any restriction to the portfolio of mitigation options requires greater emissions reductions over the short term in order to compensate for the loss of emissions reduction potential in the long term. The gap between present pledges and the GEA pathways is therefore small only if one combines both the most optimistic assumptions about pledges with the most optimistic assumptions for the full portfolio of all mitigation options. The likelihood of the gap actually being small is thus rather low, especially if one considers the history of technology failure as well as the past performances of some countries in terms of emissions reductions.

The Price of CO_2

Figure 17.39 also shows, for each of several groups of GEA pathways, the CO_2 price that would need to be introduced globally to achieve the required reductions in emissions by 2020. According to this study's estimates, CO_2 prices would need to be on the order of US$15–45 per tonne of CO_2 to keep emissions in 2020 between 2005 and 2010 levels. As discussed in Section 17.3.5, however, higher carbon prices will need to be complemented by regulation and technology standards to mobilize the required investments and to act against, for example, rebound effects or barriers to implementation. In addition, the stringency of the mitigation policies needs to increase over time, leading to CO_2 prices increasing at about the pace of the discount rate (5%/year in the present analysis). In the most stringent emissions pathways, emissions need to drop to below the level of 2000 by 2020. The global CO_2 price corresponding to such stringent reductions is above US$110/tonne of CO_2, a value comparable with average gasoline taxes in Western Europe today.

17.5.1.3   Emissions Mitigation in the Energy Sector

As discussed earlier in this section, the objective to limit temperature change to below 2°C with a likelihood greater than 50% translates into stringent emissions reduction targets for virtually all GHG-emitting sectors.

The abatement of GHG emissions can be achieved through a wide portfolio of measures in the energy, industry, agriculture, and forestry sectors, which are the principal sources of emissions and thus of global warming. Measures to reduce CO_2 emissions range from structural changes to the energy system and the replacement of carbon-intensive fossil fuels with cleaner alternatives on the supply side (such as a switch from coal power generation to the enhanced use of nuclear and renewable energy) to demand-side measures geared toward energy conservation and efficiency improvements. In addition, CCS provides an "add-on" end-of-pipe approach for the decarbonization of hydrocarbon fuels. Other important options for GHG emissions reductions encompass the enhancement of forest sinks through afforestation and reforestation activities, as well as non-CO_2 emissions reductions in the agricultural sector.

Section 17.3.2.2 provides a comprehensive discussion of specific measures to improve efficiency on the demand side. In addition, structural changes on the energy supply side of the GEA pathways are illustrated in detail in Section 17.3.3. This section primarily explores the GHG emissions implications of those transformations, with a specific focus on the resultant pace of the decarbonization of energy supply as well as on the demand-side sectors (industry, residential and commercial, and transport).

At present, roughly 50% (14 GtCO$_2$) of global CO_2 emissions from energy are due to supply-side conversion processes, including electricity and heat generation and refining, but also losses during the transmission and distribution of fuels (Figure 17.40). The other half of emissions come from the direct use of fossil fuels in the end-use sectors: industry (5.6 GtCO$_2$), transportation (6.1 GtCO$_2$), and residential and commercial (3.6 GtCO$_2$).[43] In addition, about 2.2 GtCO$_2$ are contained in industry feedstocks and about 2.7 GtCO$_2$-eq. of non-CO_2 GHGs are emitted by the energy and industrial sectors (for example, methane from coal extraction and long-lived gases such as sulfur hexafluoride and hydrofluorocarbons in industrial processes).

The stringent climate objective of the GEA pathways requires cutting CO_2 emissions from energy and industrial sources by about half in 2050 from 2000 levels (the full range across pathways is 30–70%). The bulk of these emissions reductions are achieved through decarbonization of supply, reducing its share of energy-related emissions from 50% today to between about 25–45% by 2050 (with exception of pathways assuming limited intermittent renewables assessed in Section 17.3). However, integration of supply and demand remains essential, since one of the main reasons for the comparatively rapid decarbonization of supply is the increasing quality and flexibility of fuels demanded by consumers (e.g., electricity). Higher fuel quality requires more elaborate conversion processes and thus permits decarbonization through both fuel switching (e.g., from coal power plants to renewable power) and end-of-pipe (CCS) solutions. The latter option is economic only in large centralized systems and is thus not applicable in the context of dispersed and heterogeneous demand-side sources (except for some industrial applications, such as CCS from cement production, which is considered in this analysis).

The enormous speed of supply-side decarbonization in the GEA pathways is also illustrated by the build-up rates of low-carbon power plants (nuclear, renewable, or fossil power plants with CCS), which reach a share of around 75–98% of global power generation by 2050. By comparison, the low-carbon share of primary energy increases to (still impressive but lower) shares of about 65–85% over the same period (see Section 17.3.4.3 for further details).

The decarbonization of the demand side is equally ambitious, although by mid-century significant amounts of fossil fuels continue to play a role in the final energy mix in most of the GEA pathways. Emissions reductions on the demand side are primarily due to fuel switching away from direct use of fossil fuels, as well as increased efficiency of end-use devices.[44] At the aggregate global level, emissions from the end-use sectors are reduced by about 45% in most of the GEA pathways by 2050 compared with 2000 (Figure 17.40). These reductions are achieved despite increases in energy services levels. The low-carbon share of final energy fuels for services thus needs to increase significantly, from about 30% today to 60–70% by 2050 (see Section 17.3.4.3 for further details). The GEA-Efficiency pathways, which aim at limiting demand as one of the principal measures to attain the GEA sustainability targets, show more flexibility with respect to the rate of decarbonization and structural changes and are thus obviously at the lower bound of the ranges for low-carbon shares (for final and primary energy as well as electricity shares). Efforts to reduce emissions differ significantly across regions and are generally higher in today's industrialized world than in the developing world (see Figure 17.40 and the GEA web database, at www.globalenergyassessment.org, for further regional detail).

17.5.1.4   Regional Perspectives and Equity Issues

Achieving emissions reductions, especially such drastic ones as those depicted by the GEA pathways, is a formidable task, considering that developing countries require increases in energy services and other activities that result in GHG emissions. The salient questions are how such reductions might be achieved and by whom, and what the effects (economic, distributive, etc.) might be. In other words, how is the burden of global emissions reduction going to be shared, and what might be the criteria for such burden sharing?

Of crucial importance in this context is the large disparity between industrialized and developing countries. The former are responsible for about 40% of global energy-related emissions but account for only 20% of the world population. The more industrialized countries are also responsible (some more than others) for the bulk of the historical increase in anthropogenic GHG concentrations. Conversely, developing countries will become more important contributors to GHG emissions in the future, almost independent of how high or low global GHG emissions actually turn out to be (Grubler and Nakicenovic, 1994; Riahi et al., 2007).

Regional disparities with respect to today's per capita emissions are illustrated in Figure 17.41. Differences between regions up to an order

---

43   Estimates from the industry sector include also process emissions (e.g., from cement production) and emissions related to nonenergy feedstocks (e.g., asphalt and lubricants).

44   The latter effect of reduced energy demand decreases emissions throughout the system and thus also contributes to the supply-side emissions reductions discussed above in this section.

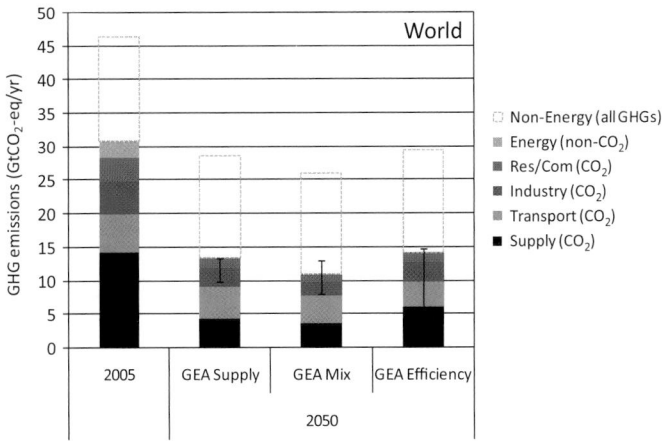

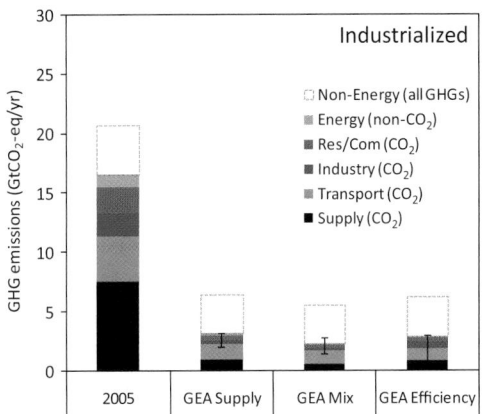

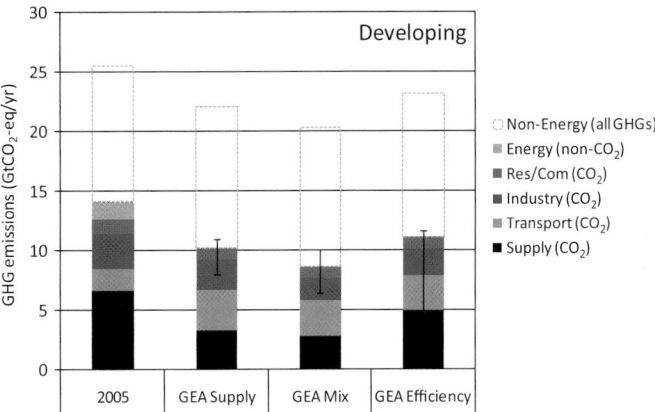

Figure 17.40 | GHG emissions from energy supply and from demand-side sectors in 2005 and in the three illustrative GEA pathways in 2050. Dashed lines indicate additional GHG emissions from the non-energy sector. Error bars show the range across all GEA pathways within each pathway group.

of magnitude are seen: all the high-income regions of the industrialized world are significantly above the world average, whereas most developing regions are considerably below the average.

The results of the GEA transition pathways clearly indicate the need for emissions reductions across virtually all regions in order to halve global per capita emissions by 2050 (Figure 17.41). Given the stringency of the GEA target and the implied magnitude of emissions reductions, today's industrialized countries must contribute proportionally greater reductions in per capita emissions. In the aggregate, this results in more equitable per capita emissions distributions in 2050 than today. Nevertheless, as Figure 17.41 illustrates, some of today's most emissions-intensive regions (e.g., North America and the former Soviet Union) continue to emit more than the world average in 2050, while the poorest regions (e.g., South Asia and sub-Saharan Africa) stay considerably below the world average.

It is important to emphasize that the vast majority of model-based mitigation analyses employ a cost-effectiveness approach. This means that emissions reductions are implemented globally when and where they are most cost-effective. The GEA scenarios are no exception in this respect. The GEA pathways address the question of when and how to spatially allocate, for

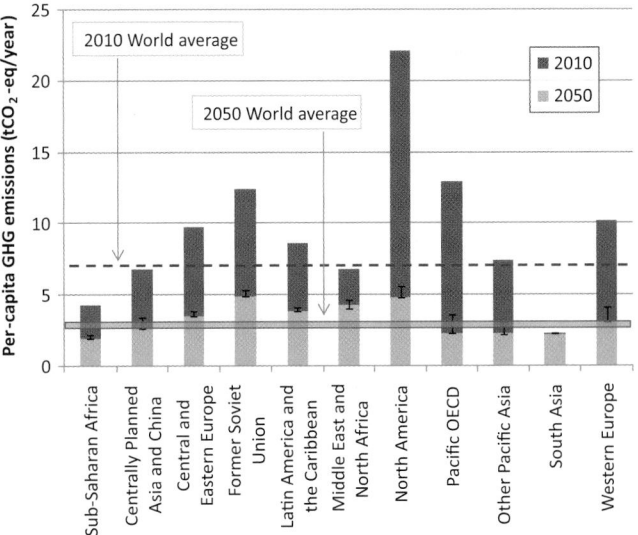

Figure 17.41 | Regional per capita GHG emissions. The full height of each bar represents per capita emissions in the indicated region in 2010, and the lower section per capita emissions in the illustrative GEA pathways by 2050. The upper section thus indicates per capita emissions reductions between 2010 and 2050. The dashed horizontal line denotes world average per capita emissions in 2010 and the solid line the range across the three illustrative GEA pathways in 2050.

example, scarce investments to meet the overall mitigation objective cost-effectively, but they do not explicitly quantify who pays for those reductions. That will depend specifically on international agreements about regional emissions entitlements and agreed-upon equity principles.

The regional allocation of future emissions reductions following equity principles may differ significantly under the cost-effectiveness approach. Generally, it is argued that given their historical responsibility and greater affluence, today's industrialized countries should take the lead in reducing emissions (Article 3.1 of the United Nations Framework Convention on Climate Change). Numerous studies have analyzed the regional emissions allocations or requirements for emissions reductions and time of participation in the international climate change regime. (For summaries see the IPCC AR4, Chapter 13, or Berk and den Elzen, 2001; Blanchard, 2002; Winkler et al., 2002; Criqui et al., 2003; Nakicenovic and Riahi, 2003; Bollen et al., 2004; Böhringer and Welsch, 2004; Groenenberg et al., 2004; Böhringer and Löschel, 2005; den Elzen and Meinshausen, 2005; den Elzen and Lucas, 2005; den Elzen et al., 2005; Höhne et al., 2005; Michaelowa et al., 2005; Höhne, 2006; and Persson et al., 2006). A large variety of system designs for allocating emissions allowances or permits has been analyzed, including contraction and convergence of per capita emissions, multistage approaches, and triptych (sectoral) and intensity targets.

A discussion of all the different proposals in the literature is beyond the scope of this chapter. Instead, this section addresses two central questions linked to the equity dimension of the GEA transition pathways. The first question is to what extent developing countries could delay their participation in emissions reduction, considering the stringency of the climate target of the GEA pathways. Here the analysis relies on findings from a recent modeling intercomparison project of the Energy Modeling Forum (EMF22), in which both modeling teams of the GEA (MESSAGE and IMAGE) were involved. The second question concerns the financial transfers that might be needed to create appropriate incentives in the developing world to join international climate agreements.

The Effect of Delayed Participation

A study by the Energy Modeling Forum (Clarke et al., 2009) investigated the effect of delayed participation of key regions in the developing world on the attainability and costs of a range of climate stabilization targets. Eleven of the leading integrated assessment modeling teams participated in the study. They jointly explored 10 alternative policy cases, assuming either full participation or a delay of the developing world and Russia in joining the international emissions mitigation regime.

Specifically, for the delayed participation scenarios, it was assumed that Brazil, Russia, India, and China do not start emissions reduction efforts until 2030, and other developing countries until 2050. The study also explored alternative emissions trajectories by differentiating between targets for CO_2-equivalent concentrations that may temporarily "overshoot" and targets that do not allow for overshoot. The summary of the EMF22 attainability analysis is presented and compared with the results of the GEA pathways in Figure 17.42.

The EMF22 results clearly indicate that whether delayed participation has any implications for the attainability of the target depends strongly on the ambitiousness, and thus the stringency, of the objective. The majority of the modeling frameworks, for example, found that although delayed participation by the developing world has significant implications for overall costs, targets above 550 ppm CO_2-eq. are still attainable. For more stringent target levels such as those adopted in the GEA (450 ppm CO_2-eq.), however, 12 out of 14 scenarios were rejected, since they were found to be infeasible under the assumption of delayed participation (see the category "Overshoot: Delay" in the right panel of Figure 17.42).

Delays by the major emitting countries of the developing world in joining a comprehensive international emissions mitigation regime would thus make attainment of the GEA objective, to limit temperature change to below 2°C with a probability greater than 50%, very unlikely. Full but differentiated participation in reduction efforts by the developing world, on the other hand, significantly increases the chance of success.

Transfers under Contraction-and-Convergence Assumptions

This section explores the implications of an illustrative burden-sharing scheme for the allocation of future emissions rights and applies it to the GEA pathways. This burden-sharing scheme is referred to in the literature as a "contraction and convergence" scheme (see, e.g., den Elzen and van Vuuren, 2007). In essence, under such a scheme, all regions need to converge to a common per capita emissions entitlement by a specified date (2050). For regions with per capita emissions above the world average, this implies reductions (hence the term "contraction") until the convergence criterion is fulfilled, but starting from very different initial conditions. For regions with per capita emissions below the world average, emissions can rise initially until they reach the world average. Thereafter, these regions also need to contract to the specified convergence level. The resulting emissions projections from the allocation scheme differ from the original GEA pathways, which assume that reductions take place where they are most cost-effective.

Figure 17.43 contrasts the difference between emissions entitlements and cost-effective emissions reductions of the original GEA pathways. In the aggregate, an equal allocation of per capita emissions by 2050 results in comparatively higher reduction needs in industrialized regions. This entails higher mitigation costs for these regions and creates an incentive to buy emissions permits on, for example, a global market. In addition to reducing total mitigation cost, trading emissions entitlements would have the co-benefit of generating revenue for developing regions. Developing countries with emissions below the world per capita average by 2050 therefore have an incentive to sell permits.

Figure 17.44 shows cumulative energy system expenditure for the original (cost-effectiveness based) GEA pathways, as well as the additional costs that would accrue for industrialized countries if the contraction-and-convergence target were achieved domestically. For developing countries this translates, of course, into lower energy system costs, due

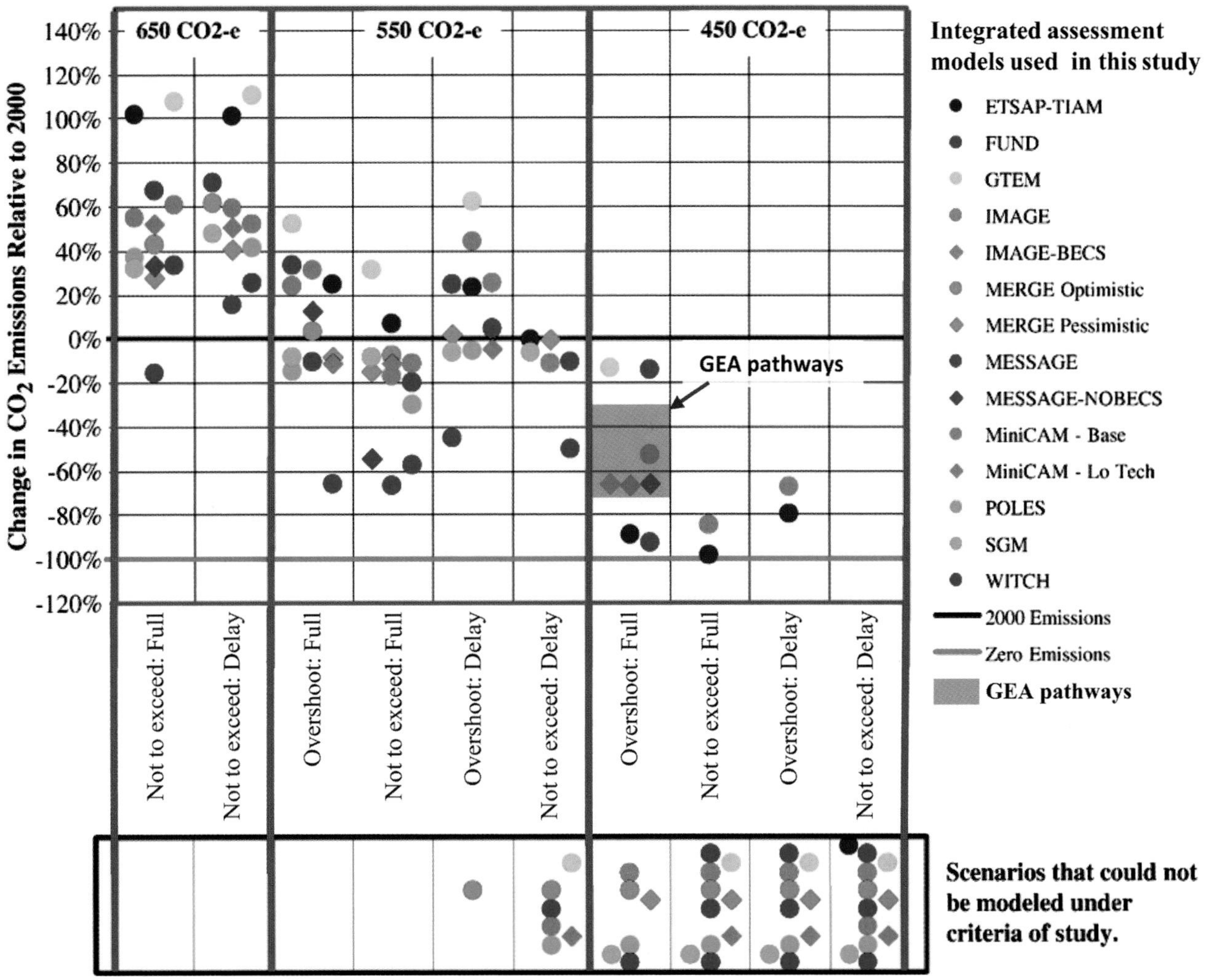

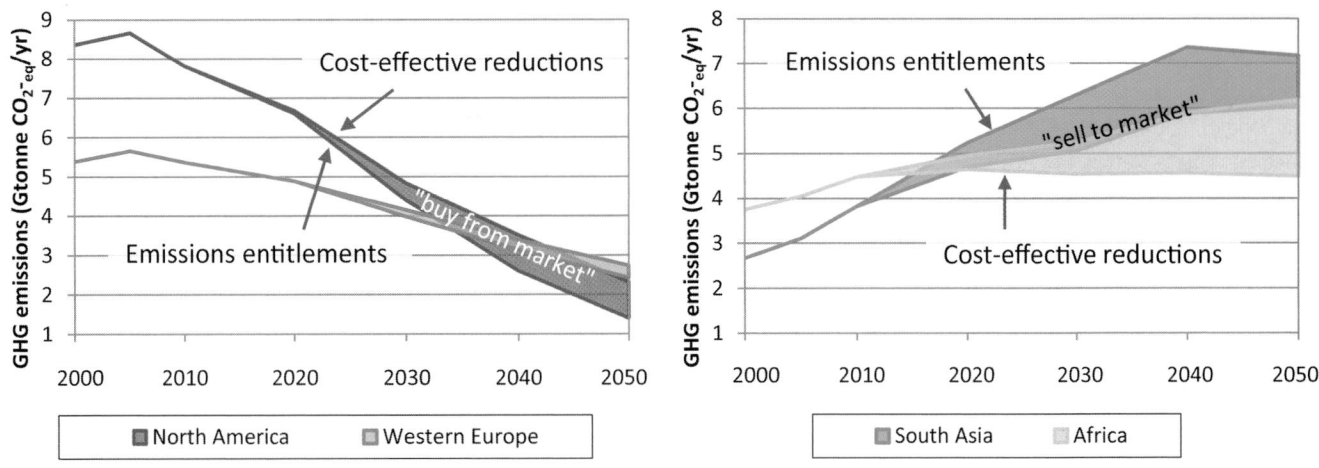

Figure 17.42 | Carbon emissions reductions from industrial and fossil sources in 2050 from 10 integrated assessment models. Dots outside the figure range indicate scenarios that were found to be infeasible under the specified criteria.

Figure 17.43 | Projected GHG emissions in the case of contraction-and-convergence allocation of emissions entitlements compared with cost-effective emissions reductions in the GEA-Mix illustrative pathway. Shaded areas in the left panel show the resulting demand for permit trade in North America and Western Europe. Shaded areas in the right panel show the resulting emissions surplus for permit sales in South Asia and sub-Saharan Africa.

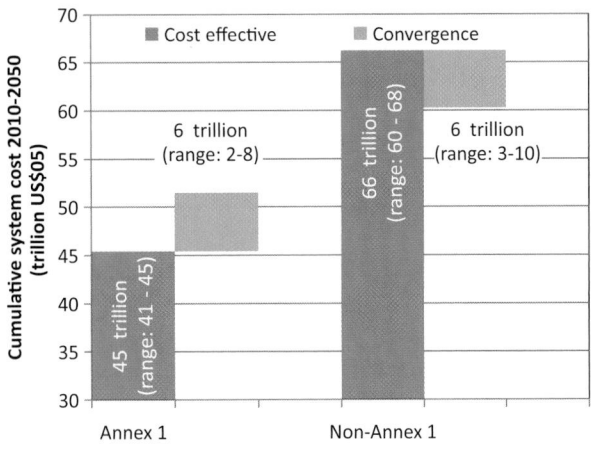

Figure 17.44 | Cumulative energy system costs of the GEA-Mix pathway in Annex 1 and non-Annex 1 regions (left bar in each pair). The right bar in each pair shows the effect on costs in the case where Annex 1 (industrialized) countries achieve the contraction-and-convergence pathway through domestic reductions only. Stated ranges are the corresponding values for the GEA-Efficiency and GEA-Supply pathways.

to their targets being relaxed, than if industrialized countries were to achieve all reductions domestically.

The increase in costs to industrialized countries to achieve the additional emissions reductions domestically approximates the region's willingness to buy permits outside the region (and thus to invest elsewhere to achieve the reductions at lower cost). This study estimates that this willingness to pay might be around US$2 trillion to US$8 trillion over the course of next 40 years, or about 3–12% of the total energy system costs of the developing world. This corresponds on average to between US$50 billion and US$210 billion/year. Although small in comparison with total energy system costs, this is a large sum compared, for example, with total official development assistance in 2000, which amounted to some US$53 billion (UNCTAD, 2003).

The estimates of financial transfers in this section are only illustrative, because they are derived from this particular contraction-and-convergence allocation scheme. Other allocation rules may yield different results. Science cannot answer the question of what particular ethical model should be used to determine the amount of international burden sharing appropriate to the challenges of mitigation and adaptation to a changing climate. Rather, this is a task for international negotiation and political decision making. Considering, however, the urgent need for rapid and globally comprehensive emissions reductions, as the GEA pathways make clear, decisions need to be made sooner rather than later, so that the 2°C target does not become out of reach.

17.5.2   Air Pollution

Pollution control is an essential component of sustainable development, as good air quality is a fundamental aspect of quality of life. Local air quality is directly linked to health, as discussed in detail in Chapter 3. As discussed in Section 17.4, household air pollution due to lack of access

to modern cooking has serious health consequences; hence, improving the quality of fuels through policies on energy access is essential. Both ambient air quality in cities and air quality within rural and urban homes are major contributors to local health. In addition, a number of air pollutants have other environmental impacts, such as acidification and eutrophication as well as damage to vegetation, as discussed in Chapter 3. In this section, the focus is on the health implications of various policy packages that include increasingly stringent air quality control policies.

17.5.2.1   Air Quality Policies

Varying levels of stringency of air quality legislation are examined here in combination with a selection of other policies sampled from the GEA scenario space described in the earlier sections on energy efficiency (Section 17.3.2), energy access (Section 17.4), and climate change (Section 17.5.1). The objective is to cover a wide range of air pollution outcomes and to analyze in detail the implications of different policy packages in terms of their health benefits. This section thus explores both future pollutant levels in the absence of further improvements in air quality legislation and GEA pathways that address all challenges simultaneously.

The assessment builds upon the MESSAGE energy model as the primary tool for deriving detailed, sector-based estimates of various pollutant gases. In addition, MESSAGE is linked to the GAINS air quality model (Amann et al., 2008) to represent different levels of air quality legislation until 2030[45] (for further details see Rafaj et al., 2010, and Rao et al., forthcoming). Regional emissions estimates for 2005 are based on historical and current inventories as described in Granier et al. (2010) and Lamarque et al. (2010). A number of air pollutants and GHGs have been downscaled to spatially explicit levels for 0.5-degree resolution (see Riahi et al., 2011 for methodology). To estimate the impacts of the spatially explicit emissions, atmospheric concentrations of particulate matter, aerosols, and ozone were derived using the TM5 model (Dentener et al., 2006; Stevenson, 2006; Kinne et al., 2006; Textor et al., 2007; Bergamaschi et al., 2007). TM5 includes contributions from (i) primary PM2.5 (particulate matter <2.5 μm in diameter) released from anthropogenic sources, (ii) secondary inorganic aerosols formed from anthropogenic emissions of sulfur dioxide, nitrogen oxides, and ammonia, and (iii) particulate matter from natural sources (soil dust, sea salt, biogenic sources). Table 17.17 describes in detail the background of the chosen policy packages and the types of air pollutants, sectors, and spatial scales covered by them.

The policies driving each of these scenarios and their relevance for air pollution outcomes are discussed in more detail below:

---

45   Although the focus of this section is on the shorter-term pollution estimates until 2030, emissions pathways are represented until 2100 based on assumptions of future improvements in emissions factors as described in Rafaj et al. (2010) and Rao et al. (2012).

Table 17.17 | Policy matrix and coverage.

Policy package	Policies			
	Air pollution	**Climate change**	**Energy efficiency**	**Energy access**
FLE	No improvement in air quality legislations beyond 2005	No climate change policy	Annual energy intensity reduction of 1.5% until 2050	No specific energy access policy; slow improvement in quality of cooking fuels
CLE1	All current and planned air quality legislations until 2030	No climate change policy	Annual energy intensity reduction of 1.5% until 2050	No energy access policy; medium improvement in quality of cooking fuels
CLE2	All current and planned air quality legislations until 2030	Limit on temperature change to 2°C in 2100	Annual energy intensity reduction of 2.6% until 2050	Moderate energy access policy
SLE1	Stringent air quality legislations globally	Limit on temperature change to 2°C in 2100	Annual energy intensity reduction of 2.6% until 2050	Moderate energy access policy
SLE2	Stringent air quality legislations globally	Limit on temperature change to 2°C in 2100	Annual energy intensity reduction of 2.6% until 2050	Policies to ensure global access to clean energy by 2030

Note: Sectors included in all policy packages are power plants, industry (combustion and process), road transport, international shipping and aviation, agricultural waste burning, biomass burning (deforestation, savannah burning, and vegetation fires). GHGs and air pollutants gridded include methane, sulfur dioxide, nitrogen oxides, carbon monoxide, volatile organic compounds, black carbon, organic carbon, and PM2.5; gridding is based on spatial allocation maps (using the dataset described in Lamarque et al. (2010) and methods from Riahi et al. (2011)).

- *No sustainability policies (FLE)*: This policy package assumes that no specific policies on sustainability are implemented. There is no change in future air pollution policies relative to 2005. Energy demand in this scenario is higher than in the GEA-Supply illustrative scenario, as no climate change policies are implemented, and therefore no feedback on energy demand from such policies is assumed. There is also no implementation of policies on improving energy access, although increasing economic growth leads to a slow decline in the use of solid fuels for cooking and heating in developing regions. As a result of the "frozen legislation" (FLE) assumptions, pollution levels in this scenario are the highest among the scenarios described.

- *Moderate air pollution policies (CLE1)*: This scenario is identical to the FLE case in terms of energy structure and lack of specific policies on climate change and energy access. However, it assumes full implementation of all current and planned air pollution legislation (CLE) worldwide until 2030. (See Table 17.18 for details of the types of measures undertaken.) Thus, this scenario provides a measure of the impact of current and planned air pollution policies in the absence of any specific climate or energy access policy.

- *Moderate air pollution, stringent climate, and moderate energy access policies (CLE2)*: This scenario is based on the illustrative scenario of the GEA-Efficiency pathways group described earlier in this chapter in terms of energy demand and use and the implementation of a stringent climate policy corresponding to a global temperature target of 2°C maximum warming. In addition, it assumes a moderate energy access policy, corresponding to availability of microfinance and a 20% fuel subsidy (as described in Section 17.4.1.2), as well as full implementation of all current and planned air quality legislation until 2030 as in the previous scenario. Thus, this scenario explicitly provides an indication of the multiple benefits of combining moderate policies on climate change, energy access, and air pollution.

- *Stringent air pollution, stringent climate, and moderate energy access policies (SLE1)*: This scenario differs from the previous one in that it assumes global implementation of extremely stringent pollution policies until 2030 (see Table 17.18 for details). These policies are much more aggressive than the currently planned legislation assumed in the previous two cases, but are less aggressive than the so-called maximum feasible reduction (MFR) level, which describes the technological frontier in terms of possible air quality control strategies by 2030 (for further details on CLE and MFR, see Amann et al., 2004).

- *Stringent air pollution, stringent climate, and universal energy access policies (SLE2)*: This is a variant of the previous scenario that includes in addition the universal access policy described in Section 17.4, and investigates specifically how stringent policies on energy access in developing regions, combined with stringent air pollution legislation, can affect emissions levels and associated health impacts.

Table 17.18 describes in detail the types of air pollution control technologies and policies adopted in the CLE and SLE cases. The information is derived and summarized from a number of GAINS-related publications including Cofala et al. (2007) and Kupiainen and Klimont (2004).

Policy Impacts on Pollutant Emissions
Anthropogenic sources are major contributors to outdoor air pollution, with the energy system alone contributing around 60% of PM2.5 emissions in 2005. A number of policies to control air pollution have been implemented, especially in the industrialized countries, in the past two decades: global air pollution control costs in 2005 are estimated at US$195 billion. However, more than 80% of the world's population is estimated to be exposed to PM2.5 concentrations exceeding WHO air quality standards (annual mean) of 10 μg/m^3 in 2005 (see Rao et al., 2012, for details). Future air pollution levels will depend on the future development of the energy system and the types of policies that are

Table 17.18 | Policies and measures for air pollution control.

	Transport	Industry and power plants	International shipping	Other
Current legislation (CLE)				
Sulfur dioxide (SO_2)	OECD: Directives on the sulfur content in liquid fuels; Non-OECD: National legislation on the sulfur content in liquid fuels	OECD: Emission standards for new plants from the Large Combustion Plant Directive (LCPD) (OJ 1988) Non-OECD: increased use of low-sulfur coal, increasing penetration of flue gas desulfurization (FGD) after 2005 in new and existing plants	MARPOL Annex VI regulations	Reduction in gas flaring, reduction in agricultural waste burning
Nitrogen oxides (NO_x)	OECD: Emission controls for vehicles and off-road sources up to the EURO-IV/ EURO-V standard Non-OECD: National emission standards equivalent to approximately EURO III-IV standards (vary by region)	OECD: Emission standards for new plants and emission ceilings for existing plants from the LCPD (OJ 1988). National emission standards on stationary sources– if stricter than in the LCPD Non-OECD: Primary measures for controlling of NO_x	Revised MARPOL Annex VI regulations	Reduction in gas flaring, reduction in agricultural waste burning
Carbon monoxide (CO)	As above for NO_x			Reduction in gas flaring, reduction in agricultural waste burning
Volatile organic compounds (VOC)	End-of-pipe measures as described above for NO_x	Solvent Directive of the EU (COM(96)538, 1997); 1999 UNECE Gothenburg Protocol to Abate Acidification, Eutrophication and Ground-level Ozone		Reduction in gas flaring, reduction in agricultural waste burning
Ammonia (NH_3)		End-of-pipe controls in industry (fertilizer manufacturing)		Substitution of urea fertilizers
PM2.5[1]		EU and national legislation on power plants and industrial sources limiting stack concentrations of PM		Reduction in gas flaring, reduction in agricultural waste burning
Stringent legislation (SLE)				
SO_2	As in CLE	High-efficiency flue gases desulfurization (FGD) on existing and new large boilers Use of low-sulfur fuels and simple FGD techniques for smaller combustion sectors High-efficiency controls on process emission sources	Revised MARPOL Annex VI and NO_x Technical Code 2008	Cessation of gas flaring, reduction in agricultural waste burning
NO_x	As in CLE	Selective catalytic reduction at large plants in industry and in the power sector Combustion modifications for smaller sources in industry and in the residential and commercial sectors High-efficiency controls on process emission sources	Revised MARPOL Annex VI and NO_x Technical Code 2008	Cessation of gas flaring, reduction in agricultural waste burning
CO	As in CLE			Cessation of gas flaring, reduction in agricultural waste burning
VOC	As in CLE	Regular monitoring, flaring, as well as control of the evaporative loses from storage Solvent use: full use of potential for substitution with low-solvent products in both "do it yourself" and industrial applications, modification of application methods and introduction of solvent management plans		Cessation of gas flaring, reduction in agricultural waste burning
NH_3		End-of-pipe controls in industry (fertilizer manufacturing)		Substitution of urea fertilizers, rapid incorporation of solid manure, low nitrogen feed and biofiltration
PM2.5 (including BC and OC)		High-efficiency electrostatic precipitators, fabric filters, new boiler types, filters, good practices	Revised MARPOL Annex VI regulations	Good practices in agriculture production, ban on agricultural waste burning

1 Legislation is for PM2.5 only, but black carbon and organic carbon emissions can be expected also to decline as a result.

implemented. The impacts of specific policies described in Table 17.17 on pollutant emissions in 2030 are examined below.

The absence of significant future legislation on air quality, combined with a lack of policies on energy efficiency and energy access (the FLE scenario), is seen to lead to a significant increase in all categories of emissions to more than 30% above 2005 levels and added deterioration in air quality, with 90% of the world's population exposed to PM2.5 concentration levels above WHO air quality standards – an increase of 10 percentage points compared with 2005.

Currently planned air quality legislation (the CLE1 scenario) is seen to curb the growth of emissions, especially in OECD countries. However, emissions continue to increase in non-OECD countries because of the overall high energy demand and very little or nonexistent air quality legislation in many countries (e.g., in Africa). Sulfur dioxide emissions decrease globally by only 2% in 2030 compared with 2005, in spite of a 30% decrease in OECD countries. Nitrogen oxide emissions increase globally to 115 Mt, a 15% increase over 2005 levels, again mainly due to increasing emissions – in particular, from the transportation and power sectors – in non-OECD countries, particularly in Asia. Globally, PM2.5 emissions decrease by around 2–3%, mainly from shifts in cooking fuels in the residential sector, currently the largest source of emissions (around 50% of the total, almost 90% of which is in non-OECD countries), as well as assumed legislation that establishes stronger controls on power plants, industry, and road transport. More than 80% of the world's population continues to be exposed to levels above the mandated WHO standards, the same as in 2005. This clearly indicates that, even if currently legislated air pollution control policies were globally implemented, only modest declines in pollutants would be expected. This occurs mainly because of increasing growth in emissions in developing countries in spite of the significant technological shifts that can be expected in many parts of the world in the next two decades.[46]

Emissions decline when air pollution policies are combined with additional policies on climate change, energy access, and energy efficiency. The effects of such combined policies are determined by the stringency of the individual policies assumed. A policy package of currently legislated air quality controls, together with policies on climate change, energy access, and energy efficiency (the CLE2 scenario), results in emissions reductions on the order of 50% for sulfur dioxide (SO$_2$), 35% for nitrogen oxides (NO$_x$), and 30% for PM2.5. Most of these reductions (up to 80%) occur in non-OECD countries, thus indicating that the co-benefits of combined policies are the highest there. Comparing the panels of Figure 17.45, transport and industrial sectors in particular are seen to be the most important sources of

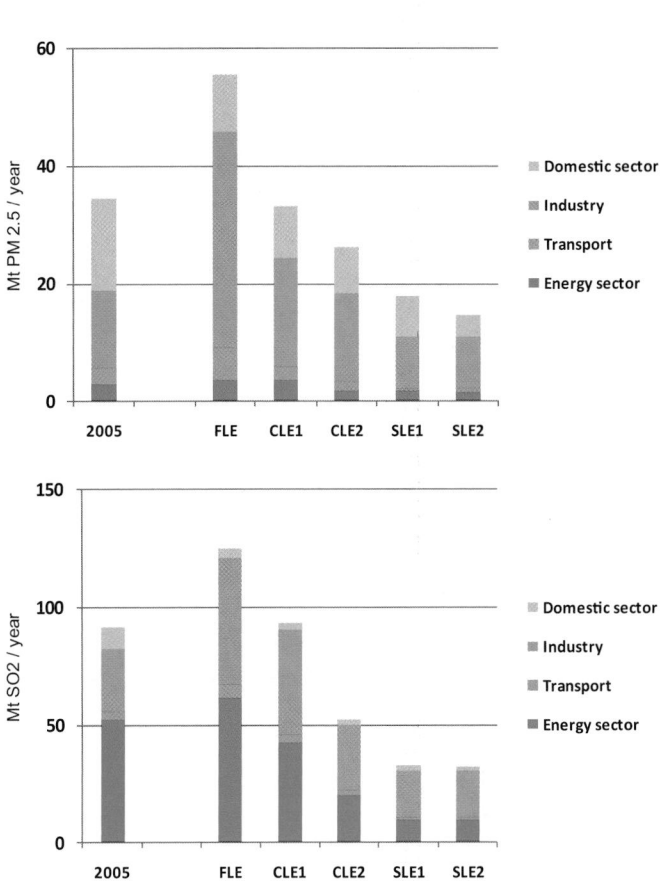

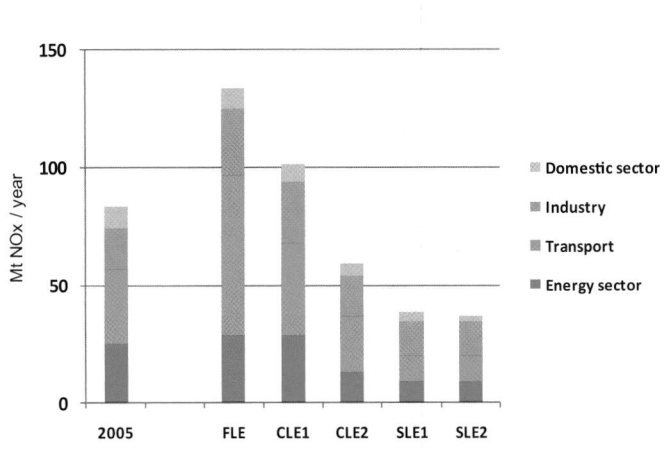

Figure 17.45 | Global energy-related pollutant emissions by sector in 2005 and under alternative policy packages in 2030.

reductions (a 28% reduction in NO$_x$ and a 35% reduction in PM2.5), as these sectors offer significant opportunities for combined policies that can tap the co-benefits of GHG mitigation and air pollution control.[47] In the residential sector, moderately stringent policies on access to modern energy forms in developing countries have a significant impact on pollutant emissions (a 60% reduction in SO$_2$ and 30–40%

---

46 Emissions from international shipping, however, show a significant decline (80% reduction in SO$_2$ and 20% reduction in NO$_x$) despite increasing fuel use in this sector. This is because of the stringent international policies that are expected to govern this sector.

47 NO$_x$ emissions from the power sector, although decreasing in the short term, may increase in the longer term because of the increase of overall electricity demand.

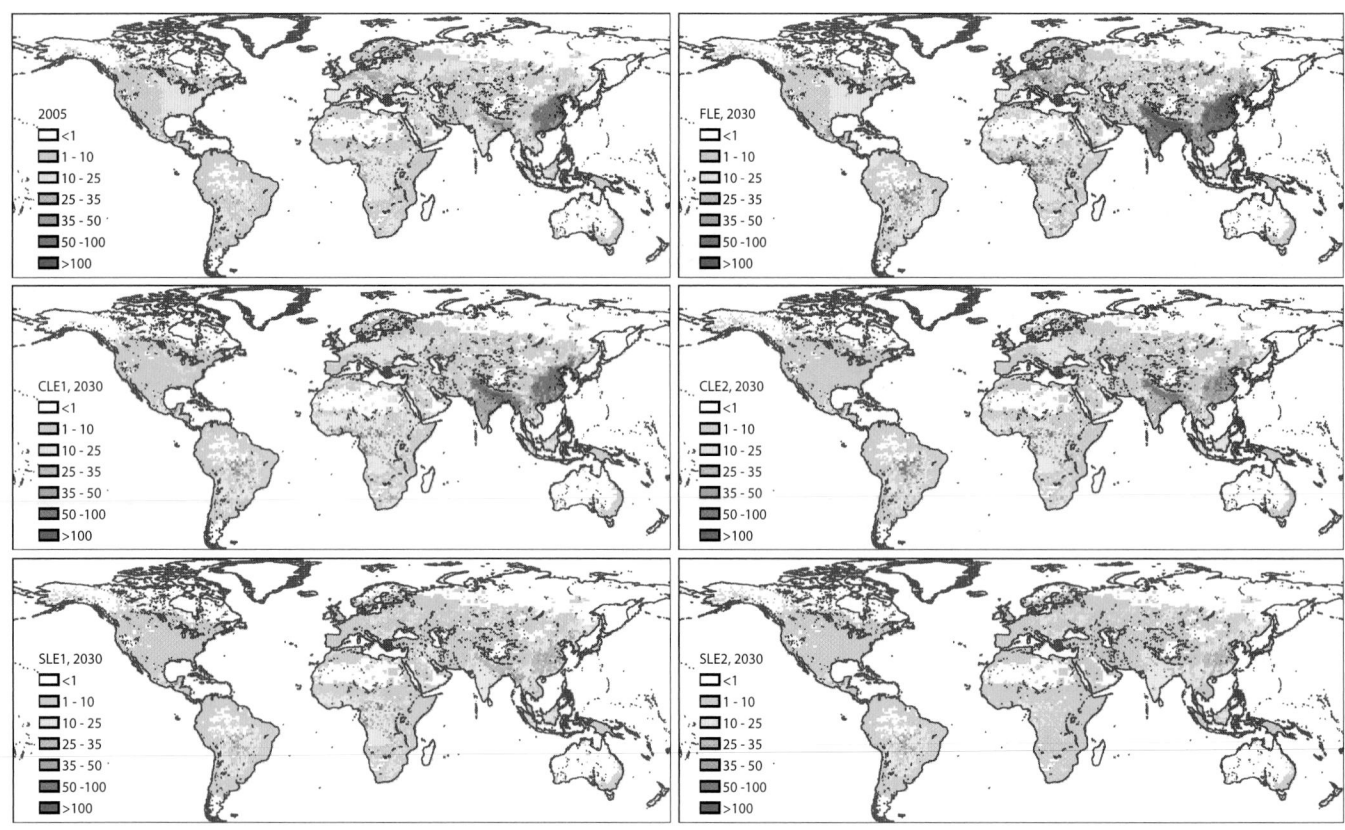

Figure 17.46 | Geographic distribution of anthropogenic PM2.5 concentrations in 2005 and under alternative policy packages in 2030 in μg/m³.

reductions in NOx and PM2.5). The pollution control costs of CLE2 are around 12% lower than those of CLE1. However, a CLE2 policy package still results in 70% of the world's population at levels beyond WHO's air quality guidelines in 2030, indicating that more stringent policies will be needed if further improvements are required.

Increasing the stringency of air quality legislation (the SLE1 scenario) leads to significant reductions across air pollutants by more than 50% (see Table 17.18 for details on controls), especially in sectors such as transport, where stricter controls yield large benefits. The annual air pollution control costs of such a scenario in 2030 are estimated at 50% lower than for the CLE1 policy package, thus implying significant co-benefits of combined policies. Around 60% of the world's population is still exposed to levels beyond WHO's air quality guidelines in 2030, but fewer than 5% are above the WHO-mandated tier I levels of 35 μg/m³ PM2.5 concentrations (Figure 17.47). Maximum benefits accrue when, in addition to stringent air quality controls, there is also a universal energy access policy that ensures clean energy globally by 2030 (the SLE2 scenario). This highlights that compliance with stringent air quality standards in developing countries cannot be achieved with only increasing the stringency of outdoor air pollution controls but will require in addition, controlling for household air pollution through access to modern cooking. This results in an overall emissions reduction of 50% in 2030 compared

with 2005 levels, 100% of the world's population below WHO-mandated tier I levels, and more than 50% of the population at levels below WHO air quality guidelines of 10 μg/m³ PM2.5 concentrations. In addition to PM2.5, there are also significant differences across the scenarios for SO_2 and other pollutant emissions. The resulting spatial emissions patterns of PM2.5 across the different scenarios are illustrated in Figure 17.46.

17.5.2.2   Health-Related Impacts

Outdoor Air Pollution

This section presents estimates of global health impacts attributable to outdoor air pollution based on implementing the various policy packages discussed in earlier sub-sections. Results presented are based on combining estimated PM2.5 concentrations with WHO (2008) data on mortality and DALYs and risk rates (RRs) detailed in Cohen et al. (2004)[48] (see Box 17.4 and Table 17.19 for comparison with alternative health impact methodology used in this study). In 2005, outdoor air pollution is estimated to result in 2.75 million deaths or 23 million DALYs lost globally, which

---

48   Both urban and rural populations are considered here.

Box 17.4 | Alternative Methodology for Calculating Health Impacts

In addition to the standard approach for calculating DALYs, an alternative methodology is also applied in the outdoor air pollution related health impact estimations. This methodology, which is based on Mechler et al. (2002) and used in World Energy Outlook 2009 (IEA, 2009a), calculates absolute changes in life expectancy based on modification of survival functions for population over 30 years of age. The end point is the statistical years of life lost (YOLLs) over the remaining lifetime of the entire population. The main features in this approach include:

Risk rates are assumed to be linear from 5 $\mu g/m^3$ until 200 $\mu g/m^3$; all cause risk rates are used as opposed to a cause specific one (RR of 1.04 is used for developing countries and RR of 1.06 for developed countries); and baseline mortality data are derived from UN Department of Economic and Social Affairs (UN DESA, 2009) and include all non-accidental mortality, i.e., not distinguished by cause of death. Country-specific life tables were used to reflect different mortality levels.

The YOLLs calculated from this approach are available in Table 17.19 (and Table 17.25 in Section 7).

Table 17.19 | Health impacts of outdoor air pollution in millions of DALYs (millions of population integrated YOLLs).

Region	2005	2030				
		FLE	CLE1	CLE2	SLE1	SLE2
World	23 (3865)	40 (12292)	33.6 (7366)	23 (4891)	14.4 (2947)	13.4 (2375)
OECD and Reform Countries	4.3 (867)	5.3 (1525)	2.2 (717)	0.95 (456)	0.5 (322)	0.5 (322)
Middle East and Africa	1.5 (265)	4.3 (1043)	3.1 (744)	1.8 (560)	1.1 (443)	0.9 (232)
Latin America and Caribbean	0.3 (91)	0.6 (234)	0.3 (137)	0.2 (122)	0.1 (97)	0.1 (97)
Asia	17 (2643)	30 (9490)	28 (5768)	20.2 (3753)	12.7 (2085)	11.8 (1725)

Source: Chapter 9 and 17.

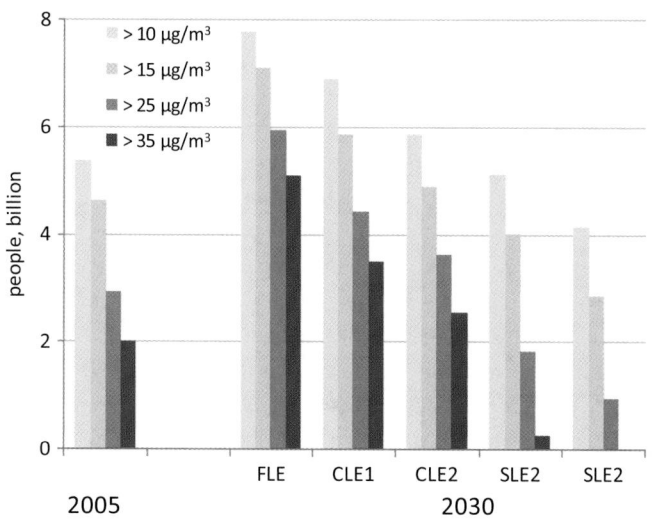

Figure 17.47 | Global population exposed to fine particulate matter concentrations exceeding WHO air quality targets in 2005 and under alternative policy packages in 2030. The rightmost bar indicates the compliance with WHO Tier 1 level of 35 $\mu g/m^3$ PM2.5, representing the global target of the GEA pathways for the environmental and health objective by 2030. Bars to the left indicate a comparison to lower (more stringent) WHO Tier levels with the leftmost bar corresponding to the lowest WHO AQG of 10 $\mu g/m^3$.

represents around 5% of all deaths, 2% of all DALYs and around 12% of the total burden that can be attributed to cardiovascular, respiratory, and lung cancer (for further discussion see Rao et al., 2012). More than 70% of this burden is felt in Asia alone.

Failure to implement further air pollution control policies beyond 2005 levels (the FLE scenario) is seen to result in a global increase of close to 50% in DALYs (and deaths) in 2030 as compared to 2005 (shown in Table 17.19), indicating that the implementation of air pollution policies is an absolute must for controlling the health-related impacts of air pollution in the future. However, an air pollution control-only policy, as in the CLE1 scenario, still leads to an increase in health impacts by more than 30% between 2005 and 2030[49] with the share of the outdoor air pollution related in the total burden increasing slightly from 2005 levels. This is mainly due to the large increases in emissions in many developing regions, particularly South Asia and Africa, where currently legislated policies do not lead to emissions declines in the future, as discussed earlier. In addition, a growing population in these regions means that the future population over 30 years of age at risk for air pollution

---

49   This is comparable to findings in the *World Energy Outlook 2009* of the IEA (2009a), which estimates a 70% growth of emissions in selected regions for the baseline scenario in spite of current air quality legislation.

Table 17.20 | Outdoor and household health impacts in 2005 and in 2030 for the CLE1 and SLE2 scenarios (in millions of DALYs).

Region	2005		2030			
			CLE1		SLE2	
	Outdoor (impacts from solid fuel use in households)*	Household	Outdoor (impacts from solid fuel use in households)*	Household	Outdoor	Household
World	22.6 (4.5–6)	41.6	33 (3–3.7)	~24	13.4	0
OECD	2.4		1.2		0.2	0
REFS	1.9		0.9		0.2	
Middle East and N. Africa	0.6		1.4		0.5	0
South Asia	7.0 (2.1–2.7)	13.8	12.8 (1.8–1.9)	8	5.7	0
Pacific Asia	1.1 (0.2–0.3)	3.9	2 (0.1–0.2)	3.5	0.6	0
Sub-Saharan Africa	0.9 (0.2–0.4)	18.6	1.9 (0.3–0.6)	10	0.5	0
Centrally Planned Asia	8.4 (2–2.6)	4.6	12.8 (0.8–1)	<2.6[1]	5.6	0
Latin America and the Caribbean	0.3 (0.02–0.05)	0.8	0.3 (0.01–0.03)	<0.4[1]	0.1	0

* Indicated in parenthesis are estimated outdoor health impacts attributable to use of solid fuels (biomass and coal) in households. Note that these estimates are based on interpolations between scenarios and are only indicative. Range represents the inherent uncertainty in calculating these impacts.

1 Explicit energy access scenarios for these regions have not been constructed. The numbers of DALYs lost for these regions for the CLE1 scenario are therefore estimated by extrapolating trends from the other regions.

will also be larger than today's, leading to increases in health impacts in most developing countries.

The combination of currently legislated air quality policies with other policies, as in the CLE2 scenario, helps to slow the health-related impacts, with 1.2 million avoided deaths in 2030 and a reduction of more than 7 million DALYs compared with the air pollution-only CLE1 scenario. While more stringent air pollution policies (as in the SLE1 scenario) yield further reductions in deaths and DALYs, including universal energy access policies for 2030 that directly affect outdoor air pollution through cleaner fuels, the SLE2 policy case yields maximum health benefits by 2030, corresponding to 2.6 million avoided deaths or a reduction of 20 million DALYs compared to the CLE1 policy case. The air pollution-related burden in 2030 (1.2 million deaths and 13 million DALYs) reduces to less than 2% of total deaths, 1% of total DALYs and around 5% of deaths and DALYs that can be attributed to cardiovascular, respiratory, and lung cancer. Thus, maximum benefits in terms of meeting environmental aims and reducing the health-related impacts of outdoor air pollution will require not only an increase in the stringency of air quality controls globally, but also an integration of a wider spectrum of policy concerns.

Comparison with Household Air Pollution

This section compares the impacts of policies to control outdoor air pollution that were explained earlier with specific policy scenarios, such as fuel subsidies and microfinance options, that facilitate access to cleaner cooking fuels (LPG) and thereby limit household air pollution (see Section 17.4.1).

Estimates of the current health impacts of household pollution are based on the effects of solid fuel dependence today, whereas future estimates are based on the detailed access scenarios described in Section 17.4.1.2 and account for forecasted demographic change and trends in background disease and mortality levels as estimated by the WHO. The methodology is described in detail in Rao et al., (forthcoming).[50] In 2005, total deaths attributed to solid fuel combustion in traditional stoves were about 2.2 million,[51] and more than 41.6 million DALYs were lost, with the impacts felt mainly by women and children. Although substantial uncertainty is associated with these estimates, policies that improve access to modern cooking have the potential to avert between 0.6 million and 1.8 million premature deaths, on average, every year until 2030, in the three regions of sub-Saharan Africa, South Asia, and Pacific Asia. These include between 0.4 million and 0.6 million deaths per year of children below the age of five. Deaths attributable to acute lower respiratory infection (ALRI) among children under five are seen to decline between 2005 and 2030 even in the absence of any access policies, but deaths due to chronic obstructive pulmonary disease (COPD) and ischemic heart disease (IHD) in adults

---

50 In contrast to outdoor effects, which are quantified only for the population older than 30 years, estimates of health impacts from household pollution include the effects on children. The present study includes impacts due to acute lower respiratory infections (ALRI) in young children, chronic obstructive pulmonary disease (COPD) in adults and ischemic heart disease (IHD) in adults. In addition, we also estimate the incidence of lung disease in adults due to the combustion of coal in homes.

51 About 1.6 million deaths were in South and Pacific Asia and sub-Saharan Africa, regions where the lack of access to modern energy carriers is the most acute and for which explicit energy access scenarios have been assessed in Section 17.4.

are expected to increase during the same period. These trends are in line with those reported by Bailis et al. (2005), who find that the observed decline in childhood ALRI mortality over time is a result of additional factors,[52] whereas the upward trend in adult incidence of COPD is mainly due to population aging. Alternatively, in the absence of any new policies to enhance access to modern cooking fuels or devices, it is estimated that in 2030 there could still be over 24 million DALYs lost due to household air pollution. See Rao et al., (forthcoming) for details on disease-specific impacts.

Table 17.20 lists the health impacts in DALYs from outdoor and household air pollution for 2005 and for an air pollution-only policy (CLE1) compared with a combined policy (SLE2). There are significant health benefits from a combination of stringent outdoor air pollution policies and a policy that ensures universal access to clean cooking by 2030. These are especially effective in developing countries that face the dual problems of outdoor air pollution due to a growing motorized fleet combined with household pollution from poor quality cooking fuels and devices. Thus, such policies can have multiple benefits both for human well-being and for the environment (as highlighted in Chapter 19 and Section 17.4), including major health gains.

17.6   Energy Security

17.6.1   Introduction

Energy security has been a major concern for energy systems for decades, and therefore needs to be addressed in the transition pathways presented in this chapter. As reviewed in detail in Chapter 5, energy security has multiple dimensions, which are not easily combined into a holistic concept or single indicator. Therefore, the concept of energy security is not used as a quantitative target or a technical modeling constraint in the GEA pathway scenarios described here. Instead, this section draws on the conceptual and quantitative framework developed in Chapter 5 to illustrate the implications of the GEA transition pathways for energy security. Section 17.7 then extends the discussion to the multiple benefits of the transition pathways, considering especially the synergistic effects between other energy development objectives and energy trade.[53]

Chapter 5 summarized the present main energy security concerns as follows:

- *Oil and transport:* volatility in the global oil market coupled with the geographic concentration of oil production; rapidly increasing

demand under potentially constrained production capacities; growing dependence of an increasing number of countries on imported oil from ever fewer producing countries, with low-income countries often facing unaffordable costs of imports; and the dominance of oil in the transportation sector, where easily-available substitutes are lacking.

- *Natural gas:* dependence of a number of countries on imported natural gas, often procured from a single supplier and delivered through a limited number of potentially vulnerable routes and infrastructure.

- *Electricity:* vulnerability of electricity systems associated with low diversity of power generation options, aging infrastructure, inadequate generation capacity, and rapid demand growth.

- *Energy export revenue:* volatility and uncertain sustainability of energy export revenue ("energy demand" security) in countries where energy is a vital economic sector.

- *Total primary energy supply vulnerabilities:* overall energy vulnerability of a number of individual countries that face several of the above concerns simultaneously.

Chapter 5 established a framework for analyzing the energy security-related vulnerabilities of energy systems associated with fuels, end-use sectors (including electricity as a carrier), and individual countries. For each of these three subsystems, that chapter identified three dimensions of energy security concerns: sovereignty (the degree of control that national governments have over energy systems), resilience (the ability of energy systems to respond to disruptions), and robustness (the risks related to the physical state of energy resources and infrastructure). This section adopts the same framework, but considers different energy subsystems to reflect game-changing developments in the transition pathways. The section analyzes not only the globally traded fuels that dominate today (oil, gas, and coal) but also those of the future (biofuels and hydrogen). The main energy end-use sectors (transportation, industry, residential and commercial) and electricity generation are also analyzed. Finally, since the modeling frameworks do not provide detail on individual countries, the analysis is applied at the world and regional level.

Table 17.21 summarizes the energy security perspectives and indicators analyzed in this section. The analysis relates to sovereignty and resilience concerns. The robustness concerns could not be addressed at this aggregated level.

This analytical framework is applied to the three illustrative GEA transition pathways (GEA-Supply, GEA-Mix, and GEA-Efficiency) but considers both of the transportation cases for each, because the transportation sector is a key policy concern, given the limited alternatives to the use

---

52 These include increased coverage and efficacy of pneumonia case management using antibiotics; increased awareness and practice of breastfeeding, which increases child immunity and survival; and other economic and technological factors.

53 The energy security analysis of the GEA pathways relies on the MESSAGE interpretations of the GEA pahtways, including the alternative transportation sector setups and reduced supply-side portfolios.

Table 17.21 | Indicators for analyzingenergy security across subsystems and security perspectives.

Subsystems	Security perspectives	
	Sovereignty	Resilience
Upstream		
Fuels (total fuel supply and globally traded fuels)	Volume and intensity of trade (by fuel and total)	Global diversity of the primary energy system; dominance of a single fuel
Downstream		
Carriers (electricity) and end uses (transport, residential and commercial, industry)	Reliance of carrier or end-use sector on insecure fuels	Diversity of fuels used in the carrier or end-use sector
Regional	Import dependency (and export flows)	Diversity of primary energy supply in the region

Table 17.22 | Characteristics of globally traded fuels in 2050 compared with oil in 2005.

Characteristic	Oil, 2005	Gas, 2050	Coal, 2050	Biofuels, 2050
Predominance (% of primary energy supply)	34	19–22	8–10	6–8
Trade volume (EJ/year)	83	35–54	10–20	7–14
Geographic concentration of production (diversity index)[1]	1.0	0.9–1.0	1.3–1.4	1.2–1.5

1 See text for definition of the diversity index.

of oil in internal combustion vehicles. The analysis focuses on changes between now and 2050, because this is the longest time horizon under which energy security concerns are considered in present policies. The analysis first considers the energy security concerns associated with fuels, then examines the future vulnerabilities of end-use sectors, and concludes with an examination of the energy security of individual GEA regions.

17.6.2   Fuels

Under the GEA transition pathways, the vulnerabilities of globally traded fuels in the aggregate as well as of individual fuels decrease over time in terms of both sovereignty and resilience.

A proxy measure of the sovereignty aspects of energy security is global trade in energy. Absolute volumes of traded energy and the share of traded energy in overall energy use (the latter referred to here as "trade intensity") indicate the extent to which regions rely on fuels produced in other regions, raising sovereignty concerns.

Figure 17.48 shows both trade volumes and trade intensity under the six GEA pathways considered here. The global aggregate energy trade volumes among the 11 GEA regions (estimated at some 104 EJ in 2005) peak in 2030 or 2040, and trade intensities peak in 2020 or 2030. Thereafter both indicators decline, so that by 2050 the intensity of trade is lower than at present but the absolute amount of trade remains higher than present values under the GEA-Supply scenarios. The decline in absolute trade volumes after 2030 is most pronounced in the Advanced Transport GEA-Mix and -Efficiency pathways and least pronounced in the high-demand, supply-dominated GEA transition pathways.

Concerning individual fuels, the analysis shows the following trends. In all the GEA pathways considered, oil is phased out in the long term. It accounts for between 9% and 15% of global primary energy supply by 2050 and declines to less than 1% by the end of the century

(Table 17.22). As a result, trade volumes of oil for all pathways peak at about 100 EJ (compared with approximately 83 EJ today[54]) between 2020 and 2030 and decline thereafter.

Present energy security concerns associated with oil drastically diminish in the GEA pathways because of their comparatively modest demand growth, which is due to efficiency improvements and a more diversified supply mix. No other fuel assumes a dominant role similar to that which oil plays today, accounting for 36% of primary energy supply worldwide.[55] Moreover, no "new oil" emerges in the global energy arena. Table 17.22 summarizes the characteristics of the globally traded fuels in 2050 as compared with oil today. Figure 17.49 shows the shares of different fuels in global primary energy supply, indicating a more diversified supply portfolio. Figure 17.50 illustrates how trade volumes and the geographic concentration of production of oil, gas, and coal change across the GEA pathways.[56] Biofuels, hydrogen, and electricity are traded in much smaller volumes (a maximum of 50 EJ for hydrogen in the GEA-Supply pathway with Advanced Transportation) and with greater geographic diversity of producers than is the case with oil today.

At the same time, by 2050 natural gas trade exhibits some of the characteristics of oil trade today under certain pathways. By 2050 gas accounts for about 20% of primary energy and 36–51 EJ of trade per year (compared with oil's current 83 EJ). Additionally, gas production stays at its current level of geographic concentration until about 2050, which is comparable to the geographic concentration of oil production, and becomes even more concentrated than current oil production under most pathways thereafter, as shown by its decreasing Shannon-Wiener diversity index in Figure 17.50. Although natural gas is a potentially more risky fuel, the overall resilience of energy systems, as measured

---

54 BP (2009) estimates a total volume of country-to-country oil trade of some 110 EJ in 2005. Thus, the GEA interregional model representation covers some 75% of actual oil trade flows at the country level.

55 Although electricity comes to dominate final energy use, it is not, strictly speaking, a "fuel," as it is produced from a variety of sources. Moreover, global trade in electricity is minimal in all pathways, never accounting for more than 2% of total electricity supply.

56 The geographic concentration of production is measured by the Shannon-Wiener Diversity Index (SWDI), described below.

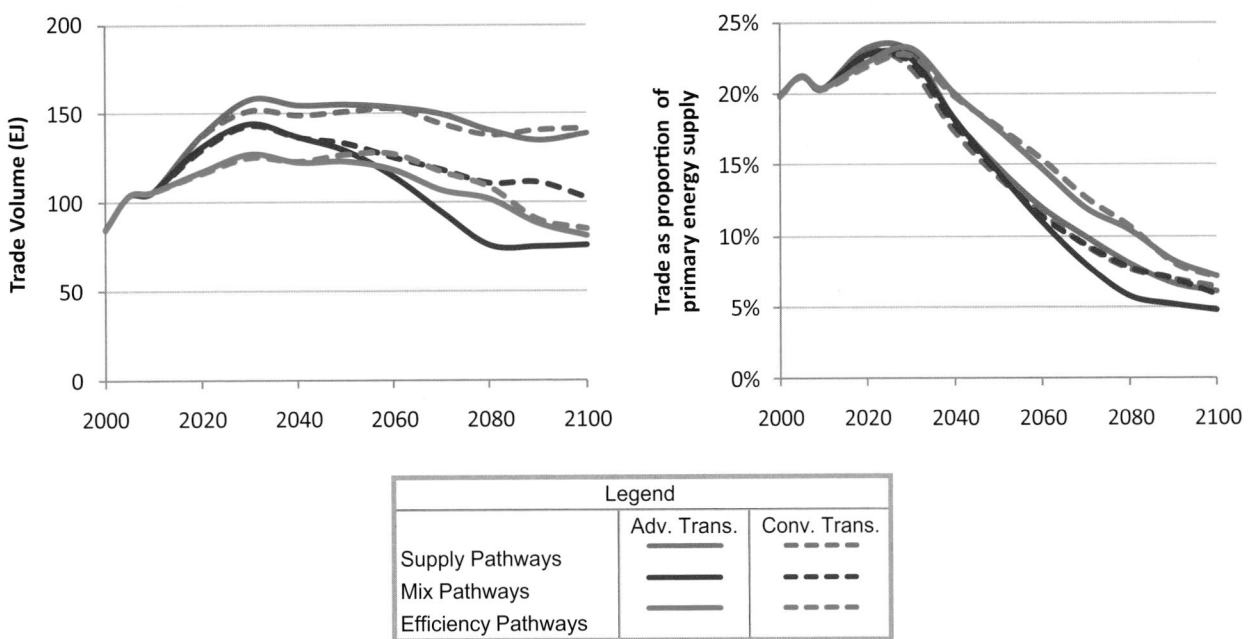

Figure 17.48 | Volume of energy trade and trade intensity in the illustrative GEA pathways. Each of the transportation variants of the three GEA transition pathways is represented.

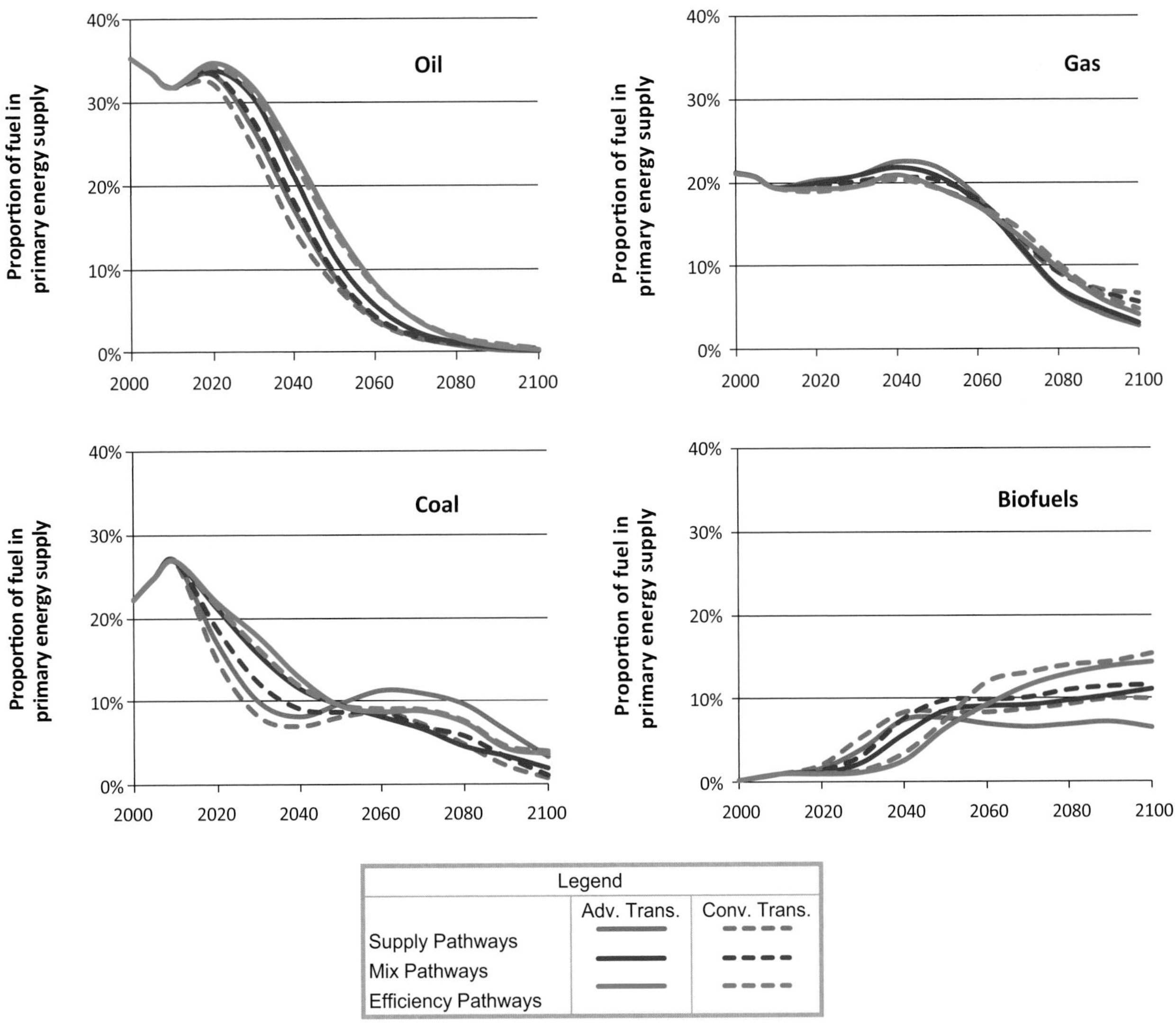

Figure 17.49 | Contributions of fossil fuels and biofuels to global primary energy supply in the GEA pathways (unrestricted portfolios for advanced and conventional transport).

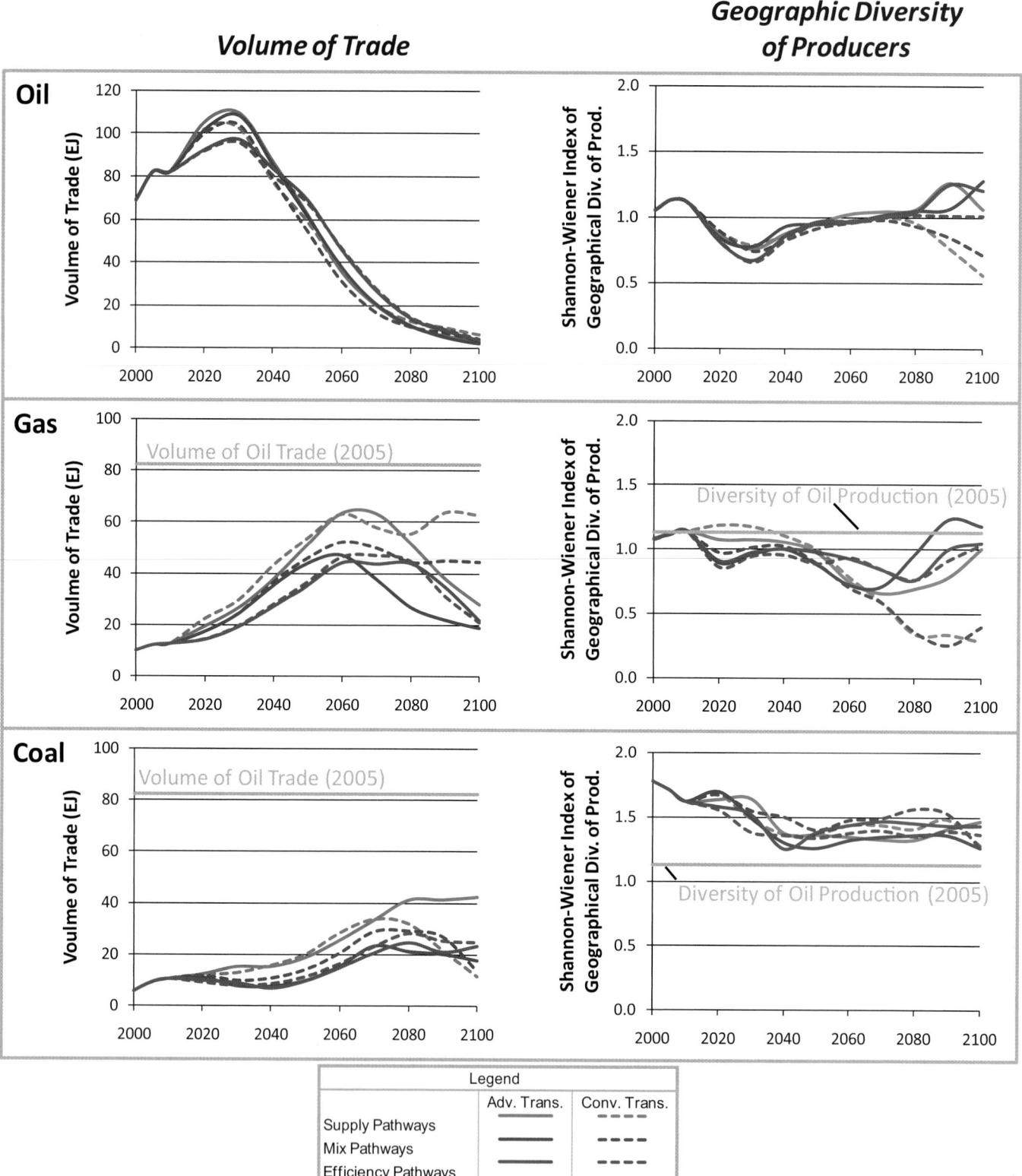

Figure 17.50 | Volume of trade and geographic diversity of production of globally traded fuels in the GEA pathways (unrestricted portfolios for advanced and conventional transport).

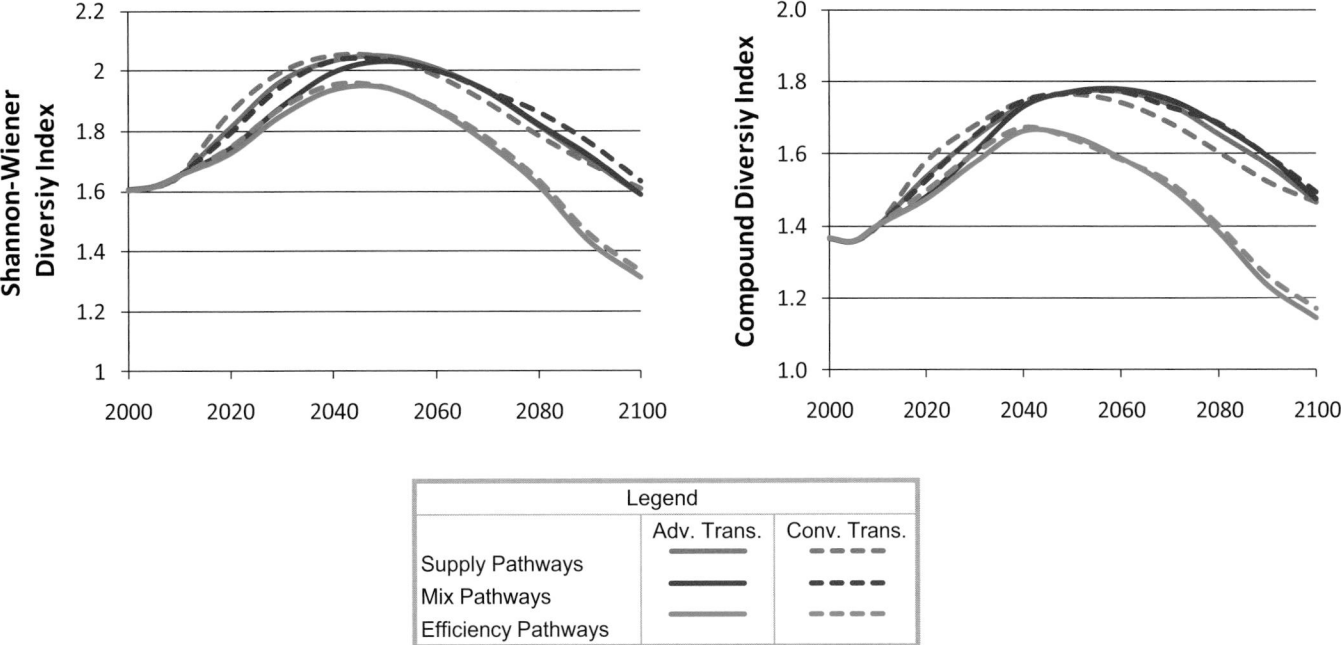

Legend		
	Adv. Trans.	Conv. Trans.
Supply Pathways	———	– – – –
Mix Pathways	———	– – – –
Efficiency Pathways	———	– – – –

Figure 17.51 | Primary energy supply diversification in the GEA pathways (unrestricted portfolios for advanced and conventional transport). See text for definitions of the indexes.

by the diversity of primary energy supply, increases under all transition pathways.

The Shannon-Wiener diversity index (SWDI; see Shannon and Weaver, 1963) is frequently applied as a measure of energy security of supply (see, e.g., Jansen et al., 2004; APERC, 2007) and electricity generation (Stirling, 1994). The index is calculated as follows:

$$SWDI = -\sum_i (p_i * \ln(p_i))$$ (1)

where p_i is the share of primary energy i in total primary energy supply. In the GEA pathways, the global SWDI rises (supply diversification increases) from the current level of 1.6 to 2.0 by 2050, before falling to between 1.3 and 1.6 in the latter half of the century.[57]

Measures of global energy trade (reflecting sovereignty concerns) and diversity (reflecting resilience concerns) can be aggregated into a single index called a compound SWDI. This compound indicator differs from the simple SWDI in that it does not count globally traded fuels as contributing to the overall diversity of primary energy supply.[58] It is calculated by excluding the imported energy in a nation's or region's diversity index:

$$Compound\ SWDI = -\sum_i (1 - m_i) * (p_i \ln(p_i))$$ (2)

where p_i is again the share of primary energy resource i in total primary energy supply, and m_i is the share of primary energy resource i that is supplied by net imports.

This indicator shows a trend similar to that of the simple SWDI (Figure 17.51), increasing from 1.4 in 2000 to about 1.7 in 2050 before falling slightly to about 1.5 in the second half of the century.[59]

17.6.3   End-Use Sectors

The diversity of fuels used in a sector is generally considered an indicator of the resilience of energy supply for the sector. Sectoral diversity indexes are shown in Figure 17.52. The increase in diversity in the transportation sector is particularly pronounced in the GEA pathways, whereas the improvement is more gradual in the other end-use sectors and in electricity generation.

Although this pattern of increasing diversity of the energy mix in individual end-use sectors is relatively homogeneous across regions, there is some regional variation. In some regions, the rise in diversity is more rapid and pronounced, whereas in others, individual fuels (particularly gas) come to dominate certain sectors in certain periods. These regional deviations are discussed in the next section.

57 It is important to note that the diversity index strongly depends on the primary energy accounting convention used. In GEA, a consistent substitution-equivalent accounting of primary energy is applied across all chapters to ensure comparability. Under this accounting convention, diversity indicators drop in the latter half of the century with the strong decarbonization of the energy system. Thus, it is also important to consider the sectoral or end-use diversity indices, discussed below.

58 This index was first used in Jansen et al. (2004) as a measure of long-term energy security.

59 See also the electronic appendix to this chapter for further illustrative examples of diversity indicators for specific regions.

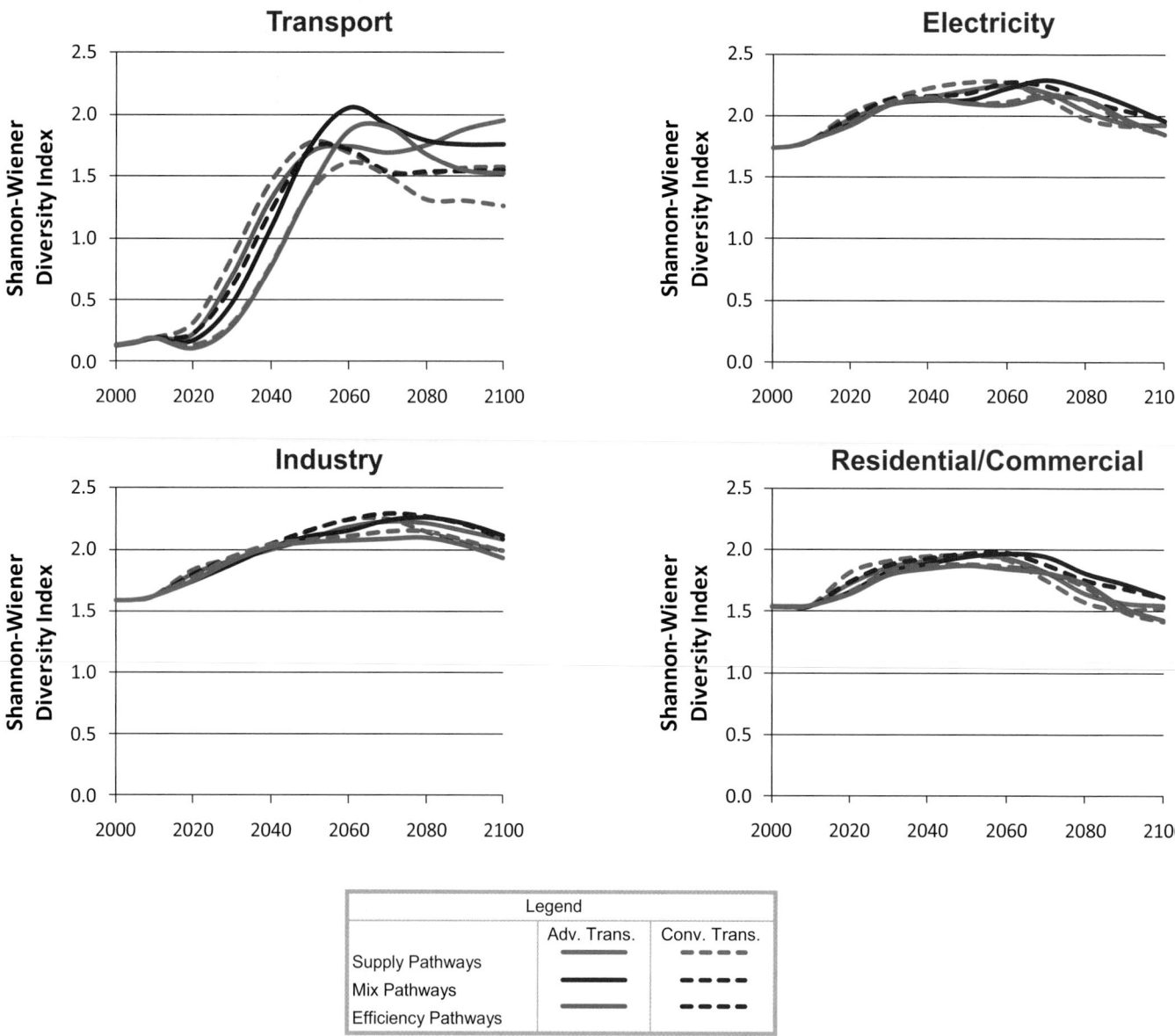

Figure 17.52 | Diversification of end-use sectors and of electricity generation in the GEA pathways (unrestricted portfolios for advanced and conventional transport).

17.6.4 Regions

At present, energy security concerns vary across countries. In the GEA pathways, different regions also face different energy security trends and challenges.

17.6.4.1 Regional Import Dependency and Energy Export Volume

Different regions fare differently in terms of sovereignty concerns related to energy trade. Fewer regions are net energy importers under the GEA pathways than today; across all pathways, the number of regions with low import dependency rises from five in 2000 to between seven and eight in 2050 and between nine and 11 in 2100 (Table 17.23).

The flip side of the decrease in energy imports is a fall in energy exports for certain regions. Energy exports provide vital revenue for a number of countries, and rapid and profound declines in such revenue could adversely affect energy-exporting regions. (Chapter 5 conceptualizes energy exports as a "vital energy service.") This drop in export volumes may be partly mitigated, however, by rising energy prices.

The most important energy-exporting region today is the Middle East and North Africa (MEA), with net energy exports of over 52 EJ in 2005, followed by the Former Soviet Union (FSU), which exported about 24 EJ in that year; Latin America and the Caribbean (LAC) and sub-Saharan Africa (AFR) each exported some 11–13 EJ.

Because of the declining share of oil in the global energy mix, MEA experiences the largest decline in energy export volumes. The region's

Table 17.23 | Import dependency of GEA regions in the GEA pathways.

Imports as share of primary energy supply	No. of regions		
	2005	2050	2100
Low[1] (<16%)	5 (AFR, CPA, FSU, LAC, MEA)	7–8 (AFR, CPA, FSU, LAC, MEA, NAM, PAO, PAS[2])	9–11 (AFR, CPA, FSU, EEU, LAC, MEA, NAM, PAO, PAS,[3] SAS,[4] WEU)
Medium (16–34%)	3 (NAM, PAS, SAS)	1–3 (EEU, PAS,[2] SAS, WEU[5])	0–2 (PAS,[3] SAS,[4])
High (>34%)	3 (EEU, PAO, WEU)	0–1 (WEU[5])	0

Note: AFR (Africa); CPA (Centrally Planned Asia); FSU (Former Soviet Union); EEU (Eastern Europe); LAC (Latin America and the Caribbean); MEA (Middle East and North Africa); NAM (North America); PAO (Pacific OECD); PAS (Pacific Asia); SAS (South Asia); WEU (Western Europe).

1 Includes net energy exporters.

2 Has medium import dependency (~25%) under the GEA-Efficiency pathways.

3 Has medium import dependency (~32%) under the GEA-Efficiency pathways.

4 Has medium import dependency (17–20%) under the GEA-Mix and GEA-Efficiency pathways with Conventional Transportation.

5 Has medium import dependency (17–30%) under the GEA-Mix pathways and GEA-Supply pathways with Advanced Transportation.

exports peak at between 80 and 90 EJ/year in 2030 before falling to 43–55 EJ/year in 2050 (slightly higher for the GEA-Efficiency pathways) and to less than 15 EJ/year in the second half of the century. Similarly, LAC's exports approximately double to about 20 EJ/year between now and 2030 or 2040 before dropping to between 7 and 10 EJ/year by the end of the century. Exports from AFR also drop, although not as profoundly, stabilizing in some pathways at approximately the present levels in the second half of the century.

The major export "winner" under all the pathways is FSU, which experiences a dramatic rise in its energy exports due to the increasing demand for gas. The region's energy exports rise from the 2005 level of some 24 EJ/year to between 47 and 52 EJ/year in 2050, and then experience a continued rise through 2070 or 2080 to between 50 and 65 EJ/year. Three other regions –Centrally Planned Asia (dominated by China), North America, and Pacific OECD – also see a rise in export volumes over all the GEA pathway analysed, although their increases are not as pronounced and differ across pathways and time horizons.

17.6.4.2   Regional Resilience and Diversity

As noted above, global energy supply becomes more diverse in the GEA pathways both as a whole and within individual end-use sectors, especially the transport sector. This increase in diversity is also observed across all GEA regions (Table 17.24), with a generally smaller

rise in MEA (a consequence of the region's exceptional oil and gas resource endowment).

The fuel diversity of electricity generation increases in all pathways in five out of the 11 GEA regions. In five other regions, diversity increases in at least some pathways. Only in Western Europe does electricity production diversity decrease compared with the current (relatively high) level.

In general, the GEA regions may be divided into three broad groups. The first group includes such industrialized regions as the Pacific OECD, Latin America and the Caribbean, North America, and Western Europe, which generally follow the global trends with respect to fuels and end-use sectors. Transitions in their energy systems are primarily driven by global factors, including the switch away from fossil fuels, increases in efficiency, and the diversification of transport technologies. Since all these transitions generally improve energy security by increasing resilience and sovereignty, energy security in these regions also improves significantly.

The second group includes Sub-Saharan Africa and Centrally Planned Asia, in which the global energy transitions provide a context for massive growth in regional energy systems. The expansion of energy systems in Sub-Saharan Africa to extend energy access to all, and in Centrally Planned Asia (dominated by China) to keep up with rapidly growing economies, results in dramatically altered configurations of energy systems, leapfrogging the inherited energy systems inertia of the industrialized world. As a result, many energy security indicators in these regions improve much more rapidly and dramatically than in the rest of the world, as their energy systems become more diverse and more reliant on regional rather than global resources.

The third group includes those regions that, because of their geography and either fossil fuel resource endowments (the Former Soviet Union, the Middle East and North Africa) or resource scarcity (Eastern Europe and South Asia), have more limited options for radical systemic change. The diversity of energy supply, especially in specific sectors in these regions, may be below the global average. For example, their transportation and electricity sectors may become dominated by natural gas, a fuel of choice in the middle of the century.

17.6.5   Conclusions on Energy Security

Under the GEA pathways, energy security improves in the world as a whole and in the majority of regions. The diversity of energy sources increases, whereas the volume and the intensity of trade decline in most pathways. No individual fuel is likely to cause energy security concerns similar to those caused by oil at present. The one exception is natural gas, which, as a transition fuel, experiences growth to some 30% of global primary energy supply (compared with oil's 36% share today) in 2050, with increasing trade flows and a decrease in the diversity of production.

Table 17.24 | Regional trends in diversity and import dependency, 2005 and 2050.

Import Dependency*			Diversity						Compound diversity index	
			Electricity		Transport		Primay Energy Supply			
Region	2005	2050	2005	2050	2005	2050	2005	2050	2005	2050
AFR	−8%	−7%	1.01	1.64–1.89	0.07	1.24–1.73	1.40	1.66–1.83	1.4	1.51–1.73
CPA	4%	8%–14%	0.74	1.72–1.85	0.21	1.06–1.77	1.17	1.87–1.97	1.05	1.56–1.73
EEU	36%	28%–34%	1.2	1.43–1.60	0.16	1.38–1.76	1.53	1.86–1.94	0.99	1.32–1.50
FSU	−54%	−56%	1.4	1.34–1.51	0.42	1.36–1.70	1.37	1.68–1.97	1.37	1.68–1.97
LAM	−34%	−13%	1.26	1.38–1.68	0.31	1.41–1.73	1.44	1.71–1.94	1.44	1.71–1.94
MEA	−187%	−46%	1	1.12–1.47	0.03	1.09–1.23	0.95	1.22–1.56	0.87	1.17–1.49
NAM	21%	2%–8%	1.46	1.55–1.80	0.08	1.28–1.76	1.54	1.87–2.04	1.34	1.71–1.91
PAO	41%	−22%	1.55	1.64–1.95	0.09	1.46–1.79	1.48	1.87–2.07	0.99	1.70–1.81
PAS	28%	10%–28%	1.47	1.53–1.88	0.02	1.28–1.72	1.50	1.96–2.04	1.2	1.45–1.74
SAS	20%	29%–32%	1.22	1.68–1.82	0.11	0.92–1.55	1.46	1.69–1.84	1.16	1.03–1.25
WEU	40%	31%–36%	1.64	1.52–1.73	0.16	1.48–1.75	1.61	1.84–1.93	1.1	1.27–1.34
World	*20%*	*13%–16%*	*1.54*	*1.79–1.92*	*0.15*	1.38–1.77	*1.62*	*1.94–2.05*	*1.36*	*1.64–1.77*

LEGEND	Import Dependency*	Diversity		Compound diversity	
	low (<16%)	High (>1.5)		High (>1.5)	
	medium (16–34%)	Medium (1.0–1.5)		Medium (1.0–1.5)	
	high (>34%)	Low (<1.0)		Low (<1.0)	

* Import dependency values are reported as negative if a region is a net energy exporter. World import dependency is the proportion of primary energy that is traded.

In the GEA pathways, individual end-use sectors generally use a more diverse mix of energy sources than today. The transportation sector, presently associated with major energy security concerns, achieves diversity similar to that in other end-use sectors. No end-use sector relies on a single fuel to the extent that transport relies on oil today.

Each of the 11 world regions generally follows the global trend toward improved energy security. Some experience a more rapid and pronounced increase in diversity and self-sufficiency of their energy systems. At the same time, some regions with more limited energy options may experience continued reliance on particular fuels (primarily natural gas) in specific sectors (transportation and electricity generation) under certain pathways.

The next section examines, among other issues, the impact of the other energy objectives for energy security, including global energy trade and primary energy source diversity.

17.7 Synergies and Multiple Benefits of Achieving Different Energy Objectives Simultaneously

The previous sections have illustrated a variety of energy futures in which the objectives of climate change mitigation, air quality, health,

access, and security could be achieved simultaneously. These pathways show that a dramatic transformation of the energy system is technically possible, and that if society truly values sustainability across all dimensions, such a transformation is indeed necessary. Transitions of this kind would likely lead to enormous synergies between objectives – synergies that are, at the moment, not fully understood by decisionmakers, or often overlooked, because the analysis is complex and requires an integrated, holistic perspective.

This section builds upon the main findings of an analysis conducted at IIASA in support of the GEA (McCollum et al., 2011), which attempts to illuminate the major synergies and, to a lesser extent, the trade-offs among the various energy objectives and the requisite policy choices and outcomes. In so doing, the analysis takes a slightly different approach from the core illustrative GEA pathways described so far in the chapter. Here the GEA-Mix scenario is used as a starting point for generating a wide array of scenarios that attempt to cover a large portion of the full scenario space across several different dimensions. Within this space, many of the scenarios are unsustainable by GEA standards, as each meets (or fails to meet) the different energy objectives to varying degrees. The analysis uses these less stringent scenarios as counterfactuals and for comparison purposes, in order to show how certain objectives and policy choices push in the same direction, while others are in conflict.

Table 17.25 | Indicators for climate change, pollution and health, and energy security and levels of satisfaction within the weak-intermediate-stringent framework.

Fulfillment	Climate Change [probability of staying within 2°C warming limit]	Pollution and Health [million DALYs (YOLLs), 2030]	Energy security [compound diversity indicator, 2030]
Weak	<20%	>33 (7300)	<1.40
Intermediate	20–50%	15–33 (2700 – 7300)	1.40–1.50
Stringent	>50%	<15 (2700)	>1.50

17.7.1   Characterization of the Full Scenario Space

In this section, the feasible scenario space is represented by several hundred distinct scenario pathways. These scenarios stretch the potential development of the energy system in several dimensions, each fulfilling the individual GEA objectives with respect to climate change, air pollution and health, and energy security to varying levels of satisfaction. For instance, some scenarios push climate change mitigation while ignoring security and air pollution, whereas other scenarios prioritize security only while ignoring the climate objective. Notably, the access objective is taken as a given in this analysis, as all scenarios have been developed to meet the access targets of the GEA, including even the corresponding counterfactual (baseline) scenario.[60] This simplification was made because energy access, compared with other objectives, has the lowest impact on energy use and GHG emissions (see Section 17.4). For further methodological details on how the full scenario ensemble was developed, see the electronic appendix.

As discussed earlier in this chapter, satisfaction of each of the individual GEA objectives can be measured in their own unique way: climate change in terms of the probability of limiting global temperature rise to 2°C (Section 17.5.1), pollution and health impacts in terms of DALYs (Section 17.5.2), and energy security in terms of a compound diversity indicator (Section 17.6). The use of such different metrics, although necessary given the far-ranging impacts of the energy system, tends to complicate the comparison of scenarios that meet certain objectives but not others. For this reason, this section adopts a simple framework to describe the scenario space across all three objectives; at the same time, it allows for ready comparison with the previous discussions in this chapter. The framework, summarized in Table 17.25, defines three levels of satisfaction – Weak, Intermediate, and Stringent – for each of the

three energy objectives. Specific numerical ranges are given for what constitutes each of these levels in terms of the relevant indicators. (Note that health impacts are also presented in terms of YOLLs, the methodology for which is described in Section 17.5.2.) Importantly, within a given scenario, the fulfillment of each objective is independent of the fulfillment of another (except for some important synergies, discussed later in this section). Therefore, a given scenario could, for example, fulfill the climate objective at the Weak level while at the same time satisfy the pollution and health objective and the energy security objective at the Intermediate level. By sharp contrast, all of the core GEA pathways described up to this point in the chapter (GEA-Efficiency, GEA-Mix, and GEA-Supply, along with their variants) have been designed to fulfill all of the objectives simultaneously at the Stringent level. In fact, the minimum allowable indicator values corresponding to the Stringent level are derived from the originally stated targets of the GEA (see Cluster I of the report and Section 17.2.3).

Figure 17.53 illustrates the full scenario space across all three dimensions: climate, pollution and health, and energy security. The degree to which each scenario (or rather, class of scenarios) fulfills the individual objectives is indicated in the figure by the shaded Weak, Intermediate, and Stringent regions. For instance, the top panel illustrates ranges of GHG emissions trajectories for all scenarios in the large ensemble that correspond to probabilities of reaching the 2°C target. The baseline scenario, which assumes no new climate, pollution and health, or energy security policies, sees the largest growth in emissions throughout the century and is therefore at the upper bound of the Weak region. Annual emissions in the baseline scenario climb from 49 GtCO$_2$-eq. in 2010 to 84 GtCO$_2$-eq. in 2050.[61] Emissions then peak near 100 Gt in the later part of the century. All other scenarios achieve emissions reductions compared with the baseline, and hence have comparatively higher probabilities of meeting the 2°C target. In the most stringent climate scenarios (lower bound on the Stringent region), emissions in 2050 are just 18.6 Gt. As discussed more fully in Section 17.5.1, reaching the 2°C target with greater than 50% probability (Stringent region) requires that emissions peak in 2020 at levels only marginally higher than today and then be reduced significantly in the decades that follow. If, however, the climate objective is of lower priority (i.e., if probabilities of meeting the 2°C target at less than 50% are acceptable), the permissible peak in emissions could certainly be greater and could even be delayed far beyond 2020. In the case of such weak and intermediate fulfillment of the climate objective, emissions reductions in the middle to late part of the century would not need to be nearly as drastic. For example, annual GHG emissions in 2050 for the Intermediate region (corresponding to a 20–50% probability of meeting the target) range from levels approximately the same as today to levels up to 45% lower. Comparing the latter case with the former, the emissions peak must occur almost two

---

60   Importantly, the baseline scenario referred to here differs from that discussed elsewhere in the chapter. Here, the baseline corresponds to a variation of the GEA-Mix pathway (thus including intermediate efficiency focus to limit energy demand), in which the policy constraints are relaxed to business-as-usual conditions. The counterfactual referred to in other sections builds upon the GEA-Supply storyline, and a corresponding baseline depicting future developments in the absence of any of GEA sustainability policies at levels of relatively higher demand.

61   Note that these GHG estimates include all well-mixed Kyoto greenhouse gases (CO$_2$, methane, nitrous oxide, sulfur hexafluoride, tetrafluoromethane, and halocarbons).

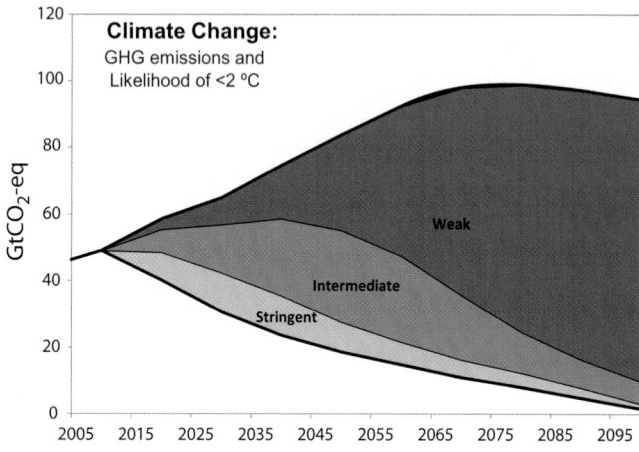

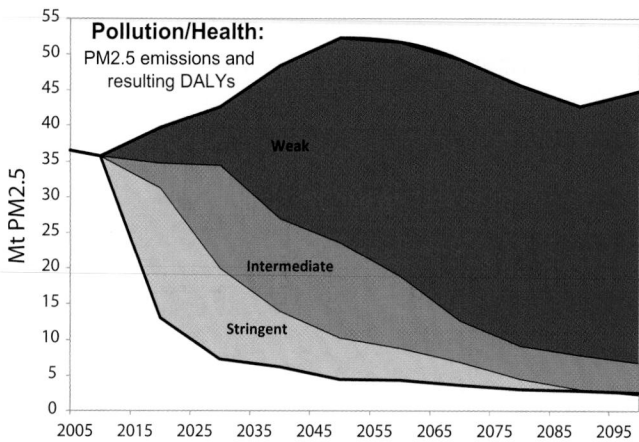

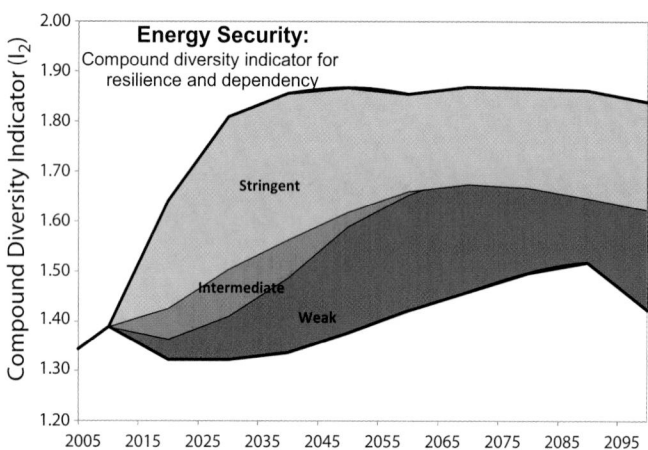

Figure 17.53 | Trajectories for global GHG emissions, PM2.5 emissions, and the compound energy diversity indicator for the full scenario ensemble. See Table 17.25 for definitions of Weak, Intermediate, and Stringent fulfillment of objectives.

decades earlier in order to preserve the feasibility of achieving the 2°C target with near 50% probability.

The middle panel of Figure 17.53 illustrates the full space of the scenario ensemble in the combined air pollution and health dimension by showing PM2.5 emissions trajectories and resulting DALYs. Particulate matter is chosen as a representative pollutant for this discussion because,

as discussed in Section 17.5.2 and Chapter 4, of all types of air pollutant emissions, PM2.5 causes some of the most serious impacts on human health.[62] The emissions trajectories shown in the figure correspond to multiple pathways for energy system development under different portfolios of air pollution control policies. These policy packages are described more fully in Section 17.5.2, where further details on the assumed types of controls are provided. Moreover, whereas that section focuses in detail on the impacts of the different air pollution policies in terms of health and other environmental benefits, this section specifically examines the economic implications of combined policies. In Figure 17.53, the shaded Weak, Intermediate, and Stringent regions correspond to DALYs at the global level (the aggregate of all world regions) that would be expected in 2030 by following the ranges of PM2.5 emissions trajectories shown. The important point here is that by making a more concerted effort to control air pollution throughout the world over the next two decades, especially in the densely populated urban centers of rapidly developing countries, the collective health of the global population can be significantly improved and DALYs can be reduced quite substantially. And although these reductions might be achieved by more stringent pollution control policies and measures (i.e., end-of-pipe technologies), they may also be achieved, to some extent, through decarbonization of the energy system in response to strong climate policy. The latter point touches upon an important synergy between the climate objective and the air pollution and health objective that, although not immediately evident in Figure 17.53, is discussed in more detail later in this section. In short, by driving the energy system toward zero-carbon, emissions-free technologies, stringent climate (and indeed energy security) policies can play an important role in reducing air pollutant emissions, even under an otherwise weak pollution policy regime. In other words, fulfillment of the pollution and health objective at the Weak, Intermediate, or Stringent level depends on measures for both pollution and climate control.

The scenarios also cover a broad space in the energy security dimension, as illustrated by the bottom panel of Figure 17.53. This analysis measures energy security using the compound diversity indicator introduced in Section 17.6 (see also Chapter 5). This indicator takes into account the diversity of primary energy resources at the global level, as well as where those resources are sourced – that is, whether from imports or domestic production. The diversity indicator rises with increasing diversity of the energy system but falls at higher levels of import dependency (see further details in Section 17.6 and the electronic appendix). In this sense, the higher the diversity indicator for a given country or region, the more secure its energy system. Figure 17.53 shows how global energy system diversity develops over time in all of the scenarios of the full ensemble, with the Weak, Intermediate, and Stringent regions grouping together scenarios that fulfill the security objective to a similar degree, as outlined in Table 17.25. The lower bound of the Weak region

---

62 Note that in addition to PM 2.5, each scenario of the large ensemble possesses unique emissions trajectories for sulfur dioxide, nitrogen oxides, volatile organic compounds, carbon monoxide, black carbon, organic carbon, and ammonia.

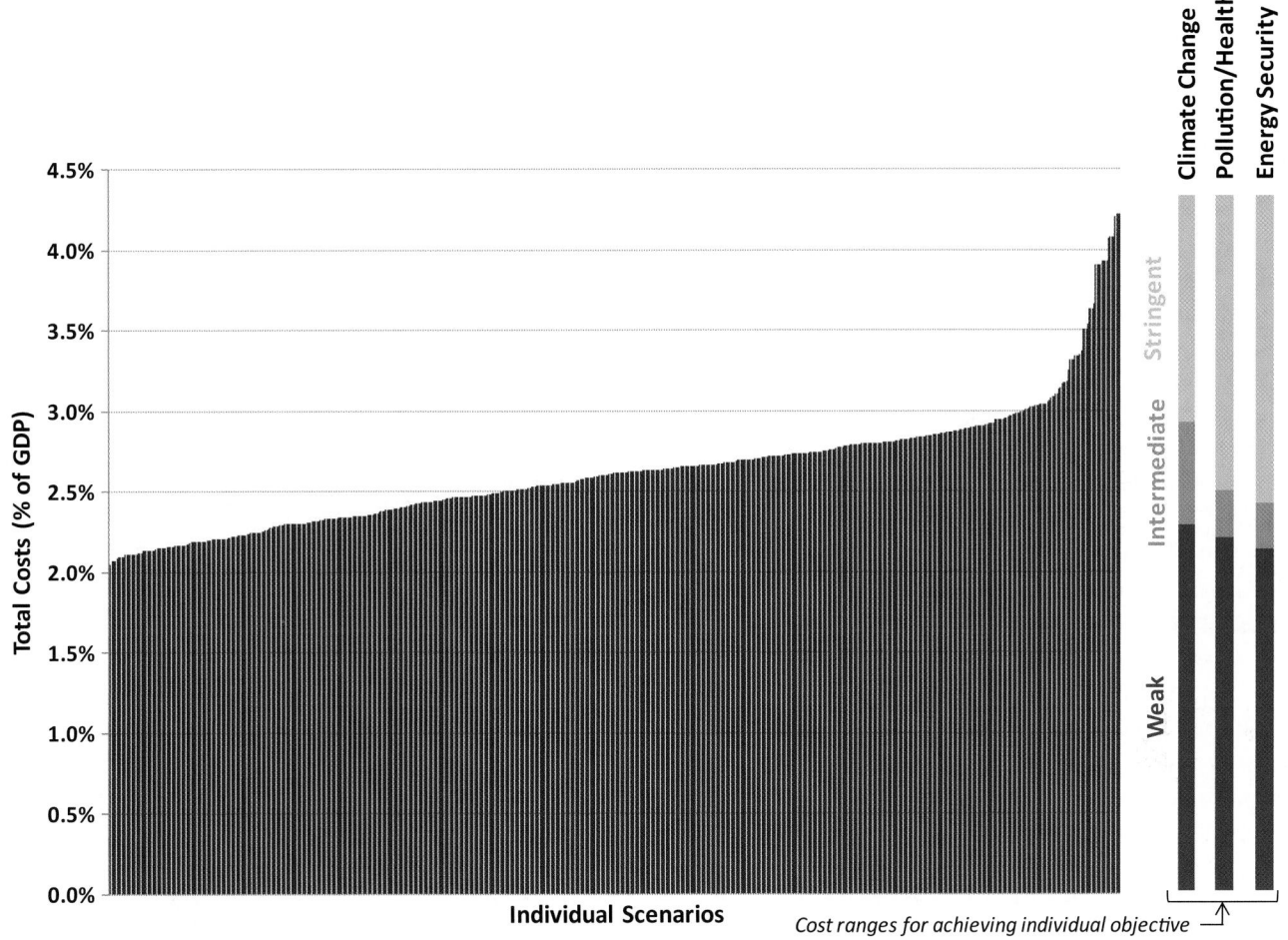

Figure 17.54 | Cumulative discounted total energy system costs for all scenarios in the full ensemble (2010–2050). Bars at right illustrate the ranges of total cost that correspond to Weak, Intermediate, and Stringent fulfillment of the climate, pollution and health, and energy security objectives.

is represented by the baseline scenario, which is obviously one of the least desirable in terms of diversity. (Nor does the baseline meet any of the other sustainability targets of the GEA, lying within the Weak region in all cases.) Compared with the baseline, virtually every other scenario, whether motivated by security or by climate policy, achieves a greater diversification of the global energy mix over time. As discussed more fully in Section 17.6, fulfilling the GEA targets for near-term energy security (the Stringent region in the figure) necessitates a global energy system that transitions to a broader portfolio of energy sources over the coming decades, while at the same time individual countries and regions (e.g., North America) come to rely less on imported energy commodities and more on domestic supplies. However, given the combination of the dominance of fossil energy in today's energy mix and the uneven distribution of fossil resource deposits around the globe, increasing energy diversity, and thus security, essentially requires that countries and regions move away from fossil energy and instead toward renewable energy sources such as biomass, wind, solar, and geothermal. Indeed, this is what emerges from the illustrative GEA pathways described previously in this chapter, as well as from the scenarios represented by

the Stringent and Intermediate regions in Figure 17.53. Section 17.7.2.2 discusses this point further.

Because the individual scenarios in the ensemble vary so greatly along the dimensions of climate change, pollution and health, and energy security, total energy system costs naturally span a fairly wide range as well. This is illustrated in Figure 17.54, where each bar represents the costs of a single scenario, and the scenarios are sorted in order of increasing costs. Included in these costs is the cumulative sum between 2010 and 2050 (discounted at 5% annually) of energy system investments (including supply and demand as well as climate change mitigation, energy security, and pollution control investments), operation and maintenance, fuel, and nonenergy mitigation costs.[63] Total system costs for each scenario are then related to the cumulative discounted sum of global GDP over the same time period. The least costly scenario in the ensemble is the baseline, since it assumes no climate change mitigation, pollution control, or energy security policies other than what is already planned over the next

_____

63 For the investment intensity of GDP, see also Section 17.3.5.

few years. Fulfillment of the GEA objectives (to any level of satisfaction) then adds to energy system costs to a certain degree. If one thinks of the multiple objectives as societal targets that the energy system should attempt to satisfy (i.e., scenario inputs), then total costs are an embodiment (i.e., scenario outputs) of the system-wide transformations that must take place in order to meet those objectives (e.g., increased utilization of advanced technologies and alternative fuels). The resulting total cost of a given scenario depends entirely on how far it goes toward satisfying each individual objective, as shown by the bars on the right side of Figure 17.54, which illustrate the ranges of scenarios, from a cost perspective, that correspond to Weak, Intermediate, and Stringent fulfillment of the climate, pollution and health, and energy security objectives. The least costly scenarios – those yielding little or no improvement in the objectives, such as the baseline – lie within the Weak region, whereas scenarios that achieve one or all of the objectives at the Intermediate or the Stringent level obviously incur costs in the middle or the upper end of the range, respectively. Notably, total costs range from 3.1–4.2% of GDP for the class of scenarios that achieves stringent fulfillment of all three objectives simultaneously. By comparison, energy system costs in the counterfactual baseline are about 2.1% of GDP over the same time period.

An important caveat to the cost analysis shown here is that it performs only a partial economic accounting. The analysis attempts to capture multiple benefits in terms of avoided or reduced costs for climate change mitigation, energy security, and pollution control. However, given the inherent difficulties in valuing human life in the economic sense, and given the vast uncertainties with respect to the economic valuation of, for example, climate-related damages, the analysis does not attempt to value other benefits of pursuing these three objectives (for a discussion of other benefits see Chapters 3, 4, and 5). For instance, the analysis does not consider the avoided costs of climate change (e.g., more frequent extreme weather events, impacts on global agriculture and food production), nor does it capture the avoided costs of adaptation to climate change (e.g., construction of sea walls, relocation of coastal populations). Similarly, the benefits accruing from reduced health expenditure and increased life expectancies have not been quantified here. Hence, the conclusions on multiple economic benefits presented in this section relate to "mitigation" costs only; they would become larger if other benefits were assigned an economic value as well.

17.7.2   Synergies between Objectives

The discussions above have already begun to show the inherent synergies, and to a lesser extent the trade-offs, among the various energy objectives and how these complex interdependencies can be illuminated through analysis of a large ensemble of possible energy futures. Among energy planners and decision makers, however, these relationships are not well enough understood. Cost trade-offs are obviously the more familiar: the greater society's aspiration for achieving the energy objectives, the larger the costs for the energy system. However, for such questions as, "How much extra might it cost to achieve each additional objective?" and "How can costs be reduced by pursuing multiple

objectives?" the answers are much less clear. The discussion that follows highlights the main findings of one of the few attempts in the scenario literature to explore the important relationships among climate change mitigation, energy security, and reduced air pollution and health impacts (for further reading, see van Vuuren et al., 2006; Cofala et al., 2009; Cofala et al., 2010; Bollen et al., 2010; McCollum et al., 2011).[64]

17.7.2.1   Climate Change Mitigation and Pollution and Health

Section 17.5.1 discussed in detail how decarbonization of the global energy system, combined with energy and conservation efforts, may be instrumental in limiting climate change to safer levels. This section takes the analysis a step further, showing that climate change mitigation can also help to reduce air pollutant emissions and their corresponding impacts on human health. Put more directly, climate change mitigation can be an important entry point for achieving society's pollution- and health-related goals. This is illustrated clearly in Figure 17.55, which relates global PM2.5 emissions in the near term (to 2030) to the probability of staying below a 2°C maximum temperature rise over the course of the century. Each data point in the figure represents values for a single scenario in the ensemble. The specific combination of pollution and climate policy stringency is what distinguishes the scenarios from one another. In particular, the different levels of air pollution control policy, indicated by the varying shapes of the data points, correspond to the scenario assumptions discussed previously in the pollution section (FLE, CLE, and SLE; see Section 17.5.2).

What one first notices in Figure 17.55 is that as the energy system is decarbonized and increasing shares of zero-carbon, pollution-free technologies are utilized, the probability of meeting the 2°C target increases, and pollutant emissions are significantly reduced. Moreover, the spread between the pollution control levels narrows as climate change mitigation becomes more of a priority. (The shaded areas in the figure help to illuminate this effect.) This last point is important, as it shows how the impacts of pollution control policy are much less variable as zero-carbon technologies penetrate the market and fossil technologies are forced out. This result stems from pollution control being applicable to fewer technologies (e.g., power plants, factories, vehicles) when there is less fossil energy in the system. A final observation is that climate change mitigation measures alone can yield pollutant emissions reductions on the order of currently planned legislation for pollution control.

Figure 17.55 also illustrates the extent to which each scenario fulfills the climate and pollution and health objectives, utilizing the Weak-Intermediate-Stringent framework discussed in Section 17.7.1.

---

64   As a supplement to the GEA, an interactive web-based scenario development tool has been developed at IIASA, which allows members of the public to improve their understanding of how different policy choices (i.e., prioritization of certain objectives above or below others) could potentially impact the development of the global energy system over the next several decades, in terms of resources, technologies, fuels, investments and the corresponding impacts on human health and the environment. To experiment with the Multi-criteria Analysis tool, see www.iiasa.ac.at/web-apps/ene/GeaMCA.

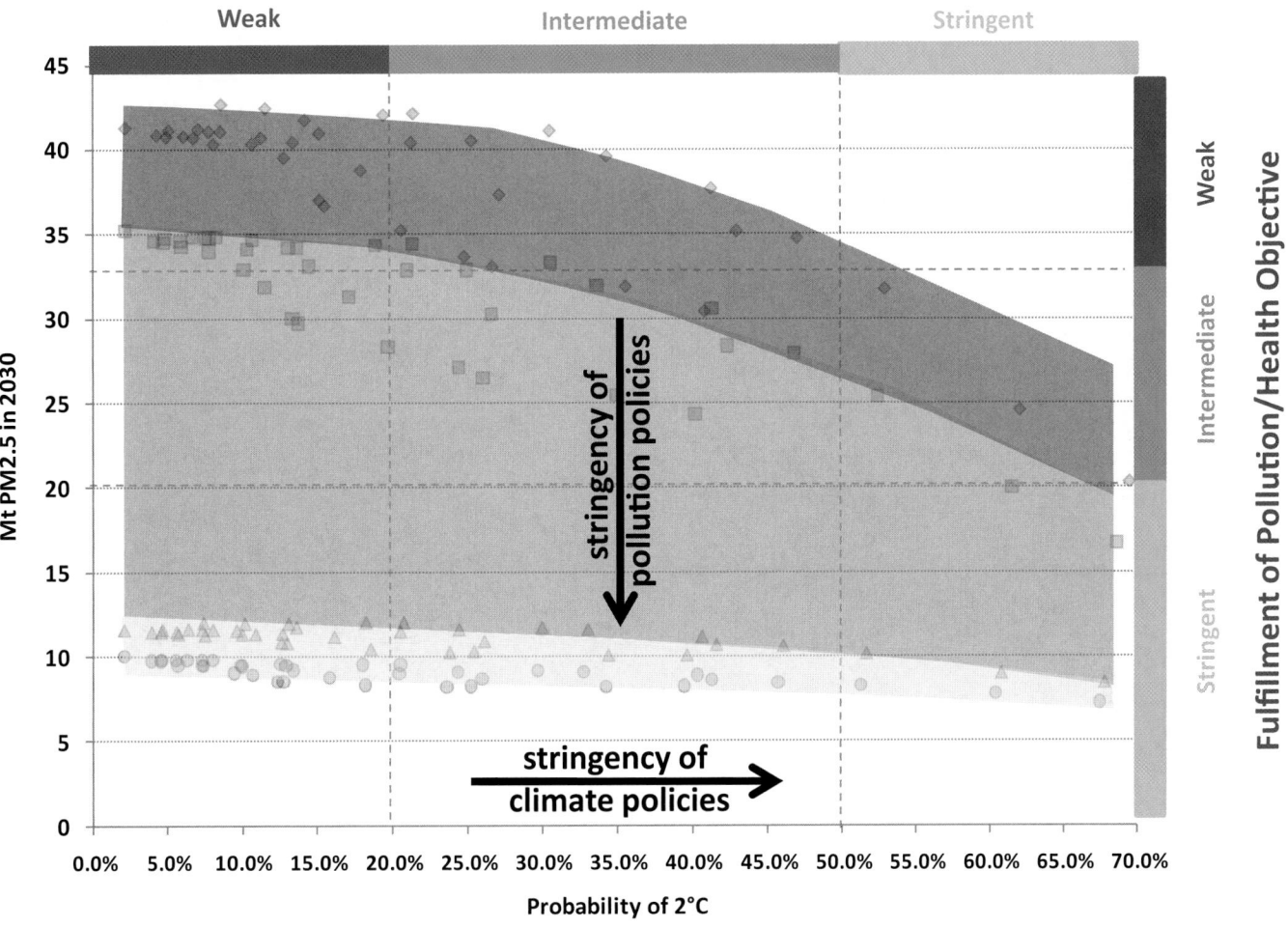

Figure 17.55 | Synergies between climate change mitigation and near-term reduction of PM2.5 emissions.

Because the core illustrative GEA pathways are designed to simultaneously satisfy all objectives at the Stringent level, they would be found in the lower-right corner of Figure 17.55. All other scenarios shown in the figure are unsustainable from the perspective of the GEA, as they satisfy the climate objective and the pollution and health objective at some other combination of levels (e.g., Weak on climate, Intermediate on pollution and health). Interestingly, the upper-right corner of the figure (corresponding to scenarios that would be Stringent on climate but Weak on pollution and health) contains not a single scenario, a result that again highlights how climate change mitigation can be an important entry point for achieving society's pollution- and health-related goals. In other words, strong climate change mitigation measures alone can yield pollutant emissions reductions that are as great as, or even greater than, currently planned pollution control legislation would likely yield in the absence of climate policy (i.e., through end-of-pipe pollution control technologies only), thereby allowing the pollution and health objective to be satisfied at the Intermediate level at a minimum. The opposite case (i.e., Weak on climate, Stringent on pollution and health) does not necessarily lead to the same conclusion, however; pollution control on its own is not likely to lead to dramatic reductions in GHG emissions. That being said, reducing key air pollutant emissions, namely, those

that cause warming (black carbon and the ozone precursors methane, nitrogen oxides, carbon monoxide, and volatile organic compounds), may be able to play a modest role in mitigating climate change. The climate feedbacks of air pollution are rather complex, and although the scenarios in the large ensemble shown here assume across-the-board reductions in all pollutants, one could certainly envision control strategies in which some specific pollutants are reduced proportionally more than others (e.g., warming components are reduced more than cooling components, namely, sulfur dioxide and organic carbon), in an effort to preserve the overall cooling effect of aerosols and, thus, to produce a net gain for the climate, or to at least remain radiant energy-neutral (Cofala et al., 2009; Ramanathan and Xu, 2010).

Reducing global air pollution levels, whether through pollution control or climate policy, or both, will necessarily lead to additional energy system costs – an important trade-off that relates to policy choices and the resulting direction of the energy system. However, given the enormous co-benefits between pollution and climate policy, achieving society's pollution and health objectives through climate change mitigation as an entry point has the potential to significantly reduce the added costs of pollution control. This is illustrated in Figure 17.56, which plots pollution

Fulfillment of Climate Objective

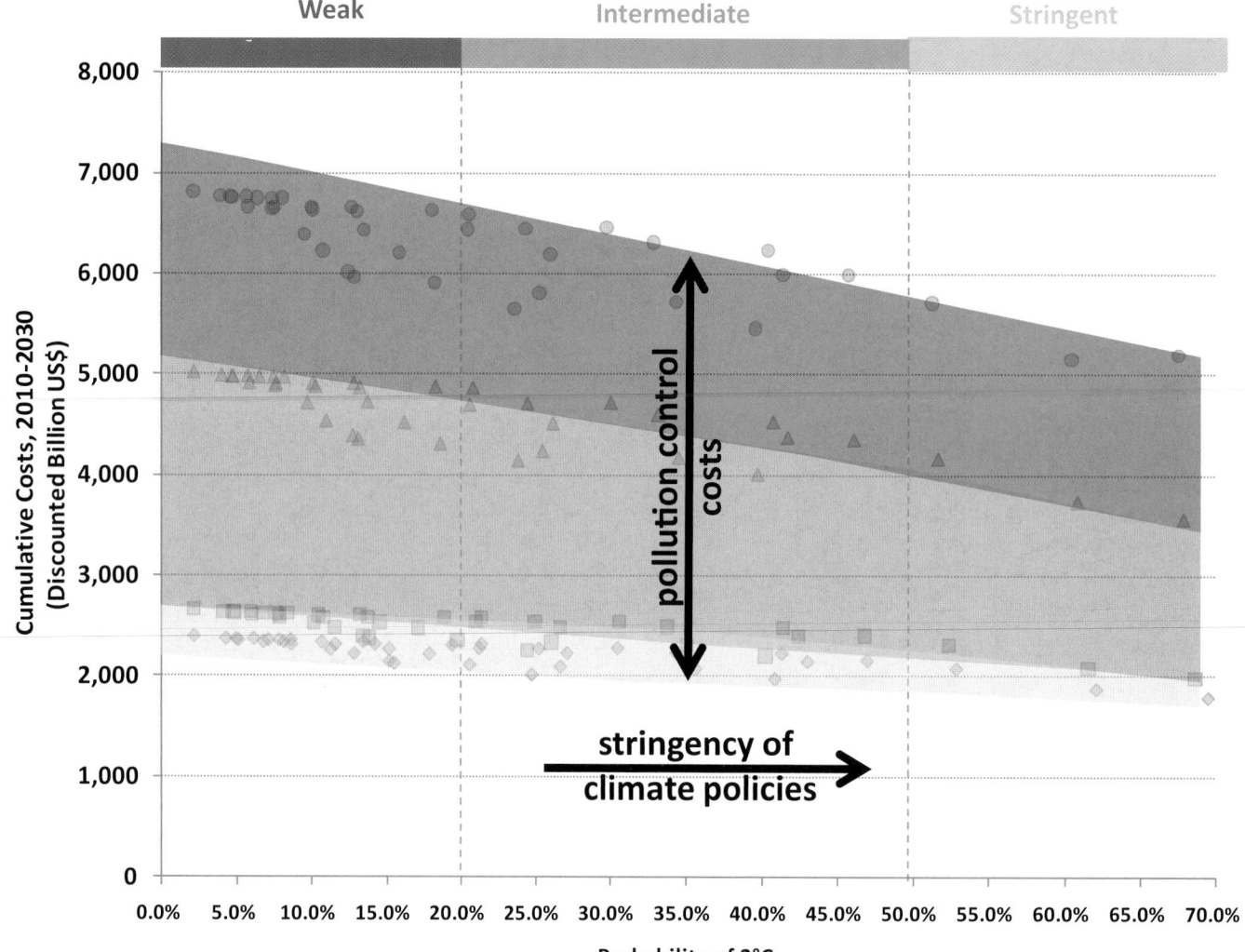

Figure 17.56 | Synergies between pollution control costs and climate change mitigation. The vertical axis represents the added costs of pollution control for each scenario in the ensemble.

control costs (relative to all other energy system costs) for each scenario in the ensemble. The data points toward the right side of the figure, particularly in the middle-right portion, are some of the most interesting, as these represent scenarios that fulfill both the climate objective and the pollution and health objective (see Figure 17.55) at the Stringent level, yet their added costs of pollution control are not much higher than in the baseline scenario (lower-left corner of the figure).

A closer look at three select scenarios of the ensemble provides a more detailed understanding of the climate-pollution-cost relationship. These three scenarios, shown in Figure 17.57, each fulfill the pollution and health objective at the Stringent level (consistent with the three illustrative sustainable GEA pathways); however, they do this by pursuing the climate objective to a greater or lesser degree (Weak, Intermediate, or Stringent fulfillment). The Weak Climate scenario represents baseline energy system development under a more stringent air pollution policy framework than would likely be realized

in a typical business-as-usual future. Such a policy adds a significant US$830 billion to total annual costs in 2030, compared with US$1630 billion for all other energy system costs (including both investments and operation and maintenance) in the same year.[65] Then, as the stringency of climate policy increases, the added costs of pollution control decrease substantially, especially in the Stringent Climate scenario, where control costs are US$470 billion less than in the Weak Climate scenario, a 57% reduction.[66] This striking result, which corroborates findings from other studies (e.g., Amann et al., 2009 for Europe), shows that a significant portion of climate change mitigation costs

---

65 Note the uncertainties of pollution control costs in absence of climate policies. For instance, pollution control costs in the so-called CLE1 scenario in Section 17.5.2 amount to about US$600 billion in 2030. Differences are due to alternative baselines used in the two sections.

66 Generally, pollution control costs of scenarios reaching the Stringent fulfillment level are on the order of US$200 billion to US$350 billion in 2030.

can be compensated for by reduced pollution control requirements, while at the same time still allowing for the stringent fulfillment of society's pollution- and health-related targets. Furthermore, the multiple benefits of climate change mitigation also show up as avoided damage costs for the impacts of air pollutant emissions on human health, though it should be clearly stated that the synergies analysis described in this section has not attempted such an estimation.

Another noteworthy observation from Figure 17.57 relates to which sectors contribute most to the added costs of pollution control. In the Weak Climate scenario, all sectors require significant amounts of investment, with the energy conversion sector and the residential and commercial end-use sectors being responsible for the bulk of the costs. In the Stringent Climate scenario, however, end-of-pipe pollution control requirements decrease substantially in all sectors.

In sum, when viewed from a holistic and integrated perspective, the combined costs of climate change mitigation and pollution control come at a significantly reduced total energy bill if the benefits of pollution reduction are properly figured into the calculation of GHG abatement strategies (see also Nemet et al., 2010). The design of cost-effective future policies, therefore, would benefit by integrating holistic portfolios of measures that address both pollution and climate objectives simultaneously. This is, of course, no simple task, given that in many countries air pollution and climate change are dealt with by separate policy institutions. For this reason the enormous co-benefits of the two objectives are often overlooked, and the costs of reaching each objective individually are often overstated (Amann, 2009). In terms of the technology mix, a robust finding of the analyses summarized in this chapter is that a key strategy for meeting both climate and pollution and health objectives is to increase the utilization

of efficiency measures as well as zero-carbon, pollution-free energy technologies, such as nuclear and renewable energy.

17.7.2.2   Climate Change Mitigation and Energy Security

The previous discussion has shown that early deployment of zero-carbon technologies can help to achieve both near-term pollution and long-term climate targets. In addition, this analysis finds that there are important synergies between decarbonization and energy security, yet another key near-term objective. In short, as countries and regions invest more heavily in renewables in an effort to decarbonize their economies, they will by extension reduce their need to import globally traded fossil energy commodities such as coal, oil, and natural gas. Because renewables (biomass, hydro, wind, solar, and geothermal) can potentially be produced almost entirely domestically (or at least regionally within a cluster of like-minded countries), they are from a dependency perspective inherently secure resources. Moreover, increased utilization of renewables and nuclear energy tends to diversify the energy resource mix away from one that relies so heavily on fossil energy. Thus, decarbonization of the energy system can simultaneously reduce import dependence and increase energy diversity, both of which are key indicators of a more secure energy supply (see Chapter 5 on energy security). In fact, the results of this analysis indicate that the most "secure" scenario, from the perspective of both diversity and trade, is one in which all regions pursue very stringent policies that promote both climate change mitigation *and* reduced import dependence.

Figure 17.58 illustrates the relationship between the climate and security objectives by showing global primary energy diversity and dependence in 2030 (measured in terms of the compound SWDI, introduced in Section 17.6.2) as a function of the probability of staying below the 2°C warming target. The third dimension captures several alternative policy levels representing the varying stringency of efforts to limit import dependency by individual world regions; these levels are grouped together by the shaded areas. Note that all the scenarios are identical with respect to the stringency of air pollution legislation that is assumed. Figure 17.59 focuses on costs, plotting the probability of meeting the 2°C target against cumulative total global policy costs as a share of global GDP between 2010 and 2030. Total policy costs, calculated relative to the baseline scenario, attempt to capture the added costs of energy security, climate change mitigation, and air pollution control policies.[67]

The double effects of decarbonization and reduced import dependence are quite clear from Figures 17.58 and 17.59. As regions pursue strategies to mitigate climate change or enact policies and procurement strategies that prioritize domestic supplies over imports, the diversity of

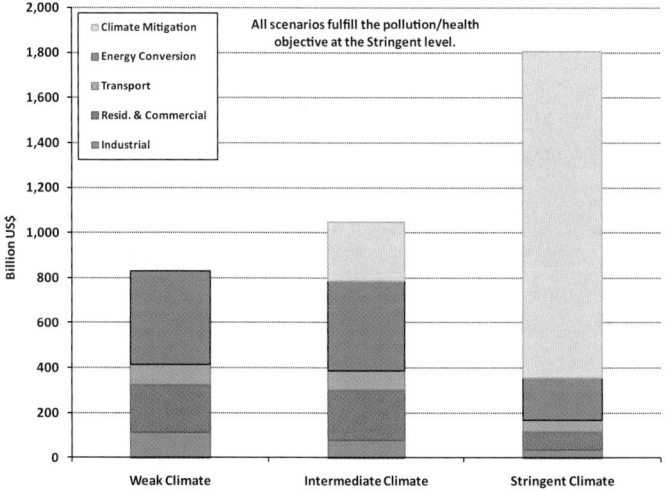

Figure 17.57 | Global annual pollution control and climate change mitigation costs for Weak, Intermediate, and Stringent climate policy scenarios in 2030. The Stringent Climate scenario achieves the 2°C target with a comparatively high probability of >60% and thus represents an upper-bound estimate for climate change mitigation costs and pollution control co-benefits across the GEA pathways.

---

67   Costs include energy system investments, pollution control investments, operation and maintenance, fuel, nonenergy mitigation, and demand reduction (i.e., the macroeconomic response).

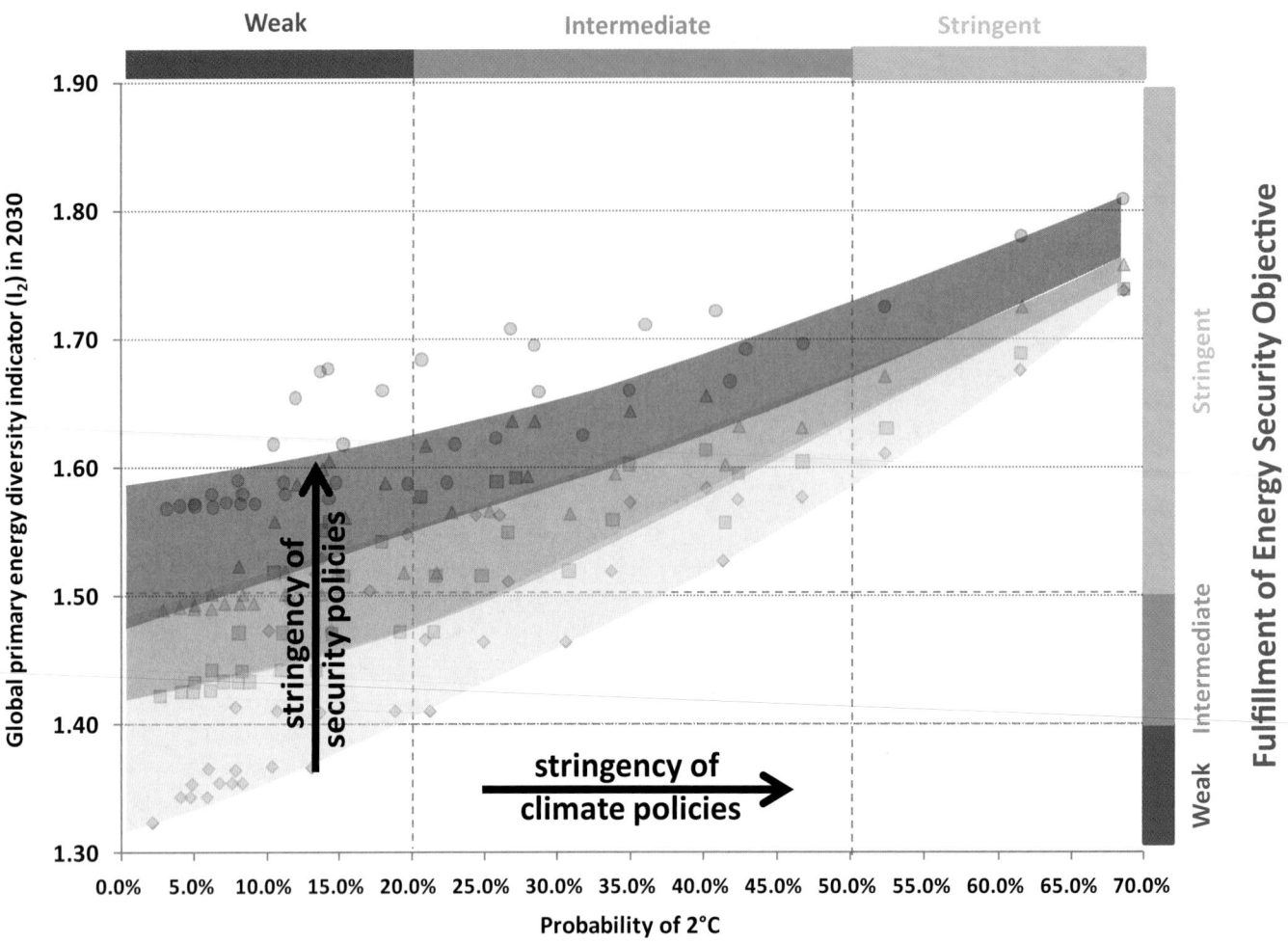

Figure 17.58 | Synergies between near-term energy security policies and climate change mitigation.

their energy resource mix is likely to increase. Naturally, pushing both the climate and security objectives adds to total energy system costs; yet, as in the relationship between climate change mitigation and pollution control, at higher levels of decarbonization, the costs of security are significantly reduced, highlighting the synergies and multiple benefits of the two objectives. As Figure 17.59 illustrates, when climate change is of relatively low priority (the Weak climate region), security costs can increase total system costs by as much as 0.2 percentage points. Conversely, under Stringent climate policies, the added costs of security approach zero.

Figure 17.60 takes a deeper look into the climate-security-cost relationship by summarizing energy security costs for three alternative pairs of scenarios.[68] The scenarios in each pair fulfill the climate objective to the same degree (Weak, Intermediate, or Stringent). What distinguishes them is the level at which the two scenarios in a given pair satisfy the energy security objective; hence, the difference in their costs represents the added costs

of security. For instance, under a Weak Climate regime, as envisioned in a business-as-usual future, this cost premium, in terms of globally aggregated annual energy system investments, is approximately US$160 billion in 2030. By comparison, under an Intermediate or a Stringent Climate regime, the added costs of security decline significantly, to just US$64 billion and US$28 billion/year, respectively (reductions of 61% and 84% compared with the Weak Climate case). As evidenced by Figure 17.60, security policy, applied at the level of individual countries and groups of countries, primarily spurs additional investments in end-use efficiency and electricity generation, while at the same time lower the global investment requirements for upstream energy extraction (coal mining and oil production). The security co-benefits that stem from climate change mitigation are then largely attributed to the reduced need for extra "security investments", since climate policy promotes energy efficiency and conservation and the increased utilization of domestically produced, low-carbon energy sources. Of course, climate policy itself also adds to the total energy bill, as is shown separately in Figure 17.60 for comparison. Climate change mitigation costs are clearly quite substantial, although it is important to note that the cost accounting for climate policy is more comprehensive than that shown for security, which captures

---

68   Each scenario incorporates the same assumptions for the stringency of air pollution legislation (CLE level). See the pollution section for more information.

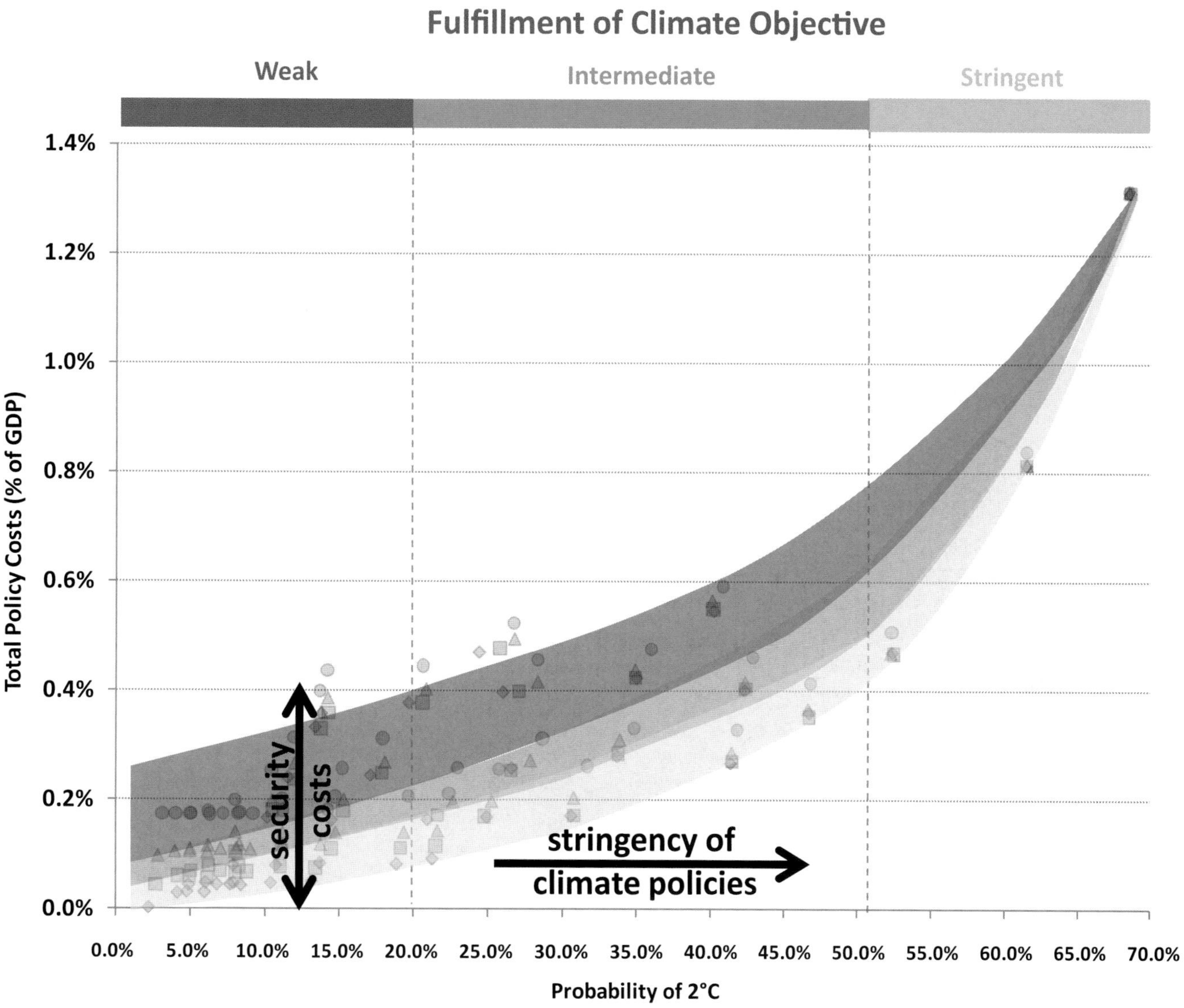

Fulfillment of Climate Objective

Weak   Intermediate   Stringent

Figure 17.59 | Total global policy costs of simultaneously achieving energy security and climate objectives to varying degrees. Cumulative discounted costs from 2010 to 2030, relative to baseline.

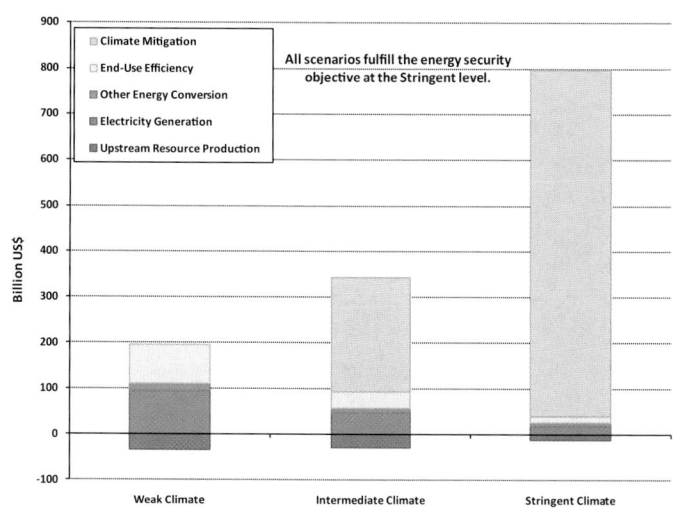

Figure 17.60 | Global annual energy security investment and climate change mitigation costs for Weak, Intermediate, and Stringent climate policy scenarios in 2030.

only investments. In the figure, climate change mitigation costs refer to all costs beyond those motivated by security policy, including investments in low-carbon technologies and their associated variable costs, as well as the costs associated with demand reduction (energy efficiency investments and conservation efforts) and nonenergy GHG mitigation measures. The bottom line is that, as with the climate-pollution-cost relationship, when viewed from a holistic and integrated perspective, the combined costs of climate change mitigation and energy security come at a significantly reduced total energy bill when the benefits of security are properly figured into the calculation of GHG abatement strategies.

17.7.3   Conclusion

The energy system of the future could potentially develop in a number of different directions, depending on how society and its decision makers prioritize various worthwhile energy objectives, including, but not limited to, climate change mitigation, energy security, and air pollution and human health. These objectives are generally discussed in the context of different time frames (security and pollution and health in the near term, climate in the medium to long term). For this reason, they frequently compete for attention in the policy world. An added challenge is that in many countries, separate policy institutions are responsible for dealing with each of the multiple objectives. As a result, the important synergies between them are not well enough understood, or are simply overlooked, and the costs of reaching each objective individually are often overstated.

In short, by adopting a holistic and integrated perspective that addresses all of the objectives simultaneously, the analysis described in this section clearly indicates that cost-effective climate-pollution-security policies are likely to lead to substantial co-benefits, in terms of costs avoided and the achievement of societal objectives for sustainability. First, fulfillment of near-term pollution and health goals at the Stringent level is greatly furthered by climate change mitigation. Under stringent climate policy scenarios, for instance, globally aggregated DALYs can be reduced by up to 22 million in 2030. At the same time, stringent climate policy can help to further the energy security goals of individual countries and regions by promoting the increased utilization of domestically available renewable energy sources that are both more dependable and more resilient than imports of fossil energy commodities. Such a strategy would lead to the diversification of a given region's supply mix, a widely acknowledged approach for achieving security. Both of these findings illustrate how climate change mitigation can be an important entry point for achieving society's pollution- and health-related goals. Moreover, the combined costs of climate change mitigation, energy security, and air pollution control come at a significantly reduced total energy bill if the multiple benefits of each are properly accounted for in the calculation of total energy system costs. For instance, the total added costs of pollution control at the global level are cut significantly (by up to US$500 billion annually in 2030 compared to a baseline scenario) as the stringency of climate policy increases and the utilization of zero-carbon, pollution-free (thus, pollution control-free) technologies rises. Similarly, security costs also decrease substantially under increasingly aggressive levels of decarbonization, and in scenarios with very stringent climate policies, the added costs of security actually approach zero (translating to an annual cost savings of more than US$130 billion in 2030). Although steps taken to mitigate climate change will themselves add to total energy system costs compared with a baseline scenario (a key trade-off), these climate costs will be substantially compensated for by the corresponding cost reductions for pollution control and energy security (key synergies).

Other economic benefits of rapidly decarbonizing the energy system are the reduced need for subsidies into carbon-intensive petroleum products and coal. Following the IEA (2009b) and Coady et al. (2010) subsidies from these fuels amount at present to about US$132 billion to US$240 billion/year. Just 15% of this total is spent directly for the poor who have limited access to clean energy. As noted in Section 17.4, subsidies for the poor must be increased in order to achieve universal access. GHG mitigation in the GEA pathways would, however, at the same time reduce consumption of carbon-intensive fossil fuels by the rest of the population, leading to a reduction in the need for subsidies for oil products and coal on the order of US$70 billion to US$130 billion/year by 2050 compared with today.

Many other benefits of the energy transformation have not been assigned economic values in detail here but are important to account for as well. As illustrated in this section and earlier, in Section 17.5.2, the health benefits of the transformation can be significant. In addition, pollution control reduces damages to vegetation and may result in significant benefits for land productivity by avoiding eutrophication and acidification (see Chapter 3). As discussed in Chapter 19 and in Section 17.4, universal access to electricity and clean cooking not only leads to significant health benefits, but also increases the productivity of the poorest, thus contributing to well-being and more equitable economic growth. In addition, limiting the global temperature rise to less than 2°C compared with preindustrial times reduces the risks for a number of different types of climate impacts, summarized by five main reasons for concern (Smith et al., 2009; see also Chapter 3): (i) the risk to unique or threatened systems; (ii) the risk of increases in extreme weather; (iii) the distribution of impacts (and the disparities of those impacts, given that some regions, countries, and populations may face greater harm from climate change); (iv) aggregate damages (assessing comprehensive measures of impacts through efforts to aggregate into a single metric, e.g., monetary damages); and (v) the risk of large-scale discontinuities (e.g., possible tipping points associated with very large impacts such as deglaciation of the West Antarctic or the Greenland ice sheet). Finally, rapid decarbonization, which leads to a stronger reliance on efficiency and zero-carbon energy (e.g., renewables), may create new job opportunities and thus provide additional economic benefits.

Realizing the multiple benefits of the energy transformation requires, however, a holistic and integrated approach that addresses a diverse set of objectives simultaneously. Although the GEA pathways have shown that such a transformation is in principle technically possible, the task remains extremely ambitious and will require rapid introduction of policies and fundamental political changes that lead to concerted and coordinated efforts to integrate global concerns, such as climate change, into local and national policy priorities such as health and pollution, access to clean energy, and energy security.

References

Amann, M., R. Cabala, J. Cofala, C. Heyes, Z. Klimont, W. Schö pp, L. Tarrason, D. Simpson, P. Wind and J. E. Jonson, 2004: *The "Current Legislation" and the "Maximum Technically Feasible Reduction" cases for the CAFE baseline emission projections*. CAFE Working Group on Target Setting and Policy Advice. Laxenburg, IIASA 39

Amann, M., L. Höglund Isaksson, W. Winiwarter, A. Tohka, F. Wagner, W. Schö pp, I. Bertok and C. Heyes, 2008: *Emission scenarios for non-CO₂ greenhouse gases in the EU-27. Mitigation potentials and costs in 2020*. IIASA, Laxenburg.

Amann, M., 2009: Air pollutants and greenhouse gases – Options and benefits from co-control. In *Air Pollution & Climate Change – Two Sides of the Same Coin?* H. Pleijel, P. E. Karlsson and D. Simpson, (eds.), Swedish Environmental Protection Agency, Stockholm pp.99–108.

Amann, M., I. Bertok, J. Borken-Kleefeld, J. Cofala, C. Heyes, L. Hoeglund Isaksson, Z. Klimont, P. Purohit, P. Rafaj, W. Schoepp, G. Toth, F. Wagner and W. Winiwarter, 2009: *Potentials and Costs for Greenhouse Gas Mitigation in Annex I Countries: Methodology Proceedings of the National Academy of Sciences of the United States of America*, International Institute for Applied Systems Analysis (IIASA), Laxenburg.

Andersen, P. H., J. A. Mathews and M. Rask, 2009: Integrating private transport into renewable energy policy: The strategy of creating intelligent recharging grids for electric vehicles. *Energy Policy*, **37**(7):2481–2486.

APERC, 2007: *A Quest for Energy Security in the 21st Century – Resources and Constraints*. Institute of Energy Economics, Asia Pacific Energy Research Centre (APERC).

Baciocchi, R., G. Storti and M. Mazzotti, 2006: Process design and energy requirements for the capture of carbon dioxide from air. *Chemical Engineering and Processing*, **45**(12):1047–1058.

Bailis, R., M. Ezzati and D. M. Kammen, 2005: Mortality and greenhouse gas impacts of biomass and petroleum energy futures in Africa. *Science*, **308**(5718):98–103.

Benjamin, R., 2007: Principles for interregional transmission expansion. *The Electricity Journal*, **20**(8):36–47.

Bergamaschi, P., C. Frankenberg, J. F. Meirink, M. Krol, F. Dentener, T. Wagner, U. Platt, J. O. Kaplan, S. Körner, M. Heimann, E. J. Dlugokencky and A. Goede, 2007: Satellite chartography of atmospheric methane from SCIAMACHY on board ENVISAT: 2. Evaluation based on inverse model simulations *Journal of Geophysical Research Atmospheres*, **112**(D02304):26.

Berk, M. M. and M. G. J. den Elzen, 2001: Options for differentiation of future commitments in climate policy: how to realise timely participation to meet stringent climate goals? *Climate Policy*, **1**(4):465–480.

Birol, F. and J. H. Keppler, 2000: Prices, technology development and the rebound effect. *Energy Policy*, **28**(6–7):457–469.

Blanchard, O., 2002: Scenarios for differentiating commitments. In *Options for protecting the climate*. K. A. Baumert, O. Blanchard, S. Llose and J. F. Perkaus, (eds.), World Resources Institute, WRI, Washington DC.

Böhringer, C. and H. Welsch, 2004: Contraction and Convergence of carbon emissions: an intertemporal multi-region CGE analysis. *Journal of Policy Modeling*, **26**:21–39.

Böhringer, C. and A. Löschel, 2005: Climate Policy Beyond Kyoto: Quo Vadis? A Computable General Equilibrium Analysis Based on Expert Judgments. *KYKLOS*, **58**(4):467–493.

Bollen, J., C, A. J. G. Manders and P. J. J. Veenendaal, 2004: *How much does a 30% emission reduction cost? Macroeconomic effects of post-Kyoto climate policy in 2020*. Netherlands Bureau for Economic Policy Analysis, The Hague.

Bollen, J., C., S. Hers and v. d. B. Zwaan, 2010: An integrated assessment of climate change, air pollution, and energy security policy. *Energy Policy*, **38**(8):4021–4030.

Bouwman, L., T. Kram and K. Klein-Goldewijk, 2006: *Integrated Modelling of Global Environmental Change. An Overview of IMAGE 2.4. Netherlands Environmental Assessment Agency*, Bilthoven.

BP, 2009: *Statistical Review of World Energy*. BP, London, UK.

Bradshaw, J. and T. Dance, 2005: Mapping geological storage prospectivity of CO2 for the world's sedimentary basins and regional source to sink matching. *Proceedings of the 7th International Conference on Greenhouse Gas Control Technologies*.

Carpenter, S. and P. Pingali, 2005: *Millennium Ecosystem Assessment – Scenarios Assessment*. Island Press, Washington DC.

Clarke, L., J. Edmonds, V. Krey, R. Richels, S. Rose and M. Tavoni, 2009: International climate policy architectures: Overview of the EMF 22 international scenarios. *Energy Economics*, **31**(Supplement 2):S64-S81.

Coady, D., R. Gillingham, R. Ossowski, J. Piotrowski, S. Tareq and J. Tyson, 2010: *Petroleum Product Subsidies: Costly, Inequitable, and Rising*. International Monetary Fund (IMF).

Cofala, J., M. Amann, Z. Klimont, K. Kupiainen and L. Hoeglund Isaksson, 2007: Scenarios of global anthropogenic emissions of air pollutants and methane until 2030. *Atmospheric Environment*, **41**(38):8486–8499.

Cofala, J., P. Rafaj, W. Schoepp, Z. Klimont and M. Amann, 2009: *Emissions of Air Pollutants for the World Energy Outlook 2009 Energy Scenarios*. IIASA, Laxenburg.

Cofala, J., M. Amann, W. Asman, I. Bertok, C. Heyes, L. Hoeglund Isaksson, Z. Klimont, W. Schoepp and F. Wagner, 2010: Integrated assessment of air pollution and greenhouse gases mitigation in Europe. *Archives of Environmental Protection*, **36**(1):29–39.

Cohen, A., R. Anderson, Bart Ostro, K. D. Pandey, M. Krzyzanowski, N. Künzli, K. Gutschmidt, A. C. Pope III, I. Romieu, J. M. Samet and K. R. Smith, 2004: Urban Air Pollution. In *Comparative quantification of health risks:global and regional burden of disease attributable to selected major risk factors*. E. M. e. al, (ed.), World Health Organization, Geneva pp.1353–1434.

Criqui, P., A. Kitous, M. M. Berk, M. G. J. den Elzen, B. Eickhout, P. Lucas, D. P. van Vuuren, N. Kouvaritakis and D. van Regemorter, 2003: *Greenhouse gas reduction pathways in the UNFCCC Process upto 2025 – Technical Report*. CNRS-IEPE, Grenoble, France.

De, S. and M. Assadi, 2009: Impact of cofiring biomass with coal in power plants – A techno-economic assessment. *Biomass and Bioenergy*, **33**(2):283–293.

Delucchi, M. A. and M. Z. Jacobson, 2011: Providing all global energy with wind, water, and solar power, Part II: Reliability, system and transmission costs, and policies. *Energy Policy*.

den Elzen, M. G. J. and P. Lucas, 2005: The FAIR model: a tool to analyse environmental and costs implications of climate regimes. *Environmental Modeling and Assessment*, **10**(2):115–134.

den Elzen, M. G. J., P. Lucas and D. P. Van Vuuren, 2005: Abatement costs of post-Kyoto climate regimes. *Energy Policy*, **33**(16):2138–2151.

den Elzen, M. G. J. and M. Meinshausen, 2005: *Meeting the EU 2°C climate target: global and regional emission implications*. Netherlands Environmental Assessment Agency (MNP), Bilthoven, the Netherlands.

den Elzen, M. G. J. and D. P. van Vuuren, 2007: Peaking profiles for achieving long-term temperature targets with more likelihood at lower costs. *Proceedings of the National Academy of Sciences of the United States of America*, **104**(46):17931–17936.

Dentener, F., J. Drevet, J. F. Lamarque, I. Bey, B. Eickhout, A. M. Fiore, D. Hauglustaine, L. W. Horowitz, M. Krol, U. C. Kulshrestha, M. Lawrence, C. Galy-Lacaux, S. Rast, D. Shindell, D. Stevenson, T. Van Noije, C. Atherton, N. Bell, D. Bergman, T. Butler, J. Cofala, B. Collins, R. Doherty, K. Ellingsen, J. Galloway, M. Gauss, V. Montanaro, J. F. Müller, G. Pitari, J. Rodriguez, M. Sanderson, S. Strahan, M. Schultz, F. Solmon, K. Sudo, S. Szopa and O. Wild, 2006: Nitrogen and sulphur deposition on regional and global scales: a multi-model evaluation. *Global Biogeochemical Cycles*, **GB4003**:21.

Dornburg, V. and A. P. C. Faaij, 2005: Cost and CO2-Emission Reduction of Biomass Cascading: Methodological Aspects and Case Study of SRF Poplar. *Climatic Change*, **71**(3):373–408.

Dornburg, V., D. van Vuuren, G. van de Ven, H. Langeveld, M. Meeusen, M. Banse, M. van Oorschot, J. Ros, G. van den Born, H. Aiking, M. Londo, H. Mozaffarian, P. Verweij, E. Lysen and A. Faaij, 2010: Bioenergy revisited: Key factors in global potentials of bioenergy. *Energy Environment Science*, **3**(3):258–267.

Edenhofer, O., C. Carraro, J.-C. Hourcade, K. Neuhoff, G. Luderer, C. Flachsland, M. Jakob, A. Popp, J. Steckel, J. Strohschein, N. Bauer, S. Brunner, M. Leimbach, H. Lotze-Campen, V. Bosetti, E. d. Cian, M. Tavoni, O. Sassi, H. Waisman, R. Crassous-Doerfler, S. Monjon, S. Dröge, H. v. Essen, P. d. Río and A. Türk, 2009: *The Economics of Decarbonization – Report of the RECIPE project.*, Potsdam Institute for Climate Impact Research, Potsdam.

Edenhofer, O., B. Knopf, T. Barker, L. Baumstark, E. Bellevrat, B. Chateau, P. Criqui, M. Isaac, A. Kitous, S. Kypreos, M. Leimbach, K. Lessmann, B. Magne, Å. Scrieciu, H. Turton and D. P. Van Vuuren, 2010a: The economics of low stabilization: Model comparison of mitigation strategies and costs. *Energy Journal*, **31**(Special):223–241.

Edenhofer, O., B. Knopf, T. Barker, L. Baumstark, E. Bellevrat, B. Chateau, P. Criqui, M. Isaac, A. Kitous, S. Kypreos, M. Leimbach, K. Lessmann, B. Magne, Å. Scrieciu, H. Turton and D. P. Van Vuuren, 2010b: The economics of low stabilization: Model comparison of mitigation strategies and costs. *Energy Journal* **31**(Special):223–241.

Ekholm, T., V. Krey, S. Pachauri and K. Riahi, 2010: Determinants of household energy consumption in India. *Energy Policy*, **38**(10):5696–5707.

European Commission, 2007: *Limiting Global Climate Change to 2 degrees Celsius: The way ahead for 2020 and beyond.* Communication from the Commission to the Council, the European Parliament, the European Economic and Social Committee and the Committee of the Regions, Brussels.

Fischedick, M., R. Schaeffer, A. Adedoyin, M. Akai, T. Bruckner, L. Clarke, V. Krey, I. Savolainen, S. Teske, D. Ürge-Vorsatz, R. Wright, 2011: Mitigation Potential and Costs. In *IPCC Special Report on Renewable Energy Sources and Climate Change Mitigation.* O. Edenhofer, R. Pichs-Madruga, Y. Sokona, K. Seyboth, P. Matschoss, S. Kadner, T. Zwickel, P. Eickemeier, G. Hansen, S. Schlömer, C. von Stechow (eds), Cambridge University Press, Cambridge, United Kingdom and New York, NY, USA.

Fisher, B. S., N. Nakicenovic, K. Alfsen, J. Corfee Morlot, F. de la Chesnaye, J.-C. Hourcade, K. Jiang, M. Kainuma, E. La Rovere, A. Matysek, A. Rana, K. Riahi, R. Richels, S. Rose, D. van Vuuren and R. Warren, 2007: Issues related to mitigation in the long term context. In *Climate Change 2007: Mitigation. Contribution of Working Group III to the Fourth Assessment Report of the Intergovernmental Panel on Climate Change.* B. Metz, O. R. Davidson, P. R. Bosch, R. Dave and L. A. Meyer, (eds.), Cambridge University Press, Cambridge, UK pp.169–250.

Forest, C. E., P. H. Stone, A. P. Sokolov, M. R. Allen and M. D. Webster, 2002: Quantifying uncertainties in climate system properties with the use of recent climate observations. *Science*, **295**(5552):113–117.

Føyn, T. H. Y., K. Karlsson, O. Balyk and P. E. Grohnheit, 2011: A global renewable energy system: A modelling exercise in ETSAP/TIAM. *Applied Energy*, **88**(2):526–534.

Gaunt, C. T., 2005 : Meeting electrification's social objectives in South Africa, and implications for developing countries. *Energy Policy*, **33**(10):1309–1317.

Gilli, P. V., N. Nakicenovic and R. Kurz 1995: *First- and second-law efficiencies of the global and regional energy systems.* Preeedings of the 16th World Energy Congress, Tokyo, Japan.

Goldemberg, J., 2007: Ethanol for a Sustainable Energy Future. *Science*, **315**(5813):808–810.

Granier, C., B. Bessagnet, T. Bond, A. D'Angiola, H. G. v. d. Gon, G. Frost, A. Heil, M. Kainuma, J. Kaiser, S. Kinne, Z. Klimont, S. Kloster, J. F. Lamarque, C. Liousse, T. Matsui, F. Meleux, A. Mieville, T. Ohara, J. C. Raut, K. Riahi, M. Schultz, S. Smith, A. M. Thomson, J. v. Aardenne, G. v. d. Werf and D. v. Vuuren, 2010: Evolution of anthropogenic and biomass burning emissions at global and regional scales during the 1980–2010 period. *Climatic Change*, **Submitted, under review.**

Greening, L., D. L. Greene and C. Difiglio, 2000: Energy efficiency and consumption – the rebound effect – a survey. *Energy Policy*, **28**(6–7):389–401.

Groenenberg, H., K. Blok and J. P. van der Sluijs, 2004: Global Triptych: a bottom-up approach for the differentiation of commitments under the Climate Convention. *Climate Policy*, **4**:153–175.

Gross, R., P. Heptonstall, M. Leach, D. Anderson, T. Green and J. Skea, 2007: Renewables and the grid: understanding intermittency. *Energy*, **160**(1):31–41.

Grubler, A. and N. Nakicenovic, 1994: *International burden sharing in greenhouse gas reduction.* International Institute for Applied Systems Analysis (IIASA), Laxenburg, Austria.

Grubler, A., 2010: The costs of the French nuclear scale-up: A case of negative learning by doing. *Energy Policy*, **38**(9):5174–5188.

Grubler, A. and K. Riahi, 2010: Do governments have the right mix in their energy R&D portfolios? *Carbon Management*, **1**(1):79–87.

Hanley, N., P. G. McGregor, J. K. Swales and K. Turner, 2009: Do increases in energy efficiency improve environmental quality and sustainability? *Ecological Economics*, **68**(3):692–709.

Hendriks, C., W. Graus and F. van Bergen, 2004: *Global carbon dioxide storage potential and costs.* Ecofys, Utrecht.

Hirschberg, S., C. Bauer, P. Burgherr, R. Dones, W. Schenler, T. Bachmann and D. Gallego Carrera, 2007: *Environmental, Economic and Social Criteria and Indicators for Sustainability Assessment of Energy Technologies.* New Energy Externalities Developments for Sustainability (NEEDS), Rome, Italy.

Höhne, N., D. Phylipsen, S. Ullrich and K. Blok, 2005: *Options for the second commitment period of the Kyoto Protocol, research report for the German Federal Environmental Agency.* ECOFYS Gmbh, Berlin.

Höhne, N., 2006: *What is Next After the Kyoto Protocol? Assessment of Options for International Climate Policy Post 2012.* Techne Press, Amsterdam.

Holttinen, H., P. Meibom, A. Orths, F. van Hulle, B. Lange, A. Tiedemann, M. O'Malley, J. Perik, B. Ummels, J. Tande, A. Estanqueiro, M. Matos, E. Gomez, L. Soder, G. Strbac, A. Shakoor, J. Smith and M. Milligan, 2009: *Design and operation of power systems with large amounts of wind power: Phase one 2006–2008.* VTT Technical Research Centre of Finland, Espoo, Finland.

IEA, 2006: *World Energy Outlook 2006.* Hrsg., (ed.) International Energy Agency (IEA) of the Organisation for Economic Co-operation and Development (OECD), Paris, France.

IEA, 2007: *Tracking Industrial Energy Efficiency and CO2 Emissions.* International Energy Agency (IEA), OECD, Paris, France.

IEA, 2009a: *World Energy Outlook 2009.* International Energy Agency (IEA), Paris, France.

IEA, 2009b: *Energy Technology Transitions for Industry.* International Energy Agency (IEA), Paris, France.

IEA, 2010: *Energy Technology Perspectives 2010.* International Energy Agency (IEA), Paris, France. International Energy Agency IEA, 2009: *Energy Balances of Non-OECD Countries*

IPCC, 2005: *Special Report on CO_2 capture and storage.* B. Metz, O. R. Davidson, H. de Coninck and L. M. Meyer, (eds.), Intergovernmental Panel on Climate Change (IPCC), Cambridge, UK.

IPCC, 2007: *Climate Change 2007: Impacts, Adaptation and Vulnerability Working Group II contribution to the Fourth Assessment Report of the Intergovernmental Panel on Climate Change (IPCC).* Intergovernmental Panel on Climate Change (IPCC), Cambridge, UK.

Jacobson, M. Z. and M. A. Delucchi, 2011: Providing all global energy with wind, water, and solar power, Part I: Technologies, energy resources, quantities and areas of infrastructure, and materials. *Energy Policy,* 39: 1154–1169.

Jansen, J. C., W. G. van Arkel and M. G. Boots, 2004: *Designing indicators of long-term energy supply security.* ECN, Petten, Netherlands.

Keith, D., M. Ha-Duong and J. Stolaroff, 2006: Climate strategy with CO2 capture from the air. *Climatic Change,* 74(1–3):17–45.

Kemmler, A., 2007: Factors influencing household access to electricity in India. *Energy for Sustainable Development,* 11(4):13–20.

Keppo, I., B. C. O'Neill and K. Riahi, 2007: Probabilistic temperature change projections and energy system implications of greenhouse gas emission scenarios. *Technological Forecasting and Social Change,* 74(7):936–961.

Kinne, S., M. Schulz, C. Textor, S. Guibert, Y. Balkanski, S. E. Bauer, T. Berntsen, T. F. Berglen, O. Boucher, M. Chin, W. Collins, F. Dentener, T. Diehl, R. Easter, J. Feichter, D. Fillmore, S. Ghan, P. Ginoux, S. Gong, A. Grini, J. Hendricks, M. Herzog, L. Horowitz, I. Isaksen, T. Iversen, A. Kirkevåg, S. Kloster, D. Koch, J. E. Kristjansson, M. Krol, A. Lauer, J. F. Lamarque, G. Lesins, X. Liu, U. Lohmann, V. Montanaro, G. Myhre, J. Penner, G. Pitari, S. Reddy, O. Seland, P. Stier, T. Takemura, and X. Tie, 2006: An AeroCom initial assessment – optical properties in aerosol component modules of global models. *Atmos. Chem. Phys.,* 6:1815–1834.

Krewitt, W., S. Teske, S. Simon, T. Pregger, W. Graus, E. Blomen, S. Schmid and O. Schäfer, 2009: Energy [R]evolution 2008 – a sustainable world energy perspective. *Energy Policy,* 37(12):5764–5775.

Krey, V. and K. Riahi, 2009: Implications of delayed participation and technology failure for the feasibility, costs, and likelihood of staying below temperature targets – greenhouse gas mitigation scenarios for the 21st century. *Energy Economics,* 31(Supplement 2):S94-S106.

Krey, V. and L. Clarke, 2011: Role of renewable energy in climate mitigation: a synthesis of recent scenarios. *Climate Policy,* 11(4): 1131–1158.

Kruyt, B., D. P. van Vuuren, H. J. M. de Vries and H. Groenenberg, 2009 : Indicators for energy security. *Energy Policy,* 37(6):2166–2181.

Kupiainen, K. and Z. Klimont, 2004: *Primary emissions of submicron and carbonaceous particles in Europe and the potential for their control.* International Institute for Applied Systems Analysis Interim Report, Laxenburg, Austria.

Lamarque, J. F., T. C. Bond, V. Eyring, C. Granier, A. Heil, Z. Klimont, D. Lee, C. Liousse, A. Mieville, B. Owen, M. G. Schultz, D. Shindell, S. J. Smith, E. Stehfest, J. Van Aardenne, O. R. Cooper, M. Kainuma, N. Mahowald, J. R. McConnell, V. Naik, K. Riahi and D. P. Van Vuuren, 2010: Historical (1850–2000) gridded anthropogenic and biomass burning emissions of reactive gases and aerosols: Methodology and application. *Atmospheric Chemistry and Physics Discussions,* 10(2):4963–5019.

LBST, 2008: *European Hydrogen Energy Roadmap* Hyways.

Leggett, J., W. Pepper and R. J. Swart, 1992: Emissions Scenarios for the IPCC: an Update. In *Climate Change 1992. The Supplementary Report to the IPCC Scientific Assessment.* J. T. Houghton, B. A. Callander and S. K. Varney, (eds.), Cambridge University Press, Cambridge pp.71–95.

Luderer, G., V. Bosetti, J. Steckel, H. Waisman, N. Bauer, E. Decian, M. Leimbach, O. Sassi and M. Tavoni, 2009: *The Economics of Decarbonization – Results from the RECIPE model intercomparison.* RECIPE Working Paper, Potsdam Institute for Climate Impact Research, Potsdam.

McCollum, D., Krey, V., Riahi K., 2011: An integrated approach to energy sustainability. *Nature Climate Change,* 1(9): 428–429.

Mechler, R., M. Amann and W. Schoepp, 2002: *A Methodology to Estimate Changes in Statistical Life Expectancy Due to the Control of Particulate Matter in Air Pollution.* International Institute for Applied Systems Analysis (IIASA), Laxenburg, Austria.

Meinshausen, M., 2006: What Does a 2°C Target Mean for Greenhouse Gas Concentrations? A Brief Analysis Based on Multi-Gas Emission Pathways and Several Climate Sensitivity Uncertainty Estimates. In *Avoiding Dangerous Climate Change.* H. J. Schellnhuber, W. Cramer, N. Nakicenovic, T. Wigley and G. Yohe, (eds.), Cambridge University Press, Cambridge, UK.

Meinshausen, M., N. Meinshausen, W. Hare, S. C. B. Raper, K. Frieler, R. Knutti, D. J. Frame and M. R. Allen, 2009: Greenhouse-gas emission targets for limiting global warming to 2 degrees C. *Nature,* 458(7242):1158–1162.

Messner, S. and M. Strubegger, 1995: *User's guide for MESSAGE III.* IIASA, Laxenburg, Austria.

Messner, S. and L. Schrattenholzer, 2000: Linking an Energy Supply Model with a Macroeconomic Model and Solving It Interactively. *Energy,* 25:267–282.

Michaelowa, A., K. Tangen and H. Hasselknippe, 2005: Issues and Options for the Post-2012 Climate Architecture – An Overview. *International Environmental Agreements,* 5(1):5–24.

Midilli, A., M. Ay, I. Dincer and M. A. Rosen, 2005: On hydrogen and hydrogen energy strategies: I: current status and needs. *Renewable and Sustainable Energy Reviews,* 9(3):255–271.

Milligan, M., D. J. Lew, D. Corbus, P. Piwko, N. Miller, K. Clark, G. Jordan, L. Freeman, B. Zavadil and M. Schuerger, 2009: *Large-Scale Wind Integration Studies in the United States: Preliminary Results.* Bremen, Germany.

Modi, V., S. McDade, D. Lallement and J. Saghir, 2005: *Energy Services for the Millennium Development Goals.* Energy Sector Management Assistance

Programme (ESMAP), United Nations Development Programme (UNDP), UN Millennium Project, and World Bank, New York.

Nakicenovic, N., A. Grübler, A. Inaba, S. Messner, S. Nilsson, Y. Nishimura, H.-H. Rogner, A. Schäfer, L. Schrattenholzer, M. Strubegger, J. Swisher, D. Victor and D. Wilson, 1993: Long-term strategies for mitigating global warming. *Energy – The International Journal*, **18**(5):401–609.

Nakicenovic, N., A. Grübler, H. Ishitani, T. Johansson, G. Marland, J.-R. Moreira and H.-H. Rogner, 1996: Energy primer. In *Climate Change 1995: Impacts, Adaptations and Mitigation of Climate Change: Scientific-Technical Analyses*. R. T. Watson, M. C. Zinyowera and R. H. Moss, (eds.), Contribution of Working Group II to the Second Assessment Report of the Intergovernmental Panel on Climate Change, Cambridge University Press, Cambridge and New York pp.77–92.

Nakicenovic, N., A. Grübler and A. McDonald, 1998: *Global energy perspectives.* Cambridge University Press, Cambridge, UK.

Nakicenovic, N. and R. Swart, (eds.), 2000: *IPCC Special Report on Emissions Scenarios*. Cambridge University Press, Cambridge.

Nakicenovic, N. and K. Riahi, 2003: *Model runs with MESSAGE in the Context of the Further Development of the Kyoto-Protocol*. IIASA, WBGU – German Advisory Council on Global Change, Berlin.

Nakicenovic, N., P. Kolp, K. Riahi, M. Kainuma and T. Hanaoka, 2006: Assessment of emissions scenarios revisited. *Environmental Economics and Policy Studies*, **7**(3):137–173.

NATURALHY, 2010: *Preparing for the hydrogen economy by using existing natural gas systems as a catalyst*. Final publishable activity report, N.V. Nederlandse Gasunie.

Nemet, G. F., T. Holloway and P. Meier, 2010: Implications of incorporating air quality co benefits into climate change policymaking. *Environmental Research Letters*, **5**(1):1–9.

O'Neill, B. C., Dalton, M., Fuchs, R., Jiang, L., Pachauri, S., Zigovad, K., 2010a: Global demographic trends and future carbon emissions. *Proceedings of the National Academy of Sciences*, **107**(41):17521–17526.

O'Neill, B. C., K. Riahi, and I. Keppo, 2010b: Mitigation implications of mid-century targets that preserve long-term climate policy options. *Proceedings of the National Academy of Sciences*, **107**(3):1011–1016.

Pachauri, S., Y. Nagai and K. Riahi, Options for and Impacts of Achieving the Household Energy Access Challenge by 2030. *Forthcoming*.

Pachauri, S., 2007: *An Energy Analysis of Household Consumption – Changing Patterns of Direct and Indirect Use in India. Springer*, Dordrecht, Netherlands.

Pandey, R., 2002: Energy policy modelling: agenda for developing countries. *Energy Policy*, **30**(2):97–106.

Pepper, W., J. Leggett, R. Swart, J. Watson, J. Edmonds and I. Mintzer, 1992: *Emission scenarios for the IPCC. An update: assumptions, methodology and results*. IPCC, Geneva, Switzerland.

Persson, T. A., C. Azar and K. Lindgren, 2006: Allocation of CO2 emission permits – economic incentives for emission reductions in developing countries. *Energy Policy*, **In Press.**

Planning Commission, 2006: *Integrated Energy Policy, Report of the Expert Committee*. Kirit Parikh Expert Committee on Integrated Energy Policy, Government of India, Planning Commission, New Delhi.

Price, L. and A. McKane, 2009: *Policies and Measures to realize Industrial Energy Efficiency and mitigate Climate Change*. UN Energy (UNIDO, LBNL, IAEA), Vienna.

Rafaj, P., S. Rao, Z. Klimont, P. Kolp and W. Schöpp, 2010: *Emissions of air pollutants implied by global long-term energy scenarios*. International Institute for Applied Systems Analysis (IIASA), Laxenburg, Austria.

Rajagopal, D. and D. Zilberman, 2007: *Review of Environmental, Economic and Policy Aspects of Biofuels*. Sustainable Rural and Urban Development Team, Development Research Group, The World Bank, Washington, DC.

Ramanathan, V. and Y. Xu, 2010: The Copenhagen Accord for limiting global warming: Criteria, constraints, and available avenues. *Proceedings of the National Academy of Sciences*, **107**(18):8055–8062.

Rao, S., V. Chirkov, F. Dentener, R. Van Dingenen, S. Pachauri, P. Purohit, M. Amann, C. Heyes, P. Kinney, P. Kolp, Z. Klimont, K. Riahi and W. Schoepp, 2012: Environmental modeling and methods for – estimation of the global health impacts of air pollution. *Environmental Modeling and Assessment* (2012): 1–10.

Riahi, K. and R. A. Roehrl, 2000: Greenhouse gas emissions in a dynamics-as-usual scenario of economic and energy development. *Technological Forecasting and Social Change*, **63**(2–3):175–205.

Riahi, K., A. Grübler and N. Nakicenovic, 2007: Scenarios of long-term socio-economic and environmental development under climate stabilization. *Technological Forecasting and Social Change*, **74**(7):887–935.

Riahi, K., S. Rao, V. Krey, C. Cho, V. Chirkov, G. Fischer, G. Kindermann, N. Nakicenovic and P. Rafaj, 2011: RCP 8.5 – A scenario of comparatively high greenhouse gas emissions. *Climatic Change*, **109**: 33–57.

Robinson, M. S., 1996: Addressing some key questions on finance and poverty. *Journal of International Development*, **8**(2):153–161.

Roehrl, R. A. and K. Riahi, 2000: Technology dynamics and greenhouse gas emissions mitigation: A cost assessment. *Technological Forecasting and Social Change*, **63**(2–3):231–261.

Rogelj, J., J. Nabel, C. Chen, W. Hare, K. Markmann, M. Meinshausen, M. Schaeffer, K. MacEy and N. Höhne, 2010: Copenhagen Accord pledges are paltry. *Nature*, **464**(7292):1126–1128.

Rogner, H. H., 1997: An Assessment of World Hydrocarbon Resources. *Annual Review of Energy and the Environment*, **22**(1):217–262.

Saygin, D., M. Patel and D. Gielen, 2010: *Global benchmarking for the industrial sector-first steps in application and analysis of competitiveness*. UNIDO, in Preparation, Vienna.

Scheepers, M. J. J., A. J. Seebregts, J. J. de Jong and J. M. Maters, 2007: *EU Standards for Energy Security of Supply – Updates on the Crisis Capability Index and the Supply/Demand Index Quantification for EU-27* Energy Research Centre of the Netherlands (ECN), Clingendael International Energy Programme (CIEP), Petten.

Schwartz, P., 1991: *The Art of the Longview: Three Global Scenarios to 2005. Doubleday Publications*, New York, NY.

Shannon, C. E. and W. Weaver, 1963: *The mathematical theory of communication.* University of Illinois Press, Urbana.

Sherif, S. A., F. Barbir and T. N. Veziroglu, 2005: Wind energy and the hydrogen economy – review of the technology. *Solar Energy*, **78**(5):647–660.

Smith, J. B., S. H. Schneider, M. Oppenheimerd, G. W. Yohee, W. Haref, M. D. Mastrandreac, A. Patwardhang, I. Burtonh, J. Corfee-Morloti, C. H. D. Magadzaj, H.-M. Füsself, A. B. Pittockk, A. Rahmanl, A. Suarezm and J.-P. v. Yperselen, 2009: Assessing dangerous climate change through an update of the Intergovernmental Panel on Climate Change (IPCC) "reasons for concern".

Proceedings of the National Academy of Sciences of the United States of America, **106**(11):4133–4137.

Smith, J. C., M. R. Milligan, D. E.A. and B. Parsons, 2007: Utility wind integration and operating impact state of the art. *IEEE Transactions on Power Systems*, **22**(2):900–908.

Sovacool, B. and M. A. Brown, 2010: Competing dimensions of energy security: An international perspective. *Annual Review of Environment and Resources*, **35**(1):77–108.

Sovacool, B. K., 2009: Rejecting renewables: The socio-technical impediments to renewable electricity in the United States. *Energy Policy*, **37**(11):4500–4513.

Stevenson, D. S. e. a., 2006: Multi-model ensemble simulations of present-day and near-future tropospheric ozone. *Journal of Geophysical Research*, **111**(D08301):23.

Stirling, A., 1994: Diversity and ignorance in electricity supply investment: Addressing the solution rather than the problem. *Energy Policy*, **22**(3):195–216.

Swider, D. J., L. Beurskens, S. Davidson, J. Twidell, J. Pyrko, W. Prüggler, H. Auer, K. Vertin and R. Skema, 2008: Conditions and costs for renewables electricity grid connection: Examples in Europe. *Renewable Energy*, **33**(8):1832–1842.

Teske, S., T. Pregger, S. Simon, T. Naegler, W. Graus and C. Lins, 2010: Energy [R] evolution 2010 – a sustainable world energy outlook. *Energy Efficiency*:1–25.

Textor, C., M. Schulz, S. Guibert, S. Kinne, Y. Balkanski, S. Bauer, T. Berntsen, T. Berglen, O. Boucher, M. Chin, F. Dentener, T. Diehl, J. Feichter, D. Fillmore, P. Ginoux, S. Gong, A. Grini, J. Hendricks, L. Horowitz, P. Huang, I. S. A. Isaksen, T. Iversen, S. Kloster, D. Koch, A. Kirkevåg, J. E. Kristjansson, M. Krol, A. Lauer, J. F. Lamarque, X. Liu, V. Montanaro, G. Myhre, J. E. Penner, G. Pitari, M. S. Reddy, Ø. Seland, P. Stier, T. Takemura, and X. Tie, 2007 : The effect of harmonized emissions on aerosol properties in global models – an AeroCom experiment. *Atmospheric Chemistry and Physics*, **7**:4489–4501.

UN DESA, 2009: *World Population Prospects: The 2008 Revision Database*. Working Paper No. ESA/P/WP.210, United Nations Department of Economic and Social Affairs (UN DESA), New York.

UNCTAD, 2003: *World Investment Report*. United Nations, New York and Geneva.

UNDP and WHO, 2009: The Energy Access Situation in Developing Countries: A Review Focusing on the Least Developed Countries and Sub-Saharan Africa. United Nations Development Programme (UNDP), New York, NY, USA and the World Health Organization (WHO), Geneva, Switzerland.

UNEP, 2010: *The Emissions Gap Report Are the Copenhagen Accord Pledges Sufficient to Limit Global Warming to 2° C or 1.5° C?* United Nations Environment Programme (UNEP), Nairobi, Kenya.

United Nations Conference of the Parties, 2009: Copenhagen Accord FCCC/CP/2009/L.7. *Conference of the Parties*, 7–18 December 2009, Copenhagen.

Vajjhala, S. P. and P. S. Fischbeck, 2007: Quantifying siting difficulty: A case study of U.S. transmission line siting. *Energy Policy*, **35**(1):650–671

van der Zwaan, B. and K. Smekens, 2009: CO$_2$ capture and storage with leakage in an energy-climate model. *Environmental Modeling and Assessment*, **14**(2):135–148.

van Ruijven, B., F. Urban, R. Benders, H. Moll, J. van der Sluijs, B. de Vries and D. van Vuuren, 2008: Modeling energy and development: An evaluation of models and concepts. *World Development*, **36**(12):2801–2821.

van Vliet, O. P. R., A. S. Brouwer, T. Kuramochi, M. van den Broek and A. P. C. Faaij, 2011: Energy use, cost and CO$_2$ emissions of electric cars. *Journal of Power Sources*, **196**(4): 2298–2310.

van Vliet, O. P. R., T. Kruithof and A. P. C. Faaij, 2010: Techno-economic comparison of series hybrid, fuel cell and regular cars. *Journal of Power Sources*, **195**(19):6570–6585.

van Vuuren, D. and K. Riahi, 2011: The relationship between short-term emissions and long-term concentration targets. *Climatic Change*, **104**(3–4):793–801.

van Vuuren, D. P., J. Cofala, H. E. Eerens, R. Oostenrijk, C. Heyes, Z. Klimont, M. G. J. den Elzen and M. Amann, 2006: Exploring the ancillary benefits of the Kyoto Protocol for air pollution in Europe. *Energy Policy*, **34**(4):444–460.

van Vuuren, D. P., J. van Vliet and E. Stehfest, 2009: Future bio-energy potential under various natural constraints. *Energy Policy*, **37**(11):4220–4230.

van Vuuren, D. P., E. Bellevrat, A. Kitous and M. Isaac, 2010: Bio-energy use and low stabilization scenarios. *The Energy Journal*, **31**(Special):192–222.

Vera, I. and L. Langlois, 2007: Energy indicators for sustainable development *Energy*, **32**(6):875–882.

Verbruggen, A. and M. Al Marchohi, 2010: Views on peak oil and its relation to climate change policy. *Energy Policy*, **38**(10):5572–5581.

von Weizsäcker, E. U., A. B. Lovins and L. H. Lovins, 1997: *Factor Four: Doubling Wealth – Halving Resource Use.* Earthscan, London.

WBCSD/IEA, 2009: *Cement Technology Roadmap 2009: Carbon emissions reductions up to 2050*. World Business Council for Sustainable Development (WBCSD)/ International Energy Agency (IEA), Geneva.

WHO, 2008: *The Global Burden of Disease: 2004 update*. World Health Organization (WHO).

Wigley, T. M. L. and S. C. B. Raper, 2001: Interpretation of high projections for global-mean warming. *Science*, **293**(5529):451–454.

Wilson, C., 2009: *Meta-analysis of unit and industry level scaling dynamics in energy technologies and climate change mitigation scenarios*. International Institute for Applied Systems Analysis (IIASA), Laxenburg, Austria.

Winkler, H., R. Spalding-Fecher and L. Tyani, 2002: Comparing developing countries under potential carbon allocation schemes. *Climate Policy*, **2**(4):303–318 (316).

Wise, M., K. Calvin, A. Thomson, L. Clarke, B. Bond-Lamberty, R. Sands, S. J. Smith, A. Janetos and J. Edmonds, 2009: Implication of limiting CO$_2$ concentration for land use and energy. *Science*, **29**(324:5931):1183 – 1186.

World Bank, 2008: *The Welfare Impact of Rural Electrification: A Reassessment of the Costs and Benefits*. World Bank, Washington D.C.

Yang, C., 2008: Hydrogen and electricity: Parallels, interactions, and convergence. *International Journal of Hydrogen Energy*, **33**(8):1977–1994.

Zeman, F., 2007: Energy and material balance of CO$_2$ capture from ambient air. *Environmental Science & Technology*, **41**(21):7558–7563

Zomers, A., 2001: *Rural Electrification, Utilities' Chafe or Challenge?* Faculty of Technology and Management, University of Twente, Enschede.

18

Urban Energy Systems

Convening Lead Author (CLA)
Arnulf Grubler (International Institute for Applied Systems Analysis, Austria and Yale University, USA)

Lead Authors (LA)
Xuemei Bai (Australian National University)
Thomas Buettner (United Nations Department of Economic and Social Affairs)
Shobhakar Dhakal (Global Carbon Project and National Institute for Environmental Studies, Japan)
David J. Fisk (Imperial College London, UK)
Toshiaki Ichinose (National Institute for Environmental Studies, Japan)
James E. Keirstead (Imperial College London, UK)
Gerd Sammer (University of Natural Resources and Applied Life Sciences, Austria)
David Satterthwaite (International Institute for Environment and Development, UK)
Niels B. Schulz (International Institute for Applied Systems Analysis, Austria and Imperial College London, UK)
Nilay Shah (Imperial College London, UK)
Julia Steinberger (The Institute of Social Ecology, Austria and University of Leeds, UK)
Helga Weisz (Potsdam Institute for Climate Impact Research, Germany)

Contributing Authors (CA) *including contributors to GEA City Energy Data Base
Gilbert Ahamer* (University of Graz, Austria)
Timothy Baynes* (Commonwealth Scientific and Industrial Research Organisation, Australia)
Daniel Curtis* (Oxford University Centre for the Environment, UK)
Michael Doherty (Commonwealth Scientific and Industrial Research Organisation, Australia)
Nick Eyre* (Oxford University Centre for the Environment, UK)
Junichi Fujino* (National Institute for Environmental Studies, Japan)
Keisuke Hanaki (University of Tokyo, Japan)
Mikiko Kainuma* (National Institute for Environmental Studies, Japan)
Shinji Kaneko (Hiroshima University, Japan)
Manfred Lenzen (University of Sydney, Australia)
Jacqui Meyers (Commonwealth Scientific and Industrial Research Organisation, Australia)
Hitomi Nakanishi (University of Canberra, Australia)
Victoria Novikova* (Oxford University Centre for the Environment, UK)
Krishnan S. Rajan (International Institute of Information Technology, India)
Seongwon Seo* (Commonwealth Scientific and Industrial Research Organisation, Australia)
Ram M. Shrestha* (Asian Institute of Technology, Thailand)
Priyadarshi R. Shukla* (Indian Institute of Management)
Alice Sverdlik (International Institute for Environment and Development, UK)

Review Editor
Jayant Sathaye (Lawrence Berkeley National Laboratory, USA)

Contents

Executive Summary

More than 50% of the global population already lives in urban settlements and urban areas are projected to absorb almost all the global population growth to 2050, amounting to some additional three billion people. Over the next decades the increase in rural population in many developing countries will be overshadowed by population flows to cities. Rural populations globally are expected to peak at a level of 3.5 billion people by around 2020 and decline thereafter, albeit with heterogeneous regional trends. This adds urgency in addressing rural energy access, but our common future will be predominantly urban. Most of urban growth will continue to occur in small- to medium-sized urban centers. Growth in these smaller cities poses serious policy challenges, especially in the developing world. In small cities, data and information to guide policy are largely absent, local resources to tackle development challenges are limited, and governance and institutional capacities are weak, requiring serious efforts in capacity building, novel applications of remote sensing, information, and decision support techniques, and new institutional partnerships. While 'megacities' with more than 10 million inhabitants have distinctive challenges, their contribution to global urban growth will remain comparatively small.

Energy-wise, the world is already predominantly urban. This assessment estimates that between 60–80% of final energy use globally is urban, with a central estimate of 75%. Applying national energy (or GHG inventory) reporting formats to the urban scale and to urban administrative boundaries is often referred to as a 'production' accounting approach and underlies the above GEA estimate. This contrasts to a 'consumption' accounting approach that pro-rates associated energy uses per unit of urban consumer expenditures, thus allocating energy uses to urban consumers irrespective of the form energy is used (direct or embodied energy) or its location (within or outside a city's administrative boundary). Available consumption-based energy accounts for cities are too limited (estimates exist for only a handful of megacities) to allow generalization but it is highly likely that urban energy use under a consumption accounting approach approximates the urban share in the world GDP, estimated by this assessment to be some 80%. There is great heterogeneity in urban energy-use patterns as revealed by this assessment drawing together a novel urban energy use data set. In many developing countries, urban dwellers use substantially *more* final energy per capita than their rural compatriots. This primarily reflects their much higher average urban incomes. Conversely, in many industrialized countries per capita final energy use of city dwellers is often *lower* than the national average, which reflects the effects of compact urban form, settlement types (multi- versus single-family dwellings) and availability and/or practicability of public transport infrastructure systems compared with those in the suburban or rural sprawl. The few available data, however, suggest that urban energy use in high-income countries is not substantially different from the national average when using a consumption-based accounting approach, whereas in low-income countries the urban-rural energy difference is likely to be even larger under this alternative energy accounting method. This Assessment concludes that both accounting methodologies provide complementary, valuable information to inform urban policy decisions. However, because of complexities in systems boundaries and accounting, urban studies need to adhere to high standards in terms of clarity and documentation of the terminology, methodology, and documentation of underlying energy data used.

Addressing Urban Challenges

Rapid migration rates and natural population growth in cities can overwhelm the provision of basic urban services, particularly for the poorest urban dwellers. Several hundred million urbanites in low- and middle-income nations lack access to electricity and are unable to afford cleaner, safer fuels which results in significant adverse consequences for human health and local air quality. Most are in low-income nations in Southeast Asia and sub-Saharan Africa. Innovations have reduced access costs – for instance, rising tariffs with low prices for 'lifeline' consumption, pay-as-you-use meters, and standard 'boards' that remove the need for individualized household wiring—but urban energy access also faces political and institutional obstacles. A large part of the urban population that lack clean energy and electricity (and other basic services like water, sanitation, and transport) live in informal settlements. It is mostly in situations where the often antagonistic relationship between local government and the inhabitants of informal settlements has changed, through widespread public support to upgrade 'slums' and squatters, that clean energy and electricity and other public services have reached the urban poor.

Housing, water supply and sanitation infrastructure, energy, and transport services are the key sustainability challenges to accommodate some three billion additional urban dwellers in the decades to come, especially in low-income countries. Informal settlements will often be one of the transitional forms of settlement for many of these new urban dwellers and will require a much more proactive, anticipatory policy approach, especially with respect to the location of informal settlements and subsequent infrastructure connections and upgrading programs. Energy-wise, low-cost and fast implementation options will take precedence over 'grand' new urban designs that require unrealistically large sustained capital provision over long periods. In low-income countries access to clean cooking fuels and electricity, as well as pro-poor transport policies, which include safer use of roads by non-motorized modes (walking and bicycling) and making public transport choices available need to be ranked high on the urban policy agenda.

From all the major determinants of urban energy use – climate, position in the global economy, consumption patterns, quality of built environment, urban form and density (including transport systems), and urban energy systems and their integration – only the final three are amenable to policymaking by city administrations, at least partially. Therefore, both in terms of leverage and potentials, energy policy at the urban scale needs to focus above all on *demand management* with a focus on energy efficient buildings, structuring urban form and density conductive to energy efficient housing forms, high-quality public transport services, and to urban energy systems integration. This demand-side focus at the urban scale represents a paradigm shift compared to the traditional, more supply-side energy policy focus at the national scale.

Systemic characteristics of urban energy use are generally more important determinants of the efficiency of urban energy use than those of individual consumers or of technological artifacts. For instance, the share of high occupancy public and/ or non-motorized transport modes in urban mobility is a more important determinant of urban transport energy use than the efficiency of the urban vehicle fleet (be it buses or hybrid automobiles). Denser, multifamily dwellings in compact settlement forms with a corresponding higher share of non-automobile mobility (even without thermal retrofit) can use less *total energy* than low-density, single-family 'Passivhaus'-standard (or even 'active,' net energy generating) homes in dispersed suburbs deploying two hybrid automobiles for work commutes and daily family chores. Evidently, urban policies need to address both systemic and individual characteristics in urban energy use, but their different long-term leverage effects should structure policy attention and perseverance. In terms of urban energy-demand management, the quality of the built environment (buildings efficiency) and urban form and density that, to a large degree, structure urban transport energy use are roughly of equal importance. Also, energy-systems integration (cogeneration, waste heat cascading) can give substantial efficiency gains, but ranks second after buildings efficiency and urban form and density, and associated transport efficiency measures, as shown both by empirical cross-city comparisons and modeling studies commissioned by this assessment. The potential for energy-efficiency improvements in urban areas remains enormous, as indicated by corresponding urban exergy analyses reviewed in this Assessment and that suggest urban energy-use efficiency is generally less than 20% of the thermodynamic efficiency frontier; representing an improvement potential of more than a factor five. Conversely, the potential of supply-side measures within the immediate spatial and functional confines of urban systems is very limited, especially for renewable energies. *Locally harvested* renewables can, at best, provide 1% of the energy needs of a megacity and a few percentage points in smaller, low-density cities because of the mismatch between (high) urban energy demand density and (low) renewable energy supply densities at the local level.

Urban Policy Choices and Priorities

The historical evolutionary processes that govern urban growth have largely been *path dependent* with variation that played out differently over time and space. Cities that evolved along alternative pathways have alternative density levels from high-density 'Asian' (e.g., Tokyo, Shanghai, Mumbai) and 'old European' (e.g., London, Madrid, Warsaw) to low density 'new frontier' (e.g., Los Angeles, Brasilia, Melbourne) pathways, each of which have different structural options available to improve energy efficiency and optimize urban energy and transport systems in terms of sustainability criteria. Despite this diversity, two important generalizations can be drawn.

First, the implications of urban density on the requirements of urban energy systems are that they need to be basically *pollution free*, as otherwise even relatively clean energy forms can quickly overwhelm the assimilative capacity of urban environments. This especially applies to the million, decentralized energy end-use combustion devices (stoves, heating systems, vehicles) for which end-of-pipe pollution control is often not an option. Thus, in the long-term all end-use energy fuels burnt in urban areas need to be of zero-emission quality. This requires energy vectors from remote 'clean' plants, as exemplified by electricity or (possibly) hydrogen. Natural gas plays the role of the transitional fuel of choice in many urban areas. This 'zero-emission' requirement for urban energy transcends the customary sustainability divide between fossil and renewable energies, as even 'carbon-neutral' biofuels when used by millions of automobiles in an urban environment will produce unacceptable levels of NO_x or O_3 pollution. The observed significant improvements in urban air quality caused by the elimination of traditional air pollutants, such as soot, particles, and SO_2, in cities of high-income countries are a powerful illustration that cities act as innovation centers and hubs for environmental improvements that can lead to a sustainability transition path. First signs of progress in traditional air pollutants are evident in low-income countries as well, as illustrated by the recent decline in the emissions of some traditional pollutants in Asian megacities discussed in this assessment. Nonetheless, an exceedingly high fraction of urban dwellers worldwide are still exposed to high levels of urban air pollution, especially total suspended particles (TSPs), with fine particle emissions (PM-10s) continuing their upward trend. A wide portfolio of policy options is available, ranging from regulatory instruments such as mandated fuel choice ('smokeless zone' regulations), air-pollution standards, regulation of large point-source emissions and vehicle exhaust standards, to market-based instruments (or hybrid) approaches that incentivize technological change. An important feature of these regulatory approaches is *dynamic target setting* to reflect changing technology options and to counter the consequences of urban growth and potential consumer 'take-back' effects. Air pollution is also the area of urban environmental policymaking where the most significant co-benefits of policies can be realized: improving access to clean cooking fuels, for example, improves human health and lowers traditional pollutant emissions, and also has (through reduced black carbon emissions) significant net global warming co-benefits.

Second, the literature repeatedly identifies important size and density thresholds as useful guides for urban planning. The importance of these urban thresholds extends to specialized urban infrastructures, such as underground (metro) transport networks that are, as a rule, economically (in terms of potential customers and users) not feasible below a threshold population size of less than one million. It also extends to energy (e.g., cogeneration-based district heating and cooling) and public transport networks, whose feasibility are framed by a gross[1] population density threshold between 50 and 150 inhabitants/hectare (ha) (5000–15,000 people/km²). Such density levels of 50–150 inhabitants/ ha certainly do *not* imply the need for high-rise buildings, as they can be achieved by compact building structures and designs, both traditional and new, including town or terraced houses, while still allowing for open public (parks) or private (courtyard) spaces – but they do not allow for unlimited (aboveground parking) spaces for private vehicles. Zoning and parking regulation, combined with public transport policies and policies that promote non-motorized transport modes and walkability thus constitute the essential 'building blocks' of urban energy efficiency and sustainability 'policy packages.'

---

1   i.e., a minimum density level over the entire settlement area that comprises residential zones of higher density with low density green spaces.

The wide variation in urban transport choices observed in the modal split in different cities illustrates that urban mobility patterns are not *ex ante* given, but rather result from specific choices of individuals and decision makers. Urban transport choices can be modified, if both a strong determination for a sustainable urban transport policy *and* a corresponding wide public acceptance of the overall goals of such a policy exist. Restrictive measures that limit individual mobility by automobiles need to be complemented by proactive policies that enhance the attractiveness of non-motorized and public transport choices, and 'soft' policy measures (e.g., fees, tariffs) also need to be complemented by 'hard' (i.e., infrastructural investment) measures. Investments in public transport systems need to find an appropriate balance between improvements that are less capital intensive and faster to implement, and radical solutions. Bus-based Rapid Transit systems (BRT) with own dedicated lanes are, therefore, a much more attractive option for many cities in low-income countries than are capital-intensive subway or light-rail systems, even though, in the long term, the latter offer the possibility of higher passenger fluxes and greater energy efficiency. Often there is no contradiction between incremental versus radical public transport policy options: for instance, BRT can also be considered as a transitional infrastructure strategy to secure public transport 'rights of way,' which offer subsequent possibilities for infrastructure upgrades, for example in putting light rail systems in BRT lanes. Many of the new urban settlements being formed meet the required density targets for attractive and efficient public transport systems. The policy issue is to exploit the available advantages before a 'lock-in' into a private automobile-dependent 'vicious' development cycle becomes entrenched.

A common characteristic of sustainable urban energy system options and policies is that they are usually systemic: for example, the integration of land-use and urban transport planning that extends beyond traditional administrative boundaries; the increasing integration of urban resource streams, including water, wastes, and energy, that can further both resource (e.g., heat) recovery and improve environmental performance. This view of a more integrated (and often also more decentralized) urban infrastructures also offers possibilities to improve the resilience and security of urban energy systems. And yet, this systemic perspective reveals a new kind of 'governance paradox.' Whereas the largest policy leverages are from systemic approaches and policy integration, these policies are also the most difficult to implement and require that policy fragmentation and uncoordinated, dispersed decision making be overcome. The urban governance paradox is compounded by weak institutional capacities, especially in small- to medium-sized cities that are the focus of projected urban growth, as well as from the legacies of market deregulation and privatization that have made integrated urban planning and coordinated energy, transport, and other infrastructural policy approaches more difficult to design and yet more difficult to implement. However, there are good reasons for (cautionary) optimism. Urban areas will continue to act as innovation centers for experimentation and as diffusion nodes for the introduction of new systems and individual technological options by providing critical niche market sizes in the needed transition toward more sustainable urban energy systems.

18.1 Introduction

18.1.1 Preamble

Towns and cities and their increasing coalescence into urban agglomerations are now the dominant form of spatial organization in which people live, economies operate, technologies are generated and used, corresponding demands for *energy services* arise and environmental sustainability is critically defined. The decade 2000–2010 marked an important watershed in human history: for the first time more than 50% of the global population are urban dwellers and estimates indicate that already some three-quarters of global (direct) final energy use is urban, with corresponding primary energy and carbon dioxide (CO_2) emissions probably being comparable.

Given the robust trends toward a convergence of much of the developing world to levels of urbanization already found in the developed world, the energy and sustainability challenges of equitable access to clean-energy services, of energy security, and of environmental compatibility at local through global scales cannot be addressed without explicit consideration of urban energy systems and their specific sustainability challenges and opportunities. The future development of the demand side for energy cannot be described without understanding changes at the level of the urban settlements. Research shows how the properties of urban areas across the world are scalable, revealing distinct patterns. Just as it is possible 'to fail to see the forest for the trees,' it is possible 'to fail to see the city for the buildings.' Hence the GEA includes a specific chapter on urbanization and urban energy use.

Paradoxes in conventional analysis abound. A single urban agglomeration, such as greater Tokyo, generates more gross domestic product (GDP) than the venerable pioneer country of the Industrial Revolution – the current United Kingdom. And yet, our statistical reporting systems almost exclusively focus on nation states, as represented by Systems of National Accounts, Energy Balances, or similar reporting standards. In fact, as detailed throughout this chapter, the difficulty of finding data at the urban level starts with the very definition of urban areas and urban populations. The study of urban energy must take this diversity and uncertainty into account. Accordingly, this chapter draws attention to the limits of comparability and policy guidance of existing studies that suffer from inconsistent or unclear system boundaries and accounting methodologies, and concludes with the need to develop clear methods and guidelines for a range of complementary urban energy accounting tools to guide policy. The energy-accounting system boundary issue thus adds to the conundrum of different territorial and administrative boundaries that range from core cities through metropolitan agglomerations to transnational urban corridors.

In this chapter the terminology used is of an 'urban system' when describing the urban phenomenon from a *functional* perspective in addition to the traditional territorial or administrative perspective. Thus, an *urban energy system* comprises all components related to the use and provision of energy services associated with a functional urban system, irrespective where the associated energy use and conversion are located in space, such as power plants and transport fuel requirements both locally and internationally (airports, ports). The full urban energy system entails both energy flows proper (fuels, 'direct' energy flows) and 'embodied' energy (energy used in the production of goods and provision of services *imported* into but also *exported* from an urban system). The functional perspective of urban energy systems highlights that urban locations and their growth (urbanization) are not only the locations of people and economic activities in space, but also include the types of activities they pursue and the infrastructural and functional framing conditions (service functions) urban agglomerations provide. Functional characteristics increasingly define urban areas and need to be reflected in urban energy systems analysis, and thus need to combine both 'production' and 'consumption' perspectives. From a spatial perspective, this chapter also extends – within the limits of available data – the traditional discussion of cities as defined by political and/or administrative boundaries toward urban agglomerations, including 'periurban' and larger metropolitan areas, through urban 'clusters' or 'corridors' to Doxiadis' (Doxiadis and Papaioannou, 1974) 'ecumenopolis.'

The future development of urban energy systems is characterized by specific challenges and opportunities. The high density of population, economic activities, and resulting energy use severely limit an obvious sustainable energy choice: In many larger cities locally harnessed renewables can provide for only some *one per cent* of urban energy use which implies large-scale *imports* of renewable energies generated elsewhere, much like in the currently dominating fossil energy systems. The diversity of activities and energy uses characteristic of urban systems opens numerous opportunities for intelligent energy management and 'recycling' (e.g., electricity-heat cogeneration and 'heat cascading', in which different energy end uses can 'feed' on waste energy flows from energy conversion and industrial applications). Both *diversity and density* (at least above a critical threshold value of some 50–150 inhabitants/ha) can be considered as key *strategic assets* of urban areas that help to use energy more efficiently by energy-systems integration, compact energy-efficient housing, and co-location of activities that can help to minimize transport distances and automobile dependence. The provision of transport services via environmentally friendly, high-quality urban public transport systems is a unique option for cities, generally not available in low-density sub-urban or rural contexts.

The vital urban infrastructures all depend on energy: water supply, treatment and waste water disposal, transport and communication systems, complex webs of food and material supplies, the resulting disposal of wastes, and, of course, energy supply itself. Many urban infrastructures have shown great adaptability, but catastrophes like hurricane Katrina, or the 9/11 attacks on New York, show that urban systems and, by extension, their populations are vulnerable. Urban infrastructures are almost always considered in isolation, but this ignores their interdependence, and common vulnerabilities, as well as their potential synergies and efficiency gains. This highlights the importance of improved planning, but this will require new institutional frameworks and the inclusion of different stakeholders to address the complex coordination issues across sectors and across spatial scales.

Urban agglomerations are dominant in terms of production, consumption, and associated energy use (irrespective of where these take place physically). But they are also unique centers of human capital, ingenuity and innovation, financial resources, and local decision-making processes, which are all 'human' resources that can be harnessed for a sustainability transition. While global and national policy frameworks are clearly needed, ultimately all implementation is *place-based* and requires local formal and informal supportive frameworks. To promote more sustainable development, cities may thus be the right scale for an 'intermediate' (even mediating) actor level between the individual and national and transnational initiatives. The urban scale is also the appropriate one to identify and realize many options in promoting energy efficiency that may not always be apparent at higher levels of policymaking.

Cities thus could become *the* innovation centers in developing and implementing solutions in the sustainability transition, a perspective that well merits their explicit consideration in an assessment like GEA.

18.1.1   Objectives and Approach

Given the above, the broad objective is to address urban energy issues in an increasingly urbanizing world from a *systemic* perspective that focuses on the specific energy challenges and opportunities represented by urban settlements.

The specific objectives of this chapter are to perform first a global assessment to establish the order of magnitude of urban energy systems and their drivers, then develop some generalized hypotheses to understand urban energy use, its differences and dynamics, and finally test these hypotheses through case studies at the local and regional levels, drawing on specific examples of individual cities.

This chapter addresses the systemic and structural interlinkages of urban systems and how these interact within and outside traditional territorial urban system boundaries. It adds to the information and knowledge of the sectorial GEA chapters (buildings, industry, transport) by addressing the *integrated* issues specific and unique to urban systems. Sectoral perspectives are therefore addressed here only to the extent that they contain an explicit urban specificity, e.g., (public) urban transport systems, or urban energy cogeneration systems.

Urbanization is a multidimensional phenomenon that can be described from demographic, land use, or economics perspectives. Empirical data are well developed for demographics (through regular population censuses) and for land-use perspectives (through remote sensing data). Conversely, there is a paucity of widely available and comparable economic data at the urban scale. Systems of National Accounts

that underlie much of the available economic data were developed and continue to be used predominantly at the national scale, with comparatively few applications at the urban scale, which explains the significant gap between urban economic data[2] and the largely theoretical discussion of urbanization in the economics literature. The literature on urban land use and urban land-cover changes, while most valuable for describing a physical dimension of urbanization, is of limited use in an energy assessment despite the richness of quantitative data available. After all, it is not the square kilometers of urban extent that can explain urban energy use, but only the linkage of urban land use with demographic and economic data and characteristics as reflected through urban form and population density, infrastructure endowments, level and structure of economic activities, lifestyles of city dwellers, among other factors. Therefore, *demographics* (population) is adopted quite naturally as a fundamental driver and core metric to discuss urbanization and urban energy use, drawing on the urban land-use change and economics literature only to the degree necessary to understand urban energy use and its variation through derived metrics centered on population, like population density, or per capita incomes and expenditures, in addition to more narrow disciplinary land-use and economic metrics, such as urban extents/form or economic structure.

As comprehensive energy information and accounts at the urban scale are extremely limited, developing a robust assessment storyline from the bottom-up alone is challenging. Therefore in this analyses a mixed approach of both top-down and bottom-up perspectives is utilized, combining estimates derived from 'downscaling' or remote sensing approaches with bottom-up statistical information where available.

18.1.2   Roadmap

After the introductory section (18.1), Section 18.2 gives an overview of the urbanization phenomenon and the contexts for understanding urban energy use. Its focus is on the demographic dimension of urbanization, both from historical and futures (scenarios) perspectives, and on the drivers and patterns of urbanization dynamics. It also addresses the specifics of urban energy systems and elaborates on the importance of system boundaries.

Section 18.3 analyzes current global, regional, and city-specific urban energy use, and comprises new 'top down' estimates of urban energy from a global perspective, complemented by 'bottom up' urban-scale energy-use data collected through the GEA Chapter 18 City Energy Data Base.

Urban energy use and systems as a whole are accompanied and shaped by multiple challenges, discussed in detail in Section 18.4. These include

---

2   With a paucity of 'official' urban scale economic data reporting, this assessment also relies heavily on estimates rather than statistics, with data derived either from 'grey literature' expert estimates (e.g., Hawksworth et al., 2007) or spatially explicit 'downscaling' exercises (e.g., Grubler et al., 2007).

issues of energy access, energy demand densities, and associated key environmental externalities, as well as energy supply constraints, which include reliability and security.

Section 18.5 outlines opportunities and response options to the urban energy challenges. Key drivers of urban energy use, policy players and main policy leverages, and some of the key infrastructure issues linked to specifics of urban systems, such as urban transport, urban energy infrastructure planning, design and implementation, and urban air-quality management are discussed. Section 18.5 also includes novel model simulations to illustrate the respective impacts of policy interventions along three main opportunity areas: urban form and density (and their influence on transport energy demand), buildings efficiency, and urban energy systems integration and optimization.

Section 18.6 summarizes the key messages and conclusions. The research, data, and information needs identified are summarized at the end of the each section rather than in a separate section at the end of the chapter.

18.2 An Overview of Urbanization

18.2.1 Overview of Urbanization, Past and Scenario Trends

18.2.1.1 The Multiple Dimensions of Urbanization

The process of urbanization involves multiple dimensions, characterized by different theories, methodologies, and literatures that follow four distinct disciplinary perspectives: demography, geography, economics, and sociology.

The resulting effects of the scale and concentration of human activity in urbanized areas are the focus of research in various disciplines. Economists often emphasize the benefits of scale of larger labor markets in cities and agglomeration effects of clustering of various industrial and service activities (Krugman, 1991; Fujita et al., 1999; UNIDO, 2009; World Bank, 2009). Climatologists discuss the consequence of urbanization on albedo changes, radiation balances, and weather patterns (Kalnay and Cai, 2003; IAUC and WMO, 2006; Souch and Grimmond, 2006). Transport planners are concerned with avoiding negative externalities of urban density, such as traffic congestion. Environmental researchers study typical patterns of the generation and distribution of pollutants in urban centers and the exposure of target populations to such hazards (McGranahan et al., 2001; McGranahan and Marcotullio, 2007). Social scientists are investigating particular urban social structures and challenges, and urban cultural modes of creativity and innovation that result from the immediate proximity of many million people that can exchange, cooperate and profit from high degrees of specialization. Understanding the consequences of urbanization on energy use in general, however, is an area of research that has attained surprisingly

little attention in empirical studies, given the relevance of urban areas for overall energy demand (IEA, 2008), their particular vulnerability to energy supply disruption, and their potential for energy savings and climate-change mitigation.

The economic, geographic, and sociological perspectives of urbanization are discussed in greater detail as driving forces of urban energy demand (see Section 18.5.2). Following these different disciplinary perspectives, four complementary concepts describe the process of urbanization:

- The demographer's approach emphasizes population. To a demographer, urbanization is the process by which a rural population becomes urban. That is, the *population* is becoming urbanized by an increase in the proportion of the population classified as urban.

- In the geographer's approach, a defined geographic area gradually loses the characteristics associated with rural areas (e.g., dominance of agricultural land uses) and gains characteristics associated with urban areas (e.g., built-up land, and high density of buildings and technological infrastructures) – the *region* is becoming urbanized).

- The economist's approach is the process of economic structural change away from primary economic activities (agriculture, forestry, mining, etc.) toward manufacturing and services (secondary and tertiary economic activities). This usually involves spatial concentration and co-location of economic actors[3] that profit from agglomeration externalities: the *economy* is becoming urbanized.

- In the sociologist's approach, individuals move from rural to urban areas and take on urban characteristics: individuals and their aggregate (i.e., *society*) are becoming urbanized.

The demographic study of urbanization is among the oldest research traditions and also the most quantitative, including scenario projections into the future. Along with economics it is also the dimension with the closest direct causal link to urban energy use. This being the reason why in the subsequent discussion the demographic perspective of urbanization is highlighted. The economic, geographical, and sociological perspectives of urbanization in turn are discussed in greater detail as driving forces of urban energy demand (see Section 18.5.2 below).

Table 18.1 illustrates these multiple dimensions of urbanization at the global level for the year 2000, the latest common year for which the various indicator data sets available. With the exception of remotely sensed urban land areas (that turn out to display the largest uncertainty range of a factor of ten) and population for which data are available directly at

---

3   In economics this is referred to as agglomeration externalities that arise from economies of scale and economies of scope effects. See Chapter 24 for a more detailed discussion and exposition of these concepts.

Table 18.1 | Various indicators of urbanization for the year 2000 at the global level in absolute amounts and as percentage urban with associated uncertainty ranges derived from the literature or estimated in this study.

Indicator		Value	Uncertainty Range	References for Uncertainty Range
Area[1]	(1000 km²)	2929	313–3524	Schneider et al., 2009
	% of total	*2.2*	*0.2–2.7*	range of globcover-grump data
Population[2]	(million)	2855	2650–3150	Uchida and Nelson, 2008
	% of total	*47*	*44–52*	Size threshold: 50,000–100,000
GDP (MER US2005$)[1]	(billion)	32008		
	% of total	*81*		not available
Final energy use[1]	(EJ)	239	176–246	this assessment
	% of total	*76*	*56–78*	(see Section 18.3.1)
Light luminosity[1,3]	(million NLIS)	33		
	% of total	*57*	*50–82*	Chapter 18 estimate
Internet routers[1,4]	(number in 1000)	8592		
	% of total	*96*	*73–97*	Chapter 18 estimate

Notes – MER: Market Exchange Rates; NLIS: Light Luminosity Intensity Sum (Index)

Sources: (1) Grubler et al., 2007; (2) UN DESA, 2010; (3) NOAA, 2008; (4) Crovella, 2007 and Lakhina et al., 2003.

the urban scale, all other indicators represent derived estimates, combining spatially explicit data sets of urban extents consistent with UN urban population statistics with other spatially explicit data of human activity.

18.2.1.2 Urbanization in a Historical and Demographic Context

People have lived in cities for millennia. Cities and their associated urban agglomerations have and continue to play key roles as centers of government, production, and trade, as well as knowledge, innovation, and productivity growth (UN, 2008). Yet during most of the past, the number of city dwellers remained exceedingly small – the majority of the population continued to live in rural areas and worked in agriculture.

Before 1800, large cities (by contemporary standards) being exceedingly rare, with Chang'an and Baghdad being likely the first examples of cities that approached one million inhabitants (Chandler and Fox, 1974; Chandler, 1987). This started to change dramatically with the (rapid) improvement of agricultural productivity, which freed people for new industrial jobs and migration into cities (the Industrial Revolution). Many cities grew to much larger sizes; rural settlements were transformed into cities, and new cities were founded. The rate of urban population

growth (the twin result of natural growth in urban populations plus net in-migration into cities) exceeded significantly that of the overall population and resulted in a secular trend toward *urbanization* (i.e., an increase in the rate or percentage of a population living in urban areas). Estimates suggest that by 1900–1920, between 12 and 14% of the global population could be classified as urban. Because of their lead in industrialization and agricultural productivity growth, the more developed countries had an urbanization rate of some 30% by the 1920s, and developing countries of only some 6% (UN DESA, 1973). Urbanization trends have accelerated ever since.

The UN reconstruction of global population trends from 1950 onwards (UN, 2008) documents a historically unprecedented global population growth in the second half of the 20th century, peaking at about 2% annually between 1965 and 1970. Global population growth rates have declined since, and current projections indicate a leveling off of global population growth toward the third quarter of the 21st century.

In 55 years, the world's population grew by about 4 billion people to approximately 6.5 billion in 2005. A defining feature of this growth is its diversity across regions, especially between more- and less-developed ones. Having entered the demographic transition earlier, OECD90 countries and countries undergoing economic reform (Eastern Europe and the former USSR) increased their population between 1950 and 2005 by a comparatively modest 62% and 51%, respectively. All other GEA regions, by contrast, more than doubled their populations. The fastest population growth occurred in the Middle Eastern and African regions, where the population more than quadrupled between 1950 and 2005. The population in Latin America and the Caribbean more than tripled, and in Asia, already the most populous region of the world, in 2005 the population was 2.7 times larger than that in 1950.

18.2.1.3 Past Urbanization Trends

The inevitable uncertainty in urban population and urbanization levels does not affect an assessment of the dynamics of the urbanization process, provided the system of national definitional criteria does not change too often over time (occasional definitional changes of individual countries do not markedly affect global or regional trends).

While populations in all regions grew over the past half century, urban populations grew even faster (Table 18.2). Between 1950 and 2005, the UN DESA (2010) estimate suggests that the Middle Eastern and African urban population multiplied more than ten-fold, increasing from 44 million to 478 million. Asia registered in 2005 an urban population almost six-times higher than it had in 1950 (175 million versus 1.3 billion), and the urban population of Latin America and the Caribbean increased more than five times, from 68 million to about 429 million people (UN DESA, 2010). By 2005 about half of the global population (3.2 out of 6.5 billion) was urban. By the time of the publication of the GEA report (2012), more than half of global population will be urban.

Table 18.2 | Urban population (millions) and as a percentage of world total urban population, and total population (in millions) for the five GEA regions and the world since 1950. Countries undergoing economic reform = Eastern Europe and former USSR.

GEA Region	Urban Population				Total Population	
	1950	2005	1950	2005	1950	2005
	in Millions		in percent of world		in Millions	
OECD90	340	730	46.6	23.1	593	963
Countries undergoing Econ. Reform	102	254	14.0	8.0	269	404
Asia	175	1276	24.0	40.3	1237	3478
Middle East and Africa	44	478	6.0	15.1	266	1115
Latin American and the Caribbean	68	429	9.4	13.5	165	552
World	**729**	**3167**	**100**	**100**	**2529**	**6512**

Source: UN DESA, 2010.

18.2.1.4   Urbanization Levels

As the urban population grew, its relative weight also increased. In 1950, just about 30% of the world's population of 2.5 billion people lived in urban areas, while the vast majority still lived in rural areas. Even for the more advanced OECD90 region, in 1950, 57% of its population lived in urban areas, while a substantial proportion of its population (43%) lived in rural areas. All other major regions were predominantly rural. Latin America and the Caribbean, and the countries of Eastern Europe and the former USSR currently undergoing economic (transition) reform already had high urbanization levels (41% and 38%, respectively) in 1950. Populations in the Asian and the Middle Eastern and African regions were still predominately rural, with just about 14% and 16% of their populations living in urban areas, respectively.

By 2005, the world had become almost 50% urban, but with significant differences across the regions. In the OECD90 and Latin America and the Caribbean countries, three out of four people lived in urban settlements; and in the Asian and the Middle Eastern and African regions about two out of five people were urban dwellers.

18.2.1.5   An Urban Future

Trends in population aging and urbanization are inevitably important to the immediate and medium-term demographic future. Population growth is still a persistent element of demographic trends in developing countries. For some developed countries, however, population growth has not only tapered off, but may be reversed with population decline as a defining feature.

Significant negative population growth is currently expected to persist for countries undergoing economic reform (i.e., the countries of the former Soviet Union/USSR and of Eastern Europe), with their population projected to shrink by about 39 million people: from 404 million in 2005 to 365 million in 2050. All other GEA regions are projected to increase their

populations between 2005 and 2050 (UN DESA, 2010). There is, however, great diversity not only between GEA regions, but also, and more significantly, within the regions and between countries. There are also different causes for the continued population growth within most GEA regions. Populations in the OECD region, which have experienced low or even very low fertility during recent decades, are expected to increase slightly, mainly because of net gains from international migration. For example, Western Europe is projected to lose about 1.4 million people between 2005 and 2050, when net migration gains are included. In the absence of that gain, in 2050 Western Europe would have about 15 million less inhabitants than in 2005 (UN DESA, 2010).

Population growth in the other GEA regions is caused by very different factors: momentum that stems from young populations and above-replacement fertility. The latter is increasingly less important because of a continued fertility decline. Combined, these factors generate the addition of 1.1 billion people to Asia, 1.2 billion to the Middle Eastern and African region, and 173 million to Latin America and the Caribbean to 2050.

Against this background, the growth of the global urban population is estimated to range from some additional 2.7 to 3.2 billion people, depending on the projected urbanization rate of growth (see Box 18.1). The generally larger growth of the projected urban population not only means that urban settlements will absorb all of the population growth between 2005 and 2050, but also that there will be a sizable redistribution of rural populations to urban areas. In addition, urban growth will be predominantly a phenomenon of the less-developed regions.

Figure 18.1 summarizes the GEA scenario results on the global level, and Figure 18.2 summarizes the five GEA world regions in terms of urban and rural populations. The continued growth of the world's urban population is set against a different growth path of the rural population.

Globally, rural population growth is projected to come to a halt around 2020, when it will reach a peak at about 3.5 billion people. This figure

Box 18.1 | Urbanization Projections Methodology

I. The UN World Urbanization Prospects

The UN World Urbanization Prospects (WUP) provides biannual estimates and projections of core demographic indicators of urbanization for all 229 countries or areas of the world and for the most populous urban settlements. The data include time series of total populations by urban and rural residence for the period 1950 to 2050, and of populations in 590 large urban settlements with 750,000 and more inhabitants for the period 1950 to 2025. In total, the entire WUP database contains estimates and projections for 4501 urban locations, covering two-thirds of the global urban population.

The WUP use a projection method that is described and discussed extensively (UN DESA, 1971; 1974; Ledent, 1980; UN DESA, 1980; Ledent, 1982; Rogers, 1982; Ledent, 1986; O'Neill et al., 2001; National Research Council, 2003; Bocquier, 2005; O'Neill and Scherbov, 2006). Such long-term projections include many uncertainties, reflected in the three urbanization scenarios discussed here. The method uses the growth difference between urban and rural populations or between city populations and total urban populations to model the dynamics of the urbanization process, an approach that is equivalent to a logistic curve fit (UN DESA, 1974). Starting with the most recent observed growth rates, future trends are calculated by taking into account the empirical observations that the pace of urbanization slows down as the proportion of the urban population increases (a characteristic feature of logistic growth processes beyond the 50% level of its asymptote). Consequently, the observed growth differentials are adjusted such that they approach a hypothetical norm obtained from past empirical observations (UN DESA, 1980). A distinguishing characteristic of the WUP is that only one central projection is performed in which the ultimate upper level of the urbanization process is not constrained; that is, all countries could ultimately converge to an urbanization rate of up to 100%.

II. Other Urbanization Projections

There are no real independent urbanization projections to the UN Urbanization Prospects as alternative scenarios invariably use the UN historical and current data as model inputs and also deploy a comparable methodological framework. The literature, however, reports scenario variants on the UN projection by either extending the time horizon beyond 2050 (the end-year reported by the UN projections) and/or alternatively relaxing the convergence assumption toward a 100% urbanization rate. For an OECD study on the climate vulnerability of port cities, Nicholls et al. (2008) extend the UN projections to 2100 and then apply a constant growth fraction to port cities at the national level to determine future port-city populations exposed to climate-change risk. In an integrated scenario exercise that explores future uncertainty in GHG emissions, Grubler et al. (2007) also extend the UN urbanization projection to 2100 and develop two additional scenario variants in which the asymptotic urbanization level is varied to explore the implications of lower urbanization. These three urbanization-rate scenarios were then combined with three alternative total population-growth scenarios (low, medium, and high) to determine the uncertainty range of future urban populations. This alternative scenario method estimated the uncertainty in the level of world urban population to range from 4.7 to 10.5 billion people by 2100, and from 5.6 to 7 billion compared to the 6.3 billion projected by the latest UN Urbanization Prospects projection for 2050 (UN DESA, 2010).

III. GEA Urbanization Scenarios

Methods and qualitative and quantitative assumptions that underlie the GEA transition pathways are described in detail in Chapter 17. In collaboration, this chapter and Chapter 17 also explore the issue of urbanization scenarios. In this context, two main features of the GEA scenario approach stand out: (1) their focus on energy issues and (2) their normative nature; that is, their objective to describe alternative, but successful, pathways toward an energy sustainability transition. As a result, the scenarios are based on a single central demographic-economic development scenario (e.g., based on the most recent UN medium population projection). The scenarios also describe pathways in which current widespread economic and energy disadvantages of poor rural populations are addressed successfully, for example with universal energy access achieved by ca. 2030. As a result, the qualitative scenario storylines describe developments in which, arguably, the pressure for rural to urban migration is significantly relieved, which suggests their reflection in the GEA urbanization scenarios. Therefore, three urbanization scenario variants based on a single medium population projection were developed. One scenario (GEA-supply) uses the UN Urbanization Prospects projection directly as input, whereas the other (lower) scenarios (GEA-mix and GEA-efficiency) adopt the methodology outlined in Grubler et al. (2007) using a model recalibrated to the most recent (2010) UN urbanization data. The resulting scenario differences in projected rural and urban populations bracket the potential impact of successful rural development policies that relieve urban migration pressures (GEA-efficiency) compared to largely unaltered patterns in rural and/or urban locational advantages (GEA-supply) and illustrate to policymakers the potential effects of altered policies that change the locational advantage of rural versus urban places.

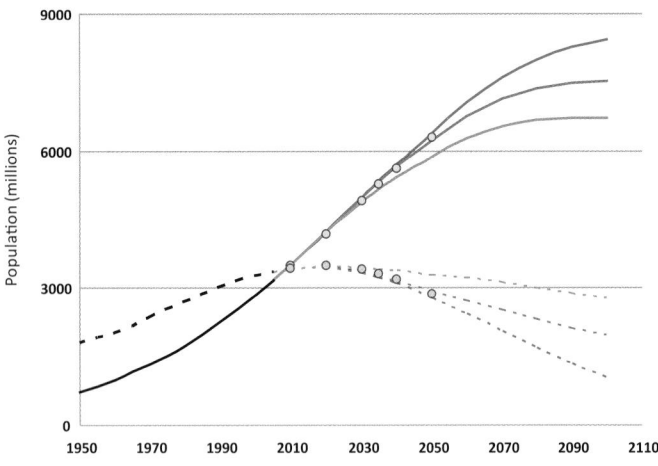

Figure 18.1 | Global urban (solid lines) and rural (dashed lines) population (in million). Historical trends 1950–2005 and scenarios to 2100 combining a medium population growth scenario (see also Chapter 17) with three alternative scenarios of urbanization rate growth (red, green, and blue, based on Grubler et al., 2007) and comparison to the most recent UN Urbanization Prospects projection (yellow and orange circles). Source: adapted from UN DESA, 2010.

is unaffected by the uncertainty of the urbanization scenarios and is a major, robust conclusion emerging from this assessment. The finding also adds urgency to corresponding efforts to improve energy access for the rural poor because if energy does not reach them soon, they will have an additional incentive to seek access in urban areas. After entering a path of negative growth, the rural population is projected to range from 2.8 to 3.3 billion by 2050 and to decline even further thereafter. However, the global picture masks stark regional differences: the Middle Eastern and African region will likely experience rural population growth until 2050 (or at the latest to 2080 in the GEA-efficiency scenario), while for all other regions the beginning of the decline the scenatio in rural populations is imminent or is already occurring (in the OECD90 region and in countries undergoing economic reform).

In 2050, the world is projected to be 70% urban with a comparatively narrow scenario uncertainty range of 64–70%. Latin America and the Caribbean, as well as the OECD90 countries, are expected to approach 90% urban, about the level of urbanization of the United Kingdom or Australia today.[4] By 2100, world urbanization levels could range from 71% to 89% with a corresponding urban population from 6.7 to 8.4 billion people. Thus, even in the lowest scenario, the world urban population in 2050 will be larger than the entire global population today.

18.2.1.6   Heterogeneity in Urban Growth

Patterns of urbanization are heterogeneous, including settlements of rapid growth, slower growth, and even cases of declining cities (Box 18.2). Urbanization is often equated with the growth of megacities. These vast, often crowded and complex, metropolitan areas with populations of 10 million or more are highly visible, and epitomize the challenges and problems of a rapidly urbanizing world, with pervasive poverty, slums, stressed infrastructures, etc. However, the reality of urbanization, both in terms of current settlement sizes (Table 18.3) and as historical and projected growth trends, is dominated by cities of smaller size (Figure 18.3). By 2005, just 19 cities worldwide met the megacity criteria (>10 million inhabitants) and their 284 million residents accounted for only about 9% of the global urban population. Conversely, some two billion people lived in cities with less than one million inhabitants, which corresponded to 63% of the world's urban population. Some one billion people or 37% of the urban population lived in smaller cities, less than 100,000 inhabitants,[5] in the year 2005.

The virtual absence of smaller cities and towns in statistical reporting, data collation, and modelling poses a serious problem and adds substantial uncertainty to any statement on urbanization trends. However, this should not lead to the erroneous conclusion not to focus analytical and policy attention on those cities that dominate the global urbanization phenomenon, most with <100,000 inhabitants. These numerous smaller cities pose a triple challenge for policymaking: data are largely absent, local resources to tackle problems with urban growth are limited, and governance and institutional capacities to implement policies for more sustainable urban growth are thin.

18.2.2   Urbanization Dynamics

18.2.2.1   Introduction

Modern towns and cities are generally recognized in the urban studies' literature as complex, largely self-organizing systems (Allen, 1997; Amaral and Ottino, 2004; Batty, 2005) that exhibit important agglomeration effects (National Research Council, 2003; World Bank, 2009). Cities are complex in the formal sense that recording their detailed behavior has vast information requirements. They are self-organizing in that interactions between partially informed citizens and between citizens and the city's 'hardware' form self-reinforcing patterns of land use and allocation of time. What was 'town planning' in the early 20th century is now seen more realistically as the facilitation of spatial patterns trying to emerge naturally (Hall, 2002).

Urban settlements show extraordinary resilience. Long-standing settlements reinvent their *raison d'être* many, many times. Large cities usually

---

4   Projected urban populations by 2050 at the country level (UN DESA, 2010) suggest that the two countries with by far the largest urban populations by 2050 will be China and India with some 1 and 0.9 billion, respectively. Ten countries are projected to have urban populations above 100 million by 2050 and, with the exception of the US (370 million), all of these are currently developing countries (projections by 2050 in millions: China, 1040 million; India, 875; Nigeria, 220; Brazil, 200; Pakistan, 200; Indonesia, 190; Bangladesh, 125; Mexico, 110; Philippines, 100).

5   Calculated mainly as a residual to the estimates of total urban population.

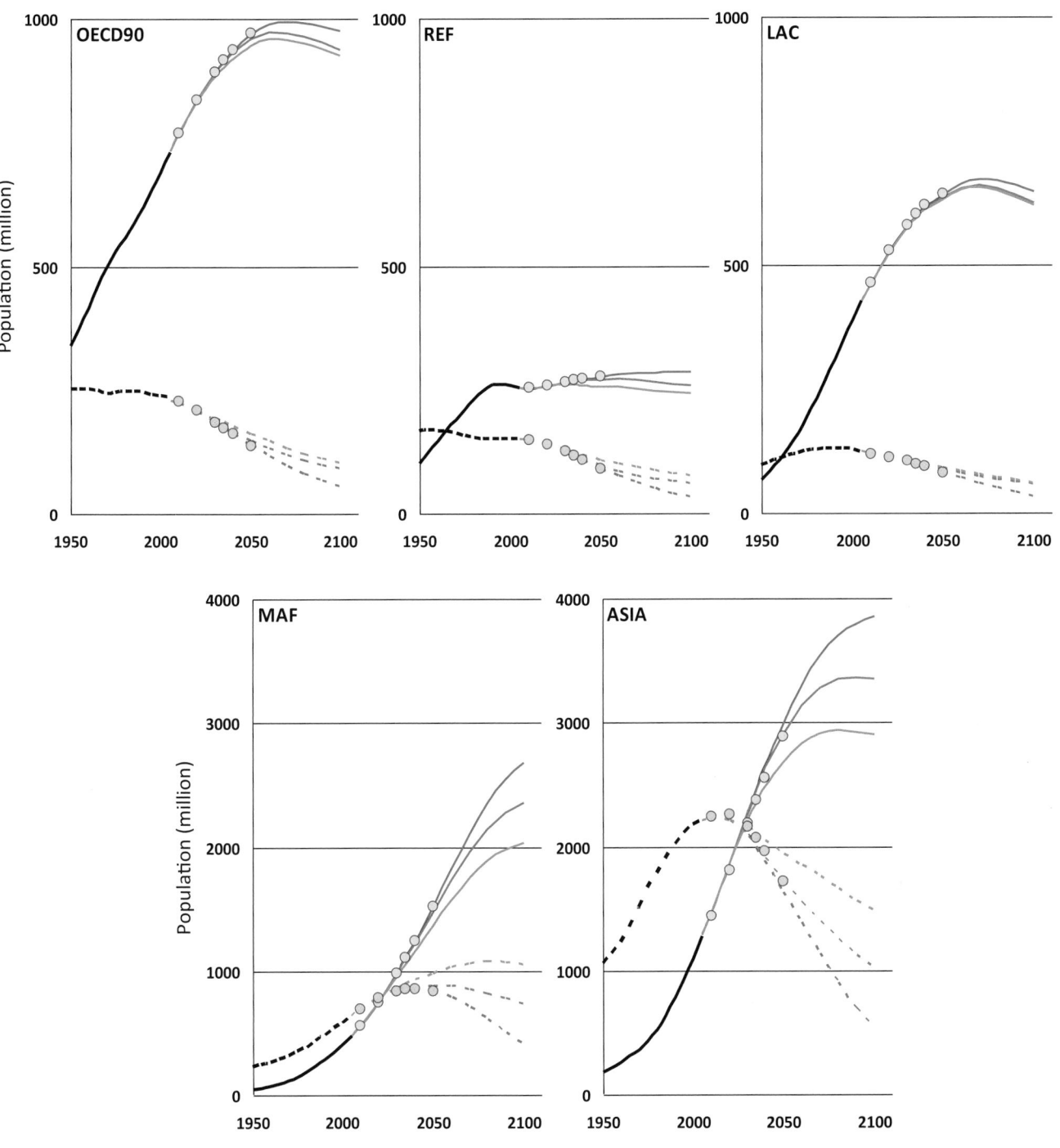

Figure 18.2 | Regional scenarios of urban and rural population (millions) for a medium population growth scenario (see also Chapter 17). Historical trends 1950–2005 and scenarios to 2100 for the five GEA regions (in millions). Regions: OECD90, OECD countries as per 1990 membership; REF: Countries undergoing economic reform, i.e., the reforming economies of Eastern Europe and the ex-USSR; LAC: Latin America and the Caribbean; MAF: Middle East and Africa; ASIA: developing economies of Asia. Urban population, solid lines; rural population, dashed lines. UN Projections (yellow and orange circles) and three alternative scenarios (red: GEA-supply, blue: GEA-mix, and green: GEA-efficiency).

confront constraints to existing paradigms first and have to develop innovative solutions (like freshwater-supply networks, and the use of clean secondary fuels) that are then taken up by innovation diffusion to smaller settlements. Energy *demand* considerations often serve as a shaping factor in settlements. Urban settlements, at least in times of peace, are more easily served by transport when in a valley, on a river, or on the seacoast. Differentiation of rental value takes place downwind of the pollution of energy-intensive industries. Prevalent solar and wind conditions often account for orientation patterns of streets and squares. In contrast, energy *supply* considerations have taken a secondary role in determining urban *form*. Indeed, the reverse has been true – urban form selects the energy mix. Large 19th century settlements needed coal because of the

Box 18.2 | Shrinking Cities

Urban growth is not a law of nature, as cities have stagnated in size over long periods of time, or even shrunk, and may well do so in the future. In addition, as national and international production and trade patterns change, some cities may lose their economic or strategic advantage and so shrink or even be abandoned entirely, for which history also provides ample examples. In the United States and Europe, many of the great 19th and early 20th century steel, textile, and mining centers and ports have lost economic importance and population (Cunningham-Sabot and Fol, 2009; Wiechmann, 2009); so too have some of the major manufacturing cities – for instance Detroit as a center of motor vehicle production. Also, in various high-income nations, from the 1970s there appeared to be a reversal of long-established urbanization trends nationally or within some regions with a net migration from large to small urban centers or from urban to rural areas. This was labeled counterurbanization, although much of it is more accurately described as demetropolitanization because it was population shifts from large metropolitan centers or central cities to smaller urban centers or suburbs or commuter communities.

Few systematic analyses have been performed on contemporary shrinking cities (UN HABITAT, 2008). Based on the latest assessment by the UN (UN DESA, 2010), of the 3552 cities with 100,000 and more inhabitants in 2005, 392 experienced a combined population loss of 10.4 million people between 1990 and 2005. While this seems a rather modest figure given the 3.2 billion urban dwellers in 2005, but for some cities population decline was substantial: a few hundred thousand inhabitants. The decline in urban population can, in some cases, be far from gradual but very fast, as in the eastern German city of Hoyerswerda (Pearce, 2010).

In the 80s, it had a population of 75,000 and the highest birth rate in East Germany. Today, the town's population has halved. It has gone from being Germany's fastest-growing town to its fastest-shrinking one. The biggest age groups are in their 60s and 70s, and the town's former birth clinic is an old people's home. Its population pyramid is upturned – more like a mushroom cloud.

Fearing decay, vandalism, and costs, high-rise apartment buildings are now being torn down. Cities also shrink as a reaction to an often painful economic restructuring process, as observed in countries of Eastern Europe and the former USSR undergoing economic reform. Most OECD90 countries have had persistently low fertility, well below replacement levels, for extended periods of time. If low fertility rates continue, overall population decline will become a reality. Shrinking cities amidst a growing or stable national population could then be replaced by a regime of national and city populations shrinking simultaneously. How this will affect urbanization remains one of the biggest challenges for understanding long-term urbanization trends.

The energy implications also remain an important area to be explored. Shrinking urban populations yield reductions in urban energy use. Such trends are, however, likely to be counterbalanced by the effects of potentially larger residential floor space available for the remaining population as well as increases in transport energy use associated with lower population density. The economic viability of urban transport and energy infrastructures in shrinking cities may also be challenged, but on this aspect no data or studies are available currently.

transportation problems associated with taking wood into the dense center. Later, they needed the coal processed into town gas to provide better quality services, like lighting. Urban settlements developed in inhospitable environments once they could depend on the provision of high-grade energy to run mechanical and electrical building services (Banham, 1969). Over the long perspective of future energy-infrastructure planning and population growth, the self-organizing properties of urban space need to be taken seriously. An historical novelty would be a reconfiguration of the urban form to reflect constraints on energy supply.

18.2.2.2   Inter-Urban Complexity

Some of the most compelling evidence that complexity theory is at work is that settlements within the range of a mobile population as ranked by population size approximately conform to Zipf's rule – size is inversely proportional to rank (Zipf, 1949; see Figure 18.4). This is true at least to within the indeterminacy of the definition of an urban population. Each urban center is then in a dynamic equilibrium with the larger urban system and, although urban settlements may present themselves as 'self-contained,' they are far from it. The implication is that at the margin, urban settlements show no total economies or diseconomies of scale. If they did, they would simply disperse to a uniform size or accumulate in a single megatropolis. This conclusion may not apply to any specific economic activity (e.g., the location of energy-intensive industries or finance centers). Large cities and small towns do different things, but large and small appear to show no special advantages in the economy as a whole. The implications of the urban rank-size rule for this assessment suggest that any undue overemphasis on the top end of the distribution curve (i.e., an exclusive focus on 'megacities') is certainly not

Table 18.3 | Population in urban locations by city size class in 2005.

Size class	Total City population	Proportion of total urban population
	2005	
	Millions	%
Total urban population	3,167	100
<100,000	1069	34
100,000–1,000,000	932	29
1,000,000–5,000,000	673	21
5,000,000–10,000,000	209	7
>10,000,000	284	9

Source: UN DESA, 2010.

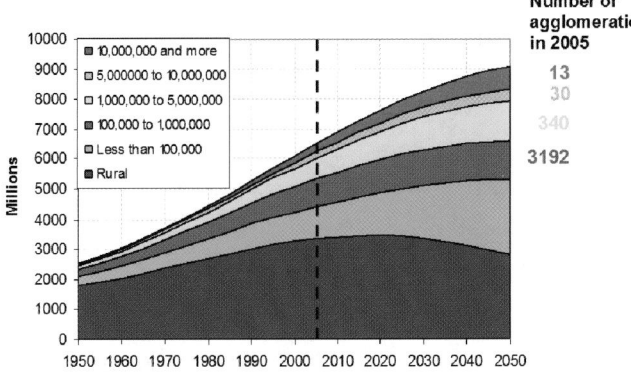

Figure 18.3 | Population by residence and settlement type (millions). Historical (1950–2005) and projection data (to 2050) (for 2005 statistics, see Table 18.3). The number of agglomerations within each size class are also shown for reference. The total number of urban settlements/agglomerations in the size class below 100,000 inhabitants remains unknown. Source: adapted from UN DESA, 2010.

warranted, as smaller cities constitute the majority of the continuum of the rank-size distribution curve.

At present, few data or studies are available that explore the implications of the rank-size rule for urban energy use. Bettencourt et al. (2007) identified economies of scale effects for urban energy infrastructures for gasoline distribution and electricity distribution grids (cables); i.e., large cities (in terms of population size) have proportionately smaller energy infrastructures than smaller cities, at least in Germany and the United States. In terms of energy use, the sparse available evidence is mixed: transport energy use (gasoline sales) seems to be somewhat lower in larger cities in the United States, which appears plausible in view of the impacts of higher population density on lower automobile dependency. Conversely, electricity use appears to grow somewhat overproportionally with city size (United States, Germany, and China). However, it is unclear whether this is a genuine urban scale effect or simply reflects fuel substitution effects where larger and denser cities exhibit higher preferences for clean, grid-dependent energy carriers (more electricity and/or gas and less oil and/or coal) compared to

smaller cities. This remains an important area for future urban energy research.

It now seems likely that the power-law size distribution results from many co-existing stochastic processes (Sornette and Cont, 1997) with a similar statistical effect to that of a long tail power law. The only constraint on candidate processes is that the stochastic growth retains a long tail power-law distribution on average as the national population grows. In fact, very simple growth processes can generate power laws. For example, Ijiri and Simon (1975) obtained a power-law tail by assuming that annual population growth is randomly distributed across settlements in proportion to their size. The key to maintaining a power-law distribution over time is not the modeling of the largest settlement, but the model mechanisms for creating and retiring the smallest.

The robustness of the Zipf law means mechanisms that populate the large number of smaller settlements play a more important role in urbanization dynamics in most countries than the growth of the few largest cities, as amply confirmed by urbanization statistics (see Section 18.2.1.6), which adds to the conclusion herein on the urgent need to study smaller sized cities in terms of their energy and environmental implications.

18.2.2.3   Intra-Urban Complexity

While the smallest settlement is, to some extent, a matter of statistical definition, at its most basic level there is an economic and physical limit to the size of settlement that can manage its own water, sewage, power, and administration collectively, at a few thousand people. This reasoning about the dynamics of growth replicates within the urban settlement itself (Batty, 2008). Quanta of growth similar in size to the minimum urban settlement appear in the history of large cities, marked out by physically distinct districts. There is a subtle interplay between the technical quanta defined by the constraints of urban engineering and the economic and social aspirations of citizens. It is possible to think of a district as a cellular automaton that can be flipped in status according to rules that relate the status of adjoining districts. This dynamic suggests that up to some rate-of-event threshold the city remains heterogeneous, with districts changing their designation from time to time. Above some co-existence threshold, groups of districts start to cluster together in one class and the designations segregate (Schelling, 1969), which can act as a precursor to the formation of slums and ghettos as development capital is diverted elsewhere. This phase change can be seeded by an infrastructure change (an investment in public space or a waste incinerator). Cellular automata models can reproduce many of the key social spatial statistics of urban settlements (Batty, 2005).

The attachment of new zones and transformation of others reflect urban-area attempts to gain the advantages of agglomeration while

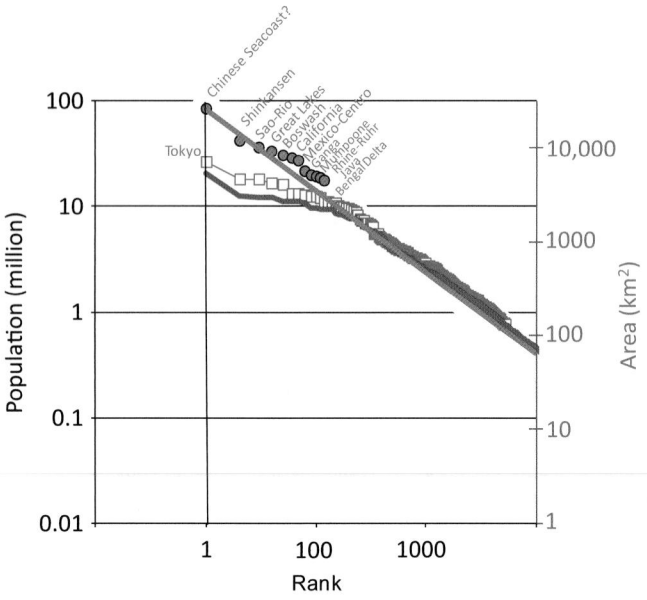

Figure 18.4 | Rank-size distribution of global cities by population. Note the irregularity in the distribution at the largest cities when defined by their administrative boundaries. Conversely, by aggregating individual cities to 'urban clusters' or 'corridors' the regularity of the global city rank-size distribution is maintained. The designation of an emerging urban cluster along the Chinese seacoast is entirely speculative. The currently largest urban cluster is found in Japan along the Shinkansen corridor, with some 70 million inhabitants. The rank-size distribution of global cities not only extends to their populations (red for cities and magenta for urban agglomerations), but also to their area (land cover; green). Source: Grubler et al., 2007, based on Marchetti, 1994; city area distribution from Yamagata, 2010, based on Kinoshita et al., 2008.

internally separating discordant functions (Fujita et al., 1999). Zoning, whether imposed or emergent, is then a natural property of an urban settlement. The dynamics of these processes have long been recognized as complex (Forrester, 1969). Indeed, modern urban planning frequently uses techniques from statistical mechanics to find the most likely aggregate properties from the nearly random effect of the thousands of citizen choices (Wilson, 2000). Administrative boundaries do not always follow the expansion of an urban area and care needs to be taken, for example, as to whether an urban settlement's quoted energy statistics refer to the current political or to the practical spatial reality.

Urban expansion can be seen as a process with land as an input, in effect making the town or city a 'space machine' reorganizing land use and connections (Hillier, 1999). Indeed, only a limited number of stable connectivity patterns seem to emerge for expanding urban spaces. This may, in part, reflect that the existing connection patterns have a very strong influence on the optimal connections within a new zone or district. Such an assumption underlies much of the modeling with cellular automata applied to understanding urban growth. Preferred joining rules are a common property in large complex systems that grow while remaining unconditionally stable (Fisk and Kerhervéa, 2006).

One constraint on the network is that expansion needs to occur such that the urban settlement remains socially and economically viable at all stages of growth. It may, in part, explain why urban areas adopt one of only a few configurations. The 'idealized' European city physically expands by absorbing smaller satellite centers and creating annular transport networks, to form a scale-free connectivity that enables access on foot to public transport. Low-income, low-rent areas cluster around those areas where transport to work is cheap.[6] In contrast, the 'idealized' North American grid city expands along its major axes and maintains a uniformity that avoids overstressing its original center, again using beltways to relieve congestion at intersections from clipping journeys. Although grid cities are superficially uniform, partitioning of rental value is often triggered by environmental factors (downwind, downriver, across the tracks) and agglomeration by local economies of scale. Both these urban configurations are characterized historically by capital investment in the city that matched the influx of population. Conversely, in many developing countries the migration rates from rural areas currently run ahead of capital formation, creating serious issues of energy access and infrastructure development (see Section 18.4.1).

18.2.2.4   Time and the City

While urban geography is naturally expressed as space, an urban settlement might equally be thought of as a complex time machine. Since the urban rationale is to bring specified people together at the same place at the same time to get things done, motion within the area needs to meet socially determined constraints in absolute time. Zahavi's rule (Zahavi, 1974) notes the rough constancy, in many countries over many decades, of the time spent traveling during a working day. This presumably reflects the constancy of many shared social norms, such as when work starts and ends and when entertainment begins. Since different modes of travel involve different amounts of physical work by the traveler, the social norm also sets upper limits on how tired the traveler can be on arrival (Kölbl and Helbing, 2003). The dynamics of land-use changes then 'solve' the consequent set of time constraints. Thinking of a city as a vast set of service connections self-consistently feasible within given absolute time windows is a complementary dynamics paradigm with some useful explicative power.

Transport improvements (or the prospect of them) tend to induce land-use changes and investment, and not save traveling time (as reviewed in Metz, 2008). In urban areas the local density of travelers on the move directly affects the average speed of travel and also indirectly decides safe speed limits. A city with high values of speed averaged over complete journeys can support journeys of longer length, within social

---

6   In cities of developing countries, economic pressures to relocate slums that enjoy a transport locational advantage are therefore high. Conversely, locating the poor far outside the city with poor (and expensive) transport access (as was the case in apartheid South Africa) further disadvantages them socially and economically.

norms, and so individual service points sweep larger areas of clients. Increased average speed supports larger agglomerations (e.g., large retail shopping malls) and so gains economies of scale. Higher speeds can give greater separation between incompatible land uses without breaking scheduling norms.

As a consequence of land-use adjustments, high-speed cities have longer average journey lengths, but this means their growth dynamics can relax to lower densities with greater local accumulation. An hour or so traveling time represents the effective limit to the radius of commercial activity of an urban settlement. The resulting dependency of transport-energy use on urban density is roughly linear when plotted against the mean interpersonal distance interpreted from two-dimensional urban density (Kenworthy and Newman, 1990), as might be expected from a largely traveler-interaction effect. Since the total start-to-finish journey time is the key determinant of location there are densities below which only individual motorized transport can provide a plausible service. By the same token, as urban density increases the interaction between travelers becomes increasingly significant and energy use ceases to fall as energy is wasted in congestion and traffic queues. Very high density centers offer little advantage to transportation because congestion is so severe (especially true in cities in which both the mass transit and private transit systems become embroiled in the same gridlock). Urban areas in this state tend to spill outwards and create a self-organizing critical density consistent with the underlying social norms.

18.2.2.5   Economic Complexity

The new insights of complex system 'econophysics' complement rather than displace traditional urban economics. The timescales and spatial statistics employed tend to be different. Ecophysics focuses on changes over long timescales and wide spatial averages, while traditional models the short-term dynamics that manage the scarcity and surplus of individual space. The overlap is the understanding of the dynamics of rents, both in the form of local taxation and 'ground' rents. These influence income distribution and the dynamics of investments and also change the economics of urban- versus rural-based activities (Irwin et al., 2009). Differences between urban settlement dynamics in different societies may also reflect institutional differences (e.g., as reviewed by Diamond, 2004).

18.2.2.6   Future Dynamics

The energy demand and energy mix of urban settlements normally follows rather than leads urban expansion (Seto and Shepherd, 2009). However, the historical generalities of urban growth hide the continual adoption and invention within the settlements themselves. Even new transport technologies that made such a fundamental contribution to urban form usually arose from solving a problem caused by an existing technology. So projecting the urban future on the basis of past trends is

dangerous if attention is not paid to changes in detailed mechanisms. Indeed, some signs indicate that this new urbanized century might be different in character from the urban-rural mix that preceded it.

For example, the simplicity of Zipf's law is now retained for the world's largest cities only if the associated metropolitan region is treated with the city as one (see Figure 18.4 above). If airline connectivity is a new paradigm of interurban transport, the connectivity of the largest cities appears not to have caught up with expectations from a natural extension of the power law to describe airline connectivity of smaller towns (Guimerà et al., 2005). Indeed, scaling arguments (Kühnert et al., 2006) suggest that the frequency with which settlements need to find paradigm-shifting innovations must increase as they grow. Since breakthroughs are, by definition, unforeseeable there is a future scenario in which the growth of the world's largest metropolitan areas falters.

Urbanization forms part of the demographic transition and is expected to deliver a stabilization and then decline in world population around the middle of this century. As a consequence urban settlements may face retractions in some unspecified form. How built form relaxes to new configurations poses a fascinating question to a new understanding of urban structure.

18.2.3   Specifics of Urban Energy Systems

In principle, urban systems are not fundamentally different from other energy systems in that they need both to satisfy a suite of energy-service demands and to mobilize a portfolio of technological options and resources. However, urban systems also have distinguishing characteristic that sets them apart:

* A high density of population, activities, and the resulting energy use and pollution (see Section 18.4.2 for a more detailed discussion).

* A high degree of openness in terms of exchanges of flows of information, people, and resources, including energy.

* A high concentration of economic and human capital resources that can be mobilized to institute innovation and transitional change.

Urban areas are characterized by high spatial densities of energy use, which correspond to their high population concentrations, and by low levels of energy harnessed and extracted: cities are loci of resource management, processing, trade, and use, rather than of resource extraction or energy generation. All settlements depend on a hinterland of agriculture, forestry, mining, and drilling; in the present fossil-fuel economy, this hinterland has a global reach. Indeed, the same may be true of many rural areas, which, in developed countries, often focus on a few crops. These specialized rural areas also require imports of energy, goods, and services, which often may be on the same scale, per capita, as those of urban areas. Indeed, spatial division of labor is a characteristic of modern societies, with consequences for local energy use and policies. In industrialized countries both urban and rural areas depend heavily on energy-intensive industry, which may be located in or outside cities. If heavy industry is located outside urban areas, urban energy use may apparently be lower than rural energy needs, even though urban dwellers also consume the products from industrial activities.

High levels of energy demand open possibilities to reap significant *economies-of-scale* effects in energy systems, in supply as well as in transport and distribution. (It is not a coincidence that, historically, many major, large hydropower resources were developed to supply electricity to large urban agglomerations, from Niagara Falls in the Unites States to Iguacu (Itaipu) in Brazil.) Simultaneous to cities being loci of a wide diversity of activities, significant *economies of scope* are possible. The wide range of different energy applications from high-temperature industrial processes down to low-temperature residential home heating, and even to the energy provision of greenhouses, allows the maximization of energy efficiency through better source-sink matching of energy flows in the system, be it through conventional cogeneration schemes or through more complex waste-heat 'cascading.' These can, however, only be realized if diverse energy uses in a city are sufficiently mixed and co-located to allow these concepts of 'industrial ecology' to be the implemented. On the negative side of density, typical urban agglomeration externalities are important: low transport efficiencies through congestion and high pollution densities add urgency to environmental improvement measures. Retrofitting a high-density built environment can also incur many transaction costs that would not apply in low-density developments. However, it is no coincidence that cities have always been the first innovation centers for environmental improvements (Tarr, 2001; 2005).

The latter perspective is perhaps the most fundamental for the transformation of energy systems. Urban agglomerations are *the* major centers and hubs for technological and social innovation (Kühnert et al., 2006), as they both dispose of and mobilize formidable resources in terms of human, innovation, and financial capital. Bringing these transformations to fruition may ultimately be of greater long-term environmental significance than any short-term environmental policies. So, energy and environmental policies in an urban context may have substantial leverage in inducing further much-needed innovation in the core, where such activities take place.

18.2.4   City Walls and Urban Hinterlands: The Importance of System Boundaries

18.2.4.1   Introduction

Measuring the energy use of cities is no easy task, compounded by the absence of widely agreed measurement concepts and data-reporting formats. And yet for reasons of scientific enquiry, policy guidance, and political negotiation the important issue of 'attribution' needs to be addressed. The seemingly easy question of "How large is the energy use (or associated GHG emissions) of a city and what can be done to reduce it?" can vary enormously as a function of alternative geographic and system boundaries chosen. Therefore, this section reviews the different issues that must be addressed in urban energy assessments and aims to clarify the various concepts and definitions and help make their differences more transparent. In a modification of an old adage that only what get's measured gets controlled, this chapter postulates that only what is measured correctly and transparently at an urban scale is useful for policy guidance.

As a simple example of the importance of boundaries in urban energy assessments consider the issue of the administrative/territorial boundary chosen for defining a given city. Barles (2009) studied the fossil-fuel use of Paris, its suburbs, and the larger Parisian metropolitan region. The per capita fossil-fuel use was lowest in the city of Paris, and increased as the region considered expanded. This phenomenon is caused by a combination of inherent and apparent factors: the inherent factor is the lower transportation energy required by areas of higher population density (in central Paris with its formidable Metro system, compared to its suburbs); the apparent effect is the changing of the system boundary to encompass more energy-intensive industrial activities located beyond the city center.

Generally, urban-energy assessments must be oriented either to physical, and hence local, energy flows (a 'territorial' or 'production' perspective) or, if trade effects are included, follow economic exchanges linked to energy use (a 'consumption' perspective). The joint consideration of the production and consumption perspectives is most likely to yield a full assessment of urban energy,[7] albeit to date the literature and data base for such a comprehensive perspectives is extremely limited.

In assessing a variety of local and upstream contributions to the urban metabolism, it is sometimes tempting to aggregate the disparate elements into one indicator. For instance, the ecological footprint (see Rees and Wackernagel, 1996) is increasingly used to describe the impact of urban resource use (e.g., in London, Barcelona, and Vancouver). The ecological footprint of an urban area is invariably larger than the surface area of the city itself, which leads to the facile (but erroneous) conclusion that the city is 'unsustainable.' Cities are part of an exchange process, whereby they produce manufactured goods and services while depending on a hinterland for their supplies – and the existence of this hinterland cannot of itself be unsustainable. Better indicators are the relative magnitude of resource use and emissions (per capita, household, or income), compared to rural or other urban populations, or environmental limits, like carbon accumulation in the atmosphere. Since the ecological footprint is, in any case, driven

---

7   The terms 'direct' and 'indirect' energy are avoided here, since they have a different meanings in each of the approaches considered. Instead, the terms 'final,' 'primary,' 'upstream,' and 'embodied' energy are used here.

Table 18.4 | Overview of urban energy-accounting frameworks.

Approaches	Data basis	Definition of energy users	Position along energy chain	Upstream or embodied energy	Territorial/Production or Consumption approach
Final energy	Physical	Final user (energetic)	Final	Not included, can be added using typical conversion efficiencies between primary to final energy for different energy carriers	Territorial Example: Regional energy statistics, Baynes and Bai, 2009
Regional energy metabolism	Physical	Region	Combination of final, secondary and primary	Not included, but can be added using typical conversion efficiencies	Territorial example: Schulz, 2007
Regional economic activity	Economic (physical extensions)	Final demand (economic)	Total Primary Energy Supply	Includes upstream & embodied energy of goods and services, no sectoral differentiation	Territorial Example: Dhakal, 2009
Energy Input-Output	Economic (physical extensions)	Final demand (economic)	Total Primary Energy Supply	Includes upstream & embodied energy of goods and services, no sectoral differentiation	Consumption Examples: Weisz et al., 2012; Wiedenhofer et al., 2011

by fossil-fuel carbon emissions, studying urban energy use directly appears to be a more constructive approach.

18.2.4.2   Energy Accounting Methods

Table 18.4 classifies the various methods used to estimate urban energy use. This classification utilizes two main criteria: the basis of the data and the definition of energy users. The first two methods are based on physical flows, and as such produce 'territorial' or 'production' oriented energy balances; the other two focus on economic flows, and are 'consumption' oriented. The economic-based models (regional economic activity and economic I-O (input-output) approaches) are further distinguished by the level of sectorial detail.

The 'final energy' method uses physical data, such as energy statistics from utilities or fuel sales, as the data basis. Users are defined as energetic end users within the city boundaries. Energetic end users are the "consumers" of final energy (such as electricity, heat, gasoline, or heating fuels). It is important that both producers and consumers (i.e., firms and households) use final energy. By disaggregating the final energy use by sector, one can differentiate between residential, commercial, and industrial uses. These sectorial accounts of urban final energy use also allow comparisons with national level data or data of other cities and can serve as a useful guide, e.g., for energy-efficiency 'benchmarking' that can guide policy.

The direct final energy account can further be extended by estimating the *upstream energy* requirements needed to provide the final energy, using e.g., lifecycle analysis. The upstream energy is the primary or secondary energy use *linked to the final energy* utilized by the end users within a city. Depending on the estimation method, the upstream energy may or may not include the energy required to extract and transport the primary energy itself. For clarity, it is crucial to specify which type of upstream energy is considered: secondary, primary, or primary, including energy for extraction activities themselves.

To avoid confusion with other terms used for upstream energy, this section reserves the term 'upstream' for energy-linked direct energy flows (mobilized outside the geographic city boundary, but linked to a city's final energy use), and the term 'embodied' for goods and services linked to indirect energy flows (i.e., energy embodied in resources, materials, and goods traded).

Final and upstream energy uses are not the total primary energy required by urban activities, since they do not include the energy needed to produce goods and services *imported* into the city. Conversely, the final and upstream energy uses of a city also include energy uses to manufacture goods and provision of services *exported* from the city (and consumed in other cities or in rural areas). Care therefore needs to be taken to avoid double counting of energy flows, for example by adding imported 'embodied' energy flows to the final and upstream energy uses of a city, but ignoring the (final and upstream) energy embodied in goods and services produced in a city, but exported for consumption elsewhere. This double counting is averted in I-O methods but can be a problem in approaches that rely on lifecycle analysis (Ramaswami et al., 2008; Hillman and Ramaswami, 2010).

The difference between the final and upstream energy use can be considerable. Take the example of electricity: typically, for 1 GJ of electricity consumed in a city, up to 3 GJ of primary energy in the form of coal must be burned in a conventional steam power plant. (This ratio is substantially lower for combined-cycle power plants fired by natural gas, which illustrates the need to consider the *actual* urban energy system characteristics rather than aggregate 'upstream' adjustment factors.) Heating fuels, such as gas and fuel oil, also have upstream energy use through their extraction and transport. According to Kennedy et al. (2009), who applied this method to the GHG emissions of 10 global cities, the lifecycle GHG emissions associated with urban fuel use are between 9% and 25% higher than their local emissions. This approach is also that followed by the Harmonized Emissions Analysis Tool of the International Council for Local Environmental Initiatives (ICLEI, 2009), although since the software is proprietary (and not transparent),

Table 18.5 | Energy and material flows of selected cities showing the importance of energy flows in the total metabolism of cities.*

City	Cape Town	Geneva	Hong Kong	Paris Petite Couronne	Singapore
Reference	Gasson (2007)	Faist (2003)	Newcombe et al. (1978)	Barles (2009)	Schulz (2005)
Population (millions)	3	0.4	3.9	6.3	4.1
Year	2000	2000	1971	2003	2000–2003
Energy (GJ/cap/year)					
Primary	40		72		258
Regional (city)	22	92	43		103
Domestic material consumption (tonnes/cap/year)					
Total		7.7		4.6	29.7
Fossils		1.8		2.1	5.2
Biomass		1.0			0.3
Construction minerals		4.8			22.5
Industrial minerals and ores		0.1			1.7[1]
Water	110	151	100		112

* Domestic material consumption = domestic extraction + imports – exports, which in the urban case is dominated by imports.

1 includes other products

geographically limited, and often applied only to municipal energy use, it is not clear how reliable or significant the results are.

The 'Regional Energy Metabolism' method also uses physical data, but defines the user as a geographic region (in this case, boundary of the urban area): thus, all energy flows that cross the regional boundary are measured. This method is based on a socio-metabolic understanding of the city, including energy, materials, water, important waste flows, infrastructure stocks, and other relevant physical attributes of a city. Indeed, a complete urban metabolism would include other energy flows, such as the calorific value of physical goods entering and leaving the city and biomass for human (and animal) nutrition, which are not included in other approaches. This method includes all the energy that is imported into the city, regardless of its conversion stage. In contrast to the final energy method, the primary energy used by the energy sector *within the city* is included in the regional energy metabolism. Thus, the location of a power plant inside or outside a city's boundaries has a large influence on the measurement of its energy use in this method. As a result, the regional energy metabolism of a city is always larger than its final energy. The mixture of primary and final energy, however, makes it difficult to compare this method's results with national level data and those of other cities.

So far, the urban metabolism approach has been applied to material flows rather than energy flows for Cape Town (Gasson, 2007), Geneva (Faist et al., 2003), Hong Kong (Newcombe et al., 1978), Paris (Barles, 2009), and Singapore (Schulz, 2006; 2010b; see also Chapter 1). Table 18.5 summarizes some findings from urban metabolism studies. Energy is extremely important in the urban metabolism: energy carriers represent a significant share (a quarter to a half) of domestic material consumption. Moreover, other materials consumed (biomass, construction and industrial minerals, and ores) also represent important flows of embodied energy.

The 'Regional Economic Activity' approach uses the urban final demand aggregate (defined in *economic* (monetary) terms, that is as gross regional product (GRP)) and a national or regional energy-to-GDP relationship to estimate the city-specific energy use. This method was used by Dhakal (2009) for major Chinese urban areas (see Section 18.5.2), and is the only possible approach when no city-specific energy data are available. However, the limitation of this method is that it ignores all city-specific drivers of urban energy use except income. The national or regional energy-to-GDP ratio usually uses total primary energy supply as the energy variable. This method is thus fully comparable with statistics on the national level.

'Energy I-O' accounting approaches are based on national economic I-O tables, which measure (usually in monetary terms) all sales and purchases of goods and services among the producing sectors and to final demand. These tables can be extended to account for physical energy flows or emissions. Based on I-O tables, the 'embodied' energy (i.e., the energy used throughout the whole production chain to produce the final goods and services) can be calculated. This approach allows allocation of energy use to specific sectors of economic activity. Using consumer-expenditure surveys for urban areas, I-O studies have assessed the energy use of Indian urban and rural populations (Pachauri et al., 2004; Pachauri, 2007), Brazilian urban households (Cohen et al., 2005), Sydney's energy use (Lenzen et al., 2004), as well as for postal district resolution maps of Australia (Dey et al., 2007). Illustrative results on direct versus embodied energy use in Asian megacities based on I-O analyses are given in Figure 18.5.

First, it is important to recognize how embodied energy flows compare with direct energy flows, particularly for high-income cities such as Tokyo. Second, the dynamics of energy flows are characterized by the

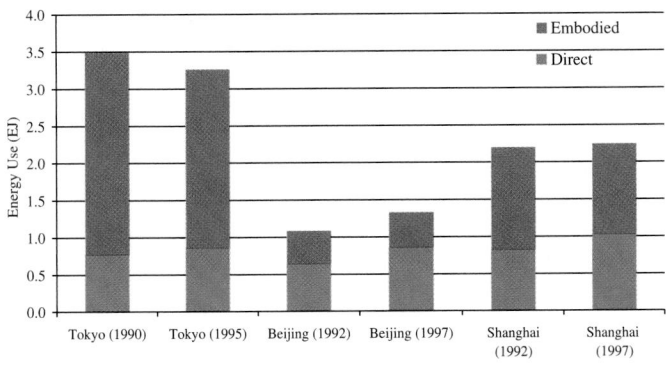

Figure 18.5 | Estimates of direct (on-site) versus embodied (via imports of embodied energy in goods and services) energy use of Asian megacities. Source: Dhakal, 2004, based on Keneko et al., 2003.

growth of direct energy flows (income growth leading to buildings and transport energy demand growth), whereas embodied energy growth remains comparatively flat. In these examples the income effect on direct energy use is larger than changes in trade flows into the city over the limited time period considered in the studies.

I-O avoids the thorny issue of double counting, as only the final demand (i.e., household consumption – and that of government, even though this is rarely reported) serves as the allocation unit of all direct (including upstream) and indirect energy flows, whereas industrial and service-sector energy uses are treated as intermediary uses, allocated to final consumption (in a city and elsewhere) based on expenditure levels and structure. Hence, this approach is often referred to as a 'consumption' approach, as opposed to the 'production' approach that focuses on apparent energy use. The I-O based 'consumption' approach can thus differentiate between cities with different incomes and final consumption patterns, but cannot capture differences in industrial and service energy use, not to mention differences in energy systems (e.g., degree of cogeneration) between different cities. The issue of the representativeness of national I-O tables for cities with different consumption structures, even within identical expenditure categories, and with different (usually higher) price levels has not been addressed by systematic studies. National I-O analyses do not provide information on embodied energy flows in commodities and goods traded internationally, which requires the use of multiregional I-O models, whose data quality is often weak. Studies performed for Norway (Peters et al., 2004), the United Kingdom (Wiedmann et al., 2007), and the United States (Weber and Matthews, 2008) show that a significant fraction of energy use can be attributed to imports, especially in industrial economies (Peters and Hertwich, 2008). The application of the I-O method to developing countries is, however, not straightforward, since the large informal sector is absent from the official I-O tables, which themselves exist in up-to-date versions only for a few developing countries. It has also been shown that uncertainties are very large for the GTAP (Global Trade Analysis Project) database used in multiregional I-O models (Weber and Matthews, 2008).

From the I-O studies, the largest categories in urban household energy use are housing, electricity, transportation, and food. Of these, food and electricity have the largest upstream and embodied energy "content." Thermal electricity generation involves heat losses up to two-thirds during transformation, plus transmission and distribution losses. Food has both nutritional and embodied energy components: a diet of 3000 kcal/day corresponds to 4.5 GJ/year in nutritional energy per person. The commercial primary energy required to produce food ranges from 2.5 to 4.0 GJ/capita for Indian urban households and from 6 to 30 GJ/household for Brazilian urban households (where the ranges correspond to low and high income brackets) to around 40 GJ/capita for European households (Vringer and Blok, 1995; Pachauri, 2004; Cohen et al., 2005). The majority of the energy embodied in food production is consumed outside city boundaries.

Comparison of energy-accounting frameworks

To compare and contrast the results from different accounting approaches Chapter 18 initiated a collaboration among various research teams to provide a quantitative illustration, applying two different methodologies ('final energy' and 'energy I-O') for two different cities: London and Melbourne. Recent (partial) results for Beijing are also included for comparison. These two approaches are often contrasted at the national level for energy or GHG, where they are generally known as 'production'[8] or 'territorial' (for the 'final energy' method), and 'consumption' (for the 'energy I-O' method) (see, for instance, Peters, 2008).

For the Melbourne study, Manfred Lenzen and his group at the University of Sydney used environmentally extended I-O methods coupled with household expenditure surveys to map direct and embodied GHG emissions of household consumption (Dey et al., 2007). This method was adapted to provide results in terms of primary energy use for the city of Melbourne (Australia). Baynes and Bai (2009) scaled state data down to the urban level, focusing on the (direct) final energy use of Melbourne city. The London study compares final energy use from official statistics (UKDECC, 2010) and results from a multiregional, environmentally extended I-O analysis with explicit representation of the household consumption vectors for the Greater London Authority. The yet unpublished study is based on a method of Minx et al., (2009) and was carried out by the Technical University of Berlin, the Potsdam Institute for Climate Impact Research, and the Stockholm Environment Institute.[9] The Beijing study (Arvesen et al., 2010) only considered household energy use (and hence misses the large industrial and service sector energy use) and combined both the final energy method (with additional approximate fuel-specific estimates of upstream conversion energy needs) with an I-O approach.

---

8   The term 'production' accounting as applied to household energy use is somewhat misleading, but to not introduce further terminological complexity, the term (well established in the national scale literature) is retained here.

9   Weisz et al., 2012.

Table 18.6 | Primary energy use for two different energy-accounting approaches for three cities for which (partial) data are available: Melbourne, London, and Beijing. All values are expressed in GJ/capita (permanent) resident population. Dashes (-) indicate categories of energy use that cannot be compared directly between the two different accounting methods.

| Primary energy GJ/capita for: | Melbourne 2001 | | Greater London 2004 | | Beijing 2007 Household energy only | |
| | Pop.: 3.2 million | | Pop.: 7.6 million | | Pop.: 12.1 million | |
	Prod. acc.	Cons. acc.	Prod. acc.	Cons. acc.	Prod. acc.	Cons. acc.
Residential heating	22	12	28	–	9	–
Residential electricity	28	30	17	–	9	–
Residential housing (heating + electricity)	50	42	45	35	18	11
Private cars	33–41	27	10	–	7	7
Nonresidential use	197	–	56	–	n.a.	–
Household consumption of goods and services	–	116	–	108	–	34
Total	279–288	184	111	143	25	52

Pop = population; Prod. acc. = production accounting; Cons. acc. = consumption accounting; n.a. = not available.

The results are summarized in Table 18.6. To enable comparison, the position along the energy chain has to be the same, so the primary energy equivalent of final energy is estimated where detailed statistics are unavailable (using standard conversion efficiency factors from Kennedy et al., 2010). To correct for different sizes of the cities, all values are expressed in GJ/capita.

The two methods cover different types of energy flows, some of which can be compared, and others cannot (denoted by dashes in Table 18.6). Energy I-O ('consumption accounting') focuses on the energy use of households within the city boundary, directly and embodied in the purchase of goods and services. A direct comparison with the territorial final energy ('production accounting') method is only possible, therefore, for the energy directly purchased by households: for residential housing (heating and electricity) and (in the case of Melbourne and Beijing) private transportation. The final energy method also measures urban nonresidential energy (for industry and commercial activities, as well a non-private transportation), which the energy I-O method does not cover. Conversely, the energy embodied in the household purchase of goods and services is not covered by final energy method.

The most interesting result lies in the (first ever) quantification of the differences between the two different accounting methods. As expected, the consumption-based accounting method yields much higher energy use for London (+30%)[10] and Beijing (+100%). Conversely, Melbourne's territorial energy use is significantly higher than the energy used directly

and indirectly by its households. This is not because the Melbourne households consume less energy: in fact, in total, they consume almost one-third more than the London households, mainly through private transportation (cars). Instead, it is because Melbourne's nonresidential energy use is almost quadruple that of London's, on a per capita basis. Melbourne is still a major industrial production center, and this industrial activity results in more energy per capita than that of household consumption. London, in contrast, has very little industrial activity, with services dominating its economic activities, and so household consumption is larger than the territorial production-account energy use. The importance of industrial energy use is also illustrated in the case of Beijing. Total secondary energy use (all sectors) in Beijing in 2007 was some 145 GJ/capita (Beijing Government, 2010), i.e., three times larger than the total direct plus embodied estimated household energy use reported in Table 18.6. Taking upstream energy conversion losses into account, the primary energy use (the energy level directly comparable to the other cities in Table 18.6) of Beijing is approximately 200 GJ/capita, i.e., in the same ballpark as Melbourne or London (production accounting). The major difference is that average per capita income in Beijing is with 10,000 *purchasing power parity* dollars (PPP$) per capita, approximately a factor four lower than that of Melbourne and a factor of six lower than London's, suggesting the twin importance of economic structure and efficiency of energy end-use as determinants of urban energy-use levels, with the latter offering a substantial potential for improvement.

The above results confirm that production accounting of energy reflects the economic structure of urban areas, and their role in the international division of labor, whereas consumption accounting energy reflects a mixture of local conditions (climate and transit infrastructure) and expenditure levels (income and lifestyle effects). This exercise also demonstrates the power of applying and comparing different methods at the urban level. By showcasing the differences in production and consumption accounting of energy, the potential role of local policy measures (e.g., transport) versus broader consumption measures (consumption

10 CO$_2$ accounts for London (Hersey et al., 2009) suggest that the differences between the two accounting methods could be 100% (some 45 versus 90 million tonnes CO$_2$ for the production versus the consumption accounting, respectively), a difference that appears very large in view of the results form the energy comparison reported here. I-O techniques are used here (Table 18.3.6) whereas the London CO$_2$ study used a lifecycle assessment approach to estimate consumption based CO$_2$ emissions (but the method and data have not been published), so there might be an additional methodological explanation for the large differences in estimates of the energy and CO$_2$ consumption-accounting 'footprint' of London.

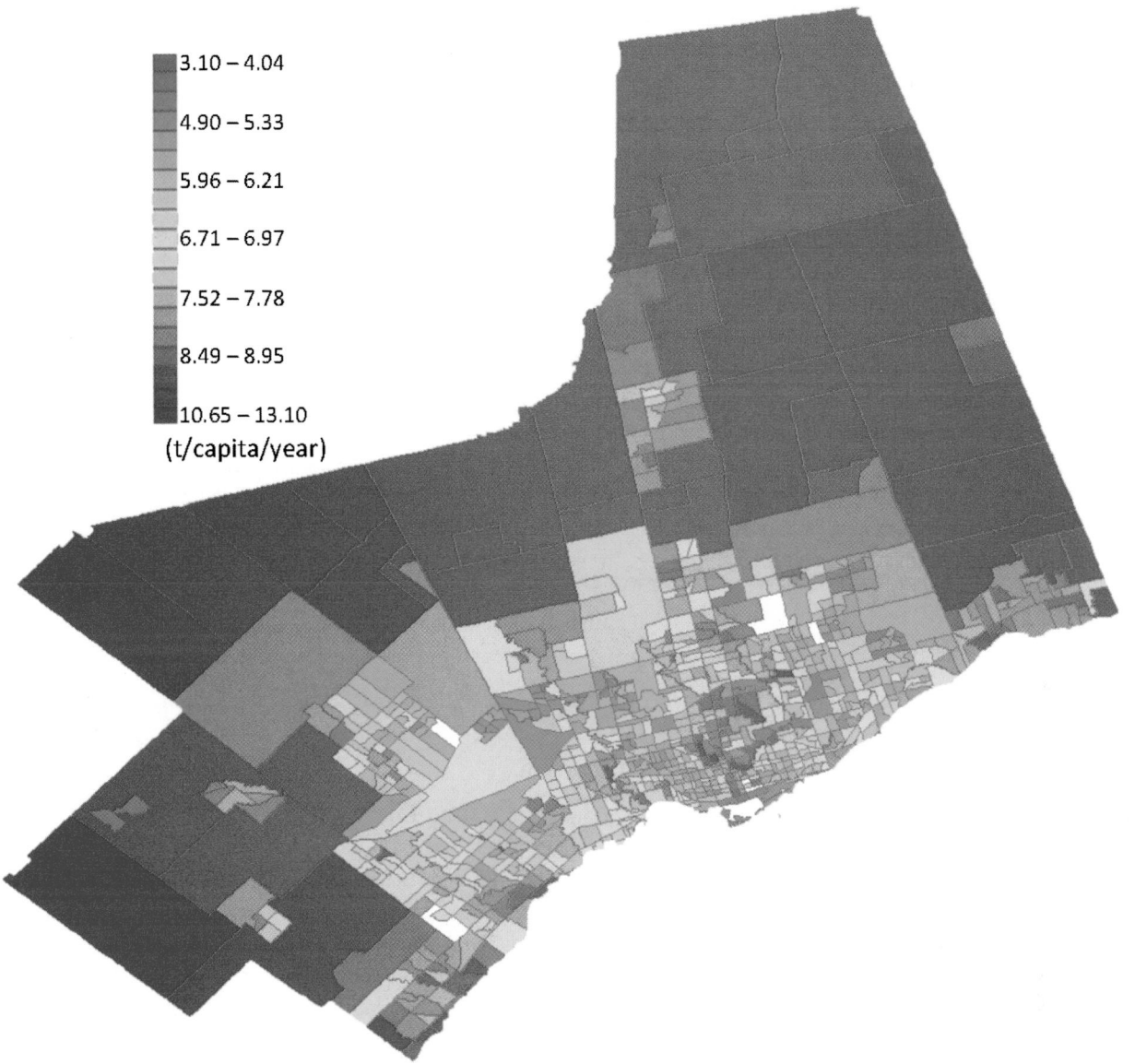

3.10 – 4.04

4.90 – 5.33

5.96 – 6.21

6.71 – 6.97

7.52 – 7.78

8.49 – 8.95

10.65 – 13.10

(t/capita/year)

Figure 18.6 | Total GHG emissions from Toronto (tonnes CO_2-equivalent/capita/year). High-resolution images as well as maps for various energy-demand subcategories (residential, transport, etc.) are available from VandeWeghe and Kennedy, 2007.

reduction, or low energy/emissions supply chain management) can be made explicit.

The comparison of the two methods indicates two policy avenues. Local policy priorities should focus on housing, transit, and industrial energy savings. But these must imperatively be complemented by shifts and reductions in the energy embodied in household purchases of goods and services to avoid that savings at the regional level are offset by increased energy demand from the consumption of goods and services, which occurs somewhere else in the world. Such a policy agenda clearly goes beyond the local level and needs to be addressed on multiple scales.

18.2.4.3   Spatially Explicit Urban Energy Accounts

Spatially explicit energy-use studies may be the key to understanding the influence of urban form and periurban and urban specificities. For Sydney, Lenzen et al. (2004) disaggregated total primary energy use in 14 areas, and followed this up with a GHG emissions atlas of Australia at the postal district level (Dey et al., 2007). For Toronto, VandeWeghe and Kennedy (2007) derived spatially explicit direct-energy use based on transportation and energy-expenditure surveys (see Figure 18.6). Andrews (2008) analyzed direct energy use of several districts in New Jersey, ranging from rural to urban. A comprehensive spatial GHG account, including discussions of uncertainties, was recently completed for the city of Lviv, Ukraine. An innovative study used the Vulcan emissions atlas to compare transportation and building emissions in urban, periurban, and rural counties of the United States (Parshall et al., 2010).

The Sydney and Toronto studies found higher energy and emissions in the outer, low-density suburbs. The US study also found a 'threshold' effect in per capita urban transportation energy compared to more rural counties. In Sydney and Toronto, building fuel use was higher in the center city. The causes for higher building energy in the center city could be the age and quality of the housing stock, the presence of an energy-intensive central business district, and higher incomes in those areas. In the Sydney study, where economic information is available, building energy use is highly correlated with income, but less correlated with population density. In Sydney, central districts tended to have higher incomes than the outer suburbs, a trend absent in the New Jersey study. Since these studies are of industrialized countries and automobile-based urban areas of North America and Australia, their results may not apply to urban areas more generally.

18.2.4.4   Recommendations for Urban Energy Assessments

The indeterminacy in defining 'urban energy' should not be misinterpreted as a flaw in the urban systems perspective. It reflects that the data are approaching the actual final-decision level at which the purpose of the decision, to some degree, can resolve many of the statistical and data ambiguities. For example, administrative boundaries and a production perspective are appropriate system boundaries if the decisions are to be undertaken by local administrations. Final energy-use data remain an essential and useful tool for analysis of energy efficiency and for crafting policies for improved efficiency. Conversely, a 'consumption' perspective on urban energy and GHG use helps to raise awareness that, ultimately, urban energy and GHG management cannot be relegated to an energy optimization task, but equally involve changing lifestyles and consumption patterns.

As urban energy statistics have a vital role in allowing agents to benchmark, it is essential to be sure that the methods are comparable and that 'gaming' is not taking place. Each of the methods described above can produce results that allow benchmarking and comparisons, as long as the method and data sources are described clearly, and consistent data, sectorial definitions, and system boundaries are applied and spelled out clearly. The method should also be as transparent and as open as possible, to guarantee reproducibility and fact checking. Moreover, examples of energy assessments that only account for some sectors are not measurements of urban energy. The sectorial distribution of energy use (residential, commercial, industrial, administrative) as well as purpose (transportation, heating) are essential complementary elements for informed analysis and decision making and should be an integral part of urban energy reporting.

Urban energy and GHG statistics should provide a basis for policy formulation, investment decisions, and further action toward climate

protection. Therefore, it is essential that their origin and data quality is made transparent and methodologies are comparable. Suggestions to improve terminology are provided in this chapter. It is far too common to read 'the energy use of this city is X Joules,' without any qualification what type of energy (final or primary, including upstream or embodied energy) is referred to. There is also a rich field in enhancing the usefulness of urban energy accounts by expanded information on energy quality (e.g., separating heat demand by low, medium, and high temperature regimes), which can form the basis of extended energy efficiency studies, for example in the form of exergy analysis (see Box 18.6).

City energy assessments should also include clear definitions of the system boundary used. Currently, many urban energy assessments, in effect, arbitrarily choose the system boundary to reduce the reported energy use or GHG emissions, for instance by claiming their electricity comes from different sources than the average regional mix, or by excluding certain energy uses that are, nonetheless, central to the very functioning of cities, such as airports or a large tourist population. Arbitrary, or ill-defined, system boundaries defy the very purpose of urban energy assessments: to guide public and private sector policies and decisions and to allow comparability and credibility of the entire process.

18.3   Urban Energy Use

18.3.1   Current Urban Energy Use (Global and Regional)

How large is the urban fraction of global energy use? This seemingly simple question is hard to answer as, contrary to the data for countries, no comprehensive statistical compilation of urban energy use data exists. With 50% of the world population being urban, a range of (largely ballpark) estimates put the urban energy share between two-thirds to three-quarters of global energy use, but such global estimates have, until recently, not been supported by more detailed assessments. This Section reviews the two detailed assessments of urban energy use available to date: the estimate of the IEA published in its 2008 *World Energy Outlook* (IEA, 2008) as well as an estimate developed by a team of researchers at the International Institute for Applied Systems Analysis (IIASA) for this study.

In the absence of detailed, comprehensive urban energy-use statistics, two analytical approaches were pursued to derive global (and regional) urban energy use estimates. One technique, which might be labeled 'upscaling,' uses a limited number of national or regional estimates of urban energy use and then extrapolates these results to the global level. This is the approach followed by the IEA (2008) study that estimated (direct) urban energy use at the primary energy level. The second approach adopts 'downscaling' techniques in which national level statistics are 'downscaled' to the grid-cell level, and then combined with geographic information system (GIS)-based data sets on

Box 18.3 | Urban Energy Data: Measurement and Quality Issues

For urban energy data and assessments two major issues need to be spelled out in a clear and transparent way: (1) system boundaries, and (2) data availability and quality issues.

(1)   Within the discussion of system boundaries two issues need to be considered:

(a)   What is the spatial or functional definition of the urban system under consideration? Does the city definition refer to the core city alone, or does the assessment include the larger metropolitan area? Does the system definition include recognition of bunker fuels[11] (transport fuels used outside of the spatial system boundary, e.g., in national and international territory) or not? Does it consider also the embodied energy associated with the use of material resources and goods other than energy carriers, or not?

(b)   What is the energy system considered? Is primary or final energy reported, and to what extent is a lifecycle perspective for the fuel provision followed (e.g., upstream energy conversion losses and associated emissions, or the costs of exploration, drilling, transporting, and refining fuels before import into the urban system are included or omitted in the analysis)?

(2)   Quality and availability of energy data: are actual statistics used or extrapolated/downscaled data? Does the assessment include noncommercial energy?[12] Which spatial and temporal resolution was considered to calculate the fuel mix for electricity provision? Are differences in technology, efficiency, etc., of power plants and other energy conversion processes recognized?

In an ideal world, urban energy reporting should adopt as wide systems boundaries and complementary accounting frameworks as is reasonably possible and available data allow.

When narrower system boundaries are adopted, a simplified sensitivity analysis of the effects of inclusion of omitted system components can help to put reported numbers into a proper perspective (i.e., complementing final energy accounts with estimates of corresponding primary energy needs, or production-based accounts by estimates of consumption-based accounts based on national I-O tables).

Incomplete reporting (e.g., of only municipal energy use) should be avoided as only a comprehensive sectoral perspective of all urban energy uses can reveal the full potential for policy intervention and assure comparability across different urban energy accounts.

Finally, data disclosure and documentation of assumptions and methods are a 'must.' Particularly, the area of urban GHG inventories is replete with examples of glossy policy briefs that do not allow the reproducibility of the numbers presented (not to mention unreported uncertainty ranges). Transparency and data disclosure are not only key from the perspective of scientific integrity, quality, and reproducibility, but they are also the key for well-informed policy choices. A comparable effort to the standardization of energy and GHG accounts at the national scale along the OECD/IPCC model is long overdue for the urban scale as well.

urban extents is used to derive spatially explicit estimates of urban energy use. These are then aggregated to the national, regional, and global levels. This approach underlies the IIASA study that estimated urban energy use at the level of (direct) final energy, also reported in this section.

In the 2008 *World Energy Outlook* (IEA, 2008), a separate chapter is devoted to urban energy use and contains estimates of 2006 base-year urban primary energy use data and a reference scenario projection to 2030. Detailed urban energy use assessments were first commissioned for a limited number of countries and regions (China,

---

11   Bunker fuels, i.e. energy used for (international) air transport and shipping, can be a substantial fraction of urban energy use. In 'world cities,' like London and New York, air and maritime bunker fuels can account for about one-third of the direct final energy use, suggesting the importance of their inclusion in urban energy accounts.

12   For many cities in developing countries, noncommercial energy forms can account for a substantial fraction of urban energy use (one-third to half). Its reporting is therefore not only key for a comprehensive urban energy accounting, but also yields important information on the potential of fuel substitution with rising urban incomes and hence future energy infrastructure needs.

the United States, the European Union, and Australasia (i.e., Australia and New Zealand)). In these regions urban energy use is estimated to range from 69% (European Union) to 80% (United States) of the primary energy use of these regions, which reflects their high degree of urbanization. For China the estimate is 75%, despite a comparatively lower urbanization rate (41% compared to 81% in the United States), but is explained by the substantially higher urban energy use in Chinese cities compared to the national average because of higher urban incomes and industrial activities. The results of the 'upscaling' of these four regional sets of data to the global level are not reported separately by region by IEA, so only global totals are discussed here.

The IEA (2008) estimates urban primary energy use at the global level to amount to some 330 EJ in the year 2006, or 67% of world primary energy use. Using an average global primary-to-final energy conversion efficiency of about 69%, the estimate translates to 230 EJ urban final energy use worldwide, which is in good agreement with the IIASA study results reported below. Estimates are also provided by major primary energy source and for electricity, assuming that the primary energy mix of cities is the same as at the national or regional average. This assumption is problematic, especially for countries of low-income, low-urbanization, particularly in Asia and Africa, where available data suggest that urban energy use structures are, in fact, very different from rural and national averages. Urban energy use is invariably characterized by much higher shares of grid-dependent energy carriers (electricity and gas) and by much lower reliance on traditional biomass fuels. This simplifying assumption in the IEA (2008) study also diminishes the plausibility of the study's scenario projections by primary energy carrier to 2030, where total urban primary energy use is projected to grow by some 56% from 2006 to 2030. In the IEA reference scenario almost 90% of global energy growth to 2030 is projected to result from urban energy use.

The IIASA study follows a different approach. Drawing on methods and data sets (see Grubler et al., 2007) developed for spatially explicit GHG emission scenarios, the IIASA study used spatially explicit GIS data sets of urban extents, constrained to be consistent with the latest UN WUP statistics (see Section 18.2.2 above) for the year 2005 as initial input. In a subsequent step, national level final energy use data by fuel (traditional biomass and electricity) as well as by end-use activity (primary, light and heavy manufacturing industries, households, and transportation) were downscaled to the grid-cell level in proportion to available spatially explicit activity variables (population, GDP, light luminosity, etc.) under a range of plausible algorithms (hence the study provides central as well as minima/maxima estimates to illustrate uncertainty). The scenarios of individual final energy use categories were then aggregated per individual grid-cell and overlaid with the urban extent map to derive the total estimated (direct) final energy use (including noncommercial traditional biomass fuels) in urban areas. Table 18.7 summarizes the results for the 11 GEA regions and five GEA world regions, as well as for the global total.

Globally, urban final energy use in the IIASA study is estimated to range from 180 to 250 EJ with a central estimate of 240 EJ, or between 56% and 78% (central estimate, 76%) of total final energy. So, in terms of final energy use (as opposed to primary energy use reported in the IEA study), cities use 240 EJ, or some three-quarters, of final energy worldwide. The absolute amounts are in good agreement with the IEA (2008) study discussed above, at least globally.[13] Readers should not be confused by the somewhat higher urban percentage (76%) of urban *final energy* use of the IIASA study when compared to the 67% estimate of the IEA for *primary energy* use. As discussed above, the assumed identity in urban fuel and energy mix with national and/or regional averages in the IEA study underestimates the level of high-quality, processed-energy forms in urban areas that entail correspondingly higher upstream energy-conversion losses. If this simplifying assumption in the IEA calculations could be relaxed, the corresponding urban primary energy estimate would become higher and much closer to the three-quarter benchmark of the IIASA study.

These observations are corroborated by commercial final energy use in urban areas, i.e., excluding traditional biomass use. For industrialized countries, estimates of urban commercial fuel use are identical to the totals reported in Table 18.7. Major differences exist, however, for some developing regions. For sub-Saharan Africa, estimates suggest that 4 EJ, or some 80% of all commercial energy use, can be classified as urban (compared to 8 EJ and 54% for total final energy, see Table 18.7). Differences for South Asia are also noticeable: 8 EJ, or 71% of final commercial energy, are classified as urban, compared to 10 EJ and 51% for total final energy. Differences for the other developing GEA regions are comparatively minor, as little noncommercial energy continues to be used in cities. The higher urban share in commercial energy results both from higher urban incomes and better urban energy access and infrastructure endowments, particularly the much higher degrees of electrification in urban areas.

Nonetheless, despite some uncertainties[14] (see Table 18.7), both the IEA and the IIASA estimates confirm a highly policy-relevant finding: While some 50% of the world's population is urban, *urban energy already dominates global energy use*, which means that the energy sustainability challenges need to be solved predominantly for urban systems.

---

13   The lack of available IEA regional estimates limits the possibilities for a more detailed comparison, but in the reported four IEA regions, urban energy use is within the respective minima/maxima regional values of the IIASA study.

14   The main source of uncertainty for the ranges reported in Table 18.7 is the fuzziness in delineating urban areas and population and hence the attribution of national energy use to the urban category. Conversely, the uncertainty in energy statistics is comparatively small, with the main uncertainty source being the lack of reliable data on urban noncommercial (traditional biomass) energy use, particularly in Africa.

Table 18.7 | Estimates of urban (direct) final energy use (including traditional biomass) for the GEA regions and the world in 2005 (in EJ and % of total final energy). See text for a discussion of urban commercial energy use and its corresponding (somewhat higher) urban share.

GEA Regions		Central estimate	%	min	%	max	%
NAM	North America	63	86%	51	69%	64	87%
PAO	Pacific OECD	14	78%	11	59%	16	92%
WEU	Western Europe	40	81%	31	64%	41	83%
EEU	Eastern Europe	6	72%	4	51%	6	72%
FSU	Former USSR	20	78%	14	54%	20	78%
AFR	Sub-Saharan Africa	8	54%	5	35%	10	71%
LAM	Latin America	17	85%	16	77%	18	89%
MEA	North Africa & Middle East	15	84%	10	58%	15	86%
CPA	China & Central Pacific Asia	32	65%	19	40%	31	65%
PAS	Pacific Asia	15	75%	10	51%	16	77%
SAS	South Asia	10	51%	5	29%	10	51%
OECD90	NAM+PAO+WEU	117	83%	92	66%	121	86%
REF	EEU+FSU	26	76%	18	54%	25	76%
MAF	AFR+MEA	23	71%	15	47%	25	79%
LAC	LAM	17	85%	16	77%	18	89%
ASIA	CPA+PAS+SAS	57	64%	35	40%	57	64%
WORLD		**240**	**76%**	**176**	**56%**	**246**	**78%**

18.3.2 GEA City Energy Data and Analysis

18.3.2.1 The GEA City Energy Data Base

An effort to compile a database with literature values of energy use on the urban scale was conducted as part of this assessment to improve understanding of the variation in energy demand of urban areas (for an example of such analyses, see Steemers, 2003).[15] The study, therefore, chose a cross-sectorial approach to compare as large a number of urban areas from as wide range of regional settings, geographies, sizes, and functions as possible, with minimal definitional constraints with respect to urban system boundaries so as to maximize data availability. In terms of energy-use, data are reported at the level of *total (direct) final energy use*, as this level of analysis creates the least ambiguity in terms of energy accounting and is also the indicator most widely available and comparable among case studies. (Accounting for primary energy equivalents or GHG emissions requires assumptions on boundary definitions, conversion factors, and efficiencies, etc., which introduce additional uncertainties in the comparisons.) Given the extreme paucity of consumption-based estimates of urban energy use (e.g., via I-O techniques), the decision to focus the database on a production approach was also straightforward.

18.3.2.2 Data Coverage

Three categories of urban statistical data were brought together in the GEA City Energy Data Base from a variety of sources: population statistics (UN, 2008), energy statistics (e.g., Dhakal, 2009; Kennedy et al., 2009; Kennedy et al., 2010), and economic statistics on gross regional economic output (or GRP, which is the urban-scale equivalent of national GDP) converted into a common 2005 denominator in purchasing power parity (PPP expressed in International\$ – Int$_{2005}$\$) terms, including Eurostat (2008) and PriceWaterhouseCoopers (2007). While population statistics are routinely collected at various levels of spatial resolution, this is rarely the case for both economic and energy data. Coherent data sets were, nonetheless, found for 200 urban units, of which 132 were from UNFCC Annex 1 (i.e., industrialized) countries and 68 were urban areas located in non-Annex-1 (i.e., developing) countries. Details on data-source limitations, as well as further statistical analyses, are reported in a GEA Chapter 18 Working Paper (Schulz, 2010a).

18.3.2.3 Analysis

Comparisons of urban scale and national scale data
This section compares data on energy use per capita, (urban) per capita income (GRP/GDP), and energy intensity of GRP/GDP at the urban scale with their national scale metrics.

---

15 A more complete GEA Chapter 18 working paper on the GEA city energy data base and its data analysis is posted on the GEA website, www.globalenergyassessment.org.

Table 18.8 | Comparison of per capita urban final energy (GJ/capita), GDP (1000 Int$_{2005}$$/capita) and energy intensity (MJ/Int$) statistics (number of observations and indicator values) at the urban and national levels. Data cover 200 urban areas, of which 132 were located in Annex-1 countries. Reported data refer approximately to the year 2000, albeit different city studies report different base years. Average and standard deviations (SD) are presented for three sample groups: 'lower,' all those cities in which urban indicators are below the respective national averages; 'higher,' indicators are higher than the national average; and 'Total,' indicators for all cities in the sample taken together.

| | | count (# of cities) higher/lower than national average | | | statistical values in GEA city energy data base | | | | | |
| | | | | | average | SD | average | SD | average | SD |
		Energy/cap.	GDP/cap.	Energy/GDP	GJ/cap.		GDP/cap.		MJ/$	
World	lower	134	94	124	85.5	30.9	29863	12629	3.1	1.1
	higher	66	106	76	126.4	100	17075	11685	9.6	1.3
	total	200	200	200	99	65.4	25643	13695	5.3	5.2
UNFCC Annex-1 countries	lower	107	67	93	93.5	23.2	33093	10274	3	0.9
	higher	25	65	39	171.9	132.3	29018	9677	6.4	4.6
	total	132	132	132	107.6	66.3	32360	10256	3.6	2.5
Non Annex-1 countries	lower	27	27	31	40.3	31	11453	8329	3.9	1.9
	higher	41	41	37	98.7	60.5	9792	4711	11.5	8
	total	68	68	68	79.5	59.2	10337	6114	9	7.5
Asia w/o Japan	lower	13	15	18	55.2	41.1	14538	14112	4	1.9
	higher	37	35	32	87.9	37.8	9938	4905	10.1	5.6
	total	50	50	50	83.3	41.3	10580	6851	9.3	5.6
Latin America, Middle East, Africa	lower	14	12	13	42.8	36.1	10121	4146	4.7	3.8
	higher	4	6	5	198.5	130.1	8446	2127	24.5	15.2
	total	18	18	18	70.5	89.7	9757	3933	8.4	11

Table 18.8 presents the overall results and a regional breakdown by status regarding Annex-1 versus non-Annex-1 assignation and geographic regions. Demographic data at the national scale are derived from UNDESA (2008; 2010), national energy statistics from IEA energy balances (IEA, 2010a; 2010b), and economic data from the International Monetary Fund (IMF, 2010).

Per capita energy use

An initial observation is that almost two out of three urban areas have a *lower than national average (direct) final energy use on a per capita basis*. This trend is more pronounced (107 out of 132) among Annex-I countries, which are overrepresented in this sample (132 out of 200). The main reason for this is the effect of urban economic structures (a higher share of less energy-intensive service activities compared to national averages) and, to a lesser degree, the effect of urban density on lower transport energy use (more public transport and soft mobility modes compared to national averages that reflect rural automobile dependence).

In non-Annex-I urban areas the reverse pattern is observed, with more than two out of three urban areas having *higher per capita final energy use* compared to their respective national averages. Among non-Annex-I countries there is pronounced *regional heterogeneity*: Africa and Latin America share the prevalence of *lower than national average urban per capita final energy use*, in contrast to Asia where urban per capita final energy use is predominantly (37 out of 50 cases) *higher than*

the national average. These patterns reflect primarily the different influences of income with the urban-rural income differential, being particularly pronounced in Asia as reflected in a corresponding higher per capita final energy use gradient.

To explain these differences requires further analysis, but preliminary findings suggest that differences in levels of incomes and in economic structure (degree of service versus industry orientation of urban economies) are likely to be the main explanatory variables. In general, the number of observations in rapidly growing economies of non-OECD Asia is much larger in the sample than those of Latin America and Africa (50 vs. 18), illustrating the need for improved energy information in urban settlements in these regions particularly.

Figures 18.7 and 18.8 summarize this statistical analysis, showing all the city observations as a cumulative plot (over population) sorted by decreasing per capita final energy use. The color code indicates whether a city is above (red) or below (blue) its respective national average. The inverse per capita energy use pattern of cities in Annex-I versus non-Annex-I countries is clear from this comparison. On average, the *lower energy-use cities in Annex-I countries have a final energy use that is one-third lower than the Annex I national average*. For non-Annex-I countries the relationship is inverse: *most non-Annex-I cities have higher (about twice) per capita final energy use* than their respective national averages, being in the same ballpark as the lower energy use city sample in the Annex-I countries (at some 100 GJ/capita).

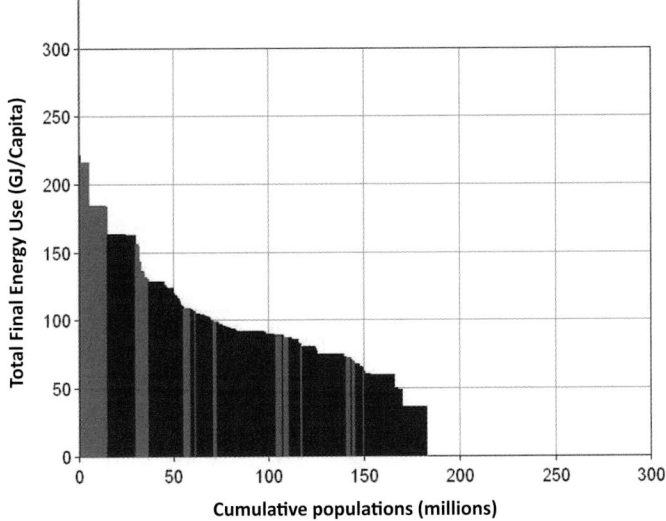

Figure 18.7 | Per capita (direct) final energy use (GJ) versus cumulative population (millions) in urban areas (n =132) of Annex-1 (industrialized) countries. Red indicates urban areas with per capita TFC *above* the national average. Blue indicates per capita final energy *below* the national average.

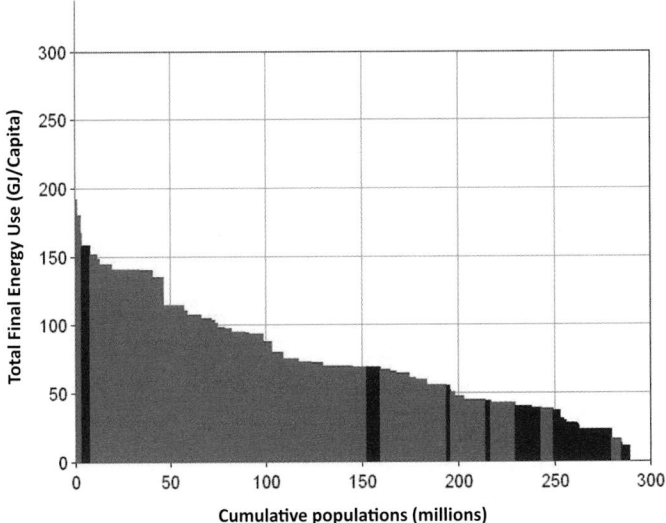

Figure 18.8 | Per capita (direct) final energy use (GJ) versus cumulative population (millions) in urban areas (n = 68) of non-Annex-1 (developing) countries. Red indicates urban areas with per capita TFC *above* the national average. Blue indicates per capita final energy *below* the national average.

These conclusions only refer to the (direct) final energy use metric adopted for the comparative analysis of our sample of 200 urban areas.

Evidence suggests that for Annex-I country cities the lower final energy use is likely to hold only for the production-accounting approach adopted for this comparison. Adding 'embodied' energy use (corrected for net trade of imports and exports of manufactured goods and services from and to urban economies) is likely to weaken the conclusion of a

lower urban energy footprint in cities of Annex-I countries compared to the national average (see Section 18.2.4 above) as lower (direct) final urban energy use is likely to be (largely) compensated by higher 'embodied energy' use associated with higher urban incomes. And yet, the lower (direct) final energy use of many urban compared to rural areas in Annex-I countries illustrates well the *urban comparative advantage for a sustainability transition*: urban areas with their corresponding more energy-efficient compact settlement structures and lesser (energy-intensive) automobile dependence and greater potential for efficiency improvements through energy 'recycling' (i.e., cogeneration and heat cascading) have larger efficiency-leverage potentials compared to those of rural areas. The challenge to reduce the energy and environmental footprint from (over)consumption (i.e., embodied energy) is not unique to urban dwellers as it applies equally to rural ones in Annex-I countries.

The situation of cities in non-Annex-I countries, particularly in Asia, is markedly different. Compared to rural areas, cities not only have higher (direct) final energy use, they also have generally much higher incomes. Thus, the urban-rural gradient in terms of per capita (direct) final energy use is amplified yet more by higher urban incomes, which further increases the rural-urban energy gradient when considering the 'embodied' energy use associated with consumption. Given the dynamics of urbanization trends (see Section 18.2.1) it is thus fair to conclude that the sustainability 'hot spot' in the decades to come will reside particularly in the rapidly growing cities of non-Annex-I countries, especially in Asia.

Per capita income

Regarding per capita income the data sample reveals much more heterogeneity than the popular conceptions of invariably rich urbanites.[16] Almost half of the urban areas in our sample had per capita GRP/GDP values below the national average. Again, the patterns diverge between Annex-I countries (where this trend is driven by the large number of relatively deprived smaller UK urban centers in the data sample, but also by such prominent examples like the capital of Germany, Berlin) and non-Annex-I countries.

In non-Annex-I urban areas the majority showed above national average per capita GRP/GDP values. In Asia, two out of three urban areas had GRP/GDP above the national per capita average. In Africa and Latin America, just over one-third of the urban areas had GRP/GDP values that exceeded the national per capita average, but two-thirds ranked below it.

Energy Intensity

Regarding energy intensity of urban GRP/GDP, the majority of urban settlements studied showed lower than national average energy intensities,

---

16   GRP data is provided only by a limited number of statistical offices or other sources. They differ in methodology and are not always strictly comparable. A more detailed discussion of economic measurement issues at the urban scale is beyond the scope of this energy assessment.

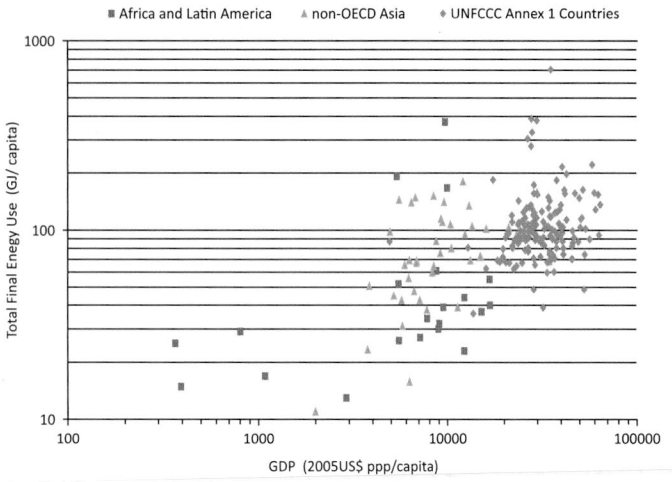

Figure 18.9 | Comparison of urban total final use (TFC) and urban income (GRP/GDP at PPP in Int$_{2005}$\$) per capita for cities in Annex-I and non-Annex-I countries.

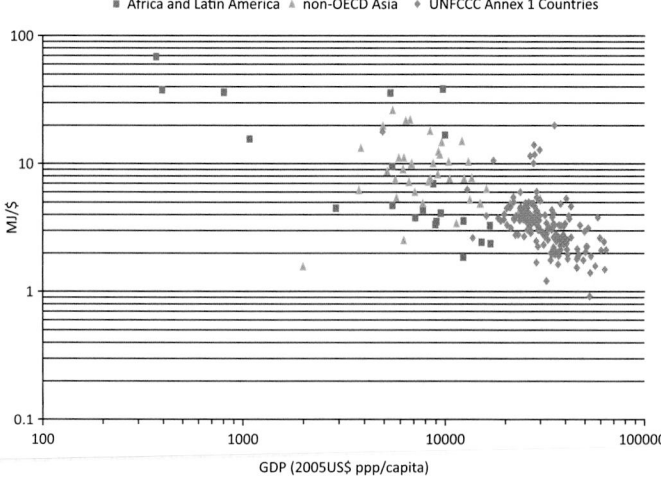

Figure 18.10 | Comparison of urban energy intensity and urban per capita income (at purchasing power parities) for cities in Annex-I and non-Annex-I countries.

which indicates the dominance of less energy-intensive tertiary sector activities in urban areas. In Annex-I countries more than two out of three settlements show lower than national level energy intensities of GDP. In the non-Annex-I countries the general trend is almost balanced, with just a few more cases of urban energy intensity that exceed the national average values. Again, the Asian urban areas show a very different pattern to those from the Latin America, Middle East, and Africa regions. Three out of four urban areas in non-OECD Asia have energy intensities that exceed the national average, while four out of five of the African, Middle East, and Latin American urban cases have the same pattern as OECD countries, with urban area energy intensities below their respective national average.

General observations

For the non-Annex-I urban areas at least three general patterns of energy use can be discerned.

One is the lower end, with final energy use under 30 GJ/capita. Per capita income is mostly less than 5000 PPP\$/capita and energy intensity is, in many cases, also quite low, below 5 MJ/\$. This low energy intensity in low-income, non-Annex-I cities does not necessarily suggest highly efficient energy systems, but rather different consumption structures (particularly lower private transport energy use). In all likelihood, the low energy intensities may also reflect an underreporting of noncommercial, traditional biofuels used by low-income urban households.

The medium range of per capita final energy use in non-Annex-I cities is from 30–100 GJ/capita and coincides with a wide range of incomes and energy intensities.

Heavy industrial urban areas show yet higher per capita final energy use of up to 350 GJ/year and over a highly variable range of income levels. In practically all urban settlements of the third group of non-Annex-I countries energy intensity is above 10 MJ/\$ (up to 39 MJ/\$ in the sample).

Patterns for Annex-I cities are markedly different. The Annex-I city panel in general appears more coherent in final energy use patterns. First, the correlation between higher urban incomes and higher final energy use tends to weaken significantly, with richer cities not necessarily using more (direct) final energy on a per capita basis (but highly likely to use much more embodied energy compared to poorer cities). Second, there is a strong and inverse correlation between urban incomes and energy intensity, with the latter falling with rising urban incomes. Only three out of the 118 Annex-I urban areas show energy-intensity values above 10 MJ/\$ and the vast majority are below 5 MJ/\$.

Variable correlations

Figure 18.9 presents the overall positive correlation between per capita incomes and (direct) final energy use. The general positive correlation, familiar from national level analyses, is also found at the urban scale. However, the variation appears to be larger at the urban level than at the national scale, which suggests a much broader spread for path-dependent urban development trajectories. Also, within the panel of Annex-I countries there is a large variation in energy use per capita (with some urban areas using more than 600 GJ/capita final energy). A proposed 'turning' (or saturation) point cannot be identified at a statistical significant level in this data set, which covers GRP/GDP ranges up to about 80,000 Int$_{2005}$\$/capita (in PPP terms), despite a visible weakening of the income-energy use link for high-income cities.

Figure 18.10 presents the relation of GRP per capita and energy intensity of GRP. Trends in non-OECD Asian cities come closest to the often proposed 'hill' (Goldemberg, 1991), with a peak in energy intensity at about 10,000 PPP\$/capita and a pronounced decline in energy intensity at higher per capita incomes. At lower income ranges, however, the data in our sample are sparse, and often also exclude the dominant noncommercial traditional biofuels, so the above findings are consistent with the observation of a 'hill' in the development of *commercial* energy intensity (Goldemberg, 1991) against a background of continuously falling *total* (including noncommercial) energy intensities with rising incomes (see Nakicenovic et al., 1998) as evident in the cities of Africa and Latin America in Figure 18.10.

18.4   The Urban Energy Challenges

The urban energy challenges are embedded within overall social, economic, and environmental development challenges and their numerous interdependencies and linkages. This energy assessment focuses on the interdependencies that bear directly on urban energy. The linkages between development and energy are most straightforward in the area of the literature on 'energy poverty,' energy access, and adequate housing and transport access for the urban poor. Hence the discussion begins with a discussion of energy access and poverty within an urban context (Section 18.4.1). The discussion then moves on to the nexus of urban energy use and urban environmental quality and the challenges imposed by the high densities of energy demand and the corresponding need for efficient and low-emission energy systems (Section 18.4.2). Lastly, Section 18.4.3 discusses the challenges for urban energy infrastructures, including reliability and security.

Other urban development challenges with more indirect implications on energy, such as urban transport, land-use, and density planning, are discussed Section 18.5.

18.4.1   Energy Access and Housing for the Urban Poor

18.4.1.1   Introduction

In the development literature, energy is not generally recognized as one of the basic needs (Pachauri et al., 2004), although it is in discussions of poverty in high-income nations (where it is referred to as 'energy poverty,' see Boardman, 1993; Buzar, 2006). One reason for this absence is that one of the main indicators of 'energy poverty' in high-income nations is the substantial proportion of household income spent on fuels and electricity (typically more than 10%). However, this is not an appropriate measure for much of the urban population in low-income and some middle-income nations because their incomes are so low in relation to the costs of food and necessities other than food that their energy use is very low. This is both in the energy used within their homes (lighting, cooking, and, where relevant, space heating and appliances), in the energy implications of the transport modes they use, and, for those who are self-employed, in the energy used in their livelihoods. Thus, the main indicator for their 'energy poverty' is in the inadequacies of the energy they can afford and in the poor quality of the energy sources they use (Boardman, 1993; Buzar, 2006) which in this assessment is referred to as "energy access." Such individuals or households also have so few consumer goods that their individual embodied energy is also low.

Thus, for nations with a proportion of fuel use from noncommercial fuels and where low-income urban households keep energy expenditures down by using dirty fuels (including wastes) or cutting fuel use, the proportion of income spent on energy is not a good indicator of poverty. In addition, an analysis of poverty in relation to energy should also consider the time and effort used to obtain needed fuels, the health implications (including those that arise from indoor air pollution and the risks of fire and burns), and the quality and convenience of the fuels used to meet daily needs (i.e., in space heating or cooling, and for hot water). Pachauri et al. (2004) suggest that, ideally, the analysis of the adequacy of energy should include primary energy use, end-use energy (especially electricity), useful energy (e.g., whether the primary or end-use energy delivers the energy needed), and the quality and adequacy of energy services for households (including transport). However, data are often only available for the first two of these. Moving out of poverty involves shifts away from the dirtier and less convenient fuels[17] and obtaining access to electricity, as well as keeping down total monetary expenditures on energy.[18]

Thus, the two most common implications of poverty in regard to energy use among urban populations in low- and most middle-income nations are, first, use of the cheapest fuels and energy-using equipment (including stoves, which bring disadvantages, especially in regard to indoor air pollution, inefficient fuel combustion, and convenience) and, second, no access to electricity. Low-income households may also limit the number of meals (in extreme circumstances to one a day) to save money both on food and cooking fuel. Poverty is also evident in the lack of space heating within cold climates or seasons – although this is difficult to measure as expenditure surveys cannot identify what consumers forgo. Owners of home-based enterprises often make significant energy purchases. Urban poor households often face much higher risks of burns and scalds for household members (especially children) and of accidental fires, underpinned by a combination of extreme overcrowding (often three or more people to each room), unsafe fires or stoves, the absence of electricity for lighting (candles and kerosene lights are used), housing built of flammable materials, high-density settlements, a lack of firebreaks, and no emergency services, including fire services (Hardoy et al., 2001; Pelling and Wisner, 2009). All the above are also often associated with homes and livelihoods in informal settlements – which helps explain the lack of electricity (with electricity utilities unwilling or not allowed to operate there), the poor-quality housing, and the lack of provision for fire-prevention and emergency services.

---

17   This includes a shift to 'clean' fuels – clean in the sense of minimizing raw pollution and health impacts for the users – for instance, with electricity and gas or energy derived from renewable energy sources being 'clean' and coal and raw biomass being 'dirty' (how dirty these are depends on the technology used in the home). Kerosene and charcoal fall between these two extremes. The term 'clean fuels' is ambiguous in that it is used to mean different things – for instance, for fuels or energy sources that have low or no CO_2 emissions rather than lower health impacts for users. In addition, electricity at the point of use may be 'clean,' but it often comes from coal-fired power stations that have high CO_2 emissions and often high levels of pollution.

18   For a more detailed discussion see also Chapter 4.

Table 18.9 | The housing submarkets used by low-income urban dwellers and their energy-use implications.

Housing type	Energy implications in the home	Energy implications for transport
Rooms rented in tenements	Typically one room per household; often electricity available, but usually too expensive to use for cooking and space heating	Usually close to sources of livelihood or demand for casual work (hence this type of accommodation is in demand)
Cheap boarding houses/dormitories (including hot beds)	Very low energy use; no provision for cooking?	As above
Informal settlement 1: squatter settlements (in many cities these house 30–60% of the entire population)	In low-income nations, usually reliance on dirtier fuels and lack of electricity; in many middle-income nations less so; for many households, part of fuel/electricity expenditure is for livelihoods in the home; illegal electricity connections may be common; often high risks from accidental fires	Many in peripheral locations, which implies high transport costs in time and money; better located squatter settlements often become expensive through informal rental or sale
Informal settlement 2: housing in illegal subdivision	More expensive than illegal land occupation, but less at risk from eviction and often with more provision of infrastructure (including electricity) or at least more possibilities of provision as the land occupation is not illegal	Many in peripheral locations which implies high transport costs; in large cities, the cheapest illegal subdivisions can imply several hours traveling a day to and from sources of income
Accommodation at the workplace	Common for single men in some cities; extent not known and includes apprentices	
Pavement dwellers and those who sleep in open or public spaces	Very low incomes so very low fuel use	Walk to work

Source: Hardoy and Satterthwaite, 1989; Yapi-Diahou, 1995; Harms, 1997; Mitlin, 1997; Mwangi, 1997; Bhan, 2009.

However, it is important to also consider the cost burdens of energy to low-income households who have access to electricity and who use cleaner, more convenient fuels. For instance, in cities such as Mumbai (India), low-income households who move from informal to formal housing benefit from access to electricity, but often find it difficult to pay the bills. Here, there are more parallels with what the literature refers to as 'energy poverty' in high-income nations. Buzar (2006) notes that increasing numbers of households in former communist states in Eastern and Central Europe[19] face difficulties in affording energy, in part because of significant energy-price increases as subsidies are removed, and in part because of the failure of the state to develop safety nets to protect low-income groups. This leaves many families with no option but to cut back on energy purchases, a problem further aggravated by cold climates and the poor energy efficiency of the building stock.

18.4.1.2   Housing Quality and Location

Around 800 million urban dwellers in low- and middle-income nations live in poor-quality, overcrowded housing with inadequate provision for basic services (UN HABITAT, 2003; 2008). A taxonomy of their housing submarkets with associated energy implications is given in Table 18.9.

Low-income groups in urban areas face limited choices in renting, buying, or building accommodation that they can afford and so have to make trade-offs between a good location, housing size and quality, infrastructure and service provision, and secure tenure (see references

Table 18.9). Good locations in relation to income-earning opportunities mean that transport expenditures can be kept down and more central locations usually have more possibilities of infrastructure and service provision. But they are also more expensive and generally have less possibility of space and of keeping down housing costs through illegally occupying land and self-built homes. At their most extreme, to obtain central locations, low-income groups live in shacks built on pavements or waste dumps or in small rooms with more than three people to a room or share beds (so a single person pays to sleep in a bed in a shared dormitory with each bed serving two or three people over a 24-hour period, known as hot beds).

One of the most extreme examples of this are the tens of thousands of pavement dwellers in Mumbai, where the choice to live on the pavement (and usually with low lean-to shacks too small to sleep in) is from a combination of their very low incomes, the central location of where they earn their incomes (they walk to work), and the impossibility of affording transport costs from less central locations (SPARC, 1990). Another example are households in Indore (India) who choose to live on land sites adjacent to small rivers that flood regularly. These have economic advantages because they are close to jobs or to markets for the goods the households produce or collect (many earn a living collecting waste). The land is cheap and, because it is public land, the residents are less likely to be evicted. These sites have social advantages because they are close to health services, schools, electricity, and water, and there are strong family, kinship, and community ties with other inhabitants (Stephens et al., 1996).

Rented accommodation or land on which houses can be built is cheaper in more peripheral locations and often more distant from income-earning opportunities – the cheaper the cost and the greater the possibility of

---

19 This section does not cover high-income nations and low- and middle-income nations that were formerly part of COMECON (termed countries undergoing economic reform in GEA).

Table 18.10 | The proportion and number of the urban population that lacks electricity and access to 'modern fuels' in developing countries, least-developed countries, and sub-Saharan Africa.

Percentage and number of the urban population	Developing countries	Least-developed countries	Sub-Saharan Africa
Lacking access to electricity	10% (226 million)	56% (116 million)	46% (124 million)
Lacking access to modern fuels	30% (679 million)	63% (130 million)	58% (156 million)

Source: UNDP and WHO (2009).[20] Statistics on the urban population are drawn from UN Population Division (UN DESA, 2008) and are for 2005. The dates for the statistics on access to electricity and modern fuels vary by country, with most being between 2003 and 2007.

building a home illegally (and so avoid paying a full rent, which is often among the main reasons why distant informal settlements develop). But this means high time- and monetary-transport costs, and it is difficult to establish the high transport costs for those living in peripheral locations because most of the data on the proportion of income spent on transport are averages for cities. In addition, it is likely that many household surveys under-represent those who live in informal settlements – for instance, a lack of formal addresses and maps makes it difficult to include their inhabitants in surveys or those responsible for collecting data fear to work in informal settlements (for an example of this, see Sabry, 2009). Peripheral locations also constrain the inhabitants' access to economic opportunities, as many locations are too distant or too expensive to commute to.

18.4.1.3 Urban Populations and Energy use in Low- and Middle-Income Nations

There are some general statistics on the forms of energy use for urban populations – for instance, in what fuels (and mix of fuels) are used and whether or not they have access to electricity (Table 18.10). However, there are no general statistics on how fuel use and access to electricity vary within nations' urban populations or within cities by income group. In part, this is because many 'energy' statistics for individuals or households are only available for national populations. Where these are disaggregated, it is often only as averages for 'urban populations' when there are very large differences between different urban centers and between different income groups within each urban center. In part, this is because the documentation of 'energy' provision deficiencies has not been given the same level of attention as, say, deficiencies in provision for water and sanitation. The only exception is the very considerable documentation on the health impacts of pollution from the use of 'dirty' fuels (and other factors, including poor ventilation and inefficient stoves), although much of this literature is for rural households and perhaps underestimates the extent of this problem among urban poor households.

Table 18.10 highlights the very considerable proportion of the urban population in low- and middle-income nations that lacks access to modern fuels – 30% (which implies close to 700 million urban dwellers). A higher proportion has access to electricity, but about half the urban population within the least-developed nations and within sub-Saharan Africa lack access to electricity. In sub-Saharan Africa alone, this implies that around 120 million urban dwellers lack access to electricity. For all developing countries taken together some 230 million urban residents lack such access.

Particular studies suggest that it is common for urban poor households in Africa and Asia to use a mix of fuels – for instance, different fuels for different kinds of food and fuel-switching at certain times of year when fuel prices or household incomes change (see Pachauri and Jiang, 2008 for China and Meikle and North, 2005 for Arusha). Regional and seasonal differences may be significant, and households are also influenced by subsidies and incentives, fuel availability, and cultural preferences. Policymakers rarely understand these complexities.

The available data on energy use by low-income urban dwellers range from very large numbers who use little or no fossil fuels and electricity (i.e., wood, dung, straw, and charcoal) through those who use kerosene and coal or coal-based fuels to those who use gas (bottled or piped) and electricity. For electricity, in some nations almost all urban households (including low-income households) have electricity and in others only a very small proportion of urban households have electricity.

Available studies also give examples of the scale of the differentials in energy used between high-income and low-income households within particular urban centers; some show that these can vary by a factor of 10 or more, but of course the scale of the differentials depends, in part, on how 'the urban population' is divided for this comparison (e.g., differentials will be greater if the richest and poorest deciles are compared instead of the richest and poorest quartiles).

The two nations with the world's largest urban populations are China and India (by 2010 these accounted for more than a quarter of the world's urban population). In India, fossil-fuel based energy sources increasingly dominate the energy mix of urban households, although biomass (including firewood and dung) continues to be used, especially by the lowest income groups. In addition, during the mid-1990s there was an evident rise in the use of liquefied petroleum gas (LPG) and

---

20 This source is inconsistent in how it reports some of the figures for access to electricity; the figures above for the least-developed countries and sub-Saharan Africa are from Figure 3, but the accompanying text (page 12) says that 46% of the urban population of least-developed countries and 56% of urban dwellers in sub-Saharan Africa lack electricity access. The report does not specify where its population figures come from, although it lists the UN Population Division's *World Population Prospects: the 2006 revision* in its sources.

electricity among urban households (Pachauri and Jiang, 2008). In China, among urban households there has been a shift away from the direct use of coal to gas and electricity (although coal is still important for a significant proportion of urban households). Energy use among urban households declined from 1985 to 2002 (from 9 GJ/capita to around 5 GJ/capita), because of a shift to more-efficient fuels (Pachauri and Jiang, 2008). Also, dependence on coal may not be reduced if coal-fired power stations are an important part of meeting the consequent rising demand for electricity.

Table 18.10 shows how most (90%) of the urban population in 'developing countries' had access to electricity and 70% had access to modern fuels (mostly gas) in 2007 – but also how the picture on energy access for urban populations was very different for the least-developed countries and for sub-Saharan Africa. When considering fuel use among urban poor households in all low- and middle-income nations, this varies from, at one extreme, continued reliance on fuelwood, charcoal, and waste materials (and no electricity) through to greater use of solid or liquid fossil fuels (coal and/or kerosene, often called transition fuels) and a proportion of households with electricity, and on to the use of cleaner fuels (LPG or connection to gas) and electricity.

This diversity in the forms of energy used is also likely to be present between urban poor households within most (but see above). The only obvious characteristics that all urban poor households share is limited purchasing power for energy (for all uses) and a desire to keep costs down, so their fuel use and fuel-energy mix is much influenced by the price and availability of different fuels and electricity. Having access to electricity at prices that low-income households can afford obviously represents a major advantage – for lighting and for key appliances (including fridges and, where needed, fans) and for the reduced risk of accidental fires. However, they will keep electricity use down (for instance, where it is expensive in comparison to other fuels, they may not use it for cooking or space heating) unless there is no better alternative (or they have made illegal connections to power grids that keep costs down).[21] Having gas for cooking and hot water (and, where needed, space heating) has great advantages of convenience and of low generation of indoor air pollution, but in many urban contexts it is only available as LPG canisters (and so less convenient and more expensive than gas piped to the home). This often makes it too expensive for large sections of the urban population.

Among the low-income households in urban centers in the lowest-income nations, fuel use is dominated by charcoal, firewood, or organic wastes (e.g., dung). The more access to fuels is commercialized, the less fuel is used. In many small urban centers in low-income nations, it may be that certain fuels (wood, dung, agricultural wastes, etc.) are cheap and that a proportion of the urban population can gather fuel rather

---

[21] Care is needed here: illegal connections are often not providing electricity free as the connection is through another household that charges for this or the occupants are tenants and have to pay the landlord extra for electricity.

Table 18.11 | The main fuels used for cooking in urban areas in developing countries, least-developed countries, and sub-Saharan Africa (in percent of urban population using particular fuels).

Percentage of the urban population	Developing countries	Least-developed countries	Sub-Saharan Africa
Using wood, charcoal, and dung for cooking fuels	18	68	54
Using coal for cooking	8	3	2
Using kerosene for cooking	7	4	20
Using gas for cooking	57	20	11
Using electricity for cooking	6	4	11

Source: UNDP and WHO, 2009.

than pay for it – but probably the larger the city, the greater the commercialization of all fuels. Also, the very limited space within the homes of most low-income urban households – especially those that live in central areas with, in many cases, less than 1 m²/person – limits the capacity to store bulky solid fuels.

Fuel use for cooking

Table 18.11 shows the contrast between the proportion of the urban population using wood, charcoal, and dung in developing countries (less than one-fifth of households), in the least developed nations (two-thirds of urban households), and in sub-Saharan Africa (more than a half). In developing countries close to two-thirds of the urban population use gas or electricity for cooking; for the least-developed nations and sub-Saharan Africa, this is less than one-quarter. There are large differences in this within the least-developed nations and in sub-Saharan Africa. For instance, for many of these nations only a small percentage of the urban population has access to electricity.

For most nations with per capita GDPs under $1100, 85% or more of their urban population use wood and charcoal for cooking – and all these nations are in sub-Saharan Africa, except Haiti. For nations with per capita GDPs above $14,000 virtually all urban households do not use wood or charcoal. For nations with per capita GDPs of $1100–4000, the variations in the percentage of the urban population that use wood or charcoal are very large (UNDP and WHO, 2009).

Households select fuels for food preparation for reasons that include cost, availability, convenience, type of food, and cooking equipment, as illustrated by a study in Ibadan (Nigeria). Kerosene was the major cooking fuel for low- and middle-income households until subsidies on petroleum products were withdrawn in 1986. As a result of the increased kerosene and cooking-gas prices, surveyed households in 1993 had begun to use fuelwood, sawdust, and other cheaper energy sources. A follow-up in 1999 discovered that households had switched back to kerosene, while also reducing the frequency of cooking, eating cold leftovers, and substituting less nutritious but faster-cooking foods (Adelekan and Jerome, 2006). A study of energy use in an informal settlement in the Cape Peninsular in South Africa showed how households that had legal electricity connections and meters could access 50 kWh/month free basic electricity, which encouraged them to cook with electricity rather than paraffin (Cowan, 2008).

Low-income urban households often cook with solid fuels that pose serious health threats to household members from indoor air pollution, especially for those with the longest exposure (see Chapter 4 for details). Among urban populations in many sub-Saharan African nations, wood and charcoal are still the most widely used cooking fuels (see, for instance, Ouedraogo, 2006 on Ouagadougou; Boadi and Kuitunen, 2005 on Accra; Kyokutamba, 2004 on Uganda; and van der Plas and Abdel-Hamid, 2005 on N'Djaména).

This reliance on charcoal by large sections of the population of major (and often rapidly growing) cities generated concerns regarding its contribution to deforestation, although a detailed study in several African nations in the late 1980s found very little evidence of this (Leach and Mearns, 1989) and a more recent review suggests that fuelwood is seldom a primary source of forest removal, although "in some of the areas where charcoal production is concentrated, this may be the case" (Arnold et al., 2006).

Urban dwellers in India are shifting to cleaner cooking fuels, although the shift between 1983 and 1999 was most evident among higher income groups. In 2000, less than 40% of the bottom two urban deciles cooked with clean fuels. And among the poorest urban groups, adoption of clean cooking fuels hardly increased from 1983 to 2000 (Viswanathan and Kavi Kumar, 2005). LPG and kerosene are highly subsidized in India, but nonpoor groups are the main beneficiaries and many low-income urban residents continue to cook with dirtier energy sources (Gangopadhyay et al., 2005; Pohekar et al., 2005).

Cooking with LPG is common in the Philippines, but poor urban households also buy kerosene or biomass fuels to keep costs down. In a survey of two low-income districts in metro-Manila, LPG was the main cooking fuel in 75% of households (APPROTECH, 2005). However, as LPG prices increased in 2004, low-income groups also began to cook with kerosene, fuelwood, or charcoal. Although residents intended to reduce expenditures, they still paid higher unit prices because they could only afford to purchase small quantities (APPROTECH, 2005).

High expenditures on energy

A considerable range of national and city studies and studies of particular settlements show how expenditures on fuels for household use are consistently burdensome for low-income households (but may not show up as high expenditures or high proportions of income spent on fuel, as discussed above). Examples of high expenditures on energy are:

- In Guatemala, cooking and lighting took up about 10% of household expenditures for the three poorest urban deciles in 2000 (ESMAP and UNDP, 2003).

- In Thailand, slum dwellers spent about 16% of their monthly income on energy (cooking, electricity, transport) in Bangkok and about 26% in Khon Kaen. Households in these slums with incomes below the poverty line spent 29% of total household income on energy in Khon Kaen and 18.5% in Bangkok – mainly because of the high cost of electricity (Shrestha et al., 2008).

- In Ethiopia, fuel and power took 11% of expenditure among urban poor (Kebede et al., 2002).

- In Sana'a, the capital of Yemen, the bottom two deciles spent over 10% of their incomes on electricity alone (ESMAP and UNDP, 2005).

- In Kibera, Nairobi's largest informal settlement, for over 100 households surveyed energy expenditures reached 20–40% of monthly incomes (Karekezi et al., 2008).

- In Rio de Janeiro (Brazil), many households in surveys in informal settlements were spending 15–25% of their incomes on energy (WEC, 2006).

- In Cairo (Egypt), households with incomes at the lower poverty line spent 8–20% of their income on electricity (Sabry, 2009).

Low-income households who obtain electricity through shared electricity meters can be charged higher rates because of rising block tariffs (examples in Kumasi (Ghana), Mumbai, and an informal settlement in South Africa are given by Devas and Korboe, 2000 and Cowan, 2008).

Space heating

Data on heating expenditures are limited, but it is clear that where space heating is needed, low-income urban dwellers can face high costs to keep warm. For instance, surveys in 1999 found that low-income city-dwellers in Armenia, Moldova, and the Kyrgyz Republic devoted 5–10% of their household incomes to heating (Wu et al., 2004). Poor households may also heat their homes with inefficient, polluting fuels to reduce expenditures. During the winter of 2002, Tbilisi's poor households who were not on the gas network resorted to using wood for heating and cooking (ESMAP, 2007). Wood prices were cheaper than those of other fuels, except natural gas. In Buenos Aires' peripheral settlements of Villa Fiorito and Budge, the average household relies on charcoal for space heating and cooking, with space heating taking up nearly 13% of household annual net energy use (Bravo et al., 2008). In the heart of South Africa's coal-mining country, residents of Vosman Township rely on coal for space heating, water heating, ironing, and cooking (Balmer, 2007). Even in the United Kingdom, four million households were deemed to live in fuel poverty in 2007 (defined by spending 10% or more of income on maintaining an adequate level of warmth) (UKDECC, 2010).

In China, coal is a key heating fuel for the poor, particularly in cold northern cities where heating may take up as much as 40% of households total energy needs (Pachauri and Jiang, 2008). Although data are not specifically available on coal use for heating, national surveys indicate that 65% of the poorest urban households utilize coal (Pachauri and Jiang, 2008). Coal-using urban residents are exposed to extremely high levels of indoor air pollution (Mestl et al., 2007).

Lighting and access to electricity

There is a clear association between the percentage of the urban population with electricity and a nation's per capita GDP (UNDP and WHO, 2009). Almost all nations with GDPs per capita of US$6000 or more have 95–100% of their urban population with electricity. For nations with per capita GDPs below US$3000, there is a quite consistent picture of rising proportions of urban households with electricity, with some variation. There is more variation between US$3000 and US$6000.

However, were the sample frames for the urban households interviewed in the surveys from which this data comes rigorous in including the needed proportion of households that are in informal or illegal settlements? For instance, half of Kenya's urban population is said to have access to electricity in 2003, yet a survey in 1998 of informal settlements in Nairobi (which house half of Nairobi's population) found that only 17.8% had electricity (APHRC, 2002).

In most middle-income nations and some low-income nations, most of the urban population has access to electricity. By 2002, there was near-universal access to power in Caracas, Buenos Aires, and Rio de Janeiro (WEC, 2006). India's household surveys in 2004–2005 found that 91% of urban households used electricity; for Chinese city dwellers, household surveys reported that 96% used electricity in 2001 (Pachauri and Jiang, 2008). Many nations, including Colombia, Dominican Republic, Egypt, Indonesia, Jordan, Pakistan, and Ukraine, report that more than 98% of their urban population has electricity (UNDP and WHO, 2009). These positive developments illustrate that the proximity to existing energy infrastructure in urban areas enables rapid progress when dedicated policies of connecting urban poor are in place. Barriers of low income and limited grid extensions therefore can be overcome.

Thus, many cities and national urban populations have a high proportion of urban poor households with access to electricity. For instance, a study of energy-use patterns in slums in Bangkok and Khon Kaen in Thailand found almost 100% with electricity connections (Shrestha et al., 2008), although in Bangkok 32% of households were connected through their neighbors (Shrestha et al., 2008). Almost all 'slum' dwellers in Cairo have electricity connections (Sabry, 2009). In Mexico in 2000, access to electricity was enjoyed by 91–97% of the lowest-quartile households in cities along the US border (Peña, 2005). National surveys in 2001 found that over 80% of Pakistan's poorest urban deciles had electricity (ESMAP, 2006).

Access to electricity is not only an issue of quantities. Quality of service in terms of regularity, reliability and the duration of provision are of equal importance.

Table 18.12 | The cost per household (in current US$) of providing electricity in different cities.

City	Cost per household (US$)
Ahmedabad	114
Manila	154
Rio de Janeiro	226
Salvador	350
Cape Town	417

Source: USAID, 2004.

The costs of electricity access for the urban poor are generally low (Table 18.12). Nonetheless, some caution is needed in using the figures in Table 18.12 because it is not clear whether these are just the cost of extending electricity to these households or also include other costs, such as the costs of extending overhead lines and upgrading the power-generation system (USAID, 2004).

A study of the costs of different 'slum' upgrading programs in Brazil showed that the provision of electricity and lighting was 1–3% of total costs, although these were comprehensive upgrading programs that included provision of water and sewer connections for each house, and building homes for those that had to be rehoused (Abiko et al., 2007). The costs would be higher as a proportion of total costs within a more minimalist upgrading program – for instance, one that only provided communal water provision and drainage and not piped water and sewer connections to each household.

Further discussion on energy access issues beyond electricity is contained in Chapter 4.

18.4.1.4   Transport

When choosing where to live, low-income individuals or households have to make trade-offs between good locations for access to income-earning opportunities, to housing quality, size, and tenure, and to infrastructure and services. In most cities in low- and middle-income nations, a significant proportion of low-income groups live in peripheral locations because it is cheaper (and often less crowded) or there are more possibilities of obtaining land on which to build housing (although usually illegally). But peripheral locations usually mean high monetary and time costs in traveling to and from work and services. Thus, transportation costs can eclipse household spending on cooking, heating, and lighting.

Various studies of transport use and expenditures in cities or of urban poor communities show that public transport costs represent a significant part of total household expenditure. For instance, for the inhabitants of eight informal settlements in Cairo transport costs were a major burden. Many such settlements on the outskirts are not adequately served by the public bus network or the metro. Many inhabitants have to use more expensive privately operated microbuses for part of the journey and a high proportion have to change to other buses or the metro for their journey. High travel costs were one reason why few children went to secondary school (Sabry, 2009). Other examples include:

- In Karachi, interviews with 108 transport users who lived in one central and four peripheral neighborhoods found that half were spending 10% or more of their income on transport (Urban Resource Centre, 2001).

- In Bandung City (Indonesia), interviews with a sample of 145 *kampong* residents found that nearly 7% of their monthly income was devoted to transport costs (Permana et al., 2008).

- In Buenos Aires, a 2002 survey found that the bottom quintile walked to work for 53% of their journeys, but they still spent over 30% of their family incomes on public transit (Carruthers et al., 2005).

- In Sao Paulo, a 2003 survey found that low-income groups spent 18–30% of their incomes on travel (Carruthers et al., 2005). Wealthy residents spent just 7% of their incomes on transport, but were able to travel far more frequently. The number of trips completed by Sao Paulo's poor was less than one-third of those completed by the highest-income residents.

- In Salvador (Brazil), low-income residents often live in the urban periphery and a survey of over 500 households in the poor neighborhoods of Plataforma and Calavera found that transport expenditures averaged 25% of monthly expenditures (Winrock International, 2005).

Thus, it is common in cities for low-income groups to face high transport expenses that curtail their travel possibilities and leave them with onerous journeys, often on foot. Transport costs also limit livelihood opportunities for low-income groups that live in peripheral locations, as the cost and time to reach parts of the city are too high. A 2003 survey in Wuhan, China, showed how prohibitively high transit costs resulted in the poor rejecting jobs far from their homes (Carruthers et al., 2005).

Some studies show how many low-income groups walk long distances to keep their transport expenditures down (see, for instance, Huq et al., 1996 for various cities in Bangladesh, and Barter, 1999 for central Bombay/Mumbai and Jakarta). So, while such individuals may pay little for transport costs, they 'pay' through long journey times and extra physical effort. In the survey of Wuhan, China (Carruthers et al., 2005), the bottom quintile reported walking for almost half of their journeys, while 27% of their travel was by public transit and 22% by cycling.

Marginalized neighborhoods may not be served by public transit, and low-income women can face particular challenges in accessing secure, efficient transportation (Watkiss et al., 2000). Informal buses have proliferated in many cities, and can help alleviate transport shortages (Zhou, 2000). However, in this unregulated sector vehicles are usually old and

Table 18.13 | Grouping households in India by the amount of energy they use and the energy services available to them (average household of five persons) in Watt-years (1 Wyr = 31.55 MJ).

Energy services of households	Useful energy use per capita (Wyr)
Associated with extreme poverty: up to one warm meal a day, a kerosene lamp, possibly a little hot water	<15 W
Associated with poverty: 1–2 warm meals per day, a few kerosene lamps or one electric bulb, some hot water	15–30
Associated with above the poverty line: two warm meals a day, hot water and lighting, some small electrical appliances for groups with electricity, possibly a scooter	30–60
Associated with a comfortable level of well-being: two or more warm meals a day, hot water, lighting, some space heating, for groups with electricity possibly some space cooling and electric appliances, possibly a scooter or an automobile	60+

Source: Pachauri et al., 2004.

overcrowded, accidents are common, and customers are vulnerable to rising or erratic fares.

18.4.1.5   Differentials within Urban Populations

Various studies show how it is common for low-income urban households in low- and middle-income nations with electricity connections to use 20–50 kWh/month (see Kulkarni et al., 1994; Karekezi et al., 2008). This is a small fraction of average household use in the United States (640–1329 kWh/month depending on the region) or Europe (341 kWh/month). So it is likely that differentials of the order of 100 or more are present between the world's wealthiest and least wealthy households with electricity. Pachauri et al. (2004) considered how the amount of energy used and the quality of energy services available varied by income-group (Table 18.13).

18.4.1.6   Summary

Development literature focuses principally on provision of water and sufficient food, not on clean energy and electricity. Several hundred million urban dwellers in low- and middle-income nations lack access to electricity and are unable to afford cleaner, safer fuels such as gas or LPG (or even kerosene). Most are in low-income nations in Asia and sub-Saharan Africa. In many such nations, more than half the urban population still rely on nonfossil fuel cooking fuels with obvious consequences for health problems (and large health burdens) and for the time needed to obtain fuels. In many low-income nations, more than half the urban population also lacks access to electricity, even though urban population concentrations lower unit costs for providing electricity and gas (or LPG gas distribution). For a large part of the urban population that lacks clean fuels and electricity, the reasons are not that they cannot afford these but that they face political or institutional obstacles to accessing them. Most live in informal settlements with no legal addresses and legal tenure, where they are also denied paved roads and provision for water, sanitation, and drainage, and often healthcare, schools, and emergency services.

A high proportion of urban dwellers in low- and middle-income nations find it difficult to afford their 'energy bills' (for fuel and, where available, electricity and expenditure on transport); these often take 10–15% of household income and for many a higher proportion.

However, in many middle-income nations (and all high-income nations) nearly all low-income urban dwellers have legal electricity connections and can afford clean fuels. The shift to clean fuels and the availability of electricity bring many advantages in terms of health, convenience, and time saved in accessing and using energy. If the cost of a legal connection to electricity and to natural gas supplies can be afforded and supplies are affordable and reliable, no time is needed to purchase or gather fuels, and then carry home solid or liquid fuels or LPG cylinders. Also, no space in the home is needed to store these. Reliable electricity supplies also bring many other obvious advantages – reliable, cheap, and safe lighting at night, the possibility of fridges, televisions, and electric fans, support for home enterprises, and a very large reduction in fire risk.

The costs of connection to an electricity grid and the use of electricity can be burdensome for low-income groups, but innovations have reduced these costs – for instance, rising tariffs with low prices for 'lifeline' electricity use (or in South Africa no charge for up to 50 kWh/month), pay-as-you-use meters, and standard 'boards' that remove the need for household wiring.

Climate change implications may pose problems for low-income urban dwellers obtaining electricity and clean fuels. However, the shift from dirty fuels to clean fuels produces a lower than expected contribution to global warming because of the inefficiencies in how dirty fuels are consumed and in the reduced contribution of fuel use to black soot aerosols. Thus, a shift to clean fossil fuels leads to major improvements in the global impacts associated with non-CO_2 emissions. In addition, current differentials in electricity use per household or in CO_2 emissions per household are likely to vary by a factor of at least 100 between the wealthiest households and the least-wealthy households.

The constraints on supporting the shift to clean fuels and providing all urban households with electricity are less in energy policy and far more in government policy and daily practice in regard to those who live in informal settlements and work in the informal economy. A large part of the population that lacks clean energy and electricity also lacks reliable piped water supplies and good provision for sanitation and drainage. They often lack access to schools and healthcare. Governments often ignore them, even though their settlements house 30–60% of the city population, most of its low-wage labor force, and many of its enterprises.

It is mostly in nations where relationships between local government and the inhabitants of these informal settlements are not antagonistic, with widespread public support for 'slum' and squatter upgrading, that clean energy and electricity reaches urban poor groups.

18.4.2   Energy Demand and Air Pollution Densities, including Heat Island Effects

18.4.2.1   Introduction

The concepts of energy demand and pollution densities refer to the *quantity of energy* used and/or produced or the pollution emitted *per unit of land*. Their common denominator and driver is urban population density. Despite being of fundamental importance in an urban context, the literature on energy or pollution densities is surprisingly limited, apart from that on urban heat island effects, reviewed in detail in Section 18.4.2.4 below.

18.4.2.2   Urban Energy Demand Densities

This Section illustrates the concept of energy supply and demand densities both generally and by drawing from contrasting examples of two high-density megacities (Tokyo and London), as well as a small, low-density city (Osnabrück, Germany). A brief discussion of associated policy issues follows.

The classic reference on energy demand and supply densities remains Smil (1991), from which Figure 18.11 is adopted in modified form.

The customary unit for energy densities is Watts per square meter (W/m²), referring to a continuous use of the power of one Watt over a year.

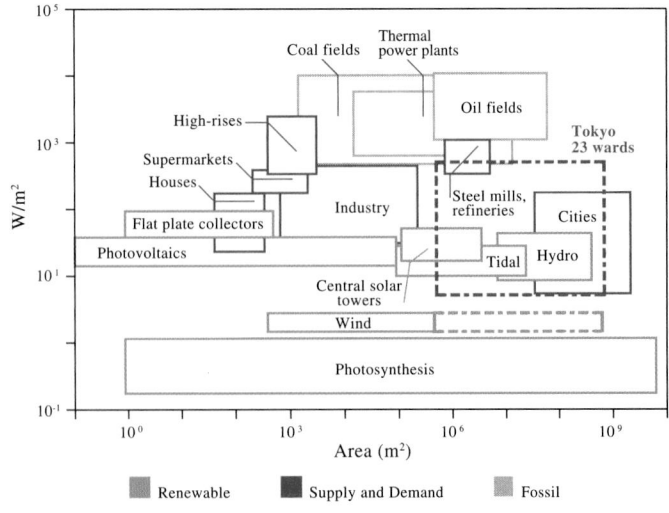

Figure 18.11 | Energy densities of energy supply from fossil (gray) and renewable sources (green) versus density of energy demand (red) for typical settings, in W/m² and m² area. Source: modified from Smil, 1991.

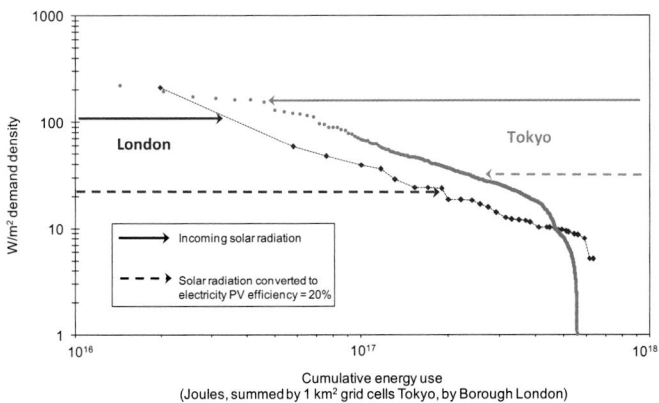

Figure 18.12 | Energy-demand densities (W/m²) for London (33 boroughs) and Tokyo (1 km² grid cells, 23 wards) versus cumulative energy use of these spatial entities (in Joules). For comparison the energy flux of incoming solar radiation (W/m²) and the electricity that could be generated (assuming photovoltaics (PVs) with a conversion efficiency of 20%) is also shown. Source: Dhakal et al., 2003; UKDECC, 2010.

Within an urban context particularly, energy demand densities are of significance. The twin influences of high population and high income mean that the spatial density of energy demand of cities typically ranges from 10–100 W/m², a range exemplified by cities such as Curitiba (Brazil) and Tokyo. Energy-demand densities in smaller portions of urban areas can approach values of 1000 W/m², as in sub-sections of the 23 wards of central Tokyo (Dhakal et al., 2003).

The significance of urban energy-demand density arises in three areas. First, the higher the energy density, the larger the impact of emissions, either as air pollutants or as waste-heat releases. Second, from an energy-demand perspective, high energy densities suggest opportunities for waste-heat recycling and economic provision of clean district heating and cooling services. Third, energy-demand densities are significant constraints for the provision of energy services through renewable energies, which (with the exception of geothermal) typically range from 0.1 to 1 W/m² and thus yield a significant mismatch between demand and supply at the urban scale.

From an energy-systems perspective it is important that the prevailing high energy-demand densities characteristic of urban areas are much in line with those of fossil fuel infrastructures and conversion devices (Grubler, 2004). The general mismatch between (high) urban energy demand and (low) renewable energy supply densities is shown with actual energy demand data for London and Tokyo in Figure 18.12.

The typical order of magnitude of energy use of a megacity is in the order of an exa-Joule (10^{18} J), a unit normally reserved reporting the energy use of entire countries. The (direct final) energy use of Tokyo's 23 wards is estimated to be about 0.6 EJ and that of the larger Tokyo Metropolitan area as 0.8 EJ (Tokyo Metropolitan Government, 2006), compared to 0.6 EJ for London (33 boroughs) and 0.8 EJ for New York City[22] (Kennedy et al., 2009). Energy-demand densities in Tokyo and London typically span a range from a few W/m² to >200 W/m², as in the

City of London or in the top 25 grid cells (i.e., top 25 km²) of the Tokyo wards that use close to 18% of Tokyo's total final energy. Such high energy-demand densities are comparable to the entirety of the solar influx, which equals 157 W/m² in Tokyo and 109 W/m² in London. Mean energy densities, 28.5 W/m² for the Tokyo 23 wards (621 km²) and 27.4 W/m² for Inner London (319 km²), are similar. (For Greater London with its larger size (1572 km²), lower population densities, and greater extent of green areas, energy densities are naturally lower, at 13 W/m².)

Assuming that all the incoming solar radiation could be converted for human energy use (e.g., to electricity with 20% efficient PV panels), the maximum renewable energy supply density would range from 22 (London) to 31 (Tokyo) W/m² in line with average demand densities in the two cities, but only under the assumption that the entire city area could be covered by PV panels! Even assuming an upper bound of potential PV area availability (roofs, etc.), the results from a low-density urban area (Osnabrück, Germany, see below) of 2% of the city area, solar energy could provide a maximum of between 0.4 (London) and 0.6 W/m² (Tokyo), which would cover between 2% (Tokyo's 23 wards) and 1–3% (Inner to Greater London) of urban energy use in the two cities. *Local renewables* can therefore only supply urban energy in niche markets (e.g., low-density residential housing), but can provide *less than 1%* only of a megacity's energy needs.[23]

Given that local renewables in large cities are at best marginal niche options (because of the density mismatch between energy demand and supply), what is their potential in small, low-density cities? Using aerial survey techniques, Ludwig et al. (2008) performed a comprehensive assessment of suitable application of rooftop solar PVs for Osnabrück (Figure 18.13). Osnabrück, with an area of 120 km² and a population of 272,000 (a density of 23 people/ha) is characterized by an incoming solar radiation of 983 kWh/m² (112 W/m²). In the study, all suitable roof areas of some 70,000 buildings in the city were assessed (considering optimal inclination as well as shadowing by adjacent buildings) and the results published for local residents in a database per individual dwelling.

---

22 Final energy use within the city limits and excluding bunker fuels (aviation, shipping). The latter are reported to be 0.28 EJ (0.2 EJ aviation fuel and 0.08 EJ marine bunkers) for New York City compared to 0.76 EJ final energy use in 2005 (Kennedy et al., 2010). For London, aviation fuel also accounted for some 0.2 EJ for the year 2000 (Mayor of London, 2004).

23 This mind experiment considers a highly efficient conversion route of solar energy via high-efficiency PVs (with 20% net conversion efficiency). Assuming biomass as an alternative reduces the energy yield by a factor of up to 20, as the average conversion efficiency of solar energy via photosynthesis is only around 1%. Conversely, considering solar hot-water collectors (with a maximum efficiency approaching 100% of incoming solar energy) also does not change drastically the conclusion of the extremely limited local renewable potentials in high density cities, as solar hot water typically provides only a few percent of energy demand (hot water accounts for 2% of final energy demand in Europe (Eurostat, 1988)). Even if this were provided entirely by solar energy where feasible (in low- to medium-density housing, as high-rise buildings do not offer sufficiently large roof areas) the yield is less than 1% of energy demand in a densely populated large city.

Figure 18.13 | Example of assessing local renewable potentials: roof area (left panel) and suitable roof-area identification for solar PV applications (right panel) for the city of Osnabrück, Germany. Red: roof area well suited for PV; orange: suitable; yellow, only conditional suitability for PV applications; grey: shadowed roof area (unsuitable). Source: modified from Ludwig et al., 2008.

The study identified a total of two million m² of suitable PV roof area for Osnabrück (corresponding to 1.6% of the city area), which if used completely for PV applications could provide some 249 million KWh of electricity, or about the entire *residential* electricity demand of the city (235 million kWh) or up to 26% of the total electricity demand of Osnabrück (940 million kWh). It is of particular interest to interpret the Osnabrück results in terms of their corresponding renewable energy supply density, which adds up to some 0.2 W/m² and can be considered a realistic upper bound of the local renewable energy potential for low-density urban areas. In the example of Osnabrück, local renewables could provide some 3.3 GJ/capita or 2% of the average German per capita final energy use of 154 GJ/capita.

This example shows an important trade-off between population density, transport energy demand, and the potential for local renewables. Generally, the areas available for harvesting local renewable energy flows are higher for a *lower* population density in an urban area. Osnabrück, with a population density of 23 inhabitants/ha and a high proportion of low-density residential housing (single-family homes) evidently offers larger potentials to harness solar energy compared to a megacity with population densities of 130 people/ha and predominantly high-rise buildings (as for Tokyo's 23 wards). However, this higher potential for harnessing local renewables at lower population densities is at odds with the potential to lower the dependence on energy-intensive individual transport modes (automobile usage) in urban areas via public transport. Public transport systems require relatively high population densities to offer an attractive and economically viable alternative to private automobiles, with the minimum population density threshold required typically above 50–100 inhabitants/ha (see Section 18.5.3). In terms of energy, there is thus an inherent trade-off between urban form, transport choices, and the potential of harnessing local renewable energy flows. Put simply, the positive energy implications of an 'active' building (e.g., a 'Passivhaus' standard energy-efficient home with PV panels on the roof that produce electricity both for its own use and for

the grid) can quickly turn negative if the building is situated in a low-density, suburban setting with a high automobile dependence.

Therefore, if renewable energies are increasingly to supply the urban energy needs on a large scale, the resulting needs for conversion and long-distance transport, as well as very large energy 'catchment' areas (the 'energy footprint' of cities), needs to be taken into account.

In an attempt to quantify the implications of the energy supply and demand-density mismatch, IIASA researchers used spatially explicit energy-demand estimates for Europe to calculate related energy-demand density zones (Figure 18.14). The study found that about 21% of final energy demand in Western Europe is below the supply density threshold of 1 W/m², a characteristic upper bound for locally harvested renewable energy flows. The corresponding value for Eastern Europe is somewhat higher, with 34% of energy demand below 1 W/m². Nonetheless, in all densely populated, highly urbanized regions, the majority of renewable energy supply has to come from areas of low population and energy-demand densities, where renewable energy flows can be harnessed and transported to the urban energy-use centers, which represents a formidable infrastructure challenge.

The findings of the IIASA study are also confirmed by a detailed, spatially explicit assessment of solar electricity (PV) potentials for all of Western Europe by Scholz (2010; 2011) (see also Chapter 11).

The Scholz study identified a total solar (rooftop)[24] PV generation potential of 638 TWh (equivalent to 2.3 EJ, or some 40% of the residential

---

24 Adding also building facades to the potential PV areas does not change the results significantly. In a study of solar PV potentials considering the entire building envelope Gutschner et al., (2001) estimated a total electricity potential of 600 TWh for a sample of 10 European countries, which is good agreement to the Scholz study (638 TWh). Facades were estimated to add another 25% to the rooftop PV potentials by Gutschner et al. (2001).

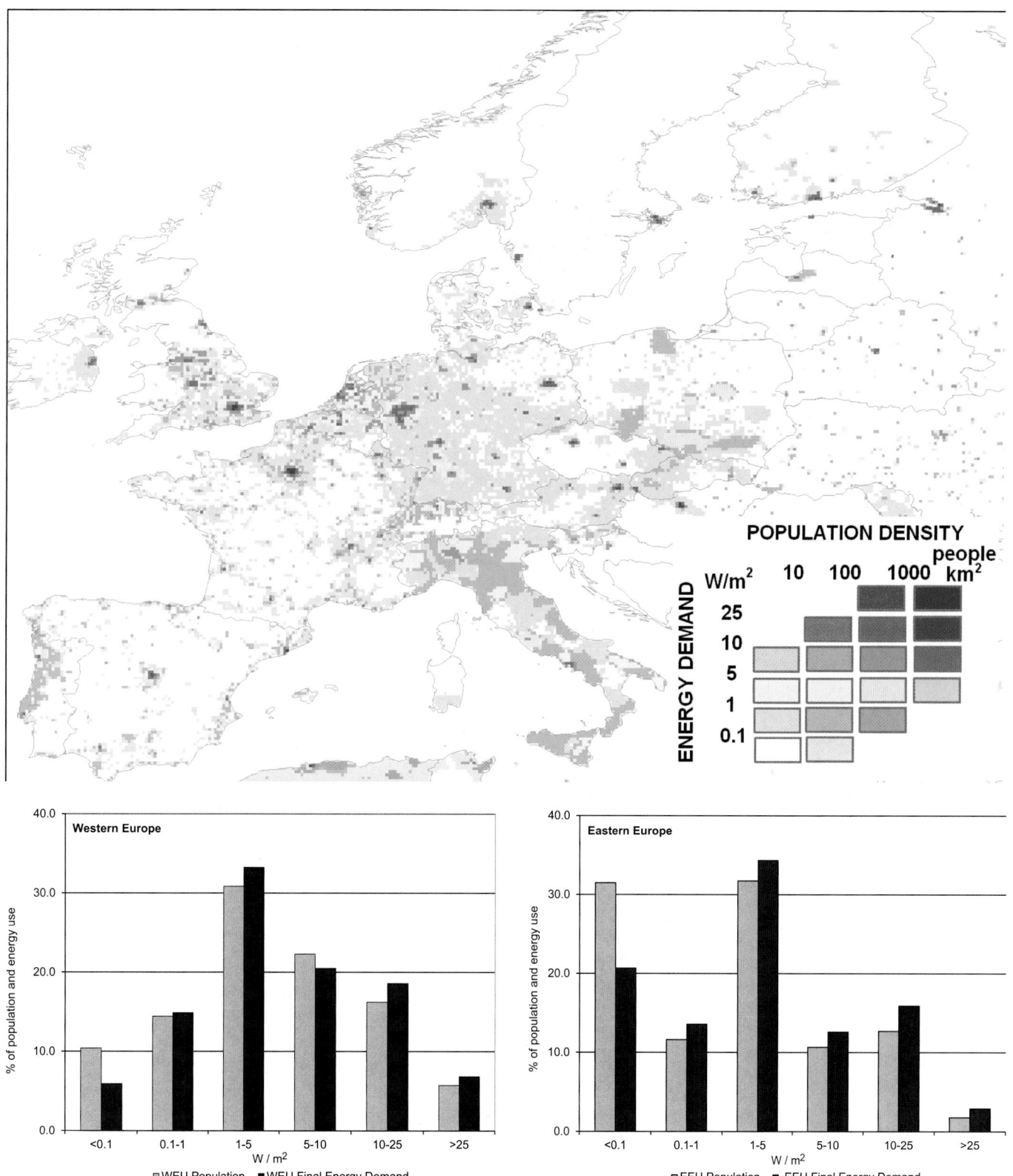

Figure 18.14 | Top: Spatially explicit energy demand densities in Europe (W/m²): Blue and white areas indicate where local renewables can satisfy low-density energy demand (<0.5–1 W/m²). Yellow, red and brown colors denote energy demand densities above 1, 5, 10, and 25 W/m² respectively.

Bottom: Distribution of population (grey) and final energy demand (black) (in percent) as a function of energy demand density classes in W/m² for Western Europe (left panel) and Eastern Europe (right panel). Only 21% (Western Europe) and 34% (Eastern Europe) of energy demand is below an energy demand density of 1 W/m² amenable to full provision by locally available renewable energy flows. The high energy densities of cities require vast energy 'hinterlands' that can be 100–200 times larger than the territorial footprint of cities proper requiring long-distance transport of renewable energies. Source: IIASA calculations commissioned by Chapter 18.

Table 18.14 | Global exposure equivalents to particulate emissions. Note, in particular, the continued dominance in developing countries of indoor air pollution from traditional biomass cook stoves compared to the urban outdoor air pollution exposure.

Group of Nations	Concentrations (µg/m³)		Exposures (GEE)[a]		
	Indoor	Outdoor	Indoor	Outdoor	Total
Developed					
Urban	100	70	5	<1	6
Rural	60	40	1	<1	1
Developing					
Urban	255	278	19	7	26
Rural	551	93	62	5	67
Total			87	13	100

a GEE = Global Exposure Equivalent

Source: adapted from Smith, 1993.

electricity demand in Western Europe, and 23% of the total electricity demand in the region). 637 TWh (99.8%) of that solar PV potential is below a maximum energy supply density of 0.5 W/m², and 563 TWh (88.2%) below a energy supply density level of 0.2 W/m² (Scholz, 2011). Renewable energy supply densities in urban areas are therefore maximum in the range of 0.2 to 0.5 W/m² which are thus between 2 to 5 percent of characteristic urban energy demand densities of 10 W/m².

18.4.2.3   Pollution Densities

A corollary of energy densities is that of pollution density. High population density also leads to high *exposure*[25] density to pollution risks.

However, at least for traditional air pollutants such as particulates, urban pollution exposures also need to be seen in context, as only approximately one-third of the global pollution exposure is urban, whereas two-thirds are rural, because of the dominance in global particulate pollution exposure of indoor air pollution in rural households of developing countries (Table 18.14 and Chapter 4). Smith (1993) developed the concept of global exposure equivalent (GEE), which represents a renormalized index of the global summation of pollution exposure (pollution concentration times population exposed) calculated for a range of human environments. According to Smith (1993), global human exposure to traditional pollutants is dominated by indoor air pollution in rural and urban households in developing countries as a result of the continued use of traditional biomass for cooking.

For more modern forms of pollution (sulfur and nitrogen oxides (SO_x and NO_x) and ozone (O_3)), the corresponding GEEs have not been

calculated, but it is highly likely that the respective role of indoor versus outdoor air pollution as the main source of a population's pollution exposure risk is reversed; that is outdoor air pollution and urban settings comprise the dominant form of pollution exposure. As an example, consider emissions of sulfur dioxide (SO_2). The 'hotspot' of sulfur emissions and pollution, which has for decades been the 'black triangle' (the coal-rich border area of Poland, the Czech Republic, and East Germany) in Europe, was remediated by successful European sulfur-emission reduction policies. The current sulfur-emission hotspot is now in China (Figure 18.15), where high elevated levels of sulfur emissions particularly affect the urban populations and triggered policy responses (see also Section 18.5.5 below).

From an environmental perspective high urban energy demand and the resulting pollution densities hold two important implications. First, energy use usually involves heat losses at well above ambient temperatures and high densities of urban energy use also imply high densities of urban waste-heat releases. These combined with the (high) thermal mass of buildings in densely built-up urban land give rise to the 'urban heat island effect' (see below) in which urban mean temperatures are several degrees higher than those of surrounding hinterlands.

Second, fuel choice becomes of paramount importance: pollution-intensive fuels (biomass or coal) used at the high demand densities of urban areas quickly result in unacceptably high levels of pollution concentration (such as the London 'killer smog' of 1952 or the current air-quality situation in many cities, especially in the developing world). Even low-pollution fuels, such as natural gas, can quickly overwhelm the pollution dissipative capacity of urban environments. So, high energy-demand density requires zero-emission fuels: electricity and perhaps, in the long run, hydrogen.

18.4.2.4   Urban Heat Island Effects

Formation of urban heat islands

Urban heat islands describe the frequently observed pattern of urban air temperatures that exceed those of neighboring, more rural areas. In temperature maps, which delineate neighborhoods of similar temperature with contour lines ('isotemperatures'), urban areas stand out as 'islands' that form 'heat domes.' For example, Figure 18.16 shows these for Tokyo, the city for which most literature on the heat island effect is available.

Urban temperatures typically peak some hours after midday, but the absolute temperature difference against rural areas can be even larger during the night under cloud-free conditions. Heat islands are facilitated in climatic situations of low air movement. Wind otherwise disperses temperature plumes. Heat islands are similar to air-pollution concentrations and they can be enhanced by local topography and climate patterns that prevent mixing of the boundary layer. In terms of average temperature difference, urban areas are often 1–3°C warmer than the

---

25 Exposure risk: product of population × pollution level × exposure time of population to pollution.

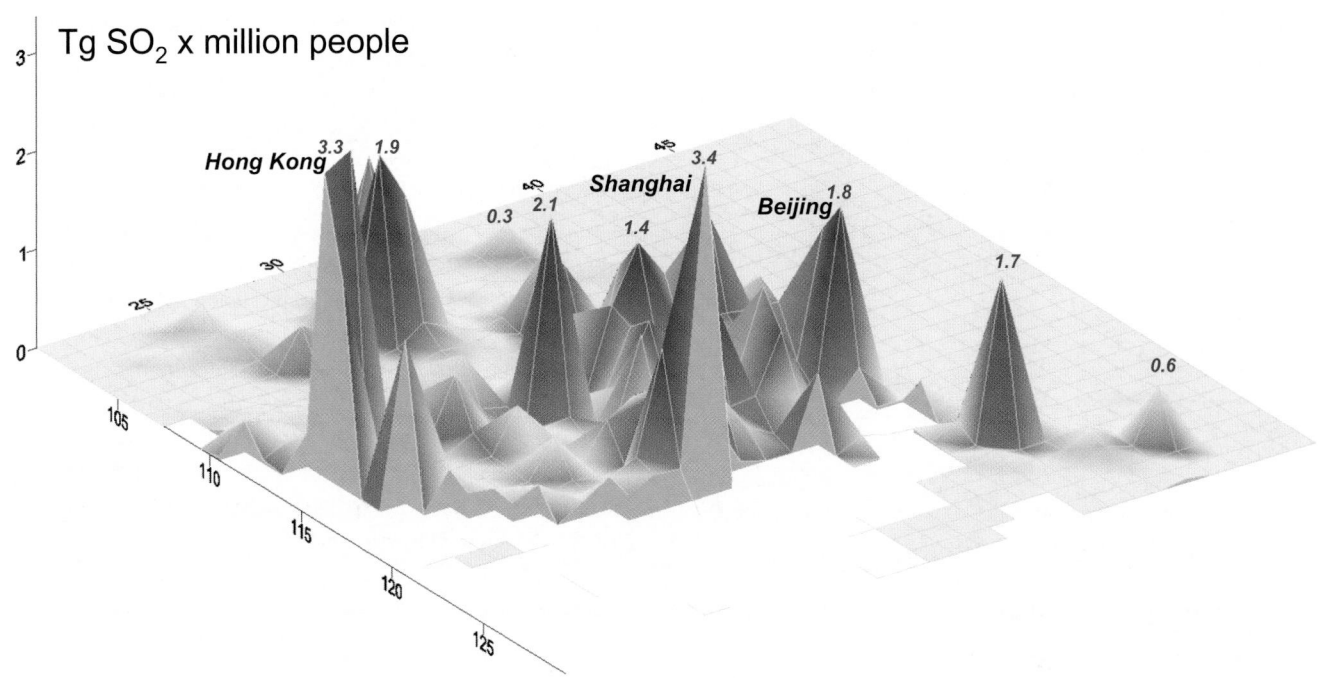

Figure 18.15 | Human exposure to sulfur emissions (population × emissions in million x Tg SO$_2$, z-axis) in China (2000), based on an analysis of gridded socioeconomic and emission data. (Units on x, and y-axis refer to geographical longitude and latitude). Note the high pollution exposure in major urban areas of China. Source: IPCC RCP scenario database (IIASA, 2010).

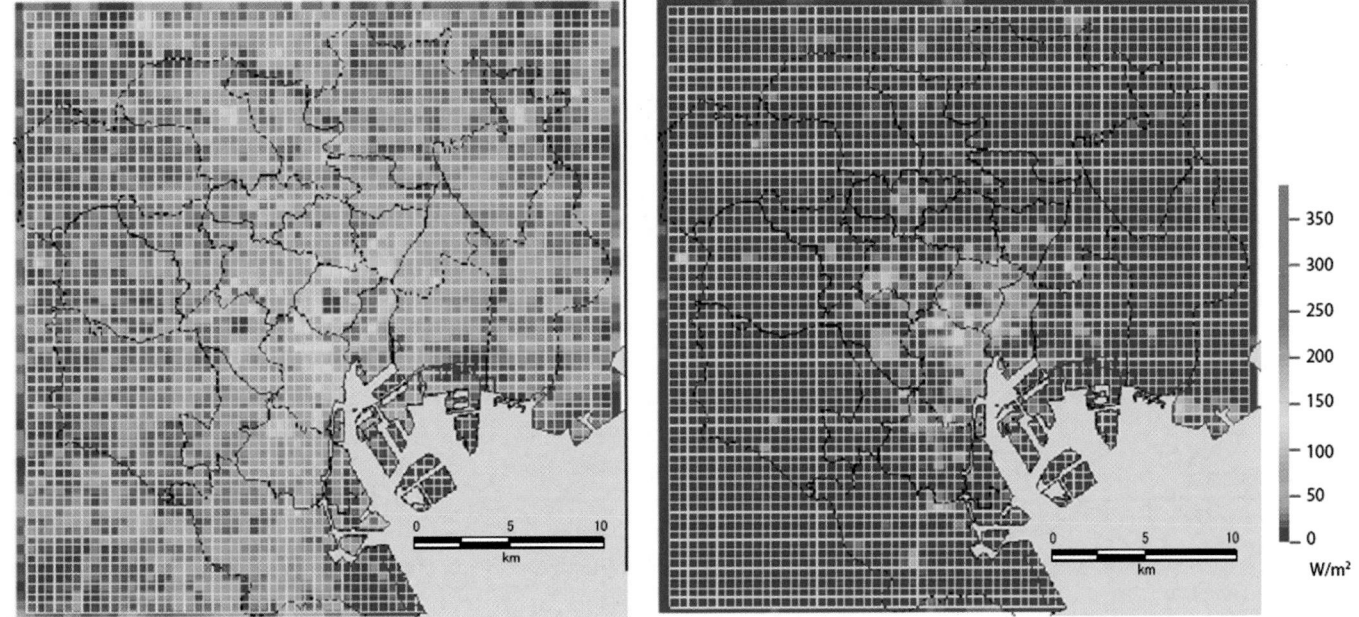

Figure 18.16 | Sensible (left) and latent (right) anthropogenic heat[26] emission in Tokyo (W/m^2). Source: Ichinose, 2008.

surrounding air, and at individual locations in calm and clear nights temperature differences can exceed 12°C (Klysik and Fortuniak, 1999). With

increasing energy use, the extent of urban heat island effects increases (Figure 18.17), which results in local warming.

Heat islands are, among other factors, caused by urban energy use through anthropogenic heat release. Without planning or intervention strategies there is a risk of maladaptation feedbacks, in which heat island countermeasures trigger increasing energy use, which amplify

---

26 Sensible heat flux: air is heated directly by the heated ground surface. Latent heat flux: evaporation from wet ground surface or from cooling towers settled on top of buildings and evapotranspiration from vegetation. This type of energy exchange does not change air temperature. Its energy is consumed in the phase change from water to moisture.

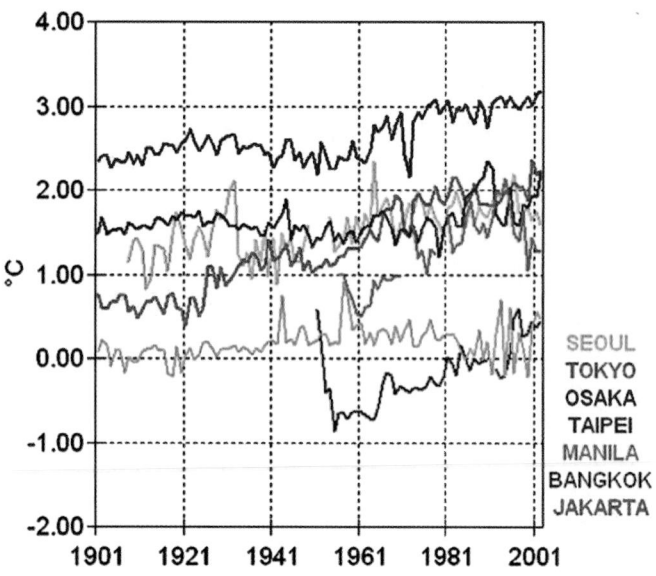

Figure 18.17 | Estimated urban heat island intensity in large Asian cities. Source: Kataoka et al., 2009.

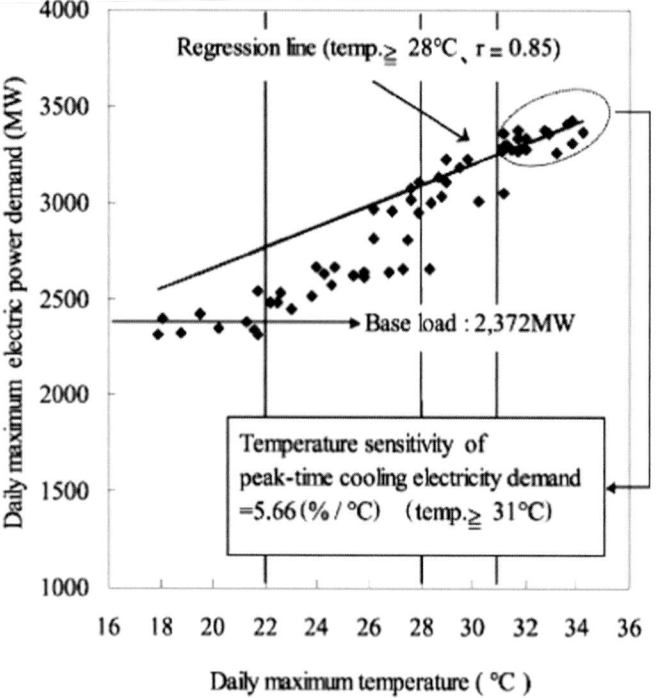

Figure 18.18 | Daily electric power demand and maximum temperature in Tokyo, June to August 1998. Source: Kikegawa et al., 2003.

the heat island (Figure 18.18). The increasing use of air-conditioning equipment in buildings and automobiles (which dump waste heat into the atmosphere) are one such example (Kikegawa et al., 2003; Crutzen, 2004).

Urban heat islands also affect the energy system directly. Elevated environmental temperatures reduce the efficiency of thermal power plants.

Also, the availability of cooling water for thermal or industrial plants can be reduced as water bodies warm up.

A range of factors contribute to the formation of urban heat islands and their relative contribution varies among urban areas (Seto and Shepherd, 2009):

a) The geographic context defines the natural radiation balance across seasons, temperature, precipitation patterns, topography, and exposure to prevailing wind patterns. In humid climate regions, the natural vegetation is often forest dominated. Urban surface-temperature variation is then, at least partly, moderated by the latent-heat transfer through evapotranspiration of adjacent vegetation (Oke, 1987). Parks with water bodies and extensive vegetation cover reduce heat island formation. In more arid climate regions with low vegetation cover, daily temperature variation is more pronounced and heat island formation is more likely.

b) The replacement of natural vegetation with artificial surface materials for buildings, squares, and transport infrastructure results in more incoming radiation being stored during daytime, particularly if materials are dark, such as bitumen and asphalt. The albedo changes and differences in specific heat capacity of construction material result in more incoming energy being accumulated in surface material during daytime, which is later emitted as infrared radiation (Taha, 1997). Thermal insulation of buildings can reduce their specific heat capacity drastically.

c) Urbanization modifies the local hydrology. Natural groundwater recharge is typically prevented by the extensive use of impermeable surface materials. Rain gutters, drainage systems, sewers, and canals channel precipitation rapidly away to avert urban flooding. Additionally, the extraction of water from local wells and for construction projects often lowers water tables, which results in lower water availability for the remaining vegetation, and consequently lower evapotranspiration and associated cooling through latent heat transfer.

d) Also, structural characteristics of the urban form (Weng et al., 2004) affect the efficiency of heat loss through radiation and convection. Narrow street canyons with a limited sky view prevent heat loss through direct radiation upward. The urban layout and orientation of street corridors in relation to prevailing wind patterns affect the efficiency of heat transfer through convection and boundary layer mixing. These factors can be addressed through planning regulations. In Germany, for example, the concept of urban ventilation pathways ('Luftleitbahnen') aims to maintain radial corridors of cold winds to reach urban centers.

e) The metropolitan area size (extent) amplifies the magnitude of the urban heat island effect. For some rapidly growing urban areas, such as Los Angeles or Kobe, a continuous increase in urban heat island temperature of up to 0.5°C/decade occurred over the past 60 years. This trend is partly amplified by changes in energy-use patterns (Böhm, 1998) and needs to be compared against general trends of global surface temperature warming of about 0.7°C over the past century (Oke, 2006; IPCC, 2007; Kataoka et al., 2009) (see Figure 18.17).

f) As energy demand is concentrated particularly in urban centers, the consequential release of anthropogenic heat is similarly dense in these areas. Industrial and service-sector activity, residential housing, and transport functions are typically clustered in close proximity. Electricity use and combustion processes in buildings and vehicles, for heating and cooling, lighting, or motion, all result in vast quantities of waste heat being released (Rosenfeld et al., 1995). To a small extent, the metabolic activity of biological body functions of the human population also contributes to this.[27] Global average estimates attribute a resulting climate forcing of about 0.028 W/m^2 to anthropogenic heat release (technical and biogenic). In North America and Europe these figures are estimated to be higher, at +0.39 and +0.68 W/m^2, respectively (Flanner, 2009). For the Ruhr area in Germany, average anthropogenic heat-related forcing values of 20 W/m^2 were calculated by Block et al. (2004). At higher spatial and temporal resolution, the values are much larger, often between 20 and 100 W/m^2. Numerical simulations for heat discharge of individual neighborhoods in the Tokyo metropolitan region, for example, indicate radiative forcing values of up to 700 W/m^2 during the day and in summer time (Dhakal et al., 2003). However, urban heat islands do not always increase urban energy demand. In higher latitudes the resulting reduction in heating demand in winter can more than compensate the additional cooling energy demand in summer. Integrated climate-energy system models increasingly aim to capture such effects (Kanda, 2006; Oleson et al., 2010).

Mitigation

Strategies for heat-island mitigation include behavioral and technological solutions. They can provide various co-benefits, including energy savings, peak-load reduction, air-quality improvements, and beneficial health, psychological, and socioeconomic effects.

Building design and layout allow solar gains of houses in summer to be minimized and increased passive gains during winter (e.g., in Passivhaus designs). Reduction in cooling demand can also be achieved through the use of deciduous vegetation for shading (including vertical greening of facades) or the application of mechanical shades, shutters, or 'smart

glass windows,' with modified transmission properties of heat and light on demand. Albedo changes and the use of reflective paint and surface material on roofs and transport infrastructure is another particularly cost-effective mechanism to reduce heat absorption in urban areas. Improving the insulation of the building stock to prevent the warming and storage of solar influx in material of high specific heat capacity, such as concrete, is another measure, just like the expansion of shading structures (vegetation or textile) in general. The shading of parking lots, for example, not only provides thermal comfort, but also reduces emission of volatile organic compounds (significant precursors of low-level O_3 formation) from parked vehicles.

Active cooling via enhanced evapotranspiration can be induced through ponds or fountains, green roofs, and tree planting, or through the generation of artificial mist (microdroplets of water) to create local cooling clouds. In preparation for urban heat waves, the city of London considered the need to prepare 'cooling shelters' for vulnerable or elderly population, as the typical UK housing stock is not equipped with air-conditioning (City of London, 2007).

Changing the timing of activity patterns to avoid the hottest hours of the day was a traditional response to hot climates. In peak-load management programs some of this rationale is revived. Operators of office space in New York City are given price incentives to start air-conditioning in the early hours of the day to reduce demand during peak hours of electricity demand (Bloomberg, 2007). While the primary motivation was to reduce peak power demand in a grid of constrained capacity, this measure also reduces the peak in waste-heat emission in the early afternoon hours. While probably not the most suitable for dense, high-rise developments like Manhattan, solar cooling devices that provide a maximum cooling output at periods of peaking outdoor temperatures appear to be suitable for low-density cities.

In 2005, the Japanese environmental ministry started a campaign titled 'Cool biz' to restrain the use of air-conditioning units to an indoor temperature of 28°C. The campaign aimed to loosen the strict dress code of full-suit, tie, and long-sleeved shirts for office workers during the hot season (June 1 to September 30) and promoted a more heat-tolerant dress code of short-sleeved shirts without ties (Pedersen, 2007).

18.4.3 Supply Constraints, including Reliability and Security

Cities depend on extended energy networks and failures can occur on a regional and national scale. On August 14, 2003, a cascading outage of transmission and generation facilities in the North American Eastern Interconnection resulted in a blackout of most of New York State as well as parts of Pennsylvania, Ohio, Michigan, and Ontario, Canada. On September 23, 2003, nearly four million customers lost power in eastern Denmark and southern Sweden following a cascading outage that struck Scandinavia. Days later, a cascading outage between

---

27 Assuming about 100 W of biological energy use per person and maximum population densities of 40,000/km² in some cities in developing countries, this factor can contribute up to 4 W/m² additional forcing. Typical urban population densities are lower.

Italy and the rest of central Europe left most of Italy in darkness on September 28. These major blackouts are among the worst power-system failures in the past few decades. They had a profound effect on power-system philosophy because these networks were some of the world's most sophisticated power-generation distribution systems. In particular, the US failure was promoted by an early underlying failure in software used to control networks, which meant the scale of the emerging problem was recognized too late to protect the cascading failure.

Energy efficiency and resilience are not coincidental outcomes. Thus, a low-energy settlement might be even more sensitive to disruptions in supply than a settlement with some slack in its energy system. A dense city with no power for its elevators may be in a worse state than a low-density urban settlement without power. Renewable power sources based on wind or solar alter the reliability profile. As they are of a much smaller unit size, they do not induce major dropouts, as happens when a large nuclear power station needs to come offline very rapidly. Conversely, the variability in available power they supply may require them to be shadowed by a rapid-response plant. They may also exacerbate failure cascades because of switching out for self-protection when the power system is stressed heavily.

The increasing dependency on power even for simple clerical work, let alone for critical functions like hospitals, means that stand-by power supplies could be an increasing feature in urban systems. One suggestion (Patterson, 2009) is that it is possible for the local distributed power generation to become dominant and the national distribution systems only handle back-up. This is already effectively the case for dwellings that use microgenerators for power and heat. Another suggestion is that more sophisticated metering and tariffs could incentivize the extension of demand-side load management from large facilities of 'interruptible supply' at the microscale. In line with the efficiency-resilience argument it is expected that vulnerability to societal interruption is higher in countries with generally very secure supplies in which the economy has sought an equilibrium that assumes secure power supplies than in countries with frequent brownouts and blackouts in which the economy has adjusted to coping with the risk.

The winter of 2008/2009 in Europe showed the vulnerability of gas-supply networks to urban areas. Gas can be stored both in the mains and in dedicated storage facilities, but gas is currently supplied directly to consumers and power generation, and so indirectly to mass-transit systems. Thus, a failure of supply pressure has wide implications. Coastal cities can increase their robustness with liquid natural gas (LNG) terminals, but LNG is a world-trade product and may come at a high price in a regional emergency.

Liberalization of gas and electricity market pressures gives the lowest prices to consumers, but the effect is to incentivize producers to 'sweat' their existing assets (e.g., Drukker, 2000). It may then become necessary to introduce further complexity into tariffs to incentivize investment and reflect the value to the consumer of lost load. Undercapitalization of energy networks, many of which were built in the 1960s, remains a real risk over the next 20 years in developed world systems.

Possibly the greatest vulnerability in an urban context is the supply of transport fuels. The weakness of the low-density settlement is its complete dependence on oil at a density below around 30–40 persons/ha (Levinson and Kumar, 1997). There are ways to save oil quickly (IEA, 2005), largely based on increasing load factors, but there is a mounting recognition of the advantages of diversifying the transport energy vector away from oil.

18.5   Urban Public and Private Sector Opportunities and Responses

18.5.1   Introduction

18.5.1.1   Concepts of Sustainable Cities and Designs

The term 'sustainable city' dates from the 1990s. The term 'sustainable development' is usually dated from the World Commission on Environment and Development Report (WCED, 1987), which devotes a chapter to 'the urban challenge.' The WCED concern was principally about issues of the urban poor in rapidly growing large cities of the South. The term sustainable development is now used more frequently in the narrower context of the need to protect the environment that underpins social and economic capital. For this reason the term 'sustainable cities' is more often associated with civic initiatives in cities of the North, addressing what is perceived as the unsustainable impact of their citizen's lifestyles, especially the generation of large volumes of waste and GHG emissions. It is largely coincident with the earlier idea of an 'ecocity.' Ecocities essentially try to contain their 'ecological footprint' (Andersson, 2006; Jabareen, 2006; Kenworthy, 2006; Pickett et al., 2008). This focus means that some projects are not always more broadly sustainable, especially as economic units.

Attempts to achieve an optimal 'sustainable urban system' in new settlements invariably resorts to some form of spatial organization. This may be provided by a city authority, but it could equally be the covenants imposed by a land developer. The intention is to gain from bringing the various strands of urban activity together into a more integrated whole. For example, reducing urban traffic noise through less need to travel by automobile and the use of quiet road surfaces or electric vehicles enables citizens to keep windows open in summer. This provides the opportunity to avoid mechanical ventilation by recovering the opportunity for natural ventilation. 'Sustainable' urban configurations are often expressed in terms of optimal residential densities linked to low-profile transport networks. This fairly crude metric is often employed in zoning regulations. The optimal configuration then seeks to avoid a very high density with highly congested services and a very low density automobile-dependent networks. This configuration is expected to induce a stronger sense of community by providing some local retail and commercial space with local interaction, itself reducing the need for automobile travel.

Energy implications of 'sustainable cities' arise naturally from their move away from automobile dependency. But overlaid is an attempt to exploit the area of the city as a source of renewable energy. At the current state of technology, most 'zero-carbon' developments are essentially *net* zero carbon, and rely on the existence of a market for surplus renewable electricity at some time of the day and the ability to import electricity at others. The installed PV capacity for the proposed new Masdar city in Abu Dhabi is nearly 200 MW within its city boundary of 6 km². Wind is less likely to be exploited in a sustainable city, except as an aid to natural ventilation, because of the ground-effect drag in urban environments. For the purist to import green electricity is wrong. While wind generators sometimes appear in iconic building structures, drag resistance of the urban surface makes them a less compelling investment than wind generators in more exposed localities. However, while the proposed ecocity at Dongtan outside Shanghai uses local agricultural wastes as an energy supply, it is not obvious how this differs from using waste collected from a wider area.

Waste is a particular issue for all cities. Cities of the ancient world produced little nonbiodegradable waste, in contrast to the large volume of solid waste that large urban settlements currently need to dispose of. 'Sustainable city' discourses thus focused on reducing waste sent to land fill by providing recycling and incineration facilities. In ironic contrast, the large cities of the developing world already have an informal economy in the periurban areas that pick clean the waste of the formal urban centers. Sustainable cities frequently try to capture waste heat from electricity generation by locating combined heat and power (CHP) sets within the city or encouraging microgeneration. For large urban areas this strategy can only maintain urban air quality if other measures designed to reduce air pollution are successful – which emphasizes the importance of treating issues and systems holistically (Box 18.4).

Box 18.4 | Zero-Carbon Cities

Planners are exploring new urban paradigms in which urban energy use is not entailed by urban form, but in part defines the urban master plan. Dongtan near Shanghai and Masdar in the Arab Emirates are two recent high-profile design studies that exemplify this.

Both Masdar and Dongtan represent a new development strategy whereby the creation of the city forms its own economic rationale. This reflects the delocalization of some classes of economic activity and hence the ability to bring these together in a desirable location. This is the theory behind the creation of Dubai as a global financial sector from scratch. Less ambitiously, Dongtan was to be a service center outside of Shanghai. Masdar was to be a knowledge center that specialized in renewable energy technologies. The master plans of both complexes are polycentric so that 'quarters' or districts are often defined by a locally dominant economic or social function. Whether the loss of monocentricity is significant is hard to tell. The cellular nature of both development plans enables stable modular growth in uncertain economic times. Both designs have a substantial external energy supply from local renewable resources and so 'to first order' are formally 'zero carbon.' The designs themselves sought to reduce energy demand substantially without affecting service delivery.

Dongtan was designed as a 'zero-carbon' development. The initial settlement study was for 80,000 inhabitants, but with expansion possibilities to over half a million. Its external supply of energy is regional biomass waste from rice production, along with solar power panels. In delivered energy terms, it is an all-electric city. Electric vehicles provide all motorized transport within the city. These can either be conventional battery-powered or fuel-cell powered vehicles that use hydrogen, but the short range is less important in Dongtan because the city is spatially organized to reduce distances for essential travel. Conventional transportation to the outside world has to be parked at the city boundary. The master design features not only low-energy concepts, but also self-sufficiency in water and waste recycling. A particular feature was the exposure of system synergies. For example, by switching to quiet clean electric vehicles, it became possible to revert to natural ventilation and day lighting for housing and offices. The estimated 'eco-foot print' for Dongtan was 1.9 ha/capita, only slightly above the WWF target (Cherry, 2007).

Masdar faces a notably different climate to Dongtan, dominated by cooling load. This demand is reduced by recourse to traditional Arabic architectural approaches that protect building facades and access routes from the sun, with narrow access spaces that still provide daylight penetration into occupied spaces. Like Dongtan, Masdar is a zero-carbon urban development. The scheme's energy supply is provided by a large solar-power energy park. Indeed, Masdar's economic rationale is as an international center in advanced renewable energy technology. Its first districts are already established and it hosts the International Renewable Energy Agency supported by over 130 nations. Industrial zoning places production facilities at the periphery of the neighborhood complex, where more conventional freight transport has access. Novel electric personal transport aims to shuttle residents between centers. As with the Dongtan design, there are no private vehicles within the city. The planned scope of the city is around 6 km².

Both Dongtan and Masdar were designed as showcases of a new vision of future master planning focused on low environmental impact, especially of energy. The focus on zero carbon and sustainability reflect the kind of private investment that the developments were intended to attract. The common theme is the value of integrated design in reducing the overall impact of urban processes. It is yet to be demonstrated whether less well-endowed and less well-organized new urban settlements can achieve similar impressive potentials. Given Masdar's initial investments of well above US$20 billion, the sheer magnitude of the investments need for housing some three billion additional urban dwellers in radical new zero-carbon city designs is staggering: well above US$1000 trillion, or some 20 years of current world GDP (Kluy, 2010)!

Whereas the Masdar project has completed at least its first phase, the Dongtan design has been set aside for the moment, although a number of new, perhaps less ambitious, settlements are under construction around Shanghai (Larson, 2009). In many ways the Dongtan design exercise has served as a valuable learning experience in 'holistic' master planning and is widely influential. As important as both projects are as showcases and experiments in new thinking and planning, the actual urban development reality is rather one of incremental, continuous change within an existing urban fabric that needs to incorporate the lessons learned from bold 'greenfield' ecocity design experiences.

The 'sustainable city' as currently conceived is not without its critics. The discourse is often so concerned with environmental factors that the local robustness of social and economic capital is unwisely taken for granted. However, examples that have been partially implemented show reductions in final energy use against normal benchmarks of 10–15% without any substantial changes in lifestyle norms. While sustainable urban form can often require significant capital 'up front' and is thus a serious obstacle under capital constraints, a possibly more fundamental issue is *institutional*. Is it possible that stakeholders within an urban context who are used to working independently can find an institutional structure where they can work together in an integrated manner. This may be easily done at the master planning stage, at which broad-brush issues are under the control of a single land-use planner, but to maintain the integrity of the master plan can be a challenge. If the problem can be solved, it would be a major disruptive technology to conventional urban planning and development and for which prototype design software is already in existence (Keirstead et al., 2009).

Summary

Urban planning measures have the potential to be very powerful methods of integrating urban services that minimize urban energy use and other ecological impacts. Nonetheless, many of the recent design exercises in 'sustainable cities' are for premium urban centers. The technology may be transferable with further development, but the systems are presently too expensive to treat them as realistic new paradigms for the urban built environment in the decades to come.

18.5.1.2   Overview of Main Policy Instruments of Relevance for Urban Energy Systems

The policy players

Policy instruments that apply to urban settlements are generally exercised through several layers of government, often as many as five or six. In theory, this plurality is to ensure that policy powers have an appropriate geographic reach for the issue at hand, and to ensure that destructive competitiveness between settlements does not undermine the quality of specific policy interventions (e.g., Baumol and Oates, 1975). Parallel arguments are deployed to explain the distribution of tax raising, tax collection, and spending powers at different levels of government. The system is extremely effective, but it can be prone to problems of coordination and regional politics. The universal tendency is to decentralize responsibility without decentralizing resources. Urban administrations are more often the delivery agent rather than the tax raiser and this frequently leads to accusations by lower tiers of government of 'underfunding' by higher levels. From time-to-time many of the cities, even in the richest nations, operate close to bankruptcy, which can limit their ability to obtain capital for projects. Generalizations are otherwise dangerous. Urban settlement patterns of governance are, in part, an accretion of local history.

The degree to which public policy is a meaningful term in an urban energy context varies greatly from the highly organized urban society of Singapore to the current chaos of Mogadishu. The issues important to energy use, such as regulation of construction standards for stationary infrastructure, are found at all levels. Control of transport provision also occurs at all levels, although land planning, infrastructure standards, and transport can be dealt with by distinct silos within several layers of public administration. Also, landowners, both private and public, often hold important powers through ownership rights that help define the urban form and its physical emergent properties. Houston is possibly the extreme with no state zoning, but relies largely on land covenants. Other relevant powers can reside with the public or private bodies that provide the utility services.

Energy use is currently an 'emergent' property of a complex urban system. Current governance structures were not designed specifically to manage energy outcomes. In any long-term perspective that embraces a very uncertain future in which energy and related environmental issues become important, it seems likely that greater clarity and effectiveness of governance-relevant structures is highly desirable. Increasingly, major cities in the world have developed 'energy plans' or 'energy strategies' (e.g., Mayor of London, 2004).

The policy instruments

Policy instruments are conveniently arranged in a hierarchy of leverage effects, with land-use planning at the base and control of infrastructure use at the apex. Sometimes the whole hierarchy is delivered at one stroke, as in a major rapid expansion of an economic zone, and energy optimization or 'zero carbon' has featured in a number of recent expansions (e.g., Masdar (Biello, 2008), Dongtan (Normile, 2008), and Incheon (Kim and Gallent, 1998)). These large enterprises are usually led by a development corporation or similar entity with powers to integrate the various s tiers of provision and exceptional access to capital. While impressive in concept, to devise a master plan durable against changes in external circumstances over decades is no mean feat (as the unrealization of the Dongtan project illustrates).

Normally, land-use planning is less ambitious and sets aggregate parameters for zones, such as permitted functions, density, or maximum building height, that guide rather than direct public and private investment. This is the pragmatic solution for the incremental redevelopment of an area. This means that upgrading and refurbishment frequently takes place in patchworks such as '22@' in Barcelona or 'Thames gateway' in London, usually within the framework of a local or regional government 'master plan.' This approach is potentially very powerful for realizing some of the advantages of economies of scale in low-energy or low-carbon technologies, but there is as yet relatively little experience as to how to exercise it effectively. There are some impressive examples in European towns, like Malmo and Freiburg, and US towns like Portland or Davis. These examples influenced recent thinking in urban design, but are apparently not impressive enough to induce widespread replication at current (low) energy or 'carbon' prices.

In many jurisdictions the planning authority has the power to apply conditions to new developments, such as mandatory connection to a district-heating scheme. These conditions can be overwritten by other energy-policy objectives. For example, in Europe a development might be required to install a renewable energy source, but energy competition policy would prohibit the imposition of an additional requirement for its exclusive use by tenants. Planning control can be effective in eliminating the most excessive energy use and giving some certainty to capital investment. However, because so much of the final energy use is delivered by instruments at lower levels in the hierarchy, planning control can seldom deliver very low-energy solutions on its own. Where a low-energy solution collides with other environmental factors, such as appearance or noise, it can militate against it.

The next level 'down' from planning control conventionally contains instruments that relate to the economic framework of the urban settlement. Since the 1980s the trend has been for local services to be provided by the private sector within a regulatory framework. Energy prices are usually regulated (or subsidized) at the national level, and not always in a manner conducive to good outcomes. Removing energy subsidies is theoretically a low-hanging fruit. In practice, the objections of those who benefit from them make such action politically contentious. However, lower tiers of government still have a number of economic instruments at their disposal, especially for transport. Thus, parking charges and more ambitious road-user charging can be used to favor or subsidize mass-transit systems. As these instruments are often redistributive, even when effective, their application can require considerable political skill. The London Congestion charge is a case in point (Taylor, 2004). More broadly, the price of land and stationary infrastructure is an important factor in all that goes on in urban settlements. Property taxes and taxes on sales can incentivize upgrading of the energy efficiency of the building stock. Where land ownership is heterogeneous, local government can facilitate the roll out of refurbishment programs, often working with large property developers. Programs of this kind can make up a substantial part of the work of 'energy-service companies' or ESCOMs (Dayton et al., 1998). The lowest tier of policy instruments relates to individual components and their direct use. Construction standards when coupled with standards for energy-control provision have a long history of application. The diffusion of new technologies is *inter alia* influenced by social networks (Fisk, 2008) and local networks, of which local government is a part and can accelerate take-up.

The provision of a mass-transport system is an important option when travel densities are sufficiently high to provide savings over private transport. However, success depends, in part, on delivering a service that is universal rather than just for the most disadvantaged. That, in turn, implies a system with sufficient coordination to ensure feasible journey times for a wide range of journeys.

Finally, urban administrations have an important role to play in administration during energy-security events (IEA, 2005).

18.5.2   Drivers of Urban Energy Use and Main Policy Leverages

18.5.2.1   Introduction

This section synthesizes existing knowledge of the main drivers of urban energy use and related policy considerations. Traditionally, comparisons and analyses of energy use and the drivers of differences are carried out at the national level. In comparison, research on the factors that determine urban energy use is still in its early stages, severely hampered by the limited availability of comparable city-level data.

Keeping the above caveats in mind, the factors that determine urban energy use can be classified into a few major groups: *natural environment* (geographic location, climate, and resource endowments), *socioeconomic characteristics* of a city (household characteristics, economic structure and dynamics, demography), *national/international urban function and integration* (i.e., the specific roles different cities play in the national and global division of labor, from production and a consumption perspectives), *urban energy systems characteristics including governance and access* (i.e., the structure and governance of the urban energy supply system and its characteristics), and last, but certainly not least, *urban form* (including the built urban environment, transportation infrastructure, and density and functional integration or separation of urban activities).

These factors do not work in isolation, but rather are linked and exhibit feedback behavior, which prohibits simple linear relations with aggregated energy use. The interaction between the driving factors may change from city to city – moreover, many of the factors are dynamic and path dependent, i.e., are contingent on historical development. There is, however, one factor that underpins all these determinants in a complex and nondeterministic way: the *history* of a city. The location of a city and the initial layout of its urban form are determined historically: witness the difference between sprawling North American cities that developed in the age of the automobile and older, compact European cities that developed their cores in the Middle Ages. Likewise, the economic activities of a city often stem from historical functions, whether as a major harbor, like Cape Town and Rotterdam, an industrial center, like Beijing now and Manchester historically, or a market and exchange center, like London, New York, and Singapore. These historical legacies may have long-term implications on urban energy use. However, there are also cases in which relatively rapid changes in the historical layout and/or the economic role of a city occur. This can be the result of war, natural disasters, or rapid socioeconomic transitions, such as industrialization or deindustrialization. Examples are Tokyo after World War II, Beijing in the past decade as transformed by China's accelerated transition from an agrarian to an industrial society, or many Eastern European cities after the fall of the iron curtain in 1989 and the subsequent economic restructuring from a centrally planned toward a market economy.

18.5.2.2   Geography, Climate, and Resource Endowments

Climate is an important factor in determining final energy use, especially for heating and cooling demands. Its influence on energy use can be measured through the metrics of heating and cooling degree days, which, in combination with the thermal quality of buildings and settings for indoor temperature, determine energy use. Urban energy demand is, in principle, not markedly different in its climate dependence than that in nonurban settings or national averages, but it is structured by the influence of other variables, such as urban form (e.g., higher settlement densities lead to smaller per capita residential floor areas), access to specific heating fuels, or income (e.g., more affluent urban households

use more air conditioning), that can amplify or dampen the effect of climate variations on urban energy demand.

National studies illustrate the quantitative impact of climate variables on energy demand. For example, Schipper (2004) reports differences in space-heating energy use (measured as useful energy) normalized to heating degree days and square meters living space for seven industrial countries. This analysis reveals substantial ranges from 50 kJ/m²/degree-day for Australia to 250 kJ/m²/degree-day for the United States in the early 1970s, and from 60 (Australia) to 160 kJ/m²/degree-day for Germany in the mid 1990s. Assuming a residential floor space of 100 m², a difference of 1500 heating degree-days, which is characteristic between northern (Denmark) and southern (Greece) Europe, translates into a variation in residential energy demand between 9 and 24 GJ, which is significant compared to a typical European household residential energy use of some 60 GJ, but nonetheless only constitutes between 9% and 24% of the typical 100 GJ/capita western European total urban final energy use. Conversely, little is known on the differences in the demand for thermal comfort as reflected in ambient indoor temperatures. A case study carried out for Metro Manila indicated that people in the highest income brackets have much lower indoor room temperature setting preferences, which leads to an increased air-conditioning demand (Sahakian and Steinberger, 2010).

The relationship between climate and urban energy use is a two-way street: climate not only influences urban energy demand, but urban areas also influence their local climate through the 'urban heat island' effect (see Section 18.4.2.4). This effect can reduce the heat demand during winter, but also enhance the need for cooling in the summer, especially in warm and humid climates. Studies show increases in the summer time cooling load in tropical and midlatitude cities (Dhakal et al, 2003). A series of studies on California show that a 0.5°C increase in temperature causes a 1.5–3% increase in peak electricity demand (Akbari et al., 1990; 1997).

To a certain extent cities inherit the resource dependencies of their respective countries, which explains, for instance, the continued use of coal in urban areas in countries endowed with large coal resources. The connection to national energy systems and their dependence on the resource base is especially pronounced for power generation, since cities often draw electricity from the national grid. In some cases, urban power plants are designed to use local resources, such as hydropower, geothermal, or wastes, but these potential resources are usually extremely limited in urban areas and provide only a small contribution to the high energy demand associated with high urban population and income densities. On the distribution and end-use side, district heating and cooling infrastructures, which allow large economies of scale, cogeneration, and energy-efficient 'cascading' schemes, are specific urban-efficiency assets, but only economically possible when the density of demand is above a threshold that warrants the investment.

18.5.2.3   Socioeconomic characteristics

The positive correlation between income and (final) energy use is long established in the traditional energy literature, especially for analyses at the national level. For the household level, correlations between income and energy use have been shown for the Netherlands (Vringer and Blok, 1995), India (Pachauri and Spreng, 2002), Brazilian cities (Cohen et al., 2005), Denmark (Wier et al., 2001), and Japan (Lenzen et al., 2006), with similar results for GHG emissions in Australia (Dey et al., 2007) and CO_2 emissions in the United States (Weber and Matthews, 2008). For Sydney, Lenzen et al. (2004) showed that urban household energy increases with household expenditure, and that most of this increase results from the energy embodied by goods and services, since direct final energy use, in contrast, increases only slowly with expenditure (albeit from high baseline levels).

Based on a production approach, urban per capita energy use is very often lower than nonurban energy use or the national average, particularly for postindustrial, service-sector oriented cities in the OECD countries (see Section 18.2.4; Brown et al., 2008; Parshall et al., 2010).

Figures 18.19 and 18.20 show the urban income-energy relationship from a production perspective. The GRP/resident is plotted in a cross-sectional analysis against energy use for a sample of Chinese cities (Figure 18.19). Figure 18.20 complements the Chinese cross-sectional analysis by a longitudinal analysis for six megacities. For both cases, income and energy increase together, albeit along distinctly different trajectories, which illustrates *path dependency*. Income is therefore far from the sole determinant of the level of energy use: for instance, Beijing and Shanghai have a higher average energy use than Tokyo, despite a lower per capita income.

In addition to income, demographic factors play a role in determining urban energy use (Liu et al., 2003; O'Neill et al., 2010). For instance, studies suggest that household size, that is the number of people living in one household, plays a role in energy use: above two people per household, economies of scale can reduce the energy used per capita (see above). This phenomenon is observed in India (Pachauri, 2004), Sydney (Lenzen et al., 2004), the United States (Weber and Matthews, 2008), and Denmark and Brazil (Lenzen et al., 2006). In Japan, in contrast, larger household sizes correlate with slightly larger energy use (Lenzen et al., 2006). Urban populations often have significantly smaller household sizes than rural populations because of smaller families and a larger generation gap, as well as smaller dwellings, and so shelter for extended families or many generations under the same roof is less likely.

The evidence for age is mixed. In Sydney, increasing age is correlated with higher residential but lower transportation energy use (Lenzen et al., 2004). Larivière and Lafrance (1999) found a positive correlation between residential electricity use with age for Canadian cities. At this point, not enough is known regarding the influence of age to make any general statement, much less predictions, applicable to cities with very diverse age pyramids that range from young and growing, to old and declining populations.

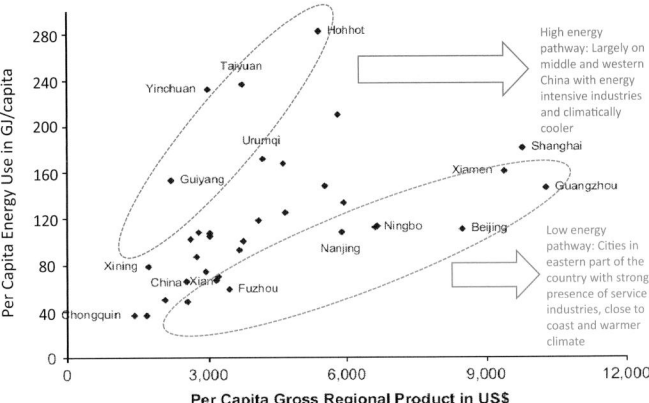

Figure 18.19 | Per capita energy use versus income for a sample of Chinese cities for the year 2006, illustrating path dependency. Per capita Gross Regional Product is expressed in $US_{2006}\$$ calculated using market exchange rates, MER. Source: Dhakal, 2009.

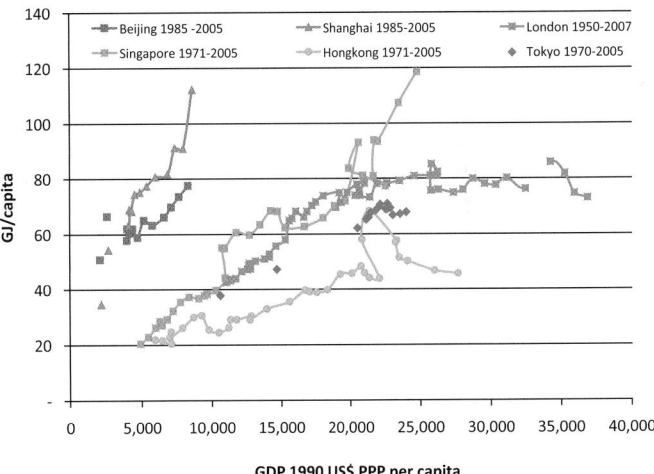

Figure 18.20 | Longitudinal trends in final energy (GJ) versus income (at PPP, in $Int_{1990}\$$)[28] per capita for six megacities. Note the path-dependent behavior. Source: Schulz, 2010a.

18.5.2.4   Role of the City in the National or Global Economy

A city's function in regional, national, and international economies has a strong bearing on its energy signature when measured from a production perspective. In the extreme case of Singapore (Box 18.5), a major center for oil refining and petrochemical production and a major international transport hub, the energy use associated with international trade in oil products, shipping, and air transport (usually subsumed[29] under 'apparent consumption' of the city's primary energy use) is four times larger than the direct primary energy use of Singapore and more than eight times larger than the final energy use of the city.

---

28   For comparison: per capita GDP (in PPP terms) in 2005 and in $Int\$_{2005}$ are: Beijing: 9238, Hongkong: 34574, London: 53145, Shanghai: 9584, Singapore: 29810, and Tokyo: 33714. (Note that a change in base year for the PPP metric changes the relative position of urban incomes in a non-proportional way.)

29   International bunker fuels are an important exception that, by simple definition, are excluded in national energy-use balances and the resulting emission inventories.

Box 18.5 | Singapore: The Importance of Trade

The case of Singapore illustrates the intricacies of energy (and emissions) accounting in trade-oriented cities that import primary energy, such as crude oil, re-export processed energy (fuels), energy-intensive products (petrochemicals), refuel ships and aircraft (bunker fuels), and import and export numerous other products and services that all 'embody' energy (see Figure 18.21). In terms of energy or CO_2-emission accounting, this extreme example amply illustrates the limitations of applying current inventory methodologies developed for national applications to the extremely open economies of cities. New, internationally agreed accounting standards are needed, as otherwise the risk of either misinforming policy or drawing arbitrary system boundaries is significant. There is a risk of 'defining away' energy use and emissions (e.g., international bunker fuels for aircraft and ships) associated with the inherent functioning of spatially defined entities (cities, city states, even small national open economies) whose interdependencies and energy/ emissions integration into the international economy provide for their very *raison d'être* and therefore need to be included in energy and GHG emission inventories.

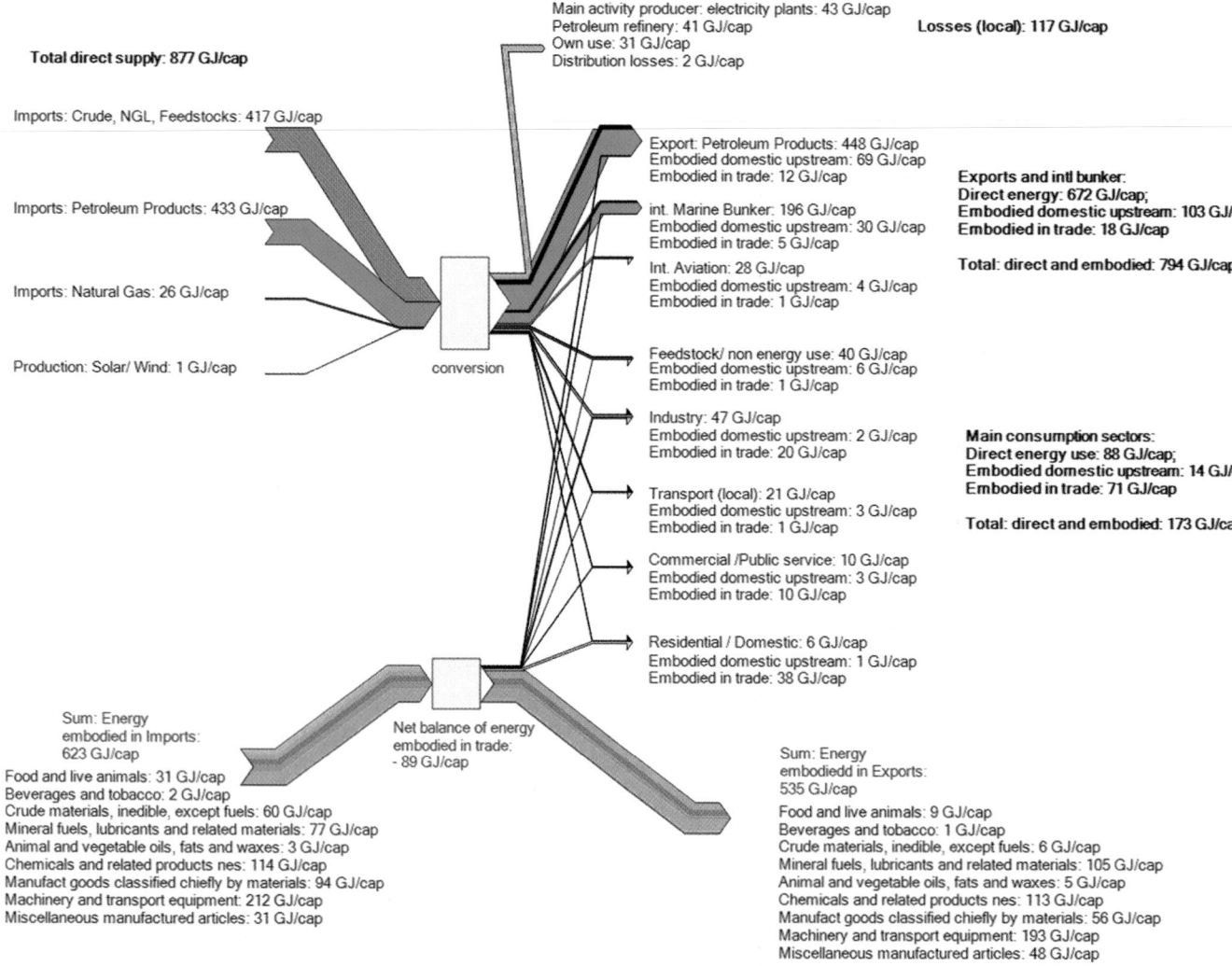

Figure 18.21 | Full per capita energy accounting for both direct and embodied energy flows of a large urban trade city, Singapore (in GJ/capita). Domestic direct and embodied energy use is 173 GJ/capita, but is dwarfed by the total energy imports to the city of 1490 GJ/capita. Total energy re-exports (direct and embodied) are 1225 GJ/capita. Source: Schulz, 2007; 2010b.

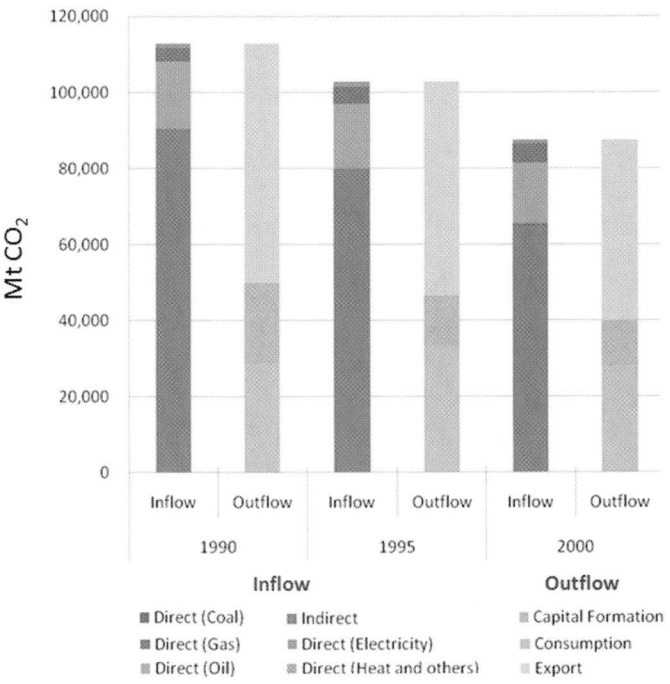

Figure 18.22 | CO$_2$ balance of Tokyo 1990, 1995, and 2000 using I-O analyses, in million tonnes CO$_2$. Source: Estimates by Shinji Kaneko and Shobhakar Dhakal.

That urban areas are usually in an intense process of energy exchange (imported and exported) with surrounding markets is again shown dramatically in Figure 18.22 for Tokyo in terms of CO$_2$ emissions. Emissions attributable to the direct and indirect energy and resource uses of Tokyo ('inflows') are balanced with the final consumption categories of these inputs as well as exports ('outflows'). Embodied energy and emission flows have gained increasing importance, comprising some 80% of Tokyo's 'inflows' and still slightly half of its 'outflows,' which illustrates Tokyo's embeddedness in the global economy. The Tokyo example also indicates the importance of energy and emissions embodied in maintaining and expanding the physical infrastructures and capital goods (reported under capital formation in Figure 18.22). Generally, private households only account for a small fraction of total capital formation (dominated by government, industry, and commerce). A consumption-based accounting that only uses household expenditures (as frequently done) therefore misses these important embodied energy and emission flows.

The 35 largest cities in China (China's key industrialization and economic drivers) are responsible for 40% of the nation's GDP and contribute overproportionally to national commercial energy use (Dhakal, 2009). Cities often specialize in certain types of manufacturing, commercial, or administrative functions. Some urban areas are also large transport hubs, such as London for air transit, or Cape Town and Rotterdam for shipping, that adds significantly to urban energy use, and is too often omitted from urban energy and GHG accounts. For instance, London's twin functions as a major international airport hub and as a global city result in an energy use from air transport that corresponds to one-third of London's total (direct) final energy use (Mayor of London, 2004).

A service-based economy can generate the same income with less energy than an economy based on the production of goods, which is one reason city per capita energy use in advanced, service-oriented economies is lower than national averages. This is also why Shanghai and Beijing have higher energy use per capita than Tokyo (see Figure 18.20 above), despite their lower GDP/capita.

If the economic activities located within a city determine its local energy use, its economic transactions with other areas entail energy use in those areas. Any product or service bought or sold entails energy use, and for service-oriented cities it may well be that the energy used, indirectly, through their economic transactions is larger than the energy used locally by their services industry. This phenomenon was shown at the level of urban household expenditures: rich households consume more energy indirectly than they do on housing, utilities, and local transit (Lenzen et al., 2004).

In addition to economic globalization, cultural globalization encourages urban upper and middle classes to adopt consumption patterns from global elites. Globalization-influenced urban development tends to favor private automobile-based individual transport modes and suburban sprawl for those who can afford it. Foreign direct investments (FDIs) and trade agreements affect the location and technology of manufacturing and commercial activities and labor reorganization (Romero Lankao et al., 2005). In China, individual cities compete with each other to attract FDIs and compromise their local environmental conditions and tax policies (Dhakal and Schipper, 2005). This type of intranational competition also occurs in other countries, such as Vietnam and India.

18.5.2.5   Energy Systems Characteristics: Governance, Access, and Cogeneration

The organization of energy markets and their controls at the urban level also influence urban energy use. Alternative organizational forms, such as state or municipal monopolies, cartels, or free-markets, impact access, affordability, and the possibility of implementing energy-saving policies. Localized energy monopolies may work closely with urban governments to further local policies, whereas free-market structures often challenge the enactment of environmental or social policies, such as renewable mandates, or the possibility of performance contracting. New York City requires (because of energy security and reliability concerns) 80% of electricity-generating capacity to be located within its territory; this means that the ability to influence the energy system is different to that in other cases. Vienna city owns its respective electricity, gas, and district-heating utility companies, and thus may have greater influence compared to cities with completely privatized and deregulated utilities. In Chinese cities, where energy companies are state-owned enterprises, the city government policies can exert strong influence on the suppliers, albeit less on the energy demand side. Many industrialized cities have put in place City Climate Actions Plans, which are expected to reduce or dampen energy use or promote shifts to renewables in the coming decades, but their success will depend on the links between city government and local energy providers. In many cities across the world, the local government is hardly able to influence the energy-supply side (because of jurisdictional and capacity limits), but may be in a position to address demand-side energy issues.

In developing countries, urban populations generally have higher levels of access to commercial energy forms than rural populations. This affects the efficiency and the intensity of the environmental impacts of energy use (Pachauri, 2004; Pachauri and Jiang, 2008): rural populations consume (often self-collected) fuels such as fuelwood, biomass, and coal; urban populations consume commercial and cleaner energy forms: electricity, oil, and gas. Owing to the low level of efficiency of biomass use, the quantity of primary energy use per capita may be similar in urban and rural settings (Pachauri and Jiang, 2008), but the different fuel structure in urban, higher income settings provides for much higher levels of energy service provision. In this sense, urban populations benefit from the high efficiency of energy-service delivery of modern fuels and distribution systems, such as electricity, gas, or bottled LPG. Access to commercial energy is much less an issue in industrialized or industrializing countries, which already have electrification levels at 100% (IEA, 2002) and where gas-distribution networks connect a majority of urban households.

Many European countries also have a long tradition of urban district heating (and more recently of district cooling) networks that either use district heating plants or CHP energy systems. CHPs, in particular, offer potential energy-efficiency gains as waste heat from electricity generation can be used for low- and medium-temperature heat demands in urban areas, with steam-driven chillers that also provide cooling energy. Traditionally, such centralized systems are capital intensive and only economic in higher density urban settings that provide for sufficiently high demand loads to warrant the investments. The recent advance in more decentralized energy solutions, including microgrids, allows such systems to be extended to lower density urban settings. Typically, cities with significant energy cogeneration have primary energy needs that can be 10–20% lower compared to systems in which all energy demands are provided by separate, individual conversion devices.

A key issue for the improved efficiency of urban energy systems is therefore an optimal matching between the various energy-demand categories and forms to energy-conversion processes and flows, usually achieved by exergy analysis (see Box 18.6).

Box 18.6 | Urban Exergy Analysis: Efficiency – How Far to Go?

An analysis of the efficiency of urban energy systems is far from a trivial task, but it is fundamental to identify options and priorities for improved efficiency in energy use. With respect to the system boundaries of the analysis, should the analysis extend to final energy (the usual level of market transactions in the energy field), to the level of useful energy, or to energy services? Should only simply energy outputs/inputs relationships be considered in defining efficiency (referred to as First Law analysis in the literature, after the First Law of Thermodynamics) or the analysis be extended to consider quality differences in energy forms (which energy form is most adequate for delivering a particular task) and efficiency, not in absolute terms (as in First Law analysis), but in relation to what thermodynamically represents an upper bound of energy conversion efficiency (as no conversion process that operates under real-world conditions can achieve 100% efficiency)? The latter concept is referred to in the literature as Second Law (after the Second Law of Thermodynamics), or *exergy* analysis (e.g., Rosen, 1992).

The literature (e.g., Nakicenovic et al., 1990; Gilli et al., 1995) identifies the value of both types of analyses (First and Second Law analysis), but also concludes that Second Law analysis enables us to extend the system boundaries to include also energy *service efficiency* (which cannot be captured in First Law analysis as it lacks a common energy denominator) and important quality characteristics of different energy forms and their adequacy to deliver a particular energy service. Therefore, an illustration of the value of exergy analysis to assess the efficiency of urban energy systems is provided here using the example of Vienna, which is compared to a few fast-track European urban-exergy analyses obtained from various research groups.

The energy system of the city of Vienna is characterized by a number of unique features. First is that the city generates much of its electricity needs within the city itself with the use of resulting waste heat through a district-heating network (recently also extended to a district-cooling network). As a result, the corresponding First Law efficiencies of Vienna's energy system are very high: 85% of secondary energy is delivered as final energy and about 50% can be used as useful energy to provide the energy service needs of the city (see Figure 18.23). The impact of cogeneration on the city's energy needs is also noticeable: without cogeneration Vienna's secondary energy use would be some 13% higher. The high First Law efficiencies suggest limited improvement potentials. However, this is not the case as revealed by a Second Law analysis of Vienna's energy system, which shows the efficiency between secondary and useful exergy is only some 17%. This suggests significant improvement potentials, for example via heat-cascading schemes that better match the exergetic quality of energy carriers with the required temperature regime of energy end-uses.

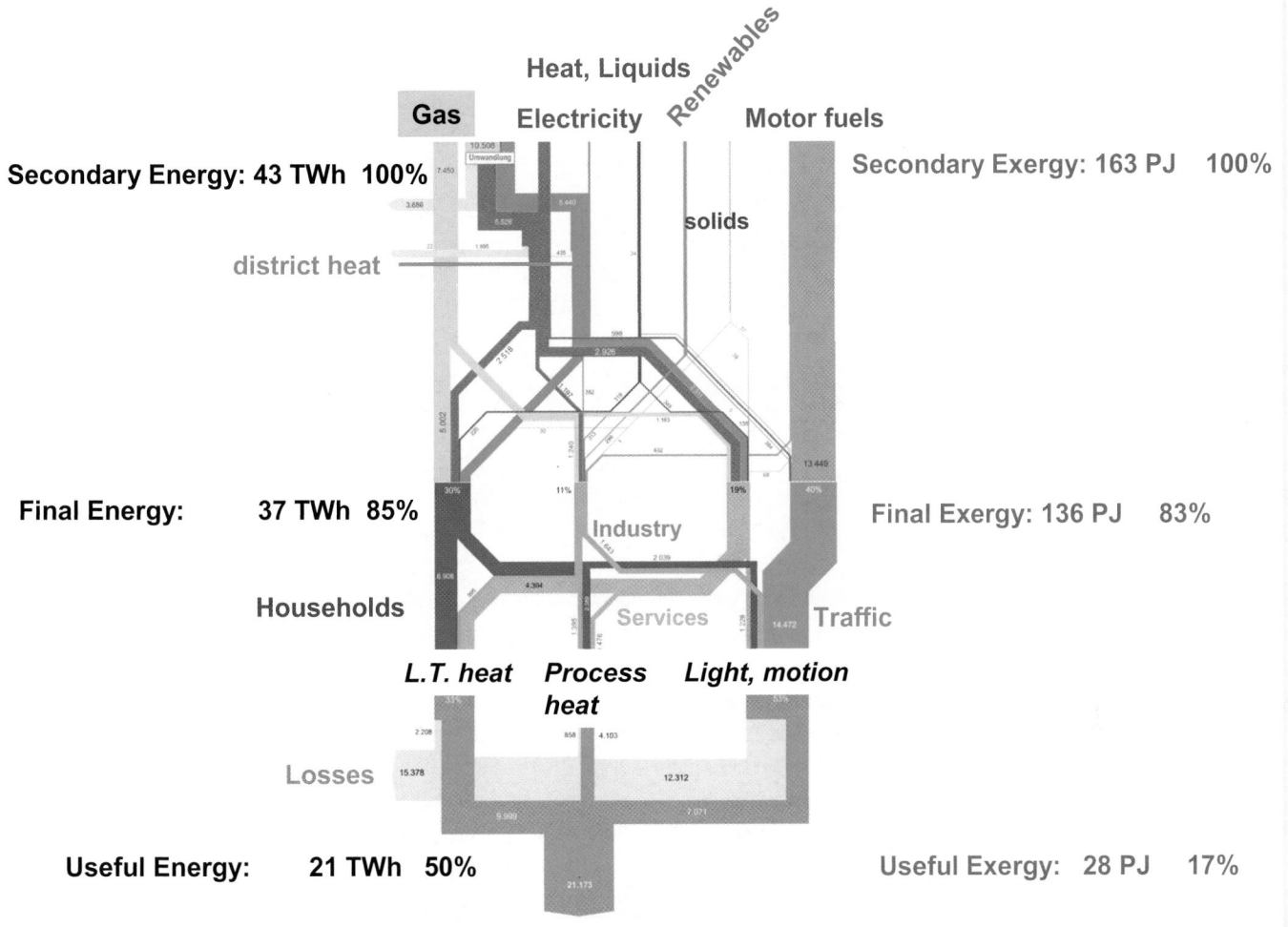

Figure 18.23 | Energy and exergy flows in the City of Vienna in 2007 between secondary and useful energy/exergy. Source: Energie Wien, 2009, (approximate) exergy efficiencies based on Gilli et al., 1996.

This assessment obtained the results of similar exergy analyses for Geneva, Switzerland (Giradin and Favrat, 2010), the Swedish city of Malmo, and London (Fisk, 2010). The results of the comparison in terms of the efficiency of useful exergy to that of secondary and primary exergy are summarized in Table 18.15.

Table 18.15 | Comparison of the efficiency of useful exergy to that of secondary and primary exergy.

	Useful exergy as % of:	
	secondary	primary
Geneva (CH)	23.2	15.5
Vienna (A)	17.2	
Malmö (S)	21.2	12.7
London (UK)	11.3	6.2
trad. Mexican village	5.7	

The results confirm earlier conclusions that, thermodynamically, urban energy systems could, in theory, be improved vastly, perhaps by as much as a factor of 20 (a similar order of magnitude as suggested by Nakicenovic et al. (1990) for OECD countries), and thus leave ample opportunity to realize feasible measures under real-world conditions and constraints that might deliver an improvement by at least a factor two. That modern urban energy systems are – despite their comparatively low exergy efficiencies – vastly more efficient (by a factor of 2–4) than traditional rural energy systems is also shown in the Masera and Dutt (1991) analysis of a traditional Mexican village with 2400 inhabitants using mostly preindustrial energy forms and conversion technologies (draft animals and fuelwood) for the provision of their energy services yielding an exergetic efficiency of only some 6%, compared to 11–23% for modern urban energy systems and uses.

18.5.2.6   The Urban Form: The Built Urban Environment and its Functions[30]

The built environment

The built urban environment comprises the totality of the urban building stock: residential, commercial, administrative, and industrial buildings, their thermal quality and spatial distribution (for a detailed exposition of building energy use, see Chapter 10). It also includes built urban infrastructures for transport, energy, water, and sewage. This environment is one of the key components for understanding the special characteristics of urban energy use as compared to rural, economy-wide, or global patterns. The unique concentration and overall scale of the built urban environment allow both economies of scale and economies of scope to occur, and thus provide options for energy-efficiency gains.

Building design

The design and thermal integrity (e.g., insulation levels) of buildings are essential for the amount of energy intensity (energy/m²) needed for heating and cooling. Reducing the energy associated with heating has been a strong focus in northern European countries, but midlatitude countries have to attempt a design a balance between heating and cooling energy demands. In many cases, newer buildings have better thermal standards, but in some cases they are poorly adapted to their climate (e.g., European- and US-style villas and apartment buildings in tropical climates, which do not have adequate shade and ventilation). Old buildings may suffer from lack of renovations, or renovations that do not apply the best possible standards. The influence of building technology on the energy used for space heating is huge: a Passivhaus standard requires that energy use for space heating be no more than 15 kWh/m² floor area per year; for low-energy houses the corresponding number is around 50 kWh/m², whereas poor thermal insulation may cause energy use for space heating of 200–400 kWh/m² in mid-European latitudes.[31]

The energy involved in the maintenance and replacement of components over a building's life should also be taken into account in assessing the energy performance of a building. For a 50+ year lifetime of office buildings, the embodied energy in construction materials plus the energy needed for decommissioning is estimated to range from 2.5 to 5 years of the building's lifetime operational energy use (Cole and Kernan, 1996; Scheuer et al., 2003; Treberspurg, 2005), with a typical value of embodied energy being between 5% and 10% of direct, operational energy needs of buildings. Including single and multifamily houses somewhat expands this range. The detailed literature review of Sartori and Hestnes (2007) reports a range from 4 to 15% embodied energy in total lifecycle energy use of buildings. Only in extremely low energy-use buildings (e.g., Passivhaus-standard or even below), with their extremely low operational energy use, does embodied energy play a somewhat greater role, reaching between 25% (Sartori and Hestnes, 2007) and 29% (Treberspurg, 2005): typical values of 20–30 kWh/m² building floor area/year of embodied energy compare to 50–60 kWh/m² building floor area/year for operational energy (heating plus electricity).

Type of buildings and uses

Next to the energy characteristics of an individual building, also the mix of building types and their density are important determinants of urban energy use.

The specificities of the urban built environment are usually a large existing stock, which requires renovation and maintenance, and new buildings in growing cities. The improvement in building stock to lower heating and cooling demands is counterbalanced by the increase in surfaces necessary to house new populations in growing cities, along with the demand of inhabitants for larger and larger apartments – even as the average household size decreases. Residential floor space per capita is known to be strongly correlated with income (e.g., Schipper, 2004; Hu et al., 2010). National averages in industrial countries range from 30 m²/person in Japan to 50 in Canada, 55 in Norway, and 80 in the United States (Schipper, 2004; US DOE, 2005). Typically, urban residential floor space per capita is lower than the national averages (to a degree counterbalanced by smaller household size), particularly for high-density cities with their corresponding high land and dwelling prices, but comprehensive statistics are lacking. For urban China, Hu et al. (2010) estimate 5 m²/person in 1990 and approximately 25 m²/person in 2007.

Newton et al. (2000) evaluated and modeled the energy performance of two 'typical' dwelling types – detached houses and apartments – across a range of climatic zones in Australia. Two main conclusions were drawn: (1) annual heating and cooling energy and embodied energy per unit area were similar for apartments and detached houses; (2) per person, however, the lifecycle energy of apartments was significantly less (10–30%) than that of detached houses in all circumstances, because the area occupied per person was much less. Norman et al. (2006) used a lifecycle analysis approach to assess residential energy use and GHG emissions, contrasting 'typical' inner-urban, high-density and outer-urban, low-density residential developments in Toronto. They found that that the energy embodied in the buildings themselves was 1.5 times higher in low-density areas than that in high-density areas on a per capita basis, but was 1.25 times higher in high-density areas than that in low-density areas on a per unit living area basis. Salat and Morterol (2006) compared 18th century, 19th century, and modernist urban areas in Paris, assessing five factors in relation to CO_2 emissions for heating: (1) the efficiency of urban form in relation to compactness; (2) a building's envelope performance; (3) heating equipment type, age, and efficiency; (4) inhabitant behavior; and (5) type of energy used. Salat and Morterol (2006) asserted that an efficiency factor of up to 20 could be achieved

---

30   A working paper on urban form and morphology contains a more extended discussion and is accessible at www.globalenergyassessment.org.

31   See http://energieberatung.ibs-hlk.de/

from the worst-performing to the best-performing urban morphology by taking these five factors into account. Salat and Guesne (2008) investigated a greater range of morphologies in Paris and found that when considering heating energy, the less dense the area, the greater the energy required for heating (see also Ratti et al., 2005).

Urban form and functions

Urbanization patterns affect the extent and location of urban activities and impact the accompanying choice of infrastructures. Newton (2000) summarized key alternative urban forms or 'archetypal urban geometries,' namely the dispersed city, the compact city, the edge city, the corridor city, and the fringe city. The merits of dispersed and compact cities ('suburban spread' versus 'urban densification') have been debated since the 19th century and a strong divide exists between the 'decentrist' (the dispersed city model) and 'centrist' (the compact city model) advocates (Brehny, 1986).

Nonetheless, one the most important characteristic of cities is density. Overall, a certain density threshold is the most important necessary (although not sufficient) condition to allow efficient and economically viable public transit (see Section 18.5.3 below). In addition, in a dense environment distribution networks are shorter, infrastructure is more compact, and district-heating and -cooling systems become feasible. Unconventional energy sources, such as sewage and waste heat, are also more accessible. High density may thus help curb urban energy use (Rickaby, 1991; Banister, 1992; Ewing and Cervero, 2001; Holden and Norland, 2005).

Most importantly, a compact city brings the location of urban activities closer. In the context of transportation, from cross-city comparisons it is well established that higher urban densities are associated with less automobile dependency and thus less transport energy demand per capita (Newman and Kenworthy, 1989; Kenworthy and Newman, 1990; Newman and Kenworthy, 1991; Brown et al., 2008; Kennedy et al., 2009). Intracity studies for Sydney (Lenzen et al., 2004), Toronto (VandeWeghe and Kennedy, 2007), and New Jersey (Andrews, 2008) also show that denser neighborhoods have lower per capita transportation energy needs. As a result, in many low-density cities, per capita energy use has grown at approximately the same rate as that in suburban areas (sprawl) (Baynes and Bai, 2009).

In many of the less-compact cities, transportation[32] by automobile is the biggest contributor to energy use (Newman and Kenworthy, 1999). The data suggest that cities with a density of 30–40 people/ha or greater developed a less automobile-based urban transport system with typical density thresholds for viable public transport systems given as above 50–100 people/ha (see Section 18.5.3). On average, residents who live at a distance of 15 km from an urban center use more than twice the transport energy compared to residents living 5 km from the center

(Stead and Williams, 2000). Nijkamp and Rienstra (1996) note that the private automobile has brought low-density living within the reach of large groups of upper and lower middle-class families. Moreover, correlations between automobile ownership and income suggest that more affluent automobile owners have a higher propensity to travel longer distances by energy-consumptive modes (Banister et al., 1997).

Diversity of function may also play a role in managing urban transport demand (Cervero and Kockelman, 1997). When strict zoning is enforced so that residential areas are separated from commercial, education, services, and work areas, private transportation is maximized. Mixed land uses and concepts of self-containment are important in reducing energy use in transport. Nevertheless, local jobs and local facilities must be suitable for local residents, otherwise long-distance, energy-intensive movements will continue (Banister et al., 1997). This coordination of land-use and transportation policies is termed transit-oriented development. The idea of location efficiency emphasizes the accessibility of opportunities, rather than how mobile one must be to find them (Doi et al., 2008); this is a central concept in recent approaches to transit-oriented development and other forms of sustainable urban development.

Also, urban density is an indicator of *potential* energy savings, especially in transportation. If infrastructure is inadequate to support the volume of traffic flow the resulting congestion can lead to higher energy use, even in high-density, built-up areas. For energy efficincy potentials of urban densities to be realized, a chain of interdependent, appropriate infrastructure, technical, and consumption decisions must be made. The correct level of public transit infrastructure requires large up-front investment and maintenance, from light rail to subways, trams, or dedicated bus routes. Adopting public transit also requires appropriate consumer behavior. In many North American cities, public transit is associated with lower economic status, and thus avoided by most people who can afford to drive, which reinforces the initial perception. A contrary example is Tokyo where the per capita energy use is smaller than in many East Asian megacities; one of the key reasons for this is the efficient rail-based public transport in Tokyo (Dhakal, 2004; 2009).

Another important energy implication of the urban form is the choice of urban energy-supply systems. District-heating and -cooling infrastructures, which allow large economies of scale and efficiency gains through cogeneration, are only possible when the density of demand is high enough to warrant the capital-intensive investment, unless such systems are mandated (and costs added to land prices). Compact urban form may also play a role in the energy used for buildings. Apartment buildings generate economies of scale compared to single-family homes, but apartment buildings may compromise decentralized low-energy design practices, such as natural lighting, ventilation, and decentralized use of PVs. Another important influence of density is at the personal consumption level. Apartment size per person tends to decrease with population density (with Hong Kong and Manhattan representing extreme examples). Effectively, the high competition for central urban space creates rents that contract floor space. However, in cities without

---

32 For a more in-depth discussion of transport energy use and its drivers see also Chapter 9 of GEA.

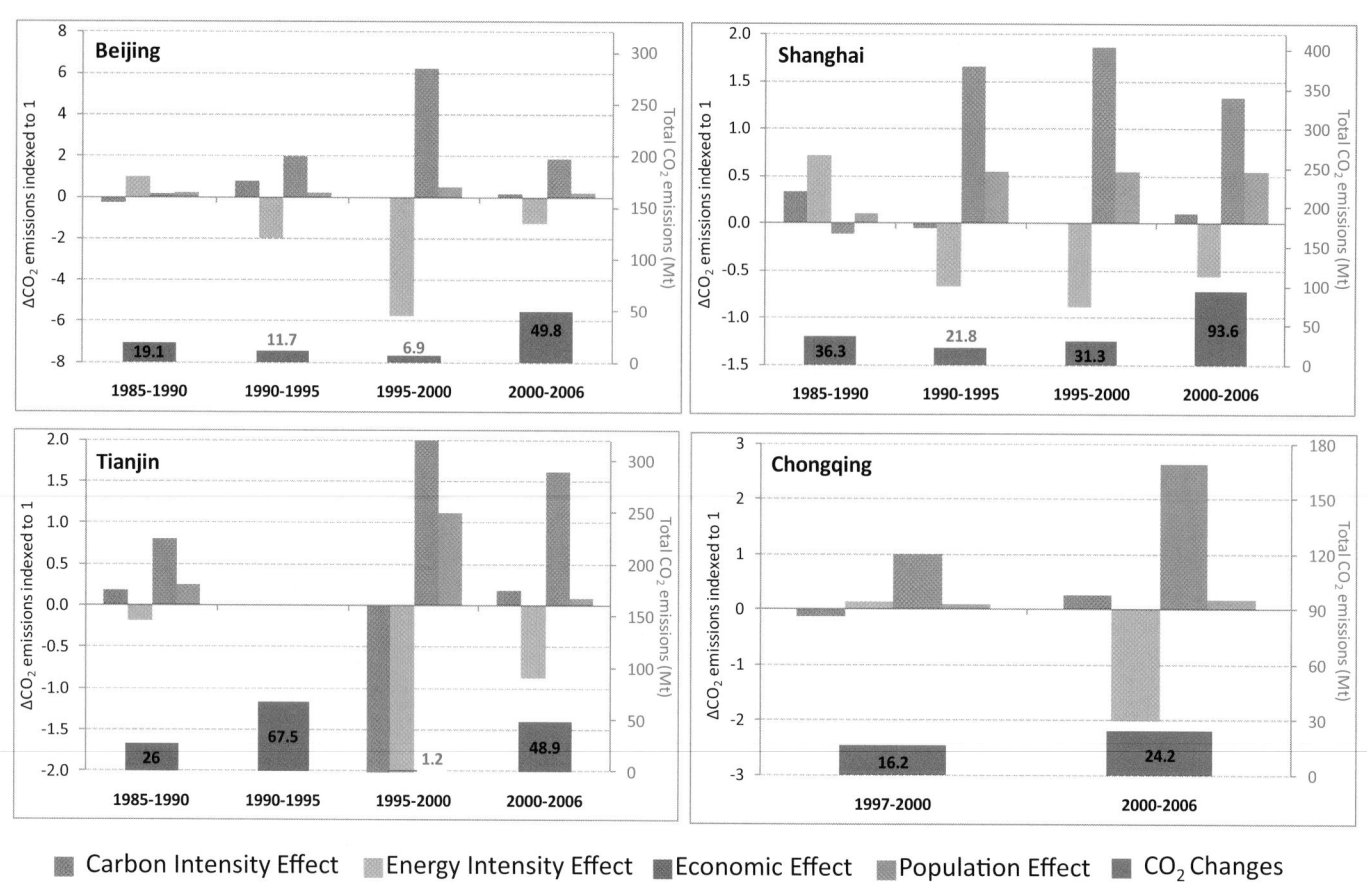

Figure 18.24 | Contribution of factors (indexed, left axis) to the changes in energy-related CO_2 emissions (million tonnes CO_2, right axis) for four Chinese cities. Source: Dhakal, 2009.

sufficient low-rental housing, even the smallest apartments can be out of reach for the poorer populations, who are forced to live in distant suburbs with poor transit connections. In many cities, suburbanization is also caused by industrial relocation from urban cores and the unplanned settlement of migrants and urban poor in the urban periphery.

More compact cities, however, may require special management to avoid the ill-effects of congestion and higher concentrations of local pollution (e.g., see Jenks et al., 1996). Urban heat island effects, for instance, may be exacerbated in dense urban cores. There may be a trade-off between the transport energy savings achieved with higher urban density versus the higher energy use of high-rise buildings. There are also trade-offs between urban density, dwelling type, block size, and the ecosystem services provided by vegetation. Both theoretically and empirically, it is by no means clear that there is an ideal urban form and morphology that can maximize energy performance and satisfy all other sustainability criteria.

18.5.2.7   Relative importance of the drivers of urban energy

No study so far has investigated the relative importance of all the factors known to influence urban energy use as described above. Existent

approaches rather contrast energy and/or CO_2 emissions with such macrodrivers as population, income, and technology, and thus follow the classic IPAT decomposition approach.[33] Such decomposition analysis has, for example, been carried out for several Chinese cities (Dhakal, 2009), where the relative changes in urban CO_2 emissions are decomposed into the factors population change, income change (measured as GDP/capita), and two technology factors: the carbon intensity of the energy system (measured as CO_2 emissions per unit of primary energy demand) and energy intensity (measured as primary energy demand per unit of GDP) for several periods of time. Although the relative contribution of these factors varies across cities and time periods, overall income is shown to be the most important driving factor for increases in carbon emissions (by far outpacing population growth), and improvements in energy efficiency to be the most important counterbalancing factor. The net result is, in all cases, an increase in carbon emissions, which indicates that economic growth has, to date, outpaced technology and efficiency gains (Figure 18.24).

Earlier work by Dhakal and Hanaki (2002) and Dhakal et al. (2003) for Tokyo using 1970–1998 data and for Seoul using 1990–1997 data also

---

33   IPAT: Impacts = Population × Affluence × Technology. For a history and discussion of the concept see Chertow, 2000.

shows that the income effect was primarily responsible for the majority of energy-related CO_2 emissions growth in Tokyo and Seoul in their respective high growth periods, that is 1970–1990 for Tokyo and 1990–1997 for Seoul. The analysis also showed that, despite an economic recession, energy-related CO_2 emissions continued to grow in Tokyo in 1990–1998, largely because of a drastic decline in the energy-intensity improvement rate (often observed in periods of economic growth stagnation or recession caused by the slower rate of capital turnover and hence the slowing introduction rate of more energy-efficient technologies and practices).

18.5.3   Transportation Systems[34]

18.5.3.1   Urban Travel Behavior

Introduction

To a large extent, traveling and mobility are a means to an end; they are necessary to enable people to fulfill essential functions (e.g., living, working, gaining education, acquiring necessary supplies, and relaxing) in the most suitable places. The situation is similar for the production of goods and services: fragmentation of the production process of goods and services implies different stages of the production chain, provided by different, more specialized enterprises at different sites, which improves the efficiency of the production at all. This causes travel demand, which is measured in different ways depending on the reasons for travel:

Frequency of trip making

This metric is a basic indicator for the degree of mobility and concurrent degrees of economic development and lifestyles. In urban settings the mean value of this figure ranges from 2.2 trips/day for people in low-income cities of developing countries (Padam and Singh, 2001) to up to 4.0 trips/day in industrialized countries (Hu and Reuscher, 2004; Sammer et al., 2009). It is expected that, in future, the frequency of trip making will rise slightly, particularly in developing countries. Generally, the frequency of trips is higher in cities than in rural areas because of higher degrees of specialization and division of labor and the higher number of attractive destinations with good accessibility, characteristic of urban areas.

Distance traveled per day

This metric reflects the transport modes used and the spatial structure and settlement density (NRC, 2009). The lower the settlement density and the more automobile-oriented an area, the longer the distances traveled per day. Globally, the mean values of these distances vary considerably, from 10 to 60 km/person/day in developing and industrialized countries, respectively (Salomon et al., 1993). In rural areas of developed countries the average distanced traveled per day is *higher* than that in cities. In developing countries the opposite is true: in rural areas distances traveled are generally very low (significantly below 10 km/person) and are significantly higher in urban areas because of both higher urban incomes and better availability of urban infrastructures and transport options. There is a close positive correlation between the distance traveled per day and transport energy use (Pischinger et al., 1997).

Travel time budget per day

This metric is an indication for the time spent traveling. For long periods this figure was comparatively stable with a small tendency to increase; it is currently on average around 70 minutes per person and day and 90 minutes per mobile person and day in countries with high motorization and automobile orientation (Hu and Reuscher, 2004; Joly, 2004). Assuming that, in the long run, working hours will decrease through continued productivity gains, this figure is expected to increase somewhat in future.

Traffic and mobility surveys help to determine the travel behavior, which is typical for various sections of the population and for different settings. Caution is, however, advised when comparing different mobility surveys because of differences in survey coverage and methods (e.g., does the survey include non-motorized modes?) and differences in sampled populations (commuters to work versus total population) and other methodological intricacies (e.g., weekday vs surveys that also include weekends). Therefore, there remain serious data gaps for comparable and up-to-date mobility surveys across a wide range of settlements and with comprehensive geographic coverage. The next sections summarize the current state of knowledge, with special emphasis on survey comparability rather than on how recent the survey date is. Despite important data limitations, some robust generic patterns can be discerned.

Modal split of cities

The modal split is an expression for the shares of different transport modes in the overall travel demand – usually measured as shares in total number of trips performed. It is a reasonable proxy for the evaluation of the environmental soundness of an urban transport system and its associated energy use. Higher shares of 'ecomobility' transport modes (including non-motorized modes like walking and cycling, as well as public transport) and corresponding smaller shares of private motorized transport imply a more environmentally friendly and energy-efficient transport system. The modal split can be conveniently depicted in a triangular diagram (Figures 18.25 and 18.26) that plots the respective shares of non-motorized modes (walking and cycling), public transport, and private motorized transport (automobiles, two-wheelers, etc.), respectively. In cities or towns with a modal-split point close to the center of the triangle, the three transport modes have fairly similar shares. In Figure 18.25, public transport dominates in the case of modal-split points close to the top of the triangle (e.g., New Delhi in 1994), private motorized transport is dominant in the lower right-hand corner (e.g., Chicago in 2001), and non-motorized transport in the lower left-hand corner of the triangle (e.g., Lackhnow in 1964). This

---

34 See also Chapter 9 for a review of transport energy.

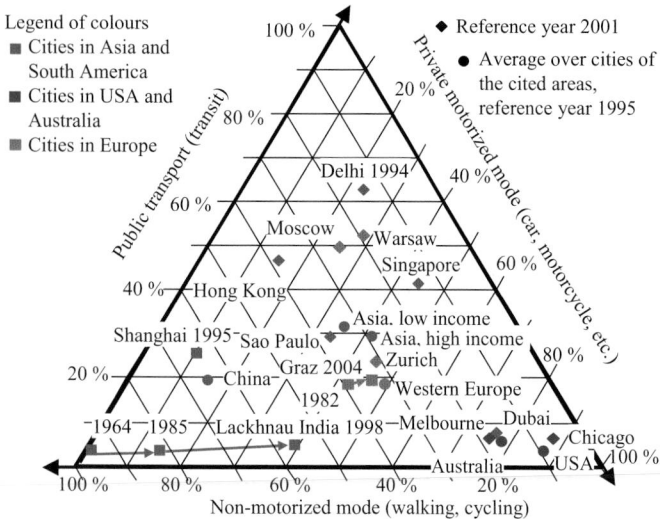

Figure 18.25 | Modal split for cities and towns on all continents (shares in trips). Source: Kenworthy and Laube, 2001; Padam and Singh, 2001; Zhou and Sperling, 2001.

representation enables the display of both cross-sectional and longitudinal observations.

Private motorized transport has a very high share between 88% and 79% in cities in the United States and Australia, where ecomobility has only a very low share of 12–21%. Thus, in these cities and towns transport systems depend strongly on private automobiles and so on fossil fuels. Cities and towns in Western Europe and well-developed Asian economies are roughly in the center of the diagram. There motorized private transport holds a share between 36% and 50%, that is transport systems are less automobile-oriented, and ecomobility holds a remarkably high share of 10–50%. In the cities and towns of China private motorized transport has thus far been fairly insignificant with a share of about 15%, but it is growing rapidly in importance.

Figure 18.25 indicates that with growing economic development and increasing incomes, the share of private motorized transport increases while the share of ecomobility decreases. This development is illustrated well by the longitudinal time trends of the modal split of the city of Lackhnau in India over the period 1964 to 1998. The United States and Australia seem to have reached a saturation level in this development; Western Europe and Asia have still some potential to develop further in this direction, unless they take countermeasures. This development does not follow any laws of nature: it is the result of transport policies and human behavior.

While Figure 18.25 shows the average modal split of cities and towns, Figure 18.26 shows the modal split for *commuter transport* in economically highly developed cities on different continents. These cities are divided in two groups: those with less and those with more than one million inhabitants.

For some selected cities the development of the modal split over time is also indicated, where comparable data are available. From the modal split for these cities and towns it is obvious that the range for individual elements is quite large, despite similar economic conditions and a high automobile-ownership rate. For example, non-motorized transport ranges from 4–50%. For walking as a separate mode, the range is from 4–21%, while cycling ranges from 0–39%. The range for public transport is also high, from 5–63%, while private motorized transport ranges from 24–80%. Thus, a high share of private motorized transport in towns and cities and an extensive use of automobiles are not an unavoidable outcome for cities in high-income countries, which illustrates the importance of urban form and transport policy choices.

The modal split of commuter transport (Figure 18.26) also shows a significant difference compared with all-travel purposes (Figure 18.25). The share of non-motorized modes is significantly lower and that of public transport higher, which is caused mainly by the longer journeys for working trips. There is no significant difference between large (over one million inhabitants) cities and smaller towns. The development of the commuting modal split over the long term shows a clear trend: motorized modes gain shares at the expense of non-motorized modes.

A considerable range for the modal split and the associated energy use can thus be observed in towns, cities, and urban agglomerations of industrialized countries, a range influenced by several factors: lifestyles, awareness of environmentally friendly and energy-saving travel behavior, objectives of urban transport policy and the willingness to implement such policies, spatial structures and settlement density, fuel prices, parking management, provision and operation of transport infrastructures for walking, cycling, public, and automobile traffic, pricing of the various transport modes (fees for both moving and stationary traffic, public transport prices), and priorities accorded to the various transport modes. Counter examples are easily found to the myth that a modern and economically viable city with a high quality of life can only be achieved with automobile-oriented private transport modes (Newman and Kenworthy, 1999; Directorate-General for Energy, 2007; Susilo and Stead, 2008).

Modal split and energy use

There exists a specific characteristic amount of energy use for every mode of transport, usually measured in terms of liters of fuel or in megajoules per passenger-kilometer traveled. The specific energy use of each mode is determined by the respective vehicle technology (e.g., size and/or weight of vehicle, engine efficiency, etc.), the vehicle occupancy rate (passengers per vehicle), and traffic conditions (degree of congestion). Overall, the specific energy use is highest for private automobile transport (Table 18.16 and Figure 18.27).

A robust finding, illustrated in Table 18.16, is that the choice of transport modes (i.e., the modal split) is a more important determinant of overall transport energy use compared to the specific energy

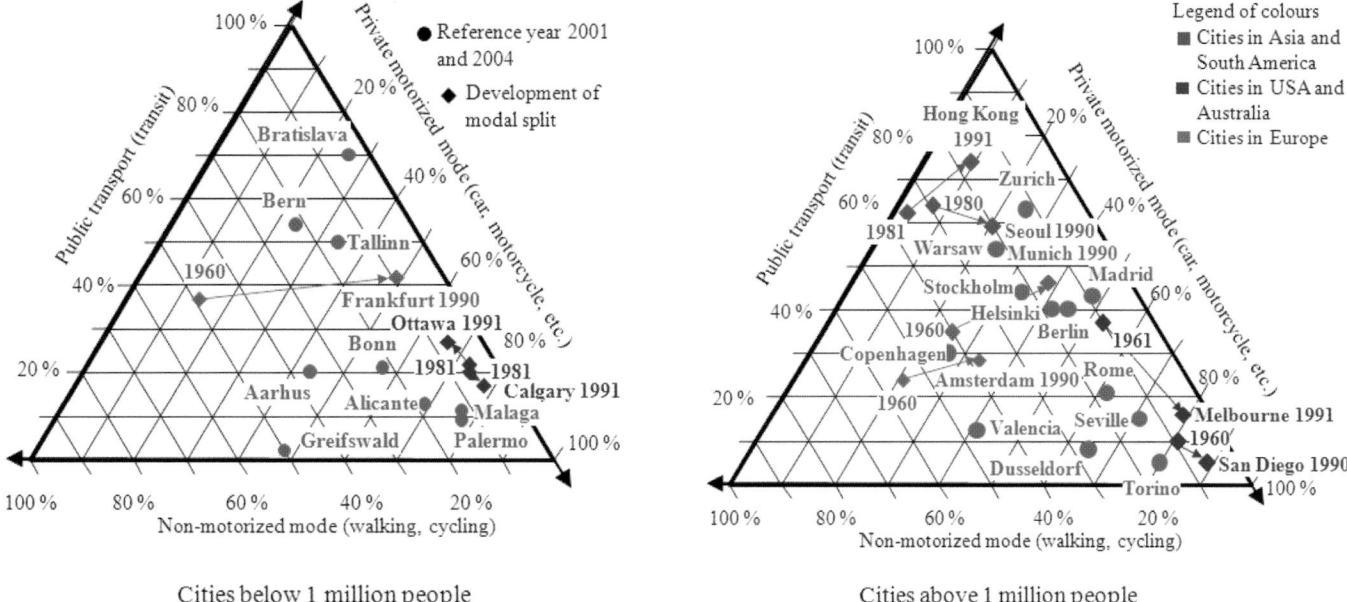

Figure 18.26 | Modal split of journeys to work in medium-sized towns with a population below one million people (left panel) and in cities with a population above one million people (right panel) in high-income economies for reference years 2001 and 2004 and selected time trends since 1960. Sources: Vivier, 2006; Steingrube and Boerdlein, 2009; Urban Audit, 2009; Wapedia, 2009.

Table 18.16 | Primary energy use per passenger-kilometer traveled for different modes, characteristic ranges for reference year 2005. Calculation based on Pischinger et al., 1997.

	Energy use per passenger-kilometer traveled	
	Fuels (liter/passenger-kilometer)	Energy (MJ/passenger-kilometer)
Private automobile traffic	0.050–0.075 (100%)	1.65–2.45 (100%)
Private motorbike	0.028–0.038 (55%)	0.92–1.25 (55%)
Public bus	0.009–0.013 (20%)	0.32–0.40 (20%)
Electric railways and public transport	0.002–0.004 (5%)	0.53–0.65 (35%)

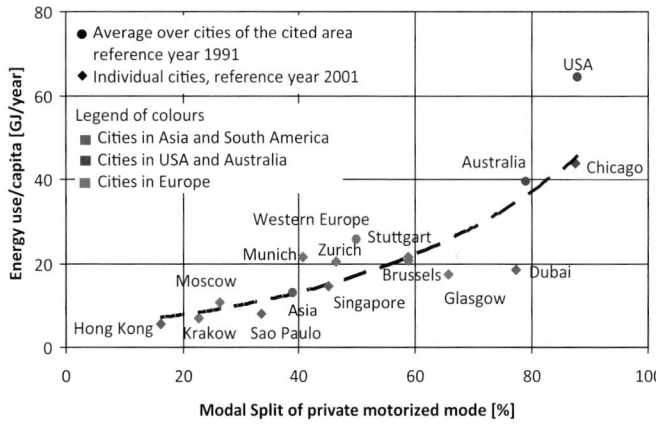

Figure 18.27 | Average energy use per capita in transport (average for countries and regions in 1991 and for selected global cities in 2001) versus share of private motorized transport in modal split. Source: data based on Kenworthy et al., 1999; Kenworthy and Laube, 2001; Vivier, 2006.

use of individual transport technologies. Put simply, using an (even energy-inefficient old) public bus is a more energy-efficient mode of transport than using a 'cutting-edge,' energy-efficient hybrid private automobile if the occupancy rate of the public bus is high (and that of the car is low). Policywise, this implies that if the objective is to minimize energy use, towns and cities should attempt to achieve a position for their modal split that favors 'ecomobility,' that is non-motorized and public transport, rather than incremental improvements in vehicle efficiency or biofuel supply for individual passenger automobiles, even if the latter may constitute important complementary options for sustainable transport planning in the interim before policy measures in urban form, traffic planning (especially for non-motorized mobility), and improved public transport systems take a long-term effect.

18.5.3.2   Specifics of Urban Transport Systems

Compared to rural and long-distance traffic, urban transport systems have some unique characteristics. Towns, cities, and urban agglomerations are places of high population and facility density. Their communication needs are high. This leads to a high density of travel demand within a limited space. Therefore transport modes are needed that require little space while offering high performance (Table 18.17). Pedestrians need the smallest amount of space, followed by public transport, cycling, shared taxis, and private motorized transport. The space required by the different transport modes and their capacity also depends critically on

Table 18.17 | Comparison of characteristic capacities of modes (assuming free-flowing traffic), space required with typical occupancy rates at peak traffic hours, infrastructure costs in urban settings, and maximum accepted distance for daily trips (ÖVG, 2009; Sammer et al., 2009).

	Non-motorized mode		Motorized mode			
	Walking	Cycling	Motorbike, two wheeler	Automobile	Shared taxi	Public transport
Capacity of a 3 m lane (person/h)	3600–4000	3600–4000	4300–5000 (max. 7200–8500)	2300–3000* (max. 9500–12,000)	5000–9000	Bus: 8000–16,000 Tram: 18,000–24,000 Underground: 30,000–60,000
Space required per person (m²/person)	0.7–0.8	6.0–7.5	13.0–15.0	21.0–28.0	7.0–12.5	Bus: 1.25–2.5 Tram: 1.7–2.3 Underground: 0.75–1.50
Infrastructure investment costs for the space required per person (€/person)**	50–150	50–150	1500–3000	Urban road: 2500–5000 Urban motorway: 50,000–200,000	1250–2500	Bus: 200–500 Tram: 2500–7000 Underground: 15,000–60,000
Accepted distance for daily trips (km)	1–2	5–10	10–20	Practically unlimited	10–20	Practically unlimited
Type of mobility service	Door-to-door service, Temporarily unrestricted service					No door-to-door service, Scheduled service

* Average occupancy rate of 1.2 people/automobile at peak hours. **1 Euro = 1.245 US$ in 2005.

occupancy rates (load factors). The figures provided in Table 18.17 refer to peak-hour traffic, which means full occupation of public transport and about 1.2 people per automobile in private motorized transport, characteristic for average load factors during peak commuting times in high-income, high automobile-ownership cities.

With respect to energy use, emissions, required space, and noise, public transport is significantly superior to private motorized transport, if a sufficiently high occupancy rate of public transport modes can be assured. From the transport user's point of view, public transport has certain disadvantages compared to private motorized transport that can only partly be compensated for. Such disadvantages are the limited spatial and temporal availability because of a scheduled service with specific stops. To make public transport attractive, a dense public transport network and a high service frequency with short intervals is necessary. A headway of service of less than 10 minutes needs to be achieved so that passengers are able to use public transport spontaneously. To make access and egress attractive, the catchment area of a public transport stop should span not more than 300–500 meters. This can only be guaranteed if the densities of development of residential areas in cities (including traffic areas) are considerably higher than roughly 100 people/ha (10,000 people/km²) and densities of development within the whole urban area are higher than 50 people/ha (5000 people/km²). The large amount of space required for road areas and the low density housing types (e.g., single family dwellings) that preordain private

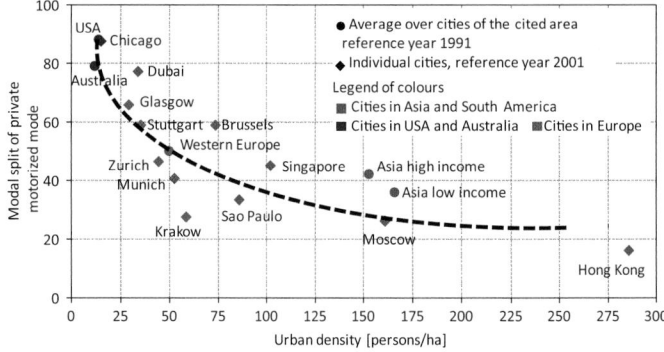

Figure 18.28 | Relation of urban density and share of private motorized transport modes (calculated from total mobility, including non-motorized modes) for individual cities and regional average cities. Source: Kenworthy et al., 1999; Kenworthy and Laube, 2001; Vivier, 2006.

automobile use mean that automobile-dependent urban settings typically have population densities far below these critical threshold levels (Figure 18.28) making attractive public transport not viable. An increasing share of private motorized transport modes causes a progressive increase of urban space and a progressive decrease of urban density. Since their density of development is unsuitable for offering attractive and economically viable public transport systems, automobile-dependent cities are Modd split of private molorized modes 'locked in' into high energy use and emissions.

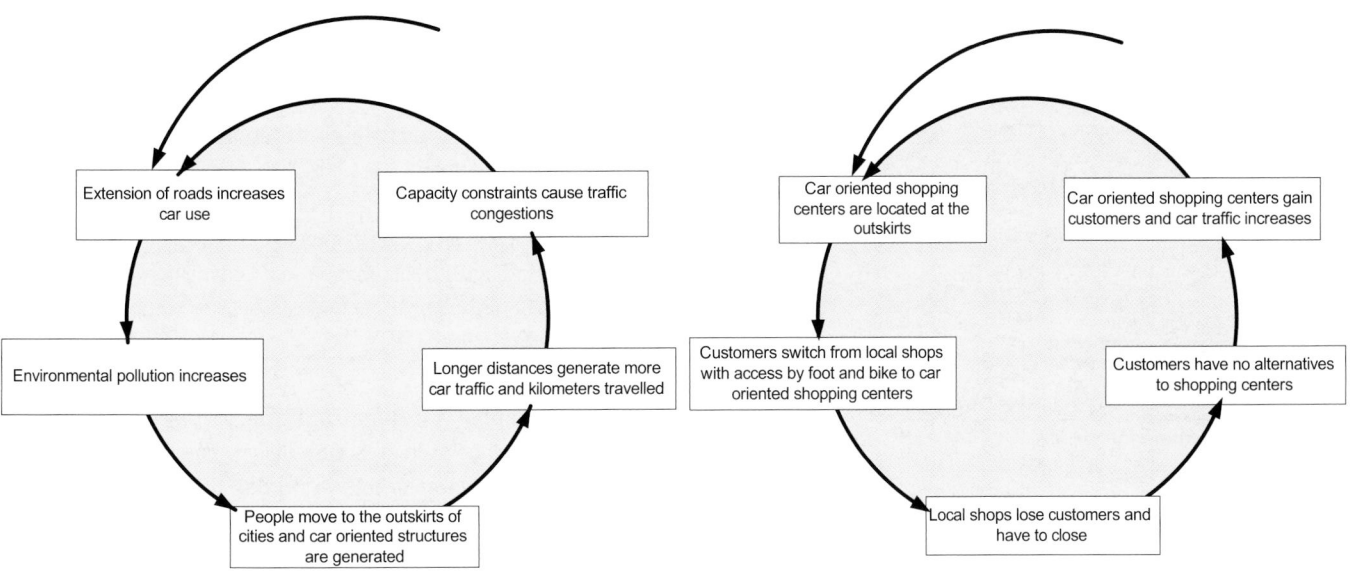

Figure 18.29 | Dynamic negative feedback effects between urban sprawl, automobile traffic, the quality of the environment, and viability of neighborhood stores in urban areas. Source: Sammer et al., 2009.

In terms of spatial planning, clear objectives with respect to sustainable urban-transport systems can be derived: stores for items of daily use need to be in locations that can be reached by non-motorized transport modes. The catchment area of such stores needs to be characterized by a sufficient density of use and compact settlement structures to make these stores economically viable. Thus, a concentration of stores for daily needs in shopping centers in automobile-oriented locations at the city perimeter should be avoided. The growth of decentralized shopping and other facilities leads to a negative feedback loop that further reinforces automobile dependence: fewer neighborhood stores, rising demand for shopping centers, increasing use of automobiles, and yet fewer customers for local stores (Figure 18.29). A similar negative feedback loop can be observed in the cause-effect chain from road-infrastructure development: the generation of more automobile traffic leads to environmental damages, lower urban quality of life, migration to suburbs, and then to yet further generation of more automobile traffic. Contrary to popular conception, any additional supply of urban road infrastructure (a frequent popular response to congestion by dense automobile traffic) inevitably also increases demand and entails more use of automobiles, ultimately simply shifting the point of congestion to higher levels of road and automobile traffic densities (Sammer et al., 2009).

Since trip lengths in local traffic are considerably shorter than those in long-distance traffic, high transport capacities at more moderate speed but with high traveling comfort are most important for short-distance, that is urban, traffic.

From an economic point of view, private motorized transport and public transport do not compete on an equal-level playing field, since the external costs of the two transport modes differ considerably. Table 18.18 offers an illustrative comparison of the external costs of road traffic and rail transport in Austria and Germany, where available data and estimation methods allow for one of the few 'apples-to-apples' comparisons possible in the extremely heterogeneous external-costs literature. Compared to rail transport, road traffic generally causes five to six times more external costs. The overall results for Austria and Germany are quite similar, but there are significant differences in some of the cost components, mostly caused by different assumptions with respect to external cost rates per externality category. (Thus, even in this case a fully consistent comparison of external costs is not possible. In the relevant literature (e.g., UNITE, 2003; Maibach et al., 2008), external cost estimates from transport vary considerably, which suggests the difficulty in arriving at widely agreed consensus values useful for policymakers.) The external costs caused by private automobile uses have to be paid for by other people. As far as the health costs of noise, emissions, and accidents are concerned, the people who finance the healthcare system have to pay for them, irrespective of whether they use an automobile frequently, infrequently, or not at all. The resulting costs of GHG emissions caused by fossil-fuel use in automobiles have to be paid by future generations that will face the consequences of climate change. From the perspective of sustainable transport systems and from an economic perspective of a fair competition among different transport modes, the externalization of costs needs to be overcome by *internalization*. Polluters should bear the resulting external costs by making them pay suitable fees or taxes. In reality this translates into an economic effect equivalent to a fuel price increase of at least a factor of 2–3, which does not look politically feasible at present. This suggests a more incremental policy strategy of a gradual phase in of external costs.

Table 18.18 | Estimates of external costs of road traffic and rail transport for passengers and freight in Austria and Germany, reference year 2005 (Pischinger et al., 1997; Sammer, 2009a) in €2005 cents per person- and tonne-km (1 Euro = 1.245US$ in 2005). For comparison, average total automobile operating costs (including taxes) per person-kilometer (paid by the driver) are about 24 €-cent per person-kilometer, including fuel cost of 6.5 €-cent (Austria, reference year 2005). External costs are thus at least up to twice the direct fuel costs (which, as mobility surveys show, dominate individual transport mode choices as few automobile drivers are aware of the total direct automobile mobility costs, including depreciation, maintenance, and operation).

External costs in €-cent per person-kilometer and ton-kilometer	Austria		Germany	
	Road traffic	Rail transport	Road traffic	Rail transport
Environment costs (caused by greenhouse and exhaust gas emissions, noise)	8.6	0.6	3.3	?
Accidents	3.1	1.5	5.1	?
External costs, Total	11.7	2.1	9.4	2.0

18.5.3.3   Strategies and Frameworks for Effective Policy Measures in Cities

This section summarizes the strategies, basic conditions, and measures to consider in policies for the effective reduction of urban transport energy use. Issues of vehicle technology and alternative fuels and the basic conditions for their use are not addressed here as they are covered in Chapter 9.

Strategies to encourage energy-efficient travel behavior need to be based on a systematic approach that takes into account the interaction of human behavior and travel demand with transport infrastructure supply, vehicle technologies, the environment, and financing.

The starting point is human decision-making which has a significant impact on urban transport systems and their energy use. Eight types of decisions by transport users have a crucial impact on urban transport systems and their effect upon the ecology, economy, and society. These decisions are usually taken in some sequence, with feedbacks taking place. Every decision taken has an impact upon subsequent decisions:

(1) **Selection of the place of residence and place of work.** The closer these two locations, the better equipped with local stores and other facilities, the more ecomobility transport choices are possible, mean less frequent motorized private transport and lower transport energy needs. To make this possible, compact settlement structures are needed; a minimum aggregate urban settlement density of well above 50 inhabitant/ha (5000 inhabitants/km²) should be the policy objective.

(2) **Selection of the availability of vehicles.** Ownership, availability, and types of vehicles (bicycle, motorbike, automobiles with different drive technologies and fuel use, season tickets for public

transport, membership in an automobile-sharing organization, etc.) have a significant impact on travel behavior and energy use.

(3) **Trip generation choice.** This is to decide whether some physical distance needs to be covered (trip) or whether an activity can be done at the place of residence or handled with the help of telecommunication technologies: good facilities at the location, for example broadband information and communication technology infrastructures and/or a garden, can help to reduce the need to travel any distances.

(4) **Decision about the time of travel.** Flexibility regarding the time of travel helps to save time, money, and energy, and to use resources in an environment-friendly way, because traffic jams and overcrowding of public transport can be avoided. This implies that wherever possible 'oversynchronization' of social activities should be avoided, for example through 'stacked' timing of school and of workplace operating hours..

(5) **Choice of destination.** A good retail, work, and school infrastructure close to the place of residence and a compact settlement structure help to avoid long commuting and shopping trips and the need to use motorized transport.

(6) **Modal choice.** Compact developments with good public (e.g., schools) and private (e.g., retail) infrastructures at the place of origin and destination, as well as good connections between the two with an attractive ecomobility offer (walking, cycling, and public transport) or a suitable offer of intermodal transport (bike-and-ride, park-and-ride, park-and-drive, park-and-bike, etc.) help to avoid unnecessary automobile trips. The quality of door-to-door connections is crucial for modal choices.

(7) **Route choice from origin to destination.** The concept of environmental zones (i.e., protected zones for residential areas, etc.) in connection with a hierarchical road network for motorized private transport and transport of goods structured according to the principles of traffic calming improves the quality of the urban environment. It helps to avoid through traffic of motor vehicles in protected areas and supports a high quality of housing and traffic safety.

(8) **Travel behavior reflection.** Reflection about previous travel behavior analyses the appropriateness of the realized choices and suggests a possible revision of the previous travel behavior in regular intervals. Global positioning system technology in connection with new traffic-information technologies will soon make it possible to check automatically whether decisions about traffic behavior are appropriate for the specified objectives (time requirements, environmental friendliness, cost of transport, etc.) or whether alternatives are preferable. This kind of individual

mobility-information system can lead to more sensitivity and awareness of transport-related decisions.

Principles and frameworks to reduce the fossil energy use in transport

- Traditional 'supply-side' measures that focus only on private vehicle infrastructure ("more and better roads") are insufficient to solve urban transport problems and risk significant consumer 'take-back' effects (i.e., induce additional automobile mobility rather than reduce congestion).

- A new fact-based systemic decision-making culture is needed for urban transport policy. Policymakers must be prepared to suggest and implement unpopular but necessary measures (e.g., internalization of external costs). It is essential that mere reactive and adaptive planning by individual measures without consideration of systemic effects and feedbacks (e.g., demand responses), such as the often ill-considered simple extension of road networks in urban areas ('adaptive planning'), is replaced by systemic and goal-oriented planning.

- As long as the true cost of various transport modes, particularly motorized private transport and public transport, are not transparent to the users through the lack of internalization of external costs, an unfair competition between transport modes is perpetuated. The 'push-and-pull principle' of restricting automobile traffic while simultaneously enhancing ecomobility modes can help to compensate for the lack of internalization of external costs.

- Cost-effectiveness criteria should be used to select policy measures to reduce fossil-fuel use and to make effective the use of limited available financial resources.

- An effective program to reduce urban transport energy use can only work as an integrated concept that takes all modes of transport into account. For the concept to be successful, regularly supervised and adapted quantitative objectives for the reduction of energy use are essential.

- To safeguard political acceptance, the development and implementation of the program needs to be supported by some suitable measures to shape stakeholders' ideas and guarantee their participation and engagement (Kelly et al., 2004).

18.5.3.4 Effective measures to reduce fossil energy use in urban transport

- **Spatial planning.** The creation of compact settlement structures with a sufficient settlement density well above 50 people/ha is essential to offer attractive non-motorized and public transport options. To this end, regulatory measures alone are insufficient; some market-based policy instruments are needed also, such as charging infrastructure development cost to users in areas of lower density.

- **Integrated planning concepts to save energy.** Since no single measure will reduce urban transport energy use, successful programs, including whole-bundle measures, need to be planned at national, regional, and local levels and implemented in a continuous process. Coordination and harmonization of transport policy across all levels is necessary.

- **Internalization of external costs.** In principle, an internalization of external costs is essential for all transport modes to make the polluters pay for the cost they cause. This is an excellent market-based instrument to reduce fossil-fuel use, which can be achieved by fuel surcharges, but also by parking fees and road pricing. The latter policy option is most effective if linked to the distance traveled and the environmental damage caused.

- **Environment-oriented road pricing.** Reduced fees for automobile pools or a variable and dynamic kind of road pricing, depending on utilization rates and congestion status (Supernak, 2005), create additional incentives to save energy. If road tolls are used only in urban areas or certain areas within cities, there is a high risk that undesired side-effects might occur, such as shifting routes to side streets, moving stores and other facilities to the outskirts where no fees are levied, and, in the long run, even a relocation of companies to areas outside cities (Sammer, 2009b), which would lead to undesirable urban sprawl.

- **Parking management schemes.** Measures to limit parking in cities and towns that cover all densely populated areas are highly effective – such measures include parking fees on public streets. These are particularly effective if they are graded depending on environmental friendliness or energy use of automobiles (Graz, 2010). Since there is considerably more private than public parking space in many towns and cities, it is recommended to include large private automobile parks (e.g., of industrial enterprises) within the parking fee scheme and combine this with levies on parking spaces for their operators (Sammer et al., 2007). In areas with a well-developed public transport system, land-use and zoning planning should fix an upper limit for available parking slots instead of (the customary) minimum number of mandated parking spaces (Sammer et al., 2005).

- **Attractive public transport.** The transport user wants a qualitatively high door-to-door mobility service, and does not care which operator is responsible for different parts of the system. Therefore the good cooperation of all public transport operators, an integrated ticketing and timetable system, and some overall responsibility for the integrated public transport system are essential for an attractive system. There is an obvious need to improve public transport with respect to intermodality, more efficient and rationalized operations, intermodal connections, information, and marketing. To compensate for some disadvantages, public transport should be given priority over motorized private transport by the provision of high-occupancy-vehicle (HOV) lanes and priority treatment at crossings with the help of suitable traffic lights. To make public transport more attractive, good synergy effects can be achieved by linking the system to non-motorized traffic (bike-and-ride, etc.) and by combining the measures already mentioned with restrictions of motorized private transport under a 'push-and-pull' policy strategy.

- **Cost-effectiveness of public transport.** When choosing types of public transport a careful check of costs and effectiveness is needed, since investment and operating costs of various systems differ considerably. In general, buses tend to be considerably cheaper than trams or suburban railways, which require tracks, and underground track-bound systems are more expensive than above-ground systems. Experiences show that Bus Rapid Transit (BRT) systems are very efficient public transport systems that provide a very attractive service at a reasonable cost level for industrialized and developing cities (e.g., Curitiba, Bogota). Renovating existing transport systems is most cost effective, for example by eliminating existing obstructions by automobile traffic or by giving public transport priority over motorized private transport with the help of HOV lanes and traffic lights.

- **Non-motorized traffic (walking and cycling).** Measures to encourage and support non-motorized traffic are a highly efficient way to save energy, particularly if they are combined with support of public transport and restrictions for automobile traffic. Suitable measures are the extension of an area-wide network of walkways and cycling routes and making them more attractive to use, more places (public and private) to leave bicycles, more information, marketing, and measures to shape people's ideas, bicycle renting, permission to transport bicycles in the public transport systems, company bicycles, etc. (Meschik, 2008).

- **Access restrictions for motorized private transport.** Environment-oriented access restrictions for automobiles in city centers are highly effective measures to save fossil fuel, for example the temporally limited or unlimited prohibition of access for certain types of vehicles, as in environmental zones to keep the air clean (Umweltzone Berlin, 2009).

- **Access contingents for automobiles with combustion engines.** This means that egress is restricted by an area-wide traffic light system in such a way that only so many automobiles are allowed to pass so that no congestion will be caused at the next crossing regulated by traffic lights. Outside these areas with such traffic-light

management, additional space needs to be provided for automobile traffic, and also HOV lanes and park-and-ride facilities.

- **Extension of road networks.** If the goal is to reduce fossil-fuel use, to extend the existing road infrastructure, as is frequently done in urban areas, cannot be recommended. Every extension of existing capacities induces an increase in motorized private transport, and further substitution away from other, more environmentally friendly transport modes.

- **Highway corridor management.** This concept aims to optimize traffic flow by suitable measures, such as HOV lanes, high occupancy-and-toll lanes by variable pricing, ramp-metering, information about suitable times, etc. Such management systems should help to avoid congestion and time loss because of congestion. It is doubtful that they would help to reduce energy use significantly in the long term. At present, some 'Integrated Corridor Management Projects' (RITA, 2009) are being conducted in the United States; they try to include public transport, for example by using emergency lanes as bus lanes.

- **Mobility management at the enterprise level.** This is a highly efficient tool. It includes many types of incentives to make individuals do without automobiles. Suitable measures are the support of automobile pools by providing incentives for their use (guaranteed home transport if the usual transport is missed, preferential parking slots, financial incentives), 'job tickets' for public transport (employers subsidize season tickets or provide them free of charge), repair and shower facilities for cyclists, etc. The effect can be increased by providing HOV lanes or by making large companies develop mobility-management plans (ICARO, 1999). In-company mobility management leads to a three-way win-win-win situation because employers, employees, and the general public all benefit. For mobility management at the company level to be successful, a permanent management process must be professionally run.

- **Voluntary programs to change travel behavior.** Such programs quite efficiently raise awareness of the impacts of energy-saving traffic behavior. They have the potential to reduce the use of fossil fuels of the target population in urban areas by 5–10%. Suitable ecomobility alternatives must be available for such programs to work. Currently, some attempts are being made to combine such programs with energy-saving measures in private households (Brög and Ker, 2009; DIALOG, 2010).

These generic policy options to reduce urban transport energy use are applicable to cities in industrialized and developing countries alike.

18.5.3.5 Summary and Conclusions for Urban Transport

- **Modal split.** The existing modal split of cities is one of the key determinants of urban transport energy use and also a good indicator for the progress toward improved sustainability of urban transport systems. A wide range of modal split and thus energy-use patterns can be observed across different urban settings. This variation in urban mobility is not *ex ante* given, but rather results from deliberate choices of individuals and of decision makers.

- **Public acceptance.** The factors that determine urban transport choices can be changed if a strong determination for sustainable traffic policy and a corresponding wide public acceptance of the overall goals of such a policy exists. This wider acceptance of the overarching goals is also the condition to implement some individual measures (e.g., traffic calming, parking fees, etc.) that often face public opposition.

- **Energy and environmental characteristics of transport supply.** An urban transport system has specific characteristics. For noise, emissions, energy use, and economic costs, non-motorized modes of transport are superior to motorized ones, and public transport is significantly superior to private motorized transport, provided a sufficiently high occupancy rate in public transport can be achieved which in turn is contingent on minimum density thresholds.

- **Precondition for attractive public transport:** To make public transport attractive, a dense public transport network and a high service frequency with short intervals is necessary. This can only be guaranteed if the densities of development of residential areas in cities are considerably higher than roughly 100 people/ha and densities of development within the whole urban area are more than 50 people/ha. The large amount of space required for road areas and the low-density housing types (e.g., single-family dwellings) that preordain private automobile uses mean automobile-dependent urban settings typically have population densities far below these critical threshold levels.

- **Strategies for more efficient energy use.** A remarkable number of suitable strategies and measures are known to reduce urban transport energy use. A very effective strategy is the promotion of high urban densities in combination with active promotion of a high-quality supply of non-motorized and public transport options combined with a restrictive automobile policy. Soft measures have generally a very high potential and high cost-effectiveness to reduce energy use, but politically they are the most difficult to sell. Transport fuel use can be reduced most effectively through well-coordinated bundles of policies and measures.

- **Internalization of external cost.** The key measure is the internalization of external costs of motorized private transport combined with the provision of a high-quality alternative offers of public and non-motorized transport. To be able to internalize external costs, political acceptance of this unpopular measure is essential.

18.5.4   Urban Energy Systems Planning, Design, and Implementation

18.5.4.1   Introduction

The energy-system elements and networks of a city reflect myriads of 'local optimizations.' The networks have thus evolved over time, but seldom exploit the opportunities for broader optimizations with other networks or urban form. They are consequently not usually resource efficient when viewed in aggregate. Systems and integration methods are now becoming available with the potential for reductions in direct primary-energy use by 30–50% without other significant physical impacts, except the advantages of reductions in externalities of energy use such as air pollution.

18.5.4.2   Modeling Urban Systems: The SynCity Model

Given the complexity of a city's energy and transport systems, it is not surprising that, to date, detailed holistic analyses of the interplay between urban form, a city's built environment, energy-demand characteristics, and its transport and energy systems have not been attempted. Bottom-up assessments of energy efficiency improvement potentials in different sectors have been developed for many cities to inform policy choices, but interactions, both in terms of potential synergies or trade-offs, cannot be explored by such compartmentalized approaches.

New computational modeling frameworks and access to new data sources promise to overcome these barriers. The relationship between key parameters, such as population density and energy, may be obscured in the real world by differences in other factors, such as wealth and income. To explore these interactions under comparable *ceteris paribus* conditions, this assessment commissioned illustrative model simulations with one such modeling framework (see Box 18.7), the results of which are reported here.[35]

In these examples, the synthetic city (SynCity) is an urban settlement for 20,000 people in a service-orientated local economy, in a moderate climate and with natural gas (and oil) as the primary fuels. Five SynCities were explored with the energy model optimizer. At one extreme, the optimization is constrained as a low-density city, fed from a power grid, with modest building-fabric energy performance. This city is taken as characteristic of one that has evolved in an economy in which resources are relatively inexpensive, such as the United States. The optimizer is left to choose the location of fixed infrastructures to minimize transport costs. The city at the other extreme is optimized with only a constraint on the lower bound to population density. It is comparable with an economy that is resource efficient, such as Japan. The location of housing and commerce, and the choice of whether to use embedded generation of power are left to the optimizer. Three intermediate cases are considered based on an intermediate density and imposed mononuclear layout (e.g., the United Kingdom).

---

35   Fuller detail is given in a GEA Chapter 18 working paper posted on the GEA website, www.globalenergyassessment.org.

Box 18.7 | SynCity Modeling Tool Kit

SynCity is a software platform for the integrated assessment and optimization of urban energy systems, developed at Imperial College London and supported by funding from BP. The goal of the tool kit is to bring together state-of-the-art optimization and simulation models so that urban energy use at different stages of a city's design can be examined within a single platform.

Three layered models within the system are used here:

- a *layout model*, which determines the optimal configuration of buildings, service provision, and transportation networks;
- an *agent-activity model*, which simulates the activities of heterogeneous agents that act within a specified urban layout to determine temporal and spatial patterns of resource demand; and
- a *resource technology network (RTN) model*, which determines the optimal configuration of energy-conversion technologies and supply networks.

The *layout model* is a mixed-integer linear programming model that seeks to satisfy urban demands for housing and activity provision, while minimizing energy demand from buildings and transport. Users specify average visit rates for each activity type and the model determines the optimal location for housing, commercial buildings and activities, and transport networks.

The *agent-activity model* is a simulation model designed to estimate the resource demands of a population that lives within a particular city layout. Briefly, the model operates as follows. First, it creates a synthetic population of individual agents with random characteristics, such as gender and education. Agents are grouped into household ensembles and assigned to jobs and dwellings. The model then loops over 16 indicative time periods that represent two seasons (summer and winter), two day types (weekday and weekend), and four time intervals during the day. For each interval, a probabilistic four-step transport model is used whereby citizens select an activity, an activity provider, a transport mode, and a travel route. The agents then move around the city and perform their planned activities, which results in spatially and temporally explicit demands for different end-use energy resources, such as electricity or heat.

The *RTN model* is also a mixed-integer linear programming model to determine the optimal configuration of energy-supply technologies to meet a given pattern of demand. The objective is to minimize the total cost of the energy-supply system that comprises the annualized cost of capital equipment (e.g., boilers, turbines, and distribution networks) and the annual cost of imported resources necessary to operate the system (e.g., supplies of gas and electricity). Users specify the full suite of possible technologies at the outset and the model returns the lowest cost system configuration.

Further details on the methodology are given in Keirstead et al., 2009.

The increased density in a compact urban layout means that individual dwellings are smaller and have less external wall area per dwelling, which results in reduced heating demands (and also in a saving of one-third, as for high-standard fabric implemented for all buildings, but at lower densities). So, efficient building design and urban density and form both yield comparable energy-demand reductions in the simulations. This highlights the importance of considering both policy options simultaneously, as to avoid the risk that the efficiency improvements of building structures with better insulation can easily be compensated by a shift toward less-compact settlement patterns.

Conversely, the construction of large houses in a low-density sparse layout increases these heat losses (one-third increase in primary energy) and also substantially increases transport energy use. This 'suburbanization' scenario, a worst-case scenario in the simulations, results in an almost three-fold increase in energy use compared to

the optimized solution. That is, building a Passivhaus standard single-family home in a low density (sub)urban area would not lower energy use substantially compared to remaining in a much less-efficient home located in a more compact urban setting (e.g., a 19th century townhouse located close to education, leisure, and shopping facilities) with its corresponding lower individual transport needs. This is clearly shown by the alternative simulated urban layouts of the model runs (Figure 18.30).

The results for each simulated city type are summarized in Table 18.19 and Figure 18.31. Numerical values are indexed to the annual primary energy use of the sparse city design (144 GJ/capita in the simulation). 'Upstream' energy is energy used at power stations to supply grid electricity to the city. 'Delivered' energy is the energy delivered to stationary infrastructure, including CHP plants, and the final end-users (i.e., a combination of final and secondary energy). The total of delivered energy,

Table 18.19 | Primary energy use of five alternative urban designs for a town of 20,000 inhabitants. Results are indexed with sparse city =100.

Type	Building fabric	Density limit	Electrical power	Layout	Transport	Delivered	Upstream	Total
Sparse	Medium	USA	Grid	Optimized	26	57	17	100
Distributed generation	Medium	UK	CHP	Mononuclear	17	52	0	69
Efficient buildings	High	UK	Grid	Mononuclear	17	28	12	57
Compact layout	Medium	Japan	Grid	Optimized	12	28	10	50
Optimized	High	Japan	Optimized	Optimized	12	31	0	43

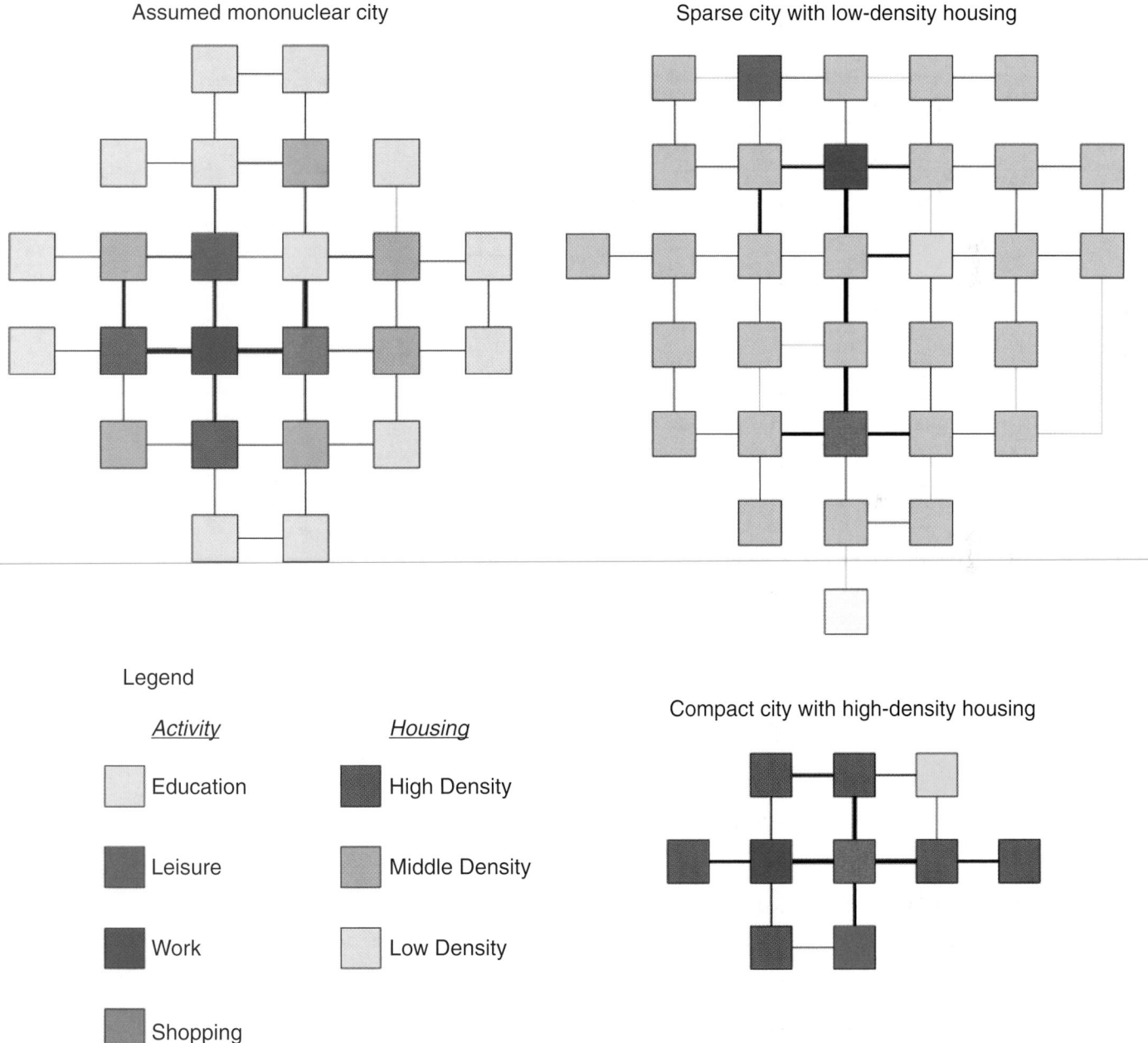

Figure 18.30 | Urban layouts, from left to right: the assumed mononuclear city, a compact city with high-density housing, and a sparse city with low-density housing. In each figure, the colored cells represent activity provision: green for leisure, blue for work, pink for shopping, and yellow for education. The gray cells represent housing at different densities, and the labels indicate the density in dwellings/ha. The black lines that connect the cells indicate road connections and indicative traffic flows.

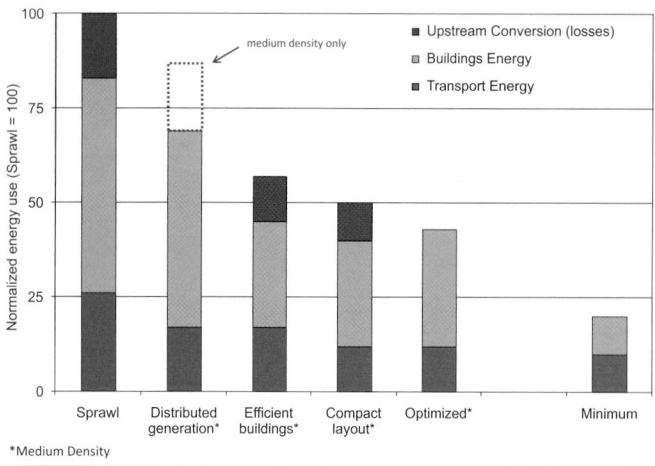

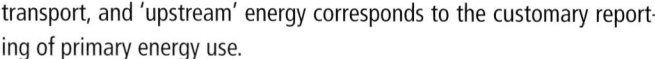

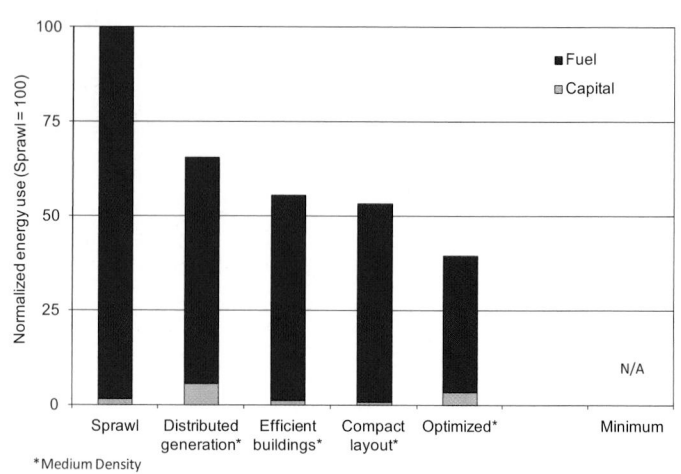

Figure 18.31 | Energy use for five alternative urban designs by major energy level and type. See Table 18.19 for definitions of the five simulations. The "Minimum" urban energy use estimates refers to implementation of the most efficient building designs (see Chapter 10) and transport options (see Chapter 9) available and which could not be considered in the scenario simulations.

Figure 18.32 | Total lifecycle costs (capital plus fuel) of the five city designs indexed to sprawl city = 100 (see Table 18.19 for definitions of the five simulations).

transport, and 'upstream' energy corresponds to the customary reporting of primary energy use.

The model results generally support the interpretation of comparative city data analyses. First, 'upstream' energy loss in power generation represents 20% of primary energy where the grid is used. To ignore this contribution and focus only on delivered energy misses important upstream implications of energy choices for power. Second, a SynCity with low resource efficiency is likely to consume about twice as much primary energy as one designed for high resource efficiency. Both transport and primary energy for heating and power are reduced by about the same proportion. This, in part, reflects that low-density cities not only require higher speeds of travel over longer distances, but that buildings tend to occupy larger areas with consequently more exposed surface for the same standard of construction. The effects of urban planning and differences in fabric standards are comparable and should be considered together with upstream consequences.

To address these three policy fields simultaneously is also of prime importance for the economic viability of cogeneration and district-heating systems. Energy-efficient, single-family homes located in low-density suburban settings are unlikely to yield the head-load densities required to install capital-intensive cogeneration systems combined with local district-heating grids (not to mention large centralized cogeneration and distribution systems, although these may allow room for other technologies, such as PV or ground-source heat pumps). To test the consequences, these city models were rerun to minimize whole lifecycle costs. The layout model (see Box 18.7) optimizes transport costs, but it is instructive to see what the resource technology model makes of

the stationary energy-service costs. The results in terms of total lifecycle costs are summarized in Figure 18.32 using UK electricity and natural gas costs as example. The discount rate used in the simulations was 6% in real terms.

The results encouragingly follow a similar pattern to that of primary-energy minimization, in part because capital costs remain a small part of the annuitized energy costs. However, minimizing only capital cost biases the outcomes away from the minimum energy solution. This emphasizes the importance of proper finance and pricing systems in bringing about optimal solutions for utilities and customers combined.

Summary

This brief analysis analyzed three ways in which cities can improve the efficiency of their energy systems: improving the quality of the built environment, increasing the density of the urban layout, and using integrated, distributed energy systems, such as CHP. A few general conclusions can be drawn from this case study:

- Cutting urban energy use by half is possible through integrated approaches that address the quality of the built environment (buildings efficiency), urban form and density, and urban energy-systems optimization (cogeneration).

- Final energy use is not a sufficient indicator of energy system performance. In cogeneration systems in particular, this metric may show an increase in delivered fuel use that masks upstream conversion and distribution losses. This effect also occurs in bioenergy-based supply systems. Primary energy use should, therefore, be the basis of scenario comparisons.

- Annual energy-system costs (i.e., the costs of energy conversion and distribution), but not demand-side measures, such as increased building efficiency or urban layout, are dominated by fuel costs. However, these costs are distributed differently between stakeholders in each of the scenarios. In current practice, most of the capital and fuel costs are paid by end consumers, whereas in a distributed energy system much more of the cost is borne by energy utilities. This suggests that to achieve overall system efficiency, policymakers should design markets that help utilities implement distributed energy installations despite their unique capital and fuel-cost structures.

Increased urban density and improved building efficiency deliver primary energy use and carbon-emission savings of about one-third each; distributed energy systems provide approximately 10% primary energy and carbon emissions saving. This indicates the importance of urban planning measures. These decisions – for example, on the building energy-performance standards or on the location of infrastructure – are difficult to change in retrofit and can lead to significant increases in energy use; in the cases studied here, urban sprawl led to a one-third increase in primary energy use. Efficient distributed-energy systems, to a certain extent, can be retrofitted into existing urban forms, but they too can benefit from long-sighted urban planning by encouraging sufficient demand density and by reducing the costs of the network infrastructure.

The above simulations suggest that the effects on energy demand of urban form and density, and that of the energy efficiency characteristics of technologies, processes, and practices (e.g., buildings) are of comparable magnitude, i.e., are comparable size 'mitigation wedges' to paraphrase a concept developed by Pacala and Socolow (2004). Conversely, the impact of narrow energy systems optimization (e.g., through cogeneration of renewables) is much smaller.

18.1.2.1   Urban Energy Systems Planning Design – Review

A holistic view of urban energy systems is multifaceted and relates to the analysis of city lifecycle, technologies, systems modeling, and optimization.

Existing versus planned layouts

The problem of planning improved urban energy systems is very different for existing and for new urban areas. The advantage of new developments (e.g., ecocities and zero-carbon cities planned in places such as China and the Middle East) is that holistic planning tools can be employed to integrate the design of the urban form, the built environment, the transportation infrastructure, and the energy system. Opportunities for resource efficiency – such as optimized transport, material, and energy cascading, demand heterogeneity, and robust network design – can be explored ahead of implementation.

Existing cities, however, are captives to their history and struggle to escape path dependency without massive capital investment. Nevertheless, relatively short turnover times for many technologies provide opportunities to improve in efficiency and integrated design. Large-scale upgrades of the built environment have been demonstrated,

for example in Germany, where new technologies at the building and building-cluster level are being employed and new ideas are emerging for new uses of existing infrastructure (e.g., using a natural gas network to transmit bioderived gas in cities, see also Chapter 10).

Energy-related technologies and systems
Demand-side management and reduction
The primary focus of efficiency in most cities is on the built environment (dwellings and commercial properties) and transport. In the built environment, there is considerable inertia and irrational behavior given the relatively low or even negative cost of GHG abatement through refurbishment. This suggests that there is a strong argument for prescription and regulation regarding building standards.

Supply side
The supply side has seen much innovation in resource-efficient network design, where integrated thinking pervades the design of new ecocities – each of which is sympathetic to its hinterland and optimized for its local climate. The integration of water and energy systems is coming to the fore in supply-side planning. In these planning applications there is an important role for systems modeling and Life Cycle Assements (LCA) in support of holistic rather than piecemeal perspectives to avoid burden shifting. Given that all city systems are subject to uncertainty and change, there is also a need for option-based design techniques that allow city growth and technological evolution and that avoid strongly path-dependent solutions.

Role of real-time systems
There is a long history of promising designs related to urban energy systems that have not delivered the expected performance. For new, resource-efficient city designs, enough of the budget and expertise must focus on the postdesign, operational phase. There are fascinating opportunities for ubiquitous sensing and computing to embed sensing and distributed 'intelligent' and autonomous control (inspired by, for example, cybernetic modeling, which indicates how local control rules can result in system solutions close to those optimized centrally). This enables effective management of real-time performance, and perhaps eventually will ensure that resource use is minimized in operation. The interactions between systems and citizens can be augmented through real-time pricing of energy and virtual energy markets, real-time displays of household, large building, neighborhood, city resource-use profiles, and personalized decision-support services.

18.5.5   Urban Air Quality Management

18.5.5.1   Air Pollution Trends

Energy and air pollution are closely linked because urban outdoor[36] air pollution is primarily a result of fuel burning in power generation,

---

36   Residential energy use, e.g., for cooking, can result in high levels of indoor air pollution and result in health impacts (see Chapters 4 and 19).

industries for domestic production and exports,[37] transport, and commercial and residential sectors. Low-quality fuels, such as coal, biomass, and high-sulfur diesel, emit more air pollutants than cleaner energy sources. From the urban energy-usage perspective, the literature shows that high urban density tends to be associated with lower per capita energy uses (see Section 18.5.2), which reduces air-pollution problems somewhat. However, the trade-off at this density reflects on the issue of air-pollution control. High density makes air-pollution control more urgent and requires better management systems, especially in the rapidly growing dense and large Asian megacities.

Historically, the concentrations of pollutants such as SO_2 and TSPs, which mainly result from industrial production systems were concentrated in cities, have declined in industrialized cities. In the United States, the average national SO_2 level at 147 sites in 2007 was 0.0038 parts per million (ppm), which is 68% lower than 0.0118 ppm in 1980, while the National Ambient Air Quality Standard remains at 0.03 ppm (USEPA, 2009). In Japan, SO_2 average levels have declined from 0.06 ppm in 1967 to 0.017 ppm in 1980, and further by 50% in 1980–2005 alone (MOE, 2005). A key component of industrial air-pollution control mechanism was the development and deployment of end-of-pipe technologies (such as flue-gas desulfurization and particulate removal), and the introduction of cleaner fuels under stringent air-quality legislations. The 'pollute first and clean up later' paradigm had actually worked for many industrial cities, but one generation suffered acute air-pollution problems. Currently, the SO_x and TSP pollutants in industrialized countries are no longer an issue for energy use in cities, because urban energy systems are dominated by electricity or other cleaner fuels, and emissions from large point sources are tightly controlled.

In developing countries, SO_x and TSPs have shown a decreasing trend in recent years (Figure 18.33) for key Asian cities. SO_2 is reasonably within WHO limits in these cities (except Beijing), while TSPs remain far higher than the WHO limits. Today, cities in developing countries can learn from the experiences of the industrialized countries and have access to the technologies developed previously to tackle SO_x and TSP. In contrast to the developed countries, industrial relocation and FDIs constitute a key aspect of air-pollution control in the developing countries, with both positive and negative effects. At the same time, relocation of existing dirty industries of cities from populated areas to less populated area or outside the cities has contributed to reductions in industrial air pollutants. Given that pollution-control technologies are readily available, the industrial air pollution in developing countries today largely results from either an inability to pay for the technology or inherent institutional, policy, and market problems.

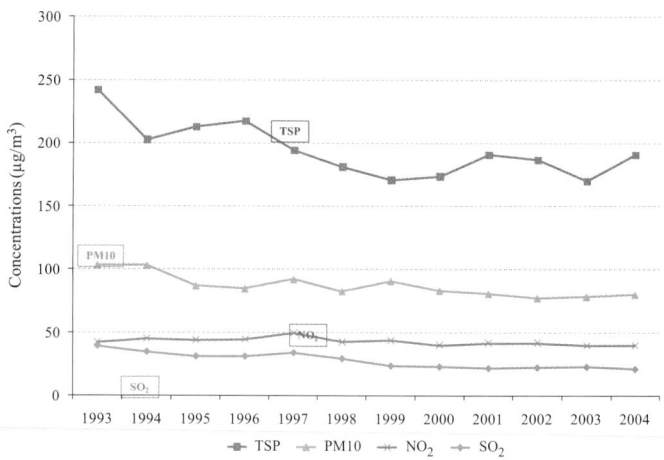

Figure 18.33 | Trends of major criteria air pollutants (1993–2004) for selected Asian cities. Source: Benchmarking Study on Air Quality Management Capability of selected Asian cities. Clean Air Initiatives in cooperation with UNEP and WHP (cf. Schwela et al., 2006). Cities included in the study: Bangkok, Beijing, Busan, Colombo, Dhaka, Delhi, Hanoi, Ho Chi Minh, Hong Kong, Jakarta, Kathmandu, Kolkata, Mumbai, Manila, Seoul, Shanghai, Singapore, Surabaya, Taipei, and Tokyo.

While industrial air pollutants are falling, the challenges to control pollutants from mobile sources, as a result of automobile dependency of cities, in particular for particulate matter (PM), NO_x, and O_3, are increasing. Even in the cities of industrialized nations, to reduce the levels of NO_x, O_3, and fine particles is proving a challenge. In the United States, monitored data show that the average levels of O_3 in 269 sites (0.078 ppm in 2007 – a reduction of 21% from 0.10 in 1980 – eight-hour average) slightly exceed National Air Quality Standards (USEPA, 2009). A recent report by the American Lung Association (ALA, 2009) shows that 125 million people (42%) in the United States live in counties that have unhealthy levels of either O_3 or particle pollution (PM-2.5). In Japan, the compliance rate of monitoring sites for O_3 is extremely low – merely 0.2% in 2004 (MOE, 2005).

PM-10 is one of the key public health issues in the cities of many developing countries, where their levels are many times higher than the WHO or USEPA. Only 160 million people in cities worldwide are breathing clean air, more than one billion need improved urban air quality, and for 740 million urban air quality is above the minimum WHO limits. While Figure 18.33 show that TSP and PM-10 generally have decreased in Asian cities, ambient concentration levels remain, nonetheless, above WHO, USEPA, and EU limits in numerous cities worldwide (Figure 18.34 and Table 18.20; See also Chapter 4).

A city-by-city analysis (Figure 18.35) further shows that PM-10 is lowest in Singapore at 30 µg/m³, which meets 50 µg/m³ standards set by USEPA (Singapore adopts USEPA standards), but fails if compared with the WHO 2005 updated guideline of 20 µg/m³. PM-10 values in all of the 20 Asian cities reviewed fail to meet WHO standards. Of 17 cities that monitor NO_x data, only eight meet the WHO guidelines, which indicates

---

37 As discussed in Section 18.6.2, global market integration and economic structural change can result in relocations of industrial activity and hence exercises an additional effect on local pollution.

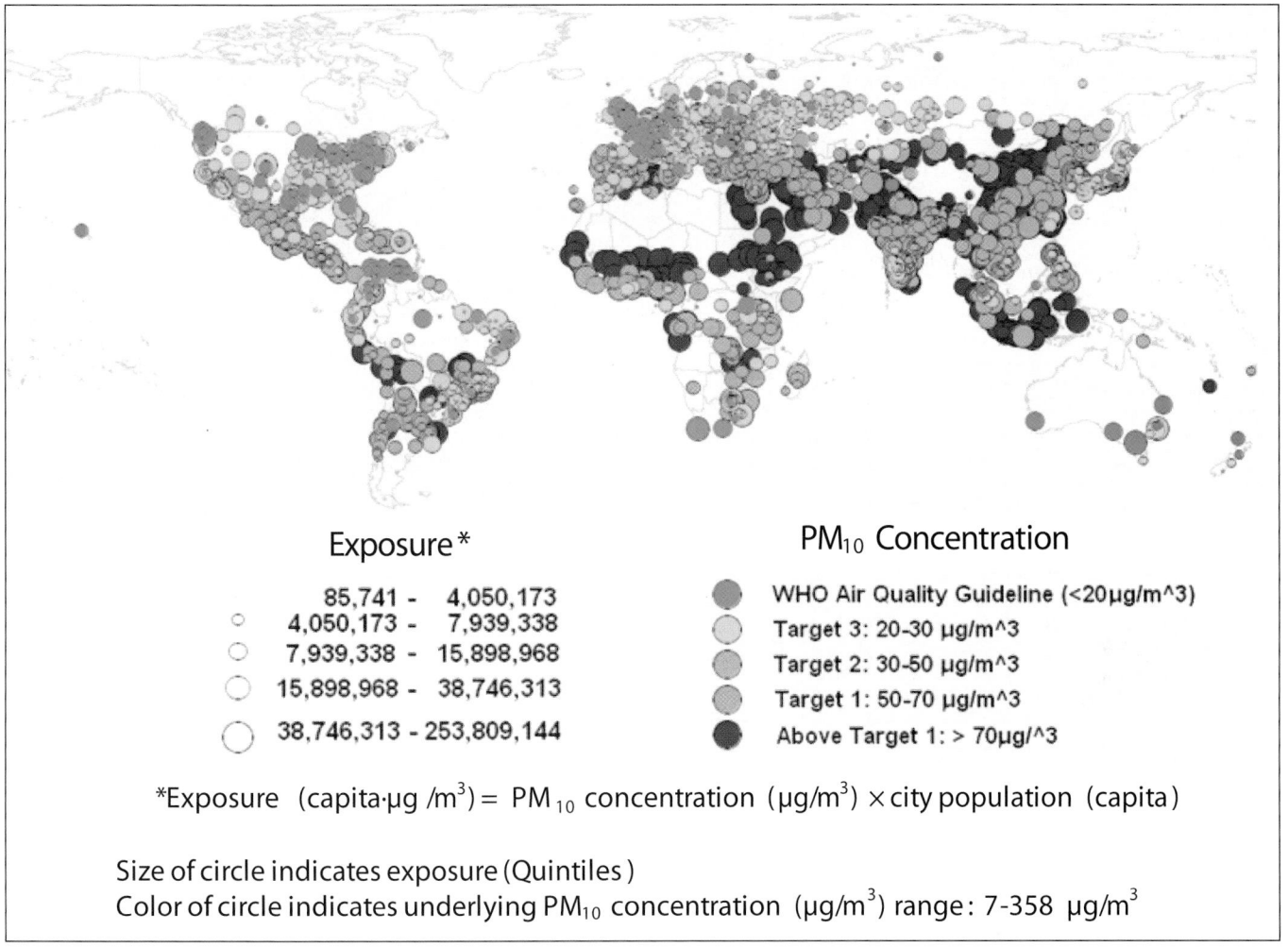

Exposure*

85,741 -	4,050,173
4,050,173 -	7,939,338
7,939,338 -	15,898,968
15,898,968 -	38,746,313
38,746,313 -	253,809,144

PM_{10} Concentration

WHO Air Quality Guideline (<20µg/m^3)
Target 3: 20-30 µg/m^3
Target 2: 30-50 µg/m^3
Target 1: 50-70 µg/m^3
Above Target 1: > 70µg/^3

*Exposure $(capita \cdot \mu g /m^3) = PM_{10}$ concentration $(\mu g/m^3) \times$ city population $(capita)$

Size of circle indicates exposure (Quintiles)
Color of circle indicates underlying PM_{10} concentration $(\mu g/m^3)$ range: 7-358 $\mu g/m^3$

Figure 18.34 | Human risk exposure to PM-10 pollution in 3200 cities worldwide (For a numerical summary see Table 18.20). Source: Doll, 2009; Doll and Pachauri, 2010, based on World Bank data.

that NO_x pollution is of special concern for Asia megacities. The reason is the rising number of private automobiles in cities of the developing world as a result of rising affordability, rising mobility demand, and slow development of public transport infrastructures. Over the years, the fuel efficiencies of new automobiles have improved considerably, the sulfur content of fuels are on a constant decline, and the emission standards have tightened (Figure 18.36), but the high volume of automobile travel demand has far overwhelmed vehicle-efficiency gains and the impacts from cleaner fuels. This confirms that urban air-quality management especially needs to address transport-related emissions from a much more systemic perspective, including transport policies that influence the urban modal split toward a reduced automobile dependence in addition to traditional vehicle efficiency, and exhaust emission and fuel standards measures.

While the majority of air pollution is associated with energy use, in many cases other sources also play an important role. Natural factors, such as dust and fine sand particles, flow across the boundary between the natural and anthropogenic sources, and also contribute to urban air pollution. The role of transboundary air pollution is particularly important for SO_x.

18.5.5.2   Examples of Air Pollution Control Measures and Urban Energy

A wide variety of air-pollution control policies and measures are in place globally. Some are system-wide and comprehensive measures, while others address a specific sector or technology depending on the prevailing sources of the air pollutants. Some of these measures are regulatory, while others are technological, managerial, or a mixture. Here a few representative examples are illustrated that touch a broad range of such measures, namely legislation, market, court rulings, and technology. Each has different implications on urban energy systems.

United Kingdom 'smoke control area' regulation

The United Kingdom started air-pollution control with a strictly source-control approach. It gradually shifted to a complex, but integrated, and risk management effects-based approach (Longhurst et al., 2009). Intense pollution from domestic coal use persisted in the United Kingdom until the 1950s and 1960s. Heightened concerns after London's Great Smog episode of December 1952 led to the introduction of The Clean Air Act of 1956 as an emergency measure. Significantly,

Table 18.20 | Number of cities and residing population categorized by ambient PM-10 WHO air quality standards (ACQ = WHO air quality guidelines met (less than 20 µg/m³), for definition of Target 1 to 3 concentration standards see Figure18.34) for a sample 3200 cities globally and by three regions (ALM = Africa, Middle East, Latin America and the Caribbean).

Global	# of Cities	Population (millions)
ACQ	446	164
Target 3	809	385
Target 2	777	409
Target 1	362	260
Above Target 1	803	739
Annex-1	# of Cities	Population (millions)
ACQ	325	121
Target 3	610	314
Target 2	371	183
Target 1	51	41
Above Target 1	26	12
ALM	# of Cities	Population (millions)
ACQ	115	41
Target 3	160	60
Target 2	228	126
Target 1	132	103
Above Target 1	205	160
Asia	# of Cities	Population (millions)
ACQ	6	2
Target 3	39	11
Target 2	178	101
Target 1	179	116
Above Target 1	572	567

Source: Doll and Pachauri, 2010 based on World Bank data.

some key industrial cities had already taken pre-emptive action, but the politics of a national measure proved more difficult. The Clean Air Act enabled local authorities to control pollution by declaring Smoke Control Areas ('smokeless zones' in which the burning of coal was banned) to whole or part of the district. Various measures were also used to ease compliance with the regulation, such as subsidies for furnace switching. In this regime, each local authority publicized the fuels that could be used and a list of exempt appliances. The Clean Air Act was further extended in 1968 to address the question of unevenness in the implementation, because in wealthier cities it was progressive while in other cities the implementation was less than that desired. The major feature of this regime was to induce a shift from coal to electricity, natural gas, and other cleaner forms of energy and implied a major transformation in energy esnd-use patterns and systems. However, with increasing levels of transport-related pollution in UK cities, NO_x concentration levels can be high and close to the statutory limits. This, in turn, may restrict further expansion of CHP systems in urban areas.

The regional clean air incentives market

The California South Coast Air Quality Management District (SCAQMD) has used a market-based system since January 1994, known as RECLAIM (Regional Clean Air Incentives Market), to reduce system-wide air pollution. SCAQMD covers 27,125 km² (10,473 square miles). At the launching of RECLAIM, it was expected to reduce the cost of achieving the same emissions reduction through a traditional command and control approach by some 40% (Harrison, 2004). Under this system, each facility participating in the RECLAIM program (facilities that emit 4 tonnes/year or more of NO_x and/or SO_x in 1990 or any later years) are allocated RECLAIM trading credits (RTCs) equal to their annual emission limits for SO_x and NO_x. The facilities must meet these allocated emissions limits. If a facility can reduce emissions more than required, they can sell surplus trading credits to the credit market, which can be bought by facilities that could not meet their own emission-reduction targets. The system is designed in such a way that the total allowable emission cap was reduced over time until 2003 after which it remained stable. It requires facilities to cut their emissions by certain amounts each year. By 2003, the program anticipated to reduce the total emission load for NO_x and SO_x in SCAQMD by 70% and 60%, respectively (Anderson and Morgenstern, 2009). In January 2005, RECLAIM decided to reduce cumulatively the emissions further, by 7.7 tonnes/day or about 20% by 2011 (Figure 18.37).[38]

Introduction of compressed natural gas vehicles in New Delhi

To address high concentrations of air pollutants in Delhi, the Supreme Court of India, acting on public interests litigation filed at the behest of civil society, directed the government of the National Capital Territory of Delhi and other authorities in Delhi to act on mitigating air pollution through specific technological interventions. The legal deliberations began as early as 1990, but the key court decisions were made in 1998. The cornerstone was the conversion of all diesel public transportation into compressed natural gas (CNG) vehicles, which began implementation in April 2001. Along with the implementation of these directives, authorities enacted a series of other measures, such as expansion of the scope of CNG coverage, scrapping of older vehicles, improving fuel quality, implementing stringent emission standards for vehicles, and improvements in the infrastructures. The implementation involved a series of policy instruments, such as penalties for noncomplying vehicles, sales-tax exemption, interest subsidy on loans to replace three wheelers, making CNG retrofitting kits available, expansion of CNG fueling stations, and others. The literature shows an ambiguous picture of the impact of CNG conversion on air pollution, but generally agrees that CNG conversion was one of the several key factors and that it triggered other measures that led to improvements in the air pollution of Delhi (Goyal and Sidhartha, 2003; Kathuria, 2004; Jalihal and Reddy, 2006; Ravindra et al., 2006; Kandlikar, 2007; UNEP, 2009).

---

38 For further updates see www.aqmd.gov/RECLAIM/index.htm.

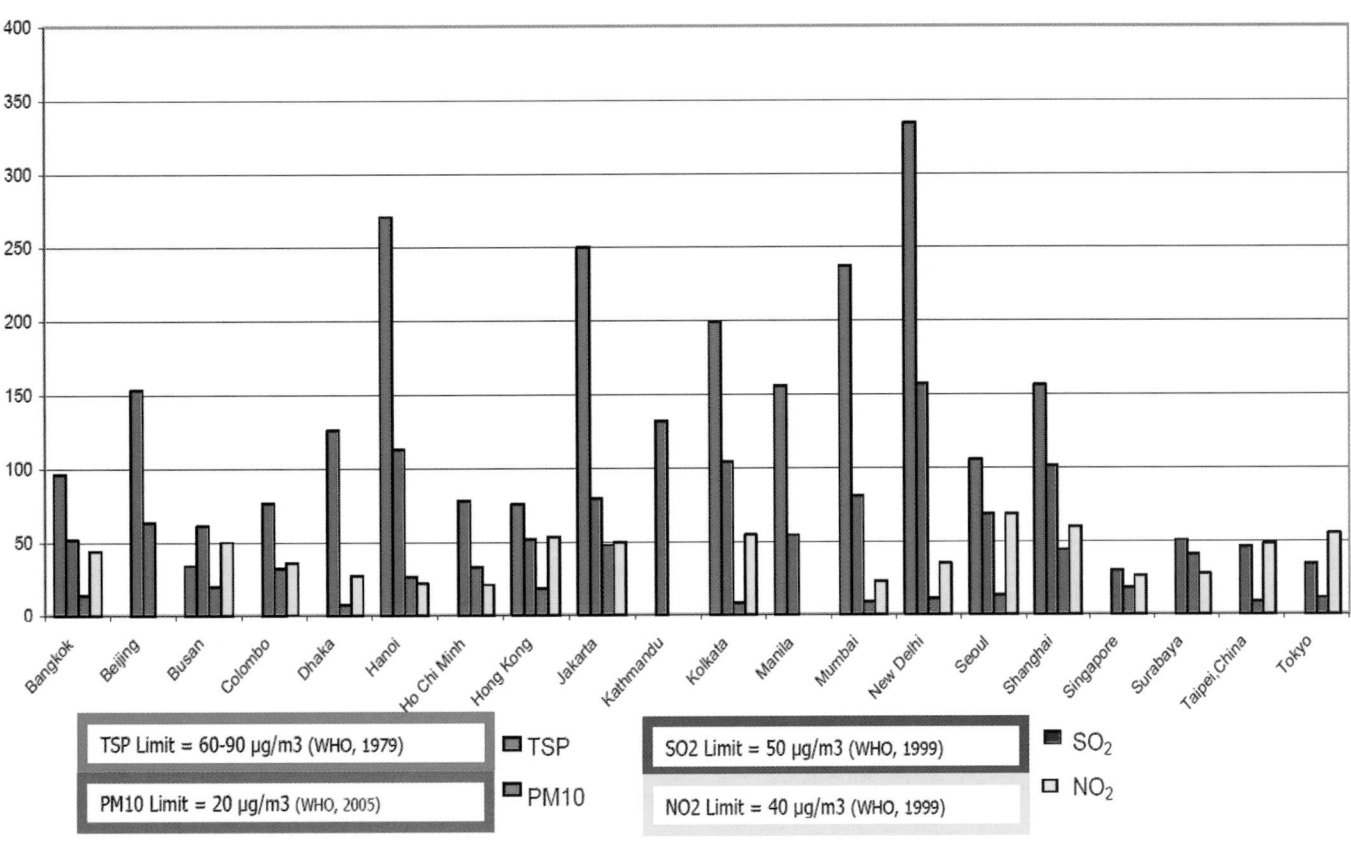

TSP Limit = 60-90 µg/m3 (WHO, 1979) ■ TSP   SO2 Limit = 50 µg/m3 (WHO, 1999) ■ SO₂

PM10 Limit = 20 µg/m3 (WHO, 2005) ▦ PM10   NO2 Limit = 40 µg/m3 (WHO, 1999) □ NO₂

Figure 18.35 | Urban concentrations of air pollutants (µg/m³) in Asian cities for 2005. Source: adapted from Schwela et al., 2006.

Country	95	96	97	98	99	00	01	02	03	04	05	06	07	08	09	10	11	12	13	14	15	16	17	18
European Union	E1	Euro 2					Euro 3				Euro 4			Euro 5					Euro 6					
Hong Kong, China	Euro 1		Euro 2				Euro 3					Euro 4			Euro 5									
South Korea												Euro 4			Euro 5									
China a							Euro 1			Euro 2		Euro 3			Euro 4									
China c							Euro 1		Euro 2			Euro 3		Euro 4				Euro 5						
Taipei, China							US Tier 1								US Tier 2 Bin 7 e									
Singapore a	Euro 1						Euro 2																	
Singapore b	Euro 1						Euro 2					Euro 4												
India							Euro 1				Euro 2					Euro 3								
India d			E1	Euro 2						Euro 3						Euro 4								
Thailand	Euro 1						Euro 2			Euro 3									Euro 4					
Malaysia		Euro 1													Euro 2				Euro 4					
Philippines									Euro 1			Euro 2									Euro 4			
Vietnam							Euro 2																	E4
Indonesia											Euro 2													
Bangladesh a											Euro 2													
Bangladesh b											Euro 1													
Pakistan															Euro 2 a		Euro 2 b							
Sri Lanka									Euro 1															
Nepal						Euro 1																		

*The level of adoption vary by country but most are based on the Euro emission standards

Italics – under discussion; a – gasoline; b – Diesel; c – Beijing [Euro 1 (Jan 1999); Euro 2 (Aug 2002); Euro 3 (2005); Euro 4 (1 Mar 2008); Euro 5 (2012)], Shanghai [Euro 1 (2000); Euro 2 (Mar 2003); Euro 3 (2007); Euro 4 (2010)] and Guangzhou [Euro 1 (Jan 2000); Euro 2 (Jul 2004); Euro 3 (Sep-Oct 2006); Euro 4 (2010)]; d – Delhi, Mumbai, Kolkata, Chennai, Hyderabad, Bangalore, Lucknow, Kanpur, Agra, Surat, Ahmedabad, Pune and Sholapur; e – US Tier 2 Bin 7 is equivalent to Euro 4 emissions standards

Figure 18.36 | Overview of vehicle emissions standards, Europe versus Asia. Source: CAI-Asia, 2009.

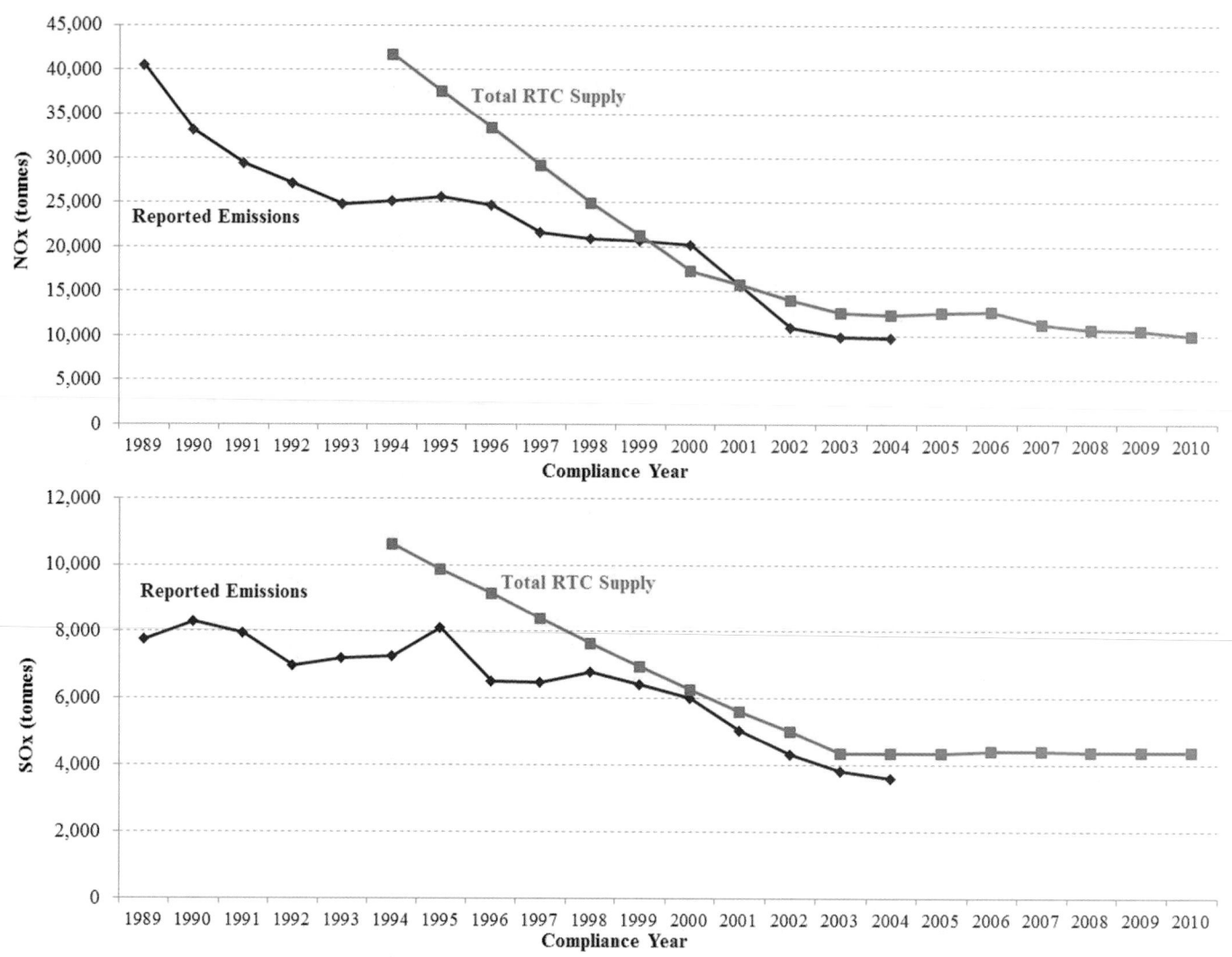

Figure 18.37 | RECLAIM's impacts on emissions in South Coast Area of California on NO$_x$ (top) and SO$_x$ (bottom) emissions versus allowable emission cap (RTCs) (in tonnes). Source: Anderson and Morgenstern, 2009.

18.5.5.3 Key Policy Issues

The ways and methods adopted to mitigate urban air-pollution impacts in urban energy systems differ widely. In Delhi, deliberate CNG introduction to reduce air pollution created a new urban energy supply, demand, and infrastructure, largely following a technology strategy. The ban on coal-fired boilers in Chinese cities such as Beijing and Shanghai, led to greater use of electricity and natural gas (Dhakal, 2004). As an example of a regulatory approach to urban air pollution, it largely follows the historical United Kingdom 'smokeless zone' regulatory model. City energy-system decisions in China are influenced by air-pollution mitigation, public transport improvements, and energy-security concerns (Dhakal, 2009). The will to control PM-10 prompted many cities in Asia, America, and Latin America to move progressively toward discouraging diesel. Europe is moving more on the path of dieselization with stricter control of the sulfur content of diesel combined with particulate filters in automobiles.

Despite improvements in both vehicle technology and fuel quality, the high growth rates in private automobile ownership and usage with rising income is proving a challenge to the control PM-10, suspended particulate matter, and NO$_x$ pollutants in cities of developing countries (Dhakal and Schipper, 2005). The key policy challenges for air pollution that have direct bearing on urban energy systems are (adapted from Schwela et al., 2006):

- Comprehensive assessment of the effectiveness of different options is needed in cities, but requires adequate institutional capacity, which remains comparatively weak in many large cities (Table 18.21), not to mention smaller ones. Often problems are addressed on a piecemeal basis without considering the complete system and thus rebound effects are prevalent.

- Development of more reliable inventories of air-pollution emissions is essential. Cities are not regularly updating their inventories

Table 18.21 | Classification of urban air-quality management capacity in Asian cities.

Capability Classification	Cities
Excellent I	Hong Kong, Singapore, Tapei, Tokyo
Excellent II	Bankgkok, Seoul, Shanghai
Good I	Beijing, Busan
Good II	New Delhi
Moderate I	Ho Chi Minh City, Jakarta, Kolkata, Metro Manila, Mumbai
Moderate II	Colombo
Limited I	Hanoi, Surabaya
Limited II	Dhaka, Kathmandu
Minimal	–

Source: Schwela et al., 2006.

and often there are serious ambiguities in emission volume and sources data.

- The need to adopt more stringent vehicle-emission standards is evident. The pace of adopting new emission standards in the face of rapidly rising private transportation is very slow. In addition, a reasonable global harmonization of air-quality and technology standards is needed. Currently, decision makers are torn between the Euro Standards and the USEPA standards, which affects technology choice and fuel regimes differently.

- Introducing cleaner fuel more actively for motor vehicles, industries, and power plants is necessary.

- Transport polices that affect urban mobility choices need to complement vehicle- and fuel-specific policy measures. In their absence, any air-quality improvements are likely to be quickly overwhelmed by continued motorized transport growth.

- Despite good policies on technology and fuel, inadequate emphasis on inspection and maintenance of systems remains one of the key challenges. Much of the existing air-pollution problems can be addressed by simple *implementation and stricter enforcement* of existing legislations, standards, and inspection and monitoring regimes for air quality.

- For transboundary air-pollution issues, such as acid rain and black carbon (emerging as key problems, particularly in Asia), regional approaches and regimes are needed, but such regional coordination emerges only very slowly in many world regions.

- To harmonize many environmental issues within common policy responses, estimation of the co-benefits of air-pollution management with respect to human health, urban energy system improvement, energy security, climate change mitigation, and ecosystems in general is essential. In developing countries such a co-benefit approach can help devise limited resources more efficiently, and

also broaden the technology and financial resources available for air-pollution control.

18.6 Summary and Conclusion

18.6.1 An Urbanizing World

The world is already predominantly urban, with the urban environment housing more than 50% of global population and accounting for even larger shares in economic and energy activities. Almost all future population growth of some three billion people to 2050 will be absorbed by urban areas. This urban growth is the combined result of natural increases in urban populations plus migration from rural to urban areas such that the increase in rural population in many developing countries will be overshadowed by population flows to the cities.

In contrast, rural populations globally will peak at around 2020 at a level of 3.5 billion people and decline thereafter. This global result masks heterogeneity in regional trends: whereas rural populations in Asia are projected to decline rapidly after 2020, the African rural population will continue to grow at least to 2040, before also declining.

Patterns of urban population growth have been and will remain heterogeneous. Most of the growth will continue to occur in small- to medium-sized urban centers, which explains the remarkable robustness of the distribution of city-size classes over time and across different regions. Growth in small cities poses serious policy challenges, especially in the developing world. In small cities, data and information to guide policy are largely absent, local resources to tackle development challenges are limited, and governance and institutional capacities are weak. Despite much public attention, the contribution of 'megacities' to global urban-population growth will remain comparatively small.

Shrinking cities in the developed world are an increasing phenomenon in urban dynamics, and could continue as below-replacement fertility levels outstrip increased longevity and so lead to declining populations in almost all high-income countries (and potentially in low-income countries in the long-term). The impacts of population contraction on urbanization remain a major unknown.

Cities of the future will have significantly older populations. Cities in developing countries will approach the age structures already prevalent in some cities of the industrialized world, with more elderly than young residents. This urban aging effect is likely to be mitigated temporarily by continued rural-urban migration as migrants comprise predominantly younger and more enterprising age cohorts, both nationally and internationally. Conversely, this demographic pattern suggests that aging will also be significant for rural settlements in low- and high-income countries alike. 'Graying' rural villages are probably the logical counterpoint of continued urbanization combined with a continued unfolding of the demographic transition worldwide.

18.6.2   Urban Energy Use

The urban share in current world-energy use varies as a function of system boundaries in terms of spatial scales (cities versus agglomerations), energy-systems definition (final commercial, total final, and total primary energy), and the boundary drawn to account for embodied energy in a city's goods and services, both imported and exported. The direct transfer of national energy (or GHG emissions) reporting formats to the urban scale is often referred to as a 'production' approach, and contrasts to a 'consumption' accounting approach that pro-rates associated energy uses (or GHG emissions) per unit of expenditure of urban consumer expenditures, thus accounting for energy uses irrespective of their form (direct or embodied energy) or location (within or outside a city's administrative boundary). Both approaches provide valuable information and should be used as complementary tools to inform urban policy decisions. However, to be useful, urban studies need to adhere to much higher standards in terms of clarity and documentation of the terminology, methodology, and underlying data used. To improve comparability, this assessment recommends specifically that all accounts based on the consumption approach (which are data- and time-intensive to prepare, and so exist only for a very limited set of megacities) be complemented by corresponding production-based energy accounts (which are much simpler and easier to determine). To ensure reproducibility, this assessment also recommends explicitly that no urban GHG-emission inventory be published without the underlying energy data used in the assessment.

Available estimates of current urban energy use based on a production approach (direct final energy use, or primary energy use, i.e., including pro-rated upstream energy sector conversion losses) suggest that urban energy use accounts for between 60% and 80% of global energy use. Total energy use is therefore already predominantly urban. Mirroring the growing importance of urban areas in demographic and economic development, urban energy use will continue to grow further as a fraction of total global energy needs. This implies that energy sustainability challenges need to be addressed and solved primarily by action in urban settings.

There is great heterogeneity in urban energy-use patterns, especially when manufacturing and transport energy uses are included. In many developing countries, urban dwellers use substantially *more* energy than their rural compatriots, which primarily reflects higher urban incomes. Conversely, in many industrialized countries per capita urban final energy use (i.e., based on a production-accounting approach) is often substantially *lower* than the national average, which reflects the effects of compact urban form, settlement types (multi- versus single-family dwellings) and availability and/or practicability of public transport infrastructure systems compared with those in the suburban or rural sprawl. The few available data, however, suggest that urban energy use in high-income countries is not substantially different from national averages using a consumption-based accounting approach that also includes energy embodied in imports. So, the effects of lowered direct final energy use through a more service-oriented urban economy, urban form and density, and resulting lower transport energy use are largely compensated by higher embodied energy use associated with higher urban incomes in high-income countries. For low-income countries, available data are too sparse to allow a similar comparison. However, it is highly likely that, because of the much higher income differential between urban and rural populations in low-income countries, their urban energy use is significantly higher on a per capita basis compared to national averages in a consumption-based accounting framework as well. Levels and structure (access to electricity, clean fuels for households, private motorized transport) of urban energy use in low-income countries are therefore a powerful leading indicator of future developments to come with rising urbanization and income growth in the developing world.

Drivers of urban energy use include geography and climate, resource availability, socioeconomic characteristics, degree of integration into the national and global economy (imports/exports), and urban form and density. Not all of these can be influenced by local governance and decision making. Priorities for urban energy and sustainability policies, therefore, should focus where local decision making and funding also provides the largest leverage effects: urban form and density (which are important macrodeterminants of urban structures, activity patterns, and hence energy use), the quality of the built environment (energy-efficient buildings in particular), urban transport policy (particularly the promotion of energy efficient and 'eco'-friendly public transport and non-motorized mobility options), and improvements in urban energy systems through cogeneration or waste-heat recycling schemes, where feasible. Local action, however, also requires local capacities and responsibilities in addressing urban energy and environmental problems, including a mediating role among the multiple stakeholders characteristic of decentralized urban decision making.

Conversely, the promotion of local solar or wind renewables will, at best, have a marginal impact on the overall energy use of larger cities (typically <1%)[39] because of the significant energy-density mismatch between (high) urban energy use and (low) renewable energy flows per unit land area available in urban areas. Smaller cities, however, could provide more avenues to integrate renewable energy into urban energy systems than large cities. Cities could also play an important role as consumers of renewable energies, creating niche market impulses as well as potentially exerting leverage on the application of sustainable social and ecological production criteria for their renewable energy suppliers.

Nonetheless, urban energy and climate policy should recognize that the most productive local decisions and policies influence the *efficiency* of urban energy use that is the demand side of the energy system, rather than its supply side.

---

39 Important exceptions include utilization of urban wastes and, where available, geothermal resources, both of which are characterized by a high energy density.

18.6.3 Facing the Challenges

18.6.3.1 Urban Poverty

Several hundred million urban dwellers in low- and middle-income nations lack access to electricity and are unable to afford the cleaner, safer fuels. Most are in low-income nations in southeast Asia and sub-Saharan Africa. In many low-income nations, more than half the urban population still rely on charcoal, fuelwood, straw, dung, or wastes for cooking, with significant adverse consequences for human health and urban air quality. A large part of the poor urban population that lacks clean fuels and electricity not only cannot afford these, but also faces political or institutional obstacles to accessing them.

In many middle-income and all high-income nations, nearly all low-income urban dwellers have legal electricity connections and access to clean fuels. The shift to clean fuels and the availability of a reliable electricity supply bring many advantages in terms of health, convenience, and time saved in accessing and using energy.

The costs of connection to an electricity grid and the use of electricity can be beyond the reach of low-income groups, but innovations have reduced these costs – for instance, rising tariffs with low prices for 'lifeline' electricity use, pay-as-you-use meters, and standard 'boards' that remove the need for individualized household wiring.

The constraints on supporting the shift to clean fuels and providing all urban households with electricity are less to do with energy policy than with authorities' handling of issues of informal settlements. A large part of the population that lack clean energy and electricity live in informal settlements. It is mostly in nations where this antagonistic relationship between local government and the inhabitants of such settlements has changed, through widespread public support to upgrade 'slums' and squatters, that clean energy and electricity reaches urban poor groups.

Housing, infrastructure, energy, and transport services are the key sustainability challenges to accommodate some three billion additional urban dwellers in the decades to come, especially in low-income countries. Informal settlements will be one of the transitional forms of settlement for many of these new urban dwellers and will require a much more proactive, anticipatory policy approach, especially with respect to the location of informal settlements and subsequent infrastructure connections and upgrading programs.

Energy-wise, low-cost and fast implementation options will take precedence over 'grand' new urban designs that require unrealistic capital provision over long periods. In low-income countries access to clean cooking fuels and electricity, as well as pro-poor transport policies, which include safer use of roads by non-motorized modes (walking and bicycling) and making public transport choices available (e.g., through BRT systems) need to receive more attention.

18.6.3.2 Livable Cities: Urban Density and Form

Urban density and form are not only important determinants for the functionality and quality of life in cities, but also for their energy use. Historically, the diversity of activities and ensuing economic and social opportunities that are the major forces of attraction to urban settings were provided by high density and co-location (mixed land-uses) of a diversity of activities that maximize the 'activity zone' of urban dwellers while minimizing transport needs. This urban history contrasts with decades of trends toward lower urban densities, which include widespread urban sprawl, and even 'ex-urban' developments.

It is widely agreed that there is no theoretical or practical argument for defining a universal 'optimal' form or density for a city. In theory (and often also in practice) higher densities increase the *economies of scope* (i.e., of activity variety) in a city. These are the main locational attractions of urban places in terms of potential number of jobs, breadth and variety of specialized trades and economic activities, along with cultural and many other attractions, usually summarized as *positive agglomeration externalities* that also extend to urban infrastructures (e.g., communication and transport networks). Conversely, higher densities can entail negative externalities as well, such as congestion, high land prices that limit the quality of residential living space for urban dwellers, or environmental problems (noise, air pollution).

Nonetheless, empirical data strongly suggest that the *net balance* of these positive and negative agglomeration externalities remained stable for extended periods of time, as illustrated by the remarkable stability of the rank-size distribution of cities (with dominating positive or negative net agglomeration externalities the growth of the *ensemble*[40] of larger cities should be above or below that of smaller cities, which is not the case). The historical evolutionary processes that govern urban growth have played out differently over time and space, which results in *path dependency*. Cities that evolved along alternative pathways have alternative density levels from high-density 'Asian' (e.g., Tokyo, Shanghai, Mumbai) and 'old European' (e.g., London, Madrid, Warsaw) to low density 'new frontier' (e.g., Los Angeles, Brasilia, Melbourne) pathways, each of which have different structural options available to improve energy efficiency and optimize urban energy and transport systems in terms of sustainability criteria.

Despite this diversity, two important generalizations can be drawn. First, the implications of urban density on the requirements of urban energy systems are that they need to be basically *pollution free*, as otherwise even relatively clean energy forms can quickly overwhelm the assimilative capacity of urban environments. This especially applies to the million, decentralized energy end-use devices (stoves, heating systems, vehicles) for which end-of-pipe pollution control is often not an option. Thus, in

---

40 Evidently, *individual* cities can forge ahead or fall behind the overall distributional pattern of aggregate uniform urban growth rates as outlined by the rank-size rule.

the long-term all end-use energy fuels consumed in urban areas need to be of zero-emission quality, as exemplified by electricity or (eventually) hydrogen, with natural gas as the transitional fuel of choice in urban areas. (Evidently, pollution levels also need to be minimized to the maximum technologically feasible at the point of production of these fuels.) This 'zero-emission' requirement for urban energy transcends the customary sustainability divide between fossil and renewable energies, as even 'carbon-neutral' biofuels when used by millions of automobiles in an urban environment will produce unacceptable levels of NO_x or O_3 pollution.

Second, the literature and above discussion repeatedly has identified important size and density thresholds that are useful guides for urban planning. The importance of these urban thresholds extends to specialized urban infrastructures, such as underground (metro) transport networks that are, as a rule, economically (in terms of potential customers and users) not feasible below a threshold population size of less than one million. It also extends to energy (e.g., cogeneration-based district heating and cooling) and public transport networks, whose feasibility (both for highly centralized and decentralized, distributed 'meso'-grids) are framed by a robust gross[41] density threshold between 50 and 150 inhabitants/ha (5000–15,000 people/km²). Such density levels of 50–150 inhabitants/ha certainly do *not* imply the need for high-rise buildings, as they can be achieved by compact building structures and designs, both traditional and new, including town or terraced houses, while still allowing for open public (parks) or private (courtyard) spaces – but they do preclude unlimited (aboveground parking) spaces for private vehicles. Zoning and parking regulation, combined with public transport policies and policies that promote non-motorized transport modes and walkability thus constitute the essential 'building blocks' of urban energy efficiency and sustainability 'policy packages.'

18.6.3.3 Mobile Cities: Urban Transport

Urban transport is a key policy concern, both for its high visibility (potential opposition to 'top-down' policies) and its crucial importance to the very functionality of cities. Two fundamental observations need to guide urban transport policies.

First, on a sustainability metric there is a clear contradiction between growing private motorized transport and growing energy use and pollution. This contradiction relates not primarily to the technological artifacts *per se* (automobiles or scooters), but rather to the *organizational form* of their *usage* as privately owned vehicles with correspondingly low occupancy rates (and thus high energy/emissions per unit service delivered). Well-designed taxis or automobile-sharing schemes are excellent examples that the (selective) use of automobiles as modes

of individual transport when needed (as opposed to their preordained use, despite congestion, simply because of a lack of alternatives) can be reconciled with the prerogatives of an energy-efficient city. Conversely, along the same sustainability metric, non-motorized mobility (walking and bicycling) and public transport schemes that function well and have high occupancy rates are the options of choice for urban mobility and should receive corresponding priority.

Second, the fundamental interrelations between demand and supply often create a 'vicious circle' for urban transport planning: more automobiles lead to congestion, which improved urban road infrastructures aim to alleviate. But more (road infrastructure) supply *induces* yet more demand (individual mobility and land-use changes) in an ever-spiraling 'rat race' of 'supply following demand' growth. A new public policy paradigm for urban transport needs to break this cycle of rebound effects through integrated urban energy transport policies that deal with both demand and supply.

The wide variation in urban transport choices observed in the modal split across cities illustrates that urban mobility patterns are not *ex ante* given, but rather result from deliberate choices of individuals and decision makers. Urban transport choices can be changed, if both a strong determination for sustainable transport policy *and* a corresponding wide public acceptance of the overall goals of such a policy exist. This wider acceptance of the overarching goals is also required to implement some individual measures (e.g., traffic calming, pricing schemes (for roads and parking, etc.)) that often face lobbying opposition. Restrictive measures that limit individual mobility by automobiles need to be complemented by proactive policies that enhance the attractiveness of non-motorized and public transport choices, and 'soft' policy measures (e.g., fees, tariffs) also need to be complemented by 'hard' (i.e., infrastructural investment) measures. The overriding goal is to turn the often automobile-dependent 'vicious' policy cycle into a 'virtuous' cycle that favors non-motorized and public transport choices.

One key measure in this context will be the progressive internalization of external costs of motorized private transport along with the provision of high-quality alternative public and non-motorized transport. Estimates for Europe suggest that these external costs are at least in the order of 6–10 cents per passenger-km (see Table 18.16), which when fully internalized would double private motorized transport costs. Comparable estimates for low-income countries are not available, but given their generally much higher road-accident rates, external costs are likely to be even larger. However, accompanying measures and strong leadership will be needed to increase the political acceptability of this invariably unpopular measure. Recent experiences with the introduction of road prices and congestion charges in cities across a wide political spectrum suggest that such policy approaches are both feasible and have the ability to alter urban transport behavior.

Attractive public transport systems require a dense public transport network and a high service frequency with short intervals, which are only

---

41   i.e. a minimum density level over the entire settlement area that comprises residential zones of higher density with low density green spaces.

feasible with a minimum threshold of urban density. A rule-of-thumb goal might be to have only urban settlements within easy walking (<500 m) access to a viable public transport service. Investments in public transport systems need to find an appropriate balance between improvements that are less capital intensive and faster to implement, and radical solutions. BRT systems are, therefore, a much more attractive option for many cities in low-income countries than are capital-intensive subway systems, even though, in the long term, the latter offer the possibility of higher passenger fluxes and greater energy efficiency. Often there is no contradiction between incremental versus radical public transport policy options: for instance, BRT can also be considered as a transitional infrastructure strategy to secure public transport 'rights of way,' which offer subsequent possibilities for infrastructure upgrades, for example in putting light rail systems in BRT lanes. Many of the new urban settlements being formed easily meet these density targets. The policy issue is to exploit the advantages before 'lock-in' into private transportation takes hold.

18.6.3.4   Efficient Cities: Doing More with Less Energy

From all the major determinants of urban energy use – climate, integration into the global economy, consumption patterns, quality of built environment, urban form and density (including transport systems), and urban energy systems and their integration – only the final three are amenable to an urban policymaking context and therefore should receive priority.

Systemic characteristics of urban energy use are generally more important determinants of the efficiency of urban energy use than those of individual consumers or of technological artifacts. For instance, the share of high occupancy public and/or non-motorized transport modes in urban mobility is a more important determinant of urban transport energy use than the efficiency of the urban vehicle fleet (be it buses or hybrid automobiles). Denser, multifamily dwellings in compact settlement forms with a corresponding higher share of nonautomobile mobility (even without thermal retrofit) use less *total energy* than low-density, single-family 'Passivhaus'-standard (or even 'active,' net energy generating) homes in dispersed suburbs with two hybrid automobiles parked in the garage and subsequently used for work commutes and daily family chores. Evidently, urban policies need to address both systemic and individual characteristics in urban energy use, but their different long-term leverage effects should structure policy attention and perseverance.

In terms of urban energy-demand management, the quality of the built environment (buildings efficiency) and urban form and density that, to a large degree, structure urban transport energy use are roughly of equal importance. Also, energy-systems integration (cogeneration, heat cascading) can give substantial efficiency gains, but ranks second after buildings efficiency and urban form and density, and associated transport efficiency measures, as shown both by empirical cross-city comparisons and modeling studies reviewed in this chapter.

The potential for energy-efficiency improvements in urban areas remains enormous, as indicated by corresponding urban exergy analyses that suggest urban energy-use efficiency is generally less than 20% of the thermodynamic efficiency frontier; this suggests an improvement potential of more than a factor five. Implementing efficiency improvements (including systemic measures) should therefore receive highest priority. In the built environment, there is considerable inertia and irrational behavior, given the relatively low or even negative cost of GHG abatement through refurbishment of buildings. There is therefore a strong argument for prescription and regulation regarding building standards.

Conversely, the potential of supply-side measures within the immediate spatial and functional confines of urban systems is very limited, especially for renewable energies. *Locally harvested* renewables can, at best, provide 1% of the energy needs of a megacity and a few percentage points in smaller, low-density cities because of the mismatch between (high) urban energy demand and (low) renewable energy supply densities. Without ambitious efficiency gains, the corresponding 'energy footprint' of cities that import large-scale, centralized renewable energies (biofuels for electricity) will be vast and at risk of producing 'collateral' damages caused by large-scale land conversions (e.g., soil carbon perturbations and albedo changes), and competition over food and water. Given that all city systems are subject to uncertainty and change, there is also a need for option-based design techniques that allow for city growth and technological evolution, and that avoid strongly path-dependent solutions and 'lock-in' into urban energy-supply systems based on current- or near-term renewable options that, ultimately, will be superseded by third and fourth generation renewable supply technology systems. Improved urban energy-efficiency leverages supply-side flexibility and resilience, and thus adds further powerful arguments for strategies that focus on urban energy-efficiency improvements.

Finally, *energy security* is a re-emerging issue for many cities because energy security declined over recent decades and security concerns need to be integrated increasingly into urban energy policies and sustainability transition analysis. Better efficiency and improved energy systems integration will also benefit urban energy security, although a further assessment of energy security is beyond the scope of this urbanization chapter (as dealt with in the energy security Chapter 5).

18.6.3.5   Clean Cities: Air Pollution

To a degree, the observed significant improvements in urban air quality caused by the elimination of traditional air pollutants, such as soot, particles, and SO_2, in cities of high-income countries are a powerful illustration that cities act as innovation centers and hubs for environmental improvements that lead to a sustainability transition path.

The first signs of progress in these traditional air pollutants are evident in countries of lower income as well, and are illustrated by the recent decline in the emissions of some traditional pollutants in Asian megacities. Nonetheless, an exceedingly high fraction of urban dwellers worldwide are still exposed to high levels of urban air pollution, especially TSPs with fine particle emissions (PM-10s) continuing their upward trend.

A wide portfolio of policy options is available, ranging from regulatory instruments such as mandated fuel choice ('smokeless zone' regulations), air-pollution standards, regulation of large point-source emissions and vehicle exhaust standards, to market-based instruments (or hybrid) approaches that incentivize technological change. An important feature of these regulatory or market-based approaches is *dynamic target setting* to reflect changing technology options and to counter the consequences of urban growth and potential consumer 'take-back' effects. The tested experiences in a diversity of settings from the United Kingdom through California to New Delhi provide valuable lessons for policy learning. However, an institutional locus on the exchange of policy lessons and capacity building (including pollution monitoring), particularly in small- and medium-sized cities in low-income countries, remains sorely lacking.

Air pollution is also the area of urban environmental policy making where the most significant co-benefits of policies can be realized: improving access to clean cooking fuels, for example, improves human health and lowers traditional pollutant emissions, and also has (through reduced black carbon emissions) net global warming co-benefits. (See also Chapters 4 and 17 on illustrations of co-benefits). Similarly, improved energy efficiency and public transport options (e.g., New Delhi's transition to CNG buses) can also yield co-benefits on a variety of fronts (lower energy and transport costs for the poor, cleaner air, improved urban functionality) and should therefore be higher on the policy agenda than more single-purpose policy measures (e.g., renewable portfolio standards for urban electricity supply). Examples of the management of urban heat island effects also illustrate well the potential significant co-benefits between mitigation and adaptation measures.

Realization of the significant potential co-benefits, however, requires more holistic policy approaches that integrate urban land-use, transport, and energy policies with the more traditional air-pollution policy frameworks. Cities in low-income countries, where the growth trends of urbanization and air pollutants are the most pronounced (to the extent that they will determine global trends), are also where institutional capacities and information needs are largest. These must be improved to be able to reap the multiple co-benefits of more integrated urban policies. A renewed effort to improve measurement and monitoring as well as planning and modeling of urban environmental quality, with a particular focus on urban energy and urban transport policy, is urgently needed. This needs to be coupled with the development of institutional capacity for the design, implementation, and enforcement of policies and plans.

18.6.4   Policy Leverages, Priorities, and Paradoxes

The highest impacts of urban policy decisions are in the areas where policies can affect local decision making and prevent or unblock spatial irreversibilities or technological 'lock-in,' or to steer away from critical thresholds.

Examples include preventing the further development of low-density, suburban housing and of shopping malls, or promotion of the co-location of high energy supply and demand centers within a city that enable cogeneration and waste-heat recycling for heating and cooling purposes (e.g., in business districts). Buildings energy-performance standards are also a prime example of policy interventions that need to be implemented as early as possible to reap long-term benefits in terms of reduced energy use and improved urban environmental quality. New technologies, like smaller micro- and mesogrids are particularly attractive options that are also suitable for deregulated market environments. The literature on urban energy use, particularly with respect to transport, also identifies a critical 'threshold' of between 50 and 150 inhabitants/ha below which public transport (or energy cogeneration) options become economically infeasible, which thus leads to overproportional increases in energy use (e.g., longer trips using private automobiles). To avoid such critical thresholds being crossed should be a high priority for urban administrations.

Given capital constraints, it is entirely unrealistic to expect 'grand' new urban 'ecodesigns' to play any significant role in integrating some three billion additional urban dwellers to 2050 into the physical, economic, and social fabric of cities. Building cities for these three billion new urban citizens along the Masdar (Abu Dhabi) model would require an investment to the tune of well above US$1000 trillion, or some 20 years of current world GDP![42] The role of such new, daring urban designs is less a template for development, but rather a 'learning laboratory' to develop and test approaches, especially to low-cost options for sustainable urban growth in low-income countries and to retrofit and adapt existing urban structures and systems across the globe.

To address urban energy sustainability challenges will also require a new paradigm for drawing systems or ecosystems boundaries that extend the traditional place-based approach (e.g., based on administrative boundaries or ecosystems such as regional watersheds or air-quality districts). Sustainability criteria need to be defined on the basis of the functional interdependence among different systems, which are not necessarily in geographic proximity to each other.

---

42 This extreme estimate does not suggest that energy efficient cities are prohibitively expensive. A wide range of measures in building retrofits and low-cost new energy efficient housing as well as in public transport policies can result in significant reductions of energy use at modest investment levels. For estimations of the investment needs of the GEA transition pathways, see Chapter 17.

System analytical and extended LCA methods are increasingly available to address the question of the social, economic, and environmental sustainability of urban energy systems that almost exclusively rely on imports. However, clear methodological guidelines and strategies to overcome the formidable data challenges are needed, a responsibility that resides within the scientific community, but that requires support and a dedicated long-term approach for funding and capacity building.

A common characteristic of sustainable urban energy system options and policies is that they are usually systemic: for example, the integration of land-use and urban transport planning that extends beyond traditional administrative boundaries; the increasing integration of urban resource streams, including water, wastes, and energy, that can further both resource (e.g., heat) recovery and improve environmental performance; or the reconfiguration of urban energy systems toward a higher integration of supply and end-use (e.g., via micro- and mesogrids) that enable step changes in efficiency, for example, through cogeneration and energy cascading. This view of more integrated and more decentralized urban infrastructures also offers possibilities to improve the resilience and security of urban energy systems.

And yet this systemic perspective reveals a new kind of 'governance paradox.' Whereas the largest policy leverages are from systemic approaches and policy integration, these policies are also the most difficult to implement and require that policy fragmentation and uncoordinated, dispersed decision making be overcome. This governance paradox is compounded by weak institutional capacities, especially in small- to medium-sized cities that are the 'backbone' of urban growth, as well as by the legacies of market deregulation and privatization that have made integrated urban planning and energy, transport, and other infrastructural policy approaches more difficult to design and yet more difficult to implement.

However, there are good reasons for (cautionary) optimism. Urban areas will continue to act as innovation centers for experimentation and as diffusion nodes for the introduction of new systems and individual technological options (Bai et al., 2010) by providing critical niche market sizes in the needed transition toward more sustainable urban energy systems. The task ahead is to leverage fully this innovation potential of cities and to scale-up successful experiments into transformative changes in energy systems. Individual and collective learning, transfer of knowledge, and sharing experiences and information across cities and among stakeholders will, as always, be key objectives to which this chapter hopes to contribute.

References

Abiko, A., L. Cardoso, R. Rinaldelli and H. Haga, 2007: Basic costs of slum upgrading in Brazil. Global Urban Development Magazine, Global Urban Development, Washington, DC, USA.

Adelekan, I. and A. Jerome, 2006: Dynamics of household energy consumption in a traditional African city, Ibadan. *The Environmentalist*, 26(2):99–110.

ALA, 2009: *State of the Air – 2008*. American Lung Association (ALA), New York, NY, USA.

Allen, P. M., 1997: *Cities and Regions as Self-Organizing Systems: Models of Complexity* Routledge, London, UK.

Amaral, L. A. N. and J. M. Ottino, 2004: Complex networks: Augmenting the framework for the study of complex systems. *The European Physical Journal B – Condensed Matter and Complex Systems*, 38(2):147–162.

Anderson, R. C. and R. D. Morgenstern, 2009: Marginal abatement cost estimates for non-CO2 greenhouse gases: lessons from RECLAIM. *Climate Policy*, 9:40–55.

Andersson, E., 2006: Urban landscapes and sustainable cities. *Ecology and Society*, 11(1):34.

Andrews, C. J., 2008: Greenhouse gas emissions along the rural-urban gradient. *Journal of Environmental Planning and Management*, 51(6):847–870.

APHRC, 2002: *Population and Health Dynamics in Nairobi's Informal Settlements*. Report of the Nairobi Cross-sectional Slums Survey (NCSS) 2000, African Population and Health Research Center (APHRC), Nairobi, Kenya.

APPROTECH, 2005: *Enabling Urban Poor Livelihood Policy Making: Understanding the Role of Energy Services in the Philippines*. National Workshop Reports, The Asian Alliance of Appropriate Technology Practitioners, Inc. (APPROTECH), Manila, Phillipines.

Arnold, J. E. M., G. Köhlin and R. Persson, 2006: Woodfuels, livelihoods, and policy interventions: Changing Perspectives. *World Development*, 34(3):596–611.

Arvesen, A., J. Liu and E. G. Hertwich, 2010: Energy Cost of Living and Associated Pollution for Beijing Residents. *Journal of Industrial Ecology*, 14(6):890–901.

Bai, X. M., B. Roberts and J. Chen, 2010: Urban sustainability experiments in Asia: Patterns and pathways. Environmental science & Policy, 13(4):312–325.

Balmer, M., 2007: Household coal use in an urban township in South Africa. *Journal of Energy in Southern Africa*, 18(3):27–32.

Banham, R., 1969: *Architecture of the Well-tempered Environment*. University of Chicago Press, Chicago, IL, USA.

Banister, D., 1992: Energy Use, Transport and Settlement Patterns. In *Sustainable Development and Urban Form*. M. J. Breheny, (ed.), Pion Ltd., London, UK.

Banister, D., S. Watson and C. Wood, 1997: Sustainable cities: transport, energy, and urban form. *Environment and Planning B: Planning and Design*, 24(1):125–143.

Barles, S., 2009: Urban Metabolism of Paris and Its Region. *Journal of Industrial Ecology*, 13(6):898–913.

Barter, P. A., 1999: Transport and urban poverty in Asia. A brief introduction to the key issues. *Regional Development Dialogue*, 20(1):143–163.

Batty, M., 2005: *Cities and Complexity: Understanding Cities with Cellular Automata, Agent-Based Models and Fractals*. The MIT Press, Cambridge, MA, USA.

Batty, M., 2008: The size, scale, and shape of cities. *Science*, 319(5864):769–771.

Baumol, W. J. and W. E. Oates, 1975: *The Theory of Environmental Policy: Externalities, Public Outlays, and the Quality of Life*. Prentice-Hall, Englewood Cliffs, NJ, USA.

Baynes, T. and X. Bai, 2009: *Trajectories of change: Melbourne's population, urban development, energy supply and use from 1960–2006*. International Institute for Applied Systems Analysis (IIASA), Laxenburg, Austria.

Beijing Government, 2010: *Energy Statistics Online*. Available at www.ebeijing. gov.cn/feature_2/Statistics/ (accessed 10 December, 2010).

Bettencourt, L. M. A., J. Lobo, D. Helbing, C. Kühnert and G. B. West, 2007: Growth, innovation, scaling, and the pace of life in cities. *Proceedings of the National Academy of Sciences*, 104(17):7301–7306.

Bhan, G., 2009: "This is no longer the city I once knew". Evictions, the urban poor and the right to the city in millennial Delhi. *Environment and Urbanization*, 21(1):127–142.

Biello, D., 2008: Eco-cities of the future. *Scientific American*, (Earth 3.0):6.

Block, A., K. Keuler and E. Schaller, 2004: Impacts of anthropogenic heat on regional climate patterns. *Geophysical Research Letters*, 31(12):L12211.

Bloomberg, M. R., 2007: *plaNYC – a greener, greater New York*. The City of New York, New York.

Boadi, K. O. and M. Kuitunen, 2005: Environment, wealth, inequality and the burden of disease in the Accra metropolitan area, Ghana. *International Journal of Environmental Health Research*, 15(3):193–206.

Boardman, B., 1993: *Fuel Poverty: From Cold Homes to Affordable Warmth*. John Wiley & Sons Ltd., London, UK.

Bocquier, P., 2005: World Urbanization Prospects: an alternative to the UN model of projection compatible with the mobility transition theory. *Demographic Research*, 12 (9):197–236.

Böhm, R., 1998: Urban Bias in Temperature Time Series – a Case Study for the City of Vienna, Austria. *Climatic Change*, 38(1):113–128.

Bravo, G., R. Kozulj and R. Landaveri, 2008: Energy access in urban and peri-urban Buenos Aires. *Energy for Sustainable Development*, 12(4):56–72.

Brehny, M., 1986: Centrists, Decentrists and Compromisers: Views on the Future of Urban Form. In *The Compact City. A Sustainable Urban Form?* M. Jenks, E. Burton and K. Williams, (eds.), Spon Press, Chapman and Hall, London, UK.

Brög, W. and I. Ker, 2009: Myths, (Mis)perceptions and reality in measuring voluntary behavioural changes. In *Transport Survey Methods – Keeping Up With a Changing World*. P. Bonnel, M. Lee-Gosselin, J. Zmud and J. L. Madre, (eds.), Emerald, Bingley, UK.

Brown, M. A., F. Southworth and A. Sarzynski, 2008: *Shrinking the carbon footprint of metropolitan America*. Metropolitan Policy Program, Brookings, Washington, DC.

Buzar, S., 2006: Estimating the extent of domestic energy deprivation through household expenditure surveys. *CEA Journal of Economics*, 1(2):6–19.

CAI-Asia, 2009: *Clean Air Initiative for Asian Cities – Annual Report 2009*. Clean Air Initiative for Asian Cities (CAI-Asia), Manila, Phillippines.

Carruthers, R., M. Dick and A. Saurkar, 2005: *Affordability of public transport in developing countries*. Transport Paper 3 World Bank, Washington, DC, USA.

Cervero, R. and K. Kockelman, 1997: Travel demand and the 3Ds: Density, diversity, and design. *Transportation Research Part D: Transport and Environment*, 2(3):199–219.

Chandler, T. and G. Fox, 1974: *Three Thousand years of Urban Growth*. Academic Press, New York, NY, USA.

Chandler, T., 1987: *Four Thousand Years of Urban Growth: A Historical Census*. Edwin Mellen Press, New York, NY, USA.

Cherry, S., 2007: *How to Build a Green City*. Available at spectrum.ieee.org/energy/environment/how-to-build-a-green-city/0 (accessed 30 December, 2010).

Chertow, M. R., 2000: The IPAT Equation and Its Variants. *Journal of Industrial Ecology*, **4**(4):13–29.

City of London, 2007: *Rising to the Challenge – The City of London Corporation's Climate Adaptation Strategy*. City of London Corporation, London, UK.

Cohen, C., M. Lenzen and R. Schaeffer, 2005: Energy requirements of households in Brazil. *Energy Policy*, **33**(4):555–562.

Cole, R. J. and P. C. Kernan, 1996: Life-cycle energy use in office buildings. *Building and Environment*, **31**(4):307–317.

Cowan, B., 2008: *Identification and Demonstration of Selected Energy Best Practices for Low-Income Urban Communitites in South Africa*. Alleviation of Poverty through the Provision of Local Energy Services (APPLES). Project No. EIE-04–168. Deliverable No. 17, Energy Research Centre, University of Cape Town and the Intelligent Energy Europe programme of the European Commission (EC), Cape Town, South Africa.

Crovella, M., 2007: *Personal Communication on Data Published in Lakhina et al. 2003*. Boston University, Boston, MA, USA.

Crutzen, P. J., 2004: New Directions: The growing urban heat and pollution island effect – impact on chemistry and climate. *Atmospheric Environment*, **38**(21):3539–3540.

Cunningham-Sabot, E. and S. Fol, 2009: Shrinking cities in France and Great Britain: A silent process? In *The Future of Shrinking Cities: Problems, Patterns and Strategies of Urban Transformation in a Global Context* K. Pallagst, J. Aber and I. Audirac, (eds.), Center for Global Metropolitan Studies, Institute of Urban and Regional Development, Berkeley, CA, USA pp.17–28.

Dayton, D., C. Goldman and S. Pickle, 1998: *The Energy Services Company (ESCO) Industry: Industry and Market Trends*. LBNL-41925, Lawrence Berkeley National Laboratory, Berkeley, CA, USA.

Devas, N. and D. Korboe, 2000: City Governance and Poverty: the Case of Kumasi. *Environment and Urbanization*, **12**(1):123–136.

Dey, C., C. Berger, B. Foran, M. Foran, R. Joske, M. Lenzen and R. Wood, 2007: An Australian environmental atlas: household environmental pressure from consumption. In *Water, Wind, Art and Debate: how environmental concerns impact on disciplinary research*. G. Birch, (ed.), Sydney University Press, Sydney, Australia pp.280–315.

Dhakal, S. and K. Hanaki, 2002: Improvement of urban thermal environment by managing heat discharge sources and surface modification in Tokyo. *Energy and Buildings*, **34**(1):13–23.

Dhakal, S., K. Hanaki and A. Hiramatsu, 2003: Estimation of heat discharges by residential buildings in Tokyo. *Energy Conversion and Management*, **44**(9):1487–1499.

Dhakal, S., 2004: *Urban Energy Use and Greenhouse Gas Emissions in Asian Megacities: Policies for a Sustainable Future*. H. Imura, (ed.) Institute for Global Environmental strategies (IGES), Kitakyushu, Japan.

Dhakal, S. and L. Schipper, 2005: Transport and environment in Asian cities: Reshaping the issues and opportunities into a holistic framework. *International Review for Environmental Strategies*, **5**(2):399–424.

Dhakal, S., 2009: Urban Energy Use and Carbon Emissions from Cities in China and Policy Implications. *Energy Policy*, **37**(11):4208–4219.

DIALOG, 2010: *Individuelle Motivation zum klimaschonenden Umgang mit Energie im Verkehr und im Haushalt (Individual motivation for climate protecting energy use in transport and household)*. G. Sammer, R. Hössinger and J. Stark, (eds.), Klima & Energie Fonds, Institute for Transport Studies, University of Natural Resources and Applied Life Sciences, Vienna, Austria.

Diamond, J., 2004: The Wealth of Nations. *Nature*, **429**(6992):616–617.

Directorate-General for Energy, 2007: *Green Paper on Urban Mobiilty: Towards a New Culture for Urban Mobility*. Directorate-General for Energy, Brussels, Belgium.

Doi, K., M. Kii and H. Nakanishi, 2008: An integrated evaluation method of accessibility, quality of life, and social interaction. *Environment and Planning B: Planning and Design*, **35**(6):1098–1116.

Doll, C. N. H., 2009: *Spatial Analysis of the World Bank's Global Air Pollution data Set*. IR-09–033, International Institute for Applied Systems Analysis (IIASA), Laxenburg, Austria.

Doll, C. N. H. and S. Pachauri, 2010: Estimating rural populations without access to electricity in developing countries through night-time light satellite imagery. *Energy Policy*, **38**(10):5661–5670.

Doxiadis, C. A. and J. G. Papaioannou, 1974: Ecumenopolis, The Inevitable City of the Future. *Environmental Conservation*, **2**(4):315–317.

Drukker, C., 2000: Economic Consequences of Electricity Deregulation: A Case Study of San Diego Gas & Electric in a Deregulated Electricity Market. *California Western Law Review*, **361**:291.

Energie Wien, 2009: *Energieflussbild Wien 2007*. Available at www.wien.gv.at/wirtschaft/eu-strategie/energie/zahlen/energieverbrauch.html (accessed 30 December, 2010).

ESMAP and UNDP, 2003: *Household Fuel Use and Fuel Switching in Guatemala*. Report No. 27274, Energy Sector Management Assistance Programme (ESMAP) of the World Bank and the United Nations Development Programme (UNDP), Washington, DC, USA.

ESMAP and UNDP, 2005: *Power Sector Reform in Africa: Assessing Impact on Poor People*. Report No. 306/05, Energy Sector Management Assistance Programme (ESMAP) of the World Bank and United Nations Development Programme (UNDP), Washington, DC, USA.

ESMAP, 2006: *Pakistan: Household Use of Commercial Energy*. Report No. 320/06, M. Kojima, (ed.) Energy Sector Management and Assistance Programme (ESMAP) of the World Bank, Washington, DC, USA.

ESMAP, 2007: *Meeting the Energy Needs of the Urban Poor: Lessons from Electrification Practitioners*. Technical Paper No. 118/07, Energy Sector Management Assistance Program (ESMAP) of the World Bank, Washington, DC, USA.

Eurostat, 1988: *Useful Energy Balances*. Statistical Office of the European Communities (Eurostat), Luxembourg.

Eurostat, 2008: *Gross domestic product (GDP) at current market prices at NUTS level 3*. Eurostat. Available at http://epp.eurostat.ec.europa.eu/extraction/evalight/EVAlight.jsp?A=1&language=en&root=/theme1/reg/reg_E3gdp.

Ewing, R. and R. Cervero, 2001: Travel and the Built Environment: A Synthesis. *Transportation Research Record: Journal of the Transportation Research Board*, **1780**(1):87–114.

Faist, M., R. Frischknecht, L. RCornaglia and S. Rubli, 2003: *Métabolisme des Activités Économiques du Canton de Genève – Phase 1*. Groupe de Travail Interdépartemental Ecosite, Geneva, Switzerland.

Fisk, D., 2008: What are the risk-related barriers to, and opportunities for, innovation from a business perspective in the UK, in the context of energy management in the built environment? *Energy Policy*, **36**(12):4615–4617.

Fisk, D., 2010: *Exergy analyses and Sankey diagrams for the cities of London and Malmö*. Imperial College London, London, UK.

Fisk, D. J. and J. Kerhervéa, 2006: Complexity as a cause of unsustainability *Ecological Complexity*, **3**(4):336–343.

Flanner, M. G., 2009: Integrating anthropogenic heat flux with global climate models. *Geophysical Research Letters*, **36**(2):L02801.

Forrester, J. W., 1969: *Urban Dynamics*. Pegasus Communications, Inc., Waltham, MA.

Fujita, M., P. Krugman and A. J. Venables, 1999: *The spatial economy: cities, regions and international trade*. MIT Press, Cambridge, MA, USA.

Gangopadhyay, S., B. Ramaswami and W. Wadhwa, 2005: Reducing subsidies on household fuels in India: how will it affect the poor? *Energy Policy*, **33**(18):2326–2336.

Gasson, B., 2007: Aspects of stocks and flows in the built environment of Cape Town. *Workshop on Analysing Stocks and Flows of the Built Urban Environment*, Norwegian University of Science and Technology (NTNU), Trondheim, Norway.

Gilli, P. V., N. Nakicenovic and R. Kurz 1995: First and Second Law Efficiencies of the Global and Regional Energy Systems *PS 3.1.16, 16th congress of the World Energy Congress (WEC)*, Tokyo, Japan.

Gilli, P. V., N. Nakicenovic and R. Kurz 1996: *First- and second-law efficiencies of the global and regional energy systems*. RR-96–2, International Institute for Applied Systems Analysis (IIASA), Laxenburg, Austria.

Giradin, L. and D. Favrat, 2010: *Bilan Énergétique/Exergétique de Genève pour 2005*. Unpublished Manuscript, École Polytechnique Fédérale de Lausanne (EPL), Lausanne, Switzerland.

Goldemberg, J., 1991: "Leap-frogging": A new energy policy for developing countries. *WEC Journal*,(December):27–30.

Goyal, P. and Sidhartha, 2003: Present scenario of air quality in Delhi: a case study of CNG implementation *Atmospheric Environment*, **37**(38):5423–5431.

Graz, 2010: *Stadt Portal des Landeshauptstadt Graz*. Available at www.graz.at/cms/ziel/245559/DE/ (accessed December 30, 2010).

Grubler, A., 2004: Transitions in Energy Use. In *Encyclopedia of Energy*. C. J. Cleveland, (ed.), Elsevier Science, Amsterdam, the Netherlands, Vol. 6, pp.163–177.

Grubler, A., B. O'Neill, K. Riahi, V. Chirkov, A. Goujon, P. Kolp, I. Prommer, S. Scherbov and E. Slentoe, 2007: Regional, national, and spatially explicit scenarios of demographic and economic change based on SRES. *Technological Forecasting and Social Change*, **74**(7):980–1029.

Guimerà, R., S. Mossa, A. Turtschi and L. A. N. Amaral, 2005: The worldwide air transportation network: Anomalous centrality, community structure, and cities' global roles. *Proceedings of the National Academy of Sciences of the United States of America*, **102**(22):7794–7799.

Gutschner, M., S. Nowak, D. Ruoss, P. Toggweiler and T. Schoen, 2001: *Potential for Building Integrated Photovoltaics*. Report IEA PVPS T7–4–2001 (Summary), International Energy Agency (IEA) of the Organisation for Economic Cooperation and Development (OECD), Paris, France.

Hall, P., 2002: *Cities of Tomorrow*. Wiley-Blackwell, Oxford, UK.

Hardoy, J. E. and D. Satterthwaite, 1989: *Squatter Citizen: Life in the Urban Third World*. Earthscan, London, UK.

Hardoy, J. E., D. Mitlin and D. Satterthwaite, 2001: *Environmental Problems in an Urbanizing World*. Earthscan, London, UK.

Harms, H., 1997: To live in the city centre: housing and tenants in central neighbourhoods of Latin American cities *Environment and Urbanization*, **9**(2):191–212.

Harrison, D., 2004: Ex post evaluation of the RECLAIM emissions trading programmes for the Los Angeles air basin. In *Tradable Permits: Policy Evaluation, Design and Reform*, Organization for Economic Cooperation and Development, (OECD), Paris, France pp.192

Hawksworth, J., T. Hoehn and M. Gyles, 2007: *Which are the Largest City Economies in the World and how might this change by 2020?* PricewaterhouseCoopers UK, London, UK.

Hersey, J., N. Lazarus, T. Chance, S. Riddlestone, P. Head, A. Gurney and S. Heath, 2009: *Capital Consumption: The Transition to Sustainable Consumption and Production in London*. BioRegional and London Sustainable Development Commission, London, UK.

Hillier, B., 1999: *Space is the Machine – A Configurational Theory of Architecture*. Cambrdige University Press, Cambridge, UK.

Hillman, T. and A. Ramaswami, 2010: Greenhouse gas emission footprints and energy use benchmarks for eight U.S. cities. *Environmental Science & Technology*, **44**(6):1902–1910.

Holden, E. and I. T. Norland, 2005: Three challenges for the compact city as a sustainable urban form: Household consumption of energy and transport in eight residential areas in the greater Oslo region. *Urban Studies*, **42**(12):2145–2166.

Hu, M., H. Bergsdal, E. van der Voet, G. Huppes and D. B. Müller, 2010: Dynamics of urban and rural housing stocks in China. *Building Research & Information*, **38**(3):301–317.

Hu, P. S. and T. R. Reuscher, 2004: *Summary of Travel Trends – 2001 National Household Travel Survey*. Federal Highway Administration, United States Department of Transportation, Washington, DC, USA.

Huq, A. T., M. Zahurul and B. Uddin, 1996: Transport and the Urban Poor. In *The Urban Poor in Bangladesh*. N. Islam, (ed.), Centre for Urban Studies, Dhaka, Bangladesh pp.123

IAUC and WMO, 2006: *Preprints: Sixth International Conference on Urban Climate*. International Association for Urban Climate (IAUC), the World Meteorological Organisation (WMO) and Göteborg University, Göteborg, Sweden.

ICARO, 1999: *Increase of car Occupancy through Innovative Measures and Technical Instruments*. G. Sammer, (ed.) European Commission under the 4th Framework Program, ICARO Consortium, Vienna, Austria.

Ichinose, T., 2008: Anthropogenic Heat and Urban Heat Islands: A Feedback System. Newsletter on Urban Heat Island Counter Measures, Committee on the Global Environment, Architectural Institute of Japan, November 2008, Tokyo, Japan.

ICLEI, 2009: *Harmonized Emissions Analysis Tool*. International Council of Local Environment Initiatives (ICLEI), Oakland, CA, USA.

IEA, 2002: *World Energy Outlook 2002*. International Energy Agency (IEA) of the Organisation of Economic Co-Operation and Development (OECD), Paris, France.

IEA, 2005: *Saving Oil in a Hurry*. International Energy Agency (IEA) of the Organisation for Economic Cooperation and Development (OECD), Paris, France.

IEA, 2008: *World Energy Outlook 2008*. International Energy Agency (IEA) of the Organisation for Economic Cooperation and Development (OECD), Paris, France.

IEA, 2010a: *Energy Balances of Non-OECD Countries*. International Energy Agency (IEA) of the Organization for Economic Cooperation and Development (OECD), Paris, France.

IEA, 2010b: *Energy Balances of OCED Countries*. International Energy Agency (IEA) of the Organization for Economic Cooperation and Development (OECD), Paris, France.

IIASA, 2010: *IPCC Reference Concentration Pathways (RCP) Scenario Database.* International Institute for Applied Systems Analysis (IIASA), Vienna, Austria. Available at http://www.iiasa.ac.at/web-apps/tnt/RcpDb/dsd?Action=htmlpage&page=welcome.

Ijiri, Y. and H. A. Simon, 1975: Some distributions associated with Bose Einstein statistics. *Proceedings of the National Academy of Sciences of the United States of America*, **72**(5):1654–1657.

IMF, 2010: *International Financial Statistics 2010.* CD-ROM Edition, International Monetary Fund (IMF), Washington, DC, USA.

IPCC, 2007: *Climate Change 2007. The Physical Science Basis. Working Group I Contribution to the Fourth Assessment Report.* Intergovernmental Panel on Climate Change (IPCC), Cambridge University Press, Cambridge, UK.

Irwin, E., K. Bell, N. Bockstael, D. Newburn, M. Partridge and J. J. Wu, 2009: The economics of urban-rural space. *Annual Review of Resource Economics*, **1**:435–459.

Jabareen, Y. R., 2006: Sustainable Urban Forms. *Journal of Planning Education and Research*, **26**(1):38–52.

Jalihal, S. A. and T. S. Reddy, 2006: Assessment of the impact of improvement measures on air quality: Case study of Delhi. *Journal of Transportation Engineering*, **132**(6):482–488.

Jenks, M., E. Burton and K. e. Williams, 1996: *The Compact City. A Sustainable Urban Form?*, Spon Press, Chapman and Hall, London, UK.

Joly, I., 2004: The link between travel time budget and speed: A key relationship for urban space-time dynamics. *European Transport Conference*, Strasbourg, France.

Kalnay, E. and M. Cai, 2003: Impact of urbanization and land-use change on climate. *Nature*, **423**(6939):528–531.

Kanda, M., 2006: Progress in the scale modeling of urban climate: Review. *Theoretical and Applied Climatology*, **84**(1):23–33.

Kandlikar, M., 2007: Air pollution at a hotspot location in Delhi: Detecting trends, seasonal cycles and oscillations. *Atmospheric Environment*, **41**(28):5934–5947.

Karekezi, S., J. Kimani and O. Onguru, 2008: Energy access among the urban poor in Kenya. *Energy for Sustainable Development*, **12**(4):38–48.

Kataoka, K., F. Matsumoto, T. Ichinose and M. Taniguchi, 2009: Urban warming trends in several large Asian cities over the last 100 years. *Science of the Total Environment*, **407**(9):3112–3119.

Kathuria, V., 2004: Impact of CNG on Vehicular Pollution in Delhi: A Note. *Transportation Research Part D: Transport and Environment*, **9**(5):409–417.

Kebede, B., A. Bekele and E. Kedir, 2002: Can the urban poor afford modern energy? The case of Ethiopia. *Energy Policy*, **30**(11–12):1029–1045.

Keirstead, J., N. Samsatli and N. Shah, 2009: Syncity: An Integrated Tool Kit for Urban Energy Systems Modelling. *Fifth Urban Research Symposium*, 29–30 June, World Bank, Marseilles, France.

Kelly, J., P. Jones, F. Barta, R. Hoessinger, A. Witte, A. Wolf, K. Beckmann, A. Costain, E. Erl, T. Grosvenor, A. Keuc, J. Machin, J. Raffaillac, G. Sammer and V. Williams, 2004: Volume 1: Concepts and Tools; Volume 2: Fac Sheets. In *Successful Transport Decision-Making: A Project Management and Stakeholder Engagement Handbook*, European Commission under the 5th Framework Program, GUIDEMAPS Consortium, Brussels, Belgium.

Keneko, S., H. Nakayama and L. Wu, 2003: Comparative Study on Indirect Energy Demand Supply and Corresponding CO₂ εμισσιονσ ιν Ασιαν Μεγα– χιτιεσ *Proceedings of International Workshop on Policy Integration Towards Sustainable Urban Energy Use of Cities in Asia*, Institute for Global Environmental Strategies, Honolulu, HI, USA.

Kennedy, C., J. Steinberger, B. Gasson, Y. Hansen, T. Hillman, M. Havránek, D. Pataki, A. Phdungsilp, A. Ramaswami and G. V. Mendez, 2009: Greenhouse Gas Emissions from Global Cities. *Environmental Science & Technology*, **43**(19):7297–7302.

Kennedy, C., J. Steinberger, B. Gasson, Y. Hansen, T. Hillman, M. Havránek, D. Pataki, A. Phdungsilp, A. Ramaswami and G. V. Mendez, 2010: Methodology for Inventorying Greenhouse Gas Emissions from Global Cities. *Energy Policy*, **38**(9):4828–4837.

Kenworthy, J., F. B. Laube, P. Newman, P. A. Barter, T. Raad, C. Poboon and B. Guia, 1999: *An International Sourcebook of Automobile Dependence in Cities 1960–1990.* University Press of Colorado, Boulder, CO, USA.

Kenworthy, J. and F. B. Laube, 2001: *Millenium Database for Sustainable Transport.* International Union of Public Transport (UITP) Brussels, Belgium.

Kenworthy, J. R. and P. W. G. Newman, 1990: Cities and Transport Energy – Lessons from A Global Survey. *Ekistics – the Problems and Science of Human Settlements*, **57**(344–45):258–268.

Kenworthy, J. R., 2006: The eco-city: ten key transport and planning dimensions for sustainable city development. *Environment and Urbanization*, **18**(1):67–85.

Kikegawa, Y., Y. Genchi, H. Yoshikado and H. Kondo, 2003: Development of a numerical simulation system toward comprehensive assessments of urban warming countermeasures including their impacts upon the urban buildings' energy-demands. *Applied Energy*, **76**(4):449–466.

Kim, K. S. and N. Gallent, 1998: Regulating industrial growth in the South Korean Capital region. *Cities*, **15**(1):1–11.

Kinoshita, T., E. Kato, K. Iwao and Y. Yamagata, 2008: Investigating the rank-size relationship of urban areas using land cover maps. *Geophys. Res. Lett.*, **35**(17):L17405.

Kluy, A., 2010: *Wenn das Öl zu Ende geht.* Die Welt 25, Available at http://www.welt.de/die-welt/kultur/article6777536/Wenn-das-Oel-zu-Ende-geht.html.

Klysik, K. and K. Fortuniak, 1999: Temporal and spatial characteristics of the urban heat island of Lódz, Poland. *Atmospheric Environment*, **33**(24–25):3885–3895.

Kölbl, R. and D. Helbing, 2003: Energy laws in human travel behaviour. *New Journal of Physics*, **5**:48.41–48.12.

Krugman, P., 1991: Increasing Returns and Economic Geography. *The Journal of Political Economy*, **99**(3):483–499.

Kühnert, C., D. Helbing and G. B. West, 2006: Scaling laws in urban supply networks. *Physica A: Statistical Mechanics and its Applications*, **363**(1):96–103.

Kulkarni, A., G. Sant and J. G. Krishnayya, 1994: Urbanization in search of energy in three Indian cities. *Energy*, **19**(5):549–560.

Kyokutamba, J., 2004: Uganda. In *Energy services for the urban poor in Africa: Issues and policy implications.* B. Kebede and I. Dube, (eds.), Zed Books, London, UK pp.231–278.

Lakhina, A., J. W. Byers, M. Crovella and I. Matta, 2003: On the geographic location of Internet resources. *IEEE Journal on Selected Areas in Communications*, **21**(6):934–948.

Larivière, I. and G. Lafrance, 1999: Modelling the Electricity Consumption of Cities: Effect of Urban Density. *Energy Economics*, **21**(1):53–66.

Larson, C., 2009: China's Grand Plans for Eco-Cities now lie abandoned. Yale Environment 360, Yale School of Forestry & Environmental Studies, 6 April, New Haven, CT, USA.

Leach, G. and R. Mearns, 1989: *Beyond the Woodfuel Crisis – People, Land and Trees in Africa.* Earthscan London, UK.

Ledent, J., 1980: *Comparative Dynamics of three demographic models of urbanization.* IIASA, Laxenburg, Austria.

Ledent, J., 1982: Rural-Urban Migration, Urbanization, and Economic Development. *Economic Development and Cultural Change*, **30**(3):507–538.

Ledent, J., 1986: A Model of Urbanization with Nonlinear Migration Flows. *International Regional Science Review*, **10**(3):221–242.

Lenzen, M., C. Dey and B. Foran, 2004: Energy requirements of Sydney households. *Ecological Economics*, **49**(3):375–399.

Lenzen, M., M. Wier, C. Cohen, H. Hayami, S. Pachauri and R. Schaeffer, 2006: A comparative multivariate analysis of household energy requirements in Australia, Brazil, Denmark, India and Japan. *Energy*, **31**(2–3):181–207.

Levinson, D. M. and A. Kumar, 1997: Density and the Journey to Work. *Growth and Change*, **28**(2):147–172.

Liu, J., G. C. Daily, P. R. Ehrlich and G. W. Luck, 2003: Effects of household dynamics on resource consumption and biodiversity. *Nature*, **421**(6922):530–533.

Longhurst, J. W. S., J. G. Irwin, T. J. Chatterton, E. T. Hayes, N. S. Leksmono and J. K. Symons, 2009: The development of effects-based air quality management regimes. *Atmospheric Environment*, **43**(1):64–78.

Ludwig, D., M. Klärle and S. Lanig, 2008: Automatisierte Standortanalyse für die Solarnutzung auf Dachflächen über hochaufgelöste Laserscanningdaten. In *Angewandte Geoinformatik 2008 – Beiträge zum 20. AGIT-Symposium Salzburg*. J. Strobl, T. Blaschke and J. Grieser, (eds.), Wichmann, Heidelberg, Germany pp.466–475.

Maibach, M., C. Schreyer, D. Sutter, H. P. van Essen, B. H. Boon, R. Smokers, A. Schroten, C. Doll, B. Pawloska and M. Bak, 2008: *Handbook on Estimation of External Costs in the Transport Sector*. Internalisation Measures and Policies for All External Cost of Transport (IMPACT), CE Delft, Delft, the Netherlands.

Marchetti, C., 1994: Anthropological invariants in travel behavior. *Technological Forecasting and Social Change*, **47**(1):75–88.

Masera, O. R. and G. S. Dutt, 1991: A thermodynamic analysis of energy needs: A case study in A Mexican village. *Energy*, **16**(4):763–769.

Mayor of London, 2004: *Greenlight to Clean Power: The Mayor's Energy Strategy.* Greater London Authority, London, UK.

McGranahan, G., P. Jacobi, J. Songsore, C. Surjadi and M. Kjellén, 2001: *Citizens at Risk: From Urban Sanitation to Sustainable Cities.* Earthscan, London, UK.

McGranahan, G. and P. J. Marcotullio, (eds.), 2007: *Urban Transitions and the Spatial Displacement of Environmental Burdens. Scaling Urban Environmental Challenges: From the Local to Global and Back*. Earthscan, London, UK.

Meikle, S. and P. North, 2005: *A study of the impact of energy use on poor women and girls' livelihoods in Arusha, Tanzania.* R8321, UK Department for International Development (DFID), London, UK.

Meschik, M., 2008: *Planungshandbuch Radverkehr (Planning Handbook of Bicycle Traffic).* Springer, Vienna, Austria and New York, NY, USA.

Mestl, H. E. S., K. Aunan, H. M. Seip, S. Wang, Y. Zhao and D. Zhang, 2007: Urban and rural exposure to indoor air pollution from domestic biomass and coal burning across China. *Science of the Total Environment*, **377**(1):12–26.

Metz, D., 2008: *The Limits to Travel: How Far will you Go?*, Earthscan, London, UK.

Minx, J. C., T. Wiedmann, R. Wood, G. P. Peters, M. Lenzen, A. Owen, K. Scott, J. Barrett, K. Hubacek, G. Baiocchi, A. Paul, E. Dawkins, J. Briggs, D. Guan, S. Suh and F. Ackerman, 2009: Input-Output Analysis and Carbon Footprigint: An Overview of Applications. *Economic Systems Research*, **21**(3):187–216.

Mitlin, D., 1997: Tenants: addressing needs, increasing options. *Environment and Urbanization*, **9**(2):17–212.

MOE, 2005: *2004 Status of Air Pollution*. Ministry of the Environment (MOE), Government of Japan, Tokyo.

Mwangi, I. K., 1997: The nature of rental housing in Kenya. *Environment and Urbanization*, **9**(2):141–159.

Nakicenovic, N., A. Gruebler, L. Bodda and P.-V. Gilli, 1990: *Technological Progress, Structural Change and Efficient Energy Use: Trends Worldwide and in Austria*. Available (accessed 20 May, 2010).

Nakicenovic, N., A. Grübler and A. McDonald, 1998: *Global Energy Perspectives*. Cambridge University Press, Cambridge, UK.

National Research Council, 2003: *Cities Transformed: Demographic Change and Its Implications in the Developing World.* The National Academies Press, Washington, DC, USA.

Newcombe, K., J. D. Kalma and A. R. Aston, 1978: The Metabolism of a City: The Case of Hong Kong. *Ambio*, **7**(1):3–15.

Newman, P. and J. Kenworthy, 1991: *Cities and Automobile Dependence: An international source book.* Avebury, Aldershot, UK.

Newman, P. W. G. and J. R. Kenworthy, 1989: Gasoline Consumption and Cities – A Comparison of U.S. Cities with a Global Survey. *Journal of the American Planning Association*, **55**(1):24–37.

Newman, P. W. G. and J. R. Kenworthy, 1999: *Sustainability and Cities: Overcoming Automobile Dependence.* Island Press, Washington, DC, USA.

Newton, P., S. Tucker and M. Ambrose, 2000: Housing form, energy use and greenhouse gas emissions. In *Achieving Sustainable Urban Form*. K. Williams, E. Burton and M. Jenks, (eds.), Routledge, London, UK pp.74–84.

Nicholls, R. J., S. Hanson, C. Herweijer, N. Patmore, S. Hallegatte, J. Corfee-Morlot, J. Château and R. Muir-Wood, 2008: *Ranking Port Cities with High Exposure and Vulnerability to Climate Extremes*. OECD Environment Working Papers No. 1, Organisation for Economic Cooperation and Development (OECD), Paris, France.

Nijkamp, P. and S. A. Rienstra, 1996: Sustainable transport in a compact city. In *The Compact City. A Sustainable Urban Form?* M. E. Jenks, E. Burton and K. Williams, (eds.), Spon Press, Chapman and Hall, London, UK pp.190–199.

NOAA, 2008: *Version 4 DMSP-OLS Nighttime Lights Series*. National Geophysical Data Center, National Oceanic and Atmospheric Administration (NOAA). Available at http://www.ngdc.noaa.gov/dmsp/downloadV4composites.html.

Norman, J., H. L. MacLean, M. Asce and C. A. Kennedy, 2006: Comparing high and low residential density: life-cycle analysis of energy use and greenhouse gas emissions. *Journal of Urban Planning and Development*, **132**(1):10–21.

Normile, D., 2008: China's Living Laboratory in Urbanization. *Science*, **319**(5864):740–743.

NRC, 2009: *Driving and the Built Environment: The Effects of Compact Development on Motorized Travel, Energy Use, and CO2 Emissions.* United States National Research Council (NRC), Washington, DC, USA.

O'Neill, B., D. Balk, M. Ezra and M. Brickman, 2001: A Guide to Global Population Projections. *Demographic Research*, **4**:203–288.

O'Neill, B. and S. Scherbov, 2006: *Interpreting UN Urbanization Projections Using a Multi-state Model*. International Instifite for Applied Systems Analysis, Laxenburg, Austria.

O'Neill, B. C., M. Dalton, R. Fuchs, L. Jiang, S. Pachauri and K. Zigova, 2010: Global demographic trends and future carbon emissions. *Proceedings of the National Academy of Sciences*, **107**(41):17521–17526.

Oke, T. R., 1987: *Boundary Layer Climates*. Routledge, New York, NY, USA.

Oke, T. R., 2006: *Initial Guidance to Obtain Representative Meteorological Observations at Urban Sites*. Instruments and Observing Methods Report No. 81, World Meteorological Organisation, Geneva, Switzerland.

Oleson, K. W., G. B. Bonan, J. Feddema and T. Jackson, 2010: An examination of urban heat island characteristics in a global climate model. *International Journal of Climatology*:n/a-n/a.

Ouedraogo, B., 2006: Household energy preferences for cooking in urban Ouagadougou, Burkina Faso. *Energy Policy*, **34**(18):3787–3795.

ÖVG, 2009: *Handbuch Öffentlicher Verkehr, Schwerpunkt Österreich*. Bohmann Druck und Verlag, Vienna, Austria.

Pacala, S. and R. Socolow, 2004: Stabilization wedges: Solving the climate problem for the next 50 years with current technologies. *Science*, **305**(5686): 968–972.

Pachauri, S. and D. Spreng, 2002: Direct and indirect energy requirements of households in India. *Energy Policy*, **30**(6):511–523.

Pachauri, S., 2004: An analysis of cross-sectional variations in total household energy requirements in India using micro survey data. *Energy Policy*, **32**(15):1723–1735.

Pachauri, S., A. Mueller, A. Kemmler and D. Spreng, 2004: On Measuring Energy poverty in Indian Households. *World Development*, **32**(12):2083–2104.

Pachauri, S., 2007: *An Energy Analysis of Household Consumption: Changing Patterns of Direct and Indirect Use in India*. Springer-Verlag, Berlin, Germany.

Pachauri, S. and L. Jiang, 2008: The household energy transition in India and China. *Energy Policy*, **36**(11):4022–4035 [2008].

Padam, S. and S. K. Singh, 2001: *Urbanization and Urban Transport in India: The Sketch for a Policy*. Central Institute of Road Transport, Pune, India.

Parshall, L., K. Gurney, S. A. Hammer, D. Mendoza, Y. Zhou and S. Geethakumar, 2010: Modeling energy consumption and CO2 emissions at the urban scale: Methodological challenges and insights from the United States. *Energy Policy*, **38**(9):4765–4782.

Patterson, W., 2009: *Keeping the Lights On*. Earthscan, London, UK.

Pearce, F., 2010: *The Population Crash*. The Guardian. London, UK, Available at http://www.guardian.co.uk/world/2010/feb/01/population-crash-fred-pearce.

Pedersen, P. D., 2007: *Human Development Report 2007/2008 – Mitigation Country Studies: Japan*. Occasional Paper No. 2007/56, Human Development Report Office, United Nations Development Programme (UNDP), New York, NY, USA.

Pelling, M. and B. Wisner, (eds.), 2009: *Disaster Risk Reduction: Cases from Urban Africa*. Earthscan, London, UK.

Peña, S., 2005: Recent developments in urban marginality along Mexico's northern border. *Habit International*, **29**(2):285–301.

Permana, A. S., R. Perera and S. Kumar, 2008: Understanding energy consumption pattern of households in different urban development forms: A comparative study in Bandung City, Indonesia. *Energy Policy*, **36**(11):4287–4297.

Peters, G., T. Briceno and E. Hertwich, 2004: *Pollution Embodied in Norwegian Comsumption*. Working Paper No. 6, Industrial Ecology Programme, Norwegian University of Science and Technology (NTNU), Trondheim, Norway.

Peters, G. P., 2008: From Production-based to Consumption-based National Emission Inventories. *Ecological Economics*, **65**(1):13–23.

Peters, G. P. and E. G. Hertwich, 2008: CO$_2$ Embodied in International Trade with Implications for Global Climate Policy. *Environmental Science & Technology*, **42**(5):1401–1407.

Pickett, S. T. A., M. L. Cadenasso, J. M. Grove, C. H. Nilon, R. V. Pouyat, W. C. Zipperer and R. Costanza, 2008: Urban Ecological Systems: Linking Terrestrial Ecological, Physical, and Socioeconomic Components of Metropolitan Areas. In *Urban Ecology*. J. M. Marzluff, E. Shulenberger, W. Endlicher, M. Alberti, G. Bradley, C. Ryan, U. Simon and C. ZumBrunnen, (eds.), Springer US, New York, NY, USA pp.99–122.

Pischinger, R., S. Hausberger, C. Sudy, J. Meinhart, G. Sammer, O. Thaller, F. Schneider and M. Stiglbauer, 1997: *Volkswirtschaftliche Kosten-Wirksamkeitsanalyse von Maßnahmen zur Reduktion der CO2-Emissionen in Österreich*. Bundesministeriums für Umwelt, Jugend und Familie and the Akademie für Umwelt und Energie, Graz, Vienna and Linz, Austria.

Pohekar, S. D., D. Kumar and M. Ramachandran, 2005: Dissemination of cooking energy alternatives in India – a review. *Renewable and Sustainable Energy Reviews*, **9**(4):379–393.

PriceWaterhouseCoopers, 2007: *UK Economic Outlook 2007*. PriceWaterhouseCoopers LLP, London.

Ramaswami, A., T. Hillman, B. Janson, M. Reiner and G. Thomas, 2008: A Demand-Centered, Hybrid Life-Cycle Methodology for City-Scale Greenhouse Gas Inventories. *Environmental Science & Technology*, **42**(17):6455–6461.

Ratti, C., N. Baker and K. Steemers, 2005: Energy consumption and Urban Texture *Energy and Buildings*, **37**(7):762–776.

Ravindra, K., E. Wauters, S. K. Tyagi, S. Mor and R. van Grieken, 2006: Assessment of Air Quality After the Implementation of Compressed Natural Gas (CNG) as Fuel in Public Transport in Delhi, India. *Environmental Monitoring and Assessment*, **115**(1–3):405–417.

Rees, W. and M. Wackernagel, 1996: Urban ecological footprints: why cities cannot be sustainable – and why they are a key to sustainability. *Environmental Impact Assessment Review*, **16**(4–6):223–248.

Rickaby, P. A., 1991: Energy and urban development in an archetypal English town. *Environment and Planning B: Planning and Design*, **18**(2):153–175.

RITA, 2009: *Integrated Corridor Management*. Available at www.its.dot.gov/icms/index.htm (accessed 30 December, 2009).

Rogers, A., 1982: Sources of Urban Population Growth and Urbanization, 1950–2000: A Demographic Accounting. *Economic Development and Cultural Change*, **30**(3):483–506.

Romero Lankao, P., H. López Villafranco, C. Rosas Huerta, G. Günther and Z. Correa Armenta, 2005: *Can Cities Reduce Global Warming? Urban Development and the Carbon Cycle in Latin America*. IAI, UAM-X, IHDP, GCP, Mexico City, Mexico.

Rosen, M. A., 1992: Evaluation of energy utilization efficiency in Canada using energy and exergy analyses. *Energy*, **17**(4):339–350.

Rosenfeld, A. H., H. Akbari, S. Bretz, B. L. Fishman, D. M. Kurn, D. Sailor and H. Taha, 1995: Mitigation of urban heat islands: materials, utility programs, updates. *Energy and Buildings*, **22**(3):255–265.

Sabry, S., 2009: *Poverty Lines in Greater Cairo: Under-estimating and Misrepresenting Poverty*. International Institute for Environment and Development (IIED), London, UK.

Sahakian, M. and J. K. Steinberger, 2010: Energy Reduction Through a Deeper Understanding of Household Consumption: Staying Cool in Metro Manila. *Journal of Industrial Ecology*, **forthcoming publication**.

Salat, S. and A. Mertorol, 2006: *Factor 20: A multiplying method for dividing by 20 the carbon energy footprint of cities: the urban morphology factor*. Urban Morphologies Laboratory, CSTB (French Scientific Centre for Building Research) and ENSMP (Ecole Nationale Superieure des Mines de Paris), Paris, France.

Salat, S. and C. Guesne, 2008: *Energy and carbon efficiency of urban morphologies. The case of Paris*. Urban Morphologies Laboratory, CSTB (French Scientific Centre for Building Research and ENSMP (Ecole Nationale Superieure des Mines de Paris), Paris, France.

Salomon, I., P. Bovy and J. P. Orefeuil, 1993: *A Billion Trips a Day: Tradition and Transition in European Travel Patterns*. Springer, New York, NY, USA.

Sammer, G., J. Stark, R. Klementschitz, G. Weber, G. Stöglehner, G. Dittrich and L. Bittner, 2005: *Instrumente zur Steuerung des Stellplatzangebotes für den Zielverkehr*. Final Report, Parts 1 and 2, In-Stella Consortium, University of Natural Resources and Applied Life Sciences, Vienna, Austria.

Sammer, G., R. Klementschitz and J. Stark, 2007: A parking management scheme for private car parks – a promising approach to mitigate congestion on urban roads? *23rd World Road Congress*, Paris, France.

Sammer, G., 2009a: Konjunkturprogramme für Verkehr – Chancen für neue Wege im Verkehr? *17. Bad Kreuznacher Verkehrssymposium Kohle für den Verkehr! Auswirkungen der Wirtschaftskrise auf die Mobilität*, Bad Kreuznach, Germany.

Sammer, G., 2009b: Non-Negligible Side Effects of traffic Management. In *Travel Demand Management and Road User Pricing; Susscess, Failure and Feasibility*. W. Saleh and G. Sammer, (eds.), Ashgate, Surrey, UK pp.268.

Sammer, G., W. J. Berger, B. Kohla, M. Meschik and J. Stark, 2009: *Schriftliche Unterlagen Vekehrsplanung und Mobility*. Lehrveranstaltung 856 102, Institut für Vekehrswesen, Universität für Bodenkultur, Vienna, USA.

Sartori, I. and A. G. Hestnes, 2007: Energy use in the life cycle of conventional and low-energy buildings: A review article. *Energy and Buildings*, **39**(3):249–257.

Schelling, T. C., 1969: Models of Segregation. *The American Economic Review*, **59**(2):488–493.

Scheuer, C., G. A. Keoleian and P. Reppe, 2003: Life cycle energy and environmental performance of a new university building: modeling challenges and design implications. *Energy and Buildings*, **35**(10):1049–1064.

Schipper, L., 2004: International Comparisons of Energy End Use: Benefits and Risks. In *Encyclopedia of Energy*. C. J. Cleveland, (ed.), Elsevier, Amsterdam, the Netherlands, Vol. 3, pp.529–555.

Schneider, A., M. A. Friedl and D. Potere, 2009: A new map of global urban extent from MODIS satellite data *Environmental Research Letters*, **4**(4):044003.

Scholz, Y., 2010: *Möglichkeiten und Grenzen der Integration Vershiedener Regenerativer Energiequellen zu einer 100% Regenerativen Stromversorgung der Bundesrepublik Deutschland bis zum Jahr 2050*. Endbericht für den Sachverständigenrat für Umweltfragen, German Aerospace Centre, Cologne, Germany.

Scholz, Y., 2011: *Data based on "Renewable Energy Based Electricity Supply at Low Costs – Setup and Application of the REMix model for Europe*. German Aerospace Centre, Stuttgart, Germany.

Schulz, N. B., 2005: Contributions of Material and Energy Flow Accounting to Urban Ecosystems Analysis. UNU-IAS Working Paper No. 136, United Nations University, Institute for Advanced Studies, Yokohama, Japan.

Schulz, N. B., 2006: *Socio-economic Development and Society's Metabolism in Singapore*. UNU-IAS Working Paper No. 148, United Nations University, Institute of Advanced Studies, Yokohama, Japan.

Schulz, N. B., 2007: The direct material inputs into Singapore's development. *Journal of Industrial Ecology*, **11**(2):117–131.

Schulz, N. B., 2010a: *Urban Energy consumption Database and Estimations of Urban Energy Intensities*. International Institute for Applied Systems Analysis (IIASA), Laxenburg, Austria.

Schulz, N. B., 2010b: Delving into the carbon footprints of Singapore – Comparing direct and indirect greenhouse gas emissions of a small and open economic system. *Energy Policy*, **38**(9):4848–4855.

Schwela, D., G. Haq, C. Huizenga, W. Han, H. Fabian and M. Ajero, 2006: *Urban Air Pollution in Asian Cities: Status, Challenges and Management*. Earthscan, London, UK.

Seto, K. C. and J. M. Shepherd, 2009: Global urban land-use trends and climate impacts. *Current Opinion in Environmental Sustainability*, **1**(1):89–95.

Shrestha, R. M., S. Kumar, S. Martin and A. Dhakal, 2008: Modern Energy Use by the Urban Poor in Thailand: a Study of Slum Households in Two Cities. *Energy for Sustainable Development*, **12**(4):5–13.

Smil, V., 1991: *General Energetics. Energy in the Biosphere and Civilizations*. John Wiley & Sons, New York, NY, USA.

Smith, K. R., 1993: Fuel Combustion, Air Pollution Exposure, and Health: The Situation in Developing Countries. *Annual Review of Energy and the Environment*, **18**(1):529–566.

Sornette, D. and R. Cont, 1997: Convergent multiplicative processes repelled from zero: Power laws and truncated power laws. *Journal de Physique I*, **7**:431–444.

Souch, C. and S. Grimmond, 2006: Applied Climatology: Urban Climate. *Progress in Physical Geography*, **30**(2):270–279.

SPARC, 1990: The Society for the Promotion of Area Resource Centres (SPARC): developing new NGO lines. *Environment and Urbanization*, **2**(1):91–104.

Stead, D. and J. e. a. Williams, 2000: Land use, Transport and People: Identifying the Connections. In *Achieving Sustainable Urban Form*. K. Williams, E. Burton and M. Jenks, (eds.), Spon Press, Taylor and Francis, London, UK.

Steemers, K., 2003: Energy and the city: density, buildings and transport. *Energy and Buildings*, **35**(1):3–14.

Steingrube, W. and R. Boerdlein, 2009: *Greifswald – die Fahrradstadt, Ergebnisse der Befragung zur Verkehrsmittel der Greifswalder Bevölkerung 2009 (Greifswald – The Bicycle City, Results of a survey for the modal split of the Greifswald population)*. Institut für Geographie und Geologie der Universität Greifswald, Greifswald, Germany.

Stephens, C., P. Rajesh and S. Lewin, 1996: *This is My Beautiful Home: Risk Perceptions towards Flooding and Environment in Low Income Urban Communities: A Case Study in Indore, India*. London School of Hygiene and Tropical Medicine, London, UK.

Supernak, J., 2005: HOT Lanes on Interstate 15 in San Diego: Technology, Impacts and Equity Issues. *PIARC Seminar on Road Pricing with Emphasis on Financing, Regulation and Equity*, Cancun, Mexico.

Susilo, Y. and O. Stead, 2008: Targeting TDM Policies based on Individual transport Emissions. *4th International Symposium on Travel Demand Management*, Vienna.

Taha, H., 1997: Urban Climate and Heat Islands: Albedo, Evapotranspiration and Anthropogenic Heat. *Energy and Buildings*,(25):99–103.

Tarr, J., 2001: Urban history and environmental history in the United States: Complementary and overlapping fields. In *Environmental Problems in European*

Cities of the 19th and 20th Century. C. Bernhardt, (ed.), Waxmann, Muenster, New York, Munich, Berlin pp.25–39.

Tarr, J., 2005: The city and technology. In *A Companion to American Technology.* C. Pursell, (ed.), Blackwell Publishing, New York pp.97–112.

Taylor, B. D., 2004: The politics of congestion mitigation. *Transport Policy*, **11**(3):299–302.

Tokyo Metropolitan Government, 2006: *Environmental White paper 2006.* The Japanese Bureau of Environment, Tokyo, Japan.

Treberspurg, M., 2005: Nachhaltige und zukunftssichere Architektur durch ressourcenorientiertes Bauen. *Österreichische Wasser- und Abfallwirtschaft*, **57**(7):111–117.

Uchida, H. and A. Nelson, 2008: *Agglomeration Index: Towards a New Measaure of Urban Concentration.* World Development Report Background Paper, World Bank, Washington, DC.

UKDECC, 2010: *Annual Report on Fuel Poverty Statistics 2009.* United Kingdom Department of Energy and Climate Change (UKDECC), London, UK.

Umweltzone Berlin, 2009: *Aktuelles Umweltzone.* Available at www.berlin.de/ Umweltzone (accessed 26 January, 2010).

UN, 2008: Report of the Secretary-General: World population monitoring, focusing on population distribution, urbanization, internal migration and development. *United Nations (UN) Commission on Population and Development 41st Session*, 7–11 April, United Nations, New York, NY, USA.

UN DESA, 1971: *Some simple methods for urban and rural population forecasts.* United Nations Department of Economic and Social Affairs (UN DESA), New York, NY, USA.

UN DESA, 1973: *The determinants and consequences of population trends: new summary of findings on interaction of demographic, economic and social factors.* United Nations Department of Economic and Social Affairs (UN DESA), New York, NY, USA.

UN DESA, 1974: *Manual VIII: Methods for Projections of Urban and Rural Population.* Population Division, United Nations Department of Economic and Social Affairs (UN DESA), New York, NY, USA.

UN DESA, 1980: *Patterns of Urban and Rural Population Growth.* Population Studies, Population Division, United Nations Department of Economic and Social Affairs, New York, NY, USA.

UN DESA, 2008: *World Urbanization Prospects: The 2007 Revision. Highlights.* Population Division, United Nations Department of Economic and Social Affairs (UN DESA), New York, NY, USA.

UN DESA, 2010: *World Urbanization Prospects: The 2009 Revision. Highlights.* Population Division, United Nations Department of Economic and Social Affairs (UN DESA), New York, NY, USA.

UN HABITAT, 2003: *The Challenge of Slums. Global Report on Human Settlements 2003.* United Nations Human Settlements Programme (UN-HABITAT), Earthscan, London, UK.

UN DESA, 2008: *State of the World's Cities 2008/2009. Harmonious Cities.* United Nations Human Settlements Programme (UN-HABITAT), Earthscan, London, UK.

UNDP and WHO, 2009: *The Energy Access Situation in Developing Countries: A Review Focusing on the Least Developed Countries and Sub-Saharan Africa.* The Energy Access Situation in Developing Countries: A Review Focusing on the Least Developed Countries and Sub-Saharan Africa, New York, NY, USA.

UNEP, 2009: *CNG conversion: Learning from New Delhi.* Available at ekh.unep. org/?q=node/1737 (accessed 2 November, 2009).

UNIDO, 2009: *Breaking In and Moving Up: New Industrial Challenges for the Bottom Billion and the Middle-Income Countries.* Industrial Development Report 2009, United Nations Industrial Development Organization (UNIDO), Vienna, Austria.

UNITE, 2003: *Unification of accounts and marginal costs for Transport Efficiency.* European Commission under the Transport RTD Programme of the 5th Framework Programme, the University of Leeds, Leeds, UK.

Urban Audit, 2009: Available at www.urbanaudit.org (accessed 30 December, 2009).

Urban Resource Centre, 2001: Urban poverty and transport: a case study from Karachi. *Environment and Urbanization*, **13**(1):223–233.

US DOE, 2005: *Residential Energy consumption Survey.* Available at eia.doe.gov/ emeu/recs/recs2005/hc2005_tables/detailed_tables2005.html (accessed 19 September, 2010).

USAID, 2004: *Innovative Approaches to Slum Electrification.* Bureau for Economic Growth, Agriculture and Trade; United States Agency for International Development (USAID), Washington, DC, USA.

USEPA, 2009: *National Trends in Sulfur Dioxide Levels.* Available at www.epa.gov/ airtrends/sulfur.html (accessed 11 May, 2009).

van der Plas, R. J. and M. A. Abdel-Hamid, 2005: Can the woodfuel supply in sub-Saharan Africa be sustainable? The case of N'Djaména, Chad. *Energy Policy*, **33**(3):297–306.

VandeWeghe, J. R. and C. Kennedy, 2007: A Spatial Analysis of Residential Greenhouse Gas Emissions in the Toronto Census Metropolitan Area. *Journal of Industrial Ecology*, **11**(2):133–144.

Viswanathan, B. and K. S. Kavi Kumar, 2005: Cooking fuel use patterns in India: 1983–2000. *Energy Policy*, **33**(8):1021–1036.

Vivier, J., 2006: *Mobility in Cities Database, Better Mobility for People Worldwide, Analysis and Recommendations.* International Association of Public Transport (UITP), Brussels, Belgium.

Vringer, K. and K. Blok , 1995: The direct and indirect energy requirements of households in the Netherlands. *Energy Policy*, **23**(10):893–910.

Wapedia, 2009, Available at wapedia.mobi/en/Modal_share (accessed December 30, 2009, 2009).

Watkiss, P., C. Brand, N. Hinds, M. Holland, C. Marx and W. Stephenson, 2000: *Urban Energy-Use: Guidance on Reducing Environmental Impacts.* DFID KaR Project R7369, United Kingdom Department for International Development (DFID), Abingdon, UK.

Weber, C. L. and H. S. Matthews, 2008: Quantifying the global and distributional aspects of American household carbon footprint. *Ecological Economics*, **66**(2–3):379–391.

WEC, 2006: *Alleviating Urban Energy poverty in Latin America.* World Energy Council (WEC), London, UK.

Weisz, H., J. Minx, P. P. Pichler and K. Feng, 2012: Direct and Indirect Household Energy Use in the UK and the Role of Income and Scale. *Energy Policy*, **in preparation**.

Weng, Q., D. Lu and J. Schubring, 2004: Estimation of land surface temperature-vegetation abundance relationship for urban heat island studies. *Remote Sensing of Environment*, **89**(4):467–483.

Wiechmann, T., 2009: Conversion strategies under uncertainty in post-socialist shrinking cities: the example of Dresden in Eastern Germany. In *The Future of Shrinking Cities: Problems, Patterns and Strategies of Urban Transformation in a Global Context.* K. Pallagst, J. Aber and I. Audirac, (eds.), Center for Global

Metropolitan Studies, Institute of Urban and Regional Development, Berkeley, CA, USA pp.5–15.

Wiedenhofer, D., M. Lezen and J. K. Steinberger, 2011: Spatial and socio-economic drivers of diret and indirect household energy consumption in Australia. In *Urban Consumption*. P. W. Newton, (ed.), CSIRO Publishing, Collingwood, Australia, pp. 251–266.

Wiedmann, T., R. Wood, M. Lenzen, J. Minx, D. Guan and J. Barrett, 2007: *Development of an Embedded Carbon Emissions Indicator – Producing a Time Series of Input-Output Tables and Embedded Carbon Dioxide Emissions for the UK by Using a MRIO Data Optimisation System*. Report to the UK Department for Environment, Food and Rural Affairs by Stockholm Environment Institute at the University of York and Centre for Integrated Sustainability Analysis at the University of Sydney, DEFRA, London, UK.

Wier, M., M. Lenzen, J. Munksgaard and S. Smed, 2001: Effects of Household Consumption Patterns on CO2 Requirements. *Economic Systems Research*, 13:259–274.

Wilson, A. G., 2000: *Complex spatial systems: the modelling foundations of urban and regional analysis.* Prentice Hall, Upper Saddle River, NJ, USA.

Winrock International, 2005: *Enabling Urban Poor Livelihoods Policy Making: Understanding the Role of Energy Services – Country Report for the Brazil Project*. DFID KaR Project R8348, Winrock International and the United Kingdom Department for International Development (DFID), Salvador, Brazil.

World Bank, 2009: *The World Development Report 2009: Reshaping Economic Geography.* The World Bank, Washington, DC, USA.

Yamagata, Y., 2010: *Personal communication of detailed data set of Kinoshita et al., 2008*. National Institute for Environmental Studies (NIES), Tsukuba, Japan.

Yapi-Diahou, A., 1995: The informal housing sector in the metropolis of Abidjan, Ivory Coast. *Environment and Urbanization*, **7**(2):11–30.

Zahavi, Y., 1974: *Traveltime Budgets and Mobility in Urban Areas*. United States Department of Transportation, Washington DC, USA.

Zhou, H. and D. Sperling, 2001: *Transportation in Developing Countries. Greenhouse Gas Scenarios for Shanghai, China*. The Pew Center, Arlington, VA, USA.

Zhou, P., 2000: Energy Use in Urban Transport in Africa. In *Energy Environment Linkages in African Cities*, United Nations Centre for Settlements (Habitat) and the United Nations Environment Programme (UNEP), Nairobi, Kenya pp.100–111.

Zipf, G., 1949: *Human Behavior and the Principle of Least-Effort: An Introduction to Human Ecology*. Addison-Wesley, Cambridge, MA, USA.

19 Energy Access for Development

Convening Lead Authors (CLA)
Shonali Pachauri (International Institute for Applied Systems Analysis, Austria)
Abeeku Brew-Hammond (Kwame Nkrumah University of Science and Technology, Ghana)

Lead Authors (LA)
Douglas F. Barnes (Energy for Development, USA)
Daniel H. Bouille (Bariloche Foundation, Argentina)
Stephen Gitonga (United Nations Development Programme)
Vijay Modi (Columbia University, USA)
Gisela Prasad (University of Cape Town, South Africa)
Amitav Rath (Policy Research International Inc., Canada)
Hisham Zerriffi (University of British Columbia, Canada)

Contributing Authors (CA)
Touria Dafrallah (Environment and Development Action in the Third World, Senegal)
Conrado Heruela (United Nations Environment Programme)
Francis Kemausuor (Kwame Nkrumah University of Science and Technology, Ghana)
Reza Kowsari (University of British Columbia, Canada)
Yu Nagai (Vienna University of Technology, Austria)
Kamal Rijal (United Nations Development Programme)
Minoru Takada (United Nations Development Programme)
Njeri Wamukonya[†] (formerly United Nations Environment Programme)

Review Editor
Jayant Sathaye (Lawrence Berkeley National Laboratory, USA)

Contents

Executive Summary

Key Challenges

- A quarter of humanity today lives without access to any electricity and almost one-half still depends on solid fuels such as unprocessed biomass, coal, or charcoal for its thermal needs. These people continue to suffer a multitude of impacts detrimental to their welfare. Most live in rural villages and urban slums in developing nations. Access to affordable modern energy carriers is a necessary, but insufficient step toward alleviating poverty and enabling the expansion of local economies.

- Even among populations with physical access to electricity and modern fuels, a lack of affordability and reliable supplies limits the extent to which a transition to using these can occur. Those who can afford the improved energy carriers may still not be able to afford the upfront costs of connections or the conversion technology or equipment that makes that energy useful.

- Beyond the obvious uses of energy for lighting, cooking, heating, and basic home appliances, uses for purposes that might bring economic development to an area are slow to emerge without institutional mechanisms in place that are conducive to fostering entrepreneurial activity and uses of energy for activities that can generate income. Without the expansion of energy uses to activities that generate income, the economic returns to energy providers are likely to remain unattractive in poor and dispersed rural markets.

- Significant success has been achieved with small pilot projects to improve energy access in some rural areas and among poor communities in urban areas. But subsequently, less thought is focused on how to scale-up from these small pilot and demonstration projects to market development and meeting the needs of the larger population.

Key Messages

1. While the scale of the challenge is tremendous, almost universal access to energy, both electricity and clean cooking for all, is achievable by 2030. As estimated in Chapter 17, this will require global investments to the tune of US$36–41 billion annually, a small fraction of the total energy infrastructural investments required by 2030, and may have a negligible or even negative impact on greenhouse gas (GHG) emissions. Immediate benefits from improved health for millions of people will result and the socioeconomic benefits from improved energy access will extend well beyond those to the current generation.

2. Electrification rates have been more rapid and far in excess of population growth rates in many countries and regions such as East Asia (including China) and Latin America. Between 1990 and 2008, almost two billion people got connected to electricity globally. This provides a basis to believe that electrifying the remaining 1.4 billion people without electricity by 2030 is feasible.

3. The progress with providing clean cooking services globally has been rather dismal over the last decades, with the numbers of people dependent on solid fuels rising in the rural regions of most developing countries and the percentage of rural populations dependent on solid fuels virtually unchanged over the last decade. This suggests that transitioning the global population to clean cooking fuels by 2030 will not be feasible and for some populations, a transition to improved stoves will be necessary to improve their cooking experience. This will require significant advances to be made in rapid diffusion of low-cost, high performing and standardized stoves and more sustainable management and practices along the entire biomass value chain.

4. While the challenge is considerable, the experiences and approaches followed in countries that have been successful in achieving improved access provide important lessons that can be applied elsewhere to achieve universal access by 2030. This will require leveraging funding from public and private sources, both for necessary investments at the macro level, and for meeting costs for low-income households at the micro level. Creative financing mechanisms and transparent cost and price structures will be key to achieving the required scale-up and quick roll-out of solutions to improve access.

5. No single solution fits all in improving access to energy for development. Programs aimed at increasing access must be cognizant of local needs, resources, and existing institutional arrangements and capabilities.

6. Supportive policy and institutional frameworks need to be created that encourage private sector participation, as well as replicability and the scale-up and scale-out of successful programs.

7. Diverse sources of energy supply (fossil and renewable), a wide portfolio of technologies, and a variety of institutional and innovative business and energy service delivery models that are adapted to local circumstances and allow for sustainable replication, deliver social benefits, and generate wealth for the community are required to meet the challenge.

8. An enabling environment shaped by sustained government commitment and enhanced capacity building at all levels is paramount to ensuring access targets are met.

9. Complementary development programs and enhancement of market infrastructure are needed to ensure sustained economic growth and steady employment and income generation for the poor, in order to provide them with a means to pay for improved energy services.

Structure and Roadmap of Chapter 19

Chapter 19 builds strongly on the concerns raised in Chapter 2 ("Energy and Society"), Chapter 3 ("Energy and Environment"), and Chapter 4 ("Energy and Health"). The Chapter is structured as follows (see Figure 19.1). Section 19.1 presents a brief overview of the linkages between energy access, human wellbeing, and the environment. It also lays the foundation for understanding why energy access is essential for poverty reduction and development. Section 19.2 assesses the historical efforts and trends and the current global status of access to electricity, clean cooking/heating fuels and stoves, and modern energy carriers for income generating activities. The following section, 19.3, provides a more differentiated and nuanced analysis of regional efforts and strategies to improve energy access in households and the status of access in each region. Finally, Section 19.4 concludes with lessons learnt and implications for the way forward. The role of policy and institutional issues is also discussed briefly in this chapter; however, a deeper discussion of this is left to Chapter 23, which focuses specifically on policies and measures for expanding energy access.

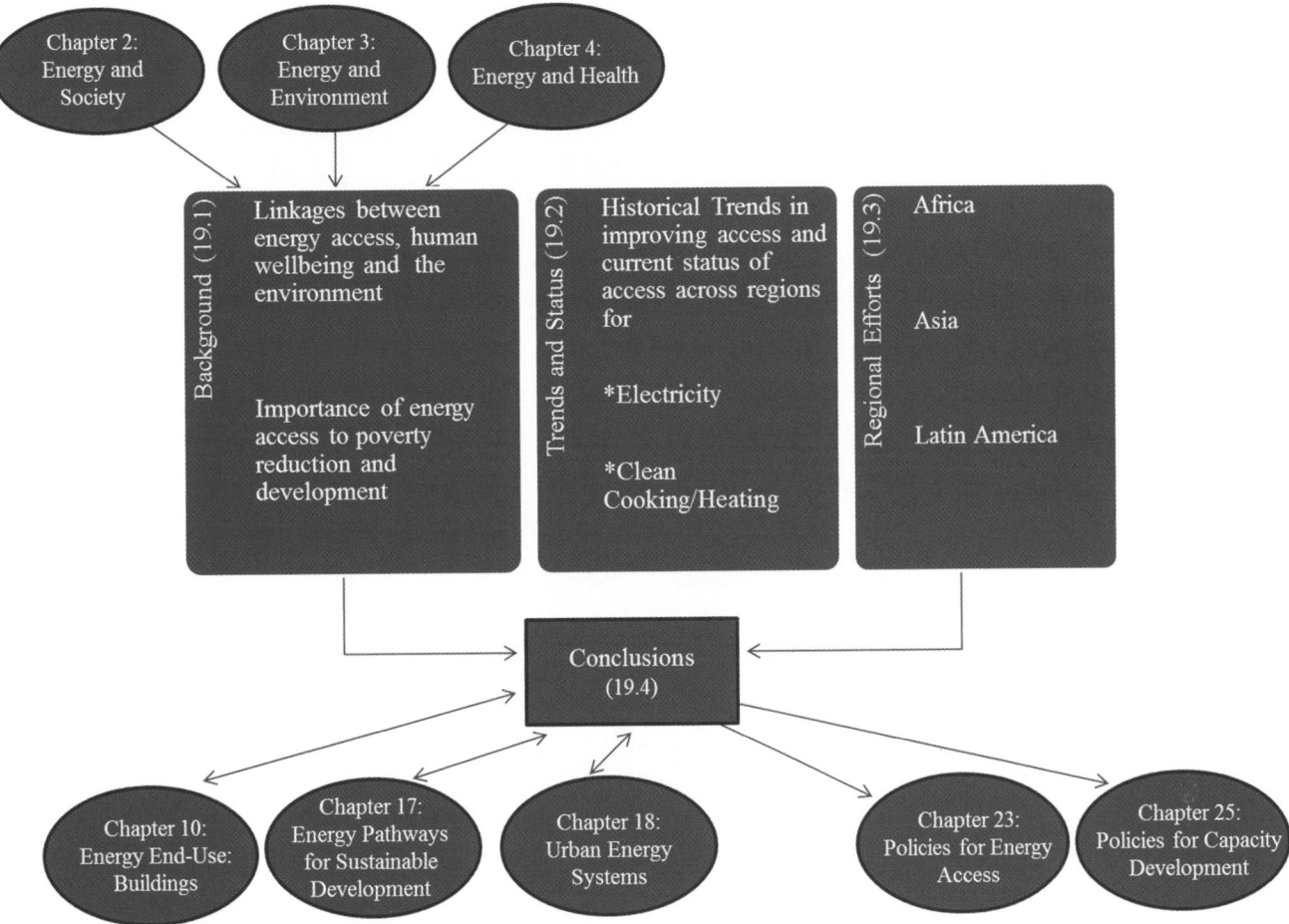

Figure 19.1 | Roadmap of Chapter 19 and its linkages to other chapters.

Box 19.1 | Definitions and Dimensions of Access

The simplest definition of universal access to modern energy is the physical availability of electricity and modern energy carriers and improved end-use devices such as cook stoves at affordable prices for all. A target of energy access for all by 2030, set by the United Nations Secretary-General's Advisory Group on Energy and Climate Change (AGECC, 2010), recommends access to be provided in accordance with this basic definition so as to enhance services such as lighting, cooking, heating, and motive power for populations in developing countries. This is no longer only a moral imperative, but also socially prudent and an economic necessity to enable the almost one and a half billion people living without any electricity and over three billion dependent primarily on solid fuels to lift themselves out of poverty and improve their living conditions. Providing access to improved energy carriers is clearly a necessary, but insufficient condition for overall poverty alleviation and socioeconomic growth. Alleviating poverty, in its totality, clearly also requires improving the earnings of the poor by providing them with more sustainable livelihood opportunities through encouraging the use of energy in activities that can generate income. This requires defining access in a much broader sense and would require making available reliable and adequate qualities and quantities of energy and the associated technologies at affordable costs in a manner that is socially acceptable and environmentally sound so as to meet basic human needs and for activities that are income generating and could empower growth and development.

Such a broader definition of access includes several elements and dimensions, including quality, reliability, adequacy, affordability, acceptability, and environmental soundness. Unfortunately, national level indicators and statistics to measure and monitor these various dimensions of access are extremely scarce, particularly for the least developed countries and regions where the issue is the most pressing.

Within GEA, we abstain from defining any global quantitative thresholds for the minimum amount of energy needed to meet basic needs. This is because basic needs are normative and vary significantly between countries and regions depending on climate, social customs and norms, and other region and society-specific factors. We are, however, mindful of the fact that some national governments have defined basic or lifeline energy entitlements for their poorest citizens. In most cases, these fall within the range of 20–50 kWh of final electricity per household per month to meet basic lighting, communication and entertainment needs, and the equivalent of 6–15 kg of LPG per household per month for cooking. Heating requirements, being seasonal, are often approximated as the equivalent of 15–30 kWh useful energy per square meter of living space, annually. Defining lifeline entitlements in final energy units of course also has the limitation that efficiencies of end-use appliances and equipment are not accounted for.

Clearly, such entitlements fall far below what is required for purposes that can generate income to empower growth and development. Ensuring adequate amounts of energy to achieve this end may require defining some average energy equity thresholds and estimating not only direct energy needs, but indirect or embodied energy requirements as well. This is a much more complex endeavor and would require taking account of national economic and energy system structures and characteristics, as the same amount of energy can provide a wide range of energy services. Previous efforts at quantifying such equity thresholds provide estimates in useful energy terms in the range of 1–2 kW per capita to meet basic needs and much more (Goldemberg et al., 1985; Imboden and Voegelin, 2000).

19.1 Introduction

19.1.1 Background

Energy deeply influences people's lives and is an engine for social development and economic growth. Over the centuries, energy has helped transform societies and has underpinned human development. Energy contributes to fulfilling the most basic human needs, including nutrition, warmth, and light. Furthermore, there is ample evidence that access to reliable, efficient, affordable, and safe energy carriers can directly affect productivity, income, and health, and can enhance gender equity, education, and access to other infrastructure services. However, energy use patterns, in terms of both quality and quantity, are highly inequitably distributed on all sides of the development divide – North and South, rich and poor, men and women, rural and urban. This inequity in energy access and use compromises human welfare and has adverse impacts on the environment. The lack of access to reliable, affordable, and modern energy carriers,[1] particularly in rural areas of developing countries, is a major challenge faced by over one-third of humanity even today. This presents a major impediment to growth and compromises progress toward sustainable development. Providing access to electricity and modern energy carriers and/or devices[2] to all populations is not only a moral obligation, but is also necessary for improving living conditions and may provide economic returns in the long run that far exceed the costs involved.

The world still faces the task of satisfying the demand for energy services of a vast majority of its population to meet basic needs for lighting, cooking, and heating, and for use in activities that can generate income. Recognizing the centrality of improving energy access for the poor, the international community has been increasingly active in discussing the setting of a global energy access target. Several governments and regional bodies have already set national targets to improve access. Building on these, the United Nations Secretary-General's Advisory Group on Energy and Climate Change (AGECC, 2010) recommends ensuring universal energy access by 2030. Meeting such a target requires the provision of affordable[3] electricity and modern fuels and improved end-use devices by 2030 to all who currently lack access. To some, this may appear unattainable, but the technologies and examples of successful policies to achieve this already exist. The challenge to meet such an access target is greater, but can have even more significance for the rural populations of the world. Before the beginning of 2008, rural areas contained more than half the world's population (UN, 2008), with nearly 90% of the rural population, close to three billion, living in developing countries. However, this half of the population still consumes only a small fraction of total global fossil fuels and electricity, and rural energy issues remain largely overlooked in national energy and developmental plans.

The use of unprocessed solid fuels, both commercial and noncommercial, on the other hand, is predominantly concentrated in rural areas of developing countries, and particularly among the poor. This remains the primary source of fuel for cooking and heating for most of the rural population in the developing world and many urban residents as well. Globally, over three billion people rely on solid fuels, largely biomass (wood and residues), charcoal, and coal for cooking and heating (UNDP and WHO, 2009). The use of biomass is both arduous and time consuming in its harvest, transport, and use and is associated with negative environmental consequences. The majority of biomass harvest is carried out by women and children and its negative impacts on them have been discussed in Chapter 2. Negative health impacts from this traditional use of biomass and other solid fuels, discussed in Chapter 3, include those due to household pollution, which accounts for an estimated almost two million deaths/year, with a higher percentage of these being women and children in developing countries (UNDP and WHO, 2009). Furthermore, the burning of wood fuels contributes to climate change through emissions of GHGs, as discussed in Chapter 4. When the biomass burnt is not sustainably harvested, the use of these fuels has the added disadvantage of no longer being CO_2-neutral. As mentioned in Chapter 4, recent evidence also shows that the climate impacts of Black Carbon are larger than previous estimates suggested, particularly for the melting of arctic and glacial ice (Ramanathan and Carmichael, 2008). Black Carbon, or soot, is a byproduct of the combustion of fossil fuel, biofuel, or biomass, including wood waste and agricultural green waste (Grieshop et al., 2009).

Access to cleaner and more efficient end-use devices, processed biomass and/or more efficient fuels can alleviate the public health, welfare, and environmental concerns associated with traditional solid fuel use discussed in the preceding paragraph. In addition, such access can address the home heating needs of those who live in colder climates. In many societies, women and girls bear the disproportionate burden of fuel gathering, home care, and cooking, and hence the provision of more efficient and safer fuels and technologies can also contribute to reducing gender inequities in health and time burdens.[4] The performance of improved biomass cook stoves is wide ranging. However, throughout this assessment, when we refer to advanced stoves we imply stoves that have proven efficiency close to that of LPG stoves and the ability to reduce emissions that are health damaging.

---

1   Modern energy carriers in this chapter refer to electricity (grid or off-grid, both renewable and fossil-based) and liquid and gaseous fuels such as liquefied petroleum gas (LPG), biogas, ethanol, and natural gas.

2   Modern devices refer to improved cook stoves (ICS) that meet a minimum efficiency and emissions standard and have a performance that matches that of LPG stoves. Access to such ICS are assumed to improve cooking energy service for the poor and are included in the universal clean cooking target discussed in Chapter 2 and analyzed in Chapter 17 on future scenarios for improving access.

3   "Affordable" is evaluated in terms of the current spending on energy services and household purchasing power.

4   Truly enhancing equity for women and children will require significant additional social and cultural change and development. However, a full discussion of these issues is beyond the scope of this assessment.

In addition to lacking access to modern fuels and devices, about a quarter of the world's population also still lacks access to any electricity (UNDP and WHO, 2009). Over 85% of those lacking access live in rural and peri-urban areas (Derdevet and Caubet, 2007). In essence, four out of five people without electricity live in rural areas in developing countries, predominantly in the least developed countries of South Asia and sub-Saharan Africa (UNDP and WHO, 2009). Thus even today, the services that electricity makes possible, from basic lighting and telecommunications with mobile phones to computer-controlled agro-industrial processing, remain outside the reach of many people in the developing world. The communication revolution was first unleashed by radio, television, and computers and has now advanced with mobile telephony and the internet. These tools have become essential for individual empowerment, enterprise development, and the functioning of social infrastructure. Access to such modern media, efficient lighting, and other labor-saving devices are impossible without electricity.

It is widely recognized that improvements in access to more efficient energy carriers, both electricity and fuels, can have huge impacts on the lives of people, particularly the poorest in the developing world (WHO, 2006; Kanagawa and Nakata, 2007; World Bank, 2008a; Hiremath et al., 2009; Khandker et al., 2009). Chapter 2 highlights the multitude of benefits that improvements in energy access can make possible for the poor. In addition to the social benefits, energy is also essential for improving productivity, which is crucial for bringing the rural poor out of subsistence activities. Irrigation pumps, processing capability, storage, and access to markets and market information is not possible without adequate energy services, increasingly enabled through electricity and mechanical power. The growth and facilitation of enterprises is also intimately linked to access to energy. Communal services such as schools and health centers also require energy. In addition, the existing base of technological infrastructure provided by standardized motor fuels is an enabler of lowered transportation costs, which allows the movement of goods and people. Expanding access to better quality energy for the poor and unserved therefore remains a major developmental and environmental challenge for the world, particularly in the case of the Least Developed Countries (LDCs).[5] National energy policies and poverty reduction strategy papers in these countries very often either neglect energy completely or focus solely on electrification. They neither reflect adequately the energy-poverty nexus nor include targets and timelines to meet the energy needs of the poor. Often there is also a misalignment between national priorities and budgetary allocations for rural energy, resulting in a lack of coherence between strategies and plans and program implementation on the ground (UNDP, 2007a).

---

5   We follow the UNDP classification of Least Developed Countries and Developing Countries. Please refer to UNDP and WHO (2009) for further details of the countries included.

19.1.2   The Poverty-Energy Relationship

Access to electricity and modern energy carriers that help fulfill energy service needs is of key importance in future efforts at poverty reduction and development, both in rural and urban areas. Chapter 2 includes a detailed discussion of the role of energy in achieving the Millennium Development Goals (MDG). In addition, Chapter 18 deals with the challenges and issues surrounding the lack of adequate and affordable energy access in urban centers. This chapter focuses more specifically on assessing the nature of the access challenge both globally and regionally, and reviewing the progress made to date in improving access in developing countries and regions.

We distinguish between rural and urban areas because the issues related to rural poverty are fundamentally distinct from those of urban poverty and the challenges related to providing access to energy for the rural poor differ from those in urban areas (for further discussion on this topic, see also Chapters 2 and 17). About 75% of the developing world's poor currently live in rural areas, with some marked regional differences (Chen and Ravallion, 2007; Ravallion et al., 2007). Analyses for very different countries, like Brazil, Ecuador, Thailand, Malawi, and Viet Nam, show that poverty rates tend to be higher in remote rural areas than in more accessible areas, and poverty is deeper and more severe in remote areas. While the numbers of rural poor are declining globally, poverty rates in rural areas remain very high, particularly in some regions, and the energy problems of the rural poor persist. The numbers of urban poor, on the other hand, have grown during the last few decades. Indeed, it has been argued that urbanization has helped reduce overall poverty, except that this is shown to be true more for rural poverty than for urban poverty (Ravallion et al., 2007; World Bank, 2008b).

Poverty is linked not only to deprivation of income, but also a lack of access to resources and assets, social networks, voice, and power (UNDP, 2010). Poverty, particularly in rural areas, is often accompanied by both a lack of access to electric power services and an extreme dependence on unprocessed biomass, coal, or charcoal for basic uses such as cooking, water and home heating, as well as a lack of adequate and appropriate energy carriers for use in activities that are income generating. There tends to be a two-way causal relationship between the lack of access to adequate, affordable, and appropriate energy forms and poverty. This has often been termed the "energy-poverty nexus" (UNDP, 2006; Masud et al., 2007) or the "vicious cycle of energy poverty" (WHO, 2006). The cycle is considered vicious because households that lack access to appropriate energy are often trapped in a vortex of deprivation. The lack of energy, in addition to insufficient access to other key services and assets, affects productivity, time budgets, opportunities for income generation, and more generally, the ability to improve living conditions. The low productivity and livelihood opportunities, in turn, result in low earnings and no or little surplus cash for these people. This contributes to the poor remaining poor and consequently, also energy poor, since they cannot afford to pay for improved energy services

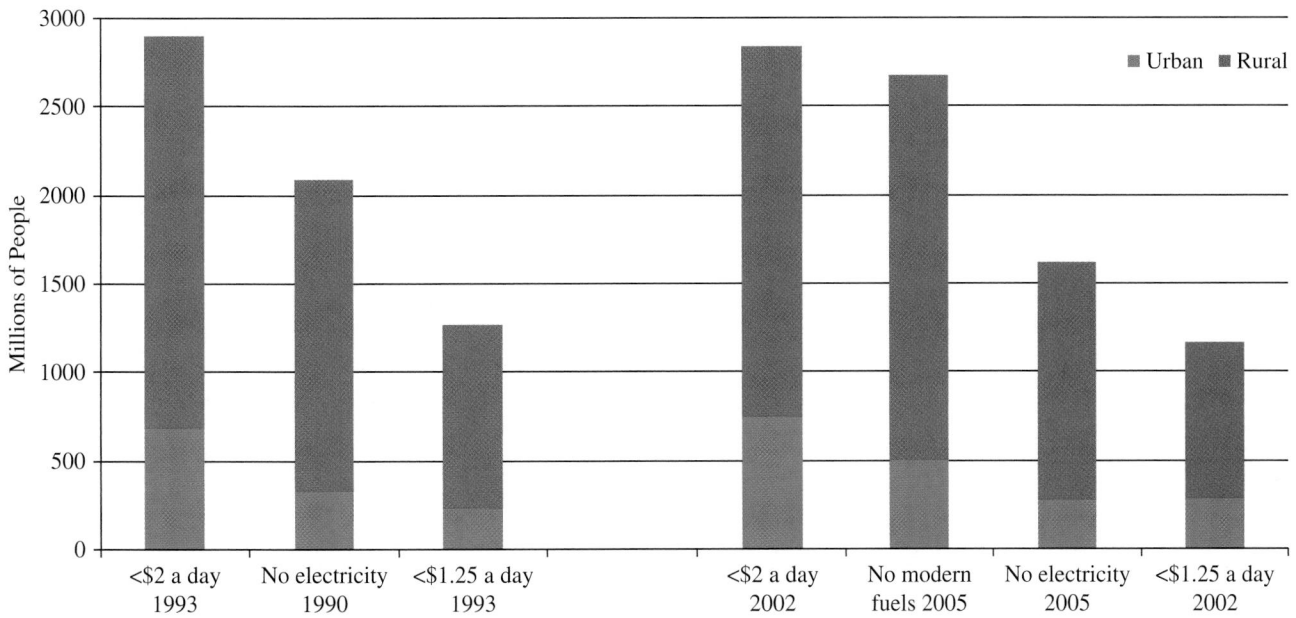

Figure 19.2 | People living in poverty and with lack of access to electricity and modern fuels. Source: data from IEA, 2002; 2007; Ravallion et al., 2007.

(often neither the fuels nor the equipment). Thus the problem of poverty remains closely intertwined with a lack of energy. This is also evident from looking at the data on the incidence of poverty and lack of access to electricity and more efficient liquid or gaseous fuels, termed "modern fuels" (see Figure 19.2).

Paradoxically, the communities that are the poorest in terms of access to energy are also often the most vulnerable and unable to cope with the threats of climate change. This is because these communities are often those most dependent on their local ecosystems for their livelihoods and energy needs. The Intergovernmental Panel on Climate Change described Africa, the world's poorest region, as "the continent most vulnerable to the impacts of projected change because widespread poverty limits adaptation capabilities" (IPCC, 2007). As these ecosystems are increasingly affected by climate change, the communities dependent on them will need other energy options and livelihood opportunities (Johnson and Lambe, 2009). Diversifying the energy sources available to these communities could thus also be an important means of enhancing the adaptive capacity of these regions.

19.1.3   The Role of Energy Access in Poverty Reduction and Rural Development[6]

Energy can reduce poverty and enable development in direct and indirect ways (Cabraal et al., 2005). Chapter 2 provides a detailed discussion

___
6   This section draws heavily on Chapter 2 and Chapter 6. Energy access alone is insufficient for development. Several other factors are essential to the development process. A fuller discussion of all these enabling factors is, however, beyond the scope of this assessment.

of the role energy can play in meeting the MDGs, more specifically in reducing poverty and improving literacy, health, gender equity, and community services, as well as how energy contributes to other positive social and environmental outcomes. All of these improvements in welfare constitute an improvement in social and human capital, which in turn can enhance the potential for higher income generation. Energy can also have a more direct influence on income in a variety of ways, such as making possible labor-saving mechanization, freeing up time, and increasing the length of productive hours in a day. The provision of energy itself is necessary, but insufficient to achieve these positive developmental benefits. To reap the largest positive impact, additional efforts and institutional mechanisms conducive to fostering entrepreneurial activity and uses of energy for activities that can generate income are also required.

Mechanical power provides energy services for productive uses and basic processing in many different rural livelihood activities undertaken in enterprises, farms, mines, workshops, forests, wells, and river crossings, to name a few. These energy services are fundamental to rural livelihoods and to the efficient transformation of natural resources into vital products and services, which results in wealth creation for producers and affordable prices for consumers.

Historically, progress in reducing rural poverty in many countries has gone hand in hand with agricultural development (World Bank, 2008b). While economists have struggled to disentangle the multiplicity of factors that enable agricultural growth, studies from Asia show that irrigation has played a prominent role. In particular, groundwater irrigation can be an important means of securing access to water as it allows farmers greater control over the amount and timing of irrigation. However, groundwater

---

Box 19.2 | The Role of Energy in Powering Rural Development in India

Literature providing evidence to support the hypothesis that agricultural growth has a strong impact on poverty reduction is well developed in the context of India, a major player in the Green Revolution (Ahluwalia, 1978). In India, groundwater irrigation was predominantly enabled by motorized pumps, either electric pumps energized through highly subsidized grid power or diesel pumps energized through low cost diesel fuel (the diesel is also subsidized). Barnes et al., (2002) report that irrigation led to income gains from 45–80% for farms of varying sizes.

More recent evidence from Gujarat state in India has also demonstrated the importance of access to electricity to rural businesses. A new scheme called *Jyotigram* was implemented in Gujarat to ensure reliable 24/7 metered supply for any nonagricultural use. Initial evidence suggests that this change from an unreliable 12 to 18 hour supply has increased rural prosperity through increased nonfarm activity as well as access to electricity for social infrastructure and communications (Shah and Varma, 2008).

---

irrigation requires energy to lift water. For growing cereal crops, this is difficult to achieve without mechanized pumps. Access to mechanized water pumps can increase incomes in multiple ways, such as:

- improving yields due to reduced risk from rainfall variability;

- facilitating a switch from single- to multi-cropping and more remunerative cash crops; and

- increasing the willingness of farmers to invest in fertilizer, improved seeds, and other farming technologies which further increase agricultural productivity, as the risk of crop failures is reduced.

The positive impact of irrigation for agricultural development and its contribution to food security, income generation, and poverty reduction has not been uniform across different regions. In general, unless the cost of irrigation is a small fraction of the value of the food produced, the enabling developmental outcomes are unlikely to materialize. Tube well irrigated agriculture in India uses on average close to 1000 kWh of electricity per hectare of irrigated land (Srivastava, 2004). With electricity supply to Indian farmers being highly subsidized, the costs associated with such mechanized irrigation remain low. While this has resulted in a litany of environmental problems associated with the over-pumping of groundwater in certain regions of India, the improvements in irrigation have been critical in creating the surpluses in food production, which in turn have enabled a transition to a more diversified economy, higher incomes, and now, with an increasing emphasis on female education, may also be helping drive a demographic transition.

In many sub-Saharan African countries with poor existing electricity grid infrastructure and very low rural demand densities, the cost associated with providing electric power for mechanized water pumps tends to be much higher. Moreover, with the exception of a few countries such as South Africa, Nigeria, and Tanzania, which have ample coal and/or natural gas resources, mechanized irrigation in the region remains dependent upon diesel pumps or grids that rely on heavy fuel oil-fired generators. In either case, the effective cost of generation (at least at the margin) is often much higher than US¢10/kWh. Indeed it is not unusual for costs of power and equipment to add up to as much as US¢30–40/kWh, thus making the cost of energy prohibitively expensive for higher food production in these regions (Modi, 2010). Reliable and reasonably priced energy is an essential ingredient for many aspects of improved or value-added agricultural and post-harvest processes, and is pivotal to enabling development and lifting millions out of poverty. Enhanced productivity in agro/food processing, artisanal activities, and microenterprises has the potential to boost economic development and improve livelihoods. In areas where electricity grids are unable to reach populations, the availability of decentralized mechanical power is particularly important for increasing the social and economic opportunities of the poor. Ironically, despite the importance of this energy service, there exists little data on mechanical power in developing countries.

Access to mechanical power can help increase efficiencies and effectiveness in production, thus raising income levels, which is an important factor for graduating from subsistence production (Box 19.2). A survey carried out in 2005 by the United Nations Development Programme (UNDP) in the villages of Sikasso and Mopti in Mali, showed that women earn additional revenue averaging US$68/year through access to mechanical power from multifunctional energy platform services. Taking into account their expenses, this translates into an average US$0.32/day, or US$44/year of additional income. The cost-benefit ratio is estimated as at least 1:2.5, given that the intervention cost is between US$80–90 per direct client (i.e., woman user) and that the minimum lifespan of a platform's engine is five years. In a country where the average gap between the dollar-a-day international poverty line and the mean income of the poor is US$0.37/day, the additional income is a significant step towards poverty reduction (UNDP, 2005).

Improvements in energy access, in addition to having a positive impact on agricultural production, processing and marketing, can also

---

Box 19.3 | "Sol de Vida": Empowering women through Solar Technologies in Costa Rica

Building solar cookers has achieved more than just providing alternative energy sources in the Guanacaste region of Costa Rica. The project has built and emphasized links with women's empowerment by creating new organizations led by women. Empowering women to take actions on their own, particularly regarding environment and livelihood issues, is a central goal of the program. So far, ten such community organizations have been created.

The solar oven promoted by Sol de Vida has been refined over the years to meet the specific needs of Central American families and continues to evolve to work under local conditions. The stove is basically a wooden box set inside another box, surrounded by insulation. The oven is covered by two panes of glass through which sunlight passes to heat the oven to an average temperature of 150°C. The stove can be built with US$100–150 worth of locally obtainable materials.

This project illustrates how a new form of energy use can be fully integrated into the lifestyle of a community. Use of the solar cookers is sustained because women build the stoves themselves. Women who learn how to build these stoves can then teach others to duplicate them at the same low cost.

Casa del Sol also creates locally adapted models of solar-powered stoves, water pumps, water heaters, and crop dryers. Educational programs at the Casa del Sol also help improve knowledge about these technologies, some of which can be reproduced locally. In fact, they have designed a parabolic solar stove which they hope to export.

Source: GEF and UNDP, 2003

---

catalyze a diversification of the rural economy into off-farm activities. Energy access can enable households to engage in a more diverse range of income-generating activities and contribute to the development of home enterprises, rural businesses, and cottage industry. The provision of adequate and affordable thermal, mechanical, and/or electrical energy is crucial for the development of rural entrepreneurship and microenterprises that often provide a significant proportion of off-farm employment opportunities in developing countries. Households are less likely to have a nonfarm enterprise and also have a lower income share from such activities if they live in a location that is more remote, has lower quality roads, lacks access to electricity, and suffers from frequent electricity blackouts. Evidence from rural Indonesia suggests that improvements in village-level infrastructure between 1993 and 2000 were associated with increases in the share of households having nonfarm enterprises (Gibson and Olivia, 2010). Dependable, reasonably priced energy access can contribute to the development and maintenance of small and medium enterprises in several ways. Mechanization and equipment upgrades can transform labor-intensive, low-production enterprises into high value-added operations, increase operating hours, and promote communication. Other benefits of improved access to energy in small enterprises are better efficiency and quality of work, better working environment, and a more attractive and secure environment for customers. In many instances, rural enterprises, especially home-based ones, are run by women. Reaching this segment of the population can serve the dual purpose of improving incomes and gender equity in these communities (Box 19.3). Promoting uses of energy that can enhance income, for both agricultural and off-farm activities, can work directly and effectively in enabling rural economic development. Recent evidence from South Africa suggests that electrification significantly raised female employment within five years. Several pieces of evidence suggest that household electrification raised employment by releasing women from home production and enabling microenterprises in South Africa (Dinkelman, 2010).

19.1.4   The Nature of the Access Challenge

Chapter 17 explores a number of interesting future scenarios for the global energy system to the year 2100. Scenarios have also been developed for household access to electricity and clean cooking until 2030 across key regions of the developing world. As a background to developing these access scenarios, this chapter provides an assessment of the historical progress of improving access. The two key indicators of relevance to this chapter, for which scenarios have been developed in GEA, are as follows:

- "People without access to electricity," which refers to populations that have no access to electricity; and

- "People without access to modern fuels or stoves," which refers to populations relying on traditional inefficient cooking devices (excluding improved solid fuel and clean-burning kerosene stoves) and solid fuels, including unprocessed biomass, charcoal, and coal.

As an input to the GEA scenario development process, a set of projections were generated for the simplest forms of these indicators for which data was available, namely populations with no access to electricity at the household level and populations relying on unprocessed biomass, coal, or charcoal for cooking and heating.

19.1.4.1   Electricity

Globally, less than 68% of the rural population has access to electricity (IEA, 2010b). Two-thirds of the global population lacking electricity access are located in sub-Saharan Africa and South Asia. The region with the lowest electrification level is sub-Saharan Africa, where only 11% of the rural population has access to electricity (UNDP and WHO, 2009). Over 600 million people, more than a third of all those without access to electricity in the world, live in South Asia.

It has been estimated that over 1.2 billion people globally will still lack electricity access in 2030 (IEA, 2010b) without the implementation of any new policies in addition to those already announced in 2010. Electricity for all by 2030 will therefore clearly not be achievable if global events are to unfold in line with current estimated projections (Figure 19.3).

A look at the historical progress with electrification reveals a mixed picture (Figure 19.3). Between 1970 and 1990, the total population without electricity access increased because population growth largely outstripped the pace of electrification in most regions of the world. Between 1990 and 2010, there was a decrease in the global population without electricity as the pace of electrification accelerated in certain countries like China and regions such as Latin America. Scenario analysis carried out in Chapter 17 for the three regions of South and Pacific Asia and sub-Saharan Africa also indicates that between 30–40% of the rural populations in these regions will continue to remain unelectrified in 2030 without additional policies and measures to accelerate access

to electricity. In sub-Saharan Africa over 70% of the rural population will remain unelectrified by 2030 without additional new policies. However, if global events were to unfold differently such that a fast track to providing electricity for all by 2030, or not much later into the future, is targeted, various policy interventions would be needed to accelerate the provision of electricity supply to households through a combination of grid and off-grid options over the next twenty years. This would require a steep acceleration in the rate of connecting new households in South Asia and sub-Saharan Africa, at a pace similar to what occurred in East Asia/China in the 1980s and 1990s. In effect, almost 20 million new households would have to be connected every year between 2010 and 2030 in order to meet the global target.

19.1.4.2   Solid Fuel Dependence

Current projections on the number of people dependent on solid fuels for cooking and heating differ in some ways from those for electricity. First, there are three developing regions of major concern, including East Asia/China in addition to South Asia/India and sub-Saharan Africa. Second, populations with no access to clean cooking fuels have continued to increase over the last decade, except in the case of China. The monotonic increases in people dependent on solid fuels in sub-Saharan Africa and the rest of Asia (excluding China) are very worrying indeed. It is estimated that almost three billion people will not improve their energy situation for cooking and heating by 2030, if current trends continue (Modi et al., 2006; IEA, 2010b). The GEA scenario analysis presented in Chapter 17 also indicates that the numbers of people dependent on solid fuels is projected to remain almost unchanged until 2030 if no new policies beyond those already in place by 2010 are implemented (Figure 19.4).

Given these trends, encouraging the use of improved stoves might be an additional way of increasing the efficiency and sustainability of the use

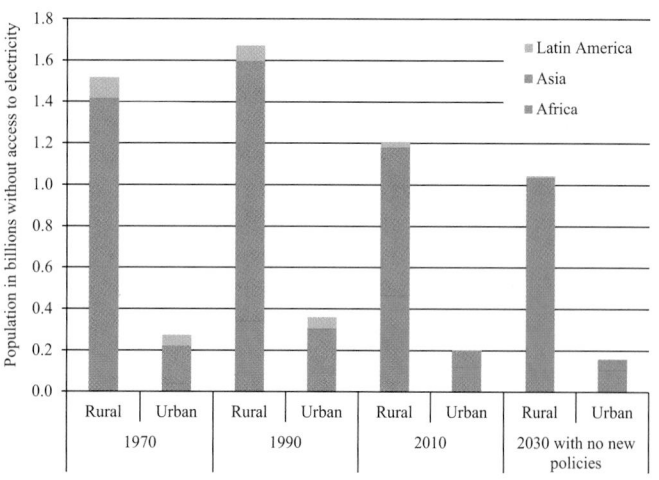

Figure 19.3 | Population without access to electricity in households in developing regions. Source: data from World Bank, 1996; IEA, 2002; UN, 2008; UNDP and WHO, 2009; IEA, 2010b.

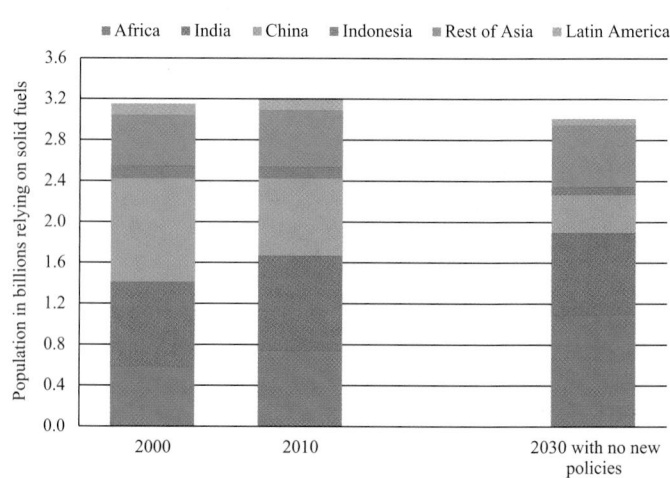

Figure 19.4 | Population dependent on solid fuels in households in developing regions. Sources: IEA, 2002; UN, 2008; UNDP and WHO, 2009; IEA, 2010b.

of biomass. Improved stoves vary widely in terms of performance, efficiency, and emissions. Better quality control and standards are clearly needed to regulate the improved stoves market. However, even relatively simple and inexpensive artisan-produced improved stoves can reduce the amount of fuel needed for cooking by as much as 20–35% (GVEP, 2009). Future scenarios for improving access to the energy services needed for cooking and heating thus consider both shifts and transitions from the use of solid fuels to more efficient liquid or gaseous fuels and the wider dissemination of improved stoves to the poorest households who are likely to continue to depend on biomass fuels in the shorter term.

Given this rather gloomy picture, is there any realistic basis for envisioning a fast track approach to providing universal access by 2030, as far as clean cooking and heating services are concerned? What possible developments at the global, regional, national, and subnational levels, and what possible policy options, could lead to deep cuts in the numbers of people dependent on solid fuels? What would be the constituting elements of such a paradigm shift and what are some of the experiences to date that would point in this direction? The following sections will review the experiences to date and provide answers to some of these key questions. Specific policies needed to achieve such desirable future scenarios will be discussed in Chapter 23, which deals specifically with policies and measures for energy access.

19.2 Past Efforts and Current Status

19.2.1 Access to Electricity

19.2.1.1 Historical Experience and Current Status

Electricity was first commercially supplied to the public in the mid-19th century in the United Kingdom, and thereafter spread quickly throughout Europe and the United States (Smil, 2005). As many nations attained independence in the 20th century, providing electricity access to their population was considered a prerequisite to modernization and progress and therefore accorded priority by the governments of these countries early on in their development. The political and social pressure to expand electrification was high in these nations, but the financial resources for doing so were often limited. As a consequence, electrification was pursued with uneven ambition and success. The historical model of pursuing electrification through a centralized energy system made it possible to benefit from economies of scale and to supply electricity to a mass market in many industrialized nations. However, this very paradigm, emphasizing the extension of a centralized grid network, also hampered a more rapid spread of electricity infrastructure to remote rural and low population density regions in many developing countries where it was not economical. In most countries, industrial and urban customers were the first to be supplied and the electrification of rural areas lagged behind.

The pace at which electrification occurred historically has been very different across nations (Figure 19.5). While Mexico took almost 90 years to electrify most of its population, Thailand achieved this in essentially a period of 20 years. While a number of factors are responsible for the uneven rate at which electrification occurred across different nations, the historical evidence supports the view that, given the commitment, an appropriate level of investments and appropriate institutional mechanisms, fast tracking the provision of electricity access is possible.

The more recent experience with electrification improvements across regions continues to remain very uneven, but provides a basis for hope. Between 1970 and 1990, over a billion people gained access to electricity, more than half of these in China alone. Between 1990 and 2008, almost two billion people gained access to electricity (Figure 19.6). In Latin America, North Africa, the Middle East, and East Asia, the pace of electrification outstripped the rate of growth of the population by a large margin, so that access significantly improved. In South Asia, the progress has been more uneven. However, in the period since 1990, the pace has increased. In sub-Saharan Africa, the rate at which new electricity connections have been provided over the last four decades has been consistently lower than the rate of population growth. This has been particularly true in rural areas.

Unfortunately, the region that faces the lowest rate of electrification today and the greatest challenge in increasing access, particularly among its rural population – sub-Saharan Africa – is also the region where the rural population density in areas without light is among the lowest in the world (Figure 19.7). This has implications for future options for expanding electricity access to these areas. Clearly, a diversity of electrification solutions is needed to increase access and centralized grid electrification alone may not be the optimal choice in all cases.

While access to electricity has been successfully extended to almost two billion people in the past 20 years, the overall picture is more complex. There is a dynamic associated with getting connected, staying connected, and increasing consumption in a situation of constrained supply. Many poor households that are connected face challenges in staying connected and increasing consumption beyond minimum levels due to poor quality, inadequate supply, and unaffordable connection costs and tariffs (PRAYAS, 2010). Per capita levels of consumption in rural areas in developing countries remain extremely low and in the case of some countries, have even declined because the growth in electricity supply to these regions has not been able to keep pace with rate of population increase and growth in demand.

Variations in the level of electricity consumption across nations, across rural and urban residents, and within rural and urban areas across income levels and different segments of the population, are very large. Just providing electricity connections does not ensure adequate access if the reliability and quality of supply remains exceedingly irregular or households just cannot afford sufficient amounts because of their exceedingly

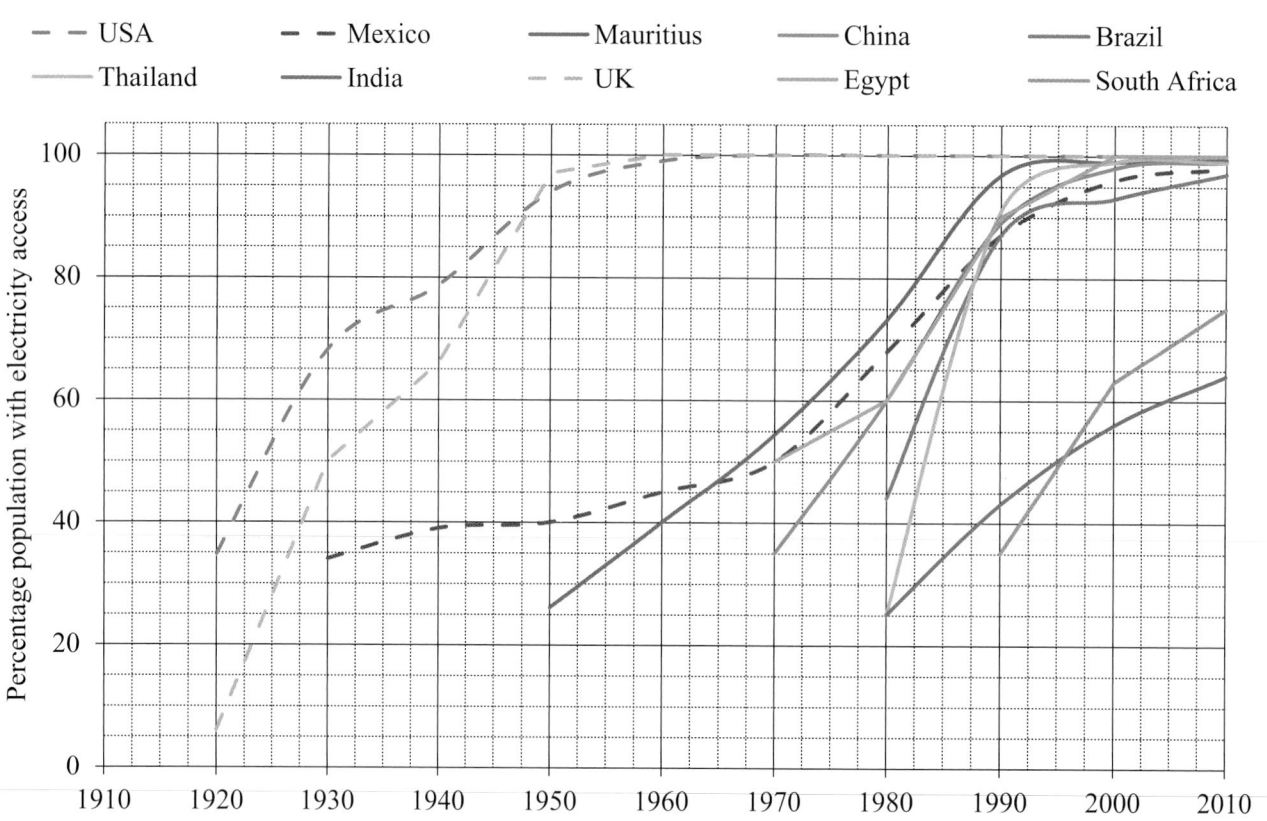

Figure 19.5 | Historical experience with household electrification in select countries. Source: Byatt, 1979; Goldemberg et al., 2004; Shrestha et al., 2004; UNDESA, 2004; Karekezi et al., 2005; Pan et al., 2006; Pachauri, 2007; UNDESA, 2007; Bekker et al., 2008; Collins, 2009.

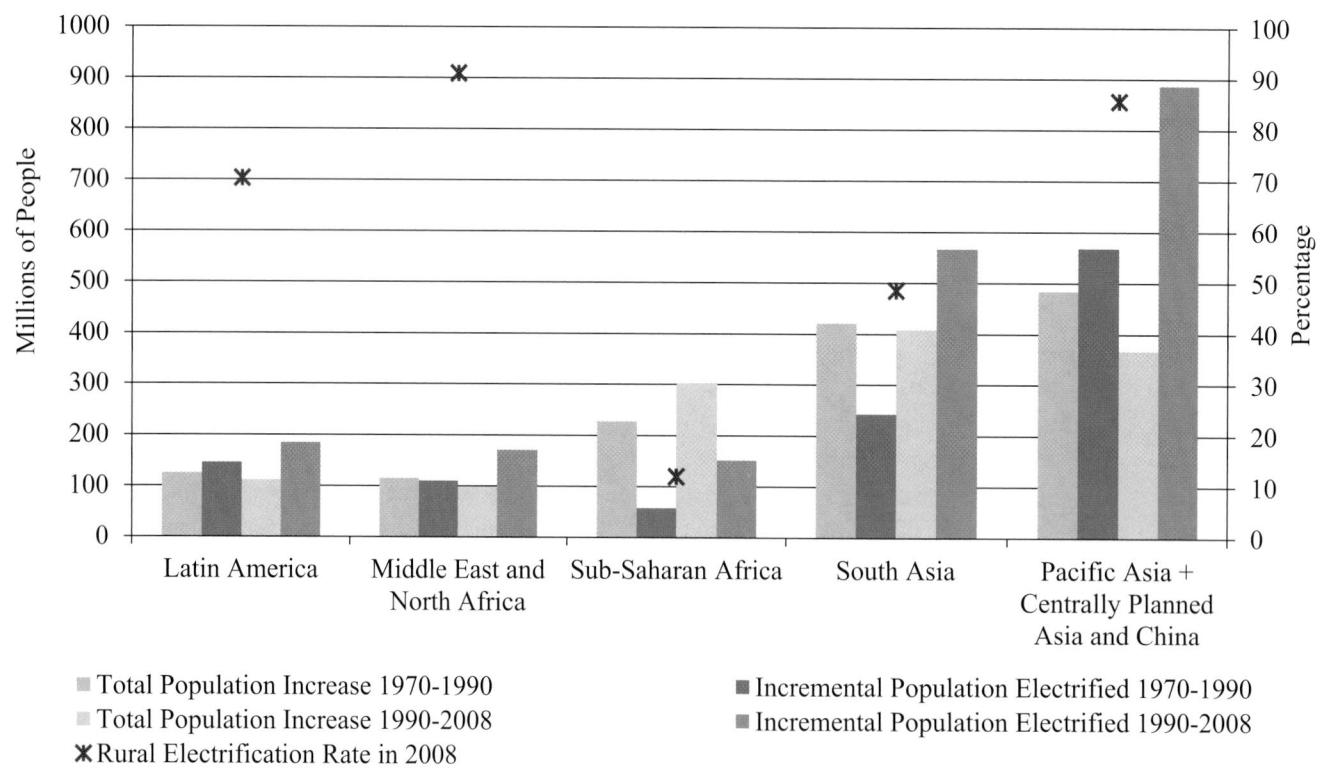

Figure 19.6 | Change in population and electrified population by region between 1970 and 2008. Source: data from World Bank, 1996; IEA, 2002; UNDP and WHO, 2009; IEA, 2010b.

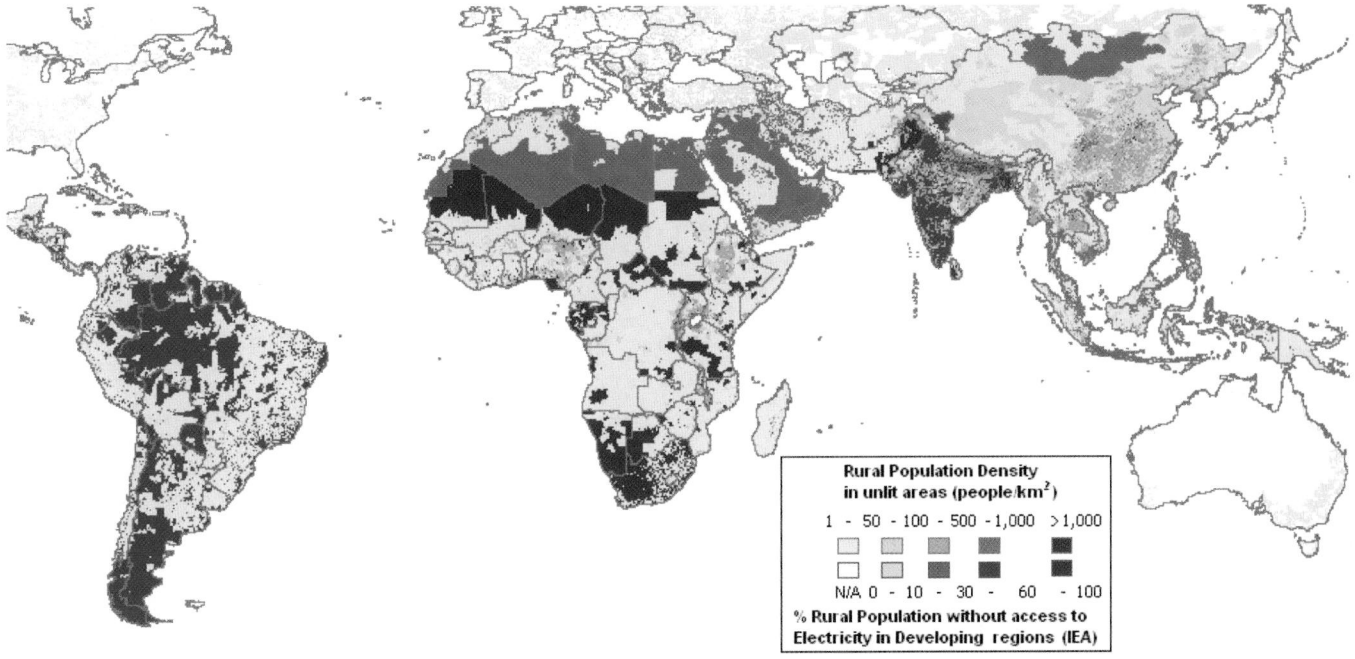

Figure 19.7 | Rural population density in unlit areas overlaid with regional estimates of population without access to electricity. Source: Doll and Pachauri, 2010.

low incomes and purchasing power. Figure 19.8 depicts changes in the mean residential electricity consumption per electrified inhabitant for major regions of the world. The average residential sector electricity consumption per electrified inhabitant is lowest in South Asia, even though a much larger proportion of the population has access to electricity in South Asia than in sub-Saharan Africa. The higher mean residential electricity consumption in sub-Saharan Africa is a consequence of the relatively high electricity consumption in South Africa. Excluding South Africa from sub-Saharan Africa would result in a lower average consumption for that region as a whole. South Africa has been relatively successful in improving access and increasing consumption of electricity even in low-income households through innovative financing and tariff schemes for providing lifeline electricity entitlements to the poorest consumers. However, issues relating to adequate metering and monitoring need to be resolved before such schemes can be implemented on a wider scale (see Box 19.8 for a fuller discussion of South Africa's lifeline electricity entitlement policy).

19.2.1.2 Social and Efficiency Benefits from Improved Energy Service through Electrification

Immediate applications of electricity in newly electrified households are for lighting and appliances, communications, and entertainment. Among community needs, public/street lighting, refrigeration, health centers and schools, piped water, communication, and the like are often cited. As already mentioned above in Section 19.1.3, electrification also benefits productive enterprises and agricultural activities. Electrification has the potential to be particularly beneficial to women, as their daily

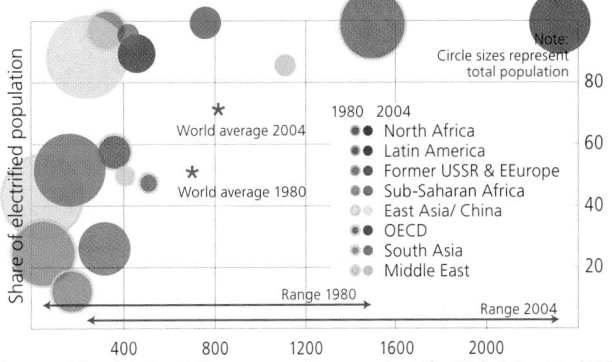

Figure 19.8 | Changes in residential electricity consumption and access. Source: Pachauri, 2007.

drudgery is reduced, their safety is enhanced, and the availability of lighting allows them to spend more time on leisure or productive activities. Electrification can also influence social capital and civil society. Well-lit streets, illuminated buildings, and systems of mass transit all increase mobility, giving citizens the ability to participate in community activities. These and other multiple benefits that are made possible through access to electricity have also already been described in Chapter 2.

Households also benefit from the use of many types of appliances that use electricity. There is a clear progression in terms of the energy services enjoyed by those connected to electricity. The first use is for lighting and entertainment. Thereafter, a wide array of benefits are potentially available – from security, comfort, and convenience to education, health, and home productivity –made possible by appliances such as electric lamps, radios, televisions, computers, refrigerators, fans, stoves, and electric pumps. In 2008, the Independent Evaluation Group of the World Bank (World Bank, 2008a) confirmed the findings of earlier World Bank work that valued the benefits of household lighting at US$5–16/month and the added benefits of entertainment, time savings, education, and home productivity at US$20–30/month (World Bank, 2002). These amounts are much higher than the US$2–5/month that a household typically pays for electricity service. However, even these low payments are often beyond the reach of cash-stripped, poor households and cannot be afforded despite the large potential benefits and high value attached to these services.

Access to electric lighting can even save households money through efficiency gains. Estimates of the effective cost of electricity for home lighting, computed on the basis of the cost of an equivalent amount of electricity needed to deliver the same amount of lighting as that provided by kerosene, are as high as US$3–4/kWh. Similar high costs are involved when one computes the cost of electricity obtained from batteries that are poorly charged and discharged. Surveys for Millennium Villages in Africa show that nearly half of households surveyed spend about US$5/month on such poor substitutes for electric lighting (Figure 19.9). This is because the substitutes to electricity used for lighting, such as kerosene or candles, are extremely inefficient (Figure 19.10). In addition to being highly inefficient, the use of kerosene and candles for lighting are associated with fire and poisoning hazards. These have been quantified for the case of South Africa in Spalding-Fecher (2005).

19.2.1.3   A Multitrack Approach for Future Electrification

Traditionally, the centralized model for electrification has been followed in most nations. To improve the status of access in the future, multiple tracks should be explored. Development strategies should consider innovations in the development and deployment of economically accessible distributed energy sources. Many renewable energy strategies look for land and natural resource conditions available in less developed landscapes where rural communities reside. The potential for providing meaningful livelihood options related to distributed energy supply and usage may provide a broader set of development pathways than those currently envisioned. Development of strategies along a more holistic framework that takes account of the socioecological system and incorporates development goals is needed.

While the reasons for continued lack of access to electricity are complex, there are two main reasons why many poor and rural households are still not connected to a source of electricity supply. The first is that a connection is not possible due to distance from a source of supply. The second is that that even though the grid may pass through the community, some households cannot afford the cost of electricity installation. If the connection of low-income households is to be made financially

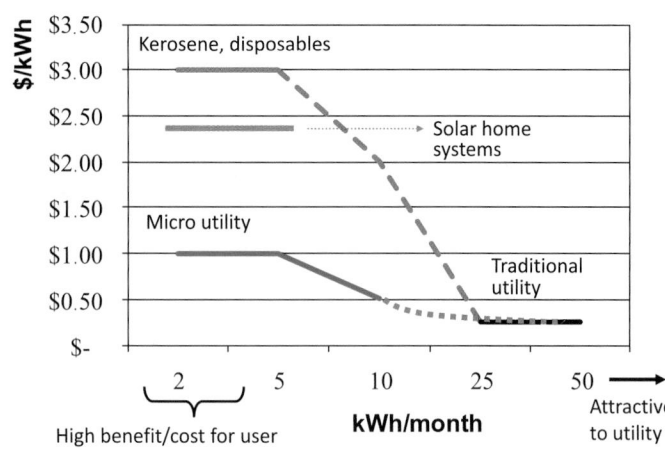

Figure 19.9 | Effective costs for lighting services. Source: Modi, 2010.

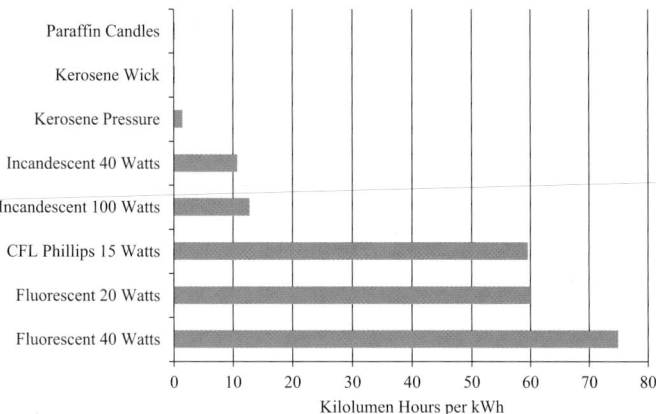

Figure 19.10 | Relative efficiency of different sources of lighting. Source: World Bank, 2010b.

viable to utilities, then special approaches are required to address the problem of low revenue caused by very low levels of consumption, and the highly dispersed and low density of consumers, particularly in rural areas. In addition, the connection and supply costs of providing electricity to low-income households need to be reduced by adopting least-cost options.

Success in the continued expansion of electricity access means adapting programs to local contexts and country environments. The past has witnessed strong advocates for centralized grid approaches to rural electrification, as well as more decentralized off-grid approaches. More recently, countries have adopted strategies that include both grid and off-grid approaches executed by various types of institutions, including public and private companies and large and small nongovernmental or microfinance organizations. Such a multitrack approach is based on the costs of supply, expected electricity demand, and development impacts of the project. The rationale is based on the difference in the cost of supply in areas with different socioeconomic and geographic profiles. Studies have shown how the cost-effectiveness boundary between grid electricity service and off-grid solar photovoltaic (PV) service changes as load density and village distance from the grid change (Cabraal et al., 1996). The implications of this are that the approach to electricity expansion must match the demand characteristics of population types. In addition, from an equity perspective, people in remote areas for which the cost of grid supply is high should not have to wait to gain access to grid electricity service when less costly alternatives could be made available with appropriate incentive policies.

Main Grid Electrification

The costs of conventional grid-based rural electrification vary greatly, both among and within countries. Local material and labor costs, terrain, and materials and construction standards can all have a major effect on the overall construction and maintenance costs. Typical figures quoted for marginal cost of grid electrification, including generation, transmission, and distribution, are in the range of US$0.10–0.20/kWh (Malik and Al-Zubeidi, 2006; Eberhard et al., 2008), with costs in rural areas typically higher than in urban ones. In many cases, the high initial costs of grid electrification can be held down by using design standards suitable for areas with less demand. Most rural consumers use about 0.5 kWh/day, much less than the minimum electricity connections typical of developing country utilities (Barnes et al., 1997). The high cost of wiring installation by utilities can be lowered by simplifying wiring codes to encourage lower electricity consumption levels. Other cost-cutting strategies include using cheaper utility poles and involving local people in construction and maintenance.

Although their institutional forms vary, as a general rule successful grid-extension programs require financially and technically strong utilities (Barnes, 2007). To ensure sustainability, distribution companies must address the issue of increased technical losses and low revenues in creative ways. The Tunisian Electricity and Gas Company (STEG), for example, reduced the capital costs of rural grid extension by shifting engineering standards and using capital subsidies provided by the government (Cecelski et al., 2007). By adopting a MALT (*mise à la terre*) design, a blend of three-phase backbone and single-phase network distribution, financing costs were reduced 20–30%. Making this technical design decision was not easy for STEG, which faced opposition from many of its own engineers accustomed to serving high-demand urban areas. This case demonstrates how careful and critical analysis of design assumptions and implementation practices reveals the potential for significant cost savings and thus more attractive financing (Cecelski et al., 2007; STEG, 2010).

Extending the grid to rural industries or commercial consumers can also promote economic growth while increasing revenue that can be used to maintain lower prices for residential and other rural consumers (Cabraal et al., 2005). Giving priority to major load centers and productive facilities can also help improve financial viability.

Microgrid Electrification

Off-grid electricity is necessary in some instances because the expansion of grid electricity will require decades to reach remote populations. In the short and medium term, the only way to reach many remote households without electricity may be through single household systems and small electricity providers using both renewable and conventional energy sources. Although these approaches to electricity provision may sound straightforward, in practice they have been difficult to implement. Decentralized, isolated distribution systems have been common in remote population centers for many decades. In most developing countries, they predate the establishment of main grid systems. The marginal costs of such systems are about US¢20–60/kWh. But diesel generators in remote locations can be hard to maintain and expensive to operate because of the high cost of spare parts and fuel. Micro hydropower systems have lower operating costs but involve higher capital costs for the systems and civil works to channel the water. Most other types of microgrids have similar cost levels.

Successful service delivery in remote locations via microgrids often involves specialized government agencies that perform an enabling role in support of private and community-based operators. For example, five years ago the Cambodian government created the Rural Electrification Fund to support small, private-sector operators of grid systems in rural areas. These indigenous operators had developed in rural and market towns, but faced significant investment constraints to expansion. To date, some 140 minigrid operators have been licensed under the new program. The Fund board, which provides overall guidance and policy oversight, comprises both public- and private-sector nominees. The Fund secretariat, which is responsible for operations, includes technical, finance, and administrative units. To ensure safety, quality, and service standards, Fund support is available only to qualified operators that are licensed by the Electricity Authority of Cambodia, the country's regulatory agency, for a minimum five-year period. The Fund has been fully operational since April 2007, and projects involving approximately 23,000 connections have been approved.

Renewable Energy and Household Systems

For countries endowed with the necessary natural resources, solar, wind, and pico and micro hydropower systems offer attractive options. The marginal costs of electricity generated by such systems are about US$0.50–1.00/kWh. Off-grid projects in such countries have taken advantage of private-sector institutions, nongovernmental organizations, and microfinance institutions that operate in rural areas. These programs can provide electricity to people in remote areas where main grid electrification is prohibitively expensive owing to the high capital cost of extending electricity lines.

The best off-grid models typically combine private-sector organizations (e.g., private entrepreneurs in Kenya have sold more than

200,000 PV systems to households that lack access to grid electricity), donor agencies, local communities, and national utilities supported by a strong energy agency whose role is to promote off-grid electrification. For example, Bangladesh's Rural Electrification and Renewable Energy Project combines main grid financing for rural electric cooperatives administered by the Rural Electrification Board and an off-grid component administered by the Infrastructure Development Company Limited, a public financial institution. The project offers participating organizations – nongovernmental and microfinance institutions, including the Grameen Bank, municipalities, and private-sector institutions – both credit and grants with which to cover about 20% of the purchase cost of solar home systems. Each participating organization signs an agreement with the Infrastructure Development Company for this blend of loans and grants, which lowers the cost for customers (Asaduzzaman et al., 2010; IDCOL, 2010). Under this program, more than 150,000 rural off-grid consumers are receiving electricity. Similar programs supported by the World Bank are functioning in more than 30 countries, including China, Sri Lanka, and Bolivia. Under these programs, funds have been committed to support services to 1.3 million consumers (seven million people). Currently, about half of these households receive electricity.

Off-grid household programs in Bangladesh and Sri Lanka demonstrate that it is possible to implement large-scale, off-grid projects that complement strong grid electrification programs. Off-grid projects in both countries have taken advantage of private-sector institutions, nongovernmental organizations, and microfinance institutions that operate in rural areas. However, this has also been backed by strong centralized institutional support.

19.2.1.4   Principles for Success

Countries must discover solutions consistent with their geography and natural resources; demographics; and socioeconomic, cultural, and political realities. Rural electrification is a dynamic, problem-solving process. Problems change as programs evolve, but certain underlying principles guide successful programs (Barnes, 2007). Governments' sustained commitment must be reflected in effective institutional structures that exhibit a high degree of operating autonomy and accountability, strong management, and dynamic leadership with the capacity to motivate and train staff. Successful programs also require effective prioritization and planning. Clear criteria based on market research are required for prioritizing areas to supply. Key factors include capital investment costs, level of local contributions, numbers and density of consumers, institutional capacities, and likely demand. For off-grid systems, reducing construction and operating costs as well as sustained financing is also vital. Sustained financing will require increased and more effective use of both domestic and external funding sources. When cost recovery is pursued, most other program elements fall into place.

Summarizing some of the key lessons and principles for successful electrification as discussed in Krupp (2007) and Barnes (2007), the following are key:

- sustained financing;
- metering and payment for cost recovery;
- local buy-in and training;
- flexible and adequate institutional arrangements;
- independent regulation; and
- lower-cost options for supplying power.

In spite of the many challenges to rural electrification, many countries have been successful in providing electricity to their rural areas. In Thailand, well over 90% of the rural population has electricity access. In Costa Rica, cooperatives and the government power utility provide electricity to nearly 100% of the rural population. In Tunisia, over 90% of rural households are already supplied. Studying countries like these and others there appear to be certain factors critical to the successful implementation of rural electrification. Strong and sustained political commitment, intensive financial support and clear earmarking of funds by the government, along with the establishment of clear planning criteria for rural electrification, are paramount to success (Barnes, 2007).

19.2.2   Access to Modern Fuels and Technologies for Cooking and Heating

Households in developing countries, particularly those in rural areas, largely rely on a range of fuels rather than electricity to meet their cooking and heating needs. Electricity use for cooking remains rare in most developing-country households because of high service and appliance costs associated with the use of electricity, limited availability, and relatively low incomes levels. Hierarchies in household energy services are quite common. Almost always, cooking and heating are the first functions fulfilled, followed by lighting and then entertainment. For the poorest people in developing countries, cooking (and space heating in particularly cold climates) can account for up to 90% of the total volume of energy used (WEC and FAO, 1999). Cooking is an energy service that is often associated with strong and highly specific fuel and appliance preferences. In addition, cooking is often only one of a range of services that are delivered from a stove or a fire. For example, coal and wood stoves serve multiple functions, including cooking, space heating, water heating, lighting, and social focus (van Horen et al., 1993). The multifunctionality of some stoves is one of the key reasons why households are at times averse to substituting their old stoves for newer more efficient and cleaner technologies.

Households use a variety of fuels for cooking and heating purposes, such as wood, dung, crop wastes, charcoal, kerosene, LPG, coal, and electricity, to name the most common. Households use the different fuels in different combinations, depending on the needs they satisfy, their availabilities, and the socioeconomic circumstances of the households. Fuels are chosen for their cost effectiveness, ease of access, and perceived efficacy in performing specific tasks. Fuel use patterns may differ at different times of the year. Tastes and cultural preferences also play an important role in decisions concerning which fuels are used, particularly in rural households.

The use of more efficient liquid or gaseous fuels and efficient appliances for heating homes and cooking food has significant social and environmental benefits and yet access to these among most of the developing world remains extremely limited (Figure 19.11 and 19.12). In sub-Saharan Africa only 16% of people use modern fuels as their primary cooking fuel. The level of reliance on modern fuels is lower than in

any other geographic region, and is comparable to that in the average for all LDCs (10%). Overall, some 41% of people in developing countries have access to different types of modern fuels for cooking. Across all developing countries, almost one-third of people (33%) use gaseous fuels (including natural gas, LPG, and biogas) as their primary cooking fuel (Figure 19.11). Use of gas is much less common in the LDCs and sub-Saharan Africa, where only 7% and 4%, respectively, of the population rely on gas as their main cooking fuel.

About 75% of people living in rural areas use traditional biomass for cooking, primarily wood, while 65% of those living in urban areas rely on modern fuels, especially gas (UNDP and WHO, 2009). Of those who rely on solid fuels, roughly 800 million people (only 27%) are estimated to use improved cooking stoves, most of which belong to households situated in China and Brazil. There has recently been an increased effort to bridge the gap between the inefficient use of traditional fuels such as wood, straw, and dung by promoting the use of improved stoves. Stove programs around the world have had an uneven history, but there are some recent developments involving more durable efficient biomass stoves that are encouraging for the future.

The environmental consequences of unprocessed biomass, charcoal, and coal use – first put before the international community several decades ago as the "other energy crisis" (Eckholm, 1975) – involve household air pollution and degradation of local and global commons. Cooking and heating with biomass fuels on open fires or traditional stoves results in high levels of health-damaging pollutants and has been associated with numerous respiratory problems, thus contributing to global mortality and morbidity. Fuel collection can lead to a deterioration of the local

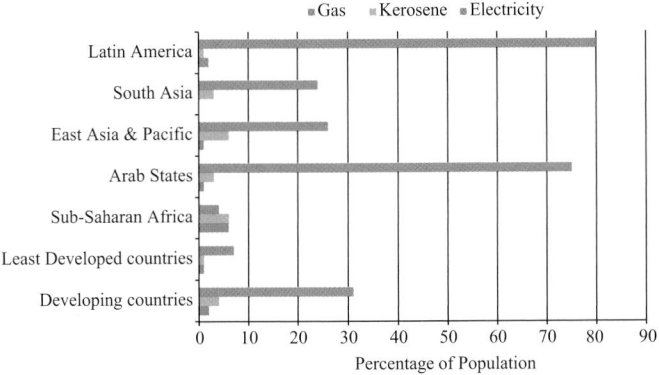

Figure 19.11 | Share of population using non-solid fuels. Source: UNDP and WHO, 2009.

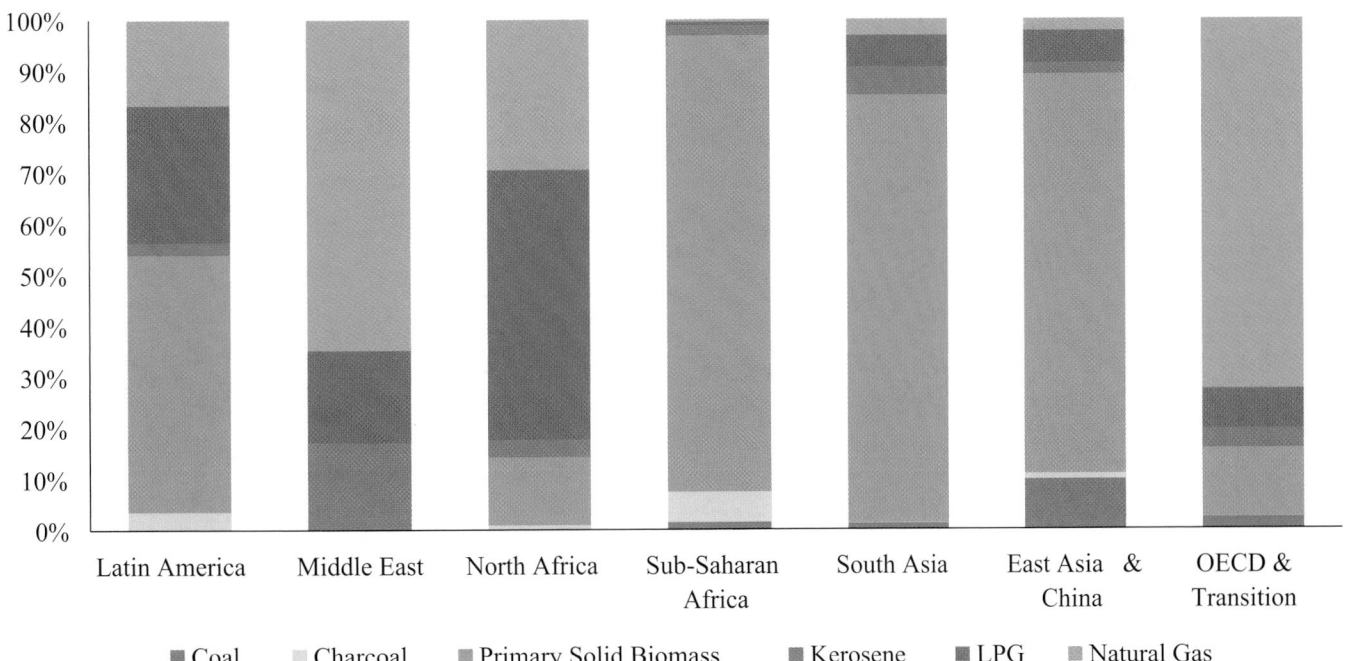

Figure 19.12 | Structure of final residential fuels use across regions in 2006. Source: Based on IEA, 2008a; b.

environment and depletion of biomass, meaning ever-longer walks to collect fuel. In Haiti, for example, the overall decline in forested areas resulting from charcoal production for urban use is well documented (Stevenson, 1989). The production and use of energy for cooking and heating in developing countries contributes to threats to human health and quality of life, affects the local ecological balance and biodiversity as demand for traditional fuels outstrips supply, and alters the climate that we live in through pollutant emissions, such as those of black carbon (Bice et al., 2008).

High dependence on traditional biomass and coal leaves many people in developing countries with few options for improving their lives. In the recent past, the health costs of the continued dependence on traditional solid fuels have been very high, particularly for many developing countries. Current estimates indicates that annually, almost two million worldwide deaths from pneumonia, chronic lung disease, and lung cancer are associated with exposure to household air pollution resulting from cooking with biomass and coal, and 99% of these deaths occur in developing countries (UNDP and WHO, 2009). The poor devote a large portion of another important asset – their time – to cooking and heating energy-related activities; women and young girls can spend in excess of six hours per day (see Figure 19.13) gathering fuelwood and water, cooking, and agro-processing (UNDP, 2007b).

Studies on understanding the factors determining fuel choices and drivers of fuels usage in developing countries are limited (e.g., Heltberg, 2004; Ouedraogo, 2006; Pachauri and Jiang, 2008; Ekholm et al., 2010). Data on fuel choices and consumption and how these factors have changed over time also remains very sparse, particularly for rural areas and in the poorest countries. Progress with expanding access to modern fuels and technologies for cooking and heating in developing countries over the past 25 years has been rather dismal. A recent review of World Bank lending for improving energy access over the period 2000–2008 also shockingly concludes that only about 1% of the total lending was dedicated to promoting a transition to more modern cooking fuels or

clean cooking devices (World Bank, 2010a). More efficient and cleaner fuels and improved stoves to meet people's most basic cooking needs are still out of the reach for the majority of populations living in developing countries, especially those in rural areas. The widespread diffusion of improved and clean cooking stoves has also yet to happen even though new designs of clean stoves are being piloted around the world. Much remains to be done on the fuel-stove package to make it available as a clean cooking option. An astounding three billion people in developing countries primarily rely on coal and traditional biomass such as wood, charcoal, and dung for their cooking and heating needs, with little or no access to more efficient, modern forms of energy (UNDP and WHO, 2009). In other words, almost half of humanity still uses traditional biomass or coal, with about 2.7 billion relying on traditional biomass alone (IEA, 2010b). Access to improved cooking stoves is also very limited. The IEA estimates that people will continue to rely on these solid fuels for the next few decades, and those relying on traditional biomass alone will increase to 2.8 billion by 2030 (IEA, 2010b) in the absence of any new policies beyond those already in place in 2010. The majority of these people will remain concentrated in rural areas of the LDCs in sub-Saharan Africa and South Asia. This places a huge burden on the economies of these countries and stifles efforts toward achieving the MDGs and poverty reduction for these populations (Modi et al., 2006).

Data on the use of biomass and other solid fuels, both in terms of the percentage of population dependent on these fuels and the actual consumption levels, remains exceedingly scarce, particularly for those countries that remain the most dependent on these fuels. Internationally comparable statistics over time are hard to find. However, over the last decade, there has been a slight decline in the number of people dependent on biomass alone, largely on account of a significant reduction in the number of people using biomass in China (Figure 19.14). Overall, little change is observed in the dependence on solid fuels, particularly among households in rural areas.

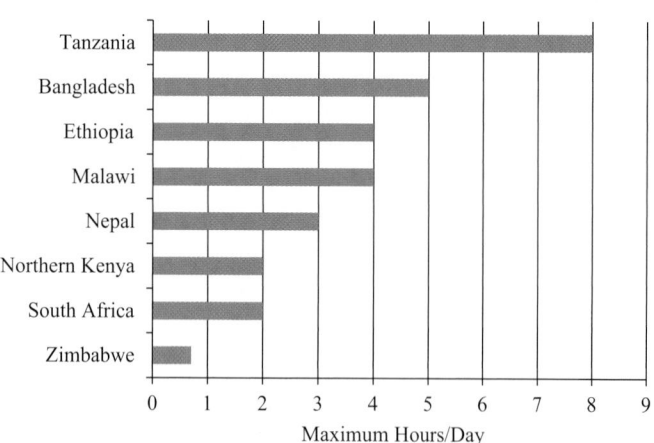

Figure 19.13 | Selected data on time spent in wood collection. Source: Practical Action, 2010.

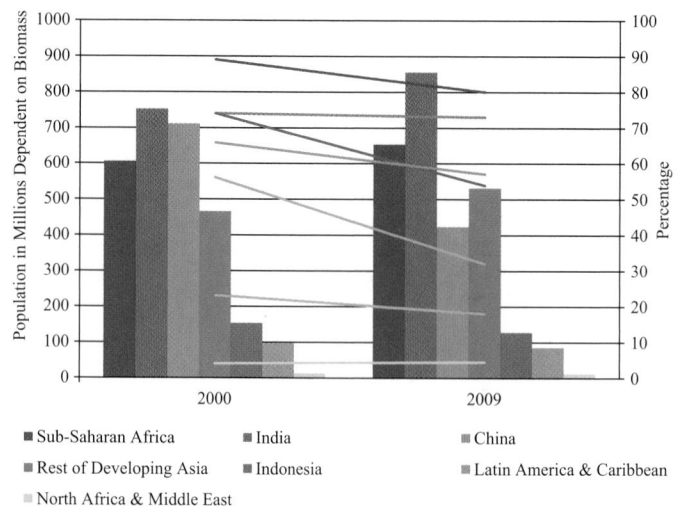

Figure 19.14 | Changes in the number of people relying on biomass. Source: IEA, 2002; 2007; UNDP and WHO, 2009.

Cooking is the major energy end-use among poor families in developing countries. Because most staple foods must be processed, conserved, and/or cooked, and thus require some form of heating before consumption, access to affordable, clean cooking fuels and equipment are among the most basic energy needs of the world's poor.

The importance of gender issues in understanding transitions in the use of cooking fuels and/or stoves cannot be overemphasized. Women tend to have limited control over and access to productive assets and income in many developing countries, especially in the poorest households. They often have little say in how much can be spent on fuels or new stoves. While not all women would embrace new fuels and stoves if given the choice, often a proposed change has to benefit the man of the house to have a chance of being adopted (Lambrou and Piana, 2006). In addition, women are disproportionately burdened by the drudgery and poor health impacts associated with the use of solid fuels. Gender issues related to energy have often been overlooked due to a lack of gender-specific data in the energy sector (Parikh, 1995; ENERGIA, 2010). However, it is well recognized that greater attention to the energy needs and concerns of women in developing countries can improve the effectiveness of energy policies and projects (Clancy, 2000).

The estimates of the numbers of people using improved cook stoves (UNDP and WHO, 2009) suggest that programs to disseminate such stoves have not had much of an impact. This is despite the fact that examples of improved cook stoves abound (see Table A19.1 in the appendix and Chapter 10 for more on improved biomass cook stoves). However, the characteristics of the stoves, the level of efficiency gains, and the level of sophistication and mode of manufacturing all vary enormously. Developing programs aimed at the sustainable dissemination of improved cooking stoves must overcome a number of barriers, including issues that relate to the maintenance and replacement of free-of-charge or self-produced stoves. Recently, there has been a surge of social entrepreneurs entering the energy field and a drive toward market models by implementing institutions and social investors, which may result in larger scale-up (GVEP, 2009). Product developers (e.g., ENVIROFIT), donors, and social investors are shifting the focus for design and diffusion of clean burning, off-the-shelf cook stoves toward a market-based approach. How successful these market approaches will be in reaching the poorest consumers remains to be seen.

Heating and cooking often go hand in hand in developing countries, particularly in cold mountainous areas where household energy is primarily used for space heating. Space heating is also common in tropical countries that have very warm weather during the day and cold temperatures at night. The need for space heating is often higher among the poor because their houses are not necessarily constructed well enough to preserve heat. Space heating is extremely important for improving health and livelihoods, especially for the young and the elderly.

19.2.2.1   Expanding Access to Modern Fuels and Technologies: Status and Implications

Energy for Cooking: There are a number of ways in which those dependent upon biomass can benefit from switching to cleaner burning cooking fuels or stoves. Practical solutions also include smokeless biomass cooking devices fitted with chimneys or hoods and switching fuels from biomass to biogas, kerosene, LPG, ethanol, or electricity. It is important to consider all of these options when evaluating suitable intervention for a specific target population. The availability and access to fuels, ease of handling, and affordability are all important determinants in the selection of a particular intervention in a given context. Contexts may be defined by factors such as urban/rural settings, domestic situation, institutional setting (schools, hospitals, prisons, etc.), poor/rich clients, and biomass-rich or -degraded areas.

Energy for Heating: Charcoal is the most popular biomass fuel for space heating in biomass stoves in peri-urban households because it emits less smoke and the thermal efficiency of charcoal stoves is relatively high. Firewood is widely used in rural homes where cooking and heating are the two major uses of fuel. Coal is widely used in China. Other fuels and energy sources such as electricity, kerosene, and LPG are not widely used in most developing country households because of limitations in affordability, availability, and accessibility.

In colder rural areas of the world (including high altitude areas within the tropics), space heating is often required during the winter months. Vast quantities of energy are used to achieve this. Not all buildings in rural areas of developing countries are designed to conserve heat. Firewood is the main fuel used for space heating in rural areas, followed by charcoal, which is the primary heating fuel for relatively few people. Providing affordable warmth to the peri-urban poor is among the main challenges facing many developing countries. The peri-urban poor often use a mix of fuels depending on availability, accessibility, and affordability. Charcoal is an important fuel for space heating in peri-urban areas; carbon monoxide poisoning is occasionally reported in peri-urban areas during the cold seasons because of use of charcoal in unventilated houses. Firewood is also used in biomass stoves with chimneys but is not a popular source of energy for space heating in peri-urban areas. Where electricity is available, small space electric heaters are used, albeit sparingly and only when extremely necessary, such as when there is a newborn baby or an elderly person in need. One option in addressing the space heating problem is housing policies – including housing finance – that take energy efficiency and improved stoves into account.

19.2.3   Access to Energy for Income-Generating Activities

The poor, especially those living in rural and remote areas of developing countries, face limited or total lack of access to reliable modern energy for activities that can generate income, commonly referred to as productive end-uses. It should be noted that the distinction between "productive uses" and "consumptive uses" is by no means clear cut. However, we refer to "productive uses" of energy as those that involve the utilization of energy, both electric and non-electric, in the forms of heat or mechanical energy, for activities that enhance income and welfare. The problem of access to energy services for productive end-uses in developing countries is far greater in rural areas than in peri-urban and urban areas (Bates et al., 2009).

To spur sustainable development in rural areas and indeed, nationally, improved rural productivity is needed. Productive end-uses can be enhanced by reducing the problems associated with dependence on solid fuels (traditional biomass and coal) in rural households and expanding access to electricity and other modern energy carriers and technologies and mechanical power. It can be accelerated by scaling up distribution of cleaner energy-conversion devices (e.g., improved stoves, gasifiers, and kilns); cleaner fuels (e.g., biogas, ethanol, and LPG); and decentralized energy options (e.g., wind, biomass micro distilleries, micro hydropower, and solar energy) that have shown some level of success (UNDP, 2006). For it to succeed, massive mainstreaming of energy issues into development planning, mobilization of finance for scaling up of energy for productive end-uses, and extensive building of national and local capacity to deliver modern energy for specific productive end-uses is needed.

To reflect the full extent of the energy access for productive end-uses in developing countries, it is important to highlight the role of mechanical power[7] in improving the lives of the poor in developing countries. Energy for mechanical power is obtained from electricity and nonelectric sources and is used for daily livelihood activities including agroprocessing, artisanal activities, and small and micro enterprises. Over the past century, technological advances have helped reduce the drudgery of human labor through the widespread use of mechanical power. Mechanical power is perhaps second only to cooking when it comes to the sorts of energy services poor people need most. Mechanical power is critical to enhancing the productive end-uses of labor and poverty alleviation.

Challenge one: With respect to access to mechanical power for productive uses, a desk survey by UNDP indicates that mechanical power is not included in most policy debates at the country level. Despite its importance in expanding access to energy services, little data exists on productive end-uses of energy or mechanical power in developing countries. Governments lack vision, time-bound targets, or data concerning the contribution of mechanical power to general human development. Even where there are targets, programs to upscale initiatives are lacking. National decision makers often fail to recognize the significance

Table 19.1 | Mechanical power is least documented by LDC countries that provide data on access to modern energy (baseline and target).

		National	Rural	Urban
Fuels for cooking/ heating	Baseline	48 (96%)	17 (34%)	15 (30%)
	Target	5 (10%)	2 (4%)	2 (4%)
Mechanical Power	Baseline	0 (0%)	2 (4%)	0 (0%)
	Target	0 (0%)	2 (4%)	3 (6%)
Electricity	Baseline	48 (96%)	38 (76%)	35 (70%)
	Target	22 (44%)	16 (32%)	9 (18%)

* The 50 LDCs are used in the calculation. (%) indicates the percentage of LDCs that provides data on access to modern energy.
* Source: UNDP and WHO 2009.

of productive end-uses through the provision of energy especially by motive power. For example, Table 19.1 shows that although five out of the 50 LDCs have a national target on access to modern fuels, and 22 have targets on access to electricity, none of them has a specific national target on access to motive power.

The lack of data is a challenge, particularly considering that the majority of developing countries' rural and peri-urban populations depend on mechanical power for activities in households, agriculture, or small and micro enterprises. For instance, small commercial enterprises and agriculture depend on motorized and nonmotorized mechanical power. Devices such as treadle pumps, ram pumps, floating pumps, wind-powered water pumping, hydro-powered carpentry, or agricultural processing enterprises such as corn threshers, are all important mechanical power devices that are commonly used in developing countries. Diesel or electric motors for mechanical power are becoming increasingly important for providing mechanical power in most developing countries. Mechanical power is one of the quickest ways that energy is used for productive end-uses, mainly in agriculture and forestry and at the small enterprise level. In terms of targets for access, national plans are lacking in all areas of energy access, with cooking fuels much more poorly represented than electricity, and mechanical power not even registering. The lack of data in this regard is a serious barrier to setting such targets. What is needed is detailed data on ownership and use of machinery and power availability at the farm level, as well as studies that link this with yields and productivity improvements (Figure 19.15).

Small and microenterprises (SMEs), many belonging to the informal sector, have become integral in many developing country economies. Studies in India have shown that SMEs enable rural households to generate nonfarm income, which can largely contribute to poverty reduction (Lanjouw and Shariff, 2002). In general, SMEs purchase (rather than harvest or collect) their energy, including electricity, LPG, kerosene, firewood, charcoal, etc. This is true even in rural areas. Moreover, despite energy being one of the significant factors for most microenterprises, there is a knowledge gap on how much energy is being used; neither is it systematically documented what role energy

---

7   Mechanical Power is defined here as the effective outcome of transforming different forms of energy sources (e.g., wind, hydro, fossil fuels, etc.) to kinetic energy (to cause motion).

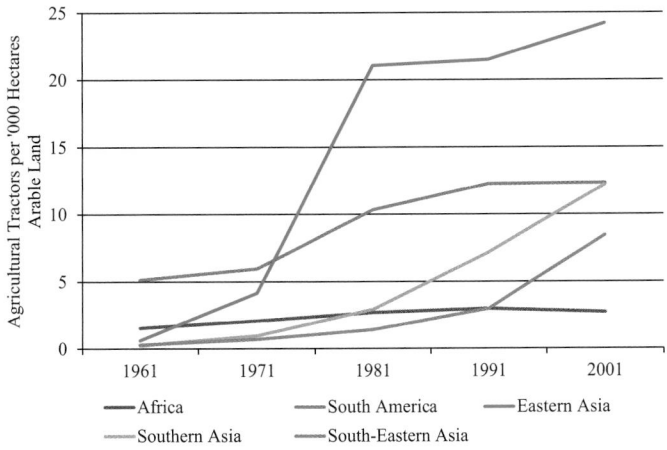

Figure 19.15 | Growth in agricultural mechanization. Source: FAOSTAT, 2010

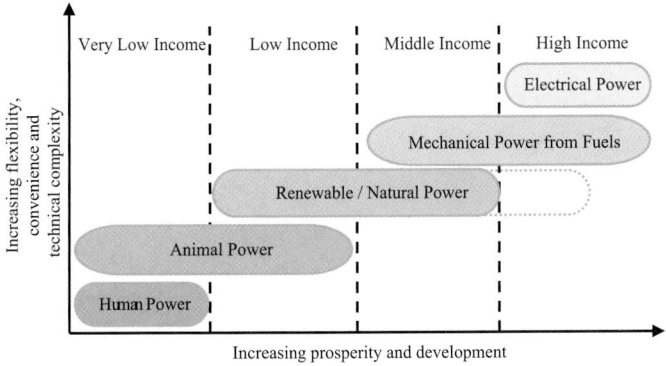

Figure 19.16 | Mechanical power ladder. Source: adapted from WHO, 2006; UNDP, 2009a.

plays in diversifying production and expanding employment opportunities by microenterprises, both in urban and rural areas (Clancy and Dutta, 2005).

Challenge two: The second challenge is the lack of improvement in technology or form of use in impoverished regions of the world today. Since the industrial revolution, access to advanced forms of mechanical power has defined the pace of human development and advancement, and shaped development in various ways in different parts of the world. However, the gap between technologies used for mechanical power in developed countries and in the rural areas of developing countries is increasing. Despite technological improvements, those without access to modern energy still depend on less efficient and effective versions of mechanical power that use human, animal, or unimproved motorized equipments to meet their energy needs, resulting in low efficiencies and limited productivity. The implications for the poor include lower incomes, increased drudgery, and a continued dependence on subsistence production practices. Access to motive power has remained an important driver of livelihood activities in impoverished regions of the world. Figure 19.16 depicts this challenge (UNDP, 2009a).

19.2.4   Institutional Development and Financing Mechanisms for Scale-up of Access

In order to meet the ambitious energy access targets discussed in Chapter 2 and highlighted in Section 19.1.1, new ways to generate additional sources of funding will be needed and funds made available will have to be allocated effectively and efficiently. In the following sections we discuss the need for innovative institutional approaches and financial mechanisms for scale-up of access initiatives and activities.

19.2.4.1   Need for Innovative Institutional Approaches

Massive diffusion of new technologies for meeting thermal energy (e.g., cooking), motive power, and electricity needs is necessary to meet the grand challenges of improving access laid out in previous sections. While electricity grids have expanded and programs have been put in place to spread distributed generation technologies and cleaner cook stoves and fuels, these efforts have often been plagued by numerous problems and the scale of expansion is barely sufficient to keep pace with increasing demand and even population growth, as in the case of many sub-Saharan countries. In many cases, the barrier is in the policies and institutional arrangements. While it may not always be clear what is needed for a given region, what is clear is the need for change to meet these grand challenges. It is also clear that some key lessons can be learned from mistakes of the past.

The circumstances in developing countries militate that energy pathways, especially for rural and peri-urban areas, be dissimilar to those followed by developed countries. This requires innovation and experimentation on both technological and institutional levels. In Ethiopia, for example, the use of inappropriate institutional structures was found to be one of the key factors hindering wide dissemination of modern energy services to rural areas (Habtetsion and Tsighe, 2002). Lack of appreciation of such approaches at a policy level is curtailing progress, as many policymakers tend to follow conventional approaches without taking into account contextual differences. It is critical that policy makers realize that the markets in developing countries are quite different from those that have emerged in the industrialized world. For example, the impetus to rationalize prices and engage in reforms of the electricity sector may actually be stronger in developing countries due to the poor state of utility finances in many of these countries, the lack of easy access to capital for system expansion, and the resulting inability to maintain existing systems or meet latent demand (e.g., the Indian utilities). However, at the same time, electricity reforms[8] could have negative impacts on low-income households that are already facing hardships quite different from those in the industrialized countries.

---

8   The role and impact of electricity and wider energy sector reforms on the poor is dealt with in greater detail in Chapter 2.

Box 19.4 | Issues with Measuring and Monitoring Access

Identifying progress in providing access and the extent of electrification depends heavily on the official definitions adopted and the measurement units used. Foley, in his 1990 paper, states that the definition of rural electrification, in particular, varies considerably across countries. In one country "rural" may include provincial towns with a population up to 50,000 while in another it may refer to small farming villages and surrounding areas. The unit of measurement also matters. The initial focus of Indian rural electrification was on "village electrification," rather than household electrification, with this being very loosely defined. As a consequence, villages were often deemed "electrified" without a single household having been connected. Changes in the definition of "electrified village" also impacted the measurement of electrification over time (Pachauri and Mueller, 2008).

The source of electricity supply can also matter. In many countries, official electrification rates refer to connections to grid electricity alone. Thus in Cambodia, where the government has not had a strong involvement in rural electrification, official electrification rates (defined only in terms of connections to a central power grid) remain very low (15%). However, according to Zerriffi (2007), the number of households with access to at least a minimal amount of electricity (e.g., enough to power a light bulb and maybe a small television) is extremely high (50% of the households have a television and an estimated 85–90% have a light bulb). Their electricity comes primarily from rural electricity entrepreneurs that run diesel-based microgrids, battery charging stations, or a combination of the two.

Measuring electricity access can also be complicated by the issue of unauthorized or illegal connections and consumption. In some countries, such as in India, transmission and distribution losses in the electricity sector can be as high as 20–30%. A large part of these losses are attributable to electricity thefts or pilferage by poor households who illegally tap electricity lines (Tongia, 2003). Official access numbers often do not account for these people.

Ideally, the quality and reliability of supply should also be a part of any access measure, but often, well-documented data and indicators measuring quality remain lacking. However, in many developing countries, particularly the least developed, the duration, reliability, and quality (measured in variability of voltage) of electricity supply remain highly irregular.

The importance of institutional factors for the implementation and sustainability of access efforts for poor and rural communities has most recently been reiterated in a World Bank review of its own rural electrification projects (World Bank, 2008a). The review of different country experiences supports the view that there is no superior institutional model. Public, private, and cooperative approaches have led to both success and failure. These models are also not necessarily mutually exclusive. What is important is the choice and strengthening of a framework that takes advantage of the country's strengths and considers the nature of their specific challenges (World Bank, 2010b). Learning these lessons has not been easy. Field surveys undertaken in Thailand show that despite flaws in the implementation of heavily subsidized solar home systems (SHS) and PV battery charging systems in rural areas, the inefficient policy continued over a 15-year period at a cost of over US$11 million (1984–2001), and by the time of the analysis, 60% of the systems were no longer operational (Green, 2004). In Viet Nam, a survey undertaken in 2006 found that 80% of SHS in a project in Vientiane Province were not working properly due to technical hitches, in some cases reducing the power available to the households to about 30 minutes a day (World Bank, 2008a). This implies the need for continuous monitoring and strategic corrections to programmatic activities (Box 19.4). However, funding is not always allocated for the necessary monitoring and evaluation activities. Furthermore, path dependencies and institutional inertia can make mid-course corrections difficult to implement (Annecke, 2008).

The lack of infrastructure and access to energy creates further economic problems at both the micro and macro level for countries. In the recent past, many sub-Saharan countries have faced sporadic and critical power shortages forcing them to resort to emergency power access methods. Over 23 countries have experienced crises, including an electricity crisis in the largest power system in the region: South Africa. This obviously undermined the well-publicized success story of rural and peri-urban electrification in South Africa. The general response to the crises has been largely dominated by the installation of high-cost diesel generators that have eroded savings made by governments or imposed additional financial burdens on those forced to install their own private generators. The consequences have included extremely high tariffs and crippled economies. In general, the low quality and reliability of centralized power, as well as issues of accessing centralized power systems, have created significant opportunities for the use of decentralized generation to meet local energy needs. However, significant financial and institutional barriers exist to the effective use of decentralized technologies, particularly for rural electrification (Zerriffi, 2010).

The provision of energy services accessible to all is often considered to be part of the social contract between governments and their people. For example, in the electricity industry, the social contract was an exchange in which the government regulates the industry and guarantees its financial viability while ensuring protection of the poor and the environment (Heller et al., 2003; Chaurey et al., 2004). However, there have been some fundamental problems with this model. It has generally relied on a combination of centralized organizations, particularly ministries, and heavy subsidies to reduce the costs to the end users. This has not always been financially viable and can exclude options that might better meet social needs. Therefore, new and innovative institutional frameworks are necessary. This may involve a greater role for the private sector, community groups, consumer organizations, and other alternatives to the centralized model of energy service delivery. However, this puts a greater emphasis on the need to rationalize the financing of rural energy efforts. The result will be an increasingly decentralized and heterogeneous approach to rural energy delivery and an emphasis on rationalizing the finances of rural energy efforts (Zerriffi, 2010).

19.2.4.2   New Financing Mechanisms

New financing mechanisms are needed for every scale of energy intervention, from large-scale infrastructure investments by both the public and private sectors to local entrepreneurs and right down to the individual household level. Mobilizing local finance is crucial for sustainability, especially taking into account that Official Development Assistance is decreasing and, in addition, it is mainly driven by donor interests (Hansen and Rand, 2006). For example, in 2006 the Kenyan utility, KenGen, offered 659 million shares to the public at 11.90 K. Shillings (~ US$0.15) with additional shares offered to KenGen employees at the same price. The goal was to raise around US$110 million. By the end of the public offering, it was over-subscribed by US$200 million (UPDEA, 2009). Notably, the investors originated from all walks of life and areas of the country, demonstrating the availability of local money and confidence in the local market.

There are a number of ways to overcome the problems of cost, affordability, and access to financial resources that do not rely entirely upon subsidies. In the case of standalone technologies, the first is to reduce the total amount of capital required by reducing the size of systems (e.g., lower wattage PV systems or smaller LPG canisters) (Cabraal et al., 1998; Barnes and Halpern, 2000; Martinot et al., 2002). Another solution to the capital cost problem for the consumers are rental models or fee-for-service models. This saves the household from having to raise enough capital to purchase the technology outright, and dealers can presumably improve their buying power and access different credit facilities (Barnes and Floor, 1996; Cabraal et al., 1998; Barnes and Halpern, 2000). A third option is to use a fee-for-service model in which the consumer only buys energy and not a technology. In the case of electricity, this would include microgrid systems or battery charging stations. These can be run by a local entrepreneur, the local government, a cooperative, or an NGO. The capital cost problem still remains for the provider of the service, however. This is an area of active institutional experimentation with various approaches for incentivizing existing financial institutions to enter into this market as well as setting up new financial

arrangements. This can include revolving capital funds and dedicated loan programs.

For the private sector at the local level, one way to address this deficiency is through what are called Market Facilitation Organizations. These are "public-private entities that support the growth of particular markets through a variety of means" (Martinot et al., 2002), ranging from more intangible benefits – such as access to information and networking – to technical support and financing.

One solution to the rural finance problem that has proven successful in a number of nonenergy areas and is now being applied to energy is the presence of microcredit lending agencies (Martinot et al., 2002; Armendáriz and Morduch, 2005). For example, Grameen Shakti has been successfully providing credit for the purchase of solar home systems in Bangladesh (Biswas et al., 2004; Uddin et al., 2006). The challenge with microcredit is that the sums may be too small for some energy purposes and at the household level, energy purchases may not lead directly to increased income, often a requirement for microlending.

Reforming the way in which energy access activities are financed and sustainably operated has potentially serious social consequences. For example, the implications of subsidy reform are that rural and peri-urban electricity consumers may be served with lower levels of energy service than their urban counterparts and by local actors rather than large government or private utilities. However, this does not absolve centralized governments of their responsibilities nor does it call for a complete removal of the international donor community from solving the problem. Some form of lifeline subsidy is needed at a minimum for low-income households.

There are conditions under which cross-subsidies could be implemented while minimizing the economic damage. Such cross-subsidies, if kept to a modest level, can be effective. Many institutional innovations may be implemented through local-scale actors, making traditional cross-subsidies for energy supply more difficult. If energy subsidies are desired, then new mechanisms may be needed. One option would be to provide subsidies directly to the end-users as an energy subsidy rather than being tied to a particular end-use or technology (Howells et al., 2006). Consumers can then make decisions based upon their energy needs and the availability of different options for meeting those needs. This would remove what is essentially a societal and political decision from affecting the functioning of the energy sector. A second option would be to create transfers among the electricity service providers either directly or via the government. This would depend on the particular institutional arrangements in each country. It could include partnerships between small actors such as NGOs, cooperatives, and small entrepreneurs and the utilities within a regulated concession model.

These new institutional arrangements may require a different role for both higher level government agencies and international donors (both official donor aid and NGO aid). This is necessary in order to maintain

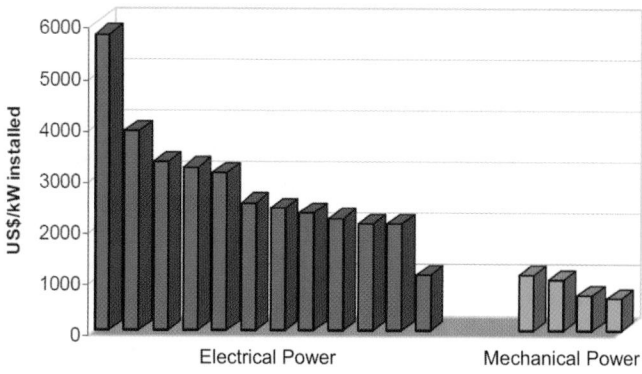

Figure 19.17 | Various costs per kW installed for rural hydro schemes in Africa, Asia, and Latin America. Source: based on UNDP, 2009a.

the social contract that previously existed, though perhaps in a different form. The key is to find a role for nonlocal actors to meet certain societal goals without destroying the market segmentation, local needs-based decision-making and possible contributions to greater rural development that comes with smaller scale solutions. Some principles, such as supporting energy planning, making investment capital available, creating incentives for commercial lending, promotional campaigns, and technical assistance have already been identified for certain markets. Centralized agencies can also aid in coordination, reduce their conflicting mandates and programs, and help create sorely lacking local institutional and organizational capacity (Bird, 1994; Cabraal et al., 1998; Radulovic, 2005; Srivastava and Rehman, 2006).

A number of development agencies are helping to facilitate microcredit schemes or small scale financing options by assisting the private sector and providing the interface between poor communities, energy providers, and private capital (UNDP, 2009b). The financial commitment of communities and entrepreneurs, together with a commercial approach, offers more guarantees for sustainability and poverty alleviation. Cost reduction, though not specific to productive end-uses of energy or mechanical power, is achieved mainly by addressing nonenergy barriers that hinder access to financing (e.g., policies, institutions), community participation, and enabling costs to be reduced by employing a local labor force and/or utilizing other community-owned assets, such as land, as collateral to secure loans. Cost reduction is also achieved by advocating for the incorporation of productive end-uses into national and international energy programs, budget allocations, strategies, and declarations, and finally, by implementing financing initiatives to scale up productive end-uses alongside other energy options in new and existing financing windows.

Financing prospects are linked to the productive end-uses ranging from financing for provision of social services, to income generation activities (commercial), to motive power for value-addition activities in rural areas (Bates et al., 2009).

For *social services*, it is likely that subsidies and grants from governments and international donors, in collaboration with relevant government ministries and NGOs, will remain a key mechanism for funding productive end-uses and mechanical power installations for basic services. For this category of services (e.g., water pumping for drinking water or sanitation), substantial incentives (government grants, support from projects/programs) remain necessary to reach the poorest. Indeed, income generated from social services is generally extremely low and is not sufficient to pay back the up-front investment. Nevertheless, tariffs in line with the beneficiaries' willingness or ability to pay should be set in order to ensure that the maintenance costs (labor force, spare parts) are covered and costs are shared between public funding, community contributions, and, where possible, private finance.

For *income-generating activities* (for instance, grain milling or manufacturing), soft and/or commercial loans, coupled in some instances with small subsidies, are instrumental in creating thriving businesses. The success story of microhydro in Nepal is mainly based on an implicit strategy aimed at prioritizing microhydro for productive end-uses through mechanical power and income-generating activities. Experiences highlighted in Khennas and Barnett (2000), based on case studies from five countries in Latin America, Africa, and Asia, illustrate the relatively low financial barriers to enter the microhydro business aimed at end-uses supplied by mechanical power. Despite interest rates of up to 17%, hundreds of schemes were developed on a sustainable basis in Nepal by small entrepreneurs.

For *enterprise-based productive end-uses* and mechanical power initiatives, there is a range of sources of funding already in existence that are potentially appropriate, based on commercial or semi-commercial loans, including AREED (African Rural Energy Enterprise Development)[9] in Africa. The Government of Senegal, for example, has used the AREED approach to develop its national program delivery in rural areas. For stand-alone productive end-uses and mechanical power systems at the farm or household level, financing and microlending models have been developed, such as that of the Grameen Bank of Bangladesh. This microlending agency has over 1000 branches and two million members and disseminates energy systems through a nonprofit rural energy company, Grameen Shakti. Loans are made after a small down payment and, while the model was initially developed for solar PV systems, it is extending into other sectors that include productive end-uses such as treadle pumps.

A survey of UNDP projects that expands access to modern energy at the local level indicates that the average cost per beneficiary for providing mechanical power by use of multifunctional platforms/equipment attached to stationery engines is US$24. Despite the relatively low cost (see Figure 19.17), there are inherent bottlenecks related to financing access to mechanical power (Bates et al., 2009).

---

9   AREED offers rural energy entrepreneurs in sub-Saharan Africa a combination of enterprise development services and start-up financing. The program allows entrepreneurs to structure their companies for growth and, by mainstreaming local financial partners, makes eventual investments possible through loans.

For *decentralized productive end-uses* and mechanical power systems, such as community water supply or shared milling resources, additional financing options are considered, drawing from existing experience in revolving funds for microhydro. Loans are given to institutions involving local government and the community, often with the management and operation of schemes headed by trained local enterprises.

The promotion of productive uses of energy with the objective to stimulate economic development should go hand in hand with other activities and instruments to support the establishment and/or development of enterprises. This requires cooperation with many actors and provision of other conditions for entrepreneurship and business (e.g., the availability of easy credit). Energy per se will not lead to the establishment of new enterprises and the alleviation of poverty. It is linked with factors of rural development, market demand and access, infrastructure, and entrepreneurship. Productive-use development must be based on demand-pull rather than technology-push.

19.3 Improvements in Household Access to Modern Energy: Regional Efforts and Status

In the following sections, the issue of household access to electricity and modern cooking/heating is assessed from a regional perspective. The focus is on the regions where the lack of access is most acute, including Africa, Asia, and Latin America. Data and information available for these three regions varies widely. For this reason the three regions are dealt with differently, although the discussion adheres to a common framework as much as possible. The objectives in discussing the following regions in more detail are to understand past trends and efforts, assess the current situation regarding access, and draw lessons from each region that might be applicable for other regions or individual nations. We do not include a deeper discussion of policies for access here, as Chapter 23 provides in-depth coverage of that issue.

19.3.1 Africa

Africa is home to about 15% of the world's population and 22% of its land. It also hosts an adequate share of energy reserves, but these remain largely unused. From 1997 to 2007 African economies grew at a steady average rate of 5.4% (World Bank, 2007) and the percentage of Africans living on US$1.25 a day decreased from 58% in 1996 to 50% in the first quarter of 2009 (World Bank, 2009). Nevertheless, Africa remains the continent with the lowest electrification level, and a third of all people in the world without access to electricity live in this region. Even today, only 11% of the rural population in sub-Saharan Africa has access to electricity and the majority of households cook with wood and charcoal over open fires or on inefficient stoves (UNDP and WHO, 2009). The use of traditional cooking fuels is highest in rural areas

(93%), but it is still very significant (about 70%) in urban households (Banerjee et al., 2009).

A detailed account of the energy access situation and efforts in North Africa, West and Central Africa, and Eastern and Southern Africa is presented in the following sections. Compared to Latin America and Asia, there are fewer studies and summaries on energy access in Africa. For this reason, data and case studies from individual countries have been widely used.

19.3.1.1 North Africa

North Africa, unlike sub-Saharan Africa, has made relatively good progress with respect to the provision of modern energy for the majority of people in the subregion.

Access to Electricity

The disparities in access to electricity between North and sub-Saharan Africa cover urban and rural populations. The rural electrification rate is around 98% in North Africa (IEA, 2010b). In urban areas of North Africa, electricity access is almost universal.

Algeria, Libya, and Egypt, the three oil and gas producing countries in North Africa, accelerated their electrification efforts over the last 30 to 40 years and achieved universal electricity access (Figure 19.18). Tunisia and Morocco, the two oil importing countries, pursued ambitious electrification drives that led to universal access in Tunisia and to 97% electricity access in Morocco. Mauritania is the exception in this region with national electrification levels as low as 30% and rural access levels at 2%. Mauritania is more similar to sub-Saharan Africa than other North African countries. Morocco's Global Rural Electrification Programme (PERG) (Box 19.5 and Fig 19.19) shows how political will, supported by technical and financial plans and capacity, lead to a steady increase in rural electrification levels from 18% in 1995 to 95% in 2008.

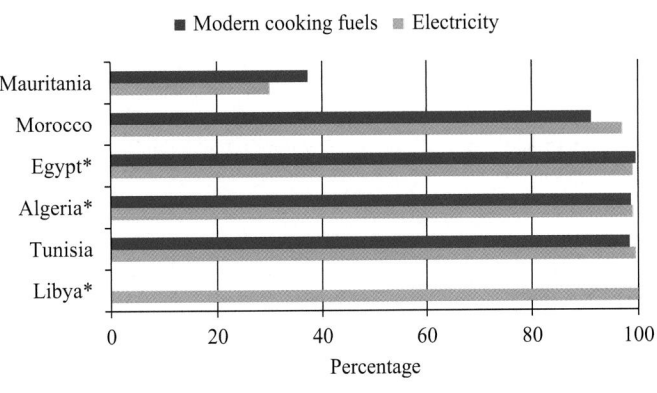

Figure 19.18 | Access to electricity and modern cooking fuels in North Africa. Source: UNDP and WHO, 2009. Note: Countries indicated with * are oil exporting.

Box 19.5 | Rural Electrification in Morocco, PERG Program

The Global Rural Electrification Program (PERG) was launched in 1996 with the aim to achieve complete rural electrification by 2010. This target date was reviewed to achieve the electrification objective in 2007, through the expansion of the electrification pace from 1000 to more than 1500 villages/year. Starting from 18% in 1995, a rural electrification level of 95% was achieved by the end of September 2008 (Figure 19.20), as a consequence of the ambitious PERG and its associated budgetary provision.

The PERG is based on the following three "global" principles:

- Territorial: it aims to provide electricity to all rural households, in all communities;

- Technical: it aims to integrate all the available electrification techniques (grid extension and decentralized power generation) to meet the electricity needs of each household and within feasible technoeconomic conditions;

- Financial: it integrates all financial resources that can be used for the rural electrification nationwide, under the PERG global financial mechanism. This financial mechanism involves three contributing partners, namely the Electricity Utility (ONE), the local authorities, and the end users/beneficiaries. The PERG budget amounted to about 20 billion Moroccan Dirham (US$2 billion).

The ONE contributes 55% of the financial cost through a fund raised from a 2% levy on grid electricity sales (35% of the electrification cost) and its own contribution of 20%. Local authorities co-finance 20% of the program costs: either 2085 MAD (~US$200) per household or 500 MAD (~US$48) per household/year over five years. Households contribute 25% of the electrification cost: either 2500 MAD (~US$240) per household or 40 MAD (~US$ 4) per month over seven years.

Decentralized electrification is being implemented mainly through solar PV installations, targeting 150,000 remote households.

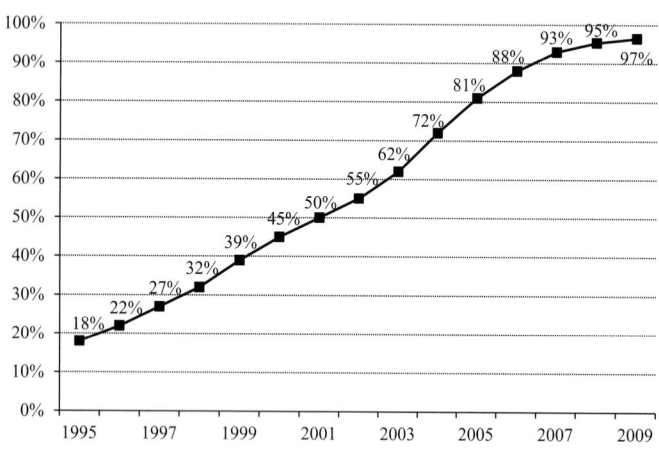

Figure 19.19 | Evolution of the rural electrification rate in Morocco 1995–2009. Source: ONE, 2009.

Access to Modern Cooking and Heating Fuels in North Africa

North Africa, including the Saharan desert, is generally a region with low rainfall. Access to traditional biomass and sustainably harvested wood resources for household cooking and heating needs has been a major challenge in the region, particularly in rural areas. North African governments have addressed the problem and successfully replaced traditional biomass with modern fuels by making access to LPG and

natural gas a policy priority. Access levels in Egypt, Algeria, and Tunisia range from 98% to almost 100%. Morocco is slightly below that and Mauritania lags behind with only 37% of households having access to LPG (Figure 19.18).

The three oil and gas producing countries – Algeria, Egypt, and Libya – used their own local resources. Morocco and Tunisia import gas from their neighbors. Most of the gas used in households is LPG, but access to natural gas for household uses is also quite well developed in the North African gas-producing countries and is progressing or emerging in the countries where gas pipelines are passing through to feed the European gas markets (Morocco and Tunisia).

In Egypt, natural gas consumption reached 25.39 Mtoe in 2003/2004. 1.7% of the total consumption is used in the residential and commercial sectors. The country has been trying to improve the availability of natural gas for residential customers by allocating service areas to several private companies since the beginning of 1998 (Drück et al., 2007).

In Tunisia, gas pipes pass through the country from the Algerian border to Italy, over an onshore length of more than 320 km. The Electricity and Gas Utility has a natural gas network that presently covers the coastal cities from Gabes to Tunis. There are plans to extend the gas network to include the Djerba and Bizerte areas (MEDREC and GNESD, 2009).

In Morocco, a natural gas market is emerging after the implementation of the Algeria-EU gas pipeline crossing the country. The first applications are for power generation, with the prospect of opening this market to improve household access to modern energy. However, about 30% of the population is still dependent on biomass (Mchirgui and Kanzari, 2006). To improve household access to LPG, some governments, such as Morocco, have been subsidizing gas bottled in small (e.g., 3 kg and 12 kg) canisters.

Access to Domestic Hot Water

Tunisia and Morocco showcase successful experiences of solar thermal applications, especially for solar water heating.

The Tunisian program, PROSOL, is supported by the electricity and gas utility (Box 19.6) and shows the importance of implementing appropriate financial mechanisms to sustain a quality dissemination of solar water heaters (SWH).

In Morocco, the SWH program (PROMASOL), supported by the Global Environment Facility, targets the installation of a capacity of 400,000 m^2 of solar collectors by 2010. According to the Renewable Energy Centre, the objective of one million m^2 of collectors is set for the year 2020.

The program is based on awareness-raising and communication, quality equipment, and after sale maintenance, as well as adequate financial mechanisms for households (leasing approach) and the tertiary/services sector.

In Egypt, the introduction of SWH technology to the national market started in 1980 with the import of 1000 home solar water heaters. In the same year, the first private local manufacturing company started and since then, SWH systems are manufactured in the country. In the mid-eighties a law was passed to promote the technology, which made the installation of solar water heaters compulsory for residential buildings in new satellite towns. Unfortunately this law did not have a lasting effect. Major obstacles included a lack of execution by the local authorities and the often poor quality of the SWH heaters, which gave the technology a bad reputation (Drück et al., 2007). More than 500,000 m^2 of solar collectors have been installed (end 2004), particularly in the new cities and tourist villages resorts. About 200,000 families are using SWH systems in Egypt. Tourist resorts and hotels are considered to be the main customers in this market. The distribution of the installed SWH systems shows that 40% of the total capacity is installed in new cities while 14%, 24%, and 14%, respectively, are installed in old cities, tourist villages, and government and public enterprises (Drück et al., 2007).

Box 19.6 | Large-scale Dissemination of Solar Water Heaters in Tunisia: PROSOL Program

Within the framework of its strategy to develop renewable energy, the Tunisian government decided to implement a program of massive dissemination of SWHs in the residential sector. This program, called "PROSOL TUNISIA" (Solar promotion for Tunisia), was launched in February 2005 and targets three types of SWH: 200-, 300-, and 500-liter capacity.

This program benefits from institutional and financial backups to promote the SWH market development. The financial support covers the following main incentives under the PROSOL:

* 20% subsidy of the SWH cost, provided by the government through the National Fund of Energy with a maximum of 100 Tunisian Dinar/m^2 (TD/m^2);

* A complementary subsidy of about 80 TD of the 300 liters SWH cost, supported by the Italian government through MEDREC Funds;

* A loan mechanism to finance the remaining cost of a SWH, granted over a period of five years, and paid through the electricity bill of the utility (STEG);

* The reduction of the interest rate, using UNEP funds during 2005 (MEDREP Program).

Therefore, the end user needs to provide only 10% cash contribution toward the SWH cost.

The installed capacity of SWH decreased to about 8000 m^2 in 2003 from around 18,000 m^2 in 2001. The PROSOL Tunisia Program aimed at reaching 225,000 m^2 SWH over the period 2005–2008; 500,000 m^2 SWH in 2009; and finally, 540,000 m^2 to achieve an installed capacity of 740,000 m^2 in 2011.

Source: MEDREC and GNESD, 2009

Lessons Learnt from North Africa

North African countries have succeeded in providing electricity access ranging from 97% to 100% of their populations. Access to modern cooking fuels is also very high and varies from 91% in Morocco to almost 100% in Egypt. The following lessons from the success of the programs in this region may be useful to other regions, particularly sub-Saharan Africa:

- The political will to implement rural electrification programs is a key driver to improve access to electricity in rural areas.

- The availability of fossil fuels resources in some countries helped in implementing early strategies for improving access to electricity and natural gas.

- The adoption of adequate financial mechanisms, involving cross subsidy, fee-for-service, stimulates large-scale access to electricity.

- Mobilization of decentralized access to electricity through renewable energy resources helps speed up access to electricity in remote villages.

- Subsidies stimulate LPG penetration and improve access to modern cooking fuels, especially when combined with the reinforcement of the LPG filling units and distribution network.

- Household access to natural gas is secured in gas-producing countries and facilitated by the gas pipelines passing through nonproducing countries.

- There is significant potential for South-South cooperation around energy access, between North African and sub-Saharan countries. Morocco and Senegal are cooperating in rural electrification through the implementation of the electrification concession in North Senegal, based on successful experiences in Morocco.

19.3.1.2   West and Central Africa

The energy access situation in West and Central Africa[10] compares poorly to North Africa and is similar to Eastern and Southern Africa (minus South Africa). Biomass energy forms the bulk of energy supply in the two regions, contributing more than 81% of final energy (GNESD, 2007) with related environmental consequences. At a national level, in countries such as Liberia, Chad, and Togo, more than 95% of the population

relies on traditional biomass for cooking and heating. Access to modern energy for cooking is very low in most countries. Access to LPG remains low, but there has been encouraging progress made in Senegal and to a lesser extent in Ghana, where a range of government policy and fiscal interventions have helped scale-up access to LPG.

Access to Electricity

According to the West African Power Pool (Diallo, 2009), only 30% of the population in West Africa has access to electricity, with 53% having access in urban areas and 7.5% in rural areas. The recent unprecedented escalation of oil prices has had a devastating effect on the economies in the region. Some countries, such as Ghana, Nigeria, Cameroon, Cote d'Ivoire, and Senegal, have household electricity access levels above 35% (IEA, 2009), with Ghana often regarded as a role model in these two regions with an access level of about 54%. At the bottom end of the scale are countries like Burundi, Chad, and Rwanda, which have access levels of 5% or below. In most rural areas, where the poor are mostly found, household access to electricity is lower than 1%. The low access levels of the poor are due in part to the high level of poverty of local communities and the underdevelopment of the electricity supply infrastructure (Sokona et al., 2004). There is insufficient grid coverage in most of the major load centers. Where available, national electricity grids are bedeviled with intermittent power supply, with frequent blackouts and sometimes power rationing as generating capacities fail to match growing populations and consumption levels (Brew-Hammond and Kemausuor, 2007). Also, the deterioration of distribution infrastructure has led to supply bottlenecks and higher technical losses comparable to other parts of sub-Saharan Africa, where inefficiencies in collection of revenues and distribution losses amount to 1.9% of GDP (Foster and Briceño-Garmendia, 2010). In Ghana, for instance, technical and commercial losses account for about 25% of supplied electricity (Energy Commission Ghana, 2008).

Several plans have been outlined in a number of countries to increase access to electricity. About 54% of the total population of Ghana has access to electricity, achieved largely through the National Electrification Scheme described in Box 19.7. The government is hoping to achieve universal access by 2020. In 2008, only 47% of the Nigerian population had access to electricity, mainly in urban areas (UNDP and WHO, 2009). The Government of Nigeria has committed resources to improve the access situation. There is an increased drive toward regional approaches in addressing the region's developmental challenges. The energy sector is spearheading this initiative, as demonstrated by the ongoing regional projects such as the West African Power Pool (WAPP) and the West African Gas Pipeline (WAGP).

In Central Africa, only 3.5% of Chad's population has access to electricity and only 30% of households in Cameroon have access to electricity (UNDP and WHO, 2009). In 2007, the Cameroon government signed an accord with the European Union worth about US$16.2 million to facilitate access to electricity in some rural areas in the country. The

---

10 The United Nations' definition of West Africa comprises all countries forming the ECOWAS, namely, Benin, Burkina Faso, Cape Verde, Cote d'Ivoire, Gambia, Ghana, Guinea, Guinea-Bissau, Liberia, Mali, Niger, Nigeria, Senegal, Sierra Leone and Togo. For the purpose of this review, Central Africa comprises all countries forming CEMAC: Cameroon, Central African Republic, Congo, Gabon, Equatorial Guinea and Chad.

Box 19.7 | Electricity Access Scale-up in Ghana

Ghana increased its electrification levels from 23% in 1985 to 54% in 2005. In 1985, only 250 out of about 4202 towns and cities in Ghana, in five of the then nine regions, had access to the national electricity grid. The national electrification drive in Ghana started in 1985 with the preparation of a project by the Volta River Authority to extend the 161 kV National Grid northward to reach all the administrative regions of Ghana under a project captioned the Northern Electrification and System Reinforcement Project (NESRP). The total project cost was estimated at US$150 million.

The Volta River Authority completed the definition of the project and obtained financial support for the first phase from the African Development Bank in 1987. Thereafter, several other multilateral agencies joined in quick succession to provide support for the entire scope of the project.

Implementation of the NESRP project was successful in every respect: financial, technical, and social. Within three years of project commencement, the national grid supplied electricity to all the regional capitals except the Upper West regional capital. The construction, commissioning, and testing of the 600 kilometers of high voltage (161 kV) lines and associated substations had been completed within budget and ahead of schedule.

The impact of the achievement of the NESRP objective by the Volta River Authority spurred the preparation of a plan called the National Electrification Scheme (NES) which was issued in 1990. The goal of the NES was to provide within a 30-year timeframe, electricity access to about 4200 settlements with populations of 500 or more. The NES was pursued through various discrete projects. Prominent among these were the Northern Electrification Project and the Self-Help Electrification Project (SHEP).

The SHEP was a nationwide scheme that was introduced as a policy framework under which communities could advance their electrification projects ahead of the dates indicated in the NES by meeting agreed criteria for community contributions to the project implementation. The SHEP aimed to connect to the national grid ahead of their respective scheduled dates any communities that:

- were within 20 km of an existing 33 kV or 11 kV network;

- had procured low-voltage poles for the network within the community; and

- had a certain minimum number of premises wired and ready to receive power.

The Government's obligation was to provide the conductors, transformers, pole-top, and other materials and assume responsibility for the construction work required to make the connection.

The achievements of the electrification drive (i.e., connecting 2350 communities in just ten years after the launch of the NES plan, reaching 40% of all communities with population exceeding 500 in 2000, and achieving an electrification rate of 54% by the end of 2005) were impressive. The NES was reviewed in 2010. It was estimated that the level of electrification stood at close to 70% at that time and the government has recommitted itself to building on the successes of the last two decades to achieve universal electrification by 2020.

Cameroonian government co-financed 50% of the projects and some 128 villages are expected to benefit.

Access to Fuels

Millions of households in West and Central Africa lack access to modern cooking fuels. For most of these households, energy from biomass (mainly fuelwood, charcoal, bagasse, and animal and agricultural waste) is the main fuel source for cooking, even with its attendant environmental and health hazards. In many countries in these regions, including Liberia, Guinea, Mali, Sierra Leone, Niger, Togo, and Chad, more than 95% of the population relies on traditional biomass for cooking and heating (Figure 19.20). Only in Gabon, Cape Verde, and Senegal does more than 40% of the population have access to modern cooking fuels. In most of these countries, biomass energy accounts for over 80% of

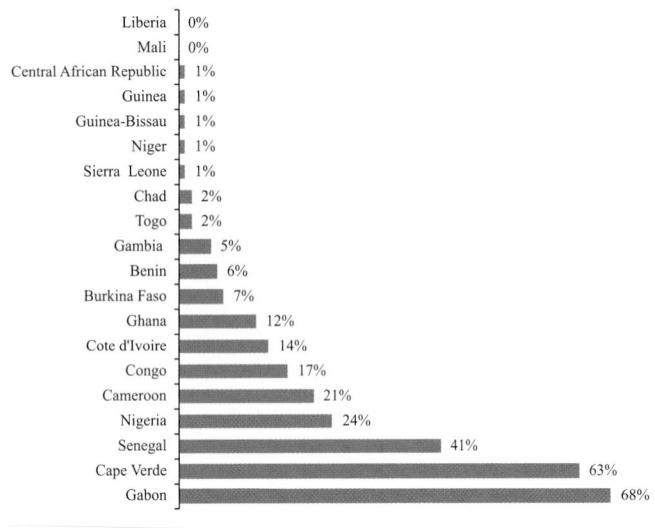

Figure 19.20 | Share of population with access to modern fuels in West and Central Africa. Source: data from UNDP and WHO, 2009.

total energy used (Hagan, 2006). The bulk of modern energy fuels used in these two regions comes from LPG.

The numbers of people relying on traditional biomass for cooking in these regions is projected to increase consistently over the next 20 years (Modi et al., 2006; IEA, 2010b). Bearing in mind the negative effects of these projections, several countries are making efforts to reverse this trend for the better. But while many countries have tried, only Gabon, Cape Verde, Senegal, have succeeded in implementing far-reaching programs to substitute woodfuels for cooking with LPG. Nigeria and Cameroon follow in that order, but the access levels are very low compared to the LPG potentials from the oil and gas industries in these two countries. Interestingly, the same countries that have somewhat higher access to electricity also have higher access to modern fuels for cooking. Ghana, and to a lesser extent Cote d'Ivoire, seem to deviate slightly from this trend. For example, Ghana's electricity access rate of over 54% compares poorly with its 12% access to modern fuels. There has been a recent drive in Ghana to focus attention on modern fuels and an ambitious target of increasing the population using LPG to 50% by 2015 has been set but it remains to be seen how this will materialize given the frequent LPG shortages in the country.

Participation of Regional Economic Communities in Energy Access Scale-up and Lessons Learnt

Following recommendations from the New Partnership for Africa's Development (NEPAD) in 2002 to the Regional Economic Communities, the Economic Community for West African States (ECOWAS) and Communauté Economique et Monétaire de l'Afrique Centrale (CEMAC) proposed some very ambitious targets. The regional organizations have developed strategies or action plans, such as the ECOWAS/UEMOA

Table 19.2 | Specific Energy Access Targets by ECOWAS and CEMAC for 2015.

	ECOWAS	CEMAC
Modern energy for cooking	100%	80%
Modern energy / electricity for basic needs in urban an peri-urban areas	100%	50%
Electricity for rural households	36%	35%
Electricity for schools, clinics and community centers	60%	56%
Mechanical power for productive uses in rural areas	60%	-

Table 19.3 | Barriers for LPG access for households and LPG suppliers.

	Demand Side	Supply side
Accessibility	Rural Household Access to LPG – Local LP Gas supply to households – Smaller cylinders – Full local sales and services – Fast and convenient refills – Plenty of cylinders frequently refilled	LP Gas Supplier Access to Rural Households – Dispersed customers – Long supply chains – Lack of rural supply infrastructure – Need for local agents and sub agents – Higher investment, risk and maintenance cost
Affordability	Low LP Gas Appliance Prices – Ability to pay – Need for credit – Small quantities – Low cost of appliance to switch to LP Gas – Access to credit – Below poverty line households	Higher Margins to Support Higher Rural Supply Cost – Low LP Gas prices – Small margins – Economic viability – Need for subsidies for market entry appliances – Need to reduce overheads – Special third party financial support
Acceptability	Rural Household View of LP Gas – Low or zero cost fuel alternatives such as wood – Higher fuel cost – Safety and proper usage – Cooking major usage – Government friendly household energy policy – Zero rated taxes – Many competing suppliers	Attractiveness of Rural Markets to Suppliers – Added cost of competitive marketing – Low margins – Costly user education – Small volumes of LP Gas – Government policy favoring other fuels such as paraffin/ natural gas – VAT on LP Gas sales – Exclusive supply territories

Source: WLPGA, 2005.

White Paper on Energy Access (ECOWAS, 2006) and the CEMAC Action Plan for Promotion of Energy Access (CEMAC, 2006) with assistance from UNDP and the EU Energy Initiative Partnership Dialogue Facility. Whereas CEMAC is aiming for 80% access to modern fuels for cooking by 2015, ECOWAS is hoping to achieve 100% access to modern fuels by 2015, with between 50% and 70% being provided through LPG and the rest through improved fuelwood cook stoves (Table 19.2). Judging from progress made so far, it is going to be very difficult to achieve these targets and they may need to be revised with greater emphasis placed on developing capacity.

The penetration of LPG, particularly in rural areas, has been very slow and access levels remain very low. The World LPG Association identified several key barriers to the accessibility, affordability, and acceptability of LPG, in particular for rural households and suppliers (Table 19.3). The World LPG Association believes that increasing access to LPG would require a concerted effort by industry and government to address all aspects of the energy puzzle, including developing local resources, financing, building capacity in local energy entrepreneurs, developing joint marketing campaigns, and increasing public awareness, with the right mix of policy changes, dissolution of market barriers, and responsible investment (WLPGA, 2005). Senegal is one country that managed to address all the barriers successfully and LPG access levels are 41% overall, with 74% in urban areas and 12% in rural areas (UNDP and WHO, 2009).[11]

Efforts that succeed in integrating productive uses and income generation activities into energy access initiatives may well turn out to be the deciding factor in improving access to households in this region. So far, private sector participation in energy access scale-up has been abysmal, as policies and tariffs have not been favorable for encouraging the private sector to venture into power production. Countries in the two regions should be learning from the experiences of Senegal – and countries outside the regions, such as Botswana and Brazil – to explore sustainable ways of increasing access to LPG, emerging biofuels, and improved cook stoves. Ultimately, a major shift is needed in the current access trajectory if realistic increases in energy access are to be achieved by 2030, which is the reference date for most forecasts by the Regional Economic Communities, the IEA, and the World Bank.

19.3.1.3   Eastern and Southern Africa

In most countries in Eastern and Southern Africa[12] energy use overall has risen and governments and utilities have made efforts to increase generation, transmission, and distribution capacities, but the progress made has been too slow to keep pace with population growth. The energy supply and use situation is generally similar to West and Central Africa (see above), but national rates of deforestation are much heavier in Eastern Africa, and in many regions, adequate supplies of wood and charcoal fuels are an issue. The cost of these fuels, when purchased, has been rising steadily. In most countries, traditional biomass still plays the major role (up to 80%) in energy supply. Some governments have implemented projects and policies to make modern cooking fuels (LPG and kerosene) more easily

available in rural and peri-urban areas (e.g., Botswana, Lesotho, and South Africa) and private companies sell them more widely now. In other countries, deregulation of LPG and kerosene made them more expensive for the end-user, leading to a decline in kerosene use. Many African countries grow sugarcane and there is a rising interest in producing ethanol for the transport sector and as a household cooking fuel. In Eastern and Southern Africa, ethanol gelfuel – ethanol with a gelling agent – was introduced, starting with the Millennium Gelfuel Project, but wider dissemination did not follow. The efficiency of technology using ethanol gelfuel compared with LPG and kerosene was also found to be significantly lower. The retail price of ethanol gel fuel would have to be well below that of kerosene and LPG to make ethanol gelfuel competitive.

Improved cook stoves have received much attention and the Kenyan Ceramic Jiko, an improved charcoal stove, is distributed to over eight million customers across Africa, from Senegal to Ethiopia and South Africa, and has become an African success story (AFREPREN/FWD, 2009). Biogas plants have been introduced on a project basis in the region. The biogas scheme in Rwanda that integrates agriculture and energy appears to be one of the more successful ones.

Access to Electricity

Levels of household access to electricity in Eastern and Southern Africa range from a low of about 6% in Rwanda to 100% in Mauritius. The countries with the highest access are Mauritius (100%) and South Africa (73%). The two countries are the two middle-income economies in the regions. Electricity supply is not always the largest limiting factor to improved access, and barriers sometimes lie in the lack of national infrastructure. The Democratic Republic of Congo (DRC), Lesotho, and Mozambique are exporting electricity through the Southern African Power Pool to other countries in the region, while their national electrification levels are less than 15%.

In Southern Africa, four countries – Botswana, Mauritius, South Africa, and Zimbabwe – have successfully extended electrification to rural areas using different approaches. In Botswana, the utility connects households on a cost-recovery basis and customers can apply for loans for their electricity connection. Monitoring the implementation and impact of the rural electrification policy, the Energy, Environment, Computer and Geophysical Applications Group (EECG, 2004) found that if the upfront payment and monthly repayments are small and extended over longer periods, the uptake of connections increases significantly. South Africa highly subsidises electricity to low-income households. Under the National Electrification Programme, access to electricity is very affordable even for the urban and rural poor. In addition, the Free Basic Electricity allocates 50kWh/month free of charge to poor households (Box 19.8). In Zimbabwe, the Rural Electrification Agency targets rural growth centres where local government infrastructure such as agricultural extension, health services, schools, and police stations are concentrated. Local councils facilitate enterprise development and lease stands to medium and small enterprises that provide services including automotive, electrical, electronic and

---

11   See Chapter 23 for further discussion on the success of transitioning to LPG as a cooking fuel in Senegal.

12   For this study, Eastern Africa includes Burundi, Djibouti, Eritrea, Ethiopia, Kenya, Rwanda, Tanzania, Somalia, Sudan, and Uganda. Southern Africa includes the 15 countries of the Southern African Development Community (SADC): Angola, Botswana, Democratic Republic of Congo (DRC), Lesotho, Madagascar, Malawi, Mauritius, Mozambique, Namibia, Seychelles, South Africa, Swaziland, Tanzania, Zambia, and Zimbabwe.

Box 19.8 | South African National Electrification Programme and the Free Basic Electricity

Through the government's National Electrification Programme, electricity connections in South Africa grew from 36 % of households in 1995 to over 70% in 2008. Electricity to low-income households is subsidised, making access affordable for the poor. The blanket roll out, in which whole areas are provided with electricity supply so all potential customers are served, not only customers applying and paying, significantly reduces cost. Technological innovations such as prepayment meters further reduce costs.

While the National Electrification Programme facilitated access to electricity, the poor did not automatically benefit from being connected. Often they could not afford to use the electricity and consumption levels among the newly connected households remained low. In 2003, the government introduced the Free Basic Electricity so the poor could benefit from the huge investments in national electrification.

The example of Cape Town illustrates how the subsidised connection works. The municipality of Cape Town charges ZAR 225 (approximately US$29) for a subsidised connection of 40 ampere to recognised areas of informal housing. If the new customer cannot pay the connection fee upfront, the amount is charged to their prepayment account and deducted gradually each time an electricity purchase is made, at the rate of 20% of the purchase. No interest is charged on the advanced connection fee. If customers use less than 450 kWh/month they are eligible for a lifeline tariff divided into three blocks. The first block up to 50 kWh/month is free, the second block from 51–150 kWh at ZAR cents 58.11/kWh, and the third block from 151–450 kWh is charged at ZAR cents 70.47/kWh. This compares to the domestic consumption tariff without subsidy of ZAR cents 93.32/kWh.

Why are some excluded from subsidised access?

Informal houses built on land not approved for electrification (flood plains, road reserves, power-line servitudes, private land, etc) cannot get a metered electricity connection and have to rely on extension cords to neighbouring houses that are metered. This generally costs twice as much as electricity from metered access.

general repairs, welding and spray painting, milling, carpentry, secretarial, and general retail services. In Mauritius, the rural electrification program started well before the country gained its independence in 1968 from the British. Since that time, the Central Electricity Board has been the only electric utility responsible for generating, transmitting, and distributing electricity in Mauritius and it had connected all households by the year 2000.

Over half a million households in Africa use PV systems for lighting and communication (AFREPREN/FWD, 2009). In Kenya, about 150,000 solar systems have been distributed through the market. Other countries, such as South Africa, started subsidised programs in the late 1990s. The number of systems distributed was far below the target figure. Financial and technical barriers and the expectation of the recipients, particularly those who live near the national grid and hoped to get a grid connection, seem to have been the major problems. In Botswana, the distribution of solar systems has been stepped up and is bringing light and communication to remote villages and dispersed cattle posts.

Access to Fuels

Fuelwood is still very widely used in Eastern and Southern Africa and 80% of the population in 13 of the 24 countries use fuelwood for cooking. In six countries, households are almost entirely dependent on woodfuels: Burundi (99.3), Madagascar (99.1%), Malawi (98.6%), Rwanda (98.6%), Somalia (99.1%), and Tanzania (96.6%). Out of these, Burundi, Malawi, and Rwanda are land-locked countries without any fossil fuel resources. In contrast, 80% of the population in four countries have access to modern fuels (electricity, gas, and kerosene). These are Djibouti (86.1%), Mauritius (95.8%), Seychelles (>95%), and South Africa (83.2%). The first three countries in the latter set have small populations below 1.3 million and make up only a small proportion of the sub-Saharan African population. Gas is a major cooking fuel in only three countries: Angola (51.9%), Botswana (45.8%), and Mauritius (91%) (Figure 19.21).

Not only a high proportion of households, but also a significant number of industries in Eastern and Southern Africa depend on fuelwood and charcoal for their energy needs. The poorest use fuelwood and as incomes rise charcoal is the preferred fuel. Charcoal is also the preferred fuel in urban areas. In densely populated areas and particularly around major cities, fuelwood and charcoal are becoming scarce and overharvesting contributes to forest and soil degradation. Woodfuels (fuelwood and charcoal) have become a lucrative trade and are a major source of income for many households. The poor are employed along the entire value chain, from the rural woodcutters and charcoal

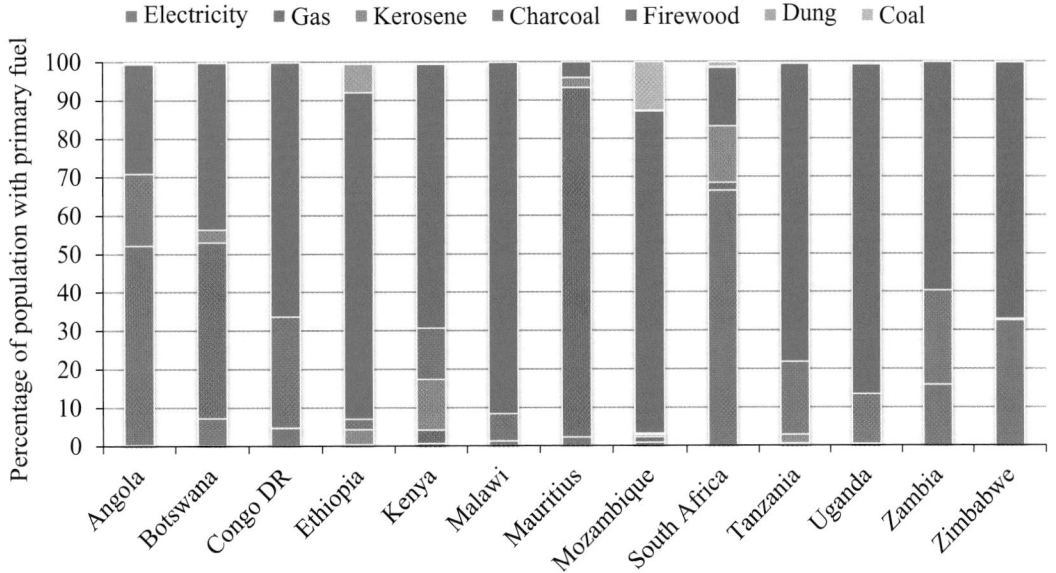

Figure 19.21 | Fuels used for cooking by country in Eastern and Southern Africa. Source: UNDP and WHO, 2009.

producers to the transporters and the urban distributors. The growing wood and charcoal markets are in many cases seen as a threat to forest and woodland resources, but if they are regarded as an opportunity to create employment for the rural poor and provide affordable energy they could provide an important contribution to poverty alleviation. This would require that forest management and the charcoal trade are well and transparently regulated and the implementation of the regulation is in the interest of all stakeholders. It is estimated that the woodfuel trade in Malawi and Rwanda is about 2% of GDP and governments could collect substantial revenues from a regulated woodfuel industry. This income could contribute to making the markets sustainable.

Generally, woodfuel policies and strategies need to be better integrated and address the following objectives: to provide a sustainable woodfuel supply for the majority of sub-Saharan Africans, to protect the environment, and to approach the fuelwood chain as an opportunity for poverty alleviation and job creation.

The traditional three-stone cook stove is still used in many parts of Eastern and Southern Africa. At the same time, improved cook stoves are locally designed and manufactured to reduce heat loss, decrease indoor air pollution, increase combustion efficiency, and improve heat transfer (AFREPREN/FWD, 2009). The best known example is the Kenya Ceramic Jiko, which is disseminated in Kenya and other African countries. The dissemination level for improved woodfuel stoves (Table 19.4) is low in relation to the number of people using woodfuels for cooking. The barriers may be limited local production levels, acceptance, and affordability of the improved cook stoves.

Table 19.4 | Dissemination of improved woodfuel cook stoves in Eastern and Southern Africa.

Country	Number disseminated
Botswana	1500
Eritrea	50,000
Ethiopia	3,010,000
Kenya	3,136,739
Malawi	3700
South Africa	1, 250,000
Sudan	100,000
Tanzania	54,000
Uganda	170 000
Zambia	4082
Zimbabwe	20,880

Source: AFREPREN/FWD, 2009.

19.3.2   Asia and Pacific

With about 60% of the world's population, the Asia and Pacific region comprises the largest of the global regions. A very high level of the population in the region still lives in poverty. Although economic development has resulted in rapid urbanization and changed the composition of the population, about two-thirds of the region's population still lives in rural areas. Furthermore, these rural areas are home to more than three-quarters of the poor in the region, who are distinguished by some of the lowest levels of per capita energy use in the world. Huge variations within Asia are evident in the levels of access to both electricity and modern fuels and more efficient devices. Progress with

Table 19.5 | Per capita energy use and percentage users by energy sources for China.

		Urban 1992	Urban 1996	Urban 1999	Urban 2001	Rural 1999
Coal	MJ	3245	2313	2085	2356	1843
	%	47.5	32.3	27	28.8	38
LPG	MJ	541	734	845	805	40
	%	45.1	53.6	56.7	56	28
Piped natural gas (urban only)/ Biomass (rural only)	MJ	892	1400	1421	1464	6214*
	%	21.3	30.8	33	34.7	62
Electricity	MJ	1445	2357	3182	3774	84
	%	93.8	92.5	94.8	96.2	97
Total	MJ	6122	6805	7544	8398	8181

* In rural areas this represents biomass use as no piped natural gas is used in rural households.

Source: Pachauri and Jiang, 2008.

electrification has also been extremely uneven across the region in the past. The problem of provision of modern fuels and/or devices for thermal energy needs, however, remains a larger challenge for Asia. For most countries in the region, the majority of the rural population relies on unprocessed biomass or coal for most of their cooking and heating needs. Over half of the total global population relying on biomass lives in China and India alone.

A detailed account of the energy access situation and efforts in the two largest Asian nations, China and India, and that for the other developing Asian countries is presented in the following sections.

19.3.2.1 China

China is an example of a country that has achieved significant success in improving the access to electricity for its rural population and in the dissemination of clean cook stoves. In overall terms, total residential energy demand increased little in China over the last couple of decades, because of a transition from inefficient to efficient fuels. However, significant changes in the pattern of residential energy occurred, largely on account of shifts in the choices of energy used in urban households. Within the rural sector, relatively little change in the patterns of household energy use took place (Figure 19.22 and Table 19.5). What has been significant is the access to modern energy among rural households in the country. However, actual consumption amounts of the modern energy sources remain very low in rural households.

Access to Electricity

China has achieved enormous success in electrifying its population.[13] During the 1980s and 1990s, almost the entire population was electrified, with over 900 million rural inhabitants gaining access to electricity (Peng and Pan, 2006). Currently, it is estimated that only about 1% of the total population remains without access to electricity. Strong government commitment was important to achieving the current status of

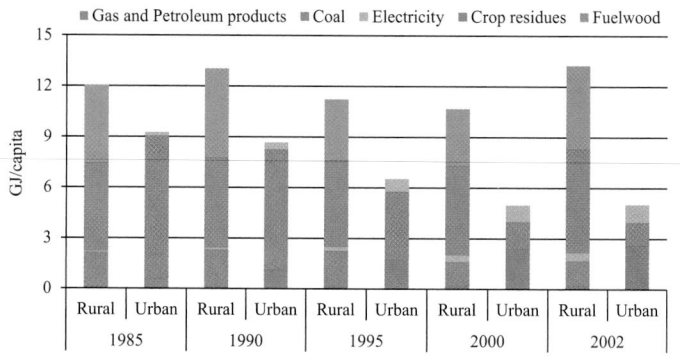

Figure 19.22 | Average household energy demand by energy type in China 1985–2002. Source: Pachauri and Jiang, 2008.

electrification in China. The latest statistics from the National Energy Administration suggest that in 2008, two million rural households still lacked electricity in China, which represents some nine to 10 million people (IEA, 2010a). Through the deployment of decentralized power systems, the government aims to supply about 10 million people with electricity by the end of 2020. The government expects, however, that by 2020 universal access will still not be achieved (the last customer will not be connected). At present, most areas without electricity are located in western regions and islands in the eastern coastal areas, far away from the grid. Most of these areas are rich in renewable energy resources (hydropower, solar, and wind energy), which can practically and economically provide electrification to remote regions. Lessons from the electrification programs in China point to certain key factors that were responsible for its success. These include strong government commitment, technological flexibility, a sense of ownership for the electrification solutions among remote communities not served by the grid, and the inclusion of the private sector in the implementation of electrification programs (IEA, 2010b).

Access to Modern Fuels and Stoves

Biomass and coal continue to be key sources of cooking and heating energy among rural households in China. Today, over 700 million

---

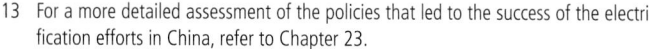

13 For a more detailed assessment of the policies that led to the success of the electrification efforts in China, refer to Chapter 23.

Table 19.6 | Shifts in the percentage of population using different sources of household energy in India.

	1983		1987–1988		1993–1994		1999–2000		2004–2005	
	Rural	Urban	Rural	Urban	Rural	Urban	Rural	Urban	Rural	Urban
LPG	0	9	1	20	2	33	6	47	12	61
Coal/coke	3	21	3	14	2	8	2	5	2	5
Electricity	15	58	24	67	36	77	47	84	54	91
Kerosene	95	92	96	88	95	83	96	75	91	55
Fuelwood	86	61	89	50	88	42	88	35	88	35
Dung	53	27	56	24	53	18	52	12	46	10

Source: Pachauri and Jiang, 2008.

inhabitants of China continue to rely on solid fuels. Although gas, oil, and electricity consumption in rural areas has increased over the last 25 years, it still remains very low, and much lower than in urban households in aggregate and per capita terms. The total consumption of biomass has remained high in rural areas because of population growth and the relatively slow transition away from the use of this fuel. Urban households in China use a larger share of modern energy. Looking at changes over time among rural households, although per capita biomass use gradually declined during most of the 1990s and coal use was moderately substituted by modern energy, no significant transition in energy use patterns occurred. In urban households, in contrast, a significant shift away from biomass and coal has taken place over the last twenty years. Without any new policies, it is projected that about a quarter of the total population will still rely on solid fuels in 2030.

The Chinese have also made significant efforts in improved dissemination of cook stoves. Between 1982 and 1999, the Chinese National Improved Stoves Program disseminated 180 million improved biomass stoves (Zhang and Smith, 2007). While it is difficult to know how many of these stoves are still in use, the recent UNDP and WHO study (2009) suggests that many still are. The Chinese ICS program, the largest and arguably most successful in the world, relied on rural private stove companies for its success. Main features of the program included:

- stove adopters paying the full cost of material and labor (about US$10);

- government-provided support to producers through designs for stove construction, training, administration, and promotion support; basically there was an indirect subsidy to pay for the costs of stove-making enterprises;

- establishment of local energy offices to provide training, service, installation support, and program monitoring;

- fostering the development of self-sustaining rural energy enterprises that manufactured, installed, and serviced the stoves; and

- an unprecedented scale of rural energy intervention.

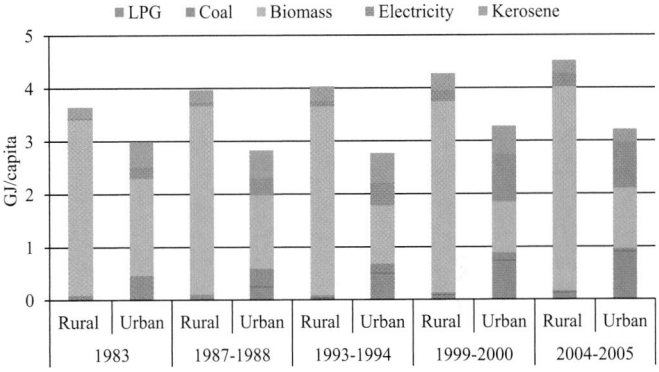

Figure 19.23 | Average household energy use by energy type in India, 1983–2005. Source: Pachauri and Jiang, 2008.

19.3.2.2 India

Even today about 40% of India's rural population lacks access to electricity and almost 80% rely on unprocessed biomass for their cooking and heating needs (Pachauri and Jiang, 2008). Several programs were implemented in the past to improve access; however, the focus was largely on improving access in urban centers and in fertile agricultural belts in rural areas.

Rural households did not witness any striking changes in their patterns of energy use. Biomass use per capita increased in absolute terms, but only slightly between 1983 and 2005. The total amounts and the proportions of commercial energy used in rural households continue to remain very low. In urban households, a much more rapid substitution of biomass by commercial fuels and electricity is evident. Biomass consumption per capita declined, and this decline resulted in a decrease in total per capita household energy demand in urban households between 1983 and 1993–1994 and between 1999–2000 and 2004–2005. However, during the mid-1990s, rise in LPG and electricity consumption among urban households drove up per capita energy use (Table 19.6 and Figure 19.23).

Access to Electricity

India today hosts the world's largest population without access to electricity. Traditionally, village electrification was used as an indicator of the

Box 19.9 | Achievements of the Rajiv Gandhi Grameen Vidyutikaran Yojana

The Rajiv Gandhi Grameen Vidyutikaran Yojana program, with a total estimated budget of over US$5 billion, is one of the most ambitious to date in India. It aims at electrifying all unelectrified villages, electrifying all households in electrified villages, and providing free electricity connections to all below-the-poverty line (BTL) households. In all, the program aims to electrify about 115 thousand unelectrified villages and connect 23.4 million BPL households. Close to 90% of the funds committed have been disbursed and the table below provides an overview of the programs achievements to date.

	Electrification of Unelectrified Villages	Intensive Electrification of Electrified Villages	Connection to BPL Households
Total Numbers (%)	103,402 (87.8%)	248,553 (69.7%)	18,912,729 (76.5%)

Source: India Ministry of Power, 2011

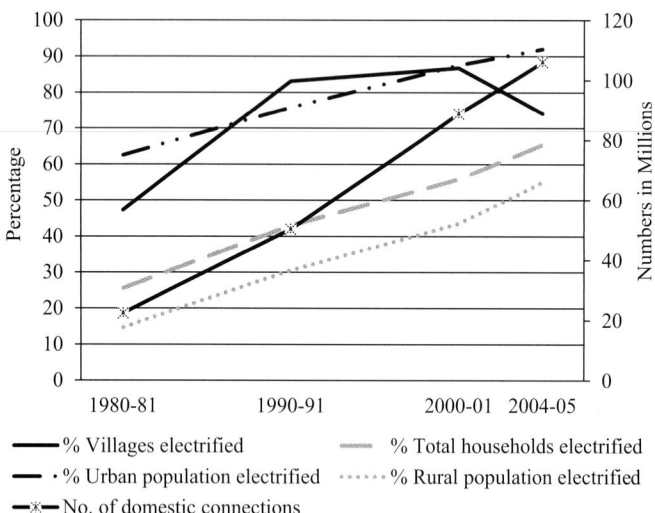

Figure 19.24 | Progress with electrification in India according to different indicators. Source: Pachauri and Mueller, 2008.

extent of rural electrification, but this did not provide an accurate picture of the actual use of electricity among households and also changed over time as a consequence of changes in the definition of "electrified village" (see Figure 19.24). Historical progress on electrification in India when measured in terms of household access has been rather poor, but the status varies significantly across states and regions.

The Indian government has recently redoubled its efforts at initiating policy reforms and new programs for accelerating electrification. The Rural Electrification Policy 2006 aimed at the provision of electricity access to all households by 2009 and a minimum lifeline consumption of one unit per household/day as a merit good by the year 2012. The National Electrification Policy 2005, which preceded this by a year, targeted total village electrification by 2010 and total household electrification by 2012 (India Ministry of Power, 2005). The main program through which universal access objectives of the Electricity Policy are being implemented is the Rajiv Gandhi Grameen Vidyutikaran

Yojana, launched in April 2005. A large effort towards grid extension and strengthening of the rural electricity infrastructure has been initiated through the Rajiv Gandhi Grameen Vidyutikaran Yojana. The government's Rural Electrification Policy 2006 also specifies, among other things, guidelines for decentralized distributed generation. The Ministry of New and Renewable Energy has also initiated a new remote village electrification program. Significant progress in providing access has been achieved in certain regions and states of the country. However, in other states, particularly among rural households, large fractions of the population are still in the dark. The ambitious targets for 2009, 2010 have not been met and future target for 2012 will also likely not be met given current trends (Pachauri and Mueller, 2008).

Access to Modern Fuels and Stoves

In 1980–1981, over 90% of energy used by households in India was from biomass sources. While this share declined to just over 80% in 2000–2001, the actual quantity of biomass consumed increased continuously over the entire period (Pachauri and Jiang, 2008). As observed in Table 19.6, major changes are evident in the percentage of persons using different energy types across rural and urban households over this period. The percentage of population using LPG increased from 9% to 61% in urban areas. However, in rural households the uptake of LPG was much slower and even in 2004–2005, only 12% of the rural population used this fuel. Thus, for the majority of the rural population even today, biomass remains the main source of cooking and heating fuel. Over 800 million people – 75% of the rural households and 22% of the urban households – rely on solid fuels in India today. Without any additional new policies, the number of people relying on solid fuels is projected to increase or at best remain unchanged by 2030.

In order to improve the access to improved end-use devices, especially among rural households who are likely to depend on biomass for the foreseeable future, the Ministry of New and Renewable Energy of the Government of India has launched a new initiative on biomass cook stoves, with the primary aim of enhancing the availability of clean and efficient energy for the energy deficient and poorer sections of society.

The new initiative is based on the recognition that cook stove technology has improved considerably in the past few years. But further advances are still possible and, indeed essential. The aim is to achieve the quality of energy services from cook stoves comparable to that from other clean energy sources such as LPG. Under this initiative, a series of pilot-scale projects are envisaged using several existing commercially-available and better cook stoves and different grades of processed biomass fuels. The goal of the program is to sell 150 million stoves in 10 years.

The Indian government is not alone in its effort to expand Indians' access to cleaner biomass cook stoves. International donors such as the Shell Foundation are increasing their support of cook stove programs that show potential for economic sustainability and scalability. Corporations such as Royal Philips Electronics, First Energy (formerly a BP company), and Bosch-Siemens, are developing cleaner cook stoves that can be customized for cooking needs around the world. Companies that can manufacture stoves include Envirofit, StoveTec, First Energy, WorldStove, and HELPS International. The Shell Foundation has invested US$3.5 million in Envirofit to support its program to sell 5–7 million stoves in seven states in India in the next five years and is investing several million dollars in a public awareness campaign. So far most companies are marketing only to consumers who can afford to pay about US$20 for a stove, which excludes the very poor. But Envirofit plans to launch a new model that customers can purchase through monthly payments to a microfinance company. The users will be required to pay about US$1 a month for around a year. Another payment option that stove companies, including Philips, may pursue is to seek carbon credits for cleaner stoves (Adler, 2010).

19.3.2.3   Other Developing Asian Countries

As mentioned above, Asia is a very diverse region. In this section we discuss the state of access in developing nations of Asia other than India and China. It is also one of the world's most dynamic regions, and as such, the region has made relatively rapid progress towards socioeconomic development. Even so, a very high level of the region's population lives in poverty, the highest in the world. As early as the 1970s, countries in the region had already adopted energy development programs that aimed to alleviate poverty. These energy development programs focused mostly on rural electrification because the majority of the poor people in these countries reside in rural areas. The view that rural electrification would be enough to generate rural development and lead to poverty alleviation was widely held when these programs were initiated.

Access to Electricity
About 14% of the total population without electricity globally is living in other developing nations in Asia. In the 1970s, nationwide electrification programs, which were generally based on grid-extension systems, were implemented with uneven success in several developing countries in the region, such as Bangladesh, Indonesia, Malaysia, Philippines, and Thailand. While some of these countries have been fairly successful in providing access to electricity to their rural populations, most of the rural population

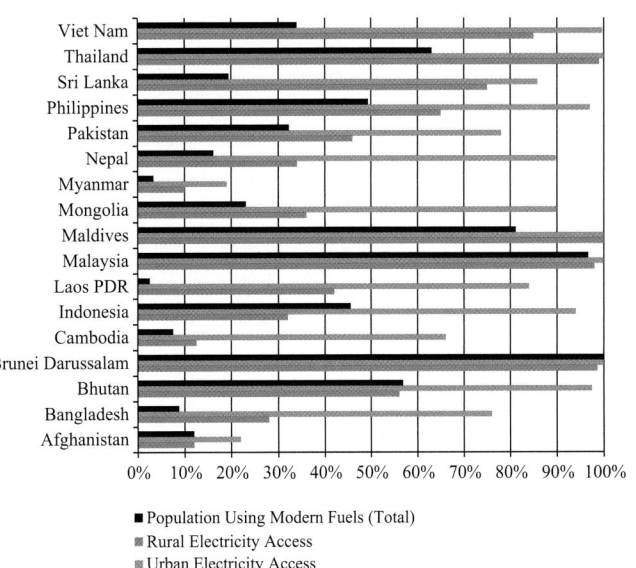

Figure 19.25 | Status of electrification and dependence on solid fuels in the Rest of Developing Asia. Source: UNDP and WHO, 2009.

in South Asia, other than in Sri Lanka and about 70 million Indonesians, are estimated to be unconnected to electricity (Figure 19.25).

More recently, Viet Nam and Laos have been extremely successful in improving the access to electricity for their populations. The World Bank and the Asian Development Bank have been particularly instrumental in improving access in these two countries.

Recent initiatives in the region have also focused on off-grid solutions to electrification and promoting renewable energy technologies. For instance, The Grameen Shakti Solar Home Systems Program in Bangladesh sells SHSs on credit. It is implemented in several districts of the country and initially targeted the installation of 8000 systems in three years. Since the systems are expensive, Grameen Shakti, part of the microfinancing institution Grameen Bank, has introduced soft-financing systems for the customer. Grameen Shakti encourages PV users to venture into income generating activities using their PV systems such as charging cellular phones; provision of light to post-harvest processing facilities, small enterprises, household-based livelihood activities and clinics so these can extend operations to early evening hours (thus increasing daily income); and power for radio/television repair shops (Shakti, 2010).

Access to Modern Fuels and Stoves
The problem of provision of modern energy for thermal energy needs remains a larger challenge for Asia. For most countries in the region, the majority of the rural population relies on unprocessed biomass or coal for most of its cooking and heating needs. Bringing about shifts in the energy-related behaviors of millions of households requires strong policies and large investments. Though complex, the problem of solid fuel use is not insoluble. Programs to reduce solid fuel use have been successful, most visibly the Chinese National Improved Stove Program mentioned above. Within other countries in the region, relatively little change in the dependence of rural populations on traditional solid fuels has taken place.

Box 19.10 | Enterprise Approach to Improved Cook Stove Programs in Asia

The Sri Lankan Anagi stove was promoted by encouraging artisans to enter business as mass producers of stoves that they market through normal market mechanisms. Village potters were trained to manufacture the models made of clay within strict standards. General distributors (wholesale buyers) visit the production centers to buy the stoves in bulk. Producers thus have their regular buyers. Stoves purchased are distributed to retail shops spread over a radius of about 200 km. Small producers living in isolated areas sell their products directly in the village. Today, about 300,000 stoves are annually produced by 120 rural potters trained by an NGO (IDEA) scattered in 14 districts of the country. The Anagi ICS has become one of the most widespread pottery items in village grocery stores.

In Cambodia, the Cambodia Fuelwood Saving Project provides technical and business development training to stove entrepreneurs to produce improved cook stoves. For each ICS, the producers invest an average of US$0.50 more than for a traditional stove. The Cambodia Fuelwood Saving Project has also created the Improved Cook Stove Producers and Distributors Association of Cambodia, which facilitates sectoral development, quality assurance, and the long-term sustainability of ICS dissemination.

In the Nepal program, capacity building inputs are provided by the government agency Alternative Energy Promotion Centre, which supported the participation of the informal private sector in cooperation with the Center for Rural Technology and other NGOs. Stove artisans (many of them women) have been trained and are now paid for their services constructing and maintaining stoves in their villages. The constructions of the improved stoves are exclusively carried out by these trained, predominantly women, masons. In 2005, there were about 1700 such trained technicians.

A UNDP survey of poor communities in the Asia Pacific region clearly indicates that rising oil prices left the poor with few choices other than to cut back on their consumption of oil products or, for uses that cannot be avoided, to bear the higher prices and look elsewhere in their household budgets to find the additional money (UNDP, 2007c). Since the urban poor rely more on oil products like kerosene and LPG, they are worse off than their rural counterparts, who are either biomass users or have the biomass option to fall back upon. The rural poor, however, are more vulnerable to higher lighting fuel prices, especially in unelectrified villages but also in electrified villages subject to frequent supply disruptions. Improved cook stove initiatives have been seen to be an important complement to improve cooking energy services for the poor and militate against rising oil prices.

19.3.3   Latin America and the Caribbean

The levels of access to electricity and clean cooking services in Latin America and the Caribbean (LAC), as shown in Section 1 of this chapter, are much higher than those in the other developing regions. Close to 90% of the region's population have access to electricity, compared to 62% for South Asia and around 28.5% for sub-Saharan Africa. (World Bank, 2010a). Access levels for clean cooking (and heating) services follow a similar pattern, with Latin America leading the way followed by Asia and then sub-Saharan Africa. Nevertheless, in Latin America and the Caribbean, approximately 200 million people currently live under the poverty line, and approximately 133 million live in urban areas and 67 million in rural areas. Seventy-two million are in absolute poverty (50% in urban areas and 50% in rural areas).

Table 19.7 shows that approximately 21.5 million people are estimated to have no access to electricity in the sample of the 14 most populated countries in the region (excluding Mexico). The largest numbers of people without access to electricity are concentrated in Peru and Brazil (over 7 million each); Bolivia, Guatemala, and Honduras (over two million each); and Nicaragua (over 1.5 million). Table 19.7 also highlights that in most countries, over 70% of the people without access to electric service are poor.

19.3.3.1   South America

The lowest energy use rates for modern energies are invariably in nations with the lowest Human Development Index (HDI) rankings. The correlation between access to modern energies, per capita energy demand, and HDI rankings are not exclusive to this region and similar correlations are observed worldwide (Fig 19.26).

In general, in all LAC countries the poor use less energy than the other social strata but they spend a higher proportion of their income on energy than the non-poor. Additionally, lower energy use by the poor is reflected in the differences in the level of access to equipment between income quintiles.

However, illegal connections, especially to electricity, could reflect very high consumption levels for services like cooking or heating. Energy access policies should be combined with efficient use of energy programs in closer coordination with public utilities.

Table 19.7 | Estimates of population without access to electricity in a sample of Latin American countries.

Country	Poor population without electric service (thousand)	Non poor population without electric service (thousand)	Total population without electric service (thousand)	% poor in total population without electric service	Country share % in total population without electric service
Argentina	57	91	148	38%	0.5%
Bolivia	2904	708	3611	80%	12.2%
Brazil	5123	2753	7875	65%	26.7%
Chile	62	168	231	27%	0.8%
Colombia	420	956	1376	31%	4.7%
Costa Rica	34	18	52	66%	0.2%
Ecuador	51	15	66	77%	0.2%
El Salvador	751	191	942	80%	3.2%
Guatemala	2569	687	3256	79%	11.0%
Honduras	2272	210	2482	92%	8.4%
Nicaragua	1377	219	1596	86%	5.4%
Paraguay	510	75	585	87%	2.0%
Peru	5264	1982	7245	73%	24.6%
Venezuela	16	19	35	46%	0.1%
Total estimate	21,410	8092	29,501	73%	100.0%

Source: ECLAC et al., 2010.

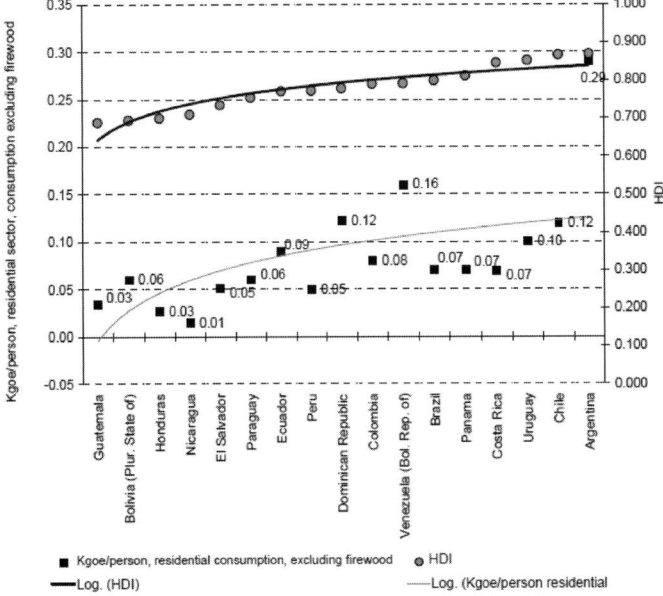

Figure 19.26 | Relationship between residential energy use and HDI in LAC nations. Source: ECLAC et al., 2010.

Access to Electricity

The process of rapid urbanization in South America and domestic migration from rural to urban areas has been accompanied by an increasing need for energy, but has also made it easier to provide electricity access. In effect, a high density concentration of potential consumers has meant lower costs to expand distribution systems. Greater access to employment

for the new inhabitants of the cities has been accompanied by higher payment capacity from an important share of urban consumers. The combination of these different issues has provided an important opportunity for cross subsidies to facilitate access for the poor population.

The information at the national-average level conceals large differences between urban and rural areas (Figure 19.27). For instance, countries with a relatively higher development level have problems of extreme poverty more serious than those of relatively lower development.

Even in countries with significant electricity access, poor households generally lack basic electrical equipment to benefit from energy services and have very limited access to communication and information technologies in comparison with the upper income groups within the same country. As shown in Figure 19.28 below, the percentage of upper income population (q5) with access to electric and communication equipment, relative to the percentage of lower income (q1) with access to the same type of equipment, can be 10 to 40 times as large in countries like Brazil, Uruguay, and Paraguay, which already have more than 90% electrical coverage. These data highlight the fundamental role of providing equipment in order to achieve effective access to energy services as coverage is extended to the poor sectors of the population.

Access to Fuels

The consumption of firewood in households drastically reduced from the 1970s to mid-1990s. Since then, firewood consumption has remained stable or even grown in some cases as shown in Figure 19.29. This

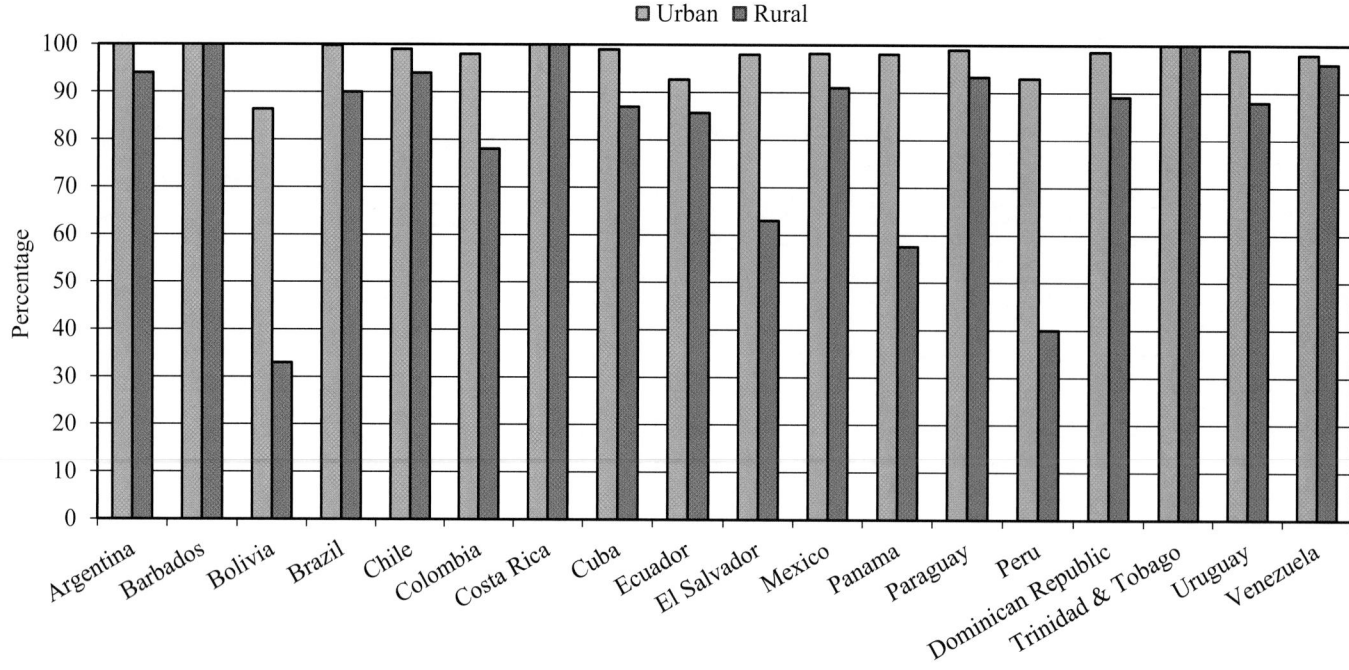

Figure 19.27 | Access to electricity in urban and rural areas of Latin America. Source: OLADE, 2008.

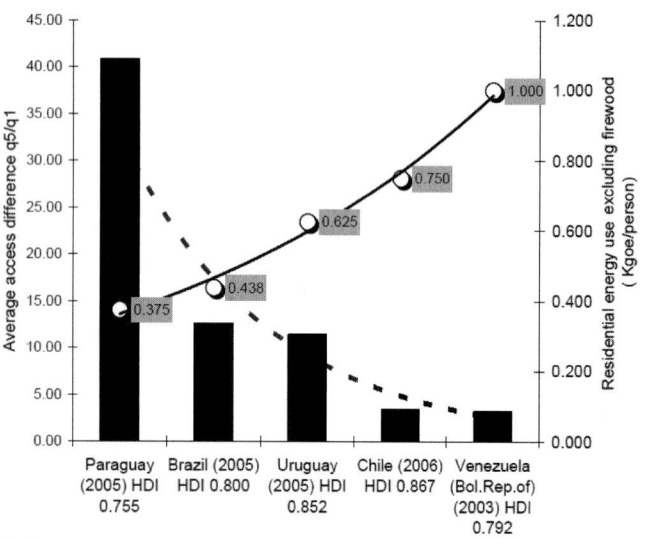

Figure 19.28 | Differences in access and use across income quintiles in select countries of Latin America. Source: ECLAC et al., 2010.

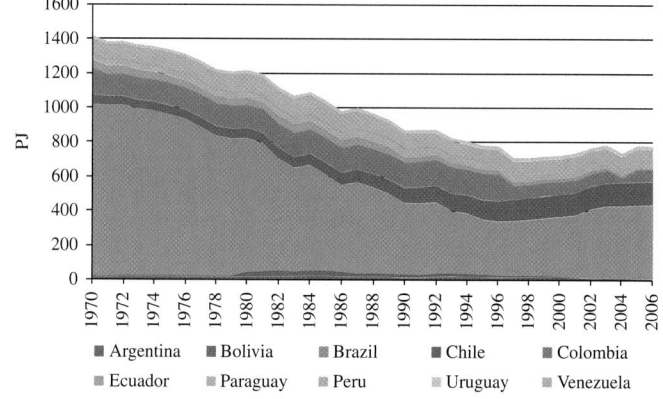

Figure 19.29 | Evolution of firewood consumption in the South American household sector. Source: based on estimates with data from OLADE, 2008.

phenomenon has been the result of rural-urban migration processes more than the introduction of end-use technologies for firewood savings. For instance, the diffusion of improved fuelwood stoves in South America is low compared to other developing regions like China and India.

In some countries, like Brazil, Chile, and Uruguay, the daily consumption of firewood per inhabitant has been growing systematically. Therefore,

despite the rural-urban migration processes and the resulting reduction of the rural population, the total consumption per inhabitant has increased. This situation represents many simultaneous realities, as in the case of Brazil where, in spite of a decreasing trend in the per capita consumption, the energy balances have shown a significant growth in firewood consumption since 1996. In the case of Chile, the consumption of firewood per rural inhabitant has also grown systematically. This may be the result of both the introduction of sustainable-use firewood programs and the impact of better living conditions of the rural population vis-à-vis the lack of commercial energy products that compete with firewood.

Use of LPG or natural gas is a clear indicator of increasing income levels and appears to be most preferred fuel among the rich, as shown for

Table 19.8 | Energy sources used for cooking according to income quintiles.

Argentina	Q1	Q2	Q3	Q4	Q5
Natural Gas	56.6%	57.8%	66.9%	76.4%	89.7%
LPG	40.2%	40.0%	32.3%	23.4%	10.3%
Brazil	Q1	Q2	Q3	Q4	Q5
LPG	43.5%	52.4%	58.9%	62.9%	72.2%
Firewood	47.5%	40.3%	37.6%	33.3%	26.2%
Charcoal	8.7%	7.0%	3.0%	3.3%	1.2%

Source: based on ECLAC et al., 2010 and OLADE, 2008.

Box 19.11 | The Brazilian Experience with LPG

In Brazil, 98% of households (including 93% of rural households) have access to LPG – a situation attributable to a government policy that has promoted the development of an LPG delivery infrastructure in all regions, including rural area, and subsidies to LPG users (Jannuzzi and Sanga, 2004; Lucon et al., 2004). Until the late 1990s, the rise in LPG use was accompanied by a sharp decline in residential wood consumption.

During the period 1973–2001, retail LPG prices were set at the same level in all regions and the average level of the subsidy amounted to 18% of the retail price. In May 2001, end-user prices were liberalized, as part of a process of deregulating the petroleum sector. At the same time, the government introduced an Auxílio-Gás ("gas assistance") program to enable qualifying low-income households to purchase LPG. Qualifying families were those with incomes less than half the minimum wage (an average daily per capita income of US$0.34/day in 2003). The total program cost in 2002 was about half that of the LPG price subsidization. This program now forms part of the Bolsa Família, by far the largest conditional cash transfer program in the developing world. Recent LPG price increases, however, appear to have led to a reversal of the trend toward lower residential biomass consumption.

Source: IEA, 2006

Brazil and Argentina in Table 19.8. In the case of Argentina, the availability of natural gas has resulted in the replacement of LPG. The lack of availability of natural gas in Brazil has meant that LPG is the preferred fuel for the rich.

19.3.3.2   Central America

The population growth of Central America has been accompanied by rural-urban migration processes, which are generally the consequence of poverty and scarce work opportunities in the rural areas. In 2005, the number of urban inhabitants reached 27.8 million people (57% of the total population) and rural inhabitants were at 21.2 million (43%).

At the beginning of the 1990s, 60% of the total population in Central America lived below the poverty line and 73.7% of the poor lived in rural areas. By 2001, the percentage of the population living below the poverty line had decreased to 50.8%, with 33.6% of the poor living in urban areas and 67.9% in rural areas. Despite this percentage reduction, the absolute number of poor people increased due to the high population growth of the region. The countries with less inequality in income distribution are Costa Rica and El Salvador, while Honduras and Nicaragua are the poorest and least developed countries of Central America.

Access to Electricity

The poorest populations in Central America generally have low access to electricity, as shown in Figure 19.30 for different income levels and various countries in the subregion. Honduras and Nicaragua show the highest levels of population without access to electric power, with 80% of the rural population in these two countries falling into the first quintile.

Even though significant progress may be noted in the level of electrification for every country in Central America, there are still approximately eight million people that do not have access to electricity, most of them in Nicaragua, Honduras, and Guatemala, and poor families in rural areas are generally not connected to the electricity grid (Serebrisky, 2007). Also, electricity accounts for close to, or more than, 10% of household expenditures for the poorest populations.

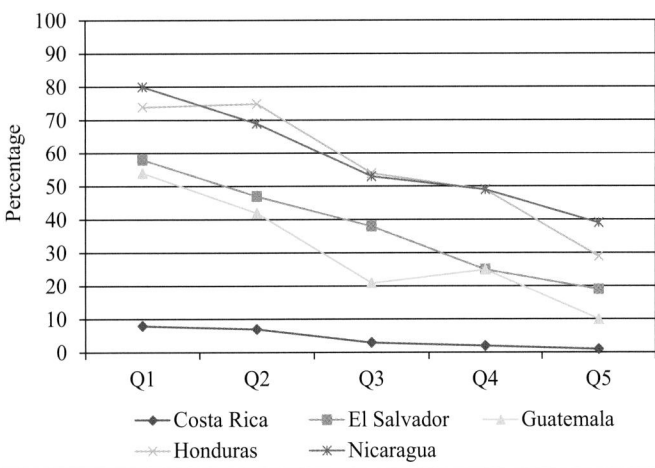

Figure 19.30 | Population without access to electricity in rural areas of Central America, by income level. CEPAL, 2011.

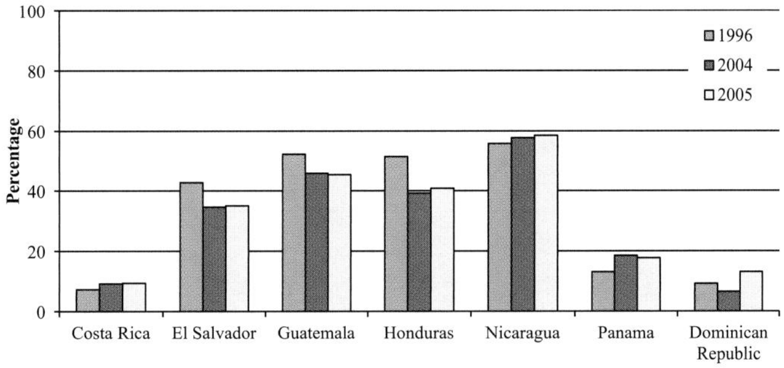

Figure 19.31 | Share of firewood in final energy demand (1996, 2004, and 2005). Source: based on OLADE, 2008.

Access to Fuels

Several LAC countries have implemented subsidies on LPG, considered the fuel of the poor. The following countries, in particular, have LPG subsidies: Argentina, Bolivia, Brazil, Colombia, Cuba, Ecuador, El Salvador, Haiti, Dominican Republic, and Venezuela (OLADE, 2008).

The poorest populations in Central America as a whole also largely depend on biomass (mainly in the form of firewood), which in 2006 represented 83% of the energy used to meet cooking and heating needs. Energy sources for cooking are mainly firewood, electric power, and LPG. Modern energy sources like LPG for cooking and heating have increased, but firewood is still the most widely used in terms of percentages.

There are significant differences in the types of energy for cooking across the subregion, as shown in Figure 19.31, with the following key features:

- electric power (high percentage both in urban and rural areas) – Costa Rica;

- LPG (most widely used) – Dominican Republic;

- LPG and firewood (both widely used) – El Salvador; and

- firewood (highly used) – Nicaragua, Honduras, and Guatemala.

Firewood thus plays an important role in final energy demand in Guatemala, Nicaragua, Honduras, and El Salvador, while the importance of firewood is relatively low in Costa Rica, Dominican Republic, and Panama. In general, LPG (or electricity) does not fully replace firewood in rural areas, but the different fuels are used complementarily; in some cases, firewood is used for cooking and LPG is used for water heating or precooked meal heating (Díaz, 2008).

For the whole subregion, firewood consumption accounted for 37% of the total energy supply and 83% of household energy supply in 2005. In terms of per capita consumption, Nicaragua and Guatemala are

Box 19.12 | The Justa Stove, Guatemala

A fuel-efficient stove called the Justa stove has been developed by Trees, Water and People (a charity working in Central America), the Aprovecho Research Center, the Honduran Association for Development, and Doña Justa Nuñez, a Honduran woman who helped design her namesake stove. The stove uses up to 70% less fuel than the open wood fires and because the design and materials used are simple, it can be made locally, using local materials, and adapted to meet local needs. Because it uses less fuel, the Justa stove decreases deforestation.

The Justa stove is relatively simple in design and can be made easily by local people in a day or less using locally available materials. The new owners of the stove have to contribute materials to the building of the stove. This gives them a personal investment in the stove, making them more likely to take good care of it.

Each stove saves 7.5 tonnes of CO_2 over a 7.5 year period. Each stove saves an average of 1 tonne of CO_2 emissions/year and 78 cubic meters of firewood over a 7.5 year period.

Source: Stoves Online, 2010

Table 19.9 | Electricity demand profile for the Caribbean region, selected countries.

Selected countries	Installed Power Capacity (MW)	Access to electricity (Total)	Access to electricity (Rural)	Per capita consumption (kWh)
Barbados	210	98%	n/d	1941
Cuba	5430	95%	87%	2321
Dominican Republic	5518	96%	89%	168
Grenada	32	82%	n/d	53
Guyana	308	82%	n/d	1220
Haiti	244	34%	n/d	341
Jamaica	854	95%	n/d	4769
Suriname	389	97%	n/d	1941
Trinidad & Tobago	1425	92%	n/d	2321

the countries with the highest firewood consumption; Honduras, El Salvador, and Panama are at an intermediate level, while Costa Rica and Dominican Republic show the lowest per capita consumption levels. The highest consumption is by lower income families in rural areas. Firewood consumption per capita in Costa Rica, Nicaragua, and Panama has tended to increase.

Sustainable energy for cooking has been pursued by means of different approaches: the adoption of new technologies like improved stoves, the introduction of modern fuels in rural areas to substitute for or complement the use of firewood, and access to electric power through energization programs in these rural areas.

19.3.3.3   The Caribbean

The Caribbean subregion faces huge challenges arising from modern globalization, declining competitiveness, trade liberalization and eroding preferences, the rising cost of imported fuel, the revolution in information technology, and very high vulnerability to natural disasters. Additionally, very high debt has placed seven Caribbean countries among the 10 most indebted countries in the world. The region is also heavily dependent on fossil fuel combustion, with petroleum products accounting for an estimated 93% of commercial energy use. The islands of the Caribbean are predominantly net energy importers, with the exception of Trinidad and Tobago.

Access to Electricity

With the exception of Cuba, Haiti, Dominican Republic, and Jamaica, the Caribbean countries are generally very small island states with populations of around one million or less and with a very important share of rural population. As shown in Table 19.9, yearly electricity consumption for these countries is well below 5000 kWh. However, data on access levels are difficult to find.

19.4   Conclusions for the Way Forward

Access to affordable modern forms of energy for populations currently without is a necessary albeit insufficient step toward poverty alleviation and the achievement of the MDGs. Providing universal access to electricity and modern fuels is not just a moral imperative. A growing body of knowledge shows that it also fosters significant social benefits and environmental improvements. It can also bring significant economic returns, particularly if policies and programs encourage the productive uses of energy to create new employment and income-generating activities through more conducive institutional mechanisms. A greater focus on scaling up pilot and demonstration projects to larger populations is also needed. The assessment of past policies and programs to improve access across different regions of the world carried out in this chapter point to the need for a paradigm shift in the approach to energy planning to meet the energy needs of the poor. An explicit focus is required on energy services. This should include a comprehensive demand-side analysis of the energy needs of poor people to support their livelihood functions, taking into account their particular constraints and opportunities. Current supply-side approaches that simply take as their starting point the provision of modern energy carriers such as electricity, petroleum, or gas, or equipment of a particular type (solar technology, improved cook stoves, biogas) are not sufficient to reap the full potential of social and economic improvements that follow from improved energy access.

A first step to achieving the paradigm shift needed in energy planning for the poor is establishing effective data collection systems based on accepted definitions and indicators of access to measure progress towards energy access targets or goals. The review included in this chapter points to significant data gaps regarding the existing energy access and use patterns in the poorest regions and for the poorest communities. Indicators that adequately assess the energy needs and describe the living conditions of such communities are required Consistent measurement frameworks and regular data collection systems on assessing

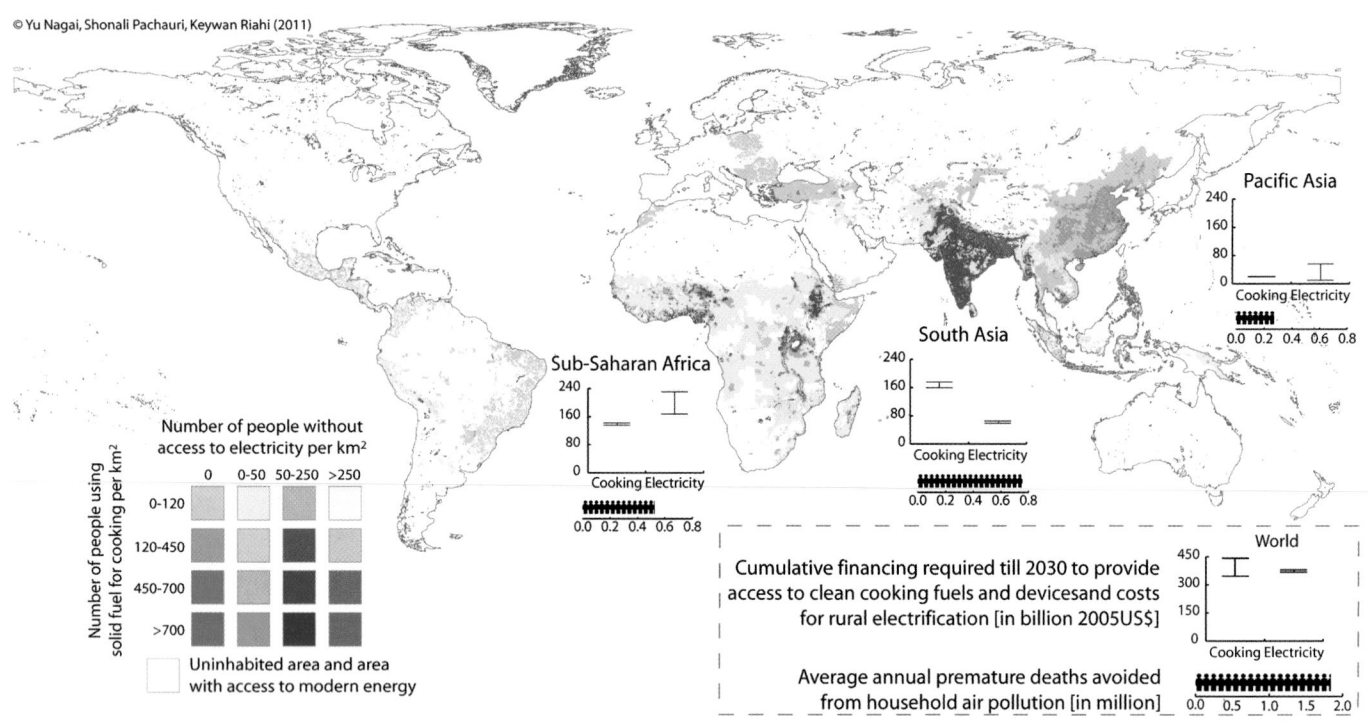

Figure 19.32 | Density of population lacking access to modern energy in 2005 and costs and health benefits of achieving a universal clean cooking and electrification goal by 2030. Colored areas show densities of people per km² without access to electricity and those that use solid fuels for cooking, e.g., dark blue and brown areas are where people do not have access to electricity and cook predominately by solid fuels. Cumulative investment requirements between 2010 and 2030 are shown for three GEA world regions and for the globe as a whole in billion 2005 US$. Also shown is the estimated population that would die prematurely from household air pollution if universal access is not achieved by 2030 (in million).

the energy situation of the poor are still lacking in many nations. In addition, the evaluation of many energy access programs and projects often fail to provide a comprehensive assessment of the impacts. An increase in the evidence base of the positive and significant impacts of such policies and projects can be instrumental in increasing efforts to enhance access activities globally.

Experience to date has resulted in a number of lessons that we must keep in mind when designing policies and programs to improve energy access in the future. Access programs and efforts are more likely to succeed if communities have an adequate understanding rather than act as passive recipients. Those designing and implementing programs need to understand that encouraging uses of energy for income generation, if built into the design of the access programs, is likely to improve sustainability. Programs that have built-in components for community training on operation and maintenance, as well as follow up with providers also have a higher likelihood of success. The extent of government commitment in creating an enabling environment is also paramount to the success of all policies. Improved access to capital that can help secure adequate financial resources, and market development that puts the customer at the center, are important for the successful scale-up of activities. Chapter 23 provides a more in-depth assessment of the full range of policies that are needed to achieve the ambitious energy access targets discussed in this chapter.

The GEA access scenarios[14] explore global strategies toward universal access to affordable and modern sources of energy by 2030. Specifically, the target calls for the provision of electricity and clean cooking fuels, including distribution of improved end-use devices to all those who currently lack access. Achieving the access goals creates multiple benefits for broader development goals, including increased productivity and decreased household air pollution and land degradation. The GEA scenarios indicate that such ambitious targets are feasible as long as financial support for dedicated access policies is provided. With respect to policies for energy access to clean cooking, the assessment suggests that fuel subsidies alone would be neither sufficient nor cost-effective in terms of achieving ambitious energy access objectives. Financial mechanisms, such as microcredit or capital grants, will need to complement subsidies to make critical end-use devices such as clean cooking stoves and connection costs affordable to the poor. The GEA scenarios estimate that the total costs for providing clean cooking services are between US$17–22 billion/year until 2030, with the difference of about US$4.7 billion/year an estimate of the capital cost associated with stove purchases that could be either met through public grants or microfinance options. While the subsidy component of this cost is substantial, it represents less than 5% of present day global fossil fuel subsidies.

---

14   Please refer to Chapter 17 for a detailed description of the GEA access scenario results.

Scenarios regarding future electrification across world regions vary tremendously, as the base level of electrification across regions is already significantly different and costs for providing grid access, which is dependent on population density, also vary enormously. The GEA electricity access scenarios, described in greater detail in Chapter 17, suggest that the investments required for additional electricity generation, operation and maintenance of plants, and rural grid expansion to reach the almost universal access target by 2030 would be between US$18.4–19 billion/year.

This suggests that the total cost for providing almost universal access to electricity and clean cooking by 2030 is between US$36–41 billion/year. There are two underlying reasons for the large range in the cost estimates derived from the GEA access scenarios. First, in the case of the estimates for providing clean cooking, the range in the estimates reflects whether the costs of new LPG stoves are assumed to be included or not. If these costs are assumed to be met through microfinance institutions, they are not included in the total cost estimate. However, if the cost of stoves is met from public grants, they are included in the estimate. The range in the cost estimates for electrification stems from differences in the modeling approaches used. A large range of estimates for the costs of providing access is also evident from a review of the literature. Typically, the global estimates range between US$30–40 billion/year (see Bazilian et al., 2010 for a recent review), though much higher estimates also exist specially for electrification. The wide range in global estimates parallel the wide range in the costs of providing access across different regions. The investment gap varies tremendously by region (see Figure 19.32). Chapter 23 provides a more in-depth and detailed discussion of the sources of funding that will need to be tapped to meet this investment gap and the kind of regional policies that will be needed for enhancing energy access. Spending on policies and measures to achieve access goals by 2030 will improve the welfare of those benefiting in several ways. Health impacts from improved household air quality have been quantified in Chapter 17. Access policies will result in averting between 0.6 and 1.8 million premature deaths, on average, every year until 2030 and saving about 24 million Disability Adjusted Life Years annually. Additional benefits that are likely to be substantial include time savings for women and children and the potential for improved livelihood opportunities.

Appendix A

Table A19.1. | Cookstove Project Details

Region: ASIA

ID	Stove Name	Stove Description	Stove Construction	Fuel Type	Typical Efficiency	Price/Cost	Number of stoves distributed	Project Date	Project Location	Financing	Reference(s)
1	Oorja	Metal stove	Manufactured	Pellet	NA	US$15 (INR 675) – 2010	65,000 families	2006	India	Full cost	Smith, 2007; Sundar, 2007
2	Envirofit G-3300	Metal (patented alloy) stove	Manufactured	Fuelwood	NA	US$10–40 – 2009	60,000	2008–2009	India	Full cost	Ritch, 2008; Envirofit, 2010
3	Envirofit B-1200	Ceramic liner stove									
5	Astra Chulha (2 and 3 pots)	Chulha models	trained masons, artisans	Fuelwood, dung cakes, agri waste	30–45% (lab test)	US$1.8–3.3 (two pot), US$4 (3 pots) – 2003	1,522,195	1984–2003 (2)	India	Max US$1.6 subsidy	Westhoff and Germann, 1995; Jagadish, 2004
6	Down-draft stoves (9 models)	Down-draft stoves, single mouth portable, single	Artisans, HH members with training	Fuel wood	NA	US$0.7 (w/o chimney), US$43 (w/ chimney), 2010	180,421	1994–2001	Bangladesh	Full cost	Khan, 2002; Hossain, 2003
8	Nada	Improved Chuhla stove	Potters, trained masons	Mustard stalk, Fuelwood, dung cake, agri residues	NA	3.2 US3 – 1993	NA	1986 – ongoing	India	78% subsidized	Westhoff and Germann, 1995
	Sahyog Chuhla				NA		NA	NA		65% subsidized	Westhoff and Germann, 1995
	Laxmi				22%	US$2.2 – 2.8 – 1993	105,000	1988–1993		US$1.6 paid in kind	Westhoff and Germann, 1995
	Sugham				28.20%						Westhoff and Germann, 1995
9	ARAVALI-U Chulha	Improved Chuhla Stove	Potters, HH members	Fuelwood, straw, dung cake, agri residues	25.30%	US$2.2 – 2.8 – 1993	20,000	1991–1993	India	50% subsidized or US$1.6 in-kind	Westhoff and Germann, 1995
	DOACHHI Chulha				20.20%	US$1.8 – 2.6 – 1993	500,000	1990–1993			Westhoff and Germann, 1995
	TNAU Chulha				26%	US$1.9 – 1993	NA	1993–1993		75% subsidized or US$2.4 in-kind	Westhoff and Germann, 1995
10	3-pot stove	Mud stove	Trained artisans	Straw and twigs	15–18%	US$1.8 – 1994	Several thousands	1988–1993 (ongoing)	Vietnam	Full cost	Westhoff and Germann, 1995
11	Silkalan	Mud stove, rice-husk stove	Artisans, clay potters, manufactured	Fuelwood	25.2% (Water boiling test)	US$7.4 – 1993	> 10,000	1987 – 1993	Philippines	Full cost	Westhoff and Germann, 1995

ID	Stove Name	Stove Description	Stove Construction	Fuel Type	Typical Efficiency	Price/Cost	Number of stoves distributed	Project Date	Project Location	Financing	Reference(s)
12	Improved Tamang Stove		Local blacksmith, welder, trained person	Fuelwood, agri residue	24.70%	US$2.0–3.0 – 1993	5,000	1986 – 1993	Nepal	40 – 90% subsidized	Westhoff and Germann, 1995
13	Tungku SAE	Clay stove	Potters	Fuelwood	12%, kitchen performance test	US$0.5 – 1993	50,000	1983 – 1999	Indonesia	Full cost	Westhoff and Germann, 1995
14	Coal Briquettes Stove	Clay stove	Workshops, crafters	Coal briquettes	40%	NA	300,000	1980 – 1994 (ongoing)	China	NA	Westhoff and Germann, 1995
15	Anagi	Clay stove	Local potters	Fuelwood, residues	17.40%	US$1.1 – 1.5 – 1994	500,000	1987 – 1997	China	NA	Westhoff and Germann, 1995
16	Priagni Chulha	Single-pot, portable stoves	Manufactured, small scale industry	Firewood, dung cakes, agri residues	26% HU, wood	US$3.6 – 6.0 – 1993	3,000,000	1983 – 1994	India	Full cost	Westhoff and Germann, 1995; Mäkelä, 2008
	Harsha Chulha				24.8% HU, wood	US$4.6 – 1993	100	NA			
	Tara Chulha				26% HU, wood	NA	NA	1991 – 1994			
17	Bekely-Darfur stove	Metal stove	Manufactured/ hand made	Fuelwood	NA	US$20.0 (15 in Sudan) – 2006	5,000	2006	Sudan	Subsidized (US$5 – 7.5)	Galitsky et al., 2006; Amrose et al., 2008
18	Kenya Ceramic Jiko (KCJ)	Charcoal burning metal ceramic stove	Artisan, manufactured	Charcoal	Thermal 30%	280 to 600 ksh	250,000 – 780,000 different reports	1982–1986, to 1995	Kenya	NA	Boiling Point, 1989; Westhoff and Germann, 1995; Kammen, 2001;
						NA	54,000	NA	Tanzania	NA	Ingwe, 2007; GTZ and HERA, 2009
						NA	52,000	NA	Uganda	NA	
						NA	45,000	NA	Ethiopia	NA	
						NA	30,000	NA	Rwanda	NA	
						NA	28,400	NA	Sudan	NA	
						NA	21,000	NA	Zimbabwe	NA	
						NA	20,500	NA	Burundi	NA	
						NA	15,400	NA	Somalia	NA	
20	Rocket Lorena	2 pot rest stove (earth, grass, water)	Artisan built	Fuelwood	NA	4.0 € – 2006	211,220	1999–2005	Uganda	Full cost	GTZ, 2006; Komuhangi, 2006; Habermehl, 2007
21	Shielded Fire	Rocket stove	Artisan built	Fuelwood	NA	US$3.0 – 9.0–2008	291,900	2004–2008	Uganda	NA	Ingwe, 2007; GTZ, 2008; GTZ and HERA, 2009
	One-pot rocket mud stove					US$1.5 – 2.0 – 2008	35,000	2006–208	Kenya	NA	

ASIA (cont.)

AFRICA

Continued next Page →

ID	Stove Name	Stove Description	Stove Construction	Fuel Type	Typical Efficiency	Price/Cost	Number of stoves distributed	Project Date	Project Location	Financing	Reference(s)
22	Albarka/Troi Pierres Ameliore	Improved 3 stove	Trained masons, artisans, HH	Fuelwood	26–28% lab test	US$0.7 – 1.4 – 2003	12,642	1989 – 1992	Niger	Full Cost	Westhoff and Germann, 1995
							120,000	1990 – 1993	Rwanda		
							150,000	1991–1992	Burkina Faso		
23	Ambo stove (Ambo Metat)	Mud stove (3 sizes)	Mud technicians, trained masons	Fuelwood	20 – 25 %	US$4.0 – 6.0 – 1993	150	150	Ethiopia		Westhoff and Germann, 1995
24	Maendeleo	Mud stove (2 sizes)	Artisans, trained HH members	Fuelwood	24 – 30% thermal	US$0.6 – 0.9 – 1993	200,000	1985 – 1994	Kenya	Full cost	Westhoff and Germann, 1995
	Maendeleo portable	Metal Stove	Workshops, crafters							Free Distribution	
	Maendeleo liner (kubi mbili)	Metal cladded firewood stove	Manufactured	Fuelwood	NA	KES250-300, US$3.50–4.20 -	290,000	2006 – ongoing	Kenya	Fullcost	Ingwe, 2007
25	Chingwa	Multi-pot stove	Local trained artisans	Fuelwood	33% thermal	US$15.4 (Z$100) – 1993	15,000	1988 – 2001	Zimbabwe	50% subsidized	Westhoff and Germann, 1995; Maya et al., 2001
26	Ouaga Ceramique	Cermic stove, clay stove	Potters	Fuelwood	30%	1,000 (FCFA) 1994	5,700	1983–1992 (ongoing)	Burkina Faso	NA *subsidized*	Westhoff and Germann, 1995
27	Tso-tso stove	Metal stove	Manufactured, sheet metal workshops	Smallpieces of wood, briquettes	23% (PHU)	US$25.4 – 1993	40,000 (18,000 refugee camps)	1986 – 1994 (ongoing)	Zimbabwe	NA, 100% subsidized	Westhoff and Germann, 1995
28	Katinbe Njamdi	Metal stove – in various sizes	Manufactured, sheet metal worker	Fuelwood	20 – 30%	1,300–5,500 FCFA – 1993	NA	1989 – 1993	Cameroon	NA (full cost0	Westhoff and Germann, 1995
29	Metal stove	metal stove	Manufactured, trained tinsmiths	Fuelwood, peat	28%	US$4.2 – 1993	25,000	1993	Rwanda	Refugee camps 100% subsidized, fami?	Westhoff and Germann, 1995
30	Teliman	Metal Stove	Manufactured, trained smiths	Fuelwood	26–30%	600–6000 FCFA – 1994	40,000	1988–1994	Mali	Full cost	Westhoff and Germann, 1995
	Mai Sauki					550–950 FCFA – 1994	34,800	1989–1992	Niger		
	NA					NA	36,200	1984–1992	Burkina Faso		
31	Modified CETA stove	Mud rocket stove	trained artisans	Fuelwood, Briquette, Agri-residue	22–26%	US$30.0 – 1994	1,500	1994	Nicaragua	85% Subsidized	Westhoff and Germann, 1995; Álvarez et al., 2004
32	Finlandia	Rocket stove	Household members	Fuelwood, agri-residue	30%	US$15 (SVC130) – 1994	500	1992 – 1997	El Salvador	60% subsidized	Westhoff and Germann, 1995; OAS, 2000; Álvarez et al., 2004

AFRICA (cont.)

LATIN AND SOUTH AMERICA

LATIN AND SOUTH AMERICA (cont.)

ID	Stove Name	Stove Description	Stove Construction	Fuel Type	Typical Efficiency	Price/Cost	Number of stoves distributed	Project Date	Project Location	Financing	Reference(s)
33	Justa Rocket stove	Brick built rocket stove	Artisan built	Fuelwood	NA	US$60.0 – 2005	4,000	1998 – 2005	Honduras	70% Subsidized	Wheldon, 2005
					NA	NA	13,000	since 1999	Latin America	NA	
34	Tezulutlan improved stove	Metal-plancha stove	Artisan built/ manufactured parts	Fuelwood	NA	US$64.0 – 2002	4,129	1998–2001	Guatemala	60% subsidized	Álvarez et al., 2004
36	INTERVIDA-type stove	Metal-plancha stove	Artisan built/ manufactured parts	Fuelwood	NA	US$88.0 – 2002	9,000	1998–2001	Guatemala	70% Subsidized	Álvarez et al., 2004
37	SIF improved stove	Metal-plancha stove	Artisan built/ manufactured parts	Fuelwood	NA	US$149.0 – 2002	90,000	1996–2001	Guatemala	90% Subsidized	Álvarez et al., 2004
38	Improved stove	Based on Indian Damrhu Chulha	Artisans trained by ICAITI	Fuelwood, charcoal, briquettes, corncobs	22%	US$7.0 – 1993	1,500	1992–1993	El Salvador	NA	Westhoff and Germann, 1995
39	FOBLOCO		HH members trained	Fuelwood	0.85 KG/ Mcal	US$15.0 – 20.0 – 1994	900	1988–1992, 1996	Paraguay	NA	Westhoff and Germann, 1995
	SEAG complete stove				NA	US$79.0 – 1994					
40	Chefina		Technicians, trained masons	Fuelwood	NA	US$52.6 – 70.2 – 1994	800	1986 – 1994 (ongoing)	Guatemala	NA	Westhoff and Germann, 1995
41	Improved stove	Stove as Asphan	Built by Ngo	Fuelwood	NA	US$74.3 – 1994	600	1991–1994 (Ongoing)	El Salvador	100% subsidized	Westhoff and Germann, 1995
42	Improved Lorena	With chimney, one piece construction	trained buildres	Fuelwood	NA	US$36.0 – 1994	1,200	1976–1994 (ongoing)	Dominican Republic	55% subsidized	Westhoff and Germann, 1995
43	Improved household stove		trained HH members	Fuelwood, carob, sapote (marmalade?)	NA	US$32.0 – 1994	50	1993 – 1996	Peru	Special group subsidy	Westhoff and Germann, 1995
44	Lorena improved stove	Various models	Trained Artisans, bricklayers, HH members	Fuelwood	NA	US$45.7 – 1994	300	1991–1994	El Salvador	40% subsidized	Westhoff and Germann, 1995
						US$21.3 – 25.5 – 1994	176	199,019,911,993	Bolivia	US$17.2 in-kind	
						US$20.0 – 1994	NA (Schools & restaurant)	NA	Peru	NA	
						US$20.0 – 1994	NA	NA	Bolivia	NA	

Continued next Page →

1451

	ID	Stove Name	Stove Description	Stove Construction	Fuel Type	Typical Efficiency	Price/Cost	Number of stoves distributed	Project Date	Project Location	Financing	Reference(s)
LATIN AND SOUTH AMERICA (cont.)	45	Fogon Cilindrico Mejorado	Improved Cylindrical Stove	Artisans, project	Fuelwood	NA	US$70.0 – 1994	2,500	1992–1994 (ongoing)	Ecuador	71.5% subsidized	Westhoff and Germann, 1995
	46	Villanueva		Builders	Fuelwood	NA	US$100.0 – 1994	500	1991–1994 (Ongoing)	Venezuela	NA	Westhoff and Germann, 1995
	47	Ecotec Rural		Trained potters,	Fuelwood, agri-residue	NA	US$22.0 – 1994	3,500	1992–1993	Guatemala	NA (full cost)	Westhoff and Germann, 1995
	48	Cocina Mejorada	Improved stove	Artisans, potters	Fuelwood, dung cakes, agri waste		US$1.5 – 1994	200	1992 – 1993	Bolivia	100% subsidized	Westhoff and Germann, 1995
	49	Onil Stove	NA	Manufactured, locally installed	Fuelwood	NA	US$150–2010	5,000	2003–2010	Guatemala	Help program 100% subsidized	Helps International, 2010; Onilstove, 2010
INTERNATIONAL	50	Inkawasi	Based on rocket elbow stove desings	Manufactured, built by artisans	Fuelwood, dung, shrubs, etc	28.19%	36 eur – 2008	14,000	2005 – 2007	Peru	NA	Klingshirn, 2006; GTZ and HERA, 2007
	51	Stovetec	Portable rocket stove	Manufactured	Fuelwood, charcoal, briquettes, corncobs	NA	US$3.0 – 12.0 FOB US – 2010	200	1999	South Africa	Full Cost	MacCarty et al, 2008; StoveTec, 2010
								500	NA	South Africa		
								70,000	2008–2010	International		

References to Appendix

Álvarez, D., C. Palma and M. Tay, 2004: *Evaluation of Improved Stove Programs in Guatelmala: Final Report of Project Case Studies*. ESMAP Technical Paper 060, Energy Sector Management Assistance Programme (ESMAP), United Nations Development Programme (UNDP), The World Bank Group, New York.

Amrose, S., G. T. Kisch, C. Kirubi, J. Woo and A. Gadgil, 2008: *Development and Testing of the Berkeley Darfur Stove*. LBNL-116E, Ernest Orlando Lawrence Berkeley National Laboratory, University of California, Berkeley, CA. Envirofit, 2010: Products: Cookstoves. Envirofit, Date Accessed: 2010, http://www.envirofit.org/products.html.

Galitsky, C., A. Gadgil, M. Jacobs and Y.-M. Lee, 2006: *Fuel Efficient Stoves for Darfur Camps of Internally Displaced Persons. Report of Field Trip to North and South Darfur, Nov. 16 – Dec. 17, 2005*. LBNL – 59540, Ernest Orlando Lawrence Berkeley National Laboratory, University of California, Berkeley, CA.

GTZ, 2006: *Scaling up Household Energy – Uganda*. Factsheet, Division Environment and Infrastructure, German Agency for Technical Cooperation (GTZ), Eschborn.

GTZ and HERA, 2007: *Inkawasi Stove*. Factsheet, German Agency for Technical Cooperation (GTZ) and Household Energy Programme (HERA), Eschborn. http://www.hedon.info/docs/CEC_inkawasi-stove_peru_factsheet-2008.pdf.

GTZ, 2008: *One-Pot Rocket Mud Stove*. Energy Advisory Project, German Agency for Technical Cooperation (GTZ) and the Ugandan Ministry of Energy and Mineral Development, Eschborn.

GTZ and HERA, 2009: *Impacts of Stove Project in Kenya*. German Agency for Technical Cooperation (GTZ) and Household Energy Programme (HERA), Eschborn.

Habermehl, H., 2007: *Economic Evaluation of the Improved Household Cooking Stove Dissemination Programme in Uganda*. German Agency for Technical Cooperation (GTZ) and the Household Energy Programme (HERA), Eschborn. Helps International, 2010: The ONIL Stove. Date Accessed: 2011, 1 March, http://www.helpsinternational.com/programs/stove.php.

Hossain, M. M. G., 2003: Improved Cookstove and Biogas Programmes in Bangladesh. *Energy for Sustainable Development*, 7(2): 97–100.

Ingwe, A., 2007: Rocket Mud Stoves in Kenya. *Boiling Point*, (53): 6–8.

Jagadish, K. S., 2004: The Development and Dissemination of Efficient domestic Cook Stoves and Other Devices in Karnataka. *Current Science*, 87(7): 926–931.

Kammen, D. M., 2001: Research, Development and Commercialization of the Kenya Ceramic Jiko and other Improved Biomass Stoves in Africa. Solutions Site Case Study, http://www.solutions-site.org/cat2_sol60.htm.

Khan, A. H. M. R., 2002: Development & Dissemination of Improved Stoves in Bangladesh. In *National Workshop on Promotion of Renewable Energy, Energy Efficiency and Greenhouse Gas Abatement PREGA)*. Dhaka.

Klingshirn, A., 2006: The Inkawasi Stove: A Success Story in the Peruvian Andes. *Boiling Point*, 52: 20–21.

Komuhangi, R., 2006: Mass Dissemination of Rocket Lorena Stoves in Uganda. *Boiling Point*, (52).

MacCarty, N., D. Still, D. Ogle and T. Drouin, 2008: *Assessing Cook Stove Performance: Field and Lab Studies of Three Rocket Stoves Comparing the Open Fire and Traditional Stoves in Tamil Nadu, India on Measures of Time to Cook, Fuel Use, Total Emissions, and Indoor Air Pollution*. Aprovecho Research Center, Cottage Grove, OR.

Mäkelä, S., 2008: *Firewood-Saving Stoves: A Review of Stove Models Based on the Documenation on the Internet*. Liana, Finland. http://www.elisanet.fi/esoini/Liana_docs/Firewood-saving_stoves_review_by_Liana.pdf.

Maya, R. S., A. P. Mhlanga, N. Nziramasanga and M. Mutyasira, 2001: *Implemenation of Renewable Energy Technologies – Opportunities and Barriers*. Case Study, Southern Centre for Energy and Environment Zimbabwe, United Nations Environment Programme (UNEP) Collaborating Centre on Energy and Environment, Roskilde. Onilstove, 2010: The ONIL Stove. Date Accessed: 2011, 1 March, http://www.onilstove.com/.

Ritch, E., 2008: Envirofit Ramps Clean-Cooking Line for India. Cleantech Group, Date Accessed: 2011, 1 March, http://cleantech.com/news/3779/envirofit-clean-cookstoves-india.

Smith, A., 2007: Growth and Innovation Lights Up India. *The BP Magazine*, (4): 5–11. StoveTec, 2010: StoveTec. Date Accessed: 2011, 1 March, http://www.stovetec.net/us/index.php.

The Hindu, June 7, 2007: Sundar, S., "A Stove and a Smokeless Kitchen." http://hindu.com/2007/06/05/stories/2007060505830500.htm.

Westhoff, B. and D. Germann, 1995: *Stove Images: A Documentation of Improved and Traditional Stoves in Africa, Asia and Latin America*. Sozietät für Entwicklungsplanung GmbH (SfE), Directorate General for Development, Commission of the European Communities, Frankfurt am Main.

Wheldon, A., 2005: *Fuel-Efficient Stoves for Rural and Urban Households*. The Ashden Awards for Sustainable Energy, Trees, Water and People (TWP), Fort Collins, CO.

References

Adler, T., 2010: Better Burning, Better Breathing: Improving Health with Cleaner Cook Stoves. *Environmental Health Perspectives*, **118**(3).

AFREPREN/FWD, 2009: *Renewable Energy for Africa – Potential, Markets and Strategies*. Renewable Energy Policy Network for the 21st Century (REN21), Deutsche Gesellschaft für Technische Zusammenarbeit (GTZ), German Federal Ministry for Economic Cooperation and Development (BMZ), AFREPREN/'FWD, Nairobi.

AGECC, 2010: *Energy for a Sustainable Future: Summary Report and Recommendations*. The UN Secretary-General's Advisory Group on Energy and Climate Change (AGECC), New York.

Ahluwalia, M. S., 1978: Rural Poverty and Agricultural Performance in India. *Journal of Development Studies*, **14**(3):298–323.

Álvarez, D., C. Palma and M. Tay, 2004: *Evaluation of Improved Stove Programs in Guatelmala: Final Report of Project Case Studies*. ESMAP Technical Paper 060, Energy Sector Management Assistance Programme (ESMAP), United Nations Development Programme (UNDP), The World Bank Group, New York.

Amrose, S., G. T. Kisch, C. Kirubi, J. Woo and A. Gadgil, 2008: *Development and Testing of the Berkeley Darfur Stove*. LBNL-116E, Ernest Orlando Lawrence Berkeley National Laboratory, University of California, Berkeley, CA.

Annecke, W., 2008: Monitoring and Evaluation of Energy for Development: The good, the Bad and the Questionable in M&E Practice. *Energy Policy*, **36**:2839–2845.

Armendáriz, B. and J. Morduch, 2005: *The Economics of Microfinance*. MIT Press, Cambridge, MA.

Asaduzzaman, M., D. F. Barnes and S. R. Khandker, 2010: *Restoring Balance: Bangladesh's Rural Energy Realities*. Energy Sector Management Assistance Program, The World Bank, Washinton, DC.

Banerjee, S., A. Diallo, V. Foster and Q. Wodon, 2009: *Trends in Household Coverage of Modern Infrastructure Services in Africa*. Policy Research Working Paper 4880, The World Bank, Washington, DC.

Barnes, D. F. and W. M. Floor, 1996: Rural Energy in Developing Countries: A Challenge for Economic Development. *Annual Review of Energy and the Environment*, **21**(1):497–530.

Barnes, D. F., R. Van Der Plus and W. Floor, 1997: Tackling the Rural Energy Problem in Developing Countries. *Finance and Development* **34**(2):11–15.

Barnes, D. F. and J. Halpern, 2000: *Subsidies and Sustainable Rural Energy Services: Can we Create Incentives without Distorting Markets?* Technical Paper 010, Energy Sector Management Assistance Programme, The World Bank, Washinton, DC.

Barnes, D. F., H. M. Peskin and K. Fitzgerald, 2002: *The Benefits of Rural Electrification in India: Implications fo Education, Household Lighting, and Irrigation*. Draft paper preprared for South Asia Energy and Infrastructure, The World Bank, Washington, DC.

Barnes, D. F., 2007: Meeting the Challenge of Rural Electrification. In *The Challenge of Rural Electrification: Strategies for Developing Countries*, RFF Press, Washington, DC pp.313–327.

Bates, L., S. Hunt, S. Khennas and N. Sastrawinata, 2009: *Expanding Energy Access in Developing Countries: The Role of Mechanical Power*. Practical Action Publishing and the UN Development Programme, New York.

Bazilian, M., P. Nussbaumer, E. Haites, M. Levi, M. Howells and K. K. Yumkella, 2010: Understanding the Scale of Investment for Universal Energy Access. *Geopolitics of Energy*, **32**(10–11):19–40.

Bekker, B., A. Eberhard, T. Gaunt and A. Marquard, 2008: South Africa's rapid electrification programme: Policy, institutional, planning, financing and technical innovations. *Energy Policy*, **36**:3115–3127.

Bice, K., A. Eil, B. Habib, P. Heijmans, R. Kopp, J. Nogues, F. Norcross, M. Sweitzer-Hamilton and A. Whitworth, 2008: *Black Carbon: A Review and Policy Recommendations*. Woodrow Wilson School of Public and International Affairs, Princeton, NJ.

Bird, R., 1994: *Decentralizing Infrastructure: For Good or for Ill?* Policy Research Working Paper 1258. Background Paper for the 1994 World Development Report, The World Bank, Washington, DC.

Biswas, W. K., M. Diesendorf and P. Bryce, 2004: Can photovoltaic technologies help attain sustainable rural development in Bangladesh? *Energy Policy*, **32**(10):1199–1207.

Boiling Point, 1989: Stove Profiles: Kenya Ceramic Jiko. Extracts from "Improved Wood, Waste and Charcoal Burning Stoves". *Boiling Point*, (17).

Brew-Hammond, A. and F. Kemausuor, (eds.), 2007: *Energy Crisis in Ghana: Drought, Technology or Policy?* University Press, Kumasi.

Byatt, I. C. R., 1979: *The British Electrical Industry 1875–1914*. Clarendon Press, Oxford.

Cabraal, A., M. Cosgrove-Davies and L. Schaeffer, 1996: Best practices for photovoltaic household electrification programs: lessons from experiences in selected countries. *World Bank Technical Paper*, **324**.

Cabraal, A. R., M. Cosgrove-Davies and L. Schaeffer, 1998: Accelerating Sustainable Photovoltaic Market Development. *Progress in Photovoltaics: Research and Applications*, **6**(5):297–306.

Cabraal, A. R., D. F. Barnes and S. G. Agarwal, 2005: Productive Uses of Energy for Rural Development. *Annual Review of Environment and Resources*, **30**(1):117–144.

Cecelski, E., J. Dunkerly, A. Ounali and M. Aissa, 2007: Electricity and Multisector Development in Rural Tunisia. In *The Challenge of Rural Electrification*. D. F. Barnes, (ed.), RFF Press, Washington, DC pp.163–197.

CEMAC, 2006: *Action Plan for the Promotion of Access to Energy in the CEMAC Region*. Summary and Final Report, Communauté Economique et Monétaire de L'Afrique Centrale (CEMAC), Bangui. CEPAL, 2011, Available at www.eclac.org/ (accessed 2 March, 2011).

Chaurey, A., M. Ranganathan and P. Mohanty, 2004: Electricity access for geographically disadvantaged rural communities – technology and policy insights. *Energy Policy*, **32**(15):1693–1705.

Chen, S. and M. Ravallion, 2007: Absolute Poverty Measures for the Developing World, 1981–2004. *Proceedings of the National Academy of Sciences*, **104**(43):16757–16762.

Clancy, J. S., 2000: How is Gender Relevant to Sustainable Energy Policies. In *Sustainable Energy Strategies: Materials for Decision Makers*, UN Development Programme (UNDP), New York.

Clancy, J. S. and S. Dutta, 2005: Women and Productive Uses of Energy: Some Light on a Shadowy Area. *UNDP Meeting on Productive uses of Renewable Energy*, 9–11 May, Bangkok.

Collins, M. J. B. W. J., 2009: Did Improvements in Household Technology Cause the Baby Boom? Evidence from Electrification, Appliance Diffusion, and the Amish. NBER Working Papers 14641.

Derdevet, M. and M. Caubet, 2007: Electricity for All: Frameworks for Implementation. *Special Report on Selected Side Events at the Fifteenth Session* of the Commission on Sustainable Development (CSD-15), International Institute for Sustainable Development and the UN Development Programme, New York.

Diallo, A., 2009: Stratégie du WAPP en Matière de Sécuriteé d'Approvisionnement en Énergie des Etats Membres de la CEDEAO. *Ad Hoc Experts Meeting on Energy Security in West Africa*, Lome.

Díaz, E., 2008: *Impact of Reducing Indoor Air Pollution on Women's Health*. RESPIRE Guatemale – Randomised Exposure Study of Pollution Indoors and Respiratory Effecs, Department of Public Health and Primary Health Care, University of Bergen, Bergen.

Dinkelman, T., 2010: *The Effects of Rural Electrification on Employment: New Evidence from South Africa*. Working Papers 1255, Woodrow Wilson School of Public and International Affairs, Princeton University, Princeton, NJ.

Doll, C. N. H. and S. Pachauri, 2010: Estimating Rural Populations without Access to Electricity in Developing Countries through Night-time Light Satellite Imagery. *Energy Policy*, 38(10):5661–5670.

Drück, H., S. Fischer, A. Al Taher, T. Núñez, J. Kochikowski and M. Rommel, 2007: *Potential Analysis for a New Generation of Solar Thermal Systems in the Southern Mediterranean Countries*. SOLATERM Project Report, European Commission 6th Framework Programme for Research and Technological Development, Brussels.

Eberhard, A., V. Foster, C. Briceño-Garmendia, F. Ouedraogo, D. Camos and M. Shkaratan, 2008: *Underpowered: The State of the Power Sector in Sub-Saharan Africa*. Background Paper 6, Africa Infrastructure Country Diagnostic.

Eckholm, E. P., 1975: *The Other Energy Crisis: Firewood*. Worldwatch Institute, Washington, DC.

ECLAC, Club de Madrid and UNDP, 2010: *Contribution of Energy Services to the Millennium Development Goals and to poverty alleviation in Latin America and the Caribbean*. Project Document, R. Kozulj, (ed.) Economic Comission for Latin America and the Caribbean (ECLAC), Club de Madrid and the United Nationsl Development Programme (UNDP), Santiago.

ECOWAS, 2006: *White Paper for a Regional Policy Geared towards increasing access to energy services for rural and peri-urban populations in order to achieve the Millennium Development Goals*. Economic Community of West-African States (ECOWAS), Abuja.

EECG, 2004: *Energy Use, Energy Supply, Sector Reform and the Poor in Botswana*. Reports 1 & 2, Energy, Environment, Computer and Geophysical Applications (EECG), Gaborone.

Ekholm, T., V. Krey, S. Pachauri and K. Riahi, 2010: Determinants of Household Energy Consumption in India. *Energy Policy*, 38(10):5696–5707.

ENERGIA, 2010: *International Network on Gender and Sustainable Energy*. Available at www.energia.org/ (accessed 30 December, 2010).

Energy Commission Ghana, 2008: *Ghana Energy Statistics 2007*. Energy Commission, Ministry of Energy, Accra.

Envirofit, 2010: *Products: Cookstoves*. Available at www.envirofit.org/products.html (accessed 1 April, 2010).

FAOSTAT, 2010: *ResourceSTAT*. Available at faostat.fao.org/default.aspx (accessed 30 December, 2010).

Foster, V. and C. Briceño-Garmendia, (eds.), 2010: *Africa's Infrastructure: A Time for Transformation*. African Development Bank, Tunis.

Galitsky, C., A. Gadgil, M. Jacobs and Y.-M. Lee, 2006: *Fuel Efficient Stoves for Darfur Camps of Internally Displaced Persons. Report of Field Trip to North and South Darfur, Nov. 16 – Dec. 17, 2005*. LBNL – 59540, Ernest Orlando Lawrence Berkeley National Laboratory, University of California, Berkeley, CA.

GEF and UNDP, 2003: *"Sol de Vida:" Improving Women's Lives Through Solar Cooking, Costa Rica*. The GEF Small Grants Programme, Global Environment Facility (GEF) and United Nations Development Program (UNDP), New York.

Gibson, J. and S. Olivia, 2010: The Effect of Infrastructure Access and Quality on Non-Farm Enterprises in Rural Indonesia. *World Development*, 38(5):717–726.

GNESD, 2007: Workshop Summary Report. *African Regional Workshop: Renewable Energy and Poverty Reduction in Africa: Best Practices for Productive Use and Job Creation*, 21–23 March, Global Network on Energy for Sustainable Development (GNESD), Dakar.

Goldemberg, J., T. B. Johansson, A. K. N. Reddy and R. H. Williams, 1985: Basic Needs and Much More with One Kilowatt per Capita. *Ambio*, 14(4/5):190–200.

Goldemberg, J., E. L. La Rovere and S. T. Coelho, 2004: Expanding Access to Electricity in Brazil. *Energy for Sustainable Development*, VIII(4):86–94.

Green, D., 2004: Thailand's solar white elephants: an analysis of 15 years of solar battery charging programmes in northern Thailand. *Energy Policy*, 32(6):747–760.

Grieshop, A. P., C. C. O. Reynolds, M. Kandlikar and H. Dowlatabadi, 2009: A black-carbon mitigation wedge. *Nature Geoscience*, 2(8):533–534.

GTZ, 2006: *Scaling up Household Energy – Uganda*. Factsheet, Division Environment and Infrastructure, German Agency for Technical Cooperation (GTZ), Eschborn.

GTZ and HERA, 2007: *Inkawasi Stove*. Factsheet, German Agency for Technical Cooperation (GTZ) and Household Energy Programme (HERA), Eschborn.

GTZ, 2008: *One-Pot Rocket Mud Stove*. Energy Advisory Project, German Agency for Technical Cooperation (GTZ) and the Ugandan Ministry of Energy and Mineral Development, Eschborn.

GTZ and HERA, 2009: *Impacts of Stove Project in Kenya*. German Agency for Technical Cooperation (GTZ) and Household Energy Programme (HERA), Eschborn.

GVEP, 2009: *Cookstoves and Markets: Experiences, Successes and Opportunities*. K. Rai and J. McDonald, (eds.), Global Village Energy Partnership (GVEP), London.

Habermehl, H., 2007: *Economic Evaluation of the Improved Household Cooking Stove Dissemination Programme in Uganda*. German Agency for Technical Cooperation (GTZ) and the Household Energy Programme (HERA), Eschborn.

Habtetsion, S. and Z. Tsighe, 2002: The Energy Sector in Eritrea: Institutional and Policy Options for Improving Rural Energy Services. *Energy Policy*, 30(11–12):1107–1118.

Hagan, E. B., 2006: *Scoping Study on Biomass in Africa: West Africa Sub-Region*. Final Report to the African Development Bank, African Development Bank, Tunis.

Hansen, H. and J. Rand, 2006: On the Causal Links Between FDI and Growth in Developing Countries. *World Economy*, 29(1):21–41.

Heller, T. C., H. I. Tijong and D. G. Victor, 2003: *Electricity Restructuring and the Social Contract*. Working Paper #15, Program on Energy and Sustainable Development, Stanford University, Stanford, CA.

Helps International, 2010: *The ONIL Stove*. Available at www.helpsinternational.com/programs/stove.php (accessed 1 March, 2011).

Heltberg, R., 2004: Fuel switching: Evidence from Eight Developing Countries. *Energy Economics*, 26(5):869–887.

Hiremath, R. B., B. Kumar, P. Balachandra, N. H. Ravindranath and B. N. Raghunandan, 2009: Decentralised Renewable Energy: Scope, Relevance and Applications in the Indian Context. *Energy for Sustainable Development*, 13(1):4–10.

Hossain, M. M. G., 2003: Improved Cookstove and Biogas Programmes in Bangladesh. *Energy for Sustainable Development*, 7(2):97–100.

Howells, M., D. G. Victor, T. Gaunt, R. J. Elias and T. Alfstad, 2006: Beyond Free Electricity: The Costs of Electric Cooking in Poor Households and a Market-friendly Alternative. *Energy Policy*, **34**(17):3351–3358.

IDCOL, 2010: *Infrastructure Development Company Limited*. Available at www.idcol.org/ (accessed 30 December, 2010).

IEA, 2002: *World Energy Outlook 2002*. International Energy Agency (IEA) of the Organisation of Economic Co-Operation and Development (OECD), Paris.

IEA, 2006: *World Energy Outlook 2006*. International Energy Agency (IEA) of the Organisation of Economic Co-Operation and Development (OECD), Paris.

IEA, 2007: *World Energy Outlook: China and India Insights*. International Energy Agency (IEA) of the Organisation of Economic Co-Operation and Development (OECD), Paris.

IEA, 2008a: *Energy Balances of Non-OECD Countries*. International Energy Agency (IEA) of the Organization for Economic Cooperation and Development (OECD), Paris.

IEA, 2008b: *Energy Balances of OCED Countries*. International Energy Agency (IEA) of the Organization for Economic Cooperation and Development (OECD), Paris.

IEA, 2009: *World Energy Outlook 2009*. International Energy Agency (IEA) of the Organisation of Economic Co-Operation and Development (OECD), Paris.

IEA, 2010a: *Comparative Study on Rural Electrification Policies in Emerging Economies*. A. Niez, (ed.) International Energy Agency (IEA) of the Organisation of Economic Co-Operation and Development (OECD), Paris.

IEA, 2010b: *World Energy Outlook 2010*. International Energy Agency (IEA) of the Organisation of Economic Co-Operation and Development (OECD), Paris.

Imboden, D. and R. C. Voegelin, 2000: *Die 2000 Watt-Gesellschaft. Der Mondflug des 21 Jahrhunderts*. ETH Bulletin 276, Swiss Federal Institute of Technology, Zürich.

India Ministry of Power, 2005: *National Electricity Policy*. Ministry of Poiwer, Government of India.

India Ministry of Power, 2011: RGGVY at a Gilance. Ministry of Power, Government of India.

Ingwe, A., 2007: Rocket Mud Stoves in Kenya. *Boiling Point*,(**53**):6–8.

IPCC, 2007: *Climate Change 2007: Synthesis Report, Fourth Assessment Report (AR4) of the IPCC*. Cambridge University Press, Cambridge, UK.

Jagadish, K. S., 2004: The Development and Dissemination of Efficient domestic Cook Stoves and Other Devices in Karnataka. *Current Science*, **87**(7):926–931.

Jannuzzi, G. M. and G. A. Sanga, 2004: LPG Subsidies in Brazil: An Estimate. *Energy for Sustainable Development*, **8**(3):127–129.

Johnson, F. X. and F. Lambe, 2009: *Energy Access, Climate and Development*. Stockholm Environment Institute, Commission on Climate Change and Development, Stockholm.

Kammen, D. M., 2001: *Research, Development and Commercialization of the Kenya Ceramic Jiko and other Improved Biomass Stoves in Africa*. Available at www.solutions-site.org/cat2_sol60.htm (accessed 23 September, 2010).

Kanagawa, M. and T. Nakata, 2007: Analysis of the Energy Access Improvement and its Socio-economic Impacts in Rural Areas of Developing Countries. *Ecological Economics*, **62**(2):319–329.

Karekezi, S., L. Majoro, K. John and A. Wambile, 2005: *Ring-fencing Funds for the Electrification of the Poor: Lessons for Eastern Africa: Sub-Regional "Energy Access" Study of East Africa* African Energy Policy Research Network (AFREPREN), Nairobi.

Khan, A. H. M. R., 2002: Development & Dissemination of Improved Stoves in Bangladesh. *National Workshop on Promotion of Renewable Energy, Energy Efficiency and Greenhouse Gas Abatement PREGA)*, Dhaka.

Khandker, S. R., D. F. Barnes and H. A. Samad, 2009: *Welfare Impacts of Rural Electrification: A Case Study from Bangladesh*. Policy Research Working Paper 4859, Development Research Group, The World Bank Washington, DC.

Khennas, S. and A. Barnett, 2000: *Best Practices for Sustaianble Development of Micro-Hydro Power in Developing Countries*. Final Synthesis Report, Contract R7215, UK Department for International Development and the World Bank, London.

Klingshirn, A., 2006: The Inkawasi Stove: A Success Story in the Peruvian Andes. *Boiling Point*, **52**:20–21.

Komuhangi, R., 2006: Mass Dissemination of Rocket Lorena Stoves in Uganda. *Boiling Point*,(52).

Krupp, C., 2007: *Electrifying Rural Areas: Extending Electricity Infrastructure And Services In Developing Countries*. PBRC Occasional Paper Series, Pacific Basin Research Centre.

Lambrou, Y. and G. Piana, 2006: *Energy and Gender in Rural Sustainable Development*. Food and Agriculture Organization (FAO), Rome.

Lanjouw, P. and A. Shariff, 2002: *Rural Non-Farm Employment in India: Access, Income and Poverty Impact*. National Council of Applied Economic Research, New Dehli.

Lucon, O., S. T. Coelho and J. Goldemberg, 2004: LPG in Brazil: lessons and challenges. *Energy for Sustainable Development*, **8**(3):82–90.

MacCarty, N., D. Still, D. Ogle and T. Drouin, 2008: *Assessing Cook Stove Performance: Field and Lab Studies of Three Rocket Stoves Comparing the Open Fire and Traditional Stoves in Tamil Nadu, India on Measures of Time to Cook, Fuel Use, Total Emissions, and Indoor Air Pollution*. Aprovecho Research Center, Cottage Grove, OR.

Mäkelä, S., 2008: *Firewood-Saving Stoves: A Review of Stove Models Based on the Documenation on the Internet*. Liana, Finland.

Malik, A. S. and S. Al-Zubeidi, 2006: Electricity tariffs based on long-run marginal costs for central grid system of Oman. *Energy*, **31**(12):1703–1714.

Martinot, E., A. Chaurey, D. Lew, J. R. Moreira and N. Wamukonya, 2002: Renewable Energy Markets in Developing Countries. *Annual Review of Energy and the Environment*, **27**(1):309–348.

Masud, J., D. Sharan and B. N. Lohani, 2007: *Energy for All: Addressing the Energy, Environment, and Poverty Nexus in Asia*. Asian Development Bank, Manilla.

Maya, R. S., A. P. Mhlanga, N. Nziramasanga and M. Mutyasira, 2001: *Implemenation of Renewable Energy Technologies – Opportunities and Barriers*. United Nations Environment Programme (UNEP) Collaborating Centre on Energy and Environment, Roskilde.

Mchirgui, R. and A. Kanzari, 2006: *Sustainable Use, Supply of Biomass and Production in Africa*. Draft Regional Report for North Africa, African Development Bank, Tunis.

MEDREC and GNESD, 2009: *Study on the Energy Security in Tunisia*. Second Draft Report, Mediterranean Renewable Energy Centre (MEDREC) & the Global Network on Energy for Sustainable Development (GNESD), New York.

Modi, V., S. McDade, D. Lallement and J. Saghir, 2006: *Energy Services for the Millennium Development Goals*. Energy Sector Management Assisstance Programme, UN Development Programme, UN Millennium Project, and the World Bank, New York.

Modi, V., 2010: *Personal Communication*.

OAS, 2000: *Energy Conservation and Environmental Protection Program Summary*. Organization of American States (OAS), Finnish International Development Agency, Washington, DC.

OLADE, 2008: *Energy-Economic Information System (SIEE)*. Latin American Energy Organization (OLADE), Quito, Equador.

ONE, 2009: *Rapport d'Activités*. Office National de l'Electricité (ONE), Kingdom of Morocco, Rabat.

Onilstove, 2010: *The ONIL Stove*. Available at www.onilstove.com/ (accessed 1 March, 2011).

Ouedraogo, B., 2006: Household Energy Preferences for Cooking in Urban Ouagadougou, Burkina Faso. *Energy Policy*, **34**(18):3787–3795.

Pachauri, S., 2007: *An Energy Analysis of Household Consumption: Changing Patterns of Direct and Indirect use in India*. Springer-Verlag, Berlin.

Pachauri, S. and L. Jiang, 2008: The Household Energy Transition in India and China. *Energy Policy*, **36**(11):4022–4035.

Pachauri, S. and A. Mueller, 2008: An Analysis of Electricity Access and Use in Indian Households: A Regional Decomposition of Consumption for 1980 – 2005. *Proceedings of the 31st International IAEE Conference*, Istanbul.

Pan, J., W. Peng, M. Li, X. Wu, L. Wan, H. Zerriffi, B. Elias, C. Zhang and D. Victor, 2006: *Rural Electrification in China 1950–2004: Historical processes and key driving forces*. Program on Energy and Sustainable Development (PESD), Stanford University, Stanford.

Parikh, J. K., 1995: Gender issues in energy policy. *Energy Policy*, **23**(9):745–754.

Peng, W. and J. Pan, 2006: Rural Electrification in China: History and Institution. *China & World Economy*, **14**(1):71–84.

Practical Action, 2010: *Poor People's Energy Outlook 2010*. Rugby, UK.

PRAYAS, 2010: *Electricity for All: Ten Ideas towards Turning Rhetoric into Reality*. Discussion Paper, Prayas Energy Group, PRAYAS, Pune, India.

Radulovic, V., 2005: Are New Institutional Economics Enough? Promoting Photovoltaics in India's Agricultural Sector. *Energy Policy*, **33**(14):1883–1899.

Ramanathan, V. and G. Carmichael, 2008: Global and Regional Climate Changes due to Black Carbon. *Nature Geoscience*, **1**(4):221–227.

Ravallion, M., S. Chen and P. Sangraula, 2007: *New Evidence on the Urbanization of Global Poverty*. Policy Research Working Paper 4199, Development Research Group, The World Bank, Washington, DC.

Ritch, E., 2008: *Envirofit Ramps Clean-Cooking Line for India*. Available at cleantech.com/news/3779/envirofit-clean-cookstoves-india (accessed 1 March, 2011).

Serebrisky, T., 2007: *Subsidios: Diseño, Implementación e Impactos*. The World Bank, Washington, DC.

Shah, T. and S. Varma, 2008: Co-Management of Electricity and Groundwater: An Assessment of Gujarat's Jyotirgram Scheme. *Economic and Political Weekly*, **43**(07):59–66.

Shakti, G., 2010: *Solar Energy Programme*. Available at www.lged-rein.org/index.php (accessed 30 December, 2010).

Shrestha, R. M., S. Kumar, S. Sharma and M. J. Todoc, 2004: Institutional Reforms and Electricity Access: Lessons from Bangladesh and Thailand. *Energy for Sustainable Development*, **VIII**(4):41–53.

Smil, V., 2005: *Creating the Twentieth Century: Technical Innovations of 1867 – 1914 and their Lasting Impact*. Oxford University Press, New York.

Smith, A., 2007: Growth and Innovation Lights Up India. *The BP Magazine*,(4):5–11.

Sokona, Y., S. Sarr and S. Wade, 2004: *Energy Services for the Poor in West Africa Sub-Regional "Energy Access" Study of West Africa* Prepared for "Energy Access" Working Group, Environnement et Développement du Tiers Monde

& Global Network on Energy for Sustainable Development, UN Environment Programme, New York.

Spalding-Fecher, R., 2005: Health Benefits of Electrification in Developing Countries: a Quantitative Assessment in South Africa. *Energy for Sustainable Development*, **IX**(1):53–62.

Srivastava, L. and I. H. Rehman, 2006: Energy for sustainable development in India: Linkages and strategic direction. *Energy Policy*, **34**(5):643–654.

Srivastava, N. S. L., 2004: Farm Power Sources, their Availability and Future Requirements to Sustain Agricultural Production. In *Status of Farm Mechanization in India*, Department of Agriculture and Cooperation, Ministry of Agriculture, Government of India.

STEG, 2010: *Société Tunisienne de l'Électricité et du Gaz (STEG)*. Available at www.steg.com.tn/en/accueil.html (accessed 30 December, 2010).

Stevenson, G. G., 1989: The Production, Distribution and Consumption of Fuelwood in Haiti. *The Journal of Developing Areas*, **24**:59–76.

Stoves Online, 2010: *Justa Stove Project*. Available at www.stovesonline.co.uk/justa-stove.html (accessed 30 December, 2010).

StoveTec, 2010: *StoveTec*. Available at www.stovetec.net/us/index.php (accessed 1 March, 2011).

Sundar, S., 2007: *A Stove and a Smokeless Kitchen*. The Hindu, Available at http://hindu.com/2007/06/05/stories/2007060505830500.htm.

Tongia, R., 2003: *Political Economy of Indian Power Sector Reforms*. Working Paper 4. Stanford, Program on Energy and Sustainable Development, Stanford University.

Uddin, S. N., R. Taplin and X. Yu, 2006: Advancement of renewables in Bangladesh and Thailand: Policy intervention and institutional settings. *Natural Resources Forum*, **30**(3):177–187.

UN, 2008: *World Urbanization Prospects: The 2007 Revision*. Population Division, UN Department of Economic and social Affairs, New York.

UNDESA, 2004: *Sustainable Energy Consumption in Africa*. United Nations Department of Economic and Social Affairs (UNDESA), New York, USA.

UNDESA, 2007: *Energy Indicators for Sustainable Development: Country Studies on Brazil, Cuba, Lithuania, Mexico, Russian Federation, Slovakia and Thailand*. UN Department of Economic and Social Affairs.

UNDP, 2005: *Achieving the Millennium Development Goals: The Role of Energy Services*. United Nations Development Programme (UNDP), New York.

UNDP, 2006: *Expanding Access to Modern Energy Services: Replicating, Scaling Up and Mainstreaming at the Local Level*. Lessons from Community-based Energy Initiatives, S. Gitonga and E. Clemens, (eds.), United Nations Development Programme (UNDP), New York.

UNDP, 2007a: *Energizing Poverty Reduction: A Review of the Energy-Poverty Nexus in Poverty Reduction Strategy Papers*. United Nations Development Programme (UNDP), New York.

UNDP, 2007b: *Gender Mainstreaming: A Key driver of Development in Environment and Energy*. Training Manual, I. Havet, F. Braun and B. Gocht, (eds.), Environment and Energy Group, United Nations Development Programme (UNDP), New York.

UNDP, 2007c: *Overcoming Vulnerability to Rising Oil Prices; Options for Asia and the Pacific*. Regional Energy Programme for Poverty Reduction, United Nations Development Programme, Bangkok.

UNDP, 2009a: *Expanding energy access in Developing countries: The role of mechanical Power*.

UNDP, 2009b: *Bringing Small Scale Finance to the Poor for Modern Energy Services: What is the Role of Government? Experiences from Burkina Faso, Kenya, Nepal and Tanzania*. United Nations Development Programme (UNDP), New York.

UNDP, 2010: *The Real Wealth of Nations: Pathyways to Human Development*. Human Development Report 2010, J. Klugman, (ed.) United Nations Development Programme (UNDP), New York.

UNDP and WHO, 2009: *The Energy Access Situation in Developing Countries: A Review Focusing on the Least Developed Countries and Sub-Saharan Africa*. The Energy Access Situation in Developing Countries: A Review Focusing on the Least Developed Countries and Sub-Saharan Africa, New York.

UPDEA, 2009: *Best practices: Power Sector Reform in Africa*. Union of Producers, Transporters and Distributors of Electric Power in Africa (UPDEA), Abidjan.

van Horen, C., A. Eberhard, H. Trollip and S. Thorne, 1993: Energy, environment and urban poverty in South Africa. *Energy Policy*, **21**(5):623–639.

WEC and FAO, 1999: *The Challenge of Rural Energy Poverty in Developing Countries*. World Energy Council (WEC) and the Food and Agriculture Organisation (FAO), London.

Westhoff, B. and D. Germann, 1995: *Stove Images: A Documentation of Improved and Traditional Stoves in Africa, Asia and Latin America*. Sozietät für Entwicklungsplanung GmbH (SfE), Directorate General for Development, Commission of the European Communities, Frankfurt am Main.

Wheldon, A., 2005: *Fuel-Efficient Stoves for Rural and Urban Households*. The Ashden Awards for Sustainable Energy, Trees, Water and People (TWP), Fort Collins, CO.

WHO, 2006: *Fuel for Life: Household Energy and Health*. World Health Organization (WHO), Geneva.

WLPGA, 2005: *Developing Rural Markets for LP Gas: Key Barriers and Success Factors*. World Liquid Petroleum Gas Association (WLPGA), Paris.

World Bank, 1996: *Rural Energy and Development Improving Energy Supplies for 2 Billion People*. A World Bank Best Practice Paper, The World Bank, Washington, DC.

World Bank, 2002: *Rural Electrification and Development in the Philippines: Measuing the Social and Economic Benefits*. Report No. 255/02, Energy Sector management Assistance Programme (ESMAP) of the World Bank, Washington, DC.

World Bank, 2007: *Africa Achieving Healthy And Steady Growth Rate*. Available at go.worldbank.org/J1M295NU60 (accessed 30 November, 2009).

World Bank, 2008a: *The Welfare Impact of Rural Electrification: A Reassessment of the Costs and Benefits*. Independent Evaluation Group, The World Bank, Washington, DC.

World Bank, 2008b: *World Development Report: Agriculture and Development*. The World Bank, Washington, DC.

World Bank, 2009: *Africa Regional Brief*. Available at go.worldbank.org/3IGKDWFTG1 (accessed 2 December, 2009).

World Bank, 2010a: *Addressing the Electricity Access Gap*. Background Paper for the World Bank Group Energy Strategy, The World Bank, Washington DC.

World Bank, 2010b: *Modernizing Energy Services for the Poor: A World Bank Investment Review – Fiscal 2000–08*. Energy Sector Management Assistance Program (ESMAP) of the World Bank, Washington, DC.

Zerriffi, H., 2007: *Making Small Work: Business Models for Electrifying the World*. Working Paper No. 63, Program on Energy and Sustainable Development, Stanford University, Stanford, CA.

Zerriffi, H., 2010: *Rural Electrification: Strategies for Distributed Generation*. Springer-Verlag, Berlin.

Zhang, J. and K. R. Smith, 2007: Household Air Pollution from Coal and Biomass Fuels in China: Measurements, Health Impacts and Interventions. *Environmental Health Perspectives*, **115**(6):848–855.

20

Land and Water: Linkages to Bioenergy

Convening Lead Author (CLA)
Suani T. Coelho (National Reference Center on Biomass, University of São Paulo, Brazil)

Lead Authors (LA)
Olivia Agbenyega (Kwame Nkrumah University of Science and Technology, Ghana)
Astrid Agostini (Food and Agriculture Organization, Italy)
Karl-Heinz Erb (Klagenfurt University, Austria)
Helmut Haberl (Klagenfurt University, Austria)
Monique Hoogwijk (Ecofys, the Netherlands)
Rattan Lal (The Ohio State University, USA)
Oswaldo Lucon (São Paulo State Environment Agency, Brazil)
Omar Masera (National Autonomous University of Mexico)
José Roberto Moreira (Biomass Users Network, Brazil)

Contributing Authors (CA)
Gunilla Björklund (Uppsala University, Sweden),
Fridolin Krausmann (Klagenfurt University, Austria),
Siwa Msangi (International Food Policy Research Institute, USA),
Christoph Plutzar (Klagenfurt University, Austria)

Review Editor
Rik Leemans (Wageningen University, the Netherlands)

Contents

Executive Summary

Sustainably managing limited resources, such as productive land areas and available freshwater, will be one of the world's most pressing challenges in the coming years. Population increases and economic growth will significantly influence humanity's future demand for land and water for different uses. In particular, changes in food and energy use will have substantial environmental impacts. They will also influence each other in many ways. At the same time, the production of food and energy, and the water resources they require, will be affected by global climate change. Sustainability issues arising from competition and synergies between future production of bioenergy and food, and related water use, are highly important in this context.

Population growth is one of the factors contributing to increased demand for land and water. While the world's population has approximately doubled since the 1960s, global economic activity has increased approximately 40 fold. Since growth in incomes is strongly correlated with increased consumption of animal-derived food (meat, milk, eggs), the combination of population increases and economic growth will likely result in increased feed and food production. This will drive up pressures on land and water resources if not counteracted by innovations that reduce land and water use. Social inequities are increasing as well, with both very rich and very poor populations often practicing 'inefficient' methods of using land and water.

Considering the importance of these issues, in particular the need to achieve the Millennium Development Goals (MDGs) and beyond (see Chapter 2), the objective of this chapter is to assess and discuss major trade-offs related to the different uses of land and water, in particular related to future energy systems, and to discuss the corresponding sustainability issues.

With respect to land use, this chapter aims to evaluate land availability, including land for bioenergy production. It discusses competition for land from different uses (food, fuel, timber, etc.), environmental impacts, and implications for natural resources (e.g., biodiversity, atmosphere, water), as well as social factors such as food security, health, and incomes.

With respect to water use, the chapter discusses water demand for different kinds of uses, with a special emphasis on water for bioenergy crops in potential competition with water needed for food production. The multiple uses of water considered include human consumption, hydro and thermal power generation, manufacturing, agriculture, water supply security, and bioenergy. Environmental, social and strategic factors, and potential trade-offs, are examined. Competition between food and energy crops may not always be over 'the same water.' Depending on the type of feedstock, it is possible to cultivate energy crops in areas where conventional food production is not feasible due to water constraints.

This chapter confirms major studies suggesting that global land and water resources will be sufficient to adequately nourish a world population of 9–10 billion people in 2050. However, the potential to additionally produce crops for bioenergy will depend on future changes in food systems (including diets), population growth, and agricultural technologies to improve crop yields and livestock feeding efficiencies, institutional arrangements, climate conditions and area demand for biodiversity conservation. Recent studies suggest that the technical potential for bioenergy production is uncertain due to these factors, which are difficult to predict. Studies mentioned in IPCC (2011) indicate that the technical bioenergy potential may reach up to 500 EJ/yr, while others found much lower potentials. This chapter only discusses the potential of dedicated energy crops.[1] Considering sustainability constraints related to possible competing land demands (food, feed and fiber production, biodiversity conservation, etc.), problems posed by possible

---

1   Dedicated (bio)energy crops are "[f]ast growing species whose biomass yields are dedicated to the production of more immediately usable energy forms, such as liquid fuels or electricity" (Sartori et al., 2006). Dedicated energy crops refer to any crop that is grown for the primary purpose to obtain bioenergy, e.g., sugarcane, switchgrass, or Salix. These crops may, of course, also be grown for other purposes. See also SECO, 2008.

deforestation, and water availability, three studies analyzed in this chapter (van Vuuren et al., 2009; Erb et al., 2009; and WBGU, 2009) found global bioenergy crop potentials of 44–133 EJ/yr in 2050. Another study, Dornburg et al. (2010), concludes that for energy crops (dedicated energy crops) they are 120 EJ/yr. So the range considered in this chapter is 44–133 EJ/yr for 2050.

For a full evaluation of all other bioenergy potentials except dedicated bioenergy crops, see Chapter 7. Chapters 11 and 12 discuss technologies and utilization pathways, including biofuels and power generation from biomass.

Reaching the stated technical bioenergy potential depends on many important factors (land and water availability, feedbacks between food, livestock and energy systems, climate change, etc.). Land allocation for different purposes will have to be monitored and managed carefully through adequate policies to minimize competition between food, bioenergy and fiber markets. Adequate policies to introduce bioenergy plantations to avoid adverse social, economic and ecological effects are also needed, since best-practice examples suggest that such plantations could be sustainable if based on sound strategies. Monitoring, managing and enforcement are required to ensure sustainability of bioenergy production. Policies for sustainable land (and water) use must include agro-economic-environmental zoning and planning, in order to consider specific environmental conditions of each region, like those already existing, for example, in Brazil.

The impact of increased biofuel use on food prices has been debated widely and is also covered here. Adequate food production for the world's growing population strongly depends on future dietary choices as well as the development of agricultural technologies that increase agricultural yields and livestock feeding efficiency through enhancements in crop management and/or genetic modifications, as well as institutional changes. The impacts of biofuel production on agricultural markets and food systems will be minimized if adequate policies for biofuels are in place to ensure sustainable production, prioritize the diversification of technologies and fuels, and identify different options for the future, based on adequate environmental zoning and sustainable policies. This must occur through public intergovernmental policies that govern and regulate markets and stimulate the adoption of efficient technologies. These policies include biofuel sustainability schemes based on certification schemes. Integrated optimization of food and bioenergy production – for example, through the use of by-products and residues, or through optimization of land allocation (zoning) – can help to mitigate possible adverse effects.

Climate change can have a substantial effect on land-use systems, but its impacts are still imperfectly understood. In subtropical and tropical regions, changes in climate and rainfall may change the agricultural suitability of a region significantly. Temperature increases may lead to a shift of some crops and agricultural areas to regions with more temperate climates, or with higher levels of soil moisture and rainfall. In general, crop productivity in the tropics may decline even with a slight increase in local temperatures (1–2°C). An increase in vulnerability of food production due to climate change may also increase the risk of hunger to a large number of people in the world, mainly in poor countries, which are most vulnerable to the effects of global warming and the least prepared to deal with its impacts (IPAM, 2002).

The chapter's main conclusion on water trade-offs is that the increasing stress on freshwater resources brought about by ever-rising demand, due mainly to population growth, is of serious concern. As population increases and development requires additional allocations of groundwater and surface water for the domestic, agriculture, energy and industrial sectors, the pressure on water resources will continue to intensify, leading to further tensions and conflicts among users, and degradation and pollution of the environment. Contamination of rivers, depletion of aquifers and increased utilization for multiple purposes are challenges commonly found in regions where demand exceeds supply and where water management is poorly conducted. Indicators such as water footprints are important to understand these important issues.

Water scarcity leads to competition between different uses, with competing demands from: cities and rural areas; rich and poor people; arid lands and wetlands; public and private sectors; infrastructure and natural environments; mainstream and marginal groups; and local stakeholders and centralized authorities. Water conflicts can arise in water-stressed areas among local communities, and countries, because sharing a very limited and essential resource is extremely difficult.

The lack of adequate legal instruments exacerbates already difficult conditions. In the absence of clear and well-established rules and regulations, severe tensions tend to dominate, and political and economic power can play an excessive role, leading to inequitable allocation of water. A well-developed system of Integrated Water Resources Management, including adequate institutional set-up and a good governance system, is needed in river basins or confined regions where demand exceeds existing supply.

Climate change is expected to account for about 20% of the global increase in physical water scarcity – and countries that already suffer from water shortages will be hit the hardest. This would include African countries, where water is a limiting factor for agricultural food production and also essential for income generation.

20.1   Introduction

World population growth, changing diets and increasing urbanization result in a surging demand for products and services that require land and water as significant inputs. In connection with these changes, increased consumption of fibers (including wood) and bioenergy are likely to lead to a considerable growth in humanity's need for biomass from agriculture and forestry. How this will affect global land and water systems will depend on innovations in agricultural technology as well as future changes in social, economic, political and legal factors that affect land use, such as land tenure, property rights, subsidies, and markets for land and agricultural products, water access and bioenergy (Global Land Project, 2005; Turner et al., 2007).

Moreover, demand for water is likely to be driven up by changes in infrastructure (households, buildings, schools, hospitals, etc.), increased requirements for basic sanitation and drinking water, energy sector uses (e.g., for hydroelectricity or cooling water) and other factors. With increasing purchasing power, consumer preferences and behavior will affect the dynamic interplay between production, demand and consumption, and the efficiency of the whole system. Consumers continue to intervene in the impacts on the water cycle through their preferences for, and use of, various goods and services.

In order to address this complex and intertwined set of challenges, a separate chapter was proposed for a comprehensive discussion of the trade-offs and synergies involved in the use of land and water, particularly as they might affect future energy systems. In presenting this discussion, Chapter 20 addresses critical issues related to environmental and social sustainability, and their policy implications. This chapter discusses only regional and global potentials for primary bioenergy from dedicated energy crops in the year 2050, taking into account sustainability constraints and interactions with other sectors such as food supply and agriculture. Chapter 7 discusses the main concepts and potentials related to all types of primary energy resources, including all types of bioenergy (residues, animal manures, municipal solid wastes, and forestry residues) except energy crops (these potentials are taken from this chapter). Chapters 11 and 12 discuss technologies and utilization pathways, including biofuels and power generation from biomass.

The term 'bioenergy' here denotes all kinds of biomass feedstocks that can be used to produce energy carriers or heat from biomass (excluding nutritional energy for humans and livestock). This ranges from combustion of any solid biomass, including municipal and rural solid waste for heat and/or electricity production to recently introduced technologies such as production of liquid or gaseous fuels for use in vehicles from sugarcane, corn, wheat or oil crops (rape, oil palm, etc.) or other biomass. This chapter only discusses primary biomass potentials (i.e., it estimates the total amount of plant biomass (feedstock) that might become available for energy conversion processes) from energy crops; issues of total biomass energy potential (including residues and biomass from forestry) are evaluated in Chapter 7.

There are concerns related to limited availability of suitable land and/or water resources for bioenergy production. This chapter addresses this controversy and discusses recent studies estimating land areas that will be available for bioenergy as well as their productivity potential. It suggests that despite several possible constraints, significant energy crop potentials could be realized if appropriate policies are implemented (IPCC, 2000; Goldemberg et al., 2008; WBGU, 2008; Erb et al., 2009; van Vuuren et al., 2009; Haberl et al., 2010; 2011).

With respect to land, the main aim of this chapter is to discuss issues of land availability for all uses and land suitability for bioenergy production. This discussion is based on the 'food first' principle, i.e., the assumption that the provision of adequate food supplies must be guaranteed when evaluating bioenergy production options (e.g., Sims et al., 2007).

Limits to freshwater availability are also highly relevant in the context of biomass production and demand, as water is a critical input affecting the primary production of terrestrial ecosystems, including agro-ecosystems. In addition, water availability is critical for many other aspects of energy systems, including hydropower, cooling water, etc. Countries have mainly controlled water by the supply-side management approach, balancing supply and demand in an increasingly precarious way, since exploitation and water consumption indices do not take into account either ecological water demand or the spatial and temporal variability of supply and demand (EC and IPTS, undated).

The following considerations are important when discussing trade-offs and synergies related to land availability:

- The productivity of land areas depends on climate and soil conditions, nutrient inputs (e.g., fertilizers), water availability (including for irrigation when needed), and many other natural and socio-economic factors. Net primary production (NPP, i.e., biomass production per unit area and year) and the production of usable plant parts (e.g., grain in the case of cereals) depend on natural conditions and management alike. Even assuming that there is enough land available for all end uses (Berndes et al., 2003; Goldemberg, 2009), the area available for bioenergy crops, as well as for food, feed and fiber, depends on agricultural yield levels.

- Sustainability of land use in agriculture and forestry is a critical issue. While increases in yields achieved through agricultural innovation can help to save land and thereby to reduce environmental problems (Burney et al., 2010), agricultural intensification has also created a host of social and environmental problems such as nutrient leaching, soil degradation, toxic effects of pesticides, and many more (IAASTD, 2009). Therefore, it is necessary to mitigate any negative environmental effects of the agricultural intensification intended to produce yield growth. When used appropriately, land can generate more than one type of product (such as food, feed, energy, or materials) or service (including protection of the soil, wastewater treatment, recreation, or nature protection) – an observation usually denoted as 'multifunctionality' or multiple use (Börjesson, 1999; Londo, 2002; Lewandowski et al., 2003; McCarney et al., 2008). Appropriate land management can reduce trade-offs or even turn them into synergies and enhance biomass production, in a win-win situation.

- Integrated optimization of biomass utilization chains – for example, the use of harvest residues or by-products of production processes, a strategy sometimes referred to as "cascade utilization of biomass" (Haberl and Geissler, 2000; WBGU, 2008) or 'integrated food energy systems" (IFES) (Bogdanski et al., 2010) – together with adequate policies, as discussed in this chapter, can help avoid or at least mitigate trade-offs between food and production of energy carriers. It can sometimes even create synergies (e.g., if agricultural by-products can be used for energy provision). This strategy can produce substantial amounts of biomass feedstock at low costs. Constraints to implementing "cascade" or IFES systems, and possible ways to overcome these, are discussed in Bogdanski et al. (2010).

The competition between production of food, biomass for energy, biomass for other uses and use of land for non-production functions (e.g., settlements, nature protection) can influence the amounts of different crops produced, their production costs, their prices, and their environmental impacts. How this competition plays out, however, depends on policy frameworks and the crops used. For example, production of biomass from perennial lignocellulosic crops (second-generation bioenergy crops) can involve the use of land not suitable for cultivation of food crops – for example, degraded and marginal land – and might provide the possibility of combining biomass production with food production or nature protection. On the other hand, some of these areas are currently used by herders for livestock rearing, often in subsistence systems – and this might create new, different trade-offs. Technical innovation in agriculture and biomass conversion can greatly influence trade-offs and synergies between different land uses. Socioeconomic development processes, such as transitions from subsistence to market-based production can play a similarly important role, including the benefits related to the creation of rural jobs through the production of bioenergy in developing countries. Adequate policies are required to achieve synergies where possible and to minimize potential conflicts and adverse affects.

Moreover, population density plays an important role: in regions with low population density, such as many areas in, for example, Africa and South America, bioenergy production with first-generation technologies (such as sugarcane in Brazil) may result in little, if any, competition with food and other end uses if based on well-designed policies (Goldemberg et al., 2008). In other regions conflicts may arise, also through indirect effects from increasing the total demands on land for biomass products. This could become a serious issue with bioenergy and agricultural production at much higher levels than today.

Central to the debate on water issues and water scarcity are water demands and end-use patterns. Water for meeting basic needs, drinking and general household use, though comparatively small in terms of volume, needs to be readily available. The inexorable growth of cities, concentrating large numbers of people in small areas, exacerbates this challenge locally.

River ecosystems, and the fish and other species living in them, of course need continued running water. However, large dams used mainly for hydropower can destroy both river and terrestrial ecosystems due to impounding, flow regulation, and fragmentation of landscapes. Water is also used to produce energy in medium and small hydropower installations, and for cooling thermal power stations. Distribution of water to different sectors, including for energy purposes and for industry, is to be decided by a water management scheme.

An important use of water is for food production. As the world population continues to increase, a growing number of people will require water for cultivation of food, fiber and industrial crops, and for livestock and fish. It is estimated that crop production to feed the growing population will need to nearly double during the next 50 years. The two main factors driving food demand are population growth and dietary change, which is strongly dependent on economic growth. With rising incomes and continuing urbanization, diets move towards consumption of more animal products, fats and sugar, as well as to a greater variety of foods. Shifts in consumption are expected between different cereal crops, and away from cereals towards livestock, fish products and high-value crops that consume more water (UN-Water/FAO, 2007).

In this context, the objectives of this chapter are to assess and to discuss major trade-offs related to the different uses of land and water, in particular related to future energy systems, land availability, and land for bioenergy. It also examines competition for land from different uses (food, fuel, timber, etc.), environmental impacts, and implications for natural resources (e.g., biodiversity, atmosphere, water), as well as social factors such as food security, health, and incomes. With respect to water use, the chapter discusses water demand for different kinds of uses, with a special emphasis on the increasing need for water for bioenergy crops in competition with the increasing need for water for food production.

In relation to water use, the chapter evaluates the multiple uses of water (human consumption, hydro and thermal power generation, manufacturing, agriculture, water security, bioenergy, etc.), as well as environmental, social and strategic issues, and potential trade-offs. Competition between food and energy crops may not always be over 'the same water.' Depending on the type of feedstock, it is possible to cultivate energy crops in areas where conventional food production is not feasible due to water constraints, i.e., the 'water footprints' are of a different character (Lundqvist et al., 2008).

20.2   Trade-offs in Land Use

This section discusses the different uses of land and the trade-offs related to availability of suitable land areas, and derives conclusions for adequate policies. The discussion of trade-offs starts with an evaluation of current land use, as a basis for analyzing perspectives for future land availability for bioenergy crops, which depends on future cropland areas and yields, livestock, and critical demand components such as food. Sustainability issues related to environmental, social and economic aspects are critical to this discussion.

20.2.1   Current Land Use and Land Availability

The global land-use dataset on a five minute grid (approximately 10x10 km at the equator) summarized in Table 20.1 has the following features:

- Reproduction of national land-use statistics (as reported by the United Nations Food and Agriculture Organization (FAO)) for cropland and forestry at the country level.

- Five land-use classes covering the Earth's total terrestrial area in the form of percent-per-grid cell layers for urban and infrastructure land, cropland, grazing land, forestry, and areas without land use (free of double-counting).

- Spatial patterns derived from thematic GIS maps based on remote sensing.

- Extensive statistical and cross-checks against other, independent datasets such as MODIS, CORINE and others (see Erb et al., 2007).

- Consistency with a geographically explicit database on net primary production (NPP) and its human use (Haberl et al., 2007) as well as with national-level biomass balances and feed balances (Krausmann et al., 2008). This consistency allows analysis of the impacts of allocating additional land for bioenergy production on all other socio-economic biomass flows, above all feed, fiber and food supply.

Most assessments are based on a "land balance" approach: total cultivable area is identified, and then areas already being cultivated are subtracted. Young (1999) argues that this approach has several inherent shortcomings, such as: (1) overestimation of cultivable land (e.g., failure to adequately account for uncultivable land such as hills, rock, outcrops, minor water bodies, etc.); (2) underestimation of land already cultivated (up to 50% in some assessments, in particular in sub-Saharan Africa); and (3) inadequate accounting of land demand for purposes other than cultivation (e.g., grazing or settlements), conservation and ecological services, and urban development. Erb et al. (2009) suggest that it is not realistic to assume that mowing and livestock grazing are confined to "permanent pastures," as reported in FAO statistics. According to Table 20.1, about 76% of the world's land surface is used more or less intensively, with around 15 million km^2 being cropland.

Currently unused areas (24%) include: almost completely unproductive land (aboveground biomass productivity below 0.04 kg/m^2/yr), currently

unused grasslands and scrublands (mostly remote and with low productivity), and the world's last remaining pristine forests. Except for pristine forests, currently unused lands are unlikely to be suitable for providing significant additional areas for cultivation in the future. With regard to use of forests, however, there are studies showing that converting pristine forests to bioenergy production would have long carbon payback times and would, therefore, not contribute to the mitigation of climate change over the next decades (WBGU, 2008; Searchinger et al., 2008). Negative environmental impacts from deforestation can be avoided by the introduction of adequate policies and enforcement measures, as discussed in Goldemberg et al. (2008). For production of bioenergy without deforestation, the areas available are those classified in Table 20.1 as cropland and grazing land.

In the case of cropland, it is possible to plant bioenergy crops on land currently lying fallow, or on cropland that becomes available if yield increases surpass the growth in demand for food, feed and fiber, thereby freeing up area for energy crops. Much larger land potentials can be mobilized on land classified as "grazing area" in Table 20.1. This category is a "remainder" category, i.e., it comprises all land not classified as infrastructure, cropland, forestry, or unused land. Land classified as "grazing land" in Table 20.1 includes a large variety of land, ranging from highly productive grasslands intensively used for grazing or mowing to barely productive, very extensively used land dominated by shrubs, more or less bare areas, and other vegetation. This category also includes degraded lands (except degraded cropland or forests), abandoned farmland, and other very extensively used lands.

Up to a certain point (which depends on the respective regional roughage demands of livestock), the use of the land in this category could be used to grow bioenergy crops. This would allow the production of bioenergy without deforestation and with few, if any, repercussions on the livestock sector (see Goldemberg, 2009; Goldemberg et al., 2008; Goldemberg and Guardabassi, 2009; Macedo et al., 2008; Nassar, 2009). In many cases, especially in degraded areas, use of such land for bioenergy production could have a favorable greenhouse gas (GHG) balance due to carbon sequestration in the soil, as discussed in Soares et al. (2009).

Urban areas and infrastructure are bound to grow due to population growth and increasing wealth, taking up sizeable areas of high-quality land, some of which is currently used as cropland. Expanding the amount of cropland available would almost exclusively mean bringing land into cultivation that is currently used for grazing or forestry. Expanding into land classified in Table 20.1 as "grazing land" is possible where this land is used extensively, i.e., where livestock density can be increased without causing degradation. This is possible in many regions where livestock densities are sufficiently low (see Erb et al., 2009). A comparison of the grazing land areas of quality classes 1 and 2 in Table 20.3 with livestock densities estimated by FAO (2006) suggests that much, but by far not all, of the available high-quality grazing land is used for intensive livestock rearing.

It is a difficult task to assess the area and spatial distribution of land that could be suitable as cropland but is at present not used for crops.

Table 20.1 | Global land use in 2000.

	(1) Infra-structure	(2) Cropland	(3) Used forests	(4) Grazing land total	(4.1) Grazing class 1	(4.2) Grazing class 2	(4.3) Grazing class 3	(4.4) Grazing class 4	(5) Unused land total	(5.1) Unused Forests	(5.2) Unused shrubs etc.	(5.3) Non-productive, snow	(6) Area total[a]
							(1000 km²)						
USA	280	1782	2598	3473	853	390	524	1706	1044	383	504	157	9178
CAN	57	458	2143	1000	170	298	131	402	5673	2032	2249	1392	9331
WEU	203	1129	1475	1480	529	235	295	421	152	21	120	11	4440
EEU	60	463	370	264	158	40	54	12	1	0	1	0	1159
FSU	240	2086	7458	7053	1052	705	1578	3718	4777	1826	2671	280	21,614
NAF	20	451	245	2013	363	156	165	1328	5256	5	59	5192	7984
EAF	13	283	382	2418	978	176	323	941	158	3	28	127	3254
WCA	34	900	3382	4125	1039	1172	610	1304	2925	139	608	2178	11,367
SAF	59	431	1873	3971	832	1058	600	1482	524	5	130	389	6859
MEE	24	347	59	1721	3	1	138	1578	3018	0	16	3002	5169
CHN	95	1496	1633	3991	891	425	1669	1005	2135	1	323	1811	9351
OEA	11	167	498	1386	190	76	188	932	349	7	77	264	2411
IND	75	1698	640	653	227	101	163	163	80	1	7	73	3147
OSA	28	437	163	803	100	32	365	306	478	1	8	469	1908
JPN	38	48	245	38	25	12	0	1	25	3	22	0	394
OCN	22	531	862	3390	407	298	899	1786	3107	264	2539	305	7913
PAS	35	831	2192	1148	531	431	52	135	111	31	80	0	4317
LAC	64	1685	8733	7932	3062	1749	489	2633	1880	1446	178	256	20,295
Other	0	1	5	21	4	1	0	15	257	0	0	257	284
TOTAL	1358	15,224	34,956	46,880	11,414	7356	8243	19,868	31,950	6168	9620	16,163	130,375

[a] Excluding Greenland, Antarctica and inland water bodies.

Note that (1) + (2) + (3) + (4) + (5) = (6). Moreover, (4.1) + (4.2) + (4.3) + (4.4) = (4) and (5.1) + (5.2) + (5.3) = (5). Differences are due to rounding.

Source: Erb et al., 2007.

The following factors need to be taken into account when calculating area potentials for agriculture and energy crops:

- Data on settlements and related infrastructure land (and projections out to 2050) should consider rural infrastructure areas required to support cropland existing now or assumed to exist in 2050 (Table 20.2, Erb et al., 2007; Erb et al., 2009).

- Land under forestry and currently unused land (wilderness, including unused forests) are increasingly accounted separately (see, e.g., Table 20.1), which enables the exclusion of forest areas from the assessment of agriculture and bioenergy crop potentials – a plausible procedure because clearing forests for agriculture and bioenergy production results in large GHG emissions (WBGU, 2008).

- Biomass balances of feed supply and livestock production allow an evaluation of whether or not grazing land remaining after conversion to grow bioenergy crops can support the required livestock feed demand (Erb et al., 2009). An assessment of grazing land quality (Erb et al., 2007) helps to identify grazing areas suitable

for bioenergy crops (Erb et al., 2009), as well as for agricultural crops.

Table 20.2 compares current infrastructure and cropland areas with estimates of quality classes of grazing lands based on (a) the assessment of global land use in Table 20.1; (b) an assessment of cropland suitability (Ramankutty et al., 2002); and (c) assessments of cropland suitability from the Global Agro-Ecological Zoning (GAEZ) maps (FAO and IIASA, 2000). The classification of grazing land quality was based on its NPP as well as on land cover information. For example, bare areas and shrub lands were assumed to be less suitable for grazing than areas with herbaceous vegetation (Erb et al., 2007).

The GAEZ study (FAO and IIASA, 2000; Fischer et al., 2002) combined soil, terrain and climate characteristics with crop production requirements. It estimates suitability for crop production at three input levels (low, medium, and high). About 30% of the earth's land surface, except for Antarctica and Greenland (a bit more than 40 million km²), was found to be at least moderately suitable for rain-fed crop production (Bruinsma, 2009).

Table 20.2 | Comparison of current (2000) infrastructure and cropland areas with two assessments of cropland suitability and current grazing areas.

	Erb et al. 2007		Ramankutty et al. 2002		GAEZ: Climate, soil, terrain & slope constraints combined			GAEZ: Suitability for rainfed cultivation – maximizing technology				GAEZ: Suitability for rainfed cultivation max. tech., without forest areas				Erb et al. 2007	
	Infra-structure 2000	Plus cropland 2000	Cropland suitability >0.7	Cropland suitability >0.5	No or (very) few constraints	plus partly with constraints	plus freq. severe constraints	Very high + high	Plus good	Plus medium	plus moderate	Very high + high	Plus good	Plus medium	plus moderate	Grazing class 1	Plus grazing class 2
USA	280	2,062	2,156	2,982	792	2,596	3,495	1,987	2,732	3,296	4,061	1,262	1,724	2,068	2,656	853	1,243
CAN	57	514	277	539	68	356	692	156	347	695	1,039	73	159	382	607	170	468
WEU	203	1,333	1,444	2,120	384	1,246	2,256	506	897	1,434	2,009	482	834	1,291	1,770	529	764
EEU	60	523	537	798	132	530	933	290	556	754	938	272	516	675	809	158	198
FSU	240	2,326	2,038	3,577	590	2,664	3,914	1,569	2,342	3,345	4,207	1,184	1,773	2,490	3,140	1,052	1,757
NAF	20	470	438	755	60	455	820	632	838	997	1,187	628	834	992	1,181	363	519
EAF	13	296	443	728	155	678	1,237	274	531	840	1,168	267	510	797	1,095	978	1,154
WCA	34	935	321	1,408	76	1,601	3,050	2,981	4,175	5,175	6,313	2,034	2,966	3,752	4,568	1,039	2,211
SAF	59	490	395	1,295	156	1,228	2,854	1,223	2,043	2,920	3,808	1,178	1,964	2,791	3,620	832	1,890
MEE	24	371	33	84	45	155	396	8	20	58	122	8	20	58	122	3	4
CHN	95	1,592	1,915	3,420	175	973	2,137	638	927	1,267	1,771	615	881	1,185	1,636	891	1,317
OEA	11	178	136	266	9	129	484	169	235	308	381	131	173	223	262	190	266
IND	75	1,773	840	1,919	351	985	1,727	1,308	1,718	2,040	2,293	1,284	1,678	1,979	2,202	227	328
OSA	28	465	191	353	24	120	426	167	192	218	261	161	183	204	242	100	132
JPN	38	86	234	299	40	116	223	28	58	89	136	22	41	57	77	25	36
OCN	22	554	397	1,070	90	467	1,210	268	510	960	1,562	233	429	807	1,324	407	705
PAS	35	866	659	1,225	106	914	2,202	445	697	1,038	1,453	385	549	734	935	531	962
LAC	64	1,750	3,727	5,584	651	3,343	7,312	4,264	6,529	8,741	11,016	2,438	3,652	4,823	5,987	3,062	4,811
TOTAL	1,358	16,584	16,181	28,422	3,904	18,556	35,368	16,913	25,347	34,175	43,725	12,657	18,886	25,308	32,233	11,410	18,765

Bold numbers indicate land potentials, i.e., in regions in which more land of a certain class is available or suitable according to the respective criterion than was used for cropland and infrastructure in 2000.

"Gross availability" of land is impressive, but it is important to bear in mind that not all of that land is really available (Young, 1999). Much of this land is either already used for other purposes (e.g., infrastructure or grazing), protected for reasons of nature/biodiversity conservation, or under forests (Bruinsma, 2009). Nachtergaele and George (2009) estimate that protected areas may cover some 2 million km², and forests some 8 million km² of that land. Further qualifications must also be made. Land suitability can only be assessed in a meaningful way for a particular crop. Land identified as potentially suitable may be prone to physical or economic constraints, such as ecological fragility, low fertility, toxic compounds in the soil, high incidence of disease, or lack of infrastructure. These factors reduce productivity and profitability (Bruinsma, 2009) and must be taken into account in the elaboration of agro-economic-environmental zoning, following the example of Brazil (see Box 20.5).

Land availability is very unevenly distributed among regions. A large percentage of the suitable land in developing countries, for any agricultural crops, including bioenergy, is located in Latin America and sub-Saharan Africa. South Asia and the Near East/North Africa have lower spare land capacity (Bruinsma, 2009).

Global infrastructure and cropland covered 16.6 million km² in 2000 – an area almost equivalent to the estimated cropland with a suitability index higher than 0.7 (Ramankutty et al., 2002). Ramankutty's cropland suitability index is calculated using climate indicators (growing degree days and water availability) and soil indicators (basically, soil pH and carbon) to estimate the probability that a grid cell possesses the physical characteristics for rain-fed cultivation.

In some regions, cropland expansion may face challenges and require costly investments, e.g., in irrigation technologies or other measures of land improvement. In other regions, considerable areas with a cropland suitability index higher than 0.7 are not yet used as cropland or infrastructure. In these regions, cropland expansion can be assumed to be less costly. Globally, only about 57% of the land with a cropland suitability index over 0.5 is already used for infrastructure and cropland. Table 20.2 shows, however, that for two regions, MEE and OSA, current infrastructure and cropland areas already exceed the land with a suitability index higher than 0.5. In these regions, quite poor land is currently used as cropland.

The GAEZ assessment of climate, soil, terrain and slope constraints suggests that the global area of land with no, very few and few constraints is smaller than the area already used for cropland plus infrastructure. However, including land classified as "partly with constraints" results in an additional area which is around the same as the area of land already used for cropland and infrastructure. Cropland expansion potentials seem to prevail in some regions, most notably in sub-Saharan Africa and Latin America and the Caribbean, where agricultural yields could also be improved significantly (Somerville et al., 2010). Moreover, changes in human diets towards less consumption of animal products would allow a reduction in cropped areas compared to a business-as-usual scenario (Aiking et al., 2006; Erb et al., 2009; Dornburg et al., 2010).

GAEZ also provides suitability estimates for rain-fed agriculture with improved technology that differentiate between potentials on currently forested land and potentials restricted to non-forested land. Here the assessment that excludes forests is most relevant (see Table 20.2). Some analyses raise the question of how well suited land potentials labeled "suitable" in this assessment really are for large-scale, intensive cultivation – in particular in tropical regions where land degradation resulting from inappropriate agricultural practices would be a widespread problem (Stocking, 2003). However, other studies (such as Somerville et al., 2010) suggest that there may be a significant potential to improve crop yields in these regions (as also discussed in FAO (2010a) for Tanzania).

Some authors (including, e.g., Showers, 2006) consider that assessments from Ramankutty et al. (2002) and the GAEZ are based on limited sets of data that were extrapolated and applied to large areas. It is recommended that in-depth regional studies should be carried out to avoid overestimating the cropland expansion potential in regions such as sub-Saharan Africa, where transfer of European cultivation techniques has caused large-scale soil erosion, and European agricultural practices often have been unsuccessful.

Much of the cultivable land in sub-Saharan Africa and Latin America is under valuable forests or in protected areas, and these regions are also those where the largest potentials for arable land are found. Ramankutty et al. (2002) argue that tropical soils could potentially lose fertility rapidly if taken into cultivation, and are highly vulnerable to climate-change impacts. IAASTD (2009) estimates that only 7% of the cultivable areas in sub-Saharan Africa, and only 12% of those in Latin America and the Caribbean, are devoid of more or less severe soil constraints that limit sustainable and profitable production. In fact, agricultural yields mainly in Africa are very low, half of those in the United States (Somerville et al., 2010), and some studies suggest large potentials to increase these yields, and also to make more area available (UNCTAD, 2009; UNF, 2008).

A comparison of area potentials from Ramankutty et al. (2002) and GAEZ with the area listed under "grazing classes 1 and 2" (last two columns of Table 20.2) shows that regions with much land in the grazing land class 1 (best-suited grazing area) are mostly also those in which there are large potentials for cropland expansion. These results are found by both Ramankutty and the GAEZ, whereas those regions with little cropland expansion potential also have small areas of high-quality grazing land. Because the assessment of grazing land quality by Erb et al. (2007) is consistent with data on grazing intensity (Haberl et al., 2007), it can be used to calculate potentials to intensify grazing in order to make land available for the additional cultivation of crops, including bioenergy crops.

The studies mentioned in this review indicate that the availability of land for cultivation of agriculture and bioenergy crops depends mostly on the following factors:

- The most important factor is the intensity with which this land is used today for other purposes, in particular grazing. Calculating livestock feed balances based on national-level livestock data from the FAO (e.g., Wirsenius 2003a; 2003b; Haberl et al., 2007; Krausmann et al., 2008) allows us to approximate the intensity with which grazing areas are currently used for feed production (e.g., Erb et al., 2009). Together with other studies mentioned above, the analysis has shown that grazing areas are used with very different intensities across the globe, suggesting that increased feed production through improved management of grazing areas could make considerable areas available for bioenergy production.

- Other important constraints include: the need to set aside valuable areas for biodiversity/nature conservation; limited water availability; and lack of infrastructure such as roads (i.e., limited accessibility). In some regions with poor soil quality, high levels of investment might be needed to allow cultivation. More important, however, is how the availability of land will change in the future due to changes in demand for products from land, and impacts of climate change.

20.2.2 Competing Future Demands for Land

Basically, land is used by humans for at least three core functions (Dunlap and Catton, 2002):

- resource supply, i.e., the provision of raw materials or energy needed for production and consumption processes, including non-renewable resources such as fossil fuels, minerals and other materials extracted from geological deposits, and renewable ones such as biomass or water diverted from current biogeochemical cycles, ultimately driven by an influx of solar energy;

- waste absorption, as well as buffering and regulating capacities of ecosystems; and

- space occupied for human infrastructures, including housing, gardening and recreational areas, as well as industrial and transport facilities.

Most human uses of land are dependent upon the land's biological productivity, i.e., its NPP per unit area and year. Many land uses involve harvesting parts of the actual or accumulated NPP in the form of biomass derived through agricultural or forestry activities (Haberl et al., 2004). At the same time, human land use often alters the land's productivity (Haberl et al., 2007). In some regions, especially sub-Saharan Africa, current land-use practices result in low yields. Implementation of adequate technologies could help to raise agricultural yields considerably (Somerville et al., 2010; FAO, 2010a; Dornburg et al., 2010).

Table 20.3 | Global average per capita supply and demand of primary biomass.

	Per capita flows of dry-matter biomass (kg/cap/yr)	Percentage of total(%)
Supply		
Harvest of primary crops	473	24%
Harvest of fodder crops	94	5%
Harvest by-products (residues)	485	24%
Grazed biomass (from grazing land)	634	32%
Wood harvest (excluding residues)	321	16%
Total supply	2007	100%
Total Supply	**2007**	**100%**
Food	248	12%
Market feed	152	8%
Seed	15	1%
Other uses	233	12%
Non-market feed, including grazing	1004	50%
Firewood	197	10%
Industrial wood	125	6%
Wastes, losses	33	2%
Total Demand	**2007**	**100%**

Source: Krausmann et al., 2008.

As shown in Table 20.1, biomass production through agriculture and forestry takes up by far the largest area. Many buffering and/or regulating services of ecosystems are to some extent compatible with productive functions (see discussion of "multifunctional land use" in the Introduction). Global biomass balances (Krausmann et al., 2008) show that most of the biomass is used for food and feed, fiber, and other uses, whereas the amount of biomass directly used for bioenergy production (as firewood) is relatively small.

Firewood reported in Table 20.3 amounted to about 22 EJ/yr globally in 2000. More than half of global biomass supply is used as feed for livestock (see Table 20.3).[2] Reuse and recycling of biomass (i.e., "cascade" or IFES utilization) already plays an important role; agricultural residues and by-products are used as feed inputs and for bioenergy production (e.g., sawdust, bark, residues from paper production, etc.). These "cascadic flows" and some underreported flows (e.g., collection of firewood on non-forested land not reported in FAO data and, therefore, also missing from Table 20.3) contributed approximately half of global bioenergy production (45±10 EJ/yr) in the year 2000. Because conversion of forests to bioenergy and agricultural crop plantations would result in a large carbon debt and poor GHG emission performance (and bioenergy options in forestry are discussed in Chapter 7), we here focus on farmland, i.e., cropland and grazing areas.

---

2   It is worth noting that, in fact, a significant part of industrial wood is used as energy (for instance, black liquor is about 50% of wood consumed in the cellulosic pulping industry).

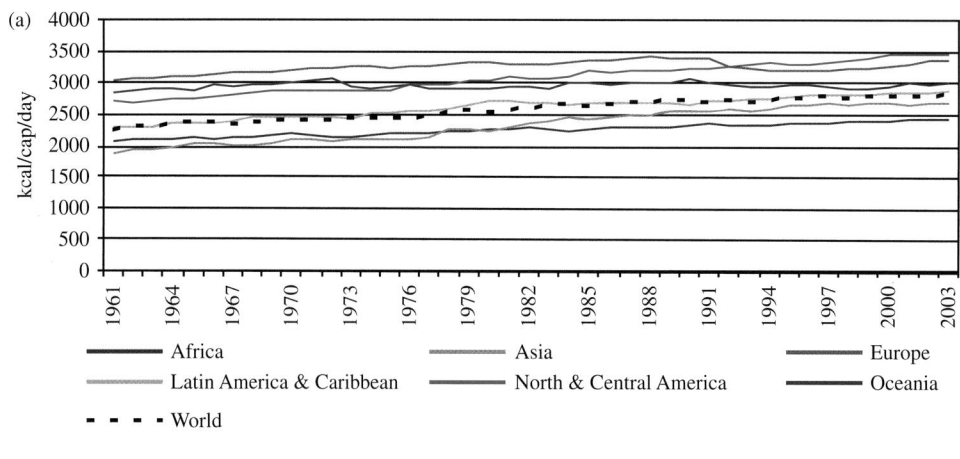

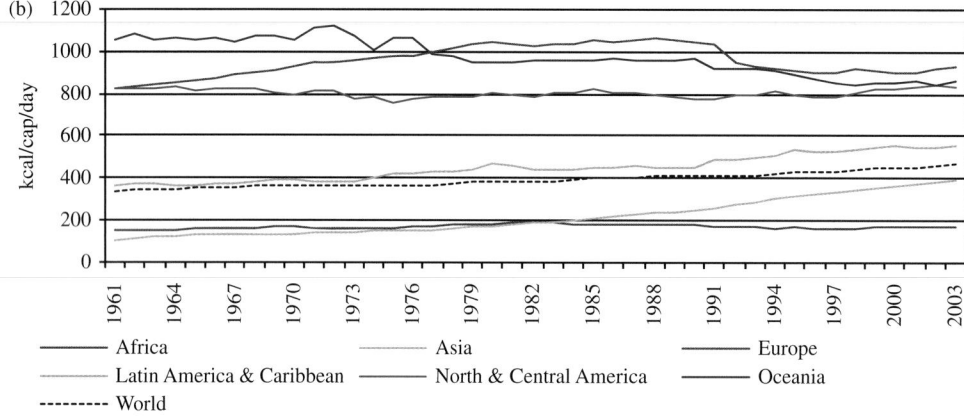

Figure 20.1 | Total food consumption (a) and consumption of animal products (b) 1961–2003. Source: based on data from FAO, 2009a.

As shown in Table 20.3, the food system accounts for a significant amount of the biomass used, as well as the area needed for producing that biomass. Three sets of factors will dominate in shaping the future trajectory of the area required globally for food production:

- The volume and composition of global food demand, which in turn depend on population growth and changes in diets: There is a strong correlation between income levels and the volume and composition of food consumed: food consumption rises with income. The proportions of animal products, sugars and fats rise with income, while consumption of cereals, pulses and roots drops with income (Erb et al., 2009). Figure 20.1 shows global trajectories of per capita food consumption 1961–2003.

- Yield levels on agricultural areas (cropland and grazing land): The total volume of crops, forage and by-products produced is the product of the area used times the yield per unit area and year. Yields are highly variable both between regions and across time (see Figure 20.2).

- Feeding efficiencies in the livestock sector: The relationship between feed input and product output (e.g., meat, milk, eggs) is highly variable between different livestock rearing systems. Cross-country

analyses as well as longitudinal data indicate how large these differences are (see Figure 20.3).

Figure 20.1 shows that the total amount of food calories consumed per capita and year is rising continuously in almost all regions. The only exceptions are Europe and Oceania after 1990, where food intake has more or less stabilized at a high level after 1990. The consumption of animal products seems to have stabilized in Europe, North and Central America and Oceania at between 800 and 1000 kilocalories per capita and day (kcal/cap/day); it is rising throughout the developing world, with the exception of Africa, where it has remained almost constant at a low level of 200 kcal/cap/day.

Improvements in agricultural technology have helped to increase yield levels across the globe considerably, while sometimes also resulting in undesired environmental consequences such as soil degradation, water pollution, and others (IAASTD, 2009), when increases in yields are based on intensive high-input monocultures. As Figure 20.2 shows, average global cereals yields grew from 1.35 metric tonnes per hectare and year (t/ha/yr) in 1961 to 3.54 t/ha/yr in 2008. However, yield growth has progressed at varying rates, and has led to considerably different yield levels across the globe. In the lowest-yielding region, yields have remained almost constant,

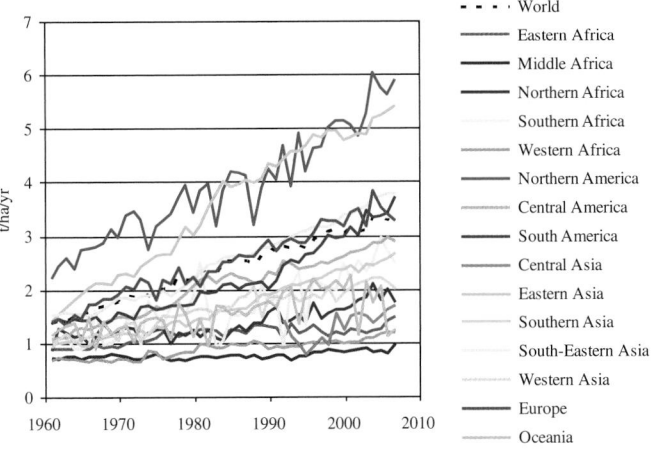

Figure 20.2 | Cereals yields in the world. Source: based on FAO, 2009a.

Legend:
- - - World
— Eastern Africa
— Middle Africa
— Northern Africa
— Southern Africa
— Western Africa
— Northern America
— Central America
— South America
— Central Asia
— Eastern Asia
— Southern Asia
— South-Eastern Asia
— Western Asia
— Europe
— Oceania

Table 20.4 | Changes in cattle numbers and area dedicated to cattle in São Paulo State, 2001–2008.

	2001	2002	2003	2004	2005	2006	2007	2008
Number of cattle (million)	13.15	13.46	13.76	13.77	14.07	13.75	12.20	11.95
Pasture area (1000 km²)	102.9	101.0	101.1	101.2	100.1	97.1	91.2	76.4
Density (number/km²)	128	133	136	136	141	142	134	156

Source: Coelho et al., 2008.

below 1 t/ha/yr, whereas yields have reached more than six times that level in the highest-yielding region, North America. In some regions where yields are currently low, especially sub-Saharan Africa, implementation of adequate technologies could help to raise agricultural yields considerably (Somerville et al., 2010; FAO, 2010a; Dornburg et al., 2010).

Feeding efficiencies in the livestock sector are also highly variable across time and space (Figure 20.3). The differences are particularly strong for grazers (cattle, buffalo, sheep, goats); less so for pigs and poultry. The amount of biomass required per unit of output and, therefore, also the area needed per unit of output depend strongly on the respective livestock rearing systems.

One main reason for the large differences in feeding efficiencies is that livestock is used in a multifunctional manner in subsistence systems. Besides producing animal-based food, livestock plays a big role as a work force (draught animals), is important for the nutrient cycle (through use of dung as fertilizer), and has important social functions, e.g., in rituals, as status symbols, as buffers for times of poor food supply, etc. (Harris, 1987; Krausmann, 2004; Wildenberg, 2005).

Market-oriented systems can be relatively "inefficient" in terms of their feed balance if area is abundant and other inputs (e.g., labor) are more costly and hence more important optimization criteria (Erb et al., 2009). Considering ecological objectives, animal welfare and product quality criteria in livestock rearing may also reduce feeding efficiencies compared to intensive, indoor-housed rearing systems, although this effect should not be over-emphasized. Compared to subsistence livestock raising, and even some existing market-oriented but feed-inefficient systems, modern, optimized organic and humane livestock systems offer large gains in terms of feeding efficiency (Erb et al., 2009).

Using less land area in livestock production by increasing grazing intensities is possible. For example, in São Paulo State, Brazil (Goldemberg, 2009), cattle density heightened in the last decade, thereby increasing area for food/bioenergy crops. Soares et al. (2009) show that the overall

balance on GHG emissions is positive, despite the increases in intensive animal husbandry and the corresponding replacement of cattle areas by sugarcane crops.

The most important uses of biomass for other uses than food, feed or bioenergy materials are pulp and paper, construction materials and chemicals, most of which come from the forestry sector, as discussed in Chapter 7. In different regions, other products – for example, cut flowers – can be important. The chemical industry could boost its use of biomass in the future, as bulk chemicals from biomass have a large potential to be substituted for fossil-fuel-based feedstocks. At present, the amounts of biomass (and related land) used for this purpose are low, and future projections still indicate a limited demand for land for that purpose. However, estimating land demands for future chemical production must also take into account the production of chemicals in bio-refineries where transport fuel and electricity can be co-generated from biomass.

Furthermore, some bio-based bulk chemicals (i.e., plastics) are often used for waste-to-energy generation in industrialized countries (see Figure 20.4).

FAO estimates that global agricultural production would have to be increased by 70% to feed the global world population expected in 2050, meeting a food supply target of 3130 kcal per capita and day (Bruinsma, 2009). This considerably exceeds global average food supply for 2000, which was 2790 kcal/cap/day (FAO, 2005), but might still leave approximately 4% of developing-country populations chronically undernourished if current patterns of inequality of food distribution persist. On the other hand, the International Food Policy Research Institute (IFPRI, 2009) indicates that daily per capita calorie availability in developing countries in 2000 was 2694 kcal, and that scenarios in these countries for 2050 could reach 2896 kcal/cap/day, if there were no climate change effects, with the largest increase (13.8%) in East Asia and the Pacific. However, there are gains for the average consumers in all countries – 3.7% in Latin America, 5.9% in sub-Saharan Africa, and 9.7% in South Asia. Taking into account climate change, calorie availability in 2050 is lower than those numbers; it actually declines relative to 2000 levels throughout the world (IFPRI, 2009).

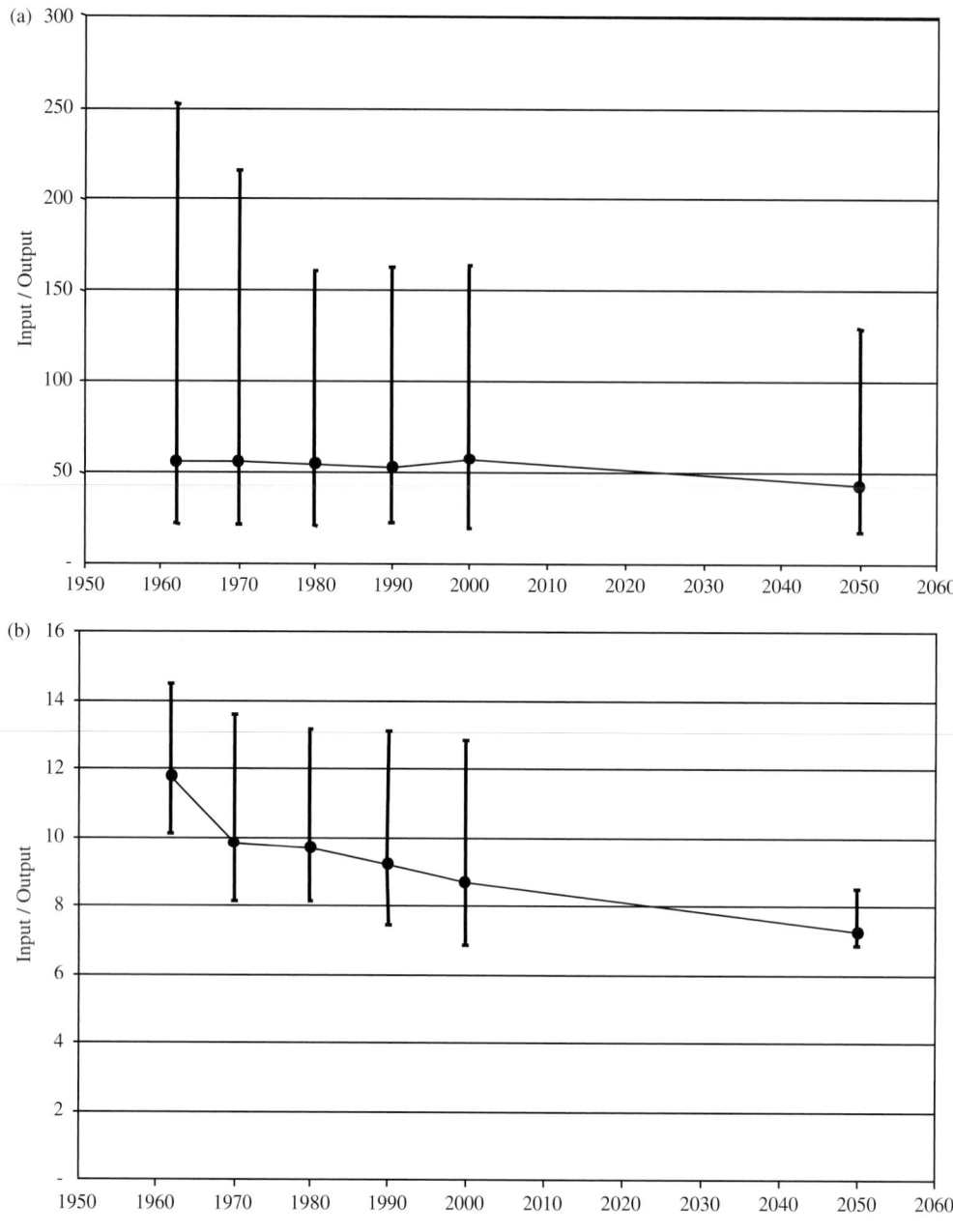

Figure 20.3 | Global feed efficiencies (feed input per unit of animal product output, both measured as dry-matter biomass): (a) Grazers (cattle and buffalo, sheep, goats); (b) Non-grazers (pigs, poultry), 1961–2000. Source: Haberl et al., 2011. Dots represent the weighted global average, whiskers the variability between the 11 regions covered in that study.

The share of animal products in people's diet would strongly affect the amount of primary biomass and area required to meet global food demand. If diets shift towards less protein from animal products, the global demand for cropland and grazing lands can be much lower than in a business-as-usual scenario (e.g., see Erb et al., 2009; Dornburg et al., 2010).

According to the "Trend" (business-as-usual) scenario of Erb et al. (2009) – based on FAO (2006) – higher yields and increased cropping intensity are expected to contribute 90% of the growth in crop

production by 2050 (80% in developing countries), with the remainder coming from land expansion. Arable land would expand by around 9% compared to 2000 in the global total. Cropland expansion would be largest in sub-Saharan Africa, Latin America and the Caribbean, and Oceania/Australia. Global cropland area would reach 16.6 million km² in 2050 in such a scenario.

In this scenario (Erb et al., 2009), growth of cropland areas in developing countries was assumed to be 12% (1.2 million km²), almost all in Africa and Latin America, which is partly offset by a decline of some

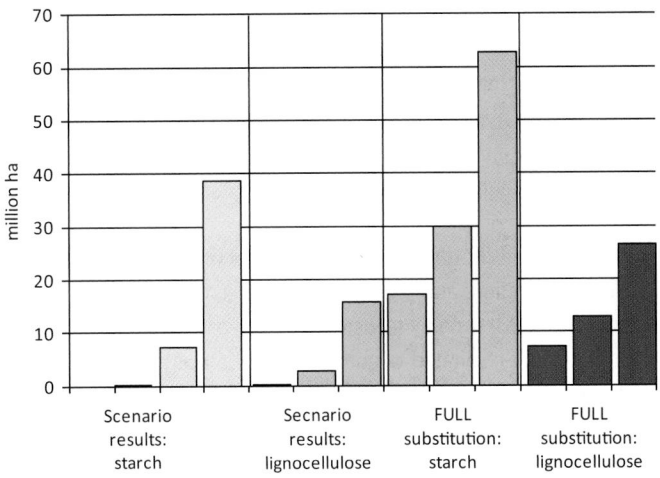

Figure 20.4 | Total land used by 2050 for the selected organic chemicals according to the three scenarios (low, medium and high) and for full substitution of selected chemicals, i.e., 100% bio-based chemical production. Source: Dornburg et al., 2008.

0.5 million km² (8%) in developed countries. Land equipped for irrigation would increase by 0.32 million km² (11 %), which is assumed to take place almost exclusively in developing countries. Water withdrawals for irrigation are forecast to increase by 11% between 2006 and 2050. It is forecast that crop yields would rise at a slower pace than in the past. Annual growth rates would halve to 0.8% per year compared to historical growth rates.

According to Cassman (1999) and Peng et al. (2000), many options to achieve yield gains have already been discovered, and further increases seem unlikely in some areas due to physiological limits. For example, these authors argue that further improvements in harvest indices that seek to increase the share of the desired product (e.g., grain) at the expense of supporting tissues such as leaves and stems (straw) seem unlikely for many cultivars because of physiological limits. Harvest indices of the most advanced rice cultivars are already around 0.50–0.55. It would seem unlikely that this can be increased substantially.

Tilman et al. (2002) argue that a continuation of past yield increases seems unlikely, because most of the best-quality farmland is already being used. According to these authors, rates of yield increases are already declining (e.g., rice in Southeast Asia), and yields have leveled off (e.g., rice in Japan, Korea, and China) as they approach limits set by soil and climate. Cassman (1999) argues that soil degradation and depletion of nutrient stocks in soils is an additional challenge. Also, a more widespread adoption of less intensive agricultural technologies (e.g., organic farming) could result in lower rates of yield growth or even declines in yields in regions where intensive conventional cultivation methods are common (Erb et al., 2009). On the other hand, Somerville et al. (2010) suggests that high investments can benefit developing countries, mainly in Africa, where countries present the lowest agricultural rates worldwide.

Improvement of management practices could help to maintain growth in yields, mostly due to improved stress tolerance, avoidance of nutrient and water shortages, improvements in pest control, etc. Some scenarios even foresee higher yield increases than the FAO (e.g., IAASTD, 2009). For example, the "Global Orchestration" scenario analyzed in the Millennium Ecosystem Assessment (2005) assumes that yields in 2050 could be, on average, 9% higher than those forecast by the FAO (2006), if world agriculture is pushed towards strong intensification.

Dornburg et al. (2010) mention other studies (Evans, 1998; Smil, 2000) suggesting that sufficient food – even for around 10 billion people – could be produced, provided that crop yields can be further improved by enhanced crop management and/or genetic modifications. It is also stressed that the large variability in regional climate and hydrology necessitates a detailed analysis of the biophysical possibilities for crop production. In any case, substantial investments will be indispensable for maintaining growth in crop yields (Khan et al., 2009), and additional ones are needed to avoid economic constraints that would prevent the realization of such technical yield potentials (Koning and van Ittersum, 2009; Somerville et al., 2010).

In many developing countries, especially in semi-arid and arid regions, yields are often far below those obtained in industrialized economies (IAASTD, 2009; IFPRI, 2009). In many regions, average increases in productivity in recent years have been only moderate, especially in Africa (FAO, 2009a; Somerville et al., 2010). Adoption, implementation and enforcement of adequate policies to foster environmentally, socially and economically sustainable yield increases could, therefore, offer large benefits (IAASTD, 2009). Recent studies call for more research efforts dedicated to "sustainable intensification," i.e., management practices and technologies that allow further yield increases but minimize adverse environmental, social or economic effects (Godfray et al., 2010).

World food systems may be affected by changes in temperature and precipitation (mean values and variability) and the atmospheric carbon dioxide (CO_2) concentration. All three factors could have substantial effects on agricultural yields. At present, however, there are substantial knowledge gaps with respect to underlying processes (e.g., downscaling of global climate scenarios to regional or local levels) and also how they will affect crop growth in the field, under real-life conditions. This depends on factors that are difficult to predict, including responses from farmers.

IFPRI (2009) analyzed climate change effects on crop production and CO_2 fertilization effects and predicted negative impacts, mainly in sub-Saharan Africa and South Asia. In fact, some regions were expected to be adversely affected by climate change, in particular in tropical regions (Cerri et al., 2007; IFPRI, 2009), but regional differentiation of climate change effects is important (IFPRI, 2009). Beyond 2050, climate change implications on crop yields and production were forecast to be severe on the global scale, with or without the CO_2 fertilization effect (IFPRI, 2009).

A recent study (Müller et al., 2010) showed that the effect of climate change on agricultural yields is highly uncertain and strongly depends on the CO_2 fertilization effect, which is poorly understood and could interact with management decisions of farmers. The study was based on simulations using the dynamic global vegetation model LPJmL. It considered three different emission scenarios implemented in five different Global Circulation Models. LPJmL was run with the CO_2 fertilization effect switched on and off, to reflect scientific uncertainty. The study found that crop yields could decrease by 13% or rise by 22% in 2050, compared to the levels forecast to prevail without climate change, depending on scenario assumptions. Using these results, Erb et al. (2009) and Haberl et al. (2011) showed that climate change impacts on agricultural crop yields would, at any given level of global food demand, result in considerable changes in the area available for the cultivation of energy crops; adverse impacts would reduce, and positive impacts would increase, the area available for bioenergy crops.

Case studies show that continuous investments in the improvement of productive technologies (i.e., Brazilian sugar cane) can mitigate or even offset adverse impacts of climate (change) on yields (von Braun, 2007). IFPRI (2009) concludes that even without climate change, greater investments in agricultural knowledge (for all different end uses) are needed to meet the demands of the future world population in 2050, mainly in developing countries, and argues that improved agricultural productivity can be an important mechanism for alleviating poverty indirectly by creating jobs and lowering food prices.

This discussion leads to the conclusion that the availability of productive areas for the cultivation of bioenergy crops in the future will strongly depend on:

- Total future food demand, which is in turn influenced by population numbers, per capita food calorie intake and the fraction of animal-based products consumed: all of this depends on future income levels and a host of other socioeconomic, political and cultural factors.

- Yield levels on farmland, in particular cropland yields: most studies agree that yield increases will be able to meet a substantial fraction of future global food demand so that the growth of cropland area required could be low, perhaps only 5–10%. But significant social, economic, environmental and technological issues related to future yield growth remain to be solved through appropriate research and technology development. Climate change could have significant impacts on yields, which could positively or negatively influence the availability of area for energy crop cultivation.

- Feeding efficiencies and many other critical issues related to the livestock sector: technological and other changes in livestock rearing are likely to contribute to increases in the output of animal-based food per unit of feed intake, but the extent of this efficiency growth, as well as its possible costs in terms of environmental impacts, product quality, and animal welfare, is at present imperfectly understood.

Comparison of livestock densities across world regions suggest that livestock densities could be increased, in some regions by large margins, thereby making substantial areas available for bioenergy crops. Appropriate management will be crucial to avoid adverse environmental and socioeconomic effects, in particular where subsistence economies might be affected.

- Area required for other purposes, including infrastructure, biodiversity conservation, and production of biogenic materials for various purposes, most notably feedstocks for the chemical industry, which seems to be of limited quantitative importance but could become more important in the future for integrated multipurpose use of biomass ("bio-refinery").

Note that conversion of forests to bioenergy crops was excluded from that discussion due to the fact that this would entail a large carbon debt, i.e., very unfavorable GHG emissions per unit of energy produced over many years if not decades. Bioenergy potentials from forestry, residues, manures and wastes are discussed separately in Chapter 7.

20.2.3   Area and energy potentials from dedicated bioenergy crops in 2050

The global potential availability of biomass for energy has been assessed in various studies. Many of these studies primarily or exclusively focus on energy crops, and the potentials estimated vary significantly. Dornburg et al. (2010), Hoogwijk et al. (2003) and Berndes et al. (2003) identify methodological differences, critical parameter assumptions and varying system boundaries that are chiefly responsible for the differences in the estimated potentials. Important parameter assumptions determining the technical potential for energy crops are:

- restrictions on land available for energy crops;

- relevant factors for future development of land use, such as population growth, diets, international food trade and technology changes, in particular with regard to crop yields, and feeding efficiencies in animal husbandry;

- future productivity of energy crops;

- agricultural commodities markets; and

- sustainability restrictions on the growth of biomass.

Table 20.5 summarizes the main features of recent studies estimating the energy potentials from energy crops. The considerable differences in the estimates result from the following factors:

- all studies use aggregate modeling approaches regarding future developments of yield and land use;

- only a few studies account for possible future land-use changes and associated uncertainty by using an scenario approach analyzing different futures (Smeets et al., 2007; Hoogwijk, 2004; van Vuuren et al., 2009; WBGU, 2009; Erb et al., 2009; Dornburg et al., 2010); and

- only a few studies explicitly consider restrictions arising from environmental and social impacts of bioenergy production (e.g., land degradation, loss of biodiversity, competition with food, and water limitations) or present spatially explicit data on land-use and bioenergy potentials (WBGU, 2009; van Vuuren et al., 2009; Erb et al., 2009; Goldemberg and Guardabassi, 2009; Dornburg et al., 2010).

According to the present review, a general tendency appears to emerge that the more recent studies show lower estimated bioenergy potentials than earlier ones. The reasons for this include the following: newer studies consider environmental constraints (e.g., carbon payback time and biodiversity conservation in WBGU, 2008); constraints on the suitability of areas for bioenergy production have become more apparent (e.g., WBGU, 2008; Dornburg et al., 2010); many areas assumed to be available for bioenergy production are already used for grazing (Erb et al., 2009); and new research has demonstrated that previous studies overestimated yields of bioenergy crops, often by 100% or more (Johnston et al., 2009).

Recently, one study (Smeets et al., 2007) suggested a very high bioenergy potential from energy crops, with an upper range that even exceeds the theoretical potential for bioenergy production discussed in Chapter 7. This very high potential resulted from a large land area (36 million km^2, i.e., 28% of the earth's land surface excluding Greenland and Antarctica) multiplied by a high productivity estimate (34 MJ/m^2/yr). For comparison, the current global average aboveground NPP is 9.5 MJ/m^2/yr.

Based on this review, four studies (Erb et al., 2009; van Vuuren et al., 2009; WBGU, 2009; Dornburg et al., 2010) were selected as the basis for this assessment. They were selected because they generally reflect essential environmental constraints such as biodiversity conservation, water scarcity, land quality/suitability, land degradation, and food supply (i.e., future cropland needs and interrelations with feed demand of livestock), although with different methods (see Table 20.5). Moreover, sufficient data allowing calculation of bioenergy potentials on the level of the 18 GEA regions were only available for three of these studies. The results derived from these three studies are then discussed in the light of a larger body of literature (including Dornburg et al., 2010).

As can be seen in the table above, the assessment of the area that would be available for dedicated energy crops varies between different studies. This survey summarized in Table 20.5 shows that yield expectations of bioenergy plantations also differ widely, approximately 1–21 kg/m^2/yr. Differences in yields of bioenergy plantations largely

result from assumptions on land suitability, choice of bioenergy crop (yields of lignocellulosic crops and perennial grasses are higher than those of food crops), and management (e.g., fertilizer input) (Harberl et al., 2010).

Table 20.6 reports the energy crop areas for three studies, and Table 20.7 the bioenergy potentials found in the four recent studies selected. These potential estimates were derived as follows:

- Van Vuuren et al. (2009) used the integrated modeling framework IMAGE and the energy model TIMER (which is a part of the IMAGE framework) to calculate available area for energy crops and related bioenergy potentials in 2050. Only abandoned agricultural land (according to an approach by Hoogwijk et al. (2005)) and natural grasslands were assumed to be available for bioenergy production, thereby assuming global accessibility factors of 75% (abandoned farmland) and 50% (natural grassland). IMAGE sub-models on land use were used to simulate land required for food production, driven by demand for food and timber, and climate change. Calculations proceeded at the level of grid cells (0.5x0.5°). Water scarcity, land degradation (based on the International Soil Reference and Information Centre's (ISRIC, 1991) *Global Assessment of Human-induced Soil Degradation* – GLASOD) and biodiversity/nature reserve areas were considered in various scenario calculations ranging from "no restrictions" to "strict criteria."

- Erb et al. (2009) followed a "food-first" approach. Assumptions on future diets (four assumptions), cropland yields (three assumptions: FAO, fully organic, intermediate), cropland expansion (+9%, +19%) and livestock feeding efficiencies (conventional, humane, organic) were derived from FAO and other studies and databases. A biomass-balance model was used to identify combinations of factors ("scenarios") that were "feasible," i.e., provided sufficient food. The model closes the balance between biomass supply (harvest of primary crops and grazing) and biomass demand (food and fiber). For scenarios classified as "feasible," the area available for bioenergy crops was calculated by assuming that grazing intensity would be maximized and, if existent, all cropland area not required for food or fiber production could be used for bioenergy. The model calculates bioenergy potentials at three levels: primary bioenergy crops on cropland not needed for food supply, primary bioenergy crop potentials on grazing areas of the highest-quality class (which is assumed to be intensified to its limits), and residue potentials. As the latter are discussed separately (see Chapter 7), Table 20.7 shows only the potentials for primary bioenergy crops. This study considered growth in infrastructure areas and assumed that there would be no deforestation for bioenergy. The MIN scenario assumes the richest diet (which is only feasible with the most intensive technology and highest yield levels), whereas the MAX scenario combines the lowest food demand with the highest possible agricultural yields and livestock feed efficiencies. The "FAO world" scenario was based on FAO (2006).

Table 20.5 | Overview of recent studies on technical potentials of biomass from energy crops

Reference	Type of potential	Regions	Time frame	(Sustainability) constraints	Land use types	Land area used [mio. Km²]	Productivity [tonnes dry matter/ha/yr]	Potential of energy crops [EJ/yr]
van Vuuren et al., 2009	Technical	Global	2050	Biodiversity, Food security, Soil degradation, Water scarcity	Abandoned agricultural land (75%) Grassland (25%)	13	Depending on land suitability and climate factors 1.0–3.2 kg dry matter/m²/yr	120–300 EJ/yr (unconstrained) 65–115 EJ/yr (constrained)
WBGU, 2008	Technical	Global	2050	Biodiversity, C balance, Deforestation, Degraded land, Food security, Water scarcity	Land suitable for bioenergy cultivation according to the crop functional types in the model, considering sustainability	2.4–5.0	7.5–12.6 t/ha/yr	34–120 EJ/yr
Campbell et al., 2008	Technical	Global	2000 (not clearly mentioned)	Agricultural lands, Ecosystems, Food security, Releasing carbon stored in forests, Water scarcity	Abandoned agricultural land (100%)	3.9–4.7	4.3 t/ha/y (AGB)	32–41 EJ/yr (AGB)
Field et al., 2008	Technical	Global	2050	Biodiversity, Food security, Ecosystems, Deforestation	Abandoned agricultural land (100%)	3.9	3.2 tC/ha/yr	27 EJ/yr (AGB)
Dornburg et al., 2010	Technical	Global	2050	Land for food excluded Various assumptions on (non-) exclusion of degraded and protected land	Not explicitly specified	Not specified	Not specified	Energy crops: 120 EJ/yr
Smeets et al., 2007	Technical	11 World regions	2050	Biodiversity, Deforestation, Food security	Surplus agricultural land (100%)	7.3–35.9	16–21 odt(oven dry tonnes)/ha/yr	215–1272 EJ/yr
Hoogwijk et al., 2005	Technical	11 World regions	2050–2100	Biodiversity, Food security	Abandoned agricultural land (100%) Remaining land not for food or material procution (10–50 %) Extensive grassland	Abandoned: 0.6–1.5 Rest land 0.3–1.4	Depending on land suitability and climate factors	Abandoned: 130–400 EJ/yr Rest land 235–240 EJ/yr Total: 300–650 EJ/yr
Erb et al., 2009	Technical	11 World regions	2050	Excluded: Land for food and feed, forestry and unproductive land	Cropland not needed for food and fiber supply Intensification of grazing land	2.3–9.9 depending on food and feed demand (44 scenarios)	Equal to potential (cropland) or actual (grazing land) NPP	Bioenergy crops: 28–128 EJ/yr Residues: 21–36 EJ/yr

- The German Advisory Council for Global Environmental Change (WBGU, 2009) considered two assumptions on future land requirements for food production, one in which the current cropland area was held constant, and one in which an additional demand for cropland area of 1.2 million km² was assumed. The other constraint was area requirements for nature protection (biodiversity hotspots, nature conservation areas, and wetlands) and exclusion of areas with carbon payback times exceeding 10 years. The study considered the impact of future climate change, including changes in CO_2 levels. Calculations were performed using the LPJmL dynamic global vegetation model that is able to simulate natural and agricultural vegetation (Bondeau et al., 2007). The MIN scenario assumes the highest area requirement for food and nature conservation, the MAX scenario the lowest. "Intermed" is the arithmetic mean of all other combinations.

- Dornburg et al. (2010) developed a sensitivity analysis, using existing modeling tools, to quantify key uncertainties regarding

biomass potentials and demand. For the sensitivity analysis, the integrated assessment model (IMAGE) was applied, using the reference scenario of the Organisation for Economic Co-operation and Development (OECD) Environmental Outlook as a baseline (OECD, 2008a). This baseline is a "medium-development" scenario in terms of changes in population, economic development, and agricultural productivity. According to the study, to assess the potential impact of water scarcity on bioenergy potentials, the maps of biomass potentials were overlaid with those of water stress as calculated by the Water Gap model. The Water Gap model uses an index in which a value of 0.2 and higher is defined as moderate water scarcity, while values above 0.4 are defined as severe water scarcity. For all the calculations, rain-fed production conditions were assumed. In order to estimate the impact of degraded land use on biomass potentials, data from the GLASOD database that classified land worldwide in terms of soil degradation was used.

Table 20.6 | Areas assumed to be used to grow bioenergy crops in three recent bioenergy potential studies.

	vanVuuren et al.		Erb et al.			WBGU		
	No criteria	Strict criteria	MIN	MAX	FAO world	MIN	MAX	Intermed.
	(1000 km²)		(1000 km²)			(1000 km²)		
USA	242	303	0	1089	676	177	402	344
CAN	193	2	0	254	157	55	186	132
WEU	88	61	21	237	80	222	579	471
EEU	29	34	60	266	205	68	91	86
FSU	58	13	582	1167	935	168	426	360
NAF	0	0	113	282	189	59	86	73
EAF	40	9	310	740	507	88	168	131
WCA	462	99	324	858	556	142	194	176
SAF	184	1	262	652	436	271	385	333
MEE	1	139	0	13	6	11	15	13
CHN	93	148	52	602	161	210	516	413
OEA	142	170	70	156	109	18	54	40
IND	3	87	0	116	51	282	318	300
OSA	9	49	0	30	13	18	31	26
JPN	25	1	1	18	5	0	0	0
OCN	258	43	2	249	120	147	486	326
PAS	4	6	243	558	428	39	217	142
LAC	711	168	398	2622	1415	480	1062	791
Total	**2545**	**1334**	**2318**	**9912**	**6047**	**2454**	**5215**	**4156**

Table 20.6 shows that the area that could be available for bioenergy crops in 2050 ranges from 1.3–9.9 million km², which is about 1–8% of the earth's total land surface excluding Antarctica and Greenland. Erb et al. (2009) found the highest land area availability of all three studies (up to 9.9 million km²), because this study did not exclude areas for nature/biodiversity conservation. The study by van Vuuren et al. (2009) found the lowest land potentials (1.3–2.5 million km²), while those of WBGU are intermediate.

Maps of bioenergy crop areas found to be available in the first three studies are shown in Figure 20.5. These three studies agree that large areas might become available in sub-Saharan Africa and South America, and that areas in cold climates will not contribute significantly to global bioenergy production. Nevertheless, significant discrepancies with regard to extent and spatial patterns prevail in other regions of the world. For example, van Vuuren et al. (2009) and WBGU (2008) find substantial areas in Northern Australia not included in the dataset by Erb et al. (2009). In contrast to van Vuuren et al. (2009), WBGU (2008) and Erb et al. (2009) identify significant potentials in eastern China, India and Southeast Asia. In the United States and Eurasia, patterns differ due to the different assumptions of each of the studies. Erb et al. (2009) and WBGU (2008), for instance, assume that areas for energy crops might be available in the US Corn Belt, whereas van Vuuren et al. (2009) identified suitable areas in western parts of the United States. For Europe,

Erb et al. (2009) and WBGU (2008) assume substantial area potentials in Eastern Europe, whereas van Vuuren et al. (2009) identified largest potentials in Northern Germany, the United Kingdom, and Ireland. In conclusion, further work on spatial patterns of possible areas for energy crops would be helpful to identify more robust estimates of suitable areas for energy crops. A refinement of the WBGU (2008) work – which basically confirmed the results of the WBGU study – was recently published (Beringer et al., 2011).

The results displayed in Table 20.6 are in line with a recent study by the IEA (2010) which suggested that 2.5–8 million km² could be available globally for bioenergy crops if constraints such as exclusion of forested land, valuable or protected habitats, etc. are properly accounted for (see also FAO, 2008a). The figures presented in Table 20.6 are a downward revision of earlier estimates that suggested area availabilities of 12.8 million km² (IPCC, 2007) or even as much as 37 million km² (WMO, 2006). The main reason is that new studies consider constraints stemming from livestock farming, nature protection and GHG emissions (carbon payback time) that have previously not sufficiently been considered.

The estimates of the global technical potential of primary bioenergy production (i.e., dedicated bioenergy plants) presented in Table 20.7 were derived from data reported in the same three studies. Table 20.7 summarizes the findings of the three studies and gives ranges for the

a) van Vuuren et al., 2009

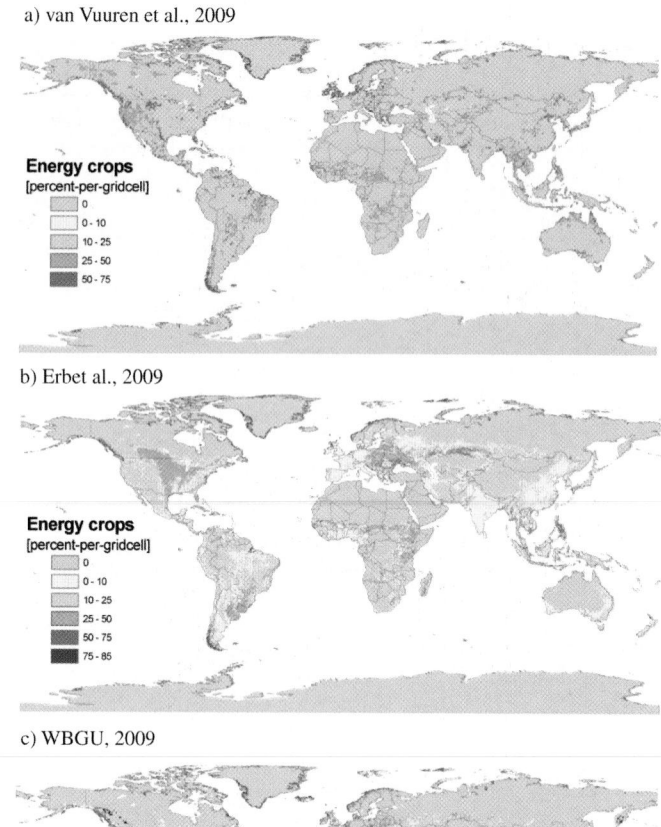

b) Erbet al., 2009

c) WBGU, 2009

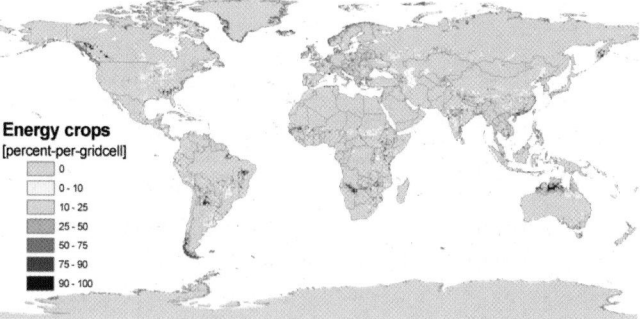

Figure 20.5 | Maps of the areas found to be available in the first three studies used in this assessment: (a) constrained scenario of van Vuuren et al., 2009; (b) TREND scenario based on Erb et al., 2009; and (c) scenario 2 (cropland high, conservation area low) based on WBGU, 2009.

bioenergy potential from primary bioenergy crops in 2050. These potentials were calculated as primary biomass supply potentials, i.e., more or less as the entire amount of aboveground biomass produced by bioenergy plants, multiplied by the gross calorific value of the biomass (18.5 MJ/kg in the study of Erb et al. (2009) and 19.0 MJ/kg in the study by WBGU (2008)). The comparison of these three studies leads to the following conclusions:

- A likely range of future primary bioenergy crop potentials in such studies for 2050 is 44–133 EJ/yr. Factors that would reduce the potential are high food demand (in terms of quantity and share of animal products), low agricultural yields in food production, low feeding efficiencies, large area requirements for nature conservation, and low energy crop yields. Factors that could help to increase the potential are low food demand and diets using fewer

animal products, high yields and feeding efficiencies, and low area requirements for nature conservation. Climate change could also affect this potential both directly and indirectly, i.e., by influencing yields of energy crops and by influencing yield (and, therefore, area requirements) of other crops. Moreover, these findings suggest that there may be trade-offs between different environmental considerations such as conserving ecosystems and biodiversity, reducing agricultural pressures of agriculture, animal welfare and water issues, as well as the production of renewable energy from biomass. While there has been progress in better understanding these feedbacks, some of them are at present incompletely understood.

- The studies agree that the largest bioenergy crop potentials are located in Latin America and the Caribbean (LAC) and in Western and Central Africa (WCA). Substantial potentials were also found in the United States, the Former Soviet Union (FSU), and Australia, New Zealand and other Oceania (OCN). There are, however, some differences in the regional distribution of total potentials that result from the differences in methodology. Regional patterns should, therefore, not be over-interpreted.

- Despite the differences in energy crop areas, the results are similar. Van Vuuren et al. (2009) assumed the highest yields and the lowest area availability, whereas Erb et al. (2009) found larger area availability but assumed the lowest energy crop yields. WBGU (2009) used one of the most advanced process-based plant growth models (LPJmL) that incorporated plant functional traits of woody and herbaceous (C4 grass) bioenergy plants, which suggested yield potentials between the two other studies.

Considering sustainability constraints related to possible competing land demands (food, feed and fiber production, biodiversity conservation, etc.), problems posed by possible deforestation, and water availability, the first three studies analyzed in this chapter (van Vuuren et al., 2009; Erb et al., 2009; and WBGU, 2009) found global bioenergy crop potentials of 44–133 EJ/yr in 2050. The fourth one, Dornburg et al. (2010), concluded that for bioenergy crops (dedicated energy crops) they are 120 EJ/yr. Thus the range considered in this chapter is 44–133 EJ/yr for 2050.

20.2.4   Bioenergy scenarios for 2050: diets, agricultural technology and climate change

Previous sections of this chapter suggest that there are strong links between diets, agricultural technology and yield changes of food and bioenergy crops resulting from climate change. Three of the studies discussed (Erb et al., 2009; van Vuuren et al., 2009; and Dornburg et al., 2010) also analyzed possible feedbacks between diets, agricultural technology, and climate change (see Section 20.2.3. above). The model used by Erb et al. (2009) calculates primary bioenergy supply potentials

Table 20.7 | Bioenergy potentials by region from dedicated bioenergy crops from the first three studies.

[EJ/yr]	Van Vuuren et al.				Erb et al.[10]			WBGU[10]			Mean[10]		
	Strict criteria	Mild criteria	No criteria	Mean strict	MIN[1]	MAX[2]	FAO world[3]	MIN[4]	MAX[5]	Intermed[6]	MIN[7]	MAX[8]	Best guess[9]
USA	8	19	29	14	0	13	8	4	13	9	4	18	10
CAN	4	4	4	4	0	3	2	1	3	2	2	3	3
WEU	3	5	7	4	0	3	1	3	13	8	2	8	4
EEU	1	3	4	2	0	3	2	1	1	1	1	3	2
FSU	2	4	5	3	5	12	9	2	9	6	3	9	6
NAF	0	0	0	0	1	3	2	1	1	1	1	2	1
EAF	1	3	3	2	3	8	5	1	3	2	2	5	3
WCA	11	14	16	12	5	12	8	1	4	3	6	11	8
SAF	3	5	5	4	3	8	5	3	8	5	3	7	5
MEE	0	0	3	0	0	0	0	0	0	0	0	1	0
CHN	3	6	10	4	0	6	2	5	13	10	3	10	5
OEA	4	7	10	5	1	2	1	1	2	1	2	5	3
IND	0	1	3	1	0	1	0	2	4	3	1	3	1
OSA	0	1	1	1	0	0	0	0	1	1	0	1	1
JPN	0	1	3	1	0	1	0	2	4	3	1	3	1
OCN	6	9	10	7	0	3	1	2	9	5	3	7	5
PAS	0	1	1	0	3	9	7	1	11	4	1	7	4
LAC	18	30	34	24	5	40	22	10	29	17	11	34	21
TOTAL	**65**	**113**	**146**	**88**	**28**	**128**	**77**	**38**	**124**	**78**	**44**	**133**	**81**

[1] Richest diet, intensive agriculture, 20% cropland expansion
[2] Most modest diet, intensive agriculture
[3] Trend scenario, based on FAO *World Agriculture towards 2030/2050*, current diet trajectory
[4] Maximum constraints, no irrigation
[5] Minimum constraints, irrigation
[6] Arithmetic mean of four scenarios with intermediate constraints, with and without irrigation
[7] Arithmetic mean of the smallest potential of the three studies
[8] Arithmetic mean of the highest potential of the three studies
[9] Arithmetic mean of "mean strict", "FAO world", and "intermediate" potentials in the three studies
[10] Sums might not add due to rounding.

in 2050 depending on assumptions on diets, cropland yields, cropland expansion, and feeding efficiencies of livestock (see Figure 20.6). The model was calibrated with a comprehensive global NPP, land-use and biomass-use database (Erb et al., 2007; Haberl et al., 2007; Krausmann et al., 2008) for 2000. The FAO report *World Agriculture: towards 2030/2050* and the "best guess" UN population forecast (Bruinsma, 2003; FAO, 2006) were used as baseline ("Trend" scenario). Assumptions for changes in diets, livestock feeding efficiency, cropland yields and cropland expansion were exogenously fed into the model based on literature reviews documented in Erb et al. (2009). The impact of climate change on cropland yields was evaluated using LPJmL runs (Müller et al., 2010).

Figure 20.6 shows how changes in diets and cropland yields may affect the global bioenergy potential of energy crops in 2050.

According to Erb et al. (2009), the main assumptions underlying these calculations were:

- **Diets:** The "Trend" diet was derived for each world region by assuming that all countries in each of the 11 regions distinguished in that study would attain a level of calorie supply and animal product consumption similar to the richest country in the region. Results were similar to those derived by the FAO with a completely different methodology. The "Rich" diet assumed a global convergence to current US and European diet patterns, but did not assume that all regions would actually reach those high levels until 2050. The "Moderate" diet assumed the same per capita level of calorie supply as the "Trend" diet but a lower share of animal products. The "Frugal" diet assumed that the global per capita level of calorie supply would remain constant around the 2000 values and only

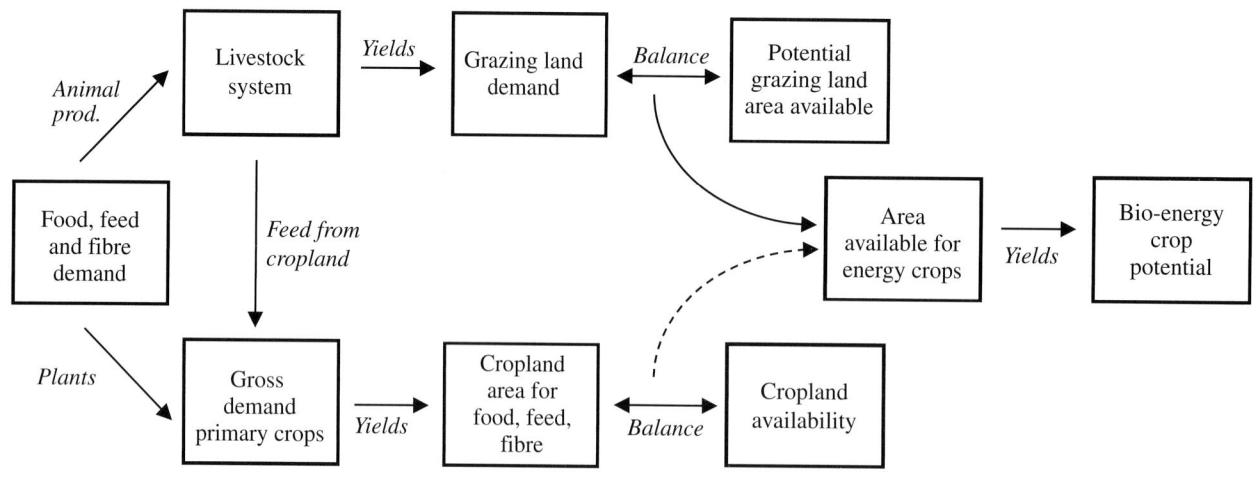

Figure 20.6 | Model structure of the land-use/biomass-balance model. Source: based on Erb et al., 2009.

20% of calories would be from animal products globally. While such a diet is nutritionally sufficient on an average basis, it would result in widespread malnutrition if current patterns of inequality of food supply persist.

- **Cropland yields:** The "Trend" assumption reproduces cropland yields as forecast by the FAO (2007a). The "Organic" yields were derived by assuming that all cropland would be cultivated according to IFOAM standards of organic agriculture. While this would imply substantial yield reductions compared to intensive conventional farming practices, it would also allow significant growth of yields in regions currently dominated by traditional low-input agriculture. The "Intermediate" yields are the mean between "Organic" and "Trend" yields, reflecting a trajectory where yield growth is constrained by environmental considerations. The "High" yields assumption was not part of the original Erb et al. (2009) study. In this case, the highest yield growth trajectory ("Global Orchestration") of the Millennium Ecosystem Assessment (2005) was adopted; yields were on average 9% higher than those forecast by the FAO (2007a).

- **Feeding efficiency:** The study contrasted conventional intensive indoor-housed feeding efficiencies with feeding efficiencies achieved if animal welfare standards or the even stricter standards of organic agriculture are adopted (here only the latter are reported; the "humane" assumptions were between conventional and organic efficiencies).

- **Cropland expansion:** The "Trend" assumption was taken from the FAO's *World Agriculture: towards 2030/2050* and assumed that global cropland area would grow by 9% between 2000 and 2050 (FAO, 2006). This was contrasted with a "Massive expansion" assumption, where growth of cropland areas between 2000 and 2050 was 19% – double the growth assumed by the FAO. Note that the assumption on cropland expansion had little influence on the bioenergy crop potential, because the study calculated the additional area that could be designated to

grow bioenergy crops if sufficient grazing area were available to meet the projected level of roughage demand (see Figure 20.6 above).

The results reported in Figure 20.7 show that diet has a strong effect on bioenergy crop potentials. The "Rich" diet leaves little space for bioenergy plantations, while the "Frugal" diet (which could only be adopted without widespread malnutrition if food distribution were egalitarian) allows for large bioenergy crop potentials. As one moves to the poorer diets, the range between the lowest and highest potential also increases. This is because the "Frugal" diet can be easily provided if the most intensive technologies (cropland yields, feeding efficiency) are adopted. However, such a combination might seem particularly unlikely. It is interesting that in the case of the "Frugal" diet, substantial energy crop potentials exist even if "Organic" yields and feeding efficiencies are assumed.

Changes in the assumptions on food crop yields also have a substantial effect on the bioenergy crop potential, as higher yield levels obviously leave more space for bioenergy plantations, assuming all other factors remain the same.

However, note the perhaps unexpected result that the lowest bioenergy potential estimate found in any of the scenarios assumes "High" yields. The reason for this is that the "Rich" diet can only be provided if "High" or at least "Trend" yields are assumed, and this diet leaves very little space available for bioenergy plantations due to the high roughage demand, irrespective of yield levels of food crops. The Erb et al. (2009) study also analyzed the possible effect of climate change on the bioenergy potential. It found that the energy crop potential under "Trend" assumptions on all parameters would be 77 EJ/yr (see Table 20.7) if no changes in yield levels due to climate change are assumed. If the CO_2 fertilization were switched off in LPJmL, however, yields were lower and the energy crop potential dropped by 18% to 63 EJ/yr, while it rose to 120 EJ/yr (+56%) if the CO_2 fertilization effect were switched on. This

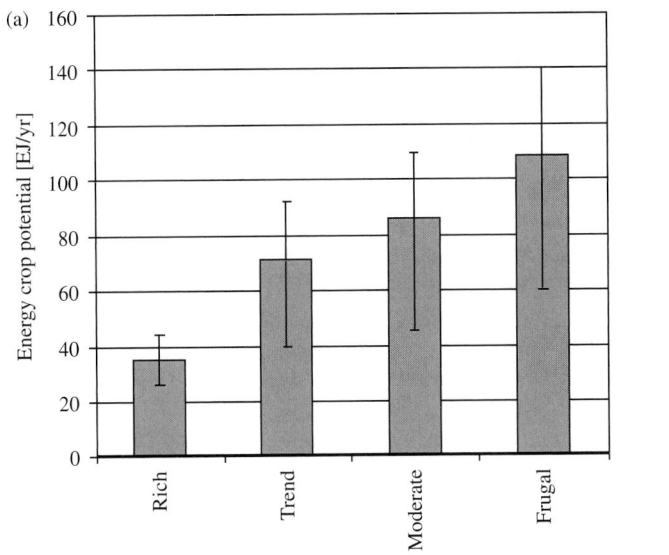

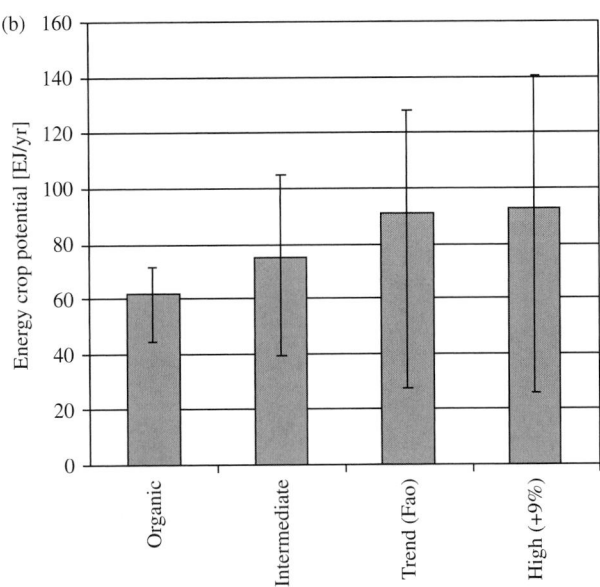

Figure 20.7 | Dependency of the global energy crop potential in 2050 on (a) changes in diets and (b) yields of food crops. Grey bars show the mean of all "feasible" scenarios (i.e., all scenarios that would deliver enough food for the respective diets), whiskers the range between the lowest and the highest scenario. Source: Erb et al., 2009 and additional calculations – see text.

suggests that the possible effect of climate change on yields introduces considerable uncertainty in estimates of global bioenergy potentials, in particular due to the indirect effect on food crops.

Van Vuuren et al. (2009) and Dornburg et al. (2010) also analyzed feedbacks between food demand, agricultural technology, and bioenergy crop potentials. In quantitative terms, comparing their results to those of the Erb et al. (2009) study, they found higher bioenergy crop potentials, mostly due to higher yields (see above), but they found the same basic dependencies of the bioenergy potential on diets and agricultural technology. It can be concluded that future bioenergy crop potentials strongly depend on diets and agricultural technology. Beyond trade-offs between food and energy, trade-offs between environmental quality goals and bioenergy potentials are also relevant. If, however, growth in yields could be reconciled with environmental quality goals (e.g., in terms of soil degradation, water pollution, biodiversity impacts, etc.) through sustainable high-yield practices or technologies, this would result in a major breakthrough in terms of food and energy supply (IAASTD, 2009; Godfray et al., 2010).

In conclusion, it can be stressed that bioenergy potential worldwide can be significant, even considering environmental restrictions to protect fragile ecosystems. However, strong investments are needed (IFPRI, 2009), mainly in capacity-building (to allow the implementation of efficient agricultural/industrial technologies, adequate policies and enforcement related to the environmental and social factors) and also to increase agricultural yields, which is fundamental to allow the implementation of the higher bioenergy potentials.

20.2.5   Bioenergy and land-use change: lessons from regional case studies

The notion of land use refers to a set of human actions – for example, arrangements, activities, and inputs – aimed at using land areas for human purposes. Land use usually results in changes in land cover, ranging from subtle effects to far-reaching alterations, including change from one land-cover type, e.g., forest, to another, such as cropland or grazing land (Lambin and Geist, 2006). The term "land use" also encompasses the social and economic purposes for which land is managed (e.g., grazing, timber extraction, or conservation). Human and natural factors in terrestrial systems are strongly linked, as captured in the recently coined notion of "land systems," conceptualized as coupled human-environment (or socio-ecological) systems in which socioeconomic and natural factors are inextricably intertwined (Global Land Project, 2005; Turner et al., 2007).

Land-use change can influence surface albedo, evapotranspiration, sources and sinks of GHGs, or other properties of the climate system, and may thus have a radiative forcing effect and/or other impacts on climate, locally or globally (Baede, 2007). These environmental effects are discussed below in Section 20.4 on sustainability. At the same time, land use and land-use change are socioeconomic processes that are influenced by a host of cultural, political, legal, economic and social factors and can have substantial repercussions on humans. This subsection draws from a selection of case studies that are intended to exemplify how these drivers and feedbacks can interact in cases related to bioenergy and land-use change.

When discussing the issue of land-use change relating to agricultural and bioenergy crops, it is important to consider not only direct effects, i.e., effects caused by establishing plantations, but also indirect effects that could result from an expansion of agricultural and bioenergy crops, e.g., displacement effects between different crops. Indirect effects – indirect land-use change (ILUC) – can be more challenging than direct effects in terms of availability of evidence and the possibility of clearly establishing causal relations. For example, deforestation could result from an expansion of land under feedstock crops and the displacement of food crops from higher-value lands (Cotula et al., 2008). On the other hand, other studies (Nassar, 2009) concluded that there is no significant evidence for ILUC from bioenergy crops.

With regard to supporting rural development, new and profitable land-use systems can provide better opportunities and long-term security for farmers and employees, plus – if processing facilities are near to farms – value-addition possibilities for profits in rural areas (Cotula et al., 2008). For countries with favorable endowments of land, labor and trade conditions, biofuels and bioenergy offer an opportunity to develop new export markets and improve the trade balance (Cotula et al., 2008). However, the FAO (2008a) provides a list of 22 developing countries that are especially vulnerable to the negative effects of bioenergy production due to a combination of high levels of chronic hunger (more than 30% undernourishment) while being highly dependent on imports of petroleum products (100% in most countries) and, in many cases, on imports of major grains (rice, wheat, and maize) for domestic consumption. Countries such as Eritrea, Niger, Comoros, Botswana, Haiti, and Liberia are especially vulnerable due to a very high level of all three risk factors (FAO, 2008a).

Almost all developing countries show strong interest in implementing bioenergy production, both liquid biofuels for transportation and solid biomass/biogas for power production (GNESD, 2010). In Africa, the preliminary conclusions from the Cogen for Africa project (AFREPREN/FWD, 2009), being developed under the coordination of AFREPREN, funded by the Global Environmental Facility/United Nations Environment Programme (GEF/UNEP) and the African Development Bank (AfDB) and aiming to implement efficient biomass-based cogeneration technologies in sub-Saharan countries, show strong interest from these countries to increase sugarcane plantations in the region, not only to produce sugar but also ethanol from molasses (a by-product from sugar production). There is also a high interest in improving agricultural productivity in the region, showing that food production can be increased together with biofuel and bioenergy production. It is important also to notice the main objective of the project – electricity production from biomass (sugarcane bagasse) to increase energy access in the region in a sustainable way.[3]

Also for Latin American countries, biofuels appear an interesting option from the experience in Brazil, without competing with other end uses

(GNESD, 2010). Many studies suggest that land potentials for agricultural crops and bioenergy in Latin America are substantial. However, competing demands for land may exist, mainly for agriculture, livestock production, and forestry. The production of agricultural and bioenergy crops has recently emerged as a contentious issue in some countries, either due to the potential direct competition between bioenergy and food crops, such as the use of maize for ethanol production in Mexico, or through direct or indirect expansion of the agriculture frontier over forests, such as soybean expansion in the Amazon (mainly to produce animal feed to export) and the Chaco Region in South America. On the other hand, there are several studies (Goldemberg, Coelho and Guardabassi, 2008; Goldemberg, 2008; Goldemberg and Guardabassi, 2009) showing positive results for Brazil and also presenting the benefits for developing countries when sustainable bioenergy production occurs, such as job generation in rural areas and local investments allowing significant development in developing countries. This could continue. The current liquid biofuel production in Latin America could be doubled sustainably based on first-generation feedstocks (Arias Chalico et al., 2009). Using second-generation feedstocks could further increase the potential. Improving productivity was found to be highly important (Pistonesi et al., 2008; Dornburg et al., 2010).

Sparovek et al. (2009) collected evidence suggesting that expansion of ethanol in Brazil from 1996 to 2006 did not contribute to direct deforestation in the traditional agricultural regions where most of the expansion took place. Their results show that sugarcane expansion did result in shrinking pasture areas and cattle head counts in these areas, as well as stronger economic growth. They could not exclude the possibility that the cattle migrated elsewhere, possibly resulting in deforestation in the Amazon. However, as mentioned above, more recent experience of sugarcane expansion in the State of São Paulo occurred without such impact, and other studies (Nassar et al., 2009) concluded that there is no evidence for such ILUC in the Brazilian sugarcane sector.

In another paper, Sparovek et al. (2007) showed that, if based on sound strategic plans, sugarcane ethanol production in Brazil could be extended in a manner that adequately considers social and environmental concerns. In their view, it would be necessary to integrate sugarcane production areas with existing land-use systems. They concluded that their development model could guarantee substantial expansion of production without resulting in displacement of extensive livestock production to remote areas, i.e., into tropical rainforests. The recent agro-environmental zoning for sugarcane both in São Paulo State and in Brazil (see Box 20.5) contributed to the achievement of these goals.

In Africa, biofuel production projects in low-income countries are often motivated by the seemingly large availability of land to grow feedstock crops. Somerville et al. (2010) claims that this would be the continent with the Earth's largest under-utilized land resources suitable to grow bioenergy crops.

On the other hand, the beginning of a biofuel boom in these countries has also raised concerns about potential environmental and/or

---

3   Coelho, S. T. Personal information by UNEP in field visits.

socioeconomic pressures. Much of Africa's land resources are characterized by soils and climate that might limit crop production, either due to low general suitability for energy crops or due to decreased potential as a result of land degradation. Bekunda et al. (2009) estimate that only 6–11% of the soils in Africa are devoid of serious constraints to effective management, with about 34% presenting medium or low potential, i.e., at least one major constraint to agriculture, and 55% altogether unsuitable for agriculture (Bekunda et al., 2009).

Food crops may be produced on lands that are less fertile but still suitable for farming, with current users having to relocate to other lands. This shift of farmers from food or cash crops to feedstocks may be voluntary in some situations, e.g., if bioenergy crop plantations offer favorable economic opportunities to farmers. For example, small-scale jatropha projects implemented in Mali have involved a shift from cotton to jatropha; this has been attributed to falling cotton prices and increases in the perceived (monetary and non-monetary) values of jatropha (Cotula et al., 2008). On the other hand, the opposite may occur. In 2011, sugarcane ethanol producers in Brazil decided to produce more sugar than ethanol, considering the high prices of sugar in the international market.[4]

There are concerns about indirect effects associated with large-scale cultivation of biofuel crops, which may include significant negative impacts on land access by local groups. For example, a multimillion dollar jatropha project in the Kisarawe district of Tanzania has been reported to involve the acquisition of 90 km[2] of land and the clearing of 11 villages, which, according to the 2002 population census, are home to 11,277 people. Approximately, US$632,400 was set aside to compensate 2840 households (African Press Agency, 2007).

However, the FAO (2010a) Bioenergy and Food Security (BEFS)[5] study for Tanzania shows that such problems can be avoided when adequate planning and agro-ecological zoning are in place. The study concludes that biofuel developments could provide an important vehicle through which to revitalize agriculture by bringing a variety of investments to increase productivity. The report shows that there are areas potentially suitable for bioenergy production, excluding those that are environmentally protected or under alternative uses. The technically viable and most competitive smallholder-integrated production chains were considered. This analysis has shown that "the dividends from investing in biofuels can have positive impacts on poverty reduction and growth." This case study is the first of several focusing on African countries in the context of the BEFS project, which has the aim of strengthening developing

countries' technical understanding of how best to mitigate the impact of bioenergy development on food security.

In Asian countries such as India, some authors have argued that substantial land areas are available for biofuel production. However, several critical issues that needed to be addressed were identified, including the costs of inputs to grow the biofuels on "wastelands," the growing demand for food production, and the social implications of converting areas currently allocated to food production into bioenergy crop plantations. Bekunda et al. (2009) recently analyzed a scenario in which one quarter of the total area of "wastelands" assumed to exist in India (i.e., 104,000 km[2]) would be converted to jatropha plantations with an average yield of 1.5 t/ha/yr of oil. They suggest that the lands would require significant inputs of nutrients and the adoption of soil and water conservation measures to realize such yield levels.

On the other hand, a recent report from The Energy and Resources Institute (TERI, 2010) discusses the potential for bioenergy in India, mainly using agricultural wastes and dedicated energy plantations (in degraded lands and wastelands) and indicates that there are 496,000 km[2] of available area categorized under "wasteland," as estimated by the Department of Land Resources. Out of this total, almost 66% falls into the classification of "wasteland suitable for land conversion," with almost 40% (129,600 km[2]) of this land in the categories of under-utilized/degraded forest land, degraded pastures and degraded land under plantation crops. The report concludes that these offer the highest potential for being converted into land for dedicated energy plantations. It also mentions that there is a large amount of available biomass in rural areas, and its usage in traditional forms causes negative social and economic impacts on rural households. In the proposed scenario, putting the available biomass to productive use would be a good strategy for the sustainable development of rural areas. Finally, the report concludes that bioenergy can contribute to rural development and poverty alleviation.

Some countries have witnessed protests against large-scale land transfers for biofuel production, indicating public concern over the implications of biofuels for land use (Cotula et al., 2008). For example, for Uganda, Cotula et al. (2008) report that there was strong public opposition to a planned allocation of national forest reserves in Bugala and Mabira to foreign plantation companies to establish oil palm and sugarcane plantations. However, conclusions from the recent Cogen for Africa project show that there is now environmental legislation in place (not only in Uganda but also in other sub-Saharan countries) to avoid such problems. Even the financial support of the AfDB is assured only when adequate Environmental Impact Assessments are developed.[6]

There are doubts about the concept of "idle" or "abandoned" land (Dufey et al., 2007; Cotula et al., 2008). In most situations, lands perceived to be

---

4   See www.conab.gov.br.

5   The BEFS project is funded by the United Nations FAO and the Government of Germany. Under the project, the FAO has developed a quantitative and qualitative framework to analyze the interplay between bioenergy and food security. The BEFS Analytical Framework provides tools that permit policymakers to make informed decisions with respect to bioenergy.

6   Coelho, S. T. Personal information by AfDB in field visits, invited by UNEP.

"idle," "under-utilized," "marginal," or "abandoned" by governments and large private operators provide a vital basis for the livelihoods of poorer and vulnerable groups, through crop farming, herding, and gathering of wild products. Further, seemingly "abandoned" land often provides important subsistence functions in times of stress to vulnerable households. Hence, the promotion of biomass production on degraded lands must avoid competition with these other land uses. These studies also claimed that other issues may cause or increase land-use conflicts, including poor enforcement of laws on land-use planning, particularly when large profits are at stake, as in the case of the expansion of oil palm plantations on native forests or even forest reserves.

Competing demands for land in Africa are primarily for agriculture and forestry, with the production of biomass for energy an emerging issue. Recent efforts have increasingly been aimed at identifying land areas for feedstock production that reduce competition with production of food and other biomass-based products. Policy suggestions have included the planting of biofuel crops on "marginal" and "idle" lands rather than prime agricultural land. Rural development and poverty alleviation from the implementation of bioenergy programs in developing countries have also been extensively discussed.

Some calculations indicate that apparent land availability differs from region to region. This implies the need for detailed regional and national assessments of the amount of land available, the quality of land for producing biofuels, potential conflicts with (or displacement of) land for food production, and the potential to increase food insecurity (Bekunda et al., 2009). Other studies (Martínez-Alier, 2002; Vanwey, 2009) also argue that subsistence agriculture is often strongly influenced by cultural or other social values that cannot be expressed in monetary terms, and that their benefits cannot be estimated using conventional methods such as cost-benefit analysis.

The European Union (EU) biofuels directive (discussed later in this chapter) is further providing incentives for the use of "degraded" lands for feedstock production. The assumption is that biofuel production will not compete with agricultural production on prime lands. Some governments have already taken steps to identify "idle" land and to allocate it for commercial biofuel production. Some governments have claimed that significant land areas are under-utilized and available for biofuel production. For instance, the Government of Mozambique has stated that only 9% of the country's 360,000 km² of arable land is currently in use, and that there is the possibility of bringing into production an additional 412,000 km² of marginal land currently not being used (Namburete, 2006, cited in Cotula et al., 2008). In fact it has expressed a strong interest in the production of biofuels in the country in a sustainable way.[7]

Another recent study (Müller et al., 2007) also suggested that sub-Saharan Africa could have substantial resources in terms of suitable land and exploitable water to expand areas for agricultural production, including bioenergy production. This conclusion is confirmed by the results of the Cogen for Africa project.

Another important issue is related to women's land rights; some studies argue that they risk being eroded by large-scale biofuel expansion, due to existing gender inequalities. In Kenya, for example, despite providing 70% of agricultural labor, women only own 1% of the land they farm (DFID, 2007). This is replicated across the developing world, with only 5% of women farmers owning their land (IUCN, 2007; Cotula et al., 2008). However, it must be noted that this is a problem for the agricultural sector as a whole and not only for biofuels.

The complexities of women's involvement are recognized in Mali, for instance, where small-scale jatropha cultivation to meet local energy needs has been promoted by both government authorities and development agencies. The Ministry of Mines, Energy and Water is implementing a US$1.6 million Programme National de Valorisation Energétique de la Plante Pourghère to promote the use of jatropha for rural electrification, conversion of vehicles to biofuels, and poverty reduction amongst rural women (Cotula et al., 2008).

The implementation of adequate social policies and enforcement can contribute to reducing these problems. For example, sugarcane plantations in Brazil, mainly in Ribeirao Preto, São Paulo State, allow significant improvements on gender and social issues.[8] Existing statistics from Uniao da Industria de Cana de Açúcar (ÚNICA, 2010) show that in this region inequalities in gender are smaller and social aspects such as strict labor legislation are very much taken into consideration.

Bioenergy produced on currently grazed lands can have large-scale impacts on livestock-rearing subsistence farmers. These may be positive or negative, depending on the implementation strategy. Large-scale bioenergy plantations owned and operated by international, vertically integrated cooperatives tend not to benefit the local farming communities where the biomass is produced, as most of the revenue is generated in the production stage that involves sophisticated biochemical conversion technologies (Sagar and Kartha, 2007). On the other hand, small-scale, locally owned and operated plants, together with sustainability certification systems, might help ensure that benefits accrue to the local farming communities (Lewandowski and Faaij, 2006).

Soyka et al. (2007), focusing on Indonesia, raised the question of whether it was possible to increase tropical biofuel production without increasing tropical deforestation. However, Wicke et al. (2011) argued that palm oil crops are not the main reason for deforestation in Malaysia and Indonesia, and that there are several other factors involved, as discussed in Section 20.4. Figure 20.8 illustrates this discussion.

---

7   Coelho, S. T. Personal information from field visit to Mozambique invited by UNCTAD, 2010.

8   Ribeirao Preto, the most developed region in the rural areas of the country, has a local economy based almost exclusively on sugarcane crops.

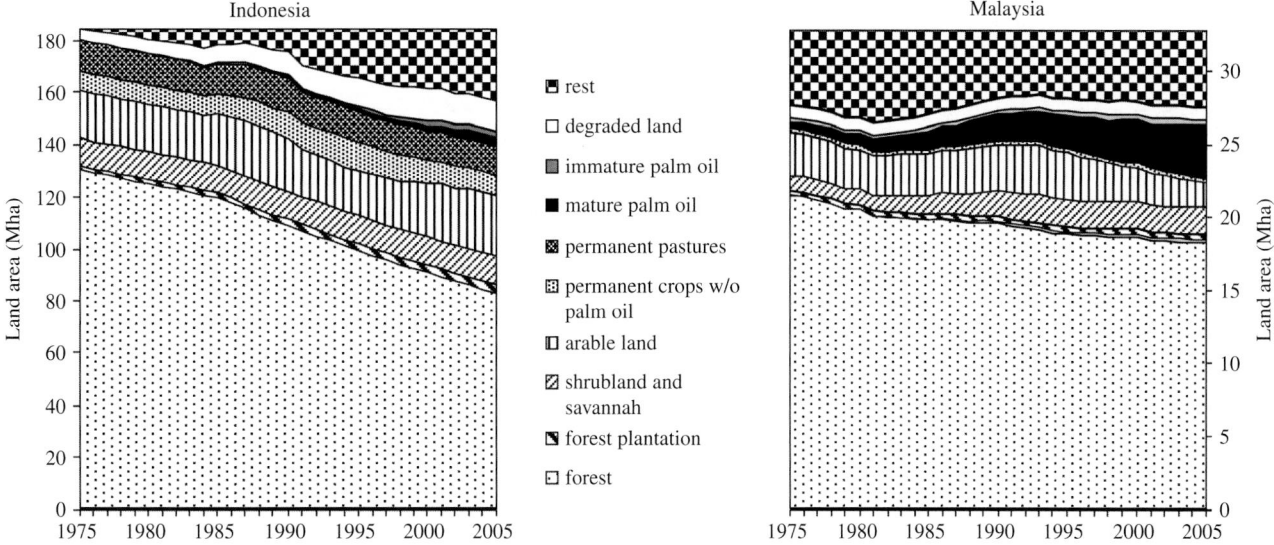

Figure 20.8 | Land-use change in Indonesia (left) and in Malaysia (right). Source: Wicke et al., 2011.

In fact, according to Wicke et al. (2011), it was found for Indonesia that "there are many, interrelated causes and underlying drivers that are responsible for this land-use change (…). Palm oil alone cannot explain the large loss in forest cover but rather a web of interrelated direct causes (including oil palm production expansion) and underlying drivers are responsible. Important direct causes were logging, palm oil expansion and other agricultural production and forest fires, while underlying drivers were found to be population growth, agriculture and forestry prices, economic growth and policy and institutional factors."

For Malaysia, Wicke et al. (2011) show that "the most important causes of land use change vary per region: In Sabah and Sarawak the most important causes have been timber extraction and shifting cultivation, while in Peninsular Malaysia, and in recent years increasingly in Sabah, forest cover has been affected most by conversion to agriculture, mainly oil palm production." The study also concludes that "additional forested land and peat land are not necessarily required for most projections of oil palm production expansion to be feasible. This is because yield improvements can largely reduce land requirements while also large amounts of degraded land exist in Indonesia. (…). As in Indonesia, yield improvements are also an important component of allowing potentially sustainable expansion. In Malaysia, yield improvements in the short term in both countries are mainly possible by applying fertilizer and other inputs more appropriately (and) practicing good harvesting standards (…)."

Moreover, according to the Malaysian Palm Oil Board, protected areas were defined in the country to preserve biodiversity and native forests.

Considering all these issues, Wicke et al. (2011) conclude that adequate policies and enforcement can minimize land-use conflicts due to the implementation of bioenergy crops, especially in regions where

motivations of small farmers ("smallholders") include strong components not usually considered in classical agro-economic toolboxes and dominant development models.

The FAO (2010a) also argues that "biofuel developments could play a pivotal role in promoting rural development through increased local employment and energy supply. Implementing bioenergy production can result in improvements or a worsening in the food security conditions depending on the bioenergy pathway chosen."

In conclusion, these case studies suggest that land allocation to different uses, including agricultural and bioenergy crop production, involves sensitive socioeconomic issues related to food security, land tenure, land-use rights, subsistence versus market economy, cultural values, etc. Careful consideration of these issues, and implementation of sensible policies and legal frameworks, will, therefore, be key for the public acceptance of bioenergy crop plantation projects and will have an important impact on whether they will be perceived as beneficial development opportunities or as "land grab" (Friis and Reenberg, 2010).

20.2.6 Economic Effects of Land Use Competition

The discussion of fuel versus food is a long one and quite controversial. In this chapter some recent studies are presented to contribute to the discussion. A number of studies have looked at the linkages between the growth of biofuel production and the dynamics of food prices – both within the 2007–2008 time period and in the follow-up to the 2009–2010 financial market crisis and recovery period. A number of papers that emerged in the immediate wake of the 2007–2008 food price spike were strongly of the opinion that growth in biofuel production was the major contributor to food

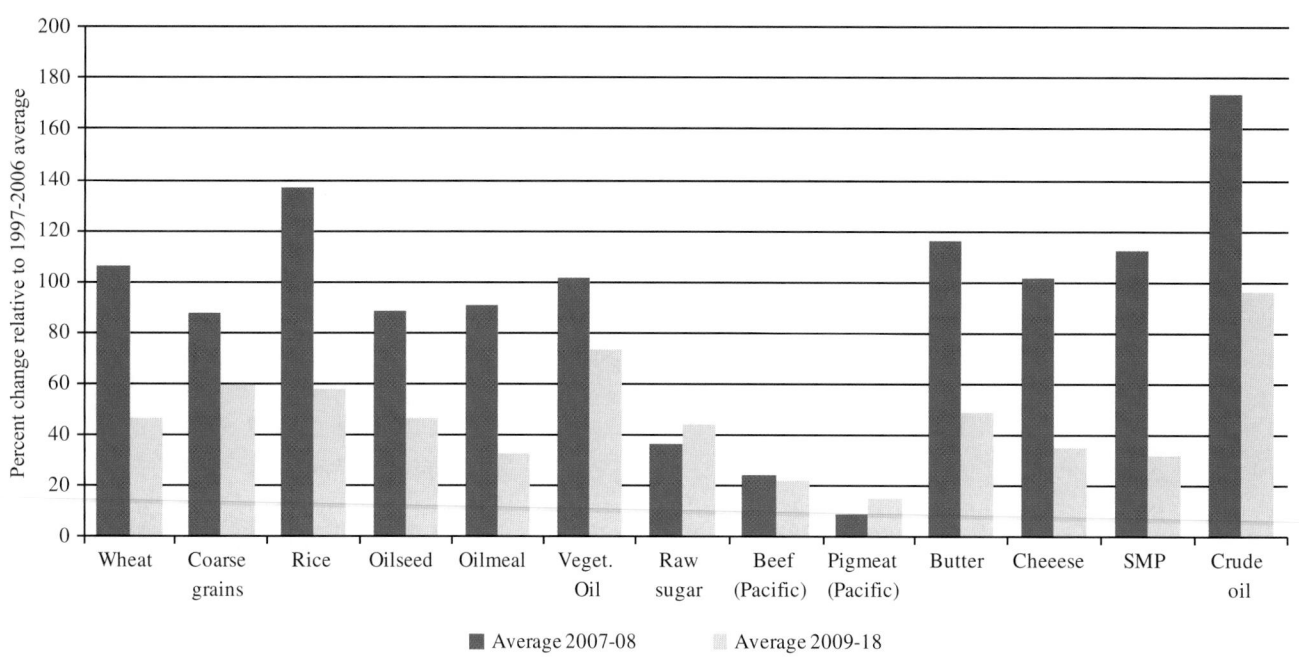

figure caption

Figure 20.9 | Nominal commodity prices projected 15–60% higher than 1997–2006 but substantially lower than in the 2007–2008 peak. Source: OECD/FAO, 2009 (p19).

price increases (Runge and Senauer, 2007; OECD, 2008b; Mitchell, 2008; von Braun, 2007). In contrast, other authors were more cautious about the estimated impact of biofuels, and placed more weight on the macroeconomic factors such as exchange rates, grain storage policies and possible market speculation that could have played a role in the food market price dynamics during that period (Trostle, 2008; Abbot et al., 2008). Besides food price impacts, some authors also critiqued the economic efficiency of biofuel policies in terms of the distortions they place on markets, as in the case of the US corn ethanol program, where mandates, blending targets and import tariffs, combine in a way that both raises commodity prices and may even encourage the consumption of fossil fuel under certain circumstances (de Gorter and Just, 2007).

In the past, according to the FAO food price indices, nominal prices of agricultural commodities fluctuated but in a medium-term perspective were mostly stable or declining from 1990–2007 (see Figure 20.9). Deflated values suggest that food prices were fluctuating around a continuous long-term downwards trajectory in real terms in the last decades, at least until 2007.

Starting in 2007, many agricultural prices increased substantially and remained high throughout most of 2008, falling sharply afterwards, most probably due to a reduction in oil prices (Faaij, 2009; Goldemberg, 2009) and reduced demand for many commodities resulting from the global financial crisis. Sugar seemed to be an exception; its price peaked in 2006 and then again in 2010 but was low during the period of high agricultural prices in 2007/2008.

These events triggered a debate on the question whether this was just another, only a bit stronger, price fluctuation that would not change the

long-term price trajectory, or whether it signaled a structural break, i.e., a long-term change in the trajectory of agricultural prices in an upward direction (FAO, 2009a; 2009b; OECD/FAO, 2009).

More recently, the OECD and FAO (OECD/FAO, 2009) showed that "despite the significant impact of the global financial crisis and economic downturn on all sectors of the economy, agriculture is expected to be relatively better off, as a result of the recent period of relatively high incomes and a relatively income-inelastic demand for food." The report also concludes that biofuels can influence agricultural prices but in general because "energy and agricultural prices have become much more interdependent with industrialized farming (…). Crude oil prices are highly volatile (…). The crude oil price (…) assumed for the baseline is about 60% higher than the 1997–2006 average in real terms, moderately increasing to US$70 per barrel by the end of projection period. If crude oil prices increased to the US$90 to US$100 per barrel level used in last year's (report), agricultural prices would be significantly higher; with the largest impact on crops, driven mainly by reduced crop production with higher input costs, but also increased feedstock demand for biofuels."

Figure 20.10c illustrates projections for the period 2009–2018. The study adjusted the prices for inflation, which are, in real terms, expected on average to be much below 2007–2008 average peak levels. Results show that "the crops expected to undergo the largest fall in real prices, compared to their 2007–08 average, are: rice, wheat, butter, cheese and skim milk powder. However, over the outlook period, real prices of products other than beef and pig meat are expected to be above their average 1997–2006 levels. In real terms, the average crude oil price assumption for the next decade is substantially below its 2007–08 peak, and remains well above, by around 60%, the 1997–2006 average level."

a)

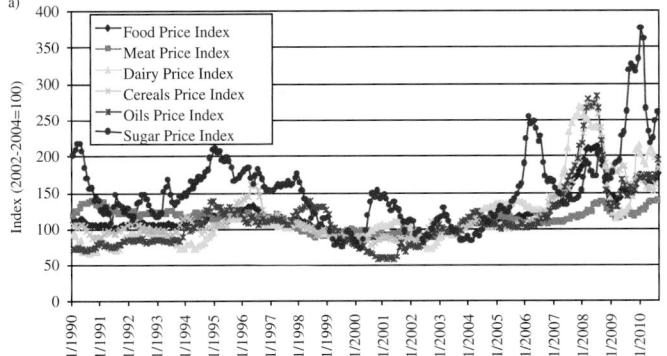

b)

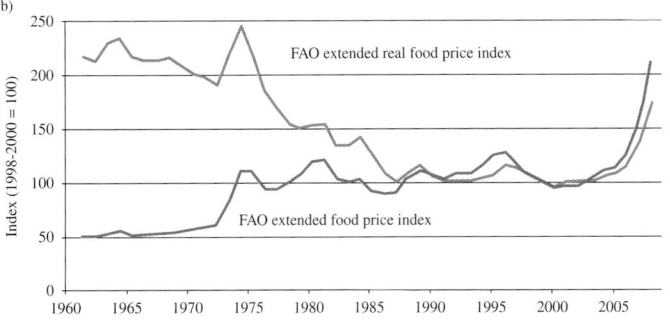

c)

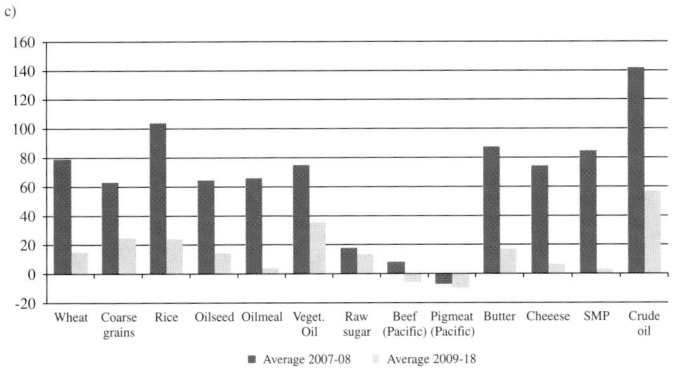

Figure 20.10 | (a) FAO food price index (nominal; monthly values, indexed to 2002/2004 = 100) for the period 01/1990–08/2010. Source: FAO, 2009c. (b) Nominal (yellow) and real FAO food price index 1961–2008. Source: FAO, 2009c. (c) Real crop prices to fall from peaks but to remain above 1997–2006 average. Source: OECD/FAO, 2009 (p19).

The list of potential explanations for the price rally in 2007 and 2008 is long and includes: low yields due to unfavorable weather (droughts); strong policies promoting biofuel production in the United States and Europe; high oil prices;[9] increasing food demand arising from strong economic growth in developing countries, in particular China and India; inflows of speculative funds; and low stocks of many agricultural commodities (OECD/FAO, 2007; 2009; FAO, 2009b; Trostle, 2008). In addition, depreciation of the value of the US dollar compared to other currencies seems to have been one of the reasons leading to the strong increases in prices. The price hikes were less dramatic when expressed in other

---

9   The oil price affects the prices of agricultural commodities because of the energy needed in agricultural production, both directly (e.g., tractors) and indirectly (e.g., fertilizer production).

currencies. However, for the United States and all other countries whose local currencies are pegged to or weaker than the US dollar, the depreciation of the dollar did in fact increase the costs of procuring food (FAO, 2009b; OECD/FAO, 2009).

Estimates of the effects of the growing production of biofuels on food prices vary, depending on the time period considered and the food price indicator chosen. The effect of increased biofuels production on the price of commodities was much larger for those commodities that are used for both food and fuel production such as maize (for food and for corn ethanol in the United States) than the impact on food price indices that consider a broader range of commodities and processed foods.

In its *World Development Report* 2008, the World Bank (2008) argued that rapid expansion of biofuels had strongly contributed to the price hike. For example, the World Bank claimed that the massive use of maize for ethanol production in the United States had reduced maize stocks and had largely caused the surge in maize prices in 2006 and 2007. While a different report by the World Bank (Mitchell, 2008) concluded that US and European biofuel policies were responsible for most of the food price increase, a later study also published by the World Bank concluded that the effect of biofuel policies had been over-estimated in that earlier study (Baffes and Haniotis, 2010).

The OECD (2008b) concludes that while biofuel policies did contribute to the increases in food prices in 2007 and 2008, they would result in rather limited price increases in the medium term, i.e., for the period from 2013 to 2017. For example, the OECD calculated that current EU and US policies would drive up prices of coarse grains by about 5% compared to the baseline. Vegetable oils would be more strongly affected; their prices were forecast by the OECD (2008b) to increase by 15%.

The FAO (2009b) and OECD/FAO (2009) discuss the 2007–2008 price hikes and distinguished short-term effects and long-term trends. With respect to short-term fluctuations in 2007–2008, the FAO concluded that growing consumption in India and China is unlikely to have contributed to the price hike, because, in those years, their growth rates were below that of the aggregate of all other countries and their trade patterns did not show sudden changes. (However, it was argued that changes in their consumption patterns could be relevant over the medium to long term.) This FAO study also concluded that the rapid growth in the use of maize for ethanol in the United States and oil crops for biodiesel in the EU, caused by the introduction of respective blending targets and subsidies, played a major role in influencing prices. In 2007, 12% of the global maize production and 25% of the global rapeseed production were used as feedstock for biofuels, and these shares had been growing rapidly. This coincided with bad harvests in 2005 and 2006 and hence record-low stocks.

With respect to speculation, the FAO mentioned that it was not clear whether speculation was driving prices, or whether it was attracted by prices that were increasing anyway. Speculation might have resulted in increased volatility, according to the FAO study. In conclusion, the FAO argued that no factor in isolation could explain the price hikes, only their combination and coincidence. With respect to possible structural effects that would change price trajectories in the coming decades, the FAO concluded that biofuels, and indeed any energy technology requiring significant amounts of highly fertile lands and, therefore, competing with food production for land resources, would influence food markets.

The FAO also reckoned that biofuels are at present not competitive with fossil fuels, with the exception of ethanol from Brazilian sugarcane. Therefore, their impact on agricultural markets would remain limited to the extent to which they were subsidized, i.e., their impact on agricultural markets would remain limited except under extraordinary circumstances, such as a rapid introduction through policy interventions in response to weather or other effects. As soon as oil prices climb to a level where biofuels become economically competitive, however, agricultural and energy markets would become linked in a new way: "[a]s energy markets are huge relative to agricultural markets, demand from the biofuel sector could *in principle* absorb any additional production in crops usable as feed stocks so the energy market would effectively set a floor price for agricultural products. It would also set a ceiling on agricultural prices at the point where they have risen so much that biofuel production is no longer competitive. It would be energy demands rather than food demands that would set agricultural product prices and agricultural product prices would be tied to *energy* prices. Clearly, this would be a major departure from how agricultural product prices have been determined in the past" (FAO, 2009c).

Other authors have drawn similar conclusions. For example, Müller et al. (2007) concluded that competition for land would increase and prices of essentially all crops would rise if a growing range of crops became competitive as feedstock for biofuel production. Moreover, other agricultural products that could be used as substitutes for other, non-renewable resources would in this case become increasingly competitive, thereby giving farmers greater flexibility to switch between different crops, e.g., between food and energy crops.

On the other hand, the OECD/FAO (2009) stressed that after the (2007–2008) high-price crisis there has again been evidence of the rapid responsiveness of global agriculture. High international commodity prices have transmitted signals to farmers to allocate more resources and increase agricultural production. However, not all farmers responded similarly, as high world prices are not transmitted to local producers in many instances. A decomposition of the response of farmers by economic region reveals that output expansion in developed countries amounted to over 13%, but developing countries together could only muster a 2% increase in their cereal production. This lack of response from a large part of the world shows the need for policy reform and additional investment in productive agriculture, particularly in many developing countries. Structural problems are likely to persist, especially for the Least Developed Countries, limiting their capacity to produce. Other studies (Goldemberg, 2009; Faaij, 2009) also concluded

that the increase in food prices was mainly caused by the rising oil price and poor food distribution in developing countries.

Also, more recent FAO estimates, derived from simulations with the OECD/FAO Aglink-Cosimo model, suggest that keeping biofuel production at 2007 levels would reduce the price of maize in 2017 by 12% and that of vegetable oil by 15% compared to the respective baseline projections (FAO, 2009c).

In brief, as raised by Dornburg et al. (2010), "it is claimed that biofuels will lead to famine, deplete water resources and destroy biodiversity and soils. (…) Biofuels are often regarded as the cause of the dramatic increases in food prices that have occurred over the past few years." In fact, "[b]iofuel developments have local, national, regional and global impacts across interlinked social, environmental and economic domains" (FAO, 2010a).

The recent increase in food prices exacerbated such criticism. However, the report from OECD/FAO (2009) mentioned above showed that agricultural markets saw a reduction in commodity prices in 2009 after their rapid rise during 2006–2008, and perspectives for future prices are not so negative. "Looking forward, real commodity prices over the 2009–2018 period are projected to remain at, or above, the 1997–2006 average. An expected economic recovery, renewed food demand growth from developing countries and the emerging biofuel markets are the key drivers underpinning agricultural commodity prices and markets over the medium term."

As more data became available in the aftermath of the 2007–2008 food price spike, and more analysis was able to be carried out on emerging market-level price, production and consumption data, other authors began to contribute additional insight into the role that biofuels could have played in driving global and regional food prices. The analysis of Heady and Fan (2010) maintained that the influence of biofuels was still strong for markets such as maize, whereas the dynamics for other grain markets such as rice were related to trade policies and national commodity prices in individual countries. Gilbert (2010) is among those authors who argued against the premise that biofuels played a major part in the food price spikes of 2007–2008, and that speculation in agricultural commodity futures had a much stronger role. Baffes and Haniotis (2010) – following on the earlier analysis of food price impacts by Ivanic and Martin (2008) – took a more middle-of-the-road position and argue that even though the role of biofuels may not have been as strong as was suggested in the earlier literature, there does exist a strong link between energy and food prices. They, like other authors, also point to the fact that agricultural commodities are becoming a large part of financial investment portfolios, and point to trends of variability in historic prices of commodities.

Zhang et al. (2010) analyzed time-series prices on fuels and agricultural commodities, and results indicate no direct long-term price relations between fuel and agricultural commodity prices. In fact rising sugar prices appear to be the leading cause of price inflation in other agricultural commodities; however, the study concludes that, with decentralized competitive agricultural markets, this sugar price impact is transitory.

The estimation of the impact of increased biofuel production on food prices has mostly been done with the help of country- or global-level economic multi-market equilibrium models that are either partial or fully comprehensive in terms of the economy-wide linkages that connect the supply and demand of agricultural commodities to important economic sectors. The nature of the particular model determines how closely the supply and demand dynamics of the agricultural food commodities themselves are linked to important input markets such as fertilizer, labor, and energy – especially biofuels and their interactions with crude oil prices. The particular structure and underlying assumptions of these models, by themselves, can have a considerable effect on the estimates of increased biofuel production on price impacts. A comprehensive meta-analysis of the differences in modeling the impacts of biofuels on land-use change (Edwards et al., 2010) showed evidence of systematic differences across types of modeling approaches, especially when moving from "partial-equilibrium" models that focus mostly on agricultural markets towards those "computable general-equilibrium" (CGE) models that consider the deeper complexities of linkages across many sectors of the economy.

These types of global CGE models have been used to demonstrate both the environmental as well as the price impact of increased biofuel production in more recent studies. The study carried out by al-Riffai et al. (2010) to evaluate the environmental implications of the EU Renewable Fuels Directive, for example, used a global CGE model with linkages to land availability to illustrate the implications of increased OECD biofuel production on GHG emissions.

Besides showing the relative "superiority" (in terms of environmental impacts and GHG emissions) of Brazilian sugarcane over US-based maize ethanol, the authors also show that liberalizing world (and especially US) trade policy to reduce tariff barriers against biofuel imports has the effect of reducing grain prices for both food use and livestock feed.

A recent study by the World Bank (Timilsina et al., 2010) illustrates the effect of increased biofuel production on land use, food prices, and poverty, and argues that moving from a scenario where OECD blending mandates and biofuel production targets are announced at current levels and are further enhanced by scaling up production capacity and doubling blending targets and biofuel production results in a near doubling of modest impacts on food prices. They also show some impacts on the supply of livestock products, sugar, and grains, as a result of the expansion of biofuels production – although the overall effect on the supply of food is relatively small (on the order of 0.2% lower than the baseline case). Even though they argue that their results are similar to those of al-Riffai et al. (2010) and others, this further illustrates the influence that model structure has on the simulated impacts of increased biofuel production.

The additional flexibility that these models have built into their structure, to capture the land-use dynamics and competition between crop, livestock and forestry cover, may understate the price impacts through their tendency to freely adjust land supply in response to relative price changes across commodities. This points to trade-offs that are inherent to different modeling approaches that try to either capture the environmental impacts of biofuels or the commodity market interactions (Nassar et al., 2011).

20.2.7 Bioenergy and Food Security

Food security (see Box 20.1) must be addressed in line with the world's commitment to eradicate extreme poverty and hunger as part of the Millennium Development Goals (MDGs) (UN, 2000). In fact, some food

Box 20.1 | Food Security

"Food security exists when all people, at all times, have physical, social and economic access to sufficient amounts of safe and nutritious food that meets their dietary needs and food preferences for an active and healthy life. There are four dimensions to food security: availability, access, stability and utilization. Availability of adequate food supplies refers to the capacity of an agro-ecological system to meet overall demand for food (including animal products, livelihoods and how producers respond to markets). Access to food refers to the ability of households to economically access food (or livelihoods), defined in terms of enough purchasing power or access to sufficient resources (entitlements). Stability refers to the time dimension of food security. Stability of food supplies refers to those situations in which populations are vulnerable to either temporarily or permanently losing access to resources, factor inputs, social capital or livelihoods due to extreme weather events, economic or market failure, civil conflict or environmental degradation, and increasingly, conflict over natural resources. Utilization of food refers to peoples' ability to absorb nutrients and is closely linked to health and nutrition factors, such as access to clean water, sanitation and medical services."

Source: Faaij, 2008.

price fluctuations will always occur. Therefore, to protect poor people, and food security generally, adequate policies have to be implemented, such as adequate agro-environmental zoning together with economic and environmental legislation.

The linkages between bioenergy and food security are complex. Food availability can be threatened if land, water and other resources are diverted from food to biofuel production (Dornburg et al., 2010). Competition for resources is reduced if biofuels are produced from non-edible crops and if the biofuel crops are cultivated on land that would not be utilized for food production in the foreseeable future. Crop selection, farming practices and yield growth patterns can have significant implications for potential impacts of biofuel growth on food availability. Food access is determined by the prices of food and the income levels of the poorest segments of society depending upon food purchases to meet their dietary needs. A significant number of people produce less food than they consume and in some cases may face an immediate negative impact in response to rising commodity prices (FAO, 2008c; 2009b; 2009d; OECD/FAO, 2009).

As mentioned by Dornburg et al. (2010), biofuel production is considered by some as the reason for rises in food prices, mainly grain prices. Poorer households spend a greater percentage of their income on food, particularly on staples, than richer ones, and are thus disproportionately affected by rising food prices. However, this question is much more complex (see discussion in Section 20.2.6.)

There are also opportunities for biofuels to contribute to greater food security. Poor farmers could benefit financially from selling commodities at higher prices. At the national level, high prices offer development opportunities for countries with significant agricultural resources and potential, such as Tanzania (Müller et al., 2007; FAO, 2010a). Greater demand for biofuels can boost incomes by revitalizing agriculture, providing new employment opportunities, and increasing access to modern energy, which can increase household welfare and contribute to rural development (FAO, 2008a; OECD/FAO, 2009).

Higher prices can also provide incentives for intensification, leading to increased food production, and improved livelihoods as long as production methods are sustainable (FAO, 2008a). Establishment of large-scale biofuel production systems could also provide benefits in the form of employment – mainly jobs in rural areas, skills development, and secondary industry (Cotula et al., 2008).

Attention must be given to developing and Least Developed Countries, since the FAO (2009d) warned that high prices for consumers do not necessarily mean high prices for poor producers in developing countries and have so far not triggered a positive supply response by smallholders there. The FAO study reported that access to means of production and assets such as land is critical for small farmers to be able to harness positive effects of price increases – if this cannot be guaranteed, large landholders are likely to benefit most from high food prices. However, the recent study from the OECD/FAO (2009) showed perspectives for agricultural commodity production increasingly shifting away from developed countries towards developing regions.

The impacts of bioenergy developments on food security depend on many factors that are country- and case-specific. Examples of these factors include the type of biomass used, the type of energy carrier produced, the type of land used for biomass production, and developments in agricultural management and in the global food markets (Box 20.2).

Box 20.2 | Preliminary insights from FAO's Bioenergy and Food Security project[10]

"Bioenergy, and particularly liquid biofuels, have been promoted as a means to enhance energy independence, promote rural development and reduce greenhouse-gas emissions. In principle there are many benefits offered by bioenergy developments but these need to be balanced against the impacts on food security and the environment. While there has been a rush by many governments to develop bioenergy alternatives to fossil fuels this has often been done in the absence of a wider understanding of the full costs and benefits of bioenergy."

Preliminary results from three countries (Peru, Tanzania, and Thailand) participating in FAO's Bioenergy and Food Security (BEFS) project suggest that bioenergy developments, if managed correctly and sustainably, can serve as an opportunity to reduce poverty and increase energy access in a number of developing countries.

The success of bioenergy developments depends most importantly on the management strategy of the investments. Biofuel developments that safeguard food security are possible, if a gradual expansion process is adopted that assesses and manages possible

---

10 The Bioenergy and Food Security (BEFS) project has been developed by the Food and Agricultural Organization of the United Nations (FAO) with funding from the German Federal Ministry of Food, Agriculture and Consumer Protection (BMELV). It was set up to assess how bioenergy developments could be implemented without hindering food security. Available at www.fao.org/bioenergy/.

negative impacts on the most vulnerable people and on the natural resource base. Land – especially large tracts of land – is analyzed to be allocated in line with transparent procedures, assessing land availability and considering projected food production needs, suitability for proposed feed stocks, and competing claims on the land. Particular attention must be paid to vulnerable groups that depend on the land for their livelihoods.

The BFES report for Peru (FAO, 2010b) shows that this country "has witnessed strong agricultural growth which has reduced rural poverty but at a much slower rate than urban poverty. Consequently, urban-rural inequalities have widened."

Overall, according to the conclusions from the project, poor people in both urban and rural areas must be directly involved in the bioenergy production chain. Rural households can benefit from price increases for crops used for both food and energy. The analysis in the case of Peru for the main food staples shows that, for example, increases in the price of maize are detrimental to the urban poor but can increase the welfare of the rural poor.

The report also stresses that "bioenergy and especially biofuel developments, in principal, hold much promise for improving agricultural growth for the benefit of the poor. However, while a mandate has been already set in Peru, feedstock production for liquid biofuel can have serious consequences on food production because they compete for the same resources. Thus, an important question is whether the mandate can be met without compromising the food security status of Peru." It concludes that the three most important policies for Peru are: to increase the knowledge base on bioenergy; to improve strategies and actions on bioenergy; and to enhance decision-making and dissemination.

Preliminary results from Tanzania indicate that countries should look beyond the most established bioenergy production chains and carefully consider feedstock and technology selection to suit the natural resource endowments, biophysical conditions, domestic capacity (including human skills), infrastructure, and economic set-up of the country. Opportunities to improve the economic viability for biofuels are closely linked to developments in the agricultural sector.

More specifically, the assessment suggests that ethanol production from cassava is a more attractive option for Tanzania than other crops, such as sugarcane. The biophysical conditions suitable for growing cassava are more widely present within the country when compared to the limited areas suitable for other crops. Additionally, cassava represents the least-cost feedstock for ethanol production in Tanzania, as it is already widely produced in all areas of Tanzania, and local farmers are familiar with the crop. Cassava ethanol production can be competitive if out-grower schemes are adopted in connection with a core plantation that can guarantee the continued optimal supplies of feedstock required by the processing plant. To ensure a viable and sustainable domestic ethanol industry based on cassava, productivity must be improved, so that the energy balance is lower; adoption of appropriate sustainable management practices, such as conservation agriculture, could considerably increase productivity. Moreover, given poor transport infrastructure and to overcome rapid post-harvest deterioration of the cassava – i.e., starch reduction – it is recommended that the cassava is sun dried into chips at the farm to extend the shelf life. The sun-dried cassava chips can then be collected in centralized sites near the area of production, and the less bulky material can be transported to an ethanol processing plant.

Analysis of the impact of rising cassava prices on the welfare of the poorest quintiles shows differences between regions and household types; while some vulnerable households may be affected negatively, there may be benefits to most people

In general, developing the biofuel industry in Tanzania will require infrastructure development, including efficient and reliable transportation networks to support the connection of the various production components along the supply chain and to facilitate processing, which requires provision of potable and industrial water (agro-processing and biofuel plants) and electrification. Biofuel processing can also be enhanced by producing value-added co-products from by-products to benefit the economies of production and to generate added-value inputs to meet local needs, i.e., biofertilizers, electricity. A sustainable biofuels industry that provides longer-term prospects for economic growth and poverty reduction can be created if the choice of technology is suited to the technical capacity available in the country.

Source: FAO, 2010a; 2010b.

Risks to food security indeed need to be considered at the local level. Large-scale allocation of land for biofuel production without adequate policies may lead to the eviction of vulnerable people, and poor people may also lose access to land, water, and other resources. In a review of large-scale land deals in sub-Saharan Africa, Cotula et al. (2009) observed that whereas there are important opportunities related to large-scale investments in land that can create employment and rural development, many countries do not have sufficient mechanisms to protect local rights and take account of local interests, livelihoods, and welfare. Insecure local land rights, inaccessible registration procedures, vaguely defined productive use requirements and insufficiently developed implementation clauses for social commitments in land contracts, as well as legislative gaps, often undermine the position of vulnerable local people. The review calls for: a careful assessment of local land uses and claims; securing land rights for rural communities; involving local people in negotiations; and proceeding with land acquisition only after their free, prior and informed consent.

On the other hand, driven by technological progress and investments in agricultural research, the rapid output growth in developing countries is particularly remarkable. Nearly 70% of incremental production came from higher yields, approximately 10% from higher cropping intensities, and only about 20% from increasing the area of land used (Müller et al., 2007).

Müller et al. (2007) also claim that, in regions where crops do not benefit from technical progress, farmers remained poor and little progress was achieved in reducing undernourishment, but in general, growth in agricultural production helped to reduce hunger and poverty. In fact, it is noted that the food production situation in developing countries has been promising; the data indicate that over the past 30 years production of food has nearly tripled. Food supply per person has increased by more than 50% in developing countries. However, this increase in gross production could result in a further increase in food consumption, but difficulties mainly related to food transportation and distribution among poor people did not allow hunger to be reduced significantly. This is mainly due to the fact that that these problems do not allow poor people to have access to food, even with the increase in food production.

A recent review on integrated food-energy systems discusses possibilities for, and constraints regarding, large-scale implementation of biofuels versus food (Bogdanski et al., 2010). If policies that integrate bioenergy farming with food and feed farming are implemented, there is a potential to decrease local food shortages and increase the incomes of the world's poorest people (Müller et al., 2007; Sparovek et al., 2007).

20.3   Water Use

Water is a natural resource necessary for human survival and important for economic activities. It is used mainly for household purposes (drinking water, sanitation), industrial purposes (e.g., food processing, mining), energy generation (in hydropower systems and for cooling towers in nuclear and thermoelectric power plants) and, most important, agriculture (food production, feed, bioenergy crops). The increasing stress on freshwater resources brought about by ever-rising demand, mainly resulting from population growth, is of serious concern (see Figure 20.11).

As the world population increases, and development requires increased allocations of groundwater and surface water for the domestic, agriculture, energy and industrial sectors, the pressure on water resources intensifies, leading to tensions and conflicts among users, and degradation and pollution of the environment. Contamination of rivers, depletion of aquifers and increased utilization for multiple purposes are challenges commonly found in regions where demand exceeds supply and where water management is poorly conducted. Indicators such as water footprints are important to understand these important issues.

Increased competition among different water uses due to growing demand is already happening. Water scarcity induces competition for water between users, between economic sectors, and between countries and regions sharing a common resource, as is the case for international rivers. Many different interests are at stake, and equitable solutions must be found regarding the competing demands between: cities and rural areas; rich and poor people; arid lands and wetlands; public and private sectors; infrastructure and natural environments; mainstream and marginal groups; and local stakeholders and centralized authorities.

Water conflicts can arise in water-stressed areas among local communities and between countries because sharing a very limited and essential resource is extremely difficult. The lack of adequate legal instruments exacerbates already difficult conditions. In the absence of clear and well-established rules and regulations, severe tensions tend to dominate, and power can play an excessive role, leading to an inequitable allocation of water. A greater focus is needed on the peaceful sharing and management of water at both international and local levels (UN-Water/FAO, 2007). A well-developed system of Integrated Water Resources Management, including adequate institutional set-up and a good governance system, is needed in river basins or confined regions where demand exceeds existing supply.

Climate change is expected to account for about 20% of the global increase in physical water scarcity – and countries that already suffer from water shortages will be hit the hardest (UN-Water/FAO, 2007). This would include African countries, where water is a limiting factor for agriculture food production and also essential for income generation. Examples of export cultures in water-stressed regions are flower farms in Kenya, Zambia, and Uganda, which are extremely dependent on continued access to water in critical basins and trying to limit costs by using innovative rainwater harvesting techniques (FAO, 2005; Belwal and Chala, 2008; Orr and Chapagain, 2007; Asea and Kaija, 2000).

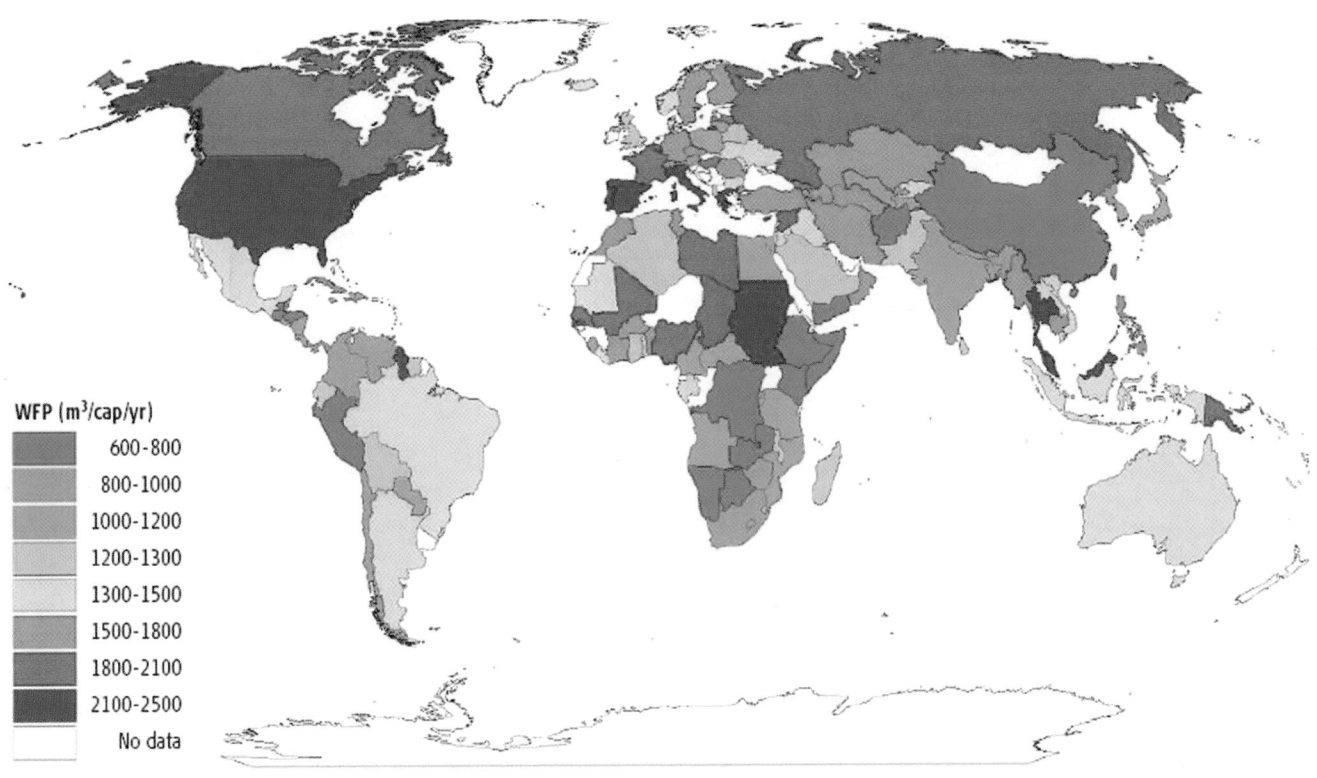

Figure 20.11 | National water footprints around the world, 2004. Source: UNESCO, 2006. Used by permission of UNESCO.

20.3.1   Water Withdrawal

In 2000, around 3800 km³ of freshwater were withdrawn from all over the world, twice the volume withdrawn 50 years earlier (WCD, 2000). Two thirds of the freshwater on Earth is frozen in glaciers and polar ice, and most of the remaining freshwater is groundwater (WCD, 2000). If all the freshwater resources on the planet were equally shared among the world's population, there would be 5000–6000 m³ of water per capita per year. However, the world's freshwater resources are distributed very unevenly, as is the world's population, and a large number of people in the low latitudes are experiencing water scarcity (see Figure 20.12).

The most dramatic water scarcity conditions are in parts of Asia and Africa. The emerging economies of India and China have regions where the climate is sub-humid or arid. Some areas depend on unsustainable pumping of groundwater aquifers and diversion of water flows for irrigation and human consumption, which can cause conflicts between communities. Unsustainable water use also occurs in developed countries, such as the mid-western parts of the United States, where the High Plains (Ogallala) aquifer has declined more than 100 feet (~30 m), and over 150 feet (~45 m) in some places. This resource underlies parts of eight states – Colorado, Kansas, Nebraska, New Mexico, Oklahoma, South Dakota, Texas, and Wyoming. The area overlying the High Plains aquifer is one of the major agricultural regions in the world. Water-level declines began in parts of the High Plains aquifer soon after the beginning of extensive groundwater irrigation (McGuire, 2007a; 2007b).

The areas of most severe water scarcity are those where high population densities converge with low availability of freshwater. Here, people not only experience a physical water scarcity, where evapotranspiration exceeds precipitation, but also a social water scarcity, where the accessible amount of water per person has decreased below 1000 m³/person/yr. Many countries are already well below the threshold value. Jordan is an extreme case, with less than 200 m³/person/yr (FAO, 2007b).

Growing scarcity and competition for water represent major threats to future advances in poverty alleviation, especially in rural areas. In semi-arid regions, increasing numbers of the rural poor are coming to see entitlement and access to water for food production, livestock and domestic purposes as more critical than access to primary health care and education. For poor people, an economic inability to cope with increased competition for water (being unable to afford rising prices) may result in economic water scarcity. Economic scarcity is caused by a lack of investment in water, or a lack of human capacity to meet the demand for water. Much of the scarcity is due to how institutions function – for example, favoring one group over another and not hearing the voices of various groups, especially women. Symptoms of economic water scarcity include scant infrastructure development, either small- or large-scale, so that people have trouble getting enough water for agriculture or drinking. Yet, even where infrastructure exists, the distribution of water may be inequitable, reflecting social distortions among different extracts of the population. Much of sub-Saharan Africa is characterized

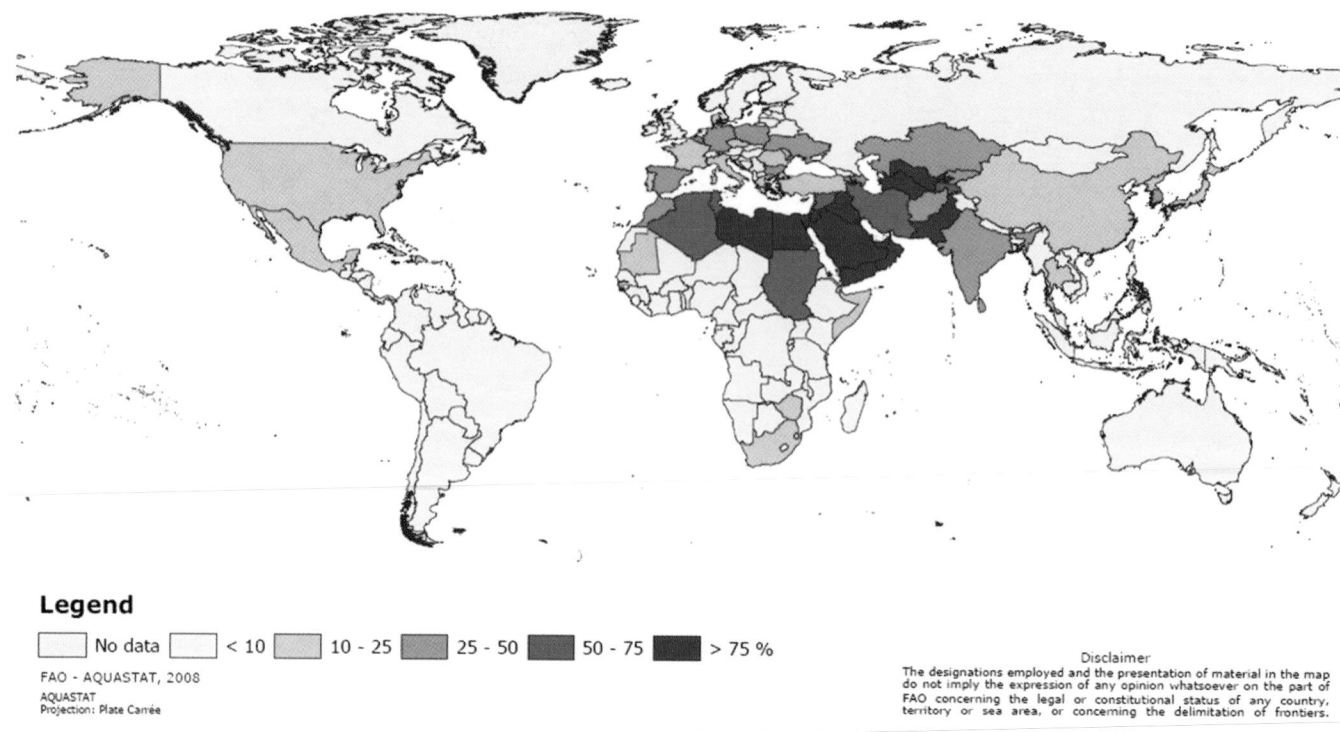

Legend

| | No data | | < 10 | | 10 - 25 | | 25 - 50 | | 50 - 75 | | > 75 % |

FAO - AQUASTAT, 2008
AQUASTAT
Projection: Plate Carrée

Disclaimer
The designations employed and the presentation of material in the map
do not imply the expression of any opinion whatsoever on the part of
FAO concerning the legal or constitutional status of any country,
territory or sea area, or concerning the delimitation of frontiers.

Figure 20.12 | Water resources withdrawn as a share of renewable water resources[11]. Source: FAO-AQUASTAT, 2008.

by economic scarcity, so further water development could do much to reduce poverty.

Physical scarcity occurs when there is not enough water to meet all demands, including environmental flows. Arid regions are most often associated with physical water scarcity, but water scarcity also appears where water is apparently abundant, when water resources are over-committed to various users due to overdevelopment of hydraulic infra-structure, most often for irrigation. In such cases, there simply is not enough water to meet both human demands and natural flow needs to supply the local environment. Symptoms of physical water scarcity are severe environmental degradation, declining groundwater, and water allocations that favor some groups over others (IWMI, 2007a).

20.3.2   Urbanization

In 2000, 75% of the total annual withdrawal of water (3532 km³) was used for agriculture, but this proportion could decrease to 58% in 2050

(IWMI, 2007a). Growing urbanization may result in tensions and conflicts for water, as large cities need to look for water supply from increasingly distant sources. One example is the city of Johannesburg, South Africa, which is largely supported by water transferred from the Orange-Senqu River basin. Farmers, who have been accustomed to liberal supplies with virtually no fees, now have to face reduced supplies and must often pay for water services (Lundqvist et al., 2007).

In some countries, governments can decide on the allocation of water from one sector to another, depending on how the water rights are defined. In other countries, where land and water rights are private, urban water service providers can buy water from the surrounding landowners, compensating them for reduced irrigation withdrawals or for giving up farming altogether. These examples are so far primarily from developed countries such as North America and Australia. Over-appropriation of ground- and surface water will continue to increase, due to increased food production and increased use of water by indus-try and cities, with the consequence that freshwater ecosystems will deteriorate. Cities and industries may secure their water supply at the expense of other sectors, such as agriculture and other rural uses, since water brings much higher economic returns for industrial and urban use (Lundqvist et al., 2008).

Urbanization is an important factor that influences agricultural markets. It is expected that the current rural population will remain more or less stable until 2030 and that population growth will be confined to urban

---

11   Renewable water resources refer to the long-term average annual flow of rivers (sur-face water) and recharge of aquifers (groundwater) generated from precipitation; they are computed on the basis of the water cycle. Non-renewable water resources refer to groundwater bodies (deep aquifers) that have a negligible rate of recharge on the human time-scale and thus can be considered as non-renewable. While renewable water resources are expressed in flows, non-renewable water resources have to be expressed in quantity (stock) (FAO-AQUASTAT, 2008).

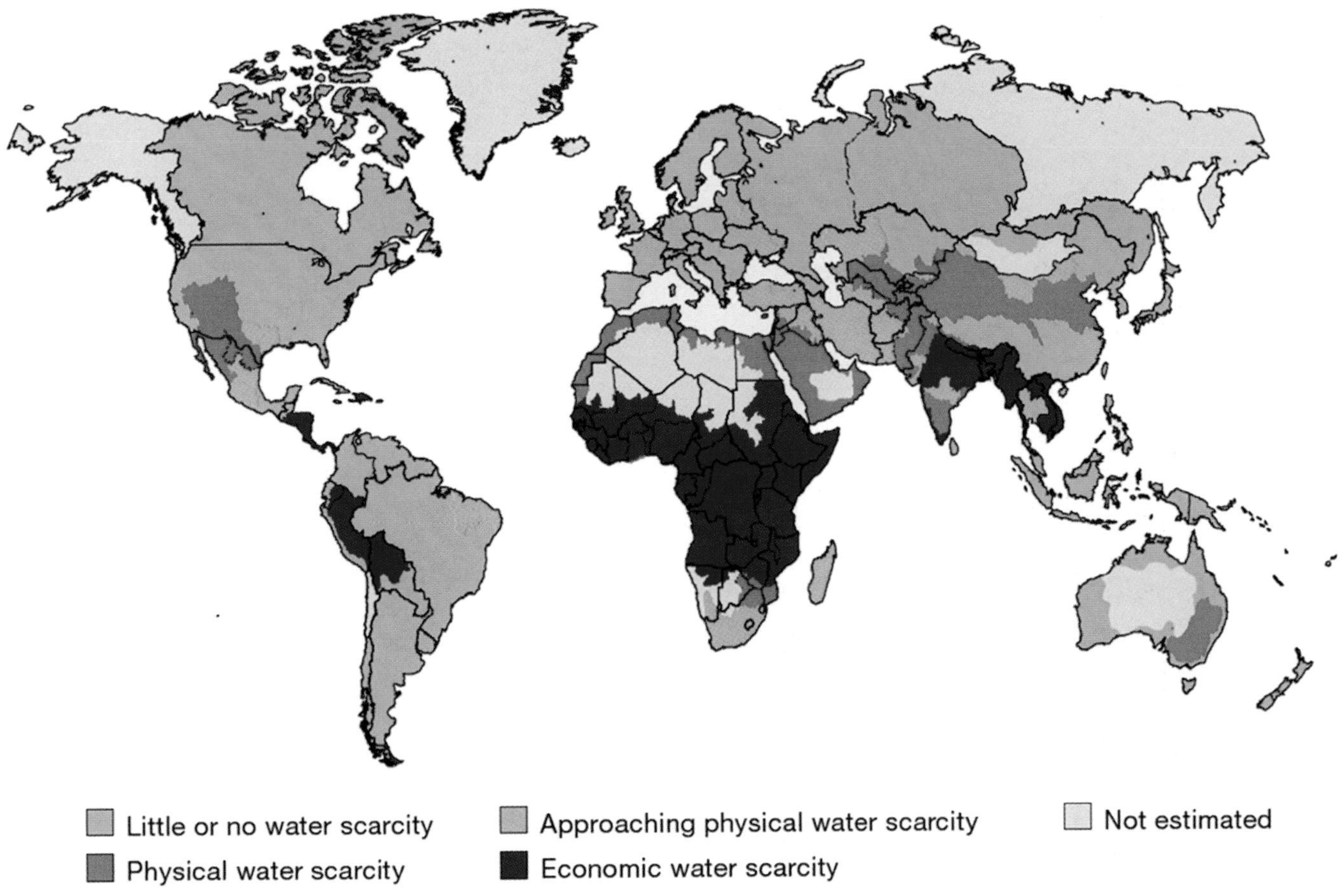
Little or no water scarcity
Physical water scarcity
Approaching physical water scarcity
Economic water scarcity
Not estimated

Figure 20.13 | Map of global water scarcity. Source: IWMI, 2007a.

areas. Urbanization has major impacts on markets due to high population density and infrastructure growth. Consumers become closely integrated into the international food markets, resulting in more food trade and changes in diets, with a shift from traditional vegetable foods to a greater demand for meat, which requires more water (Müller et al., 2007; Dufey et al., 2007, cited in Cotula et al., 2008).

In water-scarce regions the increasing reuse of wastewater after treatment processes such as deep-well pumping and large-scale desalination, would increase energy use in the water sector, thus generating more GHG emissions, unless "clean energy" options are used to generate the necessary energy input.

Possible conflicts between climate change adaptation and mitigation measures might arise with regard to water resources, such as biofuel crops that replace fossil fuels but require more water inputs, or water for cooling bioelectricity thermal plants. This indicates the importance of integrated land and water management strategies for river basins, to ensure the optimal allocation of scarce natural resources, including water. Also, both mitigation and adaptation have to be evaluated at the same time, with explicit trade-offs, to optimize economic investments while fostering sustainable development. Adaptation to changing hydrological regimes and water availability will also require continuous additional energy input (Bates et al., 2008).

20.3.3 Water Use in Agriculture

Agriculture accounts for more than 70% of the world's total water use. Its share drops in countries that import food and have a developed and diverse economy, but rises in some countries where agriculture is the primary economic activity (see Figure 20.14). Countries with water scarcity can be said to "import" virtual water when they import crops that have high water requirements.

Production of biomass for energy purposes can, in some cases, expand into areas where conventional food production is not feasible due to water constraints. This is the case with jatropha and sweat sorghum in India, where about 13 million hectares of "wasteland" are being earmarked for cultivation (FAO, 2008c).

Hydropower may offer positive environmental impacts related to water: about 75% of water reservoirs in the world were built for irrigation, flood control, and urban water supply schemes. Many could have small hydropower generation retrofits added without additional

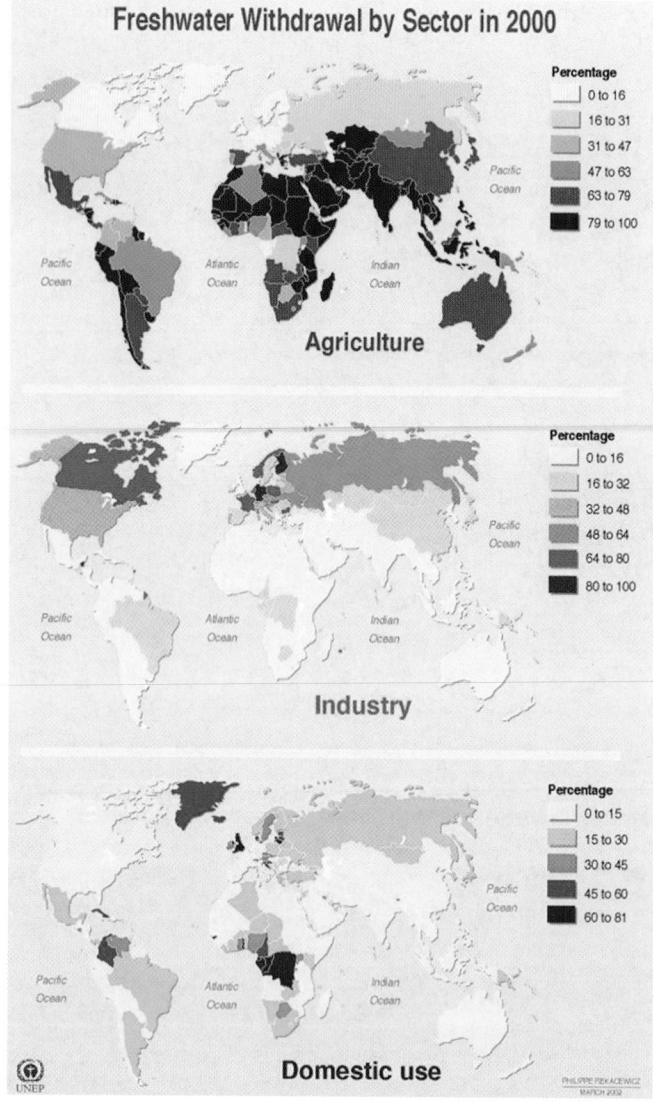

Freshwater Withdrawal by Sector in 2000

Agriculture

Percentage
- 0 to 16
- 16 to 31
- 31 to 47
- 47 to 63
- 63 to 79
- 79 to 100

Industry

Percentage
- 0 to 16
- 16 to 32
- 32 to 48
- 48 to 64
- 64 to 80
- 80 to 100

Domestic use

Percentage
- 0 to 15
- 15 to 30
- 30 to 45
- 45 to 60
- 60 to 81

Figure 20.14 | Global freshwater withdrawal: agricultural, industrial and domestic use. Source: UNEP/GRID-Arendal, 2002.

pronounced water scarcity. The growth of modern "conventional" irrigation since 1900 has been characterized by large water projects that harnessed rivers through the construction of diversion structures and canal systems. Since 1950, the spread of such technology has accelerated through state-sponsored, large-scale irrigation schemes, including large dams for water storage and regulation (sometimes including hydropower generation). Irrigated areas increased from 40 million hectares in 1900 to 100 million hectares by 1950 and to 271 million by 1998. Water from dams supports 30–40% of this area, with the remainder supplied by direct river abstraction, groundwater and traditional water harvesting systems.

In some areas, improved conditions for increased agricultural yields and better economic growth have resulted from subsidized infrastructure (less than full cost recovery), better agricultural practices, and increased access to electricity for pumping. Irrigated agriculture has contributed to growth in agricultural production worldwide, although inefficient use of water, inadequate maintenance of physical systems, and institutional and other problems have often led to poor performance and return. Emphasis on large-scale irrigation schemes facilitated consolidation of land and brought prosperity for farmers with access to irrigation facilities and economic markets. There are major multiplier effects produced by successful large irrigation schemes. However, economic support to rainfed agriculture has been limited, even though such systems supported more than 80% of farmers in developing countries, particularly in Africa. As a result, there has been a widening income gap between farmers in irrigated and rain-fed areas. Even within large-scale irrigation systems, there are inequities that lead to the marginalization of smallholders. The regional economic and development context for agriculture differs notably between industrial and developing countries. In the former, agriculture tends to be capital-intensive, with large, highly mechanized holdings requiring minimal labor. In contrast, agriculture in developing countries, particularly in Africa, supports hundreds of millions of smallholder cultivators who depend on the land for subsistence, livelihoods, and food security. These farmers generally do not have access to support mechanisms or economic resources to risk growing high-value crops in volatile market conditions. The low productivity of land and labor of many subsistence cultivators is also symptomatic of the absence of support and widespread neglect of their agriculture and irrigation systems. The extent of irrigated area in Africa is currently high on the political agenda for some countries. With less than one-third of the continent's potential under irrigation, opportunities exist for investing further in water for agriculture. This depends on access to land as well as water, and possibilities of compensating for the high rate of evapotranspiration (WCD, 2000).

Markets, commodity selection, ownership, land tenure, water storage for reliable supply, and international agreements on water allocations within river basins are all key factors in unlocking this potential. Many see the importance of reducing the food import bill borne by some African countries. Many also see the potential for boosting household incomes by creating labor opportunities. However, others recognize that, even in an optimistic scenario in which every hectare created two new

environmental impacts. The many benefits of hydroelectricity, including irrigation and water supply resource creation, rapid response to grid demand fluctuations due to peaks or intermittent renewables, recreational lakes, and food control, as well as the negative aspects, need to be evaluated for any given development. About 18% of the world's croplands now receive supplementary water through irrigation. Expanding this area (where water reserves allow), or using more effective irrigation measures, can enhance carbon storage in soils through enhanced yields and residue returns. However, some of these gains may be offset by CO_2 from energy used to deliver the water (Bates et al, 2008).

Irrigated agriculture provides a direct source of income generation and secure livelihoods for hundreds of millions of the rural poor in developing countries because of the food, income options and indirect benefits it generates. However, widespread irrigation practices may result in jobs, the significant uplifting of 60 million households would be insufficient to make major inroads into the extreme poverty that pervades the continent. A dual approach is gaining ground, one in which improvements in rain-fed food production for the very many vulnerable African farmers take place alongside the pursuit of viable irrigation opportunities (UN-Water/FAO, 2007).

Increasing water's productivity is an effective means of intensifying agricultural production and reducing environmental degradation. However, water productivity (efficiency) gains may be difficult to realize, and there are misperceptions about the actual scope for increasing physical water productivity. Much of the potential gain in physical water productivity has already been met in high-productivity regions. In addition, the amount of waste in irrigation is less than commonly perceived, especially because of reuse of water locally or downstream – farmers thirsty for water do not let it flow easily down the drain. Meanwhile, major gains and breakthroughs in agriculture that would reduce water requirements, such as those in the past from breeding and biotechnology, are much less likely to continue occurring at the same rate. Finally, a water productivity gain by one user may cause a loss or other damage to another user – an upstream gain may be offset by a loss in fisheries, or the gain may put more agrochemicals into the environment. Increasing economic water productivity would allow users to get more value per unit of water, either through switching to higher-value agricultural uses or by reducing the costs of production. Integrated approaches are important for increasing the value of water, and the number of jobs per drop (e.g., by better integrating crops and livestock in irrigated and rain-fed systems, or using irrigation water for household uses and small industries as well as agriculture). Higher physical water productivity and economic water productivity reduce poverty in two ways. First, targeted interventions enable poor people or marginal producers to gain access to water or to use water more productively for nutrition and income generation. Second, the multiplier effects on food security, employment, and incomes can benefit poor people. To reduce hunger and poverty, such programs must, however, ensure that the gains reach poor people, especially poor rural women, and are not captured by wealthier or more powerful users (IWMI, 2007a).

20.3.4   Water for Industry

Industry is also a water-demanding sector. Industrial use of water increases with country income, going from 10% of a country's total water use for low- and middle-income countries to 59% for high-income countries, where higher industrialization rates raise demand for water even though many processes are in themselves becoming less water-demanding. World water withdrawals for industry represent 22% of total water use. Pollution of freshwater supplies by industry is also a factor to consider in water management practices and policies. This is the case of both organic pollution (including from the food industry) and the inorganic one (e.g., chemicals, metal processing, and others). In developing countries, 70% of industrial wastes are dumped untreated into

waters, where they pollute the usable water supply. The current use of clean water for the dilution and transport of wastes is not sustainable. However, the total rate of water withdrawn (and returned) by industry worldwide is slowing, although water consumption for industry is still increasing. Aquatic ecosystems and species are deteriorating rapidly in many areas, and this is having an immediate impact on the livelihoods of some of the world's most vulnerable human communities by reducing protein sources for food, availability of clean water, and potential for income generation (UNESCO, 2003; 2009).

20.3.5   Energy Production, Biofuels and Water Use

Energy sources utilize water in different ways: for cooling towers in thermoelectric plants, for reservoirs in hydroelectricity production, and for irrigated biofuel crops. Despite the limited literature, an effort to assess the impacts of energy technologies on water resources from a life cycle perspective was conducted by the recent *IPCC Special Report on Renewable Energy* (Sathaye et al., 2011). Thermal power plants, hydropower and bioenergy are vulnerable to water scarcity, exhibiting risks of increased competition. Operational water consumption for different technologies in the US electricity sector is summarized in Figure 20.15.

Problems with cooling water availability (because of reduced quantity or higher water temperature) could disrupt energy supplies by adversely affecting electricity generation in thermal and nuclear power plants (Bates et al., 2008). Figures 20.16 and 20.17 compare water withdrawal across fuel cycles of electricity-generation options based on US data, except for the wind cycle, which uses information from Denmark. For thermoelectric fuel cycles, the life-cycle water withdrawal ties closely to the operational cooling type: on-site cooling water use dominates the life-cycle water withdrawal, while the dry cooling method, more expensive and energy-intensive, is an exception to the overall rule.

As conventional oil supplies become scarce and extraction costs increase, production of unconventional oil will become more economically attractive (although with higher water use and environmental costs). Mining and upgrading of oil shale and oil sands require the availability of abundant water. Technologies for recovering tar sands include open cast (surface) mining, where the deposits are shallow enough, or injection of steam into wells in situ to reduce the viscosity of the oil prior to extraction. The mining process uses about four liters of water to produce one liter of oil but produces a refinable product. Mining of oil sands leaves behind large quantities of pollutants and areas of disturbed land (Bates et al., 2008).

Most renewable energy technologies require little or no water for cooling. However, hydropower plants use water directly to generate power, diverting water from rivers through turbines, via an intake at the dam. In some cases, water is diverted outside the stream for up to several

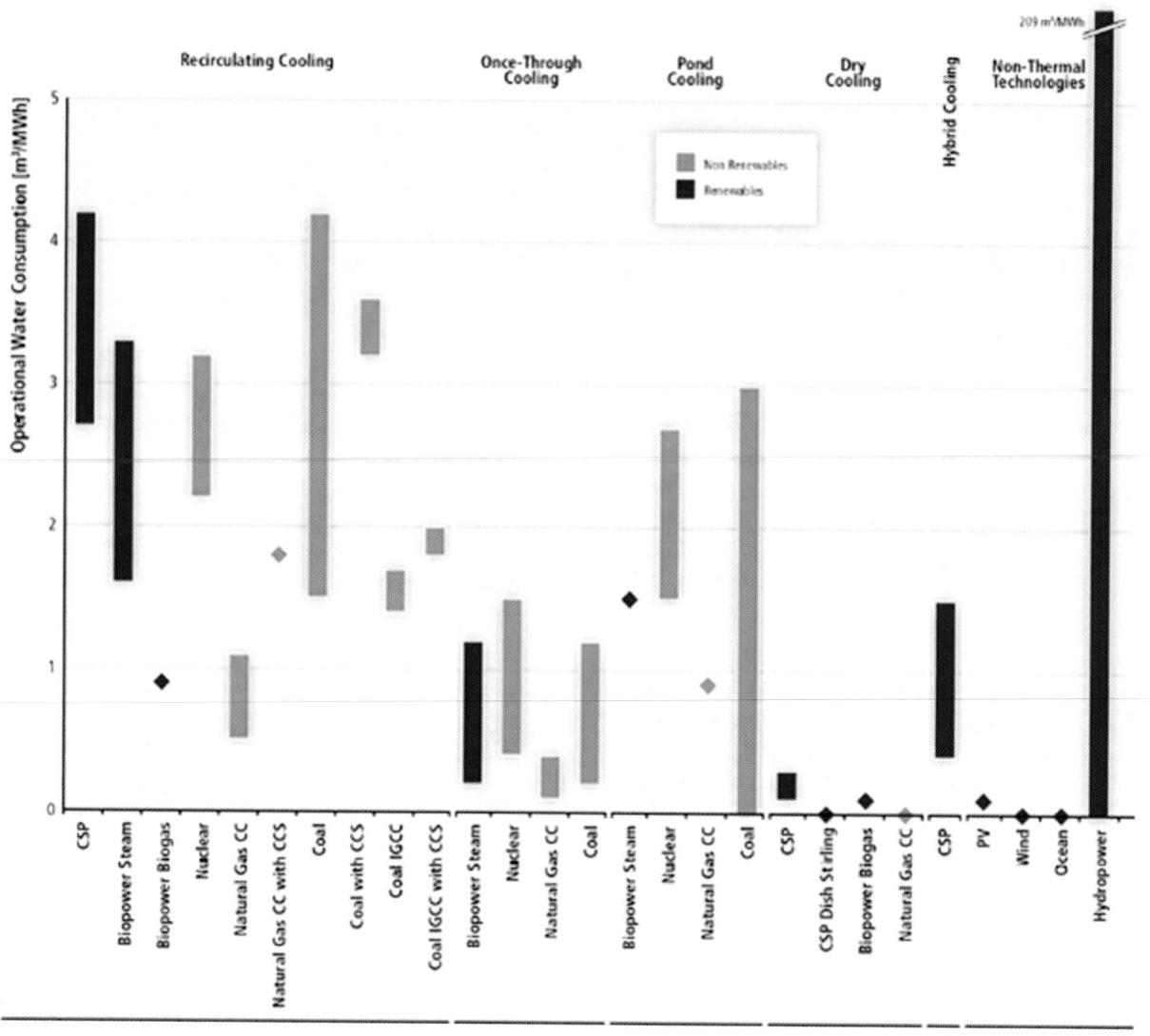

Figure 20.15 | Rates of operational water consumption by thermal and non-thermal electricity-generating technologies in the US (m³/MWh). Source: Sathaye et al., 2011.

Bars represent absolute ranges from available literature, diamonds single estimates; n represents the number of estimates reported in the sources. Upper values for hydropower result from few studies measuring gross evaporation values, and may not be representative. Notes: CSP: concentrating solar power; CCS: carbon capture and storage; IGCC: integrated gasification combined cycle; CC: combined cycle; PV: photovoltaic.

miles before being returned. Human intervention – through interbasin transfers, dams, and water withdrawals for irrigation – has fragmented 60% of the world's rivers. Since 98% of the water used in hydropower plants is returned to its source, distinctions are made between use (when water is ultimately discharged back into the original water body, although in some cases with chemical or physical alteration) and consumption (where water is lost, typically through evaporation).

The use of water to generate power at hydropower facilities imposes unique, and by no means insignificant, ecological impacts, including:

- impacts on terrestrial ecosystems and biodiversity;

- associated GHG emissions;

- altered downstream flows affecting aquatic ecosystems and biodiversity;

- altered natural flood cycles affecting downstream floodplains;

- impacts of dams on fisheries in the upstream, reservoir and downstream areas;

- enhancement of ecosystems through reservoir creation and other means; and

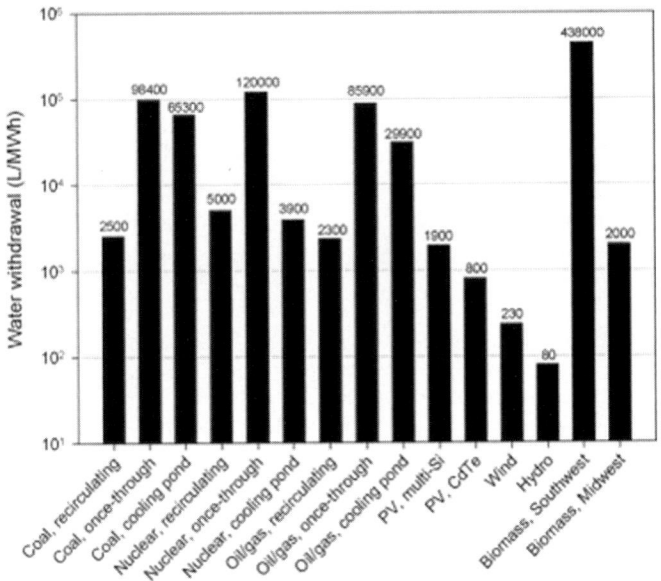

Figure 20.16 | Comparison of water withdrawal across fuel cycles, using US data, except for a Danish case for wind cycle. Source: Fthenakis and Kim, 2010.

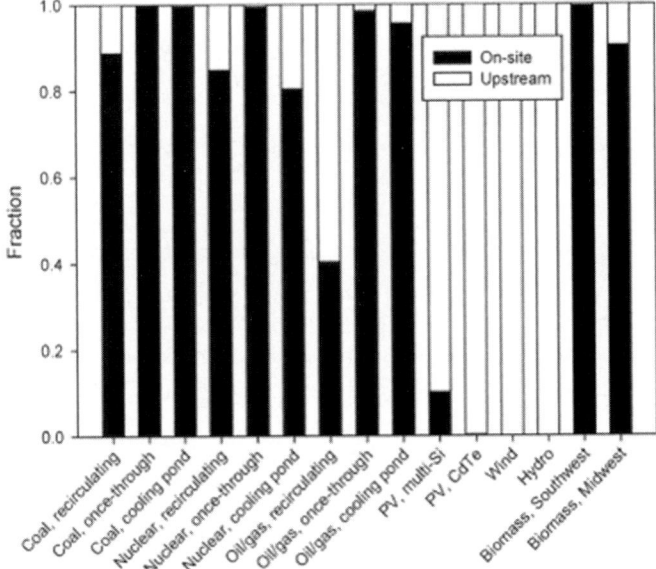

Figure 20.17 | Breakdown of water withdrawals based on the water-use stage. Source: Fthenakis and Kim, 2010.

- cumulative impacts of a series of dams on a river system (WCD, 2000).[12]

Solutions to these problems lie both in more environmentally friendly energy resources, and in more efficient energy end uses (Power Scorecard, 2000). Although there are large regional differences in the extent of hydropower development, hydrological changes will directly

affect the potential output of hydroelectric facilities – both those currently existing and possible future projects. A reduction in hydroelectric power is anticipated where and when river flows are expected to decline (Bates et al., 2008).

Geothermal fields of natural steam are rare; most produce a mixture of steam and hot water, requiring systems to separate out the hot water. Sustainability concerns relating to land subsidence, heat extraction rates exceeding natural replenishment, chemical pollution of waterways (e.g., with arsenic), and associated CO_2 emissions have resulted in some geothermal power plant permits being declined. These problems could be partly overcome by re-injection techniques and deeper drilling technology. However, expanding geothermal exploration means in the end a claim on available water resources (Bates et al., 2008).

Although bioenergy can produce climate change mitigation benefits by replacing fossil fuel use, large-scale agricultural production of biofuels could, in some cases, intensify water use, thereby reducing stream flow or groundwater reserves.

High-productivity, evergreen, deep-rooted bioenergy plantations generally have a higher water use than the land cover they replace. Some practices may also affect water quality through enhanced leaching of pesticides and nutrients. Land management practices implemented for climate change mitigation may also have different impacts on water resources.

Many of the practices advocated for soil carbon conservation – reduced tillage, more vegetative cover, greater use of perennial crops – also prevent erosion, yielding possible benefits for improved water and air quality. These practices may also have other potential adverse effects, at least in some regions or conditions. Possible effects include enhanced contamination of groundwater with nutrients or pesticides via leaching when reduced tillage practices are used. However, these possible negative effects have not been widely confirmed or quantified, and the extent to which they may offset the environmental benefits of carbon sequestration is uncertain. The group of practices known as agriculture intensification has environmental benefits that include erosion control, water conservation, improved water quality, and reduced siltation of reservoirs and waterways (Bates et al., 2008).

Total water demand for bioenergy production (prospective of about 150 EJ/year) from plantations in 2050 ranges between 4000–12,000 km³/year, depending on how much energy can be derived per unit of transpired water. Even if reduction of losses and improvements in agricultural and water productivity are accounted for, meeting the future food demand will most likely require the addition of new cropland (– 20–45% of real expansion from today's 1.5 billion ha, depending on the degree of improvements in land area productivity). The annual cropland expansion needed is estimated to be 0.48% on average between 2010–2045 (IWMI, 2007b). While the benefits of bioenergy range from

12 The European Water Framework Directive utilizes a similar classification of impacts (European Commission Environment, 2011).

reduced GHG emissions to renewability and energy independence, increased biofuel production can lead to trade-offs across other ecosystem services if not well addressed, including those derived from water. Besides decreased food supply, other trade-offs include poor water quality associated with increases in aggregate fertilizer use, nutrient runoff, and erosion (Bennett, 2008). The growing of crops for biofuel in semi-arid areas on marginal or degraded lands may require some irrigation, particularly during hot and dry seasons. Further, the processing of feedstocks into biofuels can use large quantities of water, mainly for washing plants and seeds and for cooling.

Globally, irrigation is not likely to be a major water source for biofuel production, although locally there can be water constraints in some areas where food and other non-food crops compete for the scarce available resources. In some cases, irrigated production of biofuel feedstocks has a considerable impact on local water resource balances: the Awash, Limpopo, Maputo, Nile and São Francisco river basins are cases in point. While the potential for expansion of irrigated areas may appear high in some regions on the basis of water resources and land, water quality as well as quantity may be affected. Converting pastures or woodlands into maize fields, for example, may exacerbate problems such as soil erosion, sedimentation and excess nutrient (nitrogen and phosphorous) runoff into surface waters, and infiltration into groundwater from increased use of fertilizer (FAO, 2008c).

Table 20.8 summarizes the main energy and food crops in the world today. With adequate land use management, there are enough rain-fed areas for these crops to coexist. Sugarcane ethanol is a special case, due to its significant potential to provide energy throughout large parts of the world utilizing rain-fed land (see Figure 20.18). Water use for industrial sugarcane ethanol production has improved from 5 to 1.9 m^3/t of cane (1 tonne of cane crushed produces 80–100 liters of ethanol) in the period 1997–2004. Vinasse, the distillery water sludge, is used as a fertilizer, returning part of the water used back to the soil (Elia Neto, 2005).

In terms of water use, studies in other parts of the world have shown that production of energy crops for biofuel production can have substantial impacts on water demand, especially if irrigation is used for their production (Jumbe et al., 2007). The bulk of sub-Saharan Africa's agricultural production occurs under rain-fed conditions, posing substantial risks to farmers and investors because of erratic rainfall patterns and temporal and spatial variability.

Frequent short dry spells during the growing season reduce yields and, in addition, have an indirect impact, as farmers are less likely to invest in inputs and land management. Biofuel production could be affected by these existing conditions, as unreliable rainfall and highly variable yields would also pose challenges for biofuel feedstock production.

On the other hand, new market opportunities through bioenergy production may trigger additional investments in water for agriculture to raise yields and reduce these fluctuations. The water situation in sub-Saharan Africa is also affected by the transboundary nature of most river basins, which flow through several countries. The negotiation and enforcement of water rights and international water treaties is a difficult process requiring a strong institutional infrastructure. Feedstock production requires substantial amounts of water; to produce the biomass for one liter of biofuel, crops evaporate an average 2500–3500 liters of water. Much of this water can be obtained from rainfall, where rainfall is abundant and reliable.

Compared to other regions in the world, the level of water resources development for irrigation in sub-Saharan Africa is low. Only 4% of agricultural production originates from irrigated areas, and only one-sixth of the irrigation potential in sub-Saharan Africa has been realized. In many areas of this region, water scarcity for crop production is caused by the lack of infrastructure to tap into water resources rather than an actual physical shortage (IWMI, 2007b). Development costs for irrigation are high, averaging US$6000/ha in sub-Saharan Africa compared to US$1500/ha in South Asia. It is unlikely that farmers and governments will be able to afford large investments in irrigation for biofuel feedstock, unless significant returns are obtained. Water use by other feedstock crops needs to be studied more systematically across the agro-climatic zones, especially when such crops are to be established in large-scale plantations (Benkunda et al., 2009).

20.3.6   Water Security

Although concepts of food security, energy security and access to natural resources have been widely discussed, it is only recently that the relationship between the environment and water security is becoming more widely acknowledged, mainly due to its political dimensions. It was on the agenda in the process leading up to the UN Conference on Environment and Development, held in Rio de Janeiro, Brazil, in 1992, and is reflected in the Rio Principles. Sharing water resources is an issue that particularly reflects a link between environmental degradation and resource competition resulting in conflicts. Freshwater scarcity is not just a direct cause of insecurity; it is also an indirect security threat, through its potential for causing conflicts.

A key spark for water-related tensions is a mismatch between population levels and the freshwater available for different purposes. This can produce water conflicts at a local level, such as between tribes over grazing rights or the property rights over wells, as has happened in Ethiopia. Tensions can also arise due to construction of a dam or canal which impacts water availability for riparian states, as was experienced during the 1990s by Turkey, Syria and Iraq over Turkey's Greater Anatolia Project on the Euphrates-Tigris River Basin (Coskun, 2007).

Global conflicts over water threaten geopolitical stability. An example is the Baglihar 450 MW dam, near the politically fragile Pakistan-India border along Kashmir's Chenab River. The plant will supply hydroelectric

Table 20.8 | Crops and their requirements.

Crop type	Soil requirements	Water requirements	Nutrient Requirements	Efficiency	Climate	Others
Cereals (wheat, rye, barley, oat, triticale)	less disruptive; very constant yield; straw removal affects humus balance; wheat in a wide range of soils but medium textures	wheat is the cereal with the highest water demand (450–650 mm depending on climate and length of growing period)	medium; wheat has higher fertilizer demand but good uptake	wheat on good soils and rye on poor soils for food; triticale for energy purposes	wheat: moderate climate, tropics away from the equator (winter crop)	highly developed crops; knowledge widespread among farmers
Jatropha	undemanding	both irrigated and rain-fed, wide range (200–1200 mm/yr)	adapted to low-fertility sites and alkaline soils; better yields with fertilizers	grows on marginal land, restores eroded areas, improves soil	tropical and subtropical	very resistant against pests and pathogens; toxic seeds after oil extraction
Maize	low requirements (soils well aerated and well drained); susceptible to water logging.	water demanding (500–800 mm/yr)	fertility demands for grain maize are relatively high	high water efficiency, but often irrigated (water deficits cause losses)	from temperate to tropic, mean daily temperatures above 15°C and frost-free	increased erosion risk; monoculture negative on humus balance (requires crop rotation); high pesticide use
Miscanthus (woody, perennial grass)	good water supply, not saturated brown soils with humus	crucial during main growing seasons	relatively low	high growth rates, cheap	warmer climates, fairly cold-tolerant	risk of invasive species; difficult to rehabilitate land for other uses due to deep root structure
Oil Palm	many soil types (flat, rich, deep), good drainage (tolerate floods)	demanding in excess of 2000mm, even distributed throughout the year	relatively low	highest-yielding vegetable oil crop per hectare in the world	tropical and subtropical; humid 5h/day minimum sunlight; temperature 22–32°C, basically low altitude (< 400m)	fire for land clearing, tropical forest conversion with loss of biodiversity and conflicts of land ownership with local communities, pollution by improper use of agrochemicals (little control), oil mill effluent dumping, soil erosion from land clearing
Poplar	deep, moist (highly flood tolerant)	high; irrigation might be needed	high, with good uptake	high establishing costs, not easily propagated from cuttings; short rotation crop (fast growing tree)	arctic to temperate	resistant to pests and diseases
Potato	deep, moist, well-drained and friable	500–700 mm, depending on climate; irrigation required for climatic negative balances	relatively high (under moderate demand there are late growth and soil erosion risks)	very productive	18–20°C mean; sensitive to changes; cool crop but grows well in warm conditions if water is sufficient	rotation crop (with maize, beans and alfalfa); pesticide-intensive
Rapeseed	mild, deep loamy, well-drained	600 mm/yr minimum	similar to wheat	rotation crop, poor productivity but most grown energy crop in Europe (well-known to farmers and policy supported)	very temperature-sensitive, best 15–20°C.	very intensive culture; high input of pesticides and herbicides
Rice	not demanding (permeable layer and good drainage)	very high, flooded fields	high input of fertilizers	very labor-intensive, different farming systems (rain-fed lowland, upland or dry land, irrigated, deepwater or flood-prone)	high, constant temperatures (13°C–40°C; optimum around 30°C).	high level of landscape alteration for surface infrastructure needed to move water about; anaerobic conditions in underlying soils; has unique impacts on carbon and nutrient cycling

Table 20.8 | (contd.)

Crop type	Soil requirements	Water requirements	Nutrient Requirements	Efficiency	Climate	Others
Sorghum	Deep, medium-textured, well-aerated and well-drained soils; tolerant to short periods of waterlogging, moderately tolerant to soil salinity	drought-resistant; high water supply flexibility (water holding capacity of soil very important, needs progressively less water as roots reach deeper)	high in nitrogen, low in phosphorus and potassium requirements	tropical origin, adapted to southern Europe with irrigation, high yields (less than maize), machinery can be used	over 25°C; some varieties adapted to lower temperatures; needs a lot of sunlight	numerous diseases; not competitive at the beginning
Soybean	wide range of soils, optimum growth in moist alluvial with a good organic content; best on high water capacity, good structure, loose soil	relatively high (450–700 mm/ season)	atmospheric nitrogen-fixing crop		warm conditions in tropics, subtropics and temperate climates; relatively resistant to low and very high temperatures	increased risk of erosion compared to corn
Sugar beet	wide range of soils with medium- to slightly heavy- textured, well-drained preferred; tolerant to salinity	550–750 mm/growing period; tolerant to water deficits	high demand		different climates, high sugar yields with night temperatures 15–20°C and day temperatures 20–25°C during the latter part of growing period	heavy machinery and harvested mass lead to soil compaction; can provide nesting habitat and shelter in autumn; soil erosion risks; various pesticide treatments to eliminate weeds
Sugarcane	deep soils, well-aerated, groundwater table more than 1.5 to 2.0 m below the surface	1500–1800 mm/yr, high water requirement, evenly distributed through the growing season	high nitrogen and potassium needs, low phosphate requirements; at maturity nitrogen content of soil must be as low as possible for a good sugar recovery	very high efficiency in conversion of solar energy into biomass	tropical or subtropical climate; germination 32–38°C; active growth at 20°C; for ripening lower temperatures are desirable for enrichment of sucrose	pest management very important
Sunflower	wide range of soils, deep with good water holding capacity; row crop, leaves bare soil until late spring	600–1000 mm, depending on climate and length of total growing period; water-efficient crop but often irrigated as better growth	moderate demand; good fertilizer uptake	intermediate crop for maize plantations	from arid under irrigation to temperate under rain-fed conditions; susceptible to frost; needs full sun	various pesticide treatments to combat pests
Switch grass	diverse growing conditions, from prairies to arid or marsh; conserve soil and improve its quality	tolerant to floods or drought; irrigation only necessary in very hot or very dry climates; because of deep roots, groundwater abstraction is possible	little nutrient requirements	machinery can be used; high yields in marginal or erosive land	warm season plant, good resistance to dry summer months	very little pesticide use
Willow	permanent crop, good soil cover, deep rooting and leafy canopy reduces soil erosion and prevents saturation of the land during heavy rainfall; can grow on land that is too wet for other crops	substantial quantities of water (600 mm rainfall), suffers reduced growth in dry conditions or dry years	significant nutrient uptake but good uptake also	short rotation coppice; easy and relatively inexpensive to plant	can tolerate very low temperatures in winter, but frost in late spring or early autumn will damage the top shoots	level of pest and pathogen unacceptable for food crops can be accepted here; already used in commercial or near-commercial applications; can take up heavy metals (phytoremediation); weed competition is critical; riparian buffer strips; depletion of soil nutrients from frequent and repeated harvesting; rust is an important disease; very competitive, hence no or little pesticides applications necessary

Source: UNEP, 2007; UN-Energy, 2007.

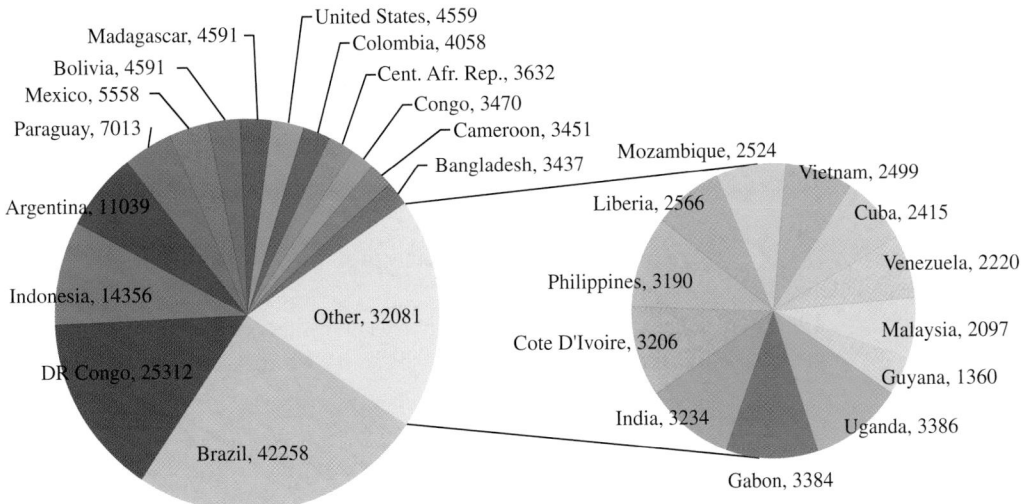

Figure 20.18 | Very suitable and suitable potential land area for sugarcane crops in major 26 countries producers, considering high-, medium- and low-input efforts, with forest area excluded, by country (units in thousand hectares). Total area in these 26 countries is 169 million hectares. Total area in all producers is 191 million hectares, equivalent to 1.46% of all land area. Source: based on data from FAO, 2009a.

power to northern India, but as the dam reaches full capacity, Pakistan is seeking compensation for what it views as a "gross violation" of an international treaty as a result of "stolen" water (Pope, 2008). Around 40% of the world's population lives in watersheds shared between nations, with more than 260 shared river basins of social and economic importance.

Basins with potential for political stresses in the coming years include: the Ganges-Brahmaputra, Han, Incomati, Kunene, Kura-Araks, Lake Chad, La Plata, Lempa, Limpopo, Mekong, Ob (Ertis), Okavango, Orange, Salween, Senegal, Tumen, and Zambezi (Wolf et al., 2003). There are complex linkages among energy, water, agriculture, and environmental degradation issues. Also linked are water-related matters affecting humanitarianism, human health, poverty reduction, economic development, environmental protection and conservation, stability, and geopolitical security. Management of water resources can be a catalyst for cross-border cooperation or a trigger for socioeconomic instability. An instrument establishing agreements over a shared water source among different stakeholders is thus an important, anticipatory structure (Peterson and Pozner, 2008).

The FAO (2008c) outlines some conditions to ensure institutional development to promote water security, as follows:

- Set conditions for more flexible and responsive service-oriented water management.

- Develop tools for water-related conflict resolution and prevention at local and district levels.

- Develop and implement economic and financial trade instruments to remove distortions in water allocation.

- Document and quantify current patterns of water use and water entitlements.

- Develop transparent water allocation mechanisms to protect water use rights while providing greater flexibility to respond to scarcity under anticipated patterns of climate change.

- Develop innovative insurance products.

Further suggestions involve better integration of water resources management, agriculture and food security databases, with much closer monitoring of irrigated and rain-fed production and clearer distinctions between the sources of supply (rainfall, surface water, and groundwater). This effort should consider specific food staples, notably rice and wheat, as well as the productivity of water-dependent aquatic environments (FAO, 2008c).

When using a wider definition of water security, such as the one used by the United Nations Educational, Scientific and Cultural Organization (UNESCO) and several other UN agencies within UN-Water, a broader approach needs to be applied to ensure this: "Water security involves protection of vulnerable water systems, protection against water related hazards such as floods and droughts, sustainable development of water resources and safeguarding access to water functions and services."

20.4   Sustainability Factors

20.4.1   Sustainable development: policies and measures

The concept of sustainable development – a significant one – aims to reconcile environmental, economic and social goals, which became more important after the World Commission on Environment and Development (1987). It recognizes the multiple challenges resulting from a declining quality of the environment, increasing resource needs of mankind, and the need to reduce malnourishment, poor hygiene, insufficient water supply and other problems related to extreme poverty, without destruction of the natural resources that are the basis for life.

The influential report *Our Common Future* (the "Brundtland Report") defined sustainable development as that which "meets the needs of the present without compromising the ability of future generations to meet their own needs" (World Commission on Environment and Development, 1987), i.e., as a universal, intra- and intergenerational human rights concept.

The 1992 United Nations Conference on Environment and Development (UNCED or Rio 92) in Rio de Janeiro, Brazil, was a major milestone in the international discussion on sustainable development. At the conference, the United Nations Framework Convention on Climate Change (UNFCCC), the Convention on Biological Diversity and Agenda 21 were proposed, and the two treaties were also opened for signatures. It was also decided to initiate negotiation of what became the UN Convention to Combat Desertification. The Climate Conference at Kyoto, Japan, in 1997 resulted in the Kyoto Protocol, a multilateral environmental agreement to curb GHG emissions worldwide. By 2000, the MDGs aimed at reducing global poverty and eliminating social exclusion by 2015. Two years later, energy (both energy efficiency and renewable energy) became a central point at the World Summit on Sustainable Development (WSSD or Rio+10) in Johannesburg.

A fairer and cleaner energy future includes the need to reduce negative social and environmental impacts. Sustainable energy also includes the goal to establish security of supply, which means both guaranteeing access in the short term and preserving resources for the long term. These concepts are closely related to geopolitics, trade liberalization and protectionism, land and water usage, and, not rarely, to military concerns.

Most policies that aim at a more sustainable energy supply rest upon four main pillars (although some of them may be more or less emphasized): more efficient use of energy, especially at the point of end use; increased utilization of renewable energy as a substitute for non-renewable energy resources; accelerated development and deployment of new energy technologies – particularly next-generation fossil fuel technologies that produce near-zero harmful emissions and open up opportunities for CO_2 sequestration; and biosequestration of carbon in terrestrial ecosystems, including soils and biota (see Box 20.4).

Achieving energy equity is another important task, providing access to modern energy to the 2 billion people that still live with only traditional solid fuels (Johansson and Turkenburg, 2004).

Sustainable development is thus a basic principle to be pursued in key policy areas directly or indirectly related to energy: technology research, development and transfer; information exchange; capacity-building;

adequate financing; ambitious environmental protection goals; removal of harmful subsidies; and shifting away from business-as-usual patterns of production and consumption.

20.4.2   Sustainable Water Use

This chapter focuses on the following aspects of sustainable water use:

- environmental issues resulting from erosion; excess use of fertilizers and agrochemicals; inadequate management of water resources; inadequate wastewater treatment and disposal; and the need for increased biodiversity protection, solid and hazardous wastes management, and protection against contamination of soils;

- resource base maintenance (freshwater protection, allocation, and sustainable use);

- social concerns (including issues related to employment and access to water, particularly for women and vulnerable groups); and

- economic issues (including weighing present versus future uses).

Table 20.9 summarizes impacts from energy on water systems.

The consumptive use of water (i.e., evapotranspiration) in energy crop production changes according to different bioenergy systems. A large variation in evapotranspiration is attributed to: varying water productivity among energy crops, related to crop type, soil, and climate, and agronomic practice, including water productivity modification options such as changing sowing date and plant density, supplemental irrigation, and microclimate manipulation; variations in the share of the aboveground biomass that is usable as feedstock in electricity/fuels production; and different conversion efficiencies of technology options available for electricity/fuels production.

Irrigation is assumed to be around 15% of the energy crop's water use calculated. If the average efficiency in irrigation water supply is 50%, then about 1175–3525 km³ of additional water would have to be withdrawn in 2050. Compared to the present withdrawal for irrigation, estimated at roughly 2600 km³/yr, clearly such additional blue water demand for energy crop irrigation implies tough challenges. Rain-fed energy crop production could potentially lead to similar impacts, by redirecting runoff water to evapotranspiration and thereby affecting downstream blue water availability and quality. Establishment of bioenergy plantations can also lead to increased evapotranspiration, especially if tree crops replace shallow-rooted grasses, herbs, or food crops. It is not possible to make general statements about the impact in terms of water depletion of expanding energy crop production, since the net change of evapotranspiration is uncertain and depends on site-specific circumstances, including the current land use that it aims to replace (Box 20.3; Figure 20.19) (Lundqvist et al., 2007).

Table 20.9 | Positive (+) and negative (-) impacts of energy on water (Lucon, 2010, own elaboration). Water indicative footprints (WF) of different energy sources (from Gerbens et al., 2008 and 2009).

Energy Source	Bioenergy	Solar	Geothermal Energy	Hydro Power	Marine Energy	Wind	Oil	Gas	Coal	Nuclear
Water usage and/or pollution	WF 24–143m^3/GJ. Potentially high water demand, especially in irrigated crops. Specific water stress for cooling towers (-).	WF 0.3m^3/GJ for solar thermal panels. Some toxic wastes from PV. Specific water stress for concentrated solar plant cooling towers. (-)	WF n.a. Risks of water contamination. Specific water stress for plant cooling towers.	WF 22 m^3/GJ according to source, but extremely variable. High impacts in water bodies and regimes (-/+)	WF n.a.	WF ~0.0 m^3/GJ.	WF 1.1 m^3/GJ. Risks of severe, large scale water contamination (-)	WF 0.1 m^3/GJ. Specific water stress for plant cooling towers.	WF 0.2 m^3/GJ. Risks of water contamination, leaching and acid rain effects. Specific water stress for plant cooling towers. (-)	WF 0.1 m^3/GJ. But potential highly toxic contamination in case of nuclear accidents, weapon proliferation tests etc. Specific water stress for plant cooling towers. (-)

Source: Gerbens-Leenes et al., 2009.

Box 20.3 | Water Pressures and Increases in Food and Bioenergy Demand. Implications of Economic Growth and Options for Decoupling

Long-term models out to 2045, as developed by Lundqvist et al. (2007), present scenarios on potential water demand, given that diets may change substantially as countries attain a higher income, and that the use of bioenergy is likely to increase. With current global levels of productivity and efficiency, the consumptive use of water to cater for growth in food demand is estimated to increase by some 50% in the period 2000–2045, from about 7000 km^3/yr to about 10,600 km^3/yr. Estimating the water requirements for the production of energy crops is more difficult, but crop expansion is a reasonable measure to adopt. Higher demands for food, energy and other goods and services are inevitable, but it is possible to achieve a decoupling effect, i.e., more income without a corresponding increase in the pressure on resources. While water productivity improvements on the order of 25% are considered feasible and realistic, reductions in losses and wastage in the food chain can be halved from the present 30%.

By adopting these decoupling measures, it is possible and desirable to meet expected future food requirements in 2045 with about the same water requirement as in 2005. In the search for decoupling opportunities between GDP growth and pressure on water and land resources, it is logical to pay attention to both production and consumption dynamics. Consumption is a critical driver, considering increasing purchasing power, preferences, and behavior. There are major challenges in balancing finite freshwater resources, including competition for different uses in the near future, and the need to stimulate improvements in efficiency. Improvements in yields and water productivity, as well as expanding the agricultural area, cannot by themselves solve the problem. Blue water sources (rivers, lakes, and groundwater) will not be enough to meet the increasing demands for food and biofuel. Recommended strategies include: better utilization of a larger fraction of the rainfall (i.e., a green water strategy); improved yields and water productivity, as well as expanding the agricultural area; and reduced consumption through changes in consumer preferences and behavior, and by achieving a more efficient food chain, primarily in terms of reduced losses and wastage "from field to table." These changes will require planning and systematic implementation over decades.

Source: Lundqvist et al., 2007.

Water use for the growth of bioenergy crops may cause environmental and social problems. Global water demand has been growing by 10% per decade, adding to water stress in many regions. Agriculture accounted in 2008 for close to 70% of all freshwater use (UNEP, 2009), of which 2% was used for bioethanol production (WBGU, 2008).

Bioenergy production may need 70–400 times as much water per unit of energy as other primary energy carriers, excluding hydropower, and ranges from 24–143 m^3/GJ (Gerbens-Leenes et al., 2009). The amount of water per unit of energy is highly dependent on the crops used, the efficiency of the cropping system, and the local hydrological and soil

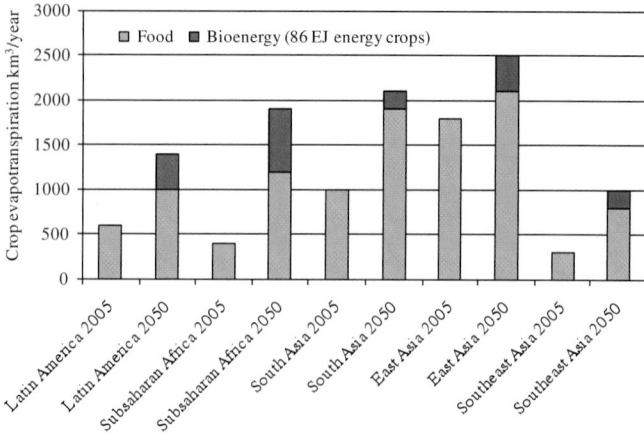

Figure 20.19 | Estimated water requirements for food today and hypothetical water requirements for food and bioenergy around year 2050 with lignocellulosic crops, plant breeding and improved agronomic practices, assuming in the model that ligno-cellulosic crops will mainly be used for with average water use efficiency of 2.5 kg biomass per m³ of evapotranspiration. Source: adapted from SIWI, 2008.

Table 20.10 | Indicative area requirements of different primary energy sources.

	Indicative area requirement (km²/PJ/yr)	Potential sustainability issues
Bioenergy crops	50–500	Impacts on C balance, ecosystems, soils and food systems can be positive or negative, depending on management
Bioenergy residues	Almost no additional area	No or little additional area required if residues or wastes can be used; possible impacts of removal of residues on soil fertility and the soil C balance need to be considered
Solar energy	1–6	Land needed for infrastructure; excess heat can be used for grain drying
Geothermal energy	≈10	Land required for infrastructure and transmission
Hydropower	<1–100	Impacts are highly site-specific and include positive (e.g., irrigation, flood control) as well as negative aspects (e.g., biodiversity and ecosystems, resettlement during construction)
Wind power	1–32	Land for wind power plants plus transmission, affects landscapes; rotors may kill birds
Oil	<1	Land required for infrastructure and transport
Natural gas	<1	Land required for infrastructure and transport
Coal	≈1	Land required for infrastructure and transport, soil contamination
Nuclear energy	<1	Land for infrastructure and transmission; potentially much larger land areas contaminated in case of an accident

Source: calculated based on Tampier (2002), except bioenergy.

conditions. Bioenergy crop plantations in marginal areas may, however, alleviate or reduce other water-related problems such as water-related erosion (Berndes, 2008).

Beyond matching supply and demand for human uses, it has become clear that management approaches need to consider also the ecological water demand and the spatial and temporal variability of supply and demand (EC and IPTS, undated).

20.4.3 Land-related sustainability issues

Energy provision almost always requires land area, at least for infra-structure, except perhaps for energy derived from the ocean (e.g., wave energy). Table 20.10 compiles indicative estimates of the area required for a yearly energy flow of one PJ for different primary energy sources. It shows that the land area required for fossil fuels and nuclear power is small compared to that needed for some renewable energies, par-ticularly bioenergy and hydropower. In most cases, land is needed more or less exclusively to host the infrastructure needed; in the case of coal additional area is required for open-cast mining.

Land areas required for hydropower plants are often small, especially for run-of-river plants, but may be substantial in the case of storage power stations or in regions with flat topography. If poorly designed, hydropower plants can produce substantial negative environmental and social effects including impacts on riverine ecosystems and biodiversity, on hydrological resources, on fisheries, and on local livelihoods, par-ticularly if dam construction requires resettlement of local populations (Goldsmith and Hildyard, 1984; Trussell et al., 1992; WCD, 2000). Plans to build large dams have, therefore, sometimes encountered strong pub-lic opposition. Hydropower plants can also have positive environmental or socioeconomic effects, including irrigation or flood control, and are

indeed sometimes constructed primarily for such purposes. These issues are discussed in detail in the World Commission on Dams (2010).

Bioenergy requires by far the largest land area per unit of energy and year; as a result, concerns about competition for land, and land-use related sustainability issues involving energy, have received most attention with regard to bioenergy. The remainder of this subsection, therefore, focuses on bioenergy. Bioenergy is produced either by planting dedicated bioen-ergy crops or using by-products, residues and wastes from agriculture, forestry, food processing, and other economic activities (see Chapter 7). The environmental effects of these pathways are, however, fundamentally different (Cherubini et al., 2009) and are, therefore, discussed separately.

20.4.3.1 Environmental effects of bioenergy crop plantations

Creating and maintaining energy crop plantations can have positive and/or negative environmental, social and economic impacts, depend-ing on the respective regional conditions, the specific crop or technol-ogy, and the implementation and management of bioenergy projects (Goldemberg at al., 2008; Dewulf and Langenhove, 2006). For example, growing energy crops can increase the demand for agricultural prod-ucts, create income and jobs in the agricultural sector, and provide clean, renewable energy. It can also have unintended adverse effects, however, e.g., on agricultural prices or food security, as discussed in Section 20.2.7. Proper implementation of bioenergy crop plantation projects is a key factor for their sustainability in terms of social, economic and regional development goals (see Sections 20.2.5 and 20.2.6). This section focuses on issues of ecological sustainability related to land demand of bioenergy crop plantations.

Tackling the question of how much bioenergy can be produced without harming the environment (European Environment Agency, 2006) and without creating adverse social and economic impacts (Section 20.2.6) requires a systems approach that takes into account the relevant interactions between various land uses and socioeconomic functions of biomass (food, fiber/materials, and energy). Some of these issues are particular interactions between food and bioenergy systems and the significance of constraints such as the conservation of ecologically valuable areas (e.g., biodiversity hotspots, wetlands, or forests) and areas under various kinds of nature protection.

From a systems perspective, two types of effects need to be considered:

- **Direct effects** refer to the environmental effects, positive or negative, of the energy crop plantation itself, i.e., processes happening on the area covered by the plantation. These effects are species-specific as well as site-specific and depend strongly on the former state of the area, in particular its former use by humans.

- **Indirect effects** are the effects of creating energy crop plantations on the global land system as a whole. These system-level effects, usually called ILUC effects, depend not only on the amount of bioenergy to be produced and on the area required for that, but also on the development of many other components of the global land system such as food demand, food crop yields, the livestock system, and many other factors. For example, depending on food demand and food crop yields, establishment of bioenergy plantations on land currently used for food production may imply that the food has to be produced somewhere else. If that is the case, and if there are environmental effects (e.g., increased GHG emissions) resulting from the creation of this additional food crop area, they have to be attributed to the establishment of the energy crop plantation. On the other hand, Nassar et al. (2011) concluded that there is no evidence that biofuels present ILUC effects in the case of Brazilian sugarcane ethanol.

Perennial grasses such as switchgrass, *Miscanthus* and short-rotation coppice are generally thought to be ecologically less demanding than the food crops used for first-generation biofuels – in terms of impacts on soils, soil erosion, biodiversity, nutrient leaching, pesticide application, etc. (Cherubini et al., 2009; Rogner et al., 2000). It must be noticed, however, that the example of Brazil with sugarcane-based ethanol shows that first-generation biofuels can be produced in a sustainable way when adequate policies are in place (Goldemberg at al., 2008; also see Box 20.5 on Biofuels Zoning in Brazil).

Environmental impacts associated with the creation of new plantations strongly depend on the prior state and use of the land. If cropland currently used for food or feed production is converted to bioenergy plantations using perennial grasses or short-rotation coppice, the direct environmental effects can be expected to be benign, as those energy crops generally have fewer detrimental impacts than food crops.

Globally, however, with growth of the world population and changes in diets towards more demanding, i.e., animal-based, foods, as most projections assume, the area required to grow food or feed crops will increase rather than shrink. Land-use change is, therefore, a central environmental issue associated with the expanding use of bioenergy crops (Sagar and Kartha, 2007; Firbank, 2008). Once considered a local environmental issue, land-use change has become recognized as a pervasive driver of global environmental change that affects ecosystems, biodiversity, the water balance, and the biogeochemical cycles of carbon, nitrogen, and many other substances and compounds (Foley et al., 2005; Lambin and Geist, 2006). On the other hand, other recent studies (Goldemberg, 2009; Goldemberg et al., 2008) conclude that the environmental impacts of bioenergy crops such as sugarcane in Brazil can be controlled. The FAO (2009b; 2009d) also discusses these issues in the above-mentioned study cases of Peru in Latin America and Tanzania in Africa, showing that these issues can be addressed with adequate policies. These issues have been discussed deeply in Sections 20.2.5. and 20.2.7. of this chapter.

The Millennium Ecosystem Assessment (2005) concluded that land and water use by humans has already reduced the ability of many ecosystems to build up resilience and deliver vital environmental services not related to biomass production, such as buffering capacity, water retention capacity, and self-regulation, among others. A comprehensive discussion of such effects is beyond the scope of this chapter (but see IAASTD, 2009; Lambin and Geist, 2006). Regarding water, the focus here is on two main water use impacts: effects on the GHG balance, in particular the carbon balance; and effects on biodiversity (see Subsection 20.5.3.3 below). The overall GHG balance of bioenergy includes three components: GHG emissions resulting from production of agricultural inputs (e.g., fertilizers, diesel for tractors, etc.) and from conversion processes from feedstocks to final energy (e.g., liquid fuels); GHG emissions from direct land use or land-use change effects; and GHG emissions from ILUC (UNEP, 2009).

The first component is well covered by established life-cycle-based methods. High-quality databases exist to estimate these effects, and data uncertainties are relatively small. It should be noted, however, that allocation rules (i.e., how emissions are allocated to each product if one process yields more than one output, which is common in bioenergy production process chains) are a methodological challenge here and need to be considered when interpreting results (e.g., Zah et al., 2007; WBGU 2009; UNEP, 2009; Fritsche et al., 2010; Macedo et al., 2008).

These issues are covered in Chapter 11, together with some recent estimates of GHG emissions related to land-use change or ILUC, and will not be further discussed here. This section focuses on fundamental issues related to GHG emissions resulting from land use and land-use change. Comprehensive information on GHG emissions per unit of energy are given in Chapter 11 (see also UNEP, 2009; Fritsche et al., 2010; Macedo et al., 2008; Nassar, 2009). First, general issues related to land use (change) and the GHG balance of ecosystems are discussed, followed by the issue of direct versus indirect effects.

Land use and land-use change directly affect the exchange of GHGs between terrestrial ecosystems and the atmosphere. In order to better understand the carbon balance of processes related to land use (change), it is important to distinguish between stocks and flows. A stock is a volume of carbon present in a system at any given point in time, while a flow is the amount of carbon moved from one compartment (e.g., the soil or the phytomass of plants) to another (e.g., the atmosphere) within a defined period of time, usually one year. With reference to terrestrial ecosystems, the most important stocks are carbon in the soil and in phytomass (see Box 20.4).

For carbon, the global terrestrial gross primary productivity (GPP) is 123 GtC/yr, of which 60 GtC/yr is returned to the atmosphere through plant respiration. Of the remaining net primary productivity (NPP) of 63 GtC/yr, 53 GtC/yr is returned through heterotrophic metabolism. The remaining 10 GtC/yr is the net ecosystem productivity (NEP). A large portion of NEP is lost because of land use, biotic stresses, fires, and other disturbances. The fraction remaining after these losses, called net biome productivity (NBP), ranges between 0.3 and 5 GtC/yr (Jansson et al., 2010).

Photosynthesis removes carbon from the atmosphere, but a considerable proportion flows back into the atmosphere through respiration of plants and heterotrophic organisms (animals, microorganisms, fungi). If there is no land-use change involved, the situation is relatively simple: if photosynthesis is greater than respiration, then the system acts as a carbon sink; otherwise, it is a carbon source. As yearly flows are large, difficult to measure, and often almost balanced, most methods used to measure the net carbon balance of ecosystems over time determine the net flows into or out of the system as the difference between stocks at different points in time. Ecosystem theory generally assumes that ecosystems gradually build up carbon stocks in vegetation and soil as they mature until they reach an equilibrium (climax) point where photosynthesis equals respiration and stocks are stable (Houghton, 1995; Schimel et al., 2001).

Table 20.11 shows that almost all carbon is stored in the soil-organic carbon (SOC) component in croplands, temperate grasslands, and tundra ecosystems. Overall, stocks are highest in forests and lowest in croplands.

The data in Table 20.11 show that deforestation, as expected, results in a massive loss of carbon to the atmosphere because large stocks in the phytomass are released in a short time. Belowground processes are more complex and much slower. If the system is used as cropland, SOC

Table 20.11 | Typical values for carbon stocks of vegetation units per unit area. Data reported for forests refer to natural or near-natural conditions except for croplands; Carbon stocks in managed forests are substantially smaller.

Vegetation unit	Phytomass (kgC/m²)	Soil organic carbon (SOC) (kgC/m²)	Total (mean) (kgC/m²)
Tropical forests	12 – 19	~12	~28
Temperate forests	5.7 – 13	~9.6	~19
Boreal forests	4.2 – 6.4	~34	~40
Tropical savannas	2.9 – 3.0	~12	~15
Tundra	0.35–0.7	~13	~14
Temperate grasslands	0.4–0.7	~24	~24
Croplands	0.2 – 0.3	~8.0	~8

Source: Watson et al., 2000; Saugier et al., 2001.

mostly declines, but in grasslands, SOC can be even larger than in forests, in particular if they are used extensively. Soil-related processes can continue over long periods of time until a new equilibrium is reached. These complexities are important here, because they are the underlying reason why it is very difficult to derive general conclusions from site-specific studies of the carbon balance of different crops. Further detailed information on this issue is shown in Box 20.4.

The carbon balance of crops, including energy crops, strongly depends on the prior state (and particular use) of the land. If, for example, land is deforested for bioenergy, then there is a large release of carbon before energy crops are produced. This "carbon debt" (Fargione et al., 2008) may be quite large, so GHG emissions of bioenergy crops from that source can exceed GHG emissions from fossil-fuel-based counterparts for years, decades, or even longer (Gibbs et al., 2008; UNEP, 2009). On the other hand, if perennial bioenergy crops are planted on degraded pasture lands, they may build up carbon stocks in the soil and act as strong carbon sinks while producing bioenergy (Goldemberg, 2009; Goldemberg et al., 2008; Goldemberg and Guardabassi, 2009; Macedo et al., 2008; Tilman et al., 2006). Moreover, the GHG balance also depends on a host of other factors, such as land-use intensity and management, crop type (perennial plants mostly build up larger carbon stocks in the soil and in phytomass than annual plants), climate, soil, landform (slope, etc.), and many more. Further information on these issues is given in Box 20.4 below.

Considering the time horizon adds further complexity. The usual assumption was that biomass production was carbon neutral except for land-use-related effects, because the CO_2 resulting from its oxidation had previously been absorbed by the plant through photosynthesis. This neglects the following intricacy:

If biomass is harvested and combusted, then the carbon is released immediately, often within one year. However, if biomass is not harvested and is left in the ecosystem, it may take years, decades or even centuries until the CO_2 returns to the atmosphere; indeed it might even be

Box 20.4 | Carbon Sequestration in Soils

World soils constitute the third largest global carbon (C) pool, after the oceanic pool at 38,400 Pg and the geological pool at 4230 Pg. The soil C pool, estimated at 2500 Pg to a depth of 1 meter, is 3.2 times the atmospheric pool of 780 Pg, and four times the biotic pool (trees, vegetation, etc.) of 620 Pg (Batjes, 1996; Eswaran et al., 1993; Falkowski et al., 2000; Lal, 2004; Pacala and Socolow, 2004; Lal, 2010).

The loss of the soil's C pool from 1850 to 2000 is estimated at 78 ± 12 Pg (Lal, 1999). Until the 1940s and 1950s, more CO_2-C was emitted from terrestrial ecosystems than from fossil fuel combustion. During 2009, 8 Pg CO_2-C/yr was emitted by fossil fuel combustion, and about 1.6 Pg CO_2-C/yr by land-use conversion. Anthropogenic activities that lead to CO_2 (and CH_4) emissions include deforestation, biomass burning, drainage of wetlands, soil tillage, excessive grazing, and extractive farming practices.

Soil processes that lead to gaseous emissions are: accelerated soil erosion, breakdown of structural aggregates, and decomposition/mineralization of soil organic matter. Two other soil processes that accentuate gaseous efflux from soil are: methanogenesis that leads to emission of CH_4 under anaerobic conditions, and nitrification/denitrification that lead to emission of N_2O.

Transfer of atmospheric CO_2 into other pools with a relatively longer residence time is called C sequestration. Thus, C sequestration in terrestrial biosphere (i.e., soils and biota) is a natural process based on photosynthesis of atmospheric CO_2 followed by humification of biomass into stable humic substances and organo-mineral complexes.

Important practices of soil organic carbon (SOC) sequestration are afforestation/reforestation, wetland restoration, establishment of a Conservation Reserve Program (CRP) or land retirement, restoration of eroded/degraded soils, no-till farming, cover cropping, application of biochar, and adoption of integrated nutrient management practices. The strategy is to create a positive ecosystem C budget. The rate of soil C sequestration ranges from negative or zero under arid and hot climates to about 2 Mg/ha/yr under cool and humid climates.

The global potential of soil C sequestration is estimated at 0.4–1.2 Pg C/yr in cropland soils (Lal, 2004) or about 1 Pg C/yr by converting 1.5 billion hectares from plow till into no-till farming (Pacala and Socolow, 2004). Improved management of grasslands (range land, pasture land) has an additional potential of C sequestration of about 1 Pg C/yr. Restoration of degraded and desertified soils has an additional potential of ~1 Pg C/yr. Restoration of peatlands (wetlands) is another important strategy for C sequestration. Adoption of these practices can be promoted through payments to land managers for ecosystem services.

Modern biofuels (e.g., ethanol, methanol, methane, and triglyceride oils) are an important alternative to fossil fuels. These are also called "green gold" fuels, because their feedstocks can be grown on farmland over and over again (Vorholz, 2006). Commonly used feedstocks of ethanol include corn grains (Pimentel, 2003) and sugarcane (Goldemberg et al., 2008). However, crop residues are also being considered as a source of cellulosic ethanol (Somerville, 2006). There are serious concerns about the use of corn grains for ethanol with regard to competition for food (Mufson, 2008), energy balance, and carbon foot print (Pimentel, 2003; Oliveira et al., 2005; Lal, 2008).

While the practical issues on cellulosic ethanol production are being addressed (Ragauskas et al., 2006), the impressive progress of ethanol production from sugarcane in Brazil needs to be objectively assessed, especially with regard to soil C dynamics. The agricultural input required for sugarcane production releases 2.27 Mg CO_2/ha/yr. Other sources of CO_2 emissions from ethanol production result from the pre-harvest burning of sugarcane and from the decomposition of vinasse. Total CO_2 emissions from ethanol production from sugarcane in Brazil are 3.31 Mg/ha/yr (Oliveira et al., 2005) compared with 5.03 Mg/ha/yr for corn-based ethanol in the United States (West and Marland, 2002; Shapouri et al., 2002).

The soil C budget for sugarcane production in São Paul State and elsewhere has not been widely studied. Plowing and other soil disturbances have negative impacts on the soil C pool by increasing the mineralization of soil organic matter and risks of soil erosion. However, the soil C pool can be increased by adopting reduced tillage, eliminating pre-harvest burning, and recycling wastes of the ethanol processing.

Source: Lal, 2005; 2006; 2010.

captured for geological time spans under certain circumstances (e.g., when carbon is stored belowground in wetlands under anaerobic conditions). Therefore, bioenergy production can only be considered "carbon neutral" in a strict sense if it results from additional plant growth that would not have occurred in the absence of bioenergy production and use (e.g., if the productivity of degraded land is increased over the previous state), if biomass is used that would be oxidized rapidly if left in the ecosystem, or if bioenergy use reduces biomass use in other sectors, e.g., if food or fiber consumption falls due to rising prices of agricultural products. These issues are currently being discussed intensively, and this debate might result in new accounting rules for GHG emissions of bioenergy (Bird et al., 2010; Pingoud et al., 2010; Searchinger et al., 2009; Searchinger, 2010; Goldemberg, 2009; Goldemberg and Guardabassi, 2009).

The above-discussed considerations generally apply to land use and land-use change, whether the land is used for food or energy crops or other purposes. Any production of biomass – no matter whether it is produced for food, as raw material, or for energy – requires land and may (or may not) result in land-use change. As long as changes in the production and consumption of biomass are studied in a systemic way – for example, by looking at changes in food, fiber and bioenergy demand over time – in a systems model that integrates plant production, all land-use effects of biomass conversion are direct, and no such thing as an "indirect effect" exists. The same applies for agriculture and industry, as well as final consumption.

ILUC effects need to be considered, however, when environmental effects of single resources, such as bioenergy crops, are to be evaluated without explicit consideration of the entire land system. Of course, the evaluation of ILUC effects involves uncertainties (Nassar et al., 2009; 2011) and can only yield approximate results, based on ceteris paribus assumptions.

In conclusion, appropriate planning and implementation of bioenergy crop plantations is of key importance. If bioenergy crop plantations can be situated on lands with low initial carbon stocks, such as degraded lands, or use perennial crops such as perennial grasses or short-rotation coppice and appropriate management technologies such as no-till agriculture, their GHG balance can be negative, or at least their GHG emissions can be low. If, on the other hand, inappropriate locations are combined with poor choices of crops and unfavorable management, GHG emissions can be very large and even exceed the fossil baseline. Therefore, adequate policies are needed to ensure that appropriate areas are used for bioenergy crops. The choice of crops should follow environmental zoning approaches, such as those currently used in Brazil (see Box 20.5 below) and in the Brazilian states of Minas Gerais and São Paulo (Joly et al., 2010).

The United Nations Department of Economic and Social Affairs (UNDESA) provides a range of recommendations that specifically address sustainability aspects of bioenergy (UNDESA, 2007). These include the use of energy crops that are suitable under local conditions and are able to grow on marginal and arid lands with limited inputs. UNDESA also favors energy crops that have a variety of by-products that would create additional income for farmers. Energy crops that are easily propagated and allow intercropping would also be important. Bioenergy should also benefit identified marginalized groups in a community.

In conclusion, adequate policies that are appropriately enforced can allow the reduction of negative, and the promotion of positive, environmental impacts of bioenergy crops.

20.4.3.2   Environmental and social effects of using by-products, residues, and wastes

Bioenergy production from agricultural by-products, residues and wastes presents advantages, as it does not: require additional land or land-use change; compete with food and fiber production; affect agricultural and food prices; or require many additional scarce inputs, such as freshwater (Berndes, 2008). Using biomass residues may help alleviate energy shortages and create employment opportunities (such as production of bioenergy crops); moreover, use of municipal solid waste (MSW) may reduce landfill requirements. However, there may also be negative environmental effects, depending on the respective biomass flow, as well as on technology and management.

The agricultural residue straw can deliver substantial amounts of energy. However, straw plays a vital role for soil fertility, soil carbon pools, and the mitigation of water and wind erosion (Lal, 2005; Lal, 2006; Wilhelm et al., 2007). WBGU (2008) assumes that about half of all crop residues could be used to produce bioenergy without compromising soil fertility. Removal of crop residues for energy purposes could also affect the GHG balance of cropping systems. There are currently studies under way evaluating how much of the residues much be left on the soil to avoid that (Hassuani et al., 2005).

The removal of biomass from forests, including forest residues, may affect forest ecosystems because the coarse woody debris is essential for biodiversity and ecosystem functioning (Harmon et al., 1986; Krajick, 2001; Shifley et al., 2006) and forest conservation objectives. The use of fuel wood and forestry residues should be jointly optimized. This can be managed through adequate capacity-building with local people, as is happening in remote villages in India (CENBIO, 2002).

Well-managed use of animal manures for biogas production can have significant positive environmental and social impacts. It reduces methane (CH_4) emissions,[13] while returning most plant nutrients and parts of the carbon back to the soil, thereby mitigating land degradation

---

13 Conversion of animal manures in biogas plants and subsequent application of the residues as fertilizer reduce methane (CH_4) compared to the storage and direct application of manures (Bhattacharya et al., 1997; Clemens et al., 2005).

Box 20.5 | Environmental Zoning in Brazil

Brazil began its large-scale sugarcane ethanol program, PROALCOOL, in 1975, when oil prices raised with the world oil crisis. Since then, many developments towards a sustainable production system have been achieved. These have resulted in important increases in both agricultural and industrial productivity (more than 3%/year, Goldemberg et al, 2008); as a consequence, production costs fell rapidly, making ethanol economically competitive with gasoline. At the same time, improvements in social and environmental legislation have been achieved, both at the federal and state level. Due to the expansion of sugarcane production in recent years, concerns about the direct impacts of land-use change led federal and state governments to adopt policies to determine suitable areas for this crop.

The state of Minas Gerais was the pioneer in this process and launched its economic-environmental zoning in 2007. The zoning is based on social, economic and environmental information that shows regional characteristics, potentialities, and vulnerabilities. It is an orienting tool that can support policymakers and entrepreneurs of different sectors (World Bank, 2011). In the state of São Paulo an agro-environmental zoning was based on studies related to soil and climate restrictions, topography, water availability, air quality, and existence of protected areas and biodiversity conservation areas; this assessment was the basis for a voluntary scheme with sugarcane producers, the Agroenvironmental Protocol. The text stipulates a set of measures to be followed, anticipating the legal deadlines for the elimination of sugarcane harvest burning and immediately halting burning practices in any sugarcane harvests located in expansion areas. It furthermore targets the protection and recovery of riparian forests and water springs in sugarcane farms, controls erosion and content water runoffs, implements water conservation plans, stipulates the proper management of agrochemicals, and encourages reduction in air pollution and solid wastes from industrial processes (Lucon and Goldemberg, 2010; SMA, 2011).

The federal government launched, in September 2009, the national agro-ecological zoning for sugarcane and, in 2010, for palm oil. This zoning identified the areas where sugarcane crop expansion can take place, and forbids sugarcane cultivation in 92.5% of national territory, including the Amazon Forest, Pantanal wetlands, and other native biomes. It identified 64 million hectares that comply with environmental and productivity requirements. The zoning was an intense program involving dozens of institutions and researchers of agricultural and environmental issues. In this process maps were produced showing soils, climate and rainfall, and topography. Land was classified and delimited by determining the areas of highest yield potential in detail (1:250,000), based on minimum productivity, with respect for the environmental regulations and which areas should be preserved, as well as trying to reduce competition with areas devoted to food production. According to these studies, there are in Brazil about 650,000 km^2 available for sugarcane and 300,000 km^2 for palm, without undesirable impacts (EMBRAPA, 2011a; 2011b).

and helping to maintain soil fertility (Rajabapaiah et al., 1993; Stinner et al., 2008). Moreover, energy from biogas can help to substitute for traditional biomass energy, which has tremendously negative health and environmental effects and currently contributes to millions of premature deaths due to respiratory diseases resulting from indoor pollution (Jaccard, 2005).

Using MSW for energy purposes lowers CH_4 emissions from waste deposits (landfills). Since waste disposal in landfills has other environmental impacts when it is not adequately controlled and enforced, such as soil and underground contamination, there can be local opposition to the construction of landfills near communities. Therefore, in some cases, incineration of MSW may provide an interesting solution for both solid waste disposal and power production.

However, effective air pollution control technology is needed to avoid large emissions of toxic pollutants such as dioxins and furans. Tight air pollution regulation that vigorously enforces the use of the most

advanced abatement technologies to reduce toxic emissions is required to minimize possible negative environmental effects from the combustion of MSW (McKay, 2002).

20.4.3.3 Biodiversity

Different trade-offs on biodiversity may result from bioenergy production, depending on the adequacy of the policies in place, and their enforcement.

On a local scale, biodiversity effects depend on crops, former land uses, and management practices. Local biodiversity may benefit from biomass crops when intensive agricultural practices are replaced by low-intensity biomass production systems. In general, first-generation European agricultural crops have more negative impacts at the local level than both mixed cropping systems and bioenergy crops in developing countries (the Brazilian experience is an example of this; Goldemberg et al.,

2008), as well as second-generation perennials and woody crops (GLOBIO, 2010).[14]

Clearing valuable ecosystems such as native rainforests and wetlands to make an expansion of the agricultural sector possible would result in large losses of natural biodiversity. If feedstock production comes predominantly from large monoculture cultivation without appropriate environmental controls, it could reduce both the number of species and, with a growing specialization on particularly suitable varieties, even the number of varieties grown. These practices could result in a greater vulnerability of the agricultural sector to non-standard crop-growing conditions such as extreme weather patterns, pests, and diseases (Müller et al., 2007). The threat to wild biodiversity from agricultural crops and bioenergy growth is associated primarily with land-use change, as already mentioned. When areas such as natural forests are converted for feedstock production, the loss of biodiversity is significant, even if an expansion of crop land is a temporary phenomenon. A further concern is the introduction of invasive species for biofuel production. Agricultural biodiversity could be affected by large-scale monocropping practices and the introduction of genetically modified materials (FAO, 2008b).

On the other hand, the conservation and/or recuperation of native forests areas and other biomes as well as fauna corridors – such as riparian forests and the so-called environmental protection areas – are fundamental for any agricultural crop and, when introduced inside large-scale plantations, can contribute to preserving or rebuilding biodiversity (Goldemberg et al., 2008; EMBRAPA, 2008).

20.4.4   Challenges

Due to the uncertainties in future oil prices and energy policies, it is still too early to understand the dimensions of the markets for biofuels sufficiently to develop a realistic scenario that quantifies the trade-offs between increased biofuel production and the provision of food (Müller et al., 2007). However, there are positive perspectives for food prices in the future, as discussed in OECD/FAO (2009).

In the long term, tropical countries will likely play an increasingly important role in feedstock production, due to favorable biophysical conditions and generally lower costs of land and labor, so long as suitable trade arrangements and stable conditions for investment prevail (Cotula et al., 2008). Considerable improvements in land and water productivity are

---

14   The origins of the present GLOBIO model (GLOBIO3) go back to the GLOBIO2 model and the Natural Capital Index-IMAGE framework (NCI-IMAGE). Both approaches are universal, policy-oriented frameworks on the impact of human activities on biodiversity. To meet the challenge of evaluating the global targets on biodiversity, an international consortium, made up of the UNEP World Conservation Monitoring Centre, UNEP GRID-Arendal and the Netherlands Environmental Assessment Agency joined forces in 2003. They combined the GLOBIO2 and the NCI-IMAGE approach into a new Global Biodiversity Model: GLOBIO3. This model has been in use since 2005 (Globio, 2010).

possible and can be significant. The example of achieving a significant increase in sugarcane productivity in Brazil shows that this is an objective that can be achieved if adequate investments are made (al-Riffai et al., 2010). In tropical farming systems, currently producing around the global average for developing countries, one can reach 2 t/ha for cereals.

In Africa, seven countries (Tanzania, Côte d'Ivoire, Burkina Faso, Ghana, Guinea, Mali, and Senegal) are participating in a UNDP Multifunctional Platform project that tackles lack of access to electricity and rural women's poverty through the provision of simple multipurpose diesel engines that are able to run on jatropha oil (Cotula et al., 2008). This project, as well as the BEFS project (FAO, 2010a; 2010b) for Peru and Tanzania (see Box 20.2), illustrates how tropical countries can implement bioenergy production if the needed overall conditions can be achieved.

As discussed in the sections above, the main challenge regarding land use is related to the increase in agricultural productivity in developing countries (mainly African countries) and the implementation of adequate policies such as agro-environmental zoning to allow the better use of land for each purpose considering the existing trade-offs. Socio-environmental assessments are fundamental for various options to meet future demands for food and biofuels.

In terms of water, challenges related to water trade-offs include the implementation of adequate strategies for water productivity improvements. This requires substantial efforts from authorities and different development agencies. A combination of incentives and sanctions are required to overcome social inertia and to demonstrate that it is a viable option. A considerable decoupling could be achieved over a period of a few decades. Technologies are generally known; the challenge is to invest in human and institutional capacity for adequate policies in all sectors, including environmental and social ones.

If improvements on the food consumption side are combined with those in water productivity on the production side, the estimated water need for food production in 2045 is estimated to be 6470 km^3/yr. Estimates for water need for bioenergy in 2050 vary considerably (4000–12,000 km^3/yr), but even the lower limit of such consumptive water requirements to produce biofuels is quite significant.

Increasing water productivity holds the key to future water scarcity challenges. Without further improvements in water productivity or major shifts in production patterns, or a more advanced allocation system within an Integrated Water Resources Management system, the amount of water needed for agriculture, industrial and domestic activities will increase by 60–90% by 2050, depending on population, incomes, and assumptions about water requirements for the environment. In agriculture alone, the total volume of water needed for crop production would be 11,000–13,500 km^3, almost double the 7130 km^3 of today (FAO, 2007). Water resources will be stressed even more from increased demand for biofuel. By 2030, world energy demand will increase by

50%, and two-thirds of this demand will come from developing countries. However, there is scope for an accelerated increase in water productivity. Water productivity in agriculture has increased steadily in recent decades, largely because of an increase in crop yields, and the potential exists for even more increases. However, the pace of such increases will vary according to the type of policies and investments put in place, with substantial variations in the impact on the environment and livelihoods of rural populations. A systematic approach to agricultural water productivity requires actions at all levels, from crops to irrigation schemes, and involving national and international economic systems, including the trade in agricultural products. It calls for an informed discussion on the scope for improved water productivity in order to ameliorate intersectoral competition for water resources and optimize social and economic outcomes (UN-Water/FAO, 2007).

"Virtual water" is the water consumed in the production process of an agricultural or industrial product (Allan, 1998). It is particularly important to water-scarce countries in their efforts to secure water for different sectors. Water for bioenergy production will become increasingly important. Taking virtual water into account in the trade balance may allow for water-scarce countries to import more water-consuming crops and produce lower water-consuming ones domestically. This may also include production of biofuels.

If improved water productivity is achieved through a more intensive application of fertilizers and other chemicals, improvements in a quantitative sense may lead to a deterioration of quality. Climate change is also a highly relevant issue but difficult to model in terms of water. Taking into account that water scarcity is already severe in many parts of the world and that growth of GDP and population and distribution are not in harmony with access to productive land and abundant water resources, the role of trade will be even more important. The virtual water perspective under those circumstances is even more important, as are the connections between production and consumption (Lundqvist et al., 2007).

With regard to hydropower, the construction of large dams often has extensive impacts on downstream areas of the river basin, including the terrestrial and aquatic ecosystems. Ecosystem restoration is a necessary measure to be addressed, and the impacts of decommissioning are complex and site-specific. The World Commission on Dams Knowledge Base demonstrates that in many cases large dams have resulted in:

- loss of forests and wildlife habitat, loss of species and degradation of upstream river basin areas due to inundation of the reservoir area, or of downstream river basin areas due to damming and subsequent loss of water for ecosystem maintenance;

- GHG emissions from reservoirs due to the anaerobic decomposition of vegetation and carbon inflows from the basin;

- loss of aquatic biodiversity, upstream and downstream fisheries and the services of downstream floodplains, wetlands and riverine, estuarine and adjacent marine ecosystems;

- creation of productive fringing wetland ecosystems with fish and waterfowl habitat opportunities in some reservoirs; and

- cumulative impacts on water quality, natural flooding, and species composition where a number of dams are sited on the same river.

The ecosystem impacts are more negative than positive and have led, in many cases, to irreversible loss of species and ecosystems. Efforts to date to mitigate the ecosystem impacts of large dams have met with limited success, owing to the lack of attention given to anticipating and avoiding impacts, the poor quality and uncertainty of predictions, the difficulty of coping with all impacts, and the only partial implementation and success of mitigation measures.

In brief, aiming to reduce the competition among the different end uses for land and water, the main challenges are related to the need for investments not only in technological aspects allowing the introduction of efficient and sustainable methods, but mainly in capacity-building in technical, economic, environmental, social, political and regulatory sectors related to this issues.

20.4.5   Environmental certification

Sustainability standards are frequently proposed for the processes of environmental permitting. Based on an agreed definition of sustainable development, specific criteria and provisions are formulated, either locally (according to community priorities and expectations) or externally (based on requirements of external markets). Different organizations have developed sustainability criteria and tools, e.g., the International Labour Organization (ILO) for acceptable labor conditions, the World Wildlife Fund for ecological factors, the World Bank for financial results, and the OECD and the UN for development policymaking and information (Lewandowski and Faaij, 2006).

Applying sustainability criteria to an environmental permitting process can be done either on a case-by-case basis or following zoning plans. The example of sugarcane zoning in São Paulo, Brazil (Joly et al., 2010), is described in Box 20.5. Several GIS datasets (e.g., climate, soil potentials, water availability and vulnerability, biodiversity protection and connectivity) led to a map with different restrictions on licensing enterprises, and subsequently to federal zoning for sugarcane and oil palm plantations in Brazil.

Considering land use, and specifically the case of biofuels, several initiatives seek to establish certification and sustainability standards. Some of these initiatives overlap, and they are all broadly consistent in their principles. Different systems have been developed by the Forest Stewardship Council (FSC, 2011), the European Retailers Produce Working Group (GLOBAL GAP, 2011), and Linking Environment and Farming (LEAF, 2011). Some schemes have stronger interfaces with bioenergy: the Roundtable on Sustainable Palm Oil (RSPO, 2011), the Roundtable on Responsible Soy Association (RTRS, 2011), and the São Paulo State Green Ethanol

Program (Lucon and Goldemberg, 2010; SMA, 2011). National policies in the EU are also supporting the assessment of sustainability and certification systems for biofuels (Ecofys, 2006). An important initiative is the Roundtable on Sustainable Biofuels (EPFL, 2007), which is developing certification criteria, led by Lausanne University and supported by the Swiss government. This initiative is currently facilitating agreement on a comprehensive set of principles for sustainable biofuels, including principles on respect for land and water rights, the socioeconomic development of communities, and food security. UNEP has also an initiative aiming to support wide adoption of sustainability criteria, avoiding breaching trade rules of the World Trade Organization (UNEP, 2011).

There are different systems and methods aiming at ensuring sustainability of bioenergy, most of which fall into the following categories:

- Demand-side, voluntary, consumer-oriented, bottom-up, providing a green label for "better products," appealing to individual perceptions, and usually covering good social practices (e.g., fair labor, small producers, from poorer regions) and organic/environmental standards (e.g., products less carbon-intensive, no deforestation in the production process, etc.).

- Demand-side, mandatory, top-down, sanitary measures and/or other requirements generally applied to imports; covering some key topics and products (e.g., absence of genetically modified organisms or proscribed substances and quality standards for a given biofuel commodity).

- Demand-side sustainability criteria, top-down general principles applied to a category of goods and services, such as biofuels, covering a broad range of topics, in many cases inspirational but also with the intention of becoming mandatory by law.

- Supply-side sustainability criteria, producer-oriented, generally voluntary schemes promoted by producer associations and/or governments, applied to main (in most cases few) topics of higher socio-environmental concern beyond law enforcement (e.g., lifecycle GHG assessments).

- Supply-side, voluntary, recognized Environmental Management Systems based on continuous improvement spirals, as in the case of the ISO 14000 series of quality standards.

An interesting example of a certification scheme already being applied is the one introduced by the Swedish company SEKAB. The company delivers about 90% of all ethanol in Sweden for E85 and ED95 (ethanol for heavy vehicles). SEKAB announced in June 2008 that it would buy certified sustainable ethanol from four Brazilian groups, in what the company said was the first deal of its kind. SEKAB said it worked with the Brazilian producers to develop sustainable and verifiable criteria for the entire life cycle of the ethanol, taking into consideration environmental, climate and social perspectives. SEKAB said the criteria are in

line with demands highlighted in the ongoing processes being led by organizations such as the UN, EU, ILO and a number of NGOs. There are requirements concerning working conditions, labor laws, and wages – and zero tolerance for child labor, non-organized working conditions (slave labor), and the destruction of rain forests. Harvesting is to be at least 30% mechanized today, increasing to 100% by 2014, and an independent international verification company will audit all production units twice a year to ensure that the established criteria are met. Criteria will gradually be developed over the coming years and synchronized with international regulations when these are in place (Lucon, 2010).

It can be observed that there has been a proliferation of certification schemes; this is a positive development, demonstrating awareness among governments, citizens, consumers and producers of the risks and challenges involved in expanding biofuel production, as mentioned by UNCTAD (2008). The inclusion of land rights criteria in some private certification schemes is also welcome. It is too early, however, to see whether they will have a real impact. The EU and government schemes, which are potentially far more influential, have not addressed land issues – in effect, giving license to European companies to ignore principles of prior informed consent in land allocation for large-scale biofuel crop cultivation (Cotula et al., 2008).

In fact, the sustainability of biofuels is a key question, and certification criteria can be used to answer it. However, as discussed in UNCTAD (2008), there are a number of issues to be considered. The same types of sustainability standards should be considered for other energy sources, and also when comparing fossil fuels with biomass energy. Sustainability standards and certification schemes, to be implemented by developing countries, mainly Least Developed Countries, need strong capacity-building measures. Also important is the question of an increase in biofuel production costs due to certification, and who could cover it. In addition, UNCTAD (2008) raised the point that certification should not be used as protectionism to farmers in industrialized countries.

Recently the United States and the EU established regulations to stimulate the use of bioenergy, in particular biofuels, and to ensure the production and use of biofuels in a sustainable manner.

The EU has two major pieces of legislation: the Renewable Energy Directive (RED) and the Fuel Quality Directive (FQD). All member states were required to implement both directives by December 2010. RED requires a minimum of 10% renewable energy fuels in transport by 2020. It also commits the EU to report by the end of 2010 on the impact of ILUC on GHG emissions from biofuels and ways to minimize that impact. FQD's target is for fuel suppliers to reduce life-cycle GHG emissions by at least 65% by weight across all transport fuels by 2020.

Minimum compliance with the EU policies includes the following: reduce GHG emissions by 50% as of 1 January 2017 and by 60% for biofuels and bioliquids produced as of 1 January 2018; biofuels may not be made from raw material obtained from land with high biodiversity values, high carbon stock or land that was peat land in or after January 2008; and biofuel feedstock grown in the EU must meet the EU's "cross-compliance" requirement. However, it must be noted that such requirements are only for liquid biofuels and do not include solid biomass (Georgescu, 2010).

The US Renewable Fuel Standard (RFS) program came into force on March 26, 2010, as required by the Energy Independence and Security Act (EISA, 2007). The program's objective is to increase the volume of renewable fuel blended into gasoline and other transportation fuels. Renewable fuels include ethanol, biodiesel, and other motor vehicle fuels made from renewable sources. All producers of gasoline to be used in the United States are obliged to comply with the annual Renewable Volume Obligation as determined by the US Environment Protection Agency (US EPA). For 2009, the RFS was 10.21%. Renewable fuel must reduce GHG emissions by 20% in life-cycle terms when compared to average transportation fuels in 2005. Similarly, biomass-based diesel and advanced biofuels must achieve a 50% reduction, and cellulosic biofuels a 60% reduction. The US EPA considers Brazilian sugarcane ethanol adequate in terms of carbon emissions, which opens significant opportunities for other countries also producing sugarcane ethanol (US EPA, 2010).

20.4   Conclusions and Policy Proposals

Whether for income generation or for local energy self-sufficiency, large-scale and small-scale biofuels production can co-exist and even work together in synergy to maximize positive outcomes for rural development. Existing experiences and practices should be disseminated to document successful experiences and to analyze the conditions that make them possible, mainly the spread of costs and benefits among local land users, investors, and government, Also, it is important to consider the extent to which such experiences can be replicated elsewhere (Cotula et al., 2008).

Overall improvements in food production may be achieved by reducing the losses and waste in the food chain from production to consumption. These losses occur in harvest, transportation, transactions, storage, handling, processing, and wholesale and retail sales, not to mention changed dietary habits (more proteins, i.e., closer to the top of the food chain) and overconsumption (more food intake than necessary). As discussed earlier in this chapter, adequate funding and capacity-building are key factors to achieve such objectives.

Decoupling and mitigating water competition between food and energy crop production is also possible (see Lundqvist et al., 2007). It is not axiomatic that expanding energy crop production leads to negative consequences relative to land, water, and other resources. Properly located, designed and managed biomass plantations can provide positive benefits, such as low water consumption and adequate choice of crops

allowing production and transportation. There are options for decoupling future water needs for food and bioenergy production from increased food and bioenergy demands, through adequate planning and zoning. Effective institutions and strong capacity-building at all levels are needed for integrated land and water resources management, allowing the introduction of adequate policies to regulate the trade-offs in their use.

Land use policies should include: protecting small-scale farmers from loss of land due to pressure from large-scale producers; respect and protection of land tenure rights; use of "informed decision-making" and full participation of stakeholders when determining land-use changes; and assessing existing land-use policies in light of potential expanded bioenergy use (UNIDO, 2008).

For food, consideration should be given to annual production of main commodities, demand, prices, and trade, both for irrigated and rain-fed production. Food demand is a function of price, income, and population. There are different areas and yield functions for rain-fed and irrigated crops; crop area and yield functions include water availability as a variable.

Conflicts between food and bioenergy can be avoided through agro-economic-ecological zoning (as presented in Box 20.5), allowing adequate use of land for each purpose. The potential impact of bioenergy production on food prices is discussed in several studies and is quite controversial, but it is clear that such adverse effects can be mitigated by appropriate policies that aim at an integrated optimization of food and bioenergy production. Adequate capacity-building is needed, mainly in developing countries, to allow the implementation and enforcement of adequate policies, such as those to regulate biomass-intensive energy systems and hydropower (limitation, water availability, cooling tower, irrigation), together with agro-ecological-economical zoning to define potentials and best aptitudes for land use, as well as imposing environmental limits in harvested areas (related, for example, to conservation units, water use, monoculture, use of pesticides, burning practices).

In the case of trade-offs in water use, it is fundamental to ensure water for all, and this requires knowledge about science, ecology, and economics, as well as ethics and international cooperation. The trade-offs in water use involve large quantities and different activities, which may conflict with hydropower, thermoelectric or bioenergy production. Global freshwater distribution is unbalanced geographically, and supply is frequently affected by contamination, depletion, and increased competition for multiple purposes. The rapid increase in water demand (due to growing populations, incomes, and unsustainable consumption) must be viewed in the context of climate change, the effects of which are difficult to predict. Expanded exploitation of water resources is only a short-term option for addressing scarcity. In the long term, only ambitious policies, heavy investments in water conservation, and adequate and integrated management of multiple uses can be solutions to the long-foreseen crisis (Rogers, 2008).

Water use alternatives need to clearly consider: improvements in the efficiency and productivity of existing irrigation systems before planning and implementing new ones; adaptation and expansion of local and traditional water management solutions; more coordinated management of surface and groundwater resources; and improvement of the productivity of rain-fed agriculture.

Efforts to promote sustainable water management practices have primarily focused on the agricultural sector as the largest consumer of freshwater. Governments have several objectives in deciding the nature and extent of inputs in agriculture. These include achieving food security, generating employment, alleviating poverty, and producing export crops to earn foreign exchange. Irrigation represents one of the inputs to enhance livelihoods and achieve economic objectives in the agricultural sector with subsequent benefits for rural development. Just as strategies and approaches to rural development are context-specific, there are numerous and diverse alternatives to agricultural development and irrigation that need to be examined. The diversity relates to scale, level of technology, performance, and appropriateness to the local cultural and socioeconomic setting.

A number of policy, institutional and regulatory factors hinder the emergence and widespread use of an appropriate mix of options that would respond to different development needs, sustain a viable agricultural sector, provide irrigation, and offer livelihood opportunities to large populations. Appropriate policy options include:

- support for innovation, modernization, adaptation, maintenance, and extension of traditional irrigation and agricultural systems;

- protection or restoration of natural functioning of deltas, floodplains, and catchments in order to sustain and enhance the productivity of traditional systems in these areas;

- transferring management to decentralized bodies, local governments, and community groups for recovering tariffs and maintenance;

- agricultural support measures, mutually reinforcing and developing intersectoral linkages in the local economy to spur rural development;

- reducing transaction costs and risks for smallholder farmers in developing countries; and

- expanding access to international markets by reducing barriers and introducing supportive domestic policies (without trade distortions such as the tariff and non-tariff barriers to OECD markets).

Government policies and institutions play an important role in the promotion of particular water and agricultural/bioenergy appropriation technologies and methods. Each method has different implications for food production, food security at the local and national levels, and the distribution of costs and benefits.

The business-as-usual policy option entails continuing along the path taken so far. Each country would proceed in setting and revising policy frameworks in line with national interests, taking into account international implications of policy decisions only where these are compatible with domestic priorities (FAO, 2008b).

Certification can play an important role, mainly in the case of biofuel production, but without an internationally agreed standard, the desire expressed by many governments to start certifying sustainable biofuels may face serious obstacles, not least under international trade law considerations (FAO, 2008b).

In the short term (about 25 years), carbon sequestration in terrestrial ecosystems (notably in degraded and desertified lands through restoration and afforestation) is a prudent strategy. In the long term (over 50 years), utilizing carbon-neutral fuel sources (biofuels, solar, hydro, wind, geothermal) is the best option (Lal, 2010).

Clearer definitions of concepts of idle, under-utilized, barren, unproductive, degraded, abandoned and marginal lands (depending on the country context) are required to avoid allocation of lands on which local user groups depend for livelihoods. Similarly, productive use requirements in countries in which security of land tenure depends on active use need to be clarified so as to minimize abuse (Cotula et al., 2008).

Governments need to develop robust safeguards in procedures to allocate land to large-scale agriculture and biofuel feedstock production where they are lacking and – even more importantly – to implement these effectively. Safeguards include clear procedures and standards for local consultation and attainment of prior informed consent, mechanisms for appeal and arbitration, and periodic review. Safeguards should be applicable across agricultural and other land-use sectors, rather than only specifically to biofuels (Cotula et al., 2008).

Although bioenergy can have significant positive impacts on rural areas through the creation of jobs, the FAO (2008a) considers that an integrated approach to social protection should be adopted for rural households, combining traditional transfers (social safety nets) and policies that enable smallholders to respond quickly to the market opportunities created by higher prices.

In the very short term, however, the supply response to higher price incentives, especially by smallholders, may be limited by their lack of access to essential inputs such as seeds and fertilizers. In these cases, social protection measures, including the distribution of seeds and fertilizers, directly or through a system of vouchers and "smart subsidies," may be an appropriate short-term response. If implemented effectively, such a program will increase the income of small producers and may reduce price increases in local markets, thereby contributing to improvements in the nutritional status of net food-buying families.

References

Abbot, P. C., C. Hurt, and W. E. Tyner, 2008: *What's Driving Food Prices?* Issue Report, Farm Foundation, Oak Brook, IL, USA.

AFREPREN/FWD, 2009: *Cogen for Africa Initiative – Project Brief*. AFREPREN/FWD, United Nations Environment Programme (UNEP), Nairobi, Kenya.

African Press Agency, 2007: *Thousands of Tanzanian peasants to be displaced for biofuels farm*. African Press Agency, 12 August 2007. pacbiofuel.blogspot. com/2007/08/ pbn-thousands-of-tanzanian-peasants to.html (accessed 19 March 2011).

Aiking, H., J. de Boer and J. Vereijken, (eds.), 2006: Sustainable Protein Production and Consumption: Pigs or Peas? Envrionment & Policy, Vol. 45. Springer, Dordrecht, the Netherlands.

Allan, J. A., 1998: Virtual Water: A Strategic Resource Global Solutions to Regional Deficits. *Ground Water*, 36(4):545–546.

al-Riffai, P., B. Dimaranan, and D. Laborde, 2010: *Global Trade and Environmental Impact Study of the EU Biofuels Mandate*. Report for the European Commission, DG TRADE, ATLASS Consortium, Brussels, Belgium.

Araujo, A., Q. Quesada-Aguilar, L. Aguilar, A. Athanas and N. McCormick, 2009: *Gender and Bioenergy*. Factsheet, International Union for the Conservation of Nature (IUCN), Gland, Switzerland.

Arias Chalico, T., M. G. García Burgos, and G. Guerrero Pacheco, 2009: *Mexico, Task 2.1: Feedstock production in Latin America, Biofuels Assessment on Technical Opportunities and Research Needs for Latin America*. BioTop Project No: FP7–213320, BioTop RTD-cooperation, Munich, Germany.

Asea, P. K. and D. Kaija, 2000: *Impact of the flower industry in Uganda*. Working Paper – WP 148. International Labor Office, Geneva, Switzerland.

Baede, A. P. M., 2007: Annex I – Glossary. In *Climate Change 2007 – The Physical Science Basis*. Contribution of Working Group I to the Fourth Assessment Report of the Intergovernmental Panel on Climate Change (IPCC). S. Solomon, D. Qin, M. Manning, Z. Chen, M. Marquis, K. B. Averyt, M. Tignor and H. L. Miller, (eds.), Cambridge University Press, Cambridge, UK and New York, NY, USA.

Baffes, J. and T. Haniotis, 2010: *Placing the 2006/08 commodity price boom into perspective*. Policy Research Working Paper 5371, World Bank, Washington, DC, USA.

Bates, B. C., Z. W. Kundzewicz, S. Wu, and J. P. Palutikof (eds), 2008: *Climate Change and Water*. Technical Paper VI of Intergovernmental Panel on Climate Change (IPCC), Geneva, Switzerland, 210 pp.

Bekunda, M., C. A. Palm, C. de Fraiture, P. Leadley, L. Maene, L. A. Martinelli, J. McNeely, M. Otto, N. H. Ravindranath, R. L. Victoria, H. Watson, and J. Woods, 2009: Biofuels in developing countries. In *Biofuels: Environmental Consequences and Interactions with Changing Land Use*. R. W. Howarth and S. Bringezu (eds.), Proceedings of the Scientific Committee on Problems of the Environment (SCOPE), International Biofuels Project Rapid Assessment, 22–25 September 2008, Gummersbach, Germany and Cornell University, Ithaca, NY, pp.249–269.

Belwal, R. and M. Chala, 2008: Catalysts and barriers to cut flower export: A case study of Ethiopian floriculture industry. *International Journal of Emerging Markets*, 3(2): 216–235.

Bennett, K., 2008: *Food or Fuel? The Bioenergy Dilemma*. www.wri.org/ stories/2008/08/food-or-fuel-the-bioenergy-dilemma (accessed November 19, 2010).

Beringer, T., W. Lucht, and S. Schaphoff, 2011: Bioenergy production potential of global biomass plantations under environmental and agricultural constraints. *GCB Bioenergy*, 3(4):299–312.

Berndes, G., 2008: *Water demand for global bioenergy production: trends, risks and opportunities*. Wissenschaftlicher Beirat der Bundesregierung Globale Umweltveränderungen (WBGU), Externe Expertise für das WBGU-Hauptgutachten "Zukunftsfähige Bioenergie," Berlin, Germany.

Berndes, G., M. Hoogwijk and R. van den Broek, 2003: The contribution of biomass in the future global energy supply: a review of 17 studies. *Biomass and Bioenergy*, 25(1):1–28.

Bhattacharya, S. C., J. M. Thomas and P. Abdul Salam, 1997: Greenhouse gas emissions and the mitigation potential of using animal wastes in Asia. *Energy*, 22(11):1079–1085.

Bird, D. N., N. Pena, H. Schwaiger and G. Zanchi, 2010: *Review of existing methods for carbon accounting*. CIFOR Occasional Paper, 978–602–8693–27–1, Center for International Forestry Research (CIFOR), Bogor, Indonesia.

Bogdanski, A., Dubois, O., Jamieson, C., Krell, R., 2010: *Making integrated Food-Energy Systems work for people and climate*. Environment and Natural Resources Management Working Paper 45, Food and Agriculture Organization (FAO), Rome, Italy.

Börjesson, P., 1999: Environmental effects of energy crop cultivation in Sweden—II: Economic valuation. *Biomass and Bioenergy*, 16(2):155–170.

Bondeau, A., P. C. Smith, S. Zaehle, S. Schaphoff, W. Lucht, W. Cramer, D. Gerten, H. Lotze-Campen, C. Mueller, M. Reichstein and B. Smith, 2007: Modelling the role of agriculture for the 20th century global terrestrial carbon balance. *Global Change Biology*, 13(3):679–706.

Bruinsma, J., 2009: *The Resource Outlook to 2050. By how much do land, water use and crop yields need to increase by 2050?* Tehcnical Paper, Expert Meeting on "How to Feed the World in 2050." Food and Agriculture Organization (FAO), Rome, Italy.

Cassman, K. G., 1999: Ecological intensification of cereal production systems: Yield potential, soil quality, and precision agriculture. *Proceedings of the National Academy of Sciences*, 96(11):5952–5959.

CENBIO, 2002: *Estado da Arte da Gaseificação – Comparação entre tecnologias de gaseificação de biomassa existentes no brasil e no exterior e formação de recursos humanos na região norte*. Centro Nacional de Referência em Biomassa, São Paulo, Brazil.

Cerri, C. E. P., G. Sparovek, M. Bernoux, W. E. Easterling, J. M. Melillo and C. C. Cerri, 2007: Tropical agriculture and global warming: impacts and mitigation options. *Scientia Agricola*, 64(1):83–99.

Cherubini, F., N. D. Bird, A. Cowie, G. Jungmeier, B. Schlamadinger and S. Woess-Gallasch, 2009: Energy- and greenhouse gas-based LCA of biofuel and bioenergy systems: Key issues, ranges and recommendations. Resources, *Conservation and Recycling*, 53(8):434–447.

Clemens, J., Triborn, M., Weiland, P., Amon, B., 2005: Mitigation of greenhouse gas emissions by anaerobic digestion of cattle slurry. Agriculture, *Ecosystems & Environment*, 112 (2–3), 171–177.

Coelho, S. T., B. A. Lora, M. B. C. A. Monteiro, 2008: A Expansão da Cultura Canarieira no Estado de São Paulo. In: VI CBPE – Congresso. Brasileiro de Planejamento Energetico, 2008, Salvador. Anais do VI CBPE, 2008.

Coskun, B. B., 2007: *More than water wars: Water and international security*. NATO Review, Winter 2007. www.nato.int/docu/review/2007/issue4/english/analysis5. html (accessed 30 September, 2010).

Cotula, L., N. Dyer, and S. Vermeulen, 2008: *Fuelling Exclusion? The Biofuels Boom and Poor People's Access to Land*. International Institute for Environment and Development (IIED), London, UK.

de Gorter, H. and D. R. Just, 2007: *The Economics of U.S. Ethanol Import Tariffs With a Consumption Mandate and Tax Credit*. Department of Applied Economics and Management Working Paper # 2007–21, Cornell University, Cornell University, Ithaca, NY.

Dewulf, J., Langenhove, H.v., 2006. *Renewables-Based Technology – Sustainability Assessment*. John Wiley & Sons Ltd., Chichester, UK.

DFID, 2007: *DFID Annual Report 2007: Development on the Record*. Department for International Development (DFID), London, UK.

Dornburg, V., B. G. Hermann and M. K. Patel, 2008: Scenario Projections for Future Market Potentials of Biobased Bulk Chemicals. *Environmental Science & Technology*, 42(7):2261–2267.

Dornburg, V., D. van Vuuren, G. van de Ven, H. Langevel, M. Meeusen, M. Banse, M. van Oorschot, J. Ros, G. van den Born, H. Aiking, M. Londo, H. Mozaffarian, P. Verweij, E. Lyseng, and A. Faaij, 2010: Bioenergy revisited: key factors in global potentials of bioenergy. *Energy and Environmental Science*, 3:258–267.

Dufey, A., S. Vermeulen, and W. Vorley, 2007: *Biofuels: strategic choices for commodity dependent developing countries*, Common Fund for Commodities, Amsterdam, the Netherlands.

Dunlap, R. E. and W. R. Catton Jr., 2002: Which Function(s) of the Environment Do We Study? A Comparison of Environmental and Natural Resource Sociology. *Society and Natural Resources*, 15(3):239–249.

EC and IPTS, undated: *Executive Summary: Towards a Sustainable/Strategic Management of Water Resources: Evaluation of Present Policies and Orientations for the Future*. European Commission (EC)'s General Directorate XVI (Regional policy and Cohesion) and the Institute for Prospective Technological Studies (IPTS) of the EC's Joint Research Centre. www.fao.org/iccd/object/doc/sustwater.htm (accessed 22 March 2011).

Ecofys, 2006: D*raft Technical Guidance for sustainability reporting under the Renewable Transport Fuel's Obligation.* Low Carbon Vehicle Partnership, London, UK.

Edwards, R., D. Mulligan, and L. Marelli, 2010*: Indirect Land Use Change from increased biofuels demand: Comparison of model and results for marginal biofuels production from different feedstocks*. Report No. JRC 59771, Joint Research Center, Institute for Energy (JRC-IE). European Commission, Luxembourg.

Elia Neto, A., 2005: Impact on the water supply. In *Sugar Cane's Energy: twelve studies on Brazilian sugar cane agribusiness and its sustainability*. I. C. Macedo, (ed.), UNICA, São Paulo, Brazil.

EMBRAPA, 2008: *Sustentabilidade Agrícola e Biodiversidade (Agricultural Sustainability and Biodiversity)*. Empresa Brasileira de Pesquisa Agropecuária (EMBRAPA). www.biodiversidade.cnpm.embrapa.br/ambiental/resumo.html (accessed 4 March 2011).

EMBRAPA, 2011a: *Zoneamento da Cana-de-Açúcar (Sugarcane Zoning)*. Empresa Brasileira de Pesquisa Agropecuária (EMBRAPA). www.cnps.embrapa.br/zoneamento_cana_de_acucar/ (accessed 4 March 2011).

EMBRAPA, 2011b: *Zoneamento do Dendê (Palm Zoning)*. Empresa Brasileira de Pesquisa Agropecuária (EMBRAPA). www.cnps.embrapa.br/zoneamento_dende (accessed 4 March 2011).

EPFL, 2007: *Roundtable on Sustainable Biofuels*. École Polytechnique Fédérale de Lausanne (EPFL), Lausanne, Switzerland.

Erb, K.-H., V. Gaube, F. Krausmann, C. Plutzar, A. Bondeau, and H. Haberl, 2007: A comprehensive global 5 min resolution land-use dataset for the year 2000 consistent with national census data. *Journal of Land Use Science*, 2(3):191–224.

Erb, K.-H., H. Haberl, F. Krausmann, C. Lauk, C. Plutzar, J. K. Steinberger, C. Müller, A. Bondeau, K. Waha, and G. Pollack, 2009: *Eating the planet: Feeding and fuelling the world sustainably, fairly and humanely – a scoping study*. Report commissioned by Compassion in World Farming and Friends of the Earth, UK. Institute of Social Ecology, Vienna, Austria and Potsdam Institute for Climate Impact Research (PIK), Potsdam, Germany.

European Environment Agency, 2006: *How much bioenergy can Europe produce without harming the environment?*, EEA Report No 7/2006. European Environment Agency (EEA), Copenhagen, Denmark.

Evans, L. T., 1998: *Feeding the ten billion: Plants and population growth*. Cambridge University Press, Cambridge, UK, and New York, NY.

Faaij, A. P. C., 2008: *Bioenergy and global food security. Externe Expertise für das WBGU-Hauptgutachten "Welt im Wandel: Zukunftsfähige Bioenergie und nachhaltige Landnutzung"*, Wissenschaftlicher Beirat der Bundesregierung Globale Umweltveränderungen (WBGU), Berlin, Germany.

Faaij, A., 2009: Global Biomass Potentials. In *International Conference Biomass in Future Landscapes – Sustainable Use of Biomass and Spatial Development*, German Biomass Research Centre (DBFZ) and Agricultural Landscape Resaerch (ZALF), 31 March 2009, Berlin, Germany.

FAO, 2005: *Irrigation in Africa in figures*. Aquastat FAO Water Report no. 29. Food and Agriculture Organizatoin (FAO), Rome, Italy.

FAO, 2006: *World Agriculture: towards 2030/2050*. Interim Report. Food and Agriculture Organizatoin (FAO), Rome, Italy.

FAO, 2007a: *Gridded livestock of the world*. Food and Agriculture Organizatoin (FAO), Rome, Italy.

FAO, 2007b: *Physical and Economic Water Scarcity*. Food and Agriculture Organizatoin (FAO), Rome, Italy. www.fao.org/nr/water/art/2007/scarcity.html.

FAO, 2008a: *Soaring Food Prices: Facts, Perspectives, Impacts And Actions Required*. High-Level Conference on World Food Security: The Challenges of Climate Change and Bioenergy, Rome, 3–5 June 2008, HLC/08/INF/1. Food and Agriculture Organizatoin (FAO), Rome, Italy.

FAO, 2008b: *Bioenergy, Food Security and Sustainability – Towards an International Framework*. High-Level Conference on World Food Security: The Challenges of Climate Change and Bioenergy, Rome, 3–5 June 2008, HLC/08/INF/3. Food and Agriculture Organizatoin (FAO), Rome, Italy.

FAO, 2008c: *The State of Food and Agriculture – Biofuels: prospects, risks and opportunities*. Food and Agriculture Organizatoin (FAO), Rome, Italy.

FAO, 2009a: FAOSTAT Statistics. Food and Agriculture Organizatoin (FAO), Rome, Italy. faostat.fao.org/site/573/default.aspx#ancor (accessed 13 April 2011).

FAO, 2009b: *The State of Food and Agriculture 2009 – Livestock in the balance*. Food and Agriculture Organizatoin (FAO), Rome, Italy.

FAO, 2009c: *The State of Agricultural Commodity market: High food prices and the food crisis – experiences and lessons learned*. Food and Agriculture Organizatoin (FAO), Rome, Italy.

FAO, 2010a: *Bioenergy and Food Security – The BEFS analysis for Tanzania*. The Bioenergy and Food Security (BEFS) Project. Food and Agriculture Organizatoin (FAO), Rome, Italy.

FAO, 2010b: *Bioenergy and Food Security – The BEFS Analysis for Peru: Supporting the policy machinery in Peru*. The Bioenergy and Food Security Project. Food and Agriculture Organizatoin (FAO), Rome, Italy.

FAO and IIASA, 2000: *Global Agro-Ecological Zones – GAEZ.* Food and Agriculture Organizatoin (FAO), Rome, Italy, and the International Institute for Applied Systems Analysis (IIASA), Laxenburg, Austria. www.fao.org/landandwater/agll/gaez/index.htm (accessed 12 June 2011).

FAO-AQUASTAT, 2008: *Proportion of renewable water resources withdrawn: MDG Water Indicator.* www.fao.org/nr/water/aquastat/globalmaps/index.stm.

Firbank, L., 2008: Assessing the Ecological Impacts of Bioenergy Projects. BioEnergy Research, 1(1):12–19.

Fischer, G., H. van Velthuizen, M. Shah, F. Nachtergaele, 2002: *Global Agro-ecological Assessment for Agriculture in the 21st Century: Methodology and Results.* IIASA Research Report RR-02–002. International Institute for Applied Systems Analysis (IIASA), Laxenburg, Austria.

Foley, J. A., R. DeFries, G. P. Asner, C. Barford, G. Bonan, S. R. Carpenter, F. S. Chapin, M. T. Coe, G. C. Daily, H. K. Gibbs, J. H. Helkowski, T. Holloway, E. A. Howard, C. J. Kucharik, C. Monfreda, J. A. Patz, I. C. Prentice, N. Ramankutty and P. K. Snyder, 2005: Global Consequences of Land Use. *Science*, **309**(5734):570–574.

Friis, C. and A. Reenberg, 2010: *Land Grap in Africa: Emerging Land System Drivers in a Teleconnected World.* GLP Report No. 1, The Global Land Project (GLP) International Project Office (IPO), Copenhagen, Denmark.

Fritsche, U., 2010: GHG emissions of future relevant biomass conversion pathways. In *Proceedings of the 16th European Biomass Conference*, 2–6 June 2010, Valencia, Spain.

Fritsche, U. R., R. E. H. Sims and A. Monti, 2010: Direct and indirect land-use competition issues for energy crops and their sustainable production – an overview. *Biofuels, Bioproducts and Biorefining*, **4**(6):692–704.

FSC, 2011: Forest Stewardship Council (FSC). www.fsc.org/ (accessed 23 October 2011).

Fthenakis, V. and H. C. Kim, 2010: Life Cycle uses of water in U.S. electricity generation. *Renewable and Sustainable Energy Reviews*, **14**(7):2039–2048.

Georgescu, A., 2010: Update on the European Renewable Energy Directive, 2009/28/ECEU. In *Presentation at Final Conference of BioTop Project*, July 2010, DG Energy, European Commission, Brussels, Belgium.

Gerbens-Leenes, W., A. Hoekstra and T. van der Meer, 2008: The Water Footprint of Energy Consumption: an Assessment of Water Requirements of Primary Energy Carriers. *ISESCO Science and Technology Vision*, **4**(5):38–42.

Gerbens-Leenes, W., A.Y. Hoekstra, and T. H. van der Meer, 2009: The water footprint of Bioenergy. *Proceedings of the National Academy of Sciences*, **106**(25):10219–10223.

Gibbs, H. K., M. Johnston, J. A. Foley, T. Holloway, C. Monfreda, N. Ramankutty, and D. Zaks, 2008: Carbon payback times for crop-based biofuel expansion in the tropics: the effects of changing yield and technology. *Environmental Research Letters*, **3**(3):034001.

Gilbert, C. L., 2010: How to Understand High Food Prices. *Journal of Agricultural Economics*, **61**(2):398–425.

GLOBAL GAP, 2011: *European Retailers Produce Working Group.* www.eurepgap.org/Languages/English/about.html. (accessed 12 March 2011).

Global Land Project, 2005: *Science Plan and Implementation Strategy.* IGBP Report No. 53/IHDP Report No. 19. International Geosphere-Biosphere Programme (IGBP) Secretariat, Stockholm, Sweden, p.64.

GLOBIO, 2010: *Modelling Human Impacts on Biodiversity.* www.globio.info/what-is-globio/history -of-globio (accessed 23 October 2010).

GNESD, 2010: *Bioenergy Theme: Summary Report.* Global Network for Sustainable Development (GNESD), Roskilde, Denmark.

Godfray, H. C. J., J. R. Beddington, I. R. Crute, L. Haddad, D. Lawrence, J. F. Muir, J. Pretty, S. Robinson, S. M. Thomas and C. Toulmin, 2010: Food Security: The Challenge of Feeding 9 Billion People. *Science*, **327**(5967):812–818.

Goldemberg, J., 2008: The Brazilian biofuels industry. *Biotechnology for Biofuels*, **1**(1):1–7.

Goldemberg, J. 2009: The Brazilian Experience with Biofuels. *Innovations Journal*, **4** (Fall):91–107.

Goldemberg, J. and P. Guardabassi, 2009: Are biofuels a feasible option? *Energy Policy*, **37**(1):10–14.

Goldemberg, J., S. T. Coelho, and P. Guardabassi, 2008: The sustainability of ethanol production from sugarcane. **Energy Policy**, **36**:2086–2097.

Goldsmith, E. and N. Hildyard, 1984: Politics of Damming. *Ecologist*, **14**(5/6):221–231.

Haberl, H., and S. Geissler, 2000: Cascade utilization of biomass: Strategies for a more efficient use of a scarce resource. *Ecological Engineering*, **16**(SUPPL. 1):111–121.

Haberl, H., N. B. Schulz, C. Plutzar, K. H. Erb, F. Krausmann, W. Loibl, D. Moser, N. Sauberer, H. Weisz, H. G. Zechmeister and P. Zulka, 2004: Human appropriation of net primary production and species diversity in agricultural landscapes. *Agriculture, Ecosystems & Environment*, **102**(2):213–218.

Haberl, H., K. H. Erb, F. Krausmann, V. Gaube, A. Bondeau, C. Plutzar, S. Gingrich, W. Lucht and M. Fischer-Kowalski, 2007: Quantifying and mapping the human appropriation of net primary production in earth's terrestrial ecosystems. *Proceedings of the National Academy of Sciences*, **104**(31):12942–12947.

Haberl, H., T. Beringer, S. C. Bhattacharya, K.-H. Erb and M. Hoogwijk, 2010: The global technical potential of bio-energy in 2050 considering sustainability constraints. *Current Opinion in Environmental Sustainability*, **2**(5–6):394–403.

Haberl, H., K.-H. Erb, F. Krausmann, A. Bondeau, C. Lauk, C. Müller, A. Plutzar and J. K. Steinberger, 2011: Global bioenergy potentials from agricultural land in 2050: Sensitivity to climate change, diets and yields. *Biomass and Bioenergy*, **35**(12), 4753–4769.

Harmon, M. E., J. F. Franklin, F. J. Swanson, P. Sollins, S. V. Gregory, J. D. Lattin, N. H. Anderson, S. P. Cline, N. G. Aumen, J. R. Sedell, G. W. Lienkaemper, K. Cromack Jr and K. W. Cummins, 1986: Ecology of Coarse Woody Debris in Temperate Ecosystems. In *Advances in Ecological Research*. A. MacFadyen and E. D. Ford, (eds.), Academic Press, Vol. 15, pp.133–302.

Harris, P. S., 1987: *Grassland Survey and Integrated Pasture Development in the High mountain Region of Bhutan*. TCP/BHU/4505[A] Food and Agriculture Organization (FAO), Rome, Italy.

Hassuani, S. J., M. R. L. V. Leal, and I. d. C. Macedo (eds.), 2005: *Biomass Power generation: sugar cane bagasse and trash*. Caminhos para Sustentabilidade series. United Nations Development Programme (UNDP) and Centro de Tecnologia Canavieira (CTC), Piracicaba, Brazil.

Hoogwijk, M., A. Faaij, R. van den Broek, G. Berndes, D. Gielen and W. Turkenburg, 2003: Exploration of the ranges of the global potential of biomass for energy. *Biomass and Bioenergy*, **25**(2):119–133.

Hoogwijk, M., B. de Vries and W. Turkenburg, 2004: Assessment of the global and regional geographical, technical and economic potential of onshore wind energy. *Energy Economics*, **26**(5):889–919.

Hoogwijk, M., A. Faaij, B. Eickhout, B. de Vries and W. Turkenburg, 2005: Potential of biomass energy out to 2100, for four IPCC SRES land-use scenarios. *Biomass and Bioenergy*, **29**(4):225–257.

Houghton, R.A., 1995: Land-use change and the carbon cycle. *Global Change Biology*, **1**(2): 275–287.

IAASTD, 2009: *Agriculture at a Crossroads.* International Assessment of Agricultural Knowledge, Science and Technology for Development (IAASTD) Global Report. Island Press, Washington, DC, USA.

IEA, 2010: *Sustainable Production of Second-Generation Biofuels – Potential and perspectives in major economies and developing countries.* Information Paper, A. Eisentraut, (ed.) International Energy Agency (IEA) of the Organization for Economic Co-operation and Development, Paris, France.

IFPRI, 2009: *Food Policy Report. Climate Change. Impact on Agriculture and Costs of Adaptation.* Food Policy Report, G. C. Nelson, M. W. Rosegrant, J. Koo, R. Robertson, T. Sulser, Z. Tingju, C. Ringler, S. Msangi, A. Palazzo, M. Batka, M. Magalhaes, R. Valmonte-Santos, M. Ewing and D. Lee, (eds.), International Food Policy Research Institute (IFPRI), Washington, DC, USA.

IPAM, 2002: *Perguntas e Respostas Sobre Mudanças Climáticas.* Instituto de Pesquisa Ambiental da Amazônia (IPAM), Belém, Brazil.

IPCC, 2000: *Land Use, Land-Use Change and Forestry.* A Special Report of the Intergovernmental Panel on Climate Change (IPCC), R. T. Watson, I. R. Noble, B. Bolin, N. H. Ravindranath, D. J. Verardo and D. J. Dokken, (eds.), Cambridge University Press, Cambridge, UK.

IPCC, 2007: *Climate Change 2007: Impacts, Adaptation and vulnerability.* Contribution of Working Group II to the Fourth Assessment Reprot of the Intergovernmental Panel of Climate Change. M. L. Parry, O. F. Canziani, J. P. Palutikof, P. J. van der Linden and C. E. Hanson, (eds.), Cambridge University Press, Cambridge, UK and New York, NY, USA.

IPCC, 2011: *IPCC Special Report on Renewable Energy Sources and Climate Change Mitigation.* O. Edenhofer, R. Pichs-Madruga, Y. Sokona, P. Matschoss and K. Seyboth, (eds.), Cambridge University Press, Cambridge, UK and New York, NY, USA.

ISRIC, 1991: *Global Assessment of Human-induced Soil Degradation (GLASOD).* International Soil Reference and Information Centre (ISRIC) World Soil Information, Wageningen, the Netherlands.

Ivanic, M. and W. Martin, 2008: Implications of higher global food prices for poverty in low-income countries. *Agricultural Economics*, **39**:405–416.

IWMI, 2007a: *Water for Food, Water for Life: A Comprehensive Assessment of Water Management in Agriculture.* International Water Management Institute (IWMI), Colombo, Sri Lanka and Earthscan, London.

IWMI, 2007b: *International Water Management Institute Annual Report 2007– 2008.* International Water Management Institute (IWMI), Colombo, Sri Lanka.

Jaccard, M., 2005: *Sustainable Fossil Fuels: The Unusual Suspect in the Quest for Clean and Enduring Energy.* Cambridge University Press, Cambridge, UK and New York, NY, USA.

Jansson, C., S. T. Wullschleger, U. C. Kalluri and G. A. Tuskan, 2010: Photosequestration: carbon bio-sequestration by plant and the prospect of genetic engineering. *Bioscience*, **60**:685–696.

Johansson, T. B. and W. Turkenburg, 2004: Policies for renewable energy in the European Union and its member states: an overview. *Energy for Sustainable Development*, **8**(1):5–24.

Johnston, M., J. A. Foley, T. Holloway, C. J. Kucharik and C. Monfreda, 2009: Resetting global expectations from agricultural biofuels. *Environmental Research Letters*, **4**(1):014004.

Joly, C. A., R. R. Rodrigues, J. P. Metzger, C. F. B. Haddad, L. M. Verdade, M. C. Oliveira and V. S. Bolzani, 2010: Biodiversity Conservation Research, Training, and Policy in São Paulo. *Science*, **328**(5984):1358–1359.

Jumbe, C., F. Msiska and L. Mhango, 2007: *Report on National Policies on Biofuels Sector Development in Sub-Saharan Africa.* Final Working Draft from Food, Agriculture and Natural Resources Policy Analysis Network (FANRPAN) to WIP Renewable Energies, Germany, for Compete Competence Platform on Energy Crop and Agroforestry Systems for Arid and Semi-arid Ecosystems – Africa.

Khan, S., M. A. Khan, M. A. Hanjra and J. Mu, 2009: Pathways to reduce the environmental footprints of water and energy inputs in food production. *Food Policy*, **34**(2):141–149.

Koning, N. and M. K. van Ittersum, 2009: Will the world have enough to eat? *Current Opinion in Environmental Sustainability*, **1**(1):77–82.

Krajick, K., 2001: Defending Deadwood. *Science*, **293**(5535):1579–1581.

Krausmann, F., 2004: Milk, manure, and muscle power. Livestock and the transformation of preindustrial agriculture in Central Europe. *Human Ecology*, **32**(6):735–772.

Krausmann, F., K.-H. Erb, S. Gingrich, C. Lauk and H. Haberl, 2008: Global patterns of socioeconomic biomass flows in the year 2000: A comprehensive assessment of supply, consumption and constraints. *Ecological Economics*, **65**(3):471–487.

Lal, R., 2005: World crop residues production and implications of its use as a biofuel. *Environment International*, **31**(4):575–584.

Lal, R., 2006: Soil and environmental implications of using crop residues as biofuel feedstock. *International Sugar Journal*, **108**:161–167.

Lal, R., 2010: Managing Soils and Ecosystems for Mitigating Anthropogenic Carbon Emissions and Advancing Global Food Security. *BioScience*, **60**(9):708–721.

Lambin, E. F. and H. Geist, (eds.), 2006: *Land-Use and Land-Cover Change: Local processes and global impacts.* Springer Berlin, Germany.

LEAF, 2011: Linking Environment and Farming (LEAF). www.assuredcrops.co.uk/leafuk/organisation/Default.asp?id=4030338 (accessed 10 December, 2010).

Lewandowski, I. and A. P. C. Faaij, 2006: Steps towards the development of a certification system for sustainable bio-energy trade. *Biomass and Bioenergy*, **30**(2):83–104.

Lewandowski, I., M. Londo, U. Schmidt, and A. P. C. Faaij, 2003: Energiepflanzen und Landnutzungsfunktionen – Ein Überblick zu Kombinationsmöglichkeiten, biophysikalischen Effekten und ökonomischen Ansätzen. *In Mitteilungen der Gesellschaft für Pflanzenbauwissenschaften Band 15.* D. Kauter, A. Kämpf, W. Claupein and W. Diepenbrock (eds). Verlag Günter Heimbach, Stuttgart, Germany, pp.110–113.

Londo, M., 2002: *Energy farming in multiple land use – An opportunity for energy crop introduction in the Netherlands*, PhD thesis, Dept. of Science, Technology and Society, Utrecht University, Utrecht, the Netherlands.

Lucon, O., 2010: *WP4: Sustainability of biofuels production in Latin America.* BioTop Project. www.top-biofuel.org (accessed 26 November 2010).

Lucon, O. and J. Goldemberg , 2010: São Paulo – The "Other" Brazil: Different Pathways on Climate Change for State and Federal Governments. *Journal of Environment and Development*, **19**(3):335–357.

Lundqvist, J., J. Barron, G. Berndes, A. Berntell, M. Falkenmark, L. Karlberg, and. Rockström, 2007: Water pressure and increases in food and bioenergy demand.

Implications of economic growth and options for decoupling. In *Scenarios on economic growth and research development: Background report to the Swedish Environmental Advisory Council Memorandum* **2007**(1):55–152.

Lundqvist, J., C. de Fraiture, and D. Molden, 2008: *Saving Water: From Field to Fork – Curbing Losses and Wastage in the Food Chain*. SIWI Policy Brief. Stockholm International Water Institute (SIWI), Stockholm, Sweden.

Macedo, I. C., J. E. A. Seabra and J. E. A. R. Silva, 2008: Greenhouse gases emissions in the production and use of ethanol from sugarcane in Brazil: The 2005/2006 averages and a prediction for 2020. *Biomass and Bioenergy*, **32**(7):582–595.

Martínez-Alier, J., 2002: Ecological debt and property rights on carbon sinks and reservoirs. *Capitalism, Nature, Socialism*, **13**(1):115–119.

McGuire, V. L., 2007a: *Changes in Water Levels and Storage in the High Plains Aquifer, Predevelopment to 2005*. USGS Fact Sheet 2007–3029, Ground-Water Resources Program, United States Geological Survey (USGS), Washington, DC, USA.

McGuire, V. L., 2007b: *Water-level changes in the High Plains aquifer, predevelopment to 2005 and 2003 to 2005*. US Geological Survey Scientific Investigations Report 2006–5324, 7pp. pubs.usgs.gov/sir/2006/5324/ (accessed 4 April 2011).

McKay, G., 2002: Dioxin characterization, formation and minimization during municipal solid waste (MSW) incineration: review. *Chemical Engineering Journal*, **86**(3):343–368.

Millennium Ecosystem Assessment, 2005: *Ecosystems and Human Well-Being: Our Human Planet*. Summary for Decision Makers. Island Press, Washington, DC, USA.

Mitchell, D., 2008: *A Note on Rising Food Prices*. Policy Research Working Paper, WPS 4682. Development Prospects Group, The World Bank, Washington, DC, USA.

MPOB, 2011: *Oil Palm & The Environment. Malaysian Palm Oil Board (MPOB)*. http://www.mpob.gov.my/index.php?option=com_content&view=article&id=520%3Aachievements&catid=131%3Aenvironment&lang=en#Introduction (accessed 14 November 2010).

Müller, A., J. Schmidhuber, J. Hoogeveen, and P. Steduto, 2007: Some insights in the effect of growing bio-energy demand on global food security and natural resources. In *Paper presented at the International Conference on Linkages between Energy and Water Management for Agriculture in Developing Countries*, 28–31 January 2007, Hyderabad, India.

Müller, C., A. Bondeau, A. Popp, K. Waha and M. Fader, 2010: *Development and Climate Change – Climate Change Impacts on Agricultural Yields*. Background Note to the World Development Report 2010, Potsdam Institute for Climate Impact Research (PIK), Potsdam, Germany.

Nachtergaele, F. and H. George, 2009: *How much land is available for agriculture?* Unpublished paper. Food and Agriculture Organization (FAO), Rome, Italy.

Namburete, H. E. S., 2006: Mozambique biofuels. In *Presentation at the African Green Revolution Conference*, Oslo, Norway, 31 August – 2 September 2006.

Nassar, A. M., 2009: Brazil as an Agricultural and Agroenergy Superpower. In *Brazil as an Economic Superpower? Understanding Brazil's Changing Role in the Global Economy*. L. Brainard and L. Martinez-Diaz, (eds.), Brookings Institution Press, Washington, DC.

Nassar, A. M., L. Harfuch, M. M. R. Moreira, L. C. Bachion, and L. B. Antoniazzi, 2009: *Impacts on Land Use and GHG Emissions from a Shock on Brazilian Sugarcane Ethanol Exports to the United States Using the Brazilian Land Use Model (BLUM)*. Report to the US Environmental Protection Agency regarding the Proposed Changes to the Renewable Fuel Standard Program. Institute for International Trade Negotiations (ICONE), São Paulo, Brazil.

Nassar, A. M., L. Harfuch, L. C. Bachion, and M. M. R. Moreira, 2011: Biofuels and land-use changes: Searching for the top model. *Focus*, **1**:224–232.

OECD, 2008a: *OECD Environmental Outlook to 2030*. Organisation for Economic Co-operation and Development (OECD), Paris, France.

OECD, 2008b: *Rising Food Prices: Causes and Consequences*. Organisation for Economic Co-operation and Development (OECD), Paris, France.

OECD/FAO, 2007: *OECD-FAO Agricultural Outlook 2007–2016*. OECD Publishing, Paris, France.

OECD/FAO, 2009: *OECD-FAO Agricultural Outlook 2009*. OECD Publishing, Paris, France.

Orr, S., and A. Chapagain, 2007: *African air-freight of fresh produce: is transport of 'virtual' water causing drought?* Fresh Perspectives, Issue 5, International Institute for Environment and Development/ Natural Resources Institute, London, UK.

Peng, S., R. C. Laza, R. M. Visperas, A. L. Sanico, K. G. Cassman and G. S. Khush, 2000: Grain Yield of Rice Cultivars and Lines Developed in the Philippines since 1966. *Crop Science*, **40**(2):307–314.

Peterson, E. and R. Pozner, 2008: *Global water futures – a roadmap for future U.S. policy*. Center for Strategic and International Studies, Washington, DC.

Pingoud, K., A. Cowie, N. Bird, L. Gustavsson, S. Rüter, R. Sathre, S. Soimakallio, A. Türk, S. Woess-Gallasch, 2010: Bioenergy: Counting on Incentives. *Science*, **327**(5970):1199–1200.

Pistonesi, H., G. Nadal, V. Bravo, and D. Bouille, 2008: *Aporte de los biocombustibles a la sustentabilidad del desarrollo en América Latina y el Caribe: Elementos para la formulación de políticas públicas*. Comisión Económica para América Latina y el Caribe (CEPAL), United Nations (UN), Santiago de Chile.

Pleskett, L., R. Slater, C. Stevens and A. Dufey, 2007: *Biofuels, Agriculture and Poverty Reduction*. Natural Resource Perspectives, Nr. 107, Overseas Development Institute (ODI), London, UK.

Pope, C. T., 2008: *U.S. needs integrated international water strategy to avert conflict, foster cooperation, says new CSIS report*. Circle of Blue. www.circleofblue.org/waternews/world/policy-think-tank-calls-for-water-bureau-in-washington/ (accessed 5 May 2011).

Power Scorecard, 2000: *Consumption of Water Resources*. Pace University, White Plains, NY. www.powerscorecard.org/issue_detail.cfm?issue_id=5 (accessed 20 November 2010).

Rajabapaiah, P., S. Jayakumar, and A. K. N. Reddy, 1993: Biogas electricity – the Pura village case study. In *Renewable Energy, Sources for Fuels and Electricity*. T. B. Johansson, H. Kelly, A. K. N. Reddy, and R. H. Williams (eds.), Island Press, Washington, DC, pp.787–815.

Ramankutty, N., J. A. Foley, J. Norman, and K. McSweeney, 2002: The global distribution of cultivable lands: current patterns and sensitivity to possible climate change. *Global Ecology and Biogeography*, **11**(5):377–392.

Rogers, P., 2008: Facing the Freshwater Crisis. *Scientific American*, **299**(2008):46–53.

Rogner, H.-H., Barthel, F., Cabrera, M., Faaij, A., Giroux, M., Hall, D.O., Kagramanian, V., Kononov, S., Lefevre, T., Moreira, R., Nötstaller, R., Odell, P., Taylor, M., 2000: Energy Resources. In: *World Energy Assessment*, J. Goldemberg, (ed.), United Nations Development Programme (UNDP), World Energy Council, New York, pp. 135–172.

RSPO, 2011: *Roundtable on Sustainable Palm Oil (RSPO)*. www.rspo.org/ (accessed 16 March 2011).

RTRS, 2011: *Roundtable on Responsible Soy Association (RTRS)*. www.responsible-soy.org/ (accessed 5 May 2011).

Runge, C. F. and B. Senauer, 2007: How Biofuels Could Starve the Poor. *Foreign Affairs*, **86**(3).

Sagar, A. D. and S. Kartha, 2007: Bioenergy and sustainable development? *Annual Review of Environmental Resources*, **32**:131–167.

Sartori, F., R. Lal, M. H. Ebinger and D. J. Parrish, 2006: Potential Soil Carbon Sequestration and CO2 Offset by Dedicated Energy Crops in the USA. *Critical Reviews in Plant Sciences*, **25**(5):441–472.

Sathaye, J., O. Lucon, A. Rahman, J. Christensen, F. Denton, J. Fujino, G. Heath, S. Kadner, M. Mirza, H. Rudnick, A. Schlaepfer, and A. Shmakin, 2011: Renewable Energy in the Context of Sustainable Development. In *IPCC Special Report on Renewable Energy Sources and Climate Change Mitigation*. O. Edenhofer, R. Pichs-Madruga, Y. Sokona, K. Seyboth, P. Matschoss, S. Kadner, T. Zwickel, P. Eickemeier, G. Hansen, S. Schlömer and C. von Stechow (eds.), Cambridge University Press, Cambridge, UK and New York, NY, USA.

Saugier, B., J. Roy and H. A. Mooney, 2001: Estimations of global terrestrial productivity: converging toward a single number? In *Terrestrial Global Productivity*. J. Roy, B. Saugier and H. A. Mooney, (eds.), Academic Press, San Diego, CA, USA.

Schimel, D. S., J. I. House, K. A. Hibbard, P. Bousquet, P. Ciais, P. Peylin, B. H. Braswell, M. J. Apps, D. Baker, A. Bondeau, J. Canadell, G. Churkina, W. Cramer, A. S. Denning, C. B. Field, P. Friedlingstein, C. Goodale, M. Heimann, R. A. Houghton, J. M. Melillo, B. Moore Iii, D. Murdiyarso, I. Noble, S. W. Pacala, I. C. Prentice, M. R. Raupach, P. J. Rayner, R. J. Scholes, W. L. Steffen and C. Wirth, 2001: Recent patterns and mechanisms of carbon exchange by terrestrial ecosystems. *Nature*, **414**(6860):169–172.

Searchinger, T., R. Heimlich, et al., 2008: Use of U.S. croplands for biofuels increases greenhouse gases through emissions from land-use change. *Science*, **319**(5867): 1238–1240.

Searchinger, T. D., S. P. Hamburg, J. Melillo, W. Chameides, P. Havlik, D. M. Kammen, G. E. Likens, R. N. Lubowski, M. Obersteiner, M. Oppenheimer, G. P. Robertson, W. H. Schlesinger, G. D. Tilman, 2009: Fixing a critical climate accounting error. *Science*, **326**(5952):527–528.

Searchinger, T. D., 2010: Biofuels and the need for additional carbon. *Environmental Research Letters*, **5**(2):024007.

SECO, 2008: *Texas Renewable Energy Resource Assessment*. Texas State Energy Conservation Office (SECO), prepared by Frontier Associates, LLC, Austin, Texas.

Shifley, S.R., F. R. Thompson, III, W. D. Dijak, M. A. Larson, and J. J. Millspaugh, 2006: Simulated effects of forest management alternatives on landscape structure and habitat suitability in the Midwestern United States. *Forest Ecology and Management*, **229**(1–3):361–377.

Showers, K. B., 2006: A History of African Soil: Perceptions, Use and Abuse. In *Soils and Societies. Perspectives from environmental history*. J. R. McNeill and V. Winiwarter (eds.), The White Horse Press, Cambridge, UK, pp.118–176.

SIWI, 2008: *Saving Water: From Field to Fork. Curbing losses and wastage in the food chain*. Stockholm International Water Institute (SIWI), Stockholm, Sweden.

SMA, 2011: *Etanol Verde*, São Paulo State Environment Secretariat, Brazil: homologa. ambiente.sp.gov.br/etanolverde/english.asp (accessed 6 June 2011).

Smeets, E. M. W., A. P. C. Faaij, I. M. Lewandowski and W. C. Turkenburg, 2007: A bottom-up assessment and review of global bio-energy potentials to 2050. *Progress in Energy and Combustion Science*, **33**(1):56–106.

Smil, V., 2000: *Feeding the world: A challenge for the twenty-first century*. MIT Press, Cambridge, MA.

Soares, L. H. B., B. J. R. Alves, S. Urquiaga, and R. M. Boddey, 2009: Mitigação das Emissões de Gases de Efeito Estufa pelo Uso de Etanol de Cana-de-açúcar Produzido no Brasil. *Circular Técnica*, **27**(2009).

Somerville, C., 2006: The Billion-Ton Biofuels Vision. *Science*, **312**(5778):1277.

Somerville, C., H. Youngs, C. Taylor, S. C. Davis, S. P. Long, 2010: Feedstocks for lignocellulosic biofuels. *Science*, **329**:790–792.

Soyka, T., C. Palmer, and S. Engel, 2007: *The Impacts of Tropical Biofuel Production on Land-use: The case of Indonesia. Conference on International Agricultural Research for Development*, Tropentag 2007. University of Kassel-Witzenhausen and University of Göttingen, Germany, 9–11 October 2007.

Sparovek, G., G. Berndes, A. Egeskog, F. L. M. de Freitas, S. Gustafsson and J. Hansson, 2007: Sugarcane ethanol production in Brazil: An expansion model sensitive to socioeconomic and environmental concerns. Biofuels, *Bioproducts and Biorefining*, **1**(4):270–282.

Sparovek, G., A. Barretto, G. Berndes, S. Martins, and R. Maule, 2009: Environmental, land-use and economic implications of Brazilian sugarcane expansion 1996–2006. *Mitigation and Adaptation Strategies for Global Change*, **14**:285–298.

Stinner, W., K. Möller, and G. Leithold, 2008: Effects of biogas digestion of clover/grass-leys, cover crops and crop residues on nitrogen cycle and crop yield in organic stockless farming systems. *European Journal of Agronomy*, **29**:125–134.

Tampier, M., 2002: *Promoting Green Power in Canada*. Pollution-Probe, Toronto, ON, Canada.

TERI, 2010: *Bioenergy for rural development and poverty alleviation*. Project Report No. 2008DG10. The Energy and Resources Institute (TERI), New Delhi.

Tilman, D., K. G. Cassman, P. A. Matson, R. Naylor and S. Polasky, 2002: Agricultural sustainability and intensive production practices. *Nature*, **418**(6898):671–677.

Timilsina, G, J. C. Beghin, D. van der Mensbrugghe, and S. Mevel, 2010: *The impacts of biofuel targets on land use and food supply: A Global CGE Assessment*. Policy Research Working Paper 5513. World Bank, Washington, DC.

Trostle, R., 2008: *Global Agricultural Supply and Demand: Factors Contributing to the Recent Increase in Food Commodity Prices*. US Department of Agriculture, Washington, DC.

Trussell, D., E. Goldsmith, and N. Hildyard, 1992: *The Social and Environmental Effects of Large Dams, Volume III: A Review of the Literature*. Wadebridge Ecological Centre, Camelford, Cornwall, UK.

UNCTAD, 2008: *Making Certification Work for Sustainable Development: The case of Biofuels*. United Nations Conference on Trade and Development (UNCTAD), New York, NY, and Geneva, Switzerland.

UNCTAD, 2009: *The Biofuels Market: Current Situation and Alternative Scenarios*. Report provided by Global Bioenergy Partnership. United Nations Conference on Trade and Development (UNCTAD), New York, NY, and Geneva, Switzerland.

UNDESA, 2007: *Small-Scale Production and Use of Liquid biofuels in Sub-Saharan Africa: Perspectives for Sustainable Development*. Energy and Transport Branch, Division of Sustainable Development. United Nations Department of Economic and Social Affairs (UNCTAD), New York, NY, and Geneva, Switzerland.

UN-Energy, 2007: *Sustainable Bioenergy: A Framework for Decision Makers.* UN-Energy, New York, NY, USA.

UNEP, 2007: *Compilations of existing certification schemes, policy measures, ongoing initiatives and crops used for bioenergy.* Working Paper, Working group on developing sustainability criteria and standards for the cultivation of biomass used for biofuels, Ministry of Nutrition and Rural Affairs of Baden Württemberg, Germany, DaimlerChrysler, and the United Nations Environment Programme (UNEP), Paris, France.

UNEP, 2009: *Assessing Biofuels – Towards Sustainable Production and Use of Resources.* United Nations Environment Programme (UNEP), Division of Technology, Industry and Economics, International Panel for Sustainable Resource Management, Nairobi, Kenya.

UNEP, 2011: *Policies, Markets and other Tools to help ensure Sustainability.* United Nations Environment Programme (UNEP), Division of Technology, Industry and Economics, Nairobi, Kenya. www.uneptie.org/energy/bioenergy/issues/policies.htm (accessed 20 October 2010).

UNEP/GRID-Arendal, 2002: *Freshwater withdrawal in agriculture, industry and domestic use*, UNEP/GRID-Arendal Maps and Graphics Library, maps.grida.no/go/graphic/freshwater_withdrawal_in_agriculture_industry_and_domestic_use (accessed 25 October 2010).

UNESCO, 2003: *Water for People, Water for Life: 1st UN World Water Development Report.* United Nations Educational, Scientific and Cultural Organization (UNESCO), Paris, France. www.unesco.org/water/wwap/wwdr/wwdr1/ (accessed 20 October 2010).

UNESCO, 2006: *Water, a shared responsibility – The United Nations World Water Development Report 2*, Map 11.2 National Water Footprints around the World. Chapagain A. K. and Hoekstra A. Y., p.391, © UNESCO-WWAP 2006.

UNESCO, 2009: *Water in a Changing World: 3rd UN World Water Development Report.* United Nations Educational, Scientific and Cultural Organization (UNESCO), Paris, France. www.unesco.org/water/wwap/wwdr/wwdr3/ (accessed 20 October 2010).

UNF, 2008: *Sustainable Bioenergy Deveopment in UEMOA Member Countries.* United Nations Foundation (UNF) Energy and Security Group, International Center for Trade (ICTSD), and The Hub for Rural Development in West and Central Africa. New York, NY, USA.

ÚNICA, 2010: *Sustainability Report.* Uniao da Industria de Cana de Açúcar (ÚNICA), São Paulo, Brazil. 128pp.

UNIDO, 2008: *Bioenergy Strategy: Sustainable Industrial Conversion and Productive Use of Bioenergy.* United Nations Industrial Development Organization (UNIDO), Vienna, Austria.

UN-Water/FAO, 2007: *Coping with water scarcity – Challenge of the twenty-first century.* UN-Water and Food and Agriculture Organization (FAO), Rome, Italy.

US EPA, 2010: *Federal Register*, Vol. 75, No. 58, Part II, 40 CFR Part 80, p.14711, United States Environmental Protection Agency (US EPA), Washington, DC, USA.

van Vuuren, D. P., J. Van Vliet, and E. Stehfest, 2009: Future bio-energy potential under various natural constraints. *Energy Policy*, **37**(11):4220–4230.

Vanwey, L., 2009: Social and distributional impacts of biofuel production. In Biofuels: Environmental Consequences and Interactions with Changing Land Use. In *Proceedings of the Scientific Committee on Problems of the Environment (SCOPE)*. R. W. Howarth and S. Bringezu (eds.), International Biofuels Project

Rapid Assessment, 22–25 September 2008, Gummersbach, Germany. Cornell University, Ithaca, NY, pp.205–214.

von Braun, J., 2007: *The World Food Situation: New Driving Forces and Required Actions*. Food Policy Report, International Food Policy Research Institute (IFPRI), Washington, DC.

Watson, R. T., I. R. Noble, B. Bolin, N. H. Ravindranath, D. J. Verardo, and D. J. Dokken, 2000: *Land Use, Land-Use Change, and Forestry*. A Special Report of the IPCC. Cambridge University Press, Cambridge, UK.

WBGU, 2008: *Welt im Wandel. Zukunftsfähige Bioenergie und nachhaltige Landnutzung.* Wissenschaftlicher Beirat der Bundesregierung Globale Umweltveränderungen (WBGU), Berlin, Germany.

WBGU, 2009: *Kassensturz für den Weltklimavertrag – Der Budgetansatz.* Wissenschaftlicher Beirat der Bundesregierung Globale Umweltveränderungen (WBGU), Berlin, Germany.

WCD, 2000: *The Report of the World Commission on Dams.* World Commission on Dams (WCD), Nairobi, Kenya.

Wicke, B., R. Sikkema, V. Dornburg, A. Faaij, 2011: Exploring land use changes and the role of palm oil production in Indonesia and Malaysia. *Land Use Policy*, **28**(1): 193–206.

Wildenberg, M. Ecology, Rituals and System-Dynamics, 2005: *An attempt to model the Socio-Ecological System of Trinket Island*. [80], 1–185. 2005. IFF Social Ecology. Social Ecology Working Paper, Vienna, Austria.

Wilhelm, W. W., J. M. F. Johnson, D. L. Karlen, and D. T. Lightle, 2007: Corn Stover to Sustain Soil Organic Carbon Further Constrains Biomass Supply. *Agronomy Journal*, **99**:1665–1667.

Wirsenius, S., 2003a: Efficiencies and biomass appropriation of food commodities on global and regional levels. *Agricultural Systems*, **77**(3):219–255.

Wirsenius, S., 2003b: The Biomass Metabolism of the Food System – A Model-Based Survey of the Global and Regional Turnover of Food Biomass. *Journal of Industrial Ecology*, **7**(1):47–80.

WMO, 2006: *Greenhouse gas bulletin: the state of greenhouse gases in atmosphere using global observation up to December 2004.* World Meteorological Organization (WMO), Geneva, Switzerland.

Wolf, A. T., S. B. Yoffe, and M. Giordano, 2003: International waters: identifying basins at risk. *Water Policy*, **5**(1):29–60.

World Bank, 2008: *Global Purchasing Power Parities and Real Expenditures.* World Bank, Washington, DC.

World Bank, 2011: *Program Information Document – Concept Stage.* Minas Gerais Partnership III DPL/PBG, P121590, Development Policy Lending, World Bank, Washington, DC, USA.

World Commission on Environment and Development, 1987: *Our Common Future.* Oxford University Press, Oxford, UK.

Young, A., 1999: Is there Really Spare Land? A Critique of Estimates of Available Cultivable Land in Developing Countries. Environment, *Development and Sustainability*, **1**(1):3–18.

Zah, R., H. Böni, M. Gauch, R. Hischier, M. Lehman and P. Wäger, 2007: *Ökobilanz von Energieprodukten: Ökologische Bewertung von Biotreibstoffen*. EMPA, St. Gallen, Switzerland.

Zhang, Z., L. Lohr, C. Escalante, and M. Wetzstein, 2010: Food versus fuel: What do prices tell us? *Energy Policy*, **38**:445–451.

21

Lifestyles, Well-Being and Energy

Convening Lead Author (CLA)
Joyashree Roy (Jadavpur University, India)

Lead Authors (LA)
Anne-Maree Dowd (Commonwealth Scientific and Industrial Research Organisation, Australia)
Adrian Muller (University of Zurich & Swiss Federal Institute of Technology Zurich, Switzerland)
Shamik Pal (Institute of Engineering & Management, India)
Ndola Prata (University of California, Berkeley, USA)

Review Editor
Sylvie Lemmet (United Nations Environment Programme)

Contents

Executive Summary

One of the objectives of the Global Energy Assessment (GEA) is to assess the means through which the potential negative economic, social and environmental impacts from energy use can be mitigated or eliminated, either by increasing the efficiency of energy use or by switching to primary energy sources and carriers. A large set of factors influence ultimate energy use beyond those related to income and affluence. These include non-economic and non-technological drivers such as behavior, lifestyle, culture, religion and the desire for improved well-being.

This chapter focuses on these underlying drivers and explores how they could influence energy use and choice of energy sources while maintaining desired levels of affluence or income. It reviews the factors that determine how socio-economic indicators of affluence and other non-technological drivers may translate into demand for energy services (for definition of energy services, see Chapter 1) and at the interventions, policies and measures (such as taxes, infrastructure, building codes, and access to information) that could modify or change lifestyles and preferences.

In addition to the consumption of goods and services and their quality, the chapter also focuses on two elements of lifestyle choices that have significant implications for energy use: diet and mobility (household energy use, another key element, is discussed in Chapter 10 while transport is discussed in Chapter 9). In general, income is often a common driver across these choices. However, modest decoupling between income and energy use can be observed at the aggregate level in many jurisdictions and time periods, by which energy use increases at a lower rate than income. This can be observed for many industrial countries but may not yield the same outcome in most developing countries for several reasons, especially when total energy accounts for significantly inefficient non-commercial energy use. It is therefore useful to look beyond income alone and explore underlying lifestyle choices and their energy implications. Behavioral change requires both knowledge contributions to change attitudes and policy implementation to provide incentives for action.

Concepts of well-being have important implications on how energy services demand and energy use are assessed. Applications of economics conventionally assume that knowledge about resource scarcity is reasonably good and reasonably widely understood. More complex notions of well-being are better able to take other factors into account even in situations of decreasing energy services.

Notions of well-being that are based only on material consumption of goods and services implicitly assume that resources needed for production of these goods and services are abundant and that either a technology can be invented to make them more productive or that new energy resources can always be found to replace depleting resources. This underlying assumption of high substitutability between different resources fails to capture more complex, multidimensional notions and differing characteristics of energy services that may actually hinder substitutability. This is of particular importance when addressing basic needs for a decent good life and the energy services demand to meet those. The assumption of substitutability is important, as wrong assumptions about it can, for example, lead to overestimations of projected well-being when the aggregate quantity of energy use is considered but not its qualitative composition.

The potential for increased sustainability that lies in strong decoupling of well-being from material consumption can be captured by notions of well-being that are not based on material consumption alone, as reflected in the gross domestic product (GDP) per capita indicator. This offers new policy options of potentially high leverage. A reduction in energy services demand does not necessarily reduce well-being and lifestyle changes can deliver some win-win options (e.g., walking and cycling) that reduce demand for energy services with the same or even improved levels of personal health and social well-being.

Life style changes are an effective and powerful approach to addressing sustainability issues, as they can provide multiple benefits like improved health, low fossil fuel based mobility, lower emissions, and nutrition, without reducing socio-economic status.

21.1   Introduction

A popular framework that can be used to understand the link between energy systems to emissions and the environment is the I=PAT framework detailed in equation below:

Impact (e.g., from CO_2 emissions) = Population × Affluence (GDP/capita) × Technology (kWh/GDP) × Emissions (e.g., from tCO_2/kWh)

Total energy use and its impact are magnified by population and affluence and are influenced by technology. They link emissions and environmental impacts from energy systems to the underlying drivers of population and affluence and the elements of energy systems – that is, the primary energy sources and supply-side and end-use conversion technologies required to meet demands for energy services (Ehrlich and Holdren, 1971; Hubacek et al., 2007).

In general, personal disposable income is often a common indicator for affluence and energy service demand. However, it can be observed that the link between income and energy use is weakening at the aggregate level in many jurisdictions and time periods. Weakening of the income-energy coupling means that energy use increases at a lower rate than income. While real global gross domestic product (GDP) increased by 1.6 times from 1990 to 2007, for example, total primary energy use increased by 1.4 times. The modest trend towards decoupling can be observed for many industrial countries but may not yield the same outcome in most developing countries for several reasons. It is therefore useful to look beyond income alone and explore underlying lifestyle choices and their energy implications. GEA is concerned with the means through which emissions and other negative impacts may be decoupled from affluence or income, either by increasing the efficiency of energy use or by switching to primary energy sources that have a reduced environmental footprint. Nevertheless, a large set of other factors influence ultimate energy use beyond these two key ones, including non-economic and non-technological drivers such as lifestyle, culture, religion, desire for improved well-being and behavior.

Conventional economic models assume that preferences are determined individually and do not change. This chapter focuses on the many underlying drivers of preference and explores how lifestyle choices can influence energy use while maintaining desired levels of well-being. It looks at the factors that determine how indicators of well-being and other non-technological drivers may translate into demand for energy services and at the interventions and policies that could modify or change lifestyles and preferences.

In addition, the chapter explores a more fundamental question of whether defining affluence in purely economic terms is an appropriate measure for well-being. Thus it explains the individual and social goals underlying energy use. To the extent that affluence or income is only one of the determinants of well-being, there may be further opportunities to decouple energy use from human well-being. The chapter includes a brief review of the emerging literature on well-being and on indicators and metrics that go beyond affluence in conceptualizing development. Thus, this chapter is about the potential for increased sustainability through changes in lifestyles and the resulting change in demand for energy services.

By focusing on consumers, a new perspective on energy system management is added that recognizes the importance of the psychological, sociological and cultural determinants of demand. Such an approach thus augments and complements the more common technological and conventional economic approaches used in the analysis of energy supply and end-use.

A focus on the consumption basket also allows the exploration of options for replacing one energy service with another. For example, in telecommuting the demand for mobility may be replaced by an increased use of electricity for information and communication appliances, perhaps leading to a reduced use of primary energy. In this view, the elements of the consumption basket of energy services are not independent, but their levels are assessed jointly based on relative costs and benefits given the prices, income, tastes and preferences, and social fabric within which a consumer functions.

21.2   Lifestyles and Energy Services Demand

Often the individual demand for energy services is expressed or assessed only at larger levels of aggregation, such as the household, community, or region. Converting to individual demand is only possible via averages (e.g., per capita average energy use to provide street lighting in a village). It is also clear that not all energy service use can be traced back to private individual consumption decisions (e.g., public/governmental provision of infrastructure services such as street lighting).

Consumer activities influencing directly the energy use (e.g., housing and private transport) account for more than 43.0% of total energy use (see Chapters 9 and 10, as well as Chapter 1, Table 1.2). Figure 21.1 shows the energy use of US citizens in 1997 for different energy services when accounted for from a life cycle perspective. Since many of the individual decisions will entail indirect energy uses, life cycle analysis can shed light on important use categories that are not directly accounted for or paid for by individuals. Figure 21.1 indicates consumer activities influencing directly energy use accounts for a lesser share in total energy compared to factors influencing indirectly the energy use. Housing operation uses the most energy among all consumer activities which is mainly contributed by the consumption of utilities (electricity/natural gas/fuel oil and other fuels/telephone, and water and other public services). For direct influences, personal travel is the most energy intensive, much of which is caused by short distance travel by automobiles and trucks.

Understanding the linkages between lifestyles and energy services demand offers insight into how changes in lifestyles may contribute to

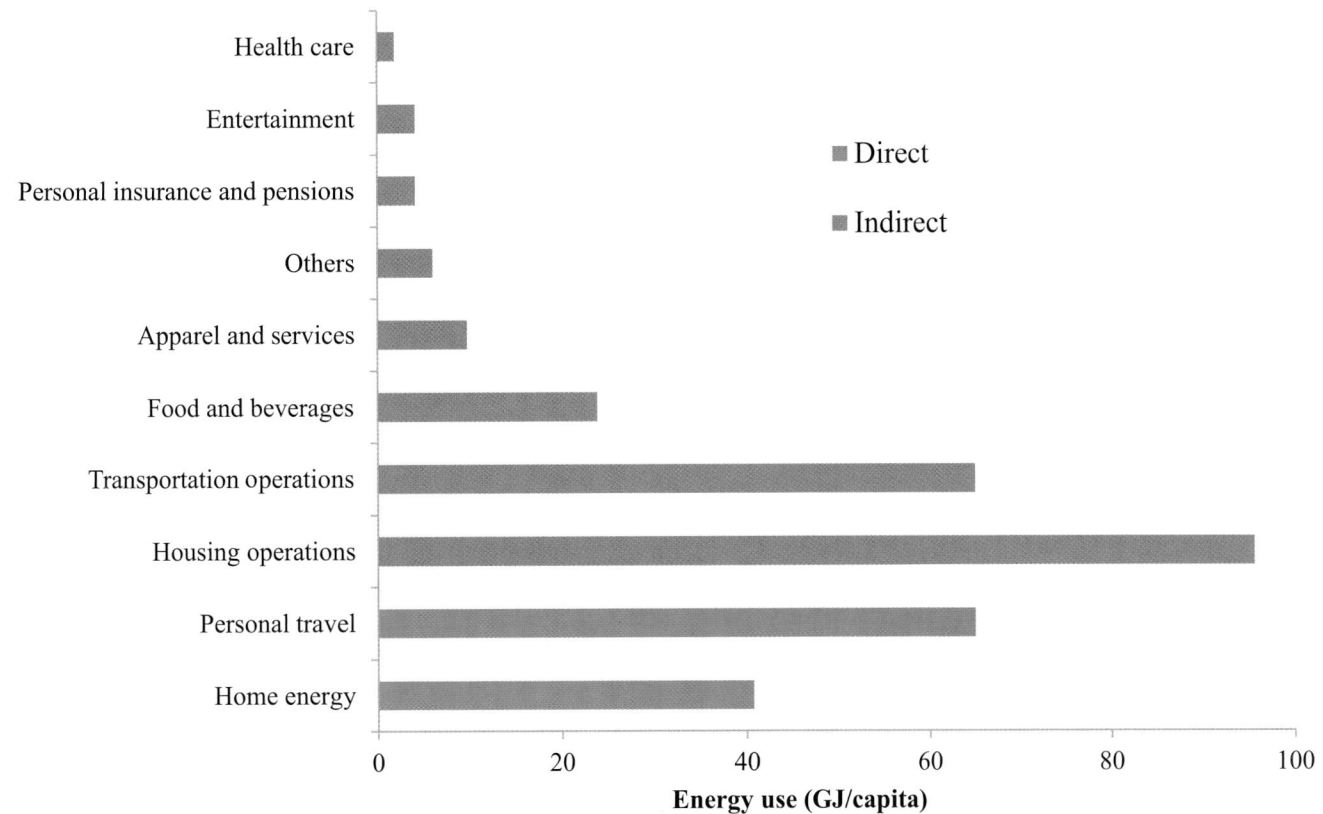

Figure 21.1 | Direct and indirect energy use by consumer activities in 1997. Source: Bin and Dowlatabadi, 2005.
Note: Home energy includes: space heating, other appliances and lighting, water heating, refrigeration and air conditioning. Transport operations include: vehicle purchases (net outlay) (cars and trucks, new/cars and trucks, used) gasoline and motor oil other vehicle expenses (vehicle finance charges, maintenance and repair, vehicle insurance, rent, lease, licenses, public transportation etc.). Housing operations include: shelter, utilities, and public services (e.g., electricity, natural gas, fuel oil and other fuels, telephone, water etc.) as well as housekeeping supplies, household furnishings and equipment (e.g., household textiles, furniture, floor coverings, major appliances, small appliances as well as miscellaneous houseware and household equipment). Personal travel includes: short and long distance travel by automobiles, trucks, air, etc.

reducing total energy use and is all the more relevant in today's internationally connected world through trade relations. Here, it is crucial to introduce the notion of well-being as the ultimate goal by which to judge supply-side or policy interventions in improving human well-being or sustaining a high level of well-being in a long-term, sustainable development context. Changes in lifestyles and energy services demand need to be assessed in relation to alternative concepts of well-being than those directly related to high economic attainments and spending. Referring to well-being also means addressing not only absolute reductions in energy services, which might be an issue for high-income consumers and regions, but also absolute increases in energy services, which are still necessary for a huge percentage of the population in developing countries.

Consider the case of private car mobility. The same level of the service (expressed as passenger-km per year or per capita km driven per year) may be associated with different levels of energy use, based on factors such as vehicle efficiency, age distribution of the vehicle stock, driving characteristics, and so on. Consequently, interventions that improve efficiency allow the reduction of energy use while maintaining the same level of consumption of the service.

From a policy perspective, however, it is not just the absolute level of the service that is important but rather the generation of utility from the consumption of the service or the production of goods and services where mobility serves as a factor input. Interventions that allow the same level of utility or welfare for reduced levels of mobility are also important from an energy and impact perspective. Such modifications of demand may often involve behavioral or lifestyle choices such as walking or cycling instead of driving for short shopping trips. At the same time, it is important to recognize that such choices may have benefits quite independent of their energy or environmental outcomes – for example, the positive effect on individual health due to increased physical activity. This example refers to industrial countries, however; the situation could be different in developing countries, where absolute levels of motorized personal mobility are very low and therefore an increase would clearly improve well-being.

Addressing changes in mobility services and technologies or in an energy carrier in a lifestyle context sharpens the vision for the full range of possible options for intervention, as the change in mobility and increase in well-being need not necessarily be based on private cars. Furthermore, an energy services and lifestyle approach emphasizes the fact that mobility can be framed differently, namely via "accessibility" of certain goods and services (see also Chapter 9), which further broadens the set of possible options for action to improve well-being, in particular if the need to gain access to these goods and services is a determinant of well-being.

Consumer energy services demand is determined by needs and preferences for multiple services, which in turn depend on, in addition to income level and prices, home country characteristics, dwelling area and type, job and leisure activities, diet preferences, cultural context, religion, etc. A certain lifestyle is characterized by a bundle of these determinants combined with a more or less explicitly framed worldview, a set of values and convictions, preferences, and behaviors. It is embedded in some social context of identity and meaning. Thus, lifestyle can be considered as an organizing concept, making explicit that human beings live a life well beyond the mono-dimensional life of an economic agent responding only to income and price variables.

The importance of characterizing lifestyles is brought out in studies by Christensen (1997) and Alfredsson (2004) of households in the United States and Sweden. Assuming floor area of the dwelling as a proxy for overall affluence levels, for families of the same size and with similar floor areas, Christensen (1997) showed the wide variation in final energy use due to choices on mobility, diet, and space heating. For example, people living near their workplace or commuting by public transport and with a dietary choice of less than the US-average amount of meat and using renewable energy-based heating systems could reduce energy use by 88% compared to those travelling long distance and eating meat more than average level, while using fossil fuel based systems. Diet and mobility choices matter not only for energy use, but also for disease burden and public health. Particular dietary choices, e.g., above average meat consumption, have been associated with higher incidence of many diseases (Mann, 2000; Fung et al., 2004; Walker et al., 2005; Caspari et al., 2009; Sinha et al., 2009). Findings from Edwards and Tsouros (2006) identify an ongoing decline in physical activity across all age groups during the past several decades. This is largely due to the mechanization of work and daily tasks, the increased use of cars, increases in sedentary work, the use of labor-saving devices, and an increase in leisure pursuits not involving physical activity (such as watching television and using a computer). Physical inactivity causes an estimated 600,000 deaths per year in Europe and leads to a loss of 5.3 million years of healthy life expectancy per year due to premature mortality and disability (Edwards and Tsouros, 2006).

The Swedish study (Alfredsson, 2004) also showed that dietary choice, mode of transport[1], fuel type, and appliance choice for indoor comfort – all changes in pattern of consumption – generally reduce energy usage though may not reduce level of energy use significantly. If energy use reduction of the full green consumption basket is taken, it turns out to be less than the added value of reductions that can be estimated from each of the green end-use-specific lifestyle changes. This is because some benefits of a partial greening of lifestyle (such as dietary choice) by end-use type is taken back by the rebound effect of an enhanced level of energy services (such as mobility) of another end-use activity or efficient mobility service may be taken back by changing lifestyle in favor of international tourism.[2] However, studies have shown that this take back effect is high in societies with more unmet demand (Roy, 2000).

Assessments of lifestyles often focus on consumption and leisure time activities and how they contribute to creating identity and meaning (Geißler, 2002). The concept of lifestyle is thus also studied in connection with "social class" and status (Wind and Green, 1974; Sobel, 1981). Social dimensions and individual lifestyles are not disjoint; lifestyles are partly the consequence of deliberate individual choices and partly determined by social contexts as well as physical and economic boundary conditions (Harrison and Davies, 1998). These lifestyle types may often be grouped along the two dimensions of "social status" (lower, middle, higher) and "basic values."

A large amount of literature assesses lifestyle as a matter of combining different activities defining consumption patterns without regard to values, convictions, and social context (Herendeen and Tanaka, 1976; Rees, 1995; Vringer and Blok, 1995; Daly, 1996; Duchin, 1998; Biesiot and Noorman, 1999; Perrels and Weber, 2000; Pachauri and Spreng, 2002; Lenzen et al., 2004; Cohen et al., 2005). There, lifestyle categories are, for example, defined by expenditure types (Minx et al., 2009) and or through correlation in a narrow operational sense between level and pattern of consumption and socioeconomic parameters such as age, sex, gender, education, occupation, or income.

Households are an important target group in any energy conservation discussions. A study involving over 300 households in the Netherlands (Benders et al., 2006) indicated that direct energy needs – energy for space heating, electricity, and motor fuel – along with indirect energy requirements, which is defined as that needed for production, distribution, and waste disposal of consumer goods and services (e.g., production of food), amounted to as much as 80% of total energy flows. But the relationship between affluence, household characteristics, and energy use is complex. Household energy use and embodied energy through consumption (building energy use aspects) are discussed in Chapter 10.

---

1   Tourism is growing rapidly in Sweden with very distant places being increasingly popular. A Swedish family's carbon budget is totally dominated by a trip to one of these distant places.

2   The rebound effect is discussed in detail in Chapter 22.

The remainder of this section describes in greater detail two other elements of lifestyle choices that have significant implications for energy use: diet and mobility.

21.2.1   Diet

Dietary choices and delivery systems can lead to considerable variations in energy service demand and resultant energy intensity of food baskets. In high income countries food accounts for 17–18% of indirect household energy use (Vringer and Blok, 1995; Reinders et al., 2003; Cohen et al., 2005).

While satisfying the nutritional needs of humans accounts for a significant portion of total primary energy demand, the individual energy components of the food supply chain that fall in each sector are relatively small, such as cooking and refrigeration (the most important energy end-uses in buildings related to satisfaction of nutritional needs, along with the provision of hot water), and represent only a few percent of building energy use. Food systems cross the boundaries of many sectors and thus affect many chapters in the GEA: buildings (Chapter 10), transport (Chapter 9), and industry (Chapter 8), as well as rural issues (Chapter 19). Therefore it is especially important to consider energy services related to nutrition. Dietary choices depend on habits but also reflect acquired tastes over time. So net benefits and costs assessment of a dietary choice will vary over time and will change with changing acquired tastes. This section examines the implication of dietary choices (including cooking) for total energy used for the provision of nutrition. In low per capita income countries as per capita income rises, consumers will shift some consumption away from lower value cereals to higher value livestock products. In developed countries, where incomes and livestock product consumption are already high, consumers are expected to make relatively small adjustments between food consumption groups with changes in income levels (Cranfield et al 1998).

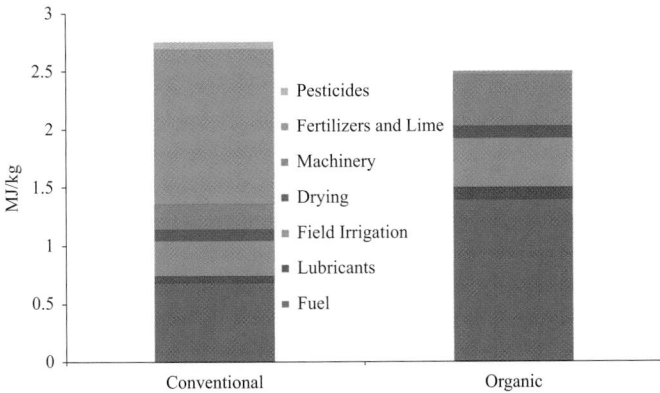

Figure 21.2 | Structure of energy inputs for conventional and organic farming systems. Source: Based on Jorgensen et al., 2005.

The total amount of energy required to provide human nutrition consists of the energy needed for agricultural production; land use and land cover change; the energy embodied in the raw food materials themselves and in the agrochemicals; the energy needed to transport the food components, water and equipment needed for food production; the industrial energy use related to food processing; and the energy used for refrigeration and cooking.

The embodied energy in the raw materials used for food directly correlates with the structure and nature of the daily diet. For instance, producing 1 kcal of grain and animal proteins requires about 2.2 kcal and 25.0 kcal of fossil energy, respectively (Pimentel et al., 2003). Thus the difference between the energy inputs for plant- and meat-based meals may exceed a factor of 10. Furthermore, the food produced by conventional agricultural methods requires a total higher energy input than organically produced food. Conventional agriculture production utilizes more overall energy than organic systems due to heavy reliance on energy intensive fertilizers, chemicals, and concentrated feed, which organic farmers forego (Ziesemer 2007). Figure 21.2 shows in some specific crops and practices in some countries in organic farming direct fossil fuel use might be higher because of relative high machine use for weeding and animal manure spreading compared to pesticides and synthetic fertilizers in conventional practices but overall energy use is less in organic farming. But in the same study (Jorgensen et al., 2005) it is shown indirect energy use is higher in conventional farming because of synthetic nitrogen fertilizer use. It is not only food production but also post-harvest practices and food delivery in which energy use needs to be examined. Both agricultural systems use separate but parallel systems, and only a few studies include data on packaging, delivery and so on. Studies (e.g., Bertilsson et al., 2008) have shown that choice of data, country context and their representation, accounting for energy use can misrepresent fuel use data across conventional and organic systems. A diet consisting mainly of locally produced, seasonal foodstuffs is significantly less energy-intensive than one based on "globalized" (and thus shipped) ingredients or produced in greenhouses. Thus an equally nutritious but plant-based diet using local, seasonal ingredients may require less total life cycle energy input than a meat-based diet using globally produced and not necessarily seasonal ingredients.

The literature on the impact of a dietary choices on energy and carbon footprint (such as Coley et al., 1998; Lenzen, 1998; Kramer et al., 1999; Carlsson-Kayama et al., 2003; Wahlander, 2004; Eshel and Martin, 2005; Baroni et al., 2006; Wrieden et al., 2007; Collins and Fairchild, 2007) focuses on the link between nutritional choices and energy use as well as ecological footprints of different diet types, mostly in industrialized countries. A key topic is the issue of meat consumption.

For many consumers, meat consumption is considered as a superior good – one that increases with rising income levels. People in

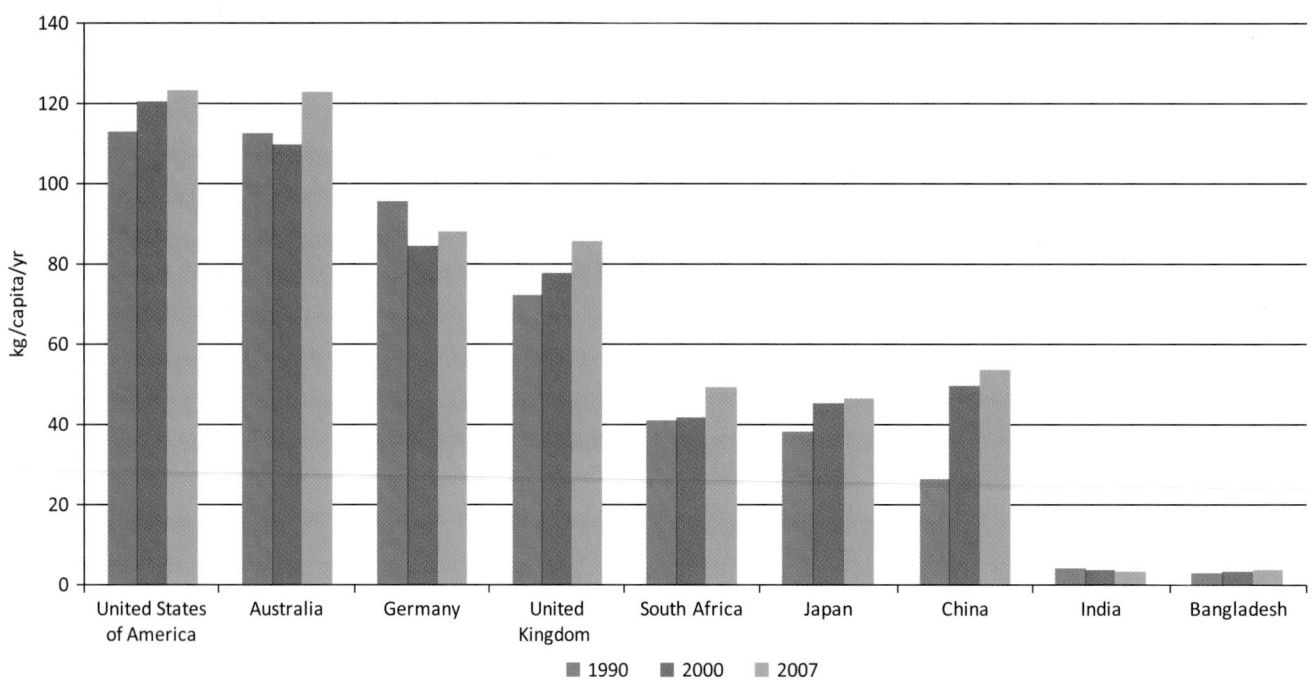

Figure 21.3 | Per capita meat consumption across regions. Source: Prepared using data from FAOSTAT, 2008.

low-income countries spend a greater portion of their budget on food, and when their income rises they increase their expenditures on different food items to a greater extent than people in wealthier countries, with the greatest increase being on higher-value food items such as dairy and meat (Regmi et al., 2001). The amount of food and foodstuffs that is not consumed but is thrown away or wasted through food preparation also increases with income, and this represents an increasing amount of embodied energy and other environmental pressures.

Population growth along with economic development has increased purchasing power, causing a demand not only for more food but also for different food varieties. Studies on human nutrition worldwide have indicated a nutrition transition toward more-affluent food consumption patterns (Gerbens-Leenes et al., 2010). Meat consumption is increasing globally, of which approximately 24% is beef. Beef consumption increases with income (Fiala, 2006). Per capita meat consumption varies widely across regions currently (see Figure 21.3) but is expected to rise with urbanization and income increases. FAO statistics show beef consumption has also the same pattern across regions. In China, while consumption change in proportion to income, change is lower for traditional pig meat consumption (0.15), for beef rate of change in consumption is more than the change in income (1.56) and for goat and poultry meat it is near proportional 0.88 to 1.05 (Masuda and Goldsmith, 2010). Proportional change would occur when value of ratio of change is equal to one i.e., rate of change in consumption is the same as rate of change in income. Urbanization

and per capita income growth have contributed to these. In India a 1994 national food survey in 32 cities indicated that 74% of urban households were non-vegetarian and that meat consumption rose in 1980–90 at 4–8% (Landes et al., 2004). In Mexico, chicken consumption is expected to rise more than 64% in the next two decades. While household consumption of chicken meat for the higher-income group can be expected to rise by 0.18% for 1% change in total household expenditure, for the lowest ten percent of households arranged in order of expenditure the change in chicken meat can be as high as 1.65% for 1% change in total household expenditure (Salazar et al., 2005). Total meat consumption in the United States is the highest and, as in most of the world (except in Germany), has been growing at a steady rate (however, at lesser rate than, e.g., United Kingdom) for a number of years.

The specific energy use (SEU) in processed meat products is very high (Table 21.1). Ruminant meat production uses about 25 kg of plant protein to produce 1 kg protein as meat, whereas for pigs this relation is about 10 kg for 1 kg meat and for poultry is around 5 kg. This is in particular the case if animals are fed with concentrate feed and are not reared in pasture-based systems; however, the latter is an option for ruminants only but not for pigs and poultry (Smil, 2002). Over the past few decades, energy needs for transport, storage, and processing has increased, especially for meat products. A study in France, the United Kingdom, the Netherlands, and Germany indicated a significant increase (14–48%) in the energy use per tonne of product over the last 15 years (Ramírez et al., 2006).

Table 21.1 | Direct specific energy use and emissions for various processed meat products in select European countries.

Product	whole and chilled (MJ/t dress carcass weight)* SEU	whole and frozen (MJ/t dress carcass weight)* SEU	cut up, deboned, and chilled(MJ/t dress carcass weight)* SEU	cut up, deboned, and frozen (MJ/t dress carcass weight)* SEU	kgCO$_2$-eq/kg of meat type**
Beef, veal, and sheep	1390	2110	2146	2866	34.6–17.4
Pork	2093	3128	2849	3884	6.35
Poultry	3096	4258–5518	3852	5014–6274	4.57

Source: *Ramírez et al., 2006; ** Srivastava, 2008.

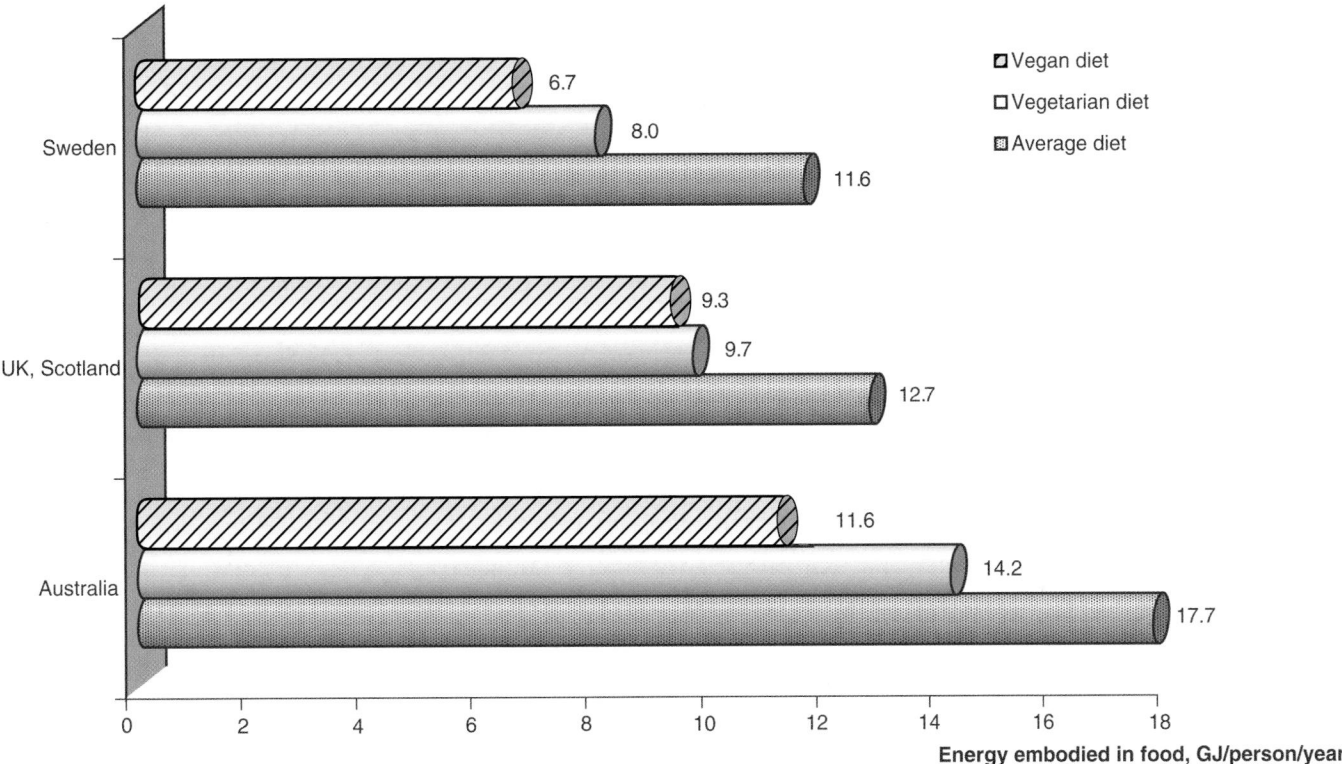

Figure 21.4 | Comparison of embodied energy in average, vegetarian, and vegan diets in Sweden (1999), Scotland (1995), and Australia (1994). Calories of fish, meat, and other animal products consumed in the average diets were replaced by a proportional increase in consumption of other food products in the vegetarian and vegan diets. Source: Coley et al., 1998; Lenzen, 1998; Wallen et al., 2004; Wrieden et al., 2007.

Figure 21.4 compares embodied energy in average, vegetarian, and vegan diets characterized with the same caloric value for Sweden, Scotland, and Australia. Embodied energy is sum total of energy use from cradle to grave of a product, i.e., energy used directly for production, transportation, marketing, as well as disposal and dismantling. There is a significant difference between energy inputs for these three diets. The difference between average and vegan diets equals 19.6 GJ/yr for a four-person family in Sweden, 24.4 GJ/yr for a similar family in Australia, and 13.8 GJ/yr in the United Kingdom – that is about 5.44 MWh/yr, 6.19 MWh/yr, and 3.02 MWh/yr respectively.

These figures are similar to the average annual electricity use per household in these countries. However, the figure also attests that other determining factors that are related to the location may influence total nutritional energy use more than the dietary choice alone. For instance, a vegan diet in Australia is associated with as much life-cycle energy use as a standard Swedish diet.

The embodied transport energy in food is determined by shipping distances, transport mode, and vehicle efficiency (Saunders, 2008). With increased globalization, the length and complexity of food supply chains

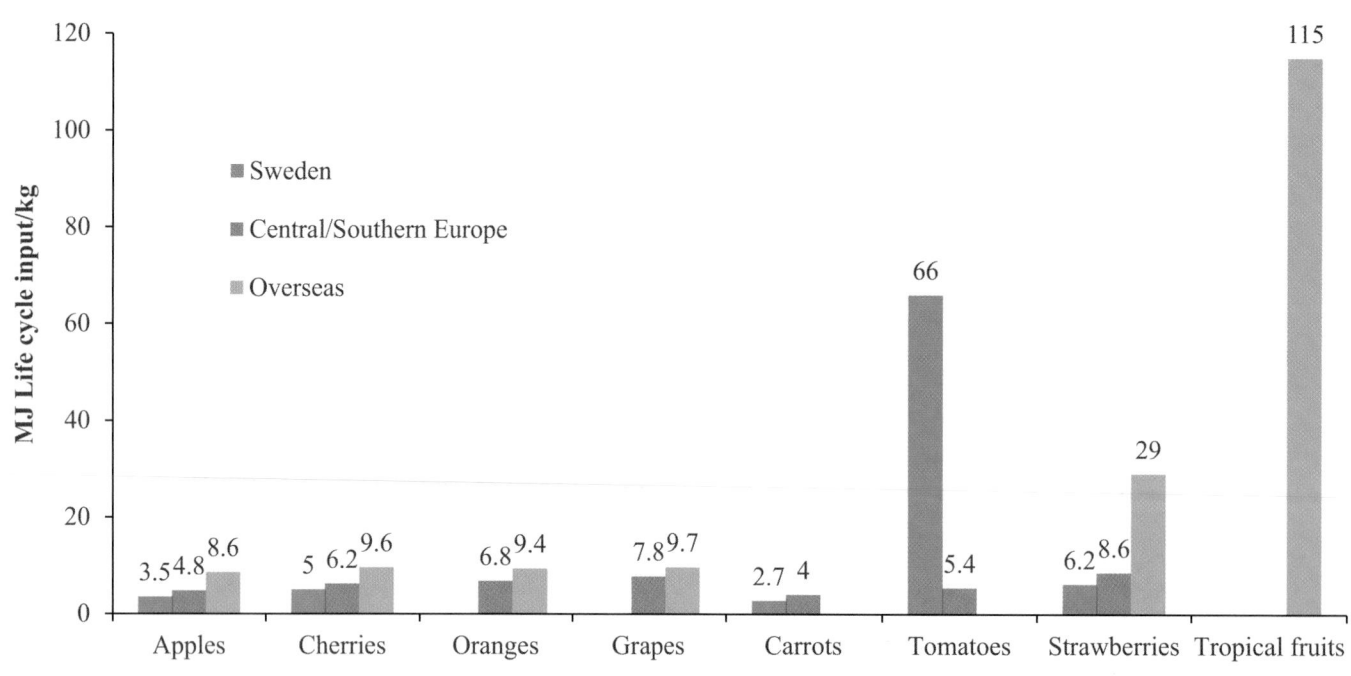

Figure 21.5 | Life-cycle energy inputs associated with local, regional, and overseas sources of food production for Sweden in 1999. Note: Tomatoes produced in Sweden were grown in a greenhouse while those in Southern Europe were open-grown; strawberries grown in the Middle East and tropical fruits were air-freighted. Source: Carlsson-Kanyama et al., 2003, in Saunders, 2008.

keeps growing (EC, 1999; Garnett, 2003), which results in increasing "food miles." Comparison with Table 21.1 suggests that two-thirds of total energy use is indirect.

Figure 21.5 compares the life-cycle energy input of fresh vegetable and fruits produced locally (in Sweden), regionally (in the European Union), and overseas. The figure illustrates that locally produced food is less energy-intensive than food shipped from a distance: the life-cycle energy input of Swedish produce differs from the overseas fruits and berries' input by factors of two (for apples, cherries) and five (for strawberries). Fresh food produced in a different country of the same region apparently requires a less transport-intensive supply system. For instance, Eastern or Southern European apples, cherries, and strawberries brought to Sweden require 1.2–1.4 times higher energy input than analogous Swedish products. In addition, it is interesting to compare the energy intensity of different production methods: for example, Swedish tomatoes grown in a greenhouse result in 66 MJ/kg of life-cycle energy input per kg, while open-grown Southern European tomatoes take 5.4 MJ/kg (Saunders, 2008), including the shipping energy to Sweden.

Another important driver of increased energy use in food production is the decrease of seasonally restricted consumption and the corresponding increasing demand for heated greenhouses or transport services. Increasing cooling demand, the globalization of the food system with corresponding transport distances, and the growing importance of processed convenience food and eating out are also important drivers.

21.2.2 Mobility[3]

The energy implications of lifestyle choice are closely related to certain mobility choice patterns. Vehicular ownership does not always mean high energy use, as the latter depends on mobility. With tourism demand rising, long distance mobility patterns are also increasing, especially air travel. The vehicle population in Beijing and Shanghai is about one-tenth that of Tokyo, while their total fuel use is one-third to one-half as high because of lower fuel economy and more miles driven. Both these components are very high in the United States.

Mobility per se is not always a goal in itself. It is about accessibility to various activities (home, work) and also for some to go places for individual recreation, or to the sources of services and products (medical appointments, shopping, etc.; see also Chapter 9). Passenger-kilometers traveled are increasing worldwide, with the United States alone accounting for 65% of total global passenger-kilometers traveled.

---

3   Please see Chapters 9 and 18 for a more detailed discussion.

Individual and household decisions on the choice of private space, suburban or city life, income, recreational demand, social status, conspicuous consumption, and so on all determine the level of mobility services demanded and the mode of transport chosen. Social theorists have also described how the disappearance of norms and economic limitations has gradually removed the restriction of individual choices and made "individual lifestyles" an appropriate way to describe differences in worldviews and consumption (Giddens, 1996; Beck, 1997; Jensen, 2008; Bauman, 1998). In societies with more value for individual space and freedom, private car mobility becomes of primary importance. The number of passenger cars is still increasing in industrial countries (World Bank, various years).

Despite increasingly more efficient engines and some attempts to promote smaller and lighter cars, motor power and the number of electronic appliances in automobiles continue to increase – with correspondingly increasing energy demand (Chapter 9). This is partly due to the aggressive marketing strategy of automobile production and distribution companies. Similar developments of increased energy services can be observed in public transport.

People in many countries will continue to depend on private cars for a long time. In countries such as the United States, widely spread-out urban living areas with low population density and inadequate public transport infrastructure necessitate private car mobility. Good practices in human settlement design – more densely populated areas; bicycle lanes; bans on parking in overcrowded areas, etc. – could counteract this development, but these are largely not replicated in newly growing and highly populated urban areas in developing countries. Furthermore, the private car seems convenient, as it is perceived to allow maximal personal freedom. So drivers systematically disregard the inconvenience of congestion and the value of using their time for other activities while they are on private transport.

21.3   Determinants of Lifestyle Choices

In many health disorders such as obesity, lifestyle choice itself is considered an extrinsic factor, as opposed to genetic factors which are intrinsic factors. Being an extrinsic factor, lifestyle choice is often times considered to be a decision variable and amenable to change. However, an in-depth analysis shows lifestyle choice itself gets determined by a host of other factors, some of which can be changed by individual decision (e.g., choice among available diet options, acquired dietary pattern) and some needs more landscape level intervention (e.g. culture, value, social norms) or macro policy (e.g., infrastructure design) intervention. Homogeneity in infrastructure and human settlement design has led to a convergence in energy service demand across various cultures and geography. Infrastructure design choices are top-down decisions and are beyond individuals. In many energy-supply system designs, the management level of

consumer demand is taken as a given based on the assumption that preferences have an intrinsic character. But a lifestyle and consumption-based approach takes into consideration various extrinsic drivers for preference which changes over time. Consumer demand for energy is determined by, among other things, infrastructure design as well. For example, it is impossible to sit on a veranda in the cool of the evening if a mechanically air-conditioned home is built without one (Shove, 2003). Individuals cannot choose to switch off their own heating or cooling services if a building is centrally heated and air-conditioned.

A number of economic (e.g., the market price for land) and institutional factors (e.g., zoning) influence land use. Factor often seeming to be left out of individual decisions are human health and environmental co-benefits. Traditional walking, bicycling, jogging, and natural green spaces are getting taken over by energy-guzzling "modern" energy-intensive health services and highly irrigated green spaces; small traditional retail stores are getting replaced by air-conditioned shopping malls. Approaches based on supply-side economics or partial end-use activities fail to take into consideration the lock-in effect of infrastructure design, which constrains demand at very high levels and leaves no flexibility to generate behavioral responses. People's choice not to switch off standby power to reduce phantom load (Roy and Pal, 2009), for instance, may be due to lack of easy access to power switches in a house.

A variety of internal and external factors influence consumer choice in areas of importance for energy use. There is, for example, evidence that religion influences consumer attitudes and behavior in general (Delener, 1994; Pettinger et al., 2004). A number of studies demonstrate that religion influences eating habits (Mennell et al., 1992; Steenkamp, 1993; Steptoe and Pollard, 1995; Shatenstein and Ghadirian, 1998; Asp 1999; Mullen et al., 2000; Blackwell et al., 2001). In many societies, religion even plays one of the most influential roles in food choice (Musaiger, 1993; Dindyal, et al., 2004).

The impact of religion on food consumption obviously depends on the religion itself. Several religions forbid certain food, such as pork and meat that has not been ritually slaughtered in Judaism and Islam, or pork and beef in Hinduism and Buddhism (Sack, 2001). Although religions may impose strict dietary laws, the number of people following them may vary considerably. For instance, it is estimated that 90% of Buddhists and Hindus (Dindyal, 2003) and 75% of Muslims compared with only 16% of Jews in the United States strictly follow their religious dietary laws (Hussaini, 1993). Due to the largely vegetarian diet of Hindus, changes in these percentages can be of considerable relevance for energy demand as energy input difference between plant- and meat-based meals may exceed a factor of 10 (see Section 21.2.1). Evidence shows that embodied energy use in an average diet which contains both vegetarian and non-vegetarian items can be almost 1.3 to 1.5 times higher than a vegetarian diet (see Figure 21.4). Differences in adherence to religious dietary prescription pertain, among other factors, to social structures, such as origin, immigration, and generation differences (Limage, 2000; Saint-Blancat, 2004; Ababou, 2005). At the same time, factors such as income convergence, markets, technology, media, and trade are leading to a homogenization in dietary patterns, thereby indicating a trend toward higher energy demand that may not be necessarily healthy.

Faiers et al. (2007) list the relevant theories and models that have been developed to explain these factors, building on a review by Jackson (2005). Further insights into the significance of these factors for consumer purchases of "green" products is provided by Ozaki and Sevastyanova (2011), who analyzed the purchase motivations of UK buyers of the Toyota Prius hybrid. They suggest that the various motivational factors could be grouped into five clusters: financial benefits; hybrid cars as exemplars of "environmentalism;" compliance with the norms of the community; attractiveness of new technology; and independence from oil producers by reducing petrol use. While their study underscores the importance of financial factors, the effects of social pressure and the aesthetic, experiential, and practical values associated with hybrid cars were also found to be important. This multidimensionality of purchase motivations highlights the importance of social, cultural, and perceptual factors in lifestyle choices, in addition to purely economic (or benefit-cost) calculations.

A cross-cultural analysis of household energy use behavior in Japan and Norway found certain similarities and differences in energy use patterns (Wilhite et al., 1996). People in both countries have good information about how much energy goes where in the home, but they exhibit an almost total lack of interest in energy efficiency and concern about the environment when shopping for appliances. In Japan, the bathing routine is extremely important to the Japanese lifestyle and at the same time very energy-intensive. Norwegians heat most of their living area most of the time, while the Japanese traditionally heat only where they are in the home and when they are there. Part of the explanation for this is culture and part is climate. When it is very cold outside, as it often is in the Norwegian winter, it is both physically and psychologically comforting to have it very warm inside (Wilhite et al., 1996).

Lyons et al. (2007) maintain that Ireland's consumption has not developed as in other European countries with increasing or high incomes but instead remains similar to that of Greece. In essence, as the rise in wealth and incomes has been very quick, consumers have not yet adjusted their spending habits and there is still room for convergence in consumption patterns. Although the Irish no longer have low incomes, they still behave, to a certain extent, as if they had.

Of particular relevance for energy services demand is the rural-urban divide especially in developing world. On average, rural incomes and expenditure levels are significantly lower than in urban areas. While a rural household's consumption pattern is biased toward food items, an urban household's expenditures have bigger shares in services (Ojha et al., 2008). Rural and urban lifestyles are influenced by the huge diversity in settlement design, infrastructure availability, and service provisions. Rural areas are more often served by decentralized systems and generally exhibit less motorized mobility and fewer modern amenities. Unless the energy-saving potential of lifestyle change is accounted for, homogenization in urbanization trends and human settlement design in newly developing areas will end up creating high-energy-using spaces. So infrastructure design needs to be diversified as well as culture-specific.

Identifying consumer preferences is a key challenge facing manufacturers and service providers. The media has been used as a powerful mass-communication tool for creating awareness and shaping public opinions. Two glaring examples of media effect on lifestyles are the successes of anti-smoking campaigns and campaigns, true across countries today, to use seat belts in cars. Wakefield et al. (1998) conclude that media can shape and reflect societal values.

21.4 Influencing Preferences and Consumer Behavior

Dietz et al. (2009) have estimated that behavioral changes involving the adoption and altered use of currently available in-home and personal transportation technologies in the United States could reduce carbon emissions by as much as 7.4%. This would require a range of nonregulatory interventions at multiple levels (individuals, communities, businesses), including interventions to address barriers to behavioral change (information, appeals, incentives) and the use of social marketing that combine mass media appeals with participatory, community-based approaches.

Many interventions to promote household efficiency do not succeed, at least to the extent expected, due to a failure to understand how people think about and make decisions regarding energy efficiency (Vandenbergh et al., 2010). For example, as Attari et al. (2010) demonstrate, householders systematically underestimate the potential for energy savings and as a result may conclude that the energy-saving actions may yield inadequate economic benefits. This happens as a result of the self-education that individuals use in making quantitative judgments about risk. Dietz (2010) suggests that linking the risk perception literature with the social psychological literature on consumption might improve the understanding of decision making regarding environmentally significant consumption.

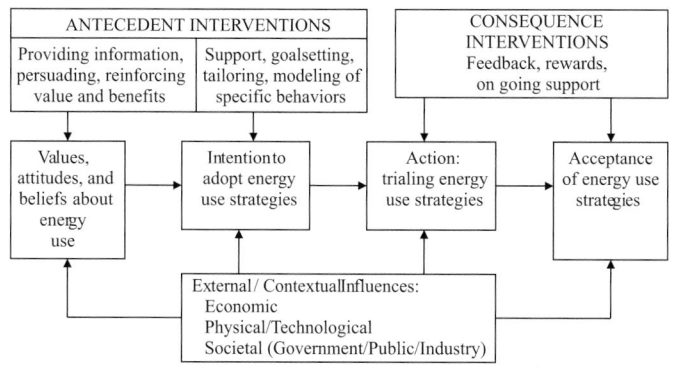

┌───┐   ┌─────────────────────┐
│ ANTECEDENT INTERVENTIONS   │   │   CONSEQUENCE   │
├──────────────────┬──────────────────────┤   │   INTERVENTIONS   │
│ Providing   │ Support, goalsetting,│   │ Feedback, rewards, │
│ information,   │ tailoring, modeling of│   │ on going support   │
│ persuading,   │ specific behaviors   │   │   │
│ reinforcing   │   │   │   │
│ value and   │   │   │   │
│ benefits   │   │   │   │
└──────────────────┴──────────────────────┘   └─────────────────────┘

Figure 21.6 | A model for behavioral change for energy use. Source: Gardner and Ashworth, 2008.

21.4.1   Information, Knowledge, and Education

Behavioral change requires both knowledge contributions to change attitudes and policy implementation to provide incentives for action. In order to achieve constructive behavioral change toward energy use, one conceptual model (adapted from Gardner and Ashworth, 2006) indicates the potential for knowledge-based interventions to increase the intention to adopt energy use strategies. Such interventions also require an understanding of factors that can influence attitudinal and behavioral change, such as situation, habits, and experience (Vaughan, 1977). The processes used to accomplish change in people's attitudes and behaviors can include persuasion or reason – or a combination of these (Cooper and Hogg, 2002).

Several models and perspectives on how to change behaviors and attitudes are available, including cognitive dissonance theory (Oskamp, 2000), the theory of planned behavior (Ajzen, 1989), the theory of reasoned action (Ajzen and Fishbein, 1980), and the theory of consumer uptake and societal acceptance (Niemeyer, 2004). Figure 21.6 (adapted from Gardner and Ashworth, 2006) presents a synthesis of many of these models and approaches. One of the key issues that needs to be accounted for in any behavioral change model is the effects of external and contextual influences. These elements cannot be controlled for, yet they can have a lasting impact on behavior.

One key element that is not included in the model but that is essential to address when specifically engaging on consumption is the issue of threat. Engaging people on reducing energy use requires a constructive response from the community to actual or perceived threats. Moser and Dilling (2007) raise caution about using fear to change behaviors. Many studies have shown that fear may change attitudes and other forms of expression but it does not necessarily assist active engagement or the changing of behaviors (Ruiter et al., 2001; Moser and Dilling, 2007). Based on a summary of research (e.g., Das et al., 2003; Ruiter el al., 2004; Moser and Dilling, 2007), achieving a productive behavioral and attitudinal outcome requires community members to:

- feel personally vulnerable to the risk;

- have useful and very specific information about possible precautionary actions;

- positively appraise their own ability (self-efficacy) to carry out the action;

- feel the suggested action will effectively solve the problem (response efficacy);

- believe the cost associated with taking precautionary action is low or acceptable;

- view the reward for not taking the action as unappealing; and

- tend to consciously and carefully process threat information (i.e., engage in central/systematic processing as opposed to peripheral/heuristic information processing).

Ten years of experience at the Akatu Institute for Conscious Consumption in Brazil have indicated that it is necessary to show individual consumers or small groups of consumers that any one person or a small group of people can have an enormous impact (Worldwatch Institute, 2010: 107). This helps persuade consumers that even an individual can have a strong positive contribution toward sustainability by changing individual consumption habits – and more so if each individual acts as a mobilizer in society for new ways of consuming that could have a much lower impact, especially on nature. The environmental movement in the last 20–30 years has focused on this aspect of sensitizing individual consumers of their responsibilities. With networking's technological advance, virtual communities are becoming very popular and have a global presence, and they can be a good means to reach out to a community. Consumers are seen as being increasingly important in the design and implementation of public policy and decisions about the delivery of services (Entwistle and Martin, 2005) and also as peer monitoring groups.

Neighborhood efforts to reduce energy use work better with realistic instructions and assessments of the threat, diagnosis and strategies, social support, people's sense of personal control over their circumstances, low-cost alternatives, and regular feedback that allows people to see that they are moving in the right direction (Morse and Doberneck, 1995; Groopman, 2004). One example of an initiative aimed at increasing knowledge among citizens and supporting concrete action is the Energymark initiative of the Commonwealth Scientific and Industrial Research Organisation in Australia. (See Box 21.1.) It shows clearly that the holistic approach to energy service demand by targeting consumers rather than any end-use component can deliver behavioral change and reduce demand for energy services.

Box 21.1 | The Energymark Initiative, Australia

The Energymark process brings together small groups of people, meeting at their own pace, to discuss energy technologies and climate change. Individuals read scientific factsheets and share their thoughts, barriers, challenges, anecdotes, and first-hand experiences. There are two benefits of this process. First, it ensures a coordinated approach to researching public perceptions of climate change and energy technologies across Australia in order to generate insights and provide an empirical benchmark for other researchers. Second, engaging the public in this way ensures the information is more likely to be translated into action by individuals because they can relate to the concepts, discuss them openly, and change their behaviors accordingly.

In total, the Energymark trials had 1713 participants from various regions of Australia, including a wide range of age groups and balanced gender representation. Following the meetings, there was a significant change in participants' knowledge, attitudes, and intended and actual behaviors. For how long these are sustained can be judged by follow up programmes. A carbon footprint calculator provided evidence that the average carbon footprint was reduced by 20.5% due to involvement in the Energymark process. The groups took, on average, 8.5 months to achieve this. Based on the costing and abatement achieved, the program could save in total 7452 tonnes of CO_2 (i.e., 4.35 tonnes per participant) in 8.5 months at an initial investment of approximately US\$500,000 (US\$250,000 committed to operational expenses related to conducting the trial and US\$250,000 for writing of the materials and establishing communication systems). Once the operational systems have been established, the cost associated with conducting more Energymark groups would be reduced. From the trial, the overall investment was 7.4US¢ to save 1 kWh through behavioral change, which is compatible with the current rate of electricity supply in Australia of around 8–16US¢/kWh.

21.4.2   The Role of Policies[4]

Detailed prescriptions and laws regulating individual consumption decisions could be supplementary to strengthen informational impact and serve as proxies for the beneficial reduction effects of carbon pricing and other instruments based on economic efficiency. Such prescriptions and laws can increase both technical and economic efficiency, while admittedly not reaching the first-best optimum. They could also be designed in an equitable way. They would, however, be very paternalistic and would interfere with individual freedom in an unacceptable way for a liberal society if implemented in due strength to trigger significant effects. Questions would arise about whether driving heavy private cars, eating meat, or traveling by airplane and private jet should be prohibited or drastically restricted.

Finally, there is a striking lack of political will, which may be the most important problem for the implementation of effective energy policy. Many energy policy instruments have exceptions, and existing gasoline taxes in almost all countries in varying degrees are way below the levels necessary for significant reductions, as can be derived from the research on price elasticities of demand for energy-intensive goods and services (see, e.g., Smith et al., 1995; Enevoldsen et al., 2007; Sterner, 2007). Stringent energy policy depends on the presence of societies that are willing to take on this burden (or chance), politicians who are willing to stand for such actions, and people who are willing to support such politicians and policies. This also limits the

effectiveness of information provision as an energy policy instrument to target individual consumers, as can be seen from the mismatch of the current level of information and the actions taken by individuals. Both costs and benefits associated with any action has important role to play. The extent to which promotional efforts succeed and the degree to which people are willing to "curb their consumption levels for the greater good of the community" (Brown and Cameron, 2000) depend on the existence of an underlying bedrock of environmental commitment.

Additional intervention options are built around the insights mentioned in the previous sections: they target energy services demand and not energy use; they account for lifestyle and well-being aspects – which can become a hindrance as well as support for reduced energy services demand. Ultimately, these alternative policy instruments need to build and develop a strong commitment to the environment, as the classical policy instruments alone do not work to deliver full potential.

Alternative policy options are found in the context of sufficiency strategies, which make the levels of energy services themselves a topic for discussion and aim at lowering those. Sufficiency thus addresses the "level" of output (or consumption) per se – and not in relation to the inputs (as technical efficiency does). It asks whether an activity needs to be performed at all (excess meat consumption, multiple car ownership, or extraordinarily high mobility service demand) and not whether it is performed "efficiently." A combination of existing and additional instruments is most promising for reducing the externalities of energy use.

---

4   Policy issues are discussed in more details in Cluster IV (Chapters 22–25).

Table 21.2 | Well-being, lifestyle change, and energy savings potentials

Lifestyle Change to Enhance Well-being	Consumption Category Intervention Point	Barrier	Energy-Savings Potential	Policy/Action (individual and public)/Regulation	Co-benefits
Less wasteful electricity usage behavior	Phantom load on electricity supply	Infrastructure design	10% of average energy bill[1]	Manual switching off of standby power points Informational campaigns	Freeing electricity for redistribution or savings in fossil fuel use and resulting emission
Healthy dietary choice with prescribed healthy meat consumption[2]	Energy embodied in excess per capita meat consumption	Marketing strategy by meat processing industries, diet gurus	1.4% of global primary energy use	Informational campaigns with better health advisory based on upper limit for meat consumption	GHG mitigation strategy; freeing up water, land, health equity through allowing consumption to rise in below-average consumption countries
Healthy lifestyle practice like walking /bicycling/more public transport usage and maximum of one car per household//kilometers driven per household	Fossil fuel using transport mode	Distorted market price (land price driven by real estate market), policy failures like government subsidy for grazing lands, wrong evaluating criterion, like conventionally measured GNP	1.3% of global primary energy use[3]	Informational campaign on reduced mobility and ill health effects Less time allocated for watching TV etc. Infrastructure design Congestion-free walkway, parks/open spaces with greenery to replace energy-guzzling gyms	Larger global mobility with same fossil fuel use

[1] Roy and Pal, 2009.

[2] Limit of lean meat consumption per year recommended by American Heart Association is per capita per year 62.6 kg (Caspari et al., 2009). Currently FAO statistics show 54 countries have per capita annual consumption higher than the standard ranging between 63.2 (Mexico)-122.79 (USA).

[3] Estimated using information on energy use and alternative green lifestyles in Christensen, (1997).

The current debate on sustainable energy systems is largely dominated by technical and economic efficiency and clean energy strategies, while sufficiency plays a minor role only. Nevertheless, many official reports dealing with the question of how to reduce emissions and energy acknowledge the crucial importance of changes in lifestyles and consumption patterns (Duchin, 1998; OECD, 1998; Lundgren, 1999; Alfredsson, 2002; ECEEE, 2006; Kaenzig and Jolliet, 2006; IPCC, 2007; DEFRA 2008). This may be due to the fact that in most cases it is not clear what is meant by changed lifestyles or consumption patterns or how to bring about those changes.

It is not only individuals who have to be targeted. Lifestyles with certain levels of comfort and high energy services demand have not necessarily evolved from individual choices. The accomplishment of comfort is entwined with fashion and property development and design. It is increasingly difficult to buy a new car that does not come readily equipped with air-conditioning. Working only at the level of individuals thus may not lead to the transformational change in lifestyle needed for a more sustainable society. So change in top-down decisions in infrastructure design, as described earlier, and educating communities (as in the Energymark Initiative) might be needed.

Table 21.2 shows various alternative additional entry points to gain energy-saving potentials from the demand side to add flexibility to supply options through lifestyle change that are win-win options so far as the goal of sustaining well-being is concerned.

21.4.3   Role of Actors

National and international agencies: government, private, communities and civil societies should be encouraging the proliferation of regional (climate-sensitive) understandings of comfort and the development of a corresponding variety of local socio-technical regimes. The definition of comfort should be the subject of explicit discussion and debate. Ecologists (Diamond, 1997; Millennium Ecosystem Assessment, 2005) are revisiting the values of local ecosystems and trying to put higher value to niches in local systems. Multiple stakeholders are seen as being increasingly important in the design and implementation of public policy and decisions about the delivery of services (Entwistle and Martin, 2005) and also as a peer monitoring group. Stakeholders have been described as those who have an interest in a particular decision, either as individuals or representatives of a group. This includes people who influence a decision, or can influence it, as well as those affected by it. The community of stakeholders can become active with regard to some policy goals. Information campaigns can be embedded in several contexts, some examples are in the Energymark example described earlier.

Corporations and all kinds of organizations can also drive information campaigns. Corporation/organizations promote lifestyles that penetrate society through their own direct consumption structure, product labeling, and purchase policies. Labeling schemes (OECD, 2009) are an important initiative, as labels provide comparable information for consumers to make informed purchase decisions. Green Labels in

Europe aim at increasing greener procurement through the promotion of energy-efficient products and awareness-raising. Examples are the Swan label for over 50 different products, the Group for Energy Efficient Appliances label for home and office appliances, and the Blue Angel label for environmentally friendly consumer goods and services (OECD, 2009), as well as Energy Star labels for energy-efficient home appliances in India (BEE, 2011). Marketing strategies play a significant role, for instance, by selecting a limited range of products that are "approved" by a company as energy-efficient. Good practices, such as encouraging employees to use carpools, bicycles, or walking for health reasons and observing Environment Day[5], Earth Day[6], Biodiversity Day[7], and so on, have the potential to change behaviors. Such labeling has led to new product development as well.

In the past, environmental disagreements between corporations and environmental groups were addressed through long, drawn-out conflicts in the news media, open-ended lengthy administrative processes in government agencies, or costly litigation (Piasecki and Asmus, 1990). Today, a more cooperative spirit can often be observed. Emerging corporate citizenship practices benefit both businesses and the communities in which they operate. Corporate environmental citizenship activities often include donations and gifts for environmental programs and incentives for employees to work with community groups on natural resource conservation and protection. Clearly, information campaigns need also be led by civil society, educational institutions, and families. Schools have a mandate to motivate and inform students. This needs to be part of school curriculum which can involve parents' participation as well otherwise there will be counter information with the possibility of undermining the actions otherwise initiated. A holistic approach can only reinforce positive actions by diverse groups of actors.

21.5   Beyond Affluence to Well-being

Well-being denotes quality of life. Numerous indices have emerged to help measure well-being through various changes in quality of life. These are being used to make comparisons across groups, nations, and time. They also help determine whether these changes are sustainable, and they evaluate policies.

For a long time, and still predominantly today, human well-being tends to be associated largely with increased levels of consumption of products and services, a rising income and the purchasing power to support this increased consumption, and a rise in energy use. In this respect, income is the measure of wealth dividing the "rich" from the "poor." Consequently, conventionally gross national product (GNP) or gross

domestic product (GDP) is the measure of choice and one that has dominated the literature.

This dominant economic notion has been challenged periodically over the years. One of the early challenges came during the 1970s, with the rise of public awareness about environmental and sustainability concerns. Serious doubts were raised as to whether traditional economic growth models were adequate in view of these new issues. This led to a broader discussion and debate about indicators of well-being. As a measure of aggregate economic activity, conventionally defined GNP or GDP does not capture social well-being, social inequities, or the depletion of natural resources. As a result of this debate, several new approaches emerged that have tried to include other aspects of well-being (Hanley et al., 1997; Hamilton and Clemens, 1999; Dasgupta, 2001).

Many of the new emerging approaches put greater emphasis on issues of poverty and provide new definitions of wealth, new ways to treat environmental degradation, and new indices to measure the quality of life (e.g., Dasgupta, 1975; IUCN, 1980; Norgaard, 1984; El Sarafy, 1989; Daly, 1990; Daly and Cobb, 1990; UNDP, 1990; Dasgupta and Maler, 1991; Commons and Perrings, 1992; Dasgupta and Maler, 1995; Hanley et al., 1997; Sen, 1999; Aggarwal et al., 2001; Dasgupta, 2001).

Measures of well-being can be classified (Dasgupta, 2001) according to whether they are based on constituents (outputs) of well-being such as health, happiness, or freedom or on the determinants (inputs) of well-being such as expenditures on food, nutrition, clothing, potable water, shelter, and access to knowledge or information. Economic literature has tended to focus on the latter, while moral philosophers and many in the social sciences have focused on the constituents of well-being. Newer indices have tried to combine both the determinants and the constituent elements of well-being.

Some recent proposals focus on the happiness index (Namgyal and Wangchuk, 1998). But many commentators, despite accepting the importance of happiness, have questioned these as being subjective, unobservable, and difficult to quantify (Kahneman et al., 1997). Theoretical and empirical research (Dasgupta, 1995, 1997; Frey and Stutzer, 1999, 2000; Narayan et al., 2000) has tried to establish more objectively the links between states of happiness and wealth or the lack therefore, employment, and the command of natural resources. Notwithstanding their deficiencies, some of these indices can be useful as supplementary tools to indicate potential entry points for policy interventions.

In response to the shortcomings of conventionally measured GNP and its tendency to measure aggregate "economic activity" rather than "social well-being," the United Nations Development Programme introduced the Human Development Index (HDI) (UNDP, 1990; 2010). Besides conventionally measured GNP (as a proxy for purchasing power based well-being), the HDI includes adult literacy and life expectancy at birth as additional crucial aspects of well-being. Despite the shortcomings of this index – it does not measure inter-temporal well-being and is still

---

5   http://www.unep.org/wed/

6   http://www.earthday.org/earth-day-history-movement

7   http://www.cbd.int/idb

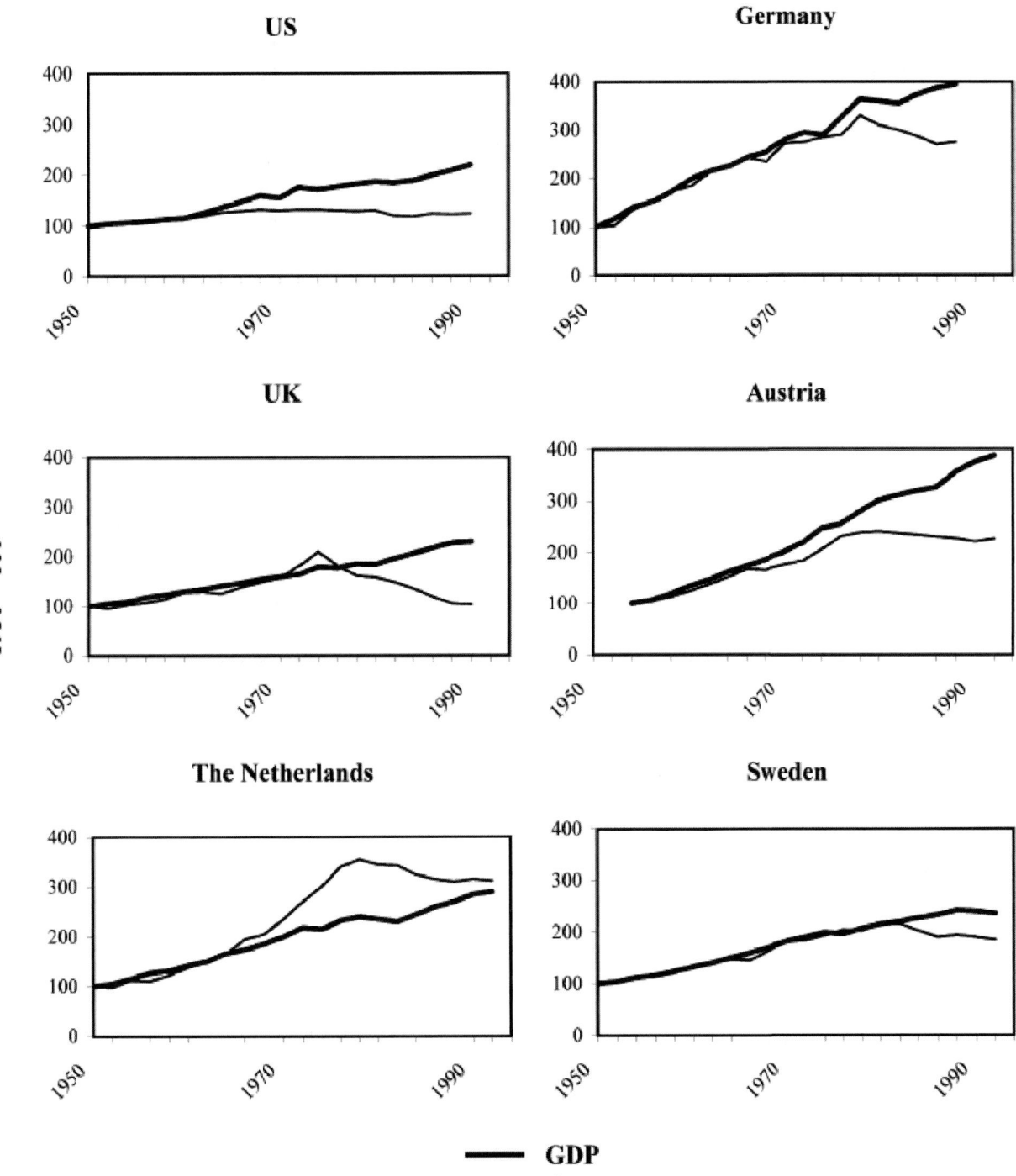

Figure 21.7 | Comparison of macro indicator: GDP and ISEW for the United States, Germany, United Kingdom, Austria, the Netherlands, and Sweden. Source: Lawn 2003.

too limited in terms of determinants of well-being – many consider the HDI a step in the right direction.

Regarding the use of consumption level as an index of well-being, studies have shown that in poor countries and communities, indices of consumption can serve as determinants of well-being and happiness.

For rich countries and communities, however, the inclusion of current consumption as a measure of well-being may not be as useful or appropriate because current consumption may not be a contributing factor to happiness of people who have much more than the basic necessities (Esterlin, 1974). Empirical research also shows that the correlation between income and individual well-being becomes weak beyond a certain level of wealth and that further increases of economic wealth do not necessarily increase individual well-being. Thus the need for more than one index to keep track of various constituents and determinants of well-being depends on the situation being considered (Nussbaum, 2000). However, the fact remains any composite measure can be non-robust to changes in weights on the components.

The concepts of well-being just described have important implications on how energy services demand and energy use are assessed. A focus on conventionally defined GNP/GDP alone is biased toward viewing changes and reductions in energy services demand as decreases in well-being. More complex notions of well-being are better able to take other factors into account even in situations of decreasing energy services.

Notions of well-being based only on consumption implicitly assume that there is high substitutability between different energy services, while more complex, multidimensional notions can capture differing characteristics of energy services that may actually hinder substitutability. This is of particular importance when addressing basic needs for a decent good life and the energy services demand to meet those. The assumption of substitutability is important, as assumptions about it can, for example, lead to overestimations of well-being when the aggregate quantity of energy use is considered but not its qualitative composition (Pachauri et al., 2004).

In the long run, however, transformational change is possible by changing the social value system and through the adoption of a better quality of life, as indicated by a Well-being Index in addition to, and other than, conventionally measured GNP, HDI or similar such new indices. Over the years a number of different indices have been developed to measure and compare the benefits and costs of growth. The first of these was Index of Sustainable Economic Welfare (ISEW). Over this time, the ISEW has been given a variety of different names and theoretical underpinning – for example, a Genuine Progress Indicator or GPI and a Sustainable Net Benefit Index or SNBI (Lawn, 2003). Empirical studies in the context of developed economies have demonstrated that there may be a threshold level for growth beyond which growth of macroeconomic system is not beneficial to human well-being. Divergence between the ISEW and conventional measures of GDP beyond a point is clearly shown in Fig 21.7. Beyond threshold, or peak, as shown by ISEW, SNBI and GPI, the suggestion is goal of these countries need to move away from growth objective to sustainable development goals. This peak is not observable in GDP measure which is the rising curve. These studies strongly support the need for refinement of new welfare & well-being measures. It also highlights the need to better link energy use with such alternative measures.

Finally, the potential for increased sustainability that lies in strong decoupling of well-being from consumption can be captured by notions of well-being that are not based on consumption. This offers new policy options of potentially high leverage. A reduction in energy services and energy use demand does not necessarily reduce well-being. And lifestyle changes can deliver some win-win options that reduce demand for energy services with the same or even improved levels of personal and social well-being.

Changes in preferences and attitudes of individuals are essential in long-term sustainable lifestyle change. In working toward sustainable energy futures, the role of the individual is crucial. "Sufficiency" is a key term in this discussion. Sufficiency directly addresses the individual, and its concepts of well-being and lifestyle aim at a thorough discussion of societal values (Muller, 2009). Leading a "sufficient" life means leading a life of moderation and prudence. This means developing awareness of the consequences of actions beyond monetized cost-benefit analysis and of markets beyond local or regional scope. Moderation would curb further growth in energy use with the corresponding externalities, including harm to others.

Numerous successful changes in lifestyles and acceptance of values such as "sufficiency" have taken place since World War II. Sufficiency notions have provided impetus for discussion and debate on society's values and the role of its citizens (Harvey, 1996; Goodman, 2010). Some of these successes provide a rich ground for exploring new measures and policies to encourage lifestyle changes in key areas of energy use in the near future as well as the development of new and more appropriate indices.

The literature on alternative indices of well-being is not conclusive enough operationally to replace conventionally measured GNP – the current index of economic well-being. There is a methodological and information gap on the energy outcome of alternative lifestyles at global and regional scales. Alternative consumption-based approaches need to be compared with supply-side-based approaches in terms of the implications for primary energy use. In public discussions, sufficiency is often related to renunciation. This nexus is always made on the basis of the current consumption level and lifestyle. Instead, efficiency-led strategies need to be tied to sufficiency to become a widely used strategy that changes values and notions of well-being.

More research is needed to assess the macroeconomic effects of large-scale switches to more-sufficient lifestyles over various time scales in the future. A key issue is how a large reduction in consumption could be absorbed by the economy without generating large unemployment. Simultaneous redistribution of labor could offer a solution, but further analysis of this is clearly needed.

References

Ababou, M., 2005: The Impact of Age, Generation and Sex Variables on Religious Beliefs and Practices in Morocco. *Social Compass*, **52**(1): 31–44.

Aggarwal, R., S. Netanyahu and C. Romano, 2001: Access to Natural Resources and the Fertility Decision of Women: The Case of South Africa. *Environment and Development Economics*, **6**(02): 209–236.

Ajzon, I. and M. Fishbein, 1980: *Understanding Attitudes and Predicting Social Behavior*. Prentice Hall, Englewood Cliffs, NJ, USA.

Alfredsson, E. C., 2002: *Green Consumption Energy Use and Carbon Dioxide Emission*. Doctoral Thesis, Department of Social and Economic Geography, Spatial Modelling Centre, Umeå University, Umeå, Sweden.

Alfredsson, E. C., 2004: "Green" Consumption – No Solution for Climate Change. *Energy*, **29**(4): 513–524.

Asp, E. H., 1999: Factors Affecting Food Decisions Made by Individual Consumers. *Food Policy*, **24**(2–3): 287–294.

Attari, S. Z., M. L. DeKay, C. I. Davidson and W. Bruine de Bruin, 2010: Public Perceptions of Energy Consumption and Savings. *Proceedings of the National Academy of Sciences*, **107**(37): 16054–16059.

Baroni, L., L. Cenci, M. Tettamanti and M. Berati, 2006: Evaluating the Environmental Impact of Various Dietary Patterns Combined with Different Food Production Systems. *Eur J Clin Nutr*, **61**(2): 279–286.

Bauman, Z., 1998: *Work, Consumerism and the New Poor*. Open University Press, Berkshire, UK.

Beck, U., 1997: *Risikoamfundet – På Vej Mod En Ny Modernitet*. Hans Reitzel, Copenhagen, Denmark.

BEE, 2011: *National Energy Labeling Programme*. http://www.beeindia.in/ (accessed 28 August, 2011).

Benders, R. M. J., R. Kok, H. C. Moll, G. Wiersma and K. J. Noorman, 2006: New Approaches for Household Energy Conservation – in Search of Personal Household Energy Budgets and Energy Reduction Options. *Energy Policy*, **34**(18): 3612–3622.

Bertilsson, G., H. Kirchmann and L. Bergström, 2008: Energy Analysis of Organic and Conventional Agricultural Systems. In *Organic Crop Production – Ambitions and Limitations*. H. Kirchmann and L. Bergström (eds.), Springer, Dordrecht, the Netherlands, pp. 173–188.

Biesiot, W. and K. J. Noorman, 1999: Energy Requirements of Household Consumption: A Case Study of the Netherlands. *Ecological Economics*, **28**(3): 367–383.

Bin, S. and H. Dowlatabadi, 2005: Consumer Lifestyle Approach to US Energy Use and the Related CO$_2$ Emissions. *Energy Policy*, **33**(2): 197–208.

Blackwell, R. D., P. W. Miniard and J. F. Engel, 2001: *Consumer Behavior*. Harcourt Inc., Orlando, FL, USA.

Brown, P. M. and L. D. Cameron, 2000: What Can Be Done to Reduce Overconsumption? *Ecological Economics*, **32**(1): 27–41.

Carlsson-Kanyama, A., M. P. Ekström and H. Shanahan, 2003: Food and Life Cycle Energy Inputs: Consequences of Diet and Ways to Increase Efficiency. *Ecological Economics*, **44**(2–3): 293–307.

Caspari, C., M. Christodoulou, J. Nganga and M. Ricci, 2009: *Implications of Global Trends in Eating Habits for Climate Change, Health and Natural Resources*. DG-IPOL-STOA-ET(2009)424735, Science and Technology Options Assessment (STOA), Directorate General for Internal Policies, European Parliament, Brussels, Belgium.

Christensen, P., 1997: Different Lifestyles and Their Impact on the Environment. *Sustainable Development*, **5**(1): 30–35.

Cohen, C., M. Lenzen and R. Schaeffer, 2005: Energy Requirements of Households in Brazil. *Energy Policy*, **33**(4): 555–562.

Coley, D. A., E. Goodliffe and J. Macdiarmid, 1998: The Embodied Energy of Food: The Role of Diet. *Energy Policy*, **26**(6): 455–459.

Collins, A. and R. Fairchild, 2007: Sustainable Food Consumption at a Sub-National Level: An Ecological Footprint, Nutritional and Economic Analysis. *Journal of Environmental Policy & Planning*, **9**(1): 5–30.

Common, M. and C. Perrings, 1992: Towards an Ecological Economics of Sustainability. *Ecological Economics*, **6**(1): 7–34.

Cooper, J. and M. A. Hogg, 2002: Dissonance Arousal and the Collective Self. In *The Social Self: Cognitive, Interpersonal and Intergroup Perspectives*. J. P. Forgas and K. D. Williams (eds.), Psychology Press, New York, NY USA.

Cranfield, J. A. L., T. W. Hertel, J. S. Eales and P. V. Preckel, 1998: Changes in the Structure of Global Food Demand. *American Journal of Agricultural Economics*, **80**(5): 1042–1050.

Daly, H. E., 1990: Toward Some Operational Principles of Sustainable Development. *Ecological Economics*, **2**(1): 1–6.

Daly, H. E. and J. B. Cobb Jr., 1990: *For the Common Good: Redirecting the Economy toward Community, the Environment and a Sustainable Future*. Green Print, London, UK.

Daly, H. E., 1996: Consumption: Value Added, Physical Transformation and Welface. In *Getting Down to Earth: Practical Applications of Ecological Economics*. R. Costanze, O. Segura and J. Martinez-Alier (eds.), Island Press, Washington, D.C., USA, pp. 49–59.

Das, E. H. H. J., J. B. F. de Wit and W. Stroebe, 2003: Fear Appeals Motivate Acceptance of Action Recommendations: Evidence for a Positive Bias in the Processing of Persuasive Messages. *Personality and Social Psychology Bulletin*, **29**(5): 650–664.

Dasgupta, A. K., 1975: *The Economics of Austerity*. Oxford University Press, New Dehli, India.

Dasgupta, P. and K.-G. Mäler, 1991: *The Environment and Emerging Development Issues. Proceedings of the World Bank Annual Conference on Development Economics 1990*, Supplement to the World Bank Economic Review and the World Bank Research Observer.

Dasgupta, P., 1995: *An Inquiry into Well-Being and Destitution*. Clarendon Press, Oxford, UK.

Dasgupta, P. and K.-G. Maler, 1995: Povert, Institutions and the Environmental Resource Base. In *Handbook of Development Economics, Vol. Iii(a)*. J. Berhrman and T. N. Srinivasan (eds.), North Holland, Amsterdam, the Netherlands.

Dasgupta, P., 1997: Nutritional Status, the Capacity for Work, and Poverty Traps. *Journal of Econometrics*, **77**(1): 5–37.

Dasgupta, P., 2001: *Human Well-Being and the Natural Environment*. Oxford University Press, Oxford, UK.

DEFRA, 2008: *A Framework for Pro-Environmental Behaviours*. Department for Environment, Food and Rural Affairs (DEFRA), London, UK.

Deleder, N., 1994: Religious Contrasts in Consumer Decision Behaviour Patterns – Their Dimensions and Marketing Implications. *European Journal of Marketing*, **28**(5): 36–53.

Diamond, J., 1997: *Guns, Germs and Steel: The Fate of Human Societies*. W.W.Norton & Company, New York, NY, USA.

Dietz, T., G. T. Gardner, J. Gilligan, P. C. Stern and M. P. Vandenbergh, 2009: Household Actions Can Provide a Behavioral Wedge to Rapidly Reduce Us Carbon Emissions. *Proceedings of the National Academy of Sciences*, **106**(44): 18452–18456.

Dietz, T., 2010: Narrowing the US Energy Efficiency Gap. *Proceedings of the National Academy of Sciences*, **107**(37): 16007–16008.

Dindyal, S. and S. Dindyal, 2004: How Personal Factors, Including Culture and Ethnicity, Affect the Choices and Selection of Food We Make. *The Internet Journal of Third World Medicine*, **1**(2): 27–33.

Duchin, F., 1998: *Structural Economics: Measuring Changes in Technology, Lifestyle and the Environment*. Island Press, Washington, D.C., USA.

EC, 1999: *Council Regulation (Ec) No 1257/1999 on Support for Rural Development from the Eagff*. Official Journal of the European Communities L160 No 1257/1999, European Comission (EC), Brussels, Belgium.

ECEEE, 2006: *The European Commission Green Paper on Energy Efficiency or Doing More with Less*. European Council for an Energy Efficient Economy (ECEEE), Stockholm, Sweden.

Edwards, P. and A. Tsouros, 2006: *Promoting Physical Activity and Active Living in Urban Environments – the Role of Local Governments*. The Solid Facts, World Health Organization (WHO), Geneva, Switzerland.

Ehrlich, P.R. and J.P. Holdren, 1971: Impact of Population Growth. *Science*, **171**(3977): 1212–1217.

El Sarafy, S., 1989: The Proper Calculation of Income from Depletable Natural Resources. In *Environmental Accounting for Sustainable Development: A Undp & World Bank Symposium*. Y. Ahmed and S. SEl Sarafy (eds.), World Bank, Washington, D.C., USA.

Enevoldsen, M. K., A. V. Ryelund and M. S. Andersen, 2007: Decoupling of Industrial Energy Consumption and CO_2-Emissions in Energy-Intensive Industries in Scandinavia. *Energy Economics*, **29**(4): 665–692.

Entwistle, T. and S. Martin, 2005: From Competition to Collaboration in Public Service Delivery: A New Agenda for Research. *Public Administration*, **83**(1): 233–242.

Eshel, G. and P.A. Martin, 2006: Diet, Energy and Global Warming. *Earth Interactions*, **10**(9).

Esterlin, R. A., 1974: Does Economic Growth Improve the Human Lot? Some Empirical Evidence. In *Nations and Households in Economic Growth: Essays in Honor of Moses Abramowitz*. P. A. Davis and M. Reder (eds.), Academic Press, New York, NY, USA.

Faiers, A., M. Cook and C. Neame, 2007: Towards a Contemporary Approach for Understanding Consumer Behaviour in the Context of Domestic Energy Use. *Energy Policy*, **35**(8): 4381–4390.

FAOSTAT, 2008: *Consumption: Livestock and Fish Primary Equivalent*. (accessed 1 December, 2008).

Fiala, N., 2006: *Estimates of US within Product Demand Elasticities for Meat*. Economic Research Service and United States Department of Agriculture, Washington, D.C., USA.

Frey, B. S. and A. Stutzer, 1999: Measuring Preferences by Subjective Well-Being. *Journal of Institutional and Theoretical Economics*, **155**(4): 755–778.

Frey, B. S. and A. Stutzer, 2000: Happiness, Economy and Institutions. *The Economic Journal*, **110**(466): 918–938.

Fung, T. T., M. J. Stampfer, J. E. Manson, K. M. Rexrode, W. C. Willett and F. B. Hu, 2004: Prospective Study of Major Dietary Patterns and Stroke Risk in Women. *Stroke*, **35**(9): 2014–2019.

Gardner, J. and P. Ashworth, 2006: *The Intelligent Grid Project: A Review of Relevant Literature*. Commonwealth Scientific and Industrial Research Organisation (CSIRO), Canberra, Australia.

Gardner, J. and P. Ashworth, 2008: Towards the Intelligent Grid – a Review of the Literature. In *Urban Energy Transition: From Fossil Fuel to Renewable Power*. P. Droege (ed.), Elsevier, Oxford, UK.

Geißler, R., 2002: *Die Sozialstruktur Deutschlands – Die Gesselschaftiliche Entwicklung Vor Und Nach Der Vereinigung*. Westdeutscher Verlag, Wiesbaden, Germany.

Gerbens-Leenes, P. W., S. Nonhebel and M. S. Krol, 2010: Food Consumption Patterns and Economic Growth. Increasing Affluence and the Use of Natural Resources. *Appetite*, **55**(3): 597–608.

Giddens, A., 1996: *Modernitet Og Selvidentitet*. Hans Reitzil, Copenhagen, Denmark.

Goodman, J., 2010: *High-Tech Growth and Low-Tech Sufficiency – Silences and Possibilities*. Draft working paper, Climate Actoin Research Group (CARG) Conference and Friends of the Earth, Australia, Canberra, Australia.

Groopman, J., 2004: *The Anatomy of Hope: How People Prevail in the Face of Illness*. Random House, New York, NY, USA.

Hamilton, K. and M. Clemens, 1999: Genuine Savings Rates in Developing Countries. *The World Bank Economic Review*, **13**(2): 333–356.

Hanley, N., J. F. Shogren and B. White, 1997: *Environmental Economics – in Theory and Practice*. Oxford University Press, Oxford, UK.

Harrison, C. and G. Davies, 1998: *Lifestyles and the Environment*. Environment & Sustainability Desk Study, The Economic and Social Research Council (ESRC), Swindon, UK.

Harvey, D., 1996: *Justice, Nature and the Politics of Difference*. Blackwell, Oxford, UK.

Hassaini, M. M., 1993: *Islamic Dietary Concepts and Practices*. Islamic Food & Nutrition Council of America, Park Ridge, IL, USA.

Herendeen, R. and J. Tanaka, 1976: Energy Cost of Living. *Energy*, **1**(2): 165–178.

Hubacek, K., and D. Guan and A. Barua, 2007: Changing lifestyle and consumption patterns in developing countries: A scenario analysis for china and India. *Futures* **35**(9): 1084–1096.

IUCN, 1980: *World Conservation Strategy: Living Resource Conservation for Sustainable Development*. United Nations Environment Programme (UNEP), World Wildlife Fund (WWF) and the International Union for Conservation of Nature (IUCN), Gland, Switzerland.

IPCC, 2007: Summary for Policy Makers. In *Climate Change 2007: Mitigation. Contribution of Working Group III to the Fourth Assessment Report of the Intergovernmental Panel on Climate Change (Ipcc)*. B. Metz, O. R. Davidson, P. R. Bosch, R. Dave and L. A. Meyer (eds.), Cambridge University Press, Cambridge, UK and New York, NY, USA.

Jackson, T., 2005: *Motivating Sustainable Consumption – a Review of Evidence on Consumer Behaviour and Behavioural Change*. Report Preapared for the Sustainable Development Research Network, Centre for Environmental Strategy, University of Surrey, Surrey, UK.

Jensen, J. O., 2008: Measuring Consumption in Households: Interpretations and Strategies. *Ecological Economics*, **68**(1–2): 353–361.

Jørgensen, U., T. Dalgaard and E. S. Kristensen, 2005: Biomass Energy in Organic Farming – the Potential Role of Short Rotation Coppice. *Biomass and Bioenergy*, **28**(2): 237–248.

Kaenzig, J. and O. Jolliet, 2006: *Umweltbewusster Konsum: Schlüsselentscheide, Akteure Und Konsummodelle*. Umwelt-Wissen Nr. 0616, Bundesamt für Umwelt (BAFU), Bern, Switzerland.

Kahneman, D., P. P. Wakker and R. Sarin, 1997: Back to Bentham? Explorations of Experienced Utility. *The Quarterly Journal of Economics*, **112**(2): 375–406.

Kramer, K. J., H. C. Moll, S. Nonhebel and H. C. Wilting, 1999: Greenhouse Gas Emissions Related to Dutch Food Consumption – Foundations and Extensions. *Energy Policy*, **27**(4): 203–216.

Landes, M., S. Persaud and J. Dyck, 2004: *India's Poultry Sector – Development and Prospects*. Agriculture and Trade Report No. WRS-04–03, Market and Trade Economics Division, Economic Research Service, United States Department of Agriculture (USDA), Washington, D.C., USA.

Lawn, P. A., 2003: A Theoretical Foundation to Support the Index of Sustainable Economic Welfare (ISEW), Genuine Progress Indicator (GPI), and Other Related Indexes. *Ecological Economics*, **44**(1): 105–118.

Lenzen, M., 1998: Energy and Greenhouse Gas Cost of Living for Australia During 1993/94. *Energy*, **23**(6): 497–516.

Lenzen, M., C. Dey and B. Foran, 2004: Energy Requirements of Sydney Households. *Ecological Economics*, **49**(3): 375–399.

Limage, L. J., 2000: Education and Muslim Identity: The Case of France. *Comparative Education*, **36**(1): 73–94.

Lundgren, L. J., Ed. 1999: *Livssstil Och Miljö, Värderingar, Val, Vanor. Livsstil Och Miljö*. Swedish Environmental Protection Agency, Stockholm, Sweden.

Lyons, S., K. Mayor and R. S. J. Tol, 2007: *Convergence of Consumption Patterns During Macroeconomic Transition – a Model of Demand in Ireland and the OECD*. Working Paper No. 205, Economic and Social Research Institute, Dublin, Ireland.

Mann, N., 2000: Dietary Lean Red Meat and Human Evolution. *European Journal of Nutrition*, **39**(2): 71–79.

Masuda, T. and P. D. Goldsmith, 2010: *China's Meat Consumption: An Income Elasticity Analysis and Long-Term Projections*. Poster prepared for presentation at the Agricultural & Applied Economics Association Joint Annual Meeting, Denver, CO, USA.

Mennell, S., A. Murcott and A. H. Van Otterloo, 1992: *The Sociology of Food: Eating, Diet and Culture*. Sage Publicatoins Ltd., London, UK.

Millenium Ecosystem Assessment, 2005: *Ecosystems and Human Well-Being – Synthesis*. Island Press, Washington, D.C., USA.

Minx, J. C., T. Wiedmann, R. Wood, G. P. Peters, M. Lenzen, A. Owen, K. Scott, J. Barrett, K. Hubacek, G. Baiocchi, A. Paul, E. Dawkins, J. Briggs, D. Guan, S. Suh and F. Ackerman, 2009: Input–Output Analysis and Carbon Footprinting: An Overview of Applications. *Economic Systems Research*, **21**(3): 187–216.

Morse, J. M. and B. Doberneck, 1995: Delineating the Concept of Hope. *Journal of Nursing Scholarship*, **27**(4): 277–285.

Moser, S. C. and D. Dilling, 2007: Toward the Social Tipping Point: Creating a Climate for Change. In *Creating a Climate for Change – Communicating Climate Change and Facilitating Social Change*. S. C. Moser and D. Dilling (eds.), Cambridge University Press, Cambridge, UK.

Mullen, K., R. Williams and K. Hunt, 2000: Irish Descent, Religion and Food Consumption in the West of Scotland. *Appetite*, **34**(1): 47–54.

Muller, A., 2009: Sufficiency – Does Energy Consumption Become a Moral Issue? *IOP Conference Series: Earth and Environmental Science*, **6**(26): 262003.

Musaiger, A. O., 1993: Socio-Cultural and Economic Factors Affecting Food Consumption Patterns in the Arab Countries. *The Journal of the Royal Society for the Promotion of Health*, **113**(2): 68–74.

Namgyal, T. S. and T. Wangchuck, 1998: Measuring Gross National Happiness: A Predictive Model for Quantifying Social and Environmental Well-Being in Bhutan. *Sherub Doenme*, **4**(1&2): 1–24.

Narayan, D., R. Patel, K. Schafft, A. Rademacher and S. Koch-Schulte, 2000: *Voices of the Poor: Can Anyone Hear Us?* Pusblished for the World Bank, Oxford University Press, New York, NY, USA.

Niemeyer, S., 2004: Deliberation in the Wilderness: Displacing Symbolic Politics. *Environmental Politics*, **13**(2): 347–372.

Norgaard, R. B., 1984: Co-Evolutionary Development Potential. *Land Economics*, **60**(2): 160–173.

Nussbaum, M., 2000: *Women and Human Development*. Cambridge University Press, Cambridge, UK.

OECD, 1998: *Towards Sustainable Consumption Patterns – a Progress Report on Member Country Initiatives*. Organisation for Economic Co-operation and Development (OECD), Paris, France.

OECD, 2009: *The Role of Consumers and Corporations in Tackling Climate Change. Consumer Empowerment and Responsible Business Conduct*, Ethical Investment Research Services (EIRIS) for the Organisation for Economic Co-operation and Development (OECD), Paris, France.

Ojha, V. P., B. Pal, S. Pohit and J. Roy, 2008: *Social Accounting Matrix for India*. Working Paper no. GCP-JU-WP-2008–10–1, Global Change Programme, Jadavpur University, Jadavpur, India.

Oskamp, S., 2000: A Sustainable Future for Humanity? How Can Psychology Help? *American Psychologist*, **55**(5): 496–508.

Ozaki, R. and K. Sevastyanova, 2011: Going Hybrid: An Analysis of Consumer Purchase Motivations. *Energy Policy*, **39**(5): 2217–2227.

Pachauri, S. and D. Spreng, 2002: Direct and Indirect Energy Requirements of Households in India. *Energy Policy*, **30**(6): 511–523.

Pachauri, S., 2004: An Analysis of Cross-Sectional Variations in Total Household Energy Requirements in India Using Micro Survey Data. *Energy Policy*, **32**(15): 1723–1735.

Perrels, A. and C. L. Weber, 2000: *Modelling Impacts of Lifestyle on Energy Demand and Related Emissions*. VATT-Discussion Papers, Government Institute for Economic Research (VATT), Helsinki, Finland.

Pettinger, C., M. Holdsworth and M. Gerber, 2004: Psycho-Social Influences on Food Choice in Southern France and Central England. *Appetite*, **42**(3): 307–316.

Piasecki, B., and P. Asmus, 1990: In Search of Environmental Excellence: moving beyond blame. Simon and Schuster, New York, NY, USA.

Pimentel, D. and M. Pimentel, 2003: Sustainability of Meat-Based and Plant-Based Diets and the Environment. *The American Journal of Clinical Nutrition*, **78**(3): 660S-663S.

Ramírez, C. A., M. Patel and K. Blok, 2006: How Much Energy to Process One Pound of Meat? A Comparison of Energy Use and Specific Energy Consumption in the Meat Industry of Four European Countries. *Energy*, **31**(12): 2047–2063.

Rees, W. E., 1995: *Reducing the Ecological Footprint of Consumption*. Workshop on Policy measures for Changing Consumption Patterns, Seoul, South Korea.

Regmi, A., M. S. Deepak, J. L. Seale Jr. and J. Bernstain, 2001: Cross-Country Analysis of Food Consumption Patterns. In *Changing Strcuture of Global Food Consumption and Trade*. A. Regmi (ed.), Economic Research Service, United States Department of Agriculture Washington D.C., USA, pp. 14–22.

Reinders, A. H. M. E., K. Vringer and K. Blok, 2003: The Direct and Indirect Energy Requirement of Households in the European Union. *Energy Policy*, **31**(2): 139–153.

Roy, J., 2000: The Rebound Effect: Some Empirical Evidence from India. *Energy Policy*, **28**(6–7): 433–438.

Roy, J. and S. Pal, 2009: Lifestyles and Climate Change: Link Awaiting Activation. *Current Opinion in Environmental Sustainability*, **1**(2): 192–200.

Ruiter, R. A. C., C. Abraham and G. Kok, 2001: Scary Warnings and Rational Precautions: A Review of the Psychology of Fear Appeals. *Psychology & Health*, **16**(6): 613–630.

Ruiter, R. A. C., B. Verplanken, D. De Cremer and G. Kok, 2004: Danger and Fear Control in Response to Fear Appeals: The Role of Need for Cognition. *Basic and Applied Social Psychology*, **26**(1): 13–24.

Sack, D., 2001: *Whitebread Protestants – Food and Religion in American Culturew*. Palgrave, New York, NY, USA.

Saint-Blancat, C., 2004: La Transmission De L'islam Auprès Des Nouvelles Générations De La Diaspora. *Social Compass*, **51**(2): 235–247.

Salazar, A., S. Mohanty and J. Malaga, 2005: 2025 Vision for Mexican Chicken Consumption. *International Journal of Poultry Science*, **4**(5): 292–295.

Saunders, H. D., 2008: Fuel Conserving (and Using) Production Functions. *Energy Economics*, **30**(5): 2184–2235.

Sen, A. K., 1999: *Development as Freedom*. Oxford University Press, Oxford, UK.

Shatenstein, B. and P. Ghadirian, 1998: Influences on Diet, Health Behaviours and Their Outcome in Select Ethnocultural and Religious Groups. *Nutrition*, **14**(2): 223–230.

Shove, E., 2003: *Changing Human Behaviour and Lifestyle: A Challenge for Sustainable Consumption?* Department of Sociology, University of Lancaster, Lancaster, UK.

Sinha, R., A. J. Cross, B. I. Graubard, M. F. Leitzmann and A. Schatzkin, 2009: Meat Intake and Mortality: A Prospective Study of over Half a Million People. *Arch Intern Med*, **169**(6): 562–571.

Smil, V., 2002: Eating Meat: Evolution, Patterns, and Consequences. *Population and Development Review*, **28**(4): 599–639.

Smith, C., S. Hall and N. Mabey, 1995: Econometric Modelling of International Carbon Tax Regimes. *Energy Economics*, **17**(2): 133–146.

Sobel, M. E., 1981: *Lifestyle and Social Structure – Conceptions, Definitions and Analyses*. Academic Press, London, UK.

Srivastava, M., 2008: *Carbon Saved by Reducing Meat Consumption*. http://www.dothegreenthing.com/wiki/display/WIKI/Eat+less+meat+and+poultry (accessed 25 November, 2011).

Steenkamp, J.-B., 1993: Foor Consumption Behavior. In *European Advances in Consumer Research*. W. F. Van Raaij and G. J. Bamossy (eds.), Association for Consumer Research, Provo, UT, USA, 1, pp. 401–409.

Steptoe, A., T. M. Pollard and J. Wardle, 1995: Development of a Measure of the Motives Underlying the Selection of Food: The Food Choice Questionnaire. *Appetite*, **25**(3): 267–284.

Sterner, T., 2007: Fuel Taxes: An Important Instrument for Climate Policy. *Energy Policy*, **35**(6): 3194–3202.

UNDP, 1990: *Human Development Report 1990*. United Nations Development Programme (UNDP), New York, NY, USA.

UNDP, 2010: *Human Development Report 2010*. United Nations Development Programme (UNDP), New York, NY, USA.

Vandenbergh, M. P., P. C. Stern, G. T. Gardner, T. Dietz and J. Gilligan, 2010: Implmenting the Behavioral Wedge: Designing and Adopting Effective Carbon Emissions Reduction Programs. *Environmental Law Reporter*, **40**: 10547.

Vaughan, G. M., 1977: Personality and Small Group Behavior. In *Handbook of Meodern Personality Theory*. R. B. Cattell and R. M. Dreger (eds.), Academic Press, London, UK.

Vringer, K. and K. Blok, 1995: The Direct and Indirect Energy Requirements of Households in the Netherlands. *Energy Policy*, **23**(10): 893–910.

Wahlander, J., 2004: *Förutsättningar För En Minskning Av Växthusgasutsläppen Frånjordbruket*. Rapport 2004:1, Jordbruks Verket, Stockholm, Sweden.

Wakefield, M. A. and F. J. Chaloupka, 1998: Improving the Measurement and Use of Tobacco Control "Inputs". *Tobacco Control*, **7**(4): 333–335.

Walker, P., P. Rhubart-Berg, S. McKenzie, K. Kelling and R. S. Lawrence, 2005: Public Health Implications of Meat Production and Consumption. *Public Health Nutrition*, **8**(04): 348–356.

Wallén, A., N. Brandt and R. Wennersten, 2004: Does the Swedish Consumer's Choice of Food Influence Greenhouse Gas Emissions? *Environmental Science Policy*, **7**(6): 525–535.

Weber, C. L. and H. S. Matthews, 2008: Food-Miles and the Relative Climate Impacts of Food Choices in the United States. *Environmental Science & Technology*, **42**(10): 3508–3513.

Wilhite, H., H. Nakagami, T. Masuda, Y. Yamaga and H. Haneda, 1996: A Cross-Cultural Analysis of Household Energy Use Behaviour in Japan and Norway. *Energy Policy*, **24**(9): 795–803.

Wind, Y. and P. Green, 1974: Some Conceptual, Measurement and Analytical Problems in Lifestyle Research. In *Life Style and Psychographics*. W. D. Wells (ed.), American Marketgin Association, Chicago, pp. 97–126.

World Bank, various years: *The Little Green Data Book*. World Bank, Washington, D.C., USA.

Worldwatch Institute, 2010: *The State of the World 2010 – Transforming Cultures from Consumerism to Sustainability*. W.W. Norton & Company, New York, NY, USA.

Wrieden, W. L., A. S. Anderson, P. J. Longbottom, K. Valentine, M. Stead, M. Caraher, T. Lang, B. Gray and E. Dowler, 2007: The Impact of a Community-Based Food Skills Intervention on Cooking Confidence, Food Preparation Methods and Dietary Choices – an Exploratory Trial. *Public Health Nutrition*, **10**(02): 203–211.

Ziesemer, J., 2007: *Energy Use in Organic Food Systems*. Natural Resources Management and Environmental Department, Food and Argiculture Organisation (FAO), Rome, Italy.

CLUSTER **4**

Chapter 22–25

22

Policies for Energy System Transformations: Objectives and Instruments

Convening Lead Author (CLA)
Mark Jaccard (Simon Fraser University, Canada)

Lead Authors (LA)
Lawrence Agbemabiese (United Nations Environment Programme)
Christian Azar (Chalmers University of Technology, Sweden)
Adilson de Oliveira (Federal University of Rio de Janeiro, Brazil)
Carolyn Fischer (Resources for the Future, USA)
Brian Fisher (BAEconomics, Australia)
Alison Hughes (University of Cape Town, South Africa)
Michael Ohadi (University of Maryland, USA)
Kenji Yamaji (University of Tokyo, Japan)
Xiliang Zhang (Tsinghua University, China)

Contributing Authors (CA)
Igor Bashmakov (Center for Energy Efficiency, Russia)
Sabine Schnittger (BAEconomics, Australia)
Julie Tran (British Columbia Utilities Commission, Canada)
David Victor (University of California, San Diego, USA)
Charlie Wilson (Tyndall Centre for Climate Change Research, UK)

Review Editors
Mohan Munasinghe (Munasinghe Institute for Development, Sri Lanka and University of Manchester, UK)
Ian Johnson (Club of Rome, Switzerland)

Contents

Executive Summary

The Global Energy Assessment (GEA) emphasizes the importance of energy to all societies, which explains a longstanding tendency for governments to be closely involved in the energy sector. The nature and extent of this involvement – the degree and types of energy-related policies – depends on a government's ideological orientation, the particular energy resource endowment in its jurisdiction, the development level of its economy, and specific concerns of its society with respect to energy access, energy security, and the environmental and human health impacts of energy supply and use.

In every country, energy's critical role for the goal of sustainable development is widely acknowledged. This means that energy-related policies need to be assessed in terms of performance with respect to the social, economic, and environmental dimensions that are encompassed by the concept of sustainable development. Ideally, energy-related policies will make advances with respect to all three of these critical sustainability dimensions. But frequently policymakers are faced with difficult trade-offs in which improvement in one dimension is at the cost of another. Thus, the first goal of energy-related policy design should be to seek win-win opportunities for simultaneously advancing social, economic, and environmental goals. When this is not possible, the goal should be to apply decision-support mechanisms that integrate diverse social objectives and values into the policy design process, such as the application of multi-criteria analysis as described by Munasinghe (1992; 2009).

Cluster I of GEA presents social, economic, and environmental dimensions of sustainable development as related to energy. These include social goals (Chapter 2), environmental protection (Chapter 3), human health (Chapter 4), energy security (Chapter 5), and economic development (Chapter 6). GEA establishes specific normative goals in these areas, chosen to reflect broad societal aspirations with respect to (1) alleviation of poverty, including universal access to modern forms of energy; (2) improved human health, including quality of life indicators; and (3) environmental sustainability, including indicators of biodiversity, water quality, and air quality.

Cluster II assesses the potential contribution of energy to these normative goals with detailed analyses of the existing and potential energy system. This includes energy efficiency-reducing energy use, while sustaining and even increasing energy services. It also includes specific energy supply options associated with socially acceptable applications of renewable energy, safe uses of nuclear power, and cleaner uses of fossil fuels. Many of the chapters within this cluster describe detailed policies with respect to, for example, energy efficient buildings or the safe use of nuclear power or protection of human health.

Cluster III provides an integrating analysis of these energy options. Of particular note is the exploration in Chapter 17 of a set of scenarios or pathways showing the potential evolution of the energy system on a sustainable development trajectory. With reduced energy use – i.e., greater energy efficiency, conserver lifestyles, urban form changes, reductions in material and water use that cause reduced energy use – potentially making a great contribution, Chapter 17 includes one pathway in which the efficiency option is pushed particularly aggressively. This scenario is enhanced by the multiple co-benefits associated with less energy supply and use. A much-expanded role for renewable energy is likewise associated with multiple benefits in many jurisdictions, such as improved security of supply from diverse domestic sources and reduced greenhouse gas (GHG) emissions. Thus, a special emphasis on renewable energy plays a key role in some energy system pathways.

Finally, Cluster IV of GEA is focused on policy, the means by which humanity might realize these pathways of energy for sustainable development – pathways that require a rapid and dramatic transition of the energy system when seen from a global perspective. Cluster IV is comprised of four chapters; this introductory chapter (Chapter 22) surveys the spectrum of energy-related policy objectives and policy measures, including examples of government involvement in the energy sector to reach both general and specific objectives of energy for sustainable development.

Text boxes in this chapter provide examples of real world policy efforts to achieve energy-related objectives. The chapter includes key policies that must be applied to meet the goals of sustainable development. It also presents specific policies necessary to achieve the sustainable development energy pathways of Chapter 17. Some of these policies are generic, and could be applied in widely different contexts, while others are more focused.

This sets the stage for subsequent chapters that focus on three specific policy challenges, namely: (1) extending energy access in the developing world (Chapter 23); (2) stimulating energy system innovation (Chapter 24); and (3) building human and institutional capacity for energy transition, especially in developing countries (Chapter 25). Finally, it should be noted that most chapters of GEA include some discussion of policies, in some cases quite detailed.

22.1   Introduction

For the purposes of policy analysis and design, the energy sector has specific policy challenges. These represent a further refinement and specification of social, economic, and environmental sustainability dimensions. These challenges reflect issues emerging throughout GEA. Some of them are only addressed via policy analysis and mechanisms presented here in Chapter 22. Market power in the energy sector, especially due to natural monopoly conditions in network industries like electricity, is one particular concern. Another is the challenge of effectively managing wealth associated with valuable resource endowments, especially from oil and natural gas. Yet another concern is the challenge of the energy system's contribution to GHG emissions and the risk of climate change. While energy-related emissions of GHGs are just one of many environmental impacts and risks associated with the energy system, the threat of climate change requires a global effort. Thus, preventing climate destabilization is a public welfare problem on a global scale and the transition of the energy system is a critical component in addressing this challenge.

Eight specific energy-related policy concerns, identified below, are addressed in this chapter.

1.   Because energy is such a critical input to the modern economy and development of society, *energy access* – at an affordable level to meet basic human needs and to offer opportunities for social and economic development – is a key objective. This is especially the case for developing economies and for energy services that benefit those at the lowest income levels. Because these segments of society are least able to afford energy systems on their own, often governments intervene to provide them publicly or through subsidies for an adequate level of service. Chapter 23 addresses this topic in detail.

2.   One mistake of past policy efforts aimed at improving energy access was to assume that technology adoption and capital formation were mainly linear processes of innovation and investment that depended little on the surrounding social, cultural, technical, financial, and institutional environment. Research into past failures, however, has indicated that an "enabling environment" is critical, and this has led to an increasing policy emphasis on *developing energy-related social capacities*. This refers to improving human, technical, financial, social, organizational, and institutional capacities, which will be critical in fostering a major transition of the energy system from its current character to one in which energy access is widespread and human and environmental impacts are dramatically reduced from today's situation. Chapter 25 addresses this topic in detail.

3.   One indicator of successful industrial economies is that they have organized their energy systems to ensure access to affordable energy for almost all members of society. But energy access does not necessarily ensure *energy security*, the assurance that this access will not be overly vulnerable to major technological problems, geopolitical conflict, terrorist sabotage, or other significant types of disruptions. In this chapter the key components of energy security – as described in Chapter 5 – are summarized, and policies to foster social and economic aspects of sustainable development are presented.

4.   The production and delivery of energy is capital intensive, requiring large investments in finding, extracting, developing, processing, transporting, and retailing. Capital-intensive activities are frequently associated with economies-of-scale, meaning that a large facility or company can provide energy at lower cost than several smaller companies. In certain sectors, like electricity, these economies-of-scale foster "natural monopolies" (utilities), where vesting a single firm with monopoly service provision is the lowest cost option for society. However, firms that operate as monopolies lack the incentive to provide goods at marginal cost as compared to competitive industries. Therefore, governments may employ policies or different types of controls – e.g., a regulated corporation – to *manage energy-related market power*. In some other energy sectors, like the petroleum industry, strong economies-of-scale mean that a few companies may be able to dominate various aspects of the market. Some degree of policy intervention may again be desirable, although unlike a natural monopoly this may involve efforts to reduce market power rather than to countenance the establishment of a regulated or state-owned monopoly.

5.   Governments face special challenges when *managing valuable energy resource endowments* and especially the substantial revenues they can generate. For countries that have few exports other than energy, government tends to play a special role in managing these exports and allocating the large earnings. The experiences with government management of these resources have varied widely. For some countries, highly valued energy resources, like oil and natural gas, have been a boon for national treasuries and a stimulus for economic growth. But such endowments can cause substantial challenges, as evidenced by mismanagement of resource revenues, negative effects on other sectors of the economy, and even rent-seeking corruption within government and industry. In the extreme, a "resource curse" can occur – the paradoxical outcome in which a country with a rich resource endowment remains poor, in part because of mismanagement of that endowment.

6.   With the expansion of the world's economy, the need has grown dramatically to *reduce environmental and human health impacts* of the energy system. Energy use affects humans and the planet's ecosystems at all scales, from uncontrolled indoor air emissions when combusting solid fuels for domestic cooking and heating in unventilated areas, to rising sea levels, extreme weather events, and new epidemiological threats from climate change. Public

policy has a critical role to play in ensuring that the impacts and risks from energy supply and end-use are reduced, in some cases dramatically. Chapter 3 disscusses the relationship between energy and the environment while Chapter 4 addresses energy and health in further detail.

7. If humanity is to achieve significant progress in pursuit of objectives like improved energy access and security, or reduced environmental impacts, it will need to *accelerate the rate of innovation, development, and dissemination of desirable energy-related technologies*. Governments can pursue policies that aim to overcome barriers to such technological transitions, ranging from public support for research and development (R&D), to programs that assist the initial market penetration of targeted technologies, to educational and institutional developments that foster an enabling environment for adopting and sustaining new technologies. Chapter 24 addresses this topic in detail.

8. The increasing globalization of the world's economic system is equally reflected in its energy system. Oil has long been traded on a global basis and this is now the case for coal and natural gas. Some regions also trade electricity. At the same time, the environmental impacts of energy use are also becoming global, especially the risk of climate change, caused in large part by the production and combustion of fossil fuels. These international dimensions of many energy-related challenges require a much greater effort to improve *coordination and implementation of international energy-related policies*, a necessity that has been recognized during the past few decades by major energy sector participants.

These eight energy-related policy goals present governments with a complex array of issues. But the policy challenge is complicated further by the need for consistency between energy-related policies, on the one hand, and the host of other government policies focused on goals like poverty alleviation, economic development, education, health, national security, and macro-economic stability, on the other. Policy consistency is a great challenge and it always should be a major focus of the design and evaluation of policy options.

In addressing these energy-related policy goals, governments have various instruments available to them. While the potential options are numerous, in a generic sense, policy is usually manifested as:

- direct public ownership or control;

- regulations and standards;

- information, education, and public engagement to promote voluntary actions;

- financial charges, such as taxes and fees; and

- subsidies, such as grants, low-interest loans, and rebates.

In choosing among these policy options, governments can rely on key evaluative criteria that policy analysts conventionally apply when assessing policies (Hahn and Stavins, 1992). These criteria are used to assess the ability of different policy options to meet their goals in a number of different ways:

- *effectiveness* – the ability of a policy to achieve the intended objectives;

- *economic efficiency* – the ability of a policy to achieve objectives at the lowest possible cost to society;

- *administrative feasibility* – the ability of a policy to avoid imposing a functional burden on government that thwarts successful implementation, such as through bureaucratic ineffectiveness or excessive information and monitoring requirements;

- *equity* – the effect of a policy on income distribution and on disadvantaged groups within society;

- *political acceptability* – the extent to which a policy can garner sufficient political support to be enacted and effectively sustained;

- *policy robustness* – the ability of a policy to perform well under highly uncertain and widely contrasted futures; and

- *policy consistency* – the extent to which a policy works in concert and not in conflict with other policies.

Policy effectiveness is often more challenging than it appears to non-experts. One key reason is that governments and public decision-making processes are sometimes portrayed in a simple way that fails to recognize factors that can cause policy impairment and policy failure. In fact, the list of factors is quite large. Small groups who are negatively affected by a policy, which otherwise has a net social benefit, can block the policy if they have strong leadership or include powerful interests. This kind of challenge may impede the development of some forms of renewable energy, for example. Powerful groups in society may gain excessive influence over the bureaucracy or politicians and may act more in their special interests rather than in the broader social interest.

Once implemented, a policy may shift the incentives in society such that people act in ways that counteract a policy's intent; this is sometimes called "moral hazard" in policy design. Corruption of politicians or bureaucrats is always a risk, though hopefully less so in more open societies. Policies that seem good on the surface could have very high transaction costs that thus impede their successful implementation. Finally, if the benefits of a public good are accessible to all, there is a risk of free-riders, agents who do not contribute to the costs yet share in the benefits, which erodes the social will to provide the public good. All of these factors must be considered when designing and comparing the effectiveness of policy options.

In the same vein, other policy evaluation criteria each present their own specific challenges. Economic efficiency is difficult to achieve when those who may be negatively affected by a policy pressure government to reduce the costs they would face under an efficient outcome. Careful policy design, however, can balance the equity objective and the efficiency objective by various redistributive mechanisms that reduce inequitable outcomes without blunting the economic efficiency requirement for clear pricing signals to suppliers and consumers.

When comparing or combining policy options, designers of energy-related policy need to assess performance against these criteria. Because no single policy is likely to outperform all others in all of these criteria, some form of multi-criteria decision analysis is an essential component of the policy development process. The outcome of this analysis will depend on the relative weights that decision makers, and society as a whole, place on each of the criteria. In some circumstances equity concerns may have an especially strong weighting, while in others it may be environmental effectiveness or economic efficiency. For a given energy policy objective, decision analysis is useful to analyze the multiplicity of policy options and of policy evaluative criteria and help the decision maker assess the trade-offs associated with one or a package of policies.

The following list provides a general description of some key trade-offs that policymakers must navigate when designing and evaluating policies for sustainable energy objectives.

- Policymakers must find an appropriate balance, in a given circumstance and setting, between the role of public authorities, on the one hand, and the delegation of responsibility for decisions about land-use, technology, and behavior to firms and households acting in a market context, on the other. Thus, in some cases, governments may wish to promote and support – with R&D, subsidies, and focused regulation – a specific technology or fuel in order to ensure its rapid dissemination. In other cases, governments may try to leave it to the market to determine which technology or fuel will actually emerge, especially when government is faced with great uncertainty about future technological evolution, costs, and preferences.

- Policymakers must find an appropriate balance between policies that are going to drive rapid change and policies that have a good chance of political acceptance. Ideally, a policy does well on both counts, but often some degree of trade-off is required.

- While policies should be as simple as possible from an administrative and bureaucratic perspective, it is important that they include sufficient sophistication in terms of experimental design, monitoring,

and hindsight evaluation so that their effectiveness can be assessed and their design improved over time.

- Policy assessments are necessary to help navigate choices about policy design when there are vested interests making strong claims about evidence that may or may not be supported by rigorous research. Examples of such complex choices include a tax on greenhouse gas emissions versus cap-and-trade systems, or fixed feed-in tariffs that guarantee a price for renewable electricity versus renewable portfolio standards that guarantee a market share.

- Policymakers must find an appropriate balance between the benefits to current versus future generations from valuable energy resource endowments. The outcome of this trade-off analysis could vary considerably, depending on: (1) the current level of well-being in the country or region; (2) the expected longevity of the resource endowment; (3) the likely future value relative to the present; and (4) the capacity of a government to ensure benefits to future generations through saving and reinvesting current returns from resource exploitation.

- Consideration must also be given to the stringency of the targets set by a policy, such as the level of greenhouse gas emission abatement. Policymakers have to decide whether the future costs and risks of not taking action – e.g., on climate change – outweigh the costs and risks to society of immediate action. These decisions will involve an assessment of the trade-off between the current and future welfare of society based, in part, on the risk of a negative outcome in the future.

Within the context of all these factors, another influence on policy choice is the ideological expectation of the role of government in the energy system. Decisions in this regard often hinge on a larger political debate around control over the economy. Should governments play a strong role, or are they willing to keep a distance and vest outcomes in individual decisions made by the myriad of actors in energy markets? The answer to this depends on institutions, goals, values, and resource endowments that can vary dramatically by country and region. Policies that work for highly industrialized countries, with a long tradition of market economies and relatively effective public institutions, may be inappropriate – or at least merit substantial modification – in developing economies. Countries that place great concern on the assured delivery of particular quantities of energy may be less willing to trust markets than are countries where competitively priced energy and market allocation has been the norm for decades. Countries where much of the population has no access to modern energy commodities may be less willing to assume that deregulated markets can meet the needs of their citizens if they have failed to do so thus far.

These kinds of decisions about the role of government and public policy in the energy sector vary by country, but also by energy service and commodity. Some countries, for example, have been wary about allowing market forces to govern commodities that the government considers strategic, even as they embrace markets in other parts of the economy. For most of the past few decades, the French government has played a dominant role in its electricity sector and in development of nuclear power, while playing a less dominant role in petroleum and natural gas. Energy policy in practice, therefore, requires that each country and region find the policy mix that best meets its goals and particular circumstances.

The following sections of this chapter include a detailed discussion of each of the eight energy-related policy goals described above. Each discussion includes descriptions and examples of policies available to further the goal, including case studies of real-world policies in different jurisdictions that present both successes and frustrations. Because three of these energy-related policy goals have separate chapters dedicated specifically to them – energy access, technology innovation, and capacity development (Chapters 23, 24, and 25, respectively) – the treatment here is cursory. Finally, it should be noted that most chapters of GEA include some discussion of policies, in some cases quite detailed.

22.2　　Part I: Eight Energy-related Policy Goals

22.2.1　　Increase Energy Access

Access to energy is one of the most urgent objectives for sustainable energy policy over the coming decades, which is why GEA dedicates Chapter 23 to this issue. Our treatment here is therefore limited to a brief description of the key issues for policy design.

About one quarter of humanity lives without access to electricity and more than a third has no access to liquid and gaseous fuels. The Millennium Development Goals for social and economic advancement will not be achieved without expanded access to clean and affordable energy. Access to energy is intimately linked with industrial productivity, communications, mobility, comfort, and other benefits that are key contributors to economic and social development. Access to electricity, in particular, contributes to higher levels of education, increased access to information, improved health care, and more effective institutions and communication networks. All of these improvements play a role in the alleviation of poverty and development of civil society (World Bank, 2000; UNDP and World Bank, 2005).

While markets can play an important role, universal access to energy is seen as a public responsibility because the benefits of its provision surpass those directly captured by individuals in their everyday participation in markets, extending to the wellbeing of all of society. During the twentieth century, governments in today's developed countries recognized the importance of access to electricity and used a combination of state enterprises and public subsidies

to develop electricity production and distribution systems. Rural electrification subsidy programs were particularly important in the widespread electrification that occurred over just a few decades, largely through the activities of public corporations or publicly supported cooperatives.

While the key focus of these subsidies was often expansion of the grid into poorer regions and non-urban areas, there were also policies to ensure that a minimum amount of electricity would be affordable to low-income households. For this, utilities developed electricity tariffs such as lifeline rates: a low initial price for base levels of electricity consumption, with higher prices for any additional consumption.

Similar policies have been pursued by developing countries for providing access to electricity. These governments and utilities usually fund expansion of the electric grid into underserved areas and provide special rates for low-income households and farmers.

Unfortunately, in many countries, expansion of electricity access has not kept pace with population growth, diminishing the likelihood of achieving the Millennium Development Goals for energy access. There are multiple reasons for this failure. First, the magnitude of the objective is truly daunting, requiring a rate of investment and electric capacity expansion that far exceeds what was even achieved in the most successful of industrialized countries over the past century. Second, in some cases, publicly operated utilities and state enterprises have performed poorly, notably in terms of making bad investments, exercising poor management practices, providing unjustifiable cross-subsidies between customer classes, and countenancing corruption. Third, in some cases, private investment, whether local or foreign, has been misdirected or even detrimental to local socio-economic development.

Energy access policies may perform best when they find a balance between mobilizing market forces to provide funds for expansion of electricity generation and distribution systems, and involvement of governments and non-government organizations to ensure social development and access for society's poorest and most isolated members – access that would not occur if the sector were simply left to private markets. Moreover, it is essential to link energy access with policies that improve access to other government social programs such as education, health services, financial resources, and modern technologies. Only in this way can improving access to energy contribute effectively to full economic and social development.

Chapter 23 details policies that, among other things, seek to combine policy design and governance with social and market forces in ways that effectively advance access to energy. While certain types of policies can be applicable in many different cultures and environments, at different levels of technological adoption and economic development, it is often the case that policies need to be tailored to these specific conditions.

This is why Chapter 23 includes an array of policies and a degree of regional disaggregation in its policy prescriptions.

Examples of focused policies that combine a balanced role for government and markets include the following:

- For extension of electricity access into rural areas, a process that involves competitive bidding for state or utility subsidies can improve the chance that these will be used to maximum effect. For rural electrification, Chile developed a program in which subsidies were allocated on the basis of the maximum grid extension that would be achieved for a given amount of money. A similar program exists in Argentina called the Proyecto de Energías Renovables en Mercados Rurales (PERMER). These programs were intended to provide electrification to very isolated communities.

- Modest public and non-government support can help small rural businesses and rural households increase their access to energy investment financing. A frequently cited example is the Grameen Bank of Bangladesh, whose energy division, called Grameen Shakti, lends money for energy-related projects, such as solar home systems that involve installing photovoltaic solar panels to provide electricity in locations without access to an electric grid.

- Innovative institutional arrangements may be able to improve the efficiency with which electricity firms operate in developing countries, especially where the traditional, centrally-managed state electric utility lacks support within the local community. Unpaid bills and electricity theft are particular problems that have been reduced

by establishing greater local control over management of the local distribution system. An example is rural electric cooperatives in Bangladesh, called Palli Biddyut Samitee.

When considering energy access, electricity is often the focus of attention. But there are also dramatic benefits from access to clean cooking and heating fuels for the poorest people in the world. Poor indoor air quality from open combustion of solid fuels is still a major source of human morbidity and mortality, especially among women and children who tend to have the highest level of exposure to these emissions (Chapter 4). Today, governments in many developing countries have programs that support in some way – e.g., education, technology subsidies, technical training – the shift from household combustion of solid fuels in open fires to cleaner and more efficient use of commercial fuels, like kerosene and liquid petroleum gases (LPGs, such as propane and butane). To this end, governments may directly subsidize cleaner gaseous and liquid fuels or the acquisition of stoves that use them. Or, government might directly subsidize the acquisition of stoves that still use solid fuels like coal and biomass, but without deleterious emissions that effect indoor air quality.

Thus, many African countries have programs to make LPGs and kerosene more easily available in rural and peri-urban areas and also to distribute improved charcoal burning stoves. If successful, such programs have the combined benefits of reducing time spent gathering biomass, reducing the environmental impacts of excessive exploitation of biomass, and improving indoor air quality with major benefits for human health, especially for women and children.

Box 22.1 | Free Basic Electricity Program in South Africa – Alison Hughes

Free Basic Electricity (FBE) was introduced in South Africa in 2003 to complement an aggressive electrification program (refer to Chapter 19). The policy was implemented after it was realized that poor households were using less than 50 kWh of electricity each month (DME, 2003a). FBE allows poor households with a legal connection to use 50 kWh of electricity each month at no cost. It has the effect of reducing the cost of a kWh of electricity and providing a safety net for consumers.

A challenge with the FBE policy lies in identifying recipients of the allocation. FBE is implemented by distributors, mainly through a blanket allocation to households consuming less than a certain quantity of kWh each month (DME, 2003a).

Pilot studies undertaken after FBE was first introduced have shown that the subsidy typically raised electricity use for lighting and other uses, allowed a more continuous use of electricity by households, and lowered the household energy bill. In this way, it can be seen as very effective. However, on average, the use of electricity increased by less than the subsidy amount of 50 kWh (DME, 2003b), which implies that poorer households continued to use alternative fuels for thermal services.

Criticism of the implementation of the policy is that it excludes many low-income households because families living in backyard shacks or whose homes are situated on land not zoned for settlement cannot receive the subsidy. It can also be argued that the policy should be extended to allow households access to units of any type of energy. This would benefit households by allowing them to purchase a fuel of their choice for cooking at a lower cost to society (Howells et al., 2005).

Box 22.2 | Bagasse-based Cogeneration in Mauritius – Stephen Karekezi

A clearly defined government policy on the use of bagasse, a by-product of the sugar industry, for electricity generation has been instrumental in the successful implementation of the cogeneration program in Mauritius. Plans and policies have been worked out over the last decade for the sugar industry in general. First, in 1985, the Sugar Sector Package Deal Act (1985) encouraged the production of bagasse for the generation of electricity. The Sugar Industry Efficiency Act (1988) provided tax incentives for investments in generation of electricity and encouraged small planters to provide bagasse for electricity generation. Three years later, the Bagasse Energy Development Programme (BEDP) for the sugar industry was initiated. In 1994, the Mauritian government abolished the sugar export duty, an additional incentive to the industry. A year later, foreign exchange controls were removed and the centralization of the sugar industry was accelerated. These and other measures are summarized in the following table.

Table 22.1 | History of Bagasse Policy in Mauritius.

Year	Policy initiatives	Key objectives/Areas of focus
1985	Sugar Sector Action Plan	– Bagasse energy policy evoked
1988	Sugar Industry Efficiency Act	– Tax free revenue from sales of bagasse and electricity – Export duty rebate on bagasse savings for firm power production – Capital allowance on investment in bagasse energy
1991	Bagasse Energy Development Programme	– Diversification of energy base – Reduction of reliance on imported fuel – Modernization of sugar factories – Enhanced environmental benefits
1997	Blue Print on the Centralization of Cane Milling Activities	– Facilitated closure of small mills with concurrent increase in capacities and investment in bagasse energy
2001	Sugar Sector Strategic Plan	– Enhanced energy efficiency in milling – Decreased number and increased capacity of mills – Favored investment in cogeneration units
2005	Roadmap for the Mauritius Sugarcane Industry for the 21st Century	– Reduction in the number of mills to six with a cogeneration plant annexed to each plant
2007	Multi-annual Adaptation Strategy	– Reduction from 11 factories to four major milling factories with coal/bagasse cogeneration plants (Belle Vue, FUEL, Medine, and Savannah) – Bio-ethanol production for the transport fuel markets. Spirits/rum and pharmaceutical products, e.g., aspirin – Commissioning of four 42 MW plants and one 35 MW plant operating at 82 bars - Promotion of the use of cane field residues as combustibles in bagasse/coal power plants to replace coal

As a result of consistent policy development and commitment to bagasse energy development in Mauritius, the installed capacity of cogeneration power has increased over the years. In 1998, close to 25% of the country's electricity was generated from the sugar industry, largely using bagasse. By 2001, of the total electricity supply in the country, 40% (half of it from bagasse) was electricity generated from sugar estates. It is estimated that modest capital investments, combined with judicious equipment selection, modifications of sugar manufacturing processes to reduce energy use in manufactured sugar, and proper planning could yield a 13-fold increase in the amount of electricity generated from sugar factories and sold to the national Mauritius power utility.

Bagasse cogeneration has delivered a number of benefits, including reduced dependence on imported oil, diversification in electricity generation, and improved efficiency in the power sector in general. It is available 100% of the time as long as bagasse production is in place, thus enhancing Mauritius' energy security. Bagasse, as a waste product, can lead to environmental problems such as fire hazards and methane emissions, which are considered potent greenhouse gases, if it is not disposed of properly. Thus, its use for power generation delivers significant local environmental as well as climate benefits. In addition, carbon dioxide produced by bagasse-based cogeneration is minimal, so it is considered a carbon-neutral option.

Cogeneration in Mauritius benefits all stakeholders through a wide variety of innovative revenue-sharing measures. The cogeneration industry has worked closely with the Government of Mauritius to ensure that substantial benefits flow to all key stakeholders of the sugar economy, including the smallholder sugar farmer. The equitable revenue sharing policies that are in place in Mauritius provide a model for emulation in ongoing and planned modern biomass energy projects in Africa. By sharing revenue with stakeholders and the small-scale farmer, the cogeneration industry was able to convince the government – which is very attentive to the needs of the small-scale farmers as a major source of votes – to extend supportive policies and tax incentives to cogeneration investments.

Box 22.3 | Government Assistance Programs for Shifting from Traditional to Advanced Technology: Solar Water Heating in Barbados and Brazil – Lawrence Agbemabiese

Government assistance programs are needed to catalyze transitions to cleaner energy technologies. This is particularly true of developing countries where poorly developed market mechanisms consistently fail to bring critical financial and economic systems into proper alignment with new technological windows of opportunity (Kaufman and Milton, 2005). A brief comparative assessment of solar water heating (SWH) initiatives in Barbados and Brazil lends credence to the view that government assistance programs are critical for replicating successful models of clean energy transitions – beyond the current handful of developing countries where this has happened.

Barbados boasts one of the highest per capita rates of SWH system ownership in the world, with more than 35,000 installed systems serving close to 40% of all households. A very active government SWH incentive program, sustained over a long period of time, is one of the main factors responsible for this achievement. In 1974 "an informed Prime Minister ... seeking ways to reduce oil dependency" of the country presided over the promulgation of the Fiscal Incentives Act (Perlack and Hinds, 2003). The Act exempted SWH raw materials, such as tanks and collectors, from the 20% import duty, effectively lowering the cost of installing a system by 5 to 10%. Simultaneously, the Act placed a 30% consumption tax on conventional electric water heaters. In 1980, an income tax amendment provided for the deduction of the full cost of a SWH installation. Though suspended briefly in 1993 as part of structural reforms, the deduction was reinstated in 1996, allowing homeowners to deduct up to 3500 Barbados Dollars (US$1750) per year to cover solar water heaters, among other home ownership costs (Perlack and Hinds, 2003). These tax incentives were paralleled by a government policy of procuring SWH systems for public buildings and publicly funded projects. This combination of incentives and programs achieved the desired result of triggering and accelerating demand for SWH in the country.

In contrast, the absence of such active and direct government support for the SWH industry in Brazil is associated with a relatively low rate of penetration of SWH systems in its energy market. The government does offer some tax incentives to households that purchase and install renewable energy technology, but these incentives are insufficient to bring SWH systems within the reach of the majority of Brazilian households, especially in poorer districts. With 20 years of experience and a large network of manufacturers, distributors, and retailers, Brazil has significant opportunities to dramatically increase the rate of SWH penetration. Compared to the case of Barbados, what is missing in Brazil appears to be the right set of end-user financing solutions backed by deliberate government assistance programs targeting poorer households.

Barbados, and a growing number of developing countries where solar thermal systems are becoming commonplace, share one thing in common: a history of strong political support, expressed through consistent market-transformative policies favoring technologies tailored to local needs. For solar water heaters, the more effective measures have included (re)structuring building and construction codes specifically designed to create or expand markets, building manufacturing capacity of local enterprises for certain components, and creating new financial incentives for suppliers and consumers.

While there are no doubt other factors at work, it is hard to deny the role government policy, including direct assistance programs, have played in some developing countries in the transition to cleaner energy technologies.

22.2.2   Develop Capacity for Energy Transitions

Whether the focus is the provision of electricity or access to cleaner fuels, a critical issue is the extent to which a given jurisdiction has the social, cultural, technical, financial, and institutional environment to develop, adopt, and sustain desired technologies and energy forms. This is why GEA devotes Chapter 25 to the issue of capacity development. Our treatment here is limited to a brief discussion of key issues for policy design.

Evaluations of failed efforts to rapidly expand energy access in developing countries suggest that these efforts foundered in part because of inadequate attention to capacity building in both developed and developing countries. Key elements of capacity development include general education, technical training, trade and cooperative associations that support an array of energy technologies and services, effective financial institutions that support a diversity of corporate sizes and organizational arrangements, a trusted legal system, innovator rights, and protection for investors. Addressing these fundamental necessities will lay the foundation for a transition to a more sustainable global energy system.

Societal efforts to shift the energy system to a different technological trajectory can benefit from a systems perspective that conceptualizes energy systems as socially embedded and historically shaped by the habits, practices, and norms of the actors within them. As new renewable energy technologies emerge in niche markets or are introduced in new environments,[1] earlier institutional frameworks and standards can raise barriers that slow the process of energy transition, as illustrated by case studies of distributed energy systems, off-grid energy solutions, and current practices in patenting (Geels, 2002; Martinot and Birner, 2005; Jacobsson and Lauber, 2006; Bergek et al., 2008).

The long term nature of energy transitions implies that capacities at the actor and systems level will change over time. Change in capacity requires feedback, flexibility, and complementarity in the design of strategies and policies for energy transitions. It demands an openness to new approaches in the formulation of legal and institutional frameworks that stimulate the development and diffusion of renewable and cleaner energy technologies, and closer attention to building capacities needed for continuous learning and innovation. Capacities that enable users, innovators, and policymakers to access knowledge and information, evaluate choices, build coalitions, and limit the negative impacts of change are critical components in an energy transition process.

To assist in this process, Chapter 25 introduces the "capacity matrix" as a tool for conceptualizing capacity development from a broad, systemic perspective. This perspective looks at habits and practices of the actors, existing norms, policies, and standards, as well as technical skills and access to information that impact on energy transitions. The capacity matrix is used to highlight the range of capacities needed in technologies, such as smart grids, small hydro for off-grid environments, wind power, and biodiesel from algae. The matrix is applied to analysis of capacity building in case studies throughout Chapter 25.

Policies for capacity development thus emphasize the dynamics of change in energy transitions, based on the capabilities, habits, and practices of all actors affecting the energy system. This is true not only in the case of helping developing countries accelerate their adoption of clean, efficient energy supply and use technologies, but also in the case of helping industrialized countries with the profound transition that is required for a more sustainable energy system. To this end, Chapter 25 includes a degree of regional disaggregation in recognition of the fact that there is no one-size-fits-all strategy for capacity development, but rather that efforts must be tailored to specific attributes of a given continent, region, country, and even sub-national locales.

22.2.3   Enhance Energy Security

While energy access emphasizes the provision of energy to individual households and communities, energy security is primarily a national-level concern focused on the uninterrupted provision of nationally vital energy services at affordable prices. As argued in Chapter 5, the transport and electricity sectors are key for all countries, whereas residential heating and industrial energy are important for many. Current energy security strategies are often focused on ensuring uninterrupted supply of forms of energy appropriate for these few sectors: liquid fuels for transport, primary energy sources and infrastructure for electricity generation, and various forms of energy for the residential and industrial sectors.

While some energy analysts argue that markets can provide sufficient energy security, others point out that in many contexts long-term security is a public good that can be "undersupplied," even by perfectly functioning markets. For example, in the absence of government regulations, markets often do not ensure adequate investments in spare generation capacity, overall system reliability in the face of extreme natural events, or may tend to favor procuring energy from politically less-reliable suppliers.

Chapter 5 distinguishes between different types of energy security concerns:

- systemic risks inherent in the design and operation of energy systems, including availability of resources, reliability of energy infrastructure, and rapid growth in energy demand;

- risks associated with control over energy resources and threats of hostile actions; and

- unexpected risks which more resilient systems are better able to withstand.

---

1   The concept of "newness," as used here, includes technologies that are new to a country, region, or user, though they may not be new to the world.

Governments can address these security and reliability concerns with a number of instruments and strategies. To address systemic risks, system design and operation should focus on reducing the likelihood of major system failures. Because this type of risk is internal to the system, governments have a large role to play in ensuring that their own domestic energy system is reliable. In this sense, these are domestically-managed risks. An electric grid, for example, should be designed to have reliable reserve capacity, regular maintenance of existing facilities, and an economically justified level of back-up generation and transmission systems to reduce the chance of system failure.

Energy demand management and forecasting is an important component of ensuring that energy supply expands at a pace that matches the growth in demand. Because energy supply facilities frequently require large investments and a long time to complete, energy forecasts should extend at least a decade or two into the future. While individual energy companies – e.g., electric, oil, natural gas – are usually active in forecasting, government should also play a role in providing a forecasting framework with assumptions about its own investments and energy needs, plus a means of coordinating the forecasts of these different entities. An aggressive energy efficiency strategy can also slow the pace of demand growth, which eases timing pressures on new supply investments.

In the case of non-electric energy demand, various requirements for domestic stockpiling of energy supplies – e.g., storage facilities for oil, natural gas, and coal – and for easier access to energy imports in the case of a failure of one component of the delivery infrastructure – such as multiple pipelines, transmission interconnects, and port facilities – can all contribute to reducing structured reliability risks.

Longer-term systemic risks are often associated with the availability, accessibility, and acceptability of primary energy resources on the national, regional, or global levels, and may be addressed by switching to more abundant, accessible, and acceptable energy resources. In some cases such a switch would mean from non-renewable fossil fuels to renewable energy sources, but this depends on the relative abundance of resources in a given location. In all changes, the decision of government to foster a particular energy source should only be taken after a risk-based economic assessment of such a strategy. In some cases, the costs of reducing structural risks are not justified by the benefits they may provide.

With respect to risks from foreign actions, some governments pursue energy self-sufficiency in order to increase energy security by fostering development of domestic energy supplies. This is an approach that European and the United States governments have debated from time to time (Kalicki and Goldwyn, 2005). But efforts of these countries pale in comparison to the commitments made in the policy that Brazil has pursued since the late 1970s, when it launched a strategy of substituting domestically produced vehicle fuel from sugar cane for oil imports,

which it perceived as too costly and volatile. Again, such a decision should be supported by risk-based economic analysis. While this was an implicit aspect of the Brazilian decision, a more explicit analysis can make the policy decision more apparent. Of course, policies, such as that of Brazil should be assessed on the full range of potential implications, including, in this case, macroeconomic effects and environmental impacts at local and global levels.

Energy self-sufficiency can be exercised at the sub-national or local level as well. Thus, many cities throughout the world, in both developed and developing countries, are pursuing decentralized energy strategies that involve greater energy efficiency – in buildings, land use, and infrastructure – with more local supply of energy. The latter may entail, for example, small-scale, urban cogeneration of power and heat for local distribution, on the one hand, and on-site solar power for electricity, water heating, and space heating, on the other. While this strategy is generally not intended to eliminate risks from external disruptions, it can reduce the degree of vulnerability to such risks.

Another strategy for addressing risks from foreign origins is through increasing national control over imported energy resources. For example, Chinese national energy companies pursue acquisition of oil-related assets and companies around the world as a way to increase leverage over the international oil market and secure oil supplies to China. The United States maintains a large military presence in the Persian Gulf region and off the coast of West Africa to minimize the risks of hostile actions against oil production and trade. Needless to say, such ambitious strategies are only available to a handful of major economies and their net benefits are not certain.

Another strategy of increasing control over energy imports is "multi-sourcing" energy supply to minimize dependence on a single source or a particular region, and/or long-term, fixed price supply contracts to protect supply sources from price instability. For example, various pipelines (Nabucco, Nord Stream, etc.) proposed and constructed to link natural gas deposits in Russia and Central Asia to European markets are intended to increase the diversity of supply routes and energy exporters, thus minimizing the consequences of potential disruptions of any one export source or transit route.

Various international institutional arrangements may also increase energy security. These range from long-term bilateral supply contracts – especially prominent in the Eurasian natural gas market – to multi-national institutions and protocols for coordinated action. The latter include the Organization of the Petroleum Exporting Countries (OPEC), which aims to stabilize oil markets and returns to oil producing countries, the International Energy Agency (IEA), established in 1974 by the Organisation for Economic Co-operation and Development (OECD) countries to coordinate oil supply strategies, and the European Energy Charter Treaty, which focuses on diverse forms of energy traded in Eurasia.

Box 22.4 | Strategic Oil Reserves – David Victor

Over the three decades since the Arab oil embargo of the early 1970s, most of the large oil-consuming nations have accumulated substantial strategic oil reserves. The United States alone has spent nearly US$50 billion in today's money to build and maintain a huge strategic stockpile of crude oil. Other large oil importers – notably in Europe and Japan – have also spent heavily to accrue their own reserves. Large new oil consumers, notably China, are in the early stages of building strategic reserves.

In theory, a well-coordinated system of oil caches can provide a buffer against harmful shocks to the world oil market, which makes them important tools of economic policy as well as elements of an effective foreign policy. In theory, oil importers can use their reserves to prevent exporters from brandishing the oil weapon when markets are tight while also making their economies less vulnerable to trouble along critical supply routes, such as the straits of Hormuz, through which about one-third of all the world's oil exports travel.

Because oil is a fungible global commodity, making effective use of oil reserves requires international coordination. That logic inspired the creation of the International Energy Agency (IEA), an arm of the OECD, in 1974. IEA is a forum for governments to discuss energy policy and, in crisis, to coordinate release of strategic oil and other emergency measures.

IEA members have drawn up contingency plans to release strategic oil at critical times, such as the eve of the 1991 Gulf War; in anticipation of the calendar rolling over to 2000 and causing unknown computer glitches that could affect energy supplies; shortly after September 11; and in the aftermath of Hurricanes Katrina and Rita in 2005. IEA members actually collectively released strategic oil reserves twice: two and a half million barrels per day in 1991 and two million barrels per day for 30 days in 2005. Like most deterrents, strategic oil reserves are rarely used in practice.

Historically, all IEA members have been drawn from the ranks of OECD membership. But the rise of emerging economies as large oil consumers – notably China, which is building its own oil reserve – is forcing IEA members to find new more flexible ways to engage countries outside the OECD.

Today's oil market is very different from the one that existed when governments created oil reserves and had more direct control over the quantity and price of oil. Today, oil prices arise through trading on complex financial markets, and this new market has led to calls for changing the systems for managing oil reserves and treating them akin to the array of other financial instruments that governments manipulate as part of economic policy.

Source: Victor and Eskreis Winkler, 2008.

Finally, robustness and resilience concerns are addressed through preparing for supply disruptions in order to minimize their impacts. All IEA members stockpile oil to protect against short-term supply disruptions. Natural gas is also stored in an increasing number of European states. Oil importers, such as European countries and Japan, levy high taxes on oil, which serves to decrease the oil intensity of their economies as a cushion against the macro-economic effects of oil price volatility.

22.2.4   Manage Energy-related Market Power

The conversion and delivery of energy is capital intensive, requiring huge investments in finding, extracting, developing, processing, transporting, and retailing. Capital-intensive sectors of the economy are often associated with a degree of market power, meaning that one or a few firms have considerable influence over the market. Market control by one firm is a monopoly. Market power held by several firms in concert is an oligopoly. Capital-intensive activities are frequently associated with economies-of-scale. For instance, a large facility or company can provide energy at lower cost than several smaller companies. In certain cases, these economies-of-scale foster natural monopolies (utilities), where vesting a single firm with monopoly service provision is the lowest cost option for society, resulting in that service being provided by either a public monopoly or a regulated private monopoly (Berg and Tschirhart, 1988).

With grid networks, as in the delivery of electricity, natural gas, and district heat, natural monopoly conditions are common. In many other settings that are potentially competitive, such as oil and natural gas extraction, high capital intensity and strategic political and economic considerations sometimes lead governments to countenance market dominance by a few large firms or even just one state-owned firm. Likewise, industrial activities that exhibit strong economies-of-scope – meaning that a firm producing one good or service is well-positioned to be the low-cost producer of related goods or services – can result in market power for just a few corporations. In sum, sectors that exhibit significant economies-of-scale, economies-of-scope, and oligopolistic conditions create difficult decisions for governments about the level of public intervention and effort in establishing market competition in society's best interest.

22.2.4.1   Electricity Sector

For most of the last century, conditions that favored monopolies were assumed to exist in various components of the electricity sector. Governments have responded in various ways.

In some cases, governments created publicly owned electric utilities. This strategy often reflected the fact that governments were the only entities willing to bear the risks associated with the construction of electricity networks in remote regions and linking distant sources of supply and demand. This strategy has also rested on the assumption that state-owned corporations would operate with the interests of the public in mind, although there has always been a concern that such corporations can become powerful entities unto themselves, unresponsive to broader public goals and perhaps poorly managed.

In some cases governments have allowed private ownership by a single, monopoly utility, one that is regulated by government or an independent utilities commission to set tariffs and approve major investments. Under the conventional utility commission, cost-of-service regulatory approach, the commission conducts public hearings in which the utility must justify, sometimes in great detail, the prudency of its investments. If approved, the commission will allow the utility to set rates that should, if the firm is effectively operated, enable it to recover its costs plus a return on investment that reflects the risks it faces.

These systems have generally worked where applied, notably in the United States and a few other power markets such as Hong Kong. But regulated private utilities are sometimes wary of taking risks with new technology, and regulators face difficulty in getting the information they need to play an effective role (Viscusi et al., 2005). Concerns have also been raised that this regulatory approach leads to overinvestment in basic infrastructure but lacks incentives for making improvements to operating efficiency (Stigler, 1971). As a consequence, utility regulators are moving toward performance-based ratemaking, an approach that does not limit the rate of return between rate hearings. This encourages the utility to continuously pursue profit maximizing efficiency gains, which the regulator will only translate into lower customer rates after

a considerable period of profit-taking by the utility – five years or perhaps longer.

Governments might redesign the network to separate parts of the energy supply chain that are amenable to competition from those where a monopoly is optimal. In the 1990s, most governments that tried to apply market forces did so by restructuring (Newberry, 1999). They unbundled the generation of electricity and often also the sale of electricity and gas to final consumers (where markets might be able to operate satisfactorily) from transmission and distribution (where competitive markets are often impractical).

The last two decades of attempts to restructure electric power systems, accompanied in some countries by a shift from central state ownership and planning toward a greater role for private entrepreneurship, is a useful period for exploring the larger questions of how energy systems and economies can be organized. The mixed experiences with electricity sector reform over the last two decades have shown the dangers, in some cases, of allowing assumptions about the inherent performance of markets to dominate real-world evidence about the special characteristics of a complicated industrial sector like electricity.

As part of separating generation from other components of the electricity system, restructuring advocates have called for the unbundling of generation assets into enough individual entities with separate ownership – private firms, municipal governments, cooperatives, and perhaps some retention of state ownership – that electricity supply competition might be possible. At the same time, the grid system would remain a monopoly in most cases; that is, independently regulated if owned by private firms. This model of electricity reform would comprise the following elements (Joskow, 2006):

- vertical separation of competitive market segments from regulated segments;

- horizontal restructuring of generation to ensure effective competition;

- horizontal integration of transmission and network operations and designation of a single system operator;

- creation of spot energy and operating reserve markets for real-time system balancing;

- unbundling of retail tariffs to separate competitive from regulated costs and prices;

- creation of independent agencies for regulating the network prices and services, including access; and

- establishment of transition mechanisms to deal with unforeseen challenges of reform.

The electricity system in England and Wales is considered to have been very close to this textbook model, and the reforms there have functioned fairly well according to most – but not all – independent experts (Joskow, 2006). However, reform efforts in some jurisdictions have been associated with major problems in terms of supply reliability and price stability. California's electricity crisis of 2000–2001 stands out as an extreme example. In California, wholesale prices were manipulated upward by some electricity traders in the spot market while local distribution utilities were not allowed to increase retail rates correspondingly. The traders were able to bid high prices for providing power, as they took advantage of supply shortfalls related to an exceptional number of units shut down for maintenance and low hydropower supply with low water conditions on the west coast of North America restricting supply. High wholesale prices combined with retail prices capped by regulators caused a rapid increase in debt of distribution utilities, and the supply shortages caused localized blackouts. While the California electricity crisis had many root causes, one was poor policy design that made it nearly impossible to enter into long-term contracts for power, which were more competitive than the short-term spot market, and the reluctance of federal regulators to intervene in the marketplace to stop short-run supply and price manipulation (Wolak, 2003).

There is still no widespread agreement about the optimal form for the electricity market. In any case, opposing ideologies and the diversity of electricity systems in different jurisdictions will ensure continued diversity of market designs. Nonetheless, some issues and lessons are generally recognized. In particular, the electricity sector has characteristics that impede and even prevent development of the degree of competition found within many more conventional commodities. There will thus continue to be an important role for the independent regulation of transmission systems, and even a role for the regulation of investment, pricing, and operation of some parts of the generation system. This regulation is necessary to:

- prevent spot market manipulation;

- ensure adequate short-run reserve capacity;

- ensure adequate investment in new generation supplies;

- determine optimal transmission capacity and operation;

- ensure transmission access for all eligible suppliers, including small-scale renewable energy; and

- promote coordination between electricity systems to maximize efficiency and reliability.

Policies for managing the electricity sector have generally focused, as does the above discussion, on the potential for integrating some degree of competition into electricity generation and retailing in large grid systems. However, in developing countries, and even in rural areas of developed countries, there are areas where isolated mini-grids are the norm and will, in fact, become more important in the future. Proper management of these systems is also necessary and requires different management strategies (Victor and Heller, 2007).

22.2.4.2   Oil Sector

There is a close relationship between the policy response to natural monopoly markets and the policy response to other forms of market power. Many of the market concerns in energy supply arise not only when monopolies reign, but when competition is imperfect, often because of the very large size – and capital commitment – in the energy industry, and also because many energy services are priced in global markets where it is difficult to ensure true competition. OPEC members agree to constrain their collective output in order to stabilize international oil prices at levels above those of pure competition. Because they collude overtly, this type of oligopoly is commonly referred to as a cartel. More conventionally, oligopoly exists in markets where a few firms dominate, with opportunities to collectively behave like a monopoly even without the overt collusion practiced by OPEC members.

There are many other settings in which such collusion could arise and governments organize to oversee the market and restore competitive outcomes. Oil refining and distribution have often been seen as activities with sufficient economies-of-scale that an oligopoly would develop in place of more aggressive competition. Anti-trust action can help ensure a competitive marketplace. Where that is not possible or is ineffective, governments may regulate prices to approximate those of relatively competitive markets, although other considerations, such as national economic performance, may also be important.

In the oil sector at the global scale, there has been a significant shift toward state ownership of oil companies. Indeed, today, most of the world's largest oil companies are owned by governments and trace their origins to decisions by governments to assert control over oil revenues by creating state-controlled oil companies, in many cases because they did not trust private enterprise, much of it foreign-owned. The experience to date is that these national oil companies vary widely in performance and strategy, and they also vary enormously in ability to actually meet the goal of government to exert greater control over the oil sector. National oil companies are also discussed in the following section on the management of valuable energy resource endowments.

Box 22.5 | Electricity Market Reform in Australia – Brian Fisher

In Australia before the mid-1990s, the electricity supply chain, comprising generation, transmission, distribution, and retailing, was owned and managed either by government monopolies spanning the entire chain, or by monopolies at the generation and transmission stages linked to monopoly businesses operating in the distribution and retailing sectors, usually with a franchised "catchment" of customers.

The ownership and management structure provided a protected environment for the industry and there were various indications that electricity was not being provided at least cost. A government-commissioned report released in 1991 cited "poor investment decisions" and "gross overstaffing," as well as reserve plant margins well above international standards, as indicators of inefficiency within the industry (Industry Commission, 1991). This and other factors spurred a wave of disaggregation, privatization, and corporatization, as well as the establishment of an interconnected market in parts of Australia and the introduction of retail contestability for the majority of consumers.

In the southern and eastern states of Australia – where networks are now interconnected – a central initiative of reform was the creation of a common electricity market. The market was managed by NEMMCO (now the Australian Energy Market Operator [AEMO]), a company established in 1996 and owned by the relevant state and territory governments. Broadly, NEMMCO resolves the supply schedules of generators with the demand schedules of purchasers to create a spot price for electricity. It then issues dispatch orders to generators indicating how much they are to produce, taking account of capacity constraints and likely transmission losses. In terms of retailing, even the smallest consumers of electricity in areas covered by the common electricity market have a choice of retailer, with the exception of those in Tasmania, where full contestability was expected by 2010. Western Australia and the Northern Territory are too far from the integrated market in eastern Australia to take part, but Western Australia has operated its own wholesale electricity market for the south-west of the state since 2006.

Most elements of transmission and distribution have not been opened to competition on the grounds that these stages have "natural monopoly" characteristics. Each state has a monopoly transmission business, and has either adopted or retained a framework whereby distribution within a region is generally handled by a single entity. Some entities have interests in both distribution and retailing.

Average labor productivity across Australia in the generation sector had more than doubled by the time the majority of deregulation reforms were underway or completed in the late 1990s. There were also significant reductions in the wholesale price of electricity. Modeling by the Australian Bureau of Agricultural and Resource Economics suggested that annual gross domestic product would be US$2.9 billion higher (in 2008 dollars) in 2010 than it would have been had electricity market reform not been undertaken (Short et al., 2001).

Box 22.6 | South Africa's Institutional Failures and the Power Crisis – Alison Hughes

There is widespread consensus in South Africa that December 2007 heralded an era of crisis in the South African electricity industry. December and the following months were characterized by rolling blackouts. For the first time in South African history, large users such as gold mines lost their electricity supply for extended periods. The underlying cause of the crisis is an inadequate reserve margin. After a long period of excess capacity, the reserve margin was allowed to drop to 6% in 2007 from 31% in 1994. While there are periods where load shedding is not necessary, it will be many years before an adequate reserve margin is restored and electricity supply stabilizes.

The reasons behind the crisis are complex. Short- and medium-term causes can be traced back to a decision to restructure the electricity sector in 1998, which involved breaking up the state monopoly utility Eskom into competing generation companies. At the same time, government wanted to encourage independent power producers to enter the market, and placed a moratorium on Eskom building new generating capacity. A new institutional framework was prepared, but in 2004, government performed a policy u-turn and announced the scrapping of the restructuring program. The rationale was two-fold: (1) government wanted state-owned enterprises to play a larger role in national development; and (2) no private investment was forthcoming for new capacity, since neither a market system nor a framework for concluding long-term power purchase agreements was put in place. Eskom was given the go-ahead to bring a new peaking plant online and mothballed coal plants are being returned to service, but no new generation units were scheduled for operation before 2012.

Another underlying cause behind the crisis was the long-term prevalence of ultra-low electricity prices, which were below long-run marginal cost and fell in real terms during the 1990s. South Africa has always relied on low-cost energy to attract energy-intensive industry and foreign investors, and has pursued a long-term policy of low energy prices. The national regulator continued this trend by keeping electricity price increases close to, or below, inflation, reflecting the average cost of Eskom producing the electricity. The pricing policy was in part due to significant over-build of generating capacity by Eskom in the 1980s, partly funded through an indirect subsidy from the country's Reserve Bank. Through the pricing policy, demand-side management and energy efficiency were dis-incentivized, load grew faster, and new investment in generating capacity was deterred.

22.2.5   Manage Valuable Energy Resource Endowments

In some countries, the natural endowment in valuable energy resources, especially conventional oil and natural gas, are of such a magnitude that energy is one of, if not the, most important sectors in the economy. As experience has shown, however, a spectacular energy resource endowment can be both a blessing and a curse. On the one hand, the revenues from the resource can allow governments to provide high levels of social services like education, health care, and welfare. On the other hand, the resource windfall can produce so many challenges, as mentioned earlier, that analysts sometimes refer to this endowment as a resource curse (Collier, 2007).

Sarraf and Jiwanji (2001) note several factors leading to resource curse type outcomes. Higher returns to investment and labor in the resource sector can cause capital shortages and wage inflation that harms other sectors of the economy, even though these other sectors would otherwise have wealth-generating potential long after the demise of a non-renewable resource like oil and natural gas. This resource dependency may be inconsequential if the resource is slated to last a century or more. However, it is extremely important in cases where the resource will be largely depleted within just a few decades.

Resource price volatility can cause great uncertainty in government revenues. Periods of high prices tend to be associated with expansion of government expenditures and borrowing, while periods of low prices are associated with cutbacks in government activity and the accumulation of debt.

Periods of high resource prices can also be associated with the rapid expansion of the resource sector itself, which can overheat an economy, in either industrialized or developing countries, and affect the competition for labor and capital with other sectors – referred to as "Dutch disease" – and also create pressure to develop the resource

more quickly than should occur from a sound technological perspective. In addition, rapid development and export puts an upward pressure on the exchange rate, which, in turn, makes it more difficult for other sectors of the economy to compete, both internationally and domestically. Developing a resource too quickly can lead to wasteful resource exploitation. An example would be to develop oil and gas reservoirs at accelerated rates that result in lower recovery of the resource than could have been achieved by development at a slower pace.

The wealth generated by a valuable energy resource is an inducement to corruption, where unscrupulous public officials and representatives of domestic and foreign firms vie for resource rents (Leite and Weidmann, 1999). This activity can be so widespread that the distribution of resource wealth to the public at large is far below what it should be, with perhaps even a negligible benefit. Even without pervasive corruption as an outcome, it is relatively easy for ineffective governments to garner political support by distributing some of the rents to key interests and thereby avoid more rigorous oversight of governing institutions (Sala-i-Martin and Subramanian, 2003). Although it need not be the case, a society with a resource windfall can actually be worse off in terms of its economic health and the strength of its social fabric than it otherwise would have been without the resource.

Governments that have successfully handled resource endowments have done so by developing various policy instruments and strategies to address the special challenges of managing energy and mineral resource wealth to the benefit of current and future generations. Some lessons are outlined below.

Governments should not use the resource rents for excessive energy subsidies to domestic consumers, but instead maximize the return from resource assets, in both domestic and international markets. Thus, domestic energy would be sold at international prices. The rents from the domestic and export sales of the resource can then be distributed to citizens as improved public services and infrastructure or even as direct resource rent payments, which is more economically efficient – and usually more equitable – than a strategy to return the rents via large subsidies to domestic energy users.

Governments can pursue some degree of economic diversification, but should only do so where there are sound prospects for the sustained development of alternative activities. There are many unfortunate examples of governments using windfall resource revenues to subsidize economic development initiatives that have no hope of standing on their own once the subsidy is later withdrawn. But there are also examples of governments helping innovative domestic firms to develop niche markets in activities that are related to resource extraction, which can be marketed internationally, or in completely new industrial activities where the country can have a comparative advantage once it passes the initial start-up hurdles.

Governments should establish mechanisms that reserve a significant share of the revenue from the resource for the future. Obviously this strategy is more affordable for rich countries – for example, Norway and its oil and gas resource revenues – or for countries with a huge resource endowment relative to the size of their population, such as the United Arab Emirates with its large oil revenues relative to its population size. The revenue that is set aside should be invested cautiously in trusts or sovereign wealth funds that are diversified to reduce the risks of wealth depletion via exchange rate shifts or mis-investment.

Governments should establish mechanisms that allow them to access some of the accumulated surplus when resource prices are low. For example, a stabilization fund would not be money set aside for future generations, but would instead be a fund that grew during times of relatively high resource prices but which government could use to sustain key levels of services during times of low resource prices. This would reduce volatility in government expenditures that would otherwise have resulted from volatility in resource prices. However, while this approach seems sound, it is difficult to apply in practice. It is impossible to know

in advance when prices are above or below their long-run average, since technological change and/or resource depletion could mean that the long-run price trend in the future will be very different in an upward or downward direction. Only with hindsight can governments determine the extent to which their strategy was providing a proper counter-cyclical balance of building up and drawing down surpluses from the resource.

Governments should create mechanisms that provide transparency in all transactions involving government officials and resource firms in order to minimize the risks of resource-motivated corruption. This objective can be advanced by legislating open accounting systems that enable watchdog organizations like Transparency International to put pressure on large, multi-national oil and gas companies to "publish what they pay" in dealings with national oil companies and governments.

Governments can create mechanisms by which part of resource revenues are returned directly to communities located in the resource areas, in part to compensate for negative effects of resource development, but also to give communities a stake in the efficient extraction of the resource and to maximize the revenues it is capable of generating.

Many oil-rich countries capture much of the economic rent associated with this high-value endowment by having national oil companies play a key role in the sector. The rents appear as profits for these companies. However, a well-designed system of resource rent taxation is important, even in cases where public ownership is predominant, because this provides transparency as to the economic value of the resource. Resource rent taxation should also not be limited to oil. Coal, natural gas, and even favorable hydropower sites can all be associated with substantial resource rents. Countries with a long experience of resource rent taxation tend to have evolved toward rent collection mechanisms that combine ex ante and ex poste obligations. In other words, companies submit competitive ex ante bids for rights or access to the resource – mineral rights, exploration rights, hydropower sites, wind farm sites – and then, during production, also pay a percentage of total revenue or net revenue.

Box 22.7 | Oil Funds – Carolyn Fischer

The Norwegian experience with oil funds is often held up as a best practice for the way a country should manage the revenues it generates from the production of oil. In Norway, the vast majority of resource revenues are transferred to a Petroleum Fund, which is used to generate income and diversify risk by investing exclusively in foreign bonds and equity. As a small, open economy, Norway would not be able to absorb the oil revenues into its own economy, so it has chosen to maintain its existing tax structure and save the wealth for future needs, such as funding the social security system. This strategy contrasts with that of Alaska, which uses most of its oil revenues to fund state government expenditures and keep taxes low. The rest is invested in the Alaska Permanent Fund, which generates dividend income for all citizens of the state.

For developing countries, current needs are arguably more pressing than future ones. High-return strategies then involve investing in human capital and critical public infrastructure, rather than equities, with the caution not to invest beyond the absorptive capacity of the economy or to protect unsustainable businesses. Botswana, a major diamond exporter, represents a success story in this sphere, where the primary mechanism of revenue management is not an explicit savings fund or allocation scheme, but rather a solid approach to budgeting. The multi-year National Development Plan process aims to stabilize government spending growth and to prioritize spending. The focus has been to expand essential public services and infrastructure – e.g., electricity, water, roads, police, health care, and education – and to provide credit to state-owned enterprises, which, in turn, have made commercial loans. At the same time, the government has accumulated international reserves and earmarked budget surpluses for stability spending in leaner years, and it has managed liquidity and the exchange rate to avoid real appreciation.

In other countries, development benefits of resource extraction have been elusive. In response, some companies have engaged directly in development and compensation programs to improve community relations. Overall, good governance seems a necessary component for resource riches to become a broad-based economic blessing, and not a curse.

Source: Fischer, 2007.

Box 22.8 | Resource Rent Taxation – Brian Fisher

The concept of economic rent is outlined in many places, but in this context the definition given by Stiglitz (1996) is most pertinent. Under competitive conditions, the economic rent accruing to a mining firm is the difference between its revenue and its costs, where costs include a "normal" return to capital – the minimum return needed to hold the capital in the mining activity. Economic rents may persist in the mining industry because there is not a perfectly elastic supply of non-renewable resources.

Brown (1948) proposed a tax that is calculated as a fixed proportion of net cash flow each year, where net cash flow is defined as the difference between revenue and total costs. Total costs include all capital expenditure during the particular year but exclude interest payments. In years when "rent" calculated in this way is positive, the government receives a fixed proportion of that rent. In years when the mining company incurs a loss, the government would rebate a fixed proportion of that loss.

The characteristic that distinguishes the resource rent tax proposed by Garnaut and Ross (1975) from a Brown tax is that there is no provision for the government to pay a rebate on losses. Instead, losses may be carried forward, and increased by a "threshold" rate of interest, until they can be deducted from future profits. Taxation is then triggered when the net cash flow from the project is positive. In cases where future profits are insufficient to offset past losses, then the private firm will bear those losses.

The rate of resource rent tax applied in the Australian petroleum sector, for example, is 40%. Although many of the profit-based royalty systems in place around the world are not good approximations to a pure resource rent tax, it is interesting to note that in most cases the tax rate is set below 15% (Otto et al., 2006).

22.2.6   Reduce Environmental and Human Health Impacts

As discussed in Chapters 3 and 4, humanity is increasingly aware of the impacts and risks to the environment and human health from energy supply and use, including indoor air pollution, local and regional air pollution, pollution in fresh water and oceans, direct damage to landscapes and ecosystems, and the major transformation of the earth's climate, with its own implications for weather, ecosystems, coastlines, human economies, and human health. In assessing these effects, some analysts find it helpful to distinguish between "extreme event risks," on the one hand, and "ongoing impacts and risks," on the other.

Extreme event risk refers to events that have an extremely low probability of occurrence but could have an extremely severe impact. Examples include a large oil spill from an ocean tanker, a major refinery explosion or gas leak, or severe radiation exposure from a nuclear accident. Policymaking for extreme event risks can be especially challenging because people tend to focus on the severe consequences of the event itself and forget to consider the very small likelihood of the event (Bier et al., 1999). Nuclear power experts, for example, often express frustration with the public's fixation on the potential human impacts of a nuclear accident while seemingly accepting a much higher ongoing rate of human harm from coal combustion (Matysek and Fisher, 2008).

Ongoing impacts and risks, like the risk of coal mining accidents, refer to operational occurrences that are fairly well understood and accepted as part of the regular impact of a particular activity. The amount of land alienated by an open pit coal mine is easily known. The impact of acid emissions on natural systems, infrastructure, and buildings is quite well understood. Scientific understanding of the impact of greenhouse gas emissions on the earth's climate has improved significantly over the past few decades, although there is still considerable uncertainty about the magnitude of temperature changes associated with different GHG atmospheric concentrations. In fact, because of this uncertainty, GHG emissions present both ongoing impacts and well-understood risks, but also an extreme event risk – runaway global warming. This extreme event has a low probability, but it is not low enough for an increasing number of experts and policymakers (Stern, 2006; Weitzman, 2009).

Ongoing impacts and risks from a particular activity can affect land, air, and water, although some effects are concentrated in one particular medium. A key challenge is balancing and mitigating these effects, as no energy source is completely free of environmental impacts.

On land, impacts and risks are associated with the exploration, extraction, processing, and transportation activities in fossil fuel mining and production, uranium mining and production, and renewable forms of energy like biomass, hydropower, wind, and solar. Specifically, these impacts include open pit coal mines, the well-dotted landscape of an oil and gas producing region, the land required for energy biomass plantations in agriculture or forestry, the land flooded for major hydropower dams, and land required for transmission lines and other networks. As humanity increasingly turns toward renewable forms of energy, land use conflicts arise also for the siting of wind power and run-of-the-river hydropower facilities.

In the air, emissions from the combustion of fossil fuel products and biomass have local, regional, and global impacts. Incomplete combustion of refined petroleum products and biomass leads to indoor emissions of carbon monoxide, black carbon, and volatile organic compounds, in addition to methane leaks, all of which can impact human health. Improved combustion will reduce these emissions and normally also reduce GHG emissions, since these are either direct or indirect GHGs.

In water, there are threats from fossil fuels to ground water, rivers, lakes, and oceans from urban runoff, urban sewage, permitted industrial effluents, and industrial accidents, including oil spills in fresh water and oceans. Discharge of water used for cooling in thermal power plants, including nuclear plants, leads to thermal pollution of waterways. Air emissions from coal-fired generation can lead to acidification of lakes and elevated mercury levels in fish and other aquatic life. Hydropower development, including small-scale, run-of-the-river hydropower, can disrupt rivers and the life systems they support.

The diversity of these threats calls for a multiplicity of policy instruments, the choice of which depends on which medium is under threat

and whether the source of the threat is local, regional, or global (Kemp, 1997; Jaffe et al., 2002; Harrington et al., 2004; Newell, 2008; Wei and Rose, 2009). The introduction of this chapter laid out a broad list of government policy types. This range is narrowed somewhat to the following list for discussing options for energy-related environment policy. These include:

1. information programs that inform firms and households of the environmental benefits and perhaps personal financial benefits from certain types of investments and behavior, such as energy efficiency, an approach that extends to sweeping campaigns of public education and engagement;

2. financial penalties (emissions pricing) that discourage rather than prohibit emissions;

3. subsidies that promote changes in investment and behavior by firms and households;

4. regulations that prohibit certain activities, technologies, or energy forms, or that regulate in ways that provide some degree of market incentives for innovation and adoption of more sustainable energy technologies, such as tradable quota obligations or tradable emission permits; and

5. direct actions by governments and their agencies at all levels to fund public sector R&D, upgrade public buildings, facilities, and equipment, improve infrastructure, and develop social capital, such as education, training, etc.

22.2.6.1   Information Programs and Public Engagement

Information programs promote environmentally beneficial energy choices by using both moral suasion and financial self-interest arguments. Moral suasion arguments might focus on global or local environmental benefits for current or future generations. Financial self-interest arguments might convince firms to promote "greenness" as a way of gaining or sustaining market share. Or, with better information about energy use rates and costs, firms and households might realize that some investments, like energy efficiency and conservation, may provide a financial gain over the long term. The challenge for information programs is to compete with all other sources of marketing information, much of which ignores the environmental and even long-term financial benefits of certain energy supply and demand choices and most of which now tries to convince consumers that all of their choices are "green."

One of the challenges is policymakers may assume that information programs can be effective on their own in stimulating a voluntary shift toward profoundly different technologies. Governments in OECD countries have relied to a large extent on information programs to incite actions by firms and households to make choices that affect environmental performance, but these have been considerably less effective than hoped for (Karamanos, 2001; Khanna, 2001). One of the lessons, in the case of climate policy, is that information on its own – without supporting emissions pricing and regulations of technologies and fuels – is unlikely to drive technological transformation toward lower carbon emissions. Recognition of this reality has shifted the thinking of policy advisors toward finding ways of combining policy options so that information programs work in concert with other, more compulsory types of policies. At the same time, some advocates of this approach emphasize the need for wide-ranging campaigns that engage key interest groups and sectors of the public – for instance, youth, rural inhabitants, etc. – in attempts at profound perceptual and behavioral change, and to increase the acceptability of stronger policy instruments, such as carbon taxes or cap-and-trade systems.

22.2.6.2   Emissions Pricing

Policies like emissions taxes, tradable emissions permit prices, or other types of financial charges seek to provide a price signal that reflects to some degree the value of the impacts and risks to the environment and human health from various types of emissions and effluents. Taxes and other charges can also be applied to specific technologies, such as inefficient vehicles or batteries, where the revenues can be used to fund recycling and the disposal of toxic materials.

Decades of price elasticity studies by economists support the argument that emission taxes can be an effective policy for environmental improvement. Economists also point out that emissions pricing is the most economically efficient policy and that this approach also scores well in terms of other policy criteria, such as administrative feasibility. Yet it is politically difficult for policymakers to implement the level of emissions pricing that would cause the kind of price changes necessary to drive profound technological change in the necessary time period. In some cases, therefore, emissions pricing may need to work in concert with other policy options.

Sweden deserves attention in this context. Emissions of carbon dioxide have dropped by 8% over the years 1990–2007, while GDP grew over 40%, largely as a result of the introduction of carbon taxes in 1990 that, among other things, drove changes in fuel use for district heating away from oil and toward woody biomass.

22.2.6.3   Subsidies

This policy approach was first discussed in the section on energy access, where it was noted that governments have subsidized: (1) electric grid extensions to provide access to electricity for rural and remote areas; (2) the prices of electricity and fuels, such as kerosene, as poverty reduction schemes; and (3) certain technologies like low emission

stoves. Subsidies are also applied to energy supply innovation (R&D), commercial development and dissemination of favored technologies, and energy forms for strategic or economic development reasons, such as nuclear power, new fossil fuel production technologies like offshore oil and unconventional oil, and renewables like hydropower and, more recently, wind and biofuels.

With the growing concern for climate change, however, the rationale for some types of subsidies has been questioned. The effect of research and production subsidies on the international price of a key commodity like oil is difficult to estimate. But since high-cost oil production technologies, like the tar sands in Canada, have received such subsidies in the past, there is reason to presume these have had a downward effect on the international price of oil and therefore an upward influence on GHG emissions. In contrast, the effect of subsidies to fossil fuel consumption is easier to estimate. These subsidies existed in many countries a few decades ago, and were especially notable in the countries of eastern Europe and the former Soviet Union. Today, such subsidies remain in other non-OECD countries – such as OPEC members – to provide lower prices, especially for gasoline and diesel. They also exist in non-OPEC developing countries to help certain customer groups, such as the subsidized price of electricity for farmers in India. The IEA (2010) estimates that, globally, the subsidies to fossil fuel consumption exceeded US$300 billion in 2009, and that reductions in these subsidies could lower fossil fuel consumption and associated GHG emissions by 5%.

While energy access is still seen as a legitimate reason for targeted and temporary energy use subsidies, there is now wider recognition that such subsidies should be aligned with policies for GHG abatement. This means that subsidies for fossil fuel consumption should be phased out and subsidies might only be allowed when they either foster non-emitting forms of energy or energy efficiency.

Indeed, subsidy programs for energy efficiency have been significant in OECD countries, especially in the electric sector. Electric utilities in the United States alone spent over US$20 billion on electricity efficiency in the two decades from 1985 to 2005. However, as with information programs, hindsight evaluations of subsidy programs for energy efficiency have increasingly challenged the efficacy of this type of non-compulsory policy. In the case of energy efficiency, for example, it can appear on the surface that each high efficiency device acquired with the help of a government or utility subsidy contributes to a more efficient energy system. However, this can be deceiving.

First, a certain number of higher efficiency devices are acquired even in jurisdictions without energy efficiency subsidies, as some technologies naturally evolve toward more efficient models. Anyone who would have purchased an efficient device anyway, but nonetheless receives a subsidy, does not actually increase the efficiency of the energy system from what it otherwise would have been – a business-as-usual trajectory. These subsidy recipients are therefore called "free riders," and their participation rate must be subtracted before the true effectiveness of the subsidy policy is known. Economists and policy analysts also use the term "adverse selection" to explain this problem.

Second, the acquisition of a more efficient device may provoke a rebound effect, which is a feedback between improvements in energy efficiency and the demand for energy services. Researchers have described various components of the rebound effect (Sorrell et al., 2009). To simplify the discussion, these are compressed into direct and indirect rebound effects.

With the *direct rebound effect*, energy efficiency investments lower the energy operating costs of energy-using buildings and equipment, which may lead to greater use. This response may occur even if the efficiency investment is not profitable; it is simply a demand response to lower operating costs. If, however, the efficiency investment is also profitable, then it increases net income. Expenditure of this additional income may cause increases in demand for the original energy service. Thus, profitable efficiency investments can induce substitution and income components of the direct rebound effect.

The *indirect rebound effect* refers to second-order developments induced by energy efficiency. The literature describes various components of indirect rebound. For simplicity, we focus on three:[2] (1) by lowering energy demand from what it otherwise would be, energy efficiency should decrease the price of energy, which could increase demand for some energy services; (2) profitable energy efficiency investments increase income, which could increase the demand not just for the original energy service, as noted above, but also for other energy services; and (3) innovation and adoption of energy-efficient technologies for a given energy service could provide spillover effects that increase the demand for other uses of energy, some of which could be completely new.

This third form of indirect rebound warrants elaboration. In essence, improvements in the productivity of a factor of production are normally associated with increases in demand for that factor. Thus, labor productivity gains are associated with increased demand for labor. Fouquet and Pearson (2006) provide an energy example by tracing the historical relationship between productivity gains in lighting services and its demand. They note that in the United Kingdom since 1800, the cost of lighting services fell to one-three-thousandth its initial level while per capita use grew by 6500 times. Although this correlation does not ensure causality, there is a strong likelihood that the dramatic drop in the cost of lighting that resulted from efficiency gains played a not-insubstantial role in the rising demand for lighting services over time. Thus, improvements in the efficiency of basic lighting technologies may accelerate the emergence of and demand for related energy services like decorative lighting and security lighting, just as improvements in

the efficiency of refrigeration technologies may accelerate the emergence of related energy services like wine coolers, water coolers, portable freezers, and desktop fridges.

While all researchers can agree with the concept of rebound, there has been considerable debate about its significance and its empirical measurement can be difficult and controversial. The reason is that changes in the cost of energy services caused by energy efficiency would be just one of many factors determining the demand for energy services during a given period. The Fouquet and Pearson (2006) study, noted above, demonstrates a correlation between lighting productivity gains and the demand for lighting. These productivity gains no doubt had a role in increased per capita demand for lighting in the 200 years since 1800 in the United Kingdom. But incomes rose dramatically during this time also, raising the question of the relative importance of rising incomes versus falling costs due to productivity gains in causing the increased demand for lighting. Some research papers or special issues of journals have tried to assess the magnitude of direct and indirect rebound effects (Sorrell et al., 2009; Schipper, 2000). In general, these surveys have given inconclusive results, but suggest that some energy end-uses are more prone to rebound effects than others, and that direct rebound effects are not particularly large for most energy services, but that indirect rebound effects could be very large over long time periods, including the development of new, but related, energy services.

Research into free riders and rebound effects with energy efficiency policies suggests that these can be quite substantial, indicating that subsidy programs too may need to be combined with other more compulsory policies like emissions pricing and technology and fuel regulations (Loughran and Kulick, 2004; Gillingham et al., 2006; Arimura et al., 2009). Finally, subsidy programs have administrative costs and these too need to be considered in the cost-effectiveness calculation, although increasingly utilities and even some governments are now doing this.

Short-term subsidies also suffer from the question of timing. Since energy-using devices are durable goods, firms and consumers can accelerate or delay purchases they would make anyway to coincide with the timing of the subsidy. This shifting behavior increases the cost relative to the benefits of one-off subsidy programs, like the recent "cash-for-clunkers" program in the United States and other countries. Even subsidies that are renewed can suffer from the perception of intermittency. In the case of the United States, production tax credit for renewable electricity generation, uncertainty about policy renewal is compounded by uncertainty about profits, since positive income is required to take advantage of the tax credit.

Given these issues with subsidies when used on their own, one strategy that policymakers are turning to is a combination of subsidies and taxes. For example, some jurisdictions have experimented with vehicle fee bate schemes, in which high emission or high fuel consumption vehicles pay a tax and this revenue is used to provide a subsidy to low emission or low fuel consumption vehicles. The subsidy, while still subject

---

2   The literature is not consistent. Some of the indirect components described here are sometimes described as direct rebound effects. And sometimes, the income effect on a given energy service, which is ascribed above to direct rebound, is considered to be part of the indirect rebound effect.

to free-riders, helps with political acceptability, while the tax may have more of an effective role.

Even without help from taxes, subsidies can be effective if substantial and well targeted. One of the most compelling recent examples is the "feed-in electricity tariff" that has enabled several European countries to rapidly increase the role of renewables like wind power in the generation of electricity. This tariff guarantees a higher price for electricity from renewables and therefore provides stable revenue projections that help independent power producers secure financing and adequate capital.[3]

Subsidies are also a means by which industrialized countries can transfer funds to developing countries in order to assist in the transformation toward a more sustainable energy system. Foreign assistance in the form of low interest loans to develop fossil fuel, nuclear, and hydro-power resources has been provided for decades by governments and international agencies like the World Bank. With the growing concern for climate change, however, new mechanisms have been created, such as the Global Environment Facility and the Clean Development Mechanism, the latter as part of the Framework Convention on Climate Change. These types of subsidies are discussed later in Section 22.2.8 on sustainable energy policy coordination at the international level.

22.2.6.4   Regulations

In some cases, regulations as a policy instrument can be strongly prescriptive in nature, such as an outright ban on a technology or resource use. Concern for the risks of an extreme event, like a nuclear accident, has led some governments to prohibit the development of nuclear power. The potential for nuclear weapons proliferation has prompted some of the world's major powers to use trade threats and technology embargoes to hinder the development of nuclear power by other countries. Concern for the loss of land and harm to migratory fish from large hydropower developments has led some governments to ban such projects, and even to dismantle some existing hydro dams.

Most countries regulate to some degree particulate and gas emissions from the combustion of fuels, including both fossil fuel products and biomass. Energy efficiency standards are applied in most countries to regulate appliances, buildings, vehicles, and industrial equipment. Regulations that affect the safety of oil tankers, petroleum refineries, and transmission lines are also commonplace.

Regulations are sometimes applied in flexible ways in order to reduce the cost of compliance by allowing exchanges between those subject to the regulation. One type of market-oriented regulation is an emissions cap-and-trade policy. A cap is set, with penalties for non-compliance, as

with other regulations. The cap is allocated – freely or by auction – in the form of permits or allowances and, because these are tradable, the trading price provides a price signal to emitters, just like an emissions tax. Each emitter has the option of reducing emissions, and selling surplus allowances to other emitters, or not reducing emissions and instead purchasing the needed allowances from someone else.

There is considerable debate about the relative merits of cap-and-trade versus emissions taxes. Cap-and-trade provides greater certainty about the emissions outcome of the policy. This is seen as desirable from the perspective of ensuring the realization of an environmental objective. In contrast, emissions taxes provide greater certainty about the effect of the policy on energy prices faced by firms and consumers. This is seen as desirable from an investment and economic efficiency perspective. Emissions taxes are likely to be much more stable than emissions prices that emerge from the market in which emissions permits are traded. In reality, however, it is possible to design cap-and-trade systems with a price floor and price ceiling and to allow banking and borrowing of emissions permits in order to reduce the tendency for emissions price volatility.

Another type of market-oriented regulation is an obligation to achieve a minimum market share of a particular good or production process. An example is a low-emission standard for vehicles that obliges producers to achieve a minimum market share for the sale of a desired type of vehicle, with an allowance for producers to trade among themselves to achieve the aggregate minimum sales requirement. Similarly, a renewable portfolio standard in electricity sets minimum market shares for renewable electricity generation, with credits for producers to trade among themselves in achieving the aggregate target. Effectively, these policies combine a subsidy for renewable energy with a tax on non-renewable energy generators.

22.2.6.5   Direct Action by Government

While the policies discussed thus far mainly involve governments trying to induce alternative behavior and technology choice by firms and households, there are also opportunities for governments to take direct action. Senior levels of government invest in R&D and can establish funding criteria that emphasize environmental and human health objectives, especially with respect to energy-supply and -use technologies. All levels of government own buildings, vehicles, and equipment, and they can require specific technological choices in terms of environmental and human health impacts. Governments can even use their purchasing power to influence the commercialization choices of manufacturers. Thus, public procurement strategies can make major equipment acquisitions contingent upon the ability of one or more competing manufacturers to produce new technologies that meet aggressive targets for energy efficiency or low environmental impact.

Local governments determine land use through planning, zoning, and building, permitting authority provision of public transportation options

---

3   It is important to note, however, that subsidies come from somewhere. If they are provided by government, they are generated mostly through taxes elsewhere in the economy. Conversely, they may be provided as cross-subsidies from other customers or industry.

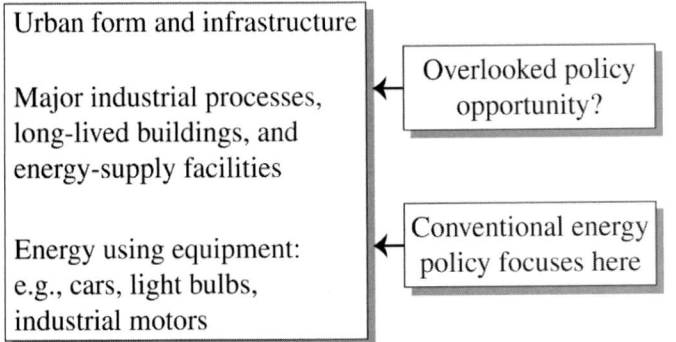

Urban form and infrastructure	←	Overlooked policy opportunity?
Major industrial processes, long-lived buildings, and energy-supply facilities		
Energy using equipment: e.g., cars, light bulbs, industrial motors	←	Conventional energy policy focuses here

Figure 22.1 | Hierarchy of energy-related decision-making. Source: Jaccard et al., 1997.

within and between urban centers, responsibility to help energy utilities site transmission and other network lines, and urban and regional control over zoning and development. An example in this area is the rising interest in community energy management, an effort by local government especially to coordinate regulations and land use decisions affecting building and facility characteristics, siting of buildings, and infrastructure investments in ways that minimize the use of energy, maximize synergies between systems, and provide opportunities for the greater use of energy forms that are more environmentally benign.

Figure 22.1 shows a hierarchy of energy-related decisions. While energy policy has traditionally focused on the buildings, industrial facilities, and equipment in the middle and lower part of the hierarchy, it is increasingly recognized that there is considerable opportunity for municipal and

sometimes more senior levels of governments to influence the evolution of urban form through land-use zoning, development permitting, siting requirements, building codes, and infrastructure investment in public transportation, district energy, and even urban liquid and solid waste collection and disposal systems. Decisions at higher levels of the hierarchy are more often taken by government, given its ultimate control over land use and major infrastructure. Also, decisions at higher levels tend to have longer-term impacts. The rate of turnover of urban form is much slower than that of vehicles or the equipment used inside buildings. This makes policies affecting the top of the hierarchy all the more urgent, given the rapid transformations that are required for a more sustainable energy system.

A final consideration is that many environmental impacts and risks cross international borders, which can make it extremely difficult to create effective policy because this requires that many countries coordinate their objectives and efforts to achieve multi-national or global public good. Depending on their level of economic development and resource endowments, countries could have very different perspectives. China is less interested in reducing coal use in order to reduce GHG emissions than is a country such as Japan, which, unlike China, is poorly endowed in terms of coal resources. The challenge of achieving a coordinated international effort with respect to the environmental impacts of energy supply and use arose in the 1980s with the cross-border challenges of acid rain and the control of acid emissions. Europe and North America made headway on this challenge, but the focus has now shifted to the need for international coordination of GHG emissions control. This special problem is addressed in Section 22.2.8 below on energy policy coordination at the international level.

Box 22.9 | Key Policies for GHG Abatement: Cap-and-trade Versus Carbon Tax – Mark Jaccard

When addressing an environmental externality like GHG emissions, economists have a preference for policies that provide a single price for GHG emissions from all activities in the economy. Thus, economists' two favorite policies are either a tax on emissions or the allocation of rights to a fixed quantity of emissions. In the latter case, tradable emission allowances are allocated by auction or a political process, perhaps reflecting historical emissions and other criteria. In any given period, those who have reduced emissions enough to hold a surplus of allowances can sell them to those who find it cheaper to purchase these instead of fully reducing emissions to match allocated allowances. Borrowing and banking of some allowances may also be permitted. The combined effect is that the market for tradable allowances, if it covers all emissions in the economy, would provide an economy-wide price signal in the same way as would an economy-wide GHG emissions tax.

The two policies differ in that the emissions charge provides price certainty while the emissions cap provides quantity certainty. The former gives investors confidence about the future price of emissions, but it leaves uncertain for government the actual emission reductions that will be induced by a given tax level. The latter gives certainty to government about the level of future emissions, but the price of tradable permits will be uncertain.

There are vigorous debates about the relative merits of these competing approaches (Weitzman, 1974). In practice, however, they may not end up that different. The level of emission taxes is uncertain to the extent that governments will inevitably adjust taxes as they observe the reductions induced by a given level of tax. And emission caps may be implemented in conjunction with price floor and price ceiling strategies that increase confidence about the price while reducing certainty about the future level of emissions. Also, governments may apply caps to industry but taxes to other sectors of the economy.

While taxes and caps are usually discussed in terms of national or regional application – a regional example being Europe – the global nature of the climate risk has led to debate about which approach is best from a global perspective. The Kyoto Protocol is based on the idea that countries can negotiate national caps and then establish allowance-trading mechanisms. Critics of this approach claim that it will be less difficult and more effective to apply harmonized carbon taxes around the globe rather than trying to extend restrictive caps from the Annex I countries of Kyoto to all countries on the planet (Nordhaus, 2007).

While taxes and caps are largely seen as policy substitutes, there is a fairly widespread belief that dramatic GHG reduction by mid-century requires the combination of a tax or cap policy with additional policies that accelerate the rate of technological change. One such policy is the allocation of large public subsidies to the initial developments of new capital-intensive technologies, such as the first full-scale CCS systems, the prototypes of new nuclear power systems, and technological innovations that provide reliable energy storage for renewable energy. Another policy is the use of regulations to guarantee market shares for new and emerging technologies that otherwise have higher costs than conventional, GHG-emitting technologies. An example is the renewable portfolio standard, which guarantees a growing share of the electricity generation market for renewables.

Box 22.10 | Information as a Policy to Influence Sustainable Energy Choices – Charlie Wilson

Information policies have been widely used to promote the adoption of sustainable energy technologies as part of broader behavior change strategies. Through the 1980s and 1990s, for example, United States utilities spent over US$20 billion providing information and incentives in response to a perceived shortfall in both households' and firms' knowledge of widely available and cost-effective investment opportunities.

Information-based approaches to promoting energy technologies remain widely used today, but have been improved by a wealth of empirical findings on the "who," "how," and "what" of information provision.

Lessons on *who* should provide information and *how* they should provide it are inter-related. Information disseminated from person to person, particularly if such people are trusted peers or social role models, is more effective at changing behavior than information spread through mass media channels. The perceived trustworthiness and credibility of the information provider is also important. So too is consistency. Clearly inconsistent policies weaken the credibility and thereby effectiveness of information – e.g., policies to reduce energy prices vs. information policies on cost-effective, energy-efficient technologies.

The *what* of information provision relates to form and content. "Folk" behavioral models that describe how people actually think about energy clearly demonstrate the importance of information that is simple and salient; i.e., it stands out in some way and is easily comparable. Efficiency ratings on product labels can be a good example of this. Targeted or otherwise personally-tailored information is also effective. In contrast, information that is technical, detailed, factual, and comprehensive is often glazed over, or interpreted subjectively and selectively, often to support pre-existing beliefs.

Another important role for information is to provide feedback on behaviors undertaken or technologies adopted. The transition underway from aggregated monthly utility bills to real-time energy use monitors enabled by smart metering will greatly improve the value of information as feedback on energy-related behavior: consider the analogy of being informed of your total food expenditure once a month, as opposed to receiving an itemized bill each time you shop.

Well-designed information policies may successfully raise awareness and support positive attitudes toward sustainable energy technologies. But the ultimate success in changing behavior is limited by contextual factors. High investment costs, coupled with limited access to capital and strong consumer preferences for immediate rather than delayed financial benefits, are a common example. Other contextual constraints include: regulations (e.g., planning guidelines); economic incentives (e.g., falling energy prices, sales taxes); social norms (e.g., larger homes); habits and routines (e.g., daily washing and cleaning); and community governance traditions (e.g., resisting outside developers).

The stronger these contextual constraints, the weaker the effectiveness of information policies. In general, therefore, information policies are effective only as part of a broad and consistent multi-pronged policy framework to promote the diffusion of sustainable energy technologies. As an example, information policies are a useful rapid response to a window of opportunity presented by some external shock, such as rapid oil price rises or supply disruptions.

Paradoxically, the best form of information provision results from peoples' actual experiences, and only comes with widespread diffusion. The communication of positive experiences through networks of families, friends, and peers is the most effective means of supporting social learning, increasing familiarity, forming favorable attitudes, and reducing the perceived risks of sustainable energy technologies.

Sources: Dietz and Stern, 2002; Owens and Driffill, 2008.

Box 22.11 | Community Energy Management and Transportation in Curitiba, Brazil – Mark Jaccard

While the rapid urbanization in developing countries presents monumental challenges, it also provides unique opportunities. A myriad of incremental, and seemingly unimportant, decisions about urban land use and infrastructure taken today will profoundly determine the ability of tomorrow's burgeoning urban centers to achieve sustainable energy systems. Curitiba, Brazil, provides an example of how effective planning can have a positive impact on a community's development, particularly in terms of energy use for transportation.

A city of over two million, Curitiba has, since the 1970s, channeled growth along five axes radiating from the city center. Each axis has a bus expressway and parallel roads for vehicles. Land use zoning has concentrated high-density development to the five axes, especially centered on interchange bus terminals that are located about every two kilometers along each axis. Passengers from lower density areas take feeder buses to these terminals, where they transfer to the express buses for travel to the city center. Costing about 1/200th per kilometer of a conventional subway system, the bus expressway nonetheless achieves comparable performance in terms of ridership and travel times. While Curitiba has a high rate of car ownership for Brazil, almost 75% of commuters use buses, resulting in 25% lower vehicle fuel consumption than similar Brazilian cities. Reduced fuel consumption contributes to the city's relatively low level of urban air pollution, and reduced vehicle use for commuting fosters a more pedestrian-oriented city center. The express bus system is operated primarily by private companies under guidelines from, and in partnership with, the municipal government.

Box 22.12 | Masdar, CCS, and Enhanced Oil Recovery – Michael Ohadi

The United Arab Emirates (UAE) is the world's second largest emitter of greenhouse gas per capita. In 2006, the Government of Abu Dhabi launched an initiative, the so-called Masdar initiative, with a primary goal to promote advanced and clean energy supply and substantial reduction in the CO_2 emissions to the environment.

The Abu Dhabi Government is pumping billions of dollars into the clean energy Masdar initiative with the double aim of capturing CO_2 emissions from major sources of CO_2 production and injecting it into oil reservoirs for enhanced oil recovery purposes. The long-term goal is to prepare the world's third-largest crude exporter country with economic diversification and a future less dependent on a supply of crude oil.

CCS technology refers to the capture of carbon at the source, then compressing and liquefying and finally transporting it by pipelines to safe and permanent storage in geological formations. The nature and size of the reservoirs of Abu Dhabi, as well as the short distance between CO_2 emission sources and oil reservoirs, created an opportunity for a reliable and technically feasible CCS project.

The UAE embarked on the Masdar CCS project and the CO_2 emission reduction program in 2007. The project aims to slash the emirate's CO_2 output by about one-third by 2020. In its first phase, the project aims by the end of 2012 to capture five million tonnes of CO_2 from power plants and industrial facilities and to transport the CO_2 to oilfields for enhanced oil recovery (EOR) applications. Injecting CO_2 in the oil reservoirs will maintain underground pressures, thus resulting in enhanced recovery of the crude oil. EOR is receiving more attention in recent years with further maturing of the oil reservoirs in the country. The world average of oil recovery is estimated at 35%, which in essence means 65% of recoverable oil is being left behind if it is not recovered through advanced technologies, such as use of injected CO_2 and other enhanced oil recovery techniques. Currently natural gas is being injected in some of the Abu Dhabi reservoirs for enhanced oil recovery purposes. It is estimated that over one billion standard cubic feet of natural gas per day is currently being injected into oilfields for enhanced oil recovery that otherwise can be used for power generation or petrochemicals.

The Masdar CCS project has already launched a pilot plant project in which CO_2 is captured from a source and is being injected in one of the oil reservoirs. Data on CO_2 diffusion and its impact on the enhanced oil recovery of the reservoir are being collected. The project involves close collaboration between the Masdar Institute, Abu Dhabi National Oil Company and its subsidiaries, the Petroleum Institute, and other academic and industrial collaborators from around the world.

The project is being developed over a multiphase road map, with the feasibility study undertaken in 2007 and the Front End Engineering Design (FEED) phase begun in August 2008. The fully developed CCS project will use CO_2 from power stations, refineries, and other industrial sources. The following emitters are planned in the first phase of the project: a major power plant, an aluminum smelter, and a steel plant. The combined phase one capacity is five million tonnes of CO_2 per year with an approximately 300-kilometer CO_2 pipeline network to carry the CO_2 to the injection site(s).

Masdar's CCS activities have considerable potential to be expanded in Middle Eastern countries where significant carbon capture potential can be located under the Kyoto Protocol. The Masdar initiative can serve as a role model for similar developments in other Middle Eastern, as well as other regional and industrialized, countries. It is an example of a clean and peaceful energy initiative that has multiple win-win objectives, including utilization of CO_2 for enhanced oil recovery to free up more oil while liberating substantial amounts of natural gas for power production and other purposes, and finally economic diversification and job creation opportunities.

22.2.7　Accelerate the Rate of Energy-related Technological Change

Innovation and technological change are essential contributors to a sustainable energy future. Energy technology innovation is addressed in detail in Chapter 24 and, as noted above, the broader social and institutional capacity on which technological change is predicated is covered in Chapter 25. Below is a summary of the context, rationale, and criteria for policy to foster innovations that are critically needed to address the joint challenges of global environmental change and the energy service needs of all of humanity.

22.2.7.1　Characteristics of Innovation and Technological Change

In its most general sense, technology is a system of means to particular ends (e.g., Grubler, 1998). Technologies comprise both physical creations – plants, equipment, devices – and information and knowledge systems – know-how, skills, experiences. Technology is thus a specific form of knowledge that can be embodied (an appliance) or disembodied (know-how). The respective importance of these two forms changes over the life cycle of a technology from invention through research, development, and demonstration to niche market applications and, ultimately, pervasive diffusion. Through this lifecycle from innovation to deployment, technological knowledge needs to be created, developed, and applied. It can also depreciate or be lost if not actively managed. Innovation policy is therefore fundamentally concerned with stimulating and managing this process of knowledge generation, application, dissemination, and feedback, and thus needs to embrace a systemic perspective (e.g., Carlsson et al., 2002). This includes the continuous feedback between different stages of a technology's lifecycle, which typically characterizes successful innovation processes. Technologies are also inherently dynamic, and the innovation process is characterized by high degrees of uncertainty. This requires adaptive policy approaches that recognize the dynamics of technological change, and allow for experimentation, foster diversity, and accommodate failures.

22.2.7.2　The Rationale for Public Technology Policy

Policy intervention in the innovation process is primarily justified by the public good characteristics of knowledge and thus the potential positive spillovers beyond the innovating agent. Additional sources of market failure exacerbate this public good problem (Jaffe et al., 2005).

New knowledge can be expensive to generate, but cheap to imitate. As a result a "free-rider" problem exists in which private firms under-invest in innovation, as the benefits cannot be fully captured. Without intervention, these knowledge spillovers result in under-allocation of resources to innovation. This problem applies to all technologies, but is exacerbated for energy technologies due to the high capital costs and development lead times, which magnify the risks for innovators, and the potentially limited returns in regulated markets. Literature related to innovation has repeatedly found that social benefits from innovation exceed the benefits accruing to private innovating firms several-fold: social rates of return are persistently and significantly higher than the private returns from innovation (Freeman, 1994). Policy intervention is needed to correct this market failure, particularly as the limited data available on private-sector R&D investment suggests a decline over the last 20 years against an ever-rising backdrop of public concerns on energy security, access, and environmental impacts.

Other sources of market failure increase the need for government intervention. Positive interaction effects may exist in cases where both under-pricing of the environment (lack of pricing or regulation to reflect negative externality costs) and under-pricing of knowledge (positive externalities of information) make the financial attractiveness of innovation investment much lower than its true value to society. Increasing returns, due to network effects and economies-of-scale of incumbent technologies, delay the deployment of nascent technologies, and possibly lead to under-investment at earlier stages in the innovation process. Information asymmetries between technology producers and technology adopters make deployment sub-optimal in early stages when technologies are unproven. Principal-agent problems – for example, between owners and occupants of buildings needing thermal retrofit to improve efficiency – hinder deployment, even after technologies prove to be reliable.

However, successful innovation depends on much more than overcoming market failures. A systemic view on innovation highlights the fact that innovation entails a series of collective, embedded processes involving many actors, networks, and institutional settings that can fail at many different levels and need to be addressed by policy intervention (Hekkert et al., 2007).

22.2.7.3   Policies for Innovation

Policies for innovation can: (1) directly target the innovation process; (2) support the innovation system; or (3) unintentionally impact innovation while targeting an unrelated concern.

Direct policies for innovation vary according to the target and timing during the innovation process. Policy is needed at each stage of this process. The role of government is most evident at the earliest stage

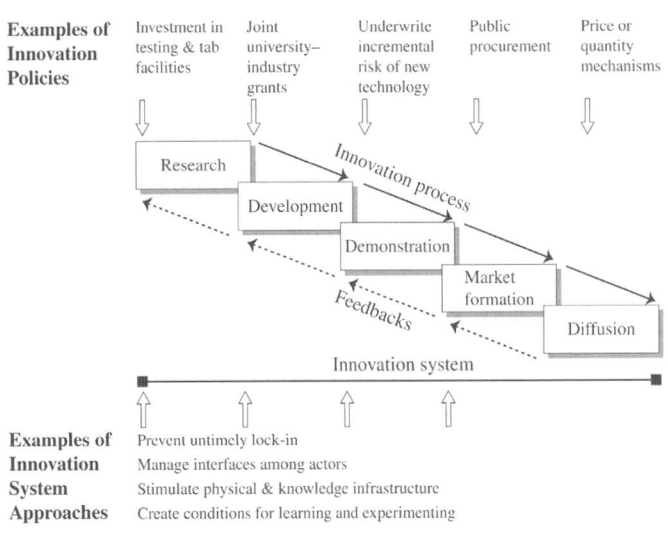

Figure 22.2 | Overview of policies for innovation. Source: developed by GEA authors.

of basic science and research. Together with the private sector, governments are also engines of applied energy R&D. But governments also need to play an important role in leveraging private sector investment at the early commercialization stages by supporting demonstration activities to reduce risks and market formation to underwrite demand.

This innovation process is situated within an overarching system comprising the actors, institutions, and networks involved in developing and commercializing a technology (see Chapter 24 for details). Innovation policies must also target the successful functioning of the innovation system (see Figure 22.2 for examples).

Policies on issues such as education, taxes and subsidies, and market regulation can all exert an important, yet indirect, influence over both the supply of, and demand for, innovations. This reinforces the need for consistency not just between direct innovation policies but also between the broader regulatory and institutional environments for innovation.

Policies supporting the supply of innovations or the development of technologies include investments in R&D, intellectual property protection, laboratory and testing infrastructure, training and skills, university–industry collaborations, formal and informal mechanisms of knowledge exchange, technology roadmaps to guide the direction of innovation, and financial incentives such as tax credits for private investments. Not all innovation, however, derives from formal R&D activities. Problem solving and incremental improvements in existing technologies are also of importance and can be stimulated and supported by public sector policies that lead to the creation of outreach, extension, and technical support programs. Policies supporting the demand for innovations as commercialized technologies include demonstration projects, public procurement, market niche creation (supply obligations), market incentives via changes in relative prices (via environmental taxes or feed-in tariffs), standards and regulations, direct financial support, education, and marketing. These "supply push" and "demand pull" policies are complements and not substitutes. Innovation success stories are typically characterized by comprehensive and consistent policy support through the entire innovation process. Although these criteria can be generalized, particular innovation policies must take into account specific local conditions or be otherwise tailored to the technological or market characteristics of an innovation.

22.2.7.4   Principles for Innovation Policy Design

Selecting innovation policies is necessarily context-specific. However, the literature (Grubler et al., 1999; Norberg-Bohm, 2000; Nemet, 2009) and the sample of energy technology case studies reviewed in Chapter 24 points to a number of guiding principles or policy criteria that are considered generic to all technology domains and adoption environments. These are:

Innovation policies need to be aligned and consistent. This should be true both within a given innovation system and between different innovation systems to maximize spillover benefits. Alignment implies a set of policies that are co-ordinated and free of contradictions. A prime counter-example is the current emphasis of public support for low carbon innovation alongside the much larger subsidization of fossil fuel technologies.

Aligning incentives for energy technology innovation is aided by an explicit and systemic innovation strategy. Such a strategy will inevitably face trade-offs between policy objectives, as is the case with standards and incentives to promote energy efficiency while maintaining downward pressure on energy prices.

Innovation policy frameworks and supporting institutions need to be stable and independent. This is the means of effectively managing the process of technology knowledge generation, application, and maintenance. Independence avoids the resistance to change and learning of the vested interests, which grow up around incumbent technologies. Policy objectives and instruments require continual re-assessment and potential adjustment toward long-term transformative goals.

Stability wards against the eventual loss of knowledge. Erratic policy support and signals can lead to rapid knowledge depreciation. Governments play an essential role in managing expectations of the demand for innovation by providing policy signals that are credible and consistent over a multi-year period to reduce the uncertainty of private sector investments reliant on distant payoffs. Policy instability acts as a barrier to commercial innovation, and can accelerate knowledge depreciation. Stability does not preclude policy dynamism and flexibility to respond to new information. But it does, however, mean patient and consistent goals backed up by predictable funding support for various stages of the innovation lifecycle.

Superior policy requires a commitment to good data. Innovation policy needs to be founded on a clear understanding of how the innovation system operates, what it requires in terms of inputs, and how to assess its effectiveness in terms of outputs. Understanding and assessing innovation systems requires far better data on innovation activities than are currently available. For example, technology-specific private sector R&D data are almost entirely lacking. Systematic and comprehensive information disclosure as a condition of policy support can help redress the current scarcity of reliable innovation outcome data. As an example, subsidies for demonstration projects and niche market deployment could be contingent on documentation and public disclosure of both successes and failures in order to facilitate wider learning.

Policy should support the different stages of innovation, and the feedbacks between them. Traditionally, innovation policies have targeted specific stages of the innovation process while neglecting the essential feedbacks between them. As an example, publicly-funded testing and demonstration facilities for new technologies can ensure disclosure of unbiased performance benchmarks to guide technology R&D. (For example, see the case study on wind power in Chapter 24.) Feedback from market deployment experience can also benefit technology design and manufacturing quality assurance. (See the case study on solar photovoltaics [PV] in Chapter 24.) Increasingly, policies to support information sharing and knowledge feedbacks need to consider the globalized nature of technology markets.

Innovation policies should facilitate widespread experimentation. Successful diffusion is underpinned by knowledge generated through the development, design, construction, and operation of a technology. Inherent uncertainties make this a process of experimentation. Experimentation leads to incremental improvements and learning-related cost reductions, and also underpins the success of increasing unit sizes or overall production volume to capture economies-of-scale. Indeed, the capital intensity, risk, and opportunity costs of many large-scale energy technologies should orient experimentation toward relatively small-scale versions of technologies in a diverse portfolio to reduce the consequence of failure. By comparison, narrowly-targeted policy support for single design large-scale demonstration projects – such as breeder reactors – is high risk.

Experimentation can, and also perhaps should be, multifarious, involving an array of different actors, forms, and stages of the technology's lifecycle. Governments should intervene to manage the natural commercial tendency to rapidly hone in on a dominant design that confers market advantages and potential cost reductions.

Innovation policies need to focus on portfolios rather than single technologies. The magnitude and array of energy system challenges, combined with inherent uncertainty in the innovation process, requires a broad portfolio of technologies. Such portfolios need to balance each technology's option value – or social benefits in case of successful diffusion – and risk, in terms of both innovation failure and investment.

These portfolio decisions will require a combination of expertise from public and private sector innovation actors. Formal scientific tools for innovation portfolio design and analysis are increasingly available, but need further development as well as an institutional home for application.

Generic principles for innovation portfolio design include: (1) incorporate options from across the entire energy system, especially including the traditionally under-researched and under-funded energy end-use technologies, and avoid an overemphasis on any single, "magic bullet" option; (2) manage resource constraints by focusing on granular, less capital-intensive technologies such as end-use innovations and smaller scale supply options; (3) link large-scale, capital intensive, and high risk innovations into global innovation portfolios that enable international coordination and cooperation; (4) avoid pre-empting the outcome of decentralized market-based technology experimentation while counteracting private sector biases toward early selection of a dominant and rapidly scalable technological solution; and (5) promote technologies that show good prospects for gains from learning-by-doing in application and economies-of-scale in production.

22.2.8 Coordinate and Implement International Energy-related Policies

Globalization of the world's economic system is associated, as is to be expected, with the globalization of its energy system. Globalization can have benefits, but it can also present major challenges.

First, globalization of the world's energy system has not had a significant benefit for the poorest people on the planet. The rate of foreign direct investment in the energy sectors in developing countries remains at only a fraction of the level needed if most people in the world are to have access to electricity and modern fuels over the next few decades. Governments have tried to make gains in this direction, in part through collective funding mechanisms like the World Bank and the International Monetary Fund, and in part through encouraging the establishment of domestic environments more favorable to foreign investment. But these strategies, thus far, have not provided the investment funds that are needed. The key issue, from a policy strategy perspective, is to find a way to make globalization work for the poorest people in the developing world.

Oil's price has been set internationally for decades, but now this is also increasingly the case for coal, natural gas, and sometimes even electricity. Globalization of the world's energy system can contribute to increased energy security, widening the options for any given country to meet its energy needs. But globalization of energy trade can also have the reverse effect, especially as individual governments lose the ability to ensure energy supplies for domestic markets at relatively stable prices. For example, while individual governments cannot prevent oil price spikes, like that of 2008, working in concert they may be able to reduce some of the energy price volatility that has occurred in the past.

Finally, when it comes to international energy policy coordination, the issue of climate change is becoming increasingly critical. Addressing the risk of climate change through GHG emissions reductions is a collective public good that can only be achieved with collective action. Most countries, perhaps starting especially with the world's largest emitters and most powerful countries, must work together to successfully reduce emissions. Thus far, the international community has put a great deal of effort into widespread negotiations, largely through the United Nations sponsored Framework Convention on Climate Change, but this has not yet produced truly significant action by those countries who have made commitments, and not yet induced enough significant emitters, like the United States, China, and India, to agree to substantial emissions cuts.

While some see globalization as largely a force for improvement in the world, others see it as the cause of problems and argue that policies should try to prevent or restrain globalization. Still others argue that globalization is inevitable, albeit undesirable, so the only alternative is to work within this reality and not resist it. Finally, some believe that with careful policy strategies, the forces of globalization can play a critical role in addressing the very challenges it presents. More of this latter perspective is discussed in the description of globalization-related policy instruments and strategies.

If globalization is to significantly increase energy access in developing countries, these countries and international agencies need to improve the climate for foreign investment in energy supply while at the same time ensuring that such investment meets the need for clean, affordable energy. Thus, international institutions like the World Bank need to keep reforming the criteria by which they support energy policies, although some significant progress in this direction has already been made in recent years.

If globalization is to enhance rather than reduce energy security in developing and developed countries, significant changes in international energy security regimes will be required. The current focus of energy security institutions such as the IEA and the OPEC on short-term fossil fuel supply stability is both increasingly ineffective – as demonstrated by recent price volatility – and untenable in the face of persistent systemic challenges associated with rapid demand growth and increasing geographic concentration, if not outright physical scarcity, of petroleum resources. Moreover, stability, inherently preferred by energy security regimes, is at odds with the rapid and radical change expected of energy systems to meet sustainability objectives. The future international energy security regimes should therefore be able to overcome these limitations by expanding focus away from the supply-side and fossil fuel orientation of today. They should also be able to provide for a meaningful dialogue between energy importers and energy exporters, including such emerging major energy users as India and China and such major energy suppliers as Russia.

Another key area for international coordination and cooperation is to enhance security measures that prevent the diversion of the peaceful uses of nuclear energy to the production of nuclear weapons by governments or terrorists. While much progress has been made in this direction over the past decades, the challenge becomes all the greater as the number of countries with nuclear capabilities increases (for further information, see Chapter 14).

Of increasingly dominant importance is that two decades of efforts have thus far failed to result in an agreement on the establishment of a global architecture for concerted international action on GHG emissions to address the risk of climate change. Many of these efforts have focused on the creation of international agreements to reduce emissions, notably the 1992 Framework Convention on Climate Change and the 1997 Kyoto Protocol. This top-down approach seeks to achieve widespread agreement among almost all countries, whether they be big or small contributors. In the long run, participation by virtually all countries will be required, if only to prevent the disintegration of agreements by a growing number of free-riders (Barrett, 2003; Aldy and Stavins, 2007). But as a starting point for reaching an international agreement with meaningful commitments and effective policies, some have argued that it may be easier to limit negotiations to a small number of countries that account for most emissions or that have the financial ability to contribute significantly to the global costs of reduction (Victor, 2001). Perhaps in a bottom-up approach, a group of 20 major emitters, like the United States, China, India, and some European countries, or the European Union as a whole, should first try to negotiate targets and policies that could form the basis for an eventual international architecture that other countries would gradually join. Other issues that need to be addressed include the effect of institutions like the World Trade Organization on efforts to achieve a meaningful global effort to reduce GHG emissions.

22.2.9   Key Energy-focused Policies[4]

For the eight previously delineated energy-related policy goals to be achieved, energy services, technologies, and operating practices of the global energy system must evolve in specific directions. Certain outcomes need to occur in terms of energy efficiency, renewable energy, fossil fuels, and nuclear power. These key energy-related outcomes represent the means for achieving the energy-related goals. A survey of the various chapters of GEA, and especially Chapter 17 on scenarios for a sustainable global energy system, suggests four outcomes of particular importance:

- The global energy system needs to become much less energy-intensive for a given level of energy services, i.e., much more energy efficient.

- The global energy system needs to rapidly shift toward renewable sources of energy, used with minimal harm to the environment and humans, such as the need for food production.

- Wherever fossil fuels are used, the environmental and human impacts and risks from their use need to be minimized.

- Wherever nuclear power is used, its energy-related and weapons-related risks to the environment and humans must be minimized.

In this section, generic policies are outlined that offer a means of achieving sustainable energy. In Section 22.3.1, some of the key policies that are likely to be effective in achieving these outcomes are presented. Additionally, Chapters 23, 24, and 25 focus especially on critical areas of policies for energy access, policies for capacity development for energy transition, and policies for accelerating energy-related technological change. The policies in Chapter 22 have been selected because: (1) they have broad applicability in terms of the urgent need for rapid technological and behavioral change; and (2) they have either already demonstrated effectiveness in real-world applications or they appear to perform well against many of the key policy evaluative criteria listed in the chapter. Many of the other chapters discuss policies, especially the ones focused on specific technology and resource options.

For advancing each of these four means to sustainable energy – efficiency, renewable energy, clean fossil fuels, and safe nuclear power – a critical policy challenge is to ensure that valuable innovations successfully navigate the hazardous steps from invention, to commercialization, to widespread market penetration. Thus, while policy support for basic and applied R&D is needed, policy support is just as important in the demonstration and initial commercial phases; when additional support can help lower the production and operating costs of new technologies and establish, or help to establish, market niches that will stimulate investor interest in the technology. Finally, market pull policies are needed to carry the product through to widespread dissemination. The policies listed under each of the four means to sustainable energy, in the following Section 22.2.9.1, cover the various phases of this technology life cycle.

22.2.9.1   Reducing Energy-intensity of Energy Services

As noted throughout GEA, much of humanity requires a dramatic expansion of energy services. For sustainability, this expansion must coincide with a significant reduction in the energy intensity of these services. Even with such a reduction, the global energy system is likely to grow significantly by mid-century as the global population and the per capita energy demand of the billions of people with minimal energy services continues to grow.

A reduction in energy intensity can occur in basically four ways. First, even without a technological change it is sometimes possible to reduce

---

4   See also Chapters 8–14.

energy use while retaining an energy service at a given level. For example, it is possible to design urban form and land use so that total or average commuting distances are decreased.

Second, there may be cases where people can be helped to find ways in which their quality of life can be sustained even while their demand for certain energy services actually declines. An example would be encouraging people to adopt indoor thermal comfort zones ranging from 18–26°C, with people living in hotter climates allowing warmer temperatures and people in cooler climates allowing cooler temperatures instead of keeping temperatures at a fixed level. Another example would be reduced consumption of meat, since less energy is required to produce most non-meat foods. For an increasing number of people in rich countries, reducing meat consumption is a dietary change associated with improved health.

Third, a shift to more energy-efficient technologies will reduce energy use per capita, everything else held equal. Of course, there is a challenge with more efficient technologies in that they tend to have higher up-front costs and energy service demands will increase (the rebound effect) as the operating costs of providing these services fall. Well-designed policies may be able to reduce this likelihood, which usually involve price changes that reinforce the goal of reducing energy use.

Fourth, a restructuring of the economy toward services, more recycling of materials, and cascading uses of energy, and away from material- and energy-intensive inputs and final products can all help to reduce energy use for a given level of energy services.

The following prominent policies can contribute to these four ways of reducing energy intensity:

- *GHG emissions pricing* in the form of carbon taxes and/or cap-and-trade schemes that raise the price of energy and make it more profitable to reduce energy use;

- requirements that prescribe *community energy management* by municipal authorities, involving coordinated efforts to reduce energy use via land-use zoning decisions, development permits, and siting requirements, and the link to public transportation and other infrastructure planning and development – including an overall goal of reducing the need for private vehicle use, especially in urban settings (transportation demand management);

- *education programs* by governments, utilities, non-governmental organizations (NGOs), industry, the media, academics, and others to foster a conserver lifestyle with respect to energy and material throughputs;

- government *energy efficiency information and subsidy programs* for all types of equipment, appliances, products, and buildings. When efficiency programs involve subsidies, they must be carefully designed

and monitored to reduce the likelihood of high free-rider rates, even though some degree of free-riding is inevitable. Revolving green funds for energy efficiency investment can provide a subsidy in the form of low-interest capital that is mostly recaptured through payments by the recipient as their energy bills are reduced through efficiency;

- in the case of electric, gas, heat, water, and sewer utilities, efficiency programs administered by the utilities themselves, called *utility demand-side management*, that might be motivated and applied relatively efficiently by regulations that entail tradable mechanisms, such as tradable energy efficiency performance standards or certified amounts of energy saved (referred to in Europe sometimes as "white certificates");

- *building code* changes that, over a specified time period, set increasingly stringent requirements for new and retrofitted structures in terms of energy use and emissions. This should affect all buildings and structures in residential, commercial, institutional, and industrial sectors;

- *efficiency regulations* that set an aggressive but reasonable phase-out schedule for the market availability of less energy-efficient industrial equipment, household appliances, transportation vehicles, and other equipment;

- *professional training* of architects, engineers, designers, and installation contractors in methods and installation of energy-efficient devices;

- use of *public procurement* and co-operative procurement of energy-using devices as opportunities to create a first market for energy-efficient products and building concepts, or increase the market volume for the best already in the market;

- *funding and encouragement of research, development, and demonstration* of more energy-efficient solutions;

- regulations that prohibit the sale of products and equipment that use energy when not in operation (called *standby power or phantom power regulations*);

- regulations that *limit the use of packaging* in retail sales that has substantial energy implications in terms the cumulative effects of production, distribution, and transportation of goods;

- in addition to increasing the retail price of various forms of energy to reflect environmental costs, utilities should be mandated by government or by independent regulators to establish tariffs that price marginal energy use, such as peak electricity or the last units consumed in a billing period, at its marginal cost of production. This form of *non-linear pricing*, also called marginal cost pricing and time-of-use pricing, tends to provide a further incentive for reduced energy use, especially in jurisdictions where the long-run cost of new supply is higher than the average present costs of operating the electricity system. For residential and commercial customers, this might be associated with the widespread distribution of rates that rise with greater energy use per billing period and/or rates that reflect peak and off-peak costs of service. And where the ability of low income customers to pay is also a concern, non-linear pricing can be combined with lifeline rates as a form of cross-subsidy between customers; and

- governments, especially at the municipal level, provide *transit pricing options*, such as a subsidized annual transit pass, that motivate commuters to rely on public transit for commuting and many other mobility needs.

In order to ensure that these policies are meeting the desired goals, governments must set targets for reducing energy intensity, assess the amount of energy saved through each policy, and monitor the progress toward meeting these targets.

22.2.9.2   Accelerating the Growth of Renewables

Although depletion of low-cost fossil fuel resources may lead to rising average prices for oil, natural gas, and perhaps even coal over the next decades, one cannot assume that energy markets will, as a consequence, rapidly shift the global energy system to renewables from its current 80--85% fossil fuel dependence. First, as long as a relatively plentiful resource like coal can still use the atmosphere as a free waste receptacle for GHG emissions, this resource will remain highly competitive with renewables for the production of electricity and even vehicle fuels. Second, incumbent technologies, such as those using fossil fuels, benefit from path dependence, in which society's technological, social, and institutional capacity is geared toward the continued success of that technological path. Third, in the same vein, renewables face many barriers to massive scale-up, some of which relate to overcoming the path dependence in favor of fossil fuels, others of which relate to the specific characteristics of renewables – namely, that frequent low-energy density implies significant claims on the land base and that their intermittency requires additional, costly investments for energy storage.

For these reasons, policies to induce a rapid scale-up of renewable energy must: (1) ensure that renewables are economically attractive for investors; and (2) ensure that non-cost barriers to renewables are reduced. Policies to price GHG emissions such as taxes and cap-and-tradable permits increase the financial attractiveness of renewable energy. These policies are constrained, however, in that it is politically difficult in developed countries, and even more so in developing countries, to burden energy users with rapid energy price increases in a relatively short timeframe. This explains why jurisdictions have also provided targeted subsidies (feed-in tariffs) and market share regulations (renewable portfolio standards) to support renewables, in addition to GHG pricing initiatives. As for the non-price barriers to renewables, policymakers

are increasingly realizing that these are also very important and have started to focus policy efforts on issues like regional land-use, including planning for siting hydropower, wind, and solar facilities; urban land-use, including solar access laws, siting of buildings, and building codes for solar hot water and passive solar heating and cooling; utility regulation, for transmission expansion and connection rights; and even technical training and education. Below are some key policies.

- *GHG emissions pricing* via carbon taxes and/or cap-and-trade schemes provide the clearest means of fostering renewables since these will raise the relative cost of GHG emitting forms of energy. Revenues from emissions pricing can be recycled to firms and households as income tax cuts or additional government program expenditures, but they could also be used to support emerging renewables technologies.

- *Feed-in tariffs* for renewables guarantee prospective independent power producers that all power they generate will be purchased by the grid at a fixed price. If the price is adequate, this provides a powerful inducement to equity and debt financing and explains why jurisdictions with feed-in tariffs have mostly had success in rapidly expanding renewables-based electricity generation. Feed-in tariffs may also be tailored to different renewable technologies according to their costs.

- *Market share mandates* provide a guaranteed market share for renewable energy, but they do not guarantee a price. They operate by creating confidence for investors in a steady and growing market for renewable energy. In the case of electricity, such mandates are commonly known as Renewable Portfolio Standards (RPS). Most market share standards offer the same credits to all eligible technologies, which tends to benefit the most commercially competitive technology, such as wind or corn-based ethanol, while providing no support to emerging technologies. In part for this reason, many renewables advocates argue that the feed-in tariff is a more effective policy. However, it is possible to create "tranches" for minimum shares for different forms of renewables or different types of technologies – such as a minimum market share for solar PV electric and a minimum market share for solar thermal electric.

- *Biofuel mandates* set a minimum content for fuels from biomass in gasoline, diesel, and perhaps eventually other fuels like heating oil and jet fuel.

- *Performance standards*, or intensity standards, set a maximum emissions content per unit of energy. These standards can also be tradable, leading to an average emissions content. An example in the renewable fuels context is low carbon fuel standards. While market share mandates may be effective in expanding renewables, they do not distinguish among the carbon, or other performance, of alternative fuels, be they renewable or non-renewable; if the ultimate goal is reducing emissions or emissions intensity, tradable performance standards are more effective. However, it may be noted that measuring the carbon performance of renewable fuels remains controversial.

- *Regional land-use and water-use planning* is important for scale-up of some renewables because of the potential land-use conflicts that renewables introduce. This might include river basin planning with hydropower, coastal zone planning with wind farms, land-use planning for concentrated solar power and biomass energy, and water-use planning for biomass energy. It is important that society understand that all energy options involve impacts and risks and to this end local communities must have meaningful participation in both regional planning and energy planning.

- *Electric transmission regulation and infrastructure planning* must focus on maximizing the opportunities for renewables. This includes transmission expansion that is tailored to the location of renewable resources and grid connection rules that address the particular needs of renewables.

- *Urban land-use zoning and infrastructure planning* are critical in removing barriers to decentralized renewables, via solar access rights, wind turbine siting rights, grid connection rights, and other key laws and policies.

- *Zero- and low-emission vehicle standards* can mandate a rising market share for vehicles fuelled by alternatives to end-use combustion of gasoline and diesel, such as biofuels, hydrogen, and electricity, all produced with zero- or low-life cycle GHG emissions.

- *Zero- and low-emission building standards* can mandate a rising market share for buildings that rely exclusively or almost exclusively on passive and active renewables and waste heat for heating, cooling, and ventilation.

- *Electric distribution utilities* must facilitate decentralized renewables electricity production via net metering, smart grids, and other policies.

- *Information programs, education, and other forms of capacity development* have a role to play in accelerating the growth of renewable sources of energy.

Similar to policies designed to reduce energy intensity, governments must set targets for the growth of renewable energy, assess the amount of energy supplied through each policy, and monitor the progress toward meeting these targets.

22.2.9.3　　Transitioning to Cleaner Uses of Fossil Fuels

The global energy system is dominated almost everywhere by fossil fuels, but the global endowment of fossil fuels is uneven. Some countries have negligible fossil fuel resources and are thus prime candidates for a transition to renewables and – possibly – nuclear, and some, like Sweden, France, and Brazil, are already moving down this path. However, countries with rich fossil fuel endowments are less willing to consider abandoning this high energy-density resource and are thus increasingly interested in technological developments that may enable continued use of fossil fuels, albeit with low or zero emissions of pollutants and especially GHGs.

This latter step requires converting fossil fuels into electricity, heat, and perhaps hydrogen while capturing and permanently storing the carbon, mostly in the form of CO_2, probably in geological formations deep in the earth. Carbon capture and storage (CCS) is expensive and will not occur without policies that price GHG emissions. Again, because of political constraints on GHG pricing, the zero- or low-emission use of fossil fuels will require additional support. Some of the following leading policies have already been enacted on an experimental basis, but these efforts would need to be intensified significantly over the next decades to realize a dramatic shift to low- and zero-emission uses of fossil fuels.

- Governments, as noted above, can implement *GHG emissions pricing* via carbon taxes and/or cap-and-trade systems. These will reduce the use of fossil fuels but also foster commercialization efforts with CCS in some jurisdictions, enabling societies to assess with less uncertainty the viability and acceptability of this option.

- Governments can *reduce all subsidies to conventional combustion of fossil fuels, which will improve the relative competitiveness of renewables and nuclear alternatives*. This includes fuel price subsidies, subsidies to private vehicle use for urban and inter-urban travel (untolled roads), especially if fuelled by gasoline and diesel from fossil fuels, and a host of subsidies to industrial, commercial, institutional, and other combustion uses of fossil fuels.

- Governments can provide demonstration and commercialization subsidies, allocated via competitive bidding perhaps, for the initial, high-risk investments in CCS.

- Some cap-and-trade systems for emissions pricing allow firms to pay into a *technology fund* for those emissions that exceed permits. These funds can be used alone, or in concert with other government subsidies, to support early CCS projects.

- Governments can offer to *pay above-market rates* for electricity, heat, or hydrogen from projects that produce these from fossil fuels with CCS. This would be similar to the feed-in tariff for renewables.

- Governments or utility regulators could *ban construction of new coal-fired electricity plants* that lack CCS or that are not "CCS ready." Some jurisdictions in developed countries have explicitly adopted this policy. Some utility regulators in the United States and elsewhere have implicitly adopted the policy by disallowing coal-fired plants under the basis that they pose financial risks for ratepayers from future increases in GHG emissions prices.

- *Emissions performance standards* can give electricity generators the option to reduce emissions per unit of electricity either by a growing share of renewables and/or nuclear or by a growing reliance on CCS.

- *Life cycle carbon fuel mandates* can give fuel retailers the option of reducing the net GHG emissions from fuel use by increasing the content of biofuels or perhaps by including some hydrogen that is produced from fossil fuels with CCS.

- *Land-use planning* can facilitate socially and environmentally acceptable siting of underground carbon storage and CO_2 pipelines. There is also a need for land-use planning to safeguard against potential impacts of carbon storage on other uses for the subterranean, such as geothermal energy, or at least consider a balance between the possible uses.

- Governments need to *legally clarify geological rights* to underground pore spaces for CO_2 storage.

- Governments need to *establish short- and long-term liabilities and risk management and monitoring responsibilities* at CO_2 storage sites and on CO_2 pipeline right-of-ways.

22.2.9.4 Fostering the Safer Use of Nuclear Power

Views on the value and risks of nuclear power differ greatly and are often polarized. Some see nuclear power as a risky technology whose potential harm far exceeds any possible benefits it might provide, such as low GHG emissions. These perceived threats from nuclear power include catastrophic accidents at nuclear plants, either through operational failures or terrorist attacks, inability to safely transport and permanently store radioactive wastes, and the exploitation of civilian nuclear expertise for the proliferation of nuclear weapons.

Depending on the severity of these concerns about nuclear power, its regulatory burden for design, permitting, operation, and decommissioning can be such that nuclear power is a high-cost option for electricity generation. However, where local, national, and/or international public policy is able to allay these concerns, then nuclear power can be a competitive energy option. But everything hinges on risk preferences among the public and decision-makers, particularly with respect to trading off the extreme event risks of nuclear power with the ongoing impacts and risks of its alternatives. The following policies, therefore, focus on how to ensure safe use of nuclear power that is both real and perceived.

- At the international level, governments and the nuclear industry need to continue to improve the *mechanisms for monitoring and*

controlling the use of nuclear power and the reprocessing of nuclear fuel to prevent the acquisition of expertise and materials for nuclear weapons production.

- Governments need to collaborate in the *establishment of permanent storage sites* for radioactive materials.

- *By facilitating collaborative investments*, governments can help the nuclear industry settle on two or three dominant designs that have the best chance of achieving regulatory approval and should thus reduce regulatory costs, which have been very high in jurisdictions like the United States.

22.3 Part II: Policy Portfolios for Sustainable Energy

22.3.1 Implementing Policies for Sustainable Energy

So far, this chapter has presented a summary of policies required to achieve the sustainable energy pathways of GEA. While these pathways are described in detail in Chapter 17, policies for sustainable energy are spread throughout GEA. This section provides further guidance on policies that are essential in some form, depending on the jurisdiction and its particular goals and challenges.

Throughout GEA, the issue has been raised that there is an urgent need to identify and promulgate key criteria that can be used to select the type and intensity of sustainable energy policies that are required to make rapid progress toward universal access to energy and a rapid transition to a cleaner, less risky global energy system.

There are two important characteristics to note about energy-related policymaking. First is the importance of understanding policymaking as an organic, iterative process. Policymaking must be a continuous learning exercise since one can never predict precisely how a policy will work. Thus, policymakers need to think experimentally when designing and implementing policies. This might mean, for example, deliberately creating real-world variations by implementing different policies or policy variants in different locations within a given jurisdiction in order to generate more useful data on policy effectiveness. To this end, the steps of policy monitoring, policy evaluation, and policy redesign are essential to good policymaking.

The second is the critical role of institutions and governance. Without effective institutions and governance practices, policies have little chance of success. Assessments of policy design and policy effectiveness cannot be divorced from efforts to improve governance. This includes: (1) increasing transparency and public involvement to reduce the risks of corruption; (2) developing higher standards for education; (3) monitoring and evaluation of the civil service; (4) ensuring effective, independent controls by the legal system; and (5) strengthening the policymaking functions of democratic processes in various ways.

This section outlines policies that can achieve the eight goals for sustainable energy policy discussed earlier in this chapter. Again, these are:

1. increase energy access;
2. develop capacities for energy transitions;
3. enhance energy security;
4. manage energy-related market power;
5. manage valuable energy resource endowments;
6. reduce environmental and human health impacts;
7. accelerate the rate of energy-related technological change; and
8. coordinate and implement international energy-related policies.

Each of the goals demands its own unique policies, but policies must be implemented in a coordinated, integrated, and mutually consistent fashion. Chapter 17 presented pathways for achieving these goals in terms of the contributions from different actions, such as increased energy efficiency, switching to more desirable fuels and technologies, better prevention and control of pollutants, and stimulating major investments to increase energy access and improve energy security. The pathways include: (1) one that strongly emphasizes energy efficiency; (2) one that focuses on a cleaner, more sustainable energy supply mix; and (3) one that is a mix of these two contrasting strategies.

The policies presented here have been selected to ensure achievement of the GEA pathways. A dominant focus is the substantial, coordinated effort required for making major economy-wide advances in energy efficiency. This includes focused, somewhat aggressive policies. Such policies are needed to induce rapid innovation, tighten efficiency regulations in energy supply and demand, and increase energy prices. The policies must create a culture of conservation among consumers and firms, change land-use zoning to increase urban density, and integrate mixed land-uses so that transportation service demands decline while low energy-using transportation modes flourish.

Achievement of greater energy efficiency, alongside a rapid increase in the use of renewable energy, cannot be pursued in isolation. These goals must be intrinsically linked to broader social and economic development goals on the one hand and to environmental protection and restoration goals on the other. Providing adequate access to modern forms of energy is a critical contributor to social and economic development that can substantially improve the standard of living of the most disadvantaged people on earth. At the same time, a rapid transition to an energy system with negligible greenhouse gas emissions is critical for preventing destabilization of the earth's climate system.

At the policy level, a similar relationship exists. Energy-focused policies must not be conceived and implemented in an isolated fashion, but instead coordinated and integrated with non-energy policies for socio-economic development and environmental protection. These latter include policies that foster sustainable urban areas, preserve forested land and biodiversity, reduce poverty, provide environmentally acceptable transportation of goods and people, ensure vibrant communities, and improve human health.

An emphasis on policy integration is a common theme throughout GEA. This explains why the Millennium Development Goals are a key consideration in the design of policies for energy access, capacity development, clean energy for households, and emissions reduction at the urban and regional levels. It also explains why broader development goals that push for accelerating the rate of energy technology innovation are a key consideration in the design of policies.

While virtually every chapter in GEA provides specific policy suggestions, this chapter and the following three (Chapters 23–25) focus especially on issues related to policy design and implementation. The section that follows combines policies from these chapters with policies found throughout GEA to produce a portfolio of policies associated with the sustainable energy pathways of Chapter 17.

To this end, the following is divided into three sections. This section, Section 22.3.1, provides a portfolio of key energy policies to address the eight major energy policy goals identified in GEA. Section 22.3.2 provides a set of policies focused especially on energy efficiency. They explain how additional policies could intensify energy efficiency efforts in order to meet the aggressive targets of the most ambitious efficiency-focused pathways. Section 22.3.3 summarizes the links between policy objectives, policy types, and investments required for the rapid transformation of the global energy system. This transformation necessitates a rapid increase in energy access in the developing world, improvements in energy security in all energy systems, and, finally, the adoption of energy technologies with much lower environmental and human impacts.

This section does not recapitulate the long list of policies contained throughout GEA, which is a very large list with, at times, specific recommendations for specific situations from social, cultural, technological, and geographic considerations. Instead, what is provided is a summary of the salient policies associated with each of the key goals, in general, and a rather precise set of policies necessary to achieve the ambitious energy efficiency improvements described in Chapter 17, in particular. Thus, these policies are not distinguished by level of economic development, type of political system, institutional arrangements, or other critical factors in real-world policy design, implementation, and review.

For case studies, literature references, and arguments to defend specific policy designs, readers can refer to the policy chapters (Chapters 22–25), as well as to specific policy discussions in individual chapters throughout GEA, depending on their specific area of interest.

22.3.1.1   Increase Energy Access

As explained in Chapter 23, because increasing access to modern forms of clean, affordable, and efficient energy is an issue of equal importance to both developed and developing countries, in urban slums as well as rural hinterlands, there is no single set of policies that can increase energy access. The following portfolio of policies was put together with developing countries particularly in mind, where access to energy tends to be lower than in developed countries. While these policies are energy-focused, their effectiveness depends on coordination with broader policies in pursuit of social and economic development – investments in education, human health, poverty reduction, social and cultural support, sound institutions, and effective government.

- Competitive bidding to allocate funds to private, public, or cooperative grid owners or managers to extend high voltage transmission lines to unserved regions or communities and to extend low voltage distribution lines to unserved households via low hook-up fees. Funding could come from government or utilities.

- Competitive bidding to allocate funds to private, public, or cooperative enterprises to provide electricity services in isolated communities or regions that are too remote for connection to the grid. This may involve subsidized installation of household devices, like solar arrays, or larger community-based systems that may use any type of renewable energy or perhaps low-carbon fossil fuels and may cogenerate electricity with heating and/or cooling services.

- Lifeline rates that ensure a basic low price for an initial quantity of electricity, with higher rates for additional amounts.

- Electric utility tariffs that provide cross subsidies from one rate class of electricity customers to another to ensure electricity access for the poorest customers. Utility funds could be augmented by public sources of revenue.

- Public funding (grants, micro-financing) for the acquisition of stoves, heaters, and other devices that use clean-burning gaseous and liquid fuels or that can combust solid fuels (coal, charcoal, biomass) without deleterious emissions that adversely affect indoor air quality or the air quality in sensitive urban and rural airsheds.

22.3.1.2   Develop Capacities for Energy Transitions

Energy capacity development does not achieve energy transitions via a few simple policies (see Chapter 25). It requires a holistic approach that recognizes the complex interconnection of capabilities, habits, and norms of actors at all levels of the energy system. This is why Chapter 25 introduces the capacity matrix as a tool for conceptualizing how capacity development from a broad systemic perspective can play a critical role in energy transitions. It is also why there is

considerable regional disaggregation in Chapter 25 – to reflect differences in cultural norms, technical skills, and access to information, not just between developed and developing countries, but also depending on the specific attributes of a given continent, region, country, and even sub-national locales.

- An energy capacity strategy must be tailored to the specific characteristics of a given location if it is to succeed in provoking a rapid transition of the energy system toward a more sustainable path. While this strategy would address basic needs for education and training, it should also be adapted to the cultural norms and practices of that location.

- Developed countries need to improve mechanisms for supporting capacity development in developing countries, including financial support, technical training, sharing of industry, trade, institutional experiences, and public education.

22.3.1.3   Enhance Energy Security

Policies by different levels of government and different institutions all have a role to play in helping reduce or respond to energy security concerns. Energy security concerns are categorized in Chapter 5 as structural, sovereignty, and resilience. Policies to address systemic risk emphasize improved management of energy system design and operation. Policies to address risks from foreign sources emphasize avoiding or mitigating external control of energy supply or hostile actions that might disrupt supply from external sources. Finally, resilience or robustness of an energy system is a critical attribute for improving energy security prospects.

While energy security is often presented as a focus of industrialized countries, where a high level of reliability is essential for modern communications and production systems, it is also of great importance for developing countries. Micro-enterprises that lack back-up self-generation are particularly vulnerable to blackouts that are a daily occurrence in many parts of the developing world. The recommendations below are designed to address these various energy security risks.

- Electric utilities should make timely supply additions and provide regular system maintenance to ensure adequate reserve capacity and back-up generation to reduce the chance of unprovoked system failure.

- Regional cooperation can play a vital role in ensuring energy security in a sustainable manner. Energy trading can help meet energy demand while maximizing scarce natural resources in the sub-region. By utilizing different peak times of neighboring countries, regional power trade can reduce the need to build new power plants in each country. Transmission interconnections that enhance reliability by allowing for exchanges between contiguous grids should be developed up to the point where their marginal potential benefits in terms of risk reduction equal their marginal costs of provision.

- Electric system operators should provide tariffs and foster technologies that reward generators and consumers who can modulate their electricity flows in to and out of the grid in immediate response to signals from the operator.

- Diversity of primary energy supply should be pursued when the costs of additional diversity are likely to be below the expected benefits depending on an assessment of contingent reliability risks. This desired diversity may be by energy form (reduced reliance on oil or some other form of energy) or by supplier (reduced reliance on supplies from one region or country).

- National governments could develop alone, or in concert with other governments, mechanisms to stockpile energy resources that have significant risks of supply disruption or price volatility so that they can release these into the market (domestic or international) at times of scarcity, thereby reducing price volatility and supply insecurity.

- During periods of unusually high energy prices, governments may decide to protect the most vulnerable users from extreme energy price increases, but they should be careful not to completely constrain prices for energy suppliers and users as this would mute market responses (increased supply, decreased demand) that are normally triggered by high prices and that eventually lead to a new equilibrium of secure supply at prices that may be acceptable even if somewhat higher than before.

- When it can be shown to increase energy security at a reasonable cost, governments could push for an increased share of local energy supply through incentives such as tax advantages, regulatory flexibility for municipalities to encourage distributed and decentralized energy supply, and grants and loans for feasibility studies and capital costs related to district energy systems, as well as support for knowledge and technical capacity building.

22.3.1.4   Manage Energy-related Market Power

Significant economies-of-scale in some parts of the energy system – transmission and distribution of electricity and natural gas, oil pipelines, district heat networks – mean that market power can be socially desirable from an economic efficiency perspective. In such natural monopoly cases, the policy challenge is to regulate public or private monopolies so that their market power does not distort upstream or downstream phases of the industry where competition may be effective, such as oil and gas production, electricity generation, retail energy commodity markets, and energy service markets. In other parts of the energy system, economies-of-scale do not create natural monopoly

conditions, but they make it likely that the industry will be dominated by a few large firms. The petroleum industry is especially known for oligopoly forms of market power, which may call for various policies to protect the public interest.

- Whether transmission and distribution monopolies are publicly or privately owned, oversight by an independent monopoly regulator is likely to be in society's interest. The regulator should have expertise in economics, engineering, accounting, and environment and the authority to review for prudent utility decisions about investments, rate levels, rate design, and associated environmental and social trade-offs.

- Government and/or the monopoly regulator should ensure that monopoly power in one phase of an industry does not confer power over other levels of the industry, especially if competition is desired for these levels. This requires a separation of grid operation and regulation from electricity generation activities, even if this separation is more functional than corporate and even if some aspects of electricity generation are still treated as a natural monopoly.

- Government and/or the monopoly regulator could require the monopoly to play a role in delivering energy efficiency programs and increasing public awareness of economic, environmental, and social considerations in business and household choices of technology and lifestyle that affect energy use. Alternatively, another entity could be given responsibility for energy efficiency.

- If governments pursue competition in electricity generation they need to understand the special nature of electricity, in particular the need to balance all demand and supply on the grid instantaneously. An authority should be created to administer the competitive market, regulate electricity supply and demand, and prevent the short-term exercise of market power in ways that cause dramatic price spikes to the detriment of some customers.

- Government and/or the monopoly regulator need to make sure that the electric monopoly has the capability and authority to plan and operate the grid in an economically, socially, and environmentally sustainable manner. This capacity increases in importance with the need for dramatic transformation of electricity supply – for instance, many small-scale, decentralized generators, some of them located in urban areas, some in isolated regions that are favorable to renewable electricity generation – and electricity demand, such as smart meters that communicate between private energy-using devices and the grid operator, and electricity generation downstream of the meter in private buildings.

- In non-monopoly cases of market power, as sometimes occurs in electricity generation, oil and gas industries, anti-trust regulatory policies must be strong enough to ensure that a few large firms cannot: (1) distort markets in ways that reduce long-run energy security;

(2) allow industry to capture excess profits from consumers via price inflation; or (3) interfere in some way with effective governance. Where the firms are associated with a particularly valuable energy resource like petroleum and natural gas, policies are also required to effectively manage this public endowment (as outlined in the next Section 22.3.2) and provide a competitive environment to promote private sector investment.

- Market power sometimes equates to political power – the ability to influence public policies in favor of a particular industry or firm. Subsidies are one manifestation of this power. While subsidies are sometimes a valuable policy tool for advancing economic and social objectives, governments must be vigilant over time that subsidies not work at cross purposes to other objectives and do not become entrenched long after their justification has passed.

- The policy portfolio should seek to eliminate or substantially reduce subsidies, such as tax breaks and royalty reductions, which work at cross-purposes to overall policy goals. For example, if the intention is to reduce reliance on fossil fuels, subsidies to oil and gas exploration and development should be eliminated, as should subsidies to the coal mining industry.

22.3.1.5   Manage Valuable Energy Resource Endowments

Depending on market conditions and international prices, petroleum and natural gas can provide massive revenue streams in countries that are particularly well endowed with these premium sources of energy. While these lucrative rents could be a short- and long-run benefit to society, this is not always the case. Sometimes the resulting rent-seeking behavior contributes to corruption and poor governance – the resource curse – and the boom and bust cycle of commodity prices swings the economy from periods of excess wealth and waste to periods of severe recession. Policies are needed to manage the revenues in ways that maximize the long-run benefits from valuable resource endowments and prepare the economy for the time when these resources are depleted or lose their value. Also, the influx of foreign investment capital pushes up the value of the currency, threatening the survival of other domestic industries. Policies are needed to help these other domestic industries survive if they have the potential to provide sustainable benefits to the economy.

- Policies should control the rate of resource exploitation so that its demand for labor and capital (with resulting inflationary effects) does not irreparably undermine other sectors of the economy (e.g., renewable resources, technology, and services) that otherwise have good prospects of providing value long after the resource is depleted.

- Policies should maximize the collection of economic rents (surplus profits) from the resource for present and future generations. Rents can be captured via royalties and income taxes on private firms and/or by

a significant role for state-owned energy companies. In the latter case, management incentives and transparency are necessary to ensure that these companies are efficiently operated and accountable.

- Some part of resource rents could be streamed into sovereign wealth funds that are invested domestically and abroad in an effort to maximize the stream of expected benefits to future generations. This means a balanced portfolio of riskier and safer investments. Foreign investment with these funds can also help reduce upward pressure on the value of the currency, thus helping other domestic industries stay competitive. A poorer country will want to invest more of the rents domestically in infrastructure, education, and social services for the present generation, as this is likely to offer good potential for benefiting present and future generations. But resource rents should rarely be used to subsidize domestic energy prices, as this only dissipates the rent in inefficient current energy use.

- Some of the resource rents could be used to help the economy smooth out the boom-bust cycle of commodity prices. But such efforts should be limited and cautious, since it is impossible to know whether future resource prices will be higher or lower than current levels. These rents can also be used to sustain an economy once the resource is depleted.

- Policies should ensure transparency of resource rent flows and the budgets of government ministries, state-owned operations, and foreign resource companies in order to minimize the opportunities for rent dissipation through corruption. An example is to use principles from the extractive industries transparency initiative.

- Collected resource rents can also be used to repair environmental damage and can be allocated to communities most negatively affected by resource exploitation.

22.3.1.6   Reduce Environmental and Human Health Impacts

Energy supply and use involves a wide range of risks and impacts on humans and the environment. Energy use plays a large role in poor urban air quality and poor indoor air quality from open burning of solid fuels, which negatively affects human health. Energy extraction and processing can have substantial and sometimes devastating effects on the landscape and on soil and water resources. And, increasingly, the role of the current energy system in increasing atmospheric concentrations of greenhouse gases will cause rapid destabilization of the planet's climate, weather, water resources, and ecosystems, as well as human land-use and their social and economic systems.

While some of these energy system effects are known and fairly predictable, others present risks with very high uncertainties, which complicates and hinders the development of collective actions to reduce environmental and human risks. A sustainable energy policy portfolio would, at a minimum, include the following elements.

- A mix of regulations, information programs, and subsidies are needed to stimulate the rapid adoption of household energy using devices (stoves, heaters) that have virtually zero indoor emissions. While fuel subsidies may be used in a limited fashion, it is generally preferable to provide subsidies and low-cost financing for the acquisition of equipment, such as efficient, zero-indoor-emission stoves (using gaseous, liquid, or solid fuels) and efficient electric light bulbs to replace the indoor combustion of kerosene and other fuels for lighting.

- Urban air quality must be protected or restored by regulations on emissions from fuel combustion in buildings, industry, and vehicles in urban settings. Some regulations may be highly prescriptive (specifying combustion technologies such as the use of catalytic converters, restricting certain fuels, or requiring connection to district heating and cooling systems), while others may focus on the absorptive capacity of a given urban airshed for a particular pollutant. The latter case could involve the establishment of airshed emission limits, perhaps involving a cap-and-trade system. Or, it could involve the establishment of emission taxes of some kind.

- As with urban air quality, regional air quality must be protected by technology and/or emissions regulations or by direct emissions pricing. Again, some form of cap that is directly related to absorptive capacity has the best chance of meeting environmental objectives. A cap-and-trade system can, however, be difficult to achieve, depending on the complexity of emissions sources and the administrative capacity of the institution that will administer and oversee the trading system.

- Extractive activities and land/water uses – e.g., coal mines, oil and gas fields, hydropower dams, diversions and reservoirs, nuclear plants, storage sites for radioactive wastes and captured carbon dioxide, wind farms, solar electricity farms – should be subject to a permitting and regulatory framework that assesses their benefits against a precautionary consideration of their impacts and risks. In some cases, this may result in regulation and controls placed on the activity in order to reduce impacts and risks and, in some cases, in the complete rejection of the activity.

- Policies are needed to accelerate the development of renewable forms of electricity generation. These include possibly a feed-in-tariff that provides a minimum price for electricity from renewable sources and/or a renewable portfolio standard that sets a minimum but growing market share for renewable electricity. In order to maximize the effectiveness of a feed-in-tariff, it should identify the source of funding and have a clear sunset provision that pushes project developers to lower the cost of power.

- Policies fostering energy from biomass should recognize the trade-off between biomass-for-food, biomass-for-fuel, and the conservation of biodiversity. Subsidies to corn-based ethanol have encouraged inefficient production of biofuel with greater than necessary pressures on food prices. Better policies would encourage the use of biomass residues and only the most efficient biomass feedstocks and conversion processes.

- Policies to foster the safe development of nuclear power can help reduce the reliance on fossil fuels and associated greenhouse gas emissions. But development of nuclear power requires an integration of policies for developing technological and institutional capacities for producing nuclear power in a given country. It also requires ongoing development of existing international institutions and processes for ensuring that nuclear technology is not diverted to weapons production.

The following policies focus on key, specific sources of greenhouse gas emissions, namely fossil fuel use in industry, buildings, the transportation sector, and the generation of electricity.

- GHG pricing (via cap-and-trade and/or carbon tax) provides the essential major long-run mechanism by which energy systems gradually shift toward low carbon emission technologies, fuels, and activities. However, policies are also required that drive initial development of new technologies that must pass scale-up thresholds to the point where experience and learning drive down costs and accelerate adoption. Major public subsidies may drive early development of carbon capture and storage from large point-sources of GHGs, such as coal-fired electricity plants, oil sands plants, coal-to-liquids plants, and some industrial plants. Another approach is to apply niche market regulations that require specific sectors to gradually increase the percentage of carbon they process that is captured and stored, called a carbon capture and storage performance standard. The goal of such an approach is to create incentives for the pursuit of an increasing number of profitable CCS projects.

- Governments must use targeted policies to swiftly overcome a lack of regulatory clarity and challenging financial barriers if they want to succeed in CCS deployment on a commercial-scale. Key policies include: (1) allocating public funds to a number of different industries to share in the initial, high-risk investments associated with the first commercial CCS projects; and (2) developing a regulatory environment conducive to CCS in terms of property rights and liabilities. For instance, governments or utility regulators could prohibit construction of new coal-fired electricity plants that lack CCS or that are not CCS-ready. Such a policy approach would occur initially in developed countries with a lagged application, along with funding support, and in developing countries as part of an international agreement. Furthermore, the creation of CCS-specific measurement and crediting protocols will be key to ensuring that CCS projects are as tradable or valuable under national GHG regulatory frameworks as other qualifying emission reduction options.

- Governments need to develop property rights and regulatory policies that clarify short- and long-term management responsibilities and liabilities for geological storage of carbon dioxide, just as they have established these for the storage of other undesired gaseous and liquid by-products.

- Regulatory policies (e.g., land-use zoning, building codes, development permitting, and local emission standards) are needed to drive a shift toward low- and zero-emission buildings. In some cases this can be done in concert with low- and zero-emission decentralized energy supply. New urban developments should be required – initially most stringently in developed countries – to be low- and zero-emission of local air pollutants and GHG emissions via the local production of energy from solar, wind, biomass, and other sources and through import of energy from zero-emission external sources. These requirements could be gradually phased in to drive the retrofit of existing buildings and redevelopment of existing urban areas.

- Vehicle emission standard policies can require manufacturers and retailers to achieve minimum levels of sales of low- and zero-emission vehicles. This can start at modest percentages initially to allow the establishment of a niche market that enables learning and experience to lower production costs, but it should be enabled to grow substantially over the next two decades toward market domination. Alternatively, vehicle emission standard policies can set average emission standards for new vehicles.

- Similar niche market emission standards should be applied to other modes of transport, namely buses, trucks, off-road vehicles and mobile equipment, boats, trains, and airplanes.

22.3.1.7   Accelerate the Rate of Energy-related Technological Change

Some of the previous sections include policies that would help to drive rapid technological change. These include an economy-wide GHG emissions price that changes the incentives for innovation of new technologies and for their adoption. Also promoted are certain types of regulations, such as renewable portfolio standards and other niche market regulations for low- and zero-emission vehicles, which also help motivate industry to innovate and market technologies with desirable attributes from an environmental or social perspective.

In this section, it is reiterated that policymakers should ensure that a sustainable energy policy portfolio consistently applies an innovation system perspective, which will be further emphasized in Chapter 24.

- A sustainable energy policy portfolio should combine supply-push and demand-pull policies to ensure that adequate resources are available for innovation. This means, for example, subsidies, tax credits, and patents law on the supply side. On the demand side, polices should encourage sufficient market demand to move innovations from the new technology phase to significant initial levels of early adoption, and eventually toward industrial competition. The portfolio should recognize the importance of innovation spillovers, ensuring that technological breakthroughs can be widely applied. It should also ensure that patent law does not hinder rapid innovation.

- Subsidies might be used in some cases to bridge the gap between development and commercialization. But this should be done only when such subsidies are expected to lead to lower costs for technologies that will have a good chance of success in the future energy system in which market prices have been corrected to reflect environmental and social externality costs. From the outset it should be clear that such subsidies are temporary via the use of sunset clauses or other termination provisions.

- A policy portfolio must try to find an appropriate balance between picking winners and letting the market decide. The vehicle emission standard provides an example in that it picks a winner by requiring that vehicles be low-emission or zero-emission, but gives flexibility to innovators and market adopters by not specifying whether such vehicles are driven by electricity, biofuels, or hydrogen. Likewise, the renewable portfolio standard picks a winner by forcing the market growth of renewable electricity, but it lets innovators and the market decide by not specifying which renewables should dominate.

22.3.1.8   Coordinate and Implement International Energy-related Policies

Globalization of the world's economic system goes hand in hand with globalization of its energy system. Prices for energy commodities are increasingly set internationally, starting initially with crude oil but now extending to natural gas, coal, refined petroleum products, and bio-fuels. Even electricity prices are increasingly influenced at the international level within European and North American markets. Political disputes between countries (Israel and Arab OPEC countries; Russia and Ukraine) can threaten energy security and energy prices in far-distant consuming countries. At the same time, the environmental impacts of energy use have become global, especially with the key role of fossil fuel combustion in rising GHG emissions. Finally, international energy policies, as well as international aid, should bear in mind the urgent need for energy access and clean energy in developing countries.

- A sustainable energy policy portfolio must include effective mechanisms by which developed countries can assure a rapid transfer of financial, technological, and institutional resources to developing countries for energy system development and transformation. Key international and multi-national institutions, such as the World Bank, the International Monetary Fund, divisions of the United Nations, and development banks such as the Asian Development Bank, African Development Bank, and Inter-American Development Bank, must ensure that their policies contribute in a consistent and effective way to all of the key energy policy goals of GEA (access, security, managing market power, environment, etc.).

- A sustainable energy policy portfolio must also include mechanisms by which countries can coordinate efforts to reduce energy price volatility (although it cannot be eliminated). These include diversifying suppliers, stockpiling energy, establishing back-up electricity and fuel supply systems, developing inter-regional electricity and natural gas trade, and discouraging the use of energy as a political weapon.

- At the international level, governments and the nuclear industry need to continue to improve their mechanisms for monitoring and controlling the use of nuclear power and the reprocessing of nuclear fuel to prevent acquisition of expertise and materials for nuclear weapons production. For instance, proliferation resistance could be increased through policies aimed at phasing out reprocessing as rapidly as possible, and placing enrichment plants under multinational management and restricting them to stable regions. Governments need to collaborate in the establishment of regional storage sites for radioactive materials. Design of the repositories should be subject to international standards and oversight to ensure that the import of spent fuel from richer countries to poorer countries would not create undue environmental hazards (see Chapter 14).

- A sustainable energy policy portfolio must include one or more mechanisms to ensure that global concentrations of GHGs in the atmosphere do not reach levels that are highly risky for human and environmental health. These mechanisms could be, for instance, country-specific targets for GHG emissions established through international negotiations based on almost universal agreement. Or, the mechanisms may be targets and obligations negotiated by individual sectors, such as specific requirements for coal-fired electricity plants to gradually incorporate CCS, or for the airline industry to gradually increase the bio-fuel share of jet fuel, or for the vehicle industry to have a rising share of zero-emission vehicles worldwide. As indicated, international coordination is necessary to mitigate emission leakage and other issues that may arise from only a subset of countries being covered by these policies.

22.3.2   Policy Portfolio Focused on Energy Efficiency

Chapter 17 presents scenarios that are distinguished in part by the degree to which energy efficiency contributes to a sustainable energy system. In part three of this chapter, the focus is on presenting the type and intensity of policy effort required to realize a global sustainable energy pathway through dramatically lower levels of energy use.

Since this is not a precise policy modeling exercise, the policies here are not calibrated to the intensity levels that would exactly match the scenarios in Chapter 17. What is presented here, instead, are descriptions of policies that can accelerate energy efficiency trends, along with suggestions on the intensity with which such policies would need to be applied in order to achieve the most ambitious energy efficiency scenario.

It is important to note, however, that achieving an aggressive energy efficiency scenario is not just a question of turning up the intensity on policy levers. If societies are to achieve development paths in which efficient technologies are rapidly innovated and adopted, while at the same time there is widespread acceptance of a conserver lifestyle, there must be social acceptance of energy policies that drive not only technological change, but also changes in some key behavioral expectations. One major example is the ownership and rate of use of private vehicles. Such a level of change is more likely to succeed if it is fostered by the development of an enabling environment. This requires the creation of a broadly shared vision that would permeate a given society's institutions, infrastructure, education, technical capacities, financial and market conditions, laws, regulations, and social norms.

The rebound effect is a critical consideration when assessing the potential for energy efficiency policies to dramatically reduce energy use. Efficiency regulations on their own are likely to increase the rebound effect – because of lower operating costs and efficiency innovations that stimulate new energy devices and services – unless they are compensated by policies that increase energy prices to reflect environmental and social externalities and that shift energy rate structures (in the case of tariffs set by electric, natural gas, and district heating utilities) to reflect the full long-run cost of new supplies. These kinds of policy changes will be challenging to enact. In the description below of ambitious energy efficiency policies, the emphasis is on those that offer the best prospects of delivering this major transformation in the relatively short period of a few decades.

There are a number of ways of presenting energy efficiency policies – by level of government, by type of policy, by source of energy, by intensity of the efficiency effort, or by end-use sector. In this section, energy efficiency policies are organized by end-use sector. They also include upstream policies affecting individual energy supply systems for electricity, fuels, and heat. Thus, this portfolio of energy efficiency policies has the following sections: industry; appliances and devices (office equipment, personal products); buildings; urban form; transportation; and agriculture.

As with the previous section, the policies in this section represent only a subset of the possible policies in a sustainable energy policy portfolio. A more detailed description of policies, with many case studies of successes and failures, is provided by the other policy chapters and indeed in most other chapters of GEA. Any particular government should collect key information on the use of energy in industry, transport, commercial, and residential sectors if it is to develop effective policies for its own energy efficiency pathway.

In all policy cases discussed below, it is assumed that energy forms with GHG emissions associated with them, at the point of use or during production – as in the case of electricity generation – will experience rising prices over the coming decades due to economy-wide GHG emissions pricing policies, i.e., tax or cap-and-trade. In the same vein, regional and local air and water pollutants are assumed to face rising costs through externality pricing policies and regulations that affect the cost of emissions and effluents.

Most analysts argue that pricing policies are the most efficient and effective way to drive much of the investment and behavioral change needed for a major transformation of the energy system. Others argue, however, that pricing policies alone are insufficient to achieve the desired policy objectives. They contend that complementary regulations, subsidies, public investments and other policies are needed to realize a rapid and profound transformation of technologies, fuels, buildings, urban form, infrastructure, and industrial plants.

Thus, depending on one's perspective, the policies listed below can be seen as potential complements to pricing policies, intended to accelerate efficiency actions that would be stimulated by rising prices. Complementary policies must be designed and implemented carefully, however, to reduce the risk that they will have a negligible effect or even negative unintended consequences.

22.3.2.1   Energy Efficiency Policies for Industry[5]

Development of GHG emissions pricing and other policies that internalize the costs of environmental harm and impact on humans will provide additional incentive for industry to adopt more efficient energy-using technologies. Additional policies may accelerate the adoption of high efficiency technologies, but they must be applied carefully. In some cases, their incremental effect will be zero once prices have been corrected to reflect environmental harm. In some cases, they may actually work at cross purposes with price-adjusting policies.

- Governments and utilities can design utility tariffs to ensure that the marginal tariffs facing customers reflect the true marginal costs of providing them with energy. Depending on the characteristics of the energy supply system, this involves time-of-use rates and/or inverted block rates in which marginal tariffs reflect an appropriate weighting of short- and long-run marginal costs.

- Governments can set regulations that are updated (for example, every five years) to prohibit market access to the least efficient third or half of technologies for a given service, such as motors, fans, conveyors, blowers, boilers, cogenerators, and process-specific equipment.

- Governments can implement "golden carrot" policies and/or niche market regulations that use financial incentives or regulation, or a combination of both, to encourage industry to develop and adopt technologies that are more efficient than the current stock.

- Governments can establish efficiency performance standards for major industries. Or governments can mandate processes that require new investments to achieve a best available technology efficiency standard. Also, phased retrofit of already installed process technologies can be required.

- Governments can provide free audits for smaller firms that find it too costly to inform themselves of energy efficiency options and audits for a fee for medium and larger firms.

- Governments can require firms of a certain size to have full-time energy managers responsible for identifying efficiency opportunities.

- Governments and utilities can provide up-front funds or financing arrangements to ensure that capital constraints are not a barrier to energy efficiency investments.

- Local governments can plan and regulate the location of industrial activities in order to maximize opportunities for waste heat transfer and other synergies between industries to reduce material and energy use.

- National and international industry associations can ensure that industries throughout the world are informed of the most energy efficient processes and technologies.

- Some modifications to patent laws may help improve innovation and a relatively quick transfer of energy efficient technology for industries in different countries.

22.3.2.2   Energy Efficiency Policies for Appliances and Devices[6]

While energy intensive industries are normally very sensitive to energy price – and therefore its technology choices will be affected to a considerable extent by emissions pricing policies – the same is not always true to the same degree with commercial and residential energy users. One reason is the difficulty and cost in getting adequate information about the efficiency of alternative equipment relative to the potential savings for a small-scale energy user. Another reason is the challenge for small consumers in demonstrating to financial institutions the lower operating cost benefits resulting from up-front investments in energy efficiency. In contrast, large industries are continually engaged in explaining the benefits and costs of their investments to financial organizations. A third

---

5   See also Chapter 8.

6   See also Chapter 10.

reason is the split incentive that often exists between the landlord, who pays for appliances and some devices, and the commercial or residential customer, who pays the energy bills. The following policies seek to address these challenges.

- Governments and utilities can ensure that marginal tariffs facing customers reflect the marginal costs of providing them with energy. Depending on the characteristics of the energy supply system, this involves time-of-use rates and/or inverted block rates in which marginal tariffs reflect an appropriate weighting of short- and long-run marginal costs. With the falling cost of meters, such tariffs are now possible for even the smallest customers.

- Governments can set regulations that are updated every five years to prohibit market access to the least efficient third or half of appliances and devices. Alternatively, these regulations can ban or mandate the phase out of certain types of less efficient technologies, such as incandescent lights.

- Governments can set regulations that prohibit appliances and devices that use electricity while not in use (turned-off, sleeping, or in standby mode).

- Governments can require efficiency rating labels on appliances and devices to raise public awareness about operating costs of such devices relative to average efficiency.

- Governments and industry associations can work to reach internationally sanctioned appliance and device standards to prevent inefficient products being phased out in industrialized countries from being transferred to developing countries, with the consequence of locking such countries into a regime of inefficient energy use and high operating costs.

- Governments can implement golden carrot policies and/or niche market regulations that use regulation or financial incentives (or a combination of both) to require manufacturers to develop and market appliances and devices that push the envelope to yet higher levels of efficiency than have thus far been experienced. This can perhaps be done initially with a subset of consumers.

- Governments and utilities can provide subsidies for early adopters of high efficiency appliances and devices, although such programs should only reward the adoption of technologies that are unlikely to have otherwise been adopted, because free-riders are a well known challenge to subsidy policy effectiveness.

- Governments and utility regulators can ensure that utilities and energy service companies have an incentive to play a key role in promoting energy efficiency among customers of all classes, particularly small-scale customers who lack resources and knowledge to access energy efficiency. This involves initiatives such as: (1) decoupling utility profits

from sales; and (2) creating tradable "white certificates" by which utilities can trade energy efficiency credits among themselves as they compete to achieve energy efficiency targets set by regulators.

22.3.2.3　Energy Efficiency Policies for Buildings[7]

Buildings use approximately one-third of all the energy used globally and are responsible for about one-third of the total energy-related emissions of GHGs (including emissions from electricity production). Because the building stock turns over slowly and because energy efficiency is much less costly if it is built in at the time of design and construction, policies usually distinguish between those affecting new buildings and those affecting existing buildings.

- For new buildings, governments can establish a set schedule for tightening over time the energy efficiency standards for the building shell; heating, ventilation, and air conditioning (HVAC) system; and domestic water heating systems to reach very low levels of energy use. The phase-in period for the strictest efficiency standards could be shorter in developed countries (a decade) than in developing countries (two or three decades in urban areas), although country-, region- and city-specific criteria could lead to very different time-frames. The phase-in period for the efficiency requirements and their stringency would be tightest for the highest efficiency scenario.

- For new buildings, governments can use development charges and siting requirements to ensure that buildings are sized and located appropriately for optimizing heat gains and losses of the building shell, depending on climate and the building's heating or cooling needs, with respect to its ambient surroundings. These can also be applied, in many cases, to maximize opportunities for waste heat recovery and/or district heating supply.

- Governments can require all existing buildings to have an energy audit at the time of sale that shows prospective buyers how the building ranks on a scale showing energy efficiency performance and GHG emissions. This would include estimated future energy costs associated with rising emissions charges. Audit results should be clearly highlighted in information related to advertising and execution of sale.

- Governments can provide subsidies (tax credits, grants, low-interest loans) for retrofits in existing buildings that drive emissions to zero and/or external energy inputs to zero for a given building. In developed countries, funds can be acquired through local improvement charges and assigned to a property rather than a person. For developing countries, funds could be acquired from domestic sources and perhaps through international transfer mechanisms like the Clean Development Mechanism (CDM). For scenarios of high-energy efficiency, retrofit subsidy programs should be much more substantial.

7　See also Chapter 10.

22.3.2.4   Energy Efficiency Policies for Urban Form

The energy efficiency of the built environment can be improved substantially through urban planning that combines land-use planning, transportation management, and a synergistic approach to the provision of energy and water services (see also Chapter 18). Such a "community energy management" approach is characterized by increased urban density and especially associated with infrastructure for public transit and district heating, mixed land-uses that reduce the need for mobility and increase the attractiveness of walking and cycling, and coordinated developments that maximize the potential between and within buildings for energy cascading – the use of waste heat. This approach can reduce use and increase reuse of water and sometimes other materials, thus indirectly reducing energy use for these services. Attractive, high-density living also increases energy efficiency by reducing the ratio of external walls to floor space in residential dwellings. Community energy management policies include the following.

- Local governments can set land-use zoning and development standards to foster targeted densities and mixed uses to reduce travel and transportation needs, energy efficient buildings, and cogeneration.

- Local governments can negotiate explicit energy performance standards into permits for re-zoning and new buildings.

- Local and regional governments can foster strategic alliances between energy utilities and developers, with explicit energy planning requirements in official community plans.

- National and regional governments can legislate the requirement that energy planning be an explicit function of local and perhaps regional governments.

- Governments can provide technical and financial support for the integration of energy considerations into traditional community planning processes.

- Governments can tie infrastructure grants to the implementation of community energy management strategies.

- Governments can reduce or eliminate minimum parking stall requirements for new developments and increase local taxes on parking spaces.

- Along with reducing incentives for vehicle traffic, governments can increase the provision of public and mass transportation.

- Governments can require the coordination of energy supply and energy use planning and investment such that cogeneration and district heating systems are expanded.

- Governments can set taxes to ensure that new dispersed developments pay the full costs they cause to energy and water utilities, transportation services, urban livability, and the environment.

These community energy management policies are relevant to all countries. However, it must be recognized that developing countries face special challenges because of rapid rates of urbanization and the lack of institutional, professional, and financial capacity to perform the tasks required – i.e., planning, zoning, and regulation, as well as developing adequate infrastructure for energy, water, sewage, transportation, and other potential collective services, like district heating and cooling. Thus, in developing countries energy efficiency in urban planning would include the following:

- Policies to support local planning, control, and perhaps ownership (municipal utilities, co-operatives, local entrepreneurs) of energy and other utility networks. This has been shown to reduce line losses due to theft and thus improve the operating efficiency of the grid.

- Policies are needed to provide mechanisms for municipal financing and micro-financing of local developments and improvements to infrastructure for energy and energy-related services like transportation, water, and sewers by using efficient technologies that reduce energy and other operating costs. These policies would include support for more effective public investment, which may be leveraged and managed to increase private investment through ensuring stable and sufficient returns on capital.

22.3.2.5   Energy Efficiency Policies for Transportation[8]

A key aspect to developing effective transportation energy efficiency policies is to integrate individual policies and regulations into packages that benefit from a synergistic interaction among the components. Two types of policy portfolios can improve the overall energy efficiency of private transportation. In the case of personal transportation, the first portfolio approach consists of policies that improve vehicle fuel efficiency. The second consists of policies aimed at reducing private vehicle travel and increasing the use of public transit, walking, and cycling within cities and of trains for long distance travel. In the same vein, with the transport of goods, one approach is to make trucks, trains, boats, and planes more efficient. While another approach is to encourage mode shifting to the mode that is most energy efficient for a particular mobility need.

It should be noted that policies to improve vehicle and transport equipment efficiency can work in opposition to policies to encourage mode shifting. More efficient vehicles have lower operating costs and become even more attractive relative to alternatives. This is why vehicle efficiency policies need to be combined with policies that keep the cost

---

8   For more on energy and transport, see Chapter 9.

of vehicle use high by reflecting congestion costs (road pricing), environmental effects (emissions pricing), and urban livability (vehicle-free streets, parking charges, parking restrictions, vehicle registration fees).

- Governments can set vehicle efficiency regulations. Such regulations can apply universally to all vehicles or can be flexible by allowing different levels of efficiency, as long as vehicles collectively achieve an average level of efficiency. This latter is the case of the Corporate Average Fuel Economy (CAFE) Standards introduced in the United States in 1975.

- Governments can use financial penalties and financial incentives alone or in combination. In the latter case, a feebate policy levies a one-time or annual charge on less efficient vehicles and uses the resulting revenue to provide a one-time or annual subsidy to efficient vehicles.

- Governments could guarantee to purchase the most efficient vehicles for their own vehicle fleets, thereby providing a guaranteed market for the next generation of new vehicles.

- Local governments can implement road pricing to reduce vehicle use and urban congestion. This may include tolls on key bridges or highways, or an electronic zonal pricing system covering a major urban agglomeration. In the case of tolls on bridges and highways, a reduced or zero rate for high-occupancy vehicles encourages more passengers per vehicle and thus a more efficient vehicle use rate. Local and regional governments should make sure that private vehicles on roads face tolls that reflect the costs they incur, including environmental and human health externalities.

- Governments can require vehicle insurance companies to charge distance-based insurance rates, determined by annual odometer readings.

- Governments can rationalize traffic signals to optimize vehicle traffic flows.

- Government regulations can set efficiency standards for modes of transport of goods (trucks, trains, boats, planes). These can have design flexibility similar to that outlined above for personal vehicles.

- Government regulations and pricing can foster modal shifting toward more energy efficient modes of transport of goods, such as a shift from trucks to trains, when possible.

22.3.2.6   Energy Efficiency Policies for Agriculture

Agriculture can be energy intensive, especially the industrialized version practiced today in many parts of the world. Direct energy use in agriculture results from farm equipment, buildings, irrigation, and post-harvest processing. Indirect energy use in agriculture results from the production of fertilizers and pesticides. In the United States, about a third of all energy used in agriculture relates to commercial fertilizer and pesticide production. Potential energy savings in the agricultural sector can be achieved through changes in on-site transportation, reduced tillage, and improvements in irrigation, drying, dairy and livestock production, and horticulture, among others.

- Governments can provide information, low interest loans, and grants to foster the use of more efficient technologies and techniques in agriculture. But as with all subsidies, care must be taken not to provide support that has little effect because it pays farmers for efficiency measures they would have undertaken anyway.

- Governments can provide training for farmers and farm personnel in ways of reducing energy use.

- Governments can tighten regulations on some types of agricultural practices, such as irrigation technologies, pesticide use, and herbicide use, in ways that would indirectly reduce energy use.

- Governments and utilities can change tariff structures to encourage energy efficiency in all farming practices.

- Governments can levy taxes on fertilizers and on fuel for tractors.

22.3.2.7   Decentralized Energy Policies and Energy Efficiency

The human population is urbanizing rapidly. Cities are now responsible for two-thirds of total primary energy use. As demand for energy services continues to grow, the energy infrastructure that cities depend on will need to be expanded, upgraded, or replaced. This provides governments with opportunities to build energy efficiency or reduced energy service demands into long-lived infrastructure, plants, and equipment. This includes everything from urban form, to energy supply, to major energy end-uses.

Expanding, upgrading or replacing energy infrastructure also presents an opportunity to move toward more decentralized, low emission energy systems. Decentralization may contribute to the goal of low cost energy access with a minimum of negative social and environmental impacts. At the same time, local awareness of energy supply can increase, and that can foster a greater emphasis on energy efficiency.

Decentralized energy systems include locally-focused mini-grids that connect small-scale local supply with local demand and energy self-sufficient industries. Decentralized energy systems can reduce the need for expensive, expansive energy grids and increase the reliability of the system. It can also provide reliable energy to communities currently not connected to the grid.

Government at all levels can support the development of smaller-scale, local energy supplies. Support can come via implementation of incentives, such as tax advantages, regulatory flexibility for municipalities and utilities to encourage distributed and decentralized energy supply, and grants and low-interest loans for feasibility studies and capital costs related to development of district energy systems. Governments must also support knowledge- and technical capacity-building. Local governments should be involved in the development of decentralized energy systems because they often build, own, and operate decentralized energy systems. The following policies can contribute to decentralized energy supply:

- Development permit area guidelines can require decentralized renewable energy systems external to buildings, such as the installation of ground-source heat pump systems, to reduce total energy use.

- Tax exemptions for development projects using local renewable energy sources can be incentives to owners and developers to promote decentralized energy retrofits on buildings or neighborhood-scale initiatives. This might include solar water heaters, heat pumps, or heat recovery systems.

- Development cost charge (DCC) reductions or exemptions, conditional on inclusion of decentralized energy generation, can provide financial incentive for developers.

- Local governments can adopt a rezoning policy that encourages decentralized energy generation in new developments. Such a policy would indicate clearly which attributes will be sought by governments when making rezoning decisions, creating incentives for developers to include decentralized energy in their plans.

- Local improvement charges can promote the use and finance the installation of renewable energy systems in existing buildings and developments throughout a community. For instance, these can be used to pay for the installation of solar hot water systems in the community.

- Local governments can use service area bylaws to provide, and charge for, decentralized energy generation and services.

- Governments can enact regulations that require renewable technologies, like solar water heating or rooftop photovoltaic systems on new buildings.

- Local governments can offer or expropriate land for the construction of local energy generating plants.

- Electric utilities can adjust their grid extension policies so that communities that are beyond eligibility for grid connection receive

technical and some financial support to develop mini-grids. Public-private partnerships of various types may be effective in this area.

- Developed countries should promote the transfer of new technologies related to decentralized energy to developing countries.

- Utility regulators need to adopt flexible, lighter approaches to utility regulation so that small-scale operations that are nonetheless defined as utilities – such as the provision of energy supply to a building, a development, a group of buildings, or a remote community – do not face the typical regulatory burdens of conventionally-sized utilities.

- International and national aid agencies may want to focus energy-related aid efforts into supporting with capital and expertise the development of decentralized energy systems in developing countries. But care should be taken so that this does not come at the expense of expanding large-scale systems, if that is more cost effective.

22.3.3   Linking Policies and Investments

It is sometimes assumed that one can measure a policymaker's commitment to a particular issue by the amount of public funds they are willing to commit. In the case of energy policy, this assumption would be misleading in some instances. Good energy policy can sometimes require little or no expenditures, as in the case of effectively managing and regulating the interplay of monopolistic and competitive forces in the electricity sector or effectively managing a valuable resource endowment. However, when it comes to the rapid expansion of the energy sector necessary to provide adequate energy access to billions of un-served and under-served people, and to the much-needed transformation of the global energy system to one with a much smaller environmental impact, effective energy policies will need to stimulate a great deal of investment from the public and especially the private sector.

Chapter 22 emphasizes the need to price harmful energy-related emissions so that energy prices reflect environmental costs. Properly pricing energy to include the value of environmental harms and risks can drive massive investments in cleaner energy technologies and in energy efficiency. Chapter 24 describes a slate of policies that can create an enabling environment for profound energy technology system innovation. But these policies will be especially effective in developed countries where lack of energy access is less of an issue. In developing countries, policy must also result in a profound investment in energy supply.

Chapter 6 describes the types of investments that are required for the global energy system. Chapter 17 links the magnitude of these investments to specific pathways in terms of potential development of energy efficiency and the expansion of energy supply from renewables, nuclear

Table 22.2 | Investment needs and policy mechanisms.

	Investment (billions of US$/year)		Policy mechanisms			
	2010	**2010–2050**	**Regulation, standards**	**Externality pricing**	**Carefully designed subsidies**	**Capacity building**
Efficiency	n.a.[1]	290–800[2]	*Essential* (elimination of less efficient technologies every few years)	*Essential* (cannot achieve dramatic efficiency gains without prices that reflect full costs)	*Complement* (ineffective without price regulation, multiple instruments possible)[3]	*Essential* (expertise needed for new technologies)
Nuclear	5–40[4]	15–210	*Essential* (waste disposal regulation and, of fuel cycle, to prevent proliferation)	*Uncertain* (GHG pricing helps nuclear but prices reflecting nuclear risks would hurt)	*Uncertain* (has been important in the past, but with GHG pricing perhaps not needed)	*Desired* (need to correct the loss of expertise of recent decades)[5]
Renewables	190	260–1010	*Complement* (renewable portfolio standards can complement GHG pricing)	*Essential* (GHG pricing is key to rapid development of renewables)	*Complement* (feed-in tariff and tax credits for R&D or production can complement GHG pricing)	*Essential* (expertise needed for new technologies)
CCS	<1	0–64	*Essential* (CCS requirement for all new coal plants and phase-in with existing)	*Essential* (GHG pricing is essential, but even this is unlikely to suffice in near term)	*Complement* (would help with first plants while GHG price is still low)	*Desired* (expertise needed for new technologies)[5]
Infrastructure[6]	260	310–500	*Essential* (security regulation critical for some aspects of reliability)	*Uncertain* (neutral effect)	*Essential* (customers must pay for reliability levels they value)	*Essential* (expertise needed for new technologies)
Access[7]	n.a.	36–41	*Essential* (ensure standardization but must not hinder development)	*Uncertain* (could reduce access by increasing costs of fossil fuel products)	*Essential* (grants for grid, microfinancing for appliances, subsidies for cooking fuels)	*Essential* (create enabling environment: technical, legal, institutional, financial)

1. Global investments into efficiency improvements for the year 2010 are not available. Note, however, that the best-guess estimate from Chapter 24 for investments into energy components of demand-side devices is by comparison about 300$ billion per year. This includes, for example, investments into the engines in cars, boilers in building heating systems, and compressors, fans, and heating elements in large household appliances. Uncertainty range is between US$100 billion and US$700 billion annually for investments in components. Accounting for the full investment costs of end-use devices would increase demand-side investments by about an order of magnitude (see Chapter 24 for details).
2. Estimate includes efficiency investments at the margin only and is thus an underestimate compared with demand-side investments into energy components given for 2010 (see note 1).
3. Efficiency improvements typically require a basket of financing tools in addition to subsidies, including, for example, low- or no-interest loans or, in general, access to capital and financing, guarantee funds, third-party financing, pay-as-you-save schemes, or feebates as well as information and educational instruments such as labeling, disclosure and certification mandates and programs, training and education, and information campaigns.
4. Lower-bound estimate includes only traditional deployment investments in about 2 GW capacity additions in 2010. Upper-bound estimate includes, in addition, investments for plants under construction, fuel reprocessing, and estimated costs for capacity lifetime extensions.
5. Note the large range of required investments for CCS and nuclear in 2010–2050. Depending on the social and political acceptability of these options, capacity building may become essential for achieving the high estimate of future investments.
6. Overall electricity grid investments, including investments for operations and capacity reserves, back-up capacity, and power storage.
7. Annual costs for almost universal access by 2030 (including electricity grid connections and fuel subsidies for clean cooking fuels).

power, and perhaps fossil fuels, provided that emissions can be captured and safely stored. The following table from Chapter 17 summarizes, in a very general way, the link between policies and investment levels.

Detailed information on this table is provided in Chapter 17. Here we simply reiterate a few key points. The energy investment columns refer to investment levels needed for different pathways. Thus, a high-energy efficiency pathway would require over US$500 billion/year of new investment. In contrast, since there are pathways in which nuclear power stagnates, its annual investment could be as low as US$15 billion/year, a sharp contrast with the almost US$400 billion/year

for an energy pathway that includes a substantial role for nuclear power. In the same vein, CCS could be abandoned in some energy pathways, while in others its rapid adoption could require almost US$100 billion/year of new investment.

The policy mechanisms of the table provide an admittedly crude simplification. There are only four policy types (regulations, pricing, subsidies, capacity building) and their role and relationship to each other is evaluated in only four ways (essential, complement, desired, uncertain). In reality, there are many variations within these four policy types, and indeed, there are policies that are hybrids in combining elements of both.

Likewise, depending on how they are designed and implemented, some policies can be complements in some cases and substitutes in others.

With these caveats, the investment table nonetheless provides a summary of the amount of investment required under various energy pathways and kinds of policies needed to drive this investment. In so doing, it gives a general sense of points made throughout this chapter, namely that: (1) strong pricing and regulatory policies will be required for the major energy transitions called for in GEA; and (2) policy coordination is essential for this effort to be effective.

References

Aldy, J. and R. Stavins (eds.), 2007: Architectures for Agreement: Addressing Global Climate Change in the Post-Kyoto World. Cambridge University Press, Cambridge, UK.

Arimura, T., R. Newell, and K. Palmer, 2009: Cost-effectiveness of electricity energy efficiency programs. In *Resources for the Future*, DP 09-48, Resources for the Future, Washington, DC, USA.

Barrett, S., 2003: Environment and Statecraft: The Strategy of Environmental Treaty-Making. Oxford University Press, Oxford, UK.

Berg, S. and J. Tschirhart, 1988: Natural *Monopoly Regulation: Principles and Practice*. Cambridge University Press, Cambridge, UK.

Bergek, A., S. Jacobsson, and B. Sanden, 2008: Legitimation and development of positive externalities: two key processes in the formation phase of technological innovation systems. *Technology Analysis and Strategic Management*, **20**(5): 575–592.

Bier, V., Y. Haimes, J. Lambert, N. Matala, and R. Zimmerman, 1999: A survey of approaches for assessing and managing the risk of extremes. *Risk Analysis*, **19**(1): 83–94.

Brown, E., 1948: Business-income taxation and investment incentives. In *Income, Employment and Public Policy: Essays in Honor of Alvin H. Hansen*. Norton, New York, USA.

Carlsson, B., S. Jacobsson, M. Holmen, and A. Rickne, 2002: Innovation systems: Analytical and methodological issues. *Research Policy*, **31**(2): 233–245.

Collier, P., 2007: The Bottom Billion: Why the Poorest Countries are Failing and What Can be Done About It. Oxford University Press, Oxford, UK.

Dietz, T., and P. Stern, 2002: New Tools for Environmental Protection: Education, Information and Voluntary Measures. National Academy Press, Washington, DC, USA.

DME, 2003a: Electricity Basic Services Support Tariff (Free Basic Electricity) Policy. Department of Minerals and Energy, Pretoria, South Africa.

DME, 2003b: *Options for a Basic Electricity Support Tariff: Supplementary Report*. Department of Minerals and Energy, Pretoria, South Africa.

Fischer, C., 2007: International Experience with Benefit-Sharing Instruments for Extractive Resources. *Resources for the Future (RFF) Report*, May 2007, Resources for the Future, Washington, DC, USA.

Fouquet, R. and P. Pearson, 2006: Seven centuries of energy services: the price of use of light in the United Kingdom (1300–2000). *The Energy Journal*, **27**(1): 139–177.

Freeman, C., 1994: The economics of technical change. *Cambridge Journal of Economics*, **6**: 587–603.

Garnaut, R. and A. Ross, 1975: Uncertainty, risk aversion and the taxing of natural resource projects. *Economic Journal*, (June) **85**: 278–287.

Geels, F., 2002: Technological transitions as evolutionary reconfiguration processes: A multi-level perspective and a case-study. *Research Policy*, **31**(8/9): 1257–1274.

Gillingham, K., R. Newell, and K. Palmer, 2006: Energy efficiency policies: A retrospective examination. *Annual Review of Environment & Resources*, **31**(1): 161–192.

Grubler, A., 1998: *Technology and Global Change*. Cambridge University Press, Cambridge, UK.

Grubler, A., N. Nakicenovic, and D. Victor, 1999: Dynamics of energy technologies and global change. *Energy Policy*, **27**: 247–280.

Hahn, R. and R. Stavins, 1992: Economic incentives for environmental protection: integrating theory and practice. *American Economic Review* **82**(2): 464–468.

Harrington, W., R. Morgenstern, and T. Sterner (eds.), 2004: *Choosing Environmental Policy: Comparing Instruments and Outcomes in the United States and Europe*. Resources for the Future, Washington, DC, USA.

Hekkert, M. P., R. A. A. Suurs, S. O. Negro, S. Kuhlmann, and R.E.H.M., 2007: Functions of innovation systems: A new approach for analysing technological change. *Technological Forecasting and Social Change*, **74**(4): 413–432.

Howells, M., D. G. Victor, T. Gaunt, R. G. Elias, and T. Alfstad, 2005: Beyond free basic electricity: The cost of electric cooking in poor households and a market-friendly alternative. *Energy Policy*, **34**: 3351–3358.

IEA, 2010: *World Energy Outlook*. International Energy Agency (IEA) of the Organisation for Economic Co-operation and Development (OECD), Paris, France.

Industry Commission, 1991: *Energy Generation and Distribution*. Canberra, Australia.

Jaccard, M., L. Failing, and T. Berry, 1997: From equipment to infrastructure: community energy management and greenhouse gas emission reduction. *Energy Policy*, **25**(13): 1065–1074.

Jacobsson, S. and V. Lauber, 2006: The politics and policy of energy system transformation – explaining the German diffusion of renewable energy technology. *Energy Policy*, **34**(3): 256–76.

Jaffe, A., R. Newell, and R. Stavins, 2002: Environmental policy and technological change. *Environmental and Resource Economics*, **22**(1–2): 41–69.

Jaffe, A. B., R. G. Newell, and R. N. Stavins, 2005: A tale of two market failures: Technology and environmental policy. *Ecological Economics*, **54**: 164–174.

Joskow, P., 2006: Introduction to electricity sector liberalization: lessons learned from cross-country studies. In *Electricity Market Reform: An International Perspective*. F. Sioshansi and W. Pfaffenberger (eds.), Elsevier, Oxford, UK.

Kalicki, J. and D. Goldwyn (eds.), 2005: *Energy and Security: Toward a New Foreign Policy Strategy*. John Hopkins Press, Baltimore, MD, USA.

Karamanos, P., 2001: Voluntary environmental agreements: evolution and definition of a new environmental policy approach. *Journal of Environmental Planning and Management*, **44**:1: 67–84.

Kaufman, S. and S. Milton, 2005: *Solar Water Heating as a Climate Protection Strategy: The Role for Carbon Finance*. Green Markets International, Arlington, Massachusetts.

Kemp, R., 1997: Environmental Policy and Technical Change: A Comparison of Technological Impact of Policy Instruments. Edward Elgar, Cheltenham, UK.

Khanna, M., 2001: Non-mandatory approaches to environmental protection. *Journal of Economic Surveys*, **15**(3): 291–324.

Leite, C. and M. Weidmann, 1999: Does Mother Nature corrupt? Natural Resources, Corruption and Economic Growth. *IMF Working Paper* 99/85. International Monetary Fund, Washington, DC, USA.

Loughran, D. and J. Kulick, 2004: Demand-side management and energy efficiency in the United States. *The Energy Journal*, **25**(1): 19–43.

Martinot, E. and S. Birner, 2005: Market transformation for energy-efficient products: lessons from programs in developing countries. *Energy Policy*, **33**(14): 1765–1779.

Matysek, A. and B. Fisher, 2008: Prospects for nuclear power in Australia and New Zealand. *International Journal of Global Energy Issues*, **30**(1,2,3,4): 309–323.

Munasinghe, M., 1992: Environmental economics and sustainable development. *Environment Paper* No.3, World Bank, Washington, DC, USA.

Munasinghe, M., 2009: Sustainable Development in Practice: Sustainomics Methodology and Applications. Cambridge University Press, Cambridge, UK.

Nemet, G. F., 2009: Demand-pull, technology-push, and government-led incentives for non-incremental technical change. *Research Policy*, **38**: 700–709.

Newberry, D., 1999: Privatization, Restructuring and Regulation of Network Utilities. MIT Press, Cambridge, MA, USA.

Newell, R., 2008: *Climate Technology Deployment Policy*. Resources for the Future, Washington, DC, USA.

Norberg-Bohm, V., 2000: Creating incentives for environmentally enhancing technological change: Lessons from 30 years of U.S. energy technology policy. *Technological Forecasting and Social Change*, **65**(2): 125–148.

Nordhaus, W., 2007: To tax or not to tax: alternative approaches to slowing global warming. *Review of Environmental Economics and Policy*, **1**(1): 26–44.

Otto, J., C. Andrews, F. Cawood, M. Doggett, P. Guj, F. J. Stermole, and J. Tilton, 2006: *Mining Royalties: A Global Study of their Impact on Investors, Government, and Civil Society*. The World Bank, Washington, DC, USA.

Owens, S. and L. Driffill, 2008: How to change attitudes and behaviours in the context of energy. *Energy Policy*, **36**: 4412–4418.

Perlack, B., and W. Hinds, 2003: *Evaluation of the Barbados Solar Water Heating Experience*. Oak Ridge National Laboratory, Oak Ridge, TN, USA.

Sala-i-Martin, X. and A. Subramanian, 2003: Addressing the natural resource curse: and illustration from Nigeria. *Working Paper 9804*. National Bureau of Economic Research, Cambridge, MA, USA.

Sarraf, M. and M. Jiwanji, 2001: Beating the resource curse: The case of Botswana. *Environment Department Working Paper 83*, Environmental Economics Series. World Bank Group, Washington, DC, USA.

Schipper, L., 2000: On the rebound: the interaction of energy efficiency, energy use and economic activity. *Energy Policy*, **28**(6–7): 351–354.

Short, C., A. Swan, B. Graham, and W. Mackay-Smith, 2001: Electricity reform: the benefits and costs to Australia. *Outlook Conference 2001*, Australian Bureau of Agricultural and Resource Economics, Canberra, Australia.

Sorrell, S., J. Dimitriopolous, and M. Sommerville, 2009: Empirical estimates of direct rebound effects: a review, *Energy Policy*, **37**: 1356–1371.

Stern, N., 2006: *The Economics of Climate Change*. UK government, London.

Stigler, G., 1971: The theory of economic regulation. *The Bell Journal of Economics and Management Science*, **2**(1): 3–21.

Stiglitz, J., 1996: *Principles of Micro-Economics*. 2nd ed. W.W Norton, New York.

UNDP and World Bank, 2005: *Energy Services for the Millennium Development Goals*. United Nations Development Programme, New York, NY, USA.

Victor, D., 2001: The Collapse of the Kyoto Protocol and the Struggle to Slow Global Warming. Princeton University Press, Princeton, NJ, USA.

Victor, D. G. and S. Eskreis Winkler, 2008: In the Tank: Making the Most of Strategic Oil Reserves. *Foreign Affairs*, **87**(4): 70–83.

Victor, D. and T. Heller (eds.), 2007: The Political Economy of Power Sector Reform: The Experiences of Five Major Developing Countries. Cambridge University Press, Cambridge, UK.

Viscussi, W., J. Vernon, and J. Harrington, 2005: *The Economics of Regulation and Antitrust*. MIT Press, Cambridge, MA, USA.

Wei, D. and A. Rose, 2009: Interregional Sharing of Energy Conservation Targets in China: Efficiency and Equity, *The Energy Journal*, **30**(4): 81–111.

Weitzman, M., 1974: Prices versus quantities. *Review of Economic Studies*, **41**(4): 477–491.

Weitzman, M., 2009: On modeling and interpreting the economics of catastrophic climate change. *Review of Economics and Statistics*, **91**(1): 1–19.

Wolak, F., 2003: Diagnosing the California Energy Crisis. *The Electricity Journal*, **16**(7): 11–37

World Bank, 2000: *Energy Services for the World's Poor*. The World Bank, Washington, DC, USA.

23

Policies For Energy Access

Convening Lead Author (CLA)
Daniel H. Bouille (Bariloche Foundation, Argentina)

Lead Authors (LA)
Hugo Altomonte (Economic Commission for Latin America and the Caribbean)
Douglas F. Barnes (Energy for Development, USA)
Touria Dafrallah (Environment and Development Action in the Third World, Senegal)
Hu Gao (Energy Research Institute, China)
Hector Pistonesi (Bariloche Foundation, Argentina)
Ram M. Shrestha (Asian Institute of Technology, Thailand)
Eugene Visagie (University of Cape Town, South Africa)

Contributing Authors (CA)
Jean Acquatella (Economic Commission for Latin America and the Caribbean)
Suani T. Coelho (Brazilian Reference Center on Biomass)
Sivanappan Kumar (Asian Institute of Technology, Thailand)
Debajit Palit (The Energy and Resources Institute, India)
Gisela Prasad (University of Cape Town, South Africa)

Review Editor
Leena Srivastava (The Energy and Resources Institute, India)

Contents

Executive Summary

A number of factors contribute to the lack of access to modern forms of energy. They include low income levels, unequal income distribution, inequitable distribution of modern forms of energy, a lack of financial resources to build the necessary infrastructure, weak institutional and legal frameworks, and a lack of political commitment to the scaling up of services. An absence of specific policies oriented to poverty alleviation often explains inequitable economic growth and, consequently, inequality in access to and use of energy. In recent years, several developing countries have defined targets aimed at improving access to electricity, but many developing countries still have no modern forms of energy access targets in place that address meeting basic energy services, including modern fuels for cooking and mechanical power.

As Chapter 2 argues, developing countries require adequate access to modern energy, especially among the poor, in order to meet the Millennium Development Goals (MDGs) as well as their own national development objectives. In line with GEA objectives, Chapter 17 pathways are designed to describe transformative changes toward a more sustainable future. A specific feature of the GEA energy transition pathways is that they *simultaneously* achieve normative goals related to all major energy challenges, including environmental impacts of energy conversion and use, as well as energy security and energy access. 'Energy access' refers to those challenges clearly described in Chapter 19, which will be addressed in this chapter.

Affordable and sustainable universal access to modern forms of energy depends on the evolution of income level and income distribution. Urbanization processes and population growth are the other variables that play a key role, but both cannot be addressed and solved by energy policy alone. Clearly, without a significant growth of per capita income in developing countries, ambitious targets on access to modern forms of energy will face barriers that are both significant and hard to overcome. As such, policies to improve energy access should be part of the strategies on poverty reduction and income distribution. Isolated solutions are not effective. Without an integrated approach to facilitate the inclusion of excluded populations as a means to alleviate poverty, the intended outcomes will not follow.[1] Reaching universal access to modern forms of energy within a period of 20 years needs robust strategies, policies and measures integrated in long-term national programs with clear targets, dedicated and guaranteed funds, and an adequate institutional framework.

If widespread energy access is to be achieved in just a few decades, energy policies would need to work in concert with economic development policies by harnessing the collective efforts and investment potential of markets, international organizations, central governments, regional governments, cooperatives, and local organizations. Large amounts of funding are needed to provide a major clean-energy infrastructure and to leverage private funding through a more sound investment environment. For the poorer regions, grants would in many cases be needed to, in combination with targeted energy subsidies for the lowest-income populations. Policies would need to support local participation in developing and managing energy systems, as this approach has been shown to have the best chance of providing a stable environment for new investment and reinvestment in increased energy access.

A variety of policies and measures applied around the world have provided considerable experience and knowledge. Overall, no single institutional model or strategy can be recommended as the right one, as evidenced by the success or failures of different models for access to modern forms of energy.[2]

---

1   For example, decentralized, energy technology itself is a package comprised of "hardware and software." The "hardware" includes the energy technology and physical project components. The "software" includes community mobilization, participatory development of the energy technology itself, capacity-building for the use of production technology and scaling-up through market development.

2   The experiences of the eighties and nineties in energy structural reforms, when the proposals coming from international financing institutions were very similar, ignoring national and local circumstances, led to the demonstration and conviction that "one size coat does not fit all," specially in institutional, regulatory and property right issues. Lessons learned from success stories are very important and they will be address in different sections of the chapter, without being prescriptive.

A major shift appears to be urgently needed in the way countries approach the formulation, planning and implementation of policies designed to facilitate access to energy and, particularly, to meet the energy services of poor people. Current supply-side approaches that simply take as their starting point the provision of modern forms of energy or equipment of a particular type are less successful in reaping the full potential of social and economic improvements that follow from improved energy access. A solid knowledge of the energy services that are a priority of the target population is a crucial prerequisite to identifying the right response with energy source and technology. This often requires innovation, testing and experimentation at both technological and institutional levels. Lack of appreciation of such approaches at the policy level may curtail progress, as many policymakers may follow approaches that do not take into account the contextual differences. The integration of centralized and decentralized options of advancing universal energy access also needs to be explored more carefully.

Fuel subsidies alone will be neither sufficient nor cost-effective in terms of achieving ambitious energy access objectives. Often, financial mechanisms, such as end-user finance, would need to complement subsidies to make critical end-use devices, connection costs for network systems like electricity and natural gas or LPG (liquefied petroleum gas) cylinder costs affordable to poor people. Up-front costs of the equipment and appliances are, in many cases, the most important barrier to access. Policies that address this will make it easier for households to cover the fixed capital costs associated with a switch to cleaner fuels.

The promotion of joint actions involving local communities and authorities, NGOs and energy utilities, both in rural and urban areas, have demonstrated to be a favorable enabling environment to advance access objectives.

Reforming the way energy is financed and sustainably operated has potentially important outcomes in the efforts to reduce inequity. Some principles, such as supporting energy planning, making infrastructure investment, creating incentives for commercial lending, generate soft loans, launching promotional campaigns, and providing technical assistance, have been identified for certain markets. Centralized agencies can also assist by improving this coordination and helping to create needed local institutional and organizational capacity.

Finally, this chapter does not aim to be prescriptive in policymaking, but seeks to introduce the challenges, conditions, and key issues that should be taken into account in the quest for a policy on access to modern forms of energy.

23.1   Chapter Roadmap

Universal access to modern forms of energy is one of the most urgent objectives of energy policies in the coming decades. It is inextricably linked to improved welfare, because energy services have a direct impact on human needs, productivity, health, education, and communication. Lack of access to modern forms of energy and the related lack of access to energy services contribute to and are a consequence of poverty, constrain the delivery of social services, limit opportunities, and often erode local environmental sustainability. Universal access to modern forms of energy has clear implications for the achievement of the MDGs, and beyond. It will help address environmental challenges, guarantee adequate levels of health, increase energy security, and promote economic development.

Several factors contribute to the lack of access to modern forms of energy. They include low incomes, unequal distribution of incomes and of modern forms of energy, lack of financial resources to build the required infrastructure, weak institutional and legal frameworks, and a lack of political commitment to the scaling up of services. An absence of specific policies oriented to poverty alleviation often explains inequitable economic growth and, consequently, inequality in access to and consumption of energy. Many developing countries have no energy access targets in place, particularly for those of its components that do most to reduce poverty. In recent years, several countries have defined targets aimed at improving access to electricity. A few, however, addressed targets on meeting basic energy services, including modern fuels for cooking and mechanical power. These, of course, are crucial if the basic needs of poor people are to be met. Access to energy should also be considered within the broader objective of equity. Access to modern forms of energy contributes to this objective.

The role of the state in the energy system varies by country. What is clear, though, is that public policies are the non-transferable responsibilities of government. Governments would need to put in place feasible and effective policies with defined objectives, targets, and strategies that are appropriate to their needs and conditions, and need to apply appropriate measures and incentives to ensure their proper implementation. At the industry level, some countries consider oil and natural gas as strategic goods to be controlled and managed directly by government through public utilities, while accepting that the private sector has an important role in the power industry. In practice, energy policy requires each country and region to find a policy mix that best meets its goals and particular national and regional circumstances, needs and priorities.

In line with the Global Energy Assessment (GEA) framework and goals, this chapter takes 'universal access to electricity, liquid, and gaseous fuels (modern forms of energy) to satisfy households' energy services' as an objective for energy access policy.[3]

The key challenge for this chapter is therefore to examine the various options of policies and strategies that facilitate universal access to modern forms of energy by 2030. The terminology employed in this chapter, such as 'access,' 'modern forms of energy,' and 'rural or urban energy' is consistent with the use of these terms in GEA, and no further definitions are added here.[4]

Section 23.3 and 23.4 introduces two contextual issues. First, the social dimension of access to energy, including the relationships between energy and poverty, the macroeconomic conditions that relate to inequity and poverty, and their influence in the energy system. Second, it examines the key challenges and barriers to increasing access to modern forms of energy.

The situations, conditions, circumstances, levels of access, sociocultural contexts, history, and other factors vary between regions and countries. Developing proposals that take account of experiential learning demand consideration of regional contexts. Sections 23.6–8 provide a picture of the situation, lessons learned and policies applied in sub-Saharan Africa, Asia, and Latin America. Regional experts offer suggestions and recommendations based on their own experiences.

Sections 23.10 and 23.11 seeks to summarize key findings and lessons learnt and present recommendations to facilitate and improve access to modern forms of energy.

23.2   Introduction

Access to modern forms of energy is a key element in poverty alleviation and an indispensable component of sustainable human development. Roughly, 1.3 billion people still have no access to electricity and some 2.7 billion people rely on traditional biomass (IEA, 2011). More than 99% of people without electricity live in developing regions, of which four out of five live in rural South Asia and sub-Saharan Africa (Baker Institute, 2006).

This widespread lack of energy access makes it reasonable to conclude that the present structures and processes within the energy sector are not functioning for the benefit of poor people. Macro energy policies in developing countries tend to focus on commercial energy carriers: electricity, coal, gas, and petroleum products. Urban users are the primary beneficiaries. Although the urban poor may also benefit to some extent, the rural poor – who are the majority of poor people in most developing countries – generally do not benefit for the most part.

---

3   This assessment should not be understood as access to basic energy services, but to an adequate level of satisfaction of energy services according to the social-cultural,

economic and environmental framework. Of course access to modern forms of energy is a necessary but not a sufficient condition for poverty alleviation and development.

4   The exception is if a concept requires a particular interpretation within the framework of Chapter 23 but for others please see Chapter 1.

Developing countries require adequate access to modern energy, especially among the poor, in order to meet the MDGs as well as their own national development objectives. In line with GEA objectives, future pathways are designed to describe transformative changes toward a more sustainable future. A specific feature of the GEA energy transition pathways is that they *simultaneously* achieve normative goals related to all major energy challenges, including environmental impacts of energy conversion and use, as well as energy security and energy access. 'Energy access' refers to those challenges outlined in Chapter 19 which will be addressed in this chapter.

Grid extension to rural areas is very expensive and off-grid or renewable programs supply relatively small quantities of electricity to rural communities. Oil products, such as kerosene or LPG, depend on road transport, which can be seasonally unreliable, and transport costs may increase the fuel cost considerably. Consequently, rural households and industries rely heavily on locally supplied biomass fuels. Energy planners and policymakers in developing countries, while usually aware of this process, rarely address the roots of the problem coherently.

The situation can be explained partly by a lack of understanding of the processes in rural areas and, in particular, by a failure to appreciate the positive contributions that access to modern forms of energy can make to sustainable livelihoods. To some extent, it is also explained by the fact that some issues related to energy, such as biomass, are not the responsibility of energy ministries, because its sources are normally the responsibility of forestry and agriculture ministries. Lack of cross-ministry cooperation is frequent. Poor understanding of the dynamics of rural energy, along with inadequate institutional cooperation, can result in a weak systemic capacity at the government level.

Energy sector reforms, both globally and in developing countries, have had specific consequences for poor people. Such reforms were not oriented to social objectives having no or negative impact on access to energy.

Next sections consider the main energy access dilemmas. It discusses relationships between poverty, development, and energy access at the macro level, and between energy policies, aggregated policies, and specific objectives, such as access to modern forms of energy. It identifies the key challenges and barriers that an energy policy must negotiate.

23.3   The Development Gap: Socioeconomic, Poverty, and Inequity Context

This chapter builds strongly on the concepts of social issues, MDGs and energy (discussed in Chapter 2). It demonstrates linkages with health and energy (Chapter 4), energy, economy, and investment (Chapter 6), scenario/pathways (Chapter 17), and energy access for development (Chapter 19). Findings from these chapters have fundamental

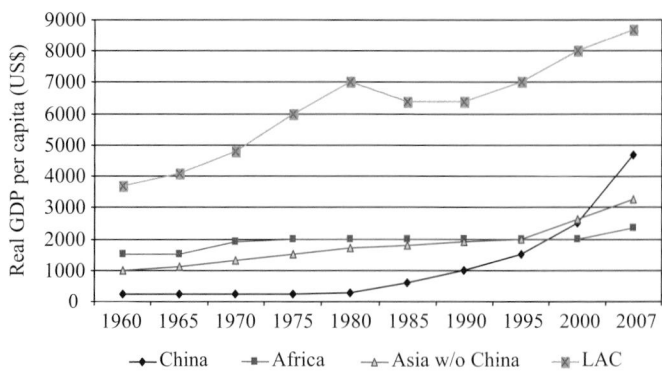

Figure 23.1 | Real per capita income of developing countries by region. Source: Based on data from UNDP Human Development Report, 2009.

implications for this chapter, which in turn strongly influences the discussions of policies, rationales (Chapter 22), and policies for capacity development (Chapter 25).

At the macroeconomic level, the link between adequate energy services – in quantity and quality – and economic development (as measured by gross domestic product (GDP)) is relatively clear, although the direction of causality may or may not always be so (Ghosh, 2002; Wolde-Rufael, 2006). A more in-depth discussion on this causality is given in Chapter 6. Generally, no socioeconomic system provides an adequate welfare environment without available, affordable, and secure energy carriers. The relationship between per capita energy use and GDP per capita is a clear indicator of the importance of energy in development. Of course, the lack of adequate energy supplies is a barrier to development.

In the 47 years, from 1960 to 2007, GDP per capita in Africa grew at less than 1% per year, in Latin America and the Caribbean (LAC) at less than 2% per year, and in Asia, excluding China, at around 2.5% (Figure 23.1). Between 1981 and 2005, sub-Saharan Africa was the only region that did not see a decline in poverty levels.[5] In absolute terms, the number of poor people in Africa has nearly doubled and, if this trend continues, by 2015 one in two of the world's poorest people will live in sub-Saharan Africa, compared with one in ten in 1980.

The weakness of the economic system[6] is a major challenge to universal access to modern forms of energy, and to achieving the MDGs. Feasible,

---

5   Global poverty has fallen sharply on the strength of China's growing prosperity over the past two decades. The proportion of the world's population living in poverty fell by half – from 52% in 1980 to 26% in 2005. In the past 20 years, poverty has been declining at 1% annually (Collier, 2007).

6   The weakness of an economic system results from all or many of the following elements: lack of a long-term vision; low competitiveness; lack of scientific and technology development, knowledge, and capacity in general; high concentration of income and welfare; low saving and investment capacity; weak physical infrastructure; unemployment; degree of informal economy; lack or inadequate management of natural resources; inadequate or inefficient institutional framework; weakness of the external sector, among others. There are five dimensions to this: knowledge, natural resources, social, economics, and politics.

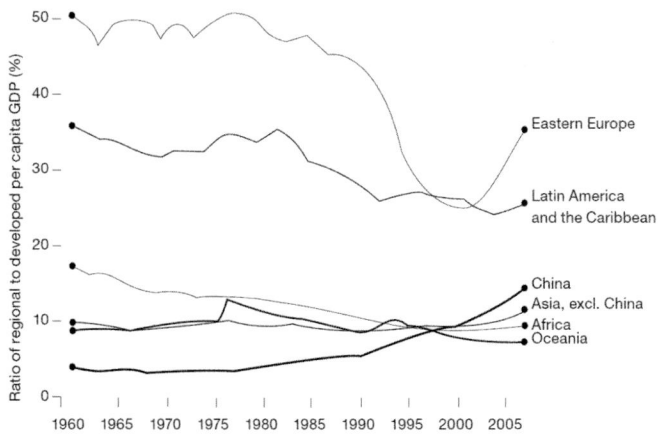

Figure 23.2 | Per capita GDP in relation to Organisation for Economic Cooperation and Development (OECD) countries. Source: UNDP Human Development Report, 2009.

affordable, and sustainable access to modern forms of energy requires an enabling macro- and microeconomic environment to guarantee such an ambitious target as universal access to modern forms of energy by 2030.

There is a clear relationship between low GDP per capita and low access to modern forms of energy. In addition, the disparity in per capita income between the richest and poorest of the world's nations has widened (Figure 23.2). According to UN Habitat (2011), nearly one billion people who live in slums lack both essential physical and social infrastructure. In this respect there has been a lack of convergence between the developed and developing world in the past 50 years, with some exceptions.

Without a significant growth in developing countries' real per capita income, ambitious targets on access to modern forms of energy will become hard to achieve. GEA pathways assume not only an important increase in GDP per capita of developing countries, but also a convergence among the developing and developed world (Chapter 17).

Improved income distribution, at both the global and domestic level, should go hand-in-hand with the creation of feasible, affordable, and universal access to modern forms of energy and cleaner cooking by 2030.

23.4 Key Challenges and Barriers to be Addressed by a Policy for Access to Modern Forms of Energy

The complex and multidimensional character of the development concept is broadly known, but qualifying it as sustainable makes the concept still more intricate. This is not just because it is hard to agree a universally acceptable and applicable definition, but because the dynamics of

a largely complex system must be qualified. These multiple dimensions are strongly related to the reality of a specific socioeconomic system, in which they interact dynamically with each other.[7]

It is evident that the current situation may not be considered as socially sustainable, particularly given extremely low incomes and the challenges in accessing the most basic services, such as health, education, housing, and energy. In respect of energy, greater sustainability demands an understanding of the access to, and coverage of, basic energy services – in terms of both quantity and quality. This should include not just household needs, but the needs of essential services, such as health, education, and drinking water.

Energy policies should be consistent with policies on employment, income, foreign affairs, institutional structure, social objectives, environment, science and technologies, regional development, financing, to name a few. General policies influence the feasibility and effectiveness of sector policies, such as energy. There must be complementarity between energy strategies and other cross-cutting policies, other sector policies, and general development policy. They cannot be developed in isolation. And to be effective, they should include specific definitions of interventions and of the system as a whole.

In consequence, it is not sufficient to address energy access as a list of strategies and measures, but also as the need to establish the analytical framework by identification, through a diagnosis, of the barriers that should be overcome for universal access to modern forms of energy.

An evaluation of the extent of the challenge should note some of the key conclusions and recommendations of a UNDP/WHO (2009) report on the energy access situation in developing countries summarized below:

- The global development community must take specific and far reaching measures to massively scale up initiatives to expand access to energy services for poor and un-served people.

- The quantity and quality of statistical information related to energy access need major improvements and further efforts to address this need

- In order to expand access to energy services, especially cooking and heating services, as well as access to mechanical power in rural and remote areas, more dedicated and broad-based efforts are needed.

- The use of targets is a key to providing a framework for tracking progress and accountability. These targets, in turn, need to be part

---

7   Complexity and, in many cases, the unique conditions of the socioeconomic systems are a key reason to avoid the proposal of prescriptions that should run overall without taking into account national circumstances and specific factors of the energy system. Prescriptions in the energy policy of the type "one size fits all" promoted during the 1980s and 1990s and still today, are the best example of the consequences of ignoring the characteristics and conditions of the target systems.

of viable energy access strategies, backed by appropriate priorities, policies and programs, and financial resources, if they are to be realized.

- If massive efforts are not made to expand the range, quality, and quantity of energy services available to poor people, many countries – particularly those with large pockets of population with low rates of access to modern forms of energy, are unlikely to achieve their development aspirations.

Although electricity use has expanded and programs are in place to spread distributed generation technologies and modern fuels, often these efforts have been plagued by numerous problems, and the scale at which the growth is occurring is barely sufficient to keep up with a universal access to electricity and modern cooking fuels target by 2030. In many cases, the barrier exists in the policies themselves, along with the institutional arrangements that impact energy services. The importance of institutional factors for implementation and sustainability, especially in rural interventions, is widely cited (e.g., World Bank, 2008a; Green Nine, 2004).

The following section examines the importance of various challenges, and how they should be considered in defining policy objectives and strategies.

23.4.1   Political Failure

In general, a policy's failure to obtain the expected results is due to one or more of the following: failure of diagnosis through lack of information, failure to identify the barriers to be overcome or the main problems that the policy should address, or inadequate strategies, instruments, and measures to address the problems and barriers. Of course, lack of political will and government commitment to prioritize investment in energy should also be considered as a failure, or a short-term view of the energy system.

Planning and policy implementation are often based on inadequate diagnosis, which means no clear identification of the problems and barriers. Consequently, strategies, instruments, and measures fail to address the objectives. A proper analysis of why the situation is as it is and how it will evolve is vital. This is the most important phase of designing and implementing an energy policy. Even with decades of experience of the analysis of human needs, the approach to addressing energy use remains based on supply and technology. In general, the wrong question is asked: how can electricity or photovoltaic (PV) panels or other technology or energy sources help to facilitate access to energy? There are other questions that would need to be asked first. Which human needs are the ones that need to be prioritized? What role can energy play in meeting these needs? What is the best combination of technology and energy to meet these priorities? What is the capacity of target users to incorporate the sources and the technology? What

other conditions would need to be put in place to ensure the proposed solutions' affordability and sustainability? How will demand change when access to modern forms of energy becomes a reality? Are the sources and technologies sufficiently flexible to address future needs? If not, what will be done? Attention would need to be given to the needs of the target populations and on the energy services to be provided.

Inadequate diagnoses are commonly related to a lack of clear understanding of the energy access situation in countries, including regional and national trends, rural/urban disparities, and the range of energy sources typically used in poor households, and are a barrier to a comprehensive situation analysis.

Biomass represents a major share of energy use by poor people in rural areas. It is collected at zero monetary cost, mainly by women and children, and therefore falls outside national energy accounts. In the formal statistical data bases, several times it becomes invisible. This invisible collection of energy source means that decision makers are rarely aware of its significance. Policies and strategies therefore fail to address the issue adequately.

Do policymakers have sufficient understanding of where to begin improving access? What improvements will be valued by poor households and communities? To answer these questions requires some understanding of how energy is obtained and used today and of future trends, both for final use and for productive activities. Also essential is knowledge of poor households' requirements for better energy services, and their willingness to pay for them.

Improving access to modern forms of energy for poor people will require greater attention on two fronts. First, policymakers and their advisers need to use as many energy data as are available to design strategies that, at least, do not close off energy options valued by poor people or distort incentives to supply and use better services. Second, they need to design policies and strategies that elicit access and energy use information effectively, to give enabling conditions for any proposed solution.

23.4.1.1   Availability and Use of Adequate Energy Access Data

Data collection on these issues was – and remains – weak. Of late, policy advisers and donor agencies have worked to improve their understanding of poor people's needs and to tailor actions to their preferences. But the data gap remains wide. Niez (2009) cites, "sound statistical data … and a clear description of the [energy services] situation" as the first precondition for successful rural energy access policies.[8]

---

8   This does not mean that the main barrier for energy access is lack of data, but in the absence of the real dimension and quantification of energy needs, the design and implementation of policies and strategies could have poor results in the actions proposed, the assessment of impacts and the effectiveness in addressing the priorities from the target population point of view.

Table 23.1 | Basic energy use in different bioclimatic zone.

Bioclimatic zone	Urban or rural	With or without access to electricity	Final energy use by household (kgoe)*	Useful energy use by household (kgoe)	Annual Electricity consumption by household (kWh)
Hot	Urban	With	243	124	1596
		Without	596	100	0
	Rural	With	658	131	1521
		Without	1069	130	0
Moderate	Urban	With	380	162	1492
		Without	706	151	0
	Rural	With	1034	170	1388
		Without	1451	174	0
Cold	Urban	With	606	262	1416
		Without	961	255	0
	Rural	With	2062	287	1353
		Without	2504	293	0

* kilograms of oil equivalent

Source: Bravo, 2004.

Available information shows that poor households rely on highly varied energy sources. They often incur real costs that are far higher than those for equivalent energy from, for example, electricity networks. Evidence suggests that poor people are, indeed, often willing to pay for more modern forms of better energy. Identification of this situation, and meeting that demand, therefore becomes a major challenge. Information as to the economic capacity to afford access to modern forms of energy is generally lacking.

A valid example is the estimation made for LAC on the quantification of basic energy services and the right to energy (Bravo, 2004):

"Any quantification of basic energy services is space specific, time related, and influenced or determined by environmental and sociocultural conditions. It is not possible to fix global, regional, or national values for basic energy services. There are three elements around any energy service: human need, appliance or equipment, and energy source.[9] In basic energy services for household consumption, only indoor energy services are included.

Basic energy services are lighting, cooking, water supply, water heating, space heating, and food conservation. The requirement for each energy service is determined by sociocultural conditions, urban, rural, or peri-urban localization, biogeographic and climatic conditions, access to energy sources, types of appliances available, building characteristics and family size, and other factors. In addition, the capacity, efficiency, and annual use of appliances determine the useful and net energy consumption.

The total basic energy use estimated for different situations are summarized in Table 23.1. This demonstrates the importance of

environmental conditions and access to an energy source, such as electricity. Of course, the structure of energy services differs by zone, urban, or rural location, and access to electricity.

Similar studies could be useful to determine the amount of basic energy services needed to assume reasonable conditions of human dignity, and should offer a figure for the right to energy. A comparison with current levels of use in useful energy could be useful to measure the extent to which the needs are met and could feed into policy and strategy development."

There is substantial variation in energy use patterns, depending on climate, local fuel resources, the country's economic history, location (urban, rural, or peri-urban), and other factors (Box 23.1). Excessive focus on supply, along with incomplete understanding of the real and priority needs of poor households, can limit proposals for positive change.

Poverty often leads households to use a mix of energy sources that are suboptimal from the economic, financial, health, and environmental perspectives. They also tend to use less modern forms of energy than they would be willing and able to if supplies were commercially available at prices that are fair such that the suppliers are able to recover their costs. Such situations are often explained by an inability to afford the up-front costs of the appliances needed to satisfy the energy services, or the costs of connecting to grids, in the case of electricity.

Households in most developing countries are getting smaller and may have fewer wage earners, which reflects factors such as higher per capita incomes, smaller family sizes, greater access to education, and increasing urbanization. People living in these smaller households are less likely to be poor and therefore more likely to have disposable income to spend on modern forms of energy. But smaller households also mean that each new electricity connection may benefit fewer people than in

---

9   The energy services beyond the basic may be different combinations of appliances and sources.

Box 23.1 | Factors Relevant to Energy Access

Understanding what type of energy carriers and end use technology are available, who uses them, how much they cost, and the benefits they provide to users, are factors to consider when assessing energy access. For instance, energy access can include measures related to:

- The quality and quantity of energy provided. Data regularly collected by statistical agencies generally do not provide detailed information on the quantity (e.g., hours of use and availability) and quality (e.g., rated voltage and frequency) of the energy services provided, although these may be available from utilities and other sources.

- Energy end-use appliances and equipment and the services they provide. Data on the availability of certain household and agricultural equipment – lighting using different energy sources, water pumps, refrigerators, and different types of stoves – are available from some international surveys, and would be useful to collect, but are beyond the scope of this report.

- Socioeconomic profiles of energy users and energy affordability. Data on income levels and geographic location of energy users are often available from statistical agencies, but other socioeconomic data on users – prices of energy services, gender-disaggregated data, and disaggregation for key sectors such as businesses, schools, and health centers – may not be.

Source: UNDP/WHO, 2009.

---

the past. This has implications for the design of programs to increase access. Understanding how this will evolve in the medium and longer terms is crucial to generating reliable market estimates.

These findings come from partial analyses in various regions, and should be examined further and verified in region and country specific contexts before policies on access to modern forms of energy are developed and implemented. The absence of a good and reliable diagnosis could be a key barrier to the success of a policy or strategy.

23.4.1.2   Coherence and Convergence between Objectives and Strategies

To ensure coherence and convergence between objectives and strategies is a necessary, but not evident, issue in policy design and implementation.

The most recent policy failures, at least in addressing the social dimension of a sustainable energy policy, are the structural reforms implemented in the past 20 years in the electricity and in other sectors of the energy system. The emblematic role given to market forces, the private sector, and the transfer of decision making to decentralized stakeholders should be assessed in relation to the results and in how they helped or did not help to increase access to modern forms of energy. Reducing the governmental role in controlling and managing a strategic good such as energy is, indeed, a key issue for important consideration in this decision.[10]

Divestiture of public utilities and energy resources, changes in the regulatory framework, openness to foreign investment, and abandonment of long-term planning were just some of the strategies proposed and implemented as a panacea to promote sustainable development.

One rationale for the reforms was that they would enable increased access to modern forms of energy and even facilitate rural electrification (Bouille and Wamukonya, 2003). Injection of new capital and the increased efficiency that would come with the need for private sector returns was viewed as essential to create the infrastructure necessary to bring electricity to rural areas and shore-up existing distribution systems.

An in-depth GNESD (2006) [11] study explains that:

> "… the reasoning behind such initiatives was simple: streamlined and restructured energy sectors, being more efficient and less costly, would widen access to energy services and produce benefits for health, education, nutrition and entrepreneurship for all. In this context, the role of government is reduced to creating an enabling environment within which private sector mechanisms develop and provide services."

The failure was largely the result of a policy vacuum as far as the needs of poor people were concerned (GNESD, 2006). The key negative impacts on poor people include reduction in electrification rates and increased

---

10 A careful reading of the fundamentals and considerations in the laws that made reform possible demonstrates very similar arguments, and proposed an *urbi et orbi* approach.

11 The main values of the GNESD report are the broad representation of countries involved, deeper analysis made by centers of excellence of the countries or regions, the use of a common methodology, and a collection of common findings. The countries included in the analysis were Kenya, Uganda, Senegal, Mali, South Africa, Zimbabwe, India, Philippines, Thailand, Bangladesh, Vietnam, China, Argentina, Peru, El Salvador, and Brazil.

Table 23.2 | Comparison of results of analysis of selected indicators for the Argentina, El Salvador, and Peru case studies.

Selected indicators	Argentina		Peru		El Salvador	
	Pre-reform	Post-reform	Pre-reform	Post-reform	Pre-reform	Post-reform
Electrification levels (%)	91	95	38	62	62	76
Annual electrification rates (%)	2.04	1.03	7.8	5.8	6.6	4.1
Average household consumption (kWh/month)	155	205	136	106	104	112
Poor household lifeline tariff (USc/kWh)	4.35	11.77	6.8	17.2	4.8	8.6/16.8[12]

Source: GNESD, 2004.

tariff levels. See, for examples, some Latin American cases (Table 23.2). This was largely due to a "one size fits all" approach, whereby whole-sale policy transfer was applied with little consideration for context, national circumstances, technical issues, or degree of development and maturity of the energy system.

The International Monetary Fund (2008) concluded:

"As of 2006, more than 80% of sub-Saharan African countries had enacted a power sector reform law, 75% had experienced private participation in power, about 66% had corporatized their state-owned utilities, more than half had established a regulator, and more than one-third had independent power producers reform programs typically followed an orthodoxy that aimed at creating competition among private electricity suppliers, but few energy markets in Africa are large enough to support the multiple suppliers needed for a competitive environment. As a result, despite reform measures, utility performance continues to be disappointing and associated hidden costs can absorb as much as 2% of GDP."

If the proposal for future strategies follows the 1990s paradigm, some lessons learned could help to avoid the same mistakes, at least in the power sector:

1. Ring fencing, or protecting, the funds for providing electrification for poor people. In several countries funds have not been protected, Kenya being the best example. Brazil and South Africa emerge as model examples of how to ring fence. In Brazil, two important measures were implemented. First, the Electricity Act has made it mandatory for all electricity distribution concessionaires to contribute to the Global Reversions Reserve (RGR – Brazil) – the national electrification fund. Second, the allocation of funds for electrification by the Reserve was predetermined by region and matches the electrification needs of specific regions.

2. Sequencing of reform. Is it better to embark on wide-scale electrification of poor populations prior to privatization, or vice versa – privatizing and thereafter launching an electrification program? In countries where wide-scale electrification

was undertaken prior to market-oriented reform such as privatization, notably South Africa, Zimbabwe, Mauritius, Thailand, and Philippines, a significant proportion of poor people gained access to electricity.

3. Explicit focus on poor people. With some exceptions, reforms in most countries examined in the Energy Access study (GNESD, 2006) did not focus explicitly on poor people. There are several ways in which reforms could ensure that poor people become a critical consideration of the reform process. Examples include Brazil, South Africa, and Bangladesh. One way to ensure that reforms explicitly focus on poor people is by enacting laws that ensure they gain access to electricity. This was the approach adopted in Brazil, which has three key laws that focus on poor people.[13]

4. Establishing dedicated institutions for poor people to have electrification is another way that reforms could ensure an explicit focus on poor people. This approach appears to have been successful in Bangladesh and South Africa. In Bangladesh, the Rural Electrification Board was established, with one of its key mandates being to widen access to electricity and to ensure poverty alleviation in rural areas (Shrestha et al., 2004a). In South Africa, the National Electrification Program was designed formally to target underprivileged groups under the apartheid regime, who constituted the majority of poor people.

5. Participation of poor people in the electrification process. Apart from ensuring an explicit focus on poor people, their involvement in the electrification process appears to be equally important. With the exception of South and Southeast Asian countries, there appears to be limited involvement of poor people in other regions covered by the Energy Access study (GNESD, 2006).

The key challenge is not to tie policies to paradigms that do not include a clear diagnosis of the national circumstances and the technical, socioeconomic, and cultural situation,[14] and to avoid

---

12 Depending on level of consumption and localization (rural or urban) the tariff range is between both figures.

13 Policy oriented to the poor during the Lula presidential period reduced the number of people living below the poverty line in Brazil by 30 million (Coutinho, 2007).

14 Proposals to introduce competition and unbundling in power systems that have less than 1000 MW of installed capacity are an example of the lack of logic in such a paradigmatic approach.

Table 23.3 | Urban and peri-urban key issues.

Category	Urban	Peri-urban
Institutional	Government representation, good infrastructure, single municipality, information systems	Absence of the state, poor infrastructure, unregulated land use, across different municipalities, poor information on land use and population growth
Land use	Regular settlements, vertical buildings, few better-off shantytowns	Many irregular settlements, many shantytowns, housing projects
Demographic	Low or negative population growth, job offers, older population	High population growth, unemployment, pressure over public infrastructure, younger population
Environment	Presence of parks, better sanitation and environmental conditions	Sanitation problems, deforestation, pollution of water sources, invasion of protected areas, landslides

Source: based on da Gama Torres, 2007.

prescriptions that appear as magic solutions without evidence of success in similar circumstances.

23.4.2 Population Dynamic: Urban and Peri-urban Access to Modern Forms of Energy

Urban energy, peri-urban energy, and rural energy make up part of the analyses of access to modern forms of energy, or the provision of adequate energy services. But is there a clear definition of each category? Is there common agreement on the meaning of each category? Which is the most challenging from the perspective of access to energy?

An urban area is characterized by a higher population density and many human features not found in areas that surround it. Urban areas may be cities, towns, or conurbations, but the term is not commonly extended to rural settlements, such as villages and hamlets. Internationally, the standard determinant of a rural area is population density. Therefore, rural areas are defined as those with low numbers of people who live on any given area of land.

However, the most challenging issue is the definition of peri-urban populations. In short, they could be considered as the transition zones, or interaction zones, where urban and rural activities are juxtaposed, and landscape features are subject to rapid modifications, induced by human activities.

For many African countries, peri-urban literally means the area around an urban settlement. It is distinctive in its diversity, with a mix of land uses and residents. It is rural in appearance, but many residents will have jobs in the nearby urban area to which they commute. Houston (2003), also cited by Buxton et al. (2008), defines the peri-urban areas on the basis of population density, employment in non-agricultural industries, and population mobility. Barr (2005) defines a similar region in terms of "rural amenity landscapes," by analyzing the relationship between rural land value and agricultural production value. Differentiation between urban and peri-urban conditions can be considered in various ways (Table 23.3).

Why are peri-urban areas important for access to energy? The challenge is twofold. First, agreement is needed on what peri-urban means. And second, population growth in developing countries is largely concentrated in peri-urban areas, which accentuates the key challenge of infrastructure development, including the energy infrastructure.

Population dynamics in developing countries show that an increasing share of the total population is found in peri-urban areas. In addition, a deficit exists in peri-urban information. The information deficit results from invisibility or the trend to ignore poverty. This refers to a situation in which policymakers are located elsewhere, where there is a lack of media coverage and focus, plus inadequate registration by urban information systems. This can lead to poorly located schools and healthcare facilities, crowded public facilities, the state's inability to regulate land use, and a lack of infrastructure such as water and energy.

Recognizing that urban, peri-urban, and rural situations imply different challenges, the peri-urban situation is the most complex. It challenges the planning of sustainable and affordable access to modern forms of energy. The multidimensional characteristics of peri-urban areas demand an integrated policy approach, including to energy access.

In relation to LPG, recent studies show different situations, all of them with the common issue of the need of intervention through public policies. There is a generalized situation of irregularity in the energy provision: LPG is not available in some places, it is expensive and the subsidized cylinder is found to be adulterated (Bariloche Foundation, 2008). LPG was found prominent in economically well-off households, however LPG had issues like affordability and delays in refill attached to it. Due to lack of awareness, LPG was perceived as a dangerous fuel. Access to biomass and kerosene was relatively easier than LPG or electricity as the latter required a valid residence proof and had higher upfront costs. Lack of awareness of the harmful effects of using conventional biomass stoves and lack of willingness to give up biomass usage due to its ease of access and the non-continuous supply of other fuels, were key factors driving continual usage of biomass even in households having other cooking fuel options. Kerosene was found to be the baseline fuel used in all households however due to factors like not-well-targeted subsidies, market distortions and need for ration cards to acquire it, and the fuel had accessibility issues (TERI, 2008). In spite of LPG not being a very common source of fuel among the urban poor population, trends indicate that its use and dissemination is steadily growing. LPG has a very high upfront cost which is normally beyond the reach of the majority of the urban poor. The overall cost of a simple cylinder with its related accessories is approximately 10 to 15 times the national per capita income. This has greatly affected the dissemination of LPG mainly among the urban poor. In addition, although the cost of refilling the LPG cylinders is normally affordable and within reach of the urban poor population, the prices are dictated by the world oil prices which fluctuate from time to time. This causes uncertainty about its use within the target group. The safety aspect and reliability (volume found in each cylinder may vary substantially pertaining to the use of LPG is also of great concern among the urban poor (Karekezi et al., 2008)). In spite of LPG not being a very common source of fuel among the urban poor population, trends indicate that its use and dissemination is steadily growing.

The key access issues for the urban poor in the slums studied in Delhi were high upfront costs, insecurity of tenure of land and the lack of recognition and permanence of many slums in Delhi. Mistargeted subsidies, market distortions, lack of accountability on part of service providers and no monitoring mechanisms were identified as the other pressing issues hindering access. Broader issues included lack of database on urban poor and their energy use patterns, unclear institutional responsibilities and lack of policies targeting clean energy access to the urban poor (TERI, 2008).

Almost 85% of the world's urban population has access to electricity. Indeed, in some parts of the developing world, namely North Africa, East Asia (including China), the Middle East, and Latin America, the level of urban-energy access is nearly universal. About 95% of population growth over the next 30 years will occur in urban areas. Thus, fast population growth and urbanization and rising demand for electricity will exert tremendous pressure on infrastructures and create strong demand for new investment. Unless appropriate steps are taken to meet that growing demand, the urban poor are certain to lose ground in access to electricity.

In urban areas, extending electricity access to poor people is a matter first and foremost of getting the policies right. The infrastructure is generally already in place in most of the world's large urban centers, except in Africa. Therefore, energy companies need to make fewer new capital investments. But even with lower capital costs and higher incomes in urban areas, poor people still often cannot afford the connection fees or monthly rates – even if conditions in these areas are more favorable in terms of distance to the network and density of population. However, in addition to low incomes, other issues are crucial for a sustainable supply, notably at the management level and also the dubious legal status of many peri-urban settlements.

Consequently, supportive policies are needed that make service expansion to the urban poor sustainable. The problem of reaching poor people in urban areas generally requires a change in the mindset of urban utilities, as serving poor populations often calls for special policies, investments, and innovative technical and financial solutions.

Urbanization growth rates are surpassing national growth rates in many countries (Box 23.2). Many urbanites are settling in peri-urban areas,

which are generally illegal. Planning for service provision, including modern forms of energy, tends to exclude such areas. Initiatives for service provision in such areas can be *ad hoc*, and have minimal impact. This is escalating poverty and crime in these areas, and retards development. At the very minimum, governments would need to provide regulatory guidance to enable access to energy services in peri-urban areas.

In recent years, several international forums have addressed the issue of peri-urban electricity problems (Rojas and Lallement, 2007). Poor people pay extremely high prices for electricity, often to illegal entrepreneurs. Often safety issues are ignored by such entrepreneurs and service levels are usually very poor. The solutions to these problems are not insurmountable. However, implementation has been lagging in many countries, and there is a need to address these issues more directly.

In addition, the lack of income may sometimes be more of a deprivation in urban than in rural areas. In the latter, poverty is usually accompanied by traditional or pre-modern ways of life. Access to energy through the use of firewood may be partially guaranteed, but in ways that are highly undesirable because of the effect on health and on the work of women and children, and also because of the correlation with other deficiencies, i.e. access to education, health services, and water.

23.4.3　Rural Energy and Electrification

Rural energy is a complex issue that encompasses a broad and diverse spectrum of resources – from petroleum fuels and coal to biomass and renewable energy – that spans multiple sectors, including forestry, electricity, and health. Many new projects are developing the technical capacity to tackle rural energy issues in all their complexity.

International agencies have sometimes taken short cuts to address this problem, and have advocated projects with a narrow technology focus. However, to focus on single technologies does nothing to develop markets, companies, or non-governmental organizations (NGOs) to support rural energy development. Such an arrangement is not conducive to solving the problem of universal access. Whatever experience is gained by a specific project cannot be applied to subsequent projects because of lack of continuity. Moreover, single efforts generally seek exceptions to regulatory policies, but are unable to change them. As a result, it is not possible to see the long-term effects of such projects in the form of greater access to quality energy services.

The development of rural energy policy institutions could provide advice and support for how to better focus the use of funds. Their role would be to promote sound policies for rural areas and innovative pilot projects. It must be understood that, for the most part, rural energy is an unprofitable business, because in many cases there are costs involved in market development that cannot be borne by private sector companies. However, by using a combination of loans and subsidies, both large- and small-scale businesses can become viable to better promote

Box 23.2 | Urban Poverty Today and Tomorrow

"Over the past half century, the world's urban population increased from around 730 million in 1950 to over 3.15 billion in 2005. Around 1 billion, or nearly one-third of the world's urban population, are now believed to be living in slum conditions. With virtually all population growth until 2030 taking place in urban areas, that number is likely to double.

In India, an alarming accompaniment to increasing population and urbanization has been the deepening of urban poverty, growth of urban slums, and the deterioration in basic service delivery. Slums in Delhi include informal settlements that are either squatter or illegal colonies where people live on undeveloped and unserved land without secure tenure or access to basic services.

In Argentina, over 13% of the population of Greater Buenos Aires lives in slums. Here, as in other cities, there is a parallel growth of rich, well-serviced neighborhoods and gated residential communities close to dense inner city or peri-urban slums that lack even the most basic of services. There is parallel growth of slums, country clubs, and closed wealthy neighborhoods (often with private security services). As slums grow even within the city, high-income housing seems to bridge these neighborhoods.

Kenya is experiencing very rapid urbanization. Over 40% of the population is urban, a figure that looks set to rise to around 50% by 2050. Around 34% of the urban population lives below the poverty line and income distribution shows a large disparity between poor people and non-poor.

In South Africa, urbanization is already ahead of the global trend. The 2001 census showed an urbanization level of over 56%, leading to major problems in terms of infrastructure, unemployment, and poverty.

In Thailand, people migrate from the countryside to urban areas in search of better employment opportunities and higher income: 81% of the dwellings in the Bangkok region house people who have migrated from other regions or slums. Most slums are in the city's core areas, but there are indications that this is changing with slums in core areas decreasing and new ones arising in urban fringe areas."

Source: GNESD, 2008.

---

a wide range of rural energy services, from LPG and grid electricity to improved stoves and tree growing. Moving forward requires a combination of research, production and delivery, support, and monitoring and evaluation, all of which must be done through a variety of businesses – including rural electric cooperatives, NGOs, private sector companies, and local municipalities – interested in serving rural energy customers.

In the case of electrification, the choice is between a large-scale infrastructure and the local delivery of energy. The current dominant development model focuses on achieving macroeconomic growth. This results in a predominance of attention to, and investment in, large-scale energy infrastructures to provide energy for growth. There is a need to redress the balance, with much more attention and investment directed toward the supply of local energy services for poverty reduction in local communities. A policy that takes account of both the infrastructure for energy development and energy access priorities is needed.

The rate of electrification in rural areas is very low, particularly in sub-Saharan Africa. This inhibits social and economic development in these areas. A combination of three drivers helps to explain this – the density and dispersion of population, the distance to the network, and the income of the population, all of which are particularly

unfavorable in these rural areas. These drivers influence the decision between grid or off-grid rural electrification. The decision-making process would need to consider what inherent limits they place on the extension of the grid, as a way to motivate the use of off-grid approaches.

In rural electrification, the key financial barriers and challenges to attract both local and foreign private companies incorporate additional barriers to the issues mentioned (ACP-EU, 2009):[15]

- Rural electrification is often not a profitable business, and there will be limited interest in such activities from a purely financial point of view, in particular since many countries require that a nationwide uniform tariff be applied.

- As it can be politically unacceptable to raise tariffs, these are often not adequate for the financial sustainability of the economic models, for both national utilities and independent producers.

---

15 Although these statements are related to Africa, they are still valid for other developing regions.

- The investment capacity of national utilities is often limited, which restricts their ability to maintain the existing network and, in the best case, to create commercially viable grid extensions. The financial situation of these utilities is often weak through a combination of factors: inadequate tariffs, low levels of revenue collection (commercial losses), technical losses, inefficiency and obsolescence of the power systems through inadequate maintenance, low level of consumption, and low rate of interconnection, among others.

- Lack of financing schemes, or poor implementation of existing schemes, dedicated to increased investments in rural access, such as cross-subsidies between urban and rural areas, special levies to benefit rural electrification, and public guarantees for loans.

- Inadequate project size. Programs are often too small to be attractive to financing institutions and, even more so, to international private investors.

- Lack of interest from local banks for rural electrification because of a lack of knowledge and of confidence in this sector. This can be explained by the fact that local financing institutions are used to working in sectors they already know and whose risks they can measure and charge for, on a short-term period, and to working with established clients. Local loans are expensive and a short-term approach to financing is not suitable for financing long-term assets.

- Lack of credit-enhancement schemes, such as various bank guarantees and co-financing instruments for private investment in rural electrification.

- Risk of unmanageable escalation of exploitation costs, because of the higher price of fossil fuel, for example. Even when these can be included in the tariff, escalation may cause a loss of customers and a reduction of receipts, and therefore inhibit planned expansion.

- The exchange risk for imported goods, which are paid for with foreign currency but with receipts in local currency.

In summary, the two main challenges for financing rural electrification are:

- How to ensure the service's long-term financial viability.

- How to divide risks among stakeholders in a sustainable manner.

The choice of short, medium, and long term outlooks is another significant consideration. In the short and medium term, the only way to reach isolated households that do not have electricity is through single-household systems and small electricity providers, using both renewable and conventional energy sources.

Off-grid household programs in Bangladesh and Sri Lanka demonstrate that it is possible to implement large-scale, off-grid projects that complement strong grid-electrification programs (Govindarajalu et al., 2008). The challenges of off-grid projects in both countries have taken advantage of private sector institutions, NGOs, and microfinance institutions that operate in rural areas. Also, they have required centralized institutional support. In Sri Lanka, financing is provided though microfinance institutions, banks, and leasing companies for renewable energy systems that are provided by the private sector and NGOs. Today, off-grid solar home systems (SHS) and village microhydropower (MH) grids provide electricity to 3% of all Sri Lankan households (World Bank, 2008b). This solution would need to be considered as short term, because SHS consist of PV modules with capacities that range from 30–60 W_p watts and therefore offer limited access.

Small grid systems vary widely, from MH to locally generated private distribution. To grow and thrive, such systems often require external technical and financial support. The challenge is, therefore, to assure the program's affordability and sustainability. Off-grid electricity has the drawback of high cost compared to that of grid electricity in urban areas, along with low financial capacity or willingness to pay for modern forms of energy in many remote or rural areas where access is lacking.

Although their institutional forms vary, successful grid-extension programs generally require financially and technically strong utilities.[16] To ensure sustainability, distribution companies must address the issue of increased technical losses and low revenues creatively, or with the introduction of cross-subsidies. In Jujuy, Argentina, for example, the utility reduced fixed costs by creating two companies, with a common management, for the grid and off-grid electricity public service. This minimized the need for government subsidies. The extension of cooperative electricity has been successful in Argentina, Bangladesh, and the Philippines.

23.4.4   Scale of Investment for Universal Access: Scenario Target

The GEA pathways carried out in Chapter 17 sought to assess options on how to achieve 'almost universal access to electricity and modern cooking fuels by 2030'.[17] This includes the diffusion of clean and efficient cooking appliances, extension of both high-voltage electricity grids and decentralized microgrids, and increased financial assistance from industrialized countries to support clean-energy infrastructures. The costs of almost universal access estimated by GEA are substantial, some US$36–41 billion/yr until 2030.

Several estimates have been made for the cost of universal access to energy services at the global, regional, and project levels (Table 23.4). In general, estimates focus on electricity – fewer data are available for fuels

---

16   See Barnes (2007), for more examples.

17   The target is "almost universal access" because reaching the remotest rural populations is exceedingly expensive and urban electrification costs are not included.

Table 23.4 | Cumulative investments to facilitate access to modern forms of energy.

Geographical focus	Goal	Cost estimates (billion US$)		Source
		Electricity	**Cooking**	
Global	Universal energy access	700 [i]	56	OECD/IEA (2010)
	Improved access to reach MDG 1	223	21 [ii]	OECD/IEA (2010)
	Universal energy access	35–40/year [iii]	39–64 [iv]	AGECC (2010)
	Universal electricity access	~55/year	1.8/year	Saghir (2010)
	Universal electricity access	35/year		IEA (2009)
	Improved access to clean cooking [v]	858		Birol (2007)
	Universal electricity access [vi]	200		World Bank Group (2006)
	Improved electricity access to reach the MDGs	665		IEA (2004)
	Universal electricity access			IEA (2003)
Regional/local				
Africa	Improved electricity access [vii]	17/year [viii]		African Development Bank (2008)
Sub-Saharan Africa	Improved energy access	6–15/year		BREW-Hammond (2010)
	Increase household electricity access to 35%	4/year		UN-Energy/Africa (2007)
East African Community (EAC)	Improved energy access [ix]	1.5	0.262	East African Community (2006) [x]
Economic Community of Central Africa States	50% electrification	1.45		CEMAC (2006)
Economic Community of West African States (ECOWAS)	60% electrification, 100% improved cooking fuels, access to mechanical power in 100% of villages	2.1	2.8	ECOWAS (2005)
South Africa	Electrificacion	US$1000 per connection [xi]		Eskom (2009), Niez (2009)
Kenya	Electrification	US$1900 per household [xii]		Parshall et al. (2009)
Botswana	Electrification	US$1100 per household [xiii]		Krishnaswamy and Stuggins (2007)
Mali	Rural electrification	US$776 per connection [xiv]		AMADEER, quoted in Foster et al. (2010), p. 199
Senegal	Increased electrification rate from 47 to 66%	0.86		ASER (2007)
Bangladesh, Cambodia, Ghana, Tanzania, and Uganda	Improved energy access in line with the MDG targets	US$13–18 per capita/year [xv]		Sachs et al. (2004)
South Asia	Universal access to LPG		449	IIASA [xvi]
Brazil	Promoting LPG access to underprivileged households		0.5 [xvii]	Jannuzzi et al. (2004)
(Unspecified)	Electrification	Above US$1200 per connection [xviii]		Practical Action (2007)

[i] Including both rural and urban grid connection, generation, transmission, and distribution; minigrid generation and distribution; off-grid generation.

[ii] Including advanced biomass stoves, LPG stoves, and biogas systems.

[iii] Based on IEA (2009b).

[iv] Improved cookstoves, 11–31; biogas, 30–40; LPG, 7–17. Includes capacity-development costs.

[v] LPG cylinders and stoves to all the people who currently still use traditional biomass.

[vi] Includes breakdown by major regions.

[vii] Reliable electric power to 90% of sub-Saharan rural population, 100% of the sub-Saharan urban population, and 100% of the both the rural and urban populations in the Northern African middle-income countries.

[viii] Considering only new generating capacity, including generation as well as transmission and distribution.

[ix] Reliable electricity for all urban and peri-urban poor; modern cooking practices for 50% of population currently using traditional cooking fuels; energy services for all schools, clinics, hospitals, and community centers; mechanical power for heating and productive uses for all communities.

[x] Including capital expenditure, programs, and loan guarantees.

[xi] The average is expected to increase as the electrification process moves to communities in more remote rural areas.

[xii] Average cost per household in a so-called realistic penetration scenario, with US$1500 and US$2615 for infilling and grid extension, respectively; based on modeling of grid extension.

[xiii] Based on project experience.

[xiv] Based on project experience from AMADER (Agence malienne pour le développement de l'énergie domestique et l'électrification rurale).

[xv] Including costs of end-use devices, fuel consumption, electrical connections, and power plants.

[xvi] Updated analysis based on the methodology described in Ekholm et al. (2010).

[xvii] Subsidies for LPG access to underprivileged households in 2003.

[xviii] New connection to electricity, based on case studies, varies from country to country, and can be as much as US$6000 in some cases.

Source: Bazilian et al. (2010) and references therein.

for caloric uses. A recent report summarized a wide range of estimates (Bazilian et al., 2010). Electrification costs range from US$5–40/capita/yr, "reflecting the large uncertainties associated with such evaluation and the sensitivity to certain assumptions." The report suggested a general underestimation of the financial effort to satisfy universal access to modern forms of energy. Most estimates consider only capital costs and do not include fuel and operation and maintenance (O&M) costs.

Several issues influence the results. The most important include the combination of grid, off-grid, and minigrid in the structure of the system, population density, urban and rural population mix, annual level of consumption per capita, and mix of generation technology and generation fuel.

For mechanical power, the available information is very poor.

The annual costs for universal electrification vary between US$12–134 billion/yr, accordingly to per capita annual consumption estimates.[18] Total estimates, including those for cooking, run from US$14–135 billion/yr. Many of the countries that require the most effort to achieve universal access are those with GDP per capita less than US$1000, a range in which lie many of the LDCs with low access to modern forms of energy.

The question is not only how much the global investment would need to be or how high the other global costs are, but where the investment has to be made, and recovered, taking into recognition the economic capacity of the target population and of the country as a whole.

Table 23.5 is a preliminary indicator of the required effort, only in terms of investment, by some developing countries if they are to satisfy universal access by 2030. Considering figures of Table 23.4 and depending on the family size, the cost of connection to electricity could be up to two times the annual income of the household.

The magnitude of the resources involved and the need to recover them during the lifetime of the investment is, perhaps, the major challenge for some developing countries.

23.4.5　Funding Gap and Financial Constraints

The great majority of people without adequate access to energy live on less than US$2/day, which makes it difficult for them to access good services, including energy services. Energy access is not without cost and the initial expenditure on electricity connections or better technologies can be high. A large funding gap in providing energy access

Table 23.5 | Population without access to electricity and Per Capita GDP – some Sub-Saharan African Countries.

Country	Share of population without electricity access (%)	GDP per capita 2008
Burundi	97	138
Liberia	97	216
Chad	97	863
Rwanda	95	465
Central Africa Republic	95	459
Sierra Leone	95	332
Gambia	92	497
Malawi	91	313
Uganda	91	455
Niger	91	391
Burkina Faso	90	578
DR Congo	89	185
Guinea-Biassau	89	264
Tanzania	89	520
Mozambique	88	477
Kenya	85	660
Ethiopia	85	657
Lesotho	84	1248
Mali	83	468
Zambia	81	436
Madagascar	81	439
Togo	80	828
Guinea	80	1224
Benin	75	216
Cameroon	71	863

Source: UNDP Human Development Report, 2010.

for poor people has not been addressed seriously by existing financial mechanisms and financing institutions.

Lack of access to (affordable) capital in many countries is a problem that exists at every scale, from national governments and large utilities through to households. Until the 1990s, most developing countries relied on the international financial institutions, the World Bank, and regional development banks for investment in the energy sector. However, during the 1980s the World Bank promoted private investment. Countries responded by reforming the sector and initially private investors moved in, especially into the larger economies. However, not only did private funds start drying up in 2000, but most investors generally avoided additional generation capacity, which contributed to power and economic crises, as evident in Brazil (Bouille and Wamukonya, 2003; Millán, 2006; Woodhouse, 2005).

The level of success in private sector financing has been rather limited. In Africa, for example, private sector financing accounted for an average

---

18　Low, urban 100 kWh/cap and rural 50 kWh/cap; medium, urban 456 kWh/cap and rural 152 kWh/cap; high, urban 456 kWh/cap and rural 360 kWh (Bravo, 2004). See also the reference for the estimation of basic energy services for LAC.

Box 23.3 | Private Activity in Energy Reaches a Record High

In 2009, there were 139 energy projects with private participation that reached financial or contractual closure in 21 low- and middle-income countries, involving investment commitments of US$58.5 billion. In addition, energy projects implemented in 1990–2008 attracted new investment of US$10 billion, bringing a total investment commitment to the energy sector of US$68.5 billion in 2009.

Private activity, however, was concentrated in just a few countries and electricity generation projects. Brazil and India accounted for 67% of investment and 43% of new projects, and for all of the growth in private activity in 2009. Electricity generation accounted for 79% of investment and 80% of new projects. 100% of the non-electricity projects were in China, while Chinese electricity projects focused on the grid and mainly BOT or BOO systems.[19]

In East Asia and the Pacific, two countries (China and the Philippines) implemented 22 new projects that represented US$7.6 billion in investments. China had seven natural gas distribution projects and five power plant projects with a total investment of US$3.1 billion. The Philippines implemented nine power plants (mainly divestitures[20]) and a concession for the national electricity transmission company, with a total investment of US$4.5 billion.

In LAC, four countries (Brazil, Chile, Costa Rica, and Peru) implemented 43 new projects that represented US$20.5 billion in investment commitments. Of these projects, 31 were for electricity generation and 12 for electricity distribution. Most projects and investments were located in Brazil: 26 power plant projects and 11 electricity transmission projects, representing US$19.4 billion investment. In addition, investment commitments to projects implemented previously in the region came to a total of US$4.6 billion.

In South Asia,[21] four countries (Bangladesh, Bhutan, India, and Pakistan) implemented 38 new projects with a total investment commitment of US$22.4 billion. Thirty-five of these projects were power plants along with three electricity transmission lines. Most activity took place in India with 23 projects and US$21.0 billion in investment. In Pakistan, 11 power plant projects reached financial closure, eight of which were emergency rental power plants. Additional investment in previously implemented projects in the region totaled US$2.2 billion.

In sub-Saharan Africa, four countries (Ethiopia, Kenya, Liberia, and Uganda) implemented five electricity generation projects that represented US$212 million in investment. Three were emergency rental power plants in Ethiopia and Kenya, while the other two were the Kakata power plant (a waste-to-energy project) in Liberia and the Buseruka hydropower plant in Uganda.

In addition to the 139 projects that reached financial or contractual closure in 2009, at least 124 projects were awarded throughout the year, but did not reach closure by December 2009. Those projects were distributed across the regions, with 41 in East Asia and the Pacific, 19 in Europe and Central Asia, 44 in LAC, 16 in South Asia, and four in sub-Saharan Africa.

Source: World Bank Group (2010).

of just US$300 million/yr over the decade 2000–2010, against a total requirement of US$4 billion/yr. The sustainability of private sector players has been minimal, with many exiting within a few years of their entry.

Overall, private sector investment in the energy sector has accounted for 15.6% of the total private sector participation in sub-Saharan Africa from 1990 to 2006 (World Bank Group, 2010). Nearly all this investment has been concentrated on national – and hence urban – needs. Attracting the private sector into rural areas has proved to be extremely difficult (see Box 23.3 for a summary of private investment in developing countries).

The conclusion from the Private Participation in Infrastructure Database of the World Bank (World Bank Group, 2010) is that there is no evidence to date on the role that the private sector could play in improving access to modern forms of energy, particularly in rural areas. Consequently, the

---

19 BOT = build, operate, and transfer; BOO = build, own, and operate. This system, in general, means that price and quantity is guaranteed by the government, and there is no risk.

20 Divestiture means that they were not new investment, only the transfer of property from the public to the private sector.

21 South Asia is one of the regions with the lowest level of access to electricity and is, at same time (according to the data of the World Bank), the region with the lowest investment from the private sector in electricity.

task of bringing in private capital and moving from rhetoric to reality remains a major public policy challenge.

A further factor that hinders access to modern forms of energy is the high up-front cost. Subsidies have been used, but it is evident that they are not always sufficient to enable poor rural and peri-urban populations to access modern forms of energy just because subsidies address energy costs, but not capital costs related to the equipment necessary to satisfy energy services.

Electrification access is also hindered by the initial connection and operational costs. Local energy entrepreneurs face a capital access problem in trying to establish businesses to meet rural energy needs. In particular, banks have a difficult time assessing the risks of these loans.

23.4.6   Capacity, Management, and Institutional Gap

Managing the process of delivering energy services for poor people requires ample public institutional capacity, specifically at the subnational and local/community levels. These include:

- analytical ability to create district-level rural energy policies and plans that are appropriate to specific locations and are, at the same time, aligned to national energy access visions, targets, and budgetary allocations;

- ability to manage financial resources transparently and accountably;

- technical capability to guide, regulate, and train non-state implementing actors (e.g. local NGOs) to initiate, deliver, and manage energy systems, energy services, energy users, entrepreneurs, and small-scale energy financing institutions; and

- ability to collect and manage data to establish a baseline and to monitor ongoing performance.

The current reality is that the majority of public institutions in developing countries have only limited capacities to handle these tasks. Moreover, a prevailing vacuum in institutional platforms for the delivery of energy services at the local level is often transferred to the national level, which makes capacity development efforts uncoordinated and often ineffective, thereby exacerbating efforts to scale up the programs.

Furthermore, without these capacities, transaction costs and operational risks increase considerably for potential actors, particularly private actors, to invest in energy activities in rural areas. This restricts their business opportunities and, at the same time, deprives rural people of access to modern forms of energy to meet basic human needs.

Institutional, systemic, and individual capacity developments – along with reinforcement of many different stakeholders' existing

capacities – are needed if the energy system is to be instrumental in bringing about sustainability. In a broad sense, capacity refers to the ability of individuals and institutions to make and implement decisions and perform functions in an effective, efficient, and sustainable manner (UNDP, 1994). This definition has three important aspects. First, it indicates that capacity is not a passive state, but is part of a continuing process. Second, it ensures that human resources, and the way in which they are utilized, are central to capacity development. Third, it requires that the overall existing context and functions of organizations be a key consideration in designing strategies for capacity development (UNCED, 1993).

A UNDP study, related to energy access in rural areas (UNDP, 2010), report on experiences from Nepal that showed that the focus needs to be on: "(1) planning, oversight, and monitoring; (2) policies and regulations; (3) situational analysis; (4) stakeholder dialogues, communication, and community mobilization; (5) setting up and enhancing institutions; (6) training program implementers and community members; and (7) implementation and management" and remarks that "developing capacities in all these areas is essential for making the scale-up of rural energy access a reality." In addition, a key conclusion is that "Upfront public investments are needed to develop national and local capacities for scaling up rural energy services delivery, and can catalyze private financing."

The challenges require innovative answers to old and new problems. They also require a search for more flexible and pragmatic strategies, approaches, tools, instruments, and action to obtain results in a new framework. The new operating environment in which energy solutions must be found suggests a new and essential role for government in terms of its responsibilities to make markets and the energy system work to satisfy, among others, the objective of universal access to modern forms of energy.

23.4.7   Gender and Energy

Inequity along gender lines is one of the main factors that drive the establishment of gender focused programs (see Chapter 2). The issue is predominantly a phenomenon of developing countries, and the gender and energy approach is justified on the basis that women's end uses of energy is different to that of men, and that providing energy to women will improve their livelihoods.

This approach has resulted in interventions that focus more on energy than on the service, and more on the woman than on her context. As such, the technological approach (UNDP, 2004), namely improved biomass stoves, ethanol stoves, or solar homes systems, taken in isolation of the development context, may achieve only marginal results.

The challenge, and the reason that this chapter does not give a particular focus to the gender perspective, is to recognize that access to modern forms of energy (and the energy problem in general) is not a household or gender-specific problem, but a development issue related to poverty and inequity. This shift acknowledges and calls for a paradigm change toward a full understanding of the macro- rather than the micro-development framework. Such a program, rather than the project-funding approach advocated and implemented by donors, could offer a good start if planned in a down–up approach in which local sector staff are involved in structuring from the outset.

Is the problem of gender in developing countries related to low income? The gender and energy approach is justified on the basis that women use energy differently from men, and that providing energy to women will improve their livelihoods. Some have also argued that energy is a basic good, implying that women are entitled to it as much as to health and education. There are arguments against energy as a basic good, but as an important input for satisfying basic needs.

The gender and energy approach focuses on the impacts suffered by women in gathering firewood. The doubt is, though, that by contextualizing and defining the energy problems from such perspectives, the approach may have masked the real issues and misdirected resources. Energy is an input to development, but an insufficient condition for development. Wamukonya remarks, "that women suffer energy problems maybe the case. That they are experiencing these problems merely because they are women is subject to debate. While there may be traditional cultural factors tying women to certain tasks, and hence curtailing employment mobility and flexibility, improvements in income levels are particularly important in determining the relationship between energy and women. In households with higher income, women can employ men or women to procure energy and where alternative modern forms of energy carriers are available, they switch to these fuels."

The challenge and remaining doubt is if the gender and energy issue and gender equity is an energy problem or a much broader socio-cultural and economic challenge. Apparently, the approach has to have a broader view than to look only for ways to substitute biomass for cooking.

The challenge, and the reason that this chapter does not go deeper on the "energy gender perspective," is to recognize that access to modern forms of energy (and the energy problem, generally) is not a household or gender-specific problem, but a development issue that is related to poverty and inequity.

This shift acknowledges and calls for a paradigm change towards a full understanding of the macro rather than the micro development framework. The program rather than the project funding approach, being advocated and implemented by donors, could offer a good start if planned in a down-up approach where local sector staff are involved in structuring from the outset.

23.4.8 Climate Change, Green Economy, and Poverty

For billions of people struggling with poverty, access to affordable energy services is of higher priority than climate change. Evidence suggests that increasing energy access to poor people would entail a small increase in the level of emissions[22].

It is expected that additional electricity will be, partially, centralized generation, partially mini-grid solutions and the remaining by isolated off-grid solutions. In the case of mini-grid and isolated off-grid, the majority should be provided by renewable.

Given that the priority objective is poverty alleviation through access to modern forms of energy, it would be more useful to look for synergies and convergence with global objectives of climate change and clean energy. Looking for a convergence and win-win actions in energy access, climate change, and poverty alleviation, GNESD has summarized some key findings in policy papers. These include:

- diversifying energy generation sources, with a wider mix of energy sources;

- promoting proven renewable energy technologies for electricity generation; and

- setting renewable energy targets in the energy mix.

Such measures could be a major contribution to reducing vulnerability to climate change and at the same time improve access to energy.

23.4.9 Decision Making under Uncertainty

Despite decades of rural energy programs, interventions, and research on rural energy, a number of gaps remain in our understanding of the dynamics of energy choice of poor households and the welfare impacts of access to modern forms of energy. This has made it more difficult to create sound public policy and to mobilize efforts that sufficiently and appropriately address the problem.

The report by UNDP/WHO (2009) remarks that "understanding what type of energy carriers and end use are available, who uses them, how much they cost, and the benefits they provide to users, are factors to consider when assessing energy access." As mentioned before, to solve such uncertainties a clear diagnosis is needed.

---

22 World Energy Outlook 2011 devoted a special chapter named "Energy for All: financing access for the poor." According to such report "achieving the Energy for All Case requires an increase in global electricity generation of 2.5% (around 840 Twh)…" "in 2030, CO_2 emissions in the Energy for All Case …are 0.7% higher than in the Baseline Scenario." The figures include LPG to replace Biomass in cookstoves.

There is a lack of information, especially based on field studies, on the quality and quantity of energy used and provided, energy end-use appliances and equipment, and the services they provide, as well as the socio-economic profile of energy users and energy affordability. Examples of this problem include:

- Costs and benefits of modern cooking fuels. Many programs and projects are justified on the basis that the benefits outweigh the costs. For clean cooking fuels, the costs include all the capital and programmatic costs, while benefits range from improved health outcomes that impact household finances to the impact on the health-care system itself. A recent set of reports and guidelines from UNDP/WHO (2009) has helped clarify how to estimate these costs and benefits, and has provided a global set of estimates. But much work remains in refining the methods and determining these values in particular circumstances.

- Ability versus willingness to pay for energy services. It is common to find projects and programs based on consumers' willingness to pay. However, the outcomes tend to demonstrate that this is a misinformed approach, as ability and willingness differ in reality. For example, the ability of poor people to pay for SHS is often based on theoretical calculations of the savings they would make by not buying kerosene. Yet reality shows that outlays on purchasing kerosene are made in small amounts and income restrictions act as a barrier to making periodically structured payments toward SHS (Green Nine, 2004).

- Opportunity costs of biomass collection. The time spent by households in collecting biomass is assumed to have an opportunity cost, because that time could be used on other activities, such as income generation or education. However, to determine the value of that opportunity cost and how it plays into households' decisions is still an active area of inquiry (Campbell et al., 1997; Arnold et al., 2003, 2006).

Therefore, estimates of benefit and potential penetration are based on theoretical or on controlled experiments and not on reality, or from taking a social[23] instead of an economic approach as the framework for estimating benefits. In many cases, inadequate knowledge and diagnosis results in poor estimations.

23.4.10 Oil Price Volatility

Crude oil prices behave much as any other commodity. They experience wide price swings in times of shortage or oversupply, through political instability, and for many other reasons (see Chapter 5). The crude-oil price cycle may extend over several years in response to changes in demand, as well as Organization of the Petroleum Exporting Countries (OPEC) and non-OPEC supply.

Since 1973, crude oil prices have swung wildly. They reached levels that few predicted and then dropped precipitously, before rising again and falling in response to global economic crises. This has a direct impact on low-income energy services.

Most rural and peri-urban populations rely heavily on kerosene or LPG. Indirectly, the prices of other goods they depend on are influenced by oil prices. Developing countries are notably more dependent on imported oil and oil products. Many countries subsidize oil to keep the products affordable.

Also, escalating and unpredictable petroleum prices have placed many countries in a dilemma on how to protect the poor communities. The costs of direct subsidies are, in many cases, unsustainable. The recent instability in the price of petroleum fuels has, in some cases, actually caused households to switch back to traditional fuels.

Oil-exporting countries like Venezuela, where the market is controlled by the public utility Petróleos de Venezuela, S.A, do not fix domestic prices and consider the opportunity cost based on international prices. A similar situation is given in Ecuador, where Petroecuador controls the domestic market.

In Nigeria, the structural reforms implemented during the 1990s, including privatization of the state oil company, increased deregulation of petroleum prices, and domestic crude-oil allocation to the Nigerian National Petroleum Corporation would be paid for at export parity with immediate effect. The objective was to attract investment from international oil companies and improve profitability. Also, using the case of Nigeria and analyzing the impact of oil-price volatility, Moser et al. (1997) arrived at the following conclusion:

> "Inflation rate depend on shocks to output and the real exchange rates. However, the findings demonstrated that fluctuations in oil prices do substantially affect the real exchange rates in Nigeria. Also, it was found out that it is not the oil price itself but rather its manifestation in real exchange rates and money supply that affects the fluctuations of aggregate economic activity proxy, the GDP. Thus, we conclude that oil price shock is an important determinant of real exchange rates and in the long run money supply, while money supply rather than oil price shocks that affects output growth in Nigeria."

This is another example of the impact of the policies of liberalization, privatization, and deregulation implemented during the 1990s (Moser et al., 1997; Onayemi, 2003; Olomola and Adejumo, 2006).

Some countries, such as Chile, China, and Indonesia, have used direct cash transfers to cushion poor households against petroleum price hikes. However, most developing countries can ill afford such measures.

---

23 Considering the benefit for the economic system as a whole, but also the direct benefit that the target population involved in the project will receive (economic benefits).

23.4.11   Final Remarks on Challenges

It is neither our intention nor possible to cover all the challenges and barriers to access to modern forms of energy.[24] National circumstances, specific conditions, drivers related to the target population, energy chains addressed (electricity or oil products and natural gas), the organization of the energy system (public, private, mix), institutional structure (policy authorities, regulatory bodies), constitutional aspects (property rights on natural resources), and availability of energy sources are just some of the challenges.

The initial challenge is to avoid defining a policy approach based on ideological preconceptions[25] or preconditions, because they leave aside or ignore potential solutions. Strategies, instruments, measures, and actions need to be the consequence and result of adequate analysis, and offer a pragmatic path toward affordable access to modern forms of energy.

It is important to be as accurate as possible in estimating costs. An underestimate gives a false notion of what is possible within a given period. Estimates and, consequently, the achievement of targets will depend on proper understanding of the relationships between the investment needed, the economic capacity of the country, the financial instruments, and the capacity of the target population to afford the costs.

A third set of challenges is found in the capacity to design and implement public policies, along with a lack of information about the energy services to be addressed in terms of quantity, quality, location, time, logic, and means.

Rapid urbanization, both historical and current, poses a further challenge in agreeing where, how, and what type of infrastructure should be developed. There are sociocultural issues related to the behavior of rural populations moving to peri-urban areas and maintaining their rural customs to satisfy energy services. The increasing and different needs of urban and rural environments, along with the prerogative to keep rural populations in rural areas, are among the challenges that would need to be part of any public policy process, along with the objective of meeting human needs.

Robust decision making in public policies is another key challenge in the quest for feasible measures and actions. The volatility of oil prices,[26]

however, brings uncertainty to the equation. Oil products are immediate, feasible, and natural substitutes for biomass and other caloric energy services. In many countries, oil is the main source of energy for power generation. Ethanol and biogas may be considered as options, but massive development in the production of such energy sources and associated appliances must be put in place immediately if they are to represent a solution for the 2.7 billion people using solid biomass as their main source of energy for cooking.

23.5   Introduction Regional Analysis

Although the aim of GEA, and of this chapter in particular, is to approach the global problem of access to modern sources of energy, the specific situations and realities of each region and country cannot be ignored. The magnitude and characteristics of the problem, the underlying reasons, the national and regional contexts, the current and historical circumstances that have influenced the situation, the policies and strategies that have contributed to solving problems (or, in some cases, to aggravating them), the socioeconomic structures, and the characteristics of energy systems, to cite only some of the many dimensions, necessarily imply that suggestions or recommendations would need to take account of the different realities and potentials. It is important to avoid the mistakes of the recent past, such as promoting particular institutional models without appropriate consideration of the peculiarities of the individual environmental, socioeconomic, and energy systems.

Section 23.4 identifies some of the principal challenges and barriers to meeting the objective of universal access to energy by 2030. It is important to prioritize a full understanding of the unique characteristics of different regions and countries.

The lack of access to modern sources of energy is dramatic in most sub-Saharan African countries and in much of South Asia and other Pacific Asia (Table 23.6). In these regions, the national and regional response capacities are different from those in LAC, where more

Table 23.6 | Electricity and human development (2008).

Concept*	Africa	Asia	LAC
Average kWh/capita	540	847	1806
Average electrification rate (%)	29	61	84
HDI high (% of population)	4	7	38
HDI medium (% of population)	48	93	62
HDI low (% of population)	48	–	–
Average GDP/capita (PPP-US$)	3101	4161	7859
GDP/capita ratio†	19	11	8

* HDI, Human Development Index; PPP, purchasing power parity.
† Relation between the GDP per capita of the richest and poorest countries in the region.

Source: Based on information from UNDP, 2011 and World Bank Group, 2011.

---

24 The energy systems have multiple dimensions as part of their own nature. Environmental, national constitution, and legal frameworks, sociocultural, economic, strategic, institutional, political, human health, security, technological, temporal, and energy reserves are the key dimensions addressed in different ways in different countries.

25 Such as "a free market is the best way for an efficient allocation of resources" or "private sector contribution is the only way to address access to modern forms of energy."

26 In many cases oil prices act as reference prices for the other energy sources (renewable and non-renewable).

Box 23.4 | Access to Modern Forms of Energy

"Access to energy services is still low in developing countries and this lack of access disproportionately affects the least-developed countries (LDCs) and sub-Saharan Africa.

- Three billion people – almost half of humanity – still relies on solid fuels: traditional biomass and coal. In LDCs and sub-Saharan Africa, more than 80% of people primarily rely on solid fuels for cooking, compared to 56% of people in developing countries as a whole.

- Two million deaths annually are associated with the indoor burning of solid fuels in unventilated kitchens. Some 44% of these deaths are children; and among adult deaths 60% are women. In LDCs and sub-Saharan Africa more than 50% of all deaths from pneumonia in children under five years and chronic lung disease and lung cancer in adults over 30 years can be attributed to solid fuel use.

- Access to improved cooking stoves is also very limited. In LDCs and sub-Saharan Africa, only 7% of people who rely on solid fuels use improved cooking stoves to help reduce indoor smoke, compared to 27% of people in developing countries as a whole.

- One-and-a-half billion people are still living in darkness – over 80% of them in South Asia and sub-Saharan Africa. More than 70% of people in LDCs and sub-Saharan Africa lack access to electricity, compared to 28% in developing countries as a whole."

Source: UNDP/WHO, 2009.

favorable macroeconomic conditions, development levels, maturity of energy systems, and contexts of regional cooperation offer a better framework in which to implement and succeed with oriented policies.

Access to electricity also varies dramatically among countries in the same region. For example, in LAC, 62% do not have access in Haiti, but only 2% lack access in Brazil. In sub-Saharan Africa, in countries such as Chad, Liberia, and Burundi, more than 95% of people lack electricity access, while only 25% are without access in South Africa, and less than 1% in Mauritius (Box 23.4).

Access to modern fuels for cooking, meanwhile, also varies dramatically among developing countries in the same region. In Asia, for instance, less than 10% of people in Bangladesh have access to modern fuels, but access is almost universal in Malaysia. In sub-Saharan Africa, less than 1% of people in Burundi, Liberia, Mali, Rwanda, Somalia, and Uganda have access to modern fuels, but 83% of people in South Africa have access.

There are significant differences in the availability, control, and management of energy sources among regions. The roles of the public and private sectors differ substantially within the same regions. Systemic, institutional, and individual capacities to implement policies and strategies are not the same. A long history of intervention by public utilities and governments in many LAC countries, for example, implies a different culture and approach to energy issues than in countries without such experience.

23.6 Africa Review: Successes, Failures, and Proposals

23.6.1 Introduction

In Africa, access to energy services varies greatly between regions, between rich and poor, and between rural and urban populations (see Chapter 19). North African countries have achieved universal access to both modern cooking fuels and electricity, with the exception of Mauritania. In sub-Saharan Africa, the situation is very different and only 17% of the population has access to modern fuels. This ranges from 0.3% in Burundi to 96% in Mauritius. At 26%, sub-Saharan Africa is the region with the lowest levels of electricity access (UNDP/WHO, 2009).

23.6.2 Access

Of the sub-Saharan African population, 26% have access to electricity, but only 6% use electricity for cooking because they cannot afford the relatively high electricity tariffs. In fact, the average power tariff of US$0.13/kWh is around twice that found in other parts of the developing world, and almost on par with that in the countries of the OECD.

To alleviate power shortages many sub-Saharan countries rely on short-term leases of diesel generators for emergency power, which leads to high average electricity costs of more than US$0.20/kWh (Eberhard and Shkaratan, 2010).

Access to electricity reflects the wide rural–urban and income divide. In sub-Saharan Africa, 71% of urban and 13% of rural residents have access, and only 4% of the lowest income quintile, as compared to 74% of the highest, have access (Banerjee et al., 2009).

From 1997 to 2007, sub-Saharan African countries invested in their infrastructures and the economy grew at about 5% per year while the power sector was growing at only 3% (Foster and Briceno-Garmendia, 2010). In 2010 sub-Saharan Africa spent US$45 billion on the power sector (just half the amount required to catch up with other developing areas) and US$30 billion of annual spending is domestically financed from the pockets of African taxpayers and consumers (Foster and Briceno-Garmendia, 2010).

The electricity generation capacity of the region with a population of 800 million is only 68 Gigawatts (GW), comparable to that of Spain with a population of 45 million, and when South Africa is not counted the total amounts to only 28 GW (Eberhard and Shkaratan, 2010). Access varies across regions and countries. North African countries, except Mauritania, have achieved universal electricity access, but in sub-Saharan Africa 561 million people, equal to 74% of the population, have no access to electricity, a figure that rises to 89% in rural areas (UNDP/WHO, 2009). The per capita consumption levels are only 457 kWh annually, on average, which reduces to 124 kWh without South Africa, compared to 1155 kWh in the developing world and 10,198 kWh in high-income countries (Eberhard and Shkaratan, 2010). When present electrification rates and population growth rates are projected to 2030, more people (654 million) will be without electricity in 2030 than in 2009 (587 million) (OECD/IEA, 2010).

Sub-Saharan Africa has adequate modern forms of energy resources (hydropower, oil and gas) for its population, but they are largely unused and 83% of the population still cook with solid biomass on open fires (UNDP/WHO, 2009). Most of the electricity generated in Africa (76%) is from thermal and/or fossil fuels (particularly from coal and oil), 22% is from hydropower, and the rest is from other sources – nuclear (South Africa) and geothermal (East Africa). All oil and gas producing countries in sub-Saharan Africa export fossil fuels. For example, in Mozambique 84% of the population still cook with solid biomass and only 12% have access to electricity, yet at the same time the country is exporting gas and electricity. Low population densities and dispersed settlement patterns in rural areas make affordable access very difficult. In Nigeria, a major oil-exporting country, 75% of the population still cook with solid biomass and have no access to modern cooking fuels.

Africa's energy situation is paradoxical in that the continent desperately needs energy for economic growth and poverty reduction, yet it is a net exporter of energy.

In addition to access to energy, both energy security and regional cooperation are among the key energy priorities in all reviewed national policy papers. Also, there is a growing interest in biofuels production and trade as an alternative option to fossil sources of energy. Efforts toward access to electricity and cooking fuels might be impacted by an additional burden of more expensive imported petrol and its derived products.

Table 23.7 | Number of people who rely on fuelwood and charcoal for cooking in assessed Sub-Saharan African countries.

Country	Population in 2006 (millions)	People who rely on fuelwood and charcoal for cooking	
		(millions)	%
Angola	16.6	15.7	95
Cameroon	18.2	14.2	78
Chad	10.5	10.2	97
Congo	3.7	2.9	80
Côte d'Ivoire	18.9	14.7	78
Equatorial Guinea	0.5	0.3	59
Gabon	1.3	0.4	33
Mozambique	21.0	16.9	80
Nigeria	144.7	93.8	65
Sudan	37.7	35.2	93
Total	273.1	204.0	75

Source: IEA, 2008.

To increase access to energy services, governments have to improve the performance of the sector in areas of governance, infrastructure, access to finance, and increasing regional trade. Also, income levels of both the rural and urban poor have to rise to make the transition from solid biomass to modern fuels and their appliances affordable.

23.6.2.1   West Africa

West African[27] countries are endowed with very significant energy potentials (oil, natural gas, uranium, hydropower, coal, renewable energy). There are major oil and gas reserves in Nigeria, Cote d'Ivoire, and Ghana. The most important reserves of oil and gas are concentrated in Nigeria. Hydropower potentials are important in Nigeria, Guinea, Ghana, Liberia, Cote d'Ivoire, and Mali. The main sources are the Niger, Senegal, and the Volta Rivers. In addition, solar resources are available and significant throughout the region and all year long.

In West Africa, electricity consumption is among the lowest in the world (on average 139 kWh/yr/capita compared with an average of 1020 kWh/yr/capita in North Africa) while the world's average is around

---

27 Africa is generally divided into five subregions: North Africa, West Africa, Central Africa, East Africa, and Southern Africa. The regions have formed economic communities and some countries are members of more than one regional community.

2400 kWh/yr/capita.[28] This low consumption is mainly a result of the low access to electricity services, especially in rural areas.

Annually, per capita electricity consumption in West Africa is slightly higher than the sub-Saharan African average. Cape Verde, Ghana, and Cote d'Ivoire have the highest electricity-access levels. They have implemented energy policies to improve access at affordable prices. The highest levels of access resulted from public policies to improve access to electricity while ensuring affordable pricing: the lifeline tariff in Ghana and subsidies in Cote d'Ivoire.

Lessons learned show that social tariff, social electrification, moderate residential tariffs, and subsidized connection were key instruments for their success. Government subsidies to LPG have been a key incentive for a large diffusion of the use of this product for cooking. Different mechanisms were used, such as cross-subsidies, specific funds, funds from the general treasury, and others.

23.6.2.2   Central Africa

In Central Africa, energy potentials are large and diverse. After Nigeria, the region has the most important oil producers in Africa – Angola, Equatorial Guinea, Gabon, Congo, and Chad. The oil reserves in Central Africa are estimated to amount to some 11.4 billion barrels representing 11% of Africa's reserves. The gas reserves are estimated to be more than 430 billion m³ (3% of Africa's reserves), and are located in Cameroon (37% of Central Africa's reserves), Congo (23%), Rwanda (12.7%), Angola (10.6%), Equatorial Guinea (8.5%), and Gabon (7.8%). However, this resource remains underexploited (CEMAC, 2006).

The region also has very important hydropower resources (1000 TWh), which is around 60% of Africa's potential. This potential is located mainly in the Equatorial zone: DR Congo ranks first with 100 GW, Cameroon is second, followed by Congo, Gabon, and Equatorial Guinea. Elsewhere in Central Africa, Rwanda has geothermal resources and there are significant methane deposits in Lake Kivu on the border between Rwanda and DR Congo.

Although the region is richly endowed with large modern forms of energy resources, wood, charcoal, and forest residues make up 70–90% of primary energy supply, and up to 95% of household energy use in some countries. A large majority of the Economic Community of Central African States (CEMAC/ECCAS/CEEAC) population uses wood energy harvested without regard for its sustainability, which is burned in unhealthy conditions.

With the exception of Gabon, where 68% of the population uses it for cooking, the use of LPG is still limited to urban areas of most of sub-Saharan Africa. In fact, this product (bottled in small to medium canisters for households and small enterprises use) is a better and cleaner fuel than wood and charcoal used for the same purposes.

Several electricity companies that have been unable to invest and keep up with growth in the demand within the localities they serve, or that have suffered conflict-related damages, now find themselves with inadequate or obsolete production and transport facilities. In the best-case scenarios, private sector companies that provide a good level of service are not able to be the driving force behind hydropower investments, because of the high investment per unit of capacity, the long-term return on the investment, and the lower rate of return of the investment.

Peri-urban electrification is, in most cases, below standard and, with the exception of Cameroon and Gabon, rural electrification has not been pursued on a significant scale. No power company has developed a pro-poor commercial culture. The overall rate of household electrification in CEMAC is less than 15%, according to Africa Development Indicators (World Bank Group, 2011).

23.6.3   The Energy Dimension in the Poverty-reduction Strategies

In general terms, a sectoral approach has been used to include the energy dimension in poverty-reduction strategy papers (PRSPs).[29] Energy has been treated as a stand-alone sector and from a supply-side perspective (power-generation systems, biomass energy production and management, electricity-grid extension, petroleum exploration), mainly under the aspect related to the macroeconomic framework enhancement or infrastructure development.

The articulation of the energy dimension with the other main axes dedicated to poverty alleviation (social and income-generating activities, human development, access to basic social services for poor and vulnerable groups, rural development, and gender equity) was not seen as very significant.

However, an awareness of this gap in considering energy for poverty alleviation as arisen and the second generation of the PRSPs engaged an interactive multistakeholders dialogue to integrate the energy dimension into the poverty-alleviation options.

Table 23.8 summarizes the energy options as considered in the PRSPs documents elaborated by selected countries in Africa (Benin, Burkina Faso, Guinea RD, Mali, Niger, Rwanda, and Senegal).

---

28 Own estimation based on information from UNDP/WHO (2009), UNDP Human Development Report (2010) and Niez (2009).

29 PRSPs: country-driven approaches to tackling poverty, which have been developed through nationwide consultations with stakeholders.

Table 23.8 | Main axes in the PRSPs.

#	Main axes in the PRSPs	Associated energy options	Benin	Burkina Faso	RD Guinea	Mali	Niger	Rwanda	Senegal
1	Macroeconomic framework and wealth creation	Power capacities and generation	X	X	X		X	X	X
		Grid extension	✓		✓			✓	✓
		Oil/gas/peat exploration	✓					✓	✓
		Energy infrastructure development	✓					✓	✓
		Private sector involvement		✓	✓			✓	✓
		Investment and financial mechanisms						✓	✓
		Energy sources diversification							✓
		Access to domestic fuels							✓
		Rural electrification						✓	✓
		Energy sector restructuring and private sector involvement							
		Power sector management		✓	✓				
2	Human and environmental capital		X		X			X	X
		Rural electrification (including renewable energy options)	✓						
3	Good governance and institutional capacities			X	X	X	X	X	
4	Job creation and development of revenue-generating activities			X	X	X	X		
		Rural electrification (including renewable energy options)		✓		X			X
		Forest management and introduction of alternative energies		✓					
5	Participative implementation		X						X
6	Rural development						X	X	
7	Infrastructure development					X	X	X	
		Improvement of access to energy					✓		
		Energy infrastructure development and rehabilitation				✓			
		Sub-regional cooperation				✓			
		Awareness raising on alternative energy forms				✓			
		Tax exempts for renewable energy				✓			
		Forests management				✓			
		Privatization				✓			
8	Private sector promotion					X	X	X	
9	Urban development						X		
10	Access to basic social services for poor and vulnerable groups and gender equity			X	X	X	X		X
		Encouragement of renewable energy uses				✓			
		Rural electrification			✓				

23.6.4   Assessment of National Energy Policies/strategies in Selected African Countries

23.6.4.1   West and Central Africa

The assessed policies[30] relate to those of selected countries such as Burkina Faso, Cote d'Ivoire, Central Africa Republic, Ghana, Liberia, Mali, Niger, Senegal, Sierra Leone, and Togo. Nearly all of these identify access to energy as an objective or a priority in their energy policies. Table 23.9 summarizes the objectives and priorities and provides common threads of the energy policies in some of the selected countries. But rarely were objectives and expected results accompanied by a set of strategies, measures, and actions to achieve the targets.

Access to energy, energy security, and regional cooperation are the key energy priorities in all the reviewed national policy papers. In some countries, efforts have been made to speed up access to clean energy forms through dedicated programs and projects. A successful initiative from Senegal in West Africa is given in Box 23.5.

23.6.4.2   East and Southern Africa

In East and Southern Africa access to energy services varies more widely than that in West and Central Africa (see Chapter 19). Countries in East and Southern Africa can be divided into three groups. The first are those that have achieved or have definite policies and targets for universal access (e.g., Mauritius, South Africa). The second group consists of countries well on their way to having and implementing policies, and actively pursuing targets for greater access (e.g., Botswana, Kenya, and Zimbabwe). The third group includes countries with very low access rates and policies that do not seem to promise greater access rates in the near future (e.g., Burundi, Malawi).

Access to finance is a major barrier to extending energy services. In East Africa, Kenya addressed the problem and successfully raised finance to improve electricity generation. In 2006 the Kenya Electricity Company raised substantial investments through a public offer (PO) on the Nairobi Stock Exchange. The PO was an unexpected success and the electricity company exceeded the targeted amount (over US$112 million) and the share offer was oversubscribed by nearly double this amount (Bhagavan, 1999).

South Africa had the political will, the financial resources, and the capacity to implement the National Electrification Programme and increase electricity access from 36% in 1995 to 75% in 2007 (Niez, 2009). The connections to poor households are very highly subsidized, which makes access affordable for poor people. Every household in an area

is provided with electricity supply, not only those customers who apply and pay, which significantly reduces cost. Other measures, such as pre-payment meters, further reduced cost. Many people could not benefit from the huge investments in electricity supply because they could not afford to use it. The government then introduced a lifeline tariff of 50 kWh free of charge for poor customers. The Free Basic Alternative Energy Tariff subsidizes energy sources such as kerosene, LPG, and renewable energy, particularly in areas not connected to the grid, but this tariff is not or is poorly implemented (Box 23.6).

In Botswana, the Rural Electrification Collective Scheme (RCS) started in 1988 is an example of adjusting conditions of supply when the initial policy does not achieve its objective. The government extends the grid to the village and customers pay for the extension to their houses. Initially uptake was very slow and it took over ten years to adapt the scheme by gradually easing payment conditions, but not the total amount, until potential customers were able to afford the smaller installments over a longer period and then electricity access substantially increased at full cost recovery.

In Zimbabwe, low take-up rates threatened the minimal returns on investment in rural electrification. The Rural Electrification Agency (REA) established in 2002 supports income-generating activities for small and medium enterprises (SMEs) in order to increase electricity demand in rural areas and stimulate small-scale commercial and industrial development. REA provides loans and delivers electrical machinery ordered by SMEs. The Rural Electrification Programme is funded by a levy on all electricity bills of 6% (in 2007) as well as government fiscal allocations. Once small enterprises had access to electricity demand went up, and the variety and use of electric machinery increased and, at the same time, the use of stand-alone generators declined.

The third group is made up of low-income countries that cannot afford the necessary additional investments to accelerate greatly their energy access rates and will have to raise more finance from external sources, move their access targets from 10 to 20 or 30 years, or use alternative low-cost technologies serving more people in the short to medium term.

In the GEA solid biomass, and in particular woodfuels, are not considered as a modern energy form. However, 80% of the population still depends on woodfuels for their energy needs in Africa. To bridge the energy gap until modern energy forms are available and affordable, there is a need to relook at the traditional woodfuels sector with a view to modernize access, use, and supply. The wood and charcoal sector must be re-evaluated because it is an economic resource from which millions of people derive jobs and income. In Malawi and Burundi – two land-locked countries – the woodfuel market contributes about 2% to GDP. Community-based woodfuel production (CBWP) has been introduced in some African countries (Madagascar, Mali, Senegal) and has proved to be a successful strategy to decentralize forest management from exclusive government control to the local level, empowering communities to

---

30 The national energy policies/strategies are reviewed based on the existing/available policy papers or other available documents that mention the principles of national energy policies.

Table 23.9 | Main focuses of national energy policies.

	Burkina Faso	Central African Republic	Ghana	Mali	Niger	Senegal	Sierra Leone	Togo
Main focuses of national energy policies								
Increasing investment and infrastructure in energy			✓			✓		✓
Enhancing security of supply and diversification of energy sources	✓	✓	✓	✓	✓	✓	✓	✓
Promoting renewable energy and energy efficiency	✓	✓	✓	✓	✓	✓	✓	✓
Managing the environment			✓	✓		✓		✓
Improving access to energy	✓	✓	✓	✓	✓	✓	✓	✓
Promoting the institutional framework	✓		✓	✓		✓	✓	✓
Enhancing energy to alleviate poverty and promote rural development	✓		✓			✓	✓	
Gender and energy								
Other energy policy focuses								
Enhancing research and development			✓	✓				
Promoting employment			✓			✓		
Exploring oil/petroleum					✓			
Developing subregional, regional, and international cooperation			✓	✓	✓	✓		✓
Capacities development		✓		✓				✓

Source: Compiled by the author, based on following documents:

Burkina Faso: Energy Policy (Source: Energy Sector Development policy paper).

Central African Republic: Energy Policy (Source: Energy Policy National Framework for Poverty Alleviation, 2003).

Ghana: Strategic National Energy Plan and Policy 2005–2020 (Source: Energy Commission of Ghana, 2005).

Mali: Energy Policy (Source: Ministry of Mines, Energy and Water, 2006).

Niger: Energy Priorities and Objectives (Source: UEMOA-BERP, 2007).

Senegal: Energy Policy (Source: Ministry of Energy, 2008).

Sierra Leone: Energy Policy (Source: Ministry of Energy and Power, 2004).

Togo: Energy Policy (Source: Ministry of Energy and Water, 2006).

Box 23.5 | Senegal LPG National Program

A national program to promote LPG use in Senegalese households was implemented in 1974. This program was developed to attenuate the effects of drought and deforestation. The program's goal was to increase LPG consumption and decrease the reliance on biomass by the most vulnerable populations. Initially, a cooking stove with an attached 2.7 kg LPG cylinder was promoted. Then, in 1983, a more solid cooking stove with a 6 kg LPG cylinder better adapted to the cooking habits and income levels was also subsidized. In addition, the Senegalese government exempted all LPG-related equipment from customs duty, and eventually subsidized the LPG itself in 1976. This program, which focused on the distribution of 2.7 and 6 kg LPG bottles (called popular gas), resulted in an annual increase in LPG consumption from 3000 tons in 1974 to nearly 140,000 tons in 2005. This represents an average annual growth rate of 10–12%. The transition from biomass to LPG was achieved gradually, particularly in urban areas. It resulted in a new domestic fuel consumption profile in urban areas characterized by the use of LPG and charcoal. The key lesson learned from the Senegalese LPG program is that the political will and adequately targeted measures are necessary to achieve large-scale access to modern forms of energy.

Source: ENDA, 2006.

Box 23.6 | Electricity Access for Poor People: a Study of South Africa and Zimbabwe, Key Findings

In both countries examined, primary data on the electrification of poor people are almost non-existent – and this forms a key limitation of this study. Although, for instance, the National Electricity Regulator in South Africa keeps track of rural electrification levels, the data are not categorized by poor and non-poor users. Because of these data limitations, the findings and conclusions of this study should not be regarded as fully conclusive.

The comparisons between South Africa and Zimbabwe indicate that the policy environment to encourage and enable the provision of energy services for poor people needs to be designed for the specific needs of the country. The reforms undertaken to enhance access to electricity realized positive outcomes, particularly under the grid-based electrification programs. In South Africa, national electrification levels more than doubled from 34% to 70% between 1994 and 2001, as they also did in Zimbabwe, growing from 20% to 42% between 1980 and 2001. The Government-funded electrification program in South Africa took a shorter time and reached a much larger proportion of the population than the program in Zimbabwe.

In an attempt to reach poor people in remote locations, both countries focused on the establishment of off-grid programs which were mostly centered on SHS powered by solar PV technology. Even if all the operational and financial problems are resolved, off-grid programs based on solar PV home systems require an urgent review as they are focused on lighting, which is not the highest priority for poor people (Davidson and Sokona, 2002). Designing energy programs for poor people must address household-cooking and water-heating needs as a priority over lighting, which would, for example, reduce dependence on fuelwood. Similar priority would need to be attached to the provision of electricity for motive power, which would support small-scale rural industries for income-generating activities, and other services such as water pumping.

In both countries, the reforms have attempted to make electricity affordable to poor people. South Africa has introduced special subsidies on electricity consumption, including some free electricity. Zimbabwe has established a rural electrification fund to subsidize rural electrification schemes.

The electricity basic services support tariff (EBSST) subsidy in South Africa, which supplies 20–50 kWh of free electricity to poor people in selected areas, seems to have realized direct benefits for poor people. It had some positive impact on poverty alleviation as it reduced electricity expenditure. The reforms in both countries have ensured the protection of funds for financing the electrification of poor people by requiring transparency and accountability, albeit in different ways. In South Africa, the National Electricity Regulator (NER) aggressively monitors and makes public the progress of the National Electrification Programme through the NER's annual reports. In

Zimbabwe, the Performance Improvement Programme includes explicit rural and urban electrification targets that the utility is obliged to meet.

In order to meet the electrification challenge in rural areas, a diverse set of technical and institutional approaches will be needed – covering large-scale grid-connected extensions and new developments, together with smaller-scale distributed energy systems using both conventional and renewable energy sources.

Strong institutions are the backbone of an efficient and effective energy sector. National policies that create the right enabling environment for investment and business-led market growth are going to be essential.

The paper recommends the following for further investigations:

- Income-differentiated electrification, both current and trend data.

- More detailed understanding, through participatory approaches, of the associated social and economic characteristics of energy-consumption patterns of poor people.

- Innovative technological approaches to reduce connection fees and distribution costs, and so reduce the overall cost of increasing access to electricity to poor people.

- Further assessment and review of the use of renewable energy, especially SHS, as a poverty-alleviation tool in off-grid electricity supply.

- Exploration of public–private management schemes that could benefit poor people. This should include an assessment of the role of independent power producers and energy service providers.

Source: Davidson and Mwakasonda, 2004.

manage their forest resources sustainably (de Miranda et al., 2010). The CBWP approach has also been successful in promoting forest rehabilitation and reducing deforestation rates, creating long-term ecological benefits. In countries where the dependence on woodfuels is very high CBWP could be part of national energy policy.

At least as a transition phase, there is an urgent need to disseminate modern woodfuel technologies more widely as part of the access to energy services agenda. Traditionally, people use woodfuels in open fires with major negative impacts on health. Modern woodfuel technologies, including gasification, save woodfuels, minimize harmful emissions, and can make sustainable use of Africa's forest resources. In addition, efficient modern charcoal kilns should be strongly supported and disseminated to improve productivity and reduce waste of forest resources.

23.6.5   Energy Strategies of Africa and Sub-regional Bodies

The New Partnership for Africa's Development (NEPAD), adopted at the Organization of African Union in Lusaka, Zambia, in July 2001,

recognizes the important role that energy plays in the development process of African countries, not only as a domestic necessity, but also as a factor of production whose cost directly affects prices of goods and services, and the competitiveness of enterprises (Zhou, 2003). In this regard, NEPAD has identified actions that need to be taken to address the critical barriers to universal access to modern energy in Africa.

NEPAD set a target for providing access to electricity for 35% of the population of Africa by 2015 and modern forms of energy for cooking, such as improved stoves or fuels like LPG, to half the population. Since then, all regional organizations have developed strategies or action plans.[31]

31 In 2002, NEPAD proposed that regional organizations, such as ECOWAS, CEMAC, or the EAC, play a key role in increasing the access to modern forms of services. NEPAD set a target for providing access to electricity for 35% of the population of Africa by 2015 and modern forms of for cooking, such as improved stoves or fuels like LPG to half the population. Since then, most of these regional organizations have developed strategies or action plans, such as the ECOWAS/UEMOA White Paper on Energy Access, the EAC Energy Access Strategy, and the CEMAC Action Plan for Promotion of Energy Access (all adopted in 2006) (Holland and Mayer-Tasch, 2007).

23.6.5.1   West and Central Africa

In West Africa, ECOWAS and the West African Power Pool (ECOWAS/ UEMOA, 2006) formulated policies to enable at least half of the rural and peri-urban population to gain access to energy services by 2015. This would give access to 36 million additional households and over 49,000 additional localities. The specific objectives are to provide access to:

- 100% of urban and peri-urban areas; in rough terms, this means doubling the current access rate.

- 36% of rural populations – where the rate in the least densely populated countries is just 1%, and for the more advanced countries is 10%.

- Moreover, 60% of the rural population will live in a locality equipped with modern basic social services – healthcare, education, drinking water, communication, and lighting. This will be achieved through either decentralized electrical facilities or grid extensions. The objective entails increasing current levels threefold.

In Central Africa the CEMAC Action Plan (CEMAC, 2006) is geared primarily to rural and peri-urban zones, and energy access development will be balanced through:

- Strong LPG dissemination in peri-urban areas (70%), increased usage in secondary towns (50%), with use rates decreasing from 35–10% depending on the size of the inhabited area.[32]

- Usage by other households of improved stoves with chimneys (proportion of households increases from urban to rural areas).

- Supplying 50% of the peri-urban population with electricity through the power grid.

- Providing 35% of rural households with grid electricity or solar kits.

- Installing a corresponding infrastructure in non-electrified villages, giving 56% of rural inhabitants access to power supplies.

A set of 11 strategic activities are included and consist of:

- coordinated development of hydropower;

- rational use of surplus biomass;

- waste from agroindustrial units and peri-urban areas;

- rural energy service projects in promotion zones;

- intensive peri-urban electrification project;

- promotion of PVs;

- optimizing the domestic fuel market;

- support for the coordinated development of the hydrocarbon market;

- elaboration of an energy charter;

- establishment of an energy access observatory; and

- technology transfer and strengthening of national value added.

23.6.5.2   East and Southern Africa

As in West Africa, in 1995 the Southern African Power Pool was created to develop electricity trade, reduce energy costs, and promote greater supply stability for the region's 12 national utilities.

The Southern African Development Community (SADC) countries signed an energy protocol which came into force in April 1998 (SADC, 1998). Key objectives of the protocol are:

- to harmonize national and regional energy policies, strategies, and programs;

- to cooperate in the development and utilization of energy and energy pooling to enhance security and reliability;

- to develop jointly the human and institutional capacity of the energy sector; and

- to promote standardization where appropriate in the energy sector.

23.6.5.3   Summary of Regional MDG-related Energy Target

Regional organizations for sub-Saharan Africa have also proposed targets for countries in their regions to adopt, as follows for 2015 as specified in the MDGs objectives.[33]

The Forum of Energy Ministers of Africa Position Paper (FEMA, 2006):

---

32 We assume a combination of measures, including substitution between sources. This means that in some areas the LPG share will increase, but in others another energy source will replace it.

33 The majority of summary is drawn from a publication by UNDP/WHO in 2009, "*The Energy Access Situation in Developing Countries: A Review Focusing on the Least Developed Countries and Sub-Saharan Africa*".

- 50% of Africans who live in rural areas and use traditional biomass for cooking would need to have access to energy services, such as improved cooking stoves, which reduce indoor air pollution, as well as efficient kerosene and gas stoves.

- 50% of urban and peri-urban poor should have access to reliable and affordable energy services for their basic energy needs, such as cooking and lighting, and productive uses such as agrocultural processing and general value addition.

- 50% of schools, clinics, and community centers should have access to modern electricity services for the provision of lighting, refrigeration, information, and communication technology.

The ECOWAS White Paper (ECOWAS/UEMOA, 2006):

- 100% of the total populations (325 million people) will have access to a modern cooking fuel.

- At least 60% of people who live in rural areas will have access to productive energy services in villages, in particular motive power to boost the productivity of economic activities.

- 66% of the population (214 million people) will have access to an individual electricity supply, or 100% of urban and peri-urban areas, 36% of rural populations, and, moreover, 60% of the rural population will live in localities with:

  - modernized basic social services – healthcare, drinking water, communication, etc.

  - access to lighting, audiovisual, and telecommunication service, etc.

  - coverage of isolated populations with decentralized approaches.

The CEMAC Action Plan for the Promotion of Energy Access (CEMAC, 2006):

- 50% of the population to have electricity access, with at least 35% of the rural population having access.

- 80% of the peri-urban and rural population to have improved access to modern fuels for cooking and heating.

The EAC Energy Access Strategy (East African Community, 2006):

- 55% of the total population in the region will have access to LPG or improved stoves and to sustained biomass supply. This is the equivalent of an additional 50% of the population that currently does not have access to modern cooking practices.

- 100% of urban and peri-urban households will be provided with an electricity service.

- 100% of the rural population will live in a locality where social service centers are equipped with energy services.

- 100% of administrative headquarters and localities with more than 3500 inhabitants will be equipped with mechanical power and heating technology.

The SADC Regional Indicative Strategic Development Plan (undated):

- 70% of rural communities will have access to electricity (by 2018), or

- 70% of rural communities will have access to modern forms of energy supplies (by 2018).

23.6.6   Key Conclusions on Policies at a Regional Level

It is evident that Africa has sufficient energy resources to fuel its own development, but so far most of the resources are undeveloped or exported. This is particularly so with oil that is sold in crude form and imported back as refined products, and yet the continent could build additional refinery capacity that can supply the bulk of Africa's needs. The energy industry is oriented to export (in oil-producing countries) and there is a lack of investment to mobilize the hydropower resources, large and small, and natural gas resources are unexploited. The hydropower resources in the DR Congo are barely exploited and are far from demand centers. There are few economies of scale in Africa, which makes the development of large energy resources unaffordable for individual countries and requires joint investments. Increasing the trade of energy, especially of oil, gas, and electricity, among African countries can significantly improve the uneven distribution of energy resources in the continent. To achieve this, capacity development and reinforcement in areas of governance, financing, and energy are necessary. Such capacity should be created at individual, institutional (individual capacity integrated with the institutions to afford its objectives), and systemic level (adequate coordination among the institutions).

Recently, regional bodies have developed a growing interest in promoting policies to improve energy access for poverty alleviation in addition to their natural regional priorities geared toward regional cooperation, infrastructure development for electricity and gas interconnections, and capacity development.

As cited above, ECOWAS, CEMAC, and SADC have developed time-bound objectives for access to modern forms of energy in their subregions in line with the MDGs horizon. This regional political will needs to be pursued and completed by action plans for the implementation and mobilization of funds.

The African Development Bank et al. (2003) has developed a strategic plan that takes poverty alleviation as one of its priorities. The Bank is updating its energy sector policy (enacted since 1994). An appropriate updated policy has to comply with national priorities and support energy security at both the macroeconomic and local levels.

South Africa, Botswana, Ghana, and Zimbabwe have implemented policies and strategies to give poor households greater access to electricity and to make the use of electricity more affordable. Emphasis on access was, in three cases, the primary focus and in their different ways the policies and strategies have achieved their objectives. When it was found that the programs started in these countries did not fully achieve their objectives, adjustments were made over time. In South Africa the Free Basic Electricity tariff was introduced, in Botswana up-front cost and the repayment rates were reduced, and in Zimbabwe the cross-subsidy was raised from 1% to 6% to pay for the program.

23.6.7   Lessons Learned

- Sub-Saharan Africa has adequate energy resources (hydropower, oil, gas, coal) to fuel its development, but they are largely unexploited. Greater regional cooperation and trade offer the least-cost option for energy development.

- The experiences of West African countries show that social tariff, social electrification, moderate residential tariffs, and subsidized connections were key instruments for their success.

- National and subregional energy policies and strategies would need to be harmonized with other relevant policies for efficient implementation and development.

- Energy policies should aim at creating conditions to support subregional and regional energy industry and market development for renewable and non-renewable energy resources.

- There are many successful African examples of addressing barriers to energy development. Learning from and upscaling these best practices will avoid costly mistakes and increase access to energy services.

- Sub-Saharan African's utilities have implemented some reforms, but more needs to be done to increase the efficiency of the power sector. Tariffs are often not cost reflective, and subsidize the affluent sector of society who could afford a connection anyway. Charging full cost to this group would save the government subsidies which might be better used to extend the grid to a poorer part of the population and advance development.

- Energy issues should be well integrated into PRSPs and other national and regional development issues. In particular, energization of rural areas achieves greater results when integrated into multisectoral rural development strategies, and programs.

- All options of energy supply, grid, minigrid, and off-grid, are valuable in their appropriate context.

23.7   Asia Review: Successes, Failures, and Proposals

23.7.1   Introduction

One-fifth of the population (some 800 million people) of the Asia-Pacific region still lack reliable access to electricity, and more than half (near two billion people) still lack access to clean cooking facilities. According to a recent report by the International Energy Agency (Niez, 2009), around 1.3 billion people globally lack access to electricity, and more than half of these are in the Asia-Pacific region. This has severe socioeconomic costs, particularly for the 641 million people in this region who live on less than US$1/day as they tend to spend a higher proportion of their income on energy.[34]

Although in some Asian and Pacific subregions the proportion of the population with access to electricity improved overall between 1990 and 2005, this growth was much less than the GDP growth rate over the same period. Growth in access to electricity and in GDP has also been much greater than the global averages (see Chapter 19). In southeast and northeast Asia, this growth was also much greater between 1990 and 2000 than that in more recent years.

By 2005, more than two million households in the region were generating electricity from stand-alone SHS.[35] However, assuming an average household size of five persons across these countries, this only equates to roughly 10 million people or only 1% of the population without access to electricity.

23.7.2   Energy Access Programs and Success Stories in Some Selected Countries

This section highlights the roles of factors that enhance access to electricity supply through successful rural electrification programs in Asia. In grid-based rural electrification programs, these factors include:

- dedicated public institutions and community organizations for rural electrification;

- self-sustainable revenue generation;

---

34   See UNESCAP, 2007 from which this section draws heavily.

35   However, they receive very limited access to electricity, as the SHS have a supply very low capacity, and supply very few energy services.

- strong non-residential customer base to cross-subsidize electricity to residential customers;

- sufficient public investment in rural electrification prior to reforms that allow private participation in electricity supply;

- economic growth;

- sound financial performance of rural electrification institutions and an ability to expand generation capacity adequately;

- involvement of stakeholders in planning and implementing rural electrification schemes;

- in some cases, the removal of regulatory barriers, e.g., house registration identity documents in Thailand.

In rural electrification programs not based on grids, the key factors to increase electricity access include innovative financing schemes to overcome the barrier of high up-front costs of isolated or decentralized electricity-generation systems (e.g., *Grameen Shakti*, see Box 23.7), SHS, and technology support.

While the figures in the next paragraphs serve as a common denominator to the problem, there exists wide disparity in rural electrification in South Asia. Sri Lanka has a rural electrification rate higher than the global average, while only 12% of the rural population in Afghanistan is connected to the grid. India, Pakistan, and Bangladesh alone constitute more than 90% of the region's population without access to electricity, with the remaining 10% in other South Asian countries.

India, Nepal, Sri Lanka, and Bangladesh have taken the lead in using off-grid technologies to create access to electrification in rural areas through a range of schemes and models. Of the region's 614 million rural people without electricity, many live in isolated communities, far from the national electricity network. These off-grid communities are generally small and dispersed, consisting of low-income households with characteristics that are economically unattractive to potential private sector energy providers, or even to government electrification programs that usually prioritize the allocation of scarce resources.

In 2008, the national electrification rate in Nepal was 64.5% with a very uneven urban–rural distribution. In urban areas, where less than 20% of the population lives, the household electrification rate is 93.1%, while the rate in rural areas is only 52.5%. It is highest in the accessible lowland regions and lowest in the mountain regions. The per capita electricity consumption is only 81 kWh/yr, one of the lowest in the world (Palit and Chaurey, 2010). In 2001, only 27% of the total population had access to grid electricity. In 1996, the Government of Nepal started a pilot electrification project under the Rural Energy Development Program (REDP) to promote modern forms of energy. Its objective was to alleviate poverty, improve livelihoods, and preserve the environment

in remote and rural parts of the country, where grid-based electrification was not expected to materialize in the near future. After the successful implementation of the pilot projects in five districts, it was extended to a further 25 districts.[36]

The REDP concept is heavily based on the decentralized and participatory decision-making process and a holistic development approach. A salient feature is its strong community mobilization process, focusing on (i) organizational development, (ii) skills enhancement, (iii) capital formation, (iv) technology promotion, (v) environmental management, and (vi) vulnerable community empowerment. Participation, transparency, consensus decision making, and inclusion of all households in the community, irrespective of class, color, creed, or gender, are the four pillars of good governance for ensuring equal ownership and equitable sharing of benefits accrued from MH systems.

By 2007, the REDP had installed 185 MH plants, with a total capacity of 2.47 MW. Together, these plants provide electricity access to more than 120,000 people for lighting and mechanical power for agroprocessing and other productive applications (Rijal et al. 2007). The main reason for REDP's success is its effective resource investment in capacity development of local stakeholders, effective community mobilization, and affordable tariff structure (GNESD, 2004).

The Energy Sector Assistance Programme (ESAP) has also been instrumental in supporting the Alternative Energy Promotion Centre to promote MH schemes of up to 100 kW. Besides loan financing available through commercial banks, there was also the provision of financial subsidies for these projects. Also, a total of 69,411 SHS were installed in the country, bettering the program target of 40,000 systems under the ESAP's first phase. The program was also successful in establishing guidelines for administering solar energy subsidies and putting in place quality assurance and monitoring systems for the solar energy projects.

Sri Lanka stands out in South Asia for its high rate of household electrification. Between 1986 and 2005, the national electrification rate improved substantially from 10.9% to 76.7%. Almost 75% of Sri Lanka's rural households are connected to the electricity grid, while another 2% are provided with basic off-grid electricity connections. In the off-grid sector, small hydropower has been the preferred option, with the first off-grid village hydropower scheme commissioned in 1992. The program resulted in a dramatic increase in the development of grid-connected and off-grid renewable energy projects, prepared and implemented by the private sector and village communities. Studies have concluded that the large-scale penetration of SHS in Sri Lanka has helped rural communities to improve their socioeconomic conditions and reduce adverse environmental impacts.

---

36 In Nepal, almost 30% of electricity supplied in the rural areas is through the off-grid route. The use of alternative energy sources for rural electrification took place because of the early realization by the Government of Nepal that the central electricity grid may not reach most rural populations (Palit and Chaurey, 2010).

With a view to enhancing rural electrification, the Energy Service Delivery Project (ESDP) was jointly initiated by the World Bank and the Government of Sri Lanka in July 1997 for a five-year period. The project aimed to create nationally coordinated programs to introduce, popularize, and consolidate alternative energy sources including village MH system. The project provides financing, including a grant portion from Global Environmental Facility, for both grid-connected schemes and off-grid connected schemes for rural electrification.

The project facilitated an accelerated development of village MH schemes in Sri Lanka, with technical backstopping from the Intermediate Technology Development Group (ITDG), a NGO. As ITDG had been promoting grid-connected small hydropower plants, off-grid MH systems, biogas systems, and small wind systems for 15 years in Sri Lanka, its expertise was helpful during the dissemination program.

The successful lessons from the ESDP are now being replicated on a larger scale under a successor program, Renewable Energy for Rural Economic Development Project, funded by the World Bank (World Bank, 2003).

In Thailand, the Provincial Electricity Authority (PEA) formulated a 25-year National Plan for accelerated rural electrification in 1977. This served as the master plan for the country's rural electrification. The PEA was able to increase electricity access to rural populations from 7% in the 1970s to 99% by 2007. The key factors behind the rapid growth in Thailand's rural electrification include the creation of a dedicated entity, the Office of Rural Electrification, by the PEA, with specific responsibilities to implement the rural electrification program, self-sustainable revenue generation, involvement of end users in planning, financing of development of distribution network, and subsidies to residential customers (Shrestha et al., 2004b). High levels of electrification were achieved in Thailand by the early 1990s. As a result, subsequent reforms in the power sector, such as private participation in power generation and tariff reform, do not seem to have affected the rural electrification in Thailand.

Bangladesh's rural electrification program was launched in 1977 at around the same time as that of Thailand. The program in Bangladesh was implemented through two-tiered institutional arrangements involving the Rural Electricity Board and rural electricity cooperatives known as *Palli Biddut Samities* (PBSs). Each consumer is a member of the PBS that serves them. The Rural Electricity Board is responsible for planning and developing the distribution network. The PBS is responsible for preparing a master plan on electrification of its members and for forecasting load growth. The PBS also manages financial and operational activities. The 70 PBSs established provide electricity to more than 40 million people living in 38,000 villages (GNESD, 2007).

Rural electrification has helped generate local employment and promoted local non-farming economic activities in Bangladesh (Barkat, 2005). This has helped to make electricity more affordable for people.

PBS has been successful in reducing system losses by 50% and improving billing collection, which stands at more than 95%, compared to national utilities (Rijal et al., 2007). According to GNESD (2007), PBS is a model to be followed by highly centralized national electricity utilities. Its success lies in effective decentralized actions, facilitation of cooperatives with subsidized finance, a revolving fund for loss-making cooperatives, subsidized power from the Bangladesh Power Development Board, performance target agreement of cooperatives with the Rural Electrification Board, and intersectoral cross subsidy provision for customers (Rijal et al., 2007; GNESD, 2007).

The rural electrification programs in Bangladesh and Thailand have a number of things in common. These include (Shrestha et al., 2004b):

* the creation of an entity with specific responsibilities to implement the rural electrification program;

* involvement of end users in the distribution network planning process;

* financing of distribution network development, the creation of the distribution networks, is funded through grants and low-interest loans from the government, as well as bilateral/multilateral agencies; and

* provision of subsidies to residential consumers of electricity.

Despite the similarities in approach, the achievements were more significant in Thailand. Only 19% of Bangladeshi households were electrified by 2000. Although the Rural Electrification Board and PBSs covered 90% of the area in Bangladesh with a basic distribution infrastructure, household connectivity is still very low. According to the IEA, the overall electrification rate in Bangladesh was 41% in 2008, with 76% of the urban population and only 28% of the rural population having access to electricity. Although the rural household electrification rate is poor, Bangladesh has recorded an impressive rural electrification performance with the help of solar PV technology, particularly SHS (Box 23.7). The solar PV program was developed by the Infrastructure Development Company Limited, Bangladesh, with the help of the World Bank.

Three factors that appear to have influenced the divergence in achievements of the rural electrification programs in Bangladesh and Thailand are financial resources, electricity generation capacity, and level of economic growth:

* Financial resources: The PEA in Thailand was able to cover its operational cost through revenue generation from the sale of electricity. This enabled the PEA to use new resources allocated for rural electrification to expand the distribution network. However, unlike the PEA, the PBSs in Bangladesh were not able to meet their operational costs. Both, however, received power from a national electricity-generating

Box 23.7 | *Grameen Shakti* Microfinance Scheme in Bangladesh

Grameen Shakti was incorporated in 1996 as part of the *Grameen* family of companies. It specialized in renewable energy, such as SHS, wind, and biogas. The main objective was to produce electricity to provide the minimum needs of electricity for lighting after dusk to enable income- generation activities to continue at night. *Grameen Shakti's* board of directors and top management had extensive experience of microcredit financing and many were founding members of the Grameen Bank. Their experience in microcredit was essential in the design of *Grameen Shakti* programs.

Grameen Shakti has various financing models (Barua, 2005).

* Mode 1: The customer has to pay 15% of the total price as a down payment during installation and the remaining 85% of the cost is paid by monthly installation within 36 months, including a 12% service charge.

* Mode 2: The customer has to pay 25% of the total price as a down payment during installation and the remaining 75% of the cost is paid by monthly installation within 24 months, with an 8% service charge.

* Mode 3: The customer has to pay 15% of the total price as a down payment during installation and the remaining 85% of the cost, including a 10% service charge, is made by 36 post-dated checks.

* Mode 4: 4% discount is given for cash purchase.

* Mode 5 (microutility): The customer has to pay 10% of the total price as a down payment during installation and the remaining cost is paid by installments within 42 months, with no service charge. Here the customer sets the system up on his/her premises, and other shop owners receive the facility of SHS in exchange for payment.

Grameen Shakti developed one of the most successful market-based programs with a social objective of popularizing SHS, as well as other renewable-energy technologies, to millions of rural villagers. *Grameen Shakti* used its Grameen Bank concept of microcredit to evolve a financial package suitable for rural people, which helped, in particular, to bring down costs. The customized pricing system based on installments helped *Grameen Shakti* to reach economies of scale with the increase in sales. Their business is centered on customer-service excellence. *Grameen Shakti* engineers pay monthly visits to households and offer their services for a small fee upon the signing of an annual maintenance agreement by clients. *Grameen Shakti* also undertakes several other activities, such as educational loans and gift schemes, which go well beyond the energy service itself and help develop trust between it and the local communities. By the end of December 2009, the total number of installations had reached 113,736 SHS.

authority at subsidized rates. But the ultimate financer of these subsidies in Bangladesh was the government, as the Bangladesh Power Development Board had been losing money and was unable to generate its own resources to pay for its operations. This was caused by combination of factors, including that the average Bangladesh Power Development Board tariff was set below the long-run marginal cost, there were high system losses (38%), and a low rate of bill recovery. The losses in Bangladesh also adversely affected the availability of financial resources from multilateral institutions, as between 1990 and 1995 they withdrew from financing the power sector in Bangladesh.

* Generation capacity: Lack of adequate supply, because the power-generation capacity expanded more slowly than the projected increase in demand. Unlike Bangladesh, generation capacity was not a barrier to expanding rural electrification in Thailand.

* Economic growth: High economic growth also implies expansion of commercial and industrial activities and therefore increased demands for electricity. The rise in the number of non-residential customers and the level of their electricity consumption provide a greater resource base for subsidizing residential consumers. The smaller, non-residential resource base was a factor that seems to have inhibited the electrification rate in Bangladesh. The per capita income growth during the rural electrification program was much faster in Thailand than in Bangladesh. At less than 2%, the poverty levels[37] in Thailand were very low, compared to 36% in Bangladesh in 2000. Bangladesh's high poverty levels imply that the consumer base with very low paying capacity is much larger in Bangladesh than in Thailand.

37 Population earning less than US1$ a day (UNDP Human Development Report, 2003).

In 1997, Indonesia promoted a household electrification program based on off-grid, stand-alone SHSs. This electrified rural areas where there were no plans to extend the government-owned utility's grid network. The Indonesian government has established a revolving fund, or grant, to implement these systems.

Users made down payments to cooperatives, and then paid monthly fees until the cost was covered. The program emphasized the formation of representative community groups (involving men and women in all stages of the electrification process), which helped to manage the electrification projects successfully.

The program promotes the use of locally made system components, which makes it cheaper and technically sustainable. This has been a success story for the off-grid electrification process, particularly for countries that lack the financial resources for grid extension in remote areas (UNEP/GNESD, 2002). The credit for this success goes to the effective management of cooperatives and the development of local manufacturing capability, which have reduced the system costs. However, as this scheme is based on a bottom-up approach, with users bearing all costs, it is limited to high-income groups only.

23.7.3   The Experience of India in Rural Electricity Access

With the largest rural population in the world, India is facing a huge electrification challenge. About 60% of rural households have access to electricity. Electricity consumption per capita is low at 543 kWh/yr/capita (Niez, 2009). There is also wide disparity in access to electricity between urban and rural populations, and also between states.

In the power generation sector, although India has considerably improved its generating capacity over the years, with installed capacity growing from 1362 MW in 1947 to 159,648 MW[38] by 2008 (Palit and Chaurey, 2010), the supply of electricity across the country currently lacks both quality and quantity. There is an extensive shortfall in supply, a poor record for outages, and high levels of technical and non-technical losses.

A number of specifically targeted schemes were launched in India to facilitate electricity access for poor people (Box 23.8). Most of these schemes were implemented by the state electricity utilities with central financial assistance disbursed through the Rural Electrification Corporation (REC).

The grid connection is the most favored approach to rural electrification for most rural households. But renewable off-grid technologies, such

as solar PV, mini-/microhydropower, biomass gasifiers, biofuel-powered generators, and small wind aerogenerators in hybrid mode are also disseminated to areas that are inaccessible to the grid, such as remote, hilly, and forested villages, islands, or hamlets that are not recognized as villages by national census records.

Off-grid technologies have been used either through the creation of local minigrids, or by disseminating household-level technology, such as solar PV, for lighting and other low-consumption activities. It is reported that off-grid capacity in India is around 13 GW, of which 10 GW is diesel and 3 GW is renewable energy (Banerjee, 2006). Off-grid power plants based on renewable energy are typically in the range 1–500 kW, and are located in independent distribution network (minigrids). Most off-grid systems have been promoted by Government of India schemes.

23.7.3.1   Rural Electrification Policy

Rural electrification is a key factor in accelerating rural development. The provision of electricity is essential for the requirements of agriculture and other important activities, including small and medium industries, *khadi* (indigenous) and village industries, cold chains, healthcare, education, and information technology. Where a grid extension is not feasible in rural areas, the policy recommends distributed generation, through either conventional or non-conventional means, as the preferred option. District committees would coordinate and review the extension of electrification in the district, review the quality of power supply and consumer satisfaction, and promote energy efficiency and conservation. The policy stresses that the state governments should actively raise awareness of electricity issues, including generation, distribution, energy conservation, and the energy efficiency and energy-water nexus.

The policy says that it is essential for energy efficiency to be promoted in rural areas through mass campaigns. It also notes that the use of inefficient and energy intensive equipment by the agricultural sector distorts the consumption pattern and results in non-optimal utilization of tariff subsidies. It recommends the use of economically viable, energy efficient farm equipment, especially irrigation-pump sets.

The main vehicle chosen to implement the policy's universal access objectives is the *Rajiv Gandhi Grameen Vidhyutikaran Yojana* (RGGVY). This was launched in April 2005 with the goal of electrifying all villages and hamlets that were without electricity, and providing access to electricity to all households within five years. The RGGVY's basic objective is to create a rural electricity infrastructure in the country in order to provide all rural households with access to electricity within a given timeframe. The RGGVY emphasizes not just village electrification, but the facilitation of rural development, employment generation, and poverty alleviation through access to electricity. This includes below-the-poverty line (BPL) households, and also caters to the needs of agriculture, small and microenterprises, cold chains, healthcare, information technology, and education.

---

38   Thermal power (coal, gas, and diesel) accounted for about 64% of total installed capacity with 102,704 MW, large hydropower for 23% with 36,863 MW, grid-connected renewables for about 10% with 15,521 MW, and the rest from nuclear power.

Box 23.8 | Policy Regimes for Rural Electrification in India

In 2002, the Gokak Committee recommended that a choice between grid connection and decentralized generation should be made on the basis of technical, managerial, and economic issues. These included distance from existing grid, load density, system losses, and load management.

The Electricity Act 2003 was enacted in June 2003. Its overall objective was to develop the electricity industry and provide electricity access to all areas. Sections 4 and 5 of the Act specifically require the Government of India to formulate appropriate and adequate policies for the supply of electricity in rural areas.

The aims of the National Electricity Policy, introduced in 2005, include:

- Access to electricity: available to all households in five years.

- Availability of power: demand to be fully met by 2012; energy and peaking shortages to be overcome and adequate spinning reserve to be made available.

- Supply of reliable and quality power to specified standards, efficiently, and at reasonable rates.

- Per capita availability of electricity to be increased to over 1000 kWh by 2012.

- Minimum lifeline consumption of 1 kWh/household per day by 2012.

- Financial turnaround and commercial viability of electricity sector.

- Protection of consumers' interests.

The National Electricity Policy also states that wherever grid-based electrification is not feasible, decentralized distributed generation facilities (either conventional or non-conventional methods of electricity generation, whichever is more suitable and economical) together with local distribution networks will be provided, so that every household can access electricity.

The scheme's scope includes the provision of a rural electricity distribution backbone (i.e., 33/11 kV substations with adequate capacity and lines in blocks), creation of a village electrification infrastructure (i.e., electrification of unelectrified villages and settlements), provision of distribution transformers of appropriate capacity in electrified villages/settlements, decentralized distributed generation and supply, and electrification of rural households below the poverty line.

The RGGVY provides a capital subsidy of up to 90%. This is disbursed through REC Limited, a nodal agency for the scheme's implementation. The other 10% is to be arranged by the project developer. It is planned to involve private developers in the program, with the subsidy to be released on an annuities basis depending on the system's performance over five years. However, electrification of BPL households is financed with a 100% capital subsidy. The guidelines say that priority shall be given to villages where grid connectivity is not foreseen in the next five years.

The remote village electrification (RVE) program of the Ministry of New and Renewable Energy (MNRE) started in 2001. It aims to provide basic lighting facilities in unelectrified census villages, regardless of whether these villages were likely to receive grid connectivity. But the scheme was subsequently modified to cover only those unelectrified census villages that are not likely to receive grid connectivity. In addition to domestic use, the scheme has the option to provide energy services for community facilities, for pumping for drinking water or irrigation, and for economic and income generating activities in the village. Like the RGGVY, central financial assistance of up to 90% of the projects' costs is provided as a grant with specific benchmarks. The balance of 10% is financed through other decentralized governmental institutions.

The Village Energy Security Programme (VESP) was conceptualized as a step toward the RVE program. It also addresses a village's total energy requirements for cooking, electricity, and motive power, and it aims to transform the largely unsustainable use of locally available biomass to

Box 23.9 | A Good Example to Follow: Village Energy Security Programme in India

Launched in 2004, the Village Energy Security Programme (VESP) is a community-based initiative that aims to provide clean, affordable energy in rural areas – home to around 70% of India's population. The focus has been on finding ways for villages – particularly those located in remote rural areas that are unlikely to be provided grid electricity in the near future – to achieve energy security based on locally available renewable energy sources (preferably biomass). The program goes beyond rural electrification to village energization. It therefore places additional emphasis on cleaner options of cooking through improved cook stoves and biogas, productive use of energy for livelihood generation, and sustaining the energy systems through captive plantation. As of December 2009, 79 test projects have been sanctioned, of which 55 were commissioned in eight different states. These test projects were undertaken in non-electrified remote villages and hamlets that are not likely to be electrified through conventional means in the immediate future. Based on a community centric approach, a one-time grant (up to 90% of the project cost) was provided to the village community for installation of energy systems capable of meeting the village community's energy demands. The community, in some cases, also provided an equity contribution (either in cash or kind) to bring in the much needed ownership required for the success of any community centric projects. Based on an assessment conducted to review the performance, impacts, and lessons of the VESP test phase, it was found that the VESP projects emerged as a vehicle to motivate the community, especially the youth, to attempt develop their skills. Local youths enhanced their skill to operate the installed power-generation systems in almost all the VESP subprojects. Innovations adopted by select Project Implementation Agencies for capacity development of the technology operators helped to improve project performance. There are mixed results for social mobilization and leadership of the Village Energy Council (VEC), with mobilization and leadership relatively better in the test projects implemented by NGOs as compared to those implemented by state departments. Revenue management is comparatively better in projects where villagers have cash income because of either existing income-generation activities or newly introduced activities after being electrified under VESP. Active involvement of Gram Panchayats in some projects helped to develop the required synergy in getting village-development funds for VESP, both toward project cost and operational expenses.

However, along with the above-mentioned best practices some shortcomings were also reported. The uptime of the projects, considering their remoteness and various other inherent technical and institutional problems, is satisfactory in some of the projects, but it is poor in most of the projects. Some of the challenges for sustainability were found to arise from less-concentrated electricity demand in the villages, low economic activity (implying a lower electricity demand), less ability to pay by the consumers, difficulty in O&M, limited technical knowledge within the VECs, and weak fuel-supply chain linkages. One or a combination of these factors leads to a low load factor and fewer hours of operation, and thereby a low capacity-utilization factor. The potential of income generation activities and productive load was also not fully exploited because of the absence of proper guidance to the VECs to initiate and execute such possibilities.

Source: Palit and Chaurey, 2010.

innovative, sustainable modern biomass energy. The objectives of the VESP are:

- To meet village energy requirements through biomass material and biomass-based conversion technology, or other renewable technologies where necessary.

- To go beyond electrification by addressing the total energy requirements, such as those required for household cooking.

- These projects would involve the installation of energy production systems: biomass gasifiers, biogas plants, plantation activities, and improved cooking stoves (*chulhas*).

23.7.3.2 Institutional and Financial Viability of Electrification

Electricity is in the concurrent list of the Indian constitution, and therefore both state and central governments have jurisdiction over it. The state governments' jurisdiction includes generation, intrastate transmission, distribution, and intrastate trading of electricity. The central government's purview includes policy formulation, generation plants catering to more than one state, interstate transmission, and interstate trading of electricity. In the rural electrification sector, the principal actors have traditionally been the state electricity utilities, because they were responsible for the distribution of electricity in the states.

Table 23.10 | Subsidies for rural electrification schemes in India.

Scheme	Target under the scheme	Subsidy vehicle	Amount of central financial assistance
RGGVY	100% household electricity access throughout India by 2012	Capital subsidy	90% grant is provided by the Government of India 10% as loan by REC to the state governments Total subsidy disbursed by September 01, 2009: INR256,790,000,000
VESP	1000 villages to be electrified within the current five-year plan	Capital subsidy Operational subsidy for first two years	90% of the total project cost Maximum CFA per household is INR20,000 10% of the total project cost
RVE	Electrification of villages and hamlets that are not likely to receive grid connectivity	Capital subsidy subject to upper limits	90% of total costs of electricity generation systems

Source: TERI, 2007.

The total costs of RGGVY schemes were estimated recently as US$13.64 billion. The Ministry of Power funds 90% of RGGVY project costs, with state governments funding the remaining 10% through either long-term loans from the REC or other financial institutions, or from their own budgets. Household-connection charges are borne by individual households. However, for unelectrified households below the poverty line, the household connections are financed wholly by capital subsidies.

The total implementation costs for the RVE program and the VESP are estimated to be US$377 million. As with the RGGVY, subsidies from MNRE cover up to 90% of the project implementation costs (Table 23.10).

23.7.3.3   India Programs: Key Lessons

- Government support is essential to the development of a successful rural electrification plan. Laws and reforms have shaped the institutional and legal framework, but without firm implementation of policies and goals, enforced through legislation, the electrification process is difficult to achieve.

- The creation of franchises for the management of local power distribution in rural settings is reported to have brought efficient billing and revenue collection, and thereby ensure stable delivery of electricity. Studies by The Energy and Resources Institute (TERI) indicate that franchises are particularly effective in managing electricity provision and cost recovery, because they are in close contact with the targeted communities. This has led to a stronger sense of ownership of the electrification process.

- The three-tier quality-monitoring mechanism established the RGGVY is reported to be ensuring proper implementation of projects, thus contributing to their efficiency and long-term sustainability. Similarly, the five-year performance warranty and annual maintenance contracts for all systems installed by the RVE program is securing proper and sustained energy supply services.

- Involving rural communities in the decision-making process has contributed substantially to the effectiveness of the off-grid electrification program, adding value to the planning process and giving communities a sense of ownership of the process.

- A study by TERI in 2008–2009 to review the off-grid and grid-connected distributed generation projects produced a number of interesting observations. Lessons learned from the study include:

- Grid-connected projects have advantages in terms of reliability and quality of supply, as the grid acts as a balancing sink or source, and also supplements power in the local area during periods of plant shutdown. The productive load, irrigation pumps, and agroprocessing can be served on demand, which is particularly significant for small, off-grid projects that are otherwise not being served at all. There also seems to be a greater community demand for grid-connected power, because of the limited hours of supply that off-grid projects provide.

- Off-grid projects encounter many sustainability challenges, because most of these projects are located in remote villages. These challenges lead to the system's lower plant load factor and low uptime. This leads to a higher generation cost, which users may not be able to afford. Users also become reluctant to pay when the plant doesn't function. Lack of payment then makes it hard for the operators to continue to run the service, creating a vicious cycle in the system that can be difficult to break out of.

23.7.4   China

China is the largest developing country and makes poverty reduction a top priority policy. Since the late 1970s, when China started opening up to the world, it has made immense progress, both economically and socially. This has been the case particularly since the early 1990s, when the country moved from a planning-based to a market-based society. Over the past 10 years, the major energy programs have been rural power-grid development, a national biogas program, a rural hydropower-based electrification program, and electrification in remote areas.

23.7.4.1   Energy Access for the Urban Poor

In general, there are no significant barriers for urban residents to access energy for any purpose. Cities of all sizes have the infrastructure to provide basic services, for example electricity, safe water, and

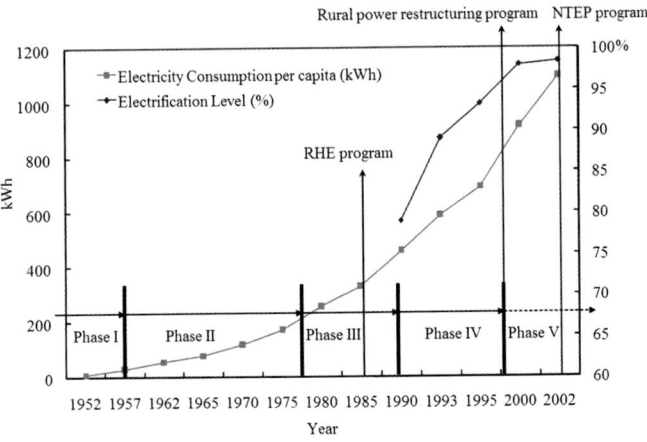

Figure 23.3 | Results of NTEP Program.

telecommunications. The household electrification level[39] in China exceeded 98.4% in 2002, and all townships[40] in China have had access to electricity since the National Township Electrification Program (NTEP) in 2005. The power-grid extension is not determined by affordability, but by the requirements of families. The Chinese government allocates a special subsidy that allows the poorest urban households to afford their basic needs, including energy.

23.7.4.2   Energy Access for Rural Populations

Energy system construction is a key part of China's development of its rural infrastructure. Rural energy should be understood in two ways. One focuses on the general idea of rural areas, while the other concentrates on hard to reach, remote, or mountainous regions. As a large country, the difference in approaches to 'general rural' and 'special rural' is highly significant.

'General rural' requires a universal energy service system. This involves the government promoting development of the energy system in line with the urban standards. This helps to improve livelihoods and the quality of life in rural areas. For 'special rural', remote areas or mountainous regions where infrastructure is difficult to access or construct, China uses local resources, such as small hydropower, wind, or solar energy, to provide clean and relatively cheap energy (in comparison to other approaches, such as grid extension). These are included in the NTEP, which uses wind turbine, solar PV panels, and other advanced technologies.

Electricity is the most difficult infrastructure to develop in rural areas. China has moved rural electrification forward through five main stages (Figure 23.3):

---

39 Household electrification level indicates the percentage of electrified households of all national households.

40 Administrative regions in China are separated by province (autonomous region), city, county, township, and village.

1. The first stage was the initial development period for rural power, from 1949 to 1957. Rural electricity generation depended largely on local resources, but local communities also developed small-scale power stations.

2. The second stage was a stable period of rural power development that lasted from 1958 to 1977. This aimed to meet the demands of agricultural production and irrigation.

3. The third stage, from 1978 to 1989, saw the rural power infrastructure develop rapidly, driven by small hydropower-based primary electrification.

4. The fourth stage, from 1989 to 1998, began as a planned development period from 1989 to 1998. From 1989 the government introduced standardized management to rural electrification construction to ensure effective development.

5. The fifth stage, which has been underway since 1998, highlights management reform of rural power and restructuring of rural grids.

23.7.5   Lessons Learned

*   Energy access needs to be closely linked with the creation of social equity, rather than taken in isolation. Energy programs for poverty reduction need to be always integrated into a wider social program, along with access to clean water, roads, and education.

*   In general, the experiences were related to social goals, meaning access to modern forms of energy for final use to improve welfare. Energy as a production input in poor and isolated rural areas in combination with other factors and opportunities could create income opportunities. For this, more and deeper analysis is needed.

*   Governments need to play an important role in achieving equity. They should, therefore, take responsibility for planning, organization, and social mobilization aimed at fulfilling social targets.

*   Developing countries seem to have no unified approaches to addressing poverty reduction and energy access. Speeding up urbanization, under an umbrella of national economic development, would help increase long-term access to energy.

*   A universal energy service demands a range of approaches, with specific services for the extreme poor. A universal energy service should not only be oriented to the consumers who can afford the cost, but actually realize a general infrastructure for all the populations.

*   Successes in disseminating solar technologies, such as those that Bangladesh, Sri Lanka, Nepal, and India have achieved, demonstrate

that off-grid programs, in association with the private sector and rural microfinance institutions, are realistic. Projects can be scaled up appropriately, with improved access to capital, development of effective and reliable after-sales service, customer-focused market development, and routine stakeholder participation.

- The Thai and Bangladesh experiences in rural electrification show that, besides the creation of a dedicated entity to implement rural electrification programs, involvement of end users in the distribution-network planning process, and a policy to subsidize low-income residential users, the electricity pricing policy would need to be such that the electricity revenue at least meets the operating costs of rural electrification. Furthermore, the economic growth of the country could provide a sound base for cross-subsidizing the residential users under rural electrification and the generation capacity of the grid system should not be a barrier to expand rural electrification.

- The Indian experience in rural electrification shows the key roles of commitment to targets, enforced through legislation, involvement of the rural communities in the decision-making process (which adds value and gives communities a sense of ownership of the process), and local management to have brought efficient billing and revenue collection, and thereby ensure the stable delivery of electricity.

- Off-grid projects encounter many sustainability challenges because most of these projects are located in remote villages. These challenges lead to the system's lower plant-load factor and low uptime. This leads to a higher generation cost, which users may not be able to afford.

23.8   Latin America Review

23.8.1   Introduction

In many Latin American and Caribbean countries the relationship between energy and poverty is either lacking or treated superficially in national development plans, energy policies, or poverty-reduction strategies. There is also little research on the linkages between access to energy services and national development goals, poverty alleviation and reduction, and environmental issues.

Despite high rates of urbanization, around 30 million people, of which 21.4 million are poor, still do not have access to electricity. Access to modern fuels for cooking is another major problem: either families do not have access to modern fuels or, if they have, it accounts for a disproportionate share of their income (ECLAC, 2009).

It is important to increase our knowledge of the relationships between poverty and energy. It is also important to show that if the goal is to provide access to modern forms of energy, energy policies cannot be approached from a macro view of the energy industry, nor should they be merely part of other issues, such as energy security, geopolitical issues,

or climate-change impacts. Energy access policies should not be considered in isolation, but integrated into development policies, regional policies, and other general policies such as health and education.

It is remarkable that several LAC countries do not have policies oriented to reducing poverty and inequity, with the exception of social assistance (from fiscal funds) to alleviate poverty. Such assistance alleviates and reduces the pressures on poor populations, but do not create conditions to reduce or eliminate poverty. Although access to modern forms of energy, like electricity, is relatively high, poverty levels and inequity also remain significant.

23.8.2   Poverty and Energy Access

Approximately 200 million people currently live below the poverty line in LAC, of which some 133 million live in urban areas and 67 million in rural areas.[41] In this region, poverty is more an urban than a rural reality, but the share of urban and rural poverty differs widely across countries and subregions. In South America, about 70% of poor people live in urban areas, whereas only 48% of poor people are urban[42] in Central America (Table 23.11)[43].

Key qualitative differences exist between policy approaches aimed at improving access to energy by poorer populations in rural and urban areas. It is an issue that would need to be considered in the rural and urban contexts, as well as in the context of poverty in general.

Approximately 21.5 million poor people have no access to electricity in the region's 14 most populated countries (excluding Mexico). The region is characterized by highly heterogeneous energy resource endowments, levels of economic development, and Human Development Index across countries, and by very large socioeconomic asymmetries between the top and bottom income groups of the population within each country (see Chapter 19).

Available data show that poor households pay a much larger share of their income for energy services. The bottom income quintile of the population pays between 5% and 16% of their median monthly income for energy, while the top income quintile pays between 0.5% and 3% of their income (Figure 23.4).

The same situation is evident in the sample of Central American countries (see Chapter 19). The difference in energy use and energy expenditure

---

41   During a decade of reforms (1990–2000), the number of poor people increased by 14% as a result of structural reforms (from nearly 210 million to 240 million).

42   Brazil alone accounts for approximately 50% and 40% of the total urban and rural poverty, respectively, in South America. In Central America, El Salvador, Guatemala, Honduras, and Nicaragua account for more than 70% of the total urban poverty, and more than 81% of the total rural poverty in that subregion.

43   See ECLAC, 2009 from where this section draws heavily.

Table 23.11 | Rural and urban poverty estimates in 2006.*

Country	Urban poverty (million people)	%	Rural poverty (million people)	%	Total poverty (estimate in million people)	%	% urban	% rural
Argentina	7.4	7.9	N/A	-	7.4	6	100	0
Bolivia	5.1	5.4	2.7	6.9	7.8	6	65	35
Brazil	46.8	49.8	15.6	39.4	62.4	47	75	25
Chile	2.0	2.1	0.3	0.7	2.2	2	88	12
Colombia	10.0	10.7	11.5	29.1	21.5	16	47	53
Chile	2.0	2.1	0.3	0.7	2.2	2	88	12
Ecuador	3.3	3.5	2.4	6.1	5.7	4	58	42
Paraguay	1.9	2.0	1.7	4.2	3.6	3	53	47
Peru	6.2	6.6	5.2	13.0	11.3	8	54	46
Uruguay	0.6	0.6	N/A	–	0.6	0	100	0
Venezuela	8.6	9.2	N/A	–	8.6	6	100	0
South America	**93.9**	**100**	**39.6**	**100**	**133.5**	**100**	**70**	**30**
Costa Rica	0.5	4.1	0.3	2.5	0.8	3	60	40
El Salvador	1.6	13.6	1.6	12.5	3.3	13	50	50
Guatemala	2.9	23.9	4.3	32.8	7.2	29	40	60
Honduras	2.0	16.3	2.9	22.3	4.9	19	40	60
Nicaragua	2.0	16.5	1.8	13.7	3.8	15	52	48
Panama	0.5	3.9	0.5	3.9	1.0	4	48	52
Dominican Republic (DR)	2.6	21.6	1.6	12.3	4.2	17	62	38
Central America + DR	**12.0**	**100**	**13.2**	**100**	**25.2**	**100**	**48**	**52**

Source: CEPAL, 2009.

* Estimates are based on reported percentages of urban and rural poor from the most recent household surveys available in each country. N/A = not available.

of the top income quintile is as much as three to six times that of the bottom quintile.

Despite the high level of urbanization and electricity coverage achieved by LAC countries, energy accessibility still remains very much an unresolved development challenge for many countries in the region. Lack of access to modern forms of energy sources disproportionately affects low-income groups in most countries, aggravating the large socioeconomic asymmetries that remain a major development hurdle for the region.

A general lack of appropriate policies is the principle bottleneck that now impedes access to modern energies in LAC. In the power market, many countries have converted their state-owned monopolies into privatized systems over the past three decades. Many are unbundled competitive markets. These competitive markets place great emphasis on short-term demand with the capacity to pay, and place a premium on existing generation and new-generation investment that have very short construction lead times and low capital intensity. In such markets, the state has largely removed itself from the electricity business. The model does not tend to encourage aggressive electrification programs in rural areas, where profits are hard to generate (Organization of American States, 2004).

Energy affordability and the need to increase energy access for poor people still features as a low priority in the political agenda of most countries. Neither is part of the regional policy agenda, which has tended to focus on the issues of energy security in the face of higher oil prices, reducing the investment gap in energy infrastructure, and addressing the regulatory gaps and barriers that persist after the energy sector reform process of the 1990s.

23.8.3 Energy Access Policies in Rural and Urban Areas

The policy challenges for rural and urban areas are different in terms of alternative models, technology options, management opportunities, energy portfolios, financial transfers, and guaranteed utilities investments. The barriers that public policies must overcome are different in nature, dimension, level, and magnitude for rural and urban areas (Box 23.10).

In both areas, there is a deficit of information to address the energy-poverty-economy-environment linkage effectively. This is a serious obstacle to policy formulation, as it restricts clear characterization of the problems (lack of systematic information and diagnosis).

In remote rural areas of Latin America, the high cost of providing a grid electricity service and LPG for cooking combined with the low payment capacity

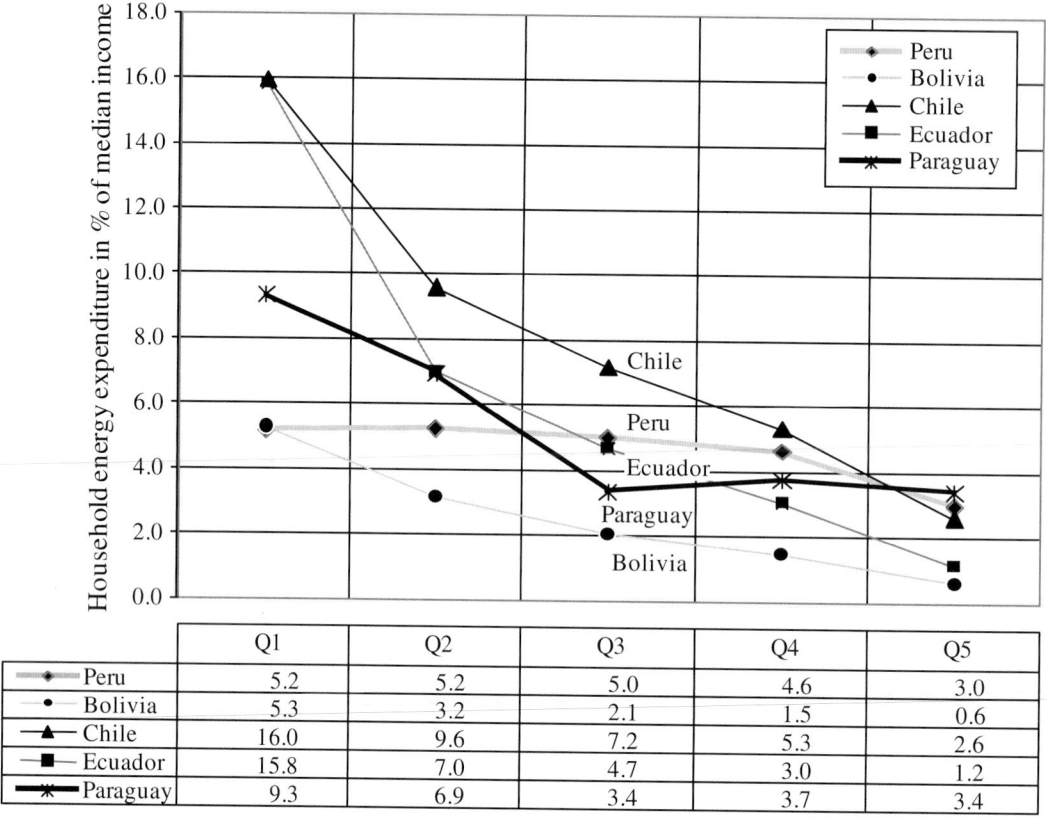

Figure 23.4 | Energy expenditure as percentage of median income per quintile (Q1 to Q5) in South America. Source: ECLAC, 2009. © ECLAC, Figure 7.

	Q1	Q2	Q3	Q4	Q5
Peru	5.2	5.2	5.0	4.6	3.0
Bolivia	5.3	3.2	2.1	1.5	0.6
Chile	16.0	9.6	7.2	5.3	2.6
Ecuador	15.8	7.0	4.7	3.0	1.2
Paraguay	9.3	6.9	3.4	3.7	3.4

and constraints on equipment availability, presents a major challenge.[44] Successful interventions are characterized by multiple policies, including:

- capacity development and technical support;

- subsidized payment and credit for equipment for off-grid solutions at both the household and community levels;

- subsidized monthly energy service bills to match the ability to pay;

- promotion of sustainable firewood use and improved stove/household ventilation programs;

- promotion of women's education and health programs; and

- promotion of community participation and integration into broader national poverty reduction.

Experience in Peru with programs that provide decentralized energy access solutions, such as single-household PV panels and batteries, shows better rates of success when families purchase the equipment and take responsibility for its maintenance through schemes adjusted to

their ability to pay. Programs in which the same type of equipment was installed for free by public agencies, without transfer of property, show higher failure rates, with the equipment often non-operational after only a few years because of lack of maintenance. This experience highlights the importance of engaging active participation and financial responsibility of the target communities, adjusted to their ability to pay, to ensure the sustainability of decentralized energy access interventions.

In poor urban settlements, insufficient income and absence of credit combine with illegal occupation of lands, inadequate equipment, clandestine connections to electric power services, and difficult access to regular fuel distribution channels for cooking, heating, and water-heating purposes (e.g., LPG service, gas networks). All these factors prevent access through regular channels, which results in low-quality, informal services and/or deprivation of basic energy needs, with strong impacts on the education and labor prospects that are critical for successful integration into the urban economy. In urban contexts, energy deprivation in combination with inadequate access to sanitary and water services might also have a disproportionate impact on women's and children's health.

Paradoxically, in urban environments, high energy use levels may be found in poor families because they often use second-hand, low-efficiency equipment acquired at low cost in informal markets. This situation highlights the importance of incorporating equipment replacement and efficient use of energy into urban energy access programs. This should

---

44 The supply of electric power through local solutions, such as small hydropower stations or PV systems through solar panels, could be an alternative to grid extension.

Box 23.10 | Barriers Faced by Energy-access Policies in Rural and Urban Contexts

Problems and barriers in urban areas

- Insufficient or irregular cash income prevents paid access to available energy services and adequate equipment.

- Clandestine connections to electricity grids – associated with accidents, fire hazard, precarious connections, and equipment failures.

- Low-efficiency, second-hand household equipment creates high consumption and waste.

- Precarious or non-existing titling of housing and land properties prevents access to public services and credit through regular channels.

- Social marginalization and insecurity discourage private and public investments to extend existing grids into informal settlements.

- Failure to apply targeted public subsidies to extend service infrastructure that addresses poor settlements.

- Failure to apply cross-subsidies and block-tariff schemes to facilitate access by low-income clients.

- Failure to integrate energy-access projects into urban poverty-reduction programs.

- Resistance to using public subsidies for infrastructure investment.

Problems and barriers in rural areas

- Geographic isolation can render uneconomic the extension of electricity-grid coverage (>$1000–1500 per additional connection).

- Limited distribution of modern fuels (e.g., LPG) for caloric uses because of geographic isolation, distance to markets, and absence of distributing enterprises.

- Insufficient income/lack of credit prevents purchase of decentralized technology equipment (e.g., PV household systems, biomass and minihydropower systems for isolated municipalities).

- Failure of public investment to build the infrastructure required for energization.

- Failure to apply targeted public subsidies to induce the purchase of decentralized energy technologies by households and communities.

- Illiteracy and low levels of education and participation in programs.

- Failure to integrate energy access into rural development and poverty- reduction programs.

- Lack of political will and low priority of disenfranchised rural communities with weak political representation.

---

include mechanisms for financing new and more efficient equipment, as well as destruction of the inefficient equipment replaced.[45, 46]

Table 23.12 lists the most common types of energy access policies found in rural and urban contexts in LAC countries. Policies used in both contexts, such as subsidization of electric tariffs and household fuels (e.g., LPG, kerosene), are listed under general policies.

23.8.4   Selected Energy Access Policy Experiences in LAC

23.8.4.1   General Fuel Subsidies

Fuel subsidies as part of an energy policy are nominally justified as a means to provide energy access to populations that are otherwise unable to afford it. However, in reality, fuel-price subsidies in LAC are driven by political decisions, and are largely appropriated by affluent sectors of the population who are the largest fuel consumers. The available evidence does not point to any discernible positive impact of these generic fuel subsidies on improving energy access for the poorer population. Despite that, it is estimated that total public expenditure on fossil fuel subsidies in LAC countries amounted to US$25 billion in 2005, of which approximately 8%, or US$2.14 billion, was for LPG. Venezuela, Ecuador, and Argentina spend the largest amount in LPG subsidies[47] (Table 23.13). Uruguay, Brazil, Peru, Jamaica, Honduras, and Chile do not use this type of subsidy. A major opportunity exists in the region to reorient public expenditure away from these highly regressive subsidies, and toward more carefully designed subsidy schemes that specifically target only the poor beneficiaries.

---

45 Equipment replacement and rational use of energy are the focus of recent proposals in Argentina, Brazil, Colombia, and Cuba, and includes new labeling for refrigerators and the retirement of older, inefficient equipment (ECLAC, 2008).

46 For a more in-depth analysis, see ECLAC, 2009.

47 Transport fuels: diesel oil is subsidized in Venezuela, Mexico, Argentina, Ecuador, and Colombia. Fuel oil is subsidized in Mexico, Venezuela, Ecuador, and Argentina.

Table 23.12 | Policies for energy access in a sample of LAC countries.

Policies for energy access	Argentina	Brazil	Bolivia	Colombia	Chile	Guatemala	Paraguay	Peru
General policies								
Electricity tariff subsidies, general or targeted to poor households	✓	✓						
LPG or kerosene subsidies, general or targeted to poor households	✓	✓	✓	✓			✓	
Promotion of private investments in energy-access solutions and/or equipment provision and service	✓	✓	✓	✓	✓	✓		✓
After-sales service and maintenance of energy-access solutions								✓
Promotion of productive and income-generating activities		✓	✓	✓	✓	✓	✓	
Improved access to basic water and sanitation services for poor households	✓	✓		✓	✓	✓	✓	✓
Rural area policies								
Rural electrification through grid extension (extending electric grid to rural off-grid populations)	✓	✓	✓	✓	✓	✓	✓	✓
Rural electrification through decentralized renewable energy sources (e.g., PV, biomass, mini hydropower)	✓	✓	✓	✓	✓	✓	✓	✓
Equipment programs for poor households in rural areas (e.g., stoves, refrigeration, lighting, pumps)		✓		✓			✓	✓
Firewood-substitution programs (LPG, kerosene, and other fuels for cooking use)			✓	✓	✓	✓		✓
Sustainable firewood programs (managed firewood use with improved stoves)		✓			✓	✓		
Capacity development, participatory planning, and local stakeholder engagement in developing energy-access solutions in rural areas		✓	✓		✓	✓	✓	✓
Gender-specific programs			✓			✓		
Urban-area policies								
Regularization of clandestine electricity-grid connections	✓	✓						
Equipment replacement and rational use of energy programs for poor households in urban areas (e.g., refrigeration, lighting)		✓						
Participatory planning and local stakeholder engagement in developing energy-access solutions in urban areas.		✓						

Table 23.13 | LPG subsidies in LAC countries (2005).

Countries	LPG subsidy expenditure in 2005 (US$ million)
Argentina	597.52
Bolivia	35.39
Colombia	168.17
Cuba	73.39
Ecuador	488.89
El Salvador	53.35
Haiti	4.16
Dominican Republic	181.07
Venezuela	531.33
LAC	**2133.27**

Source: Ríos Roca et al., 2007.

23.8.4.2 Brazil: Targeted LPG and Electricity Subsidies for Poor Families

In Brazil, LPG is the main fuel used for cooking. In the late 1990s, as part of a general revision of public expenditure, the government proceeded with the removal of general LPG subsidies. To mitigate the price impact on low-income families from the withdrawal of LPG subsidies, the government created a new scheme in 2001, *Vale Gas*, to subsidize energy supply for low-income families. A gas voucher was introduced for LPG, which provided financial assistance to consumers with a per capita income of up to half the minimum wage. This program was intended to assist approximately nine million people, more than half of whom resided in the northeastern region.

To be eligible for this subsidy, families had to prove income shortage and enlist with the government's registry of poor families. From 2003, the Lula government maintained this program, incorporating it as part of other social support mechanisms, known today as *Bolsa Família*. It has reached some 12 million families (almost 25% of all families in Brazil) at a cost of US$5.5 billion in 2008 (ESMAP/WEC, 2006).

Targeted electricity subsidies have been used to reduce electricity costs for poor consumers. Progressive rates (low rates for lower levels of consumption) were targeted at consumers who could provide evidence of financial need. In this case, the eligibility of families for subsidized rates was delegated to the utility companies, under the supervision of the National Electric Energy Agency, the regulatory entity of the Brazilian electrical system. In the Amazon region, targeted fuel subsidies are fundamental to increasing energy access. Since 1993, diesel oil has been subsidized through the *Conta de Consumo de Combustíveis* fund (Fuel Consumption Account) with resources collected from special taxes on all electricity bills for households in the interlinked system. This is an example of a cross-subsidization scheme between grid-serviced electricity consumers and off-grid diesel generator users in remote areas.

Incorporation of energy access into poverty-alleviation programs in Brazil

The lowest levels of energy access[48] are found in the north and northeast regions of Brazil, together with the lowest Human Development Indexes in Brazil. The 2000 National Census (Instituto Brasileiro de Geografia e Estatística, 2001) shows that 64% of households without access to electric lighting have a family income below two minimum wages. This figure increases to 89% for households below these minimum wage units. Since the 1960s, the Government of Brazil has invested in rural electrification programs, recognizing that energy access in isolated areas is central to reducing poverty and hunger, improving health, literacy, and education, and improving the living conditions of women and children. The federal Program for Energy Development in State and Municipalities began in December 1994 with the explicit goal to reach isolated regions that do not have grid coverage, mainly through PV systems and locally available renewable sources. The major focus was the electrification of about 50,000 schools without electricity and of water pumps in areas vulnerable to drought. This program was later incorporated by the Cardoso government (1995–2002) into the *Luz no Campo* program, run by *Eletrobras*, the state-owned electric utility. Its objective was to finance the electrification of one million new rural consumers over a period of three years through grid extension. In turn, the Lula government (2003–2010) maintained this program under the name of *Luz para Todos* as an integral part of the government poverty-alleviation policy *Bolsa Família*. Official data show that as a result of the program 1,877,362 additional households gained access to electricity during the past five years.[49]

The major features in the success of the Brazilian programs include:

- political priority and government commitment to rural electrification programs since the 1960s;

- continued public investment and strengthening of program budgets through various administrations from 1995 to 2009;

- effective mechanisms for targeting the available subsidies exclusively to poor families in need, thus ensuring highly efficient public expenditure; and

- integration of the energy access program into the government's broader policy of social support for poverty alleviation.

---

48 Habitants without access (%): Brazil = 5.5, north = 17.6, northeast = 11.1, middle-west = 3.9, south = 3.1, and southeast = 1.9 (Ministero de Minas y Energía, 2003).

49 This section draws heavily from a paper elaborated by Suani Teixeira Coelho, Patricia Guardabassi, Beatriz Lora, and José Goldemberg.

Box 23.11 | Colombia's Act 142 (Régimen Legal de Bogotá D.C., 1994)

Colombia's Act 142 of June 11th 1994 established a regime for the provision of household public services. It sets out explicit social equity criteria, such as:

- continued expansion of public services coverage through systems that offset the insufficient payment capacity of users;

- proportional tariff rates for low-income sectors according to principles of equity and solidarity;

- subsidies to people demonstrating insufficient income;

- stratification of individual households according to a common national methodology and criteria, and clear identification of the households to be provided public services at the municipality level;

- public investment support, and use of other incentive instruments, to promote utility companies in departments and at the national level.

23.8.4.3 Colombia: Massive Gas-application Program 1997–2009

The use of cross-subsidy schemes helps lower income users to pay for services and covers their basic needs.[50] In practice, higher income households, commercial, and industrial users pay a surplus on the full cost of the public service. This is defined by the Solidarity and Redistribution Fund. These funds are used to subsidize the public service cost for the lower income users. Intermediate income groups pay the full price.

These principles have guided Colombia's massive gas application program since the mid-1990s. The program started by connecting lower income users, providing natural gas for household caloric uses. The plan increased the number of households using gas from 0.5 million in 1991 to 4.3 million by July 2007, and the number of municipalities covered from 191 in 2000 to 415 in 2007. Available data show that the program has continued to target low-income users as the primary beneficiaries. Of these 4.3 million households, 53% are in low income categories (strata 1–2) and receive subsidized rates, and 85% are in the lower half of income categories (strata 1–3). Only 15% of households fall into the upper income categories (strata 4–6). It is difficult to quantify the amount of state contributions to this plan, but it is widely known that the highest cost did not derive from the subsidies to poor people, but from an income transfer made by the state company *Ecopetrol* (producer of the natural gas) to another state company, Ecogas (distributor and transporter of the natural gas, which was privatized in 2007). It is difficult to assess if this model could be replicated in other countries, as it depended on the particular institutional framework in Colombia at that time. The main lesson from this policy experience is that when there is sufficient political will and state backing, it is possible to find formulae to achieve massive gains in bringing energy services to the lower income population (ECLAC, 2009).

---

50 Official Journal: Santafé de Bogotá, D.C., Monday, July 11, 1994. CXXX N° 41.433.

23.8.4.4 The Argentine Off-grid Electrification: an Example for a Medium Developing Country

Argentina has made significant progress in its efforts to reform the power sector. While it has a relatively high overall rate of electrification (over 95%), substantial numbers of the rural population still remain without electricity services (over 25%). The Renewable Energy for Rural Markets Project (PERMER) aims to provide about 35,000 remote rural households, 1750 public services (rural schools, health posts), and 500 productive uses with electricity through provincial 'off-grid concessions' that are negotiated or bid out for minimum subsidy and regulated by independent provincial regulating agencies. The concessionaire is free to choose the least-cost technologies applied to meet its obligation to provide universal service.

This project subsidy is about 50–60% and paid partly at the time of procurement of a new lot of systems and partly against met installation targets, to balance the advantage of a direct control of outputs with manageable working-capital costs to the concessionaire. Installations, service quality, and customer satisfaction are verified *ex post* by the regulator. The monthly fees paid by the user are for O&M costs and for recovering the concessionaire's share of investment costs.

To ensure that energy services meet the local demand, as well as to attract private sector interest, market studies were conducted for all interested provinces. The existence of such a market study is one of three preconditions for becoming a PERMER participating province entitled to subsidies from the bank loan.

The business model is changing 'from grid to off-grid.' The management and technical personal of the utility involved in the project is linked to the utility in charge of the grid-connected market. Various lessons have been learned and the utility way of doing business has changed. Most of these changes are aimed at lowering the extremely high costs of operating and maintaining the very remote and dispersed systems. One common underlying element to most of these changes is an increased attention to the responsibility of local users, microenterprises, and sub-contractors for O&M, fee collection, and new installations. This approach is a promising way to improve concession models and combine a variety of important advantages:

- Reductions on O&M cost as they can replace costly field visits by visits of the subcontractor.

- The subcontractors are closer to the market, both in geographic and social terms, and can hence react directly and in a more flexible way to individual customer needs.

- The majority of users are indigenous – and so are the subcontractors. This improves communication.

- Being present locally, the subcontractor has a better grip on potential reasons for payment default.

- This local social control works both ways because subcontractors feel responsible toward 'their users' for the overall service quality.

- Some of the users pay the subcontractor with goods instead of money, which reduces default rates and increases local market efficiency.

- Additional high-value employment and income is created in local microenterprises.

- One of the subcontractors has started to offer a variety of additional services in response to local demand.

The user is the central part of any off grid system; therefore borders blur between user and utility. One of the most interesting findings of a recent study on the emerging issues of service quality is that both the perception of responsibilities and the specific O&M arrangements for the individual off-grid systems have evolved over time, depending on the specific users and technicians.

Training is crucial. A full-time sociologist has been hired to improve user training and demonstration material for SHS. A training session is given to each user during system installation. The aim is to:

- improve battery treatment so as to decrease life-cycle cost and increase user satisfaction (fewer failures);

- avoid misuse of components (e.g., short-cutting fuses and charge controllers, use of inadequate loads) so as to reduce system failures (and the related costly repair visits!);

- improve the users' understanding of their systems and energy efficiency measures (i.e., to increase 'energy culture') and the roles and responsibilities of each player so as to improve overall user satisfaction; and

- ultimately to lower the default rates.

Regarding the regulator, two important ways for cost reduction were identified:

- Adopt adequate 'off-grid service standards.' Specific regulations for rural service have to be adopted to allow for the different service levels provided by off-grid systems. Regulations were simplified to consider the off-grid situation.

- The driving force behind all of the utility's improvements to its business model is the aim to decrease the high costs of visiting remote, decentralized systems. In any remote off-grid system, the user is by definition the only one who is always at hand. User behavior therefore decides about the ultimate quality of service, and interventions from the utility side should be avoided where possible. The most interesting improvement is the integration of independent local microenterprises as 'subcontractors' in the downstream part of the rural service-delivery chain.

23.8.4.5   Bolivia and Peru: Rural Electrification in Remote Areas

In Peru, electricity coverage in rural areas has made steady gains, increasing from approximately 8% in 1993 to 30% in 2007. The Government of Peru has announced plans to further extend electricity coverage to 5.6 million people through public investments of US$1.33 billion between 2008 and 2017 (Ministerio de Energía y Minas, Republic of Perú, 2008). In Bolivia, electricity coverage of rural areas rose from 6.8% in 1976 to 28.3% in 2001 (Espinoza, 2005). Through its latest 2002–2007 rural electrification plan (*Plan Bolivia de Electrificación Rural,* PLABER, 2003) the Government of Bolivia expected to expand this coverage to 45% by 2008, investing some US$170 million to connect 200,000 additional families. The estimated cost per additional rural connection in both countries runs slightly above US$1000 per household, increasing for more remote communities.

The experience of both countries clearly shows that gains in rural electricity coverage require a strong commitment of public resources to infrastructure investments over extended periods. It also shows that the support of international cooperation agencies and regional development banks[51] has played a key role in supplementing scarce national funds to tackle successfully the large social investments demanded by rural electrification.

---

51   Such as Corporación Andina de Fomento, InterAmerican Development Bank, etc.

The development of a market for decentralized energy technologies in rural areas requires large investments in pilot projects, technical assistance, technology transfer and diffusion programs, capacity development, and the engagement of local communities. Many of these extension activities have been funded through international cooperation projects. Bolivia has made steady gains in rural electricity coverage from about 12% in 1992 to over 30% in 2005.

In both countries, rural communities show a positive willingness to pay for energy services, with many installing diesel generators despite their relative higher tariffs. Around 25% of households in remote rural areas actively seek to install PV systems, through both private and international cooperation projects. Low rural incomes severely limit the ability to pay the full cost, so a key factor in the success of these programs is the design of appropriate financing schemes. Financing schemes generally take the form of a combination of a monthly service tariff (to be paid by the user) and a one-time subsidy paid by the government to install the equipment. Depending on the scheme and the ability to pay, a portion of the equipment might also be paid by the user through a combination of microcredit, tariff payments, or other means. Ideally, a project should catalyze the development of a self-financed local system, with paying users and service providers organized through a private or local community enterprise.

Financing schemes in successful projects are generally characterized by the following features:

- Rural households assume ownership and financial responsibility for part of the cost of the equipment that is transferred to them. This creates incentives to care for and maintain the equipment owned by the household (e.g., in-house equipment, such as batteries and lights).

- Sufficient public funds are assigned to the project to subsidize the capital and installation costs of the technology as required by the rural household's ability to pay. The public agency must also ensure thorough testing and certification of the technology prior to installation.

- Monthly service tariffs paid by rural households must be set at a level that reflects both the ability to pay and the coverage of local costs. Both are necessary to ensure the system's longer term sustainability.

- A private or community enterprise is established to collect tariffs, provide maintenance and technical support, and ensure the whole system's quality of service.

23.8.5   Lessons Learned

- Results depend on a proactive role by government in the energy sector, and in many cases through its own public utilities, to accelerate universal access to modern forms of energy.

- Definition of feasible and attainable targets for sectors without access, as well as possible resource availability, along with the target group's economic, social, and environmental condition.

- Specific support, particularly in rural areas, to facilitate the mobilization of local funds to contribute to closing the funding gap. The mobilization of these resources requires appropriate mechanisms and enabling institutional and legal frameworks, not only at national, but also at local levels.

- Adequate management models are needed to guarantee long-term sustainability for rural access to electricity and modern fuels. Examples of failure were related to the absence of or inadequate management of projects for off-grid or remote settlements energization.

- A specific pro-poor regulatory framework to protect poor communities and promote access to modern forms of energy at an affordable price and tariff.

- Scale-up investments targeted to decentralized energy systems. Successful examples and expertise around the world should be leveraged.

- The scope of energy use and investment subsidies in both rural and urban areas should be defined clearly.

- Allocate funding and resources to create local capacities and promote energy literacy to ensure the effective involvement of local actors and their organizations in the energy planning and decision-making processes. Capacity and knowledge are the key elements to empower poor people to participate in the energy debate – and in the production, implementation, operation, maintenance, and use of the local energy infrastructure.

- Energy for poor people would need to included, as a specific chapter, in an integrated framework of energy projections.

- Reasonable supply horizons would need to be guaranteed by means of callable investment plans.

- Capacity development and reinforcement would also need to go hand by hand in the development of pro-poor policies.

- Deeper integration and cooperation at the regional level could facilitate access to modern forms of energy at the same time as reducing energy costs, expanding the market, increasing possibilities for projects that are not feasible at national level, sharing energy resources, and promoting technological development.

23.9   Concluding Remarks and Suggestions

A substantial amount of analytical work has been carried out on the main ways to address the challenge of access to modern forms of energy. Some recent accomplishments include:

Table 23.14 | Examples of energy access – direct and indirect assistance.

Access investment type	Investments or grants in:	Objective
Direct	Modern fuels for cooking Rural, urban, and peri-urban electricity Productive uses of energy in homes and small businesses Energy efficiency for households, communities, or small businesses Institutional development and reinforcement	modern forms of energy for households and communities modern forms of energy for new or improved productive uses and small enterprise development Development of new institutions to support energy access Improvements in energy efficiency (household or building efficiency for residential energy)
Indirect	Development of information systems and data records Improvements in policies, strategies, and technical assistance Sector studies of energy-access issues Power plants, transmission, and other infrastructure that support the development of greater energy access	Facilitate improved investment for energy access Investments in supporting infrastructure necessary to extend new or improve quality to existing households Promote economic development that can help poor people in more indirect ways Generate information on energy services for poor people Development of policies and strategies to enhance energy access

- identifying and documenting best practices in rural electrification;

- mobilizing expanded investment from both the public and private sector;

- developing frameworks to regulate new institutional arrangements for modern forms of energy provision, including private electricity distributors that serve rural and peri-urban populations;

- developing methodologies and case studies that demonstrate the benefits of targeted energy service investments for poor people;

- improved understanding through surveys and other research on how poor people meet their energy needs in the rural and peri-urban context; and

- regional strategies for scaling-up energy access, focusing mainly on electricity.

The main areas identified as a priority for addressing energy access include the expansion of rural electrification programs in many developing countries through grid or off-grid programs, a greater attention to the policy reforms necessary to address energy for the urban and peri-urban poor, and a refocusing on the problems involved in the use of traditional fuels for cooking.[52] It is also important to address the more upstream investments necessary to expand energy access to poor people. These are issues highlighted as important in the transition to higher quality fuels and appliances for poor households in developing countries.

This development of knowledge, experiences, and proposals is an important contribution to the design and implementation of policy strategies and concrete actions. However, the figures show that major efforts are still needed to make universal access a reality.

International development agencies are committed to promoting affordable energy access in developing countries, but sometimes retain many of the old biases of previous strategies and policies. An energy strategy must start from a deep analysis of the need for energy by poor people. Energy access and affordability are multidimensional and context dependent, and the appropriate responses to these are equally diverse. In several cases 'energy' is essentially taken as synonymous with 'electricity,' which does not demonstrate an understanding of the full range of energy needs, and so cannot fully address energy access.

Past investments in programs such as rural electrification and renewable energy for rural areas have yielded significant achievements in terms of progress in countries that are committed to such programs, along with new intervention models that are also replicable elsewhere. In most countries, these investments must, of course, be complemented by the development of supporting infrastructure (Table 23.14).

Current priorities demand greater strategic direction, such as:

- Direct assistance to poor people to facilitate access to modern fuels and electricity through various measures and incentives.

- Reduce costs and increase efficiency in upstream infrastructure for urban, peri-urban, and rural populations.

- Improve, promote, and implement energy efficient programs oriented to poor people.

- Promote good governance, including transparent and pro-poor regulatory mechanisms.

- Removal of other key barriers, such as up-front costs.

- Develop specific institutional frameworks with fixed targets and adequate resources, both human and financial.

- Implement international financing mechanisms, based on soft loans, oriented to develop and subsidize the necessary infrastructure to facilitate access to modern forms of energy.

---

52   For reasons that need deeper analysis, the majority of the efforts are oriented to electricity and much less so to fuels for caloric uses.

Case-specific analyses are necessary to develop appropriate solutions. Some countries face major difficulties in addressing the problem because of the low level of development and weakness of the economic system.

23.10   Key findings and lessons learned

This chapter's discussions and analyses offer a range of lessons and insights:

- Although efforts have been made to improve access to electricity and modern fuels, the energy gap – both between developed and developing countries, and between wealthier and poorer people in many countries – persists, and is even widening.

- Governments and international organizations recognize the positive effects of access to energy services. However, very few countries have yet developed a comprehensive approach to improving access with a specific focus on poor people.

- There are significant gaps in our knowledge of energy services, as well as a poor understanding of the main reasons for, and barriers to, access to modern forms of energy.

- The structural reforms implemented during the 1980s and 1990s emphasize efficiency, but ignore social issues. This has helped create a non-friendly environment for energy access.

- Priorities could include the following:

  ▷ Increase energy coverage for households and productive uses.

  ▷ Enhance generation capacity by including regional projects and regional cooperation.

  ▷ Address energy services for key public facilities, such as schools and clinics.

  ▷ Push for better achievement of basic human needs, represented by energy services, such as provision of electricity for lighting, health, education, and community services, and modern fuels and technologies for cooking, heating, and sanitary uses.

  ▷ Solutions would need to be tailored to individual country contexts, especially in relation to institutions, capacity, and energy resource availability. A mix of actions for each country would be driven by the availability of resources, plus the legal, regulatory, and policy environment, the institutional and technical capacity, the relative cost of implementation of the different solutions, and/or the sociocultural contexts.

  ▷ Capacities should be developed and reinforced at individual, institutional, and systemic levels to enable feasible and affordable policies to be designed and implemented.

- The significant gains achieved in electricity coverage, for example in India, China, and Latin America, were mainly the result of strong political commitment and public investment programs executed through public utilities with clear development mandates.

- The transference of resources between the oil productive chain and the natural gas productive chain has financed infrastructure development and expansion in several Latin American countries.

- Levels of poverty, particularly in rural areas of LDCs, require that infrastructure investments be either subsidized or wholly executed through public funds, because the low income capacity of the target users prevents full cost recovery.

- Governments must acknowledge that extending energy access further to poor people will only result from a political decision that establishes clear mandates, targets to be achieved, and commitment of the required public funds over extended periods of time (decades). There is sufficient evidence that important changes will only be possible through political will at the highest level.

- International cooperation and development banks have played a key role in providing concessionary financing and grants to supplement national funds, making the undertaking of these large investments possible. This would hopefully continue.

- Investment and O&M expenditure on modern forms of energy represent a significant proportion of total investment and GDP in LDCs. Achieving universal access in these countries by 2030 will require international support, political compromise in sustainable development, and reduction of inequality.

- Effective targeting of direct and indirect subsidies has been a key to all the successful experiences of expanding access to electricity, modern fuels, and associated technologies to poor populations. It will not be possible to avoid subsidies if the targets are to be achieved. In many countries, significant opportunities exist to apply cross-subsidization and differentiated/block-tariff schemes in energy services across income groups, to enable adequate access and regularization of services to the poorest populations.

- Facilitating access to appliances was a key component for households and, in successful programs, allowed them to benefit from increased energy use. Energy services provision should include access to modern forms of energy and access to appliances.

The main lesson learned from past experiences on the implementation of strategies and actions for access to modern forms of energy and technologies is that there is some room for original or innovative proposals.

Key basic principles or enabling conditions that appear repeatedly in documents or proposals include:

- Providing universal access to modern forms of energy is a necessity, not a luxury.

- The need for good understanding of the energy services to be addressed, along with reliable analysis.

- It is imperative to transform the paradigm. Energy activities are not just another industry, but a system with strong socioeconomic and environmental dimensions, with direct impacts on the sustainable development of any country. Energy is not a commodity, but a strategic good. It is essential for economic, social, and environmental sustainability.

- Governments must make long-term commitments, with explicit and clear public policies oriented toward poor people.

- Several good practices can be replicated elsewhere in places that experience similar circumstances (policies and regulation, capacity development, technical standards, best-available technologies, financing and implementation approach, coordinated research and development).

- The need for soft funds and financing mechanism to develop infrastructure and to provide up-front costs for the potential consumers.

- Direct government involvement in implementation, through public utilities or private or non-profit organizations (e.g., NGOs, cooperatives) or adequate public-private cooperation.

- The oriented subsidies (direct or indirect) to create enabling economic conditions.

- Access to modern forms of energy for poor people should not only be conditioned to clean energy forms but also to clean appliances.

23.11   Policy Options and Some Recommendations

To achieve the overall goal of economic growth and poverty alleviation but avoid the pitfalls of poor planning and inappropriate targets is a daunting challenge in any context. Providing universal access to modern forms of energy poses its own set of additional institutional and policy challenges, which have historically hindered efforts to increase energy affordability and are a major reason why so many people remain without access to better energy services. Policy recommendations can take the form of general ideas or guidelines. Regional and national contexts should be considered in defining strategies, instruments, and measures.

23.11.1   Diagnosis and Information

A better understanding and a clearer diagnosis of the structure and functioning of energy systems, along with the needs (energy services) to be supplied, is needed. It has often been absent in the discussion of proposals and the role of public policies.

An information system on energy use will help to identify problems and barriers, and will promote understanding of local conditions, sociocultural behavior, and a system's ability to implement actions.

Section 23.3 remarks, adequate information on needs to be satisfied and priority energy services to be provided are the necessary initial step. Identification of barriers and problems to address energy affordability is a key issue to identify, design, and implement adequate strategies and measures.

Good policies and strategies need good diagnosis. Wrong or absent diagnosis could mean inadequate proposals. Support and funds to diagnosis and information would need to be part of the strategies.

23.11.2   Reform of the Tariff, Tariff Structure, and the Subsidy Systems

Subsidies are generally justified as a response to inequality and social expectations in energy provision (Barnes and Halpern, 2000; UNEP, 2002). However, their net effect can be positive or negative depending on the intended goals of the subsidy, and the way a subsidy is implemented.

An effective tariff and subsidy regime has to be transparent and minimize administrative costs to avoid gaming of the system and to maximize the benefits that accrue to the intended recipients (UNEP, 2002). The subsidies themselves would need to have some of the following characteristics:

- **Clear mandates**. Subsidies would need to have clear mandates and be appropriately financed to ensure that the financial burden is acceptable and properly allocated, and that the opportunity costs are not too high.

- **Targeted**. Subsidies could be designed so as to reach those most in need and to ensure that resources are not wasted. Lifeline tariffs can be tied to other aid programs to ensure they are meeting the recipients' needs.

- **Phased**. Subsidies would need to be established with clear guidelines for their phasing out, such as a sunset clause or performance-based milestones.

- **Market enhancing**. Subsidies that help develop and nurture a market early in its development can be very effective, while subsidies that can undercut a growing market should be avoided.

- **Flexible**. Subsidy programs would also need to incorporate flexibility to deal with the uncertainties that are an inevitable part of making changes to institutions and markets.

- **Complemented**. It needs to be complemented with funds toward solving the first cost capital financing problem (Barnes and Halpern, 2000). Up-front costs of equipment are, usually, the key barrier.

Generally speaking, social tariff and social prices are the instruments with which to address poor people. Two conditions are key elements in the definition and application of such social tariff and prices: to be sure that all the poor people that need the subsidy are included (guarantee of inclusion) and to be sure that people that do not need the subsidy are not included (guarantee of exclusion).

Finally, for economic sustainability of the system it is necessary that average costs of the total energy provided be covered by the average tariff or by the price recovered by the provider.

23.11.3 Innovation and Guarantee of Financing Mechanisms

Financing mechanisms are needed for every scale of energy intervention. Mobilizing affordable and genuine international, regional, national, and local funds is crucial. Sustainable energy access will not be achieved by occasional or intermittent donors and actions.

Existing and new financing mechanisms could be oriented so that they have a real impact in addressing energy access for poor people. This review process needs to be inclusive, and result in the modification of procedures and mechanisms to enable small and non-conventional energy programs targeted to poor people to receive adequate funding. Likewise, the proportion of pro-poor energy investments would need to increase consistently with the magnitude of the challenge at hand. This is urgent if the funding gap is to be bridged.

The mobilization of local funds should also be considered. International financial institutions and donors can play a facilitating role. The mobilization of these resources requires appropriate mechanisms and enabling institutional and legal frameworks not just at the national level, but locally too.

23.11.4 Changes in Regulatory Structure

Current regulations often act at cross-purposes to the efficient delivery of services through new and innovative technology and institutional options, such as the use of distributed generation (ESMAP, 2001; Morgan and Zerriffi, 2002). This is the result of both regulatory structure and regulatory practice in systems that have generally been put in place with centralized utilities in mind. It is difficult for such centralized

regulatory systems to monitor effectively a large number of smaller operators. Similarly, the regulatory burdens on the smaller operators of a system designed for large utilities can be prohibitive. These regulatory problems can be handled in a variety of ways, including the creation of standardized licenses, the delegation of regulatory responsibility, flexible power quality and reliability standards, and tailoring regulations to different types of entities (Reiche et al., 2006).

23.11.5 Capacity Development and Strengthening

Capacity development can be understood as the processes of creating, mobilizing, enhancing (or upgrading), and converting the skills and expertise of institutions in the contexts required to achieve the specifically desired socioeconomic outcomes. Capacity development must be achieved through activities at the individual, institutional, and systemic levels. Capacity-development efforts at each of these levels are discrete elements of the capacity-development process (Bouille and McDade, 2002).

No single institution can have the capacity to resolve the complex governance, energy, and capacity-development challenges and their linkages to issues of equity, environmental sustainability, economic efficiency, and public sector management. The challenges are not static, but change in dimension, location, priority, and costs as the energy system evolves.

An interdisciplinary perspective and a multistakeholder approach are needed to address the multiple dimensions of access to modern forms of energy.

Resources would need to be allocated to develop local capacities and promote better knowledge of local energy. Local populations would need to be involved in energy planning and decision-making processes. Better understanding of the role of energy can contribute to the sustainability of systems and improve relations between the energy provider and the energy user. Target actors for capacity development should include government, the private productive sector, civil society, academia, consultant institutions, and the media.

23.11.6 Policy Alignment

Energy policy is part of a wider development policy and would need to be aligned with other sector policies and objectives. If these policies are misaligned, they can reduce the effectiveness of any given policy. Policy misalignments can occur when different energy policies work at cross-purposes or when government priorities that could benefit from an effective energy policy are not aligned.

While individual technology choices are made at the microlevel (that is, by an individual or small group of decision makers), these decisions are influenced by policies at the macrolevel, generally, the government (Stewart, 1987). The impact of macropolicies on technology choice is the result of changing "firm objectives, resource availability and cost, markets, and technology" (Stewart, 1987). For example, high import tariffs on technology components can significantly drive up the price of distributed systems and so act as a barrier to access. Taxes can also change the relative pricing of traditional versus modern fuels, creating price pressures even on those without access.

At a broader level, there is a need to link rural and peri-urban energy supply more closely with rural development. This would shift the focus from minimal household supply to a more comprehensive approach to energy that includes productive activities and other welfare-enhancing uses of electricity. Ideally, the linkages between energy and other policy priorities, such as health, education, and poverty alleviation, should be recognized explicitly and local solutions that address these needs be encouraged and supported. By taking a more comprehensive approach, revenues can be increased because of the higher ability and willingness to pay by productive users, and the effectiveness of investments should be higher because of the link with welfare-enhancement goals. The new mix of customers and demand allows for natural market segmentation, which improves the viability of energy supply efforts.

To be effective, policies and programs within the energy sector geared toward low-income energy services also need to be aligned with other poverty-alleviation efforts. Provision of infrastructural services, including modern forms of energy, does not change the poverty equation. When development efforts fail, improvements to electricity supplies alone have little effect on local welfare. Work done in Peru found that providing a combination of services – electricity, water, sanitation, and telephones – had a greater impact on poverty reduction than that provided by a single service. Adding a fourth service resulted in welfare improvements that were seven times greater than those delivered by the second service.

Many populations experience an inability to pay for energy services. In some cases, such as in Lao People's Democratic Republic, villages have been connected to electricity for 15–20 years, but around a quarter of households remain unconnected (Independent Evaluation Group, 2008). Where purchasing power is the underlying problem, it is critical that provision of modern forms of energy be evaluated alongside other development options. Overall, given the low returns on investment in energy services, and in particular electrification, it is essential that public resources be concentrated on the investments that have the greatest impact on development. Poor people are more likely to benefit from an access to modern forms of energy approach that creates jobs and raises income, as opposed to a simple household connection.

23.11.7   Regional Integration and Cooperation

The traditional approach of limiting energy planning and service provision to the political frontiers could be a barrier for access to modern forms of energy. There are several reasons why nation-based planning is suboptimal:

- The geography of energy supply options does not necessarily correspond to political boundaries – the cleanest and cheapest energy source may lie across national borders.

- National energy markets are often too small to justify the investments needed for particular energy supply options.

- Cross-border energy supply often provides diversification of energy source – a key component in energy security and cost reduction.

Latin America offers several examples in which integration and cooperation among countries reduce energy costs and, indirectly, facilitate access to energy.[53] Africa, for example, has very good potential for such integration based on projects or interconnection. An integrated energy market, such as in the European Union, is not necessary. An expanded regional market could offer feasible technical options to reduce costs and still maintain the integrity of national markets. Bilateral or multilateral agreement on energy projects that share costs and benefits could be an enabling action to expand markets to low-income populations. The development of energy markets on a regional basis offers significant benefits, as the linking of national petroleum and electricity industries can help mobilize private and domestic investments by expanding market size.

Major benefits are associated with regional energy integrations: improved security of supply, better resource allocation, enhanced environmental quality, and wider deployment of renewable energy resources, all of which contribute to access to modern forms of energy and sustainable energy systems.

Africa has a regional diversity that offers substantial opportunities for integration. Regional integration is increasingly being seen as a way for individual countries that suffer from structural and economic weaknesses to join the global economy. Better macroeconomic conditions will facilitate access to energy.

23.11.8   Final Remarks

Overall, and on the basis of successful experiences of increasing access to modern forms of energy, no single approach can be recommended

---

53   Close to 100% of Paraguay's population has access to electricity. This was made possible by South America's two largest binational hydroelectric power plants: Itaipú, made and financed by Brazil, and Yacyretá, implemented by Argentina. As both countries need the energy and have the capacity to implement the project, they could afford the project. Paraguay provided 50% of the river and paid 50% of the investment in energy. Both projects enabled Paraguay to have full access to electricity, a very difficult task for the country to achieve alone. A similar situation exists in Central America and in the Caribbean, where the major oil-producing countries (Venezuela and Mexico) are cooperating with small, oil-importing countries in the Caribbean. For example, Venezuela and Cuba exchange oil for medical assistance.

above the others. What is clear, however, is that the current institutional arrangements and policies have met with mixed success, at best. Reforms are needed, at global and country level, to strengthen the feasibility of energy projects for poor people, expand the range of actors involved, open up the regulatory system, and allow for innovation. Several examples demonstrate that some of the main factors for success include:

- Political will and government priority;

- Continuing support, both financial and administrative;

- Effective mechanisms for effective targeting of policy interventions; and

- Integration of energy policies with other sectoral policies, particularly those dealing with poverty alleviation where these exist.

Success depends on regional, national, and local circumstances. In some instances, such as Nepal's Rural Energy Development Program (REDP), a decentralized and participatory decision-making process and a holistic development approach is very important. This goes together with a strong community-mobilization process that focuses on:

- organizational development,

- skills enhancement,

- capital formation,

- promotion of technology,

- environmental management, and

- empowerment of vulnerable communities.

Participation, transparency, consensus decision making, and inclusion of all households in the community, irrespective of class, color, creed, or sex, are the four pillars of good governance to ensure equal ownership and equitable sharing of benefits accrued from MH systems.

Challenges and economic, sociocultural, and political barriers require more elaborated strategies and a higher global compromise to satisfy GEA objectives. The key catalyst to improving access to energy in some Latin American and Asian developing countries is political will. The examples of India, China, Argentina, Chile, or Brazil all demonstrate that if a political decision is made, the results are positive. However, this has not been the case in many sub-Saharan Africa countries, in LDCs, and in those with a very low Human Development Index. In these countries, universal access to modern forms of energy by 2030 will not be achieved with microactions and isolated measures, unless they are integrated into

a long-term national program with clear targets, dedicated and guarantee funds, adequate institutional frameworks, and robust strategies. An approach based on providing a few thousand solar lamps to rural settlements, or a PV with capacity of just a few watts, will improve the quality of life for poor rural inhabitants, but are not capable of real improvement in welfare and securing the inclusion of marginalized populations. They can be seen as intermediate actions, but should not be considered as a solution for access to modern forms of energy.

Reforming the way in which energy is financed and sustainably operated has a major potential to reduce inequality. Lifeline subsidies are needed, at the very least. There are conditions under which cross-subsidies can be effective, but this may require a different role for both higher level government agencies and the international system. Some principles, such as supporting energy planning, making investment capital available, creating incentives for commercial lending, promotional campaigns, and technical assistance, have already been identified for certain markets. Centralized agencies can also aid in coordination, in eliminating conflicts in mandates and programs, and in helping to build much needed local institutional and organization capacity.

The circumstances in developing countries militate that the energy path, especially for rural and peri-urban areas, be dissimilar to that followed by developed countries. This will require innovation and experimentation on both the technological and institutional levels. Lack of appreciation of such approaches at the policy level is curtailing progress, because many policymakers tend to follow conventional approaches without taking account of contextual differences.

We know that universal access to modern forms of energy at the household level depends on various factors, such as prioritizing energy access, long-term policy commitments by national governments to create strong institutional, regulatory, and legal frameworks, and financing from all available sources. It is important that governments facilitate support from national and international development organizations on the research, design, and development of appropriate technologies. Collecting, compiling, and sharing knowledge is equally important.

It is expected that governments will report on the progress they have made on addressing the energy access agenda in all its dimensions, including funding for small decentralized solutions, community capacity development, leverage of local indigenous financing, and achievement of national energy access targets, among others.

Last, but not least, the wide range of material analyzed shows that abundant ideas and proposals are available to address many different situations. What is absent are political decisions to implement them.

References

ACP-EU, 2009: Position paper on rural and peri-urban electrification, 2nd ACP-EU Energy Facility, *Newsletter: EuropeAid – Energy Facility*, Development and Cooperation, European Commission.

AGECC, 2010: Clean Energy Access for All – Low Carbon Energy Technologies and Poverty Alleviation, United Nations, New York, NY, USA.

African Development Bank, 2008: Clean Energy Investment Framework for Africa – The Role of the African Development Bank Group. Tunis, Tunisia.

African Development Bank, Asian Development Bank, Department for International Development – United Kingdom, Directorate-General Development – European Commission, Federal Ministry for Economic Cooperation and Development – Germany, Ministry of Foreign Affairs – Development Cooperation – The Netherlands, Organization for Economic Cooperation and Development, United Nations Development Programme, United Nations Environment Programme, and The World Bank, 2003: *Poverty and Climate Change Reducing the Vulnerability of the Poor Through Adaptation*.

Arnold, M., G. Köhlin, R. Persson, and G. Shepherd, 2003: *Fuelwood Revisited: What Has Changed in the Last Decade?* CIFOR Occasional Paper No. 39. Center for International Forest Research, Bogor, Indonesia.

Arnold, J. E. M., G. Köhlin, and R. Persson, 2006: Woodfuels, livelihoods, and policy interventions: changing perspectives. *World Development*, **34**(3):596–611.

ASER, 2007: *Costing for National Electricity Interventions to Increase Access to Energy, Health Services, and Education – Senegal Final Report*. Agence Sénégalaise d'Electrification Rurale, Energy Group, Columbia Earth Institute.

Baker Institute, 2006: *Poverty, Energy and Society*. Energy Forum. James A. Baker III Institute for Public Policy, Rice University, Houston, Texas, USA.

Banerjee, S., A. Diallo, V. Foster, and Q. Wodon, 2009: *Trends in Household Coverage of Modern Infrastructure Services in Africa*. Policy Research Working Paper 4880. World Bank, Washington, DC, USA.

Banerjee, R., 2006: Comparison of options for distributed generation in India. *Energy Policy*, **34**(1):101–111.

Bariloche Foundation, 2008: *Urban and Periurban Energy Access – UPEA II* – Global Network on Energy for Sustainable Development.

Barkat, A., 2005: Bangladesh Rural Electrification Program: a success story of poverty reduction through electricity. In *International Seminar on Nuclear War and Planetary Emergencies 32nd Session*. R. Ragani (ed.), 19–24 August 2004, The Science and Culture Series, Erice, Italy.

Barnes, D. F. and J. Halpern, 2000: *Subsidies and Sustainable Rural Energy Services: Can We Create Incentives Without Distorting Markets?* Joint UNDP/World Bank Energy Sector Management Assistance Program, Word Bank, Washington, DC, USA.

Barnes, D. (ed.), 2007: *The Challenge of Rural Electrification: Strategies for Developing Countries*. Resources for the Future, Washington, DC, USA.

Barr, N., 2005: *The Changing Social Landscape of Rural Victoria*. Department of Primary Industry, Victoria, Australia.

Barua, D. C., 2005: *Success of Grameen Shakti in the field of Renewable Energy Sector in Bangladesh.* Grameen Bank Bhaban, Dhaka, Bangladesh.

Bazilian, M., P. Nussbaumer, E. Haites, M. Levi, M. Howells and K. K. Yumkella, 2010: Understanding the Scale of Investment for Universal Energy Access. *Geopolitics of Energy*, **32**(10–11):19–40.

Bhagavan, M.R. (ed.), 1999: *Petroleum Marketing in Africa. Issues in Pricing, Taxation and Investments.* Zed Books, London, UK.

Birol, F., 2007. Energy Economics: A Place for Energy Poverty in the Agenda? *The Energy Journal*, **28**(3).

Bouille, D. and S. McDade, 2002: Chapter 6: Capacity development. In *Energy for Sustainable Development: Policy Agenda*. T. Johansson and J. Goldemberg (eds), United Nations Development Programme, New York, NY, USA.

Bouille, D. and N. Wamukonya, 2003: Power sector reforms in Latin America: a retrospective agenda. In: *Electricity Reform: Social and Environmental Challenges*. N. Wamukonya (ed.), United Nations Environment Programme, Roskilde, Denmark.

Bravo, V., 2004: *Requerimientos básicos y mínimos de energía de los pobladores urbanos y rurales pobres e indigentes de América Latina y el Caribe*, Fundación Bariloche, Argentina.

Brew-Hammond, A., 2010: Energy access in Africa: Challenges ahead. *Energy Policy*, **38**(5):2291–2301.

Buxton, M., A. Alvarez, A. Butt, S. Farrell, and D. O'Neil, 2008: *Planning Sustainable Futures for Melbourne's Peri-urban Region*. RMIT University, Melbourne, Australia.

Campbell, B., M. Luckert, and I. Scoones, 1997: Local-level valuation of savanna resources: a case study from Zimbabwe. *Economic Botany*, **51**(1):59–77.

CEMAC (Communauté Economique et Monétaire de l'Afrique Centrale), 2006: Action Plan for the Promotion of Access to Energy in the CEMAC Region: Summary of Final Report. European Union Energy Initiative, Eschborn, Germany.

Collier, P., 2007: *The Bottom Billion. Why the Poorest Countries are Failing and What Can be Done About It*. Oxford University Press, London, UK.

Coutinho, D. R., 2007: *Law and Development Policies in Brazil: Decentralization and Coordination in the Bolsa Família Program*. Fundacion Getulio Vargas, Rio de Janeiro, Brazil.

da Gama Torres, H., 2007: *Peri-Urban Growth in Latin America*. PPT presentation prepared for the UN Expert Group Meeting on Population Distribution, Urbanization, Internal Migration and Development, Cebrap, Brazil.

Davidson, O. and Y. Sokona, 2002: *A New Sustainable Energy Path for African Development: Think Bigger, Act Faster*. Energy and Development Research Centre, University of Cape Town, Cape Town, South Africa.

Davidson, O. and A. Mwakasonda, 2004: *Electricity Access to the Poor: A Study of South Africa and Zimbwabe*. University of Sierra Leone, Freetown, Sierra Leone and University of Cape Town, Cape Town, South Africa.

de Miranda, R. C., S. Sepp, E. Ceccon, S. Mann, and B. Singh., 2010: *Sustainable Production of Commercial Woodfuel: Lessons and Guidance from Two Strategies*. ESMAP, World Bank, Washington, DC, USA.

East African Community, 2006: *Regional Initiatives to Increase Energy Access: the case of the East African Community*. Gesellschaft für Systems Engineering, Vienna, Austria.

Eberhard, A. and M. Shkaratan, 2010: *Africa's Power Infrastructure*, Policy Working Paper, World Bank, Washington, DC, USA.

ECLAC, 2008: *Contribution of Energy Services to the Millennium Development Goals and Poverty Alleviation in LAC*, Fourth International Symposium on Technological Frontiers and Rural Sector Energization. Economic Commission for Latin America and the Caribbean (ECLAC), Santiago, Chile.

ECLAC, 2009: *Contribución de los Servicios Energéticos: a los Objetivos de Desarrollo del Milenio y a la Mitigación de la Pobreza en América Latina y el Caribe*, Project documents, No. 281 (LC/W.281-P/E). Economic Commission for Latin America and the Caribbean (ECLAC), Santiago, Chile.

ECOWAS, 2005: *White Paper for a Regional Policy – Geared towards increasing access to energy services for rural and periurban populations in order to achieve the Millennium Development Goals*. Economic Community of West African States.

ECOWAS and UEMOA, 2006: *White Paper for a Regional Policy Geared Towards Increasing Access to Energy Services for Rural and Peri-Urban Populations in order to Achieve The MDGs*. ECOWAS and UEMOA. Banjul, Gambia.

Ekholm, T., V. Krey, S. Pachauri, and K. Riahi, 2010: Determinants of household energy consumption in India. *Energy Policy*, 38(10):5696–5707.

ENDA, 2006: *Assessment of Capacity Building Needs in the Field of Sustainable Energy*. Energy, Environment, Development Program. Environment and Development Action in the Third World (ENDA). Non-published internal document.

Energy Commission of Ghana, 2005: *Strategic National Energy Plan and Policy 2005–2020*. Volume 1. The Republic of Ghana.

Eskom, 2009: Annual Report. www.eskom.co.za/annreport09/ar_2009/index_annual_report.htm (accessed 7 June 2010).

ESMAP, 2001: *Best Practice Manual: Promoting Decentralized Electrification Investment*. ESM248. Energy Sector Management Assistance Program (ESMAP) c/o Energy and Water –World Bank, Washington, DC, USA.

ESMAP/WEC, 2006: *Brazil: How do the Peri-Urban Poor Meet their Energy Needs: a Case Study of Caju Shantytown, Rio de Janeiro*. ESMAP Technical Paper 094. ESMAP c/o Energy and Water –World Bank, Washington, DC, USA and World Energy Council (WEC), London, UK.

Espinoza, W. C., 2005: *Diagnóstico del Sector Energético en el Área Rural de Bolivia: Proyecto: Electrificación Rural*. OLADE, CIDA, University of Calgary, Canada.

FEMA, 2006: *Energy and the Millennium Development Goals in Africa*. Position paper prepared for the UN World Summit. Forum of Energy Ministers of Africa (FEMA), Washington, DC, USA.

Foster, V. and C. Briceno-Garmendia (eds.), 2010: *Africa's Infrastructure: A Time for Transformation*. World Bank, Washington, DC, USA.

Ghosh, S., 2002: Electricity consumption and economic growth in India. *Energy Policy*, 30(2):125–129.

GNESD, 2004: *Energy Access Theme Results. Synthesis/Compilation Report*. Global Network on Energy for Sustainable Development (GNESD), Risø National Laboratory – DTU, Denmark.

GNESD, 2006: *Making power sector reform work for the poor*. Global Network on Energy for Sustainable Development (GNESD), Risø National Laboratory – DTU, Denmark.

GNESD, 2007: *Reaching the Millennium Development Goals and Beyond: Access to Modern Forms of Energy as a Prerequisite*. Global Network on Energy for Sustainable Development (GNESD), Risø National Laboratory – DTU, Denmark.

GNESD, 2008: Peri-urban access study shows common results, *GNESD News*. Global Network on Energy for Sustainable Development (GNESD), Risø National Laboratory – DTU, Denmark.

Govindarajalu, C., Elahi, R., and J. Najendran, 2008. *Electricity Beyond the Grid: Innovative Programmes in Bangladesh and Sri Lanka*. ESMAP Knowledge Exchange Series, No. 10, World Bank, Washington, DC, USA.

Green Nine, 2004: *The EU's new Constitution: Assessing the Environmental Perspective*. Green Nine, Brussels, Belgium.

Holland, R. and L. Mayer-Tasch, 2007: *Regional Approaches to Energy Access in sub-Saharan Africa*, European Union Energy Initiative – Partnership Dialogue Facility.

Houston, P., 2003: *National Audit of Peri-Urban Agriculture: Leading with diversity*. Planning Congress Adelaide, Australia.

IEA, 2003: *World Energy Investment Outlook 2002*. International Energy Agency (IEA), Organisation of Economic Co-Operation and Development (OECD), Paris, France.

IEA, 2004: *World Energy Outlook 2004*. International Energy Agency (IEA), Organisation of Economic Co-Operation and Development (OECD), Paris, France.

IEA, 2008: *World Energy Outlook 2008*. International Energy Agency (IEA), Organisation of Economic Co-Operation and Development (OECD), Paris, France.

IEA, 2009: *World Energy Outlook 2009*. International Energy Agency (IEA), Organisation of Economic Co-Operation and Development (OECD), Paris, France.

IEA, 2011: *World Energy Outlook 2011*. International Energy Agency (IEA), Organisation of Economic Co-Operation and Development (OECD), Paris, France.

Independent Evaluation Group, 2008: *The Welfare Impact of Rural Electrification: A Reassessment of the Costs and Benefit*. World Bank, Washington, DC, USA.

Instituto Brasileiro de Geografia e Estatística, 2001: Reporte 2001. www.ibge.gov.br/english/ (accessed June 12, 2011).

IISD, 2009: *The Sustainable Development Timelines*. International Institute for Sustainable Development (IISD), New York, NY, USA.

International Monetary Fund, 2008: *Africa's Energy Shortage. Africa's Power Supply Crisis: Unraveling the Paradoxes*, IMF Survey online, www.imf.org/external/pubs/ft/survey/so/2008/CAR052208C (accessed August 3, 2011).

Jannuzzi, G. M., A. Romeiro, C. Melo, F. Piacente, G. Esteves, H. Xavier, Jr., and R. D. M. Gomes, 2007: Agenda Eletrica Sustentavel 2020: Estudo de cenarios para um setor eletrico brasileiro eficiente, seguro e competitivo. *Série Técnica*. WWF-Brasil, Brasilia, Brazil.

Karekezi, S., J. Kimani, and O. Onguru, 2008: Energy access among the Urban and Peri-Urban Poor in Kenya. Draft report prepared for Global Network on Energy for Sustainable Development (GNESD), Energy Environment and Development Network for Africa (AFREPREN/FWD), Nairobi, Kenya.

Krishnaswamy, V. and G. Stuggins, 2007: *Closing the Electricity Supply-Demand Gap*. World Bank, Washington, DC, USA.

Ministerio de Obras Públicas, Servicios y Vivienda, 2007: *Programa "Electricidad para Vivir con Dignidad."* Viceministerio de Electricidad y Energías Alternativas, Bolivia.

Millán, J., 2006: *Entre el Mercado y el Estado*. Banco Interamiericano de Desarrollo, Washington, DC, USA.

Ministerio de Energía y Minas, Republic of Perú, 2008: *Plan Nacional de Electrificación Rural (PNER) Periodo 2008–2017*, Ministerio de Energía y Minas, Lima, Peru.

Ministry of Energy, 2008: *Lettre de Politique de Développement du Secteur de l'Energie*. Government of Senegal.

Ministry of Energy and Power, 2004: The Energy Policy for Sierra Leon.

Ministry of Energy and Water, 2006: *Energy Sector in Togo*. Government of Togo.

Ministry of Mines, Energy and Water, 2006: *National Energy Policy*. Government of Mali.

Morgan, M. G. and H. Zerriffi, 2002: The regulatory environment for small independent micro-grid companies. *The Electricity Journal*, 15(9):52–57.

Moser, G., S. Rogers, R. van Til, R. Kibuka, and I. Lukonga, 1997: *Nigeria: Experience with Structural Adjustment*. Occasional Paper, No. 148, International Monetary Fund, Washington, DC, USA.

Niez, A., 2009: *Comparative Study on Rural Electrification Policies in Emerging Countries*, International Energy Agency (IEA), Organisation of Economic Co-Operation and Development (OECD), Paris, France.

OECD/IEA, 2010: *Energy Poverty: How to Make Modern Energy Access Universal?* International Energy Agency (IEA), Organisation of Economic Co-Operation and Development (OECD), Paris, France.

Olomola, P. and A. Adejumo, 2006: Oil price shock and macroeconomic activities in Nigeria. *International Research Journal of Finance and Economics*, **3**:28–34.

Onayemi, T., 2003: Nigeria Oil: Prices, Politics and the People. *Nigeria Today,* www.nigeriatoday.com/nigeria_oil.htm (accessed August 3, 2011).

Organization of American States, 2004: *Policy Reform for Sustainable Energy in Latin America and the Caribbean*. Unit for Sustainable Development and Environment, Policy Series, Number 5.

Palit, D. and A. Chaurey, 2010: *Off-grid Electrification Experience in South Asia: Status and Best Practices.* Working paper series, working paper 1. OASYS – South Asia Project.

Parshall, L., D. Pillai, S. Mohan, A. Sanoh, and V. Modi, 2009: National electricity planning in settings with low pre-existing grid coverage: Development of a spatial model and case study of Kenya. *Energy Policy*, **37**(6):2395–2410.

Practical Actions, 2007: *Bridging the Funding Gap to Ensure Energy Access for the poor*. www.practicalactionconsulting.org (accessed December 2, 2010).

Régimen Legal de Bogotá D.C., 1994: Ley 142 de 1994 Nivel Nacional La Secretaría General de la Alcaldía Mayor de Bogotá D.C. Bogotá, Diario Oficial.

Reiche, K. B., B. Tenenbaum, and C. Torres de Mästle, 2006: *Promoting Electrification: Regulatory Principles and a Model Law*. Energy and Mining Sector Board Discussion Paper, Paper No. 18, Energy and Mining Sector Board, World Bank, Washington, DC, USA.

Rijal, K. N., K. Bansal, and P. D. Grover, 2007: Economics of shaft power applications in rural areas of Nepal. *International Journal for Energy Research*, **19**(4):289–308.

Ríos Roca, A., M. B. Garrón, and P. G. Cisneros, 2007: *Focalización de los subsidios a los combustibles en América Latina y el Caribe: Análisis y Propuesta*. The Latin American Energy Organization (OLADE), Quito, Ecuador.

Rojas, J. M. and D. Lallement, 2007: *Meeting the Energy Needs of the Urban Poor: Lessons from Electrification Practitioners*. ESMAP Technical Paper 118/07. World Bank, Washington, DC, USA.

Sachs, J., J. McArthur, G. Schmidt-Traub, C. Bahadur, M. Faye, and M. Kruk, 2004: *Millennium Development Goals Needs Assessments – Country Case Studies of Bangladesh, Cambodia, Ghana, Tanzania and Uganda*. Millennium Project.

SADC, 1998: *Energy Protocol*. The Southern African Development Community, Pretoria, South Africa.

Saghir, J., 2010: Energy and Development: Lessons Learned.

Shrestha, R., S. Kumar, M. J. Todoc, and S. Sharma., 2004a: *Institutional Reforms and their Impact on Rural Electrification: Case Studies in South and Southeast Asia*. GNESD, UNEP, Nairobi, Kenya.

Shrestha, R. M., S. Kumar, S. Sharma and M. J. Todoc, 2004b: Institutional Reforms and Electricity Access: Lessons from Bangladesh and Thailand. *Energy for Sustainable Development*, **VIII**(4):41–53.

Stewart, F., 1987: *Macro-Policies for Appropriate Technology in Developing Countries*. Westview Press, Boulder, CO, USA.

TERI, 2007: *Evaluation of Franchise system in selected districts of Assam, Karnataka and Madhya Pradesh*, Project Report 2006ER39. The Energy and Resources Institute (TERI), New Delhi, India.

TERI, 2008: *Supply of clean energy services to urban and peri-urban poor*. Global Network on Energy for Sustainable Development, Project Report No. 2007UD21.

UNCED, 1993: *Agenda 21 / Río Declaration on Environment and Development / Statement of Forest*. United Nations Conference on Environment and Development (UNCED), United Nations Department of Public Information, New York. www.un.org/esa/dsd/agenda21/index.shtml (accessed August 3, 2011).

UNDP, 1994: *Capacity Development: Lessons of Experience and Guiding Principles*. United Nations Development Programme (UNDP), New York, NY, USA.

UNDP, 2004: *Gender and Energy for Sustainable Development: A Toolkit and Resource Guide*. United Nations Development Programme (UNDP), New York, NY, USA.

UNDP, 2010: *UNDP and Energy Access for the Poor: Energizing the Millennium Development Goals*. United Nations Development Programme (UNDP), New York, NY, USA.

UNDP, 2011: International Human Development Indicators. hdrstats.undp.org/en/tables/default.html (accessed April 22, 2011).

UNDP Human Development Report, 2003: *Millennium Development Goals: A Compact Among Nations to End Human Poverty*. United Nations Development Programme (UNDP), New York, NY, USA.

UNDP Human Development Report, 2009: *Overcoming barriers: human mobility and development*. United Nations Development Programme (UNDP), New York, NY, USA.

UNDP Human Development Report, 2010: *The Real Wealth of Nations: Pathways to Human Development*. United Nations Development Programme (UNDP), New York, NY, USA.

UNDP/WHO, 2009: *The Energy Access Situation in Developing Countries: A Review Focusing on the Least Developed Countries and Sub-Saharan Africa*. United Nations Development Programme (UNDP), New York, NY, USA, and World Health Organization (WHO), Geneva, Switzerland.

UN-Energy/Africa, 2007: *Energy for Sustainable Development: policy options for Africa*. UN-ENERGY/Africa.

UNEP, 2002: *Reforming Energy Subsidies: An Explanatory Summary of the Issues and Challenges in Removing or Modifying Subsidies on Energy that Undermine the Pursuit of Sustainable Development*. United Nations Environment Program (UNEP), Nairobi, Kenya.

UNEP/GNESD, 2002: *Energy Access. Making Power Sector Reform Work for the Poor*. United Nations Environment Program (UNEP), Nairobi, Kenya.

UNESCAP, 2007: Fact sheet I: Energy issues related to access, social development, poverty and the Millennium Development Goals (MDGs). In Energy Security and Sustainable Development in Asian and the Pacific. United Nations Economic and Social Commission for Asia and the Pacific (UNESCAP), Bangkok, Thailand.

UN Habitat, 2011: *State of the World's Cities 2010/2011 – Cities for All: Bridging the Urban Divide*. UN Habitat, Nairobi, Kenya.

UNCTAD, 2002: *Economic Development in Africa. From Adjustment to Poverty Reduction: What is New?* UNCTAD/GDS/AFRICA/2, United Nations, New York and Geneva.

Wamukonya, N., 2002: *A critical look at gender and energy mainstreaming in Africa*. Draft paper distributed at a side event organized by UNDESA/DAW and WEDO at Prep Com III, G*ender perspectives in sustainable development.*

Wang, M., 2005: *Poverty Reduction Strategy Papers within the Human Rights Perspective*. In P. Alston, and M. Robinson (eds.), *Human Rights and Development*. OUP, Oxford, pp.447–474.

Williams, J. L., 2011: Oil Price History and Analysis (Updating). WTRG Economics. www.wtrg.com/prices.htm (accessed February 19, 2011).

Wolde-Rufael, Y., 2006: Electricity consumption and economic growth: a time series experience for 17 African countries. *Energy Policy,* **34**:1106–1114.

Woodhouse, E. J., 2005: *A Political Economy of International Infrastructure Contracting: Lessons from the IPP Experience*. Program on Energy and Sustainable Development, Stanford, CA, USA.

World Bank, 2003: *Renewable Energy Development Project*. World Bank, Washington, DC, USA.

World Bank, 2008a: *The Welfare Impact of Rural Electrification: A Reassessment of the Costs and Benefits*. Independent Evaluation Group, World Bank, Washington, DC, USA.

World Bank, 2008b: *Designing Sustainable Off-Grid Rural Electrification Projects: Principles and Practice*, World Bank, Washington, DC, USA.

World Bank Group, 2006: *An Investment Framework for Clean Energy and Development: A Progress Report*. The World Bank Group, Washington, DC, USA.

World Bank Group, 2010: *Private Participation in Infrastructure Database*. The World Bank Group, Washington, DC, USA.

World Bank Group, 2011: Africa Development Indicators. data.worldbank.org/data-catalog/africa-development-indicators (accessed March 13, 2010).

World Bank Group, 2011: World Development Indicators (WDI). data.worldbank.org/indicator (accessed April 22, 2010).

Zhou, P. P., 2003: *Taking the NEPAD Energy Initiative Forward: Regional and Sub-Regional Perspectives. Dakar, Senegal.*

Policies for the Energy Technology Innovation System (ETIS)

24

Convening Lead Author (CLA)
Arnulf Grubler (International Institute for Applied Systems Analysis, Austria and Yale University, USA)

Lead Authors (LA)
Francisco Aguayo (El Colegio de México)
Kelly Gallagher (Tufts University, USA)
Marko Hekkert (Utrecht University, the Netherlands)
Kejun JIANG (Energy Research Institute, China)
Lynn Mytelka (United Nations University-MERIT, the Netherlands)
Lena Neij (Lund University, Sweden)
Gregory Nemet (University of Wisconsin, USA)
Charlie Wilson (Tyndall Centre for Climate Change Research, UK)

Contributing Authors (CA)
Per Dannemand Andersen (Technical University of Denmark)
Leon Clarke (University of Maryland, USA)
Laura Diaz Anadon (Harvard University, USA)
Sabine Fuss (International Institute of Applied Systems Analysis, Austria)
Martin Jakob (Swiss Federal Institute of Technology, Zurich)
Daniel Kammen (University of California, Berkeley, USA)
Ruud Kempener (Harvard University, USA)
Osamu Kimura (Central Research Institute of Electric Power Industry, Japan)
Bernadette Kiss (Lund University, Sweden)
Anastasia O'Rourke (Big Room Inc., Canada)
Robert N. Schock (World Energy Council, UK and Center for Global Security Research, USA)
Paulo Teixeira de Sousa Jr. (Federal University Mato Grosso, Brazil)

Review Editor
Leena Srivastava (The Energy and Resources Institute, India)

Contents

Dedication

We dedicate this work to our families for their understanding and support for our long, collaborative journey that led to this chapter and which included both sad and joyful moments.

In loving memory to Maria, Georg, and Gerth, who departed us while we were working on this chapter.

With a warm welcome to Heidi Marie, Estelle, Henry Ian, Inés, and Alfred Victor who joined us in our collective travel towards a sustainable future.

Executive Summary

Innovation and technological change are integral to the energy system transformations described in the Global Energy Assessment (GEA) pathways. Energy technology innovations range from incremental improvements to radical breakthroughs and from technologies and infrastructure to social institutions and individual behaviors. This Executive Summary synthesizes the main policy-relevant findings of Chapter 24. Specific positive policy examples or key take-home messages are highlighted in italics.

The innovation process involves many stages – from research through to incubation, demonstration, (niche) market creation, and ultimately, widespread diffusion. Feedbacks between these stages influence progress and likely success, yet innovation outcomes are unavoidably uncertain. Innovations do not happen in isolation; interdependence and complexity are the rule under an increasingly globalized innovation system. Any emphasis on particular technologies or parts of the energy system, or technology policy that emphasizes only particular innovation stages or processes (e.g., an exclusive focus on energy supply from renewables, or an exclusive focus on Research and Development [R&D], or feed-in tariffs) is inadequate given the magnitude and multitude of challenges represented by the GEA objectives.

A first, even if incomplete, assessment of the entire global resource mobilization (investments) in both energy supply and demand-side technologies and across different innovation stages suggests current annual Research, Development & Demonstration (RD&D) investments of some US$50 billion, market formation investments (which rely on directed public policy support) of some US$150 billion, and an estimated US$1 trillion to US$5 trillion investments in mature energy supply and end-use technologies (technology diffusion). *Major developing economies like Brazil, India and above all China, have become significant players in global energy technology RD&D, with public- and private-sector investments approaching US$20 billion, or almost half of global innovation investments, which is significantly above the Organisation for Economic Co-operation and Development (OECD) countries' public-sector energy RD&D investments (US$13 billion). Important data and information gaps exist for all stages of the energy technology innovation investments outside public sector R&D funding in OECD countries, particularly in the areas of recent technology-specific private sector and non-OECD R&D expenditures, and energy end-use diffusion investments.*

Analysis of investment flows into different stages of the innovation process reveals an apparent mismatch of resource allocation and resource needs.

Early in the innovation process, public expenditure on R&D is heavily weighted toward large-scale supply-side technologies. Of an estimated US$50 billion in annual investment globally, less than US$10 billion are allocated to energy end-use technologies and energy efficiency.

Later in the innovation process, annual market (diffusion) investment in supply-side plant and infrastructure total roughly US$_{2005}$0.8 trillion, compared with a conservative estimate of some US$1–4 trillion spent on demand-side technologies. These relative proportions are, however, insufficiently reflected in market deployment investment incentives of technologies, which almost exclusively focus on supply-side options, to the detriment of energy end use in general and energy efficiency in particular foregoing also important employment and economic growth stimuli effects from end-use investments that are critical in improving energy efficiency. *The need for investment to support the widespread diffusion of efficient end-use technologies is also clearly shown in the GEA pathway analyses. The demand side generally tends to contribute more than the supply-side options to realizing the GEA goals. This apparent mismatch suggests the necessity of rebalancing public innovation expenditure and policy incentives to include smaller-scale demand-side technologies within innovation portfolios.*

Given persistent barriers to the adoption of energy-efficient technologies even when they are cost competitive on a life cycle basis, technology policies need to move toward a more integrated approach, simultaneously stimulating the *development* as well as the *adoption* of energy efficiency technologies and measures. *R&D initiatives that fail to incentivize consumers to adopt the outcomes of innovation efforts (e.g., promoting energy-efficient building designs without strengthened building codes, or Carbon Capture and Storage [CCS] development without a price on carbon) risk not only being ineffective but also precluding the market feedback and learning that are critical for continued improvements in technologies.*

Little systematic data are available for private-sector innovation inputs (including investments), particularly in developing countries. Information is patchy on innovation spillovers or transfers between technologies, between sectors, and between countries. It is also not clearly understood how fast knowledge generated by innovation investments may depreciate, although policy and investment volatility are recognized as critical factors. Technical performance and economic characteristics for technologies in the lab, in testing, and in the field are not routinely available. Innovation successes are more widely documented than innovation failures. Although some of the data constraints reflect legitimate concerns to protect intellectual property, most do not. Standardized mechanisms to collect, compile, and make data on energy technology innovation publicly available are urgently needed. The benefits of coupling these information needs to public policy support have been clearly demonstrated. *A positive policy example is provided by the early US Solar Thermal Electricity Program, which required formal, non-proprietary documentation of cost improvements resulting from public R&D support for the technology.*

The energy technology innovation system is founded on knowledge generation and flows. These are increasingly global, but this global knowledge needs to be adapted, modified, and applied to local conditions. The generation of knowledge requires independent and stable institutions to balance the competing needs and interests of the market, policy makers, and the R&D community. *The technology roadmaps and the policy regime that characterize innovation in end-use technologies in the Japanese Top Runner program are a good example of the actor coordination and knowledge exchange needed to stimulate technological innovation.*

Generated knowledge needs to spread through the innovation system. Knowledge flows and feedbacks create and strengthen links between different actors. This can take place formally or informally. Policies that are overly focused on the development of technological "hardware" should be rebalanced to support interactions and learning between actors. *The provision of test facilities in the early years of the Danish wind industry is a good example of how policy can support knowledge flows and the strengthening of collaborative links within networks of actors in an innovation system (energy companies, turbine manufacturers, local owners).*

Long-term, consistent, and credible institutions underpin investments in knowledge generation, particularly from the private sector, and consistency does not preclude learning. Knowledge institutions must be responsive to experience and adaptive to changing conditions. Although knowledge flows through international cooperation and experience sharing cannot presently be analyzed in detail, the scale of the innovation challenge emphasizes their importance alongside efforts to develop the capacity to absorb and adapt knowledge to local needs and conditions. *The current global cooperation in energy technology innovation is well illustrated by the International Energy Agency (IEA) technology cooperation programs reviewed in Section 4.4; all invariably show a sparse involvement from developing countries.*

Clear, stable, and consistent expectations about the direction and shape of the innovation system are necessary for innovation actors to commit time, money, and effort with only the uncertain promise of distant returns. To date, policy support for the innovation system has been characterized by volatility, changes in emphasis, and a lack of clarity. *The debilitating consequences on innovation outcomes of stop-go policies are well illustrated by the wind and solar water heater programs in the United States through the 1980s, as well as the large-scale (but fickle) US efforts to develop alternative liquid fuels (Synfuels). The legacy of such innovation policy failures can be long lasting. The creation of a viable and successful Brazilian ethanol industry through consistent policy support over several decades, including agricultural R&D, guaranteed ethanol purchase prices, and fuel distribution infrastructures, as well as vehicle manufacturing (flex fuel cars), is a good example of a stable, aligned, and systemic technology policy framework. It is worth noting that even in this highly successful policy example, it has taken some three decades for domestic renewable ethanol to become directly cost competitive with imported gasoline.*

Policies need also to be *aligned*. Innovation support through early research and development is undermined by an absence of support for their demonstration to potential investors and their subsequent deployment in potential markets. Policies to support innovations in low-carbon technologies are undermined by subsidies to support carbon-intensive technologies. *Fuel efficiency standards that set minimum (static) efficiency floors fail to stimulate continuous technological advances, meaning innovations in efficiency stagnate once standards are reached. As a further example of misalignment, the lack of effective policies to limit the demand for mobility mean efficiency improvements can be swamped by rising activity levels.*

Policies should support a wide range of technologies. However seductive they seem, "silver bullets" do not exist without the benefit of hindsight. Innovation policies should use a portfolio approach under a risk-hedging and "insurance policy" decision-making paradigm. Portfolios need to recognize also that innovation is inherently risky. Failures vastly outnumber successes. *Experimentation, often for prolonged periods (decades rather than years), is critical to generate the applied knowledge necessary to support the scaling up of innovations to the mass market.*

The whole energy system should be represented in innovation portfolios, not only particular groups or types of technologies; the entire suite of innovation processes should be included, not just particular stages or individual mechanisms. Less capital-intensive, smaller-scale (i.e., *granular*) technologies or projects are less of a drain on scarce resources, and failure has less serious consequences. *Granular projects and technologies with smaller scales (MW rather than GW) therefore should figure prominently in any innovation portfolio.*

Finally, public technology policy should not be beholden to incumbent interests that favor support for particular technologies that either perpetuate the lock-in of currently dominant technologies or transfer all high innovation risks of novel concepts to the public sector.

24.1　　Introduction

24.1.1　　Welcome to Chapter 24

Unlike resources found in nature, technological and social innovations are human-made resources that can be generated and expanded as a matter of social choice but come with costs and with uncertain outcomes. Energy-technology innovations not only encompass new inventions and improvements in the performance or attributes of technologies like coal gasification, solar thermal electricity, batteries, or energy-efficient windows or light bulbs, but also in how firms develop and markets and users relate to and utilize such technologies. Social innovations that result in changes in behavior of technology suppliers as well as users can therefore be just as important as improvements in technological efficiency or emissions performances of individual technological artifacts.

Innovations do not fall like manna from heaven; they need to be created through a multistage process. The stages include research, development, demonstration, market formation, and finally, the culminating pervasive diffusion of successful innovations. In the most general definition, energy technology change is the capital-embodied result of institutionalized R&D and collective learning processes[1] between developers/suppliers and users of technologies, operating within specific innovation and adoption environments that are strongly shaped by policies. This chapter therefore adopts a systemic view of an Energy Technology Innovation System (ETIS) and focuses on the particular role of policy in the energy innovation process and the functioning of ETIS.

Chapter 24 is both theoretical and deeply empirical: it provides the first ever quantitative estimate of global investments in energy-technology innovation (Appendix I), as well as a rich set of new case studies (summarized in Appendix II). These case studies trace the evolution of individual energy technologies, describe often neglected aspects of energy technology innovation, and assess the role of policies in influencing energy technology innovation. Throughout, this chapter emphasizes the importance of understanding the energy-technology innovation system in its entirety, including its many feedbacks. Because the energy-technology innovation system is complex and remains incompletely understood, readers are advised to use caution when seeking precise mathematical formulations for models or simple policy recipes. Nonetheless, despite its limitations, a systems perspective on energy technology innovation – particularly one that integrates supply and demand aspects – offers new insights that complement and improve upon traditional views and resulting fragmented technology policy approaches.

Chapter 24 provides guidance to policy makers about how to positively influence energy innovation, as well as how policy can be harmful and counterproductive. Common myths are explicitly examined. Refraining from being overly prescriptive about particular individual policy instruments,

Chapter 24 instead offers broad guidelines drawn from the case studies for improved innovation policies that recognize both the inevitable uncertainty in the innovation process and its systemic nature. The chapter concludes with research and information/data needs and summary findings. Space limitations preclude a full presentation of the 20 case studies drawn upon in Chapter 24. They are presented in one-page summaries as an appendix to this text and are available upon request.[2]

24.1.2　　Roadmap of Chapter 24

Figure 24.1 shows a roadmap of Chapter 24. After the introduction (Section 24.1), Chapter 24 moves to the assessment of ETIS, which consists of three main parts.

Section 24.2 characterizes ETIS. The review is necessarily selective, but identifies key components and themes. Features of ETIS are organized around knowledge and learning (Section 24.2.2); attributes of energy technologies and their industries and drivers of changing technology characteristics, such as economies of scale and scope (Section 24.2.3); and the functions of actors and associated institutions (Section 24.2.4). These are the distinct mechanisms of innovation described in the engineering, economics, management, and sociological literature and include knowledge accumulation (and depreciation), economies of scale and scope, and various learning processes. This part concludes with an integrative representation of ETIS and its components according to the "functions of innovation systems" literature. This emphasizes the dynamic, evolving nature of an ETIS over time (Section 24.2.5).

Section 24.3 identifies ways of assessing ETIS. The breadth of assessment metrics are reviewed in detail in the *Assessment Metric* case study

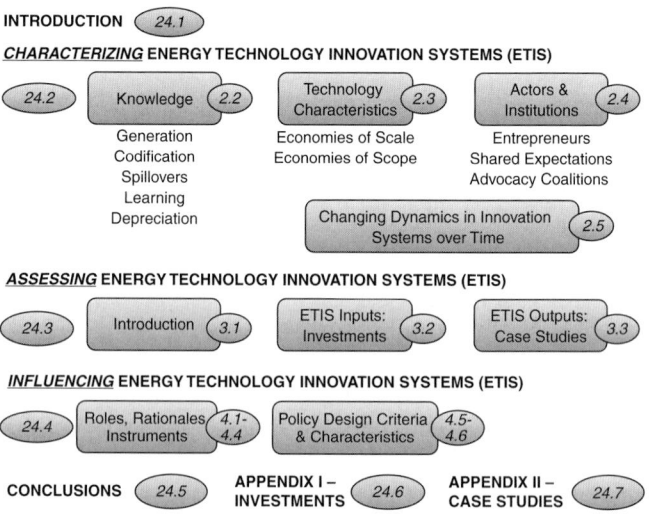

Figure 24.1 | Roadmap of Chapter 24

1　Excellent historical studies on oil-refining (Enos, 1962) and gas turbines (Watson, 2004) illustrate these processes.

2　Available at www.globalenergyassessment.org.

(see summary in Appendix II). The discussion then proceeds to a quantitative commensurate assessment of current investment inputs into ETIS, as summarized in Section 24.3.2. In order to assess ETIS in terms of outcomes (Section 24.3.3), Chapter 24 draws upon 20 case studies, which are summarized in Appendix II. Section 24.3.3 provides an overview of the case studies, their rationale, and selected illustrative examples of ETIS outputs across a variety of energy technologies.

Section 24.4 then examines the question of how to influence the direction and effective functioning of the ETIS. The policy community is a key constituency for the findings in Chapter 24. The ETIS framework presented is an integrative conceptual framework that neither can nor should be used to generate policy prescriptions. Therefore, after an overview of actors and rationales for technology policy (Sections 24.4.1 and 24.4.2), policy models and instruments (Section 24.4.3) and their increasingly international dimension (Section 24.4.4) are outlined. Chapter 24 abstracts generalizable *policy design* guidelines and criteria that should support innovation success and mitigate against innovation failure (Section 24.4.6).

Chapter 24 culminates in a discussion of the research, data, and information needs identified in this assessment (Section 24.5.1), as well as overall conclusions (Section 24.5.2).

As noted, this chapter is written to provide a practical guide for policy makers concerned with supporting the effective functioning of the ETIS in the context of the GEA objectives on climate, access, security, and health. To develop policy guidance, Chapter 24 also reviews some key characteristics and metrics of ETIS. These sections (24.2.2 to 24.2.5) are written with greater technical depth and language, and are aimed also at those in the research and business communities interested in understanding the fundamentals and mechanisms of innovation in an energy context. Readers more interested in policy aspects can move on to Section 24.3, revisiting the more technical material of Section 24.2 at a later stage. Given the range of potential audiences, considerable effort has been made to define key terms (see Table 24.1 and also the GEA Glossary), use consistent terminology, and support conceptual arguments with empirical details from the case studies.

24.1.3   Technological Change in Energy Systems

Technological and congruent institutional and social changes have been widely recognized as main drivers for long-run economic growth ever since Solow (1957), and for broader societal development as well (Freeman and Perez, 1988). In terms of causality, caution is advised as technology and institutional/social setting co-evolve, mutually depending on and cross-enhancing each other. Technological change in energy systems to a large degree determines how efficiently energy services can be provided, at what costs, and with which associated externalities. Scholars agree on the importance of technological change in past and future energy transitions (e.g., Smil, 1994; Grubler, 1998; Nakicenovic et al., 2000; Grubler, 2008; and the literature review in Halsnæs et al., 2007).

The *Grand Designs* case study (summarized in Appendix II, see also Wilson and Grubler, 2011) provides a synthesis of major patterns driving historical energy transitions and contrasts this historical perspective by examining also the scenario literature on the importance and patterns of technological change in alternative futures. The transformative power of technology arises from: (1) combinations of interrelated individual technologies (clustering) and applications of technologies outside their

Table 24.1 | List of key terms

Key Term	Definition as Used in Chapter 24
ETIS	Energy Technology Innovation System: the innovation systems approach applied to the energy system. In this approach, innovation is understood as an interactive process involving a network of firms and other economic agents that, together with the institutions and policies that influence their innovative behavior and performance, bring new products, processes, and forms of organization into economic use
Invention	origination of an idea as a technological solution to a perceived problem or need (usually codified via a patent)
Innovation	putting ideas into practice through an (iterative) process of design, testing, and improvement, including small-scale demonstration or commercial pilot projects, and culminating in the establishment of an industrial capability to manufacture a given technological innovation
Diffusion	widespread uptake of a technological innovation throughout the market of potential adopters
R&D (Research and Development)	knowledge generation by directed activities (e.g., evaluation, screening, research) aimed at developing new or improving on existing technological knowledge
Demonstration	construction of technology prototypes or pilots demonstrating technological feasibility
RD&D (Research, Development and Demonstration)	integration of the required upfront stages in a technology life cycle (invention-innovation) related terms: RDD&D (i.e., RD&D + deployment)
Market Formation	application of a technology in a specific limited market setting (or niche) by harnessing either a specific comparative advantage (e.g., PV electricity in remote areas without grid connections) or via public early deployment incentives (e.g., feed-in tariffs) related terms: market creation, niche markets, deployment, early commercialization
Learning	improved (technological) knowledge derived from production experience (learning-by-doing) and/or user experience (learning-by-using) that leads to performance improvements, including cost reductions
Knowledge Spillovers	knowledge transfer between different innovation actors and technology application fields through mechanisms such as imitation, trade, licensing, foreign direct investment, and/or movement of people

initial sector/use (spillovers); (2) the continued improvements of technology performance and costs as a result of innovation efforts and market growth (learning and economies of scale effects, among others); (3) energy end-use and technology users/consumers are particularly critical; and (4) generally, rates of capital turnover and technological change in the energy systems remain slow. These four "grand" patterns of energy technological change are addressed in more detail below.

(1) No individual technology, as important it may be, is able to transform whole energy systems that are large and complex. The importance of technology arises in particular through *clustering* (combinations of interrelated individual technologies) and *spillover* (applications outside the initial sector/use for which a technology was initially devised) effects. The concept of general purpose technologies (GPT) (e.g., Lipsey et al., 2006) captures this notion that some technologies, like steam power or electricity, find multiple applications across many sectors, industries, and energy end-uses. Technologies operate more effectively as families or as "gangs" rather than as individuals. Strong interrelatedness conditions major innovations in the energy sector to a multitude of complementary changes, including also new business and financing models, as demonstrated in the history of electric light and power (Hughes, 1983) or the emergence of oil-based individual motorized mobility with automobiles (Freeman and Perez, 1988). Once a technology is adopted, a number of related technologies, derived products, and business models become established. Improvements and knowledge about possibilities and applications accumulate, generating further learning economies as the application range grows (Watson, 2004). Combined, these processes create powerful self-reinforcing mechanisms that make it very difficult to dislodge a dominant technological regime, a fact referred to in the technology literature as "path dependency" or "technology lock-in" (e.g., Frankel, 1955; Arthur, 1988a; 1988b; 1989; Unruh, 2000). As a result, new technologies, even when economically feasible, face higher short-term adoption costs compared to established technologies (Cowan and Hulten, 1996; Unruh, 2000).

(2) Generally, when new technologies are introduced, they are initially crude, imperfect, and very expensive (Rosenberg, 1994). Incumbent technologies are generally more advanced in their respective technology life cycle and thus enjoy an associated learning and deployment advantage (Cowan, 1990). Therefore, performance (the ability to perform a particular task or deliver a novel energy service) of a new energy technology initially dominates economics as a driver of technological change and diffusion. Only after an extended period of experimentation, learning, and improvements, and the establishment of a corresponding industrial base (in many cases, profiting from standardization, mass production, and scale economies of a growing industry) do new technologies become capable of competing with existing ones on a pure cost basis. In other words, *attractive beats cheap*, at least initially. Policy intervention can short-cut this evolutionary pattern and are justified when "attractiveness" is defined by lower externalities (e.g., emissions, energy security, etc.). However, such policy interventions come at a price: either costly direct public subsidies or changed economic incentives (via levies, fees, taxes imposed on incumbent, undesirable technologies – and paid for

by consumers). There is also a risk of policy-induced premature "lock-in" in technologies that ultimately turn out to be either socially undesirable, too expensive, or risky for unregulated markets (cf. the *French Nuclear* case study in Appendix II), or pose unanticipated social/environmental challenges, e.g., land competition with food production or greenhouse gas emissions associated with fertilizer use and land-use changes in the case of first generation biofuels (Plevin et al., 2010).

(3) The history of past energy transitions highlights the critical importance of end-use services (i.e., consumers, energy *demand*). Historically, energy supply has *followed* energy demand in technology applications, and energy end-use markets have been the most important market outlets for new energy technologies (as quantified in the *Grand Designs* case study. See also Appendix I for a quantification of current energy end-use versus energy supply investments). In other words, new energy technologies generally need to find consumers (users), preferably many. This holds important implications for both modeling future energy transition scenarios and technology innovation and diffusion policies alike, where energy end-use technologies are often underrepresented.

(4) The process of technological change (from innovation to widespread diffusion) takes considerable time, usually many decades. In addition, rates of change become slower the larger the energy system or its components, and when consequences of those changes are more disruptive. A novel approach that quantifies the historical scaling dynamics of energy technologies and illustrates this conclusion is reported in the *Scaling Dynamics* case study. The historically slow rates of change of energy technologies and systems, which span from several decades up to a century (for a review, see Grubler et al., 1999), arise from four phenomena:

- Capital intensiveness: investments in energy technologies are among the most capital-intensive across industries, characterized by high upfront costs, a high degree of specificity of infrastructure, long payback periods, and strong exposure to financial risk (IEA, 2003). Capital intensiveness, therefore, ceteris paribus slows technology diffusion.

- Longevity of capital stock: the lifetime of the capital stock of energy systems in many end-use applications (buildings), conversion technologies (refineries, power plants), and above all, infrastructures (railway networks, electricity grids), is generally long compared to other industrial equipment or consumer products (Smekens et al., 2003; Worrell and Biermans, 2005). Longevity of capital stock tends to slow capital turnover and thus the diffusion speed of new technologies.

- Learning/experimentation time: extended time is required for experimentation, learning, and technology development from invention to innovation, to initial specialized niche market applications, and finally, in case of success, to pervasive adoption across many sectors, markets, and countries.

- Lastly, considerable time is also required for technology clustering and spillover effects to emerge.

Only in exceptional cases does the diffusion of new technologies proceed via a premature retiring of existing capital stock, as is the case in current cell phone markets or with information and communication technologies (ICT) in general. In view of the generally slow rates of change in large technology systems like energy, pervasive technological transformations require a long-term view, and it is better for transition initiatives to start sooner rather than later.

The above characteristics of technological change in energy systems are important for policy, as they suggest that approaches must be systemic, long-term, and cognizant of inevitable innovation uncertainties. Short-term, piecemeal efforts to stimulate innovation and speed technology diffusion are unlikely to result in the kind of major technological transformations needed to achieve more sustainable energy systems as called for throughout the GEA.

24.2 Characterizing Energy Technology Innovation Systems

24.2.1 Introduction to the Energy Technology Innovation System

24.2.1.1 From Linear Models to Innovation Systems

The evolution of technology is often conceptualized through a life cycle model that proceeds sequentially from birth (invention, innovation), to adolescence (growth), maturity (saturation), and ultimately senescence (decline due to competition by more recent innovations). Models of innovation describe the drivers and mechanisms behind this technology life cycle. These have evolved substantially and continue to evolve further. The intellectual history of innovation concepts reaches back into the nineteenth century (e.g., Marxist economic theories and their conceptualization of technological innovation). Still influential today are the theories of Joseph A. Schumpeter (1942), who emphasized the importance of radical or disruptive technological and organizational changes, the role of entrepreneurship, and competition. In contrast to Schumpeter's emphasis on radical "breakthrough" innovations, the importance of the compounded effects of numerous, smaller (incremental) innovations is also now widely recognized. Concepts formulated by Vannevar Bush in his 1945 report to the US president, *Science the Endless Frontier*, were influential on early models of innovation (Bush, 1945). These are often referred to as "linear" models. These models emphasize the role of basic,[3] largely publicly funded science in a linear innovation process from basic research to applied development, demonstration, and concluding with the diffusion process (see the upper part of Figure 24.2).

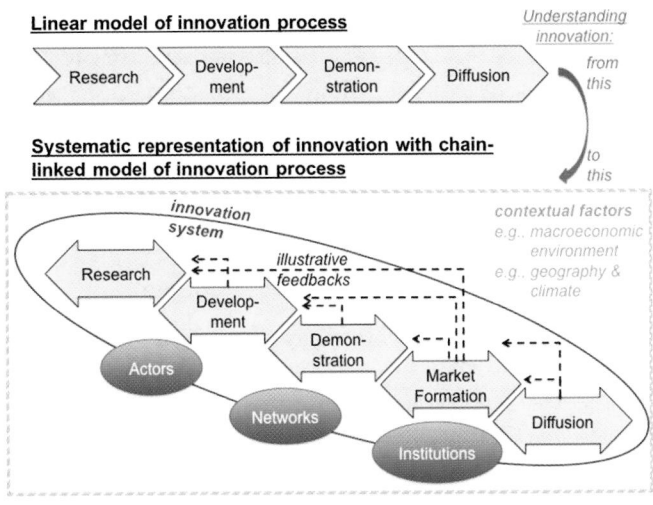

Figure 24.2 | The Evolution of Thinking on Innovation Processes.

In truth, it is well understood that the innovation process is neither linear nor unidirectional (Mowery and Rosenberg, 1979; Landau and Rosenberg, 1986; Freeman, 1994). Rather, the stages of the innovation process are linked, with feedbacks between each stage, giving rise to the term "chain-linked" model (Kline and Rosenberg, 1986; Brooks, 1995). This is illustrated in the lower part of Figure 24.2. The linear knowledge flow direction from basic science to applied technology as implied by the old "linear" model is now recognized to be more complex because it can also go in the opposite direction, with applied technologies enabling breakthroughs in basic science.[4] Likewise, research has identified the importance of knowledge spillovers and networks in collective learning processes, as there is no quasi-automatic "trickle down" from basic scientific knowledge to industrial applications of that knowledge.

Figure 24.2 represents the main modifications and additions to this "chain-linked" model of the innovation process. In this improved model there are multiple feedbacks among the different stages and their interaction, combining elements of "supply push" (forces affecting the *generation* of new knowledge) and "demand pull" (forces affecting the *demand* for innovations) (see the review in Halsnæs et al., 2007). Indeed, the stages often overlap with one another and the more interaction among the various stages, the more efficient the innovation process as offering more possibilities for learning, and knowledge and technology spillovers. And, of course, some technologies are successful without proceeding through each step in the innovation process (Grubler, 1998).

The distinction between supply-push and demand-pull has traditionally been important, especially as they imply different technology policy instruments – e.g., public R&D expenditures or incentives for private R&D as classical technology "supply" instruments versus government purchase programs, mandated quantitative portfolio standards, regulated feed-in tariffs, or subsidies as classical technology "demand"

---

3   The term "basic research" refers to study and research in pure science that aims to increase the scientific knowledge base. This type of research is often purely theoretical and has the intent of increasing the basic understanding of certain phenomena or behaviors; it does not seek to identify concrete applications of phenomena studied or to solve particular applied problems.

4   For example, satellite measurements leading to the discovery and subsequent explanation of previously unrecognized environmental problems such as stratospheric ozone depletion.

policy instruments. As argued here, from the perspective of a systemic innovation model characterized by multiple feedbacks, this technology supply-demand dichotomy is artificial to a degree. Transformative technological change generally requires the simultaneous leveraging of all innovation stages, processes, and feedbacks, and thus a combination of both supply- and demand-side technology policy instruments.

In an additional improvement over previous models, a market formation stage[5] has been added in explicit recognition of the so-called "valley of death"[6] observed in this innovation process between technology demonstration and diffusion. Many technologies fail at this or a similar hurdle between development and demonstration if they are too expensive, otherwise uncompetitive, too difficult to scale up, or lack perceived market demand. Market formation activities support new technologies that can struggle to compete with incumbent technologies that enjoy economies of scale and the learning advantages resulting from their more mature technology life cycle. In some cases, natural market niches exist that value the relative advantages of the new technology and offer a price premium. In other cases, it is important to create new niches (Kemp et al., 1998).[7]

The importance of the institutional context in which innovation occurs is also increasingly emphasized (Nelson, 1993; Geels, 2004). This points to the need for a more systemic approach to innovation, extending beyond the technology-focused "hardware" innovation process to also include analysis of actors, networks, and institutions.

Finally, the broader context of the innovation system matters. Technological, national or geographical factors affect the relative importance, roles, and relationships between components of the innovation system or the specific incentives structures in place. The concept of "national systems of innovation" (Nelson, 1993; Lundvall, 2009) describes this specificity. As a result, innovation systems for specific energy technologies vary substantively in their details, involving different sets of actors (e.g., incumbents or new entrants), interacting in different ways (e.g., research or market development), focusing on different problems (e.g., problem solving or learning by doing), and acting at different spatial scales (e.g., national or global) (Jacobsson and Lauber, 2006; Hekkert et al., 2007).

24.2.1.2 The Innovation Systems Approach

Taken together, the different elements described in the preceding section comprise the innovation systems approach used as the conceptual framework for this chapter. This is represented in the lower part of Figure 24.2. The different traditions of innovation and energy technology research outlined above (from linear to systemic) are drawn upon to support this chapter's integrative perspective. The innovation systems approach centers on the set of factors that drive and direct innovation *processes*. From a systemic perspective, innovation is understood as an interactive process involving a network of firms and other economic agents (most notably users) who, together with the institutions and policies that influence their innovation and adoption behavior and performance, bring new products, processes, and forms of organization into economic use (Nelson and Winter, 1982; Freeman and Perez, 1988; Lundvall, 1992).

The innovation systems approach emphasizes that the life cycle of a particular technology must develop in tandem with its corresponding innovation system (Jacobsson and Johnson, 2000). For new technologies that are incremental improvements to existing ones, innovation systems are already in place. For example, the development of a more efficient gas turbine occurs within a mature innovation system comprised of large firms with high R&D spending, strong networks between suppliers and users of the technology, established markets and well-aligned institutional infrastructures. In contrast, innovation systems need to be built up for radically new or disruptive innovations that strongly deviate from existing technologies and practices (van De Ven, 1993). Current examples of radical innovations in the energy domain are solar photovoltaic (PV) and electric vehicles. Innovation systems emerging around such technologies may be characterized by poorly developed markets, misaligned institutional settings, poorly structured knowledge networks, and small firms with limited resources to develop and market the new technology (Alkemade et al., 2007).

It takes time to build up an innovation system, particularly for radical innovations whose initial development typically takes place over decades (see the *Scaling Dynamics* case study). Weak or immature innovation systems may delay the progress of an innovation, or decrease the likelihood of its success (van De Ven, 1993). In the initial stages of the innovation process, only a few actors are involved in developing a new technology. Over time, other actors enter, the knowledge base starts to grow, often the legitimacy of the new technology increases, and more financial resources become available (although sometimes creating exuberant expectations that can lead to investment bubbles). Through this "formative phase," the innovation system around a new technology is built up (Jacobsson and Lauber, 2006). At a certain point, the innovation

---

5 Traditionally, market formation policies have been used in the defense (e.g., jet engines) and space sectors (e.g., photovoltaic [PV] technology) to kick-start a minimum level of technology demand for a nascent industry or technology.

6 The "valley of death" describes a situation where a successful R&D project either cannot attract funding for further development, or, once developed, cannot attract funding for large-scale demonstration of the new technology. It can also occur when the capital intensiveness of a project exceeds the financial resources of an otherwise willing investor (e.g., venture capital) or when promised public support does not materialize. For instance, the US FutureGen "clean coal" (advanced coal gasification, combined cycle, electricity generation plant combined with CCS) demonstration project was discontinued mid-stream after the US Department of Energy stopped funding due to substantial cost overruns (Rapier, 2008).

7 Examples of "natural" market niches include the first applications of PV in instruments (calculators) and toys in the Japanese electronics industry that did not need any public policy support and incentives, see the *Solar PV* case study in Appendix II. For an example of "created" niche markets, consider the case of Switzerland, where regulation requires electricity back-up systems for all public and technological infrastructures. This has created niche markets for microturbines and fuel cell applications for onsite electricity generation in hospitals, supermarkets, or cell phone towers.

system becomes large and developed enough for technology diffusion to take place during a "growth phase" (Jacobsson and Lauber, 2006).

24.2.1.3　The Energy Technology Innovation System (ETIS): What is it? Why is it needed?

ETIS is the application of this systemic perspective on innovation to energy technologies. In terms of the innovation system, this means the synthesis and analysis of data on the various stages of the innovation process; on different inputs, outputs and outcomes; on actors and institutions; and on the key innovation processes. In terms of the energy system, this means the synthesis and analysis of data on both the energy supply side and the energy demand side; on different energy technologies; and on both developed and developing countries. ETIS is thus an integrative approach that aims to comprehensively cover all the components of energy technology innovation systems, in terms of innovations, mechanisms of change and supporting policies, and energy technologies (supply and end-use), as well as in terms of geographical and actor network coverage.

Why is such a systemic approach needed? The GEA sets out clearly the magnitude of the challenge facing the global energy system. The GEA transition pathways – described in Chapter 17 – illustrate that a substantive and pervasive technological transformation in energy systems towards vastly improved efficiency and decarbonization is needed. This holds regardless of the ongoing debate over whether it is possible to improve existing technologies incrementally, with the primary challenge one of diffusion (Pacala and Socolow, 2004), or whether breakthroughs with radically new technologies are needed with the main challenge being basic and applied research (Hoffert et al., 2002; Hoffert, 2010).

It is the magnitude of the challenge that most clearly points to the need for a systemic perspective rather than a piecemeal approach focused on particular technologies (e.g., PV or CCS) or particular drivers (R&D or feed-in tariffs). This is fully supported by the accumulating body of knowledge on innovation processes and innovation histories, both successful and failed. New research carried out for this chapter adds to these findings. All point to the interrelationships and dependencies within effectively functioning innovation systems. This too necessitates a systemic approach.

ETIS has certain key characteristics which emerge repeatedly through the literature and are worth emphasizing. These include interdependence, uncertainty, complexity, and inertia. Interdependence means that different components of ETIS influence one another; moreover, the strength and direction of these influences may change. The outcomes of the innovation process are irreducibly uncertain, and it is *not* possible to ensure *ex ante* success for technology A if recipe B is followed. Complexity arises inevitably from the number and variety of innovation system components and their shifting interdependencies. This is further exacerbated by context-dependence in the application of the ETIS

framework to specific energy technologies. Inertia also arises from interdependencies, and is exacerbated by the long-lived capital stock and infrastructures in the energy system, as discussed above.

From these characteristics follow certain key implications for efforts to intervene in ETIS to support its effective functioning. Again, these emerge repeatedly in the literature and include coherence, alignment, consistency, stability, and integration. "Effective functioning" is used here in a qualitative sense. ETIS that demonstrate the full complement of drivers, mechanisms, actors, and institutions described in this chapter are more likely to be successful than ETIS that are lacking in one or more areas. Failure and success are not defined in absolute terms. Innovation system success could be interpreted most simply as widespread diffusion of new technologies and practices and when innovation benefits outweigh costs (in a large societal context). This is the ultimate outcome of interest for innovation processes in the context of energy system transformation required by the GEA objectives. Conversely, innovation system failure can be dramatic, as in a technology which fails in the "valley of death," or relative, as in a technology which diffuses slowly, to a low extent, or in a stop-start manner.

24.2.1.4　Strengths and Weaknesses of the ETIS Perspective

A systemic approach to innovation in an energy context is largely novel and challenges some established wisdoms. This is a recurring theme throughout Chapter 24 and is explicitly noted in the final policy guidance section, which directly questions certain policy myths, and in the quantitative assessment of financial inputs into ETIS presented in Appendix I. The systemic perspective necessitates an integrative analysis: from large-scale supply-side technologies to dispersed end-use technologies within the energy system and from early stage R&D through market formation to diffusion activities. Conventional data collection and analysis (as well as the formation of public and commercial institutions) has tended to focus on one piece of this puzzle. This chapter's comparative assessment makes (within the limitations of available data) commensurate what have to date largely been apples, oranges, pears, and peaches. Certain patterns emerge from this commensuration that have direct implications for the ETIS and its effective functioning. An example is an apparent mismatch between the target of innovation investments and the need for diffusion investments. This is explained and discussed at length below. Here, it suffices to note that the implications of the systemic perspective offer a challenge to prevailing practice and thinking. One example is the question of whether the technological, market, and institutional differences between the supply side and demand side of the energy system mean an integrative comparison is worthwhile or even meaningful. The ETIS perspective contends that it is, as the resulting insights are both important and potentially transformative.

Despite the strengths of the systemic perspective, its weaknesses and limitations should also be acknowledged. Though rich and detailed in certain areas, ETIS research is weaker in others, such as feedbacks between components of innovation systems. Studying innovation from a systemic perspective in the energy domain is a relatively young endeavor, with an empirical bias toward national, sustainable, and supply-side energy technologies. Policy experiments and field experience are largely still ongoing, particularly in a Northern European context that, together with Japan, provides many of the innovation histories from which the ETIS framework has been inductively derived. Studies in developing countries are particularly lacking, although this assessment begins to fill the gap with specific case studies on R&D expenditures in emerging economies, energy technology innovation in China, and lessons from solar PV market deployment in rural Kenya. Data are partial, incommensurate, or otherwise limited, as discussed below in the context of assessing ETIS (see Section 24.5.1 on data and information needs identified in this assessment). The understanding of mechanisms and linkages is incomplete. As a result, the ETIS perspective should not be interpreted as a full systemic dynamics model that can support quantitative modeling, simulation, or optimization. Rather, ETIS as developed in this chapter is a conceptual framework with the necessary generality to apply across the entire energy domain.

24.2.1.5 Empirical Basis of the ETIS Perspective

ETIS integrates current understanding of innovation processes within the energy system, their interlinkages, and the roles and influence of different actors and institutions including public policy. This systemic perspective is founded upon empirical work on technology histories such as wind power, processes such as learning, actor networks such as advocacy coalitions, social institutions such as expectations, and so on. This empirical work is covered in extensive literatures, which are referenced throughout the text. In addition, this chapter contributes a series of new empirical studies that are summarized in Appendix II, published in full in a companion volume, and referenced throughout the text. These are also summarized in Tables 24.2, 24.3, and 24.4 below and discussed further in the *Assessment Metrics* case study, summarized in Appendix II.

It is important to emphasize this empirical basis for the ETIS perspective. The various components of the ETIS described here characterize what is understood about successful innovation, as well as what may be missing

Table 24.2 | Chapter 24 case studies (innovation histories): demand-side technologies.

Short Name	Summary Description	Example of Relevance for ETIS	Chapter Section
Hybrid Cars	Development of hybrid electric vehicles in Japan, United States, and China, emphasizing the role of public policy.	Importance of policy alignment and consistency. Role of market demand and end-user preferences.	24.7.9
Solar Water Heaters	Early success and later failure of the solar water heater industry, particularly in the United States.	Lasting legacies of industry failure, including knowledge depreciation. Alignment of innovation system actors.	24.7.10
Heat Pumps	Different stages of heat pump diffusion in Sweden and Switzerland, emphasizing the role of public policy.	Interactions between supply of, and demand for, innovation. Importance of policy stability and consistency.	24.7.11
US Vehicle Efficiency	The "CAFE" standard for vehicle efficiency in the United States, and its influence on technological change.	Interaction between policy standards and changing market characteristics, including prices.	24.7.12
Japanese Efficiency	The "Top Runner" program to improve end-use efficiencies in Japan, and the role of dynamic incentives.	Flexible policies creating dynamic incentives within a clear overall strategic direction.	24.7.13

Table 24.3 | Case studies (innovation histories): supply-side technologies.

Short Name	Summary Description	Example of Relevance for ETIS	Chapter Section
Wind Power	Evolution of innovation stages and strategies in different wind power markets worldwide.	Need to integrate RD&D support with market formation. Interaction and feedback between innovation actors.	24.7.14
Solar PV	Development of solar PV in different markets worldwide, focusing on drivers of cost reduction.	Long-term R&D support complemented by market formation activities to stimulate commercial learning.	24.7.15
Kenyan PV	Market dynamics in the solar PV market in Kenya, emphasizing product quality issues.	Local institutions to set and enforce standards for quality control and assurance.	24.7.16
Solar Thermal	Early experience of solar thermal electricity in the US, and spillovers to later stage production.	Codification of knowledge. Interaction between R&D and learning to support cost reductions.	24.7.17
US Synfuels	History of US government investment in synthetic fuel production as oil substitute, and ultimate innovation system "failure."	Over-exuberant expectations in the context of changing market conditions. Public/private roles in innovation system.	24.7.18
French Nuclear	Review of pressurized water reactor (PWR) program in France, including cost escalation.	Interaction between learning effects and institutions, including standards and regulatory stability. Limitations of learning paradigm in technology cost reductions.	24.7.19
Brazilian Ethanol	History of ethanol production and developments in automotive technologies in Brazil, focusing on supporting role of policy.	Coalitions and shared expectations among innovation system actors, and interactions between related technologies.	24.7.20

Table 24.4 | Case studies (innovation processes and metrics).

Short Name	Summary Description	Example of Relevance for ETIS	Chapter Section
Grand Designs	Review of patterns, drivers, and dynamics of energy systems, both historically and in future scenarios.	Knowledge spillovers and inter-dependence between innovations creating inertia and path dependence.	24.7.1
Scaling Dynamics	Comparison of rates and extents of growth in 8 energy technologies historically.	Experimentation with small-scale units takes place during extended early formative phase.	24.7.2
Technology Portfolios	Tools to guide the selection of innovation portfolios under conditions of uncertainty.	Formal tools to support portfolio selection and diversification. Historical under-representation of end use technologies.	24.7.3
Knowledge Depreciation	Loss or obsolescence of knowledge, with examples in the context of energy innovation.	Importance of stable, sustained inputs to knowledge generation. High potential rates of knowledge depreciation.	24.7.4
Assessment Metrics	Quantitative metrics and qualitative approaches for assessing innovation.	Assessment of inputs, outputs, or outcomes, but not typically innovation systems as a whole. Data limitations.	24.7.5
Chinese R&D	Overview of R&D investments in China, with relevant institutions and mechanisms.	Substantive and growing R&D, dominated by industry but with strong government role. Supply-side technology emphasis.	24.7.6
Emerging Economies R&D	Review of RD&D investments in 6 major emerging economies, by technology type and source.	Increasing importance of emerging economies in global energy technology innovation system. Supply-side technology emphasis.	24.7.7
Venture Capital	Trends and targets of venture capital investments in energy technology innovation.	Rapidly growing venture capital investments in energy innovation. Non-fossil supply-side emphasis.	24.7.8

in cases of failed innovation. In this way, it should be treated as descriptive, induced, and interpreted *ex post* from available evidence; it should not be treated as normative. Partly for this reason, we use the ETIS perspective in the concluding section to abstract some general guidance for policy makers rather than offering more specific, universal prescriptions. As noted, the ETIS perspective is not sufficiently developed nor detailed to support a quantified system dynamics model. Nor is it necessarily appropriate to generate formalized, testable hypotheses which are replicable in both form and method across different technologies and contexts. Clearly, further work is needed in this direction. Testing, critiquing, and improving the ETIS perspective is of critical importance, given the magnitude of transformation needed in the energy system as captured by the GEA.

24.2.2 Characteristics of ETIS (I): Knowledge

24.2.2.1 Sources and Generation of Knowledge

"The most fundamental resource in the modern economy is knowledge and, accordingly, the most important process is learning" (Lundvall, 1998). Knowledge is a ubiquitous and powerful driver of technological change. Technological knowledge can be basic ("know-why") or applied ("know-how"), as well as publicly available (e.g., through scientific or engineering journals) or entirely tacit (e.g., resting with accumulated experience of a production engineer in manufacturing). Understanding the process of generation, reproduction, and diffusion of knowledge and the constraints to knowledge flows is therefore critical for innovation policy.

Knowledge is generally largely a public good. Once produced and disclosed, it is difficult to control or restrict its use. For activities organized around reward systems based on reputation and primacy of discovery, like science, this poses less of a problem (Dasgupta and

David, 1994). As more basic knowledge becomes integrated into technological solutions and into the realm of private production, the public aspect of knowledge which makes it expensive to generate, but cheap to reproduce, is generally classified as a source of market failure. This results in underinvestment in knowledge production (Arrow, 1962a). This is the traditional argument for encouraging public sector support for the generation of basic knowledge and allowing knowledge appropriability of private R&D through systems of property right protection.

Knowledge is generated at several different levels of innovation systems, through several distinct processes of knowledge exchange and transformation within and between agents and institutions. It is, therefore, a powerful source of feedback, correction, and advance in innovation systems. Basic science is, of course, a strong component of innovation in energy systems (Ausubel and Marchetti, 1997) and the disciplines that support it are numerous. However, technical change in energy systems tends to be based dominantly in engineering practices and disciplines, and the origins of major innovations have often come from outside basic energy science proper. (The best historical example is the development and application of steam engines much before the discovery of the Laws of Thermodynamics.) These specialized sources of knowledge outside classical basic science are crucial for energy technology innovation. Technology knowledge is also spawned during productive experience and a result of producer-user collaborations in technology development and of producer-producer collaborations in technology production at the manufacturing stages (von Hippel, 1988; Lundvall, 1992; see also Fridlund, 2000 for a case study on user-producer learning in the electric power sector in Sweden). Experience in production and use, as well as knowledge exchange between different types of producers and users, are important sources of specific knowledge that cannot be generated by scientific research (i.e., cannot be predicted from general principles or comparable technologies). This feedback process from application experience to redesign and engineering has been particularly important in energy technologies (see the *Wind Power*, *Solar PV*, and *Kenyan PV* case studies summarized in Appendix II) and highlights the importance of market applications as main knowledge and learning feedbacks enhancing R&D efforts that require appropriate incentives beyond R&D.

Applied energy research and development on the supply side has been oriented and guided by four main sets of goals, or "focusing devices" (Rosenberg, 1982):

- harnessing new energy sources and carriers with special qualities like energy density, abundance, transportability, but also flexibility, convertibility, and modularity;

- increasing efficiency, both in thermodynamic as well as in economic terms (cost reductions);

- improving control, security, and stability of energy conversion and delivery infrastructures; and

- improving adverse social and environmental impacts of energy systems.

Whereas the first three goals can be considered endogenous for the energy system, the goal of improved social and environmental performance has historically been triggered by regulation. R&D intensity (R&D expenditures per unit of value added) in the energy sector is strongly differentiated between energy supply (e.g., electric utilities) and energy demand (e.g., manufacturing of electricity using equipment such as TVs or computers) industries. At least in developed countries, energy supply industries have lower R&D intensity levels than the manufacturing average, similarly low as in the textile industry (see Figure 24.3). Conversely, electrical machinery, transport equipment and motor vehicles exhibit higher than average R&D intensities, although no information is available to differentiate their R&D into energy- and non-energy-related components (e.g., more fuel efficient engines versus safety improvements in motor vehicles).

24.2.2.2 Characteristics of Knowledge: Codification and Spillovers

Knowledge possesses a number of unique characteristics. It is *nonrival* (the use of knowledge by someone does not preclude its uses by someone else); *nonexhaustible* (can be used/reproduced without paying for an additional copy); *combinatorial* (knowledge is combined from different specific knowledge bases); and *cumulative* (builds on pre-existing knowledge). These features of knowledge can generate very high social rates of return and call for minimizing or even eliminating access costs to knowledge if social welfare is to be maximized (Foray, 2004). Indeed, much knowledge relevant in an industry is routinely shared and used by private actors through nonmarket mechanisms, a phenomenon referred to as knowledge "spillover."

Knowledge spillovers across sectors and across countries have been considered key engines of technological development and economic growth. The most salient sources of knowledge spillovers explored (e.g., Falvey et al., 2004) are universities and R&D centers; personnel training; scientific publications and patents; personnel movements, inter-firm turnover; formal and informal networks of scientists, engineers, and technicians; licensing and technology transfer agreements; foreign direct investment; international research collaboration and research; reverse engineering, and international trade in final, intermediate, and capital goods.

The literature is vast and not free of controversies. In general, it has been shown that knowledge spillovers positively impact growth and productivity (Coe and Helpman, 1995; Coe et al., 2009), with highly localized

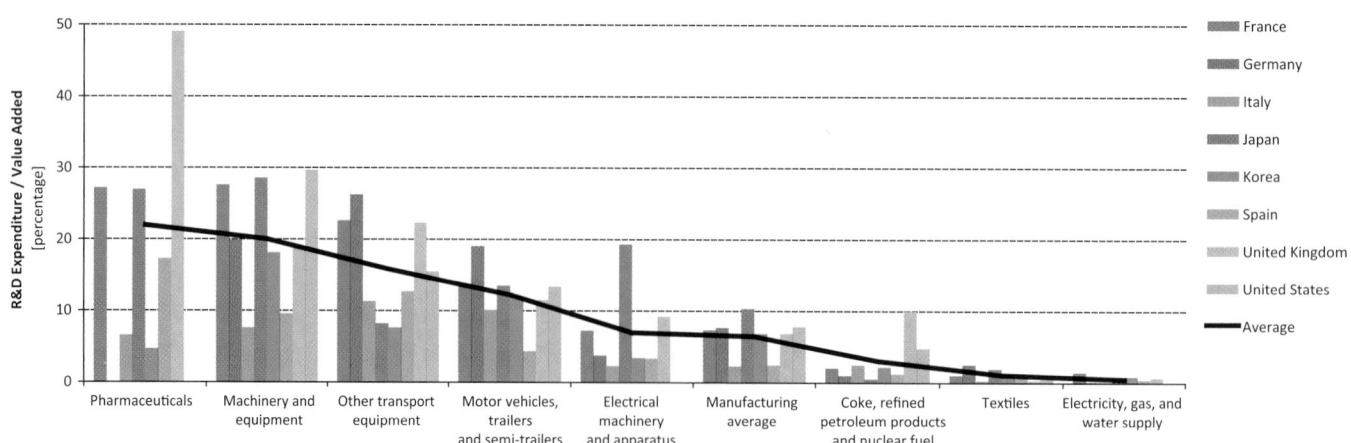

Figure 24.3 | R&D intensity (expenditures per value added, in percent) for selected sectors and OECD countries in 2002. Source: The OECD Research and Development Expenditure in Industry database and STAN Database.

geographical effects (Jaffe and Trajtenberg, 1999). It has also been shown that the extent of international knowledge spillovers depends on national R&D efforts (Mancusi, 2008; Unel, 2008). Thus, knowledge spillovers depend critically on the assimilative capacity of the recipient country. There are very few studies focusing specifically on knowledge spillovers in the energy sector (an exception being Bosetti et al., 2008). Measuring knowledge spillovers poses many challenges, for knowledge is highly heterogeneous, unobservable, and difficult to define in quantitative terms (Mohnen, 1997; van Pottelsberghe de la Potterie, 1997).

However, there are other important barriers to learning and knowledge advancement through spillovers, which are more relevant for innovation policy design. Knowledge implies more than simply information such as data and formulas; it also implies the cognitive capability to interpret, process, and articulate information. Similarly, in order to understand a problem and make sense of knowledge produced elsewhere, firms require a minimum threshold of accumulated knowledge or absorptive capacity, derived from their direct involvement in R&D, production experience, or training (Cohen and Levinthal, 1990). In many cases, it has been shown that firms engage in R&D not only to produce new knowledge, but to be able to learn knowledge produced elsewhere (Cohen and Levinthal, 1989). In other words, enforcing knowledge appropriability (i.e., knowledge access) is insufficient for knowledge generation and learning (Foray, 2004). Policies are therefore also needed to further absorptive (assimilative) capacity to facilitate learning in firms, but also in the public sector and society at large. For energy technology innovation policy, this implies that an exclusive focus on technology transfer (knowledge appropriability) is entirely insufficient to guarantee technology diffusion, as the latter critically depends on sufficient local knowledge or assimilative capacity. Local knowledge is needed to first decide which globally available technologies offer local diffusion potentials, which requires capacity building for technology assessment both in the public and private sectors. Local knowledge is also needed to import/manufacture, adapt, install, and above all *use* a new energy technology effectively (see e.g., the *French Nuclear* case study, which illustrates the importance of local technological capability in the successful scaling up of imported reactor designs, or the *Kenyan PV* case study as an illustration of the problems arising if local capacity for testing and evaluating technical characteristics [and defects] of imported PV cells are unavailable).

It also important to note that not all technological knowledge is embodied in artifacts or codified in blueprints, manuals, and sets of instructions; much knowledge remains tacit (uncodified) in the form of personal or institutional knowledge and skills. In this case, knowledge generation implies practicing and accumulation of experience (Nelson and Winter, 1982), which needs to be achieved locally and cannot be substituted by "imports." Generally, new knowledge tends to be less codified and articulated, which makes it more difficult to reproduce, memorize, recombine, and learn, which in turn makes it costly to transfer in a usable form (von Hippel, 1994). For new energy technologies, this implies above all a need to develop a

local industrial base (either in manufacturing or in using/marketing a new technology) and to promote partnerships that can further the transfer of tacit knowledge. For example, the *Wind Power* case study illustrates the importance of public testing facilities for new wind turbine designs that furthered localized learning and tacit knowledge transfer among firms.

Knowledge is also highly dependent on local conditions and environments and on market structures, and therefore tends to be dispersed by the division of labor and specialization, which raises difficult coordination and communication problems (Machlup, 1984; Smith, 2002). The public sector therefore has an important role to play in furthering formal or informal coordination mechanisms, e.g., technology development "roadmaps,"[8] and setting dynamic technology performance standards (see the Top Runner program described in the *Japanese Efficiency* case study).

Lastly, knowledge can depreciate rapidly if it is not continuously replenished. This depreciation occurs through personnel turnover and retirement, technological obsolescence, and institutional inertia (see Section 24.2.2.4).

For all these reasons, knowledge cannot simply be assumed to generate automatic "spillover" effects. A (local) enabling learning environment is as important, as is the generation of knowledge (elsewhere) in the first place.

The nature of spillovers also depends on the type of knowledge, on the type of industry, and, finally, on the life cycle stage of a technology. In the early phases of the technology life cycle, disembodied knowledge tends to be predominant. As learning experience accumulates,[9] underlying phenomena are better understood and measured, practices and procedures become increasingly codified, and hardware is adapted and developed. The more codified and embodied (in artifacts) technological knowledge is, the easier it is to capture spillovers.

Empirical studies have shown evidence on the importance of knowledge spillovers in new energy technologies, particularly in the cases of PV (Watanabe et al., 2002) and wind energy (Kamp et al., 2004; Lako, 2004).

In Watanabe et al. (2002), a virtuous cycle between R&D, knowledge spillovers, market growth, and price reduction is thoroughly illustrated for the case of the Japanese PV industry. By stimulating public and private R&D on a broad base of industrial sectors and by simultaneously creating incentives for niche market deployment,

---

8 Successful examples of technology roadmaps can be found especially in the semiconductor industry.

9 The concept of learning via knowledge spillovers is sometimes referred to as "learning-by-interacting" (Lundvall, 1992).

the Japanese Sunshine Program triggered a range of mutually enhancing positive feedback mechanisms. The technology knowledge stock increased rapidly both from proprietary R&D and knowledge spillover effects among Japanese PV R&D and manufacturing firms (even after accounting for knowledge depreciation). Next to innovation, learning contributed to important cost reductions in PV cells, which further induced increases in demand and production; market growth in turn further increased R&D expenditures, closing a positive feedback loop between knowledge generation and market development. As shown in the *Solar PV* case study, PVs also benefited importantly from inter-industry learning spillovers and scale economies, among other factors. The *Wind Power* case study confirms this conclusion: knowledge spillovers have been important sources of innovation and design improvements, including knowledge spillovers between industries (materials, aerodynamic simulations, and designs) and between manufacturers nationally (facilitated by testing stations), as well as internationally. Denmark's success involved not just R&D, but also the facilitation of feedback between users and producers of wind turbines (Garud and Karnoe, 2003). This positive feedback process, based upon inter- and intra-firm knowledge spillovers in which learning and absorptive capacity production are mutually reinforced, can potentially augment an industry's capability of knowledge production (Foray, 2004). The *Solar Thermal* case study shows how interactions between researchers and firms commercializing technologies create feedbacks so that real world problems can stimulate improvements in design. This "learning-by-doing" can be more important for technology development than R&D, and is discussed further below. The *Kenyan PV* case study shows how interactions between firms and users are important to communicate how users adapt technologies into their daily practices, thereby revealing potential quality and technology problems that in turn form the basis for further design improvements.

24.2.2.3　Knowledge Creation through Learning-by-Doing and Learning-by-Using

Processes of learning are essential for the development and introduction of new technologies and comprise a complex set of actors (*who*) and processes (*what* and *how*). Learning results in improved and standardized production processes and products, which in turn can often result in cost reductions. In order to highlight that learning processes require dedicated efforts rather than just the passage of time, resulting cost reductions are often illustrated by so-called "learning curves," i.e., curves that describe the cost development as a function of cumulative production (as a proxy for learning).[10] In reality

however, it is not the act of production per se that provides a source of learning but rather a complex set of interrelated processes that include learning at the individual and organizational scale, classical economies of scale in manufacturing, knowledge spillovers, market conditions and structure (e.g., raw material prices, degree of competition), etc. Common to all is that they exercise their impact on costs and cost reductions via accumulated production/market deployment and/or growing industry size.

The learning curve originates from observations that workers in manufacturing plants became more efficient as they produced more units (Wright, 1936; Alchian, 1963; Rapping, 1965). The roots of these microlevel observations can be traced back to early economic theories about the importance of the relationship between specialization and trade, which were based in part on individuals developing expertise over time (Smith, 1776). Drawing on the concept of learning in psychological theory, Arrow (1962b) formalized a model explaining technical change as a function of learning derived from the accumulation of experiences in production: Arrow's "learning by doing" (LbD) model. Accumulating experience in the early stages of an innovation's life cycle can be a powerful strategy both for maximizing the profitability of firms (BCG, 1972) and for the societal benefits of technology-related public policy.

In its original conception, the learning curve referred to the changes in the productivity of labor that were enabled by the experience of cumulative production within a manufacturing plant. It has since been refined. For example, Bahk and Gort (1993) make the distinction between "labor learning," "capital learning," and "organizational learning." Subsequently, the Arrow model was complemented by an analogous concept, that of "learning by using" (LbU), referring to learning effects from the perspective of technology users (e.g., plant and equipment operators or consumers as opposed to technology producers).

Often, the term "experience curve" is used in the literature to provide a more general formulation of the learning concept, including not just labor but all manufacturing costs (Conley, 1970) and aggregating entire industries rather than single plants (Dutton and Thomas, 1984) or entire technological "trajectories" rather than individual technology generations. Though somewhat different in scope, each of these concepts is based on Arrow's explanation that "learning-by-doing" provides

---

10　In the functional form of: $C_t = C_0 * X^{-\varepsilon}$ where the costs at *t* are a function of the costs at *0* multiplied by the cumulative production volume X to the exponent of the so-called learning rate ε. Thus a learning rate ε=0.2 indicates a cost reduction by 20% per doubling of cumulative production. A related concept used in the literature is the so-called progress ratio PR = $[P_0 * (2X)^{-\varepsilon}]/[P_0 * X^{-\varepsilon}]=2 - \varepsilon$. Hence a progress ratio of 0.8 (1-ε) indicates that costs are at 0.8 of their original value after a doubling

of cumulative production. A specific characteristic of learning curves is that costs are reduced by a constant percentage with each doubling of output. As the doubling of cumulative production is more rapid in the early phases of an industry compared to mature ones, cost reductions invariably are steeper in early industries and tend to decline with increasing maturity as the doubling of cumulative production volume requires ever longer time. The observed cost reduction for different technologies analyzed in the literature covers a range up to 40% cost reductions for each doubling of the total number of units produced (e.g., Argote and Epple, 1990). There are also well-documented cases in the aircraft and nuclear reactor industries that exhibit "negative" learning, i.e., costs increase rather than decrease with accumulated production volumes. (See the *French Nuclear* case study in Appendix II.)

opportunities for cost reductions and quality improvements and the following discussion will therefore refer to this phenomenon simply as "learning"[11] or "learning curve."

There have been a number of misconceptions about how to use and interpret learning curves (see Neij, 2004; Nemet, 2006) that have resulted in certain pitfalls in communicating with policy makers. For instance, a popular misinterpretation of learning (or experience) curves is that policies can simply "buy down" (e.g., Brennand, 2001) technology costs through, for example, one-sided demand-pull policies without due regard to the equally needed supply-side technology innovation policy aspects. A second pitfall of cost "buy down" policy approaches is the assumption that the extent of possible "learning" can be *ex ante* anticipated (i.e., "forecasted"), which evidently is not possible. Evidence from the descriptive case studies performed within Chapter 24 is used to inform the general insights for policy design covered in Section 24.4.

Learning phenomena and curves are frequently evoked in cost trend analysis of nonstandardized products produced globally or nationally. The resulting uncertainties and variation in estimates/projections need, however, to be assessed critically (van der Zwaan and Seebregts, 2004). Variations in calculated learning (or experience) curve parameters as a function of different levels of spatial aggregation, observational period, or the use of price data as a proxy for cost information are examples of important uncertainties. Future prices may not only be reduced as a function of "learning," but may also increase as a result of escalating input prices, quality improvements of the product, or lack of competition in quasi monopoly markets, especially under rapid demand growth. The example of wind turbines illustrates this point nicely. Whereas prices fell in line with the "learning curve concept" until the late 1990s, prices have risen since the early 2000s. A recent study reviewed cost estimates in off-shore wind projects in the United Kingdom and found a real-term cost escalation of a factor of two since 2000 to some £3000/kW, or US$5000/kW (UKERC, 2010). Similar trends have been observed in the VS as well (Bolinger and Wiser, 2012). The reasons for such cost escalation are varied and include rising material prices, much larger wind turbines, and component supply bottlenecks, among others. It also appears that policy (favorable feed-in tariffs) stimulated rapidly increasing demand growth for wind turbines and enabled opportunistic pricing strategies of wind turbine manufacturers. The rapidly growing demand led to significant supply bottlenecks, further reducing competition in this oligopolistic industry.

---

11   Often the literature refers to "experience curves" to describe cost reductions in aggregate costs over entire industries and across a whole series of technological generations and thus includes the effects of product design changes and improvements as well as manufacturing economies of scale in cost reductions, which were initially excluded in "learning curve" analyses (for a discussion of these two concepts see Nemet, 2009b). However, this terminological distinction has not become standardized and aggregate cost reductions over entire industries continue to be referred as "learning curves" as well. To add to the confusion, the ensuing cost reductions of learning processes are also defined differently in the literature, either as percent cost reductions per doubling of cumulative output (or manufactured units), or as the percent original costs remaining after a doubling of cumulative output (cf. the discussion in footnote 10).

Secondly, the driving forces of the cost reductions remain unexplained by such aggregate analyses. In order to develop scenarios of possible future technology cost trends as a guide for policy, the sources of cost reduction must be identified and analyzed separately – be they characteristics of innovations (radical or incremental), knowledge generation and spillovers, the importance of economies of scale effects, or other factors. Such detailed analyses are increasingly becoming available (see, e.g., Krawiec et al., 1980; Hall and Howell, 1985; Nemet, 2006; van den Wall Bake et al., 2009; see also the *Solar PV* and the *Brazilian Ethanol* case studies).

R&D investments or other sources of knowledge generation are now increasingly recognized as not being simple substitutes for the accumulation of actual production experience (see the discussion in Halsnæs et al., 2007). These two sources of technological learning are therefore more usefully conceptualized as complementary to each other.

24.2.2.4   Knowledge Depreciation

It is important to recognize that technological knowledge – like all knowledge – can be accumulated (learned) but equally lost (unlearned). Knowledge depreciation particularly affects settings in which knowledge remains largely tacit, residing in individuals or organizational entities (e.g., managers) and needs to be acquired again in case of staff turnover or stop-and-go production schedules. A second type of depreciation occurs as old knowledge becomes obsolete. Knowledge can depreciate because of an insufficient "recharge" (Evenson, 2002) of knowledge in cases where innovation proceeds rapidly such that old technological knowledge is no longer relevant for updated processes/techniques, but new learning cannot proceed quickly enough because of financial constraints. It is the latter type of unlearning that is of particular concern in energy technology innovation systems – when rapid rates of innovations coincide with erratic funding and policy support, for which history provides ample examples.

Both knowledge gained from experience through market deployment and knowledge gained from R&D depreciate. The *Knowledge Depreciation* case study (Appendix II) discusses the case of R&D knowledge depreciation of nuclear power (and energy efficiency) R&D in the member countries of the International Energy Agency (IEA), but there is also evidence of nuclear knowledge depreciation beyond R&D, such as in the ability to construct nuclear power plants illustrated in the recent disappointing experiences (substantial costs and construction time overruns) at the two construction sites of the European Pressurized Water Reactor (EPR). (See also the *French Nuclear* case study.)

The particular vulnerability of tacit knowledge to depreciation implies that learning by doing is prone to especially high rates of knowledge depreciation, since experience in production is less likely to be codified than research and development activities and may thus be "unlearned" rather quickly.

As an example of the first source of knowledge depreciation, it has long been recognized in the management literature that in service industries (e.g., pizza franchises) organizational and production knowledge can be lost quickly, especially under high rates of staff turnover (that may be up to 300%/year). Argote et al. (1990) and Darr et al. (1995) report knowledge depreciation rates of between 25–50% month in service industries. Such high rates of knowledge depreciation basically imply that after a year only between 0–5% of the original knowledge of an organization remains. Kim and Seo (2009) arrive at similarly high depreciation rates of some 26%/month in their analysis of Liberty ships manufacture during World War II, even if earlier studies on the same case (e.g., Thompson, 2007) estimate much lower rates of 4–6%/month.

A classic technological un-learning case discussed in the literature is that of the Lockheed L1011 Tri-Star aircraft (Argote and Epple, 1990). The experience in aircraft industries suggests a significant reduction in manufacturing costs as more production experience (output volumes) is accumulated (learning by doing). The only exception seems to be the Lockheed Tri-Star aircraft. When production resumed after an extended production halt, manufacturing costs were much higher and also did not decline, reflecting lost experience or unlearning. The reason for this knowledge depreciation was basically the same as for pizza franchises. During the production halt, the entire staff of the manufacturing plant was fired, including managers who, according to Michina (1992; 1999), are the main locus of organizational learning and knowledge accumulation in aircraft manufacturing. Benkard (2000) reports corresponding knowledge depreciation rates of typically 40%/year in aircraft manufacturing.

In innovation-intensive industries, estimates of knowledge depreciation are extremely limited. Hall (2007) provides one of the few comprehensive efforts to estimate knowledge depreciation across various industry sectors, relating the market value of firms to patent data to estimate the R&D knowledge depreciation in six US industry sectors. Hall finds knowledge depreciation to vary significantly over time and across industries, with median R&D knowledge depreciation rates of between 15%/year (drugs and instruments) to 36%/year (electrical). Watanabe et al. (2002) provide one of the few estimates for energy technologies. By constructing a knowledge stock model for the Japanese PV industry that includes both R&D by firms and knowledge spillovers from other firms (measured via patent citations), he estimates a mean PV knowledge depreciation rate of some 30%/year (Watanabe et al., 2002). This implies that without continuous recharge (R&D), an existing technology knowledge stock is reduced to some 25% of the original value after five years and to less than 5% after 10 years. Nemet (2009a) provides an illustration for the US wind turbine industry by analyzing a set of "highly cited" wind energy patents. He finds that 40% of all (cumulative) citations occur during the first five years, after which citations decline to basically zero after 25 years. This declining trend in patent citations after year five reflects their decreasing significance and can be used as a proxy for R&D knowledge depreciation, which corresponds to a rate of approximately 10%/year after the fifth year.

The available literature thus suggests typical knowledge depreciation rates of 10–40%/year in industries comparable to energy (i.e., where innovation and R&D play a significant role). Given such high rates of obsolescence of technological innovation knowledge, continuous knowledge recharge becomes extremely important. In case of erratic stop-and-go policy support for knowledge generation, e.g., through R&D, knowledge depreciation rates can outweigh knowledge recharge rates, as discussed in the *Knowledge Depreciation* case study for nuclear. The nature of knowledge generation support (stable versus erratic) is thus as important as the absolute levels of resources made available, which provides an important argument for stability and gradual expansion of inputs to ETIS over "crash" programs that may not be sustained over any significant time periods.

In their study of knowledge depreciation of professional services, Boone et al. (2008) conclude that the extremely low rates of knowledge depreciation found in engineering design firms are explained by comprehensive knowledge documentation (earlier designs are documented and kept for subsequent use), as well as much lower staff turnover rates (3%), particularly among senior engineers. As such, the study provides valuable lessons for improved knowledge management in ETIS, highlighting the importance of documentation, codification, and preservation of knowledge, as well as the need for a minimum degree of continuity in senior staff that are the living memory of organizations.[12]

24.2.3 Characteristics of ETIS (II): Economies of Scale and Scope

A discussion of the sources of technological advance is not complete without reference to the rich body of literature on sources for cost improvements beyond product and process innovation and new knowledge application discussed above, including in particular economies of scale and economies of scope. While economies of scale are recognized as particularly powerful drivers in the historical evolution of energy industries such as electric power (e.g., Lee and Loftness, 1987) or petroleum refining (Enos, 1962), economies of scope merit discussion here for their potential impact on future energy systems.

24.2.3.1 Economies of Scale

What are Economies of Scale?
Economies of scale describe reductions in average unit costs as output or production increase over the long run, assuming all factors of

---

12 It is significant that private firms practicing comprehensive reassignment (job-rotation) policies frequently lack any institutional memory of earlier corporate strategies and corresponding innovation successes and failures. For instance, a group of managers at a major oil company attended a workshop with one of this chapter's authors. At the workshop, these managers contemplated possible investment strategies into renewables, entirely unaware that the company had invested previously in solar PV and biofuels and had already sold these activities.

production are variable. They are often conflated with technical returns to scale, which describe a more than proportional increase in output for a given increase in inputs. For examples, see Table 24.5. For further discussion and theory, see Chapter 4 of Rosegger (1996).

Economies of scale can act at different levels (see Table 24.5). Cost reductions associated with scaling at the unit level of a technology (e.g., size of wind turbines) or at the level of manufacturing plants (size/output of a manufacturing plant producing wind turbines) are all important drivers of change in energy technologies.

Unit and Manufacturing Level Economies of Scale

Figure 24.4 illustrates the increasing unit scales of energy technologies over the twentieth century (for other graphical examples, see Smil, 1994; 2008). The graph on the top shows average unit capacities; the one on the bottom shows unit scale frontiers (maximum size of units produced/installed) with characteristically concentrated periods of scaling up (note the log-scale y-axis on both graphs). Increases in unit scales have been a pervasive phenomenon for energy technologies throughout the twentieth century.

Observed increases in unit capacities of energy technologies are strongly linked to falling costs driven by economies of scale, particularly for large capacity energy supply and conversion technologies. For example, over a 50-year period, beginning with World War II, the average cost/unit output of a performance optimized fluid catalytic cracking unit in a US oil refinery fell by around 4%/year on a compounded basis (close to an absolute cost reduction by a factor of seven). These continuous cost improvements were driven by capital and labor productivity gains asso-

ciated with an order of magnitude increase in unit capacity from 15,000 to 140,000 barrels/day (Enos, 1962).

Additional examples of unit-level economies of scale are presented in the *Solar PV, Solar Thermal,* and *French Nuclear* case studies. Distributed energy conversion and end-use technologies are more likely to be characterized by manufacturing scale economies. The exemplar is the car, beginning with the Model T Ford produced from 1908–1927. Bywords of Fordist manufacturing include *specialization* of machine tools, *routinization* of labor tasks, *standardization* of output, and sequencing of manufacturing along *assembly lines* (Raff, 1991). These technical, process, and organizational innovations allowed a scaling of manufacturing output from a single plant at Highland Park near Detroit, with remarkable increases in labor productivity. In a 12-month period between 1913 and 1914, the labor requirements for assembly of a single Model T fell from 12.5 to 1.5 man-hours. This, in turn, contributed to the price of a Model T falling by two-thirds over its 20-year production run (Ruttan, 2001).

Further examples and discussion of manufacturing level economies of scale are presented in several of the case studies, including *Solar PV, Hybrid Cars, Heat Pumps*, and *Brazilian Ethanol*.

Drivers of Economies and Diseconomies of Scale

Table 24.5 summarizes some common drivers of scale economies. The examples given above further emphasize the relationship between scale and technical efficiency (e.g., wind power), labor productivity (e.g., cars), and capital productivity (e.g., refineries). Demand growth and standardization are two other important factors.

Table 24.5 | Economies of scale at different levels, using wind power as an example.

Level of Scale Economy	Example of Scale Economy	Outcome of Scale Economy using Wind Power as an Example
unit level	increases in wind turbine blade length and tower height cf. **technical returns to scale**: outputs increase proportionally more than inputs, e.g., inputs = f(area), outputs = f(volume) fixed balance of system components spread	lower $/MW for larger MW wind turbines higher MWh/MW for larger MW wind turbines (no reference to costs) lower $/MW for
plant level (also facility or installation level)	over larger numbers of units operating	larger MW wind farms
manufacturing level (also production level)	capital productivity improved by spreading fixed input costs over higher output volumes	lower $/turbine for larger turbine manufacturing facilities
organizational level (also firm level)	labor productivity improved by specialization; lower cost volume purchases of capital equipment*	lower $/MW or $/turbine for larger turbine manufacturers or wind farm developers
industry level	political economic influence (rent-seeking) securing, e.g., increased subsidy or price support*	lower cost or higher revenue as contribution of industry to GDP or job increases
inter-industry level (also external or system level)**	development of enabling infrastructure (e.g., distribution networks) and institutions (e.g., forward contracts); development of complementary industries (e.g., materials or equipment suppliers)	lower cost or higher revenue as share of total electricity production increases

*Rosegger (1996) refers to these as "pecuniary" economies of scale, which are nontechnical and associated with input costs or output revenues.
** External economies of scale are one explanation for the potential benefits of both national industry clusters and transnational economic unions e.g., Henriksen et al., 2001a.

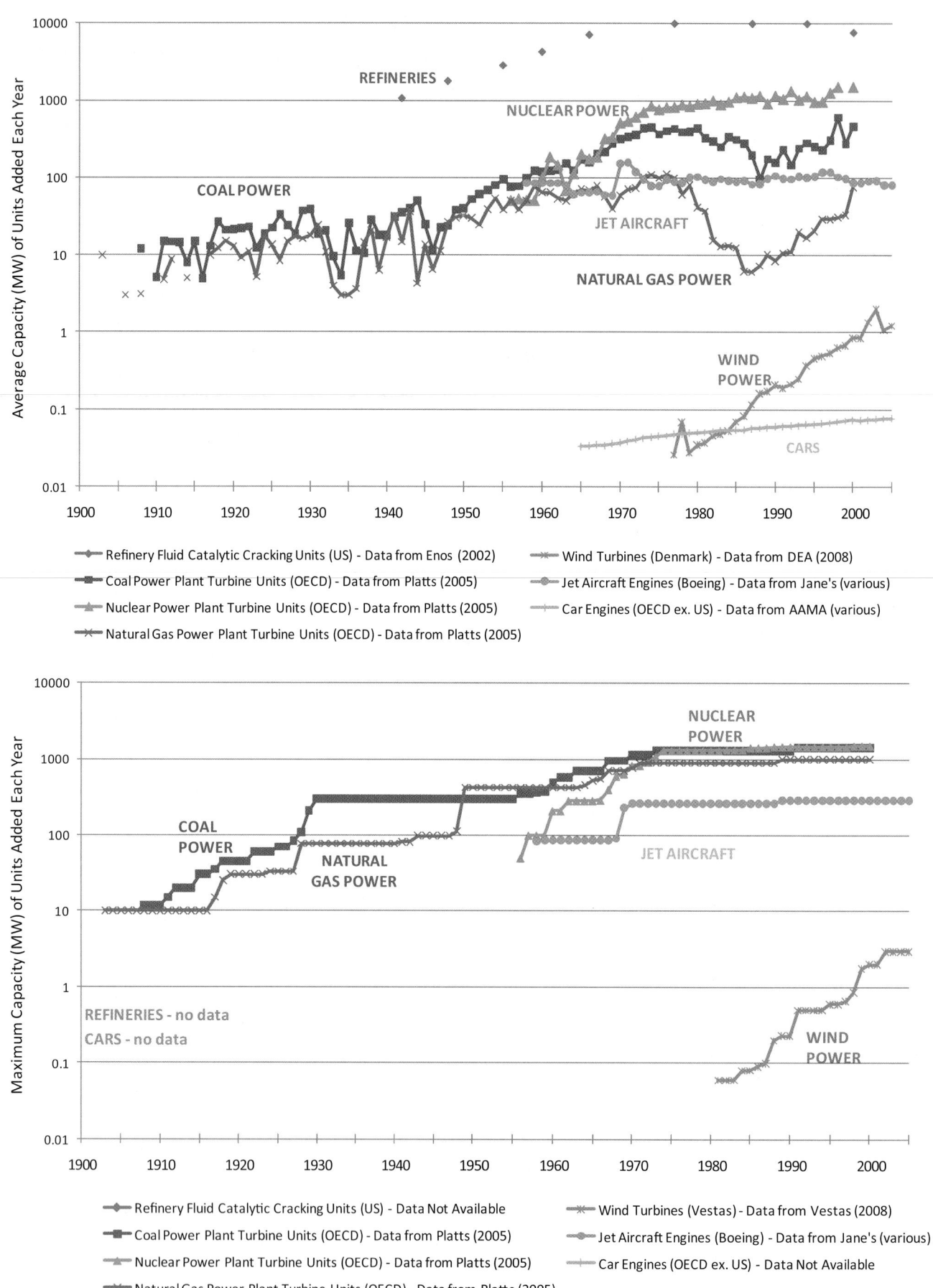

Figure 24.4 | Increasing unit of scale of energy technologies, average (top) and scale frontier (bottom). Unit scales are expressed as MW capacities and plotted on log-scale y-axes. For details, see: Wilson, 2009.

For energy technologies with perceived social benefits, consistent government policies to support market demand are needed to underwrite the scaling up of unit and/or manufacturing capacities. Conversely, stop-start market-based policies undermine manufacturers' confidence, increase the risk of investing in scaling up production, and ultimately can result in market collapse. This is clearly demonstrated in the *Solar Thermal* case study.

Alongside demand growth, technology standardization has proven important for manufacturing scale economies at the unit or plant levels. The more successful growth of nuclear power in France in comparison to the United States can be attributed in part to the standardization of reactor and plant design and to knowledge spillovers, both profiting from a well-tested US Westinghouse reactor design. Later in the French nuclear program, scaling of unit capacities from the standardized 900 MW reactor to more bespoke or "Frenchified" 1300–1500 MW reactors led to marked cost increases (see the *French Nuclear* case study).

These cost increases also demonstrate the possibilities of diseconomies of scale – not lower, but higher $/MW or $/MWh – associated with an increase in the complexity of designing, building, and operating integrated technologies close to the unit scale frontier (Rosegger, 1996). For large-scale power plants in the late 1970s, complexity led to additional borrowing costs during the construction phase due to delays (Koomey and Hultman, 2007) and foregone revenues during the operating phase due to an increase in the frequency and duration of unforced outages (Ruttan, 2001; Lovins et al., 2002). Increasingly stringent health, safety, and environmental regulations also impacted larger scale plants more severely. The transition through the 1980s and 1990s from large-scale coal and nuclear power plants to smaller scale, more flexible and less capital-intensive natural gas-fired units (better suited for increasingly deregulated markets) illustrates well that the availability of economies of scale are contingent on both technical and market factors.

Isolating Economies of Scale as a Source of Cost Reduction
Simple engineering models relate cost to scale in the form:

$$\frac{\$_t}{\$_{t-\square}} = \left(\frac{scale_t}{scale_{t-\square}} \right)^{(scale\ factor)} \qquad (1)$$

with a scale factor less than one indicating economies of scale. A study of wind turbines in Germany during the 1990s found that a scale factor of 0.84 described the observed fall in wind turbine investment costs per MW installed capacity by 11% for a doubling of turbine size from 0.3 to 0.6 MW, and by a further 18% for a subsequent unit size increase from 0.6 to 1.5 MW (Grubler, 2010).

However, observed average cost reductions during the development and commercialization of an energy technology commonly conflate unit and manufacturing level scale economies as well as learning effects. Isolating the contribution of economies of scale at different levels, therefore, requires models that disaggregate the various influences on cost.

Econometric approaches typically use a Cobb-Douglas functional form (linear in log-transformed variables) to explain unit cost as a function of scale. Models fitted to data on coal power plants built in the United States from 1960–1980 showed that a doubling of unit scale reduced the cost per unit capacity by 12–24%, controlling for learning effects, compliance with environmental regulation, and changes in productivity and input prices (Joskow and Rose, 1985; McCabe, 1996). It is worth noting that the data period to which these models were fitted describes the rapid growth phase of unit scale during which unit scale economies might be expected to be most evident (see Figure 24.4 above).

More sophisticated engineering models control for the effect of non-scale related drivers of cost reductions over time. A good example is presented in the *Solar PV* case study for the United States. Manufacturing scale economies were found to explain 43% of observed cost reductions in solar PV module cost ($/W$_{peak}$) between 1980 and 2001 (Nemet, 2006). During this period, manufacturing plant output had scaled by two orders of magnitude, from 125 kW/year to 14 MW/year.

24.2.3.2   Economies of Scope

Whereas economies of scale describe the reduction in unit costs with increasing scale of production of a standardized good/commodity, economies of scope describe the reduction in unit costs that can be achieved by producing more products jointly as opposed to individually (see Panzar and Willig, 1975; Teece, 1980, for an exposition).

The traditional manufacturing economic literature draws the following contrast: whereas economies of scale describe production processes where the focus is on quantity and the emphasis is on reducing unit costs, the focus in economies of scope is on product variety. Economies of scope are realized through the effective sharing of knowledge, facilities, equipment, and other inputs such as marketing and design services. In addition, machinery and production processes were designed to facilitate and speed-up the process of change-over between products. Economies of scope can also be realized when there are cost savings arising from byproducts in the production process.

In the energy field, the literature on economies of scope is somewhat limited (examples include Mayo, 1984; Kwon and Yun, 2003; Farsi et al., 2008; Shum and Watanabe, 2008). A concept used to describe energy plants that reap economies of scope from the production of an array of multiple energy carriers is sometimes referred to as energy "combinates." The prime example of economies of scope in energy are, therefore, cogeneration systems, i.e., the joint production of electricity and heat in power plants (combined heat and power [CHP] plants), where the waste heat can be used for heating purposes, e.g., via district heating systems. More recently, such schemes also include district cooling systems, such as using steam generated in electricity plants to drive chillers and the distribution of chilled water for air conditioning purposes.

A necessary condition for the realization of economies of scope in cogeneration systems are either an appropriate co-location of industrial and end-use applications with energy conversion facilities or the possibility to interconnect these diverse users and energy uses through district heating/cooling pipeline systems. Given the usually large scale of electricity generating plants, such schemes are generally only economically viable if generation and demand are not too distantly located (e.g., cogeneration from power plants located within larger cities) or have a sufficiently high energy demand density to justify the high investment costs of heat pipeline systems. Conversely, the advent of decentralized distributed energy systems can increase cogeneration potentials. Examples of applications include natural gas microgeneration (with waste heat recovery for heating of commercial and larger residential buildings and sales of electricity back to the grid),[13] or biomass gasification coupled to gas engines for joint production of electricity and heat.

24.2.4 Characteristics of ETIS (III): Actors and Institutions

24.2.4.1 Introduction

Innovation processes take place within an environment that consists of actors and institutions (Edquist, 2001). Collectively, this makes up the innovation system at the heart of the ETIS perspective. Models of innovation typically highlight the various stages of innovation, and the interactions and feedback loops between these phases (see Figure 24.2 above). The systemic approach emphasizes that innovation is also a collective activity involving many actors, and that innovation processes are influenced by their institutional settings and its corresponding incentive structures (Edquist and Johnson, 1997; Lundvall, 2007).

Institutions are not only organizations and formal structures like rules and regulations, but also established habits, practices, routines, and norms of the various actors within the system. In this sense, institutions can be seen as learned patterns of behavior and interaction, marked by the historical specificities of a particular system and moment in time. As such, their salience and strength may shift as conditions change. Learning and unlearning on the part of actors in the innovation system are thus essential to the evolution of a system in response to new challenges.

There are many different actors and institutions in ETIS. Many have been covered in preceding sections. An example in the section on knowledge feedbacks and spillovers are the networks of interacting and cooperating innovation system actors that enable and mediate this knowledge exchange (Carlsson and Stankiewicz, 1991). Three examples not covered previously are considered below in more detail. These are entrepreneurs and experimentation, shared expectations, and advocacy coalitions.

---

13 See for example, www.lichtblick.de/h/index.php.

Another key set of actors and institutions influencing the innovation system relate to public policy. Policies can reinforce or support broader institutional change within the innovation system with regard to learning, collaboration, risk taking, consumer preferences, and so on. While some measure of policy stability is necessary, adaptive policy making in response to feedback from policy dynamics (the interaction of policies and the traditional habits and practices of the actors) are important for stimulating innovation under conditions of uncertainty (Mytelka, 2000).

Policies to support market formation have proven important in encouraging renewable energy technologies through their early commercialization. Such policies might involve subsidies, tax incentives, regulated feed-in tariffs, procurement policies, minimum production quotas, and exemptions from regulation, among others (Raven, 2007). The *Wind Power* case study shows the importance of early market creation efforts by the Danish government for small-scale wind turbines. The *Solar Thermal* case study and the *Kenyan PV* case study illustrate the importance of niche markets for building the capacity to construct, operate, maintain, and assure quality control of a new technology.

The roles and importance of different actors and institutions varies between innovation systems, and also changes over the life cycle of an innovation. For example, a wide range of both public and private actors can be involved in mobilizing resources to support ETIS, and in developing their skills and competences as a result of this resource mobilization (Carlsson and Stankiewicz, 1991). Typically, as innovation systems increase in maturity, the importance of private actors increases (Suurs and Hekkert, 2009a). Various case studies profile the changing importance of public and private actors. In the *US Synfuels* case study, for example, a hybrid approach to public-private roles was ultimately undermined by how mobilized resources were used.

24.2.4.2 ETIS Actors and Institutions (I): Entrepreneurs

There is no such thing as an innovation system without entrepreneurs (Carlsson and Stankiewicz, 1991). Entrepreneurial risk-taking is essential to cope with the large uncertainties surrounding new combinations of technological knowledge, applications, and markets (Meijer and Hekkert, 2007). Above all, entrepreneurship is needed for bringing new technologies, products, and practices to markets. Experimentation is integral to the process of knowledge generation and learning described earlier, allowing for the evaluation of the reactions to new applications on the part of consumers, governments, competitors, and suppliers. The role of the entrepreneur is therefore to turn the potential of new knowledge, networks, and markets into concrete actions that both generate and take advantage of new business opportunities. Entrepreneurs can either be new entrants who see new market opportunities (e.g., university spin-offs) or incumbents who diversify their business strategy to take advantage of new developments.

In most innovation processes, a period of variety creation driven by experimentation takes place before a dominant design emerges and is further developed. The *Wind Power* case study shows how entrepreneurial experimentation led to a large variety of wind turbine designs from which the three-blade vertical axis turbine eventually emerged as the dominant design. The *Hybrid Cars* case study shows the importance of entrepreneurial experimentation among Japanese car manufacturers for initiating the buildup of an innovation system. The *Scaling Dynamics* case study emphasizes the importance of experimentation during the formative phases of many different energy technologies, whereas the *French Nuclear* case study illustrates the consequences of "short-cutting" extended experimentation in the interest of rapid upscaling of a dominant technological design.

24.2.4.3   ETIS Actors and Institutions (II): Shared Expectations

Innovation is always characterized by uncertainty. The expected merits of a new technology in terms of performance or costs cannot be known *ex ante* as being shaped by the agency at play in ETIS (R&D strategies and funding levels, niche market developments, extent of possible learning effects, etc.) as well as by exogenous factors (such as oil prices, as illustrated in the *US Synfuels* case study). Strategic decisions on the direction of the innovation process and the more promising technological avenues can be important to sustain innovation system development. Note, however, that trying to force technology development over short timeframes risks disappointing high initial expectations, as is demonstrated in the *US Synfuels* case study.

Shared or collective expectations are an important means of reducing uncertainty and stimulating entrepreneurial activity (van Lente and Rip, 1998; Borup et al., 2006). This is clearly demonstrated by the *Solar PV* case study. Shared expectations help guide the search of actors within the innovation system by selecting technological alternatives from the variety created by knowledge generation activities. As such this function is equivalent to the "visioning" step advanced in the technological transition literature (Smith and Stirling, 2010; see also Chapter 16 for a more detailed discussion).

Guidance functions can be provided by a variety of actors, including individual firms, actor networks, or governments, as has been the case in Japan (see *Japanese Efficiency* case study). Policies can shape changing societal preferences to reflect public policy objectives like greenhouse gas emission reduction. The *Hybrid Cars* case study clearly shows how zero-emission vehicle regulations and large-scale R&D programs affected the research direction of car manufacturers. The *Brazilian Ethanol* case study shows that strong political leadership can be another important means of guiding the search within an innovation system. Long-term technology roadmaps are another important means of establishing shared technological innovation expectations, as are credible long-term policy signals – for instance, in the form of pollution (e.g., carbon) taxes that rise over time at pre-announced, predictable rates.

24.2.4.4   ETIS Actors and Institutions (III): Advocacy Coalitions

New energy technologies face resistance from actors with interests vested in incumbent systems. To build up an innovation system, actors – usually from nongovernmental organizations (NGOs) and industry – must counteract this inertia through, for example, political lobbying and building advocacy coalitions (Sabatier, 1987; Sabatier, 1988). Public institutions may also contribute (Fligstein, 1997), as in the case of planning agencies advising regional or national governments to develop supporting policies for emerging technologies. In all such cases, innovation system actors try to convince other actors to take particular actions that they cannot conduct themselves. Nonetheless, it should be emphasized that incumbent technology systems often have great lobbying power to resist changes.

24.2.5   Changing Dynamics Over Time in Effectively Functioning ETIS

Researchers in the "functions of innovation systems" tradition have identified seven key processes in emerging innovation systems that are needed for a successful maturation through the formative phase (Jacobsson and Lauber, 2006; Bergek et al., 2008). These key processes interact strongly and all can potentially be supported by policy makers. Although networks between actors in an energy innovation system tend to be international, national policies can strongly influence how the formative phase in specific countries occurs. Table 24.6 summarizes the seven key processes and references the earlier sections in which they were discussed.

Positive interactions and feedbacks between the key processes shown in Table 24.6 are integral for the successful build up of an innovation system (Jacobsson and Bergek, 2004; Hillman et al., 2008). These positive feedback loops are referred to as "motors of innovation" (Suurs and Hekkert, 2009b).

In the first phase of innovation system development, a "science and technology push motor" is characterized by knowledge development and the creation of positive expectations about the new technology by scientists and engineers to help guide the search and stimulate funding that supports further knowledge development in and beyond R&D.

The science and technology push motor can turn into an "entrepreneurial motor" if entrepreneurial experimentation leads to knowledge exchange between the entrepreneurs and the research community. Next, a "system building motor" involves a wide range of actors involved in the development and production of a technology lobbying for market formation and an alignment of institutional structures with the needs of the new technology. If successful, this collective activity can overcome the resistance to change of incumbent actors with vested interests.

Table 24.6 | Seven key processes in innovation systems.

#	Key Process	Summary Description	Relevant Sections in Chapter 24
1	Entrepreneurial experimentation	Taking risks, creating variety, "field" testing, developing business opportunities.	Knowledge Generation; Learning; Actors (Entrepreneurs)
2	Knowledge development and exchange in networks	Generating and sharing knowledge to improve performance, learn from experience, etc.	Knowledge Generation; Knowledge Spillovers; Learning; Actors (Exchange)
3	Guidance of the search	Strategic directioning of the innovation process to reduce uncertainty.	Economies of Scale and Scope; Institutions (Expectations)
4	Market formation	Creating, protecting or supporting niches for innovations to enter the market.	Innovation Models; Niche Markets
5	Resource mobilization	Allocating financial, material and human capital to the innovation process.	Economies of Scale and Scope; Metrics and Assessment
6	Counteract resistance to change	Overcoming systemic inertia and vested interests.	Actors (Advocacy)
7	Materialization	Building up production or manufacturing capacity.	Innovation Models; Economies of Scale; Metrics and Assessment

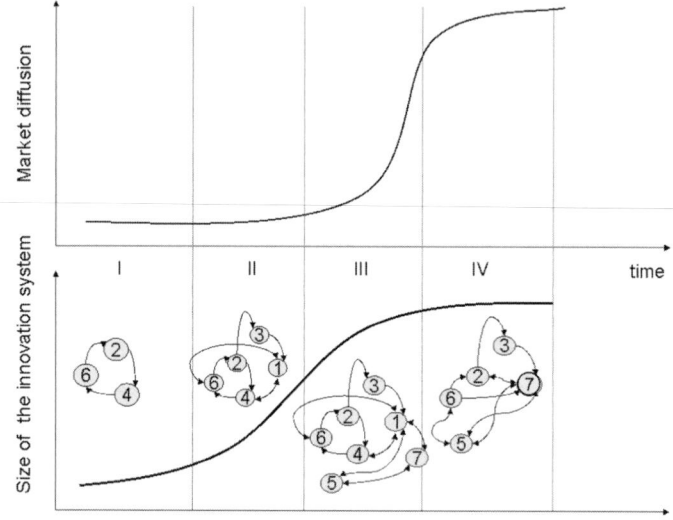

Figure 24.5 | Stylized life cycle of a new technology over time (Phases I to IV, top) and corresponding build up of the technology's innovation system and the seven key processes involved (circles) and their interactions (arrows) (bottom). See Table 24.6 for a definition of the key processes.

Finally, a "market motor" may emerge once market formation has taken place and a technology has started to diffuse with a concomitant build up of production capacity. Overcoming resistance to change and guidance of the search now become less important. The succession of these four motors of innovation is represented stylistically in Figure 24.5 (bottom panel), alongside the S-shaped curve describing a technology's life cycle (top panel). The build up of the innovation system can be measured in terms of the number of actors involved, the extent and complexity of the networks between these actors, and the specific institutions aligned with the innovation. However, such indicators are specific to individual technologies and innovation system contexts and cannot be generalized. Figure 24.5 shows how the innovation processes (represented as circles) and the interactions between them (represented as arrows) increase through the formative phase prior to the technology's diffusion into the market (Phases I-II). As market diffusion accelerates (Phase III), the innovation system grows to its maximum (Phase IV).

24.3 Assessing Energy Technology Innovation Systems

24.3.1 Introduction

There is no uniform simple metric to describe ETIS in terms of commensurate measures of needed inputs and corresponding system outputs. Unlike in macroeconomics, it is not possible to develop a simple production function of ETIS. The *Assessment Metrics* case study reviews in detail the literature on innovation metrics from which Chapter 24 draws to structure this section.

In terms of assessing the inputs to ETIS, Section 24.3.2 (see also Appendix I) provides a comprehensive overview of ETIS in terms of current (as of 2005) investments into energy technologies across the entire life cycle phases of ETIS, from knowledge generation to market formation (niche market investments), to technology diffusion (in both energy supply and end-use technologies). The choice of this metric arises from three considerations:

- *Novelty*: Such an overview assessment has to date been absent in the literature, which invariably has focused only on pieces of the entire system, like public energy R&D expenditures, or investments into renewables.

- *Commensurability*: The metric needs to be comparable across all technologies, across all ETIS activities/processes, and across different sectors and markets (regions); hence the use of the US dollar as a core metric.

- *Centrality*: Investments are a central element of ETIS, constituting a core input to knowledge generation and any embodiment of technological change. They are a key process in ETIS (resource mobilization) and also constitute an important constraint for ETIS in their own right.

The adoption of a single metric to assess current inputs into ETIS does not suggest that readers should ignore the multidimensionality of metrics proposed in the literature on ETIS (see the *Assessment Metrics* case study) but rather to allow to put the relative weight of different stages of ETIS and currently revealed preferences (in terms of resource mobilization) into a quantitative perspective. As such, Section 24.3.2 also provides a baseline against which the needed redirection of ETIS investment flows as described by the GEA transition pathways scenarios (Chapter 17), for which these numbers served as important input, can be assessed (see in particular Section 17.3.5).

The assessment in terms of outputs of ETIS is necessarily more eclectic and illustrative as there is no single common metric for describing technological change (the "output" of ETIS) across all technologies, life cycle stages, and markets. Section 24.3.3 therefore provides salient illustrative examples drawn from Chapter 24's case studies. The focus is on two core dynamic metrics: technology diffusion and costs, which evolve over time to illustrate both their dynamic nature as well as their mutual interdependence. Falling costs drive expanding market applications which, in turn, provide yet further cost reductions, e.g., through economies of scale effects in manufacturing and learning (by doing and using) processes.

24.3.2   Quantitative Assessments of Inputs (Investments)

24.3.2.1   Introduction and Overview

This section attempts a first ever quantitative overview of financial resources that constitute a fundamental input to energy technology innovation in terms of required resource mobilization. Evidently, money is not the only resource that needs to be mobilized: the development of knowledge, skills, supporting institutional settings, etc. is important too. Financial investment data, however, are more readily available than other ETIS input, output, or outcome metrics. In addition, they are a useful tool for policy makers, as budgets are a key policy tool in governments and industry alike. Finally, even if the information provided below is still relatively scarce, investments in innovation give a sense of the scale of the energy innovation enterprise.

The key messages of this section are as follows. First, there are formidable data problems associated with the need to describe energy technology innovation, which highlights important areas of future research and renewed initiatives to provide better technology-specific data for informed policy choices. In addition, consideration of institutions needed to collect and share these data at the national and international levels is badly needed.

Second, this section illustrates the increasing scale of resource mobilization across successive stages of ETIS, from research, development, and demonstration (~50 billion[14]), to market formation investments (~150

billion), and finally to the dominant diffusion investments (>1000 billion). If large-scale technological change is on the agenda, changes in the diffusion environment and associated incentives for technology adoption and diffusion – e.g., through changes in relative prices – are key in addition to developing improved technologies in the upstream stages of ETIS.

Third, this analysis reveals that the structure of current investments in ETIS is highly asymmetrical between the dominance of diffusion investments in energy end-use technologies, and their under-representation in the investments in the earlier stages of ETIS. In other words, this overview helps elucidate the relatively large support for supply-side technologies such as fossil fuels and nuclear energy in RD&D. This is difficult to reconcile with the energy innovation needed to respond to the multitude of challenges of current energy systems, ranging from energy access to energy security and climate change mitigation, all of which call for vastly improved energy end-use efficiency.

Fourth, six major emerging economies – Brazil, the Russian Federation, India, Mexico, China, and South Africa, known collectively as BRIMCS countries – now account for a significant fraction of global ETIS. However, significant regional imbalances, particularly in the support for energy RD&D, persist. The increasing globalization of ETIS in general and of energy technology RD&D in particular suggests that new mechanisms for international technology cooperation and coordination might be called for, which again raises the question of the need of an appropriate institutional (re)design, as existing institutions such as the IEA are limited in scope and membership (mostly oil-importing OECD countries).

Innovation inputs are quantified in this chapter by the associated financial resource mobilization per broad technology class and by stage of the technology life cycle. The definition of the innovation stage is straightforward, as characterized by RD&D expenditures, which are a well-defined expenditure category in macroeconomic and corporate accounts. The subsequent phase of market creation investments is defined by either relying on special funding mechanisms such as venture capital or special (government-induced) market incentives such as feed-in tariffs, production tax credits, and the like, but the definitional boundaries are necessarily more blurred. Finally, diffusion investments are those that represent commercialization of mature technologies and that need no special policy incentives to mobilize the required investment in markets. Evidently, all investments across the entire technology life cycle will always be influenced by the overall incentive environment, as characterized by relative prices, taxes, etc., i.e., by numerous nontechnology-specific policies. What differentiates market creation from diffusion investments is the degree to which investments rely on dedicated technology policy support for their early

---

14   R&D expenditures represent aggregates of national statistics, which are mostly available only in International $ (i.e., in PPP terms). When expressed in US$ (i.e. at market

exchange rates, MER), R&D expenditures would be lower by some $10 billion. As private sector R&D is significantly underreported, a global order of magnitude estimate of $50 billion energy R&D can be considered commensurate with the subsequent niche market and diffusion investment numbers that are expressed in US$ (i.e., MER-based).

Table 24.7 | Summary of current global public and private ETIS investments (in billion US$_{2005}$) by stage and type of technology application (first order estimates and ranges from the literature).

	Innovation (RD&D)		market formation		diffusion	
End-use & efficiency	>>8	1)	5	8)	300–3500	15)
Fossil fuel supply	>12	2)	>>2	9)	200–550	16)
Nuclear	>10	3)	0	10)	3–8	17)
Renewables	>12	4)	~20	11)	>20	18)
Electricity (Gen+T&D)	>>1	5)	~100	12)	450–520	16)
Other* and unspecified	>>4	6)	<15	13)	n.a.	
Total	>50	7)	<150	14)	1000–<5000	19)

Notes: * hydrogen, fuel cells, other power & storage technologies, basic energy research

1) Public RD&D 1.8 billion (IEA, 2009a; BRIMCS case study); private RD&D >>6 billion (WEC, 2001; BRIMCS case study).
2) Public RD&D 2 billion (IEA, 2009a; BRIMCS case study); private RD&D: >10 billion (WEC, 2001; BRIMCS case study).
3) Public RD&D >6.2 billion (IEA, 2009a; BRIMCS case study); private RD&D >3.4 billion (WEC, 2001; BRIMCS case study).
4) Numbers also include renewable electricity. Public RD&D (excl. electricity): 2 billion (IEA, 2009a; BRIMCS case study); private RD&D (includes electricity): 7 billion.
5) Only public RD&D (IEA, 2009a; BRIMCS case study).
6) Only public RD&D (IEA, 2009a; BRIMCS case study).
7) Lower bound estimate (rounded number)
8) NEF/SEFI, 2009 includes 2 billion asset finance p.13, plus estimated 2 billion from venture capital (based on 15 billion total VC in 2008 and assuming category proportion in cumulative VC investments over the 2002–2008 period).
9) Estimated 2 billion from venture capital only (based on 15 billion total VC in 2008 and assuming category proportion in cumulative VC investments over the 2002–2008 period).
10) Classified as mature technology and reported under diffusion investments.
11) Biomass and biofuels total of 24.8 billion (NEF/SEFI, 2009, p.13) minus 8 billion Brazilian ethanol (accounted for as diffusion investment) plus 2.4 billion estimated VC investments.
12) ~90 billion asset finance (NEF/SEFI, 2009, p.13, including wind, solar, geothermal, marine and small hydro plus estimated ~8 billion from VC).
13) Unaccounted for technology categories.
14) Rounded number, estimated market formation investments ~140 billion derived from NEF/SEFI, 2009.
15) Chapter 24 first order estimate, rounded numbers, cf. Appendix I; lower bound: central estimate of energy-using components of end-use investments (297 billion), upper bound: upper range of total end-use investments (3549 billion).
16) Source: Table 24.5 in Appendix I.
17) Estimate for 2–3 GW reactor completions per year (IAEA-PRIS, 2010) at assumed costs between 1500–2500 $/kW.
18) Source: Table 24.5 Appendix I, fuels only.
19) Rounded numbers.

market deployment. A tentative, albeit incomplete, attempt at a global overview is provided in Table 24.7. Further details are provided at Appendix I.

24.3.3   Case Study Assessments of Innovation Outputs

As part of this assessment, Chapter 24 conducted 20 case studies. This section discusses the rationale for conducting these case studies, as well as the rationale for selecting this particular set of cases. All 20 assess the innovation system and are intended to complement the quantitative overview described above. The implications for understanding energy innovation and for public policy are described in Sections 24.4.5 and 24.4.6 below. Some illustrative results from the assessment of innovation processes that emerge from the case studies are included here as well (see Section 24.3.3.3). Space limitations precluded the full presentation of all case studies in this chapter; they are summarized in

Appendix II. The full case studies will be published separately quot are also reported on the GEA Chapter 24 website.[15]

24.3.3.1   Rationale and Logic of Case Studies

The rationale for conducting case studies arises from the need to complement quantitative evaluations with richer descriptive characterizations of innovation systems. This assessment uses evidence from descriptive case studies to further illustrate the general insights for policy design covered in Section 24.4. The complexities of the dynamics of the innovation process are often ignored in quantitative models. While a growing body of work on quantitative evaluation improves understanding, their explanatory power has so far proven limited.

---

15   For more information, see www.globalenergyassessment.org.

Attempts to econometrically identify the effects of demand-pull and technology-push – e.g., Kouvaritakis et al. (2000); Watanabe et al. (2000); Miketa and Schrattenholzer (2004); Klaassen et al. (2005) – have provided limited claims in the available studies because of their sensitivity to assumptions about the depreciation of R&D knowledge stock and about the lags between policy signals and decisions to innovate. Both of these parameters have proven difficult to estimate empirically. Using the observation that most technologies tend to decline in cost over time, the notion of the "learning (or experience) curve" has been widely used to simulate the cost reductions that can be expected from programs that subsidize demand (Duke and Kammen, 1999; Wene, 2000; IEA, 2008b). However, observed discontinuities in learning rates, perhaps resulting from omitted variable bias, limit their reliability. Moreover, large dispersion in observed learning rates, even including negative rates, complicates choices of which point estimates to apply (Nemet, 2009b).

The relationship between R&D investments and technical change is even more difficult to model, in part due to the inherent stochasticity of the R&D process. One notable approach has been to measure the value of the commercialized projects that emerged from federal R&D programs (NRC, 2001). This cost-benefit valuation approach has been used to evaluate the US Department of Energy's wind and PV R&D investments. Key shortcomings in this approach center on the assumptions needed to construct a counterfactual case in which one must characterize outcomes in the event that the R&D investment was not made. Prospectively, another approach common to R&D management has been employed in which decision analytic techniques are often used to obtain the necessarily subjective judgment of experts who are most familiar with the specific technologies (Peerenboom et al., 1989; Sharpe and Keelin, 1998; Clemen and Kwit, 2001). A report by the National Research Council (NRC, 2007) recommends that the US Department of Energy adopt a process including expert elicitations. They provided prototype elicitations for carbon storage, a vehicle technologies program, and four other programs. Examples of such assessments include studies of PVs and carbon capture (Baker et al., 2009).

More generally, quantitative assessments of innovation systems may be biased toward the selection of cases for which detailed data are available. This may explain the lack of empirical work on energy end-use technologies relative to supply technologies. Also, it may limit comparisons across countries, as there may be insufficient variables for comparing the results of heterogeneous studies across countries. Finally, the reduction of the complex process of innovation to a few factors may omit important aspects of the system and contribute little in mechanistically explaining causality.

The qualitative descriptions in case studies provide an avenue for incorporating explicit considerations of the innovation system's complexities and feedbacks, which would otherwise be ignored. It is important to note that generalizing from case studies is limited by the specifics of context and technical characteristics, which are described. Selection bias is also discussed. Policy conclusions take these limitations into account.

24.3.3.2   Summary of Case Studies

The selection of case studies was based on the following criteria. First, in many cases the focus was on individual technologies. For these technology-focused studies, technologies were selected that had a dynamic aspect – for example, technically, economically, or in terms of deployment. Second, many of the cases included a situation in which public policies played an important role in affecting the process of innovation and diffusion. The case studies were particularly interested in describing the activities of governments. Third, to the extent that data availability allowed, an effort was made to include international diversity and include cross-country comparisons. Fourth, care was taken to also include from the available case studies illustrations of innovation failures or imperfections (e.g., *US Synfuels*, *Solar Thermal*, or *French Nuclear*). Finally, some case studies were conducted because they illustrate specific attributes of the innovation system, as described above. In selecting studies under this criterion, a special effort was made to evaluate ETIS characteristics that are poorly explained in the literature (e.g., *Knowledge Depreciation* or *Scaling Dynamics*).

This chapter's assessment includes 20 case studies, which can be categorized in several ways (Table 24.8). Six of the case studies explicitly address innovation in developing countries. Fifteen include some assessment of government actions affecting the innovation system. At least 12 include a discussion of knowledge feedbacks in the innovation system – or in some cases, the lack of feedbacks. The case studies can also be categorized by whether they were conducted in order to illustrate a particular aspect of the innovation process described above, or whether they were focused on evaluating the innovation system for a particular technology. The latter category was divided into cases that examined end-use technologies and those that examined supply-side technologies. Emphasis was placed on achieving a rough balance between these two technologies to address the exceptionally weak empirical basis for understanding innovation in end-use technologies. Table 24.8 categorizes the 20 case studies as (1) illustrations of specific characteristics of innovation systems; (2) energy end-use technologies; and (3) energy supply technologies.

24.3.3.3   Illustrative Examples of Innovation Outputs from Case Studies

Six examples illustrate the types of work that can be used to evaluate the outcomes of the innovations process, particularly in response

Table 24.8 | Innovation case studies conducted for Chapter 24.

Short Name	Summary Section		Developing Country	Public Policy
		1. Illustrations of Specific Characteristics of Innovation Systems		
Grand Designs	24.7.1	Grand Designs: Historical Patterns and Future Scenarios of Energy Technological Change	√	
Scaling Dynamics	24.7.2	Historical Scaling Dynamics of Energy Technologies	√	
Technology Portfolios	24.7.3	Technology Portfolios		√
Knowledge Depreciation	24.7.4	Knowledge Depreciation		√
Assessment Metrics	24.7.5	Metrics for Assessing Energy Technology Innovation	√	√
Chinese R&D	24.7.6	China Energy Technology Innovation Landscape	√	√
Emerging Economies R&D	24.7.7	Energy R&D in Emerging Economies (BRIMCS)	√	√
Venture Capital	24.7.8	Venture Capital Investments in the Energy Industry	√	
		2. Energy End-use Technologies		
Hybrid Cars	24.7.9	Hybrid Cars	√	√
Solar Water Heaters	24.7.10	Solar Water Heaters	√	√
Heat Pumps	24.7.11	Heat Pumps		√
US Vehicle Efficiency	24.7.12	Role of Standards – Example: CAFE		√
Japanese Efficiency	24.7.13	Role of Standards – Example: Japanese Top Runner Program		√
		3. Energy Supply Technologies		
Wind Power	24.7.14	Comparative Assessment of Wind Turbine Innovation and Diffusion Policies		√
Solar PV	24.7.15	Comparative Assessment of PV (European Union, Japan, United States)		√
Kenyan PV	24.7.16	Solar Innovation and Market Feedbacks: Solar Photovoltaics in Rural Kenya	√	
Solar Thermal	24.7.17	Solar Thermal Electricity		√
US Synfuels	24.7.18	The US Synthetic Fuels Program		√
French Nuclear	24.7.19	French Pressurized Water Reactor Program		√
Brazilian Ethanol	24.7.20	Ethanol in Brazil	√	√

to public policy. Common to all is the adoption of the perspective of technology diffusion, as well as costs. In terms of ETIS outputs, the dynamics of widening market applications, as well as costs and their interdependence, serve as core metrics in these six illustrative examples.

Figure 24.6 illustrates one possible outcome of ETIS in terms of accelerated diffusion of new energy technologies in later adopting regions via spillover and learning effects (*Scaling Dynamics* case study).

Figure 24.7 illustrates the response to policy, in this case the introduction of the Corporate Average Fuel Economy (CAFE) standard in the United States, in terms of diffusion of the induced technological innovations in the automotive sector (*US Vehicle Efficiency* case study).

Figure 24.8 illustrates the components in cost reductions of sugarcane production in the Brazilian ethanol industry as a twin example of an analytical opening up the "black box" of technology cost improvements and an example of technology responsiveness to an exemplary decades-long sustained public policy effort (*Brazilian Ethanol* case study).

Figure 24.9 shows cost reductions and technical improvements in early solar thermal electricity generation in the United States, from 1982–1992. A virtuous cycle (i.e., a positive, self-reinforcing feedback loop) of unfolding of ETIS came abruptly to an end with the discontinuation of policy support, illustrating the pitfalls of erratic policies and the key importance of continuous policy support (*Solar Thermal* case study).

Figure 24.10 shows the declining cost of PV associated with the Japanese subsidy program from 1994–2004 and provides the positive example of the responsiveness of ETIS outputs to a sustained and predictable policy environment (*Solar PV* case study).

Finally, Figure 24.11 summarizes the cost trends of non-fossil energy technologies analyzed in the Chapter 24 case studies. These data have been updated with most recent cost trends (2010) available in the literature for PV Si Modules (IPCC SRREN, 2011) and US onshore wind turbines (Wiser and Bolinger, 2011). Note that the summary illustrates comparative cost trends only and is not suitable for direct economic comparison of different energy technologies due to important differences between the economics of technology components (e.g., PV modules or heat pumps [only]) versus total systems installed, cost versus

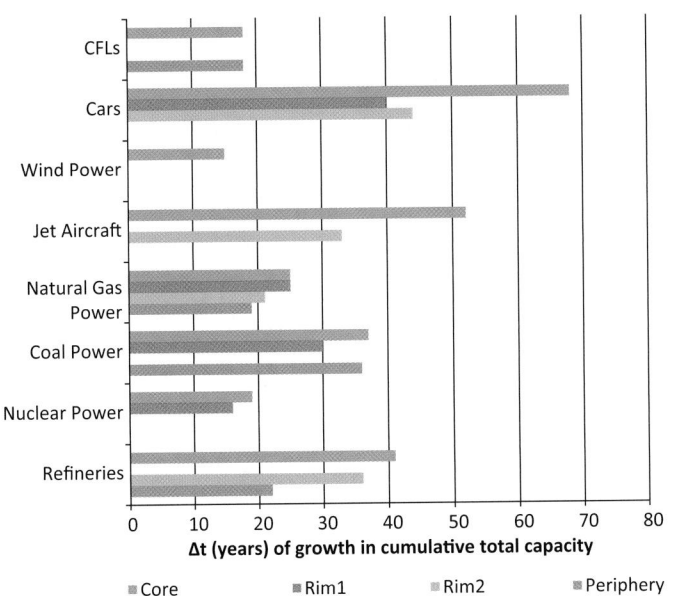

Figure 24.6 | Diffusion rate or Δt – time to grow from 10–90% of installed cumulative capacity – between innovating *Core* regions and later adopting rim (graduated between *Rim1* and *Rim2*) and *Periphery* regions for a range of energy technologies (CFLs denoting compact fluorescent light bulbs). Note that the geographical diffusion categorization is based on technology adoption rates rather than country aggregates. Source: Wilson, 2009.

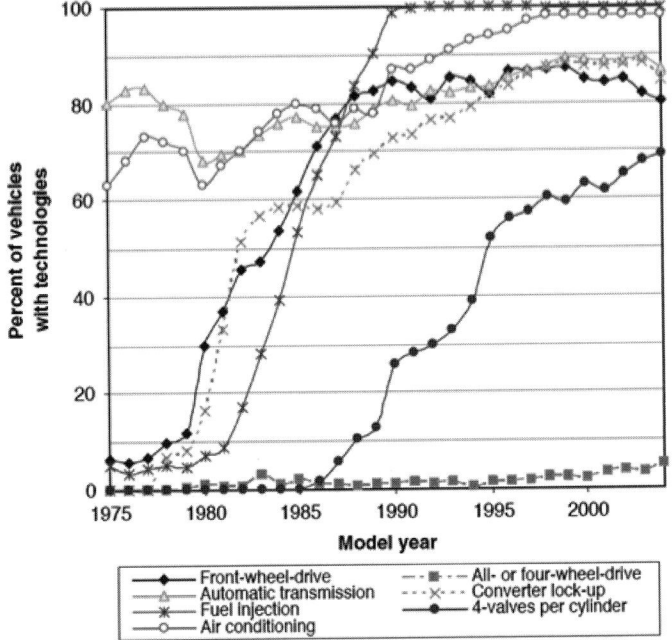

Figure 24.7 | Diffusion of automotive technologies during implementation of US Fuel Efficiency (CAFE) standards. Source: Lutsey and Sperling, 2005. © National Academy of Sciences. Reproduced with permission of the Transportation Research Board.

price data, and also differences in load factors across technologies (e.g., nuclear's electricity output per kW installed is between a factor three to five larger than that of PV or wind turbine systems). Despite a wide range in cost trend experiences across technologies, two important

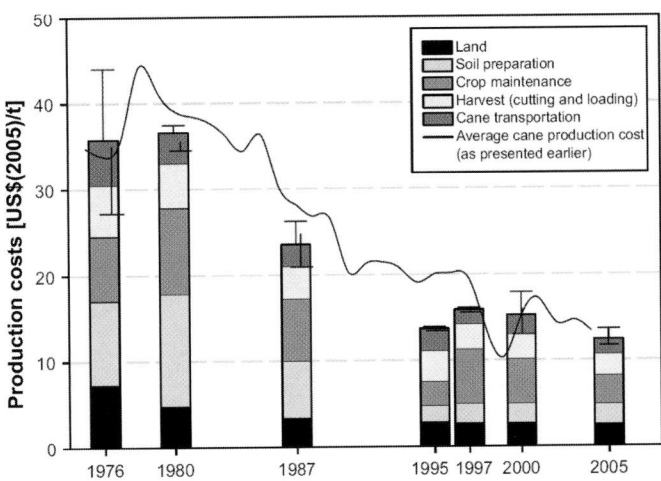

Figure 24.8 | Decomposition of cost components and their improvements for Brazilian ethanol (cane production only). Source: van den Wall Bake et al., 2009.

observations stand out: (1) there is a marked contrast between nuclear showing persistent cost escalation versus the other non-fossil technologies, that generally show declining costs/prices with accumulated market deployment experience. (2) Improvement trends are highly variable across technologies and also over time. For some technologies (e.g. wind in the United States and Europe) historical cost improvements were temporarily reversed after the year 2003/2004 suggesting possible effects of ambitious demand-pull policies in the face of manufacturing capacity constraints and rising profit margins that (along with rising commodity and raw material prices) have led to cost escalations in renewable energy technologies as well.

24.4 Energy Technology Innovation Policy

24.4.1 Public vs. Private Actors: Roles and Differences

There are a multitude of actors involved in ETIS that either can be differentiated with respect to their role in a technology's life cycle (technology research, development, or adoption, i.e., the supply of and the demand for technology innovation) or with respect to their nature as public or private sector institutions (governments, firms, associations), or individuals (entrepreneurs, consumers). Moreover, the innovation actor landscape that has traditionally be defined within a national context is becoming increasingly globalized, considering the increasing role of multinational firms and direct foreign investments as sources of technological change, and the role of multilateral institutions (World Bank, GEF) and NGOs, which are increasingly involved in ETIS.

The role of these actors can vary considerably. A firm can be a developer of a particular technological innovation while at the same time an adopter for another innovation; their respective role as actors over a technology life cycle also changes. Actors are also extremely heterogeneous in terms of their technology knowledge and competence

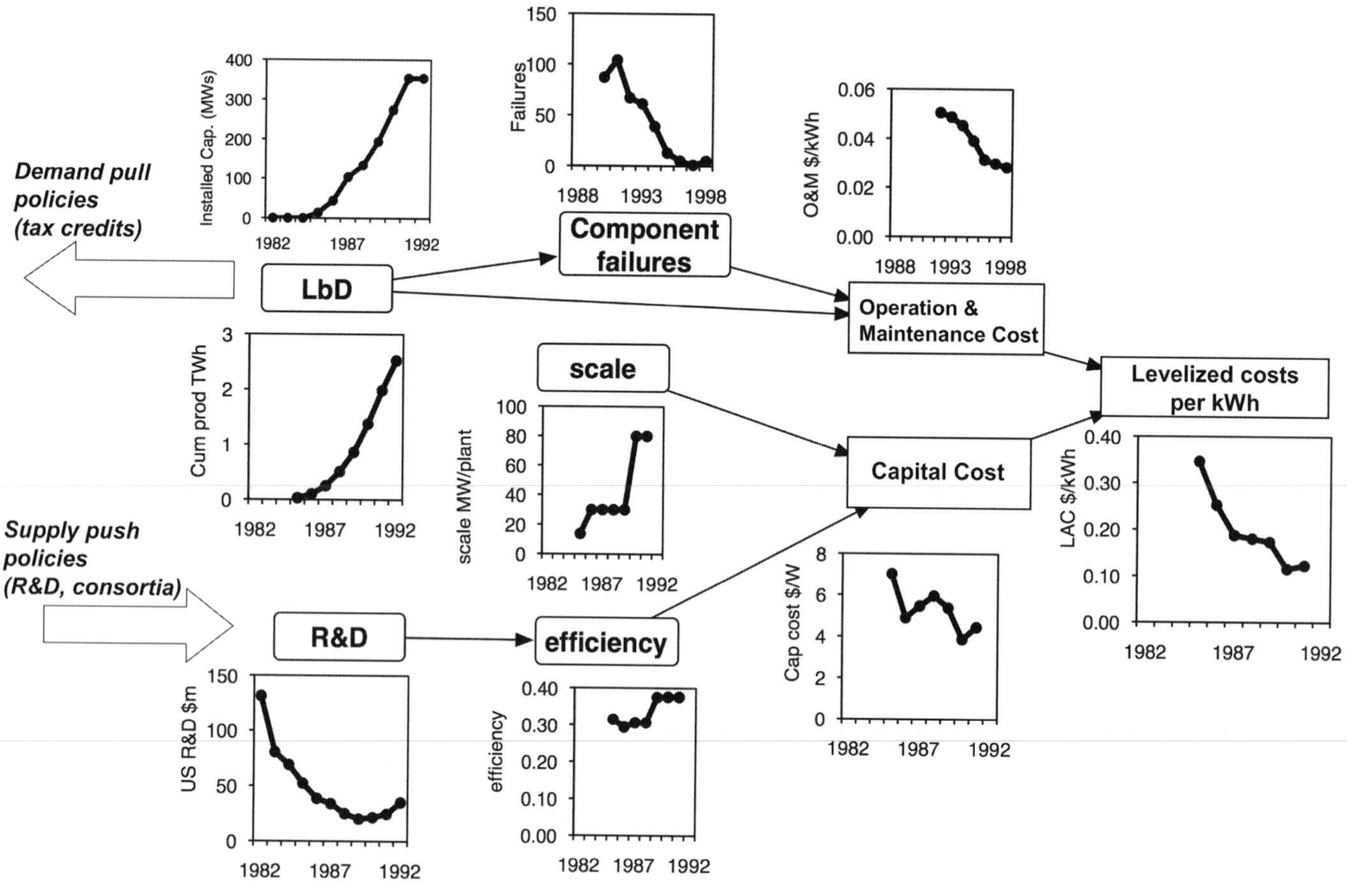

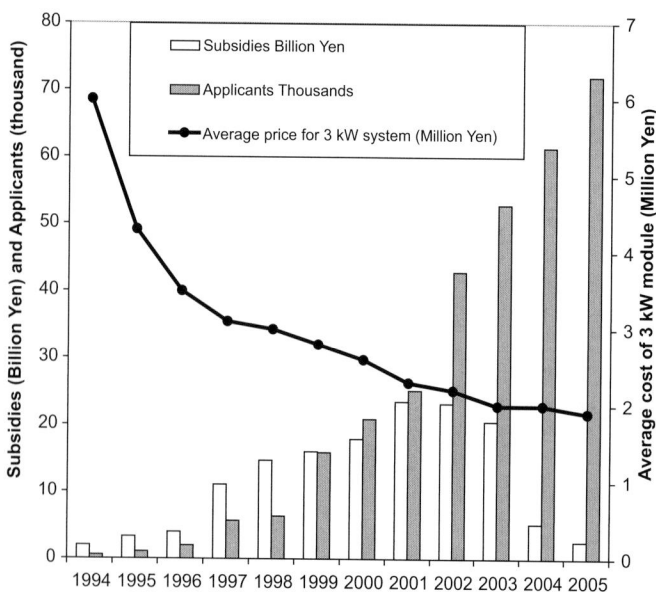

Figure 24.9 | Causal link model diagram for cost reductions and technical improvements in early solar thermal electricity generation in the United States, 1982–1992. Source: *Solar Thermal* case study. Note: R&D = Research & Development, LbD = Learning by Doing.

Figure 24.10 | Japan subsidies and applications for rooftop PV systems versus average 3 kW module price. The 2005 data correspond to specific investment costs of US$_{2005}$5816/kW and the 2.6 billion yen subsidy to US$_{2005}$23 million. Source: based on Jäger-Waldau, 2006.

base and the resources they can mobilize for innovation (development or adoption), as well as in their characteristics (e.g., different discount rates applied to energy efficiency investments).

For the purpose of this chapter, the differentiation between public and private sector actors is of particular importance. Whereas private sector actors are the main actors of technological innovation in terms of performers of R&D, technology developers, and in manufacturing and marketing of technological innovation, they cannot influence associated (knowledge or environmental) externalities nor the incentive environment in which innovation takes place. This accords a special role to the public sector with respect to technology innovation policy and is reflected in traditional areas of public policy concern (see Section 24.4.2), including public R&D funding, incentives for private R&D, the setting of technology or environmental performance standards, and the general area of economic incentives for technology adoption (e.g., via taxes or subsidies).

Lastly, the literature on institutional innovation (e.g., Ruttan, 1996) is relevant here. Evolving institutional settings can be interpreted by themselves as forms of social "techniques" or innovation that can help to overcome knowledge asymmetries or split incentives that can hinder

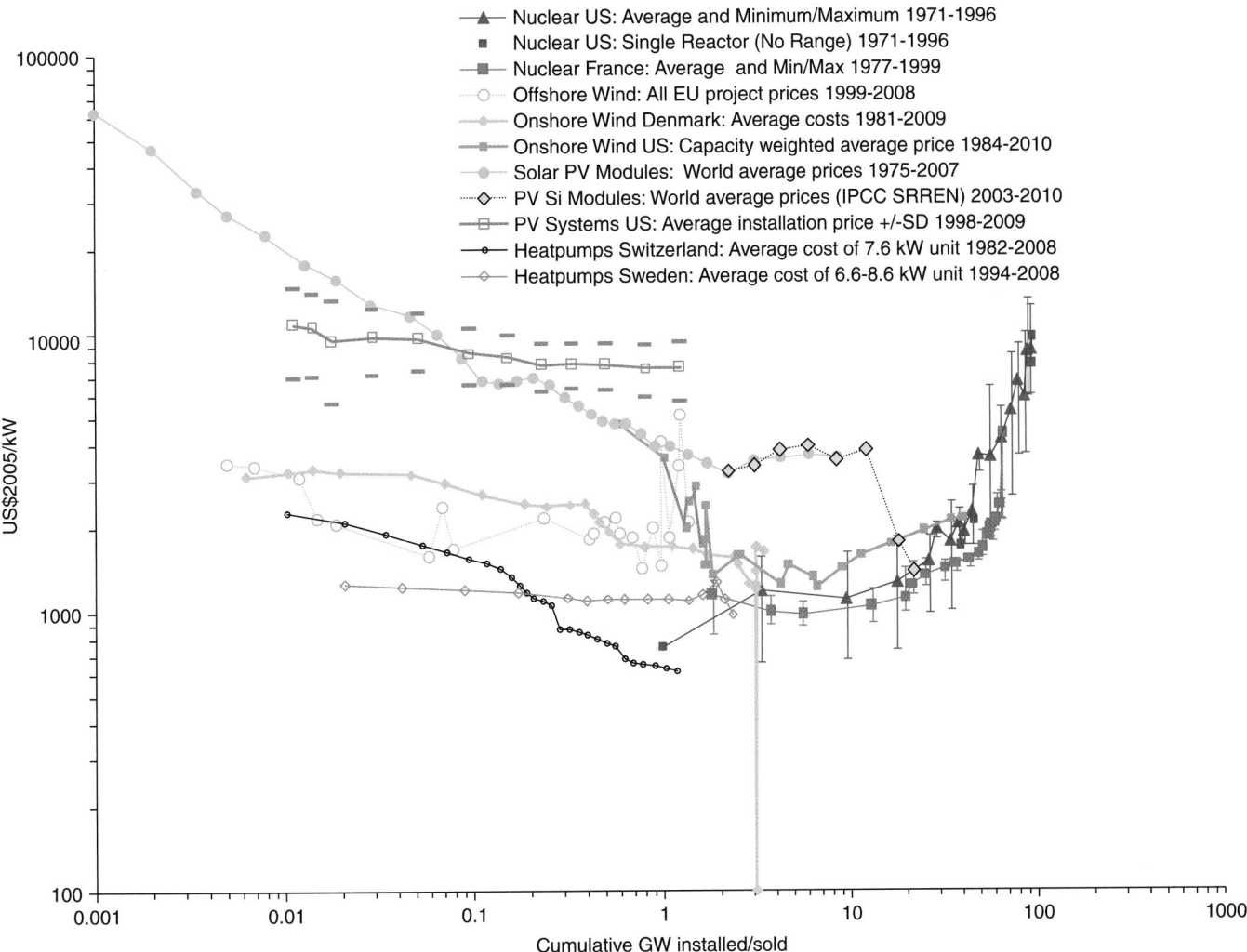

Figure 24.11 | Chapter 24 Case Studies summarized: Cost trends of selected non-fossil energy technologies (US2005/KW installed capacity) versus cumulative deployment (cumulative GW installed). Source: Chapter 24 case studies.

socially desirable technological innovation. For instance, energy service companies (ESCOMs) are an example of an institutional innovation in the form of new actors who can assist households or public sector entities, like schools, in the adoption of energy efficiency innovation.

24.4.2   Rationale for Public Policy

When considering the rationale for investments in energy technology innovation (ETI), there are two main questions to be answered: why should anyone – the government or private companies – engage in ETI? And, what is the particular rationale for government policy and investments in ETI?

A private firm would endeavor to innovate in response to a perceived need in the marketplace or to create a new product to market to the world. Such profit-maximizing behavior is obvious and important because the global energy marketplace is indeed very large. The IEA estimates that

investments in energy supply alone will cumulatively total US$22 trillion globally between 2006 and 2030 (IEA, 2007). Thriving businesses in the energy domain translates into economic development, economic growth, the maintenance and creation of jobs, high-technology domestic sales, and perhaps exports. ETI can also reduce the costs of delivering energy services to consumers, freeing them to save or spend money on other goods and services in the economy. An important question is whether or not the private sector invests sufficiently in ETI, and unfortunately the data do not yet exist to answer that question satisfactorily. Companies are not required to disclose information about their investments, and even if they did, it would be difficult to determine what fraction of the private-sector expenditures by automobile companies and other manufacturers of energy-consuming goods can be counted as efforts to improve the energy efficiency of products. It is also hard to define energy technology. Information on private-sector expenditures in the early-deployment phase is more readily available because venture capital (VC) and private equity investments, asset finance projects, and corporate finance deals are announced and tracked (e.g., Anadon et al., 2009; see also Section 24.6.2 in Appendix I).

There are reasons to believe that the private sector probably under-invests in ETI, both in comparison to historical R&D investment levels and in view of the social and environmental challenges related to the energy sector. First, what data exist indicate that investments are declining. According to one analysis of the US private sector using data from the National Science Foundation's annual survey of companies, private sector ETI investments fell approximately 20% during 1994–2004. The US electricity sector's R&D arm, the Electric Power Research Institute (EPRI), saw its budget decline by a factor of three during that time period (Nemet and Kammen, 2007). Also, companies are far more likely to invest in short-term RD&D projects that are likely to bear fruit in the near term than to invest in longer term, more fundamental R&D. This is especially true during times of broader economic turmoil or recession and energy price volatility. Such volatility leads to a "lumpy" pattern of investment on the part of the private sector, where big investments are followed by precipitous declines, and vice versa. Innovation requires sustained and steady "inputs" – people who are able to focus over the longer term on improving or inventing energy technologies, and adequate and stable resources to do so (Gallagher et al., 2006).

Turning to the particular rationales for government involvement in ETI, the first is to support, complement, and encourage the private sector's efforts because a vibrant energy sector in any economy will contribute to economic growth and prosperity. Second, energy services are fundamental human needs, and improvement of those services can better the human condition. If innovation reduces the costs of those services, consumer welfare and human well-being are improved. In poor countries, where millions still lack access to basic modern energy services provided by energy carriers such as like electricity, the government has an especially important role to play in developing appropriate technologies for rural energy users, and devising and implementing demonstration and deployment programs for better cookstoves, heating and building technologies, and so forth.

Government investment in ETI is also justified to make energy supply more reliable and secure; help the energy system emit fewer pollutants; and reduce the negative impacts of energy extraction, conversion, and use of water and land resources. In other words, market failures to protect the environment and enhance the security of a country help justify government involvement in the innovation process.

The last rationale for government policy for ETI is to overcome market barriers. Incumbent energy technologies or systems tend to have institutions, infrastructures, and policies that support them, providing barriers to entry for new technologies (sometimes called lock-in or path-dependence). There is also a famous valley of death between the invention phase of innovation and the deployment phase. This valley is really two valleys, because there are often difficulties moving from R&D to demonstration (which is expensive), and then again difficulties taking a proven technology to the marketplace during the early deployment phase. Governments can erect bridges across these valleys to reduce the barriers and speed the passage of these technologies from the lab

to the market. In sum, policy can help push and pull advanced, cleaner, and more efficient technologies into the marketplace.

24.4.3 Models and Instruments of Policy

Policies for innovation can directly target the innovation process, support the innovation system, or unintentionally impact innovation while targeting an unrelated concern.

Direct policies for innovation vary according to their target and their timing during the innovation process. Policy is needed at each stage of this process (see the top of Figure 24.12 for examples). The role of government is typically viewed as being most evident at the earliest stage of basic science and research. However, together with the private sector, governments are also engines of applied energy R&D. But governments must also play an important role in leveraging private sector investment at the early commercialization stages by supporting demonstration activities (to reduce risks) and market formation (to underwrite demand). Finally, through regulations and other policies, including tax and fiscal policies, governments also strongly influence the diffusion of energy technologies.

There is often an intermediate stage between demonstration and diffusion that can be considered a market formation or early deployment stage. Here, government can play a critical role because policies are often needed to create an initial market to ease the passage of new energy technologies into the marketplace. First-of-a-kind technologies are often more expensive, and governments can create niche markets through procurement and other policies (e.g., feed-in tariffs or technology portfolio standards) to create demand for advanced or cleaner

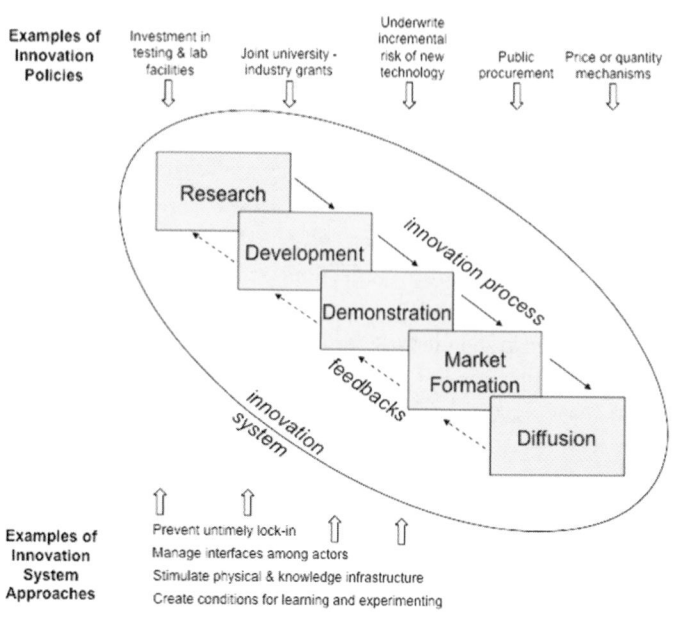

Figure 24.12 | Overview of policies for innovation systems.

energy technologies. With this support, entrepreneurs can experiment and test the market. Technological learning occurs through experience. Even after the niche market has been exploited, policy intervention may be needed to broaden and deepen the market through the elimination of market hurdles, provision of information, tax incentives, or low-interest loans. At some point, a given technology becomes competitive in the marketplace, and the government can exit the market-formation stage. For new, cleaner energy technology to be competitive in the broader market, government policies are also needed to correct for market externalities and define the rules of the game (e.g., through a carbon tax). Because there are so many market distortions, technologies cannot be assumed to freely compete in the global marketplace.

The innovation process is situated within an overarching system comprised of the actors, institutions, and networks involved in developing and commercializing a technology (see Section 24.2 for details). Innovation policies must therefore also target the smooth functioning of the innovation system (see bottom of Figure 24.12). Although government policy affects all stages of innovation, rarely do we see evidence of comprehensive government strategies to optimize the efficiency of the ETIS. Instead, government policies persistently aim at isolated components of the system, such as support for R&D without regard to which policies are needed to maximize feedbacks in the system, or which market-formation policies will be needed if and when the technologies emerge in the demonstration phase.

Policies on issues such as education, taxes and subsidies, and market regulation can exert an important but indirect influence on innovation supply and demand. This reinforces the need for consistency, not just between direct innovation policies but also between the broader regulatory and institutional environments for innovation.

Policies supporting the supply of innovations or the development of technologies include investments in R&D, intellectual property protection, laboratory and testing infrastructure, training and skills development, university-industry collaborations, formal and informal mechanisms of knowledge exchange, technology roadmaps to guide the direction of innovation, and financial incentives such as tax credits for private investments. Not all innovation, however, derives from formal research and development activities. Problem solving and incremental improvements in existing technologies are also of importance and can be stimulated and supported by public sector policies that lead to the creation of outreach, extension, and technical support programs. Policies supporting the demand for innovations as commercialized technologies include demonstration projects, public procurement, market niche creation (e.g., supply obligations), and the creation of appropriate market incentives. Market incentives may be created via changes in relative prices (e.g., environmental taxes or feed-in tariffs), standards, and regulations. These supply-push and demand-pull policies are context-specific complements rather than substitutes. Innovation success stories are typically characterized by comprehensive and consistent policy support through the entire innovation process (see Figure 24.12). Particular innovation

policies must account for specific local conditions or be otherwise tailored to the technological or market characteristics of an innovation.

24.4.4 International Dimension to Energy Technology Innovation and Policy

International energy technology spillovers and feedbacks will depend on both local and global factors, and policy at both these levels is crucial. Energy systems and technologies are highly internationalized, and knowledge and learning in the energy sector has an intrinsic international dimension. Moreover, as shown in some of the case studies and the quantitative ETIS investment analysis (Appendix I), non-OECD countries have progressively invested in and developed capabilities in earlier stages of development of new technologies (e.g., China in coal gasification, India in wind turbines, and Brazil in biofuels). The globalization of technology has the potential to significantly increase the rate of energy innovation if international feedback and learning are properly enhanced.

The majority of energy technology is diffused through private means; it routinely flows through foreign investment, licensing agreements, and international trade. Each channel implies different modes of technology transfer and, depending on the effectiveness of local learning investment, different levels of local assimilation (Cohen and Levinthal, 1989). Many of these flows are actually intra-firm technology transfer, since energy firms are counted among the largest multinational corporations in the world, with very high indexes of internationalization (measured as share of affiliate sales on total sales; see UNCTAD, 2002). But this is not exclusive to the fossil energy industries: PV and wind turbine technologies were also developed as a result of experience and learning that crossed borders. But international inter- and intra-firm transfer do not occur without the appropriate systemic incentives, including subsidies, public R&D investment, tariffs, standards, resource mobilization, and the guiding action of policies and institutions.

Incentives for technology diffusion in general include market conditions and government policy, but cleaner energy technologies require normally explicit incentives. While the existence of more advanced and cleaner energy technologies has led many to believe that latecomer countries will leapfrog to such technologies (Goldemberg, 1998), this is by no means an automatic process. Rather, the process is predicated on developing technological capabilities (absorptive capacity) and appropriate market incentives for technology diffusion.

National policy incentives, coupled with either the financial resources to buy and/or the indigenous technological capabilities to make or assimilate advanced technologies, are required in this situation. Local policy incentives and institutions are crucial to overcome adoption hurdles, especially for cleaner technologies. The development of China's automobile industry is one example. The Chinese government supported its firms to purchase automotive technology through licensing and joint venture arrangements, but for a long time failed to elicit pollution-control technology adoption due to the lack of pollution control standards until 2000. A lack of leapfrogging was observed with respect to pollution-control technology, but rapid leapfrogging occurred during the 1990s for automotive technology more generally in China (Gallagher, 2006). In the case of ethanol in Brazil, the government employed a comprehensive strategy involving standards, market incentives, RD&D investments, and human resource development to create a sugarcane-based ethanol industry (see the *Brazilian Ethanol* case study). In Mexico, opportunities were missed to cultivate a local wind industry and accompanying capability accumulation. Although the Instituto de Investigaciones Electricas (a public R&D center) had made the first steps toward developing wind turbine technology, the government failed to create incentives at the manufacturing stage, and opted for turn-key plant technology transfer (Borja-Diaz et al., 2005; Aguayo, 2008).

International policy also creates incentive frameworks for the diffusion of energy technologies. This includes treaties (e.g., Kyoto Protocol), norms (e.g., technical standards), and institutions regulating trade, finance, investment, environment, development, security, and health issues (e.g., IEA, World Trade Organization, World Bank, and United Nations Development Programme). However, most of these are not oriented specifically to foster energy technology innovation. A current research gap is the dearth of studies examining the impact of these international policies and institutions on energy innovation. The effectiveness of international schemes like the Clean Development Mechanism and offset markets as energy innovation mechanisms remains unclear.

Adoption of energy innovations tends to proceed faster in late-adopting countries, but achieves lower technological levels than in inventor countries, as shown in the *Scaling Dynamics* case study. This is true also for traditional, carbon-intensive technologies. Most developing countries are currently experiencing a rapid transition into energy- and carbon-intensive modern energy infrastructures and fossil-fuel dependent end-use technologies, most notably the private automobile. Augmenting international spillovers and knowledge flows at early phases of scaling up could enable scope and scale economies in the formative phases of technology, accelerating technology improvement and pulling new technologies from their initial niches of application.

International knowledge spillovers through government-sponsored collaboration efforts seem weak compared to what is needed to foster a significant global energy transition. Perhaps the most prominent example of international energy technology collaboration among governments is the implementing agreements of the IEA. The IEA provides support for numerous international cooperation and collaboration agreements in energy technology R&D, deployment, and information dissemination (IEA, 2010). These agreements cover a broad range of technologies, but they strongly differ in the scope, stage, and level of commitment of the R&D process. Moreover, many are not really R&D collaboration projects, but simply institutional arrangements for information exchange or standardization. While non-member countries and international organizations may participate, OECD members' presence dominates (see Figure 24.13 below). There are other examples of government-supported international collaboration (e.g., the fusion reactor project ITER, or the Carbon Sequestration Leadership Forum), but their effectiveness remains to be assessed.

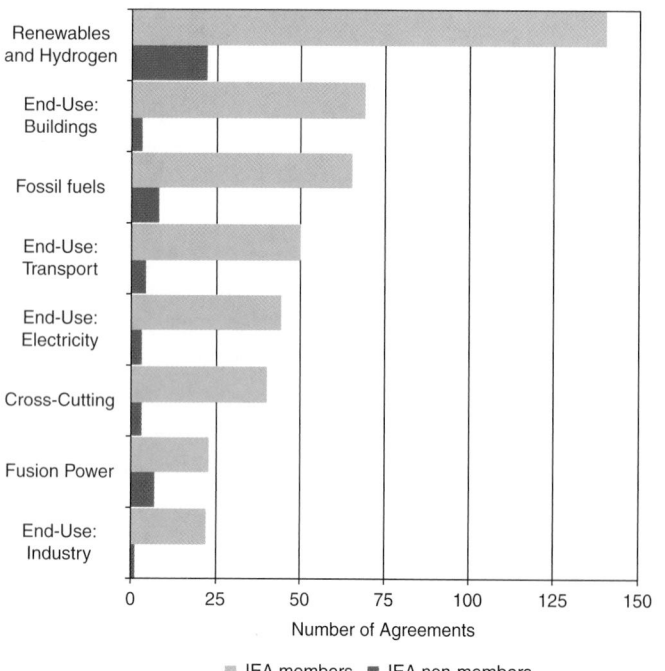
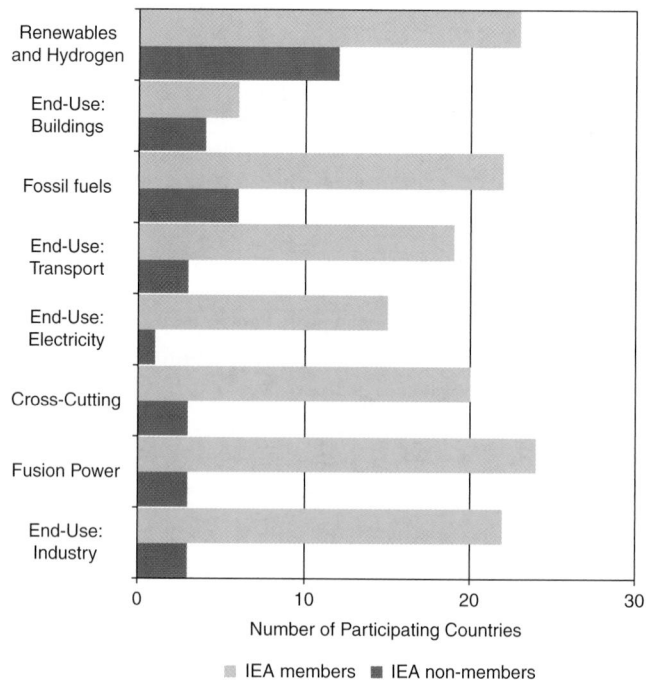

Figure 24.13 | Examples of international technology cooperation via IEA agreements, by participating countries (IEA members and non-members), number of agreements covered (left panel) and number of participating countries (right panel). Note in particular the relative sparse participation from developing countries, which is less surprising considering the exclusive OECD membership of IEA. Source: based on IEA, 2010.

Another potentially important channel for international spillovers and technology transfer is government bilateral agreements. Their real impact on technology spillovers and advancement has not been assessed and will probably remain constrained by national interest issues and disagreement. These bilateral agreements are predominantly North-North, but there are examples of North-South cooperation, and most recently South-South energy technology flows, such as Brazil's set of agreements with 12 African and Caribbean countries for transferring sugar-based ethanol technology. Given their growing importance as both energy consumers and as energy technology providers, India and China have attracted most of the attention, spurring a constellation of binational dialogue and exchange programs such as the new United States-China Clean Energy Research Center, agreed to in November 2009 by President Barack Obama and President Hu Jintao. Again, these programs' effectiveness in fostering innovation and technology transfer remain to be evaluated.

24.4.5   Policy Design Guidelines/Criteria

24.4.5.1   Introduction

The previous sections have outlined the main drivers of technological change embedded in a *systemic* conceptual perspective of the ETIS. The systemic perspective highlights that drivers and policies to stimulate technological change (innovation and diffusion) are closely interrelated. This section summarizes the main findings from the technology policy case studies of Chapter 24 that can guide the design of technology innovation and diffusion policies. In the view of the authors, these guiding principles for policy design carry more weight than the choice of particular policy instruments (including, e.g., externality pricing, preferential feed-in tariffs for emerging technologies, or various forms of subsidies or quantitative regulations such as technology performance standards) that are discussed in the various case studies. In other words, while the policy guidelines outlined below are considered generic and applicable across *all* technology fields and adoption environments, the choice of individual technology-related policy instruments needs to be tailored to technology- and locality-specific circumstances, but are invariably guided by the overarching general policy principles.

Ignoring the systemic characteristics of technological change often leads to a partial view and fragmented (even contradictory) policy frameworks. Although it is well understood that technology is fundamental to solving the energy challenges of our time, including climate change, energy security, and economic growth, what remains less clear is how to most effectively create and deploy new and improved technologies. There are no simple answers; innovation systems are highly complex and interconnected, meaning that decision makers must guard against overly simplistic responses that hide the need for flexible and broad policy approaches to meet energy innovation challenges. Included

as boxes within the following sections are several stylized examples of these simplifications – or "policy myths"– and brief explanations of how such simplification might lead policy makers to actions that will not achieve their goals.

24.4.5.2   Create Knowledge! Or: How to Enable Technological Learning while Learning about Technologies Yourself

One cannot influence the creation of technological knowledge in an effective way without knowledge of how the ETIS operates (and its institutions), the inputs it requires, and how to assess the effectiveness of innovation policies (outputs and "outcomes"). A special need in knowledge development relates to data on innovation activities themselves, which for the most part are poor, scattered, and incomplete. Informed innovation policy cannot be created in a knowledge vacuum. It needs to rely on data and appropriate metrics and indicators that can guide adaptive innovation policy design. For instance, our knowledge on technology-specific private sector energy R&D is woefully inadequate, implying that public and private sector innovation priorities risk being misaligned or even contradictory.

Research on energy innovation requires consistent, long-term, comparable, and more detailed data on innovation inputs and outputs, including information disclosure on policy programs. This information is critical for assessing not only the direction and rates of technological change and identifying needs in different areas of the innovation process, but also for evaluating society's response to energy challenges, including policies themselves. Policy makers need to communicate clearly strict quid pro quo conditions for policy support. For instance, direct subsidies on nascent technologies such as demonstration projects and niche market deployment need to be contingent on public disclosure and documentation of successes and failures in the deployment and performance of new technologies, in order to enable learning and the preservation of technology experimentation knowledge.

Decisions and choices that policy aims to influence depend on the structures in which actors are embedded. As technology systems develop, vested interests emerge, not only in the private sector and intermediate institutions but also in the policy-making realm itself. The risk of moral hazard and a poor ability to learn from mistakes can introduce rigidity and biases within the innovation system. This is why societies require reassessment and institutional learning at higher levels of the innovation system, in order to be able to learn and readjust policy objectives, priorities, and instruments. The need for independent and stable institutions that act as intermediaries between the twin vagaries of the policy and market environments, e.g., in the form of technology assessment institutions, cannot be stressed enough. Innovation policy needs institutional capacity for designing, implementing, and monitoring innovation policies, which is lacking in almost all countries, as well as at the international level.

24.4.5.3　Assure Feedbacks! Or: How to Create/Enable Knowledge Flows for Technology Learning and Spillovers

Formal and informal information feedback processes are essential for sustained and successful innovation. This is well known and well cited in the literature, yet it has proven virtually impossible to institutionalize. Even this recommendation has sometimes been seen as an excessive burden on already over-worked public officials, NGOs, and contractors who have little interest in long-term monitoring (PCAST, 1997). Government can support these essential feedbacks in a variety of ways, but can also hinder – or even block – essential information and knowledge flows.

For instance, governments can support knowledge feedback between demonstration projects and niche market applications back to R&D by providing facilities where new technology options are tested and results communicated back to developers/manufacturers. The *Wind Power* case study illustrates the success of test stations in Denmark to support knowledge feedback and quality assurance. The results were widely spread, as networking between actors (i.e., manufactures) was also supported. The test station establishment resulted in essential knowledge and technology development. However, the case study of wind energy also illustrates the negative experience in the Netherlands, where the government supported competitiveness rather than networking. In the Netherland case, the test station environment did not support essential feedbacks and information exchanges between Dutch manufactures.

Niche markets and early market deployment can also provide essential feedbacks. For many new energy technologies, early experience in production and use, including experience in operation and maintenance (O&M), have been essential for successful development because experience is fed back into R&D and design changes. For example, wind turbines developed in the early 1980s were assembled from standard components, and feedback in use and O&M were essential for the development and tailoring of specialized wind turbine components; this in turn supported the development of specialized suppliers. Moreover, high costs related to production and O&M are important drivers for feedback and improvements of products and production processes. Another example for this type of essential market feedback is provided in the *Kenyan PV* case study of PV applications in rural Kenya (see Appendix II). Problems with quality in the PV systems were only revealed through extensive market deployment (and not in earlier demonstration stages or via traditional manufacturing quality control) and led to the subsequent improvement of the technology. Governments or NGOs can assist in this feedback process by providing documentation and public disclosure of market deployment experiences with novel technologies.

Feedback from niche markets and early markets are also important for the formation of the entire innovation system, taking into account not only the technology itself, but also actors and legal and economic frameworks. Essential feedback can be provided by evaluating the process of market formation. By evaluating how the system of innovation is evolving, e.g., the development of knowledge and actor networks, ongoing policy programs can be redesigned to improve in effectiveness and efficiency.

For both technology and market development, extended feedback loops could be achieved through international cooperation and experience sharing. Reporting is essential to overcome any discontinuities (for longer or shorter periods) in the support of technology and market development. Such international knowledge exchange initiatives (e.g., through some IEA programs) remain in their infancy.

24.4.5.4　Globalize! Or: How to Devise Local Policies to Productively Harness the International Flow of Energy Technologies

Energy technologies are intrinsically international. They constantly flow in the private sector through international licensing agreements, joint ventures, direct investment, and trade. Feedbacks in the energy innovation process can and should occur across national borders. To encourage the development and deployment of advanced/new technologies, policies are often needed to create a coherent incentive structure. Local policies are also necessary to foster absorptive capacity to take advantage of technology and knowledge produced abroad.

Protecting intellectual property rights (IPR) is an important aspect of knowledge exchange, but not a sufficient condition – nor even the most important factor – for enabling the transfer of technologies and knowledge (see Box 24.1). Technology diffusion, both across industries as well as across countries, consists fundamentally of adapting existing solutions to new environments through an iterative process of knowledge exchange, revision, reconstruction, and improvement. Setting up the conditions for accessing and assimilating foreign technologies necessarily implies building a local system to produce and reproduce this knowledge.

Developing countries can access new energy technologies through external technology sources like specialized suppliers and multinational firms, or they can support indigenous development of advanced technologies by implementing a comprehensive strategy that includes policy support for human resource development, investments for RD&D, and market formation. Simply buying technologies from abroad is often insufficient because developing countries assimilate these technologies but not the related knowledge about how to adapt, reproduce, and improve upon them.

Naturally, the financial requirements for acquiring hardware, machinery, and equipment are a central aspect of international technology diffusion, especially in capital-intensive, large, and embodied energy technologies. International financial schemes and institutions play a role in the current technological lock-in to the extent that they tend to screen-out investment allocations to cleaner energy sources, local R&D efforts, and knowledge infrastructures. Local and global efforts to mobilize the appropriate financial resources and schemes must be aimed at reducing the valley-of-death transit of clean, advanced, new energy technologies to enable technology and knowledge flows across borders.

Efforts to align national policies toward more effective technology transfer mechanisms must take into account both the predominance of private channels of technology transfer, as well as the role of public investment and incentives needed to provide a level playing field for advanced new clean technologies.

Box 24.1 | How can we ensure that all regions and sectors have access to, and are using, the best technology?

Myths: *"You can just buy (transfer), whatever technology is needed."*

"If we just fix IPR issues, technology will transfer seamlessly."

"If developing countries only had strong IPRs, technologies would transfer to their countries."

"If only international IPRs were weaker, developing countries would rapidly adopt new technologies."

Technological capabilities and technology levels vary widely across regions. Given that the non-OECD regions will represent an increasing share of the global energy system, effective technology deployment within and to those regions will lay the foundation of growth that is consistent with energy-related objectives.

Patents and other Intellectual Property Rights (IPRs) instruments are not a sufficient condition for innovation or technology transfer. Technology is much more than the "blueprints" of information disclosed in a patent. There are plenty of other conditions and investment needs to be met. Income thresholds set up limits to the scale to which technologies can be applied, limiting their attractiveness. Specialized inputs and infrastructures must be timely and effectively supplied. Skills in operating and integrating complex systems need to be developed. Moreover, many innovations are not patented, and in many industries firms rely on other means for seizing technology advantages (Levin et al., 1987). This means that simply adjusting intellectual property rights will prove far from sufficient to bring about the necessary technology transfer.

The literature discussing the impact of IPR on energy technology transfer is scarce. There is a dearth of empirical or literature evidence that lacking IPR protection is a strong barrier to technology diffusion. In fact, there is emerging evidence to the contrary, i.e., other barriers such as capital costs, lack of infrastructure, lack of local policy incentives like performance standards, feed-in tariffs, and subsidies, and lack of financial resources are more important. In the case of PV and biofuels, for example, the high number of supplier firms and flexibility of sources will most likely reduce the space for monopolist practices in technology contracts. The same seems to be true in the more concentrated wind turbine industry, where developing countries' firms have developed local industries through licensing. However, IPR protections may be a barrier to industry entry for developing countries in the future (Barton, 2009), even when other entry barriers (manufacturing experience) may play a much larger role.

Patents build up incentives for innovation by providing means to control and shape technology transfer. The current global context is already one in which IPRs have been considerably strengthened by prohibiting or restricting compulsory licensing, reversing of burden of proof, and extending of patenting dimensions (Maskus, 2000). Many models on patents show that balancing a patent's dimension can actually reduce the social costs of IPRs, depending on the structure of demand (Nordhaus, 1969; Klemperer, 1990; La Manna, 1992). The resulting trend to maximize all dimensions of patents in TRIPS (Trade Related Intellectual Property Issues) and other trade related IPR frameworks have limited the scope for IPR policies to more rationally foster innovation and transfer by customized IPR systems that properly balance private and social costs.

24.4.5.5 Be Stable! Or: How to Create Policy Stability and Credible Commitments on which Innovation Depends

Governments need to create expectations for actors in the innovation system that are stable and consistent over a multi-year period. Uncertainty in expectations about future policies increases the risk of investing in innovation for energy technologies. Because externalities are pervasive in the clean energy sector – due to both knowledge spillovers and environmental externalities – these distant payoffs rely heavily on policy instruments. However, if expectations about the level or existence of these policy instruments several years in the future are uncertain, firms will discount the value of future policies and under-invest in innovation. Because technology development is in itself a risky endeavor, private sector energy companies will only respond to policies that are credible, last more than a few years, and have a reasonable degree of stability. Moreover, volatility can accelerate knowledge depreciation and loss. Technology policy can be dynamic and flexible to reflect new information, but broad goals must be consistent and funding levels for support of the various stages of the innovation life cycle need to be predictable for the private sector to engage and invest in the creation of new technological knowledge. For energy problems that cannot be solved quickly, patience and predictability are needed.

The case studies make clear the adverse effects of policy volatility and rapidly shifting priorities among policy makers, as well as successes that have resulted from a more recent shift to longer time horizons. R&D budgets have been notoriously volatile. The history of US energy R&D funding is not characterized by stable budgets, but by changes that are much larger than annual changes in economic activity and overall research spending. More than half the time, annual program budgets rose or fell by more than 10% (Nemet, 2007). Wind power, solar thermal electricity, and solar water heaters boomed in the early 1980s and then the industries were devastated by dramatic program cuts in the mid-1980s, even if partially restored soon thereafter. Innovation, job creation, and manufacturing dropped in the United States, and even 25 years later, the focus of activity on these "abandoned" technologies remains outside the country.

A more recent policy innovation has been the shift to policies that ensure stability by including time horizons that set expectations about the intensity of government activity, for example, over ten year periods. The Japanese New Sunshine Program in the 1990s set declining levels of subsidies over ten years. The California Million Solar Roofs Bill set subsidies for 10 years. Renewables obligations in many US states set levels 15–20 years in the future, usually with annual interim targets. An important cautionary note is that long term commitments like these often include clauses that allow loopholes for governments and actors to avoid meeting these commitments should compliance become more difficult than expected, for example through the ability to pay low penalties. A "safety valve" clause in cap and trade has a similar effect if not paired with a symmetric price floor. While the flexibility to change targets may have social benefits, it is important to understand the price paid in terms of reduced incentives for investment for private actors. The shift to longer time horizons for policy making has been an important development, but can also be undermined by implementation details allowing excessive flexibility in cases of nonattainment.

Box 24.2 | How quickly can we move the energy innovation system? How long of a commitment to energy innovation is needed?

Myths: *"If we throw enough money at this, we can make it happen quickly."*

"This is a man-on-the-moon project."

There is no doubt that today's energy challenges call for quick action, and increases in government funding may play a key role in the strategy for solving these problems. In framing the energy challenge, many have evoked the memories of rapid, focused projects to achieve single national goals, such as the Manhattan Project in the United States to develop nuclear weapons, or the US effort to put a man on the moon in less than 10 years.

Although there is a need to pursue energy innovation more aggressively, energy embodies a far broader range of technologies and actors than a Manhattan Project. Virtually every citizen of the globe is an energy user, and therefore has the ability to choose technologies to deploy and fuels to purchase. Energy supplies are produced and provided by a vast range of actors and there is a wide range of supply sources: fossil fuels, bioenergy, nuclear power, solar power, wind power, and others. Each of these involves multiple technology competitions and opportunities for improvements. Historically, accelerated technology deployment programs relied on "selected" single-mission driven technology winners. Meeting future energy needs likely benefits by bringing multiple, competing technology options to the market.

Further, the challenges that face the energy system over the coming century will not be met within a decade. For example, climate change research indicates that the carbon dioxide (CO_2) emissions reductions required to stabilize CO_2 concentrations will be more stringent in the longer-term than in the shorter-term and reductions must continue indefinitely. The challenge for decision makers is to develop the support for a sustained, long-term effort to enhance energy innovation.

24.4.5.6   Align Incentive Structures! Or: How to Avoid Confusing the Market

To maximize the effectiveness of ETIS, it is essential to align incentive structures and employ consistent policy signals. These alignments should be durable so that there is predictability over time for the ETIS. When there is inconsistency or lack of alignment, the efficiency and effectiveness of the system is undermined.

There are two types of alignment that must be considered: (1) alignment of incentives within a given innovation system; and (2) alignment of incentives for different innovation systems to encourage spillovers. To illustrate how to align a particular system, we can utilize the model of the growth phase of the technology life cycle. Aligned policies would include the development of an explicit strategy for supporting technologies that are invented through demonstration and testing, and formation of larger-than-niche markets to facilitate the transition of technologies across both "valleys of death" (from R&D to demonstration, and demonstration to early deployment). Government often must also establish policies that create incentives for technologies to be pulled into the marketplace. Throughout the growth phase of a technology (or set of technologies), governments need to elicit information from actors and devise mechanisms for feeding that information to other actors and other stages of the innovation process. In other words, government often must facilitate the integration of supply and demand and overcome barriers to sharing information essential to the good functioning of energy markets.

When government fails to align the incentive structures for achieving desired outcomes, contradictions emerge and perverse outcomes flourish. Three examples are helpful to illustrate these contradictions.

The *US Vehicle Efficiency* case study on the CAFE standard and the *Hybrid Cars* case studies (see Appendix II) show that the main policy incentive for the development and deployment of energy-efficient vehicle technologies in the United States is the CAFE standard. However, the United States does not impose significant fuel taxes or any fees on the purchase of inefficient vehicles (though it provides a subsidy in the form of an income tax credit for the purchase of certain advanced technology vehicles like plug-in hybrids). As a result, manufacturers are encouraged to innovate only as much as the standard implies, and consumers have virtually no incentive to drive less or to purchase more efficient vehicles. The CAFE standard thus creates a floor for minimum levels of innovation, which implies that the policy mix is incomplete. As one would expect, vehicle miles traveled have steadily grown in the United States (especially because the cost per mile of driving is lower with the fuel efficiency standards), and there was an explosion in the purchases of large passenger cars and light trucks, which had a weaker standard. If the US government added a tax to the price of gasoline, it would help create consumer demand for more fuel-efficient vehicles, thereby facilitating the integration of supply and demand.

The large government investments into RD&D of fossil, fission, fusion, efficiency, and renewable energy technologies can be charitably labeled uncoordinated, and possibly characterized as completely at odds with the much larger government subsidization of the deployment of fossil-fuel technologies. In 2008, IEA member countries invested US$14 billion in energy RD&D, and US$1.6 billion in total for fossil energy technologies (IEA, 2009b). While fossil fuel subsidies may support the fossil fuel RD&D investments, they strongly distort the market for nonfossil energy technologies. US fossil fuel subsidies alone are at the level of approximately US$10 billion/year for traditional fossil fuels, according to a recent report (Adeyeye et al., 2009). Fossil fuel subsidies in the non-OECD countries are estimated to be approximately US$170 billion/year (IEA, 2006). Globally, subsidies to fossil fuels may be on the order of US$500 billion/year, of which about US$100 billion is estimated to be provided to producers (GSI, 2009).

For an example of the lack of policy coordination, consider the United States. There is little evidence in the United States of bureaucratic coordination at the federal level. R&D strategies and decisions are largely made by the Department of Energy in the Federal Executive Branch (although appropriation decisions for R&D are made by Congress, often in contradiction to the R&D strategies set forth by the Department of Energy). Congress, however, establishes the market formation policies – the subsidies, loan guarantees, tax credits, carbon taxes, and cap-and-trade systems. Sometimes the Environmental Protection Agency establishes the market formation or deployment policies (e.g., sulfur dioxide emission trading system; performance standards for power plants), but Congress usually confers this authority. In this system, it is difficult to map out and implement an efficient innovation strategy, much less insure that feedback loops are established and maximized.

Alignment of incentives for ETI probably cannot be achieved without an explicitly designed and implemented innovation strategy for energy technologies. Even with such a strategy, multiple objectives in a country's energy policy can cause a misalignment of incentives. A common misalignment is government making RD&D investments in energy efficiency while simultaneously subsidizing the price of retail fuels. Another example of misalignment is the government encouraging RD&D investments in wind energy when local planning and zoning laws prohibit the installation of wind turbines.

24.4.5.7   Be Systemic! Or: How to Address Innovation in a Comprehensive Way

Innovation policies tend to focus on specific technologies. A narrow technology focus runs counter to the systemic view of energy technology innovation developed throughout this chapter. As well as technology-specific innovation processes, the innovation system is comprised of actors, organizations, infrastructure, and institutions. Relationships and feedbacks between these various components of the innovation system underpin the drivers and mechanisms of innovation discussed in

Box 24.3 | What mix of policy instruments would most effectively spur innovation?

Myths: *"The solution is a massive ramp up of R&D expenditures."*

"If prices are right, innovation will take care of itself."

It is well established that innovation takes place through a system of complementary actors and processes. Each or any of these processes may prove to be a choke point in the innovation system. Government-supported R&D is a core element of the innovation system, and many argue that government R&D expenditures should be increased in the face of our current energy challenges. However, government R&D is far from the only component of the energy innovation system.

Internalizing the costs associated with climate change and energy security to the feasible extent would provide more appropriate signals for technology deployment and private-sector innovation and investment activities. However, a wide range of market failures exist that prevent the private sector from investing in innovation in ways that are consistent with social needs, even if prices are right. Effective pricing is far from sufficient for a robust energy innovation system.

Decision makers find themselves in a situation where there is no single policy mechanism that will support a robust energy innovation system. The system must include a well-functioning and well-targeted government R&D program, but appropriate incentives through the pricing of externalities is also critical for producing demand signals that will induce learning and create incentives for private-sector R&D. A wide range of policy challenges remain regarding other core elements of a robust innovation system, including intellectual property rights and institutional structures to support widespread deployment of new technologies.

Section 24.2. Taken together, these components comprise the selection environment that shapes the outcomes of the innovation process.

A systemic approach to innovation policy requires a package of policy instruments that should adapt to changes and be geared toward triggering changes over time in the innovation systems. Policy packages may also differ from one innovation system to the next. Within these packages, policies must be aligned and consistent in their targets and objectives, as discussed above. Policy packages must also be broad in their coverage, supporting the successful functioning of the whole innovation system. A good example is the cooperation and knowledge feedbacks between different actors and institutions furthered by incentives for collaboration and mandates of information disclosures. Overall, the policy package needs to support knowledge development, feedback processes, and learning for the entire – or at least for the essential parts – of the innovation system. The *Wind Power* case study discusses the development of wind energy in Denmark and shows the success of a systemic approach that includes several actors (energy companies, wind turbine producers, smaller wind turbine owners) and the relationships between those actors, as well as essential institutional features (connection to the grid, spatial planning and permitting process). The *Japanese Efficiency* case study also illustrates this point nicely.

In the case of end-use technologies, actor-networks within innovation systems are often more complex, involving end-users, local authorities, wholesalers, retailers, branch organizations, consultants, installers, energy companies, architects, etc., as well as socially constructed norms, habits, routines, and values. Although these social institutions may play an important part in the success or failure of innovation processes, innovation policies tend to focus on technologies. The broader dynamic between technological change and social change is either sidelined or framed as a simple push-pull relationship with technologies driving responses in social institutions. As a result, social innovation, referring explicitly to changes in the adoption, use, and adaptation of technologies in a social and institutional context, is marginalized as a target for innovation policy. This can be attributed in part to the deep ideological, conceptual, and analytical differences between social innovation and technological innovation, which extend into the policy domain.

Myriad forms of social innovations include participatory planning processes, community-based initiatives, social learning, normative messaging on utility bills, information provision (to change attitudes), educative initiatives (to change values), supply chain alliances and pressures, new business models, and reporting and disclosure requirements. The package of instruments developed to support innovation systems should include policies targeting social innovation as well.

The interdependence of innovation outcomes with social change also highlights the limitations of innovation policy, even when designed within this systemic perspective. Even comprehensive, aligned, and stable policy packages can never guarantee successful innovation outcomes due to irreducible uncertainties.

Box 24.4 | Is innovation policy exclusively concerned with technologies?

Myths: *"The problem is technological. If we solve that, social change will follow."*

"Technological change only arises in response to social needs."

Technological and social change are interdependent. The success of technology-focused innovation processes, as well as the effectiveness of these policies, is conditional on supportive and adaptive social institutions. Participatory planning processes to foster the scaling-up and diffusion of wind power, and community-based organizations that promote energy efficient behaviors and technology adoption, are two current examples. Policies to support social innovations should be considered an integral part of an innovation system approach.

24.4.5.8   Experiment! Or: How to Stop Worrying about Failure

The formative phase of a technology's life cycle concerns the transition from development to diffusion. The dynamics of energy technologies that have successfully diffused historically indicate the importance of building out large numbers of units as a key feature of this formative phase. (A unit refers to a steam turbine in a coal or nuclear plant, a wind turbine, a photovoltaic module, and so on.) From the perspective of governments, experimentation, debugging, improvements and learning possibilities are all proportional to the number of units built during the formative phases of a technology's life cycle. Granularity is therefore a key variable determining innovation and investment risks, as well as the extent of possible experimentation, improvements, and learning favoring smaller, unit-scale, MW-scale projects over larger, "lumpy" GW-scale projects.

The demonstration and early deployment of many units is a natural feature of modular, distributed technologies like residential solar hot water or PV systems, with relatively low capital requirements per unit (even if costs per unit of energy output/capacity might be high). The unit costs of modular technologies may be driven down during the early deployment and diffusion phases by manufacturing scale economies in addition to learning effects. However, the same tendency to focus initially on building out unit numbers is also observed in the histories of large, centralized supply-side technologies. Despite the apparent availability of unit scale economies (i.e., falling unit costs at larger unit capacities), formative phases were historically characterized by only incremental increases in unit capacity over an extended, decadal period. Considerable experience was gained with smaller scale units before significant jumps in unit scale were successfully attempted (e.g., coal power, jet aircraft, and wind turbines).

The consistency of this pattern points to the importance of experimentation with many different units as a precursor both to widespread diffusion and to up-scaling (i.e., pushing technologies up the unit scale frontier). The hallmark of this approach to innovation is granularity: individual eggs in many small baskets. Note that granularity here refers to the formative phases of a given technology; design criteria for portfolios of technologies are addressed below.

Promoting substantial one-off increases in unit scale prior to numerous formative experiments in both precommercial (demonstration) and commercial contexts should therefore be considered a high-risk strategy. The comparative histories of wind power in Denmark and Germany support this cautionary note. Experimentation to generate knowledge on a technology's performance, efficiency, reliability, and other service attributes enhances the required capacity for capturing unit-scale economies. The design, construction, and operation of many different smaller scale units not only leads to incremental improvements and learning-related cost reductions, but also would appear to underpin the success of subsequent larger-scale units.

The importance of these formative experiments helps explain why the time period over which unit scale economies are captured does not appear to accelerate from early to late markets. Although large scale units may be transferable from early to late stage markets, the late stage markets need time to form the requisite institutional capacity to support these increases in unit scale. A context of international cooperation and knowledge transfer would, of course, support the aggregation of learning from experimentation processes running concurrently around the world.

Experimentation can and also perhaps should be multifarious, involving an array of different actors, forms, and stages of the technology's life cycle. In practice, many smaller scale variants of a technology may be pursued on parallel tracks by competing and heterogeneous commercial interests. This granular approach to innovation policy diversifies risk and reduces the consequence of failure.

Government's role should be to fill in the gaps by, for example:

- underwriting small-scale demonstration projects for socially robust innovations with less immediate or higher risk private returns;

- supporting variety in the early deployment phase by creating and protecting differentiated niches;

- reducing upfront capital barriers;

- managing the natural commercial tendency to rapidly confirm a dominant design that confers market advantages and potential cost benefits through scale economies; and

- avoiding over-emphasis on rapid unit scaling.

Failure is an inherent feature of a multifarious and granular portfolio of innovation experiments. Venture capitalists may build energy technology portfolios with an expected nine-in-ten failure rate, knowing that the one in 10 breakthrough will support returns for the portfolio as a whole. Accountability for taxpayer dollars and the associated political risks of funding failures (among other things) makes public innovation policies less tolerant. A shift in mindset is needed to recognize that perfect foresight and innovation are awkward bedfellows, not just in the early R&D phase, but also during the early deployment phase of experimentation with many unit numbers. Learning what does not work supports learning about what does work, and this in turn supports both diffusion and unit scaling. Building diverse portfolios of modular or smaller-scale technologies helps spread this risk of failure. Conversely, concentrating public resources on the scaling up of a particular technology (be it fusion power or GW-scale CCS projects) reduces portfolio diversity and magnifies the risk of failure.

24.4.5.9   Focus on Technology Portfolios! Or: How to Not Pick a Winner, but to be Picky on your Picks

The broad range of necessary ETIs combined with inevitable innovation uncertainty suggests that innovation policies must consider not a random collection of innovations, but rather a wide portfolio of technologies.

Innovation portfolios reflect the combination of technology options pursued within the innovation system that reflects both their respective option value (i.e., societal/ environmental/ economic benefits in the case of successful development and diffusion) as well as their associated risks (e.g., innovation failure or investment risks).

In designing innovation portfolios, a number of basic criteria need to be taken into account:

1. The portfolio needs to reflect a blend of options comprising the entire energy system and spread the investments across many technologies and projects.

2. The innovation portfolio should encompass all salient elements of the technology development cycle and all different channels of technology knowledge creation, such as R&D. demonstration, niche market deployment incentives, and market creation measures.

3. Given inevitable resource constraints, the design of diversified portfolios is more feasible when focusing on granular, less capital-intensive technologies such as end-use innovations and smaller-scale supply options. Conversely, large-scale, capital-intensive, high-risk innovations can meaningfully only be considered in global innovation portfolios. A common thread in case studies on both historical as well as current energy technologies (compare the studies on *Hybrid Cars, Solar Water Heaters, Heat Pumps, Japanese Efficiency, Wind Power,*

Box 24.5 | Are the technologies available that would be necessary to take action today?

Myths: *"No innovation is needed; all that's required is political will."*

"We can't take action now because the needed technologies are not available."

The full suite of technologies that will ultimately be deployed to address the energy challenges in the coming century is not currently available. Technology continues to advance and redoubled efforts to spur innovation will certainly lead to improvements in technologies over the coming decades. At the same time, there are numerous reasons to take action to deploy currently available technologies today: innovation systems are most effective when there is communication between technology users and developers; a range of non-technology factors are associated with the broad deployment of many technologies, and these can take years to develop; capital investments today may preclude effective action in the future; and in many applications, there exist technologies that could have dramatic near-term effects, such as end-use technologies.

Decision makers find themselves in a situation where a wide range of beneficial actions are possible today to deploy existing technology, yet not all technologies are ready for deployment. The challenge for decision makers is to implement policies that will allow some technologies to develop further before moving from the laboratory to the field, and experimentation and feedback with other technologies at a small scale. These policies should support the private sector's role in deploying still other technologies at a large scale if they are clearly proven at smaller scales.

Solar PV, and *Kenyan PV* with the *US Synfuels* and *French Nuclear* case studies) is the importance of granularity. Smaller unit scales of energy technologies (e.g., the size of a one MW wind turbine compared to 1000 times larger typical size of a one GW nuclear reactor) not only result in lower innovation failure risk[16] for a given project, but also in lower barriers to innovation adoption in case of successful demonstration. Technology policy can also profit from granularity, as enabling the spreading of innovation risks across a broader range of a multitude of smaller-scale innovations without requiring the premature selection of few capital intensive innovation projects, preempting decentralized market-based decision processes. It is not coincidental that granular, smaller-scale innovations in energy end-use technologies have been the dominant source of technological advance in energy systems historically and will likely remain so in the future.

4. In portfolio design, the inherent tension between the desirable goal of maintaining technological diversity and the equally desirable goal of improved economics through standardization and technology focus needs to be considered.

5. The core element of a broad-based innovation portfolio is that policy makers need to be pickier with their technology picks. On one side, innovation policies need to avoid preempting the outcome of decentralized market-based technology innovation, experimentation, and early market deployment decisions that are key in technology development. On the other side, public sector innovation policy legitimately needs to complement private market technology innovation biases against large-scale, investment-intensive technologies that might be crucial in addressing broader social and environmental goals. While resource limitations inevitably require focus on a few strategic technologies, there is a downside: decision makers need to be aware of the pressures to concentrate capital-intensive, high-risk innovation projects (potential innovation "lemons") in the domain of public sector innovation portfolios, implying higher risks that the public sector disproportionally shoulders. The result is an invitation for public support for high-risk innovations or bad business practices (e.g., according a quasi-monopoly to private sector-advanced technology options that capitalize on the economic rents from lavish public subsidies, without "delivering" in terms of much needed technology innovation, such as persistent cost improvements of new energy technologies).

6. Innovation portfolio design ideally includes a blend of the respective options values of technologies, both from a demand-pull and a supply-push perspective. Policy makers also need to be explicit on their underlying risk measures and criteria that enter innovation

portfolio design to enhance the legitimacy and transparency of the resulting resource allocation across options and measures.

7. The importance and complexity of these portfolio decisions will require more expertise within governments, as not all technology decisions can (or should) be made by the private sector. Fortunately, formal scientific tools for innovation portfolio design and analysis are becoming increasingly available (cf. the *Technology Portfolios* case study). These tools can help move the discussion on innovation portfolios to more rational ground, but the further development of these tools requires continued policy and funding support much like energy technologies themselves, as well as an institutional locus for their application.

24.4.6   Conclusions: Generic Characteristics for Energy Technology Innovation Policies

Energy systems consist of interconnected technologies, knowledge bases, and practices that are interdependent and operate in dynamic ways. They are simultaneously embedded into prevailing economic incentive structures and broader social contexts. To be effective, innovation policy must be conscious of these interdependencies across time, space, and actors, and act coherently. Isolated policies acting in only one realm of the market (for example, by stimulating niche markets through tariff policies under a cost buy down paradigm) will not yield strong innovation if not accompanied by coordinated and aligned support for R&D and demonstration projects. Likewise, an isolated technology push focused on R&D without regard to the economic and institutional incentive structures and barriers prevailing in energy markets will be counterproductive for innovation, effectively curtailing the essential feedback and learning from market deployment on which R&D so critically depends.

The changing nature of technology should also be reflected in policy design. Instruments and incentives have to be adapted to the particular problems, tensions, and bottlenecks that characterize each stage of a technology's innovation life cycle. Resource and financial hurdles are, for example, of a very different type in R&D phases than in demonstration and early deployment stages; as market formation advances, the need for agents' coordination also changes in nature and degree. Policies for preserving variety or accelerating standardization are extremely sensitive to the state of development of the technology.

The dynamics of technology over time require the attention of innovation policy to development times and feedback processes. Technology research and development is a time-consuming activity that requires patient funding and active networking. Policy design must therefore be sensitive to the timing of investments and returns. Since knowledge development on suitable solutions and technology improvement takes a long time – and knowledge rapidly depreciates under erratic policy signals – early and persistent policy actions are very important. Long-term planning and policies that persist before being able to reap economic,

---

16   Given equal probability of failure of two innovation projects, the one with smaller unit scale and hence lower total costs (millions as opposed to billions of dollars) results in lower innovation risk (defined as failure probability times [economic] consequences, i.e., loss of investment).

Box 24.6 | How should policy focus and resources be allocated across technologies?

Myths: *"All technological options will be needed everywhere."*

"We need a balanced portfolio."

"Technological failures are a sign that mistakes have been made."

"We'll need as much diversity as possible."

"Technology X is the answer."

There is a strong tendency for decision makers and technology advocates to focus on a single technology as being the key to meeting energy challenges – the "silver bullet" approach. This single technology focus is a result of both a lack of understanding of the complexity of the global energy system and the tendency of technology experts to support the continued expenditures on single technologies. However, research has consistently indicated that a broad and diversified portfolio is critical to provide options throughout the energy system, to allow for experimentation, and because history has shown that technological expectations are often wrong. A necessary condition for a successful innovation system is the exposure of nonproductive avenues for innovation and the ability to drop these avenues from the portfolio.

At the same time, a diversified portfolio is not necessarily an effective portfolio. There exist a wide array of possible diversified portfolios, and the need for diversity does not obviate the need for decision makers to identify avenues that are most promising and or lacking in effort and adjust the portfolio appropriately. Further, there are many avenues of investment for which increased effort is an obvious need.

The challenge for decision makers is the management of diversified portfolios. This requires institutional structures that are capable of effectively collecting and synthesizing information, robust in the face of necessary technological failures, and capable of adjusting as information changes.

social, and environmental returns are key to ETI. To be consistent in time, policies should be adaptive, flexible, and patient during the early phases of technology development to preserve technological variety, with efficiency assessments and selection of programs/projects in later stages.

The inherent, strong uncertainty that characterizes early phases of innovation calls for flexible institutional mechanisms that are able to actualize expectations regularly. However, as the technology life cycle advances and uncertainty about technical features decreases, capital-intensive investments demand long-term policy stability. Institutional design aimed at accelerating innovation must be aware of this trade-off between maintaining experimentation and technological variety and the economic drive towards standardization, and be able to switch policy priorities over time, but in a predictable, consistent manner.

Knowledge is a crucial factor in innovation, and its nature critically influences policy outcomes. Systemic approaches particularly emphasize the fact that learning processes in innovation occur at many different levels and flow in multiple directions. Knowledge develops and accumulates through a range of complementary processes and activities, which in turn condition the future absorptive capacity for new technological

knowledge. Supporting and facilitating these complementary learning processes is a crucial complement for innovation policy, unfortunately too often ignored. The quasi-public and distributed nature of knowledge, together with the strong positive feedbacks between different knowledge bases, call for adequate, timely policy support to these activities, which are easily screened out by market processes. At the same time, policy design should reflect the understanding that knowledge can become obsolete and experience can be lost. Innovation policy is incomplete if it does not adequately address the conservation of memory and a continuous renewal of the knowledge base, which is particularly threatened by stop-and-go erratic policy support.

24.5   Conclusions

24.5.1   Research, Data, and Information Needs

A number of important gaps in data, information and research were identified in the above assessment. Addressing these issues is not only of academic interest but equally critical for improved technology innovation policies, and hence is of greater societal relevance. Ten important areas are summarized here, regrouped into two broad categories: data and information, and research needs. A central theme is the need to develop better indicators and quantitative data as well as operational models and criteria to answer the core questions of technology innovation policy regarding effectiveness, i.e., what is the most appropriate policy instrument for a particular purpose, what resources are required, and what are the likely response times of the innovation system to policy interventions?

24.5.1.1   Data Needs

In terms of data, four areas stand out where the gap between data needs and availability is particularly large:

- data on innovative activities (R&D) pursued by private firms;

- data on technology specific investments, particularly in end-use technologies;

- data on knowledge spillovers across different innovation fields and at the international level including, in particular, technology-specific trade data and on joint technology development collaborations; and

- systematic and up-to-date data on performance and economic characteristics for energy technologies that are internationally comparable and widely available for technology studies and policy assessments.

24.5.1.2   Information Needs

Information needs include the following areas:

- identification of a limited set of appropriate and manageable criteria and metrics for the assessment of innovation systems in terms of inputs, outputs, and outcomes that can be matched with available or to be developed data sets;

- operational measurement models that describe knowledge depreciation in R&D and LbD processes; and

- criteria for the selection of technology specific case studies especially in a comparative context across countries and across technologies.

24.5.1.3   Research Needs

In terms of research, this assessment has identified the following areas:

- the development of conceptual models that answer the question of how measurable inputs and outputs of innovation systems relate to each other;

- the development of a "meta-theory" of technological change that enables to establish appropriate *ceteris paribus* conditions to be able to compare and assess the dynamics of change and of policy effectiveness across different technologies and development/adoption environments; and

- comparative assessments of the effectiveness of alternative policy instruments aiming at influencing individual components or the entirety of ETIS.

24.5.2   Conclusions on Energy Technology Innovation

Substantial and accelerated innovation is essential to respond to the sustainability challenges of energy systems at all levels, including the local, national, regional, all the way up to the globe scale. Further, a coordinated approach is needed that works within and between industrialized and developing nations.

Such innovations will comprise a combination of both incremental, cumulative changes and radical, discontinuous changes that can only emerge if the various innovation dimensions are nurtured simultaneously. Innovation entails technological, social, and institutional, as well as economic, driving and embedding factors that need to work hand in hand in the development, testing, and ultimate selection and adoption of new innovations.

A core message of this chapter is that the drivers of innovation, as well as the policies that support it, are complementary rather than substitutable for each other. This requires attention to fundamental innovation, or technology push, which needs to be coordinated with efforts to facilitate the expansion of the market opportunities – demand pull – that move innovations from laboratory to cost-effective deployment. As such, the energy sustainability challenge requires changes in whole innovation *systems* rather than simply more independent, individual innovations.

The synthesis of the available literature and case studies suggests that successful innovation systems and their supporting policies are characterized by three main features: alignment, consistency, and patience.

Alignment (i.e., comprehensive and contradiction-free) means that the various forces and policies that drive innovation are considered holistically and not from the perspective that any single driver can substitute for the (lack of) other drivers. Accelerated R&D in new energy technologies without economic incentives for ultimate adoption of the innovations will not yield the much needed change to redirect energy systems toward sustainability.

Likewise, consistency of policies and drivers is key. Incentive structures need to remain stable and not at odds with each other. All innovation actors (researchers, industry suppliers, and customers, the end consumers) rely on predictability and consistency of the innovation environment; otherwise the costs of taking the inevitable innovation risks become prohibitive. An important task for research is to provide a framework that can be used to examine energy and carbon outcomes of innovation policies, including the relative benefits and costs of policy tools aimed at the expansion of the low-carbon energy sector.

Finally, patience is needed. The time lags between basic and applied research, development and testing, market introduction, and ultimate diffusion and the required feedbacks to earlier innovation stages in the process are substantial. Joint expectations (or visions) need time to emerge and accommodating social and institutional settings need time to develop. The international dimension in the development and diffusion of innovations also requires patience.

Innovation and technology policies also can no longer remain fragmented, *ad hoc*, and concentrated on individual technological options. A much more strategic and long-term approach is required to harness the potentials of well-functioning ETIS. Goals and objectives, weighting of different (sometimes conflicting) objectives, strategies and implementation plans to be followed, evaluation criteria for continued reassessment, etc., all need to be formulated to involve relevant stakeholders and take account of international developments. Above all, this requires institutional innovations as, at present, corresponding institutional frameworks and learning capabilities are insufficiently developed. Successful examples such as the Japanese national system of innovation can provide inspiration, but institutional solutions need to be custom-tailored to their specific national or regional circumstances.

A paradigm for a strategic and long-term approach to ETIS is the concept of adaptive/policy learning. Strategies and policies need built-in mechanisms to assure flexibility and the ability to adjust courses of actions and policies to reflect new developments, in order to be able to react to and correct for unanticipated outcomes and surprises. There is an inherent tension between the desired criteria of flexibility/adaptability on one hand, and the equally desirable criteria of alignment, consistency, and patience on the other. These seemingly contradictory objectives can and should be reconciled, but it will involve an open institutional and policy architecture that can mobilize collective learning processes and widely shared strategic goals.

Openness implies increased sharing of data and experience and non-exclusive networks of actors, nationally as well as internationally. In short, much higher levels of cooperation and knowledge exchange are needed to address the potential tensions between national policies and an increasingly globalized ETIS landscape via formal and informal information sharing, cooperation, and coordination agreements. Existing institutional solutions such as the IEA or the International Renewable Energy Agency can serve as useful entry points, but must be expanded to become both truly international and comprehensive in terms of their energy systems perspective, in particular in moving away from the traditional energy-supply bias to include energy efficiency and conservation in a more integrated way.

As daunting as the agenda for a systemic, consistent, and aligned long-term technology policy framework may appear, improvements can be implemented gradually. An illustrative roadmap for action over time consistent with the dynamics of capital stock turnover rates in energy systems is given in Table 24.9.

Table 24.9 | Illustrative roadmap for the development of a systemic, aligned, and consistent policy framework for energy technology innovation and diffusion matching policy approaches to realistic timescales of outcomes.

Timescale of Policy Outcome	Examples of Policy Approaches
short term (e.g., to 2020) capital stock additions (some)	create, stimulate and protect market niches around performance advantages of new technologies deploy market-ready, clean technologies through credible and stable incentive mechanisms develop long-term technology innovation and market deployment strategies in a consultative process, creating "joint expectations" reduce/eliminate direct or indirect subsidies for technologies not aligned to long-term technology strategy and portfolios use "sunset" clauses for planned retirement of depreciated, inefficient, or polluting capital vintages
medium term (e.g., to 2050) capital stock additions (all), capital stock turnover (some)	expand public and private R&D investments stably in diversified portfolios designed to manage risks and corresponding with end-use needs underwrite many granular and multifarious technology demonstration and learning cycles support disclosure, interaction, and feedback between innovation system actors engage in multiple international collaborative projects to further knowledge dissemination and technology spillovers align innovation and market deployment incentives (e.g., recycling externality pricing revenues back to R&D and market deployment incentives)
long term (e.g., to 2100) capital stock additions (all), capital stock turnover (all)	set long-term targets with appropriate monitoring and enforcement mechanisms to sustain shared technology expectations maintain portfolio diversity to prevent premature lock-in or standardization set technology standards for the gradual phase out of "bridging" technologies
throughout (present-2100)	create and nurture formal and informal institutional settings for technology assessment, evaluation, portfolio design, and knowledge sharing

24.5.3   What is New?

At the end of any assessment, it is legitimate to ask: what's new? Readers will undoubtedly form their own opinions, but the (subjective) perspective of this chapter's writing team is summarized below.

As a first, the chapter provides a synthetic overview of the resource requirements (in terms of investments) of the entire energy technology innovation and diffusion process. This assessment thus includes new data, including R&D expenditures in emerging economies, a summary synthesis of diffusion investments, and novel first-order estimates of end-use technology investments for which information has been lacking to date. Notwithstanding the key role of innovation in the front end of the technology life cycle, the numbers nonetheless confirm the predominance of assuring an appropriate incentive environment for the adoption of innovations, where typically 80–90% of financial resources need to be mobilized. The numbers also confirm the critical importance of energy end-use and related technologies in ETIS that need to be better reflected in R&D and market deployment incentives and new business models.

The chapter also contains a rich set of new case studies on energy technological change, which are both novel in their scope and constitute a useful resource for research as well as policy learning. The case studies represent new, original research and are unusual for a major international assessment such as the GEA. However, the fact that they could be conducted even under extreme resource constraints points to the wide interest and collaborative spirit of the technology community, which can be harnessed further to improve knowledge exchange and learning for successes and failures, moving technology policy forward.

The chapter also opens the "black box" of technological innovation and change. It analyzes the "finer grain" underlying the change in the multitude of attributes and drivers of innovation: new knowledge, but also knowledge depreciation, economies of scale, linkages and spillovers to other sectors, phenomena of increasing returns inherent in knowledge generation and in infrastructure-intensive interconnected systems such as energy, as well as resource constraints and (relative) input prices. The new findings, while tentative, again confirm the importance of alignment and consistency under the overall umbrella of the complementarity of policies/measures/incentives rather than their substitutability. These findings add a new dimension to naive perceptions of the prevalence of either supply push (e.g., accelerated, stepped-up R&D programs) or demand pull (e.g., "cost buy down" in new technologies) that have characterized much of the literature and policy debates to date.

Finally, this chapter introduces the novel concept of "granularity." Historically, successful innovations are characterized by a prevalence of a multitude and a diversity of "small" (locally adapted) solutions to problems that occur even at a global scale, as opposed to singular, large-scale, planetary solutions (be it geo-engineering, solar power satellites, or a single design for a nuclear fusion reactor). "Granular," small scale innovations offer the potential of multiple and repeated experimentations, learning, and adaptation to diverse adoption environments. An example is the dramatic difference in industrial and managerial experience that stems from building and operating a million wind turbines, as opposed to some 1000 nuclear reactors. From that perspective, the critical innovations paving the way to energy sustainability will reside in energy end-use (e.g., efficiency) and locally adapted supply options (e.g., in smaller-scale renewables) that are in stark contrast to the prevalence of a uniform, global technology landscape that has characterized the fossil fuel age.

24.6   Appendix I: Investments into ETIS

24.6.1   RD&D Investments

R&D expenditures at the macroeconomic level are routinely collected by national and international statistical agencies (see OECD, 2007). The data usually differentiate by funding source (public versus private sectors); by R&D performing institution (government laboratories, universities, or by private firms); and, finally, by broad economic sector. Methodologies, data collection, and compilations are well established (OECD, 2002).

Energy-related or technology-specific RD&D data are not reported separately in these macroeconomic statistical frameworks, creating formidable data challenges. For a concise review of data sources, methodological issues, and limitations of energy RD&D data, see Dooley (2000). Energy- and technology-specific RD&D data are available for public sector expenditures in member countries of the IEA (IEA, 2009b), but information on non-IEA countries (Brazil, China, and India, or Russia, to name the most important ones) and especially for private sector energy RD&D are extremely fragmented and sparse. Evidence suggests that the IEA public sector energy RD&D statistics may cover only about a quarter of all energy-related RD&D globally, where private sector RD&D and increasingly non-IEA countries substantially contribute to energy RD&D. This assessment therefore includes an effort to compile national energy RD&D data on the emerging economies of BRIMCS (Brazil, Russia, India, Mexico, China, and South Africa), conducted by a team of researchers at Harvard University (Kempener et al., 2010b; see the *Emerging Economies R&D* case study summarized in Section 24.7.8) that also includes (albeit incomplete) coverage of private sector RD&D in these countries. For private sector energy RD&D in OECD countries, this chapter could only draw on the survey conducted by the World Energy Council in 2001 (WEC, 2001) for a sample of OECD countries. More recent international comparative data are simply unavailable.

24.6.1.1   Public Sector Energy RD&D

Figure 24.14 summarizes the trends in public sector energy RD&D since 1974 in IEA member countries and contrasts it with total public RD&D

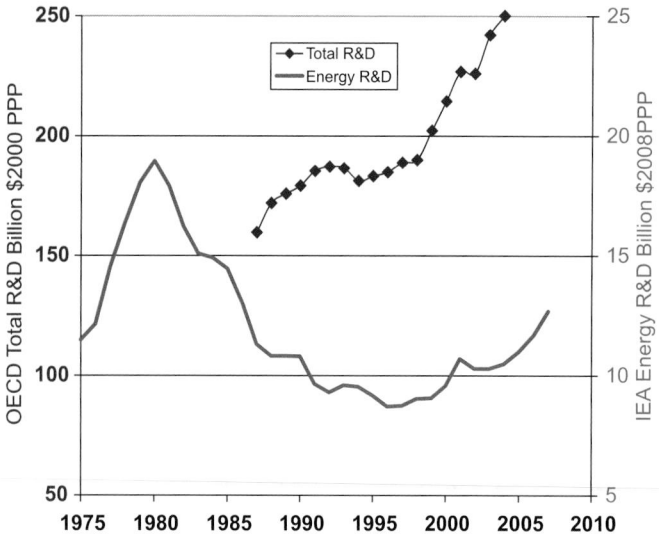

Figure 24.14 | OECD public R&D expenditures (US$_{2000}$ PPP) versus public sector energy R&DD in IEA countries (us$_{2008}$ PPP). Note the factor of 10 difference in the scale of the two expenditure categories and the opposing time trends since 1985. Source: Doornbosch and Upton, 2006 (in US$_{2000}$ PPP); IEA, 2009a (in US$_{2008}$ PPP).

expenditures. While total reported energy technology RD&D expenditures by definition include demonstration investments in addition to R&D expenditures, little detailed data are available. IEA (2009b) reports a total of some $550 million[17] at purchasing power parities (PPP) for seven countries for which such data are available. The United States represents the bulk of this figure, with approximately $444 million development expenditures, corresponding to some 4% of total public energy technology RD&D in all the IEA member countries in 2008. Hence, it is fair to say that public energy RD&D expenditures are in fact mostly R&D expenditures with little expenditure on technology demonstration proper.

A defining characteristic of energy RD&D is both its comparably small magnitude (5% of total government RD&D), as well as its boom and bust cyclical nature, characterized by rapid expansion in the wake of the oil crises of the 1970s, its subsequent collapse (with corresponding impacts on knowledge depreciation), and only gradual recovery after the year 2000. This is in stark contrast to the continually expanding overall R&D budget in IEA member countries. These trends are extensively discussed in the literature (Dooley and Runci, 2000; Doornbosch and Upton, 2006) and have been referred to repeatedly as "R&D under-investment" by researchers (e.g., Nemet and Kammen, 2007) and business executives (e.g., AEIC, 2010).

Figure 24.15 summarizes the historical evolution of IEA member country energy RD&D by broad technology class, illustrating a third area of

concern: asymmetries in public energy RD&D portfolios (see Box 24.7 on RD&D Portfolios). Total public sector RD&D in IEA member countries in 2008 amounted to some $12.7 billion (PPP). Close to $5 billion was spent on nuclear (fission and fusion), $3 billion on "other" energy technologies (hydrogen; electric power outside renewables, fossils, and nuclear; electricity transport and distribution; as well as basic energy research), and about $1.5 billion on fossil fuels and energy efficiency, respectively (for a tabular overview for IEA countries see Box 24.7. For BRIMCS countries, see overview in Section 24.7.7).

As discussed above, comparable internationally comprehensive energy RD&D statistics for non-IEA member countries are lacking. This results in the (incorrect) perception that energy RD&D and technology development is primarily performed in OECD countries. Given that this is no longer the case, enlarging global energy RD&D reporting systems remains a critical task. The Kempener et al. (2010b) energy RD&D survey on BRIMCs countries, suggests that public energy RD&D in the six BRIMCS countries amounted to some $2.7 billion dollars[18] in aggregating national data sources, compared with US$4.4 billion public energy RD&D in the United States in 2008 (IEA, 2009b). The case of BRIMCS countries also illustrates the fact that the traditional distinction between public (i.e., government) and private (privately-owned companies) sectors as sources of RD&D funding becomes increasingly blurred. Whole or partially state-owned enterprises (e.g., national oil and gas companies, utilities) constitute an important part of the energy sector in developing and emerging economies and also in many OECD countries. The RD&D expenditures in state-owned enterprises are strongly determined by national governmental policies. Combining public and semi-private energy RD&D, BRIMCS countries have a total current energy RD&D budget of some $15 billion (PPP), about equal to the entire public sector energy RD&D expenditures in IEA member countries ($13 billion, PPP) and still about half of the combined public and private energy RD&D in OECD countries (estimated here at approximately $25 billion). A commonality in the energy RD&D budgets of IEA member countries and BRIMCS countries is the dominance of fossil fuel and nuclear technologies, which currently receive some $11 billion (PPP) in total RD&D funding, or some 75% of total energy RD&D in BRIMCS countries. IEA member countries invest close to 50% of their public energy RD&D investments in these technologies (>$6 billion; $10 billion when private RD&D investments are included).

24.6.1.2   Private Sector Energy RD&D

The situation with respect to data availability of private R&D expenditures is dire. The Directorate General for Research of the European Commission states, "Despite the growing need, statistics on energy R&D expenditures in the private sector remain a problem" (EC, 2005).

---

17 2008 International $. The R&D expenditure data in this section are expressed throughout in terms of purchasing power parity (PPP) in order to improve comparability across very different economies. One drawback of PPP is, however, that simple ex post adjustments to a common base year (2005 in GEA) are not possible for international, cross-country statistics. Differences to 2005$ PPP should however be minor. PPP values are denoted here simply as "$" to differentiate them from the US$.

18 Data limitations preclude a comparison for an identical base year. Instead, the totals reported aggregate national statistical data for reporting years that vary between 2004 (China) to 2008 (Brazil, Russia, India, and South Africa). Numbers are expressed in terms of PPP.

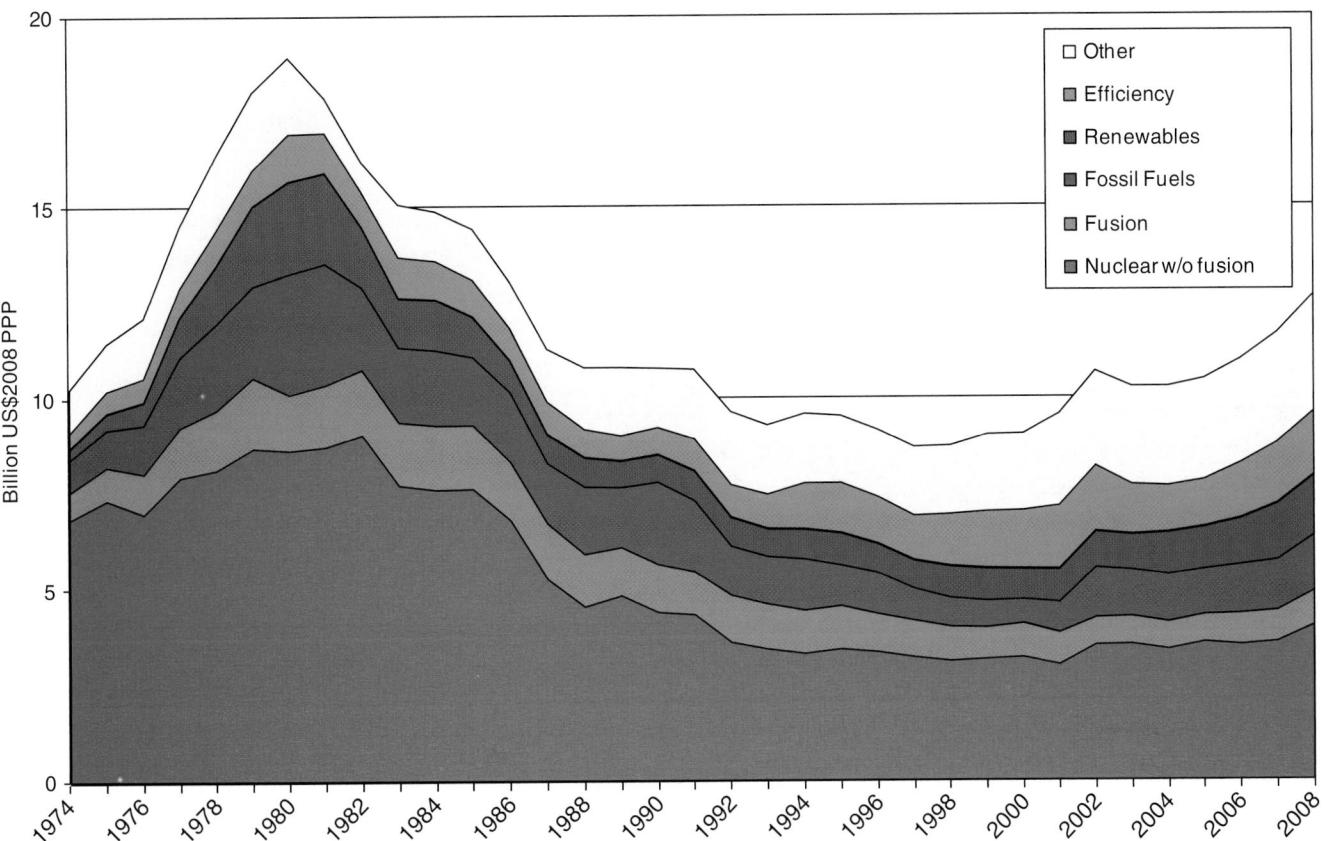

Figure 24.15 | Public sector energy RD&D in IEA member countries by major technology group 1974–2008 (in US$2000 million PPP). Source: data based on IEA, 2009a.

Macroeconomic RD&D statistics by broad economic sector are available for a sample of 19 OECD countries in the OECD Structural Analysis database (OECD, 2009). However, the latest year reported is 2002. Data for other countries are not collected and available in any internationally comparable form. In 2002, business enterprises excluding extractive industries (coal mining; oil and gas extraction) performed R&D equivalent to $433 billion (PPP in 2000$) in the sample of 19 OECD countries provided. The only sectorial breakdown available that bears directly on the energy sector are "coke, refined petroleum products and nuclear fuel" ($2.7 billion) and "electricity, gas, and water supply" ($2 billion), with an OECD total of energy-related private sector RD&D of less than $5 billion (PPP). Of course, RD&D performed in the manufacturing sector has a bearing on energy use, e.g., such as in electrical machinery ($13 billion), motor vehicles[19] ($50 billion), or aircraft ($20 billion; see NSF, 2010). How much of the RD&D performed in these sectors is energy-related remains unknown.

The only available survey of private sector RD&D specific to the energy sector is the study conducted by the World Energy Council (WEC, 2001), covering the period 1997–2000 for a sample of seven OECD countries, which are summarized in Table 24.10. In addition to the WEC survey, the most recent private sector RD&D data for the United States for 2004 (NSF, 2009) are listed for comparison.

The information available on total OECD private sector energy RD&D from 1993–2000 amounted to some $12 billion annually. The technology-specific breakdown is too incomplete and the data too old to warrant a detailed discussion. However, it is noteworthy that with the exception of Japan, private sector R&D on energy efficiency appears either to be unrecorded or, when subsumed under the "other" (or unaccounted for) category, remains extremely small outside Japan. Thus, the sparse available data suggest that also private sector energy R&D seems to follow the supply side (fossil and nuclear) over-emphasis apparent in public sector R&D in OECD countries.

For non-OECD countries, the Kempener et al. (2010b) survey on BRIMCS puts the OECD numbers in perspective (for a detailed breakdown, see the table in Section 24.7.7). The information available on investments in energy RD&D in the BRIMCS countries by the private sector and state-owned enterprises also amount to about $12 billion (PPP 2008), albeit based on more recent data (2004–2006).

---

19 The six automobile manufacturers listed among the top 25 global corporations in 2006, performed between US$4.6 billion (Honda) to US$7.5 billion (Toyota) in R&D, with a total of some US$39 billion. Other corporations, whose R&D is likely to include an important energy component are Siemens (US$6.6 billion), Samsung (US$5.9 billion) Matsushita (US$4.9 billion), Sony (US$4.6 billion), and Bosch (US$4.4 billion) (NSF, 2010). This listing alone suggests that in terms of private energy-related RD&D, energy end-use technologies are most likely to be of much greater importance than energy supply.

Box 24.7 | R&D Portfolios

How can we assess current energy technology R&D portfolios, i.e., the technologies we invest in, with the technology investments needed for an energy sustainability transition?

One way is to describe alternative futures through the scenario technique and use models to calculate the future market potential of specific energy technologies. This potential can be contrasted with public sector energy R&D spending (see the *Technology Portfolio* case study based on Grubler and Riahi (2010) and summarized in Appendix II below).

Given that the future is inherently uncertain, one needs to explore a wide range of possibilities, i.e., a reasonable number of scenarios. The results of these scenarios can be analyzed to derive "need-based" technology portfolios that can guide RD&D allocations across technology fields. A large-scale scenario study (Riahi et al., 2007) explicitly addressed the question of how the portfolio of GHG mitigation technologies changes as a function of the representation of salient uncertainties including energy demand, resource constraints, availability and costs of technologies, and the magnitude of GHG emissions constraints. For the quantification of the respective role of individual groups of technologies in the entire GHG emission reduction portfolios, the concept of mitigation "wedges" (Pacala and Socolow, 2004) was used. A mitigation wedge as defined in Riahi et al., 2007 is simply the contribution to the cumulative emissions reduction over the period 2000–2100 that a particular option provides compared to a baseline scenario. First, three baseline scenarios without GHG emissions constraints were compared to corresponding hypothetical baselines that assume a "frozen" state of technology in the year 2000 (i.e., no technological change/improvements). Then, for each baseline scenario a range of increasingly stringent GHG emissions constraint scenarios are calculated (constraints vary from as low as 450 ppm CO_2-equivalent GHG concentration by 2100 all the way up to 1390 ppmv-equivalent). Additional model sensitivity analyses then explored the impacts of the unavailability of particular technological options (e.g., nuclear or CCS). The calculated aggregated technology specific GHG mitigation "wedges" are summarized in Figure 24.16, showing mean as well as minima/maxima across all scenarios explored. The ranking of different mitigation options is quite robust across the scenarios explored, with energy efficiency and conservation being the single most important option with typically >50% contribution and nuclear with a typical 10% contribution to cumulative 2000–2100 emission reduction. The results are representative of other modeling studies, e.g., as reported by the Energy Modeling Forum (EMF-22) where the maximum share of nuclear energy ranges between 11–12% (Calvin et al., 2009) to 9–14% (Gurney et al., 2009) by the end of the 21st century.

It is also instructive to compare the calculated future mitigation potentials of technologies with RD&D expenditures, summarized for the total of all IEA countries above. A *significant mismatch in R&D*

portfolios in favor of nuclear and to the detriment of energy efficiency and conservation emerges. Nuclear received well above 50% of all cumulative (1974–2008) R&D expenditures, with energy efficiency receiving less than 10%, whereas their respective role in the GHG mitigation portfolios is exactly the inverse. To put these numbers into an absolute perspective: cumulative public R&D into energy efficiency totaled some 38 billion $_{2008}$ (in purchasing power [PPP] terms), which is lower than total cumulative expenditure into fusion energy ($41 billion PPP). Current R&D levels into renewable and CCS (which is subsumed in the "other" category above that includes *inter alia* also hydrogen, fuel cells, and basic energy research) are also much lower than a future "need-based" analysis suggests, albeit the mismatch is less striking than the one comparing energy efficiency to nuclear.

Were current energy technology R&D portfolios to represent the respective "option value" of alternative technologies in a climate-constrained world, one would have to increase current R&D into energy efficiency by at least a factor five or by some $6 billion PPP per year (thus not proposing a reduction in nuclear R&D). Given that improved energy efficiency has multiple public benefits beyond climate change (e.g., less energy use, reduced local air pollution, and lessened import dependence) even more ambitious increases in public energy R&D budgets for energy efficiency would be justified.

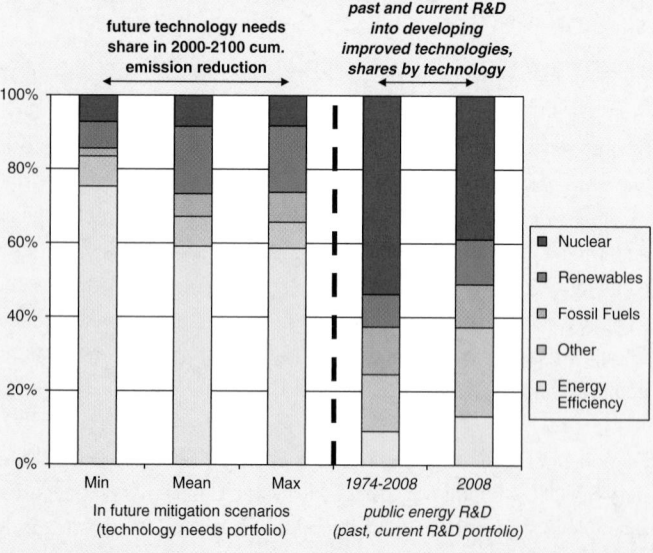

Figure 24.16 | Distribution of past (1974–2008) and current (2008) public sector energy technology R&D portfolios in member countries of the IEA (right) versus portfolios of future GHG mitigation needs (min/mean/max, left) derived from an extensive scenario uncertainty analysis. Source: adapted from Grubler and Riahi, 2010.

Table 24.10 | Private sector energy RD&D, selected OECD countries from WEC (2001) survey (in billion US$_{2001}$$^{[22]}$). Also, for the US latest available data for 2004 (NSF, 2009) are shown for comparison.

Billion US$	US	US	Japan	Korea	Sweden	France	Denmark	Spain	Total
in year	2000	2004	1997	1998	1997	1998	1993	1998	~2000
Efficiency			6.10						6.1
Fossil fuels	0.81	1.04	0.84	0.03		0.56			2.2
Nuclear	0.03	0.03	1.00			1.17			2.2
Renewables			0.29						?
Other or non-spec.	0.36	1.21		0.03		0.05			?
Total	1.20	2.28	8.6	0.06	0.1	1.78	0.08	0.08	11.9

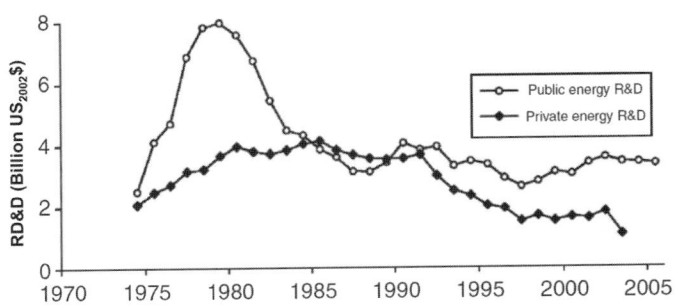

Figure 24.17 | Trends in US public and private energy RD&D (billion us$_{2002}$$^{[20]}$). Source: Nemet and Kammen, 2007.

There is evidence from the United States that private sector energy RD&D appears to follow comparable trends as public R&D (Figure 24.17), as both are influenced by rising and falling oil prices. One interpretation of these joint trends is "that the signal of commitment that a large government initiative sends to private investors outweighs any crowding-out effects associated with competition over funding or retention of scientists and engineers" (Nemet and Kammen, 2007). By analogy, the same influences also appear at work in periods of declining public R&D budgets. The available empirical evidence at present appears insufficient to support the often advanced argument of "crowding-out" effects, i.e., expanded public sector R&D would substitute (crowd-out) private sector R&D (e.g., Popp, 2006).

24.6.1.3   Total Energy RD&D

Based on the limited data available, the order of magnitude estimate of global energy RD&D amounts to some $50 billion (PPP) with some $15 billion in public sector RD&D and up to $35 billion by the private sector. About half of all energy RD&D is spent on fossil fuels and nuclear according to this assessment. The Sustainable Energy Finance Initiative's (SEFI) estimate (UNEP/SEFI/NEF, 2009[21]) of global RD&D into sustainable

energy of some $12.4 billion for the year 2005 (including $6.8 billion private and $5.6 billion public sector RD&D) is insufficiently documented to allow a more in-depth comparison, but are likely to represent an optimistic estimate.

24.6.2   Market Formation Investments

Market-formation investments include public and private investments in the early stages of technological diffusion and are sometimes also referred to as "niche market" investments. In the energy domain, these investments include government subsidies for certain technologies (e.g., feed-in tariffs or production tax credits) and public procurement. They also include private investments that may take advantage of markets created by government policies, such as renewable performance standards or price instruments like carbon taxes.

Market-formation investments in the energy sector as a whole are difficult to track, because many transactions are unreported, ways of measuring market-formation investments are not yet harmonized internationally, and efforts to track such investments are only relatively recent.

24.6.2.1   Analysis of Market Formation Sustainable Energy Investments

- Market formation investments in sustainable energy (solar, wind, biofuels, biomass and waste-to-energy, marine and small-hydro, geothermal, efficiency, and other low-carbon technologies/services) can be measured by activity in three main asset classes: venture capital/private equity; new listings on public markets; and asset finance. Figure 24.18 shows the distribution across the three asset classes with total investments across all regions growing

---

20   To convert to US$_{2005}$ $ multiply by 1.09.

21   The Sustainable Energy Finance Initiative is convened by UNEP, with participation from Bloomberg New Energy Finance (NEF): www.sefi.unep.org/.

22   These rather outdated (but only available) data are presented here mainly for illustrative purposes. In order to avoid confusion, currencies have not been converted to the more recent GEA base year of 2005.

Box 24.8 | How large are the investments in sustainable energy?

It has been reported that over US$200 billion in investments have been made into sustainable energy,[23] and in 2008, this grew to US$223 billion (UNEP/SEFI/NEF, 2009). However, as a leading indicator for the growth of sustainable energy infrastructure, the total US$223 billion is misleading because it includes a number of transactions – a full 30% – that are purely financial, such as mergers and acquisitions (M&A), refinancings, and buyouts. These transactions represent changes in ownership rather than input investments into energy technologies.

Figure 24.18 shows a breakdown of the US$223 billion in investment transactions in 2008 for sustainable energy as tracked by UNEP/SEFI/NEF,[24] and maps them against the categories of RD&D and market formation used in this chapter. By doing so, a clearer picture of the scale of input investments into the sustainable energy technology system emerges. Technology-specific (as opposed to ownership changes represented by some US$67 billion for mergers and acquisitions) investments amounted to a total of some US$$_{2008}$160 billion. RD&D (public and private) made up US$18 billion, with the bulk of the investments (US$141 billion) classified as market formation investments into sustainable energy technologies.

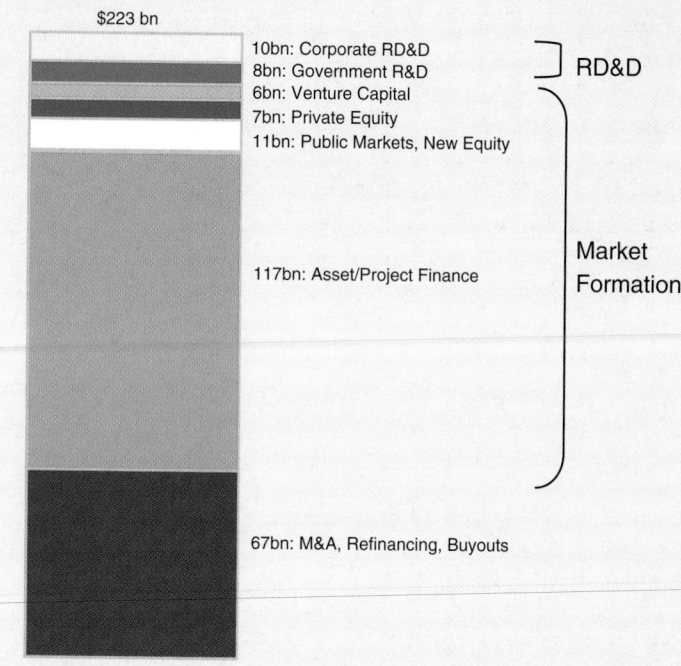

Figure 24.18 | Investments into sustainable energy in 2008. Data source: SEFI/UNEP/NEF, 2009.

---

at a compound annual growth rate (CAGR) of 63%/year between 2004 and 2008.

- The largest asset class within market formation investments is asset finance, which includes the building of new assets financed via either project finance or balance sheet/syndicated equity. While often investing in quite large and more mature technologies, asset finance investments in sustainable energy are counted here as part of "market formation" because they are highly dependent on governmental subsidies and incentives such as tax equity credits or feed-in tariffs. The amounts also include estimated investments for small-scale and residential installations of sustainable energy technologies such as biodigesters, micro-wind turbines, and solar hot water systems.

- Brazilian ethanol was excluded from market-formation investment totals, as these investments are no longer substantially supported

by government subsidies (see the *Brazilian Ethanol* case study). The net effect of doing so was to reduce the total amounts invested by US$17.2 billion for 2004–2008 and US$8 billion in 2008 across all asset classes. (These investments are included in the diffusion investment category in this assessment, summarized in Section 24.6.3 below.)

- The technology sector attracting the most investment for 2004–2008 is wind. Wind-related investments grew at an AGR of 51%/year in this period.

- Investments into energy efficiency are small (~2%).

- The regions that saw the most investment were OECD countries, notably Europe, at 45%, and North America, at 30%, of total investments for the 2004–2008 period.

---

23   "Sustainable energy," according to UNEP/SEFI/NEF, includes: solar, wind, biofuels, biomass and waste to energy, marine and small-hydro, geothermal, efficiency, and other low-carbon technologies/services. It excludes large-scale hydro (>50MW) and all nuclear power.

24   Figure 24.18 includes both project finance for large scale installations and small scale and residential installations as estimated by UNEP/SEFI/NEF.

24.6.2.2   Spotlight on Venture Capital and Private Equity (VC/PE) Energy Investments

While often characterized as investing in highly risky assets, VC/PE investors typically invest after a good deal of technology risk has already

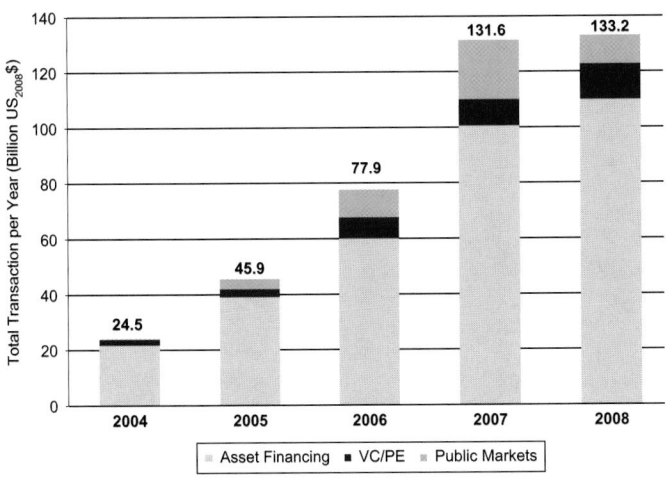

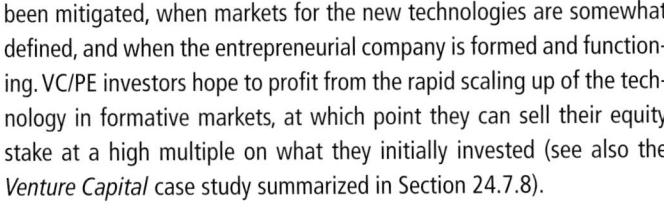

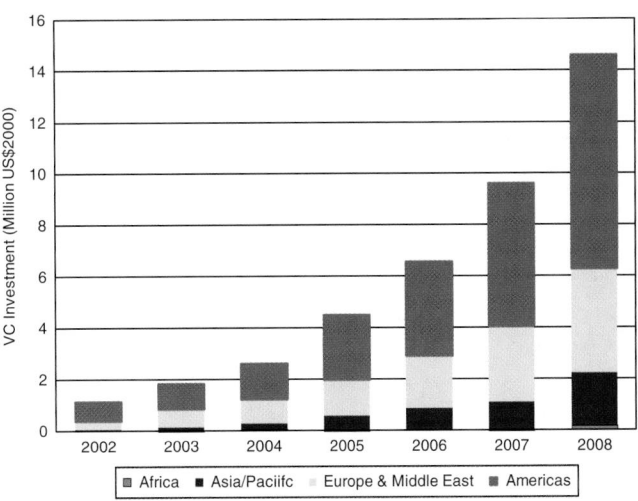

Figure 24.19 | Yearly investments in market-formation for sustainable energy technologies 2004–2008 (total transaction/year in billion US$_{2008}$). The total market formation investments in 2008 exclude US$8 billion in Brazilian ethanol (classified here as diffusion investments) from the total of US$141 billion reported in Figure 24.18. Source: O'Rourke, 2009; UNEP/SEFI/NEF, 2009; bloomberg new energy finance database (courtesy of ERD3 Project Harvard; NEF/SEFI, 2009).

Figure 24.20 | Total amounts of VC/PE in energy by year and region (in 1000 US$_{2008}$). Sources: O'Rourke, 2009; Bloomberg New Energy Finance database (courtesy of ERD3 project harvard; NEF/SEFI, 2009); Thomson Reuters Venturexpert Database (Thomson Financial, 2009).

been mitigated, when markets for the new technologies are somewhat defined, and when the entrepreneurial company is formed and functioning. VC/PE investors hope to profit from the rapid scaling up of the technology in formative markets, at which point they can sell their equity stake at a high multiple on what they initially invested (see also the *Venture Capital* case study summarized in Section 24.7.8).

Compared to coal, oil and gas, and nuclear energy, sustainable energy technologies are less mature technologies, and VC/PE capital and skill can help accelerate the transition from demonstration to market adoption.

The following investment amounts build upon the UNEP/SEFI/NEF data presented above by adding fossil-fuel technologies and installations with data gathered from the Thomson/Reuter's VentureXpert database (Thomson Financial, 2009). Total amounts and numbers of investments are likely to be higher than what is listed here because investments made by VC/PE funds into energy technology companies forming new markets are often not publicly reported, and investments made by angel (individual) investors are also not reliably reported, even if such investments serve as an important source of capital for both very new and later stage venture companies with energy technologies.

There has been a dramatic growth of investment by VC/PE investors into energy – and specifically into clean energy technologies – since the mid to late 2000s. Figure 24.20 illustrates the recent growth of VC investment into both fossil and non-fossil energy technologies following the regional disaggregation given in the original data source.

- Between 2002 and 2008, at least US$40.88 billion was invested by VC/PE investors into energy technology firms; in some 2,375 transactions.

- In 2008, the total amount of energy (fossil and non-fossil) investments made by professional VC/PE investors worldwide was US$14.6 billion. This grew from some $1.14 billion in 2002.

- The compound annual growth rate (CAGR) for the 2002–2008 period is 53%/year for the total amounts invested and 26%/year for the number of investment rounds ("deals") made.

- The bulk of the investment went into North American and European companies with non-fossil-based energy generation technologies. This uneven distribution reflects both the limited availability of data outside North America and Europe, and the limited availability of professional VC/PE in some regions of the world with other forms of financing more common such as family-firm or investments made by large corporations.

Figure 24.21 shows that the majority of investments made by VC/PE investors are in sustainable/renewable energy generation and end-use efficiency technologies over fossil fuels and power generation technologies. Specifically:

- solar energy-related technology companies have attracted the highest amounts of investments overall, representing some 30% of global energy VC for the whole period. Solar investments grew particularly rapidly between 2005–2008 in terms of numbers of deals and total amounts invested; and

- "end-use efficiency" (such as smart energy metering in buildings, demand response software systems, high efficiency lighting, etc.) also grew in the period, totaling US$5.5 billion for the period.

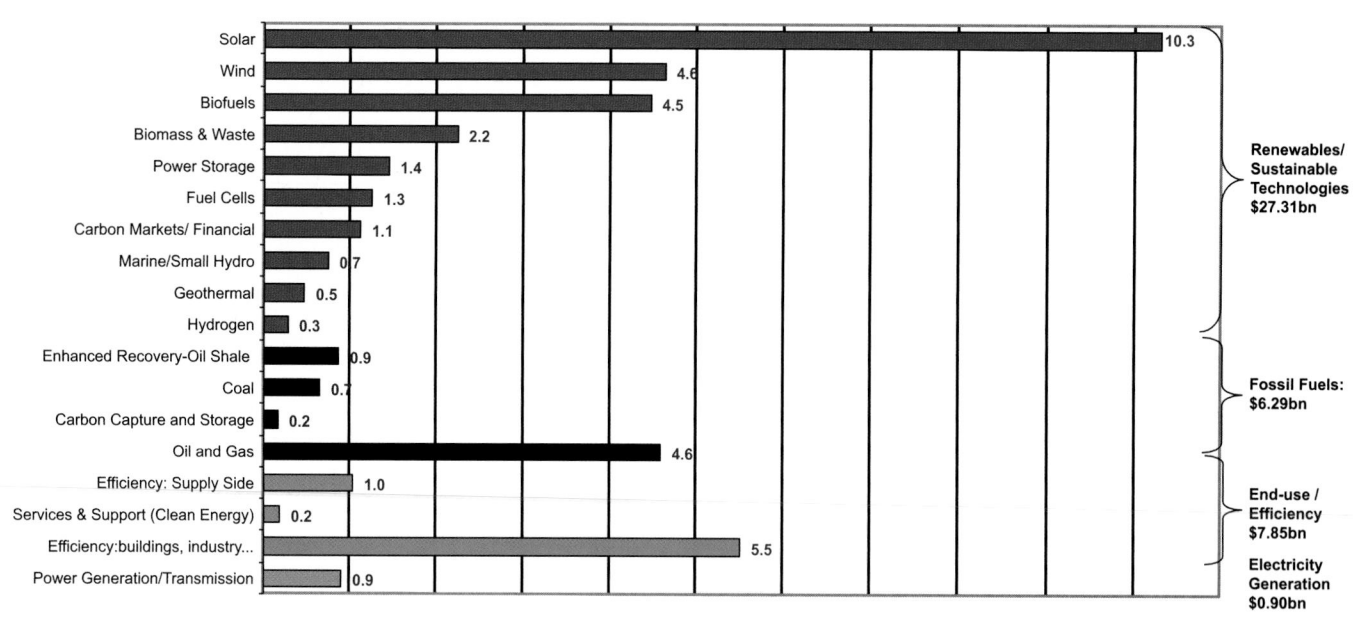

Figure 24.21 | VC/PE investments in energy technologies by technology cluster between 2002–2008 in billion US₂₀₀₈$. ("Biofuels" category excludes investments made into Brazilian ethanol companies, accounted for diffusion investments in Section 24.6.3 below.) Source: O'Rourke, 2009; Bloomberg New Energy Finance database (courtesy of ERD3 Project Harvard; NEF/SEFI, 2009); Thomson Reuters Venturexpert Data Base (Thomson Financial, 2009).

24.6.3 Diffusion Investments

Energy sector diffusion investment data are sparse and not collected systematically nationally or internationally. Modeling studies, as well as limited survey data, allow estimates of energy-supply investment levels, but energy end-use investment data are almost entirely lacking. Instead of concluding with the simple statement of data unavailability, this assessment explores some of the reasons for the lack of data and aims to provide at least some plausible estimates of orders of magnitude to provoke discussion and subsequent research in this extremely under-researched and under-reported area.

24.6.3.1 Energy Supply Investments

Data on energy supply investments are extremely limited, so the literature typically relies on model estimates (multiplying statistical data and/or estimates on capacity expansion with average technology-specific investment costs to derive total energy supply investments) or limited surveys. Energy supply modeling studies have become available since the mid-1990s in academia (e.g., Nakicenovic and Rogner, 1996; Nakićenović et al., 1998; Riahi et al., 2007), as well as from the work of the IEA, particularly the World Energy Investment Outlook (IEA, 2003); the Energy Technology Perspectives (IEA, 2006; 2008b); and the recurrent projections of IEA's World Energy Outlook (e.g., IEA, 2006; 2007; 2008a; 2009a), which also contain unique survey data on energy supply investments, particularly in the oil and gas industry. A common feature (and drawback) of all modeling studies is that energy sector

investments are not reported for their corresponding base year values, but instead as cumulative totals of the projection horizon of typically 30 years. The absence of published base year input data for energy sector investment projections not only reduces the credibility of the modeling studies, but also makes an assessment of current investment levels and structure and a comparison among the different studies an almost impossible task.[25] In the assessment below, we summarize available information by drawing on the only modeling study that has disclosed its underlying base year energy investment numbers (Riahi et al., 2007)[26] and the surveys reported in IEA's WEO (IEA, 2006; 2008a; 2009a). Because of the significant price escalation observed for energy sector investments (particularly for oil and gas since 2004), the Riahi et al. (2007) estimate (that refers to year 2000 investments and price levels) can be considered a lower bound, assuming recent price escalations will not remain permanent. Conversely, the IEA numbers can be considered as an upper-bound estimate of investments in energy supply (see Table 24.11).

Despite differences in estimated supply-side investments per category, the available data suggest a likely order of magnitude of energy-supply

---

25 Therefore wherever possible, underlying investment numbers of modeling studies should be made publicly available.

26 Numbers have been published in an interactive web-based database. Base year data refer to capacity additions and price levels for the year 2000 but were expressed in US₁₉₉₀$. These were converted to the GEA standard of US₂₀₀₅$ using the US GDP deflator multiplier of 1.4. However, despite being expressed in US₂₀₀₅$, price levels remain that of the year 2000, as energy sector-specific price deflators are not available internationally.

Table 24.11 | Range of energy supply investments in Billion US$_{2005}$$. (T&D: transport and distribution of electricity).

		LOW[1]	HIGH[2]
		2000 prices & activity	2005–07 prices & activity
FUELS			
	UPSTREAM:		
	Exploration fossil fuels	n.a.	40
	Extraction fossil fuels	180	180–360
	DOWNSTREAM†:	n.a.	100–140
	Synfuels, fossil	1	7
	Biofuels	20	n.a.
	Other	20	n.a.
TOTAL FUELS		>220	300–550**
POWER			
	Electricity generation:		
	Fossil	110	n.a.
	Non-fossil	100	n.a.
	Total	210	220–300
	T&D	>>70	?-230-?
TOTAL POWER		>500	450–520*
TOTAL SUPPLY INVESTMENTS		>720	750–840*

* Total minima/maxima ranges are not additive from (sub-)component min/max ranges.

** Mimima excludes exploration while maxima includes exploration.

† Downstream includes refining, pipelines etc.

1 Riahi et al., 2007.

2 IEA, 2006; 2008a; 2009a.

side investment of some US$_{2005}$$700 billion/year that could extend to some US$840 billion in 2007/2008, considering the higher ranges reported in the literature. Investments are dominated by electricity generation and transport and distribution (T&D), with some US$_{2005}$$500 billion. Fossil fuel supply, particularly the "upstream" component (i.e., exploration and production), accounts for US$250[27]-400 billion, mostly for oil and gas.

Renewables that figured prominently in market formation investments discussed above are minor players under the market conditions characterizing current diffusion investments. Liquid and gaseous biofuels account for US$20 billion, including US$8 billion for Brazilian ethanol (UNEP/SEFI/NEF, 2009). Large-scale hydropower (<US$100 billion for annual capacity additions of between 25–30 GW) make up a maximum of 17% of current supply-side investments.

Major uncertainties include the accounting for oil and gas exploration activities (at some US$40 billion) that are, strictly speaking, not energy technology investments. When categorized as RD&D activity for future oil/gas reserves – as is the practice by some companies – oil and gas exploration would represent the single largest RD&D spending in the energy technology field. Major differences also exist for electricity transport and distribution infrastructure investments for which only modeling study data are available and estimates differ by about a factor of three. The IEA WEO 2008 projection of average annual electricity T&D infrastructure investments of US$230 billion over the period 2007–2015 appears extremely high, and is comparable to the corresponding electricity generation capacity expansion investments. Lastly, it is interesting to note that no studies available report actual data for current investments in nuclear energy (even though nuclear figures prominently in future projections). According to IEA (2002; 2009a), installed nuclear capacity expanded by 20 GW between 1999 and 2007. IEA (2008a) reports an increase from 358 to 376 GW between 2000 and 2006, which yields an average annual net increase in nuclear capacity of between 2–3 GW, mostly in Asia where investment costs are comparatively modest at an estimated 1500–2500 $/kW (see Chapter 14). This suggests current investments of between US$3–7.5 billion/year for nuclear

27 Taking the Riahi et al. (2007) estimate of US$220 billion, complemented by not reported investment categories, the estimated grand total includes US$230 billion for fossil fuels and US$20 billion for biofuels.

reactors, which makes this the only technology in which RD&D investments *exceed* diffusion investments. (Given its technological maturity of over 40 years since its first market introduction, nuclear can reasonably not be considered a technology in its market formation stage, where such an investment pattern would be both possible and plausible.)

The assessment of the nuclear industry in terms of technological and investment risks by markets, as reflected in actual technology investments, departs markedly from the overemphasis of nuclear in public RD&D portfolios. This misalignment suggests two critical questions for technology policy. First, is the public sector energy RD&D confined to investments in innovations that ultimately find little market appeal? Alternatively, given the heavy emphasis of public RD&D on nuclear, is the public sector providing sufficiently consistent market deployment incentives so the heavily subsidized technology finds market applications?

Evidence regarding the time trend of supply-side energy investments is scarce in the literature. An intriguing empirical finding from the United States, however, shows a significant decline in energy supply-side investments as a share of sector revenues for electricity generation in the second half of the twentieth century (Figure 24.22). The declining investments (as a share of revenues) in the US electricity sector suggest a substantial thinning of resources available for capital turnover and diffusion of new technologies as a twin result of slowing demand growth and energy sector deregulation and liberalization. At present, it remains unclear if this trend is a specific phenomenon of OECD countries or of US electricity supply (an increasingly deregulated sector). However the example supports the conclusion that better current and longitudinal data on energy sector investments are needed for improved decision-making.

This assessment of diffusion investments has focused on the global level for the simple reason that regionally disaggregated investment survey data are lacking. Modeling studies suggest that current (year 2000) energy supply-side investments are distributed about 60:40 between Annex I and non-Annex I countries, as defined by the United Nations Framework Convention on Climate Change (UNFCCC, 1992). Short-term projections (e.g., to 2030 by IEA, 2009a) suggest roughly a 50:50 split between energy supply investment needs between Annex I and non-Annex I countries, for a global total of cumulative energy supply investments 2008–2030 of some US$_{2008}$25 trillion.

24.6.3.2 Energy End-use Investments

The decentralized nature of these investments by private households (and their corresponding classification as consumer expenditures rather than investments) and by firms (whose energy-specific investments go unrecorded) explains the absence of energy end-use investment numbers in the literature. The small-scale nature and formidable definitional

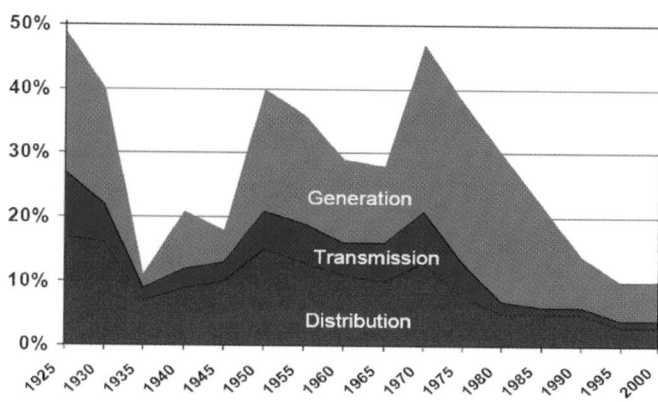

Figure 24.22 | Declining investments (as share of revenues) in the US electricity sector, 1925–2000. Source: modified from EPRI, 2003.

challenges[28] of these numbers also contributes to their absence. This lack of data, even model estimates,[29] introduces a serious challenge in both energy modeling and policy, because the potentially largest source of energy demand (and emissions) reduction is either entirely ignored or assumed to cost nothing. Customary energy and climate policy models deal with energy end-use costs by either "assuming away" missing data by exogenous (and policy independent) autonomous energy efficiency trends or by considering investment costs for the incremental component of energy end-use investments related to improved energy efficiency, which in itself provides a formidable definitional and data challenge.

To address this gap, Chapter 24 presents the first global, bottom-up estimate of total investment costs in energy end-use technologies. Volume data (production, delivery, sales, and installations) and cost estimates to approximate total investment costs in 2005 are estimated in both end-use technologies and their specific energy-using components.[30] Low and high sensitivities around central estimates are included, taking into account uncertainties in both volume and cost assumptions. The intention is to provide a first order, educated guess point of comparison with supply-side investments. Supporting data and a discussion text are posted on the GEA Chapter 24 website[31]

---

28 For instance, it is far from trivial to discern the energy component in the total investments of a new building. Depending on where the systems boundary is drawn, one could look at the heating and air conditioning system, including that part of the building structure that determines its energy use (insulation, windows). Indeed, the entire building structure may be considered.

29 Some studies include incremental energy end-use technology investments associated with additional energy efficiency gains above a typical "business as usual" scenario (e.g., IEA, 2009a). Apart from introducing additional definitional ambiguities (i.e., what constitutes incremental investments), the modeling is usually only done for a few technologies (e.g., transport), which limits its usefulness to inform policy.

30 Available data do not allow a further disaggregation into those subcomponents of investments on energy efficiency improvements, which remains an important future research task.

31 See www.globalenergyassessment.org.

to document the assumptions underlying the estimates below, solicit feedback and comments, and invite further research in this critical area (See also Wilson and Grubler, 2011).

To ensure comparability between supply-side and demand-side investments, a common definition of the unit of analysis is needed. Supply-side investments are quantified at the level of the power plant, refinery, or liquefied natural gas terminal. These are complex, integrated technological systems with energy conversion technologies at their core. These energy-converting components are configured within their corresponding technological system to provide a traded energy carrier to intermediate users (utilities, fuel distributors, pipeline, or shipping companies).

The logical demand-side analogues of these technological systems are the aircraft, vehicle, refrigerator, and home heating system. Although generally less complex, each of these technological systems similarly has an energy conversion technology at their core (i.e., the jet engine, internal combustion engine, compressor, boiler). In addition, each is configured to provide a useful service to final users.

With demand-side technologies, however, this definition of the unit of analysis is problematic. Investments in (and performance of) end-use technologies are dependent on investments in associated infrastructure such as airports, roads, and buildings. Is it meaningful to quantify the investment cost of a home heating system without quantifying the investment cost of a home and the insulation level that determine the dimensioning of the home heating system in the first place? Is the end-use technology to consider a boiler or a building?

Although the same issue exists on the supply-side, it is largely addressed by additionally quantifying investment costs in associated transmission and distribution infrastructures in policy models, as comprehensive statistics are also lacking on the supply side. The problem on the demand side is that the same approach would result in a sum of the total investment costs in all building structures,

roads, railways, ports, airports, industrial machinery, equipment, and appliances. Such an exercise would amount to a *reductio ad absurdum*.

A pragmatic pathway out of this system boundary ambiguity is to provide a range of estimates for a range of system boundaries of energy end-use technologies. An initial broader definition and data set describes end-use technologies as the smallest (or cheapest) discrete purchasable units by final consumers. This implies boilers and air conditioning units not houses, and dish washers and ovens not kitchens. A second, narrower definition and data set describes the specific energy-using components of these end-use technologies. This implies engines in cars, and light bulbs in lighting systems. Table 24.12 summarizes these distinctions for the technologies analyzed. In some cases (e.g., industrial motors, mobile heating appliances), a distinct energy-using component was not identified.

The investments in 2005 in end-use technologies are estimated to be on the order of US$1–3.5 trillion; the estimate in 2005 in the energy-using components of these end-use technologies is on the order of US$0.1–0.7 trillion. The breakdowns of these totals by technology are given in Table 24.13 and Table 24.14.

It should be emphasized that these investment cost ranges are rather *underestimates*, as many end-use technologies are omitted from the analysis. Although the principal end-use technologies in terms of the costs of their energy-using components (not the technologies themselves) are captured, investment costs in many technologies cannot be quantified. These include all propeller-based and noncommercial aircraft; helicopters; all military technologies; mass transit systems (whose costs are extremely site specific); heating and cooling systems in commercial and institutional buildings (new build and retrofits); water heaters; information and communication technologies; small appliances; other consumer electronics; and all industrial equipment and processes other than motors (e.g., blast furnaces, pulp mills, cement kilns). With the exception of industrial plants, the inclusion

Table 24.12 | Summary of technologies and components included in estimates of energy end-use technology investments.

End-Use Service	End-Use Technology	Energy-Using Component
mobility	commercial jet **aircraft**	jet **engine**
mobility	**vehicles** (cars and commercial)	internal combustion **engine**
space conditioning	**central heating systems** (boiler/furnace, ducts/pipes, radiators, controls, energy supply infrastructure network connections for new systems)	**boiler** or furnace
space conditioning	**air conditioning systems** (AC unit, ducts, controls, energy supply infrastructure network connections for new systems)	air conditioning **unit**
space conditioning	**mobile heating appliances** (e.g., portable convection / fan heaters)	same definition as technology
lighting	**lighting** (light bulb and fixture)	light **bulb**
food storage and cooking	large **household appliances** (fridges, freezers, clothes washers and dryers, dish washers, cookers)	compressors, motors, fans, heating elements
various (e.g., processing)	industrial motors	same definition as technology

Table 24.13 | Estimated investment costs in selected end-use technologies (in billion US$_{2005}$).

End-Use Technologies in 2005	low sensitivity	central estimate	high sensitivity
	in billion US$_{2005}$		
Commercial jet aircraft	12	28	50
Cars	540	758	1194
Commercial vehicles	270	427	672
Buildings (retrofits) – central heating systems	47	250	979
Buildings (new) – central heating systems	33	93	248
Mobile heating systems	2	4	5
Buildings (retrofit) – air conditioning systems	9	42	137
Buildings (new) – air conditioning systems	7	20	41
Lighting	17	38	83
Large household appliances	45	75	124
Industrial motors	2	6	16
GRAND TOTAL COSTS	**984**	**1741**	**3549**

Table 24.14 | Estimated investment costs in "energy-using components" of selected end-use technologies (in billion US$_{2005}$).

Energy-Using Components of End-Use Technologies in 2005	low sensitivity	central estimate	high sensitivity
	in billion US$_{2005}$		
Commercial jet aircraft	3	7	13
Cars	36	76	159
Commercial vehicles	27	57	119
Buildings (retrofits) – central heating systems	13	52	158
Buildings (new) – central heating systems	9	20	41
Mobile heating systems	2	4	5
Buildings (retrofit) – air conditioning systems	5	21	69
Buildings (new) – air conditioning systems	4	10	20
Lighting	12	27	59
Large household appliances	11	18	53
Industrial motors	2	6	16
GRAND TOTAL COSTS	**124**	**298**	**712**

of these categories should not substantially increase the investment cost range for energy-using components, as suggested by back-of-the-envelope sensitivity analyses; however, they would substantially increase the investment cost range for end-use technologies in their broader definition.

Given the definitional problems described above, the appropriate point of comparison for estimates of supply-side investment costs is a range spanning the narrow category of "energy-using components" at the lower end, to the broader category of "end-use technologies" at the upper end. Taking also into account the extent of end-use technologies missing from this analysis, *the range of demand-side investment costs is conservatively in the order of US$0.3–4.0 trillion.*

This compares with supply-side investment costs on the order of US$0.7 trillion/year.

Although the two ranges span the same orders of magnitude, the upper bound of demand-side investment costs is four times higher than its supply-side equivalent, recalling also that this is likely a (potentially substantial) underestimate. Interestingly, this result aligns with the IEA's estimation that demand-side investment needs exceed supply-side investment needs by a factor of 4 to 5 in the IEA climate policy scenarios (IEA, 2008b). Disaggregating the data by region shows that approximately two-thirds of the end-use investments in 2005 are in Annex I countries; the remaining one-third are in developing economies.

24.7 Appendix II: Summaries of Case Studies of Energy Technology Innovation

24.7.1 Grand Designs: Historical Patterns and Future Scenarios of Energy Technological Change[32]

The case study reviews patterns, drivers, and typical dynamics (rates of change) in energy systems from a historical as well as futures (scenario) perspective. From a historical perspective, two major energy transitions, each of which took up to a century to unfold, can be identified: the phase of growth in coal-fired steam power, and its subsequent displacement by oil and electricity-related end-uses and technologies (Figure 24.23). Similar far-reaching future transitions are also described in the scenario literature as a function of alternative assumptions on rates and direction of inventive activities and performance and cost improvements of new energy technologies.

Summary Points

- Technological and social innovations have been core drivers of historical energy transitions and remain so in future scenarios.

- Energy history and future are characterized by four "grand" patterns of technological change in energy systems:

  - clustering and spillover effects dominate over singular technologies;

  - performance dominates over costs in the early phases of technology development;

  - end-use applications dominate over energy supply; and

  - time constants of change are substantial, spanning from many decades to a century.

- There is evidence that, contrary to the historical evidence and popular conception, rates of systems transitions have substantially slowed since the 1970s.

24.7.2 Historical Scaling Dynamics of Energy Technologies[33]

The twentieth century has witnessed extensive diffusion of many supply-side and end-use energy technologies as part of a wholesale transformation of the energy system. Entire industries have grown, but so too have the size of technologies at the "unit" level (e.g.,

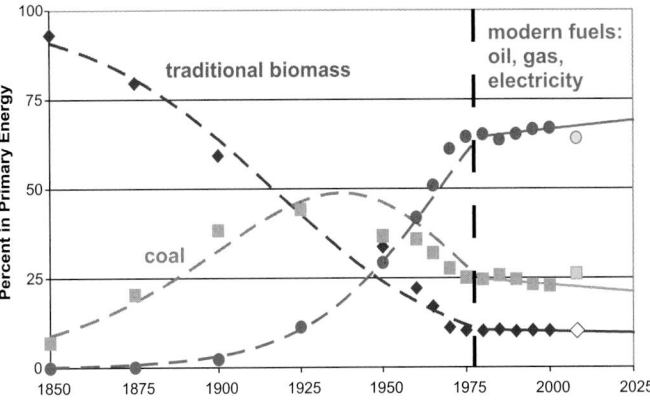

Figure 24.23 | Two "grand" transitions global energy systems measuring market shares in total primary energy use, 1850–2007 (preliminary data) of biofuels, coal, and modern energy carriers (oil, gas, electricity). Note in particular the long period of initial slow market penetration of new technologies and the significant slow-down of historical technology dynamics after 1975.

the rated capacity of a steam turbine or a car engine). Analyzing these historical growth dynamics at both the industry and unit level reveals some general patterns that appear robust across very different energy technologies. First, increases in unit size generally follow a period of experimentation with many smaller-scale units. This is particularly the case for technologies like nuclear or wind power with clear economies of scale at the unit level. Second, the extent to which an energy technology industry grows is consistently positively related to the time duration of that growth. These and other findings have important implications for policy, not least in striking a cautionary note on pushing for significant jumps in technology unit size before a formative phase of experimentation and learning with smaller-scale units has been completed.

Summary Points

- Growth of energy technology industries comprises a *formative* phase, then a *scaling* phase that precedes or is concurrent with an *industry growth* phase.

- The formative phase involves many smaller-scale, granular units with only small increases in unit size. The scaling phase sees large increases in unit sizes, particularly at the scale frontier, and a large increase in numbers of units. The industry growth phase is driven by large numbers of units at larger unit sizes and describes the maturing industry.

- Experimentation with many smaller-scale units tends to precede substantive increases in unit size. The formative phase following an energy technology's introduction into the market is an often lengthy process of testing and experimentation with many small-scale units, which allows technologies to be "debugged" through a process of "designing-by-experience" (Ruttan, 2001). Successful experimentation, improvements, and learning appear to be critically determined

---

32 Arnulf Grubler.

33 Charlie Wilson.

by the granularity and the associated smaller financial and innovation failure risks of smaller-scale unit projects. Resulting learning effects lead to cost and performance improvements, but also facilitate the subsequent capture of unit scale economies as the industry matures.

- The relationship between the extent and the time duration of an industry's growth is consistent for different energy technologies. It is intuitively obvious that the extent of growth should correlate positively with the time period over which that growth occurs (not-withstanding the many factors that affect diffusion rates). However, the consistency of that relationship for very different energy technologies is surprising. The case study also identifies an important learning externality for later adopting regions that achieve generally faster diffusion compared to leading innovation centers. Compared to historical dynamics, future scenarios reviewed were generally found to be conservative in their technology scaling and market expansion assumptions.

24.7.3   Technology Portfolios[34]

The case study reviews a range of methodologies and model-based applications that can assist policy decisions under (inevitable) technology innovation uncertainty.

One possible approach is based on scenario analysis. As an example of that method, a detailed scenario exercise performed at IIASA explores uncertainties in main scenario drivers of future greenhouse gas (GHG) emissions (energy demand, resource and technology availability, etc.) and extent of future climate constraints (represented through a range of stabilization targets). The study identifies energy efficiency and conservation as the single most important, and also the most robust, technology option across all scenarios. This result is in stark contrast to past and present public sector R&D portfolios, which continue to be dominated by nuclear R&D at the expense of energy efficiency.

A second analytical approach is based on a portfolio theory that helps capture the benefits from (technology) portfolio diversification in the framework of risk-averse decision-making. An example modeling study suggests that risk aversion leads to higher adoption rates of currently higher-cost energy technology options such as modern biomass and renewables, but also CCS. The modeling study also suggests higher short-to-medium term investments into advanced technologies under risk aversion. It is also possible to not only consider variance as a risk measure, but take into account the risk of high impact tail events. For example, a risk premium of only about 1% of total energy expenditures was found to decrease the value of the 99th percentile extreme event by more than a factor of two. Diversification thus not only reduces the

mean of risk exposure but drastically lowers the tails of extreme and undesirable outcomes.

Portfolio theory and scenario analysis can also be used in combination, where scenario analysis provides the basis for describing the uncertainty space and portfolio-based approaches then help to identify optimal risk hedging strategies and resulting technology portfolios.

Summary Points

- Formal tools, e.g., scenario analysis and portfolio theory, are increasingly available to move technology policy decisions (e.g., R&D, or early niche market investments) onto a more rational ground.

- A generic pattern from modeling studies that applies these methods to the field of energy technologies involves *portfolio diversification* and *enhanced experimentation* with earlier niche market investments, the extent of which depends on the (user specified) degree of risk aversion in addition to the underlying innovation uncertainty distributions.

- A comparison of technology portfolio scenario studies with past and current energy R&D portfolios reveals that the latter are highly biased, with energy efficiency/conservation underrepresented and nuclear R&D overrepresented in comparison to their respective option values in a climate-constrained world.

24.7.4   Knowledge Depreciation[35]

The case study first reviews the sources of knowledge depreciation that consist of knowledge lost (e.g., due to staff turnover), as well as knowledge made obsolete (e.g., due to rapid innovation). The literature of typical knowledge depreciation rates is reviewed and the limited examples related to energy technologies are discussed in more detail. Illustrative calculations show the implications of knowledge depreciation for two groups of energy technology innovations: nuclear power and energy efficiency based on public energy R&D statistics of IEA member countries (Table 24.15).

Summary Points

- Knowledge depreciation rates – characterized by high staff turnover – can be substantial, reaching 100%/year in service industries. In the energy technology field, knowledge depreciation rates range from 10%/year (wind turbines) to 30%/year (solar PVs) due to technological innovation-induced knowledge obsolescence.

---

34   Sabine Fuss and Arnulf Grubler.

35   Arnulf Grubler and Gregory Nemet.

Table 24.15 | Current (2007) and cumulative (1974–2007) R&D expenditures of IEA member countries in US$_{2007}$ billion and estimate of remaining knowledge stock, assuming an average knowledge depreciation rate of 20%/year.

	current R&D		cumulative R&D		remaining knowledge capital stock	
	10^9 US$_{2007}$\$	%	10^9 US$_{2007}$\$	%	10^9 US$_{2007}$\$	as % of cum. R&D
Energy efficiency	1.6	13.1%	38	8.8%	7	18.4%
Fossil fuels	1.4	11.5%	55	12.8%	6	10.9%
Renewables	1.5	12.3%	37	8.6%	6	16.2%
Nuclear fission	3.7	30.3%	194	45.1%	18	9.3%
Nuclear fusion	0.9	7.4%	42	9.8%	4	9.5%
Others	3.1	25.4%	64	14.9%	13	20.3%
Total	12.2	100.0%	430	100.0%	54	12.6%

Source: IEA, 2008b and authors' calculation.

- With knowledge depreciation, continuous knowledge recharge through sustained and stable R&D and niche market deployment efforts becomes critical.

- The *pathway* (stable, gradually rising) of policy support is as – if not more – important than the *absolute level* of policy support when characterized by "boom and bust" innovation investment cycles.

- Erratic policy support leads to substantially higher knowledge depreciation, even under otherwise high support levels. Estimates of the energy technology R&D knowledge stock for IEA member countries suggest that nuclear has suffered substantially more from knowledge depreciation than energy efficiency, which is characterized by much lower but more stable R&D expenditures.

Energy technology innovation is characterized by a mixed, staged involvement of both public and private resources, and an interdependent set of time-lagged processes linking innovation inputs to outputs. In other fields, econometric techniques are commonly used to evaluate outcome metrics such as net social returns on R&D. However, their application to date in the energy field has been limited due to data constraints, as well as the lack of consensus on objectives. Unanswered questions include: does successful energy technology innovation reduce energy or carbon intensity? Reduce unit costs? Reduce the cost of final service provision? Improve security, reliability, and flexibility? Increase option value?

Cross-country comparisons using both quantitative and qualitative approaches are complicated by the absence of standardized data, local specificities in energy technology innovation systems, and confounding factors such as labor cost differentials, intellectual property rights systems, and economic structure.

24.7.5   Metrics for Assessing Energy Technology Innovation[36]

Assessing the performance of energy technology innovation processes or systems is complex. There are different assessment methods or approaches – both qualitative and quantitative – as well as many different assessment metrics, which are reviewed in more detail in the case study. These metrics are proxies for innovation inputs, outputs, and outcomes that are either intangible (e.g., knowledge stock, practical problems and solutions) or tangible (e.g., scientists, laboratories, installed technologies). *Input* metrics describe financial and labor inputs to the innovation process (e.g., R&D investments). *Output* metrics describe defined products of the innovation process (e.g., cost reductions). *Outcome* metrics describe broader energy sector or economy-wide impacts of the successful diffusion of innovations into the marketplace (e.g., reduction in emissions intensity).

Summary Points

- Assessments of energy technology innovation can be qualitative, quantitative, or a combination of both. Qualitative assessments often complement the analysis of time series data by adding analytical rigor and depth.

- Assessments typically center on a particular technology within a national context, and/or focus on the effectiveness of innovation policies.

- Metrics of energy technology innovation relate to either *inputs, outputs,* or broader *outcomes* of the innovation process. With the partial exception of technological learning rates, there are no metrics that comprehensively link inputs to outputs and outcomes.

- The lack of available and reliable data on innovation inputs and outputs hampers standardized and cross-comparable assessments of energy technology innovation systems across technologies and countries.

---

36   Charlie Wilson.

Table 24.16 | Energy Technology R&D in China in 2004 (in million Yuan) by institutional actor, performer, and funding source (> symbols indicate incomplete reporting). See also Table 24.17.

R&D performed by	funded by	basic research	end-use & efficiency	fossil fuels	electric power	T&D	others & unspec.	Total
				Technology Areas				
Nat'l Program Energy Research	Government	1274					785	2059
Institutions of Higher Education	Government				744		187	931
(Public) R&D Institutions	Government			21	199			220
	Enterprises			78	127			205
	Total			99	326			425
Subtotal (R&D performed by public sector)		1274		99	1070		972	3415
R&D by Industry	Government		78	878	146	83		1185
	Enterprises		>>1873	14382	5092	2753		>>22227
	Total		>>1951	15260	5238	2836		23334
Grand Total		2548	>>1951	15359	6308	2836	972	>28700
	incl. Gov't funded	2548	78	899	1089	83	972	5669

24.7.6   China: Energy Technology Innovation Landscape[37]

The case study provides a first-time comprehensive (even if in some aspects, incomplete) overview of the entirety of the energy technology innovation landscape in China. Major programs, institutional actors, funding sources, and R&D allocation by broad technology groupings are outlined and draw on data for the year 2004 (Table 24.16). Mechanisms for setting innovation and R&D priorities as well as strategic and priority areas for China's ETIS are outlined.

Summary Points

- Energy technology R&D in China is both substantial and expanding rapidly. The survey is partially incomplete, as it does not cover R&D in the automotive and other end-use technologies (e.g., appliances) industries. Nevertheless, the survey indicates a total resource mobilization of greater than RMB29 billion in 2004 for energy technology R&D. This translates to either $3.5 or $9.4 billion, based on either market or purchasing power exchange rates, respectively, and compares to a total public energy R&D budget of the United States of US$5.6 billion in the same year.

- Despite its unique feature as a largely centrally planned economy, energy technology R&D in China (not unlike in OECD economies) is dominated by industry, which performs 88% of R&D. Largely government owned enterprises also provide for some 85% of all energy technology R&D funding in China.

- The energy technology portfolio (again, not unlike in OECD countries) is dominated by supply-side options, even considering incomplete data available for end-use technologies such as the automotive sector. Within energy supply technologies, fossil fuel-related technologies account for more than 50% of all energy R&D, followed by electric power and T&D with more than 30% of all energy R&D.

24.7.7   Energy R&D in Emerging Economies (BRIMCS)[38]

This case study provides an overview of energy RD&D expenditures of six major emerging economies referred to as BRIMCS: Brazil, the Russian Federation, India, Mexico, China, and South Africa. For comparison purposes, corresponding US data are also included. The data summarized below synthesizes a wide array of sources that have to date not been compiled in a consistent fashion. RD&D expenditures are differentiated by broad technology group as well as by funding source based on the most recent published data available in each country (Table 24.17). Funding sources include (federal) government or 100% state-owned enterprises (SOE), as well as other sources such as local governments, partially-owned SOEs, private industry investments, or NGOs.

---

37 Kejun Jiang.

38 Ruud Kempener, Laura Diaz Anadon, and Kelly Sims Gallagher.

Table 24.17 | Energy RD&D in BRIMCS countries (Million US$ at PPP).

in Million 2008 PPP $Int*	Fossil (incl. CCS)	Nuclear (incl. fusion)	Electricity, transmission, distribution & storage	Renewable energy sources	Energy Efficiency	Energy technologies (not specified)	TOTAL
United States – Gov't	659	770	319	699	525	1160	4132
United States – Other**	1162	34	no data	no data	no data	1350	2545
Brazil – Gov't	79	8	122	46	46	12	313
Brazil – Other	1167	no data	no data	no data	no data	1844	1351
Russia – Gov't	20	no data	22	14	25	45	126
Russia – Other	411	no data	no data	no data	no data	508	918
India – Gov't	106	965	35	57	no data	no data	1163
India – Other	694	no data	no data	no data	no data	no data	694
Mexico – Gov't	140	32	79	no data	no data	no data	252
Mexico – Other	0.1[1]	no data	no data	no data	263[3]	19[4]	282
China – Gov't	6755	12	no data	no data	136	4900	11803
China – Other	289	7	no data	no data	26	985	1307
South Africa – Gov't	no data	133	no data	no data	no data	9	142
South Africa – Other	164	31[2]	26	7	no data	no data	229
BRIMCS – Gov't	7100	1149	> 259	> 117	> 208	>4966	> 13799
BRIMCS – Other	2724	>> 38	>> 26	>> 7	> 289	> 1696	> 4781
BRIMCS – Grand Total	9824	>1187	> 285	> 497	> 497	> 6662	> 18580

* Data from United States, Brazil, Russia, India, China and South Africa are based on 2008, data from Mexico is from 2007.

** United States data on industry expenditure is from 2004 (NSF, 2009).

1 Based on PEMEX's fund for Scientific and Technological Research on Energy.

2 Based on total non-governmental investments into PBMR Ltd.

3 Based on 2005 R&D expenditure in car manufacturing industry (CONACYT, 2008).

4 Based on 2005 R&D expenditure in utilities sector (CONACYT, 2008).

> These cumulative values are based on data from only three to four BRIMCS countries, so actual expenditures are likely to be higher.

>> These cumulative values are based on data from two BRIMCS countries or less, so actual expenditures are expected to be much higher.

Source: Chapter 24 case studies and Kempener et al., 2010a.

Summary Points

- Public energy RD&D in BRIMCS countries is substantial and amounts to some $14 billion in PPP terms, slightly above the entirety of the public energy R&D budget of all IEA member countries combined (some $13 billion in PPP terms). Including non-governmental R&D funding sources, total R&D in BRIMCS countries is estimated to total nearly $19 billion.

- The significance of energy RD&D expenditure in BRIMCS countries challenges the traditional view that new energy technologies are predominantly developed within OECD countries and points to the need to include the BRIMCS countries in regular international statistical reporting and in a comprehensive global strategy to promote energy technology innovation.

- BRIMCS countries' energy RD&D data show that fossil fuel and nuclear energy receive the highest level of RD&D support, with renewables and energy efficiency highly underrepresented both in expenditures and statistical reporting.

24.7.8   Venture Capital in the Energy Industry[39]

Access to capital is a major enabler to the scale and speed of technological innovation as produced by entrepreneurs. Entrepreneurial firms face a multitude of risks and barriers. For those working with new technologies, technological, market, and financing risks are paramount. Venture capital (VC) is a source of capital that is willing to finance companies in this risky

---

39 Anastasia O'Rourke.

stage of the technology life cycle. By investing equity in such risky companies, VC investors are often considered "technology gatekeepers" who have helped to select and create "waves of technological innovation" that have transformed industries (Florida and Smith Jr., 1990).

The case study reviews recent trends in energy VC investments by category and region, drawing on and synthesizing a large body of literature and statistical information that has to date not been available in the public domain. There has been a dramatic growth of VC investment in clean energy technologies since the mid to late 2000s. Detailed statistical trends are presented in the VC case study.

Summary Points

- In 2008, the total amount of energy (fossil and non-fossil) investments made by professional VCs worldwide was US$15.5 billion.

- The compound annual growth rate (CAGR) for the period 2004–2008 is 22%/year of the number of investment rounds (deals) and 45%/year CAGR for the total amounts invested.

- The bulk of the investment went to North American and European companies with non-fossil-based energy generation technologies (solar; biofuels and biomass) and to storage technologies (particularly batteries). Significant investments were also made in end-use energy technologies such as smart energy metering in buildings, demand response software systems, or high-efficiency engines.

- While growth in VC investments has been dramatic, VC makes up only a comparatively small portion of all the capital employed to launch energy technologies into the market worldwide.

- Other niche-market investments in energy are needed as well. These include investments made at a very early stage by private individuals (angel investors); large company internal investments; investments in late-stage growth and private equity (primarily using debt instruments); project finance (also debt, often used to build larger-scale energy production facilities such as wind farms); and finally, investments in energy-technology firms that are listed on various public markets.

- The contribution of VCs to all private investments in energy combined sits at approximately 10%, according to data from New Energy Finance (IEA, 2009a; UNEP/SEFI/NEF, 2009).

24.7.9   Hybrid Cars[40]

This case study reviews the available literature on the development and deployment of hybrid-electric vehicles (HEVs), and particularly

---

40   Kelly Sims Gallagher.

examines the role of government in this history. Three country-specific case studies are provided for Japan, the United States, and China. Key factors in the development and deployment of HEVs are identified and discussed.

Some governments are interested in promoting HEVs and other alternatives to conventional internal combustion (IC) engines because of concerns about oil security, air pollution, and global climate change. HEVs achieve greater fuel efficiency than conventional IC vehicles, although the extent of improvement in efficiency depends greatly on the configuration of the specific HEV system.

The three case studies below show that the drivers of invention did not substantially differ among the three countries, but the policy mechanisms and incentives for deployment diverged significantly. In all three countries, the governments originally pushed harder for alternative automotive technologies other than hybrids, such as pure electric vehicles or hydrogen-fuel-cell vehicles. In some cases, private firms made R&D choices that do not appear to have been strongly influenced by public policy, other than to provoke the firms to explore fuel-efficient technologies. In other cases, government policies appear to actually have turned firms away from HEVs. Once HEVs emerged in the marketplace, however, the government response was completely different in the three countries, especially in terms of the extent to which each government was prepared to support their transition through the "early deployment" phase of innovation to facilitate widespread market diffusion.

Summary Points

- Policy for government investments in the RD&D of advanced-vehicle technologies was initially poorly coordinated, with policy for the early deployment of these technologies in all three countries. Japan and the United States reactively established policies to support the early deployment of HEVs once they were introduced to the market, but Japan implemented much more effective policies to support commercialization than the United States.

- Hybrid car consumers are clearly responsive to increased gas prices and other fiscal incentives, including sales tax reductions or exemptions and feebate schemes, and they also appear to buy hybrids out of concern for the environment or energy security.

- Leadership within firms appears to have been a major factor in explaining the relative success among the firms in developing and commercializing HEVs.

- Political and economic factors, most prominently the concern about energy security, were initially the main drivers for government technology policy and investments in advanced vehicle technologies in Japan, the United States, and China.

24.7.10   Solar Water Heaters[41]

The experience of innovation policy related to solar water heaters in the United States is a story of policy intermittency. A key general finding is that bad outcomes are often not easily forgotten, and can have substantial spillover effects on other technologies. This presents a challenge to the need to support experimentation and intelligent failures.

Solar water heaters (SWH) use a working fluid to absorb sunlight and provide heating and hot water in residential and commercial settings. The technology is currently cost effective, especially in large installations with high demand for hot water. Real-time electricity pricing is considered a potential boost to SWH. China is by far the world's largest market for solar water heaters. An important historical episode for this technology was the programs in the United States in the late-1970s and early 1980s. The biggest technical improvement was the advent in the 1970s of selective coatings, which would absorb more sunlight. Since then, the technology has been rather stable, with some improvements in lifetime and reliability. The key period of R&D investment was in the 1970s at national laboratories and universities. The subsidies in the 1980s were not monitored. There was rampant abuse of subsidies in the 1980s, and many installations leaked and caused extensive damage to structures far in excess of the cost of the water heater itself. The response was to avoid the technology for many years; the US industry went from US$1 billion/year in 1982 to US$30 million/year in the late 2000s. Hundreds of firms went out of business. As a result, much of the learning gained in the period of rapid deployment was lost. The perception of poor reliability persists and has proven difficult to overcome – bad news lasts. An exception to this general policy failure has been a program in Hawaii that makes consumer rebates contingent on an inspection that occurs one year after installation.

Summary Points

- A large, high-profile failure in the early stage of the US SWH industry has proven extremely difficult to overcome.

- For several years after this failure, the technology was not trusted.

- Experience was lost with the collapse of the industry.

- Verification is essential and not expensive.

- Inspection and verification have proven successful in Hawaii.

- Both R&D and incentives placed excessive focus on collector units rather than on system integration.

- Lack of involvement of utilities and builders appears to have hurt the industry.

24.7.11   Heat Pumps – Innovation and Diffusion Policies in Sweden and Switzerland[42]

Innovation and diffusion policies for the development and introduction of heat pumps provide an interesting case study on policy learning. Heat pumps have been supported by several countries since the 1970s as a strategy to improve energy efficiency, support energy security, reduce environmental degradation, and combat climate change. Sweden and Switzerland have been essential to the development and commercialization of heat pumps in Europe. In both counties, numerous policy incentives have lined the path of technology and market development. Early policy initiatives were poorly coordinated but supported technology development, entrepreneurial experimentation, knowledge development, the involvement of important actors, the formation of essential associations and organizations, and early market formation. The market collapse in the mid-1980s could have resulted in a total failure – but did not. The research programs continued in the 1980s, and a new set of stakeholders formed – both publicly and privately funded researchers, authorities, and institutions – and provided an important platform for further development. In the 1990s and 2000s, Sweden and Switzerland introduced more coordinated and strategic policy incentives for the development of heat pumps. The approaches were flexible and adjusted over time. The policy interventions in both counties supported essential learning, successful development and diffusion processes, and cost reductions of the heat pumps. The assessment of innovation and diffusion policies for the heat pump systems can be used to illustrate some general policy conclusions.

Summary Points

- The assessment shows the need for strategic, *long-term*, and *continuous* support. First attempts to introduce a new technology failed, and continuous support was needed to overcome initial shortcomings. Technological change takes considerable time.

- The combination of policy instruments may have to change and the government's approach should be *flexible*. The policy intervention may initially allow uncoordinated intervention to support entrepreneurial testing, but then should be developed into credible, stable, and transparent strategies that allow industry to make long-term investments.

- The policy interventions need to be *system-oriented* and consider both the development of the technology and its emerging market. In

---

41   Gregory Nemet.

42   Lena Neij, Bernadett Kiss, and Martin Jakob.

other words, R&D is important as a part of the policy strategy, but not enough.

- The assessment indicates a need for *testing and certification* processes to support technical quality and create credibility and legitimacy. R&D initiatives and subsidies require testing and certification to support a stable market development.

- The support of networking to improve strategic integration and the use of learning and to ensure feedback and spillover effects seems essential.

24.7.12   Role of Standards – The US CAFE Standard[43]

In 1975, the US government passed the Federal Automotive Fuel Efficiency Standards, which specified mandatory levels of miles per gallon of gasoline consumed for new vehicles averaged across each manufacturer's vehicle fleet. The required Corporate Average Fuel Efficiency (CAFE) standard increased from 18 miles per gallon (13 liters/100 km) in 1978 to 27.5 miles per gallon (8.6 liters/100 km) in 1985. The standard was effective in meeting its goals; actual fuel efficiency has never fallen below the government requirement. In fact, actual fuel efficiency has almost always exceeded the mandatory level. This over-compliance, combined with the collinear rise in the price of gasoline during the period of escalating standards, suggests that prices have played a role in motivating efforts to improve fuel efficiency, not the CAFE standard alone. The standard for passenger cars has remained the same since 1985, while actual efficiency has improved slightly. The standard will rise to 30 miles per gallon (7.8 liters/100 km) in 2011. Standards also exist in Europe, China, Japan, Australia, and Canada, and are well in excess of US requirements – by nearly 100% in the cases of Japan and the European Union.

The CAFE standard has affected the rate and direction of technological change in vehicles. End-use efficiency has improved almost continuously for the past 30 years. This rise in efficiency was used to accomplish different ends during the period of policy escalation (1975–1985) and after it (1986-present). In the first period, efficiency improvements were directed toward improving miles per gallon. After 1985, almost all of the efficiency improvements were used to increase other attributes, including acceleration, towing capacity, and vehicle size. Energy conversion efficiency has improved in drive trains, engines, drag, and rolling resistance. Drive train and engine energy conversion efficiency improved from 1975–1985; efficiency improved at the rate of 2–3%/year. After 1985, efficiency improvement slowed to about 1%/year, although that rate has increased since 2000. The continuity of this improvement makes it difficult to attribute to CAFE. However, consideration of what end-use characteristics these efficiency gains were used for is revealing; from

1975–1985, vehicle weight dropped by about one-third and acceleration remained the same. After 1985, when CAFE standards stopped rising, efficiency improvements were used to power increasingly heavier vehicles that could accelerate considerably faster. It needs to be noted that the CAFE standards did not apply to all road vehicles, excluding in particular light duty trucks, which incentivized a change in the composition of the road vehicle fleet towards pick-up trucks and, later on, Sport Utility Vehicles (SUVs) with much higher gasoline consumption CAFE regulated passenger cars.

Summary Points

- CAFE standards had a real effect on technological change.

- The improvement in miles per gallon was accomplished not only by a shift to lighter, less-powerful vehicles, but also by the adoption of new energy-efficient technologies.

- Attribution of these changes to the regulations, rather than to gasoline prices, is less clear since the two are so well correlated.

- Standards that apply only to parts of the technological artifacts in use risk behavioral responses from manufacturers and consumers that can go against the original intention of the efficiency regulation.

- The effectiveness of CAFE may actually have been important after 1985 when it served as a fuel economy floor in the face of persistently low gasoline prices.

24.7.13   Role of Standards – The Japanese Top Runner Program[44]

In 1998, Japan initiated a unique program – the Top Runner Approach – to improve the energy efficiency of end-use products and to develop "the world's best energy-efficient products." Under this program, the most energy-efficient product on the market during the standard-setting process sets the Top Runner Standard for all corresponding manufacturers and products.

The Program started with nine products: room air conditioners, fluorescent lighting, television sets, copying machines, computers, magnetic disk units, video cassette recorders, refrigerators, passenger vehicles, and freight vehicles. The scope was reviewed every two to three years and gradually expanded to include 21 products by 2009. It is now considered one of the major pillars of Japanese climate policy. The case study examines 12 years of the program's experience. It first reviews the structure of the Top Runner Approach and then illustrates its impacts. It

---

43 Gregory Nemet.

44 Osamu Kimura.

also discusses issues associated with the approach, and concludes with some implications.

Summary Points

- Dynamic, continuously adjusted standards have been successful in accelerating the trend of energy efficiency improvement in many end-use products, such as room air conditioners and passenger vehicles. In these cases, the standards provided a clear direction for product development by aiming at higher energy efficiency and eliminating low-efficiency products from the market.

- The case study illustrates that ambitious policies that match market conditions and technological conditions can work well to induce remarkable energy efficiency improvements. Because such conditions depend on the country and the phase of technological development, careful design and adjustment are required for effective policy making.

- Some preconditions may be necessary for success of the Top Runner Approach. One is the Japanese market structure, which is dominated by a limited number of domestic producers. Another precondition is the existence of cost-effective potentials for efficiency improvement. Last, the specifics of the Japanese systems of innovation, in which public and private sectors cooperate largely through informal networks in the dynamic standard setting and implementation process, may be a further precondition for success. When these conditions were met, the Top Runner Approach resulted in a substantial outcome in terms of efficiency gains.

- Although the achievement of the Top Runner Approach is remarkable, there remain some issues for policy consideration. The largest one is the lack of explicit consideration of potential impacts on consumers in the standard setting process. Because the approach is based on the Top Runner products on the market, price increases due to energy efficiency improvements are not explicitly considered. This might lead to product prices too high for consumers to achieve pay back within the lifetime of the product.

24.7.14 Comparative Assessment of Wind Turbine Innovation and Diffusion Policies[45]

Wind turbines have become a mainstream technology – a first choice for many when investing in new electricity generation facilities. This comparative case study addresses how governmental policy has been formulated and formed to support the wind turbine innovation process. Three innovation stages and corresponding innovation strategies are identified. First is the stage of early movers in the 1970s and early 1980s, covering pioneer countries such as Denmark, the United States, Germany, the

Netherlands, Great Britain, and Sweden. Second, the stage of booming markets in the 1990s, guided by the successful Danish innovation path of the 1980s, is described. Third is the stage of emerging markets in the 1990s and 2000s, including countries such as India, China, and Korea. Within these different periods, common key elements in governmental policy strategies can be identified as essential for a sustainable and successful innovation process.

Summary Points

- Support for diversity in technology and market formation is essential. The experience of wind turbine policy intervention shows the importance of applying a diversified technology portfolio. Moreover, the study illustrates the difficulties in foreseeing the drivers and trends of any given technology and the need to provide subsidies for implementation to many actors.

- RD&D is fundamental but not enough. In many countries, wind energy innovation was initially supported through RD&D only and the innovation process was envisioned to be linear. However, the RD&D funding alone did not bring about any commercial applications.

- To support technology innovation, quality assurance is essential. An important component of the innovation path of wind turbines was the development of a certification process.

- Support for innovator interaction and networking is essential. However, models for interaction and networking have only gradually developed over time and have been designed differently in different countries.

- Support requires a systemic approach. The case of wind energy shows that governmental policy needs to support the development of the entire innovation system, i.e., the development of the turbines and its infrastructure, but also the involvement of actors, necessary networks, and institutions.

- Support needs to be stable and continuous. The history of wind turbines development is long; it started in the 1880s. Many failures have occurred over time. The continuous support allowed knowledge creation and learning, as well as essential market formation that paved the successful innovation path of onshore wind energy and that now is the basis for the development of offshore wind energy.

24.7.15 Comparative Assessment of Photovoltaics (PV)[46]

A variety of factors, including government activities, have enabled the two order-of-magnitude reductions in the cost of PV over the past five decades. Despite this achievement, the technology remains too expensive

---

45 Lena Neij and Per Dannemand Andersen.

46 Gregory Nemet.

compared to existing electricity sources in many applications, such that widespread deployment depends on substantial future improvements. No single determinant predominantly explains the improvement to date; R&D, economies of scale, learning-by-doing, and knowledge spillovers from other technologies have all played a role in reducing system costs. Moreover, interactions among factors that enable knowledge feedbacks – for example, between demand subsidies and R&D – have also proven important.

Conversion efficiency, economies of scale, and the emergence of sequential niche markets have been important factors accounting for the impressive cost reductions in PV. Improvements in electrical conversion efficiency have been important to cost reductions, accounting for about one-third of the decline in cost over time. R&D, especially public sector R&D, has been central to this change. Deployment of PV has benefited from a sequence of niche markets where users of the technology were less price sensitive and had strong preferences for characteristics such as reliability and performance, which allowed product differentiation. Governments have played a large role in creating or enhancing these niche markets. Increasing demand for PV has reduced costs by enabling opportunities for economies of scale in manufacturing. Japan's program was especially innovative in that it took not only a long time horizon but also set a declining subsidy such that it fell to zero after the 10 years of the program. This provided not only expectations of demand, but also clear expectations of future levels of subsidy. The Renewable Energy Law in Germany in the 2000s successfully replicated many of the features of Japan's program.

Summary Points

- An array of supporting policy instruments is required: R&D, demand subsidies, etc.

- Timing matters: the question of when to switch from a focus on R&D to deployment is important.

- Much of the success of multiyear demand-side programs (in Japan and Germany) is because these programs created long-term expectations of future demand that enabled large investments in manufacturing facilities, which brought down costs through economies of scale.

- R&D support also needs a long-term commitment, whether through budgets or grants spanning multiple years, or supporting policies. Examples of supporting policies include Japan's Sunshine Program and the United States' Project Independence, which demonstrated commitment by making this area of work a serious national priority.

- Niche markets have been crucial, although they are most effective when not government supported.

- The success of new technological generations may require renewed R&D support even while markets for the existing technology are expanding.

24.7.16 Solar Innovation and Market Feedbacks: Solar PVs in Rural Kenya[47]

The solar PV market in Kenya is among the largest and most dynamic per capita in the world. Over 30,000 systems are sold each year. Much of this activity is related to the unsubsidized, purely free market sale of household solar electric systems, which account for an estimated 75% of solar equipment sales in the country. Solar is the largest source of new electrical connections in rural Kenya.

Despite this undisputed commercial success, product quality has been a significant concern (Jacobson and Kammen, 2007). Quality problems emerged first with the amorphous silicon solar modules that entered the Kenyan market in the early 1990s. This situation created a serious problem in the market, as many potential solar customers were unable to determine which brands performed well and which did not. Through independent testing in 1999, underperforming suppliers and models were identified and consequently withdrawn from the market, but quality problems resurfaced after 2004.

In response, the Kenya Bureau of Standards (KBS) moved to formulate and enforce quality standards for both amorphous and crystalline silicon PV modules. Currently, the KBS requires that import companies secure a certificate that validates their product conforms to the respective Kenyan standards prior to bringing the modules into the country. This certificate of conformity must be issued by an accredited laboratory. No such facility exists in Kenya, so this testing needs to take place in laboratories in Europe, North America, and Asia.

Summary Points

- Product quality assurance and standards constitute an important element for the diffusion of new energy technologies, particularly for decentralized systems like solar PV that are installed at residential sites by local businesses relying on imported modules.

- The recurrent emergence of quality problems in the Kenya PV market confirms that the issue of product quality control cannot be solved decisively by one-time testing efforts or focusing on the improvement of individual low-performing brands. Rather, institutional solutions that persistently require high performance for all brands are needed to ensure quality.

---

47   Daniel Kammen.

- The formulation, implementation and enforcement of product quality standards requires appropriate local institutional capacity, which needs to be developed and maintained for each market.

- The potential market feedbacks between end-users and suppliers of energy technologies can be weakened when testing stations are only available overseas.

24.7.17   Solar Thermal Electricity[48]

Solar thermal electricity (STE) production technology has improved through a combination of learning-by-doing and R&D investments. Even though R&D levels for this technology were small (even declining) relative to other energy technology programs, these programs played an important role in making full use of the knowledge gained through experience in manufacturing, installation, and operation of these facilities. R&D helped codify and document the often-tacit knowledge that accrued to operators through experience. As a result, this knowledge could be shared across firms and even across countries. It also preserved at least some of the value of the knowledge over time – especially during more than a decade of stagnation in the 1990s, when essentially no large plants were built worldwide. The rebirth of the STE industry during the 2000s indicates that at least some of this knowledge accumulated in the 1980s informs current designs and operation.

Most of our historical knowledge about technical change in STE comes from the 350 MW (mostly troughs) systems that were deployed in California in the 1980s. Since the mid-2000s, a new round of installations has begun, primarily in Spain and the southwestern United States. These installations encompass all three types of STE systems: troughs, concentrators, and dishes. By the end of 2010, total STE capacity neared 1 GW, and several additional GW (>2 in Spain, >3 in the United States) are likely to break ground in 2011. A distinguishing feature of STE, relative to other renewables, is that STE requires big, risky investments. Early investors may have to "eat between one and three US$200 million plants" before improvements enable profitable operation. A policy implication is that the scale of technology requires different types of incentives from smaller-scale renewables such as PV. Two policy instruments are driving a resurgence in STE installations over the past 10 years: (1) California's aggressive renewables obligation, possibly increasing to 33% by 2020; and (2) Spain's feed-in tariff. Complementary policies, such as loan guarantees and tax credits, have driven investment as well.

Summary Points

- Interaction between learning-by-doing and public R&D investment – even if small – was important for the substantial decline in operating costs of STE systems.

- This technology appears to be one for which higher R&D could not have substituted for deployment; learning by doing was essential.

- Initial investments required were large, and chunky, due to scale. Firms knew they would need to absorb losses on the first few plants – a classic valley of death problem that was eventually overcome by the alternative energy bubble on Wall Street in the 1980s. As a result, early plants needed both guaranteed tariffs and capital cost subsidies.

- Part of the value of the R&D investment was the formal documentation of cost improvement efforts, which was publicly available and nonproprietary.

- Technically, the 1980s California (solar energy generating system) plants have been successful. They steadily improved their performances, are still providing power two decades after they were installed, and provide the basis for newly designed plants in Spain and California. The failure of the company that built them was due to falling energy prices and consequent changes in policy rather than technical problems, illustrating the substantial market risks associated with policy intermittency.

24.7.18   The US Synthetic Fuels Program[49]

In response to the drastic oil price increase in the wake of the oil crises in 1979, subsidies for the demonstration and deployment of synfuels – a supply-side technology – received a rare confluence of support from the US government and energy experts (Deutch and Lester, 2004). The policy objective was to ameliorate the energy-security consequences and macroeconomic effects of the US dependence on imported oil by using the country's extensive coal, heavy oil, and oil shale deposits. Given estimates that synfuels would cost only US$60/barrel against the backdrop of (erroneously) projected rapid further increases in oil prices, the demonstration and deployment of synfuels production capability was seen as a major backstop or insurance policy (Deutch, 2005). The US Congress created the Synthetic Fuels Corporation (SFC) in 1980 and gave it the ambitious mandate to achieve production of 0.5 million barrels/day by 1987, and 2 million barrels/day by 1992. The goal of producing 2 million barrels/day in 1992 would have replaced over one-quarter of US crude oil and petroleum product imports, implying a scaling-up to one-quarter of the entire market in only 12 years, which was clearly unrealistic. Five and a half years later, and after expenditures of billions of dollars, the program was terminated without achieving its production goal (Gaskins and Stram, 1991).

Summary Points

- Deterministic cost-benefit analysis should not be the only measure used to make policy decisions when there is deep uncertainty

---

48 Gregory Nemet.

49 Laura Diaz Anadon, Gregory Nemet, with contributions from Bob Schock.

associated with market conditions (future oil prices) and technical feasibility. Policies aimed at providing insurance against a risk that does not materialize do not necessarily imply a policy failure.

- The mandate to meet ambitious production targets in relatively short timeframes regardless of market conditions and with little technical information was highly risky, partly because of the high expenses involved and partly because it was not accompanied by a long-term vision and corresponding innovation patience.

- Deciding when there is enough technical information to move from applied R&D to demonstration and deployment (or, in the SFC case, production) is difficult, yet crucial. This is especially the case in a situation in which the market and political environments change quickly. In large technology demonstration efforts, a good management and evaluation system is necessary to allow for timely decisions about whether and when to redefine program goals.

- Boom and bust cycles of support should be avoided, as they disrupt the innovation process, result in knowledge depreciation, and deter wider engagement in the enterprise.

- Policy backlash – the risk aversion induced by generalizing the experience of perceived failure of one program to unrelated programs – can have important and long-lasting effects.

24.7.19　The French Pressurized Water Reactor Program[50]

The case study reviews the French nuclear Pressurized Water Reactor (PWR) Program as an example of successful scaling-up of a complex and capital-intensive energy technology. Starting in the early 1970s, France built 58 PWRs with a total gross installed capacity of 66 GWe. On completion in 2000, they produced some 400 TWh/year of electricity, or close to 80% of France's electricity production. The institutional setting is characterized by a number of features. These include a high degree of standardization; external learning via the use of proven US reactor designs under a Westinghouse license; high regulatory stability; the effective absence of any public opposition; and a powerful national utility, ÉDF (Électricité de France), which acted both as *principal* and *agent* in reactor construction. The case study reviews the economics of this successful scale-up of nuclear reactor technology, identifying a significant cost escalation in real-term reactor construction costs, but also a remarkable stability in reactor operation costs. This cost stability is particularly noteworthy considering the need for load modulation in a system relying as heavily on base-load nuclear as in the case in France.

Summary Points

- Even under a most favorable institutional setting, earlier hopes of significant declines in nuclear reactor construction costs did not materialize, illustrating a case of negative learning much like the example of the United States (although cost escalation in France remained substantially below US trends).

- The case study thus demonstrates the limits of the learning paradigm: the assumption that costs invariably decrease with accumulated technology deployment. Not only do nuclear reactors across all countries with significant programs invariably exhibit negative learning (cost increase rather than decline), but the pattern is also quite variable, defying approximations by simple learning-curve models.

- While reactors' real construction costs increased steadily, their *operating* costs remained low and flat in France, as well as for many reactors elsewhere. Perhaps nuclear's "valley of death" is its inherently high investment costs and their tendency to rise beyond economically viable levels. The success of ÉDF in combining principal *and* agent in the construction process, which limited price escalation trends (especially in comparison to the US experience), could be an option worth considering for minimizing cost escalation. Conversely, this logic may suggest that competitive nuclear power is unlikely to be achieved in a private free market, which instead is tending to produce the rapid innovations that now competitively challenge nuclear power.

- The case study also provides valuable lessons for energy technology and climate policy. Cost projections of novel technologies are an inherent element in any climate change policy analysis. The case study confirms the earlier conclusion of Koomey and Hultman (2007) that projections of the future need to be grounded much more firmly within the historical observational space. Climate policy analysis must include a wider variation in cost uncertainties, as revealed by past experiences, than has previously been assumed in policy analysis and models.

24.7.20　Ethanol in Brazil[51]

Brazil's first ethanol program (PROALCOOL), launched in 1975, was a direct response to the dramatic rise in imported petroleum prices in 1973. The military government of the time saw this as a challenge to Brazil's financial stability and energy security, since the country imported 80% of the fuel used by its transport sector. Moreover, Brazil had extensive sugar plantations that were facing increased challenges to their exports from European Union trading preferences with the African, Caribbean,

---

50　Arnulf Grubler.

51　Paulo Teixeira de Sousa Jr. and Lynn Mytelka.

and Pacific Associated States – ACP countries – and the emergence of corn syrup and other close substitutes for sugar.

PROALCOOL was initially a classic import substitution policy. Subsidies were used to expand ethanol production, then in its infancy, and to induce users to shift to dedicated engines for ethanol that could handle a gasoline blend with more than 5–10% ethanol. When gasoline prices fell a few years later, those who had shifted were left paying the higher costs of ethanol, while the original problem of oil imports remained.

In this changed context, the government decided to invest in the research needed for Brazil to become a more efficient ethanol producer and thus be in a position to eventually eliminate subsidies. At the core of this process was a research partnership that brought together the Brazilian Agricultural Research Corporation and Copersucar, a cooperative of sugar mill and ethanol plant owners. Between 1975 and 2002, ethanol production increased from 0.6 to 12.6 million cubic meters and the price paid to alcohol producers dipped below that of Rotterdam gasoline prices (Goldemberg et al., 2004). By early in the new millennium, increasing yields and reduced processing costs (van den Wall Bake et al., 2009) had eliminated the need for subsidies. The decision to develop flex-fuel engines, in collaboration with foreign-owned automobile producers, strengthened the domestic auto industry and led to a dramatic shift in consumption habits and practices, thus further building the market. Introduced in 2003, flex-fuel vehicles accounted for 81% of the light vehicle registrations by 2008 (ANFAVEA, 2008).

Five policy lessons emerge from the Brazilian ethanol experience:

- First, developing a portfolio of fuel options was important for Brazilian development more broadly.

- Second, domestic research was a major contributor to the positive development outcome over the longer term.

- Third, building coalitions among interested parties helped to sustain long-term development efforts.

- Fourth, sugar cane was grown on large plantations, concentrated heavily in a single state. This left few opportunities for small holders and meant transporting ethanol elsewhere around the country with potentially negative effects on net CO_2 benefits.

- Fifth, ethanol from sugar cane tended to maintain the structure of large-scale, concentrated production. Lessons such as these have been applied in the development of the biodiesel sector. A variety of inputs have been identified for biodiesel and several of these are available in most regions and can be grown by small holders and processed and distributed locally.

References

Adeyeye, A., J. Barrett, J. Diamond, L. Goldman, J. Pendergrass and D. Schramm, 2009: *Estimating U.S. Government Subsidies to Energy Sources: 2002–2008*. Environmental Law Institute, Washington, DC.

AEIC, 2010: *A Business Plan for America's Future*. American Energy Innovation Council.

Aguayo, F., 2008: Renewable Energy and Sustainable Development in Mexico: The case of Wind Energy on the Tehuantepec Isthmus. *Mexico's Energy Future Seminar*, 22 June – 3 July, John F. Kennedy School of Government, Harvard University, Cambridge, MA.

Alchian, A., 1963: Reliability of Progress Curves in Airframe Production. *Econometrica: Journal of the Econometric Society*, **31**(4):679–693.

Alkemade, F., C. Kleinschmidt and M. Hekkert, 2007: Analysing emerging innovation systems: a functions approach to foresight. *International Journal of Foresight and Innovation Policy*, **3**(2):139–168.

Anadon, L. D., K. S. Gallagher, M. Bunn and C. Jones, 2009: Tackling U.S. Energy Challenges and Opportunities: Preliminary Policy Recommendations for Enhancing Energy Innovation in the United States. Energy Technology Innovation Policy Group, Belfer Center for Science and International Affairs, Harvard University, Cambridge, MA.

ANFAVEA, 2008: *Brazilian Autmotive Industry Yearbook*. Associacao Nacional dos Fabricantes de Veiculos Automores – Brasil (ANFAVEA), Sao Paolo.

Argote, L., S. L. Beckman and D. Epple, 1990: The Persistence and Transfer of Learning in Industrial Settings. *Management Science*, **36**(2):140–154.

Argote, L. and D. Epple, 1990: Learning Curves in Manufacturing. *Science*, **247**(4945):920–924.

Arrow, K. J., 1962a: The Economic Implications of Learning by Doing. *The Review of Economic Studies*, **29**(3):155–173.

Arrow, K. J., 1962b: Economic welfare and the allocation of resources for invention. In *The Rate and Direction of Economic Activity*. R. Nelson, (ed.), Princeton University Press, Princeton pp. 609–625.

Arthur, B. W., 1988a: Competing Technologies. In *Technical Change and Economic Theory*. G. Dosi, C. Freeman, R. Nelson, G. Silverberg and L. Soete, (eds.), Pinter Publishers, London.

Arthur, B. W., 1988b: Self-Reinforcing Mechanisms in Economics. In *The Economy as an Evolving Complex System*, Westview Press, Boulder, CO.

Arthur, B. W., 1989: Competing Technologies, Increasing Returns, and Lock-In by Historical Events. *The Economic Journal*, **99**(394):116–131.

Ausubel, J. H. and C. Marchetti, 1997: Elektron: Electrical systems in Retrospect and Prospect. In *Technological Trajectories and the Human Environment*. J. H. Ausubel and H. D. Langford, (eds.), National Academy Press, Washington, DC pp.115–140.

Bahk, B.-H. and M. Gort, 1993: Decomposing Learning by Doing in New Plants. *Journal of Political Economy*, **101**(4):561–583.

Baker, E., H. Chon and J. Keisler, 2009: Advanced solar R&D: Combining economic analysis with expert elicitations to inform climate policy. *Energy Economics*, **31**:S37–S49.

Barton, J. H., 2009: Patenting and Access to Clean Energy Technologies in Developing Countries. WIPO Magazine, World Intellectual Property Organization, March 2009.

BCG, 1972: *Perspectives on Experience*. The Boston Consulting Group, Boston, MA.

Benkard, C. L., 2000: Learning and Forgetting: The Dynamics of Aircraft Production. *American Economic Review*, **90**(4):1034–1054.

Bergek, A., S. Jacobsson, B. Carisson, S. Lindmark and A. Rickne, 2008: Analyzing the functional dynamics of technological innovation systems: A scheme of analysis. *Research Policy*, **37**(3):407–407.

Bolingar, M. and Wiser, R., 2012: Understanding wind turbine price trends in the U.S. Over the past decade. *Energy Policy*, **42**:628–641.

Boone, T., R. Ganeshan and R. L. Hicks, 2008: Learning and Knowledge Depreciation in Professional Services. *Management Science*, **54**(7):1231–1236.

Borja-Diaz, M. A., O. A. Jaramillo-Salgado and F. Mimiaga-Sosa, 2005: *Primer Documento del Proyecto Eoloeléctrico del Corredor Eólico del Istmo de Tehuantepec*. Instituto de Investigacione Eléctricas-PNUD-GEF, México.

Borup, M., N. Brown, K. Konrad and H. Van Lente, 2006: The sociology of expectations in science and technology. *Technology Analysis & Strategic Management*, **18**(3–4):285–298.

Bosetti, V., C. Carraro, E. Massetti and M. Tavoni, 2008: International energy R&D spillovers and the economics of greenhouse gas atmospheric stabilization. *Energy Economics*, **30**(6):2912–2929.

Brennand, T. P., 2001: Wind energy in China: policy options for development. *Energy for Sustainable Development*, **5**(4):84–91.

Brooks, H., 1995: What We Know and Do Not Know About Technology Transfer: Linking Knowledge to Action. In *Marshaling technology for development*, National Academy Press, Washington, DC.

Bush, V., 1945: Science The Endless Frontier. A Report to the President, United States Government Printing Office, Washington, D. C.

Calvin, K., J. Edmonds, B. Bond-Lamberty, L. Clarke, S. H. Kim, P. Kyle, S. J. Smith, A. Thomson and M. Wise, 2009: 2.6: Limiting climate change to 450 ppm CO_2 equivalent in the 21st century. *Energy Economics*, **31**(Supplement 2):S107–S120.

Carlsson, B. and R. Stankiewicz, 1991: On the nature, function and composition of technological systems. *Journal of Evolutionary Economics*, **1**(2):93–119.

Clemen, R. T. and R. C. Kwit, 2001: The Value of Decision Analysis at Eastman Kodak Company, 1990–1999. *INTERFACES*, **31**(5):74–92.

Coe, D. T. and E. Helpman, 1995: International R&D spillovers. *European Economic Review*, **39**(5):859–888.

Coe, D. T., E. Helpman and A. W. Hoffmaister, 2009: International R&D spillovers and institutions. *European Economic Review*, **53**(7):723–723.

Cohen, W. M. and D. A. Levinthal, 1989: Innovation and learning: the two faces of R&D. *The Economic Journal*, **99**(397):569–596.

Cohen, W. M. and D. A. Levinthal, 1990: Absorptive Capacity: A New Perspective on Learning and Innovation. *Administrative Science Quarterly*, **35**(1):128–152.

Conley, P., 1970: Experience curves as a planning tool. *IEEE Spectrum*, **7**(6):63–68.

Cowan, R., 1990: Nuclear power reactors: a study in technological lock-in. *The Journal of Economic History*, **50**(3):541–567.

Cowan, R. and S. Hulten, 1996: Escaping Lock-In: The Case of the Electric Vehicle. *Technological Forecasting and Social Change*, **53**(1):61–79.

Darr, E., D., L. Argote and D. Epple, 1995: The Acquisition, Transfer, and Depreciation of Knowledge in Service Organizations: Productivity in Franchises. *Management Science*, **41**(11):1750–1762.

Dasgupta, P. and P. A. David, 1994: Toward a new economics of science. *Research Policy*, **23**(5):487–521.

Deutch, J. M. and R. K. Lester, 2004: *Making technology work: applications in energy and the environment.* Cambridge University Press, Cambridge, UK.

Deutch, J. M., 2005: *What Should the Government Do to Encourage Technical Change in the Energy Sector?* MIT Joint Program on the Science and Policy of Global Change, Massachusetts Institute of Technology, Cambridge MA.

Dooley, J. J., 2000: A Short Primer on collecting and Analyzing Energy R&D Statistics. PNNL-13158, Batelle.

Dooley, J. J. and P. J. Runci, 2000: Developing nations, energy R&D, and the provision of a planetary public good: A long-term strategy for addressing climate change. *Journal of Environment & Development*, 9(3):215–239.

Doornbosch, R. and S. Upton, 2006: *DO WE HAVE THE RIGHT R&D PRIORITIES AND PROGRAMMES TO SUPPORT THE ENERGY TECHNOLOGIES OF THE FUTURE?* Round Table on Sustainable Development – SG/SD/RT(2006)1, OECD, Paris.

Duke, R. and D. M. Kammen, 1999: The economics of energy market transformation programs. *The Energy Journal*, 20(4):15–64.

Dutton, J. M. and A. Thomas, 1984: Treating Progress Functions as a Managerial Opportunity. *Academy of Management. The Academy of Management Review*, 9(2):235–248.

EC, 2005: Energy R&D Statistics in the European Research Area. European Commision, Brussels.

Edquist, C. and B. Johnson, 1997: Institution and Organizations in Systems of Innovation. In *Systems of Innovation: Technologies, Institutions and Organization.* C. Edquist, (ed.), Pinter Publishers, London pp.41–60.

Edquist, C., 2001: Innovation Systems and Innovation Policy: the state of the art. *DRUID's Nelson-Winter Conference*, Aarlborg.

Enos, J. L., 1962: Petroleum progress and profits: a history of process innovation. MIT Press, Cambridge, MA.

EPRI, 2003: Electricity Technology Roadmap – Meeting the Critical Challenges of the 21st Century. Electric Power Research Insititute (EPRI), Palo Alto, CA.

Evenson, R. E., 2002: Induced Adaptive Invention/Innovation and Productivity Convergence in Developing Countries. In *Technological Change and the Environment*. A. Grubler, N. Nakicenovic and W. D. Nordhaus, (eds.), Resources for the Future Press, Washington DC, USA pp.61–96.

Falvey, R., N. Foster and D. Greenaway, 2004: Imports, exports, knowledge spillovers and growth. *Economics Letters*, 85(2):209–213.

Farsi, M., A. Fetz and M. Filippini, 2008: Economies of Scale and Scope in Multi-Utilities. *The Energy Journal*, 29(4):123–143.

Fligstein, N., 1997: Social skill and institutional theory. *American Behavioral Scientist*, 40(4):397–406.

Florida, R. and D. F. Smith Jr., 1990: Venture capital, innovation, and economic development. *Economic Development Quarterly*, 4(4):345–361.

Foray, D., 2004: *The Economics of Knowledge.* The MIT Press, Cambridge, MA.

Frankel, M., 1955: OBSOLESCENCE AND TECHNOLOGICAL CHANGE IN A MATURING ECONOMY. *American Economic Review*, 45(3):296–320.

Freeman, C. and C. Perez, 1988: Structural Crises of Adjustment, Business Cycles and Investment Behaviour In *Technical Change and Economic Theory*. G. Dosi, C. Freeman, R. Nelson, G. Silverberg and L. Soete, (eds.), Pinter Publishers, London pp.38–66.

Freeman, C., 1994: The Economics of Technical Change. *Cambridge Journal of Economics*, 18(5):463–463.

Fridlund, M., 2000: Procuring Products and Power, Developing International Competitiveness in Swedish Electrotechnology and Electric Power. In *Public Technology Procurement and Innovation*. C. Edquist, L. Hommen and L. Sipouri, (eds.), Kluwer Academic Publishers, Dordrecht.

Gallagher, K. S., 2006: Limits to leapfrogging in energy technologies? Evidence from the Chinese automobile industry. *Energy Policy*, 34(4):383–394.

Gallagher, K. S., J. P. Holdren and A. D. Sagar, 2006: ENERGY-TECHNOLOGY INNOVATION. *Annual Review of Environment & Resources*, 31(1):193–242.

Garud, R. and P. Karnoe, 2003: Bricolage versus breakthrough: Distributed and embedded agency in technology entrepreneurship. *Research Policy*, 32(2):277–300.

Gaskins, D. and B. Stram, 1991: *A Meta Plan: A Policy Response to Global Warming*. CSIA Discussion Paper 91–3, Kennedy School of Government, Harvard University.

Geels, F. W., 2004: From sectoral systems of innovation to socio-technical systems: Insights about dynamics and change from sociology and institutional theory. *Research Policy*, 33(6–7):897–920.

Goldemberg, J., 1998: FROM PHYSICS TO DEVELOPMENT STRATEGIES. *Annual Review of energy and the environment*, 23:1–23.

Goldemberg, J., S. T. Coelho, P. M. Nastari and O. Lucon, 2004: Ethanol learning curve-the Brazilian experience. *Biomass and Bioenergy*, 26(3):301–304.

Grubler, A., 1998: *Technology and Global Change.* Cambridge University Press, Cambridge, UK.

Grubler, A., N. Nakicenovic and D. G. Victor, 1999: Dynamics of energy technologies and global change. *Energy Policy*, 27(5):247–280.

Grubler, A., 2008: *Energy Transitions*. Encyclopedia of Earth. C. J. Cleveland, (ed.), Washington, DC.

Grubler, A., 2010: The costs of the French nuclear scale-up: A case of negative learning by doing. *Energy Policy*, 38(9):5174–5188.

Grubler, A. and K. Riahi, 2010: Do governments have the right mix in their energy R&D portfolios? *Carbon Management*, 1(1):79–87.

GSI, 2009: *Kinds of subsidies, who uses them and how big they are*. www.globalsubsidies.org/en/research/kinds-subsidies-who-uses-them-and-how-big-they-are-0 (accessed 25 August, 2010).

Gurney, A., H. Ahammad and M. Ford, 2009: The economics of greenhouse gas mitigation: Insights from illustrative global abatement scenarios modelling. *Energy Economics*, 31(Supplement 2):S174–S186.

Hall, B. H., 2007: *MEASURING THE RETURNS TO R&D: THE DEPRECIATION PROBLEM*. NBER Working Paper Series, Working Paper 13473, National Bureau of Economic Research, Cambridge, MA.

Hall, G. and S. Howell, 1985: The Experience Curve from the Economist's Perspective. *Strategic Management Journal*, 6(3):197–213.

Halsnæs, K., P. Shukla, D. Ahuja, G. Akumu, R. Beale, J. Edmonds, C. Gollier, A. Grubler, M. Ha Duong, A. Markandya, M. McFarland, E. Nikitina, T. Sugiyama, A. Villavicencio and J. Zou, 2007: Framing issues. In *Climate Change* 2007: *Mitigation*. B. Metz, O. Davidson, P. Bosch, R. Dave and L. Meyer, (eds.), Contribution of Working Group III to the Fourth Assessment Report of the Intergovernmental Panel on Climate Change, Cambridge University Press, Cambridge, UK pp.117–167.

Hekkert, M. P., R. A. A. Suurs, S. O. Negro, S. Kuhlmann and R. E. H. M. Smits, 2007: Functions of innovation systems: A new approach for analysing technological change. *Technological Forecasting and Social Change*, 74(4):413–432.

Henriksen, E., K.-H. Midelfart and F. Steen, 2001a: *Economies of Scale in European Manufacturing Revisited*. CEPR Discussion Papers, Centre for Economic Policy Research, London, UK.

Henriksen, E., F. Steen and K. Ulltveit-Moe, 2001b: *Economies of Scale in European Manufacturing Revisited.* CEPR Discussion Paper no. 2896, Centre for Economic Policy Research, London.

Hillman, K. M., R. A. A. Suurs, M. P. Hekkert and B. A. Sandén, 2008: Cumulative causation in biofuels development: a critical comparison of the Netherlands and Sweden. *Technology Analysis & Strategic Management,* **20**(5):593–593.

Hoffert, M. I., K. Caldeira, G. Benford, D. R. Criswell, C. Green, H. Herzog, A. K. Jain, H. S. Kheshgi, K. S. Lackner, J. S. Lewis, H. D. Lightfoot, W. Manheimer, J. C. Mankins, M. E. Mauel, L. J. Perkins, M. E. Schlesinger, T. Volk and T. M. L. Wigley, 2002: Advanced Technology Paths to Global Climate Stability: Energy for a Greenhouse Planet. *Science,* **298**(5595):981–987.

Hoffert, M. I., 2010: Farewell to Fossil Fuels? *Science,* **329**(5997):1292–1294.

Hughes, T. P., 1983: *Networks of Power: electrification in Western society, 1880–1930.* The Johns Hopkins University Press, Baltimore and London.

IAEA-PRIS, 2010: IAEA Power Reactor Information System. www.iaea.org/programmes/a2/ (accessed 2 November 2010).

IEA, 2002: *World Energy Outlook.* International Energy Agency, Paris.

IEA, 2003: *World Energy Investment Outlook.* International Energy Agency, Paris.

IEA, 2006: *World Energy Outlook.* International Energy Agency, Paris.

IEA, 2007: *World Energy Outlook: China and India Insights.* International Energy Agency, Paris.

IEA, 2008a: *World Energy Outlook.* International Energy Agency, Paris.

IEA, 2008b: Energy Technology Perspectives: Energy Technology Perspectives to 2050. International Energy Agency – OECD, Paris.

IEA, 2009a: *World Energy Outlook.* International Energy Agency, Organization for Economic Cooperation & Development, Paris.

IEA, 2009b: *R&D Statistics.* www.iea.org/stats/rd.asp (accessed 25 August 2010).

IEA, 2010: *World Energy Outlook.* International Energy Agency, Organization for Economic Cooperation & Development, Paris.

Jacobson, A. and D. M. Kammen, 2007: Engineering, institutions, and the public interest: Evaluating product quality in the Kenyan solar photovoltaics industry. *Energy Policy,* **35**(5):2960–2960.

Jacobsson, S. and A. Johnson, 2000: The diffusion of renewable energy technology: An analytical framework and key issues for research. *Energy Policy,* **28**(9):625–640.

Jacobsson, S. and A. Bergek, 2004: Transforming the energy sector: the evolution of technological systems in renewable energy technology. *Industrial and corporate change,* **13**(5):815–849.

Jacobsson, S. and V. Lauber, 2006: The politics and policy of energy system transformation-explaining the German diffusion of renewable energy technology. *Energy Policy,* **34**(3):256–276.

Jaffe, A. B. and M. Trajtenberg, 1999: International Knowledge Flows: Evidence from Patent Citations. *Economics of Innovation & New Technology,* **8**(1–2):105–137.

Jäger-Waldau, A., 2006: *PV Status Report 2006.* European Commission, Joint Research Centre, Ispra, Italy.

Joskow, P. L. and N. L. Rose, 1985: The effects of technological change, experience, and environmental regulation on the construction cost of coal-burning generating units. *The Rand Journal of Economics,* **16**(1):1–27.

Kamp, L. M., R. Smits, E. Andriesse and D. Cornelis, 2004: Notions on learning applied to wind turbine development in the Netherlands and Denmark. *Energy Policy,* **32**(1625–1637).

Kemp, R., J. Schot and R. Hoogma, 1998: Regime shifts to sustainability through processes of niche formation: The approach of strategic niche management. *Technology Analysis & Strategic Management,* **10**(2):175–196.

Kempener, R., L. D. Anadon and J. Condor, 2010a: Governmental Energy Innovation Investments, Policies, and Institutions in the Major Emerging Economies: Brazil, Russia, India, Mexico, China, and South Africa. Energy Technology Innovation Policy Discussion Paper #2010–16, Belfer Center for Science and International Affairs, Harvard Kennedy School, Cambridge, MA.

Kempener, R., L. D. Anadon, J. Condor and J. Kenrick, 2010b: A Comparative Analysis of Energy Technology Innovation Policies in Major Emerging Economies: Brazil, Russia, India, Mexico, China and South Africa. HKS Faculty Working Paper Series: Harvard University, Cambridge, MA.

Kim, I. and H. L. Seo, 2009: Depreciation and transfer of knowledge: an empirical exploration of a shipbuilding process. *International Journal of Production Research,* **47**(7):1857–1857.

Klaassen, G., A. Miketa, K. Larsen and T. Sundqvist, 2005: The impact of R&D on innovation for wind energy in Denmark, Germany and the United Kingdom. *Ecological Economics,* **54**(2–3):227–240.

Klemperer, P., 1990: How Broad Should the Scope of Patent Protection Be? *The Rand Journal of Economics,* **21**(1):113–131.

Kline, S. J. and N. Rosenberg, 1986: An Overview of Innovation. In *The Positive Sum Strategy: Harnessing Technology for Economic Growth.* R. Landau and N. Rosenberg, (eds.), National Academy Press, Washington, DC.

Koomey, J. and N. E. Hultman, 2007: A reactor-level analysis of bus bar costs for US nuclear plants, 1970–2005. *Energy Policy,* **35**(11):5630–5630.

Kouvaritakis, N., A. Soria and S. Isoard, 2000: Modelling energy technology dynamics: methodology for adaptive expectations models with learning by doing and learning by searching. *International Journal of Global Energy Issues,* **14**(1):104–115.

Krawiec, F., J. Thornton and M. Edesess, 1980: *An Investigation of Learning and Experience Curves.* Solar Energy Research Institute, Report SERI/TR-353-459 prepared for the US Department of Energy, Golden, Colorado.

Kwon, O. S. and W.-C. Yun, 2003: Measuring economies of scope for cogeneration systems in Korea: a nonparametric approach. *Energy Economics,* **25**(4):331–338.

La Manna, M. A., 1992: Optimal patent life vs optimal patentability standards. *International Journal of Industrial Organization,* **10**(1):81–89.

Lako, P., 2004: Spillover Effects from Wind Power: Case study in the framework of the project Spillovers of climate policy. Energy Research Centre of the Netherlands, Petten, the Netherlands.

Landau, R. and N. Rosenberg, (eds.), 1986: *The Positive sum strategy: harnessing technology for economic growth.* National Academy Press, Washington D.C.

Lee, T. H. and R. L. Loftness, 1987: *Managing Electrotechnology Innovation in the USA.* Working Papper WP-87-54, Laxenburg, Austria.

Levin, R., A. Klevorick, R. Nelson and S. Winter, 1987: Appropriating the Returns from Industrial Research and Development. In *Brooking Papers on Economic Activity,* Brookings Institution Press, Washington, DC, Vol. Vol. 3, pp.783–820.

Lipsey, R. G., K. I. Carlaw and C. T. Bekar, 2006: *Economic Transformations: General Purpose Technologies and Long Term Economic Growth.* Oxford University Press, New York.

Lovins, A. B., E. Kyle Datta, T. Feiler, K. R. Rábago, J. Swisher, A. Lehmann and K. Wicker, 2002: *Small is Profitable: The Hidden Economic Benefits of Making Electrical Resources the Right Size.* Rocky Mountain Institute, Snowmass, CO.

Lundvall, B.-A., 1998: Why Study National Systems and National Styles of Innovation? *Technology Analysis & Strategic Management*, **10**(4):407–422.

Lundvall, B.-Å., (ed.) 1992: National Systems of innovation: Towards a theory of innovation and interactive learning. Pinter Publishers, London.

Lundvall, B.-Å., 2007: National Innovation Systems-Analytical Concept and Development Tool. *Industry and Innovation*, **14**(1):95–119.

Lundvall, B.-Å., 2009: Innovation as an Interactive Process: User-Producer Interaction to the National System of Innovation. *African Journal of Science, Technology, Innovation and Development*, **1**(2&3):10–34.

Lutsey, N. and D. Sperling, 2005: Energy Efficiency, Fuel Economy, and Policy Implications. *Transportation Research Record: Journal of the Transportation Research Board*, 1941:Figure 6(a), p14. Copyright, National Academy Sciences, Washinton, DC, 2005. Reproduced with permission of teh Transportation Research Board.

Machlup, F., 1984: Knowledge, its creation, distribution and economic significance: The economics of information and human capital. Princeton University Press, Princeton, NJ.

Mancusi, M. L., 2008: International spillovers and absorptive capacity: A cross-country cross-sector analysis based on patents and citations. *Journal of International Economics*, **76**(2):155–165.

Maskus, K. E., 2000: Chapter 2: A Road Map for the TRIPs ahead. In *Intellectual property rights in the global economy*, Institute for International Economics, Washington DC.

Mayo, J. W., 1984: The Technological Determinants of the U.S. Energy Industry Structure. *The Review of Economics and Statistics*, **66**(1):51–59.

McCabe, M. J., 1996: Principals, Agents, and the Learning Curve: The Case of Steam-Electric Power Plant Design and Construction. *The Journal of Industrial Economics*, **44**(4):357–375.

Meijer, I. and M. P. Hekkert, 2007: Managing Uncertainties in the Transition Towards Sustainability: Cases of Emerging Energy Technologies in The Netherlands. *Journal of Environmental Policy and Planning*, **9**(3–4):281–298.

Miketa, A. and L. Schrattenholzer, 2004: Experiments with a methodology to model the role of R&D expenditures in energy technology learning processes; first results. *Energy Policy*, **32**(15):1679–1692.

Mishina, K., 1992: *Learning by New Experiences*. Working Paper 93-084, Harvard Business School, Cambridge MA.

Mishina, K., 1999: Learning by New Experiences: Revisiting the Flying Fortress Learning Curve. In *Learning by doing in markets, firms, and* countries, N. R. Lamoreaux, D. M. G. Raff and P. Temin, Eds., The University of Chicago Press, Chicago. pp.145–184.

Mohnen, P., 1997: Introduction: Input-output analysis of interindustry R&D spillovers. *Economic Systems Research*, **9**(1):3–9.

Mowery, D. and N. Rosenberg, 1979: The influence of market demand upon innovation: a critical review of some recent empirical studies. *Research Policy*, **8**(2):102–153.

Nakicenovic, N. and H.-H. Rogner, 1996: Financing global energy perspectives to 2050. *OPEC Review*, **20**(1):1–23.

Nakicenovic, N., J. Alcamo, G. Davis, B. de Vries, J. Fenhann, S. Gaffin, K. Gregory, A. Grubler, Y. J. Tae, T. Kram, E. L. La Rovere, L. Michaelis, S. Mori, T. Morita, W. Pepper, H. Pitcher, L. Price, K. Riahi, A. Roehrl, H.-H. Rogner, A. Sankovski, M. Schlesinger, P. Shukla, S. Smith, R. Swart, S. van Rooijen, N. Victor and D. Zhou, 2000: *Special Report on Emissions Scenarios*. IPCC and Cambridge University Press, Cambridge, UK.

Nakićenović, N., A. Grubler and A. McDonald, (eds.), 1998: *Global energy: perspectives*. Cambridge University Press, Cambridge.

NEF/SEFI, 2009: *Analysis of Trend and Issues in the Financing of Renewable Energy and Energy Efficiency*. Global Trends in Sustainable Energy Investment 2009, UNEP & Basel Agency for Sustaianble Energy, New Energy Finance (NEF) / Sustainable Energy Finance Initiative (SEFI), Basel, Switzerland.

Neij, L., 2004: The development of the experience curve concept and its application in energy policy assessment. *International Journal of Energy Technology and Policy*, **2**(1–2):3–14.

Nelson, R. and S. Winter, 1982: *An Evolutionary Theory of Economic Change*. Harvard University Press, Cambridge, MA.

Nelson, R., (ed.) 1993: *National Innovation Systems: A Comparative Analysis*. Oxford University Press, New York.

Nemet, G. F., 2006: Beyond the learning curve: factors influencing cost reductions in photovoltaics. *Energy Policy*, **34**(17):3218–3232.

Nemet, G. F., 2007: Policy and Innovation in Low-carbon energy Technologies. PhD Dissertetion, University of California, Barkeley, CA.

Nemet, G. F. and D. M. Kammen, 2007: U.S. energy research and development: Declining investment, increasing need, and the feasibility of expansion. *Energy Policy*, **35**(1):746–746.

Nemet, G. F., 2009a: Demand-pull, technology-push, and government-led incentives for non-incremental technical change. *Research Policy*, **38**(5):700–700.

Nemet, G. F., 2009b: Interim monitoring of cost dynamics for publicly supported energy technologies. *Energy Policy*, **37**(3):825–825.

Nordhaus, W. D., 1969: Inventions, Growth and Welfare: a Theoretical Treatment of Technological Change. MIT Press, Cambridge, MA.

NRC, 2001: *Energy Research at DOE was it Worth it? Energy Efficiency and Fossil Energy Research 1978 to 2000*. Board on Energy and Environmental Systems, National Research Council (NRC), Washington, D. C.

NRC, 2007: Prospective Evaluation of Applied Energy Research and Development at DOE (Phase Two). National Research Council (NRC), Washington, DC.

NSF, 2009: *Research and Development in Industry: 2004*. NSF-09–301 National Science Foundation, Arlington VA.

NSF, 2010: Science and Engineering Indicators 2010. Chapter 4: Research and Development: National Trends and International Linkages. Report NSB 10-01, National Science Fundation, Arlington, VA.

O'Rourke, A. R., 2009: *The Emergence of Cleantech*. Yale University, New Haven CT.

OECD, 2002: *Frascati Manual: Proposed Standard Practice for Surveys on Research and Experimental Development*. The Measurement of Scientific and Technological Activities, Organisation for Economic Co-Operation and Development, Paris.

OECD, 2007: Science, Technology and Innovation Indicators in a Changing World: Responding to Policy Needs. *Paper discussed at the OECD Blue Sky II Forum*, Organisation for Economic Co-operation and Development, Ottawa, Canada.

OECD, 2009: *R&D Expenditure in Industry* ISIC Rev.3, OECD STAN (Structural Analysis Database), Paris.

Pacala, S. and R. Socolow, 2004: Stabilization Wedges: Solving the Climate Problem for the Next 50 Years with Current Technologies. *Science*, **305**(5686):968–972.

Panzar, J. and R. Willig, 1975: *Economies of scale and scope in Multi-Output Production*. Economics Discussion Paper No. 33, Bell Laboratories, Murray Hill, NJ.

PCAST, 1997: *Report to the President on Federal Energy Research and Development for the Challenges of the Twenty-First Century*. President's Committee of

Advisors on Science and Technology Panel on Energy Research and Development, Washington, D. C.

Peerenboom, J. P., W. A. Buehring and T. W. Joseph, 1989: Selecting a Portfolio of Environmental Programs for a Synthetic Fuels Facility. *Operations Research*, **37**(5):689–699.

Plevin, R. J., M. O'Hare, A. D. Jones, M. S. Torn and H. K. Gibbs, 2010: Greenhouse Gas Emissions from Biofuels' Indirect Land Use Change Are Uncertain but May Be Much Greater than Previously Estimated. *Environmental Science & Technology*, **44**(21):8015–8021.

Popp, D., 2006: R&D Subsidies and Climate Policy: Is There a ``Free Lunch"? *Climatic Change*, **77**(3):311–341.

Raff, D. M. G., 1991: Making Cars and Making Money in the Interwar Automobile Industry: Economies of Scale and Scope and the Manufacturing behind the Marketing. *The Business History Review*, **65**(4):721–753.

Rapier, R., 2008: *FutureGen Project Stopped*. www.consumerenergyreport.com/2008/01/31/futuregen-project-stopped/ (accessed 31 January 2008).

Rapping, L., 1965: Learning and World War 2 Production Functions. *The Review of Economics and Statistics*, **47**(1):81–86.

Riahi, K., A. Grubler and N. Nakicenovic, 2007: Scenarios of long-term socio-economic and environmental development under climate stabilization. *Technological Forecasting and Social Change*, **74**(7):887–935.

Rosegger, G., 1996: The Economics of Production and Innovation: An Industrial Perspective. Butterworth-Heinemann Ltd, Oxford, UK.

Rosenberg, N., 1982: *Inside the black box: technology and economics.* Cambridge University Press, Cambridge, UK.

Rosenberg, N., 1994: *Exploring the black box: technology, economics, and history.* Cambridge University Press, Cambridge, UK.

Ruttan, V., 2001: Technology, Growth, and Development: an induced innovation perspective. Oxford University Press, Oxford, UK.

Ruttan, V. W., 1996: Induced Innovation and Path Dependence: A Reassessment with Respect to Agricultural Development and the Environment. *Technological Forecasting and Social Change*, **53**(1):41–59.

Sabatier, P. A., 1987: Knowledge, Policy-Oriented Learning, and Policy Change. *Science Communication*, **8**(4):649–692.

Sabatier, P. A., 1988: An advocacy coalition framework of policy change and the role of policy-oriented learning therein. *Policy Sciences*, **21**(2–4):129–169.

Schumpeter, J. A., 1942: *Capitalism, Socialism and Democracy.* Harper, New York.

Sharpe, P. and T. Keelin, 1998: How SmithKline Beecham makes better resource-allocation decisions. *Harvard Business Review*, **76**(2):45–57.

Shum, K. L. and C. Watanabe, 2008: Towards a local learning (innovation) model of solar photovoltaic deployment. *Energy Policy*, **36**(2):508–508.

Smekens, K. E. L., P. Lako and A. J. Seebregts, 2003: *Technologies and technology learning, contributions to IEA's Energy Technology Perspectives*. Report ECN-C--03-046, Energy Research Centre of the Netherlands, Petten, the Netherlands.

Smil, V., 1994: *Energy in World History*. Westview Press, Boulder, CO.

Smil, V., 2008: Energy in Nature and Society: General Energetics of Complex Systems. MIT Press, Boston, MA.

Smith, A., 1776: Chapter 1: Of the Division of Labor. In *An Inquiry into the Nature of the Causes of the Wealth of Nations*, Methuen and Co., London.

Smith, A. and A. Stirling, 2010: The Politics of Social-ecological Resilience and Sustainable Socio-technical Transitions. *Ecology and Society*, **15**(1):11.

Smith, K., 2002: *What is the 'knowledge economy'? Knowledge intensity and distributed knowledge bases*. Discussion Paper Series, United Nations University, Institute for New Technologies, Maastricht, The Netherlands.

Solow, R. M., 1957: Technical Change and the Aggregate Production Function. *The Review of Economics and Statistics*, **39**(3):312–320.

Suurs, R. A. A. and M. P. Hekkert, 2009a: Competition between first and second generation technologies: Lessons from the formation of a biofuels innovation system in the Netherlands. *Energy*, **34**(5):669–679.

Suurs, R. A. A. and M. P. Hekkert, 2009b: Cumulative causation in the formation of a technological innovation system: The case of biofuels in The Netherlands. *Technological Forecasting and Social Change*, **76**(8):1003–1020.

Teece, D. J., 1980: Economies of scope and the scope of the enterprise. *Journal of Economic Behavior and Organization*, **1**(3):223–247.

Thompson, P., 2007: How Much Did the Liberty Shipbuilders Forget? *Management Science*, **53**(6):908–918.

Thomson Financial, 2009: *Thomson Reuters VentureXpert Data Base.* vx.thomsonib.com/NASApp/VxComponent/NewMain.htm (accessed 15 July 2010).

UKERC, (ed.) 2010: Great Expectations:The cost of offshore wind in UK waters – understanding the past and projecting the future. UK Energy Research Centre, London.

UNCTAD, 2002: *World Investment Report 2002: Transnational Corporations and Export Competitiveness*. Report UNCTAD/WIR/2002, United Nations Conference On Trade And Development, Geneva, Switzerland.

Unel, B., 2008: R&D spillovers through trade in a panel of OECD industries. *The Journal of International Trade & Economic Development*, **17**(1):105–133.

UNEP/SEFI/NEF, 2009: *Global Trends in Sustainable Energy Investments 2009*. UN Environment Programme; Sustainable Energy Finance Initiative; New Energy Finance, Nairobi.

UNFCCC, 1992: *Convention Text* UN Framework Convention on Climate Change (UNFCCC), IUCC, Geneva.

Unruh, G. C., 2000: Understanding carbon lock-in. *Energy Policy*, **28**(12):817–830.

van De Ven, H., 1993: The development of an infrastructure for entrepreneurship. *Journal of Business Venturing*, **8**(3):211–230.

van den Wall Bake, J. D., M. Junginger, A. Faaij, T. Poot and A. Walter, 2009: Explaining the experience curve: Cost reductions of Brazilian ethanol from sugarcane. *Biomass and Bioenergy*, **33**(4):644–658.

van der Zwaan, B. and A. Seebregts, 2004: Endogenous learning in climate-energy-economic models – an inventory of key uncertainties. *International Journal of Energy Technology and Policy*, **2**(1–2):130–141.

van Lente, H. and A. Rip, 1998: The Rise of Membrane Technology: From Rhetorics to Social Reality. *Social Studies of Science*, **28**(2):221–254.

van Pottelsberghe de la Potterie, B., 1997: Issues in assessing the effect of interindustry R&D spillovers. *Economic Systems Research*, **9**(4):331–357.

von Hippel, E., 1988: *The Sources of Innovation.* Oxford University Press, Oxford, UK.

von Hippel, E., 1994: "Sticky Information" and the Locus of Problem Solving: Implications for Innovation. *Management Science*, **40**(4):429–439.

Watanabe, C., K. Wakabayashi and T. Miyazawa, 2000: Industrial dynamism and the creation of a "virtuous cycle" between R&D, market growth and price reduction – The case of photovoltaic power generation (PV) development in Japan. *Technovation*, **20**(6):299–312.

Watanabe, C., C. Griffy-Brown, B. Zhu and A. Nagamatsu, 2002: Inter-Firm Technology Spillover and the "Virtuous Cycle" of Photovoltaic Development in Japan. In *Technological Change and the Environment*. A. Grubler, N. Nakicenovic and W. Nordhaus, (eds.), Resources for the Future, Washington. pp.127–159.

Watson, J., 2004: Selection environments, flexibility and the success of the gas turbine. *Research Policy*, **33**(8):1065–1080.

WEC, 2001: Energy Technologies for the 21st Century: Energy Research, Development and Demonstration Expenditure 1985–2000: An International Comparison. A Report by a Study Group of the World Energy Council (WEC), London.

Wene, C. O., 2000: *Experience Curves for Energy Technology Policy.* International Energy Agency, Paris.

Wilson, C., 2009: Meta-analysis of unit and industry level scaling dynamics in energy technologies and climate change mitigation scenarios. Interim Report IR-09-029, International Institute for Applied Systems Analysis, Laxenburg, Austria.

Wilson, C. and A. Grubler, 2011: A Comparative Analysis of Annual Market Investments in Energy Supply and End-use Investments. Interim Report IR-11-032. International Institute for Applied Systems Analysis, Laxenburg, Austria.

Worrell, E. and G. Biermans, 2005: Move over! Stock turnover, retrofit and industrial energy efficiency. *Energy Policy*, **33**(7):949–962.

Wright, T. P., 1936: Factors Affecting the Cost of Airplanes. *Journal of Aeronautical Sciences*, **3**:122–128.

25 Policies for Capacity Development

Convening Lead Author (CLA)
Lynn Mytelka (United Nations University-MERIT, the Netherlands)

Lead Authors (LA)
Francisco Aguayo (El Colegio de México)
Grant Boyle (McCarthy Tétrault LLP, Canada)
Sylvia Breukers (Duneworks, the Netherlands)
Gabriel de Scheemaker (Conduit Ventures Ltd., UK)
Ibrahim Abdel Gelil (Arabian Gulf University, Bahrain)
René Kemp (United Nations University-MERIT, the Netherlands)
Joachim Monkelbaan (International Centre for Trade and Sustainable Development, Switzerland)
Carolina Rossini (University of São Paulo, Brazil)
Jim Watson (University of Sussex, UK)
Rosemary Wolson (Council for Scientific and Industrial Research, South Africa)

Contributing Authors (CA)
Staffan Jacobsson (Chalmers University of Technology, Sweden)
Upendra Tripathy (Government of India)
John T. Wilbanks (The Ewing Marro Kauffman Foundation, USA)

Review Editor
Youba Sokona (United Nations Economic Commission for Africa)

Contents

Executive Summary

This chapter focuses on capacities and capacity development for energy transitions. The transitions put forward in GEA require a transformation of energy systems that demand significant changes in the way energy is supplied and used today, irrespective of whether the technologies involved are new to the world or to a country, its producers or users.

Energy transitions are, by definition, long-term, socially embedded processes in the course of which capacities at the individual, organizational, and systems levels, as well as the policies for capacity development themselves, will inevitably change. From this perspective, capacity development can no longer be seen as a simple aggregation of individual skills and competences or the introduction of a new "technology." Rather, it is a broad process of change in production and consumption patterns, knowledge, skills, organizational forms, and – most importantly – in the established practices and norms of the actors involved, or what are called informal institutions. In other words, a host of new and enhanced capacities will be needed over time. Informal institutions are reflected in a range of beliefs and boundaries that shape choices about new energy technologies. These can include engineering beliefs about what is feasible or worth attempting and boundaries that shape the processes of choice, such as lines of research to pursue, kinds of products to produce, or practices of consultation and dialogue. They also emerge as "path dependence" in contexts where earlier investments result in high sunk costs, habits and practices are entrenched, and "expert views" are shaped by earlier thinking that narrows the range of choices to established technologies and evaluation techniques.

There is abundant evidence of the slow pace at which new energy technologies are being developed and diffused. Many recent studies have pointed to the continued lack of finance as a serious impediment to the transformation of energy systems in both the North and the South. But finance alone, as this chapter illustrates, does not explain the problems encountered in scaling up projects for energy transitions in developing countries. Nor does it provide a satisfactory answer to the question raised in GEA as to why, if more energy-efficient and carbon-neutral technologies are available, they are not more widely used. Investments that involve innovative changes are by their nature costly, risky, and likely to deliver returns only in the long term. How, then, might pathways to energy transitions be built today that do not compromise change in the future, especially in poor countries that can ill afford to make the same mistakes that industrialized countries have done and then expend the resources to clean up and rebuild in the next generation? A first step is to better understand the challenges posed by informal institutions because of the way they shape choices in the development of new technologies and affect the speed of technology diffusion in both developing and industrialized countries. A second is to develop the capacities to work with and around existing habits and practices in carrying out energy transitions.

Societies and organizations differ in the extent to which established habits, practices, and norms favor interaction and dialogue or more hierarchical patterns of communication. Until recently, dialogues have not featured centrally in project planning and development. Case studies of the implementation of onshore wind power in Europe and the diffusion of solar home systems in Asia and Africa illustrate that dialogues play an important role in developing and institutionalizing capacities for energy transitions. Dialogues may not "solve" problems, but they do open channels for innovative ways to deal with them.

For dialogues to work, confidence-building measures that recognize the legitimacy of local concerns, interests, and needs, as well as take account of the informal institutions shaping the behavior of actors involved in the change process, are essential in gaining broad societal support for an energy transition. And this is the case whether it concerns the choice of renewable energies, the adoption of smart grids or improved cookstoves, a shift from individual mobility to urban public transport, or the decision to construct energy efficient buildings. At the broader national level, a framework for domestic consensus building that involves the wider community in discussions about key issues and choices and their consequences must be developed. One such approach is to create spaces for dialogue about transition pathways, processes and policies that bring together members of the public and private sector, academics, and civil society in what the Dutch have called "transition platforms." As energy transitions are long term processes, spaces for dialogue at the national level also help to legitimize feedback and facilitate adaptive policymaking over time.

The choices made in selecting energy transition pathways and the capacities that will be needed in their pursuit are also inextricably linked to broader development issues and goals. These, too, must become visible as part of a dialogue process and the tradeoffs must be made explicit. The successful implementation of wind power in Germany and Denmark and the targeted bottom-up approaches in the development of small hydro projects in China, Bhutan, Nepal, and Rwanda are examples. The latter illustrate the way mutual interests can be accommodated through dialogues in which local government bodies are brought into the process early in the development of a project and a mechanism for continued dialogue and adaptive policymaking over the life of the project is put in place.

In developing countries, a multi-goal approach can both speed the diffusion of new energy technologies and stimulate the development process. Linking to domestic market building by creating or strengthening capacities for the design, manufacture, distribution, and repair of improved cookstoves and multifunctional energy platforms and the installation of solar home systems have generated multiple benefits such as these. Distributed energy systems based on biofuels are also playing an important role in the broader development process. Developing capacities for making choices about biofuels in technical as well as in economic and financial terms and evaluating these trade-offs are of critical importance in creating such benefits and ensuring their long-term sustainability. Greater attention, for example, needs to be paid to assessing whether, when, and how to scale up the production of biofuels in a given context.

Pursuing a multi-goal approach to energy transitions involves an analysis of local resource availability and the early identification of linkages and support structures that will be needed to build local markets for these new technologies. A multi-goal perspective must then be consciously built into policies, programs, and financing mechanisms. Strengthening the capacities of governments, universities, non-governmental organizations (NGOs), and other organizations in data collection and analysis, research, and problem-solving in the relevant science, engineering and social science disciplines will be essential in supporting such policymaking processes. Developing these capabilities, however, can take many years at a time when decisions about energy transitions that involve biofuels and other clean energy technologies require attention now. This chapter thus explores the role that a variety of different forms of networking for knowledge sharing and collaborative problem solving can play in contributing to the development of domestic capacities for research, analysis, and innovation.

Open innovation practices in industrialized countries are already helping to speed up the pace of problem solving in a growing number of sectors. Users, especially companies, but also government agencies and community organizations, are becoming increasingly aware of these possibilities. The core insight of the open innovation approach is the ability to use the world outside an organization to generate internally useful knowledge in a purposeful manner. This involves the development of networks and the capacities needed to use them. Developing countries are now moving in this direction. In particular, they are strengthening their domestic research sector and linking it to the wider economy through incentives to engage in problem solving, facilitate adaptive changes in the course of energy transitions, and support the process of making choices about energy pathways. Open innovation practices were a critical factor in boosting productivity and reducing costs in the development of ethanol from sugarcane in Brazil, as well as in supporting the growing of jatropha in a multi-cropping environment in Mali and the use of jatropha oil for power and electricity generation.

Designing policies to increase the demand for and supply of clean, energy-efficient technologies, speeding their adoption in sectors such as transport, energy, buildings and building materials, and strengthening their contribution to overall development goals requires the capacity to work from a systems perspective. Bringing key actors together in task forces, panels, and platforms could play an important role here. While not yet widespread, there is evidence that universities in a growing number of countries have created interdisciplinary graduate programs in sustainable engineering that could train a new generation of intermediaries and help change the existing reward structure. But more needs to be done, and more rapidly than in the past, to meet GEA's goals.

25.1   Introduction

"We can't solve problems by using the same kind of thinking we used when we created them." – Albert Einstein

New capacities and new thinking will be needed to take the world to a more sustainable state. This report has outlined the nature and magnitude of the many challenges ahead. Included among these challenges is the need to bring cleaner energy technologies[1] within the reach of users in the developing world, in terms of both access and price (Chapter 19), and to reduce the negative health impacts of the use of traditional lighting implements, cookstoves and fuels (Chapters 2 and 4). Doing so can create new livelihood opportunities that stimulate growth and reduce poverty, a point we develop in this chapter. From a global perspective, current efforts to reduce greenhouse gas emissions are overwhelmed everywhere by the increasing use of fossil fuels. Thus, worldwide a greater emphasis must be placed on the energy end-use sectors (Chapters 8–10), in order to make use of more energy-efficient technologies and systems, as well as on the supply side, in order to dramatically increase the provision of renewable energies (Chapter 11) while modernizing conventional supply systems (Chapters 12 and 13).

There is abundant evidence of the slow pace at which new energy technologies are being developed and diffused (ESMAP, 2008). Many recent studies have pointed to the continued lack of finance as a serious impediment to the transformation of energy systems in both the North and the South. But finance alone does not explain the problems encountered in scaling-up projects for energy transitions in developing countries. Nor does it provide a satisfactory answer to the question raised in previous chapters as to why, if more energy-efficient technologies are available, they are not more widely used (Chapters 8, 9, 10, 18, and 22).[2] In both developing and industrialized countries, lack of knowledge of, and access to – these technologies are major constraints. Meeting the challenges of energy transitions will thus require more comprehensive and integrated approaches to capacity development than those based mainly on building technical and management skills. As one study on capacity and capacity development cautioned, these concepts "are so all-encompassing that practitioners have often found it difficult to make operational sense of them […] It is [thus] important to begin by asking the question, 'capacity for what?'" (DAC, 2006).

This chapter focuses on capacities and capacity development for energy transitions, with particular emphasis on the pathways explored in Chapter 17.[3] Capacity is defined here as the ability of individuals, organizations,[4] societies, and communities to make choices, perform functions,

solve problems, and set and achieve objectives in a sustainable manner. Capacity development is understood as the process through which these abilities are obtained, strengthened, adapted, maintained, or changed over time.

Energy transitions are broadly defined as those processes that involve changes to "new," "clean," "sustainable," or "energy-efficient" technologies and the formal and informal institutions required to support them. There is no assumption that energy transitions take place only at the frontier of global knowledge. Moreover, while seemingly very different, energy transitions have much in common. Each involves changes to existing patterns of production and demand that will require new knowledge, skills, organizational forms and habits, practices and norms, whether the process is taking place in a least developed, developing, emerging, or industrialized economy. Because the points of departure for energy transitions are very different across these varied contexts, it is important to develop approaches that enable us to perceive these differences and take them into account in the design of policies, programs, and processes for energy transitions over time.

The transformation of energy systems, irrespective of whether the technologies involved are new to the world or to a country, its producers or users,[5] is a long-term, socially embedded process that involves a host of other changes that extend beyond merely introducing a new technology. Although capacity development has traditionally been seen as a major concern in developing countries, given the nature and magnitude of the changes needed in energy systems around the world, attention to capacity development is now required globally.

Ultimately, the world must shift to renewable, clean, and green energy sources, zero-energy buildings, and more efficient energy supply systems. This shift will involve multiple interrelated transition processes, each of which will give rise to new capacity issues. As a result, capacity development policies and practices will have to change.

This chapter is structured as follows. Section 25.2 conceptualizes energy transitions. It briefly addresses the theoretical tenets of a transition perspective, as well as how and why it is relevant in helping to understand the levels, processes, mechanisms, actors, and institutions that affect and are affected by technological change. It also introduces the policy dimension in energy transitions. Section 25.3 reviews the capacity development literature and then links the energy transition perspective

---

1   Clean energy technologies are generally defined to include fuels such as natural gas, liquid petroleum gas (LPG), biodiesel, and bioethanol. Some also include carriers such as electricity and end-use devices, e.g., improved cookstoves (Goldemberg et al., 1985; UN AGECC, 2010; see also Chapter 19).

2   This is also the theme of the UNIDO Industrial Development Report (UNIDO, 2011).

3   GEA-Efficiency is a pathway in which energy demand is low and there is a strong emphasis on energy efficiency standards and regulations across all demand sectors. The GEA-Supply pathway focuses on the rapid up-scaling of supply-side options, a necessary condition for achieving a transition where energy demand remains high. New infrastructures and renewables are also important in this pathway. The GEA-Mix pathway is an intermediate transition pathway that emphasizes diversity of the energy supply mix with greater prominence given to local choices and resource endowments and incentives to increase energy efficiency and reduce energy use (see Chapter 17).

4   For example: government agencies, enterprises, universities, farms, and NGOs.

5   This approach to "new" energy technologies is adapted from Ernst et al., 1998.

to the concept of capacity development for energy transitions and the systems-embedded, innovative nature of this process. We argue that models of technology transfer and deployment often fail to appreciate the complexity of change processes, and we introduce the concepts of path dependence[6] and informal institutions[7] as analytical tools.

Having thus set out the conceptual framework, subsequent sections discuss specific aspects of capacity development for energy transitions in developed and developing countries. The sections are organized thematically, paying particular attention to a number of capacity issues raised in the three GEA energy transition pathways. These include energy efficiency; the development of new, clean energy technologies and their diffusion; and the choice of energy supply mix and role of local resource endowments in the development of transition strategies. A range of capacities will be needed to develop and implement these strategies in a manner that will contribute to resolving energy access issues and ensure that they make positive contributions to livelihoods and to sustainable development. The six case studies following the conclusions of this chapter amplify the discussions of these challenges.

Section 25.4 addresses the issue of path dependence and informal institutions as impediments to change. Change processes inevitably face challenges that emerge from established interests, habits, practices, and norms. The case studies in this section illustrate these problems in the adoption of energy-efficient technologies and decentralized energy systems. Section 25.5 addresses the importance of dialogue at the local level among relevant stakeholders, illustrated here in the case of wind power. It also stresses local market development and the role of feedback and flexibility at the project level in supporting the diffusion of new energy technologies. Section 25.6 focuses on distributed energy systems and how they can serve multiple goals and bring about multiple benefits at the local level. Examples include the local production of biofuels and the development of small and micro-hydro stations for lighting and stationary power. Access to knowledge and information is critical in the decision-making processes through which choices about the energy supply mix are made. Section 25.7 discusses some of the capacities that are needed to make such choices and new ways of accessing knowledge, including open and distributed innovation processes and the creation of knowledge networks. The latter are also of interest for the role they can play in reducing the brain drain from developing countries.

The "management" of energy transitions is discussed in Section 25.8, which examines the Dutch energy transition approach as an innovative model for creating strategic intelligence for policy and innovation. The Dutch model is of interest to developed and developing countries because of the potential for creative adaptation of three of its core elements: its focus on transformative change, its reliance on bottom-up

processes, and its use of "platforms" to enroll business and other non-state actors in the energy transition process. This section also looks at how oil producers are adapting to the need for change.

Section 25.9 moves to the international level, which has assumed an important role in energy-related negotiations, particularly with regard to trade and technology transfer. The section focuses on strengthening capacities for effective participation in these negotiations. Section 25.10 draws out a number of "lessons" for capacity development and policies that might stimulate and support capacity development processes for energy transitions. These take the form of pointers and issues for consideration rather than a strict list of what actions to take. Capacity development, as we acknowledge throughout, is a complex process that defies a focus on organizations and a simple management approach.

25.2   Energy Transitions: Moving Frontiers

Past energy transitions have been part of broad, systemic processes of change in patterns of production and consumption of goods and services (Smil, 1994; Fouquet and Pearson, 1998; Grubler, 2004). The shift to oil, for example, stemmed from the growing demand for gasoline to power forms of transportation that either did not exist before 1880 (motorized vehicles and airplanes) or relied on the use of wind (ships). The development of automobiles, in turn, owed much to the growing demand for mobility (Geels, 2005).[8]

During each such transition, the old and new energy regimes, based on different technologies, fuels, products, behavior and organizations, co-exist. What is sometimes viewed as a process of technology diffusion is really a more complex transformation process in which a new technological regime grows out of an existing one, with change first occurring in niche markets. The new regime does not emerge in isolation but is connected with the emergence of new technologies, expectations, skills, management systems, supplier-user relationships, ideas, and changes in the regulatory framework (Freeman and Perez, 1988). It sees new fuels and technologies replacing others after a long period of co-existence. Both the technology and social context change in a process that can be seen as co-evolution (Rip and Kemp, 1998). An energy transition thus involves a moving frontier along multiple dimensions. Moreover, these are not deterministic processes, and may fail to emerge or spread, as illustrated in Chapter 19 by the wide gaps in access to modern energy technologies between urban and rural areas of the developing world. Understanding the context in which such transitions occur is critical in bringing about a global energy transformation.

The best theoretical attempt to understand these multilayered processes of systemic change is the transition approach, which is based on a multilevel framework of niches, regimes, and landscapes (Rip and Kemp,

---

6   Path dependence is the tendency for past practices and decisions to shape present choices.

7   Informal institutions include the traditional habits, practices, and norms of actors in the system.

8   See Chapter 16 for a more extensive discussion of energy transitions in historical perspective.

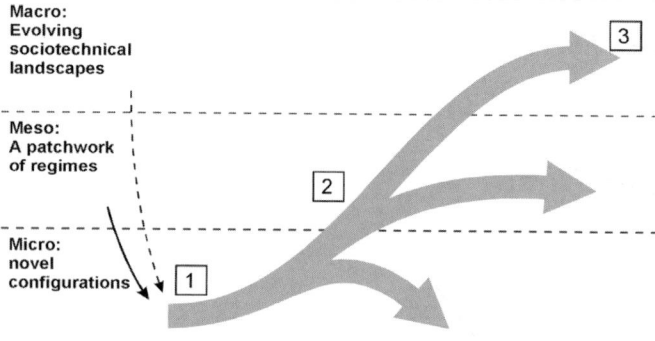

Macro:
Evolving
sociotechnical
landscapes

Meso:
A patchwork
of regimes

Micro:
novel
configurations

[1]
[2]
[3]

[1] Novelty, shaped by existing regime
[2] Evolves, is taken up, may modify regime
[3] Landscape is transformed

Figure 25.1 | The multilevel transition model. Source: Rip and Kemp, 1996.

1996; 1998; Geels, 2002; 2004; 2005; Geels and Kemp, 2012). Central to the transition framework are regimes, i.e., the dominant practices, search heuristics, outlook or paradigm, and the ensuing logic of appropriateness pertaining in a domain – a sector, policy area, or a field of science and technology – that give it stability and orientation and guide decision making (Kemp, 1994). In the case of automobiles, for example, at a high level of aggregation there is a fossil-fuel energy regime, and at a lower level of aggregation there is an internal combustion engine regime for automotive technology. Users are part of a regime that is a component of what Geels (2004) has called the "socio-technical regime." A regime is an interpretive analytical concept that invites the analyst to investigate the "deep structure" behind activities, e.g., shared beliefs, norms, standardized ways of doing things, heuristics, and rules of thumb (Geels and Kemp, 2012). Policy may be involved in promoting or regulating these socio-technical regimes.

Regimes are part of a broader context – the macro landscape of values, infrastructures, settlements, political associations, prices, lifestyles, aspirations, and concerns of people. The landscape may variously sustain a particular regime or put it under pressure. In the case of automobiles, it does both. At the micro level are niches, novelties, and alternative practices (Figure 25.1). Within this scheme, novelties (new products, technologies, or special institutional arrangements) emerge in niches, particular domains of use, constellations of actors, and a geography, all of which are part of a broader world consisting of regimes and landscapes. What happens in each niche is shaped by the external developments with which it interacts. Each of these levels has its own type of dynamics, but developments are interdependent. The diffusion of niche products is constrained by competition from existing, well-developed products and cultural practices.

Transitions are viewed as the outcome of pressures on socio-technical regimes (e.g., cultural criticism, regulations) and changing conditions, the development of alternative products and systems, elements of which are breaking out of niches because of learning economies, support from various actors (including incumbent companies, politicians, civil society) and processes of alignment (Geels, 2002).

The transition perspective has relevance for transitions to energy systems that meet the GEA goals (see Cluster I). Change may occur within an existing regime. For example, carbon reductions may be achieved through technical fixes that leave the logic of the regime intact. Change may also occur through the transformation or replacement of an existing regime with a new regime of energy carriers involving different technologies, practices, paradigms, and institutions. Here, it is important to note that the knowledge and capacity issues involved in adopting technical fixes and creating new regimes are altogether different. The introduction of the catalytic converter as a technical fix to reduce car emissions, for example, required no changes on the demand side or in the powertrain of vehicles. Neither car manufacturers nor users had to acquire special capabilities. In the case of regime-shifting change, however, such as the move to renewables, a wider range of capacity issues are involved in achieving changes in the lifestyles, habits, and practices of users and producers. Society has to create and sustain a capacity for making choices about options and configurations, skills for installation, and knowledge for product development, load management, defining appropriate policies, and adapting policies to changing circumstances.

In order to pursue a path towards an energy transition, a cluster of new technologies will need to be fostered (Jacobsson and Bergek, 2004; Stern, 2006). Energy transitions therefore involve a moving frontier that advances for different technological options at different rates and in different contexts. This has a bearing on both capacities and capacity development processes. When these new technologies come with risks, as they often do in regime-shifting change, the capacities needed to understand the risks, and to contain and manage them, become important. Special government policies and science and technology programs may be needed to stimulate development of the technologies that underpin a new regime, support their diffusion, and guarantee safety, as well as manage transition processes.

Past energy transitions in the industrialized world have included, for example, the transition from wood to coal, from coal to natural gas for power and heating, from gaslight to electric light, the creation of centralized power systems and nuclear technologies, or more recently, the development and diffusion of renewable and clean energy technologies in both developed and developing countries. In many cases, governments have played a role in encouraging transition processes by increasing public investments in science and technology programs, providing market incentives, and introducing regulations and standardization. But governments alone cannot bring about an energy transition. The following section links the energy transition perspective to the capacity development issues that arise from the systems-embedded and innovative nature of the transition process.

25.3 Capacities and Capacity Development for Energy Transitions

Capacity building and capacity development have mainly been associated with developing countries in the past, but this is no longer the case. Energy transitions involve a broad set of changes in existing patterns of production and consumption, in the knowledge and skills required, the organizational forms, and the business and governance practices, as well as other social habits, practices and norms, whether the process is taking place in a least developed, developing, emerging or industrialized economy. We thus go beyond the traditional development literature to identify a range of capacities that are of increasing relevance in energy transitions. The section begins with a brief review of the early development and "capacity building" literature, its recent evolution, and the challenges it still faces. It then explores issues of capacity development for energy transitions that emerge from the application of an innovation systems approach.

25.3.1   From Capacity Building to Capacity Development

The early development literature, as Bouille and McDade (2002) noted, referred to capacities in terms of skills, and the process of creating and/or strengthening them, as "capacity building." It assumed that relatively few local capacities existed in specific domains and these would need to be transferred[9] to developing countries mainly through foreign aid and investment. Initially, development agencies focused on providing goods, services, and foreign experts to government ministries in developing countries (UNDP, 2009a), and consequently on ensuring that public sector organizations were capable of managing large public investment programs (Lusthaus et al., 1999).

During the 1990s, a substantial and growing part of the population in developing countries remained without access to energy, although significant investments in energy infrastructure had been realized (WBG, 2003). Donors became increasingly concerned about the effectiveness of foreign aid and this led to major changes in policy, including a rethinking of development assistance practices and the process of capacity building. The policy changes began in 1991, with the publication of a report entitled, *Principles for Evaluation of Development Assistance*, by the Development Assistance Committee (DAC) of the Paris-based Organisation for Economic Co-operation and Development (OECD). Initially a conversation among donor countries about "the role of aid evaluation in the aid management process," the principles went beyond earlier conceptions of donor-recipient relationships to view development assistance as "a co-operative partnership exercise between donors and recipients" (DAC, 1991). The nature of this partnership was not defined, however, nor were prospective partners included in the discussions. In 2005, this changed when developed and developing countries met in Paris and endorsed the view that capacity development was fundamentally an endogenous process that should be led by the recipient countries, with donors playing a supporting role (OECD, 2005/2008).

The Paris Declaration on Aid Effectiveness sought to address issues of sustainable development that went beyond those of finance. This led to the emergence of a new concept, "capacity development." In contrast with "capacity building," capacity development recognized that some capacities existed everywhere upon which one could build.[10] The Paris Declaration also took a broader view of the actors involved, moving beyond individuals to include organizations and society (Fukuda-Parr et al., 2002; Bouille and McDade, 2002). The Economic Community of West African States (ECOWAS), in a white paper for a regional policy on access to energy services, agreed, advocating that "capacity-building [should...] be aimed both at private operators (local operators, investors, donors, etc.) and public actors (ministries, regulatory authorities, rural electrification agencies, etc.)" (ECOWAS, 2006). The white paper also stressed the need for linkages across countries within a region (a point to which we return in Section 25.7.3 on knowledge networks).

The capacity development approach understood capacity as an outcome and capacity development as a process. In contrast with standard assessments of individual performance or project outcomes, some of the capacity development literature came to regard participatory monitoring and evaluation at various points in the life of a project as part of a learning process. From this perspective, feedback is a desired outcome that serves not only as a means for measuring progress, but also as an opportunity for "questioning, reflecting, learning and modifying" (IDRC, 2008). More broadly still, tools for evaluative thinking, such as "outcome mapping," adopted by the International Development Research Centre (IDRC), are more attentive to "changes in the behavior, relationships, activities, or actions of the people, groups, and organizations with whom a program works directly" (Earl et al., 2001). This approach has important implications for the kinds of capacities and capacity development processes that are needed for energy transitions (Young, 2008).

The new conceptual insights notwithstanding, "[i]n recent years about a quarter of donor aid, or more than US$15 billion a year, has gone into technical co-operation, the bulk of which is ostensibly aimed at capacity development. Despite the magnitude of these inputs, evaluation results confirm that development of sustainable capacity remains one of the most difficult areas of international development practice" (DAC, 2006).

The development of sustainable capacity has remained elusive (see Box 25.1), for a number of reasons. One is the persistence of earlier thinking and practices, since "until recently capacity development was viewed mainly as a technical process, involving the simple transfer of knowledge or organizational models from North to South" (DAC, 2006). Such views share with the early development literature a conceptualization of technology as embodied in machinery and equipment, and its transfer and assimilation as automatic and context-free (Bell, 1984; Forbes and Wield, 2002). Development was often seen as a linear process "a sort of race along a fixed track, where catching-up is merely a question of relative speed" (Perez, 1988).

---

9   In the 1950s, development economists such as W. Arthur Lewis and Raul Prebisch argued that developing countries were not capturing an adequate share of the gains from technological change. The solution was to industrialize. But the scarcity of capital, skills, and technology made it imperative for these countries to attract foreign investment (Lewis, 1950; Prebisch, 1959). This view subsequently became entrenched in the development literature (see Bell, 1984; Mytelka, 1989; Forbes and Wield, 2002).

10   Although they differ in their approach to energy transition processes and the capacities needed to support them, the terms capacity building and capacity development are often used interchangeably in the literature.

Box 25.1 | "We are making progress, but not enough"

In its 2003 Human Development Report (HDR), the United Nations Development Programme (UNDP) embraced the capacity development approach and stressed the need for initiative and ownership to reside with and in developing countries. This complemented the evolving view of development assistance as a partnership, but it did not resolve the challenges involved in putting such an approach into practice.

Ways to address the many issues of ownership, inclusive partnerships and the effectiveness of aid raised in Paris in 2005 (OECD, 2005/2008) were still being worked out at a follow up meeting in Accra, Ghana, as the Accra Agenda for Action (2008) illustrated. They included the lack of capacities in developing countries to plan, manage, implement, and account for the results of policies and programs; the need to deal with issues of predictability, conditionality of funds, and the untying of aid on the donor side; and the continued difficulties in scaling up for more effective aid.

Source: Accra Agenda for Action, 2008.

During the 1980s and 1990s, researchers in developing countries began questioning the linearity and passive nature of the catch-up process. Effective absorption and diffusion of new technologies, it was found, mainly took place when accompanied by two broad sets of purposive actions:

- innovative practices, such as adapting imported technologies to suit local conditions, adjusting to new sources of inputs, altering the product mix, and resolving bottlenecks in production (Stewart, 1984; Katz, 1985; Mytelka, 1985; Westphal et al., 1985; Lall, 1992); and

- technological learning, by engaging in a process of intensive technological searching and protracted negotiations with prospective sellers or licensers of technology, by adopting a strategy focused on adaptation and upgrading of the original design and by developing in-house research capabilities (Rosenberg and Frischtak, 1985; Katz, 2004; Ernst et al., 1998; Bell, 2009).

Such purposive actions were not easy or cheap, but were often facilitated by a supportive policy environment (Bell, 1984; Kim, 1997; Nelson, 2004). These lessons were soon forgotten, however, and development practice returned to a focus on more passive approaches to technology transfer through development assistance projects and foreign direct investment, rather than on the learning and technological mastery that drove the process of catching up in the "Asian tigers" (Kim, 2004; Mytelka, 2004).

Another factor contributing to the difficulties encountered in capacity development is the continued donor focus on "skills development and individual training" (Baser and Morgan, 2008), with a particular emphasis on upgrading public sector governance (World Bank and IMF, 2004). As those who have adopted "institutional" or "systems" approaches point out, the way public agencies function is shaped as much by the training individuals receive as by their social and political context (Lusthaus et.al. 1999; Baser and Morgan, 2008; DAC, 2006). Inspired by the work of Douglass North, they define institutions as the "formal and informal rules of the game," which distinguishes institutions

from "organizations" (North, 1994). This approach goes beyond a focus on public agencies to include a wider array of organizations whose involvement in energy transition processes is critical. It also acknowledges that these actors cannot be divorced from the "structures of power and influence," or from the informal "institutions" in which they are embedded (DAC, 2006). From this perspective, training programs can go only so far in bringing about change. Instead, as the business and innovation systems literatures have shown, what often impede change in organizations are the habits, practices, routines, and norms of the actors within the organization and in the wider society (Teece, 1988),[11] a point developed in the framework presented in Section 25.3.2.

Lastly, an emerging view, more often heard than put into practice, is that the private sector should take on a more prominent role in the design, financing, and delivery of rural and/or renewables-based technologies (van de Vleuten et al., 2007). Public agencies often lack the capacity and motivation to improve rural energy conditions in developing countries and focus quasi-exclusively on urban energy needs, while expecting private actors to address rural conditions. Private firms, local or international, often lack the in-country capacity to deliver the necessary energy carriers or technologies, in part also because policies and regulation do not favor this type of solution.

25.3.2   An Innovation Systems Approach

The definition of capacity as the ability of individuals, organizations, society, and communities to make choices, perform functions, solve problems and set and achieve objectives, and of capacity development as an endogenous learning process through which these abilities are obtained, strengthened, adapted, maintained, or changed over time, owes much to the work of UNDP (2008a; 2008b; 2009d). It differs from the UNDP definition, however,

---

11   An example of how these informal institutions can slow the process of change can be found in Section 25.11, Case Study 1, on the introduction of smart grids in the UK.

in three important ways. First, as in the broader institutional and systems literature, this definition distinguishes between organizations, such as industries, farms, government ministries, hospitals, universities, non-governmental organizations (NGOs), and informal institutions. Second, it is rooted in an understanding of energy transitions as processes of innovative change that take place interactively across the multiple levels shown in Figure 25.1. Third, it relates to the growing importance of learning and innovation in the processes of development and change.

During the 1980s and 1990s, changes in production and competition in the industrialized countries drew attention to the role of knowledge and innovation in the competitiveness of firms and, by extension, of nations (OECD, 1992). Over time, the knowledge intensity of production extended beyond the high technology sectors to reshape a broad spectrum of traditional and new industries in both developed and developing countries (Mytelka, 2000). In the developing world, this change placed a premium on learning, capacity development, and innovation in the process of catch-up and keep-up, as case studies in Asia, Latin America, and Africa have shown (Ernst et al., 1998; Muchie et al., 2003; Pietrobelli and Rabellotti, 2006). This would lead to a growing interest in innovation systems thinking as a useful approach to contextualizing learning and innovation processes, first in the industrialized countries and, more recently, in the developing world.[12]

As a set of conceptual tools and frameworks, the innovation system approach is still evolving. Since its appearance in the 1980s, the approach has provided comprehensive and integrated analyses of the processes of learning and innovation in various societies and economies. It is particularly relevant to energy transitions, which are broad processes of innovation and change.

An innovation system can be seen as a network of firms and other economic agents that, together with the institutions and policies that influence their innovative behavior and performance, bring new products, new processes, and new forms of organization into economic use (Lundvall, 1992; Nelson, 1993).[13] The strength of the innovation systems approach lies in its focus on the interactions among these actors and their embeddedness in an institutional and policy context that influences their behavior and performance. Institutions in this sense are not organizations, but "sets of common habits, routines, norms, rules and established practices that regulate the relations and interactions between individuals and groups" (Edquist, 1997) and prescribe behavioral roles, constrain activity and shape expectations, or what we call here informal institutions (Hall and Taylor, 1996; Edquist, 1997; Scharpf, 1997; Storper, 1998; Hukkinen, 1999). These habits and practices are learned behavior patterns marked by the historical specificities of a particular system and

moment in time. As such, their relevance may diminish as conditions change. This is particularly likely during the course of energy transitions, which are, by definition, dynamic processes.

From a policy perspective, the strength of the innovation systems approach lies in the attention it draws to policy dynamics and the way these emerge from the interaction between policies and the habits and practices of the actors whose behavior is targeted by policy (Mytelka, 2000). Policies can be designed to reinforce desirable habits and practices, or to induce change, provided that there is some understanding of the habits and practices of local actors, and that a continuous process of feedback provides the inputs for adaptive policymaking. Learning and unlearning on the part of firms, policymakers, and other actors in an emerging or established innovation system are thus at the heart of its ability to respond to new challenges. Continuous flows of knowledge and information provide critical inputs into this process.

These insights have particular relevance for energy transitions that are long-term processes and will require adaptive policymaking to successfully carry out the broad range of interrelated changes, negotiate the multiple pathways, and manage the risks and uncertainties involved. Capacities at the actor and systems levels can be expected to differ both across socioeconomic contexts and over time in energy transition processes. By conceptualizing energy systems as socially embedded and historically shaped by the habits and practices of the actors within them, the innovation systems approach draws attention to the importance of addressing issues that arise out of these informal institutions, in the design of capacity development policies and processes. Among these are the challenges posed by path dependence,[14] i.e., the tendency for past practices and decisions to shape present choices, and lock-in, where complementary technologies or products can very quickly gain dominance in a market even if they are inferior.[15]

Path dependence in energy transitions is reflected in a range of beliefs and boundaries that shape choices about new energy technologies. These can include engineering "beliefs about what is feasible or at least worth attempting," and boundaries that shape processes of choice such as lines of research to pursue, kinds of products to produce, organizational routines and development trajectories (Teece, 1988; Mytelka, 2008).[16] Path dependence also emerges in contexts where earlier investments result in high sunk costs, habits and practices are entrenched (see Box 25.2), and "expert views" are shaped by earlier thinking that narrows the range of choices to established technologies and evaluation techniques.

---

12 In Brazil and South Africa, an innovation systems approach has been applied in research, in industrial surveys, and in policymaking. It has also been adopted by the European Union, by individual Asian countries, and by New Partnership for Africa's Development (NEPAD) for their innovation surveys.

13 The innovation system approach is introduced in Chapter 24.

14 For more on the concepts of path dependence and lock-in, see Chapter 17.

15 The classic example of lock-in is the QWERTY keyboard used in typewriters and maintained in computers, even though the placement of keys makes it inefficient (David, 1985). For further lock-in examples, see Chapter 19.

16 The business literature has recently begun to focus on "taking the bias out of meeting" and strategic plans by "gathering data, discussing analogies, and stimulating debate that together can diminish the impact of cognitive biases on critical decisions" (Lovallo and Sibony, 2010). However, they do not seem to be aware of the way in which past practices can also affect the selection of data and analogies.

Box 25.2 | Path Dependence

In 2008, the World Bank's International Finance Corporation approved a US$450 million loan for the Tata Mundra project, designed to build five coal-fired power plants in the Indian state of Gujarat. This illustrates the challenge that path dependence can pose for energy transitions. In this case, traditional norms and practices that oppose the use of subsidies and employ short-term, static cost comparisons to evaluate alternative options, provided the rationale for an energy choice that will have negative implications for India's CO_2 emissions in the long term.

The first of the Tata Mundra power plants will be commissioned in mid-2011. It is expected to emit 40% less CO_2 than existing coal-fired power plants in India, but given the lifespan of the new plants, it will contribute "23.4 million tonnes of CO_2 per year" to the environment for the next 25–30 years (IFC, 2009).

IFC vice-president Rashad Kaldany justified this decision on the grounds that, in comparison with other alternatives, "[t]his is by far the least expensive and to try to do something like either wind or solar would cost huge amounts in terms of subsidies" (Wroughton, 2008). While this may be true in the short term, taking longer-term considerations into account, such as the future costs of retrofitting a plant with carbon capture and sequestration technology, or the imposition of a carbon tax on coal-fired power plants, might have altered the choice matrix.

In 2010, the World Bank approved yet another loan for the construction of a coal-fired power plant, this time in South Africa. Once fully operational in 2015, the Medupi power station will emit 26 million tonnes of CO_2/year (Duffy, 2010). Decisions such as these, to support old, polluting technology and to promote new and renewable forms of energy only when they are commercially viable, create further disincentives to change energy demand and supply patterns and make it even more difficult to move down a cleaner energy pathway.

Investments that involve innovative changes are, however, by their nature costly, risky, and likely to deliver returns only in the long term. How, then, might pathways be built today that do not compromise change in the future, especially in poor countries that can ill afford to make the same mistakes that industrialized countries have done and then expend the resources to clean up and rebuild in the next generation? To address this challenge, two complementary approaches are discussed.

One approach involves rethinking the standard economic calculations that have so far failed to take account of social, economic, and environmental costs and their cumulative impacts over time.[17] The other approach recognizes the long time horizons required to develop industrial production capacity in new energy technologies and to build the necessary infrastructure and markets. It therefore favors developing a portfolio of new technologies and associated industries in parallel, and not in sequence (Jacobsson and Bergek, 2004). From this perspective, the most appropriate policy would be not to select the current "best," "off-the-shelf" technology, as neoclassical economists would prefer, but simultaneously to foster a broad range of technologies and associated industries (Sandén and Azar, 2005).[18] This would apply

not only to traditional leaders in technology, but also to newcomers such as Brazil, China, and India, and to other countries that today only import their capital goods. In the latter cases, the time needed to build up production capacity would, of course, be shorter than developing entirely new industries. As a major Latin American research program in the 1980s demonstrated, even importing, successfully operating, and reproducing technology requires substantial engineering efforts and may take decades rather than years to develop (Katz, 1983).[19]

A better understanding of informal institutions is also needed because of the way they shape choices in the development of new technologies and affect the speed of technology diffusion in both developing and industrialized countries. Diffusion, as we saw in the previous section, is associated with innovation in organizational routines, new mindsets, practices, and supplier-user relationships. It is thus a context-specific process involving elements of innovation (Kline and Rosenberg, 1986). Much of the literature on energy transformations, especially in

---

17 Daniel M. Kammen, the World Bank's new "clean-energy czar," reiterated this point in a recent interview (Broydo Vestel, 2010).

18 For example, F-i-T is an instrument for a broad range, while green certificate markets always select the best off the shelf.

19 The introduction, use, and diffusion of imported solar cells, for example, would require the capacities of local engineering firms to design and deliver whole systems; roof and façade manufacturers, electricians, and architects to incorporate the new technology and extend its applications; and changes in formal and informal institutions to align with the new technology. See the cases of Bangladesh and Senegal in Section 25.5.2.

developing countries, retains an earlier language – that of technology transfer – or has adopted a new language – that of "technology deployment" – both of which conceive of the process in contextually neutral and linear terms. In most developing countries, for example, technology transfer through donor-supported projects has been one of the principal ways in which "new" energy technologies have been deployed. But they have rarely led to rapid and sustainable diffusion – a problem that is widely acknowledged and most often attributed to the general lack of absorptive capacity in developing countries.[20] Absorptive capacity is only part of the problem, however. The availability of affordable energy technologies, especially for lighting and power, is important, and the habits, practices, norms, and interests of local actors play a major role.

A recent study of capacity development initiatives based on an analysis of 18 case studies,[21] suggests that donor practices also tend to reinforce path dependence in the project design and implementation phases (Baser and Morgan, 2008), and thus are likely to contribute to the slow diffusion of new energy technologies in developing countries in several ways. Current approaches to capacity analysis are overly focused "on the macro and the aggregate levels such as state building, improved governance and democratization." At the operational level, the experience of managing projects and programs has given rise to an emphasis on "prediction, targeting, control, results and accountability," and an interest primarily in "ideas that are simple, make immediate sense and can be easily integrated into what they are already doing" (Baser and Morgan, 2008).

This section has moved beyond the development literature in its understanding of the range of capacities and the types of capacity development processes that impact on energy transitions. While finance and governance remain major challenges, capacities that will strengthen the flexibility of existing systems to respond to change are critically important in undertaking and sustaining an energy transition. Among these, building the capacity for long-term, dynamic, multi-goal strategic programming and policymaking will be needed. This will not be an easy task in developing countries. It will require vastly increased access to and capacity for the analysis of a continuous stream of domestic and international information and knowledge as inputs to the policy process.

Within the policy process, capacities for individual and organizational learning must be strengthened. In both developing and developed countries, a broader definition for the identification of capacities and capacity development processes to include the learning and

unlearning of habits and practices by all actors in the "innovation" system will be needed. "Institutional" capacity development in this sense refers not only to changes in formal rules, legislation, or standards. It also calls for innovation in organizational routines, mindsets, and feedback mechanisms, as well as the creation and strengthening of new networks that enhance interactions and enable learning processes.[22] From this perspective, capacity development for energy transitions aims to embed new practices and the technologies that are part of these practices, within existing social contexts. The next section explores this process in the context of energy efficiency in diverse environments.

25.4   Path Dependence and Informal Institutions

Path dependence and informal institutions have been important contributing factors to the limited acceptance and slow diffusion of a wide variety of new energy and energy-saving technologies. This section illustrates these challenges with reference to the adoption of new and potentially more efficient energy-related practices and energy technologies in urban transport, energy services, buildings, and construction materials. While energy efficiency is often simply a palliative practice within an existing regime,[23] it can – and should – become a pathway towards a regime-altering energy transition. We therefore look at where and how this is being done, and what lessons might be drawn from these experiences for the kinds of capacities and capacity development processes that are needed.

25.4.1   Urban Transport

In the 1960s, major cities in Europe and North America began to develop and implement urban transport policies. These policies were sometimes implemented in collaboration with national governments and the private sector, and then followed decades of fluctuating oil prices and increased traffic congestion and pollution. These policies were not particularly regime-altering. They focused on the creation or extension of metro systems, the reintroduction of the electric trolley, and the development of light rail systems and, more recently, of bus rapid transit (BRT) lanes.[24] To a lesser extent, they have also included partnerships that could significantly alter urban transport technologies by accelerating the adoption of electric and hydrogen fuel-cell

---

20 For discussions of technology transfer processes and specific cases, see Mytelka, 1985; 1986; Bell and Pavitt, 1997; Kim, 1997; 2003; and Mytelka 2004.

21 Although the study does not deal specifically with new energy technologies, the cases on education, environmental action, local government support, and NGOs are relevant for issues raised in this chapter.

22 See the discussions on open innovation systems and knowledge networking in Section 25.7.

23 A contemporary example is the introduction of energy efficiency standards for automobile engines, not to discourage their use but as a means to deal with energy security issues by reducing consumption of imported gasoline.

24 For an overview of these systems, see Chapter 9.

vehicles, notably in urban bus networks.[25] There are also experiments that might lead to substantial changes in personal mobility practices, such as the Velib self-service, subscription-based, bicycle hire system launched in Paris in 2007 (C40 São Paulo Summit, 2011) and, more recently, the development of similar programs for automobiles. All of these approaches aim to reduce the use of cars in urban environments by working with existing habits of personal mobility, but around the practice of doing so through personal vehicles that clog city streets and collectively generate far more CO_2. European cities are at the forefront of these initiatives.

The first decade of the 21st century also marked an important watershed in human history when, for the first time, more than 50% of the global population are urban dwellers (see Chapter 18). Many of the world's mega-cities are located in developing countries where energy and transport policy and planning are weak and public transport systems do not meet emerging needs. The situation is particularly problematic in Africa (Davison and Sokona, 2001), but it has emerged in the new millennium as a major challenge in cities across the developing world, largely as a result of the continued growth in urban populations and in the use of automobiles.[26]

A number of strategies have been developed to deal with this problem and to attract automobile users to switch to public transport systems. Some of these strategies have emphasized energy efficiency and linked it to a process of ensuring the survival of the public bus systems. In the case of the Bangalore Metropolitan Transport Corporation (BMTC), this required significant changes in government policies and in the management of the corporation (see Box 25.3).

Box 25.3 | Bangalore Metropolitan Transport Corporation

Until August 1997, when it became an independent state-owned corporation, the Bangalore Metropolitan Transport Corporation (BMTC) was typical of India's public bus companies. Its revenues failed to cover costs, and buses were old and inefficient in their use of fuel and polluting. Morale and discipline among workers was low, strikes were frequent, a lack of managerial skills contributed to poor operational efficiency, and there was little scope for introducing changes from the bottom up.

When financial support from the Karnataka State government was not forthcoming and transport services were threatened with privatization, BMTC's management needed to develop a new strategy. They began by bringing the workforce on board. Training, more transparent management practices, and innovative labor welfare measures encouraged workers to be more attentive to maintaining their buses and ensuring compliance with pollution norms. Increasing the number of bus depots from 13 to 29 made it possible to monitor fuel consumption more closely and improve maintenance practices and operational efficiency. As revenues grew, BMTC was able to retire 1700 aged vehicles and purchase 2974 new ones that met European standards.

BMTC also worked hard to change traditional practices in its relationship with the Karnataka government, and in 2003 set up a cell to create an environmental plan for the company and develop closer working relationships with the Karnataka State Pollution Control Board. By 2007, in comparison with Mumbai, Delhi, Chennai, and Ahmedabad, the BMTC had the highest bus to population ratio, the youngest bus fleet, and the best fuel economy ratio. It was also India's only large profit-making state transport corporation. The BMTC has received many awards in the field of affordable public transport, and fuel efficiency in urban services. Perhaps the most important impact, however, was the federal government's decision to diffuse the BMTC model across India's cities.

Source: Torres-Montoya, 2007; Tripathy et al., 2008.

---

25 These began with the California Fuel Cell Partnership, launched in 1999, and a series of EU programs designed to test hydrogen fuel cell (HFC) buses and refueling stations in some member states (e.g., the CUTE and ECTOS programs). These programs have expanded through collaboration with municipalities, regional governments, and industry in the building of hydrogen refueling infrastructure across Europe and the introduction of HFC and electric HFC buses and other vehicles into commercial use. See the Hydrogen Bus Alliance (www.hydrogenbusalliance.org) and HyRaMP, the European Regions and Municipalities Partnership for Hydrogen and Fuel Cells (www.hy-ramp.eu).

26 The challenge presented by urban population growth and increased automobile use has been particularly acute in Beijing (since the Olympic Games in 2010), in New Delhi (despite the conversion of bus and taxi fleets to less polluting compressed natural gas), Bogota, and Mexico City. Over the past few years, in each of these countries, city and national governments have combined forces to develop transport, urbanization, and related strategies to overcome this problem, sometimes with external funding from, for example, the Clean Development Mechanism in the case of Bogota, and the Clean Technology Fund in the case of Mexico City.

Other strategies have focused on the issue of access to transportation and the role it can play in achieving broader development goals. Several of these initiatives have been supported by local mayors. Curitiba, capital of Brazil's Paraná state, was a pioneer in that respect. Driven by population growth, the city moved away from earlier urban planning efforts based on the automobile, to the design of a master plan, adopted in 1965, for integrated transport and urban planning. As it evolved, the plan focused on the development of a public transport system that was unique at the time – the BRT system (Lindau et al., 2010).

There are now 47 BRT systems worldwide, and more are under construction (see Chapter 9). Among developing countries, 16 cities in Latin America have built BRT systems, and the model has spread to Asia. Africa, however, lags far behind. In sub-Saharan Africa, only Johannesburg, South Africa's largest city, has thus far built the first line of its BRT system, Rea Vaya.[27] Focused on the need to overcome the legacy of apartheid, Rea Vaya links Soweto residents to Johannesburg and is interconnected with the city's other transport networks, Metrorail and the Gautrain. Rea Vaya buses are powered by fuel-efficient diesel engines that meet stringent Euro 4 emissions standards.[28] To the extent that it induces the anticipated 15% of car users to shift to public transport, Rea Vaya is expected to substantially reduce CO_2 emissions.[29]

Despite the attractiveness of these transport systems – more frequent bus services, low fares – the use of private cars continues to grow (see Chapter 9). While access to cleaner and more efficient transport systems potentially influence the behavior of urban and peri-urban residents, more is needed to change established habits and practices with respect to the choice between personal and public transport. On the whole, BRT systems, moreover, have not moved away from standard bus technology and imported petroleum-based diesel fuel. There is some tentative movement in this direction in a few larger cities – for example, in the production of biodiesel and its use in some delivery vans in Cape Town, South Africa, and the development of biodiesel in Brazil.[30] But, of more interest is the spontaneous development in the production and use of biodiesel for the more traditional minibus/matatu/tuk-tuk sector, which plays an important role in urban transport in many developing countries.

Box 25.4 | Biodiesel Initiative in Tanzania

After years in business in the United States, a Tanzanian national returned to his home country and established Mafuta Sasa Biodiesel in 2008. Tanzania consumes over two million liters of petroleum diesel daily, all of which was imported when the company was founded. Mafuta Sasa is positioning itself to replace 2% of Dar es Salaam's diesel market with biodiesel from used cooking oil that has been chemically transformed into biodiesel suitable for use in diesel engines or generators. Matatu owners, the company's principal customers, can now purchase biodiesel at a price 20% lower than that of imported petroleum diesel.

Source: Mwakilasa, 2008; Mafuta Sasa, 2011.

These two examples notwithstanding, given the continued rise in automobile use in growing urban areas, greater efforts will be needed to integrate urban planning and energy policy with other policies that affect the habits and practices of actors on both the demand and the production side (Sinha, 2003; Mutizwa-Mangiza, 2009). Only then can we move away from the car culture of the past and towards more energy-efficient and environmentally sustainable transport systems of the future.

25.4.2   Smart Grids

Problems of path dependence and informal institutions also impact on the speed with which new energy-related technologies diffuse in the industrialized world. Sunk costs in centralized grid systems and the engineering skills that maintain them or, as we show in the next section, traditional cement plants that cannot easily be converted to the production of low-energy/low-carbon cement, for example, explain some of the reluctance in adopting these technologies. In other cases, the costs of conversion and "cost-sharing" formulas may be controversial, also

---

27   Accra, the capital of Ghana, is beginning construction on its future system, an elevated monorail that will allow maximum use of existing streets and walkways.

28   The first European emissions standards date from a 1970 directive. Each amendment to this directive is numbered consecutively: Euro 1, Euro 2, and so on. Euro 4 dates from 2005.

29   Estimates of expected CO_2 savings from Rea Vaya can be found on the City of Johannesburg website (City of Johannesburg, 2011). Similar estimates exist for Bogota's TransMilenio (ESMAP, 2009) and Mexico City's new BRT lines.

30   In the city of Sao Paulo, a new project is now underway to replace diesel in buses by ethanol from sugarcane and an additive (cenbio.iee.usp.br/projetos/best.htm). The project is funded by the European Union. The city has also launched a hydrogen fuel cell demonstration project.

slowing the process of change. In the United Kingdom, for example, the regulatory formulas that governed distribution network companies until recently did not allow companies to recover the costs of investment in RD&D for network innovation (Sauter and Bauknecht, 2009). What makes these issues so intractable is the way existing habits, practices, and norms interact with the above factors.

A deeper understanding of traditional habits and practices will be required to overcome the problems associated with path dependence and lock-in that slow the processes of change towards low-carbon energy systems. Two types of lock-in are particularly relevant here. First, most energy systems are locked into a high-carbon pathway and are heavily dependent on fossil fuels (Unruh, 2000). Second, in some countries, there has been an historical lock-in to centralized energy supplies and policymaking. This second form of lock-in has its roots in the search for economies of scale in electricity generation (Foresight, 2008). However, this can be problematic if the transition to more sustainable low-carbon energy systems requires some measure of decentralization. As Chapter 15 pointed out, optimized interconnected networks making use of real time information and modern communications technology linking energy supply sub-systems provides redundancy, robustness and flexibility, that can lower overall costs. Critical elements in the new system include distributed generation and microgrids, integrated in digital smart grids that can monitor and heal themselves. Public-private partnerships are important in achieving this vision. Changes such as these will require the development of new habits, as well as interactive practices and skills.

The United Kingdom's energy system provides an apt illustration of these tensions, since it has both a highly centralized architecture and structure, and is subject to ambitious plans for a transition to a low-carbon economy. Two-thirds of the UK heating demand is met by natural gas from a centralized network (DTI, 2007). Less than 10% of heat demand is met by off-grid heat generation – i.e., by sources other than gas or electricity. Furthermore, less than 10% of the United Kingdom's electricity is supplied from renewable energy or combined heat and power (CHP) plants connected to the electricity distribution network. These levels of decentralization are much lower than those in most other northern European countries (WADE, 2006). This centralization of the United Kingdom's energy infrastructure is mirrored in policy frameworks and energy market rules. Local and regional institutions have little say in the shape of energy investment and decision making, except with respect to planning decisions (Foresight, 2008). Only very recently has a legal restriction on the ability of local authorities to sell electricity been lifted (DECC, 2010b).

This centralized approach to energy provision is not necessarily problematic in industrialized countries.[31] An analysis by the UK government

shows that there are many ways to meet the United Kingdom's ambitious target of cutting carbon emissions by 80% by 2050 (DECC, 2009). Some of these are technically possible while relying mostly on centralized electricity generation. However, many of the government's policies and plans rely on significant decentralization – including the strategy to meet the UK share of the renewable energy target, the plans for new economic incentives for local renewable heat, and the desire to have smarter grids that balance electricity supply and demand in more complex ways (DECC, 2010a). To facilitate this, some new policies and regulations have been developed and implemented. Grants for householders to install "micro-generation" in their homes, feed-in tariffs for smaller renewable energy installations, and new incentives for electricity distribution companies are just some of these.

It is too early to tell what impact these schemes will have. The experiences of the household micro-generation grants have been mixed. Frequent rule changes and adjustments created significant uncertainty for householders and installers (Bergman and Jardine, 2009). A key aim of many of them is to build new capacity. In the case of micro-generation, the installers, plumbers, and electricians who work in households need new skills, while distribution network companies need to build up capacity and skills in innovation to complement their business-as-usual system maintenance skills (see Case Study 1, Section 25.11 for more details).

25.4.3   Passive Buildings and Building Materials

Similar challenges related to path-dependent choices also confront the speed with which new energy-saving materials and the technologies for passive- or even positive-energy buildings[32] are being taken up, whether in refitting the old or building the new (see Chapter 10). "Novel cements" that use less energy and produce less CO_2 during manufacture have been given tepid reviews in industry fact sheets (Taylor and Collins, 2006). The testing of new building materials, moreover, does not figure prominently on the agenda of voluntary initiatives such as the World Business Council for Sustainable Development's Cement Sustainability Initiative, "Getting the Numbers Right," which focuses on the development of sector-specific CO_2 accounting and reporting protocols (WBCSD, 2009). In 2009, MIT researchers announced that the basic molecular structure of cement has now been decoded, opening up new opportunities for working around entrenched habits and practices. As pointed out by MIT Professor Franz-Josef Ulm, a "validated molecular model" will make it possible to manipulate the chemical structure and design new cement and concrete products for both strength and environmental qualities (Brehm, 2009). To speed these new low-carbon products to market, however, will still require funding for 2–3 years of development and testing to overcome earlier uncertainties among users and producers in the industry. Policies introducing CO_2 as a factor in pricing cement,

---

31 The centralized approach to energy provision may be more problematic in a number of developing countries, particularly where low population densities and large distances between cities and villages prevent the rolling out of a low-cost infrastructure through the extension of urban energy systems. We discuss a number of alternatives to deal with this problem in Section 25.5.2 on solar home services.

---

32 Positive-energy buildings could, for example, supply excess energy to the grid.

such as carbon taxes or caps on carbon, as well as new building codes and other energy- and environment-related legislation, would widen the market for low-carbon/low-energy materials.

The building sector has also been slow to encourage the demand for such products, and in many countries has found scant regulatory support for doing so, although a few high-profile positive-energy buildings have been built in the last few years, such as the Elithis Tower in France and the Masdar headquarters building under construction in Abu Dhabi. This is now beginning to change.

Progress is now being made in some niche areas that might point to ways of working around and with existing habits and practices in the building sector. C40 Cities, a voluntary group of mayors from among the world's largest cities, is encouraging the emergence of Mayors as champions in energy-efficient and eco-friendly buildings and urban transportation systems. Since the Mayor of Seoul made a commitment to retrofitting in 2007, 87 buildings – 45 public and 42 private – have been designated for retrofitting, and 62 of these had been completed by the end of 2009 (Oh, 2009). Holistic retrofits "can achieve 50–90% final energy savings in thermal energy use in existing buildings with cost savings typically exceeding investments," but GEA scenario work demonstrates a significant lock-in risk. If building codes are introduced universally and energy retrofits accelerate, but policies do not mandate state-of-the-art efficiency levels, substantial energy use and corresponding greenhouse gas (GHG) emissions will be locked in for many decades (see Chapter 10).

Elsewhere, innovative builders and developers are experimenting with the integration of multiple energy-saving technologies in new building construction. In the United States, with support from federal, state and municipal governments, new housing is being built that combines geothermal heat pumps, energy recovery ventilation systems, and innovative wall insulation technology using solar panels and autoclaved aerated concrete (Colaneri, 2010; McDonough, 2008).

The Leadership in Energy and Environmental Design (LEED), an internationally recognized green building certification system launched in 1998, grew out of the environmental NGO movement in the United States.[33] Hosted by the US Green Building Council in Washington, DC, LEED has brought together a broad coalition of actors including lawyers, scientists, environmentalists, NGOs, business firms, and government agencies, to define green buildings and to establish common, measurable standards. By 2010, some 400 buildings in the United States and 37 in other countries[34] had been awarded LEED certification at the platinum

level, which assigns points based on criteria such as sustainable site development, water saving, energy efficiency,[35] materials selection, and indoor environmental quality. LEED certification, especially at the platinum level, is becoming recognized as a mark of quality in green building design, construction, and retrofitting, and is helping to bring about changes in existing norms and practices in the building industry. Developing an appropriate policy and regulatory environment could further encourage this process.

In designing policies to increase interest in energy efficiency solutions on both demand and supply sides of the transport, energy, buildings, and building materials sectors, the capacity to work from a systems perspective is critical. New approaches are required to overcome the reluctance of architects, builders, and engineers to work with new materials or, as we saw in the case of smart grids, to accelerate the introduction of new distributed energy systems. Bringing key actors together on task forces, panels, and platforms[36] could play an important role here. While not yet widespread, there is evidence that universities in a growing number of countries have created interdisciplinary graduate programs in sustainable engineering that could train a new generation of users[37] and potentially help to change existing reward structures in the building profession. But more needs to be done, and more rapidly, in order to develop the capacity to integrate knowledge in new ways by, for example, establishing sustainability certificate programs that will attract current building and materials professionals, bring them up to date, and encourage the processes of networking and knowledge exchange.

Strengthening local capacities to design regulatory policies, support programs, and financial stimulus packages that correspond to domestic conditions will also be needed to accelerate the introduction and rapid diffusion of new energy technologies, as we saw in the case of smart grids. But in a long-term process such as an energy transition, developing capacities for and practices of adaptive policymaking will be essential over time as conditions change. Such adjustments may require legislation introducing mandatory review processes or sunset clauses that provide for the automatic phase-out of existing legislation.

25.5   Dialogue, Market Building, and Diffusion

Societies and organizations differ in the extent to which established habits, practices, and norms favor either interaction and dialogue or more hierarchical patterns of communication. They may be based on

---

33 This discussion of LEEDS is drawn from the websites of the US Green Building Council (www.usgbc.org), the Natural Resources Defense Council (www.nrdc.org), an environmental NGO whose chief scientist, Robert Watson, was one of the pioneers in the green building movement and founding chairman of the LEED rating system, and M. Landman's list of "LEED Platinum Certified Buildings" based on the database of certified projects (Landman, 2010).

34 These countries include Australia, Brazil, Canada, China, Germany, India, Saudi Arabia, South Korea, Sri Lanka, the United Arab Emirates, and the United Kingdom.

35 A reading of their on-line literature does not make it clear whether certification at the platinum level implies a passive plus building (www.green-buildings.com/content/781911-leed-certification-conspiracy).

36 Platforms are discussed in the Dutch transition approach (Section 25.8.3).

37 These include prestigious institutions such as Kyoto University in Japan; KTH Royal Institute of Technology in Sweden; the Indian Institute of Technology in Mumbai; the University of Cambridge in the United Kingdom; and Stanford, the University of Michigan, and MIT in the United States.

intense interactions or go-it-alone practices, accepting of criticism and the need for change, or see all criticism as negative and build little feedback and few mechanisms for introducing corrective action into project design and implementation. This section explores the importance of dialogue in developing and institutionalizing capacities for energy transitions. It illustrates this with regard to the role of dialogues in the implementation of onshore wind power in Europe and in the diffusion of solar home systems in Asia and Africa. In these cases, links to market building were particularly relevant. Finally, the section reviews the slow pace of diffusion of multifunctional energy platforms and improved cookstoves, and assesses the contributions that dialogue and market building might play in the development of sustainable solutions to these challenges. Understanding dialogues is a first step in this process.

Until recently, dialogues have not featured centrally in project planning and development. Instead, in both developed and developing countries, a common approach has involved a top-down, linear process, in which the flows of information and knowledge are driven by governments or the research or business sectors. Communication consists in the transfer of information and knowledge through the distribution of printed materials, awareness-raising campaigns, the internet, and formal training programs. As the literature now stresses, experts must recognize that they have access to only one of many "forms of knowing and valuing," and often lack sufficient knowledge of the qualities of places, problems, potential solutions, and how to make policies work effectively (Gibbons et al., 1994; Healey, 1998; Renn, 1999; DAC, 2006). Consultations involving focus groups, village-level discussions, and stakeholder meetings can give the impression of being two-way processes, but the boundaries of such consultations are pre-established and often provide limited opportunities for those consulted to express interests, needs, and preferences other than those already on the agenda (Gibbs,1997). The process often fails to reach a broad enough range of potential users to stimulate the adoption of flexible processes at the project design stage, and thus few opportunities for adaptive change are built in once a project is under way. Interpreting the results of consultation processes is also problematic if, as the German physicist Werner Heisenberg noted in relation to his work on quantum mechanics, "what we observe is not nature in itself, but nature exposed to our method of questioning" (Heisenberg, 1962).

An interactive dialogue process, in contrast, offers room for the articulation of a broad range of views, interests, preferences, and needs. Preparing for such a dialogue process involves dealing up front with issues of how to proceed, the content, where the information will come from, the form and frequency of interactions, channels of knowledge and information, and the expected outputs and outcomes of the process. Establishing the legitimacy of interactive dialogue processes also extends to the concepts of "monitoring and evaluation," and the need to transform these somewhat punitive sounding – and thus often ignored – activities so that they provide feedback and flexibility in the course of project implementation (IDRC, 2008).

25.5.1 Dialogue at the Local Level: Wind Turbines

The deployment of onshore wind farms is shaped by a broad array of national policies, but often involves relatively small-scale, decentralized, and location-dependent applications that need planning consent from local authorities. Successful implementation therefore requires numerous local decisions and investments in the wind farm siting process.

In many countries, national policies favoring wind power have come to signify a top-down imposed development model that takes little note of local land-use planning and environmental issues. They often do not encourage project developers to involve local stakeholders in the design and planning of wind projects. In the Netherlands,[38] for example, most project developers are accustomed to building large power stations and have little interaction with local stakeholders. They have failed to recognize local planning concerns, sparking opposition to wind farm project from municipalities and other local stakeholders. Dutch energy companies faced difficulties in getting wind projects built in the 1990s and this contributed to the demise of the turbine manufacturing industry.[39]

In the United Kingdom, the limited development of wind power during the 1990s was due to the lack of experience with small-scale, decentralized, and location-dependent technologies, and the dominance of large, established utilities in wind projects. Some of the first large projects were planned in windy areas, which were ecologically sensitive hilly areas where there was no industrial activity. Public concerns and reluctance among local planning authorities resulted. The tradition of grassroots initiatives to protect the landscape subsequently made local opposition to wind projects very effective.

In Germany, in contrast, national policies were more favorable, enabling the grassroots environmental movement to propose locally based wind projects in collaboration with local communities. In the state of North Rhine-Westphalia, for example, federal and state governments responded to the grassroots movement, and in the early 1990s policy support was tailored to the needs of grassroots wind project initiatives. The German feed-in tariff system guaranteed access to the grid and did not discriminate against small, independent initiatives. Thus a diverse range of actors was able to become involved in both turbine manufacturing and project development.[40] Local ownership has been helpful in mobilizing support from the local level upwards, while the involvement

---

38 The case studies in Germany, the Netherlands, and the United Kingdom summarized here are drawn from the work of Breukers and Wolsink, 2008; Agterbosch and Breukers, 2009; and Toke et al., 2008.

39 A very different type of project developer, the Dutch wind power cooperatives, adopted a much more collaborative approach, but their impact has remained marginal.

40 In the early 1990s, Denmark also introduced feed-in tariffs. These were not only stable over long periods, allowing investors to plan their projects with confidence, but were also set quite high for the first years. Although the level of subsidies was later reduced, by 2002 almost 2900 MW of generating power had been installed and a strong industry built. Denmark's main focus in the mid-2000s is on re-powering existing onshore wind farms and on developing offshore projects (UNDP/GEF, 2008).

of diverse actors at local, regional, state, and federal levels precluded early resistance, as in the United Kingdom and the Netherlands. The institutionalization of environmental concerns in society and in politics also contributed to the successful development of wind projects as a form of local energy activism. Later, when the turbine and related industries were booming, the support for wind power became broader based – wind power represented not only environmental, but also economic (e.g., employment) and industrial (manufacturing and related services) interests.

Habits and practices are not immutable, but are embedded in the broader social context. In Germany, with the rapid growth of the wind energy sector in the late 1990s and the early 2000s, locally based projects have become less evident, and projects are increasingly developed by outsiders. Although local support has waned and resistance has increased, implementation began to decline when an impressive level of capacity had already been installed. North Rhine-Westphalia has achieved more in a generally less conflicting environment than has been the case in either the United Kingdom or the Netherlands.

While some "unlearning" took place in the German context, there seems to have been some "learning" in the Dutch and British contexts. In the late 1990s, Dutch policies no longer favored energy companies over other project developers. Farmers with access to financial support were responsible for a steady rise in the implementation of locally owned projects. These farmer-led projects provide clear local economic benefits and are developed by people who are part of the local political and social contexts. A further advantage is that the farmers own the land on which the turbines can be sited. Moreover, the farmers are well organized at several levels and so have been able to enter into collective negotiations with national energy distributors, local and provincial authorities, etc.

In the United Kingdom, there have been efforts to reinvigorate a "cooperative" tradition for wind power, stemming from the acknowledgement that the projects proposed and planned by "outside" companies are not always successful in gaining local support. In addition to such "community project initiatives," some commercial project developers have adopted a strategy encouraging the early involvement of local stakeholders.[41] Although there have been successes, the majority of wind turbine projects still meet with opposition from the firmly rooted and increasingly well-organized movement to protect the landscape.

As the case of onshore wind power shows, attempts to speed up implementation of new energy projects without sufficiently addressing local concerns are unlikely to engender societal commitment and acceptance. They may result in stalemate situations where further developments

are blocked in a context of conflict and polarization.[42] One lesson is that energy transitions involving decentralized and location-dependent applications need support from the bottom up from a diversity of actors and networks. Such applications need to "fit" into the contexts for which they are proposed. This is not just a matter of compliance with formal procedures, permits and technical standards, but is very much a social-political process involving local stakeholders.

25.5.2   Market Building and the Diffusion of Solar Home Services Systems

For decades, technology transfer through donor-supported projects has been one of the main ways in which new energy technologies have been deployed in developing countries. Although many of these projects were designed to provide energy access for the poor, especially those in rural areas with no access to energy grids, widespread diffusion of these technologies has not followed. This section discusses the problems encountered in the introduction of solar home systems (SHSs) in developing countries, and how they have been addressed.

As early as the 1970s, SHSs were envisaged as off-grid energy services, but the costs and time needed to improve the technology and to build local capacity to produce spare parts and repair the systems drove many away. Although petroleum price hikes temporarily increased interest, efforts to develop SHS technology were not supported by government policies and SHSs were largely abandoned until the mid-1990s.

By this point, the technical performance of SHSs had improved, but producers still faced the challenge of ensuring consistent quality. The absence of quality standards contributed to customer dissatisfaction. Whether treadle pumps, solar water heaters or SHSs, standards and certification are needed "to increase customer confidence […] and reduce the negative perception arising from low equipment quality, unsatisfactory installation and incorrect operation and maintenance procedures" (GNESD, 2007). The prices of these systems had not come down sufficiently for most rural households, as studies in South Africa and elsewhere have shown (ERC, 2007; GNESD, 2007).

With the exception of Kenya, which has one of the most "dynamic commercial photovoltaic markets in Africa with a non-subsidized demand of 1–1.3 MWp per year that has been growing at an annual rate of 15% since the mid-90s" (GTZ, 2009), the diffusion of this technology has remained relatively slow in Africa (Hankins, 2011). Ho (2011) notes, "[i]n 2005, Kenya was home to just over 150,000 solar systems with a median size of 25 watts; coverage has since reached some 300,000

---

41 In line with this UK experience, but related to micro-hydro power development in developing countries, Rwanda has encouraged dialogues between beneficiary communities and local entrepreneurs in the development of plans for micro-hydro. See Section 25.6.2.

42 This point is not limited to onshore wind power, as is shown by a recent example of the long-term process of litigation that characterized the offshore deployment of wind turbines in the US state of Massachusetts.

households."[43] Sustained market growth in Kenya has been fuelled by a number of factors. On the demand side, these include the "(r)elatively high incomes among farmers (coffee, tea, horticulture), rural teachers, civil servants and businesses with a strong demand for consumer electronics (TVs, radios, cell phones)" (Ho, 2011) and support from NGOs and missions that provide medical and other services in remote areas that require off-grid power. On the supply side, it has been supported by the availability of photovoltaic (PV) modules in a range of sizes, local battery manufacture, an active entrepreneur class with strong connections between Nairobi and rural areas, and growing government interest. The latter was reflected in the Rural Electrification Program which, working with two foreign companies, installed PV systems in over 150 schools in remote off-grid areas between 2005 and 2009 (GTZ, 2009). With regard to capacity development, however, "(o)utside the solar companies themselves, Kenya has no organized solar energy training programs for artisans or engineers. It does have some university-level courses in alternative energy, but these are fairly basic and do not prepare 'solar engineers' per se" (GTZ, 2009).

With a view to removing barriers to the delivery and financing of SHS, the World Bank and the Global Environment Facility (GEF) initiated a series of pilot projects to provide off-grid solar energy services to rural households in 12 developing countries (Martinot et al., 2001).[44] The project tested two approaches for the delivery of SHSs – a dealer-sales approach and an energy service company (ESCO) approach – but it left open the possibility of switching from one to the other, or modifying the chosen approach once the project was underway. This flexibility was critical in developing the capacities for learning and adaptation at the project management and local levels that later proved essential in the diffusion of SHSs in a number of countries.

The range of countries in which these approaches were tested lends robustness to the findings and are supported by evidence from other projects introducing SHSs in Kenya, South Africa, Nepal (Brew-Hammond et al., 2008), and Ethiopia (ERC, 2007; UNDP, 2007). Among the main findings of these studies, three in particular stand out.

First, demonstration projects such as these need to be turned into experiments in learning for all partners, including donors, supplier firms, and clients. From the choice of approach for the delivery of SHSs to the type of financing and the focus on capacity building activities aimed at market development, learning through feedback and continuous dialogue were critical in moving the project ahead. This is particularly relevant where there are small market players who are frequently overlooked when the need for scale or minimum quality prevails. Without learning through feedback, opportunities for upgrading can also be lost as local

residents and businesses, which initially adopted solutions that were far from ideal, acquired the means to purchase a better system.

Second, the decision to build flexibility into the project design is critical. This recognizes that different habits, practices, norms, and experiences of government staff, suppliers, and users may require adaptive changes along the way. The Sri Lanka project, for example, adopted both dealer-sales and ESCO models. Early in the project, one dealer decided to shift to the ESCO model but soon found that "the costs of monthly collections among dispersed and remote rural population [were too] high. The firm did not have sufficient rural infrastructure and standing in rural communities to handle collections effectively and efficiently," and so switched back to direct sales, this time facilitated by credit from a microfinance organization (Martinot et al., 2001). The flexibility built into the project design legitimized a process of adaptation that could take place once it became clear that rural practices and norms were hampering the firm's ability to function. Since the communities' experiences with microfinance made this other option possible, the ESCO model has largely been abandoned.

Third, the emphasis on creating and building a market means that projects must focus on those capacities needed to encourage the diffusion of SHS technologies even after a project has ended (Martinot et al., 2005). Taking a broad systems perspective, these include strengthening the capacities of dealers to engage in community outreach and tailor systems to local needs, developing novel financing mechanisms, and encouraging the emergence of local installation, maintenance and repair services. In Ethiopia, for example, the Solar Energy Foundation, which won an Ashden Award in 2009, created a school for young people to train as solar technicians in Rema, Ethiopia, where 2100 SHSs were installed between 2006 and 2008. Graduates of this school have opened Solar Centers in four other areas in Ethiopia, thus stimulating the diffusion of SHS systems across the country (Ashden, 2009). In the World Bank/GEF project, working with standard-setting bodies to ensure product quality and with regulatory bodies to review pricing policies was also important.[45]

Since 1997, Sri Lanka has benefited from three energy projects funded by the World Bank, International Development Association, and the GEF. By 2008, these projects had "created a vibrant local industry of suppliers, developers, financiers, consultants, and trainers," and "some 120,000 households were using solar home systems, with 750 new installations occurring monthly. Nearly 6000 households... [were] obtaining electricity from micro-hydro minigrids that communities own, operate and manage" (World Bank, 2008). It thus appears that Sri Lanka has developed the capacities to sustain the diffusion of SHSs in the near term.

The Bangladesh case shows, perhaps more clearly, how creating a market can drive the diffusion process, as well as promote the development of local manufacturing and value chains. Initially, the government's Rural Electrification and Renewable Energy Development

---

43 In addition, observations in Kenya also show that a larger number of rural households obtain electricity from car batteries and small PV systems than from the national grid (Jacobson, 2004). This resembles the early automobile industry, where American farmers and rural dwellers used their Model T Fords not only for transport but to run corn shellers, grinders, wood saws, water pumps, and fodder and ensilage cutters (Kline and Pinch, 1996).

44 The 12 pilot projects were implemented in Argentina, Bangladesh, Cape Verde, China, Dominican Republic, India, Indonesia, Kenya, Lao PDR, Morocco, Sri Lanka, and Vietnam.

45 Examples of this can be found in Martinot et al. (2001).

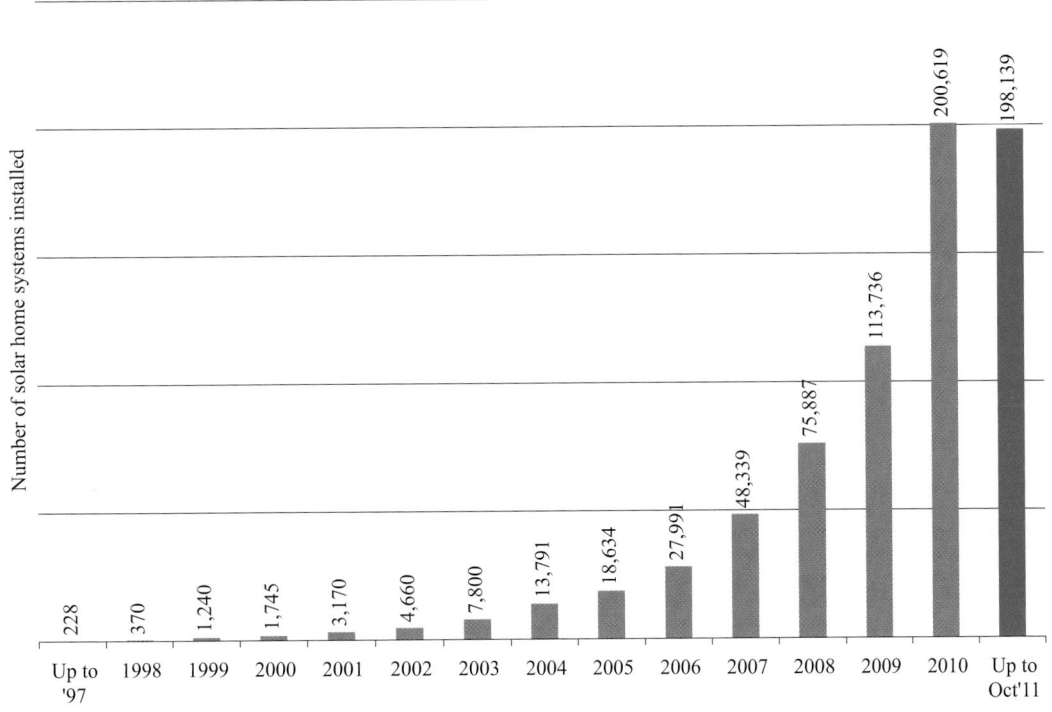

Figure 25.2 | Yearwise installation of solar home systems (SHS) in Bangladesh. Source: based on Grameen Shakti, 2011.

Project, funded by the World Bank and GEF and managed by the Infrastructure Development Company Limited (IDCOL), expected the installation of 50,000 systems by the end of 2007, but by then over 150,000 households had installed solar systems, and 7000 more were being added each month (REN21, 2008).

Three factors were important in the rapid diffusion of SHSs in Bangladesh. One was the availability of subsides from IDCOL to participating organizations that acted as distributors. This enabled these organizations to extend credit to buyers. The second was the creation in 1996 of Grameen Shakti, a not-for-profit company under the Grameen Bank that focused on bringing energy to rural areas. By 2007, it employed 1500 field staff, had trained 1000 engineers and 1000 local technicians, of which most were women, and had established a network of 390 village unit offices in Bangladesh's 64 districts. It was through these village unit offices that Grameen Shakti promoted SHSs. Soft credit through installments, which made SHSs affordable, community involvement in planning, which led to social acceptance, and effective after-sales service contributed to the rapid development of the SHS market in Bangladesh.[46] By the end of 2010, Grameen Shakti technicians had installed 518,220 Solar Home Systems and the number installed per year had risen dramatically as the capacities needed

to put these systems into place developed (Grameen Shakti, 2011; see also Figure 25.2). The third factor was the role of Rahimafrooz Batteries Ltd., a local company that turned its experience in battery manufacture to the production of rechargeable batteries for PV systems. The company has since diversified into the design of charge controllers and fluorescent lamps for local manufacture. It has supplied 25,000 complete systems to distributors and provides training and technical support. In 2006, Rahimafrooz Batteries Ltd. was awarded an Ashden Award for sustainable energy (Ashden, 2006).

In addition to the "for-profit" sector, many foundations, NGOs, and social enterprises are now playing a role in developing local capacities for the diffusion of clean energy technologies, and in stimulating the development of supportive manufacturing and repair capabilities in developing countries. In Africa, for example, the African Rural Energy Enterprise Development (AREED) network,[47] funded by UNEP, GEF, and other donors, works through NGOs in Botswana, Zambia, Tanzania, Mali, Senegal, and Ghana (Agbemabiese, 2008) to simulate and support energy transitions in rural and urban areas by providing advisory services to small- and medium-sized enterprises (SMEs) and linking these to local sources of seed financing development of energy demand-side services (Brew-Hammond and Kemausuor, 2008).

---

46 Grameen Shakti was awarded the European Solar Award in 2003, the Ashden Award for Sustainable energy in 2006, and the Right Livelihood Award "for outstanding vision and work on behalf of our planet and its people" in 2007 (Right Livelihood Award, 2011).

47 The "REED" model has been implemented in Africa (AREED), Brazil (BREED) and China (CREED); see AREED, undated.

Box 25.5 | African Rural Energy Enterprise Development (AREED)

AREED's underlying philosophy and its innovative, networked approach makes it unique. Both of these characteristics encourage dialogue and the search for innovative solutions. AREED is, in its own words, based on the powerful idea that "impoverished people can transform their lives and break out of the vicious circle of poverty when they are empowered from the ground up" with proven, powerful, practical energy technologies. At the same time, collective experience from a variety of energy initiatives is driving renewed interest in alternative delivery models. This trend is manifested in the growing support for SMEs as versatile energy service providers capable of extending access to energy, while promoting poverty reduction and development through bioenergy systems such as the production and use of plant-based oils to power multifunctional platforms and village electrification schemes.

AREED also seeks to penetrate rural markets that would otherwise be unattractive to entities motivated solely by profit. Its focus is on "social enterprises" – defined as business organizations whose surpluses are reinvested in the business or in the community to achieve social objectives.

Source: AREED, undated.

In Senegal, AREED is working through Environment and Development Action in the Third World, an NGO, to develop a multifaceted market for solar energy technology that brings together local entrepreneurs, NGOs, government agencies, and financial service providers. With rising oil prices, the interest in solar energy has increased in Senegal, encouraged by government support for SMEs and solar technology. The agency for Rural Electrification through Local Initiatives "is responsible for the development of small energy business delivery through the use of solar technology" (Drame, 2008). Nonetheless, companies dealing in solar heaters went bankrupt (Drame, 2008). When the AREED program started, several of the bankrupt companies came together to form Africaine de Maintenance et d'Équipement (AME), a company that installs and maintains solar water heating systems. AME received a loan from AREED in 2001, which was repaid in full as the company's management capacity improved. A second loan went to Prosoleil, which manufactures solar water heating systems, and a third to Motagrisol, a producer of solar-powered milling machines (Drame, 2008). Working through the AREED network, a broader base for solar energy services is being built in Senegal.

25.5.3 Diffusion: "Scaling Up" and "Scaling Out" at the Project Level

Both the "scaling up" and "scaling out" of pilot projects are important in the diffusion of new energy technologies to users in rural, off-grid areas or to the urban poor. Scaling up involves moving from small-scale demonstration projects to larger, commercially viable enterprises that might benefit from economies of scale to reduce prices and thus increase sales. This process was discussed in Section 25.4.1 in connection with a series of European programs that demonstrated, tested, and are now moving towards the commercialization of electric and hydrogen fuel-cell buses. Scaling up can also be achieved through collaboration between research and users (see Sections 25.6 and 25.7).

Scaling out involves the diffusion of new technologies to new users. This section focuses on two energy technologies that are important in rural, off-grid areas and to the urban poor: multifunctional energy platforms (MFPs)[48] and improved cookstoves. The diffusion of these technologies to micro and small enterprises, households, and communities has proven challenging. The focus of this section is on how adaptive learning and the development of local capabilities for market-building can stimulate, support, and sustain a scaling out process.

25.5.3.1   The Multifunctional Platform

MFPs provide access to mechanical power for income generation in off-grid areas. UNDP's MFP Program was built on an earlier project by International Fund for Agricultural Development (IFAD) and UNIDO. Launched by UNDP in Mali in 1994,[49] it has since been extended to Burkina Faso, Ghana, Senegal,[50] and other West African countries with the financial assistance of a growing number of donors. Its 16-year history offers many insights into the challenges still facing the diffusion of new energy technologies in developing countries.

---

48 An MFP is a source of mechanical and electrical energy meant to reduce the need for repetitive manual labor, especially that of women in rural Africa. Typically, an MFP consists of a diesel engine (8–12 hp) mounted on a chassis that can power equipment such as grinding mills, de-huskers, de-hullers, oil press equipment, water pumps, and battery chargers.

49 The launch was followed by a preparatory phase and then the full pilot phase.

50 Due to conflicting figures and the absence of reference dates in a number of UNDP publications and other reports, only the first four West African countries in which the MFP program operated are discussed here, but this covers a 13-year period in the life of the program. In addition, recent independent initiatives to set up MFPs will also be mentioned.

One of the program's main objectives was the inclusion of village women in an income-generating activity. Women, through their village associations, would thus be the owners, managers and operators of the platforms and were trained in the necessary skills. They were also responsible for mobilizing a portion of the funds needed for the MFP's housing and tools. Although in some cases, men were employed to operate an MFP, it was not until the Ghana MFP project was launched in 2005 that ownership by individuals, both men and women, became a possibility. Another objective was sustainability, often the Achilles heel of such projects once donors are no longer involved, which was achieved by training and building up a network of mechanics and providing them with motorcycles so that they could repair the platforms whenever necessary.

MFP country projects proceeded in two phases. The project began in Mali with a three year pilot phase (1996–1999), during which 48 MFPs were installed.[51] This was followed by a first scaling out phase (2000–2004), at the end of which a total of 576 MFPs had been installed in Malian villages (UNDP, undated). The pace was somewhat slower in Burkina Faso, where 25 MFPs were installed during the pilot phase (2000–2003). A lack of consistent data makes it difficult to determine how many MFPs were installed in the national diffusion phase (2005–2009).[52] A recent UNDP press release, however, reported that MFPs have been installed in 441 localities in that country (UNDP, 2010). In Senegal, 44 MFPs were installed during the pilot phase (2002–2005) (UNDP, undated). The diffusion phase did not begin until two years later, with the support of the Bill and Melinda Gates Foundation and other funding sources. The additional funding also made possible the extension of the MFP programs in Mali and Burkina Faso. The Ghana MFP project began with a pilot phase (2005–2007), during which only 32 of the 40 platforms anticipated were actually installed because the budget "did not adequately capture potential issues of commodity inflation and change-orders" (Brew-Hammond and Crole-Rees, 2004; UNDP and UNOPS, 2008). This also slowed the start of the scaling out process, for which negotiations were only completed in 2009.

Unlike the emphasis on building local markets and the adaptive changes that were introduced during the pilot phase of the Solar Home Services project (Section 25.5.2), the design and implementation of the MFP project was marked by established routines and practices that slowed the diffusion process. The early MFP projects, for example, did not support the development of independent initiatives to establish MFPs. A duly registered women's association had to request a platform. Nonetheless, such initiatives began to emerge in Mali, where "19 platforms were installed without any subsidies from the project. In these cases, the villages directly contacted private artisans who had been trained by the

project, and procured and installed the multifunctional platform on a commercial basis. These […] are not monitored by the project and no information has been gathered" (UNDP, 2004).

In contrast to the capacity development practices in the SHS cases, the MFP program also spent heavily on literacy training in addition to training in the management and operation of the MFPs.[53] This was costly but necessary in a program designed to enable rural women to assume ownership and management control of the MFP and to operate it. In its earlier phases, however, it might have contributed to a relatively narrow range of operations, such as milling and grinding, when a wider scope for income-generating activities through the operation of the platform – including sawing, welding, and water pumping – would have been possible but required a much higher financial outlay. This was noted in a review of the MFP project in Mali, which recommended measures such as increases in subsidies and help from other members of the community, such as men's associations (UNDP, 2004). The issue was raised again, however, in the Evaluation of the Ghana MFP pilot phase where the report recognized that "[t]here is a clear demand for other mechanized services at all MFPs visited during the evaluation […] More flexibility is needed here" (UNDP and UNOPS, 2008). Greater flexibility in adjusting to new interests, in accommodating the need for diversity in delivery models and in nurturing village-to-village learning and networking processes could have made a difference in the speed with which MFPs diffused.[54] This is particularly relevant where independent initiatives, such as the decision to replace imported diesel with jatropha oil,[55] have taken place with positive results in contexts as different as Maurolo, a small village in Mali, or Garalo, a much larger town in that country.[56]

Although the diffusion of multifunctional platforms has so far been slow, this is now changing. Feedback, flexibility, and adaptive learning are evident in the expansion of new services offered by MFPs in Mali, Burkina Faso, and Ghana, and the diversity of operational model approaches. In Ghana, of the 32 platforms installed, 11 are owned by individual women, 11 by groups and cooperatives (mostly women), nine by individual men, and one is run as a partnership (UNDP and UNOPS, 2008). Using MFPs as part of an expanding number of energy services such as water pumping and electricity provision is accelerating, and the capacities in engineering, finance, decision-making, and

---

51 The MFP program targeted villages with between 800 and 2000 people. This meant 2300 villages in Mali, 3200 villages in Burkina Faso, and 1400 villages in Senegal.

52 One UNDP source states that "[t]o-date 120 MFPs have been installed," but the document itself has no date (UNDP, undated).

53 A small number of men – five out of 30 trainees per literacy class and per MFP village – were also trained.

54 Deployment of energy platforms was based on a donor/government selection process and was not supportive of independent initiatives to redesign or install platforms. Recent research has shown that farmer-to-farmer learning processes can speed up the diffusion of new technologies in developing countries. They provide a tool for small-scale farmers that enables them to gain insight into their own performance in farm management, and to learn by comparing their performance with that of colleague farmers (Sinja et al., 2004; Taweekul et al., 2009).

55 Jatropha oil is a vegetable oil from the *Jatropha curcas* plant. When the plant's seeds are crushed, it produces a high-quality biodiesel.

56 Maurolo uses jatropha oil in its MFP; Garalo is using it for the generation of electricity (see Section 25.6.1; KITE, 2004).

monitoring are being developed. The diffusion process is also moving beyond the established MFP Programs in West Africa to East Africa, where discussions on the need for MFPs is underway in Tanzania (Caniels and Romijn, 2010) and two MFPs have been installed in Uganda (EPA, 2010a; 2010b). Changes are also taking place as the program opens to newcomer initiatives in the development of jatropha as a replacement for diesel by commercial enterprises such as Mali biocarbonate (Mali Biocarburant, 2010). Columbia University, working with Engineers without Boarders (USA), has developed and tested an engine modification kit and are working with a local NGO to test it further in Uganda (EPA, 2008).

25.5.3.2 Cookstoves

Improved cookstoves are another important addition to the portfolio of "new" energy technologies that serve multiple purposes (see also Chapters 4, 10, and 19). They can raise the level of energy efficiency and save on the amount and cost of wood and charcoal. Their design, production, repair, and use can also stimulate local learning and innovation and contribute to market creation and income-generating activities, as well as reduce damage to the environment and to the health of the women who are the primary users. The many possible benefits from this one "new" technology have made it the focus of donor attention for decades.

During the 1970s and early 1980s, international donor agencies introduced nearly 100 stove projects and programs. A review of these efforts by the Canadian International Development Research Centre (IDRC), however, showed that widespread diffusion was not taking place (Krugmann, 1988). The Energy Sector Management Assistance Program also reported on this in more detail in the mid-1990s, after which several World Bank projects were reoriented (Barnes et al., 1994). Nonetheless, this problem has persisted into the new millennium. Chapter 19 provides a table of 52 "improved" cookstoves that were developed from the mid-1980s to 2009 and distributed in Africa, Asia, and Latin America. Only three of these models – two in India and one in Africa – succeeded in achieving a distribution of over one million stoves. In the following section, we look briefly at some of the factors that contributed to the past failure and the emerging success of these models.

Two of the most important factors contributing to the failure of improved cookstove projects stem from the fact that they were designed in the North and transferred to the South. The former led to a choice of materials and associated production costs that tended to raise the price of stoves beyond local household income possibilities. It also led to inappropriate designs for users, mainly women, whose inputs were rarely sought. It also limited income-generating opportunities from the growth of local production and the development of skilled artisans for maintenance and repair.

Among the exceptions to this practice was the Kenyan jiko, a charcoal stove that was built and sold through conventional artisanal market channels. But even more important in understanding the speed and extent of the diffusion of jikos, in the absence of subsidies or special programs, were a number of design features that helped to build the market, provide opportunities for the development of local supply chains that supported the diffusion process, keep costs of the jiko low and prices within an affordable range for rural households. The stoves, for example, were made by local artisans and the insulation provided by ceramic liners was manufactured in local factories. When the liners wear out, artisans can either replace them or take the old jiko as part of the purchase price for a new one. The artisan then replaces the liner, refurbishes the stove, and resells it. Ceramic jikos are also much more energy efficient, and their cost varies with size and quality and can be bought for prices as low as US$2. According to a study by the German organization GTZ, between 1997 and 2001 about 1.6 million Kenyan ceramic jikos were sold, and "it became a model for a number of other stoves like the Lakech stove in Ethiopia" (Brew-Hammond et al., 2008).

Local design, manufacture, and repair were also important in the development and diffusion of improved mud chulhas with a chimney in India and a major factor in the reduction of smoke pollution in the homes of users. As many as 28 million improved chulhas, including replacement stoves, were sold from 1983–1998. While the chulhas have been praised for their positive impact on health, questions have been raised about their efficiency, particularly when the chimneys are replaced by the women themselves or by untrained artisans, with considerable loss in energy efficiency (Kishore and Ramana, 2002). Further research to upgrade the chulhas by lengthening the time between rebuilding mud walls and chimney and training programs for local artisans in repair and upgrading of the stoves would have improved their overall energy impact as well as contributing to local livelihoods.

More recently, the emergence of social entrepreneurship and microfinance in the Indian energy sector has led to the development of lower cost, more efficient stoves that emit little smoke and are able to reach the bottom-of-the-pyramid markets. These new start-ups, however, are struggling to fund additional research and larger-scale production that would help to reduce costs and prices (GVEP, 2009). They also face new competitors entering the market with "high-tech" wood and agricultural residue stoves designed in the North with the financial support of international foundations.

In the past few years, new stakeholders have become interested in stoves, including commercial European and US firms, with a view to capture carbon funds and reduce deforestation at the same time. Some completely different stove models ("second generation") were the result. These included scientifically designed models, such as those developed by Envirofit. These stoves look modern and guarantee good performance in terms of fuel efficiency and emissions. They are also not cheap and can be several times more expensive than traditional stoves. Contrary to traditionally improved stoves, second-generation stoves are produced on a large scale in China[57] or in the industrialized world and imported into the developing countries. The role of traditional stove makers, as we shall see below, is minimal.

Envirofit, a nonprofit organization based in the United States, received donations and institutional support from the Shell Foundation that helped "to fund product development and early stage product commercialization" (Envirofit, 2010). The company was able to work closely with Colorado State University's Engines and Energy Conversion Laboratory to develop improved cookstoves based on new materials and processes for which patents are pending. Like the jiko, these cookstoves burn wood and residues, as opposed to new fuels such as liquid petroleum gas (LPG), but are more energy efficient than traditional stoves. Envirofit's business plan is based on high-volume, centralized manufacturing and global supply chain management that is expected to generate "strong margins" through their sales channels in developing countries (Envirofit, 2010). With a price tag of US$35, the diffusion of such stoves would require significant subsidies from international donors. In May 2008, Envirofit launched their cookstove business in India. Since then, the Shell Foundation "has helped them to develop market strategies which focus on rural villages across southwest India … One of the challenges was to make the stoves affordable to users" (Gomes, 2009).[58]

However, stoves from Envirofit are designed to use firewood and residues, and not charcoal, whereas the main urban market for cooking energy, particularly in Africa, is for charcoal. Charcoal is an efficient and convenient fuel for the end-user, but if production efficiencies are taken into account, it is no longer a very efficient fuel. Charcoal-using households use 30–100% more energy for cooking than firewood-using households, depending on the efficiencies of the stoves used (GTZ, 2010). Unless trees for charcoal making come from farmers' fields or plantations, deforestation may accelerate. Charcoal is a quickly growing commodity in urban areas, and it is also a lower cost cooking fuel than any of the modern alternatives such as kerosene, LPG, and electricity. Since towns are expanding rapidly across the developing world, its consumption is increasing rapidly. To meet broader development goals, strengthening capacities for the design and manufacture of improved cookstoves for charcoal and the development of the sustainable production of charcoal in developing countries could be of importance. E&CO, Inc. is doing just this.

A not-for-profit investment corporation founded in 1994, E&CO supports small and medium-sized enterprises involved in supplying clean energy technologies. It works closely with these local entrepreneurs and with local partners to support research, development, and production in developing countries. In collaboration with AREED, which provides key business support services as well as seed funding to local enterprises in the

energy sector, E&CO provided a first loan of US$70,000 to Toyola Energy, a small start-up company making improved cookstoves in Ghana. Having trained in energy-efficient stove manufacturing under the Ghana household energy project in 2003, the proprietor, Suraj Wahab, was making a small number of charcoal stoves that "reduce the amount of charcoal needed for cooking by around 35% resulting in significant household savings" (EnergyAccess, 2010). By 2010, Toyola had produced and sold more than 50,000 stoves to households across Ghana and was the recipient of two further loans from E&CO, each in the amount of US$100,000.

This section took the discussion of informal institutions a step further in its identification of capacities needed for the development and diffusion of "new" renewable energy technologies such as wind, solar power, and woodfuels. These have an important role to play in meeting the energy needs, sustainable development goals, and economic and environmental objectives of communities in developed and developing countries. Critical factors in the success of such ventures are dialogue in the initial phases of project development and the inclusion of mechanisms for feedback and flexibility in project design. Dialogue between users and producers also contributes to the kind of signaling that creates awareness of opportunities for widening local markets and speeding up the process of technology diffusion.

For local entrepreneurs in developing countries to take advantage of these new flows of information, however, there will be a need to strengthen capacities for the local design, manufacture, repair, and distribution of new energy-related technologies, such as improved SHSs, cookstoves, and MFPs. As these cases have illustrated, however, the creation of new types of funding and the capacities to access these funds through stronger business support systems are also needed to make such potential benefits a reality.

25.6   Distributed Energy Systems and Development: A Multi-Goal Perspective

This section expands on the discussion above by looking at a variety of biofuels and the capacities and capacity development processes that could enable their local production and use in the generation of electricity and in building an evolving portfolio of clean energy technologies for the transport sector. It also explores the development of small and village-level hydro for on- and off-grid applications. Decentralized production of biofuels and distributed energy systems could contribute to the realization of development goals, while simultaneously moving towards an energy transition that includes smallholders and the poor.

25.6.1   Energy Access, Income Generation and Inclusion: Making Choices about Biofuels

Biofuels have had a checkered history. They are alternatively praised as a means to provide energy access, energy security, and reduced

---

57   Aprovecho Research Center, located in Cottage Grove, Oregon, is the designer and developer of improved "Rocket Cook Stoves," which it disseminates through its "not-just-for-profit entity" StoveTec, created in 2008. The stoves are mass produced in Shengzhou, China (ARC, 2011).

58   Richard Gomes is communications and business manager, Shell Foundation UK.

greenhouse gas emissions, and criticized as contributing to the destruction of tropical forests and competing with food crops for the use of agricultural land. As the world's population grows and climate change affects agricultural production, making choices about biofuels will require new skills, information, and capacities to analyze trade-offs and design, monitor, manage, and enforce policies that ensure sustainability (see Chapter 20).

By 2005, with oil prices again rising, interest worldwide has been attracted to liquid biofuels for the transport sector (see ESMAP, 2005b; 2010 for further information).[59] Although electric vehicles are also making a comeback in urban environments, levels of electricity output are insufficient to meet their widespread use in Europe in the near term[60] and are likely to be a niche or longer-term option, particularly in the developing world. Biofuels are also being developed for energy generation, thus creating new options for distributed energy in off-grid environments.[61] From an even broader, but longer-term development perspective, biofuels can open up opportunities in developing countries for what the OECD, in a recent study, has termed the "bioeconomy"[62] (OECD, 2009). Strengthening local research capacities, participating in collaborative research and development (R&D) efforts, and coordinating policies across government ministries are key elements in laying the foundation for the long-term development of the bioeconomy and the multiple options it will offer in the future (ESMAP, 2005a; OECD, 2009). In the past, this would have been dismissed as unrealistic, even as a long-term goal for developing countries. However, the recent development of biofuels from sugarcane, jatropha, and algae suggest that building a bioeconomy in the developing world is not an altogether unrealistic proposition, as the following examples illustrate. Evidence of the role that local R&D capacities can play in stimulating the growth of the bioeconomy in the developing world is already emerging.

In Brazil, for example, sugarcane-based ethanol was the first biofuel to be commercialized successfully in a developing country and perhaps in the world.[63] A long-term research effort that brought government, growers, and ethanol producers into the process was a major factor in bringing down prices and eliminating subsidies. It is important to emphasize here that land-use issues were not of major importance in Brazil, where sugar plantations were already in place and options for sugar exports were in decline.[64] The use of sugarcane as the feedstock for ethanol in Brazil has contributed to its overall positive climate benefit (UNEP, 2009). Most sugarcane plantations, however, were concentrated initially in the north-east and south-east regions of the country[65] and the need to transport ethanol around the country resulted in some negative effects on net CO_2 benefits. The plantation system, moreover, largely excluded smallholders from the biofuels sector.

In the development of biodiesel production the Brazilian government sought to include smallholders and the poor. A variety of inputs were identified for biodiesel and several of these are available in most regions of Brazil. They can be grown by smallholders and processed and distributed locally. Government policies have been put in place to encourage the purchase of feedstock for biodiesel from smallholders (Teixera et al., 2008). In keeping with earlier successful practices in the development of higher-yielding sugarcane varieties for ethanol production, the research that has made possible the development of crop varieties and production techniques for biodiesel was subsidized by the government and has been supported by both the public and private sectors.

Biodiesel from palm oil is an example of how this new approach is leading to a rethinking of the role that decentralization can play in strengthening smallholder production and in generating externalities. In the Amazon region of Brazil, oil palm trees are long-lived and abundant. Producing biodiesel from the palm oil of existing trees is an option with multiple benefits in areas where choices are few and the damage to forests is already great.[66] Palm oil biodiesel production would help resolve the problem of electrification in the Amazon region, which has the lowest electrification rate in the country (Coelho et al., 2005). Since 2002, the Brazilian Reference Center on Biomass (CENBIO), in collaboration with other research institutes, has been working to develop and test a rural electrification model for the Amazon based on the use of local palm oil (Coelho et al., 2003). This could result in "a quite significant improvement in [greenhouse gas] emissions compared to conventional

---

59 In 2010, the US Department of Energy awarded US$78 million to the National Alliance for Advanced Biofuels and Bioproducts and the National Advanced Biofuels Consortium.

60 See McKinsey and Company, 2010, a study prepared for 28 companies from the automobile, oil and gas, utilities, industrial gas, equipment car manufacturers, wind and electrolyser companies, one NGO and two government organizations. The report emphasizes the importance of developing a range of technologies that will ensure the long-term sustainability of mobility in Europe and in this connection, stresses the need to reassess the role of fuel-cell electric vehicles in the light of recent technological breakthroughs.

61 Taking advantage of their experience with flex-fuel engines in the transport sector, Brazil's Petrobras, in collaboration with General Electric, has upgraded an 87 MW power plant to operate using either natural gas or ethanol. General Electric was interested in this research project because in other countries it has some 770 turbines similar to those in the Brazilian plant that could be converted to run on ethanol. Both companies are members of the Model Fuels Consortium, whose goal is to accelerate adoptions of Green Fuels (GE, 2010; Petrobras, 2011).

62 The "bioeconomy" spans an array of sectors, including agriculture, health, environment, and industry.

63 For a case study of this process, see Chapter 24. Additional discussion is available in Chapter 11.

64 This resulted from a combination of the special trading arrangements concluded between the European Economic Community and its associated African, Caribbean, and Pacific states, and between the former Soviet Union and Cuba, and the production of sugar from beetroots in Europe and the development of corn syrup for use in prepared food and beverages in the United States.

65 Nowadays 80% of the alcohol production is from states in southeast (São Paulo, Parana, and Minas Gerais) and mideast states (Goias, Mato Grosso do Sul, Mato Grosso, and Tocantinhs). In Parana State sugarcane is produced by smallholders organized in cooperatives.

66 In a number of African countries, palm oil has also traditionally been used for lighting.

diesel. On the other hand, if areas that were not previously cultivated are converted to palm oil production, the net resulting balance can be dramatically negative" (UNEP, 2009).

On May 6, 2010, former President Lula da Silva announced the Program for Sustainable Production of Palm Oil on abandoned and degraded agricultural lands in the Amazon region. The program, designed to support poor farmers in the region, specifically prohibits the expansion of oil palm at the expense of native forests (Butler, 2010), although whether this can be enforced is debatable. Estimates of the production of biodiesel from harvesting palm nuts and processing the oil locally in the Amazon region put the cost of the biodiesel below that of the petroleum-based fuels currently transported to the region over land. In addition to this saving in fuel costs, the harvests and processing can provide local populations with income and potentially forestall further destruction of the rainforest to plant soybean destined for export as part of the animal feed chain.

Jatropha has been praised as a feedstock for biofuels because of its ability to grow on marginal land and in arid environments unsuitable for food crops. It is a particularly good substitute for petroleum-based diesel and kerosene, and can be used as a transport fuel or as a replacement for off-grid stationary power. Small-scale production of jatropha can also help to restore the productivity of degraded land (UNEP, 2009) and open new income-generating opportunities.

In Mali, for example, jatropha is now being grown on "marginal" and unused land to provide fuel for a local 300 kW power plant (UNEP, 2009; ACCESS SARL, 2010). In early 2007, rather than rely on imported diesel for a future off-grid generator, villagers from the commune of Garalo in Mali chose to plant jatropha.[67] Of the 440 hectares (ha) planted with jatropha, two ha were prepared as a nursery and 95% of the fields were planted by individual farmers,[68] of which 6% (25 ha) were for women (Burrel, 2008). The main purpose of the plant is to provide access to electricity for a range of energy services that include lighting, refrigeration, the use of welding equipment and agricultural processing machinery by businesses and workshops. It was also designed to reduce the cost of the village water pumping system by replacing the diesel genset then in use with electricity from the local mini-grid (Rijssenbeek and Togola, 2007).

Research and advisory services are playing an important role in the development of smallholder jatropha farms. In 2008, to assess the condition of the jatropha plants, a sample of 118 ha comprised of 100 fields in 21 of the villages, were surveyed by a local agronomist working with the International Crop Research Institute for the Semi-Arid Tropics. The capacity to form such linkages between farmers and research is critical for learning, problem solving, and the sustainability of projects that involve new technologies. Feedback from this survey, for example, has contributed to strengthening the positive outcomes of the project in two ways. First, the survey indicated that 71% of the farmers intercropped jatropha with food crops such as maize, sorghum, millet, peanut, sesame, and beans. These farmers had thus dealt with the assumed need to choose between food and fuel by developing a system that produced both. Second, the survey alerted farmers unfamiliar with jatropha cultivation to the need for wider spacing between plants if intercropping were to continue as the jatropha plants grew (Burrell, 2008).

In 2008, the power station in the Garalo township provided electricity to 230 clients, of which 198 are households, 19 are shops, restaurants, workshops, and mills, two are healthcare providers, seven are local government buildings, and four are places of worship.[69] This had increased to 251 consumers in 2010, according to ACCESS SARL, a Malian energy service provider that has been selected by the municipality to manage its power station and grid (ACCESS SARL, 2010). It is expected that the number of users will more than double over the next two or three years as the jatropha plants mature and the yields and incomes of smallholders increase. Local oil transformation is also contributing to income generation, economic growth, and poverty reduction in the broader community as access to and costs of energy services decline (Burrell, 2008).

Much of the success of this project can be attributed to strong community participation nurtured by a continuous dialogue and the availability of information and technical support from local agencies such as the Mali Folk Centre (Goertz, 2006; Burrell, 2008). This has led to considerable follow-up, experimentation, and local learning, for example, in shifting from the costly process of using plastic bags for seedlings in the nursery to the use of direct seeding in the fields. Improving methods of intercropping is another area that has benefited from local research and is serving to increase farmers' incomes.[70]

The problem with using jatropha oil for fuel arises when it is reconceptualized as an export commodity. Upscaling the smallholder model for the large-scale production of biodiesel fuel for the domestic market, and even more so for export, requires considerable rethinking.

First, the assumption that jatropha is a low-cost, low-input crop that grows virtually by itself is problematic in this new context. Recent research in India, for example, has shown that the price of jatropha oil depends upon increasing the yields, and that this in turn requires improved seeds that reach their potential when well watered and often when fertilized (Altenburg et al., 2008). The failure to reach anticipated yield levels has led to the abandonment of a number of joint ventures in India (Dogbevi, 2009a; 2009b).

---

67 The village worked closely with the Mali Folk Centre, a Malian NGO that is a member of the AREED network.

68 The remaining 5% were collective fields.

69 The data come from billing records. In addition to the clients, 45 public lights have been installed. Every client pays 500 CFA (approximately US$1) extra for these lights (Burrell, 2008).

70 Changes in billing practices are also under consideration to reduce the financial burden on customers.

Second, a larger volume of feedstock is needed. Regrouping land parcels into large-scale farms and increasing mechanization are new practices introduced as a means to increase yields. But these practices require substantial amounts of capital. They also lead to changes in the organizational model away from smallholder farming to cooperatives, in some cases, but more frequently into contract farming with multinational corporations as partners. These practices also result in the outright sale of land to local or foreign firms, resulting in the potential alienation of smallholder lands and requiring considerable attention to regulatory and other policies that currently do not exist.

Algae are yet another entrant into the biofuel chain that could contribute to clean energy technologies, the growth of the bioeconomy, and the broader process of sustainable development. In a popular (non-technical) classification for liquid fuels made from biomass, such fuels are discussed in terms of first, second, and third generations where the "different generations are distinguished primarily by the feedstocks from which they are derived and the extent to which they are (or are not) commercially developed" (Chapter 11). Algae are currently placed among third-generation biofuel fuels. However, "land use constraints, competition with food and demand for biomass with high caloric value have made marine based biomass an attractive alternative supply option over the past years [...] Promising concepts for aquatic biomass include (1) land-based open ponds for microalgae, (2) horizontal lines between off shore infrastructure, e.g., wind farms for macroalgae (seaweed), and (3) vertical lines near shore in densely used areas and nutrient-rich areas for macroalgae" (Chapter 7). Marine algae use saline water, CO_2, and non-arable land. Located near point-sources of CO_2, such as factories and coal-fired power plants, hybrid production systems[71] open opportunities for clean-up,[72] while optimizing the system cost and contamination of production (Huntley and Redalje, 2006).

Algae technologies can also provide bio-based inputs or end products for a wide array of markets. Algae biomass, for example, can be converted to end applications such as biofuels and animal feed. Hydro-treated algal oil can be mixed in any ratio with conventional diesel or aviation fuel, and can then make use of the existing infrastructure. Of the algal biomass, approximately one-third can be converted into biofuels. Dried whole algae or the remaining dried defatted algal mass can be used as a source of high-quality protein in animal feed (Kiron et al., 2009) and in food products. Although the market for protein is likely to expand rapidly due to population growth, the supply of high-quality, algae-based protein may become greater than the growth in demand (Balagopal et al., 2010). A high supply could lower feed and

food prices and change the structure of the animal feed and food industries.[73]

Microalgae are single-cell organisms that convert CO_2 and sunlight into biomass through photosynthesis. Compared with terrestrial alternatives, marine algae do this very efficiently. They have the potential to produce substantially more biomass per m^2 per day than conventional biofuel feedstocks, such as palm oil or rapeseed oil, but without the need for fresh water and minimizing the need for land, phosphates, and nitrates. Because the production of the end product can take place in a liquid medium, advanced industrial processing techniques are available that have the potential to reduce processing costs and allow scaling up to very large volumes.[74]

Production yields are directly dependent on the amount of light reaching the growing algae, so that the most efficient algae facilities will be located between latitudes 23.5°N and 23.5°S. Most development of know-how, however, is currently taking place in the infrastructure-rich United States. This opens opportunities to build capacities outside the US, particularly as the most effective, non-genetically engineered algae strains for any given commercial open pond facility are likely to be found locally. Expertise is rare anywhere at the moment, which necessitates global cooperation and the use of virtual teams that include researchers and other experts based in different locations. Developing countries could play an important role if domestic knowledge capabilities in these domains are strengthened. There are also interesting research, development, and testing opportunities in closed bio-reactor processes in developing countries, as current collaborative research between a German and a South African university illustrates.[75]

Solving the many engineering problems and market issues that new industries face requires bringing people together from different disciplines and different industries, as well as patience. This goes beyond earlier practices of consulting with researchers from different firms or industries. Instead, it opens a dialogue between disciplines that speak different languages, have different points of view, and may display different logic based on different value systems. As the discussion of open innovation systems in Section 25.7.1 shows, networks of this sort have already been put in place to speed up the process of problem solving

---

73 An algal industry with a capacity of 450 kbd (1 kbd = 1000 barrels/day) of algal oil could supply up to 15% of the crude protein needed in the compound animal feed market by 2030, thus enriching crops such as protein-poor cassava for use as animal feed and reducing the damage to land and forests from soybean monocropping.

74 Biofuels must be both profitable and sustainable. Algae do not yet meet both criteria simultaneously. An important element of cost reduction is to maximize the biomass and/or algal oil produced per m^2 per day, which would minimize the physical footprint of a facility for a given plant capacity.

75 In 2009, Jacobs University (Germany) partnered with Nelson Mandela Metropolitan University (South Africa) to expand the testing of its photobioreactor, which currently uses marine microalgae to convert CO_2 from a coal-fired power plant, breweries, refineries, and cement factories in the Eastern Cape region into organic compounds through photosynthesis (Guetali SADC, 2009).

---

71 Hybrid systems combine (expensive) photo bioreactors with (contamination-prone) open ponds.

72 The closer, the more concentrated, and the purer the CO_2, the cheaper the cost of production will be.

in a variety of new technologies. This form of cooperation should be stimulated and encouraged. Corporations that are the first to master this will be more successful in executing the associated complex integrated projects and create more value.

25.6.2   Small and Micro-Hydro: Bottom-up Solutions

Hydropower is the main source of renewably generated electricity in the world today and there is still considerable potential for exploitation, in particular in rural areas of developing countries (see Chapter 7, Section 7.6 and Chapter 11, Section 11.3). Many developing countries have tried to build small and micro hydro projects,[76] some for connection to grids and others for off-grid use. Where many projects failed is through a lack of local ownership: plants are designed and constructed by foreign firms with minimal input from local stakeholders. Once commissioned, plants are transferred to the local community, which then must assume operation and maintenance responsibilities as well as costs. This has had negative consequence for sustainability. We take a closer look at the experiences of China, Bhutan, and Rwanda here.

By the end of 2007, after 30 years of unremitting efforts, China had built more than 45,000 small hydropower (SHP) stations with a total installed capacity of about 50 gigawatts. Together, they generate 150 TWh/year and account for one-third of the total hydropower nationwide. By developing small hydropower, half of the territory of China, one-third of the counties, and over 300 million people had access to electricity in 2009 (REN21, 2009).

The large number of SHP plants in China is directly related to transmission system needs, government tax policies, and feed-in power rates. The Ministry of Water and Power also played an important role in this process by providing interest-free loans to China's rural utilities and technical assistance to ensure that locally developed plans conformed to national standards. Although large hydro plants can generate huge amounts of energy, the existing grid system prevented the efficient transmission of electricity to rural villages. By providing policy support and financial incentives, funding for renovating rural grids, especially through the Clean Development Mechanism (CDM) since 2004, and technical training, the state built pilot counties of primary rural electrification infrastructure and encouraged local governments and farmers to develop small hydropower to solve the problem of local access to electricity. This environment encouraged private companies to invest in the construction of small hydro plants, and up to 85% of the funding now comes from private sources (Zhao and Zhu, 2004). Zhejiang province is a case in point.

Severe power shortages in Zhejiang province necessitated reforms. Due to preferential government policies (a guaranteed power price, macro planning, and a standardized bidding process) and stable and high profits, the Zhejiang water resources industry has attracted the highest national amount of private investment in this sector. The national fund for grid construction was used to build the connecting infrastructure, while private funds were tapped to develop the small hydropower sources. Private investment covers 80% of installed small hydro capacity.

A wide range of investors have been involved in this process. Water resources enterprises, power corporations, individuals, overseas Chinese and shareholding corporations have cooperated in establishing corporations and limited companies. Shareholding SHP corporations, originally located in villages or counties, became provincial group corporations. This not only quickened the exploitation of hydropower resources, but also fostered water conservation, developed a wide range of local capabilities, and raised living standards.

Village-scale power, however, continues to be one of China's biggest challenges for isolated villages. Early project experience suggests that village power applications will require government assistance for some time due to economics, the challenges of management arrangements, and the need for local regulators and tariff frameworks. Most village-scale power systems managed by traditional utilities have not been successful in China. Other operating and management models are clearly needed. Even collecting tariffs high enough to cover operating and maintenance costs can be a challenge – about 1 RMB/kWh (or about US$0.10/kWh) seems to be about the maximum.

SHP development in China is decentralized and managed by local governments and people. The state encourages the development of local small hydro power resources – based on the principle of "self-construction, self-management, and self-consumption" and the policy of "those who invest should enjoy the revenue and ownership." The results have been impressive. The bottom-up development of village power schemes undertaken by local village organizations and business firms, seems to show a greater chance of success by coordinating mutual interests at the earliest stage of development, generating commitment through face-to-face communications with working people, and following a rigorous process for clarifying the interests, roles, and responsibilities of all parties.[77]

---

76 Definitions of small, micro-, and mini-hydropower vary across countries and can even include systems with a capacity of just a few megawatts. The United States India, and Brazil define microhydro as <100 kW and mini hydro as 100–1000 kW (Moreira and Poole, 1993). In China, small hydropower refers to plants with an installed capacity of up to 50 MW (Zhou et al., 2009). In Nepal, small, micro-, and mini-hydropower systems are defined respectively as plants with 1–15 MW (usually feeding into a grid), 100 kW-1 MW, sometimes stand-alone, but more often feeding into a grid. Micro installations with capacities of 5–100 kW usually provide power for a small community off-grid (AEPC and UNDP, 2009).

---

77 A bottom-up approach to rural electrification in Nepal also appears promising, although its lessons "are not being adopted and upscaled at a desired pace," despite the passage of a Local Self-Governance Act in 1999, and evidence that "local government bodies are dissatisfied with central agencies, which are perceived to be indifferent to local proposals including local program needs, objectives and modalities" (Nepal, 2007).

The Kingdom of Bhutan faced similar problems in its efforts to develop off-grid micro-hydro projects. Due to their lack of financial viability, the Bhutan Power Corporation has been reluctant to take over the management of the new micro-hydro plants (MHP) being built. When the Chendebji micro-hydro was commissioned in October 2005, an alternative model involving the community had to be developed. This community-based management model is being piloted in Chendebji and is expected to be refined and used for similar projects in the future. To ensure sustainability and diffusion of micro-hydropower in the rural areas, such projects should be coupled with the development of income-generating opportunities and a program of continuous training to maintain the skills of local plant operators (see Case Study 2, Section 25.11).

The situation in Rwanda differs from that in both Bhutan and China. A small, densely populated country, Rwanda had an available national electricity generation capacity of 71 MW in 2010 (Bensch et al., 2010). This provides access to about 10% of its 10 million people. Some 270 MHP sites have been identified, of which about 20 have been developed or are under development.[78] Grid expansion is rapid and sites that even five years ago were expected to remain more than 15 km from the national transmission grid are now close enough to be grid-connected. As a result, most MHP plants have a choice between connecting to the grid or not, based on a trade-off between incremental investments for transmission lines and incremental income from being able to sell 100% of the produced electricity.

Public funds were used to develop most of the MHP plants and operational responsibilities were, until recently, given to local communities. These were not able to manage the plants in a sustainable manner (GTZ, 2010). The national electricity company is not interested in managing these plants even when grid connected, as they are too small and associated management costs are too high compared to its larger plants. Over the past several years, a number of privately owned MHPs became operational and more are under construction. These are funded by a private sector participation program for the development of micro-hydro in Rwanda (Pigaht and van der Plas, 2009). Under this program, local entrepreneurs develop the plan for these plants jointly with beneficiary communities, secure commercial loans from local banks, and are financially supported by a government subsidy. Since plant owners have taken loans to finance part of the investment costs, they have every incentive to ensure that the plants continue to function properly. The government has now decided that management of all public MHP plants will be auctioned off to the private sector, and future MHP plants will only be established with contributions from the private sector. Although the cost of electricity from these plants is slightly higher than from 100% publicly financed plants to cover private financing costs, more funds are available for immediate development of village power systems. In addition, the sustainability of privately financed plants is likely to be higher, and plants are developed around locations with the highest interest among the local population (Pigaht and van der Plas, 2009).

Overall, there is great potential for the further development of hydropower, especially in rural areas of developing countries and with the participation of local communities and private sector players. But the successful development of this potential requires comprehensive government policies at the local, regional, and national levels. Such polices include financial support through taxes, feed-in tariffs, and soft loans; the creation of programs for technical assistance and capacity development in the design, maintenance, and management of small and micro-hydro; and changes in regulatory policies that make conditions more favorable for private sector investment in village-based micro-hydro projects. The success of targeted bottom-up approaches in China, Bhutan, Nepal and Rwanda illustrates the need for mutual interests to be accommodated through dialogues in which local government bodies are brought into the process early in the development of a project and a mechanism for continued dialogue over the life of the project is put in place.

This section has also illustrated that, in addition to enabling energy access, distributed energy systems based on biofuels could have an important role to play in the broader development process. The need to develop capacities for making choices about biofuels in technical, as well as economic and financial, terms, and in evaluating trade-offs is of critical importance in creating such benefits and ensuring their long-term sustainability. Greater attention, for example, should be paid to assessing whether, when, and how to scale up the production of biofuels in a given context. Strengthening the research and problem-solving capabilities of local universities and research institutes in agro-, bio-, and social sciences will be essential in supporting such policymaking processes. Developing these capabilities, however, can take many years at a time when decisions about energy transitions that involve biofuels require attention now. The next section explores the role that a variety of different forms of networking can play in contributing to the development of domestic capabilities for research, analysis, and innovation.

25.7   Strengthening and Sustaining Capacities for Networking and Innovation

In the last quarter of the 20th century, open and distributed innovation emerged as a response to new needs for knowledge sharing in the information technology and biopharmaceutical industries that were not being met by the existing patent system (Harhoff et al., 2007; Dechezlepretre et al., 2009). The networks and knowledge-sharing approaches that were developed helped to deal with the diversity of knowledge inputs that characterized these new technologies, enabled enterprises in developed countries to catch up with the technological frontier,[79] and contributed to more rapid problem solving that accelerated the diffusion of new technologies. Networks for knowledge sharing and collaborative research are now emerging in the field of new energy technologies.

---

78   This involves about 17 MW in plants with capacity of between 100 kW and 5 MW.

79   See the case studies of the international collaborative ventures of US firms in the telecommunications equipment industry, pharmaceuticals, biotechnology, aircraft, robotics, and automobiles (Mowery, 1988) and of European firms (Mytelka, 2001).

This section discusses the special features of open and distributed innovation approaches as they have emerged in the developed countries. It then focuses on the application of these approaches in developing countries, notably in the strengthening of university-industry linkages and in a variety of new knowledge networks that support research, analysis, and problem solving in new energy technologies, and thus speed their development and diffusion. From a multi-goal perspective, networks such as these are now also being designed to support broader development goals by reducing the brain drain from developing countries.

25.7.1   Networks for Open and Distributed Innovation

Open innovation explores the potential of an organization, i.e., a firm or university, in a networked environment (Chesbrough, 2006). A network culture provides an opportunity to connect more people, who embody a wider knowledge base, to an organization's mission.[80] The core insight of open innovation is the ability to use the world outside an institution to generate internally useful knowledge – and the core dependency of open innovation, in turn, is the need to make the flow of knowledge in and out of an institution a purposeful thing, not a random process.

A world of purposeful information flow is at odds with many of the business structures of the last 50 years, especially intellectual property rights. The idea of leveraging the sharing of knowledge as a key business strategy, rather than using intellectual property exclusively as a measure to prevent or exclude competition, is thus a novel one that has not yet resulted in the widespread adoption of models such as business-to-business networks or the collaborative research programs that are now common in the European Union.[81] There is, however, movement in this direction. The BIOS Project in Australia, for example, began with a single core patented technology for transferring genes in plants. It was developed specifically to avoid encumbrance from existing patents, so that users could develop plants without any fear of "reach through" claims affecting things that were as yet undiscovered. Now called the Initiative for Open Innovation, it offers a range of legal tools that allow the possibility for many classes of contracts other than patent licenses. These tools encourage new improvements to the underlying technology,[82] as well as improve the potential for profit-making into the covenant.

---

80 For example, to contribute to internal projects from the outside, to take a project that fails to gather internal support forward using outside funding, to generate novel projects outside and "spin into" new internal projects.

81 These date back to the European Strategic Programme for Research and Development on Information Technologies (ESPRIT) in 1983. Funded in part by the European Union, each project must be cross-national and bring together at least one enterprise with other actors (universities, users) for collaborative research and technology development (RTD). Annual competitions are held to select projects for funding and all partners have full access to knowledge and technology developed within the project (Mytelka, 2001).

82 A plant created would be freely available, but if a licensee improved the gene transfer tool itself and patented the improvement, then the improvement would have to be made available back to all other licensees of the tool (IOI, 2011).

Other projects aim to go even further towards the creation of infrastructure for open innovation by specifically encouraging patent holders to make their patent portfolio available for licensing through public license offers that are open to everyone on reasonable terms, while retaining the defensive benefits of patents. This goal is based on the understanding that many patent holders have patented inventions that could have broad or new applications in areas that they did not anticipate, but they may not have a strategy to actively license them or offer them for such uses. By making public license offers on reasonable terms, patent holders can encourage others to seek out novel uses that could have important economic or environmental benefits (see Science Commons, 2011).

Capacity development in this area involves the creation and use of standard licenses for most uses of patents outside a company's core business space. It also creates open and widespread patent "landscapes" in key areas so that risks and opportunities are well understood and public-private partnerships to ensure that vendors are in place to provide services to entrepreneurs, large companies, and universities, each of which plays a key role in the innovation ecosystem under open innovation.

The second approach, generally called distributed innovation, is most closely associated with the open source software culture reflected in the development of Linux (Feller et al., 2005). Distributed innovation is less a modification of existing practices than a truly disruptive aspect of the network culture. In terms of innovation power, it consists of a collected set of individuals whose individual actions "snap together" into a coherent group through standard technical systems and digital networks. Distributed innovation has now spread from information and communications technologies to the biosciences, pharmaceuticals, environment, agriculture, and related energy technologies. Recent studies have provided evidence that this approach is able to speed problem solving in technologies that span multiple knowledge bases and have relevance to multiple disciplines involved in new energy technologies (Lakhani et al., 2007; Lakhani and Pannetta, 2007). InnoCentive, an online network that puts "seekers" in touch with "solvers" of technology and policy problems, for example, has begun to include energy-related problems on its list (see Case Study 3, Section 25.11).

A related development is the recent emergence of informal Internet discussion groups on subjects of importance to developing countries. Their purpose is to exchange information on a certain technology or process, use the collective experience of network members to solve problems posed by network participants and to advance knowledge and understanding of these technologies to a wider audience of commercial firms, researchers, government officials, and students. Among these, two well-known examples are the Stoves Discussion Group (Stoves, 2011) and the Microhydropower Discussion Group (microhydropower.net, 2011).

Scientific publications are also playing an important role in knowledge dissemination and network creation. The push to share knowledge has increased tremendously since the Internet became commonplace. "Open Source" publications[83] co-exist with for-fee publications, and numerous government organizations, donor organizations, UN Agencies, NGOs, and universities provide open access to reports and papers prepared by their staff or consultants.

25.7.2 University-Industry Linkages

Brazil has a long history of government support for domestic research, as the development of ethanol and biodiesel illustrates (see Section 25.6.1). The technology transfer offices in Brazilian universities and the development of collaborative relationships between universities/public research institutes and the enterprise sector trace their origins to the transition from a "top-down" system to one that operates at multiple levels – municipal, regional, national, and multinational. This transition allowed for the emergence of new initiatives by new actors, especially universities and industrial associations (Lahorgue et al., 2005). Technology parks, technology transfer offices (TTOs), science parks, and incubators have proliferated since the 1990s (Coutinho, 2001). But, despite these initiatives, Brazilian manufacturing firms have relatively low innovation capacity. Their spending on innovation is mainly related to the purchase of machinery and equipment and, to a lesser extent, on research and the development of new products and processes. In contrast with other countries, moreover, Brazilian companies do not receive significant public financing for these activities (BRASIL, 1996; 2004; 2007), and about 60% of R&D activities in Brazil are funded and carried out by the government (Cruz and Mello, 2006). But this includes major research institutes in state-controlled companies such as the Brazilian oil company, Petrobras, and the research and reference centers located in a number of Brazilian universities.

Petrobras was a pioneer in the creation of knowledge networks as a vehicle for the development of offshore oil technology at a time when low oil prices created a disincentive for major companies to invest in this technology. Between the late 1960s and the early 2000s, "Petrobras massively transformed its offshore technology networks [...] from a passive learning network to an active learning network" and developed in-house research capabilities that later enabled them to engage in strategic networks that involved two-way flows of knowledge and collaborative research activities"[84] (Dantas and Bell, 2009). The reform of the power sector in the mid-1990s opened yet more opportunities for networking and collaborative research. From the start of the energy privatization process, and as a condition within the concession agreements, companies were obliged to invest 1% of

annual net revenues in energy efficiency and R&D.[85] This was complemented in 2000 by the creation of "CT-Energy," a fund to support public interest research,[86] which specifically included energy R&D and energy efficiency among its "social benefit investments."

In 2002, the Brazilian government's White Book on Science, Technology and Innovation (see Portal do Ministério da Sciência e Technologia, 2002) focused on creating "an effective science, technology and innovation national system" in Brazil. This was followed by Brazil's current policy framework, the industrial, technological, and foreign trade policy, launched in 2003, the 2004 Innovation Law (10.973/2004), and the expansion of sectoral funds, including those earmarked for renewable energy technologies that encourage and financially support a broad range of energy-related networks.

Brazil currently has a substantial network of research groups, or reference centers. They serve as recognized regional or national centers of excellence in a specific domain and are also open information networks in the field of energy technology. Among the research groups and networked centers is the award-winning GreenSolar research group, whose multidisciplinary research on solar thermal energy technologies draws together university departments of civil and mechanical engineering, electronics, control and automation, architecture, computer science, and management.[87] GreenSolar also works closely with companies such as Electrobras and the main business association in this field, and ABRAVA, the Brazilian branch of the Global Solar Thermal Energy Council. Among the reference centers are CENBIO, which undertook the research on palm oil production in the Amazon (see Section 25.6.1),[88] and the Brazilian Reference Centre on Biofuels (CERBIO), which works in collaboration with the Paraná Technology Institute to undertake research, development, and innovation on biofuels. It also evaluates the technical feasibility and social and environmental impacts of these biofuels. CERBIO was chosen by the federal government to support the national biodiesel program, which brings together some 30 institutions including universities and enterprises. In this latter capacity, CERBIO is also part of the "Brazilian Service for Technical Answers," through which specialists address technical questions in various fields.

---

83 Such as www.ashden.com, www.scidev.net, and www.scirp.org/journal/lce.

84 The concept of two-way knowledge flows in collaborative research was developed and extensively applied to analyze strategic partnerships and networks in the European Union (Mytelka, 2001).

85 This measure was intended "to avoid the risk that the new companies, mostly owned by foreign investors, would transfer all their R&D efforts elsewhere, a trend already noted by [...] Bourgeois and Jacquier-Roux, 2001" (Jannuzzi, 2005). See also www.annel.gov.br.

86 CT-Energy does not substitute for the obligation to invest, present in the concession contracts. The fund supports energy efficiency programs and initiatives that would not be considered by utilities or market agents, such as the development of energy efficiency standards, consumer training courses, promotion of events, scholarships or research grants directed to projects that will contribute to improvements in energy supply and use. The fund supports regional initiatives to develop local capacity and projects (Jannuzzi, 2005).

87 GreenSolar is based at the Pontifícia Universidade Católica in Minas Gerais, Brazil.

88 Information on Brazilian zoning for sugarcane at both federal and state levels is presented in Chapter 20.

Like Brazil, South Africa issued a white paper on science and technology in 1996 outlining the concept of a national innovation system (DST, 1996). Three years later, it had become evident that, in the context of South Africa's complex socioeconomic challenges, more would need to be done to raise investment in science and technology. This led to the development of the South African research and development strategy, which provided funding for a range of technology missions that were central to innovation and to the promotion of economic and social development (Mehlomakulu, 2008). The strategy also recommended interventions to improve the efficiency and effectiveness of the national innovation system, including the establishment of a national agency to stimulate innovation and the introduction of legislation to encourage technology transfer from public research institutions to industry. After declining for much of the 1990s, gross expenditure on R&D has been rising steadily, approaching a level of 1% of GDP,[89] more than half of which is undertaken by industry and not government, as in Brazil.

In most of the "research-intensive" universities and science councils, TTOs were set up during the late 1990s and early 2000s at the initiative of the institutions themselves, who recognized the need for and were willing to devote some resources to this activity, as occurred in Brazil. Active government involvement in promoting technology transfer from public research organizations began shortly thereafter, leading to more TTOs.

While the newer TTOs are still in the process of institutionalizing the relevant policies, procedures, and practices, the more established TTOs have now accumulated some degree of experience and expertise. Patenting and licensing, nonetheless, remain low. Various factors have contributed to this (Wolson, 2008; Sibanda, 2009). The pipeline of invention disclosures is thin. Since the number of potential disclosures is related to institutional R&D expenditure (Heher, 2005), a smaller system is likely to yield fewer disclosures. Furthermore, researchers are often reluctant to make disclosures, whether due to a lack of interest or incentive or skepticism about the academic technology transfer endeavor. A relatively small pool of researchers with interest and experience in the technology transfer process also affects the quality of disclosures and, as a consequence, they are often not ideal subject matter for commercialization. The costs of patenting are high, and many institutions do not have adequate budgets to pursue international filings. Although the number of technology transfer practitioners is increasing, capacity remains limited, as do capacity-building opportunities. Finally, local licensing opportunities are scarce, with South African technology-led companies typically accessing their technology from abroad, in the belief that local institutions are, for the most part, not a valuable source of innovation, according to a 2005 Innovation Survey (DST, 2007) or are undertaking their own research. Both the South African Technology Innovation Agency and the new Intellectual Property Act provide for financial support and

capacity-building activities, which are welcomed, and are expected to assist in overcoming some of the identified problems once they are up and running (see Case Study 4, Section 25.11).

There has also been some concern that South Africa is not capturing the benefits of its own R&D. However, while examples exist of technologies developed in South Africa that have been commercialized abroad, a recent study investigated some of these cases, but the data failed to identify any technologies that had been transferred overseas and successfully commercialized to the point that they were generating income to their current owners, without benefit to the original inventors (Pouris, 2008).

One example from the field of renewable energy clearly illustrates that commercialization outside the country is by no means synonymous with denying South Africa and South African inventors the optimal benefit of their innovations. The technology for cheap, thin-film, silicone-free photovoltaic cells was developed at the University of Johannesburg with support from the Innovation Fund, a government agency that promotes technological innovation. The university and the inventor set up a company in South Africa to commercialize the technology. The company's decision to license the technology to a German firm attracted strong criticism, on the basis that this was "unpatriotic" and with the implication that the technology was being "lost" to South Africa. But while the first plant was set up in Germany, plans are now afoot to build a South African manufacturing plant that will generate substantial investment and employment and essentially kick-start a new local industry. This is a collaboration between the German licensee and various local investors, including the company owned by the university and inventor. The local partners have also invested in the German firm, resulting in a cross-shareholding that has facilitated access to relevant expertise, know-how, suppliers, and finance that were not available within the country. Arguably, had efforts to set up a local plant been confined to South Africa, it would have taken far longer to build the necessary capacity and attract the funding required (Planting, 2006; Pouris, 2008; Reuters, 2009). While few would argue against licensing new technologies to new spin-out or start-up or existing domestic companies, this case shows that this is by no means the only way to capture value from a technology for the country.

25.7.3   Creating Knowledge Networks: South-South, North-South

Knowledge networks for joint research and problem solving, knowledge sharing, analysis, and long-term planning can contribute positively to the development and diffusion of new energy technologies in a number of ways. Such networks, for example, can potentially create confidence that attracts greater funding and might strengthen the sense of local ownership within the context of donor-supported projects. Participation in knowledge networks that strengthens local research capacities, moreover, can create incentives that reduce brain drain, thus contributing to

---

the provision of context-specific inputs that are needed for policymaking, for problem solving, and for the design, development, and diffusion of new energy technologies. These benefits are, of course, potential. In this section, we explore some of the ways in which knowledge networks that provide such benefits have emerged and evolved. The focus is on developing and least developed countries where, until recently, such opportunities have been rare.

There is much concern in the developing world about the problem of brain drain. The challenges that this presents are exacerbated by the growing need for access to a continuous stream of knowledge and information in order to make choices about energy technologies and pathways towards energy transitions. Some efforts are being made through networks of centers of excellence, such as the Global Network on Energy for Sustainable Development (see GNESD, 2009), to provide inputs for policymaking in developing countries.

One of several partnerships in the field of energy that were launched at the World Summit on Sustainable Development (WSSD) in September 2002, Global Network on Energy for Sustainable Development (GNESD) became operational in 2004. GNESD received core funding from Germany and Denmark. It has also received initial support from France, Italy, the United Kingdom, the UN Foundation, UNDP, REEEP, and UNEP. It was originally composed of 21 public and private research centers and NGOs engaged in research, with a variety of different strengths in the energy area and situated in both developed and developing countries. However, over the years GNESD has evolved into a South-South knowledge network consisting of 10 centers of excellence. Current membership spans a number of sub-regions, focusing on policy-oriented research that addresses different aspects of energy access and Millennium Development Goals achievement. Some members have worked together before in other networks.[90] A number of the members are themselves networks; consequently, GNESD's reach is quite wide.

GNESD's work focuses on energy access, energy security, renewables and bioenergy for the poor – notably those in urban and peri-urban areas – and energy policy issues facing governments in the developing world. Within this framework, a steering committee, currently composed of representatives from eight member organizations in the developing world, three from OECD countries, and two from donor organizations, meet annually to decide on the network's thematic programs. The first of these focused on energy access issues. It reviewed existing energy policies in Africa, Asia, and Latin America to identify the effects of reforms that have been carried out in the past and to recommend policy measures that can increase the possibility of bringing energy to those currently without access. One of its outcomes was a publication on how to overcome barriers and unlock potentials in renewable energy technologies (GNESD, 2007) tailored to the needs

of users, which was followed up by workshops in developing countries organized by member centers to disseminate their findings.[91] GNESD also undertakes research at the request of developing country governments. In 2010, the UN Secretary-General's Advisory Group on Energy and Climate Change recommended that "[e]xisting knowledge networks should be mobilized and new ones built [...] to accelerate the transfer of best practices (with respect to modern energy system policies and regulations)," and specifically cited GNESD as "a good example of knowledge creation and sharing on energy policy analysis" (UN AGECC, 2010).

In the sciences, the challenges of the brain drain remain acute, and the benefits derived from earlier efforts to build North-South networks in the areas of health and biosciences have not relieved this. In July 2009, the Wellcome Trust[92] announced a grant of £30 million (US$47.9 million) over five years to support seven newly created research consortia based in Africa. The grant is renewable after an independent review. The approach is new in several ways. First, the objective is to strengthen research capacities in Africa itself. The seven consortia are thus based in Africa, each one headed by an African researcher, and their research will be conducted mainly in Africa. As Dr. Alex Ezeh, executive director at the African Population and Health Research Center in Kenya, who is leading the Consortium for Advanced Research Training in Africa, noted, "(n)otwithstanding the attention it has historically received, research capacity remains very weak in Africa. Existing research capacity-strengthening programs and collaborative partnerships in Africa are largely driven by Northern academic and research institutions" (Wellcome Trust, 2009). To this, Dr. Margaret Gyapong, director of the Dodowa Health Research Center of the Ghana Health Service and head of the new Research Institute for Infectious Diseases of Poverty, one of the seven consortia, added, "Africa is losing many of its best scientists to the brain drain, partly because of the absence of coordinated institutional strategies and national research environments [...] poor integration of knowledge across disciplines, and a lack of consistent engagement with policymakers, users and community beneficiaries" (Looi, 2009). The seven consortia include 52 African institutions in 18 African countries, and have linkages to research centers in Australia, Denmark, Norway, Switzerland, the United Kingdom, and the United States. While it is still too early to evaluate its contribution to the development of research capacity in Africa; if successful, it might constitute a new approach to consortia development for energy transitions.

---

90 Interview with Daniel Bouille, Bariloche Foundation, Argentina, center representative.

91 Its second theme, renewable energy technologies, was launched in early 2005 and completed during 2006 and 2007. Two other themes were launched thereafter, one on urban and peri-urban access to energy, and the other on energy security. The former has already produced case studies of India, Senegal, Argentina, Thailand, and Kenya, and dissemination workshops have been held to discuss lessons from the needs assessment and the analysis for policies and programs for capacity building.

92 The Wellcome Trust is the United Kingdom's largest charity.

Box 25.6 | UNU Geothermal Training Programme

The United Nations University Geothermal Training Programme (UNU-GTP) is aimed at professionals employed by local organizations, public or private, in which geothermal work is already underway or starting up. Potential participants are nominated by their organizations and the selection process emphasizes direct insertion upon return (see UNU, 2011).

The GTP offers courses in nine fields related to geothermal technology and good environmental practice. The programs of individual participants are tailored to these fields so that, over time, teams are built up with the capacity to support local research, exploration, production energy, and environmental planning. This strengthens the "ownership" of geothermal resource planning in projects funded through development assistance and strengthens the set of capabilities needed for research, development, and planning in these countries.

The adoption of a regional approach in areas with large reserves of untapped geothermal potential has encouraged the development of knowledge networks that support local learning and have helped to attract investment in geothermal exploration and energy supply. The Rift Valley in East Africa illustrates this approach. It began with a gradual expansion in the number of participants from Rift Valley countries in the six-month training and research program. This was followed by the first "short course" in Africa, which brought together 33 participants from Kenya, Djibouti, Eritrea, Ethiopia, Tanzania, and Uganda, the six member countries of the African Rift Geothermal Energy Development Facility (ARGeo), a program supported by GEF, UNEP, and KfW, a German development bank (Hamlin, 2004). After successfully testing advanced seismic and drilling techniques in Kenya that will substantially reduce costs (UNEP, 2008), a five-year program of exploratory drilling in the ARGeo countries has been developed and funded (ARGeo, 2010).

A similar approach is being pursued in Central America (Fridleifsson, 2006).

In the field of new energy technologies, the role that training programs can play in supporting research, development, and diffusion of geothermal energy in developing countries has been already demonstrated by the United Nations University Geothermal Training Programme (UNU-GTP). Established in 1978 and located in Iceland, the program was designed to assist in establishing groups of specialists in selected organizations – research institutes, municipal energy utilities, and universities – in developing countries with significant geothermal potential. These groups, in turn, would create local capacity for the exploration, development, and use of geothermal energy in these countries (see UNU, 2010). Initially, the vehicle for capacity development was a six-month specialized training program that brought scientists and engineers from developing countries to Iceland. In 2005, this was complemented by "short courses" on thermal energy topics open to both specialists and decision makers from the public and private sectors. These workshops are held in the developing countries. Co-sponsored by local energy agencies, they include graduates of the UNU-GTP in their teaching staff (Georgsson, 2008).

In terms of standard assessments of project outputs and outcomes, the UNU-GTP has been very successful. Over the period 1979–2010, 452 scientists and engineers from 47 countries completed the program, including more than 75 from China, 53 from Kenya, 31 from the Philippines, 30 from El Salvador, and 27 from Ethiopia and Indonesia. Today, China is a world leader in the direct use of geothermal energy. Kenya, the Philippines, and El Salvador obtain 10–22% of their total electricity from geothermal energy, and Ethiopia has started its first geothermal power plant (see UNU, 2010).[93] What is unique about this program, however, is its many spillovers – some intended from the outset and others the product of iterative learning and adaptation processes.[94] These innovative practices have created incentives for trained personnel to return home and to stay in jobs for which they have been trained. This helps to meet the challenges of the brain drain, as well as contributing to the sustainability of geothermal research and local production once donor funding ends.

As the variety of new initiatives reviewed in this section illustrates, networking and open and distributed innovation can play an important role in providing the continuous flow of information, "research-based

---

93 Since its introduction in 2000, 25 have also graduated from the MSc program.

94 See Sections 25.3.1 and 25.3.2 for a discussion of these differences.

evidence," and analysis that governments need to make informed choices about energy technologies and pathways. However, as Simon Maxwell, director of the United Kingdom's Overseas Development Institute (ODI) points out, "engaging with policy requires more than just research skills" (Young, 2008). Researchers will need to work closely with policymakers throughout the research process, from identifying the problem to undertaking the research itself and drawing out the recommendations for policy and practice. These capabilities are not inherent in the research task, nor is collaboration of this sort the norm in policymaking practice. Working closely with researchers, bilateral and multilateral development organizations, and NGOs, ODI has developed an iterative approach to "engaging with policy" that draws on the concepts of complexity, on the tools of outcome mapping developed by the IDRC (see Section 25.3.1), and policy engagement tools developed by the ODI itself. The process of learning to work in partnership begins by mapping the political context around the policy issue, identifying the key participants and the desired changes in behavior. Developing an engagement strategy, analyzing internal capacities to effect change, and establishing the assessment tools and learning frameworks will then be needed to make alignment and learning in partnership happen (Young, 2008). As this list reveals, many new capacities, in the sense of habits, practices, norms, and values, will be developed in this process.

Participation in networks also strengthens the capacity of governments, local industry, research, and NGOs to innovate and speed up the processes of change. For this to benefit developing countries, however, a concerted effort must be made to strengthen local research capacity, foster a better understanding of open licensing practices in relation to intellectual property rights, and develop and implement policies that support networking and innovation. Innovation, as we have stressed from the outset, must be broadly conceptualized to include both the uptake of knowledge at the frontier and knowledge that is new to the user, community, or country, if not to the world. Problem-solving networks, especially in developing countries, are thus as important as those that lead to patented products and processes.

As the case studies in this section also illustrate, participation in networks and the capacities to innovate are not limited to South Africa and Brazil, but provide opportunities for a wide range of developing countries to move more rapidly towards an energy transition, as the example of geothermal energy showed.[95] A co-benefit of this process is that it can create the positive environment that encourages researchers to remain at home, thus reducing the brain drain.

25.8   Making and Managing Energy Transitions

This section focuses on the way capacities in the energy sector – notably skills, habits, practices, legal and other norms, and policies – have been built up. But, it also explores the extent to which they have been maintained over time, transferred to new applications, or progressively unlearned as they became less relevant in the course of an energy transition. We look at three very different cases from established oil and gas producers and exporters. In the first two, we focus on the build-up and erosion of R&D capabilities in Mexico's oil industry and its slow start in moving towards renewable energies and on the Abu Dhabi Economic Vision 2030, which is designed to turn the emirate into a knowledge-based leader in new and renewable energy technologies. The third case examines the approach to an energy transition process taken by the Netherlands.

25.8.1   From Oil to Clean Energy Systems: Policy Learning and Unlearning In Mexico

The discovery of giant oil deposits in Mexico during the late 1960s and early 1970s led to a production boom that turned Mexico into a major oil exporter. Over the next two decades, the Instituto Mexicano del Petróleo (IMP), the research arm of PEMEX, Mexico's National Petroleum Company, emerged as a major R&D center, with strong capabilities in exploration, development, and engineering projects, as well as flexible production planning. The IMP undertook basic and applied R&D, the training of researchers and personnel, and detailed engineering projects.[96] This would change as Mexico entered a period of economic instability.

With oil prices rising, interest rates at historically low levels, and import-substitution-led growth slowing down, the Mexican government yielded to the temptation to fuel economic growth by relying on oil revenues and external debt. This mode of development financing collapsed with the crisis of 1982. The "hands-off" model of economic liberalization that followed reoriented growth towards export-oriented industries, substantially reducing the share of oil revenues in total exports, but it did not bring about macroeconomic stability or sustained growth. The accumulation of reserves became negative in this period and after the Mexican financial crisis of 1995, a second surge in oil exports would again push annual production higher.

Since then, accelerated oil exploitation has been driven by a new investment model in which the burden of activity has been shifted to the private sector, leveraging these investments with internal debt. The adoption of these new financial schemes has had a strong negative impact on capacity building. During the crisis years, IMP's

---

95 Some of the smaller developing countries, of which Kenya (*jikos*, geothermal energy) and Cuba (modernization of thermoelectric plants, natural gas [Saenz, 2008]) are interesting examples, have forged ahead in specific energy-related areas, linking problem solving to the needs of local users.

96 "By the late 1970s, about 90% of the detail engineering of Pemex projects was developed in Mexico" (Guajardo, 2007).

capacities were not renewed, as PEMEX abandoned efforts to attract new investment and relied increasingly on technological outsourcing. The new investment model was based on build-operate-lease and build-operate-transfer operations carried out by foreign firms. In these turnkey projects, all R&D, licensing, design, engineering, and sometimes operation is undertaken by the contracting firms. The buyer acquires a ready-to-use plant (or the final product). By "buying" these capacities, this investment model eliminates the demand for local engineering and R&D skills, and therefore the learning space needed for local capacity building and the development of new enterprises and organizations. By the early 2000s, the output of Mexico's major oil fields, particularly the Cantarell field, had entered a period of decline, slightly ahead of other giant oil fields in the world (Höök et al., 2009). Nevertheless, production for export continued. Growing certainty about the nature and volume of deep-water reserves in the Gulf of Mexico convinced PEMEX officials that the company must rapidly engage in deep-water drilling and development in order to maintain the pace of extraction. This led to a relaxation of the legal constraints on subcontracting to private partners, on the grounds that PEMEX lacked technical and operational experience to exploit these offshore resources (see SENER, 2007).

Indeed, oil exploration, development, and production in deep waters require a different set of technologies and capacities. Drilling below 500 meters in the sea demands floating platforms, special drilling ships, and satellite guided systems, as well as robotized and other remote-controlled equipment and installations, all equipped to operate under extreme climatic and pressure conditions. Naturally, using these systems also requires specialized capabilities in project design and development that PEMEX clearly lacks. But this is what other firms at similar stages of technological development, such as Petrobras in Brazil, have done (Ballarin, 2002). Moreover, it is precisely what PEMEX did in the past.

The challenge goes beyond the difficulties of building efficient (and safe) deep-sea drilling platforms. Electricity generation in Mexico was opened to independent producers in 1993, and by 2008, the private share in generation had grown to 40%. The lack of incentives and a lowest-cost policy for new installed capacity meant that most capacity growth in combined-cycle gas turbines was also constructed via turnkey investment schemes. In addition to its negative impact on the development of local engineering and research capacities, this policy framework has blocked the adoption of renewable energy sources. A similar situation prevails in the development of wind energy (Borja et al., 2005), where turnkey investments are the rule. The Mexican government has since committed itself to exploiting existing, and developing new, renewable energy sources (SENER, 2006), and has recognized the role of technology R&D in this process (SENER, 2002). But in the absence of a coherent policy framework for capacity building beyond some support to basic research and market deregulation, there is little demand for detailed engineering, design, and R&D services from local institutes and firms. In the solar energy sector, for example, despite the accumulation

of important capabilities in the Instituto de Investigaciones Eléctricas (IIE),[97] the lack of an industrial incentive framework has prevented spin-offs and transfers of knowledge and technology from the public to the private sector. Moreover, no local industry supplying photovoltaic components has developed. Movement towards an energy transition will thus require a rethinking of current practices and policies that shape incentives, innovation, and the diffusion of new energy technologies.

25.8.2   Abu Dhabi's Economic Vision for 2030

In contrast with Mexico, which "unlearned" earlier positive habits of capacity development, Abu Dhabi illustrates movement towards the learning of new habits and practices as it seeks to overcome its dependence on oil and gas, and to develop domestic capacities for an energy transition. Since the discovery of petroleum in 1958, Abu Dhabi's energy sector has been the primary driver of both economic development and environmental degradation. Industrial growth in the emirate, as in much of the Arab Gulf region, has involved the construction of thermal electric power plants, aluminum smelters, cement plants, oil refineries and platforms, and petrochemical and fertilizer plants. According to the World Resources Institute, the United Arab Emirates (UAE), of which Abu Dhabi is a member and the capital, is the world's second largest CO_2 emitter on a per capita basis (WRI, 2010).

In view of its small population, the transition to an oil and gas economy also brought with it a heavy reliance on foreign technology and on foreign labor and expertise to operate it. Efforts have been made to reduce the reliance on oil by diversifying into agriculture, manufacturing, tourism, and aviation; by re-exporting commerce; and by establishing free-zone facilities and telecommunications. And yet these have tended to exacerbate the problem. The region's member states face a growing challenge in maintaining the demographic balance between nationals and expatriates. In an interview in 2008, a human resource specialist from the Dubai Municipal Government reported that expatriates occupied 99% of jobs in the private sector and 91% in the public sector out of a total 3.1 million employees in the UAE: "Going by the trend, [...] UAE nationals will account for less than four per cent" by 2020 (Ahmed, 2008).

Arab states have begun to implement a variety of schemes to encourage nationals into employment and to force private companies to hire more of them. So far, these efforts have been largely unsuccessful in reducing the reliance on expatriates and integrating nationals into the economy.

---

97 These include capabilities in resource assessment, project development, and plant and prototype design. IIE also defined the technical requirements for design and installation of PV systems with a nominal capacity of up to 30 kWp connected in parallel to the low-tension distribution network. Also, in collaboration with the Federal Electricity Commission and the Rural Development Program, the IIE provided engineering services that enabled the installation of over 40,000 solar panels (IIE, 2006).

The solution to this challenge lies primarily in the education system, which needs to prepare and equip students with not only the right skills and qualifications but also the work ethic to make them competitive in the job market. Education reforms and training initiatives are being put in place to address the skills gap and to ensure that the supply of suitably qualified employees meets the market's growing requirements, yet the outcomes of these efforts remain to be seen.

One encouraging development is the Abu Dhabi Economic Vision 2030,[98] launched in 2006, which identifies nine pillars that will form the architecture of the emirate's social, political, and economic future. These include the creation of a sustainable knowledge-based economy, for which the above education reforms and the Masdar initiative (see Case Study 5, Section 25.11) will be critical. Unlike Mexico, using the financial resources generated by the "energy-oil and gas" sector is a central element in the energy transition process envisaged in Abu Dhabi's vision of the future and its application in the Masdar initiative. To support the latter, the government has announced that renewable resources will meet 7% of the emirate's total power-generating capacity by 2020. It is estimated that this will create a market valued at US$ 6–8 billion, and will reduce CO_2 emissions by 2.4 million tonnes/year.[99]

Despite the progress made, the Masdar Initiative has not been without its problems. In October 2010, Abu Dhabi announced its intention to reduce the costs of the Masdar initiative by 15% and push the completion date from 2016 to between 2020 and 2025. As Sultan al Jaber, the chief executive of Masdar, emphasized, "The vision as a whole remains intact. No scale-back, no scale-down. There isn't a model or example anywhere around the world today that we can use. We have to come up with the answers ourselves" (*The National*, 2010). Partly in reaction to the impact of the financial crisis on property markets, this decision was also a response to lessons learned over the past several years in the design and construction of a carbon-neutral city. Concentrated solar power (CSP), which uses mirrors to reflect sunshine on to a central collection point, for example, is less effective in Abu Dhabi, where humidity and sand in the air can make the sun's rays less intense than anticipated. New approaches to CSP will have to be developed. The use of geothermal power in Abu Dhabi will also require new thinking, as the far deeper drilling required will be a more costly process than initially foreseen. These unexpected developments point to the need to make educational reform an integral part of any strategy to move from fossil-fuel-based economies to renewable energy systems and to the importance of systems thinking, interdisciplinary approaches and continuous feedback throughout a transition process.

25.8.3   The Dutch Energy Transition Approach

An interesting attempt at working towards an energy transition to a low-carbon economy is the Dutch energy transition approach. It is significant because of its focus on transformative change, its reliance on bottom-up processes, and its enrolment of business and other non-state actors in the transformation process. It grew out of deliberations between innovation researchers and policymakers in 2000, in which policymakers came to accept that a traditional policy approach based on short-term goals formulated by various ministries (in an uncoordinated way) and cost-effectiveness will not work. They also recognized that fostering radical innovation requires a different approach, one that is more forward-looking, with attention to possibilities and barriers, drawing on ideas for innovative change among market actors and innovative thinkers, with policy aligned to identified transition paths. This new way of thinking has been described as follows:

> It is clear that working on fundamental changes to the energy system can only be successful if the government adjusts its policy instrumentarium accordingly. This means that the policy for research and development, the stimulation of demonstration projects, and the (large-scale) market introduction must be brought in line with the selected transition pathways. In addition, the suggestions for new policies put forward by the platforms must be taken seriously. At this point, the government faces a major challenge, because much of the current policy was formulated based on the classic way of thinking that is characterized by a top-down approach and dominated by short-term objectives, implemented by fragmented and individually-operating departments and ministries, on which market influences do not or hardly have any effect (Dietz et al., 2008).

At the heart of the energy transition project are the activities of seven "transition platforms": new gas, green resources, chain efficiency, sustainable electricity supply, sustainable mobility, built environment, and energy-producing greenhouses. These platforms bring together individuals from the private and the public sector, academia, and civil society to identify innovative system configurations in a particular field, to develop a common ambition for an area of energy use or energy conversion, to think up pathways and programs, and to suggest transition experiments. Through these platforms, 31 transition paths have been selected.

The portfolio of transition paths contains technological innovations at different stages of development. The choice is based on technologies in which there is expertise in the Netherlands and an interest from business actors to work on them.

Based on suggestions from the transition platforms, a transition action plan was formulated in 2006 with the following goals:

- reducing CO_2 emissions by 50% by 2050 in a growing economy;
- increasing the rate of energy saving from 1.2% to 1.5–2%/year;

---

98   The Abu Dhabi Economic Vision 2030 was designed based on the advice of development experts from Norway, Ireland, and New Zealand (Abu Dhabi, 2008).

99   As part of the Masdar initiative, Abu Dhabi has also created a US$50 million development fund that will be used to provide loans to support renewable energy projects in the developing world until 2016.

Table 25.1 | The frontrunners' desk in the Netherlands.

Services for innovators	Services for policymakers
Obtain financial support from existing instruments	Make existing instruments more conducive for innovation
Contact relevant agencies and government staff	Improve policy coordination between and within ministries
Overcome legal problems and problems with permits	Encourage case-sensitive implementation of existing and new policy
Widen their network and strengthen the organizational set-up of the innovation trajectory	Encourage policy development in areas of the innovation chain that are not well covered
Provide business support and public relations help for successful market introduction	Be of service to businesses in a case-sensitive way

Source: Weterings, 2006.

- making the energy system progressively more sustainable; and
- creating new business.[100]

The platforms are intended to mobilize the interests of society and of business in innovative change and to promote learning and action. Issues of wider interest (or more strategic nature) are taken care of via a special council called the Regieorgaan Energietransitie Nederland.

The transition approach was consciously set up as a vehicle for coordinated socio-technical change and policy change. Through their involvement in the platforms, businesses are encouraged to work on and assist in the development of low-carbon innovations, including those that are not yet ready for the market. The platforms fulfill an intelligence function based on discussion, specially commissioned studies, and a coordinating function for emerging technology innovation systems.[101]

In addition to the transition platforms, an interesting policy innovation is the "front-runners' desk," created in 2004, which seeks to help innovative companies with problems they encounter, and to help policy to become more innovation friendly. Problems varied from difficulties with obtaining financial support (from government or the private sector), to problems with getting permits. Between January 2004 and March 2006, 69 companies approached the desk to discuss their problems. In 59% of cases, the problems were solved thanks to the intervention of the desk, in 12% of the cases the companies could not be helped, and in the remaining cases (29%), the desk was still dealing with the issue at the time of the evaluation (Weterings, 2006). Table 25.1 provides an overview of the functions of the front-runners' desk for innovators and policymakers.

The energy transition involves government agencies on different issues and at all levels. Some mechanism of cooperation is thus needed. Special arrangements have been set up for this task in the Netherlands, the most important of which is the Interdepartmental Project directorate Energy transition (IPE). The IPE plays an important role in "taking initiatives," "connecting and strengthening initiatives," "evaluate existing policy and to act upon the policy advice from the Regieorgaan and transition platforms," to "stimulate interdepartmental coordination" and to "make the overall transition approach more coherent" (Staatscourant, 2008). In drawing upon suggestions from the platforms, however, there is a danger of the transition process becoming a closed shop. Options outside the portfolio are at a disadvantage, but are not locked out. New initiatives may emerge outside the platforms through parliament or because certain powerful parties in society are able to secure policy support for it. This happened in the case of battery-powered electric vehicles, for which a coalition of NGOs, businesses (Essent, Better Place), finance (ING, Rabobank), and the Urgenda (a coalition for sustainability action) successfully lobbied ministers and parliament to provide special support to electric vehicles.

On the whole, policy coordination has improved in recent years. For example, battery-powered electric vehicles, hybrid electric vehicles and other low-emission vehicles are subject to special fiscal treatment. There is more cooperation between ministries and among government, business, research, and civil society. There is also more cooperation between previously separate national and regional initiatives.

Various kinds of capabilities are being created in a strategic way for technologies and policy through the Dutch transition approach. Areas for creating capabilities are being identified through the activities of the transition platforms, and they are supported by policy in a provisional manner. The government's ability to act to bring about systemic change has improved, and there is a commitment to transitional change that is absent in many countries. Indeed, the political discussion on low-carbon energy in many countries is narrowly focused on centralized choices of carbon capture and storage (CCS), and natural gas and nuclear power plants. These discussions are also taking place in the Netherlands, but more options can be explored through the energy transition approach.

---

100   The official goals in 2009 for 2020 were: 2% rate of energy saving a year, 20% share for renewable energy, and 30% reduction in CO_2 emissions.

101   As an illustration, in 2009, the platform for sustainable mobility will: (i) make recommendations for the fiscal treatment of clean vehicles; (ii) discuss the action plan on alternative mobility with leasing companies; (iii) examine how natural gas and green gas may pave the way for hydrogen; (iv) evaluate experiences with bus experiments funded in the first tender; (v) offer advice on how public transport concessions may be used for innovation; (vi) assist in the implementation of five pilots about smart grids and electric mobility; (vii) launch or stimulate pilots for sustainable biofuels (high blends and biogas) and hydrogen in five cities, in cooperation with Germany and the province of Flanders in Belgium (Kemp, 2010).

The Dutch approach is not without faults. The platforms are dominated by well-known energy companies (the insider-outsider problem). A second problem is that subsidies for energy investment are funded through general taxes, which makes those subsidies vulnerable to cuts. It is better to fund these through energy bills, as is the case in Germany, in order to create an element of continuity in energy transition policy. This takes on particular importance when governments change. Dutch clean energy policies have been characterized by major discontinuities that have hampered energy innovation (Verbong et al., 2008). Third, the liberalization agenda conflicted with the energy transition agenda. Its use in an add-on way contributed to policy incongruence (Kern and Howlett, 2009). This is not the fault of the energy transition approach, but the problem has not yet been resolved.

25.8.4   Moving Towards an Energy Transition

Oil- and gas-producing and exporting countries have faced considerable difficulty in overcoming the tendency to rely heavily on this one sector, to the detriment of greater economic diversity and flexibility. Significant new thinking and practices have emerged in these countries, although overcoming the challenges of path dependence, sunk costs, and established habits and practices to move towards an energy transition has been particularly difficult.

The broad set of initiatives discussed in this section provides pointers for working with and around habits, practices, and challenges in other environments. The Norwegian government, for example, created the Norwegian Petroleum Fund in 1990. All central government revenues from petroleum operations are deposited in this fund, all of which is invested abroad to provide a buffer between current petroleum revenues and the use of these revenues in the Norwegian economy. The government uses about 4% of its petroleum revenues annually (Norges Bank, 2004). The Norwegian energy innovation system also includes Enova SF, a public enterprise created in 2002 and owned by the Ministry of Petroleum and Energy. Enova SF advises the ministry on issues related to new renewable energies and energy efficiency, with the goal of reducing energy use. Its funding comes from a levy on the electricity distribution tariffs (Enova, undated; Borup et al., 2008). Brazil, already well along the pathway towards cleaner energy in the transport sector, is planning a development fund using oil revenues. Abu Dhabi's Masdar initiative has grown out of a similar process, and the Arab Gulf states are developing the education and research base needed for an energy transition that will involve more than simply importing technologies and expertise.

The policy dimension has also been important in moving towards energy transitions in these countries. Mexico, having moved away from a focus on building and sustaining capacity in the oil sector, is now on the brink of developing a cleaner energy strategy based on wind and solar power for which capacity development will be necessary. Some capacities have already developed within existing organizations, but policy coordination

and new policy measures that provide the incentives and support the use of this new knowledge for innovation are lacking.

The Netherlands has adopted a particularly innovative approach to energy transition thinking and practice that provides pointers for others looking to work with and around existing habits and practices in their own environments. The approach is embodied in the role of energy platforms. These platforms bring together individuals from the private and public sectors, academia, and civil society to identify and make proposals to government about policies for different energy-related initiatives. They are not just "talk shops," but spaces for dialogue that inform the participants of their diverse interests, concerns, and goals. They mobilize and legitimize a process of change and the kinds of support measures that would be needed for movement along consensual pathways. Their proposals feed directly into the policy process.

25.9   Enhancing Capacity for Effective Participation at the International Level

A wide variety of new technology, energy, and environment issues are increasingly being dealt with at the international level and across many different venues. They include trade, intellectual property, and antidumping issues at the World Trade Organization (WTO); technology transfer and the future of a multilateral climate framework at meetings of the United Nations Framework Convention on Climate Change (UNFCCC); funding opportunities through the CDM under the Kyoto Protocol; the Global Environment Facility (GEF); and other funds managed by the World Bank. Enhancing the capacity for effective participation in international negotiations has thus become an arena within which substantive knowledge about the subject under negotiation will be needed, as well as the capacity to design projects for and secure access to the many funding sources now available for adaptation to climate change, and new energy technologies and their diffusion. This will require the strengthening of capacities for networking, learning, and innovation[102] along multiple dimensions.

Developing countries soon faced the rising transaction costs of settling antidumping complaints, which were first signaled in the controversies generated by the difficulties that developing countries encountered in accessing drugs to deal with HIV/Aids and the role of the Agreement on Trade-Related Aspects of Intellectual Property Rights. Like most developing countries, Brazil initially met this new international legal challenge by hiring foreign law firms. Subsequently, a few Brazilian law firms created departments specializing in WTO negotiations, and they signed agreements with government and major law schools in Brazil to create domestic expertise in international trade law. The presence of local lawyers on law school

---

102   For example, UNDP's "Capacity development for policy makers" project seeks to strengthen the national capacity of developing countries to develop policy options for addressing climate change, as inputs to the UNFCCC negotiations.

faculties helped to form expert groups that strengthened the capacity development process (Shaffer et al., 2008).

Further experiments in capacity development began when Brazil established an advanced WTO mission in Geneva and signed agreements with Brazilian schools of law and of international relations to create a training program in Geneva – the first of its kind in Brazil. This experiment played an important role in strengthening Brazil's capacity to participate in the more informal dispute settlement negotiations in Geneva, where many such disputes were actually resolved (Shaffer et al., 2008). The virtuous cycles of learning and practice further strengthened Brazil's network connections, with some surprisingly positive and unexpected benefits.[103]

25.9.1 Funding Energy Transitions: A Capacity Development Approach

The CDM, as a development financing mechanism, marks a new innovation in international environmental law and with it the need for a new set of capacities. The CDM is an emissions offset trading system that allows countries to meet their Kyoto targets by investing in emissions reduction projects in developing countries, where the marginal costs of GHG abatement are expected to be lower than in industrialized countries. It has the added purpose of helping to promote sustainable development and technology transfer to developing countries (Doranova, 2009).

After several years in operation, the results of the CDM are mixed. The mechanism has attracted considerable investment, with more than 4500 projects currently at some stage of development and approval (UNEP-Risø, 2011). But there are also a range of problems:

- there are concerns that many of these projects would have happened without the CDM (lack of additionality or environmental integrity);

- few projects promote sustainable development;

- projects are not distributed equally (especially in Africa);

- there are high administrative burdens and transaction costs; and

- there is unequal negotiating power among project participants (see van Asselt and Gupta, 2008).

The CDM has generally not spurred innovation and is not in and of itself poised to foster any type of energy transition in developing countries.

There is also no firm evidence that the mechanism has encouraged participating companies in the North to move towards cleaner technology, rather than simply buying time. Many CDM projects are "end-of pipe" control technologies such as HFC-23 control projects,[104] which favor existing interests in pursuing the cheapest emissions reductions. Path dependence thus remains the driver, not change.

Multilateral climate negotiations are in a state of transition. This is not surprising, considering that they cover not only the respective emission cuts by different groups of countries, but also topics as wide as agriculture, deforestation, technology transfer, intellectual property rights, competitiveness, international transportation, and financial mechanisms to support climate change mitigation in developing countries. This requires a significant amount of capacity development with negotiators. The European Capacity Building Initiative (ECBI, 2011) is an initiative for sustained capacity building in support of international climate change negotiations. The ECBI promotes a more level playing field between government delegations during international climate change negotiations and aims to facilitate mutual understanding and trust, both between European and developing countries and among developing countries.

The most important topic for the foreseeable future in the climate change negotiations is financing and crediting. Moving forward approaches such as the Clean Development Fund, the existing Least Developed Country Fund, and the Special Climate Change Fund, requires specialized and substantive knowledge from all negotiators involved. Currently, most negotiators in the climate talks are career diplomats who lack sufficient knowledge of financial markets, technology development, and scientific fields relevant to climate change to lay the foundational framework for energy transitions. Negotiators too often do not have the capacity to develop a clear overview of the options on the negotiation table, let alone how they might reach strategic targets that are in the common interest.

The capacity to have a broad overview of the climate negotiations and to translate diplomatic decisions into action will be crucial in the next few years. This is especially the case now that the "collective commitment" that emerged in the Copenhagen Accord to provide "new and additional resources […] approaching US$30 billion to developing countries for adaptation, technology development and transfer and capacity building in the period 2010–12" (UNFCCC, 2009) has been confirmed at the UN Climate Change Conference in Cancún in December 2010. Developed countries also committed to a goal of jointly mobilizing US$100 billion a year by 2020, "in the context of meaningful mitigation actions and transparency on implementation" (UNFCCC, 2009). Such funds must focus on building capacity in developing countries.

---

103 An analysis of data on over 1300 antidumping investigations initiated by different countries against other WTO members showed that "those countries that are abundant in legal capacity are more likely to challenge AD duties brought against them and less likely to be targeted by AD duties in the first place. These results are especially striking in light of the fact that we control for these countries' market power, which make credible their threat to retaliate" (Busch et al., 2008).

104 The top-two renewable energy projects (after hydropower) are biomass (129) and wind (161) projects. Together, they generated 37,272,397,000 carbon credits (Certified Emission Reductions, or CERs), whereas the 18 HFC (fluoroform) projects in the CDM pipeline as of October 2009, generated 218,637,000 CERs. See UNEP-Risø, 2011.

Even if these funds were to be fully subscribed, there would still remain a vital need for capacity development to enable developing countries to participate effectively in international negotiations concerning the governance and use of these funds and to benefit from the various funding opportunities. There is also a need for the funding itself to support broader innovative capacity in low-carbon technologies in developing countries – not just a series of low-carbon hardware projects (e.g., Ockwell et al., 2009). Past practice has shown that funding mechanisms such as the CDM have had a limited impact on the development and diffusion of new energy technologies. Most developing countries do not have the capacity to design such projects or identify partners and work with them to do so. The CDM recognized this and began, rather late, to emphasize a needs assessment approach, but without the methodology and training that would make local choices possible. The CDM's own methodology, doing baseline studies and attempting to determine additionality, are meaningless in the context of least developed countries, where the capacity to undertake data collection and analysis and to evaluate choices is critical yet largely absent. Only in the past several years have UNEP and the GEF begun a series of workshops to fill the gap, as the case of Bahrain and the role of the Arabian Gulf University in that country illustrate (see Case Study 6, Section 25.11). Of particular interest is the way in which the use of local education and research institutes to train analysts in the developing world strengthens their capacity to support regional networks.

In conclusion, effective participation in the international system is critical for energy transitions. Although currently going through stages of learning, global regimes including the CDM have an important role to play in encouraging and supporting change. This cannot be done solely through financial flows. It must also provide legitimacy for citizen action. An equitable framework for legitimate engagement and negotiation in which stakeholders can participate at a substantive level is part of the global enabling environment that would ultimately enable the system to work for energy transitions.

25.10   Conclusions

Bringing cleaner, safe, and affordable energy and energy services to people, moving away from our current dependence on fossil fuels, and reducing the need for energy to the largest possible extent, are key energy challenges for society today in both developed and developing countries. Meeting these challenges will involve major changes in energy supply systems, energy-using technologies, and their associated practices. Improving trajectories based on fossil fuels and building supercritical coal-fired power plants will not enable us to meet these challenges – it will merely lead to problems in the future. Ultimately, we must shift to renewable and green energy sources, electric transport technologies, zero-energy buildings, and more efficient energy supply systems. The focus in this chapter, therefore, has not been on "policies for capacity development," in general, but on the challenges of energy

transitions and the capacities and capacity development processes that are needed to overcome them.

As energy transitions will differ across countries, and perhaps even between regions within a country, these conclusions do not attempt to provide a definitive list of what should be done. But the experiences reviewed here do point to a number of principles that might usefully be applied when thinking about capacities and capacity development policies for sustainable energy transitions in a given environment. This takes us one step beyond the classic discussion about what needs to be done, and towards an exploration of ways in which such changes might be brought about in a variety of different contexts. It also opens opportunities for reflection on the multiple benefits that might be generated for livelihoods, learning, innovation, consensus building, and development, through the capacity development process.

25.10.1   Getting There from Here: First Principles

Energy transitions are long-term, socially embedded processes that involve changes in production and consumption patterns, knowledge, skills, formal and informal institutions, and the habits and practices of the actors involved. These characteristics have major consequences for the choice of an energy transition pathway today and the capacities that might be needed to move down that pathway successfully in the future. From this perspective, capacity development can no longer be seen as a simple aggregation of individual skills and competences or the introduction of a new "technology." In the course of an energy transition, moreover, capacities at the actor and systems levels, and the policies for capacity development themselves, will inevitably change. This will require feedback and the capacity for flexible adaptation over time. How might we deal with the uncertainties and complexities generated by such a process?

25.10.1.1   Principle 1: Understanding the Challenges

All change processes inevitably face challenges that emerge from established habits, practices, norms, and interests, or what we have called "informal institutions and path dependency." These can slow both the development and the diffusion of new energy technologies. In conjunction with the "lock-in" effects of earlier investment in physical plants, supplier linkages, infrastructure and skills, and efforts to move towards energy efficiency improvements are also slowed.

Understanding these informal institutions and learning to work with and around them in a given context are important first steps in an energy transition process. This principle was illustrated in a wide variety of cases that include efforts to improve energy end-use efficiency in transport (Section 25.4.1) and buildings (Section 25.4.3), the introduction of new energy technologies and their accompanying practices in the case of smart grids (Section 25.4.2) and wind turbines (Section

25.5.1), and the diffusion of solar home systems (Section 25.5.2) and cookstoves. What emerged from these studies is the need for dialogue (Section 25.5) of many sorts and at many levels in dealing with the challenges of energy transitions.

25.10.1.2   Principle 2: Dialogues Make a Difference

Dialogues may not "solve" these problems, but they do open channels for innovative ways to deal with them. Moreover, they open channels that traditional awareness-raising programs, based on public meetings, conferences, one-off consultations, and focus groups, cannot. For dialogues to work, confidence-building measures that recognize the legitimacy of local concerns, interests, and needs is a first step in gaining broad societal support for an energy transition. And this is the case whether it concerns renewable energies, lighting services, improved cookstoves, or personal mobility in the public transport system.

Dialogues must begin early in the process of conceptualization to be effective. This is when the identification of needs and problems and the range of options available and preferences for how to deal with them are first discussed. Dialogue should not be limited to the usual stakeholders, nor should solutions be reduced to a few predetermined options. An open mind should be kept vis-à-vis nonconventional solutions. This applies to both broad policy directives and the development of programs and projects at the local level. It is at this point in the process that a framework for domestic consensus building must be developed by involving the wider community in discussions about key issues and choices, as well as their consequences. One such approach consists of creating spaces for dialogue about transition pathways, processes, and policies that bring together members of the public and private sectors, academics, and civil society in what the Dutch have called transition platforms (Section 25.8.2).

Identifying relevant stakeholders and building the capacities needed to engage meaningfully in this new practice are essential in initiating and sustaining energy transition processes. Not all stakeholders will have the same view of a problem and possible solutions. In a dialogue, it is understood that the solution to barriers may not lie in support policies for the technology but in finding the appropriate technology configuration, which will require coalition building and stakeholder learning. This opens opportunities to reconsider the proposed solution, a process that may result in its rejection or major adaptation. "Not in my backyard" resistance may be turned into constructive dialogue focused on the question: if not this, then what else?

25.10.1.3   Principle 3: Dialogue Processes Have Wide Applicability in Energy Transitions

Dialogues will be needed throughout the lifetime of a program or project in an energy transition. To make dialogues meaningful, mechanisms for feedback and flexibility will have to be built into programs and projects – the former to legitimize the use of "negative" feedback as a positive incentive for change, and the latter to create space for adaptation to take place as energy transitions evolve.

Going beyond the project level, the emergence of decentralized and distributed processes, energy-related practices, and new technologies have raised the importance of local actors in decision-making processes. Examples include wind power in Germany, the Netherlands, and the United Kingdom, agriculture-based renewables in Brazil and Mali (Section 25.6.1) and micro hydro in China, Bhutan, Nepal, and Rwanda (Section 25.6.2). In this context, dialogue can become a useful bridging mechanism for consensus building across national and local levels.

Although the notion of dialogues has been in circulation for some time in the capacity development literature, the practice of designing projects through a dialogue has not been systematically applied (Section 25.3.1). Long-standing discussion on the need for developing country ownership of donor-funded projects can also benefit from this engagement in the dialogue process. The success of such an approach, however, depends critically on the extent to which those in partner countries have access to the knowledge and information they require to develop positions of their own. It also depends upon the extent to which experts from donor agencies have learned to engage in meaningful dialogue processes.

For least developed and many developing countries, the capacities to articulate their positions in international negotiations are also an issue. The experience of Brazil in international negotiations (Section 25.9) and the development of programs such as those at the Arabian Gulf University (Section 25.9 and Case Study 6, Section 25.11) suggest ways to meet this capacity development need. Indeed, UNDP, IDRC (Section 25.3.1) and ODI (Section 25.7.3), among others, have developed tools for this purpose.

25.10.2   Making Choices about Energy Transition Pathways

Access to knowledge and information – and the capacity to use such inputs – are critical in making choices about energy transitions, whether at the level of individual actors, communities, or national governments. Most developed and emerging economies have created national agencies for this purpose. These agencies monitor international developments and regularly undertake national surveys that provide comparative information for making choices among alternatives in both the short and the longer term. With access to a wide array of information on local and international conditions, and the capacity to use such data, these countries are developing methodologies that will enable them to move from static cost comparisons to analyses of factors that might affect longer-term costs and savings (Section 25.3.2). This longer-term perspective opens up opportunities for building flexibility into strategies for energy transitions and the necessary policies for capacity development.

Many developing countries, however, lack access to continuous flows of knowledge and information on new and improved energy technologies, the policies and practices of other governments, or enterprise strategies. But these countries' governments also face the challenge of collecting local and national data on the state of education, innovation, industry, energy use, and efficiency. Surveys and studies undertaken by UNESCO, UNIDO, and New Partnership for Africa's Development (NEPAD) attest to this.[105] Strengthening the capacities of governments, universities, NGOs, and other organizations in data collection and analysis can widen the range of choice of new energy technologies to meet local needs and expectations.

25.10.2.1   Principle 4: Accessing Knowledge and Information

Problem solving can be accelerated by developing channels for access to knowledge and information. These channels consequently encourage the more rapid diffusion of new energy technologies. The beneficiaries would be manifold, going beyond local governments, research bodies, and the banking and business sectors to include agricultural extension services, energy users, and intermediaries such as urban transport and utility companies, architects, and builders. National reference centers in Brazil (Section 25.7.2) and front-runner desks in the Netherlands (Section 25.8.2) are two examples of how this concept can be adapted to different contexts.

To encourage a more diversified set of criteria for rewards, such as research grants and promotions, policies will need to be put in place that stimulate and support linkages between research and users – whether in the business, health, government, NGO, or other sectors (Section 25.7.2). Research funding will also need to be made available for this purpose. This can take a number of forms. In Brazil, national reference centers are attached to universities and are supported by university research. Two quite different approaches linking R&D to its potential users are, first, CT-Energy, a fund created in Brazil to support "public interest research," including energy R&D and energy efficiency, and, second, the integrated approach to R&D, energy industries, and investment in the Masdar initiative (see Case Study 5, Section 25.11).

25.10.2.2   Principle 5: Create Opportunities for Interdisciplinarity and Systems Thinking

The higher education system is crucial for sustaining an energy transition, yet current practices will need to change if it is to play this role. The long-term and uncertain nature of energy transition processes requires flexibility that standard discipline-based educational and training

programs do not often provide. This at least suggests there is more need for interdisciplinarity in graduate, postgraduate, continuing education, and training programs. In many countries, this will require policy changes at the national and university levels to create this flexibility now and thus strengthen the capacity for adaptive curricula changes as the relative importance of issues, technologies, and challenges change over time.

Interdisciplinarity impacts the speed with which new energy technologies are developed and diffused in disciplines where established practices mediate between the demand for and acceptance of these technologies. Examples include architecture, engineering, housing and construction, new materials, transport, urban planning (Section 25.4.3), and the biosciences (Section 25.6.1). Targeting these disciplines will enable university programs to offer a wider view of the issues and options that are currently available and to study and assess those that are on the horizon. For developing countries, interdisciplinarity could provide evidence-based research as an input into planning and policymaking processes across a wide spectrum of actors in civil society, the business sector, and government.

Managing energy transitions is another area in which universities could play an important role. They could create opportunities to update policymakers on the range of available options and strengthen the contribution of scientists and social scientists in decision-making processes. Both the rapid turnover in personnel at the middle and senior advisory and policy levels in developing countries, and the need for mobility to widen perspectives in the industrialized world, make continuous education for energy transitions imperative. Interdisciplinary masters-level programs designed to bring together scientists, and policymakers have a role to play here.

25.10.2.3   Principle 6: Use Knowledge Networks to Solve Problems during Energy Transitions

Open innovation practices and networking in industrialized countries are helping to speed up the pace of problem solving in a growing number of sectors. Users, especially companies, but also government agencies and community organizations, are becoming increasingly aware of its possibilities (see Case Study 3, Section 25.11). The core insight of this approach is the ability to use the world outside an organization to generate internally useful knowledge in a purposeful manner. This involves the development of networks and the capacities needed to use them.

Developing countries are moving in this direction now as well. In particular, they are strengthening their domestic research capacity (Section 25.7.3) and problem solving (Section 25.7.1), and they are facilitating adaptive changes in the course of energy transitions and supporting the process of making choices about energy pathways. It was a critical factor in boosting productivity and reducing costs in the development of ethanol from sugarcane in Brazil (Section 25.6.1). In Mali, with the

---

105 The lack of up-to-date statistical information in the surveys published by UN Agencies show how difficult it has been to collect comparable data at regular intervals from local ministries. The launch of innovation surveys by NEPAD has also faced a lack of trained personal and funding.

support of the local member of the AREED network, research has supported the growing of jatropha in a multi-cropping environment and the use of jatropha oil for power and electricity generation (Section 25.6.1). The UNEP-funded REED network, of which AREED is a member, has supported renewable energy enterprise development in Senegal (Section 25.5.2) and its GNESD network (Section 25.7.2). This network comprises centers of excellence in industrialized, emerging, and developing economies, undertakes joint energy policy-related research, and provides a space for knowledge sharing on the links between sustainable energy and other development and environmental priorities, as well as technology and policy options.

In the field of new energy technologies, United Nations University Geothermal Training Programme has designed its training program to deal with the problem of brain drain by strengthening capacities for local research, exploration, and the development and production of geothermal energy in East Africa. Participants in the six-month workshop in Iceland had jobs in this field in East Africa to which they would return. Bringing policymakers and practitioners from the region together in local workshops supported network building among these countries (Section 25.7.3).

25.10.3   Policies and Practices in Making and Sustaining Energy Transitions

The long-term, socially embedded nature of energy transition processes gives rise to various challenges. The choices made in selecting energy transition pathways are also inextricably linked to broader development issues and goals. These, too, must become visible as part of a dialogue process, and the tradeoffs must be made explicit. The case studies reviewed in this chapter make clear that in laying the foundation for an energy transition, it is necessary to work from a longer-term, systems perspective.

25.10.3.1   Principle 7: Energy Efficiency is a First Step towards an Energy Transition

Energy efficiency improvements can potentially bring about changes in existing habits and practices of energy use, but they do not in and of themselves constitute energy transitions. They are only one step on a path that might lead to a more transformative change. In designing energy efficiency programs, it is important to take a longer-term perspective that builds in a strategy for developing the steps to follow.

Urban public transportation provides a good example of how this might be done. Energy efficiency improvements in this sector are driven mainly by the need to reduce costs and increase fuel efficiency, especially where fuel is imported, and to address transport access issues. For instance, bus rapid transit (BRTs) systems – of which there are currently

47 around the world, including 16 in Latin America, several in Asia, and one in sub-Saharan Africa (Section 25.4.1) – provide relatively cheap and frequent bus services that widen access to employment opportunities. In some cases, they have also been designed to lure drivers away from their automobiles, thus reducing traffic congestion in urban areas, contributing to changes in current habits and practices, and reducing overall GHG emissions.

But there is a need to go beyond these objectives and develop a strategy that replaces fossil fuels with renewable energy. Examples of this can be found in programs such as the European Regions and Municipalities Partnership for Hydrogen and Fuel Cells, and in the creation of a private sector firm in Tanzania that produces biodiesel from used cooking oil, which it sells to the minibus (matatu) sector (Section 25.4.1).

25.10.3.2   Principle 8: Adopting a Multi-Goal Perspective in Energy Transition Strategies Can Generate Multiple Benefits

As the transport sector illustrates, an extensive array of multiple benefits can be envisaged as part of an energy transition process. Support for the development of biofuels in developing countries and their use in urban and rural areas is one way to create a continuous movement towards an energy transition over the long term. It also offers income-generating opportunities for rural smallholders who intercrop biofuel crops with food crops, as the examples of Garalo in Mali and the Brazilian Amazon region illustrate (Section 25.6.1). Adopting a multi-goal perspective starts with an analysis of local resource availability and the early identification of linkages and support structures that will be needed to build local markets for biofuels, and cleaner energy technologies such as cookstoves and multifunctional energy platforms. At the same time, such a perspective can help to identify alternatives to existing energy pathways. For this to take place, a multi-goal perspective must be consciously built into policies, programs, and financing mechanisms from the outset.

In both industrialized and developing countries, SMEs continue to be major drivers of innovation processes, as well as contributors to livelihoods and employment creation. Opening space and creating the support structures for the establishment of innovative SMEs in the provision of energy services has played a role in stimulating market growth and the diffusion of new energy technologies in such diverse areas as solar home systems, renewable energies, and off-grid energy access.

In creating small-scale energy solutions, knowledge of the local market is critical but needs to be complemented by networks and local organizations that help to strengthen business skills and the ability to adapt technologies to local needs, as the AREED examples show (Section 25.5). This requires users to have some domain knowledge and technical skills, as well as the habits and practices that support learning and innovation. Financing for SMEs to develop such capabilities is scarce in developing countries. New international support programs that might strengthen capabilities to meet the challenges of development and climate change were envisaged in Copenhagen and in Cancun, at the 16th meeting of the Conference of the Parties, a decision was adopted to establish a Technology Mechanism comprised of a Technology Executive Committee and the Climate Technology Centre and Network (UNFCCC, 2011).

How these new organizations will operate still needs to be defined. Ensuring that developing countries have the capacities to participate meaningfully in future negotiations on programs and practices under the Technology Mechanism will be critical to their success and sustainability. As to funding, the European Parliament acknowledges that this will be a challenge but believes it to be feasible. Members of the European Parliament's "climate parliament" have formally proposed that at least 5% of the EU budget should be devoted to ensuring a rapid transition from fossil fuels to wind, solar, and other renewables in the European Union and a further one billion euro/year be used to support renewable energy in developing countries. Such funds could support local NGOs, domestic foundations, R&D, grant-making councils, and business support structures in developing countries, and take this process of building rural and peri-urban energy solutions into a broader program of market building and SME upgrading.

25.10.3.1   Principle 9: Consciously Pursue Market-Building Strategies as a Driver of Diffusion

The development of a conscious strategy of market building in energy-related projects and programs is also important in driving the widespread diffusion of new energy technologies in developing countries. The diffusion of solar home systems in Kenya, Sri Lanka, and Bangladesh (Section 25.5.2) and micro-hydro in China, Bhutan, and Nepal, illustrate some of the challenges and successes of bottom-up processes that take a market-building approach (Section 25.6.2 and Case Study 3, Section 25.11). The absence of such a strategy and its impact on diffusion is also clear, notably in the traditional approach to improved cookstoves. The latter relies on stoves designed abroad with expensive materials and that are imported, as opposed to stoves with longer local value chains, opportunities for local repair and manufacture, and thus with lower prices (Section 25.5.3). Similarly, the early experience in the diffusion of multifunctional energy platforms lacked a built-in feedback and adaptive learning strategy that limited market development and slowed the diffusion of these platforms (Section 25.5.3 and Case Study 2, Section 25.11). This is now changing, as the case of Garalo in Mali (Section 25.6.1) demonstrates. The provision of business services and coaching to start-ups and SMEs would stimulate the diffusion of clean and more efficient energy technologies at the same time as it opened opportunities for a wide range of benefits.

25.10.3.2   Principle 10: Governments are Critical Actors in Energy Transition Processes

As the above principles illustrate, a policy aimed at fostering energy transitions will need to involve a range of government bodies at various administrative levels in the analyzing of problems and the designing and implementing of policies. No government ministry can be expected to have all of the required instruments in its "arsenal." Adopting a systems approach and a multi-goal perspective will require dialogues across ministries and the creation of inter-ministerial teams for policy integration and coordination. Capacities for dialogue and policy coordination must therefore exist or be created.

Designing for flexibility over the course of an energy transition will require access to the knowledge needed to create a portfolio of options that allow for different development possibilities. The portfolio should build on existing capacities and the interests, concerns, and needs of local actors and national actors alike. Developing countries might begin this process by identifying those capacities that are needed across many renewable energy technologies and support universities in the development of interdisciplinary programs around each knowledge nexus. This would create the flexibility to move between different technologies as conditions change.

Another task for strategic decision makers is to develop the analytical capacity to create and manage a shifting portfolio of renewable options as technological frontiers are breached, costs decline, and needs change. This includes the ability to undertake comparative assessments of the impacts of different choices and combinations of choices, as well as how these change over time. These impacts could include energy efficiency, consumer habits and practices of energy use, GHGs and other emissions, job losses and gains, and inducements offered to move towards an energy transition. Linkages to local knowledge centers and participation in wider networks will be essential if governments are to play these roles.

Government, therefore, will be one of the key actors in the transition process at local, national, and international levels. Developing and keeping a focus on transformative change, however, cannot succeed in the absence of broad-based support, continuous dialogue, and a long-term, systems-based approach to change.

25.11   Case Studies

25.11.1   Case Study 1: Path Dependence and Lock-in: Decentralized Energy in the United Kingdom

Efforts to increase the contribution from decentralized energy in the United Kingdom have been underway for over 10 years,[106] driven mainly

---

106 These efforts date back to the report of the government-industry Embedded Generation Working Group, *Report into Network Access Issues: A Consultation Document* (DTI, 2001).

by government targets to expand the use of renewable energy (including small-scale installations) – but also by a desire to explore more fully the scope for district heating, smart distribution grids, and more active participation by individuals and communities in the United Kingdom's low carbon transition (HM Government, 2009). Progress with many forms of decentralized energy has been slow. One of the reasons for this is that the existing system is highly centralized, and the United Kingdom has no recent tradition of decentralized electricity production (Foresight, 2008). This means that local institutions that understand local needs, such as local authorities, play a relatively minor role in decisions on energy investments and strategies. Another barrier is local opposition, particularly to proposals for wind farms from which many communities feel they have little to gain (Toke et al., 2008). In addition, there are economic and financial barriers. Small-scale renewable and low-carbon energy technologies tend to have high capital costs, and can in some cases face additional economic barriers such as the costs of connections to the grid.

The regulation of the electricity system provides a good illustration of the barriers to decentralization – and also how recent reforms are attempting to tackle them. Technical challenges arise because significant increases in decentralized electricity generation will have a particularly strong impact on the existing electricity grid, which was not designed for that purpose. Furthermore, decentralized generation is not just about connecting new plants to the system, but also about innovation and transformation of the system. Decentralization is part of a more general move to a more complex electricity system in the United Kingdom, which includes smart meters in homes, more sophisticated supply-demand balancing facilitated by smart grids, and the possible diffusion of plug-in electric vehicles (DECC, 2009).

Associated regulatory challenges arise because a more decentralized electricity system requires distribution network operators to play a more active role in managing their networks – and therefore need the appropriate skills and incentives to do so. The technical and regulatory impacts of decentralized generation will vary according to whether the approach taken is a piecemeal connection or a transformation of the system (Sauter and Bauknecht, 2009). In the former case, each generator is appraised with reference to the current system and is connected as a "one off." In the latter case, innovative solutions are needed for the development of generation and network infrastructure, and must work in conjunction with demand-side measures. The United Kingdom's price-based approach to regulation, for example, allows network companies to increase their charges by the retail price index minus an efficiency factor – a formula known as RPI-X. While it has been argued that this should act as a driver of innovation, it only does so if the aim of the innovation is to increase the efficiency of current network operations. As a result, innovations that would facilitate significant changes to the architecture and operation of networks have not been encouraged.

Some tentative moves to encourage innovation were introduced in 2005, but the impacts were limited (Woodman, 2006), largely because the

innovative capacity within network companies started from a very low base, and the incentives were too weak and limited. It is only recently (in April 2010) that the regulator, the Office of Gas and Electricity Markets (Ofgem) has introduced significant new incentives for innovation in the form of a Low Carbon Networks Fund (Ofgem, 2010). This is designed to provide a much stronger rationale for distribution companies to move in the direction of smarter grids that could not only include decentralized generators, but also encompass the early rollout of electric vehicles and integrate the demand side through smart meters – something that the government supports (e.g., DECC, 2009). However, it remains to be seen what impact the new Ofgem fund will have, and whether it will be enough to overcome historical barriers to decentralized energy systems.

25.11.2   Case Study 2: A Bottom-up Approach for Micro-Hydro in Bhutan

The development of a hydropower industry is recognized as the primary driving force for economic development in the Kingdom of Bhutan. The Power System Master Plan (PSMP, 2004) identified technically and economically feasible and potential hydropower sites that could provide a total of 23,760 MW of electricity. The current installed generating capacity is 1505 MW.[107]

Bhutan's rural electrification efforts are guided by "Bhutan 2020," a policy document that set a target of providing electricity to 50% of the rural population by 2012, and to all households by 2020. The policy also seeks to transform the role of the state from that of a "provider" to an "enabler." Bhutan's clear policy directives and its Electricity Act 2001, which provided for free licenses up to 500 kW, are helping to stimulate development of micro-hydropower.

The high cost of extending the grid, however, has made it difficult to connect a large percentage of the rural population to the national power grid. Based on past rural electrification projects, it is estimated that electrifying one rural household costs about US$1800. This needs to be financed by the government, and on-grid rural consumers receive substantial operational subsidies.[108]

In the past, the Bhutan Power Corporation owned, managed, operated, and maintained the country's micro-hydropower plants. However, due to the corporation's lack of financial viability, it has been reluctant to take over the management of the new micro-hydros being built. Thus, when the Chendebji micro-hydro plant was commissioned in October 2005, an alternative model involving the community had to be developed. The community-based management model now being piloted in

---

107   This includes the recently commissioned Tala hydropower project.

108   The average rural consumer utilizes less than 50 kWh/month and falls in the lowest tariff category of US$0.013/kWh. The estimated cost of supplying electricity to rural households is about US$0.10/kWh.

Chendebji is expected to be refined and the lessons learned applied in future projects.

In particular, the development of micro-hydropower should be coupled with the creation of income-generating opportunities that will enhance the economic self-reliance of rural communities. Further, by integrating micro-hydropower with the grid at higher feed-in tariffs, the sale of surplus power could improve load management and displace the need for government subsidies. Finally, training programs are needed to maintain the skills of local plant operators, and education campaigns are necessary to raise consumers' awareness of the need for energy conservation.

25.11.3 Case Study 3: How Open Innovation Systems Speed Problem Solving: InnoCentive

Most of the work on distributed innovation systems has focused on software development. In a recent study, a team of researchers looked at the value of openness in scientific problem solving in the life sciences (Lakhani et al., 2007). The study considered a diversity of disciplines, several of which are relevant in the development of biofuels and other renewable energy sources.

Unlike the open source software approach, the InnoCentive network was not a spontaneous development but an offshoot of a multinational pharmaceutical firm, Eli Lilly. Its creation recognized both the explosion of knowledge and the combinatorial nature of knowledge bases in the technologies that emerged in the last quarter of the 20th century.

InnoCentive describes itself as an "open innovation community of smart, creative people, who provide solutions to tough problems" (InnoCentive, 2011). Membership of InnoCentive is open to anyone who is attracted both by the challenge of solving problems and by the financial rewards that are attached. The notion that this is a community along the lines of the open source software community is misleading, however. InnoCentive's problem solvers rarely interact among themselves or with the seekers. Another difference lies in the "openness" of the knowledge produced by solvers, who must sign an agreement upon joining InnoCentive under which all rights to the knowledge they generate belong to the seeker.

Despite the absence of a free flow of knowledge and open access to and use of the knowledge generated by solvers – both of which are arguably among the factors that enable open innovation systems to speed the pace of innovation and diffusion – InnoCentive has served to accelerate problem solving for seekers who are willing to share some information with solvers. One study, for example, illustrates the efficacy of this approach in the case of "166 discrete scientific problems from the research laboratories of 26 firms from 10 different countries between June 2001 and January 2005" (Lakhani et al., 2007). Firms came from a variety of industries, including agrochemicals, biotechnology, chemicals,

and pharmaceuticals, and most had attempted to solve the problem within their own laboratories. Out of 166 problems, 49 (29.5%) were solved using this approach (Lakhani et al., 2007).

The outreach that such an approach permits is interesting for problem solving. "On average, 240 (sd: 195, range: 19–1058) individuals examined each detailed problem statement and 10 (sd: 14, range: 0–103) individuals submitted solutions for evaluation" (Lakhani et al., 2007). More recent data provide a higher rate of problems solved. Of the "(a) round 900 challenges that have been posted so far by 150 firms [...] (m)ore than 400 have been solved"(*The Economist*, 2009). The range of seekers and problems has already extended to the renewable energy sector (Bent, 2009).

25.11.4 Case Study 4: Creating a Policy Framework for Innovation and Technology Transfer in South Africa

The South African white paper on science and technology (DST, 1996) created a National System of Innovation based framework, for key enabling policies and strategies that would inform the strategic development of science and technology in post-apartheid South Africa. In 2002, the national R&D strategy, implemented by the Department of Science and Technology, aimed to improve the impact of the white paper by identifying interventions needed to address systemic weaknesses. One of these involved improving the protection and management of intellectual property from publicly financed research as a way to "bridge the innovation chasm."

New legislation aiming to address this issue has recently been passed. It prescribes in some detail the policies, processes, and structures that recipients of public R&D funding will have to institute in managing their intellectual property and provides for the establishment of a National Intellectual Property Management Office and Intellectual Property Fund to assist in financing the costs of obtaining intellectual property protection. The question of whether the legislation has struck the right balance between creating an enabling environment and regulating relevant activities continues to generate debate. This is of particular importance in the South African context, where more than half of R&D expenditure comes from industry. The focus of the legislation on publicly funded research could therefore potentially limit the overall impact of these interventions.

With regard to local university-industry linkages, for example, until now, the ownership of intellectual property arising out of collaborative research or industry-sponsored research at universities has been a matter of negotiation. Under the new legislation, the circumstances under which a company may own intellectual property have been significantly reduced. There is concern that this might act as a deterrent to companies sponsoring research at, or collaborating with, universities (Wolson, 2007).

The R&D strategy also recommended that a core agency be set up to stimulate technological innovation by aligning, coordinating, and providing a single point of strategic direction for a range of existing and anticipated innovation support instruments. In the first instance, this will involve consolidating several existing agencies and institutions, mainly from within the Department of Science and Technology, under the umbrella of the new Technology Innovation Agency. The new policy instruments aimed at promoting innovation and the transfer and commercialization of technology bode well for a supportive environment. It is hoped that they are implemented in a flexible, enabling, and responsive manner so as to successfully cultivate the anticipated improvements in the systems.

25.1.1   Case Study 5: The Masdar Initiative

Abu Dhabi has taken a bold step towards creating a knowledge-based economy. The Masdar initiative, driven by the Abu Dhabi Future Energy Company, is designed to be a global cooperative platform for engagement in the search for solutions to pressing issues such as energy security, climate change, and the development of human expertise in sustainability. If successful, it will also serve to position Abu Dhabi as a world-class R&D hub for new energy technologies and drive the commercialization and adoption of these and other technologies in sustainable energy, carbon management, and water conservation. In parallel with "Vision 2030," Masdar will also assist economic diversification and the development of a knowledge-based economy, while enhancing Abu Dhabi's record of environmental stewardship (see Masdar, 2011).

Masdar has five business units. Its Carbon Management Unit is developing greenhouse gas emission reduction projects, creating value by trading carbon Certified Emission Reduction certificates under the CDM of the Kyoto Protocol. The unit also develops sustainable technologies, including large-scale projects that generate sizable carbon emission reductions, focusing on CCS. This unit presumes the sustainability of the CDM and the global carbon market beyond 2012. Among its projects are a partnership with Bahrain's National Oil and Gas Authority to reduce carbon emissions derived from energy efficiency projects in the Bahraini oil industry and the development of a national carbon capture network to reduce Abu Dhabi's carbon footprint. The first phase of the network will capture around 6.5 million tonnes of CO_2 from power plants and industrial facilities for injection into oil reservoirs for enhanced oil recovery. Natural gas will be processed to hydrogen and CO_2, with the former being used for power generation and the latter re-injected into oil fields.

The Industries Unit invests globally and locally to establish a portfolio of production assets to provide Masdar access to technology and markets for the renewable energy value chain. The unit's current flagship investment is Masdar PV, which was created in April 2008 with the goal of becoming a top-three global thin-film PV manufacturer. A US$600 million phase-one investment will fund the construction of plants in Germany and Abu Dhabi with a combined annual output of 210 MW.

The Masdar Institute of Science and Technology, developed in cooperation with the Massachusetts Institute of Technology, offers masters- and doctoral-level degree programs focused on the science and engineering of advanced energy and sustainable technologies. The aim is to develop a pool of highly skilled scientists, engineers, managers, and technicians capable of accelerating the development of technology in the region and globally. The first class of 92 students, of which 13% are UAE nationals,[109] started in September 2009. A faculty of 20 academics from 11 countries, eight of whom are from Arab origins, has been recruited from internationally renowned universities.

Masdar's property development unit is currently constructing the carbon-neutral Masdar City l, soon to become home to 40,000 residents as well as research centers and companies with expertise in clean technology. The International Renewable Energy Agency is also located here, though it is uncertain whether much research will actually be conducted at its headquarters.

The utility assets and management unit aims to secure for Masdar a leading position in future energy markets by building a renewable energy portfolio through investments in companies with promising technology. The unit is positioned as a renewable energy power project developer, in bridging the gap between equipment manufacturers that lack project development capital and local utilities that lack the renewable energy economics, knowledge, and know-how. The current focus is on CSP, PV, wind, and waste-to-energy systems.

Within this framework, Masdar recently connected a 10 MW PV plant to the electricity grid, which will reduce Abu Dhabi's carbon footprint by 15,000 tonnes/year and will supply all the energy needs of the Masdar Institute of Science Technology. The plant, consisting of 87,777 panels (50% thin film and 50% crystalline silicon) is projected to generate 17,500 MWh of clean energy each year. A thin-film plant was expected to begin production in late 2010. Other new projects include the construction and operation of a 500 MW hydrogen-fired power plant[110] and a 100 MW CSP plant that will use parabolic trough technology. This is planned to be the first of many CSP plants in Abu Dhabi that will feed green power into the grid.

25.1.1   Case Study 6: Strengthening Effective Participation at the International Level: The Arabian Gulf University

The Arabian Gulf University (AGU) was established in Bahrain in 1979 and as a regional university was the first of its kind. With the founding of the Gulf Cooperation Council (GCC) in 1981, AGU became a regional

---

109   The remaining students came from 22 countries, including Algeria, Egypt, Eritrea, Germany, the Netherlands, Iceland, India, Mexico, Morocco, Nigeria, Pakistan, Palestine, Taiwan, Turkey, the United States, and the United Kingdom.

110   The project is a joint venture between BP, Rio Tinto, and Hydrogen Energy, a UK-based company that produces low-carbon hydrogen for electricity generation.

institution of the GCC with the objective of contributing to regional integration and development in the region and strengthening regional efforts in human resources development. It has since become a center for building research and training capacities on energy and the environment in the GCC countries and strengthening their participation on such issues at the international level.

AGU's unique teaching program is based on a multidisciplinary, integrated studies approach that emphasizes problem-based learning through case studies. To ensure flexibility in responding to the emerging development needs of the region, the AGU has adopted a flexible structure based on programs rather than rigid departments, focusing on technology and the environment. The latter was further strengthened by the creation of the Sheikh Zayed Al Nahyan Academic Chair of Environmental Sciences in 1998, which aims to enhance research and capacities in resource management and environmental issues. Two major milestones have been the launch of the environmental management program at AGU in 2004 and the creation of the Zayed seminar series. The seminars are particularly important as a networking mechanism across the region which, IPCC models predict, will become considerably hotter and drier over the century. Gelil (2010) observes, "If the Arab countries have failed, for decades, to join hands on most fronts, the climate change problem might be a 'golden opportunity' to start."

Five seminars have now been held. The first laid the basis for the Environmental Management Program at AGU. The others examined the potential of alternative energy in the GCC, sustainable cities, and the issues of climate change, energy, water, and food security in the Arab region. The seminars were attended by 400 participants from the GCC, of whom 57% were from the academic community, 23% from government agencies, 15% from the private sector, and the remaining 5% from NGOs, the media, etc. Although the seminars could play an important role in knowledge production and sharing, the low level of participation by governments and policymakers from the region thus far limits opportunities for joint problem solving and policy coordination.

The AGU has begun to gain credibility and name recognition that is enabling it to help strengthen the capacity of Bahrain and other Arab Gulf countries to participate in international energy negotiations, and to benefit from financial and technical assistance programs provided by GEF and other agencies. The AGU has assisted the government of Bahrain to develop its initial national communication to the UNFCCC in 2005. It subsequently was invited to lead two of the research teams that prepared the Kingdom of Bahrain's second national communication. The studies provided assessments of the vulnerability of biodiversity and natural ecosystems to climate change and of mitigation measures. In addition, the project was designed to strengthen Bahrain's human, institutional, scientific, technical, and informational capacities. Moreover, it was designed to enhance its ability to respond to the challenges of climate change and meet its obligations, as a Non-Annex I Party, towards UNFCCC, and therefore each proposed activity of the project included an element of capacity development. Staff from Bahrain's Environmental Agency participated in the research projects, managed the process, and participated in training throughout the project. At the end of the project, the staff of the agency will establish a unit to address all issues related to climate change.

The AGU is also working as a collaborating center of UNEP in the preparation of the Global Environment Outlook (GEO). AGU faculty members have contributed to the West Asia chapter in the series of GEO reports, the latest of which was GEO-4, published in 2007 (UNEP, 2007). The holder of the Zayed Academic Chair has also been a member of the team of international experts led by the International Institute for Sustainable Development, which developed the GEO Integrated Environmental Assessment training manual used by UNEP to build national and regional capacity to undertake these assessments and prepare GEO national reports. The manual has been used in training activities in a number of Arab countries, including Egypt, Jordan, Yemen, and Qatar (UNEP and IISD, 2008).

References

Abu Dhabi, 2008: *Abu Dhabi Economic Vision 2030*. The Government of Abu Dhabi, Abu Dhabi, United Arab Emirates.

ACCESS SARL, 2010: *Electrification Rural a Base de Biocarburant le Pourghere dans la commune rural de Garalo dans le Sud du Mali*. September 28, 2010, Bamako, Mali.

AEPC and UNDP, 2009: *Successful Delivery of Modern Energy Services to the Poor in Rural Areas: Lessons from Nepal*. Working Draft, United Nations Development Programme (UNDP) and Alternative Promotion Center (AEPC), Nepal.

Agbemabiese, L., 2008: Renewable energy enterprise development. In *Renewable Energy for Rural Areas in Africa: The Enterprise Development Approach*. A. Brew-Hammond and F. Kemausuor (eds.), Kwame Nkrumah University of Science and Technology (KNUST), Ghana, pp.6–14.

AGECC, 2010: *Energy for a Sustainable Future*. The UN Secretary-General's Advisory Group on Energy and Climate Change (AGECC), Summary Report and Recommendations, United Nations, New York.

Accra Agenda for Action, 2008: Endorsement Statement. *3rd High Level Forum on Aid Effectiveness*, Accra, Ghana.

Agterbosch, S. and S. Breukers, 2009: Socio-political embedding of onshore wind power in the Netherlands and North Rhine-Westphalia. *Technology Analysis & Strategic Management*, **20**(5): 633–48.

Ahmed, A., 2008: Expats make up 99% of private sector staff in UAE. *gulfnews.com*. April 7, 2008. gulfnews.com/news/gulf/uae/employment/expats-make-up-99-of-private-sector-staff-in-uae (accessed 30 May 2010).

ARC, 2011: *StoveTec*. Aprovecho Research Center (ARC). www.aprovecho.org/lab/stovetec-story (accessed May 30, 2011).

AREED, undated : *AREED II*. African Rural Energy Enterprise Development (AREED). www.areed.org (accessed May 30, 2010).

Ashden, 2006: *Rahimafrooz Renewable Energy, Bangladesh: Manufacturing and Installing Solar Home Systems*. The Ashden Awards for Sustainable Energy, London, UK www.ashdenawards.org/winners/rahimafrooz (accessed May 30, 2011).

Ashden, 2009: *2009 Ashden Awards Case Study: The Solar Energy Foundation*. The Ashden Awards for Sustainable Energy, London, UK. www.ashdenawards.org/blog/solar-energy-foundation-bbc (accessed May 30, 2010).

Balagopal, B., P. Paranikas and J. Rose, 2010: *What's Next for Alternative Energy?* Boston Consulting Group (BCG), Boston, MA, USA.

Ballarin, P., 2002: Petrobras: Estratégia e Esforço Tecnológico. *Petróleo & Gás Brasil*, Vol. 3, pp. 5–6.

Barnes, D., R. Van der Plas, K. Openshaw and K. R. Smith, 1994: *What Makes People Cook with Improved Biomass Stoves? A Comparative International Review of Stove Programs*. Technical Paper No. 242, The World Bank, Washington, DC, USA.

Baser, H. and P. Morgan, 2008: *Capacity, Change and Performance: Study Report*. Discussion Paper No. 59B, European Centre for Development Policy Management (ECDPM), Maastricht, The Netherlands.

Bell, M., 1984: "Learning" and the accumulation of industrial technological capacity in developing countries. In *Technological Capability in the Third World*. M. Fransman and K. King (eds.), Macmillan, Basingstoke, UK, pp.187–209.

Bell, M., 2009: *Innovation Capabilities and Directions of Development*. STEPS Working Paper 33, Social, Technological and Environmental Pathways to Sustainability (STEPS) Centre, Brighton.

Bell, M. and K. Pavitt, 1997: Technological accumulation and industrial growth: Contrasts between developed and developing countries. In *Technology, Globalisation and Economic Performance*. D. Archibugi and J. Michie (eds.), Cambridge University Press, Cambridge, UK, pp.83–137.

Bensch, G., J. Kluve and J. Peters, 2010: *Rural Electrification in Rwanda: An Impact Assessment Using Matching Techniques*. Ruhr Economic Papers No. 231, Rheinish-Westfälisches Institut für Wirtshaftsfortschung (RWI), Essen, Germany.

Bent, Mark, 2009: *How InnoCentive accelerates our R&D*. InnoCentive. wn.com/SunNight_Solar_CEO_How_InnoCentive_accelerates_our_R&D (accessed May 30, 2010).

Bergman, N. and C. Jardine, 2009: *Power from the People: Domestic Microgeneration and the Low Carbon Buildings Programme*. ECI Research Report No. 34, Oxford Environmental Change Institute (ECI), University of Oxford, UK.

Borja-Díaz, M. A., O. A. Jaramillo-Salgado and F. Mimiaga-Sosa, 2005: *Primer Documento del Proyecto Eoloeléctrico del Corredor Eólico del Istmo de Tehuantepec*. Instituto de Investigaciones Eléctricas-PNUD-GEF, Mexico.

Borup, M., P. D. Anderson, S. Jacobsson and A. Midtune, 2008: *Nordic Energy Innovation System Patterns of Need and Cooperation*. Nordic Energy Research, Copenhagen, Denmark.

Bouille, D. and S. McDade, 2002: Capacity development. In *Energy for Sustainable Development: A Policy Agenda*. T. B. Johansson and J. Goldemberg (eds.), United Nations Development Programme (UNDP), New York, NY, USA, pp.173–205.

Boyle, G., J. Kirton, R. Lof and T. Nayler, 2009: Transitioning from the CDM to a Clean Development Fund. *Carbon and Climate Law Review*, **1**: 14–22.

BRASIL, 1996: *Industrial Property Law*. Lei n. 9.279, May 14, 1996. www.planalto.gov.br/ccivil_03/Leis/L9279.htm (accessed July 25, 2009).

BRASIL, 2004: *On the Incentives for Innovation and Scientific and Technological Research and Other Subjects*. Lei n. 10.973, December 2, 2004. www.planalto.gov.br/ccivil_03/_ato2004–2006/2004/Lei/L10.973.htm (accessed July 25, 2009).

BRASIL, 2007: *Plano de Ação de Ciência, Tecnologia e Inovação para o Desenvolvimento Nacional 2007–2010*. www.inovacao.unicamp.br/report/integras/index.php?cod=197 (accessed July 25, 2009).

Brehm, D., 2009: Cement's basic molecular structure finally decoded. *MIT News*, September 14, 2009.

Breukers, S. and M. Wolsink, 2008: Wind power implementation in changing institutional landscapes: An international comparison. *Energy Policy*, **35**(5): 2737–2750.

Brew-Hammond, A. and A. Crole-Rees, 2004: *Reducing Rural Poverty through Increased Access to Energy Services: A Review of the Multifunctional Platform Project in Mali*. United Nations Development Programme (UNDP), New York, NY, USA.

Brew-Hammond, A. and F. Kemausuor (eds.), 2008: *Renewable Energy for Rural Areas in Africa: The Enterprise Development Approach*. Kwame Nkrumah University of Science and Technology (KNUST), Kumasi, Ghana.

Brew-Hammond, A., L. Darkwah, G. Obeng and F. Kemausour, 2008: *Renewable Energy Technology, Capacity and R&D in Africa*. Background paper for the International Conference on Renewable Energy in Africa, organized by the Government of Senegal, the African Union, BMZ and UNIDO, 16–18 April 2008, Dakar, Senegal.

Broydo Vestel, L., 2010: Q and A: The renewable energy czar. *Green: A Blog about Energy and the Environment*. *New York Times*, December 9, 2010.

Burrell, T., 2008: *Garalo Bagani Yelen, a Jatropha-fuelled rural electrification for 10,000 people in the Commune of Garalo*. Mali Folk Centre (MFC) Nyetaa, Mali.

Busch, M. L., E. Reinhardt and G. Shaffer, 2008: *Does Legal Capacity Matter? Explaining Dispute Initiation and Antidumping Actions in the WTO*. Series Issue Paper No. 4, ICTSD Project on Dispute Settlement, International Centre for Trade and Sustainable Development (ICTSD), Geneva, Switzerland.

Butler, R. A., 2010: Brazil launches major push for sustainable palm oil in the Amazon. *mongabay.com*, (accessed May 7, 2010).

C40 São Paulo Summit, 2011: *Velib – A new Paris love affair*. C40 São Paulo Summit. www.c40saopaulosummit.com/site/conteudo/index.php?in_secao=36&lang=3&in_conteudo=44 (accessed May 30, 2011).

Caniels, M. and H. Romijn, 2010: *The Jatropha Biofuels Sector in Tanzania 2004–9: Evolution Towards Sustainability?* Working Paper 10.04, Einhoven Centre for Innovation Studies (ECIS), The Netherlands, forthcoming in *Research Policy*, 2011.

City of Johannesburg, 2011: *New York transport experts to ride on Rea Vaya*. City of Johannesburg, South Africa. Press release, February 22, 2011. www.joburg.org.za/index.php?option=com_content&view=article&id=6267&catid=209&Itemid=114 (accessed February 24, 2011).

Chesbrough, H., 2006: *Open Innovation: Researching a New Paradigm*, Oxford University Press, Oxford, UK.

Coelho, S. T., O. C. da Silva, S. M. S. G. Velazquez, M. B. C. A. Monteiro and C. E. G. Silotto, 2003: *Energy from Vegetable Oil in Diesel Generators: Results of a Test Unit at Amazon Region*. Brazilian Reference Center on Biomass (CENBIO), Electrochemical and Energy Institute, University of São Paulo, Brazil.

Coelho, S. T., O. C. da Silva, A. F. L. Andrade and F. de Godoy, 2005: *Palm Oil as Fuel to Conventional Diesel Engines in the Amazon Region Isolated Communities*, Brazilian Reference Center on Biomass (CENBIO), Electrochemical and Energy Institute, University of São Paulo, Brazil.

Colaneri, K., 2010: Kearny Student, Hoboken developer receive Governor's environmental awards. *The Jersey Journal*, December 20, 2010.

Coutinho, L. N., 2001: *Protegendo as invenções: a visão de pesquisadores brasileiros sobre facilidades e dificuldades do caminho entre a bancada e o INPI*. Masters Diss., Centro Federal de Educacão Celso Suckow da Fonseca, Rio de Janeiro, Brazil.

Cruz, C. H. B. and L. Mello, 2006: *Boosting Innovation Performance in Brazil*. OECD Working Paper 532, Economics Department, Organisation for Economic Co-operation and Development (OECD), Paris, France.

DAC, 1991: *Principles for Evaluation of Development Assistance*. Development Assistance Committee (DAC), Organisation for Economic Co-operation and Development (OECD), Paris, France.

DAC, 2006: *The Challenge of Capacity Development: Working towards Good Practice*. DAC Guidelines and Reference Series. Development Assistance Committee (DAC), Organisation for Economic Co-operation and Development (OECD), Paris, France.

Dantas, E. and M. Bell, 2009: Latecomer firms and the emergence and development of knowledge networks: The case of Petrobras in Brazil. *Research Policy*, **38**(5): 829–844.

David, P., 1985: Clio and the economics of QWERTY. *American Economics Review*, **75**(2): 332–337.

Davidson, O. and Y. Sokona, 2001: Energy and sustainable development: Key issues for Africa. In *Proceedings of the African High-level Regional Meeting on Energy and Sustainable Development*. N. Wamukonya (ed.), 1–3 January 2001, United

Nations Development Programme (UNDP) Collaborating Centre on Energy Development, Risø National Laboratory, Roskilde, Denmark, pp.1–19.

DECC, 2009: *Smarter Grids: The Opportunity*. UK Department of Energy and Climate Change (DECC), London, UK.

DECC, 2010a: *Annual Energy Statement*. DECC Departmental Memorandum. UK Department of Energy and Climate Change (DECC), London, UK.

DECC, 2010b: *Huhne ends local authority power struggle*. UK Department of Energy and Climate Change (DECC). Press release, August 9, 2010, London, UK.

Dechezlepretre, A., M. Glachant, I. Hascic, N. Johnstone and Y. Meniere, 2009: *Invention and Transfer of Climate Change Mitigation Technologies on a Global Scale: A Study Drawing on Patent Data*. Final Report, Centre d'Economie Industrielle (CERNA) Mines Paris Tech, Paris, France.

Dietz, F., H. Brouwer and R. Weterings, 2008: Energy transition experiments in the Netherlands. In *Managing the Transition towards Renewable Energy: Theory and Practice from Local, Regional and Macro Perspectives*. J. van den Bergh and F. Bruinsma (eds.), Edward Elgar, Cheltenham, UK, pp. 217–244.

Dogbevi, E. K., 2009a: Update: Any lessons for Ghana in India's jatropha failure? *Ghana Business News*, 23 May 23, 2009. ghanabusinessnews.com/2009/05/23 (accessed May 30, 2010).

Dogbevi, 2009b: Another major Jatropha project suffers setback. *Ghana Business News*, July 28, 2009. www.ghanabusinessnews.com/2009/07/28 (accessed May 30, 2010).

Doranova, A., 2009: *Technology Transfer and Learning under the Kyoto Regime*. Diss., United Nations University, Maastricht Economic and Social Research Training Center on Innovation and Technology (UNU-MERIT), Maastricht, The Netherlands.

Drame, A., 2008: The AREED experience in Senegal. In *Renewable Energy for Rural Areas in Africa: The Enterprise Development Approach*. A. Brew-Hammond and F. Kemausuor (eds.), Kwame Nkrumah University of Science and Technology (KNUST), Kumasi, Ghana.

DST, 1996: *Preparing for the 21st Century*. White Paper on Science and Technology, Department of Arts, Culture, Science and Technology (DST), Government of South Africa, Pretoria, South Africa.

DST, 2006/2007: *National Survey of Research and Experimental Development*. Department of Arts, Culture, Science and Technology (DST), Government of South Africa, Pretoria, South Africa.

DST, 2007: *South African Innovation Survey 2005: Highlights*. Department of Arts, Culture, Science and Technology (DST), Government of South Africa, Pretoria, South Africa.

DTI, 2001: *Report into Network Access Issues: A Consultation Document*. Embedded Generation Working Group, Department of Trade and Industry (DTI), London, UK.

DTI, 2007: *Meeting the Energy Challenge*. Report CM 7124. Department of Trade and Industry (DTI), UK.

Duffy, J., 2010: World Bank pressured on clean energy. *International Herald Tribune*, October 11, 2010, pp.11–12.

Earl, S., F. Carden and T. Smutylo, 2001: *Outcome Mapping: Building Learning and Reflection into Development Programs*. International Development Research Centre (IDRC), Ottawa, Canada.

ECBI, 2011: *The European Capacity Building Initiative*. European Capacity Building Initiative (ECBI). www.eurocapacity.org (accessed May 30, 2011).

ECOWAS, 2006: *White Paper for a Regional Policy on Access to Energy Services for Populations in Rural and Peri-Urban Areas for Poverty Reduction in Line with Achieving the MDGs in Member States*. Economic Community of West African

States (ECOWAS), Twenty Ninth Summit of the Authority of Heads of State and Government, January 12, 2006, Niamey, Niger.

Edquist, C., 1997: Systems of innovation approaches: Their emergence and characteristics. In *Systems of Innovation: Technologies, Institutions and Organisations*. C. Edquist (ed.), Pinter, London, UK.

EnergyAccess, 2010: Toyola Case Study. *EnergyAccess*. energyaccess.wikispaces. com/Toyola+-+Case+Study (accessed August 5, 2011).

Enova SF, undated: *Funding, Scope and Objectives*. Enova SF, Royal Norwegian Ministry of Petroleum and Energy, Trondheim, Norway. www.enova.no/sitepageview.aspx?sitePageID=1001 (accessed May 30, 2010).

Envirofit, 2010: *Envirofit*. www.envirofit.org (accessed May 30, 2010).

EPA, 2008: *Multifunction Energy Platform (MFP) Pilot, Columbia University Earth Institute and Engineers without Borders*. United States Environmental Protection Agency (EPA), October 28, 2010, Washington, DC, USA.

EPA, 2010a: *Final Report: Multifunction Energy Platform (MFP) Pilot*. United States Environmental Protection Agency (EPA), Washington, DC.

EPA, 2010b: *MFP and Jatropha Program, Description*. Columbia University, May 26, 2010, New York, NY, USA.

ERC, 2007: *Electricity from Solar Home Systems in South Africa, Case Study 2*. Energy Research Centre, University of Cape Town, South Africa.

Ernst, D., L. Mytelka and T. Ganiatsos, 1998: Technological capabilities in the context of export-led growth: A conceptual framework. In *Technological Capabilities and Export Success*. D. Ernst, T. Ganiatsos and L. Mytelka (eds.), Routledge, London, UK, pp.5–45.

ESMAP, 2005a: *Advancing Bioenergy for Sustainable Development: Guidelines for Policymakers and Investors*. Report 300/05, prepared for United Nations Development Programme (UNDP) and World Bank Energy Sector Management Assistance Program (ESMAP), Washington, DC.

ESMAP, 2005b: *Potential for Biofuels for Transport in Developing Countries*, Joint United Nations Development Programme (UNDP)/World Bank Energy Sector Management Assistance Program (ESMAP), Washington, DC, USA.

ESMAP, 2008: *Accelerating Clean Energy Technology Research: Development and Deployment Lessons from Non-Energy Sectors*. Working Paper 138, P. Avalo and J. Coony (eds.), World Bank Energy Sector Management Assistance Program (ESMAP), Washington, DC, USA.

ESMAP, 2009: *Good Practices in City Energy Efficiency, Bogota, Colombia: Bus Rapid Transit for Urban Transport*. World Bank Energy Sector Management Assistance Program (ESMAP), Washington, DC, USA.

ESMAP, 2010: *Improved Cookstoves and Better Health in Bangladesh: Lessons from Household Energy and Sanitation Programs*. Final Report, World Bank Energy Sector Management Assistance Program (ESMAP), Washington, DC, USA.

Feller, J., B. Fitzgerald, S. Hissam and K. Lakhani, 2005: *Perspectives on Free and Open Source Software*. MIT Press, Cambridge, MA, USA.

Forbes, N. and D. Wield, 2002: *From Followers to Leaders: Managing Technology and Innovation in Newly Industrializing Countries*. Routledge, London, UK.

Foresight, 2008: *Powering Our Lives: Sustainable Energy Management and the Built Environment*. Government Office for Science, London, UK.

Fouquet, R. and P. J. G. Pearson, 1998: A thousand years of energy use in the United Kingdom. *The Energy Journal*, **19**(4): 1–41.

Freeman, C. and C. Perez, 1988: Structural crises of adjustment: Business cycles and investment behavior. In *Technical Change and Economic Theory*. G. Dosi, C. Freeman, R. Nelson, G. Silverberg and L. Soete (eds.), Pinter, London, UK, pp. 38–66.

Fridleifsson, I., 2006: *UNU-GTP Capacity Building and Central America*. Presented at the Workshop for Decision Makers on Geothermal Projects in Central America, November 26-December 2, 2006, organized by United Nations University Geothermal Training Programme (UNU-GTP) and LaGeo in San Salvador, El Salvador.

Fukuda-Parr, S., C. Lopes and K. Malik (eds.), 2002: *Capacity for Development: New Solutions to Old Problems*. Earthscan, London, and United Nations Development Programme (UNDP), New York, NY, USA.

GE, 2010: *Brazil energy milestone: GE, Petrobras using sugarcane-based ethanol to produce electricity*. General Electric (GE), press release, January 19, 2010, Rio de Janeiro, Brazil.

Geels, F. W., 2002: Technological transitions as evolutionary reconfiguration processes: A multi-level perspective and a case-study, *Research Policy*, **31**(–): 1257–1274.

Geels, F. W., 2004: From sectoral systems of innovation to socio-technical systems: Insights about dynamics and change from sociology and institutional theory. *Research Policy*, **33**(8–9): 897–920.

Geels, F. W., 2005: *Technological Transitions: A Co-Evolutionary and Socio-Technical Analysis*. Edward Elgar, Cheltenham, UK.

Geels, F. W. and R. Kemp, 2012: The transition perspective as a new perspective for road mobility study. In *Automobility in Transition? A Socio-Technical Analysis of Sustainable Transport*. F. W. Geels, R. Kemp, G. Dudley and G. Lyons (eds.), Routledge, New York, NY, USA, pp. 49–79.

Gelil, I. A, 2010: Towards a science-led climate policy in the Arab region. *Nature Middle East*, June 9, 2010. www.nature.com/nmiddleeast/2010/100609/full/nmiddleeast.2010.160.html (accessed May 30, 2011).

Georgsson, L., 2008: *Geothermal Energy in the World*. United Nations University Geothermal Training Programme (UNU-GTA) and Orkustofnun, Uganda Short Course, November 11, 2008, Uganda.

Gibbs, A., 1997: Focus groups. *Social Research Update 19*, University of Surrey, UK.

Gibbons, M., C. Limoges, H. Nowotny, S. Schwartzman, P. Scott and M. Trow, 1994: *The New Production of Knowledge: The Dynamics of Science and Research in Contemporary Societies*. Sage, London, UK.

GNESD, 2007: *Renewable Energy Technologies and Poverty Alleviation: Overcoming Barriers and Unlocking Potentials*. Global Network on Energy for Sustainable Development (GNESD), Roskilde, Denmark.

GNESD, 2009: *What is GNESD?* Steering Committee 2008–2009, Global Network on Energy for Sustainable Development (GNESD). www.gnesd.org (accessed May 30, 2010).

Goertz, L., 2006: *Mali Multi-functional Platform Project, Evaluation and Diagnostic Report*. Engineers without Borders (EWB) Canada and the Mali Multi-functional Platform Staff, Toronto.

Goldemberg, J., T. B. Johansson, A. K. N. Reddy and R. H. Williams, 1985: Basic needs and much more with one kilowatt per capita. *Ambio*, **14**(24): 190–200.

Gomes, R., 2009: Market solutions to combat indoor air pollution: Shell Foundation and Envirofit International. In *Cookstoves and Markets: Experiences, Successes and Opportunities*. K. Rai and J. McDonald (eds.), Global Village Energy Partnership, London, pp.7–9.

Grameen Shakti, 2011: *Energy for Development: The Grameen Shakti Model*. Proceedings of the Global Social Business Summit, Vienna, Austria.

Grubler, A., 2004: *Transitions in Energy Use*. In *Encyclopedia of Energy*. C. J. Cleveland (ed.), Elsevier; Amsterdam, pp. 163–177.

GTZ, 2009: *Target Market Analysis Kenya's Solar Energy Market*. Federal Ministry of Economics and Technology and Deutsche Gesellschaft für Technicsche Zusammenarbeit (GTZ) GmbH, Eschborn, Germany.

GTZ (GIZ), 2010: *Energising Development Partnership*. EnDev Fact Sheet, Deutsche Gesellschaft für Technicsche Zusammenarbeit (GTZ) GmbH, Eschborn, Germany.

Guajardo, G., 2007: Lecciones sobre cambio tecnológico e institucional en la investigación y desarrollo del petróleo en México. *Agenda para el desarrollo. Política Energética*, Vol. 8. J. L. Calva (ed.), UNAM-Porrúa, México DF, Mexico, pp.138–155.

Guetali SADC, 2009: *Marine algae biofuel launched in the Eastern Cape*. Guetali South African Development Community (SADC), March 23, 2009. guetalisadc. blogspot.com/2009/03/marine-algae-biofuel-pilot-project.html (accessed May 30, 2010).

GVEP, 2009: *Cookstoves and Markets: Experiences, Successes and Opportunities*. K. Rai and J. McDonald (eds.), Global Village Energy Partnership (GVEP), London,UK.

Hall, P. and R. Taylor, 1996: Political science and the three institutionalisms. *Political Studies*, **44**: 936–957.

Hamlin, T. and A. Fikre-Marium, 2004: *The African Rift Geothermal Energy Development Facility (ARGeo)*. Presentation to the UNEP and KfW Renewable Energy Conference, Bonn, German

Hankins, M., 2011: A solar strategy for Africa: International players set to expand key market. *Renewable Energy World*, January 4, 2011. www.renewableenergy-world.com/rea/news/print/article/2011/01/a-solar-strategy-for-africa (accessed May 30, 2011).

Harhoff, D., B. Hall, G. von Graevenitz, K. Hoisl and S. Wagner, 2007: *The Strategic Use of Patents and its Implications for Enterprise and Competition Policies*. EU Final Report No. ENTR/05/82, European Union (EU), Munich, Germany

Healey, P., 1998: Building institutional capacity through collaborative approaches to urban planning. *Environment and Planning*, **A30**: 1531–1546.

Heher, T., 2005: Implications of international technology transfer benchmarks for developing countries. *International Journal of Technology Management and Sustainable Development*, **4**(3): 207–225.

Heisenberg, W., 1962: *Physics and Philosophy: The Revolution in Modern Science*. Harper and Row, New York, USA.

HM Government, 2009 : *The UK Low Carbon Transition Plan: National Strategy for Climate and Energy*. Stationery Office, London, UK.

Ho, M.-W., 2010: *Eradicating Rural Poverty with Renewable Energies*. ISIS Report, 24/11, Institute of Science in Society (ISIS), London, UK.

Höök, M., R. Hirsch and K. Aleklett, 2009: Giant oil field decline rates and their influence on world oil production. *Energy Policy*, **37**(6): 2262–2272.

Hukkinen, J., 1999: *Institutions in Environmental Management: Constructing Mental Models and Sustainability*. Routledge, London, UK.

Huntley, M. E. and D. G. Redalje, 2006: CO_2 mitigation and renewable oil from photosynthetic microbes: A new appraisal. *Mitigation and Adaptation Strategies for Global Change*, **12**: 573–608.

IDRC, 2008: *Research Matters – Knowledge Translation: Bridging the 'Know-do' Gap*. International Development Research Centre (IDRC), Ottawa, and the Swiss Agency for Development and Cooperation, Canada and Switzerland.

IFC, 2009: *FAQ – Tata Mundra Project*. International Finance Corporation (IFC). www.ifc.org/ifcext/southasia.nsf/Content/TataMundra_FAQ (accessed May 30, 2011).

IIE, 2005/2006: *Sexto Informe de Labores*. Instituto de Investigaciones Eléctricas (IIE), Morelos, Mexico.

InnoCentive, 2011: *InnoCentive*. www.innocentive.com (accessed May 30, 2011).

IOI, 2011: *Global Initiative for Open Innovation Launched to Promote Patent System Transparency*. Initiative for Open Innovation (IOI). Press release, July 13, 2009, Geneva, Switzerland. www.openinnovation.org/daisy/ioi/home.html (accessed May 30, 2011).

Jacobson, A., 2004: *Connective Power: Solar and Social Change in Kenya*. Diss., Energy and Resources Group, University of California, Berkeley, USA.

Jacobsson, S. and A. Bergek, 2004: Transforming the energy sector: The evolution of technological systems in renewable energy technology. *Industrial and Corporate Change*, **13**(5): 815–849.

Jannuzzi, G. de M., 2005: Power sector reforms in Brazil and its impacts on energy efficiency and research and development activities. *Energy Policy*, **33**(13): 1753–7162.

Katz, J., 1983: Technological change in the Latin American metalworking industries: Results from a programme of case studies. *CEPAL Review*, **19**: 85–143.

Katz, J., 1985: Domestic technological innovations and dynamic comparative advantages: Further reflections on a comparative case-study programme. *International Technology Transfer Concepts, Measures, and Comparisons*. N. Rosenberg and C. Frischtak (eds.), Praeger, New York, USA, pp.127–166.

Katz, J., 2004: Market-oriented reforms, globalization and the recent transformation of Latin American innovation systems. *Oxford Development Studies*, **32**(3): 375–388.

Kemp, R., 1994: Technology and the transition to environmental sustainability: The problem of technological regime shifts. *Futures*, **26**(10): 1023–1046.

Kemp, R., 2010: The Dutch Energy Transition Approach, International Economics and Economic Policy, **7**: 291–316.

Kern, F. and M. Howlett, 2009: Implementing transition management as policy reforms: a case study of the Dutch energy sector, *Policy Sciences*, **43**: 391–408.

Kim, L., 1997: *Imitation to Innovation: The Dynamics of Korea's Technological Learning*. Harvard Business School Press, Boston, MA, USA.

Kim, L., 2003: Foreign direct investment, technology development and competitiveness: Issues and evidence. In *Competitiveness, FDI and Technological Activity in East Asia*. S. Lall and S. Urata (eds.), Edward Elgar, Cheltenham, UK, pp.12–82.

Kim, L., 2004: The multifaceted evolution of Korean technological capabilities and its implications for contemporary policy. *Oxford Development Studies*, **32**(3): 342–363.

Kiron, V., W. Phromkunthong, M. Huntely, I. Archibald, and G. de Scheemaker, 2010: Marine Microalgae: A green alternative protein source in aquatic feeds. European Aquaculture Meeting (September 2010), Porto, Portugal.

Kishore, V. V. N. and P. V. Ramana, 2002: Improved cookstoves in rural India: How improved are they? A critique of the perceived benefits from the National Programme on Improved Chulhas (NPIC). *Energy*, **27**: 470–463.

KITE, 2004: *Technical Feasibility of Jatropha Oil and Biodiesel as Fuel for Rural MFPs*. 2nd Draft Report, Prepared by A. Brew- Hammond, J. Robinson, D. Derzu and C. Ciarra, Kumasi Institute of Technology and Environment, Ghana.

Kline, R. and T. J. Pinch, 1996: Taking the black box off its wheels: The social construction of the automobile in rural America. *Technology and Culture*, **37**: 776–795.

Kline, S. J. and N. Rosenberg, 1986: *An Overview of Innovation in the Positive Sum Strategy*. R. Landau and N. Rosenberg (eds.), National Academies Press, Washington, DC, USA.

Krugmann, H., 1988: *Improved Cookstove Programs: Boon or Boondoggle?* International Development Research Centre (IRDC), Ottawa, Canada, pp. 22–23.

Lahorgue, M., M. Santos and J. Mello, 2005: *Economic Development Mission in Brazilian Universities.* Triple Helix 2005 Conference, May 17–20, 2005, Turin, Italy.

Lakhani, K. R., L. B. Jeppesen, P. A. Lohse and J. A. Panetta, 2007: *The Value of Openness in Scientific Problem Solving.* Working Paper 07–050, Harvard Business School, Boston, MA, USA.

Lakhani, K. R. and J. A. Pannetta, 2007: The principles of distributed innovation. *Innovations*, 2(3): 97–112.

Lall, S., 1992: Technological capabilities and industrialization. *World Development*, 20(2): 165–186.

Landman, M., 2010: *LEED Platinum Certified Buildings.* M. Landman Communications and Consulting, San Francisco, CA, USA. www.mlandman.com/gbuildinginfo/leedplatinum.shtml (accessed May 20, 2011).

Lewis, W. A., 1950: The industrialization of the British West Indies. *Caribbean Economic Review*, May.

Lindau, L., D. Hidalgo and D. Facchini, 2010: Curitiba, the cradle of Bus Rapid Transit. *Built Environment*, 36(3): 274–22.

Looi, M.-K., 2009: Platform for research – African Institutions Initiative. Wellcome Trust, London, July 3, 2009. www.wellcome.ac.uk/news/2009/features/wtx055738.htm (accessed May 30, 2010).

Lovallo, D. and O. Sibony, 2010: *The Case for Behavioral Strategy.* McKinsey Quarterly, March. www.mckinseyquarterly.com/The_case_for_behavioral_strategy_2551 (accessed May 30, 2011).

Lundvall, B.-A., 1992: *National Systems of Innovation: Towards a Theory of Innovation and Interactive Learning.* Pinter Publishers, London, UK.

Lusthaus, C., M.-H. Adrien and M. Perstinger, 1999: *Capacity Development: Definitions, Issues and Implications for Planning, Monitoring and Evaluation.* Occasional Paper No. 23, Montreal, Canada.

Mafuta Sasa, 2011: *Mafuta Sasa Biodiesel Limited.* Mafuta Sasa Biodiesel Limited, Dar es Salaam, Tanzania. www.mafutasasa.com/about-tanzania (accessed May 30, 2011).

Mali Biocarburant, 2010: *Company Profile: Mali Biocarburant,* Bamako, Mali. www.malibiocarburant.com/website%20example%20-%20copie/Mali_Biocarburant_SA/MBSA/MBSA_files/Depliant%20MBSA.pdf (accessed May 30, 2010).

Martinot, E. and S. Birner, 2005: Market transformation for energy-efficient products: lessons from programs in developing countries, *Energy Policy*, 33(14): 1765–1779.

Martinot, E., A. Cabraal and S. Mathur, 2001: World Bank/GEF solar home system projects: Experiences and lessons learned 199–000. *Renewable and Sustainable Energy Reviews*, 5: 39–57.

Masdar, 2011: *Masdar Initiative: About Us.* Masdar: A Mubadala Company. www.masdar.ae/en/Menu/Index.aspx?MenuID=42&mnu=Pri, (accessed May 30, 2011).

McDonough, M., 2008: *Mass Wall Building: An Achievable Approach to Green, High Performance and Net Zero Energy Buildings,* White Paper, Hoboken Brownstone Co., Hoboken, NJ.

McKinsey and Company, 2010: *A Portfolio of Power-Trains for Europe: A Fact-based Analysis. The Role of Battery Electric Vehicles, Plug-in Hybrids and Fuel Cell Electric Vehicles.* McKinsey and Company, London, UK.

Mehlomakulu, B., 2008: Hydrogen and fuel-cell technology issues for South Africa: The emerging debate. In *Making Choices About Hydrogen. Transport Issues for Developing Countries.* L. K. Mytelka and G. Boyle (eds.), United Nations University (UNU) Press, Tokyo and International Development Research Centre (IRDC), Ottawa, Canada, pp.324–345.

Microhydropower.net, 2011: *Microhydropower Discussion Group.* Wim Jonker Klunne, moderator. www.microhydropower.net (accessed May 30, 2011).

Moreira, J. R. and A. D. Poole, 1993: Hydropower and its constraints. In *Renewable Energy: Sources for Fuels and Electricity.* T. B. Johansson, H. Kelly, A. K. N. Reddy and R. H. Williams (eds.), Island Press, Washington, DC, USA, pp. 73–120.

Mowery, D. (ed.), 1988: *International Collaborative Ventures in U.S. Manufacturing,* Ballinger, Cambridge, MA, USA.

Muchie, M., P. Gammeltoft and B.-A. Lundvall (eds.), 2003: *Putting Africa First: The Making of African Innovation Systems.* Aalborg University Press, Denmark.

Mutizwa-Mangiza, N. 2009: Why urban planning systems must change. *Urban World*, 1(4):16–21.

Mwakilasa, M., 2008: *Mafuta Sasa Biodiesel Limited.* Bid Network. www.bidnetwork.org/page/144465/en (accessed May 30, 2011).

Mytelka, L. K., 1985: Stimulating effective technology transfer: The case of textiles in Africa. In *International Technology Transfer Concepts, Measures, and Comparisons.* N. Rosenberg and C. Frischtak (eds.), Praeger, New York, pp.77–126.

Mytelka, L. K., 1986: The Transfer of Technology: Myth or Reality? In *The European Community's Development Policy: The Strategies Ahead.* C. Cosgrove and J. Jamar (eds.), De Tempel, Bruges, pp. 243–28l.

Mytelka, L. K., 1989: The unfulfilled promise of African industrialization. *African Studies Review.* 32(3): 77–137.

Mytelka, L. K., 2000: Local systems of innovation in a globalized world economy. *Industry and Innovation,* 7(1)15–32.

Mytelka, L. K., 2001: Mergers, acquisitions and inter-firm technology agreements in the global learning economy. In *The Globalising Learning Economy.* D. Archibugi and B.-A. Lundvall (eds.), Oxford University Press, Oxford, UK, pp. 127–144.

Mytelka, L. K., 2004: Catching up in new wave technologies. *Oxford Development Studies,* 32(3): 389–405.

Mytelka, L. K., 2008: Hydrogen fuel cells and alternatives in the transport sector: A framework for analysis. In *Making Choices About Hydrogen Transport Issues for Developing Countries.* L. Mytelka and G. Boyle (eds.), United Nations University Press, Tokyo, Japan, and International Development Research Centre (IDRC), Ottawa, Canada, pp.5–38.

Nelson, R. (ed.), 1993: *National Innovation Systems: A Comparative Analysis.* Oxford University Press, Oxford, UK.

Nelson, R., 2004: The challenge of building an effective innovation system for catch-up. *Oxford Development Studies,* 32(3): 365–374.

Nepal, 2007: *Energy and Poverty in Nepal: Challenges and the Way Forward.* Regional Energy Programme for Poverty Reduction, United Nations Development Programme (UNDP) Regional Centre, Bangkok, Thailand.

Norges Bank, 2004: *The Long-term Investment Strategy of the Norwegian Petroleum Fund.* Address by Governor Svein Gjedrem to the Argentium Conference, September 30, 2004.

North, D. C., 1994: Institutional change: A framework of analysis. *Economic History,* 9412001, EconWPA.

OECD, 1992: *Technology and the Economy: The Key Relationships.* TEP Report, Organisation for Economic Co-operation and Development (OECD), Paris, France.

OECD, 2005/2008: *The Paris Declaration on Aid Effectiveness (2005) and the Accra Agenda for Action (2008).* Organisation for Economic Co-operation and Development (OECD), Paris, France.

OECD, 2009: *The Bioeconomy to 2030: Designing a Policy Agenda.* Organisation for Economic Co-Operation and Development (OECD), Paris, France.

Ockwell, D., J. Watson, A. Verbeken, A. Mallett and G. MacKerron, 2009: *A Blueprint for Post-2012 Technology Transfer to Developing Countries.* Policy Briefing No.5, Sussex Energy Group, University of Sussex, Sussex, UK.

Ofgem, 2010: *Low Carbon Networks Fund Governance.* Document V.3. Office of the Gas and Electricity Markets (Ofgem), London, UK.

Oh. S.-H., 2009: Seoul's bid to be the world's greenest city. *Urban World,* **1**(4): 5–7.

Perez, C., 1988: New technologies and development. In *Small Countries Facing the Technological Revolution.* C. Freeman and B.-A. Lundvall (eds.), Pinter, London, UK, pp. 85–97.

Petrobras, 2011: Petrobras joins consortium to accelerate adoption of green fuels. *Newsnet,* February 16, 2011. www.newenergyworldnetwork.com (accessed May 30, 2011).

Pietrobelli, C. and R. Rabellotti (eds.), 2006: *Upgrading to Compete: SMEs, Clusters and Value Chains in Latin America.* Harvard University Press, Cambridge, MA, USA.

Pigaht, M. and R. van der Plas, 2009: Innovative private micro-hydro power development in Rwanda. *Energy Policy,* **37**(11): 4753–4760.

Planting, S., 2006: A place in the sun. *Financial Mail,* November 3, 2006. free.financialmail.co.za/innovations/06/1103/cinn.htm_fin_mail_3_nov_2006.aplace in the sun sasha_planting (accessed May 30, 2011).

Portal do Ministério da Sciência e Technologia, 2002: *Livro Branco de Ciencia, Technologia e Inovacao.* Portal do Ministério da Sciência e Technologia, Brazil. acessibilidade.mct.gov.br/index.php/content/view/18765.html (accessed May 30, 2011).

Pouris, A., 2008: *Science-Industry Relations and the SA Innovation Chasm: Searching for Lost Technologies.* Paper prepared for the Department of Trade and Industry, South Africa.

Prebisch, R., 1959: The role of commercial policies in underdeveloped countries. *American Economic Review, Papers and Proceedings,* **49**: 251–73.

REN21, 2008: *Renewables 2007: Global Status Report.* Renewable Energy Policy Network for the 21st Century (REN21) Secretariat and Worldwatch Institute, Washington, DC, USA.

REN21, 2009: *Chinese Renewables Status Report.* Background Paper, Renewable Energy Policy Network for the 21st Century (REN21), Paris, France.

Renn, O., 1999: A model for analytic-deliberative process in risk management. *Environmental Science and Technology,* **33**: 3049–3055.

Reuters, 2009: SA's thin film solar tech at commercial stage. *Mail & Guardian* (South Africa), October 13, 2009.

Right Livelihood Award, 2011: *The Right Livelihood Award.* www.rightlivelihood.org/grameen_shakti.html (accessed May 30, 2011).

Rijssenbeek, W. and I. Togola, 2007: *Jatropha Village Power in Garalo, Mali: A New Dimension for People, Planet and Profit Actions.* Mali Folk Centre (MFC) Nyetta, Mali.

Rip, A. and R. Kemp, 1996: *Towards a Theory of Socio-technical Change.* Mimeo University of Twente, Enschede, the Netherlands.

Rip, A. and R. Kemp, 1998: Technological change. In *Human Choice and Climate Change: An International Assessment.* S. Rayner and E. L. Malone (eds.), Vol. 2, Batelle Press, Washington, DC, USA, pp. 327–400.

Rosenberg, N. and C. Frischtak, 1985: *International Technology Transfer: Concepts, Measures, and Comparisons.* Praeger, New York, USA.

Saenz, T., 2008: The path to innovation: The Cuban experience. *International Journal of Technology Management and Sustainable Development,* **7**(3): 205–221.

Sandén, B. and C. Azar, 2005: Near-term technology policies for long-term climate targets: Economy-wide versus technology-specific approaches. *Energy Policy,* **33**: 1551–1576.

Sauter, R. and D. Bauknecht, 2009: Distributed generation: Transforming the electricity network. In *Energy for the Future: A New Agenda.* I. Scrase and G. MacKerron (eds.), Palgrave Macmillan, Basingstoke, UK.

Science Commons, 2011: *Patent licenses.* Science Commons. sciencecommons.org/projects/patent-licenses (accessed May 30, 2011).

SENER, 2002: *Programa de Investigación y Desarrollo del Sector Energía 2002–2006.* Secretaría de Energía de Mexico (SENER), Mexico.

SENER, 2006: *Renewable Energies for Mexico's Development.* Secretaría de Energía de Mexico (SENER) and Gesellschaft für Internationale Zusammenarbeit (GTZ), Mexico.

SENER, 2007: *Diagnóstico de la Situación de Pemex.* Secretaría de Energía de Mexico (SENER) and Pemex, Mexico.

Scharpf, F., 1997: *Games Real Actors Play: Actor-Centered Institutionalism in Policy Research.* Westview Press, Boulder, CO.

Shaffer, G., M. R. Sanchez and B. Rosenberg, 2008: The trials of winning at the WTO: What lies behind Brazil's success. *Cornell International Law Journal,* **41**(2): 383–501.

Sibanda, M., 2009: Intellectual property, commercialization and institutional arrangements at South African publicly financed research institutions. In *The Economics of Intellectual Property in South Africa.* World Intellectual Property Organization, Geneva, pp. 113–145.

Sinha, K. C., 2003: Sustainability and urban public transportation. *Journal of Transportation Engineering,* **129**(4): 331–341.

Sinja, J., J. Karugia, M. Waithaka, D. Miano, I. Baltenweck, S. Franzel, R. Nyiakal and D. Romney, 2004: Adoption of fodder legumes technology through farmer-to-farmer extension approach. *Uganda Journal of Agricultural Sciences,* **9**: 222–226.

Smil, V. 1994: *Energy in World History.* Westview Press, Boulder, CO, USA.

Staatscourant, 2008: *Instellingsbesluit van het Regieorgaan Energietransitie.* Netherland Staatscourant 25 Feb 2008, nr. 39, p. 39.

Stern, N., 2006: *The Economics of Climate Change.* The Stern Review, Government of the United Kingdom, His Majesties Treasury, London, UK.

Stewart, F., 1984: Facilitating indigenous technical change in third world countries. In *Technological Capability in the Third World.* M. Fransman and K. King (eds.), Macmillan, Basingstoke, UK, pp. 81–94.

Storper, M., 1998: Industrial policy for latecomers: Products, conventions and learning. In *Latecomers in the Global Economy.* M. Storper, T. Thomadakis and L. Tsipouri (eds.), Routledge, London, UK, pp.13–39.

Stoves, 2011: *Improved Biomass Cooking Stoves*. Global Alliance for Clean Cookstoves. Leslie Cordes, moderator. www.bioenergylists.org (accessed May 30, 2011).

Taweekul, K., J. Caldwell, R. Yamada and A. Fujimoto, 2009: Assessment of the impact of a farmer-to-farmer learning and innovation scaling out process on technology adaptation, farm income and diversification in Northeast Thailand. *International Journal of Technology Management and Sustainable Development*, **8**(2): 129–144.

Taylor, M. G. and D. Collins, 2006: *Novel Cements: Low Energy, Low Carbon Cements*. British Cement Association (BCA), Fact Sheet 12, May 25, 2006.

Teece, D., 1988: Technological change and the nature of the firm. In *Technical Change and Economic Theory*. G. Dosi, C. Freeman, R. Nelson, G. Silverberg and L. Soete (eds.), Pinter, London, UK, pp.256–281.

Teixeira da Sousa, P., E. Dall'Ogllio, M. Sato, J. Marta, A. Brito and C. Spindola, 2008: The ethanol and biodiesel programmes in Brazil. In *Making Choices About Hydrogen Transport Issues for Developing Countries*. L. Mytelka and G. Boyle (eds.), United Nations University (UNU) Press, Tokyo, Japan, and International Development Research Centre (IDRC), Ottawa, Canada, pp.118–140.

The Economist, 2009: InnoCentive: A market for ideas. *The Economist*, September 17, 2009. www.economist.com/node/14460185 (accessed May 30, 2011).

The National, 2010: Masdar's vision remains unchanged. *The National*, October 12, 2010.

Toke, D., S. Breukers and M. Wolsink, 2008: Wind power deployment outcomes: How can we account for the differences? *Renewable and Sustainable Energy Reviews*, **12**: 1129–1147.

Torres-Montoya, M., 2007: *Innovative Public-Private Sector Cooperation in Urban Transport in India*. Paper 08–117, World Bank, Washington, DC, USA.

Tripathy, U., P. S. Sandhu and D. C. Prakash, 2008: *Turnaround Success Story of Urban Public Transport Undertaking – BMTC*. Bangalore Metropolitan Transport Corporation (BMTC), Bangalore, India.

UN AGECC, 2010: *Energy for a Sustainable Future: Summary Report and Recommendations*. Secretary-General's Advisory Group on Energy and Climate Change (AGECC), United Nations (UN), April 28, 2010, New York, USA.

UNDP, 2004: *Reducing Rural Poverty through Increased Access to Energy Services: A Review of the Multifunctional Platform Project in Mali*. United Nations Development Programme (UNDP), Bamako, Mali.

UNDP, 2007: *Energy and Poverty in Nepal: Challenges and the Way Forward*. Regional Energy Programme for Poverty Reduction, United Nations Development Programme (UNDP) Regional Centre, Bangkok, Thailand.

UNDP, 2008a: *Capacity Development Practice Note*. United Nations Development Programme (UNDP), New York, USA.

UNDP, 2008b: *Successfully Implementing, Replicating and Scaling up Rural Energy Programmes: Costs and Financing*. Powerpoint presentation, United Nations Development Programme (UNDP), New York, USA.

UNDP, 2009a: *Capacity Development: A UNDP Primer*. United Nations Development Programme (UNDP), New York, USA.

UNDP, 2009d: *Towards Sustainable Production and Use of Resources: Assessing Biofuels*. United Nations Development Programme (UNDP), New York, USA.

UNDP, 2010: Le Burkina Faso poursuit sa lute contre la pauvrete. *Centre de Press*, November 15, 2009.

UNDP, undated: *MFP Approach in West Africa Results and Potentials for expanding Access to Modern Energy Services in Burkina Faso, Ghana, Mali and Senegal*. United Nations Development Programme (UNDP), New York, USA.

UNDP, GEF, 2008: *Promotion of Wind Energy: Lessons Learned from International Experience and UNDP-GEF Projects*. United Nations Development Programme (UNDP), New York, USA.

UNDP and UNOPS, 2008: *Evaluation of the Ghana Multi-Functional Platform Pilot Project*. Evaluation Team: E. Morris, R. King and J. Winiecki. United Nations Development Program (UNDP), New York, USA.

UNEP, 2007: *Global Environment Outlook: Environment for Development (GEO-4)*. United Nations Environment Programme (UNEP), Nairobi, Kenya, and New York, USA.

UNEP, 2008: Hot Prospect-Geothermal Electricity set for Rift Valley Lift-Off in 2009. United Nations Environment Programme (UNEP), press release, December 2008. UNEP, 2009: *Towards Sustainable Production and Use of Resources: Assessing Biofuels*. International Panel for Sustainable Resource Management, United Nations Environment Programme (UNEP), Paris, France.

UNEP Risø Centre, 2011: *CDM/JI Pipeline Analysis and Database*. United Nations Environment Programme (UNEP) Risø Centre, Roskilde, Denmark. cdmpipeline. org (accessed May 30, 2011).

UNEP and IISD, 2008: *Integrated Environmental Assessment Training Manual, 2008*. United Nations Environment Programme (UNEP) and International Institute for Sustainable Development (IISD), Winnipeg, Canada.

UNFCCC, 2009: *Copenhagen Accord*. Report No. FCCC/CP/2009/L.7, United Nations Framework and Convention on Climate Change (UNFCCC), Conference of the Parties, December 7–18, 2009, Copenhagen, Denmark.

UNFCCC, 2011: *Report of the Conference of the Parties on Its Sixteenth Session* and *Addendum Part Two: Action Taken by the Conference of the Parties at Its Sixteenth Session, Contents Decisions adopted by the Conference of the Parties*. UNFCCC/CP/2010/7/Add. 1, United Nations Framework and Convention on Climate Change (UNFCCC), November 29-December 10, Cancun, Mexico.

UNIDO, 2011: *Industrial Development Report 2011*. United Nations Industrial Development Organization (UNIDO), New York, USA.

Unruh, G. C., 2000: Understanding carbon lock-in. *Energy Policy*, **28**: 817–830.

UNU, 2010: *United Nations University Geothermal Training Programme*. United Nations University (UNU). www.unugtp.is (accessed December 12, 2010).

van Asselt, H. and J. Gupta, 2008: *Stretching too Far? Developing Countries and the Role of the Flexibility Mechanisms beyond Kyoto*. Paper presented to the Conference on Climate Law in Developing Countries Post-2012: North and South Perspectives, September 28, 2008, University of Ottawa, Canada.

van de Vleuten, F., N. Stam and R. van der Plas, 2007: Putting solar home systems into perspective: What lessons are relevant? *Energy Policy*, **33**(3): 1439–1451.

Verbong, G. P. J., F. W. Geels and R. P. J. M. Raven, 2008: Multi-niche analysis of dynamics and policies in Dutch renewable energy innovation journeys (1970–2006): Hype-cycles, closed networks and technology-focused learning. *Technology Analysis and Strategic Management*, **20**(5): 55–73.

WBCSD, 2009: *The Cement Sustainability Initiative*. World Business Council for Sustainable Development (WBCSD), Geneva, Switzerland.

WBG, 2003: *Note to the Informal Board Briefing on Infrastructure*. World Bank Group (WBG), February 13, 2003.

Wellcome Trust, 2009: African institutions lead international consortia in £30 million initiative. Wellcome Trust, press release, July 2, 2009, London, UK.

Westphal, L., L. Kim and C. J. Dahlman, 1985: Reflections on the Republic of Korea's acquisition of technological capability. In N. Rosenberg and C. Frischtak (eds.), *International Technology Transfer Concepts, Measures, and Comparisons*. Praeger, New York, UK, pp.167–221.

Weterings, R., 2006: *Quick scan koplopersloket. Een evaluatie van werkwijze, output en effecten*, Competentie Centrum Transities, The Netherlands.

Wolson, R., 2007: The role of technology transfer offices in building the South African biotechnology sector: An assessment of policies, practices and impact. *Journal of Technology Transfer*, **32**: 343–365.

Wolson, R., 2008: *The Evolving Policy Landscape for Technology Transfer from Public Research Organisations in South Africa*. Masters Diss., University of Cape Town, South Africa.

Woodman, B., 2006: *Ofgem, Innovation and Distributed Generation: Recent Initiatives*. BIEE/ UK ERC Academic Conference, September 20–22, 2006, Oxford, UK.

World Bank, 2008: *Designing Sustainable Off-Grid Rural Electrification Projects: Principles and Practices*. Operational Guidance for World Bank Group Staff, The Energy and Mining Sector Board, Washington, DC, USA.

World Bank, 2009: *Energy Strategy Approach Paper*. World Bank, Washington, DC, USA.

World Bank and IMF, 2004: *Global Monitoring Report: Policies and Actions for Achieving the Millennium Development Goals and Related Outcomes*. World Bank and International Monetary Fund (IMF), Washington, DC, USA.

WRI, 2010: *Climate Analysis Indicators Tool (CAIT)*. World Resources Institute (WRI), Washington, DC, USA.

Wroughton, L., 2008: World Bank approves funds for Indian coal-fired plant. *Reuters*, 8 April 2008.

Young, J., 2008: Impact of research on policy and practice. *Capacity.org*, **35**: 4.

Zhao, J. and X. Zhu, 2004: *Private Participation in Small Hydropower Development in China – Comparison with International Communities*. UNHYDRO, Beijing.

Zhou, S., X. Zhang and J. Liu, 2009: The trend of small hydropower development in China. *Renewable Energy*, **34**: 1078–1083.

IV

Annexes I to IV

Acronyms, Abbreviations and Chemical Symbols

Acronyms, Abbreviations and Chemical Symbols

1P	proven reserves	
2P	proved plus probable reserves	
3P	proved plus probable plus possible reserves	
ABC	atmospheric brown cloud	
ABNT	Associação Brasileira de Normas Técnicas (Brazilian National Standards Organization)	
ABS	advanced biomass stoves	
AC	alternating current	
ACCESS	Australian Community Climate and Earth-System Simulator	
ACEEE	American Council for an Energy-Efficient Economy	
ACP	African, Caribbean, and Pacific states	
ADB	Asian Development Bank	
AEFP	America's Energy Future Panel	
AER	all electric range	
AEU	annual energy use	
AfDB	African Development Bank	
AFR	Sub-Saharan Africa region	
AGC-21	Advanced Gas Conversion for the 21st Century	
AGECC	United Nations Secretary-General's Advisory Group on Energy and Climate Change	
AGR	advanced gas reactor	
AGU	Arabian Gulf University (Bahrain)	
AIA	American Institute of Architects	
AIRE	Atlantic Interoperability Initiative to Reduce Emissions	
Al_2O_3	alumina	
ALRI	acute lower respiratory infection	
AME	Africaine de Maintenance et d'Équipement (solar water heating equipment and maintenance company based in Dakar)	
AMI	advanced metering infrastructure	
AMM	automatic meter management	
AMR	automated meter reading	
ANL	Argonne National Laboratory (United States)	
ANME	National Agency for Energy Management (Tunisia)	
ANSI	American National Standards Institute	
AOT40	accumulated O_3 exposure over a threshold of 40 parts per billion	
API	American Petroleum Institute	
APINA	Air Pollution Information Network for Africa	
APS	(IEA World Energy Outlook) Alternative Policy Scenario	
AQG	air quality guidelines	
AREED	African Rural Energy Enterprise Development	

ARGeo	African Rift Geothermal Energy Development Facility
ARI	acute respiratory infections
ARMZ	Atomredmetzoloto JSC (Russia)
ASHRAE	American Society of Heating, Refrigeration and Air Conditioning Engineers
ASME	American Society of Mechanical Engineers
ASN	Nuclear Safety Authority (France)
ASP	average selling price
ASPIRE	Asia and Pacific Initiative to Reduce Emissions
ASU	acid separation unit
ATR	autothermal reformer/reforming
AVR	Arbeitsgemeinschaft Versuchsreaktor (a prototype pebble bed reactor in West Germany)
BAPS	(IEA) Beyond Alternative Policy Scenario
BAT	best available technology
BAU	business-as-usual
BC	black carbon
BCIRA	British Cast Iron Research Association
BCL	brown coal liquefaction
BDCPM	brushless direct-current permanent magnet motors
BEDP	Bagasse Energy Development Programme (Mauritius)
BEE	Bureau of Energy Efficiency (India)
BEFS	Bioenergy and Food Security
BEOP	breakeven oil price
BEV	battery-powered electric vehicle
BGR	Bundesanstalt für Geowissenschaften und Rohstoffe (Federal Institute for Geosciences and Natural Resources – Germany)
BIGCC	biomass integrated gasification combined cycle
BII	biomass input index
BioCCS	bioenergy carbon capture and storage
BiPV	building-integrated photovoltaic
BMTC	Bangalore Metropolitan Transport Corporation
BOS	balance-of-systems
BP	British Petroleum
BPL	below-the-poverty line
BRIC	Brazil, Russia, India, and China
BRICS	Brazil, Russia, India, China, and South Africa
BRIMCS	Brazil, Russia, India, Mexico, China, and South Africa
BRT	bus(-based) rapid transit (system)
BTG	biomass to gasoline via methanol intermediate
BTG-RC-CCS	BTG plant with recycle synthesis design and CO_2 capture and storage

BTG-RC-V	BTG plant with recycle synthesis design and CO$_2$ venting	**CBTL2-OT-CCS**	CBTL1 plant with OT synthesis design and with CO$_2$ capture and storage
BTL	Biomass-to-Liquids, where liquids refers to Fischer-Tropsch fuels	**CBTL-OT-CCS**	CBTL plant with once-through synthesis design and with CO$_2$ capture and storage
BTL-RC-CCS	BTL plant with recycle synthesis design and CO$_2$ capture and storage	**CBTL-OT-V**	CBTL plant with once-through synthesis design and with CO$_2$ venting
BTL-RC-V	BTL plant with recycle synthesis design and CO$_2$ venting	**CBTL-RC-CCS**	CBTL plant with RC synthesis design and with CO$_2$ capture and storage
BUENAS	Bottom-up Energy Analysis System	**CBWP**	community-based woodfuel
BWR	boiling water reactor	**CC**	chilled ceiling
C	carbon	**CCGT**	combined cycle gas turbine
C$_2$F$_6$	Hexafluoroethane – CFC-116	**CCLP**	Chevron coal liquefaction
CAEP	Committee on Aviation Environmental Protection	**CCP**	CO$_2$ capture project
CAES	compressed-air energy system	**CCS**	carbon (dioxide) capture and storage
CAFE	corporate average fuel economy	**CDM**	clean development mechanism
CAGR	compound annual growth rate	**CEMAC**	Communauté Economique et Monétaire de l'Afrique Centrale (Economic and Monetary Community of Central Africa)
CANDU	Canada Deuterium Uranium reactor		
CAP	climate-active pollutant		
CARNFCHT	Committee on Assessment of Resource Needs for Fuel Cell and Hydrogen Technologies (United States)	**CEN**	European Committee for Standardization
		CENBIO	Centre Brasileiro de Referência em Biomass (Brazilian Reference Center on Biomass)
CARTA	Consortium for Advanced Research Training in Africa	**CENELEC**	The European Committee for Electrotechnical Standardization
CASBEE	Comprehensive Assessment System for Built Environment Efficiency	**CEPCI**	Chemical Engineering Plant Cost Index
		CER	certified emissions reduction (in the CDM)
CBIGCC-CCS	coal/biomass integrated gasification combined cycle with CO$_2$ capture and storage	**CERBIO**	Centro Brasileiro de Referência em Biocombustíveis (Brazilian Reference Center on Biofuels)
CBIGCC-V	coal/biomass integrated gasification combined cycle with CO$_2$ venting		
		CERM	coordinated emergency response mechanism
CBM	coal-bed methane	**CERN**	European Organization for Nuclear Research
CBTG	coal + biomass to gasoline	**CERTS**	Consortium for Electricity Reliability Solutions
CBTG1	coal + biomass to gasoline, with reduced biomass/coal input ratio (relative to CBTG)	**CF**	capacity factor
		CF$_4$	Tetrafluoromethane – CFC-14
CBTG1-PB-CCS	CBTG1 plant with partial bypass of syngas and with CO$_2$ capture and storage	**CFB**	circulating fluidized bed
		CFC	Chlorofluorocarbon
CBTG-PB-CCS	CBTG plant with partial bypass of syngas and with CO$_2$ capture and storage	**CFC-14**	Tetrafluoromethane – CF$_4$
		CFC-116	Hexafluoroethylene – C$_2$F$_6$
CBTG-RC-CCS	CBTG plant with recycle synthesis design and with CO$_2$ capture and storage	**CFL**	compact fluorescent lamps/light bulbs
		CGNPC	China Guangdong Nuclear Power Holding Company
CBTL	coal + biomass to liquids, where liquids refers to Fischer-Tropsch fuels		
		(CH$_2$)$_4$(COOH)$_2$	adipic acid
CBTL1	coal + biomass to FTL, with reduced biomass/coal input ratio (relative to CBTL)	**CH$_3$OH**	methanol
		CH$_4$	methane
CBTL1-OTA-CCS	CBTL1 plant with OT synthesis design and with aggressive CO$_2$ capture and storage	**CHP**	combined heat and power
		CIGCC	coal integrated gasification combined cycle
CBTL1-OT-CCS	CBTL1 plant with OT synthesis design and with CO$_2$ capture and storage	**CIGCC-CCS**	coal integrated gasification combined cycle with CO$_2$ capture and storage
CBTL2	coal + biomass to FTL, with reduced biomass/coal input ratio (relative to CBTL1)	**CIGCC-V**	coal integrated gasification combined cycle with CO$_2$ venting
CBTL2-OTA-CCS	CBTL plant with OT synthesis design and with aggressive CO$_2$ capture and storage	**CIGI**	Centre for International Governance Innovation
		CLE	current and planned legislation

CLEEN	Consortium for Continuous Low Energy, Emissions, and Noise (United States)	**CTL-RC-V**	CTL plant with RC synthesis island design and with CO_2 venting
ClO_2	Chlorine dioxide	**CTSL**	catalytic two-stage liquefaction
CMH	ceramic and metal halide	**DAC**	Development Assistance Committee (OECD)
CNG	compressed natural gas	**DAE**	Department of Atomic Energy (India)
CNNC	China National Nuclear Corporation	**DALY**	disability-adjusted life year
CO	carbon monoxide	**DBC**	divided blast cupola
CO_2	carbon dioxide	**DC**	direct current
CO_2-eq	carbon dioxide equivalent	**DCAP**	downstream plus upstream CO_2 capture
CO_2CRC	Cooperative Research Centre for Greenhouse Gas Technologies	**DCAP-1**	downstream plus upstream CO_2 capture (with reference to DME production from coal)
CODP	crude oil derived products	**DCAP-2**	more aggressive downstream plus upstream CO_2 capture (with DME production from coal)
COGEN	cogeneration		
Combi-Plant	Combined Power Plant (for renewable energy)	**DCAP-3**	most aggressive downstream plus upstream CO_2 capture (with DME production from coal)
COP	coefficient of performance		
COPD	chronic obstructive pulmonary disease	**DCC**	development cost charge
COPD	crude oil products displaced	**DCV**	demand-controlled ventilation
COS	carbonyl sulfide	**DECC**	Department of Energy and Climate Change (United Kingdom)
CPA	Centrally Planned Asia and China region		
CPRs	common pool resources	**DER**	distributed energy resources
CPWD	Central Public Works Department (India)	**DFID**	Department for International Development (United Kingdom)
CRA	Comparative Risk Assessment		
CREED	China Rural Energy Enterprise Development	**DG**	distributed generation
CSE	coast of saved energy	**DLR**	Deutsches Zentrum für Luft- und Raumfahrt (German Aerospace Center)
CSF	consol synthetic fuel		
CSI	Cement Sustainability Initiative	**DME**	Department of Minerals and Energy
CSIR	Council for Scientific and Industrial Research (South Africa)	**DME**	dimethyl ether
		DNI	direct normal irradiation
CSLF	Carbon Sequestration Leadership Forum	**DOAS**	dedicated outdoor air supply
CSM	Centre de Stockage de la Manche (a radioactive waste depository in France)	**DRI**	direct reduced iron
		DSM	demand-side management
CSP	concentrating/ed solar power	**DV**	displacement ventilation
CSS	cyclic steam stimulation	**E & P**	exploration and production
CTF	Clean Technology Fund	**EAF**	electric arc furnace
CTG	coal-to-gasoline via methanol intermediate	**EBRD**	European Bank for Reconstruction and Development
CTG-PB-CCS	coal-to-gasoline with partial bypass of syngas past methanol synthesis and with CO_2 capture and storage	**EC**	European Commission
		ECBI	European Capacity Building Initiative
		ECBM	enhanced coal bed methane
CTG-PB-V	coal-to-gasoline with partial bypass of syngas past methanol synthesis and with CO_2 venting	**ECOWAS**	Economic Community of West African States
		ÉDF	Électricité de France
		EDGAR	Emissions Database for Global Atmospheric Research
CTG-RC-CCS	coal-to-gasoline plant with recycle design for methanol synthesis and with CO_2 capture and storage		
		EDI	Energy Development Index
		EEA	European Environment Agency
CTL	coal-to-liquids, where liquids refers to Fischer-Tropsch fuels	**EEB**	energy efficiency in buildings
		EEDSM	Energy Efficiency and Demand-Side Management
CTL-OT-CCS	CTL plant with OT synthesis island design and with CO_2 capture and storage	**EES**	electric energy storage
		EESL	Energy Efficiency Services Limited
CTL-OT-V	CTL plant with OT synthesis island design and with CO_2 venting	**EEU**	Central and Eastern Europe region
		EGS	enhanced geothermal systems
CTL-RC-CCS	CTL plant with OT synthesis island design and with CO_2 capture and storage	**EI**	energy intensity

EIB	European Investment Bank	**FSR**	fast-spectrum reactor
EIFS	External Insulation and Finishing System	**FSU**	Former Soviet Union region
EITI	Extractive Industries Transparency Initiative	**F-T**	Fischer-Tropsch
EMBRAPA	Empresa Brasileira de Pesquisa Agropecuária	**FTL**	Fischer-Tropsch Liquids
EMF	Energy Modeling Forum	**G20**	Group of Twenty (international group of finance
ENDA-TM	Environment and Development Action in the		ministers and central bank governors from
	Third World (Senegal)		20 major economies)
EOR	enhanced oil recovery	**G7**	Group of Seven (international finance group
EPA	Environment Protection Agency		consisting of the finance ministers from seven
EPBD	Energy Performance of Buildings Directive		industrialized nations)
	(European Union)	**G8**	Group of Eight (forum for the governments of
EPC	energy performance contracting		eight of the world's largest economies)
EPFL	École Polytechnique Fédérale de Lausanne (Swiss	**GAEZ**	Global Agro-Ecological Zoning
	Federal Institute of Technology in Lausanne)	**GAINS**	Greenhouse Gas and Air Pollution Interactions
EPI	Energy Performance Indicator		and Synergies
EPR	European Pressurized Reactor	**GBTL**	natural gas + biomass-to-liquids, where liquids
EPRI	Electric Power Research Institute		refers to Fischer-Tropsch fuels
ESAP	Energy Sector Assistance Programme	**GBTL-OT-CCS**	GBTL plant with once-through synthesis design
ESCO	energy service company		and with CO_2 capture and storage
ESDP	Energy Service Delivery Project	**GBTL-OT-V**	GBTL plant with once-through synthesis design
ESP	electrostatic precipitator		and with CO_2 venting
ESPRIT	European Strategic Programme for	**GCC**	Gulf Cooperation Council
	Research and Development on Information	**GCCSI**	Global Carbon Capture and Storage Institute
	Technologies	**GCV**	gross calorific value
ETC	Enrichment Technology Company	**GDF**	Gaz de France
ETI	energy technology innovation	**GDP**	Gross Domestic Product
ETIS	Energy Technology Innovation System	**GEA**	Global Energy Assessment
EtOH	ethanol	**GECF**	Gas Exporting Countries Forum
ETS	emission trading scheme	**GEE**	global exposure equivalent
EU	European Union	**GEF**	Global Environment Facility
EUR	estimated ultimate recoverable oil	**genset**	engine-generator
EURODIF	European Gaseous Diffusion Uranium Enrichment	**GEO**	Global Environment Outlook
	Consortium	**GFAAF**	Global Framework for Aviation Alternative Fuels
EV	electric vehicle	**GHG**	greenhouse gas
EWEA	European Wind Energy Association	**GHGI**	greenhouse gas emission index
EWG	Energy Watch Group	**GIS**	geographic information system
EWM	excess winter mortality	**GLASOD**	Global Assessment of Human-induced Soil
FAO	Food and Agriculture Organization (of the		Degradation
	United Nations)	**GM**	General Motors
FBE	Free Basic Electricity (South Africa)	**GMT**	global mean surface temperature
FC	fuel cell	**GNEP**	Global Nuclear Energy Partnership
FCC	fluid catalytic cracker	**GNESD**	Global Network on Energy for Sustainable
FCV	fuel cell vehicles		Development
FDI	foreign direct investment	**GPT**	general purpose technologies
FE	fuel efficiency	**GPU**	ground power unit
FEED	Front End Engineering Design	**GRP**	gross regional product
FERC	Federal Energy Regulatory Commission	**GTAP**	Global Trade Analysis Project
FGD	flue gas desulphurization	**GTI**	Gas Technology Institute
FIREX	fast ignition experiment (Japan)	**gTKP**	Global Transport Knowledge Partnership
FIT	feed-in tariffs	**GTL**	(natural) gas-to-liquids, where liquids refers to
FLE	frozen legislation		Fischer-Tropsch fuels
FQD	Fuel Quality Directive	**GWEC**	Global Wind Energy Council

GWP	global warming potential
H	atomic hydrogen
H_2	molecular hydrogen
H_2O	water
H_2S	hydrogen sulfide
ha	hectare
HCFC	hydrochlorofluorocarbons (hydrogenated fluorocarbons)
HCO_3-	bicarbonate ion
HDD	heating degree day
HDI	Human Development Index
HDR	Human Development Report (UNDP)
HEU	highly (energy) enriched uranium (containing 20% or more 235U)
HEV	hybrid electric vehicle
HFC	hydrogen fuel cell
HFC-134a	CH_2FCF_3
HFCV	hydrogen fuel cell vehicles
HFO	heavy fuel oil
HG	mercury
hh	household
HHSFU	household solid fuel use
HHV	higher heating value
HIR	halogen infrared-reflecting
HIV/AIDS	Human Immunodeficiency Virus /Acquired Immune Deficiency Syndrome
HLW	high-level waste
HNO_3	nitric acid
HRJ	hydro-treated renewable jet
HRSG	heat recovery steam generator
HV	hybrid vehicles
HVAC	heating, ventilation, and air conditioning
HVDC	high voltage direct current
HWR	heavy-water reactor
HyRaMP	European Regions and Municipalities Partnership for Hydrogen and Fuel Cells
IAASTD	International Assessment of Agricultural Knowledge, Science and Technology for Development
IAEA	International Atomic Energy Agency
IAI	International Aluminium Institute
IARC	International Agency for Research on Cancer
IC	internal combustion
ICAO	International Civil Aviation Organization
ICE	internal combustion engine
ICEF	India-Canada Environment Facility
ICRISAT	International Crop Research Institute for the Semi-Arid Tropics
ICS	improved cook stove
ICT	information and communication technology
IDCOL	Infrastructure Development Company Limited

IDP	integrated design process
IDRC	International Development Research Centre (Canada)
IEA	International Energy Agency
IEA-OES-IA	IEA Ocean Energy Systems Implementing Agreement
IEA-PVPS	International Energy Agency-Photovoltaic Power Systems Programme
IEEE	Institute of Electrical and Electronics Engineers
IEF	International Energy Forum
IEP	International Energy Program
IFA	International Fertilizer Industry Association
IFES	Integrated food energy systems
IFPRI	International Food Policy Research Institute
IGCC	integrated (coal) gasification combined cycle
IHD	ischemic heart disease
IIASA	International Institute for Applied Systems Analysis
IIDP	Institute for Infectious Diseases of Poverty (Ghana)
IIE	Instituto de Investigaciones Eléctricas (Electric Research Institute – Mexico)
IISD	International Institute for Sustainable Development
ILO	International Labour Organization
ILUC	indirect land-use charge
IMAGE	Integrated Model to Assess the Global Environment
IMO	International Maritime Organization
IMP	Instituto Mexicano del Petróleo (Mexican Institute of Petroleum)
INMM	Institute for Nuclear Materials Management
INPO	Institute of Nuclear Power Operations (United States)
INPRO	International Project on Innovative Nuclear Reactors and Fuel Cycles (IAEA)
INSAG	International Nuclear Safety Advisory Group (IAEA)
I-O	input-output
IOC	international oil company
IOI	Initiative for Open Innovation
IP	intellectual property
IPAM	Instituto de Pesquisa Ambiental da Amazônia (Amazon Environmental Research Institute)
IPCC	Intergovernmental Panel on Climate Change
IPE	Interdepartmental Project directorate Energy transition
IPFM	International Panel on Fissile Materials
IPR	intellectual property right
IRENA	International Renewable Energy Agency
IRRE	internal rate of return on equity

IRSN	Institut de Radioprotection et de Sûreté Nucléaire (Institute of Radioprotection and Nuclear Safety – France)	**MALT**	mise à la terre
		MDC	minimum dispatch cost
		MDG	Millennium Development Goal
ISEW	Index of Sustainable Economic Welfare	**MDO**	marine diesel oil
ISG	integrated starter/generator	**MEA**	Middle East and North Africa region
ISL	in situ leaching	**MEA**	Millennium Ecosystem Assessment
ISO	Independent System Operators	**MEA**	monoethanolamine
ISO	International Organization for Standardization	**MEE**	Middle East
ISP	Integrated service provider	**MEM**	Microgrid energy management
ITER	international thermonuclear experimental reactor	**MEPS**	minimum efficiency performance standards
IUEC	International Uranium Enrichment Center	**MEPS**	minimal energy performance standards
IWMI	International Water Management Institute	**MER**	market exchange rates
JI	Joint Implementation	**MESSAGE**	Model for Energy Supply Strategy Alternatives and their General Environmental Impact
JNFL	Japan Nuclear Fuel Limited		
JODI	Joined Oil Data Initiative	**MeOH**	methanol
KAERI	Korea Atomic Energy Research Institute	**MFP**	multifunctional energy platform
KBS	Kenya Bureau of Standards	**MFR**	maximum feasible reduction
KCJ	Kenya Ceramic Jiko	**MH**	microhydropower
KiKK	Kinderkrebs in der Umgebung von Kernkraftwerken (Childhood Cancer in the Vicinity of Nuclear Power Plants – Germany)	**MHD**	magnetohydrodynamics
		MHP	micro-hydro plants
		MIC	middle-income country
LAC	Latin America and the Caribbean region	**MNRE**	Ministry of New and Renewable Energy
LbD	learning by doing	**MOE**	Ministry of the Environment
LBNL	Lawrence Berkeley National Laboratory	**MOX**	mixed oxide
LbU	learning by using	**MPI**	multidimensional poverty index
LCA	life cycle assessment	**MS**	modal shift
LCIA	life-cycle impact assessment	**MSA**	methane sulfonic acid
LCFS	low-carbon fuel standard	**MSW**	municipal solid waste
LCOE	levelized cost of electricity/energy	**MTBE**	methyl tert-butyl ether
LCOF	levelized cost of fuel	**MTG**	methanol to gasoline
LCOH	levelized cost of heat	**MURE**	Mesures d'Utilisation Rationelle de l'Énergie (Rational Use of Energy Measures)
LDC	least-developed country		
LDCF	Least Developed Country Fund (UNFCCC/GEF)	**MVA**	manufacturing value added
LDV	light-duty vehicle	**MVA**	monitoring, verification, and accounting
LED	light-emitting diode	**MVR**	mechanical vapor recompression
LEED	Leadership in Energy and Environmental Design	**N**	atomic nitrogen
LEU	low-enriched uranium (containing less than 20% 235U)	N_2	molecular nitrogen
		N_2O	nitrous oxide
LHV	lower heating value	**NAAQOs**	National Ambient Air Quality Objectives
LIC	low-income country	**NAAQS**	National Ambient Air Quality Standards
LNG	liquefied natural gas	**NAM**	North America region
LPG	liquefied petroleum gas (propane/butane mixture)	**NATCARB**	National Carbon Explorer
		NDRC	National Development and Reform Commission
LRT	light rapid transit		
LRTAP	long range transboundary air pollution	**NEA**	Nuclear Energy Agency (OECD)
LSE	liquid solvent extraction	**NEDO**	New Energy and Industrial Technology Development Organization
LUI	land-use intensity		
LVPCC	Latrobe Valley Post Combustion Capture Project	**NEF**	new energy finance
LWR	light-water reactor	**NELFA**	Natural Energy Laboratory of Hawaii Authority
LZC	low- and zero-carbon technologies	**NEPAD**	New Partnership for Africa's Development
MAGICC	Model for the Assessment of Greenhouse-gas Induced Climate Change	**NER**	net energy ratio
		NES	National Electrification Scheme

NESRP	Northern Electrification and System Reinforcement Project	**-OT-**	once-through (referring to design of synthesis island)
NETL	National Energy Technology Laboratory (United States)	**-OTA-**	once-through (referring to design of synthesis island) with aggressive CO_2 capture design
NFCRs	naturally fractured carbonate reservoirs		
NFPA	National Fire Protection Association	**OTEC**	ocean thermal energy conversion
NG	natural gas	**PACE**	property-assessed clean energy
NGCC	natural gas combined cycle	**PAF**	population attributable fraction
NGCC-CCS	natural gas combined cycle with CO_2 capture and storage	**PAH**	polycyclic aromatic hydrocarbon
		PAO	Pacific OECD region
NGCC-V	natural gas combined cycle with CO_2 vented	**PAS**	Other Pacific Asia region
NGL	natural gas liquids	**-PB-**	partial bypass (referring to design of methanol synthesis island)
NGO	non-governmental organization		
NH$_3$	ammonia	**PBS**	Palli Biddut Samities (rural electricity cooperatives in Bangladesh)
NIF	National Ignition Facility		
NiMH	nickel-metal hydride (batteries)	**PC**	personal computer
NMEEE	National Mission on Enhanced Energy Efficiency	**PC**	pulverized coal
NMT	non-motorized transport/transportation	**PCBs**	polychlorinated biphenyls
NMV	non-motorized vehicles	**PC-CCS**	pulverized coal plant with CO_2 capture and storage
NMVOC	non-methane volatile organic carbon		
NO	nitric oxide	**PCM**	phase change material
NO$_x$	nitrogen oxides (the sum of NO and NO_2) atomic oxygen	**PC-V**	pulverized coal plant with CO_2 venting
		PDF	probability density functions
NOC	national oil company	**PDVSA**	Petróleos de Venezuela Sociedad Anonima (Oils of Venezuela, the Venezuelan state-owned petroleum company)
NPC	National Petroleum Council		
NPP	net primary production		
NPP	nuclear power plant	**PE**	private equity
NPT	Treaty on the Non-Proliferation of Nuclear Weapons (Non-Proliferation Treaty)	**PEA**	Provincial Electricity Authority
		PEI	Princeton Environmental Institute
Nr	reactive nitrogen	**PEM**	polymer electrolyte membrane
NPV	net present value	**PERMER**	Proyecto de Energías Renovables en Mercados Rurales (Renewable Energy Project for the Rural Market – Argentina)
NRC	National Research Council		
NRES	new renewable energy sources		
NS	new stock		
NSG	Nuclear Suppliers Group	**Petrobas**	Petroleo Brasileiro S. A. (Brazilian energy corporation)
O$_2$	molecular oxygen		
O$_3$	ozone	**PFC**	Perfluorocarbon
O&M	operating and maintenance	**PFCHEV**	plug-in fuel cell hybrid electric vehicle
OC	organic carbon	**PFM**	public finance mechanism
ODA	official development assistance	**PHEV**	plug-in hybrid electric vehicle
ODS	ozone-depleting substances	**PIC**	products of incomplete combustion
ODI	Overseas Development Institute	**PISI**	gasoline port injection spark ignition engine
OECD	Organisation for Economic Co-operation and Development	**PM**	particulate matter
		PM$_{2.5}$	particles with an aerodynamic diameter < 2.5 μm
Ofgem	Office of Gas and Electricity Markets	**PM$_{10}$**	particles with an aerodynamic diameter < 10 μm
OLADE	Organizacion Latinoamericana de Energia (Latin America Energy Organization)	**PMU**	phasor measurement unit
		PNG	piped natural gas
OLED	organic light-emitting diode	**POP**	persistent organic pollutant
OPEC	Organization of the Petroleum Exporting Countries	**POX**	partial oxidation reforming
		ppm	parts per million
		ppbv	parts per billion by volume
OSA	Other South Asia region	**PPP**	purchasing power parities
OSART	Operation Safety Review Team (IAEA)	**PR**	performance ratio

PRGF	Partial Risk Guarantee Fund		SAS	South Asia region
PROALCOOL	National Alcohol Program (Brazil's first ethanol program)		SBS	Sick Building Syndrome
			SCCF	Special Climate Change Fund
PROCEL	Programa Nacional de Conservação de Energia Elétrica (National Electrical Energy Conservation Program – Brazil)		SCO	Shanghai Cooperation Organization
			SCR	selective catalytic reduction
			SD	standard deviation
PRSP	poverty-reduction strategy papers		SEFI	Sustainable Energy Finance Initiative
PSA	pressure swing absorption/absorbers		SEGS	solar energy generating system
PV	photovoltaic		SESAR	Single European Sky ATM Research
PWR	pressurized (light) water reactor		SEU	specific energy use
PYE	person year equivalents		SF_6	sulfur hexafluoride
R&C	residential and commercial		SFC	synthetic fuels corporation
R&D	research and development		SGE	supercritical gas extraction
R/P	reserves-to-production ratio		SHEP	Self-Help Electrification Project
RAR	reasonably assured resources		SHP	small hydropower
-RC-	recycle (referring to design of synthesis island)		SHS	solar home system
			SI	Système international d'unités (International System of Units)
RCSP	Regional Carbon Sequestration Partnerships			
RD&D	research, development, and deployment/ demonstration		SloCAT	Partnership for Sustainable Low Carbon Transport
			SLE	stringent air pollution, stringent climate, and universal energy access policies
RDD&D	research, development, demonstration, and deployment			
REA	Rural Electrification Agency		SME	small- and medium-sized enterprise
REC	Rural Electrification Corporation		SME	small and microenterprise
RED	Renewable Energy Directive		SME(S)	superconducting magnetic energy storage
REED	Rural Energy Enterprise Development		SMP	Sustainable Mobility Project (of the WBCSD)
REEEP	Renewable Energy and Energy Efficiency Partnership		SMR	steam methane reforming
			SNBI	Sustainable Net Benefit Index
REREDP	Rural Electrification and Renewable Energy Development Project (Bangladesh)		SNG	synthetic natural gas
			SO_2	sulfur dioxide
RES	renewable energy sources		SO_x	sulfur oxides
RF	radiative forcing		SOC	soil-organic carbon
RF	recoverable fraction		SODIS	solar water disinfection
RFS	renewable fuel standard		SOE	state-owned enterprises
RGGVY	Rajiv Gandhi Grameen Vidhyutikaran Yojana (program for rural electricity infrastructure and household electrification in India)		SOFC	solid oxide fuel cell
			Sofidif	Société franco-iranienne pour l'enrichissement de l'uranium par diffusion gazeuse (French-Iranian Society for the Enrichment of Uranium by Gas Diffusion)
RISI	Resource Information Systems Incorporated			
ROK	Republic of Korea			
RPR	residue to product ratio		SRES	Special Report on Emissions Scenarios (IPCC)
RPS	Renewable Portfolio Standards		SRMC	short run marginal cost of electricity
RR	risk rate		SSAC	state system accounting and control
RTE	repowering through experiment		SSTR	Steady State Tokamak Reactor
RTO	Regional Transmission Organizations		STE	solar thermal electricity
RVE	remote village electrification		STEG	Tunisian Electricity and Gas Company
S	atomic sulfur		sub PC	subcritical pulverized coal
S&L	standard and labeling		sup PC	supercritical pulverized coal
SADC	Southern African Development Community		sup PC-V	supercritical coal power plant that vents CO_2
SAGD	steam-assisted gravity drainage		SUTP	Sustainable Urban Transport Project
SAGSI	Standing Advisory Group on Safeguards Implementation (IAEA)		SUV	sports utility vehicles
			SWDI	Shannon-Wiener Diversity Index
SAPIERR	Support Action Pilot Initiative for European Regional Repositories (European Union)		SWH	solar water heaters/heating
			SWU	separative work unit

T&D	transport and distribution	US	United States
tce	tonne of coal equivalent	USGBC	United States Green Building Council
TDM	transportation demand management	USGS	United States Geological Survey
TECPAR	Paraná Technology Institute	UV	Ultraviolet
TEGs	thermal-electric generators	UV-B	ultraviolet-B radiation
TEPCO	Tokyo Electric Power Company	V2G	vehicle-to-grid
TERI	The Energy Research Institute	VAT	value added tax
TEU	twenty-foot equivalent units	VAV	variable air volume
TIA	Technology Innovation Agency	VC	venture capital
TM5	Global Chemistry-Transport Index	VCFEE	Venture Capital Fund for Energy Efficiency
TPC	total plant cost	VESP	Village Energy Security Programme
TPES	total primary energy supply	VMT	vehicle-miles traveled
TRC	Total Recordable Case Rate	VOC	volatile organic compound
TSP	total suspended particle/particulate matter	VPSA	vacuum pressure swing absorbers
TTO	technology transfer office	VVER	Vodo-Vodyanoi Energetichesky Reactor (Russian pressurized water reactor)
TTW	tank-to-wheel		
U	uranium metal	WANO	World Association of Nuclear Operators
U_3O_8	uranium oxide	WBCSD	World Business Council for Sustainable Development
UAE	United Arab Emirates		
UCAP	upstream CO_2 capture (with reference to plant designs for DME production from coal)	WBGU	Wissenschaftlicher Beirat der Bundesregierung Globale Umweltveränderungen (German Advisory Council for Global Environmental Change)
UCG	underground coal gasification		
UF_6	uranium hexafluoride		
UGS	underground (gas) storage	WCD	World Commission on Dams
UIC	underground injection control	WEA	World Energy Assessment
UITP	International Association of Public Transport	WEC	World Energy Council
UK	United Kingdom	WENRA	Western European Nuclear Regulators
UN	United Nations	WEO	World Energy Outlook
UNCTAD	United Nations Conference on Trade and Development	WEU	Western Europe region
		WGS	water gas shift
UNDESA	United Nations Department of Economic and Social Affairs	WHO	World Health Organization
		WINS	World Institute of Nuclear Security
UNDP	United Nations Development Programme	WISE	World Information Service on Nuclear Energy
UNECE	United Nations Economic Commission for Europe	WMD	weapon of mass destruction
UNEP	United Nations Environment Programme	WNA	World Nuclear Association
UNESCO	United Nations Educational, Scientific and Cultural Organization	WO PC-V	written-off pulverized coal plants and CO_2 venting
UNFCCC	United Framework Convention on Climate Change	WRI	World Resources Institute
		WSSD	World Summit on Sustainable Development
UNIDO	United Nations Industrial Development Organization	WTA	willingness to accept
		WTO	World Trade Organization
UNU-GTP	United Nations University Geothermal Training Programme	WTP	willingness to pay
		WTT	well-to-tank
UNU-MERIT	United Nations University – Maastricht Economic Research Institute on Innovation and Technology	WTW	well-to-wheel
		WUP	World Urbanization Prospects
URENCO	Uranium Enrichment Consortium	XTL-CCS	plants that co-produce electricity and Fischer-Tropsch liquids transportation fuels from X, where X is coal, biomass, or coal+biomass
US DOE	United States Department of Energy		
US EIA	United States Energy Information Administration		
US EPA	United States Environmental Protection Agency		
US NAS	United States National Academy of Sciences	YOLL	years of life lost
US NRC	United States Nuclear Regulatory Commission	ZEFI	zero-emission fuels index

Technical Guidelines: Common terms, definitions and units used in GEA

Technical Guidelines: Common terms, definitions and units used in GEA

This document provides technical guidelines for common terms, definitions, and units used in various chapters of the Global Energy Assessment (GEA). Additionally, it covers the proposed spatial resolution (definition of regional aggregates), base year values, units and definition of energy, and sectorial definitions.

Definition and aggregation of GEA Regions

Three different levels of regional aggregations are used in GEA, each of which serves for different purposes and levels of analysis.

1. **5 regions**: This aggregation is the common, consistent denominator across the different models proposed to be used in GEA. Thus, it represents the minimum level of spatial disaggregation for all analytical input/output data within GEA that involved modeling work. For instance, data for the GEA pathways are available systematically for the 5 GEA regions only. This is the highest (i.e., minimum) level of spatial disaggregation across all GEA chapters.

2. **11 Regions:** This is the spatial resolution available for all IIASA models used in GEA, which provided the backbone especially for the GEA scenario work for developing the pathways. Input data used in GEA modeling work (e.g., resource potentials, present energy use patterns, technology characteristics that are regionally different, etc.) have been provided at least at the level of these 11 world regions. The 11 regional

Table 1 | 5 Regions.

Regional acronyms	Regional definition
OECD90	UNFCCC Annex I countries
REF	Eastern Europe and Former Soviet Union
ASIA	Asia excl. OECD90 countries
MAF	Middle East and Africa
LAC	Latin America and the Caribbean

OECD90 = Includes the OECD90 countries, therefore encompassing the countries included below (11 region) in the regions WEU, NAM and PAO.

REF = Countries undergoing economic reform, i.e. countries listed under the regions EEU and FSU below (11 regions).

ASIA = The countries included in the regions SAS, PAS and CPA are aggregated into this region.

MAF = This region includes the Middle East and African countries that make up the regions AFR and MEA above.

LAC = This region includes the Latin American countries that make up the LAC region above.

aggregations represent just an approximate denominator across all models used within GEA. Data for 11 regions are available for specific scenarios and for a limited set of models only. The results of the GEA pathways are documented in detail at the interactive web-based GEA scenario database hosted by IIASA: www.iiasa.ac.at/web-apps/ene/geadb.

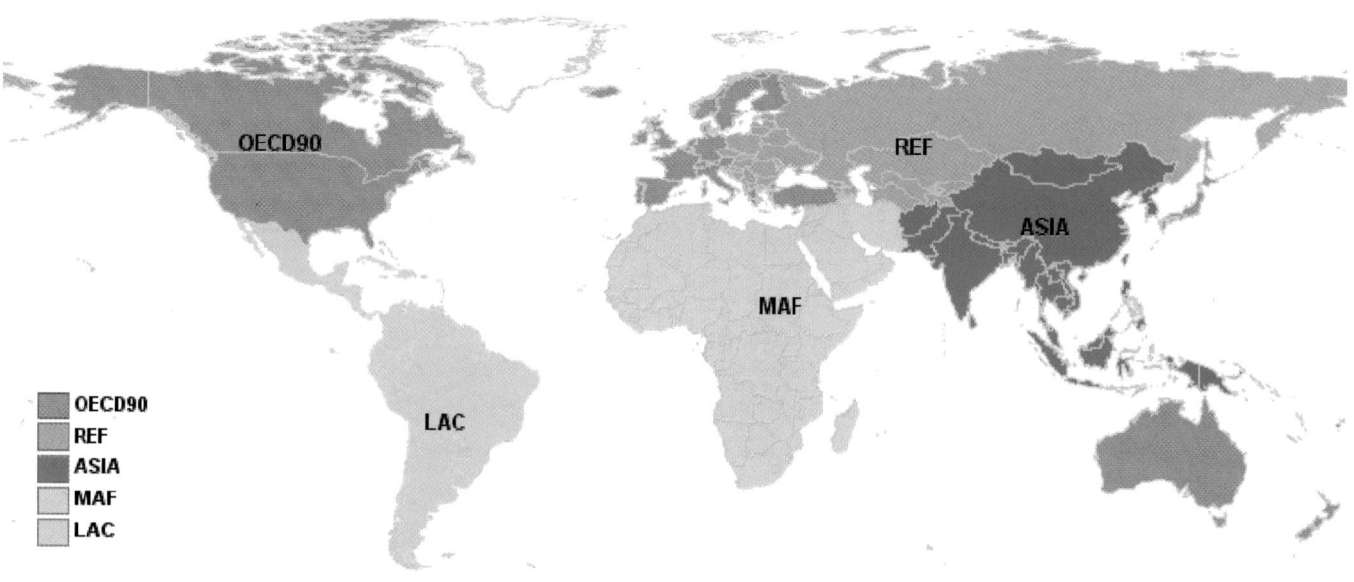

Figure 1 | Minimum GEA regional detail (5 regions).

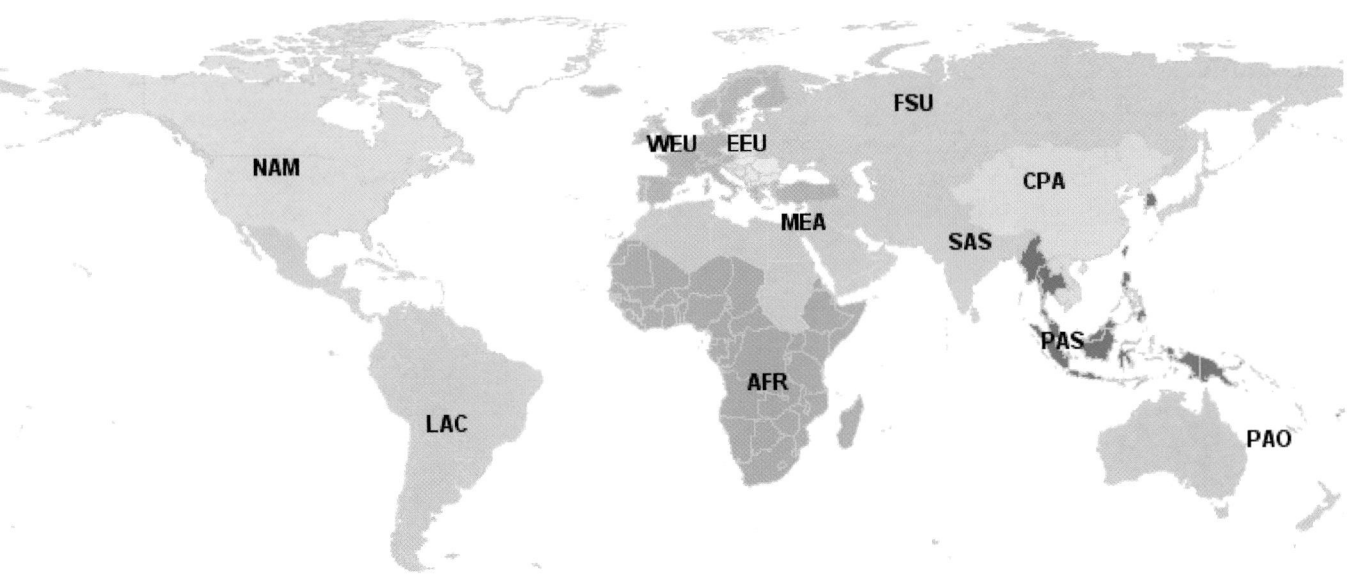

Figure 2 | GEA Scenario Model Regions (11 regions).

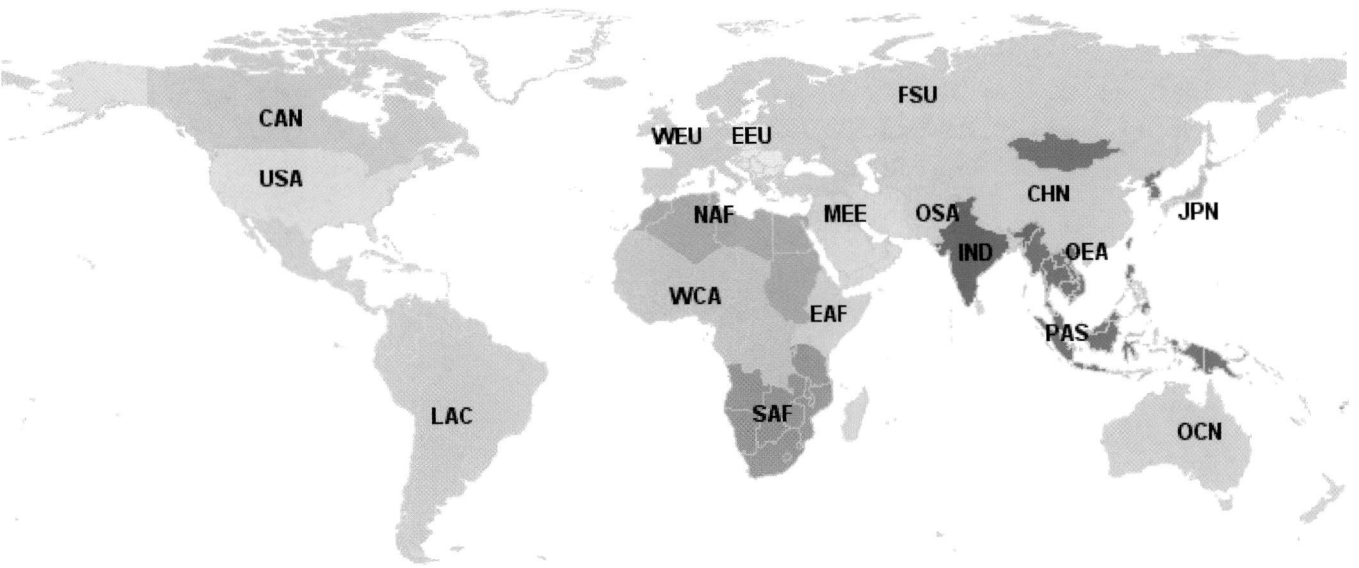

Figure 3 | Detailed GEA Regions (18 regions).

3. *18 regions*: This is the minimum spatial disaggregation needed to reconcile the regional definition differences across models used within GEA (e.g., including IMAGE models). This regional definition was used for all salient analytical work within GEA, e.g., for describing base year patterns of energy end-use and supply, energy access, resource potentials (fossil and renewable), technology deployment, policy implementation, etc. When available data did not permit the representation of all 18 regions, the 11 regions defined above was used, or alternatively at least the 5 GEA Regions. Base year (2005) proportional scaling techniques are suggested in cases where only 5-regional input data are available but finer regional detail is required.

Base Year: 2005

A common GEA base year is 2005. This is the year for which most critical energy statistics were available in time for GEA's analytical work. More recent data where provided wherever available.

Table 2 | 11 Regions.

Regional acronyms	Regional definition
NAM	North America
WEU	Western Europe
PAO	Pacific OECD
EEU	Central and Eastern Europe
FSU	Former Soviet Union
CPA	Centrally Planned Asia and China
SAS	South Asia
PAS	Other Pacific Asia
MEA	Middle East and North Africa
LAC	Latin America and the Caribbean
AFR	Sub-Saharan Africa

NAM = North America (Canada, Guam, Puerto Rico, United States of America, British Virgin Islands)

WEU = Western Europe (Andorra, Austria, Belgium, Cyprus, Denmark, Finland, France, Germany, Greece, Greenland, Holy See, Iceland, Ireland, Italy, Liechtenstein, Luxembourg, Malta, Monaco, Netherlands, Norway, Portugal, San Marino, Spain, Sweden, Switzerland, Turkey, United Kingdom)

PAO = Pacific OECD (Australia, Japan, New Zealand)

EEU = Central and Eastern Europe (Albania, Bosnia and Herzegovina, Bulgaria, Croatia, Czech Republic, Estonia, The former Yugoslav Rep. of Macedonia, Hungary, Latvia, Lithuania, Montenegro, Poland, Romania, Serbia, Slovak Republic, Slovenia)

FSU = Newly independent states of the former Soviet Union (Armenia, Azerbaijan, Belarus, Georgia, Kazakhstan, Kyrgyzstan, Republic of Moldova, Russian Federation, Tajikistan, Turkmenistan, Ukraine, Uzbekistan)

CPA = Centrally planned Asia and China (Cambodia, China (incl. Hong Kong and Macao), Korea (DPR), Laos (PDR), Mongolia, Viet Nam)

SAS = South Asia (Afghanistan, Bangladesh, Bhutan, India, Maldives, Nepal, Pakistan, Sri Lanka)

PAS = Other Pacific Asia (American Samoa, Brunei Darussalam, Fiji, French Polynesia, Kiribati, Indonesia, Malaysia, Marshall Islands, Micronesia, Myanmar, Nauru, New Caledonia, Palau, Papua, New Guinea, Philippines, Republic of Korea, Samoa, Singapore, Solomon Islands, Taiwan (China), Thailand, Timor-Leste, Tonga, Tuvalu, Vanuatu)

MEA = Middle East and North Africa (Algeria, Bahrain, Egypt (Arab Republic), Iraq, Iran (Islamic Republic), Israel, Jordan, Kuwait, Lebanon, Libya/SPLAJ, Morocco, Occupied Palestine Territory, Oman, Qatar, Saudi Arabia, South Sudan, Syria (Arab Republic), Tunisia, United Arab Emirates, Western Sahara, Yemen)

LAC = Latin America and the Caribbean (Antigua and Barbuda, Argentina, Bahamas, Barbados, Belize, Bermuda, Bolivia, Brazil, Chile, Colombia, Costa Rica, Cuba, Dominica, Dominican Republic, Ecuador, El Salvador, French Guiana, Grenada, Guadeloupe, Guatemala, Guyana, Haiti, Honduras, Jamaica, Miartinique, Mexico, Netherlands Antilles, Nicaragua, Panama, Paraguay, Peru, Saint Kitts and Nevis, Santa Lucia, Saint Vincent and the Grenadines, Suriname, Trinidad and Tobago, Uruguay, Venezuela)

AFR = Sub-Saharan Africa (Angola, Benin, Botswana, Burkina Faso, Burundi, Cameroon, Cape Verde, Central African Republic, Chad, Comoros, Cote d'Ivoire, Congo (DR), Djibouti, Equatorial Guinea, Eritrea, Ethiopia, Gabon, Gambia, Ghana, Guinea, Guinea-Bissau, Kenya, Lesotho, Liberia, Madagascar, Malawi, Mali, Mauritania, Mauritius, Mozambique, Namibia, Niger, Nigeria, Reunion, Rwanda, Sao Tome and Principe, Senegal, Seychelles, Sierra Leone, Somalia, South Africa, Saint Helena, Swaziland, Tanzania, Togo, Uganda, Zambia, Zimbabwe)

Table 3 | 18 Regions.

Regional acronyms	Regional definition
USA	United States of America
CAN	Canada
WEU	Western Europe, incl. Turkey
EEU	Central and Eastern Europe
FSU	Former Soviet Union
NAF	Northern Africa
EAF	Eastern Africa
WCA	Western and Central Africa
SAF	Southern Africa
MEE	Middle East
CHN	China
OEA	Other East Asia
IND	India
OSA	Other South Asia
JPN	Japan
PAS	Other Pacific Asia
OCN	Australia, New Zealand, and other Oceania
LAC	Latin America and the Caribbean

USA = Unites States (Guam, Puerto Rico, United States of America, British Virgin Islands)

CAN = Canada

WEU = Western Europe (Andorra, Austria, Belgium, Cyprus, Denmark, Finland, France, Germany, Greece, Greenland, Holy See, Iceland, Ireland, Italy, Liechtenstein, Luxembourg, Malta, Monaco, Netherlands, Norway, Portugal, San Marino, Spain, Sweden, Switzerland, Turkey, United Kingdom)

EEU = Central and Eastern Europe (Albania, Bosnia and Herzegovina, Bulgaria, Croatia, Czech Republic, Estonia, The former Yugoslav Rep. of Macedonia, Hungary, Latvia, Lithuania, Montenegro, Poland, Romania, Serbia, Slovak Republic, Slovenia)

FSU = Newly independent states of the former Soviet Union (Armenia, Azerbaijan, Belarus, Georgia, Kazakhstan, Kyrgyzstan, Republic of Moldova, Russian Federation, Tajikistan, Turkmenistan, Ukraine, Uzbekistan)

NAF = North Africa (Algeria, Egypt (Arab Republic), Libya/SPLAJ, Morocco, Sudan, Tunisia)

EAF = Eastern Africa (Burundi, Eritrea, Ethiopia, Kenya, Madagascar, Mauritius, Seychelles, Somalia, Uganda)

WCA = Western and Central Africa (Benin, Burkina Faso, Cameroon, Cape Verde, Central African Republic, Chad, Comoros, Cote d'Ivoire, Congo (DR), Djibouti, Equatorial Guinea, Gabon, Gambia, Ghana, Guinea, Guinea-Bissau, Liberia, Mali, Mauritania, Niger, Nigeria, Sao Tome and Principe, Senegal, Sierra Leone, Togo)

SAF = Southern Africa (Angola, Botswana, Burundi, Malawi, Mozambique, Namibia, Reunion, Rwanda, Saint Helena, South Africa, Swaziland, Tanzania, Zambia, Zimbabwe)

MEE = Middle East (Bahrain, Iraq, Iran (Islamic Republic), Israel, Jordan, Kuwait, Lebanon, Occupied Palestine Territory, Oman, Qatar, Saudi Arabia, Syria (Arab Republic), United Arab Emirates, Western Sahara, Yemen)

CHN = China (incl. Hong Kong and Macao)

OEA = Other East Asia (Cambodia, Korea (DPR), Laos (PDR), Mongolia, Viet Nam)

Economic data are expressed in 2005 US$. GDP at constant (2005) prices in US$ was calculated using the following equations:

$$\frac{\text{GDP at constant (2005)}}{\text{price in national currency}_i(t)} = \frac{\text{GDP at current price}_i(t)}{\dfrac{\text{GDP deflator}_i(t)}{\text{GDP deflator}_i(2005)}}$$

$$\frac{\text{GDP at constant (2005)}}{\text{price in US\$}_i(t)} = \frac{\dfrac{\text{GDP in constant (2005)}}{\text{price in national currency}_i(t)}}{\text{XRATE}_i(2005)}$$

Where, i = country indicator, t = year indicator, $\text{XRATE}_i(2005)$ = annual average market exchange rate in 2005 [Local currency / US$].

For example, 272 [billion 2007 Euro] GDP in Austria can be converted in 326 [billion 2005 US$] as follows:

$$262\ [2005\ \text{Euro}] = \frac{272\ [2007\ \text{Euro}]}{\left(\dfrac{112.93\ [\text{GDP deflator (2007)}]}{108.51\ [\text{GDP deflator (2005)}]}\right)}$$

$$326\ [2005\ \text{US\$}] = \frac{262\ [2005\ \text{Euro}]}{(0.803\ [\text{XRATE (2005)}])}$$

All relevant data series for each country is from the World Economic Outlook (WEO) database by IMF at: www.imf.org/external/data.htm.

Depending on the specific context, economic indicators were presented in purchasing power parity (PPP) or market exchange rates (MER). In either case, the choice of PPP or MER should be clearly stated in the text and figures. The equations above are given for market exchange rates (XRATE). For using PPPs, above XRATE needs to be replaced by the following term:

Table 3 | (cont.)

IND = India

OSA = Other South Asia (Afghanistan, Bangladesh, Bhutan, Maldives, Nepal, Pakistan, Sri Lanka)

JPN = Japan

OCN = Oceania (Australia, New Zealand)

PAS = Other Pacific Asia (American Samoa, Brunei Darussalam, Fiji, French Polynesia, Gilbert-Kiribati, Indonesia, Malaysia, Marshall Islands, Micronesia, Myanmar, Nauru, New Caledonia, Palau, Papua, New Guinea, Philippines, Republic of Korea, Singapore, Solomon Islands, Taiwan (China), Thailand, Timor-Leste, Tonga, Tuvalu, Vanuatu, Samoa)

LAC = Latin America and the Caribbean (Antigua and Barbuda, Argentina, Bahamas, Barbados, Belize, Bermuda, Bolivia, Brazil, Chile, Colombia, Costa Rica, Cuba, Dominica, Dominican Republic, Ecuador, El Salvador, French Guiana, Grenada, Guadeloupe, Guatemala, Guyana, Haiti, Honduras, Jamaica, Martinique, Mexico, Netherlands Antilles, Nicaragua, Panama, Paraguay, Peru, Saint Kitts and Nevis, Santa Lucia, Saint Vincent and the Grenadines, Suriname, Trinidad and Tobago, Uruguay, Venezuela)

Table 4 | Sample XRATE for major currencies.

Local Currency	XRATE [Local Currency / US$] in 2005
Euro	0.803
British Pound	0.549
Japanese Yen	108.9
Chinese Yuan	8.195
Canadian Dollar	1.211
Australian Dollar	1.309

$$\frac{\text{GDP at local currency } (t = 2005)}{\text{GDP-PPP in nternation \$ } (t = 2005)}$$

The relevant statistics are available from the World Development Indicators (World Bank, 2007).

Discount Rate

The discount rate represents the annual rate at which the effects of future events are reduced so as to be comparable to the effect of present events. Assumptions in regard to the discount rate adopted have important implications when considering medium- to long-term scenarios. However, there is no rate universally accepted since it is laden with value judgment. For practical reasons, GEA adopted a uniform rate of 5% throughout. Other rates are used in specific cases (e.g. from the literature) whenever possible with a clear reference to the rate used is made in such instances.

Energy Flows (Joules)

Both *final* and *primary energy* use levels are reported systematically in GEA using SI units (i.e., Joules or multiples thereof, and to allow for additional units, e.g., kWh if so desired, but always keeping the Joules common denominator). Often alternative units from original sources in the literature are shown in addition to the SI units. For the sectorial end-use chapters, only final energy is reported to avoid the risk of double counting.

Table 5 | Units based on SI.

Physical quantity	Unit	Symbol
Length	meter	m
Mass	kilogram	kg
Time	second	s
Thermodynamic temperature	Kelvin	K
Energy	Joule	J
Power	Watt	W
Temperature	Degree Celsius	°C

Table 6 | Selected SI prefix.

Prefix	Symbol	10^n
exa-	E	10^{18}
peta	P	10^{15}
tera	T	10^{12}
giga	G	10^9
mega	M	10^6
kilo	k	10^3
hector	h	10^2
deca	da	10
-	-	10^0
deci	d	10^{-1}
centi	c	10^{-2}
milli	m	10^{-3}
micro	μ	10^{-6}
nano	n	10^{-9}
pico	p	10^{-12}
femto	f	10^{-15}

Accounting of Primary Energy

A consistent methodology was used for converting non-combustible and non-fossil energy (e.g., renewable and nuclear energy for electricity generation) to primary energy equivalent across the assessment. A widely accepted method for the accounting of primary energy from these sources does not exist. Studies in the past have either used the substitution equivalent method, assuming specific efficiencies for renewable sources, or they have applied the direct equivalent method, which accounts the energy output (e.g., electricity) of above energy sources as their primary energy. The SRES (IPCC, 2000) have used the direct equivalent method, while e.g., Nakicenovic et al. (1998) and the IPCC (2007) have applied uniform conversion efficiencies across different sources (38.6% and 33%, respectively). Other assessments, such as the World Energy Assessment (UNDP, 2004) or the World Energy Outlook (IEA, 2007a), combine both methods, and assume different conversion factors for different conversion technologies (in the WEA, for instance, nuclear: 3, hydro, wind and solar: 1, geothermal: 10).

It was decided to keep the accounting methodology transparent and as simple as possible in GEA given that a widely accepted and consistent

method across different studies does not exist. As a guide for the selection of the conversion efficiency, a global average efficiency of electricity and heat generation for the year 2005 was used. This corresponds to an average efficiency of 35% for electricity generation, and 85% for heat generation (IEA, 2007b). For example, 1[EJ] of electricity generated by wind or nuclear at the secondary energy level is accounted as 2.86 [EJ] (i.e., 1/0.35) at the primary energy level. Likewise, 1[EJ] of heat generated (and used as heat) by solar or geothermal is accounted as 1.17 [EJ] at the primary energy level (i.e., 1/0.85). See Chapter 1 for further discussion of primary energy accounting methods.

Accounting of Heating Values

To maintain clarity and consistency throughout the report, the heating value of a substance, i.e., the amount of heat released during combustion of a specified amount, is expressed in J/kg.

Heating values are clearly defined in the text as either HHV (Higher Heating Value) or LHV (Lower Heating Value). It is assumed that heating values are given as LHV unless explicitly noted in the text as HHV. The difference is that HHV includes the energy of condensation of the water vapor contained in the combustion products.

Sectorial Definitions

The reporting and analysis within GEA's sectorial chapters adheres strictly to the principle of *mutual exclusiveness.*

Therefore, efficiency improvement (or carbon reduction) potentials for end-use sectors (e.g., industry) are always strictly separated from corresponding "upstream" energy systems/sectors (e.g., electricity generation). For example, emissions reductions (e.g., from the electricity generation sector) at the supply side need to be discussed/accounted in the "upstream" GEA chapter, while the end-use demand chapters should report emission/energy reduction potentials due to fuel substitution and energy conservation in that end-use sector only. Thus, sectorial assessments were not "inflated" by accounting for (often unrelated) "upstream" energy systems changes that should be reported separately. The GEA energy modeling framework enabled a rigorous and consistent accounting across sectors and the impacts of policy measures at the sector level.

References

IEA, 2007a: *World Energy Outlook 2007: China and India Insights*. International Energy Agency (IEA), Paris.

IEA, 2007b: *Energy Balances, 2007 Edition*. International Energy Agency (IEA), Paris.

IPCC, 2000: *Emissions Scenarios*. Special Report of Working Group III of the International Panel on Climate Change (IPCC), Cambridge University Press, Cambridge.

IPCC, 2007: *Fourth Assessment Report: Climate Change 2007*. International Panel on Climate Change (IPCC), Cambridge University Press, Cambridge.

Nakicenovic, N., A. Grubler and A. McDonald (eds.), 1998: *Global Energy Perspectives*. International Institute for Applied System Analysis (IIASA), World Energy Council.

UNDP, 2004: *World Energy Assessment: 2004 Update*. United Nations Development Programme (UNDP), United Nations Department of Economic and Social Affairs (UNDESA), World Energy Council.

Contributors to the Global Energy Assessment

Contributors to the Global Energy Assessment

Jean Acquatella
Economic Commission for Latin America and the Caribbean

Adeola Adenikinju
University of Ibadan, Nigeria

Lawrence Agbemabiese
United Nations Environment Programme

Olivia Agbenyega
Kwame Nkrumah University of Science and Technology, Ghana

Astrid Agostini
Food and Agriculture Organization, Italy

Francisco Aguayo
El Colegio de México

Roberto F. Aguilera
Curtin University, Australia

Gilbert Ahamer
University of Graz, Austria

John Ahearne
Sigma Xi, USA

Hugo Altomonte
Economic Commission for Latin America and the Caribbean

Markus Amann
International Institute for Applied Systems Analysis, Austria

Laura Diaz Anadon
Harvard University, USA

Per Dannemand Andersen
Technical University of Denmark

Cristina L. Archer
California State University and Stanford University, USA

Doug J. Arent
National Renewable Energy Laboratory, USA

Robert Ayres
European Institute of Business Administration, France

Christian Azar
Chalmers University of Technology, Sweden

Ines Azevedo
Carnegie Mellon University, USA

Xuemei Bai
Australian National University

Kalpana Balakrishnan
Sri Ramachandra University, India

Rangan Banerjee
Indian Institute of Technology-Bombay

Douglas F. Barnes
Energy for Development, USA

Jennie Barron
Stockholm Environment Institute, University of York, UK

Igor Bashmakov
Center for Energy Efficiency, Russia

Timothy Baynes
Commonwealth Scientific and Industrial Research Organisation, Australia

Morgan Bazilian
United Nations Industrial Development Organization

Kamel Bennaceur
Schlumberger, France

Sally M. Benson
Stanford University, USA

Ruggero Bertani
Enel Green Power S.p.A., Italy

S.C. Bhattacharya
International Energy Initiative, India

Dan Bilello
National Renewable Energy Laboratory, USA

Gunilla Björklund
Uppsala University, Sweden

Brenda Boardman
University of Oxford, UK

Daniel H. Bouille
Bariloche Foundation, Argentina

Grant Boyle
McCarthy Tétrault LLP, Canada

Sylvia Breukers
Duneworks, the Netherlands

Abeeku Brew-Hammond
Kwame Nkrumah University of Science and Technology, Ghana

Ian Bryden
University of Edinburgh, UK

Thomas Buettner
United Nations Department of Economic and Social Affairs

Stan Bull
National Renewable Energy Laboratory, USA

Matthew Bunn
Harvard University, USA

Colin Butler
Australian National University

Zoë Chafe
University of California, Berkeley, USA

Aleh Cherp
Central European University, Hungary

Helena Chum
National Renewable Energy Laboratory, USA

Leon Clarke
University of Maryland, USA

Suani T. Coelho
National Reference Center on Biomass, University of São Paulo, Brazil

Yu Cong
Energy Research Institute, China

Peter Cook
Cooperative Research Centre for Greenhouse Gas Technologies, Australia

Robert Corell
Global Environment Technology Foundation, USA

Felix Creutzig
Technical University of Berlin, Germany

Daniel Curtis
Oxford University Centre for the Environment, UK

Touria Dafrallah
Environment and Development Action in the Third World, Senegal

Ogunlade R. Davidson
Ministry of Energy, Sierra Leone

John Davison
IEA Greenhouse Gas R&D Programme, UK

Felix Dayo
Triple "E" Systems Inc., USA

Heleen de Coninck
Energy Research Centre of the Netherlands

Luiz Alberto de Melo Brettas
Brazilian National Civil Aviation Agency

Adilson de Oliveira
Federal University of Rio de Janeiro, Brazil

Gabriel de Scheemaker
Conduit Ventures Ltd., UK

Paulo Teixeira de Sousa Jr.
Federal University Mato Grosso, Brazil

Frank Dentener
Joint Research Centre, Italy

Shobhakar Dhakal
Global Carbon Project and National Institute for Environmental Studies, Japan

Anatoli Diakov
Moscow Institute of Physics and Technology, Russia

Ming Ding
Delft University, the Netherlands

Michael Doherty
Commonwealth Scientific and Industrial Research Organisation, Australia

Anne-Maree Dowd
Commonwealth Scientific and Industrial Research Organisation, Australia

Carolina Dubeux
Federal University of Rio de Janeiro, Brazil

Maurice B. Dusseault
University of Waterloo, Canada

Lisa Emberson
Stockholm Environment Institute and University of York, UK

Karl-Heinz Erb
Klagenfurt University, Austria

Nick Eyre
Oxford University, UK

Andre Faaij
Utrecht University, Netherlands

Ian Fairlie
Consultant on Radiation in the Environment, UK

Karim Farhat
Stanford University, USA

Sara Feresu
University of Zimbabwe

Maria Josefina Figueroa
Technical University of Denmark

Carolyn Fischer
Resources for the Future, USA

Brian Fisher
BAEconomics, Australia

David J. Fisk
Imperial College London, UK

Theo H. Fleisch
BP America (retired), USA

Tira Foran
Commonwealth Scientific Industrial Research Organisation, Australia

Roger Fouquet
Basque Centre for Climate Change, Spain

Junichi Fujino
National Institute for Environmental Studies, Japan

Sabine Fuss
International Institute of Applied Systems Analysis, Austria

Luc Gagnon
HydroQuébec, Canada

Kelly Gallagher
Tufts University, USA

Hu Gao
Energy Research Institute, China

Ibrahim Abdel Gelil
Arabian Gulf University, Bahrain

Dolf Gielen
United Nations Industrial Development Organization

Asmerom Gilau
Triple "E" Systems Inc., USA

Stephen Gitonga
United Nations Development Programme

Robert Goldston
Princeton Plasma Physics Laboratory, USA

Andreas Goldthau
Central European University, Hungary

Luis Gomez-Echeverri
International Institute for Applied Systems
Analysis, Austria

Peter Graham
University of New South Wales, Australia

Arnulf Grubler
International Institute for Applied Systems
Analysis, Austria and Yale University, USA

Helmut Haberl
Klagenfurt University, Austria

Richard Haeuber
United States Environmental Protection
Agency

Keisuke Hanaki
University of Tokyo, Japan

Maureen Hand
National Renewable Energy Laboratory, USA

Danny Harvey
University of Toronto, Canada

Marianne Haug
University of Hohenheim, Germany

Kebin He
Tsinghua University, China

Marko Hekkert
Utrecht University, the Netherlands

Francisco Hernandez
Lund University, Sweden

Sergio Tirado Herrero
Central European University, Hungary

Edgar Hertwich
Norwegian University of Science and
Technology

Conrado Heruela
United Nations Environment Programme

Kevin Hicks
Stockholm Environment Institute,
University of York, UK

Monique Hoogwijk
Ecofys, the Netherlands

Richard Hosier
World Bank, USA

Alison Hughes
University of Cape Town, South Africa

Larry Hughes
Dalhousie University, Canada

Jane Hupe
International Civil Aviation Organization,
Canada

Toshiaki Ichinose
National Institute for Environmental Studies,
Japan

Morna Isaac
PBL Netherlands Environmental Assessment
Agency, the Netherlands

Mark Jaccard
Simon Fraser University, Canada

Staffan Jacobsson
Chalmers University of Technology, Sweden

Jill Jäger
Sustainable Europe Research Institute,
Austria

Martin Jakob
Swiss Federal Institute of Technology Zurich

Kathryn Janda
Oxford University, UK

Gilberto Jannuzzi
University of Campinas, Brazil

Jaap Jansen
Energy Research Centre of the Netherlands

Jessica Jewell
Central European University, Hungary

Kejun Jiang
Energy Research Institute, China

Yi Jiang
Tsinghua University, China

Eberhard Jochem
Fraunhofer Institute for Systems and
Innovation Research, Germany

Thomas B. Johansson
Lund University, Sweden

Francis X. Johnson
Stockholm Environment Institute, Stockholm
University, Sweden

Arthur Johnson
Hydrate Energy International, USA

Ian Johnson
Club of Rome, Switzerland

Suzana Kahn Ribeiro
Federal University of Rio de Janeiro,
Brazil

Mikiko Kainuma
National Institute for Environmental Studies, Japan

Daniel Kammen
University of California, Berkeley, USA

Shinji Kaneko
Hiroshima University, Japan

Stephen Karekezi
AFREPREN/FWD, Kenya

Anders Karlqvist
Swedish Polar Research Secretariat

Tadahiro Katsuta
Meiji University, Japan

James E. Keirstead
Imperial College London, UK

Francis Kemausuor
Kwame Nkrumah University of Science and Technology, Ghana

René Kemp
United Nations University-MERIT, the Netherlands

Ruud Kempener
Harvard University, USA

John Kimani
AFREPREN/FWD, Kenya

Osamu Kimura
Central Research Institute of Electric Power Industry, Japan

Patrick Kinney
Columbia University, USA

Bernadette Kiss
Lund University, Sweden

Tord Kjellstrom
Umea University, Sweden

Zbigniew Klimont
International Institute for Applied Systems Analysis, Austria

Shigeki Kobayashi
Toyota Central R&D Laboratories, Japan

Peter Kolp
International Institute for Applied Systems Analysis, Austria

Christian Kornevall
World Business Council for Sustainable Development, Switzerland

Reza Kowsari
University of British Columbia, Canada

Diana Kraft
REN21, France

Fridolin Krausmann
Klagenfurt University, Austria

Wolfram Krewitt†
German Air and Space Agency

Volker Krey
International Institute for Applied Systems Analysis, Austria

Sivanappan Kumar
Asian Institute of Technology, Thailand

Rattan Lal
The Ohio State University, USA

Eric D. Larson
Princeton University and Climate Central, USA

Hans Larsen
Technical University of Denmark

Rik Leemans
Wageningen University, the Netherlands

Sylvie Lemmet
United Nations Environment Programme

Philippe Lempp
German Development Ministry, Germany

Manfred Lenzen
University of Sydney, Australia

Zheng Li
Tsinghua University, China

Vladimir Likhachev
Russian Academy of Sciences

Guangjian Liu
North China Electric Power University

Jeff Logan
National Renewable Energy Laboratory, USA

Oswaldo Lucon
São Paulo State Environment Agency, Brazil

John Lund
Geo-Heat Center, Oregon Institute of Technology, USA

Nora Lustig
Tulane University, USA

Jordan Macknick
National Renewable Energy Laboratory, USA

Mili Majumdar
The Energy and Resources Institute, India

François Maréchal
Swiss Federal Institute of Technology Lausanne, Switzerland

Omar Masera
National Autonomous University of Mexico

Denise L. Mauzerall
Princeton University, USA

Peter McCabe
Commonwealth Scientific and Industrial
Research Organisation, Australia

David McCollum
University of California, Davis, USA

Charles McCombie
Independent Consultant

Susan McDade
United Nations Development Programme

Aimee T. McKane
Lawrence Berkeley National Laboratory, USA

Thomas McKone
Lawrence Berkeley National Laboratory, USA

James E. McMahon
Lawrence Berkeley National Laboratory, USA

Anthony McMichael
Australian National University

Michael McNeil
Lawrence Berkeley National Laboratory, USA

Mark Mehos
National Renewable Energy Laboratory, USA

Tim Merrigan
National Renewable Energy Laboratory, USA

Jacqui Meyers
Commonwealth Scientific and Industrial
Research Organisation, Australia

Alan Miller
International Finance Corporation, USA

Sevastianos Mirasgedis
National Observatory of Athens, Greece

Catherine Mitchell
University of Exeter, UK

Vijay Modi
Columbia University, USA

Joachim Monkelbaan
International Centre for Trade and
Sustainable Development, Switzerland

José Roberto Moreira
Biomass Users Network, Brazil

Granger Morgan
Carnegie Mellon University, USA

Siwa Msangi
International Food Policy Research Institute,
USA

Adrian Muller
University of Zurich, Switzerland and Swiss
Federal Institute of Technology Zurich

Mohan Munasinghe
Munasinghe Institute for Development, Sri
Lanka and University of Manchester, UK

Luis Mundaca
Lund University, Sweden

Shuzo Murakami
Keio University, Japan

Iyngararasan Mylvakanam
United Nations Environment Programme

Lynn K. Mytelka
United Nations University-MERIT, the
Netherlands

Yu Nagai
Vienna University of Technology, Austria

Koji Nagano
Central Research Institute of Electric Power
Industry, Japan

Hitomi Nakanishi
University of Canberra, Australia

Nebojsa Nakicenovic
International Institute for Applied
Systems Analysis and Vienna University of
Technology, Austria

Lena Neij
Lund University, Sweden

Gregory Nemet
University of Wisconsin, USA

George L. Nicolaides
Wildcat Venture Management, USA

Hans Nilsson
FourFact, Sweden

Aleksandra Novikova
Climate Policy Initiative, German Institute for
Economic Research

Victoria Novikova
Oxford University, UK

Virginia Sonntag O'Brien
REN21, France

Anastasia O'Rourke
BigRoom Inc., Canada

Peter Odell
Erasmus University Rotterdam,
the Netherlands.

Michael Ohadi
Universitys of Maryland, USA

Marina Olshanskaya
United Nations Development Programme

Shonali Pachauri
International Institute for Applied Systems
Analysis, Austria

Saptarshi Pal
Central European University, Hungary

Shamik Pal
Institute of Engineering and Management,
India

Debajit Palit
The Energy and Resources Institute, India

Riddhi Panse
Indian Institute of Technology-Bombay

Mahesh Patankar
Independent Energy Sector Consultant

Anand Patwardhan
Indian Institute of Technology-Bombay

Ksenia Petrichenko
Central European University, Hungary

Hector Pistonesi
Bariloche Foundation, Argentina

Christoph Plutzar
Klagenfurt University, Austria

Gisela Prasad
University of Cape Town, South Africa

Ndola Prata
University of California, Berkeley, USA

Lynn Price
Lawrence Berkeley National Laboratory, USA

Pallav Purohit
International Institute for Applied Systems
Analysis, Austria

Krishnan S. Rajan
International Institute of Information
Technology, India

M.V. Ramana
Princeton University, USA

Andrea Ramirez
Utrecht University, the Netherlands

Saumya Ranjan
Indian Institute of Technology-Bombay

Anand Rao
Indian Institute of Technology-Bombay

Shilpa Rao
International Institute for Applied Systems
Analysis, Austria

Amitav Rath
Policy Research International Inc., Canada

Rob Raven
Eindhoven University of Technology, the
Netherlands

Xiangkun Ren
Shenhua Coal Liquefaction Research Center,
China

Keywan Riahi
International Institute for Applied Systems
Analysis, Austria

Kamal Rijal
United Nations Development Programme

Johan Rockström
Stockholm Environment Institute, Stockholm
University, Sweden

Hans-Holger Rogner
International Atomic Energy Agency, Austria

Mathis L. Rogner
International Institute for Applied Systems
Analysis, Austria

Marc A. Rosen
University of Ontario Institute of Technology,
Canada

Carolina Rossini
University of São Paulo, Brazil

Joyashree Roy
Jadavpur University, India

Lau Saili
International Hydropower Association, UK

Constantine Samaras
Rand Corporation, USA

Gerd Sammer
University of Natural Resources and Applied
Life Sciences, Austria

Jayant Sathaye
Lawrence Berkeley National Laboratory, USA

David Satterthwaite
International Institute for Environment and
Development, UK

Deger Saygin
Utrecht University, the Netherlands

Jules Schers
PBL Netherlands Environmental Assessment
Agency, the Netherlands

Christoph Schillings
German Aerospace Center

Jürgen Schmid
Fraunhofer Institute for Systems and
Innovation Research, Germany

Mycle Schneider
Consultant on Energy and Nuclear Policy,
France

Sabine Schnittger
BAEconomics, Australia

Robert N. Schock
World Energy Council, UK and Center for
Global Security Research, USA

Niels B. Schulz
International Institute for Applied Systems
Analysis, Austria and Imperial College
London, UK

Seongwon Seo
Commonwealth Scientific and Industrial
Research Organisation, Australia

Ali Shafiei
University of Waterloo, Canada

Nilay Shah
Imperial College London, UK

Ram M. Shrestha
Asian Institute of Technology, Thailand

Priyadarshi R. Shukla
Indian Institute of Management

Dale Simbeck
SFA Pacific, USA

Ralph Sims
Massey University, New Zealand

Wim Sinke
Energy Research Centre of the Netherlands

Kirk R. Smith
University of California, Berkeley, USA

Aaron Smith
National Renewable Energy Laboratory, USA

Adrian Smith
University of Sussex, UK

Ricardo Soares de Oliveira
Oxford University, UK

Youba Sokona
United Nations Economic Commission for
Africa

Weiwei Song
Tsinghua University, China

Benjamin Sovacool
National University of Singapore

Ashutosh Srivastava
Indian Institute of Technology-Bombay

Leena Srivastava
The Energy and Resources Institute, India

Kjartan Steen-Olsen
Norwegian University of Science and
Technology

Julia Steinberger
The Institute of Social Ecology, Austria and
University of Leeds, UK

Lars Strupeit
Lund University, Sweden

Terry Surles
Desert Research Institute, USA

Tatsujiro Suzuki
Tokyo University, Japan

Alice Sverdlik
International Institute for Environment and
Development, UK

Minoru Takada
United Nations Development Programme

Richard Taylor
International Hydropower Association, UK

Theodore Thrasher
International Civil Aviation Organization,
Canada

Robert Thresher
National Renewable Energy Laboratory, USA

Julie Tran
British Columbia Utilities Commission,
Canada

Upendra Tripathy
Government of India

Craig Turchi
National Renewable Energy Laboratory, USA

Wim C. Turkenburg
Utrecht University, the Netherlands

Neha Umarji
Indian Institute of Technology-Bombay

Diana Ürge-Vorsatz
Central European University, Hungary

Eric Usher
United Nations Environment Programme

Sergey Vakulenko
Cambridge Energy Research Associates, USA

Harry Vallack
Stockholm Environment Institute, University
of York, UK

Rita van Dingenen
Joint Research Center, Italy

Denis van Es
Energy Research Centre, South Africa

Bas van Ruijven
PBL Netherlands Environmental Assessment
Agency, the Netherlands

Wilfried van Sark
Utrecht University, the Netherlands

Oscar van Vliet
International Institute for Applied Systems
Analysis, Austria

Detlef P. van Vuuren
PBL Netherlands Environmental Assessment
Agency, the Netherlands

Geert Verbong
Eindhoven University of Technology, the
Netherlands

Preeti Verma
The Climate Group, India

David Victor
University of California, San Diego, USA

Eugene Visagie
University of Cape Town, South Africa

Frank von Hippel
Princeton University, USA

Seppo Vuori
VTT Technical Research Centre of Finland

Horst Wagner
Montan University Leoben, Austria

Rahul Walawalkar
Customized Energy Solutions, India

Njeri Wamukonya†
formerly, United Nations Environment
Programme

Jim Watson
University of Sussex, UK

Sandy Webb
Independent Consultant

Jan Weinzettel
Norwegian University of Science and
Technology

Helga Weisz
Potsdam Institute for Climate Impact
Research, Germany

John Weyant
Stanford University, USA

John T. Wilbanks
The Ewing Mario Kauffman Foundation, USA

Paul Wilkinson
London School of Hygiene and Tropical
Medicine, UK

Robert H. Williams
Princeton University, USA

Charlie Wilson
Tyndall Centre for Climate Change Research,
UK

Rosemary Wolson
Council for Scientific and Industrial
Research, South Africa

Ernst Worrell
Utrecht University, the Netherlands

Iain Wright
BP, UK

Vladimir Yakushev
Gazprom, Russia

Kenji Yamaji
University of Tokyo, Japan

Kurt Yeager
Electric Power Research Institute and Galvin
Electricity Initiative, USA

Suyuan Yu
Tsinghua University, China

Hisham Zerriffi
University of British Columbia, Canada

Qiang Zhang
Tsinghua University, China

Xiliang Zhang
Tsinghua University, China

Li Zhou
Tsinghua University, China

Ji Zou
Renmin University of China

ANNEX **IV** Reviewers of the Global Energy Assesment Report

Reviewers of the Global Energy Assesment Report

Dilip Ahuja
National Institute of Advanced Studies, India

Anas Alhajji
NGP Energy Capital Management, LLC, USA

Maria Argiri
formerly, International Energy Agency

Vicki Arroyo
Georgetown Climate Center, USA

Alan Atkisson
CEO AtKisson Inc., ISIS Academy, Balaton Group, USA

Patil Balachandra
Indian Institute of Science

Fritz Barthel
Energy and Mineral Resources, Germany

David F. Batten
Commonwealth Scientific and Industrial Research Organization, Australia

Frans Berkhout
VU University, the Netherlands

Christoph Bertram
Potsdam Institute for Climate Impact Research, Germany

Preety Bhandari
United Nations Framework Convention on Climate Change Secretariat, Germany

Kornelis Blok
Utrecht University, the Netherlands

Valentina Bosetti
Fondazione Eni Enrico Mattei, Italy

Richard A. Bradley
Consultant, USA

Elizabeth Cecelski
ENERGIA International Network on Gender and Sustainable Energy, the Netherlands

Akanksha Chaurey
Tata Energy Research Institute, India

Francisco de la Chesnaye
Electric Power Research Insitute, USA

Nikhil Desai
Independent consultant, USA

Hadi Dowlatabadi
University of British Columbia, Canada

Olivier Dubois
Food and Agriculture Organisation, Italy

Gautam S. Dutt
International Energy Initiative, Argentina

Geoff Dutton
Rutherford Appleton Laboratory, UK

James A. Edmonds
Pacific Northwest National Laboratory, USA

Wolfgang Eichhammer
Fraunhofer Institute for Systems and Innovation Research, Germany

Per Eikeland
Fridtjof Nansen Institute, Norway

Paul Epstein†
formerly, Harvard University, USA

Marianne Fay
World Bank, USA

Peter Fraenkel
Marine Current Turbines Ltd., UK

Antony Froggatt
Chatham House, UK

Bill Fulkerson
University of Tennessee, USA

Bradford Gentry
Yale University, USA

John Gibbons
National Academy of Engineering and American Academy of Arts and Sciences, USA

Michael W. Golay
Massachusetts Institute of Technology, USA

Donna L. Goodman
Earth Child Institute, USA

Charles Goodman
Southern Company (retired), USA

Paul Graham
Commonwealth Scientific and Industrial Research Organization, Australia

David L. Greene
Oak Ridge National Laboratory, USA

Don Grether
Lawrence Berkeley National Laboratory, USA

Andrei Gritsevskyii
International Atomic Energy Agency, Austria

Michael Grubb
Cambridge University, UK

Waclaw Gudowski
Royal Institute of Technology, Sweden

Eshita Gupta
Indian Statistical Institute

Pablo Gutman
World Wildlife Fund, USA

Javier Hanna
United Nations Framework Convention on Climate Change Secretariat, Germany

John Bøgild Hansen
Haldor Topsøe A/S, Denmark

Nikos Hatziargyriou
National Technical University of Athens, Greece

Marianne Haug
University of Hohenheim, Germany

Peter Haugan
University of Bergen Thormøhlensgate,
Norway

Detlev Heinemann
Oldenburg University, Germany

Peter Hennicke
Wuppertal Institute for Climate, Environment
and Energy, Germany

Vera Höfele
Wuppertal Institute for Climate, Environment
and Energy, Germany

Adonai Herrera Martinez
European Bank for Reconstruction and
Development, Germany

Mark Hopkins
United Nations Foundation, USA

Luiz Horta Noqueira
Itajubá Federal University, Brazil

Chuck Howard
CddHoward Consulting Ltd., Canada

Ernst Huenges
Helmholtz Centre Potsdam, Germany

Steven Hunt
Practical Action Consulting, UK

Hillard Huntington
Stanford University, USA

Antonina Ivanova Boncheva
Autonomous University of Southern Baja
California Sur, Mexico

Roderick Jackson
Oak Ridge National Laboratory, USA

Arnulf Jaeger-Waldau
Joint Research Centre, Italy

Michael Jefferson
London Metropolitan Business School, UK

Catrinus Jepma
University of Groningen, the Netherlands

Hongguang Jin
Chinese Academy of Sciences

Veena Joshi
Embassy of Switzerland, India

James R. Katzer
ExxonMobil (retired) and Iowa State
University, USA

Gregory Keoleian
University of of Michigan, USA

Emek Barış Kepenek
Middle East Technical University, Turkey

Ilkka Keppo
University College London, UK

Anund Killingtveit
Norwegian University of Science and
Technology

Jong-Inn Kim
Asian Development Bank, Philippines

Jonathan G. Koomey
Stanford University, USA

Sivanappan Kumar
Asian Institute of Technology, Thailand

Balesh Kumar
National Geophysical Research Institute,
India

Vello Kuuskraa
Advanced Resources International, USA

Anthony Land
Anthony Land Associates, Botswana

Melissa Lapsa
Oak Ridge National Laboratory, USA

Louis Lebel
Chiang Mai University, Thailand

Stefan Lechtenböhmer
Wuppertal Institute for Climate, Environment
and Energy, Germany

Nicolas Lefèvre-Marton
Princeton University, USA

Vladimir Likhachev
Russian Academy of Sciences

David Lobell
Stanford University, USA

Alexander Luedi
WR Lloyd Limited, China

Nestor Luna Gonzalez
Latin American Energy Organization,
Ecuador

Landis MacKellar
International Institute for Applied Systems
Analysis, Austria

Alexei A. Makarov
Russian Academy of Sciences

Maxwell Mapako
Natural Resources and Environment, South
Africa

Anil Markandya
Basque Centre for Climate Change, Spain

Gregg Marland
Appalachian State University, USA

Ajay Mathur
Government of India

Helio Mattar
Akatu Institute for Conscious Consumption,
Brazil

Doug McKay
Shell InternationaL Ltd., United Arab
Emirates

James Meadowcroft
Carleton University, Canada

Tatyana Mitrova
Russian Academy of Sciences

Arild Moe
Fridtjof Nansens Institute, Norway

Mark R. Montgomery
State University of New York-Stony Brook,
USA

Shantanu Mukherjee
United Nations Development Programme

Peter Mulder
University of Amsterdam, the Netherlands

Svend Munkejord
SINTEF Energy Research, Norway

Rogier Nijssen
Knowledge Centre Wind turbine Materials
and Constructions, the Netherlands

Lars Nilsson
Lund University, Sweden

Dong-Woon Noh
Korea Energy Economics Institute

Tor Nygaard
Norwegian University of Life Sciences

Joan Ogden
University of California Davis, USA

Dennis S. Ojima
Colorado State University, USA

Debo Oladosu
Oak Ridge National Laboratory, USA

Ralph P. Overend
Biomass and Bioenergy, Canada

Tony Owen
University College London, Australia

Karen Palmer
Resources for the Future, USA

Martin K. Patel
Utrecht University, the Netherlands

Rashmi S. Patil
Indian Institute of Technology-Bombay

Walt Patterson
Chatham House, UK

Martin Pehnt
Institute for Energy and Environmental
Research, Germany

Joachim Peinke
University of Oldenburg, Germany

Per F. Peterson
University of California Berkeley, USA

Cédric Philibert
International Energy Agency, France

Gonzalo Piernavieja Izquierdo
Canary Islands Institute of Technology, Spain

Robert Pindyck
Massachusetts Institute of Technology, USA

Luiz Pinguelli Rosa
Federal University of Rio de Janeiro, Brazil

Lawrence Pitt
University of Victoria, Canada

Maximilian Posch
National Institute for Public Health and the
Environment, the Netherlands

Graham Pugh
US Department of Energy

Tinus Pulles
Netherlands Organization for Applied
Scientific Research

Burton Richter
Stanford University, USA

Michael Rock
Bryn Mawr College, USA

Richard Alexander Roehrl
United Nations Department of Economic and
Social Affairs

Adam Rose
University of Southern California, USA

Mark Rosenberg
Queens University, Canada

Teodoro Sanchez
Practical Action, UK

Ajit Sapre
Reliance Industries Limited, India

Guido Schmidt-Traub
CDC Climat Asset Management, France

Jan Sendzimir
International Institute for Applied Systems
Analysis, Austria

Karen Seto
Yale University, USA

Evgeny Shvarts
World Wildlife Fund-Russia

Toufiq Siddiqi
East-West Center, USA

Jim Skea
UK Energy Research Centre

Ruud Smits
Utreht University, the Netherlands

Robert Socolow
Princeton University, USA

Luc Soete
United Nations University-MERIT/Maastricht
University, the Netherlands

Allen Solomon
Cape Arago Ecological Consultants, USA

Mohammad Soltanieh
Sharif University of Technology, Iran

Laszlo Somlyody
Budapest University of Technology and
Economics, Hungary

Ashok Sreenivas
Prayas Energy Group, India

Will Steffen
Australian National University

Andrew Stirling
University of Sussex, UK

Harry C. Stokes
Project Gaia, Inc., USA

Gary Stuggins
World Bank, USA

Salvador Suárez García
Canary Islands Institute of Technology, Spain

Yoshiharu Tachibana
Tokyo Electric Power Company, Japan

Anil Terway
formerly, Asian Development Bank

Jefferson Tester
Cornell University, USA

Thomas Theisohn
Learning Network on Capacity Development,
France

Stefan Thomas
Wuppertal Institute for Climate, Environment
and Energy, Germany

Victoria Thoresen
Hedmark University College, Norway

Dennis Tirpak
World Resources Institute, USA

Michael Toman
World Bank, USA

David Trimm†
formerly, Commonwealth Scientific and
Industrial Research Organization, Australia

Anthony Turhollow
Oak Ridge National Laboratory, USA

Hal Turton
Paul Scherrer Institute, Switzerland

Julio Usaola Garcia
University Carlos III de Madrid, Spain

Bob van der Zwaan
Energy research Centre of the Netherlands,
Columbia University, USA, and Johns
Hopkins University, Italy

Adriaan van Zon
United Nations University-MERIT/Maastricht
University, the Netherlands

Thyjagarajan Velumail
United Nations Development Programme

Ivan Vera
United Nations Department of Economic and
Social Affairs

Fernando Viana
Viana & Associates LLC, USA

Nadejda M. Victor
Booz Allen Hamilton, USA

Spyros Voutsinas
National Technical University of Athens,
Greece

Steve Wiel
Collaborative Labeling & Appliance
Standards Program, USA

Thomas J. Wilbanks
Oak Ridge National Laboratory, USA

Robert Williams
United Nations Industrial Development
Organization

Harald Winkler
University of Cape Town, South Africa

Anny Wong
RAND Corporation, USA

Francis D. Yamba
Centre for Energy, Environment and
Engineering Zambia

Xianli Zhu
Technical University of Denmark

Reviewer 49

Reviewer 60

Reviewer 93

Reviewer 118

Reviewer 172

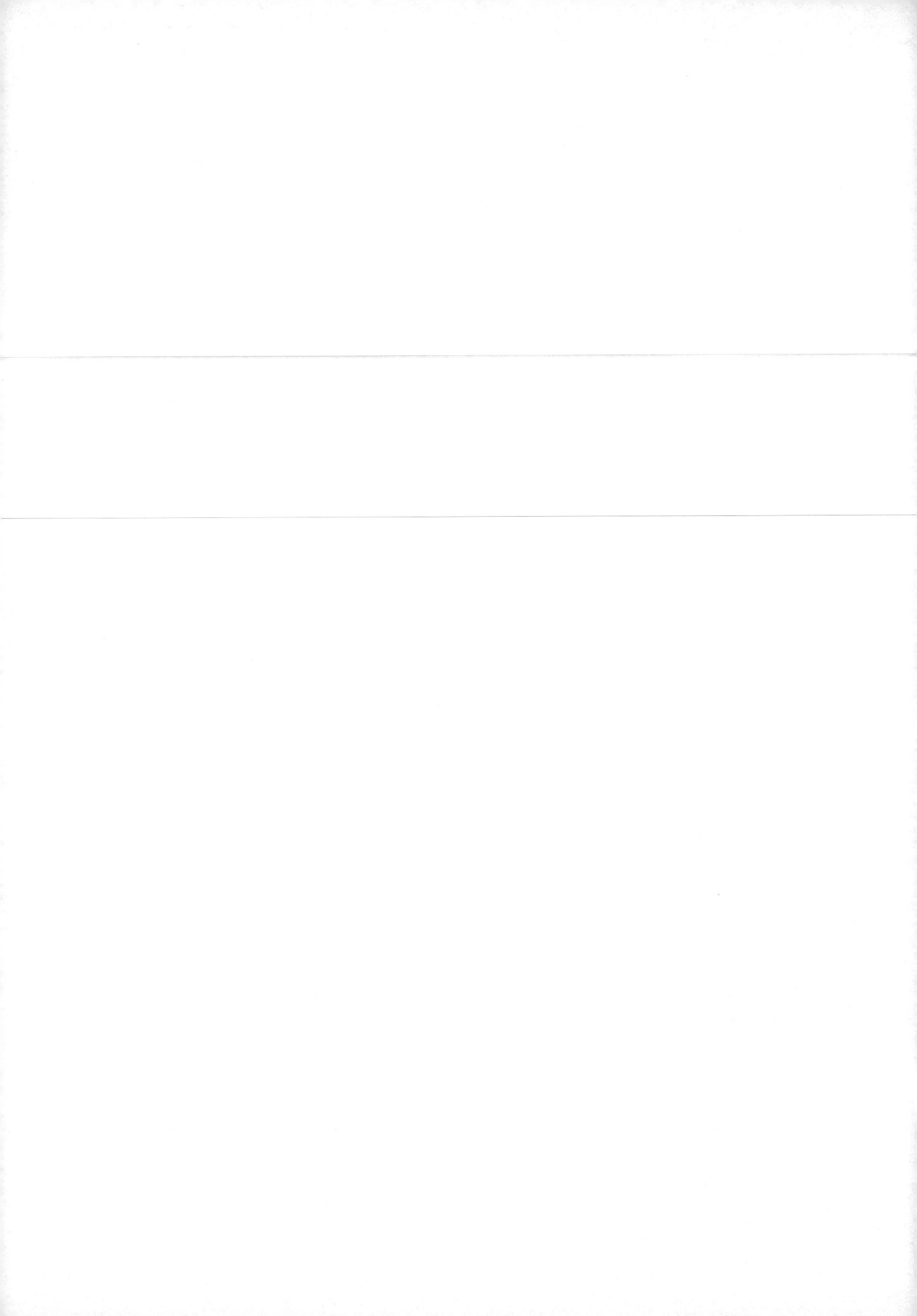

Index

Index

Page numbers for figures, tables and boxed material are shown in *italic*

Index

Index

Index

Index

Index

Index